THE UNIVERSITY OF ARIZONA SPACE SCIENCE SERIES
RICHARD P. BINZEL, GENERAL EDITOR

Asteroids III
William F. Bottke Jr., Alberto Cellino, Paolo Paolicchi,
and Richard P. Binzel, editors, 2002, 785 pages

TOM GEHRELS, GENERAL EDITOR

Origin of the Earth and Moon
R. M. Canup and K. Righter, editors, 2000, 555 pages

Protostars and Planets IV
Vincent Mannings, Alan P. Boss, and Sara S. Russell, editors, 2000, 1422 pages

Pluto and Charon
S. Alan Stern and David J. Tholen, editors, 1997, 728 pages

Venus II—Geology, Geophysics, Atmosphere, and Solar Wind Environment
S. W. Bougher, D. M. Hunten, and R. J. Phillips, editors, 1997, 1376 pages

Cosmic Winds and the Heliosphere
J. R. Jokipii, C. P. Sonett, and M. S. Giampapa, editors, 1997, 1013 pages

Neptune and Triton
Dale P. Cruikshank, editor, 1995, 1249 pages

Hazards Due to Comets and Asteroids
Tom Gehrels, editor, 1994, 1300 pages

Resources of Near-Earth Space
John S. Lewis, Mildred S. Matthews, and Mary L. Guerrieri, editors, 1993, 977 pages

Protostars and Planets III
Eugene H. Levy and Jonathan I. Lunine, editors, 1993, 1596 pages

Mars
Hugh H. Kieffer, Bruce M. Jakosky, Conway W. Snyder,
and Mildred S. Matthews, editors, 1992, 1498 pages

Solar Interior and Atmosphere
A. N. Cox, W. C. Livingston, and M. S. Matthews, editors, 1991, 1416 pages

The Sun in Time
C. P. Sonett, M. S. Giampapa, and M. S. Matthews, editors, 1991, 990 pages

Uranus
Jay T. Bergstralh, Ellis D. Miner, and Mildred S. Matthews, editors, 1991, 1076 pages

Asteroids II
Richard P. Binzel, Tom Gehrels, and Mildred S. Matthews, editors, 1989, 1258 pages

Origin and Evolution of Planetary and Satellite Atmospheres
S. K. Atreya, J. B. Pollack, and Mildred S. Matthews, editors, 1989, 1269 pages

Mercury
Faith Vilas, Clark R. Chapman, and Mildred S. Matthews, editors, 1988, 794 pages

Meteorites and the Early Solar System
John F. Kerridge and Mildred S. Matthews, editors, 1988, 1269 pages

The Galaxy and the Solar System
Roman Smoluchowski, John N. Bahcall, and Mildred S. Matthews, editors, 1986, 483 pages

Satellites
Joseph A. Burns and Mildred S. Matthews, editors, 1986, 1021 pages

Protostars and Planets II
David C. Black and Mildred S. Matthews, editors, 1985, 1293 pages

Planetary Rings
Richard Greenberg and André Brahic, editors, 1984, 784 pages

Saturn
Tom Gehrels and Mildred S. Matthews, editors, 1984, 968 pages

Venus
D. M. Hunten, L. Colin, T. M. Donahue, and V. I. Moroz, editors, 1983, 1143 pages

Satellites of Jupiter
David Morrison, editor, 1982, 972 pages

Comets
Laurel L. Wilkening, editor, 1982, 766 pages

Asteroids
Tom Gehrels, editor, 1979, 1181 pages

Protostars and Planets
Tom Gehrels, editor, 1978, 756 pages

Planetary Satellites
Joseph A. Burns, editor, 1977, 598 pages

Jupiter
Tom Gehrels, editor, 1976, 1254 pages

Planets, Stars and Nebulae, Studied with Photopolarimetry
Tom Gehrels, editor, 1974, 1133 pages

Asteroids III

Asteroids III

William F. Bottke Jr.
Alberto Cellino
Paolo Paolicchi
Richard P. Binzel

Editors

With 150 collaborating authors

Foreword by Tom Gehrels

THE UNIVERSITY OF ARIZONA PRESS
Tucson

in collaboration with

LUNAR AND PLANETARY INSTITUTE
Houston

About the front cover:

A view from the surface of a near-Earth asteroid that is encountering the Earth-Moon system. The viewer is standing on one end of an elongated, somewhat banana-shaped asteroid, looking down the long axis at the other end. Boulders and regolith can be seen in the foreground, while a crater in the middle distance can be observed across the "valley" between the two ends. At the opposite end, the asteroid curves up and away from the viewer. Numerous boulders and craters can be found over the asteroid's surface. The Moon is to the left and Earth is to the right. Painting by Bill Hartmann.

About the back cover:

A view of some of the "ponded" deposits that can be found on near-Earth asteroid (433) Eros. The *NEAR Shoemaker* spacecraft apparently landed in one such deposit. These deposits appear to be the result of deposition of finer material into depressions such as craters, after the material was sorted from the upper portion of the asteroid's regolith. Though ponds can be found at all latitudes of Eros, the largest ones occur in a narrow zone near the equator. The color of the ponded material appears to be bluer than the surrounding terrain. The regolith mobility mechanism that produced Eros' ponded deposits is unknown; possibilities include seismic shaking induced by impacts and photoelectric sorting (i.e., solar energy may electrically charge particles on Eros, enough so some can levitate off the surface and settle in a different location). MET Image 156087831-7866, 2.4°S, 179.1°W; crater is ~90 m in diameter, image acquired at 0.4 m per pixel.

The University of Arizona Press in collaboration with the Lunar and Planetary Institute

First printing

♾ This book is printed on acid-free, archival-quality paper.
Manufactured in the United States of America

07 06 05 04 03 02 6 5 4 3 2 1

Library of Congress Cataloging-in-Publication Data

Asteroids III / W.F. Bottke Jr. ... [et al.], editors.
p. cm. — (Space science series)
Includes bibliographical references and index.
ISBN 0-8165-2281-2 (cloth : alk. paper)
1. Asteroids. I. Title: Asteroids 3. II. Bottke, W. F. (William Frederick), 1966- III. Series.
QB651 .A856 20002
523.44—dc21
2002015179

British Library Cataloguing-in-Publication Data

A catalogue record for this book is available from the British Library.

Contents

5.2. Relationships

5.3. Origins

5.4. The Impact Hazard

PART VI: DATABASES

Collaborating Authors

Scientific Organizing Committee

The editors thank the following scientists for their assistance in the planning stages of this book:

M. A. Barucci
C. R. Chapman
Cl. Froeschlé
D. R. Davis
A. Fujiwara
A. W. Harris (JPL)
K. Keil
D. Lupishko
H. Y. McSween Jr.
A. Morbidelli
K. Muinonen
E. F. Tedesco
V. Zappalà

Acknowledgment of Reviewers

The editors gratefully acknowledge the following individuals, as well as several anonymous reviewers, for their time and effort in reviewing chapters for this volume:

Erik Asphaug
Olivier Barnouin-Jha
Antonella Barucci
Irina Belskaya
Greg Black
Ted Bowell
Mike Brown
Thomas H. Burbine
Joseph Burns
Schelte J. Bus
Donald Campbell
Humberto Campins
Adriana Campo Bagatin
Andrea Carusi
Alberto Cellino
Clark Chapman
Marc Chaussidon
Steven Chesley
Beth Ellen Clark
Marcello Coradini
Dale Cruikshank
Don Davis
Stan Dermott
Mario Di Martino
Deborah Domingue
Alain Doressoundiram
Elisabetta Dotto
Jack Drummond
Daniel P. Durda
Yan Fernandez
George Flynn
Akira Fujiwara
Paul Geissler
Amitabha Ghosh
Owen Gingerich
Richard Greenberg
Gerhard Hahn
Paul Hardersen
Jacques Henrard
Carl Hergenrother
Kevin Housen
Ellen Howell
Scott Hudson
David Hughes
Boris Ivanov
Dave Jewitt
Steve Kortenkamp
Jeffrey A. Larsen
Daniela Lazarro
Anne Lemaître
Stan Love
Guenter Lugmair
Francesco Marzari
Lucy McFadden
William B. McKinnon
Harry Y. McSween Jr.
Patrick Michel
Andrea Milani
Stefano Mottola
Thomas Müller
Karri Muinonen
Carl Murray
Fathi Namouni
David Nesvorný
Mike Nolan
David P. O'Brien
William J. Owen Jr.
Paolo Paolicchi
Jean-Marc Petit
Petr Pravec
Steve Price
David Rabinowitz
Michel Rapaport
Derek Richardson
Andrew Rivkin
Alan Rubin
Sho Sasaki
Hans Scholl
Joe Spitale
Duncan Steel
Robert Strom
Mark Sykes
Hiroshi Takeda
Peter Thomas
Giovanni Valsecchi
Rob Whiteley
Gareth Williams
Don Yeomans
Alexander Zaitsev
Vincenzo Zappalà
Michael Zolensky

Foreword

Asteroids were special from the beginning. *Piazzi* (1803) was not even searching for the missing planet at 2.6 AU; he was working on a new stellar catalog and indeed noted Ceres as a star on January 1, 1801, but discovered the next night that his star had moved! It was not proper for scientists to use emotional words, of course, not until *Chandrasekhar* (1987) connected the quest for beauty with the choice of good science. But you can be sure that Piazzi had a good time learning that it was not a comet, not a planet, but something new (*Cunningham,* 2001). More excitement followed when positions yielded an orbit, and when other "minor planets" were found. And so a breed of "asteroiders" was born whose latest generation is comprised of the readers, authors, referees, and editors of this book. What is it that is so captivating about the objects once contemptuously called the "vermin of the sky"? You may wish to answer that question for yourself.

The asteroiders did not mind long hours spent in cold domes or with tedious calculations. Daniel Kirkwood (1814–1895) persistently pursued collisional dynamics far ahead of its time (see *Edmondson,* 2000). Through the 1950s, the *Ephemerides of Minor Planets* were computed in Leningrad by a roomful of women, each spending all day with her abacus. Photometry was systematically askew with one magnitude too bright at the 11th and two at the 16th visual magnitude, but it was fun to correct that after the discovery of the opposition effect made new determinations consistent (*Gehrels,* 1956). The IAU meeting of 1958 had its Commission 20 members debating the value of finding more asteroids; there were so many already. But they didn't quit. By 1970 there were about two dozen asteroiders who organized a meeting, and a book was produced. To me, that was "Asteroids I" (*Gehrels,* 1971), for it became the prototype of this series. Life was less lonely among the hundred or so, but the meeting's great debate about missions ended nascent planning for *Pioneer 12* to fly by a comet and/or an asteroid (also see *Stuhlinger et al.,* 1972; *Whipple et al.,* 1973). It would have cost only 40 million dollars as the spacecraft was flight-qualified ready (i.e., the one currently hanging in the U.S. National Air and Space Museum), but its science would of course have been nowhere as great as 10 years later for Comet Halley.

Asteroiders had great times in a field that was gaining respectability. The astrophysicists looked more kindly at applications for time with their telescopes after *McCord et al.* (1970) showed that the asteroid spectra, in reflected light, are diagnostic of composition. It became gradually clear how asteroids participate in the origin of planets, stars, and life (*Mannings et al.,* 2000). Stuhlinger's ion propulsion demonstrated active targeting on Deep Space I all the way to Comet Borrelly. We are still a special breed in a corner of the barn; there is no Nobel Prize to aim for, in fact there is no asteroid prize at all — the attraction is enough. There is beauty also in the hazards, because we are taking care of them. We are working to safeguard our planet, even if the world does not seem to want to be saved. We can understand that too, because it was not until the 1980s that we heeded the warnings of our peers (*Öpik,* 1951; *Urey,* 1957, 1973).

What of the future? It looks even brighter than the past. There are ever-larger telescopes, anyone may compete for a space mission, and advanced theories are being tested with sophisticated data. New generations are better trained and more carefully selected than older ones. For the beginners, I have advice that has worked well for family and friends, which is not to simply look for a job in *Physics Today* or the AAS Job Listing, because their position descriptions are often written for someone else. You have to *become* that someone. It is your life, you decide what you are going to be, and where. Study the group you wish to join for years in advance; know everyone involved, their papers and their dreams. Visit them a year in advance so they can create your job. Stride into their doors with, "I want to work with you because you are the greatest." They know that too, so you are on the same wavelength, but then follow through with showing what you and only you can do for them.

For all of us, the greatest advance still comes from the unknown. Asteroid lightcurves and pole determinations were pioneered simply because it was fun; at NASA Headquarters, Maury Dubin thought of future missions to justify our funding. *Kowal* (1996) aimed at Planet X and found Chiron, the first of a new Centaur class. *Jewitt* (1999) believed in transneptunian objects; he did not give up for years, trying different telescopes until he found them.

I failed where Kowal and Jewitt succeeded, for lack of focus. But I do have an example in pursuit of fascination. The flight of *Sputnik* inspired the dream of a photopolarimeter on the *Voyager* Mars mission; that mission was not flown, but we were funded to prepare for a future opportunity. So, we flew *Polariscope* on a balloon at 36.6 km and proved the theoretical transmission at 2250 Å between ozone and oxygen absorptions, actually found 1.1 magnitude extinction per airmass, and obtained interstellar, Mars, and Venus polarizations. It was a tough challenge for a small university crew. It got us on *Pioneers 10* and *11* for the "Jupiter-Asteroid missions," which became the "Jupiter-Saturn missions" after the realization that precise targeting was needed to fly by anything, even in the asteroid belt (*Gehrels,* 1988).

My lack of focus punctuates the counterpoint of attraction, namely that intelligent determination is also essential. And then there is the example of a kind and kindred soul, held up for us in the Van Biesbroeck Award for unselfish service to science and colleagues. Ask your intuition what is helpful and beautiful. Then let your rigorous reason follow through.

REFERENCES

Chandrasekhar S. (1987) *Truth and Beauty, Aesthetics and Motivations in Science.* Univ. of Chicago, Chicago, Illinois. 180 pp.

Cunningham C. J. (2001) *The First Asteroid.* Star Lab Press, Surfside, Florida. 498 pp.

Edmondson F. K. (2000) Daniel Kirkwood — "Dean of American Astronomy." *Mercury, 29(3),* 27–31.

Gehrels T. (1956) Photometric studies of asteroids. V. The lightcurve and phase function of 20 Massalia. *Astrophys. J., 123,* 331–338.

Gehrels T., ed. (1971) *Physical Studies of Minor Planets.* NASA SP-267. 717 pp.

Gehrels T. (1988) *On the Glassy Sea,* pp. 111–132. American Institute of Physics, New York.

Jewitt D. (1999) Kuiper Belt objects. *Annu. Rev. Earth Planet. Sci., 27,* 287–312.

Kowal C. T. (1996) *Asteroids: Their Nature and Utilisation,* 2nd edition. Praxis, Chichester, England. 170 pp.

Mannings V., Boss A. P., and Russell S. S., eds. (2000) *Protostars and Planets IV.* Univ. of Arizona, Tucson. 1452 pp.

McCord T. B., Adams J. B., and Johnson T. V. (1970) Asteroid Vesta: Spectral reflectivity and compositional implications. Science, 168, 1445–1447.

Öpik E. (1951) Collision probabilities with the planets and the distribution of interstellar matter. *Proc. R. Irish Acad., 54A,* 165–199.
Piazzi G. (1803) Praecipuarum stellarum inerrantium positiones mediae ineunte seculo XIX ex observationibus habitis in Specula Panormitana ab anno 1792 ad annum 1802. Panormi M.DCCC.III. Typis Regiis (see also http://www.astropa.unipa.it/HISTORY/history.htm).
Stuhlinger E., Alfvén H., Arrhenius G., Bourke R., Doe B., Dwornik S., Friedlander A., Gehrels T., Guttmann C., Strangway D., and Whipple F. (1972) *Comets and Asteroids: A Strategy for Exploration.* NASA TMX-64677. 93 pp.
Whipple F., Atkins K., Arrhenius G., Brandt J., Doe B., Gehrels T., Ness N., Owen T., Rasool S. I., Stuhlinger E., and Wasson J. (1973) *The 1973 Report and Recommendations of the NASA Science Advisory Committee on Comets and Asteroids: A Program of Study.* NASA TMX-71917. 69 pp.
Urey H. C. (1957) Origin of tektites. *Nature, 179,* 556–557.
Urey H. C. (1973) Cometary collisions and geological periods. *Nature, 242,* 32–33.

Tom Gehrels
Tucson, Arizona
November 2001

Preface

"Big things have small beginnings." This phrase has stayed with the editors throughout the process involved in producing the book *Asteroids III,* not only because it provides a rationale into why studies of asteroids (and comets) are important for understanding planetary formation, but also because it describes how contributions from scientists around the world have helped produce a stunning mosaic of our current state of knowledge about the asteroids. Two hundred years after 1 Ceres was discovered, asteroids can no longer be considered mere points of light in the sky. Spacecraft missions, advanced Earth-based observation techniques, and state-of-the-art numerical models are continually revealing the detailed shapes, structures, geological properties, and orbital characteristics of these smaller denizens of our solar system. This volume brings together the latest information obtained by spacecraft, combined with astronomical observations and theoretical modeling, to present our best current understanding of asteroids and the clues they reveal for the origin and evolution of the solar system. This collective knowledge, prepared by a team of more than 100 international authorities on asteroids, includes new insights into asteroid-meteorite connections, possible relationships with comets, and the hazards posed by asteroids colliding with Earth.

The book's contents include reports on surveys based on remote observation and summaries of physical properties; results of *in situ* exploration; studies of dynamical, collisiónal, cosmochemical, and weathering evolutionary processes; and discussions of asteroid families and the relationships between asteroids and other solar system bodies. It the hope of the editors that *Asteroids III* will carry on the fine tradition established by the previous two generations of this book, in that it will stand as the definitive source on asteroids for the next decade.

We are deeply indebted to the many authors and referees for their efforts in helping us create *Asteroids III.* We also thank the members of the *Asteroids III* scientific organizing committee for helping us shape the structure and content of the book. The conference associated with the book, "Asteroids 2001: From Piazzi to the 3rd Millennium," held June 11–16, 2001, in Palermo, Italy, was simply superb. We are extremely grateful to our hosts Salvatore Serio and Giorgia Foderà of the Palermo Observatory and their staff. We also thank Carlo Blanco, the meeting's scientific organizing committee, and the presenters for the stimulating program. Papers associated with this conference are also appearing in a special issue of *Memorie della Società Astronomica Italiana* (*Journal of the Italian Astronomical Society, Vol. 73, No. 3,* 2002) and *Icarus* (*Vol. 159, No. 2,* 2002). The editors would also like to recognize the critical contributions of the *Asteroids III* production staff, namely Renée Dotson and the tireless workers at the Lunar and Planetary Institute, whose invaluable assistance and pleasant demeanor made life as easy as possible for the editors, authors, and referees.

Finally, editor William Bottke would like to thank his wife Veronica and three children, Kristina-Marie, Laura, and Julie, for putting up with him during the many long nights that were required to assemble this work. For him, it has been the experience of a lifetime. Co-editors Cellino, Paolicchi, and Binzel give no less thanks to their own asteroid families (Paola and Alice for A.C.; Alessandra for P.P.; Michelle, Steven, and Christine for R.P.B.) who have endured, encouraged, and supported this extraordinary project.

William F. Bottke Jr., Alberto Cellino, Paolo Paolicchi, and Richard P. Binzel

Part I: Introduction

An Overview of the Asteroids: The Asteroids III Perspective

William F. Bottke Jr.
Southwest Research Institute

Alberto Cellino
Astronomical Observatory of Torino

Paolo Paolicchi
University of Pisa

Richard P. Binzel
Massachusetts Institute of Technology

---

1. INTRODUCTION: WHY WE STUDY ASTEROIDS

"We are now on the threshold of a new era of asteroid studies," wrote Tom Gehrels in 1971 (*Gehrels,* 1971). These words proved quite prophetic for the three decades of physical observations and theoretical understanding that followed. As the third century of asteroid research has begun in 2001, we can once again picture ourselves standing on a new threshold. This new century begins with asteroids no longer being starlike points of light in our telescopes, but resolved worlds with distinctly measurable, sizes, shapes, and surface morphologies. Each has a unique history to unravel, a history that begins at the time of formation of our own planet and the entire solar system.

The physical nature, distribution, formation, and evolution of asteroids are fundamental to our understanding how planet formation occurred and, ultimately, why life exists on Earth. In our current solar system, asteroids (and comets) are the most direct remnants of the original building blocks that formed the planets. As such, they contain a relatively pristine record of the initial conditions that existed in our solar nebula some 4.6 G.y. ago. The asteroids that have survived since that epoch, however, have experienced numerous collisional, dynamical, and thermal events that have shaped their present-day physical and orbital properties. Interpreting this record via observations, laboratory studies of meteorites, and theoretical/numerical modeling can tell us much about how the bodies in our solar system have evolved with time. In fact, even though asteroids represent only a tiny fraction of the total mass of the terrestrial planets, their large numbers, diverse compositions, and orbital distributions provide powerful constraints for planet-formation models. For example, in the inner solar system, the orbital and physical characteristics of the asteroid belt and the surfaces of the terrestrial planets scarred by asteroid impacts can be used to narrow the range of possible starting conditions that can conceivably create a planetary system similar to our own. While the dynamical properties of the terrestrial planets (e.g., masses, heliocentric distances, spin rates, obliquities) also provide valuable constraints, many of these characteristics may have been significantly influenced by large collisions and other stochastic events, potentially making it problematic to reproduce these traits in any modeling effort. Moreover, other useful constraints, such as a planet's geochemical and isotopic characteristics, are often difficult to interpret because each of the terrestrial planets has experienced differentiation and thermal evolution with unknown starting conditions. Thus, asteroids and other small body populations may provide the key pieces of the cosmological puzzle that allow us to decipher why at least one of the planets in our solar system harbors life.

Motivated by this theme, the planetary community's decadal survey recently published a list of three central questions raised by the structure and nature of various asteroid populations (*Sykes et al.,* 2002). These questions are intimately related to the need of finding an answer to a number of long-debated problems. We repeat them here, since they help put into context the work that is presented throughout *Asteroids III.*

I. What was the compositional gradient of the asteroid belt at the time of initial protoplanetary accretion?

- What is the population and compositional structure of the main asteroid belt today?
- How do dynamics and collisions modify this structure over time?
- What are the physical properties of asteroids?
- How do surface modification processes affect our ability to determine this structure?

II. What fragments originated from the same primordial parent bodies, and what was the original distribution of those parent bodies?

- What asteroid fragments are associated dynamically, suggesting a common origin?
- What objects are geochemically linked?

III. What are the early steps in planet formation and evolution?

- What are the compositions and structures of surviving protoplanets?

- What does their cratering record combined with the cratering record of younger surfaces reveal about the primordial size distribution of objects in the asteroid belt?
- What do meteorites tell us about formation and evolution processes of these bodies? How well do they sample the asteroid belt?

A second and perhaps more practical reason to study asteroids has to do with the fact that some of these bodies are capable of striking Earth with enough energy to produce severe or possibly catastrophic damage to our civilization. Over the last two decades, it has been convincingly argued that the impact of a multikilometer asteroid or comet 65 m.y. ago led to a mass extinction event that eliminated the dinosaurs (e.g., *Alvarez et al.,* 1980). Because events like this are increasingly seen as harbingers of things to come rather than anomalies, scientists from around the world are now employing a variety of techniques (e.g., remote sensing, numerical modeling, spacecraft missions) to detect bodies near Earth and understand their physical properties.

Since the publication of *Asteroids II* some 13 years ago, the asteroid community has made tremendous progress in addressing many of the diverse issues described above. To give the reader a sense of how far we have come over this time, recall that before the 1990s, we had yet to visit our first asteroid with a spacecraft, nor had asteroid satellites been directly imaged by groundbased observations. Detailed radar investigations of the shapes and surface properties of near-Earth objects (NEOs) were in their infancy, while the number of known NEOs was only ~100. Dedicated NEO surveys like Spacewatch and LINEAR were only beginning to come on line or were years away. Public awareness that impacting asteroids and comets posed a threat to life on Earth was almost nonexistent.

Now, when comparing the table of contents between *Asteroids II* and *Asteroids III*, one is immediately struck by the essential role that charge-coupled-device (CCD) technology has had in advancing our understanding of asteroid surfaces and rotations. CCDs have enabled the number of objects for which high-quality spectra are available to grow from hundreds to thousands (chapter by Bus et al.), to push the limits of rotation studies down to small sizes where complex rotation states are evident (chapter by Pravec et al.), to expand our observational sample of outer-belt asteroids (chapter by Barucci et al.), and to explore in detail the complex mineralogy that reveals the petrologic history of the asteroids (chapters by Rivkin et al. and Gaffey et al.). Improvements in radiometric techniques and the application of larger aperture telescopes have pushed forward our capabilities and understanding of the thermal properties of both main-belt and near-Earth asteroids (NEAs) (chapter by Harris and Lagerros). Astounding advances in radar imaging reconstruction techniques, coupled with the upgraded capability of the Arecibo facility, have allowed radar studies to blossom among a new generation of researchers, as summarized in the chapter by Ostro et al. Routine observations of asteroids from space platforms (chapter by Dotto et al.) plus our first *in situ* exploration of asteroids by space probes (chapters by Cheng, Sullivan et al., Chapman, and Farquhar et al.) have most strikingly propelled our knowledge forward by allowing us to explore asteroids as real places rather than as distant points of light.

The great strides made in exploring asteroids observationally have been matched by the tremendous progress made in simulating the nature and evolution of asteroids via numerical models. The advent of inexpensive, fast workstations and increasingly sophisticated numerical algorithms have allowed fairly realistic numerical experiments to become commonplace. It is now possible to track the orbits of planetesimals and planetary embryos for millions of years while including all relevant gravitational perturbations. We can also now model (with reasonable accuracy) what happens when kilometer-sized asteroids strike one another at speeds of several kilometers per second. To check the results of these "virtual laboratories," scientists must constantly compare their results with observational constraints. If the models are wrong, they must be refined or thrown out, as appropriate. This checks-and-balances system in asteroid research has produced a much deeper understanding of small-body evolution, such that many of the questions addressed by the decadal survey can now be attacked.

Our purpose for this chapter is to convey the current key points of understanding for the asteroids. In doing so, we hope to help map out the content of this book and point the reader to literature that has connections beyond this volume. Of course, the best purpose for reviewing our current understanding is setting the context for future asteroid investigations.

2. BRIEF HISTORY OF THE PRIMORDIAL ASTEROID BELT

2.1. Formation of Main-Belt Asteroids

The largest reservoir of asteroids in the inner solar system is the main belt, located between the orbits of Mars and Jupiter. The processes by which the main belt took on its current attributes are believed to be linked to planet formation, particularly the formation of the terrestrial planets and Jupiter. The sequence of planet formation in the inner solar system, which involves the gradual coalescence of many tiny bodies into rocky planets, can be divided into four stages: (1) the accumulation of dust in the solar nebula into kilometer-sized planetesimals, (2) runaway growth of the largest planetesimals via gravitational accretion into numerous protoplanets isolated in their feeding zones; (3) oligarchic growth of protoplanets fed by planetesimals residing between their feeding zones; and (4) mutual perturbations between Moon-to-Mars-sized planetary embryos and Jupiter, causing collisions, mergers, and the dynamical excitation of small-body populations not yet accreted by the embryos (e.g., *Safronov,* 1969; *Weidenschilling,* 2000). With the discovery and knowledge of numerous stellar disks and the detection of extrasolar planets (e.g., *Marcy et al.,* 2000), our understanding of the basic processes of accretion and planet formation has seen a broad and rapid growth beyond what is covered here. (Note: For a more detailed look at these processes, the reader is directed to the many excellent review chapters found in *Protostars and Planets IV,* published in 2000 by the University of Arizona Press).

Meteorites provide the clock for timing planetesimal formation. Several chronometers provided by short-lived radionuclides (e.g., ^{53}Mn decaying to ^{53}Cr; ^{26}Al decaying to ^{26}Mg) are described in the chapter by Shukolyukov and Lugmair. According to their high-resolution chronological studies, the calcium-aluminum-rich inclusions (CAIs) found in chondritic meteorites, considered the first condensates of matter in the solar system, have an estimated formation age of ~4571 m.y. Some 2 m.y. after the formation of the CAIs, asteroids with diameters $D > 10$ km had formed in the asteroid belt. Objects that had accreted significant amounts of ^{26}Al during this time were heated as ^{26}Al decayed into ^{26}Mg. In some cases, the heat budget on these asteroids was high enough to produce aqueous alteration, metamorphism, melting, or even differentiation.

The thermal evolution of large asteroids during this epoch is the subject of the chapter by McSween et al. Their modeling work suggests that there are several variables that determine how an asteroid might be modified over time by radiogenic heat: (1) accretion time, which determines the quantity and distribution of active ^{26}Al available to a given asteroid; (2) the composition of the accreted material, which changes with increasing heliocentric distance to include more water and other volatiles; (3) the thickness and thermal conductivity of an asteroid's regolith, which may regulate its ability to eliminate heat; and (4) the asteroid's ultimate diameter. Using these results, we can begin to interpret the surface properties and compositions of asteroids in the asteroid belt. For example, McSween et al.'s results point to a possible solution for this frequently-cited main-belt enigma: (4) Vesta, a $D = 530$ km asteroid in the inner main belt, has differentiated while (1) Ceres, a much larger $D = 930$ km asteroid in the central main belt, apparently has not. It is possible that objects in the inner main belt formed much more quickly than those farther out in the main belt, such that Vesta may have accreted more active ^{26}Al than Ceres. In addition, the presumed greater availability of volatiles in the outer main belt may have helped inhibit large-scale melting events. The observational search for water and aqueous alteration on main-belt asteroids is discussed in the chapter by Rivkin et al.

As far as we know, Vesta is the only remaining differentiated object in the main belt with an intact interior structure consisting of a core, olivine-rich mantle, and a basaltic crust. Assuming that Vesta is the ultimate source of the howardites, eucrites, and diogenite (HED) meteorites, a considerable amount of information can be inferred about Vesta's history. For example, Shukolyukov and Lugmair claim that differentiation ended on Vesta by ~4565 m.y., some 6 m.y. after the formation of the CAIs. In the chapter by Keil, the geologic history of Vesta is examined in detail. One of the interesting questions raised by their work is the hypothesis that the primordial main belt may, at one time, have held additional Vesta-like objects. Their claim is supported by evidence suggesting that 108 of the 135 asteroids sampled by meteorites were partially melted or completely melted and differentiated by postaccretionary heating events (e.g., iron meteorites and stony-irons like pallasites and mesosiderites). The possible existence of these bodies raises several important questions about the evolutionary history of the primordial main belt (e.g., What happened to these Vesta-like objects? Are there any meteoritical or asteroidal remnants from this population residing in the current main belt? What can these survivors tell us?), some of which are (briefly) addressed in the next section.

Additional constraints on this early period in solar system history come from the chapter by Scott, who investigates the effects of asteroid accretion, metamorphism, melting, and collisional evolution on various meteorite samples (e.g., chondrites, differentiated meteorites, breccias). Meteorites provide strong constraints for asteroid-evolution models, since they provide a physical record of real asteroid properties and the effects of impact damage. At the present time, however, many meteorite samples do not appear consistent with *Asteroids II*-era models of how the asteroid belt evolved over time.

2.2. Dynamical Excitation of the Primordial Main Belt

To decipher the timing and enigmatic processes of planet formation, it is useful to understand how the main belt reached its current state. As described in the chapter by Petit et al., the formation of Jupiter had significant repercussions for the evolution of the primordial main belt. We list the principal changes below:

1. *Large mass depletion.* Model results suggest the primordial main belt contained ~2–10 $M_\oplus$ of material, enough to allow the asteroids to accrete on relatively short timescales. The current main belt, however, is depleted of mass, such that it only contains ~5×10^{-4} $M_\oplus$ of material. The mechanism that eliminated the mass is constrained by the presence of Vesta's basaltic crust. If the main belt were massive for too long, Vesta's crust would have been obliterated by collisions.

2. *Strong dynamical excitation.* Initially, the eccentricities e and inclinations i of asteroids within the primordial main belt were low enough that accretion could occur. The median e, i values of asteroids in the current main belt, however, are high enough that collisions produce fragmentation rather than accretion.

3. *Radial mixing of asteroid types.* Asteroid thermal models suggest that the outer main belt should contain more "primitive" objects than the more heated/processed inner belt (chapter by McSween et al.). This trend is roughly reproduced in the current orbital distribution of the taxonomic classes, with S-type asteroids dominating the inner belt, C-

type asteroids dominating the central belt, and D-/P-type asteroids dominating the outer main belt. The boundaries between these main taxonomic types, however, are not sharp; some C and D asteroids can be found in the inner main belt, while some S-type asteroids can be found in the outer main belt. While it is difficult to produce this configuration using thermal models alone, it is plausible that some process (or processes) partially mixed the taxonomic types after their formation or that conditions at any one heliocentric distance changed while asteroids were forming.

Petit et al. claim that these characteristics are byproducts of a dynamical removal mechanism associated with the late stages of planet formation. Model results show that Jupiter's formation could have produced sweeping resonances and/or the scattering and excitation of large planetary embryos within the main belt, thereby eliminating most of the bodies from the primordial main belt shortly after Jupiter reached full size (i.e., presumably some 10 m.y. after the formation of planetesimals). This same dynamical excitation event should have scattered some asteroids away from their formation location, leaving the main belt in a condition comparable to its current state.

If this scenario is valid, we can postulate that Vesta is the lone survivor of a large population of similar objects that quickly disappeared from the main belt. Only a fraction of these Vesta-like objects was shattered by impacts before they were scattered out of the main belt, leaving behind an odd assortment of scraps that was then winnowed over 4.5 G.y. of comminution and dynamical evolution. Fortunately, remnants of this lost "Flying Dutchman" population may still exist in the current main belt or in our meteorite records (e.g., basaltic asteroid 1459 Magnya? olivine-rich A-type asteroids? numerous unusual types of stony-iron and iron meteorites? other?). These objects may be compelling targets for future rendezvous and/or sample return missions.

As an aside, we point out that the dynamical excitation of the primordial main belt may have provided Earth with most of its water. Model results by *Morbidelli et al.* (2000) indicate that collisions between large outer main-belt asteroid and Earth are capable of delivering nearly all of Earth's water budget. Comets, on the other hand, probably do not contribute more than 10% (*Morbidelli et al.,* 2000; *Levison et al.,* 2001). These results are consistent with the deuterium/hydrogen (D/H) ratio of terrestrial ocean water, which matches the D/H ratio of outer main-belt material [e.g., carbonaceous chondrites (*Dauphas et al.,* 2000)] but not that of observed comets coming from the Oort Cloud (e.g., *Meier et al.,* 1998). These results suggest that the outer main belt, which has not been extensively studied nor sampled by meteorites, is a fertile ground for new research projects and spacecraft missions over the next several decades.

2.3. Trojan Asteroids

The Trojan asteroids, the second largest reservoir of asteroids in the inner solar system, are located around the L_4 and L_5 Lagrange points of Jupiter. (We use the term "asteroid" here, although the compositional differences between outer-main-belt asteroids, Trojans, and dormant/extinct comets are thought to be subtle. For this reason, terms like "asteroid" or "comet" are ambiguous, such that both could be reasonably applied when these populations are discussed.) Even though the population of Trojans is thought to be only slightly less numerous than the current main belt (see the chapter by Jedicke et al.), the Trojans frequently receive far less attention then their next-door neighbor. The reasons probably have to do with the fact that Trojans are more distant and therefore are harder to observe and they lack the distinctive spectral features seen among many main-belt asteroids. Still, as described in the chapter by Barucci et al., the physical properties of the Trojans are an interesting counterpart to the diverse main-belt asteroids. Their spectra are featureless and reddish, indicating that they are probably covered by organic molecules. Most Trojans that have been examined closely appear to belong to the D-type taxonomic class, while their rotational properties appear to be similar to main-belt objects. The origin and evolution of the Trojans, as described by Marzari et al., are probably linked to the growth and evolution of Jupiter. As the gaseous envelope collapsed and accreted onto Jupiter's core and quickly increased its mass, the libration regions near the L_4 and L_5 Lagrange points would have readily expanded, capturing planetesimals that happened to be wandering near these zones. The more Jupiter increased its mass, the more the libration amplitudes of the captured planetesimals would shrink, forcing some objects into orbits consistent with known Trojans. Eventually, the planetesimals with large libration amplitudes and shorter dynamical lifetimes escaped, leaving behind the Trojan swarms observed today. Because this mechanism had a negligible effect on the eccentricities and inclinations of the captured bodies, it is possible, although unlikely, that the orbital distribution observed among the Trojans today is primordial. A more likely scenario is that some unknown postcapture mechanism produced the high inclinations observed among the known Trojans. Several potential mechanisms capable of increasing inclination values among the Trojan asteroids are discussed in the Marzari et al. chapter.

3. PRESENT STATE OF THE MAIN BELT

As described above, the evolution of the primordial main belt was characterized by numerous impacts, thermal processing, and substantial dynamical upheaval, all which occurred over a relatively short time span (on the order of ~100 m.y.). Once this dramatic epoch ended, however, the evolution of the remaining population occurred more slowly, with the dominant physical processes being collisions, now more infrequent than before, and several dynamical mechanisms capable of modifying the orbits of asteroids over time. To understand the evolution of main-belt asteroids, we need to understand how these processes operate. [We overlook for now the possibility that the main belt was significantly modified by the Late Heavy Bombardment (LHB). The LHB refers to an event ~3.9 G.y. ago where the inner solar system was ravaged by numerous impactors (e.g., *Hartmann et al.,* 2000). Evidence suggests that most lunar basins were formed at this time. Although the LHB's origin and length are unknown, numerical studies have suggested an intriguing possibility: The LHB may have been caused by a sudden dynamical depletion of small bodies in the primordial outer solar system some 600 m.y. after the birth of the solar system (*Levison et al.,* 2001). If this scenario is valid, the LHB would have had several consequences for the main belt: (1) Projectiles from the LHB may have disrupted main-belt asteroids. (2) Gravitational interactions between ejected outer solar system bodies and Jupiter may have caused Jupiter to migrate inward, such that asteroids could have been trapped, excited, or even ejected by sweeping jovian resonances. Even if the LHB scenario described in *Levison et al.* (2001) proves to be incorrect, it is clear that the history of the main belt (and the solar system) is intimately linked to our understanding of the LHB.]

The surface of an asteroid tells the story of what has happened to the body over time. To read that story, we must document an asteroid's morphological, orbital, and spectral properties, interpret how these factors relate to one another, and place them in context. *In situ* exploration and remote sensing observations help us estimate an asteroid's shape, surface spectral signature, size, rotation rate, cratering record, and bulk density. Other asteroid properties (e.g., internal composition, internal structure, chemical history) cannot be thoroughly assessed unless we land on the asteroid or bring back well-documented samples to terrestrial laboratories. Moreover, when interpreting available observations, we should take into account possible effects of space weathering due to long exposure of the surfaces to solar wind, cosmic rays, and micrometeorite impacts (see chapter by Clark et al.).

3.1. Collisions

Collisions are the principle geologic process occurring on asteroids today. Mutual collisions between asteroids have ground down earlier populations, processing their members into smaller and smaller fragments. In addition to being comminuted, asteroids are also scarred by impacts. The nature of the size distribution of the bombarding asteroid population is such that numbers increase strongly as size decreases. For this reason, asteroids are likely to experience numerous cratering events before eventually being disrupted by a more energetic impact. The records left behind by these events yield important information about the target as well as the bombarding population.

Though our understanding of high-velocity impact physics remains incomplete, we have made significant progress over the last decade. Recall that at the time of *Asteroids II,* no asteroid had yet been imaged by spacecraft. For this reason, much of our knowledge about asteroid internal structures was deduced from observations of ancient catastrophic disruption events among large asteroids (i.e., asteroid families) and laboratory impact experiments, where centimeter-sized projectiles were fired into targets at several kilometers per second. The conventional wisdom based on these inferences was that most small asteroids were monolithic shards produced by collisions. Consequently, their reaction to cratering and catastrophic disruption events was thought to be driven by their physical strength. As images of asteroids or asteroid-like objects [e.g., Phobos, (951) Gaspra, (243) Ida, (253) Mathilde, (433) Eros] were analyzed in the 1990s; however, it became apparent that we were missing something important. For example, each of these bodies had sustained a collision energetic enough to produce a multikilometer crater. In fact, in the case of Mathilde, the largest craters were comparable to the dimensions of the asteroid itself! The only way to explain the existence of these large craters was that some unexplored aspects of impact physics were allowing these objects to escape catastrophic disruption.

As described in the chapters by Asphaug et al., and Holsapple et al., these surprising results have been investigated using numerical hydrocodes, which can model the pressures, temperatures, and energies produced by asteroid-sized impacts with reasonable accuracy, and laboratory impact experiments, which can provide groundtruth for the complicated physics occurring when real objects collide at high velocities. Their results indicate that large asteroid craters form in the "gravity-scaling" regime, where the final size of the crater is controlled by the target's gravity. In contrast, small asteroid craters form in the "strength-scaling" regime, where the final size of the crater is governed by the strength of the target's surface. A collision in the gravity regime works in the following way. Imagine a body tens of meters in diameter striking an undamaged rocklike kilometer-sized asteroid. According to hydrocode modeling, the shock front launched by the collision pulverizes material at the impact site, such that crater excavation occurs in virtually strengthless material. Since weak material does not transport energy efficiently, much of the impact energy is deposited near the impact site. In this manner, craters formed in the gravity regime can be significantly larger than suggested by extrapolation of strength-scaling laws. In fact, a powerful enough shockwave may shatter the target asteroid, effectively creating a collection of gravitationally-bound components. This behavior provides insight into why multikilometer asteroids can have such enormous craters.

Until recently, hydrocode modeling of asteroid collisions had concentrated on relatively simple experiments: The target asteroids were assumed to be initially undamaged, the composition of the projectile and target asteroids was assumed to behave like basalt or other common terrestrial rocks, and the porosity of the target material was considered negligible. As our interpretation of asteroid and meteorite data has grown more sophisticated over time, however, it has become clear that these approximations do not address the full range of asteroid configurations. As described in the chapter by Britt et al., asteroid and meteorite density trends suggest that asteroids can be divided into three general groups: (1) asteroids that are essentially solid objects; (2) asteroids with macroporosities near 20% that are probably heavily fractured, and (3) asteroids with macroporosities >30% that may be considered gravitational aggregates. (Gravitational aggregates are commonly referred to in the literature as "rubble piles." Unfortunately, this term has led to several misunderstandings between geologists and numerical modelers, each of whom use the term "rubble pile" to describe a specific asteroid attribute. To avoid future entanglements, Richardson et al.'s chapter proposes new terminology in which a "rubble pile" describes the asteroid structure one might find if a bunch of rocks were dumped off a truck, while the term "gravitational aggregate" describes any object comprised of loosely consolidated material.) Although simulating impact events on model asteroids from the last two groups is challenging, many insights have already been gleaned from recent numerical and experimental data. Descriptions of the current state of the art can be found in chapters by Asphaug et al., Britt et al., and Holsapple et al.

These results led Richardson et al. to propose a new classification scheme for asteroid interiors, one that provides a more useful means for predicting how asteroids should react to short- and long-term stresses. Richardson et al. categorized the spectrum of possible asteroid configurations using two parameters, porosity and relative tensile strength (RTS). Porosity provides some measure of how asteroids react to impact energy. For example, since high-porosity objects are thought to absorb impact energy rather than transmit it, they should be more difficult to fragment/disrupt. On the other hand, RTS can be used to describe the structure of flaws inside an asteroid. Objects with high RTS, characterized as fractured or monolithic, should resist long-term stresses like planetary tidal forces, while objects with low RTS, characterized as highly fractured or shattered, are more susceptible to those same stresses. Using these parameters, it is possible to make some general statements about how different asteroids react to impacts:

1. Collisions on monolithic or moderately fractured asteroids (high RTS, low porosity) produce compressive waves that easily reach the farside of the object. The reflected compressive wave turns into a tensile wave that can produce damage and spalls. This structure may be a good description for <100-m asteroids.

2. Fractured or shattered asteroids (moderate RTS, low porosity) contain significant numbers of faults and joints that help to suppress the tensile wave, such that the object is more difficult to disrupt. This structure may be a good description for some S-type asteroids [e.g., (243) Ida, (433) Eros, and (951) Gaspra].

3. Asteroids with rubble-pile structures or highly porous structures (low RTS, moderate to high porosity) absorb impact energy via compression, with little to no tensile wave developed in the structure. When impact energy is damped, craters may form by compaction. This structure may be a good description for C-type asteroids like (253) Mathilde.

From an evolutionary standpoint, we expect collisions to eventually fracture or shatter most asteroids, ultimately turning them into gravitational aggregates. Observational evidence and numerical results described by several chapters (Richardson et al., Asphaug et al., Pravec et al., Britt et al., and Merline et al.) suggest that such structures may persist down to objects only a few hundred meters in diameter:

Asteroid spins. As described in the chapter by Pravec et al., the rotation rates of observed kilometer-sized and smaller asteroids indicate that relatively few are spinning fast enough to be in a state of tension. Moreover, the rotation period distribution of $D > 0.15$ km asteroids abruptly truncates at $P \gtrsim 2$ h, which is where gravitational aggregates of typical asteroidal density would begin to fly apart from centrifugal forces. Since solid objects should conceivably be able to spin at nearly any rate, the data suggests that most $D > 0.15$ km bodies lack tensile strength.

Binary asteroids in near-Earth space. Doublet craters, created by the nearly simultaneous impact of objects of comparable size, have been found on all the terrestrial planets. These craters were almost certainly formed by the impact of binary asteroids (see also the chapter by Merline et al.). To explain the quantity of observed doublets (i.e., ~10% on Earth), one must infer a steady-state population of ~15–20% binary asteroids in near-Earth space (see chapters by Pravec et al. and Merline et al.). A plausible way of producing such a large quantity of binaries in that region is to have gravitational aggregates undergo tidal disruption when passing close to a terrestrial planet, with some of the shed fragments entering orbit around the remnant asteroid. If this scenario is valid, we can infer that many near-Earth asteroids are virtually strengthless.

Asteroid densities. As described in the chapter by Britt et al., many S- and C-type asteroids have higher porosities (≳20%) that the average porosity values derived from various ordinary and carbonaceous meteorites (~10%). These results suggest many asteroids possess substantial macroporosity. The nature of this macroporosity (e.g., cracks between adjacent blocks? void spaces?) is an open question. Additional information about asteroid densities can be found in the chapter by Hilton.

Giant craters. Nearly every observed asteroid (e.g., Ida, Mathilde, Vesta, Eros) contains an impact crater on its surface that is comparable to the average radius of the body itself (see chapters by Chapman and by Sullivan et al.). The fact that these crater-forming impacts did not disrupt the target body implies the interior is highly fractured.

Studies of asteroid collisions via hydrocode and laboratory experiments can also be used to understand collisional evolution in the main belt. As reported in the chapters by Jedicke et al. and Davis et al., the cumulative size frequency distribution (SFD) of observed main-belt asteroids is wavy, with "bumps" at diameter D ~ 3–4 km and 100 km. These waves were something of a surprise, because *Dohnanyi* (1969) predicted that an asteroid population in collisional equilibrium should eventually evolve to a SFD with a cumulative power law slope index of –2.5. Dohnanyi's model, however, also made a number of assumptions: (1) His model asteroids were spherical and had equal densities, (2) the response of his model asteroids to impacts was size-independent, and (3) his model population had no lower cutoff in mass. In the real main belt, none of these assumptions can be considered valid: Asteroids have a variety of different collisional outcomes depending on their physical properties and the target/impactor size, while small bodies escape the main belt via Poynting-Robertson and Yarkovsky forces (see chapters by Dermott et al. and Bottke et al.). An important factor in negating Dohnanyi's assumption of self-similarity in the collision process is the transition between strength- and gravity-dominated scaling regimes that occurs near 100–300 m. As summarized in the Davis et al. chapter, gravity tends to make larger asteroids more difficult to disrupt. Ultimately, this perturbation introduces a wavelike effect into the main belt's SFD that produces the bumps observed at D ~ 3–4 km and 100 km.

3.2. Collisional Outcomes: Ejecta, Satellites, and Families

As described in the chapter by Scheeres et al., one common side effect of asteroid collisions is the production of ejecta and regolith. Data from asteroid polarimetric observations (e.g., *Dollfus et al.,* 1989) have suggested for some time that many large asteroids have regoliths. Interestingly, regolith has also been directly observed on asteroids Gaspra, Ida, Mathilde, and Eros as well as the martian moons Phobos and Deimos. The inferred existence of deep regolith (~10–100 m) on bodies with low escape velocities was unexpected, mainly because laboratory impact experiments had predicted that small rocky asteroids should lose nearly 100% of their ejecta in any given impact event. More recent modeling efforts have shown that the excavation phase of crater formation on a rocky asteroid is dominated by low velocities, such that a significant fraction of a crater's ejecta may remain gravitationally bound to the asteroid. Stranger results are found with porous media; laboratory impact experiments suggest that collisions into asteroids with this internal structure may produce craters primarily by compaction rather than excavation and that little ejecta is produced (see chapters by Britt et al., Asphaug et al., Richardson et al., and Holsapple et al.). These distinctive physical properties may explain why Mathilde's large craters, whose boundaries are adjacent to one another, somehow managed to avoid disturbing one another during their formation. Clearly our insights into these types of bodies will remain limited until we obtain more *in situ* observations of highly porous asteroids.

The fate of low-velocity crater ejecta launched from an asteroid depends on several factors: the asteroid's size, shape, density distribution, and rotation; the trajectory and velocity of the ejecta; and the launch location. Because irregularly shaped asteroids have irregular gravitational fields, orbital dynamics near the asteroid can be surprisingly complex. For this reason, numerical modeling is typically needed to interpret an asteroid's regolith or boulder distribution. Interestingly, these simulations can often be used to identify recent large impacts. For example, the subtle color/albedo variations observed across Ida's surface appear to be associated with the relatively fresh Azzurra Crater, while most of the house-sized boulders observed on Eros appear to have been ejected from the relatively fresh Shoemaker Regio Crater (see chapters by Scheeres et al., Sullivan et al., and Chapman).

Asteroid collisions may also produce asteroid satellites. As described in the chapter by Merline et al., asteroid satellites have now been observed around many main-belt asteroids, NEOs, transneptunian objects, and at least one Trojan asteroid. Several methods have been used to discover these binaries, including direct imaging (e.g., groundbased adaptive optics, HST, spacecraft), delay-Doppler radar techniques (see chapter by Ostro et al.), and sophisticated studies of lightcurve data. Most of the asteroid satellites discovered so far are small compared with their primary, although a few [e.g., (90) Antiope, (617) Patroclus] can be considered double asteroids. Interestingly, the majority of asteroids with satellites appear to be primitive in nature. Assuming this is not a selection effect, it would suggest that the physical properties of primitive asteroids (and their reaction to impacts) play a critical role in satellite formation. More data will be needed to draw a definitive conclusion.

Several possible mechanisms for forming asteroid satellites are discussed by Merline et al. and Paolicchi et al. Although planetary tidal disruptions may be responsible for creating many of the binaries found in near-Earth space (see chapter by Richardson et al.), the most likely scenario for producing binaries in non-planet-crossing populations involves collisions. There appear to be several ways that collisions produce asteroid satellites: (1) Bodies ejected from the impact site with similar trajectories may go into orbit around one another, and (2) gravitational perturbations or collisions between ejected fragments may change their trajectories enough for some fragments to enter into stable orbits around the primary (or each other). Tidal torques between the primary and secondary may also play a role, either by stabilizing orbits or by causing the secondary to crash back into the primary.

Another byproduct of asteroid collisions is asteroid families. As described in the chapter by Zappalà et al., a family is produced when a large asteroid undergoes a catastrophic disruption, leaving behind numerous fragments with similar proper semimajor axes a, eccentricities e, and inclinations i. Proper elements are of critical use in identifying families, since they are quasi-integrals of motion (i.e., they are nearly constant with time). The chapter by Knežević et al. discusses how new synthetic theories allow proper elements (and their errors) to be computed more precisely than previous methods. The means of identifying asteroid families using proper elements is discussed in the chapter by Bendjoya and Zappalà. The two most common methods used to determine family membership are (1) the hierarchical clustering method (HCM), which requires that a given asteroid be located within a given velocity difference, or cutoff velocity, to a neighboring asteroid in proper (a, e, i); and (2) the wavelet analysis method (WAM), which determines asteroid density concentrations within a proper (a, e, i) distribution. Once a family is recognized, the ejection velocities of the family members can be computed using the (a, e, i) differences between the inferred orbit of the original parent body and the current orbits of the family members. Note that this technique assumes that the orbits of the family members (particularly the semimajor axes) have been essentially constant since the family was created. As discussed in the chapters by Nesvorný et al. and Bottke et al., however, nearly all prominent asteroid families may have been spread in (a, e, i) since their formation via resonances and the Yarkovsky effect.

Zappalà et al. also discuss the size-frequency distributions of observed asteroid family members (typically with D > 5–10 km), which tend to have power-law indexes that are steeper than those measured for nonfamily asteroids. If this trend continues to smaller asteroids, it would suggest that families dominate the main-belt population at small sizes. The size-velocity trends measured from asteroid family fragments suggest that ejection velocities for multikilometer asteroids could be several 100 m s^{-1}. If true, family breakup events could potentially flood nearby resonances with fragments, possibly enough to produce an "asteroid" shower on Earth. Observations of asteroid family members, described by Cellino et al., suggest that most family members share similar, but not identical, spectral signatures. While this helps to verify the asteroid identification methods described by Bendjoya and Zappalà, it also suggests that no observed families were derived from the catastrophic disruption of a differentiated object (i.e., no asteroid association looks like it came from the core, mantle, and crust of a Vesta-like body). For this reason, spectral signatures may be used in the future to expand the membership of asteroid families by identifying those members that have evolved away from the family cluster over time.

3.3. Asteroid Geology

With this rudimentary understanding of how asteroids react to collision events, we can examine the geology of the four asteroids visited by spacecraft: Gaspra, Ida, Mathilde, and Eros. As described in the chapters by Sullivan et al., Chapman, Cheng, and Farquhar et al., all these bodies have shapes and surface morphologies shaped by collisions. We briefly review their characteristics below:

(951) Gaspra, an S-type member of the Flora family, has an elongated shape (18.2 × 10.5 × 8.9 km) characterized by broad, flat facets and shallow concavities several kilometers across. It has been suggested that some of these concavities may be ancient craters or spalls. The depth of Gaspra's regolith is at least a few meters, though some studies suggest it could also be significantly deeper. Observed grooves on Gaspra's surface suggest the asteroid may have originated as a single collisional fragment that was fractured or shattered by one or more collisions. The albedo, color, and photometric properties across Gaspra's surface are similar. The size-frequency distribution of Gaspra's craters is surprisingly steep (differential power law slope index of –4.3) and below saturation, such that Gaspra's surface may reveal the shape of the production population's size distribution.

(243) Ida, an S-type member of the Koronis family, is elongated (29.9 × 12.7 × 9.3 km) and has an irregular surface covered by impact craters as large as 12 km. Ida appears to be billions of years old, with its surface saturated in D < 1 km craters and several large craters showing signs of significant degradation. Observations of boulders, shallow mass-wasting features, grooves, and infilled craters all suggest that Ida possesses a substantial regolith, perhaps 50–100 m deep. The grooves themselves, some 4 km long, suggest that collisions have fractured or shattered the asteroid. Ejecta blocks can be found in several locations, with their distribution suggesting they are impact products derived directly from the largest craters or that they were swept up after an impact event by the leading rotational edge of Ida. Ida also has a satellite named Dactyl that has an average radius of 0.7 km. Color data and photometric modeling suggest Ida/Dactyl have similar compositions and surface textures.

(253) Mathilde, a C-type asteroid, was encountered by the *NEAR Shoemaker* spacecraft. Although only a fraction of Mathilde's surface was imaged, a best-fit ellipsoid suggests its shape is 66 × 48 × 46 km. Mathilde's surface is dominated by several large craters with diameters that exceed the asteroid's average radius. Despite this fact, there is no evidence that these craters ever interfered with one another; no ejecta blankets, ejecta blocks, or grooves have been observed. The density of Mathilde is ~1.3 g cm^{-3}, low enough to imply that this body may be highly porous (although it is not known whether it is macroporosity or microporosity). No spectral or albedo contrasts can be seen.

(433) Eros is an S-type near-Earth asteroid that was the final destination of the *NEAR Shoemaker* spacecraft. It has a curved shape and a length of 34 km. While the largest craters on Eros are Himeros (9 km), Shoemaker (7.6 km), and Psyche (5.3 km), most of the house-sized blocks observed on Eros's surface appear to have been produced by the Shoemaker impact. The entire surface is covered by regolith, although its thickness is uncertain. An examination of Eros' surface morphology and gravitational field suggests the asteroid was once a single ejecta fragment that was fractured or shattered by subsequent impacts. One of the most curious features on Eros is its ponded deposits, flat regions that cover the bottoms of some depressions. These deposits are concentrated within 10° of the equator. The mechanism that produced the ponded deposits is unknown, although seismic shaking and electrostatic levitation are the leading candidates. Most of Eros' impact craters were produced while it was a member of the main belt, since (1) Eros has only recently evolved out of the main belt, and (2) the impact flux on main-belt asteroids (or those crossing into the main belt) is several orders of magnitude higher than

impact rates on the Moon or those near-Earth asteroids collisionally decoupled from the main belt. The size distribution of Eros' D > 100 m craters is similar to Ida, while there is a notable deficiency of D < 100 m craters. The reason why Eros has so few small craters is unknown.

We believe this list could be readily expanded to include (4) Vesta, which has been closely examined both by HST observations and laboratory studies of HED meteorites (Dotto et al., Keil, Burbine et al., Scott), and several near-Earth asteroids extensively explored by radar (Ostro et al.). The power of various remote sensing techniques has reached the point that we are now capable of doing many things that were once the purview of spacecraft alone.

3.4. Dynamical Evolution of Main-Belt Asteroids

Since the publication of *Asteroids II*, our understanding of the dynamical evolution of asteroids has undergone significant advances. Propelled by the advent of symplectic integration algorithms and inexpensive but powerful workstations, numerical models are now capable of tracking the often chaotic orbital paths taken by test bodies for tens to hundreds of millions of years while including all gravitational perturbations produced by the planets. These timescales, which are orders of magnitude longer than the best integrations available a decade ago, have allowed us to develop a much deeper comprehension for how main-belt asteroids are transported to escape hatches that can take them out of the main belt and into the inner solar system.

Our advances in this field have come in two main areas:

1. *Effects of mean-motion and secular resonances.* As described in the chapters by Nesvorný et al. and Morbidelli et al., tremendous progress has been made in our understanding of how resonances modify the eccentricities and inclinations of main-belt and planet-crossing asteroids. It has been shown that test bodies entering several powerful mean-motion resonances with Jupiter (e.g., 3:1, 4:1, 5:2) can have their eccentricities pumped up to Earth-crossing values, usually over timescales of ~1 m.y. In some cases, orbital motion inside these resonances is chaotic enough that test bodies can be pushed directly onto Sun-grazing orbits. Similarly, the ν_6 secular resonance lying along the inner edge of the main belt is now seen as one of the primary sources of near-Earth objects (NEOs). Less dramatic but also important are the narrow mean-motion resonances produced by Mars and Jupiter and the three-body resonances produced by Jupiter and Saturn. These weaker but far more numerous resonances crisscross the main belt, such that most asteroids do not have to travel very far to interact with them. In some cases, these resonances can push main-belt asteroids onto Mars-crossing orbits, although over longer timescales than those described above (e.g., 10^7 to 10^9 yr).

2. *Effects of Yarkovsky thermal forces.* Nongravitational forces have been shown to play an important role in allowing asteroidal material to escape the main belt. As described in the chapter by Bottke et al., small bodies orbiting the Sun absorb sunlight, heat up, and reradiate the thermal energy after a short delay produced by thermal inertia. This emission, while tiny, produces a force that can lead to secular changes in the object's semimajor axis. This so-called "Yarkovsky effect" compels 0.1-m to 20-km bodies to slowly spiral inward or outward as a function of their spin, orbit, and material properties. A variant of this force called YORP (Yarkovsky-O'Keefe-Radzievskii-Paddack) is also capable of modifying the spin rates of asteroids.

Prior to these advances, it was generally believed that most NEOs and meteorites escaped the main belt by being directly injected by collisions into one of several powerful resonances (e.g., the ν_6 secular resonance or the 3:1 mean-motion resonance with Jupiter). As described by Bottke et al., however, this scenario is inconsistent with observations, numerical simulations of catastrophic collisions, and the cosmic-ray exposure ages of meteorites. A better scenario for how asteroids and meteoroids are delivered from their parent bodies in the main belt to the inner solar system (and Earth) is the following: (1) An asteroid undergoes a catastrophic disruption or cratering event and ejects numerous fragments; most are not directly injected into a resonance. (2) D < 20 km fragments start drifting in semimajor axis under the Yarkovsky effect. (3) These bodies jump over or become trapped in chaotic mean-motion and secular resonances that change their eccentricity and/or inclination. (4) Asteroids drifting far enough may fall into mean-motion or secular resonances capable of pushing them onto planet-crossing orbits. From here, they become members of the Mars-crossing and/or NEO populations.

At smaller size scales, Dermott et al. describes how interplanetary dust particles (IDPs) produced by asteroid collisions drift out of the main belt via Poynting-Robertson drag, a radiation effect that causes small objects to spiral inward as they absorb energy and momentum streaming radially outward from the Sun and then reradiate this energy isotropically in their own reference frame. Millions of metric tons of dust are delivered every year to Earth by this process. The interaction between these drifting IDPs and resonances can produce interesting effects. For example, both IRAS and COBE data have detected an Earth-resonant dust ring. It is produced when asteroid dust spiraling toward Earth becomes temporarily trapped in a co-rotation resonance near 1 AU. Eventually, the eccentricities of the dust particles get high enough to escape the resonance, allowing some to strike Earth.

4. THE NEAR-EARTH-OBJECT POPULATION

The near-Earth object (NEO) population, including both asteroids and active/extinct comets, are defined as those bodies having perihelion distances $q \leq 1.3$ AU and aphelion distances $Q \geq 0.983$ AU. Subcategories of the NEO population include the Apollos ($a \geq 1.0$ AU, $q \leq 1.0167$ AU) and Atens ($a < 1.0$ AU, $Q \geq 0.983$ AU), which are on Earth-crossing orbits, and the Amors (1.0167 AU $< q \leq 1.3$ AU), which are on nearly-Earth-crossing orbits. A population inside Earth's orbit (Q < 0.983 AU) is also expected to exist (see Morbidelli et al.). Evidence from the lunar cratering record suggests that the NEO population has been in steady-state for roughly the last 3 G.y. and has been comprised of bodies ranging in size from dust-sized fragments to objects tens of kilometers in diameter. Even so, temporary NEO showers following energetic collisional events in the main belt cannot be ruled out (Zappalà et al.).

Most NEOs are believed to be fragments of main-belt asteroids that, following ejection in a collision event involving a larger asteroid millions of years ago, evolved via resonances and nongravitational forces until reaching an Earth-approaching orbit. The rest are thought to be ejected members of comet reservoirs in the outer solar system, namely the Kuiper Belt and the Oort Cloud. In much the same way that rocks and sediments in a riverbed yield information on the types of material found upstream, the NEOs (and meteorites) can reveal a great deal about the nature of the bodies found in all asteroid and comet reservoirs. The majority of attention given to NEOs, however, is unrelated to their intrinsic scientific value; instead, it concerns the fact that a collision between a multikilometer asteroid/comet and Earth could potentially wreak regional-to-global devastation on our biosphere. For this reason, it is important that we understand the NEO population and the potential threat it represents to humanity.

4.1. Detecting Near-Earth Objects

The search for NEOs is described in the chapter by Stokes et al. Over the last decade, an increasing number of dedicated surveys (e.g., Spacewatch, LINEAR, LONEOS, Catalina Sky Survey, NEAT) have come on line to scan the skies for NEOs. In that time, the number of known NEOs has jumped from roughly 100 to well over 1000. At present, it is believed that over 50% of the objects with absolute magnitude H < 18 (i.e., roughly D > 1 km objects) have been discovered out to a semimajor axis a < 7.4 AU. This significant rise in productivity stems from several factors: a political recognition of the potential threat to Earth from NEOs, which in turn led to greater resources and more dedicated telescopes; a switch from visual detection via photographic searches to more sophisticated automated searches via sensitive CCD detectors; and an increase in available computing power. The immediate goal of these search programs is to find 90% of the kilometer-sized NEOs by 2008 (i.e., it is thought that an Earth impactor with a diameter larger than about 1 km could potentially produce global catastrophic consequences for human life on Earth; see chapter by Morrison et al.). Reaching that desired level of completeness, however, may be complicated by the fact that some NEOs are more difficult to detect than others.

Once a NEO has been detected, it is important to determine its orbit as precisely as possible and then compute whether the body has any chance of striking Earth in the near future. Chapters by Bowell et al. and Milani et al. discuss in detail the technical side of these issues.

4.2. Modeling the Near-Earth-Object Population

The remarkable progress made in finding NEOs over the last decade has also been accompanied by substantial numerical and theoretical work. Together, these advances give us a much more profound understanding of the orbital and size distribution of the NEO population than we had at the time of *Asteroids II*. In particular, advances in two areas have allowed us to compensate for a paucity of direct information on the nature of the NEO population:

1. *Observational biases.* Every NEO survey is limited by several factors: the orbits, sizes, and albedos of the bodies; the capabilities of the detector (e.g., limiting magnitude, degree of sky coverage); the software applied to sift asteroids and comets from the background; and factors related to the physical location of the site (e.g., weather, distance from city lights). The effects of these components on asteroid and comet detection are known as observational bias. As explained in the chapter by Jedicke et al., once these biases are understood and mathematically modeled, it is possible to take an observed population sampled by various surveys and deduce the properties of the actual population, even though many of the objects in that population remain undetected. Jedicke et al. describe the bias-corrected absolute magnitude distribution of several small-body populations: NEOs, main belt, Trojans, Centaurs, and the transneptunian objects.

2. *Dynamical pathways taken by near-Earth objects.* To determine the orbital distribution of the NEOs, it is important to understand how NEOs travel from their source regions to the observed orbits and their ultimate demise (i.e., elimination by striking the Sun, a planet, or being ejected out of the solar system). This type of modeling can be challenging, particularly because NEOs come from numerous regions inside (or adjacent to) sources like the main asteroid belt, transneptunian region, and/or Oort Cloud. As described in chapters by Morbidelli et al. and Weissman et al., tracking the evolution of thousands of test bodies from these source regions via numerical integration has allowed us to put together a picture of where test bodies from these regions are statistically most likely to spend their time.

By combining these components together, Morbidelli et al. and Weissman et al. explain how the evolution of asteroids and comets respectively can be used to compute a reasonably accurate model of the debiased orbital and absolute magnitude distribution of the NEO population as well as a measure of the relative importance of each distinctive NEO source region.

It is also possible to derive information on the NEO populations from the crater size-frequency distribution found on the terrestrial planets. As described in the chapter by Ivanov et al., the "wavy" shape of the crater size-frequency distributions found on the Moon and Mars are similar to the size-frequency distributions inferred for the main-belt population. This relationship suggests that (1) the planet-crossing object population may be dominated by asteroids (unless cometary impactors have a comparable size distribution to that found in the main belt) and (2) the planet-crossing objects are mainly replenished by bodies diffusing out of a main-belt population in collisional equilibrium. Using crater-scaling laws and estimates of the physical parameters and crater rates on the Moon and Mars, Ivanov et al. "back out" the shape of the planet-crossing asteroid population. To reasonable accuracy, their results appear to be consistent with estimates provided by more direct observational methods.

4.3. Near-Earth-Object Shapes

In the past decade radar studies of asteroids have provided accurate information on NEO shapes, surface properties, and rotation rates/states. The observational methods and results of radar astronomy are discussed in the chapter by Ostro et al. Most of the asteroids investigated in detail by radar are NEOs, primarily because the echo's signal-to noise ratio goes as $(\text{distance})^{-4}$ (i.e., as the object approaches Earth, the radar return gets significantly stronger). If the target echoes are strong enough to get good resolution in both time delay (range) and Doppler frequency (radial velocity), it is possible to construct an accurate three-dimensional model of the object's shape as well as its precise spin state. These results indicate that asteroids come in all shapes and sizes, from featureless spherical balls to irregularly shaped objects with craggy, bumpy, and/or cratered surfaces. One main-belt object, (216) Kleopatra, even has the shape of a dog bone! Radar studies also show that some NEOs are in complex rotation states [e.g., (4179) Toutatis] and that a few have satellites (see chapters by Pravec et al. and Merline et al. for more information on these topics).

Additional information on asteroid shapes can be mined from photometric data. As described in the chapter by Kaasalainen et al., new lightcurve inversion techniques are making it possible to determine unique solutions for convex ("hull") shapes. These shape models mimic asteroid silhouettes in three dimensions, such that they provide reasonable facsimiles of asteroids like (951) Gaspra and (433) Eros. The future of this technique is very promising, particularly since it might be used to analyze nearly any asteroid that can be modeled as a triaxial ellipsoid.

4.4. Linking Meteorites and Near-Earth Objects to their Parent Bodies

An important goal of asteroid science is to link meteorites and their immediate precursors back to their parent bodies. To accomplish this task, we need to combine data from several different disciplines: dynamical modeling, spectroscopic observations, petrology, and mineralogy. It is difficult work, but enough progress has been made over the last decade that some positive connections can be described.

Spectroscopic observations tell us about composition of material on an asteroid's surface. Over the last decade, asteroid spectra have become an increasingly rich source of information on asteroid physical properties, particularly after CCD technology was incorporated into spectrographs. Understanding the detailed mineralogy derived from spectroscopic measurements has been a long-term goal and current progress and challenges are described herein by Gaffey et al. As reviewed in the chapter by Bus et al., survey results now suggest that the main-belt population contains some 26 different taxonomic types. Most of these taxonomic types can also be found within the NEO population. As discussed by Binzel et al., the NEO population appears to be, more or less, a representative sample of the taxonomic types found in the inner and central main-belt populations. These include V-type asteroids, which are thought to be derived from (4) Vesta, and E-type asteroids, which may be predominantly derived from the Hungaria asteroid region (i.e., the Hungarias are non-Mars-crossing asteroids with $1.77 < a < 2.06$ AU and $i > 15°$).

At present, roughly 80% of the NEOs with known taxonomic types are bright, S-type (or Q-type) bodies. As Binzel et al. point out, however, this value should be used carefully because dark C-type asteroids have lower albedos than S-types and thus are less numerous for a given absolute magnitude than their S-type counterparts. The fraction of observed NEOs with low albedos gradually increases as you move to larger semimajor axes, reflecting a comparable albedo gradient seen among the main-belt asteroids. NEOs residing in the Jupiter-family comet region are consistent with this trend; many are thought to be extinct comets. As discussed by both Weissman et al. and Binzel et al., extinct comet candidates generally have featureless spectra with flat to slightly increasing red slopes spanning the dynamic range between C- to D-type asteroids.

One of the major problems remaining since *Asteroids II* has to do with the mysterious source of the ordinary chondrite parent bodies. It has been suggested that ordinary chondrites, ordinary chondrite-like NEOs, and S-type asteroids are genetically related, primarily because they all appear to have similar spectral features. Evidence supporting this idea comes from observational work described in Binzel et al., where the spectra of several NEOs show a clear transition between S-type asteroid and ordinary chondrite-like spectra, and from the spectral features of (243) Ida, where ejecta associated with the relatively young Azzurra Crater appears to have spectral features that trend toward those of ordinary chondrites (see Scheeres et al., Sullivan et al., and Chapman). Still, the mechanism producing this transition has remained enigmatic. As described by Clark et al., a possible solution to this issue could be the surface modification processes commonly known as "space weathering." Space weathering is defined as any process that changes the spectroscopic properties on airless bodies. Processes that produce space weathering are impacts, solar-wind ion implantation, sputtering, and micrometeorite bombardment. Based on an analysis of lunar samples that show clear signs of space weathering, it has been suggested that space-weathering processes observed on asteroids may be caused by the deposition of condensates bearing submicroscopic Fe onto grain surfaces. The condensates are produced when the target material is vaporized, presumably by solar-wind sputtering and micrometeorite bombardment. For this process to work effectively, the target material must contain an abundance of Fe. Thus, asteroids with olivine-rich surfaces, like ordinary chondrites and S-type asteroids, are more likely to undergo optical maturation processes than those rich in pyroxene, like the HED meteorites or the basaltic crust of (4) Vesta. Other important factors may include (1) the impact energies of those micrometeorites striking the target, (2) the age of the surface, and/or (3) the ability of the object to retain regolith.

Finally, the chapter by Burbine et al. tries to put together the big picture and discuss whether various meteorite classes can be linked to specific parent objects in the main belt. For example, Burbine et al. presents the meteoritic and spectroscopic evidence (pro and con) that (1) ordinary chondrites are linked to the S-asteroids [i.e., specifically, the S(IV)-type asteroids], (2) CM chondrites were produced by the C-type asteroids, (3) iron meteorites and enstatite chondrites come from M-type asteroids, and (4) the HED meteorites come from (4) Vesta. Although several of the matches are compelling, Burbine et al. admit that most of the postulated parent bodies (with the possible exception of Vesta) do not have enough singular spectral features to conclusively link them with a meteorite class. In many cases, these issues may not be fully settled until spacecraft missions can return main-belt asteroid samples to Earth.

4.5. Impact Hazard

While many now recognize that the threat to life on Earth from impacting NEOs is real and important, it is less clear what we should do about it. From a policy perspective, one would expect that the resources devoted to this issue should be correlated with the perceived level of risk. Unfortunately, humanity has a mixed track record in dealing with the sometimes severe consequences that come solely from low-probability events (e.g., communities are often allowed to build homes on flood plains). For these reasons, the chapter by Morrison et al. reexamines the impact hazard issue and attempts to guide a path through the complicated policy issues that lie in front of us. Current estimates suggest that impacts capable of producing a global ecological catastrophe occur roughly twice per million years. The reaction to this perceived threat over the last decade has been remarkable; through hard work and persistence, scientists and issue advocates appear to have placed the impact hazard issue onto the radar screens of politicians from around the world. The outcome of this debate is still unclear, with budgetary constraints and competing priorities within the scientific community threatening to slow or even stop progress. Hopefully, by the time *Asteroids IV* is written, a much more complete picture of the asteroid and comet hazard issue will have emerged, enabling scientists, governments, and the public to deal with it effectively through international cooperation.

5. PREDICTING THE FUTURE

As the great baseball philosopher Yogi Berra reportedly said, "The hardest thing to predict is the future." With that in mind, we speculate on a few of the issues that might show up in *Asteroids IV*.

We believe it is likely that data from the least-studied portion of the main belt, namely the outer main belt and the Trojan populations, may become a major research goal as the search for the source of Earth's water budget becomes more extensive. The connection (or lack thereof) between these regions and the volatile content of the other terrestrial planets, particularly Mars, will also increase in importance. Several additional topics should also continue to stimulate inquiry among dynamical modelers, observers, and meteoriticists alike: (1) What can the current configuration of the main belt tell us about the processes that shaped planet formation? (2) What can the observed population of asteroids and/or meteorites tell us about the primordial asteroids that disappeared early in solar system history? (3) Was the main belt significantly shaped by the events producing the Late Heavy Bombardment, and/or was it an important source of impactors? (4) How have stochastic events (e.g., the breakup of a D > 100 km asteroid) changed the flux of material reaching Earth over time? (5) How do the physical properties of asteroids vary as a function of taxonomic type and how do they affect geologic processes like collisions? (6) What are the similarities and differences between outer main-belt asteroids, Trojan asteroids, and extinct comets? (7) How has the impactor flux of asteroids and comets varied with time? (8) How have asteroid and comet impacts affected the evolution of life on all the planets of the solar system?

From a numerical modeling perspective, we expect that faster computers and advanced codes will allow us to accurately simulate collisions between porous objects, such that impacts onto asteroids like Mathilde can be readily understood. They may also allow us to simultaneously track the collisional and dynamical evolution of millions of interacting bodies (with all appropriate physics included). These advances will eventually allow us to model the formation of the solar system (and main belt) without making too many oversimplifications. We also expect advances in several areas, including (but not limited to) such fundamental physics as the scattering of light by solid surfaces (see Muinonen et al.), the effect of nongravitational forces on the dynamical evolution of asteroids, and the inferred equilibrium shape of asteroids only partially dominated by gravity.

Our understanding of asteroid regolith properties may also be on the cusp of some important advances. Recent radiometric studies indicate that several small NEOs might exhibit radiometric-derived albedos that are unexpectedly high for their taxonomic class. These results, if confirmed, would suggest that the thermal and physical properties of asteroid surface regolith layers change as a function of asteroid diameter, possibly enough to have significant implications for the dynamical evolution of small bodies (e.g., Bottke et al.) and for our understanding of how asteroids scatter light (e.g., Muinonen et al.).

Spectroscopic observations of asteroids will be increasingly extended to near-infrared wavelengths owing to the development of increasingly sensitive detectors and a growing cadre of larger-aperture telescopes. These advances in capabilities will allow asteroids at smaller sizes to be observed and allow increasing sophistication in the interpretation of their mineralogic and petrologic properties. With increasing knowledge of these properties, for example within asteroid families, we can look forward to advances in our revelation of the properties of asteroid interiors. Undoubtedly, these advances will also challenge our understanding of the early solar system evolutionary processes that are recorded within these remnant planetesimals.

Finally, we anticipate that data provided by the upcoming asteroid missions described in the chapter by Farquhar et al. will overturn at least some of the ideas proposed in *Asteroids III*. It is even possible that the upcoming *Dawn* mission to Ceres and Vesta will help to reestablish these objects as planetary worlds (e.g., *Stern and Levison,* 2002). Note that the chapter by Foderà Serio et al. describes how Ceres shortly held planetary status until it was demoted for unspecified reasons. *Asteroids IV* may also contain data from the first asteroid sample return mission. If so, we should be able to make our first direct links between several meteorite classes and their parent bodies in the main belt. Regardless, our understanding of asteroid physical properties will continue to grow by leaps and bounds as a consequence of both groundbased and *in situ* observations.

Acknowledgments. The authors thank C. Chapman and M. Sykes for careful and constructive reviews of this manuscript. We also thank E. Asphaug, D. Durda, R. Jedicke, H. Levison, A. Morbidelli, D. Nesvorný, and D. Richardson for helpful comments and suggestions. W. Bottke acknowledges support for this work from NASA's Origins of Solar Systems, Planetary Geology and Geophysics, and Near-Earth Object Observations programs. P. Paolicchi acknowledges support from the Italian Education Ministry, through the National Research Program (COFIN 1998–2001). R. Binzel acknowledges support for this work from the NASA Planetary Astronomy and NSF Solar System Astronomy programs.

REFERENCES

Alvarez L. W., Alvarez W., Asaro F., and Michel H. V. (1980) Extraterrestrial cause for the Cretaceous-Tertiary extinction. *Science, 208,* 1095–1108.

Dauphas N., Robert F., and Marty B. (2000) The late asteroidal and cometary bombardment of Earth as recorded in water deuterium to protium ratio. *Icarus, 148,* 508–512.

Dohnanyi J. S. (1969) Collisional model of asteroids and their debris. *J. Geophys. Res., 74,* 2531–2554.

Dollfus A., Wolff M., Geake J. E., Dougherty L. M., and Lupishko D. F. (1989) Photopolarimetry of asteroids. In *Asteroids II* (R. P. Binzel et al., eds.), pp. 594–616. Univ. of Arizona, Tucson.

Gehrels T. (1971) Asteroid masses and densities. In *Physical Studies of Minor Planets* (T. Gehrels, ed.), p. vii. NASA SP-267, Washington, DC.

Hartmann W. K., Ryder G., Dones L., and Grinspoon D. (2000) The time-dependent intense bombardment of the primordial Earth/Moon system. In *Origin of the Earth and the Moon* (R. M. Canup and K. Righter, eds.), pp. 805–826. Univ. of Arizona, Tucson.

Levison H. F., Dones L., Chapman C. R., Stern S. A., Duncan M. J., and Zahnle K. (2001) Could the lunar "Late Heavy Bombardment" have been triggered by the formation of Uranus and Neptune? *Icarus, 151,* 286–306.

Marcy G. W., Cochran W. D., and Mayor M. (2000) Extrasolar planets around main-sequence stars. In *Protostars and Planets IV* (V. Mannings et al., eds.), pp. 1285–1311. Univ. of Arizona, Tucson.

Meier R., Owen T. C., Matthews H. E., Jewitt D. C., Bockelee-Morvan D., Biver N., Crovisier J., and Gautier D. (1998) A determination of the HDO/H_2O ratio in comet C/1995 O1 (Hale-Bopp). *Science, 279,* 842.

Morbidelli A., Chambers J., Lunine J. I., Petit J.-M., Robert F., Valsecchi G. B., and Cyr K. E. (2000) Source regions and time scales for the delivery of water to Earth. *Meteoritics & Planet. Sci., 35,* 1309–1320.

Safronov V. S. (1969) *Evolution of the Protoplanetary Cloud and Formation of the Earth and Planets.* Nauka, Moscow.

Stern S. A. and Levison H. F. (2002) Regarding the criteria for planethood and proposed classification schemes. *IAU Proceedings,* in press.

Sykes M. and 30 colleagues (2002) Exploring main belt asteroids. In *The Future of Solar System Exploration (2003–2013)* (M. Sykes, ed.), in press. ASP Conference Series 272.

Weidenschilling S. J. (2000) Formation of planetesimals and accretion of the terrestrial planets. *Space Sci. Rev., 92,* 295–310.

Giuseppe Piazzi and the Discovery of Ceres

G. Foderà Serio
Universita' di Palermo

A. Manara
Osservatorio Astronomico di Brera

P. Sicoli
Osservatorio Astronomico di Sormano

In this chapter we focus on the circumstances that led Giuseppe Piazzi (1746–1826) to discover the first asteroid, Ceres, on January 1, 1801. Through the examination of published and archival documentation, we shed light on the reaction of the astronomical community at the announcement of the discovery and on Piazzi's puzzling behavior. In the end, we briefly discuss the discoveries of Pallas, Juno, and Vesta and the theories put forward to explain their nature.

1. INTRODUCTION

Gioacchino Giuseppe Maria Ubaldo Nicolò Piazzi was born in Ponte, Valtellina, July 16, 1746, to one of the wealthiest families of the region. The penultimate of 10 sons, most of whom died as children, his parents worried about his health and for this reason quickly baptized him at home. The register of baptisms of St. Maurizio Church clearly specifies "*ob imminens vitae periculum,*" or "because of impending danger of death" (*Maineri,* 1871; *Invernizzi et al.,* 2001).

Following the tradition that encouraged younger children of wealthy and noble families to take holy orders, Giuseppe joined the Teatine order at the age of 19. We do not have firsthand documents about his early studies, but we know from documents preserved in the Archive of the Palermo Observatory that between 1770 and 1780 he was requested by his superiors to teach philosophy and mathematics in many different Italian cities, including Rome, Genoa, and Ravenna. In 1781, he was appointed to the Chair of Mathematics in the newly established Accademia dei Regi Studi of Palermo (which became the University of Palermo in 1806); a few years later, in 1787, he was named to the Chair of Astronomy even though he was not yet even an amateur astronomer. In a matter of only a few years, however, he was to become one of the most respected astronomers of his time (Fig. 1).

In March 1787, soon after he was charged with overseeing the construction of a new observatory at Palermo, Piazzi departed for a three-year stay at the major astronomical centers of Paris and London. During his travels he gained the esteem and friendship of some of the most reputed astronomers of the time, including Lalande, Messier, Mechain, Cassini, Maskelyne, and Herschel. Moreover, he succeeded in securing for the new observatory a unique instrument: the famous 5-foot circular-scale altazimuth telescope made by Jesse Ramsden of London (*Piazzi,* 1792; *Pearson,* 1829; *Chinnici et al.,* 2001). Returning to Palermo in November 1789, Piazzi was able, in a matter of months, to have the new observatory built on top of the tower of Santa Ninfa at the Royal Palace.

Fig. 1. Giuseppe Piazzi and Ceres. This oil portrait (60.5 × 73 cm) was most likely painted at the very beginning of the nineteenth century (1803?) by the Sicilian artist Giuseppe Velasco (1750–1827), a friend of Piazzi and the author of many portraits in the Palermo Observatory collection. Courtesy of the Palermo Astronomical Observatory.

Fig. 2. The Palermo Circle by Jesse Ramsden (1730–1800), the greatest of the eighteenth-century instrument makers. With this instrument, Piazzi discovered Ceres in 1801. It was completed in 1789 after almost two years of intense work. The telescope has a 7.5-cm objective lens; the altitude scale (5 feet in diameter) was read with the aid of two diametrically opposed micrometer microscopes while the azimuth scale (3 feet in diameter) was read by means of a micrometer microscope. Recently refurbished (*Brenni et al.,* 2001; *Chinnici et al.,* 2001), all the parts of the telescope are original to the time of Ceres' discovery, except for the eyepiece, which was replaced in 1855. Reproduced by permission of the Palermo Astronomical Observatory.

Encouraged by the possession of the 5-foot Palermo Circle (Fig. 2), whose accuracy was regarded to be much superior to that of any other existing instrument (*Lalande,* 1803), Piazzi centered his scientific program on the accurate measurements of stellar positions. His observational technique required that each star had to be observed for at least four nights before its position could be established. This painstaking work resulted in the publication in 1803 of his first star catalog (*Piazzi,* 1803). For this highly regarded work, he was awarded the prize for mathematics and physics at the Institut National de France, Fondation Lalande, (*Lalande,* 1804) and was elected a fellow of the Royal Society. It was while working on this catalog that Piazzi, on January 1, 1801, unexpectedly discovered Ceres, the "missing planet" between the orbits of Mars and Jupiter.

2. THE PROBLEM OF THE MISSING PLANET

It is well known that Johannes Kepler was the first to suggest in his *Mysterium Cosmographicum* (1596) the existence of a planet between Mars and Jupiter. Kepler's motivations for such a suggestion were of course very different from the factors that would prompt a modern scientist to make a cosmological hypothesis. The question Kepler asked himself was along the lines of, "Why had God been motivated mathematically to select the planetary orbits in the way He had?" Kepler's concern focused on the reasons the number, size, and motion of the circles were as they were and not otherwise. The gap between Mars and Jupiter was especially difficult to explain and Kepler tried a bold approach: "Between Jupiter and Mars I placed a new planet, and also another between Venus and Mercury . . . Yet the interposition of a single planet was not sufficient for the huge gap between Jupiter and Mars" (*Kepler*, 1596).

Throughout the centuries the question of a possible explanation of this gap was taken up by such luminaries as Newton, Kant, J. H. Lambert, David Gregory, William Whiston (Newton's successor at Cambridge), and Christian Wolff (*Hoskin,* 1999). However, the first to invoke what is today known as the Titius-Bode Law, in the form we are familiar with, was Johann Daniel Titius, in his German translation of Charles Bonnet's *Contemplation de la Nature,* first published in 1766. The relation regarding the mean distances of the planets from the Sun caught the attention of Johann Elert Bode, a professional astronomer who later became director of the Berlin Observatory. Bode was convinced of the validity of such a rule and inserted it in a footnote of the second edition of his book *Anleitung zur Kenntnis des gestirten Himmels*, published in 1772:

> This latter point seems in particular to follow from the astonishing relation which the known six planets observe in their distances from the Sun. Let the distance from the Sun to Saturn be taken as 100, then Mercury is separated by 4 such parts from the Sun. Venus is 4 + 3 = 7. The Earth 4 + 6 = 10. Mars 4 + 12 = 16. Now comes a gap in this so orderly progression. After Mars there follows a space of 4 + 24 = 28 parts, in which no planet has yet been seen. Can one believe that the Founder of the universe had left this space empty? Certainly not. From here we come to the distance of Jupiter by 4 + 48 = 52 parts, and finally to that of Saturn by 4 + 96 = 100 parts. (*Hoskin*, 1993)

Because of Bode's interest and prominence, the relation assumed a new importance, especially among a small but very determined group of German astronomers. Moreover, the 1781 discovery of Uranus, whose orbit fit well into the Titius-Bode series, was a remarkable confirmation of the relation and reinforced the belief that there must be a planet between Mars and Jupiter.

One of the most determined hunters of the missing planet was Baron Franz Xaver von Zach, astronomer of the Duke of Gotha and director of the Seeberg Observatory. Von Zach's strategy to find the planet was a very reasonable one: He limited his investigations to the zodiac and produced an accurate catalog of zodiacal stars in the hope of detecting any newcomers that would fall under his telescopic field of view. He even went as far as to try to calculate a possible orbit of the as-yet-undiscovered planet. Von Zach's efforts, however, remained unsuccessful, and he reasoned that a cooperative attack to the problem was necessary. In September 1800, he made what he called "a small astronomical tour." He traveled to Celle near Hannover, to Bremen, and to the nearby town of Lilienthal. The purpose of this tour was to organize, with von Ende, Olbers, Gildemeister, Schroeter, and Harding, a society whose aim was to scrutinize the entire zodiac down to the smallest telescopic stars. The society was established on the afternoon of September 20, 1800, and took the name of Vereinigte Astronomische Gesellschaft, usually referred to as the "Lilienthal Society." The society's members decided to divide the zodiac into 24 zones of 15° longitude and ±7°–8° latitude and to allocate by lot each zone to one astronomer. It was thus necessary to ensure the cooperation of 18 other astronomers. The 24 members of the society, who became known as the "Himmels Polizei" (or "Celestial Police") (*von Zach,* 1801a) were J. E. Bode (Berlin), J. S. G. Huth (Frankfurt/Oder), G. S. Klügel (Halle), J. A. Koch (Danzig), J. F. Wurm (Blauebeuren), F. von Ende (Celle), J. Gildemeister (Bremen), K. L. Harding (Lilienthal), W. Olbers (Bremen), J. H. Schroeter (Lilienthal), F. X. von Zach (Gotha), J. T. Bürg (Vienne), T. Bugge (Copenhagen), D. Melanderhielm (Stockholm), J. Svanberg (Uppsala), F. T. Schubert (St. Petersburg), J. C. Burckhardt (Paris), P. F. A. Mechain (Paris), C. Messier (Paris), C. Thulis (Marseille), N. Maskelyne (Greenwich), W. Herschel (Slough), B. Oriani (Milan), and G. Piazzi (Palermo). However, not all the astronomers appearing on this list actively participated in the society's program, nor were all of them immediately invited to be part of it. For example, no record has been found up to now that William Herschel was actually invited to actively participate (Hoskin, personal communication, 2001). Lalande, who had been invited, decided not to participate because he was "*occupé du travail de la méridienne*" at the time (*Delambre,* 1806) while von Zach's invitation letter to Barnaba Oriani was dated May 29, 1801, well after the discovery of Ceres. As for Piazzi, he never received an invitation. Von Zach, in the same letter to Oriani, wrote: "You and Piazzi were in the list of this astronomical Society that was established in September 1800. . . . when you will write to Piazzi, invite him in the name of the Society." Quite rude behavior on the part of the Perpetual Secretary of an astronomical society!

3. THE DISCOVERY OF CERES

Lacking an invitation, Piazzi was totally unaware of the society's plans when at about 8:00 p.m. on January 1, 1801, while working on his star catalog, he detected what he termed a tiny star in the "shoulder" of Taurus (*Piazzi,* 1801). Piazzi measured its position and, as was his custom, reobserved it the following night and found that it had moved. He thought at first it was a mistake but by January 4 he was convinced it was a new "star," possibly a comet. As was customary at the time, he alerted the press the very same day. The news of the discovery of a comet was soon published by foreign newspapers, where the news was picked up by the end of February by at least two professional astronomers: Lalande in Paris and Bode in Berlin. On January 24, Piazzi, having observed the new star for a total of 14 nights, finally decided to write to Bode and to Barnaba Oriani in Milan.

The choice of writing only to Oriani and Bode may at first appear a bit strange. Oriani was his best friend, but why Bode? And why only these two? The fact is that Piazzi was not ready to commit himself as to the nature of the new star. To Oriani, he sent the positions for January 1 and January 23 plus the information that on January 11, the motion of the star had changed from retrograde to direct. But he also added a paragraph in which he stated openly that he thought that the object he had observed might have been a new planet:

> I dare . . . to write you, impatient as I am to give you [news] . . . On the 1st of January I have observed in the shoulder of Taurus a star of the 8th magnitude which, on the following night, that is the 2nd, advanced by about 3'30" northwards and about 4' towards Aries' section. I did verify my observations on the 3rd and 4th, and found approximately the same motion. On 5, 6, 7, 8, 9, the sky was covered. I did see the star again January 10 and 11, and then on 13, 14, 17, 18, 19, 21, 22, and 23. On the first observation its R.A. was 51°47' and its northern declination was 16°8'. From 10 to 11 it turned from retrograde to direct motion, and on the observation of 23, it had R.A. 51°46', Dec. 17°8'. *I have announced this star as a comet, but since it shows no nebulosity, and moreover, since it had a slow and rather uniform motion, I surmise that it could be something better than a comet. However, I would not by any means advance publicly this conjecture. As soon as I shall have a larger number of observations, I will try to compute its elements.* (*C. A.,* 1874) (italics added)

To *Bode* (1802) instead he sent the same technical information but referred clearly to the object as a comet, adding only, as a veiled suggestion, that it did not have any "appreciable nebulosity":

> On January 1st I discovered a *comet* in Taurus, it had 51°47' of Right Ascension, and 16°8' of northern declination. On the 11th its motion, until then retrograde westwards, changed into direct motion eastwards; on the 23rd it had a right ascension of 51°46' and a northern declination of 17°8'. I shall continue to observe it and I hope to be able to observe it all along February. It is very tiny, and reaches at most a star of the 8th magnitude *without appreciable nebulosity*. Please, let me know if it has already been observed by other astronomers, for in this case I will not bother with the calculation of its orbit. (italics added)

Piazzi's ambiguity did not pass unnoticed. On the matter *Bode* (1802) later wrote: "It is absolutely incomprehensible to me why Mr. Piazzi in his letter of the same date to me calls his discovered moving star a comet, and even in some of the following letters insists in this opinion, disregarding my objections, and nevertheless in his first letter to Mr. Oriani favours its planetary nature." It is evident at this point that Piazzi decided to write to Bode, who was known for helping to create the Titius-Bode law, just to probe his reaction. Unfortunately for Piazzi, the letters reached Bode on March 20 and Oriani on April 5. On February 27, Lalande, having read in the *Journal de Paris* that a comet had been discovered in Palermo, wrote to Piazzi asking for his observations. The letter arrived at the beginning of April and Piazzi, who had not yet received any reaction either from Bode or Oriani, was obliged to send his complete set of observations to Lalande. It should be understood that Lalande was not only a good friend of Piazzi but was also the Gran Maestro of the Lodge of the Neuf Seurs; Piazzi himself was a freemason. In other words, it was very difficult for Piazzi not to answer Lalande.

Piazzi's observations were sent on April 11 to Lalande and Oriani but not to Bode. In the meantime Bode had not been idle. As soon as he received Piazzi's letter he jumped on the idea that the object might have been the famous missing planet. He quickly calculated a circular orbit on the basis of his hypothesis about the distance and period of the supposed planet, verified that Piazzi's observations were consistent with this idea, and on March 26 gave a preliminary announcement at the Prussian Academy of Sciences. Immediately afterward he alerted von Zach, who at the time was editor of the *Monatliche Correspondenz*, a monthly publication designed to quickly spread astronomical news. At their first meeting after Easter, Bode presented a new memoir at the Prussian Academy. He went as far as to announce the discovery of the new planet to the press in Hamburg, Jena, and Berlin and to name it "Juno" (*Bode,* 1802). Von Zach was in favor of the name "Hera" proposed by Duke Ernst of Saxe-Gotha 15 years before the object's discovery (*von Zach,* 1801a), and this name was at first widely accepted, at least in Germany. Oriani, writing to Piazzi on July 25, 1801, said: "I have to forewarn you that the name Hρα, or 'Hera,' that is Juno has been given to it almost universally in all Germany" (*C. A.,* 1874). Piazzi, who called his planet "Ceres Ferdinandea" in honor of the patron goddess of Sicily and of King Ferdinand of Bourbon (*Piazzi,* 1801), certainly did not agree. Writing to Oriani on August 25, 1801, he made no secret of his sentiment: "If the Germans think they have the right to name somebody else's discoveries they can call my new star the way they like: as for me I will always keep it the name of Cerere and I will be very obliged if you and your colleagues will do the same" (*C. A.,* 1874). In the end, the name Ceres was accepted by the astronomical community. "As for me," wrote von Zach to Oriani on February 25, 1802, "I shall continue to call it Ceres but I beg Piazzi to drop Ferdinandea because it is a bit too long."

The young Burckhardt, to whom Lalande had passed Piazzi's observations received on May 31, immediately set to work and by June 6 sent to von Zach the elements of a circular orbit, followed three days later by those of an elliptical orbit (*von Zach,* 1801b; *Bode,* 1802). By the end of June the astronomical community was convinced that the "star" discovered by Piazzi was a planet. But how to reobserve Ceres once it reappeared in the morning sky in order to obtain conclusive evidence? Good ephemerides were needed. Unfortunately, the best ephemerides available at that time (derived from Burckhardt's ellipse) were not accurate enough to mount an effective search. In fact, the problem with which astronomers and mathematicians were confronted was "to determine the orbit of a celestial body, without making any hypothesis, from observations covering a space neither too large nor such as to allow the special methods to be applied" (*Gauss,* 1809). Because Piazzi's observations covered only 3° and were so few (21) before Ceres became lost to the Sun in the evening sky, the "special methods" used by Burckhardt simply did not apply. Astronomers were growing desperate because during the month of August they had been looking for the planet without success, and, above all in France, they began to doubt the real existence of the new celestial body. In a letter to Oriani dated July 6, 1801, von Zach criticized Piazzi for having kept his observations secret for a long time, in effect preventing other astronomers from observing it and better understanding its true nature.

Reporting some suspicions, von Zach wrote:

> There are some astronomers who are starting to doubt the real existence of such a star, Burckhardt suspects that the observations are very wrong, it is a fact that he gave you and Bode a Declination wrong by at least half a degree, Burckhardt says that there are other errors. Now I cannot conceive, how an observer as experienced as Piazzi, provided with the best instruments, a complete Circle, and a transit telescope by Ramsden, could incur such mistakes in his meridian observations?

In another letter to Oriani dated December 18, 1801, as yet unaware of having actually observed Ceres on December 7, von Zach asked:

> What is going on with the *Ceres Ferdinandea*? Nothing has been found as yet either in France or in Germany. Peoples are starting to doubt: Already sceptics are making jokes about it. What is Devil Piazzi doing? La Lande wrote me that he [Piazzi] has changed again his observations and that he has made a new Edition of them! What does that mean? La Lande in his letter adds: *This is why I do not believe in the planet.*

In the September issue of the *Monatliche Correspondenz,* Piazzi's complete set of observations were finally published (Fig. 3). The young mathematician Carl Friedrich Gauss, at the time only 24, immediately seized on the importance of the problem. In the preface to his *Theoria motus* he wrote: "Nowhere in the annals of Astronomy do we meet with so great an opportunity, and a greater one could hardly be imagined, for showing most strikingly the value

Beobachtungen des zu Palermo d. 1 Jan. 1801 von Prof. Piazzi neu entdeckten Gestirns.

1801	Mittlere Sonnen-Zeit	Gerade Aufsteig. in Zeit	Gerade Aufsteigung in Graden	Nördl. Abweich.	Geocentrische Länge	Geocentr. Breite	Ort der Sonne + 20" Aberration	Logar. d. Distanz ⊙ ♁
	St ′ ″	St ′ ″	° ′ ″	° ′ ″	Z ° ′ ″	° ′ ″	Z ° ′ ″	
Jan. 1	8 43 27.8	3 27 11.25	51 47 48.8	15 37 43.5	1 23 22 58.3	3 6 42.1	9 11 1 30.9	9.9926156
2	8 39 4.6	3 26 53.85	51 43 27.8	15 41 5.5	1 23 19 44.3	3 2 24.9	9 12 2 28.6	9.9926317
3	8 34 53.3	3 26 38.4	51 39 36.0	15 44 31.6	1 23 16 58.6	2 58 9.9	9 13 3 26.6	9.9926324
4	8 30 42.1	3 26 23.15	51 35 47.3	15 47 57.6	1 23 14 15.5	2 53 55.6	9 14 4 24.9	9.9926418
10	8 6 15.8	3 25 32.11	51 23 1.5	16 10 32.0	1 23 7 59.1	2 29 0.6	9 20 10 17.5	9.9927641
11	8 2 17.5	3 25 29.73	51 22 26.0	……	……	……	……	……
13	7 54 26.2	3 25 30.30	51 22 34.5	16 22 49.5	1 23 10 27.6	2 16 59.7	9 23 12 13.8	9.9928490
14	7 50 31.7	3 25 31.72	51 22 55.8	16 27 5.7	1 23 12 1.2	2 12 56.7	9 24 14 13.5	9.9928809
17	……	……	……	16 40 13.0	……	……	……	……
18	7 35 11.3	3 25 55.0	51 28 45.0	……	……	……	……	……
19	7 31 28.5	3 26 8.15	51 32 2.2	16 49 16.1	1 23 25 59.2	1 53 38.1	9 29 19 53.8	9.9930607
21	7 24 2.7	3 26 34.27	51 38 34.1	16 58 35.9	1 23 34 21.3	1 46 6.0	10 1 20 40.3	9.9931434
22	7 20 21.7	3 26 49.42	51 42 21.3	17 3 18.5	1 23 39 1.8	1 42 28.1	10 2 21 32.0	9.9931886
23	7 16 43.5	3 27 6.90	51 46 43.5	17 8 5.5	1 23 44 15.7	1 38 52.1	10 3 22 22.7	9.9932348
28	6 58 51.3	3 28 54.55	52 13 38.3	17 32 54.1	1 24 15 15.7	1 21 6.9	10 8 26 20.1	9.9935062
30	6 51 52.9	3 29 48.14	52 27 2.1	17 43 11.0	1 24 30 9.0	1 14 18.0	10 10 27 46.2	9.9936332
31	6 48 25.4	3 30 17.25	52 34 18.8	17 48 21.5	1 24 38 7.3	1 10 54.6	10 11 28 28.5	9.9937007
Febr. 1	6 44 59.9	3 30 47.21	52 41 48.0	17 53 36.5	1 24 46 19.3	1 7 30.9	10 12 29 9.6	9.9937703
2	6 41 35.8	3 31 19.06	52 49 45.9	17 58 57.5	1 24 54 57.9	1 4 10.5	10 13 29 49.9	9.9938423
5	6 31 31.5	3 33 3.70	53 15 40.5	18 15 1.0	1 25 22 43.4	0 54 28.9	10 16 31 45.5	9.9940751
8	6 21 39.2	3 34 58.50	53 44 37.5	18 31 23.3	1 25 51 29.5	0 45 6.0	10 19 33 33.1	9.9943276
11	6 11 58.2	3 37 6.54	54 16 38.1	18 47 58.8	1 26 26 40.0	0 36 2.9	10 22 35 11.4	9.9945823

Fig. 3. Page showing Piazzi's complete set of observations as published in the *Monatliche Correspondenz,* September 1801, p. 280. Courtesy of the Brera Astronomical Observatory.

of this problem, than in this crisis and urgent necessity, when all hope of discovering in the heavens this planetary atom, among innumerable small stars after the lapse of nearly a year, rested solely upon a sufficient approximate knowledge of its orbit to be based upon these very few observations" (*Gauss,* 1809). In a matter of a little more than a month, he produced what was termed by von Zach a "perfect ellipse" (*Bode,* 1802), and his computed elements and ephemerides (*von Zach,* 1801c), in good agreement with modern values, provided positions quite different compared to the solutions that had been proposed up to then. Using these new ephemerides, von Zach directed his telescope on December 7 and actually observed Ceres, but bad weather in the following days prevented him from continuing his observations: "After Dec. 7 I have not had a clear sky. That day I observed many unknown stars, that I have not found in any catalogue not even in the one in folio that Bode has recently published . . . On Dec. 16th there was a break in the clouds and I had observed many little stars of 4 or 5 magnitude . . . When No. 1 had to transit across the meridian it didn't come. Great joy! I thought I had caught this coquette Ceres but the joy lasted less than a minute since I didn't see either N. 2 or N. 3. It was a light haze that hid them from me" (Von Zach to Oriani, Dec. 18, 1801). All uncertainties were swept definitively away when first von Zach on December 31, 1801, and then Olbers, two days later in Brema, not far from Gauss' estimated position, were finally and independently able to confirm the recovery of Ceres and, using Gauss' words, "restore the fugitive to the observations."

It is interesting to read what von Zach wrote to Oriani on January 14, 1802:

> I hasten to inform you, that I found the Ceres Ferdinandea on December 7 of last year. I had already published this observation in the January 1802 issue of my journal . . . without realizing then that it was the planet; but I suspected it was. On Decbr. 31, I verified the thing and my suspect star had changed its position, on January 11 I observed it for the third time (the weather here is terrible) and I had the certainty of my finding, that I have the pleasure to announce to you . . . Mr. Olbers has discovered the planet Ceres *independently* at Bremen, but later than me, on Jan. 2. I said independently since in truth he made the discovery as well as I did, since I had not sent him my observations, that I kept secret until after Jan.ry 11 when I was completely sure of my Discovery. . . . I hope that Piazzi, or you other Gentlemen Astronomers [taking advantage] of the beautiful Italian climate have found the planet before me.

A few months later, on March 28, 1802, Olbers in Bremen detected a second small body, Pallas, with features similar to those of Ceres but with an orbit that had higher eccentricity and inclination. On May 6, 1802, William Herschel presented to the Royal Society a memoir, "Observations on the two lately discovered celestial bodies," and on May 22 he sent to Piazzi a long letter in which he summarized the content of his memoir: "I say in my paper, that the interesting discoveries of Mr. Piazzi and Olbers have introduced to our acquaintance a new species of celestial bodies, with which hitherto we have not been acquainted" (*C. A.,* 1874). This is one of Herschel's strokes of genius. Out of only two cases, Ceres and Pallas, and examining carefully in his memoir the principal features of planets and comets he drew the correct conclusion and proposed for these new bodies the name of "asteroids" because of their starlike appearance. Needless to say, the astronomical community was slow in accepting the ideas offerred by *Herschel* (1802), and the term asteroid became widely accepted only early in this century.

But what about Piazzi? What was he doing while throughout Europe astronomers debated, calculated orbits, and published papers? Piazzi simply remained idle. It has often been written that immediately after his last observation on February 11 he fell ill. While this is certainly true, and it is also true that between April and May he had to work on the meridian line of the Palermo cathedral that had to be solemnly reopened at the beginning of June, the real point is that he did not want his observations to be published before he could himself calculate an orbit, and he encountered tremendous difficulties in calculating even a circular orbit. Moreover, during the entire summer Piazzi continued to call the object a comet or "new star." This behavior puzzled and even infuriated the astronomical community. For instance, the Astronomer Royal Nevil Maskelyne in the summer of 1801 wrote to an unknown correspondent the following letter:

> There is great astronomical news: Mr. Piazzi, Astronomer to the King of the two Sicilies, at Palermo, discovered a new planet the beginning of this year, and was so covetous as to keep this delicious morsel to himself for six weeks; when he was punished for his illiberality by a fit of sickness, by which means he lost the track of it; and now a german Astronomer, having got some of his observations, has calculated an orbit in our system as near as he could from such few observations, and had just informed us where

> he thinks it should be looked for in the course of the summer and autumn.
>
> It will not be so easy to recover, as the lost Cupid, when Venus said you might spy among 20 immediately by his air and complection. But this having been only a star of the 8th at first, & now for some months to come not bigger than the 10th or 12th will not be easily distinguished among 40.000 or 50.000 stars of similar appearance as it can be only known by its motion, which cannot be seen immediately but require observations of the relative position of several stars among which it is to be looked for.
>
> What a deal this imprudent Astronomer has to answer for! It is now publicly proposed, in a german publication, to all Astronomers in Europe to hunt for it. (*Howse,* 1989)

To explain Piazzi's behavior, it is useful to recall that at the time of his appointment to the chair of astronomy, he was not an astronomer but simply a 40-year-old professor of mathematics whose lectures were appreciated and considered up to date. He could hardly be considered a mathematical scholar, as he had never published. However, he rapidly became an excellent observer and made important contributions to astronomy. In addition to his two excellent star catalogs, we may remember that in 1806, well before Bessel, he was the first to detect the unusually large proper motion of 61 Cygni and to point this star out to the astronomical community as a good candidate for parallax measurements (*Foderà-Serio,* 1990). Yet Piazzi never mastered theoretical astronomy. In his letters to Oriani he repeatedly asked his friend to send him the best formulae for calculating the orbits and to furnish him with explanations in order not to be obliged to "go back to the theory of attraction."

A second, more important point is related to the difficult environment in which Piazzi had to work. The establishment in Palermo of an Astronomical Observatory had been considered futile even by some of the professors of the Accademia dei Regi Studi and, in 1795, Piazzi had lost his most powerful "protector," Viceroy Prince of Caramanico. This fact left him in the power of the "envious academics" who went so far, once Ceres had been reobserved, as to circulate the tale that "Piazzi had been discovered by Ceres" (*Angelitti,* 1927).

4. PALLAS, JUNO, AND VESTA AND SOME REMARKS ON THE ORIGIN OF THE ASTEROIDS

As already mentioned, on March 28 1802, Wilhelm Olbers at Bremen, while "observing with his telescope all the small stars in the wing of Virgo, to be sure of their position, so that he could more easily establish the position of the planet" [Ceres], detected a 7-magnitude star that "he was absolutely sure that was not there at the time of his first observations" [in January]. He "took its position; and continuing to observe it during two hours, he could see that it had changed place during this span of time" (*Lalande,* 1803). Within two days, Olbers was sure it was a new planet. He hastened to name it Pallas, and alerted Baron von Zach, who was able to observe it on April 4. In the following days he circulated his observations to the astronomical community. In this way not only was Pallas observed by many astronomers throughout Europe but the possibility that it might have been a comet, because of its hazy appearance and exceptional inclination of its orbit, was soon dismissed. It is amazing to read in the original reports how matter-of-factly the news of the discovery of another tiny planet orbiting around the Sun at about the same distance of Ceres was received by the astronomical community. For instance, von Zach, in a letter to Piazzi dated April 8, 1802 (less than two weeks after the discovery) wrote: "The star of D. Olbers, that I have had the honour to announce to you [April 5], is actually a primary Planet that revolves around the sun on a highly inclined orbit. . . . It exists then between Mars & Ceres; & undoubtedly many more planets of this kind must exist in the various spaces among the Planets; . . . It is to you, Eminent Confrère, that we owe all these discoveries, without your Ceres, no Pallas. Without Pallas no future discoveries by any of us. What a new field!" (*Piazzi,* 1802). This letter, along with many others, expresses well not only the lack of surprise with which the astronomers received the announcement of the new discovery, but also shows the community's willingness to accept that many other new bodies could exist both in orbits lying between Mars and Jupiter or "dans les differents espaces des Planétes." However, Bode's "law" remained in the mind of some astronomers as a law that could have a physical basis yet to be discovered. The first to try to save the law was the discoverer of Pallas. Beginning no later than June 1802, Olbers dealt with the problem of reconciling the existence of Pallas with the "beautiful harmonious law of planetary distances." Taking advantage of the fact that the mean distances from the Sun of Ceres and Pallas were pretty much the same, he suggested that they were fragments of a full-sized planet that had once occupied the gap between the orbits of Mars and Jupiter and had fragmented either under the action of internal forces or because of the impact of a comet: ". . . and if Ceres and Pallas were only pieces and big fragments of a pristine larger planet disrupted by internal natural forces or by the external impact of a comet?" (*von Zach,* 1802).

As a consequence, the discovery of other asteroids was expected and, in addition, their frequently observed variation in luminosity could be readily explained. In fact, as fragments of an exploded planet they were obviously "lacking roundness" and hence "in their rotation they were not always reflecting the same quantity of light" (*Oriani,* 1802). Olbers' theory seemed reasonable and was accepted by many astronomers who further reasoned that for a catastrophic explosion (initially at least) the orbits of all the fragments would have intersected in the place of the explosion and on the opposite side of the Sun. It was by observing constantly in the regions of Cetus and Virgo (where the orbits of Ceres and Pallas intersected) that Harding at Lilienthal discovered Juno on September 1, 1804.

Immediately after this discovery, Hofrath Huth, in a letter to Bode dated September 21, 1804 (*Bode,* 1804), offered a different theory: "I hope that this [planet] is not the last one that will be found between Mars and Jupiter. I think it very probable that these little planets are as old as the others and that the planetary mass in the space between Mars and Jupiter has coagulated in many little spheres, almost all of the same dimensions, at the same time in which happened the separation of the celestial fluid and the coagulation of the other planets."

On March 29, 1807, Olbers, observing at Brema in the same regions of the sky where Ceres, Pallas, and Juno had been discovered, found his second asteroid, which Gauss named Vesta. Four "positive cases" in seven years were not a representative sample, but, in the absence of evidence to the contrary, they were enough to reinforce belief in Olbers' theory. Lagrange, in a well-known paper "Sur l'Origine des Comètes" (1812), considering that "Olbers hypothesis, however extraordinary it may appear, is nevertheless not unlikely," explored the consequences of the breaking of a planet into two or more fragments under the influence of internal forces. He found that this could add the required ellipticities and inclinations to Laplace's hypothesis on the formation of the solar system.

For nearly 40 years thereafter, no additional minor planets were added to the list, until finally K. L. Hencke, after 15 years of intense dedicated work, found Astraea in 1845. What accounted for the long interval of no discoveries? It is impossible to properly address this question here. Thus we will limit ourselves to make three points that are certainly relevant.

1. At least some astronomers, misled by Olbers' theory, concentrated their searches on a limited area in the directions of the orbital intersections between Ceres and Pallas.

2. The search for little planets was not considered a relevant scientific problem in and of itself. Because of that, many professional astronomers, while dedicating time to observations of new planets once they had been found, did not engage in systematic searches for them. For example, Delambre in 1806 wrote: "We further remark that these four planets [Uranus, Ceres, Pallas, and Juno] were found while searching for something else, and conclude that the real way to deserve and to encounter such accidents is to be occupied in some grand undertaking, which in itself is of real use, and keeps us constantly on the route to such discoveries; it is, for example, to work, as M. Piazzi, to perfect and augment the stellar catalogue, observing each star repeatedly for several days: this method has the double advantage to register in the catalogue only the reliable positions, and to evidence in the long run the planets that could still be confused among the innumerable quantity of very faint stars scattered in the sky" (*Delambre,* 1806).

3. Even though astronomers were open to the possibility that more planets were yet to be discovered [e.g., see the last lines of the above quotation and von Zach's letter to Piazzi (*Piazzi,* 1802)], the lack of suitable star maps discouraged many of them from undertaking a work that, while time-consuming, could not offer any guarantee of success. It is certainly not by chance that it was only after these became available, in the last years of the 1840s, that the discovery of new asteroids followed at a regularly increasing rate.

As asteroids began to be discovered in all parts of the sky, Olbers' theory began to be questioned. In 1857, when the number of asteroids amounted to a mere 50, *Arago* (1857) wrote: "The large number of these bodies known today leads one to believe that there are other causes for their birth. The intersections of pairs of orbits of the small planets are far from being all in agreement with Olbers' hypothesis; nevertheless, the interlacing of their orbits suggests an intimate relationship between many of these bodies, and this is a curious subject of research for astronomers in the phenomena they present." In the subsequent years a number of astronomers took up the subject. Among these was Daniel Kirkwood, known for having discovered the so-called Kirkwood gaps, who hypothesized that the asteroids originated from a ring of nebular mass that was prevented from forming a planet by the attractive pull of Jupiter (*Kirkwood,* 1867), a view favored by modern astronomers.

Acknowledgments. Unless otherwise stated, all the letters quoted are preserved in the Archives of the Brera Astronomical Observatory. We wish to thank Agnese Mandrino, archivist and librarian of the Brera Observatory, for her invaluable assistance in this research. We also thank the Brera Astronomical Observatory both for the material placed at our disposal and its financial support. The research of Giorgia Foderà Serio at the University of Palermo is funded by Ministero dell'Istruzione, dell'Università e della Ricerca.

REFERENCES

Angelitti F. (1927) Per il centenario della morte dell'astronomo Giuseppe Piazzi avvenuta il 22 luglio del 1826. *Memorie della Società Astronomica Italiana, n.s., vol. III,* pp. 369–395.

Arago F. (1857) *Astronomie Populaire*. Tome quatrième. Gide, Paris. 856 pp.

Bode J. E. (1802) *Von dem neuen, zwischen Mars und Jupiter entdeckten achten Haupt planeten des Sonnensystems.* In der Himburgischen Buchhandlung, Berlin. 136 pp.

Bode J. E. (1804) Aus einem Schreiben des Hrn. Hofrath Huth aus Frankfurt an der Oder vom 21, Sept. 1804. *Astronomisches Jahrbuch für das Jahr 1807 nebst einer Sammlung der neuesten in die astronomischen Wissenschaften einschlagenden Abhandlungen, Beobachtungen und Nachrichten,* pp. 265–267

Brenni P., Chinnici I., and Foderà-Serio G. (2001) The restoration of three large instruments of the Palermo Astronomical Observatory. *Bull. Sci. Instrum., 71,* 11–16.

C. A. (1874) *Corrispondenza Astronomica fra Giuseppe Piazzi e Barnaba Oriani* (G. Cacciatore and G. V. Schiaparelli, eds.). Univ. of Hoepli, Milano. 204 pp.

Chinnici I., Foderà-Serio G., and Brenni P. (2001) The Ramsden Circle at the Palermo Astronomical Observatory. *Bull. Sci. Instrum. Soc., 71,* 2–10.

Delambre J. B. J. (1806) Planète de Harding ou Junon. In *Connaissance des Tems ou des Mouvemens Ce'lestes, a'li Usage des Astronomes et des Navigateurs pour l'An 1808,* pp. 417–427.

Foderà Serio G. (1990) Giuseppe Piazzi and the discovery of the proper motion of 61 Cygni. *J. History Astron., XXI,* 275–282.

Gauss C. F. (1809) *Theoria Motus Corporum Coelestium in Sectionibus Conicis Solem Ambientium.* F. Perthes et I. H. Besser, Hamburg. 228 pp.

Herschel W. (1802) Observations on the two lately discovered celestial bodies. *Philos. Trans. R. Soc. London, 2,* 213–232.

Hoskin M. (1993) Bode's Law and the discovery of Ceres. In *Physics of Solar and Stellar Coronae: G. S. Vaiana Memorial Symposium* (J. F. Linsky and S. Serio, eds.), pp. 36–46. Kluwer, Dordrecht.

Hoskin M., ed. (1999) *The Cambridge Concise History of Astronomy.* Cambridge Univ., Cambridge. 362 pp.

Howse D. (1989) *Nevil Maskelyne: The Seaman's Astronomer.* Cambridge Univ., Cambridge. 280 pp.

Invernizzi L., Manara A., and Sicoli P. (2001). *L'astronomo Valtellinese Giuseppe Piazzi e la scoperta di Cerere.* Fondazione Credito Valtellines, Ed. Bonazzi, Sondrio. 159 pp.

Kepler J. (1596) *Prodromus dissertationum cosmographicarum continens mysterium cosmographicum de admirabili proportione orbium celestium deque causis coelorum numeri, magnitudinis, motuumque periodicorum genuinis et propiis, demonstratum per quinque regularia corpora geometrica.* Excudebat Georgius Gruppenbachius, Tubingae. Translated by Duncan A. M. (1981) *The Secret of the Universe.* Abaris, New York. 267 pp.

Kirkwood D. (1867) *Meteoric Astronomy: A Treatise on Shooting-Stars, Fire-Balls and Aerolites.* Lippincott & Co, Philadelphia. 129 pp.

Lagrange J. L. (1812) Sur l'origine des comètes. In *Connaissance des Tems ou des Mouvemens Ce'lestes, a'li Usage des Astronomes et des Navigateurs pour l'An 1814,* pp. 211–218

Lalande J. (1803) *Bibliographie astronomique avec l'Histoire de l'Astronomie depuis 1781 jusqu'à 1802.* De l'Imprimerie de la République, Paris. 916 pp.

Lalande J. (1804) Prix adjugé à M.Piazzi, de la fondation Lalande. In *Connaissance des Tems ou des Mouvemens Ce'lestes, a'li Usage des Astronomes et des Navigateurs pour l'An XV,* pp. 454–455.

Maineri B. E. (1871) *L'Astronomo Giuseppe Piazzi Notizie Biografiche,* Tipografia già Domenico Salvi e Co., Milano. 135 pp.

Oriani B. (1802) Osservazioni del nuovo pianeta Pallade Olbersiana fatte al Settore Equatoriale. In *Ephemerides Astronomicae Anni 1803 ad Meridianum Mediolanensem supputatae ab Angelo De Cesaris,* pp. 22–34.

Pearson W. (1829) *An Introduction to Practical Astronomy, Vol. 2.* Longman, London. 708 pp.

Piazzi G. (1792) *Della Specola Astronomica dè Regj Studj di Palermo, Libri Quattro.* Dalla Reale Stamperia, Palermo. 240 pp.

Piazzi G. (1801) *Risultati delle Osservazioni della Nuova Stella scoperta il dì 1. Gennajo all'Osservatorio reale di Palermo.* Nella Reale Stamperia, Palermo. 25 pp.

Piazzi G. (1802) *Della scoperta del nuovo pianeta Cerere Ferdinandea ottavo tra i primarj del nostro sistema solare.* Nella Stamperia Reale, Palermo. 65 pp.

Piazzi G. (1803) *Praecipuarum stellarum inerrantium positiones mediae ineunte saeculo XIX ex observationibus habitis in Specula Panormitana ab anno 1792 ad annum 1802.* Typis Regiis, Palermo.

von Zach F. X. (1801a) Über einen zwischen Mars und Jupiter längst vermuteten, nun wohnscheinlich entdeckten neuen Hauptplaneten unseres Sonnen Systems. *Monatliche Correspondenz zur Beforderung der Erd- und Himmelskunde, III,* 592–623.

von Zach F. X. (1801b) Fortgetze Nachrichten über den zwichen Mars und Jupiter längst vermuteten, nun wohnscheinlich entdeckten neuen Hauptplaneten unseres Sonnen Systems. *Monatliche Correspondenz zur Beforderung der Erd- und Himmelskunde, IV,* 53–67.

von Zach F. X. (1801c) Fortgetze Nachrichten über den längst vermuteten neuen Haupt-Planeten unseres Sonnen-Systems. *Monatliche Correspondenz zur Beforderung der Erd- und Himmelskunde, IV,* 638–649.

von Zach F. X. (1802) Fortgetze Nachrichten über den neuen Haupt-Planeten unseres Sonnen-Systems, Palls Olbersiana. *Monatliche Correspondenz zur Beforderung der Erd- und Himmelskunde, VI,* 71–95.

Part II:
Remote Observations

Asteroid Orbit Computation

Edward Bowell
Lowell Observatory

Jenni Virtanen
University of Helsinki

Karri Muinonen
University of Helsinki and Astronomical Observatory of Torino

Andrea Boattini
Instituto di Astrofisica Spaziale

During the last decade, orbit computation has evolved from a deterministic pursuit to a statistical one. Its development has been spurred by the advent of the World Wide Web and by the greatly increased rate of asteroid astrometric observation. Several new solutions to the inverse problem of orbit computation have been devised, including linear, semilinear, and nonlinear methods. They have been applied to a number of problems related to asteroids' computed sky-plane or spatial uncertainty, such as recovery/precovery, identification, optimization of search strategies, and Earth collision probability. The future looks very bright. For example, the provision of milliarcsec-accuracy astrometry from spacecraft will allow the modeling of asteroid shape, spin, and surface-scattering properties. The development of deep, widefield surveys will mandate the almost complete automation of the orbit computation process, to the extent that one will be able to transform astrometric data into a desired output product, such as an ephemeris, without the intermediary of orbital elements.

1. INTRODUCTION

There has been a revolution in orbit computation over the past decade or so. It has been fueled in large part by observational efforts to find near-Earth asteroids (NEAs). For example, as *Stokes et al.* (2002) makes plain, the volume of observations has increased by almost 2 orders of magnitude, which has resulted in an even greater surge in the rate of orbit computation. In the book *Asteroids II, Bowell et al.* (1989) and *Ostro* (1989) seem to have been the only authors to refer to uncertainty in the positions of asteroids, the former in the context of recovering a newly discovered asteroid at a subsequent lunation, the latter in regard to the improvement of an asteroid's orbit when radar observations have been incorporated. Indeed, *Bowell et al.* (1989) end their chapter uncertain about the future of then-infant CCD observations. There is no reference to the need for automated orbit computation or to the rapid dissemination of data worldwide, both of which we now take for granted, and both of which have been enabled by an enormous increase in computing power and a concomitant reduction in costs. One can argue that it was the development of the World Wide Web that created a synergy between observers and orbit computers, each driving advances in the other's research, to the benefit of all. As we will describe, the development and application of powerful new techniques of orbit computation have largely kept pace.

The derivation of orbital elements from astrometric observations of asteroids is one of the oldest inversion problems in astronomy. Usually, the observational data comprise a set of right ascensions and declinations at given times, although other types of data, such as radar time delay and Doppler astrometry, may also be used. Six orbital elements, at a specified epoch, suffice to describe heliocentric motion, a common parametrization being by means of Keplerian elements a, e, i, Ω, ϖ, M (respectively, semimajor axis, eccentricity, inclination, longitude of the ascending node, argument of perihelion, and mean anomaly at a specified time). In the two-body case, only M changes with time. In the n-body case, in which account is taken of gravitational perturbations due to planets and perhaps other bodies, along with relativistic and nongravitational effects, all six elements may change with time. Most remarkably, the inversion method developed by Gauss in 1801 (e.g., see *Gauss,* 1809; *Teets and Whitehead,* 1999), in response to the loss of the first-discovered asteroid Ceres, has never really been supplanted (in fact, there exists a family of methods originating from the ideas of Gauss and elaborated by others). Gauss also introduced the method of least squares, which sowed the seeds for some of the probabilistic methods of orbit computation that have been developed over the past decade. A description of these methods and their application will be our central preoccupation in this chapter. For more detailed summaries of the development of orbit computation, readers are referred to *Danby* (1988) and *Virtanen et al.* (2001).

Hitherto, it has been customary to use the term *orbit determination* to cover the sequential processes of preliminary (initial) orbit fitting (normally using observations at three times) and differential correction (normally consisting of a least-squares fit using all the observations deemed accurate enough, planetary perturbations optionally being allowed for), independent of whether statistical or nonstatistical techniques are used. In this chapter, we largely ignore the terms orbit determination, determinacy, and indeterminacy because, in our experience, their use results in unnecessary confusion. The title of our chapter pertains to both the inverse problem of deriving the orbital element probability density from the astrometric observations and the prediction problem of applying the probability density.

It seems of doubtful use to apply statistical techniques to asteroids having an arbitrarily small number of observations. When there are two observations only, the number of observations is smaller than the number of orbital elements to be estimated. Nevertheless, techniques that are best described as statistical, and that have been developed during the past decade for inverse problems involving small numbers of observations, are superior to earlier techniques and appear to capture the overall significance of the observational uncertainty. We advise that, when there are fewer than on the order of 10 observations, the resulting orbital-element probability density must be applied with particular caution.

Note that, in this chapter, we are only peripherally concerned with advances in celestial mechanics; for example, work on integrators and solar-system dynamics is outside our purview. We concentrate on modeling the interrelations among the astrometric observations, orbital elements, and the predictions they allow.

We cannot emphasize enough how much online software and databases have contributed to the explosion of observational activity in the post-*Asteroids II* era. They have provided observers powerful tools for planning observations. The Minor Planet Center (MPC) has been a pioneer in ephemeris and targeting services (http://cfa-www.harvard.edu/cfa/ps/mpc.html). Lowell Observatory's asteroid services (http://asteroid.lowell.edu) have been most useful for main-belt asteroid observing; they are one of only two services that provide ephemeris uncertainties for non-near-Earth asteroids. The Jet Propulsion Laboratory's *Horizons* ephemeris computation system (http://ssd.jpl.nasa.gov/horizons.html) is maintained by JPL's Solar System Dynamics Group. At the University of Pisa, the Near Earth Objects–Dynamic Site (NEODyS; http://newton.dm.unipi.it/cgi-bin/neodys/neoibo) is a compilation of orbital and observational information on near-Earth asteroids. Its counterpart, ASTDyS (http://hamilton.dm.unipi.it/cgi-bin/astdys/astibo), pertains to numbered and multiapparition asteroids. Both of the University of Pisa sites rely on an orbit-computation freeware package called OrbFit (http://newton.dm.unipi.it/~asteroid/orbfit) (*Milani,* 1999), developed by a consortium of about a dozen astronomers.

We begin our review with a description of the theoretical underpinnings of the principal new techniques of orbit estimation. All of them accommodate probabilistic treatment, by which we understand fitting orbits that satisfy the observations within their estimated uncertainty. In section 3, we describe the implications of the new work as it pertains, in a broad sense, to predicting asteroid positions; in section 4, we outline some areas of future research that are likely to become prominent. We conclude with some thoughts and speculations about what might engage the interest of orbit computers a decade hence.

2. INVERSE PROBLEM

The inverse problem entails the derivation of an orbital-element probability density from observed right ascensions and declinations. Gauss' theory made use of the normal distribution of observational errors and the method of least squares. That single least-squares orbit solutions can be misleading was realized long ago. For one, "unphysical" orbital elements may result; for another, poor convergence of the differential correction process can be an indicator of large uncertainties in the orbital elements. Computer-based iterative orbit-estimation methods have made it possible to explore part of the parameter space of orbit solutions very rapidly. For example, Väisälä orbits, in which it is assumed that an asteroid is observed at perihelion (*Väisälä,* 1939), can provide a useful — though not foolproof — tool for discriminating between newly discovered NEAs and main-belt asteroids (MBAs) and for estimating their ephemeris uncertainties. Regardless of an asteroid's single orbit solution, a fundamental question to ask is, "How accurate are the orbital elements?"

At the time of the Asteroids II conference, all star catalogs contained zonal errors, sometimes amounting to an arcsec or more. Moreover, their relatively low surface densities and lack of faint star positions introduced field mapping and magnitude-dependent positional errors in asteroid astrometry. Beginning in 1997, the HIPPARCOS reference frame was used for high-density catalogs such as GSC 1.2 and USNO-A2.0 that are practically free of zonal errors. Even better catalogs, such as UCAC, are in the pipeline (e.g., see *Gauss,* 1999). For asteroid astrometry at its current best accuracy of tens of milliarcseconds, these catalogs are for most purposes error free and therefore introduce no systematic error. The remaining practical problems are related to generally poor stellar magnitudes and color indices and, as time goes by, positional degradation arising from imperfectly known stellar proper motions. If one knows, or can infer, which reference catalog was used for asteroid astrometric reductions, it is possible, to an extent, to compensate for systematic zonal errors. Such a technique was used by *Stone et al.* (2000) in their prediction of stellar occultations by (10199) Chariklo. In principle, the majority of historic astrometric data could be so corrected, though the amount of labor might be large. A new format for the submission of astrometric data, to include reference-frame and S/N ratio information, will soon be implemented by the MPC.

2.1. Inversion Using Bayesian Probabilities

In the inverse problem, the derivation of an orbital-element probability density is based on a linear relationship between observation, theory, and error: The observed position is the sum of the position computed from six orbital elements, a systematic error, and a random error. It is customary to assume that the systematic error has been corrected for or is sufficiently arbitrary to be accounted for by the random error. A probability density is assumed for the random error: Typically, based on statistics of the O-C residuals, the density is assumed to be Gaussian (accompanied by an error-covariance matrix). Using Bayes' theorem, the orbital-element probability density then follows from the aforementioned linear relationship and is conditional upon (and thus compatible with) the observations. Because of the nonlinear relationship between the orbital elements and the positions computed from them, the rigorous orbital-element probability density is non-Gaussian. The orbital-element probability density may, if desired, be multiplied by an *a priori* probability density, such as the distribution of the orbital elements of known asteroids.

Only during the past decade has it been understood that the assumption of a Gaussian distribution of errors is sometimes not applicable to asteroid and comet orbit computation. Particularly for observations made recently, accidental astrometric errors (due to observational noise) are small compared to reference-star positional errors, known to have regional biases (zonal errors), or imperfect field mapping. Thus the error distribution in O-C positional residuals appears to be non-Gaussian. Although *Muinonen and Bowell* (1993) studied non-Gaussian probability densities for the observational error — using methods based on so-called statistical inversion theory (*Lehtinen,* 1988; *Menke,* 1989; *Press et al.,* 1994) — the Gaussian density remained the most attractive statistical model for the random error. In spite of criticism of Gaussian hypotheses, alternative statistical models — aside from the work by *Muinonen and Bowell* (1993) — have not been put forward. Gaussian hypotheses are further supported by the impossibility of discriminating between systematic and random errors; indeed, if there is a large number of sources of uncorrected systematic error, the total error is essentially random, and the central limit theorem elevates carefully constructed Gaussian hypotheses above all other statistical hypotheses.

In treating orbits probabilistically, orbit computers concerned with astrodynamic applications have taken the lead over those working on the analysis of groundbased observations. For example, *Cappellari et al.* (1976) were pioneers in their analysis of spacecraft trajectories, although *Brouwer and Clemence* (1961) gave an introduction to orbital error analysis. The first application of statistical inversion theory (see references above) to asteroid orbit determination was explored by *Muinonen and Bowell* (1993), who studied non-Gaussian observational noise using an $\exp(-|x|^{\kappa})$-type probability density ($\kappa = 2$ is the Gaussian hypothesis). In certain cases, they observed noticeable differences between Gaussian and non-Gaussian modeling, but overall, the advantages of applying ad hoc non-Gaussian hypotheses remain questionable.

Carpino et al. (2002) discussed a method for assessing statistically the performance of a number of observatories that have produced large amounts of asteroid astrometric data, the goal being to use statistical characterization to improve asteroid positional determination. In their analysis, which encompassed all the numbered asteroids, they computed best-fitting orbits and corresponding O-C astrometric residuals for all the observations used. They derived some empirical functions to model the bias from each contributing observatory — sometimes almost an arcsec — as well as some non-Gaussian characteristics. These functions could be used to estimate correlation coefficients among datasets. *Bykov* (1996) has pursued a similar goal, using Laplacean orbit computation, as did *Hernius et al.* (1997), who study large numbers of multiple-apparition asteroids.

In a recent study, completing the work in *Muinonen and Bowell* (1993), *Muinonen et al.* (2001) found that nonlinear statistical techniques, in contrast to the linear approximations in section 2.2, are not automatically invariant in orbital element transformations. Interestingly, without proper regularization via a Bayesian *a priori* probability density, even using the same orbital elements at different epochs can affect the probabilistic interpretation, rendering the interpretation dependent on the epoch chosen. Muinonen et al. then put forward a simple *a priori* probability density — the square root of the determinant for the Fisher information matrix computed for given orbital elements — that guarantees invariance analogous to that of the linear approximation.

There are several practical consequences of allowing for the invariance. First, the probabilistic interpretation no longer depends on which orbital elements — that is, Keplerian, equinoctial, or Cartesian — are being used. Second, the interpretation is specifically independent of the epoch of the orbital elements. Third, the definition of the linear approximation by *Muinonen and Bowell* (1993) is revised and now includes the determinant part of the *a posteriori* probability density. Finally, the invariance principle is highly pertinent to all inverse problems in which the parameter uncertainties are expected to be substantial.

2.2. Linear Approximation

In the linear approximation (e.g., *Muinonen and Bowell,* 1993; *Muinonen et al.,* 2001), the relationship between orbital elements and sky-plane positions at the observation dates is linearized, and the regularizing *a priori* probability density is assumed constant. The resulting orbital-element probability density is Gaussian, and can be sampled using standard Monte Carlo techniques. Typically, the linear approximation works remarkably well for multiapparition MBAs, and even for single-apparition NEAs having observational arcs spanning a few months, whereas it can fail for multiapparition transneptunian objects (TNOs).

Using the rigorous statistical ranging technique (section 2.4), the validity of the linear approximation can be studied thoroughly for different orbital elements. For short-arc orbits, there is a hint that Cartesian elements, in comparison to Keplerian and equinoctial elements, offer the best linear approximation. Intuitively, this can be understood in a straightforward way. First, because of the Gaussian hypothesis for the observational error, the two transverse positional elements at a given observation date tend to be Gaussian. Second, because the transverse velocity elements are related to the difference of two sky-plane positions having Gaussian errors, the two transverse velocity elements are also almost Gaussian. The line-of-sight position and velocity elements can, however, be remarkably non-Gaussian.

Using the linear approximation, *Muinonen et al.* (1994) carried out orbital uncertainty analysis for the more than 10,000 single-apparition asteroids known at the time. Though of limited application, that analysis captured the dramatic increase of orbital uncertainties for short-arc orbits and led naturally to a study, via eigenvalues, of the covariance matrix by *Muinonen* (1996) and *Muinonen et al.* (1997). The latter found that there exists a bound, as a function of observational arc and number of observations, outside which the linear approximation can be applied, and inside which nonlinearity dominates the inversion. This notion is in agreement with the experience by B. Marsden (personal communications, early 1990s) that covariance matrices of the linear approximation can be misleading for short-arc orbits.

Chodas and Yeomans (1996, 1999), relying on the linear approximation, described orbit computation techniques that can incorporate both optical and radar astrometry, including a Monte Carlo algorithm that uses a square-root information filter. *Bielicki and Sitarski* (1991) and *Sitarski* (1998) developed random orbit-selection methods, paying particular attention to multivariate Gaussian statistics, outlier rejection, and proper weighting of observations. Their approach is based on the linear approximation, where random linear deviations are added to least-squares standard orbital elements. *Milani* (1999) discussed the linear approximation using six-dimensional confidence boundaries.

The ultimate simplification of the linear approximation is a one-dimensional line of variations along the principal eigenvector of the covariance matrix (*Muinonen,* 1996), a precursor to the more general one-dimensional curves of variation used in semilinear approximations (section 2.3). However, as pointed out by *Muinonen et al.* (1997) and *Milani* (1999), there are substantial risks in applying the line-of-variations method.

2.3. Semilinear Approximations

The problem raised by nonlinear effects has been stated several times, starting from *Muinonen and Bowell* (1993), who defined the rigorous, non-Gaussian *a posteriori* probability density and developed a Monte Carlo method for it. Their work constitutes the backbone of many of the theoretical developments over the past decade.

A cascade of one-dimensional semilinear approximations follows from the notion that the complete differential correction procedure for six orbital elements is replaced, after fixing a single orbital element [mapping parameter; for example, the semimajor axis or perihelion distance; cf. *Bowell et al.* (1993) and *Muinonen et al.* (1997)], by an incomplete procedure for five orbital elements. Varying the mapping parameter and repeating the algorithm allows one to obtain a one-dimensional, nonlinear curve of variation, along the ridge of the *a posteriori* probability density in the phase space of the orbital elements.

While the incomplete differential correction procedure has been used by several researchers over the decades, only *Milani* (1999), in what he terms the multiple-solution technique, has systematically explored its practical implementation. In Milani's technique, the mapping parameter is the step along the principal eigenvector of the covariance matrix computed in the linear approximation. However, the covariance matrix and the eigenvector are recomputed after each step, allowing efficient tracking of the probability-density ridge. Because of the nonlinearity of the inverse problem for short-arc asteroids, Milani's multiple-solution technique has turned out to be particularly successful in many of the applications described in section 3 and in *Milani et al.* (2002). The nonlinear technique is very attractive and efficient because of its simplicity.

2.4. Statistical Ranging

The linear and semilinear approximations, relying on differential correction procedures, run into severe convergence problems for short observational arcs and/or small numbers of observations. For such circumstances, the true six-dimensional orbital element probability density cannot generally be collapsed into a single dimension.

Virtanen et al. (2001) devised a completely general approach to orbit estimation that is particularly applicable to short orbital arcs, where orbit computation is almost always nonlinear (see also *Muinonen,* 1999). They term it "statistical ranging," and use Monte Carlo selection of orbits in orbital-element phase space. From two observations, angular deviations in right ascension and declination are chosen and topocentric ranges are assumed by randomly sampling the sky plane and line-of-sight positions. Then large numbers of sample orbits are computed from the pairs of heliocentric rectangular coordinates to map the *a posteriori* probability density of the orbital elements. In a follow-up paper, *Muinonen et al.* (2001) studied the invariance in transformations between different element sets and developed a Spearman rank correlation measure for the validity of the linear approximation. They were also able to accelerate the computational efficiency of the Monte Carlo method by up to two orders of magnitude. Computations for most short-arc objects — including, for example, all single-apparition TNOs — are carried out in seconds to minutes. For longer

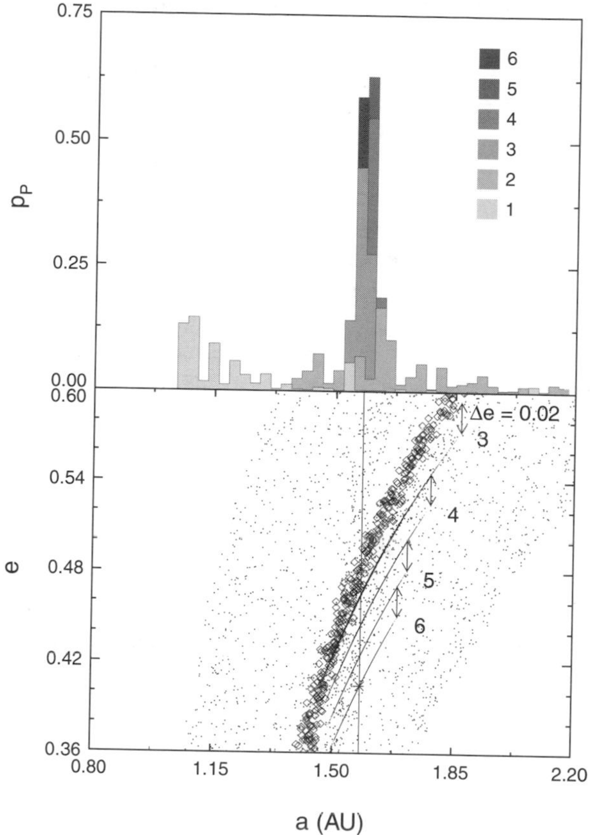

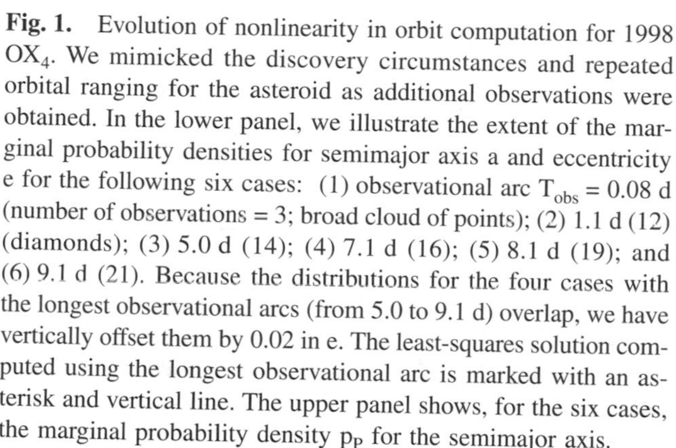

Fig. 1. Evolution of nonlinearity in orbit computation for 1998 OX_4. We mimicked the discovery circumstances and repeated orbital ranging for the asteroid as additional observations were obtained. In the lower panel, we illustrate the extent of the marginal probability densities for semimajor axis a and eccentricity e for the following six cases: (1) observational arc T_{obs} = 0.08 d (number of observations = 3; broad cloud of points); (2) 1.1 d (12) (diamonds); (3) 5.0 d (14); (4) 7.1 d (16); (5) 8.1 d (19); and (6) 9.1 d (21). Because the distributions for the four cases with the longest observational arcs (from 5.0 to 9.1 d) overlap, we have vertically offset them by 0.02 in e. The least-squares solution computed using the longest observational arc is marked with an asterisk and vertical line. The upper panel shows, for the six cases, the marginal probability density p_P for the semimajor axis.

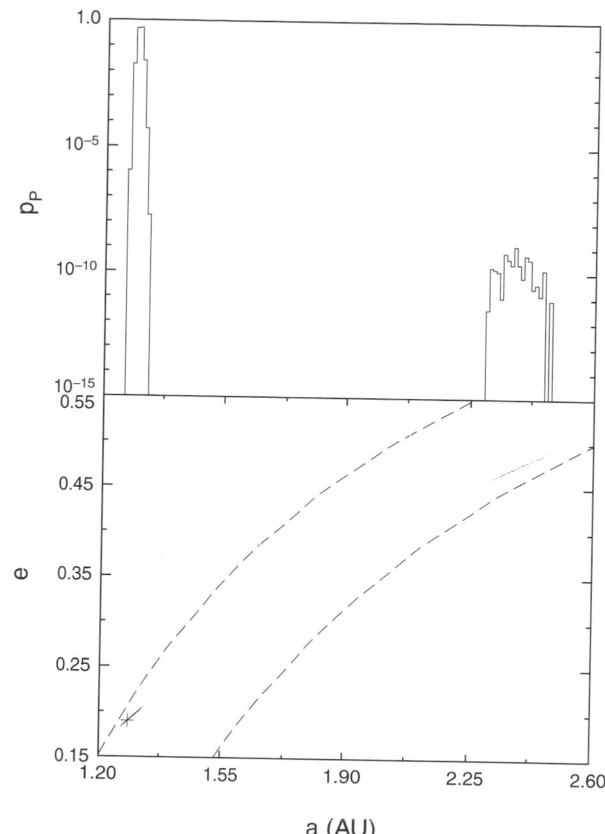

Fig. 2. Bimodal orbital solutions for 1999 XJ_{141} (format is similar to that of Fig. 1). The solutions occur in the regions around (a ~ 1.28 AU, e ~ 0.18), which contains the current orbit (cross) following identification with 1993 OM_7 (see MPEC 2000-C34), and another region around a ~ 2.3 AU. Dashed lines separating Earth-crosser and Amor orbits (upper) and Amor and Mars-crosser orbits (lower) indicate that both solution regions imply an Amor type asteroid. The upper panel shows the marginal probability density p_P for the semimajor axis.

arclengths (say, several weeks for NEOs and MBOs), mapping the six-dimensional orbital-element phase-space region is tedious. In such cases, the ranging technique is more practical for detailed studies, such as exploring the validity of the linear approximation, than routine orbit computation. The statistical ranging technique can be applied to asteroids having only two observations, providing estimates on six orbital elements based on only four data points. The intriguing mathematical questions involved will be the subject of future study.

Figures 1 and 2 show some of the results that can be obtained from statistical ranging. In Fig. 1, the evolution of possible orbits of the Apollo-type asteroid 1998 OX_4 can be followed as the observational arc gradually lengthens. Although the initial 2-h discovery arc leads to possible orbits extending to e ~ 0 and a ~ 180 AU, the probability that the asteroid is an Apollo type is as high as 74%. After being recovered one night later (second case, diamond symbols), Apollo classification is confirmed at very nearly 100% probability. One discordant observation on the second night (see also Fig. 7) is easily recognizable because of the existence of eight additional observations on the same night; it was thus omitted from the orbit computation. In the upper plot, the narrowing of the marginal probability density in a has become very pronounced after only 5 d of observation.

The asteroid 1999 XJ_{141} was discovered near 90° solar elongation on December 13, 1999, followed till December 24, and then lost. It was only the realization that the observations of 1999 XJ_{141} led to a double solution — not uncommon for asteroids discovered near quadrature — that allowed the asteroid to be linked to independently discovered 2000 BN_{19} and then 1993 OM_7, the last observed on only two nights. The statistical ranging technique very nicely resolves the orbital-elements ambiguity, as shown in Fig. 2. Applying statistical ranging to the observations made between December 13 and 24, 1999, and imposing a 1.1-arcsec rms residual shows a bimodal distribution of possible solutions (lower panel). However, as shown in the upper panel, the probability mass is overwhelmingly concentrated in the region at smaller a, leaving vanishingly small probability for the larger-a, e solutions. A continuum of orbital solutions between the two regions can be generated only by allowing unexpectedly large rms residuals of several arcseconds.

Virtanen et al. (2002) automated the statistical ranging technique for multiple asteroids, and applied statistical ranging to TNOs. Whereas TNO orbital element probability densities tend to be very complicated (in contrast to those of NEAs), the projected distributions of sky-plane uncertainties are remarkably unambiguous. Virtanen et al. found that dynamical classification of TNOs is reasonably secure for orbital arcs exceeding 1.5 yr and were able to use the known TNO population (as of February 2001) to make a statistical assessment of orbital types (see section 3.4).

2.5. Other Advances

Approaches based on the variation of topocentric range and the angle to the line of sight have been developed by *McNaught* (1999) and *Tholen and Whiteley* (2002). The former technique is not readily amenable to uncertainty estimation, whereas the latter does sample the orbital-element probability density. In essence, both are methods that, instead of using two Cartesian positions, as does the statistical ranging technique, use Cartesian position and velocity. However, the statistical ranging technique has an advantage over the position-velocity techniques: For ranging, the two angular positions involved are well fitted by every trial solution without iteration at that stage; for position-velocity techniques, there is an *a priori* requirement that the two positions be timed so as to allow for an accurate estimation of motion.

Although at opposition the apparent motion is usually a good indicator of an object's topocentric distance (*Bowell et al.*, 1990), the situation is quite different at smaller solar elongation, where confusion among the various asteroid classes is much greater and multiple orbital solutions may occur.

Parallax is a very useful tool for short-arc orbital determination. For surveys of the opposition region it can be useful for improving the first guess of a Väisälä solution when observations from two nights are available. The improvement can be significant if, from at least one night, data are distributed over an arc of several hours. In a study on searching for NEOs at small solar elongation, *Hills and Leonard* (1995) suggested discriminating NEA candidates not only using the traditional method based on their large daily motion, but also by taking advantage of the parallax resulting from observations at different locations. For this purpose they considered a baseline of 1 R (Earth radius). Such a baseline could conveniently be implemented by observing at appropriate intervals using, for example, the Hubble Space Telescope (e.g., *Evans et al.*, 1998), but it is not easily attainable by groundbased observers. *Boattini and Carusi* (1997) suggested that, even using a baseline of 0.1–0.2 R between stations, good range estimates could be made up to distances of 0.5 AU if the observations are performed near simultaneously.

Marsden (1991) modified the Gauss-Encke-Merton (GEM) method of three-observation orbit computation to deal with mathematically ill-posed cases (usually for short arcs). He extended the procedure to pertain to cases having two observations only, which has a useful application to newly discovered putative Earth-crossing asteroids, where Väisälä's method yields only aphelic orbits. More recently, *Marsden* (1999) developed the concept of *lateral orbits*, in which the true anomaly is ±90° and the object is on the *latus rectum*. According to G. V. Williams (personal communication), these methods are used at the MPC to compute the ephemeris uncertainty for newly discovered NEAs, as published at http://cfa-www.harvard.edu/iau/NEO/TheNEOPage.html. L. K. Kristensen (in preparation, 2002) looked at the accuracy of follow-up ephemerides based on short-arc orbits and found that the geocentric distance and its time derivative are the essential parameters that determine the accuracy of predicted positions.

Yeomans et al. (1987, 1992) assessed the improvement of NEA orbits when radar range and Doppler data are combined with optical astrometric observations. They found that long-arc orbits were improved only modestly, whereas short-arc orbits could be improved by several orders of magnitude, often to the extent that additional near-term optical observations would afford little further improvement. The relative weights of the optical and radar observations follow the Bayesian theory by *Muinonen and Bowell* (1993): Radar time-delay and Doppler astrometry are incorporated by defining the optical *a posteriori* probability density to be the radar *a priori* probability density. In practice, the inverse problem then culminates in the study of the joint χ^2 of the optical and radar data. Regularization, such as that described in section 2.1, is not needed because the linear approximation is usually valid when radar data are incorporated. Both OrbFit and NEODyS have recently been upgraded to account for radar astrometry, too.

Bernstein and Khushalani (2000) devised a linearized orbit-fitting procedure in which accelerations are treated as perturbations to the inertial motions of TNOs. The method produces ephemerides and uncertainty ellipses, even for short-arc orbits. They applied it to devising a strategy for computing accurate TNO orbits from a minimum number of observations, a matter of importance when one considers the expense of using large telescopes.

3. PREDICTION

Here our fundamental question is, "What is an asteroid's positional uncertainty in space or on the sky plane?" Answering this question allows observers to search for asteroids whose positions are inexactly known, and to know when it is useful or necessary (from the standpoint of orbit improvement) to secure additional observations. It allows orbit computers to identify images of asteroids in archival photographic or CCD media, forms the basis of Earth-impact probability studies, and guides spaceflight planners in pointing their instruments and in making course corrections. This type of analysis belongs to the prediction (or projection) problem of applying the orbital-element probability density derived in section 2.

3.1. Ephemeris Uncertainty, Precovery, and Recovery

In the linear approach to finding a lost asteroid, an asteroid is sought on a short segment of a curve, known as the line of variations, which is defined as the projection of the asteroid's orbit on the sky plane. It is generally computed by varying the asteroid's mean anomaly M. The extent of the line of variations that needs to be searched is readily computed using *Muinonen and Bowell*'s (1993) linear approximation and forms the basis of a number of URLs developed by Bowell and colleagues at Lowell Observatory (see http://asteroid.lowell.edu). For example, the asteroid orbit file astorb.dat contains current and future ephemeris uncertainty information for more than 127,000 asteroids, and the Web utility obs may be used to build a plot indicating when an asteroid is observable, based in part on sky location, brightness, and maximum acceptable ephemeris uncertainty.

Although a good approximation when the sky-plane uncertainty is small, the linear approach is usually inappropriate for short observational arcs or for completely lost asteroids. *Milani*'s (1999) semilinear theory identified sources of nonlinearity. The observation function, or the mapping of given orbital elements onto the sky plane, is also nonlinear. Such effects cannot be neglected if one or more close approaches to a planet occur in the interval between the two. Because the observation function is nonlinear, the image of the confidence ellipsoid will appear as an ellipse on the sky plane only if the ellipse is quite small.

Milani (1999) and *Chesley and Milani* (1999) considered semilinear methods of ephemeris prediction. They explored the validity of the linear approximation and introduced methods for propagating the orbital uncertainty using semilinear approximations. In uncertain cases, they recommended that the linear approximation be used only as a reference.

The computation of the observation function is relatively straightforward as the solution of differential equations is not involved. Thus it is possible to overcome the difficulty of nonlinearity on the sky plane by computing it at many points. A burden of this approach is that delineating the six-dimensional confidence region uniformly requires a large number of samples, many of which may result in predictions close to each other when mapped onto the two-dimensional space of the observations. *Milani* (1999) developed the concept of the *semilinear confidence boundary*, in which an algorithm is used to identify points, in the confidence ellipsoid, that map close to the boundary of the confidence region. This approach has been used for the recovery and precovery of asteroids.

Following the pioneering efforts of the Anglo-Australian Near-Earth Asteroid Survey (AANEAS) program from 1990 to 1996 (*Steel et al.,* 1997), newly discovered NEOs are now routinely sought in photographic and CCD archives. Indeed, several groups of "precoverers" are currently active. For example, the DLR-Archenhold Near Earth Objects Precovery Survey (DANEOPS) group in Germany (http://earn.dlr.de/daneops) and the team of E. Helin and K. Lawrence at the Jet Propulsion Laboratory have made significant contributions, focusing on NEOs. DANEOPS was able to precover 1999 AN_{10}, the first asteroid known to have a nonzero collision probability before its precovery (*Milani et al.,* 1999), and Helin and Lawrence precovered 1997 XF_{11}. *Boattini et al.* (2001a) started an NEA precovery program called the Arcetri NEO Precovery Program (ANEOPP) in 1999. Although most of their identifications have been made using good initial orbits, ANEOPP has successfully made use of the mathematical tools developed for OrbFit to guide precovery attempts of NEOs having poor orbits. The multiple-solution approach has proved to be a most effective search method. Other groups such as that at Lowell Observatory and amateur astronomers such as A. Lowe have conducted extensive precovery programs on asteroids in general.

Precovery activity has spawned other scientifically interesting results in the past decade. For example, the identification of the Amor-type asteroid (4015) 1979 VA with periodic Comet Wilson-Harrington (1949 III) on a plate from 1949 (*Bowell and Marsden,* 1992) showed clear evidence that the distinction between asteroids and comets is not straightforward.

Recovering completely lost asteroids (i.e., asteroids having similar probability of being located at any orbital longitude) requires different techniques. Using a semilinear approach, *Bowell et al.* (1993) attempted to recover then-lost (719) Albert by eliminating, through searches of archival photographic media, parts of the perihelion distance space where the asteroid was not seen ("negative observations"). (They would have succeeded had sufficient resources been available for the search.) Several groups are searching for 1937 UB (Hermes), observed (poorly) for less than 5 d. For example, ANEOPP is using the multiple-solution technique to examine regions of Palomar and U.K. Schmidt plates. Currently, ~10–15% of the orbital element phase space has been examined. L. D. Schmadel and J. Schubart, reporting on their work on Hermes at http://www.rzuser.uni-heidelberg.de/~s24/hermes.htm, indicate that they used a method similar to that devised by *Bowell et al.* (1993).

For asteroids having very poor orbit solutions, though not completely lost, the semilinear confidence boundary is very effective for planning a search, in particular because the recovery region is usually very elongated; even when its length is tens of degrees, the width may only be a few arcseconds. One can also compute a sequence of solutions along the major axis of the confidence region. This multiple-solution approach has recently been used to recover some

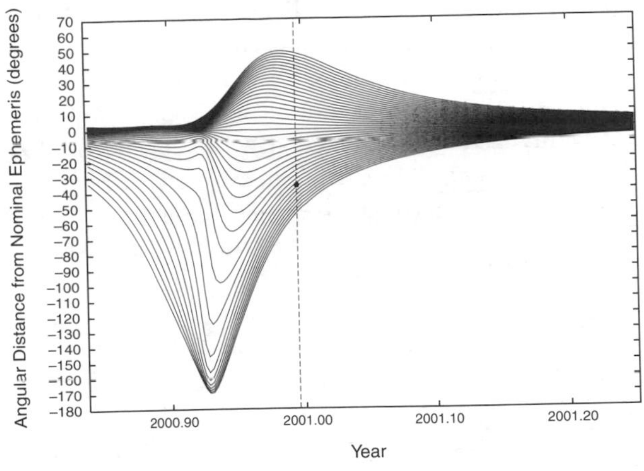

Fig. 3. Precovery of 1998 KM_3. Forty-one continuous curves, each corresponding to one of 41 orbits generated using the multiple-solution method out to ±5σ (actually, 201 orbits were used to sample the confidence region to be searched, but for clarity only one in five of them is shown). The angular distance from the nominal solution y = 0 is plotted against time. The precovery location of 1998 KM_3 is indicated by the dot.

particularly difficult targets. For example, Fig. 3 shows how recovery was achieved of 1998 KM_3, an Apollo asteroid whose sky-plane uncertainty was several tens of degrees at the time of the observations (see *Boattini et al.,* 2001b). The plot shows the angular distance from the nominal solution given by y = 0, which is orbit #21, and the confidence region is populated by 20 orbits on each side of the nominal solution (actually, 201 orbits were considered in the original computation). Note that the asymmetric shape of the confidence region is an indicator of nonlinearity. 1998 KM_3 was recovered on December 30, 2000, the asteroid being located between orbits #8 and #9.

The challenges of recovering TNOs center on accurate ephemeris prediction. Although projection of orbital uncertainties onto the sky plane results in almost linear distributions, the need to use large telescopes for follow-up observations mandates a firm understanding of possible TNO orbits. An efficient method of ephemeris uncertainty ellipse computation was presented in *Bernstein and Khushalani* (2000). However, to map the ephemeris uncertainties for very short-arc TNOs (a few days to weeks of observations) over intervals of a year or more, rigorous methods, such as statistical ranging, are appropriate. Even for longer arcs (two apparitions, say), TNO orbits may be very imperfectly known. Figure 4 shows the results of applying statistical ranging to 1998 WU_{31}, a TNO discovered in the fall of 1998 and recovered in the fall of 1999. The sky-plane uncertainty distributions are based on computations that assume 1.0-arcsec observational noise (probably an overestimate by a factor of 3–4). The correlations between ephemeris and orbital-element uncertainties — especially between right ascension and semimajor axis or eccentricity — provide a good way of reducing the sky-plane search region. One could also make use of the orbital-elements distribution of known TNOs to limit the search region even more, though such an approach could bias against successfully following up TNOs having unusual orbits.

3.2. Identification

A great deal of effort has recently been expended to establish identifications among asteroid observations, that is, trying to link observations of two independently discovered asteroids and hence showing that they are the same object. Orbit computation plays a key part in the asteroid identification process, as *Milani et al.* (1996, 2000a, 2001), *Sansaturio et al.* (1996, 1999) show in developing identification algorithms (see also http://copernico.dm.unipi.it/identifications). Such algorithms can be classified into two types, depending on the number and time distribution of the two available sets of astrometric observations.

The first type is based on a comparison of the orbital elements, as computed by a least-squares fit to the observations making up each arc, and provides so-called orbit identification methods. These methods use a cascade of filters based on identification metrics and take into account the difference in the orbits weighted by the uncertainty of the solution, as represented by the covariance matrix. Orbit pairs that pass all the filters are subjected to accurate computation by differential correction to compute an orbit that fits all or most of the observations.

In the second type of algorithm, an orbit computed for one of the arcs is used to assess the observations that make up the other arc. These constitute so-called observation attribution methods. The key to such methods lies in finding a suitable single observation, termed an attributable, that fits the observations of the second arc and at the same time contains information on both the asteroid's position on the sky and its apparent motion. As in the orbit-identification algorithm, the attribution method uses a cascade of filters that allow comparison in the space of the observations and in that of the apparent motion, and point to a selection of the pairs to be subjected to least-squares fits to all the observations.

A. Doppler and A. Gnädig (Berlin), starting in 1997, developed an automated search technique. After generating ephemerides, candidate matches are identified in surrounding regions whose sizes are dependent on the properties of the initial orbit. The candidate matches are then subjected to differential orbit correction, and, for apparently successful fits, additional identifications are sought. Doppler and Gnädig have to date made about 10,000 identifications. One of them involved 2001 KX_{76} (now numbered as 28976), which turned out to be the largest known small body after Pluto.

Applying statistical ranging to TNOs, *Virtanen et al.* (2002) showed that the identification problem can sometimes coincide with the inverse problem. They applied statistical ranging routinely to most of the multiapparition TNOs; that is, they succeed in deriving sample orbital elements linking observations at different apparitions without any preparatory work. The first and last observations used

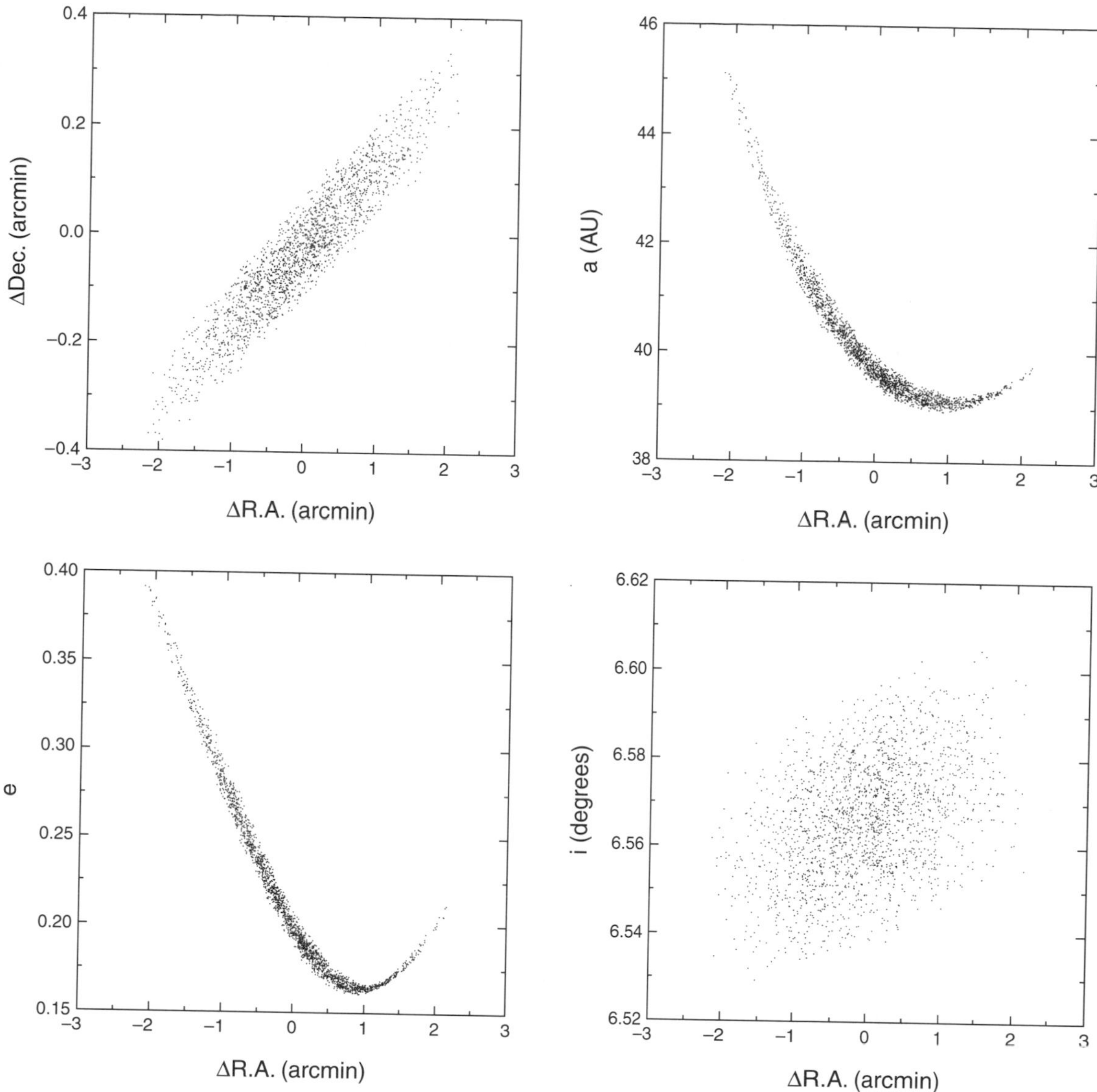

Fig. 4. Statistical ranging ephemeris prediction for the TNO 1998 WU_{31}. Sky-plane uncertainty distributions are predicted on October 11, 2000, one year after the last observation.

in the computation are sometimes separated by several years.

3.3. Collision Probability

Because this topic is the subject of another chapter in this book (see *Milani et al.,* 2002), we comment only on some aspects of orbit computation. The collision probability was mathematically defined by *Muinonen and Bowell* (1993) in their Bayesian probabilistic work on the inverse problem.

A first indicator of Earth or other planetary close approach is afforded by the value of the minimum orbital intersection distance (MOID) (*Bowell and Muinonen,* 1993; *Milani et al.,* 2000b), the closest distance between an asteroid's orbit and that of a planet. An asteroid's Earth MOID can usually be determined surprisingly accurately, even for short observational arcs. In the absence of close planetary approaches or strong mean-motion resonance, MOIDs do not change by more than 0.02 or 0.03 AU per century. Therefore, if it can be established that an asteroid's planetary MOID is larger than, say, 0.05 AU one can be sure that no planetary collision is imminent. If it is less, further calculations are necessary. As a next step, *Bonanno* (2000) derived an analytical formulation for the MOID, together with its uncertainty, which allows one to identify cases of asteroids that might have nonnegligible Earth impact probability even though the nominal values of their Earth MOIDs are not small.

To assess the possibility of an Earth impact in 2028 by asteroid 1997 XF_{11}, *Muinonen* (1999) developed the concept of the maximum likelihood collision orbit: Given the time window for the collisional study and the astrometric observations, it is possible to derive the collision orbit, or "virtual impactor" in the terminology of *Milani et al.* (2000c), that is statistically closest to the maximum likelihood orbit. A safe upper bound for the collision probability follows by assigning all the probability outside the confidence boundary of the maximum likelihood collision orbit to the collision. For 1997 XF_{11}, the rms values of the maximum likelihood collision orbit were 3.81 arcsec in right ascension and 4.06 arcsec in declination, rendering the collision probability negligible (*Muinonen,* 1998). The collision orbits derived in that study in spring 1998 were the first of their kind. Soon thereafter, collision orbits became a subject of routine computation for essentially all research groups assessing collision probability.

Whereas maximum likelihood collision orbits can be very useful for assessing primary close approaches, they do not yield the actual collision probability. Based on the technique of statistical ranging, and motivated by the approximate one-dimensional semilinear techniques of *Milani et al.* (2000c), *Muinonen et al.* (2001) developed a general six-dimensional technique to compute the collision probability for short-arc asteroids. In that technique, every sample orbital element set from ranging is accompanied by a one-dimensional line of acceptable orbital solutions. Along these lines, collision segments are derived, allowing estimation of the general collision probability (as well as the phase space of collision orbits).

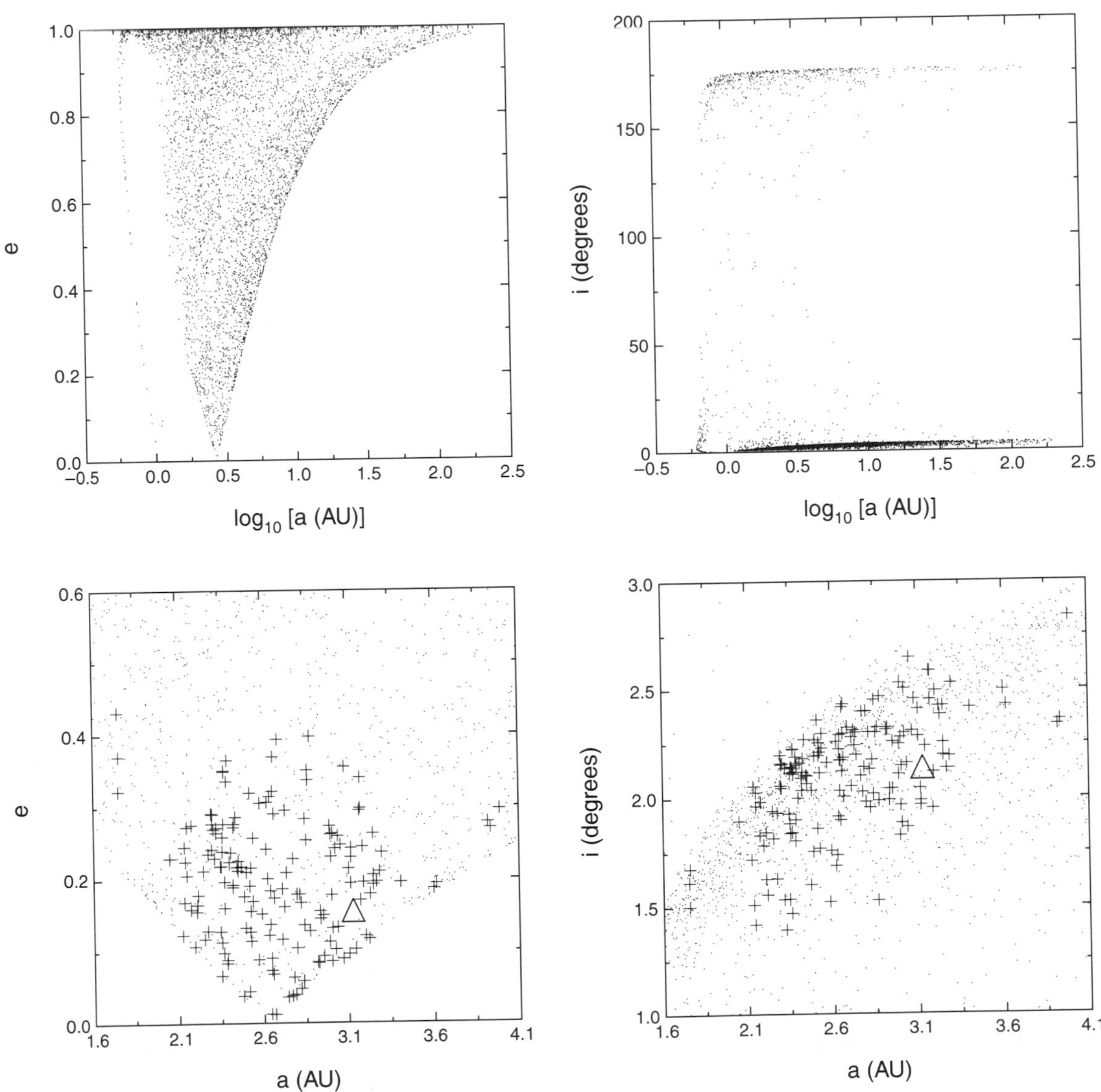

Fig. 5. Classification using statistical ranging of 1990 RM_{18}, observed over a 3-d arc. Five thousand orbits were computed assuming 1.0-arcsec observational noise. The upper panels show the extent of the *a posteriori* probability in the (a, e) and (a, i) planes. In the lower panels, the a, e, i probability density of known multiapparition asteroids has been incorporated as *a priori* information. Accurate orbital elements of 1990 RM_{18} are indicated by a triangle.

3.4. Classification

One of the motivations for developing methods of initial orbit computation is the classification of asteroid orbital type. The question is particularly relevant for Earth-approaching asteroids, which need to be recognized from the very moment of discovery to ensure follow-up. Because of the small numbers of observations, we call for special caution in the discussion of probabilities.

In his study of detection probabilities for NEAs, *Jedicke* (1996) derived probabilities that a given asteroid is an NEA based solely on its rate of motion. His approach, entirely analytical, requires modeling the orbital element and absolute magnitude distribution of the asteroid population.

The era of statistical orbit inversion has given us the means of addressing the classification problem probabilistically. The orbital element probability density can be mapped rigorously using the method of statistical ranging (*Virtanen et al.,* 2001), which leads in a straightforward way to an assessment of probabilities for different orbital types. Figure 5 illustrates the case of 1990 RM_{18} (now numbered as 10343). Using no *a priori* information, 1990 RM_{18} had a 28% probability of being an Apollo asteroid and also had nonzero (0.2%) probability of being an Aten. Incorporating the probability density of known multiapparition asteroids as *a priori* information secures its classification as an MBO with 99.9% certainty, with peaks near the widely separated Flora and Themis family regions (the asteroid's orbital elements are now known to be consistent with Themis family membership).

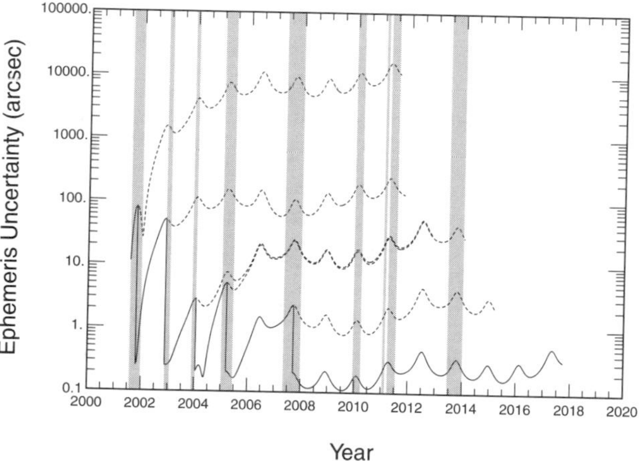

Fig. 6. Hierarchical Observing Protocol (HOP), requiring very few observations, shows how 2001 LS_8 could be numbered in 2007. Vertical bars show windows of observability from Lowell Observatory, assuming a V-band limiting magnitude of 19.0, a minimum solar elongation of 60°, a minimum galactic latitude |b| = 20°, and a southern declination limit of –30°. The continuous curve shows the time evolution of the ephemeris uncertainty, assuming linear error propagation as described by *Muinonen and Bowell* (1993), vertical drops being when additional ±1-arcsec-accuracy observations are called for. Dashed curves indicate subsequent ephemeris uncertainty evolution in the absence of follow-up observations.

Dynamical classification of TNOs has been attempted using statistical ranging by *Virtanen et al.* (2002), as described in section 2.4. Although lacking a clear definition of the various dynamical classes, their results show that classification is very uncertain for observational arcs shorter than six months. Their population analysis does not support the existence of near-circular orbits beyond 50 AU, although the existence of objects having low to moderate eccentricities is not ruled out. They call for more detailed dynamical studies to address the question of the edge of the transneptunian region.

3.5. Optimizing Observational Strategy

Bowell et al. (1997) devised a method they term Hierarchical Observing Protocol (HOP) to choose optimum times to observe asteroids so their orbits would be improved as much as possible. The method combines some of the results from *Muinonen and Bowell* (1993) and *Muinonen et al.* (1994). Of course, the result of a given optimization strategy depends on the choice of metric. HOP is predicated on minimizing the so-called leak metric, a metric based on longitude, eccentricity, and angular momentum (k) (*Muinonen and Bowell,* 1993). HOP results from four semiempirical rules are: (1) To be numberable, an asteroid's ephemeris uncertainty should be <2 arcsec for at least 10 yr into the future. (Except for TNOs, this criterion produces very similar results to the quite different algorithm used at the MPC.) (2) When a new astrometric observation has been incorporated into an orbit computation, the resulting improvement in its orbital elements leads to a proportionate improvement in the asteroid's ephemeris uncertainty. (3) Soon after discovery, ephemeris uncertainty increases approximately as the 3/2 power of the time interval since observation. (4) For multiapparition asteroids, peak ephemeris uncertainty usually occurs near opposition; sometimes (for NEAs) at conjunction and/or greatest solar elongation.

Taken together, the rules imply that, following discovery, observations should be made at geometrically increasing intervals (supplemented by observations used as a check in cases where there would be temporal outliers that, if erroneous, could unknowingly distort an orbit). Thereafter, observations should be made at times of maximum ephemeris uncertainty, depending on observing circumstances. Prioritized lists of asteroid targets may be generated, in accordance with the precepts of HOP, using http://asteroid.lowell.edu/cgi-bin/koehn/hop, and an asteroid's preferred observational cadence is amenable to automated evaluation (see http://asteroid.lowell.edu/cgi-bin/koehn/obsstrat). Figure 6 illustrates a possible observing sequence for 2001 LS_8, an MBO observed, at this writing, on 4 nights over a 15-d interval. Observations are called for in five future apparitions before the criteria for numberability are met, though those at the beginning of 2004 are likely redundant as the subsequent ephemeris uncertainty is little reduced.

In a similar vein, *Kristensen* (2001) studied the distribution of observations leading to the smallest future ephemeris uncertainty. He also determined the best time to make observations to maximally improve an orbit.

Optimizing a strategy for follow-up observations of NEAs is a more complex task than one aimed at MBOs. However, as with MBO strategies, the first requirement is that appropriate orbital accuracy be achieved to allow recovery at a future apparition. At the Spaceguard Central Node, NEAs are classified into four categories, according to observational urgency. The categories are established based on a figure of merit computed from the product of five terms, including sky-plane ephemeris uncertainty, a parametrization of the difficulty in recovering the target at the next opportunity, and a measure of the rate at which the asteroid's magnitude is increasing and solar elongation decreasing. A similar algorithm has been developed to prioritize the need for recovering NEAs at a second apparition. Additional information is given at http://spaceguard.ias.rm.cnr.it/SSystem/SSystem.html. For groundbased observations, one sometimes has just a few days to observe an NEO before it disappears into the sunlit sky. For this reason, the greatest effort in an NEO observing coordination service concerns the provision of appropriate weights for each factor involved in the computation of observational urgency. For an MBO discovered not too far from opposition, there are always at least two or three months available for follow-up observations, and the asteroid can readily be recovered within two years of discovery.

4. FUTURE RESEARCH

We comment on anticipated advances in orbit estimation techniques, and on the development of three areas: (1) the continuing acceleration of astrometric data acquisition, especially as large-area surveys go fainter and provide more accurate positions; (2) the impending advent of microarcsec-accuracy spacecraft observations; and (3) the future of some established applications.

4.1. Deep Wide-Field Surveys

It is likely that the current broad-areal-coverage NEA surveys, most of which use meter-class telescopes (see *Stokes et al.,* 2002), will gradually migrate to larger instruments. If the limiting magnitude of such surveys eventually reaches V_{lim} = 22 or fainter, one may anticipate that the discovery rate of asteroids will increase by 1–2 orders of magnitude. Moreover, larger-aperture telescopes inevitably have larger image scales, resulting in increased image sampling and more accurate astrometry.

However, further advances in moving-object detection using larger telescopes will require substantial new modeling. Some will involve optimizing the detection rate by improving detection algorithms, population distribution models (i.e., a, e, i, H, with H being the absolute magnitude), the partitioning of ecliptic/nonecliptic searching, and understanding the interplay among observational requirements of the various classes of moving objects (fixed objects, too, if telescopes are not dedicated to moving-object detection). Also, unless several larger telescopes team together in the search for moving objects — which seems unlikely in the next decade — then a given telescope will have to carry out its own moving-object follow-up observations. How this requirement drives observational strategy is a very complicated matter. Some of the considerations for future work on orbit computation are:

1. The number of visits to a region per lunation requires modeling the observational lifetimes of faint discoveries and the accuracy of their orbits, many of which need to be good enough that recovery in a subsequent apparition is assured. TNOs are easier to deal with, but the question of how to sample optimally, so as to establish the best orbits from the minimum number of observations, needs more work.

2. At faint magnitude limits, night-to-night linkage can, depending on the observing cadence, become difficult. For example, the sky-plane surface density of asteroids at V = 24 mag is expected to be several hundred/deg^2, implying a mean separation of a few arcminutes. What are the observational requirements (how many nights/lunation and at what cadence) that will lead to unambiguous linkage of all or most of the detections?

3. The possibility of detecting moving objects by shifting and coadding images taken on different nights needs to be further developed. Potentially, the minimum detectable S/N ratio could be substantially reduced, perhaps by a factor of 2, which could result in a doubling of the number of moving-object detections. Shifting and coadding has been attempted for single-night observations made during pencil-beam searches for TNOs (*Gladman et al.,* 1998), where the variance of sky-plane motion is small and predictable, but never for the cases of NEAs and MBOs, whose rates of motion change — sometimes substantially — from night to night, and, if very close, during a night. Work is needed on understanding the range of nonlinear motion to be expected in a given observational situation. Shifting and coadding has also been used for follow-up work, even in rich starfields (*Hainaut et al.,* 1994; *Boehnhardt et al.,* 1997).

4. More work needs to be done on the general problem of moving-object trail detection. The only astronomical treatment we know of is by *Milani et al.* (1996), who showed that faint asteroid trails can be detected at S/N ratios <1. However, the method is far too slow to be routinely applied.

5. Confident near-real-time assessment of an asteroid's orbital class, perhaps using the methods described in section 3.4, will be *de rigueur* for deep surveys, during which a single large telescope could detect on the order of 100 NEAs/h. Planning follow-up observations, which will frequently require large telescopes, will necessitate streamlining current methods, such as that used at the Spaceguard Central Node.

6. One can envisage the creation of a multidimensional all-sky "plot" of moving objects, which could be used to visualize objects available for observation or recovery according to chosen criteria, such as ephemeris uncertainty (equivalently, orbit improvement) or imminence of loss.

It is well known that the accuracy of ordinary CCDs as polarization detectors is no better than 0.1%. CCDs have electrical matrix detector arrays that contain thin photosensitive silicon layers. The path of incoming light depends on its polarization state, so computed astrometric position may be shifted. To the best of our knowledge, this effect has not been studied in the laboratory.

4.2. Spacebased Observations

The Global Astrometric Interferometer for Astrophysics (GAIA; http://astro.estec.esa.nl/GAIA) is an approved ESA mission that will greatly impact the observation of asteroids (*Lindegren and Perryman,* 1996). The GAIA satellite will observe the whole sky to V ≈ 20 mag down to solar elongations as small as 35°. It will generate astrometric data accurate to about 10 microarcsec for a 15th-magnitude detection, degrading to about 1 milliarcsec for the faintest stars. GAIA will observe asteroids in the Milky Way and at small solar elongations, and the extraordinary quality of the astrometric data will make a remarkable contribution to asteroid science. GAIA will also spur asteroid orbit-improvement work, including for example, detection of movement of an asteroid's photocenter with respect to its mean apparent motion and modeling in terms of surface light scattering properties, shape, and rotation; detection of and orbital characterization binary asteroids, determination of physical parameters (mass, size, density, taxonomy); and a means of testing general relativity.

ESA's *BepiColombo* mission (http://sci.esa.int/home/bepicolombo), with a launch expected in 2010, will primarily address the study of Mercury. However, from such a privileged site so close to the Sun, a small telescope mounted on the spacecraft will be used to search primarily for NEAs. The astrometric accuracy of the system will be comparable to that of current groundbased optical data, and parallactic ranging might be possible because of the spacecraft's 2.4-h polar orbit. Otherwise, use of the observations will pose a new challenge in orbit computation and identification because groundbased follow-up observations will require "triangulation" from two widely separated observing sites (GAIA faces a similar challenge). When successful, the triangulation will result in exceptionally accurate orbital elements.

4.3. Future Applications

Hoffmann (1989) and *Standish and Hellings* (1989), in their work on asteroid mass determination using the perturbative effects of asteroid-asteroid close encounters and on the orbital motion of Mars, respectively, were optimistic that radar observations would soon present an opportunity to improve the accuracy of asteroid mass estimates. Although the number of radar observations has increased greatly of late (see *Ostro et al.,* 2002), they have yet to be applied to mass determination. Instead, the accelerating accumulation of astrometric data has allowed an increased number of asteroid masses, but only a slight improvement in their accuracy (e.g., see *Hilton,* 1999; *Michalak,* 2000, 2001). Perhaps greater hope of improving mass estimates lies in increased accuracy (milliarcsec?) from spacecraft observations. In principal, if effects such as photocenter/center-of-mass offset can be well modeled, one could expect an order-of-magnitude improvement in mass determination accuracy and a great increase in the number of asteroids whose masses could usefully be measured. However, such improvement would probably require that high-accuracy observations be conducted for a decade or two.

Timing of stellar occultations provides a direct means of determining the projected size and shape of an asteroid (*Millis and Elliot,* 1979). During the past three years, the availability of HIPPARCOS and similar high-accuracy star catalogs has greatly helped improve the accuracy of predicted ground-track locations (*Dunham,* 2001). Asteroid ephemeris uncertainty is now generally small enough that initial predictions are accurate to a few tens of milliarcsec, and little, if any, last-minute astrometric refinement of asteroid orbits is necessary. However, results from the recent advances in orbit computation have not yet been applied to occultation prediction.

Recovering completely lost asteroids requires the intelligent use of observational and manpower resources. We have already mentioned the cases of (719) Albert and 1998 KM_3 (section 3.1), but it is clear that many more targets could be found, some of them possibly hazardous NEAs, among the growing number of targets. The methods in use at present, mainly concentrating on the examination of archived media, will in future be augmented by the provision of deep and wide NEA search images (section 4.1). As this happens, it will be possible to make systematic (and, one hopes, automated) searches for lost asteroids such as 1937 UB (Hermes). One way this could be accomplished is based on the technique of "position/motion," in which a family of orbits (perhaps generated via the multiple-solution or statistical-ranging methods) produces ephemeris positions and brightnesses that can be searched for target asteroids moving at predicted rates. If the motion is unusual, as it often is for NEAs, secure identification can be based on a single night's observation. The aforementioned technique appears first to have been applied to the Amor 1997 NJ_6, which was identified with 1999 WZ (*Boattini and Williams,* 1999).

4.4. Prospects in Orbit Computation

Using high-accuracy astrometric observations, modeling the angular deviation of an asteroid's photometric center from its center of mass will become increasingly important. As a first approximation, these so-called photocenter effects can be accounted for using simple ellipsoidal or more advanced convex-hull models for asteroid shapes, along with simple light-scattering modeling (see *Kaasalainen et al.,* 2002).

Long observational arcs can be of major significance for the study of relativistic effects, diurnal and seasonal Yarkovsky effects (e.g., *Vokrouhlický et al.,* 2000), radiation pressure effects (e.g., *Vokrouhlický and Milani,* 2000), and effects caused by the deviation of the photocenter from the

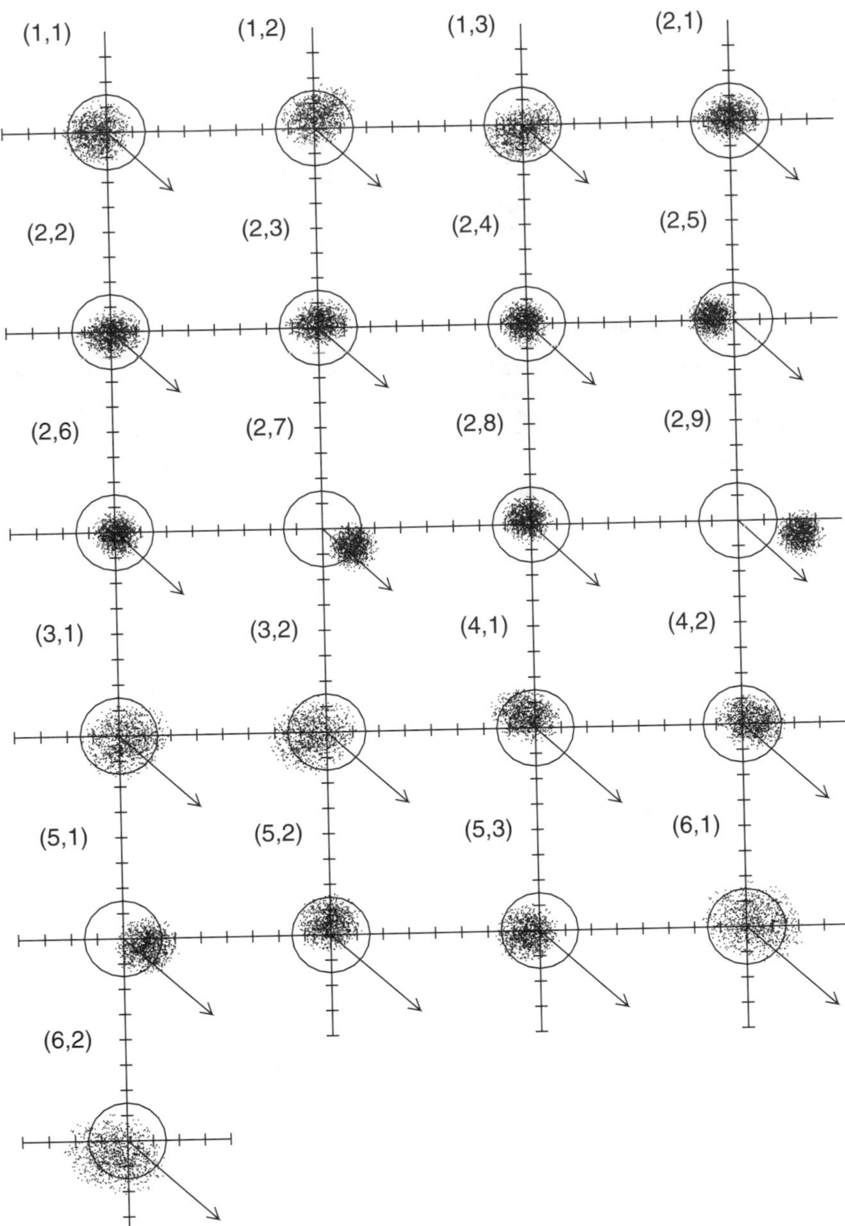

Fig. 7. Distributions of O-C residuals in right ascension (abscissae) and declination (ordinates) for 21 observations of 1998 OX_4 made between July 24 and August 4, 1998, and computed using statistical ranging in which 0.5-arcsec observational noise was assumed. Original observations are located at the origins of the 8 × 8-arcsec panels, and the surrounding circles show the applied 3σ regimes. Labels indicate the night and observation number on that night; for example, label (4,2) pertains to the second observation on night four. Three-minute motion vectors are plotted for each observation.

center of mass. In this context, understanding the absorption, scattering, and emission of radiation by asteroid surfaces is of utmost importance (see *Muinonen et al.,* 2002). Also, we envisage that the mutual gravitational attraction of asteroids could be manifest as systematic astrometric errors for long-arc asteroids and could therefore provide new opportunities for asteroid mass determination. However, it may still be too difficult to discriminate among the various effects.

The statistical-ranging technique shows great future potential and will be subject to further optimization. First, by revisiting the algorithm for selecting the two topocentric range intervals, the technique will be made more efficient in the generation of sample orbital elements in the proximity of the maximum likelihood solution. Second, by developing efficient numerical integrators, the technique will become more readily applicable to longer observational arcs. We note that the statistical-ranging technique is particularly suited to identifying TNOs, as shown by the systematic application of the method to many multiapparition orbits (*Virtanen et al.,* 2002).

Using statistical ranging, low-accuracy astrometric observations could be dealt with in a routine manner. Such observations can result from spacecraft (e.g., IRAS), from imperfect groundbased observations (e.g., the 1937 discovery observations of Hermes; see also section 3.1) and from future groundbased discovery programs optimized for discovery efficiency with the intention that follow-up observations are made to secure accurate orbital elements.

From our experience with statistical ranging, a single motion vector can be consistent with asteroid classes ranging from near-Earth to transneptunian objects. Such degeneracy can only be removed by follow-up observations. Also, near-stationary-point astrometry can lead to broad orbital-element distributions for near-Earth asteroids, whose orbits are typically considerably more constrained (see section 2.4).

Systematic errors in astrometry could be identified using statistical ranging (or other statistical methods) as illustrated by Fig. 7, which shows results for 1998 OX_4, a now-lost Apollo asteroid that has been the subject of intense scrutiny because of its possible impact with Earth (e.g., *Milani et al.,* 2000c). The distribution (not single value) of O-C residuals for each observation has been computed assuming 0.5-arcsec observational noise. The offsets between the observations and theoretical distributions could reveal the presence of timing errors, correlations between observations made by the same observatory, or errors in reference star catalogs. For 1998 OX_4, timing (along-the-track) errors cannot be identified because of the asteroid's slow rate of motion. Outliers [for example, observations (2,7) and (2,9)] are clearly revealed, suggesting a future application of the ranging technique to the weighting of observational data.

It is plausible that new techniques will allow more efficient computation of planetary collision probabilities for asteroids. Techniques based on statistical ranging can be optimized by further work on the region of phase space that points to collisions (see *Milani et al.,* 2002).

In cases of convergent differential correction, the phase space of orbital element solutions can be efficiently mapped using a Monte Carlo simulation of random observational errors (also noted by *Chesley and Milani,* 1999). The least-squares fit offers reference right ascension and declination values for each observation date. By repeatedly applying random errors to the reference values, convergent differential correction allows the computation of an arbitrarily large number of sample orbital elements that describe the orbital-element probability density. Invariance of the probabilistic analysis needs to be assured, as described in section 2.

4.5. Concluding Speculation

It has become ever more difficult to view the classical orbit computation problem as nonstatistical. Indeed, the post-*Asteroids II* era has witnessed the triumph of statistical techniques. We believe that, within a decade, classical methods of orbit computation will become largely obsolete, as the ever-continuing acceleration of computer speed should allow an equivalence between astrometric observations and orbital-element probability density. We hope to be able to transform very rapidly from the observations to end products such as ephemerides, ephemeris uncertainty, or asteroid classification, without the intermediary of orbital elements that continue to be essential in long-term dynamical studies of asteroids. One consequence could be a major change in the way the MPC passes data from observers to clients. It could be that the MPC need only maintain the observational database. All other operations, such as ephemeris generation, could consist of client-requested automatic analytical and statistical treatment of the astrometric data. An embryonic form of this concept was put forward by *Bowell* (1999).

Acknowledgments. We thank A. Doppler, A. Gnädig, and M. E. Sansaturio for information on their methods of identifying asteroids and J. Piironen for guidance regarding polarimetric effects in CCD chips. We are grateful to reviewers S. R. Chesley and A. Milani for perceptive and useful comments, and we thank M. Carpino and O. Bykov for their input on observation statistics. E.B.'s contribution has been funded by NASA Grant NAG5-4742, Lowell Observatory, and the University of Helsinki. K.M. is grateful to Osservatorio Astronomico di Torino for its kind hospitality during his sabbatical stay and to the Academy of Finland for funding his research. J.V.'s funding was provided by the University of Helsinki Science Foundation, and A.B.'s was provided by IASF-CNR and ASI.

REFERENCES

Bernstein G. and Khushalani B. (2000) Orbit fitting and uncertainties for Kuiper belt objects. *Astron. J., 120,* 3323–3332.

Bielicki M. and Sitarski G. (1991) Nongravitational motion of comet P/Swift-Gehrels. *Acta Astron., 41,* 309–323.

Boattini A. and Carusi A. (1997) Atens: Importance among near-Earth asteroids and search strategies. *Vistas in Astronomy, 41,* 527–541.

Boattini A. and Williams G. V. (1999) *Minor Planet Electronic Circular,* 1999–W27.

Boattini A., D'Abramo G., Forti G., and Gal R. (2001a) The Arcetri NEO precovery program. *Astron. Astrophys., 375,* 293–307.

Boattini A., D'Abramo G., Stokes G., and Marsden B. G. (2001b) *Minor Planet Electronic Circular,* 2001–J05.

Boehnhardt H., Babjon J., and West R. M. (1997) An optimized detection technique for faint moving objects on a star-rich background. A search for the nucleus of comet 46P/Wirtanen. *Astron. Astrophys., 320,* 642–651.

Bonanno C. (2000) An analytical approximation for the MOID and its consequences. *Astron. Astrophys., 360,* 411–416.

Bowell E. (1999) A new protocol for the operation of the Minor Planet Center. *Bull. Am. Astron. Soc., 31,* 1116.

Bowell E. and Marsden B. G. (1992) *(4015) 1979 VA = Comet Wilson-Harrington (1949 III).* IAU Circular 5585.

Bowell E. and Muinonen K. (1993) Earth-crossing asteroids and comets: Groundbased search strategies. In *Hazards Due to Comets and Asteroids* (T. Gehrels, ed.), pp. 241–258. Univ. of Arizona, Tucson.

Bowell E., Chernykh N. S., and Marsden B. G. (1989) Discovery and follow up of asteroids. In *Asteroids II* (R. P. Binzel et al., eds.), pp. 21–38. Univ. of Arizona, Tucson.

Bowell E., Skiff B. A., Wasserman L. H., and Russell K. S. (1990) Orbital information from asteroid motion vectors. In *Asteroids, Comets, Meteors III* (C.-I. Lagerkvist et al., eds.), pp. 19–24. Uppsala Universitet, Uppsala, Sweden.

Bowell E., Wasserman L. H., Muinonen K., McNaught R. H., and West R. M. (1993) A search for the lost asteroid (719) Albert. *Bull. Am. Astron. Soc., 25,* 1118.

Bowell E., Koehn B. W., Wasserman L. H., and Muinonen K. (1997) Hierarchical observing protocol (HOP) for asteroids. *Bull. Am. Astron. Soc., 29,* 974.

Brouwer D. and Clemence G. M. (1961) *Methods of Celestial*

Mechanics. Academic, New York. 598 pp.

Bykov O. P. (1996) CCD-observations of asteroids: Accuracy and the nearest perspectives for application of the Laplacian orbit determination method. In *Dynamics, Ephemerides, and Astrometry of the Solar System* (S. Ferraz-Mello et al., eds.), p. 451. IAU Symposium 172.

Cappellari J. O., Velez C. E., and Fuchs A. J. (1976) *Mathematical Theory of the Goddard Trajectory Determination System.* Goddard Space Flight Center Report X-582-76-77.

Carpino M., Milani A., and Chesley S. R. (2002) Error statistics of asteroid optical astrometric observations. *Icarus,* in press.

Chesley S. R. and Milani A. (1999) Nonlinear methods for the propagation of orbital uncertainty. Paper 99-148 in AAS/AIAA Space Flight Mechanics Meeting, 16–19 August 1999, Girdwood, Alaska.

Chodas P. W. and Yeomans D. K. (1996) The orbital motion and impact circumstances of comet Shoemaker-Levy 9. In *The Collision of Comet Shoemaker-Levy 9 and Jupiter* (K. S. Noll et al., eds.), pp. 1–30. Cambridge Univ., Cambridge.

Chodas P. W. and Yeomans D. K. (1999) Orbit determination and estimation of impact probability for near-Earth objects. Paper 99-002 in 21st Annual AAS Guidance and Control Conference, 3–7 February 1999, Breckenridge, Colorado.

Danby J. M. A. (1988) *Fundamentals of Celestial Mechanics.* Willmann-Bell, Richmond. 466 pp.

Dunham D. W. (2001) Planetary occultations for 2001. *Sky and Telescope, 101,* 116–120.

Evans R. W. and 22 colleagues (1998) Asteroid trails in Hubble Space Telescope WFPC2 images: First results. *Icarus, 131,* 261–282.

Gauss C. F. (1809) *Theory of the Motion of the Heavenly Bodies Moving about the Sun in Conic Sections.* Dover, New York. 326 pp. (1963).

Gauss F. S. (1999) Precise star positions for positioning applications. In *Proceedings of the Nautical Almanac Office Sesquicentennial Symposium* (A. D. Fiala and S. J. Dick, eds.), pp. 311–323. U.S. Naval Observatory, Washington, DC.

Gladman B., Kavelaars J. J., Nicholson P. D., Loredo T. J., and Burns J. A. (1998) Pencil-beam surveys for faint trans-Neptunian objects. *Astron. J., 116,* 2042–2054.

Hainaut O., West R. M., Smette A., and Marsden B. G. (1994) Imaging of very distant comets: Current and future limits. *Astron. Astrophys., 289,* 311–324.

Hernius O., Lagerkvist C.-I., Lindgren M., Tancredi G., and Williams G. V. (1997) UESAC — The Uppsala-ESO survey of asteroids and comets. *Astron. Astrophys., 318,* 631–638.

Hills J. G. and Leonard P. T. (1995) Earth-crossing asteroids: The last days before Earth impact. *Astron. J., 109,* 401–417.

Hilton J. L. (1999) U.S. Naval Observatory ephemerides of the largest asteroids. *Astron. J., 117,* 1077–1086.

Hoffman M. (1989) Asteroid mass determination: Present situation and perspectives. In *Asteroids II* (R. P. Binzel et al., eds.), pp. 228–239. Univ. of Arizona, Tucson.

Jedicke R. (1996) Detection of near-Earth asteroids based upon their rates of motion. *Astron. J., 111,* 970–982.

Kaasalainen M., Mottola S., and Fulchignoni M. (2002) Asteroid models from disk-integrated data. In *Asteroids III* (W. F. Bottke Jr. et al., eds.), this volume. Univ. of Arizona. Tucson.

Kristensen L. K. (2001) Optimum observing strategies. *Astron. Nachr., 322,* 47–56.

Lehtinen M. S. (1988) On statistical inversion theory. In *Theory and Applications of Inverse Problems* (H. Haario, ed.), pp. 46–57. Longman Scientific and Technical, New York.

Lindegren L. and Perryman M. A. C. (1996) GAIA: Global astrometric interferometer for astrophysics. *Astron. Astrophys. Suppl. Ser., 116,* 579–595.

Marsden B. G. (1991) The computation of orbits in indeterminate and uncertain cases. *Astron. J., 102,* 1539–1552.

Marsden B. G. (1999) Ephemerides of small bodies of the solar system. In *Proceedings of the Nautical Almanac Office Sesquicentennial Symposium* (A. D. Fiala and S. J. Dick, eds.), pp. 333–351. U.S. Naval Observatory, Washington, DC.

McNaught R. H. (1999) Unusual object 1998 DK_{36} (Letter to the Editor). *J. Brit. Astron. Assoc., 109,* 299.

Menke W. (1989) *Geophysical Data Analysis: Discrete Inverse Theory.* Academic, San Diego. 289 pp.

Michalak G. (2000) Determination of asteroid masses. I. (1) Ceres, (2) Pallas and (4) Vesta. *Astron. Astrophys., 360,* 363–374.

Michalak G. (2001) Determination of asteroid masses. II. (6) Hebe, (10) Hygiea, (15) Eunomia, (52) Europa, (88) Thisbe, (444) Gyptis, (511) Davida and (704) Interamnia. *Astron. Astrophys., 374,* 703–711.

Milani A. (1999) The asteroid identification problem I. Recovery of lost asteroids. *Icarus, 137,* 269–292.

Milani A., Villani A., and Stiavelli M. (1996) Discovery of very faint small asteroids by automated trail detection. *Earth, Moon, Planets, 72,* 257–262.

Milani A., Chesley S. R., and Valsecchi G. B. (1999) Close approaches of asteroid 1999 AN_{10}: Resonant and non-resonant returns. *Astron. Astrophys., 346,* L65–L68.

Milani A., La Spina A., Sansaturio M. E., and Chesley S. R. (2000a) The asteroid identification problem III: Proposing identifications. *Icarus, 144,* 39–53.

Milani A., Gronchi G. F., and Bonanno C. (2000b) Minimal orbital intersection distance: Computation, uncertainty, and secular evolution. *Bull. Am. Astron. Soc., 32,* 863.

Milani A., Chesley S. R., Boattini A., and Valsecchi G. B. (2000c) Virtual impactors: Search and destroy. *Icarus, 144,* 12–24.

Milani A., Sansaturio M. E., and Chesley S. R. (2001) The asteroid identification problem IV: Attributions. *Icarus, 151,* 150–159.

Milani A., Chesley S. R., Chodas P. W., and Valsecchi G. B. (2002) Asteroid close approaches: Analysis and potential impact detection. In *Asteroids III* (W. F. Bottke Jr. et al., eds.), this volume. Univ. of Arizona, Tucson.

Millis R. L. and Elliot J. L. (1979) Direct determination of asteroid diameters from occultation observations. In *Asteroids* (T. Gehrels, ed.), pp. 98–118. Univ. of Arizona, Tucson.

Muinonen K. (1996) Orbital covariance eigenproblem for asteroids and comets. *Mon. Not. R. Astron. Soc., 280,* 1235–1238.

Muinonen K. (1998) Upper bounds for the Earth-XF_{11} collision probability. *Bull. Am. Astron. Soc., 30,* 1030.

Muinonen K. (1999) Asteroid and comet encounters with the Earth: Impact hazard and collision probability. In *The Dynamics of Small Bodies in the Solar System: A Major Key to Solar System Studies* (A. E. Roy and B. A. Steves, eds.), pp. 127–158. Kluwer, Dordrecht.

Muinonen K. and Bowell E. (1993) Asteroid orbit determination using Bayesian probabilities. *Icarus, 104,* 255–279.

Muinonen K., Bowell E., and Wasserman L. H. (1994) Orbital uncertainties of single-apparition asteroids. *Planet. Space Sci., 42,* 307–313.

Muinonen K., Milani A., and Bowell E. (1997) Determination of initial eigenorbits for asteroids. In *Dynamics and Astrometry of Natural and Artificial Celestial Bodies* (I. M. Wytrzyszczak et al., eds.), pp. 191–198. Kluwer, Dordrecht.

Muinonen K., Virtanen J., and Bowell E. (2001) Collision probability for Earth-crossing asteroids using orbital ranging. *Cel. Mech. Dyn. Astron., 81,* 93–101.

Muinonen K., Piironen J., Shkuratov Yu. G., Ovcharenko A., and Clark B. E. (2002) Asteroid photometric and polarimetric phase effects. In *Asteroids III* (W. F. Bottke Jr. et al., eds.), this volume. Univ. of Arizona, Tucson.

Ostro S. J. (1989) Radar observations of asteroids. In *Asteroids II* (R. P. Binzel et al., eds.), pp. 192–212. Univ. of Arizona, Tucson.

Ostro S. J., Hudson R. S., Benner L. A. M., Giorgini J. D., Magri C., Margot J.-L., and Nolan M. C. (2002) Asteroid radar astronomy. In *Asteroids III* (W. F. Bottke Jr. et al., eds.), this volume. Univ. of Arizona, Tucson.

Press W. H., Flannery B. P., Teukolsky S. A., and Vetterling W. T. (1994) *Numerical Recipes in FORTRAN: The Art of Scientific Computing,* 2nd edition. Cambridge Univ., Cambridge. 963 pp.

Sansaturio M. E., Milani A., and Cattaneo L. (1996) Nonlinear optimization and the asteroid identification problem. In *Dynamics, Ephemerides and Astrometry of the Solar System* (S. Ferraz-Mello et al., eds.), pp. 193–198. Kluwer, Dordrecht.

Sansaturio M. E., Milani A., La Spina A., and Muinonen K. (1999) The asteroid identification problem. In *Impacts of Modern Dynamics in Astronomy* (J. Henrard and S. Ferraz-Mello, eds.), pp. 363–364. IAU Colloquium 172, Kluwer, Dordrecht.

Sitarski G. (1998) Motion of the minor planet 4179 Toutatis: Can we predict its collision with the Earth? *Acta Astron., 48,* 547–561.

Standish E. M. and Hellings R. L. (1989) A determination of the masses of Ceres, Pallas, and Vesta from their perturbations upon the orbit of Mars. *Icarus, 80,* 326–333.

Steel D. I., McNaught R. H., Garradd G. J., Asher D. J., and Russell K. S. (1997) AANEAS: A valedictory report. *Austr. J. Astron., 7,* 67–77.

Stokes G. H., Evans J. B., and Larson S. M. (2002) Near-Earth asteroid search programs. In *Asteroids III* (W. F. Bottke Jr. et al., eds.), this volume. Univ. of Arizona, Tucson.

Stone R. C., McDonald S. W., Elliot J. L., and Bowell E. (2000) 10199 Chariklo stellar occultation candidates: 1999–2005. *Astron. J., 119,* 2008–2017.

Teets D. and Whitehead K. (1999) The discovery of Ceres: How Gauss became famous. *Mathematics Magazine, 72,* 83–93.

Tholen D. and Whiteley R. J. (2002) Short arc orbit solutions and their application to near Earth objects. *Icarus,* in press.

Väisälä Y. (1939) Eine einfache Methode der Bahnbestimmung. *Mitt. Sternw. Univ. Turku, Suomalainen Tiedeakatemia, 1,* 1–32.

Virtanen J., Muinonen K., and Bowell E. (2001) Statistical ranging of asteroid orbits. *Icarus, 154,* 412–431.

Virtanen J., Muinonen K., and Bowell E. (2002) Statistical ranging of orbits for transneptunian objects. *Icarus,* in press.

Vokrouhlický D. and Milani A. (2000) Direct solar radiation pressure on the orbits of small near-Earth asteroids. *Astron. Astrophys., 362,* 746–755.

Vokrouhlický D., Milani A., and Chesley S. R. (2000) Yarkovsky effect on small near-Earth asteroids: Mathematical formulation and examples. *Icarus, 148,* 118–138.

Yeomans D. K., Chodas P. W., Keesey M. S., Ostro S. J., Chandler J. F., and Shapiro I. I. (1992) Asteroid and comet orbits using radar data. *Astron. J., 103,* 303–317.

Yeomans D. K., Ostro S. J., and Chodas P. W. (1987) Radar astrometry of near-Earth asteroids. *Astron. J., 94,* 189–200.

Near-Earth Asteroid Search Programs

Grant H. Stokes and Jenifer B. Evans
Massachusetts Institute of Technology Lincoln Laboratory

Stephen M. Larson
University of Arizona

The discovery of the potentially hazardous near-Earth asteroid (NEA) component of the minor-planet population has been enhanced by better detecting and computing technology. A government mandate to quantify the terrestrial impact hazard and to detect 90% of all NEAs larger than 1 km can be realistically addressed. The characteristics, capabilities, and strategies of the major search programs illustrate the challenges and solutions toward meeting the Spaceguard goal. This chapter reviews the historical context of early asteroid detection and of the current and anticipated search programs. It describes the search systems and discusses challenges in maximizing the NEA detection rate.

1. INTRODUCTION AND BACKGROUND

The past decade has witnessed an explosion in the number of cataloged asteroids. Of the more than 35,000 numbered asteroids discovered in the last 200 years, 62% of these were found in the past decade. Similarly, of the 1785 NEAs that are now known, 89% have been discovered in the past 10 years. During this time the total number of observations published by the Minor Planet Center (MPC) has grown from 792,439 in 1992 to 11,626,118 in 2002, and the total routinely grows by more than 300,000 per month. This dramatic increase in the observation and discovery rate has been enabled by the charge-coupled-device (CCD)-based search systems that have replaced their film-based predecessors. This chapter will give a brief historical perspective of asteroid searches and an overview of general search strategies and specific search systems in existence today. In addition, we will address the current status and issues facing the continuing search for asteroids.

2. HISTORICAL SEARCH METHODS

Ceres, the first asteroid discovered, was found in 1801 by Giuseppe Piazzi, who was in the process of using direct visual telescopic observations to construct a star catalog. Ceres was recorded as a star the first night it was noticed, but moved the next night it was observed. Thus, Ceres became the first of a new class of objects based on its motion relative to the fixed background stars. This original method of discovering asteroids by differentiating their motion from that of the stars has not changed since 1801; however, the technology available to apply that original method has evolved significantly. Search technology evolved from direct telescope observations, which led to the discovery of the first few hundred asteroids, to film-based observations in the 1890s. The first asteroid discovered by photography, (323) Brucia, was discovered by Max Wolf in 1891. The most common photographic-based search methods were either (1) to take a single long exposure and visually inspect the plate or film for trails as the asteroid moved during the exposure, or (2) to take two successive exposures and visually compare them for moving objects. The latter was accomplished either with a blink comparator to show the asteroid moving between two positions, or by viewing the two images stereoscopically, rendering the moving object as suspended above the plane of the nonmoving background. Asteroid search systems based on photography dominated the field until the mid-1980s when modern CCD- and computer-based systems emerged.

Photographic search techniques employed during the 1900–1990 era had a number of advantages over direct visual observation techniques: Large areas could be photographed with much less observer fatigue and a permanent record of the observations was maintained. Photographic systems, however, have two limitations that are overcome by the new solid-state detector technology of the 1990s. First, the sensitivity of the film is poor, with a QE of ~1% for objects having the solar spectrum, and secondly, the moving object detection process relied on humans who are subject to fatigue and nonuniform performance.

The conversion to CCD-based detection systems and computer processing of the data was first demonstrated by the Spacewatch system in 1984 (*Gehrels,* 1986). At the time, the readout rate of astronomical-quality CCD detectors was very slow (~25 Kpx/s for optimal photometry), and the computation capability that could be applied to processing the data was quite limited by today's standards. T. Gehrels and R. McMillan developed a search system that accommodated the available technology in a very clever way. They developed a drift-scan technique where the readout rate was clocked to the sidereal drift rate across the CCD. The long scans produced images with minimal pixel-to-pixel (flat-field) variation, and readout overhead was limited to ramp-up and ramp-down fields at either end. The data rate was well matched to the capabilities of the processing equipment available at the time. The use of CCDs enabled much more sensitive search systems due to their ability to achieve quantum efficiencies as high as 65% for objects having the solar spectrum, almost 2 orders of magnitude over that of film. Spacewatch achieved a faint limiting magnitude because of the long integration time achieved as a star drifts across the CCD. Thus, the Spacewatch system was very well tailored for the available telescope and technology.

Evolution to the more commonly used CCD-based "step-stare" asteroid search approach came in late 1995 with the development of the Near-Earth Asteroid Tracking (NEAT) search program (*Pravdo et al.,* 1999), a joint project between NASA's Jet Propulsion Laboratory (JPL) and the U.S. Air Force. NASA/JPL developed a 4080 × 4080 CCD camera that was mounted in the space of the usual Ebsicon camera on a 1-m U.S. Air Force GEODSS (groundbased electro-optical space surveillance) telescope. A step-stare system works much like a film-based search, with multiple images taken over an hour or two. Each image is "exposed" while the telescope tracks at the sidereal rate and is then read out to a computer before the next exposure is taken. Faster, multichannel readout rates generate a much higher data rate requiring greater computing capacity and offer more flexibility in balancing coverage and depth. The NEAT system was able to search more sky than Spacewatch by using a CCD array with 4× as many pixels and a larger angular pixel scale and by using shorter integration times and not going as deep.

Step-stare search systems were also developed in the late 1980s for use by the U.S. Air Force to find and track Earth-orbiting satellites, a process that has many elements in common with searching for asteroids. Both targets traverse the same cluttered background and are differentiated from the background based on their particular motion. Earth orbiting satellites are generally brighter, but move at a larger angular rate. The stringent military requirements for satellite search led MIT's Lincoln Laboratory to develop a specialized search system, consisting of a custom-manufactured, frame-transfer CCD and a high-capacity computer processing system on a 1-m GEODSS-type telescope (*Viggh et al.,* 1998). In March 1998, the search system started operating in an asteroid search role as the Lincoln Near-Earth Asteroid Research project (LINEAR) (*Stokes,* 2000) and has since provided 70% of the worldwide asteroid discovery rate.

3. SEARCH OBJECTIVES AND MOTIVATION

Efforts to find and catalog asteroids over the past 200 years were initially inspired by scientific curiosity and a desire to understand the structure of the solar system. More recently these efforts have been spurred by the NASA search goal and as part of the International Spaceguard effort. The NASA goal, mandated by the US Congress, is to discover 90% of all potential impactors with diameters in excess of 1 km by the year 2008 (*NASA,* 1998). One-kilometer-diameter asteroids are thought to mark the threshold size for globally catastrophic consequences in a collision, and various models indicate there are between 500 and 2100 such objects (*Morrison et al.,* 1992; *Rabinowitz et al.,* 2000; *Bottke et al.,* 2000; *Stuart,* 2000). The ability to achieve the NASA goal with existing and proposed systems can be viewed as a function of the search volume for these systems with respect to the size of the target object. Search volume is defined as the maximum range at which a system can detect an object of a certain diameter multiplied by the area of sky covered. As the diameter of the limiting object decreases, so does the maximum range at which it can be detected, and the search volume therefore decreases. Longer integration times may increase the limiting range but will decrease the area of sky covered due to limited time available for searching, thereby not necessarily increasing the volume. Improved detector and computing technology has stretched the search volume for 1-km objects to make the NASA goal potentially achievable in the future with the current smaller aperture telescopes, and discussion is shifting toward extending the inventory to smaller objects that could cause significant regional damage. For example, a catalog 90% complete to the 300-m size is one of the proposed objectives of the Large-aperture Synoptic Survey Telescope (*NRC,* 2001) recommended by the recent astronomy decadal study. The Spaceguard report (*Morrison et al.,* 1992) estimates that there are between 12,500 and 50,000 NEAs larger than 300 m. Below a few hundred meters in size, the limited search volume of existing or proposed search systems precludes near-term cataloging. Such objects will require a continuous search effort that could also address detection of potentially dangerous long-period comets not amenable to cataloging. Below ~100 m, the mass that survives entry through the atmosphere is likely too small to create widespread destruction.

Despite 200 years of searching for asteroids and the dramatic increase in discoveries and observations in the past decade, we are just now approaching the point of having found approximately one-half of the large NEAs set as the goal by NASA. This state of affairs testifies to the inherent challenges associated with finding asteroids in general and NEAs in particular. Searching for asteroids has all the typical difficulties associated with finding faint objects in a cluttered and complex environment. Specifically, the faint observed magnitudes limit the search volume; search areas and times are denied by the Sun, Moon, and weather; and the cluttered stellar background makes detection difficult, especially in the galactic plane. In fact, the instantaneous image of an asteroid observed from Earth is essentially indistinguishable from a star. Given this state of affairs, asteroid searchers have labored to achieve sustained, high search-rate operations while maintaining good sensitivity. Searchers have attempted to access as much as possible of the total volume where asteroids could reside. This leads to the obvious questions: Where do asteroids reside, where are they most easily detected, and how is the sensitivity and limited search volume best utilized?

To optimize survey efficiency, it is necessary to understand the time-dependent spatial distribution of asteroids, their rates of motion, and apparent brightnesses. This has been covered by *Bowell and Muinonen* (1994) using simulations and by *Harris* (1998). The earlier results are supported by the large quantity of new data collected in recent years. A brief summary with new plots is presented for completeness.

The apparent ecliptic rate of motion with respect to ecliptic longtitude of a typical main-belt asteroid is shown in Fig. 1 along with the rates of motion of similar asteroids with varied semimajor axes. NEAs have a wider distribution of orbital elements than main-belt asteroids and therefore display a wider distribution of apparent angular motions. As the search region moves away from opposition toward 60°, the angular rates decrease and the difference in apparent angular rates between NEAs and main-belt asteroids decrease, making them more difficult to distinguish. In addition, the favorable addition of the Earth's motion to the asteroid's motion is lost. As a result, searching at opposition has traditionally been considered the most productive method for detecting NEAs.

A second factor affecting searches is the apparent visual magnitude. The visual magnitude is a function of the solar phase angle, the distance between the asteroid and Earth, and the distance between the asteroid and the Sun. Consider the apparent magnitude of a 1-km asteroid (H = 18.0) on the ecliptic plane as its location is varied away from opposition. Figure 2 is a contour plot showing how the relative magnitude varies. The intrinsic phase function of the asteroid adds to the peak brightness at opposition, which is added to the increased brightness due to the decreased distance between the asteroid and the observer and between the asteroid and the Sun. The candle shape of Fig. 2 also describes how the relative magnitude varies as an asteroid located at opposition is varied along ecliptic latitude. The maximum brightness occurs at the ecliptic and decreases at higher declinations, much as the maximum occurs at opposition when the ecliptic latitude is held constant.

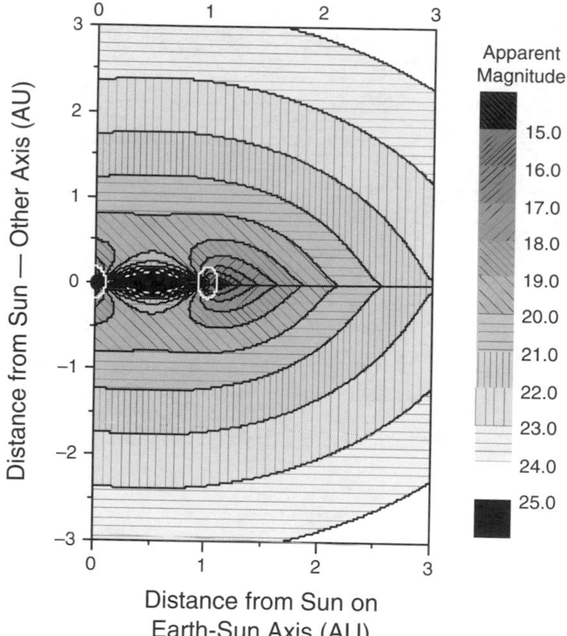

Fig. 2. Visual magnitude of an H = 18.0 asteroid as a function of its ecliptic position.

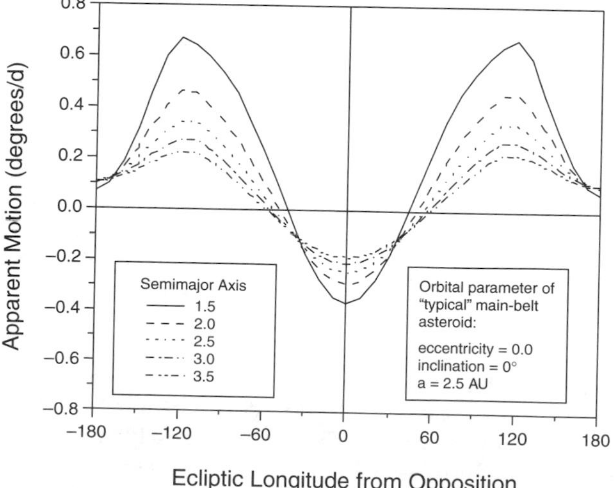

Fig. 1. Ecliptic rates of motion of asteroids.

In summary, it would appear that an observer's best chance of detecting an NEA is to search near the ecliptic at opposition for optimal rates of motion and apparent visual magnitude. Given that all asteroids, regardless of their inclination, will pass through the ecliptic twice at their nodes with each orbit, this is a reasonable approach. This is also the most common practice. A review by *Marsden* (1994) shows that 86% of all the numbered asteroids at that time were discovered within 30° opposition and 96% were discovered within 20° of the ecliptic. However, with the advent of the CCD, and the introduction of full-sky search capabilities by modern surveys, alternative search strategies need to be considered. What are the benefits of looking away from opposition and away from the ecliptic?

A contour plot showing the number of known NEAs per cubic AU as a function of ecliptic plane coordinates is shown in Fig. 3. The plot was generated by computing the percentage of time each known NEA spends in each heliocentric cylindrical bin, then summing over the bins. The current known NEA population, uncorrected for observational bias, was used for this model. The highest density of NEAs is located in the ring around the Sun that includes Earth, and the density of observable NEOs diminishes with geocentric distance. The contour plot also shows potential search directions at various solar elongations (labeled as angle from opposition). Of interest when considering vari-

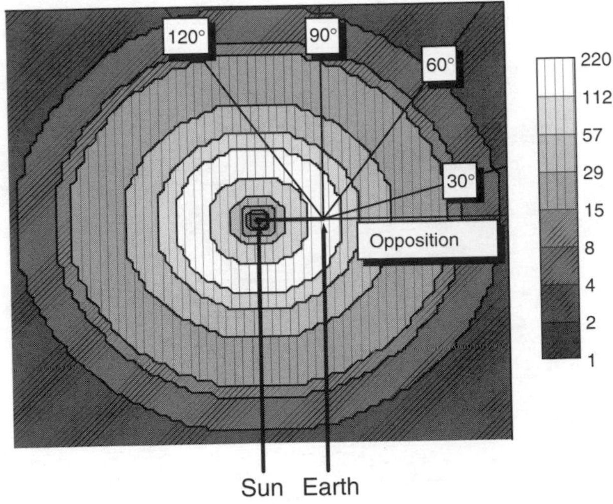

Fig. 3. Density of known NEAs in ecliptic coordinates, and potential search lines.

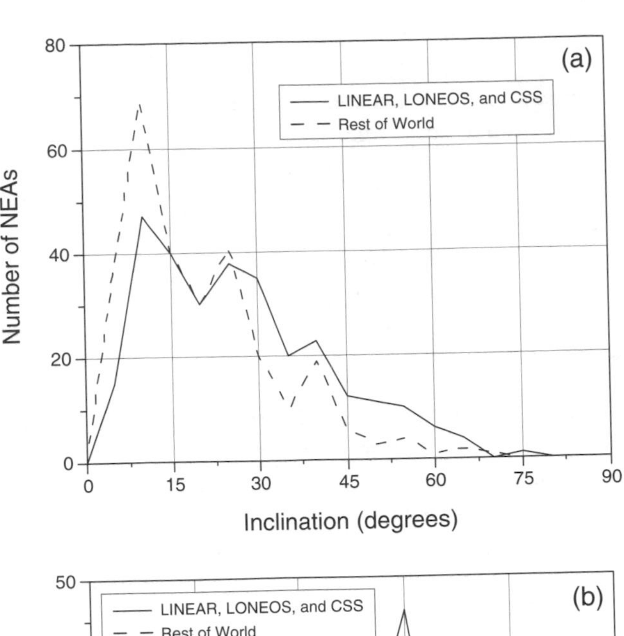

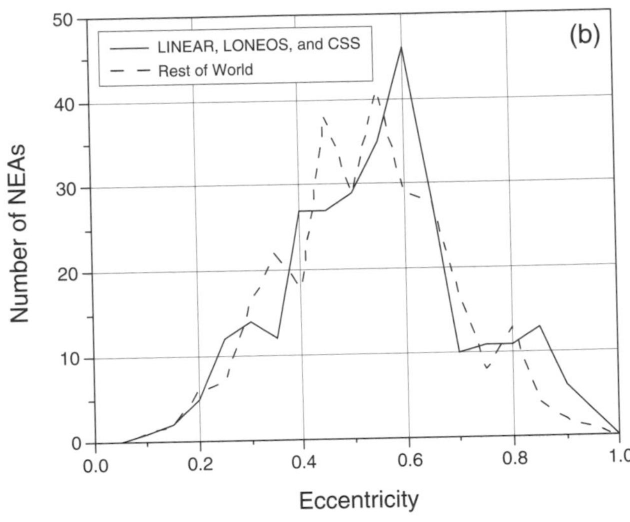

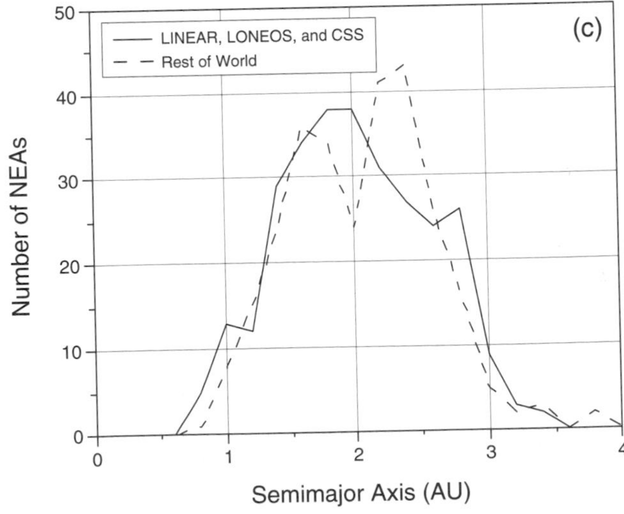

Fig. 4. **(a)** Inclination distribution for NEAs discovered by LINEAR, LONEOS, and CSS and NEAs discovered by others. **(b)** Eccentricity distribution for NEAs discovered by LINEAR, LONEOS, and CSS and NEAs discovered by others. **(c)** Semimajor axis distribution for NEAs discovered by LINEAR, LONEOS, and CSS and NEAs discovered by others.

ous search strategies is the NEO sky density as a function of detection limit. This plot shows heuristically that at the simplest level the probability of detecting NEAs actually increases away from opposition due to the longer line of sight through higher-density regions of NEOs. The negative aspects of such a search include observing through greater air mass and less favorable solar phase angles. Each of these effects should be considered, and each varies system to system with the location and sensitivity of equipment. A rigorous discussion of probabilities of detection is covered by *Jedicke* (1996) and supports a similar conclusion regarding the potential benefits of searching away from opposition: The probability of detection increases away from opposition if you can search deep enough.

Searching away from the ecliptic generally does not provide for optimal detection efficiency. For any given asteroid, it is generally easier to detect it near the ecliptic. However, NEAs on an inclined orbit are less likely to be found at small ecliptic latitudes because they reside longer at high apparent latitudes. This may justify an all-sky search strategy. In 1998, three large search programs began operation: LINEAR, Catalina Sky Survey, and LONEOS. Due to the significantly increased search capability with the introduction of these surveys, full-sky searches have become more common, primarily through these three programs. The effect has been to refine our understanding of the statistical distribution of the known population. Figure 4a shows the distribution by inclination of all known NEAs with H ≤ 18.0 discovered by LINEAR, LONEOS, and Catalina Sky Survey as compared to those discovered by other surveys more likely to be concentrating near ecliptic opposition. Note that the distribution is both flatter and has a longer tail towards higher inclinations for discoveries by these three. Figures 4b and 4c show similar plots for eccentricity and semimajor axis, with some differences amongst these parameters. This variation may indicate the current distribution for all these parameters has been biased by past search methods.

Regarding the question at the beginning of this section on the best place to search, there is no single answer. If the goal is to detect an NEA, traditional opposition ecliptic searches are fine. If the goal is to find and catalog all NEAs, given the apparent sky distribution of NEAs, an all-sky search is appropriate if it is possible. However, the details of any search strategy are determined by the individual search system's capabilities. An overview of the current major search systems is given in the following section.

4. NEAR-EARTH-OBJECT SEARCH PROGRAMS

The characteristics of the major existing NEO search surveys and a few of the planned surveys are summarized in Table 1. Their individual approach and emphases are described below in the approximate order they became operational. Additionally, Table 2 lists the total discoveries, total NEA discoveries, length of time each survey has been operating, and discovery rates. The equipment used by each survey often dictates the basic way the survey operates, with the detection algorithms and search patterns also contributing to a survey's discovery rate.

4.1. Spacewatch

The Spacewatch program, started by T. Gehrels and R. McMillan of the University of Arizona's Lunar and Planetary Laboratory in 1980, was the first NEO search program to utilize CCD imagers to look for NEOs (*Carusi et al.,* 1994). Spacewatch has had exclusive access to the Steward Observatory 0.9-m telescope (f/5.34) on Kitt Peak outside of Tucson, Arizona, since 1982 and made the first test scans in 1983. As described previously in the historical section, the program uses a drift-scanning mode in which the telescope is kept stationary while the CCD is read out at the sidereal rate of the sky (*Gehrels,* 1991). Spacewatch first used an RCA 320 × 512-pixel CCD, then upgraded to a Tektronics (now SITe) 2048 × 2048-pixel CCD with 0.027-mm pixels in 1989, and finally switched to a 2048 × 2048-pixel, thinned, backside-illuminated CCD with 0.028-mm pixels in 1992. With the current system, each pixel is 1.05 arcsec. In a drift-scan mode, a swath of sky about 0.5° in declination and 7° in right ascension is completed in about 30 min with an effective integration time (on the equator) of 143 s. This permits detection of objects as faint as V = 21.7 on good nights. Due to star trails at higher declinations moving in arcs that smear during readout, the useful declination range of the drift-scan system is limited to ±30°. To access higher declinations, Spacewatch uses a stare mode.

TABLE 1. Comparison of current near-Earth object search programs.

Program	Spacewatch	Spacewatch	NEAT	NEAT	LONEOS	LINEAR*	CSS	Bisei
Aperture (m)	0.9	1.8	1.2	1.2	0.6	1	0.4	0.5
f Number	5.3	2.7	1.9	2.5	1.9	2.2	3.0	1.9
Telescope Type	Newt.	folded	prime	Schmidt	Schmidt	folded	Schmidt	Cass.
Array Dimensions	2048 × 2048	2048 × 2048	4080 × 4080	4080 × 4080 × 3	2048 × 4096 × 2	1960 × 2560	4080 × 4080	2048 × 4096 × 2
Pixel Size (mm)	0.024	0.024	0.015	0.015	0.0135	0.024	0.015	0.015
Pixel Size (arcsec)	1	1	1.4	1	2.5	2.25	2.5	3.2
FOV (deg²)	0.3	0.3	2.5	3.8	8.3	2	8.1	3.1
Cooling (°C)	–90	–90	–20	–40	–130	–50	–30	–100
Readout Mode	drift scan	ds/stare	stare	stare	stare	stare	stare	stare
CCD Type	thin	thin	thick	thick	thin	thin	thick	thin
Exposure (sec)	150	150	20	60	45	5	60	23
Magnitude Limit	21.7V	22.6V	19.5V	21V	19.3V	19.2V	19.3R	19
Coverage (deg²/h)	7.4	7.4	174	174	400	1050	194	—
Number of Visits	3	3	3	3	4	5	4	—
Effective Coverage	2.5	2.5	58	58	100	210	48	—

*Two identical telescopes.

TABLE 2. Discovery statistics for current search programs as of February 2002.

	Spacewatch	NEAT	LONEOS	LINEAR	CSS	Bisei
Total Discoveries	48,054	12,803	12,713	157,346	4309	141
NEA Discoveries	254	158	106	925	46	1
Months of Operation	135	70	45	75	15	—
Discovery Rate	356	183	282	2098	287	—
NEA Discovery Rate	1.88	2.26	2.36	12.33	3.07	—

The Spacewatch automated detection system utilizes three visits and a default S/N limit of 2.5σ. The observer is able to visually inspect the candidate objects near real time, and based on the rate of motion, can estimate if the object is an NEA.

Spacewatch's future plans include upgrading the 0.9-m telescope with a shorter focal length primary and a mosaic of four 2048 × 4608-pixel Marconi CCDs with 0.0135-mm pixels. This will give 1-arcsec pixels, but will have 9× the field coverage of the current system and will be used in stare mode. Additionally, a 1.8-m (f/2.7) telescope has very recently come online with a fainter detection limit of V = 22.6. It has a thinned 2048 × 2048-pixel CCD with 0.024-mm, 1-arcsec pixels. The telescope altitude-azimuth mount drive system is designed to allow scanning in any direction over a wide range of rates, as well as in stare mode.

Spacewatch is the longest-running survey of the currently existing surveys, and has consistently contributed toward the cumulative knowledge of NEOs. This survey has discovered some of the smallest and nearest NEOs known (*Rabinowitz,* 1992).

4.2. Near-Earth Asteroid Tracking (NEAT)

The Near-Earth Asteroid Tracking (NEAT) program is run by JPL (E. Helin, S. Pravdo, K. Lawrence, and D. Rabinowitz) and is currently operating at two sites: one at the U.S. Air Force Maui Optical Station in Hawai'i (NEAT/MSSS), and the other at Palomar Observatory in California (NEAT/Palomar).

Beginning in 1995, this program utilized a GEODSS 1-m (f/2.2) telescope with a Loral large-format, 4080 × 4080-pixel CCD with 0.015-mm, 1.4-arcsec pixels. The telescope was shared with the USAF at Haleakala, where NEAT was able to utilize 12 nights per month centered on the new moon from December 1995 to December 1996 and 6 nights from January 1997 to February 1999. To help ensure more telescope time per month, NEAT modified the 1.2-m telescope of the U.S. Air Force Maui Space Surveillance System (MSSS), managed by the Air Force Research Laboratory, with a similar camera at the prime focus. This telescope became their operational system in February 2000. It maintains the 1.4-arcsec pixel scale and field of the GEODSS configuration, but is optically faster at f/1.9. It is capable of detecting objects to V ~ 19.5 with unfiltered exposures of 20 s.

The NEAT detection software utilizes three visits and was written to work autonomously in real time and to present small subarrays for visual inspection by scientists at JPL. The JPL-located scientists generate the nightly pointing lists for a computer in Maui that controls the sequencing of the telescope automatically once onsite operators open the telescope.

To further extend the telescope time available to NEAT, the 1.2-m (f/2.5) Oschin Schmidt telescope at Palomar Observatory has been recently upgraded for autonomous pointing, and a camera with three thick 4080 × 4080-pixel CCDs with 0.015-mm, 1.0-arcsec pixels that provide a total 3.8-deg^2 field of view (FOV). It is capable of detecting objects to V ~ 21 with 60-s exposures. Looking toward the future, a camera of 112 2400 × 600-pixel thinned CCDs with 0.013-mm, 0.9-arcsec pixels covering 9.3 deg^2 is expected to reach V ~ 21.5 and is being constructed at Yale University. It is scheduled to replace the current NEAT camera in late 2002 and JPL/NEAT and YALE/QUEST will share the telescope time thereafter.

NEAT was one of the early CCD search surveys and many of their past images have been archived at the Skymorph online archive where compressed versions of the images can be accessed to search for and measure prediscovery images of interesting objects. Skymorph is linked to the JPL Horizons site to facilitate easy identification of potentially useful imagery.

4.3. Lowell Near-Earth Object Survey (LONEOS)

The Lowell Near-Earth Object Survey (LONEOS) is run by Lowell Observatory (E. Bowell, B. Koehn) and utilizes a 0.59-m (f/1.0) modified Schmidt telescope located at the Lowell Anderson Mesa site near Flagstaff, Arizona. Two cryocooled 2048 × 4096 Marconi CCDs with 0.0135-mm, 2.4-arcsec pixels provide a total FOV of 8 deg^2. Using an unfiltered stare mode with 45-s integration, the system is capable of reaching V ~ 19.3 and can be run automatically.

The system utilizes four visits, and moving-object detection is automatic, although the observer can visually inspect candidate objects in near real time and compute short-arc Vaisala orbits to identify possible NEAs.

The LONEOS approach of using a large-format CCD with large pixels and large FOV has been successful.

Future plans include a collaboration with scientists at the U.S. Naval Observatory Flagstaff Station using a new mosaic camera on their 1.3-m telescope, which should double the current LONEOS discovery rate. Even further in the future is the potential 4-m Next Generation Lowell Telescope with a 3.2-deg^2 FOV and V ~ 22.3.

4.4. Lincoln Laboratory's Near-Earth Asteroid Research (LINEAR)

Lincoln Laboratory's Near-Earth Asteroid Research (LINEAR) program is run by the MIT Lincoln Laboratory using two experimental 1-m (f/2.2) GEODSS telescopes at the Lincoln Laboratory Experimental Test Site (ETS) near Socorro, New Mexico (*Stokes et al.,* 2000). Each telescope has a specialized low-noise, frame-transfer, eight-channel, 1960 × 2560-pixel, thinned CCD designed and fabricated by Lincoln Laboratory (*Viggh et al.,* 1998). The 0.024-mm, 2.25-arcsec pixels provide a 2-deg^2 FOV. While the system is capable of shutterless exposures of 0.1–100 s, the nominal operation mode is exposures from 5 to 30 s depending on the phase of the Moon and the time of the year. At 10-s integration, the system is capable of V ~ 19.2.

The LINEAR detection software utilizes five visits ~30 min apart and requires only three hits for a detection. This approach is effective even in dense star fields near the galactic plane. Rates of moving objects are used as a means of identifying potential NEAs, which are then verified visually. All interesting objects are automatically followed up on the subsequent observing night.

This survey approach is to cover large areas of sky by maximizing efficiency and minimizing cycle time between frames. Additionally, using two nearly identical telescopes, it is possible to cover the entire available sky each lunation and to cover some areas of the sky more often than once per lunation. These characteristics have enabled LINEAR to be the most productive of the surveys thus far.

4.5. Catalina Sky Survey (CSS)

The Catalina Sky Survey (CSS), a project of the University of Arizona's Lunar and Planetary Laboratory, utilizes the 0.42-m (f/3) Schmidt at the Catalina Observatory station north of Tucson (*Larson et al.,* 1999). The 4080 × 4080 thick, single-channel, Loral CCD has 0.015-mm, 2.5-arcsec pixels and a FOV of 8.1 deg^2. It is capable of detecting objects to R = 19.5 with 60-s integrations. Upgrades underway include a full-aperture, 0.7-m corrector and a cryogenically cooled, four-channel, thinned CCD.

The CSS detection software utilizes four visits to reduce false positives with the default S/N limit setting of 1.5σ. Sequencing is automated, freeing the observer to visually validate the candidate objects.

Future plans include developing a southern hemisphere component by modifying the 0.6-m Uppsala Schmidt at Siding Spring, Australia. The system is being modified for surveying and will utilize the same CSS detector and software. The project will have exclusive use of the telescope for surveying and will concentrate on the southern sky not reachable by the other surveys.

4.6. Bisei Spaceguard, Japan

The Bisei facility is a new facility devoted to NEO and space debris studies (*Isobe et al.,* 2000). The facility consists of a 0.5-m (f/2) Cassegrain telescope with two 2048 × 4096-pixel thinned SITe CCDs and a 1.0-m (f/3) Cassegrain telescope with 10 CCDs of the same size. These provide 9.6- and 6.8-deg^2 FOVs with 3.2- and 1-arcsec pixels respectively. At the time of this writing, the facility is preparing to come online.

Being outside of the southwestern U.S. weather pattern where most of the other surveys are located will allow for continued surveying during traditional seasonal lows.

4.7. Detection Algorithms

Automated moving-object detection is a key element of any search program. It is a nontrivial task to develop and write code to efficiently identify moving objects in a star field and generate astrometric positions to report to the Minor Planet Center. Search programs spend a significant fraction of their resources in producing and refining code to maximize detection efficiency.

Each project has independently developed its own code according to the type of data obtained and other numerous constraints such as available computer speed. It is beyond the scope of this review to describe the details of detection algorithms used by the various surveys, but we describe some basic elements that are commonly used.

In all cases, three, four, or five images of the same field are taken at appropriate intervals to allow the moving objects with the targeted range of apparent rates to move more than a few pixels. The larger the number of visits to a field, the easier it is to compensate for the effects of spurious "cosmic-ray" events and the merging with field stars. This has to be balanced against being able to cover more sky with fewer visits.

The most common detection method involves generating a list of the centroid positions of objects in the image, performing an astrometric solution using a reference catalog such as the USNO-A2 or UCAC-1 catalogs, and eliminating objects whose positions do not differ between visits by some predefined amount. The remaining objects are tested for consistent motion and are grouped as potential moving-object candidates.

Another method is to work in image space by registering images and taking differences between individual visits and a median of all the visits to eliminate the stationary field stars. This method is computationally intensive, but may be better able to differentiate moving objects from field stars. On the other hand, the added noise of the median image degrades the signal-to-noise ratio of an object.

In most cases, the candidate objects are validated visually by blinking or similar means. The ratio of false to real objects and available personnel may dictate how close to the noise floor objects are extracted from the images. It may also restrict visual validation to only the obvious NEO candidates.

4.8. Followup Observations

Maximizing survey coverage precludes making followup observations of potential NEOs that are more efficiently made with smaller field telescopes. Confirming observations on subsequent nights provide the extended arc needed to determine their orbits with sufficient precision to identify NEOs. Fortunately, there is a sizable network of amateur and professional observers who consult the Minor Planet Center's NEO Confirmation Web page each night and provide the bulk of additional astrometry needed to identify NEOs. In the past few years, the development of affordable CCD cameras, fast computers, sophisticated software, better reference star catalogs, larger telescopes, and the ability to communicate effectively over the Internet has enabled amateurs and professionals at small observatories to out-produce the professionals at major observatories of 10–15 years ago. Amateur followups have become a vital component in the NEO inventory effort down to V ~ 20. In addition to providing the necessary astrometry, this followup contribution is expanding toward obtaining better time-resolved photometry from which sizes and shapes of NEOs can be estimated.

While followup of objects to V ~ 20 are well supported by the various amateurs and professionals, followup observations of the fainter objects, such as those found by Spacewatch, have proven to be more difficult. Followup of these objects are being made more routinely with the Spacewatch 1.8-m telescope, Catalina and Mt. Lemmon 1.5-m telescopes, and the 1-m Siding Spring telescope. As the existing and proposed search systems continue to improve their sensitivity and search to fainter magnitudes, these followup limitations will need to be addressed.

5. COORDINATION OF SEARCH PROGRAMS

Asteroid searches over the past 20 years have been carried out by a number of independent groups using a variety of telescopes and detectors. Prior to the U.S. Congressional mandate, there were only three groups consistently observing. There was relatively little competition for funding and regions of the sky to search. As support to achieve the congressional mandate became available through NASA, development of detectors and efficient detection algorithms became the natural outcome of a more competitive environment. Today, with the growing number of mature survey systems, it is possible to cover the entire observable sky on a monthly basis. To optimize efficiency and minimize needless duplication, it is appropriate to consider coordination among the major search programs

An active series of discussions started in 1999 between the major U.S. survey teams to address the issue of coordination. The premise of the discussions has been that the resources available for asteroid search should be applied in such a way as to maximize the total joint productivity of the search systems. One notable result has been the e-mail distribution to all the surveys of coverage lists of fields successfully observed with estimates of typical limiting magnitudes. This is represented graphically by the Minor Planet Center (MPC) Web site and the Lowell Observatory site. An example is shown in Plate 1. This method of coordination allows each search program to plan the night's observations with an understanding of what other area programs have covered recently, and how deep the coverage was. As an added benefit, the site is available to amateur observers and other small-scale programs that want to work around the major systems.

The next phase of the coordination effort centers on finding some *a priori* method of coordinating the various cooperating search systems in a way that achieves a larger joint productivity. This objective is quite complex due to the considerable differences between the operation's concepts and tempos, capabilities and maturities. The current consensus of the discussions is that most of the survey systems are evolving quickly and have not yet demonstrated their ultimate capability. In addition, we believe that in order to achieve a level of coordination beyond that currently operating, the capabilities of each of the systems must be evaluated in a common framework and against a common standard. Considerable discussion on the subject has led to an agreement to pursue this goal on several fronts as follows:

Task 1. Maintain comparable search experience information for all searches. This will allow a better understanding of how the search programs compare with one another and will allow extraction of experience information on a common basis. This will be accomplished by defining common elements for each search program's database, which will contain look-by-look standard measures of the seeing and magnitude limit based upon star measurements.

Task 2. Develop a common understanding of the most effective search strategies to effect a search for 1-km and larger asteroids. For example, decide how to distribute observation effort across the sky for best productivity. This task was initially approached by plotting the LINEAR detections of all NEAs and all large NEAs (H > 18) relative to the opposition in ecliptic coordinates. Only the first detection during a lunation was included and detections resulting from directed followup activities were excluded. The results of these plots are shown in Figs. 5a and 5b. The plots indicate that LINEAR detects asteroids at all declinations and an all-sky search is an appropriate strategy for systems with similar capability. If search systems can achieve a limiting magnitude performance substantially better than LINEAR, the strategy question will need to be revisited.

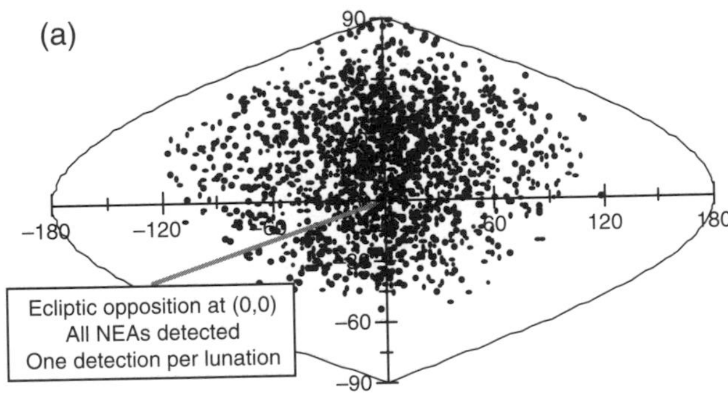

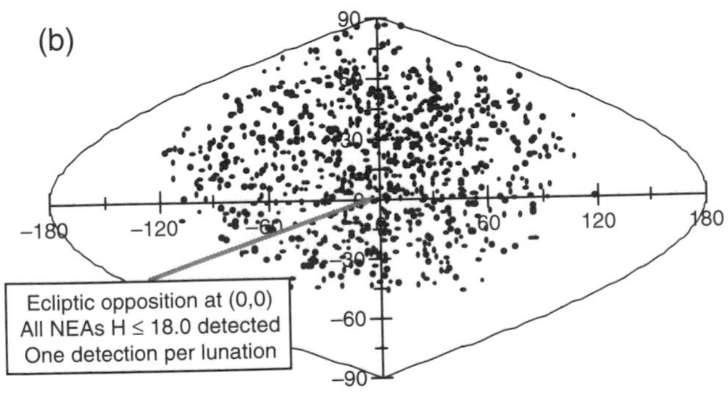

Fig. 5. **(a)** Distribution of all NEAs detected by LINEAR and plotted with respect to ecliptic opposition. **(b)** Distribution of all large NEAs (H ≤ 18) detected by LINEAR and plotted with respect to ecliptic opposition.

Task 3. Develop a common measure of capability/capacity for each search system to enable a systematic approach to coordinating the search programs. One of the most important metrics for a search effort is the volume of space that the survey can search for asteroids larger than H = 18. This volume may be calculated by calibrating, on a field-by-field basis, the depth of the search for the detection of an H = 18 object. Once a reliable calibration method is found, the volume of each field can simply be accumulated over a period of time to generate an effective search volume.

Achieving a reliable field-by-field calibration is complex due to the variable conditions under which the observing is accomplished, and the fact that considerable data-dependent processing occurs to detect moving asteroids. The calibration method chosen by the LINEAR program is discussed here as an example of such a process. The most obvious method to calibrate the magnitude performance of a system is to pick stars of known magnitude from the fields and determine how bright a star must be to achieve some standard signal-to-noise ratio in the detection system. Given that the CCD pixels are large and the integration times are short, asteroids do not streak and are indistinguishable from stars in any given frame. This process of computing signal-to-noise to determine a system's sensitivity takes into account many of the factors effecting the performance of the search, such as weather and seeing, but fails to consider the aspects of the detection algorithm that look for moving objects. In order to validate the star S/N ratio as a valid general indicator of search depth, a large set of data taken near the ecliptic was identified and the detections extracted. Objects in the MPC catalog expected to be in the LINEAR search area and with known absolute magnitudes were identified and their magnitudes corrected for distance and illumination geometry. With these inputs, a field-by-field calibration of the search's ability to detect asteroids could be accomplished by plotting the known asteroids detected and those not detected as a function of magnitude. An example of such a plot is shown in Fig. 6. Such a calibration process is possible near the ecliptic, where there are abundant known asteroids, but breaks down off the plane where there may only be a few to zero detections per field.

The moving-object calibration process described above validated the process of using the star signal-to-noise ratio as a measure of the search depth on a frame-by-frame basis. Thus, the accumulated search volume of the search may be calculated and is shown in Fig. 7 in cubic AU.

Another relevant piece of information yielded by the volume calculation is the ability to measure the evolution of a search system. Current asteroid search systems are all in various stages of an evolution toward increased maturity and capability. As search systems add capability, the rate of volume searched should grow. The second curve in Fig. 7 shows a running computation of the volume searched by LINEAR in the preceding year. Note that the system search capability evolved rapidly over 1999/2000 and appears to have leveled off in 2001. After achieving the fundamental capability of the existing telescope/detector/processing system, the variations in the plot should reflect observing experience (i.e., lunations, weather, equipment failure, staff availability, etc.). Once the fundamental capabilities of the various systems have been reached and characterized, the coordination process between the search systems in pursuit of the common goal of maximizing joint detection rate will be much easier to arrange.

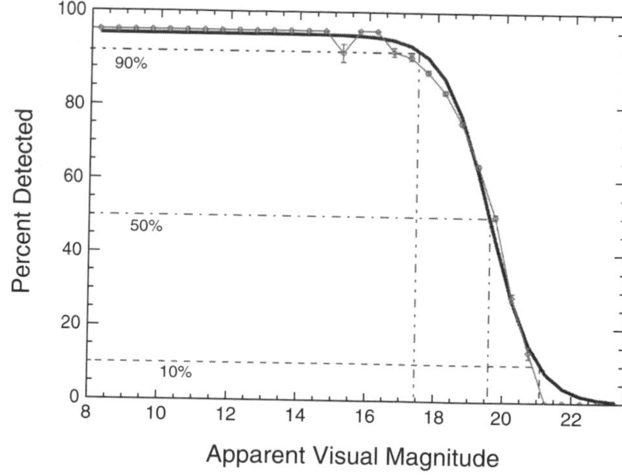

Fig. 6. A single night's detection efficiency plot from LINEAR.

6. THE FUTURE OF SEARCH TECHNOLOGY

This chapter has given an overview of the current search systems, how they came about, and how the steps taken by them, both alone and together, have furthered our general knowledge of NEAs. What does the future hold for these and yet-to-be-developed systems?

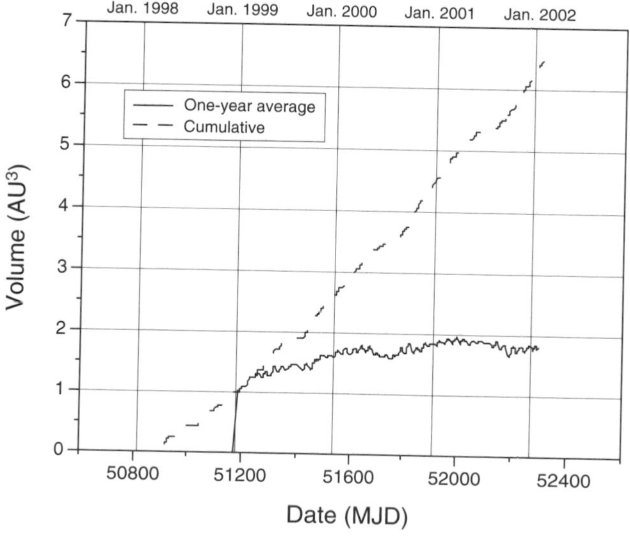

Fig. 7. Actual volume searched by the LINEAR program.

The march of Moore's Law has allowed the development of CCD detectors and the processing capacity to match to the point that there are probably no more factors of 2 in increased search performance for search systems using 1-m telescopes such as LINEAR. In designing the LINEAR system, considerable effort went into minimizing bottlenecks by matching the capabilities of each of the subsystems to work well with the rest of the system and to maximize the total system capability. It would appear that given unlimited detector mosaic sizes, the limitation of asteroid search systems is aperture size. Historically, the astronomical community has built telescopes with ever-increasing aperture, but with small FOVs compared to that needed to search a reasonable fraction of the sky; sensitivity to bore down on interesting objects has been the mainline interest rather than search. In addition, wide-FOV telescopes become progressively harder and more expensive to build as the aperture size increases. Therefore, there are very few existing wide-FOV telescopes with apertures exceeding 1 m. The very thick corrector plate on the venerable 1.2-m Palomar Schmidt is testament to the difficulties of building large-aperture, wide-FOV telescopes.

As the existing systems evolve toward maximum capacity, and when new larger aperture systems join the search in the future, the interest in coordination will grow. The first steps toward coordination have been defined and a few steps taken. The future holds more thorough system analysis and characterization for the sake of optimizing search strategies for objects of ever decreasing size. When the goal of cataloging 90% of all NEAs greater than 1 km in diameter has been reached, the bar will most likely be lowered to NEAs >300 m in diameter. If that goal is met, NEAs of 100 m still pose a serious regional threat.

REFERENCES

Bottke W. F. Jr., Jedicke R., Morbidelli A., Petit J. M., and Gladman B. (2000) Understanding the distribution of near-Earth asteroids. *Science, 288,* 2190–2194.

Bowell E. and Muinonen K. (1994) Earth-crossing asteroids and comets: Groundbased search strategies. In *Hazards Due to Comets and Astroids* (T. Gehrels, ed.), pp. 149–197. Univ. of Arizona, Tucson.

Carusi A., Gehrels T., Helin E. F., Marsden B. G., Russell K. S., Shoemaker C. S., Shoemaker E. M., and Steel D. I. (1994) Near-Earth objects: Present search programs. In *Hazards Due to Comets and Astroids* (T. Gehrels, ed.), pp. 127–147. Univ. of Arizona, Tucson.

Gehrels T. (1986) CCD scanning. In *Asteroids, Comets, and Meteors II* (C.-I. Lagerkvist et al., eds.), pp. 19–20. Uppsala Univ., Uppsala.

Gehrels T. (1991) Scanning and charge-coupled devices. *Space Sci. Rev., 58,* 347–375.

Harris A. W. (1998) Evaluation of ground-based optical surveys for near-Earth asteroids. *Planet. Space Sci., 46,* 283–290.

Isobe S., Mulherin J., Way S., Downey E., Nishimura K., Doi I., and Saotome M. (2000) A cost-effective, advanced-technology telescope system for detecting near-Earth objects and space debris. In *Proceedings of SPIE on Telescope Structures, Enclosures, Controls, Assembly/Integration/ Validation and Commissioning,* Vol. 4004.

Jedicke R. (1996) Detection of near-Earth asteroids based upon their rates of motion. *Astron. J., 111,* 970–982.

Larson S., Spahr T., Brownlee J., Hergenrother C., and McNaught R. (1999) The Catalina sky survey and southern hemisphere NEO survey. In *Proceedings of the 1999 AMOS Technical Conference,* pp. 182–186.

Marsden B. G. (1994) Asteroid and comet surveys. In *Astronomy from Wide-Field Imaging* (H. T. Mac Gillivray et al., eds.), pp. 385–399. Kluwer, Dordrecht.

Minor Planet Center (2001) *Sky Coverage Services for Observers.* (Available on line at http://scully.harvard.edu/~cgi/SkyCoverage.html.)

Morrison D. and 23 colleagues (1992) *The Spaceguard Survey: Report of the NASA International Near-Earth-Object Detection Workshop.* Jet Propulsion Laboratory, Pasadena.

NASA (1998) *Strategic Plan of NASA's Office of Space Science 1998.* U.S. Govt. Printing Office, Washington, DC.

NRC (National Research Council) Astronomy and Astrophysics Survey Committee (2001) *Astronomy and Astrophysics in the New Millennium.* National Academy Press, Washington, DC. 246 pp.

Pravdo S. H., Rabinowitz D. L., Helin E. F., Lawrence K. J., Bambery R. J., Clark, Groom S. L., Levin S., Lorre J., Shaklan S. B., Kervin P., Africano J. A., Sydney P., and Soohoo V. (1999) The Near-Earth Tracking (NEAT) program: An automated system for telescope control, wide-field imaging, and object detection. *Astron. J., 117,* 1616–1633

Rabinowitz D. L., Helin E. F., Lawrence K., and Pravdo S. (2000) A reduced estimate of the number of kilometre-sized near-Earth asteroids. *Nature, 43,* 165–166.

Rabinowitz D. L. (1992) The flux of small asteroids near the Earth. In *Comets, Asteroids, Meteors 1991* (A. W. Harris and E. Bowell, eds.), pp. 481–485. NASA, Washington, DC.

Stokes G. H., Evans J. B., Viggh E. M., Shelly F. C., and Pearce E. C. (2000) Lincoln Near-Earth Asteroid Program (LINEAR). *Icarus, 148,* 21–28.

Stuart J. S. (2000) The near-Earth asteroid population from Lincoln near-Earth asteroid research (LINEAR) data (abstract). *Bull. Am. Astron. Soc., 32,* 1023.

Viggh H. E. M., Stokes G. H., Shelly F. C., Blythe M. S., and Stuart J. S. (1998) Applying Electro-optical space surveillance technology to asteroid search and detection: The LINEAR program results. *Proceedings of the Sixth International Conference and Exposition on Engineering, Construction, and Operations in Space,* pp. 373–381.

Asteroid Close Approaches: Analysis and Potential Impact Detection

Andrea Milani
Università di Pisa

Steven R. Chesley
Jet Propulsion Laboratory

Paul W. Chodas
Jet Propulsion Laboratory

Giovanni B. Valsecchi
Istituto di Astrofisica Spaziale

Recently, several new tools and techniques have been developed to allow for robust detection and prediction of planetary encounters and potential impacts by near-Earth asteroids (NEAs). We review the recent history of impact-prediction theory and cover the classical linear techniques for analyzing encounters, which consists of precise orbit determination and propagation followed by target-plane analysis. When the linear approximation is unreliable, there are various suitable approaches for detecting and analyzing very low-probability encounters dominated by strongly nonlinear dynamics. We also describe an analytic approach that can provide valuable insight into the mechanisms responsible for most encounters. This theory is the foundation of impact monitoring systems, including those currently operational and those being developed.

1. INTRODUCTION

The last 10 years have seen tremendous progress in our ability to assess the risk of an asteroid or comet colliding with Earth. The catalyst for much of the increased interest in these near-Earth objects (NEOs) was a request by the U.S. Congress in 1990 that NASA undertake two workshop studies, one to study ways of increasing the discovery rate of these objects and another to study the technologies and options for deflecting or destroying an NEO should it be found to pose a danger to life on Earth. The report from the first of these workshops proposed an international NEO survey program called Spaceguard, borrowing the name from a similar project in Arthur C. Clarke's science fiction novel *Rendezvous with Rama* (*Morrison,* 1992). The report noted the need for development of new procedures and software to "assess the uncertainty [in Earth-object distance] for any future close approaches."

The exploding state of knowledge of the impact hazard problem in the early 1990s was well captured by an earlier monograph in this series, *Hazards Due to Comets and Asteroids* (*Gehrels et al.,* 1994). In that volume, *Bowell and Muinonen* (1994) suggest the use of the minimum orbital intersection distance (MOID) for close encounter analyses, and define the class of potentially hazardous asteroids (PHAs) as those asteroids having a MOID with respect to Earth of <0.05 AU. In the same volume, *Chodas and Yeomans* (1994) describe a system for predicting future asteroid and comet close approaches, including the computation of close-approach uncertainties and impact probabilities via linear methods. The impact probability computation via linear methods in the impact plane was introduced in the context of asteroid and comet collisions by *Chodas* (1993). The method saw an immediate application when Comet Shoemaker-Levy 9 was found to be on a collision course with Jupiter. Within a few days of the impact announcement, the impact probability was calculated to be 64% (*Yeomans and Chodas,* 1993), and it reached 95% only a week later. Somewhat different linear methods for computing collision probabilities by monitoring the distance between Earth and the asteroid uncertainty ellipsoid are described by *Muinonen and Bowell* (1993). *Chodas and Yeomans* (1996) performed an early nonlinear analysis of orbital uncertainties in an investigation of the prebreakup orbital history of Comet Shoemaker-Levy 9.

In March 1998, the problem of computing impact probabilities received a great amount of press attention because of a prediction by Brian Marsden of the Smithsonian Astrophysical Observatory that the sizable asteroid 1997 XF_{11} would make an extremely close approach to Earth in 2028, coupled with a widely misunderstood statement that a collision was "not entirely out of the question" (*Marsden,* 1999). A linear analysis of the impact probability was immediately performed by Paul W. Chodas and Donald K. Yeomans of the Jet Propulsion Laboratory (JPL), and the chance of collision in 2028 was found to be essentially zero.

When prediscovery observations from 1990 were found the next day and included in the orbital solution, they only served to confirm that there was no possibility of collision in 2028 (*Chodas and Yeomans,* 1999b; *Muinonen,* 1999).

Three months after the 1997 XF_{11} story hit the news, Marsden opened a new area of investigation by suggesting that prior to the discovery of the 1990 observations, 1997 XF_{11} had in fact a small possibility of collision in the decade or so after 2028 because the deep close approach in 2028 could alter the asteroid's orbital period to bring it back to Earth several orbits later. Linear methods were inadequate to analyze these later collision possibilities because the 2028 close approach introduced a strong nonlinearity into 1997 XF_{11}'s predicted motion. Primarily to investigate the post-2028 impact scenarios of 1997 XF_{11}, two groups independently and simultaneously developed new nonlinear methods for analysis of collision possibilities much further into the future than had been possible with the linear methods. *Chodas and Yeomans* (1999a) applied a Monte Carlo technique to sample the linear six-dimensional confidence region at the epoch of observation and then numerically integrate over the timespan of interest using the nonlinear equations. *Milani et al.* (1999, 2000b) applied the multiple-solutions approach to sample the central line of variations of the nonlinear confidence region at epoch and then numerically integrate over the timespan of interest in a similar fashion. The two methods yielded very similar results when applied to the hypothetical 1997 XF_{11} case without the prediscovery observations, with each obtaining an impact probability on the order of 10^{-5} for the year 2040 (*Chodas and Yeomans,* 1999c; *Milani et al.,* 2000b).

In early 1999, Milani et al. applied these same techniques to the case of asteroid 1999 AN_{10} while developing the theory of resonant and nonresonant returns. They also identified for the first time an impacting solution consistent with a complete set of NEA observations, although the collision probability was very small (*Milani et al.,* 1999). This result was confirmed by independent investigators. As more observations of 1999 AN_{10} became available, both the Pisa and the JPL group identified two much more likely impacting scenarios for the years 2044 and 2046, with probabilities on the order of 10^{-6}, the highest seen to that date. Observations later found on archival plates drove these probabilities to essentially zero.

In mid-1999, a second asteroid, 1998 OX_4, was determined (based on all available observations) to have some collision potential (*Milani et al.,* 2000a). Unfortunately this asteroid had become lost, because of its faintness and its location in the Milky Way, despite concerted efforts of numerous observers around the time of discovery, and there was no practical way to obtain additional observations to refine the collision hazard posed by this object. *Milani et al.* (2000a) proposed a method, based on the concept of a negative observation, for ruling out the known collision possibilities without requiring the recovery of 1998 OX_4; their approach has been successfully implemented (Andrea Boattini, personal communication, 2001). In their reassessment of the case, *Muinonen et al.* (2001) found additional possibilities of collision, generally confirming the results of *Milani et al.* (2000a). However, this case highlighted the need for *automatic collision monitoring,* since it is highly likely that, had the collision potential been recognized at the time of the discovery, the resources needed to observe the newly discovered PHA and refine its orbit would have been forthcoming. In response, the CLOMON monitoring system (*Chesley and Milani,* 2000) was developed at the University of Pisa, and it has been used to detect numerous impact possibilities among recently discovered asteroids. Importantly, all these threatening events (apart from some involving very small bodies) were eventually eliminated as a result of additional observations, which in some cases came as a direct result of CLOMON impact detection. CLOMON was, however, a prototype system with an uncertain level of completeness. At the time of this writing, more advanced monitoring systems, based on principles described in this chapter, are under independent development at the University of Pisa and at JPL.

This chapter is organized to give an overview of the ideas and theories needed to reliably detect and analyze potentially threatening Earth-asteroid encounters. The next section describes the most important tool used in encounter analysis, the target plane. In section 3, we describe the linear theory of orbit determination and encounter analysis and also discuss the limitations of the theory. In sections 4 and 5 we describe the various nonlinear sampling and analysis methods available. Section 6 presents an analytic approach that can be used to understand the chaotic nature of the problem in terms of resonant returns and keyholes. Finally, section 7 discusses a few of the more significant open problems that remain.

2. PLANETARY ENCOUNTERS

2.1. Target Planes

The primary method of studying planetary encounters, whether the encounters are with spacecraft or minor planets, continues to rely on the concept of a target plane. Here we use this term generically to encompass any of the various planes and coordinate systems that may be used to study a specific encounter of an asteroid with Earth. (For simplicity we limit this development to asteroid encounters with Earth. The extension to comet encounters, where nongravitational forces must be considered, or to encounters with other planets is straightforward.) In this sense, a target plane is simply a geocentric plane oriented normal to the asteroid's geocentric velocity vector. The point of intersection of the asteroid trajectory with the target plane provides considerable insight into the nature of an encounter, especially when uncertainty is carefully taken into account.

While the fundamental concept is rather straightforward, there are several issues we must address that can unfortunately obscure this simplicity. In general, either of two distinct planes and several coordinate systems can be used. The two available target planes pass under various names through the published literature, which is occasionally confusing, but for the purposes of this development we set forth the following usage:

Target plane or b-plane. The b-plane is the classical target plane that has been used in astrodynamics since the 1960s (e.g., *Kizner,* 1959) and by *Greenberg et al.* (1988) in the framework of Öpik's theory of close encounters (*Öpik,* 1976). It is oriented normal to the incoming asymptote of the osculating geocentric hyperbola, or, equivalently, it is oriented normal to the unperturbed relative velocity $\mathbf{v}_\infty$. The b-plane is named in reference to the so-called impact parameter b, which is the distance from the geocenter to the intercept of the asymptote on this plane, i.e., the minimum encounter distance along the unperturbed trajectory.

Modified target plane (MTP). The MTP is modified in the sense that it is oriented normal to the geocentric velocity at the point of closest approach along the actual trajectory. Rather than noting the position of the intersection of the unperturbed orbit, as is done with the b-plane, we mark the intersection of the perturbed asteroid trajectory with the target plane when using the MTP. The term "MTP" was introduced by *Milani and Valsecchi* (1999), although the concept was introduced earlier (e.g., *Chodas and Yeomans,* 1994).

The main difference between the two types of target planes arises from the fact that the deflection of the asteroid by the planet or gravitational focusing is directly indicated with the use of the MTP, while the gravitational focusing is in some sense hidden when the b-plane is used. For encounters taking place at high relative velocities, or at large distances, the deflection will be very small and the distinction between the two planes becomes negligible.

One difficulty that arises from the visibility of gravitational focusing on the MTP is that nearby trajectories are deflected by different amounts. When the deflection is significant, as is typically the case for very deep encounters or those with low relative velocities, this leads to substantial nonlinearity in the mapping from the asteroid's preencounter state onto the impact plane. Moreover, the orientation of the plane itself can change significantly with modest changes to the asteroid's trajectory, and this poses another important source of nonlinearity. Of course, for very slow encounters that lead to temporary capture of the asteroid, an asymptote does not exist, and therefore the b-plane approach fails; in such cases the MTP must be used.

This allows us to rate the utility of the b-plane and MTP according to the deflection caused by the encounter. For very low deflections, the two approaches are essentially indistinguishable, while for moderate deflections the preferred plane depends on the circumstances and purposes at hand. For relatively large deflections, the greater nonlinearity of the MTP causes the b-plane to be generally superior, but for very low relative velocities (leading to capture in elliptical orbits) only the MTP is available.

2.2. Target Plane Coordinates

Conventionally, the coordinate origin is at the geocenter, but the orientation of the coordinate system on the target plane is arbitrary. The reference system has been variously fixed by aligning the axes so that one of the nominal target plane coordinates is zero (i.e., the asteroid is placed on an axis) or by aligning a coordinate axis with either the projection of Earth's polar axis (e.g., Figs. 1 and 2) or the projection of Earth's heliocentric velocity (e.g., Fig. 3).

One of the most important objectives in a target-plane analysis is to determine whether a collision is possible and, if not, to decide how deep the encounter will be. When using the MTP we plot the position of the asteroid at the point of closest approach; hence this information is immediately available. On the other hand, with the b-plane we obtain the minimum distance of the unperturbed trajectory at its closest approach point, which is the impact parameter b. The impact parameter alone does not reveal whether the perturbed trajectory will intersect the figure of Earth, but this information can be extracted by scaling Earth's radius $r_\oplus$ according to

$$b_\oplus = r_\oplus\sqrt{1 + \frac{v_e^2}{v_\infty^2}}$$

where v_e is Earth's escape velocity

$$v_e^2 = \frac{2GM_\oplus}{r_\oplus}$$

With this we can say that a given trajectory impacts if $b < b_\oplus$, and not otherwise. An alternative to simply scaling the figure of Earth on the b-plane is to scale b while leaving the figure of Earth at $r_\oplus$. For any single orbit this is obviously equivalent, but when computing the coordinates of slightly different asteroids, each with a slightly different $\mathbf{v}_\infty$, the scaling is not uniform (*Chodas and Yeomans,* 1999c).

A convenient target-plane reference system (ξ, η, ζ) is obtained by aligning the negative ζ-axis with the projection of Earth's heliocentric velocity $\mathbf{v}_\oplus$, the positive η-axis with the geocentric velocity (i.e., normal to the b-plane), and the positive ξ-axis in such a way that the reference system is positively oriented. With this frame of reference, the target-plane coordinates (ξ, ζ) indicate the cross-track and along-track miss distances, respectively. In other words, ζ is the distance by which the asteroid is early or late for the minimum possible distance encounter. The associated early/late timing of the target-plane crossing ($\eta = 0$) is $\Delta t = \zeta/(v_\oplus \sin\theta)$, where θ is the angle between $\mathbf{v}_\infty$ and $\mathbf{v}_\oplus$. On the b-plane the ξ coordinate is the minimum distance that can be obtained by varying the timing of the encounter. This distance, known as the minimum orbital intersection distance (MOID), is equivalent to the minimum separation between the osculating ellipses, without regard to the location of the objects on their orbits. We note that this interpretation of the target-plane coordinates is valid only in the linear approximation and can break down for distant encounters (e.g., beyond several lunar distances).

Whatever target plane or coordinate frame is used, the idea that an asteroid can avoid an impact, either through being off time or by having an orbit that does not even intersect Earth, is important. To have an impact the object must have a small enough MOID and be on time for the collision. With this in mind, we can characterize an encounter quite well given only the MOID and Δt. The MOID, which is strictly a function of the osculating elements of the asteroid and Earth, can be computed in various ways, with the numerical approach of *Sitarski* (1968) being the most widely used. A recently developed semianalytic approach (*Gronchi,* 2001) can be more robust, with better handling of the cases of multiple local minima of the distance between the two orbits (so-called local MOIDs).

3. LINEAR ENCOUNTER THEORY

3.1. Least-Squares Orbit Determination

The details of the orbit determination process are not within the scope of this presentation; readers should consult *Bowell et al.* (2002). However, we do need to touch on a few of the key ideas of estimation theory to establish our notation and terminology. First, we describe the initial conditions at some epoch t_0 for the asteroid under consideration as the six-dimensional vector X of orbital elements. The observation residuals $\rho = \rho(X)$ are a function of the six orbital elements, and they form an m-dimensional vector where $m > 6$ is the number of (scalar) observations.

The weighted least-squares method of orbit determination seeks to minimize the weighted RMS of the residuals ρ, so we define the cost function as

$$Q = \frac{1}{m}\rho^T W \rho$$

where W is a square, symmetric (though not necessarily diagonal) matrix that should reflect the *a priori* RMS and correlations of the observation errors. We denote the dependence of the residuals on the elements by

$$B = \frac{\partial \rho}{\partial X}(X)$$

where B is an $m \times 6$ matrix. Then we can compute the derivative of the cost function as

$$\frac{\partial Q}{\partial X} = \frac{2}{m}\rho^T W B$$

The stationary points of the cost function Q are solutions of the system of nonlinear equations $\partial Q/\partial X = 0$, which are usually solved by some iterative procedure. The most popular is a variant of Newton's method, known in this context as differential corrections, with each iteration making the correction $X - \Delta X \rightarrow X$, where

$$\Delta X = (B^T W B)^{-1} B^T W \rho$$

This converges (normally) to the best-fitting or nominal solution X^*, where $\Delta X = 0$. We use the usual terminology for the normal matrix $C_X = B^T W B$, and covariance matrix $\Gamma_X = C_X^{-1}$. We note that these matrixes can be computed not only for the nominal solution but also for nearby values of X, a point that will be put to use in sections 4.3 and 5.3 in handling nonlinear situations.

3.2. Orbital Confidence Region

The expansion of the cost function at a point $X = X^* + \Delta X$ in a neighborhood of X^* is

$$Q(X) = Q(X^*) + \frac{1}{m}\Delta X^T C_X \Delta X + \ldots = Q(X^*) + \Delta Q(X)$$

where the dots indicate higher-order terms in ΔX plus a term containing the second derivative of ρ (see *Milani,* 1999). A confidence region is a region where the solutions are not too far from the nominal, as measured by the penalty ΔQ. We shall indicate with $Z_X(\sigma)$ the confidence ellipsoid defined by the inequality

$$m\Delta Q(X) \simeq \Delta X^T C_X \Delta X \leq \sigma^2$$

If the confidence ellipsoid $Z_X(\sigma)$, defined by the normal matrix C_X, is small enough (either because the constraint is sufficiently strong, i.e., the eigenvalues of C_X are large, or because the choice of σ is sufficiently small) then the higher-order terms in $\Delta Q(X)$ are also small and $Z_X(\sigma)$ is a good approximation of the region in the space of orbital elements where the penalty $\Delta Q < \sigma^2/m$. In such cases we say that the linear theory of estimation holds; however, when the confidence ellipsoid is larger, it can be an unreliable representation of the region containing alternative solutions still compatible with the observations.

The confidence region is generally viewed as a cloud of possible orbits centered on the nominal solution, where density is greatest, and surrounded by a diffuse boundary. This is classically represented by the multivariate Gaussian probability density, which reflects the likelihood that a given volume of the confidence region will contain the true solution; this Gaussian distribution has X^* as the mean and Γ_X as the matrix containing the variances and the covariances of the distributions of the elements (hence its name). But the use of the Gaussian probability density to describe orbital uncertainty assumes that the observational errors are Gaussian, and that W accurately reflects the observational uncertainty. Indeed, the extent of the confidence region is directly dependent on the choice of W, and therefore proper observation weighting is crucial for reliably determining the orbital confidence region. However, given the fact that the error statistics for asteroid astrometry are rarely characterized, and moreover are often not even Gaussian, it should be clear that the selection of W is very problematic. This implies that a careful probabilistic interpretation of the orbital confidence region is elusive [see *Carpino et al.* (2001) for an expanded discussion of this issue].

3.3. Linear Mapping to the Target Plane

Ultimately we wish to use the orbital confidence region to infer how close an asteroid could pass during a given close approach and, if collision is possible, to determine the probability of collision. To do this we must map the orbital uncertainty onto the target plane.

The nonlinear function F maps a given orbit X to a point Y on the target plane: $Y = F(X)$. This mapping consists of a propagation from an epoch near the observations t_0 to the time of the encounter t_1, followed by a projection onto the target plane. The Jacobian DF linearly maps orbits near the nominal to nearby points on the target plane

$$\Delta Y = DF(X^*)\ \Delta X$$

The matrix DF is practically computed by propagating numerically the orbit together with the variational equations (providing the state transition matrix), then projecting on the cross section defined by the target plane. In the linear approximation — i.e., when the confidence region is small and F is not too nonlinear — the confidence ellipsoid $Z_X(\sigma)$ in the space of orbital elements maps onto the target plane as an ellipse $Z_Y(\sigma)$ defined by

$$\Delta Y^T\ C_Y\ \Delta Y \leq \sigma^2$$

where C_Y is the normal matrix describing the uncertainty on the target plane. As is well known from the theory of Gaussian probability distributions (*Jazwinski,* 1970), the relationship between the covariance matrices of the variables X and Y can be shown by

$$C_Y^{-1} = \Gamma_Y = DF\ \Gamma_X\ DF^T$$

3.4. Linear Target Plane Analysis

Continuing under the assumption of linearity, we can characterize the target-plane ellipse Z_Y given only the nominal target-plane coordinates Y, which marks the center of the ellipse, and the associated covariance matrix Γ_Y, which indicates the size and orientation of the ellipse. The square roots of the eigenvalues of Γ_Y are the semimajor and semiminor axes of Z_Y, and the corresponding eigenvectors indicate the orientation of the associated ellipse axes.

At this point we define several terms and variables associated with the target-plane analysis that are important for interpretation. First, d denotes the distance from the asteroid to the geocenter on the target plane, i.e., $d = \|Y\|$ (on the b-plane we have $d = b$). Also, α denotes the angle between Y and the major axis of Z_Y. The minimum distance from the major axis to the geocenter is $d \sin \alpha$. The semimajor and semiminor axes of Z_Y are termed the stretching Λ and the semiwidth w, respectively, for reasons that we describe immediately.

When an encounter occurs several years or decades in the future, as is typical for the cases considered to date, the orbital uncertainty is strongly dominated by uncertainty in the anomaly. In such cases Z_Y is very long and slender and thus in the vicinity of Earth the ellipse can often be treated as a strip of constant width, which allows a convenient interpretation of the above parameters. Specifically, under this "strip approximation," when moving along the major axis, or spine, of Z_Y we are essentially changing only the anomaly of the asteroid; thus the minimum distance to the geocenter is the minimum encounter distance possible for any variation in the timing of the encounter. But this is precisely the definition of the MOID, which leads to the result that $d \sin \alpha$ is a good approximation for the MOID in such situations. It follows that the semiwidth w is essentially the uncertainty in the MOID. Furthermore, under the strip approximation, the semimajor axis of Z_Y will be closely aligned with the projection of Earth's heliocentric velocity on the target plane; i.e., it is aligned with the ζ-axis, if the reference system on the target plane was chosen in this way. Its length, which is the stretching Λ, indicates the amount by which the original ellipsoid Z_X has been stretched by the propagation F, as well as the timing uncertainty of the encounter $\Delta t = \Lambda/(v_\oplus \sin \theta)$. In some cases, this strip approximation can be inappropriate, generally because either the encounter is close to the time of the observations t_0, leading to a less-eccentric ellipse with errors not dominated by uncertainty in the time of the encounter t_1, or because the mapping F is locally highly nonlinear and interrupted returns (see section 5.4) are present on the target plane. In the former case, a strictly linear analysis of the encounter will typically be suitable, while in the latter case more sophisticated methods are needed to detect and analyze the encounter.

To compute the impact probability, one simply integrates the target-plane probability density p_{TP} over the cross-sectional area of Earth (*Chodas and Yeomans,* 1994). Within the Gaussian formalism, the bivariate probability density p_{TP} can be computed as the product of two univariate probability densities

$$p_{TP}(\sigma_1,\sigma_2) = p(\sigma_1)p(\sigma_2)$$

In fact, the probability density $p(\sigma)$ is often assumed to be Gaussian, but can also be modeled as a uniform density over the interval $|\sigma| \leq 3$, so that $p(\sigma) = 1/6$ within this interval and zero elsewhere; in the latter case the above formula is an approximation. Note that the value 3 for the limits on σ is somewhat arbitrary and based on experience; a rigorous statistical limit cannot be computed without going through a rigorous weighting procedure as described in section 3.2 and in *Carpino et al.* (2000). Although the real error distribution of asteroid astrometry is clearly not uniform, it has much stronger tails than a Gaussian distribution with formally derived standard deviations. In practice, the true probability distribution lies somewhere between the two, and either distribution will often suffice to obtain an order-of-magnitude estimate of the impact probability.

The recently proposed Palermo Technical Scale can be used to infer the significance of a potential impact event, based on the impact probability, size of the object, and time until the potential impact (*Chesley et al.,* 2001). However, when the impact probability is very low, say $\ll 10^{-9}$, it is practically indistinguishable from zero; it is therefore reasonable to say that an impact is not possible. In a strictly theoretical sense, if the probability is computed within a Gaussian formalism, it is never zero, although the values can be meaninglessly small.

3.5. Limitations of Linear Theory

The linear theory described so far has little use when searching for low-probability impact solutions. It should serve as a point of reference for applying the nonlinear approaches described in the following sections, but it is very important to use this valuable tool in a critical way, being aware of its limitations.

Nonlinearity arises from each stage in the computation procedure:

1. The least-squares fit for the elements X at epoch t_0 can be poorly constrained and so the ellipsoid $Z_X(\sigma)$ has some very long axes and is a poor approximation of the true confidence region. This form of nonlinearity is often manifested by confidence regions that appear as curved ellipsoids, or "bananoids."

2. The integral flow is nonlinear, and the longer the time of propagation the more nonlinearity accumulates. This means that even if the confidence ellipsoid were adequate at epoch t_0, the linear mapping of the covariance to a much later time t_1 might be very different from the exact propagation of the orbits in the original confidence ellipsoid, which will often appear folded, leading to interrupted returns as described by resonant return theory (sections 5.4 and 6).

3. The projection from the space of elements X onto the target plane Y is nonlinear, partly as a result of the transformation from orbital elements to Cartesian coordinates. Additionally, we note that for encounters with substantial timing uncertainty, the target-plane confidence ellipse will have a very long major axis, and the true confidence region will be actually curved by an amount corresponding to the curvature of Earth's heliocentric orbit (i.e., the true confidence region deviates from the confidence ellipse to the same extent that Earth's path deviates from a line). This problem is most severe for encounters with very large timing uncertainty or very low relative velocity.

Besides the more obvious limitation of linear theory, that the confidence region is not ellipsoidal, there is also the important problem that only the encounters experienced by the nominal orbit are even detected. And yet it is not unusual for the asteroid uncertainty to grow to a significant fraction of the entire orbit, or even wrap all the way around the orbit one or more times by the time of a potential collision, at which time the nominal orbit may be far from Earth, even on the opposite side of the Sun, and thus the linear analysis would not indicate any hazard at that time. Indeed, threatening encounters cannot even be reliably detected unless the probability of impact rises above $\sim 10^{-4}$, assuming that only encounters of the nominal orbit passing within 0.1 AU are considered.

We know that the curvature of the true confidence region, with respect to its linear approximation, is essentially the curvature of Earth's heliocentric orbit. Thus the use of the linear approximation cannot be accurate to 10^{-4} AU when used over a distance larger than $\sqrt{10^{-4}} = 10^{-2}$ AU. Thus, as a rule of thumb for Earth encounters, if d cos α is not more than several lunar distances, then the local curvature of the confidence region will have only a small effect and the value of d sin α will be a reliable estimate of the MOID to within a few Earth radii. This implies that even encounters detected along the nominal orbit cannot be reliably recognized as threatening using linear theories until the impact probability exceeds $\sim 10^{-3}$. For all these reasons, the linear analysis of the nominal trajectory is often inadequate or even erroneous, and nonlinear sampling and analysis techniques are required.

4. VIRTUAL ASTEROIDS

4.1. Discretization of the Confidence Region

When an asteroid is discovered, we cannot immediately know "the orbit" of the real object; rather, we can describe our knowledge by thinking of a swarm of virtual asteroids (VAs) with slightly different orbits all compatible with the observations. The reality of the asteroid is shared among the VAs, in the sense that only one of them is real but we do not know which one. Since the confidence region contains a continuum of orbits, each virtual asteroid is in turn representative of a small region; i.e., its orbit is also uncertain, but to a much smaller degree. This smaller uncertainty enables us to use for each VA some local algorithms, such as linearization, that would be inappropriate over the entire confidence region. Note that the nominal orbit is just one of the VAs, and is not extraordinary in this context.

The problem is then how to sample the confidence region with a limited number of VAs (in practice, a number ranging between a thousand and several tens of thousands) in such a way that, if an impact is indeed compatible with the available information, a virtual impactor (VI) is found among the selected virtual asteroids (*Milani et al.,* 2000a). Again, a VI is not an isolated collision orbit, but a representative of a small connected set in the confidence region in the initial conditions space, formed by points leading to a collision with a given planet at approximately the same time. Two main approaches have been devised to scan for VIs: the Monte Carlo methods and the line-of-variations methods.

4.2. Monte Carlo Sampling

The Monte Carlo (MC) methods directly use the probabilistic interpretation of the least-squares principle. Since the orbit-determination process yields a probabilistic distribution in the space of orbital elements, the distribution can be randomly sampled to obtain a set of equally probable virtual asteroids. They will be more dense near the nominal solution, where the probability density is maximum, and progressively less dense as the RMS of the residuals increases.

The random sampling needed to begin an MC test can be done in a simple way in the linear case, where the probability density in the orbital-elements space is well approximated by a Gaussian distribution with the mean value and covariance established by the nominal solution. In this case, a standard random number generator, providing a unit variance Gaussian distribution for a single variable, can be used to generate a random sampling according to the Gaussian distribution in the elements space (*Chodas and Yeomans,* 1999c).

In the nonlinear case, the probability density in the orbital-elements space is not given by an explicit analytical formula and, although it theoretically exists, cannot be directly computed. Thus the random sampling needs to take place in the space of observations, where the probability density is assumed to be Gaussian according to some assumed error model, although establishing such a model is a complex problem (*Carpino et al.,* 2001). The orbit determination process is then repeated, and the nominal solution for each set of modified observations is taken as a VA. Of course, this nonlinear MC procedure is computationally more expensive and should be used only when a linear MC is inappropriate. In practice, however, the cases that require the nonlinear MC treatment have very short observational arcs (typically much less than 30 days), and the time interval over which to search for impactors is orders of magnitude longer. Thus the computational load of the orbit determination is much less than the propagation, and the overhead of the nonlinear MC method (with respect to the linear MC) is not important.

4.3. Line of Variations Sampling

Whereas the MC samples consist of a scattering of unrelated points covering the entire six-dimensional confidence region, there are important computational advantages to sampling a reduced subspace (e.g., a one-dimensional continuous line) that is hopefully representative of the entire confidence region. For this purpose we can use the line of variations (LOV), which is the line of weakness of the orbit-determination solution. In the linear case, when the confidence region is well represented by the confidence ellipsoid, the LOV is simply the major axis of the ellipsoid Z_X, defined by the eigendirection V_1 associated with the largest eigenvalue λ_1 of the nominal covariance matrix $\Gamma(X^*)$, and its semilength is $\sigma_1 = \sqrt{\lambda_1}$. In the nonlinear case, for each solution X in the confidence region we can compute the covariance matrix $\Gamma(X)$ and the corresponding unit eigenvector $V_1(X)$ satisfying some orientation rule, e.g., the component of V_1 along the axis of the semimajor axis is positive. Then $\sigma_1(X)V_1(X)$ is a well-defined, smooth vector field, and the LOV is the unique solution of the ordinary differential equation

$$\frac{dX}{d\sigma} = \sigma_1(X)V_1(X)$$

with initial conditions at the nominal solution X^* for $\sigma = 0$. A numerical method for solving this equation is described by *Milani* (1999); more accurate methods have been developed for this and other applications.

It is natural to sample the LOV with VAs spaced uniformly in the cumulative probability so that each VA, representing a slice of the confidence region, has the same probability of representing the true orbit. One approach is to assume a Gaussian distribution along the LOV, which leads to short steps in the independent variable σ near zero and progressively larger steps as one moves away from the nominal solution. An alternate and more convenient approximation assumes a uniform probability density up to some value (e.g., for $-3 \leq \sigma \leq 3$); under this hypothesis the sampling will be uniform in σ.

4.4. Discussion

Our ultimate objective in this development is to investigate the entirety of the orbital confidence region to determine if there is any meaningful possibility of future collision. To this end, after computing a set of virtual asteroids, one must propagate each of them forward to the end of some timespan of interest, which in most cases extends 50–100 yr into the future. During the propagations any planetary encounters are noted for later analysis.

The most straightforward means of searching for collisions is to note the VAs that impact directly. With this very simple approach, if some of the VAs have close approaches within the radius of the planet they qualify as Virtual Impactors, and a simple estimate of the probability of impact for that asteroid over the timespan of interest would simply be the ratio of VIs to VAs. The LOV method of generating VAs is ill suited to this technique because such VAs cannot impact unless the LOV directly intersects Earth on the target plane, even though nearby off-LOV solutions may impact. However, the Monte Carlo-derived VAs do sample the entire volume of the orbital confidence region, and thus they can reveal all potential collisions with a reliability described by Poisson statistics. But this points to the main limitation of such a simple method: A VI can be reliably detected only if the impact probability is of the order of the inverse of the number of VAs used. For example, to achieve a complete scan of the confidence region down to probabilities of impact of $\sim 10^{-6}$ would require the propagation of a few million orbits, presently requiring a week or two of CPU time (for just one NEA). Thus this approach is currently too inefficient to be generally practical. For this reason it is necessary to apply some form of local analysis in the neighborhood of the close approaches detected for each VA.

The LOV approach is especially favorable for local analysis because one can exploit the fact that the LOV VAs belong to a continuous one-dimensional set. Thus it is possible to interpolate between the target plane points belonging to two consecutive VAs. As an example, if one VA passes from the nodal point before Earth, and the next consecutive one passes behind Earth, by continuity there must be a point on the LOV, corresponding to an intermediate value of σ, such that the encounter takes place at the MOID. Therefore the MOID of the VA orbit, at a time shortly before the encounter, can be used as a criterion to identify any potentially threatening encounters. In such cases some local analysis is needed to search for VI regions significantly smaller than the VA sampling scale, thus permitting searches for VIs with a probability significantly lower than the inverse of the number of VAs. Monte Carlo solutions are less well suited to interpolation because they are not ordered and are not constrained to a subspace. However, as was explained in section 3.4, the trace of the confidence region on the target plane is often a very narrow strip (e.g., see Figs. 1 and 2). In such cases this LOV interpolation can also be done using MC-derived VAs, but one must take care to ensure that the interpolation is based on dynamically similar VAs, as this is not generally clear *a priori*.

Even though interpolation along the LOV is generally very effective, we cannot be confident that the LOV approach will detect all potential collisions. If some VIs lie well off the LOV and are separated from it by some strong nonlinearity (e.g., the MOID having been changed with respect to the value along the LOV by some encounter at the other node or with another planet), then the VAs selected along the LOV may fail to indicate some potentially threatening encounters. The MC approach would, in principle, reveal such encounters. In the end, we reach the conclusion that both methods have important advantages, and a robust search procedure should use both methods in a coherent hybrid scheme to ensure efficient and reliable detection of all VIs above a given probability.

5. NONLINEAR TARGET-PLANE ANALYSIS

5.1. All Possible Encounters

Whether the confidence region in initial orbital elements is sampled with a Monte Carlo method, yielding a cloud of VAs, or with a Line Of Variations method, yielding a one-dimensional string of VAs, the samples are all numerically integrated forward to some time horizon (typically 50–100 yr), and all close approaches to perturbing bodies within some detection threshold distance are recorded. At each close approach, the time, distance, geocentric position, and velocity coordinates are noted. The close-approach threshold distance used for Earth is typically ~0.1–0.2 AU.

As noted in the previous section, a simple search for VIs among the sample VA orbits is not adequate for finding all potential collisions because this approach is likely to miss low-probability impacts. Consider, for example, a lost asteroid. Its confidence region quickly spreads along the entire orbit, which for a typical NEA has a length of ~10 AU. Since the diameter of Earth is less than 10^{-4} AU, and the number of samples is typically ~10^4 or less, it is unlikely that one of the virtual asteroids will impact directly, even if the orbits intersect and an impact is possible. There is clearly a need to use a method of local analysis in the neighborhood of a given VA to search for nearby VIs.

When ordered by time, the close approaches of all the VAs are seen to cluster around a series of encounter times — for example, when Earth passes close to the asteroid's nodal crossing point with the asteroid nearby. We call the subset of VAs that approach Earth around a given encounter time a *virtual shower*, since it is reminiscent of a stream of meteoroid particles encountering Earth. It is possible that the entire set of VAs form the shower if, for example, the encounter time is close to the initial time t_0. Often, however, the confidence region is so spread out along the orbit that only a small subset of VAs pass within the close-approach detection threshold.

A shower can be broken down into separate trails, which are subsets of closely related VAs, each subset following a qualitatively different dynamical path to the encounter. Often, a shower contains several different trails (see Table 1 in *Milani et al.,* 1999). For example, an asteroid that is lost by more than one full revolution can produce a shower containing trails in which VAs have performed different numbers of revolutions around the Sun since time t_0. Another situation occurs when VAs go through a close approach that changes their orbital periods in such a way as to bring them back for another close approach years later. Such *resonant returns* typically occur in pairs characterized by different stretching (see section 6).

When the individual approaches of a shower are plotted in the target plane, the separate trails typically reveal themselves through differing characteristic MOID values (Fig. 1a). Even if the initial confidence region is sampled with a Monte Carlo method, and the VAs are therefore off the line of variations, the confidence regions of the trails are usually highly stretched, with widths much narrower than the differences in MOIDs. As a result, it is often possible to separate the VAs of a shower into distinct trails simply by sorting on MOID (*Chodas and Yeomans,* 1999c). This method will not work, however, if the MOIDs of the trails are too similar. A more reliable method of separating the VAs of a shower into separate trails is to sort and index them according to their semimajor axis, the variation of which leads to the stretching in the target plane. This indexing can be immediately obtained with the LOV method and requires only minimal computation for MC samples. The trails can then be detected as sequences of n consecutive solutions X_i, $i = k, k + 1, \ldots, k + (n - 1)$ within a given shower (Fig. 1b).

5.2. Filtering Trails

The points of a trail are just sample points of a continuous line more or less following the LOV, and none need approach very close to the minimum possible distance for that trail. To decide if an impact is possible, we need to de-

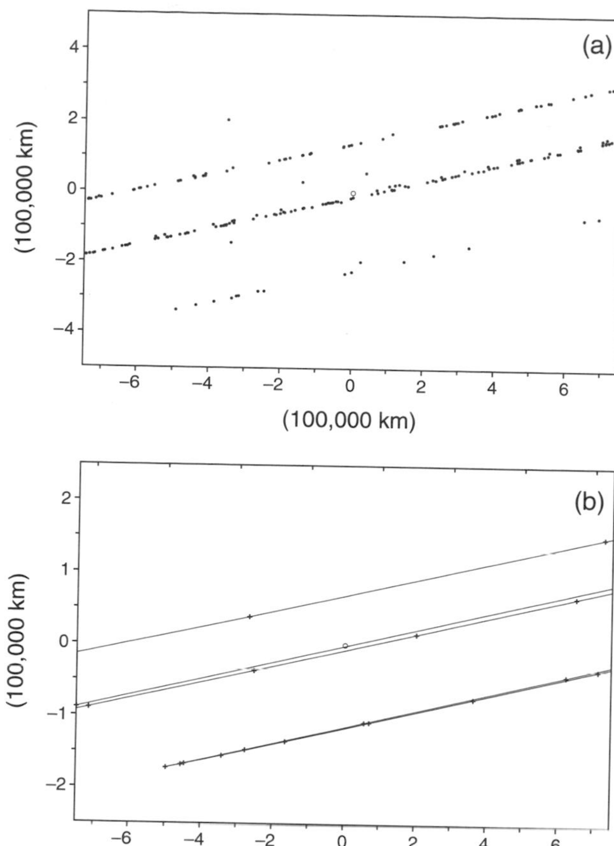

Fig. 1. The January 2046 shower from the VAs of 1998 OX_4 plotted on the b-plane with VAs derived by **(a)** MC and **(b)** LOV methods. Earth is represented to scale by the small open circle in the center of each plot. Since one of the trails clearly intersects Earth, a collision is possible. Note that the axes in this figure are not the (ξ, ζ) axes discussed in section 2.2, and the trails are not therefore aligned with the vertical axis.

termine precisely how close this trail approaches Earth. Since this new problem is computationally intensive, it is essential to apply some filter to discard immediately the trails that clearly cannot collide. Two methods are available for this: selection by MOID and linear analysis.

For each VA, we can compute the MOID, reduced by the amount of gravitational focusing (as described in section 2.2 for $b_\oplus$) to obtain the perturbed minimum approach distance $MOID_{GF}$. To determine whether an orbit will impact, the MOID should be computed based on the elements near the time of the encounter but not during the encounter (because the effects of the encounter itself will corrupt the MOID). However, lunar perturbations on Earth cause high-frequency fluctuations in the MOID, with amplitudes on the order of one Earth radius. This effect can be minimized by computing the MOID one lunar period before the encounter, but in general if $MOID_{GF} < 2r_\oplus$ then a collision for that particular trail cannot safely be ruled out, while larger values of $MOID_{GF}$ do indicate that no impact can take place due solely to timing uncertainty. But the MOID itself is uncertain due to the dispersion of the orbital elements at the time of the encounter (*Bonanno,* 2000), so in the absence of MOID uncertainty information we must use a generous safety margin, i.e., only ruling out collision when $MOID_{GF}$ is greater than, say the lunar distance.

As an alternative, we can perform a linear target-plane analysis for the closest-approaching VA found in each trail and discard the cases in which an impact is well outside of the confidence ellipse. Both methods have advantages and disadvantages. If the confidence ellipse on the target plane is wider than our safety margin, then the MOID of the VA under consideration is not representative of the values the MOID could assume for initial conditions inside the confidence region but far from the LOV. On the other hand, if the VA considered has only a shallow approach (e.g., further than 0.02 AU as discussed in section 3.5), the results of the linear analysis can be unreliable. The best solution is to use information from both methods, e.g., using the value of the MOID and allowing for the possibility of a decrease of its value by the 3σ width of the confidence ellipse.

Once a particular trail has been found to allow a very low minimum distance, i.e., too close to rule out a collision, then we need to carefully determine the actual value of this minimum. We first consider straightforward cases of trails with simple geometry and then turn to complex cases with strong nonlinearities.

5.3. Straightforward Encounters

When the VAs of a simple trail are projected onto the target plane, the ζ-coordinates of the intercept points form a monotonic sequence proceeding from one side of Earth to the other (i.e., spanning zero). To find the VA which makes the closest possible approach of a trail, several methods can be used.

One technique is a variant of Newton's method, which starts from the solution X_j with the closest approach among the VAs for the trail and applies corrections to obtain a new VA with an even closer approach. First, we compute the direction of the long axis of the target-plane confidence ellipse for X_j; we then find the point Y_{min} along this line that is closest to Earth in the target plane. The desired correction in the target plane is therefore $\Delta Y = Y_{min} - F(X_j)$. We then need to determine a change ΔX in the initial orbital elements about X_j that satisfies the linear mapping

$$DF\ \Delta X = \Delta Y$$

Since DF maps from a six-dimensional space to a two-dimensional space, there is no unique solution, but with the further constraint that the solution minimize the sum of the squares of residuals, a unique solution for ΔX can be found (*Milani et al.,* 2000a). With the new solution $X_j + \Delta X$ in hand, we numerically integrate to the encounter time and compute the new close approach distance. Since the direction of the correction ΔY is recomputed at each step, any curvature introduced by gravitational focusing is accounted for. The iterations are stopped when the close approach distances stop decreasing. This approach has proven very effective in most cases, but it is sometimes prone to divergence, especially in strongly nonlinear cases, as will be described in section 5.4.

Another method for finding the VA producing the minimum possible close approach along the LOV uses the *regula falsi* interpolation method. Starting from two consecutive VAs having target-plane ζ-coordinates with opposite signs, Y_j and Y_{j+1}, we compute the point Y_{min} closest to Earth along the line between them. We then obtain a new VA by interpolating to the corresponding point along the LOV between X_j and X_{j+1} in the space of initial orbital elements and iterate the procedure. This approach does not require the computation of the confidence ellipse in the target plane or the use of the linear mapping DF, but it does require that the trail comprise at least two VAs. This is in contrast to Newton's method above, which is more complex but can proceed from only a single point on the target plane.

Once the minimum possible distance D has been found for initial conditions along the LOV, we need to take into account the width of the confidence region on the target plane. The trace of the LOV on the target plane is a kind of spine of the confidence region, and the width can be estimated by using the minor axis w of the confidence ellipse (see section 3.4). For example, since w corresponds to 1σ, the minimum distance to the confidence region corresponding to σ = 3 is approximately D – 3w, assuming that D > 3w. The knowledge of the distance between the LOV and Earth, measured in sigmas, can be essential for evaluating the probability of impact, as detailed in section 5.5. This approximation corresponds to linearization in the neighborhood of the LOV point corresponding to the close approach at distance D, and this is adequate if D and w are small enough.

Both Newton's and the interpolation methods move along the line of variation to determine the minimum encounter distance. There is also a method developed by *Sitarski* (1999) that moves in the shortest direction to find the minimum cost Q for collision, and with this information the impact hazard can be eliminated, or at least the impact probability bounded. This method does not, however, permit a careful assessment of the impact probability since the probability density in the vicinity of the VI is unknown. Another method has been proposed by *Muinonen* (1999) to find the minimum value of the penalty ΔQ compatible with a collision, without restricting the search to the LOV. A mixed method, involving both one-dimensional and multidimensional explorations of the confidence region, has been developed by *Muinonen et al.* (2001). Additional discussion on this point is found in section 3.3 of *Bowell et al.* (2002).

The upper trails indicated in Fig. 1 are typical straightforward trails. Another excellent example of this type was the 2028 close approach of 1997 XF_{11} (based upon the 1997–1998 observations only). This case has a special feature: The confidence region is long (millions of kilometers), but the nominal solution results in a quite close approach in 2028. Thus linearization around the nominal is good enough to draw the conclusion that an impact in 2028 is not possible (Fig. 2a), despite the fact that the departure between the confidence region and the linear ellipse is evident (hundreds of kilometers) near the ends of the 3σ target-plane ellipse (Fig. 2b); this departure is greater on the MTP than on the b-plane (*Milani and Valsecchi,* 1999; *Chodas and Yeomans,* 1999c).

5.4. Complex Encounters

Some trails do not follow the simple behavior described above. When the VAs for these complex trails are projected into the target plane, the sequence of intercept points approaches Earth along the LOV, slows down its approach,

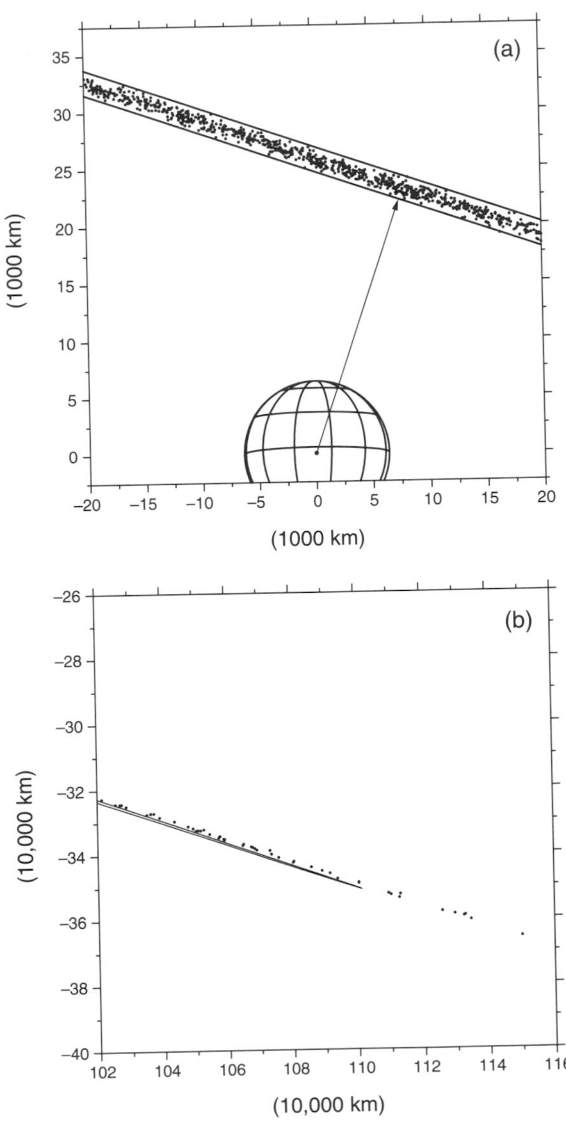

Fig. 2. The linear (3σ ellipse) and nonlinear (Monte Carlo points) confidence regions on the 2028 b-plane for the 88-d arc of 1997 XF_{11}. **(a)** Confidence region near nominal solution; **(b)** confidence region near 3σ. The full region extends over some 2,000,000 km, but is only 2000 km wide.

and then turns back and recedes in the direction from which it approached (see Fig. 7 in *Milani et al.,* 2000b). We call this behavior an interrupted return, and it arises as a result of previous close approaches, as explained in section 6. The bottom trail depicted in Fig. 1 is an interrupted return, as is clear from Fig. 1b where the trail both enters and exits the figure on the right. If the orientation of the LOV is mapped on the target plane for each VA then this reversal in direction can easily be detected.

It can be shown that for an interrupted return the derivative of Δt with respect to the σ parameter on the LOV passes through zero, and, not surprisingly, Newton's method can fail completely for these strongly nonlinear cases. Methods to identify such cases, and to perform an alternative local analysis to detect possible degenerate VIs and estimate their impact probability, are being tested. One possible approach is to use an LOV method including controls to detect interrupted returns and revert to an MC method restricted to the neighborhood of the VA near the folding point of the interrupted return. Another possibility is to use the regula falsi method along the LOV, exploiting the fact that minima of the approach distance D as a function of σ do exist, and they can be identified as zeros of the derivative $dD/d\sigma$. We must handle interrupted returns carefully because the point on the target plane at which the trail appears to stop and go back has relatively low stretching and may actually lie inside Earth's disk. This corresponds to the tangent case of section 6 and may result in a probability of impact larger that that of an ordinary trail.

5.5. Probability of Impact

If the minimum possible close approach distance of a trail is less than one Earth radius, then we know that an impact is possible, and our next task is to estimate the probability of impact P_I. In the following, we assume that we have identified the orbit X_I with the minimum possible encounter distance along the LOV as described in section 5.3. For this orbit we have σ_Λ, which is the σ distance from the nominal along the LOV, and we also have the target-plane covariance Γ_Y, from which we obtain the local stretching Λ and semiwidth w.

Now, taking X_I as nominal, we can compute the impact probability $\tilde{P}_I$ according to the linearized procedure described in section 3.4. We note that, since we are at the minimum distance point along the LOV, the assumption of local linearity is valid despite the fact that the location and properties of the trail itself may be dominated by nonlinear effects (see Fig. 2). But since X_I is not actually the nominal orbit, we must correct $\tilde{P}_I$ for the distance from the nominal to obtain

$$P_I = \tilde{P}_I \frac{p(\sigma_\Lambda)}{p(0)}$$

where we have used again the univariate probability density $p(\sigma)$ discussed in section 3.4.

When Λ is large (e.g., greater than the lunar distance), as is typical for highly stretched, far-future encounters, the strip approximation (section 3.4) permits the assumption that the probability density is a constant over the cross-section of Earth, which allows a very simple and convenient computation of the probability. If we can further assume that $w \ll r_\oplus$, then simple one-dimensional methods may be used, since the full width of the confidence strip falls onto Earth. In this case, the geometric chord length ℓ of the intersection of the trail's LOV with Earth's disk is the main parameter of interest, and P_I is computed from the cumulative probability distribution over ℓ

$$P_I = \frac{p(\sigma_\Lambda)}{\Lambda}\ell$$

On the other hand, when $w \gg r_\oplus$, fully two-dimensional methods must be used to compute impact probability; this only requires the additional consideration of the lateral distance to the geocenter σ_w so that the probability density orthogonal to the LOV can be computed. P_I is then calculated from a constant bivariate probability density integrated over the cross-section of Earth

$$P_I = \frac{p(\sigma_\Lambda)\, p(\sigma_\omega)}{\Lambda\omega}\pi r_\oplus^2$$

When $w \simeq r_\oplus$, these approximations are not suitable because $p(\sigma_w)$ cannot be assumed constant over the width of Earth: A two-dimensional probability integral needs to be computed. Even more complicated cases can be handled with suitable probabilistic formalisms, such as the ones of *Muinonen et al.* (2001).

6. RETURNS AND KEYHOLES

6.1. Resonant Returns

Virtual impactors are normally found numerically, given that a realistic physical model of their motion is quite complex; on the other hand, an exploration of the problem using analytical tools can give insight into the approximate location of VIs in elements space. Let us discuss the simple and rather common case of a VI whose impact takes place at a resonant return.

A resonant return occurs when, as a consequence of an encounter with Earth, the asteroid is perturbed into an orbit of period $P' \approx k/h$ yr, with h and k integers; then, after h revolutions of the asteroid and k revolutions of Earth, both bodies are again in the region where the first encounter occurred and a second encounter takes place (*Milani et al.,* 1999).

An analytical theory of resonant returns has been recently developed by *Valsecchi et al.* (2001). This theory treats close encounters with a suitable extension of Öpik's theory (*Öpik,* 1976), and adds a Keplerian heliocentric propagation (modified to account for the evolution of the

MOID; see section 6.2) between encounters, thus establishing a link between the outcome of an encounter and the initial conditions of the following one. The motion during an encounter with Earth is modeled by simply assuming that it takes place on a hyperbola; one of the asymptotes of the hyperbola, directed along the unperturbed geocentric encounter velocity $\mathbf{v}_\infty$, crosses the b-plane at a right angle, and the vector from Earth to the intersection point is **b**.

We can describe $\mathbf{v}_\infty$ in terms of its modulus, v_∞, and two angles, θ and ϕ; θ is the angle between $\mathbf{v}_\infty$ and Earth's heliocentric velocity $\mathbf{v}_\oplus$, while ϕ is the angle between the plane containing $\mathbf{v}_\infty$ and $\mathbf{v}_\oplus$ and the plane containing $\mathbf{v}_\oplus$ and the ecliptic pole.

We recall the (ξ, η, ζ) reference frame defined in section 2.2, where η is normal to the b-plane and ζ is oriented in the direction opposite to that of the projection of $\mathbf{v}_\oplus$ on the b-plane. Then ξ lies in both the b-plane and the plane normal to $\mathbf{v}_\oplus$ (*Greenberg et al.,* 1988). This choice of the b-plane coordinates has the nice property that ξ is simply the local MOID, while ζ is proportional to the time delay with which the asteroid "misses" the closest possible approach to Earth.

Öpik's theory then simply states that the encounter consists of the instantaneous transition, when the small body reaches the b-plane, from the preencounter velocity vector $\mathbf{v}_\infty$ to the postencounter vector $\mathbf{v}'_\infty$, such that $v'_\infty = v_\infty$, and θ' and ϕ' are simple functions of v_∞, θ, ϕ, ξ, and ζ; the angle between $\mathbf{v}_\infty$ and $\mathbf{v}'_\infty$ is the deflection γ, given by

$$\tan\frac{\gamma}{2} = \frac{c}{b}$$

where $c = GM_\oplus/v_\infty^2$. Finally, simple expressions relate a, e, i to v_∞, θ, ϕ (*Carusi et al.,* 1990), and ω, Ω, f to ξ, ζ, t_0 (*Valsecchi et al.,* 2001); t_0 is the time at which the asteroid passes at the node closer to the encounter. It is important to note that for small encounter distance a depends only on v_∞ and θ and, correspondingly, a' depends only on v_∞ and θ'.

We can formulate the condition for a resonant return to occur as a condition on θ'; for a ratio of the periods equal to k/h, we obtain a value of a', say a'_0, given by $a'_0 = \sqrt[3]{k^2/h^2}$, and a corresponding value of θ', say θ'_0.

Given θ, we can easily compute θ'_0 from ξ, ζ:

$$\cos\theta'_0 = \cos\theta\frac{b^2 - c^2}{b^2 + c^2} + \sin\theta\frac{2c\zeta}{b^2 + c^2}$$

thus obtaining the locus of points on the b-plane leading to a given resonant return. We can rearrange it in the form

$$\xi^2 + \zeta^2 - 2D\zeta + D^2 = R^2$$

that is the equation of a circle centered on the ζ-axis (*Valsecchi et al.,* 2000); R is the radius of such a circle and D is the value of the ζ-coordinate of its center, and they can be obtained from

$$R = \left|\frac{c\sin\theta'_0}{\cos\theta'_0 - \cos\theta}\right|$$

$$D = \frac{c\sin\theta}{\cos\theta'_0 - \cos\theta}$$

Actually, the expression for the locus of points on the b-plane leading to a given resonant return contains terms that are of the third order in ξ, ζ, and c (*Valsecchi et al.,* 2001), but for the purposes of the present qualitative discussion these terms can be ignored.

Figure 3a shows the arrangement of the b-plane circles corresponding to resonant returns to close encounters in 2040, 2044, and 2046 for the August 2027 encounter of asteroid 1999 AN_{10}; these correspond, respectively, to the mean-motion resonances 7:13, 10:17, and 11:19. A straight line is also drawn in the plot at $\xi = 6r_\oplus$ that represents a string of VAs having the same orbit and MOID (and thus the same ξ) and spaced in the timing of encounter with Earth (and thus different values of ζ). It is clear that in such an arrangement there are two regions for each resonance that lead to resonant returns, and it is also clear that they are located at the crossings of the straight line with the appropriate circle. Figure 3a also shows the circle associated with the 3:5 mean-motion resonance, corresponding to a resonant encounter in 2032; note that the straight line does not actually touch the circle, implying that no close encounter takes place.

6.2. Keyholes

Chodas (1999) introduced the term "keyhole" to indicate small regions of the b-plane of a specific close encounter such that, if the asteroid passes through one of them, it will hit Earth on a subsequent return. This can be generalized by requiring that the subsequent approach occurs to within a given small distance. An impact keyhole is thus just one of the possible preimages of Earth's cross section on the b-plane and is therefore tied to the specific value for the postencounter semimajor axis that allows the occurrence of the next encounter at the given date. Figure 3b depicts the two keyholes associated with the 7:13 resonance that are also shown in the top half of Fig. 3a.

The b-plane circles corresponding to a given a' address the question of the timing (i.e., the ζ-coordinate) of the subsequent encounter. However, if we use a simple Keplerian propagation between encounters, this would leave unchanged the MOID (the ξ-coordinate) of the next encounter. This would be unrealistic, as in general, the MOID varies between encounters for two main reasons: On a long timescale, secular perturbations (*Gronchi and Milani,* 2001) make it slowly evolve through the so-called Kozai cycle, or ω-cycle, while on a shorter timescale significant quasiperiodic variations are caused by planetary perturbations and, for planets with massive satellites, by the displacement of the planet with respect to the center of mass of the planet-satellite system.

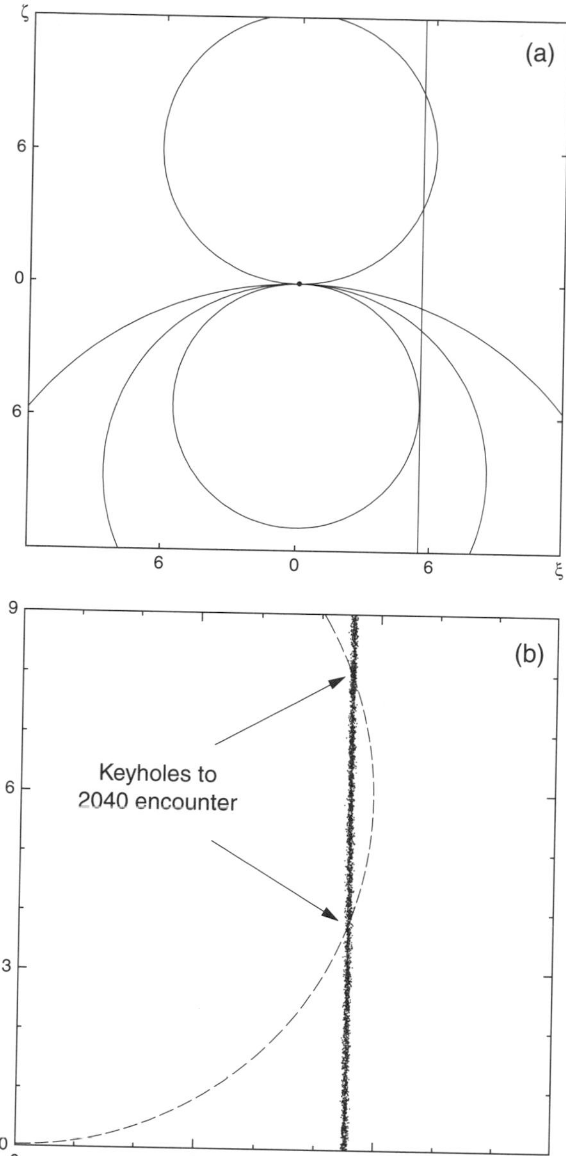

Fig. 3. Circles corresponding to various mean-motion resonances on the b-plane of the August 2027 encounter with Earth of asteroid 1999 AN_{10}. Distances are in Earth radii augmented for gravitational focusing ($b_\oplus$; see section 2.2). **(a)** Uppermost circle: 7:13 resonance; then, with the centers along the ζ-axis, from top to bottom: 3:5, 10:17, and 11:19 resonances. The vertical line at $\xi = 5.5\ b_\oplus$ represents fictitious asteroids, all with the same orbit as 1999 AN_{10} and spaced in the time of encounter with Earth. **(b)** Explicit depiction of keyholes for the 2040 returns on the 2027 b-plane of 1999 AN_{10}. The stream of Monte Carlo points corresponds to the vertical line in **(a)**. The dashes represent the best-fitting circle passing through the impacting zones.

For the purpose of obtaining the size and shape of an impact keyhole we can, however, just model the secular variation of the MOID as a linear term affecting ξ", the value of ξ at the next encounter

$$\xi'' = \xi' + \frac{d\xi}{dt}(t_0'' - t_0')$$

where t_0' and t_0'' are the times of passage at the node, on the post-first-encounter orbit, that are closest to, respectively, the first and the second encounter. We can compute the time derivative of ξ either from a suitable secular theory for crossing orbits (*Gronchi and Milani,* 2001) or by deducing it from a numerical integration. The result could then be corrected to take into account the short periodic terms, possibly by using the output of a numerical integration. Without such numerical corrections, the theory would reliably predict very close encounters (e.g., within a few thousandths of AU), but not collisions, which require accuracies an order of magnitude better.

The computation of the size and shape of an impact keyhole is described in detail in *Valsecchi et al.* (2001); here we give a qualitative discussion. Let us start from the image of Earth in the b-plane of the subsequent encounter, the one in which the impact should take place. We denote the coordinate axes in this plane as ξ", ζ" and consider the circle centered in the origin and of radius $b_\oplus$. The points on the b-plane of the first encounter that are mapped — by the Keplerian propagation plus the MOID drift — into the points of Earth's image circle in the b-plane of the second encounter constitute the Earth preimage we are looking for.

As far as the location is concerned, impact keyholes must lie close to the intersections, in the first encounter b-plane, of the circle corresponding to the suitable resonant return and the vertical line expressing the condition that the MOID be equal to

$$-\frac{d\xi}{dt}(t_0'' - t_0')$$

What about size and shape? To address this question, we can examine the matrix of partial derivatives $\partial(\xi'',\zeta'')/\partial(\xi,\zeta)$. We do this under the assumption, as in *Valsecchi et al.* (2001), that for the encounter of interest $c^2 \ll b^2$, since this condition is not too restrictive in most asteroid encounters with Earth. The matrix of partial derivatives has the following structure

$$\begin{bmatrix} \frac{\partial\xi''}{\partial\xi} & \frac{\partial\xi''}{\partial\zeta} \\ \frac{\partial\zeta''}{\partial\xi} & \frac{\partial\zeta''}{\partial\zeta} \end{bmatrix} \approx$$

$$\begin{bmatrix} 1 + \mathcal{O}\left(\frac{c}{b}\right) & \mathcal{O}\left(\frac{c}{b}\right) \\ \mathcal{O}(1) - 2hf\frac{\xi\zeta}{b^4} & 1 + \mathcal{O}\left(\frac{c}{b}\right) + hf\frac{\xi^2 - \zeta^2}{b^4} \end{bmatrix}$$

where f is a constant not depending on h.

The first row of the matrix shows that the ξ dimension of the preimage of Earth is essentially unchanged; the second row, however, gives a completely different picture. First, both derivatives grow linearly with h, the number of heliocentric revolutions the asteroid makes between the two encounters; second, the relative size of the two elements of this row depends critically on the values of ξ and ζ. Apart from the case in which $\xi \approx \zeta$, which we will discuss below, $\partial\zeta''/\partial\zeta$ can be rather large; this means that the separation in ζ of two VAs, that grows with $\partial\zeta''/\partial\zeta$ going from the first to the second encounter, can increase by a large amount. If we now go backward in time, from the second to the first encounter, the expansion in ζ becomes a contraction; this contraction affects the preimage of Earth, by squeezing it along ζ. The result of this is that the keyhole has the form of a lunar crescent that closely follows the appropriate b-plane circle, with a "width" in ξ of about $2b_\oplus$ and a maximum thickness of $\sim 2b_\oplus/(\partial\zeta''/\partial\zeta)$.

When $\xi = \zeta$, the third term in $\partial\zeta''/\partial\zeta$ becomes 0, and somewhere in the vicinity the entire derivative becomes very small, <1; this can happen if the absolute value of the MOID at the first encounter is equal to the radius of the appropriate resonant circle. As a consequence of the smallness of $\partial\zeta''/\partial\zeta$, the dimension in ζ of the keyhole can be larger than $b_\oplus$!

The possibility of anomalously large keyholes is a manifestation of a more general phenomenon that takes place whenever a string of VAs is almost tangent to, but does not cross, a resonant return circle, as in the case of the 3:5 resonant circle shown in Fig. 3a. When this is the case, what happens at the time of the resonant return is that the string of VAs "enters" the b-plane of the resonant encounter from the side of either positive or negative ζ but then, since none of the VAs quite reaches the resonant semimajor axis, at some point it has to "turn back." The turning point is obviously a point where $\partial\zeta''/\partial\zeta = 0$. This situation is frequently encountered in numerical integrations and is also known as "interrupted return," as shown in the lower trail of Fig. 1. If this was to take place with the MOID at the second encounter $<b_\oplus$, an anomalously large keyhole would occur.

As stated above, the computation of VIs requires sophisticated numerical modeling of the orbital evolution. In this context, the simplifications made in the theory just described may lead one to think that it has no practical value. However, the main advantage of the theory is the geometrical understanding of the global structure of potentially impacting solutions that it gives, and this is particularly relevant for complex or pathological cases such as interrupted returns. In this sense, it can be considered as a tool for guiding the numerical explorations and for contriving fictitious critical scenarios to be fed to automatic monitoring programs for testing.

7. CONCLUSIONS AND FUTURE WORK

The last three years have seen both an enormous increase in our ability to identify impact possibilities and the establishment of the first efficient impact-monitoring system; this progress has been in part driven by the discovery of objects in orbits with very small, but detectable, Earth-impact regions within their confidence region. The situation is now evolving toward a more systematic study of the problem, with the goal of improving the currently operating monitoring system (CLOMON) and of establishing an additional and independent one at JPL.

As a consequence of the first two years or so of operation of CLOMON, with the capability of detecting collision possibilities for newly discovered objects while they are still observable, it has become routinely possible to concentrate followup observations on potentially colliding NEAs; in fact, most NEAs larger than ~100 m in diameter discovered after the end of 1999, having impact possibilities in the confidence regions of their short-arc orbits, have been followed long enough that their confidence regions could be reduced to the point of excluding any detectable impact.

The analytic theory as it stands is still largely incomplete. Future developments should include (1) a complete treatment of nonresonant returns, i.e., returns taking place at the other node, for asteroids having either both nodes at a heliocentric distance very close to 1 AU, or the other node at a heliocentric distance very close to the orbital radius of another planet; (2) a better understanding of the complex phenomena associated with interrupted returns and anomalously large keyholes; and (3) the possibility of sequences of more than two encounters, with the last encounter of the sequence leading to an impact.

The work now in progress, the results of which should be available soon, includes the following developments and tests: (1) the Sentry system, a fully independent monitoring system at JPL, which will include "second-generation" features such as the ones mentioned below (the second-generation CLOMON-2 system is also under development); (2) improved methods to handle interrupted returns in order to sharply reduce the cases of divergent iterations in the Newton method; (3) mechanisms to implement focused or densified Monte Carlos to handle difficult cases, including interrupted returns; and (4) a new observation weighting scheme that allows the use of modified Gaussian probabilities as a reliable estimate of impact probabilities.

Note added in proof: The second-generation impact monitoring systems are now on line, both the Sentry system at JPL (http://neo.jpl.nasa.gov/risk/) and the CLOMON2 system at the University of Pisa (http://newton.dm.unipi.it/cgi-bin/neodys/neoibo?riskpage:0;main) and duplicated at the University of Valladolid (http://unicorn.eis.uva.es/cgi-bin/neodys/neoibo?riskpage:0;main).

Acknowledgments. This research was conducted at the University of Pisa and at IAS-CNR under a contract with the Italian Space Agency, and at the Jet Propulsion Laboratory, California Institute of Technology, under a contract with the National Aeronautics and Space Administration.

REFERENCES

Bonanno C. (2000) An analytical approximation for the MOID and its consequences. *Astron. Astrophys., 360,* 411–416.

Bowell E. and Muinonen K. (1994) Earth-crossing asteroids and comets: Groundbased search strategies. In *Hazards Due to Comets and Asteroids* (T. Gehrels, ed.), pp. 149–197. Univ. of Arizona, Tucson.

Bowell E., Virtanen J., Muinonen K., and Boattini A. (2002) Asteroid orbit computation. In *Asteroids III* (W. F. Bottke Jr. et al., eds.), this volume. Univ. of Arizona, Tucson.

Carpino M., Milani A., and Chesley S. R. (2001) Error statistics of asteroid optical astrometric observations. *Icarus,* in press.

Carusi A., Valsecchi G. B., and Greenberg R. (1990) Planetary close encounters: Geometry of approach and post-encounter orbital parameters. *Cel. Mech. Dyn. Astron., 49,* 111–131.

Chesley S. R. and Milani A. (2000) An automatic Earth-asteroid collision monitoring system. *Bull. Am. Astron. Soc., 32,* 0602.

Chesley S. R., Chodas P. W., Milani A., Valsecchi G. B., and Yeomans D. K. (2001) Quantifying the risk posed by potential Earth impacts. *Icarus,* in press.

Chodas P. W. (1993) Estimating the impact probability of a minor planet with the Earth. *Bull. Am. Astron. Soc., 25,* 1236.

Chodas P. W. (1999) Orbit uncertainties, keyholes, and collision probabilities. *Bull. Am. Astron. Soc., 31,* 2804.

Chodas P. W. and Yeomans D. K. (1994) Predicting close approaches of asteroids and comets to earth. In *Hazards Due to Comets and Asteroids* (T. Gehrels, ed.), pp. 241–258. Univ. of Arizona, Tucson.

Chodas P. W. and Yeomans D. K. (1996) The orbital motion and impact circumstances of Comet Shoemaker-Levy 9. In *The Collision of Comet Shoemaker-Levy 9 and Jupiter* (K. S. Noll et al., eds.), pp. 1–30. IAU Colloq. 156.

Chodas P. W. and Yeomans D. K. (1999a) Could asteroid 1997 XF_{11} collide with Earth after 2028? *Bull. Am. Astron. Soc., 31,* 703.

Chodas P. W. and Yeomans D. K. (1999b) Orbit determination and estimation of impact probability for near-Earth objects. Paper AAS 99-002, 21st Annual AAS Guidance and Control Conference, Breckenridge, Colorado.

Chodas P. W. and Yeomans D. K. (1999c) Predicting close approaches and estimating impact probabilities for near-Earth objects. Paper AAS 99–462, AAS/AIAA Astrodynamics Specialists Conference, Girdwood, Alaska.

Gehrels T., Matthews M. S., and Schumann A. M., eds. (1994) *Hazards Due to Comets and Asteroids.* Univ. of Arizona, Tucson. 1300 pp.

Greenberg R., Carusi A., and Valsecchi G. B. (1988) Outcomes of planetary close encounters: A systematic comparison of methodologies. *Icarus, 75,* 1–29.

Gronchi G. F. (2001) On the stationary points of the squared distance between two ellipses with a common focus. *SIAM J. Sci. Comp.,* in press.

Gronchi G. F. and Milani A. (2001) Proper elements for Earth-crossing asteroids. *Icarus, 152,* 58–69.

Jazwinski A. H. (1970) *Stochastic Processes and Filtering Theory.* Academic, New York. 376 pp.

Kizner W. (1959) *A Method of Describing Miss Distances for Lunar and Interplanetary Trajectories.* External Publication No. 674, Jet Propulsion Laboratory, California Institute of Technology.

Marsden B. G. (1999) A discourse on 1997 XF_{11}. *J. British Interplanet. Soc., 52,* 195–202.

Milani A. (1999) The asteroid identification problem I: Recovery of lost asteroids. *Icarus, 137,* 269–292.

Milani A. and Valsecchi G. B. (1999) The asteroid identification problem II: Target plane confidence boundaries. *Icarus, 140,* 408–423.

Milani A., Chesley S. R., and Valsecchi G. B. (1999) Close approaches of asteroid 1999 AN_{10}: Resonant and non-resonant returns. *Astron. Astrophys., 346,* L65–L68.

Milani A., Chesley S. R., Boattini A., and Valsecchi G. B. (2000a) Virtual impactors: Search and destroy. *Icarus, 145,* 12–24.

Milani A., Chesley S. R., and Valsecchi G. B. (2000b) Asteroid close encounters with Earth: Risk assessment. *Planet. Space Sci., 48,* 945–954.

Morrison D., ed. (1992) *The Spaceguard Survey: Report of the NASA International Near-Earth-Object Detection Workshop.* (Available on line at http://impact.arc.nasa.gov/reports spaceguard/)

Muinonen K. (1999) Asteroid and comet encounters with the Earth: Impact hazard and collision probability. In *The Dynamics of Small Bodies in the Solar System, A Major Key to Solar System Studies* (A. E. Roy and B. A. Steves, eds.), p. 127. NATO ASI Proc. 522, Kluwer, Dordrecht.

Muinonen K. and Bowell E. (1993) Asteroid orbit determination using Bayesian probabilities. *Icarus, 104,* 255–279.

Muinonen K., Virtanen J., and Bowell E. (2001) Collision probability for Earth-crossing asteroids using orbital ranging. *Cel. Mech. Dyn. Astron., 81,* 93–101.

Öpik E. J. (1976) *Interplanetary Encounters: Close-Range Gravitational Interactions.* Elsevier, New York. 155 pp.

Sitarski G. (1968) Approaches of the parabolic comets to the outer planets. *Acta Astron., 18,* 171–195.

Sitarski G. (1999) How to find an impact orbit for the Earth-asteroid collision. *Acta Astron., 49,* 421–431.

Valsecchi G. B., Milani A., Gronchi G. F., and Chesley S. R. (2000) The distribution of energy perturbations at planetary close encounters. *Cel. Mech. Dyn. Astron., 78,* 83–91.

Valsecchi G. B., Milani A., Gronchi G. F., and Chesley S. R. (2001) Resonant return to close approach: Analytical theory. *Astron. Astrophys.,* in press.

Yeomans D. K. and Chodas P. (1993) *Periodic Comet Shoemaker-Levy 9 (1993e).* IAU Circular 5807.

Observational Selection Effects in Asteroid Surveys and Estimates of Asteroid Population Sizes

Robert Jedicke
University of Arizona

Jeffrey Larsen
University of Arizona

Timothy Spahr
Smithsonian Astrophysical Observatory

Every imaginable asteroid survey, including physically collecting them, is limited in capability by the detector, the observer, software, and other sundry parameters. In order to correct the observed population for these selection effects the surveyors must develop a detailed understanding of the operating parameters for their entire detection system. We provide an overview of the main selection effects for any survey's asteroid detection capabilities and mathematical basis for the concepts involved in correcting for these observational biases. Furthermore, we illustrate in a series of figures some of the more interesting aspects of how selection effects affect studies of near-Earth objects (NEO). The currently known and bias-corrected absolute magnitude distributions for near-Earth, main-belt, Trojan, Centaur, and transneptunian objects are presented on the same figure and in the same functional form for ease of comparison. It represents an inventory of the major populations of minor planets, and detailed studies of these distributions (described in other chapters in this book) yield insights into (among other things) their strengths and dynamical and orbital evolution.

1. INTRODUCTION

Every observational study of the minor planets is influenced by selection effects created during the data collection process. Conclusions then drawn from the raw survey data will be incorrect unless an appropriate compensation is made by applying a mathematical bias correction factor. The bias is a complex function of many factors relating to the asteroids (orbital parameters, size, albedo), detector (telescope, CCD size, sensitivity, and photometric response), software (requirements for automatic detection, ability to operate under confusion with background stars), and observer (region selection, weather conditions, psychology), to name just a few. The interaction of all these elements is unique to every study, whether it be astrometric, radar, spectrophotometric, etc. In this chapter we discuss selection effects that can affect the observed distribution of orbital elements and absolute magnitudes and present examples of how the data can be appropriately corrected.

The true orbital element, size, and taxonomic distributions of asteroids provide a wealth of information regarding present and past conditions in the solar system. They trace the location and migration of the major planets throughout the history of the solar system, the formation locations for major and minor planets, the amount of mass present in the original asteroid belt, the cratering history of planets, and the sources of NEOs. In addition, their current dynamical structure indicates the long-term stability of the entire system. This wealth of information, which forms the basis for discussion in many other chapters of this book, depends on the unbiased (actual) distribution, size, and composition of the solar system's smaller objects.

In 1979, at the time of publication of *Asteroids* (*Gehrels,* 1979), there were ~2190 numbered objects. Ten years later, at the time *Asteroids II* (*Binzel et al.,* 1989) was published, there were ~4295. Now, by the time of this writing, there are ~27,000. The complete list of objects with reasonably good orbits is currently over 120,000 and the main belt is probably observationally complete to absolute magnitude (H) ~ 13 (based on the location of the "roll-over" in histograms of the incremental H distribution of main-belt objects). There is essentially no room for large, undiscovered populations of asteroids in the main belt. Histograms of the osculating elements for main-belt objects with H ≤ 13 now provide an excellent picture of the structure of the belt. Given the spread in apparent visual magnitude across the 1.5-AU width of the belt (nearly 2 magnitudes at opposition for objects of the same size on the inner and outer edges), the inner region is complete to H ~ 15. Interestingly enough, asteroids in this H range display visual magnitudes of about 10–12, which is too bright for some existing surveys to detect. In the vernacular of this chapter, a bias exists against the detection of these larger main-belt asteroids.

The 12 years since the publication of *Asteroids II* (*Binzel et al.,* 1989) have witnessed many important discoveries through advances in computing power and CCD technology combined with increases in funding for minor-planet discovery. Almost all astrometric positions are now provided by CCD observers and their precision/accuracy is roughly a factor of 10 better than 15 years ago. It is not uncommon for several thousand asteroids per day to be observed astrometrically. Completely new classes of asteroids (e.g., trans-neptunian objects) and a number of apparently inactive retrograde objects have been discovered in tandem with a barrage of multiwavelength spectrophotometric surveys.

Reaping the full benefits of information sown by this increase in effort, technology, and statistics requires a better understanding of the observational selection effects. This chapter introduces some of the concepts and techniques involved in correcting asteroid survey results for their intrinsic biases. The net result of the intensive effort to locate and measure asteroid properties and then correct for systematic biases is that the complete dynamical structure and physical makeup of the asteroids is slowly being determined.

Bowell et al. (1989a) suggested that the set of numbered asteroids is a relatively unbiased sample of the structure of the asteroid belt. While this is not strictly true, subsets of the numbered and unnumbered asteroids with high-quality orbital elements can provide remarkable insight into various asteroid subpopulations. Note that *Kirkwood* (1868) and *Hirayama* (1918) both made breakthroughs in the structure of the asteroid belt with remarkably few objects. Kirkwood used only 100 asteroids to identify the gaps, which now carry his name, and Hirayama used a total of 790 objects to find the 37 that delineated the Eos family.

The inherent difficulty in correcting for selection effects is that we cannot observe all asteroids simultaneously, nor can we measure all the properties that influence our bias. Some scaling factor must be derived to extend the results of any particular study to the entire asteroid population. Recent studies such as *Rabinowitz et al.* (2000), *Bottke et al.* (2000), and others have advanced our understanding of selection effects in surveys that discover asteroids. Their "bias-corrected" results allow more refined estimates of the absolute magnitude and orbit distribution of asteroids. Still, much work is required in order to gain insights on physical characteristics such as size, rotation rate, albedo, surface, and bulk properties.

An important piece of the puzzle is the size-frequency distribution, a primary constraint on modeling the evolution of the solar system's small bodies (*Davis et al.,* 2002). The goal of many asteroid surveys is to expand the envelope of measured bodies to ever-smaller sizes with the result that many different subsets of the asteroid population currently enjoy estimates of their size-frequency distribution. The results provide valuable clues to the history and physical properties of minor bodies and thus the entire solar system. In this chapter, we present the bias-corrected absolute magnitude-frequency distribution for five different populations of asteroids.

2. SELECTION EFFECTS IN ASTEROID SURVEYS

When you reconstruct a jigsaw puzzle, it is always reassuring to know the number of pieces before you start. A 2000-piece monstrosity will require roughly double the time to complete than a 1000-piece puzzle. If the puzzle is new you can be confident that all the pieces will be in the box and that they will all fit together nearly seamlessly, resulting in a satisfying state of completion.

The situation is entirely different when attempting to piece together the results of asteroid surveys into a picture of their size or orbit distribution. Vast stretches of the smallest pieces of the puzzle remain unknown and even big chunks of the puzzle (large asteroids easily within the realm of discoverability for asteroid surveys) may not have been found due to protective resonances.

Unable to create an inventory by trawling for asteroids with a gigantic net, planetary astronomers are thus hobbled by the limitations of their detectors. The smaller pieces of the asteroid puzzle, those that are intrinsically more interesting and often seem to be of the most importance in distinguishing between competing theories, are not well observed and the correction factors associated with converting their observed to the actual number are large with correspondingly huge errors. A discovery "bias" exists in favor of some types of asteroids and the prejudice often seems pathologically oriented against the planetary astronomer's goals of understanding the solar system.

In general, there are a set of M quantities $\overline{x} = (x_1,\ldots,x_M)$ associated with each asteroid and we construct a histogram of the number of observed objects as a function of those quantities $n(\overline{x})$ in bins of width $\Delta\overline{x} = (\Delta x_1,\ldots,\Delta x_M)$. From the observed set of objects $n(\overline{x})$ we need to account for the detector's properties in order to determine the actual distribution of the objects $N(\overline{x})$. In practice, it is usually necessary to accumulate observations over many of the quantities (x_i) because their values are unknown or the statistics in each bin are too small for meaningful results. For instance, a single bin in a distribution of semimajor axis (a) values encompasses asteroids possessing a wide range of eccentricities (e), inclinations (i), absolute magnitudes (H), etc.

The detection efficiency (ε) will depend on the $\overline{x}$ in a secondary manner. The ability to detect an asteroid does not depend directly on its physical or orbital parameters, only circuitously through the effect they have on the asteroid's appearance in a field of view (brightness, rate and direction of motion, etc.). The detector system mangles the actual set of parameters to produce a skewed observed distribution. For this reason it is very dangerous to use raw observations in place of a corrected set of data.

Specifically with asteroids, the set of $\overline{x}$ may comprise the orbital elements (a; e; i; Ω, longitude of ascending node; ω, longitude of perihelion; M, mean anomaly) and physical parameters such as H, slope parameter (G), albedo, type, etc., but it is rare for all these parameters to be well-measured for a specific object. Unfortunately, every one of these parameters affects a detector system's ability to detect the asteroid. The detection efficiency, when restricted to a subset of parameters, depends on the distribution of the other (unmeasured) physical and orbital parameters. In the following subsections we address detection or computational problems that influence the efficiency for identifying asteroids.

We will use a traditional nomenclature for Sun (S), Earth (E), and object (O) where $\overline{SE} = \bar{R}$, $\overline{EO} = \bar{\Delta}$, and $\overline{SO} = \bar{r}$. We orient the x-axis in the direction toward the vernal equinox and the z-axis in the direction of the north ecliptic pole. The geocentric ecliptic longitude and latitude are repesented by λ and β respectively, while the heliocentric equivalents are ℓ and b.

Many of the data-biasing effects discussed in the following subsections are interrelated and therefore difficult to treat separately. The difficulty inherent to presenting the effects should be kept in mind when contemplating the interplay of all the parameters affecting a system's ability to discover asteroids. The back-of-the-envelope calculations discussed herein only skim the surface of what a global and rigorous simulation would reveal.

2.1. Asteroid Sky-Plane Distribution

The selection of sky regions covered during an asteroid survey introduces selection bias into the orbital elements for detected asteroids. For example, the "inclination effect" (section 2.6) causes more distant classes of asteroids to appear to be clustered toward the ecliptic. Since distant asteroids are much less sensitive to the "phase effect" (section 2.7) it is, in some sense, easier to detect them farther from opposition than it is for closer populations.

The apparent sky distribution to a given limiting magnitude for a specific asteroid class spreads out dramatically in ecliptic latitude as the mean semimajor axis for the class decreases. Figures 1a–c illustrate this effect using the three types of near-Earth asteroids (NEAs) as examples. The most distant Amors (perihelion q = a(1 – e) in the range 1.017 < q ≤ 1.3 AU and $\bar{a}$ ~ 2.4) are clustered strongly toward the ecliptic, whereas the Apollos (a ≥ 1.0 AU and q ≤ 1.017 AU and $\bar{a}$ ~ 2.0) are less so. To a limiting magnitude of V = 22 both sets preferentially appear further from opposition. The sky-plane distribution of the Atens [a < 1.0 AU and aphelion Q = a(1 + e) > 0.983 AU and $\bar{a}$ ~ 0.83] shown in Fig. 1c is dramatically different — their distribution extends much higher in ecliptic latitude and is sparse toward the opposition point.

Clearly, when undertaking a survey for NEOs, the choice of survey regions dramatically impacts the relative fractions of discovered Atens, Apollos, and Amors.

2.2. Trailing Losses

Well-behaved stars in the field of view display consistent point spread functions (PSF) and are easy to detect compared to moving asteroids that leave trails during an exposure. The (typically) Gaussian-like instantaneous PSF for an asteroid is spread over many pixels as it moves across the field of view. Modern CCDs may contain square or rectangular pixels and an asteroid's apparent motion on the array may be in any orientation and display a wide range of rates. As the light from the asteroid is spread along the length of the trail each pixel receives less light than it would have had the asteroid remained stationary. The effect is dubbed a "trailing loss" and is represented by a description of the reduction in peak apparent magnitude (Δm_T) as a function of the rate of motion (ω).

Of course, no light is actually "lost" due to motion of the asteroid across the detector; it is just more difficult to detect manually and automatically. While the human eye is very effective at identifying linear structures, even with faint signals barely above the background noise level, automated detection of the "streaks" is computationally expensive — dramatically more so than locating Gaussian-like PSF sources and identifying those that have moved linearly between suc-

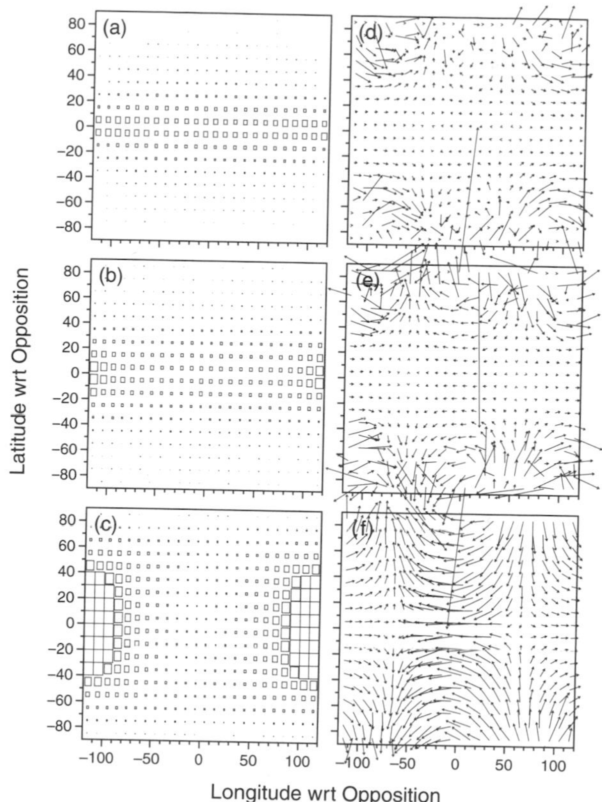

Fig. 1. **(a)–(c)** The sky-plane distribution of Amors, Apollos, and Atens respectively in the night sky relative to opposition for V < 22. The size of each box within each figure is proportional to the relative number of NEOs of each type at that location. **(d)–(f)** Mean-motion vectors for a realistic set of Amor, Apollo, and Atens respectively for V < 22. The motion vectors for Amors and Apollos are at twice the scale of the Atens. The mean rate of motion at opposition in ecliptic latitude is 0.0°/d for all three types (as expected) while the rates in longitude are –0.3°/d, –0.4°/d, and –1.4°/d respectively.

cessive images. Most of the trailed objects identified by the Spacewatch project were discovered by an observer through real-time examination of the incoming imagery.

Jedicke and Herron (1997) performed numerical integrations of the motion of a Gaussian PSF across a pixelized detection device at a variety of rates, angles, and offsets with respect to the pixel's grid. They found that trailing losses were apparent even for very small rates of motion (comparable to the FWHM of the PSF during the exposure interval). In general, losses are small for low rates and increase linearly at high rates. The exact form of the trailing loss function (Δm_T vs. ω) will depend on the system's CCD, pixel shape, PSF, etc., and must be determined for each system independently. Trailing losses can be measured directly by clocking the telescope drive in a nonsidereal fashion to trail stars of known brightness at desired rates.

The ability to detect an asteroid in a CCD image is related to its brightness and its rate of motion across the detector. These parameters are in turn related to the asteroid's orbit elements, which ties into the bias calculation. At opposition the formulae for the rate of motion of an object are relatively simple

$$\dot{\beta} = \pm\frac{\sqrt{\mu}}{r(r-1)}\sqrt{a(1-e^2)}\sin i \tag{1}$$

$$\dot{\lambda} = \frac{\sqrt{\mu}}{r-1}\left[\frac{\sqrt{a(1-e^2)}}{r}\cos i - 1\right] \tag{2}$$

as determined by *Bowell et al.* (1994a). Therefore, a detector system that has a rate-dependent efficiency automatically also has an (a, e, i) dependency.

Figures 1d–f illustrate the relative mean rates and directions of motions of NEAs as a function of their class (Amors, Apollos, Atens) and position with respect to opposition (λ_{oppn} and β). The orbital elements used to generate this figure were derived from the NEO model of *Bottke et al.* (2000). It is clear that trailing losses will be larger for all objects as the search moves away from the ecliptic and that they will be greater for Atens than for the other types of NEOs. These considerations are important when planning a survey. For example, if a search was intended to find primarily Atens, it would make sense to shorten the exposure time since increased length of exposure would not create a brighter image of the rapidly moving object.

2.3. Stationary Points

The mean-motion vectors in Fig. 1 also illustrate some locations in the sky at which objects appear to move very slowly. When this occurs their motion between images of the same region may not be sufficient to identify the objects as nonstellar. The points at which an object's rate of motion in ecliptic longitude is zero are termed the "stationary points." Using the vectors defined in section 2, the stationary points for any object are given by the condition (*Green,* 1985) that

$$\left[(\bar{r}-\bar{R})\times(\dot{\bar{r}}-\dot{\bar{R}})\right]\times\hat{z} = 0$$

The location of the stationary point depends on the object's orbital elements and, once again, implicitly ties the orbital elements to the bias through the apparent rates of motion.

Figure 1 shows the Amor stationary point at about 55°, Apollos at about 65°, and Atens closer to 70° from opposition. Clearly, the design of a survey for asteroids and the correction for the bias must take into account the location of the stationary points and the system's ability to distinguish slow rates of motion.

2.4. Magnitude Cutoff

The most basic selection effect is imposed by the magnitude restriction(s) of the surveying system. Every asteroid survey has both a faint and a bright limiting apparent magnitude. The limitation on detection of bright asteroids is sometimes surprising, usually due to the saturation of the detector (photographic film or CCD) creating a very large "blob" instead of a profile characteristic of the detector's PSF.

The contemporary standard for converting an asteroid's H and G into apparent magnitude (M) is provided by *Bowell et al.* (1989b)

$$M = H_M + 5\log\left(\frac{r\Delta}{1\ AU^2}\right) - 2.5\log\left[(1-G)\Phi_1(\alpha) + G\Phi_2(\alpha)\right] \tag{3}$$

$$0° \le \alpha \le 120°$$

where $r = |\bar{r}|$, $\Delta = |\bar{\Delta}|$, and $\Phi_{1,2}$ are functions of the phase angle (α), which is the angle between Earth and Sun as viewed from the asteroid.

Figures 2a and 2b turn equation (3) inside out and show the maximum geocentric distance (Δ) at which an object of a given H is visible for a system with fixed limiting magnitude (V_{lim}) as a function of the angular distance from opposition. The most obvious feature is simply that as V_{lim} increases it is possible to detect objects at greater distances. Second, as the asteroid moves away from opposition the distance at which it can just be detected decreases due to phase effects [e.g., shadowing, coherent backscattering, single-particle phase function (*Muinonen et al.,* 2002)]. In most cases there is a dramatic drop in the maximum distance moving from opposition to 90° elongation. The "U"-shaped functions in Fig. 2a for V_{lim} = 21 and 24 are a reflection of the changing geometry. At opposition, in the case of the V_{limit} = 24 system and H = 18 object, it is ~4.5 AU from the Sun and ~3.5 AU from Earth. Roughly 120° from opposition the same object is detectable at ~3.5 AU despite the inhospitable phase because it is only half the distance from the Sun (~2.7 AU).

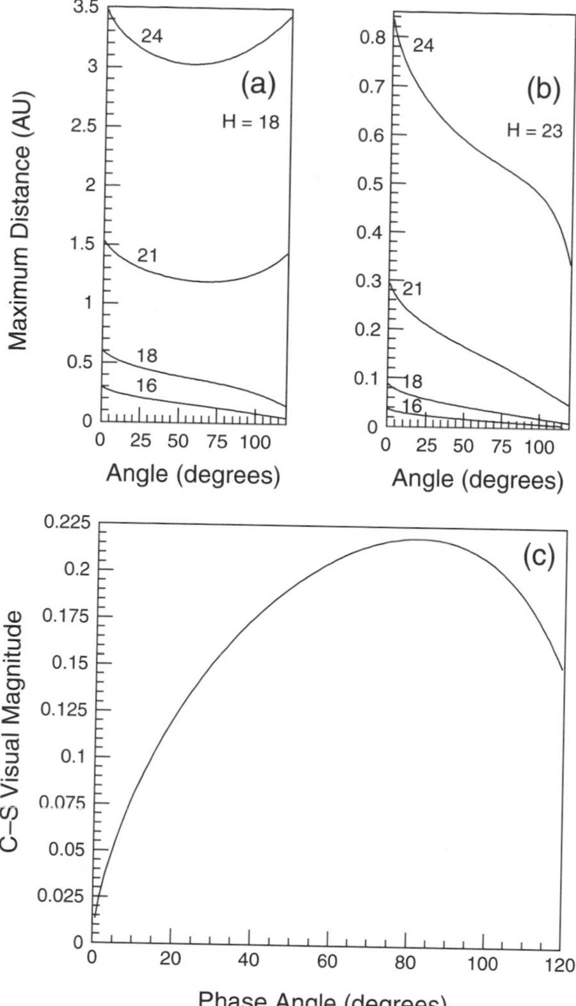

Fig. 2. **(a)** Maximum geocentric distance at which an H = 18 object is detectable for fixed limiting magnitudes of 16, 18, 21, and 24 as a function of the angular distance from opposition. **(b)** Same as in **(a)** except for H = 23. **(c)** Difference in apparent visual magnitude between a C-type and S-type asteroid of the same absolute magnitude as a function of the phase angle.

2.5. Line-of-Sight (LoS) Specificity

In any single image it is clear that only objects whose orbits can place them in the telescope's line-of-sight (LoS) are detectable. Consider the relationship for heliocentric ecliptic longitude given by

$$\tan(\ell - \Omega) = \tan(\nu + \omega)\cos i$$

where ν is the true anomaly. For the purpose of elucidation, restrict the survey's LoS to the opposition point so that $\ell = \Omega$. In this case, the observer "catches" objects as they penetrate the ecliptic at their ascending or descending node and they can only detect asteriods of a specific Ω. Furthermore, along the LoS toward opposition, the only asteroids that can be detected are those satisfying the relation tan (ν + ω) cos i = 0. Since cos(i) = 0 only for i = 90°, observing at opposition requires only finding asteroids for which ν + ω = 0. Therefore, to provide good sampling of the longitude of ascending node requires wide areal coverage in ecliptic longitude, especially in surveys that are otherwise strongly tainted by a need to search on the ecliptic. Even long-term surveys with otherwise good coverage around the ecliptic fall victim to this selection effect, as described in section 2.9.6.

2.6. Inclination Effect

The geocentric ecliptic latitude for an object is given by

$$\sin\beta = \frac{r}{\Delta}\sin(\nu + \omega)\sin i \tag{4}$$

Trivially, it says that zero-inclination objects must always be observed on the ecliptic. In this section we are interested in the maximum geocentric ecliptic latitude (β_{max}) attainable for a class of objects with a specific set of (a, e, i). Simple geometrical arguments persuade us that β_{max} is achieved when the object is at perihelion when Earth, the Sun, and the object all lie in the same plane (same heliocentric ecliptic longitude). Under these circumstances

$$\tan\beta_{max} = \frac{q\sin i}{q\cos i - R} \tag{5}$$

Letting Earth's heliocentric distance be 1 AU it is simple to determine β_{max} for a variety of asteroid classes. Hungarias have a_{min} = 1.78 AU, e_{max} = 0.18 and i_{max} = 34° (*Gradie et al.,* 1989) so β_{max} ~ 75.6° — it is impossible to find Hungarias with geocentric ecliptic latitude greater than about 76°.

The sinusoidal variation of β in equation (4) (through the dependence of the true anomaly on time) indicates that objects with nonzero inclination spend more time at high β than near the ecliptic. A survey with an explicit purpose of finding high-i objects should survey at ecliptic latitudes that maximize the probability of detecting these objects.

2.7. Phase-Angle Effects

The link between the slope parameter (G) and the phase angle (α) in equation (3) introduces a relationship between the viewing geometry and the difference in apparent magnitude of C- and S-type asteroids (which have different average G values). The effect is shown graphically in Fig. 2c, which indicates that for the same H, r, and Δ, a C-type asteroid is fainter than the S-type as a function of the phase angle [G_C = 0.15 and G_S = 0.25 following *Luu and Jewitt* (1989)]. They were concerned with the C:S ratio in the NEO population, often discovered at large phase angles due to their proximity to Earth. Since V_C-V_S increases with phase angle to a maximum of about 0.2 for α ~ 80° the C:S ratio for NEOs would be subject to a stronger bias than more distant asteroids discovered at smaller phase angles.

2.8. Color Effects

The choice of survey detector can bias results and lead to preferential discovery of certain compositional classes of asteroids as discussed in *Jedicke and Metcalfe* (1998).

Figure 3 compares the relative reflectance spectra from two different asteroids (one C-type and one S-type) obtained from the SMASS survey (*Xu et al.,* 1995) plotted with arbitrary zero points but equal scales. Since the SMASS spectra have a blue cutoff at ~4500 Å, we extrapolate the behavior of each spectra for shorter wavelengths according to *Cochran and Vilas* (1997). In each case the presented reflectances are scaled to $\lambda = 5600$ Å. Underneath the asteroid spectra we reproduce the spectral throughputs of two surveys (detector + filter only). The first is the Palomar Leiden Survey (PLS) (*van Houten et al.,* 1970), which used Kodak 103a-O photographic emulsions filtered by a Schott GG13 filter (very similar to the Wratten WG2 actually used to give the survey a "B"-like behavior). The second survey is the forthcoming Spacewatch mosaic camera, Marconi Technology CCD42-90 CCDs filtered by a Schott OG-515 filter, which approximates a "V + R + I" bandpass.

The two surveys discover asteroids using light coming from completely different portions of the spectra. When the spectra are recombined with the solar spectrum and located so as to have equal apparent V magnitudes, the PLS would see the C-type asteroid brighter than the S-type asteroid by about 0.13 mag. Using the Spacewatch mosaic, the situation is reversed. Therefore samples of asteroids compiled by each survey would contain a bias toward different classes. If the size of the magnitude difference seems small, consider that the number-magnitude distribution for asteroids increases rapidly with apparent magnitude leading to an ~30% overrepresentation of the favored asteroid type in the faintest magnitude bin for each survey.

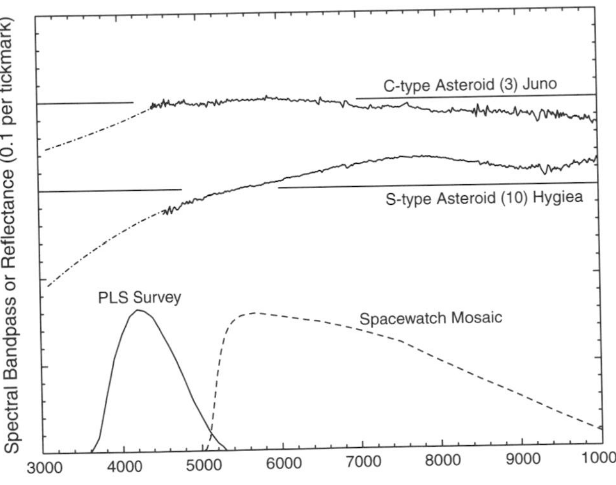

Fig. 3. Asteroid reflectance spectra (top two curves) (*Xu et al.,* 1995) and spectral bandpasses for a photographic survey (*van Houten et al.,* 1970) and the new Spacewatch mosaic CCD camera currently under construction.

2.9. Surveying Factors

This section highlights other selection effects introduced by surveying strategies, detector configuration, computing capabilities, and factors unrelated to an asteroid's physical or orbital parameters. While it may not be possible to eliminate these selection effects, a well-planned survey with good quality controls should be able to reduce their impact with a concomitant increase in the ease of data analysis. Most of these effects are introduced and treated carefully by *Bowell and Muinonen* (1994) and can be compensated for by a sufficiently robust debiasing routine.

2.9.1. Intra-image leakage. Asteroids move during an exposure, as described in section 2.2, and in the interval between exposures. Most surveys rely on successive images of a fixed portion of sky and then compare the list of objects found in each image to identify those that have moved. Some asteroids will be located near the edge of the field of view and will move off the imaging area between exposures; the likelihood of this happening is determined by the object's proximity to the edge of the image and its rates of motion. Since the rates are related to the orbital elements (e.g., equation (1)) a selection effect is introduced. The effect can be reduced with large fields of view, reduced time intervals between exposures, or by offsetting successive images by the mean displacement of asteroids in the field, but all these effects are interrelated.

2.9.2. Survey picket fence. All surveys have a limited field of view and attempt to choose search strategies that optimize their spatial coverage. The "picket fence effect" may be thought of as inter-image leakage if we extend the nomenclature of section 2.9.1. Any selection of survey regions will allow asteroids with fast or slow motions to be missed between visits to adjoining regions. Once again, these missed asteroids occupy portions of the orbital element parameter space, which subsequently become misrepresented. Careful selection of the survey regions can reduce the effect.

2.9.3. Detector picket fence. A variant of the picket fence will affect new large-mosaic CCDs planned or in operation for the next generation of NEO surveys. The gaps between the CCDs comprising the mosaic can be nonnegligible and asteroids may spend enough time within these gaps to make them undetectable. The ratio of gap size to detector size, and time between revisits to a field during a night, will bias against certain motion rates and their underlying elements.

2.9.4. New sky coverage. A two-dimensional region on the sky-plane imaged on one night corresponds to a three-dimensional volume occupied by asteroids from a variety of populations with different apparent motions and physical properties. The depth to which the volume is searched depends upon the observation's (system + conditions) limiting magnitude and an object's physical parameters. Thus, a revisit on a subsequent night to the same area will not sample the same spatial volume, nor will it see the same objects (which may have faded, brightened, or changed their rates of motion). Conversely, surveys intending to cover the sky can poorly choose revisit strategies and, due to Earth's motion, revisit the same spatial volumes for an asteroid population multiple times.

2.9.5. Detection techniques. Modern CCD surveys depend exclusively on automated motion detection software (pioneered by *Rabinowitz,* 1991) while earlier photographic surveys (e.g., *van Houten et al.,* 1970; *Helin and Shoemaker,* 1979) relied on experienced human examination of the images to identify moving objects. Every imaginable algorithm short of one that is 100% efficient will introduce a selection bias into the detections made with each system. Peak efficiencies for automated systems vary but it appears that >90% efficiency is routinely achievable for objects more than a magnitude brighter than the limiting threshold. The PLS was ~100% efficient using visual inspection at bright magnitudes.

Automated detection software will be constrained in two fundamental manners: rates of motion and brightness. Both of these are related to a moving object's orbital and physical parameters and therefore the detection software biases the observed distribution of objects. No debias mechanism is complete unless it also incorporates the effects of the detection technique. The linking process is not difficult in principle but can become computationally expensive without setting limits on rates and requiring "cuts" on the allowable variation in brightness of the object, linearity of the rates, etc. These "cuts" are chosen to balance the number of false positive detections against the rejection efficiency, and their side effect will be to impose related cuts on the types of asteroids that can be detected by the survey.

2.9.6. Seasonal and weather effects. The weather patterns and seasons at a survey location will have an effect on the detected orbital element distribution. In addition, the length of observing night combined with the Milky Way historically lead to a strong ecliptic longitude selection of discovered asteroids. Furthermore, bad seeing and extinction can systematically influence magnitude estimates unless careful photometric compensation takes place. Improperly accounting for these effects will bias H estimates downward.

These effects are nicely illustrated in Fig. 4a, as is the concept of search volume for the Spacewatch survey (see section 2.9.4). The wide gap in the $-\hat{y}$ direction points toward the galactic center while the smaller gap in the opposite direction is clearly the winter Milky Way. They avoided the confusion of stars in the galactic plane where detecting asteroids in dense star fields drastically reduces the efficiency and increases the false positive detection rate. The sparse coverage in the fourth quadrant is due to summer time surveying when nights are shorter and the southwestern United States experiences its "monsoon" period. The radial "spokes" in the set of detections are due to the monthly observing patterns where survey regions are clustered toward opposition during the new moon. Even though the data points in Fig. 4a represent a decade's worth of surveying, the repeated use of region selections still leave their imprint.

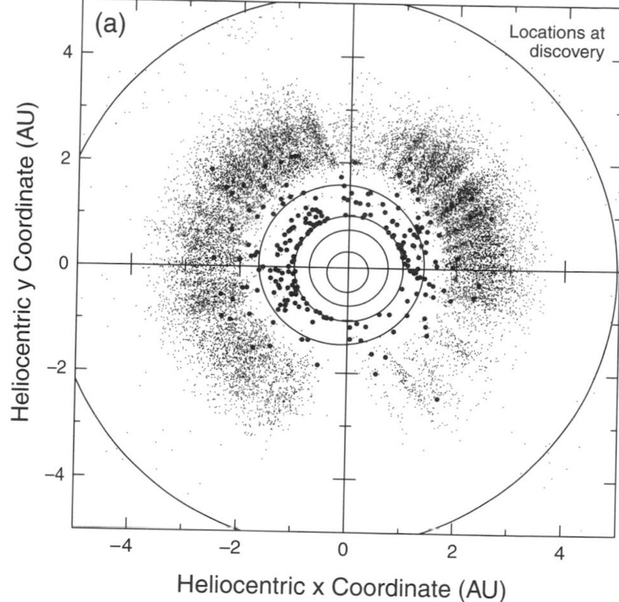

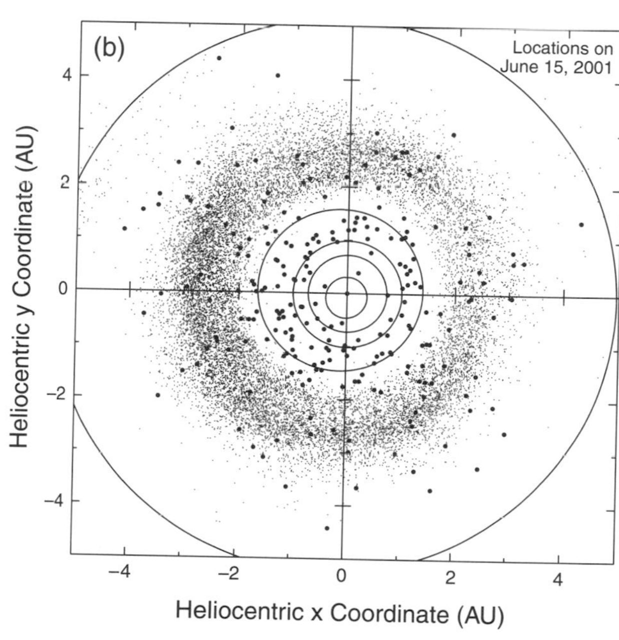

Fig. 4. **(a)** Heliocentric (x,y) location of Spacewatch objects at the time of discovery. **(b)** Heliocentric (x,y) location of all Spacewatch objects as of June 15, 2001. Main-belt asteroids and NEOs are represented by small and large points respectively. The x-axis lies along the direction to the vernal equinox. The bold circles represent the orbits of Mercury, Venus, Earth, Mars, and Jupiter respectively moving outward from the origin.

Spacewatch tends to concentrate their regions near opposition and near the ecliptic in order to maximize their rate of NEO discoveries. The effect of this selection bias is seen in Fig. 4b, which shows the location of all the Spacewatch objects as of June 15, 2001. Even though data were obtained over many years, there is still a visible depletion of asteroids in the direction toward the vernal equinox ($\hat{x}$) due to the nonrandom distribution of detected angular elements.

The effect is visible when comparing the L_4 and L_5 Trojan asteroid swarms leading and following Jupiter in its orbit respectively. In the Spacewatch data the L_4 swarm (in the second quadrant clustered around the bold circle representing Jupiter's orbit) seems much more populated than the L_5 swarm (in the first quadrant). In fact, there are 105 Trojans in the L_4 group and only 28 in the L_5 group, but these are biased raw numbers. A rough correction for the amount of sky covered in each direction provides a pseudoestimate for the L_5 group of 98 objects, very close to the 105 Trojans detected at L_4.

2.10. Bias Calculation

Consider a particular object with a set of orbital and physical parameters $\bar{x}$ and a set of N_R regions of the sky that have been searched. The efficiency for detecting the object (ε_R) in a region (R) depends upon its set of observable characteristics ($\bar{o}$) when it appears in that region (rates, magnitude, etc.), which in turn depend on the orbital and physical parameters [$\bar{o}(\bar{x})$]. Usually the interesting parameters to be debiased are not within the set of observed parameters but are related to them in some complicated nonlinear fashion. More likely than not, each region will have its own efficiency characteristics because surveys rarely search under consistent conditions and often modify strategies, exposure times, analysis software, etc. Letting $H_R(\bar{x})$ be the probability that the object with parameters $\bar{x}$ appears in a region, the total efficiency for detecting this particular object is then

$$\varepsilon_T(\bar{x}) = 1 - \prod_R \{1 - H_R(\bar{x})\, \varepsilon_R[\bar{o}(\bar{x})]\} \qquad (6)$$

We represent the probability as $H_R(\bar{x})$ to promote the idea of a multidimensional Heavyside function in the phase space of orbital parameters. Objects with $\bar{x}$ either appear in the region R or do not — the probability is either 0 or 1. It is the efficiency term that defines the probability that the object will be *detected* in the field of view.

Typically, the detections of asteroids are binned in one of the parameters (x_i) in bins of width Δx_i and the desire is to determine the actual number of objects within each bin. The bin widths should be selected based on typical errors in the measurement of each x_i; preexisting knowledge of limits, thresholds, and the time-rate of change for each parameter; and computational considerations. Letting $n(x_i)$ represent the number of objects with values of x_i in the range (x_i, $x_i + \Delta x_i$) we need to determine $\bar{\varepsilon}(x_i)$ such that $N(x_i) = n(x_i)/\bar{\varepsilon}(x_i)$ where $N(x_i)$ is the corrected/actual number of objects in the bin. The average efficiency in the bin is then

$$\bar{\varepsilon}(x_i) = \frac{1}{\Delta V} \int_{x_i}^{x_i+\Delta x_i} dy_i \prod_j^{j \neq i} \int_{x_j} dy_j\, N(\bar{y})\, \varepsilon_T(\bar{y}) \qquad (7)$$

where ΔV is the number-weighted volume of the $\bar{x}$ parameter space encompassed within the bin

$$\Delta V = \int_{x_i}^{x_i+\Delta x_i} dy_i \prod_j^{j \neq i} \int_{x_j} dy_j\, N(\bar{y}) \qquad (8)$$

The notation indicates that we integrate over the bin in x_i but must perform a multiple-integral over the full range of all the other parameters (x_j, $j \neq i$). It is important to note that we have introduced a weighting factor $N(\bar{y})$ that is none other than the actual distribution of objects as a function of the orbital and physical parameters. Here we identify the crux of the problem when correcting for selection effects: A determination of the binned average efficiency will always depend on the otherwise (and usually) unknown actual distribution of objects that is being determined.

The average efficiency defined by equation (7) is appropriate for a survey that uniquely identifies each object in the histogram. NEO survey programs (e.g., Spacewatch, LINEAR, etc.) meet this requirement since (almost) every NEO identified during the search is not repeated when accumulating a histogram of orbital parameters. On the other hand, if the objects are not necessarily unique (e.g., *Jedicke and Metcalfe,* 1998), we instead require a bias correction factor (B) such that $N(x_i) = n(x_i)/B(x_i)$. This equation provides the means of determining the bias as $B = n/N$ (note that B may be >1). Once again, the problem is that N is the quantity to be determined. But let's proceed with formalizing the determination of bias utilizing a simulation of a detector system.

The actual total number of objects in a population is simply

$$N = \int_{\bar{x}_i} N(\bar{y})\, d\bar{y}$$

and the actual observed number of objects for a survey would be

$$n = \sum_R \int_{\bar{x}_i} H_R(\bar{y})\, \varepsilon_R[\bar{o}(\bar{y})]\, N(\bar{y})\, d\bar{y}$$

The equations for the total and observed number of objects in a *simulated* population and survey (N' and n' respectively) would be identical to those for N and n except that every parameter in the integral would correspond to the simulated (primed) variable.

Most studies are interested in the corrected *distribution* of objects as a function of a subset of $\bar{x}$ if the statistics are large enough to justify spreading the results over a number of bins. It is impossible to obtain data on every single parameter applicable to the object, so most experiments "project" the observations into a limited number of subparameters. Consider the case of debiasing the distribution as a function of the parameter x_i so the actual number of objects in the range $(x_i, x_i + \Delta x_i)$ is

$$N(x_i) = \int_{x_i}^{x_i+\Delta x_i} dy_i \prod_j^{j \neq i} \int_{x_j} dy_j \, N(\bar{y})$$

The detected number of objects in the same bin would be

$$n(x_i) = \sum_R \int_{x_i}^{x_i+\Delta x_i} dy_i \prod_j^{j \neq i} \int_{x_j} dy_j \, H_R(\bar{y}) \, \varepsilon_R[\bar{o}(\bar{y})] \, N(\bar{y})$$

Once again, the equations for the simulated detector $[n'(x_i)$ and $N'(x_i)]$ are identical except for each variable being primed. The bias correction factor is then

$$B'(x_i) = n'(x_i)/N'(x_i) \tag{9}$$

Note that we could have defined bias to be the inverse of that given here such that $N = nB'$, but we use the current notation for congruence with the concept of efficiency. The equation for $B'(x_i)$ is impossible to calculate analytically except perhaps in some very restricted trivial survey that would be of little consequence.

An easily implemented technique for solving complicated equations such as equation (9) involves a Monte Carlo integrator. The technique has been described in detail elsewhere (e.g., *James,* 1980) but the general idea is simply to sample the complicated function randomly within the limits of integration. In evaluating equation (9) the Monte Carlo technique is tantamount to performing a simulation of an asteroid survey and has been performed as such (e.g., *Bowell and Muinonen,* 1994; *Harris,* 1998; *Muinonen,* 1998; *Tedesco et al.,* 2000), while others (e.g., *Rabinowitz et al.,* 1994; *Jedicke and Herron,* 1997; *Rabinowitz et al.,* 2000) have performed the simulations in order to determine the correction bias.

A model population $N'(\bar{x})$ is randomly generated and then passed through a survey simulator that should attempt to mimic every aspect of the detector's capabilities. The model population should parallel the actual distribution as closely as possible. It is probably justified to attempt an iterative approach to the evaluation of B' as the corrected population is determined. Then, for each $\bar{x}$ in the generated model, determine whether the object appears in any of the surveyed regions $[H_R(\bar{x})]$. If it did, calculate the observable parameters $(\bar{o})$ of the object in the region and the efficiency for detecting the object $[\varepsilon'_R(\bar{o}), 0 \leq \varepsilon'_R \leq 1]$. It is detected if a random number generated in the interval [0,1] is $\leq \varepsilon'_R(\bar{o})$ and the set of objects thus detected is $n'(\bar{x})$.

The estimate of the actual number of objects in a bin is then $N_{est}(x_i) = n(x_i)/B'(x_i)$. However, a simple substitution reveals that $N_{est}(x_i) = N'(x_i)\,[n(x_i)/n'(x_i)]$, which highlights an important point discussed in more detail in the next few paragraphs: The population estimate depends upon both a good input model and a good detector simulation. Together they must be able to reproduce the observed distribution $[n'(x_i) \sim n(x_i)]$, but an incorrect model coupled with an incorrect detector simulation may yield the observed distribution as well. The solution is not unique and the latter scenario will produce an invalid population estimate. In the best-case scenario the detector simulation is perfect and the generated distribution will, of necessity, be the same as the actual population.

The technique is only applicable where $B'(x_i) \neq 0$. You cannot even set a limit for the actual number of objects in a bin that contains no objects if $B' = 0$. In practice, a contiguous region in $\bar{x}$-space where $B'(x_i) \geq B'_{min}$ should be identified and explicitly quoted when using the result. A good value for B'_{min} can be determined iteratively by repeatedly applying the procedure with different values (and therefore different ranges in $\bar{x}$-space) until the debiased results are insensitive to small changes in B'_{min}. It would also be good practice to test the debiasing on fake generated data to ensure that it is possible to regenerate the fake data from the results of the simulation using the selected threshold.

If it is possible to assume that the distribution in x_i is independent of the other parameters $[N(\bar{x}) = N(x_i)\,N(x_j; j \neq i)]$ *and* the efficiency is also separable for the parameter of interest $\{\varepsilon_R[\bar{o}(\bar{x})] = \varepsilon_R[\bar{o}(x_i)]\,\varepsilon_R[\bar{o}(x_j; j \neq i)]\}$, then we can write

$$B'(x_i) = \frac{\displaystyle\sum_R \int_{x_i}^{x_i+\Delta x_i} dy_i \, \varepsilon_R\left[\bar{o}(y_i)\right] N(y_i) \prod_j^{j \neq i} \int_{x_j} dy_j \; N(y_j; j \neq i)\, H_R(\bar{y})\, \varepsilon_R\left[\bar{o}(y_j; j \neq i)\right]}{\displaystyle\int_{x_i}^{x_i+\Delta x_i} dy_i \, N(y_i) \prod_j^{j \neq i} \int_{x_j} dy_j \, N(y_j; j \neq i)}$$

Furthermore, if the experimentalist has carefully chosen their Δx_i such that the bin size is smaller than the scale at which $N(x_i)$ and $\varepsilon'_R[\bar{o}(x_i)]$ change quickly, *and* taken steps to ensure that the efficiency in the parameter is nearly the same for all regions $\{\varepsilon_R[\bar{o}(x_i)] \sim \varepsilon[\bar{o}(x_i)]\}$, then

$$B'(x_i) \sim \varepsilon'\left[\bar{o}(x_i)\right] \times \frac{\displaystyle\sum_R \prod_j^{j \neq i} \int_{x_j} dy_j \, N(y_j; j \neq i)\, H_R(\bar{y})\, \varepsilon_R\left[\bar{o}(y_j; j \neq i)\right]}{\displaystyle\prod_j^{j \neq i} \int_{x_j} dy_j \, N(y_j; j \neq i)}$$

Only when all these conditions are met is the bias for a specific parameter independent of the generated distribution in that parameter. Furthermore, the generated distributions in all the other (hidden) parameters must mirror those of the actual (yet probably unknown) distributions.

The situation seems mathematically intractable and replete with conditions that cannot be satisfied for a realistic system. But we have found that the technique is still useful;

even large variations in the hidden parameters seem to average out under suitable conditions. *Spahr* (1998) tested the method by generating bias values according to equation (9) for an asteroid survey as a function of (a, e, i, H), where he first assumed that the angular orbital elements were distributed evenly in the range [0, 2π] and then pathologically. Despite the tremendous difference in the generated distributions the final bias values differed by only ~10%. A final caveat: Spahr's tests took place on a simulated wide-area survey, which would have the effect of averaging over large changes (or errors) in the input distributions. Narrow-field surveys will require much more care in generating their input distributions for the bias calculation.

Recent attempts to debias asteroid search programs have done so using this technique on a restricted grid in (a, e, i, H) space. The bin size in each of the four parameters is chosen based on some combination of the measurement error associated with each parameter, the physics of the situation, and the available statistics in the distribution. Even though the angular orbit elements may be known, to speed up the calculation of the bias their generated Monte Carlo distribution was assumed to be flat while "reasonable" values were chosen for the physical parameters.

Bottke et al. (2000) used B' in a slightly different manner. They created a theoretical model NEO population M(a, e, i, H) depending upon four parameters (α_i; i = 1…4) and then fit M(a, e, i, H) × B'(a, e, i, H) to the observed distribution of Spacewatch NEOs n(a, e, i, H) to determine the α_i. Once the four fit parameters were known the model gave their prediction for the actual distribution of NEOs.

2.11. Bias Examples

The intimidating discussion of the previous section should be read through pragmatic glasses. Many of these surveying factors may not be known or controlled in a manner amenable to debiasing. The result is that debiasing takes into account known factors, those that are most important, and those that can be accommodated in a reasonable time frame.

Figure 5a shows the complicated form of the NEO bias for a 100% efficient system in the range $3.5 \leq V < 18.5$ and $0.3 \leq \omega < 10.0$ (°/d) with 0% efficiency outside those ranges. It was calculated for an instantaneously imaged 100 deg^2 circular region centered at opposition ($\lambda_{oppn} = \beta = 0$) at the autumnal equinox. We used the Monte Carlo method outlined in section 2.10 in which N' = 10^6 fake orbits were generated randomly and evenly within each (a, e, i, H) bin (of width 0.1 AU, 0.05, 5°, and 0.5 respectively) and all the angular elements were generated randomly and evenly in the range [0, 2π]. The bias was only calculated for bins containing NEOs — some portion of the bin must have q < 1.3 AU and Q > 1.0 AU. The bias calculated for such a simple survey is really a measure of the restrictions placed on the accessible angular elements.

In general, the detection probability increases with eccentricity for constant semimajor axis (see Fig. 5a). This is because objects with high eccentricity spend more of their time moving slowly at greater heliocentric distances and are more likely to be available in the survey's search volume.

At constant eccentricity the bias increases to a maximum and then decreases. The reason for the increasing bias is identical to the last paragraph — objects with larger semimajor axis spend more time in the search volume. However, the semimajor axis can increase only so much before the distance to the objects causes their apparent brightness to drop below the limiting magnitude of the detector system.

The bias against Atens at opposition should be expected after examining Fig. 1c, where it is clear that most Atens spend little time near the antisolar point. The importance of the Aten subpopulation and search strategies to enhance their discovery rates was discussed by *Boattini and Carusi* (1997).

Comparing Figs. 5a and 5b highlights the difference in detecting objects of different inclination. The probability of detecting an i ~ 32.5° NEO is about an order of magnitude smaller than that for i ~ 2.5° objects. Due to the high-inclination, and the *a priori* assumption that the angular elements are distributed evenly, what is happening is simply that the system can only detect objects whose longitude of the ascending node (Ω) lies close to the circular field of view on the ecliptic at opposition (see section 2.5).

In Fig. 5c the bias is shown for an object 1000× smaller (in volume) than in Fig. 5a (~1 km diameter instead of ~10 km). Many of the features that were identified in Fig. 5a are apparent in this one as well. The major change is the dramatic reduction in discoverability of the objects, especially at large semimajor axis. If the objects are distant and small they are going to be harder to detect. The straight "ridge" that existed in Fig. 5a at about 2.0 AU has become a curved "ridge" at much smaller semimajor axis for the smaller objects. Note that on the small-a side of the "ridge," where the faint-end limiting magnitude of the system has not had an effect on discoverability, the bias is identical to that for the larger objects.

The difference in bias between scanning at opposition and at $\lambda_{oppn} = 30°$ is subtle when examined on the log plots used in Figs. 5a–c. At the risk of confusing the reader, we show the ratio between the bias values at $\lambda_{oppn} = 30°$ and 0° in Figs. 5d–f with a linear vertical scale. The figures are illustrative of the impact of the rate cut, which has a dramatic effect on some regions of the (a, e) phase space and little to no effect in other regions. Obviously, the most spectacular differences occurs with combinations of (a, e) that lead to rates of motion under 0.3°/day.

Close examination of Figs. 5d–f reveals that the Amor-like asteroids have smaller bias values than the Apollos due to their stationary point lying closer to the survey region. They lie along the arc of nonzero bin values starting at a ~ 1.3 AU; the further the bin from this curve, the less effect the rate cut has on the bias.

There are "troughs" in the bias ratios for a ~ 2.0 AU and e ~ 0.8 (for a > 2.0 AU) corresponding to combinations of orbital elements, which lead to a high probability that the object will display a small rate of motion when positioned

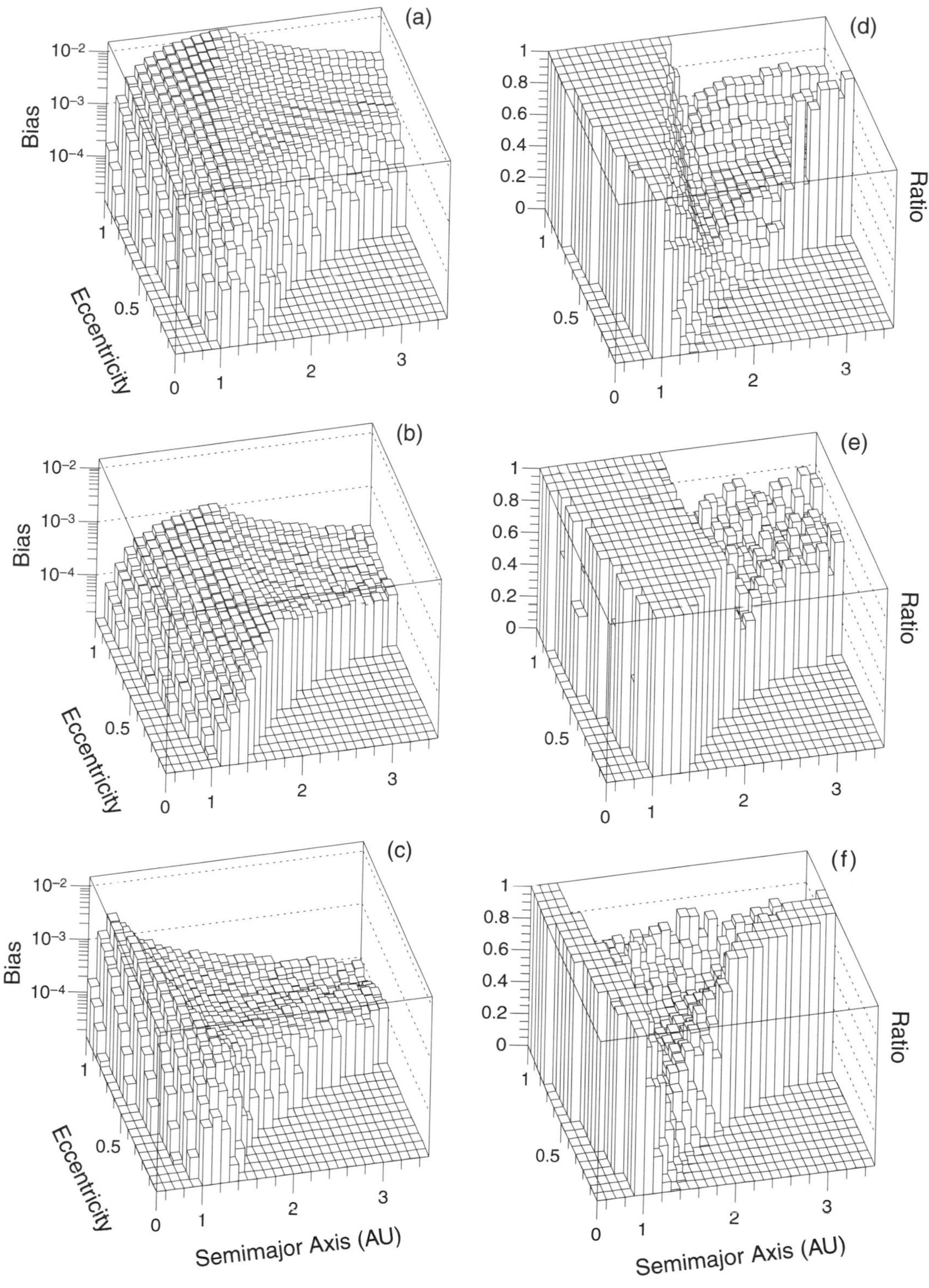

Fig. 5. **(a)–(c)** Discovery bias for NEOs in a 100-deg² circular field near opposition. **(a)** i = 2.5° and H = 13.0 (D ~ 10 km). **(b)** i = 32.5° and H = 13.0 (D ~ 10 km). **(c)** i = 2.5° and H = 18.0 (D ~ 1 km). **(d)–(f)** Ratio of the discovery bias near λ_{oppn} = 30° with respect to the bias near opposition [**(a)–(c)**]. **(d)** i = 2.5° and H =13.0 (D ~ 10 km). **(e)** i = 32.5° and H = 13.0 (D ~ 10 km). **(f)** i = 2.5° and H = 18.0 (D ~ 1 km).

in the observation area. The Amors and distant Apollos are affected adversely, while the Aten-type NEAs are not at all affected because they display unusual rates of motion even 30° from opposition.

Searching for objects of higher inclination on the ecliptic but further from opposition (see Fig. 5e) is not as adversely affected because they pass through the ecliptic with a relatively high $\dot{\beta}$, which adds quadratically with $\dot{\lambda}$. So they are more likely to possess rates greater than the detection threshold.

In Fig. 5f note the appearance of a second "plateau" in the (a, e) bias landscape for larger (a, e) hugging the q < 1.3 AU limit. These are objects that normally reside in the outer solar system but every once and a while whip quickly by Earth so their rates can be above the detection threshold.

It is perhaps more interesting to examine the bias values when scanning at (λ_{oppn}, β) = (0°, 30°) as shown in Fig. 6. There is little difference in the bias values between Figs. 6a and 6c despite the difference of 5 in absolute magnitude of the objects because they need to be relatively close to Earth before they can appear in the search volume and the reduced magnitude has little impact.

On the other hand, there are now gigantic tracts of the (a, e) phase space where B' = 0 because these objects simply can not appear at 30° ecliptic latitude as described in section 2.6. Scanning at this latitude is not an effective means of discovering low-inclination Apollos.

Detection of higher-i orbits (Fig. 6b) is mostly unaffected by searching off the ecliptic. These objects are relatively close to Earth and moving quickly so that they are assured of passing the detection thresholds.

The most important concept to understand after having examined Figs. 5 and 6 is that the probability of detecting an object depends sensitively and interdependently upon the orbital and physical parameters of the object. Meaningful distributions of these parameters can only be obtained through a careful debiasing of the data, taking into account as many parameters as possible, and the interpreter of the results should always be aware of potentially skewed distributions due to some subtle bias that was not properly corrected.

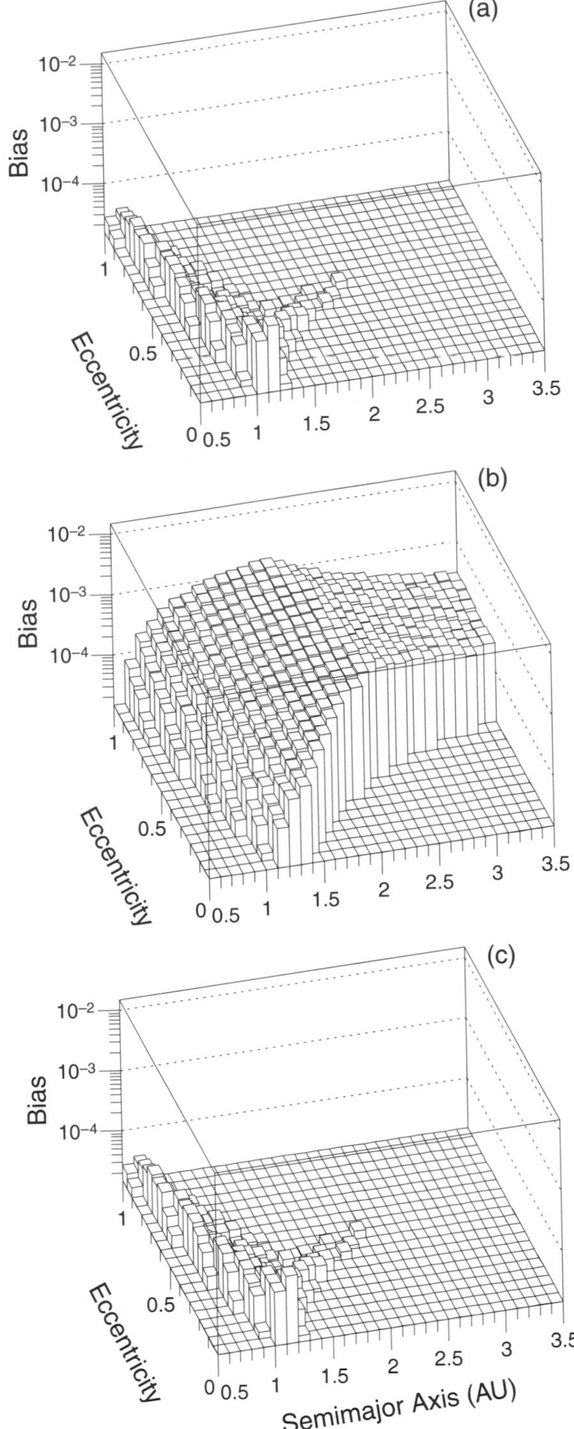

Fig. 6. Discovery bias near (λ_{oppn}, β) = (0°, 30°). **(a)** i = 2.5° and H = 13.0 (D ~ 10 km). **(b)** i = 32.5° and H = 13.0 (D ~ 10 km). **(c)** i = 2.5° and H = 18.0 (D ~ 1 km).

3. ASTEROID POPULATION SIZE DISTRIBUTIONS

The discussion of various selection effects presented in section 2 should be regarded as a brief survey of the involved process required to account for observational bias in asteroid surveys. In practice, surveying groups have not accounted for all the possible selection mechanisms, and many surveys (e.g., KBO, Centaurs) do not generate enough statistics to attempt a multidimensional debiasing of their data. In these cases they often rely on sky-plane surface densities or theoretically inferred populations to infer or fit the bias-corrected size distribution.

Figure 7 presents the differential absolute magnitude number distribution for five different important populations of asteroids. Each population is discussed separately in order of increasing heliocentric distance in the following five subsections. In particular, we discuss how and where we obtained the normalization and slopes of the debiased distributions as a function of H (indicated as lines on the

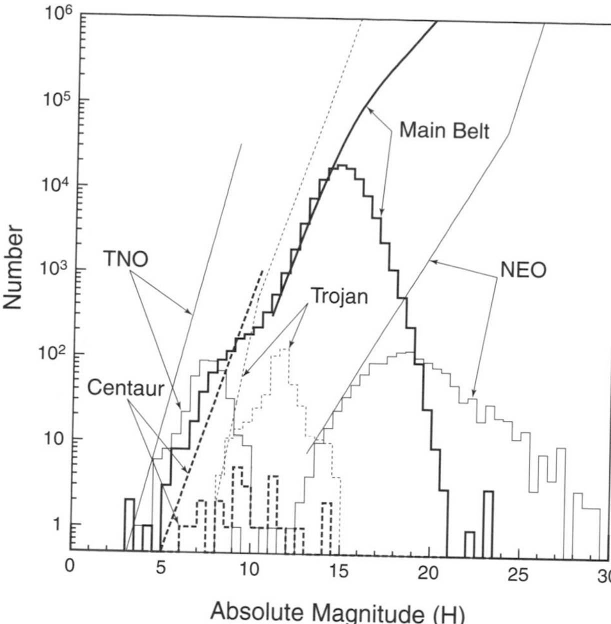

Fig. 7. Differential H distributions for five asteroid populations. Bias-corrected estimates for each population are shown as lines of various types. Each line merges toward the left of the figure with the known distribution of asteroids as of July 18, 2001, derived from the database of *Bowell et al.* (1994b).

figure). We present the data in terms of absolute magnitude (instead of diameter) because this parameter is determined unambiguously from the apparent magnitude and phase angle obtained from the orbit elements. The histograms of known asteroids all derive from the ASTORB database (*Bowell et al.,* 1994b).

In the following sections we present the debiased H distributions with a functional representation of the form

$$\frac{dn}{dH} = K\ 10^{\alpha(H - H_0)} \tag{10}$$

The cumulative number of asteroids with H < H* is then

$$N(H < H^*) = \frac{K\ 10^{\alpha(H^* - H_0)}}{\alpha \ln 10} \tag{11}$$

The parameter α is known as the "slope parameter" and is a key ingredient when modeling asteroid populations. We often needed to convert from the differential size distribution $dn/dr \propto r^{-a}$ where r is the radius of the object and a is a slope parameter to the form of equation (10). Under the assumption of a constant albedo the relationship is simply $\alpha = a/5$.

In all cases there still remain large errors on the estimated debiased number distributions — the bias correction factors are large with a small number of detections for the faintest objects. We refer the reader to the references and the references therein for a discussion of the errors on the populations and fit parameters.

3.1. Near-Earth Objects (NEOs)

The NEOs currently enjoy a beneficent funding environment and associated high rate of discovery due to the prospect and danger of impacts of large asteroids with Earth (*Gehrels,* 1994). There are currently over 1300 known NEOs spanning the range 9.4 < H < 29, found by about 55 different observatories. The population is generally considered to be "complete" to H ~ 15. The last asteroid found with H < 15 was in late 2000 despite ongoing intensive worldwide searches. There are still almost certainly a few large stragglers that have not yet been found because their orbital elements prevent frequent passage by Earth near opposition.

There are more than 50 objects with H < 15, but this is probably insufficient to constrain theoretical production estimates and expected size distributions for these objects. An attempt to correct the known distribution of all NEOs found by the entire ensemble of observatories would be a sisyphean adventure. Some attempts to debias the NEO distributions for specific surveys include *Shoemaker et al.* (1990) and *Rabinowitz et al.* (2000). The corrected distribution shown as the straight solid line on Fig. 7 comes from *Bottke et al.* (2001) (H ≤ 24) (section 2.10) and *Rabinowitz et al.* (2000) (24 < H < 31), whose results are in good agreement with each other and independent estimates. *Rabinowitz et al.* (2000) suggests that there is an upswing in the number distribution that occurs near H = 24 so their function has been normalized to equal the *Bottke et al.* (2002) result at H = 24. The debiased NEO curve in Fig. 7 is given by

$$\frac{dn}{dH} = 14.0 \times 10^{0.35 \times (H - 13.0)} \qquad H \le 24 \tag{12}$$

$$\frac{dn}{dH} = 99112 \times 10^{0.70 \times (H - 24.0)} \qquad 24 < H < 31 \tag{13}$$

which implies that there are about 1000 NEOs with diameter larger than about 1 km (corresponding to H ~ 18).

No object is currently known with an orbit entirely interior to Earth's orbit (IEO), but there is every reason to believe that such objects must exist as discussed in *Michel et al.* (2000). Because of the nature of the fitting technique implemented by *Bottke et al.* (2002), their model and predictions already incorporate the IEO population. They estimate the IEO population to be about 2% of the entire number of NEOs, implying about 20 IEOs larger than ~1 km in diameter.

3.2. Main Belt

Main-belt asteroid discoveries span over 200 years and there are currently more than 120,000 with good orbits. The discovery rate has increased dramatically over the past few years because of parasitic observations during NEO surveys (*Stokes et al.,* 2002). Based on extrapolations of the size distribution and the rate of discovery for the brightest objects, they are now arguably complete for H < 13 (see section 1) throughout the main belt (2.0 AU < a < 3.5 AU with q > 1.666 AU).

Extending the H distribution to fainter magnitudes requires debiasing the observational data, and this is best achieved using data from a single survey. This has been performed by a few surveys [e.g., PLS (*van Houten et al.,* 1970), IRAS (*Cellino et al.,* 1991), Spacewatch (*Jedicke and Metcalfe,* 1998), SDSS (*Ivezić et al.,* 2001)], but each was haunted by its own selection demons.

The PLS made an excellent attempt at debiasing their data sample of about 2000 asteroids. Unfortunately, their debiased estimates for the larger asteroids in the population lie below the currently known number of objects. The Spacewatch survey's attempt had to contend with other complicating factors including many fields covering small areas of the sky under variable conditions. They matched the set of known asteroids at large sizes, verified a downturn in the slope of the H distribution near H ~ 15.5 seen in the PLS study, and found a reduced slope parameter for the smallest objects they could observe. Recent results for the debiased main-belt asteroid H distribution obtained by the SDSS confirm the transition to a shallower slope, but extrapolations of the size distribution to H < 15.5 are about a factor of 2 below the estimates from the Spacewatch survey.

The tremendous advantage of the SDSS results over the PLS and Spacewatch debiasing attempts is their large number statistics in a small number of images during a tightly controlled observing campaign with a big telescope. For these reasons we adopt their results for the debiased main-belt H distribution shown on Fig. 7.

It is important to note that the debiased SDSS main-belt size distribution lies everywhere below the known asteroids for H < 14. The source of this discrepancy is likely a systematic error in the absolute magnitudes in the databases of known asteroids (Ž. Ivezić and M. Jurić, personal communication, 2001). Reporting of observed asteroid magnitudes is notoriously inconsistent (T. B. Spahr, personal communication, 2001). The Minor Planet Center (MPC) regularly weights magnitudes reported by different groups according to their historical photometric accuracy and performs standard conversions from B and R (for instance) to V before calculating H_v. The ASTORB database (*Bowell et al.,* 1994b) uses the reported photometry in determining the absolute magnitude. It is clear that if reported magnitudes are not exactly B or R then the calculated H_v will be systematically incorrect if there are a sufficient number of "bad" reported magnitudes.

Since the main belt's known or debiased H distribution covers 3.0 < H < 19 and includes a few slope transitions, it is impossible to fit equation (10) over the entire range. The SDSS fit their debiased results to a more complicated functional form that naturally incorporated a single slope transition, but we would like to maintain our form for the entire main belt. Table 1 provides parametric values for n, α, and

TABLE 1. Main-belt model parameters.

H	n	α	K
5.25	3	0.8519375	0.0001010
5.75	8	0	8
6.25	8	0.6547179	0.0006473
6.75	17	0.6986694	0.0003269
7.25	38	0.4527928	0.0198169
7.75	64	0.3057228	0.2734186
8.25	91	0.2108332	1.6583163
8.75	116	0.3007717	0.2708355
9.25	164	0.1046558	17.6513577
9.75	185	0.1661526	4.4379702
10.25	224	0.3573374	0.0487013
10.75	338	0.4291861	0.0082256
11.25	554	0.3079478	0.1901505
11.75	789.7390	0.5845556	0.0001069
12.25	**1547.966**	**0.5724986**	**0.0001502**
12.75	**2992.338**	**0.5554168**	**0.0002480**
13.25	**5671.776**	**0.5319492**	**0.0005074**
13.75	**10463.90**	**0.5010692**	**0.0013489**
14.25	**18630.66**	**0.4627473**	**0.0047433**
14.75	**31739.63**	**0.4186979**	**0.0211743**
15.25	**51398.54**	**0.3726901**	**0.1065190**
15.75	**78939.77**	**0.3298225**	**0.5041878**
16.25	**115400.3**	**0.2947575**	**1.8723967**
16.75	**162026.4**	**0.2699278**	**4.8785930**
17.25	**221080.1**	**0.2549590**	**8.8410969**
17.75	**296503.1**	**0.2475485**	**11.9685917**
18.25	**394278.9**	**0.2448941**	**13.3809528**

Values for the parametric representation of the H distribution. H gives the center of a 0.5-mag bin. n is the actual or predicted number of main-belt asteroids in the bin. α and K are the slope parameter and normalization respectively for the H bin such that $n = K\,10^{\alpha H}$. Numbers in bold are derived from the results of the SDSS main-belt study.

K such that $n = K\,10^{\alpha H}$ within each half-magnitude bin. The entire main belt H distribution may thus be reproduced in the range 5.0 < H < 18.5 with an intuitive continuation of the slope parameter as employed in the other population's distributions. Due to the problems with calculated absolute magnitudes as described in the last paragraph, we have arbitrarily decided to use the known distribution for H ≤ 12 and the SDSS results for H > 12.

Davis et al. (2002) discuss various models of the main belt's size distribution with emphasis on the implications for the collisional evolution of the belt. When a large asteroid is disrupted by a collision it produces a "family" of asteroids with similar orbital elements that gradually disperse in (a, e, i) space. The size distribution of the remants is more steep (larger slope parameter) than that of the background population of asteroids, and some (e.g., *Zappalà and Cellino,* 1996) suggest that the steep slope must dominate at small asteroid sizes. The size-distribution figure in *Davis et al.* (2002) shows that the debiased SDSS and Spacewatch data disagree with these family-based models, while unpublished ISO observations apparently confirm the expectations of a higher slope parameter.

3.3. Trojans

The dashed histogram in Fig. 7 shows the known H distribution for Trojan asteroids. The corrected Trojan distributions have been discussed in the literature (e.g., *Shoemaker et al.,* 1989; *van Houten et al.,* 1991), and we present results from *Jewitt et al.* (2000) as the dotted line in the figure. For this discussion we define a Trojan as an object with 4.8 AU < a < 5.2 AU and q > 4.2 AU. The Trojan population is nearly complete to about H ~ 9.0 and they propose that a transition point in the H distribution exists near H ~ 10.25.

To produce the corrected distribution we solved for K in equation (10) knowing there are 17 Trojans with H < 9.0. The transition to the shallower slope occurs at H = 10.25 and we normalized their functional form for larger H to match at the transition point. The final differential H distribution for the debiased Trojan population is then

$$\frac{dn}{dH} = 43.1 \times 10^{1.1 \times (H - 9.0)} \qquad H < 10.25 \qquad (14)$$

$$\frac{dn}{dH} = 1022 \times 10^{0.6 \times (H - 10.25)} \qquad 10.25 \le H < 16.0 \qquad (15)$$

3.4. Centaurs

There are currently only 25 objects in the region 5.5 AU < a < 30 AU with q > 5.2 AU, which we define as the Centaur regime. Of the five populations discussed in this section this set of asteroids contains by far the fewest number of known objects. The Centaurs are most likely rejects from the transneptunian region (discussed in section 3.5): fragments of asteroids knocked into this region by collisions outside the orbit of Neptune or thrown into the giant-planet zone through chaotic orbital perturbations.

Due to the scarcity of Centaurs in the sky-plane there are only a limited number of studies of their number distribution (e.g., *Jewitt et al,* 1996; *Larsen et al.,* 2001). The only attempt at the H distribution is that of *Jedicke and Heron* (1997), whose entire result depends on a single Centaur detection with the Spacewatch survey. The most likely value of α is 0.61, surprisingly close to the value for the fainter Trojans in equation (15) and consistent with the slope parameter for the transneptunians discussed in section 3.5.

We normalize the Centaur differential number H distribution using their result that with 99% confidence there are ≤3 Centaurs brighter than H = 6. Assuming that N(H ≤ 6) = 3 we find

$$\frac{dn}{dH} = 4.2 \times 10^{0.61 \times (H - 6.0)} \qquad H < 10.5 \qquad (16)$$

This relation and the known H distribution are shown on Fig. 7 as a thick dashed line and histogram respectively.

3.5. Transneptunian Objects

The semimajor axis space outside Neptune (a > 30.1 AU) is populated by the transneptunian objects (TNOs). For our purposes we further restrict them to be non-Uranus-crossing (q > 19.2 AU). Since the discovery of the first TNO (other than Pluto) by *Jewitt and Luu* (1993) they have generated a great deal of interest because of the secrets they might unveil regarding the formation of the solar system. We adopt *Trujillo et al.* (2001) for the TNO population estimate because it appears to be consistent with most other surveys. The currently known H distribution for over 400 TNOs is shown as a thin-lined histogram in Fig. 7 and the expected differential distribution

$$\frac{dn}{dH} = 70000 \times 10^{0.8 \times (H - 9.15)} \qquad H < 9.15 \qquad (17)$$

is shown as a thin solid line. dn/dH was normalized using *Trujillo et al.*'s (2001) estimate that N(D > 100 km) = 3.8 × 10^4. From the figure it is clear that the TNOs are the most populous of the five types of asteroids and also contain the most mass.

4. DISCUSSION

Exciting new observational programs promise advances in the near future and further understanding of the size and absolute magnitude distributions for asteroids of all classes. It appears that the funding environment is still favorable to asteroid surveying so that the completion level for all populations will gradually be pushed to higher H and smaller sizes. There are also some Earth- and spacebased plans for enhanced surveying of some populations.

We believe that the major problem requiring immediate attention with regard to asteroid size distributions is the resolution of the absolute magnitude problem in asteroid databases. The MPC and other international asteroid clearing houses are calculating and extrapolating orbits regularly and dealing effectively with the ever-increasing discovery and recovery rates. But the utility of these observations for size distributions without believable absolute magnitudes is questionable. This problem needs to be resolved, perhaps with some sort of quality assurance program for each observatory that reports magnitudes to the MPC.

It is estimated that ~50% of the NEO population larger than H = 18 have now been discovered and have good orbits. As mentioned above, their sizes are much less well determined and the problem is likely as bad or worse than the situation for the main belt. Better estimates of the size or H distribution of all the asteroids will improve through better debiasing techniques, and perhaps more so through obtaining better photometry. Size-distribution estimates for the main belt appear to be in contradiction with family-based models (see section 3.2), but improved datasets to smaller sizes (described below) should be available in the next few years to resolve this discrepancy. The size distributions for the other major classes of asteroids (Trojan, Centaur, TNO) will improve with time and be extended to ever smaller sizes, although probably not at the same rate as the closer objects.

The prospects for continued and increased reporting rates from various existent (e.g., LINEAR, NEAT, LONEOS, Spacewatch) and proposed NEO surveys around the world appears to be very good. Even though their main goal is the discovery of new, potentially hazardous NEOs, the parasitic discovery and recovery of other types of asteroids has led to the tremendous increase in the number of asteroids with good orbits. The SDSS has already made an impressive contribution to the distribution of asteroids in the main belt (see section 3.2) using only their commissioning data. Their dataset will eventually be about 10× larger, allowing even more detailed analysis of the gross structure of the main belt.

An extension of a PLS-like dataset to smaller sizes should be forthcoming soon. Conceived by P. Farinella, D. Davis, and B. Gladman at the Protostars and Planets IV meeting, a dataset was acquired in March 2001 by a team working out of the Planetary Science Institute (Tucson). Data was acquired over six nights in March 2001 on KPNO's 3.8-m telescope with the CCD mosaic. This survey should yield orbits of roughly 1000 main-belt asteroids to V ~ 23 — good enough to extend the size distribution to the subkilometer level in the inner main belt. Analysis should be completed in 2003.

The most ambitious Earth-based proposal is without a doubt the Large Synoptic Surveying Telescope (LSST) proposed in *Astronomy and Astrophysics in the New Millenium* (2000). This 6.5-m-class telescope with an ~3° field of view will produce a digital map of the visible sky every week to ~24th magnitude. It is likely that this project will be funded and, if it is conducive to finding asteroids, will create a revolution in surveying capabilities going more than five magnitudes deeper than any existent survey of similar sky-coverage ability.

A spacebased platform provides an excellent but expensive vantage point from which to survey asteroids. Suggestions for dedicated asteroid satellites abound but there is serious consideration for parasitic searches using two European Space Agency spacecraft: *GAIA* and *BepiColombo*. The former will be placed into orbit around Earth's L_2 point and is expected to be operational for five years. In this low-radiation environment, with the Sun, Earth, and Moon "behind" the instrument, they are expecting a high observing efficiency and anticipate a significant number of serendipitous observations of asteroids. The *BepiColombo* mission to Mercury offers the possibility of an observing platform interior to Earth's orbit, which could be extremely effective for identifying potentially hazardous NEOs and locating IEOs.

No matter how large the telescope employed for asteroid surveying, independent of its physical location, regardless of the software's ability to extract usable signal from the noisy background, the most interesting planetary science will always lie just over the horizon, pushing the boundary of what is achievable at any time. Gleaning important information on the edge of our detection capability will require an increasing awareness of selection effects and enhanced ability to correct for observational bias.

Acknowledgments. We thank Bill Bottke, Robert McMillan, and David L. Rabinowitz for valuable discussions. The two reviewers (Karri Muinonen and Alberto Cellino) provided excellent direction and suggestions for improving the structure and content for this chapter. R. Jedicke and J. Larsen are supported by grants to the Spacewatch Project from NASA (NAG5-7854 and NAG5-8095), AFOSR (F49620-00-1-0126), The David and Lucile Packard Foundation, The Steven and Michele Kirsch Foundation, John and Ilene Nitardy, and other contributors.

REFERENCES

Astronomy and Astrophysics in the New Millenium (2000) *Astronomy and Astrophysics in the New Millenium,* p. 69. National Academy, Washington, DC. (Available on line at http://books.nap.edu/books/0309070317/html/R1.html)

Binzel R. P., Gehrels T., and Matthews M. S., eds. (1989) *Asteroids II.* Univ. of Arizona, Tucson. 1258 pp.

Boattini A. and Carusi A. (1997) Atens: Importance among near-earth asteroids and search strategies. *Vistas in Astronomy, 41,* 527–541.

Bottke W. F., Jedicke R., Morbidelli A., Petit J.-M., and Gladman B. (2000) Understanding the distribution of near-Earth asteroids. *Science, 288,* 2190–2194.

Bottke W. F., Morbidelli A., Jedicke R., Petit J.-M., Levison H., Michel P., and Metcalfe T. S. (2002) Debiased orbital and size distributions of the near-Earth objects. *Icarus,* in press.

Bowell E. and Muinonen K. (1994) Earth-crossing asteroids and comets: Groundbased search strategies. In *Hazards Due to Comets and Asteroids* (T. Gehrels and M. S. Matthew, eds.), pp. 417–462. Univ. of Arizona, Tucson.

Bowell E., Chernykn N. S., and Marsden B. G. (1989a) Discovery and follow up of asteroids. In *Asteroids II* (R. Binzel et al., eds.), pp. 21–38. Univ. of Arizona, Tucson.

Bowell E., Hapke B., Domingue D., Lumme K., Peltoniemi J., and Harris A. W. (1989b) Application of photometric models to asteroids. In *Asteroids II* (R. Binzel et al., eds), pp. 524–556. Univ. of Arizona, Tucson.

Bowell E., Skiff B., Wasserman L., and Russell K. S. (1994a) Orbital information from asteroid motion vectors. In *Asteroids, Comets, Meteors III* (C. I. Lagerkvist et al., eds.), pp. 19–24. Uppsala Universitet, Uppsala.

Bowell E., Muinonen K., and Wasserman L. H. (1994b) A public-domain asteroid orbit data base. In *Asteroids, Comets, Meteors 1993,* p. 477. IAU Symposium 160.

Cellino A., Zappalà V., and Farinella P. (1991) The size distribution of main-belt asteroids from IRAS data. *Mon. Not. R. Astron. Soc., 253,* 561–574.

Cochran A. L. and Vilas F. (1997) The McDonald Observatory serendipitous UV/blue spectral survey of asteroids. *Icarus, 127,* 121.

Davis D. R., Durda D. D., Marzari F., Campo-Bagatin A., and Gil-Hutton R. (2002) Collisional evolution of small body populations. In *Asteroids III* (W. F. Bottke Jr. et al., eds.), this volume. Univ. of Arizona, Tucson.

Gehrels T., ed. (1979) *Asteroids.* Univ. of Arizona, Tucson. 1191 pp.

Gehrels T., ed. (1994) *Hazards Due to Comets and Asteroids.* Univ. of Arizona, Tucson. 1300 pp.

Gradie J. C., Chapman C. R., and Tedesco E. F. (1989) Distribution of taxonomic classes and the compositional structure of the asteroid belt. In *Asteroids II* (R. Binzel et al., eds.), pp. 316–335. Univ. of Arizona, Tucson.

Green R. M. (1985) *Spherical Astronomy.* Cambridge Univ., Cambridge. 480 pp.

Harris A. W. (1998) Evaluation of ground-based optical surveys for near-Earth asteroids. *Planet. Space Sci., 46,* 283–290.

Helin E. F. and Shoemaker E. M. (1979) Palomar planet-crossing asteroid survey. *Icarus, 31,* 415–419.

Hirayama K. (1918) Groups of asteroids probably of common origin. *Astron. J., 31,* 185–188.

Ivezić Ž. et al. (2001) Solar system objects observed in the Sloan digital sky survey commissioning data. *Astron. J., 122,* 2779–2784.

James F. (1980) Monte Carlo theory and practice. *Rep. Prog. Phys., 43,* 1145–1189.

Jedicke R. and Herron J. D. (1997) Observational constraints on the Centaur population. *Icarus, 127,* 494–507.

Jedicke R. and Metcalfe T. S. (1998) The orbital and absolute magnitude distributions of main belt asteroids. *Icarus, 131,* 245–260.

Jewitt D. and Luu J. (1993) Discovery of the candidate Kuiper belt object 1992 QB1. *Nature, 362,* 730–732.

Jewitt D., Luu J., and Chen J. (1996) The Mauna Kea-Cerro-Tololo (MKCT) Kuiper Belt and Centaur survey. *Astron. J., 112,* 1225.

Jewitt D. C., Trujillo C. A., and Luu J. X. (2000) Population and size distribution of small jovian Trojan asteroids. *Astron. J., 120,* 1140–1147.

Kirkwood D. (1868) On the nebular hypothesis, and the approximate commensurability of the planetary periods. *Mon. Not. R. Astron. Soc., 29,* 96.

Larsen J. A., Gleason A. E., Danzl N. M., Descour A. S., McMillan R. S., Gehrels T., Jedicke R., Montani J. L., and Scotti J. V. (2001) The Spacewatch wide-area survey for bright Centaurs and trans-Neptunian objects. *Astron. J., 121,* 562–579.

Luu J. and Jewitt D. (1989) On the relative numbers of C types and S types among near-Earth asteroids. *Astron. J., 98,* 1905–1911.

Michel P., Zappalà V., Cellino A., and Tanga P. (2000) Note: Estimated abundance of Atens and asteroids evolving on orbits between Earth and Sun. *Icarus, 143,* 421–424.

Muinonen K. (1998) Discovery and follow-up simulations for small Earth-crossing asteroids. *Planet. Space Sci., 46,* 291–297.

Muinonen K., Piironen J., Shkuratov Yu. G., Ovcharenko A., and Clark B. E. (2002) Asteroid photometric and polarimetric phase effects. In *Asteroids III* (W. F. Bottke et al., eds.), this volume. Univ. of Arizona, Tucson.

Rabinowitz D. L. (1991) Detection of Earth-approaching asteroids in near real time. *Astron. J., 101,* 1518–1529.

Rabinowitz D. L., Bowell E., Shoemaker E. M., and Muinonen K. (1994) The population of Earth-crossing asteroids. In *Hazards Due to Comets and Asteroids* (T. Gehrels, ed.), pp. 285–312. Univ. of Arizona, Tucson.

Rabinowitz D. L., Helin E., Lawrence K., and Pravdo S. (2000) A reduced estimate of the number of kilometre-sized near-Earth asteroids. *Nature, 403,* 165–166.

Shoemaker E. M., Shoemaker C. S., and Wolfe R. F. (1989) Trojan asteroids — Populations, dynamical structure and origin of the L4 and L5 swarms. In *Asteroids II* (R. Binzel et al., eds.), pp. 487–523. Univ. of Arizona, Tucson.

Shoemaker E. M., Wolfe R. F., and Shoemaker C. S. (1990) Asteroid and comet flux in the neighborhood of Earth. In *Global Catastrophes in Earth History* (V. L. Sharpton and P. D. Ward, eds.), pp. 335–342. Geol. Soc. Am. Spec. Paper 247.

Spahr T. B. (1998) *Determining the Orbital Element and Size Distribution of Main Belt Asteroids.* Univ. of Florida, Gainesville. 98 pp.

Stokes G. H., Evans J. B., and Larson S. M. (2002) Near-Earth asteroid search programs. In *Asteroids III* (W. F. Bottke Jr. et al., eds.), this volume. Univ. of Arizona, Tucson.

Tedesco E. F., Muinonen K., and Price S. D. (2000) Space-based infrared near-Earth asteroid survey simulation. *Planet. Space Sci., 48,* 801–816.

Trujillo C. A., Jewitt D. C., and Luu J. X. (2001) Properties of the trans-Neptunian belt: Statistics from the Canada-France-Hawaii telescope survey. *Astron. J., 122,* 457–473.

van Houten C. J., van Houten-Groeneveld I., Herget P., and Gehrels T. (1970) The Palomar-Leiden survey of faint minor planets. *Astron. Astrophys. Suppl. Ser., 2,* 339.

van Houten C. J., van Houten-Groeneveld I., Wisse-Schouten M., Bardwell C., and Green W. E. (1991) The Second Palomar-Leiden Trojan survey. *Icarus, 91,* 326–333.

Xu S., Binzel R. P., Burbine T. H., and Bus S. J. (1995) Small main-belt asteroid spectroscopic survey: Initial results. *Icarus, 115,* 1.

Zappalà V. and Cellino A. (1996) Main belt asteroids: Present and future inventory. In *Completing the Inventory of the Solar System* (T. Rettig and J. Hahn, eds.), pp. 29–44. ASP Conference Series 107.

The Comparison of Size-Frequency Distributions of Impact Craters and Asteroids and the Planetary Cratering Rate

B. A. Ivanov
Institute for Dynamics of Geospheres, Russian Academy of Sciences

G. Neukum
DLR Institute of Space Sensor Technology and Planetary Exploration and Institute of Geosciences, Free University of Berlin

W. F. Bottke Jr.
Southwest Research Institute

W. K. Hartmann
Planetary Science Institute

The well-investigated size-frequency distributions (SFD) for lunar craters can be used to estimate the SFD for projectiles that formed craters both on terrestrial planets and on asteroids. Our results suggest these distributions may have been relative stable over the past 4 G.y. The derived projectile SFD is found to have a shape that is similar to the SFD of main-belt asteroids as compared with the astronomical observations (Spacewatch asteroid data, Palomar-Leiden survey, IRAS data) and *in situ* images obtained by space missions. This result suggests that asteroids (or, more generally, collisionally evolved bodies) are the main component of the family of impactors striking the terrestrial planets.

1. INTRODUCTION

Asteroids leaving the main belt (MB) may strike the Sun, be ejected out of the solar system, or create an impact crater when striking a terrestrial planetary body (i.e., planets, their satellites, or smaller bodies like asteroids or comets). Impacts craters have been measured and counted on all terrestrial planets and several asteroids. The relative age of different surfaces can be estimated by calculating crater densities, or the number of craters of a given diameter per unit area. Thus, crater densities provide planetologists with an instrument to compare the geological ages of various planetary surfaces, provided that the impactor flux striking each surface is known and the relationship between impact energy and crater size is well understood. All things being equal, larger crater densities imply older surfaces.

These issues have been recently covered by *Neukum et al.* (2001), *Ivanov* (2001), and *Hartmann and Neukum* (2001). In addition to crater chronology, the size-frequency distribution (SFD) of impact craters may be used to estimate the "production SFD" (often named "production function") of projectiles that formed those craters. To determine the production SFD, one must assume that that surface of the body was once a blank slate, such that the craters observed today directly reflect the size spectrum of the projectiles. Once we obtain a crater-derived "projectile" SFD, we can compare it to the SFD of observed asteroids and comets. In this way, we obtain some sense of how small bodies have evolved over time. On some surfaces, this integrated history may stretch all the way back to the "dark era" of the heavy bombardment period.

The projectiles capable of forming craters on the terrestrial planets today come primarily from three populations: (1) asteroids derived from the main belt, (2) Jupiter-family comets (JFCs) derived from the Kuiper Belt, and (3) long period comets (LPC) derived from the Oort Cloud (*Morbidelli et al.,* 2002; *Weissman et al.,* 2002). Other less-important contributors include the Trojan asteroids and the Halley-type comets.

Each impactor population mentioned above has undergone a specific, and possibly unique, accretion/collisional/evolutionary history, such that their SFD are potentially quite different from one another. Moreover, these populations produce planet-crossing projectiles with characteristic orbits and physical properties, such that sorting out the importance of various impactor populations can be complicated. Fortunately, dynamical models and observational work provide some constraints. For example, *Bottke et al.* (2002a) show that asteroids, rather than JFCs or Trojans, currently provide most of the terrestrial planet impact craters coming from a < 7.4-AU orbits (i.e., over 90%). The fraction of craters formed by LPCs (active and dormant), however, is not well understood (see *Weissman et al.,* 2002). To keep things as simple as possible, we will assume in this chapter that asteroids are the dominant source of impactors on the terrestrial planets.

The largest reservoir of asteroids in the inner solar system is the asteroid belt. As described in *Bottke et al.* (2002b) and *Morbidelli et al.* (2002), collisions, the Yarkovsky effect, and numerous mean-motion and secular resonances gradually push D < 20-km asteroids out of the main belt and onto planet-crossing orbits. We call this subpopulation the planet-crossing asteroids, or PCAs. PCAs are short-lived compared to the age of the solar system, such that new main-belt bodies must steadily resupply them over time. It is plausible that the differences between the main-belt SFD and that of the PCA populations are a consequence of the mechanisms producing new PCAs. Hence, a comparison between the SFD of terrestrial planet craters and those of various asteroid populations (e.g., main-belt asteroids, PCAs) should help find new constraints for future investigations of solar system evolution.

2. LUNAR PRODUCTION FUNCTION

The Moon is an ideal test site to study cratering records, particularly since most lunar endogenic activity ended more than 3 G.y. ago [with several important exceptions; see *Hiesinger et al.* (2000)]. Thus, over the last 3 G.y. or so, impacts alone have dominated modifications to the lunar landscape. Moreover, space missions have studied the Moon extensively. Returned samples of lunar landing sites give us a unique opportunity to ascribe an age to craters and regions where accumulated impact craters have been meticulously counted (see *Stöffler and Ryder,* 2001; *Neukum et al.,* 2001). Hence, on the Moon we can estimate the cratering rate to be the number of craters of a given diameter accumulated at a given surface during a given time interval.

We start by assuming an ideal case (i.e., where a planetary surface erased by some process begins to accumulate craters). Before the crater degradation/obliteration processes change the population of these craters, the crater SFD reflects the production SFD or "the standard distribution" of the projectiles. To understand this production function, many authors have tabulated and generalized a huge amount of data on lunar crater counts. In this chapter, we will concentrate on the lunar "production" crater SFD proposed by W. Hartmann and G. Neukum (see *Neukum et al.,* 2001).

2.1. Hartmann Production Function (HPF)

To represent the crater SFD found on the terrestrial planets, Hartmann uses a log-incremental SFD representation with a standard diameter bin size. We call his result the "Hartmann production function," or HPF. The number of craters per kilometers squared is calculated for craters in the diameter bin $D_L < D < D_R$, where D_L and D_R are the left and right bin boundary and the standard bin width is $D_R/D_L = 2^{1/2}$.

The tabulated HPF is an assemblage of data selected by Hartmann to present the production function for one specific moment of time: the average time of lunar mare sur-

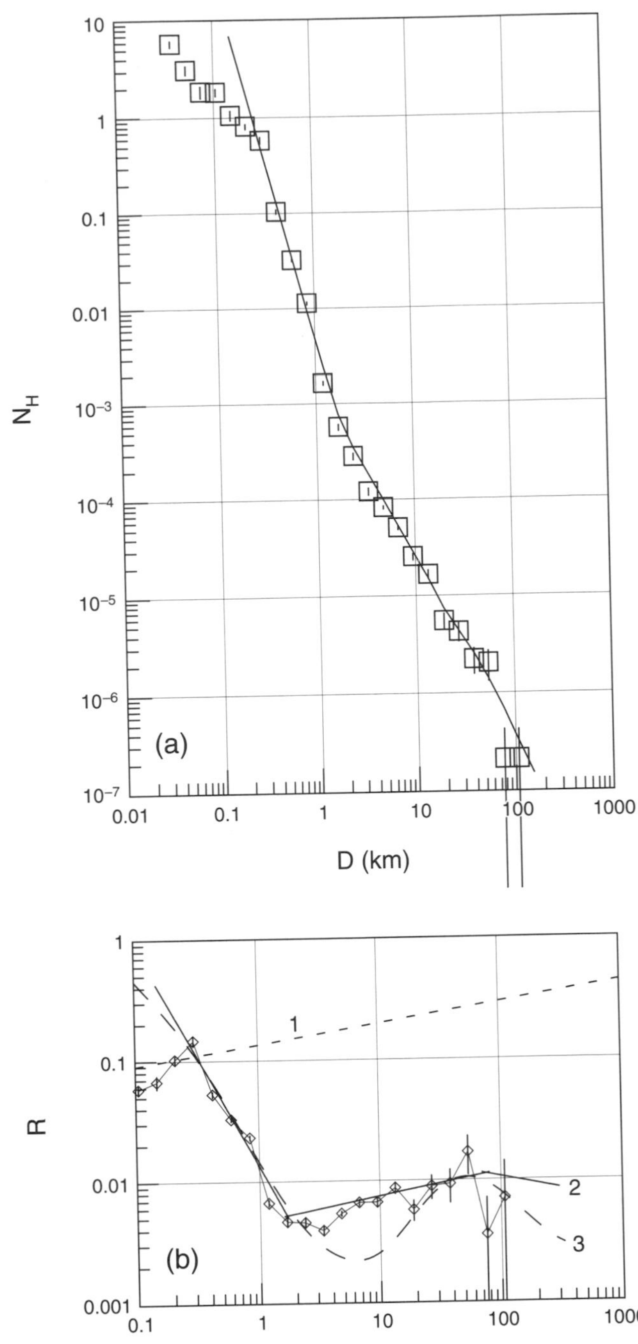

Fig. 1. **(a)** The incremental representation of the Hartmann production function (HPF). The HPF, in a direct sense, is the set of points shown in the plot. Straight lines represent the piece-wise power law fitting to the data (equation (1)). **(b)** Comparison of production functions derived by Hartmann (HPF) and Neukum (NPF) in the R plot representation. The maximum discrepancy between HPF (2) and NPF (3) (roughly a factor of 3) is observed in the diameter bins around D ~ 6 km. Below D ~ 1 km and in the diameter range of 30–100 km, the HPF and NPF give the same or similar results. Fitting the HPF to equation (3), we obtain a model age of 3.4 G.y. The NPF, which is fit to the wide range count of impact craters in the Orientale Basin, yields a model age of ~3.7 G.y. The dashed line 1 represents the approximate saturation level estimated by *Hartmann* (1995).

TABLE 1. Coefficients in equation (2).

a_i	"Old" N(D) (*Neukum,* 1983)	"New" N(D) (*Neukum et al.,* 2001)	"New" N(D) Sensibility*	R(D) for Projectiles (*Ivanov et al.,* 2001)
a_0	−3.0768	−3.0876		—
a_1	−3.6269	−3.557528	±3.8%	+1.375
a_2	+0.4366	+0.781027	±3.9%	+0.1272
a_3	+0.7935	+1.021521	±2.5%	−1.2821
a_4	+0.0865	−0.156012	±1.6%	−0.3075
a_5	−0.2649	−0.444058	±0.88%	+0.4149
a_6	−0.0664	+0.019977	±1.3%	+0.1911
a_7	+0.0379	+0.086850	±0.78%	−0.04261
a_8	+0.0106	−0.005874	±1.8%	−0.03976
a_9	−0.0022	−0.006809	±1.8%	-3.1802×10^{-3}
a_{10}	-5.18×10^{-4}	$+8.25 \times 10^{-4}$	±5.6%	$+2.799 \times 10^{-3}$
a_{11}	$+3.97 \times 10^{-5}$	$+5.54 \times 10^{-5}$	±24.1%	$+6.892 \times 10^{-4}$
a_{12}	—	—	—	$+2.614 \times 10^{-6}$
a_{13}	—	—	—	-1.416×10^{-5}
a_{14}	—	—	—	-1.191×10^{-6}

*"Sensibility" is the coefficient variation that changes the N(D) value a factor of 2 up and down.

face formation. Here the condition to have a fresh surface is satisfied by the fact that most lunar mare basalt samples have a narrow range of ages [e.g., 3.2–3.5 Ga (*Stöffler and Ryder,* 2001); note that some lava flows may be younger (see *Hiesinger et al.,* 2000)]. Hence, the age variation is represented by a factor of 1.1.

The tabulated HPF has been constructed by combining and averaging crater counts in different areas of the Moon. For this reason, it should be treated as a relatively reliable model to deduce the projectile production function. The HPF, in incremental form, takes the form of a piece-wise three-segment power law (*Hartmann,* 1995; see also *Ivanov,* 2001)

$$\log N_{2^{1/2}} = -2.616 - 3.82 \log D_L, \quad D < 1.41 \text{ km} \tag{1a}$$

$$\log N_{2^{1/2}} = -2.920 - 1.80 \log D_L, \quad 1.41 \text{ km} < D < 64 \text{ km} \tag{1b}$$

$$\log N_{2^{1/2}} = -2.198 - 2.20 \log D_L, \quad D > 64 \text{ km} \tag{1c}$$

This function is shown in Fig. 1. Hartmann's choice of power-law segments was made in the 1960s when this work was begun. Note that some selections were for historical reasons; at that time, only the shallow branch with 1.41 km < D < 64 km was well established, with the preexisting literature suggesting various laws for asteroids and meteorites that Hartmann was attempting to relate to lunar data.

2.2. Neukum Production Function (NPF)

In a series of publications, Neukum [for summaries, see *Neukum* (1983) and *Neukum and Ivanov* (1994)] proposed an analytical function to describe the cumulative SFD of lunar impact craters. He showed that the production function had been more or less stable from Nectarian to Copernican epochs (i.e., from practically more than 4 G.y. ago until now). By this time the full size spectrum of craters was known, and in contrast to the piece-wise exponential equations used for the HPF, Neukum computed a polynomial fit to the cumulative number of craters, N, per kilometers squared with diameters larger than a given value D. For the time period of 1 G.y., N(D) may be expressed (*Neukum,* 1983) as

$$\log_{10}(N) = a_0 + \sum_{n=1}^{11} a_n [\log_{10}(D)]^n \tag{2}$$

where D is in km, N is the number of craters with diameters > D per km^2 per G.y., and the coefficients a_n are given in Table 1. Equation (2) is valid for D from 0.01 km to 300 km.

Recently, the NPF was slightly reworked in the largest-crater part by careful remeasuring in the size range (*Ivanov et al.,* 1999, 2001; *Neukum et al.,* 2001). The time dependence of the a_0 coefficient is discussed in the following section 2.3.

A similar equation is used here to present the projectile SFD derived below. Coefficients for this projectile SFD are also listed in Table 1. In the projectile SFD column, the first coefficient a_0 has been set to zero for simplicity. This coefficient determines the absolute number of projectiles. The absolute value of a_0 for projectiles may be found by fitting to observational data (see Fig. 5 below).

2.3. Toward a Unified Production Function

In Fig. 1b, the NPF and HPF are shown in an R plot together with selected data for crater counts on the lunar maria and Orientale Basin. The NPF was fit to the crater counts using an assumed age. We find that both the HPF and NPF are a good match to the observational data below D ~ 1 km. However, for D > 1 km, the HPF goes well above the NPF, meeting again the NPF at crater diameters D ~ 40 km. A maximum discrepancy of a factor of 3 between HPF and NPF is observed in the diameter bins around D ~ 6 km. Below D ~ 1 km and in the 30 < D < 100 km range, the HPF and NPF give the same or similar results.

Although Fig. 1b shows that the HPF and NPF share some similarities, the discrepancy of a factor of 3 for 2 < D < 20 km craters requires further investigation. In general, one should be cautious in interpreting data in this range, particularly since different datasets (including the University of Arizona Arthur Crater Catalog from the 1960s) show somewhat different SFD curvatures. We believe that additional studies of lunar mare data will be needed to further refine the accuracy of the main production function curve.

To use a production function, one should first select a portion of the lunar surface where all the accumulated craters since the last resurfacing event can be counted. Examples of such "time slices" are (1) the Orientale Basin, which erased a large area near the base of the Imbrian stratigraphic horizon; (2) the emplacement of mare basalts (*Hartmann,* 1970; *Hartmann et al.,* 1981); (3) Eratosthenian-aged craters, which mostly have good stratigraphic dates (*Wilhelms et al.,* 1987); and (4) crater rays, which in some cases have a limited lifetime and thus mark an approximate time horizon.

Using the NPF, we show crater counts for lunar areas that differ by a factor of 100 in crater area density (Fig. 2). An examination of these "time slices" suggests that we cannot rule out the simple hypothesis that the lunar production function had a constant shape from ~4 Ga (lunar

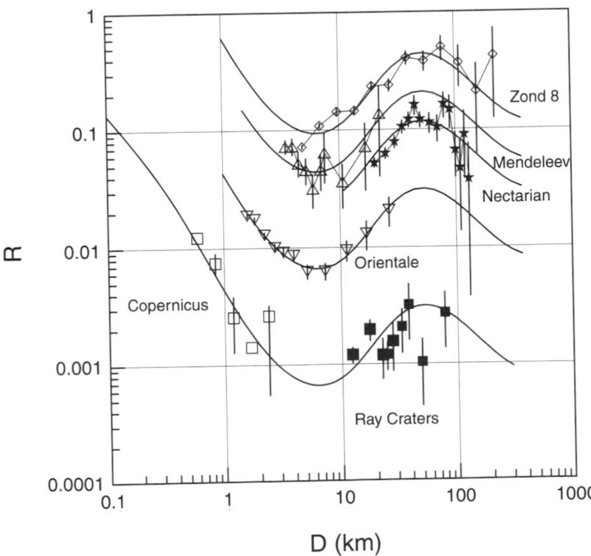

Fig. 2. R plot for several "time slices" of the lunar impact chronology fitted with the NPF curve (equation (2)). See data description in *Neukum et al.* (2001).

highland formation) to ~1 Ga (ray craters). Thus, within the limits of data accuracy, we can assume that the projectile SFD was stable over this interval. To check this hypothesis, we will compare lunar data to those cratering records found on other planets and asteroids.

Although the HPF and NPF have some differences, both assume that the general shape of the SFD striking the Moon over the last 4 G.y. was the same. A different point of view is given by R. Strom (*Strom and Neukum,* 1988; *Strom et al.,* 1992), who claims that the "modern" (postmare) production function is quite different from that produced during the epoch of the late heavy bombardment. A more extensive treatment of this subject can be found in *Strom and Neukum* (1988).

3. CRATERS ON TERRESTRIAL PLANETS

Taking the "theoretical" lunar SFD described in the previous section, we find it useful to scale it to the other terrestrial planets and compare our results with previously published measurements for selected areas on Mercury, Venus, Earth, and Mars. As discussed in more detail by *Hartmann* (1977), this procedure requires that we derive the average impact velocity for each planetary body and then determine the average projectile size needed to make a given size crater. The issue of converting between projectiles and craters via a crater-scaling law is discussed elsewhere (e.g., *Ivanov,* 2001).

3.1. Average Impact Velocity

To compute the average impact velocity of PCAs striking various terrestrial planets, we first need an estimate of the orbital distribution of the PCA population. Assuming that the PCA population is well represented by (1) observed asteroids taken from the "astorb.dat" catalog provided by E. Bowell (http://naic.edu/~nolan/astorb.html) and (2) in the case of Earth, by the debiased model population of Earth-crossing objects estimated by Rabinowitz (1993), we can compute impact velocities for various planets using an extended Öpik method (*Wetherill,* 1967). We recognize that the use of either dataset has some limitations. For (1), we assumed that the observed population of large PCAs did a reasonable job of sampling the debiased population. For (2), we assumed that the sparseness of data points in Rabinowitz's model would not affect our results.

Overall, we found that the average impact velocities of asteroids striking Earth from (1) and (2) were similar, although somewhat higher velocities were found for the debiased population of Earth-crossers. Specifically, the asteroid catalog gives the average impact velocity of 18.6 km s^{-1} (for all asteroids with absolute magnitude H < 15), while estimates from Rabinowitz's model yielded 20.2 km s^{-1}. We believe this difference is a byproduct of peculiar sampling among the H < 15 objects; model results suggest that a substantial fraction of undetected Earth-crossing asteroids with H < 15 have semimajor axes a > 2 AU, large eccentricities, and/or high orbital inclinations (*Bottke et al.,* 2000, 2002a; *Stuart,* 2001; *Morbidelli et al.,* 2002). These values

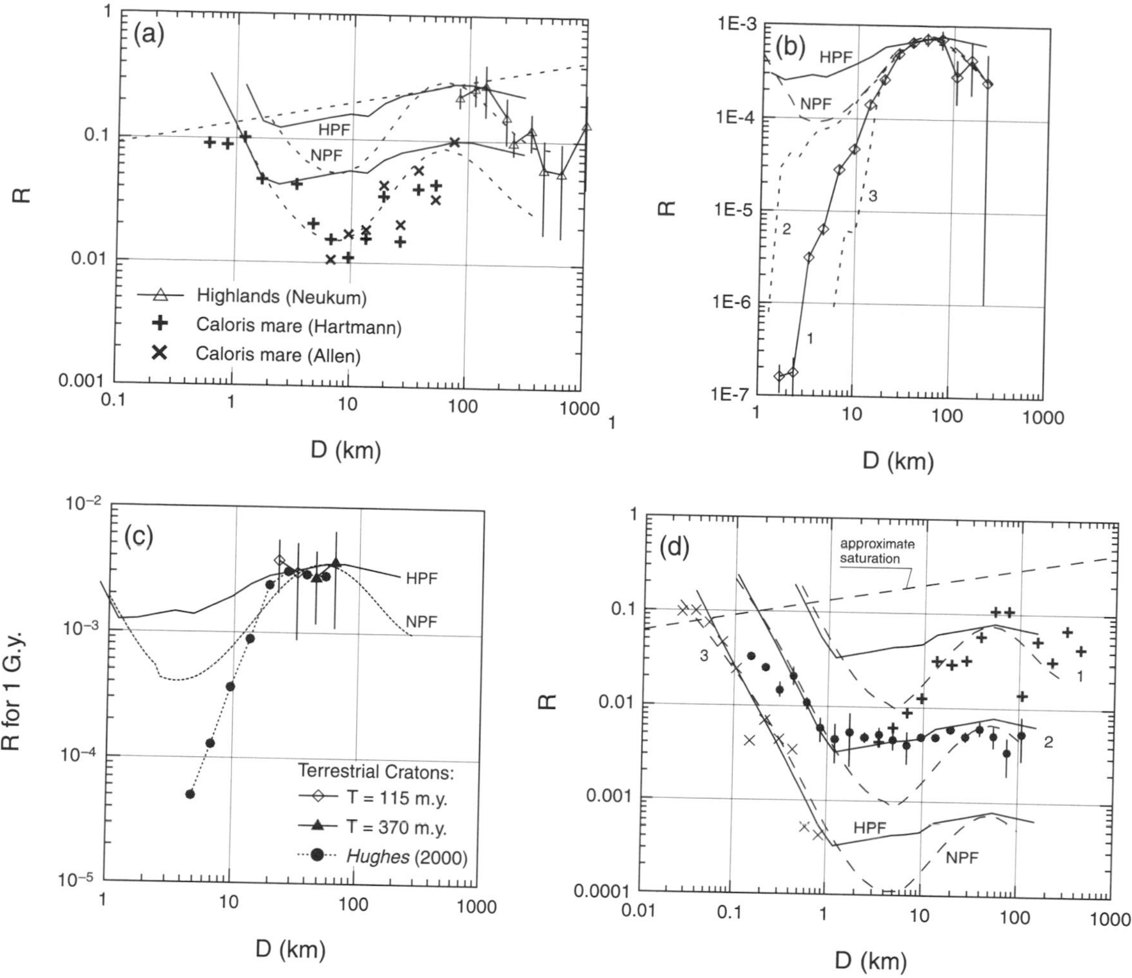

Fig. 3. **(a)** Crater counts for Mercury's highlands and Caloris mare basin compared with the "lunar analog" curves constructed with HPF and NPF. Dashed lines show an approximate saturation level after *Hartmann* (1995). Crater counts are digitized from figures in *Hartmann et al.* (1981). **(b)** R plot for the size-frequency distribution of venusian craters (1) in comparison with the lunar curve recalculated for venusian conditions with the *Schmidt and Housen* (1987) scaling law and *Croft*'s (1985) crater collapse model. Dashed curves 2 and 3 represent two models of how projectiles undergo atmospheric disintegration (*Ivanov et al.,* 1997). "Lunar analogs" for an atmosphere-less Venus are shown as HPF and NPF. **(c)** The R plot for terrestrial craters in comparison with data for cratons (North American + European). To determine the change in the impactor flux over time, the two datasets are divided by factors of 0.115 (115 m.y.) and 0.370 (370 m.y.) in order to put them at the 1-G.y. position. Black dots are crater counts provided by *Hughes* (2000). **(d)** R plot for martian craters. The crater SFD for heavily cratered terrain [1, after *Hartmann et al.* (1981)], "young plains" (2, after *Strom et al.* (1992)] and a relatively younger volcanic caldera floor (*Hartmann et al.*, 1999a). Dashed lines show an approximate saturation level after *Hartmann* (1995).

also account for gravitational focusing, which decreases with increasing encounter velocity. Hence, different model assumption about encounter velocities result in variation of impact rate comparisons for the terrestrial planets.

3.2. Projectile Sizes

Using scaling laws and estimated impact velocities, one can find the projectile SFD, dN/dD_P, for a given impact-crater SFD, dN/dD. To simplify the problem in this chapter, we test the endmember hypothesis of a purely asteroidal projectile flux onto the terrestrial planets. Having such an estimate, we keep open the problem of the cometary impact fraction in the observed crater population. For the same reason, we assume the same projectile density [2.7 g cm^{-3}, the density of typical S-type asteroids (*Britt et al.,* 2002)] for all estimates. The procedure we use to transform craters into projectiles (and vice versa) can be found in *Ivanov et al.* (2001). Briefly, the crater-scaling law derived by *Schmidt and Housen* (1987) is used to estimate the transient and simple crater diameter for a given projectile. The collapse of complex craters (e.g., *Melosh and Ivanov,* 1999) is taken into account using the semiempirical model derived by *Croft* (1985) (see also *Chapman and McKinnon,* 1986; *McKinnon et al.,* 1991).

The SFD of the projectiles is approximated in the same form as equation (2), with a polynomial of 14th degree valid for projectile diameters between ~0.25 m and ~27 km. Computing the largest projectile sizes from the largest craters, however, is more complicated. The SFD for large craters is based on the lunar basins, all of which are very old and do not appear in younger crater populations. Moreover, the basin assignment of a crater diameter D to a given basin involves interpretation of the origin of multiring structures.

Polynomial coefficients for the R plot are listed in Table 1. The estimated projectile SFD is used below to produce a model ("lunar analog") for Mercury, Venus, Earth, and Mars. This SFD, shown graphically in Fig. 3, is also compared with the recent data on the main-belt SFD (Fig. 4).

3.3. Cratering Records

In this section, we apply the lunar-based projectile SFD to each terrestrial planet, accounting for the specific impact velocity and gravity of each planetary body. Our results are summarized in Fig. 3, which shows crater counts from various regions of Mercury, Venus, Earth, and Mars compared to the "lunar analog" SFD curve.

3.3.1. Mercury. The mare surface in the Caloris Basin is one of the few areas suitable for production function measurements of small- to intermediate-sized craters on Mercury. Figure 3a compares direct measurements and calculated SFD (the "lunar analog"). The good coincidence of these data shows a definite similarity of projectile SFDs on the Moon and Mercury in the projectile diameter range from 1 to ~100 km, with a steep part for smaller craters and the "R minimum" for ~8-km craters. However, the age of Caloris Basin is comparable to the age of Orientale Basin.

3.3.2. Venus. *Magellan* data allow us to compare the lunar data averaged over the last 3 G.y. with a planetary surface with an age of ~0.5 Ga. The presence of the atmosphere may be taken into account using a model of projectile atmospheric passage [i.e., model results for both Venus and Titan can be found in *Ivanov et al.* (1997)]. The resulting comparison (Fig. 3b) suggests that venusian craters were formed by a projectile population with a similar SFD for D > 10 km (projectile diameters D_P > 2 km). The R maximum at D ~ 50–70 km exists both on the Moon (3–4 G.y.) and on Venus (~0.5 G.y.). Based on these results, we conclude that the corresponding R maximum in the projectile distribution in the range of D_P ~ 5 km is stable in time.

We point out that our model of venusian atmospheric entry and projectile destruction is based on the work described in *Ivanov et al.* (1997). *McKinnon et al.* (1997), on the other hand, present a different model of atmospheric shielding that also reproduces venusian crater counts, even though they use a simple power-law SFD for projectiles. The disagreement between the two models needs to be studied more thoroughly. Here we only note that *Ivanov et al.* (1997) derive their results from numerical simulations of stony bodies striking Venus' atmosphere, while *McKinnon et al.* (1997) use analytical estimates.

3.3.3. Earth. *Hartmann* (1965, 1966) pointed out that large terrestrial craters reflect an older population, while smaller craters are continually removed by erosion, producing an observed SFD that differs from the production function. The inspection of data from the North American and European cratons (*Grieve and Shoemaker,* 1994) suggests that it is possible to distinguish two populations of craters: (1) eight craters with diameters from 24 to 39 km, the oldest being 115 m.y. old; and (2) eight craters with diameters from 55 to 100 km, with the oldest being 370 m.y. old. The oldest age in each set gives an estimate of the accumulation time. For a proper balance between crater diameter bin width and the number of craters per bin, only two bins for each age subpopulation are used to represent the crater production rate.

We assume that craters smaller than ~20 km in the younger set and smaller than ~45 km in the older set are depleted by erosion. Figure 3c shows the R plot of terrestrial data recalculated to the reference age of 1 Ga assuming a constant crater production rate (R values are divided by 0.115 and 0.370 respcctively). The assumption of a constant crater production rate does not appear to contradict the lunar chronology when it is scaled to Earth. We find that the recent terrestrial production rate estimated by *Hughes* (2000) yields similar results to our simplified analysis (dark circles in Fig. 3c). The poor statistics for terrestrial craters cannot help us resolve the production function's shape. However, the terrestrial craters do help us constrain variations in the impact rate.

3.3.4. Mars. Several groups have investigated the martian crater record. A good summary of our knowledge from the Viking program can be found in *Strom et al.* (1992) for craters with D > 8 km, while more recent crater data obtained by *Mars Global Surveyor* is discussed by *Hartmann et al.* (1999a,b). A recent review of both datasets can be found in *Hartmann and Neukum* (2001) for craters down to D = 11 m. Here we present limited examples of the martian cratering record; in general, they show the same trends as those seen on the other terrestrial planets (Fig. 3d). It is believed that small martian craters have a high obliteration rate, such that any analysis of the data requires one to model how the crater density evolves toward equilibrium. The equilibrium-controlling processes may even vary in different regions or times on Mars. Still, the presence of a R maximum in the SFD for highlands can be readily observed. The R minimum in the SFD is less obvious, especially in crater counts published by *Strom et al.* (1992). We postpone the discussion of this important question for the future.

3.3.5. Craters on asteroids. The *Galileo* and *NEAR Shoemaker* spacecraft returned images of four asteroids: Gaspra, Ida, Mathilde, and Eros (*Belton et al.,* 1992, 1994; *Chapman et al.,* 1996a,b; *Veverka et al.,* 1997, 2000). All four bodies are covered by impact craters. Assuming that main-belt asteroids strike one another at average velocities of 5.5 km/s, it is possible to estimate the small-projectile asteroid SFD from the crater SFD (*Ivanov et al.,* 2001).

We assume that on Gaspra, the impact crater SFD represents the production function. Ida's smaller craters, on the other hand, are believed to be close to saturation (equilibrium); only the largest craters are believed to be below the saturation limit (*Chapman et al.,* 1996a). Large craters on Ida may be formed in the gravity regime (*Asphaug et al.,* 1996).

For Mathilde and Eros, we use the published impact crater SFD (*Veverka et al.,* 1997, 2000). The scaling of craters on Mathilde is not well defined due to the unusually low density (high porosity) of the target. As a first-order approximation, we use scaling parameters presented by *Schmidt and Housen* (1987) for the loosest soil.

Using the assumptions discussed above, it is possible to construct a model projectile distribution for all imaged asteroids (Fig. 4). To simplify things as much as possible, Fig. 4a shows R values for craters vs. projectile diameter, while Fig. 4b shows all asteroid cratering data (below a saturation limit) fit to a single curve. Using Eros data as a reference level, we conclude that its younger surfaces have a crater density that is a factor of 0.01 less than that of its oldest surfaces (*Veverka et al.,* 2000; *Chapman,* 2002). On Gaspra, the same ratio is ~0.15, while it is ~0.6 for Ida and Mathilde (within the accuracy limits of available scaling laws).

For comparison, the lunar-derived projectile SFD is also shown. The model results for asteroid craters demonstrate the presence of the R minimum of the projectile SFD curve. We find this minimum is in approximately the same range of projectile diameters as that seen for the near-Earth asteroids (NEAs).

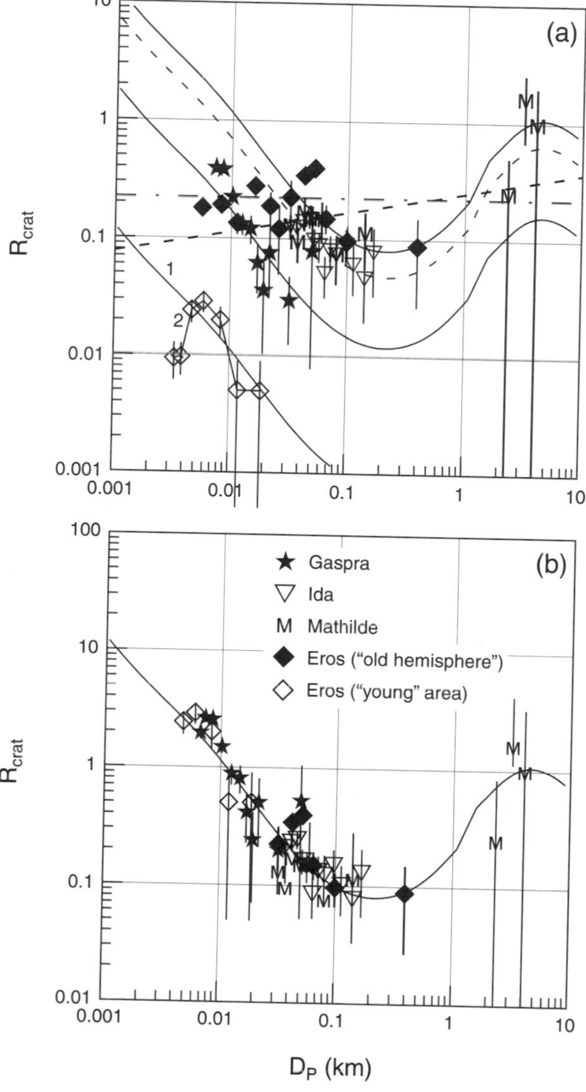

Fig. 4. The R plot for the projectile population derived from craters found on Gaspra, Ida, Mathilde, and Eros. The solid lines represent the projectile SFD derived from lunar cratering records. **(a)** R values calculated for craters plotted against their estimated projectile diameter: 1 — geometrical saturation limit; 2 — empirical saturation limit according to *Hartmann* (1995). **(b)** Selected data points from **(a)** below a saturation crater density have been shifted vertically to fit onto a single curve.

When comparing the cratering records found on the terrestrial planets and asteroids, we found that we could not reject the idea that a single projectile population formed most observed craters. Thus, it seems reasonable to assume that asteroids dominate the population of crater-forming bodies in the inner solar system. The search for deviations from this simple assumption will be left for future work. The interplanetary comparison is consistent with conclusions made by *Hartmann* (1995), namely that (1) the lunar crater record is consistent with a relatively uniform size distribution of interplanetary impactors stretching back 4 G.y. and (2) this same population has struck all the bodies of the inner solar system.

4. MAIN-BELT AND NEAR-EARTH ASTEROIDS

4.1. Main Belt

In this section, we compare our projectile SFD with direct astronomical observations of main-belt asteroids. Earth-based astronomical observations and the satellite infrared survey (IRAS) have revealed all main-belt asteroids with diameters larger than about 40 km (*van Houten et al.,* 1970; *Gradie et al.,* 1989; *Cellino et al.,* 1991). For smaller diameters, one usually assumes a power-law SFD of the form $dN/dD_P \sim D_P^{-k}$, where the value of k may vary between 2.95 [the so-called PLS-slope, after the Palomar-Leiden Survey (see *van Houten et al.,* 1970)] up to 3.5 [the so-called *Dohnanyi* slope (see *Dohnanyi,* 1969)]. K = 3.5 is a typical value for a self-similar cascade of fragments. *Davis et al.* (1994) used a geometrical average of two possible power-law distributions, the PLS distribution and estimations by *Cellino et al.* (1991), to analyze the IRAS data (Fig. 5).

Deviations from a simple power-law crater SFD, considered above, suggest that the asteroid SFD also deviates from a simple power law at diameters smaller than the completeness limit. For large bodies (~100 km in diameter) the non-power-law SFD is thought to be an intrinsic feature of the initial distribution of small bodies before the main belt accumulated (*Davis et al.,* 1985). *Hartmann et al.* (1999c) pointed out that the Yarkovsky effect may have strong influence in controlling size distribution below $D_p \sim 10$ m, and

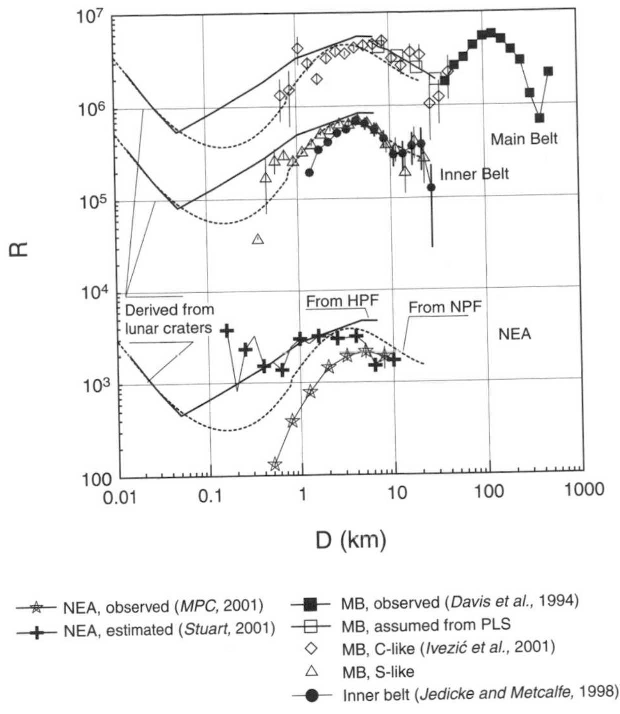

Fig. 5. R plot for main-belt asteroids according to *Davis et al.* (1994) (filled squares — observed asteroids, open squares — assumed by Palomar-Leiden survey), Spacewatch data by *Jedicke and Metcalfe* (1998) for all the inner belt (black dots error bars), Sloan digital sky survey data by *Ivezić et al.* (2001) for C-like (diamonds) and S-like (triangles) asteroids, and the projectile SFDs found using lunar craters according to HPF and NPF. Observed NEAs are shown as stars (*MPC,* 2001), while crosses are for the LINEAR survey data (*Stuart,* 2001). The R value for asteroids is calculated in the same way as that for craters: $R = D_p^3\, dN/dD_p$.

this influence may vary from one asteroid taxonomic type to another (see also *Bottke et al.,* 2002b).

Another possible mechanism for producing such deviations is based on modeling results describing impact evolution in the main belt (i.e., a "wavy" SFD) (*Campo Bagatin et al.,* 1994a,b; *Durda et al.,* 1998; *Davis et al.,* 2002). In a model proposed by *Campo Bagatin et al.* (1994a,b), "waves" seen in the main-belt SFD are produced when small particles are removed from the collisional cascade. A lack of small fragments result in an increase in the number of larger bodies that would have been destroyed by these projectiles. *Durda et al.* (1998), using numerical modeling results of catastrophic collisions (e.g., *Love and Ahrens,* 1996; *Melosh and Ryan,* 1997; *Benz and Asphaug,* 1999), found that the transition from strength scaling to gravity scaling was also capable of producing a wave in the main-belt SFD. Occurring for bodies with diameters near a few hundred meters, self-gravity helps prevent catastrophic disruption events by allowing fragments to reaccumulate with the target asteroid. As bodies get stronger via gravity, more projectiles of that size are available to disrupt larger asteroids, ultimately leading to a wave in the SFD.

Jedicke and Metcalfe (1998) published an estimate of the debiased main-belt SFD based on absolute magnitudes measured as part of the Spacewatch survey. The SFD was converted from absolute magnitudes to diameters by *Durda et al.* (1998) using albedo measurements taken from large asteroids (*Gradie et al.,* 1989). Their SFD estimates are believed to be valid for asteroid diameters large than 2 km. For comparison, the more direct IRAS SFD estimates are considered valid for D > 20 km (*Cellino et al.,* 1991).

In Fig. 5, data from direct observations (following *Davis et al.,* 1989) and estimations by *Jedicke and Metcalfe* (1998) are shown. The projectile SFD, determined above using lunar impact craters, can be compared to these data. Our analysis yields the following points: (1) The direct astronomical data show a relative R minimum at asteroid diameters D ~ 30–40 km; the depth of this minimum may vary for different semimajor axes a. This minimum corresponds to a lunar impact crater ~300 km in diameter. The lunar SFD also shows this minimum. In fact, the lunar SFD, which contains basins up to a size of ~1000 km, shows the same rising R characteristics as the main-belt SFD as it approaches its R maximum near ~100 km. (2) Asteroid diameters have a second R maximum at D ~ 4–5 km. This maximum is visible in the inner and central main belt (2.0 < a < 2.6 AU). It may also exist in the outer main belt, although the needed data is currently beyond our detection limit. The asteroidal R maximum at D ~ 4–5 km corresponds to a lunar crater diameter of ~50 km. The lunar SFD does show this maximum. (3) For asteroid diameters between 2 and 20 km, the general shape of the SFD of the inner asteroid belt and the SFD derived from lunar impact craters is identical within the error limits of available data.

Based on this evidence, we can make some predictions. Figure 5 shows that the projectile SFD derived from lunar craters has an inflection point near 50 m. We suggest that the SFD of main-belt asteroids — the main source of projectiles for cratering on terrestrial planets — may have a similar inflection point. To test this idea, we will need better and more numerous main-belt observations than are currently available.

4.2. Near-Earth Asteroids

A similar SFD is found for NEAs. Figure 6 gives a summary of several recent SFDs estimated from astronomical observations and debiased modeling work of the NEA population. These results are compared to the projectile SFD derived from lunar cratering records with HPF and NPF. The similarity between (1) crater-forming projectiles derived from 1- to 4-G.y.-old surfaces on the Moon, (2) the observed main-belt asteroid population (Fig. 5), and (3) NEAs (Fig. 6) suggests a common connection, namely that the main belt is the predominant source of both the current NEA population and those projectiles that have struck the Moon over the last several billion years. The wavelike shape of

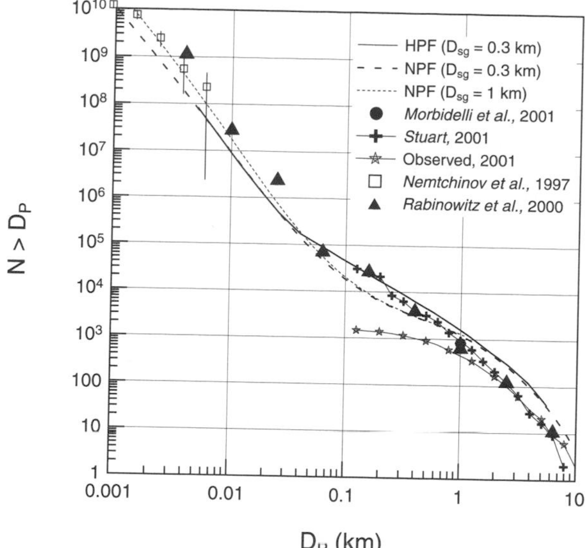

Fig. 6. Estimates of a cumulative, N > D_P, size-frequency distribution for NEAs. The solid, dashed, and dotted lines are model distributions derived from the HPF and NPF for various assumed strength-to-gravity transition diameters for lunar craters. The absolute position of these curves corresponds to the lunar chronology (equation (3)) combined with estimated average impact probability for Earth-crossing asteroids (ECAs). The number of NEAs (defined as bodies with $q < 1.3$ AU) is larger. For observed bodies with $H < 15$, the ECA to NEA ratio is ~0.57. Recent astronomical estimates by *Rabinowitz et al.* (2000), *Morbidelli et al.* (2001), and *Stuart* (2001) are generally consistent with these estimates. Satellite observations of bolides entering Earth's atmosphere (*Nemtchinov et al.,* 1997) are consistent with our results for small (D_P < 10 m) bodies, though we caution that it is problematic to convert light flashes detected in the atmosphere into projectile sizes. The average probability of ECA impacts is used to estimate the total number of projectiles and the impact (or atmospheric entrance) rate.

the main-belt SFD appears to be a byproduct of collisional processes (*Davis et al.,* 2002). The shape of the NEA population is probably an artifact of both the shape of the main-belt SFD and the subtle dynamical mechanisms that allow some main-belt asteroids to escape over long timescales [i.e., collisions coupled with the Yarkovsky effect and chaotic resonances (*Bottke et al.,* 2002b; *Morbidelli et al.,* 2002)].

5. PLANETARY CRATERING RATES

The balance between the PCA supply and depletion rates controls the abundance of PCAs and, accordingly, the terrestrial planet cratering rate (*Morbidelli et al.,* 2002). Using a conservative approach, we assume that the currently observed population of PCAs has been in steady state over the last 3 G.y. For more ancient periods (i.e., the end of the heavy bombardment period), we assume that the relative cratering rates on the terrestrial planets with respect to the Moon was the same as relative cratering rates for the currently observed PCAs. Each of these assumptions should be used with caution.

Most investigators believe that the cratering projectile flux over the last 3 G.y. was relatively constant (with possible variations within a factor of 2). Prior to 3 G.y., the PCA flux was much larger, though it was certainly decaying with time. This ancient period is commonly named "the Late Heavy Bombardment," or LHB. The LHB decay rate is still considered controversial (*Hartmann et al.,* 2000). In this section, we adopt the values favored by *Neukum* (1983). *Neukum* (1983) used lunar rock dating to calibrate the lunar crater count and thus reveal the general character of the bombardment flux decay. His work can be expressed analytically in the form

$$N(1) = 5.44 \times 10^{-14} (\exp(6.93T) - 1) + 8.38 \times 10^{-4} T \tag{3}$$

with N(1) the number of craters larger that 1 km in diameter per km^2 and T being the crater accumulation time (crater retention age) in G.y. Assuming the shape of the SFD is constant in time, equation (3) should be valid for any crater diameter with the proper numerical coefficients.

5.1. Cratering Rate Comparison

The estimated lunar cratering rate has been calibrated using returned samples that have been dated. For other planets, we can rely only upon measured impact crater densities. To translate the lunar crater chronology to other planets, one needs to take into account the differences in planetary gravity, number of planet-crossing bodies, and orbital parameters of the potential projectiles, which in turn control their collision probabilities and impact velocities. As an important example, we compare here the crater rate on Mars to that of the Moon. We discuss first the impact rate, defined as the number of impacts of bodies of the same size per unit surface area per unit of time. The Mars/Moon impact rate ratio is named "R bolide" (R_b) following *Hartmann* (1977). This factor is the basis of the system to date the martian surface — a system highly consistent with martian meteorite age data (*Hartmann,* 1999).

Gravity and target materials are vary from planet to planet. Hence the "R bolide" ratio does not correspond directly to the cratering rate ratio, R_c. Moreover, for a non-power-law projectile SFD, the cratering rate ratio depends intimately on crater diameter; note that different-sized projectiles can form craters with the same diameter on different planets, such that the projectiles may lie on completely different parts of the production SFD.

Estimates of R_b and R_c can be found in various ways. *Ivanov* (2001) followed the technique proposed for comets by *Shoemaker and Wolfe* (1982): (1) The lunar crater SFDs were used to derive the projectile SFDs; (2) the orbits and sizes of observed asteroids were used to both calibrate the projectile SFDs and compute their associated impact velocities; and (3) these values were inserted into crater-scaling laws, which were then used to determine crater-production rates on Mars and the Moon and the values R_b and R_c. The limitation of this "bottom-up" approach is that, so far, observed (and potentially biased) asteroids have been used to determine several important parameters.

The alternative "top-down" approach relies on the recent work by *Bottke et al.* (2002a) (see also *Morbidelli et al.,* 2002), who constructed a model of the debiased orbital and size distribution of the NEA and Mars-crossing asteroid populations.

5.2. R_b from Observed Asteroids

A quantitative description of the PCA flux requires a good understanding of the physical and dynamical evolution of main-belt asteroids (e.g., *Bottke et al.,* 2002a; *Morbidelli et al.,* 2002) or a method to deduce this information from the observed population. *Ivanov* (2001) used the latter approach to estimate the PCA flux. His method, originally developed by *Shoemaker and Wolfe* (1982) to estimate the population of short-period comets, was as follows: Ivanov tabulated the set of observed PCA orbits, sorted them according to perihelion distance, q, and then estimated their average collision probabilities and impact velocities as a function of q. In this way, one uses observed orbits as an assumed representative ensemble of the statistically "averaged" steady-state population of PCAs. In this chapter, we used the February 2002 list of osculating orbits in the small-body database "astorb.dat" (http://asteroid.lowell.edu).

The observed asteroid population is incomplete. Several techniques to remove observational bias exist (e.g., *Rabinowitz,* 1993, 1997; *Rabinowitz et al.,* 1994; *Jedicke and Metcalfe,* 1998; *Bottke et al.,* 2000; *Jedicke et al.,* 2002). For Mercury and Venus, we can make only rough estimates of the cratering rate from the observed asteroid population because many objects with large (a, e, i) values have yet to be detected. For Mars, however, the list of observed asteroids allows us to make a simple correction to estimate the Mars/Moon cratering-rate ratio. We use a simple approach to remove the obvious bias in the q-distribution: We sort PCAs by their absolute magnitude and assume that the brightest asteroids represent a nearly-observationally complete set of objects. For now, we assume that the population of Mars-crossers with $H < 15$ bodies is relatively complete [in close agreement with *Bottke et al.* (2000, 2002a)]. To test the completeness of the $H < 15$ observations, we compared its q distributions for $H < 12$ bodies (*Ivanov,* 2001). Both N(q) functions appear similar to one another. Note that this procedure cannot be used for NEAs, because the number of $H < 12$ asteroids is less than one. Here we assume that the q distribution of asteroids with $H < 15$ may be used to estimate N(q) function for smaller bodies.

For each target (Mars or the Moon) and projectile (PCA), the probability and impact velocity were calculated using an update of Öpik's formulas [refined by *Wetherill* (1967) for the general case of elliptic orbits both for a target and a projectile]. Öpik's method assumes that the apsides and nodes have random positions. The total collision probability (with a correction factor used to account for observational bias among small Mars-crossers) and the average impact velocity yielded the values needed to compute the cratering rate.

The average impact velocity (over all possible Mars orbit eccentricities) was found to be $\langle v_{Mars} \rangle = 9.6$ km s^{-1}. The Mars/Moon impact rate ratio R_b averaged in time is ~2 for asteroids of the same size (*Ivanov,* 2001).

5.3. R_b Estimates from Celestial Mechanics Modeling

The limitation of the "bottom-up" approach presented in the previous section is that, so far, observed (and potentially biased) asteroids have been used to determine several important parameters. In this section, we check these results using a "top-down" approach. Our computation relies on the recent work by *Bottke et al.* (2002a) (see also *Morbidelli et al.,* 2002), who constructed a model of the debiased orbital and size distribution of the NEO and intermediate Mars-crossers (IMC) populations. The latter is defined as the population of the asteroids with main-belt-like semimajor axis and inclination, and which intersect the orbit of Mars within a secular cycle of the eccentricity oscillation. Notice that the Bottke et al. model does not incorporate other populations of Mars-crossers, like the Hungarias, Phoceas, and the isotropic comets of Oort Cloud origin, but these are all expected to be secondary sources as far as cratering is concerned.

From this model, we compute the impact rates between our debiased populations and Moon/Mars using the methodology of *Bottke et al.* (1994). To account for secular variations in Mars/Moon (e, I) values over time, we compute their evolution over the next 10 m.y. using the analytical approximations provided by *Laskar* (1988). Collision probabilities and mean impact velocities are then computed between these values and a grid of test particles uniformly distributed in (a, e, i) space, with each particle representing a component of the NEA and IMC debiased populations.

Our results indicate the interval between $H < 18$ impacts on the Moon, ~11 m.y., is over 10× longer than that for Mars, while the mean impact velocities of material striking the Moon (~19 km s^{-1}) is twice that of objects striking Mars at approximately 10 km s^{-1}. Thus, using these numbers, we calculate that the ratio of $H < 18$ impactors striking Mars over that of the Moon per km^2 per year $R_b = 2.8$.

Computing R_c requires the conversion of H into projectile diameter d and crater diameter D. Since the ratio of bright to dark objects in the Mars-crossing and Earth-crossing populations are not well known, and to keep things simple, we will assume that our projectiles have the same intrinsic properties as those described previously in this chapter.

5.4. Cratering Rate Ratio

To compute R_b, one needs to take into account the differences between the average impact velocity and surface gravity of Mars and the Moon. Using the modern scaling law for simple craters, we estimate the ratio of crater diameters produced on Mars and the Moon with an impact of the same projectile to be

$$D_M/D_m = (\langle v_{Mars}\rangle/\langle v_{Moon}\rangle)^{\beta}(g_M/g_m)^{-\gamma} \quad (4)$$

where exponents $\beta = 0.43$ and $\gamma = 0.22$, according to *Schmidt and Housen* (1987), and subscripts M and m are for Mars and the Moon respectively. *Ivanov* (2001) estimated the $\langle v_{Mars}\rangle/\langle v_{Moon}\rangle$ ratio to be $(9.6/16.1) \approx 0.6$. The velocity term in equation (4) consequently yields a factor of 0.8. In Bottke's estimate from section 5.3, the ratio $\langle v_{Mars}\rangle/\langle v_{Moon}\rangle = (10/19) = 0.53$, and the velocity term in equation (4) is 0.76. The gravity term in equation (4) is the same for both models: $(3.69/1.62)^{-0.22} \approx 0.83$. The total difference in D_M/D_m ratio in two discussing models is rather small: 0.5 for Ivanov's estimates vs. 0.63 for Bottke's estimate.

If N(>D) curves on both planetary bodies are compared for the same (steep or shallow) power slope $[N(>D) \sim D^{-b}]$, the cratering rate ratio is expressed as

$$R_c = (D_M/D_m)^b R_b \quad (5)$$

For a steep branch of HPF (equation (1a)) b = 3.82 for craters below ~1.4 km in diameter. The same slope is given by the NPF (see Fig. 1b). For a shallow branch (D > 1.4 km) b = 1.8 (equation (1b)). For D < 5 km, one can ignore crater modification issues (see details in *Ivanov,* 2001). Consequently, the cratering rate ratio is estimated as $R_c = 0.97$ for b = 1.8 and $R_c = 0.43$ for b = 3.82 from *Ivanov* (2001) and $R_c = 1.22$ for b = 1.8 and $R_c = 0.49$ for b = 3.82 from Bottke's model. Hence the difference in R_b of 40% (2.8 vs. 2) is reduced to the difference in R_c of 27% for b = 1.8 and 14% for b = 3.82. We conclude that the accuracy of our "R bolide" value is formally improved for our "R crater" value.

The work above suggests that the shape of the projectile SFD striking Mars and the Moon must be understood before the martian and lunar crater rates can be compared to one another. In terms of impact rates, it appears that the proximity of Mars to the asteroid belt (more available projectiles) is partially compensated by its smaller average impact velocity and larger surface gravity. For example, although a given size projectile may strike Mars 2–4× more frequently than the Moon, it generally creates a smaller crater. Consequently, the Mars/Moon cratering rate ratio varies from 0.4 to 1.6, depending on the steepness of the projectile's SFD (*Ivanov,* 2001).

6. CONCLUSIONS

1. For the last ~4 G.y., we claim that the shape of production function on the Moon has not changed within the limits of observational accuracy for craters below at least 300 km in diameter.

2. Application of cratering scaling laws allow us to estimate the size-frequency distribution (SFD) of projectiles from the measured SFD of lunar impact craters. The estimated projectile SFD has a complex form with wavy deviations from a simple power law.

3. The estimated SFD of projectiles allows us to reproduce the impact crater SFD on the terrestrial planets. We conclude that a single projectile population formed the majority of impact craters.

4. The projectile SFD derived from the lunar crater population is similar to the SFD of asteroids in the main asteroid belt (within the limits of accuracy of available data). Within the same limits of accuracy, it appears that the contribution of comets to crater formation in the inner solar system either replicates the wavy SFD seen for asteroids or is relatively insignificant. We believe that main-belt asteroids have provided most of the crater-forming objects striking the terrestrial planets, and that these projectiles belong to a collisionally evolved family of objects.

Acknowledgments. We thank W. McKinnon and R. Strom for helpful reviews. B.I. is supported by the Russian Foundation for Basic Science (RFBR) project #01-05-64564-a. W.F.B would like to acknowledge support from NASA's Planetary Geology and Geophysics (NAG5-10331), NEO Observations (NAG6-9951), and Mars Data Analysis (NAG5-10603) programs.

REFERENCES

Asphaug E., Moore J. M., Morrison D., Benz W., Nolan M. C., and Sullivan R. J. (1996) Mechanical and geological effects of impact cratering on Ida. *Icarus, 120,* 158–184.

Belton M. J. S., Veverka J., Thomas P., Helfenstein P., Simonelli D., Chapman C. R., Davies M. E., Greeley R., Greenberg R., and Head J. W. III (1992) Galileo encounter with 951 Gaspra — first pictures of an asteroid. *Science, 257,* 1647–1652.

Belton M. J. S., Chapman C. R., Veverka J., Klaasen K. P., Harch A., Greeley R., Greenberg R., Head J. W. III, McEwen A., Morrison D., Thomas P. C., Davies M. E., Carr M. H., Neukum G., Fanale F. P., Davis D. R., Anger C., Gierasch P. J., Ingersoll A. P., and Pilcher C. B. (1994) First images of asteroid 243 Ida. *Science, 265,* 1543.

Benz W. and Asphaug E. (1999) Catastrophic disruptions revisited. *Icarus, 142,* 5–20.

Bottke W. F., Nolan M. C., Greenberg R., and Kolvoord R. A. (1994) Collisional lifetimes and impact statistics of near-Earth asteroids. In *Hazards Due to Comets and Asteroids* (T. Gehrels and M. S. Matthews, eds.), pp. 337–357. Univ. of Arizona, Tucson.

Bottke W. F., Jedicke R., Morbidelli A., Petit J. M., and Gladman B. (2000) Understanding the distribution of near-Earth asteroids. *Science, 288,* 2190–2194.

Bottke W. F., Morbidelli A., Jedicke R., Petit J. M., Levison H. F., Michel P., and Metcalfe T. S. (2002a) Debiased orbital and absolute magnitude distribution of the near Earth objects. *Icarus, 156,* 399–433.

Bottke W. F. Jr., Vokrouhlický D., Rubincam D. P., and Brož M. (2002b) The effect of Yarkovsky thermal forces on the dynamical evolution of asteroids and meteoroids. In *Asteroids III* (W. F. Bottke Jr. et al., eds.), this volume. Univ. of Arizona, Tucson.

Britt D. T., Yeomans D., Housen K., and Consolmagno G. (2002) Asteroid density, porosity, and structure. In *Asteroids III* (W. F. Bottke Jr. et al., eds.), this volume. Univ. of Arizona, Tucson.

Campo Bagatin A., Cellino A., Davis D. R., Farinella P., and Paolicchi P. (1994a) Wavy size distribution for collisional systems with a small-size cutoff. *Planet. Space Sci., 42,* 1049–1092.

Campo Bagatin A., Farinella P., and Petit J.-M. (1994b) Fragment ejection velocities and the collisional evolution of asteroids. *Planet. Space Sci., 42,* 1099–1107.

Cellino A., Zappalà V., and Farinella P. (1991) The asteroid size distribution from IRAS data. *Mon. Not. R. Astron. Soc., 253,* 561–574.

Chapman C. R. (2002) Cratering on asteroids from *Galileo* and *NEAR Shoemaker.* In *Asteroids III* (W. F. Bottke Jr. et al., eds.), this volume. Univ. of Arizona, Tucson.

Chapman C. R. and McKinnon W. B. (1986) Cratering of planetary satellites. In *Satellites* (J. A. Burns and M. S. Matthews, eds.), pp. 492–580. Univ. of Arizona, Tucson.

Chapman C. R., Ryan E. V., Merline, W. J., Neukum G., Wagner R., Thomas P. C., Veverka J., and Sullivan R. (1996a) Cratering on Ida. *Icarus, 120,* 77–86.

Chapman C. R., Veverka J., Belton M., Neukum G., and Morrison D. (1996b) Cratering on Gaspra. *Icarus, 120,* 231–245.

Croft S. K. (1985) The scaling of complex craters. *J. Geophys. Res., 90,* 828–842.

Davis D. R., Chapman C. R., Weidenschilling S. J., and Greenberg R. (1985) Collisional history of asteroids: Evidence from Vesta and the Hirayama families. *Icarus, 62,* 30–35.

Davis D., Weidenshilling S. J., Farinella P., Paolicchi P. and Binzel R. P. (1989) Asteroid collisional history: Effects on sizes and spins. In *Asteroids II* (R. P. Binzel et al., eds.), pp. 805–826. Univ. of Arizona, Tucson.

Davis D. R., Ryan E. V., and Farinella P. (1994) Asteroid collisional evolution: Results from current scaling algorithm. *Planet. Space. Sci., 43,* 599–610.

Davis D. R., Durda D. D., Marzari F., Campo Bagatin A., and Gil-Hutton R. (2002) Collisional evolution of small-body populations. In *Asteroids III* (W. F. Bottke Jr. et al., eds.), this volume. Univ. of Arizona, Tucson.

Dohnanyi J. W. (1969) Collisional model of asteroids and their debris. *J. Geophys. Res., 74,* 2531–2554.

Durda D., Greenberg R., and Jedicke R. (1998) Collisional models and scaling laws: A new interpretation of the shape of the Main-Belt asteroid distribution. *Icarus, 135,* 431–440.

Gradie J. C., Chapman C. R., and Tedesco E. W. (1989) Distribution of taxonomic classes and the compositional structure of the Asteroid Belt. In *Asteroids II* (R. P. Binzel et al., eds.), pp. 316–335. Univ. of Arizona, Tucson.

Grieve R. A. F. and Shoemaker E. M. (1994) The record of the past impacts on Earth. In *Hazards Due to Comets and Asteroids* (T. Gehrels, ed.), pp. 417–462. Univ. of Arizona, Tucson.

Hartmann W. K. (1965) Terrestrial and lunar flux of meteorites in the last two billion years. *Icarus, 4,* 157–165.

Hartmann W. K. (1966) Martian cratering. *Icarus, 5,* 565–576.

Hartmann W. K. (1970) Lunar cratering chronology. *Icarus, 13,* 299–301.

Hartmann W. K. (1977) Relative crater production rates on planets. *Icarus, 31,* 260–276.

Hartmann W. K. (1995) Planetary cratering I: Lunar highlands and tests of hypotheses on crater populations. *Meteoritics, 30,* 451–467.

Hartmann W. K. (1999) Martian cratering VI. Crater count isochrons and evidence for recent volcanism from Mars Global Surveyor. *Meteoritics & Planet. Sci., 34,* 167–177.

Hartmann W. K. and Neukum G. (2001) Crater chronology and the evolution of Mars. In *Chronology and Evolution of Mars* (R. Kallenbach et al., eds.), pp. 165–194. Kluwer, Dordrecht.

Hartmann W. K., Strom R. G., Weidenschilling S. J., Blasius K. R., Woronow A., Dence M. R., Grieve R. A. F., Diaz J., Chapman C. R., Shoemaker E. M., and Jones K. L. (1981) Chronology of planetary volcanism by comparative studies of planetary craters. In *Basaltic Volcanism on the Terrestrial Planets,* pp. 1050–1127. Pergamon, New York.

Hartmann W. K., Berman D., Esquerdo G. A., and McEwen A. (1999a) Recent martian volcanism: New evidence from Mars Global Surveyor (abstract). In *Lunar and Planetary Science XXX,* Abstract #1270. Lunar and Planetary Institute, Houston (CD-ROM).

Hartmann W. K., Malin M. M., McEwen A., Carr M., Soderblom L., Thomas P., Danielson E., James P., and Veverka J. (1999b) Evidence for recent volcanism on Mars from crater counts. *Nature, 397,* 586–589.

Hartmann W K., Farinella P., Vokrouhlický D., Weidenschilling S. J., Morbidelli A., Marzari F., Davis D. R., and Ryan E. (1999c) Reviewing the Yarkovsky effect: New light on the delivery of stone and iron meteorites from the asteroid belt. *Meteoritics & Planet. Sci., 34,* 161–167.

Hartmann W. K., Ryder G., Dones L., and Grinspoon D. (2000) The time-dependent intense bombardment of the primordial Earth/moon system. In *Origin of the Earth and Moon* (R. M. Canup and K. Righter, eds.), pp. 493–512. Univ. of Arizona, Tucson.

Hiesinger H., Jaumann R., Neukum G., and Head J. W. III (2000) Ages of mare basalts on the lunar nearside. *J. Geophys. Res., 105,* 29239–29776.

Hughes D. W. (2000) A new approach to the calculation of the cratering rate of the Earth over the last 125±20 m.y. *Mon. Not. R. Astron. Soc., 317,* 429–437.

Ivanov B. A. (2001) Mars/moon cratering rate ratio estimates. In *Chronology and Evolution of Mars* (R. Kallenbach et al., eds.), pp. 87–104. Kluwer, Dordrecht.

Ivanov B. A., Basilevsky A. T., and Neukum G. (1997) Atmospheric entry of large meteoroids: Implication to Titan. *Planet. Space Sci., 45,* 993–1007.

Ivanov B. A., Neukum G., and Wagner R. (1999) Impact craters, NEA, and Main Belt asteroids: Size-frequency distribution (abstract). In *Lunar and Planetary Science XXX,* Abstract #1583. Lunar and Planetary Institute, Houston (CD-ROM).

Ivanov B. A., Neukum G., and Wagner R. (2001) Size-frequency distributions of planetary impact craters and asteroids. In *Collisional Processes in the Solar System* (H. Rickman and M. Marov, eds.), pp. 1–34. Kluwer, Dordrecht.

Ivezić Ž., Tabachnik S., Rafikov R., Lupton R. H., Quinn T., Hammergren M., Eyer L., Chu J., Armstrong J. C., Fan X., Finlator K., Geballe T. R., Gunn J. E., Hennessy G. S., Knapp G. R., Leggett S. K., Munn J. A., Pier J. R., Rockosi C. M., Schneider D. P., Strauss M. A., Yanny B., Brinkmann J., Csabai I., Hindsley R. B., Kent S., Lamb D. Q., Margon B., McKay T. A., Smith J. A., Waddel P., York D. G., and the SDSS Collaboration (2001) Solar system objects observed in the Sloan Digital Sky Survey Commissioning Data. *Astron. J., 122,* 2749–2784.

Jedicke R. and Metcalfe T. S. (1998) The orbital absolute magnitude distributions of Main Belt asteroids. *Icarus, 131,* 245–260.

Jedicke R., Larsen J., and Spahr T. (2002) Observational selection effects in asteroid surveys and estimates of asteroid population sizes. In *Asteroids III* (W. F. Bottke Jr. et al., eds.), this volume. Univ. of Arizona, Tucson.

Laskar J. (1988) Secular evolution of the solar system over 10 million years. *Astron. Astrophys., 198,* 341–362.

Love S. and Ahrens T. J. (1996) Catastrophic impacts on gravity dominated asteroids. *Icarus, 124,* 141–155.

McKinnon W. B., Chapman C. R., and Housen K. R. (1991) Cratering of the Uranian satellites. In *Uranus* (J. T. Bergstralh et al., eds.), pp. 629–692. Univ. of Arizona, Tucson.

McKinnon W. B., Zahnle K. J., Ivanov B. A., and Melosh H. J. (1997) Cratering on Venus: Models and observations. In *Venus II* (S. W. Bougher et al., eds.), pp. 969–1014. Univ. of Arizona, Tucson.

Melosh H. J. and Ryan E. V. (1997) Note: Asteroids shattered but not dispersed. *Icarus, 129,* 562–564.

Melosh H. J. and Ivanov B. A. (1999) Impact crater collapse. *Annu. Rev. Earth Planet. Sci., 27,* 385–415.

Morbidelli A., Bottke W. F., Jedicke R., Michel P., and Tedesco E. F. (2001) NEO albedo distribution and impact hazards (abstract). *Bull. Am. Astron. Soc., 33,* 1137.

Morbidelli A., Bottke W. F. Jr., Froeschlé C. H., and Michel P. (2002) Origin and evolution of near-Earth objects. In *Asteroids III* (W. F. Bottke Jr. et al., eds.), this volume. Univ. of Arizona, Tucson.

MPC (2001) *Asteroid Observing Services.* Minor Planet Center Data File available on line at http://asteroid.lowell.edu.

Nemtchinov I. V., Svetsov V. V., Kosarev I. B., Golub' A. P., Popova O. P., Shuvalov V. V., Spalding R. E., Jacobs C., and Tagliaferri E. (1997) Assessement of kinetic energy of meteoroids detected by satellite-based light sensors. *Icarus, 130,* 259–274.

Neukum G. (1983) Meteoritenbombardement and datierung planetarer oberfluchen. Habilitation Dissertation for Faculty Membership, University of Munich. 186 pp.

Neukum G. and Ivanov B. A. (1994) Crater size distribution and impact probabilities on Earth from lunar, terrestrial-planet, and asteroid cratering data. In *Hazards Due to Comets and Asteroids* (T. Gehrels, ed.), pp. 359–416. Univ. of Arizona, Tucson.

Neukum G., Ivanov B., and Hartmann W. K. (2001) Cratering records in the inner solar system. In *Chronology and Evolution of Mars* (R. Kallenbach et al., eds.), pp. 55–86. Kluwer, Dordrecht.

Rabinowitz D. L. (1993) The size-distribution of the Earth-approaching asteroids. *Astrophys. J., 407,* 412–427.

Rabinowitz D. L. (1997) Are Main-Belt asteroids a sufficient source for the Earth-approaching asteroids? Part II. Predicted vs. observed size distribution. *Icarus, 130,* 287–295.

Rabinowitz D. L., Bowell E., Shoemaker E., and Muinonen K. (1994) The population of Earth-crossing asteroids. In *Hazards Due to Comets and Asteroids* (T. Gehrels, ed.), pp. 285–312. Univ. of Arizona, Tucson.

Rabinowitz D. L., Helin E., Lawrence K., and Pravdo S. (2000) A reduced estimate of the number of kilometer-sized near-Earth asteroids. *Nature, 403,* 165–156.

Schmidt R. M. and Housen K. R. (1987) Some recent advances in the scaling of impact and explosion cratering. *Intl. J. Impact Engng., 5,* 543–560.

Sekanina Z. and Yeomans D. K. (1984) Close encounters and collisions of comets with the earth. *Astron. J., 89,* 154–161.

Shoemaker E. M. and Wolfe R. (1982) Cratering time scales for the Galilean satellites. In *Satellites of Jupiter* (D. Morrison, ed.), pp. 277–339. Univ. of Arizona, Tucson.

Stöffler D. and Ryder G. (2001) Stratigraphy and isotope ages of lunar geologic units: Chronological standard for the inner solar system. In *Chronology and Evolution of Mars* (R. Kallenbach et al., eds.), pp. 9–54. Kluwer, Dordrecht.

Strom R. G. and Neukum G. (1988) The cratering record on Mercury and the origin of impacting objects. In *Mercury* (F. Vilas et al., eds.), pp. 336–373. Univ. of Arizona, Tucson.

Strom R. G., Croft S. K., and Barlow N. G. (1992) The Martian impact cratering record. In *Mars* (H. H. Kieffer et al., eds.), pp. 383–423. Univ. of Arizona, Tucson.

Stuart J. S. (2001) A near-Earth asteroid population estimate from the LINEAR survey. *Science, 294,* 1691–1693.

van Houten C. J., van Houten-Groeneveld I., Herget P., and Gehrels T. (1970) The Palomar-Leiden survey of faint minor planets. *Astron. Astrophys. Suppl., 2,* 339–448.

Veverka J., Thomas P., Harch A., Clark B., Bell J. F. III, Carcich B., Joseph J., Chapman C., Merline W., Robinson M., Malin M., McFadden L. A., Murchie S., Hawkins S. E. III, Farquhar R., Izenberg N., and Cheng A. (1997) NEAR's flyby of 253 Mathilde: Images of a C asteroid. *Science, 278,* 2109–2114.

Veverka J., Robinson M., Thomas P., Murchie S., Bell J. F., Izenberg N., Chapman C., Harch A., Bell M., Carcich B., Cheng A., Clark B., Domingue D., Dunham D., Farquhar R., Gaffey M. J., Hawkins E., Joseph J., Kirk R., Li H., Lucey P., Malin M., Martin P., McFadden L., Merline W. J., Miller J. K., Owen W. M., Peterson C., Prockter L., Warren J., Wellnitz D., Williams B. G., and Yeomans D. K. (2000) NEAR at Eros: Imaging and spectral results. *Science, 289,* 2088–2097.

Weissmann P. R., Bottke W. F. Jr., and Levison H. F. (2002). Evolution of comets into asteroids. In *Asteroids III* (W. F. Bottke Jr. et al., eds.), this volume. Univ. of Arizona, Tucson.

Wetherill G. W. (1967) Collisions in the asteroid belt. *J. Geophys. Res., 72,* 2429–2444.

Wilhelms D. E., McCauley J. F., and Trask N. J. (1987) *The Geologic History of the Moon.* U.S. Geol. Survey Prof. Paper 1348. 302 pp.

Asteroid Masses and Densities

James L. Hilton
U.S. Naval Observatory

Since 1989 both traditional groundbased techniques and modern spacecraft techniques have increased the number of asteroids with known masses from 4 to 24. At the same time, the shapes for 16 of these asteroids have been determined with sufficient precision to determine reliable volumes and bulk densities of these bodies. This review paper will look at the masses and densities that have been determined.

1. INTRODUCTION

Although knowledge of the masses and bulk densities of the asteroids is critical in assessing their composition, determining these quantities is a difficult task. Asteroid perturbations of the inner planets of the solar system constitute the largest insufficiently modeled perturbation of high-accuracy planetary ephemerides (*Standish,* 2000). Thus, improved knowledge of asteroid masses is required before planetary ephemerides can be improved. After 200 years of observations, masses have been determined for only 24 asteroids. The next several years should see a large increase in the number of asteroid mass and density determinations due to new methods of determining their masses.

The difficulty in determining asteroid masses lies in their small size. Even the largest main-belt asteroid, 1 Ceres, estimated to contain 30–40% of the mass of the main belt, is only 1% the mass of the Moon. Determining the mass of an asteroid requires observation of its gravitational effect on another body such as an asteroid satellite, or a perturbed body such as another asteroid or a spacecraft. Currently, 12 asteroids are known to have natural satellites (*Chapman et al.,* 1995; *Elliot et al.,* 2001; *Merline et al.,* 1999, 2000, 2001a,b, 2002; *Margot et al.,* 2001; *Veillet,* 2001). Two of these asteroids with satellites, 1998 WW_{31} and 2001 QT_{297}, are transneptunian objects. In 2000, 433 Eros became the first asteroid to be orbited by a spacecraft, *NEAR Shoemaker* (*Yeomans et al.,* 2000). Perturbations on *NEAR Shoemaker* were also used to determine the mass of 253 Mathilde (*Yeomans et al.,* 1998).

To first order, the perturbation of a test body can be estimated using the two-body ballistic particle model

$$\tan \frac{1}{2}\theta = \frac{G(m + M)}{v^2 b}$$

where θ is the angle of deflection in the center-of-mass frame of reference, m is the mass of one body, M is the mass of the other body, G is the gravitational constant, v is the relative velocity of the encounter, and b is the impact parameter (see Fig. 1).

Most asteroids have orbital planes near the ecliptic. Examination of the perturbing equation (see *Danby,* 1988, section 11.9) shows that a coplanar encounter will change only the semimajor axis and/or eccentricity, such that to a first approximation the major change to most perturbed asteroid orbits will be a change in these two elements. In addition, the perturbation is weak, so the change in orbital elements of the perturbed asteroid is small. Thus, the predominant observable for a perturbed asteroid is the cumulative change in its longitude as a function of time caused by a change in its semimajor axis.

Bulk densities are a function of only the mass and volume. If an asteroid's mass is known, determining the volume is equivalent to determining its bulk density. Except for the largest asteroids, their mean diameters give only rough approximations of their volumes because they do not have nearly spheroidal shapes.

2. EARLY MASS DETERMINATIONS

More than 150 years after the discovery of 4 Vesta in 1807, *Hertz* (1966) made the first asteroid mass determination by analyzing its perturbation on 197 Arete. Since Arete's orbital period is nearly 5/4 that of Vesta, it encounters Vesta every 18 years. The multiple encounters enhance the size of the perturbation of Vesta's mass on Arete, making the mass of Vesta easier to determine.

Between 1966 and 1989, masses were determined for three other asteroids, 1 Ceres (*Schubart,* 1970, 1971a,b, 1974; *Landgraff,* 1984, 1988; *Standish and Hellings,* 1989), 2 Pallas (*Schubart,* 1974, 1975; *Standish and Hellings,* 1989), and 10 Hygiea (*Scholl et al.,* 1987). During this time *Hertz* (1968) and *Standish and Hellings* (1989) redetermined the mass of Vesta. Aside from the masses determined by *Standish and Hellings* (1989), all these masses were determined using asteroid-asteroid perturbations.

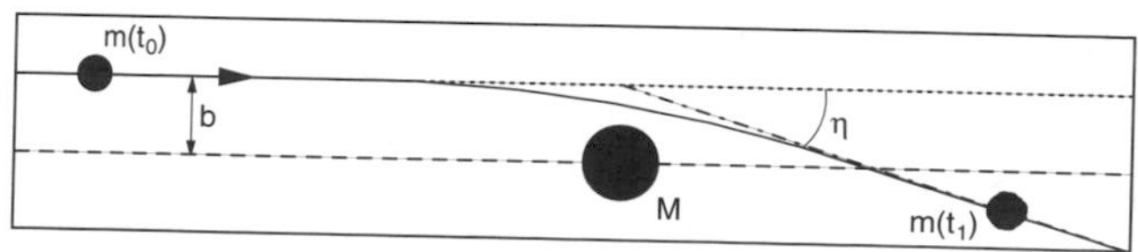

Fig. 1. Ballistic approximation of a small asteroid by a large one. From *Hilton et al.* (1996).

The masses of Ceres and Pallas were determined from their mutual perturbations, but they were not determined simultaneously. *Schubart* (1971a,b, 1974) determined Ceres' mass and subsequently determined Pallas' mass (*Schubart,* 1974, 1975). Unfortunately, Schubart's method of producing normal points from the observations of Ceres and Pallas resulted in masses for Ceres that were too high in both analyses: $(6.7 \pm 0.4) \times 10^{-10}\ M_\odot$ (*Schubart,* 1971a,b) and $(6.0 \pm 0.7) \times 10^{-10}\ M_\odot$ (*Schubart,* 1975). The masses subsequently determined for Pallas were $(1.3 \pm 0.4) \times 10^{-10}\ M_\odot$ (Schubart, 1974) and $(1.1 \pm 0.2) \times 10^{-10}\ M_\odot$ (*Schubart,* 1975).

Standish and Hellings (1989), using *Viking* lander radar ranging data from 1976 through 1981, simultaneously determined the masses of Ceres, Pallas, and Vesta from their perturbations of Mars. These ranging observations have an accuracy of 7 m over the period from 1976 through 1980 and 12 m from 1980 through 1981. The masses determined were $(5.0 \pm 0.2) \times 10^{-10}\ M_\odot$ for Ceres (a 15% decrease from Schubart's mass), $(1.4 \pm 0.2) \times 10^{-10}\ M_\odot$ for Pallas (a 30% increase), and $(1.5 \pm 0.3) \times 10^{-10}\ M_\odot$ for Vesta (a 9% increase). Thus, the masses of the three largest asteroids were known with an uncertainty of $2\text{–}3 \times 10^{-11}\ M_\odot$.

3. MODERN MASS DETERMINATIONS

Since 1989, masses have been determined for 20 additional asteroids, along with 13 new determinations for the mass of Ceres, 5 for Pallas, 7 for Vesta, and 1 for Hygiea. These recent mass determinations are summarized in Table 1.

Most of these mass determinations have been made using the classic method of observing the gravitational perturbation on a test asteroid. Although the stated uncertainties in the masses have also improved, the actual uncertainties from the classic method are more likely closer to the $2\text{–}3 \times 10^{-11}\ M_\odot$ of *Standish and Hellings* (1989).

The Standish (personal communication, 2001) and *Pitjeva* (2001) masses for Ceres, Pallas, and Vesta were determined from their perturbations of Mars measured from *Viking* lander and *Mars Pathfinder* time delay radar observations.

Eleven of the mass determinations were made using methods previously unavailable. Although there were insufficient observations to determine an orbit for its satellite, 243 Ida I (Dactyl), the mass of 243 Ida was determined with an uncertainty of 15% by *Petit et al.* (1997) based on the constraint that Dactyl's orbit is stable.

The masses of 45 Eugenia (*Merline et al.,* 1999), 90 Antiope (*Merline et al.,* 2002), and 87 Sylvia (*Margot et al.,* 2001) were determined from observations of satellites using groundbased adaptive optics. Adaptive optics observations of 762 Pulcova suggest that it is not a single body, but is made up of two components orbiting each other, nearly in contact (*Merline et al.,* 2002).

The masses of 1999 KW_4, 2000 DP_{107}, and 2000 UG_{11} (*Margot et al.,* 2001) were determined using radar time delay-Doppler. Preliminary reduction of observations of 1999 KW_4 indicate that the motion of the primary about the center of mass is observable, making it possible to determine masses for both components of this binary asteroid.

The masses of 253 Mathilde (*Yeomans et al.,* 1998) and Eros (*Yeomans et al.,* 2000) are the first two asteroid masses determined by observing the perturbation of a spacecraft in the vicinity of the asteroid.

The Konopliv et al. (personal communication, 2002) mass of Vesta was determined from its perturbation of Eros and, indirectly, the *NEAR Shoemaker* spacecraft during a close approach (0.416 AU) between Vesta and Eros. The rather large distance of the encounter demonstrates both how large the perturbations of the largest asteroids can be and the sensitivity of spacecraft radar time-delay observations.

Future possible asteroid missions, such as *Dawn* (*Russell et al.,* 2001), *Muses-C* (*Fujiwara et al.,* 2001), and *Hera* (*Sears et al.,* 2000), will return information on the masses, volumes, and densities of the asteroids they encounter. Space missions, however, are limited because of the expense required for the manufacture, launch, and monitoring of a spacecraft.

4. LIMITATIONS TO DETERMINING ASTEROID MASSES FROM THE PERTURBATIONS OF ASTEROIDS

The correlation between the *a priori* mass used for a third asteroid, such as Pallas, can significantly change the mass determined for an asteroid, such as Ceres. Figure 2 shows the mass of Ceres determined by several authors using different test asteroids. Except for *Standish and Hellings* (1989) and *Hilton* (1999), all the mass determinations since 1989 have used approximately the same *a priori* mass for Pallas. Generally, the masses determined agree with each other to within $1 \times 10^{-11}\ M_\odot$. Hilton showed that using exactly the same data but a different *a priori* mass for Pallas, a significantly different mass is obtained for Ceres. The mass determined is independent of the asteroid used as a test body, but varies linearly with the mass adopted for Pallas. In this particular case, the correlation between the masses of Ceres and Pallas is caused by the similarity in the mean motions and mean longitudes of Ceres and Pallas ($0° \leq |\lambda_{Ceres} - \lambda_{Pallas}| \leq 35°$ between 1801 and 2001).

More recent mass determinations such as *Michalak* (2001) and *Goffin* (2001) have included analyses of how changing the *a priori* mass of other asteroids in the model can change the value of the mass to be determined. Goffin's analysis shows that that the high inclination and eccentricity of Pallas is sufficient to discriminate between the its perturbation and that of Ceres on a test asteroid, in disagreement with Hilton.

Long-range encounters with multiple medium-sized asteroids (diameter ~50–150 km) induces noise in the orbit of the perturbed asteroid in a mass determination. The individual perturbations may be insignificant, but the encounters are numerous enough that their combined effect can

TABLE 1. Recent asteroid mass determinations.

Asteroid	Mass $M_\odot$	Method of Determining Mass	Reference
1 Ceres	$(4.7 \pm 0.3) \times 10^{-10}$	asteroid perturbation	*Goffin* (1991)
	$(4.80 \pm 0.08) \times 10^{-10}$	asteroid perturbation	*Sitarski et al.* (1992)
	$(4.8 \pm 0.2) \times 10^{-10}$	asteroid perturbation	*Williams* (1992)
	$(4.62 \pm 0.07) \times 10^{-10}$	asteroid perturbation	*Sitarski and Todorovic-Juchniewicz* (1992)
	$(5.0 \pm 0.2) \times 10^{-10}$	asteroid perturbation	*Viateau and Rapaport* (1995)
	$(4.71 \pm 0.09) \times 10^{-10}$	asteroid perturbation	*Carpino and Knežević* (1996)
	$(4.26 \pm 0.09) \times 10^{-10}$	asteroid perturbation	*Kuzmanoski* (1996)
	$(4.79 \pm 0.04) \times 10^{-10}$	asteroid perturbation	*Viateau and Rapaport* (1997a)
	$(4.76 \pm 0.02) \times 10^{-10}$	asteroid perturbation	*Viateau and Rapaport* (1998)
	$(4.39 \pm 0.04) \times 10^{-10}$	asteroid perturbation	*Hilton* (1999)
	$(4.70 \pm 0.04) \times 10^{-10}$	asteroid perturbation	*Michalak* (2000)
	$(4.76 \pm 0.02) \times 10^{-10}$	Mars perturbation	E. M. Standish (personal communication, 2001)
	$(4.81 \pm 0.01) \times 10^{-10}$	Mars perturbation	*Pitjeva* (2001)
2 Pallas	$(1.59 \pm 0.05) \times 10^{-10}$	asteroid perturbation	*Hilton* (1999)
	$(1.2 \pm 0.3) \times 10^{-10}$	asteroid perturbation	*Michalak* (2000)
	$(1.17 \pm 0.03) \times 10^{-10}$	asteroid perturbation	*Goffin* (2001)
	$(1.08 \pm 0.04) \times 10^{-10}$	Mars perturbation	E. M. Standish (personal communication, 2001)
	$(1.00 \pm 0.01) \times 10^{-10}$	Mars perturbation	*Pitjeva* (2001)
4 Vesta	$(1.40 \pm 0.04) \times 10^{-10}$	asteroid perturbation	*Sitarski and Todorovic-Juchniewicz* (1992)
	$(1.69 \pm 0.05) \times 10^{-10}$	asteroid perturbation	*Hilton* (1999)
	$(1.36 \pm 0.05) \times 10^{-10}$	asteroid perturbation	*Michalak* (2000)
	$(1.31 \pm 0.02) \times 10^{-10}$	asteroid perturbation	*Viateau and Rapaport* (2001)
	$(1.34 \pm 0.02) \times 10^{-10}$	Mars perturbation	E. M. Standish (personal communication, 2001)
	$(1.36 \pm 0.01) \times 10^{-10}$	Mars perturbation	*Pitjeva* (2001)
	$(1.38 \pm 0.03) \times 10^{-10}$	spacecraft perturbation*	Konopliv et al. (personal communication, 2002)
10 Hygiea	$(5.6 \pm 0.7) \times 10^{-11}$	asteroid perturbation	*Michalak* (2001)
11 Parthenope	$(2.6 \pm 0.1) \times 10^{-12}$	asteroid perturbation	*Viateau and Rapaport* (1997b)
	$(2.56 \pm 0.07) \times 10^{-12}$	asteroid perturbation	*Viateau and Rapaport* (2001)
15 Eunomia	$(4 \pm 1) \times 10^{-12}$	asteroid perturbation	*Hilton* (1997)
	$(1.2 \pm 0.4) \times 10^{-11}$	asteroid perturbation	*Michalak* (2001)
16 Psyche	$(9 \pm 3) \times 10^{-12}$	asteroid perturbation	*Viateau* (1999)
20 Massalia	$(2.44 \pm 0.4) \times 10^{-12}$	asteroid perturbation	*Bange* (1998)
45 Eugenia	$(3.0 \pm 0.1) \times 10^{-12}$	observation of satellite	*Merline et al.* (1999)
52 Europa	$(2.6 \pm 0.9) \times 10^{-11}$	asteroid perturbation	*Michalak* (2001)
87 Sylvia	$(7.6 \pm 0.6) \times 10^{-12}$	observation of satellite	*Margot et al.* (2001)
88 Thisbe	$(7 \pm 1) \times 10^{-12}$	asteroid perturbation	*Michalak* (2001)
90 Antiope	$(4.14 \pm 0.05) \times 10^{-13}$	observation of satellite	*Merline et al.* (2002)
121 Hermione	$(4.7 \pm 0.8) \times 10^{-12}$	asteroid perturbation	*Viateau* (1999)
243 Ida	$(2.2 \pm 0.3) \times 10^{-14}$	observation of satellite†	*Petit et al.* (1997)
253 Mathilde	$(5.19 \pm 0.02) \times 10^{-14}$	spacecraft perturbation	*Yeomans et al.* (1998)
433 Eros	$(3.6 \pm 0.9) \times 10^{-15}$	spacecraft perturbation	*Yeomans et al.* (2000)
444 Gyptis	$(4 \pm 2) \times 10^{-12}$	asteroid perturbation	*Michalak* (2001)
511 Davida	$(5.6 \pm 0.7) \times 10^{-11}$	asteroid perturbation	*Michalak* (2001)
704 Interamnia	$(4 \pm 2) \times 10^{-11}$	asteroid perturbation	*Landgraff* (1992)
	$(3.5 \pm 0.9) \times 10^{-11}$	asteroid perturbation	*Michalak* (2001)
762 Pulcova	$(1.28 \pm 0.02) \times 10^{-12}$	observation of satellite	*Merline et al.* (2002)
1999 KW_4	$(1.1 \pm 0.2) \times 10^{-18}$	observation of satellite	*Margot et al.* (2001)
2000 DP_{107}	$(2.2 +1.0,-0.3) \times 10^{-19}$	observation of satellite	*Margot et al.* (2001)
2000 UG_{11}	$(5 +1,-2) \times 10^{-21}$	observation of satellite	*Margot et al.* (2001)

* The Konopliv et al. mass of Vesta was determined from its perturbation of Eros and, indirectly, the *NEAR Shoemaker* spacecraft during a close approach (0.416 AU) between Vesta and Eros.

† The mass of 243 Ida was determined based on the constraint that its satellite is in a stable orbit, not from actual observation of the orbit.

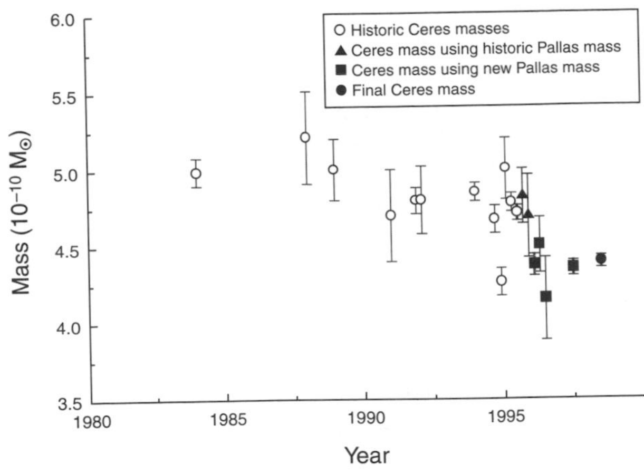

Fig. 2. Historic determinations of the mass of 1 Ceres.

be significant. An example of the effect of unmodeled perturbations can be seen in the ephemeris of Mars. The single largest source of uncertainty in Mars' motion is perturbations by unmodeled asteroids (*Standish and Williams,* 1990). As a result of these perturbations the uncertainty in Mars' position increases by about 0.01 arcsec per century. *Williams* (1984) identifies 36 asteroids with diameter ≥100 km (*Tedesco et al.,* 2002), including Ceres, that are significant perturbers of Mars. Aside from Ceres, the median Mars-asteroid distance at mean opposition for these asteroids' distance is 5.2× greater than the Ceres-asteroid distance at mean opposition. These smaller Ceres-asteroid distances at mean opposition have two consequences. First, the median force on Ceres is 27× greater than that acting on Mars. Second, the encounters take place over a much longer period of time. The median Ceres-asteroid synodic period is 11× greater than the median Mars-asteroid period. A quantitative determination of the increase in the perturbation would require a significant effort, but this qualitative analysis shows that the perturbation of Ceres is easily tens to hundreds of times greater than the perturbation of Mars. Hence, unmodeled perturbations from several medium-sized asteroids can easily dominate over the perturbation of a single large asteroid.

Ceres and Pallas made their closest approach to each other nearly 200 years ago, about the time they were discovered. Thus, determination of their masses from mutual perturbations rely critically on the oldest observations. *Michalak* (2000) made mass determinations for Ceres from its perturbations of Pallas and vice versa. These two determinations produced results similar to *Hilton* (1999). Michalak assumed that there were systematic errors in the early observations and chose not to use these determinations in his final weighted average for the masses of Ceres and Pallas. Hilton examined the oldest observations of Ceres, Pallas, and Vesta, and found no evidence for systematic errors in the observations. It is still possible that perturbations from many long-range encounters over 200 years have made the results unreliable. These observations have a RMS (O–C) ~ 3" in both right ascension and declination, so a small systematic error may not be evident.

Hilton et al. (1996) point out that 348 May should be ideal for determining the mass of Ceres even though it is subject to perturbations from several other asteroids. Although *Davis and Bender* (1977) found an encounter between May and 511 Davida, Hilton et al. failed to find this encounter, and Davis and Bender failed to find the encounter with Ceres. In both searches for encounters the authors made the assumption that perturbations were small enough that an encounter with one asteroid would not change the orbit of the perturbed asteroid enough to hide an encounter with another asteroid. Since Hilton et al. and Davis and Bender used different initial conditions for May and each missed an encounter found by the other, this assumption appears to be false.

In a followup to the work of *Hilton et al.* (1996) and *Davis and Bender* (1977), I attempted to determine the mass of 511 Davida using the software described in *Hilton* (1999). Using the previously determined mass for Ceres, both Ceres and Davida were included as perturbers of May. A search was then made for additional encounters. If one was found, the mass of the perturbing asteroid was added to the model and the process repeated. This process found encounters between May and seven asteroids (1 Ceres, 16 Psyche, 52 Europa, 87 Sylvia, 451 Patientia, 511 Davida, and 704 Interamnia). At this point, the sparse observational history of May made it impractical to determine the masses of any of these asteroids.

Is there evidence for significant noise being induced into the orbits of perturbed asteroids used in the determination of asteroid masses? *Viateau and Rapaport* (1997b, 2001) include perturbations from eight asteroids in their determination of the mass of 11 Parthenope from the perturbation of 17 Thetis. Aside from the perturbation by Vesta, most of these perturbations are small, but they are cumulatively large enough to affect the mass determined at the 1σ level. *Michalak* (2001) shows how encounters with other large asteroids affect the mass determined for a given large asteroid using a particular test asteroid.

The *Hilton* (1999) ephemeris positions of Juno have large systematic residuals with respect to the early (pre-1825) observed positions. A search for perturbers of Juno was made following the same scheme described above for May. The final model included perturbations from nine asteroids (1 Ceres, 2 Pallas, 4 Vesta, 16 Psyche, 24 Themis, 87 Sylvia, 216 Kleopatra, 511 Davida, and 704 Interamnia). However, including all these perturbing asteroids was still insufficient to remove all the systematic error.

Several authors such as *Sitarski and Todorovic-Juchniewicz* (1995), *Carpino and Kneževič* (1996), *Viateau and Rapaport* (1998), *Michalak* (2000), and *Goffin* (2001) have determined the masses of Ceres, Pallas, and Vesta by taking the weighted mean of mass determinations of several perturbed asteroids. The results for each individual determination can vary widely. For example, Goffin's determination of the mass of Pallas used 16 asteroids with individual determinations ranging from $(0.48 \pm 0.56) \times 10^{-10}$ to $(2.8 \pm 1.1) \times 10^{-10}\ M_\odot$. The unweighted mean of the mass determinations is $1.4 \times 10^{-10}\ M_\odot$ with a variance of $0.8 \times 10^{-10}\ M_\odot$. *Michalak* (2000) rejected approximately 20% of the mass determinations he made due to large residuals. *Michalak* (2001) made mass determinations of seven other asteroids, and rejected a similar proportion of the results. This rejection rate indicates the level at which the perturbation of test asteroids by other unmodeled asteroids is great enough to be significant in at least 20% of the cases. It is not clear how many of the adopted mass determinations are contaminated by smaller perturbations caused by unmodeled asteroids.

The conclusion is that the classical method of determining asteroid masses is limited by uncertainty in the masses of the large asteroids and perturbations by other, unmodeled asteroids. Examination of those authors who have made mass determinations of asteroids based on the perturbations of multiple test asteroids suggest that the actual uncertainty in the masses is on the order of $10^{-11}\ M_\odot$.

The limitation of long-range perturbations can be eliminated by reducing the time around the encounter by the test asteroid to a short enough period that the encounter truly can be treated as a two-body deflection of a ballistic particle. Reducing the observing period requires an increase in the accuracy of the observations of the perturbed body. Past mass determinations made by optical observation of a test asteroid used the detection of changes in the mean longitude of a few arcseconds over periods on the order of 50 yr. If the perturbations of other asteroids are to be ignored, then the period of observation of the test asteroid needs to be reduced to a fraction of an orbital period (i.e., a few months), requiring that the position of the test asteroid be determined to a few milliarcseconds.

The *GAIA* mission (*GAIA,* 2000) will produce single-observation accuracies that are good enough to make asteroid mass determinations by observing a perturbed body over the period of a few months. Another high accuracy source of data is radar time delay-Doppler observations. Since radar observations can determine the position of the center of mass of the perturbed asteroid with an accuracy of 1 km or better (*Ostro,* 1993), the perturbed motion of an asteroid could be detected over time periods as short as a few days. However, since it relies on a signal sent out from a radar station, this technique has a r^{-4} falloff in the return signal. The perturbed asteroids are usually only tens of kilometers in diameter, so observations require the use of the Arecibo radio telescope, which is restricted in its declination and hour angle range.

5. EARLY SIZE DETERMINATIONS

Determination of the mean diameters, and hence the volumes and bulk densities, of even the largest asteroids has taken well over a century to produce reliable results. The first attempt to determine the asteroid mean diameters was made by *Herschel* (1802). Using a projection system to measure their angular diameters, he determined 161.6 miles (260.0 km) for the diameter of Ceres and 147 miles (237 km) for Pallas. These results are nearly a factor of 4 too small for Ceres and more than a factor of 2 too small for Pallas. A similar determination by *Schröter* (1811) produced diameters of 2613 km for Ceres, 3380 km for Pallas, and 2290 km for Juno. Clearly, direct observation of the disks of the asteroids using early nineteenth century equipment and techniques produced inaccurate results. As late as 1979 the *Schubart and Matson* (1979) radius for Ceres was uncertain by 75 km, leading to an uncertainty in its bulk density of 50%.

Bruhns (1856) made the first indirect determination of the size of the asteroids using their brightnesses, a technique that requires knowledge of the asteroid's albedo. Bruhns chose an average of the albedos of Saturn, Uranus, Neptune, and the Galilean satellites, all of which have albedos much higher than that of most asteroids. Thus, the diameters he determined for 39 asteroids were too small. In particular, he found the diameter of Ceres to be 365 km; Pallas, 277 km; Juno, 180 km; and Vesta, 367 km.

Barnard (1895), using impersonal filar micrometer observations, determined the diameters of Ceres (780 ± 80 km), Pallas (490 ± 100 km), Juno (190 ± 20 km), and Vesta (390 ± 40 km). These diameters were the definitive values for the first half of the twentieth century. Micrometer determinations of the asteroid diameters tended to produce values that are significantly smaller than the current diameters. For example, *Dollfus* (1971) determined the diameters of the first four asteroids: Ceres, 770 km; Pallas, 490 km; Juno, 195 km; Vesta, 390 km. Like Barnard's diameters, they are all systematically too small. A discussion of the sources of systematic errors in micrometer measurements can be found in *de Vaucouleurs* (1964).

Hamy (1899) made the first attempt to determine the diameter of Vesta using an interferometer. The result, 390 ± 50 km, similar to that of direct measurements, is too small.

The first twentieth century determination of asteroid sizes was made by *Windorn* (1967), who estimated the albedos of 1 Ceres, 2 Pallas, 4 Vesta, and 7 Iris from their polarimetric properties. These albedos then allowed him to estimate effective diameters from their photometry.

Allen (1971) pioneered the radiometric method of determining asteroid diameters. This method compares the amount of reflected visible radiation with the amount of radiated infrared radiation. Since these two quantities depend complimentarily on the albedo, the diameter and albedo are determined simultaneously.

Early polarimetric and radiometric diameters have uncertainties of about 100 km. Thus, the density of Ceres had an uncertainty of about 15% and the densities of Vesta and Pallas had uncertainties of about 20%. The uncertainties in the diameters and masses of the asteroids contributed approximately equally to the uncertainty in the densities. Subsequent improvement in the models, and a combination of both polarimetry and radiometry techniques for the albedo estimates, has reduced the uncertainty for the diameters of the largest asteroids to about 5% (*Tedesco et al.,*

2002). However, the use of the radiometric method to determine the volume of an asteroid becomes severely limited since the shapes of the smaller asteroids depart from a spheroid. Even the largest of the S-class asteroids, 15 Eunomia, departs significantly from a spheroidal shape (see Fig. 3).

Worden and Stein (1979) determined diameters from speckle interferometry observations of Pallas and Vesta. The diameter for Vesta, 550 ± 23 km, is in good agreement with modern determinations, but the diameter for Pallas, 673 ± 55 km, is approximately 26% greater than the currently accepted value. The authors point out that the nonlinearity and nonuniformity of the photographic film they used could cause systematic errors; furthermore, these errors were more likely to affect the diameter of Pallas. Until recently, detector nonlinearity has remained the greatest obstacle to using speckle interferometry to determine asteroid sizes and shapes.

Observing stellar occultations by asteroids is another method of inferring asteroid shapes. This method has the advantage of making accurate ($\sigma \sim$ a few kilometers) determinations of the length of the chord observed. Individual chord lengths are determined by timing the length of the occultation at a known place within the path on Earth's surface. The rate of motion of the asteroid from its ephemeris allows the timing data to be converted to a length. The shape is then formed by observing multiple chords at several different places across the path. Thus, the asteroid's projected shape at a specific rotational phase and relative position can be determined from multiple chords observed during a single occultation.

Occultation observations have a severe restriction. Since the track of an asteroid occultation on the surface of Earth is only a couple hundred kilometers wide, the ephemeris for the asteroid needs an accuracy of about 1 arcsec to assure that most of the observers are within the path of the shadow. There is also a significant chance that the shadow will pass over a large body of water or be unobservable at one or more locations due to local weather conditions.

TABLE 2. Occultations of asteroids with mass determinations.

Asteroid	No. Occult. Obs.	Max. No. Chords
1 Ceres	2	13
2 Pallas	9	141
4 Vesta	2	21
10 Hygiea	5	2
11 Parthenope	1	1
15 Eunomia	3	2
45 Eugenia	1	1
121 Hermione	2	3
216 Kleopatra	3	10
433 Eros	1	10
704 Interamnia	4	11

As of February 27, 2002, the International Occultation Timing Association (IOTA) had collected data from 331 occultations of 213 asteroids (*Dunham,* 2002). These occultations cover the period from February 19, 1958, through February 26, 2002, and include several observations of asteroid occultations for which mass determinations have been made (see Table 2).

Since many chords need to be observed to determine the shape of an asteroid, a large number of observers stationed along the expected path of the asteroid's shadow are required. For example, 22 teams of observers obtained just seven chords during an occultation of SAO 85009 by Pallas (*Wasserman et al.,* 1979). Of the 331 occultations for which

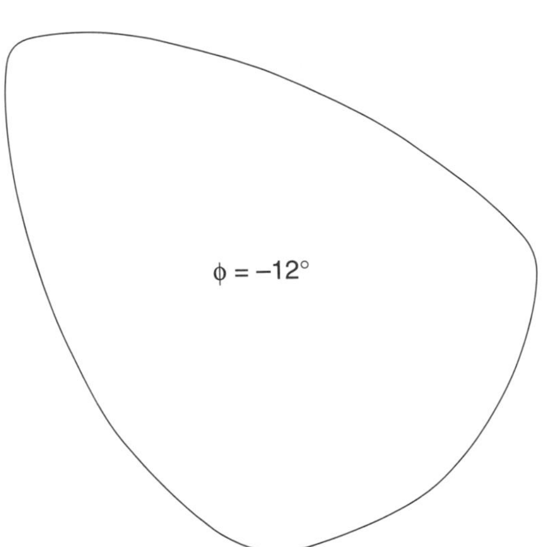

Fig. 3. The shape of 15 Eunomia at a rotation phase of –12°. From *Ostro and Connelly* (1984).

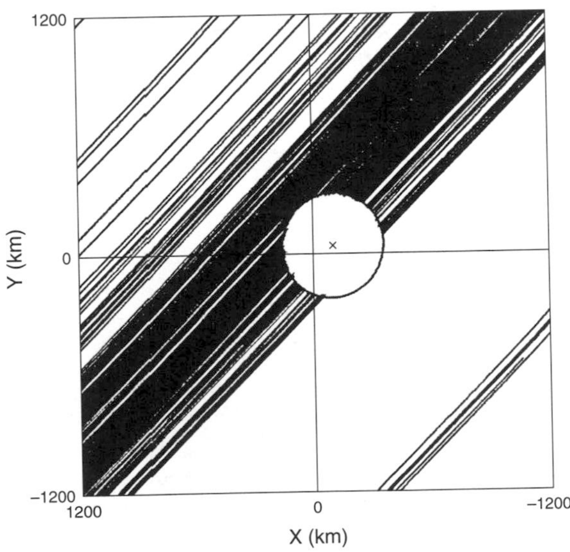

Fig. 4. The shape of 2 Pallas determined by 141 chords observed in its occultation of 1 Vulpeculae by *Dunham et al.* (1990).

IOTA has collected observations, only 56 occultations yielded five or more chords and only four occultations yielded 20 or more chords. Figure 4 shows the result from a total of 141 chords from the occultation of 1 Vulpeculae by Pallas in 1983 (*Dunham et al.,* 1990).

A single occultation gives only a two-dimensional projection of the shape of an asteroid. Thus, a single occultation cannot give enough information to reliably determine the volume for a nonspheroidal asteroid.

6. MODERN SIZE DETERMINATIONS

The determinations of the shapes of asteroids no longer have to rely on the indirect methods developed and used during most of the previous 30 years. Instead, using both spacebased observing platforms and groundbased adaptive optics, direct imaging dominates with much more accurate angular resolution.

The most direct method has been through encounters of asteroids by spacecraft. *Galileo* imaged both 951 Gaspra (*Belton,* 1994) and 243 Ida (*Belton et al.,* 1996), while *NEAR Shoemaker* imaged 253 Mathilde (*Yeomans et al.,* 1998) and 433 Eros (*Yeomans et al.,* 2000). These encounters have allowed highly accurate determinations of the size and shape of these four irregular asteroids. Integrating over the shape of the asteroids (e.g., *Yeomans et al.,* 2000) allows an accurate determination of their volumes to be derived. Combined with mass information, the densities of three of these asteroids, Ida, Mathilde, and Eros, have been determined. In the case of Eros, additional gravity field data show that its composition is probably homogeneous. Although spacecraft imaging *in situ* is highly accurate, it is also extremely expensive. Thus, this technique cannot be counted on for providing volumes for most asteroids.

Direct imaging using optical telescopes has been greatly improved recently by using instruments either in space (*Thomas et al.,* 1997; *Storrs et al.,* 1999) or using adaptive optics at groundbased observatories (*Drummond et al.,* 1998; *Merline,* 2002).

Thomas et al. (1997) have determined the size and shape of Vesta to an uncertainty of only 5 km using the Hubble Space Telescope. This gives its volume with an accuracy of 4%, which is comparable to that attainable from occultation studies. *Storrs et al.* (1999) have provided similar results for several smaller main-belt asteroids. Their results also show that even relatively large (mean diameter ~300 km) asteroids may have nonspheroidal shapes.

Drummond et al. (1999) have used adaptive optics techniques to produce groundbased determinations of the sizes and shapes of Ceres and Vesta with absolute accuracies similar to those produced from Hubble observations. *Drummond and Christou* (1994) indicates that these more recent groundbased observations do not suffer from the systematic errors of earlier ones.

An even more accurate method of asteroid size and shape determination exists in the form of radar time delay-Doppler observations. This method was developed theoretically by *Ostro et al.* (1988) and has been refined to provide

TABLE 3. Current best estimates of asteroid bulk densities.

Asteroid		Density (g cm^{-3})	Reference
1 Ceres*	*Michalak* (2000)	2.03 ± 0.05	*Parker et al.* (2002)
	E. M. Standish (personal communication, 2001)	2.06 ± 0.05	
	Pitjeva (2001)	2.08 ± 0.05	
2 Pallas*	*Goffin* (2001)	3.1 ± 0.3	*Drummond and Cocke* (1989)
	E. M. Standish (personal communication, 2001)	2.9 ± 0.3	
	Pitjeva (2001)	2.6 ± 0.2	
4 Vesta*	E. M. Standish (personal communication, 2001)	3.4 ± 0.2	*Thomas et al.* (1997)
	Pitjeva (2001)	3.5 ± 0.2	
	Konopliv et al. (personal communication, 2002)	3.5 ± 0.3	
16 Psyche		1.8 ± 0.6	*Viateau* (1999)
20 Massalia		2.7 ± 1.1	*Bange* (1998)
45 Eugenia		1.2 (+0.6,–0.3)	*Merline et al.* (1999)
87 Sylvia		1.6 ± 0.3	*Tedesco et al.* (2002)
90 Antiope		1.8 ± 0.8	*Merline et al.* (2002)
121 Hermione		1.8 ± 0.4	*Viateau* (1999)
243 Ida		2.7 ± 0.4	*Petit et al.* (1997)
253 Mathilde		1.3 ± 0.2	*Veverka et al.* (1997)
433 Eros		2.67 ± 0.03	*Yeomans et al.* (2000)
762 Pulcova		1.5 ± 0.4	*Merline et al.* (2002)
1999 KW_4		2.4 ± 0.9	*Margot et al.* (2001)
2000 DP_{107}		1.6 (+1.2,–0.9)	*Margot et al.* (2001)
2000 UG_{11}		1.5 (+0.6,–1.3)	*Margot et al.* (2001)

* Densities using each of the three most recent mass determinations are given for Ceres, Pallas, and Vesta. All three densities are computed using the same volume for comparison.

detailed shapes of asteroids such as 4769 Castalia (*Hudson and Ostro,* 1994) and 216 Kleopatra (*Ostro et al.,* 2000). This technique is limited by a r^{-4} falloff in the return signal.

Ostro and Connelly (1984) pioneered the determination of asteroid shapes from inversion of their light curves. Nowadays, higher-quality data, more powerful computers, and more sophisticated models, such as those used by *Kaasalainen et al.* (2001), could provide very good estimates of asteroid sizes and shapes.

Another future source for asteroid volumes will be GAIA. With a basic resolution of 20 milliarcsec (*GAIA,* 2000), a 50-km-diameter asteroid at 1.5 AU would subtend approximately 2 pixels. Thus, several observations over a rotation period will allow shapes to be determined with an uncertainty of about 40–50%. Although other techniques already available are more accurate, *GAIA* will be able to produce the first large-scale determination of the volumes of large- to medium-sized asteroids.

The bulk density of an asteroid is its mass per unit volume. If the mass and volume of an asteroid are known, its density can easily be derived. There are currently 16 asteroids with reliable masses and volumes for which bulk densities can be derived. The current best estimates for densities of individual asteroids is summarized in Table 3.

7. SUMMARY

The last 10 years have seen great progress in the determination of the masses of asteroids. The classical method of asteroid mass determination by observing the perturbation of a test asteroid over 30–50 years is seriously limited by asteroid interactions not included in the model used to determine the mass. Fortunately, new methods such as spacecraft astrometry, observation of asteroid satellites, and high-accuracy observations (~1 milliarcsec) of the perturbed asteroid over a few months are being developed. These methods will allow better determination of the masses of individual asteroids. Another method that has great promise is radar observations of perturbed asteroids. These observations are of such high accuracy that the perturbation of the massive asteroid can be observed over a few days or months. The greatest drawback to this technique is the r^{-4} falloff in the return signal.

Better determination of asteroid shapes and volumes have also seen major improvements over the last decade. Groundbased adaptive optics and spacecraft-based observations have become major contributors to asteroid shape determinations. Other methods such as deconvolution of asteroid lightcurves and radar image synthesis also have much to contribute as well. Knowledge of both the mass and volume of an asteroid lead directly to the bulk density.

REFERENCES

Allen D. A. (1971) The method of determining infrared diameters. In *Physical Studies of Minor Planets* (T. Gehrels, ed.), pp. 41–44. NASA SP–267, Washington, DC.

Bange J. (1998) An estimation of the mass of asteroid 20-Massalia derived from the Hipparcos Minor Planet Data. *Astron. Astrophys., 340,* L1–L4.

Barnard E. E. (1895) Micrometrical determinations of the diameters of the minor planets Ceres (1), Pallas (2), Juno (3), and Vesta (4) made with the Filar micrometer of the 36-inch equatorial of the Lick Observatory, and on the albedos of those planets. *Mon. Not. Roy. Astron. Soc., 56,* 55–63.

Belton M. J. S. (1994) Galileo encounter with 951 Gaspra First pictures of an asteroid. *Science, 257,* 1647–1652.

Belton M. J. S., Chapman C. R., Klaasen K. P., Harch A. P., Thomas P. C., Veverka J., McEwen A. S., and Pappalardo R. T. (1996) Galileo's encounter with 243 Ida: An overview of the imaging experiment. *Icarus, 120,* 1–19.

Bruhns C. C. (1856) De Planetis Minoribus inter Martem et Jovem circa Solem versantibus. Ph.D. thesis, Berlin Observatory, Berlin. 39 pp.

Carpino M. and Knežević Z. (1996) Asteroid Mass Determination: (1) Ceres. In *Dynamics, Ephemerides, and Astrometry of the Solar System* (S. Ferraz-Mello et al., eds.), p. 203. Proc. IAU Symp. 72, Kluwer, Dordrecht.

Chapman C. R. and 11 colleagues (1995) Discovery and physical properties of Dactyl, a satellite of asteroid 243 Ida. *Nature, 374,* 783–785.

Danby J. M. A. (1988) *Fundamentals of Celestial Mechanics, 2nd edition.* Willmann-Bell, Richmond, Virginia. 466 pp.

Davis D. R. and Bender D. F. (1977) Asteroid mass determination: A search for further encounter opportunities (abstract). *Bull. Am. Astron. Soc., 9,* 502–503.

de Vaucouleurs G. (1964) Geometric and photometric parameters of the terrestrial planets. *Icarus, 3,* 187–235.

Dollfus A. (1971) Diameter measurements of asteroids. In *Physical Studies of Minor Planets* (T. Gehrels, ed.), pp. 25–31. NASA SP–267, Washington, DC.

Drummond J. D. and Christou J. (1994) Images of asteroids 1 Ceres, 2 Pallas, and 4 Vesta with adaptive optics at the Starfire optical range. In *SPIE Proceedings: Adaptive Optics in Astronomy* (M. A. Ealey and F. Merkle, eds.), pp. 468–471 SPIE, Bellingham, Washington.

Drummond J. D. and Cocke W. J. (1989) Triaxial ellipsoid dimensions and rotational pole of 2 Pallas from two stellar occultations. *Icarus, 78,* 323–329.

Drummond J. D., Fugate R. Q., Christou J. C., and Hege E. K. (1998) full adaptive optics images of asteroids Ceres and Vesta; rotational poles and triaxial ellipsoid dimensions. *Icarus, 132,* 80–99.

Dunham D. W. (2002) *Observed Minor Planet Occultation Events Version of 2002 February 27.* International Occultation Timing Association, Topeka, Kansas.

Dunham D. W. and 45 colleagues (1990) The size and shape of (2) Pallas from the 1983 occultation of 1 Vulpeculae. *Astron. J., 99,* 1636–1662.

Elliot J. L., Kern S. D., Osip D. J., and Burles S. M. (2001) *2001 QT_297.* IAU Circular 7733.

Fujiwara A., Mukai T., Abe M., Yano H., Kawaguchi J., and Uesugi K. (2001) Sample return and science by Muses-C (abstract). In *Asteroids 2001 Conference,* p. 132. Osservatorio di Palermo, Palermo, Sicily.

GAIA Science Advisory Group (2000) *GAIA Composition, Formation, and Evolution of the Galaxy. Results of the Concept and Technology Study.* Draft Version 1.6.

Goffin E. (1991) The orbit of 203 Pompeja and the mass of Ceres.

Astron. Astrophys., 249, 563–568.

Goffin E. (2001) New determination of the mass of Pallas. *Astron. Astrophys., 365,* 627–630.

Hamy M. (1899) Sur la Mesure Interférentielle des Petits Diamètres. Application aux Satellites de Jupiter et a Vesta. *Bull. Astronomique, 16,* 257–274.

Herschel W. (1802) Observations on the two lately discovered celestial bodies. *Philos. Trans. R. Astron. Soc., 92,* 213–232.

Hertz H. G. (1966) *The Mass of Vesta.* IAU Circular 1983.

Hertz H. G. (1968) Mass of Vesta. *Science, 160,* 299–300.

Hilton J. L. (1997) The mass of the asteroid 15 Eunomia from observations of 1313 Berna and 1284 Latvia. *Astron. J., 114,* 402–408.

Hilton J. L. (1999) U.S. Naval Observatory ephemerides of the largest asteroids. *Astron. J., 117,* 1077–1086.

Hilton J. L., Seidelmann P. K., and Middour J. (1996) Prospects for determining asteroid masses. *Astron. J., 112,* 2319–2329.

Hudson R. S. and Ostro S. J. (1994) Shape of asteroid 4769 Castalia (1989 PB) from inversion of radar images. *Science, 263,* 940–943.

Kaasalainen M., Torppa J., and Muinonen K. (2001) Optimization methods for asteroid lightcurve inversion. II. The complete inverse problem. *Icarus, 153,* 37–51.

Kuzmanoski M. (1996) A method for asteroid mass determination. In *Dynamics, Ephemerides, and Astrometry of the Solar System* (S. Ferraz-Mello et al., eds.), p. 207. Proc. IAU Symp. 172, Kluwer, Dordrecht.

Landgraff W. (1984) *Minor Planet Circular 8665.*

Landgraff W. (1988) The mass of Ceres. *Astron. Astrophys., 191,* 161–166.

Landgraff W. (1992) A determination of the mass of (704) Interamnia from observations of (993) Moultona. In *Chaos Resonance and Collective Dynamical Phenomena in the Solar System* (S. Ferraz-Mello, ed.), pp. 179–182. Proc. IAU Symp. 152, Kluwer, Dordrecht.

Margot J.-L., Brown M. E., de Pater I., and Roe H. (2001) Discovery and characterization of three binary asteroids, 2000 DP_{107}, 2000 UG_{11}, and 87 Sylvia (abstract). In *Asteroids 2001: From Piazzi to the Third Millennium,* pp. 19–20. Osservatorio di Palermo, Palermo, Sicily.

Merline W. J. and 9 colleagues (1999) Discovery of a moon orbiting 45 Eugenia. *Nature, 401,* 565–569.

Merline W. J., Close L. M., Dumas C., Shelton J. C., Menard F., Chapman C. R., and Slater D. C. (2000) Discovery of companions to asteroids 762 Pulcova and 90 Antiope by direct imaging (abstract). *Bull. Am. Astron. Soc., 32,* 1017.

Merline W. J., Close L. M., Siegler N., Potter D., Chapman C. R., Dumas C., Menard F., and Slater D. C. (2001a) *S/2001 (617) 1.* IAU Circular 7741.

Merline W. J., Menard F., Close L. M., Dumas C., Chapman C. R., Slater D. C., Margot J.-L., and Brown M. E. (2001b) *S/2001 (22) 1.* IAU Circular 7703.

Merline W. J., Weidenschilling S. J., Durda D. D., Margot J.-L., Pravec P., and Storrs A. D. (2002) Asteroids do have satellites. In *Asteroids III* (W. F. Botke Jr. et al., eds.), this volume. Univ. of Arizona, Tucson.

Michalak G. (2000) Determination of asteroid masses: I. (1) Ceres, (2) Pallas, and (4) Vesta. *Astron. Astropyhs., 360,* 363–374.

Michalak G. (2001) Determination of asteroid masses: II. (6) Hebe, (10) Hygiea, (15) Eunomia, (52) Europa, (88) Thisbe, (444) Gyptis, (511) Davida, and (704) Interamnia. *Astron. Astrophys., 374,* 703–711.

Ostro S. J. (1993) Planetary radar astronomy. *Rev. Modern Phys., 65,* 1235–1279.

Ostro S. J. and Connelly R. (1984) Convex profile from asteroid light curves. *Icarus, 57,* 443–463.

Ostro S. J., Connelly R., and Belkora L. (1988) Asteroid shapes from radar echo spectra: A new theoretical approach. *Icarus, 73,* 15–24.

Ostro S. J., Hudson R. S., Nolan M. C., Margot J.-L., Scheeres D. J., Campbell D. B., Magri C., Giorgini J. D., and Yeomans D. K. (2000) Radar observations of asteroid 216 Kleopatra. *Science, 288,* 836–839.

Parker J. W., Stern S. A., Thomas P. C., Festou M. C., Merline W. J., Young E. F., Binzel R. P., and Lebofsky L. A. (2002) Analysis of the first disk-resolved images of Ceres from ultraviolet observations with the Hubble Space Telescope. *Astron. J., 123,* 549–557.

Petit J.-M., Durda D. D., Greenberg R., Hurford T. A., and Geissler P. E. (1997) The long-term dynamics of Dactyl's orbit. *Icarus, 130,* 177–197.

Pitjeva E. V. (2001) Progress in the determination of some astronomical constants from radiometric observations of planets and spacecraft. *Astron. Astrophys., 371,* 760–765.

Russell C. T. and 14 colleagues (2001) Dawn: A proposed Discovery mission to orbit Vesta and Ceres (abstract). In *Asteroids 2001 Conference,* p. 322. Osservatorio di Palermo, Palermo, Sicily.

Scholl H., Schmadel L. D., and Roser S. (1987) The mass of the asteroid (10) Hygiea derived from observations of (829) Academia. *Astron. Astrophys., 179,* 311–316.

Schröter J. H. (1811) *Lilienthalische Beobachtungen der neurendeckten Planeten Ceres, Pallas, und Juno, zur genauen und richtigen Kenntniss iher 1805 wahren Grössen, Atmosphären und übrigen merkwürdigen Naturverhältnisse im Sonnengbiete, etc.*

Schubart J. (1970) *The mass of Ceres.* IAU Circular 2268.

Schubart J. (1971a) The planetary masses and the orbits of the first four minor planets. *Cel. Mech., 4,* 246–249.

Schubart J. (1971b) Asteroid masses and densities. In *Physical Studies of Minor Planets* (T. Gehrels, ed.), pp. 33–39. NASA SP-267, Washington, DC.

Schubart J. (1974) The masses of the first two asteroids. *Astron. Astrophys., 30,* 289–292.

Schubart J. (1975) The mass of Pallas. *Astron. Astrophys., 39,* 147–148.

Schubart J. and Matson D. L. (1979) Masses and densities of asteroids. In *Asteroids* (T. Gehrels, ed.), pp. 84–97. Univ. of Arizona, Tucson.

Sears D. W. G., Brownlee D. D., Pieters C., Lindstrom M., Britt D., Clark B. C., Gefert L., Gorevan S., and Preble J. C. (2000) A multiple near-Earth asteroid sample return mission called Hera (abstract). In *Meteoritics & Planet. Sci., 35,* A145.

Sitarski G. and Todorovic-Juchniewicz B. (1992) Determination of the mass of (1) Ceres from perturbations on (203) Pompeja and (348) May. *Acta Astron., 42,* 139–144.

Sitarski G. and Todorovic-Juchniewicz B. (1995) Determination of the masses of Ceres and Vesta from perturbations on four asteroids. *Acta Astron., 45,* 673–677.

Standish E. M. (2000) Dynamical reference frame — Current relevance and future prospects. In *Towards Models and Constants for Sub-Microarcsecond Astrometry* (K. J. Johnston et al., eds.), pp. 120–126. Proc. IAU Coll. 180, U.S. Naval Observatory, Washington, DC.

Standish E. M. and Hellings R. W. (1989) A determination of the masses of Ceres, Pallas, and Vesta from their perturbations upon the orbit of Mars. *Icarus, 80,* 326–333.

Standish E. M. and Williams J. G. (1990) Dynamical reference frames in the planetary and Earth-Moon systems. In *Inertial Coordinate Systems on the Sky* (J. H. Lieske and V. K. Abalakin, eds.), pp. 173–182. Kluwer, Dordrecht.

Storrs A., Weiss B., Zellner B., Burleson W., Sichitiu R., Wells E., Kowal C., and Tholen D. (1999) Imaging observations of asteroids with Hubble Space Telescope. *Icarus, 137,* 260–268.

Tedesco E. F., Noah P. V., Noah M., and Price S. D. (2002) The supplemental IRAS minor planet survey. *Astron. J., 123,* 1056–1085.

Thomas P., Binzel R. P., Gaffey M. J., Zellner B. H., Storrs A. D., and Wells E. (1997) Vesta: Spin pole, size, and shape from HST images. *Icarus, 128,* 88–94.

Veillet C. (2001) *S/2000 (1998 WW_31) 1.* IAU Circular 7610.

Veverka J. W. and 16 colleagues (1997) NEAR's flyby of 253 Mathilde: Images of a C asteroid. *Science, 278,* 2109–2114.

Viateau B. (1999) Mass and density of asteroids (16) Psyche and (121) Hermione. *Astron. Astrophys., 354,* 725–731.

Viateau B. and Rapaport M. (1995) The orbit of (2) Pallas. *Astron. Astrophys. Suppl. Ser., 111,* 305–310.

Viateau B. and Rapaport M. (1997a) Improvement of the orbits of asteroids and the mass of (1) Ceres. In *Proceedings of the ESA Symposium "Hipparcos — Venice '97"* (B. Battrick, ed.), pp. 91–94. ESA, Noordwijk, The Netherlands.

Viateau B. and Rapaport M. (1997b) The Bordeaux Meridian observations of asteroids. First determination of the mass of (11) Parthenope. *Astron. Astrophys., 320,* 652–658.

Viateau B. and Rapaport M. (1998) The mass of (1) Ceres from its gravitational perturbations on the orbits of 9 asteroids. *Astron. Astrophys., 334,* 729–735.

Viateau B. and Rapaport M. (2001) Mass and density of asteroids (4) Vesta and (11) Parthenope. *Astron. Astrophys., 370,* 602–609.

Wasserman L. H. and 17 colleagues (1979) The diameter of Pallas from its occultation of SAO 85009. *Astron. J., 84,* 259–268.

Williams G. V. (1992) The mass of (1) Ceres from perturbations on (348) May. In *Asteroids, Comets, and Meteors 1991* (A. W. Harris and E. Bowell, eds.), pp. 641–643. Lunar and Planetary Institute, Houston.

Williams J. G. (1984) Determining asteroid masses from perturbations on Mars. *Icarus, 57,* 1–13.

Windorn T. (1967) Zur Photometrischen Bestimmung der Durchmesser der Kleinen Planeten. *Annalen der Universitäts-Sternwarte Wein, 27,* 109–119.

Worden S. P. and Stein M. K. (1979) Angular diameter of the asteroids Vesta and Pallas determined from speckle observations. *Astron. J., 84,* 140–142.

Yeomans D. K. and 12 colleagues (1998) Estimating the mass of asteroid 253 Mathilde from tracking data during the NEAR flyby. *Science, 278,* 2106.

Yeomans D. K. and 15 colleagues (2000) Radio science results during the NEAR-Shoemaker spacecraft rendezvous with Eros. *Science, 289,* 2085–2088.

Asteroid Rotations

P. Pravec
Astronomical Institute of the Czech Republic Academy of Sciences

A. W. Harris
Jet Propulsion Laboratory

T. Michałowski
Adam Mickiewicz University

The rotations of asteroids larger than ~40 km in diameter have a distribution close to Maxwellian that suggests that they are either original bodies of the asteroid main belt or its largest, collisionally evolved remnants. Small asteroids (0.15 < D < 10 km) show significant excesses of both slow and fast rotations, a "barrier" against spins faster than ~12 rotations per day, and some of them are binary systems on inner-planet-crossing orbits with a characteristic fast rotation of their primaries. These small asteroids are collisionally derived fragments, mostly with negligible tensile strength (rubble-pile or shattered interior structure). They mostly gained angular momentum through collisions, but noncollisional factors have also affected their spins and perhaps shapes. In the intermediate size range (10 < D < 40 km), the populations of large and small asteroids overlap. Most tiny asteroids smaller than D ≈ 0.15 km are rotating so fast that they cannot be held together by self-gravitation and therefore must be coherent bodies. They likely are single fragments of the rubble that make up larger asteroids from which the smaller ones are derived.

1. INTRODUCTION

The fundamental characteristic of asteroid rotation is the rotational angular momentum. The angular momentum vector ($\vec{L}$) as well as the inertia tensor ($\hat{I}$) are changed through collisions and other processes of asteroid evolution. They are related with the angular velocity vector ($\vec{\omega}$)

$$\vec{L} = \hat{I}\vec{\omega} \tag{1}$$

The inertia tensor is generally a symmetric tensor containing six independent components. A convenient choice of the system of coordinates in the asteroid-fixed frame gives zero nondiagonal components. The diagonal components $I_x \leq I_y \leq I_z$ are then the principal moments of inertia; the axes are called the principal inertia axes.

In a general rotation state, the spin vector $\vec{\omega}$ is not constant due to the varying moment of inertia about the instantaneous spin axis; its direction and size change on a timescale usually on the order of the rotation period. The excited rotational motion has been described by, e.g., *Samarasinha and A'Hearn* (1991) and *Kaasalainen* (2001). The complex rotation results in a stress-strain cycling within the body. Since the asteroid is not a completely rigid body, the excess rotational energy is dissipated in the asteroid's interior and the spin state asymptotically reaches the lowest energy state, which is a rotation around the principal axis of the maximum moment of inertia I_z. The energy-dissipation profile may be complex, but a reasonable estimate of the timescale τ of damping of the excited rotation to the lowest energy state of principal-axis rotation has been derived by *Burns and Safronov* (1973) assuming a low-amplitude libration

$$\tau \sim \frac{\mu Q}{\rho K_3^2 R^2 \omega^3} \tag{2}$$

where μ is the rigidity of the material composing the asteroid, Q is the quality factor (ratio of the energy contained in the oscillation to the energy lost per cycle), ρ is the bulk density of the body, K_3^2 is a dimensionless factor relating to the shape of the body with a value ranging from ~0.01 for a nearly spherical one to ~0.1 for a highly elongate or oblate one, R is the mean radius of the asteroid, and ω is the angular velocity of rotation. *Harris* (1994) estimates the parameters in equation (2) and expressed the damping timescale as

$$\tau = \frac{P^3}{C^3 D^2} \tag{3}$$

where $P = 2\pi/\omega$ is the rotation period, D is the mean diameter of the asteroid, and C is a constant of ~17 (uncertain by a factor of ~2.5) for P in hours, D in kilometers, and τ in billions (10^9) of years. For most asteroids, the damping timescale is much shorter than the characteristic timescale of events causing excitation of their rotations; all but the slowest rotators and one very small superfast rotator (2000 WL_{107}; P. Pravec et al., in preparation, 2002) have been found with rotations close to principal-axis rotation states.

Efroimsky (2001) has argued for an even shorter damping timescale than above, although observationally, the transition from principal-axis rotation to "tumbling" seems fairly consistent with the above damping timescale. The reader can find more details on the theoretical efforts to estimate the damping timescale in the chapter by *Paolicchi et al.* (2002).

With groundbased observations, we cannot directly measure the asteroid angular momentum vector. The observable parameter is the spin vector; the angular momentum can be estimated from that parameter using equation (1) with an estimate of the moment of inertia based on the asteroid shape, size, and bulk density. Among methods of derivation of the spin vector from ground based observations, the most frequently used one is lightcurve observations. [For the general analysis of disk-integrated data obtained by lightcurve observations, see *Kaasalainen et al.* (2002).] Lightcurve observations are relatively inexpensive in terms of equipment needed but require a lot of observing time (often many nights over several years) to gather enough data for a full solution of the spin vector. Nevertheless, an estimate of the period of rotation and some indication of the shape can be obtained much more easily, often from observations of only a few nights. Thus data on rotation rates, along with amplitudes of variation, are the most abundant. Most studies of rotation characteristics of asteroids are based mainly on lightcurve-derived rotation rates. Likewise, most of the results that we review and present in this chapter are based on analyses of asteroid rotation rates.

2. DISTRIBUTIONS OF ROTATION RATES

In this section, we review analyses of distributions of asteroid rotation rates and also redo some of these analyses using new data. We use the compilation of asteroid lightcurve data maintained by A. W. Harris and available on the Internet (e.g., the IAU Minor Planet Center Web site, http://cfa-www.harvard.edu/iau/lists/LightcurveDat.html). The version used here is dated March 1, 2001, and contains rotation periods and lightcurve amplitudes of 984 asteroids with quality codes Rel ≥ 2. *Harris and Young* (1983) define the Rel quality code scale used in the data tabulation. Here it suffices to note that Rel = 2 corresponds to period determinations that are accurate to ~20%, or in error at most due to some cycle ambiguity or number of extrema per rotation cycle, in rare cases perhaps even off by a factor of 2, but not more. Higher reliabilities have no significant ambiguity. Errors of the period estimates of 970 asteroids of the sample are not greater than a few tens percent, so they are suitable for statistical analyses. For another 14 objects, lower limits on their periods have been estimated and we plausibly take a value of 1.5× the lower limit as an estimate of the period in those cases. Omitting them would lead to an increase of the bias against slow rotators, which is nevertheless inevitably still present in the sample due to the lightcurve technique used. Asteroid diameter estimates are mostly from the IRAS Minor Planet Survey (*Tedesco et al.*, 1992); for objects with no other diameter estimate available, diameter is estimated from absolute magnitude and an assumed albedo typical for their taxonomic or orbital class. The uncertainty of such diameter estimates is likely within a factor of 1.5, and almost certainly within a factor of 2.

2.1. Spin Rate vs. Size

In Fig. 1, the spin rate vs. diameter is plotted using the data for 984 asteroids. The geometric mean spin rate has been computed using the "running box" method described in *Pravec and Harris* (2000). Briefly, the geometric mean spin rate $\langle f \rangle$ is computed within a box of 50 objects, shifted down in diameter by one object each time. Slowly rotating asteroids with $f < 0.16 \langle f \rangle$ are excluded from the computation, since their excess would affect the computation significantly. The computed geometric mean spin rate is thus representative of the population of the asteroids with the slow rotators excluded. The computation has been stopped at D = 0.15 km. The abrupt change of the rotation properties around this diameter does not warrant a use of the method over the boundary. We note that the choice to use the geometric rather than arithmetic mean rate has the advantage that it is less sensitive to outlying extreme values and its results are the same for P (spin period) as for f (spin rate) in the analysis. We also did the analysis with the arithmetic mean spin rate and obtained similar results, so the choice to use the geometric rather than arithmetic mean spin rate is not critical for the conclusions.

In Fig. 2, the computed $\langle f \rangle$ vs. diameter is shown in detail. Arrows at the bottom of the diagram show positions of every fiftieth object (starting from the twenty-fifth) to indicate the resolution in diameter. Points of the curve shifted by 50 objects or more are independent; thus there are only about 19 fully independent points in the curve. The formal $\pm 1\sigma$ uncertainty of the computed $\langle f \rangle$ is represented with the dashed curves; it is about 10% at any diameter.

The geometric mean spin rate is about 3.0 d^{-1} for the largest asteroids (mean D ~ 200 km) and decreases to 1.8 d^{-1} as D decreases to ~100 km. Then it increases to about 4.0 d^{-1} as D decreases down to ~10 km. The "wavy" deviations from a monotonic increase of the computed $\langle f \rangle$ between 10 and 100 km may not be real, as the peaks and troughs differ by ~2σ or less and, moreover, occur at intervals similar to the resolution in D in that size range. This suggests that statistical variations in the sample are a cause of the deviations. Only the local minimum of $\langle f \rangle$ at D ~ 40 km is marginally significant at ~3σ. Below D = 10 km, there is apparently only a small increase of $\langle f \rangle$ with decreasing diameter, but more data are needed to establish its statistical significance. Below we discuss the rotation characteristics in different size ranges.

2.2. Large Asteroids (D > 40 km)

Several researchers have estimated a lower bound diameter of the "large asteroid" group on a basis of statistical

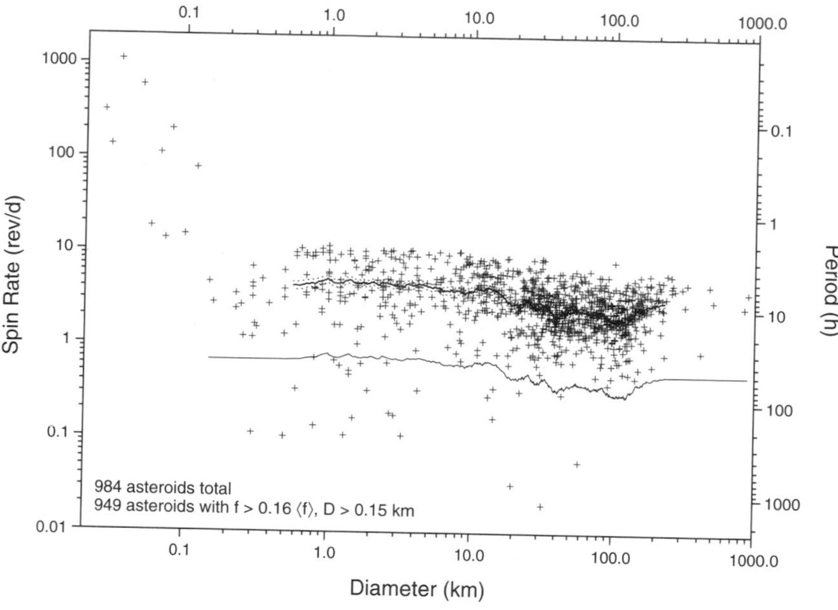

Fig. 1. Asteroids' spin rate vs. diameter plot. The bold solid curve is the geometric mean spin rate ⟨f⟩ vs. diameter; the thin solid curve is the limit f = 0.16 ⟨f⟩ used for excluding slow rotators.

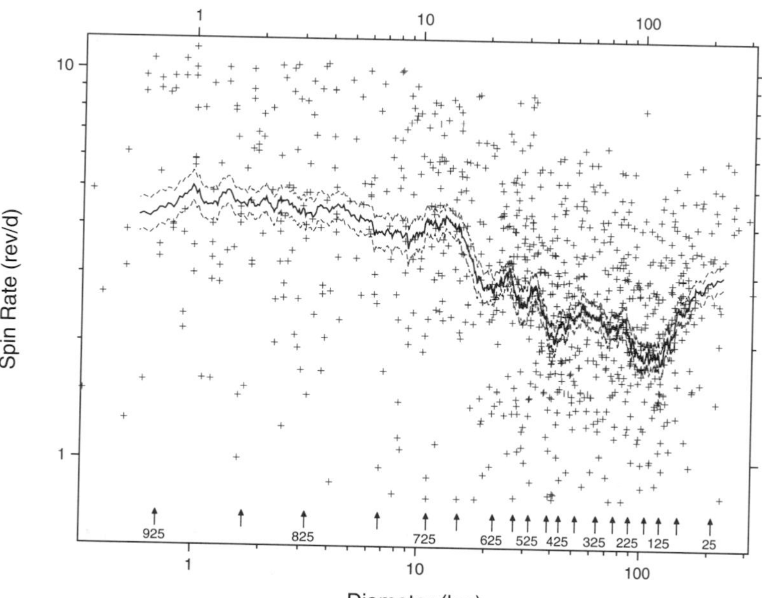

Fig. 2. The densest part of Fig. 1 shown in detail. The dashed curves represent ±1σ errors of ⟨f⟩ (solid curve) computed from variance of a sample within the running box.

analyses of the asteroid rotation properties. Recent estimates range generally from 30 to 50 km (e.g., *Fulchignoni et al.,* 1995; *Donnison and Wiper,* 1999). *Pravec and Harris* (2000) give a lower limit of 40 km based on their study, which shows that larger deviations from Maxwellian distribution (see below) start to occur in the 30–40-km-diameter range. We repeat the analysis using the larger dataset and come to the same conclusion. The occurrence of several relatively rapidly rotating asteroids and one very slow rotator at diameters just above 30 km indicates that the lower diameter limit of the large asteroid group is about 35 km with an uncertainty of several kilometers; the error being dominated by likely systematic errors of the asteroid diameter estimates themselves. For the analyses presented below, we adopt as the lower limit a diameter of 40 km, consistent with the work by *Pravec and Harris* (2000).

The Maxwellian distribution has the form

$$n(f) = \sqrt{\frac{2}{\pi}}\frac{Nf^2}{\sigma^3}\exp\left(\frac{-f^2}{2\sigma^2}\right) \quad (4)$$

where n(f) df is the fraction of objects in the range (f,f + df), N is the total number of objects, and σ is the dispersion. Instead of σ, one can as well use $\overline{f^2}$, the mean squared spin frequency of the distribution, where $\overline{f^2} = 3\sigma^2$ for the Maxwellian distribution. A distribution of spin rates of asteroids is Maxwellian if all the three components of $\overline{\omega}$ are distributed according to a Gaussian with zero mean values and equal dispersions. It is plausible to expect that this is a typical outcome for a collisionally evolved system (*Salo,* 1987). However, deviations from the Maxwellian distribution of rotation rates are expected because of, e.g., an inhomogeneous sample. In fact, it is known that the sample of asteroids is inhomogeneous; they have different masses and material properties. Nevertheless, it still makes sense to compare observed distribution with a Maxwellian: If it is close to Maxwellian, it suggests that the system is collisionally relaxed. If it is not, we must investigate the deviations in detail.

Comparisons of histograms of asteroidal spin rates with a Maxwellian distribution have been done by many researchers (e.g., *Harris and Burns,* 1979; *Farinella et al.,* 1981; *Binzel et al.,* 1989; *Fulchignoni et al.,* 1995; *Donnison and Wiper,* 1999). They generally have found that the spin-rate distribution is Maxwellian for larger asteroids but non-Maxwellian below a certain diameter. They use, however, an assumption of a constant dispersion (or, equivalently, mean spin rate) for asteroids of all sizes in each investigated sample. We know that this is an incorrect assumption; the mean spin rate varies considerably with size (as shown in Figs. 1 and 2).

Attempting to account for different sizes of asteroids, *Pravec and Harris* (2000) normalize each spin rate to the geometric mean spin rate at the given size and test that distribution for consistency with a Maxwellian. This approach should eliminate a size-dependent part of the sample inhomogeneity. They find that the distribution of spin rates of large asteroids is nearly Maxwellian, with only a small deviation among the fastest rotators. We repeat the analysis with the larger sample and the result is plotted in Fig. 3. The most extreme outlying point, at $D \approx 100$ km and $f \approx 8$ d^{-1} (see Fig. 2), is the asteroid 522 Helga (*Lagerkvist et al.,* 2001). They note an ambiguity in their solution and that Helga's period may be ~7 d^{-1}, which is not so extreme, and the determination is quite uncertain, only barely of Rel = 2. Thus, we exclude this point from a statistical test below. According to a χ^2 test, the observed distribution of spin rates of large asteroids is only marginally different from a Maxwellian distribution. The hypothesis that the distribution is Maxwellian can be rejected at the 95% confidence level, but not at the 99% level. An excess occurs at normalized spin rates from 2.2 to 2.8. Cheking it in detail, we have found that this is at least partly caused by M-type asteroids that have a mean spin rate 1.46× (formal error ±0.17) the mean spin rate of "primitive" (BCDFGPT) and "differentiated stony" (SQAEVR) asteroids. Actually, most asteroids in the excess of $f/\langle f \rangle$ from 2.2 to 2.8 for which a taxonomic class is known are M types. So, a fully sufficient explanation of the marginally significant deviation of the distribution of spin rates of large asteroids from Maxwellian is the presence of the faster-than-average-rotating population of M-type asteroids. An explanation of their greater mean spin rate is, however, uncertain. A theory by *Harris* (1979) predicts that the mean spin rate depends on a square root of the bulk density, which would be consistent with their expected properties. On the other hand, estimated bulk densities of C and S types differ significantly as well, and we see no difference between the mean spin rates of the "primitive" and the "differentiated stony" classes.

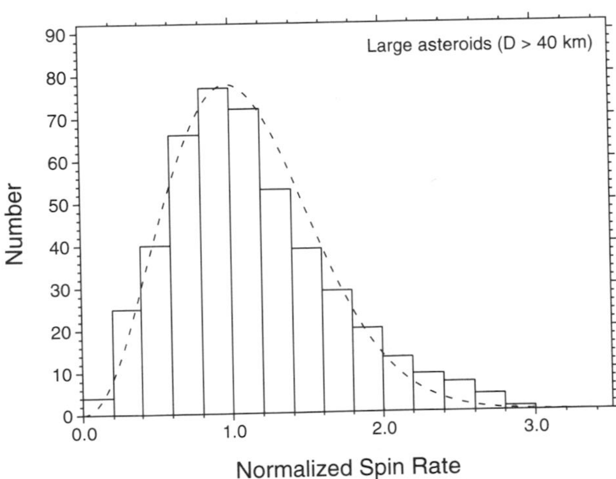

Fig. 3. Histogram of $f/\langle f \rangle$ for 460 asteroids with $D > 40$ km. The dashed curve is the corresponding Maxwellian distribution.

A plausible explanation of the minimum of $\langle f \rangle$ at D ~ 100 km is an effect proposed by *Dobrovolskis and Burns* (1984) called "angular momentum drain." They point out that if the characteristic velocity of impact ejecta is in the range of the surface escape velocity of the asteroid, then more of the ejecta coming out of a crater in a prograde sense with respect to the asteroid's rotation will escape than ejecta coming out in a retrograde sense. This asymmetric distribution of escaping ejecta produces a recoil that may lead to a slowing of the spin rate with many impacts. The surface escape velocity in the size range of the minimum in spin frequency at ~100 km diameter is ~100 m/s, plausibly in the right range corresponding to typical impact ejecta velocities. *Cellino et al.* (1990) describe a similar effect (calling it "angular momentum splash") based on the same concept and working on marginally catastropic disruptions. Tidally slowed binary asteroids cannot account for the reduction in mean spin without substantially broadening the distribution, since only a small fraction of the population can be binaries.

The large asteroid group includes 22 "jovian" and "transjovian" asteroids (Trojans, Centaurs, and transneptunian objects). Their number is too small for a thorough separate statistical analysis but the present group is indistinguishable from main-belt rotational properties. We therefore include them together with large main-belt asteroids in the analysis described above, but of course excluding them would not alter the results for the main-belt asteroids. When a significantly larger sample of these distant objects is available in the future, it may be possible to find some differences in their rotational properties.

In general, the rotational properties of asteroids larger than D = 40 km suggest that they are either original bodies of the main asteroid belt or their largest, collisionally relaxed remnants.

2.3. Small Asteroids (0.15 km < D < 10 km)

Below D ≈ 40 km, the distribution of the rotation rates is non-Maxwellian, with excesses both at the fast and slow spins. The range between 10 and 40 km is where a steep increase of the mean spin rate occurs (Fig. 2). We consider this range as a transitional region where the large and small asteroid groups overlap. Since the samples are mixed there, we do not study this region in detail. We also note that some objects in this diameter range are recognized members of dynamical (Hirayama) families. Some of the families have specific rotation distributions likely related to specific formation conditions of the families (e.g., see *Binzel et al.,* 1989); this can contribute to the non-Maxwellian distribution of spin rates in the range 10 < D < 40 km as well.

The distribution of spin rates of asteroids with 0.15 < D < 10 km is strongly non-Maxwellian (see Fig. 4). There are significant populations of both slow ($f \leq 0.8$ d^{-1}) and fast ($f \geq 7$ d^{-1}) rotators (*Pravec and Harris,* 2000). Several previous investigators have attempted to formally fit the distribution as a sum of Maxwellians, e.g., *Fulchignoni et al.* (1995) and *Donnison and Wiper* (1999). The fits by the outlying Maxwellians, particularly to the slow rotating population, are poor.

It is apparent in Fig. 1 that there is a "barrier" against rotations faster than f ≈ 12 rotations per day in the size range of ~1–10 km diameter. This is further apparent in Fig. 5, where we can also see that among the very fastest rotators, $f > 6$ d^{-1}, there is a tendency toward more spherodial shape with increasing spin rates. These characteristics suggest that most of the small asteroids are loosely bound, gravity-dominated aggregates with negligible tensile strength (*Harris,* 1996; *Pravec and Harris,* 2000). They find that the "barrier" to fast rotation is at f ≈ 11 rotations per day for nearly spherical bodies, corresponding to stability for strengthless bodies of bulk density greater than ~2.5 g/cm³. The maximum spin rate for a strengthless nonspherical body of a given density is less with increasing elongation of the shape, thus the "barrier" shifts to a slower rotation rate with increasing amplitude of lightcurve variation. In February 2001, Pravec et al. find that 1950 DA with D ≈ 1 km has a period of 2.12 h (results available at http://www.asu.cas.cz/~ppravec/). It can still be a strengthless object if its bulk density is >2.9 g/cm³, which is plausible for a silicate body with very little porosity, that is, a "shattered" body rather

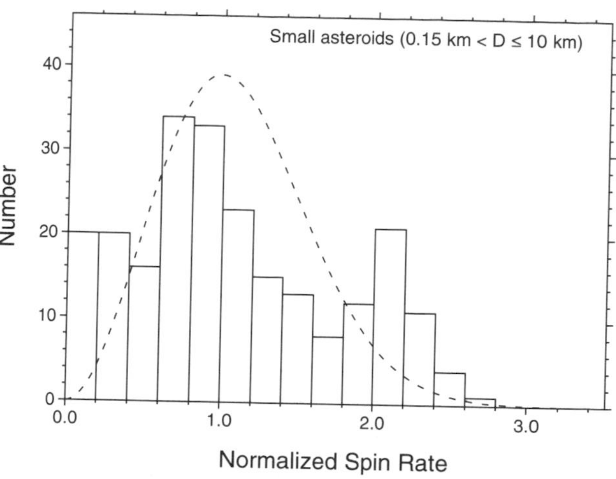

Fig. 4. Histogram of f/⟨f⟩ for 231 asteroids with 0.15 < D ≤ 10 km. The dashed curve is the Maxwellian distribution for the same number of objects. Of them, 164 are NEAs and Mars-crossers; most of the rest are inner main-belt asteroids.

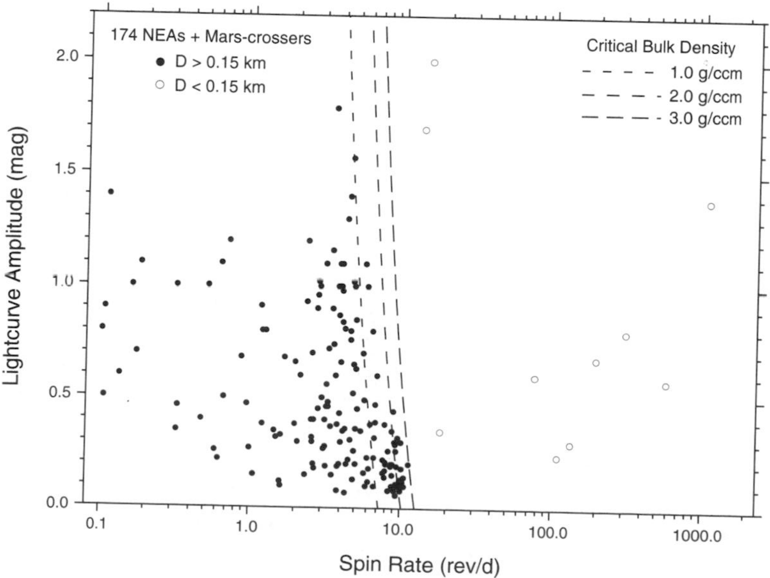

Fig. 5. The observed lightcurve amplitude vs. spin rate of near-Earth and Mars-crossing asteroids. The dashed curves are approximate upper limits of spin rates of bodies held together by self-gravitation only, with bulk densities plausible for asteroids; $f_c = \sqrt{\rho/(1 + \Delta m)}/3.3$ h (ρ in g/cm³) (*Pravec and Harris,* 2000).

than a "rubble pile" [see definitions in *Richardson et al.* (2002)] or one with substantial metal content. Further studies should refine the placement of the "barrier" around our estimate of f ≈ 11–12 rotations per day and give a better understanding of the interior structure of the fastest rotators just below the "barrier."

In October 2001, after the first draft of this chapter was written, *Pravec and Kušnirák* (2001) found that 2001 OE_{84} is a superfast rotator with a period of 29.19 min, thus it must have a nonzero tensile strength (cf. section 2.4). With its estimated size ~0.9 km, it is the first known coherent object in the size range 0.15 < D < 10 km, and the first one that breaks the "barrier." It shows that coherent bodies exist also in the small asteroids group but we cannot estimate their fraction well from the observed statistic of the one superfast rotator among a few hundred asteroids in the group. It is likely an exceptional object, maybe an unusually large "rubble" fragment of a larger asteroid released in a collision. The "barrier" against such fast rotations holds for all other observed asteroids in the group.

The excess of fast rotators at $f/\langle f \rangle \geq 1.7$ (corresponding approximately to f ≥ 7 revolutions per day in the diameter range of 1–10 km where the mean spin rate changes only slightly) lies just below the "barrier" (Fig. 4; the "barrier" in f is less obvious in this figure since it is somewhat smeared there due to the normalization to the computed mean spin rate). The cause of this excess of fast rotators is not quantitatively understood, but may be related to radiation pressure effects (*Rubincam,* 2000; *Bottke et al.,* 2002) or, for planet-crossing bodies, gravitational interactions with planets during close encounters (*Scheeres et al.,* 2000; *Richardson et al.,* 1998). The shapes of at least some of the bodies might be affected by spinup as well. Only asteroids with more spheroidal shapes or those whose shapes were so reconfigured during the spinup process could be spun up to the rates not much below the "barrier." The shape reconfiguration could even be associated with a loss of mass and angular momentum (see below).

A significant fraction of the observed population of fast rotators are actually binary near-Earth asteroids (*Merline et al.,* 2002; also *Pravec and Harris,* 2000, and references therein). Primaries of the observed and suspected binary NEAs are fast rotators with low amplitudes. It is probable that creation and/or evolution mechanisms of the binary NEAs are connected with the fast rotations of the strengthless bodies. A mechanism of tidal breakup of such asteroids during close encounters with the terrestrial planets has been suggested to create binary NEAs (*Bottke and Melosh,* 1996; *Richardson et al.,* 1998). An additional mechanism may be a rotational fission of strengthless asteroids when they are spun up by collisions or non-gravitational effects (see *Paolicchi et al.,* 2002). The reader can find more information on binary NEAs in the chapter by *Merline et al.* (2002).

The origin of the population of slow rotators is unclear. None of mechanisms proposed earlier has explained all observed characteristics of the slow rotator population or its specific members. Radiation pressure effects (*Rubincam,* 2000; *Bottke et al.,* 2002) are ineffective for objects larger than several kilometers. Despinning as a result of outgassing of now-extinct comet nuclei certainly cannot explain large slow rotators like 253 Mathilde or 288 Glauke. Tidal forces from close planetary encounters (e.g., *Scheeres et al.,* 2000) likewise cannot apply to these main-belt objects. Tidal evolution of binaries, such as the process that has led to the synchonization of Pluto and Charon with a period of ~6 d (e.g., *Farinella et al.,* 1981; *Weidenschilling et al.,* 1989) cannot explain periods longer than this, as the rate of transfer of angular momentum from the primary's spin to the secondary's orbit stalls out in the age of the solar system for rocky bodies the size of small asteroids when the satellite recedes out to only about a 5-d period orbit. Furthermore, the very slow rotators Mathilde, Glauke, and Toutatis are proven not binary objects. Recently *Harris* (2002) has fit the distribution of the excess of slow rotators to a function of the form $N(<f) \propto f$, and he offers another explanation for slow rotations — that they result from disintegration of high mass ratio (~1:5) binaries through the rapid transfer of rotational energy of the primary into the orbit of the secondary due to the irregular gravity field of the primary. While results of the model still do not fit the observed cumulative spin rate distribution that appears linearly proportional to f at slow rates, at least it is a plausible mechanism even for the largest slow rotators. Further, more detailed modeling should show if the theory can really fully explain the population of slow rotators.

In general, the rotational properties of asteroids with 0.15 < D < 10 km suggest that they are collisionally derived fragments, mostly with negligible tensile strength. Most of them are rubble piles or shattered bodies according to the definitions of *Richardson et al.* (2002). They mostly gain angular momentum through collisions (see also *Paolicchi et al.,* 2002) but are also affected by noncollisional effects. Their rubble-pile and shattered internal structures are likely a result of collisions that shattered either original parent bodies from which the small asteroids were derived or the small asteroids themselves but did not disperse them (e.g., see *Love and Ahrens,* 1996). We cannot, however, rule out the possibility that the rubble pile structure could be primordial in some cases.

2.4. Very Small Asteroids (D < 0.15 km)

Most asteroids with absolute magnitudes H > 22 (corresponding to D ≤ 0.15 km) rotate with periods of less than 2 h. In fact, all well-derived periods for asteroids below the diameter noted are shorter than that (*Steel et al,* 1997; *Ostro et al.,* 1999; *Pravec et al.,* 2000; *Whiteley et al.,* 2002; C. Hergenrother et al., personal communication, 2001); though optical and radar observations of a few asteroids suggest longer periods, they need to be confirmed by further work. The observed rotations are so fast that the bodies are in a state of tension and cannot be held together by self-gravitation. In Fig. 1, they form a distinct group in the upper left part of the diagram. In Fig. 5, they lie in the right part of the diagram, behind the curves of maximum possible spin rates of bodies held together only by self-gravitation. They could be held together, however, by very meager bonds; even the fastest known rotator, 2000 DO_8, with a period of 1.30 min and a long axis of ≈80 m, has a centrifugal acceleration at the ends of the long axis of only ≈0.26 m/s^2, and the minimum required tensile strength for it is on the order of 2×10^4 Pa, which is $\approx 10^{-3}$ less than the typical tensile strength of well-consolidated rock (see *Ostro et al.,* 1999). While they are sometimes called "monoliths," their internal structure can, however, be almost anything except "true" rubble pile. In the framework of the definitions proposed by *Richardson et al.* (2002), they can lie anywhere except the very left side in their Fig. 1.

The very small, very fast rotating asteroids with nonzero tensile strength are likely collisional fragments of larger asteroids. The transition from larger rubble-pile to smaller "monolithic" asteroids at $D \approx 0.15$ km appears surprisingly sharp. It has been proposed that this is the characteristic size of the largest "rubble" fragments that make up larger asteroids from which the smaller ones are derived (*Pravec et al.,* 2000). No known postformation spinup mechanism is so sensitive to size that it could explain the dramatic change of the rotation properties in such a narrow range around the transition diameter. The apparent truncation of the size range of the observed very fast rotators likely corresponds to the size limit of monolithic fragments from the disruption of larger asteroids. *Whiteley et al.* (2002) show that if there were "monolithic" fragments larger than the observed boundary diameter present in large asteroids, we would observe fast rotating "monoliths" also among asteroids larger than $D \approx 0.15$ km. In other words, we would observe a gradual transition from slower to faster rotations in a wider size range and not the sharp transition in the very narrow range that we see. The apparent threshold size is in fair agreement with results of hydrodynamic computations by *Love and Ahrens* (1996), *Melosh and Ryan* (1997), and *Benz and Asphaug* (1999) that show that asteroids as small as a few hundred meters are gravitationally bound, strengthless rubble piles. The recently discovered superfast rotation and therefore the coherent nature of the ~0.9-km-sized asteroid 2001 OE_{84} (see section 2.3) may be an exception to the scheme described above.

The very small asteroids likely gained fast spins in their creation by impacts on larger asteroids (*Asphaug and Scheeres,* 1999). They were generally in excited rotation states immediately after the creation. No lightcurve aperiodicities that would be indicative of tumbling have been observed in the sample of ten very small asteroids that we analyzed; it appears that they are relaxed to states close to principal-axis rotation. In a larger sample of a few tens of small superfast rotators that have been observed more recently, mostly by C. Hergenrother et al. (personal communication, 2001), there is one tumbling asteroid, 2000 WL_{107}. So, the fraction of excited rotations among superfast rotators with diameters from a few tens to a few hundred meters may be on the order of several percent; a principal axis rotation is a usual state of superfast rotators. Their damping timescales computed from equation (3) range from 10^3 to 10^8 yr, shorter than or comparable to their collisional and dynamical lifetimes of 10^7–10^8 yr (*Farinella et al.,* 1998) if they have been created in the main belt and later delivered to near-Earth orbits. All the very small super-fast rotating asteroids are near-Earth objects. There is a considerable uncertainty, however, in the product μQ (equation (2)); the value $\mu Q = 5 \times 10^{12}$ (CGS units) chosen by *Harris* (1994) may be too small for the bodies that are likely more rigid than rubble piles. Thus, their damping timescales may be actually longer by more than 1 order of magnitude than estimated above. A statistic of their principal vs. nonprincipal axis rotations that can be derived when a much larger sample of the very small asteroids rotations is available may provide a constraint on their material properties. The rotations of the very small asteroids, however, might be influenced by postformation processes. Among them, radiation pressure effects might be particularly effective for the small bodies as they are proportional to the inverse square of the diameter. On the contrary, the bare-rock surface and the fast rotation of the bodies can reduce the effect by more than an order of magnitude (*Rubincam,* 2000). The spins of the very small asteroids can also be affected by collisions. *Farinella et al.* (1998) estimate that the timescale at which a major change of the spin occurs is $\sim 3 \times 10^7$ yr for a main-belt stony asteroid with a diameter of 100 m and a rotation period of 0.5 h; this would be shorter than its collisional lifetime of $\sim 2 \times 10^8$ yr. As a result of these spin-changing mechanisms, the present rotation rates of the very small asteroids may not be relaxed rates from their initial, postformation rotations.

3. SPIN-VECTOR DISTRIBUTION

A review of lightcurve techniques used for estimation of asteroid shapes, sidereal periods, and ecliptic coordinates (λ_p, β_p) of poles is given by *Kaasalainen et al.* (2002).

Magnusson (1986, 1990) analyses the spin vectors of 20 and 30 asteroids, respectively, and concludes that prograde rotating asteroids are in a slight majority. He notes the apparent lack of poles close to the ecliptic plane and attributes this bimodality to a possible observational selection effect; an asteroid with a pole at a low ecliptic latitude may be seen nearly pole-on in some apparitions, giving an amplitude too small for an unambiguous epoch determination. Therefore, more lightcurve observations are needed for such an asteroid than are needed for asteroid with a pole far from the ecliptic plane.

Drummond et al. (1988; 1991), using results for 26 asteroids, confirm the bimodality of the observed pole distribution and concur with its possible explanation due to the observational selection effect. They note, however, that it may be just a statistical fluctuation but also consider the possibility that the observed bimodality may be real and may reflect a primordial distribution of spin rates.

At present we have available a larger sample of asteroid pole estimates. The most complete spin and shape database

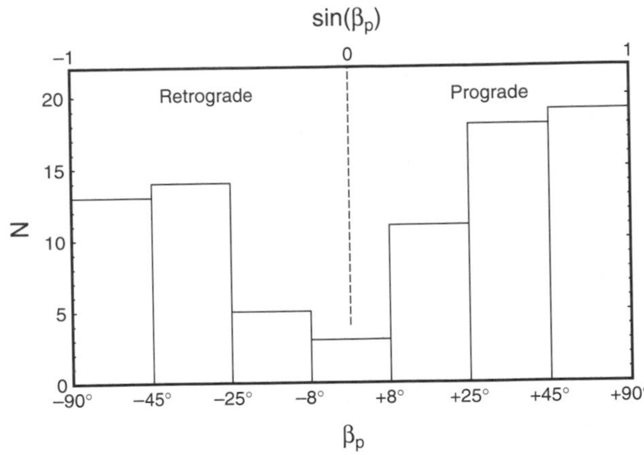

Fig. 6. Distribution of the ecliptic latitudes of north poles of rotations for 83 asteroids.

has been compiled by P. Magnusson and can be found at The Small Bodies Node of the NASA Planetary Data System (http://pdssbn.astro.umd.edu/) or at the Uppsala Observatory's Web site (http://www.astro.uu.se). The database contains results published in numerous papers up to 1995 (see references attached to the database). Additional spin-vector estimates were published by *Michałowski* (1996a,b), *Kryszczyńska et al.* (1996), *Erikson et al.* (1999), and *Michałowski et al.* (1995, 2000, 2001). The present sample contains estimates of spin vectors with senses of rotation for 83 asteroids, which are mostly large and bright asteroids well observed photometrically during at least a few apparitions.

Figure 6 shows the observed distribution of the sine of the ecliptic latitudes of the asteroids' north poles. The distribution is bimodal and not flat as would be expected for a random distribution of the spin vectors; the Kolmogorov-Smirnov test shows that it is not uniform at about 85% confidence level. The lack of asteroids poles close to the ecliptic plane is apparent. There are 48 asteroids with prograde rotations, 32 with retrograde rotations, and only 3 with poles at low ecliptic latitudes ($|\beta_p| < 8°$). The bimodality of the pole distributions, if real, is not understood yet but may be primordial.

4. CONCLUDING REMARKS

Our understanding of asteroid rotations has advanced significantly in recent years, mostly thanks to several productive observational programs that brought a wealth of new data, especially for small asteroids. A new category of very fast spinning small "monolithic" asteroids has been found. The significant population of binary systems with fast spinning primaries was found to be present among near-Earth asteroids. The population of slow rotators is better established and described now than it was 10 years ago. The study of rotation rates and amplitudes of small asteroids reveals that most of them are strengthless bodies with rubble-pile or shattered interior structure. The overall picture of rotational characteristics of asteroids and their relation to other asteroidal properties has been improved. We expect that in years to come further data from lightcurve observations and other techniques will continue supplying new information that will allow us to achieve a better understanding of many aspects of asteroid rotation characteristics and their relations to asteroid evolution processes. In the following paragraph, we mention a few directions of the work that should be particularly interesting.

It is clear that the most interesting information is likely to be gained in fields not explored well enough up to now. Observers should concentrate on inner-planet-crossing objects of all sizes. These observations are likely to lead, in addition to a general increase of the sample allowing better statistical studies, to new detections of NEA binaries and consequently to a better understanding of their characteristics, creation, and evolution; an improved description of the characteristics of the very fast rotating small "monoliths" population and the transition between "monoliths" and rubble piles around D ≈ 0.15 km; and an increase of the sample of slow rotators so that we get new data for testing theories of their creation and evolution. It would be worthwhile to study similarly sized (a few kilometers and smaller) asteroids in the main belt so as to reveal possible differences in their properties from those of NEAs that could shed more light on processes working exclusively or preferentially on objects on inner planet-crossing orbits. Studies of distant asteroids (Trojans, Centaurs, and transneptunian objects) are needed to compare their properties to the population of large main-belt asteroids. We point out the need for each observing program to be designed so as to minimize observational biases against slowly rotating, low-amplitude, and faint objects. Such biases are present in the available sample of asteroid rotations of all sizes and orbits, and every effort to reduce them will improve our knowledge of asteroid rotations and their implications. We anticipate that 10 years from now we will have a much more detailed and complex picture of the asteroid rotational properties over wide ranges of sizes and orbits and that it will allow us to reach a more comprehensive understanding of processes of asteroid evolution.

Acknowledgments. The research at Ondřejov has been supported by the Grant Agency of the Czech Republic, Grant 205-99-0255, by the Grant Agency of the Czech Academy of Sciences, Grant A3003204, and by the Space Frontier Foundation within "The Watch" program. The research at the Jet Propulsion Laboratory, California Institute of Technology, was supported under contract from NASA. T.M. was supported by Polish KBN Grant 2 P03D 007 18.

REFERENCES

Asphaug E. and Scheeres D. J. (1999) Deconstructing Castalia: Evaluating a postimpact state. *Icarus, 139,* 383–386.

Benz W. and Asphaug E. (1999) Catastrophic disruptions revisited. *Icarus, 142,* 5–20.

Binzel R. P., Farinella P., Zappalà V., and Cellino A. (1989) Asteroid rotation rates: Distributions and statistics. In *Asteroids II* (R. P. Binzel et al., eds.), pp. 416–441. Univ. of Arizona, Tucson.

Bottke W. F. and Melosh H. J. (1996) Binary asteroids and the formation of doublet craters. *Icarus, 124,* 372–391.

Bottke W. F., Vokrouhlický D., Rubincam D. P., and Brož M. (2002) The effect of Yarkovsky thermal forces on the dynamical evolution of asteroids and meteoroids. In *Asteroids III* (W. F. Bottke Jr. et al., eds.), this volume. Univ. of Arizona, Tucson.

Burns J. A. and Safronov V. S. (1973) Asteroid nutation angles. *Mon. Not. R. Astron. Soc., 165,* 403–411.

Cellino A., Zappalà V., Davis D. R., Farinella P., and Paolicchi P. (1990) Asteroid collisional evolution. I. Angular momentum splash: Despinning asteroids through catastrophic collisions. *Icarus, 87,* 391–402.

Dobrovolskis A. R. and Burns J. A. (1984) Angular momentum drain: A mechanism for despinning asteroids. *Icarus, 57,* 464–476.

Donnison J. R. and Wiper M. P. (1999) Bayesian statistical analysis of asteroid rotation rates. *Mon. Not. R. Astron. Soc., 302,* 75–80.

Drummond J. D., Weidenschilling S. J., Chapman C. R., and Davis D. R. (1988) Photometric geodesy of main-belt asteroids. II. Analysis of lightcurves for poles, periods, and shapes. *Icarus, 76,* 19–77.

Drummond J. D., Weidenschilling S. J., Chapman C. R., and Davis D. R. (1991) Photometric geodesy of main-belt asteroids. IV. An updated analysis of lightcurves for poles, periods, and shapes. *Icarus, 89,* 44–64.

Efroimsky M. (2001) Relaxation of wobbling asteroids and comets — theoretical problems, perspectives of experimental observations. *Planet. Space Sci., 49,* 937–955.

Erikson A., Berthier J., Denchev P. V., Harris A. W., Ioannou Z., Kryszczyńska A., Lagerkvist C.-I., Magnusson P., Michałowski T., Nathues A., Piironen J., Pravec P., Šarounová L., and Velichko F. (1999) Photometric observations and modelling of the asteroid 85 Io in conjunction with data from an occultation event during the 1995–96 apparition. *Planet. Space Sci., 47,* 327–330.

Farinella P., Paolicchi, P., and Zappalà V. (1981) Analysis of the spin rate distribution of asteroids. *Astron. Astrophys., 104,* 159–165.

Farinella P., Vokrouhlický D., and Hartmann W. K. (1998) Meteorite delivery via Yarkovsky orbital drift. *Icarus, 132,* 378–387.

Fulchignoni M., Barucci M. A., Di Martino M., and Dotto E. (1995) On the evolution of the asteroid spin. *Astron. Astrophys., 299,* 929–932.

Harris A. W. (1979) Asteroid rotation rates II. A theory for the collisional evolution of rotation rates. *Icarus, 40,* 145–153.

Harris A. W. (1994) Tumbling asteroids. *Icarus, 107,* 209–211.

Harris A. W. (1996) The rotation rates of very small asteroids: Evidence for "rubble pile" structure (abstract). In *Lunar and Planetary Science XXVII,* pp. 493–494. Lunar and Planetary Institute, Houston.

Harris A. W. (2002) On the slow rotation of asteroids. *Icarus, 156,* 184–190.

Harris A. W. and Burns J. A. (1979) Asteroid rotation rates I. Tabulations and analysis of rates, pole positions and shapes. *Icarus, 40,* 115–144.

Harris A. W. and Young J. W. (1983) Asteroid rotation IV. 1979 observations. *Icarus, 54,* 59–109.

Kaasalainen M. (2001) Interpretation of lightcurves of precessing asteroids. *Astron. Astrophys., 376,* 302–309.

Kaasalainen M., Mottola S., and Fulchignoni M. (2002) Asteroid models from disk-integrated data. In *Asteroids III* (W. F. Bottke Jr. et al., eds.), this volume. Univ. of Arizona, Tucson.

Kryszczyńska A., Colas F., Berthier J., Michałowski T., and Pych W. (1996) CCD photometry of seven asteroids: New spin axis and shape determinations. *Icarus, 124,* 134–140.

Lagerkvist C.-I., Erikson A., Lahulla F., Di Martino M., Nathues A., and Dahlgren M. (2001) A study of Cybele asteroids. I. Spin properties of ten asteroids. *Icarus, 149,* 191–197.

Love S. G. and Ahrens T. J. (1996) Catastrophic impacts on gravity dominated asteroids. *Icarus, 124,* 141–155.

Magnusson P. (1986) Distribution of spin axes and senses of rotation for 20 large asteroids. *Icarus, 68,* 1–39.

Magnusson P. (1990) Spin vectors of 22 large asteroids. *Icarus, 85,* 229–240.

Melosh H. J. and Ryan E. V. (1997) Asteroids: Shattered but not dispersed. *Icarus, 129,* 562–564.

Merline W. J., Weidenschilling S. J., Durda D. D., Margot J.-L., Pravec P., and Storrs A. D. (2002) Asteroids *do* have satellites. In *Asteroids III* (W. F. Bottke Jr. et al., eds.), this volume. Univ. of Arizona, Tucson.

Michałowski T. (1996a) Pole and shape determination for 12 asteroids. *Icarus, 123,* 456–462.

Michałowski T. (1996b) A new model of the asteroid 532 Herculina. *Astron. Astrophys., 309,* 970–978.

Michałowski T., Velichko F. P., Di Martino M., Krugly Yu. N., Kalashnikov V. G., Shevchenko V. G., Birch P. V., Sears W. D., Denchev P., and Kwiatkowski T. (1995) Models of four asteroids: 17 Thetis, 52 Europa, 532 Herculina, and 704 Interamnia. *Icarus, 118,* 292–301.

Michałowski T., Pych W., Berthier J., Kryszczyńska A., Kwiatkowski T., Boussuge J., Fauvaud S., Denchev P., and Baranowski R. (2000) CCD photometry, spin and shape models of five asteroids: 225, 360, 416, 516, and 1223. *Astron. Astrophys. Suppl., 146,* 471–479.

Michałowski T., Pych W., Kwiatkowski T., Kryszczyńska A., Pravec P., Borczyk W., Erikson A., Wisniewski W., Colas F., and Berthier J. (2001) CCD photometry, spin and shape model of the asteroid 1572 Posnania. *Astron. Astrophys., 371,* 748–752.

Ostro S. J., Pravec P., Benner L. A. M., Hudson R. S., Šarounová L., Hicks M. D., Rabinowitz D. L., Scotti J. V., Tholen D. J., Wolf M., Jurgens R. F., Thomas M. L., Giorgini J. D., Chodas P. W., Yeomans D. K., Rose R., Frye R., Rosema K. D., Winkler R., and Slade M. A. (1999) Radar and optical observations of asteroid 1998 KY_{26}. *Science, 285,* 557–559.

Paolicchi P., Burns J. A., and Weidenschilling S. J. (2002) Side effects of collisions: Spin rate changes, tumbling rotation states, and binary asteroids. In *Asteroids III* (W. F. Bottke Jr. et al., eds.), this volume. Univ. of Arizona, Tucson.

Pravec P. and Harris A. W. (2000) Fast and slow rotation of asteroids. *Icarus, 148,* 12–20.

Pravec P. and Kušnirák P. (2001) *2001 OE_{84}*. IAU Circular 7735.

Pravec P., Hergenrother C., Whiteley R., Šarounová L., Kušnirák P., and Wolf M. (2000) Fast rotating asteroids 1999 TY_2, 1999 SF_{10}, and 1998 WB_2. *Icarus, 147,* 477–486.

Richardson D. C., Bottke W. F., and Love S. G. (1998) Tidal distortion and disruption of Earth-crossing asteroids. *Icarus, 134,* 47–76.

Richardson D. C., Leinhardt Z. M., Melosh H. J., Bottke W. F. Jr., and Asphaug E. (2002) Gravitational aggregates: Evidence and evolution. In *Asteroids III* (W. F. Bottke Jr. et al., eds.), this volume. Univ. of Arizona, Tucson.

Rubincam D. P. (2000) Radiative spin-up and spin-down of small asteroids. *Icarus, 148,* 2–11.

Salo H. (1987) Numerical simulations of collisions between rotating particles. *Icarus, 70,* 37–51.

Samarasinha N. H. and A'Hearn M. F. (1991) Observational and dynamical constraints on the rotation of comet P/Halley. *Icarus, 93,* 194–225.

Scheeres D. J., Ostro S. J., Werner R. A., Asphaug E., and Hudson R. S. (2000) Effects of gravitational interactions on asteroid spin states. *Icarus, 147,* 106–118.

Steel D. I., McNaught R. H., Garradd G. J., Asher D. J., and Taylor A. D. (1997) Near-Earth asteroid 1995 HM: A highly-elongated monolith rotating under tension? *Planet. Space Sci., 45,* 1091–1098.

Tedesco E. F., Veeder G. J., Fowler J. W., and Chillemi J. R. (1992) *The IRAS Minor Planet Survey.* Philips Laboratory Report PL-TR-92-2049, Hanscom AFB, Massachusetts.

Weidenschilling S. J., Paolicchi P., and Zappalà V. (1989) Do asteroids have satellites? In *Asteroids II* (R. P. Binzel et al., eds.), pp. 643–658. Univ. of Arizona, Tucson.

Whiteley R. J., Tholen D. J., and Hergenrother C. W. (2002) Lightcurve analysis of 4 new monolithic fast-rotating asteroids. *Icarus, 157,* 139–154.

Asteroid Photometric and Polarimetric Phase Effects

Karri Muinonen
University of Helsinki and Astronomical Observatory of Torino

Jukka Piironen
University of Helsinki

Yurij G. Shkuratov and Andrej Ovcharenko
Kharkov National University

Beth E. Clark
Cornell University

We review progress in the photometric and polarimetric observations of asteroids, including theoretical and experimental advances in interpreting scattering and absorption of light by asteroid surfaces. Photometric phase effects, such as the opposition effect, the overall phase function, and the so-called amplitude-phase relationship, provide insight into local physical characteristics of the surface, e.g., surface roughness and porosity, and single-particle geometrical and optical properties. Polarimetric phase effects, the negative degree of linear polarization in particular, unveil single-particle properties beyond those derivable from photometry. In order to make progress in the interpretation of the observations, it is obligatory to consider the full vector nature of light, as dictated by Maxwell's equations. We review the primary physical mechanisms for the opposition effect and the negative linear polarization. The coherent backscattering mechanism has been established to contribute to the observed opposition effects and negative linear polarizations. The traditional shadowing mechanism explanation of the opposition effect has been studied via extensive computer simulations. Laboratory photometric and polarimetric experiments with regolith-simulating samples provide an invaluable test bench for theoretical modeling. We review the progress in laboratory work, emphasizing the need for experiments with well-documented samples. We show simultaneous heuristic coherent-backscattering modeling of asteroid photometric and polarimetric phase effects for different asteroid classes. The recent findings imply enhanced future utilization of the phase effects for the benefit of asteroid science.

1. INTRODUCTION

Photometric observations for large numbers of asteroids indicate an opposition effect, a nonlinear increase of disk-integrated brightness at small solar phase angles, the angle between the Sun and the observer as seen from the asteroid (Fig. 1a). Vast sets of asteroid polarimetric observations show negative polarization, a peculiar degree of linear polarization $(I_\perp - I_\parallel)/(I_\perp + I_\parallel)$ for unpolarized incident sunlight: At small phase angles, the disk-integrated brightness component $I_\parallel$ with the electric vector parallel to the scattering plane defined by the Sun, the asteroid, and the observer predominates over the perpendicular component $I_\perp$ (Fig. 1b).

The photometric and polarimetric phase effects of asteroids have a colorful history, with important observational advances in recent years. B. Lyot discovered the negative polarization of (1) Ceres and (4) Vesta in 1934 (*Dollfus et al.,* 1989), while *Gehrels* (1956) discovered the opposition effect for the S-class asteroid (20) Massalia. In the late 1980s, *Harris et al.* (1989b) recorded narrow opposition effects for the bright E class asteroids (44) Nysa and (64) Angelina, a discovery of paramount significance for the overall theoretical interpretation of opposition phenomena. In the 1990s, *Rosenbush et al.* (2002) made polarimetric observations of (64) Angelina, obtaining evidence for a sharp polarization feature similar to that observed for Saturn's rings by *Lyot* (1929). In our terminology, we do not distinguish between the opposition effect and "opposition spike" (narrow opposition effect) and the negative polarization surge and "polarization opposition effect" (sharp negative polarization surge).

During the past two decades, we have seen a crucial advance in theories for the physical causes of the opposition phenomena: The coherent backscattering mechanism (CBM) has been established to contribute to the photometry and polarimetry of asteroids at small solar phase angles (*Shkuratov,* 1985; *Muinonen,* 1989; *Hapke,* 1990; *Mishchenko and Dlugach,* 1993), challenging the traditional shadowing-mechanism (SM) interpretations. CBM is a multiple-scattering mechanism for scattering orders higher than the second (inclusive), whereas SM is a first-order multiple-scattering mechanism. There are several reviews on the mechanisms for the backscattering phenomena of atmosphereless

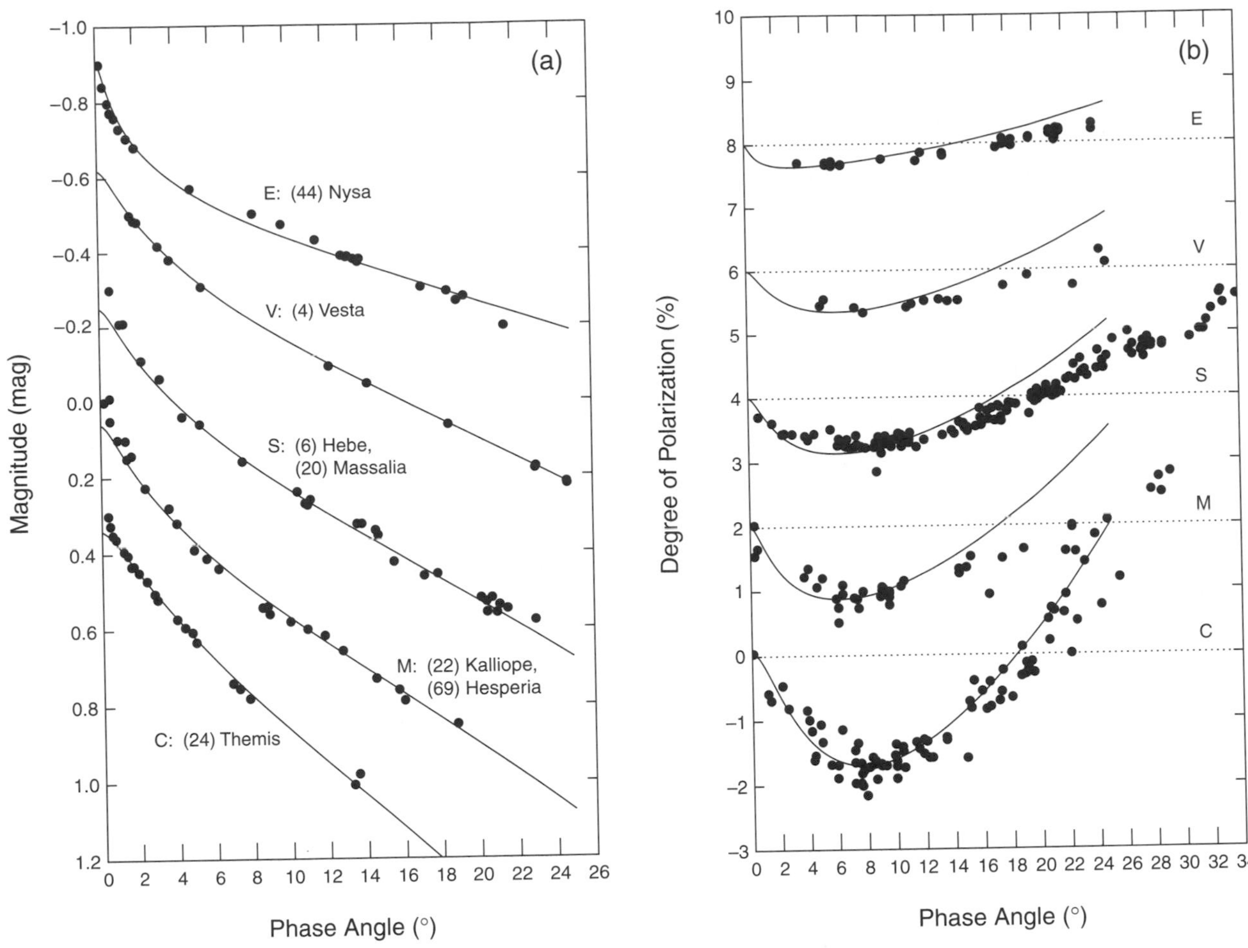

Fig. 1. **(a)** Opposition effect and **(b)** negative linear polarization observations for C-, M-, S-, V-, and E-class asteroids. For illustration, the opposition effects are presented on a relative magnitude scale, and the negative polarizations of M-, S-, V-, and E-class asteroids have been shifted upward by 2, 4, 6, and 8 vertical units respectively. The solid lines illustrate results from heuristic theoretical modeling (section 4). The polarimetric observations are from the Small Bodies Node of the Planetary Data System (http://pdssbn.astro.umd.edu/sbnhtml). The photometric observations are: (44) Nysa, *Harris et al.* (1989b); (4) Vesta, *Gehrels* (1967); (6) Hebe, *Gehrels and Taylor* (1977); (20) Massalia, *Gehrels* (1956); (22) Kalliope, *Scaltriti et al.* (1978); (69) Hesperia, *Poutanen et al.* (1985); and (24) Themis, *Harris et al.* (1989a) (see section 4 for further information on observations and theoretical modeling).

solar system bodies (e.g., *Muinonen,* 1994; *Shkuratov et al.,* 1994), and we refer the reader to these reviews for a more detailed history of, e.g., CBM and SM.

There are no exact electromagnetic solutions for light scattering by what asteroid surfaces are presumably composed of, i.e., close-packed random media of inhomogeneous particles large compared to the wavelength. The related direct problem of computing scattering characteristics for media with well-specified physical properties continues to pose a major challenge. In the inverse problem of deriving information about the physical properties of asteroid surfaces based on the observational data available, it is mandatory to make use of simplified scattering models. For example, the most popular photometric models (e.g., *Hapke,* 1986; *Lumme and Bowell,* 1981) for the inverse problem account for the single-particle albedo and phase function, the volume density (or fraction) of particles, and the roughness of the interface between the scattering medium and the free space. We note that Hapke's model has continued to be the subject of critical debate (e.g., *Mishchenko,* 1994; *Hapke,* 1996), a debate that we cannot enter into in detail. It is widely agreed that separate analyses of disk-integrated photometric or polarimetric phase effects provide two ill-posed inverse problems. However, analyzing the two datasets simultaneously may remove some of the ambiguities.

CBM for the opposition effect is described in Fig. 2a for second-order scattering. An incident electromagnetic plane wave (solid and dashed lines; wavelength λ and wavenumber $k = 2\pi/\lambda$) interacts with two scatterers A and B, which are of the order of the wavelength to hundreds of wavelengths apart, and propagates to the observer to the left. The two scattered wave components due to two opposite propagation directions between the scatterers interfere constructively in the conical directions defined by rotating the light source

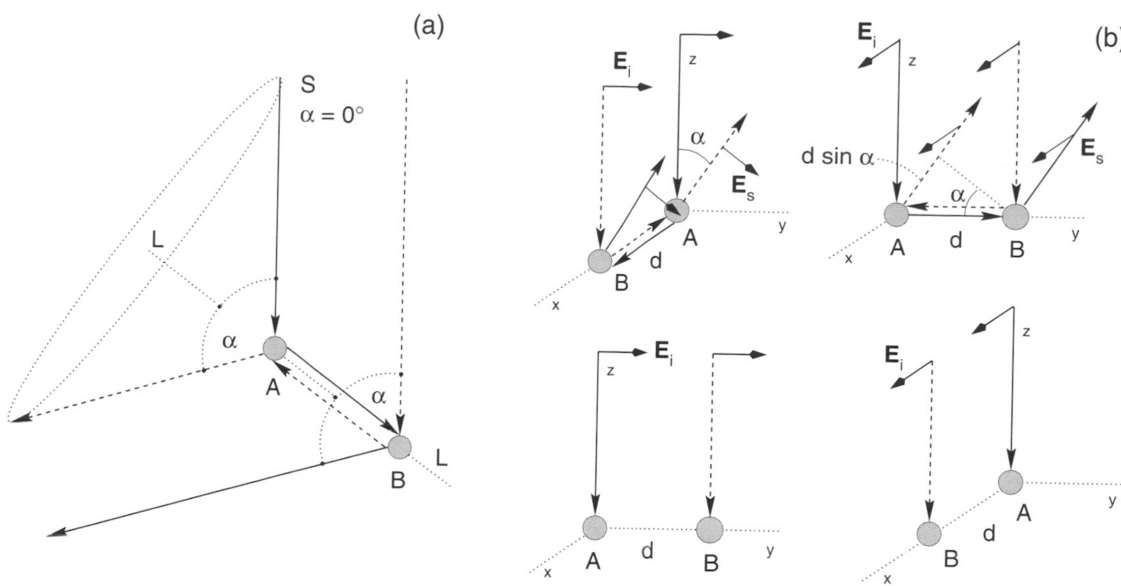

Fig. 2. **(a)** Coherent backscattering mechanism (CBM) for the opposition effect. The multiply scattering electromagnetic wave components propagating in opposite directions between the scatterers (solid and dashed lines) interfere constructively in conical directions about the axis L, always including the exact backscattering direction (phase angle α = 0°). In other directions, the interference is arbitrary depending on the wavelength, and the distance and orientation of the first and last scattering elements. **(b)** Illustration of CBM for the negative linear polarization for unpolarized incident light. In the yz-plane in the scattering geometry leading to positive polarization, the interference depends on the phase difference kd sin α (upper right panel; k is the wave number), whereas the interference is always constructive in the geometry causing negative polarization (upper left panel). Interaction with the polarization vector parallel to the line connecting the elements is typically suppressed (two lower panels) as compared to interaction with the perpendicular polarization vector. Averaging over scatterer locations will result in an opposition effect and net negative polarization (see text).

direction S about the axis L joining the two end scatterers. We illustrate a scattering direction on the cone precisely opposite to the light source direction. Thus, the exact backward direction (phase angle α = 0°) is on the constructive-interference cone for arbitrary locations of the two scatterers, whereas in other directions, interference varies from constructive to destructive. Three-dimensional averaging over scatterer locations results in a backscattering enhancement with decreasing angular width for increasing order of interactions, because the average distance between the end scatterers is larger for higher orders of interactions. The scattering processes can be caused by any disorder or irregularity in the medium.

CBM for the negative degree of linear polarization is explained for second-order scattering in Fig. 2b. The incident radiation is unpolarized by definition, which requires the derivation and proper averaging of the Stokes vectors corresponding to the scattered electromagnetic fields ($\mathbf{E}_s$) for two linear polarization states of an incident plane wave ($\mathbf{E}_i$). In Fig. 2b, incident polarizations parallel and perpendicular to the scattering plane (here referred to as the yz-plane) are treated in the two leftmost and two rightmost panels respectively.

Consequently, in Fig. 2b, an incident electromagnetic plane wave interacts with two scatterers A and B at a distance d from one another aligned either on the x-axis or the y-axis, while the observer is in the yz-plane. For the present geometries, the constructive interference cones of Fig. 2a reduce to the yz- and xz-planes, depending on the alignment of the scatterers. Since first-order scattering is typically positively polarized (e.g., Rayleigh scattering and Fresnel reflection), the scatterers sufficiently far away from each other ($kd = 2\pi d/\lambda \gg 1$) interact predominantly with the electric field vector perpendicular to the plane defined by the source and the scatterers (two upper panels), while interaction with the electric field vector parallel to that plane is suppressed (two lower panels). The observer in the yz-plane will detect negative polarization from the geometry in the upper left panel of Fig. 2, and positive polarization from the geometry in the upper right panel. However, the positive polarization suffers from the phase difference kd sin α, whereas the phase difference for the negative polarization is zero for all phase angles. Averaging over scatterer locations will result in negative polarization near the backward direction. Scattering orders higher than the second experience similar preferential interaction geometries, and contribute to negative polarization. As above for the opposition effect, the contributions from increasing orders of scattering manifest themselves at decreasing phase angles.

In an asteroid's surface, SM is relevant for essentially all length scales much larger than the wavelength of incident light, and is the most striking first-order multiple-scattering effect. We can distinguish between shadowing by the rough interface between the regolith and the free space and shadowing by the internal geometric structure of the regolith. In practice, it can be difficult if not impossible to discriminate between the two shadowing contributions in disk-integrated photometric data. In both cases, SM is due to the fact that a ray of light penetrating into the scattering medium and incident on a certain particle can always emerge back along the path of incidence, whereas in other directions, the emerging ray can be blocked by other particles. The internal SM depends mainly on the volume density of the scattering medium, whereas the interfacial SM depends mainly on surface roughness (here standard deviation of the interfacial slopes). Modern modelings of the internal SM and interfacial SM are offered by, e.g., *Peltoniemi and Lumme* (1992) and *Lumme et al.* (1990) respectively.

We summarize the central questions raised by the observations of photometric and polarimetric phase dependences for asteroids as follows: (1) What are the physical mechanisms responsible for the observed phenomena? (2) What possible knowledge can be derived from the observations? (3) What observations are needed to draw conclusions about the physical mechanisms? (4) What future astronomical observations will serve to progress asteroid science? We note the fact that photometric and polarimetric observations can yield information on the physical characteristics of asteroid surfaces, but not necessarily on asteroid interiors.

The books *Asteroids* and *Asteroids II* include excellent reviews of the photometric, polarimetric, spectroscopic, mineralogical, and petrological advances through the late 1980s. *Bowell et al.* (1989) were already aware of CBM (or interference). In this chapter, we concentrate on photometric and polarimetric phase effects of asteroids. [A companion chapter (*M. Kaasalainen et al.,* 2002) focuses on the derivation of spin state, global shape, and semiempirical scattering parameters from photometric lightcurves of asteroids.] Section 2 summarizes the observational progress in groundbased photometry and polarimetry and spacecraft photometry (there is no spacecraft polarimetry on asteroids). In section 3 we review recent advances in studies of CBM and SM. Recent experimental findings are reviewed with emphasis on joint photometric and polarimetric measurements of samples relevant for asteroid research. In section 4, we present and discuss some heuristic modeling of the photometric and polarimetric phase effects. We close with future prospects in section 5.

2. OBSERVATIONAL PROGRESS

2.1. Groundbased Photometry

A vast number of asteroid photometric lightcurves valuable for phase curve studies have been published by *Harris and Young* (1989), *Harris et al.* (1989a,b), and *Wisniewski et al.* (1997). Among others, these observations are included with references in the *Asteroid Photometric Catalogue (APC)* by *Lagerkvist et al.* (2001), nearly 10,000 lightcurves are included.

The H,G magnitude system of asteroids (*Bowell et al.,* 1989) has been quite successful in general, playing a major role in asteroid science. However, it suffers from certain known drawbacks (e.g., *Lagerkvist and Magnusson,* 1990). The H,G system fails to fit the narrow opposition effects of E-class asteroids (*Harris et al.,* 1989b). Furthermore, it shows poor fits to the phase curves of certain dark asteroids: A linear fit sometimes yields smaller rms errors (e.g., *Piironen et al.,* 1994; *Shevchenko et al.,* 1996). *Verbiscer and Veverka* (1995) also analyzed the H,G magnitude system using Hapke's photometric model, and obtained photometric fits that were as good as those of the H,G system. However, we note that Hapke's photometric model has five parameters, whereas there are only two parameters in the H,G system.

Striking dependence of the photometric phase curve on the asteroid illumination and observation geometry in different apparitions was detected for certain well-observed asteroids by *Piironen et al.* (1994) and *Shevchenko et al.* (1996), yielding apparition-dependent H,G values. For (64) Angelina, such apparition dependence had already been discovered by *Poutanen* (1983). For near-Earth objects, it was observed by *Pravec et al.* (1997), and was observed for (83) Beatrix by *Krugly et al.* (1994). The phenomenon derives from variations in the physical characteristics of the surface or from global slope effects of a very irregular shape. These observations indicate that one has to be careful in analyses of photometric phase curves pertaining to different apparitions. *M. Kaasalainen et al.* (2001) suggest ways to overcome the underlying ambiguities by reducing the phase curves to so-called reference illumination and observation geometries. In addition, the empirical reduction techniques by *Muinonen et al.* (2002a) allow one to concentrate on the opposition-effect enhancement factor and angular width that can be independent of the apparition geometry.

Some of the most extensive observational campaigns of asteroids in recent years have been presented by *Mottola et al.* (1997) and *Magnusson et al.* (1996). The near-Earth asteroid (6489) Golevka (1991 JX) was observed photometrically, radiometrically, and with radar (*Mottola et al.,* 1997). Results showed that this strangely-shaped 300-m asteroid is one of the brightest ever observed (geometric albedo near 0.6). A similar campaign compiled for the near-Earth asteroid (1620) Geographos showed that this asteroid shows little variation in the physical characteristics of its surface; however, its strong lightcurve variations indicate global surface shape variations (*Magnusson et al.,* 1996). These two campaigns showed unequivocally how valuable knowledge of asteroids can be gathered via concerted efforts of several observing sites operating at complementary wavelengths.

Lagerkvist and Magnusson (1990) studied phase curves of asteroids with the help of the H,G magnitude system. Their conclusion was that the S-, C-, and M-class asteroids are well separated into different classes by their G values. Since the early 1990s, the number of asteroids for which phase curves were measured down to subdegree phase angles has increased up to about 20 and thus more than doubled, thanks to long-term observational programs by

Shevchenko et al. (1996, 1997, 2002) and *Piironen et al.* (1994). The main result from these studies is that the small-phase-angle behavior of asteroids is found to depend primarily on geometric albedo. The program by Shevchenko et al. has been devoted to observations of asteroids of various compositions at the smallest possible phase angles. As a result, detailed phase curves down to phase angles of 0.3° or less have been obtained in the V band for asteroids of P, C, G, M, S, and E classes.

Particularly interesting is the analysis of the available observational data on asteroid opposition effects by *Belskaya and Shevchenko* (2000). This survey shows unequivocally that the opposition effect differs for different asteroid taxonomic classes. This is an indication of the importance of composition in determining scattering behavior. *Belskaya and Shevchenko* (2000) found that a ratio of intensities at the phase angles of 0.3° and 5° vs. asteroid geometric albedo reveals a non-monotonic dependence of the opposition effect on albedo. The amplitude of the opposition effect decreases both for dark- and high-albedo asteroids. The largest amplitude occurs with moderate-albedo asteroids of S and M classes, which show pronounced opposition effects starting at phase angles of 5°–7°. The phase angle of 0.3° was chosen for the opposition effect estimation because, at smaller phase angles, very few observations are available. High-albedo E-class asteroids are characterized by steep and narrow opposition phenomena starting at phase angles of about 3° (*Harris et al.,* 1989b). Asteroids with similar geometric albedos show practically identical phase curves. The differences between individual phase curves for high- and moderate-albedo asteroids are on the order of the observational scatter (typically 0.02–0.03 mag).

Low-albedo asteroids show greater diversity in their phase curves. G-class asteroids tend to exhibit wide opposition effects starting at about 6°. C-class asteroids tend to show narrower and shallower effects. P- and F-class asteroids tend to show practically linear phase curves down to phase angles of about 2° (*Belskaya and Shevchenko,* 2000). Overall, results show that low-albedo asteroids with larger U-B colors tend to have stronger opposition effects.

Zappalà et al. (1990) modeled the variation in the amplitude of rotational lightcurves with solar phase angle to produce what is called the amplitude-phase relationship. They applied their model to observations of C-, M-, E-, and S-class asteroids and found that the dependence of the rotational lightcurve amplitude with solar phase is similar for C-, M-, and E-class asteroids, and is steeper for S-class asteroids. In both cases, the lightcurve amplitude increases with increasing phase angle. They found evidence that, as a first approximation, the amplitude-phase relationship was linear within 0°–30° phase angle, challenging the nonlinear dependences showing up in a similar study by *Karttunen and Bowell* (1989). Furthermore, *Karttunen and Bowell* did not report differences in the amplitude-phase relationships for S- and C-class asteroids. *Zappalà et al.* (1990) acknowledge that the discrepancies between the two studies may be due to the small observation set to which they applied their model. However, if it can be shown that the amplitude-phase relationship does map taxonomic class, then this has implications for the similarity of surface characteristics (such as porosity and roughness) within asteroid classes.

In general, extrapolation of photometric phase curves to zero phase angle (or any phase angle) depends on the assumed approximating function, and may lead to inappropriate estimates of the amplitude of the opposition effect. The very same problem is inherent in the estimation of absolute H magnitudes for near-Earth objects from large-phase-angle brightness estimates, and may thus affect the estimated numbers of such objects (see *Jedicke et al.,* 2002).

2.2. Groundbased Polarimetry

Polarimetric observations of asteroids have continued in the years following the publication of *Asteroids II*, adding significantly to the extensive campaign by *Zellner et al.* (1974), *Zellner and Gradie* (1976), and *Le Bertre and Zellner* (1980). Maximum positive polarizations have been reached for (1685) Toro by *Kiselev et al.* (1990) and for (4179) Toutatis by *Mukai et al.* (1997). The polarimetric data are collected at the Small Bodies Node of the Planetary Data System (http://pdssbn.astro.umd.edu/sbnhtml/) (*Lupishko and Vasilyev,* 1997).

Rosenbush et al. (1997) have found evidence that Io, Europa, and Ganymede — Galilean satellites of Jupiter — show negative polarization surges with sharp features at very small phase angles. They review some of the asteroid polarimetric observations, pointing out disagreements in data obtained by different research groups. *Rosenbush et al.* (2002) then provide evidence for the existence of a similar sharp feature in the negative polarization surge of (64) Angelina.

Broglia and Manara (1990, 1992, 1994) and *Broglia et al.* (1994) carried out a polarimetric study of a few large asteroids, searching for polarimetric changes with rotational phase. Such variations were found for most of the objects studied, in particular for (6) Hebe. Polarimetric observations of (6) Hebe by *Migliorini et al.* (1997) confirmed the polarization lightcurve. Because the same study found no changes in spectra at different rotational phases, it is inferred that the regolith structural variations cause the polarization light curve in this case. *Mukai et al.* (1997) found a change of polarization with rotational phase in their study of (4179) Toutatis. They concluded that this S-class asteroid exhibited substantial surface variegations.

Clark et al. (2001) provide a brief review of Earth-based observations of asteroid (433) Eros. One of the intriguing results is that although the surface of Eros exhibits globally different types of terrains, polarimetric observations in the published literature indicate no variations greater than 1 part in 40 (*Zellner and Gradie,* 1976).

Based on the polarimetric observations accumulated since late 1970s, *Lupishko and Mohamed* (1996) updated the empirical correlation rules between first the geometric albedo and the polarimetric slope at the inversion angle, and second the geometric albedo and the minimum polarization. They provided an example case where the polarimetrically determined albedo appeared to be more realistic than the radiometrically determined one. *Cellino et al.* (1999) carried out polarimetric observations of small asteroids to obtain the polarimetric slopes and/or minimum polarizations and thus the geometric albedos that allowed them to assess the accuracy of IRAS geometric albedos for small asteroids. They saw some indications for an overestimation of these IRAS albedos.

The most considerable sets of asteroid polarimetric observations in recent years come from *Lupishko* (1999), *Lupishko and Kiselev* (1995), *Lupishko and Vasilyev* (1997), *Lupishko and Di Martino* (1998), and *Lupishko et al.* (1988, 1991, 1994, 1995, 1999), as well as *Chernova et al.* (1991, 1994). *Kiselev et al.* (1990, 1994, 1996, 1999) published polarization observations of several asteroids, including the near-Earth asteroids (1685) Toro, (1036) Ganymed, (1627) Ivar, and (2100) Ra-Shalom. In the first place, the significance of these observations lies in their quantity and rich phase angle coverage.

It is well known that the main-belt asteroid (4) Vesta shows variations of the negative-polarization degree over its surface (*Degewij et al.,* 1979; *Lupishko et al.,* 1988). New UBVRI polarimetric observations of Vesta (*Lupishko et al.,* 1999) during its rotation period confirmed the variations, showing a record high relative variation of 0.24% in the V-band degree of polarization, inversely correlated with the asteroid lightcurve. In addition, for the first time for asteroids, the observations suggested a variation of polarization-plane position angle with Vesta's rotation, maximum in the U band (8°) and minimum (2.5°) in the I band. During Vesta's rotation period, the polarization degree and position angle vary along a closed cycle, indicating a periodic change connected to Vesta's rotation. The results may be explained by the presence of orderly oriented linear features on the asteroid surface (*Lupishko et al.,* 1999), such as grooves and slopes, related to the giant 460-km crater on Vesta, recently detected by the Hubble Space Telescope (*Zellner et al.,* 1997).

The UBVRI-polarimetric observations of the Apollo asteroid (4179) Toutatis (*Lupishko et al.,* 1995) in the phase angle range of 15.8°–51.4° suggested that the Stokes parameter U (corresponding to the polarizations with the plane of vibration oriented at an angle of 45° and 135° to the scattering plane) differs from zero, when the Stokes parameter Q (corresponding to the polarizations with the plane of vibration parallel to the scattering plane and perpendicular to it) is close to zero. The corresponding values of the position angle of the polarization plane in the proper coordinate system, measured from the axis perpendicular to the scattering plane, also differs from 0° or 90° and is equal to about 45°. This would indicate an aspect of Toutatis' polarization not connected with the scattering plane. Possible causes could derive from the surface heterogeneity and complex shape. However, these observations are not confirmed by the simultaneous observations by *Mukai et al.* (1997). Furthermore, *Rosenbush et al.* (1997) emphasize that there are controversies over whether or not the Stokes U-parameter variations are real.

Kiselev et al. (1994) and *Lupishko et al.* (1995) noticed that, for S-class asteroids, the absolute value of negative polarization measured at the phase angle of 10° increases with wavelength, while the positive polarization (at 40°–90°) displays a clear decrease with wavelength. This indicates an inversion of spectral dependence of S-class asteroid polarization (*Lupishko and Kiselev,* 1995; *Lupishko and Di Martino,* 1998). In contrast, the negative polarization degree of low-albedo asteroids (1) Ceres, (704) Interamnia, and others (*Belskaya et al.,* 1987; *Lupishko et al.,* 1994) decreases with wavelength. This raises the question of whether the positive polarization degree of these asteroids increases with wavelength in a way that is the reverse of that for S-class asteroids. Measurements of polarization for Interamnia at phase angles of 10.6° (negative polarization) and 22.1° (positive, since the zero-crossing angle equals 15.7°) show that the spectral dependence of its polarization is transformed in accordance with the law described above. However, the spectral dependence of positive polarization of C-asteroids (2100) Ra-Shalom at the phase angle of 60° (*Kiselev et al.,* 1999) and (1580) Betulia at 39° (*Tedesco et al.,* 1978) reveals a small reduction of polarization with increasing wavelength.

2.3. Spacecraft Photometry

Since the time of *Asteroids II*, disk-resolved observations by two spacecraft, *Galileo* and *NEAR Shoemaker*, have dramatically increased our understanding of asteroid spectrophotometric properties. The *Galileo* spacecraft flew by asteroids (951) Gaspra and (243) Ida in the early 1990s, obtaining images and spectroscopy. *Helfenstein et al.* (1994, 1996) have performed extensive analyses of these observations based on Hapke's photometric model. Generalized, these analyses provide a detailed examination of the photometric behavior of S-class asteroids. In addition, the Helfenstein et al. work on Gaspra and Ida provided the first quantitative analyses of the distribution and magnitude of asteroid surface color variations at broad-band wavelengths in the visible. These analyses showed that asteroid surface color variations occur on small and large spatial scales, and can be linked with surface processes, such as cratering.

The first of the *NEAR Shoemaker* mission targets, C-class asteroid (253) Mathilde, has been photometrically analyzed by *Clark et al.* (1999). Looking at photometric properties at a broadband wavelength of 0.7 μm, Clark et al. showed that Mathilde is remarkably uniform in albedo and other photometric properties on spatial scales of 1 km or more. The main *NEAR Shoemaker* mission target, S asteroid (433) Eros, has also been analyzed in terms of its photometric and spectrophotometric properties across a wide wavelength range of 0.55–2.4 μm, and a wide phase angle range of 1.2° to 111° (*Clark et al.,* 2002; *Domingue et al.,* 2002; *Veverka et al.,* 1999, 2000).

While most spacecraft studies of asteroid photometric properties have used broadband visible wavelengths, one study in particular has examined narrow-band multiwavelength observations of (433) Eros. Such a multiwavelength approach is useful in isolating model parameter effects because some purely geometric scattering model parameters should not be wavelength-dependent. Specifically, near-infrared spectrometer observations (0.8–2.4 μm) of Eros have been compared with the Hapke photometric model at phase angles ranging from 1.2° to 111.0° and at spatial resolutions of 1.25–5.5 km/spectrum (*Clark et al.,* 2002). A 15% increase in the 1-μm band depth was observed at high phase angles. In contrast, only a 5% increase in continuum slope from 1.5 to 2.4 μm and essentially no difference in the 2-μm band depth were observed at higher phase angles. These contrasting phase effects imply that phase-dependent differences must be accounted for in the computation of parametric measurements of 1- and 2-μm band areas. Whole-disk phase curves derived from the models indicate that Eros exhibits phase reddening of 8–12% across the phase angle range of 0°–100° (Fig. 3). Phase reddening is most severe for wavelengths inside the 1.0- and 2.0-μm silicate absorption features. These results agree in both sense and magnitude with the laboratory study of silicate materials performed by *Gradie et al.* (1980). It is interesting that the phase reddening effects are not accompanied by variations in the single-particle phase functions, implying that single-scattering albedo, and hence multiple scattering, are more important controls of phase reddening than is the single particle phase function.

In the *Clark et al.* (2002) study, two of five Hapke model parameters exhibited a notable wavelength dependence: (1) The single-scattering albedo mimics the spectrum of Eros, and (2) there is a decrease in angular width of the opposition surge with increasing wavelength from 0.8 to 1.6 μm. Such opposition surge behavior is not adequately modeled with the shadow-hiding Hapke model, consistent with coherent backscattering phenomena near zero phase.

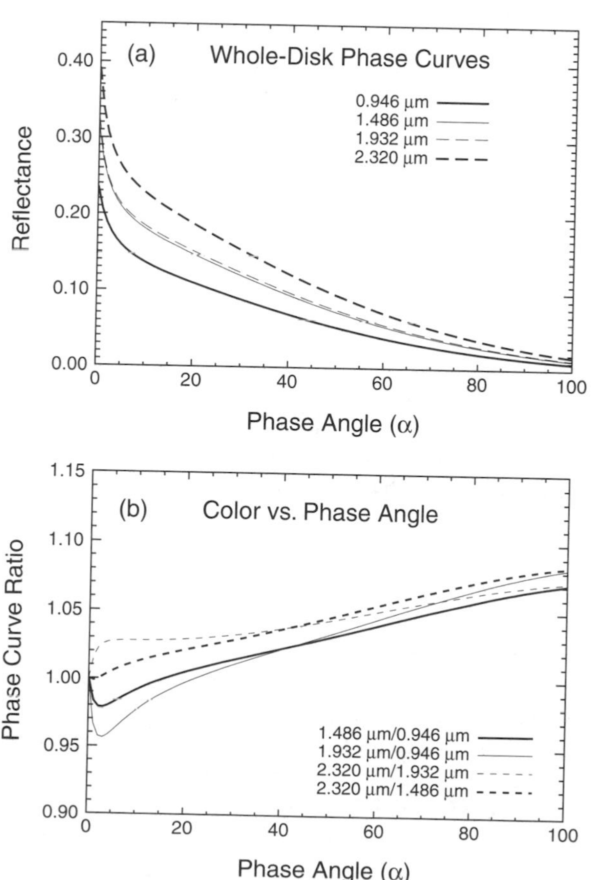

Fig. 3. Disk-integrated multiwavelength opposition effects of asteroid Eros as derived from the disk-resolved observations of *NEAR Shoemaker*.

3. THEORETICAL AND EXPERIMENTAL ADVANCES

Theoretical advances in asteroid photometry and polarimetry concern the fundamental physical understanding of the opposition effect and negative polarization. These advances have been supported by numerous laboratory measurements.

3.1. Coherent Backscattering Mechanism

Based on both theoretical and experimental findings, *Shkuratov* (1985) suggested that the coherent backscattering mechanism is relevant for the opposition effect and negative polarization of atmosphereless solar system bodies. *Muinonen* (1989, 1990) solved the electromagnetic scattering problem of two interacting dipole particles analytically and showed that the coherent backscattering opposition effect is accompanied by negative polarization. He offered a physical explanation independent of the work by *Shkuratov* (1985). Using rigorous T-matrix computations, *Mishchenko* (1996b) showed unequivocally that CBM played a role in light scattering by two interacting spherical particles (cf. *Muinonen,* 1989). *Lumme et al.* (1997) saw tentative clues of CBM in their application of the discrete-dipole approximation to scattering by dense clusters of spherical particles.

Furthermore, *Muinonen et al.* (1991) showed unequivocally that the opposition effect and negative polarization follow naturally when the electromagnetic scattering problem of a dipole particle above a plane-parallel interface of solid optically isotropic and homogeneous medium is treated exactly. Based on the resulting shallow and narrow negative polarizations, they suggested that interactions among the small-scale inhomogeneities are more important than those between small-scale inhomogeneities and surface elements. It is interesting that the morphology of the observed asteroid opposition effects and negative polarizations is already apparent in their example computations. *Muinonen and Lumme* (1991) (see also *Shkuratov,* 1989; *Videen,* 2002) showed that second-order external Fresnel reflection from two spherical particles is large compared to the wavelength revealed signatures of the opposition effect and negative polarization. However, the intensity component due to external Fresnel reflections is only a fraction of the total intensity scattered by an asteroid's surface.

Shkuratov and his colleagues continued their theoretical investigations of the opposition effect and negative polarization (e.g., *Shkuratov,* 1991; *Zubko et al.,* 2002). In addition, computer modeling of ray tracing in particulate media, including simulations in the opposition phenomena context, began (e.g., *Zubko et al.,* 2002; *Shkuratov et al.,* 2002). *Hapke* (1990, 1993) continued theoretical work on scattering of light, with CBM receiving particular attention. *Kolokolova et al.* (1993) provided statistical evidence that CBM was the primary contributor to the negative polarization for media of subwavelength-sized scatterers.

Mishchenko and Dlugach (1993) and *Mishchenko* (1996a) concentrated on modeling CBM for discrete random media of scatterers including polarization effects. The former study offered advanced CBM modeling for the opposition effect of E-class asteroids (see also *Shkuratov and Muinonen,* 1992), while the latter study described extensive computations of various multiple scattering parameters at the exact backscattering direction using the reciprocity principle of electromagnetic scattering. *Mishchenko et al.* (2000) made use of the analytical theories by *Ozrin* (1992) and *Amic et al.* (1997) to compute coherent backscattering by conservative media of Rayleigh scatterers. They were able to offer reference results, that is, accurate predictions for the values of the amplitude and width of the opposition effect and the shape and depth of the negative polarization surge.

Tishkovets (2002) and *Tishkovets et al.* (2002) studied CBM for layer-type and semi-infinite discrete random media of absorbing scatterers including multiple scattering up to the second order. They theorized that the considerable width of the negative polarization surge as compared to the photometric surge is due to the longitudinal component of the electromagnetic near-fields in the discrete scattering medium.

Very recently, *Muinonen* (2002; see also *Muinonen et al.,* 2002b) put forward a computational algorithm for coherent backscattering by absorbing and scattering media. The algorithm mimics radiative transfer but, for each multiple scattering event, it additionally computes the coherent backscattering contribution. The algorithm is currently under further development, and the first results are in accordance with the findings by *Muinonen et al.* (1991), extending and not disagreeing with the reference results by *Mishchenko et al.* (2000).

3.2. Shadowing Mechanism

In their ray-tracing simulations for close-packed random media of spherical particles, *Peltoniemi and Lumme* (1992) noted that the radiative transfer models by *Hapke* (1986) and *Lumme and Bowell* (1981) can be severely in error for close-packed random media of spherical particles. In particular, the error can be considerable, even a factor of several, for grazing emergence of rays.

Stankevich et al. (1999) carried out computer simulations of shadowing for semi-infinite plane-parallel media of close-packed spherical particles. They emphasized that it is nontrivial to generate semi-infinite close-packed media of constant volume density. *Muinonen et al.* (2001) studied SM for clusters of opaque spherical particles. By increasing the number of constituent particles, they revealed the gradual increase of the opposition effect due to shadowing, thus providing supporting theoretical evidence for an SM opposition effect. However, for realistic particle volume fractions, the angular widths of the opposition effects were typically larger than those observed for asteroids.

In his Ph.D. dissertation, *Peltoniemi* (1993a) completed an extensive geometric optics study of scattering of light by planetary regoliths. *Peltoniemi* (1993b) and *Stankevich et al.* (2002) offered three-dimensional Gaussian modeling for the planetary regolith geometry. For such random media, Stankevich et al. studied shadowing in orders higher than the first, concluding that those shadowing contributions are negligible as compared to the first-order one [in agreement with *Esposito* (1979)].

Shepard and Campbell (1998) utilized a fractal surface model in the simulation of light scattering from rough surfaces. They compared their model to Hapke's and discussed the meaning of roughness in problems concerning scattering of light. *Hillier* (1997) put forward a double-layer radiative transfer model to explain the opposition effect using only SM. He concluded that it may not be necessary after all to invoke CBM to explain the observations. The possibility of an additional enhancement of SM was considered by *Shkuratov and Helfenstein* (2001). They hypothesized that a fractallike, hierarchical structure of asteroid surfaces could enhance the SM contribution to the opposition effect.

3.3. New Laboratory Results

Photometric studies by *Oetking* (1966) continue to deserve attention due to the completeness of the choice of samples and intriguing variety of results. The results of this study have been very useful in planning new laboratory experiments of light scattering.

Extensive laboratory photometric and polarimetric studies were carried out by *Shkuratov et al.* (1992) and *Shkuratov and Opanasenko* (1992) at the phase-angle range of 2°–50°. They showed, in particular, that the depth and the inversion angle of negative polarization depend strongly on the microscopic optical inhomogeneity of the surfaces, e.g., a mixture of smoked MgO and soot gives a minimum polarization down to –3.5% and an inversion angle reaching up to 35° (*Shkuratov,* 1987).

Laboratory measurements have been rare at phase angles <1°. Recently, simultaneous polarimetric and photometric measurements were carried out at phase angles covering 0.2°–4° for unpolarized incident light (*Shkuratov et al.,* 1999, 2002; *Ovcharenko and Shkuratov,* 2000). The instrument was calibrated with a comparable photometer at the Jet Propulsion Laboratory (*Nelson et al.,* 1999). Samples with controlled structural characteristics and albedos were studied. Several important results for testing theoretical models were obtained in the laboratory studies by *Shkuratov et al.* (2002).

First, the volume density of the scattering medium can dramatically influence the negative polarization. Figure 4 shows data for coatings of smoked MgO before and after compressing. Due to compression, the angle of minimum polarization shifts from the small phase angle of 1° up to the phase angle range of 5°–10°, changing the negative surge from asymmetric to regular and symmetric. The width of the opposition effect increases along with the compression.

Second, a strong particle-size dependence of the negative polarization for powdered dielectric surfaces was found for particle-size separates of Al_2O_3 studied by *Nelson et al.* (2000). Data for the separates of 0.1, 0.5, 1.0, 3.2, and 12.1 μm are shown in Fig. 4. These separates have almost the same high albedo. For grain sizes larger than 1 μm, the depth of the negative polarization does not exceed 0.25%. The smallest fractions of 0.1 and 0.5 μm reveal similar dependences with the minimum near 1.6° and amplitude of 0.8%. The polarization of the 1.0 μm fraction is located between the fine and coarse powdered samples. Figure 4 demonstrates the systematic evolution of polarimetric phase dependences along with particle size. As for the photometric data, the phase functions of the samples of 0.1, 0.5, 1.0, and 12.1 μm are fairly inert to the particle size. The measurements of Al_2O_3 particle-size separates thus provide an example where the negative polarization is a more sensitive particle-size discriminator than the opposition effect. Third, *Ovcharenko and Shkuratov* (2000) found a MgO-like surge of negative polarization for soot.

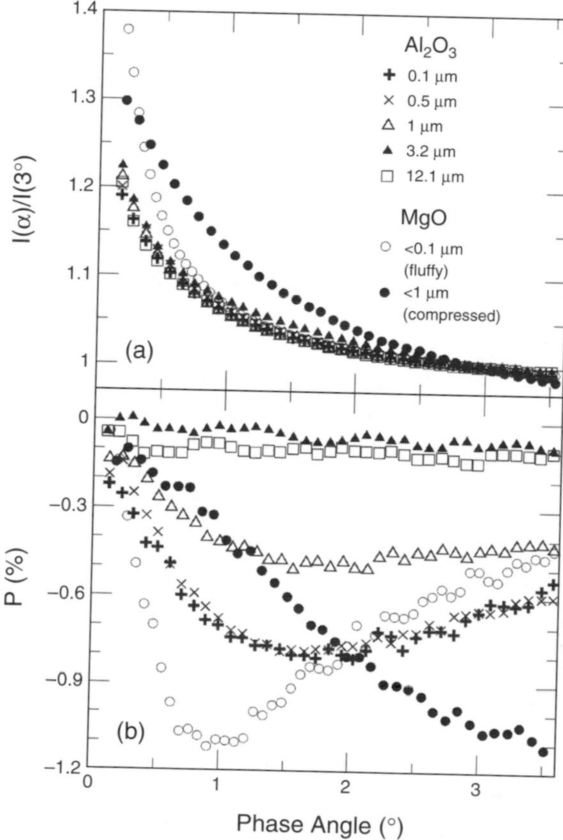

Fig. 4. Phase functions of brightness and polarization degree for plane-parallel coatings of smoked MgO before and after compression, as well as for particle-size separates of Al_2O_3 powder (*Shkuratov et al.*, 2002), in the nadir-imaging geometry.

Hapke et al. (1993, 1998) put forward a hypothesis that experimental measurements for the phase dependences of the so-called linear and circular polarization ratios could discriminate between CBM and SM contributions to the opposition effect of the Moon. In the former work, they concluded that CBM was the primary cause of the opposition effects of atmosphereless solar system bodies, but in the latter, they found evidence for SM contributions. Further theoretical and experimental studies are needed to assess these interesting analyses. *Nelson et al.* (1999, 2000) published small-phase-angle measurements of Al_2O_3 powders. They confirmed a dependence of opposition effect on the size of particles. The maximum opposition effect was seen when the particle sizes were a few times the wavelength. *Shepard and Arvidson* (1999) studied basalt samples and suggested that the opposition surge and negative polarization found in pulverized samples is caused by CBM. Also relevant for asteroid studies, the AMIE microcamera [J-L. Josset, PI; cf. *Muinonen et al.* (2002c)] onboard the ESA SMART-1 mission to the Moon (launch in February 1, 2003) should soon contribute to our general understanding of the opposition effect.

Capaccioni et al. (1990) published numerous laboratory phase curves of terrestrial rocks and meteorites. The laboratory phase curves at moderately large phase angles are qualitatively comparable to the observed phase curves of asteroids. *Piironen et al.* (1998) measured single-particle albedos of meteorites (for particle sizes from a few tens of micrometers upwards) using the device described in *Sasse et al.* (1996). Their results indicated fairly high single-particle albedos that did not show a strong dependence on particle size. *Kamei and Nakamura* (2002) have carried out extensive photometric measurements of various meteorites and other powdered materials. Their estimates of the single-particle albedos were in agreement with those measured by *Piironen et al.* (1998).

McGuire and Hapke (1995) studied scattering of light by large particles in the laboratory. They concluded that large particles are mostly backward scattering (excluding forward diffraction), which is a logical result because their analog particle sizes were on the order of 1 cm. Recently, *Volten et al.* (2001) carried out extensive scattering matrix measurements for mineral aerosol particles, for example, and were able to obtain similar scattering matrices from ray-optics simulations with Gaussian random particles, i.e., stochastically shaped particles described by the lognormal distribution with given mean (radius), standard deviation, and correlation function (e.g., *Muinonen,* 2000). The measurements were then studied further by *Nousiainen et al.* (2002), who were able to improve the ray-optics analyses by *Volten et al.* (2001) and interpret the measurements using a modified ray-optics approximation, where the Gaussian particles are assumed to consist of an optically isotropic and homogeneous medium with both specular and Lambertian surface elements (in addition, Lambertian internal interfaces were assumed). Long-term experimental work on scattering by snow (*Piironen et al.,* 2000) has now been extended by *S. Kaasalainen et al.* (2002) to experimental laboratory work on backscattering characteristics of various regolith-type samples.

4. DISCUSSION

We now return to asteroid photometric and polarimetric observations and theoretical modeling in Fig. 1. As for the photometric observations in Fig. 1a, there are several important details. The observations of (44) Nysa are referred to as the mean light level of the lightcurve, including a correction of 0.0007 mag/° subtracted from the observations for phase angles before the opposition and added to the observations after the opposition [*Harris et al.* (1989b); correction attributed to changing illumination and observation geometry]. The observations of (4) Vesta are referred to as the mean light level (*Gehrels,* 1967), and so are the observations of (20) Massalia (*Gehrels,* 1956) and (24) Themis [*Harris et al.* (1989a), using Fourier analysis]. The observations of (6) Hebe are referred to as the maximum light level (*Gehrels and Taylor,* 1977), and combined from several apparitions using shifts based on an optimum fit using the H,G magnitude system (*Bowell et al.,* 1989); naturally, the shifts should be recomputed here, but for the current purpose of illustration, the H,G shifted dataset will suffice [it overlaps nicely with the data for (20) Massalia]. The observations of (22) Kalliope represent the maximum light level (*Scaltriti et al.,* 1978), as well as the observations of (69) Hesperia (*Poutanen et al.,* 1985). Additionally, the phase curve pairs for S-class asteroids and M-class asteroids have been combined graphically by introducing essentially arbitrary shifts along the magnitude axis. This simple combination technique is appropriate for the simplified theoretical modeling that follows.

The observed photometric and polarimetric phase effects in Fig. 1 provide an immediate constraint on the possible physical mechanisms involved: Laws of physics dictate that the first derivatives at the zero phase angle must vanish. The observations indicate that this vanishing must take place at very small phase angles not covered by the data. To produce such sharp features, the underlying physical mechanism must extend over a large number of wavelengths of incident light. Both SM and CBM are then eligible to explain the photometric observations at small phase angles (Fig. 1a), but SM suffers from the drawback of not being proven to cause negative polarization (Fig. 1b). Furthermore, SM would seem to require unrealistically small particle volume densities to produce sufficiently narrow opposition effects (*Muinonen et al.,* 2001). We thus note that, on one hand, there is so far no theoretical, observational, or experimental evidence against CBM as the mechanism for the opposition effect and negative polarization surge, whereas on the other hand, there is no compelling evidence for SM providing negative polarization.

Figure 1 shows heuristic, perhaps provocative modeling of the opposition effect and negative polarization of C-, M-, S-, V-, and E-class asteroids using Monte Carlo computations for coherent backscattering (*Muinonen,* 2002; *Muinonen et al.,* 2002b). The numerical technique relies on the reciprocity principle for electromagnetic scattering, allowing a renormalization of the coherent backscattering contributions at each multiple-scattering event. The technique is currently available for finite/semi-infinite plane-parallel media and spherical media of scatterers, as well as for semi-infinite plane-parallel media embedded in a semi-infinite optically isotropic and homogeneous host medium. At the present, numerical studies have begun, and further refinements and applications are expected in the future.

Coherent backscattering is here computed for a plane wave (wavelength $\lambda = 2\pi/10$ μm) normally incident on a semi-infinite medium of discrete Rayleigh scatterers with single-scattering albedos of 0.9 and medium mean-free-path length of $\ell = 1, 2, 3, \ldots, 50$ μm. For Rayleigh scatterers, the single-scattering albedo chosen roughly results in a maximum opposition effect amplitude as well as a maximum depth of the negative polarization surge. Each computation used 5×10^4 rays.

In Fig. 1, an ad hoc exponential weighting scheme of $\exp(-\ell/L)$ has been applied to the numerical results in all cases with L = 1.0, 2.0, 2.0, 2.0, and 6.0 μm for the C-, M-, S-, V-, and E-class curves respectively. The weighting scheme derives from the idea that, rather than modeling the random scattering medium using a single mean free path only, we allow a distribution of mean free paths and thus physical characteristics on asteroid surfaces. Additionally, the CBM computations have been multiplied by a linear function $1 - K\alpha$ (e.g., mimicking SM), where K = 0.021/°, 0.017/°, 0.015/°, 0.013/°, and 0.009/° for C-, M-, S-, V-, and E-class curves respectively. Finally, because of the excess depth of the negative polarization surges computed (see below), for comparison with the polarimetric observations, they have been divided by the scaling "fudge factors" of 1.5, 2.3, 3.0, 4.0, and 7.2 in a fashion similar to the comparison of computations and experimental data in *Mishchenko et al.* (2000).

We point out that, as evident in Fig. 1, we cannot obtain excellent fits for both the photometric and polarimetric phase effects: There are problems in simultaneously producing the amplitude of the opposition effect and the width of the negative polarization surge. We have chosen to treat the photometric and polarimetric observations in a fair way by showing equally good or poor fits for both sets of observations. Based on the present computations, we note that the asymmetric polarization surge implied for E-class asteroids is in overall agreement with the observations by *Rosenbush et al.* (2002) and that the best fits are obtained for C-class asteroids (with the smallest polarization scaling factor). The most severe fitting problems are encountered in the photometry and polarimetry of S- and M-class asteroids.

We note that multiple Rayleigh scattering results in negative polarization surges that are too deep and too narrow to fully explain the current observations. We can predict that better agreement with both photometric and polarimetric data can be obtained by incorporating multipole scatterers of degrees higher than the second (Rayleigh scatterers are pure electric dipole scatterers). Such multipole scatterers provide more forward-scattering and less positively polarizing scattering phase matrices, in agreement with what is needed to improve the present modeling. Such phase matrices can exhibit both negative and positive polarization, and their interplay with CBM and the observed negative polarization surges are topics of future research.

Furthermore, there are two primary physical mechanisms that could extend the negative polarization surge toward larger phase angles without affecting the opposition effect significantly at smaller phase angles: (1) diffraction connected to SM and affecting the CBM contributions from well inside a densely packed discrete medium, (2) divergence connected to CBM contributions from inhomogeneities inside otherwise optically isotropic and homogeneous particles of the scattering medium. The second mechanism has already been seen to widen the negative polarization surges (*Muinonen,* 2002).

It is probable that SM is contributing in a moderately nonlinear way to the photometric phase effect, thus improving the agreement between the observations and theoretical computations. The polarimetric scaling factors offer at least a clue that SM cannot be solely responsible for the poorness of the current fits: Shadowing contributions should be smaller for brighter asteroids, whereas the scaling factors are larger for brighter asteroids. One more caveat is still due. We have made use of a single value for the single-scattering albedo, whereas albedo obviously varies among the asteroid classes presently studied. In effect, the scaling factors probably reflect the current use of a fixed single-scattering albedo. Consistent with earlier analyses, the linear photometric slopes (parameter K above) become steeper with darker asteroids whose photometric phase effects, in agreement with decreased multiple scattering, are more strongly affected by internal and interfacial SM.

5. FUTURE PROSPECTS

5.1. Theory and Experiments

The coherent backscattering and shadowing mechanisms appear to be primarily responsible for the observed phase effects (question 1 in section 1). Based on the present discussions, we can indicate a direction for further theoretical studies of the coherent backscattering mechanism: For scattering models to be used in inverse light-scattering problems of asteroids, the coherent backscattering contributions could be computed with the help of electromagnetic field expansions in terms of vector spherical harmonics. Such field expansions yield scattering phase matrices that account for general multipole radiation. How such expansions are related to the physical properties of the scattering media is a fundamental problem of physics that is ripe for development.

We predict that further theoretical and computational progress in light scattering by natural particles will allow a construction of extensive electronic databases of scattering models. Such models can be rapidly applied in the interpretation of asteroid observations, mimicking the readiness of analytical theoretical models such as the Lumme-Bowell and Hapke models, for example. In addition, theoretical models can, in part, be applied in the interpretation of radar, X-ray, and spectroscopic observations of asteroids. During the next decade, we envisage an update to the current H,G magnitude system of asteroids. This update will involve new parameters, new definitions, and our new understanding of the physics involved.

The discussions in section 4 give hints about the possible knowledge (question 2 in section 1) that can be derived from the photometric and polarimetric observations of asteroids. Indeed, it can be promising that the simple modeling was not fully successful: All the observational data together appear to put constraints on the physical characteristics of the single scatterers themselves and the media composed of them.

In laboratory experiments, the physical properties of materials relevant for asteroid studies, such as meteorites, need to be quantitatively measured. Such characterization should include both geometrical and optical properties of the particle surfaces as well as their interiors. In particular, the source of absorption in such materials needs to be assessed. Is the absorption of light due to the optically homogeneous and isotropic host matter, or due to the disorder embedded in it? What are the typical distributions of complex refractive indices? From mineralogical and spectroscopic studies (see *Gaffey et al.,* 2002), we know that these questions are complex and highly wavelength-dependent. It is a challenge for future studies to explain the curious color effects at small phase angles in Fig. 3.

It is probable that wavelength-sized void scatterers are inherent in media of close-packed particles with complicated internal structures. It is intriguing that such void scatterers have single-scattering albedos equal to unity, being potential contributors to coherent backscattering mechanism and the overall brightness in dark media consisting of small particles (e.g., soot).

Additional experiments are needed for coherent backscattering and shadowing by meteorite and other naturally occurring materials. Near-future ICAPS light-scattering experiments in microgravity on board the International Space Station can offer valuable insight (*Blum et al.,* 1999). Where does the main contribution of coherent backscattering come from: from the internal or surface structures of single particles that are large compared to the wavelength, or from interactions between small wavelength-scale particles? Is there a way to distinguish between these two contributions?

How much does diffraction affect the shadowing mechanism? Such experiments will profit from simultaneous theoretical computations.

5.2. Observations

While the recent past has been eye-opening in terms of theoretical advances, we envisage future focus on the observations and the inverse problems they provide. The inverse problems benefit from sophisticated scattering models that should be available soon. Empirical analyses of photometric and polarimetric phase effects could result in viable means for physical classification of asteroids.

For revising the magnitude system for asteroids (question 4 in section 1), it would be important to obtain photometric observations for near-Earth asteroids of various classes in several wavelength bands at wide phase angle ranges, spanning from very large to very small phase angles. The photometric observations could be accompanied by polarimetric observations in order to develop a suitable polarimetric system for asteroids, and contribute to our understanding of the physical mechanisms involved (question 1 in section 1). It is, however, important to realize that developing the magnitude system requires particular caution. As near-Earth asteroids are small, their surface characteristics can differ from those of the generally larger main-belt asteroids, e.g., with small particles missing entirely, giving rise to substantially differing phase curves even for asteroids belonging to the same taxonomic classes.

In the near future, asteroid astrometry and photometry will become intertwingled by, for example, asteroid photocenters differing from asteroid barycenters, and the solar electromagnetic radiation pressure and Yarkovsky thermal emission forces acting on the asteroids (see *Bowell et al.,* 2002). Thus, a full utilization of highly accurate astrometric observations, for example, those planned for the ESA cornerstone mission GAIA (launch around 2011), will benefit from enhanced photometric observation campaigns of large numbers of asteroids (and vice versa).

To move forward with theoretical modeling, we recommend small-phase-angle polarimetric and photometric observations of dark asteroids (question 3 in section 1). Are there indications of sharp but suppressed parts of the opposition effect? Do we see a saturation or even a decrease of the amplitude of negative polarization? Are there sharp features at small phase angles in the polarimetry of S- or M-class asteroids? Narrow-band multiwavelength polarimetric observations would help interpret the color effects in Fig. 3.

As for bright asteroids, we recommend continued polarimetric observations of Angelina and Nysa in order to decisively determine their polarization characteristics. Are there apparition dependences in the negative polarization characteristics? Whether the Stokes parameter U is typically nonzero at the zero-crossing angle of polarization requires further observational work. Reaching an observational consensus on issues like this is crucial for the overall credibility of modern asteroid polarimetry. In essence, we call for a calibration campaign of asteroid polarimetry.

A major breakthrough is still pending in the observationally challenging circular polarimetry of asteroids. It could yield valuable information about an asteroid's spin state, shape, and microscopic physical properties beyond photometry and linear polarimetry.

Acknowledgments. We are grateful to D. Domingue and I. Belskaya for extensive constructive reviews of an early version of the chapter, and A. Cellino for his guidance with the chapter contents. A. Harris (JPL) offered valuable help with the photometric observations, while V. Rosenbush, D. Lupishko, and N. Kiselev helped us review the state of the art in polarimetric observations. We appreciate theoretical remarks by K. Lumme and J. Peltoniemi, experimental comments by A. Nakamura, and observational contributions by V. Shevchenko. G. Videen offered much-respected cricitism on the figures explaining CBM. Additional comments were offered by M. Kaasalainen, T. Nousiainen, and S. Kaasalainen. K.M. is grateful to the Astronomical Observatory of Torino for hospitality during his sabbatical stay and for the Academy of Finland for financial support. This research is supported in part by INTAS Grant No. 2000-0792.

REFERENCES

Amic E., Luck J. M., and Nieuwenhuizen Th. M. (1997) Multiple Rayleigh scattering of electromagnetic waves. *J. Phys. I, 7,* 445–483.

Belskaya I. N. and Shevchenko V. G. (2000) Opposition effect of asteroids. *Icarus, 147,* 94–105.

Belskaya I. N., Lupishko D. F., and Shakhovskoj N. M. (1987) Negative-polarization spectra for five asteroids. *Sov. Astron. Lett., 13,* 219–220.

Blum J., Cabane M., Henning T., Holländer W., Levasseur-Regourd A.-C., Lumme K., Marijnissen J., Muinonen K., Poppe T., Prodi F., Wagner P., and Worms J.-C. (1999) Research with small particles onboard the ISS. In *2nd European Symposium on Utilisation of the International Space Station* (A. Wilson, ed.), pp. 285–289. ESA SP-433.

Bowell E., Hapke B., Dominque D., Lumme K., Peltoniemi J. I., and Harris A. W. (1989) Application of photometric models to asteroids. In *Asteroids II* (R. P. Binzel et al., eds.), pp. 524–556. Univ. of Arizona, Tucson.

Bowell E., Virtanen J., Muinonen K., and Boattini A. (2002) Asteroid orbit computation. In *Asteroids III* (W. F. Bottke Jr. et al., eds.), this volume. Univ. of Arizona, Tucson.

Broglia P. and Manara A. (1990) A search for rotational variations in the optical polarization of 3 Juno and 7 Iris. *Astron. Astrophys., 237,* 256–258.

Broglia P. and Manara A. (1992) A study of the polarimetric lightcurve of the asteroid 16 Psyche. *Astron. Astrophys., 257,* 770–772.

Broglia P. and Manara A. (1994) Polarimetric observations of 51 Nemausa during its 1991 apparition. *Astron. Astrophys., 281,* 576–578.

Broglia P., and Manara A., and Farinella P. (1994) Polarimetric observations of (6) Hebe. *Icarus, 109,* 204–209.

Capaccioni F., Cerroni P., Barucci M. A., and Fulchignoni M. (1990) Phase curves of meteorites and terrestrial rocks: Laboratory measurements and application to asteroids. *Icarus, 83,* 325–348.

Cellino A., Gil Hutton R., Tedesco E. F., Di Martino M., and Brunini A. (1999) Polarimetric observations of small asteroids: Preliminary results. *Icarus, 138,* 129–140.

Chernova G. P., Lupishko D. F., Shevchenko V. G., Kiselev N. N., and Saljes R. (1991) Photometry and polarimetry of asteroid 47 Aglaja. *Kin. Phys. Neb. Tel., 7,* 20–26 (in Russian).

Chernova G. P., Lupishko D. F., and Shevchenko V. G. (1994) Photometry and polarimetry of asteroid 24 Themis. *Kin. Phys. Neb. Tel., 10,* 45–49.

Clark B. E., Veverka J., Helfenstein P., Thomas P. C., Bell J. F. III, Harch A., Robinson M. S., Murchie S. L., McFadden L. A., and Chapman C. R. (1999) NEAR photometry of asteroid 253 Mathilde. *Icarus, 140,* 53–65.

Clark B. E., Lucey P., Helfenstein P., Bell J. F. III, Peterson C., Veverka J., McConnochie T., Robinson M., Bussey B., Murchie S., Izenberg N., and Chapman C. (2001) Space weathering on Eros: Constraints from albedo and spectral measurements of Psyche crater. *Meteoritics & Planet. Sci., 36,* 1617–1638.

Clark B. E., Helfenstein P., Bell J. F. III, Peterson C., Veverka J., Izenberg N. I., Domingue D., Wellnitz D., and McFadden L. (2002) NEAR infrared spectrometer photometry of asteroid 433 Eros. *Icarus, 155,* 189–204.

Degewij J., Tedesco E. F., and Zellner B. (1979) Albedo and color contrasts on asteroid surfaces. *Icarus, 40,* 364–374.

Dollfus A., Wolff M., Geake J. E., Lupishko D. F., and Dougherty L. M. (1989) Photopolarimetry of asteroids. In *Asteroids II* (R. P. Binzel et al., eds.), pp. 594–616. Univ. of Arizona, Tucson.

Domingue D. L., Robinson M., Carcich B., Joseph J., Thomas P., and Clark B. E. (2002) Disk-integrated photometry of 433 Eros. *Icarus, 155,* 205–219.

Esposito L. W. (1979) Extensions to the classical calculation of the effect of mutual shadowing in diffuse reflection. *Icarus, 39,* 69–80.

Gaffey M. J., Cloutis E. A., Kelley M. S., and Reed K. L. (2002) Mineralogy of asteroids. In *Asteroids III* (W. F. Bottke Jr. et al., eds.), this volume. Univ. of Arizona, Tucson.

Gehrels T. (1956) Photometric studies of asteroids. V. The lightcurve and phase function of 20 Massalia. *Astrophys. J., 123,* 331–338.

Gehrels T. (1967) Minor planets. I. The rotation of Vesta. Photometric studies of asteroids. *Astron. J., 72,* 929–938.

Gehrels T. and Taylor R. C. (1977) Minor planets and related objects. XXII. Phase functions for (6) Hebe. *Astron. J., 82,* 229–237.

Gradie J., Veverka J., and Buratti B. (1980) The effects of scattering geometry on the spectrophotometric properties of powdered material. *Proc. Lunar Planet. Sci. Conf. 11th,* pp. 799–815.

Hapke B. (1986) Bidirectional reflectance spectroscopy: 4. The extinction coefficient and the opposition effect. *Icarus, 67,* 264–280.

Hapke B. (1990) Coherent backscatter and the radar characteristics of outer planet satellites. *Icarus, 88,* 407–417.

Hapke B. (1993) *Theory of Reflectance and Emittance Spectroscopy.* Cambridge Univ., New York. 450 pp.

Hapke B. (1996) Are planetary regolith particles back scattering? Response to a paper by M. Mishchenko. *J. Quant. Spectr. Rad. Transfer, 55,* 837–848.

Hapke B. W., Nelson R. M., and Smythe W. D. (1993) The opposition effect of the Moon: The contribution of coherent backscatter. *Science, 260,* 509–511.

Hapke B., Nelson R., and Smythe W. (1998) The opposition effect of the Moon: Coherent backscatter and shadow hiding. *Icarus, 133,* 89–97.

Harris A. W. and Young J. W. (1989) Asteroid lightcurve observations from 1979–1981. *Icarus, 81,* 314–364.

Harris A. W., Young J. W., Bowell E., Martin L. J., Millis R. L., Poutanen M., Scaltriti F., Zappalà V., Schober H.-J., Debehogne H., and Zeigler K. W. (1989a) Photoelectric observations of asteroids 3, 24, 60, 261, and 863. *Icarus, 77,* 171–186.

Harris A. W., Young J. W., Contreiras L., Dockweiler T., Belkora L., Salo H., Harris W. D., Bowell E., Poutanen M., Binzel R. P., Tholen D. J., and Wang S. (1989b) Phase relations of high albedo asteroids: The unusual opposition brightening of 44 Nysa and 64 Angelina. *Icarus, 81,* 365–374.

Helfenstein P., Veverka J., Thomas P. C., Simonelli D. P., Lee P., Klaasen K., Johnson T. V., Breneman H., Head J. W., Murchie S., Fanale F., Robinson M., Clark B., Granahan J., Garbeil H., McEwen A. S., Kirk R. L., Davies M., Neukum G., Mottola S., Wagner R., Belton M., Chapman C., and Pilcher C. (1994) Galileo photometry of asteroid 951 Gaspra. *Icarus, 107,* 37–60.

Helfenstein P., Veverka J., Thomas P. C., Simonelli D. P., Klaasen K., Johnson T. V., Fanale F., Granahan J., McEwen A. S., Belton M., and Chapman C. (1996) Galileo photometry of asteroid 243 Ida. *Icarus, 120,* 48–65.

Hillier J. K. (1997) Shadow-hiding opposition surge for a two-layer surface. *Icarus, 128,* 15–27.

Jedicke R., Larsen J., and Spahr T. (2002) Observational selection effects in asteroid surveys. In *Asteroids III* (W. F. Bottke Jr. et al., eds.), this volume. Univ. of Arizona, Tucson.

Kaasalainen M., Torppa J., and Muinonen K. (2001) Optimization methods for asteroid lightcurve inversion. II. The complete inverse problem. *Icarus, 153,* 37–51.

Kaasalainen M., Mottola S., and Fulchignoni M. (2002) Asteroid models from disk-integrated data. In *Asteroids III* (W. F. Bottke Jr. et al., eds.), this volume. Univ. of Arizona, Tucson.

Kaasalainen S., Piironen J., Muinonen K., Karttunen H., Peltoniemi J. I., and Näränen J. (2002) Experiments of backscattering from regolith-type samples. In *Electromagnetic and Light Scattering by Non-Spherical Particles, Gainesville 2002* (B. Gustafson et al., eds.), pp. 143–146. Army Research Laboratory, Adelphi, Maryland.

Kamei A. and Nakamura A. M. (2002) Laboratory study of bidirectional reflectance of powdered surfaces: On the asymmetry parameter of asteroid photometric data. *Icarus,* in press.

Karttunen H. and Bowell E. (1989) Modelling asteroid brightness variations II. The uninterpretability of light curves and phase curves. *Astron. Astrophys., 208,* 320–326.

Kiselev N. N., Lupishko D. F., Chernova G. P., and Shkuratov Yu. G. (1990) Polarimetry of asteroid 1685 Toro. *Kin. Phys. Neb. Tel., 6,* 77–82 (in Russian).

Kiselev N. N., Chernova G. P., and Lupishko D. F. (1994) Polarimetry of asteroids 1036 Ganymed and 1627 Ivar. *Kin. Phys. Neb. Tel., 10,* 35–39 (in Russian).

Kiselev N. N., Shakhovskoy N. M., and Efimov Yu. S. (1996) On the polarization opposition effect of E-type asteroid 64 Angelina. *Icarus, 120,* 408–411.

Kiselev N. N., Rosenbush V. K., and Jockers K. (1999) Polarimetry of asteroid 2100 Ra-Shalom at large phase angle. *Icarus, 140,* 464–466.

Kolokolova L. O., Mishchenko M. I., and Wolff M. (1993) On the negative polarization of light scattered by subwavelength regolith grains. *Mon. Not. R. Astron. Soc., 260,* 550–552.

Krugly Yu. N., Shevchenko V. G., Velichko F. P., Bowell E., Piironen J., Kwiatkowski T., Kryszczyńska A., and Michałowski T. (1994) Asteroid 83 Beatrix — photometry and model. *Astron. Astrophys. Suppl. Ser., 108,* 143–149.

Lagerkvist C.-I. and Magnusson P. (1990) Analysis of asteroid lightcurves. II. Phase curves in a generalized HG-system. *Astron. Astrophys. Suppl. Ser., 78,* 519–532.

Lagerkvist C.-I., Piironen J., and Erikson A., eds. (2001) *Asteroid Photometric Catalogue,* 5th edition. Astronomical Observatory, Uppsala University, Sweden. 201 pp.

Le Bertre T. and Zellner B. (1980) Surface texture of Vesta from optical polarimetry. *Icarus, 43,* 172–180.

Lumme K. and Bowell E. (1981) Radiative transfer in the surfaces of atmosphereless bodies. I. Theory. *Astron. J., 86,* 1694–1704.

Lumme K., Peltoniemi J. I., and Irvine W. M. (1990) Diffuse reflection in stochastically bounded semi-infinite media. *Trans. Theory Stat. Phys., 19,* 317–332.

Lumme K., Rahola J., and Hovenier J. W. (1997) Light scattering by dense clusters of spheres. *Icarus, 126,* 455–469.

Lupishko D. F. (1999) Photometry and polarimetry of asteroids: Results of observations and data analysis. Doctor's thesis, Kharkov National University. 259 pp.

Lupishko D. F. and Di Martino M. (1998) Physical properties of near-Earth asteroids. *Planet. Space Sci., 46,* 47–74.

Lupishko D. F. and Kiselev N. N. (1995) Inversion effect of spectral dependences of asteroid polarization (abstract). *Bull. Am. Astron. Soc., 27,* 1064.

Lupishko D. F. and Mohamed R. A. (1996) A new calibration of the polarimetric albedo scale of asteroids. *Icarus, 119,* 209–213.

Lupishko D. F. and Vasilyev S. V. (1997) Asteroid polarimetric database. *Kin. Phys. Neb. Tel., 13,* 17–23.

Lupishko D. F., Belskaya I. N., and Kvaratshelia O. I. (1988) Polarimetry of Vesta in 1986 opposition. *Astron. Vestnik, XXII:2,* 142–146.

Lupishko D. F., Efimov Yu. S., and Shakhovskoy N. M. (1991) Ceres' peculiar polarization (abstract). In *23rd Annual Meeting of the DPS of the AAS,* p. 184. Palo Alto, California.

Lupishko D. F., Kiselev N. N., Chernova G. P., Shakhovskoj N. M., and Vasilyev S. V. (1994) Polarization phase dependences of asteroids 55 Pandora and 704 Interamnia. *Kin. Phys. Neb. Tel., 10,* 40–44.

Lupishko D. F., Vasilyev S. V., Efimov Yu. S., and Shakhovskoj N. M. (1995) UBVRI-polarimetry of asteroid 4179 Toutatis. *Icarus, 113,* 200–205.

Lupishko D. F., Efimov Yu. S., and Shakhovskoj N. M. (1999) Position angle variations of the polarization plane of asteroid 4 Vesta. *Solar System Res., 33,* 45–48.

Lyot B. (1929) Recherches sur la polarisation de la lumière des planètes et de quelques substances terrestres. *Ann. Obs. Paris, 8(1),* 1–161.

Magnusson P., Dahlgren M., Barucci M. A., Jorda L., Binzel R. P., Slivan S. M., Blanco C., Riccioli D., Buratti B. J., Colas F., Berthier J., De Angelis G., Di Martino M., Dotto E., Drummond J. D., Fink U., Hicks M., Grundy W., Wisniewski, W., Gaftonyuk N. M., Geyer E. H., Bauer T., Hoffmann M., Ivanova V., Komitov B., Donchev Z., Denchev P., Krugly Yu. N., Velichko F. P., Chiorny V. G., Lupishko D. F., Shevchenko V. G., Kwiatkowski T., Kryszczyńska A., Lahulla J. F., Licandro J., Mendez O., Mottola S., Erikson A., Ostro S. J., Pravec P., Pych W., Tholen D. J., Whiteley R., Wild W. J., Wolf M., and Šarounová L. (1996) Photometric observations and modeling of asteroid 1620 Geographos. *Icarus, 123,* 227–244.

McGuire A. F. and Hapke B. W. (1995) An experimental study of light scattering by large, irregular particles. *Icarus, 113,* 134–155.

Migliorini F., Manara A., Scaltriti F., Farinella P., Cellino A., and Di Martino M. (1997) Surface properties of (6) Hebe: A possible parent body of ordinary chondrites. *Icarus, 128,* 104–113.

Mishchenko M. I. (1994) Asymmetry parameters of the phase function for densely packed scattering grains. *J. Quant. Spectr. Rad. Transfer, 52,* 95–110.

Mishchenko M. I. (1996a) Diffuse and coherent backscattering by discrete random media — I. Radar reflectivity, polarization ratios, and enhancement factors for a half-space of polydisperse, nonabsorbing and absorbing spherical particles. *J. Quant. Spectr. Rad. Transfer, 56,* 673–702.

Mishchenko M. I. (1996b) Coherent backscattering by two-sphere clusters. *Opt. Lett., 21,* 623–625.

Mishchenko M. I. and Dlugach J. M. (1993) Coherent backscatter and the opposition effect for E-type asteroids. *Planet. Space Sci., 41,* 173–181.

Mishchenko M. I., Luck J.-M., and Nieuwenhuizen T. M. (2000) Full angular profile of the coherent polarization opposition effect. *J. Opt. Soc. Am., A17,* 888–891.

Mottola S., Erikson A., Harris A. W., Hahn G., Neukum G., Buie M. W., Sears W. D., Harris A. W., Tholen D. J., Whiteley R. J., Magnusson P., Piironen J., Kwiatkowski T., Borczyk W., Howell E. S., Hicks M. D., Fevig R., Krugly Yu. N., Velichko F. P., Chiorny V. G., Gaftonyuk N. M., Di Martino M., Pravec P., Šarounová L., Wolf M., Worman W., Davies J. K., Schober H.-J., and Pych W. (1997) Physical model of near-earth asteroid 6489 Golevka (1991 JX) from optical and infrared observations. *Astron. J., 114,* 1234–1245.

Muinonen K. (1989) Electromagnetic scattering by two interacting dipoles. In *Proc. 1989 URSI Electromagnetic Theory Symposium,* pp. 428–430. Stockholm, Sweden.

Muinonen K. (1990) Light scattering by inhomogeneous media: Backward enhancement and reversal of polarization. Ph.D. thesis, University of Helsinki.

Muinonen K. (1994) Coherent backscattering by solar system dust particles. In *Asteroids, Comets and Meteors 1993* (A. Milani et al., eds.), pp. 271–296. Kluwer, Dordrecht.

Muinonen K. (2000) Light scattering by stochastically shaped particles. In *Light Scattering by Nonspherical Particles: Theory, Measurements, and Applications* (M. I. Mishchenko et al., eds.), pp. 323–352. Academic, San Diego.

Muinonen K. (2002) Coherent backscattering by absorbing and scattering media. In *Electromagnetic and Light Scattering by Non-Spherical Particles, Gainesville 2002* (B. Gustafson et al., eds.), pp. 223–226. Army Research Laboratory, Adelphi, Maryland.

Muinonen K. and Lumme K. (1991) Light scattering by solar system dust: The opposition effect and the reversal of linear polarization. In *Origin and Evolution of Dust in the Solar System* (A.-C. Levasseur-Regourd and H. Hasekawa, eds.), pp. 159–162. IAU Colloquium 126, Kluwer, Dordrecht.

Muinonen K. O., Sihvola A. H., Lindell I. V., and Lumme K. A. (1991) Scattering by a small object close to an interface II. Study of backscattering. *J. Opt. Soc. Am., A8,* 477–482.

Muinonen K., Stankevich D., Shkuratov Yu. G., Kaasalainen S., and Piironen J. (2001) Shadowing effect in clusters of opaque spherical particles. *J. Quant. Spectr. Rad. Transfer, 70,* 787–810.

Muinonen K., Piironen J., Kaasalainen S., and Cellino A. (2002a) Asteroid photometric and polarimetric phase curves: Empirical modeling. *Mem. S. A. It.,* in press.

Muinonen K., Videen G., Zubko E., and Shkuratov Yu. (2002b) Numerical techniques for coherent backscattering by random media. In *Cosmic Dust and Its Optics: Bratislava Contributions* (G. Videen and M. Kocifaj, eds.), pp. 261–282. Kluwer, Dordrecht.

Muinonen K., Shkuratov Yu. G., Ovcharenko A., Piironen J., Stankevich D., Miloslavskaya O., Kaasalainen S., and Josset J.-L. (2002c) The SMART-1 AMIE experiment: Implication to the lunar opposition effect. *Planet. Space Sci.,* in press.

Mukai T., Iwata T., Kikuchi S., Hirata R., Matsumura M., Nakamura Y., Narusawa S., Okazaki A., Seki M., and Hayashi K. (1997) Polarimetric observations of 4179 Toutatis in 1992/1993. *Icarus, 127,* 452–460.

Nelson R., Hapke B., Smythe W., Shkuratov Yu., Ovcharenko A., and Stankevich D. (1999) The reflectance phase curves at very small phase angle: A comparative study of two goniometers (abstract). In *Lunar and Planetary Science XXX,* Abstract #2068. Lunar and Planetary Institute, Houston (CD-ROM).

Nelson R. M., Hapke B. W., Smythe W. D., and Spilker L. J. (2000) The opposition effect in simulated planetary regoliths. Reflectance and circular polarization ratio change at small phase angle. *Icarus, 147,* 545–558.

Nousiainen T., Muinonen K., and Räisänen P. (2002) Scattering of light by large Saharan dust particles in a modified ray-optics approximation. *J. Geophys. Res.,* in press.

Oetking P. (1966) Photometric studies of diffusely reflecting surfaces with applications to the brightness of the Moon. *J. Geophys. Res., 71,* 2505–2513.

Ovcharenko A. A. and Shkuratov Yu. G. (2000) Weak-localization effect for light backscattered by surfaces with a complex structure. *Opt. Spectrosc., 88,* 253–259.

Ozrin V. D. (1992) Exact solution for coherent backscattering of polarized light from a random medium of Rayleigh scatterers. *Waves Random Med., 2,* 141–164.

Peltoniemi J. I. (1993a) Light scattering in planetary regoliths and cloudy atmospheres. Ph.D. thesis, University of Helsinki, Finland.

Peltoniemi J. I. (1993b) Radiative transfer in stochastically inhomogeneous media. *J. Quant. Spectr. Rad. Transfer, 50,* 655–671.

Peltoniemi J. I. and Lumme K. (1992) Light scattering by closely packed particulate media. *J. Opt. Soc. Am., A9,* 1320–1326.

Piironen J., Bowell E., Erikson A., and Magnusson P. (1994) Photometry of eleven asteroids at small phase angles. *Astron. Astrophys. Suppl. Ser., 106,* 587–595.

Piironen J., Muinonen K., Nousiainen T., Sasse C., Roth S., and Peltoniemi J. I. (1998) Albedo measurements on meteorite particles. *Planet. Space Sci., 46,* 937–943.

Piironen J., Muinonen K., Keränen S., Karttunen H., and Peltoniemi J. I. (2000) Backscattering of light by snow: Field measurements. In *Observing Land From Space: Science, Customers and Technology, Advances in Global Change Research 4* (M. M. Verstraete et al., eds.), pp. 219–229. Kluwer, Dordrecht.

Poutanen M. (1983) UBV photometry of asteroid 64 Angelina. In *Asteroids, Comets, Meteors* (C.-I. Lagerkvist and H. Rickman, eds.), pp. 45–48. Uppsala Universitet, Uppsala.

Poutanen M., Bowell E., Martin L. J., and Thompson D. T. (1985) Photoelectric photometry of (69) Hesperia. *Astron. Astrophys. Suppl. Ser., 61,* 291–297.

Pravec P., Wolf M., Šarounová L., Mottola S., Erikson A., Hahn G., Harris A. W., and Young J. W. (1997) The near-Earth objects follow-up program. II. Results for 8 asteroids from 1982 to 1995. *Icarus, 130,* 275–286.

Rosenbush V. K., Avramchuk V. V., Rosenbush A. E., and Mishchenko M. I. (1997) Polarization properties of the Galilean satellites of Jupiter: Observations and preliminary analysis. *Astrophys. J., 487,* 402–414.

Rosenbush V. K., Kiselev N. N., Avramchuk V. V., Jockers K., Shakhovskoj N. M., and Efimov Yu. S. (2002) Coherent effects in light scattering from atmosphereless solar system bodies. In *Electromagnetic and Light Scattering by Non-Spherical Particles, Gainesville 2002* (B. Gustafson et al., eds.), pp. 279–282. Army Research Laboratory, Adelphi, Maryland.

Sasse C., Muinonen K., Piironen J., and Dröse G. (1996) Albedo measurements on single particles. *J. Quant. Spectr. Rad. Transfer, 55,* 673–681.

Scaltriti F., Zappalà V., and Stanzel R. (1978) Lightcurves, phase function and pole of the asteroid 22 Kalliope. *Icarus, 34,* 93–98.

Shepard M. K. and Arvidson R. E. (1999) The opposition surge and photopolarimetry of fresh and coated basalts. *Icarus, 141,* 172–178.

Shepard M. K. and Campbell B. A. (1998) Shadows on a planetary surface and implications for photometric roughness. *Icarus, 134,* 279–291.

Shevchenko V. G., Chiorny V. G., Kalashnikov A. V., Krugly Yu. N., Mohamed R. A., and Velichko F. P. (1996) Magnitude-phase dependences for three asteroids. *Astron. Astrophys. Suppl. Ser., 115,* 475–479.

Shevchenko V. G., Belskaya I. N., Chiorny V. G., Piironen J., Erikson A., Neukum G., and Mohamed R. (1997) Asteroid observations at low phase angles. I. 50 Virginia, 91 Aegina, and 102 Miriam. *Planet. Space Sci., 45,* 1615–1623.

Shevchenko V. G., Belskaya I. N., Krugly Yu. N., Chiorny V. G., and Gaftonyuk N. M. (2002) Asteroid observations at low phase angles. II. 5 Astraea, 75 Eurynome, 77 Frigga, 105 Artemis, 119 Althaea, 124 Alkeste, and 201 Penelope. *Icarus,* in press.

Shkuratov Yu. G. (1985) On the origin of the opposition effect and negative polarization for cosmic bodies with solid surface. *Astron. Circular No. 1400,* pp. 3–6. Sternberg State Astronomical Institute, Moscow.

Shkuratov Yu. G. (1987) Negative polarization of sunlight scattered from celestial bodies: Interpretation of the wavelength dependence. *Sov. Astron. Lett., 13,* 182–183.

Shkuratov Yu. G. (1989) A new mechanism of the negative polarization of light scattered by the surfaces of atmosphereless celestial bodies. *Astron. Vestnik, 23,* 176–180.

Shkuratov Yu. (1991) An interference model of the negative polarization of light scattered by atmosphereless celestial bodies. *Solar System Res., 25,* 134–142.

Shkuratov Yu. G. and Helfenstein P. (2001) The opposition effect and the quasi-fractal structure of regolith: Theory. *Icarus, 152,* 96–116.

Shkuratov Yu. G. and Muinonen K. (1992) Interpreting asteroid photometry and polarimetry using a model of shadowing and coherent backscattering. In *Asteroids, Comets, Meteors 1991* (A. W. Harris and E. Bowell, eds.), pp. 549–552. Flagstaff, Arizona.

Shkuratov Yu. G. and Opanasenko N. V. (1992) Polarimetric and photometric properties of the Moon: Telescope observation and laboratory simulation. 2. The positive polarization. *Icarus, 99,* 468–484.

Shkuratov Yu. G., Opanasenko N. V., and Kreslavsky M. A. (1992) Polarimetric and photometric properties of the Moon: Telescopic observations and laboratory simulations. 1. The negative polarization. *Icarus, 95,* 283–299.

Shkuratov Yu. G., Muinonen K., Bowell E., Lumme K., Peltoniemi J. I., Kreslavsky M. A., Stankevich D. G., Tishkovets V. P., Opanasenko N. V., and Melkumova L. Y. (1994) A critical review of theoretical models for the negative polarization of light scattered by atmosphereless solar system bodies. *Earth, Moon, Planets, 65,* 201–246.

Shkuratov Yu. G., Kreslavsky M. A., Ovcharenko A. A., Stankevich D. G., Zubko E. S., Pieters C., and Arnold G. (1999) Opposition effect from Clementine data and mechanisms of backscatter. *Icarus, 141,* 132–155.

Shkuratov Yu., Ovcharenko A., Zubko E., Stankevich D., Miloslavskaya O., Muinonen K., Piironen J., Nelson R., Smythe W., Rosenbush V., and Helfenstein P. (2002) The opposition effect and negative polarization of structural analogs of planetary regoliths. *Icarus,* in press.

Stankevich D. G., Shkuratov Yu. G., and Muinonen K. (1999) Shadow-hiding effect in inhomogeneous and layered particulate media. *J. Quant. Spectr. Rad. Transfer, 63,* 445–458.

Stankevich D., Grynko Ye., Shkuratov Yu., and Muinonen K. (2002) A computer simulation of multiple scattering of light rays in systems of opaque particles. *J. Quant. Spectr. Rad. Transfer,* in press.

Tedesco E., Drummond J., Gandy M., Birch P., Nikoloff I., and Zellner B. (1978) 1580 Betulia: An unusual asteroid with an extraordinary lightcurve. *Icarus, 35,* 344–359.

Tishkovets V. P. (2002) Multiple scattering of light by a layer of discrete random medium: backscattering. *J. Quant. Spectr. Rad. Transfer, 72,* 123–137.

Tishkovets V. P., Litvinov P. V., and Lyubchenko M. V. (2002) Coherent opposition effects for semi-infinite discrete random medium in the double-scattering approximation. *J. Quant. Spectr. Rad. Transfer, 72,* 803–811.

Verbiscer A. J. and Veverka J. (1995) Interpretation of the IAU two-parameter magnitude system for asteroids in terms of Hapke photometric theory. *Icarus, 115,* 369–373.

Veverka J., Thomas, P. C., Bell J. F. III, Bell M., Carcich B., Clark B., Harch A., Joseph J., Martin P., Robinson M., Murchie S., Izenberg N., Hawkins E., Warren J., Farquhar R., Cheng A., Dunham D., Chapman C., Merline W. J., MacFadden L., Wellnitz D., Malin M., Owen W. M. Jr., Miller J. K., Williams B. G., and Yeomans D. K. (1999) Imaging of asteroid 433 Eros during NEAR's flyby recoinnaisance. *Science, 285,* 562–564.

Veverka J., Robinson M., Thomas P., Murchie S., Bell J. F. III, Izenberg N., Chapman C., Harch A., Bell M., Carcich B., Cheng A., Clark B., Domingue D., Dunham D., Farquhar R., Caffey M. J., Hawkins E., Joseph J., Kirk R., Li H., Lucey P., Malin M., Martin P., MacFadden L., Merline W. J., Miller J. K., Owen W. M. Jr., Peterson C., Prockter L., Warren J., Wellnitz D., Williams B. G., and Yeomans D. K. (2000) NEAR at Eros: Imaging and spectral results. *Science, 289,* 2088–2097.

Videen G. (2002) Polarization opposition effect and second-order ray-tracing. *Appl. Opt.,* in press.

Volten H., Muñoz O., de Haan J. F., Vassen W., Hovenier J., Muinonen K., and Nousiainen T. (2001) Scattering matrices of mineral aerosol particles at 441.6 nm and 632.8 nm. *J. Geophys. Res., 106,* 17375–17401.

Wisniewski W. Z., Michałowski T. M., Harris A. W., and McMillan R. S. (1997) Photometric observations of 125 asteroids. *Icarus, 126,* 395–449.

Zappalà V., Cellino A., Barucci A. M., Fulchignoni M., and Lupishko D. F. (1990) An analysis of the amplitude-phase relationship among asteroids. *Astron. Astrophys., 231,* 548–560.

Zellner B. and Gradie J. (1976) Minor planets and related objects. XX. Polarimetric evidence for the albedos and compositions of 94 asteroids. *Astron. J., 81,* 262–280.

Zellner B., Gehrels T., and Gradie J. (1974) Minor planets and related objects. XVI. Polarimetric diameters. *Astron. J., 79,* 1100–1110.

Zellner B. H., Albrecht R., Binzel R. P., Gaffey M. J., Thomas P. C., Storrs A. D., and Wells E. N. (1997) Hubble Space Telescope images of asteroid 4 Vesta in 1994. *Icarus, 128,* 83–87.

Zubko Eu., Ovcharenko A., and Shkuratov Yu. G. (2002) Polarimetric weak-localization effect in scattering of natural light in the region of small phase angles. *Opt. Spectrosc., 92,* 443–448.

Asteroid Models from Disk-integrated Data

Mikko Kaasalainen
University of Helsinki

Stefano Mottola
DLR German Aerospace Center

Marcello Fulchignoni
Observatoire de Paris

Photometric observations are the primary source of information on the bulk of the asteroid population. The macroscopic shapes, rotational states, and scattering parameters of asteroids can be deduced from photometric measurements of total brightness in different viewing/illumination geometries. Shape classes include convex, nonconvex, and binary objects, while rotational states include relaxed rotation, precessing motion, and mutually orbiting configurations. We discuss the role of lightcurve inversion in light of results from other sources.

1. INTRODUCTION

Disk-resolved images can be obtained of only a limited number of the small atmosphereless bodies of our solar system. This is why disk-integrated photometry, i.e., measurements of total brightness, is, and will remain, a major source of information on these objects. The total brightness of an object as a function of time (as the viewing/illumination geometry changes) is called its lightcurve; the parameters to be determined from lightcurve observations are the object's shape, its rotational state, and the scattering properties of its surface.

The problem of lightcurve inversion has been studied for nearly a century; however, restricted choices of observing geometry and scattering law in the early analytical studies have caused erroneously negative preconceptions about the uniqueness and stability of the solution (*Russell,* 1906). Partly because of this, and partly because of the relative scarcity of accurate lightcurve observations, the shape models used in lightcurve inversion until recently have been extremely simple — typically triaxial ellipsoids or their modifications — while the rotational states have usually been heuristically determined from certain characteristics of the lightcurves. Such traditional methods are well described in, e.g., *Magnusson et al.* (1989, 1996), *Barucci et al.* (1992), *Kwiatkowski* (1995), and *Binzel et al.* (1993); they are practical if the data are very limited. We encourage the reader to study the *Asteroids II* chapter by *Magnusson et al.* (1989): The methods described there are especially suitable for getting the first pole-direction estimate for an asteroid not yet well observed.

The view taken in this chapter is that lightcurve inversion is a typical representative of a problem category known as inverse problems. These are solved with modern deconvolution methods and optimization techniques. By an asteroid model we mean a general physical model not based on a shape given *a priori.* The model must also utilize all available (photometric) data in the analysis, not just, e.g., amplitudes or epochs of lightcurve features. The first general analysis of the inverse problem, including uniqueness theorems for convex shapes, is offered in *Kaasalainen et al.* (1992a,b). A robust and generally applicable inversion scheme is developed in *Kaasalainen and Torppa* (2001) and *M. Kaasalainen et al.* (2001). Models for precessing and binary systems have recently been developed as well (*Kaasalainen,* 2001; *Mottola and Lahulla,* 2000; *Pravec and Hahn,* 1997).

A number of spacecraft and radar images of asteroids have been obtained during the last decade (see *Chapman,* 2002; *Cheng,* 2002; *Ostro et al.,* 2002; and references therein). The most conspicuous feature in these is the existence of substantial irregularities: sharp edges, completely asymmetric shapes, globally significant valleys and indentations, and large craters. The number and accuracy of groundbased lightcurve measurements have also increased significantly, so it is important to know how much information can really be obtained from photometry in principle and how well the results can be expected to describe the real objects.

In section 2, we discuss the fundamental role of lightcurve inversion in light of (and compared with) the results obtained with other groundbased and *in situ* methods during the last decade. In section 3, we address some generic concepts of lightcurve analysis, while section 4 is devoted to the discussion of convex inversion and the corresponding stability properties. We discuss the methods and limitations of nonconvex inversion in section 5, while section 6 is concerned with the role of the light-scattering model of the surface material. The determination of the rotational state is the subject of section 7, and binary models are discussed in section 8. Finally, in section 9, we list some conclusions and topics for future work. Readers who want to form a quick overview of the subject can skip sections 4–6 on first reading.

2. GROUND TRUTH AND THE ROLE OF LIGHTCURVE INVERSION

The shapes of asteroids result from (1) the accretion process in the swarm of planetesimals or (2) the (repeated) fragmentation of a parent body caused by a catastrophic collision, followed in some cases by (3) a possible gravitational reaccumulation into a "rubble pile" structure. Single fragments are mostly irregular in shape, while the remnants of the original planetesimals and large rubble piles can mostly be expected to have more regular shapes, close to rotational equilibrium shapes, i.e., biaxial and triaxial ellipsoids (*Farinella et al.,* 1981; *Holsapple et al.,* 2002; *Richardson et al.,* 2002; *Scheeres et al.,* 2002). Continuous exposure to the space environment (crater and regolith formation, micro-impact erosion, and bombardment by charged particles) affects the surfaces of asteroids by changing their global morphology as well as their macroscopic and microscopic photometric properties.

Modeling the shapes and global surface properties of asteroids from disk-integrated photometric data necessitates robust and flexible inversion methods. The foremost objective in establishing the degree of reliability is the mathematical assurance of the uniqueness and stability of the solution, as discussed in the sections below. The reassuring knowledge that we are on firm ground instead of thin ice is important, but it is also necessary to check the adopted inversion method against reality. This can be done if the shape and surface properties of the body that originated a given set of lightcurves are known. This "calibration" has been explored both theoretically and through laboratory simulations (e.g., *Barucci and Fulchignoni,* 1985; *M. Kaasalainen et al.,* 2001).

The images obtained by spacecraft, orbiting observatories, radar observations, or adaptive optics show that the larger asteroids (Ceres, Vesta) have relatively regular global shapes, close to a spheroid or equilibrium ellipsoid, while the medium- and small-sized objects have more irregular shapes, characterized by sharp edges, concavities, and peculiar features that dominate the global morphology (e.g., the saddle-shaped feature Himeros on Eros). The shapes and topography of such closely observed objects can be represented with accurate numerical models (*Simonelli et al.,* 1993) such as the one shown in Fig. 1. The obtained models allow us to calculate the volume of the body and estimate its density, provided that a mass value is available from mutual perturbations of asteroidal orbits, spacecraft trajectories, or the presence of a satellite (*Hilton,* 2002). The knowledge of the shapes of these bodies is essential for modeling their internal structure, if other data (spectrometry, altimetry, gravimetry, mineralogical composition, etc.) are available. Moreover, such cartography can be used to interpret an asteroid's surface features in terms of its evolution.

The availability of the disk-resolved dataset on the asteroids 951 Gaspra (*Thomas et al.,* 1994), 243 Ida (*Thomas et al.,* 1996), and 433 Eros (*Veverka et al.,* 2000), taken during the *Galileo* and *NEAR Shoemaker* missions, represents the ground truth that allows us to understand the limits of the adopted inversion methods. All these objects have been well observed photometrically; the corresponding detailed inversion results are discussed in *M. Kaasalainen et al.* (2001). The global triaxial dimensions of the lightcurve inversion and probe models agree within 5–10% for each of the three bodies, and the global-scale details of the two shape models resemble each other closely. The spin-vector directions and rotation rates also completely agree (within a few degrees) with the spacecraft data for Eros and Gaspra; the restricted observation geometries for Ida allow two possible pole solutions, one of which closely agrees with the correct pole. Lightcurve inversion results also closely agree (*M. Kaasalainen et al.,* 2001) with the radar-based models for 1620 Geographos (*Hudson and Ostro,* 1999) and 6489 Golevka (*Hudson et al.* 2000).

Traditional pole determination methods and their variants have also produced good estimates of the pole directions and rotation periods for most of the "test-case" asteroids above (see the corresponding references above and references therein). These estimates are particularly useful as initial values that can be refined with the techniques discussed in this chapter. Traditional pole determination schemes are often unable to resolve pole ambiguities (section 7), and irregularly shaped asteroids may confuse such methods [cf. the analysis of 6489 Golevka by *Mottola et al.* (1997b)].

Figure 1 shows the topographic model of 951 Gaspra, based on *Galileo* data, while Fig. 2 shows the corresponding model from lightcurve inversion at the same viewing geometry (30°N of the equator and 60° off the long axis of the body). The two figures well demonstrate three important points (each discussed in more detail in the sections below): (1) Lightcurve inversion provides a good model of the global shape of the object. (2) The stable (most reliable) shape

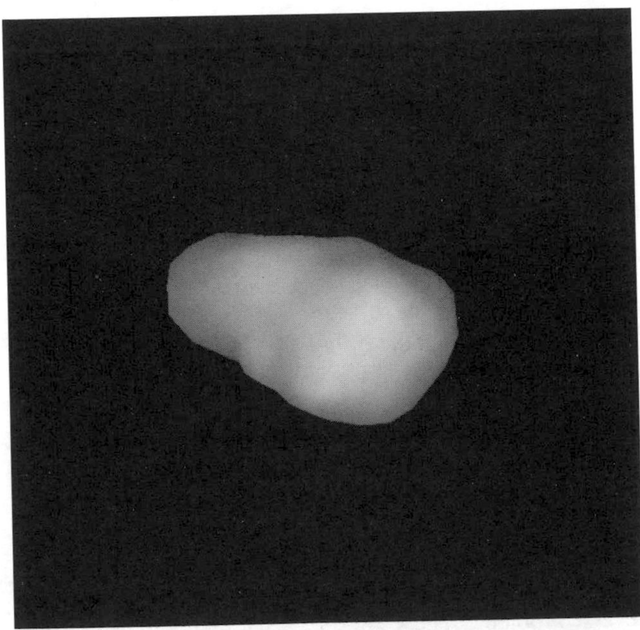

Fig. 1. A view of the topographic model of 951 Gaspra, based on *Galileo* data.

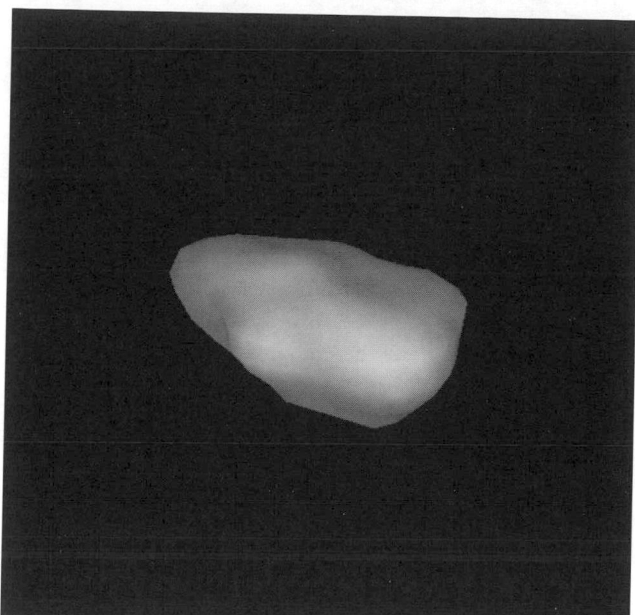

Fig. 2. A view of the lightcurve inversion model of 951 Gaspra at the same geometry as Fig. 1.

description obtained from the lightcurves can be thought of as the convex shape best mimicking the silhouette of the body in all viewing directions. (3) Detailed topographic/nonconvex features can seldom be confirmed using disk-integrated photometric data alone.

Besides satisfying the explorer's curiosity, knowledge of asteroid shapes, spin states, and photometric properties offers further insights. The shape resolution from lightcurve inversion is often sufficient to imply the nature of the body's structure, especially if the absolute size can be estimated from the albedo and/or a thermal infrared model. A more general goal is to deduce the evolutionary history and common physical properties of, e.g., an asteroid family from the individual shapes, surface characteristics, and spin states of its members. Shape statistics can be established for suitably large sample groups using, e.g., the Gaussian random sphere (*Muinonen,* 1998).

Cartography-like precision is limited to the few asteroid targets of space missions and, to a lesser extent, to those resolved by especially powerful observational techniques. The number of asteroids that will be explored by space missions in the next decades will remain small (*Farquhar et al.,* 2002) even though the results are expected to increase our knowledge significantly, and the radar and detailed adaptive optics/orbiting telescope observations will concern a limited number of asteroids (some very large ones and several near-Earth asteroids) in comparison with the total asteroid population. Moreover, due to the long observation spans, the potential accuracy of the lightcurve-based sidereal rotation period is better than that of any other method. Thus, groundbased photometric data will continuously be a fundamental source of information for the shapes, rotational states, and surface properties of asteroids.

3. DIRECT VS. INVERSE PROBLEM

The total brightness L of an object can be written as $L = L(\mathcal{R}, \mathcal{S}, \mathcal{D}; \mathcal{O}, \mathcal{A}; t)$, where $\mathcal{R}$, $\mathcal{S}$, and $\mathcal{D}$ are the sets of parameters describing the shape, scattering properties, and rotational state of the object respectively. The orbital parameters describing the motion of the observer and the asteroid are given by $\mathcal{O}$ and $\mathcal{A}$, so the evolution of L in time t, i.e., the asteroid's lightcurve, is completely determined by the five parameter sets.

Lightcurves produced by arbitrary objects can be computed numerically using a ray-tracing code. Second-order scattering can usually be neglected, so it is straightforward to write a multipurpose procedure that checks which parts of the surface are visible to both Earth and the Sun. To this end, the surface is best given as a polyhedron with small triangles (on the order of 1000) as facets; precomputed lists of possible ray-blockers for each facet make the computation fast. This tessellation can be applied separately to each object of a group; thus binary objects can be handled with the same code (suitably modified to include time dependence if the rotation is nonsynchronous).

Once it is known that a surface patch ds is both visible and illuminated, its contribution dL to the total brightness is given by (omitting irrelevant scale factors such as the squares of distances)

$$dL = S(\mu, \mu_0)\varpi \, ds \tag{1}$$

where ϖ and S are albedo and the scattering law (in a simple form here; more arguments can naturally be included, as in the full Hapke model), $\mu = \mathbf{E} \cdot \mathbf{n}$ and $\mu_0 = \mathbf{E}_0 \cdot \mathbf{n}$, where $\mathbf{E}$ and $\mathbf{E}_0$ are, respectively, unit vectors toward the observer (Earth) and the Sun, and $\mathbf{n}$ is the surface unit normal. Lambert's law, for example, is $S_L = \mu\mu_0$, while the Lommel-Seeliger law is $S_{LS} = S_L/(\mu + \mu_0)$. Note that L is given in intensity units rather than magnitudes. The separate expression ϖ for albedo as a scale factor representing the local darkness of the surface is intuitively obvious for sufficiently large surface patches. The complete physical description of the small-scale structure represented in the scattering law S, with the albedo of the surface material as one parameter, is much more complicated. It should be stressed that "albedo" is, indeed, a somewhat vaguely defined concept, and our ϖ is not the Bond, the geometric, or the single-scattering albedo [often denoted by w or ϖ_0; see *Bowell et al.* (1989)].

The direct problem of lightcurve computation is straightforward: Just add together all the relevant dL contributions from equation (1) to obtain the total brightness L at any given geometries. However, the inverse problem is obviously difficult: The effects of $\mathcal{R}$, $\mathcal{S}$, and $\mathcal{D}$ are combined in L, and now the sets should be disentangled. A general solution can only be found when all parameters are determined simultaneously. Our aim is thus to minimize

$$\chi^2 = \|\mathbf{L}_{obs} - \mathbf{L}(\mathcal{R}, \mathcal{S}, \mathcal{D})\|^2 \tag{2}$$

where $\mathbf{L}_{obs}$ and $\mathbf{L}$ are vectors containing the observed and modeled brightnesses respectively at the observation epochs.

An important aspect of the scattering law is its strong dependence on the solar phase angle $\alpha = \text{acos}\,(\mathbf{E} \cdot \mathbf{E}_0)$. If this cannot be modeled accurately, fitting absolute brightnesses is useless. On the other hand, absolute brightnesses of observations are often not known accurately enough in any case. To obtain proper line-over-points fits, the only possibility is usually to regard some or all of the photometric data as relative. In this case we minimize

$$\chi^2_{rel} = \sum_i \left\| \frac{\mathbf{L}^{(i)}_{obs}}{\bar{L}^{(i)}_{obs}} - \frac{\mathbf{L}^{(i)}}{\bar{L}^{(i)}} \right\|^2 \tag{3}$$

where, through the average brightnesses $\bar{L}^{(i)}$ of each lightcurve sequence i, both the observed and the model lightcurves are renormalized to mean brightnesses of unity. This corresponds to leaving an offset magnitude for each lightcurve as a free parameter to be determined; equation (3) is more advantageous as it discards all scale factors and thus keeps the number of free parameters as low as possible. Relative photometry is slightly more insensitive to the flattening of the body shape in the direction of the rotation axis than absolute photometry (for example, a bare spheroid can obviously be mistaken for a sphere if only relative photometry is available). This naturally emphasizes the role of good polar-aspect coverage.

Using equation (3) reflects a natural phenomenon: It is primarily the shapes of the lightcurves that are strongly connected with the pole, the period, and the shape of the asteroid. The absolute brightnesses are principally connected with the scattering properties, thus one can actually decouple one set of parameters from the rest to some extent.

4. CONVEX MODELS

4.1. Representations of a Convex Surface

The inverse problem becomes analytically tractable if it is assumed that the body can be modeled with a convex shape. Such a shape is computationally easiest to give in the form of a convex polyhedron. Define the matrix M by

$$M_{ij} = S_j(\mu^{(ij)}, \mu_0^{(ij)})\varpi_j \tag{4}$$

where S_j and ϖ_j are the scattering law and albedo at the facet j; $\mu^{(ij)} = \mathbf{E}_i \cdot \mathbf{n}_j$ and $\mu_0^{(ij)} = \mathbf{E}_{0i} \cdot \mathbf{n}_j$ for the observation i (in the asteroid's frame of reference); and $\mathbf{n}_j$ is the chosen surface outward normal of the facet j. If either $\mu^{(ij)}$ or $\mu_0^{(ij)}$ is less than or equal to zero, M_{ij} vanishes. From equation (1) we immediately have

$$\mathbf{L} = \mathrm{M}\mathbf{s} \tag{5}$$

where the vector $\mathbf{s}$ contains the areas of the facets of the polyhedron. These must obviously be positive; the easiest way to guarantee this is to represent each s_j exponentially, the optimization parameter being now the exponent a_j

$$s_j = \exp(a_j) \tag{6}$$

Since the number of fitted parameters must be large (on the order of 1000) to make sure that the result does not depend on the directions of the surface normals, the conjugate-gradient method (see, e.g., *Press et al.,* 1992) is efficient for minimizing $\chi^2_{(rel)}$. Once the areas of the facets are known, the vertices of the facets can be obtained straightforwardly by iteratively solving the so-called Minkowski problem (*Kaasalainen and Torppa,* 2001; *Lamberg,* 1993; *Lamberg and Kaasalainen,* 2001).

It is also possible to use the more general concept of curvature function given as a smooth function series (*Kaasalainen and Torppa,* 2001); this function automatically determines the areas of the polyhedron facets at any given resolution. The optimization parameters are now the coefficients of the series. Since the number of the parameters to be solved for is not large (typically from, say, 40–100), it is advantageous to use the Levenberg-Marquardt optimization scheme (*Press et al.,* 1992). This very fast and robustly converging approach is especially suitable for obtaining the dynamical parameters $\mathcal{D}$ (section 7). For a typical lightcurve set (i.e., on the order of 1000 brightness measurements), one minimization run is done in seconds. The complete analysis of a target typically requires a few tens of runs started at different initial values for the rotation and scattering parameters (see sections 6 and 7).

4.2. Albedo Variegation

Obviously $\mathbf{s}$ in equation (5) may be taken to represent the unknown facet albedo or its product with the facet area (depending on the scattering law, more complicated formulations can be considered as well but they are not relevant to the discussion). Shape parameters can thus be separated from those of albedo only by using suitable constraints.

Simulations and inversion results (*Cellino et al.,* 1989; *Kaasalainen et al.,* 2002b) have shown that significant albedo variegation is seldom required even for drastic lightcurve irregularities, while spacecraft images and simple physical considerations imply that asteroid surfaces are not likely to be covered with extensive high-contrast albedo "paintings." All the lightcurves of the "test-case" asteroids mentioned in section 2 could be completely explained with the asteroids' global shapes. The slight albedo markings visible to the eye in the disk-resolved probe images are negligible in the disk-integrated sense. Thus we have good reason to attribute lightcurve variation to shape as much as possible and to invoke albedo variegation only when necessary.

One of the advantages of convex modeling is that the result contains a straightforward indicator of the probable strength of albedo variegation on the asteroid's surface (*Kaasalainen and Torppa,* 2001). This quantity can be represented in terms of a "residual area": The effect of the asymmetry of albedo variegation on the surface is equivalent to that of a surface patch the size of the residual area. If this area is negligible, the lightcurve features are in all probability caused by the shape.

If the residual area is larger than, say, 1% of the total surface area, we must use the above-mentioned principle of shape domination to obtain a plausible shape/albedo result. This can be given straightforwardly in the form of a regularizing function that, e.g., describes a simple one-spot model over a freely adjustable shape (*Kaasalainen et al.,* 2002b). This spot corresponds to the observed residual area. Another possibility is to use physical *a priori* assumptions. For example, asteroids large enough to be molded mostly by self-gravitation can be expected to be quite spheroidal, if the lightcurve variations are not large. In such cases the shape can be approximated as the best-fit ellipsoid, and the albedo map can then be "painted" on it (an early application of this approach is given for 4 Vesta in *Cellino et al.,* 1987).

4.3. Uniqueness and Stability

An important fact is that the global shape resulting from the separation of albedo asymmetry is close to the correct one as long as the albedo markings on the surface are not abnormally bright and/or extensive. This can best be understood via the convex formulation: The overall shape of a convex body may change very little even if the areas of individual facets change considerably. This property could be called Minkowski stability in recognition of the rather subtle mathematical theorems on convex bodies by Minkowski (e.g., *Lamberg and Kaasalainen,* 2001, and references therein). It is precisely this property that makes convex inversion so robust; it is also the reason why it is much safer to attribute brightness changes to shape rather than albedo. This stability also means that the inversion result is *not* very sensitive to random noise in observations. In fact, the result is insensitive even to the (realistic) choice of the light-scattering model of the surface (section 6).

The convex shape result is not exactly the convex hull of the body but a shape close to it, "trimmed" such that its shadowing properties best mimic those of the original, often strongly nonconvex shape. The nonconvexities can be seen as deviations from this basic shape. Minkowski stability means that the local curvature may be trimmed in various ways at individual locations on the surface without influencing the global shape of the convex model very much. In addition to providing this stability, only the convex formulation admits uniqueness theorems (*Kaasalainen et al.,* 1992a; *Kaasalainen and Torppa,* 2001), which is invaluable for the reliability of the inverse solution.

The global characteristics of even quite strongly nonconvex bodies can well be recognized from the convex representation. This is illustrated by Figs. 3 and 4; Fig. 3 portrays the radar-based model of 6489 Golevka viewed at 20° below the equator, while Fig. 4 shows the lightcurve-based convex model at the same geometry. The convex model fitted the observations within the noise, i.e., no nonconvexities were required to explain the lightcurve features. This is quite typical of groundbased observations.

5. NONCONVEX MODELS

Lightcurve observations can usually be well explained with a convex model even when the data are produced by a shape known to have large concave features. Thus we can conclude that lightcurves seldom carry detailed information

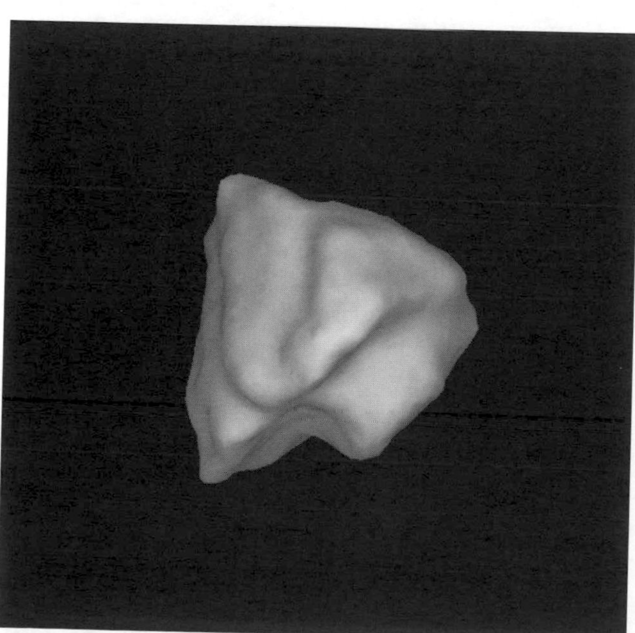

Fig. 3. A view of the radar-based model of 6489 Golevka (courtesy of Scott Hudson).

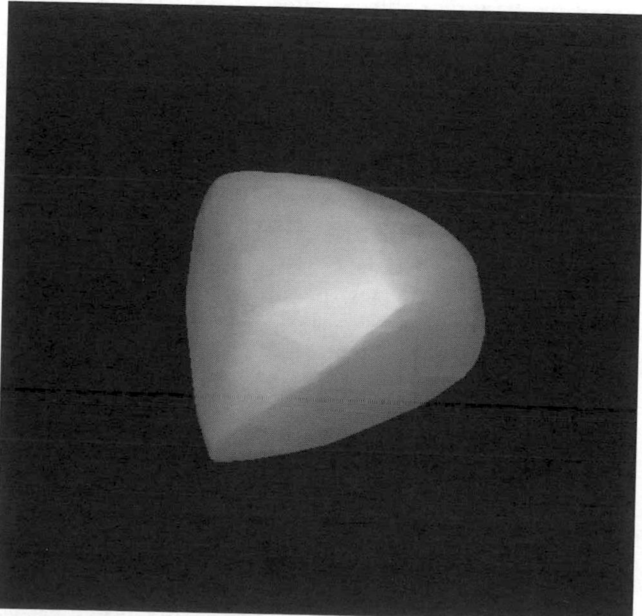

Fig. 4. A view of the convex lightcurve inversion model of 6489 Golevka at the same geometry as Fig. 3.

on nonconvex features. This is mostly due to the fact that solar phase angles have to be very large to cause striking shadowing effects. Thus the signature of nonconvex features is usually drowned in the noise at low and intermediate solar phase angles. Another significant factor is that the light-scattering properties of the surface can seldom be modeled very accurately.

From the technical point of view, the main complications in nonconvex modeling are that all uniqueness theorems are lost, and the parameter space is usually peppered with local minima especially if there is noise in the data. A practical approach is to use a short function series for the radii of the vertices of a triangulated surface in given directions. The trial lightcurves are now computed using equation (1) and a fast ray-tracing algorithm; the resulting χ^2 is minimized with Levenberg-Marquardt, statistical, or other suitable algorithms. The model of Fig. 2 is an example of this method.

Nonconvex models usually do not really reach lower χ^2s than convex ones. Minkowski stability no longer applies to them; thus, even in theory, reliable general nonconvex inversion requires highly accurate observations, very favorable observation geometries, and an accurate scattering model. Smoothness regularization is essential, and it is also useful to regularize the rotation axis to align with the axis of the maximum moment of inertia (cf. *Pravec et al.,* 2002). Fortunately, large flat areas on the convex solution already indicate the presence and locations of major nonconvex features.

The convex solution can indeed be seen as the "basic" shape of the object: Various nonconvex perturbations give very similar fits, but the unperturbed shape is stable. We know that asteroids are not convex. However, due to its inherent instability, a nonconvex model can be regarded as reliable only if it fits the data clearly better than a convex one. In other words, the convex formulation provides us with a good diagnostic tool: If we happen to stumble upon one of the very rare objects whose lightcurve data cannot be fitted well with a convex model, we can be quite certain that there must be considerable global concavities on the surface. Contact binaries with a distinctly double-lobed appearance are good candidates for such a class of objects.

6. SCATTERING MODELS AND PHASE CURVES

Accurate, universal few-parameter descriptions of the scattering properties of the surface material are hard to come by. The existing models, such as those by Hapke or Lumme and Bowell (see *Bowell et al.,* 1989), are still inadequate and known to produce ambiguous and unrealistic parameter values in inverse problems. A typical example of a scattering phenomenon not yet properly modeled is the opposition effect, i.e., the brightening near-zero phase angle caused by coherent backscattering and shadowing (e.g., *Muinonen et al.,* 2002).

For lightcurve inversion, the scattering law must be simple; too many parameters and possibilities cause instability and unrealistic results. Also, it is often easier (at least in the first analysis) to simply express the general photometric properties of the surface rather than try to obtain detailed physical parameters. A useful scattering model for this purpose is

$$\begin{aligned} S(\mu, \mu_0, \alpha) &= f(\alpha)[S_{LS}(\mu, \mu_0) + c\, S_L(\mu, \mu_0)] \\ &= f(\alpha)\mu\mu_0\left(\frac{1}{\mu + \mu_0} + c\right) \end{aligned} \tag{7}$$

which combines single (Lommel-Seeliger term S_{LS}) and multiple scattering (Lambert term S_L) with a weight factor c for the latter. For the sake of convenient inversion, the phase function $f(\alpha)$ is taken to multiply the sum of the single and multiple scattering terms. Such a simplification can be justified by the ambiguity in what should be called single or multiple scattering in a medium consisting of small particles and their aggregates. In this formulation $f(\alpha)$ can be determined afterward from a set of scale factors (obtained by dividing the average observed brightness by the corresponding model brightness) for each lightcurve while it does not have to be known when solving for the other parameters. An exponential and linear model is a versatile choice for this purpose (*S. Kaasalainen et al.,* 2001; *Muinonen et al.,* 2002)

$$f(\alpha) = a \exp\left(-\frac{\alpha}{d}\right) + k\alpha + 1 \tag{8}$$

where a and d are the amplitude and scale length of the opposition effect, and k is the overall slope of the phase curve (with the linear part at zero phase angle normalized to unity).

The parameters of any given scattering model can be directly incorporated in the optimization procedure. Hapke, Lumme-Bowell, and the above scattering model all give quite similar results for the shape and the rotational state (*M. Kaasalainen et al.,* 2001). Inversion should obviously be performed with at least two different scattering laws to establish a rough error estimate. Once the scattering characteristics, shape, and the rotation period and pole are known for the asteroid, one can compute "proper" reference phase curves as functions of solar phase angle in, e.g., an equatorial illumination and viewing geometry. Such well-defined phase curves (see Fig. 5), defined as the intensity time average of the model lightcurve as a function of α, are naturally more reliable than badly defined curve fits obtained more or less directly from observed magnitudes. In this way one can also define other characteristics such as the amplitude-phase relationship (*Zappalà et al.,* 1990). Reference phase curves are a stable means of representing an object's global photometric characteristics; as Fig. 5 shows, different scattering laws produce very similar phase curves for 433 Eros. The phase curves from the Hapke model (dotted line) and equations (7) and (8) (solid line) differ slightly only at the

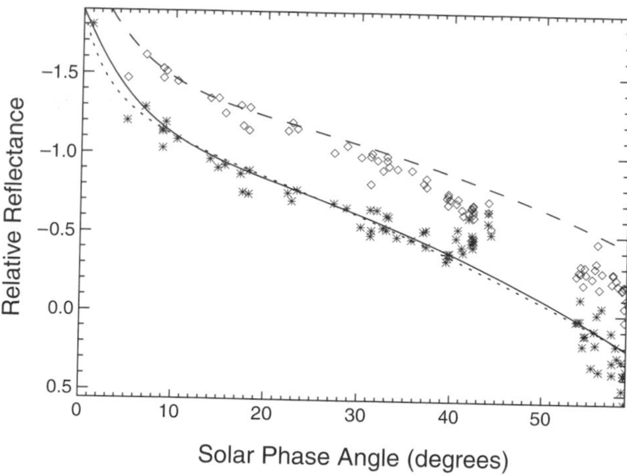

Fig. 5. The reference phase curve for 433 Eros obtained with the exponential-linear phase function (8) (solid line), together with the phase curve obtained with the best-fit Hapke parameters (dotted line), and the computed maxima at the equatorial geometry (dashed line). Observed average brightnesses are plotted with asterisks, and the observed maxima with diamonds (note the displacing effect of nonequatorial geometries).

low solar phase angles, where there were not enough observations to determine the phase function in equation (8) accurately. The difference between the computed maxima (dashed line) and the phase curve describes the amplitude-phase relationship.

7. ASTEROID ROTATIONS AND PHOTOMETRIC DATA

7.1. Single-Period Lightcurves

The great majority of asteroids are principal-axis rotators exhibiting single-period lightcurves, whereas tumbling bodies and binary systems produce multiple-period lightcurves. An asteroid in the relaxed state of principal-axis rotation revolves about its angular momentum vector **M** coinciding with the asteroid's axis of maximum moment of inertia, so the rotational state is fully described by three parameters: the ecliptic latitude β and longitude λ of the direction of the rotation axis (pole) and the rotation speed ω. The sidereal period is $P = 2\pi/\omega$. The asteroid always rotates in the positive direction about the pole, i.e., ω is always positive. Prograde and retrograde motion are indicated by positive and negative β respectively.

A rough estimate of the current synodic period is easily obtained by observing a lightcurve long enough to see a repeated pattern; an improved estimate can be obtained by fitting a Fourier series to a few consecutive lightcurves, if the orbital motion is slow compared to the rotational one. The often-used method of building a composite lightcurve from the data of neighboring observing nights should, however, be practiced only with extreme caution. When data over long periods of time and various observing geometries are used, the only reliable way of estimating the sidereal period is to include it in the general pole and shape analysis.

Let $\mathbf{r}_{ecl}$ denote a vector in the ecliptic coordinate frame where the origin is translated to the asteroid. This vector transforms to the vector $\mathbf{r}_{ast}$ in the asteroid's own frame (where z-axis is aligned with the rotation axis) by the rotation sequence

$$\mathbf{r}_{ast} = R_z[\phi_0 + \omega(t - t_0)]\ R_y(90° - \beta)\ R_z(\lambda)\ \mathbf{r}_{ecl} \qquad (9)$$

where t is the time and $R_i(\alpha)$ is the rotation matrix corresponding to the rotation of the coordinate frame through angle α in the positive direction about the i-axis. The angle ϕ_0 and the epoch t_0 can be chosen at will.

The directions **E** and $\mathbf{E}_0$ of Earth and the Sun as seen from the asteroid are now simple functions of β, λ, and ω, so these parameters can be included in equations (2) or (3). Obviously there are several local minima in χ^2, so multiple initial values for the pole and the period must be applied. The case of the pole is the simplest. A standard choice is to use a few directions in each octant of the celestial sphere as starting points; such a grid will usually cover all the local minima. One can also use the pole estimates of possible previous models. [For methods of obtaining simple first estimates of rotation parameters, see *Magnusson et al.* (1989, 1996).]

If the lightcurve set covers many years and apparitions, the period space is filled with densely packed local minima. The smallest separation ΔP of the local minima when plotting χ^2 as a function of the trial period P is roughly given by

$$\Delta P \approx \frac{1}{2}\frac{P^2}{T} \qquad (10)$$

where $T = \max(|t - t_0|)$ within the lightcurve set. This derives from the phenomenon that if P is changed by ΔP, the minima and maxima of the model lightcurve at $t_0 \pm T$ will undergo a phase shift of 180°; for the typical double-sinusoidal lightcurves, they will then be at roughly the same places as with P. One should thus use initial periods less than ΔP apart from each other covering the whole of the interval within which P can be expected to lie. Lightcurve inversion uses all apparitions and observations for the period estimate; thus, the more apparitions that are available, the more pronounced the correct local minimum of the period is.

Error estimates for the pole and the period are not completely obvious. Formal errors (such as the square roots of the elements of the covariance matrix) should not be reported; they are worthless in practice since the effects of random noise are usually negligible compared to systematic and model errors. A practical and robust way of estimating the errors is to perform a series of optimizations with different light-scattering models and initial parameter values. The results with χ^2 close to the best one describe the parameter distribution; this gives at least a lower bound to the error.

If the solar phase angle were always exactly zero, it would be impossible to determine the object's sense of rotation from disk-integrated data since both **M** and –**M** would give equally good solutions. Similarly, if the object always moved exactly in the plane of the ecliptic, both **M** and the reflection of –**M** in the ecliptic plane (i.e., two λs differing by 180°) would be acceptable solutions. Many asteroids often come close to one or both of these limiting cases, resulting in two to four possible solution regions. In these cases it is particularly important to perform a simultaneous solution of both the shape and rotational state, because even if the resulting χ^2s of the different solutions turn out not to be sufficiently different, the shape solution can still betray the wrong pole by its more unrealistic appearance.

7.2. Multiple-Period Lightcurves

Recently, mainly due to the establishment of systematic photometric monitoring programs of NEAs (*Mottola et al.,* 1995; *Pravec et al.,* 1998a; *Erikson et al.,* 2000), the presence of multiple-period lightcurves has been detected. Such observations, when the effects of the relative motion between the target and observer can be neglected or compensated for, can be subjected to standard power-spectrum and multidimensional Fourier analysis (*Press et al.,* 1992). With these methods, the probable underlying frequencies of the data can be singled out and given initial estimates.

A Fourier series representation can also be used for checking the quality of the data and highlighting the possible patterns against noise. In general, the brightness L of any bona fide doubly periodic lightcurve (extension to further dimensions being obvious) can be expanded as

$$L = \sum_{ij} a_{ij} \cos(i\omega_1 t + j\omega_2 t) + b_{ij} \sin(i\omega_1 t + j\omega_2 t) \quad (11)$$

where $\omega_n = 2\pi/P_n$, i = 0, 1, 2, …, and j = …, –1, 0, 1, … (only nonnegative integers when i = 0). Note that several i, j combinations in coefficients are present in principle even for low orders. Thus the sometimes-used approximative expansion with two one-dimensional Fourier series (*Pravec and Hahn,* 1997) is more suitable for binaries than for complex rotators, because the lightcurves of the former have an additive two-component physical origin. As will be discussed below, it is precisely this difference in the physical origins that usually allows one to distinguish the lightcurves of the two classes from each other.

A total of eight parameters are needed for a complete description of the motion of the asteroid's inertia ellipsoid in force-free precession. These parameters define the size and direction of the angular momentum vector; the angles of rotation, precession, and nutation; and the relative axes of the inertia ellipsoid. The body shape is described in the coordinate frame defined by the inertia ellipsoid, so the shape/scattering parameter part of the problem is exactly the same as in relaxed rotation, and the dynamical parameters can be included in the analysis in the same manner as in equation (9) (*Kaasalainen,* 2001).

The tumbling motion is governed by two periods: an exact one, P_ψ, for the rotation about an extremal axis of the inertia ellipsoid, and an average one, P_ϕ, for the precession about **M**. The nutation period is exactly one-half the rotation period. The (main) peaks of the power spectrum are located at frequencies that are linear combinations of f_ψ and f_ϕ, where f = 1/P. Prominent peaks are usually found at $2f_\phi$ and $2(f_\phi \pm f_\psi)$; other low harmonics and combinations are typically seen as well. The factor 2 is the same shape-related phenomenon as in double-sinusoidal ordinary lightcurves. The combination of the frequencies is perhaps best understood when one considers the apparent frequency in the limiting case of principal-axis rotation: The two frequencies are still present, but they are only seen as the single frequency $f_\phi \pm f_\psi$. When the body is slightly tilted, an additional spectrum peak corresponding to the precession frequency will make its appearance.

Internal stresses in a tumbling asteroid cause it to lose rotational energy and thus reach the relaxed state of principal-axis rotation in a relatively short time interval (*Pravec et al.,* 2002). This makes small NEAs the natural candidates for observed precessing rotators: As the body size decreases, relaxation time increases while the time intervals between significant collisions inducing precession decrease (*Paolicchi et al.,* 2002). Similarly, because the lifetime of tumbling motion increases as the rate of rotation decreases, we are more likely to observe slow complex rotators (such as 4179 Toutatis) than fast ones.

8. BINARY ASTEROIDS

The existence of binary asteroid systems has long been postulated on the basis of theoretical considerations (e.g., *Weidenschilling et al.,* 1989), and their existence was confirmed at the turn of the millennium with such techniques as photometric observations, adaptive optics, and radar [for the last two, see *Merline et al.* (2002) and *Ostro et al.* (2002)]. Photometrically, the clearest indication of a binary system involves two periods. It has been suggested that particular lightcurve features could represent the signature of binary structures where the two bodies are in contact or separated but remain in synchronous rotation with their mutual orbital motion (*Van Flandern et al.,* 1979). However, especially for contact binaries such interpretations can usually be made only indirectly (*Kaasalainen et al.,* 2002a) because of the small photometric information content on nonconvexities as discussed in section 5.

In a binary system, the presence of a companion body causes occultation-eclipse events under certain viewing and illumination geometries. If the central-body rotation is not synchronous with the orbital period of the secondary body, an observer would possibly detect two additive components in the lightcurves (in intensity space), one related to the rotation of the primary and one to the orbital motion of the secondary. Under certain observing conditions, both precessing and binary systems can produce lightcurves that are qualitatively similar and whose correct identification requires accurate photometry and good coverage.

The key to the identification of binaries is the additive nature of the two components in the binary system. If we subtract from the lightcurve the periodic component corresponding to the rotation of the primary, and we are left with the unequivocal signature of eclipse-occultation events, we can positively identify the asteroid as a binary system. A number of NEAs have been interpreted as binary systems on the basis of their lightcurves; among the first were 1994 AW1 (*Pravec and Hahn,* 1997), 3671 Dionysus (*Mottola et al.,* 1997a), and 1991 VH (*Pravec et al.,* 1998b).

Among the best-observed cases of binary systems whose binary nature and dynamical model has been inferred by photometric measurements alone is 1996 FG3 (*Mottola and Lahulla,* 2000; *Pravec et al.,* 2000). Figure 6a shows a lightcurve of 1996 FG3. The solid line represents a Fourier fit of the rotational lightcurve of the primary. Figure 6b shows the lightcurve of 1996 FG3 with the rotational component of the primary removed. The flat-bottomed minimum corresponds to a total superior eclipse, in which the secondary is totally occulted by the primary and/or its shadow. The deeper minimum corresponds to an inferior eclipse event (the secondary in front of the primary). Its amplitude is larger because, under these observing conditions, both the satellite and its shadow obscure the central body for a certain period of time. The unocculted lightcurve between eclipses displays a gentle curvature. This feature is interpreted as the result of an elongated and synchronous secondary, which produces a lightcurve with the same period as the orbital one and with maximum brightness at maximum satellite elongation.

A major reward of the modeling of binary systems is that it is possible to estimate the average bulk density of the bodies without knowing their absolute sizes. Using the scale-free expressions for the volumes of the model shapes and their sizes vs. orbital separation, density follows directly from Kepler's third law, whereas the definition of the masses would require the absolute size scale. Knowledge of the density of asteroids is of critical importance for understanding their origin and nature, and the characterization of the orbital motion of a satellite around a central body is the only way to determine this quantity from groundbased measurements.

Obviously, general lightcurve inversion of nonsynchronous binaries is even more complicated and ambiguous than nonconvex inversion. However, the detection of mutual event features in lightcurves places strong constraints on the geometry of the system, and by making reasonable assumptions one can use a simplified approach to infer the basic properties of the system. For example, the assumptions used in the case of 1996 FG3 (*Mottola and Lahulla,* 2000) are that (1) the central body and its satellite have the same photometric properties, (2) the satellite revolves round the primary in a circular orbit confined to the primary's equatorial plane, and (3) the secondary rotation is synchronous with its orbital period. These assumptions are introduced to limit the number of free parameters in the minimization problem, while still allowing a reasonable solution. In particular, the second assumption is justified by the timing of the occultation events, which implies a very small deviation from a circular orbit, whereas the third assumption is justified by the observed periodicity of the lightcurve of the secondary body.

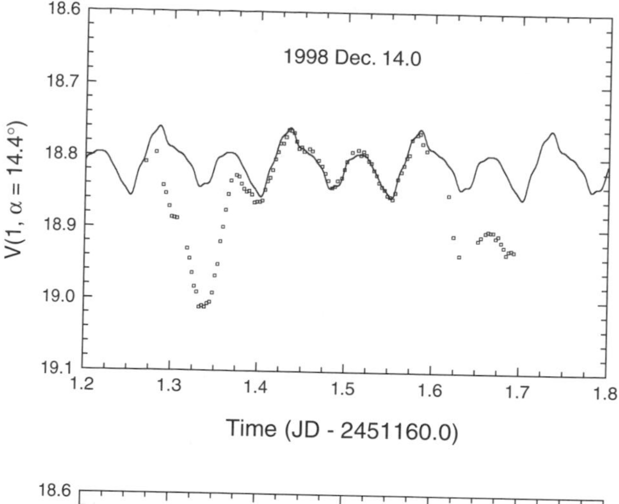

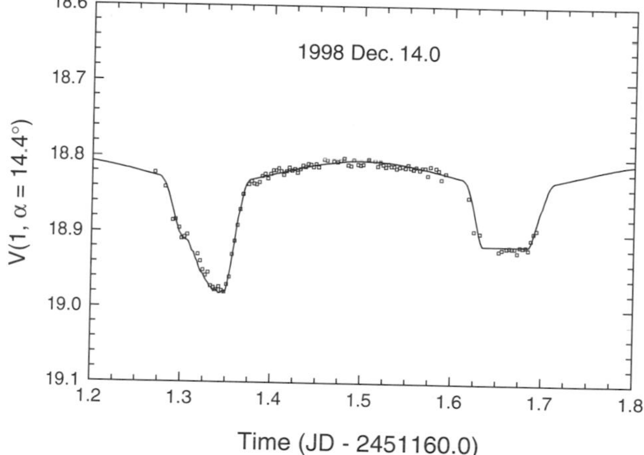

Fig. 6. **(a)** A lightcurve of the binary asteroid 1996 FG3. **(b)** The same lightcurve as in **(a)**, but with the rotational component of the binary removed. The occultation lightcurve for the best model solution is shown as a solid line.

Both the lightcurve and the dynamical characteristics of the binary system depend only on the relative sizes and separation of the bodies, so all geometric quantities are left scale-free. In the first analysis (in which lightcurves are not available at several observing geometries), both the central object and its satellite can be treated as triaxial ellipsoids approximated by polyhedra with triangular facets; eclipses and occultations are taken into account by means of ray-tracing techniques. The rotational lightcurve amplitude of the central body constrains its A and B semiaxes, while C is left as a free parameter. In addition, the measurement of the amplitude of the lightcurve of the secondary allows one to express the a and b satellite semiaxes as a function of C. The c satellite semiaxis is not constrained by the lightcurves in this analysis. As its influence on the uncertainty of the density determination is rather small, one can assume b:c = 1 in the modeling procedure. In the case of 1996 FG3 the scattering law is the five-parameter Hapke model with average parameters for C-type asteroids. To find the model that best describes the binary system, a grid search is performed over the parameter space (defined by the central body semiaxis C, the orbital radius R, and the ecliptic longitude and latitude λ, β of the orbital plane normal) to find the solution that minimizes the residuals to the observed lightcurves. The occultation lightcurve for the best-model solution is shown in Fig. 6b as a solid line.

9. CONCLUSIONS AND FUTURE WORK

Lightcurve inversion is an efficient deconvolution technique whose resolving capacity lies, roughly speaking, between space telescope and radar capabilities. The particular advantage of photometric observations is the wide distance range of analyzable objects, extending from near-Earth to main-belt asteroids. The rotational state as well as the characteristic light-scattering properties and large-scale shape features of an object can be well inferred from good-quality data by lightcurve inversion. The global shape of the target can usually be best described with a convex model, as lightcurves seldom carry information on small shape details or nonconvex features.

The solution of the inverse problem is detailed and stable if accurate measurements made at various observing geometries are available; the key principle is thus to conduct well-planned observational campaigns. Well-equipped amateur observers and the development of automatic telescopes should be of considerable assistance in this. Near-Earth asteroids are particularly rewarding targets, since a comprehensive model of an NEA can often be constructed after one suitable apparition, i.e., an observation span of only a few months.

The requirements of various observing geometries are not very stringent. Almost all asteroids to which traditional triaxial ellipsoid methods can be applied can be analyzed with the general method as well. The main objective is to observe the target at as many ecliptic longitudes (and latitudes) and solar phase angles as possible. Thus, a preliminary model of a main-belt asteroid can be built from two suitable apparitions, while three or four apparitions are usually sufficient for the construction of a good model (*Kaasalainen et al.,* 2002b). While accuracy is desired, even substantial random noise in lightcurves does not preclude their use for modeling purposes; also, relative photometry is sufficient for shape and rotation analysis.

There are several aspects of photometric analysis that should be investigated further; we briefly list here but a few. Though no unique theorems can be shown for binary asteroids, dynamical constraints or other types of regularization can be used to provide more detailed solutions. Binaries with more complex rotational states are also possible. Theoretical models of light scattering are still not adequate; neither is it clear how well the physical scattering parameters of the surface can really be determined from disk-integrated photometric data. Such parameters include not only the characteristics of the surface regolith particles, but also statistical topographic variations on small size scales. Lightcurve observations are not restricted to brightness measurements at visual wavelengths. Observations in the IR, together with theoretical models of thermal emission (*Harris and Lagerros,* 2002), can be particularly valuable in deducing the albedo variegation and composition of the surface.

Perhaps the most interesting future prospect is the use of complementary data simultaneously with photometry in asteroid modeling. Examples of such data include interferometric observations (speckle or other types), timings of stellar occultations, and precise astrometric measurements (from orbiting instruments) for which the photocenter of the target does not coincide with its center of mass. The full Stokes vector of light also includes the state of polarization, but so far not many polarimetric observations of asteroid surfaces have been carried out (*Cellino et al.,* 1999), and theoretical modeling hardly exists. Complementary data will undoubtedly prove to be very valuable in modeling details and removing possible ambiguities.

REFERENCES

Barucci M. A. and Fulchignoni M. (1985) Numerical and laboratory simulations of photometric properties of asteroids. In *Asteroids, Comets, Meteors II* (C.-I. Lagerkvist and H. Rickman, eds.), pp. 45–54. Uppsala Universitet Repocentralen, HSC, Uppsala.

Barucci M. A., Cellino A., De Sanctis C., Fulchignoni M., Lumme K., Zappalà V., and Magnusson P. (1992) Ground-based Gaspra modelling: Comparison with the first Galileo image. *Astron. Astrophys., 266,* 385–394.

Binzel R. P. and 18 colleagues (1993) Asteroid 243 Ida: Ground-based photometry and a pre-Galileo physical model. *Icarus, 105,* 310–325.

Bowell E., Hapke B., Domingue D., Lumme K., Peltoniemi J., and Harris A. W. (1989) Application of photometric models to asteroids. In *Asteroids II* (R. P. Binzel et al., eds.), pp. 524–556. Univ. of Arizona, Tucson.

Cellino A., Zappalà V., Di Martino M., Farinella P., and Paolicchi P. (1987) Flattening, pole, and albedo features of 4 Vesta from photometric data. *Icarus, 70,* 546–565.

Cellino A., Zappalà V., and Farinella P. (1989) Asteroid shapes and lightcurve morphology. *Icarus, 78,* 298–310.

Cellino A., Gil Hutton R., Tedesco E. F., Di Martino M., and Brunini A. (1999) Polarimetric observations of small asteroids: Preliminary results. *Icarus, 138,* 129–140.

Chapman C. (2002) Cratering on asteroids from *Galileo* and *NEAR Shoemaker.* In *Asteroids III* (W. F. Bottke Jr. et al., eds.), this volume. Univ. of Arizona, Tucson.

Cheng A. F. (2002) Near Earth Asteroid Rendezvous: Mission summary. In *Asteroids III* (W. F. Bottke Jr. et al., eds.), this volume. Univ. of Arizona, Tucson.

Erikson A., Mottola S., Lagerros J. S. V., Lindgren M., Piironen J., Oja T., Hahn G., Lagerkvist C.-I., Harris A. W., Nathues A., and Neukum G. (2000) The near-Earth objects follow-up program III. Lightcurves for 12 objects from 1992 and 1995. *Icarus, 147,* 487–497.

Farinella P., Paolicchi P., Tedesco E. F., and Zappalà V. (1981) Triaxial equilibrium ellipsoids among the asteroids? *Icarus, 46,* 114–123.

Farquhar R., Kawaguchi J., Russell C., Schwehm G., Veverka J., and Yeomans D. (2002) Spacecraft exploration of asteroids: The 2001 perspective. In *Asteroids III* (W. F. Bottke Jr. et al., eds.), this volume. Univ. of Arizona, Tucson.

Harris A. W. and Lagerros J. S. V. (2002) Asteroids in thermal infrared. In *Asteroids III* (W. F. Bottke Jr. et al., eds.), this volume. Univ. of Arizona, Tucson.

Hilton J. (2002) Asteroid masses and densities. In *Asteroids III* (W. F. Bottke Jr. et al., eds.), this volume. Univ. of Arizona, Tucson.

Holsapple K., Giblin I., Housen K., Nakamura A., and Ryan E. (2002) Asteroid impacts: Laboratory experiments and scaling laws. In *Asteroids III* (W. F. Bottke Jr. et al., eds.), this volume. Univ. of Arizona, Tucson.

Hudson R. S. and Ostro S. J. (1999) Physical model of asteroid 1620 Geographos from radar and optical data. *Icarus, 140,* 369–378.

Hudson R. S. and 26 colleagues (2000) Radar observations and physical model of asteroid 6489 Golevka. *Icarus, 148,* 37–51.

Kaasalainen M. (2001) Interpretation of lightcurves of precessing asteroids. *Astron. Astrophys., 376,* 302–309.

Kaasalainen M. and Torppa J. (2001) Optimization methods for asteroid lightcurve inversion. I. Shape determination. *Icarus, 153,* 24–36.

Kaasalainen M., Lamberg L., Lumme K., and Bowell E. (1992a) Interpretation of lightcurves of atmosphereless bodies. I. General theory and new inversion schemes. *Astron. Astrophys., 259,* 318–332.

Kaasalainen M., Lamberg L., and Lumme K. (1992b) Interpretation of lightcurves of atmosphereless bodies. II. Practical aspects of inversion. *Astron. Astrophys., 259,* 333–340.

Kaasalainen M., Torppa J., and Muinonen K. (2001) Optimization methods for asteroid lightcurve inversion. II. The complete inverse problem. *Icarus, 153,* 37–51.

Kaasalainen M., Torppa J., and Piironen J. (2002a) Binary structures among large asteroids. *Astron. Astrophys., 383,* L19–L22.

Kaasalainen M., Torppa J., and Piironen J. (2002b) Models of twenty asteroids from photometric data. *Icarus,* in press.

Kaasalainen S., Muinonen K., and Piironen J. (2001) A comparative study of the opposition effect of icy solar system objects. *J. Quant. Spect. Rad. Transf., 70,* 529–543.

Kwiatkowski T. (1995) Sidereal period, pole and shape of asteroid 1620 Geographos. *Astron. Astrophys., 294,* 274–277.

Lamberg L. (1993) On the Minkowski problem and the lightcurve operator. Academia Scientiarum Fennica, Series A, I. Mathematica dissertationes 87, Univ. of Helsinki.

Lamberg L. and Kaasalainen M. (2001) Numerical solution of the Minkowski problem. *J. Comp. Appl. Math., 137,* 213–227.

Magnusson P., Barucci M. A., Drummond J., Lumme K., Ostro S. J., Surdej J., Taylor R. C., and Zappalà V. (1989) Determination of pole orientations and shapes of asteroids. In *Asteroids II* (R. P. Binzel et al., eds.), pp. 67–97. Univ. of Arizona, Tucson.

Magnusson P. and 46 colleagues (1996) Photometric observations and modeling of asteroid 1620 Geographos. *Icarus, 123,* 227–244.

Merline W. J., Weidenschilling S. J., Durda D. D., Margot J.-L., Pravec P., and Storrs A. D. (2002) Asteroids *do* have satellites. In *Asteroids III* (W. F. Bottke Jr. et al., eds.), this volume. Univ. of Arizona, Tucson.

Mottola S. and Lahulla F. (2000) Mutual eclipse events in asteroidal binary system 1996 FG3: Observations and a numerical model. *Icarus, 146,* 556–567.

Mottola S., De Angelis G., Di Martino M., Erikson A., Hahn G., and Neukum G. (1995) The near-Earth objects follow-up program: First results. *Icarus, 117,* 62–70.

Mottola S., Hahn G., Pravec P., and Šarounová L. (1997a) *S/1997 (3671) 1.* IAU Circular 6680.

Mottola S. and 27 colleagues (1997b) Physical model of near-Earth asteroid 6489 Golevka (1991 JX) from optical and infrared observations. *Astron. J., 114,* 1234–1245.

Muinonen K. (1998) Introducing the Gaussian shape hypothesis for asteroids and comets. *Astron. Astrophys., 332,* 1087–1098.

Muinonen K., Piironen J., Shkuratov Yu. G., Ovcharenko A., and Clark B. E. (2002) Asteroid photometric and polarimetric phase effects. In *Asteroids III* (W. F. Bottke Jr. et al., eds.), this volume. Univ. of Arizona, Tucson.

Ostro S. J., Hudson R. S., Benner L. A. M., Giorgini J. D., Magri C., Margot J.-L., and Nolan M. C. (2002) Asteroid radar astronomy. In *Asteroids III* (W. F. Bottke Jr. et al., eds.), this volume. Univ. of Arizona, Tucson.

Paolicchi P., Burns J. A., and Weidenschilling S. J. (2002) Side effects of collisions: Spin rate changes, tumbling rotation states, and binary asteroids. In *Asteroids III* (W. F. Bottke Jr. et al., eds.), this volume. Univ. of Arizona, Tucson.

Pravec P. and Hahn G. (1997) Two-period lightcurve of 1994 AW1: Indication of a binary asteroid? *Icarus, 127,* 431–440.

Pravec P., Wolf M., and Šarounová L. (1998a) Lightcurves of 26 near-Earth asteroids. *Icarus, 136,* 124–153.

Pravec P., Wolf M., and Šarounová L. (1998b) Occultation/eclipse events in binary asteroid 1991 VH. *Icarus, 133,* 79–88.

Pravec P., Šarounová L., Rabinowitz D. L., Hicks M. D., Wolf M., Krugly Y. N., Velichko F. P., Shevchenko V. G., Chiorny V. G., Gaftonyuk N. M., and Genevier G. (2000) Two-period lightcurves of 1996 FG3, 1998 PG, and (5407) 1992 AX: One probable and two possible binary asteroids. *Icarus, 146,* 190–203.

Pravec P., Harris A. W., and Michałowski T. (2002) Asteroid rotations. In *Asteroids III* (W. F. Bottke Jr. et al., eds.), this volume. Univ. of Arizona, Tucson.

Press W. H., Teukolsky S. A., Vetterling W. T., and Flannery B. P. (1992) *Numerical Recipes in Fortran: The Art of Scientific Computing,* 2nd edition. Cambridge Univ., Cambridge. 963 pp.

Richardson D. C., Leinhardt Z. M., Melosh H. J., Bottke W. F. Jr., and Asphaug E. (2002) Gravitational aggregates: Evidence and evolution. In *Asteroids III* (W. F. Bottke Jr. et al., eds.), this volume. Univ. of Arizona, Tucson.

Russell H. N. (1906) On the light-variations of asteroids and satellites. *Astrophys. J., 24,* 1–18.

Scheeres D. J., Durda D. D., and Geissler P. E. (2002) The fate of asteroid ejecta. In *Asteroids III* (W. F. Bottke Jr. et al., eds.), this volume. Univ. of Arizona, Tucson.

Simonelli D., Thomas P. C., Carcich B. T., and Veverka J. (1993) The generation and use of numerical shape models for irregular solar system objects. *Icarus, 103,* 49–61.

Thomas P. C., Veverka J., Simonelli D., Helfenstein P., Carcich

B., Belton M. J. S., Davies M. E., and Chapman C. (1994) The shape of Gaspra. *Icarus, 107,* 23–36.

Thomas P. C., Belton M. J. S., Carcich B., Chapman C., Davies M. E., Sullivan R., and Veverka J. (1996) The shape of Ida. *Icarus, 120,* 20–32.

Van Flandern T. C, Tedesco E. F., and Binzel R. P. (1979) Satellites of asteroids. In *Asteroids* (T. Gehrels, ed.), pp. 443–465. Univ. of Arizona, Tucson.

Veverka J. and 32 colleagues (2000) NEAR at Eros: Imaging and spectral results. *Science, 289,* 2088–2097.

Weidenschilling S. J., Paolicchi P., and Zappalà V. (1989) Do asteroids have satellites? In *Asteroids II* (R. P. Binzel et al., eds.), pp. 643–658. Univ. of Arizona, Tucson.

Zappalà V., Cellino A., Barucci M. A., Fulchignoni M., and Lupishko D. (1990) An analysis of the amplitude-phase relationship among asteroids. *Astron. Astrophys., 231,* 548–560.

Asteroid Radar Astronomy

Steven J. Ostro
Jet Propulsion Laboratory, California Institute of Technology

R. Scott Hudson
Washington State University

Lance A. M. Benner and Jon D. Giorgini
Jet Propulsion Laboratory, California Institute of Technology

Christopher Magri
University of Maine at Farmington

Jean-Luc Margot
California Institute of Technology

Michael C. Nolan
Arecibo Observatory

Radar is a uniquely powerful source of information about the physical properties and orbits of asteroids. Measurements of the distribution of echo power in time delay (range) and Doppler frequency (radial velocity) produce two-dimensional images that can provide spatial resolution as fine as a decameter if the echoes are strong enough. With adequate orientational coverage, such images can be used to construct detailed three-dimensional models, define the rotation state precisely, and constrain the object's internal density distribution. As of May 2002, radar signatures have been measured for 75 main-belt asteroids (MBAs) and 105 near-Earth asteroids (NEAs). We summarize specific results for radar-detected asteroids, which span 4 orders of magnitude in diameter and rotation period. Radar has revealed both stony and metallic objects, principal-axis and complex rotators, smooth and extremely rough surfaces, objects that must be monolithic and objects that probably are not, spheroids and highly elongated shapes, contact-binary shapes, and binary systems. Radar also has expanded accurate orbit-prediction intervals for NEAs by as much as several centuries.

1. INTRODUCTION

One of the goals of this book is to outline developments since *Asteroids II* (*Binzel et al.,* 1989), which was completed in 1988. The subsequent 13 years have seen critical developments in technical aspects of asteroid radar astronomy, including increases in sensitivity and versatility of telescopes, the evolution and optimization of observational techniques, and the invention of analytical methods to optimize extraction of information from radar images. Observations commensurate with those developments have raised the number of radar-detected asteroids from 52 (19 NEAs + 33 MBAs) in mid-1988 (*Ostro,* 1989) to 180 (105 NEAs + 75 MBAs) in May 2002 and have produced an enormous body of information about the physical properties of asteroids. An equally significant development since *Asteroids II* is an increase in the number of persons lead-authoring asteroid radar papers at an average rate of one person per year.

The following sections outline the technical developments and observational highlights of the past 13 years, summarize the major conclusions drawn from the explosive increase in data, and discuss current problems and challenges to be faced during the coming decade. See *Ostro* (2002a) for a list of radar-detected asteroids, *Ostro* (1993, 2002b) for reviews of planetary radar principles and techniques, *Ostro* (1994) for a discussion of radar's role in NEO hazard identification and mitigation, and *Harmon et al.* (1999) for a review of radar observations of comets.

2. STRATEGIES, TELESCOPES, AND TECHNICAL DEVELOPMENTS

2.1. Telescopes and Observing Strategies

The basic strategy of an asteroid radar experiment is to measure the distribution of echo power in time delay and Doppler frequency, usually in the opposite sense (OC) of circular polarization as transmitted as well as in the same sense (SC) as a function of the object's orientation and plane-of-sky (POS) direction. SC echo would be absent in mirrorlike backscattering from surface elements for which size and radius of curvature are huge compared to the wavelength, but would become increasingly significant if there is near-surface "roughness" at scales near the wavelength or any kind of multiple scattering. Hence SC/OC is a measure of near-surface structural complexity (see section 3).

The achievable delay/Doppler resolution depends first of all on the echo's signal-to-noise ratio (SNR), the ratio of echo power to the root-mean-square receiver noise. The SNR depends primarily on the target's distance R_{tar}, diameter D_{tar}, and rotation period P; the telescope's effective area A and transmitter power P_{tx}; and the integration time Δt

$$SNR \sim R_{tar}^{-4} D_{tar}^{3/2} A^2 P_{tx} P^{1/2} \Delta t^{1/2} \quad (1)$$

The integration time needed to achieve any given signal-to-noise ratio for a given target increases as $R_{tar}^8 A^{-4}$. This is why the 305-m Arecibo telescope and the 70-m Goldstone antenna (DSS-14) are almost entirely responsible for the history of asteroid radar, why we try to observe asteroids at their closest approaches to Earth, and why observations of extremely close NEAs are especially lucrative.

A major structural upgrading of the Arecibo telescope and modernization of its computer hardware and software have made it an order of magnitude more sensitive and much more versatile than it was a decade ago. Transmitter upgrades and installation of a quasioptical transmit/receive switch on DSS-14 have increased its effectiveness for NEA radar astronomy by reducing the switching time from ~20 s to ~1 s. Arecibo can see almost twice as far as Goldstone, but Goldstone's greater steerability gives it access to twice as much sky and lets it track objects at least 3× longer than Arecibo. For very close targets, transmit/receive switching and transmitter on/off cycling has been avoided by transmitting continuously from DSS-14 and receiving continuously with DSS-13, a 34-m antenna 22 km from DSS-14 and connected to it by a fiber-optics cable. The complementarity of Arecibo and Goldstone has been exploited frequently.

Given unlimited echo strength, the delay resolution is limited by the rate at which signals produced by available transmitter amplifier tubes (klystrons) can be modulated, currently about 20 MHz, which corresponds to 0.05 μs (7.5 m of range), compared to 2 μs (300 m) when *Asteroids II* was written. The transmitted signal travels the round-trip distance to the target and the echo is measured using what is effectively an extremely sensitive voltmeter, the output of which is sampled and digitized. The received voltage is optionally divided into time-delay cells, and then a T_{coh}-long coherent time series of voltage samples within a given time-delay cell is Fourier-transformed to produce a spectrum of the echo power within that cell with a resolution of $\Delta f = T_{coh}^{-1}$. Because of the intrinsically noiselike nature of the voltage samples (*Jenkins and Watts,* 1968), the SNR of a single-power spectrum will generally be <3 for even the strongest radar echo. The fractional noise (dominated by thermal noise for weak echoes or by "self-noise" for strong ones) in a sum of N such "looks" is $N^{-1/2}$, so it is desirable to sum many looks. Even when the echo strength overwhelms the thermal noise of the receiver, one sometimes chooses to increase N in order to reduce the self-noise, consequently sacrificing frequency resolution. Thus, a delay-Doppler image is intrinsically a time exposure, combining coherent integration in the Fourier analysis with incoherent integration in the sum of independent looks. The optimum Δf for an image with a given tolerable level of noise will depend in part on the echo bandwidth, and hence on the target's size and spin state (see section 2.3).

At radio frequencies, the phase of the electromagnetic field is maintained all the way to the samplers, so the Fourier analysis and summing of looks can be done in software after the fact, allowing SNR to be traded for frequency resolution in the data reduction rather than during the data acquisition. However, measurement of range information requires coding of the transmitted signal, so the range resolution must be chosen before beginning the experiment.

During an observation, one can remove the Doppler frequency shift ν introduced by the radial motion of the target by tuning the frequency of either the transmitter or the receiver according to a Doppler-prediction ephemeris, with the goal of ensuring that the frequency corresponding to echoes from the target's center of mass (COM) is zero in the coordinate system of the acquired data. The Doppler shift varies as the target moves and as the Earth rotates, and it must be adjusted many times per second. One also uses the prediction ephemerides to slew the time base for sampling the echoes in order to register sequential samples of echoes from any given range cell on the target. In practice, there will always be a nonzero error $\Delta\nu_{eph}$ in the Doppler-prediction ephemeris, which is equivalent to a nonzero rate of change in the delay-ephemeris' error $\Delta\tau_{eph}$ and hence in the rate of image smearing in delay

$$d\Delta\tau_{eph}(t)/dt = -\Delta\nu_{eph}(t)/F_{tx} \quad (2)$$

where F_{tx} is the transmitter carrier frequency (2380 MHz for Arecibo; originally 8495 MHz for Goldstone but changed to 8510 MHz in September 1991 and then to 8560 MHz in March 1999). Thus a reasonably accurate Doppler ephemeris is a prerequisite for imaging with fine range resolution.

Main-belt asteroids that enter the current radar detectability windows usually have many decades of accumulated optical astrometry, so preradar pointing uncertainties typically are on the order of 1 arcsec, Doppler uncertainties are small compared to the intrinsic frequency dispersion of the echoes, and delay uncertainties are typically of the same order as the object's diameter. For MBA experiments to date, echo strength has been the only factor that has limited obtainable delay-Doppler resolution.

For NEAs, the accuracy of the ephemerides is often a major concern, and for newly discovered objects it is critical, because ephemeris accuracy decays away from the interval spanned by astrometric data. Often, unless follow-up optical astrometry is obtained between the date of a discovery announcement and the date when an initial radar observation is

TABLE 1. Residuals for past near-Earth-asteroid recoveries.

Object	Recovery Date	O	R	O/R
1989 PB (4769 Castalia)	May 1990	24"	0.4"	60
1991 AQ	September 1994	57°	0.1°	380
1986 DA (6178)	October 1994	56"	0.9"	60
1991 JX (6489 Golevka)	March 1995	3600"	4.6"	780
1986 JK (14827)	April 2000	114°	0.1°	910

Here O represents the residual (i.e., the observed position minus the predicted position) for an orbit solution incorporating only optical astrometry. R represents the residual for an orbit solution using radar as well as optical. O/R is the ratio of residuals for the two cases.

attempted, the pointing uncertainty will be large compared to the ~2-arcmin widths of the Arecibo and Goldstone radar beams. In practice, to avoid an intolerable sacrifice of sensitivity, pointing should be good to at least 15 arcsec. Once echoes have been detected, even coarse Doppler astrometry is adequate to shrink the orbit uncertainties enough to ensure sufficient pointing accuracy throughout the discovery apparition.

2.2. Radar Astrometry

Radar reconnaissance of a new NEA generally proceeds from detection and Doppler-ephemeris refinement using a continuous-wave (cw, or monochromatic) waveform to delay-ephemeris refinement using a time-modulated waveform (generally a binary-phase-coded waveform; J. K. Harmon, in preparation, 2002) with a fairly coarse delay-resolution cell (called a baud), to the finest-baud and finest-Δf imaging supported by the echo strength. In optimizing a setup's tradeoffs between time resolution, spatial resolution, and noise level, one must consider the accuracy of the delay-Doppler ephemeris and whatever is known about the target's size and spin state. Since NEA radar windows are short and telescope time is precious and difficult to obtain on short notice, rapid refinement of orbits and ephemerides using radar astrometry is critical. Installation of onsite software at Goldstone and Arecibo has permitted radar detection of several newly found NEAs within 12 h of the discovery announcement (1999 TY2, 2001 AV43, and 2001 FR85, all at Arecibo) and has dramatically sped up the progression from cw detection to high-resolution imaging (only 15 min for 2001 FR85).

Almost all radar astrometry (*Giorgini,* 2002; see also *Ostro et al.,* 1991a, and *Yeomans et al.,* 1992) reports the time delay and/or Doppler frequency (at a given transmitter frequency) corresponding to echoes from the target's center of mass received at a specific UTC epoch and at a particular telescope reference point; the transmitting telescope's reference point also is specified. For most antennas, the reference point is the intersection of the elevation and azimuth axes.

As an example of radar astrometry, imaging of Toutatis (*Ostro et al.,* 1995a) using transmissions from DSS-14 and reception at DSS-13 yielded estimates of the 8510-MHz Doppler frequency, 248335.943 ± 0.04 Hz, and time delay, 26.056497203 s ± 0.23 μs, corresponding to echoes from the asteroid's center of mass received at December 6, 1992, 16:40:00 UTC. These points' residuals with respect to the most recent Toutatis orbit are −0.098 Hz and 0.268 μs, so their fractional precision is 10^{-8} for the delay and 10^{-6} for the Doppler.

The fine fractional precision of radar astrometry plus its orthogonality to optical angle measurements make it powerful for refining orbits. A single radar detection secures the orbit well enough to prevent "loss" of a newly discovered asteroid (*Yeomans et al.,* 1987). Table 1 lists residuals at the first post-discovery-apparition recovery of several NEAs, for an orbit using just optical data and also for an orbit using both radar and optical data. Table 2 demonstrates how radar shrinks the sky area that must be searched for a given probability of recovering several NEAs observed only during their discovery apparition.

With radar astrometry, the length of the interval over which an asteroid's orbit can be calculated with a given level of accuracy can be increased by decades or centuries even for multiapparition NEAs. Let us give two examples, defining a "reliable" prediction interval as one that encompasses all those approaches to within 0.1 AU from Earth for which the 3σ uncertainty in the date of closest approach is less than 10 d. Then for Toutatis, observed with radar during the last three of its five optically observed apparitions, the optical-only interval is 1353–2262 and the radar + optical interval is 1353–2532; uncertainty associated with the very close Earth approach in 1353 precludes reliable identification of earlier close Earth approaches. For the single-apparition object 2001 CP36, the optical-only interval is 1989–2004 and the radar + optical interval is 1718–2225.

Ephemeris uncertainties are strategically important at every stage of an asteroid radar experiment and normally are calculated whenever new radar or optical astrometry warrants orbit refinement. Thus, every astrometric measurement lets one assess the accuracy of both the observing ephemeris and the uncertainties that had been calculated for it. Similarly, postfit residuals (observations minus the values calculated from the orbit solution) let one assess the accuracy both of the astrometry and the uncertainty quoted for it (Table 3).

TABLE 2. Search areas for future near-Earth-asteroid recoveries.

Object	Most Favorable Earth-based Recovery Date	Data Span (d)	Astrometry			Gap (yr)	3σ Search Area (arcsec²)		
			Optical	Doppler	Delay		O	R	O/R
1990 OS	November 2003	13	26	2	0	13	4.8E8	4.5E6	106
2000 EH26	July 2005	140	49	4	2	4	25772	675	38
1998 ML14	August 2013	225	234	6	6	15	1.9E7	3.8E5	49
2001 AV43	November 2013	38	42	1	0	12	8.9E7	1.8E6	49
1998 KY26	May 2024	11	207	2	2	26	14568	168	87
1999 TY2	October 2064	5	110	1	0	65	2.3E7	1.0E6	23
2001 FR85	March 2081	7	36	3	1	80	1.6E7	1.7E4	956

Each of these objects was observed optically over a short timespan and also was a radar target of opportunity, resulting in the listed numbers of optical (RA + DEC), Doppler, and delay measurements. On the right, we give the total sky area for the 3σ orbit-determination uncertainties mapped onto the sky at the next favorable Earth-based recovery date (which we define as the next time when the apparent visual magnitude exceeds 20 during reasonable sky-brightness conditions) for both an optical-data-only (O) orbit solution and a radar + optical-data (R) orbit solution. The O/R ratio conveys how a handful of radar measurements can reduce sky search areas for an object with minimal optical followup astrometry. Dynamical peculiarities unique to each object, such as the number of planetary close approaches, affect these results. For example, 1999 TY2 is unusual in that its 23° inclination to the ecliptic reduces the effects of in-plane gravitational perturbations, which shrinks mapped uncertainties.

TABLE 3. Radar astrometry residuals.

Antennas		Normalized Doppler Residuals			
TX	RCV	mean	σ	RMS	N
Arecibo	Arecibo	−0.119	0.4547	0.4679	94
DSS 14	DSS 14	−0.160	0.7915	0.8041	113
DSS 14	DSS 13	−0.455	1.095	1.144	12
Haystack	Haystack	−0.452	0.1552	0.4654	2
DSS 14	Evpatoria	0.1629	0.5998	0.6052	18
Sites combined (Doppler):		0.01415	0.6991	0.6978	239

Antennas		Normalized Delay Residuals			
TX	RCV	mean	σ	RMS	N
Arecibo	Arecibo	0.00044	0.6919	0.6868	68
DSS 14	DSS 14	−0.2510	0.6949	0.7351	88
DSS 14	DSS 13	−0.4635	2.154	2.127	14
Sites combined (delay):		−0.1679	0.9041	0.9170	170
Doppler/delay combined:		−0.06153	0.7949	0.7963	409

Statistics for normalized Doppler and delay postfit residuals (r_i, the measurement minus that predicted by a weighted-least-squares estimate of the orbit from all optical and radar astrometry, normalized by the measurement uncertainty assigned by the observer) obtained from 1968 through March 2001. Here RMS = $[\Sigma(r_i^2)/N]^{1/2}$ and the standard deviation (σ) equals $\{\Sigma[(r_i - \bar{r}^2]/(N - 1)\}^{1/2}$, with N the number of observations and $\bar{r}$ the mean residual. Arecibo has historically produced the lowest-noise, least-biased astrometry, followed by DSS-14 (Goldstone). Most of the DSS-13 (Goldstone) astrometry is from the December 1996 Toutatis campaign. Evpatoria results are from the Golevka experiment in 1995. Haystack results are from observations published of asteroid 1566 Icarus in 1968.

When ephemeris uncertainties become smaller than the intrinsic delay-Doppler dispersion of a target's echoes, the challenge becomes to locate the target's COM in the delay-Doppler image plane. The accuracy of this process rests on how well one knows the target's size and shape. Thus, orbit refinement is tightly coupled to determination of physical properties: Radar measurements that produce new information about a target's size, shape, rotation, or surface properties generally have astrometric value, and vice versa.

2.3. Radar Estimation of Shapes and Spin States

Interpretation of radar images is complicated by the geometry of the delay-Doppler projection (Fig. 1). Constant-

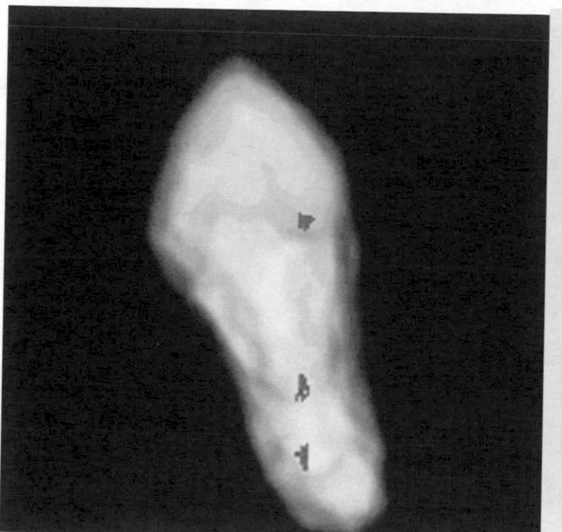
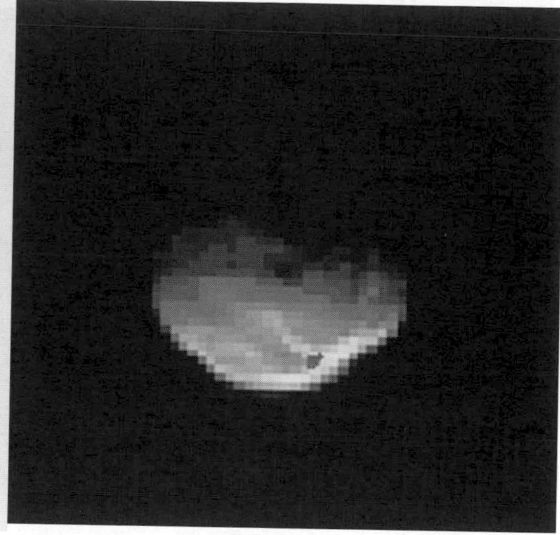

Fig. 1. Geometry of delay-Doppler images. The left frame is a plane-of-sky view of the low-resolution radar model of Toutatis (*Hudson and Ostro,* 1995). Planes of constant time delay (range) are parallel to the plane of the picture. Planes of constant Doppler frequency (line-of-sight velocity) are normal to that plane and aligned vertically. The three dots lie on a line defined by the intersection of a single constant-delay plane and a single constant-Doppler plane. The right frame is a synthesized radar image of the model, arranged so delay increases from bottom to top and Doppler increases from left to right. The three highlighted points on the asteroid model return echo at the single highlighted point in the delay-Doppler image: a three-to-one mapping.

delay planes are normal to the radar line of sight; for a rotating rigid body, constant-Doppler planes are parallel to both the line of sight and the target's apparent spin vector. These two orthogonal sets of parallel planes divide the three-dimensional target into three-dimensional resolution cells in a manner analogous to the way one cuts a potato into strips with rectangular cross sections. In optical imaging, the sets of planes that cut the target are parallel to the line of sight, and we see only the end of each cell that faces us. Thus one knows *a priori* that each point in an optical picture corresponds to a single point on the asteroid's surface — a one-to-one mapping. However, for delay-Doppler imaging, it may be possible for the radar to see both ends of a three-dimensional resolution cell. For very irregular objects, the radar may even see surface elements that lie inside the cell between these ends, e.g., if the cell slices through the sides of a concavity. Thus a delay-Doppler image is generally a many-to-one mapping that contains a form of global aliasing referred to as the north/south ambiguity: One cannot know, *a priori,* which (or even how many) points on the surface contributed echo power to a given pixel.

Moreover, the length equivalent of frequency in an image depends on the asteroid's apparent spin vector $\mathbf{W}_{app}$, as follows: Let us ignore second-order terms (*Ostro,* 1993, and references therein) and parallax, and assume that the COM is a constant distance from the radar. The Doppler frequency of an echo from a point $\mathbf{r}$ with respect to the COM is

$$\nu = (\mathbf{W}_{app} \times \mathbf{r}) \cdot \mathbf{e}/(\lambda/2) \quad (3)$$

where the unit vector $\mathbf{e}$ points from the COM to the radar. The radial velocity equivalent of 1 Hz is half a wavelength (λ) per second. The target's apparent spin vector satisfies

$$\mathbf{W}_{app} = \mathbf{W} + \mathbf{W}_{sky} \quad (4)$$

where $\mathbf{W}$ is the intrinsic spin vector and $\mathbf{W}_{sky}$ is the contribution from the target's apparent plane-of-sky (POS) motion in the frame of the radar telescope. Useful conversion factors are (*Ostro et al.,* 1995a)

$$\text{Goldstone (8560 MHz)}: \quad \text{km/Hz} = 87.2/|\mathbf{W}_{app} \times \mathbf{e}| = P/(4.13 \cos\delta) \quad (5)$$

$$\text{Arecibo (2380 MHz)}: \quad \text{km/Hz} = 312/|\mathbf{W}_{app} \times \mathbf{e}| = P/(1.15 \cos\delta) \quad (6)$$

where $\mathbf{W}_{app}$ is in degrees per day,

$$P = 360/|\mathbf{W}_{app}| \quad (7)$$

is the instantaneous, apparent spin period in days, and

$$\delta = \cos^{-1}(|\mathbf{W}_{app} \times \mathbf{e}|/|\mathbf{W}_{app}|) \quad (8)$$

is the instantaneous, apparent, target-centered declination of the radar.

An asteroid's echo bandwidth is given by

$$B = 4\pi D \cos\delta/\lambda P \quad (9)$$

where D is the breadth, measured normal to the line of sight,

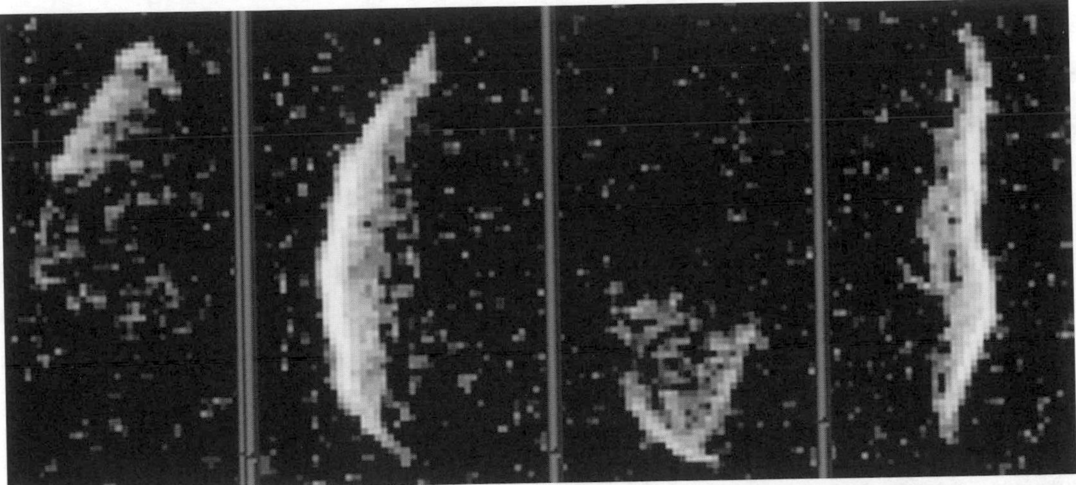

Fig. 2. Images of Geographos, taken with the radar close to the asteroid's equatorial plane (*Ostro et al.,* 1996). The four frames were taken with the radar illumination coming from the top, left, bottom, and right. When superposed, the images define the object's pole-on silhouette.

of the asteroid's pole-on silhouette. D can also be visualized as the asteroid's POS extent normal to the projected apparent spin vector. With B in Hz, we can write

$$\text{Goldstone (8560 MHz):} \quad B = 99.7\ D\cos\delta/(24\ P) \tag{10}$$

$$\text{Arecibo (2380 MHz):} \quad B = 27.7\ D\cos\delta/(24\ P) \tag{11}$$

For most asteroids, **W** is constant and parallel to both the angular momentum vector **L** and the maximum-moment principal axis of inertia (the asteroid's "short axis"). For such principal-axis (PA) rotators, two constant angles in an inertial frame fix the direction of **W** and the third of these Euler angles gives the rotation phase. However, for a nonprincipal-axis (NPA) rotator (*Hudson and Ostro,* 1995), **W** is not parallel to **L** and is not constant in inertial or body-fixed coordinates. Rather, the body-fixed inertia ellipsoid precesses about **L** while **W** executes a periodic motion in the body, its direction defining a closed curve on the inertia ellipsoid. (All three Euler angles are functions of time.) Given principal moments of inertia $I_s \geq I_m \geq I_l$ for the short, intermediate and long axes, NPA rotation is defined by eight parameters: two moment-of-inertia ratios, three initial values for the Euler angles, and initial values for each component of **W**. That is, the determinations of spin state and shape are tightly coupled. For a PA rotator, the coupling is loose, and specification of the spin state is trivial. Given Doppler-only images (i.e., cw spectra) that thoroughly sample more than half a rotation, estimates of echo-edge frequencies can be inverted to estimate the shape of the convex hull on a PA rotator's pole-on silhouette (*Ostro et al.,* 1988) as well as the COM echo frequency.

For delay-Doppler images, *Hudson* (1993) showed that a delay-Doppler image sequence can be inverted using constrained least squares to yield estimates of the target's shape, spin state, and radar-scattering properties, as well as the delay-Doppler trajectory of the COM. This is possible because each surface location has a unique delay-Doppler trajectory as the target rotates, unless the view is equatorial. In that case, the north-south ambiguity cannot be broken, and at best a three-dimensional shape model would convey attributes of the object's shape in a nonunique manner. Ironically, an image sequence taken from within the target's equatorial plane provides unambiguous measurement of the target's pole-on silhouette (Fig. 2; *Ostro et al.,* 1995b, 1996).

The accuracy of radar-based shape reconstruction depends on the echoes' strength and orientational coverage as well as the target's shape and spin state, in a manner that has been calibrated by extensive numerical experiments as well as by laboratory simulations using laser radar and clay models (*Andrews et al.,* 1995). Fortuitously, the Castalia, Toutatis, Geographos, and Golevka imaging experiments during 1989–1996 provided comprehensive experience with nonequatorial and equatorial viewing geometry, with PA and NPA rotators, and with the merits of incorporating optical lightcurves in the inversion (see sections 4.1–4.4).

In inverting radar images, one usually uses a polyhedral shape model with enough vertices to ensure reconstruction of the most detailed structure revealed in the images, while employing penalty functions to suppress structural features not required by the data. A typical approach is to estimate the free parameters **x** by minimizing an objective function

$$\Phi(\mathbf{x}) = \chi^2(\mathbf{x}) + \Sigma\ \beta_i\ \gamma_i(\mathbf{x}) \tag{12}$$

where penalty function $\gamma_i(\mathbf{x})$ has weight β_i and $\chi^2(\mathbf{x})$ is the weighted sum of squared residuals between delay-Doppler image pixel values and the corresponding values predicted by the physical model. Penalty functions that may be useful in certain circumstances include dynamical functions that force PA rotation or internal-density homogeneity as well as structural functions that suppress concavities. By examining how $\chi^2(\mathbf{x})$ and the distribution of residuals vary as a function of the β_i, one can assess the strength of evidence for such characteristics as complex rotation, nonuniform internal density, shape bifurcation, and exotic topography.

Reconstruction of shapes and spin states of NPA rotators like Toutatis is extremely difficult because of the nature of the coupling between the eight spin-state parameters and the shape parameters. NPA rotations can lead to enhanced orientational coverage, but they also tend to be extremely slow (as expected theoretically; *Harris,* 1994), so many days of radar observations are needed to obtain enough orientational and POS coverage to constrain all the spin parameters.

Asteroids for which radar-based shape constraints have been published are listed in Table 4 in order of decreasing SNR of the data. Of the several asteroids imaged at useful resolutions by spacecraft, Eros (*Veverka et al.,* 2000) is the only one for which radar-derived shape constraints are available. Goldstone echo spectra (Doppler-only "one-dimensional images") were obtained in 1975 by *Jurgens and Goldstein* (1976) with a nearly equatorial view and a total signal-to-noise ratio (SNR) of only 70σ. Hence those ancient radar data contain much less useful shape information compared to the other datasets listed in Table 4. *Ostro et al.* (1990b) used spectral-edge frequencies to estimate the hull H on the asteroid's pole-on silhouette S, and *Mitchell et al.* (1998) applied *Hudson*'s (1993) shape reconstruction methodology to all echo spectral elements (~15× more data points) to estimate S. The radar-derived estimates of H (Fig. 3) and S reproduce the object's pole-on shape characteristics within the associated uncertainty intervals (*Ostro et al.,* 2000b), lending confidence to shape reconstructions based on superior datasets.

Accurate shape models of small NEAs open the door to a wide variety of theoretical investigations that are central to understanding the nature, origin, and evolution of these objects but previously have been impossible or have used simplistic models (spheres or ellipsoids). For example, with detailed models of real objects, it is possible to explore the evolution and stability of close orbits (*Scheeres et al.,* 1996, 1998, 2000) with direct application to the design of spacecraft rendezvous and landing missions, studies of retention and redistribution of impact ejecta, and questions about the origins and lifetimes of asteroidal satellites. Given information or a realistic assumption about the internal density distribution, one can use a shape model to estimate the distribution of gravitational slopes, which can elucidate characteristics

TABLE 4. Radar shape constraints.

Asteroid	SNR	Delay, Doppler Cells	Max. |δ|	Sky Arc	Shape Constraint	Reference
4179 Toutatis	10000	40,50	80°	127°	3D: 1600	*Hudson and Ostro* (1995)
	10000	160,200	80°	127°	3D: 20000	R. S. Hudson et al. (personal communication, 2001)
6489 Golevka	2000	25,15	66°	94°	3D: 1024, H	*Hudson et al.* (2000a)
1620 Geographos	2000	68,55	10°	5°	H,S	*Ostro et al.* (1995b, 1996)
					3D: 512	*Hudson and Ostro* (1999)
4769 Castalia	1000	4,10	35°	1°	H	*Ostro et al.* (1990a)
					3D: 167	*Hudson and Ostro* (1994)
1998 ML14	1000	20,25	50°	33°	3D: 512	*Ostro et al.* (2001a)
7822 (1991 CS)	800	0,50	?	40°	H	*Benner et al.* (1999a)
216 Kleopatra	200	15,30	60°	3°	3D: 256	*Ostro et al.* (2000a)
1862 Apollo	200	12,0	10°	1°	H	*Ostro et al.* (2002)
1998 KY26	100	1,14	50°	54°	3D: 124	*Ostro et al.* (1999a)
2063 Bacchus	80	8,6	20°	30°	3D: 256	*Benner et al.* (1999b)
6178 1986 DA	80	2,20	?	1°	H	*Ostro et al.* (1991b)
433 Eros	70	0,30	10°	8°	E	*Jurgens and Goldstein* (1976)
					H	*Ostro et al.* (1990b)
					3D: 128, S	*Mitchell et al.* (1998)
2100 Ra-Shalom	70	0,30	?	8°	H	*Shepard et al.* (2000)
					E	*Ostro et al.* (1983)
1685 Toro	50	0,11	?	13°	H	*Ostro and Connelly* (1984)
1620 Ivar	50	5,10	?	7°	H	*Ostro et al.* (1990c)

Radar-based shape constraints include the pole-on silhouette (S), the hull (H) on that silhouette, triaxial ellipsoid models (E), and three-dimensional shape models (3D: number of shape parameters) reconstructed using *Hudson*'s (1993) technique. Here SNR is the approximate signal-to-noise ratio of an optimally filtered sum of all echoes. The next column gives the maximum number of delay and Doppler resolution cells placed on the target, the maximum asteroid-centered declination |δ| achieved during the observations, and the plane-of-sky arc spanned. Existing radar data can support estimation of three-dimensional shape models for 6178 (1986 DA), 7822 (1991 CS), 4197 (1982 TA), 7482 (1994 PC1), 6037 (1988 EG), 1036 Ganymede, 10115 (1992 SK), 1999 JM8, 1999 RQ36, 4486 Mithra, 2100 Ra-Shalom, 2000 RD53, 2000 DP107, 2000 UG11, 4183 Cuno, 2000 YA, 2000 XK47, 2000 YF29, 23187 (2000 PN9), 29075 (1950 DA), 25143 (1998 SF36), 2000 EE104, 1999 KW4, 2000 PH5, 22771 (1999 CU3), 1998 ST27, 33342 (1998 WT24), and 4660 Nereus.

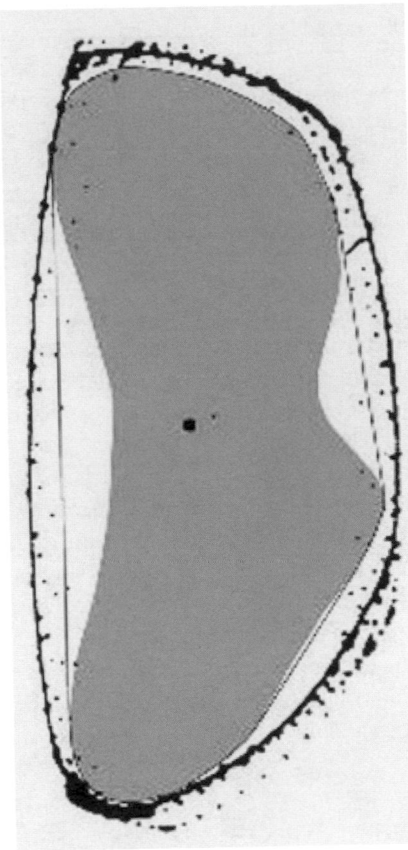

Fig. 3. Eros' pole-on silhouette and its convex hull H compared to the *Ostro et al.* (1990b) estimate of the hull (outermost solid curve) and its associated uncertainty (dots). The radar-derived lower bound on H's maximum breadth was 4% larger than the true value and the radar-derived lower bound on H's minimum breadth was 0.7% too large.

of the asteroid's surface and interior. A shape model also allows realistic investigation of the effects of collisions in various energy regimes on the object's rotation state, surface topography, regolith, and internal structure (*Asphaug et al.,* 1998). In principle, thorough thermal-infrared observations and modeling of an object for which a radar-derived shape model exists would likely provide fruitful insight into the surface's thermal properties.

3. OVERVIEW OF ASTEROID RADAR PROPERTIES

3.1. Polarization Ratio and Radar Albedo

The circular polarization ratio, μ_C = SC/OC, can easily be determined upon detection of an asteroid's echoes. Unlike radar cross sections (and hence radar albedos), which are the products of factors that suffer from systematic uncertainties generally on the order of 10%, an estimate of SC/OC is contaminated almost entirely by statistical error from receiver noise. SC/OC is a measure of the near-surface structural complexity at scales near the wavelength. An asteroid's SC/OC can be taken as a crude estimate of its rock coverage, that is, the fraction of the surface area covered by roughly wavelength-sized rocks.

Constraints on surface properties become increasingly ambiguous and model-dependent as SC/OC increases from near zero, where single reflections from smooth surface facets dominate the scattering process (e.g., Ceres and Pallas; *Ostro et al.,* 1985), to values of a few tens of percent or larger, indicating increasing contributions from single reflections from rough surfaces and/or multiple scattering. 2101 Adonis, 1992 QN (*Benner et al.,* 1997), 2000 EE104 (*Howell et al.,* 2001), and 33342 (1998 WT24) are the extreme examples, with SC/OC near unity. The merits of current interpretations of radar properties (albedo, polarization ratio, and scattering law; e.g., *Mitchell et al.,* 1995; *Magri et al.,* 2001) should be clarified upon spacecraft reconnaissance of asteroids for which detailed radar-derived physical models are available.

For SC/OC near zero, the OC radar albedo ($\hat{\sigma}_{OC}$, equal to the OC radar cross section divided by the target's projected area) is a first-order estimate of the Fresnel power reflection coefficient R, which for most homogeneous materials that are candidates for asteroid surfaces (solid metal is an exception) has been found empirically to depend simply on bulk density d_{bulk} (*Magri et al.,* 2001; *Garvin et al.,* 1985; *Olhoeft and Strangway,* 1975)

$$R = \tanh^2(d_{bulk}/6.4 \text{ g cm}^{-3}) \tag{13}$$

Using the radar properties of Eros and the *NEAR Shoemaker* determination of Eros' L-chondritic composition, *Magri et al.* (2001) have calibrated radar constraints on asteroids' near-surface regolith porosity p, solid-rock density d_{solid}, and bulk density $d_{bulk} = (1 - p)\, d_{solid}$.

3.2. Comparison of Main-Belt Asteroids and Near-Earth Asteroids

Magri et al. (1999) presented histograms of the distributions of estimates of SC/OC ratio and OC albedos for 37 MBAs, and Benner (2002) lists corresponding data for NEAs. The SC/OC distribution for NEAs is nonnormal, so nonparametric tests must be used to compare the NEA and MBA distributions. The Mann-Whitney rank-sum test indicates at a high confidence level that these two distributions have different medians; furthermore, the same conclusion holds when C-class objects alone are considered and when S-class objects alone are considered. Levene's test implies that NEAs also exhibit a wider range of polarization ratios than do MBAs, both for the full samples and for S-class targets alone. No such difference in variances can be firmly inferred for C-class objects alone, although we note that only four known C-class NEAs are in the sample. Evidently, most NEA surfaces are moderately or highly complex at decimeter scales, while MBAs are smooth or else moderately rock-covered.

The OC albedo distribution for 19 NEAs with reasonable albedo estimates is statistically similar to that for the 35 MBAs available to *Magri et al.* (1999). The same conclusion holds if we compare the distributions for the quantity ($\hat{\sigma}_{OC} - \hat{\sigma}_{SC}$), which represents an attempt to correct for diffuse scattering on the assumption that such scattering produces randomly polarized radiation (*Magri et al.,* 1999; *Shepard et al.,* 2000). If we instead use the quantity ($\hat{\sigma}_{OC} - 2\hat{\sigma}_{SC}$), relevant if diffusely scattered echo power has SC/OC = 1:2 (as is measured for the Moon and inner planets; *Harmon and Ostro,* 1985), then S-class NEAs and S-class MBAs have different medians, although equal medians still cannot be ruled out when all taxonomic classes are considered.

We conclude that the surfaces of NEAs and MBAs probably differ less strongly in bulk density than in fractional rock coverage. However, we stress the following caveats: First, NEA albedo estimates often are highly uncertain, since very few of the target diameters are well known. Second, in order to obtain Fresnel reflectivity and hence near-surface bulk density, one must correct the OC albedo for the effects of diffuse scattering from wavelength-scale structures, a correction which is large and uncertain for NEAs with high SC/OC.

3.3. Comparison of Main-Belt-Asteroid Classes

Statistical analyses of disk-integrated properties of the 37 MBAs observed during 1980–1995 (*Magri et al.,* 1999) indicate that M-class MBAs have higher radar albedos and a wider range of albedos than do MBAs from the other taxonomic classes; there is no evidence that C and S MBAs have different albedo distributions; and there is some suggestion, worthy of future study, that primitive B, F, G, and P MBAs are not as radar-bright as C and S MBAs. There is no statistically significant evidence that different taxonomic classes of MBAs have different polarization ratio distributions, despite suggestions to the contrary based on visual inspection of these distributions. The similarity between the C and S MBA albedo distributions implies similar near-surface regolith bulk densities. The hypothesis of ordinary chondritic composition for the S-class MBAs is reasonably consistent with the radar data, provided that these asteroids have typical lunar porosities. Nevertheless, it is possible that some targets have high-porosity regoliths of stony-iron composition. The M-class MBA sample apparently contains both metallic objects (such as 216 Kleopatra) and stony objects (possibly hydrated, W-class objects; *Rivkin et al.,* 2000).

The high-OC radar albedo (0.31 ± 0.06) of the largest M-class asteroid, 16 Psyche, implies a moderately high near-surface bulk density of 2.8 (+0.5, −0.6) g cm^{-3} and hence, if the porosity is about 50%, a grain density within ~20% of 5.6 g cm^{-3}. The inference of surface bulk density relies on a lab-experimental relationship between microwave reflectivity and bulk density, a relationship which would not apply if Psyche has a solid metal surface; however, it seems hard to believe that such a large object would not retain any fine impact debris on its surface. *Magri et al.* (1999; see also *Ostro et al.,* 1985) conclude that Psyche is more likely to be a metal-rich object than, say, an enstatite chondrite analog. It is not clear how to reconcile Psyche's large radar reflectivity with the bulk density, 1.8 ± 0.6 g cm^{-3}, implied by the value of Psyche's mass determined astrometrically by *Viateau* (2000).

4. RESULTS FOR SELECTED ASTEROIDS

4.1. 4769 Castalia (1989 PB)

A sequence of delay-Doppler images of Castalia (*Ostro et al.,* 1990a) obtained at Arecibo two weeks after its August 1989 discovery reveal it to consist of two kilometer-sized lobes in contact. Least-squares estimation of Castalia's three-dimensional shape from the radar images supports the hypothesis that Castalia is a contact-binary asteroid formed from a gentle collision of the two lobes and also constrains the object's surface morphology and pole direction (*Hudson and Ostro,* 1994). Analysis of Castalia lightcurves using the radar-derived shape refined the available pole constraints and yielded estimates of average Hapke parameters (*Hudson et al.,* 1997).

4.2. 1620 Geographos

Observations of Geographos (*Ostro et al.,* 1995b, 1996) with a nearly equatorial view yielded several hundred images with ~100-m resolution. The pole-on silhouette's extreme dimensions are in a ratio, 2.76 ± 0.21, that establishes Geographos as the most elongated solar system object imaged so far (Fig. 2). The images show craters as well as indications of other sorts of large-scale topographic relief, including a prominent central indentation. Protuberances at the asteroid's ends may be related to the pattern of ejecta removal and deposition caused by the asteroid's gravity field, or perhaps to tidal effects during a close planetary encounter (*Bottke et al.,* 1999).

4.3. 4179 Toutatis

Imaging of Toutatis during its close approaches in 1992 (*Ostro et al.,* 1995a), 1996 (*Ostro et al.,* 1999b), and 2000 achieved resolutions as fine as 125 ns (19 m in range) and 8.3 mHz (0.15 mm/s in radial velocity), placing hundreds to thousands of pixels on the asteroid. Inversion of a low-resolution subset of the 1992 images (*Hudson and Ostro,* 1995) produced a comprehensive physical model (Fig. 4), including estimates of the asteroid's shape and moment-of-inertia ratios, initial conditions for the asteroid's spin and orientation, the radar-scattering properties of the surface, and the delay-Doppler trajectory of the center of mass. Lightcurves synthesized using that model provide a good fit to optical photometry that spans phase angles from 0.2° to 121.4° (*Spencer et al.,* 1995), and incorporation of the lightcurves in the modeling process slightly refines the spin state (*Hudson and Ostro,* 1998).

Toutatis is rotating in a long-axis mode characterized by periods of 5.4 d (rotation about the long axis) and 7.4 d (average for long-axis precession about the angular momentum vector); see *Scheeres et al.* (1998) for a detailed description

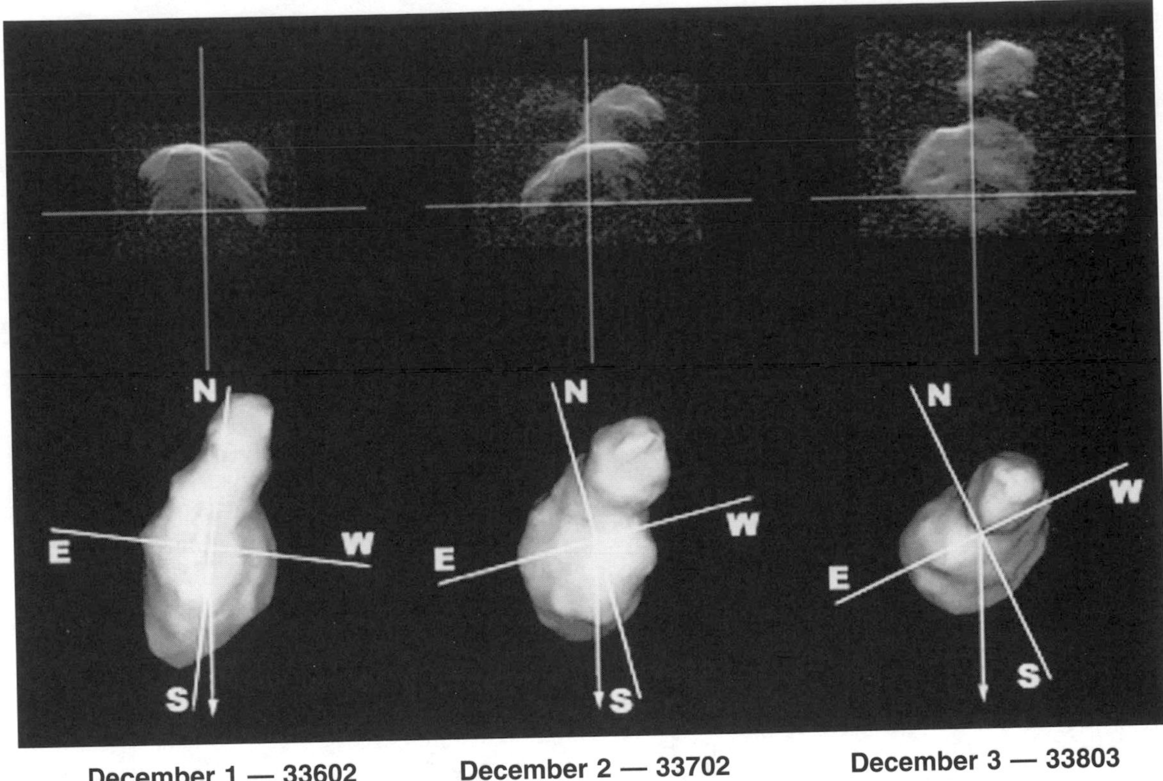

Fig. 4. High-resolution delay-Doppler images from each of three observation dates in 1996 and the corresponding plane-of-sky (POS) appearance of *Hudson and Ostro*'s (1995) physical model. The cross hairs are 5 km long and centered on Toutatis' center of mass (COM). The radar images are plotted with time delay (range) increasing from top to bottom and Doppler frequency (radial velocity) increasing from left to right. In the POS images, the model is rendered with a Lambertian scattering law, with the viewer and the illumination source colocated. The cross hairs are aligned north-south and east-west on the plane of the sky. In each POS frame, the arrow radiating from the COM shows the POS projection of the instantaneous spin vector.

of the spin state. The asteroid's principal moments of inertia are in ratios within 3% of 3.19 and 3.01, and the inertia tensor is indistinguishable from that of a homogeneous body; such information has yet to be determined for any other asteroid except Eros and probably is impossible to acquire in a fast spacecraft flyby. Dimensions along the principal axes are within 0.10 km of 1.92, 2.40, and 4.60 km. The asteroid's centimeter-to-decameter surface characteristics are strikingly uniform. The disk-integrated, 3.5-cm circular polarization (SC/OC) ratio averages 0.29 ± 0.01 and is independent of orientation at the several percent level. SC/OC = 0.25 ± 0.02 at 6 cm (*Zaitsev et al.*, 1993). The 3.5-cm OC radar albedo averages 0.24 ± 0.03; it depends on orientation, as expected from the asteroid's angular scattering behavior (limb-darkening slightly more than Lambertian). The OC albedo of a sphere with Toutatis' radar properties would be 0.21, or 3× the lunar value. The radar properties and available nonradar constraints are consistent with Toutatis' surface having a smooth component that is at least one-third covered by rocks at least as large as a decimeter. If this S-class asteroid is mineralogically similar to stony-iron meteorites, then the smooth surface component probably is regolith with porosity resembling that of lunar soil. If the mineralogy is ordinary chondritic, then the smooth surface component may be nearly solid with not much more than a centimeter of overlying regolith.

The highest-resolution radar images obtained in 1992 and 1996 have been used to construct a Toutatis model consisting of 39,996 triangular facets of roughly equal area, defined by the locations of 20,000 vertices (R. S. Hudson et al., personal communication, 2001). The average spatial resolution of the model is approximately 34 m, significantly finer than the *Hudson and Ostro* (1995) model (1600 vertices, resolution 84 m). The high-resolution model (Fig. 5) reveals complex linear features as well as circular crater-like structures down to the resolution limit. The noncrater-like features may be the manifestation of complex interior configurations involving monolithic fragments with various sizes and shapes, presumably due to collisions in various energy regimes.

Given the intensity and outcome of the Toutatis investigations, it may be appropriate to think of that experiment as a flyby of Toutatis by an Earthborne package of radar sensors. A spacecraft flyby would be hundreds of times more expensive than the incremental cost of the radar investigation and would not have been able to determine the asteroid's shape and spin state nearly as well. Obviously, the ratio of cost and risk to science return for a flyby or rendez-

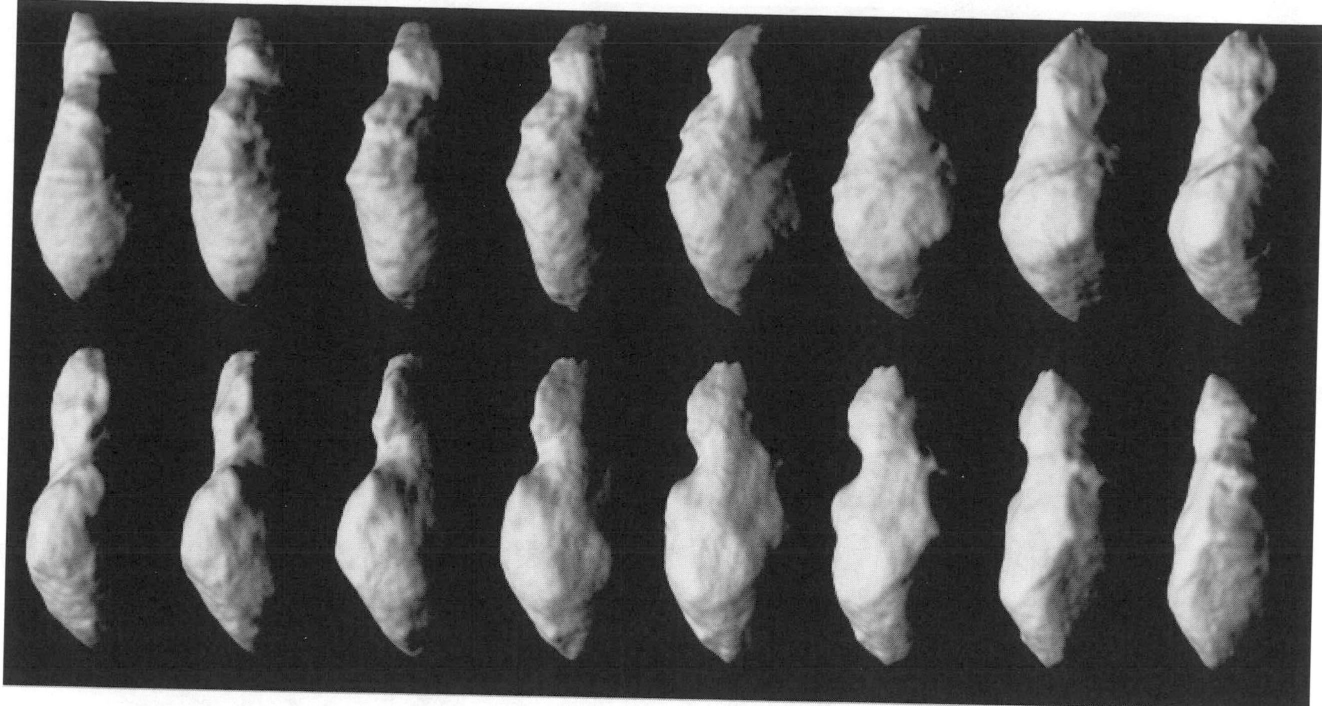

Fig. 5. The high-resolution Toutatis model (R. S. Hudson et al., personal communication, 2001) rendered at 22.5° intervals of rotation about the long (x) axis, with illumination coming from within the y-z plane at a 70° angle from the viewer's line of sight.

vous mission (let alone a sample return or a piloted mission) would be dramatically reduced if Toutatis-class radar reconnaissance could be performed on the target in advance.

Toutatis' radar-refined orbit has weighted root-mean-square residuals of 0.98 arcsec, 1.8 mm s^{-1} in radial velocity, and 73 m in range. Integration of that orbit into the past and future (*Ostro et al.,* 1999b) shows that Toutatis' pattern of close approaches to Venus, Earth, and Mars is highly asymmetric about the current epoch. The probability of the orbit intersecting the Earth is zero for at least the next six centuries. Toutatis will make its closest planetary approach since at least 1353 and until at least 2562 on September 29, 2004, when the closest COM-to-COM separation of Earth and Toutatis will be 1,549,834 ± 10 km (4.0 lunar distances).

4.4. 6489 Golevka

Goldstone 8510-MHz radar images of Golevka in June 1995 and physical modeling (*Hudson et al.,* 2000a) reveal a half-kilometer object (Plate 2) whose shape is extraordinarily angular, with flat sides, sharp edges and corners, and peculiar concavities. Extremely large gravitational slopes in some areas indicate the presence of exposed, solid, monolithic rock. This asteroid, the first subkilometer object studied in this much detail, is more likely to be a collision fragment than an unconsolidated rubble pile.

The Golevka experiment led to several technical milestones. After three radar-refined generations of orbit solutions during June 3–8, the uncertainty ellipse for the delay-Doppler location of the COM was "inside" the asteroid. The ephemeris used on June 9, when the asteroid made its closest approach (0.034 AU), was later found to be accurate to 0.01 Hz (a radial velocity of 0.2 mm/s, or about the speed of the tip of a large clock's minute hand). The preliminary model generated on that date was qualitatively almost indistinguishable from the one published by *Hudson et al.* (2000a). June 13–15 saw the first intercontinental radar astronomy observations, consisting of Goldstone cw transmissions and reception of Golevka's echoes with the Evpatoria (Ukraine) 70-m antenna on each date (*Zaitsev et al.,* 1997) and reception of echoes with the Kashima (Japan) 34-m antenna on June 15 (*Koyama et al.,* 2001). [In 1992, observations with the Evpatoria antenna transmitting and the Effelsberg 100-m antenna receiving had yielded echoes from Toutatis (*Zaitsev et al.,* 1993). In 1999, Golevka echoes from Arecibo transmissions were detected at the Deep Space Network's 70-m DSS-63 antenna near Madrid. In 2001, the Medecina (Italy) 32-m antenna detected echoes from 33342 (1998 WT24) using Goldstone and then Evpatoria transmissions.]

4.5. 4486 Mithra

Images of Mithra reveal a double-lobed object, apparently more severely bifurcated than any other nonbinary NEA imaged to date (*Ostro et al.,* 2000c). The bandwidth of the echoes is consistently very narrow, implying some combination of very slow rotation (evident from the barely noticeable variation in the appearance of images over several hours) and a radar line of sight not far from the apparent spin vector at any time during the experiment; the

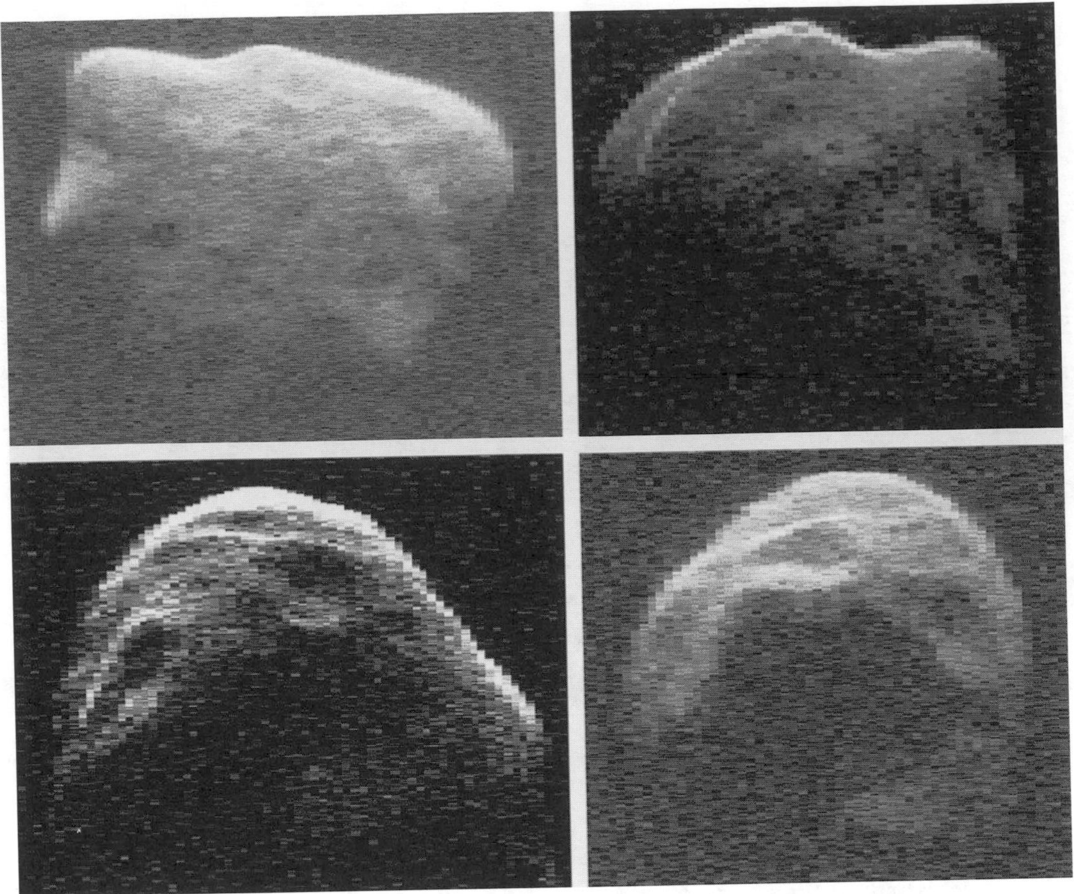

Fig. 6. Radar images of 1999 JM8 (*Benner et al.,* 2002a) from Arecibo (top left and bottom right frames) and Goldstone. Radar illumination is from the top. The vertical resolution is 15 m for Arecibo, 38 m in the top Goldstone image, and 19 m in the bottom Goldstone image. The horizontal resolutions depend on the asteroid's spin state, which is not yet known.

radar-observed sky arc was only about 35°. No simple periodicity in the day-to-day image sequence is evident, so nonprincipal-axis rotation is suggested. The alignment of the two lobes is almost parallel to the projected, apparent, instantaneous spin vector in some images but almost perpendicular to it in others, providing additional evidence for a very unusual spin state.

4.6. 1999 JM8

Observations by *Benner et al.* (2002) during this object's discovery apparition (its closest approach until at least 3000) reveal an irregularly shaped object with a maximum dimension that exceeds 5 km and numerous craterlike features with diameters from 100 m to more than 1 km (Fig. 6). 1999 JM8 is in a slow NPA rotation state that may not be very "far" from a PA state. The images provide the strongest evidence to date for a circular polarization ratio feature on an asteroid.

4.7. 1998 KY26

Radar and optical observations (*Ostro et al.,* 1999a) of this NEA shortly after its discovery reveal a slightly elongated, possibly lumpy spheroid with a diameter of about 30 m, a composition analogous to carbonaceous chondritic meteorites, and a rotation period of 10.7 min, which is too rapid for 1998 KY26 to consist of multiple components bound together just by their mutual gravitational attraction. On the other hand, E. Asphaug (personal communication, 1999) has noted that the required tensile strength is very small, comparable to that of snow. 1998 KY26 has the lowest rendezvous ΔV (the total change in velocity required to leave low Earth orbit and rendezvous with another body) of any object with a well-known orbit.

4.8. 216 Kleopatra

Radar observations of this M-class asteroid (*Ostro et al.,* 2000a) reveal a dog-bone shape with overall dimensions of 217 × 94 × 81 km (±25%). The object's high radar albedo, low SC/OC, and gentle gravitational slopes are consistent with a regolith having a metallic composition and a porosity comparable to that of lunar soil. *Ostro et al.* (2000a) argue that Kleopatra's shape is probably the outcome of an exotic sequence of collisional events, and that much of its interior may have an unconsolidated rubble-pile structure.

4.9. 288 Glauke

The combination of Arecibo radar echoes and available visible/infrared data indicates that Glauke is an S-class object slightly smaller and less elongated than 243 Ida, with radar surface properties near the average for S asteroids in the main belt, and in an extraordinarily slow (~50-d) rotation state (*Ostro et al.,* 2001b).

4.10. 1999 GU3

Optical and radar observations (*Pravec et al.,* 2000a) reveal that 1999 GU3 is a subkilometer-sized object with an apparent rotation period of 9 d, low visual and radar albedos, and colors more consistent with the ordinary chondrites than the vast majority of main-belt asteroids.

4.11. 25143 (1998 SF36)

Extensive Arecibo and Goldstone observations of this object were conducted in March–April 2001 in response to its selection as the target of the joint Japanese/NASA MUSES-C sample-return mission. A first look at the images reveals a somewhat ellipsoidal, but asymmetrical, object with overall dimensions of roughly 0.6 × 0.3 km. The surface's small-scale roughness is comparable to Eros', but the surface density appears lower.

4.12. 2063 Bacchus and 3908 Nyx (1980 PA)

Delay-Doppler images of Bacchus (*Benner et al.,* 1999b) reveal an asymmetrical, bifurcated shape with radar-derived dimensions (1.1 × 0.5 × 0.5 km) and optical brightness that imply radar and optical albedos among the highest measured for any asteroid, consistent with, for example, a regolith-free basaltic surface. The surface of the basaltic (V-class) object 3908 Nyx is unusually rough at centimeter-to-decimeter scales (*Benner et al.,* 2002b). One or both of these objects may be fragments of Vesta (*Binzel and Xu,* 1993).

4.13. 7 Iris, 9 Metis, 654 Zelinda, and 12 Victoria

Radar spectra for these large MBAs show evidence for prominent topography: large, flat regions on Iris, Metis, and Zelinda, and a nonconvex, possibly bifurcated shape for Victoria (*Mitchell et al.,* 1995).

4.14. Spheroids

1999 RQ36 (*Hudson et al.,* 2000b), 7822 (1991 CS) (*Benner et al.,* 1999a), 2100 Ra-Shalom (*Shepard et al.,* 2000), 1998 ML14 (*Ostro et al.,* 2001a), 29075 (1950 DA) (*Giorgini et al.,* 2002), and 6037 (1988 EG) have nearly circular pole-on silhouettes. Very strong, well-resolved images of 1999 RQ36 reveal a nearly featureless spheroid. Inversion of images of 1998 ML14 reveals a 1-km spheroid with several-hundred-meter protrusions on one side.

4.15. Very Small Near-Earth Asteroids

As of March 2002, NEAs with H > 21 (corresponding to diameters of ~0.2 km or less) constitute 24/103 or 23% of all radar-detected NEAs (Table 5). Objects in this size range are likely to comprise an increasing percentage of radar-detected NEAs, due to the relatively fast rate at which they are being discovered. Some of these asteroids are comparable in size to boulders seen on the surface of Eros (*Veverka et al.,* 2001). Most of these objects appear to have rotation periods no longer than an hour, but at least one object, 2001 EC16, is a very slow rotator (J. L. Margot et al., in preparation, 2001). Existing radar systems typically resolve the smallest radar detectable asteroids into only a few range pixels or less, so much of the leverage for shape modeling will have to come from Doppler resolution.

4.16. Binary Near-Earth Asteroids

Goldstone and Arecibo delay-Doppler observations during September 2000 to October 2001 revealed that 2000 DP107, 2000 UG11, 1999 KW4, 1998 ST27, and 2002 BM26 are binary systems. For 2000 DP107 (*Margot et al.,* 2002), images show separations up to at least 1 km between the components, which have different sizes and rotation states. Estimates of the diameters based on range extents are 800 m and 300 m for the primary and secondary respectively. Preliminary fits to delay-Doppler data indicate an orbital period of 1.7 d and a semimajor axis of 2.6 km (uncertainties ~10%). These parameters imply that the density of the primary is about 1.7 g cm^{-3}. For 2000 UG11 (*Nolan et al.,* 2001), preliminary estimates of average diameters, based on range extents at 15-m resolution, are 230 and 100 m. The components' maximum separation is more than 300 m and its orbital period is 19 ± 1.5 h. For 1999 KW4, very thorough, high-SNR, decameter-resolution delay-Doppler images should characterize the components and their dynamics in detail. The images (e.g., Fig. 7) show separations up to at least 2 km between the components, whose sizes differ by a factor of about 3 (*Benner et al.,* 2001a; S. Ostro et al., in preparation, 2002). Initial analysis suggests that the density of the primary is between 1.8 and 3.0 g cm^{-3}. For 1998 ST27, the images show separations up to at least 4 km between the components, whose sizes differ by a factor of at least 3 (*Benner et al.,* 2001b).

5. CHALLENGES FOR THE COMING DECADE

5.1 Telescope Time

Both Arecibo and Goldstone are heavily oversubscribed (as foreseen; *Ostro,* 1994). Additionally, numerous NEA opportunities at Goldstone are lost due to airspace clearance protocols that require all transmissions to be approved by military and civilian authorities. During the past few years, Arecibo was used for radar astronomy 8% of the time, of which more than half was for asteroids. DSS-14

TABLE 5. Very small radar-detected near-Earth asteroids.

Asteroid	H (mag)	D_{radar} (m)	D_C (m)	D_S (m)	Period (h)	SC/OC	Reference*
1998 BY7	21.5	NA	300	150	<1.3	0.4 ± 0.04	Ostro et al.
1998 KY26	25.5	30	48	24	0.18	0.5 ± 0.1	*Ostro et al.* (1999a)
1999 FN19	22.4	110	190	95	<11	0.22 ± 0.01	Benner et al.
1999 NW2	23.1	NA	150	72	<4.2	0.35 ± 0.02	Benner et al.
1999 TN13	23.6	NA	120	57	<1.3		Benner et al.
1999 TY2	23.3	NA	130	66	0.12		*Pravec et al.* (2000b)
2000 EH26	21.2	120	260	130	<33		*Benner et al.* (2001a)
2000 EW70	21.1	360	360	180	<33		Margot et al.
2000 LF3	21.6	NA	300	150	<1.0	<0.2	Nolan et al.
2000 PH5	22.3	120	170	90	0.29	0.29 ± 0.02	Margot et al.
2000 UK11	25.0	<60	52	26	~0.05		*Nolan et al.* (2001)
2000 YA	23.7	120	110	52	<1.3		*Benner et al.* (2001a)
2001 AV43	24.3	NA	72	36	?		Nolan et al.
2001 BF10	22.3	NA	170	83	<0.4		Howell et al.
2001 CP36	23.7	90	110	55	slow?	~1.0	Nolan et al.
2001 EC16	22.2	150	200	100	slow	~0.2	*Margot et al.* (2001)
2001 FR85	24.5	?	79	40	?		Nolan et al.
2001 JV1	21.3	?	320	160	<29		Ostro et al.
2001 SP263	25.8	?	42	21	?		Nolan et al.
2001 UP	25.7	?	44	22	?		Nolan et al.
2001 WM15	25.0	?	60	30	?		Nolan et al.
2001 XX4	21.9	?	250	130	?		Nolan et al.
2001 YP3	21.9	?	250	130	?		Nolan et al.
2002 AV	20.7	?	380	190	?		Ostro et al.
2002 FD6	22.3	110	210	110	?		Benner et al.

*References without dates are in preparation.

Objects with H > 21 (suggesting diameters less than about 200 m) are listed along with constraints on diameter, rotation period, and circular polarization ratio. D_{radar} is twice the echoes' range extent except for 1998 KY26, for which a shape model is available. For some objects, range-resolved data are not available (NA). D_C and D_S are diameters corresponding to typical optical geometric albedos p_v for the C (p_v = 0.05) and S (p_v = 0.20) classes, calculated using the equation log p_v = 6.259 – log D – 0.4 H, from *Bowell et al.* (1989). For objects other than 1998 KY26 and 1999 TY2, we combined D_C with the echo bandwidth to set an upper bound on the period. For 1999 NW2, photometric colors (M. Hicks, personal communication, 1999) suggest an SQ classification, so we combined D_S with the echo bandwidth to set an upper bound on the period.

was used for radar astronomy 6% of the time, of which less than half was for asteroids. These percentages are unlikely to increase much, and they certainly cannot keep up with the NEA discovery rate. Even now, only about half of asteroid radar time requests are allocated. NPA rotators are generally very slow, so many days of observations are required for a unique spin-state solution to be tractable. Thus, limitations on telescope time allocation undermine characterization of objects in the most interesting spin states. Perhaps it is time to propose a dedicated asteroid radar with 10× the sensitivity of Arecibo, whose ~$200 million cost would be comparable to that of a Discovery mission (*Ostro,* 1997).

5.2. Multiantenna Observations and Interferometry

The potential of the recently completed 100-m Greenbank Telescope (GBT, part of the National Radio Astronomy Observatory, NRAO) as the receiving antenna in "bistatic" radar experiments is both promising and challenging.

The Arecibo-GBT configuration, which already has yielded a detection of 2000 EC16 (J. L. Margot et al., in preparation, 2001), will permit more lucrative observations of very close and/or narrowband (very small or slowly rotating) NEAs than would be possible with Arecibo monostatically. Goldstone-GBT observations will enjoy a tenfold increase in sensitivity over the DSS-14-to-DSS-13 configuration. However, apart from the logistics of securing antenna time on two institution's instruments, establishment of an unbiased system for accurate delay astrometry will be more difficult than it was for the DSS-14-to-DSS-13 system, which has a fiber-optics linkage. Similar comments apply to 2380-MHz Arecibo-Goldstone observations, which have been conducted several times; observations involving Arecibo and the 25-m VLBA antenna on St. Croix, which have been used for bistatic imaging of the Moon; and 8560-MHz Goldstone-Arecibo observations, which are planned. This chapter was finalized right after completion of a three-week

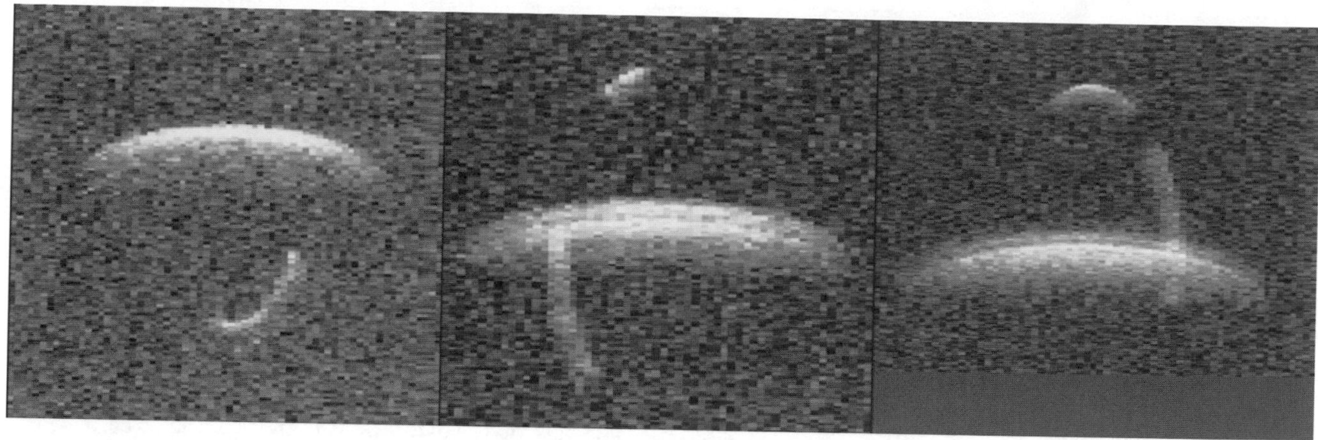

Fig. 7. Several-hour delay-Doppler time exposures of radar echoes from binary asteroid 1999 KW4 (S. Ostro et al., in preparation, 2002). Distance from Earth increases toward the bottom, and speed from Earth increases toward the left. The motion of the secondary (smaller) component about the primary (larger) component is clockwise. Gaps in the trail are due to breaks in the data-taking. The primary appears much wider than the secondary because it is a few times bigger and is rotating much faster. Although the components have the same speeds along the radar line of sight and the same distances from the radar where their echoes overlap, their positions in space are never the same. The components orbit a common center of mass, and each component's average distance from that point is inversely proportional to its mass. The motion of the relatively massive primary is much less obvious than the motion of the secondary, but can be seen in the double appearance of the primary's top edge in the two time exposures that follow the secondary from in front of the primary to behind it. These Goldstone (8560-MHz, 3.5-cm) images have overall extents of 37.5 μs × 67 Hz (5.6 km × 1.2 m s^{-1}).

experiment on 38071 (1999 GU3) incorporating Arecibo-GBT and Goldstone-GBT delay-Doppler imaging (L. A. M. Benner et al., in preparation, 2002).

Aperture-synthesis observations using transmission from Goldstone and reception at the 27-antenna Very Large Array (VLA) achieved marginal resolution of 324 Bamberga and 7 Iris (*de Pater et al.,* 1994) and helped to constrain their pole directions. Goldstone-VLA detections of Toutatis by the same authors and of Golevka by *Hudson et al.* (2000) provided POS astrometry that is roughly comparable in accuracy with the best traditional optical astrometry, that is, when Hipparcos-based reference catalogs are used in the reductions.

Apart from the Goldstone-VLA work, all useful asteroid radar imaging has relied on delay/Doppler resolution. In principle, interferometric methods can make radar images with very fine angular resolution, similar to optical pictures and free from the ambiguities that afflict delay-Doppler images. The VLA's longest baseline is 36.4 km and its finest achievable angular resolution at 8560 MHz is 0.24 arcsec. Enlargement of the array could shrink the resolution by nearly an order of magnitude. The Very Long Baseline Array (VLBA) has a maximum baseline of 8611 km and in principle can achieve a resolution at 8560 MHz of 0.85 milliarcsec. Arecibo-VLBA echoes from 2000 EW70 (G. J. Black et al., in preparation, 2001) have been detected, but actual VLBA radar images of asteroids have yet to be constructed for any object. *Koyama et al.* (2001) attempted interferometric observations of 4197 1982 TA using Goldstone transmissions and reception at the Kashima and Usuda antennas in Japan, but analysis of the echoes was inconclusive. Single baseline interferometry has been attempted with Golevka (Arecibo-Madrid), 1999 RQ36 (Arecibo-Goldstone), and 1999 KW4 (Arecibo-Goldstone); see *Margot and Nolan* (1999). Although interferometric techniques are logistically daunting, their promise is considerable. Perhaps the most tantalizing experiment would be a multiday sequence of simultaneous delay-Doppler images and radar interferometric images of a comet nucleus and coma at very high echo strength.

5.3. Data Interpretation

Reconstruction of physical models from delay-Doppler images has proved to be relatively straightforward for uniform rotators but horribly difficult for NPA rotators because of the size of the parameter space. This estimation problem demands serious attention.

Vokrouhlický et al. (2000) have suggested that the existence of precise radar astrometry of certain asteroids at multiple apparitions might reveal orbit perturbations resulting from the Yarkovsky effect, a subtle nongravitational phenomenon related to anisotropic thermal emission from the asteroid's surface. They list several opportunities for acquiring such data during the next few decades. The associated covariance space, especially the coupling between constraints on the Yarkovsky effect and constraints on asteroid thermal behavior, indicates that definitive conclusions will require modeling of the effect using the actual shape and spin state of any candidate target, with a suitably parameterized thermal model.

Integration of the radar + optical orbit of 29075 (1950 DA) (*Giorgini et al.,* 2001) has indicated a very close approach to Earth (with a small possibility of impact) in March 2880 that was not evident in solutions based solely on initial optical data. The associated uncertainties depend primarily on the Yarkovsky acceleration, which depends on the object's shape, spin state, and thermal properties. This situation dramatizes the fundamental coupling between the physical properties of NEAs and long-term prediction of their trajectories, as well as the value of radar astrometry for orbit determination.

The radar discovery of NEA binary systems presents the challenge of extracting physical and dynamical properties from the echoes. Delay-Doppler images are taken in a coordinate system tied to the planetary ephemerides, so one can determine the orbits of the components relative to the system's center of mass as well as relative to each other. Thus the components' mass and density ratios as well as their total mass can be derived. Optimum estimation procedures need to be developed and supported by appropriate covariance studies. Ideally, one would merge the results with both shape estimation and heliocentric orbit refinement procedures to construct a single, comprehensive estimator for all the relevant physical and dynamical parameters for single and multiple bodies.

REFERENCES

Andrews A. K., Hudson R. S., and Psaltis D. (1995) Optical-radar imaging of scale models for studies in asteroid astronomy. *Optics Lett., 20,* 2327–2329.

Asphaug E., Ostro S. J., Hudson R. S., Scheeres D. J., and Benz W. (1998) Disruption of kilometre-sized asteroids by energetic collisions. *Nature, 393,* 437–440.

Benner L. A. M. (2002) Summaries of asteroid radar properties [on line]. California Institute of Technology, Pasadena [cited June 1, 2002]. Available on line at http://echo.jpl.nasa.gov/~lance/asteroid_radar_properties.html.

Benner L. A. M., Ostro S. J., Giorgini J. D., Jurgens R. F., Mitchell D. L., Rose R., Rosema K. D., Slade M. A., Winkler R., Yeomans D. K., Campbell D. B., Chandler J. F., and Shapiro I. I. (1997) Radar detection of near-Earth asteroids 2062 Aten, 2101 Adonis, 3103 Eger, 4544 Xanthus, and 1992 QN. *Icarus, 130,* 296–312.

Benner L. A. M., Ostro S. J., Rosema K. D., Giorgini J. D., Choate D., Jurgens R. F., Rose R., Slade M. A., Thomas M. L., Winkler R., and Yeomans D. K. (1999a) Radar observations of asteroid 7822 (1991 CS). *Icarus, 137,* 247–259.

Benner L. A. M., Hudson R. S., Ostro S. J., Rosema K. D., Giorgini J. D., Yeomans D. K., Jurgens R. F., Mitchell D. L., Winkler R., Rose R., Slade M. A., Thomas M. L., and Pravec P. (1999b) Radar observations of asteroid 2063 Bacchus. *Icarus, 139,* 309–327.

Benner L. A. M., Ostro S. J., Giorgini J. D., Jurgens R. F., Margot J. L., and Nolan M. C. (2001a) *1999 KW4.* IAU Circular 7632.

Benner L. A. M., Ostro S. J., Giorgini J. D., Jurgens R. F., Margot J. L., and Nolan M. C. (2001b) *1998 ST27.* IAU Circular 7730.

Benner L. A. M., Ostro S. J., Nolan M. C., Margot J.-L., Giorgini J. D., Hudson R. S., Jurgens R. F., Slade M. A., Howell E. S., Campbell D. B., and Yeomans D. K. (2002a) Radar observations of asteroid 1999 JM8. *Meteoritics & Planet. Sci., 37,* 779–792.

Benner L. A. M., Ostro S. J., Hudson R. S., Rosema K. D., Jurgens R. F., Yeomans D. K., Campbell D. B., Chandler J. F., and Shapiro I. I. (2002b) Radar observations of asteroid 3908 Nyx. *Icarus, 158,* in press.

Binzel R. P. and Xu S. (1993) Chips off of asteroid 4 Vesta — Evidence for the parent body of basaltic achondrite. *Science, 260,* 186–191.

Binzel R. P., Gehrels T., and Matthews M. S., eds. (1989) *Asteroids II.* Univ. of Arizona, Tucson. 1258 pp.

Bottke W. F. Jr., Richardson D. C., and Love S. G. (1999) 1620 Geographos and 433 Eros: Shaped by planetary tides? *Astron. J., 117,* 1921–1928.

Bowell E., Hapke B., Domingue D., Lumme K., Peltoniemi J., and Harris A. W. (1989) Application of photometric models to asteroids. In *Asteroids II* (R. P. Binzel et al., eds.), pp. 524–556. Univ. of Arizona, Tucson.

de Pater I., Palmer P., Mitchell D. L., Ostro S. J., Yeomans D. K., and Snyder L. E. (1994) Radar aperture synthesis observations of asteroids. *Icarus, 111,* 489–502.

Garvin J. B., Head J. W., Pettengill G. H. and Zisk S. H. (1985) Venus global radar reflectivity and correlations with elevation. *J. Geophys. Res., 90,* 6859–6871.

Giorgini J. D. (2002) Small-Body Astrometric Radar Observations [on line]. California Institute of Technology, Pasadena [cited June 1, 2002]. Available on line at http://ssd.jpl.nasa.gov/radar_data.html.

Giorgini J. D., Ostro S. J., Benner L. A. M., Chodas P. W., Chesley S. R., Hudson R. S., Nolan M. C., Klemola A. R., Standish E. M., Jurgens R. F., Rose R., Chamberlin A. B., Yeomans D. K., and Margot J.-L (2002) Asteroid 1950 DA's encounter with Earth in 2880: Physical limits of collision probability prediction. *Science, 296,* 132–136.

Harmon J. K. and Ostro S. J. (1985) Mars: Dual-polarization radar observations with extended coverage. *Icarus, 62,* 110–128.

Harmon J. K., Campbell D. B., Ostro S. J., and Nolan M. C. (1999) Radar observations of comets. *Planet. Space Sci., 47,* 1409–1422.

Harris A. W. (1994) Tumbling asteroids. *Icarus, 107,* 209–211.

Howell E. S., Nolan M. C., DeRemer L., and Margot J.-L. (2001) Arecibo radar observations of near-Earth asteroid 2000 EE104 (abstract). *Bull. Am. Astron. Soc., 33,* 1153.

Hudson S. (1993) Three-dimensional reconstruction of asteroids from radar observations. *Remote Sens. Rev., 8,* 195–203.

Hudson R. S. and Ostro S. J. (1994) Shape of asteroid 4769 Castalia (1989 PB) from inversion of radar images. *Science, 263,* 940–943.

Hudson R. S. and Ostro S. J. (1995) Shape and non-principal-axis spin state of asteroid 4179 Toutatis from radar images. *Science, 270,* 84–86.

Hudson R. S. and Ostro S. J. (1998) Photometric properties of asteroid 4179 Toutatis from lightcurves and a radar-derived physical model. *Icarus, 135,* 451–457.

Hudson R. S. and Ostro S. J. (1999) Physical model of asteroid 1620 Geographos from radar and optical data. *Icarus, 140,* 369–378.

Hudson R. S., Ostro S. J., and Harris A. W. (1997) Constraints on spin state and Hapke parameters of asteroid 4769 Castalia using lightcurves and a radar-derived shape model. *Icarus, 130,* 165–176.

Hudson R. S., Ostro S. J., Jurgens R. F., Rosema K. D., Giorgini J. D., Winkler R., Rose R., Choate D., Cormier R. A., Franck C. R., Frye R., Howard D., Kelley D., Littlefair R., Slade

M. A., Benner L. A. M., Thomas M. L., Mitchell D. L., Chodas P. W., Yeomans D. K., Scheeres D. J., Palmer P., Zaitsev A., Koyama Y., Nakamura A., Harris A. W., and Meshkov M. N. (2000a) Radar observations and physical modeling of asteroid 6489 Golevka. *Icarus, 148,* 37–51.

Hudson R. S., Ostro S. J., and Benner L. A. M. (2000b) Recent delay-Doppler radar asteroid modeling results: 1999 RQ36 and craters on Toutatis (abstract). *Bull. Am. Astron. Soc., 32,* 1001.

Jenkins G. M. and Watts D. G. (1968) *Spectral Analysis and Its Applications.* Holden-Day, San Francisco. 525 pp.

Jurgens R. F. and Goldstein R. M. (1976) Radar observations at 3.5 and 12.6 cm wavelength of asteroid 433 Eros. *Icarus, 28,* 1–15.

Koyama Y., Nakajima J., Sekido M., Yoshikawa M., Nakamura A. M., Hirabayashi H., Okada T., Abe M., Nishibori T., Fuse T., Ostro S. J., Choate D., Cormier R. A., Winkler R., Jurgens R. F., Giorgini J. D., Yeomans D. K., Slade M. A., and Zaitsev A. L. (2001) Radar observations of near Earth asteroids 6489 Golevka and 4197 (1982 TA). *J. Comm. Res. Lab., 48,* 143–150.

Magri C., Ostro S. J., Rosema K. D., Mitchell D. L., Campbell D. B., Chandler J. F., Shapiro I. I., Giorgini J. D., and Yeomans D. K. (1999) Mainbelt asteroids: Results of Arecibo and Goldstone radar observations of 37 objects during 1980–1995. *Icarus, 140,* 379–407.

Magri C., Consolmagno G. J., Ostro S. J., Benner L. A. M., and Beeney B. R. (2001) Radar constraints on asteroid regolith compositions using 433 Eros as ground truth. *Meteoritics & Planet. Sci., 36,* 1697–1709.

Margot J.-L. and Nolan M. C. (1999) Radar interferometric imaging of near-Earth asteroids (abstract). *Bull. Am. Astron. Soc., 31,* 2002.

Margot J. L., Nolan M. C., Benner L. A. M., Ostro S. J., Jurgens R. F., Slade M. A., Giorgini J. D., and Campbell D. B. (2000) *Satellites of Minor Planets.* IAU Circular 7503.

Margot J. L., Nolan M. C., Benner L. A. M., Ostro S. J., Jurgens R. F., Giorgini J. D., Slade M. A., and Campbell D. B. (2002) Binary asteroids in the near-Earth object population. *Science, 296,* 1445–1448.

Mitchell D. L., Ostro S. J., Rosema K. D., Hudson R. S., Campbell D. B., Chandler J. F., and Shapiro I. I. (1995) Radar observations of asteroids 7 Iris, 9 Metis, 12 Victoria, 216 Kleopatra, and 654 Zelinda. *Icarus, 118,* 105–131.

Mitchell D. L., Ostro S. J., Hudson R. S., Rosema K. D., Campbell D. B., Velez R., Chandler J. F., Shapiro I. I., Giorgini J. D., and Yeomans D. K. (1996) Radar observations of asteroids 1 Ceres, 2 Pallas, and 4 Vesta. *Icarus, 124,* 113–133.

Mitchell D. L., Hudson R. S., Ostro S. J., and Rosema K. D. (1998) Shape of asteroid 433 Eros from inversion of Goldstone radar Doppler spectra. *Icarus, 131,* 4–14.

Nolan M. C., Margot J.-L., Howell E. S., Benner L. A. M., Ostro S. J., Jurgens R. F., Giorgini J. D., and Campbell D. B. (2001) Radar observations of near-Earth asteroids 2000 UG11 and 2000 UK11 (abstract). In *Lunar and Planetary Science XXXII,* Abstract #2055. Lunar and Planetary Institute, Houston (CD-ROM).

Olhoeft G. R. and Strangway D. W. (1975) Electrical properties of the first 100 meters of the moon. *Earth Planet. Sci. Lett., 24,* 394–404.

Ostro S. J. (1989) Radar observations of asteroids. In *Asteroids II* (R. P. Binzel et al., eds.), pp. 192–212. Univ. of Arizona, Tucson.

Ostro S. J. (1993) Planetary radar astronomy. *Rev. Mod. Phys., 65,* 1235–1279.

Ostro S. J. (1994) The role of groundbased radar in near-Earth object hazard identification and mitigation. In *Hazards Due to Comets and Asteroids* (T. Gehrels, ed.), pp. 259–282. Univ. of Arizona, Tucson.

Ostro S. J. (1997) Radar reconnaissance of near-Earth objects at the dawn of the next millennium. In *Near-Earth Objects: The United Nations International Conference* (J. Remo, ed.), pp. 118–139. Annals of the New York Academy of Sciences, Vol. 822.

Ostro S. J. (2002a) Radar-Detected Asteroids [on line]. California Institute of Technology, Pasadena [cited June 1, 2002]. Available on line at http://echo.jpl.nasa.gov/asteroids/index.html.

Ostro S. J. (2002b) Planetary radar astronomy. In *The Encyclopedia of Physical Science and Technology, 3rd Edition, Vol. 12* (R. A. Meyers, ed.), pp. 295–328. Academic, New York.

Ostro S. J. and Connelly R. (1984) Convex profiles from asteroid lightcurves. *Icarus, 57,* 443–463.

Ostro S. J., Campbell D. B., and Shapiro I. I. (1983) Radar observations of asteroid 1685 Toro. *Astron. J., 88,* 565–576.

Ostro S. J., Campbell D. B., and Shapiro I. I. (1985) Mainbelt asteroids: Dual-polarization radar observations. *Science, 229,* 442–446.

Ostro S. J., Connelly R., and Belkora L. (1988) Asteroid shapes from radar echo spectra: A new theoretical approach. *Icarus, 73,* 15–24.

Ostro S. J., Chandler J. F., Hine A. A., Shapiro I. I., Rosema K. D., and Yeomans D. K. (1990a) Radar images of asteroid 1989 PB. *Science, 248,* 1523–1528.

Ostro S. J., Rosema K. D., and Jurgens R. F. (1990b) The shape of Eros. *Icarus, 84,* 334–351.

Ostro S. J., Campbell D. B., Hine A. A., Shapiro I. I., Chandler J. F., Werner C. L., and Rosema K. D. (1990c) Radar images of asteroid 1627 Ivar. *Astron. J., 99,* 2012–2018.

Ostro S. J., Campbell D. B., Chandler J. F., Shapiro I. I., Hine A. A., Velez R., Jurgens R. F., Rosema K. D., Winkler R., and Yeomans D. K. (1991a) Asteroid radar astrometry. *Astron. J., 102,* 1490–1502.

Ostro S. J., Campbell D. B., Chandler J. F., Hine A. A., Hudson R. S., Rosema K. D., and Shapiro I. I. (1991b) Asteroid 1986 DA: Radar evidence for a metallic composition. *Science, 252,* 1399–1404.

Ostro S. J., Hudson R. S., Jurgens R. F., Rosema K. D., Winkler R., Howard D., Rose R., Slade M. A., Yeomans D. K., Giorgini J. D., Campbell D. B., Perillat P., Chandler J. F., and Shapiro I. I. (1995a) Radar images of asteroid 4179 Toutatis. *Science, 270,* 80–83.

Ostro S. J., Rosema K. D., Hudson R. S., Jurgens R. F., Giorgini J. D., Winkler R., Yeomans D. K., Choate D., Rose R., Slade M. A., Howard S. D., and Mitchell D. L. (1995b) Extreme elongation of asteroid 1620 Geographos from radar images. *Nature, 375,* 474–477.

Ostro S. J., Jurgens R. F., Rosema K. D., Hudson R. S., Giorgini J. D., Winkler R., Yeomans D. K., Choate D., Rose R., Slade M. A., Howard S. D., Scheeres D. J., and Mitchell D. L. (1996) Radar observations of asteroid 1620 Geographos. *Icarus, 121,* 44–66.

Ostro S. J., Pravec P., Benner L. A. M., Hudson R. S., Šarounová L., Hicks M. D., Rabinowitz D. L., Scotti J. V., Tholen D. J., Wolf M., Jurgens R. F., Thomas M. L., Giorgini J. D., Chodas P. W., Yeomans D. K., Rose R., Frye R., Rosema K. D., Winkler R., and Slade M. A. (1999a) Radar and optical observations of asteroid 1998 KY26. *Science, 285,* 557–559.

Ostro S. J., Hudson R. S., Rosema K. D., Giorgini J. D., Jurgens R. F., Yeomans D. K., Chodas P. W., Winkler R., Rose R., Choate D., Cormier R. A., Kelley D., Littlefair R., Benner L. A. M., Thomas M. L., and Slade M. A. (1999b) Asteroid 4179 Toutatis: 1996 radar observations. *Icarus, 137,* 122–139.

Ostro S. J., Hudson R. S., Nolan M. C., Margot J.-L., Scheeres D. J., Campbell D. B., Magri C., Giorgini J. D., and Yeomans D. K. (2000a) Radar observations of asteroid 216 Kleopatra. *Science, 288,* 836–839.

Ostro S. J., Hudson R. S., and Thomas P. C. (2000b) Evaluation of radar constraints on the shape of Eros (abstract). *Eos Trans. AGU, 81,* F771.

Ostro S. J., Hudson R. S., Benner L. A. M., Nolan M. C., Margot J.-L., Giorgini J. D., Jurgens R. F., Rose R., Yeomans D. K. (2000c) Radar observations of asteroid 4486 Mithra (abstract). *Bull. Am. Astron. Soc., 32,* 1003.

Ostro S. J., Hudson R. S., Benner L. A. M., Nolan M. C., Giorgini J. D., Scheeres D. J., Jurgens R. F., and Rose R. (2001a) Radar observations of asteroid 1998 ML14. *Meteoritics & Planet. Sci., 36,* 1225–1236.

Ostro S. J., Nolan M. C., Margot J.-L., Magri C., Harris A. W., and Giorgini J. D. (2001b) Radar observations of asteroid 288 Glauke. *Icarus, 152,* 201–204.

Ostro S. J., Rosema K. D., Campbell D. B., and Shapiro I. I. (2002) Radar observations of asteroid 1862 Apollo. *Icarus, 156,* 580–583.

Pravec P., Šarounová L., Benner L. A. M., Ostro S. J., Hicks M. D., Jurgens R. F., Giorgini J. D., Slade M. A. Yeomans, D. K., Rabinowitz D. L., Krugly Y. N., and Wolf M. (2000a) Slowly rotating asteroid 1999 GU3. *Icarus, 148,* 589–593.

Pravec P., Hergenrother C., Whiteley R., Šarounová L., Kušnirák P., and Wolf M. (2000b) Fast rotating asteroids 1999 TY2, 1999 SF10, and 1998 WB2. *Icarus, 147,* 477–486.

Rivkin A. S., Howell E. S., Lebofsky L. A., Clark B. E., and Britt D. T. (2000) The nature of M-class asteroids from 3-mm observations. *Icarus, 145,* 351–368.

Scheeres D. J., Ostro S. J., Hudson R. S., and Werner R. A. (1996) Orbits close to asteroid 4769 Castalia. *Icarus, 121,* 67–87.

Scheeres D. J., Ostro S. J., Hudson R. S., Suzuki S., and de Jong E. (1998) Dynamics of orbits close to asteroid 4179 Toutatis. *Icarus, 132,* 53–79.

Scheeres D. J., Ostro S. J., Werner R. A., Asphaug E., and Hudson R. S. (2000) Effects of gravitational interactions on asteroid spin states. *Icarus, 147,* 106–118.

Shepard M. K., Benner L. A. M., Ostro S. J., Harris A. W., Rosema K. D., Shapiro I. I., Chandler J. F., and Campbell D. B. (2000) Radar observations of 2100 Ra-Shalom. *Icarus, 147,* 520–529.

Spencer J. R., Akimov L. A., Angeli C., Angelini P., Barucci M. A., Birch P., Blanco C., Buie M. W., Caruso A., Chiornij V. G., Colas F., Dentshev P., Dorokhov N. I., de Sanctis M. C., Dotto E., Ezhkova O. B., Fulchignoni M., Green S., Harris A. W., Howell E. S., Hudecek T., Kalashnikov A. V., Kobelev V. V., Korobova Z. B., Koshkin N. I., Kozhevnikov V. P., Krugly Y. N., Lazzaro D., Lecacheux J., MacConnell J., Mel'Nikov S. Y., Michałowski T., Mueller B. E. A., Nakamura T., Neese C., Nolan M. C., Osborn W., Pravec P., Riccioli D., Shevchenko V. S., Shevchenko V. G., Tholen D. J., Velichko F. P., Venditti C., Venditti R., Wisniewski W., Young J., and Zellner B. (1995) The lightcurve of 4179 Toutatis: Evidence for complex rotation. *Icarus, 117,* 71–89.

Veverka J., Robinson M., Thomas P., Murchie S., Bell J. F., Izenberg N., Chapman C., Harch A., Bell M., Carcich B., Cheng A., Clark B., Domingue D., Dunham D., Farquhar R., Gaffey M. J., Hawkins E., Joseph J., Kirk R., Li H., Lucey P., Malin M., Martin P., McFadden L., Merline W. J., Miller J. K., Owen W. M., Peterson C., Prockter L., Warren J., Wellnitz D., Williams B. G., Yeomans D. K. (2000) NEAR at Eros: Imaging and spectral results. *Science, 289,* 2088–2097.

Veverka J., Thomas P. C., Robinson M., Murchie S., Chapman C., Bell M., Harch A., Merline W. J., Bell J. F., Bussey B., Carcich B., Cheng A., Clark B., Domingue D., Dunham D., Farquhar R., Gaffey M. J., Hawkins E., Izenberg N., Joseph J., Kirk R., Li H., Lucey P., Malin M., McFadden L., Miller J. K., Owen W. M., Peterson C., Prockter L., Warren J., Wellnitz D., Williams B. G., and Yeomans D. K. (2001) Imaging of small-scale features of 433 Eros from NEAR: Evidence for a complex regolith. *Science, 292,* 484–488.

Viateau B. (2000) Mass and density of asteroids (16) Psyche and (121) Hermione. *Astron. Astrophys., 354,* 725–731.

Vokrouhlický D., Milani A., and Chesley S. R. (2000) Yarkovsky effect on small near-Earth asteroids: Mathematical formulation and examples. *Icarus, 148,* 118–138.

Yeomans D. K., Ostro S. J., and Chodas P. W. (1987) Radar astrometry of near-Earth asteroids. *Astron. J., 94,* 189–200.

Yeomans D. K., Chodas P. W., Keesey M. S., Ostro S. J., Chandler J. F., and Shapiro I. I. (1992) Asteroid and comet orbits using radar data. *Astron. J., 103,* 303–317.

Zaitsev A. L., Sokolsky A. G., Rzhiga O. R., Vyshlov A. S., Krivtsov A. P., and Shubin V. A. (1993) Radar investigation of asteroid 4179 Toutatis with 6 cm waves. *Radiotekhnika Elektronika, 38,* 1842–1850 (in Russian). Translation: Zaytsev A. L., Sokol'skiy A. G., Rzhiga O. R., Vyshlov A. S., Krivtsov A. P., and Shubin V. A. (1993) Radar investigations of the asteroid 4179 Tautatis at wavelength of 6 cm. *J. Comm. Tech. Electronics, 38:16,* 135–143 (in English, by Scripta Technica, Inc., 1994).

Zaitsev A. L., Ostro S. J., Ignatov S P., Yeomans D. K., Petrenko A. G., Choate D., Margorin O. K., Cormier R. A., Mardyshkin V. V., Winkler R., Rghiga O. N., Jurgens R. F., Shubin V. A., Giorgini J. D., Krivtsov A. P., Rosema K. D., Koluka Y. F., Slade M. A., Gavrik A. L., Andreev V. B., Ivanov D. V., Peshin P. S., Koyama Y., Yoshikava M., Nakamura A. (1997) Intercontinental bistatic radar observations of 6489 Golevka (1991 JX). *Planet. Space Sci., 45,* 771–778.

Visible-Wavelength Spectroscopy of Asteroids

Schelte J. Bus
University of Hawai'i

Faith Vilas
NASA Johnson Space Center

M. Antonietta Barucci
Observatoire de Paris–Meudon

Since first becoming available for astronomical research in the early 1980s, charge-coupled-device (CCD) spectrographs have had a profound impact on our ability to measure the spectral-reflectance properties of asteroids. High signal-to-noise, low-resolution spectra, covering the visible-wavelength region from 0.4 to 1.0 μm, are now routinely obtained for asteroids much fainter than were measured during the Eight-Color Asteroid Survey. By recording the entire spectral range in a single exposure, some of the difficulties associated with multifilter photometry, arising from the inherent rotation of asteroids or from temporal variations in sky conditions, can be avoided. Studies involving CCD spectroscopy have resulted in the discovery of several absorption features in the spectra of asteroids and have provided new insights into the compositional nature of asteroid surfaces. Spectral surveys have also helped to refine our understanding of the orbital distributions of asteroid classes. We discuss the practical aspects of asteroid spectroscopy, focusing on observing procedures, data reduction techniques, and potential sources for uncertainty in the reduced spectra. We also review some of the applications of asteroid spectroscopy, and discuss how these observations have impacted the structure of asteroid taxonomy.

1. INTRODUCTION

A fundamental characteristic of an asteroid is its color. Incidental sunlight is either scattered or absorbed by mineral grains on an asteroid's surface; depending on the optical properties of these grains, the fraction of light that is reflected to Earth can vary as a function of wavelength. Based on his microphotometric measurements of photographic spectra, *Bobrovnikoff* (1929) first reported variations in the color of asteroids. Bobrovnikoff noted differences in the spectral continua of 12 asteroids, measured over the wavelength interval of 0.39–0.47 μm, and even monitored changes in the color of Vesta over time, which he was able to correlate with the asteroid's rotation. However, because of the limitations of photographic spectroscopy, very little progress was made in the study of asteroid colors until the advent of photoelectric detectors. A systematic investigation of the broadband UBV colors of asteroids was begun in the mid-1950s, leading to the identification of two distinct populations of objects based on their spectral reflectance properties (*Wood and Kuiper,* 1963).

The true spectral nature of asteroids started to become apparent as programs of narrow-band spectrophotometry were initiated in the late 1960s (*McCord et al.,* 1970; *Chapman et al.,* 1971). These narrow-band observations sampled the entire visible spectrum, from 0.3 to 1.1 μm, with higher spectral resolution than before. Besides the diversity in color that was observed, these data helped define some of the key characteristics that describe asteroid spectra over the visible spectrum, including strong absorption bands in the UV and near 1 μm (e.g., *Chapman and Gaffey,* 1979). These data also provided the basis for developing the first rigorous asteroid taxonomy (*Chapman et al.,* 1975) and allowed for more thorough characterizations of asteroid surface materials in the context of mineralogy (e.g., *Gaffey and McCord,* 1979).

Spectrophotometric studies of asteroids peaked in the mid-1980s with the completion of the Eight-Color Asteroid Survey (ECAS) (*Zellner et al.,* 1985). Using specially chosen filters with bandpasses targeting the major features contained in asteroid spectra, this survey sampled nearly 600 asteroids and formed the basis for the Tholen taxonomy (*Tholen,* 1984). At about the same time, a new generation of astronomical spectrographs was being developed that incorporated charge-coupled-device (CCD) cameras to record the spectra. Originally designed as an electronic analog to magnetic bubble memory, the introduction of CCDs as optical detectors helped revolutionize many areas of observational astronomy (*Janesick and Elliott,* 1992). The high sensitivity and stable, two-dimensional format of a CCD makes it ideal for use in spectroscopy. The use of long-slit spectrographs in the study of asteroids has many advantages over traditional filter photometry. These spectrographs provide the capability of imaging much, if not all, of the visible spectrum in a single exposure, with spectral resolutions that are much higher than possible with standard photometry. This essentially eliminates the potential for uncertainties that can arise due to photometric variability tied to an asteroid's rotation. Along with the asteroid spectrum, the background sky is measured simultaneously along the spectrograph slit, allowing for much more precise fitting and removal of the sky than was ever before possible. Unlike spectrophotometry, which demands that measurements be made under clear skies, spectroscopy does not require photometric conditions, and thus a larger fraction of the available telescope time can be utilized.

The primary objective in reflectance spectroscopy is simple: to determine the ratio of reflected sunlight to incident sunlight as a function of wavelength. There are several factors, however, that add to the complexity of asteroid spectroscopy and contribute to uncertainties in the observations. First, the fact that the Sun cannot be directly measured with a CCD spectrograph requires the use of other solar-like stars that act as spectral proxies. Any differences between the spectra of these solar analogs and the actual solar spectrum can result in artifacts in the reduced asteroid spectrum. Because the sunlight reflected from an asteroid only interacts with grains on the very top surface, the measured spectrum represents only the optical properties of those grains. It is generally assumed that this upper surface material is representative of the bulk composition of the asteroid, but other properties of these grains, including their temperature, size distribution, and the effects of long-term exposure to space, can also affect their spectral properties. Since the disks of asteroids are unresolved as viewed from Earth, the measured spectrum represents the reflectance characteristics averaged over the entire illuminated disk and can be affected by changes in viewing geometry. Before the light reflected from the asteroid reaches Earth's surface, it passes through our atmosphere, where it can be scattered and absorbed. Some of the most difficult aspects of reducing asteroid spectra involve correcting for these atmospheric effects. Finally, the characteristics of both the spectrograph and detector must be understood, so that any instrumental offsets that might be present in the measurements can be calibrated and removed.

2. INSTRUMENTATION

A CCD is a two-dimensional, solid-state detector that is sensitive to visible-wavelength light. It is composed of an array of closely spaced capacitors, commonly referred to as picture elements or pixels. When a positive potential is placed across this array, a potential well is formed at the location of each pixel. These potential wells collect and hold electrons that are freed when incoming photons interact with the silicon atoms that make up the detector substrate. At the completion of an exposure, the collected charge is read out by manipulating the voltages between the rows and columns of the array in a systematic way, stepping the charge across the detector pixel by pixel. The output signal is then amplified and converted into counts, or analog-digital units (ADUs), to make up the resulting image.

The advantages of CCDs over older two-dimensional detectors are numerous. They are dimensionally very stable and typically have high sensitivities (quantum efficiencies or "QE") over the visible wavelengths; they also typically have low read noise and a large, linear dynamic range. While the use of CCDs in astronomical imaging evolved rapidly in the mid- to late 1970s, the widespread application of CCDs to astronomical spectroscopy did not occur for nearly a decade. This delay can mainly be attributed to the time required for the design and fabrication of new long-slit spectrographs that could take full advantage of the sensitivity and large format offered by CCDs.

Long-slit CCD spectrographs are capable of low- to medium-resolution spectroscopy (R ~ 100–5000, where R is defined as $\lambda/d\lambda$) while recording spatial detail along a one-dimensional cut on the sky defined by the length and orientation of the spectrograph slit. For asteroids, this long-slit capability means that a sample of the surrounding sky is measured simultaneously with the asteroid, as shown in Fig. 1a, allowing for a much more precise determination and removal of the background sky signal. To obtain the highest possible signal-to-noise (S/N) spectra for faint asteroids, observations are usually made at very low spectral resolutions, so that the flux from an asteroid is spread over the fewest number of pixels along the dispersion direction. This low spectral resolution can be achieved with spectrographs utilizing either reflection gratings or grisms (a combined grating and prism) as the dispersing element. Grism spectrographs are particularly powerful in that they are capable of imaging the entire visible spectrum in one exposure, although the proper selection and use of blocking filters is necessary to avoid the overlap of spectral orders (*Xu et al.*, 1995). The total wavelength coverage that can be achieved with any spectrograph depends strongly on the throughput of its optics and the QE of the CCD. Though some systems have been optimized for observations in the UV (e.g., *Cochran and Vilas,* 1997), the sensitivity of many CCD spectrographs falls off sharply short of 0.4 μm, such that the resulting asteroid spectra do not extend as far into the UV as those measured by filter photometry.

3. OBSERVATIONS AND DATA REDUCTION

The observing strategies used in asteroid spectroscopy will vary depending on the telescope and instrumentation used and the goals of the particular observing program. In general, a sequence of observations includes several spectral images of the target asteroid and of one or more solar analog stars (*Hardorp,* 1978). Also, a series of calibration images are taken that include bias frames, flatfield images, and an arc lamp image for wavelength calibration.

Each asteroid and standard star observation is made by centering the object at a prescribed location along the slit, so that the maximum amount of light from the object passes through the slit. Because asteroids are typically moving against the background sky, offset rates are applied to the nominal sidereal track rate of the telescope (if possible) to improve the guiding. Many spectrographs are equipped with cameras that view the slit plane, allowing the sky region immediately outside the slit to be continuously monitored.

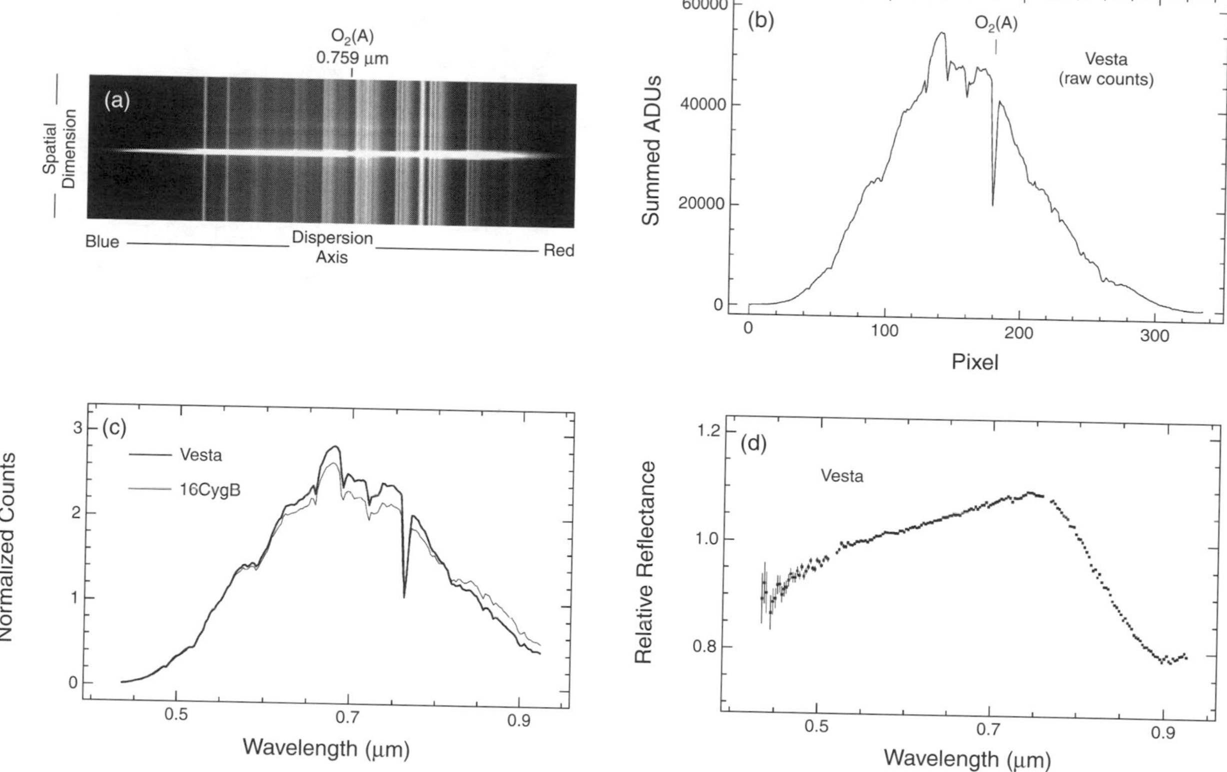

Fig. 1. Four steps in the reduction of asteroid spectra. **(a)** Spectral image of Vesta, taken on April 18, 1996, at the Michigan-Dartmouth-MIT Observatory, which has been bias and flat-field corrected. Bright vertical stripes (along the spatial dimension) are night-sky emission lines. The strong O_2(A) atmospheric absorption band is labeled. **(b)** Extracted one-dimensional spectrum obtained by summing the object counts (ADUs) along each column within the extraction aperture and subtracting the fitted sky level. The falloff in sensitivity at both ends of the spectrum is clearly seen. **(c)** Spectrum for both Vesta and the solar-analog star 16CygB after each has been calibrated for dispersion (wavelength correction) and atmospheric extinction. To simplify the comparison, the counts in both spectra have been normalized to 1.0 at 0.55 μm. **(d)** Ratio spectrum (Vesta divided by solar analog), normalized at 0.55 μm. The formal error bars are very small except at the blue end of the spectrum.

By watching for spillover of the asteroid image outside the slit during an exposure, small corrections to the guiding can be made. Exposure times for the faintest targets are usually limited to 10–20 min, because of the accumulation of cosmic-ray events on the detector. Taking several images of each target asteroid and solar analog is always advisable. Besides helping to improve the overall S/N, multiple images allow for more complete identification and correction of discrepant pixels in the individual images and for the monitoring of any systematic offsets in the extracted spectra.

Bias frames are taken by setting the exposure to 0 s and keeping the shutter closed. The result is a readout of the bias level, the positive offset in the mean signal level of an image that results from the bias voltage applied across the CCD during each exposure. These bias images are used to create a residual bias structure frame that contains any low spatial-frequency variations in the bias level that might exist across the CCD. Flatfield images are taken to map the wavelength-dependent pixel-to-pixel variations in sensitivity across the CCD. These images are obtained by uniformly illuminating the spectrograph slit with a light source whose color closely approximates that of the nighttime sky. In spectroscopy, the flatfield correction accounts not only for variations in sensitivity across the CCD but also for spatial variations in the throughput of the spectrograph slit and optics.

The procedures for reducing and calibrating CCD spectra of asteroids were first outlined by *Vilas and Smith* (1985). Similar descriptions of the reduction process are given by *Luu and Jewitt* (1990), *Sawyer* (1991), *Xu et al.* (1995), *Bus and Binzel* (2002a), and others. Though the details may vary depending on the instrumentation and observing strategies used by each observing program, the fundamental steps used to reduce asteroid spectra remain the same. These steps can be summarized as four basic procedures:

1. *Preprocessing of the CCD images.* This involves the calibration and correction of systematic effects associated with the instrument and detector. Two principal corrections are included: the subtraction of a bias level followed by the division of a flat field. Secondary corrections may include subtraction of a dark current and identification and removal of cosmic-ray events. Correcting for the bias offset is usually done by first subtracting an average bias level derived from an overscan region contained in each image and then subtracting the residual structure frame.

A nightly flatfield image is created by averaging the individual flatfield frames taken throughout the night. This average flat is often fitted and ratioed to a high-order, two-dimensional surface, resulting in a flatfield image with an average value along the dispersion axis of ~1.0. When this normalized flat is then divided into each spectral image, small pixel-to-pixel variations in sensitivity are corrected without significantly changing the level of counts (ADUs) in the data. In very low-resolution spectroscopy, where spectra covering the entire sensitivity range of the detector are recorded in a single exposure, division by a flat field can sometimes lead to increased noise at the ends of the spectra. In this case, the flatfield correction is often omitted, and all object spectra are carefully positioned to lie on the same rows of the CCD. This allows for much of the variation in pixel-to-pixel sensitivity along the asteroid spectrum to be divided out in the ratio to the solar-analog spectrum.

2. *Extraction of one-dimensional spectra.* The spatial information contained in the two-dimensional image is collapsed to form a one-dimensional spectrum. First, a line is fitted that traces the peak flux along the length of the two-dimensional spectral image. Centered along this trace, an aperture is defined that has a fixed width and whose length contains the entire object spectrum. The width of this aperture is carefully chosen to contain most of the object flux while minimizing the amount of background sky that is included. Along each column of the image, the sky level is fitted and subtracted. The remaining object flux contained within the aperture is then summed along each column. The resulting spectrum represents the sky-subtracted counts of the object in ADUs as a function of row number, as shown in Fig. 1b.

3. *Calibration of the extracted spectra.* Wavelength and extinction corrections are now applied to each extracted spectrum. The wavelength, or dispersion correction is determined by measuring the column pixel positions for spectral lines recorded in the arc lamp image and fitting these to their known wavelength values. In applying the dispersion correction, the object spectrum is resampled, so that the channels of the corrected spectrum are uniformly spaced in wavelength, as shown in Fig. 1c. It is important that the flux is correctly partitioned between the new spectral channels, and that the total flux in the spectrum is conserved during this resampling.

A correction for atmospheric extinction is needed since the observations of asteroids and standard stars are typically obtained through different airmasses. Atmospheric extinction arises from a combination of Rayleigh scattering, ozone absorption, and extinction due to aerosols, and it is strongest at short wavelengths (*Hayes and Latham,* 1975). A nightly extinction model, giving extinction per airmass as a function of wavelength (spectral channel), can be created if observations of standard stars are obtained at several different airmasses throughout the night. Because extinction can vary with direction in the sky, calibration stars must also be observed over a range in azimuth. The calculated extinction coefficients are then used to correct the flux in each channel of the standard star spectrum to match the airmass of the asteroid observation (*Vilas and Smith,* 1985). A simpler though somewhat less precise approach to correcting for atmospheric extinction involves the use of mean extinction coefficients that have been previously measured for the observing site. The use of mean extinction coefficients works well if the observations of both the asteroids and standard stars are restricted to relatively low airmasses (*Bus,* 1999).

4. *Normalization to a solar-analog star.* The calibrated asteroid spectrum is finally divided by that of a standard solar-analog star. To facilitate the comparison of spectra for different objects, the ratioed asteroid spectrum is then normalized to unity at a standard wavelength, usually 0.55 µm (the center of the photometric V band) as shown in Figs. 1c and 1d or 0.70 µm (corresponding to the R band). An uncertainty for each channel can be calculated based on the gain and readnoise of the CCD, and the object and sky counts for both the asteroid and standard star spectra assuming Poisson statistics.

Residual features that do not divide out fully may be present in the normalized spectrum. These artifacts usually arise from temporal variations in the depths of atmospheric absorption bands or from the misalignment of stellar absorption lines due to small errors in the wavelength calibration. The latter is usually caused by instrument flexure or variations in the positioning of objects in the slit and can be minimized by applying small pixel shifts between the asteroid and solar analog spectra. For observations obtained using a wide slit, changes in seeing will alter the effective spectral resolution, resulting in a potential mismatch in line widths between the asteroid and solar analog spectra and leading to residual stellar line features in the normalized spectrum. Residual atmospheric features can be more difficult to correct. One method for removing these artifacts, described by *Kelley and Gaffey* (2000), finds combinations of standard star observations taken throughout the night for which the mean, when divided into the asteroid spectrum, produces the cleanest result. Over the visible wavelength region, residual features associated with either atmospheric absorption bands or stellar lines are typically narrow, and in most cases, their presence has little impact on the interpretation of broader features contained in the asteroid spectrum. Often the points associated with these residual features can simply be omitted from the final spectrum.

Beyond the formal uncertainties that can be attached to each channel of a reduced spectrum, there is the potential for systematic offsets affecting the overall shapes of spectra, potentially impacting how these data are interpreted. The average slope of an asteroid spectrum is the characteristic that seems most sensitive to these systematic effects. While slow variations in spectral slope may be indicative of spectral heterogeneity over the surface of a rotating asteroid, sporadic fluctuations in slope are sometimes observed when a rapid sequence of spectra is taken. These fluctuations are more likely to be instrumental in nature and are most easily explained as light being preferentially lost from either the blue or red ends of the extracted spectra. Possible sources

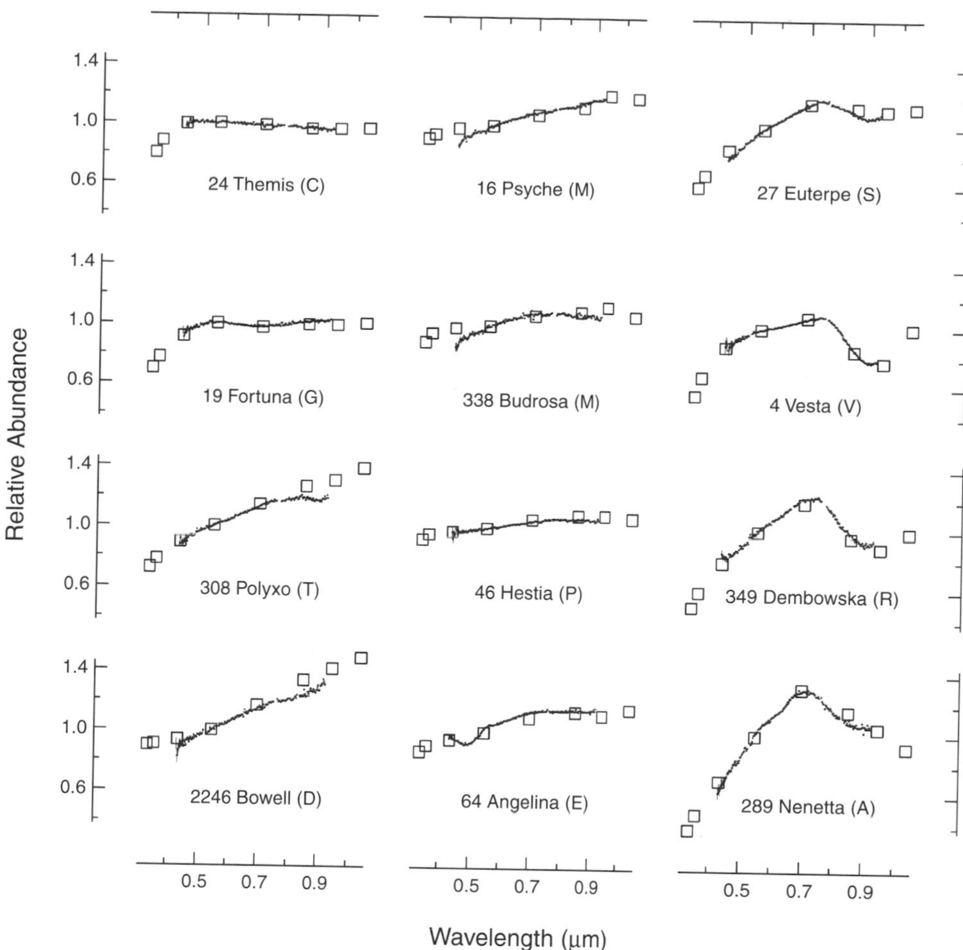

Fig. 2. A comparison between the Eight-Color Asteroid Survey (ECAS) colors (open squares) and SMASSII spectra (closed dots with error bars) for selected asteroids in various Tholen taxonomic classes. Those asteroids along the middle column (16 Psyche, 338 Budrosa, 46 Hestia, and 64 Angelina) show the range of spectral diversity among the Tholen X-class (E, M, and P classes, which can only be separated based on albedo) asteroids observed in the SMASSII data. Minor absorption features that are visible include the broad 0.7-μm band in the spectrum of 19 Fortuna, a 0.49-μm band in the spectrum of 64 Angelina, and a subtle 0.65-μm feature in the A-type spectrum of 289 Nenetta.

for error in the measured spectral slopes of asteroids include differential refraction (*Filippenko,* 1982) and fluctuations in seeing. Experience gained during the SMASSII survey (*Bus,* 1999) shows that while large fluctuations in spectral slope are relatively uncommon, they are more likely to be encountered under poor observing conditions that include high humidity, low-level clouds, and variable seeing.

Other likely causes for systematic offsets in asteroid spectra revolve around the methods used in correcting for atmospheric extinction, and the choice of solar analog stars. As the number of programs devoted to asteroid spectroscopy increases, the potential for systematic offsets between the various datasets is becoming more apparent. Systematic offsets in slope and in the depths and positions of bands are being routinely described when the spectra of objects observed by multiple programs are compared (*Vilas and Smith,* 1985; *Sawyer,* 1991; *Burbine,* 1999). Because of the number of variables involved, including instrumentation, observing strategies, and reduction and calibration techniques, the source of these systematic offsets can be hard to identify. This can make the intercomparison of results obtained by different observing programs problematic, and it also underscores the significance of large-scale asteroid spectroscopy programs in which the observations and data reduction are carried out in the most internally consistent manner possible.

4. SPECTRAL FEATURES

The existence of absorption features in the visible-wavelength spectra of asteroids can be explained by crystal field theory and charge-transfer mechanisms (*Burns,* 1970, 1981). These absorption bands are diagnostic of particular rock-forming minerals and provide a powerful means for probing the composition of asteroid surfaces. Thorough reviews of the analyses and interpretations of asteroid spectra, based on mineralogical considerations, are presented by *Gaffey et al.* (1989, 1993a) and *Pieters and McFadden* (1994).

There are three principal features in asteroid spectra over the visible wavelengths that have remained fundamental to the description and classification of asteroids since the mid-1970s. These are (1) the presence or absence of a UV absorption feature due to strong Fe^{2+} intervalence charge-transfer transitions; (2) the slope of the spectrum longward of 0.55 µm, the magnitude of which depends of the presence or absence of reddening agents such as Fe-Ni metal or organics; and (3) the presence or absence of a silicate absorption feature longward of 0.7 µm and for which the band minimum is typically centered near 1 µm. These features can be clearly seen in Fig. 2, where comparisons between ECAS colors and SMASSII spectra are plotted for asteroids belonging to various taxonomic classes as defined by *Tholen* (1984).

The increased sensitivity and spectral resolution afforded by CCD spectroscopy has also led to the identification of several weaker absorption bands in asteroid spectra. The depths of these bands range from 1% to 5% below the continuum level, and require high S/N observations for detection. Observations with somewhat higher spectral resolutions (R ~ 300–500) are often used in the study of these features to ensure the accurate calibration and removal of solar absorption lines and telluric water bands in the normalized spectra. Descriptions of the minor absorption bands recognized thus far are given below, arranged by taxonomic class:

A-type asteroids. A minor absorption band, centered near 0.65 µm, is observed in the spectrum of 289 Nenetta. Laboratory spectra of olivines show the presence of this band is correlated with low forsterite:fayalite (Mg:Fe) ratios (*Sunshine et al.,* 1998).

C-type asteroids. In a spectroscopic study of primitive asteroids, *Vilas and Gaffey* (1989) report the detection of a broad absorption band centered near 0.7 µm with depths of up to 5% below the continuum. This feature is attributed to the presence of phyllosilicates formed by aqueous alteration processes. The 0.7-µm feature was studied in detail by *Sawyer* (1991) and is clearly seen in the SMASSII spectrum of 19 Fortuna plotted in Fig. 2. Much weaker bands have been identified, centered at 0.43 µm (*Vilas et al.,* 1993) and at 0.60–0.65 and 0.80–0.90 µm (*Vilas et al.,* 1994), that are also associated with Fe oxide minerals produced by the aqueous alteration of anhydrous silicates.

S-type asteroids. Two minor features are identified, centered near 0.60 and 0.67 µm, that are consistent with small amounts of either oxidized Fe-Ni metal or spinel-group minerals in the surface regolith (*Hiroi et al.,* 1996).

V-type asteroids. A weak aborption band centered near 0.506 µm is observed in the spectrum of 4 Vesta (*Cochran and Vilas,* 1998), as illustrated in Fig. 3, and has been identified in the spectra of six other V-type asteroids belonging to the Vesta family (*Vilas et al.,* 2000). This feature is consistent with the presence of augite (a high-Ca form of pyroxene) in freshly exposed surface materials.

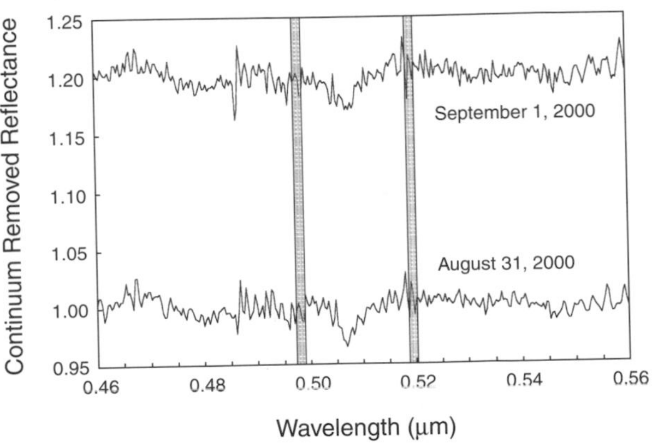

Fig. 3. A magnified section of the Vesta spectrum, measured on two consecutive nights and reduced as described by *Cochran and Vilas* (1997). The spectral continuum levels have been fitted and removed, and the September 1 spectrum has been offset by 0.20 for clarity. Vertical bars show the extent of the 0.506-µm Fe^{2+} pyroxene band that, because of its shape, indicates the presence of augite, a Ca-rich form of pyroxene. The strength of this feature suggests that some of the material on the surface of Vesta is relatively fresh, possibly excavated from Vesta's interior during a more recent cratering event (*Vilas et al.,* 2000).

X-type asteroids. A relatively strong absorption band centered near 0.49 µm and a much weaker feature centered near 0.60 µm are observed in the spectrum of 64 Angelina, as seen in Fig. 2. More than 20 additional X-type asteroids have been identified with similar but substantially weaker features (*Bus,* 1999). *Burbine et al.* (1998) suggest that these features are consistent with the presence of the Fe sulfide mineral troilite.

5. EFFECTS OF SURFACE PROPERTIES ON ASTEROID SPECTRA

Composition is the dominant factor contributing to the optical properties of an asteroid surface. However, other characteristics of the surface material can affect these optical properties and can lead to significant variations in the measured reflectance spectrum of the asteroid. Four factors are currently recognized that have the potential for altering the observed spectra:

1. *Phase reddening.* The reddening of reflectance spectra with increased phase angle has been occasionally noted in groundbased spectrophotometric observations of asteroids (*Millis et al.,* 1976; *Bowell and Lumme,* 1979; *Murchie and Pieters,* 1996) and has been observed in laboratory measurements of meteorite and mineral powders (*Gradie et al.,* 1980; *Gradie and Veverka,* 1986). The best quantitative assessment of phase reddening on an asteroid has been made recently using results from the *NEAR Shoemaker* Near-Infrared Spectrometer observations of Eros (*Clark et al.,* 2002), which revealed a reddening of the spectral slope by 8–12% over the phase angle range of 0°–100°. To compare the colors of near-Earth asteroids with those

of asteroids near the 3:1 resonance, *Luu and Jewitt* (1990) applied nominal corrections for phase-reddening to values of the normalized reflectivity gradient (slope), S'.

2. *Space weathering.* The term "space weathering" has been used to describe the darkening and reddening of a planetary surface over time and has been applied to asteroids to help explain some of the spectral mismatches between asteroids and meteorites (e.g., *Chapman,* 1996). The processes responsible for space weathering have remained poorly defined until recently. *Moroz et al.* (1996) showed that quick melting and recrystallization of mafic materials, induced by short laser pulses, had the effect of darkening this material and increasing its spectral slope. *Yamada et al.* (1999) showed that the degree of alteration due to laser irradiation depends on mineralogy, with olivine exhibiting larger spectral variations than pyroxenes. The most recent investigations by *Pieters et al.* (2000) and *Sasaki et al.* (2001) point to the production of nanophase Fe on the surfaces of grains as the underlying cause of these alterations in spectral reflectance.

3. *Particle size.* The presence of particulate regolith on asteroid surfaces has been long known from polarimetric studies (e.g., *Dollfus et al.,* 1989) and has recently been confirmed through high-resolution spacecraft imaging of Eros (*Veverka et al.,* 2001). The size distribution of particles making up this regolith plays a significant role in the optical properties of the surface and can affect the observed spectral slope and band depths (e.g., *Johnson and Fanale,* 1973). Because of the relatively solid link between Vesta and the HED meteorites (*Binzel and Xu,* 1993), laboratory spectra of HED meteorite powders have been used to constrain the distribution of particle sizes present on the surfaces of Vesta and members of the Vesta family (*Hiroi et al.,* 1994, 1995; *Burbine et al.,* 2001).

4. *Temperature.* Depending on albedo and rotational parameters, the surface temperatures of asteroids vary from 120 K at the distance of the Jupiter Trojans to over 300 K for near-Earth asteroids (*Hinrichs et al.,* 1999). The shapes of spectral bands associated with olivines and pyroxenes are sensitive to temperature (e.g., *Roush,* 1984; *Singer and Roush,* 1985), which can affect the mineralogical interpretations of asteroid spectra (*Lucey et al.,* 1998; *Moroz et al.,* 2000).

6. TAXONOMY

Asteroid taxonomy has continued to evolve ever since *Chapman et al.* (1975) proposed the first classification system based on spectral reflectance properties. This evolution in taxonomy is a natural process, spurred on by the introduction of new and larger asteroid datasets and the availability of numerous classification algorithms. The history of asteroid classification has been well documented, first by *Bowell et al.* (1978) and more recently by *Tholen and Barucci* (1989), and will not be repeated here. Instead, we focus on the impact that CCD spectroscopy has had on the philosophy behind asteroid taxonomy and describe the feature-based taxonomy developed as part of the SMASSII survey (*Bus,* 1999; *Bus and Binzel,* 2002b).

Multivariate analysis techniques, such as cluster analysis, are commonly used in the derivation of taxonomic systems. For a classification to be successful, the objects being classified must cluster together in groups that are well separated in some parameter space. Based on the combined Eight-Color Asteroid Survey (ECAS) (*Zellner et al.,* 1985) and IRAS albedo (*Veeder et al.,* 1989, *Tedesco et al.,* 1992) datasets, four taxonomies have thus far been proposed: (1) the *Tholen* (1984) taxonomy, derived using a minimum spanning tree clustering algorithm; (2) the *Barucci et al.* (1987) taxonomy, derived using G-mode analysis; (3) the *Tedesco et al.* (1989) three-parameter taxonomy derived based on the visual identification of groupings in a parameter space defined by two asteroid colors and the IRAS albedos; and (4) the taxonomy of *Howell et al.* (1994) that

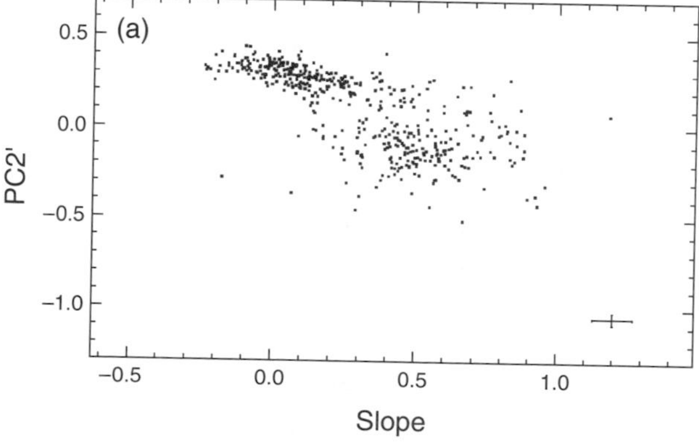

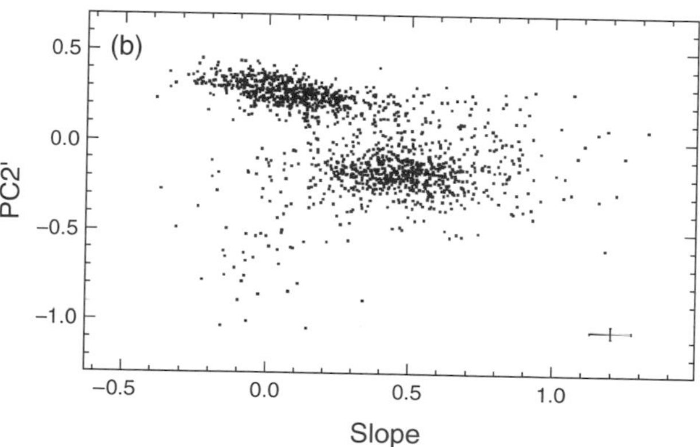

Fig. 4. Plots showing the distribution of the spectral components Slope and PC2' for SMASSII asteroids. The Slope parameter is defined as the average fitted slope across the entire spectrum, while PC2' is strongly correlated with the depth of the 1-μm band, with lower values of PC2' corresponding to increased band depth. Error bars shown in the lower right corner of each plot represent the 1σ uncertainty in the component score for an object observed only once. **(a)** Plot of only the brighter SMASSII asteroids also observed during ECAS. **(b)** Plot of all SMASSII asteroids. The increased dispersion of points plotted is indicative of the greater diversity in spectral properties observed as smaller asteroids are measured.

TABLE 1. Summary of taxonomic classes.

Tholen Class	Barucci Class	Howell Class	SMASSII Class	Description
B, F	B0–B3	B + F	B	Linear, featureless spectrum with bluish to neutral slope
			Cb	Similar to B spectrum with neutral to slight reddish slope
G	G0	Cv, Cx	Cg	Strong absorption (λ < 0.55 μm), and featureless with neutral to reddish slope (λ > 0.55 μm)
			Cgh	Similar to Cg spectrum, with shallow absorption centered near 0.7 μm
C	C0		C	Weak to medium absorption (λ < 0.55 μm), neutral to slightly reddish and featureless (λ > 0.55 μm)
			Ch	Similar to C spectrum, with shallow absorption centered near 0.7 μm
E, M, P		E. M, P	Xc	Slightly reddish spectrum, featureless except for broad convex curvature centered near 0.7 μm
	E0, M0		Xk	Similar to Xc spectrum, but redder slope (λ < 0.7 μm), and generally flat (λ > 0.7 μm)
			X	Generally featureless, linear spectrum with slight to moderate reddish slope
			Xe	Overall slope slight to moderately red, concave absorption feature centered near 0.5 μm, with occasional secondary absorption centered near 0.6 μm
T	D0–D3	T	T	Moderately reddish slope (λ < 0.75 μm), and generally flat (λ > 0.85 μm)
D		D	D	Relatively featureless spectrum with very steep red slope
S	S0–S3	K	K	Moderately steep red slope (λ < 0.75 μm) and flat to slightly bluish (λ > 0.75 μm)
		So	L	Very steep red slope (λ < 0.75 μm) and flat to slightly bluish (λ > 0.75 μm)
			Ld	Similar to L spectrum, but steeper red slope (λ < 0.75 μm)
			Sa	Similar to S spectrum, but with steeper slope (λ < 0.7 μm)
			Sl	Similar to S spectrum, but with steeper slope (λ < 0.7 μm) and a shallower absorption (λ > 0.75 μm)
		S	S	Moderately steep, reddish slope (λ < 0.7 μm), and a moderate to deep absorption band (λ > 0.75 μm)
			Sr	Similar to S spectrum, but with very steep red slope (λ < 0.7 μm) and a deeper absorption (λ > 0.75 μm)
		Sp	Sk	Similar to S spectrum, but with shallower reddish slope (λ < 0.7 μm) and a shallower absorption (λ > 0.75 μm)
			Sq	Similar to S spectrum, but with shallower reddish slope (λ < 0.7 μm)
Q	—	—	Q	Moderately steep red slope (λ < 0.7 μm) and a deep, very rounded absorption feature (λ > 0.75 μm)

TABLE 1. (continued).

<table>
<tr><th>Tholen Class</th><th>Barucci Class</th><th>Howell Class</th><th>SMASSII Class</th><th>Description</th></tr>
<tr><td>A</td><td>A0</td><td>A</td><td>A</td><td>Very steep to extremely steep red slope ($\lambda < 0.75$ μm) and a moderately deep absorption ($\lambda > 0.75$ μm). Reflectance maximum or 1 μm feature usually more rounded than in S-type spectrum</td></tr>
<tr><td>R</td><td rowspan="2">V0</td><td>R</td><td>R</td><td>Very steep red slope ($\lambda < 0.7$ μm) and a deep absorption feature ($\lambda > 0.75$ μm). Reflectance maximum more sharply peaked than in S-type spectra</td></tr>
<tr><td>V</td><td>V</td><td>V</td><td>Moderate to very steep red slope ($\lambda < 0.7$ μm) with an extremely deep absorption band ($\lambda > 0.75$ μm)</td></tr>
<tr><td>—</td><td>—</td><td>—</td><td>O</td><td>Moderately red slope ($\lambda < 0.55$ μm), then less steep ($0.55 < \lambda < 0.7$ μm). Deep absorption ($\lambda > 0.75$ μm)</td></tr>
</table>

implemented artificial neural networks to separately analyze the ECAS data and the combined data from ECAS and the 52-color asteroid survey (*Bell et al.,* 1988). The taxonomies resulting from these four studies are fundamentally similar, suggesting that statistically significant boundaries exist between clusters of objects in the combined ECAS/IRAS dataset and that the classification results are relatively insensitive to the choice of clustering methods used.

Initial analysis of the SMASSII spectra (*Bus,* 1999) revealed a conspicuous absence of gaps separating spectral types, with the only significant void being that separating the asteroids whose spectra contain a 1-μm silicate band from those that do not. The spectral component plots shown in Fig. 4 indicate an overall increase in spectral diversity, as smaller asteroids are included. The apparent continuum of objects in spectral component space depicted in Fig. 4b is verified by the continua in spectral features observed in asteroid families (e.g., *Doressoundiram et al.,* 1998; *Lazzaro et al.,* 1999) and between previously defined taxonomic classes (*Binzel et al.,* 1996). Various attempts to classify the SMASSII spectra using clustering techniques were unsuccessful, usually resulting in the identification of two very large classes, and a number of small groupings consisting of outlying objects. These initial results prompted a reassessment of the goals of asteroid classification and ultimately led to the development of a new feature-based taxonomy.

Development of the SMASSII taxonomy was based on five principles: (1) The established framework provided by the Tholen taxonomy was utilized in an attempt to preserve the historic structure and spirit of past asteroid taxonomies. (2) The SMASSII classes were defined solely on the presence (or absence) of absorption features contained in the visible-wavelength spectra. (3) The classes were arranged in a way that reflects the spectral continuum revealed by the SMASSII data. (4) To properly parameterize the various spectral features, different analytical and multivariate analysis techniques were used in the classification of the SMASSII asteroids. Because some spectral features are very subtle and not easily parameterized, the classification procedures allowed for visual inspection of the data and for some class labels to be assigned based on human judgment. (5) When possible, the sizes (scale-lengths) and boundaries of the taxonomic classes were defined based on the spectral variance observed in natural groupings among the asteroids, such as dynamical families.

The SMASSII taxonomy consists of 26 classes. Several of the smaller classes defined by Tholen, such as the A, Q, R, V, D, and T classes remain unchanged in the SMASSII system. Larger associations such as the S and C "complexes" have been subdivided. The X-type asteroids were described by Tholen as being spectrally degenerate, and could only be subdivided into the E, M, and P classes based on albedo. However, the SMASSII spectra reveal subtle variations among the members of the X complex, allowing these objects to be divided into four classes based solely on spectral features. Descriptions of the SMASSII classes are given in Table 1, along with comparisons to previous (*Tholen,* 1984; *Barucci et al.,* 1987; *Howell et al.,* 1994) taxonomies. The continuum in spectral properties represented by the SMASSII taxonomy is depicted in Fig. 5.

One drawback of the SMASSII taxonomy is that applying this system to newly observed asteroids can be cumbersome. Various descriptions of the SMASSII spectral classes such as those provided in Table 1 and elsewhere (*Bus,* 1999; *Bus and Binzel* 2002b) allow for the classification of individual objects. However, a more automated approach, such as that provided by a neural network (*Bus and Binzel,* 2000) is required if this taxonomy is to be readily applied to large datasets. An important factor to consider is the uncertainty associated with taxonomic labels. The classification assigned to an asteroid is only as good as the observational data. If subsequent observations of an asteroid reveal variations in its spectrum, whether due to compositional heterogeneity over the surface of the asteroid, variations in viewing geometry, or systematic offsets in the observations themselves, the taxonomic label may change.

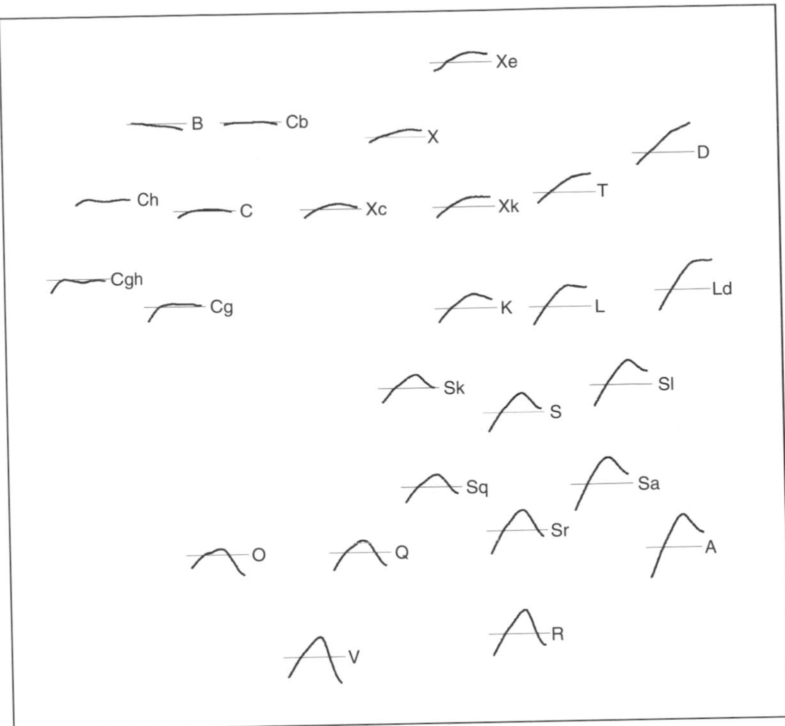

Fig. 5. Diagram showing all 26 SMASSII taxonomic classes, from *Bus and Binzel* (2002b). The spectra are arranged in a pattern that approximates the location of each class in spectral component space. The average spectral slope increases from left to right, and the depth of the 1-µm silicate absorption band generally increases from top to bottom.

When this occurs, we should not feel compelled to decide which label is "correct," but should rather accept these distinct labels as a consequence of our growing knowledge about that object.

7. NEAR-INFRARED SPECTROSCOPY

Just as CCDs have revolutionized visible-wavelength spectroscopy, the introduction of new generation near-infrared array detectors (*Hodapp,* 2000) is providing the capability of obtaining high-quality spectra of asteroids out to 2.5 µm and beyond. A new low- to medium-resolution NIR spectrograph and imager, called SpeX, is now in use at the NASA Infrared Telescope Facility (IRTF) on Mauna Kea (*Rayner et al.,* 1998). SpeX is capable of low-resolution spectroscopy (R ~ 50–250) over the wavelength range of 0.8–2.5 µm, while in a higher-resolution (R ≤ 2500) cross-dispersed mode, the spectral coverage is extended to 5.5 µm. SpeX is just one of a number of NIR spectrographs being designed and built at observatories around the world. Because the NIR region (~1–4 µm) contains absorption bands that are fundamental to studies of mineralogy (*Gaffey et al.,* 1989), these instruments could have an even greater impact on asteroid studies than CCD spectrographs.

The observing strategies and data-reduction techniques used in measuring the NIR spectra of asteroids are similar to those described earlier for CCD spectroscopy. The calibration of these data is complicated, however, by an increase in the number and relative strengths of night-sky emission lines and by the presence of strong telluric absorption bands due to atmospheric H_2O and CO_2. The depths of these absorption bands are sensitive to the amount of precipitable water in the atmosphere above the observer and can vary rapidly as sky conditions change. As discussed earlier, one method for minimizing the effects of telluric absorption bands is based on combining numerous solar-like star spectra and finding an average in which the absorption band depths most closely match those in the asteroid spectrum (*Kelley and Gaffey,* 2000). Another approach is based on nightly observations of stars belonging to the spectral class A0. The spectra of these stars are nearly featureless and can therefore be used to determine the profiles and relative strengths of absorption bands caused by Earth's atmosphere. By scaling these telluric features to match the band depths in each asteroid and solar-star spectrum, the presence of residual atmospheric features in a normalized asteroid spectrum can be minimized. A similar method uses a model of atmospheric transmission like that produced by the ATRAN software package (*Lord,* 1992). Using this algorithm, the amount of precipitable water can be varied, allowing the structure of atmospheric absorption features to be fitted and removed from each observation prior to dividing the asteroid spectrum by that of the solar-like star.

Due to the strengths and complex structure of the major telluric bands, proper wavelength calibration of both the asteroid and standard star spectra is crucial. Minor offsets in wavelength between the spectra can result in significant residuals in the normalized asteroid spectrum. These wavelength offsets are usually measured in small fractions of a pixel and are commonly associated with instrument flexure. These offsets can be corrected by shifting one of the spectra with respect to the other, where the magnitude and direction of the shift is usually determined by cross-correlating the absorption features contained in the spectra.

8. DISCUSSION

The widespread availability of CCD spectrographs has resulted in a wealth of spectroscopic data on asteroids. Long-term observing programs, in which the same instrumentation, observing strategies, and reduction techniques are used, are particularly valuable, as these are capable of producing large, internally consistent sets of asteroid spectra. The largest spectroscopic surveys to date include (1) a survey of low-albedo asteroids by *Sawyer* (1991), which contains observations of 115 asteroids; (2) the first phase of the Small Main-belt Asteroid Spectroscopic Survey (SMASSI) with 316 asteroids observed (*Xu et al.,* 1995); (3) SMASSII, with 1447 objects observed (*Bus and Binzel,* 2002a); and (4) the Small Solar System Objects Spectroscopic Survey (S^3OS^2) by *Lazzaro et al.* (2001), an ongoing survey that has already produced measurements for 800 asteroids. The combined efforts of these surveys along with several smaller studies have produced spectral measurements for an estimated 3000 individual asteroids. Even so, there are tens of thousands of additional main-belt asteroids that could become bright enough to be within reach of a 4-m class telescope, but which currently lack any spectroscopic measurements.

The applications of spectroscopy to studies of the origin and evolution of the asteroids are numerous. As the number of asteroids with measured spectra increases, bias-corrected models of the compositional structure of the asteroid belt can be refined (*Gradie et al.,* 1989; *Bus,* 1999), and unusual occurrences of objects can be identified. One such example is the V-type asteroid 1459 Magnya, an outer main-belt object (at 3.15 AU) whose basaltic nature was recently identified from spectral observations (*Lazzaro et al.,* 2000). All other known V-class asteroids are associated with the Vesta family (*Binzel and Xu,* 1993) and are confined to the inner main belt between 2.2 and 2.5 AU. Anomalous spectral types are also bound to be discovered, like that measured for asteroid 3628 Boznemcova (*Binzel et al.,* 1993), which led to the creation of a new taxonomic O class.

Recent spectroscopic studies of dynamical asteroid families have revealed strong similarities between the members of each family (e.g., *Doressoundiram et al.,* 1998; *Lazzaro et al.,* 1999; *Bus,* 1999). These findings have helped confirm the genetic reality of asteroid families and can be used to place constraints on collisional and dynamical models of family formation, as discussed by *Cellino et al.* (2002). Asteroid families provide unique opportunities to peer into the interiors of once-larger parent bodies and to search for spectral (compositional) heterogeneity among the members. Rotationally resolved spectra of individual asteroids can also be searched for evidence of spectral variation over their surfaces (*Murchie and Pieters,* 1996; *Cochran and Vilas,* 1998; *Mothé-Diniz et al.,* 2000; *Howell et al.,* 2001).

Close approaches by near-Earth objects (NEOs) provide opportunities to measure the spectral properties of very small bodies, many with diameters much smaller than 1 km. Spectral studies of NEOs are crucial for understanding their compositional distributions and for tracing origins of these objects to source regions in the main belt (*Binzel et al.,* 1996; *Hammergren,* 1998; *Hicks et al.,* 1998) and within the comet population (e.g., *Chamberlin et al.,* 1996). A fundamental goal of asteroid spectroscopy is to establish links between meteorites and the asteroids (or asteroid classes) from which they were derived. These identifications allow meteorites to be placed into a geologic context while constraining the composition and thermal histories of the asteroidal parent bodies (*Lipschutz et al.,* 1989; *Pieters and McFadden,* 1994). While some credible links have been identified between specific meteorite and asteroid classes (see *Burbine et al.,* 2002), much work remains before a consistent picture of the asteroid belt emerges that is based on both asteroid and meteorite evidence (*Bell et al.,* 1989; *Cellino,* 2000).

By combining results from visible and NIR spectroscopy, questions regarding mineralogy can be more fully addressed. The spectral interval between 0.7 and 2.5 μm is particularly important to studies of silicate minerals, such as pyroxenes, olivines and plagioclase, due to fundamental absorption bands centered near 1 and 2 μm (e.g., *Burns,* 1970; *Adams,* 1974, 1975; *Cloutis et al.,* 1986). The ability to obtain high-quality spectra over this interval is prompting the use of more advanced spectral analysis techniques, like the Modified Gaussian Model (MGM) (*Sunshine et al.,* 1990), to deconvolve individual absorption bands associated with the different silicate mineral phases. Reliable estimates of silicate mineralogy are key to constraining the thermal histories of silicate-rich asteroids and are helping to clarify the petrologic nature of the S-type asteroids (*Gaffey et al.,* 1993b; *Sunshine et al.,* 2002). The spectral window between 2.5 and 3.5 μm is significant to studies of hydrated minerals on asteroids, due to absorption bands centered near 3 μm that are associated with bound water and structural OH found in hydrated silicates (e.g., *Lebofsky et al.,* 1981; *Jones et al.,* 1990; *King et al.,* 1992). Together with visible-wavelength observations of the 0.7-μm phyllosilicate feature (*Vilas and Gaffey,* 1989; *Vilas and Sykes,* 1996; *Barucci et al.,* 1998), measurements of the 3-μm absorption band can be used to map the spatial extent of aqueous alteration on surfaces of asteroids and to constrain the heating mechanisms that led to this alteration, as discussed in *Rivkin et al.* (2002). With the addition of NIR spectra, asteroid taxonomy will also likely benefit as new classification schemes will ultimately be developed that are more representative of mineralogy.

REFERENCES

Adams J. B. (1974) Visible and near-infrared diffuse reflectance spectra of pyroxenes as applied to remote sensing of solid objects in the solar system. *J. Geophys. Res., 79,* 4829–4836.

Adams J. B. (1975) Interpretation of visible and near-infrared diffuse reflectance spectra of pyroxenes and other rock-forming minerals. In *Infrared and Raman Spectroscopy of Lunar and Terrestrial Minerals* (C. Karr, ed.), pp. 91–116. Academic, New York.

Barucci M. A., Capria M. T., Coradini A., and Fulchignoni M. (1987) Classification of asteroids using G-mode analysis.

Icarus, 72, 304–324.

Barucci M. A., Doressoundiram A., Fulchignoni M., Florczak M., Lazzarin M., Angeli C., and Lazzaro D. (1998) Search for aqueously altered materials on asteroids. *Icarus, 132,* 388–396.

Bell J. F., Owensby P. D., Hawke B. R., and Gaffey M. J. (1988) The 52-color asteroid survey: Final results and interpretation (abstract). In *Lunar and Planetary Science XIX,* pp. 57–58. Lunar and Planetary Institute, Houston.

Bell J. F., Davis D. R, Hartmann W. K., and Gaffey M. J. (1989) Asteroids: The big picture. In *Asteroids II* (R. P. Binzel et al., eds.), pp. 921–945. Univ. of Arizona, Tucson.

Binzel R. P. and Xu S. (1993) Chips off of asteroid 4 Vesta: Evidence for the parent body of basaltic achondrite meteorites. *Science, 260,* 186–191.

Binzel R. P., Xu S., Bus S. J., Skrutskie M. F., Meyer M. R., Knezek P., and Barker E. S. (1993) Discovery of a main-belt asteroid resembling ordinary chondrite meteorites. *Science, 262,* 1541–1543.

Binzel R. P., Bus S. J., Burbinc T. H., and Sunshine J. M. (1996) Spectral properties of near-Earth asteroids: Evidence for sources of ordinary chondrite meteorites. *Science, 273,* 946–948.

Bobrovnikoff N. T. (1929) The spectra of minor planets. *Lick Obs. Bull., 14,* 18–27.

Bowell E. and Lumme K. (1979) Colorimetry and magnitudes of asteroids. In *Asteroids* (T. Gehrels, ed.), pp. 132–169. Univ. of Arizona, Tucson.

Bowell E., Chapman C. R., Gradie J. C., Morrison D., and Zellner B. (1978) Taxonomy of asteroids. *Icarus, 35,* 313–335.

Burbine T. H. (1999) Forging asteroid-meteorite relationships through reflectance spectroscopy. Ph.D. thesis, Massachusetts Institute of Technology, Cambridge.

Burbine T. H., Cloutis E. A., Bus S. J., Meibom A., and Binzel R. P. (1998) The detection of troilite (FeS) on the surfaces of E-class asteroids (abstract). *Bull. Am. Astron. Soc., 30,* 1025–1026.

Burbine T. H., Buchanan P. C., Binzel R. P., Bus S. J., Hiroi T., Hinrichs J. L., Meibom A., and McCoy T. J. (2001) Vesta, Vestoids, and the howardite, eucrite, diogenite group: Relationships and the origin of spectral features. *Meteoritics & Planet. Sci., 36,* 761–781.

Burbine T. H., McCoy T. J., Meibom A., Gladman B., and Keil K. (2002) Meteoritic parent bodies: Their number and identification. In *Asteroids III* (W. F. Bottke Jr. et al., eds.), this volume. Univ. of Arizona, Tucson.

Burns R. G. (1970) *Mineralogical Applications of Crystal Field Theory.* Cambridge Univ., New York. 224 pp.

Burns R. G. (1981) Intervalence transitions in mixed-valence mainerals of iron and titanium. *Annu. Rev. Earth Planet. Sci., 9,* 345–383.

Bus S. J. (1999) Compositional structure in the asteroid belt: Results of a spectroscopic survey. Ph.D. thesis, Massachusetts Institute of Technology, Cambridge.

Bus S. J. and Binzel R. P. (2000) A neural network simulation of the SMASSII asteroid taxonomy (abstract). *Bull. Am. Astron. Soc., 32,* 1004.

Bus S. J. and Binzel R. P. (2002a) Phase II of the Small Main-Belt Asteroid Spectroscopic Survey: The observations. *Icarus, 158,* 106–145.

Bus S. J. and Binzel R. P. (2002b) Phase II of the Small Main-Belt Asteroid Spectroscopic Survey: A feature-based taxonomy. *Icarus, 158,* 146–177.

Cellino A. (2000) Minor bodies: Spectral gradients and relationships with meteorites. *Space Sci. Rev., 92,* 397–412.

Cellino A., Bus S. J., Doressoundiram A., and Lazzaro D. (2002) Spectroscopic properties of asteroid families. In *Asteroids III* (W. F. Bottke Jr. et al., eds.), this volume. Univ. of Arizona, Tucson.

Chamberlin A. B., McFadden L. A., Schulz R., Schleicher D. G., and Bus S. J. (1996) 4015 Wilson-Harrington, 2201 Oljato, and 3200 Phaethon: Search for CN emission. *Icarus, 119,* 173–181.

Chapman C. R. (1996) S-type asteroids, ordinary chondrites, and space weathering: The evidence from Galileo's fly-bys of Gaspra and Ida. *Meteoritics, 31,* 699–725.

Chapman C. R. and Gaffey M. J. (1979) Reflectance spectra for 277 asteroids. In *Asteroids* (T. Gehrels, ed.), pp. 655–687. Univ. of Arizona, Tucson.

Chapman C. R., Johnson T. V., and McCord T. B. (1971) A review of spectrophotometric studies of asteroids. In *Physical Studies of Minor Planets* (T. Gehrels, ed.), pp. 51–65. NASA SP-267, Washington, DC.

Chapman C. R., Morrison D., and Zellner B. (1975) Surface properties of asteroids: A synthesis of polarimetry, radiometry, and spectrophotometry. *Icarus, 25,* 104–130.

Clark B. E., Helfenstein P., Bell J. F. III, Peterson C., Veverka J., Izenberg N. I., Domingue D., Wellnitz D., and McFadden L. (2002) NEAR Infrared Spectrometer photometry of asteroid 433 Eros. *Icarus, 155,* 189–204.

Cloutis E. A., Gaffey M. J., Jackowski T. L., and Reed K. L. (1986) Calibrations of phase abundance, composition, and particle size distribution for olivine-orthopyroxene mixtures from reflectance spectra. *J. Geophys. Res., 91,* 11641–11653.

Cochran A. L. and Vilas F. (1997) The McDonald Observatory serendipitous UV/blue spectral survey of asteroids. *Icarus, 127,* 121–129.

Cochran A. L. and Vilas F. (1998) The changing spectrum of Vesta: Rotationally resolved spectroscopy of pyroxene on the surface. *Icarus, 134,* 207–212.

Dollfus A., Wolff M., Geake J. E., Lupishko D. F., and Dougherty L. M. (1989) Photopolarimetry of asteroids. In *Asteroids II* (R. P. Binzel et al., eds.), pp. 594–616. Univ. of Arizona, Tucson.

Doressoundiram A., Barucci M. A., Fulchignoni M., and Florczak M. (1998) Eos family: A spectroscopic study. *Icarus, 131,* 15–31.

Filippenko A. V. (1982) The importance of atmospheric differential refraction in spectroscopy. *Publ. Astron. Soc. Pac., 94,* 715–721.

Gaffey M. J. and McCord T. B. (1979) Mineralogical and petrological characterization of asteroid surface materials. In *Asteroids* (T. Gehrels, ed.), pp. 688–723. Univ. of Arizona, Tucson.

Gaffey M. J., Bell J. F., and Cruikshank D. P. (1989) Reflectance spectroscopy and asteroid surface mineralogy. In *Asteroids II* (R. P. Binzel et al., eds.), pp. 98–127. Univ. of Arizona, Tucson.

Gaffey M. J., Burbine T. H., and Binzel R. P. (1993a) Asteroid spectroscopy: Progress and perspectives. *Meteoritics, 28,* 161–187.

Gaffey M. J., Bell J. F., Brown R. H., Burbine T. H., Piatek J. L., Reed K. L., and Chaky D. A. (1993b) Mineralogical variations within the S-type asteroid class. *Icarus, 106,* 573–602.

Gradie J. and Veverka J. (1986) The wavelength dependence of phase coefficients. *Icarus, 66,* 455–467.

Gradie J., Veverka J., and Buratti B. (1980) The effects of scattering geometry on the spectrophotometric properties of powdered material. *Proc. Lunar Planet. Sci. Conf. 11th,* pp. 799–815.

Gradie J. C., Chapman C. R., and Tedesco E. F. (1989) Distribution of taxonomic classes and the compositional structure of the asteroid belt. In *Asteroids II* (R. P. Binzel et al., eds.), pp. 316–335. Univ. of Arizona, Tucson.

Hammergren M. (1998) The composition of near-Earth objects. Ph.D. thesis, University of Washington, Seattle.

Hardorp J. (1978) The sun among the stars. I–A search for solar spectral analogs. *Astron. Astrophys., 63,* 383–390.

Hayes D. S. and Latham D. W. (1975) A rediscussion of the atmospheric extinction and the absolute spectral-energy distribution of Vega. *Astrophys. J., 197,* 593–601.

Hicks M. D., Fink U., and Grundy W. M. (1998) The unusual spectra of 15 near-Earth asteroids and extinct comet candidates. *Icarus, 133,* 69–78.

Hinrichs J. L., Lucey P. G., Robinson M. S., Meibom A., and Krot A. N. (1999) Implications of temperature-dependent near-IR spectral properties of common minerals and meteorites for remote sensing of asteroids. *Geophys. Res. Lett., 26,* 1661–1664.

Hiroi T., Pieters C. M., and Takeda H. (1994) Grain size of the surface regolith of asteroid 4 Vesta estimated from its reflectance spectrum in comparison with HED meteorites. *Meteoritics, 29,* 394–396.

Hiroi T., Binzel R. P., Sunshine J. M., Pieters C. M., and Takeda H. (1995) Grain sizes and mineral compositions of surface regoliths of Vesta-like asteroids. *Icarus, 115,* 374–386.

Hiroi T., Vilas F., and Sunshine J. M. (1996) Discovery and analysis of minor absorption bands in S-asteroid visible reflectance spectra. *Icarus, 119,* 202–208.

Hodapp K. W. (2000) Near-infrared detector arrays: Current state of the art. *Proc. SPIE, 4008,* 1228–1239.

Howell E. S., Merényi E., and Lebofsky L. A. (1994) Classification of asteroid spectra using a neural network. *J. Geophys. Res., 99,* 10847–10865.

Howell E. S., Rivkin A. S., Vilas F., and Soderberg A. M. (2001) Aqueous alteration in low albedo asteroids (abstract). In *Lunar and Planetary Science XXXII,* Abstract #2058. Lunar and Planetary Institute, Houston (CD-ROM).

Janesick J. and Elliott T. (1992) History and advancements of large area array scientific CCD imagers. In *Astronomical CCD Observing and Reduction Techniques* (S. B. Howell, ed.), pp. 1–67. ASP Conference Series 23.

Johnson T. V. and Fanale F. P. (1973) Optical properties of carbonaceous chondrites and their relationship to asteroids. *J. Geophys. Res., 78,* 8507–8518.

Jones T. D., Lebofsky L. A., Lewis J. S., and Marley M. S. (1990) The composition and origin of the C, P, and D asteroids: Water as a tracer of thermal evolution in the outer belt. *Icarus, 88,* 172–192.

Kelley M. S. and Gaffey M. J. (2000) 9 Metis and 113 Amalthea: A genetic asteroid pair. *Icarus, 144,* 27–38.

King T. V. V., Clark R. N., Calvin W. M., Sherman D. M., and Brown R. H. (1992) Evidence for ammonium-bearing minerals on Ceres. *Science, 255,* 1551–1553.

Lazzaro D., Mothé-Diniz T., Carvano J. M., Angeli C. A., Betzler A. S., Florczak M., Cellino A., Di Martino M., Doressoundiram A., Barucci M. A., Dotto E., and Bendjoya P. (1999) The Eunomia family: A visible spectroscopic survey. *Icarus, 142,* 445–453.

Lazzaro D., Michtchenko T., Carvano J. M., Binzel R. P., Bus S. J., Burbine T. H., Mothé-Diniz T., Florczak M., Angeli C. A., and Harris A. W. (2000) Discovery of a basaltic asteroid in the outer main belt. *Science, 288,* 2033–2035.

Lazzaro D., Carvano J. M., Mothé-Diniz T., Angeli C., and Florczak M. (2001) S3OS2: A visible spectroscopic survey of around 800 asteroids (abstract). In *Asteroids 2001: From Piazzi to the 3rd Millennium,* p. 174. Osservatorio di Palermo, Sicily.

Lebofsky L. A., Feierberg M. A., Tokunaga A. T., Larson H. P., and Johnson J. R. (1981) The 1.7- to 4.2-μm spectrum of asteroid 1 Ceres: Evidence for structural water in clay minerals. *Icarus, 48,* 453–459.

Lipschutz M. E., Gaffey M. J., and Pellas P. (1989) Meteoritic parent bodies: Nature, number, size and relation to present-day asteroids. In *Asteroids II* (R. P. Binzel et al., eds.), pp. 740–777. Univ. of Arizona, Tucson.

Lord S. D. (1992) *A New Software Tool for Computing Earth's Atmospheric Transmission of Near- and Far-Infrared Radiation.* NASA TM-103957. NASA Ames Research Center, Moffett Field, California.

Lucey P. G., Keil K., and Whitely R. (1998) The influence of temperature on the spectra of the A-asteroids and implications for their silicate chemistry. *J. Geophys. Res., 103,* 5865–5871.

Luu J. X. and Jewitt D. C. (1990) Charge-coupled device spectra of asteroids. I. Near-Earth and 3:1 resonance asteroids. *Astron. J., 99,* 1985–2011.

McCord T. B., Adams J. B., and Johnson T. V. (1970) Asteroid Vesta: Spectral reflectivity and compositional implications. *Science, 168,* 1445–1447.

Millis R. L., Bowell E., and Thompson D. T. (1976) UBV photometry of asteroid 433 Eros. *Icarus, 28,* 53–67.

Moroz L. V., Fisenko A. V., Semjonova L. F., Pieters C. M., and Korotaeva N. N. (1996) Optical effects of regolith processes on S-asteroids as simulated by laser shots on ordinary chondrite and other mafic materials. *Icarus, 122,* 366–382.

Moroz L., Schade U., and Wasch R. (2000) Reflectance spectra of olivine-orthopyroxene-bearing assemblages at decreased temperatures: Implications for remote sensing of asteroids. *Icarus, 147,* 79–93.

Mothé-Diniz T., Lazzaro D., Carvano J. M., and Florczak M. (2000) Rotationally resolved spectra of some S-type asteroids. *Icarus, 148,* 494–507.

Murchie S. L. and Pieters C. M. (1996) Spectral properties and rotational spectral heterogeneity of 433 Eros. *J. Geophys. Res., 101,* 2201–2214.

Pieters C. M. and McFadden L. A. (1994) Meteorite and asteroid reflectance spectroscopy: Clues to early solar system processes. *Annu. Rev. Earth Planet. Sci., 22,* 457–497.

Pieters C. M., Taylor L. A., Noble S. K., Keller L. P., Hapke B., Morris R. V., Allen C. C., McKay D. S., and Wentworth S. (2000) Space weathering on airless bodies: Resolving a mystery with lunar samples. *Meteoritcs & Planet. Sci., 35,* 1101–1107.

Rayner J. T., Toomey D. W., Onaka P. M., Denault A. J., Stahlberger W. E., Watanabe D. Y., and Wang S.-I. (1998) SpeX: A medium-resolution IR spectrograph for IRTF. *Proc. SPIE, 3354,* 468–479.

Rivkin A. S., Howell E. S., Vilas F., and Lebofsky L. A. (2002) Hydrated minerals on asteroids: The astronomical record. In *Asteroids III* (W. F. Bottke Jr. et al., eds.), this volume. Univ. of Arizona, Tucson.

Roush T. L. (1984) Effects of temperature on remotely sensed mafic mineral absorption features. M.S. thesis, University of Hawai'i, Honolulu.

Sasaki S., Nakamura K., Hamabe Y., Kurahashi E., and Hiroi T. (2001) Production of iron nanoparticles by laser irradiation in a simulation of lunar-like space weathering. *Nature, 410,* 555–557.

Sawyer S. R. (1991) A high resolution CCD spectroscopic survey of low albedo main belt asteroids. Ph.D. thesis, University of Texas, Austin.

Singer R. B. and Roush T. L. (1985) Effects of temperature on remotely sensed mineral absorption features. *J. Geophys. Res., 90,* 12434–12444.

Sunshine J. M., Pieters C. M., and Pratt S. F. (1990) Deconvolution of mineral absorption bands: An improved approach. *J. Geophys. Res., 95,* 6955–6966.

Sunshine J. M., Binzel R. P., Burbine T. H., and Bus S. J. (1998) Is asteroid 289 Nenetta compositionally analogous to the Brachinite meteorites? (abstract). In *Lunar and Planetary Science XXIX,* Abstract #1430. Lunar and Planetary Institute, Houston (CD-ROM).

Sunshine J. M, Bus S. J., Burbine T. H., McCoy T. J., and Binzel R. P. (2002) Unambiguous spectral evidence for high- (and low-) calcium pyroxene in asteroids and meteorites (abstract). In *Lunar and Planetary Science XXXIII,* Abstract #1356. Lunar and Planetary Institute, Houston (CD-ROM).

Tedesco E. F., Williams J. G., Matson D. L., Veeder G. J., Gradie J. C., and Lebofsky L. A. (1989) A three-parameter asteroid taxonomy. *Astron. J., 97,* 580–606.

Tedesco E., Veeder G., Fowler J., and Chillemi J. (1992) *The IRAS Minor Planet Survey*. Phillips Lab PL-TR-92-2049, Hanscom Air Force Base, Massachusetts.

Tholen D. J. (1984) Asteroid taxonomy from cluster analysis of photometry. Ph.D. thesis, University of Arizona, Tucson.

Tholen D. J. and Barucci M. A. (1989) Asteroid taxonomy. In *Asteroids II* (R. P. Binzel et al., eds.), pp. 298–315. Univ. of Arizona, Tucson.

Veeder G. J., Tedesco E. F. and Matson D. L. (1989) Asteroid results from the IRAS survey. In *Asteroids II* (R. P. Binzel et al., eds.), pp. 282–289. Univ. of Arizona, Tucson.

Veverka J., Thomas P. C., Robinson M., Murchie S., Chapman C., Bell M., Harch A., Merline W. J., Bell J. F., Bussey B., Carcich B., Cheng A., Clark B., Domingue D., Dunham D., Farquhar R., Gaffey M. J., Hawkins E., Izenberg N., Joseph J., Kirk R., Li H., Lucey P., Malin M., McFadden L., Miller J. K., Owen W. M., Peterson C., Prockter L., Warren J., Wellnitz D., Williams B. G., and Yeomans D. K. (2001) Imaging of small-scale features on 433 Eros from NEAR: Evidence for a complex regolith. *Science, 292,* 484–488.

Vilas F. and Gaffey M. J. (1989) Phyllosilicate absorption features in main-belt and outer-belt asteroid reflectance spectra. *Science, 246,* 790–792.

Vilas F. and Smith B. A. (1985) Reflectance spectrophotometry (~0.5–1.0 μm) of outer-belt asteroids: Implications for primitive, organic solar system material. *Icarus, 64,* 503–516.

Vilas F. and Sykes M. V. (1996) Are low-albedo asteroids thermally metamorphosed? *Icarus, 124,* 483–489.

Vilas F., Larson S. M., Hatch E. C., and Jarvis K. S. (1993) CCD reflectance spectra of selected asteroids. II. Low-albedo asteroid spectra and data extraction techniques. *Icarus, 105,* 67–78.

Vilas F., Jarvis K. S., and Gaffey M. J. (1994) Iron alteration minerals in the visible and near-infrared spectra of low-albedo asteroids. *Icarus, 109,* 274–283.

Vilas F., Cochran A. L., and Jarvis K. S. (2000) Vesta and the Vestoids: A new rock group? *Icarus, 147,* 119–128.

Wood J. H. and Kuiper G. P. (1963) Photometric studies of asteroids. *Astrophys. J., 137,* 1279–1285.

Xu S., Binzel R. P., Burbine T. H., and Bus S. J. (1995) Small Main-belt Asteroid Spectroscopic Survey: Initial results. *Icarus, 115,* 1–35.

Yamada M., Sasaki S., Nagahara H., Fujiwara A., Hasegawa S., Yano H., Hiroi T., Ohashi H., and Otake H. (1999) Simulation of space weathering of planet-forming materials: Nanosecond pulse laser irradiation and proton implantation on olivine and pyroxene samples. *Earth Planets Space, 51,* 1255–1265.

Zellner B., Tholen D. J., and Tedesco E. F. (1985) The eight-color asteroid survey: Results for 589 minor planets. *Icarus, 61,* 355–416.

Mineralogy of Asteroids

M. J. Gaffey
University of North Dakota

E. A. Cloutis
University of Winnipeg

M. S. Kelley
NASA Johnson Space Center and Georgia Southern University

K. L. Reed
The Space Development Institute

The past decade has seen a significant expansion both in the interpretive methodologies used to extract mineralogical information from asteroid spectra and other remote-sensing data and in the number of asteroids for which mineralogical characterizations exist. Robust mineralogical characterizations now exist for more than 40 asteroids, an order a magnitude increase since *Asteroids II* was published. Such characterizations have allowed significant progress to be made in the identification of meteorite parent bodies. Although considerable progress has been made, most asteroid spectra have still only been analyzed by relatively ambiguous curve-matching techniques. Where appropriate and feasible, such data should be subjected to a quantitative analysis based on diagnostic mineralogical spectral features. The present paper reviews the recent advances in interpretive methodologies and outlines procedures for their application.

1. INTRODUCTION

In the beginning (ca 1801), asteroids were points of light in the sky. The primary measurements were positional, and the primary questions concerned the nature of their orbits. Despite some early attempts to determine their composition (e.g., *Watson,* 1941), the next major efforts in asteroid science occurred in the 1950s with the development of instrumental photometry, the measurement of asteroid lightcurves, and investigations of rotational periods, body shapes, and pole orientations. The first true compositional investigations of asteroids began in 1970 with the study of asteroid 4 Vesta (*McCord et al.,* 1970) supported by spectral studies of minerals (*Burns,* 1970a,b; *Adams,* 1974,1975) and meteorites (*Salisbury et al.,* 1975; *Gaffey,* 1976). A parallel effort developed an asteroid taxonomy that was suggestive of — but not diagnostic of — composition (*Tholen,* 1984; *Tholen and Barucci,* 1989).

In the past decade, spacecraft flyby (951 Gaspra, 243 Ida, 253 Mathilde) and rendezvous (433 Eros) missions as well as high-resolution Hubble Space Telescope, radar, and adaptive-optics observations have removed asteroids from the "unresolved point source" class. This has allowed the surface morphology and shape of these bodies to be studied.

In a similar fashion, advances in meteorite science and remote-sensing capabilities have opened the early history of individual asteroids and their parent bodies to sophisticated investigation. The central questions of current asteroid studies focus on geologic issues related to the original compositions of asteroidal parent bodies and the chemical and thermal processes that altered and modified the original planetesimals. Based on the small size of the planetesimals and on meteorite chronologies, it is known that all significant chemical processes that affected these minor planets were essentially complete within the first 0.5% of solar-system history. Asteroids represent the sole surviving *in situ* population of early inner solar-system planetesimals, bodies from which the terrestrial planets subsequently accreted. Material in the terrestrial planets has undergone substantial modification since the time of planetary accretion. Hence, asteroids provide our only *in situ* record of conditions and processes in the inner (~1.8–3.5 AU) portions of the late solar nebula and earliest solar system.

The meteorites provide detailed temporal resolution of events in the late solar nebula and earliest solar system, but the spatial context for that information is poorly constrained. Asteroid compositional studies, by elucidating relationships between individual asteroids and particular meteorite types, provide a spatial context for the detailed meteoritic data. Moreover, asteroid studies can identify types of materials not sampled in the terrestrial meteorite collections. The major goal of current asteroid investigations is to better understand the conditions and processes that prevailed in the late solar nebula and early solar system.

It is no longer possible to view asteroid studies as isolated from meteoritic and geologic constraints. One major corollary of this reality is that interpretations of asteroid compositions must be consistent with the nature of meteorites and with the geologic processes that produced or were active within the parent bodies. If different techniques give conflicting answers for an individual body, at best only one can be correct. And if one must choose between competing interpretations, one should select the most geologically and meteoritically plausible option. If one invokes an esoteric asteroid assemblage, the onus is on the investigator to justify it geologically. (For example, if one were to interpret a 50-km asteroid as being composed of kaolinite, one must be able to describe plausible geologic or meteorite parent body conditions for producing a 50-km mass of kaolinite!) Interpretations without geologic or meteoritic support must be considered as suspect until plausible geologic formation processes can be defined. A solid grounding in meteoritics and geology should now be considered as a prerequisite to the analysis of asteroid compositions.

The meteorites also constrain how we think about the asteroids. It is now clearly evident that the meteorite collection is an incomplete and biased sample of its asteroidal source region(s). For example, at least half of the mineral assemblages that geochemical considerations indicate should be formed with the known meteorite types are not represented in our collections. There are >55 types of asteroid-core material present as magmatic iron meteorites in our collections. The corresponding mantle and crust lithologies are essentially absent from these collections. The meteorites, nevertheless, provide strong constraints on the sampled portion of the asteroid belt. *Keil* (2000) has shown that ~135 individual meteorite parent bodies have contributed samples to our meteorite collections. Of these ~135 parent bodies, 80% have experienced strong heating that produced partial to complete melting. Thus, in the meteorite source regions (presumably primarily the inner belt), igneous asteroids are the rule rather than the exception. *Keil* (2000) also notes that the 20% of meteorite parent bodies that did not experience at least partial melting did experience heating to some degree.

Thus in thinking about at least the inner asteroid belt, one should consider that a large portion of the bodies we see probably either experienced strong heating or are fragments of parent bodies that experienced strong heating. This is the reverse of an older view (e.g., *Anders,* 1978) that most meteorite parent bodies, and by inference asteroids, are primarily primitive (i.e., chondritic) bodies.

Several of the major scientific issues that could benefit most significantly from sophisticated mineralogical characterizations of asteroids to be properly addressed include (1) constraining the conditions, chemical properties, and processes in the inner solar nebula as it evolved into the solar system; (2) independently confirming the dynamical pathways and mechanisms that deliver asteroidal bodies and their meteoritic fragments into Earth-crossing orbits; (3) characterizing additional types of materials present in the asteroid belt but not sampled in the meteorite collections; and (4) defining the chemical and thermal evolution of the planetesimals prior to accretion into the terrestrial planets.

2. BEFORE THE ANALYSIS CAN PROCEED

Mineralogical characterizations of asteroids — often somewhat inaccurately termed "compositional" characterizations — rely primarily on the identification and quantitative analysis of mineralogically diagnostic features in the spectrum of a target body. However, before such features can be identified and used for mineralogical characterizations, the spectrum must be properly reduced and calibrated. Inadequate reduction and calibration can introduce spurious features into the spectrum and/or suppress or significantly distort real features in the spectrum.

2.1. Atmospheric Extinction Corrections

Spectra begin their existence as measurements of photon flux vs. wavelength. These measurements may involve discrete filters, variable filters where bandpass is a function of position, interferometers, or dispersion of the spectrum by prism or grating. These raw flux measurements are converted to reflectance or emission spectra by ratioing the flux measurements at each wavelength for the target object (e.g., an asteroid) to a reference standard measured with the same system. For reflectance spectra this can be expressed schematically as

$$\text{Reflectance}_\lambda\ (\text{object}) = \text{Flux}_\lambda\ (\text{object})/\text{Flux}_\lambda\ (\text{standard}) \tag{1a}$$

This ratioing procedure removes the wavelength-dependent instrumental sensitivity function. If the standard object (typically a star) has a Sun-like wavelength-dependent flux distribution, the resulting ratio is a true relative reflectance spectrum (i.e., relative fraction of light reflected as a function of wavelength). If the standard does not have a Sun-like flux distribution, the object:standard ratio needs to be corrected at each wavelength for this difference. This is shown schematically by

$$\text{Reflectance}_\lambda\ (\text{obj}) = \text{Flux}_\lambda\ (\text{obj})/\text{Flux}_\lambda\ (\text{star}) \times \text{Flux}_\lambda\ (\text{star})/\text{Flux}_\lambda\ (\text{Sun}) \tag{1b}$$

For groundbased telescopic observations, this conceptually simple procedure is complicated by the transmission function of the terrestrial atmosphere. Equations (1a) and (1b) are strictly true only if both the object and the standard are observed simultaneously through the same atmospheric path. This is almost never the case. There are several methods of overcoming this problem. The simplest and most commonly used approach is to divide the object flux spectrum by a standard star flux spectrum, where the standard star observation is chosen to have an airmass (atmospheric path length) similar to the object observation and to have been

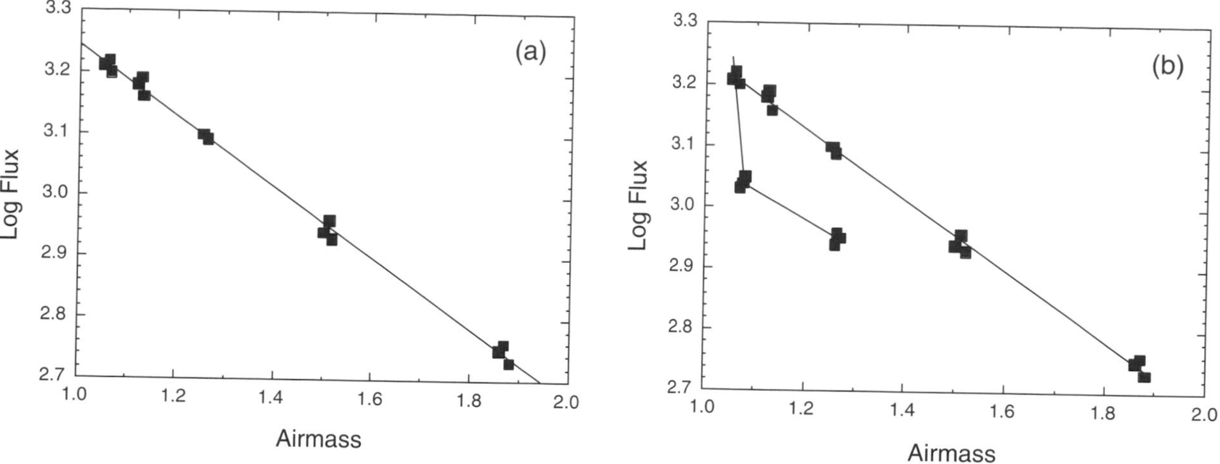

Fig. 1. **(a)** A schematic extinction curve for a particular wavelength computed from the measured fluxes at that wavelength for a standard star observed over a range of airmasses during a single night with stable and uniform atmospheric transmission. The absorption is assumed to follow a Beer-Lambert law, where the log of the flux is linear with the airmass. Airmass is calculated as the secant of the angle away from the zenith. **(b)** A schematic extinction curve where there is either an east-west asymmetry in the sky or the atmospheric transmission changed about the time the standard star crossed the meridian.

observed within a short time interval of the object observation. There are no strict criteria on what constitutes "similar" and "short." One can also extrapolate the standard star fluxes to the airmass of the object observation using standard extinction coefficients (atmospheric flux attenuation per unit airmass as a function of wavelength) for the particular observatory. Operationally, a ratio is acceptable when the atmospheric absorption features are "minimized." This may involve the division of an individual observation by a number of standard star observations until an acceptable ratio is generated. Clearly this is most easily accomplished for spectral regions where the extinction coefficient is small (i.e., no strong atmospheric absorptions), but still involves a significant subjective selection of "best."

A more rigorous means of removing the effects of differential atmospheric transmission involves computation of the extinction coefficients for each night or portion of a night. This is shown schematically in Fig. 1a. The slope and intercept of the linear least-squares fit can be used to compute the effective flux of that standard star at any airmass in the appropriate interval. The set of slopes and intercepts for all wavelengths of the observations is termed a "starpack." The slopes in a starpack should mimic the atmospheric transmission as a function of wavelength. Figure 1a shows the fit to standard flux measurements at a particular wavelength when the atmospheric transmission is stable over time and symmetric about the zenith. More typically, neither of these conditions is met, and the result is shown in Fig. 1b. For example, at Mauna Kea Observatory, there is commonly an upwind/downwind asymmetry in the extinction. This is due to the presence of a very thin (usually imperceptible to the eye) orographic haze formed over the peak and extending some distance downwind. The pattern shown in Fig. 1b is the result of extra absorption in the eastern sky (downwind) as the object rises toward the meridian. When the object passes the western edge of the cloud, the flux rises. As the object moves westward to higher airmass, the regular extinction slope is seen. As would be expected, the effect is most pronounced at the wavelengths of the atmospheric water features.

2.2. Special Instrumental Considerations

Additionally, most instrumental systems used for asteroidal observations have their own special "quirks" that can affect final reflectance spectra. Since asteroid spectral studies often push the capabilities of telescopic instruments, it is important to be sensitive to the presence of such special data-reduction requirements, which may be of little or no significance to classical astronomers. Although space limitations do not permit a full discussion of such quirks in the variety of instruments that have been used to obtain asteroid spectra, previous examples include the beam inequality and coincidence counting effects in the two-beam photometer system (*Chapman and Gaffey,* 1979), lightcurve-induced spectral slopes in CVF and filter-photometer systems (e.g., *Gaffey et al.,* 1992; *Gaffey,* 1997), and channel shifts in CCD and other array detectors as evidenced by the mismatch of the narrow atmospheric O_2 absorption line at 0.76 μm in unedited spectra (e.g., *Xu et al.,* 1995; *Vilas and Smith,* 1985; *Vilas and McFadden,* 1992).

The recent advent of moderate-resolution ($\lambda/\Delta\lambda \sim 100$–500) asteroid observations with NIR array detectors has produced a particularly keen need to detect and compensate for potential subtle instrumental artifacts in the resulting spectra. At these resolutions, small shifts in the placement of the dispersed spectrum onto the detector (due to unavoidable instrument flexure) produce significant artifacts in the reflectance spectrum in regions where the measured flux is changing rapidly with wavelength. This is shown in Fig. 2.

Even a small shift in the location of the spectrum on the detector (1 channel or ~1/500 of the length of the dispersed spectrum) produces significant structure in the resulting ratio as shown by Fig. 2b. The effect is most pronounced where the spectral flux changes most rapidly from channel to channel, such as the short wavelength edge of the 1.4-µm atmospheric water-vapor band. In that region, a distinct spike is observed, which is negative (below the background curve) when the channel shift of the "numerator" spectrum is positive relative to the "denominator" spectrum and positive (above the background curve) when the channel shift is negative. If spikes at a few channels around the steepest portions of the flux curves were the sole effect, those points could simply be deleted with little loss of data.

However, examination of Fig. 2b shows that the effect of such a shift is much more pernicious. For example, consider the region around 1.4 µm, where the ratio is ~10% high. In an asteroid:star ratio (reflectance spectrum), the negative spike near 1.34 µm could be readily detected and deleted. However, the ratio would still overestimate the reflectance by ~10%, decreasing to 0% between ~1.36 and 1.5 µm. Since this is a region where the plagioclase feldspar feature is observed in eucrite meteorites and Vesta-type spectra, the presence of this effect would distort the feldspar feature, either increasing or decreasing its apparent intensity depending on the relative direction of the channel shift.

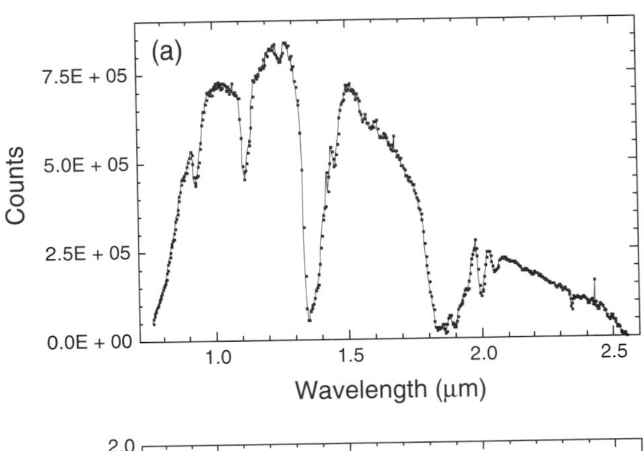

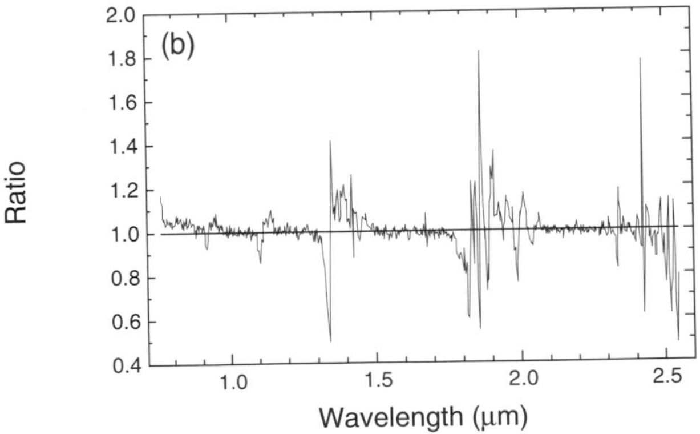

Fig. 2. **(a)** Spectral flux distribution (instrumental counts vs. wavelength) for standard star BS5996 observed with a NIR-array spectrometer (the SpeX Instrument operating in low resolution or asteroid mode at the NASA Infrared Telescope Facility at Mauna Kea Observatory). **(b)** Ratio of flux for this observation offset by +1 channel (~1/500 the length of the spectrum on the array detector) relative to the original data. Without the channel offset the ratio would lie on the heavy line at 1.0.

From the point of view of characterizing asteroid mineralogy, the effect is most severe on Band II, the ~2-µm pyroxene-absorption feature present in varying intensities in the spectra of most chondrites, basaltic achondrites, ureilites, SNCs, lunar samples, and S-, V-, and R-type asteroids. This feature is generally centered between 1.8 and 2.1 µm. Again referring to Fig. 2b, one effect of a +1 channel shift is the introduction of a high-frequency noise component, degrading the quality of the ratio spectrum between 1.8 and 2.1 µm. But more importantly, the ratio curve will be up to 15% low between 1.75 and 1.80 µm and up to 15% high between 1.9 and 2.02 µm. This would have the effect of shifting the effective band center toward a shorter wavelength, resulting in an erroneous determination of pyroxene composition. The opposite effect would be observed if the channel shift was in the opposite direction (i.e., –1 channel). The magnitude of the change in the apparent spectrum is generally comparable to — and often greater than — the intensity of Band II in S-type spectra. The actual magnitude of these effects will depend on the magnitude of the channel shift. Unless detected and corrected, this effect has the potential to significantly distort the position, shape, and intensity of the absorption feature (and hence, the interpreted mineralogy).

3. INTERPRETIVE METHODOLOGIES FOR THE MINERALOGICAL ANALYSIS OF ASTEROID SPECTRA

3.1. Taxonomy

Taxonomy is the classification of a group of subjects (objects, phenomena, organisms, etc.) into classes based on shared measured properties. The utility of any taxonomy is a function of the relevance of the properties employed in the classification. In the case of the asteroid taxonomic classes, the parameters are observational (e.g., spectral slope, color, albedo, etc.) but in most cases are not particularly diagnostic of either mineralogy or composition. It is generally safe to assume that objects in different taxonomic classes are physically or compositionally different from each other, although the nature of the difference is generally not well constrained. However, the converse is not true. It is not safe to assume that members of a taxonomic class have similar compositions, mineralogies, or genetic histories.

The reliance on asteroid taxonomy, while useful for suggesting both similarities and differences between asteroids, is seductive and can actually impede geological interpretation of the asteroid belt. Assuming that such classifications imply some mineralogical or petrogenetic similarities among all (or most) members of a class may not be warranted and can be counterproductive (e.g., *Gaffey et al.,* 1993b). For example, the early debate over the S-asteroid class was largely focused on whether this group consisted of differentiated or primitive objects. Recent interpretations of individual S-asteroid spectra indicate that this class includes both differentiated and primitive members (*Gaffey et al.,* 1993a, and references therein; *Gaffey and Gilbert,* 1998; *Kelley and Gaffey,* 2000). The S class has now been subdivided into a number of mineralogical subclasses, which may be further subdivided as new data and interpretations are generated. We have seen a similar expansion in the overall number of asteroid classes over the years as the quantity and quality of observational data have improved. At some point the knowledge of individual asteroids becomes sufficiently detailed that their class becomes largely redundant, just as the classification "mammal" becomes largely redundant when one is investigating the relationships between different members of the "cat" family.

3.2. Curve Matching

For many years, analysis of asteroid reflectance spectra relied on a comparison to laboratory reflectance spectra of meteorites (e.g., *Chapman and Salisbury,* 1973; *Chapman,* 1976). This approach provided many important insights into plausible surface mineralogies for many asteroids, but had a number of limitations and could lead to incorrect interpretations or an unnecessarily restricted range of possible mineralogies. Limitations of this approach include (1) the fact that it generally cannot provide robust insights into surface mineralogies of asteroids that have no spectrally characterized meteorite analogs, (2) the lack of spectral reflectance data for terrestrially unweathered samples of many meteorite classes, and (3) the spectral variations associated with changes in grain size, viewing geometry, and temperature (e.g., *Adams and Filice,* 1967; *Egan et al.,* 1973; *Singer and Roush,* 1985; *Gradie and Veverka,* 1986; *Lucey et al.,* 1998). Moreover, extrinsic effects such as space weathering can modify the spectral shape and obscure potential genetic links to meteorites. Nevertheless, curve matching is useful for identifying the possible reasons asteroid spectra differ from possible meteorite analog spectra.

A similar spectral shape between an asteroid and a comparison sample (e.g., a meteorite or simulant) may suggest that they are composed of similar materials, but alternative interpretations are often possible. Initially, there is the issue of incompleteness of the comparison sample set. Curve matching will select the closest match to the asteroid spectrum from among the sample set. However, if — as is almost always the case — the sample set is not exhaustive, the match may not be meaningful.

The second issue with curve matching is exactly what is meant by "similar." For example, a numerical algorithm can be used to estimate the mean deviation between an asteroid spectrum and a suite of comparison spectra in order to select the sample or mixing model that most closely matches the asteroid. This has been done by a number of investigators (e.g., *Hiroi et al.,* 1993; *Clark,* 1995). However, the match is generally to the entire available spectral curve. Such a "full-spectrum" match does not address the issue of which portions of the spectrum are most significant. For example, the position and shape of the centers of the absorption features must be matched very closely to have any confidence in the result. By contrast, even very poor matches to the spectral curve outside the absorption features may be of little or no significance. Thus "curve matching" can only provide a reasonably credible result when a wavelength-dependent weighting function is applied to account for the relative importance of the match in different spectral intervals. To date, no such wavelength-weighted matching procedure has been used, and thus all curve matching results should be viewed with at least some skepticism.

3.3. Mineralogical Analysis of Spectra

Mineralogical characterizations of asteroids — and investigations of possible genetic linkages to meteorites — involve analysis of spectral parameters that are diagnostic of the presence and composition of particular mineral species. Not all minerals — nor even all meteoritic minerals — have diagnostic absorption features in the visible and NIR (VNIR) spectral regions where most asteroid data have been obtained. Fortunately, a number of the most abundant and important meteoritic minerals do exhibit such diagnostic features. The most important set are crystal field absorptions arising from the presence of transition metal ions [most commonly bivalent iron/Fe^{2+}/Fe(II)] located in specific crystallographic sites in mafic (Mg- and Fe-bearing) silicate minerals. Mafic minerals (e.g., olivine, pyroxene, certain Fe-phyllosilicates, etc.) are the most abundant phases in all chondrites and in most achondrites. They are also present as the major silicate phases in stony irons and as inclusions in a significant number of iron-meteorite types. A mineral is a particular composition (or range of compositions in the case of a solid-solution series) formed into a particular crystallographic structure. The wavelength position, width, and intensity of these crystal-field absorptions are controlled by structure and composition that are directly related to the fundamental definition of a mineral. The quantum-mechanics-based theory behind these features is summarized by *Burns* (1970a,b, 1993). *Adams* (1974, 1975), *Salisbury et al.* (1975), *Gaffey* (1976), *King and Ridley* (1987), *Gaffey et al.* (1989), *Cloutis and Gaffey* (1991a,b), and *Calvin and King* (1997) show examples of the reflectance spectra of meteorites and the minerals of which they are made.

3.3.1. Mafic mineral compositions. The relationship between diagnostic spectral parameters (especially absorption-band center positions) for reflectance spectra of a number of mafic minerals, particularly pyroxene and olivine, was first outlined by *Adams* (1974, 1975) and revisited by *King and Ridley* (1987) and *Cloutis and Gaffey* (1991a). All these spectral measurements were made at room temperature. In equation form, the relationships between absorption-band center and pyroxene composition (molar Ca content [Wo] and molar-Fe content [Fs]) shown in those papers can be expressed as

$$Wo\ (\pm 3) = 347.9 \times \text{BI Center}\ (\mu m) - 313.6 \quad (Fs < 10;\ Wo_{\sim 5-35}\ \text{excluded}) \tag{2a}$$

$$Wo\ (\pm 3) = 456.2 \times \text{BI Center}\ (\mu m) - 416.9 \quad (Fs = 10\text{–}25;\ Wo_{\sim 10-25}\ \text{excluded}) \tag{2b}$$

$$Wo\ (\pm 4) = 418.9 \times \text{BI Center}\ (\mu m) - 380.9 \quad (Fs = 25\text{–}50) \tag{2c}$$

$$Fs\ (\pm 5) = 268.2 \times \text{BII Center}\ (\mu m) - 483.7 \quad (Wo < 11) \tag{3a}$$

$$Fs\ (\pm 5) = 57.5 \times \text{BII Center}\ (\mu m) - 72.7 \quad (Wo = 11\text{–}30,\ Fs_{<25}\ \text{excluded}) \tag{3b}$$

$$Fs\ (\pm 4) = -12.9 \times \text{BII Center}\ (\mu m) + 45.9 \quad (Wo = 30\text{–}45) \tag{3c}$$

$$Fs\ (\pm 4) = -118.0 \times \text{BII Center}\ (\mu m) + 278.5 \quad (Wo > 45) \tag{3d}$$

Note that the "excluded" ranges in the Wo calibrations (equations (2a) and (2b)) and the Fs calibration (equation (3b)) represent compositions not present among natural minerals (*Deer et al.,* 1963, p. 5).

No single equation is applicable to the full range of pyroxene compositions. The values in parentheses after each equation indicate the range of compositions to which that equation applies. The uncertainty indicated for each determination is the mean of the differences (rounded up to the next whole number) between the predicted and measured compositions among the set of samples used to establish each calibration.

Since neither the Ca nor Fe content is known initially, these equations are used in an iterative fashion to derive the pyroxene composition. The following example illustrates this process. The Band I and Band II positions of a pyroxene-dominated asteroid spectrum are 0.937 and 1.914 μm respectively. The dominance of pyroxene can be established by a high band-area ratio (see below) or an unbroadened 1-μm absorption feature. From equation (2a), the 0.937-μm Band I center corresponds to Wo_{12}. This value indicates that among the equation (3) options, equation (3b) (appropriate for Wo_{11-30}) should be used. Plugging the Band II center wavelength into equation (3b) gives a value of 37 (Fs_{37}). However, the initial Wo determination was made with equation (2a), which is appropriate only for $Fs_{<10}$. Based on the Fs determination, equation (2c) should be used and gives a value of 11.6 (Wo_{12}). This is consistent with the use of equation (3b), so the process has converged. The possible alternate computations are:

Equation (3a) and Band II → Fs_{29} ⇒
Equation (2c) and Band I → Wo_{12} ⇒
Equation (3b) and Band II → Fs_{37}

Equation (3b) and Band II → Fs_{37} ⇒
Equation (2c) and Band I → Wo_{12}

Equation (3c) and Band II → Fs_{21} ⇒
Equation (2b) and Band I → Wo_{11} ⇒
Equation (3b) and Band II → Fs_{37}

Equation (3c) and Band II → Fs_{21} ⇒
Equation (2b) and Band I → Wo_{11} ⇒
Equation (3a) and Band II → Fs_{30}

Equation (3d) and Band II → Fs_{52} ⇒
Equation (2c) and Band I → Wo_{12} ⇒
Equation (3b) and Band II → Fs_{37}

Equation (2b) and Band I → Wo_{11} ⇒
Equation (3a) and Band II → Fs_{30} ⇒
Equation (2c) and Band I → Wo_{12}

Equation (2b) and Band I → Wo_{11} ⇒
Equation (3b) and Band II → Fs_{37}

Equation (2c) and Band I → Wo_{12} ⇒
Equation (3b) and Band II → Fs_{37}

The results are relatively insensitive to the pathway chosen. All converge on Wo_{11-12} and Fs_{37} or Fs_{30}. The Wo content near 11 mol% produces the ambiguity in the Fs content, since it allows selection of either equation (3a) or (3b) to determine the Fe content.

The relationship between absorption-band position and olivine composition was explored by *King and Ridley* (1987). Unfortunately, the band shift with composition in olivine is small and band position provides a relatively poor constraint on olivine composition.

3.3.2. Mafic mineral abundances. *Cloutis et al.* (1986) developed a calibration between abundance in olivine-orthopyroxene (Ol-Opx) mixtures and the ratio of areas for the 1- and 2-μm absorption bands (Band II/Band I area; Band Area Ratio; BAR). Figure 3 shows a typical spectrum with the individual band areas indicated. (All these spectral measurements were made at room temperature.) For the interval of 10–90% orthopyroxene, this relationship can be expressed as:

$$Opx/(Opx + Ol) = (0.417 \times BAR) + 0.052 \tag{4}$$

It should be emphasized that this relationship applies directly only to olivine-orthopyroxene mixtures. An increasing error will be introduced by the presence an increasing clinopyroxene component in the mixture.

The Band I center position of an olivine-pyroxene mixture is a function of both the pyroxene composition (equations (2) and (3) above) and the olivine-pyroxene abundance ratio. Using the Band I center and BAR calibrations of *Cloutis et al.* (1986), a correction factor can be applied to the measured Band I center to remove the effect of olivine. Figure 4 shows the displacement in the Band I center as a function of BAR. When this correction factor is applied to

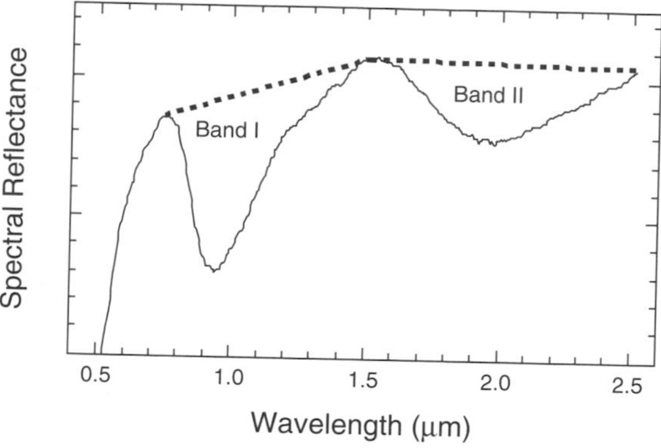

Fig. 3. The spectral reflectance curve of an olivine-pyroxene dominated meteorite sample (average LL3 chondrite). Band I and Band II are indicated. The band areas are the areas between the spectral curve and linear continua (shown as dashed lines) fitted tangent to the spectral curve at the edge of the absorption feature. The band area ratio (BAR) is the ratio of the area of Band II to that of Band I (BII/BI). Band centers are calculated by dividing out the linear continuum (to remove the effects of spectral slope) and fitting a polynomial function to the wavelength interval around the minimum in the ratioed curve.

the Band I center for olivine-orthopyroxene mixtures, the resulting Band I (corrected) and Band II values should plot along the pyroxene trend line shown by *Adams* (1975) and *Cloutis and Gaffey* (1991a). This provides a test of the validity of derived band positions. It also allows a "first-cut" determination of the calcic (Wo) content of the pyroxene.

The weak plagioclase feature near 1.25 μm in plagioclase-pyroxene mixtures (e.g., the eucrite members of the HED meteorite group) can be used to determine the plagioclase-pyroxene abundance ratio. *Gaffey et al.* (1989) describes the preliminary quantative calibration outlined by *McFadden and Gaffey* (1978).

3.3.3. One or two pyroxenes? Depending on their petrogenesis, anhydrous mafic rock assemblages (i.e., those dominated by Fe- and Mg-bearing silicates, such as olivine and pyroxene) commonly contain either a pyroxene of a single composition or two coexisting pyroxene compositions related by an equilibration constant dependent upon the geologic processes that formed the assemblage. The calibrations of 1- and 2-μm band positions to pyroxene composition [*Adams* (1974); *Cloutis and Gaffey* (1991a); equations (2) and (3)] work well for spectra from samples that contains a single pyroxene. However, the calibrations break down when a significant amount of a second pyroxene component is present.

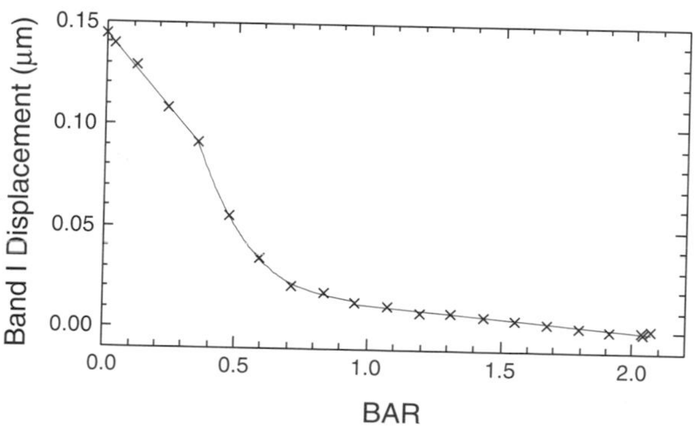

Fig. 4. The displacement in the Band I center in olivine-orthopyroxene mixtures relative to the Band I center for the pyroxene phase as a function of the band area ratio (BAR). BAR is inversely proportional to olivine abundance.

TABLE 1. Comparison of predicted and observed Band II centers for ordinary chondrites.

	H Chondrites	L Chondrites	LL Chondrites
Pyroxene composition*	$Fs_{17.2}$	$Fs_{21.3}$	$Fs_{24.1}$
Expected Band II center	1.868 μm	1.883 μm	1.893 μm
Observed Band II center	1.931 μm	1.944 μm	1.976 μm
Difference (actual — expected)	+0.063 μm	+0.061 μm	+0.083 μm

**Gomes and Keil* (1980).

An example of this discrepancy is evident when the above calibrations are applied to the Band II positions for the ordinary chondrites. These positions (e.g., *Gaffey,* 1976) are longward of the expected wavelengths based on the orthopyroxene compositions commonly cited for ordinary chondrites. This is shown on Table 1.

This discrepancy is not caused by the olivine component, since the spectra of olivine:orthopyroxene mixtures do not show this shift. In ordinary chondrites, low-Ca orthopyroxene is the dominant pyroxene phase, but there is a small but spectrally significant high-Ca pyroxene (diopside/augite) component. In the older meteorite literature, the high-Ca pyroxene component was not generally distinguished (or distinguishable) because of its small grain size and/or its presence only as very fine exsolution lamellae in orthopyroxene. As a result, often only the dominant orthopyroxene component was reported. *Brearley and Jones* (1998, p. 3-231) note that high-Ca pyroxene (augite) is the third most abundant silicate mineral component in the matrix of unequilibrated ordinary chondrites. *McSween et al.* (1991) include high-Ca pyroxene (diopside) in their normative calculations of ordinary chondrite mineralogy. Their calculations indicate that high-Ca pyroxene accounts for approximately 12%, 17%, and 19% of the pyroxene in average H-, L-, and LL-chondrites, respectively.

The high-Ca component in these chondrites has a composition of ~$Fs_{10}Wo_{45}$. The Band I and II positions for such a pyroxene are approximately 1.03 μm and 2.30 μm. On Fig. 5, the *Adams* (1974) and *Cloutis and Gaffey* (1991a) band positions are plotted for opx and cpx (small open symbols); the average band positions for H, L, and LL chondrites (large solid symbols); the expected band positions for the low-Ca (opx) in these chondrites based on the calibrations discussed above (large open symbols); the band position for the high-Ca pyroxene in these meteorites (large square with enclosed X); and the Band I position corrected

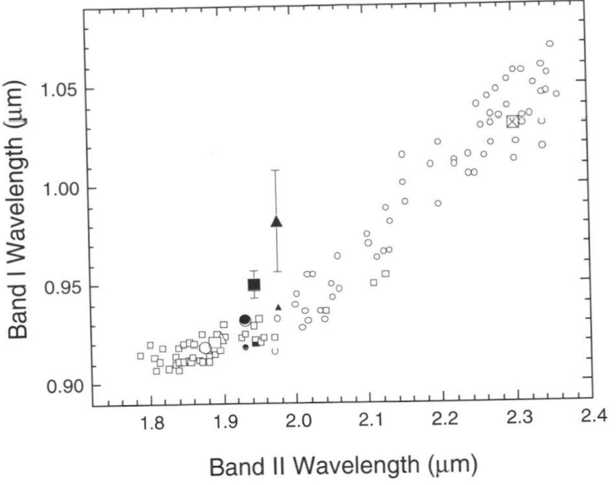

Fig. 5. Plot of Band I center vs. Band II center for low- and high-Ca pyroxenes from *Adams* (1974) and *Cloutis and Gaffey* (1991a) (small open squares and circles respectively). The average measured band positions for H, L, and LL chondrites (large solid symbols: circle, square, and triangle respectively) are compared to the "olivine corrected" band positions (small solid symbols) and the positions of the low-Ca pyroxene components (large open symbols) for these chondrites. The position of the high-Ca pyroxene component of these chondrites is plotted as square with an inscribed "X."

for olivine using the relationship shown in Fig. 3 (small solid symbols).

The ordinary chondrites are longward of their expected wavelengths based on their commonly cited pyroxene compositions, as shown in Table 1. The effect of the second (high-Ca) pyroxene (~$Fs_{10}Wo_{45}$) on the Band II position in ordinary chondrites appears to be an almost-linear function of the relative abundance of the two pyroxenes. The comparison between the fraction of the pyroxene that is high-Ca from average normative mineralogy (*McSween et al.,* 1991) and the abundance calculated from the Band II position assuming a linear relationship is summarized in Table 2. The agreement is sufficiently good to support the assumption that the Band II position is approximately a linear function of the abundances of the low-Ca and high-Ca pyroxene components in such assemblages.

Sunshine et al. (2002) have raised the intriguing possibility of S-type asteroids with silicates consisting solely of low-Ca orthopyroxene and high-Ca clinopyroxene. They identified two such asteroids (17 Thetis and 847 Agnia) that have BAR values normally indicating a substantial olivine

TABLE 2. Comparison of normative and spectrally derived high-Ca pyroxene abundance.

	H Chondrites	L Chondrites	LL Chondrites
High-Ca/total pyroxene (normative)	12%	17%	19%
High-Ca/total pyroxene (spectra)	13%	14%	20%

component. Obviously the calibration (equation (4)) for pyroxene abundance in an olivine-orthopyroxene assemblage would be inappropriate for such assemblages. Therefore it is important to establish the applicability of this calibration before simply applying it to an S asteroid. Fortunately, there is a procedure for doing so.

Figure 4 shows the displacement off of the pyroxene trend (Fig. 5) as a function of the BAR value in olivine-orthopyroxene mixtures. If the displacement derived from the BAR value is subtracted from the Band I position for an olivine-orthopyroxene assemblage, the resulting point will fall on the pyroxene trend. The same procedure for a mixture of Ca-poor orthopyroxene and calcic clinopyroxene will result in a point below the pyroxene trend line. Thus the band center positions and the BAR value for an S-type spectrum can be used to establish whether or not equation (4) is applicable to that spectrum.

3.3.4. "Featureless" spectra. Analysis of asteroid spectra will not necessarily provide a complete inventory of the minerals that reside on the surface. Spectral analysis is only capable of detecting the presence of spectrally dominant minerals, i.e., those that possess unique or distinguishable spectral properties. These minerals may or may not be the most abundant. The minerals that are spectrally dominant will depend on the other phases that may be present. As an example, opaque materials such as macromolecular organics and magnetite, present in carbonaceous chondrites at the few weight percent level, tend to dominate the spectral signature of these meteorites, in spite of the fact that other minerals, such as olivine and layer lattice silicates are much more abundant (e.g., *Cloutis,* 1996; *Cloutis and Gaffey,* 1994). This spectral dominance is attributable to the fact that these opaque components are fine-grained, dispersed throughout the meteorites, and very effective absorbers of incident light (Fig. 6a). As a result, most carbonaceous chondrite VNIR spectra exhibit low reflectance and very subdued or undetectable absorption bands, even when their predominant olivine or layer lattice silicate constituents exhibit strong features (e.g., *Calvin and King,* 1997). *Clark and Lucey* (1984) demonstrated a similar suppression of strong water-ice features with the addition of a very small carbon black component to the ice.

Meteoritic metal, whether present as part of the original lithology or arising from reduction of ferrous Fe by a reducing agent (e.g., C) at elevated temperatures (e.g., *Walker and Grove,* 1993) or by the impact melting of minerals (*Hapke,* 2001) is an example of a material with no well-defined absorption features in the visible and NIR spectral regions. The presence of this material in many asteroid reflectance spectra is inferred by the way it modifies the spectra of mafic silicates with which it is associated. Figure 6b shows the reflectance spectra of pure olivine, meteoritic metal, and a 50:50 intimate mixture of this olivine and meteoritic metal. The spectrum of the mixture is very similar to that of the metal, particularly in terms of overall slope, which increases toward longer wavelengths. This spectral shape, commonly referred to as red-sloped, is characteristic of many S-class asteroid spectra (*Gaffey et al.,* 1993a,b). This

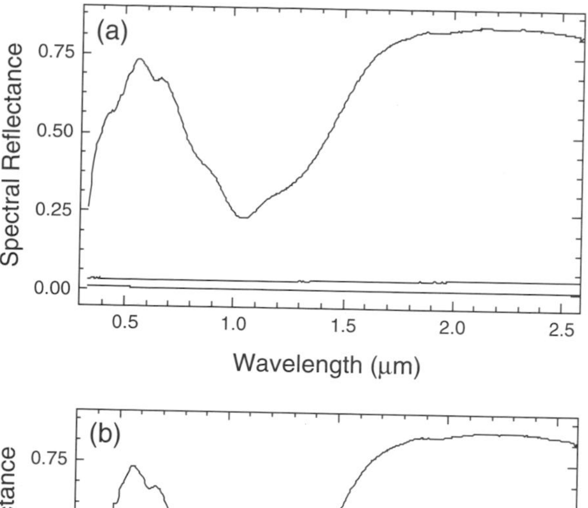

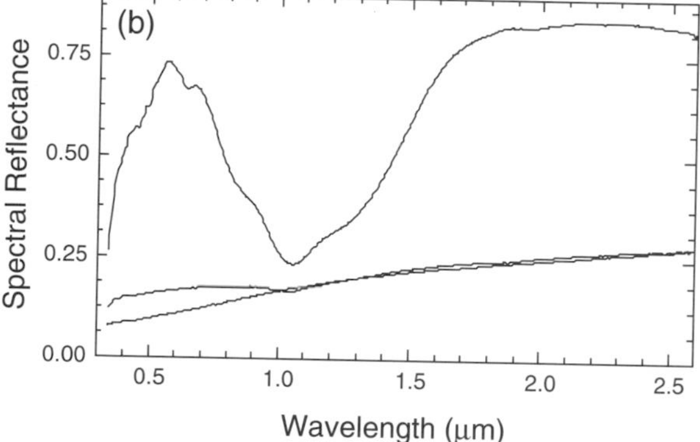

Fig. 6. **(a)** Reflectance spectra of 45–90-μm grain size olivine (upper curve), <0.021-μm size C lampblack (lower curve), and an intimate mixture of 98 wt% olivine and 2 wt% C lampblack (middle curve). Note: Bottom axis is –0.05. **(b)** Reflectance spectra of 45–90-μm grain size olivine (upper curve), 45–90-μm size metal from the Odessa iron meteorite (lower curve at 0.5 μm), and an intimate mixture of 50 wt% olivine and 50 wt% metal (middle curve at 0.5 μm).

effect, in conjunction with the common occurrence of metal in most meteorite classes, and the presence of meteorites with similar metal-silicate assemblages collectively suggest that red-sloped asteroid spectra that exhibit mafic-silicate absorption bands are indicative of mafic silicate plus metal assemblages. It should be noted, however, that the spectra of iron meteorites exhibit a range of slopes and spectral curvatures, from strongly reddened to weakly reddened and from nearly linear to sharply curved (*Gaffey,* 1976; *Britt and Pieters,* 1988). Until the nature of these spectral variations are better understood, one should be wary of asserting that NiFe metal on asteroid surfaces should introduce a linear reddening component or that failure to exhibit a linear reddened spectrum is evidence against a NiFe metal component on asteroid surfaces.

As noted above, intimate mixtures of various meteoritically relevant dark or opaque materials with mafic silicates can alter our ability to successfully derive mafic silicate abundances and compositions. Studies of the effects of opaque phases include meteoritic metal, magnetite, amorphous C (lampblack), graphite, coal tar extract, bitumen, silicon carbide, cementite, and wustite (*Cloutis et al.,* 1990b,c). In all cases, the presence of red-sloped accessory phases causes the wavelength position of the reflectance minimum of the mafic silicate absorption band in the 1-μm region to shift to shorter wavelengths, with the magnitude of the shift increasing with increasing content of the red-sloped phase; shifts of up to 0.080 μm have been produced. However, when a straight-line continuum (tangent to the reflectance spectrum on either side of this absorption band) is divided out of the spectrum, the wavelength position (center) of the absorption band is restored to its original wavelength position to within ±0.005 μm or less. This allows the composition of the mafic silicates to be retrieved even in the presence of abundant (many tens of percent) red-sloped phases such as meteoritic Fe. Spectrally flat opaque phases generally have less of an effect on band positions. Band areas and depths, on the other hand, are more strongly affected by the presence of nonmafic components, either red sloped or spectrally flat (e.g., *Moroz and Arnold,* 1999).

In some cases, the presence of a particular mineral is determined by a process of elimination. Most E-class asteroids are characterized by high reflectance (>40%) and relatively featureless spectra (*Zellner,* 1975; *Gaffey et al.,* 1992). These characteristics severely limit the range of potential surface mineralogies, since there are few geologically plausible meteoritic or asteroidal materials with high reflectance and featureless reflectance spectra in the 0.3–2.5-μm region (e.g., *Kelley and Gaffey,* 2002). When combined with the presence of meteorites with similar spectral properties (aubrites) and other factors, this strongly suggests the surfaces of these asteroids are dominated (both spectrally and compositionally) by an Fe-free or Fe-poor enstatite.

3.4. "Space Weathering" and Mineralogical Interpretations of Asteroid Spectra

One of the complicating factors in the interpretation of asteroid spectra has been the issue of space weathering and its potential to modify the spectrum (e.g., *Clark et al.,* 2002). "Space weathering" is the catchall term for the process or processes that modify the optical properties of surfaces of atmosphereless bodies exposed to the space environment. The importance of space weathering on the Moon was well documented by the examination of the first lunar samples returned by the Apollo missions (e.g., *Adams and McCord,* 1971). The spectra of lunar soils exhibited systematically lower albedos, weaker mineral absorption features, and redder spectral slopes than the bedrock lithologies from which they were derived (e.g., *Pieters et al.,* 1993). (In the jargon of lunar and asteroidal spectrophotometry, "red," "reddish," and "redder" all refer to spectra whose reflectance increases with increasing wavelength across the visible and/or NIR interval. Such samples are seldom — if ever — visually red as perceived by the human eye.)

Based on the detailed study of lunar soil samples, a physical model for space weathering has been outlined in a series of recent papers (*Pieters et al.,* 2000; *Noble et al.,* 2001; *Taylor et al.,* 2001). The mechanism involves the presence of tiny (4–30 nm) metallic-Fe grains (nanophase metallic Fe, or $npFe^0$) in thin layers or patinas of glass preferentially coating the surfaces of the finest size fraction of the lunar soil. These grain coats are produced either by condensation of soil minerals vaporized by micrometeorite impacts into the soil and/or by deposition of atoms sputtered off of soil grains by solar-wind particles (e.g., *Hapke,* 2000). Those investigators make a cogent and persuasive case for the proposed mechanism as the process that alters the optical and spectral properties of the lunar soil. However, it was also concluded that this mechanism was important on asteroid surfaces and that it potentially allowed large numbers of ordinary chondrite parent bodies to be present in the asteroid belt (*Pieters et al.,* 2000; *Hapke,* 2000).

The issue of space weathering has been an area of considerable contention among asteroid scientists (e.g., *Gaffey et al.,* 1993a; *Chapman,* 1996; *Clark et al.,* 2002, and references therein). The issue arises from an apparent paradox. Although ordinary chondrites account for the major fraction (~75%) of meteorites falling to Earth, it was recognized very early that few — if any — main-belt asteroids spectrally resembled ordinary chondrites. To explain this discrepancy, space weathering was invoked to optically alter the surface of what were expected to be common ordinary chondrite-type asteroids so as to produce the spectral mismatch. Thus asteroidal space weathering was invoked to reconcile an expectation (that ordinary chondrites were from a common type of asteroid) with an observation (that few — if any — asteroids actually exhibited ordinary chondritelike spectra).

There is clear evidence from telescopic observations for optical alteration of asteroidal surfaces (e.g., *Gaffey et al.,* 1993c). The *Galileo* flyby of 243 Ida and the *NEAR Shoemaker* rendezvous with 433 Eros have reinforced that conclusion, but have also complicated the situation, since the alteration differs between the two asteroids and both differ from the Moon. In the case of the Moon, increased weathering produces weakened absorption features, lowered albedos, and significant color changes. For Ida, there are significant color variations but virtually no albedo variations. Eros exhibits the opposite pattern, with virtually no color variations and large albedo variations. The "space weathering" process(es) occurring on asteroid surfaces are not simply a weakened form of lunar-style space weathering. Moreover, the type of weathering process can clearly vary between otherwise grossly similar bodies (i.e., Eros and Ida are both S-type asteroids).

The real issue is the degree to which "space weathering" affects the interpretation of asteroidal spectra. In spectral-matching methodology, such spectral modification is potentially devastating since it makes the spectral properties of the surficial layer of an asteroid different from that of the substrate and can therefore produce a spurious result if such effects are not considered. The use of spectral parameters (e.g., band centers, BARs) as outlined in previous sections of this chapter appears to mitigate space-weathering effects (e.g., *Gaffey,* 2001; *Hapke,* 2001). Based on preliminary analysis of lunar samples, *Gaffey* (2001) notes that the diagnostic parameters are essentially unchanged until a high degree of weathering is reached, a level generally beyond that proposed or observed on asteroid surfaces. So while "space weathering" can alter asteroid surfaces, making the spectra redder and weakening the absorption features, it does not appear to significantly impact the mineralogical interpretations of those spectra. This is consistent with the generally similar results of the analysis of the NIR spectra (affected by space weathering) and the X-ray/γ-ray spectra (unaffected by space weathering) from the *NEAR Shoemaker* observations of Eros (*McCoy et al.,* 2001).

3.5. Importance of Laboratory Spectral Studies

While the field of asteroid research has made many important advances in determining mineralogies for a number of asteroids, particularly from analysis of reflectance spectra (e.g., *McCord et al.,* 1970; *Gaffey et al.,* 1993a; *Gaffey and Gilbert,* 1998; etc.), the fact remains that for a distressingly large number of asteroids, surface mineralogies are poorly constrained, and even when mineralogies can be approximated, widely different interpretations of parent-body history and evolution can often be accommodated by available data. Some mafic-silicate-dominated assemblages, such as 433 Eros (*Veverka et al.,* 2000) are consistent with either partially differentiated (i.e., primitive achondrites analogous to lodranites, winonaites, and acapulcoites) or undifferentiated assemblages (i.e., ordinary chondrites). Even when high-quality spectral data are available and supplemented by elemental-abundance information, as is the case for Eros, this uncertainty may still not be resolvable (*Trombka et al.,* 2000; *Veverka et al.,* 2000; *McCoy et al.,* 2001). However, interpretation of the observational spectral data in light of relevant laboratory spectral data and/or rotationally resolved spectral observations can be used to resolve the differentiated nature of an S-class asteroid (*Gaffey,* 1984).

To a significant extent the paucity of detailed mineralogic interpretations and the ambiguity associated with many interpretations arises from the limitations in or the lack of appropriate interpretive calibrations. Such calibrations are largely derived from laboratory spectral studies of meteorites and mineral mixtures intended to simulate asteroid lithologies. For example, the need for a systematic spectral study of olivine, low-Ca pyroxenes, and high-Ca pyroxenes was noted above. Similarly, the spectral behavior of NiFe metal needs to be better understood. The behavior of NiFe metal in assemblages depends on the petrologic associations. For example, the metal in undifferentiated (i.e., ordinary chondrite) meteorites is spectrally neutral (*Gaffey,* 1986), while Fe from differentiated meteorites is invariably red-sloped when measured (*Gaffey,* 1976; *Britt and Pieters,* 1988; *Cloutis et al.,* 1990a, and references therein). It has been speculated that the metal grains in chondrites are coated with an optically thick alteration layer that suppresses the spectral signature of the metal (e.g., *Gaffey,* 1986), but an actual spectroscopic/petrologic study is needed to understand this effect.

In addition, gas-rich and shock-blackened meteorites, inferred to have been exposed at the surface of their parent bodies at some point in their histories, do not show evidence for spectral reddening, as would be expected for many of the space-weathering mechanisms that have been proposed (*Britt and Pieters,* 1991a,b; *Keil et al.,* 1992). Instead, these "space-weathering" effects generally result in reductions in overall reflectance and mafic-silicate absorption-band depths. One important result to note is that the reflectance spectra of many S-class asteroids have been successfully reproduced by numerical mixing models using various combinations of meteoritic metal plus mafic silicates (e.g., *Hiroi et al.,* 1993). It would be very valuable to confirm the validity of such mixing models with spectral measurements of corresponding physical mixtures.

3.6. Nature of the 3-μm Absorption Feature

The reported presence of an absorption feature in the 3-μm region of many asteroid spectra (e.g., *Rivkin et al.,* 2000, and references therein) has received much attention recently because of its potential importance as an indicator of "hydration." This wavelength interval is of importance for asteroid research because the fundamental O-H stretching bands of H_2O/OH occur in this region. Characteristics of absorption features in this region, such as wavelength position, the number and shapes of the absorption bands, and their relative intensities, can be used to distinguish a range of geological materials as well as provide information on the speciation of H_2O/OH (*McMillan and Remmele,* 1986; *Clark et al.,* 1990; *S. J. Gaffey et al.,* 1993).

For some asteroid types, a 3-μm "water" feature is expected based on the presence of hydrated mineral species in their best meteorite analogs. The presence of such a feature in many low-albedo asteroid types (e.g., C-, B-, and G-types) is important supporting evidence for the validity of metamorphic types 1 and 2 carbonaceous chondrites as analogs to the surface material of these asteroids. Asteroids 1 Ceres and 2 Pallas are examples of objects with strong 3-μm features in their spectra (e.g., *Jones et al.,* 1990; *Rivkin et al.,* 2002). Thus this is an important wavelength region for providing critical insights into asteroid surface compositions. However, 3-μm features have also been reported in asteroids whose most plausible meteoritic analogs are anhydrous assemblages (e.g., *Rivkin et al.,* 1995, 2000, 2002). This has led to a reevaluation of the significance of the 3-μm feature in asteroid spectra.

There are observational difficulties (e.g., telluric water vapor absorptions, low reflectance) associated with obtaining reliable spectra in this wavelength region. The simple presence of a 3-μm absorption band has often been interpreted as indicative of "hydration" or widespread aqueous alteration (e.g., *Merényi et al.,* 1997; *Rivkin et al.,* 1995, 2000, 2001; *Howell et al.,* 2001). However, a number of plausible alternative explanations are also possible. These include materials normally considered to be anhydrous (such as mafic silicates) containing structural OH (*Miller et al.,* 1987; *Ingrin et al.,* 1989; *Skogby and Rossman,* 1989; *Skogby et al.,* 1990), the presence of fluid inclusions (*Roedder,* 1984; *Zolensky et al.,* 1999a,b), the presence of xenolithic hydrous meteorite components on asteroid surfaces from impacts (e.g., *Lovering,* 1962; *Binns,* 1969; *Wilkening,*1977; *Neal and Lipschutz,* 1981; *Keil,* 1982; *Nozette and Wilkening,* 1982; *Rubin and Keil,* 1983; *Sears et al.,* 1983; *Verkouteren and Lipschutz,* 1983; *Yanai et al.,* 1983; *Bunch and Rajan,* 1988; *Lipschutz et al.,* 1988; *Scott et al.,* 1989; *Pellas,* 1991; *Zolensky et al.,* 1991, 1993; *Buchanan et al.,* 1993a,b), and solar-wind-implanted H (*Starukhina,* 2001), or the presence of troilite, which also exhibits an absorption feature in the 3-μm region (*Vaughan and Craig,* 1978). All these processes can give rise to absorption features in the 3-μm region in otherwise "anhydrous" lithologies. Band depths in the 3-μm region of asteroid spectra, which range up to 50% (and up to 88% in one case) (*Larson et al.,* 1983; *Merényi et al.,* 1997; *Rivkin et al.,* 2000, 2002) are consistent with all of these alternative mechanisms.

Available meteorite reflectance spectra, including a variety of falls and finds (*Miyamoto et al.,* 1989; *Miyamoto,* 1991; *Salisbury et al.,* 1991; *Miyamoto and Zolensky,* 1994), indicate that band depth in the 3 μm region in meteorite spectra, including recent falls, can range up to 30% for H_2O/OH contents as low as 0.1 wt%. Reflectance and transmittance spectra of "anhydrous" meteorites such as achondrites, E- chondrites, and ordinary chondrites (including finds and recent falls) normally exhibit absorption bands in the 3-μm region attributable to O-H stretching vibrations (*Sandford,* 1984; *Miyamoto,* 1987, 1990, 1991; *Wagner et al.,* 1988; *Miyamoto et al.,* 1989; *Salisbury et al.,* 1991; *Miyamoto and Zolensky,* 1994; *Bishop et al.,* 1998a,b). The presence of 3-μm absorption features due to adsorbed atmospheric water in the spectra of otherwise anhydrous samples simply emphasizes the minute quantity of water required to produce a 3-μm absorption feature in a spectrum.

Numerous meteorite regolith breccias are mixtures of anhydrous (achondritic) and hydrous (e.g., carbonaceous chondrite) clasts (e.g., *Lovering,* 1962; *Binns,* 1969; *Wilkening,* 1977; *Neal and Lipschutz,* 1981; *Keil,* 1982; *Nozette and Wilkening,* 1982; *Rubin and Keil,* 1983; *Sears et al.,* 1983; *Verkouteren and Lipschutz,* 1983; *Yanai et al.,* 1983; *Bunch and Rajan,* 1988; *Lipschutz et al.,* 1988; *Scott et al.,* 1989; *Pellas,* 1991; *Zolensky et al.,* 1991, 1993; *Buchanan et al.,* 1993a,b). The asteroid belt is a dynamic environment and a perception of asteroid surfaces as being simple, pristine assemblages is oversimplified. Mixing of different materials and incorporation of widely different assemblages into asteroid regoliths is probably the norm rather than the exception. Thus we should expect some anhydrous (differentiated) asteroids to show evidence for a 3-μm "water" absorption band due to this dynamical mixing. *Starukhina* (2001) has discussed the formation of hydroxyl groups by interaction of solar-wind protons with O atoms in regolith particles. Those calculations indicate that reasonable concentrations of solar-wind H-produced hydroxyl groups in a regolith could produce the 3-μm band depths seen in many asteroids.

This discussion emphasizes the difficulties that can arise in the interpretation of a 3-μm spectral feature and the importance of carrying out such interpretations in a solid geologic and meteoritic context. As observational data for the 3-μm region improve (e.g., *Rivkin et al.,* 2001) the fine structure in this wavelength region can be used to distinguish between the various interpretations listed above and identify the specific "water-bearing" species (e.g., *Bishop et al.,* 1994).

3.7. Identification of Minerals on the Basis of Multiple Spectral Features

Identifying specific minerals on asteroid surfaces from reflectance spectra must be approached cautiously. Our confidence in mineralogical interpretations increases as the number of unique or diagnostic absorption features (not necessarily the number of absorption bands) increases.

As an example, the surface mineralogies of the A-class asteroids are generally interpreted as indicating olivine-rich assemblages (*Cruikshank and Hartmann,* 1984). This is based on the presence of a single absorption feature near 1.05 μm. A number of other minerals also exhibit an absorption band in this region, notably pyroxene (*Cloutis and Gaffey,* 1991a). However, the "shape" of this feature (i.e., evidence for side lobes near 0.85 and 1.2 μm), the lack of associated absorption features in the 2-μm region, the widespread presence of olivine in the inner solar system, and a number of olivine-rich meteorites in our collections, present strong collective evidence for the correctness of this interpretation. The complete surface mineralogies of these asteroids are still uncertain, however. The asteroid spectra do not match the reflectance spectra of pure olivine in all details, particularly in terms of overall slope. A number of possible factors may account for this spectral mismatch, including the presence of metal, chromite, temperature differences between laboratory and asteroid spectra (*Lucey et al.,* 1998; *Moroz et al.,* 2000), or "space weathering."

Interpretations of asteroid reflectance spectra must also be internally consistent. As an example, assigning a 3-μm absorption feature to a specific hydrated phase has certain implications. These include the expected presence of accompanying H_2O/OH overtone and combination bands in the 1.4- and/or 1.9-μm regions, which should persist even when hydrated minerals are exposed to the space environment (*Bishop and Pieters,* 1995). These overtone bands are generally an order of magnitude less intense than the 3-μm O-H stretching fundamental band and the depth of these overtone/combination bands can be further reduced by the presence of opaque minerals, or if they lie in a region of low overall reflectance; thus they may not always be detectable. However, if the asteroid albedo is moderate or high (e.g., >15%), a hydrated assemblage should exhibit detectable 1.4- and 1.9-μm features. If these hydrated minerals are also Fe-bearing, we would also expect additional absorption bands in the 0.4–0.9-μm region to be present (e.g., *Vilas et al.,* 1994).

4. MINERALOGICAL CHARACTERIZATIONS OF ASTEROIDS

Over the past two decades there has been a slow but steady improvement in both the quality of asteroid spectral data and in the interpretive methodologies and calibrations used to analyze that data. It is useful to summarize the present status of detailed mineralogical interpretations of asteroids as well as the advances that have been made in such interpretations in the 13 years since the *Asteroids II* volume was published. The mineralogical characterizations of specific asteroids are listed in Table 3.

4.1. S-type Asteroids

The analysis of S-asteroid survey spectra by *Gaffey et al.* (1993a) made the first detailed application of the phase-abundance calibration of *Cloutis et al.* (1986). The results showed the diversity of lithologies present within this single taxonomic type. *Gaffey et al.* (1993a) identified seven mineralogical subtypes of the S-taxon, where the silicate assemblages ranged from nearly monominerallic olivine [subtype S(I)] through basaltic silicates [subtype S(VII)]. Of the seven subtypes, only one [subtype S(IV)] did not require igneous processing to produce the observed assemblage. *Gaffey et al.* (1993a) concluded that at least 75% — and probably >90% — of the S-type asteroids had experienced at least partial melting within their parent bodies. This result strongly contradicted the commonly held notion (e.g., *Anders,* 1978; *Wetherill and Chapman,* 1988) that most S-asteroids must be undifferentiated ordinary chondrites (modified by space weathering) in order to account for the preponderance of ordinary chondrites among meteorite falls. However, this result was consistent with the evidence that a large majority of the meteorite parent bodies experienced high temperatures and underwent at least partial differentiation (e.g., *Keil,* 2000).

4.2. 6 Hebe

Of particular interest was subtype S(IV), which could (but was not required to) include undifferentiated assemblages such as ordinary chondrites. This significantly narrowed the field of candidates as potential ordinary chondrite parent bodies. Combined with improved understanding of meteorite delivery mechanisms (e.g., *Farinella et al.,* 1993a,b; *Morbidelli et al.,* 1994; *Migliorini et al.,* 1997), these results allowed *Gaffey and Gilbert* (1998) to identify asteroid 6 Hebe as the probable parent body of both the H chondrites and the IIE iron meteorites. There was also significant skepticism concerning the "H-chondrite + metal" assemblage invoked in the paper (e.g., *Keil,* 2000). Therefore, the fall of the Portales Valley meteorite was particularly timely. Portales Valley consists of masses of H6 chondrite either

TABLE 3. Mineralogically characterized asteroids.

Asteroid	Class	Diameter	a (AU)	Family	Predicted Mineralogy (Meteorite Affinity)*	Reference
3 Juno	S(IV)	244	2.670	N	Px/(Ol + Px) ~ 0.30 (possible L chondrite)	1†
4 Vesta	V	~570 × 460	2.362	Y	Pyroxene-plagioclase basalt (HED meteorites)	2
Vesta Family	V/J	—	—	Y	Range of HED lithologies	3,19,20
6 Hebe	S(IV)	185	2.426	N	Px/(Ol + Px) ~ 0.4 and NiFe metal (H chon, IIE irons)	4
7 Iris	S(IV)	203	2.386	N	Px/(Ol + Px) ~ 0.30, Px ~ $Fs_{42}Wo_7$ or Cpx-bearing (possible L chondrite)	1†
8 Flora	S(III)	141	2.201	Y	Px/(Ol + Px) ~ 0.25, NiFe metal, Fe-rich Px and/or Ca-rich Px	5
9 Metis	S(I)	168–210	2.386	Y	Px/(Ol + Px) ~ 0.12 and NiFe metal	6
11 Parthenope	S(IV)	162	2.452	N	Px/(Ol + Px) ~ 0.34	1†
12 Victoria	S(II)?	117	2.334	Y	Px/(Ol + Px) ≤ 0.14 and NiFe metal	1†
15 Eunomia	S(III)	272	2.644	Y	Ol + NiFe + (low-Ca Px) to high-Fe and Ca Px basalt	7
16 Psyche	M	264	2.922	N	NiFe metal	8
17 Thetis	S	93	2.469	N	Px/(Ol + Px) ~ 0.54 or low-Ca Px: Calcic Px ~ 60:40 and Ol <15%	22†
18 Melpomene	S(V)	148	2.296	N	Px/(Ol + Px) ~ 0.47	1†
20 Massalia	S(VI)	151	2.408	Y	Px/(Ol + Px) ~ 0.66, Px ~ $Fs_{26}Wo_4$	1†
25 Phocaea	S(IV)	78	2.400	N	Px/(Ol + Px) ~ 0.30	1†
26 Proserpina	S(II)	99	2.656	N	Px/(Ol + Px) ~ 0.30	1†
27 Euterpe	S(IV)		2.347	N	Px/(Ol + Px) ~ 0.25	1†
37 Fides	S(V)	112	2.642	Y	Px/(Ol + Px) ~ 0.72 ± 0.17	1†
39 Laetitia	S(II)	159	2.769	Y	Px/(Ol + Px) ~ 0.26	1†
40 Harmonia	S(VII)	111	2.267	N	Px/(Ol + Px) ~ 0.89 (0.68–1.0), Px ~ $Fs_{27}Wo_9$	1†
42 Isis	S(I)	107	2.441	Y	Px/(Ol + Px) ~ 0.0 and NiFe metal(?)	1†
43 Ariadne	S(III)	65	2.203	Y	Px/(Ol + Px) ~ 0.13–0.20	1†
44 Nysa	E	73	2.422	Y	Nearly Fe-free enstatite (<Fs_1)	9
63 Ausonia	S(II–III)	108	2.395	Y	Px/(Ol + Px) ~ 0.27 and NiFe metal(?)	1†
67 Asia	S(IV)	60	2.421	N	Px/(Ol + Px) ~ 0.46	1†
68 Leto	S(II)	127	2.782	Y	Px/(Ol + Px) ~ 0.18	1†
80 Sappho	S(IV)	82	2.296	N	Px/(Ol + Px) ~ 0.30	1†
82 Alkmene	S(VI)	64	2.765	N	Px/(Ol + Px) ~ 0.61 ± 0.16; Px ~ $Fs_{23}Wo_2$	1†
113 Amalthea	S(I–II)	48	2.376	Y	Px/(Ol + Px) ~ 0.13, Px ~ $Fs_{34}Wo_{14}$ and NiFe metal (?)	6,10
115 Thyra	S(III–IV)	84	2.380	Y	Opx + Ca Px assemblage indicated	1‡
215 Kleopatra	M	217 × 94 × 81	2.767	N	NiFe metal	11
246 Asporina	A	64	2.695	N	Olivine + NiFe metal(?) (possible pallasite)	12,13
289 Nenetta	A	42	2.874	N	Olivine + NiFe metal(?); Px/(Ol + Px) < 0.05 (possible pallasite)	1†,13
349 Dembowska	R	140	2.925	Y	Px/(Ol + Px) ~ 0.45; Px ~ $Fs_{25}Wo_{10}$	14
354 Eleonora	S(I)	162	2.796	N	Px/(Ol + Px) ~ 0.05	1†
387 Aquitania	S	106	2.742	Proposed	Spinel (in abundant CAIs?) (CV or CO chondrite)	15
433 Eros	S(IV)	39 × 13 × 13	1.458	N	Px/(Ol + Px) ~ 0.36; Px ~ $Fs_{42}Wo_{18}$ or Cpx-bearing (possible LL chondrite)	16,21
446 Aeternitas	A	43	2.788	Y	Olivine + NiFe metal (?); Px/(Ol + Px) < 0.05 (possible pallasite)	1†,12,13
532 Herculina	S(III)	231	2.772	N	Opx + Ca Px assemblage indicated	1‡
584 Semiramis	S(IV)	56	2.374	Y	Px/(Ol + Px) ~ 0.26 + NiFe metal(?); Px ~ $Fs_{42}Wo_{16}$	1†
674 Rachele	S(VII)	101	2.924	N	Px/(Ol + Px) ~ 0.68–1.0	1†
847 Agnia	S	32	2.783	Y	Low-Ca Px: Calcic Px ~ 60:40; Ol < 15%	22
863 Benkoela	A	32	3.200	N	Olivine + NiFe metal (?) (possible pallasite)	13
980 Anacostia	S	89	2.741	Proposed	Spinel (in abundant CAIs?) (CV or CO chondrites)	15
1036 Ganymed	S(VI–VII)	41	2.662	N	Px/(Ol + Px) ~ 0.64–0.82; Px ~ $Fs_{34}Wo_4$	1†
3103 Eger	E	1.5	1.406	N	Iron-free enstatite	17
1986 DA	M	2.3	2.811	N	NiFe metal	18

*Except where otherwise noted, "Px" indicates a low-Ca orthopyroxene.
†Mineral abundances and compositions using spectral parameters from *Gaffey et al.* (1993a) and equations (2)–(4) and Fig. 4 from the present paper.
‡Olivine-poor or olivine-free two-pyroxene assemblage indicated by test described in text.

References: (1) *Gaffey et al.* (1993a); (2) *Gaffey* (1997); (3) *Binzel and Xu* (1993); (4) *Gaffey and Gilbert* (1998); (5) *Gaffey* (1984); (6) *Kelley and Gaffey* (2000); (7) *Reed et al.* (1997); (8) *Ostro et al.* (1985); (9) *Gaffey et al.* (1993b); (10) *Gaffey* (2002); (11) *Ostro et al.* (2000); (12) *Hiroi et al.* (1993); (13) *Lucey et al.* (1998); (14) *Abell and Gaffey* (2000); (15) *Burbine et al.* (1992); (16) *McFadden et al.* (2001); (17) *Gaffey et al.* (1992); (18) *Ostro et al.* (1991); (19) *Hiroi et al.* (1995); (20) *Vilas et al.* (2000); (21) *McCoy et al.* (2001); (22) *Sunshine et al.* (2002).

embedded in a NiFe metal matrix or crosscut by large (centimeter-scale) metal veins (e.g., *Kring et al.,* 1999; *Pinault et al.,* 1999; *Rubin and Ulff-Møller,* 1999; *Ruzika et al.,* 1999). Although there are still questions concerning the exact mechanism(s) that melted the metal and mixed the H-chondrite masses and the molten metal, there can no longer be any doubt that assemblages actually exist like that invoked for the Hebe surface by *Gaffey and Gilbert* (1998).

It was of course fortuitous that the identification of Hebe as the probable parent body of the H chondrites and the

IIE iron meteorites was followed so closely by the fall of the Portales Valley meteorite. In the search for meteorite parent bodies, one seldom has such a strong piece of supporting evidence provided in such a timely manner. The linkage of a major meteorite group (~32% of meteorite falls) to specific location in the asteroid belt now allows us to place these meteorites — and the detailed chemical and chronological information that they provide — into a spatial context in the early solar system, and in turn, to place temporal and chemical constraints on that specific location in the asteroid belt.

4.3. Spinel-bearing S-type Asteroids

Two spectrally anomalous bodies (387 Aquitania and 980 Anacostia) were identified during the analysis of S-asteroid survey data. Analysis of these two spectra (*Burbine et al.,* 1992) indicated a spectrally significant spinel component, and it was proposed that the spinel was associated with a high abundance of Allende-type white inclusions, indicating accretion at an early stage of nebular cooling. Because of similarities in orbital elements, it was further suggested that these two mineralogically distinctive bodies were fragments of a single parent body, i.e., a "compositionally detected" family. Such a genetic family would be far below the detection limit of any dynamical family identification system yet presented.

4.4. 4 Vesta

Gaffey (1997) used rotational spectral variations to map the large-scale lithologic variations across the surface of Vesta. The lithologies range from diogenite to eucrite, suggesting a largely intact crust (eucrites) with lower crustal units (diogenites) exposed — presumably by impact structures — in several regions. Hubble Space Telescope multicolor imagery (*Binzel et al.,* 1997) shows appropriate color variations across the surface corresponding to the spectrally derived lithologic map.

4.5. Vesta Family Objects

An extensive investigation of small objects in the Vesta family (*Binzel and Xu,* 1993) shows spectra very similar to that of Vesta itself. (Some of these bodies are not members of the dynamical Vesta family, but their distinctive spectra strongly indicate they are members of the "true" or genetic Vesta family.) Although the CCD spectral coverage was inadequate to obtain actual mineralogical characterizations, the close similarity to the very distinctive Vesta spectrum leads to a robust conclusion that the family members are HED assemblages. Apparent differences in the portion of the 1-μm feature available in the CCD spectral interval were noted and attributed to different HED lithologies.

Hiroi et al. (1995) use a curve-fitting routine to define several parameters [band center (actually band minimum), band depth, and band width] of the partially sampled 1-μm band in the CCD spectra of a number of Vesta family objects. They show a range in band positions twice the range of band position they measured in a set of HED meteorites [but very comparable to the range of band positions for HEDs shown by *Gaffey* (1997)], consistent with these bodies sampling a range of HED lithologies. However, the incomplete and variable coverage of the 1-μm band and the use of band minimum instead of band center make a more detailed comparison between individual family members difficult.

Vilas et al. (2000) identify a narrow feature centered at 0.5065 μm suggestive of a high-Ca pyroxene component (augite) that varied between different Vesta family members. This is consistent with different HED lithologies, since eucrites have more Ca-rich pyroxenes than diogenites.

Burbine et al. (2000) interpret ~0.44 to ~1.65 μm spectra of vestoids, both within and outside the dynamically defined Vesta family. Based primarily on the wavelength position of the reflectance maximum near 1.4 μm, they conclude that all of these spectra were consistent with surface compositions similar to eucrites and howardites, and none of them were similar to diogenites. Extension of vestoid spectral observations to 2.5 μm should allow testing of the presence or absence of diogenite-type assemblages and would provide robust constraints on excavation depth within the crust of the parent body, Vesta.

Kelley et al. (2001b, 2002) present the first detailed mineralogical analysis of a vestoid (1929 Kollaa) with spectral coverage extended to 2.5 μm. They conclude that the mineralogy of Kollaa is consistent with an HED-type assemblage as expected, and more specifically that it indicates a cumulate eucrite-type assemblage. Such an assemblage would form deep in the eucritic upper layer of the Vesta crust.

The Vesta family emphasizes the need to specify the type of "family" being discussed. *Farinella et al.* (1992) drew attention to the inherent ambiguities in the use of the term "family" when referring to asteroid groupings derived from different criteria. They suggest that the term "family" should be reserved for groupings passing both statistical and taxonomic/spectroscopic tests that should be genetic associations. These can be presumed to have a common origin in a single parent body. Based on spectral criterion, we would argue that all of the vestoids (identified by their characteristic spectra) in the Vesta region should be considered members of the Vesta family. The distinction between vestoids within and outside the dynamically defined Vesta family is relevant only so far as it provides a potentially powerful dataset to refine our understanding of the initial distribution and/or the dynamical evolution of bodies within a genetic family.

4.6. 9 Metis

Kelley and Gaffey (2000) show that the surface assemblage of Metis was dominantly olivine-rich mafic silicates [ol/(ol + px) ~ 0.88] and NiFe metal. Based on the similarity of orbital elements and sharing an uncommon S-asteroid mineral assemblage, they conclude that Metis and 113 Amalthea were genetically related and were the largest surviving fragments of a disrupted parent body (asteroid family).

4.7. 15 Eunomia

Reed et al. (1997) use simultaneous visible and thermal IR lightcurve observations of Eunomia to derive the body shape and albedo variability of this asteroid. They show that the NIR spectrum of Eunomia varies substantially on opposite ends of this "almond-shaped" body. The spectrum of the "pointy" end indicates abundant olivine and metal with lesser pyroxene, predominately low-Ca. The "rounded" end is more basaltic and contains abundant high-Fe and high-Ca pyroxene. Eunomia appears to expose a crust-to-core cross-section of its parent body.

4.8. 3103 Eger (1982 BB)

Based on a high albedo and a relatively featureless spectrum, near-Earth asteroid (NEA) 3103 Eger was identified as Fe-free enstatite (*Gaffey et al.,* 1992). Based upon the uniqueness of this assemblage in the NEA population and meteorite fall patterns, it was concluded that Eger is the probable parent body of the enstatite achondrite meteorites (aubrites) and is itself derived from the Hungaria region of the innermost asteroid belt.

4.9. Iron Alteration Minerals and Phyllosilicates in Dark Asteroids

Weak visible region features in CCD spectra of low albedo asteroids have been detected and confirmed by Vilas and coworkers (*Vilas and Gaffey,* 1989; *Vilas et al.,* 1993; *Vilas et al.,* 1994). These features are indicative of a number of minerals (e.g., the clay mineral, antigorite) and Fe-alteration products consistent with the mineralogy present in various CI1 and CM2 carbonaceous chondrites. Although considerable work remains, especially in the spectral study of the carbonaceous chondrite mineral phases, it is evident that this type of investigation is the key to unraveling the undoubtedly complex mineralogy of the dark asteroids.

5. IMPLICATIONS FOR ASTEROID FAMILY STUDIES

5.1. Types of Asteroid Families

One area of asteroid research in which detailed mineralogical characterizations are at the forefront is the study of asteroid families. Clusters of asteroids in similar orbits, or "families," were first recognized nearly a century ago (*Hirayama,* 1918). Depending on the dynamical and statistical or subjective criteria used, the number of proposed families ranges from about 6 to more than 100 (e.g., *Hirayama,* 1933; *Brouwer,* 1951; *Arnold,* 1969; *Kozai,* 1979; *Williams,* 1989, 1992; *Zappalà et al.,* 1990, 1994, 1995). From the outset, it was suggested that most asteroid families resulted from cataclysmic collisions between asteroids resulting in the breakup of individual parent bodies.

In the present paper, we explicitly define an "asteroid family" (aka "genetic family") as the fragments of a partially or completely disrupted parent body. This is an idealized definition that may not be testable in many situations. Thus, families can include a large number of fragments of their parent body, or families may be depleted to the point where only a few fragments remain. Accordingly, there are different types of asteroid families based on the criteria used to establish or study the family (e.g., dynamical families, taxonomic families, mineralogical families, etc.).

Dynamical families are clusters of asteroids identified in proper orbital-element space (e.g., *Zappalà et al.,* 1990, 1994, 1995; *Williams,* 1992). There are significant disagreements on the total number of such families, their memberships, and the criteria and techniques used to identify them. However, it is also highly probable that statistically well-defined dynamical families are true genetic families.

Physical studies of families to test whether their members are "truly" related began with studies of their broadband colors (*Gradie and Zellner,* 1977). Investigations based on visible wavelength spectral properties and taxonomy use dynamical families as a starting point for making assessments of the "reality" of families (e.g., *Chapman,* 1987; *Bell,* 1989; *Chapman et al.,* 1989; *Granahan,* 1993; *Bus,* 1999). The tacit assumption is that a family generated by collisional breakup of a parent body will be composed of fragments of one taxonomic type or of compatible taxonomic types. These studies have been used to identify probable interlopers in families.

In addition, the goal of detailed mineralogic studies is to recognize genetic ("true" or "real") families, and such studies similarly use dynamical families as a starting point. This approach makes assessments of family membership based on the presence of mineralogical assemblages consistent with derivation from a single parent body. A goal of such studies of genetic asteroid families is to reveal uniquely preserved information that may help to unlock the secrets of the timing of particular events in the early solar system and its subsequent history. The presence and nature of collisionally produced genetic asteroid families can provide important constraints on the processes involved in the disruptions of large (~100–1000 km) bodies, the collisional lifetime of asteroids as a function of size and composition, the thermal history and internal compositional structure of their parent bodies, and the rate of orbital diffusion of families.

Historically, most tests of asteroid family membership have relied on taxonomy over visible wavelengths (e.g., *Chapman,* 1987; *Bell,* 1989; *Chapman et al.,* 1989; *Granahan,* 1993). The work of *Bus* (1999) and *Bus et al.* (1996) combine both visible wavelength spectra and spatial location (within each grouping) to examine the likely membership and extent of a number of small families, where good consistency between spectral characteristics and location is found in each case. Further studies are now extending family member spectral measurements into the NIR out to ~1.6 μm (e.g., *Carvano et al.,* 2001; *Burbine et al.,* 2000), where the consistency of spectral characteristics remains strong. It is important to emphasize that the similarity of observational characteristics does not "prove" a family relationship, nor does taxonomic dissimilarity "disprove" a family relationship. The next progressive step for investigating the "reality" of dynamical or suggested families is the study of their detailed mineralogical and compositional properties. Such studies can more rigorously test whether a potential family is compatible with derivation from a common parent body and can establish whether it is a probable genetic family.

It is important to realize the implications of the large number of magmatic iron meteorite types (e.g., *Keil,* 2000) for asteroid family studies. The large number of iron meteorite parent bodies constrains the minimum number of parent-body disruption events (family-forming events) that have occurred during solar-system history. Currently there are ~25 recognized dynamical families that appear to have been produced by collision disruption and partial-to-complete dispersal of their parent bodies (e.g., *Marzari et al.,* 1999). Most of the potential dynamical families surveyed by *Bus* (1999) were considered valid because they were spectrally similar and spectrally distinct from the background objects. Spectrally homogeneous families such as these are very improbable sources of iron meteorites. Instead, magmatic iron meteorites must represent excavated core material from fully (or extensively) differentiated bodies. The families produced by the disruption of such differentiated bodies should consist of a variety of lithologies (basalts and HED type assemblages, dunites, pallasite-type assemblages, and Fe) and spectral types (V, A, S, M). Such families appear absent among the recognized dynamical families. Thus, most of the >55 families that released the >55 types of magmatic iron meteorites must have been depleted and/or dispersed so that they no longer provide dynamical grouping identifiable by current search techniques. It remains an open question as to whether any of the differentiated material other than the high-strength Fe core samples still exists in a recognizable form in the asteroid belt today. *Burbine et al.* (1996) propose a scenario in which the lower-strength differentiated materials may have been "battered to bits" and entirely or almost entirely removed from the asteroid belt. This model is testable by pushing detailed mineralogical studies to smaller and smaller main-belt asteroids.

If they are detectable at all based on orbital criteria, such depleted and dispersed families could resemble many of the smaller and generally not statistically significant Williams families (*Williams,* 1989, 1992). It has been demonstrated that a detailed compositional study of such small families combining groundbased visible and NIR (~0.4–2.5 μm) spectral data is feasible and can show where probable genetic relationships exist (*Kelley and Gaffey,* 1996, 2000, 2002; *Kelley et al.,* 1994).

Until recently, no dynamical asteroid family had sufficient spectral coverage of its membership to adequately permit compositional testing of its genetic reality. The Family Asteroid Compositional Evaluation Survey (FACES) was created to fill gaps in existing family asteroid spectroscopic databases and to obtain new data on additional family members. To date, the ongoing observational phase of this survey has collected visible and NIR spectroscopic data on more than 150 main-belt asteroids and NEAs. Appropriate data covering the volume or mass majority of several asteroid families have now been assembled.

5.2. Estimating Mass Loss from Asteroid Families and from the Asteroid Belt

When a genetic family originally forms from disruption of its parent body, the mass fractions of different layers (lithologies) within the parent (e.g., NiFe core, olivine mantle, basaltic crust, etc., in a differentiated body) are a function of the chondritic type that made up the raw material of the parent body (e.g., *Gaffey et al.,* 1993a). Since mineralogical investigation of the members of a family can identify the original bulk compositions, the original relative abundance of the different lithologies can be determined. In a highly evolved family, the composition and size of the larger family members can be used to place a lower limit on the parent-body size and hence establish a lower limit on the mass lost from the family (e.g., *Kelley and Gaffey,* 2000). Since even the most evolved families are certainly younger than the solar system, and since the collisional evolution of asteroids is essentially independent of whether or not they are members of families, the mass-depletion factors of the highly evolved families place a lower limit on the mass depletion of the asteroid belt as a whole.

A case in point is the *Williams* (1992; also personal communication, 1992) Metis family. Asteroids 9 Metis and 113 Amalthea represent 98% of the remaining volume of the dynamical Metis family. *Gaffey* (2002) shows that 113 Amalthea is a silicate (olivine + pyroxene) fragment of a parent body that most likely underwent significant differentiation (thermal processing) and represents an uncommon S subtype. *Kelley and Gaffey* (2000) demonstrate that 9 Metis is likely to be the remnant core of its parent body. Metis appears to be metal-rich, but still retains a significant silicate signature matching that of Amalthea. Therefore, this pair of asteroids passes the genetic test: Their silicate assemblages are very similar and quite distinct from most other S asteroids, and they have similar orbital elements. The silicate-mineral abundances (olivine-to-pyroxene ratio) for these asteroids are within the range of abundances that can be derived from igneous processing of CV/CO-, H-, L-, and LL-chondrite meteorite-type parent materials. The ranges of silicate-to-metal ratios of these meteorite groups are well known (e.g., *Brearley and Jones,* 1998). Since Metis appears to preserve the core-mantle boundary, its diameter was used to represent a *minimum* core size (metal fraction) for the parent body. It was then possible to calculate a range of silicate fractions for each of the possible meteorite starting compositions, and hence a range of parent-body diameters for each group. The minimum parent-body diameters ranged from ~330 km to ~490 km across the suite of meteorite analogs. Once the diameter of the parent body was constrained, so is the volume ($v = 4\pi r^3/3$). The volume of material remaining in the family was then calculated using the best diameters available for the individual asteroids. Consequently, a percentage of the remaining volume was determined (total remaining volume/parent body volume) for each of the possible starting groups. This resulted in a range of remaining volume of only 4–14%. Therefore, it appears that a *minimum* of 86–96% of the original volume of the Metis parent body has been lost over time, or has yet to be identified in the background asteroid population.

Similar exercises are being performed for additional asteroid families in the FACES database. The better the compositional evaluations of individual family members are, the tighter the resulting parent-body constraints will be. Since any genetic family is by definition younger than the asteroid belt, the highest depletion observed among asteroid families sets the lower limit on the overall depletion of the asteroid belt. Current results suggest an initial asteroid belt much more massive than that presently observed, almost certainly >100× as massive and probably >1000× as massive. Additional studies of small (i.e., highly evolved) genetic asteroid families will further constrain this limit.

5.3. Maria Family

Space does not permit a detailed discussion of the mineralogical issues of individual asteroid familes, so the Maria family is used as an example. This family is located adjacent to the 3:1 mean motion resonance with Jupiter and could be an important contributor to the terrestrial-meteorite flux. G. Wetherill (personal communication, 1991) suggests that this family is a good dynamical candidate for an ordinary chondrite source. The survey spectra of several members of this family appear very similar to one another. This would be expected for an undifferentiated, chondritic parent body. Based upon a dynamical investigation, *Zappalà et al.* (1997) identify the Maria family as the most promising potential source of the two largest NEAs, including 433 Eros, which was the target of NASA's *NEAR Shoemaker* mission. Mineralogical investigation of this family would shed light on the important issues of the source of ordinary chondrites and the rate of orbital diffusion adjacent to a major resonance, and it should allow us to test the proposed main-belt source region for 433 Eros. There is now excellent spectral coverage and mineralogical information for Eros (e.g., *Veverka et al.,* 2000; *Kelley et al.,* 2001a; *McFadden et al.,* 2001; *McCoy et al.,* 2001). To date, however, a rigorous mineralogical assessment of the Maria Family and its potential relationship to Eros has not been done.

5.4. Asteroid Families as Probes of Heliocentric Variations in Asteroid Histories

The variation in asteroid composition with heliocentric distance is some function of the compositional gradient in the solar nebula and the nature and intensity of the early transient heat source. From meteorites, which are naturally delivered samples from a limited set of asteroids, it is well established that an intense heating event took place during the first 2–3 m.y. (~0.05%) of inner solar-system history. Because asteroids suggest a strong heliocentric gradient in this thermal event, there is an implication that it was due to electromagnetic induction heating during a solar T-Tauri episode (e.g., *Herbert et al.,* 1991). The alternate transient heat source, a short-lived radionuclide such as ^{26}Al or ^{60}Fe, would imply either a strong enrichment of these isotopes in the inner nebula or an accretionary wave taking 4–5 m.y. to propagate outward through the inner solar nebula (e.g., *Grimm and McSween,* 1993).

Characterization of the postaccretionary temperatures attained within asteroid parent bodies, derived from their mineralogy and as a function of heliocentric distance, would define the spatial gradient in this early heat source (*Hardersen and Gaffey,* 2001). For a T-Tauri episode, this thermal gradient constrains the intensity and duration of the presolar outflow and can be used to estimate the total mass lost from the protosolar object. For the short-lived radionuclide option, the thermal gradient would constrain the nebular elemental heterogeneity and/or timing of the planetesimal accretion as a function of heliocentric distance. Based on the concentration of S(IV) asteroids near the 3:1 jovian resonance at 2.5 AU (*Gaffey et al.,* 1993a), the thermal evolution of these asteroids appears to have been retarded by the formation of Jupiter. This should provide a time constraint on the formation of Jupiter once the thermal histories of these asteroids are understood.

Asteroid families provide one of the best means of constraining these thermal and/or compositional gradients. For families passing a genetic test ("true" or "real" families), comparison of the mineralogical compositions of a representative sample of family members allows the degree of internal differentiation to be well defined. This in turn tightly constrains the temperature history and original composition of the specific parent body at the heliocentric location of the family.

6. EXISTING NEEDS, FUTURE DIRECTIONS

There are a number of areas that require further research in order to improve our ability to analyze the reflectance spectra of asteroids. Advances in observational capabilities are now outstripping the interpretive methodologies needed to analyze such data. Some of the more urgent requirements include:

1. Understanding how and to what extent space weathering affects the reflectance spectra of asteroids and our ability to extract diagnostic spectral parameters from such spectra.
2. Developing methods to reliably derive absorption-band areas for mafic silicates in metallic Fe-bearing assemblages.
3. Establishing the optimum wavelength regions and spectral resolutions for detecting the presence and quantifying the abundance and composition of specific minerals.
4. Quantifying the effects of temperature and vacuum on reflectance spectra of various minerals.
5. Defining the effects of very fine-grained and dispersed opaque minerals (e.g., magnetite, troilite) on mafic-silicate reflectance spectra.
6. Expanding the spectral-compositional database for minerals relevant to asteroids, especially low-albedo aster-

oids (e.g., Fe-bearing clays such as those in carbonaceous chondrites, sulfates, etc.).

7. Gaining a more rigorous understanding of the properties that affect the slope and linearity of iron-meteorite spectra.

8. Establishing a quantitative calibration for determining the presence and composition of high Ca-pyroxene in mixtures of olivine, low-Ca pyroxene and high-Ca pyroxene.

Acknowledgments. The authors are grateful for the comments of the editor (A. Cellino) and three reviewers (T. Burbine, R. Binzel, and an anonymous reviewer) whose comments helped to improve this chapter. Support for M.J.G. to carry out this effort came from NASA Planetary Geology and Geophysics Grant NAG5-10345, NASA Exobiology Program Grant NAG5-7598, and NSF Planetary Astronomy Grant AST-9318674. The work of M.S.K. was supported by the NRC Associateships Program. M.J.G., M.S.K., and K.L.R. were visiting astronomers at the Infrared Telescope Facility, operated by the University of Hawai'i under contract to the National Aeronautics and Space Administration.

REFERENCES

Abell P. A. and Gaffey M. J. (2000) Probable geologic composition, thermal history, and meteorite affinities for mainbelt asteroid 349 Dembowska (abstract). In *Lunar and Planetary Science XXXI,* Abstract #1291. Lunar and Planetary Institute, Houston (CD-ROM).

Adams J. B. (1974) Visible and near-infrared diffuse reflectance spectra of pyroxenes as applied to remote sensing of solid objects in the solar system. *J. Geophys. Res., 79,* 4829–4836.

Adams J. B. (1975) Interpretation of visible and near-infrared diffuse reflectance spectra of pyroxenes and other rock-forming minerals. In *Infrared and Raman Spectroscopy of Lunar and Terrestrial Minerals* (C. Karr, ed.), pp. 91–116. Academic, New York.

Adams J. B. and Filice A. L. (1967) Spectral reflectance 0.4 to 2.0 microns of silicate rock powders. *J. Geophys. Res., 72,* 5705–5715.

Adams J. B. and McCord T. B. (1971) Optical properties of mineral separates, glass and anorthositic fragments from Apollo mare samples. *Proc. Lunar Sci. Conf. 2nd,* pp. 2183–2195.

Anders E. (1978) Most stony meteorites come from the asteroid belt. In *Asteroids: An Exploration Assessment* (D. Morrison and W. C. Wells, eds.), pp. 57–78. NASA CP-2053.

Arnold J. R. (1969) Asteroid familes and "jet streams." *Astron. J., 74,* 1235–1242.

Bell J. F. (1989) Mineralogical clues to the origins of asteroid dynamical families. *Icarus, 78,* 426–440.

Binns R. A. (1969) A chondritic inclusion of unique type in the Cumberland Falls meteorite. In *Meteorite Research* (P. Millman, ed.), pp. 696–704. Reidel, Dordrecht.

Binzel R. P. and Xu S. (1993) Chips off of asteroid 4 Vesta: Evidence for the parent body of basaltic achondrite meteorites. *Science, 260,* 186–191.

Binzel R. P., Gaffey M. J., Thomas P. C., Zellner B., Storrs A. D., and Wells E. N. (1997) Geologic mapping of Vesta from 1994 Hubble Space Telescope images. *Icarus, 128,* 95–103.

Bishop J. L. and Pieters C. M. (1995) Low-temperature and low atmospheric pressure infrared reflectance spectroscopy of Mars soil analog materials. *J. Geophys. Res., 100,* 5369–5379.

Bishop J. L., Pieters C. M., and Edwards J. O. (1994) Infrared spectroscopic analysis on the nature of water in montmorillonite. *Clays Clay Minerals, 42,* 702–716.

Bishop J. L., Mustard J. F., Pieters C. M., and Hiroi T. (1998a) Recognition of minor constituents in reflectance spectra of Allan Hills 84001 chips and the importance for remote sensing of Mars. *Meteoritics & Planet. Sci., 33,* 693–698.

Bishop J. L., Pieters C. M., Hiroi T., and Mustard J. F. (1998b) Spectroscopic analysis of Martian meteorite Allan Hills 84001 powder and applications for spectral identification of minerals and other soil components on Mars. *Meteoritics & Planet. Sci., 33,* 699–707.

Brearley A. J. and Jones R. H. (1998) 3. Chondritic meteorites. In *Reviews in Mineralogy, Vol. 36: Planetary Materials* (J. J. Papike, ed.), pp. 3-1 to 3-398. Mineralogical Society of America, Washington, DC.

Britt D. T. and Pieters C. M. (1988) Bidirectional reflectance properties of iron-nickel meteorites. *Proc. Lunar Planet. Sci. Conf. 18th,* pp. 503–512.

Britt D. T. and Pieters C. M. (1991a) Black ordinary chondrites: An analysis of abundance and fall frequency. *Meteoritics, 26,* 279–285.

Britt D. T. and Pieters C. M. (1991b) The bidirectional reflectance spectra of five gas-rich ordinary chondrites (abstract). In *Lunar and Planetary Science XXII,* pp. 139–140. Lunar and Planetary Institute, Houston.

Brouwer D. (1951) Secular variation of the orbital elements of the minor planets. *Astron. J., 56,* 9–32.

Buchanan P. C., Zolensky M. E., and Reid A. M. (1993a) Carbonaceous chondrite clasts in the howardites Bholghati and EET87513. *Meteoritics, 28,* 659–682.

Buchanan P. C., Zolensky M. E., Reid A. M., and Barrett R. A. (1993b) EET87513 clast N: A CM2 fragment in an HED polymict breccia (abstract). In *Lunar and Planetary Science XXIV,* pp. 209–210. Lunar and Planetary Institute, Houston.

Bunch T. E. and Rajan R. S. (1988) Meteorite regolithic breccias. In *Meteorites and the Early Solar System* (J. F. Kerridge and M. S. Matthews, eds.), pp. 144–164. Univ. of Arizona, Tucson.

Burbine T. H., Gaffey M. J., and Bell J. F. (1992) S-asteroids 387 Aquitania and 908 Anacostia: Possible fragments of the breakup of a spinel-bearing parent body with CO3/CV3 affinities. *Meteoritics, 27,* 424–434.

Burbine T. H., Meibom A., and Binzel R. P. (1996) Mantle material in the main belt: Battered to bits? *Meteoritics & Planet. Sci., 31,* 607–620.

Burbine T. H., Buchanan P. C., Binzel R. P., Bus S. J., Hiroi T., Hinrichs J. L., Meibom A., and McCoy T. J. (2000) Vesta, vestoids, and the howardite, eucrite, diogenite group: Relationships and the origin of spectral differences. *Meteoritics & Planet. Sci., 36,* 761–781.

Burns R. G. (1970a) *Mineralogical Applications of Crystal Field Theory.* Cambridge Univ., Cambridge. 224 pp.

Burns R. G. (1970b) Crystal field spectra and evidence of cation ordering in olivine minerals. *Am. Mineral., 55,* 1608–1632.

Burns R. G. (1993) *Mineralogical Applications of Crystal Field Theory,* 2nd edition. Cambridge Univ., Cambridge. 551 pp.

Bus S. J. (1999) Compositional structure in the asteroid belt: Results of a spectroscopic survey. Ph.D. thesis, Massachusetts Institute of Technology, Cambridge. 67 pp.

Bus S. J., Binzel R. P., and Burbine T. H. (1996) Asteroid families: Myth or reality. *Bull. Am. Astron. Soc., 28,* 1097.

Calvin W. M. and King T. V. V. (1997) Spectral characteristics of iron-bearing phyllosilicates: Comparison to Orgueil (CI1), Murchison and Murray (CM2). *Meteoritics & Planet. Sci., 32,* 693–701.

Carvano J. M., Lazzaro D., Mothé-Diniz T., Angeli C. A., and

Florczak M. (2001) Spectroscopic survey of the Hungaria and Phocaea dynamical groups. *Icarus, 149,* 173–189.

Chapman C. R. (1976) Asteroids as meteorite parent-bodies: The astronomical perspective. *Geochim. Cosmochim. Acta, 40,* 701–719.

Chapman C. R. (1987) Distributions of asteroid compositional types with solar distance, body diameter, and family membership. *Meteoritics, 22,* 353–354.

Chapman C. R. (1996) S-type asteroids, ordinary chondrites, and space weathering: The evidence from Galileo's fly-bys of Gaspra and Ida. *Meteoritics & Planet. Sci., 31,* 699–725.

Chapman C. R. and Gaffey M. J. (1979) Reflectance spectra for 277 asteroids. In *Asteroids* (T. Gehrels and M. S. Matthews, eds.), pp. 655–687. Univ. of Arizona, Tucson.

Chapman C. R. and Salisbury J. W. (1973) Comparisons of meteorite and asteroid spectral reflectivities. *Icarus, 19,* 507–522.

Chapman C. R., Paolicchi P., Zappalà V., Binzel R. P., and Bell J. F. (1989) Asteroid families: Physical properties and evolution. In *Asteroids II* (R. P. Binzel et al., eds.), pp. 386–415. Univ. of Arizona, Tucson.

Clark B. E. (1995) Spectral mixing models of S-type asteroids. *J. Geophys. Res., 100,* 14443–14456.

Clark B. E., Hapke B., Pieters C., and Britt D. (2002) Asteroid space weathering and regolith evolution. In *Asteroids III* (W. F. Bottke Jr. et al., eds.), this volume. Univ. of Arizona, Tucson.

Clark R. N. and Lucey P. G. (1984) Spectral properties of ice-particulate mixtures and implications for remote sensing: I. Intimate mixtures. *J. Geophys. Res., 89,* 6341–6348.

Clark R. N., King T. V. V., Klejwa M., Swayze G. A., and Vergo N. (1990) High spectral resolution reflectance spectroscopy of minerals. *J. Geophys. Res., 95,* 12653–12680.

Cloutis E. A. (1996) Spectral properties of hydrocarbon-bearing geological materials (abstract). In *Lunar and Planetary Science XXVII,* pp. 237–238. Lunar and Planetary Institute, Houston.

Cloutis E. A. and Gaffey M. J. (1991a) Pyroxene spectroscopy revisited: Spectral-compositional correlations and relationship to geothermometry. *J. Geophys. Res., 96,* 22809–22826.

Cloutis E. A. and Gaffey M. J. (1991b) Spectral-compositional variations in the constituent minerals of mafic and ultramafic assemblages and remote sensing implications. *Earth Moon Planets, 53,* 11–53.

Cloutis E. A. and Gaffey M. J. (1994) An X-ray diffraction and reflectance spectroscopy study of iron sulphides (abstract). In *Lunar and Planetary Science XXV,* pp. 273–274. Lunar and Planetary Institute, Houston.

Cloutis E. A., Gaffey M. J., Jackowski T. L., and Reed K. L. (1986) Calibration of phase abundance, composition, and particle size distribution for olivine-orthopyroxene mixtures from reflectance spectra. *J. Geophys. Res., 91,* 11641–11653.

Cloutis E. A., Gaffey M. J., Smith D. G. W., and Lambert R. St. J. (1990a) Reflectance spectra of "featureless" materials and the surface mineralogies of M- and E-class asteroids. *J. Geophys. Res., 95,* 281–293.

Cloutis E. A., Gaffey M. J., Smith D. G. W., and Lambert R. St. J. (1990b) Reflectance spectra of mafic silicate-opaque assemblages with applications to meteorite spectra. *Icarus, 84,* 315–333.

Cloutis E. A., Gaffey M. J., Smith D. G. W., and Lambert R. St. J. (1990c) Metal silicate mixtures: Spectral properties and applications to asteroid taxonomy. *J. Geophys. Res., 95,* 8323–8338.

Cruikshank D. P. and Hartmann W. K. (1984) The meteorite-asteroid connection: Two olivine-rich asteroids. *Science, 223,* 281–283.

Deer W. A., Howie R. A., and Zussman J. (1963) *Rock-Forming Minerals, Vol. 2: Chain Silicates.* Longman, London.

Egan W. G., Veverka J., Noland M., and Hilgeman T. (1973) Photometric and polarimetric properties of the Bruderheim chondritic meteorite. *Icarus, 19,* 358–371.

Farinella P., Davis D. R., Cellino A. and Zappalà V. (1992) From asteroid clusters to families: A proposal for a new nomenclature. In *Asteroids, Comets, Meteors 1991* (A. W. Harris and E. Lowell, eds.), pp. 165–166. Lunar and Planetary Institute, Houston.

Farinella P., Gonczi R., Froeschlé Ch., and Froeschlé C. (1993a) The injection of asteroid fragments into resonances. *Icarus, 101,* 174–187.

Farinella P., Froeschlé C., and Gonczi R. (1993b) Meteorites from the asteroid 6 Hebe. *Cel. Mech. Dyn. Astron., 56,* 287–305.

Gaffey M. J. (1976) Spectral reflectance characteristics of the meteorite classes. *J. Geophys. Res., 81,* 905–920.

Gaffey M. J. (1984) Rotational spectral variations of asteroid (8) Flora: Implications for the nature of the S-type asteroids and for the parent bodies of the ordinary chondrites. *Icarus, 60,* 83–114.

Gaffey M. J. (1986) The spectral and physical properties of metal in meteoritic assemblages: Implications for asteroid surface materials. *Icarus, 66,* 468–486.

Gaffey M. J. (1997) Surface lithologic heterogeneity of asteroid 4 Vesta. *Icarus, 127,* 130–157.

Gaffey M. J. (2001) Asteroids: Does space weathering matter? (abstract). In *Lunar and Planetary Science XXXII,* Abstract #1587. Lunar and Planetary Institute, Houston (CD-ROM).

Gaffey M. J. (2002) Asteroid 113 Amalthea: Nature, origin and meteorite affinities of an S(I) assemblage. *Icarus,* submitted.

Gaffey M. J. and Gilbert S. L. (1998) Asteroid 6 Hebe: The probable parent body of the H-type ordinary chondrites and the IIE iron meteorites. *Meteoritics & Planet. Sci., 33,* 1281–1295.

Gaffey M. J., Bell J. F., and Cruikshank D. P. (1989) Reflectance spectroscopy and asteroid surface mineralogy. In *Asteroids II* (R. P. Binzel et al., eds.), pp. 98–127. Univ. of Arizona, Tucson.

Gaffey M. J., Reed K. L., and Kelley M. S. (1992) Relationship of E-type Apollo asteroid 3103 (1982 BB) to the enstatite achondrite meteorites and the Hungaria asteroids. *Icarus, 100,* 95–109.

Gaffey M. J., Bell J. F., Brown R. H., Burbine T. H., Piatek J. L., Reed K. L., and Chaky D. A. (1993a) Mineralogical variations within the S-type asteroid class. *Icarus, 106,* 573–602.

Gaffey M.J., Burbine T. H., and Binzel R. P. (1993b) Asteroid spectroscopy and the meteorite connection: Progress and perspectives. *Meteoritics, 28,* 161–187.

Gaffey M. J., Bell J. F., Brown R. H., Burbine T. H., Piatek J. L., Reed K. L., and Chaky D. A. (1993c) Spectral evidence of size dependent space weathering processes on asteroid surfaces (abstract). In *Lunar and Planetary Science XXIV,* pp. 515–516.

Gaffey S. J., McFadden L. A., Nash D., and Pieters C. M. (1993) Ultraviolet, visible, and near-infrared reflectance spectroscopy: Laboratory spectra of geologic materials. In *Remote Geochemical Analysis: Elemental and Mieralogical Composition* (C. M. Pieters and P. A. J. Englert, eds.), pp. 43–77. Cambridge Univ., Cambridge.

Gomes C. B. and Keil K. (1980) *Brazilian Stone Meteorites.* Univ. of New Mexico, Albuquerque. 162 pp.

Gradie J. and Veverka J. (1986) The wavelength dependence of phase coefficients. *Icarus, 66,* 455–467.

Gradie J. C. and Zellner B. (1977) Asteroid families: Observational evidence for common origins. *Science, 197,* 254–255.

Granahan J. C. (1993) Investigation of asteroid family geology. Ph.D. dissertation, Univ. of Hawai'i, Honolulu. 187 pp.

Grimm R. E. and McSween H. Y. Jr. (1993) Heliocentric zoning of the asteroid belt by aluminum-26 heating. *Science, 259,* 653–655.

Hapke B. (2000) Space weathering in the asteroid belt (abstract). In *Lunar and Planetary Science XXXI*, Abstract #1087. Lunar and Planetary Institute, Houston (CD-ROM).

Hapke B. (2001) Space weathering from Mercury to the asteroid belt. *J. Geophys. Res., 106,* 10039–10073.

Hardersen P. S. and M. J. Gaffey (2001) Unraveling the thermal structure of the asteroid belt from meteoritic and asteroidal evidence (abstract). In *Lunar and Planetary Science XXXII*, Abstract #1103. Lunar and Planetary Institute, Houston (CD-ROM).

Herbert F., Sonett C. P., and Gaffey M. J. (1991) Protoplanetary thermal metamorphism: The protostellar wind electromagnetic induction hypothesis. In *The Sun in Time* (C. P. Sonett et al., eds.), pp. 710–739. Univ. of Arizona, Tucson.

Hirayama K. (1918) Groups of asteroids probably of common origin. *Proc. Phys. Math. Soc. Japan, Ser. 2, No. 9,* 354–361.

Hirayama K. (1933) Present state of the families of asteroids. *Proc. Imp. Acad. Japan, 9,* 482–485.

Hiroi T., Bell J. F., Takeda H., and Pieters C. M. (1993) Modeling of S-type asteroid spectra using primitive achondrites and iron meteorites. *Icarus, 102,* 107–116.

Hiroi T., Binzel R. P., Sunshine J. M., Pieters C. M., and Takeda H. (1995) Grain sizes and mineral compositions of surface regoliths of Vesta-like asteroids. *Icarus, 115,* 374–386.

Howell E. S., Rivkin A. S., Vilas F., and Soderberg A. M. (2001) Aqueous alteration in low albedo asteroids (abstract). In *Lunar and Planetary Science XXXII*, Abstract #2058. Lunar and Planetary Institute, Houston (CD-ROM).

Ingrin J., Latrous K., Doukhan J. C., and Doukhan N. (1989) Water in diopside: An electron microscopy and infrared spectroscopy study. *Eur. J. Mineral., 1,* 327–341.

Jones T. D., Lebofsky L. A., Lewis J. S., and Marley M. S. (1990) The composition and origin of the C, P, and D asteroids: Water as a tracer of thermal evolution in the outer belt. *Icarus, 88,* 172–192.

Keil K. (1982) Composition and origin of chondritic breccias. In *Workshop on Lunar Breccias and Soils and Their Meteoritic Analogs,* pp. 65–83. LPI Tech. Rpt. 82-02, Lunar and Planetary Institute, Houston.

Keil K. (2000) Thermal alteration of asteroids: Evidence from meteorites. *Planet. Space Sci., 48,* 887–903.

Keil K., Bell J. F., and Britt D. T. (1992) Reflection spectra of shocked ordinary chondrites and their relationship to asteroids. *Icarus, 98,* 43–53.

Kelley M. S. and Gaffey M. J. (1996) A genetic study of the Ceres (Williams #67) asteroid family. *Bull. Am. Astron. Soc., 28,* 1097.

Kelley M. S. and Gaffey M. J. (2000) 9 Metis and 113 Amalthea: A genetic asteroid pair. *Icarus, 144,* 27–38.

Kelley M. S. and Gaffey M. J. (2002) High-albedo asteroid 434 Hungaria: Spectrum, composition and genetic connections. *Meteoritics & Planet. Sci.,* in press.

Kelley M. S., Gaffey M. J., and Williams J. G. (1994) Compositional evidence in favor of a genetic link between the Nysa and Hertha asteroid families (abstract). In *Lunar and Planetary Science XXV,* pp. 689–690. Lunar and Planetary Institute, Houston.

Kelley M. S., Gaffey M. J., Vilas F., and Hardersen P. S. (2001a) Recent, ground-based, near-infrared spectral observations of asteroid 433 Eros (abstract). In *Lunar and Planetary Science XXXII*, Abstract #2112. Lunar and Planetary Institute, Houston (CD-ROM).

Kelley M. S., Vilas F., Gaffey M. J., and Abell P. A. (2001b) The first confirmation of a common origin for a small v-class asteroid with 4 Vesta and the HED meteorites. GSA Annual Meeting, Boston, Abstract #20247.

Kelley M. S., Vilas F., Gaffey M. J., and Abell P. A. (2002) Compositional evidence for a common origin of mainbelt asteroid 1929 Kollaa with 4 Vesta and the HED meteorites. *Meteoritics & Planet. Sci.,* in press.

King T. V. V. and Ridley I. W. (1987) Relation of the spectroscopic reflectance of olivine to mineral chemistry and some remote sensing implications. *J. Geophys. Res., 92,* 11457–11469.

Kozai Y. (1979) The dynamical evolution of the Hirayama family. In *Asteroids* (T. Gehrels, ed.), pp. 334–358. Univ. of Arizona, Tucson.

Kring D. A., Hill D. H., Gleason J. D., Britt D. T., Consolmagno G. J., Farmer M., Wilson S., and Haag R. (1999) Portales Valley: A meteorite sample of the brecciated and metal-veined floor of an impact crater on an H-chondrite asteroid. *Meteoritics & Planet. Sci., 34,* 663–669.

Larson H. P., Feierberg M. A., and Lebofsky L. A. (1983) The composition of asteroid 2 Pallas and its relation to primitive meteorites. *Icarus, 56,* 398–408.

Lipschutz M. E., Verkouteren R. M., Sears D. W. G., Hasan F. A., Prinz M., Weisberg M. K., Nehru C. E., Delaney J. S., Grossman L., and Boily M. (1988) Cumberland Falls chondritic inclusions: III. Consortium study of relationship to inclusions in Allan Hills 78113 aubrite. *Geochim. Cosmochim. Acta, 52,* 1835–1848.

Lovering J. F. (1962) The evolution of the meteorites — evidence for the co-existence of chondritic, achondritic and iron meteorites in a typical parent meteorite body. In *Researches on Meteorites* (C. B. Moore, ed.), pp. 179–197. Wiley, New York.

Lucey P. G., Keil K., and Whitely R. (1998) The influence of temperature on the spectra of the A-asteroids and implications for their silicate chemistry. *J. Geophys. Res., 103,* 5865–5871.

Marzari F., Farinella P., and Davis D. R. (1999) Origin, aging, and death of asteroid families. *Icarus, 142,* 63–77.

McCord T. B., Adams J. B., and Johnson T. V. (1970) Asteroid Vesta: Spectral reflectivity and compositional implications. *Science, 168,* 1445–1447.

McCoy T. J., Burbine T. H., McFadden L. A., Starr R. D., Gaffey M. J., Nittler L. R., Evans L. G., Izenberg N., Lucey P. G., Trombka J. I., Bell J. F. III, Clark B. E., Clark P. E., Squyers S. W., Chapman C. R., Boynton W. W., and Veverka J. (2001) The composition of 433 Eros: A mineralogical-chemical synthesis. *Meteoritics & Planet. Sci., 36,* 1661–1672.

McFadden L. A. and Gaffey M. J. (1978) Calibration of quantitative mineral abundances determined from meteorite reflection spectra and applications to solar system objects. *Meteoritics, 13,* 556–557.

McFadden L. A., Wellnitz D. D., Schnaubelt M., Gaffey M. J., Bell J. F. III, Izenberg N., Murchie S., and Chapman C. R. (2001) Mineralogical interpretation of reflectance spectra of Eros from NEAR near-Infrared spectrometer low phase flyby. *Meteoritics & Planet. Sci., 36,* 1711–1726.

McMillan P. F. and Remmele R. L. Jr. (1986) Hydroxyl sites in SiO_2 glass: A note on infrared and Raman spectra. *Am. Mineral., 71,* 772–778.

McSween H. Y. Jr., Bennett M. E. III., and Jarosewich E.(1991) The mineralogy of ordinary chondrites and implications for asteroid spectrophotometry. *Icarus, 90,* 107–116.

Merényi E., Howell E. S., Rivkin A. S., and Lebofsky L. A. (1997) Prediction of water in asteroids from spectral data shortward of 3 µm. *Icarus, 129,* 421–439.

Migliorini F., Manara A., Scaltriti F., Farinella P., Cellino A., and Di Martino M. (1997) Surface properties of (6) Hebe: A possible parent body of ordinary chondrites. *Icarus, 128,* 104–113.

Miller G. H., Rossman G. R., and Harlow G. E. (1987) The natural occurrence of hydroxide in olivine. *Phys. Chem. Minerals, 14,* 461–472.

Miyamoto M. (1987) Diffuse reflectance from 0.25 µm to 25 µm of the Yamato-691 enstatite chondrite. *Mem. NIPR Spec. Issue 46,* pp. 123–130. National Institute of Polar Research, Tokyo.

Miyamoto M. (1990) Differences in the degree of weathering between Antarctic and non-Antarctic ordinary chondrites: Infrared spectroscopy. *Workshop on Differences Between Antarctic and Non-Antarctic Meteorites,* pp. 68–71. LPI Tech. Rpt. 90-01, Lunar and Planetary Institute, Houston.

Miyamoto M. (1991) Differences in the degree of weathering between Antarctic and non-Antarctic meteorites inferred from infrared diffuse reflectance spectra. *Geochim. Cosmochim. Acta, 55,* 89–98.

Miyamoto M. and Zolensky M. E. (1994) Infrared diffuse reflectance spectra of carbonaceous chondrites: Amount of hydrous minerals. *Meteoritics, 29,* 849–853.

Miyamoto M., Kojima H., and Yanai K. (1989) Weathering of some Antarctic meteorites: Infrared spectroscopy. *Proc. NIPR Symp. Antarc. Meteorites, 2,* pp. 296–302. National Institute of Polar Research, Tokyo.

Morbidelli A., Gonczi R., Froeschlé Ch., and Farinella P. (1994) Delivery of meteorites through the ν_6 secular resonance. *Astron. Astrophys., 282,* 955–979.

Moroz L. and Arnold G. (1999) Influence of neutral components on relative band contrasts in reflectance spectra of intimate mixtures: Implications for remote sensing.1. Nonlinear mixing modeling. *J. Geophys. Res., 104,* 14109–14121.

Moroz L., Schade U., and Wasch R. (2000) Reflectance spectra of olivine-orthopyroxene-bearing assemblages at decreased temperatures: Implications for remote sensing of asteroids. *Icarus, 147,* 79–93.

Neal C. W. and Lipschutz M. E. (1981) Cumberland Falls chondritic inclusions: Mineralogy/petrology of a forsterite chondrite suite. *Geochim. Cosmochim. Acta, 45,* 2091–2107.

Noble S. K., Pieters C. M., Taylor L. A., Morris R. V., Allen C. C., McKay D. S., and Keller L. P. (2001) The optical properties of the finest fraction of lunar soil: Implications for space weathering. *Meteoritics & Planet. Sci., 36,* 31–42.

Nozette S. and Wilkening L. L. (1982) Evidence for aqueous alteration in a carbonaceous xenolith from the Plainview (H5) chondrite. *Geochim. Cosmochim. Acta, 46,* 557–563.

Ostro S. J., Campbell D. B., and Shapiro I. I. (1985) Mainbelt asteroids: Dual-polarization radar observations. *Science, 229,* 442–446.

Ostro S. J., Campbell D. B., Chandler J. F., Hine A. A., Hudson R. S., Rosema K. D., and Shapiro I. I. (1991) Asteroid 1986 DA: Radar evidence for a metallic composition. *Science, 252,* 1399–1404.

Ostro S. J., Hudson R. S., Nolan M. C., Margot J.-L., Scheeres D. J., Campbell D. B., Magri C., Giorgini J. D., and Yeomans D. K. (2000) Radar observations of asteroid 216 Kleopatra. *Science, 288,* 836–839.

Pellas P. (1991) Exotic clasts in meteoritic breccias (abstract). *Meteoritics, 26,* 384.

Pieters C. M., Fischer E. M., Rode O., and Basu A. (1993) Optical effects of space weathering: The role of the finest fraction. *J. Geophys. Res., 98,* 20817–20824.

Pieters C. M., Taylor L. A., Noble S. K., Keller L. P., Hapke B., Morris R. V., Allen C., McKay D. S., and Wentworth S. (2000) Space weathering on airless bodies: Resolving a mystery with lunar samples. *Meteoritics & Planet. Sci., 35,* 1101–1107.

Pinault L. J., Scott E. D. R, Bogard D. D., and Keil K. (1999) Extraordinary properties of the metal-veined, H6 Portales Valley chondrite: Evidence for internal heating versus shock-melting origins (abstract). In *Lunar and Planetary Science XXX,* Abstract #2048. Lunar and Planetary Institute, Houston (CD-ROM).

Reed K. L., Gaffey M. J., and Lebofsky L. A. (1997) Shape and albedo variations of asteroid 15 Eunomia. *Icarus, 125,* 445–454.

Rivkin A. S., Howell E. S., Britt D. T., Lebofsky L. A., Nolan M. C., and Branston D. D. (1995) 3-µm spectrophotometric survey of M- and E-class asteroids. *Icarus, 117,* 90–100.

Rivkin A. S., Howell E. S., Lebofsky L. A., Clark B. E., and Britt D. T. (2000) The nature of M-class asteroids from 3-µm observations. *Icarus, 145,* 351–368.

Rivkin A. S., Davies J. K., Clark B. E., Trilling D. E., and Brown R. H. (2001) Aqueous alteration on S asteroid 6 Hebe? (abstract). In *Lunar and Planetary Science XXXII,* Abstract #1723. Lunar and Planetary Institute, Houston (CD-ROM).

Rivkin A. S., Howell E. S., Vilas F., and Lebofsky L. A. (2002) Hydrated minerals on asteroids: The astronomical record. In *Asteroids III* (W. F. Bottke Jr. et al., eds.), this volume. Univ. of Arizona, Tucson.

Roedder E., ed. (1984) *Reviews in Mineralogy, Vol. 12: Fluid Inclusions.* Mineralogical Society of America, Washington, DC.

Rubin A. E. and Keil K. (1983) Mineralogy and petrology of the Abee enstatite chondrite breccia and its dark inclusions. *Earth Planet. Sci. Lett., 62,* 118–131.

Rubin A. E. and Ulff-Møller F. (1999) The Portales Valley meteorite breccia: Evidence for impact-induced metamorphism of an ordinary chondrite (abstract). In *Lunar and Planetary Science XXX,* Abstract #1618. Lunar and Planetary Institute, Houston (CD-ROM).

Ruzicka A., Snyder G. A., Prinz M., and Taylor L. A. (1999) Portales Valley: A new metal-phosphate-rich meteorite with affinities to Netschaëvo and H-group chondrites (abstract). In *Lunar and Planetary Science XXX,* Abstract #1645. Lunar and Planetary Institute, Houston (CD-ROM).

Salisbury J. W., Hunt G. R., and Lenhoff C. J. (1975) Visible and near-infrared spectra: X. Stony meteorites. *Mod. Geol., 5,* 115–126.

Salisbury J. W., D'Aria D. M., and Jarosewich E. (1991) Midinfrared (2.5–13.5 µm) reflectance spectra of powdered stony meteorites. *Icarus, 92,* 280–297.

Sandford S. A. (1984) Infrared transmission spectra from 2.5 to 25 µm of various meteorite classes. *Icarus, 60,* 115–126.

Scott E. R. D., Taylor G. J., Newsom H. E., Herbert F., and Zolensky M. (1989) Chemical, thermal and impact processing of asteroids. In *Asteroids II* (R. P. Binzel et al., eds.), pp. 701–739. Univ. of Arizona, Tucson.

Sears D. W., Kallemeyn G. W., and Wasson J. T. (1983) Composition and origin of clasts and inclusions in the Abee enstatite chondrite breccia. *Earth Planet. Sci. Lett., 62,* 180–192.

Singer R. B. and Roush T. L. (1985) Effects of temperature on remotely sensed mineral absorption features. *J. Geophys. Res., 90,* 12434–12444.

Skogby H. and Rossman G. R. (1989) OH^- in pyroxene: An experimental study of incorporation mechanisms and stability. *Am. Mineral., 74,* 1059–1069.

Skogby H., Bell D. R., and Rossman G. R. (1990) Hydroxide in pyroxene: Variations in the natural environment. *Am. Mineral., 75,* 764–774.

Starukhina L. (2001) Water detection on atmosphereless celestial bodies: Alternative explanations of the observations. *J. Geophys. Res., 106,* 14701–14710.

Sunshine J. M., Bus S. J., Burbine T. H., McCoy T. J., and Binzel R. P. (2002) Unambiguous spectral evidence for high- (and low-) calcium pyroxene in asteroids and meteorites (abstract). In *Lunar and Planetary Science XXXIII,* Abstract #1356. Lunar and Planetary Institute, Houston (CD-ROM).

Taylor L. A., Pieters C. M., Keller L. P., Morris R. V., McKay D. S., Patchen A., and Wentworth S. (2001) The effects of space weathering on Apollo 17 mare soils: Petrographic and chemical characterization. *Meteoritics & Planet. Sci., 36,* 285–299.

Tholen D. J. (1984) Asteroid taxonomy from cluster analysis of photometry. Ph.D. thesis, Univ. of Arizona, Tucson. 150 pp.

Tholen D. J. and Barucci M. A. (1989) Asteroid taxonomy. In *Asteroids II* (R. P. Binzel et al., eds.), pp. 298–315. Univ. of Arizona, Tucson.

Trombka J. I., Squyres S. W., Brückner J., Boynton W. V., Reedy R. C., McCoy T. J., Gorenstein P., Evans L. G., Arnold J. R., Starr R. D., Nittler L. R., Murphy M. E., Mikheeva I., McNutt R. L. Jr., McClanahan T. P., McCartney E., Goldsten J. O., Gold R. E., Floyd S. R., Clark P. E., Burbine T. H., Bhangoo J. S., Bailey S. H., and Petaev M. (2000) The elemental composition of asteroid 433 Eros: Results from the NEAR-Shoemaker X-ray spectrometer. *Science, 289,* 2101–2105.

Vaughan D. J. and Craig J. R. (1978) *Mineral Chemistry of Metal Sulfides.* Cambridge Univ., Cambridge.

Verkouteren R. M. and Lipschutz M. E. (1983) Cumberland Falls chondritic inclusions — II. Trace element contents of forsterite chondrites and meteorites of similar redox state. *Geochim. Cosmochim. Acta, 47,* 1625–1633.

Veverka J., Robinson M., Thomas P., Murchie S., Bell J. F. III, Izenberg N., Chapman C., Harch A., Bell M., Carcich B., Cheng A., Clark B., Domingue D., Dunham D., Farquhar R., Gaffey M. J., Hawkins E., Joseph J., Kirk R., Li H., Lucey P., Malin M., Martin P., McFadden L., Merline M. J., Miller J. K., Owen W. M. Jr., Peterson C., Prockter L., Warren J., Wellnitz D., Williams B. G., and Yeomans D. K. (2000) NEAR at Eros: Imaging and spectral results. *Science, 289,* 2088–2097.

Vilas F. and Gaffey M. J. (1989) Identification of phyllosilicate absorption features in main-belt and outer-belt asteroid reflectance spectra. *Science, 246,* 790–792.

Vilas F. and McFadden L. A. (1992) CCD reflectance spectra of selected asteroids. I. Presentation and data analysis considerations. *Icarus, 100,* 85–94.

Vilas F. and Smith B. A. (1985) Reflectance spectrophotometry (0.5–1.0 μm) of outer-belt asteroids: Implications for primitive, organic solar system material. *Icarus, 64,* 503–516.

Vilas F., Hatch E. C., Larson S. M., Sawyer S. R., and Gaffey M. J. (1993) Ferric iron in primitive asteroids: A 0.43-μm absorption feature. *Icarus, 102,* 225–231.

Vilas F., Jarvis K. S., and Gaffey M. J. (1994) Iron alteration minerals in the visible and near-infrared spectra of low-albedo asteroids. *Icarus, 109,* 274–283.

Vilas F., Cochran A. L., and Jarvis K. S. (2000) Vesta and the Vestoids: A new rock group? *Icarus, 147,* 119–128.

Wagner C., Arnold G., and Wäsch R. (1988) The infrared transmission spectrum of the Salzwedel meteorite. *Meteoritics, 23,* 93–94.

Walker D. and Grove T. (1993) Ureilite smelting. *Meteoritics, 28,* 629–636.

Watson F. G. (1941) *Between the Planets.* Harvard Books on Astronomy, Blakiston, Philadelphia. 222 pp.

Wetherill G. W. and Chapman C. R. (1988) Asteroids and meteorites. In *Meteorites and the Early Solar System* (J. F. Kerridge and M. S. Matthews, eds), pp. 35–67. Univ. of Arizona, Tucson.

Wilkening L. L. (1977) Meteorites in meteorites: Evidence for mixing among the asteroids. In *Comets, Asteroids, Meteorites: Interrelations, Evolution and Origins* (A. H. Delsemme, ed.), pp. 389–396. Univ. of Toledo, Toledo.

Williams J. G. (1989) Asteroid family identifications and proper elements. In *Asteroids II* (R. P. Binzel et al., eds.), pp. 1034–1072. Univ. of Arizona, Tucson.

Williams J. G. (1992) Asteroid families — An initial search. *Icarus, 96,* 251–280.

Xu S., Binzel R. P., Burbine T. H., and Bus S. J. (1995) Small main-belt asteroid spectroscopic survey: Initial results. *Icarus, 115,* 1–35.

Yanai K., Matsumoto Y., and Kojima H. (1983) A Brachina-like inclusion in the Yamato-75097 L6 chondrite: A preliminary examination. *Proc. 8th Symp. Antarc. Meteorites, NIPR Spec. Issue 30,* pp. 29–35. National Institute of Polar Research, Tokyo.

Zappalà V., Cellino A., Farinella P., and Knežević Z. (1990) Asteroid families. I. Identification by hierarchical clustering and reliability assessment. *Astron. J., 100,* 2030–2046.

Zappalà V., Cellino A., Farinella P., and Milani A. (1994) Asteroid families: II. Extension to unnumbered and multiopposition asteroids. *Astron. J., 107,* 772–801.

Zappalà V., Bendjoya Ph., Cellino A. , Farinella P., and Froeschlé C. (1995) Asteroid families: Search of a 12,487-asteroid sample using two different clustering techniques. *Icarus, 116,* 291–314.

Zappalà V., Cellino A., Di Martino M., Migliorini F., and Paolicchi P. (1997) Maria's family: Physical structure and possible implications for the origin of giant NEAs. *Icarus, 129,* 1–20.

Zellner B. (1975) 44 Nysa: An iron-depleted asteroid. *Astrophys. J. Lett., 198,* L45–L47.

Zolensky M. E., Barrett R. A., and Ivanov A. V. (1991) Mineralogy and matrix composition of CI clasts in the chondritic breccia Kaidun (abstract). In *Lunar and Planetary Science XXII,* pp. 1565–1566. Lunar and Planetary Institute, Houston.

Zolensky M. E., Weisberg M. K., Barrett R. A., and Prinz M. (1993) Mineralogy of dark clasts in primitive versus differentiated meteorites (abstract). In *Lunar and Planetary Science XXIV,* pp. 1583–1584. Lunar and Planetary Institute, Houston.

Zolensky M. E., Bodnar R. J., Gibson E. K. Jr., Nyquist L. E., Reese Y., Shih C.-Y., and Wiesmann H. (1999a) Asteroidal water within fluid inclusion-bearing halite in an H5 chondrite, Monahans (1998). *Science, 285,* 1377–1379.

Zolensky M. E., Bodnar R. J., and Rubin A. E. (1999b) Asteroidal water within fluid inclusion-bearing halite in ordinary chondrites (abstract). *Meteoritics & Planet. Sci., 34,* A124.

Asteroids in the Thermal Infrared

Alan W. Harris
DLR Institute of Space Sensor Technology and Planetary Exploration, Berlin

Johan S. V. Lagerros
Uppsala University

The importance of the thermal-infrared spectral region for investigations of asteroids lies traditionally in the dependence of the thermal emission of an asteroid on its visual albedo and size. Knowledge of albedos and sizes is crucial in many areas of asteroid research, such as mineralogy and taxonomy, the size-frequency distribution of families and populations of asteroids (e.g., near-Earth asteroids), and the relationship between asteroids in the outer solar system and comets. However, the rapid increase in the availability of computing power over the past decade has cleared the way for the development of sophisticated thermophysical models of asteroids with interesting new areas of application. Recent progress in the thermal modeling of asteroids and observational work in the thermal infrared are described. The use of thermal models for the physical characterization of asteroids and other applications is discussed.

1. INTRODUCTION

Observations in the thermal infrared began to play an important role in the physical characterization of asteroids in the 1970s, at a time when developments in detector technology and telescope design first enabled routine observational work to be carried out in this difficult spectral region. In general, thermal emission from asteroids dominates reflected solar radiation at wavelengths longer than about 5 μm. Observations from the ground are limited by atmospheric absorption to various windows in the range 5–20 μm. This range is generally adequate for measuring the thermal continuum spectrum around the emission peak of objects out to, and within, the main belt. However, the thermal continua of more distant asteroids, such as Centaurs and trans-Neptunian objects, peak at longer wavelengths and reliable observations often require airborne or orbiting facilities, although some useful measurements of the thermal emission of such objects have been made from the ground in the thermal infrared and at submillimeter and millimeter wavelengths.

Two fundamental parameters characterizing an asteroid are its visual albedo and size. In general, the visual albedos of main-belt asteroids range from ≤0.1 for taxonomic classes such as B, C, D, and P to 0.5 and higher for some members of the E class. Classes such as M, Q, R, S, and V are associated with intermediate values of albedo. Photometry of an asteroid in the optical region alone provides information only on the product $D^2 \times$ albedo, where D is the object's diameter. However, in contrast to the effect on its optical brightness, increasing the albedo of an asteroid would cause its thermal emission to *decrease* because a lighter surface absorbs less solar energy and thus has a lower equilibrium temperature. So by combining the physical relationships governing the observed scattered and emitted radiation of an asteroid, it is possible to determine both the size of the object and its albedo. This is the principle on which the vast majority of the size and albedo determinations of asteroids are based, including those in the Infrared Astronomical Satellite Minor Planet Survey (IMPS) (*Tedesco,* 1992), which represents a major milestone in the cataloging of asteroid diameters and albedos.

In order to interpret observations of the thermal emission of an asteroid in terms of physical parameters a thermal model is required. The most commonly adopted approach is that embodied in so-called standard thermal models, in which a simple temperature distribution on a smooth spherical surface is assumed, which falls from a maximum at the subsolar point to zero at the terminator. This produces good results if the asteroid in question has a small thermal inertia, rotates slowly, is observed at a small solar phase angle, and is not heavily cratered or irregularly shaped. A related simple model can be used to describe the alternative extreme case of a fast-rotating spherical asteroid with high thermal inertia. In practice, as discussed and illustrated in sections 3 and 5, simple thermal models can be refined to extend their usefulness to more realistic cases. In Fig. 1 the dependence of the wavelength of the thermal emission peak on distance from the Sun is illustrated in terms of these simple thermal models. In Fig. 2 model fluxes vs. heliocentric distance are plotted for a 100-km asteroid observed at the wavelength of the thermal emission peak and at 20 μm.

In recent years much attention has been given to near-Earth asteroids (NEAs), mainly as a result of increased awareness of the impact hazard. Currently, the rate of discovery of new NEAs is far outstripping the rate at which this important population of asteroids can be characterized in terms of physical parameters. The application of thermal-infrared techniques to NEAs presents special problems due to the wide range of solar phase angles at which NEAs are

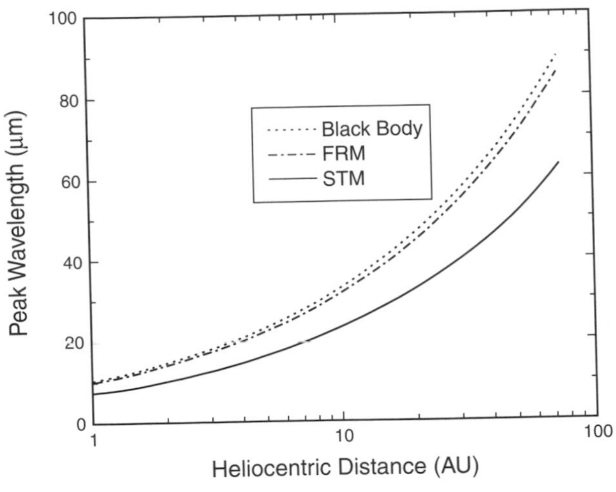

Fig. 1. Dependence of the wavelength of the peak of the thermal emission (in units of W m^{-2} μm^{-1}) on heliocentric distance, calculated for a black body and for asteroids with $p_v = 0.1$, G = 0.15, $\varepsilon = 0.9$ and the characteristics assumed in the standard thermal and fast rotating models (STM and FRM; see section 3). (When comparing this plot with similar data note that the wavelength of the thermal emission peak depends on whether units of W m^{-2} μm^{-1} or W m^{-2} Hz^{-1} are used.)

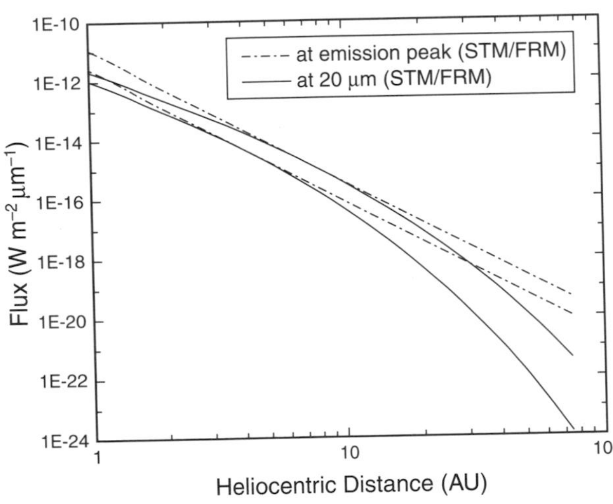

Fig. 2. Dependence of observed thermal flux on heliocentric distance as calculated with the standard thermal and fast rotating models (STM and FRM; see section 3) at the wavelength of the emission peak (dash-dot lines) and at a wavelength of 20 μm (continuous curves). The flux scale is valid for an asteroid with diameter = 100 km, $p_v = 0.1$, G = 0.15, $\varepsilon = 0.9$, $\alpha = 0°$, and geocentric distance = heliocentric distance. A rule-of-thumb value for the minimum detectable (5σ) flux with a large groundbased telescope, such as the Keck or Gemini, at 20 μm in 1 h of on-source integration time is 3.8×10^{-17} W m^{-2} μm^{-1} (=5 mJy); the Space Infrared Telescope Facility (SIRTF) is expected to better this by some 2 orders of magnitude. Note that W m^{-2} μm^{-1} can be converted to mJy via $F(mJy) = 3.336 \times 10^{14} \lambda^2 F(W\ m^{-2}\ \mu m^{-1})$, where λ is wavelength in μm (1 mJy = 10^{-29} W m^{-2} Hz^{-1}).

observed and the fact that, compared to observed main-belt asteroids, NEAs are small, irregular bodies that may in many cases have predominantly regolith-free, rocky surfaces and therefore relatively high thermal inertias (section 3).

For a number of applications, such as the use of asteroids as infrared calibration sources or detailed comparison of remotely sensed physical properties of particular objects with parameters derived from flyby/rendezvous missions or radar imaging, the accuracy achievable with simple thermal models is often inadequate and more sophisticated thermophysical models are required (section 4). Thermophysical models have been used successfully in the calibration of, and analysis of data from, the Infrared Space Observatory (ISO), which was operational from November 1995 until April 1998. ISO provided spectroscopic, photometric, imaging, and polarimetric data in the wavelength range 2–240 μm of more than 40 asteroids.

2. SOME BASIC CONCEPTS

If the effects of thermal inertia and surface roughness are negligible, the thermal emission from any point on an asteroid's surface can be considered to be in instantaneous equilibrium with the solar radiation absorbed at that point, which is the basic assumption underlying standard thermal models. In the case of a spherical asteroid the total absorbed solar radiation, S_{abs}, is simply

$$S_{abs} = \pi \frac{D^2}{4} S(1 - A) \tag{1}$$

where D is its diameter, S is the solar flux at the asteroid, and A is the bolometric Bond albedo. The bolometric Bond albedo refers to the total scattered solar energy in all directions and at all wavelengths, ratioed to the incident energy. The observationally more relevant and widely quoted albedo is the visual geometric albedo, p_v, which refers to sunlight in the visible region only (i.e., at a wavelength of 0.56 μm) reflected back toward the source, compared to that from a perfect (Lambertian) diffusely reflecting surface. Since the Sun's spectral energy distribution peaks in the visible region and the dependence of asteroid albedos on wavelength is normally small, it is usual to assume that

$$A = A_v = q\, p_v \tag{2}$$

where q is the phase integral. This allows the physically important parameter A to be linked directly to p_v. In the standard H, G magnitude system described by *Bowell et al.* (1989), in which H is the absolute magnitude and G is the slope parameter,

$$q = 0.290 + 0.684\ G \tag{3}$$

and H is defined as the V-band magnitude of an asteroid at mean brightness reduced to a distance of 1 AU from the Sun and Earth and zero solar phase angle. The relation between D, H, and p_v, which follows from the definitions of those parameters and the solar constant, can be written

$$D = \frac{1329}{\sqrt{p_v}} 10^{-\frac{H}{5}} \quad (4)$$

(e.g., *Fowler and Chillemi,* 1992). In the H, G system absolute magnitudes and albedos include the opposition effect, which is a brightness enhancement, or steepening of the phase curve, at small solar phase angles ($\alpha < 10°$). Therefore, as a rule, H values are numerically slightly lower (brighter) than V(1,0) magnitudes quoted in work published before 1990. It is important to remember that adoption of the H, G magnitude system also affects albedos: They are increased by typically 30%, so caution is called for when comparing albedos quoted in post-1990 publications with those in earlier work. Furthermore, since G is a function of albedo, the difference between V(1,0) and H, and the increase in albedo, are both albedo-dependent.

For detailed definitions and further discussion and references see, e.g., *Hansen* (1977), *Morrison and Lebofsky* (1979), *Lebofsky and Spencer* (1989), and *Hapke* (1993).

3. USE AND LIMITATIONS OF SIMPLE THERMAL MODELS

The basis of the asteroid diameter and albedo determinations in the IRAS Minor Planet Survey, and much of the later work in this field, is the refined "standard" thermal model of *Lebofsky et al.* (1986), which we refer to here as the STM. Details of the STM and related simple models are discussed in earlier chapters of the *Asteroids* series and references therein. The basis of the STM is the assumption of instantaneous equilibrium between insolation and thermal emission and a simple temperature distribution on a smooth spherical surface of the form

$$T(\varphi) = T(0) \cos^{1/4} \varphi \quad (5)$$

where φ is the angular distance from the subsolar point. T(0), the maximum (subsolar) temperature, is given by

$$T(0) = [(1 - A)S/(\eta\ \varepsilon\ \sigma)]^{1/4} \quad (6)$$

(cf. equation (1)), where ε is the emissivity and σ the Stefan-Boltzmann constant. The temperature on the nightside ($\varphi >$ 90°) is assumed to be zero, which is a reasonable assumption at small phase angles where the dayside flux dominates. The parameter η is called the "beaming parameter" because it was introduced as a means of modifying the model temperature distribution to take account of the observed enhancement of thermal emission at small solar phase angles, α, due to surface roughness, known as the beaming effect (see section 4). In practice, η can be thought of as a normalization or calibration parameter that allows a first-order correction for the effects of beaming, thermal inertia, and rotation. For a perfectly smooth sphere with zero thermal inertia η has the value 1.0; introducing surface roughness, i.e., beaming, causes η to decrease. In general, thermal inertia and rotation reduce the subsolar temperature and therefore have the opposite effect on η (*Spencer et al.,* 1989; *Spencer,* 1990). In the case of large main-belt asteroids the dominant effect is normally beaming. In the STM of *Lebofsky et al.* (1986) η is set at 0.756, the value that gives the correct diameters (as derived from occultation observations) of the asteroids 1 Ceres and 2 Pallas from photometric measurements at 10 μm. To enable its use at nonzero solar phase angles the STM also includes a (wavelength-independent) empirical thermal-infrared phase coefficient of 0.01 mag/deg, which is based on observations of main-belt asteroids at phase angles $\alpha \leq 30°$. Note that this treatment of the phase correction does not take into account the "evening/morning effect," in which the observed thermal emission of asteroids can be slightly different from that predicted on the basis of the empirical phase coefficient, depending on a combination of thermal inertia, rotation vector, and the side of opposition they are observed on (see *Lebofsky and Spencer,* 1989; *Spencer,* 1990; and references therein). However, the data necessary to correct for this effect are rarely available.

The use of thermal models to derive asteroid diameters and albedos is illustrated in Fig. 3. Note that the accuracy of albedo values derived via thermal models depends strongly on the accuracy of the adopted H value (cf. equation (4) and Fig. 3). It often happens, especially in connection with IMPS

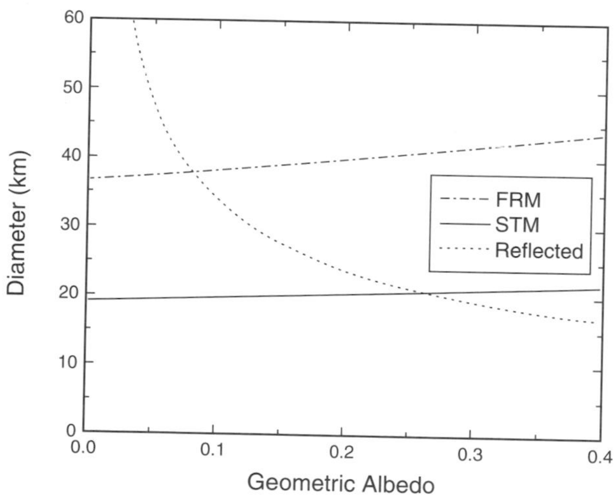

Fig. 3. Diameter/geometric albedo dependencies for a 10-μm flux measurement and an absolute magnitude, H(max) = 10.47, of 433 Eros at lightcurve maximum (see *Harris and Davies,* 1999, for details). The dash-dot and continuous curves represent the diameter/albedo relationships resulting from the condition of equilibrium between insolation and thermal emission embodied in the FRM and STM respectively (equation (6)). The dotted curve represents the diameter/albedo relationship implied by the adopted absolute magnitude, H (equation (4)). The points of intersection of the curves give the corresponding solutions for diameter and albedo.

TABLE 1. Comparison of IRAS MPS diameters with diameters from occultation measurements for main-belt asteroids with lightcurve amplitudes of 0.15 mag or less.

Asteroid	D (km) IRAS MPS	D (km) Occ.	(IRAS-Occ.)/Occ. (%)	NS	Lightcurve Amplitude (mag)
4 Vesta	468 ± 27	531	−12	1	0.12
51 Nemausa	148 ± 3	137 ± 8	+8	6	0.10–0.14
65 Cybele	237 ± 4	230 ± 16	+3	6	0.04–0.12
85 Io	155 ± 4	178	−13	3	0.15
93 Minerva	141 ± 4	171 ± 1	−18	2	0.04–0.10
105 Artemis	119 ± 3	103	+16	3	0.15
106 Dione	147 ± 3	147 ± 15	0	6	0.10
144 Vibilia	142 ± 3	178 ± 11	−20	4	0.13
324 Bamberga	229 ± 7	228 ± 9	0	2	0.07
444 Gyptis	160 ± 13	161	−1	4	0.15
471 Papagena	134 ± 5	127 ± 8	+6	4	0.11–0.13
704 Interamnia	317 ± 5	329	−4	10	0.03–0.11

The column headed NS gives the number of IRAS sightings used in the diameter determinations. The errors are as given in the data sources quoted. Data from the compilation of *Dunham et al.* (2001) with quality codes 3 and below have been excluded. For the asteroids 4, 105, 444, 704, Dunham et al. quote two diameters, a × b, based on elliptical fits to the observed chords; in these cases the occultation diameters quoted here are $(a \times b)^{0.5}$. In the case of 4 Vesta, Dunham et al. report two quality code 4 occultation measurements, namely 501 km and 561 km; the value given in the table is the mean of these, which agrees well with the result of *Thomas et al.* (1997) (529 ± 10 km) from imaging with the Hubble Space Telescope.

albedos, that more accurate H values become available at a later stage, necessitating corrections to the original solutions for albedo and diameter. *Harris and Harris* (1997) have devised a convenient, approximate method for recalculating albedos and diameters given new, improved H values, which avoids recourse to detailed thermal-model calculations.

Despite its simplicity, the STM has proved to be successful in the determination of diameters and albedos of main-belt asteroids. In Table 1 STM-derived asteroid diameters from the IRAS survey are compared with values derived from occultation measurements. The occultation diameters are taken from the online compilation of *Dunham et al.* (2001) and the IRAS diameters from the IRAS Minor Planet Survey Final Report (*Tedesco,* 1992). Due to the different times and aspects of the IRAS and occultation measurements, meaningful comparison of diameter measurements becomes very difficult for significantly nonspheroidal asteroids having large-amplitude lightcurves. For this reason the comparison in Table 1 has been limited to asteroids with peak-to-peak lightcurve amplitudes of 0.15 mag or less (*Lagerkvist et al.,* 1989; *Harris,* 2001).

The RMS fractional difference between the IMPS diameters and diameters derived from occultation measurements in Table 1 is 11%. No large systematic discrepancy is evident between the two sets of data: The mean difference is −3% (if the three asteroids with only one or two IRAS sightings are excluded, this falls to −0.6%). Despite the success of the STM in the IMPS, users of the model should always be aware of the limitations inherent in its generally unrealistic assumptions.

It is clear from the work of *Veeder et al.* (1989) that the limits of applicability of the STM are reached in the study of near-Earth asteroids (NEAs). Veeder et al. derived albedos and diameters for 22 NEAs from their 10-μm photometry and found that the albedos derived on the basis of the STM for five of these objects lie well above the range expected for their spectral classes. Veeder et al. assumed that this discrepancy is due to some small asteroids having relatively high-thermal-inertia surfaces, resulting from the lack of an insulating layer of regolith, and the failure of the STM to describe their thermal characteristics adequately. The first study in which a standard thermal model was found to give results seriously inconsistent with those obtained with other techniques is that of *Lebofsky et al.* (1978), who had to introduce a nonstandard, fast-rotating/high-thermal-inertia model to obtain a diameter and an albedo for the C-type NEA 1580 Betulia consistent with those obtained from radar and visual-polarimetry studies.

If we consider the simple temperature distribution of the STM, peaking at the subsolar point, but now assume a large thermal inertia and rapid rotation with the Sun in the equatorial plane, the temperature distribution becomes smoothed out in longitude. The limit of a longitude-independent temperature distribution corresponds to the fast rotating model (FRM), as described by *Lebofsky and Spencer* (1989). It can be shown that the maximum temperature in this case, i.e., that at the equator, is given by equation (6), with π substituted for η. It follows that the maximum temperature in the FRM is lower than the subsolar temperature in the STM, other parameters being equal. In the FRM the temperature decreases toward the poles as in equation (5), with φ in this case representing asteroidal latitude. In the FRM, 50% of the thermal emission originates on the nightside, whereas in the STM the nightside emission is zero. For an aid in visualizing the temperature distributions of the STM and FRM, see Fig. 4 of *Lebofsky and Spencer* (1989).

Use of the FRM instead of the STM for 1580 Betulia and the five "nonstandard" cases reported by *Veeder et al.* (1989) results in albedos largely consistent with the spectral classes of these NEAs. However, the work of Veeder et al. raises fundamental questions regarding the applicability of thermal-infrared techniques to the derivation of albedos and diameters of NEAs. For example, how do we know which model to apply to thermal-infrared photometry of a newly discovered NEA? The physical characteristics of NEAs are largely unknown and, in general, probably lie somewhere between those assumed in the extreme cases of the STM and FRM. *Harris* (1998) has proposed a solution in the form of a modified STM (the near-Earth asteroid thermal model, NEATM), which utilizes the information in the flux distribution around the thermal emission peak to determine the optimum value of η in each case.

In general the STM and FRM predict very different flux distributions, with the FRM emission peaking at a longer wavelength (see Fig. 1). In the STM the beaming parameter, η (cf. equation (6)), is fixed at a value of 0.756, and in the FRM there is assumed to be no beaming effect (*Lebofsky and Spencer,* 1989) and equation (6) holds with η replaced by π. In contrast, the NEATM allows η to vary to provide the best fit to the observed flux distribution. In effect, the model temperature distribution in the NEATM is modified to force consistency with the observed apparent color temperature of the asteroid, which depends on thermal inertia, surface roughness, and spin vector. In addition, the NEATM differs from the STM in its treatment of the infrared phase effect. NEAs are often observed at phase angles much larger than 30°, i.e., beyond the range in which the empirical phase coefficient used with the STM was derived and is known to apply. Furthermore, it is not clear to what extent the surfaces of relatively small, irregularly shaped NEAs give rise to a beaming effect similar to that observed in the case of large main-belt asteroids. There is an urgent need for observations of the thermal-infrared phase effect in the case of NEAs; since such data are not yet available, the solar phase angle is accounted for in the NEATM by integrating the thermal flux over that portion of the sunlit surface of the object (assumed spherical) visible to the observer. This treatment takes account of the wavelength dependence of the phase effect, in contrast to that of the STM (for further details see *Harris,* 1998).

Future observations of NEAs over a wide range of phase angle may allow empirical phase curves to be constructed that could be used to improve the application of simple thermal models to these objects.

In Table 2 NEA albedos and diameters calculated using the STM, NEATM, and FRM are compared. Comparison diameters, estimated from radar observations or from the *NEAR Shoemaker* mission in the case of 433 Eros, and spectral types are also given. When comparing results in this table with those of *Veeder et al.* (1989) note that the p_v values given here are on the H, G magnitude system and must be reduced by some 30% for comparison with the p_v values derived by Veeder et al. Furthermore, some updates to the H values and geometry assumed by Veeder et al. have been made.

It is clear from Table 2 that in at least nine cases the STM albedos are much higher than would be expected from the spectral types. Furthermore, the STM diameters for 1580 Betulia, 2100 Ra-Shalom, and 6489 Golevka are too small compared to the constraints applied by radar results. Likewise, in some eight cases the FRM gives albedos that are below the ranges normally associated with the corresponding spectral classes and/or diameters that are too large. Overall, the NEATM appears to give the most consistent agreement with the spectral types and radar sizes; its failure in the cases of 1580 Betulia and 6489 Golevka is probably due to the lack of spectral data for model fitting, necessitating the use of a default value for η.

A few cases in Table 2 merit individual discussion:

433 Eros. Thanks to the *NEAR Shoemaker* mission very reliable size and albedo information is now available for this object, which makes it a useful target for testing asteroid thermal models. However, it is one of the largest NEAs and therefore its physical characteristics with regard to the parameters of importance for thermal modeling may not be representative of the population as a whole. In particular, it has a substantial regolith and a boulder-strewn, heavily cratered surface. The results in Table 2 from the STM and the NEATM (but not the FRM) are satisfactory in this case. It is interesting to note that the *NEAR Shoemaker* size (D_{eff} (max) = 20.6 km) is slightly smaller than that derived from earlier radar observations and thermal-infrared observations and modeling. The discrepancy is probably due to the very elongated and irregular shape of Eros (cf. section 4.1): *Morrison* (1976) obtained D_{eff} = 22 ± 2 km while *Lebofsky and Rieke* (1979) found D_{eff} = 25 ± 1.5 km. Lebofsky and Rieke pointed out that their result is in good agreement with the radar results of *Jurgens and Goldstein* (1976), which indicate D_{eff} = 24.2 km. Note that *Mitchell et al.* (1998) have performed a more sophisticated analysis of the radar data obtained during the close approach of Eros in 1975 and generated shape models that are in good agreement with the results from the *NEAR Shoemaker* mission.

3671 Dionysus. This asteroid is suspected of being binary and may be a rubble pile (*Mottola et al.,* 1997b; *Pravec et al.,* 1998). On the basis of optical spectra alone, S. J. Bus (personal communication, 2000) suggests a C classification, for which an albedo of $p_v \leq 0.1$ would be expected on the basis of main-belt asteroid taxonomy. The albedos quoted in Table 2 are well above this range. The model giving the lowest albedo (p_v = 0.16) is the NEATM, but this value is based on an extremely large value of η. *Harris and Davies* (1999) favored the results of the FRM in this case due to the uncertainties associated with possible significant thermal emission from the nightside of a rapidly rotating object (period = 2.7 h), which would lead to underestimation of the albedo by the NEATM. However, if Dionysus really

TABLE 2. Diameters and albedos of near-Earth asteroids from simple thermal models.

Near-Earth Asteroid	α (°)	D_{eff} (km) STM	NEATM	FRM	Radar or s/c	η	p_v STM	NEATM	FRM	Type
433 Eros (lc max.)	10	20.5	23.6	36.2	20.6	1.05	0.27	0.20	0.09	S
	31	21.0	23.6	36.6		1.07	0.26	0.21	0.09	
(lc min.)	10	11.8	14.3	21.0	14.1	1.15	0.32	0.22	0.10	
1566 Icarus	93–100	0.88	1.27	1.05	1 – 4	d	0.70	0.33	0.49	Q ?
1580 Betulia	10–15	3.3	3.9	5.7	≥5.4	d	0.24	0.17	0.08	C
1627 Ivar	5	7.4	6.9	13.6	8.5 ± 3	0.67	0.22	0.26	0.07	S
1862 Apollo	35	1.2	1.4	1.9	~1.2	1.15	0.35	0.26	0.15	Q
1866 Sisyphus	35	7.5	8.9	13.1	—	1.14	0.20	0.14	0.07	S
1915 Quetzalcoatl	29	0.34	0.40	0.56	—	d	0.42	0.31	0.16	S
1980 Tezcatlipoca	63	4.5	6.7	6.8	—	1.54	0.31	0.14	0.14	S
2100 Ra-Shalom	42	1.7	2.5	2.6	≥2.4	1.80	0.26	0.13	0.11	C
2201 Oljato	105	1.3	2.1	1.5	—	d	0.63	0.24	0.49	S,E ?
3200 Phaethon	48	3.6	5.1	5.6	—	1.60	0.22	0.11	0.09	F
3551 Verenia	50	0.75	0.87	1.08	—	d	0.72	0.53	0.35	V
3554 Amun	16	1.8	2.1	2.9	—	d	0.23	0.17	0.09	M
3671 Dionysus	58	0.86	1.5	1.10	—	3.1	0.52	0.16	0.31	C ?
3757 1982 XB	31	0.33	0.39	0.52	—	d	0.46	0.34	0.19	S
4055 Magellan	14	2.5	3.0	4.5	—	d	0.34	0.24	0.11	V
6053 1993 BW3	51	2.7	3.1	3.9	—	d	0.25	0.18	0.11	Q,R ?
6178 1986 DA	31	2.0	2.3	3.2	2.3 ± 0.6	d	0.19	0.14	0.07	M
6489 Golevka	89	0.22	0.29	0.26	0.53 ± 0.03	d	1.08	0.63	0.78	Q,V ?
14402 1991 DB	36	0.52	0.56	0.79	—	0.98	0.18	0.16	0.08	B,C ?

In those cases in which data at only one wavelength are available (e.g., data from *Veeder et al.,* 1989), or spectral data are inadequate for model fitting, a default value of η of 1.2 is used in the NEATM (denoted by "d" in the η column). The diameter, D_{eff}, is the effective diameter, i.e., the diameter of a sphere of equivalent projected area. For 433 Eros results are given for lightcurve maximum and minimum; the comparison effective diameters are from the results of the *NEAR Shoemaker* mission (P. Thomas, personal communication, 2000; T. Kwiatkowski, personal communication, 2000). The albedo from *NEAR Shoemaker* is 0.25 ± 0.06 (*Veverka et al.,* 2000). Lightcurve effects have been taken into account where corresponding optical photometry is available; in all other cases included here the lightcurve amplitude is known to be small (≤0.3 mag). The formal errors in D_{eff} and p_v are much smaller than the modeling uncertainties (cf. Table 1). Original data sources:

433: see *Harris and Davies* (1999).
1566: radar size from *Pettengill et al.* (1969). *Mahapatra et al.* (1999) suggest a diameter based on their radar observations of D_{eff} = 0.8 km. However, given the adopted H value of 16.3, this size implies an extremely high albedo of $p_v > 0.7$. Taxonomic type suggested by *Hicks et al.* (1998).
1627: observations by Binzel, Harris, Delbó, and Davies (in preparation).
1866: observations by Harris and Davies (in preparation).
1980: lightcurve maximum, *Harris and Davies* (1999).
2100: lightcurve maximum, *Harris et al.* (1998); radar size from *Shepard et al.* (2000).
2201: E type suggested by *Hicks et al.* (1998).
3671: *Harris and Davies* (1999); tentative taxonomic type from S. J. Bus (personal communication, 2000).
6053: *Pravec et al.* (1997); taxonomic type suggested by *Hicks et al.* (1998).
6178: radar size from *Ostro et al.* (1991).
6489: radar size from *Hudson et al.* (2000); taxonomic type suggested by *Hicks et al.* (1998).
14402: observations by Binzel, Harris, Delbó, and Davies (in preparation); lightcurve amplitude ~0.1 mag (P. Pravec, personal communication, 2000); tentative taxonomic type from R. Whiteley (personal communication, 2001).
All other objects: see *Harris* (1998).

has $p_v \leq 0.1$, then the error must be in the opposite sense. This would imply that the measured flux from this object is much too low, even compared to the prediction of the NEATM. It appears that a very irregular shape, which would invalidate the assumption of sphericity in the thermal models and might give rise to shadowing, is unlikely to be the explanation since the lightcurve amplitude is only 0.14 mag. The NEATM albedo, $p_v = 0.16$, for (14402) 1991 DB, apparently a B/C type, poses a similar dilemma. However, as discussed in section 5.2, it may be unwise in the case of NEAs to draw conclusions based on the spectral type/albedo correlations of main-belt asteroids.

6489 Golevka. Thermal-model calculations based on observations made at very large solar phase angles are relatively unreliable; this is particularly true if the object has a very irregular shape, as in the case of Golevka. Furthermore, at the time of the thermal-infrared observations the object was almost pole-on. With only one 10-μm measurement available in this case, spectral fitting is not possible and the NEATM results in Table 2 are based on the default value of η = 1.2. *Mottola et al.* (1997a) attempted to take the extreme geometry into account in their model calculations, which were based on modified versions of the STM and FRM. However, in this case none of the models gives results consistent with the radar size of *Hudson et al.* (2000). Since a shape model of Golevka is now available from the radar observations, this object presents an opportunity for realistic thermophysical modeling of a small irregular asteroid to probe the limitations of the simple models.

In general, the derivation of sizes and albedos of NEAs on the basis of simple thermal models is subject to greater uncertainty than in the case of main-belt and distant asteroids. An important task for the future is to investigate the phase effect in the case of small, irregular asteroids, both observationally and via thermophysical modeling. Although the available dataset is still small, there is some evidence that the η value derived from spectral fitting, which is related to the apparent color temperature, increases with solar phase angle, α (see Table 2). If so, this would confirm that a more sophisticated treatment of the phase effect is required in the STM and NEATM and that caution should be exercised in interpreting large values of η in terms of high thermal inertia alone.

4. THERMOPHYSICAL MODELING

While the STM and its derivatives are very useful for investigations such as those described above (see also section 5), simple models have obvious limitations when it comes to detailed physical interpretation of high-quality observational data or the prediction of accurate thermal-infrared fluxes from asteroids for calibration and other purposes. Thus there have been a number of attempts at more sophisticated modeling of the thermophysical processes involved. However, before reviewing this field we first consider the complexity of the problem.

Sunlight heats the surface of a rotating, possibly irregularly shaped asteroid. Scattered sunlight and thermal emission from other parts of the surface contribute to the local radiation field. The surface may be rough and irregular on all scales, giving rise to complex shadow patterns. Radiation penetrates the porous surface material, and there are radiative and conductive heat transfer processes within the surface. The effective emissivity and the state of polarization of the thermal emission leaving the surface depend on the porosity, the particle size distribution and the wavelength-dependent refractive index.

4.1. Shape and Spin State

The STM assumes asteroids to be spherical, which is the simplest assumption in the absence of shape information. However, asteroids are often elongated and irregularly shaped, and generate a thermal lightcurve as they rotate. Correcting for the shape and the rotational phase is in many cases more important than taking account of the thermophysical processes discussed below. Ellipsoidal shape models of asteroids can be derived by inverting groundbased visual lightcurves (*Kaasalainen et al.,* 2002). Higher-order estimates of three-dimensional shapes exist for a small set of objects based on imaging with spacecraft, space telescopes, and radar (see the relevant chapters in this book). Spin vectors can often be derived from such investigations. Rotation periods can be derived from lightcurve studies covering a few apparitions to remarkable precision. By combining the results of such efforts, it is now possible to accurately predict the orientation and projected area at any time of a significant number of objects (see, for example, *Müller and Lagerros,* 1998). To predict thermal fluxes from these objects, a simple approach is to scale the STM flux by the projected area to obtain a synthetic thermal lightcurve. For more accurate work, however, systematic effects arising from the correlation between curvature and temperature and the highly nonlinear Planck function have to be considered (*Brown,* 1985; *Lagerros,* 1996a). To be more precise, the model thermal flux is computed using

$$F_\lambda = \frac{1}{\Delta^2} \oint (I_{em} + I_{ref} + I_{sc})\, \mu\, dS \qquad (7)$$

where μ is the direction cosine of the surface element dS, Δ is the distance to the observer, and the intensities I are those of the emitted, reflected, and multiply scattered radiation respectively. Points not visible to the observer are taken into account by letting $\mu = 0$. The contribution from reflected sunlight is normally completely negligible in the thermal infrared. *Lagerros* (1997) modeled multiply scattered radiation and concluded that due to large-scale shape features it can, in most cases, be ignored.

4.2. Albedo and Emissivity

Albedo and emissivity enter the thermophysical modeling by determining the heating and cooling rates, and thereby the surface temperature. At this stage the appropriate energy-balance equation uses quantities averaged over wavelength and direction. In the next step, the directional- and wavelength-dependent albedo and emissivity are required to determine the observed reflected and emitted radiation. The study of wavelength-dependent emissivity features may turn out to be of great importance for asteroid mineralogy. Thermal-infrared spectra obtained in early investigations are essentially gray (*Gillett and Merrill,* 1975; *Feierberg and Witteborn,* 1983; *Green et al.,* 1985) but showed a feature around 10 μm that was thought to be either due to silicates or of atmospheric origin. More recently, airborne observations (*Cohen et al.,* 1998), and especially observations with ISO (*Dotto et al.,* 2002, and references therein), have revealed identifiable features for the first time.

However, the mineralogical interpretation is still an open issue and much remains to be done in bringing together laboratory, observational, and theoretical results in this area.

In principle, compositional variations around the surface of a body can be detected by comparing visual and thermal lightcurves. Dark spots in the visual would be warm and bright in the thermal infrared and would therefore give rise to a phase shift in the lightcurves. An interesting example of roughly anticorrelated thermal and visible lightcurves is given by *Lellouch et al.* (2000), who analysed ISO far-infrared data of the Pluto-Charon system on the basis of a thermal model that takes into account thermal inertia and beaming. The method has been applied to asteroids (*Lebofsky et al.,* 1988) but difficulties arise in separating spots from shape effects (*Lagerros,* 1997).

4.3. Heat Conduction

Heat conduction in the regolith introduces thermal inertia, defined by

$$\Gamma = \sqrt{k\rho c} \tag{8}$$

where k is the thermal conductivity, ρ the density, and c the heat capacity. In the STM Γ = 0 and the surface is in instantaneous equilibrium with the solar radiation. A nonzero thermal inertia lowers the temperature contrast around the body and introduces a time lag in the diurnal temperature variation (Fig. 4, upper panel). *Müller and Lagerros* (1998) derived thermal inertias between 5 and 25 J m^{-2} $s^{-0.5}$ K^{-1} for a few of the largest main-belt asteroids. These values are compatible with the values for the Moon (Γ ~ 50) and Mercury (Γ ~ 80) (*Spencer et al.,* 1989) if values of k, ρ, and c measured for lunar soil (*Keihm,* 1984) are assumed, with k and c scaled for the lower temperatures of the main-belt asteroids. In comparison, Γ is around 2500 for solid rock, which illustrates the dramatic effect of the porosity of the regolith. A number of near-Earth asteroids appear to have relatively high thermal inertias (*Tedesco and Gradie,* 1987; *Veeder et al.*, 1989; *Harris et al.*, 1998; *Harris and Davies,* 1999), presumably because these small bodies have insufficient gravity to retain impact ejecta and build up significant regoliths.

The thermal skin depth (*Wesselink,* 1948; *Spencer et al.,* 1989; *Lagerros,* 1996a), defined by

$$l_s = \sqrt{\frac{k}{\rho c \omega}} \tag{9}$$

in which ω is the angular rotation rate, is very useful for characterizing the scale length of the diurnal heat wave. Typical values of l_s for asteroids lie in the range 10^{-3}–10^{-2} m, many orders of magnitude smaller than the objects in question. Thus the heat diffusion process is highly localized on a diurnal timescale, implying that the application of simple one-dimensional vertical heat transfer models is valid.

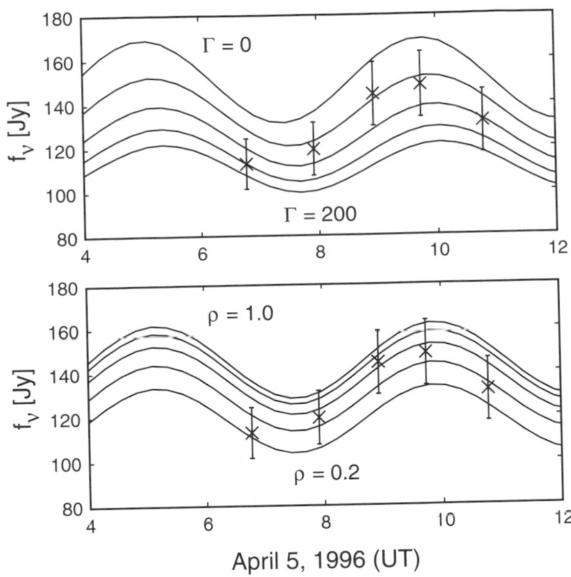

Fig. 4. Observed and modeled thermal lightcurves of 532 Herculina at a wavelength of 20 μm (*Müller and Lagerros,* 1998). Model curves are computed for thermal inertias (upper panel) Γ = 0, 50, ..., 200, and r.m.s. surface roughness slopes (lower panel) ρ = 0.2, 0.4, ..., 1.

A practical tool for characterizing the solutions of the vertical heat transfer problem is the dimensionless thermal parameter given by

$$\Theta = \frac{\Gamma\sqrt{\omega}}{\varepsilon\,\sigma\,T_{ss}^3} \tag{10}$$

where T_{ss} is the STM subsolar temperature (T(0) in equation (6)). Θ = 0 corresponds to the STM, in which zero thermal inertia and no rotation are assumed, whereas Θ → ∞ corresponds to the "fast rotator" (FRM) case. The thermal parameter is the result of transforming the heat diffusion problem into a dimensionless form (*Spencer et al.,* 1989). The parameter Θ is very useful in thermophysical modeling since heat transfer calculations can be expressed in terms of Θ alone (*Lagerros,* 1996a).

4.4. Surface Roughness

There is now ample evidence that asteroid surfaces are generally heavily cratered and rough on all scales. In combination with the lack of an atmosphere and small thermal skin depths, surface roughness gives rise to substantial temperature contrast, even on small scales, and produces the so-called beaming effect in which thermal emission is enhanced in the solar direction. The beaming effect is probably mainly due to the effect of the walls of subsolar craters, which cause the crater floors to receive more radiation than their surroundings and therefore present higher brightness temperatures to an observer at a small phase angle. In fact, such an observer would preferentially see relatively warm, sunward-facing surface elements over the whole disk. Thus, in effect, thermal emission is "beamed" in the sunward direction (see, for example, *Hansen,* 1977; *Spencer,* 1990; and references therein). As described in section 3, the beaming parameter, η, is used in the STM to adjust the surface temperature to take into account this and other effects. At small phase angles beaming increases both the mean flux level and the amplitude of thermal lightcurves (Fig. 4, lower panel), while at large phase angles conservation of energy dictates that observers see less thermal emission as a result of the beaming effect.

A number of surface roughness models have been used to describe the beaming effect on the basis of multiple scattering in spherical craters (*Buhl et al.,* 1968; *Winter and Krupp,* 1971; *Hansen,* 1977; *Spencer,* 1990; *Lagerros,* 1998; *Emery et al.,* 1998), parabolic craters (*Vogler et al.,* 1991; *Johnson et al.,* 1993), and stochastic surfaces (*Jämsä et al.,* 1993). Whereas η in the STM is a disk-integrated constant, detailed investigations show that the beaming effect actually depends on the degree of roughness, wavelength, albedo, emissivity, and the viewing and illumination geometry. In particular, thermophysical modeling of the beaming effect could aid in the interpretation of observations of NEAs, which are often made at large phase angles.

While most studies support the assumption that surface roughness is the main cause of the beaming effect, a possible alternative explanation is radiative transfer processes in the regolith. However, the modeling of *Hapke* (1996) indicates that such processes can contribute only about 20% of the observed effect. Radiative heat transfer in the regolith has also been modeled by *Henderson and Jakosky* (1994, 1997), who have investigated, for example, solid-state greenhouse effects.

4.5. Polarization

Observations have been made of the polarization of the thermal emission from the Moon, 1 Ceres, Io, and Mercury (*Heiles and Drake,* 1963; *Clegg and Carter,* 1970; *Johnson et al.,* 1983; *Goguen and Sinton,* 1985; *Mitchell and De Pater,* 1994). Polarization in the case of thermal emission is due to the refraction of the radiation leaving the surface. Although the effect is small in disk-integrated data, the state of polarization imposes constraints on the dielectric properties of the surface material, surface roughness, thermal inertia, porosity, and spin vector orientation (*Henderson et al.,* 1992; *Redman et al.,* 1995; *Lagerros,* 1996b; *Lagerros et al.,* 1999) (see also *Dotto et al.,* 2002).

4.6. Applications of Thermophysical Models

As is evident from the above discussion, thermophysical models have important applications in the study of the physical properties of asteroids, from the determination of accurate diameters and albedos of main-belt asteroids and the thermal inertia and surface roughness of asteroids with well-known shapes and sizes, to the study of spectral features in the thermal infrared. Here we briefly describe two further interesting applications of thermophysical models.

4.6.1. Calibration standards. Provided their fluxes can be predicted with sufficient precision, asteroids can act as convenient standard sources throughout the thermal infrared, a region of the spectrum lacking in suitable celestial calibration standards. As calibration sources, asteroids have the advantages of wide availability over the sky and a broad range of flux levels. With the advent of accurate thermophysical modeling the use of asteroids for this purpose has recently become feasible, as has been demonstrated in the case of the ISO project (*Müller and Lagerros,* 1998). The ISO project used 10 main-belt asteroids as calibration standards. The accuracy achievable with thermophysical modeling was better than 10% with the four primary standards (Ceres, Pallas, Vesta, Herculina) and mostly better than 15% with the remaining six asteroids used as secondary standards (see *Dotto et al.,* 2002).

4.6.2. The Yarkovsky effect. As described above, the temperature distribution over the surface of an asteroid is normally very asymmetric. For example, if the thermal inertia and rotation rate are small, or if the rotation axis is aligned with the Sun, there is a prominent peak in the surface temperature distribution at the subsolar point (equation (5)). Due to the momentum of the thermal photons, there is a net reactive force associated with asymmetric thermal emission that can significantly influence the long-term evolution of an object's orbit. This phenomenon has become known as the Yarkovsky effect (see *Bottke et al.,* 2002). In general there are two components of the Yarkovsky effect, one dependent on rotation, which acts perpendicular to the rotation axis, and the other dependent on the object's orbital motion, which acts along the rotation axis (*Rubincam,* 1995; *Farinella et al.,* 1998; *Vokrouhlický et al.,* 2000). Thermal inertia plays a crucial role in determining the overall net perturbation.

Interest in the Yarkovsky effect has increased recently with the realization that it may help to explain the delivery of small fragments (≤100 m) from the main belt into Earth-crossing orbits by causing them to drift into main-belt resonant orbits (*Farinella et al.,* 1998). Furthermore, thanks to high-precision radar astrometry of near-Earth asteroids, it may soon become possible to detect the orbital drift caused by the Yarkovsky effect. Apart from challenging measurements of minute changes in orbital parameters, confirmation of Yarkovsky-induced orbital drift requires sophisticated thermophysical modeling to facilitate accurate prediction of the effect. In most cases, lack of information on shape and thermal properties precludes calculation of the perturbation to the level of accuracy required. However, for one or two NEAs, for which accurate shape models are available or preliminary estimates suggest a relatively large effect (e.g., 1998 KY_{26} and 6489 Golevka), it may be possible to predict the magnitude of the perturbation to an accuracy commensurate with that of radar astrometry (*Vokrouhlický et al.,*

2000). The Yarkovsky effect may become an important new area of application for thermophysical models.

5. PHYSICAL CHARACTERIZATION OF ASTEROIDS

5.1. Distant Asteroids

Thanks to the improving capabilities of groundbased telescopes and the availability of orbiting infrared observatories, it is now possible to probe more distant and cooler asteroids in the thermal infrared. An important question is the relationship between distant asteroids, such as trans-Neptunian objects (TNOs) and Centaurs, and comets. Thermal-infrared investigations of such objects reveal their albedos, which are an important indicator of surface characteristics and possible similarities to cometary nuclei.

The detection of TNOs in the thermal infrared is particularly challenging. The thermal emission of an object at a heliocentric distance of 30 AU peaks at a wavelength of 40 μm or longer (see Fig. 1), in a region of the spectrum inaccessible from the ground. On the basis of the STM the corresponding flux at 40 μm from an object with a diameter of 100 km is 2.6×10^{-18} W m^{-2} $μm^{-1}$ (1.4 mJy) at the Earth (see Fig. 2). *Thomas et al.* (2000) report a 2.7 σ detection of the TNO 1993 SC of 11.5 ± 4.2 mJy at 90 μm with the Infrared Space Observatory. On the basis of a standard thermal model they estimated a geometric albedo of p_v ~ 0.02 and a diameter of around 328 km. The low albedo is in line with expectations for cometary nuclei. *Jewitt et al.* (2001) have made the first groundbased thermal observation of a TNO, albeit in the submillimeter region. Using the James Clerk Maxwell Telescope on Mauna Kea, Hawai'i, Jewitt et al. observed 20000 Varuna at a wavelength of 850 μm and derived a red geometric albedo of p_R ~ 0.07 and a diameter of about 900 km on the basis of a standard thermal model. The albedo is larger than previously assumed for TNOs but it is uncertain by about a factor of 2, mainly due to lack of knowledge of the lightcurve phase at the time of the observations.

Centaurs have orbits between those of Jupiter and Neptune (i.e., between 5 AU and 30 AU, although the aphelion distances of some exceed 30 AU) and may be objects evolving dynamically from the Kuiper Belt toward becoming short-period comets. A few Centaurs, including 2060 Chiron, actually display cometary activity. *Campins et al.* (1994) interpreted their 10- and 20-μm observations of 2060 Chiron, performed with the NASA 3-m Infrared Telescope Facility and the 4.5-m Multiple Mirror Telescope, in terms of the STM and reported an albedo of about 0.14 and a diameter of some 180 km. Interestingly, the albedo is much higher than would be expected for a cometary nucleus, but some residual contamination from Chiron's coma cannot be ruled out. A similar size was obtained by *Altenhoff and Stumpff* (1995) from thermal observations at 1.2 mm. Other Centaurs for which albedos and sizes have been determined from thermal-infrared observations and application of standard thermal models are 5145 Pholus and 10199 Chariklo. In both cases observations were made at 20 μm with the 3.8-m UK Infrared Telescope. The albedos reported for both objects are p_v ~ 0.045 (*Davies et al.,* 1996; *Jewitt and Kalas,* 1998), a value similar to that found for cometary nuclei, such as Comet P/Halley (e.g., *Keller,* 1990, and references therein). A slightly higher albedo for Chariklo, p_v = 0.055 ± 0.008, was reported by *Altenhoff et al.* (2001) from thermal observations at 1.2 mm.

A few asteroids have retrograde orbits, a characteristic suggestive of a cometary link. One of the first such objects to be discovered is (20461) 1999 LD_{31}, which has perihelion and aphelion distances of 2.38 AU and 46.5 AU respectively and an orbital period of 121 yr. In the lists of the Minor Planet Center, 1999 LD_{31} is an unusual object that does not fit into any existing category. Despite the similarity of its orbit to that of a Halley-type comet, there have been no reports from observers of a comet-like appearance or unusual brightness variability during its perihelion passage. Spectrophotometric observations in the 5–20-μm range with the 10-m Keck I telescope, interpreted on the basis of the STM and similar thermal models (Fig. 5), indicate that 1999 LD_{31} has p_v = 0.03 ± 0.01 and a diameter of some 14 km (*Harris et al.,* 2001). The visible-infrared colors are consistent with those of a D-type asteroid. The perihelion distance is smaller than that of some short-period comets, thus it seems possible that 1999 LD_{31} is an example of a dormant or extinct comet. Thermal-infrared Keck observations of further objects of this type, interpreted on the basis of a standard thermal model, have been reported by *Fernández et al.* (2001); in all cases the albedos are similar to that found for 1999 LD_{31}.

5.2. Near-Earth Asteroids

Could a significant fraction of the NEA population consist of cometary nuclei depleted of volatiles? Current observational evidence from reflection spectra and albedos suggests not. Even the FRM albedos in Table 2 are generally higher than would be expected for cometary nuclei. However, the fraction of the NEA population that has been sampled is still very small and observation selection effects may significantly bias the observed sample against dark objects. A major problem is our ignorance of what a "dead comet" actually looks like. A few NEAs that have optical reflection spectra typical of C-type asteroids appear to have albedos much higher than would be expected from comparison with main-belt C-type asteroids (see Table 2). If, for whatever reason, exhausted cometary nuclei can also have brighter-than-expected surfaces, there may be many examples lurking unrecognized in the NEA population. It appears that S-type NEAs may also have higher albedos on average, than S-types in the main belt (see Table 2 and *Binzel et al.,* 2002). Whether relatively high albedos in the case of NEAs are a real phenomenon indicative, for example, of recently exposed, relatively unweathered surfaces, or other special surface characteristics associated with

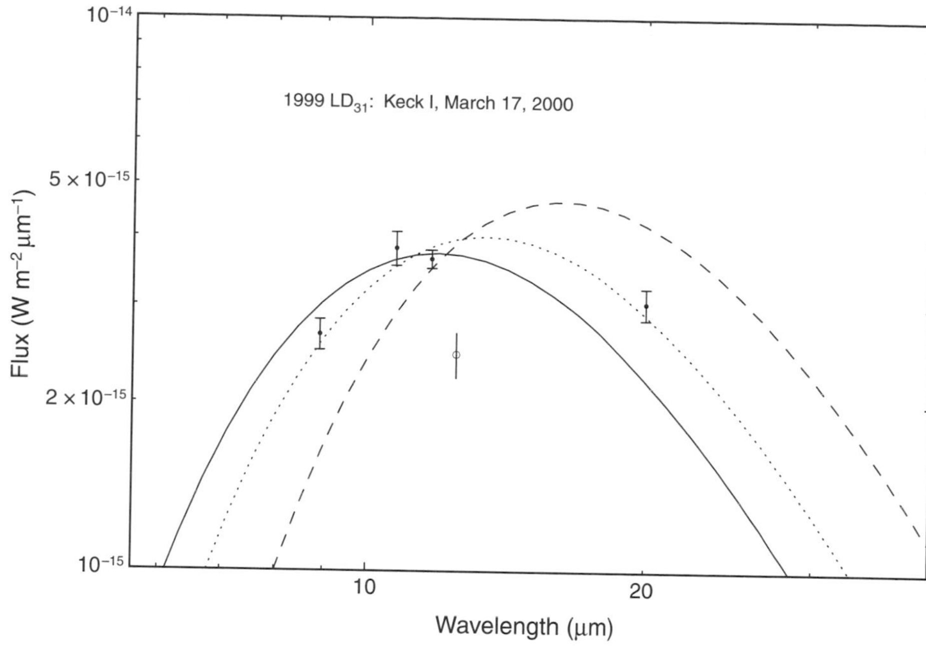

Fig. 5. Model fits to thermal-infrared fluxes of the retrograde asteroid 1999 LD_{31} from observations with the Keck I telescope (*Harris et al.,* 2001). The curves represent the STM (continuous), the FRM (dashed), and the NEATM (dotted). The fit provided by NEATM with η = 1.22 is better than that of the STM, which adopts η = 0.756 (for model fitting the apparently anomalous measurement at 12.5 μm was disregarded). The STM gives a slightly smaller diameter and a slightly larger albedo than the NEATM. The geometry at the time of the observations was: heliocentric distance = 2.696 AU, geocentric distance = 1.873 AU, α = 14.3°.

NEAs or very small asteroids in general, or are the result of inadequacies in the thermal models, is not clear at present. Note, however, that a relatively high average albedo for S-type NEAs would be consistent with the trend to ordinary-chondrite-type reflection spectra with decreasing size observed in the NEA population, which is also attributed to a lack of space weathering of relatively young surfaces (*Binzel et al.,* 2002).

Confirmation of the existence of NEAs with anomalously high albedos may come from radar observations. We note that some anomalies in the spectral type/albedo correspondence of NEAs are already apparent from radar data. The NEAs 1999 GU_3, 6489 Golevka, and 2063 Bacchus all have similar Q-type optical spectra or colors (*Pravec et al.,* 2000; *Hicks et al.,* 1998); however, their albedos, p_v, derived from radar estimates of their sizes and absolute visual magnitudes, could hardly be more diverse: 0.08, 0.15, and 0.56 respectively (*Pravec et al.,* 2000; *Hudson et al.,* 2000; *Benner et al.,* 1999). Recent radar observations of 1566 Icarus indicate a diameter of 0.8 km or less (*Mahapatra et al.,* 1999), which would imply an extremely high value for p_v of more than 0.7. Clearly results such as these need to be verified but if similar surprises turn up frequently in investigations of NEAs, it will become clear that we still have a long way to go in understanding the physical nature of the NEA population.

6. SUMMARY

The STM has proved to be very successful in the determination of diameters and albedos of main-belt asteroids. Standard thermal models are also applicable to investigations of distant asteroids, such as Centaurs and trans-Neptunian objects, and have revealed that such objects, like cometary nuclei, are characterized by very low albedos. With some modifications the STM can be used to provide insight into the physical properties of near-Earth asteroids, although care must be exercised when applying simple models to irregularly shaped objects observed at high solar phase angles. Some C-type NEAs appear to have anomalously high albedos, a potentially important result for the taxonomy and surface properties of NEAs but one that requires confirmation. For detailed interpretation of high-quality thermal-infrared data of individual asteroids, and applications such as the calibration of space instrumentation in the thermal infrared, more sophisticated thermophysical models are required.

The thermophysical models discussed to date in the literature cover a wide range of physical processes. However, the construction of a unified, self-consistent model that takes all relevant effects into account, and can accurately predict and interpret thermal radiation in terms of the physical properties of asteroids, is still a task for the future. A potentially important application of thermophysical modeling is in the prediction of the orbital evolution of small asteroids perturbed by the Yarkovsky effect. Many of the effects discussed in this chapter are also relevant to the interpretation of optical data, where there is a similar need to unify available models. It is important to remember that uncertainties in the treatment of scattered sunlight also limit the accuracy achievable with thermal modeling. The ultimate general-purpose model would be a seamless integration of the physics of scattered and thermally emitted radiation.

Acknowledgments. We thank the reviewers, E. F. Tedesco and T. G. Müller, for their constructive comments. A.W.H. wishes to acknowledge many lively and thought-provoking discussions on the thermal modeling of asteroids with M. Delbó.

REFERENCES

Altenhoff W. J. and Stumpff P. (1995) Size estimate of "asteroid" 2060 Chiron from 250 GHz measurements. *Astron. Astrophys., 293,* L41–L42.

Altenhoff W. J., Menten K. M., and Bertoldi F. (2001) Size determination of the Centaur Chariklo from millimeter-wavelength bolometer observations. *Astron. Astrophys., 366,* L9–L12.

Benner L. A. M., Hudson R. S., Ostro S. J., Rosema K. D., Giorgini J. D., Yeomans D. K., Jurgens R. F., Mitchell D. L., Winkler R., Rose R., Slade M. A., Thomas M. L., and Pravec P. (1999) Radar observations of asteroid 2063 Bacchus. *Icarus, 139,* 309–327.

Bottke W. F. Jr., Vokrouhlický D., Rubincam D. P., and Broz M. (2002) Dynamical evolution of asteroids and meteoroids using the Yarkovsky effect. In *Asteroids III* (W. F. Bottke Jr. et al., eds.), this volume. Univ. of Arizona, Tucson.

Bowell E., Hapke B., Domingue D., Lumme K., Peltoniemi J., and Harris A. W. (1989) Application of photometric models to asteroids. In *Asteroids II* (R. P. Binzel et al., eds.), pp. 524–556. Univ. of Arizona, Tucson.

Brown R. H. (1985) Ellipsoidal geometry in asteroid thermal models: The standard radiometric model. *Icarus, 64,* 53–63.

Buhl D., Welch W. J., and Rea D. G. (1968) Reradiation and thermal emission from illuminated craters on the lunar surface. *J. Geophys. Res., 73,* 5281–5295.

Campins H., Telesco C. M., Osip D. J., Rieke G. H., Rieke M. J., and Schulz B. (1994) The color temperature of (2060) Chiron: a warm and small nucleus. *Astron. J., 108,* 2318–2322.

Clegg P. E. and Carter B. S. (1970) Measurements of lunar radiation in the wavelength range centred at 1.2 mm. *Mon. Not. R. Astron. Soc., 148,* 261–274.

Cohen M., Witteborn F. C., Roush T., Bregman J., and Wooden D. (1998) Spectral irradiance calibration in the infrared. VIII. 5–14 micron spectroscopy of the asteroids Ceres, Vesta, and Pallas. *Astron. J., 115,* 1671–1679.

Davies J. K., Tholen D. J., and Ballantyne D. R. (1996) Infrared observations of distant asteroids. In *Completing the Inventory of the Solar System* (T. W. Rettig and J. M. Hahn, eds.), pp. 97–105. ASP Conf. Series 107.

Dotto E., Barucci M. A., Müller T. G., Storrs A. D., and Tanga P. (2002) Observations from orbiting platforms. In *Asteroids III* (W. F. Bottke Jr. et al., eds.), this volume. Univ. of Arizona, Tucson.

Dunham D. W. and colleagues (2001) International Occultation and Timing Association Web site, version 21 May 2001 (www.anomalies.com/iotaweb/index.htm).

Emery J. P., Sprague A. L., Witteborn F. C., Colwell J. E., Kozlowski R. W. H., and Wooden D. H. (1998) Mercury: Thermal modeling and mid-infrared (5–12 μm) observations. *Icarus, 136,* 104–123.

Farinella P., Vokrouhlický D., and Hartmann W. K. (1998) Meteorite delivery via Yarkovsky orbital drift. *Icarus, 132,* 378–387.

Feierberg M. A. and Witteborn F. C. (1983) Detection of silicate emission features in 8- to 13-μm spectra of main belt asteroids. *Icarus, 56,* 393–397.

Fernández Y. R., Jewitt D. C., and Sheppard S. S. (2001) Low albedos among extinct comet candidates. *Astrophys. J. Lett., 553,* L197–L200.

Fowler J. W. and Chillemi J. R. (1992) IRAS asteroid data processing. In *The IRAS Minor Planet Survey* (E. F. Tedesco, ed.), pp. 17–43. Tech. Rpt. PL-TR-92-2049, Phillips Laboratory, Hanscom Air Force Base, Massachusetts.

Gillett F. C. and Merrill K. M. (1975) 7.5–13.5 micron spectra of Ceres and Vesta. *Icarus, 26,* 358–360.

Goguen J. D. and Sinton W. M. (1985) Characterization of Io's volcanic activity by infrared polarimetry. *Science, 230,* 65–69.

Green S. F., Eaton N., Aitken D. K., Roche P. F., and Meadows A. J. (1985) 8- to 13-μm spectra of asteroids. *Icarus, 62,* 282–288.

Hansen O. L. (1977) An explication of the radiometric method for size and albedo determination. *Icarus, 31,* 456–482.

Hapke B. (1993) *Theory of Reflectance and Emittance Spectroscopy.* Cambridge Univ., Cambridge. 455 pp.

Hapke B. (1996) Applications of an energy transfer model to three problems in planetary regoliths: The solid-state greenhouse, thermal beaming, and emittance spectra. *J. Geophys. Res., 101,* 16833–16840.

Harris A. W. (1998) A thermal model for near-Earth asteroids. *Icarus, 131,* 291–301.

Harris A. W. (2001) Asteroid lightcurve data file, updated March 1, 2001 (cfa-www.harvard.edu/iau/lists/LightcurveDat.html).

Harris A. W. and Davies J. K. (1999) Physical characteristics of near-Earth asteroids from thermal infrared spectrophotometry. *Icarus, 142,* 464–475.

Harris A. W. and Harris A. W. (1997) On the revision of radiometric albedos and diameters of asteroids. *Icarus, 126,* 450–454.

Harris A. W., Davies J. K., and Green S. F. (1998) Thermal infrared spectrophotometry of the near-Earth asteroids 2100 Ra-Shalom and 1991 EE. *Icarus, 135,* 441–450.

Harris A. W., Delbó M., Binzel R., Davies J. K., Roberts J., Tholen D. J., and Whiteley R. (2001) Visible to thermal-infrared spectrophotometry of a possible inactive cometary nucleus. *Icarus, 153,* 332–337.

Heiles C. E. and Drake F. D. (1963) The polarization and intensity of thermal radiation from a planetary surface. *Icarus, 2,* 281–292.

Henderson B. G. and Jakosky B. M. (1994) Near-surface thermal gradients and their effects on mid-infrared emission spectra of planetary surfaces. *J. Geophys. Res., 99,* 19063–19073.

Henderson B. G. and Jakosky B. M. (1997) Near-surface thermal gradients and mid-IR emission spectra: A new model including scattering and application to real data. *J. Geophys. Res., 102,* 6567–6580.

Henderson B. G., Jakosky B. M., and Randall C. E. (1992) A Monte Carlo model of polarized thermal emission from particulate planetary surfaces. *Icarus, 99,* 51–62.

Hicks M. D., Fink U., and Grundy W. M. (1998) The unusual spectra of 15 near-Earth asteroids and extinct cometary candidates. *Icarus, 133,* 69–78.

Hudson R. S. and 26 colleagues (2000) Radar observations and physical model of asteroid 6489 Golevka. *Icarus, 148,* 37–51.

Jämsä S., Peltoniemi J. I., and Lumme K. (1993) Thermal emission from a rough surface: Ray optics approach. *Astron. Astrophys., 271,* 319–325.

Jewitt D. and Kalas P. (1998) Thermal observations of Centaur 1997 CU26. *Astrophys. J. Lett., 499,* L103–L106.

Jewitt D., Aussel H., and Evans A. (2001) The size and albedo of the Kuiper-belt object (20000) Varuna. *Nature, 411,* 446–447.

Johnson P. E., Kemp J. C., Lebofsky M. J., and Rieke G. H. (1983) 10 μm polarimetry of Ceres. *Icarus, 56,* 381–392.

Johnson P. E., Vogler K. J., and Gardner J. P. (1993) The effect of surface roughness on lunar thermal emission spectra. *J. Geophys. Res., 98,* 20825–20829.

Jurgens R. F. and Goldstein R. M. (1976) Radar observations at 3.5 and 12.6 cm wavelength of asteroid 433 Eros. *Icarus, 28,* 1–15.

Kaasalainen M., Mottola S., and Fulchignoni M. (2002) Asteroid models from disk-integrated data. In *Asteroids III* (W. F. Bottke Jr. et al., eds.), this volume. Univ. of Arizona, Tucson.

Keihm S. J. (1984) Interpretation of the lunar microwave brightness temperature spectrum: Feasibility of orbital heat flow mapping. *Icarus, 60,* 568–589.

Keller H. U. (1990) Surface features and activity of the nucleus of Comet Halley. In *Comet Halley Investigations, Results, Interpretations, Volume 2* (J. W. Mason, ed.), pp. 133–145. Ellis Horwood, Chichester.

Lagerkvist C.-I., Harris A. W., and Zappalà V. (1989) Asteroid lightcurve parameters. In *Asteroids II* (R. P. Binzel et al., eds.), pp. 1162–1179. Univ. of Arizona, Tucson.

Lagerros J. S. V. (1996a) Thermal physics of asteroids I: Effects of shape, heat conduction and beaming. *Astron. Astrophys., 310,* 1011–1020.

Lagerros J. S. V. (1996b) Thermal physics of asteroids II: Polarization of the thermal microwave emission from asteroids. *Astron. Astrophys., 315,* 625–632.

Lagerros J. S. V. (1997) Thermal physics of asteroids III. Irregular shapes and albedo variegations. *Astron. Astrophys., 325,* 1226–1236.

Lagerros J. S. V. (1998) Thermal physics of asteroids IV. Thermal infrared beaming. *Astron. Astrophys., 332,* 1123–1132.

Lagerros J. S. V., Müller T. G., Klaas U., and Erikson A. (1999) ISOPHOT polarization measurements of the asteroids (6) Hebe and (9) Metis at 25 μm. *Icarus, 142,* 454–463.

Lebofsky L. A. and Rieke G. H. (1979) Thermal properties of 433 Eros. *Icarus, 40,* 297–308.

Lebofsky L. A. and Spencer J. R. (1989) Radiometry and thermal modeling of asteroids. In *Asteroids II* (R. P. Binzel et al., eds.), pp. 128–147. Univ. of Arizona, Tucson.

Lebofsky L. A., Veeder G. J., Lebofsky M. J., and Matson D. L. (1978) Visual and radiometric photometry of 1580 Betulia. *Icarus, 35,* 336–343.

Lebofsky L. A., Sykes M. V., Tedesco E. F., Veeder G. J., Matson D. L., Brown R. H., Gradie J. C., Feierberg M. A., and Rudy R. J. (1986) A refined "standard" thermal model for asteroids based on observations of 1 Ceres and 2 Pallas. *Icarus, 68,* 239–251.

Lebofsky L. A., Greenberg R., Tedesco E. F., and Veeder G. J. (1988) Infrared lightcurves of asteroids 532 Herculina and 45 Eugenia: Proof of the absence of significant albedo markings. *Icarus, 75,* 518–526.

Lellouch E., Laureijs R., Schmitt B., Quirico E., de Bergh C., Crovisier J., and Coustenis A. (2000) Pluto's non-isothermal surface. *Icarus, 147,* 220–250.

Mahapatra P. R., Ostro S. J., Benner L. A. M., Rosema K. D., Jurgens R. F., Winkler R., Rose R., Giorgini J. D., Yeomans D. K., and Slade M. A. (1999) Recent radar observations of asteroid 1566 Icarus. *Planet. Space Sci., 47,* 987–995.

Mitchell D. L. and De Pater I. (1994) Microwave imaging of Mercury's thermal emission at wavelengths from 0.3 to 20.5 cm. *Icarus, 110,* 2–32.

Mitchell D. L., Hudson R. S., Ostro S. J., and Rosema K. D. (1998) Shape of asteroid 433 Eros from inversion of Goldstone radar Doppler spectra. *Icarus, 131,* 4–14.

Morrison D. (1976) The diameter and thermal inertia of 433 Eros. *Icarus, 28,* 125–132.

Morrison D. and Lebofsky L. A. (1979) Radiometry of asteroids. In *Asteroids* (T. Gehrels, ed.), pp. 184–205. Univ. of Arizona, Tucson.

Mottola S. and 27 colleagues (1997a) Physical model of near-Earth asteroid 6489 Golevka (1991 JX) from optical and infrared observations. *Astron. J., 114,* 1234–1245.

Mottola S., Hahn G., Pravec P., and Šarounová L. (1997b) S/1997 (3671) 1. *IAU Circ. 6680.*

Müller T. G. and Lagerros J. S. V. (1998) Asteroids as far-infrared photometric standards for ISOPHOT. *Astron. Astrophys., 338,* 340–352.

Ostro S. J., Campbell D. B., Chandler J. F., Hine A. A., Hudson R. S., Rosema K. D., and Shapiro I. I. (1991) Asteroid 1986 DA: Radar evidence for a metallic composition. *Science, 252,* 1399–1404.

Pettengill G. H., Shapiro I. I., Ash M. E., Ingalls R. P., Rainville L. P., Smith W. B., and Stone M. L. (1969) Radar observations of Icarus. *Icarus, 10,* 432–435.

Pravec P., Wolf M., and Šarounová L. (1998) Occultation/eclipse events in binary asteroid 1991 VH. *Icarus, 133,* 79–88.

Pravec P., Šarounová L., Benner L. A. M., Ostro S. J., Hicks M. D., Jurgens R. F., Giorgini J. D., Slade M. A., Yeomans D. K., Rabinowitz D. L., Krugly Y. N., and Wolf M. (2000) Slowly rotating asteroid 1999 GU_3. *Icarus, 148,* 589–593.

Pravec P., Wolf M., Šarounová L., Harris A. W., and Davies J. K. (1997) Spin vector, shape, and size of the Amor asteroid (6053) 1993 BW_3. *Icarus, 127,* 441–451.

Redman R. O., Feldman P. A., Pollanen M. D., Balam D. D., and Tatum J. B. (1995) Flux density estimates at millimeter wavelengths of asteroids near opposition from 1996 to 2005. *Astron. J., 109,* 2869–2879.

Rubincam D. P. (1995) Asteroid orbit evolution due to thermal drag. *J. Geophys. Res., 100,* 1585–1594.

Shepard M. K., Benner L. A. M., Ostro S. J., Harris A. W., Rosema K. D., Shapiro I. I., and Chandler J. F. (2000) Radar observations of asteroid 2100 Ra-Shalom. *Icarus, 147,* 520–529.

Spencer J. R. (1990) A rough-surface thermophysical model of airless planets. *Icarus, 83,* 27–38.

Spencer J. R., Lebofsky L. A., and Sykes M. V. (1989) Systematic biases in radiometric diameter determinations. *Icarus, 78,* 337–354.

Tedesco E. F., ed. (1992) *The IRAS Minor Planet Survey.* Tech. Rpt. PL-TR-92-2049. Phillips Laboratory, Hanscom Air Force Base, Massachusetts (online version available at pdssbn.astro.umd.edu/sbnhtml/asteroids/physical_param.html).

Tedesco E. F. and Gradie J. (1987) Discovery of M class objects among the near-Earth asteroid population. *Astron. J., 93,* 738–746.

Thomas N., Eggers S., Ip W.-H., Lichtenberg G., Fitzsimmons A., Jorda L., Keller H. U., Williams I. P., Hahn G., and Rauer H. (2000) Observations of the trans-Neptunian objects 1993 SC and 1996 TL_{66} with the Infrared Space Observatory. *Astrophys. J., 534,* 446–455.

Thomas P. C., Binzel R. P., Gaffey M. J., Zellner B. H., Storrs A. D., and Wells E. (1997) Vesta: Spin pole, size, and shape from HST images. *Icarus, 128,* 88–94.

Veeder G. J., Hanner M. S., Matson D. L., Tedesco E. F., Lebofsky L. A., and Tokunaga A. T. (1989) Radiometry of near-Earth asteroids. *Astron. J., 97,* 1211–1219.

Veverka J. and 32 colleagues (2000) NEAR at Eros: imaging and spectral results. *Science, 289,* 2088–2097.

Vogler K. J., Johnson P. E., and Shorthill R. W. (1991) Modeling the non-grey-body thermal emission from the full moon. *Icarus, 92,* 80–93.

Vokrouhlický D., Milani A., and Chesley S. R. (2000) Yarkovsky effect on small near-Earth asteroids: Mathematical formulation and examples. *Icarus, 148,* 118–138.

Wesselink A. J. (1948) Heat conductivity and nature of the lunar surface material. *Bull. Astron. Inst. Neth., 10,* 351–363.

Winter D. F. and Krupp J. A. (1971) Directional characteristics of infrared emission from the moon. *The Moon, 2,* 279–292.

Observations from Orbiting Platforms

E. Dotto
Istituto Nazionale di Astrofisica Osservatorio Astronomico di Torino

M. A. Barucci
Observatoire de Paris

T. G. Müller
Max-Planck-Institut für Extraterrestrische Physik and ISO Data Centre

A. D. Storrs
Towson University

P. Tanga
Istituto Nazionale di Astrofisica Osservatorio Astronomico di Torino and Observatoire de Nice

Orbiting platforms provide the opportunity to observe asteroids without limitation by Earth's atmosphere. Several Earth-orbiting observatories have been successfully operated in the last decade, obtaining unique results on asteroid physical properties. These include the high-resolution mapping of the surface of 4 Vesta and the first spectra of asteroids in the far-infrared wavelength range. In the near future other space platforms and orbiting observatories are planned. Some of them are particularly promising for asteroid science and should considerably improve our knowledge of the dynamical and physical properties of asteroids.

1. INTRODUCTION

In the last few decades the use of space platforms has opened up new frontiers in the study of physical properties of asteroids by overcoming the limits imposed by Earth's atmosphere and taking advantage of the use of new technologies.

Earth-orbiting satellites have the advantage of observing out of the terrestrial atmosphere; this allows them to be in operation 24 h per day, every day, without moonlight and/or weather limitations. Space observations are not affected by atmospheric absorptions and/or emissions. Also, because of the absence of turbulence, they are not limited by the atmospheric seeing and they are diffraction-limited in a wide wavelength range.

Starting with IRAS and continuing through IUE, ISO, MSX, and HST, the spectral range investigated has been considerably enlarged and the sample of known asteroid diameters and albedos has been greatly increased. This has led to significant advances in our understanding of the physical properties of asteroids, especially larger ones.

The InfraRed Astronomical Satellite (IRAS), which operated from January to November 1983, was a joint project of the United States, United Kingdom, and Netherlands. It performed an unbiased survey of more than 96% of the sky at four infrared bands (12, 25, 60, and 100 μm), allowing determination of the albedos and the diameters for about 1800 asteroids. The results have been widely presented and discussed in the IRAS Minor Planet Survey (*Tedesco et al.,* 1992) and the Supplemental IRAS Minor Planet Survey (*Tedesco et al.,* 2002). This survey has been very important in the new assessment of the asteroid population: The asteroid taxonomy by *Barucci et al.* (1987), its recent extension (*Fulchignoni et al.,* 2000), and an extended study of the size distribution of main-belt asteroids (*Cellino et al.,* 1991) are just a few examples of the impact factor of this survey. IRAS albedo values, together with color indexes determined by the ECAS asteroid survey (*Zellner et al.,* 1985), have provided important input for classifying asteroids by a multivariate statistical method (*Barucci et al.,* 1987). On the basis of the IRAS dataset on asteroid albedos and diameters, *Cellino et al.* (1991) found that the size distribution of asteroids cannot be fitted by a single-exponent power law: Below a critical value of diameter of ~150 km, the size distribution resembles a power law with index $\alpha \sim 1$, while for larger objects the index rises to $\alpha \sim 3$. Moreover, the size distribution changes completely when different semimajor axis regions are considered and when families are taken into account. On the basis of this result, the size distribution of larger bodies seems to be little altered by collisional evolution, while at smaller sizes the apparent discrepancy between the size distributions of family members and nonfamily objects seems to be a subject of research that is interesting and not yet fully understood.

The International Ultraviolet Explorer (IUE) was launched in January 1978, and remained operational until September 1996. During its lifetime IUE acquired ultraviolet spectra (0.23–0.33 μm) of about 50 asteroids between 1978 and 1995. *Roettger and Buratti* (1994) determined the geometric albedo of 45 objects at 0.267 μm and obtained the first ultraviolet phase curves of asteroids. Long-exposure spectra obtained by *A'Hearn and Feldman* (1992) suggested the presence of water escaping from the surface of 1 Ceres. Statistically significant OH emission has been detected in an exposure off the northern limb of the asteroid after perihelion, whereas it was not detected in an exposure off the southern limb of the object before perihelion. The authors argued that the presence of a northern polar cap that changes with time (accumulating frost during winter and dissipating during summer) is compatible with the obtained results. Other results obtained by *Festou et al.* (1991) for 4 Vesta are mentioned in section 4.

The Hipparcos Space Astrometry Mission was an ESA project for measuring positions, distances, motions, brightness, and colors of stars. Between 1989 and 1993 it observed 48 asteroids and pinpointed more than 100,000 stars with an accuracy 200× better than ever before. These data have very good accuracy (~0.01 arcsec) and allowed significant improvement in the orbital parameters of observed asteroids (*Hestroffer et al.,* 1998). The analysis of the gravitational perturbations on the orbits of the 48 minor planets observed by Hipparcos allowed the determination of the largest asteroids' masses (*Viateau and Rapaport,* 1998; *Bange,* 1998).

The Midcourse Space Experiment (MSX) (http://sd-www.jhuapl.edu) was launched in 1996 and carried out observations in a wide wavelength range from ultraviolet to midinfrared (*Mill et al.,* 1994). Three hundred seventy-five main-belt asteroids have been observed by *Price et al.* (2001). *Tedesco et al.* (2001a) presented the results of the MSX Infrared Minor Planet Survey (MIMPS), consisting of the orbital elements of 26,791 asteroids and 332 sightings of 169 different asteroids. Among these were 31 asteroids that were not included in the IRAS observations.

In recent years the most important space platforms for the observation of asteroids have been the Infrared Space Observatory (ISO) and the Hubble Space Telescope (HST). ISO provided a very large sample of infrared data for asteroids, while HST obtained some impressive results in the construction of the geologic maps of some of the biggest asteroids. ISO terminated its operative phase in April 1998. HST is still operational. Since their results over the last decade have provided a significant improvement on the knowledge of the physicochemical nature of asteroids, we devote the rest of this chapter to a description of their characteristics and their principal results on asteroid science.

2. INFRARED SPACE OBSERVATORY

The ISO (*Kessler et al.,* 1996; *Kessler,* 1999), equipped with four sophisticated and versatile scientific instruments, provided astronomers with a facility of unprecedented sensitivity for an exploration of the universe at infrared wavelengths range 2–240 μm. The satellite was a great technical and scientific success, with most of its subsystems operating far better than the specifications and its scientific results impacting practically all fields of astronomy. At a wavelength of 12 μm, ISO was 1000× more sensitive and had 100× better angular resolution than its predecessor, the all-sky-surveying IRAS. During its routine operational phase (from February 4, 1996, to April 8, 1998), ISO made over 26,450 scientific and approximately 4000 calibration observations, ranging from objects in our own solar system right out to the most distant extragalactic sources. In addition to the dedicated observations, ISO also obtained parallel data while other instruments were prime, and serendipitous data during slews of the satellite.

The ISO's operational orbit had a period of just below 24 h, an apogee height of 70,600 km and a perigee height of 1000 km. The lower parts of this orbit were inside Earth's van Allen belts, which reduced the observing time per orbit to about 16 h. The accuracy of the pointing system was at the arcsecond level, with an absolute pointing error of 1.4" and a short-term jitter of less than 0.5". The tracking of solar system objects with ISO was limited to objects with apparent velocities of less than 120"/h. Geometric constraints limited ISO observations to about 15% of the sky at any one time.

The cryogenic system enabled ISO to observe for nearly 29 months. The combination of superfluid and normal He cooled the infrared detectors, the scientific instruments, and parts of the telescope to temperatures of 2–4 K.

2.1. Infrared Space Observatory (ISO) Observations of Asteroids

The ISO performed spectroscopic, photometric, imaging, and polarimetric measurements of ~40 different asteroids at infrared wavelengths between 2 and 240 μm. All the asteroid observations are summarized in Table 1. More details on the programs, including the scientific abstracts of the proposals, can be found in the ISO Data Archive (http://www.iso.vilspa.esa.es).

Several instruments and observing modes were used to observe asteroids. The ISOCAM instrument (*Cesarsky et al.,* 1996; *Cesarsky,* 1999) performed observations of asteroids in two different modes designated as CAM01 (general observation) and CAM04 (spectrophotometry). The long wavelength spectrometer, or LWS (*Clegg et al.,* 1996, 1999), which covered the wavelength range from 43 to 196.7 μm, observed asteroids in mode LWS01 (grating range scan) and LWS02 (grating line scan). The ISOPHOT instrument (*Lemke et al.,* 1996; *Lemke and Klaas,* 1999) with its three subinstruments (PHT-C, PHT-P and PHT-S) observed asteroids in five different modes: photometry in single pointing and staring raster modes (PHT03, PHT22), photometry in scanning/mapping mode (PHT32), spectrophotometry (PHT40), and polarimetry (PHT50). Several asteroids have been observed in "many OBSMODEs" for calibration purposes (*Laureijs et al.,* 2001). The short wavelength spectrometer, or SWS (*de Graauw et al.,* 1996; *de*

TABLE 1. ISO and HST asteroid observations.

Asteroid	ISO Program	ISO Instrument	HST Program	HST Instrument
1 Ceres	THENCRE GALSAT	PHT03 PHT40	1268 Albrecht	FOC (UV) imaging*
	ASALAMA AST_MIN	SWS06	4545 A'Hearn	FOS spectroscopy*
	HLARSON ASTEROID	PHT40 SWS06	5175 Albrecht	FOC (UV) imaging
	LWS_CAL	LWS01 LWS02 LWS99	5842 Stern	FOC (UV) imaging
	BSCHULZ PHT_CAL	many OBSMODEs	8583 Storrs	WFPC2 (unaber.) imaging
	SWS_CAL	SWS06 SWS99		
2 Pallas	THENCRE GALSAT	PHT03 PHT40	8583 Storrs	WFPC2 (unaber.) imaging
	ASALAMA AST_MIN	SWS06		
	HLARSON ASTEROID	PHT40 SWS06		
	LWS_CAL	LWS01 LWS02 LWS99		
	BSCHULZ PHT_CAL	many OBSMODEs		
	SWS_CAL	SWS06 SWS99		
3 Juno	THENCRE GALSAT	PHT03 PHT40		
	BSCHULZ PHT_CAL	many OBSMODEs		
	SWS_CAL	SWS06 SWS99		
4 Vesta	THENCRE GALSAT	PHT03 PHT40	5489 Zellner	WF/PC1 (aber.) imaging
	BESCHULZ VESTALC1	PHT03 PHT22 LWS02	5175 Albrecht	FOC (UV) imaging
	LWS_CAL	LWS01 LWS02 LWS99	6481 Zellner	WFPC2 (unaber.) imaging
	BSCHULZ PHT_CAL	many OBSMODEs	7433 McCarthy	NICMOS (IR) imaging
	SWS_CAL	SWS06 SWS99		
5 Astraea			8583 Storrs	WFPC2 (unaber.) imaging
6 Hebe	TMUELLER AST_POL1/2	PHT50	8583 Storrs	WFPC2 (unaber.) imaging
7 Iris			8583 Storrs	WFPC2 (unaber.) imaging
8 Flora			6559 Storrs	WFPC2 (unaber.) imaging
9 Metis	TMUELLER AST_POL1/2	PHT50	4521 Zellner	WF/PC1 (aber.) imaging
10 Hygiea	THENCRE GALSAT	PHT03 PHT40	6559 Storrs	WFPC2 (unaber.) imaging
	MBARUCCI ASTEROID	PHT03 PHT40		
	MBARUCCI ASTEROI2	SWS06		
	CAM_CAL	CAM01 CAM04		
	LWS_CAL	LWS01 LWS02 LWS99		
	BSCHULZ PHT_CAL	many OBSMODEs		
	SWS_CAL	SWS06 SWS99		
11 Parthenope			6559 Storrs	WFPC2 (unaber.) imaging
13 Egeria	HLARSON ASTEROID	PHT40 SWS06	8583 Storrs	WFPC2 (unaber.) imaging
14 Irene			8583 Storrs	WFPC2 (unaber.) imaging
15 Eunomia			7488 Zappalà	FGS
			8583 Storrs	WFPC2 (unaber.) imaging
18 Melpomene			4764 Weiss	WF/PC1 (aber.) imaging
			5238 Westphal	WF/PC1 (aber.) imaging
			8583 Storrs	WFPC2 (unaber.) imaging
19 Fortuna			4521 Zellner	WF/PC1 (aber.) imaging
20 Massalia	CAM_CAL	CAM01 CAM04	8583 Storrs	WFPC2 (unaber.) imaging
29 Amphitrite			6559 Storrs	WFPC2 (unaber.) imaging
38 Leda			8583 Storrs	WFPC2 (unaber.) imaging
43 Ariadne			7488 Zappalà	FGS
44 Nysa			7488 Zappalà	FGS
			8583 Storrs	WFPC2 (unaber.) imaging
45 Eugenia			8583 Storrs	WFPC2 (unaber.) imaging
46 Hestia	CAM_CAL	CAM01 CAM04	8583 Storrs	WFPC2 (unaber.) imaging
49 Pales			8583 Storrs	WFPC2 (unaber.) imaging
51 Nemausa			8583 Storrs	WFPC2 (unaber.) imaging
52 Europa	THENCRE GALSAT	PHT03 PHT40	8583 Storrs	WFPC2 (unaber.) imaging
	ASALAMA AST_MIN	SWS06		
54 Alexandra	BSCHULZ PHT_CAL	many OBSMODEs	6559 Storrs	WFPC2 (unaber.) imaging
56 Melete	CAM_CAL	CAM01 CAM04		
63 Ausonia			7488 Zappalà	FGS
65 Cybele	THENCRE GALSAT	PHT03 PHT40	8583 Storrs	WFPC2 (unaber.) imaging
	CAM_CAL	CAM01 CAM04		
	BSCHULZ PHT_CAL	many OBSMODEs		

TABLE 1. (continued).

Asteroid	ISO Program	ISO Instrument	HST Program	HST Instrument
77 Frigga	MBARUCCI ASTEROID	PHT03 PHT40		
87 Sylvia			8583 Storrs	WFPC2 (unaber.) imaging
89 Julia			6559 Storrs	WFPC2 (unaber.) imaging
93 Minerva			8583 Storrs	WFPC2 (unaber.) imaging
106 Dione	BSCHULZ PHT_CAL	many OBSMODEs	8583 Storrs	WFPC2 (unaber.) imaging
107 Camilla			8583 Storrs	WFPC2 (unaber.) imaging
109 Felicitas			4521 Zellner	WF/PC1 (aber.) imaging
111 Ate			3744 Zellner	FOS spectroscopy*
114 Kassandra	MBARUCCI ASTEROID	PHT03 PHT40		
	MBARUCCI ASTEROI2	SWS06		
121 Hermione			8583 Storrs	WFPC2 (unaber.) imaging
130 Elektra			8583 Storrs	WFPC2 (unaber.) imaging
144 Vibilia			6559 Storrs	WFPC2 (unaber.) imaging
146 Lucina			4764 Weiss	WF/PC1 (aber.) imaging
150 Nuwa	CAM_CAL	CAM01 CAM04		
182 Elsa			4784 Schenk	FOS spectroscopy*
216 Kleopatra			4764 Weiss	WF/PC1 (aber.) imaging
			7488 Zappalà	FGS
224 Oceana			4784 Schenk	FOS spectroscopy
243 Ida			5633 Zellner	WFPC2 (unaber.) imaging
288 Glauke			8583 Storrs	WFPC2 (unaber.) imaging
308 Polyxo	MBARUCCI ASTEROID	PHT03 PHT40		
	MBARUCCI ASTEROI2	SWS06		
	AFITZSIM DTYPE	PHT40		
313 Chaldaea	BSCHULZ PHT_CAL	many OBSMODEs		
336 Lacadiera	AFITZSIM DTYPE	PHT40		
375 Ursula			8583 Storrs	WFPC2 (unaber.) imaging
433 Eros			4669 Whipple	FGS*
434 Hungaria			4521 Zellner	WF/PC1 (aber.) imaging
444 Gyptis			8583 Storrs	WFPC2 (unaber.) imaging
498 Tokio	AFITZSIM DTYPE	PHT40		
511 Davida	MBARUCCI ASTEROID	PHT03 PHT40		
532 Herculina	BSCHULZ PHT_CAL	many OBSMODEs	4764 Weiss	WF/PC1 (aber.) imaging
			5238 Westphal	WF/PC1 (aber.) imaging
624 Hektor	MBARUCCI ASTEROID	PHT03 PHT40	4764 Weiss	WF/PC1 (aber.) imaging
	MBARUCCI ASTEROI2	SWS06	7488 Zappalà	FGS
674 Rachele			4521 Zellner	WF/PC1 (aber.) imaging
702 Alauda			3744 Zellner	FOS spectroscopy
709 Fringilla			4521 Zellner	WF/PC1 (aber.) imaging
			8583 Storrs	WFPC2 (unaber.) imaging
804 Hispania	CAM_CAL	CAM01 CAM04		
899 Jokaste			4784 Schenk	FOS spectroscopy
911 Agamemnon	MBARUCCI ASTEROID	PHT03 PHT40		
914 Palisana	MBARUCCI ASTEROID	PHT03 PHT40		
944 Hidalgo			4784 Schenk	FOS spectroscopy*
1137 Raissa			4521 Zellner	WF/PC1 (aber.) imaging*
1144 Oda			3744 Zellner	FOS spectroscopy
1172 Aneas	MBARUCCI ASTEROID	PHT03 PHT40		
1220 Crocus			6559 Storrs	WFPC2 (unaber.) imaging
1437 Diomedes	MBARUCCI ASTEROID	PHT03 PHT40		
1980 Tezcatlipoca	AHARRIS ASTEROID	PHT03 PHT32		
		PHT40 CAM01		
2062 Aten	CAM_CAL	CAM01 CAM04		
2201 Oljato			4784 Schenk	FOS spectroscopy*
2411 Zellner			3153 Buie	WF/PC1 (aber.) imaging*
2703 Rodari	CBARBIER ROSAST	PHT03		
3200 Phaethon	AHARRIS ASTEROID	PHT03 PHT32	7884 Campins	NICMOS (IR) imaging
		PHT40 CAM01		and spectroscopy

TABLE 1. (continued).

Asteroid	ISO Program	ISO Instrument	HST Program	HST Instrument
3361 Orpheus			5483 Storrs	WF/PC1 (aber.) imaging*
3671 Dionysus	AHARRIS ASTEROID	PHT03 PHT32 PHT40 CAM01		
3840 Mimistrobell	CBARBIER ROSAST	PHT03		
	MFULCHIG AROSETTA	PHT03 PHT40		
4015 Wilson-Harrington	AHARRIS ASTEROID	PHT03 PHT32 PHT40 CAM01	2432 Zellner	WF/PC1 (aber.) imaging
			7884 Campins	NICMOS (IR) imaging and spectroscopy
4179 Toutatis	AHARRIS ASTEROID	PHT03 PHT32 PHT40 CAM01	4748 Noll	WF/PC1 (aber.) imaging

*Bad data (missed target, too weak, saturated).

Graauw, 1999), covered the wavelength range from 2.38 to 45.2 μm. SWS observations of asteroids were taken in mode SWS06 (grating wavelength range scan). The ISO modes 99 (LWS99 and SWS99) included a manual setting of the instrument commands. Data taken with these modes are nonstandard, used mainly for special calibration purposes.

The ISO observations underwent careful data processing and calibration. This data processing includes several corrections due to the properties of the detector, instrumental effects by electronics and optics, and external effects such as cosmic rays. The effects and their handling in the data processing and calibration procedures are described in the ISO Handbooks (2001) (http://www.iso.vilspa.esa.es).

Photometric and spectroscopic uncertainties were caused mainly by the difficulties in modeling the instrument behavior in space. Detector transients, nonlinearities, and frequent glitches from high-energy particles limited the final accuracy. In the relatively unexplored far-infrared wavelength range the structured and bright infrared background, along with uncertainties from celestial calibration standards, influenced the final data quality. The standards included stars with confidence in the absolute flux level of about 3% in the 2.5- to 35-μm wavelength range and better than 10% at longer wavelengths up to 300 μm (*Cohen et al.,* 1999). Asteroids were used in the far-infrared between 35 and 200 μm as photometric and spectroscopic references. Based on *Müller and Lagerros* (1998), the accuracy of the predicted asteroid fluxes is better than 10% in the wavelength range from 24 to 200 μm for a few large and well-known asteroids. The outer planets Neptune and Uranus were used for calibration at the longer wavelengths in the high flux density range up to 1000 Jy. The model predictions are based on *Griffin and Orton* (1993) and are accurate to approximately 5% for the continuum flux.

Among the ISO data on asteroids there are (1) complete spectra from 2 to 200 μm of bright objects with spectral resolutions of 1000–2000 in the SWS range and about 200 in the LWS range, (2) imaging and photometric asteroid measurements with sensitivities 1000× higher than IRAS at 12 μm, and (3) the first polarimetric measurements of the disk-integrated thermal emission of asteroids at 25 μm.

2.2. Disk-integrated Polarized Thermal Emission of Asteroids

Depending on wavelength, refractive index, and viewing geometry, the thermal-infrared emission from asteroids originates from some depth below the surface. The radiation propagates in the porous regolith toward the surface, where it is refracted according to Fresnel's equations. As a consequence, the thermal emission becomes polarized. Due to symmetry, however, there can be no net polarization in the disk-integrated flux from a circular symmetric temperature distribution. But the nonzero thermal inertia together with the rotation of the object cause an asymmetry in the temperature distribution. The effect increases with phase angle and for elongated asteroids. The contribution of subsurface emission to the total emission is wavelength dependent, with the largest contribution at peak wavelength and beyond.

Observing the polarized thermal emission is an unexplored and potentially very promising technique for studies of the surface regolith and its influence on the thermal emission. Additionally, it facilitates the interpretation of infrared spectra of asteroids. Asteroids 6 Hebe and 9 Metis have been observed at a wavelength of 25 μm with ISOPHOT (*Lagerros et al.,* 1999). Due to the observing geometry and the elongated shapes of both objects, the temperature distribution on the surface was expected to be asymmetric. Consequently it was expected to measure polarized thermal emission, which could then be attributed to surface properties. Unfortunately, for this technique the ISO targets were limited due to strong visibility and brightness constraints.

The scientific interpretation was done on the basis of the polarization code of the ThermoPhysical Model (TPM) developed by *Lagerros* (1996a,b, 1997, 1998). Critical for polarization analysis is the modeling of the small-scale surface roughness, which was approximated by hemispherical segment craters covering a smooth surface. Analytical solutions were used for the multiple scattered solar and thermally emitted radiation inside the craters. A higher surface roughness generally lowers the polarization due to a randomizing of the scattering planes. The predicted degree of linear polarization increases with higher refractive index and with higher absorption coefficient of the surface material.

The model flux densities for 6 Hebe and 9 Metis were in good agreement with the photometric results. No linear polarization was detected, but useful constraints on the properties of the regolith of these objects have been obtained (*Lagerros et al.,* 1999) on the basis of the obtained upper limits and the extended model. The derived detection limits restricted the possible parameter space in surface roughness, refractive index, and thermal inertia. For Hebe the observations were inconclusive since they coincided with a minimum in the polarization curve. The Metis observations favored a low refractive index and high surface roughness.

2.3. Fundamental Thermal Emission Parameters of Main-Belt Asteroids

The thermophysical emission aspects of several asteroids were analysed on the basis of a large uniform pre-ISO database of about 700 individual observations, in the wavelength range 7–200 µm. The TPM developed by *Lagerros* (1996a, 1997, 1998) (see preceding section) was applied to investigate surfaces roughness, heat conduction, and the emissivity over the thermal wavelength range (*Müller and Lagerros,* 1998). The model included aspects of observing geometry, illumination conditions, shape, and spin vector and, if available, physical size and the albedo [more details about the thermal models are reported and discussed in *Harris and Lagerros* (2002)].

The investigations indicated very low levels of heat conduction, expressed in thermal inertias between 5 and 25 J m^{-2} $s^{-0.5}$ K^{-1}. Based on the Ceres observations alone, a thermal inertia of $\Gamma = 15$ J m^{-2} $s^{-0.5}$ K^{-1} was found. This value is much smaller than the lunar value of 50 J m^{-2} $s^{-0.5}$ K^{-1} (*Spencer et al.,* 1989). This is mainly due to the differences between the environment of the asteroid regolith (lower temperature, lower density) and the lunar environment.

Infrared observations show that the thermal emission is directed more strongly toward the solar direction at the expense of the emission at larger phase angles. This beaming effect is largest at midinfrared wavelengths. For example, the 10-µm flux is 20–40% higher than predicted for a smooth Lambertian surface. The thermal infrared beaming parameters for Ceres have been determined together with the thermal inertia, from a least-squares method including all observational infrared data. The derived values are f = 0.6 (fraction of surface covered by craters) and ρ = 0.7 (r.m.s. of the surface slopes). *Lagerros* (1998) compared these values with the default beaming parameter ($\eta = 0.756$) of the Standard Thermal Model (STM) (*Lebofsky and Spencer,* 1989), derived by *Lebofsky et al.* (1986) from observations of 1 Ceres and 2 Pallas. A nice agreement has been found when heat conduction is neglected. With the thermal inertia taken into account, a rougher surface is needed in order to produce the same amount of beaming.

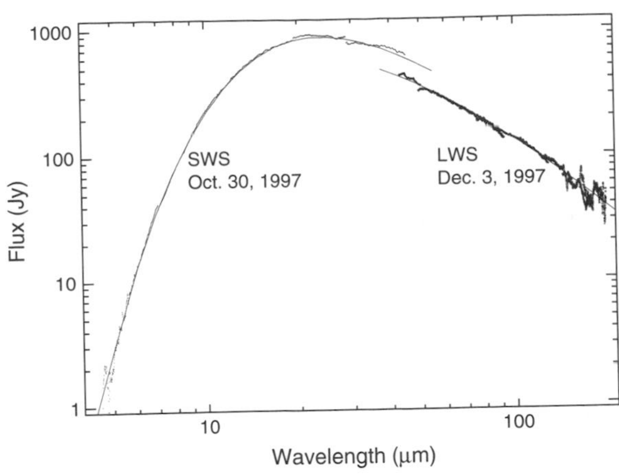

Fig. 1. The asteroid 1 Ceres observed by ISO's LWS and SWS. LWS and SWS observations have been taken at different epochs with different flux level for Ceres. The TPM predictions are overplotted as solid lines.

Figure 1 shows SWS and LWS observations of the asteroid 1 Ceres from two different epochs. The corresponding TPM predictions are overplotted.

Further comparisons with ISO observations of asteroids belonging to different taxonomic classes show that the thermal parameters of Ceres are also valid for other main-belt asteroids (*Müller and Lagerros,* 2002). This means that the TPM parameters allow a reliable prediction of the thermal emission for main-belt asteroids when the shape and H-G values are known. The TPM and the derived thermal parameters are also useful in understanding thermal asteroid spectra where the wavelength-dependent emissivity and the beaming effect are clearly visible. The thermal inertia together with the spin state cause different fluxes before and after opposition due to the morning/afternoon effect. Here again, the TPM allows combined analysis and interpretation of observations. Note that the model does not provide a perfect reproduction of the spectrum of an asteroid. Limitations are introduced by the necessity to model all the physical characteristics of the body with the minimum number of free parameters. The agreement on all flux levels over the full thermal emission spectrum of different asteroids at different epochs confirms that this kind of thermal model is plausible.

2.4. Asteroid Size-Frequency Distribution

During the ISO mission six deep maps of a 12-arcmin square field on the ecliptic were obtained through the ISO 12-µm "IRAS" filter (ISOCAM LW10 band) (*Tedesco and Desert,* 1999). These regions were sampled to an 8-µm equivalent depth of ~0.6 mJy at S/N ratio of 5. Thus, most of the asteroids detected by ISO were more than 200× fainter than those detected by IRAS, whose limiting sensitivity was about 150 mJy at 12 µm.

The ISO survey results (160 ± 40 asteroids per square degree) were modeled with the Statistical Asteroid Model (*Tedesco et al.,* 2001b). This resulted in a value of ~160 ± 10 asteroids per square degree with diameters greater then ~1.7 km. The ISO data seem to suggest that the actual number of kilometer-sized asteroids is significantly greater than expected by some size-frequency distribution models (e.g., *Farinella et al.,* 1992), and in reasonable agreement with the Statistical Asteroid Model.

2.5. Asteroids Serendipitously Seen by the Infrared Space Observatory (ISO)

Many surveys and large observing programs have been conducted by ISO. Surveys close to the ecliptic plane, performed by ISOCAM and ISOPHOT, include many asteroids (*Müller,* 2001). The detection of asteroids at thermal wavelengths has many applications. For well-known objects thermal model predictions can be tested against the measured infrared brightness. In all cases radiometric diameters and albedos can be determined and compared to direct size determinations and/or IRAS results, if available. The midinfrared ISOCAM observations allow further studies of beaming and surface structure properties, while the far-infrared measurements give clues about the as-yet poorly known emissivity behavior of asteroids.

A first analysis of the approximately 40,000 CAM parallel observations between 6 and 15 μm revealed infrared fluxes of about 50 asteroids, most of them not seen by IRAS (T. G. Müller, personal communication, 2002). Many additional objects are expected to be included in other CAM surveys, with some of the deep observations being sensitive enough to detect 1-km-diameter objects in the asteroid midbelt.

ISOPHOT performed observations at 170 μm while the satellite was slewing from one target to the next one. This so-called ISOPHOT Serendipity Survey covered approximately 15% of the sky with a limiting sensitivity of 1 Jy.

The ISO database can now be systematically searched for solar system objects with known orbits. The flux determination allows thermophysical investigations and the derivation of surface parameters. But ISO also saw moving targets where so far no counterpart in the Minor Planet Database is known. For these objects only a statistical analysis is possible at the moment, at least as long as the orbits and the visual brightnesses are not known.

2.6. Spectroscopic and Spectrophotometric Results

The interpretation of spectral features in mid- and far-infrared is not easy, as asteroid surfaces are composed by mixtures of minerals whose absorption features are combined following nonlinear paths. This means that the spectral features of a mineral present even at a few percent level on the surface of an asteroid can dominate the asteroid spectrum. Moreover, asteroid spectra are affected not only by the surface composition but also by several unknown physical parameters, such as density, mineralogy, particle size, packing, and other effects (*Logan et al.,* 1975).

Spectroscopic and spectrophotometric observations of asteroids have been carried out by ISO with PHT-P (filters from 3.29 to 100 μm), PHT-C (filters from 50 to 200 μm), PHT-S (low-resolution spectroscopy in the ranges 2.5–4.9 μm and 5.8–11.6 μm), SWS (high-resolution spectroscopy in the range 2.38–45.2 μm), and LWS (spectroscopy in the range 43–196.7 μm with a resolution of about 200). STM and TPM (see preceding section) have been applied to the data. Subsolar and black-body temperatures have been computed for several observed asteroids (*Barucci et al.,* 1997; *Dotto et al.,* 1999, 2000). Further ISOPHOT data allowed the determination of albedo and diameter of some near-Earth asteroids (*Harris and Davies,* 1999; *Harris and Lagerros,* 2002). *Heras et al.* (2000) published SWS rotationally resolved spectra of 4 Vesta in the wavelength range between 19.4 and 27.6 μm. A short discussion of these results is reported in section 4. *Dotto et al.* (2000) presented and discussed the spectra obtained by PHT-S for five bright asteroids: 1 Ceres, 2 Pallas, 3 Juno, 4 Vesta, and 52 Europa. Since these objects are among the brightest and best-known main-belt asteroids, with spin vector, shape, and size computed with a good precision, their thermal continuum has been modeled using TPM. The spectral features that remain after division for the modeled thermal continuum have been analyzed and discussed in terms of surface composition of the objects.

The main features observable in the midinfrared spectral range, which are diagnostic of the mineralogical composition of the surface of the observed asteroids, can be categorized into three classes: reststrahlen, Christiansen, and transparency features. Reststrahlen bands are fundamental stretching and bending vibration bands that in the case of silicates occur in the 8–25-μm region. The Christiansen feature is associated with the principal molecular vibration band, where the refractive index changes rapidly, and occurs at a wavelength that for silicates is just short of the Si-O stretching vibration bands. A volume scattering feature of fine particulates is the transparency feature, which forms an emissivity trough between the fundamental stretching and bending vibration bands of silicates. In the spectral region where the absorption coefficient decreases, grains become more transparent and volume scattering comes to dominate the scattering process as the particle size is reduced. A detailed description of these spectral features is given in several papers (e.g., *Salisbury,* 1993).

In order to investigate the surface composition of the five brightest asteroids observed by ISO, *Dotto et al.* (2000) compared the observed spectra with the emissivity of meteorites and minerals available in literature and new laboratory spectra of a selected sample of minerals. The thermal emissivities of all the five asteroids show a strong signature around 10–11 μm, suggesting the presence of silicates on the surface of these bodies. As an example, the observed emissivity of the C-type asteroid 52 Europa and of pyroxenes and olivines obtained by laboratory experiments is

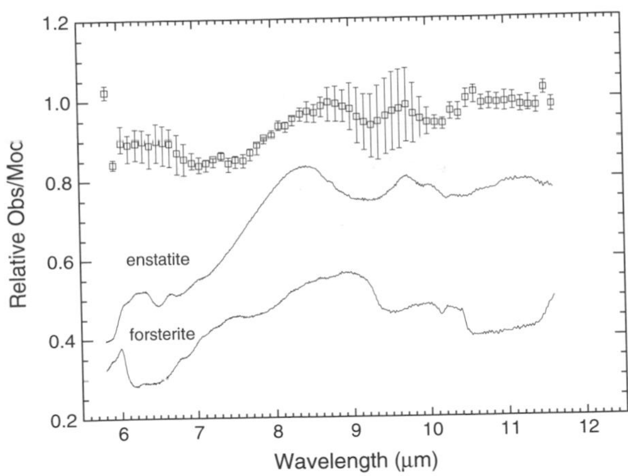

Fig. 2. Relative PHT-S OBS/MOD of 52 Europa compared with the emissivity of a sample of olivine (forsterite) and a sample of pyroxene (enstatite) obtained by laboratory measurements. The spectra are vertically offset for clarity (from *Dotto et al.,* 2000).

shown in Fig. 2. The spectral behavior between 8 and 11 μm suggests the presence of a mixture of pyroxenes and olivines on the surface of 52 Europa. In particular, the 8.8-μm maximum in the spectrum of 52 Europa seems to be consistent with the Christiansen peak of olivine, which occurs at a distinctively long wavelength (*Salisbury et al.,* 1991a). At longer wavelengths, olivine and pyroxene exhibit two major reststrahlen bands separated by a band gap that seem to be consistent with the ISO spectrum of Europa even though the error bars in this region are high.

3. HUBBLE SPACE TELESCOPE

Hubble Space Telescope (HST) is a cooperative program of the National Aeronautics and Space Administration (NASA) and the European Space Agency (ESA). It is a 2.4-m UV/VIS/NIR telescope in a 96-min orbit (about 600 km above ground level). Operating since 1990, HST has provided, up to February 2002, images of more than 60 asteroids (Table 1). HST is a queue-scheduled robot, with little real-time interactive capability. It has an array of instruments that can be (and have been) changed out during refurbishment visits by astronauts. A complete presentation of the up-to-date information on the instrument complement and capabilities is given at the Space Telescope Science Institute (STScI) Web site (http://www.stsci.edu) or the Web site of the European Coordinating Facility (ECF) (http://www.stecf.org). The "HST Primer" provides a good overview of the history and current capabilities of the observatory. In the following sections some general characteristics are given.

3.1. Hubble Space Telescope (HST) Observations of Asteroids

Several HST instruments have been used to observe asteroids. The Faint Object Camera (FOC), built by the European Space Agency, is so sensitive that objects brighter than 21st magnitude must be dimmed by the camera's filter systems to avoid saturating the detectors. With a broadband filter, the brightest object that can be accurately measured is 20th magnitude. The Wide Field Planetary Camera (WFPC2), which replaced the original Wide Field/Planetary Camera (WF/PC1) in December 1993, is a spare instrument developed in 1985 by the Jet Propulsion Laboratory in Pasadena, California. The "heart" of WFPC2 consists of an L-shaped trio of wide-field CCD sensors and a smaller, high-resolution ("planetary") camera tucked in the square's remaining corner. The relay mirrors in WFPC2 are designed to correct for the spherically aberrated primary mirror of the observatory. The Faint Object Spectrograph (FOS), which was one of the original four axial instruments on HST removed during the second servicing mission in February 1997, was used to make spectroscopic observations of astrophysical sources from the near-ultraviolet to the near-infrared (0.115–0.8 μm). The Near Infrared Camera and Multi-Object Spectrometer (NICMOS) provides the capability for infrared imaging and spectroscopic observations of astronomical targets between 0.8 and 2.5 μm. NICMOS exhausted its onboard cryogen load in January 1999. It is hoped that the installation of the NICMOS Cooling System in servicing mission 3B, performed in March 2002, will allow the restoration of at least some of the capability. More details are reported at the STScI Web site. The Fine Guidance Sensors (FGSs) are used to point the telescope and track moving targets, but can also be used as interferometers to determine the size and separation of multiple systems of stars or asteroids.

3.2. Image Restoration

Image restoration requires high signal-to-noise images and a good knowledge of the point spread function (PSF), and only if these are available can the effects of the PSF be at least partially removed from the original image. Some of the most popular techniques to do this are the Richardson-Lucy formulation (*Richardson,* 1972; *Lucy,* 1974) and the Maximum Entropy formulation (*Gull and Daniell,* 1978; *Wu,* 1994). These have been used quite successfully to "clean up" the aberrated HST images and to improve the spatial resolution when the S/N ratio was adequate. These algorithms work quite well in the standard astronomical case of point sources with perhaps a slowly varying background, but in the planetary case of a bright disk with a sharp limb and features on the disk, they can cause trouble. The problem is that the iterative algorithms do a good job of fitting the limb (where most of the digital information is) but overshoot at the edge, causing a "ringing" effect.

A new algorithm ["MISTRAL" (cf. *Conan et al.,* 2000)] gets around this problem. In contrast to the "blind" deconvolution programs that keep information about the real image and the PSF separately, this method is "myopic" in allowing some information exchange. HST has a well-characterized and stable PSF, however, so this aspect of the algorithm is not vital to the reconstruction of planetary images. What is of great interest is that the input parameters can be varied to eliminate the "ringing" effects at the edge of the restoration.

Storrs et al. (2000) used this method to restore WF/PC1 (aberrated) images of asteroid 216 Kleopatra to show the rotation of the asteroid during the time of the observations.

3.3. Hubble Space Telescope (HST) Results

More than 20 observational programs have generated asteroidal data in the HST archive (Table 1); however, some of these are only useful for engineering or missed the target entirely. The study of asteroids with HST has centered mainly on imaging programs, searching for companion objects and surface features, although some searches for cometary emissions (e.g., comae, OH) have been done as well.

All HST data is available through the HST data archive, which can be accessed through the STScI Web site cited above. Care should be taken in the analysis of data from the archive, particularly data from early in the program. The "HST Data Handbook" (also available from the STScI Web site) is an invaluable guide to understanding HST data.

3.3.1. Imaging. Many asteroid observations have been performed by HST using WF/PC1 or the new WFPC2. The method for many of these programs was to image the object with the Planetary Camera using four or five filters. These are chosen to define the blue continuum slope, the width and depth of the 1-μm "silicate feature," and the 0.7-μm hydration feature.

Binzel et al. (1997) and *Storrs et al.* (1998) used image ratios to study the mineralogical variation on the asteroid surfaces. For 4 Vesta, *Binzel et al.* (1997) had enough data to make albedo maps in each filter, which could then be ratioed. These maps are discussed in section 4.

Storrs et al. (1998) attempted a broad imaging analysis, looking for surface variegation in asteroids 8 Flora, 10 Hygiea, 11 Parthenope, 29 Amphitrite, 54 Alexandra, 89 Julia, 144 Vibilia, and 1220 Crocus. Because only one HST orbit was devoted to each asteroid (16 were used for the studies of 4 Vesta described in section 4), only one hemisphere of each of these bodies was imaged. This raises the possibility of phase effects affecting the mineralogical map. No surface variegation was reported, and this was confirmed by a reanalysis of the data presented by *Storrs et al.* (1999a). These papers used the maximum entropy image restoration method, however, which causes some loss of information due to edge effects (see section 3.2).

Evans et al. (1998) report an unusual use of serendipitous asteroidal observations with HST. As part of a program to monitor the performance of WFPC2, they noted the paths of 96 asteroids that were in the field of view of long exposures of fixed targets. HST's orbital motion during these long exposures causes the trails to be curved and an analysis of the size and shape of the curve gives a good estimate of the distance to the asteroid. Most of their objects are too faint to turn up in groundbased surveys, and are probably only a few kilometers in diameter. Evans et al. report a significantly shallower slope to the population distribution ($N \propto H^{0.25}$ vs. $N \propto H^{0.46}$ from *Tedesco et al.,* 2001b) for these small objects than is seen for larger bodies, similar to the result of *Cellino et al.* (1991) discussed in section 1. The small numbers in this sample may not be significant, though.

One of the major projects *Zellner et al.* (1989) expected to be addressed by HST was the search for asteroidal companions. The absence of atmospheric effects on HST images allows diffraction limited operation over a very large field of view. The results to date of searches for companion objects to asteroids are discussed in *Merline et al.* (2002).

The best-known asteroidal companion is Dactyl, orbiting 243 Ida. It was discovered by the *Galileo* spacecraft during its flyby and HST observations were made shortly thereafter, to check for companions farther from Ida and to constrain Dactyl's orbit. No companions were observed, but the observations put constraints on the orbit and thus the mass and bulk density of Ida as published by *Belton et al.* (1995, 1996).

Storrs et al. (1999b) carried out a wide program to look for asteroidal companions. No binary systems were detected.

A limiting case for an asteroidal companion is when the companion is of comparable size to the primary and orbits it very closely, perhaps even touching it. This contact binary situation has been suggested for a variety of asteroids. *Noll et al.* (1995) report HST images of 4179 Toutatis. Their observation was made at a maximum in the asteroid lightcurve yet it shows no extension, indicating that the elongated object [as seen in radar delay-doppler images (*Hudson and Ostro,* 1995)] must have been nearly end-on to Earth at the time.

Recently, observations of 49 Pales showed no companion. The image is elongated, however, and subsequent processing may reveal whether or not this object is a contact binary, as suggested by *Tedesco* (1979). Observations during the same program confirmed the recently discovered companion of 87 Sylvia (IAUC 7588), providing visible colors for the companion object that appear slightly less red than the primary object (IAUC 7590). A companion of 107 Camilla (see *Merline et al.,* 2002) has also been detected.

3.3.2. Fine Guidance Sensor results. The Fine Guidance Sensors (FGS) are used to point HST to the extreme degree of accuracy necessary for long exposures of faint targets. In normal operation, two of the FGSs are used for spacecraft attitude control. The third FGS is free to carry out astrometric and photometric observations, including (1) measuring the relative positions of sources to a precision of a few milliarcseconds, (2) measuring the separations and magnitude differences of binary stars, and (3) measuring stellar angular diameters. Technical details are available in the FGS handbook (available at http://www.stsci.edu). Observations with FGS, in TRANSfer mode, are potentially the most sensitive way of picking up close companions of asteroids. The technique was pioneered by a program targeting 433 Eros. The technical implementation of such observations is daunting, however, and only one program has been successful so far (*Tanga et al.,* 1999). The photometric measurements of the beams resulting from interference, repeated for different values of inclination of the incoming wavefront, provide the response function of the instrument. This produces a characteristic "S-curve" whose features are related to the shape and the size of the target, as projected on the sky plane (Fig. 3). In the case of asteroid observations, the S-curve was sampled with a stepsize of 1–1.5 milliarcsec, over a total length corresponding to 2 arcsec.

In order to retrieve shape and size parameters, the S-curve observed by FGS is compared with a calibration response curve obtained by the observation of pointlike stars. The simplest model, supported by various theoretical arguments about the equilibrium figures of asteroids (see, e.g., *Leone et al.,* 1984), is to assume that the observed bodies are composed of one (or two) triaxial ellipsoids. Systematic fit residuals can give hints about other effects (complex shape, albedo variations) not taken into account. The success of the observation, in terms of shape reconstruction and measurement, depends upon (1) a careful choice of the observing epoch, together with a good knowledge of the spin axis orientation; (2) a compromise between the number of scans that have to be averaged (depending upon magnitude) and the spin rate of the asteroid; and (3) a sufficiently long observation, at several rotational phases. The result of the fit process constrains the model parameters (in this case, the semiaxes of the ellipsoids and their on-sky orientation), with a level of uncertainty determined by the geometry of the observation. The extent of the model can be constrained with a typical 1-σ deviation that can be as low as 1 milliarcsec, an exceedingly good value corresponding to about 1 km at $\Delta = 1.2$ AU.

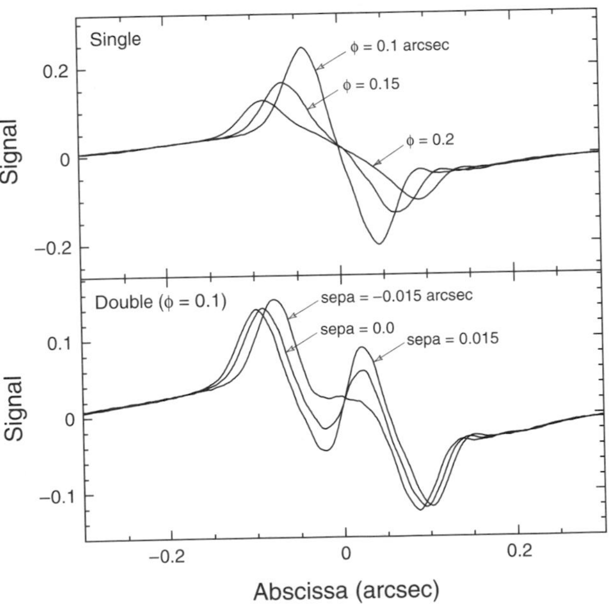

Fig. 3. Synthetic FGS S-curves obtained by convolution of the response curve obtained on a pointlike source with the brightness distribution of a single (upper panel) and double (lower panel) object. Each curve represents the signal as measured along one of the FGS axes. The single body is represented by a uniform disk of size Φ. The double one is composed by two uniform disks (placed along the axis) at different separations (sepa).

Even if an accurate calibration of the absolute scale of the FGS has not been performed, the engineering specifications of the instrument guarantee an error smaller than 10% (*Tanga et al.,* 2001). This characteristic makes the FGS one of the most precise tools available for measurement of asteroid diameters and shapes. Due to sensitivity limitations, however, the FGS gives the best results with bodies whose brightness is not lower than V ~ 13. This requires an average of 4–8 scans in order to reach a S/N ratio better than 10. A larger number of averaged response functions allows in principle the observation of fainter bodies, but it implies that the orientation of the asteroid has not changed significantly during each sequence of scans.

The limit for useful data is probably represented by 624 Hektor, which at magnitude V ~ 14 required about 15 min of observations to reach, by scan averaging, a S/N ratio around 3.

In the case of 63 Ausonia, using average pole coordinates derived from those reported in literature (http://www.astro.uu.se/), it has been shown that the scan shape can be exceedingly well fit at all epochs using an ellipsoid of sizes 151 × 66 × 66 km, with a residual uncertainty of the order of 1 km on axis a and b. The third axis is more poorly constrained due to the nearly pole-on geometry. Furthermore, since the projected on-sky shape is observed, the residual ambiguity on the pole longitude is easily resolved (Hestroffer et al., personal communication, 2002). Of course, given the high level of precision reached, the orientation and amount of the phase effect must be taken into account. Theory based on shape stability criteria for fluid bodies predicted a possible binary object (*Leone et al.,* 1984; *Cellino et al.,* 1985), while Hestroffer et al. (personal communication, 2002) find a single but very elongated shape.

Another interesting case is that of 216 Kleopatra. Despite being another candidate binary, this asteroid was revealed to be a very elongated — but probably single — body by means of radar observations (*Ostro et al.,* 2000). FGS data strongly support this result, suggesting that 216 Kleopatra can be modeled as a shape made up of two ellipsoids, aligned along their semimajor axis and slightly superimposed (Fig. 4). The best fit is obtained with ellipsoids of size 152 × 75 × 36 km and 143 × 70 × 51 km, having a center-to-center separation of 125 km (*Tanga et al.,* 2001).

3.3.3. Spectrographic observations. Only a limited amount of spectrographic work has been done on asteroids with HST. This is primarily because of the small apertures of the spectrographic instruments and the relatively poorly known ephemerides of most asteroids. To acquire an object in an HST spectrograph, its position must be known to within 2 arcsec, as the acquisition is completely robotic.

A good solar spectrum is necessary for determining the reflectance spectrum of asteroids. This has proven to be more problematical than originally thought. The "solar type" spectra used in calibrating NICMOS and other instruments are not solar in detail — the stars were selected for their broad color match to the Sun. For a solar-type star to be faint enough to be a good standard for the sensitive HST instruments, however, it must be far enough away that its

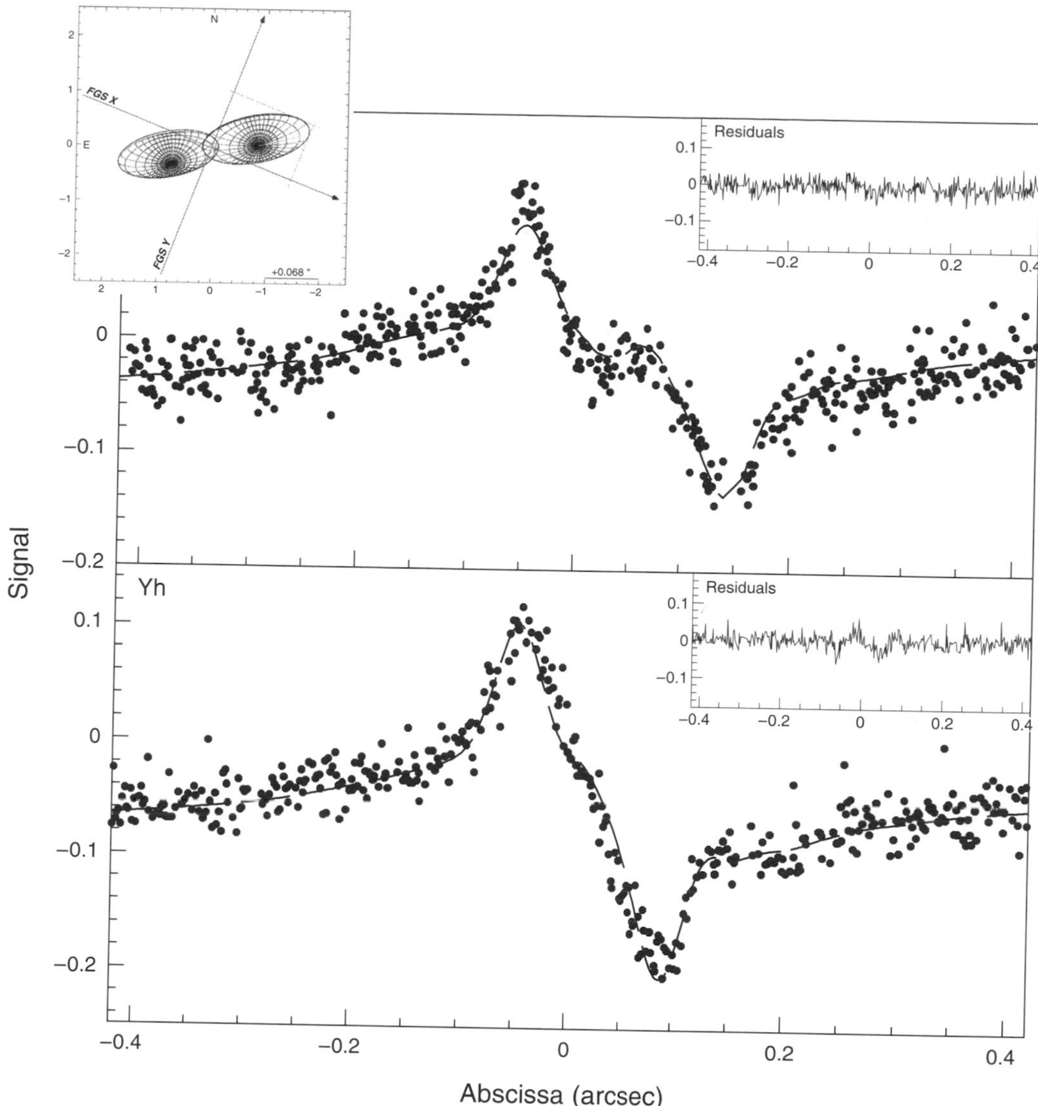

Fig. 4. An example of the response curve obtained in the case of 216 Kleopatra (dots) and model fit (line). The larger separation of the two lobes along the FGS-X direction clearly shows a signature in the center similar to that of a double body (Fig. 3). Small residuals appear on the FGS-Y, and are visible over several scans, thus hinting of the presence of albedo or shape effects. On the left is the model obtained for 216 Kleopatra and its on-sky orientation at the beginning of the observation. The orientation of the FGS-X and FGS-Y axis is also plotted.

spectrum is significantly reddened. Thus the stars discussed in *Colina et al.* (1996) are generally of an earlier type than the Sun, and are reddened to solar colors. This confusion, coupled with potential instrumental artifacts not removed in the standard processing, leads to the recommendation that the observers use a real solar analog spectrum observed by the instrument in the configuration in which they have taken their data. These can be obtained from the HST data archive (http://archive.stsci.edu/).

One program attempted to observe the region just off the poles of 1 Ceres. Unfortunately, limitations in HST target acquisition and tracking capabilities allowed too much scattered light in these spectra. No results were published. Several programs looked for comae and/or surface variegation in asteroids. A preliminary analysis is reported by *Campins et al.* (1999), indicating the presence of some hydrated minerals on 3200 Phaethon.

4. THE SPECIAL CASES OF 4 VESTA AND 10 HYGIEA

Two large asteroids have been thoroughly observed by both ISO and HST: 4 Vesta and 10 Hygiea.

4 Vesta is the third largest asteroid, with an IRAS diameter of 468 km and an albedo $p_V = 0.42$ (*Tedesco et al.,* 1992). Its history is discussed in *Keil* (2002). Here we will point out the results obtained on the basis of observations carried out by IUE, HST, and ISO. Great interest in the

geology of Vesta followed the discovery that howardite, eucrite, and diogenite (HED) meteorites could be samples excavated from its surface. These meteorites have spectral features common to Vesta and to a cluster of small (<10 km) V-type asteroids extending from the region surrounding Vesta to the edge of the 3:1 resonance at 2.5 AU (*Binzel and Xu,* 1993). Impacts may have excavated enough crustal materials to form this widely scattered family of "vestoids": Some of them approach the chaotic region associated with the 3:1 and ν_6 resonances from where fragments can be rapidly transferred to Earth-crossing orbits (*Migliorini et al.,* 1997).

Festou et al. (1991) obtained simultaneous visible (near B filter) and UV (0.17–0.32 µm range) observations of Vesta with IUE. They found at all rotational phases small-amplitude colored features interpreted as part of a progressive extended color change that renders the hemisphere before the lightcurve maximum bluer than the other one.

The most impressive asteroidal work done with HST has been the observation of 4 Vesta. A trio of papers by *Zellner et al.* (1997), *Thomas et al.* (1997a), and *Binzel et al.* (1997) introduced a systematic analysis of Wide Field Planetary Camera data that showed a small dark spot followed by a generally darker region that moved across the disk as the asteroid rotated. The spin pole, size, and shape of the asteroid were constrained, and images across the visible wavelength region and into the near-infrared were analyzed to show mineralogical variegation. Interestingly, this variegation did not closely match the albedo variation (compare the upper and lower panels of Plate 3): The 0.439-µm map shows a dark spot and a dark region that follows it in rotation, while the mineralogical map shows that these are part of a larger region with a depressed visible continuum and a fairly wide 1-µm silicate band. The data came from HST observations carried out with the original (aberrated) HST optics, which smearing when coupled with the changes in resolution over the factor of 2.4 wavelength range may account for the lack of a definitive correlation between albedo and mineralogy. Coupled with the time of observation (the subobserver latitude was +20°), albedo and compositional maps were made of only the region between +50° and –10° latitude. Observations of 4 Vesta at perihelion were made using WFPC2. This correction, coupled with a more equatorward subobserver latitude, gave us the first good view of the southern hemisphere, reported by *Thomas et al.* (1997b). On the basis of these images, *Thomas et al.* (1997a) derived a shape fit by an ellipsoid with semiaxes of 289, 280, 229 (±5) km. Deconvolution of these images, as well as shape fitting to the raw images, showed a large mountain on the south pole of the asteroid, surrounded by a ring of mountains over halfway to the equator from there (see Plate 7). This was interpreted as a huge crater (diameter 460 km) with a large central peak [6 km above the reference spheroid used by *Thomas et al.* (1997b) or about 18 km above the bottom of the crater]. A few smaller depressions in the topography may be craters as well. *Thomas et al.* (1997b) interpret trends in the brightness ratios across the rim and into the southern crater as showing increased Ca content deeper in the crater. On the western hemisphere the HST data indicate a predominance of eucritelike assemblage, with the spectra characteristic of single pyroxene with a modest Ca component. On the eastern hemisphere the spectra indicate the presence of a diogenite-like component (a low-Ca pyroxene) and an olivine component (*Binzel et al.,* 1997). The discovery of substantial impact excavation on Vesta (the crater excavated about 1% of the volume of 4 Vesta) is consistent with the idea for the origin of basaltic achondrite HED meteorites. There are thus convincing observational and dynamical arguments to suggest that Vesta is the actual parent body for the suite of these meteorites, which represent about 6% of all meteorites falling on Earth. As such, Vesta would represent one of the few known solar system bodies for which actual rock samples are available in terrestrial laboratories.

Vesta was also observed by ISO with SWS (*Heras et al.,* 2000) and PHT-S (*Dotto et al.,* 2000). *Heras et al.* (2000) analyzed rotationally resolved SWS spectra between 19.4 and 27.6 µm and identified spectral features associated with the presence of olivine and pyroxene silicate on the surface of this object. They concluded that olivines are dominant on the eastern hemisphere, while pyroxenes, which are evident at every rotational aspect, become dominant on the western hemisphere. *Dotto et al.* (2000) conducted a detailed investigation of the spectral features of the 5.8–11.6-µm PHT-S spectrum. To model the thermal continuum they used the TPM described earlier. The comparison between the obtained spectrum and laboratory spectra of known minerals and meteorites shows that the observed structure around 9.1 µm seems to be compatible with the presence of olivines on the surface of Vesta. This result is compatible with the data obtained by *Gaffey* (1997), who suggested that the surface of Vesta is composed by a dark howardite or polymict eucrite assemblage. Bright regions on one hemisphere would be produced by impacts that exposed brighter diogenite and olivine.

Asteroid 10 Hygiea is the fourth biggest asteroid with an IRAS diameter of 407 km and an albedo p_V = 0.07 (*Tedesco et al.,* 1992). Hygiea rotates around its principal axis in a retrograde sense with a period of 27.63 ± 0.02 h (*Michalowski et al.,* 1991; *Erikson and Magnusson,* 1993; *López-González and Rodríguez,* 2000). *Ragazzoni et al.* (2000) found a shape with an average diameter of 444 ± 35 km and a semimajor axis ratio a/b = 1.11 using speckle interferometry. Hygiea is classified as a C-type asteroid. Aqueous alteration products have been detected on its surface by *Jones et al.* (1990) and *Vilas* (1994) based on absorption features at 3.0 µm and 0.7 µm. Other observations in the visible region, performed at different epochs, did not detected the 0.7-µm absorption feature (*Bus,* 1999; *Fornasier et al.,* 1999). These results may imply composition variation in the surface, not confirmed by *Mothé-Diniz et al.* (2001).

HST observations of Hygiea have been presented and discussed by *Storrs et al.* (1999a). They saw only one side

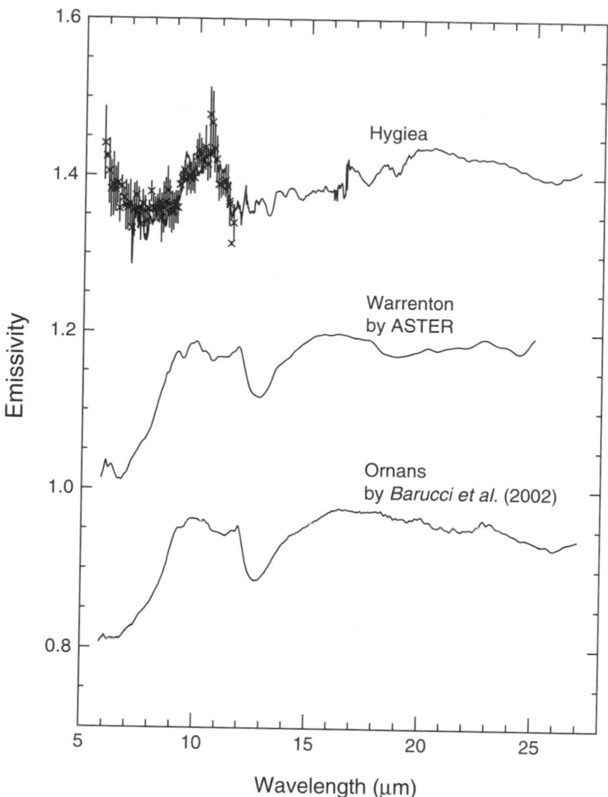

Fig. 5. Comparison between PHT-S and SWS spectra of 10 Hygiea and laboratory spectra of Ornans (by *Barucci et al.,* 2002) and Warrenton (by the ASTER database at http://speclib.jpl.nasa.gov). The spectra are vertically offset for clarity (from *Barucci et al.,* 2002).

of the asteroid (during one HST visibility period of about 45 min). They observed no surface markings, and did not detect any companion down to a limit seven magnitudes fainter than the primary asteroid. The WFPC2 images were circular, and some significant brightness changes during the sequence of exposures suggest rotation in agreement with the published rotational solutions. Further analysis of the data is in progress.

The ISO spectroscopic observations of Hygiea have been performed with PHT-S and SWS up to 11.6 and 45 μm respectively (*Barucci et al.,* 2002). TPM has been used to model the thermal continuum. There were no matches between the observed spectra and those of minerals obtained in the laboratory. The comparison with meteorite spectra shows a similarity to spectra of the carbonaceous chondrite meteorites. This confirms previous work in which C-type asteroids were associated with this type of meteorite. The comparison with all the available laboratory spectra (*Barucci et al.,* 2002) shows some similarities between Hygiea and Ornans and Warrenton (CO3 meteorites) at small grain size (Fig. 5). In this spectral range the most diagnostic feature is the Christiansen feature. As shown by *Salisbury et al.* (1991b), the Christiansen feature of carbonaceous chondrites occurs at a longer wavelength than for the other meteorite types. From the comparison of the Hygiea spectrum with the meteorite emissivities it has been found that the feature of Hygiea at about 9.3 μm seems to be consistent with the Christiansen peak of Ornans and Warrenton. The analogy with these meteorites is also supported by the comparison of the transparency features around 13 and at 26 μm.

These CO3 meteorites seem to show the presence of aqueous alteration processes (*Zolensky and McSween,* 1988), even if the possible products of aqueous alteration within CO chondrites have received limited attention. In fact, although present in these meteorite classes, they are not as pervasive as in the CIs and CMs.

If 10 Hygiea is really compatible with CO meteorites, this would imply that it is a "primitive" object that has undergone some metamorphism.

5. CONCLUSIONS AND FUTURE WORK

In the last decade orbiting platforms have provided important information about the physical properties of asteroids. The data on asteroids collected by ISO, because of their uniqueness and ISO's wide spectroscopic possibilities, have provided important information on the asteroid surface composition. A large sample of infrared data of asteroids up to 200 μm has been collected and several diameters and albedos have been computed.

Important results have been obtained by HST, even if many of the possibilities mentioned by *Zellner et al.* (1989) have not yet been accomplished. For example, the high-resolution mapping discussed by *Zellner et al.* has only been done for 4 Vesta. Programs are currently underway to study the largest main-belt asteroids and transneptunian objects. Several instruments that may be useful for asteroidal observations (e.g., the Advanced Camera for Surveys, the Wide Field Camera 3, and the NICMOS Cooling System) are planned for installation in the near future. In particular, the Wide Field Camera 3, scheduled to replace WFPC2 in 2003, will have both CCD detectors for the UV and visible wavelength region, as well as infrared sensitive detectors to give it coverage from 0.2 to 1.8 μm. Combination studies with both the visible and infrared channels will allow complete coverage in both imaging and low-resolution spectroscopy over the entire 1-μm "silicate" feature, and part of the 2-μm feature.

Orbiting platforms, whose advantage is the ability to observe outside the terrestrial atmosphere, gave us access to a wide range of wavelengths and provided the possibility of better analyzing the physical characteristics of asteroids. The identification of minerals on the surface of an asteroid is made easier by the analysis of their spectral features occurring at different wavelengths. Although this analysis is neither easy nor unique and further uncertainties are introduced by thermal models and laboratory experiments, the increase in the amount of collected data obliges us to improve our techniques to interpret the obtained information, and this is a challenging task for solving the puzzle of physical properties of asteroids.

Several new space platform projects (e.g., SIRTF, ASTRO-F, FIRST-Herschel, NGST, and GAIA) are planned to be operational within the near future and will allow us to update our knowledge of the dynamical and physical properties of asteroids. The Space Infrared Telescope Facility (SIRTF) (http://ssc.ipac.caltech.edu), which is the final element in NASA's "Great Observatories" program, is scheduled for launch in December 2002. It will perform spectroscopy and radiometry of small bodies of the solar system from the asteroid main belt, through the Trojan clouds, to the Kuiper Disk and comets. The Herschel Space Observatory (http://astro.estec.esa.nl/), the European Space Agency's fourth "Cornerstone Mission," will perform photometry and spectroscopy in the 60–670-μm range. Its launch is scheduled for early 2007 and it is planned as a three-year mission. The Next Generation Space Telescope (NGST) (http://www.ngst.nasa.gov/) is scheduled to be launched in 2009 in a L_2 orbit to replace HST. It will be a 6-m-class telescope equipped with cameras and spectrographs working between 0.6 and 28 μm. A very promising future space platform is the ESA space astrometry mission GAIA (http://astro.estec.esa.nl/), which is scheduled to be launched around 2010–2012 and will be operation for five years. Present estimates are that GAIA will detect between 10^5 and 10^6 asteroids, compared with about 20,000 currently known. The study of the physical properties of a wide sample of asteroids, including masses, densities, sizes, shapes, and taxonomic classes, is scheduled (*de Boer et al.,* 2000).

Acknowledgments. The authors thank S. D. Price and P. S. Hardersen for their careful reviews. They also wish to acknowledge D. Hestroffer (IMCCE, Paris) for his contribution to the illustration of the FGS section, useful discussions, and his work in FGS data analysis.

REFERENCES

A'Hearn M. F. and Feldman P. D. (1992) Water vaporization on Ceres. *Icarus, 98,* 54–60.

Bange J. (1998) An estimation of the mass of asteroid 20-Massalia derived from the HIPPARCOS minor planets data. *Astron. Astrophys., 340,* L1–L4.

Barucci M. A., Capria M. T., Coradini A., and Fulchignoni M. (1987) Classification of asteroids using G-mode analysis. *Icarus, 72,* 304–324.

Barucci M. A., Crovisier J., Doressoundiram A., Encrenaz T., Knacke R. F., Lellouche E., Fulchignoni M., Dotto E., and De Sanctis M. C. (1997) First ISO results on asteroids. In *First ISO Workshop on Analytical Spectroscopy* (A. M. Heras et al., eds.), p. 251. ESA SP-419.

Barucci M. A., Dotto E., Brucato J. R., Müller T. G., Morris P., Doressoundiram A., Fulchignoni M., De Sanctis M. C., Owen T., Crovisier J., Le Bras A., Colangeli L., and Mennella V. (2002) 10 Hygiea: ISO infrared observations. *Icarus, 156,* 202–210.

Belton M., Chapman C., Thomas P., Davies M., Greenberg R., Klaasen K., Byrnes D., D'Amario L., Synnott S., Merline W., Petit J.-M., Storrs A., and Zellner B. (1995) The bulk density of asteroid 243 Ida from Dactyl's orbit. *Nature, 374,* 785–788.

Belton M. J. S., Mueller B. E. A., D'Amario L. A., Byrnes D. V., Klaasen K. P., Synnott S., Breneman H., Johnson T. V., Thomas P. C., Veverka J., Harch A. P., Davies M. E., Merline W. J., Chapman C. R., Davis D., Denk T., Neukum G., Petit J.-M., Greenberg R., Storrs A., and Zellner B. (1996) The discovery and orbit of 1993 (243)1 Dactyl. *Icarus, 120,* 185–199.

Binzel R. P. and Xu S. (1993) Chips off of asteroid 4 Vesta — Evidence for the parent body of basaltic achondrite meteorites. *Science, 260,* 186–191.

Binzel R. P., Gaffey M. J., Thomas P. C., Zellner B. H., Storrs A. D. and Wells E. N. (1997) Geologic mapping of Vesta from 1994 Hubble Space Telescope images. *Icarus, 128,* 95–103.

Bus B. (1999) Compositional structure in the asteroid belt: Results of a spectroscopic survey. Ph.D. thesis, Massachusetts Institute of Technology, Cambridge..

Campins H., McCarthy D., Kern S., Weaver H. A., and Brown R. H. (1999) The 1–2.5 micron spectrum of 3200 Phaethon observed with HST's NICMOS. *Bull. Am Astron. Soc., 31,* 1121.

Cellino A., Pannunzio R., Zappalà V., Farinella P., and Paolicchi P. (1985) Do we observe light curves of binary asteroids? *Astron. Astrophys., 144,* 355–362.

Cellino A., Zappalà V., and Farinella P. (1991) The size distribution of main-belt asteroids from IRAS data. *Mon. Not. R. Astron. Soc., 253,* 561–574.

Cesarsky C. J. (1999) ISOCAM: First assessment after the end of the mission. In *The Universe as Seen by ISO* (P. Cox and M. F. Kessler, eds.), pp. 45–49. ESA SP-427.

Cesarsky C. J. and 65 colleagues (1996) ISOCAM in flight. *Astron. Astrophys., 315,* L32–L37.

Clegg P. E. and 62 colleagues (1996) The ISO long wavelength spectrometer. *Astron. Astrophys., 315,* L38.

Clegg P. E. and the LWS Consortium (1999) The ISO long wavelength spectrometer: Description, performance and highlights. In *The Universe as Seen by ISO* (P. Cox and M. F. Kessler, eds.), pp. 39–43. ESA SP-427.

Cohen M., Walker R. G., Carter B., Hammersley P., Kidger M., and Noguchi K. (1999) Spectral irradiance calibration in the infrared. X. A self-consistent radiometric all-sky network of absolutely calibrated stellar spectra. *Astron. J., 117,* 1864–1889.

Colina L., Bohlin R. C., and Castelli F. (1996) The 0.12–2.5 micron absolute flux distribution of the Sun for comparison with solar analog stars. *Astron. J., 112,* 307–315.

Conan J.-M., Fusco T., Mugnier L. M., and Marchis F. (2000) MISTRAL: Myopic deconvolution method applied to ADONIS and to simulated VLT-NAOS images. *ESO Messenger, 99,* 38–45.

de Boer K. S. and the GAIA Science Advisor Group (2000) *GAIA: Composition, Formation and Evolution of the Galaxy. Report on the Concept and Technology Study.* ESA-SCI(2000)4.

de Graauw Th. (1999) Summary of ISO SWS performance and science highlights. In *The Universe as Seen by ISO* (P. Cox and M. F. Kessler, eds.), pp. 31–37. ESA SP-427.

de Graauw Th. and 61 colleagues (1996) Observing with the ISO Short Wavelength Spectrometer. *Astron. Astrophys., 315,* L49–L54.

Dotto E., Barucci M. A., Crovisier J., Doressoundiram A., Encrenaz Th., Fulchignoni M., Knacke R., Lellouch E., Morris P. W., Müller T. G., and Owen T. (1999) ISO observations of asteroids. In *The Universe as Seen by ISO* (P. Cox and M. F.

Kessler, eds.), pp. 165–168. ESA SP-427.

Dotto E., Müller T. G., Barucci M. A., Encrenaz Th., Knacke R. F., Lellouch E., Doressoundiram A., Crovisier J., Brucato J. R., Colangeli L., and Mennella V. (2000) ISO results on bright Main Belt asteroids: PHT-S observations. *Astron. Astrophys., 358,* 1133–1141.

Erikson A. and Magnusson P. (1993) Pole determinations of asteroids. *Icarus, 103,* 62–66.

Evans R. W., Stapelfeldt K. R., Peters D. P., Trauger J. T., Padgett D. L., Ballester G. E., Burrows C. J., Clarke J. T., Crisp D., Gallagher J. S., Griffiths R. E., Grillmair C., Hester J. J., Hoessel J. G., Holtzmann J., Krist J., McMaster M., Meadows V., Mould J. R., Ostrander E., Sahai R., Scowen P. A., Watson A. M., and Westphal J. (1998) Asteroid trails in Hubble Space Telescope. *Icarus, 131,* 261–282.

Farinella P., Davis D. R., Cellino A., and Zappalà V. (1992) The collisional lifetime of asteroid 951 Gaspra. *Astron. Astrophys., 257,* 329.

Festou M. C., Stern S. A., and Tozzi G. P. (1991) Asteroid 4 Vesta — Simultaneous visible and ultraviolet IUE observations. *Icarus, 94,* 218–231.

Fornasier S., Lazzarin M., Barbieri C., and Barucci M. A. (1999) Spectroscopic comparison of aqueous altered asteroids with CM2 carbonaceous chondrite meteorites. *Astron. Astrophys. Suppl. Ser., 135,* 65–73.

Fulchignoni M., Birlan M., and Barucci M. A. (2000) The extension of the G-mode asteroid taxonomy. *Icarus, 146,* 204–212.

Gaffey M. J. (1997) Surface lithologic heterogeneity of asteroid 4 Vesta. *Icarus, 127,* 130–157.

Griffin M. J. and Orton G. S. (1993) The near-millimeter brightness temperature spectra of Uranus and Neptune. *Icarus, 105,* 537.

Gull F. and Daniell G. J. (1978) Image reconstruction from incomplete and noisy data. *Nature, 272,* 686–690.

Harris A. W. and Davies J. K. (1999) Physical characteristics of near-Earth asteroids from thermal infrared spectrophotometry. *Icarus, 142,* 464–475.

Harris A. W. and Lagerros J. S. V. (2002) Asteroids in the thermal infrared. In *Asteroids III* (W. F. Bottke Jr. et al., eds.), this volume. Univ. of Arizona, Tucson.

Heras A. M., Morris P. W., Vandenbussche B., and Müller T. G. (2000) Asteroid 4 Vesta as seen with the ISO short wavelength spectrometer. In *Thermal Emission Spectroscopy and Analysis of Dust, Disks, and Regoliths* (M. L. Sitko et al., eds.), pp. 205–212. ASP Conference Series 196.

Hestroffer D., Morando B., Hog E., Kovalevsky J., Lindegren L., and Mignard F. (1998) The HIPPARCOS solar system objects catalogues. *Astron. Astrophys., 334,* 325–336.

Hudson R. S. and Ostro S. J. (1995) Shape and non-principal axis spin state of asteroid 4179 Toutatis. *Science, 270,* 84–86.

Jones T. D., Lebofsky L. A., Lewis J. S., and Marley M. S. (1990) The composition and origin of the C, P, and D asteroids: Water as a tracer of thermal evolution in the outer belt. *Icarus, 88,* 172–192.

Keil K. (2002) Geological history of asteroid 4 Vesta: The "smallest terrestrial planet." In *Asteroids III* (W. F. Bottke Jr. et al., eds.), this volume. Univ. of Arizona, Tucson.

Kessler M. F. (1999) The ISO mission: Past and future. In *The Universe as Seen by ISO* (P. Cox and M. F. Kessler, eds.), pp. 23–29. ESA SP-427.

Kessler M. F., Steinz J. A., Anderegg M. E., Clavel J., Drechsel G., Estaria P., Faelker J., Riedinger J. R., Robson A., Taylor B. G., and Ximenez de Ferran S. (1996) The Infrared Space Observatory (ISO) mission. *Astron. Astrophys., 315,* L27–L31.

Lagerros J. S. V. (1996a) Thermal physics of asteroids. I. Effects of shape, heat conduction and beaming. *Astron. Astrophys., 310,* 1011–1020.

Lagerros J. S. V. (1996b) Thermal physics of asteroids. II. Polarization of the thermal microwave emission from asteroids. *Astron. Astrophys., 315,* 625.

Lagerros J. S. V. (1997) Thermal physics of asteroids. III. Irregular shapes and albedo variegations. *Astron. Astrophys., 325,* 1226–1236.

Lagerros J. S. V. (1998) Thermal physics of asteroids. IV. Thermal infrared beaming. *Astron. Astrophys., 332,* 1123–1132.

Lagerros J. S. V., Müller T. G., Klaas U., and Erikson A. (1999) ISOPHOT polarization measurements of the asteroids (6) Hebe and (9) Metis at 25 micron. *Icarus, 142,* 454.

Laureijs R. J., Klaas U., Richards P. J., Schulz B., and Abraham P. (2001) *The ISO Handbook Volume V: PHT — The Imaging Photo-Polarimeter.* SAI-99-069/Dc, Version 1.2, July 2001 (available on line at http://www.iso.vilspa.esa.es).

Lebofsky L. A. and Spencer J. R. (1989) Radiometry and thermal modelling of asteroids. In *Asteroids II* (R. P. Binzel et al., eds.), pp. 128–147. Univ. of Arizona, Tucson.

Lebofsky L. A., Sykes M. V., Tedesco E. F., Veeder G. J., Matson D. L., Brown R. H., Gradie J. C., Feierberg M. A., and Rudy R. J. (1986) A refined 'standard' thermal model for asteroids based on observations of 1 Ceres and 2 Pallas. *Icarus, 68,* 239–251.

Lemke D. and Klaas U. (1999) ISOPHOT — Performance, results and outlook. In *The Universe as Seen by ISO* (P. Cox and M. F. Kessler, eds.), pp. 51–60. ESA SP-427.

Lemke D. and 48 colleagues (1996) ISOPHOT — capabilities and performance. *Astron. Astrophys., 315,* L64–L70.

Leone G., Paolicchi P., Farinella P., and Zappalà V. (1984) Equilibrium models of binary asteroids. *Astron. Astrophys., 140,* 265–272.

Logan L. M., Hunt G. R., and Salisbury J. W. (1975) The use of mid-infrared spectroscopy in remote sensing of space targets. In *Infrared and Raman Spectroscopy of Lunar and Terrestrial Minerals* (C. Karr Jr., ed.), pp. 117–142. Academic, New York.

López-González M. J. and Rodríguez E. (2000) Lightcurves of 10 Hygiea, 241 Germania and 509 Iolanda. *Astron. Astrophys. Suppl. Ser., 145,* 255–261.

Lucy L. B. (1974) An iterative technique for the rectification of observed distributions. *Astron. J., 79,* 745–754.

Merline W. J., Weidenschilling S. J., Durda D. D., Margot J.-L., Pravec P., and Storrs A. D. (2002) Asteroids *do* have satellites. In *Asteroids III* (W. F. Bottke Jr. et al., eds.), this volume. Univ. of Arizona, Tucson.

Michałowski T., Velichko F. P., Lindgren M., Oja T., Lagerkvist C.-I., and Magnusson P. (1991) The spin vector of asteriod 10 Hygiea. *Astron. Astrophys. Suppl. Ser., 91,* 53–59.

Migliorini F., Morbidelli A., Zappalà V., Gladman B. J., Bailey M. E., and Cellino A. (1997) Vesta fragments from ν_6 and 3:1 resonances: Implications for V-type NEAs and HED meteorites. *Meteoritics & Planet. Sci., 32,* 903–916.

Mill J. D., O'Neil R. R., Price S., Romick G. J., Uy O. M., Gaposchkin E. M., Light G. C., Moore W. W., Murdock T. L., and Stair A. T. (1994) Midcourse space experiment: Introduction to the spacecraft, instruments, and scientific objectives. *J. Spacecraft Rockets, 31,* 900–907.

Mothé-Diniz T., Di Martino M., Bendjoya Ph., Doressoundiram

A., and Migliorini F. (2001) Rotationally resolved spectra of 10 Hygiea and a spectroscopic study of the Hygiea family. *Icarus, 152,* 117–126.

Müller T. G. (2001) Asteroids in the infrared — Serendipitous observations with ISO. *Planet. Space Sci., 49,* 787–791.

Müller T. G. and Lagerros J. S. V. (1998) Asteroids as far-infrared photometric standards for ISOPHOT. *Astron. Astrophys., 338,* 340–352.

Müller T. G. and Lagerros J. S. V. (2002) Asteroids as calibrations standards in the thermal infrared for space observatories. *Astron. Astrophys., 381,* 324–339.

Noll K. S., Weaver H. A., Storrs A. D., and Zellner B. (1995) Imaging of asteroid 4179 Toutatis with Hubble Space Telescope. *Icarus, 113,* 353–359.

Ostro S. J., Hudson R. S., Nolan M. C., Margot J.-L., Scheeres D. J., Campbell D. B., Magri C., Giorgini J. D., and Yeomans D. K. (2000) Radar observations of asteroid 216 Kleopatra. *Science, 288,* 836–839.

Price S. D., Egan M. P., Walker R., Noah P., Tedesco E. F., Carey S. J., Mizuno D., Kuchar T. A., Murdock T., Barker E., and Jayaraman S. (2001) MSX images and photometry of small solar system objects. *Bull. Am. Astron. Soc., 33,* 41.23.

Ragazzoni R., Baruffolo A., Marchetti E., Ghedina A., Farinato J., and Niero T. (2000) Speckle interferometry measurements of the asteroids 10-Hygiea and 15-Eunomia. *Astron. Astrophys., 354,* 315–320.

Richardson W. H. (1972) Bayesian-based iterative method of image restoration. *J. Opt. Soc. Am., 62,* 55–59.

Roettger E. E. and Buratti B. J. (1994) Ultraviolet spectra and geometric albedos of 45 asteroids. *Icarus, 112,* 496–512.

Salisbury J. W. (1993) Mid-infrared spectroscopy: Laboratory data. In *Remote Geochemical Analysis: Elemental and Mineralogical Composition* (C. M. Pieters and P. A. J. Englert, eds.), pp. 79–98. Cambridge Univ., New York.

Salisbury J. W., Walter L. S., Vergo N., and D'Aria D. (1991a) *Infrared (2.1–25 μm) Spectra of Minerals*. Johns Hopkins Univ., Baltimore.

Salisbury J. W., D'Aria D. M., and Jarosewich E. (1991b) Mid-infrared (2.5–13.5 microns) reflectance spectra of powdered stony meteorites. *Icarus, 92,* 280–297.

Spencer J. R., Lebofsky L. A., and Sykes M. V. (1989) Systematic biases in radiometric diameter determinations. *Icarus, 78,* 337–354.

Storrs A., Wells E., Stern A., and Zellner B. (1998) Surface heterogeneity of asteroids: HST WFPC2 images 1996–1997. *Bull. Am. Astron. Soc., 30,* 1026.

Storrs A., Wells E., Zellner B., Stern A., and Durda D. (1999a) Imaging observations of asteroids with HST. *Bull. Am. Astron. Soc., 31,* 1089.

Storrs A., Weiss B., Zellner B., Burleson W., Sichitiu R., Wells E., Kowal C., and Tholen D. (1999b) Imaging observations of asteroids with Hubble Space Telescope. *Icarus, 137,* 260–268.

Storrs A. D., Dunne C., Conan J.-M., and Mugnier L. (2000) HST imaging observations of asteroid 216 Kleopatra. *Bull. Am. Astron. Soc., 32,* 1487.

Tanga P., Loreggia D., Hestroffer D., Lattanzi M., Cellino A., Guglielmetti F., Di Martino M., and Zappalà V. (1999) Direct measurements of asteroid sizes and duplicity search by the HST FGS interferometer. *Bull. Am. Astron. Soc., 31,* 20.03.

Tanga P., Hestroffer D., Berthier J., Cellino A., Lattanzi M., Di Martino M., and Zappalà V. (2001) HST/FGS observations of the asteroid 216 Kleopatra. *Icarus, 153,* 451–454.

Tedesco E. F. (1979) Binary asteroids — Evidence for their existence from lightcurves. *Science, 203,* 905–907.

Tedesco E. F. and Desert F.-X. (1999) The ISO faint asteroid survey. *Bull. Am. Astron. Soc., 31,* 12.04.

Tedesco E. F., Veeder G. J., Fowler J. W., and Chillemi J. R. (1992) *The IRAS Minor Planet Survey.* Phillips Laboratory PL-TR-92-2049 (available from the National Technical Information Service).

Tedesco E. F., Price S. D., and Egan M. P. (2001a) MIMPS. *Bull. Am. Astron. Soc., 33,* 41.24.

Tedesco E. F., Cellino A., Zappalà V., Egan M. P., and Price S. D. (2001b) SAM. In *Asteroids 2001 Conference,* pp. 196–197. Osservatorio di Palermo, Sicily.

Tedesco E. F., Noah P. V., Noah M., and Price S. D. (2002) The Supplemental IRAS Minor Planet Survey. *Astronom. J., 123,* in press.

Thomas P. C., Binzel R. P., Gaffey M. J., Zellner B. H., Storrs A. D., and Wells E. (1997a) Vesta: Spin pole, size, and shape from HST images. *Icarus, 128,* 88–94.

Thomas P. C., Binzel R. P., Gaffey M. J., Storrs A. D., Wells E. N., and Zellner B. H. (1997b) Impact excavation on asteroid 4 Vesta: Hubble Space Telescope results. *Science, 277,* 1492–1495.

Viateau B. and Rapaport M. (1998) The mass of (1) Ceres from its gravitational perturbations on the orbits of 9 asteroids. *Astron. Astrophys., 334,* 729–735.

Vilas F. (1994) A cheaper, faster, better way to detect water of hydration on solar system bodies. *Icarus, 111,* 456–467.

Wu N. (1994) Model updating in the MEM algorithm. In *The Restoration of HST Images and Spectra* (R. J. Hanisch and R. L. White, eds.), pp. 58–63. Space Telescope Science Institute, Baltimore.

Zellner B., Tholen D. J., and Tedesco E. F. (1985) The eight-color asteroid survey — Results for 589 minor planets. *Icarus, 61,* 355–416.

Zellner B., Wells E. N., Chapman C. R., and Cruikshank D. P. (1989) Asteroid observations with the Hubble Space Telescope and the Space Infrared Telescope Facility. In *Asteroids II* (R. P. Binzel et al., eds.), pp. 949–969. Univ. of Arizona, Tucson.

Zellner B. H., Albrecht R., Binzel R. P., Gaffey M. J., Thomas P. C., Storrs A. D., and Wells E. N. (1997) Hubble Space Telescope images of asteroid 4 Vesta in 1994. *Icarus, 128,* 83–87.

Zolensky M. and McSween H. Y. (1988) Aqueous alteration. In *Meteorites* (J. F. Kerridge and M. S. Matthews, eds.), pp. 114–143. Univ. of Arizona, Tucson.

Hydrated Minerals on Asteroids: The Astronomical Record

A. S. Rivkin
Massachusetts Institute of Technology

E. S. Howell
Arecibo Observatory

F. Vilas
NASA Johnson Space Center

L. A. Lebofsky
University of Arizona

Knowledge of the hydrated mineral inventory on the asteroids is important for deducing the origin of Earth's water, interpreting the meteorite record, and unraveling the processes occurring during the earliest times in solar system history. Reflectance spectroscopy shows absorption features in both the 0.6–0.8 and 2.5–3.5-μm regions, which are diagnostic of or associated with hydrated minerals. Observations in those regions show that hydrated minerals are common in the mid-asteroid belt, and can be found in unexpected spectral groupings as well. Asteroid groups formerly associated with mineralogies assumed to have high-temperature formation, such as M- and E-class asteroids, have been observed to have hydration features in their reflectance spectra. Some asteroids have apparently been heated to several hundred degrees Celsius, enough to destroy some fraction of their phyllosilicates. Others have rotational variation suggesting that heating was uneven. We summarize this work, and present the astronomical evidence for water- and hydroxyl-bearing minerals on asteroids.

1. INTRODUCTION

Extraterrestrial water and water-bearing minerals are of great importance both for understanding the formation and evolution of the solar system and for supporting future human activities in space. The presence of water is thought to be one of the necessary conditions for the formation of life as we know it. Furthermore, the long-term survival of human-staffed bases on other planets is critically dependent upon the existence and exploitation of an easily reached water source. Current theories for the origin of Earth's ocean require a contribution from both asteroids and comets, although the relative importances of the asteroidal and cometary fractions is still under investigation (*Delsemme,* 2000; *Morbidelli et al.,* 2000). Asteroids are the primary source of meteorites, many of which show evidence of an early heating episode and varying degrees of aqueous alteration (e.g., *DuFresne and Anders,* 1962). The identification and characterization of hydrated minerals (defined in this chapter as any mineral containing H_2O or OH) among the asteroids is important for understanding a wide range of solar system formation and evolutionary processes.

1.1. Water in the Solar Nebula and Solar System

Because O is the third most abundant element in the solar system (after H and He), its stable H compound is expected to be common. Indeed, water is found throughout the outer solar system on satellites (*Clark and McCord,* 1980; *Clark et al.,* 1984), Kuiper Belt Objects (KBOs) (*Brown et al.,* 1997), and comets (*Bregman et al.,* 1988; *Brooke et al.,* 1989) as ice, and on the planets as vapor (*Larson et al.,* 1975; *Encrenaz et al.,* 1999). It is also found bound into minerals (as well as at the polar caps) on Mars (*Sinton,* 1967; *Blaney,* 1991; *Bell and Crisp,* 1991), Europa (*McCord et al.,* 1999), and some asteroids (*Lebofsky,* 1978; *Feierberg et al.,* 1981), and of course as vapor, liquid, ice, and bound into minerals here on Earth.

The heliocentric distance at which the temperature in the solar nebula allowed water ice to condense is sometimes called the "snow line" (or "ice line" or "dew line"). This distance falls within the current asteroid belt, though the exact distance depends upon the specific model parameters used and the nebular age considered (*Cyr,* 1998). Once condensed, ice grains may then have migrated inward toward the Sun through gas drag (*Cyr et al.,* 1998). This would potentially allow water to be available for accretion into planetesimals located throughout much of the asteroid belt, even if the snow line were relatively distant.

Current hypotheses postulate asteroids began as mixtures of water ice and anhydrous silicates. A heating event early in solar system history was responsible for creating magma on some bodies (like 4 Vesta), and melting the ice and driving aqueous alteration on others (such as 1 Ceres and 2 Pallas), but apparently was too weak to do either in some cases (e.g., Trojan asteroids). Both electrical induction heating caused by a solar-wind plasma flow and heating by ^{26}Al decay have been postulated as the mechanisms that generated the high temperatures that caused the probable differentiation of some of the asteroids (*Herbert and Sonnett,* 1979; *Grimm and McSween,* 1993) that produced the iron meteorites. For the carbonaceous chondrites, the heating event was apparently sufficient to melt the internal ice and drive aqueous alteration reactions, creating hydrated minerals. For most carbonaceous chondrites, the heating did not pass the point where hydrated minerals would begin to dehydrate, although the CV meteorites show evidence of dehydration (*Krot et al.,* 1997).

1.2. Hydrated Minerals

Hydrated minerals include both silicates and nonsilicates in the scope of this review. Phyllosilicates (or "clay minerals") are commonly found on Earth as weathering products of rocks or in hydrothermal systems. Nonsilicate hydrated minerals include such species as the oxides brucite and goethite, the carbonate hydromagnesite, and the sulfide tochilinite, each of which is known in the meteorite collection (*Rubin,* 1996). Although a full discussion of the petrogenesis and classification of hydrated minerals is beyond the scope of this paper, we note that formation of hydrated minerals, particularly clay minerals, occurs rapidly and easily in environments where anhydrous rock and water are together. Serpentine-group, smectite-group, and chlorite-group minerals are the phyllosilicates most common in meteorites (*Rubin,* 1996), as detailed in section 2. The different clay groups have different numbers of OH groups between the two SiO_2 layers. In some cases in addition to the structural OH, water molecules are adsorbed between the silicate layers. Depending on the cations involved, the layers can range from being only weakly held together by the water molecules, or strongly held by shared O atoms. Phyllosilicates range from serpentine with two OH groups for each silicon to clays such as talc, which have one OH for each silicon. Micas have two OH groups for each three silicon atoms, but are not generally seen in meteorites. *Deer et al.* (1962) give complete compositional and structural details of these minerals.

Studies of hydrated minerals on asteroids are important both for what they tell us about the specific body under study and for insights into the population as a whole. The suite of hydrated minerals present on an asteroid provides clues as to which meteorite groups are appropriate analogs. The distribution of hydrated minerals in the asteroid belt touches upon questions of the homogeneity of the solar nebula, the heat sources present, and how much mixing of planetesimals occurred. Because absorption features from hydrated minerals are often the only ones present in reflectance spectra of low-albedo asteroids, they provide one of the only means of determining compositions for these bodies by remote sensing.

Knowledge of the presence or absence of hydrated minerals is also critical to the interpretation of asteroid densities. The most important factors in meteorite densities are the amount of metal and the amount of hydrated minerals (*Consolmagno and Britt,* 2002). The meteorites with the lowest densities are pervasively aqueously altered, like Tagish Lake and Orgueil. Knowledge of the hydration state of an asteroid, therefore, can be used for inferring its porosity and internal structure.

Hydrated minerals give rise to absorption features throughout the visible and near-IR, as reviewed in *Gaffey et al.* (1993b). However, many of these features are not observable using remote sensing techniques because of telluric spectral absorption bands (detailed below). Two spectral regions have been the focus for hydrated mineral studies on asteroids: the 3-µm region spanning 2.4–3.6 µm, and the visible between 0.4–0.9 µm. Since the publication of *Asteroids II,* much fruitful work has been done in the former region, and study of the latter region for hydrated minerals has grown from practically nil. The highlights of research in each area are detailed below. Section 2 touches upon the hydrated minerals present in the meteorite collection, current theories about their creation, and spectral studies of hydrated minerals in meteorites. Astronomical observations, including possible complexities in data collection and analysis and the techniques used to counter these complexities, appear in section 3. Interpretations of the astronomical observations, separated by spectral classes, make up section 4. Section 5 discusses the correlations found between the visible and IR observations of hydrated minerals on asteroids, as well as issues of rotational variability and possible size effects. Section 6 concerns alternatives to hydrated minerals that have been proposed as explanations for the 3-µm absorption band. Future work and open questions are discussed in section 7. Finally, the major conclusions we can draw about asteroids from observations of hydrated minerals are presented in section 8.

2. HYDRATED MINERALS IN METEORITES

2.1. Provenance of Hydrated Minerals in Meteorites

In the meteorite collection, hydrated minerals are found mostly among the carbonaceous chondrites of metamorphic grades 1 and 2, associated with the CI, CM, and CR carbonaceous chondrite types. These meteorites have mineralogies indicating low levels of metamorphism (<1200°C). and evidence for aqueous alteration (*Sears and Dodd,* 1988; *Rubin,* 1996). The CI, CM, and CR chondrites also are distinct from other chondrites in their O-isotopic signatures and oxidation state, which is related to the formation of hydrated minerals during aqueous alteration. The CM and CI carbonaceous chondrites typically contain 5–15% H_2O/OH by weight, some more than 20% (*Salisbury et al.,* 1991a; *Takaoka et al.,* 2001; *Burbine et al.,* 2002).

The CI chondrites are composed almost entirely (≥90 vol%) of fine-grained phyllosilicates, including serpentines, though other hydrous and hydroxylated minerals are also present (*Browning et al.,* 1996). The CM chondrites also contain abundant phyllosilicates and other hydroxylated minerals including saponite, talc, and the sulfide tochilinite. The CR chondrites also contain matrix phyllosilicates in the form of Fe-rich serpentine-group minerals, saponite, and chlorite-group minerals (with Ca-carbonates and magnetite associated with the phyllosilicates) (*Rubin,* 1996). There is evidence of aqueous alteration and some subsequent dehydration in the CV chondrites (*Krot et al.,* 1997). Although not normally considered hydrated, some ordinary chondrites also have phyllosilicates (like the LL chondrite Semarkona), or show evidence of aqueous processes (the H chondrites Zag and Monahans) (*Hutchison et al.,* 1997; *Zolensky et al.,* 2000). To date, no evidence of hydrated minerals has been found in achondrites.

2.2. Nebula, Parent-Body, or Preaccretional?

Thermodynamical arguments suggest that phyllosilicates could not have been formed in the solar nebula since the kinetics of the formation reaction are too slow (*Prinn and Fegley,* 1989). This has been challenged by *Bose* (1993), who found that the activation energy used by Prinn and Fegley is too large by a factor of 2, and that the timescale using the correct activation energy is well within the lifetime of the solar nebula. Recent work by *Rietmeijer and Nuth* (2000) shows that gas-to-solid condensation in the MgO-FeO/Fe_2O_3-SiO_2 system results in Si-rich saponite dehydroxylates, which are metastable. They propose that in the solar nebula dust grains of this composition would be formed and be highly reactive during both dry thermal and aqueous alteration, ensuring that they would eventually be altered.

There is some observational evidence that silicate dust found in rims around grains was hydrated before it was incorporated into the carbonaceous chondrites (*Metzler et al.,* 1992). However, textural evidence from these meteorites shows that the majority of the hydrated minerals formed on parent bodies in the presence of liquid water. It is still uncertain whether those parent bodies were the size of the asteroids we see today or were still accreting into larger bodies (e.g., *Bischoff,* 1998). A more thorough discussion of aqueous alteration processes on asteroids as understood through meteorite studies is provided in *Zolensky and McSween* (1988) and current research topics can be found in *Zolensky et al.* (1997).

2.3. Fluid Inclusions

Hydrated minerals are most abundant on CI and CM carbonaceous chondrites, but occur to a lesser extent in other chondrites as well. Recently, fluid inclusions have been discovered in Monahans and Zag, both H ordinary chondrites (*Zolensky et al.,* 1999, 2000). Early analysis shows that the fluid is a brine, and has prompted a search for fluid inclusions on other meteorites that could have been overlooked. *Saylor et al.* (2001) searched 19 carbonaceous chondrites and found fluid inclusions on 5, with possible inclusions on several others. The three CM2 meteorites with inclusions were moderately to highly aqueously altered. Meteorites both more and less aqueously altered were not found to have inclusions, which may mean that a narrow range of conditions are necessary for fluid inclusions to form and be retained. Although this finding is of tremendous importance for analysis of direct samples of asteroidal water, fluid inclusions are not likely to be detected by remote sensing. The 3-μm band detected by remote sensing arises only from (at most) the outer few tens of micrometers of the surface material, where the presence of fluid inclusions is unlikely, and only a minor contribution at best compared to the presence of phyllosilicates in the regolith.

2.4. Meteorite Spectroscopy in the 3-μm Region

Meteorites have come under increasing scrutiny in the 3-μm spectral region in the last decade. *Salisbury et al.* (1991a) devised an empirical relation between the adsorbed/telluric-derived water content of (nominally anhydrous) minerals and their band depth in the 3-μm region, as measured in air. *Miyamoto and Zolensky* (1994) used spectra of heated samples of powdered carbonaceous chondrites, and showed that the integrated intensity of the 3-μm band (found by numerically integrating the area of the absorption feature in a continuum-removed reflectance spectrum) was closely correlated to the observed H content of carbonaceous chondrites. *Sato et al.* (1997), in turn, correlated the reflectances at 2.90 and 3.20 μm divided by the reflectance at 2.53 μm to the integrated intensity. This set of empirical relation allows the H/Si ratio to be determined for remote asteroid observations, assuming the meteorite powders studied by *Miyamoto and Zolensky* (1994) and *Sato et al.* (1997) are similar to asteroid regoliths. The H/Si ratio for 433 Eros was determined in this way by *Rivkin and Clark* (2001), who found an upper limit of 3% H_2O for that body. Caution should be taken, however, as the integrated intensities of the same meteorite can have a relatively large scatter (*Rivkin,* 1997).

Calvin and King (1997) measured the spectra of Fe-bearing phyllosilicates including serpentines and berthierines, and showed that simple linear combinations of Fe- and Mg-phyllosilicates closely approximate the spectra of CM and CI chondrites, in particular the shape of the 3-μm band.

Most asteroids have shallower band depths than the common hydrated meteorites. *Hiroi et al.* (1996) found that the best matches to the spectra of large C-class asteroids in the 3-μm region were unusual thermally metamorphosed CI/CM chondrites. This has been interpreted as evidence for heating events on the largest C asteroids, like 1 Ceres and 2 Pallas, with peak temperatures up to 500°–600°C, as detailed further in section 5. This work assumes that "space weathering" effects in the 3-μm region and/or on low-albedo asteroids are negligible. However, as mentioned in section 7.2 [and in further detail in *Clark et al.* (2002)], the effect of "space weathering" on low-albedo objects is currently an open question.

3. ASTRONOMICAL OBSERVATIONS

3.1. Infrared Wavelength Observations

Water and hydroxyl-bearing minerals give rise to diagnostic absorptions throughout the IR. A 3-μm absorption feature is typically caused by a combination of the very strong OH-radical absorption feature and the very strong first overtone of the 6-μm H_2O fundamental. Adsorbed water has a symmetric stretch mode giving rise to an absorption feature at 3.1 μm and an antisymmetric stretch feature at 2.9 μm. Structural hydroxyl (OH) that has been incorporated into mineral lattices produces a stretch absorption at 2.7 μm. The depth of the fundamental H_2O/OH feature depends on many parameters besides abundance of hydrated material, including particle size, albedo, and temperature. However, water abundances as low as fractions of a weight percent are observable using 3-μm spectroscopy (*Salisbury et al.,* 1991a). Laboratory studies have shown that higher-order overtones at shorter wavelengths (1.4 and 1.9 μm) are masked more easily by opaque minerals than the 3-μm feature (*Clark,* 1981), and they are not always seen even in freshly fallen, phyllosilicate-bearing meteorites (*Hiroi et al.,* 2001a). In addition, observations of these overtone bands on asteroids may be difficult or impossible even in high-albedo mineral assemblages; reflectance spectra of Mars analogs containing hydrated minerals shows that the 1.9-μm band greatly decreases in strength as the atmospheric pressure decreases, suggesting it will not be observed on the atmosphereless asteroids (*Bishop and Pieters,* 1995).

The 3-μm absorption feature in minerals is shifted in wavelength relative to the telluric absorption (see Fig. 1), and is much broader than the atmospheric band, though the center of the OH band at 2.7 μm is not observable from Earth.

At wavelengths near 3 μm, thermal flux from objects in the asteroid belt can become a significant fraction of the total flux and must be modeled and removed to give a true idea of the reflectance spectrum. The specifics of how much thermal flux is emitted depends critically on variables such as solar distance at the time of observation as well as fixed values such as diameter and albedo; however, for most main-belt low-albedo asteroids the thermal contribution is 1–10% of the total flux at 3.1–3.5 μm. A "standard thermal model" (STM) for asteroids was formulated by *Lebofsky et al.* (1986) for use with 3-μm observations. Thermophysical models including values for thermal inertia and rotation rate have been developed as well as refinements to the STM for near-Earth asteroids. Although a thorough review of asteroid thermal models is beyond the scope of this chapter, *Lebofsky and Spencer* (1989) gives a review of the state of the field as of 1989, and *Harris and Lagerros* (2002) updates advancements since that time. Because the uncertainty

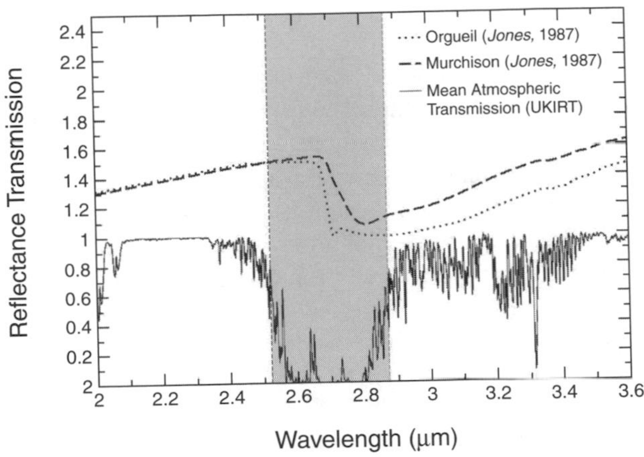

Fig. 1. A comparison of atmospheric transmission and laboratory meteorite spectra: Although the 2.52–2.85-μm region has less than 50% transmission, and the 2.56–2.82-μm region less than 10%, the meteorite absorption features are sufficiently wide that they are easily detectable in areas of greater transmission. The shaded area has less than 50% transmission. The meteorite spectra, from *Jones* (1989), have been offset by 0.5 units for clarity. The atmospheric data are from the UKIRT Web site using the program IRTRANS4, assuming an altitude of 4200 m, an airmass of 1.0, and 1.2 mm precipitable water.

in the thermal contribution is typically much smaller than the observational uncertainties at ~2.95 μm, where band depths are typically calculated, the uncertainties in the thermal flux are typically neglected in band depth measurements. The thermal models do affect band shape beyond ~3.1 μm, and typically the preferred thermal model is one that gives a continuum slope at the long-wavelength end of the 3-μm band consistent with the continuum slope shortward of 2.5 μm.

There are other species that can give rise to absorption features near 3 μm, but all are ices that are not stable on asteroidal surfaces (carbon dioxide, methane, ammonia). The NH_3^+ ion gives rise to a shallow feature near 3.07 μm, and has been reported in an ammoniated phyllosilicate on Ceres by *King et al.* (1992), as detailed below. Macromolecular organic solids have absorption features near 3.4 μm. Although fully expected on outer-belt asteroidal surfaces, no confirmed organic feature has yet been seen (*Dumas et al.,* 1998), although methanol or a product of methanol has been found in the spectrum of the Centaur 5145 Pholus (*Cruikshank et al.,* 1998). A search for an overtone of CN-bearing organic material at 2.1–2.4 μm on outer-belt asteroids yielded no detections (*Howell,* 1995). An absorption feature due to an Al-OH stretch near 2.2–2.3 μm commonly seen in terrestrial clays has not to date been seen in asteroids or meteorites.

3.2. Visible Wavelength Observations

Spectra of main-belt C- and G-class asteroids and some C-, G-, and P-class asteroids in the Cybeles (mean semi-

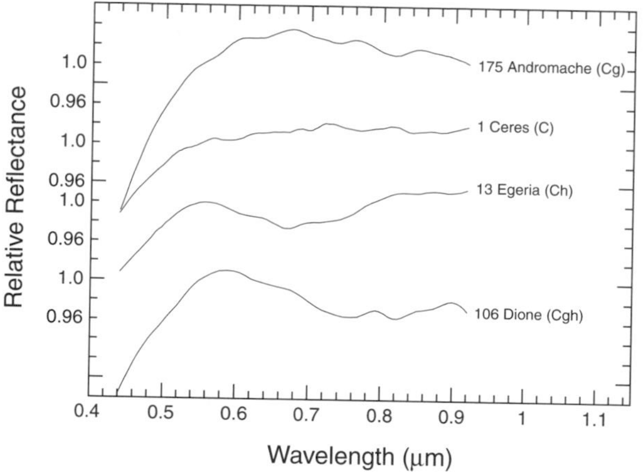

Fig. 2. The spectra shown are taken as part of the SMASS II survey (*Bus,* 1999). The Tholen C and G classes are reorganized primarily into the C, Cg, Ch, and Cgh classes. The "g" is appended to those objects whose spectra turn down sharply at the 0.44-μm end. The "h" is appended to those objects whose spectra show the broad iron charge-transfer feature near 0.7 μm. Some objects have both of these features. The iron features near 0.43 μm and 0.65 μm are narrower, and do not show clearly at this scale.

major axis a ~ 3.4 AU) show an absorption feature centered near 0.7 μm with a width of roughly 0.25 μm, attributed to an $Fe^{2+} \rightarrow Fe^{3+}$ charge transfer transition in oxidized Fe found in phyllosilicates (*Vilas and Gaffey,* 1989). Figure 2 contains examples of this feature. Independent research by *Sawyer* (1991) showed that this feature is very common among the main-belt low-albedo asteroids. This absorption feature and absorption features seen in the laboratory spectra of CM2 carbonaceous chondrite meteorites and terrestrial phyllosilicates are similar in band position and strength, further supporting the connection between the C-class asteroids and carbonaceous chondritic material formed through the aqueous alteration process. *King and Clark* (1997) caution that a feature near 0.7 μm is found in many different minerals, and that its presence alone is not diagnostic for phyllosilicates. The vast majority of minerals with a 0.7-μm band are Fe- and OH-bearing silicates, however. As detailed in section 5, joint studies at both 0.7 and 3 μm show that in low-albedo asteroids (the C, B, G, F, and P classes) at least, the correlation between these features is quite good. Since CCD visible spectra are relatively easy to obtain with moderate-sized telescopes, many efforts have been made to observe this feature and determine the hydration state of asteroids (*Barucci et al.,* 1994, 1998; *Fitzsimmons et al.,* 1994; *Dahlgren and Lagerkvist,* 1995; *Dahlgren et al.,* 1997; *Lazzarin et al.* 1995; *Carvano et al.,* 2001; *Jewitt and Luu,* 1990). *Fornasier et al.* (1999) found 65% of the mostly C-class objects they observed to have the 0.7-μm feature, in good agreement with the fraction of C asteroids with the 3-μm feature (Table 1). It is unclear at this time why later visible surveys (as well as the 3-μm surveys) disagree with the quoted results of *Sawyer* (1991).

Thirteen C-, P-, and G-class (and 1 K-class) asteroids observed in the UV/blue spectral region show an absorption feature at 0.43 μm attributed to an Fe^{3+} spin-forbidden transition in Fe alteration minerals similar to jarosite [$KFe_3(SO_4)_2(OH)_6$] (*Vilas et al.,* 1993; *Cochran and Vilas,* 1997). A strong correlation between the 0.43-μm feature and the 0.7-μm feature was apparent in these asteroids, although none of the asteroids was observed in the UV spectral region on more than one night in the original 1993 work. The fea-

TABLE 1. Asteroids observed at 3 μm and 0.7 μm, with the fraction showing absorption features for each wavelength region.

	3 μm			0.7 μm		
Tholen Class	Observed	With Band*	Percent	Observed	With Band	Percent
C	32	20	63	45	20	44
B	1	1	100	1	0	0
G	5	5	100	6	6	100
F	5	1	20	5	2	40
T	4	3	75	5	0	0
D/P	20	3	15	22	4	18
K/L	5	2	40	1	0	0
S	24	1	4	4	0	0
E	6	4	67	1	0	0
M/W	27	10	37	23	1	4

*Uncertainties vary, so the band depth limits are not uniform in these data.

At both wavelengths, variability in this feature might be present on some asteroids, so the number and percentage showing the absorption feature is a lower limit. Note that the B asteroid is 2 Pallas, which may not be representative of the B class as a whole. This table includes data from *Lebofsky* (1980), *Feierberg et al.* (1985), *Jones et al.* (1990), *Rivkin et al.* (1995), *Howell* (1995), *Rivkin* (1997), *Merényi et al.* (1997), and *Jarvis et al.* (1999), as well as further unpublished data from Rivkin and Howell. For those asteroids with multiple designations (e.g., CF, XC), the first letter was taken as the class. The K and L classes are from *Bus* (1999), the W class from *Rivkin et al.* (1995).

TABLE 2. Observed absorption features associated with hydrated minerals on asteroids.

Wavelength (μm)	Width (μm)	Transition	Reference
<0.4	>0.1	$Fe^{2+} \rightarrow Fe^{3+}$ intervalence charge	e.g., *Gaffey and McCord* (1979)
0.43	0.02	6A1 → 4A1,4E(G) Fe^{3+} spin-forbidden as in jarosite	*Vilas et al.* (1993)
0.60–0.65	0.12	6A1 → 4T2(G) Fe^{3+} in Fe alteration minerals	*Vilas et al.* (1994)
0.7	0.3	$Fe^{2+} \rightarrow Fe^{3+}$ in phyllosilicates	*Vilas and Gaffey* (1989)
0.80–0.90	0.08	6A1 → 4T1(G) Fe^{3+} in Fe alteration minerals	*Vilas et al.* (1994)
3.0	>0.7	structural hydroxyl (OH) interlayer and adsorbed H_2O	*Lebofsky* (1978, 1980)
3.07	0.2	H_2O ice NH_4-bearing saponite	*Lebofsky et al.* (1981) *King et al.* (1992)

ture is commonly found in the C and G classes, and absent in the D class, in agreement with results from the 3-μm region (K. S. Jarvis, personal communication, 2001). Because of interference by nearby absorption lines in solar-analog standard stars, however, full use and exploration of this absorption feature has been problematic (*Jarvis et al.*, 1998).

Spectra of five low-albedo F-, C-, P-, and D-class asteroids show weak absorption features centered near 0.60–0.65 μm and near 0.8–0.9 μm (*Vilas et al.*, 1994). These features have been tentatively identified as due to ferric Fe absorptions in Fe alteration minerals such as the oxyhydroxide goethite [a-Fe_3^+O(OH)], Fe oxide hematite (Fe_2O_3), or Fe sulfate jarosite, although none of these minerals has been identified in the terrestrial meteorite collection. Spectra of low-albedo asteroids showing the 0.7-μm feature sometimes also have smaller absorption features near 0.6 and 0.9 μm superimposed on the 0.7-μm feature (*Vilas et al.*, 1994). The possibility that these features are the Fe alteration features seen in conjunction with the phyllosilicate absorption was proposed (*Vilas et al.*, 1994), but spinel-group minerals are also a possibility (*Hiroi and Vilas,* 1996). The presence and interpretation of these features remain open questions pending future work with extended wavelength coverage to the near-IR, and perhaps a dedicated survey.

Table 2 lists the absorption features associated with hydrated minerals that have been identified to date in the spectra of asteroids and their current mineralogical interpretations. In the visible/near-IR spectral region (0.3–2.5 μm), these are weak absorption features having typical depths of 1–5% of the background continuum. The 3-μm feature typically has a depth of 15–30% or more in low-albedo asteroids. The widest features (>0.3 μm) can be detected at smaller depths, but care must be taken in determination of the continuum. The 3-μm feature often extends to 4 μm in laboratory spectra, but due to the uncertainty in the thermal emission in the case of asteroids closer to the Sun than 3 AU, the band continuum is usually measured between 2.5 and 3.35 (or 3.5) μm. The band depth is measured at 2.95 μm relative to this continuum.

4. RESULTS FOR ASTEROID CLASSES

The hydrated mineral features on asteroids appear to have one of two general band shapes, corresponding roughly to the rounded H_2O-like absorption feature, and the "checkmark" OH-like absorption feature (see Fig. 3). In general (and as detailed below), hydrated minerals on low-albedo asteroids tend to have the checkmark shape, while on high-albedo asteroids they have a rounded shape, when band shapes can be determined.

Table 1 summarizes the results for the major asteroid classes. As can be seen, there are large differences in the fraction of hydrated members from class to class. The specifics for each class will be discussed below. We note, however, that because of the possibility of surface variegation (discussed further in section 5), the number of hydrated asteroids should be considered a lower limit. Because the Tholen asteroid taxonomy (*Tholen and Barucci,* 1989) was the most widely used one during the period in which this work has been done, we will use the Tholen taxonomy in most cases below. We note that interpretations of asteroid populations are necessarily influenced by the choice of classification schemes, and that labeling a potentially diverse group of mineralogies with a similarly diverse set of formation conditions with a single letter should be undertaken with care.

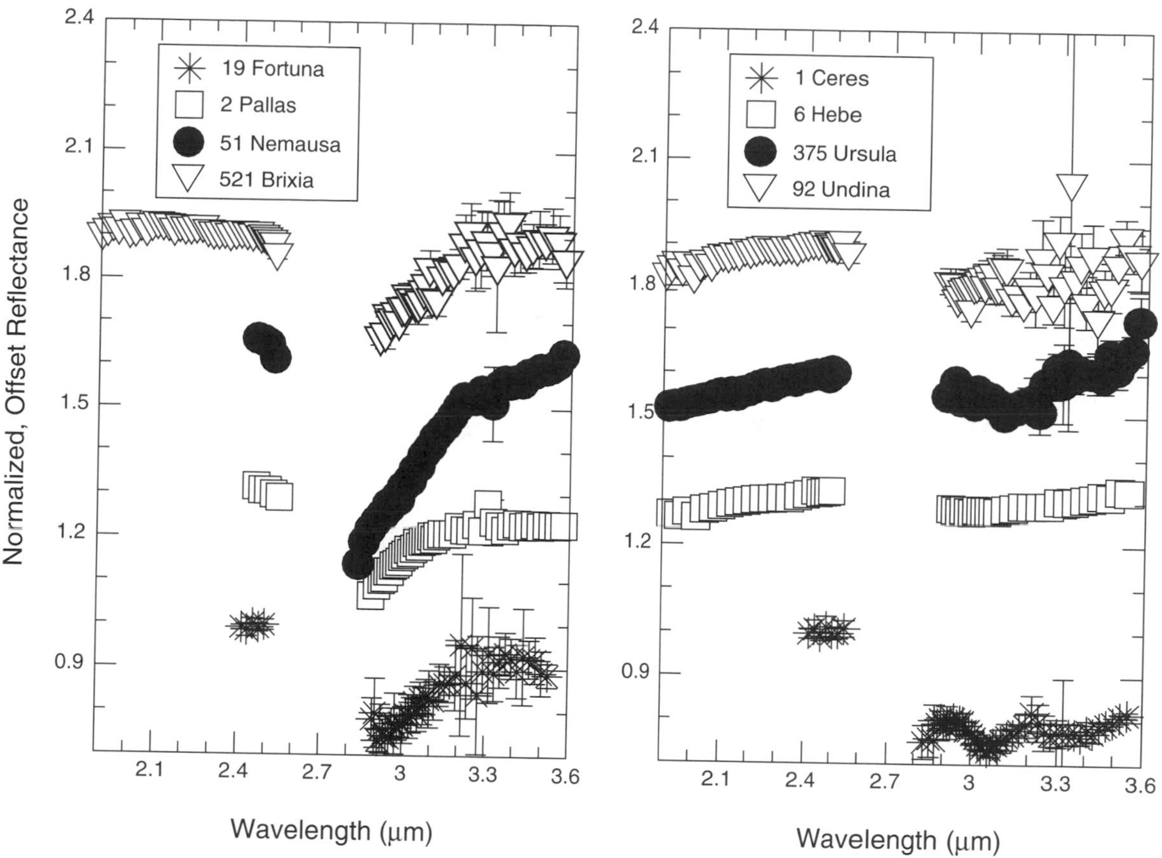

Fig. 3. "Checkmark" vs. "round" shaped features. The asteroids on the left all exhibit "checkmark"-shaped absorption features at 3 μm: Their reflectance rises monotonically in a nearly linear fashion. Those on the right have a more rounded feature. Most C-class asteroids with 3-μm absorptions tend to have the checkmark-shaped feature. Higher-albedo objects with features have rounded ones, though shallow band depths make this difficult to determine at times. Ceres and Ursula are notable exceptions to this rule of thumb. Rounded features are reminiscent of H_2O-dominated minerals, while the checkmark shapes are more similar to OH-dominated ones. Thermal corrections have been applied to these data using the standard thermal model (STM). These data were obtained by A. Rivkin and J. Davies at UKIRT, and the manuscript is in preparation. Error bars are included in this graph, although for most of the spectral region they are smaller than the points.

4.1. Low-Albedo Objects

4.1.1. C-class and related asteroids. *Jones et al.* (1990) observed 19 low-albedo objects, mostly of the C class and subclasses. They determined that the fraction of hydrated asteroids decreased with increasing semimajor axis from the middle of the asteroid belt outward. This was interpreted as evidence that the asteroids were originally composed of mixtures of ice and anhydrous silicates, and that hydrated minerals were formed through an aqueous alteration event rather than in the nebula. In this interpretation, the midbelt hydrated asteroids were heated enough to melt the internal ice and form phyllosilicates while the outer belt objects never achieved a temperature high enough to melt ice. This paradigm has continued to this day.

The C, B, G, and F asteroids have been proposed as comprising an alteration sequence (*Bell et al.,* 1989, and references therein). These subclasses all have different fractions of hydrated members, from the ubiquitously hydrated G asteroids (all six observed have a feature) to the mostly anhydrous F asteroids (one of five hydrated). This supports the idea of an alteration sequence, where the G asteroids were heated to the point where melted ice could cause pervasive aqueous alteration, and the F asteroids were heated further to the point that the hydrated minerals were destroyed. However, *Sawyer* (1991) interprets the F asteroids as unaltered rather than heated and dehydrated, since dehydration and recrystallization would be expected to raise the albedos of F asteroids beyond the observed values. The unaltered interpretation of Sawyer assumes an electromagnetic-induction heating scenario, with the exteriors of the parent bodies heated more than the interiors. With the relatively small sample size of F asteroids, the interpretation of F asteroids remains unsettled although their association with C asteroids seems secure.

The largest asteroid, 1 Ceres, was observed by *Lebofsky et al.* (1981) to have an absorption feature at 3.07 μm within the broad hydrated mineral feature. This was interpreted as water ice, which was shown to be marginally stable near Ceres' poles (*Lebofsky et al.,* 1981). Later work by *King et al.* (1992) suggested that this subfeature was due to NH_3^+-bearing phyllosilicates. *Rivkin* (1997), using a combination of first-look ISO data and UKIRT observations, found ammoniated smectite to be a good fit for the spectrum of Ceres in the 2.4–2.6-μm region, including the 2.6–2.8-μm area observed from ISO and unobservable from the ground (Fig. 4). Ammonium phyllosilicates are relatively rare on Earth, found in hydrothermal deposits. *Lewis and Prinn* (1984) considered the possibility of NH_3^+-rich brines on asteroids, which could be the origin of these minerals on Ceres. However, observations of OH near the sunlit pole of Ceres by *A'Hearn and Feldman* (1992) are more consistent with the presence of water ice. The discovery of a Ceres-like band shape on 375 Ursula suggests that the mineralogy may not be unique to Ceres (Rivkin et al., in preparation, 2002).

In general, most low-albedo asteroids with absorption features have band shapes indicative of OH-dominated minerals rather than H_2O-dominated ones, seen for example in Fig. 5, a fit to the spectrum of 13 Egeria from *Rivkin* (1997). Both types of minerals have been seen in meteorites. The explanation for this may be that the H_2O in these minerals may be unstable in the vacuum of space while the structural OH can remain; but then why do the higher-albedo objects show a H_2O-like feature? Another possibility is that the water-rich meteorites may come from a relatively restricted area on its parent body.

The 3-μm band depths for C-class asteroids are not as deep as those found in the laboratory for their putative analogs, the hydrated carbonaceous chondrite meteorites. In some cases, band depths are less by a factor of 3 or more. Differences in determining the continuum and choosing where to measure the band depth may account for some of the discrepancy, although real differences are also present in some cases. One explanation that has been suggested for this is that the asteroids have experienced some degree of thermal processing, driving off water and destroying hydrated minerals (*Hiroi et al.,* 1996). Another possibility is that "space weathering" processes similar to those found at shorter wavelengths could be at work (*Clark et al.,* 2002). How micrometeorite impacts might affect the hydrated mineral inventory of an aqueously altered body will no doubt be the subject of future work. We may speculate that the energy of impact might go to liberating water before it can create the nanophase Fe seen in lunar soils and believed to be present on chondritic, anhydrous surfaces (*Noble et al.,* 2001; *Sasaki et al.,* 2001). As a result, the reddening and darkening effects seen on lunar surfaces and laboratory simulations might not occur on hydrated bodies until they are dehydrated (which might never happen). Alternately, we note that "space weathering" effects due to nanophase Fe coatings are not currently expected to be very important on low-albedo objects like the C-class asteroids (*Pieters et al.,* 2000). This is because they are most effective when coating transparent minerals like olivine and pyroxene; while nanophase Fe has a lower albedo than olivine and pyroxene and small amounts can radically change a spectrum, it has a higher albedo than the opaque minerals present in the C, P, D, etc., asteroids and will have little effect. A final possibility is that the meteorites we have collected are from a small subset of the C-class (and related) asteroids, and the differences in spectra that we see should be taken at face value. *Burbine* (1998) proposed that the G-class asteroids could contain the parent bodies of the CM meteorites. In-

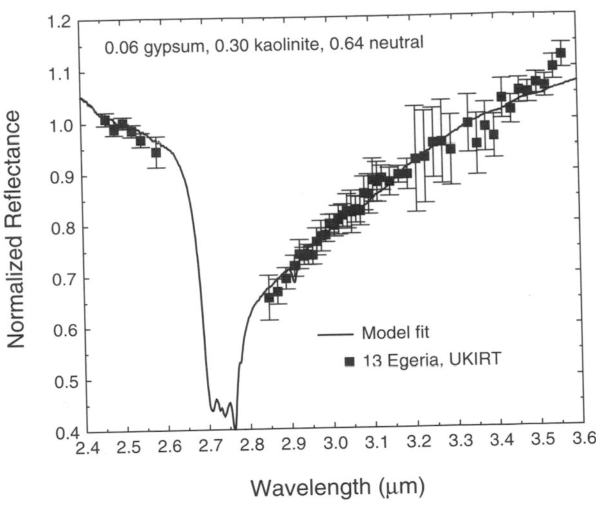

Fig. 5. Fit to 13 Egeria using minerals from the *Salisbury et al.* (1991b) spectral library. A nonlinear mixing code using the Hapke equations was used in this fit (*Clark et al.,* 1993, 2001). Figure taken from *Rivkin* (1997).

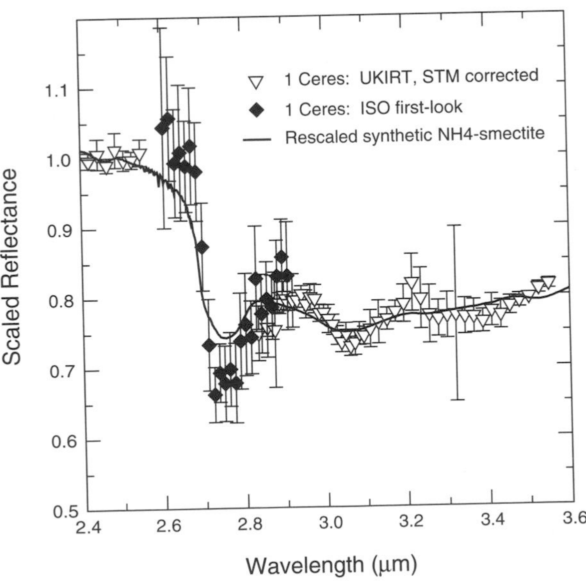

Fig. 4. First-look ISO data for Ceres, plus CGS4 UKIRT data [both from *Rivkin* (1997), vs. ammoniated smectite from T. King (personal communication, 1997)]. The smectite spectrum has been mathematically mixed with a spectrally neutral material to match Ceres' band depth. The match is quite good, including in the region visible only from space.

terestingly, these asteroids have the deepest 3-μm bands, and *Rivkin and Davies* (2002) have calculated that the G-class asteroids 13 Egeria and 106 Dione, as well as the CU-class asteroid 51 Nemausa, have water contents consistent with the average CM chondrite values, which range from 6.5–12% (*Jarosewich,* 1990).

Hydrated minerals and internal water ice have been invoked as a means to account for the lower-than-expected densities of asteroids (*Veverka et al.,* 1997; *Merline et al.,* 2000) and the martian satellites (*Avanesov et al.,* 1991). However, the lack of a 3-μm absorption feature on 253 Mathilde does not support abundant internal ice or hydrated minerals on that body (*Rivkin et al.,* 1997a), suggesting porosity is the sole cause of its low density. Similarly, the lack of absorption features on Phobos and Deimos (*Murchie and Erard,* 1996; *Rivkin et al.,* 2002) suggests that porosity is the main cause of their low densities. This topic is covered in more detail in *Britt et al.* (2002).

4.1.2. D-class and P-class asteroids. The D-class and P-class asteroids dominate the outer belt and Trojan clouds. Observations in the 3-μm region by *Jones et al.* (1990), *Lebofsky et al.* (1990), and *Howell* (1995) found no hydrated asteroids of these classes beyond 2.9 AU. Asteroid 336 Lacadiera, interestingly, shows evidence of variation in its 3-μm band (*Howell,* 1995), and is the D-class asteroid with the smallest semimajor axis in the main belt by far (2.25 AU). Asteroid 50 Virginia, classed X by Tholen, was found to have an albedo of 0.05, based on combined visible and thermal-IR observations, and is thus a P asteroid. A 3-μm band was seen on Virginia, making it the only hydrated P asteroid known. Its semimajor axis is 2.65 AU, again much closer than most P-class asteroids. In the Bus taxonomy, 336 Lacadiera is Xk, and 50 Virginia is Ch, suggesting they are distinct from the outer-belt P and D asteroids. Further work by *Emery and Brown* (2001) also found no 3-μm absorptions on Trojan asteroids.

A detailed study of the Trojan asteroid 624 Hektor by *Cruikshank et al.* (2001) found no 3-μm feature, but showed that up to 40% serpentine (representing ~6% water) could still be present but masked by low-albedo constituents (they use elemental C). Therefore, they argue, the Trojan asteroids could have relatively abundant hydrated minerals, contrary to the interpretations of *Jones et al.* (1990). *Cruikshank et al.* (2001) further note that the trend of decreasing 3-μm band strength with heliocentric distance could instead be due to increasing abundance of elemental C. Both Cruikshank et al. and *Jones et al.* (1990) interpret the Trojans as potentially ice-rich bodies, with near-surface ice masked by low-albedo constituents and/or devolatilized by impacts.

Studies of the Tagish Lake carbonaceous chondrite show a reflectance spectrum quite close to the D-class asteroids in the 0.3–2.5-μm region, and it has been proposed as an analog for those asteroids (*Hiroi et al.,* 2001b). However, Tagish Lake has a significant absorption feature in the 3-μm region (~20%) that is absent in the D asteroids. Hiroi et al. propose that this mismatch may be relatively minor due to large observational uncertainties in the asteroid spectra, and that any mismatches that do exist could be due to dehydration of the D-asteroid regoliths through micrometeorite bombardment. This idea is further discussed in section 7 for asteroids in general. Tagish Lake has a very low albedo (~3%), with 4–5% C and a matrix dominated by phyllosilicates (*Roots et al.,* 2000). The model used by *Cruikshank et al.* (2001) for 624 Hektor had 20% amorphous C. Perhaps the regolith of Trojan asteroids becomes particularly concentrated in low-albedo contituents. Or, Tagish Lake could be a fragment of an inner-belt D asteroid, for which hydration features have been seen (such as 336 Lacadiera).

4.2. Medium- and High-Albedo Objects

4.2.1. E-class and M-class (and W-class) asteroids. *Jones et al.* (1990), in the course of their studies of outer-belt asteroids, discovered 3-μm absorptions on the M-class asteroids 55 Pandora and 92 Undina, which was completely unexpected on bodies presumed to be analogous to iron meteorites. Surveys of 27 M-class asteroids by *Rivkin et al.* (1995) and *Rivkin et al.* (2000) confirmed the presence of this feature on 10 of them, including both 92 Undina and 55 Pandora. They separated the hydrated members of this group to form the "W class." The 0.7-μm feature correlated with the 3-μm hydration feature (see below) has also been seen on some of these bodies [e.g., 135 Hertha and 201 Penelope (*Busarev and Krugly Yu,* 1995; *Howell et al.,* 2001a)], reinforcing the hydrated mineral interpretation. In general, however, the 0.7-μm feature has not been seen on W-class asteroids.

Rivkin et al. (2000) found the 3-μm feature to be correlated with size in a sample of 27 M and W asteroids, with the larger bodies (diameter >65 km) likely to have the feature (75% with the feature) and the smaller bodies unlikely to have it (10% with the feature). On the basis of these observations as well as polarimetric (*Belskaya and Lagerkvist,* 1996), radar (*Ostro et al.,* 1993; *Magri et al.,* 1999), and meteorite cooling data (*Haack et al.,* 1990), *Rivkin et al.* (2000) argued that many of the M-class asteroids were not in fact analogous to iron meteorites, but to the primitive enstatite chondrites. Although enstatite chondrites with hydrated minerals are not currently known in the meteorite collection, there is no cosmochemical objection to their existence.

E asteroids have also been interpreted as igneous objects, akin to the aubrite (enstatite achondrite) meteorites. They dominate the innermost part of the asteroid belt in the Hungaria region. They are rare in the asteroid belt as a whole. *Rivkin et al.* (1995) and *Rivkin* (1997) reported on six E-class asteroids, and found the largest four of them to show an absorption feature at 3 μm. Aqueous alteration scenarios for aubrites are theoretically possible, though they are not seen in the meteorite collection. None of the survey objects were from the Hungaria region, which still may be the source of the aubrite meteorites known on Earth.

As mentioned in section 6, *Cloutis and Burbine* (1999) and *Fornaiser and Lazzarin* (2001) suggested that a feature found near 0.5 μm in some E asteroids may be due to troilite or other sulfides, and that the mineral responsible may also be responsible for the 3-μm feature in E-class and perhaps M-class asteroids. However, the group of asteroids showing the visible absorption include both those with and without a 3-µm feature, and asteroids with the 3-µm feature include those both with and without the visible feature. Therefore, the two absorptions appear uncorrelated with each other in E-class asteroids.

With the deluge of new visible-region data, as well as taxonomies that show a continuum of spectral types, the notion of hydrated M and E asteroids is somewhat less shocking. The work of *Bus* (1999) shows that asteroids populate the spectral space between the classical C and X classes, and that there are bodies with spectra that are transitional between the two. Using an artificial neural network to classify asteroids based on their 0.3–2.5-µm spectra, *Howell et al.* (1994b) showed that some nominally M-class asteroids had spectra that were actually more similar to those of C asteroids. Similarly, recent findings suggest that the albedos of common taxonomic types have a much larger range than previously considered, at least among the near-Earth population (*Harris,* 2001). Given the blurring of taxonomic lines compared to the time of the *Asteroids II* book, it is perhaps not surprising that hydrated minerals are being found on asteroids previously considered to be devoid of them (the M asteroids, for example). Furthermore, given the evolving understanding of meteorite delivery (*Bottke et al.,* 2002), and the realization that the meteorite collection may be dominated by a relatively small number of parent bodies, the finding that some asteroid mineralogies are not represented as meteorites is not as bothersome as it may have been a decade ago.

4.2.2. S-class asteroids. The S-class asteroids have been the subject of great controversy throughout the decade since *Asteroids II* was published. Given that the two main interpretations of S-asteroid mineralogy are either entirely anhydrous (stony-iron/primitive achondrites) or largely anhydrous (ordinary chondrite meteorites), these asteroids have been expected to show no absorption bands due to hydrated minerals. A compilation of the existing S-asteroid spectra by *Rivkin et al.* (1997b) found this to be true. Further observations were motivated by the realization that some OC meteorites do have aqueous alteration products, and in an effort to generate longer-wavelength data for use in spectral mixing models, a spectrophotometric survey of 17 S-class asteroids at 3 µm was made by Rivkin from 1998 to 1999. The asteroids surveyed include 7 S(IV) asteroids, which are those that are most closely related to ordinary chondrites based on their 1–2.5-µm spectrum (*Gaffey et al.,* 1993a). Although these data are still undergoing final analysis, preliminary results show that any absorption features have upper limits on band depth of 5–7%.

Eaton et al. (1983) studied 11 asteroids from 3 to 4 µm, of several classes, including 4 S asteroids. Because the spectra do not extend shortward of 2.5 µm, it is difficult to scale these data to existing near-IR data, and these data are not included in Table 1. However, of the S asteroids observed by Eaton et al., only 8 Flora was interpreted as possibly having a 3-µm absorption band.

The S-class asteroid 6 Hebe has been proposed as the parent body of the H-chondrite meteorites (*Farinella et al.,* 1993; *Migliorini et al.,* 1997; *Gaffey and Gilbert,* 1998). A 3-µm absorption feature of 5 ± 2% was discovered on this body by *Rivkin et al.* (2001), with evidence of rotational variation. The discovery of aqueous alteration products in H-chondrite meteorites (*Zolensky et al.,* 2000), and their absence in stony-iron meteorites and other achondrites proposed as S asteroid analogs by *Bell et al.* (1989) (among others), makes this absorption feature an indication that at least the last heating event on Hebe did not melt the surface, and that at least one S-class asteroid is more akin to the primitive meteorites than the igneous ones. Observations of Hebe included in *Rivkin* (1997) were interpreted as evidence of an anhydrous surface, although these data have somewhat larger uncertainties than the *Rivkin et al.* (2001) observations, and are consistent with them.

The Near-Infrared Mapping Spectrometer on the *Galileo* spacecraft obtained data in the 3-µm region for 951 Gaspra and 243 Ida (and its satellite Dactyl). These data have not yet been fully analyzed, but preliminary results show no absorption feature [Fanale and Granahan, reported in *Chapman* (1996)].

4.2.3. K-class and L-class asteroids. The K-class asteroids have spectra intermediate between S and C asteroids, and have been proposed as the CV/CO parent bodies (*Bell,* 1988; *Burbine et al.,* 2001). Most of them are members of the Eos family and may have a common origin in a collisional event. The L class was created by *Bus* (1999) as similar to K asteroids, with a steeper UV slope. Both hydrated and anhydrous members of the K and L classes have been seen (*Rivkin et al.,* 1998; *Rivkin,* 1997), supporting the proposed relationship between these bodies and the CV/CO chondrites. The CV/CO chondrites have some hydrated members as well as some that show evidence of dehydration by subsequent heating (*Krot et al.,* 1997; *Rubin,* 1996).

5. CONNECTIONS BETWEEN AND VARIABILITY IN THE 3-µm AND 0.7-µm ABSORPTIONS

5.1. Correlation of 0.7-µm and 3-µm Bands

Vilas (1994) found an 80% correlation between the 3-µm band and the 0.7-µm charge transfer band. The sample of objects was relatively small (27), and the observations at the two different wavelengths had been made many years apart in most cases. In a more recent compilation, 34 of 51 asteroids (or 67%) observed at both wavelengths had consistent bands (*Howell et al.,* 2001b). Nevertheless, this provides a useful tool for predicting the likelihood of the hydration state. Based on the presence of the 0.7-µm feature, an empirical algorithm for predicting the presence of water of hydration data from the Eight-Color Asteroid Survey (ECAS) (*Zellner et al.,* 1985) was developed and applied to the ECAS photometry of asteroids and outer planet satellites (*Vilas,* 1994). The percentage of objects in low-albedo, outer main-belt asteroid classes that test positively for water of hydration increases from P → B → C → G class and correlates linearly with the increasing mean albedos of those objects testing positively (*Vilas,* 1994). This suggests that the aqueous alteration sequence in the solar system ranges from the P-class asteroids, representing the least-altered objects created at temperatures attained at the onset of aqueous alteration, to the G-class asteroids, representing the upper range of the alteration sequence among low-albedo asteroids. Although this scenario is an oversimplification, it provides a framework for further investigation. The F-class asteroids do not as a group seem to be hydrated, based on both visible and 3-μm spectra. The G-class asteroids are all hydrated, and have both 3-μm and 0.7-μm bands. If these are the most extensively altered, the 0.7-μm band might be expected to be absent, with all the Fe^{2+} having been converted to Fe^{3+}.

The M-class asteroids with a 3-μm absorption band do not generally have the 0.7-μm band. Along with the lack of an FeO absorption edge at 0.3 μm, this may be evidence for a different composition for those M-class objects with a 3-μm absorption, hence the W class (*Rivkin et al.,* 2000). Alternatively, these objects could represent a mixture of metals with phyllosilicates that have been significantly altered to the extent that Fe^{2+} has been leached out of the phyllosilicates and sequestered as Fe^{3+} in opaque phases such as magnetite or Fe sulfides. This allows the albedo to increase as the particle size of the opaque grains increases, reducing the grain surface area that can absorb light so that light is reflected not absorbed. This process also reduces the presence of oxidized Fe, thus removing the 0.7-μm absorption feature (*Vilas,* 1994).

The inconsistencies in the observations prompted an attempt to examine the relationship between these bands more closely (*Howell et al.,* 2001a). Rotationally resolved visible spectra were obtained of a number of asteroids that did not have concurrent bands at 3 μm and 0.7 μm. Several possibilities could account for these bands not being concurrent: If the hydrated silicates are unevenly distributed across the asteroid surface, then perhaps rotational variability explains the discrepancy. Only Fe-bearing hydrated silicates exhibit the 0.7-μm band, so the presence of the 3-μm band alone might occur on Fe-poor objects. The 3-μm band is stronger, so it could be detected on some low-albedo objects in cases where the weaker 0.7-μm band cannot be seen. However, the 0.7-μm band should not be present when the 3-μm band is absent, unless it can also arise from anhydrous minerals. While some high-albedo salts have a 3-μm feature, the asteroids in the Howell et al. study are in taxonomic classes with low albedos, which puts strong constraints on the abundance of these minerals.

5.2. Thermal Alteration and Hydrated Minerals

Hydrated silicates can be used as very sensitive tracers of thermal history (*Hiroi et al.,* 1996). In addition to altering olivine and pyroxene to form hydrated silicates (e.g., serpentine), the aqueous alteration process produces oxidized Fe that has absorption bands in the visible and UV spectral regions. Moderate subsequent heating can alter the depth or eliminate some or all of these bands. *Hiroi et al.* (1996) finds that Murchison (CM2) material exhibits a strong UV band due to FeO, 0.7-μm band from Fe^{2+}-Fe^{3+} charge transfer, and 3-μm band due to H_2O/OH when heated less than 400°C. Between 400° and 600°C, the 0.7-μm band weakens and disappears, and the 3-μm band gets shallower. At temperatures above 600°C, the 3-μm band disappears as the minerals are completely dehydrated (Fig. 6). Asteroid 511 Davida has been observed extensively, and a 3-μm band has been seen with variable depths over at least 25% of the rotation period. However, at a similar sub-Earth latitude, the 0.7-μm band was not seen at any rotation phase. A mild heating episode, occurring after the aqueous alteration, with temperatures reaching 400°–600°C, can explain these observations. Another possibility is that all the Fe^{2+} has been converted to Fe^{3+}, which would account for the 0.7-μm band being absent on an object where the 3-μm band is seen.

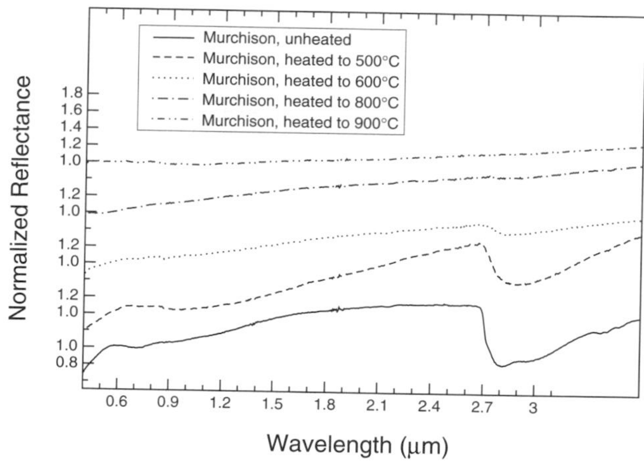

Fig. 6. Spectra of the CM meteorite Murchison, taken after heating to different temperatures. Mild heating drives off adsorbed water, but leaves the structural water and OH. Further heating dehydrates and destroys the matrix phyllosilicates, with the 3-μm band disappearing as a result. These data are from *Hiroi et al.* (1996).

5.3. Rotational Variation

Although the visible 0.7-μm band indicates hydration when it is present, when it is not seen no firm conclusions can be drawn about the hydration state. In a sample of 51 asteroids, 31% of those that did not have the 0.7-μm band did have a 3-μm band indicating hydrated minerals (*Howell et al.,* 2001a). Rotational variation is common, and several observations at different rotation phases are necessary to accurately determine if hydrated minerals are present. Examples of rotational variability in 0.7-μm band depth can be seen in Fig. 7.

Vilas and Sykes (1996) predict that there should be more surface compositional diversity among smaller-diameter

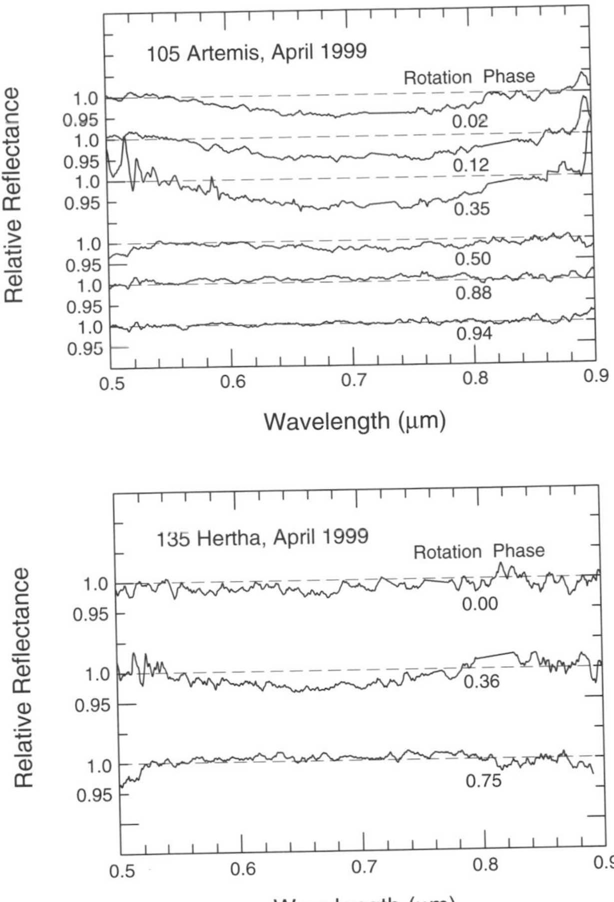

Fig. 7. Visible spectra were obtained at different rotational phases using the 2.1-m telescope at McDonald Observatory. The spectra show changes in the absorption band at 0.7 μm as the asteroid rotates. The upper plot shows that 105 Artemis has an absorption feature on one hemisphere, but not on the other. The lower plot shows a feature at one rotation phase of 135 Hertha. These variations are interpreted as localized areas of hydrated silicates exposed on the surface. This inhomogeneity could be in the regolith, or the underlying material, or both.

low-albedo asteroids, while the larger-diameter low-albedo asteroids should be individually homogeneous. However, rotationally resolved spectral observations indicate that up to 45% of asteroids have both hydrated and anhydrous surface regions in all size ranges. Asteroid 10 Hygiea (410 km diameter) has at least two distinct surface regions that show the 0.7-μm band, and other distinct regions that show only the 3-μm band. Still other areas have neither band. These observations cannot be explained by the same type of thermal metamorphism described above for 511 Davida, unless material from a wider range of depths is now exposed on the surface of 10 Hygiea. Spectra of asteroid 444 Gyptis (C) on opposite sides of the body show that one side of the asteroid has aqueously altered material, while the other side has no evidence of aqueous alteration products (*Thibault et al.,* 1995).

5.4. Putting It All Together in a Thermal Evolution Scenario

The distribution and diversity of aqueous alteration end products on asteroids in principle tells us about the original composition and the thermal evolution of the asteroids. However, the interaction and interdependence of the many different processes involved make disentangling them a challenge.

Grimm and McSween (1989) propose that radioactive heating is sufficient to aqueously alter asteroid parent bodies, provided that accretion occurs quickly enough. *Cohen and Coker* (2000) agree that ^{26}Al can produce sufficient heat for aqueous alteration, if accretion is well underway by 3 m.y. after nebula collapse, which agrees with dynamical models (*Wetherill,* 1990; *Weidenschilling and Davis,* 2001). While not ruled out, the electrical induction model for heating of asteroids to produce aqueous alteration (*Herbert and Sonett,* 1979) is no longer strongly supported, based on an improved understanding of the T-Tauri phase of solar-type stars (*Grimm and McSween,* 1993). Under both proposed heating mechanisms, most asteroids with diameters above 20 km, between 2.6 and 3.5 AU (where most of the C-class asteroids are found), would be heated to the point of water mobilization. Larger-diameter asteroids (up to diameters of 100 km) would undergo increasingly greater heating, with the interiors of the asteroids reaching higher temperatures than the surfaces (*Herbert and Sonett,* 1979; *Grimm and McSween,* 1993). The formation of Jupiter, however, resulted in asteroid orbits being pumped up in eccentricity and inclination, creating a collisionally disruptive environment (e.g., *Davis et al.,* 1989). Small asteroids observed today are representative of the interiors of larger (but not the largest) asteroids and have likely experienced multiple disruptive events. Asteroids having diameters greater than 100 km would also have been shattered, but remain gravitationally bound "rubble piles" with material churned by multiple collisions (*Davis et al.,* 1989). The spectral characteristics of the larger-diameter asteroids would represent the combination of thermally metamorphosed and aqueously altered material. This scenario predicts the larger asteroids to have more hydrated minerals than the smaller asteroids, since the latter are remnants of heated asteroid interiors, while the surfaces of larger asteroids have always been near the cooler exteriors of their parent bodies. This implicitly assumes similar-sized asteroids had similar starting materials as well as thermal and collisional histories.

The prediction of a size-band depth correlation was tested by applying the results of the algorithm for identifying the 0.7-μm feature (*Vilas,* 1994) to ECAS photometry of 153 low-albedo asteroids, binned by diameter (*Howell et al.,* 2001a). A trend of fewer low-albedo asteroids having the 0.7-μm feature at smaller sizes is consistent with the hypothesis of *Vilas and Sykes* (1996) and the scenario outlined above (Table 3).

However, if both 0.7-μm and 3-μm bands are used, the hydrated fraction is nearly constant with size. The large

TABLE 3. Percentage of Tholen CBFG asteroids testing positively for the 0.7-μm feature divided by diameter.

Diameter Range (km)	Number with 0.7-μm Feature	Total	Ratio
0–50	9	29	0.31
50–100	11	40	0.28
100–150	32	57	0.56
150–200	12	20	0.60
200–250	3	7	0.43

number of smaller objects observed as part of the SMASS II survey greatly adds to the number of asteroids below 20 km observed spectrally. However, these are observations at a single rotation phase, and should be viewed as a lower limit on the number of objects with hydration on some part of the surface, which is consistent with an *increase* in the hydrated fraction at smaller sizes. Additional study of smaller asteroids is required for a better understanding of how the surface material represents the interior of larger objects, now broken into fragments. A wide diversity seems to exist among asteroids of similar size, which may require a more complete integration of alteration mechanisms, regolith evolution, and collisional history in order to reach a complete understanding of hydrated surface mineralogy.

6. PROPOSED ALTERNATIVE INTERPRETATIONS FOR HYDRATED MINERAL ABSORPTION FEATURES

The interpretation that the 3-μm and 0.7-μm absorption features on asteroids are due to hydrated minerals has been challenged in some quarters. Their presence on the E- and M-class asteroids in particular has led some workers to propose alternative explanations that can maintain the interpretation of these asteroid classes as differentiated objects, with metallic Fe and no FeO present.

6.1. Hydroxyl Creation Through Solar-Wind Interaction

Starukhina (2001) has proposed that solar-wind-implanted H may react with silicates (or other O-bearing minerals) in asteroid regoliths, resulting in OH groups that are detectable spectroscopically but are not indicative of hydrated minerals and that did not originate through aqueous alteration. She calculates that OH created by the solar wind could explain all positive observations of 3-μm absorption bands on asteroids (although she notes that the calculated values for solar-wind-induced band strengths are upper limits and may overestimate the effect). However, this model and these calculations make some predictions that are not borne out: The calculated saturation time for solar-wind H near the equators of main-belt asteroids is a few hundred years, yet, the fraction of asteroids with detectable 3-μm absorption features is far from unity. The calculated band depth for solar-wind-induced OH increases with increasing continuum reflectance, yet no such correspondence is seen in the asteroids. The predicted absorption depth for 4 Vesta is 20% or greater (assuming saturation), but the upper limit on its band depth is 1% (*Lebofsky*, 1980). As instruments and techniques advance, however, and as observational uncertainties in the 3-μm region will more routinely approach the 1% level, the creation of OH through solar-wind interactions will need to be considered in interpreting weak 3-μm absorptions.

6.2. Troilite

Cloutis and Burbine (1999) and *Fornaisier and Lazzarin* (2001) suggested that a feature found near 0.5 μm in some E asteroids may be due to troilite (FeS) or other sulfides, and that the mineral responsible may also be responsible for the 3-μm feature in E-class and perhaps M-class asteroids. If true, this would potentially allow S contents of these asteroids to be determined, and would strengthen the interpretation of these bodies as differentiated. *Cloutis and Burbine* (1999) presented spectra of synthetic and natural troilite and pyrrhotite, which showed evidence of a 3-μm absorption feature. However, this feature when seen in troilite is likely due to laboratory contamination by atmospheric water. Because of the high concentration of water in Earth's atmosphere and its rapid reaction time with unoxidized material at room temperature, special precautions must be taken in the laboratory to ensure that any absorption features seen are due to the nonterrestrial properties of the sample of interest, particularly in powdered samples. *Salisbury et al.* (1991a) show that adsorbed water, either as a result of terrestrial weathering or simply from exposure of powdered samples to air, can contain up to 2% H_2O by weight. The 3-μm absorption bands shown in *Cloutis and Burbine* (1991b) are similar in depth and shape to those seen on quartz and barite in *Salisbury et al.* (1991b), and attributed to adsorbed water by those authors.

Furthermore, the group of E-class asteroids showing the visible absorption include both those with and without a 3-μm feature, and asteroids with the 3-μm feature include those both with and without the visible feature (*Rivkin and Howell,* 2001). Therefore, the two absorptions appear uncorrelated with each other in E-class asteroids.

6.3. Is the Observation of Water/Hydroxyl on Asteroids Meaningful?

As mentioned above, the strength of the 3-μm absorption is such that it can be detected at small concentrations, particularly when it is not mixed with very low-albedo constituents (*Clark,* 1983). The band strength is enhanced for small particle sizes, which may make a small amount in an asteroid regolith detectable (*Salisbury et al.,* 1991a). *Cloutis* (2001) argued that because the 3-μm band is seen in nominally anhydrous materials on Earth, it is not diagnostic for hydration on the asteroids or Mars when small concentrations are indicated. However, this study used powdered laboratory samples measured in air as a comparison, which is subject to adsorbed water as described above. Cloutis notes that adsorbed water from Earth's atmosphere may be responsible for the water feature in these nominally anhydrous laboratory materials. In the hard vacuum of space, with billions of years of temperature cycling, such loosely bound water can be expected to be lost. In addition, in the laboratory, one can measure band depths at 2.8 μm, which for asteroids is a wavelength that is not observable from the ground since telluric water vapor makes the atmosphere opaque. Of necessity, the band depths at 2.95–3.0 μm that are measured on astronomical objects are less than would be seen at 2.8 μm from a spaceborne instrument. In comparing astronomical spectra to laboratory measurements, caution must be used to ensure that the band depth refers to the same spectral region, relative to the same continuum.

7. OPEN QUESTIONS AND FUTURE WORK

7.1. Telescopic Work

The advent of larger telescopes and more sensitive instruments open up a larger number of possible targets for observational work. It is possible now to observe objects with diameters of ~10 km or smaller in the main belt and look for the 0.7-μm band, while 3-μm observations can reach 30-km low-albedo objects. This affords the opportunity to study trends of hydration state with size, following up on the necessarily tentative results presented in section 5. Near-Earth asteroids (NEAs) can also be studied in greater detail. At the time of this writing, only three NEAs have been observed at 3 μm [433 Eros (*Rivkin and Clark,* 2001), 4179 Toutatis (*Howell et al.,* 1994a), and 1036 Ganymed (Rivkin, unpublished data, 1999)], all of which are S asteroids, and all of which are anhydrous. The relative paucity of C asteroids in the NEA population has made observing them difficult, but they are sufficiently bright to make them feasible targets for observations at 0.7 μm. Increased instrumental sensitivities and larger telescopes also make it possible to do studies of dynamical/collisional families. This work has been initiated for the 0.7-μm feature, with work by *Bus* (1999) for the Chloris and Dora families, *Florczak et al.* (1999) for 36 members of the Themis family, and *Monthé-Diniz et al.* (2001) for 10 objects in the Hygiea family. *Bus* (1999) found the Chloris and Dora families to both be quite homogeneous in terms of hydration states, while the others found some differences between members of these much larger families. These types of studies should be extended to the 3-μm region. Similarities and differences among band depths and band shapes in a single family would help determine the nature and homogeneity of the aqueous alteration process, and could help determine any dependence upon depth in the parent body.

7.2. "Space Weathering" and Impact Processes on Hydrated Asteroids

The effect of regolith processes on determining the olivine and pyroxene abundances on an airless surface is an area of much current research. These "space weathering" effects are described in detail in *Clark et al.* (2002). We note that these studies are of great importance to studies of hydrated minerals. If the spectral effects are greater at visible wavelengths than in the IR, as it appears currently, the relative strengths of the 0.7- and 3-μm features may be usable as a maturity metric for hydrated asteroids. If these processes can cause diminished band depths for the hydration bands, attempts to accurately determine water contents will be complicated, but lower limits can still be found.

The effect of large impacts on the volatile budget of asteroids is still unknown. Impacts may dehydrate hydrated target material, although inefficient energy coupling may leave most of the ejecta largely unaffected. Furthermore, the most-heated fraction of ejecta may also be the fraction least likely to be retained on an asteroid. The presence of phyllosilicates in interplanetary dust particles (IDPs) (*Rietmeijer and MacKinnon,* 1985; *Zolensky and Kieller,* 1991; *Keller and Zolensky,* 1991) suggests that hydrated minerals can survive not only impact and ejection but millions of years in vacuum, followed by atmospheric entry. Experiments designed to study the spectral changes in phyllosilicates due to impacts have been attempted. *Boslough et al.* (1980) took spectra of shocked and unshocked nontronite and serpentine, and found that the absorption features due to OH and H_2O decreased after shocking. However, the postshock features were still quite strong, and the shock technique was not truly analogous to an impact. *Thibault et al.* (1997) performed impact experiments on serpentine and obtained reflectance spectra of serpentine pre- and postimpact, finding that the 1.9-μm H_2O feature was gone, but the 1.4-μm OH feature (and presumably the 2.7-μm fundamental) was largely unchanged. Obviously, hydrated minerals can survive family-forming impacts on an asteroid-wide scale, as shown by the prevalence of Ch and Cgh asteroids (which have the 0.7-μm feature) in the Dora and Chloris family (*Bus,* 1999).

As better data and analysis help determine the identities and relative abundances of specific hydrated minerals present on asteroidal surfaces, and we begin to understand how the surface constituents reflect internal compositions, aqueous alteration models will necessarily benefit. At this writing, these models have focused almost entirely on the meteorite record, without benefit of much in the way of asteroidal constraints. Future telescopic work will sample a much larger range of parent bodies and further limit the formation conditions of hydrated minerals.

8. THE BIG PICTURE, IN WATERCOLORS

The presence of 3-μm and 0.7-μm absorption features on some (but not all) asteroids tells us that hydrated minerals can persist on the surface of airless bodies, presumably since early in solar system history. Whatever regolith maturation may be occurring on 1 Ceres, 2 Pallas, 13 Egeria, and other large asteroids does not destroy the signature of the hydrated minerals present. The first-order correlation of this feature with taxonomic classes is similarly important. If it were found on all asteroids, in all classes, it would suggest an exogenic process (like micrometeorite impacts) brought in the hydrated minerals. However, because the S asteroids as a group are very unlikely to have hydrated minerals compared to the C-class asteroids, we can infer that the hydrated mineralogy of the asteroid classes is an intrinsic compositional difference. The same is true even at subclass levels, where nearly every one of the G-class asteroids has a 3-μm absorption feature, while only one of the F-class asteroids does despite the same sample size for each class. The large-scale generation of OH by solar-wind-generated reactions on all these asteroids can be rejected by the same reasoning, as detailed in section 6.

The depth of the 3-μm absorption feature also makes any exogenic origin unlikely. Take for example the hydrated M-class (W-class) asteroids, which have band depths on the order of 5–10%. This is in constrast to asteroids like 16 Psyche, 45 Eugenia, and 185 Eunike, which all have upper limits of 1–2% for 3-μm band depth. If the feature on the hydrated M-class asteroids were due to carbonaceous micrometeorites, these micrometeorites would need to comprise at least several percent of the regolith to give an absorption feature of this depth, assuming a 100% band strength for the carbonaceous micrometeorites. Yet these micrometeorites would have had to avoid the asteroids, which show no feature.

In summary, we conclude the following:

1. Almost every taxon has both hydrated and anhydrous members based on their 3-μm spectra. For some taxa (e.g., Ms), this probably is the result of disparate mineralogies having similar spectra in the 0.3–2.5-μm region. For others (e.g., Cs), both hydrated and anhydrous versions of the same mineralogy are probably present. The fraction of hydrated members is different in the different classes, ranging from ~65% for the C class to ~5% for the S class.

2. Broadly speaking, there appear to be two kinds of 3-μm band shapes. A "checkmark" shape is much more common in our sample among low-albedo asteroids than a rounded shape. A rounded shape is more common among the higher-albedo hydrated asteroids. Ceres has a modified rounded shape, as does 375 Ursula, implying that the hydrated mineral assemblage on Ceres may be rare but is not unique.

3. The presence of a 0.7-μm feature appears to be sufficient to identify the presence of hydrated minerals. The lack of such a feature does *not* appear to be sufficient to rule out the presence of hydrated minerals (since the 3-μm feature is more sensitive, and some higher-albedo asteroids do not appear to have 0.7-μm features).

4. Used in tandem, the 0.7- and 3-μm absorption features may constrain the maximum heating a body has experienced, although that heating may have been strongly nonuniform.

5. Rotational variation is common if not ubiquitous in larger-diameter asteroids, but it is uncertain if this is a remnant of formation or the result of thermal or impact evolution. Rotational variation also appears to explain past discrepancies between low-albedo asteroid hydration states as determined from 0.7- and 3-μm observations.

Acknowledgments. This work was partially supported by the NASA Planetary Geology and Geophysics and Planetary Astronomy programs. Reviews by S. J. Bus and D. Cruikshank greatly improved this manuscript. Thanks to the NASA IRTF, UKIRT, and McDonald Observatory, who have generously granted telescope time to perform many of the studies summarized here. The ADS abstract database aided the literature search enormously.

REFERENCES

A'Hearn M. F. and Feldman P. D. (1992) Water vaporization on Ceres. *Icarus, 98,* 54–60.

Avanesov G., Zhukov B., Ziman Y., Kostenko V., Kuzmin A., Muravev V., Fedotov V., Bonev B., Mishev D., Petkov D., Krumov A., Simeonov S., Boycheva V., Uzunov Y., Weide G. G., Halmann D., Possel W., Head J., Murchie S., Schkuratov Y. G., Berghanel R., Danz M., Mangoldt T., Pihan U., Weidlich U., Lumme K., Muinonen K., Peltoniemi J., Duxbury T., Murray B., Herkenhoff K., Fanale F., Irvine W., and Smith B. (1991) Results of TV imaging of Phobos — experiment VSK-Fregat. *Planet. Space Sci., 39,* 281–295.

Barucci M. A., Lazzarin M., Owen T., Barbieri C., and Fulchignoni M. (1994) Near-infrared spectroscopy of dark asteroids. *Icarus, 110,* 287–291.

Barucci M. A., Doressoundiram A., Fulchignoni M., Florczak M., Lazzarin M., Angeli C., and Lazzaro D. (1998) Search for aqueously altered materials on asteroids. *Icarus, 132,* 388–396.

Bell J. F. (1988) A probable asteroidal parent body for the CV or CO chondrites. *Meteoritics, 23,* 256–257.

Bell J. F. and Crisp D. A. (1991) Near-infrared (1.3–4.0 micron) imaging spectroscopy of Mars (abstract). In *Lunar and Planetary Science XXII,* pp. 73–74. Lunar and Planetary Institute, Houston.

Bell J. F., Davis D. R., Hartmann W. K., and Gaffey M. J. (1989) The big picture. In *Asteroids II* (R. P. Binzel et al., eds.), pp. 921–945. Univ. of Arizona, Tucson.

Belskaya I. N. and Lagerkvist C.-I. (1996) Physical properties of M-class asteroids. *Planet. Space Sci., 44,* 783–794.

Bischoff A. (1998) Aqueous alteration of carbonaceous chondrites: Evidence for preaccretionary alteration. A review. *Meteoritics & Planet. Sci., 33,* 1113–1122.

Bishop J. L. and Pieters C. M. (1995) Low-temperature and low atmospheric pressure infrared reflectance spectroscopy of Mars soil analog materials. *J. Geophys. Res., 100,* 5369–5379.

Blaney D. L. (1991) Infrared imaging of Mars between 2.4 microns and 5.0 microns (abstract). In *Lunar and Planetary Science XXII,* pp. 111–112. Lunar and Planetary Institute, Houston.

Bose K. (1993) Equilibrium and kinetic studies in the system MgO-SiO_2-H_2O. Ph.D. dissertation, University of Arizona, Tucson.

Boslough M. B., Weldon R. J., and Ahrens T. J. (1980) Release of water from hydrous minerals due to impact (abstract). In *Lunar and Planetary Science XI,* pp. 97–99. Lunar and Planetary Institute, Houston.

Bottke W. F. Jr., Vokrouhlický D., Rubincam D. P., and Brož M. (2002) The effect of Yarkovsky thermal forces on the dynamical evolution of asteroids and meteoroids. In *Asteroids III* (W. F. Bottke Jr. et al., eds.), this volume. Univ. of Arizona, Tucson.

Bregman J. D., Tielens A. G. G. M., Witteborn F. C., Rank D. M., and Wooden D. (1988) 3 micron spectrophotometry of Comet Halley: Evidence for water ice. *Astrophys. J., 334,* 1044–1048.

Britt D. T., Yeomans D., Housen K, and Consolmagno G. (2002) Asteroid density, porosity, and structure. In *Asteroids III* (W. F. Bottke Jr. et al., eds.), this volume. Univ. of Arizona, Tucson.

Brooke T. Y., Knacke R. F., Owen T. C., and Tokunaga A. T. (1989) Spectroscopy of emission features near 3 microns in Comet Wilson (1986l). *Astrophys. J., 336,* 971–978.

Brown R. H., Cruikshank D. P., Pendleton Y. J., and Veeder G. J. (1997) Surface composition of Kuiper belt object 1993 SC. *Science, 276,* 937–939.

Browning L. B., McSween H. Y., and Zolensky M. E. (1996) Correlated alteration effects in CM carbonaceous chondrites. *Geochim. Cosmochim. Acta, 60,* 2621–2633.

Burbine T. H. (1998) Could G-class asteroids be the parent bodies of the CM chondrites? *Meteoritics & Planet. Sci., 33,* 253–258.

Burbine T. H., Binzel R. P., Bus S. J., and Clark B. E. (2001) K asteroids and CO3/CV3 chondrites. *Meteoritics & Planet. Sci., 36,* 245–253.

Burbine T. H., McCoy T. J., Meibom A., Gladman B., and Keil K. (2002) Meteoritic parent bodies: Their number and identification. In *Asteroids III* (W. F. Bottke Jr. et al., eds.), this volume. Univ. of Arizona, Tucson.

Bus S. J. (1999) Compositional structure in the asteroid belt: Results of a spectroscopic survey. Ph.D. dissertation, Massachusetts Institute of Technology, Cambridge.

Busarev V. V. and Krugly Yu. N. (1995) A spot of hydrated silicates on the M asteroid 201 Penelope? (abstract). In *Lunar and Planetary Science XXVI,* pp. 197–198. Lunar and Planetary Institute, Houston.

Calvin W. M. and King T. V. V. (1997) Spectral characteristics of iron-bearing phyllosilicates: Comparison to Orgueil (CI1), Murchison and Murray (CM2). *Meteoritics & Planet. Sci., 32,* 693–701.

Carvano J. M., Lazzaro D., Mothé-Diniz T., Angeli C. A., and Florczak M. (2001) Spectroscopic survey of the Hungaria and Phocaea dynamical groups. *Icarus, 149,* 173–189.

Chapman C. R. (1996) S-type asteroids, ordinary chondrites, and space weathering: The evidence from Galileo's fly-bys of Gaspra and Ida. *Meteoritics & Planet. Sci., 31,* 699–725.

Clark B. E., Lucey P. G., Bell J. F., and Fanale F. P. (1993) Spectral mixing models of S-type asteroids (abstract). In *Lunar and Planetary Science XXIV,* pp. 303–304. Lunar and Planetary Institute, Houston.

Clark B. E., Lucey P., Helfenstein P., Bell J. F., Peterson C., Veverka J., McConnochie T., Robinson M. S., Bussey B., Murchie S. L., Izenberg N. I., and Chapman C. R. (2001) Space weathering on Eros: Constraints from albedo and spectral measurements of Psyche crater. *Meteoritics & Planet. Sci., 36,* 1617–1637.

Clark B. E., Hapke B., Pieters C., and Britt D. (2002) Asteroid space weathering and regolith evolution. In *Asteroids III* (W. F. Bottke Jr. et al., eds.), this volume. Univ. of Arizona, Tucson.

Clark R. N. (1981) The spectral reflectance of water-mineral mixtures at low temperatures. *J. Geophys. Res., 86,* 3074–3086.

Clark R. N. (1983) Spectral properties of mixtures of montmorillonite and dark carbon grains: Implications for remote sensing minerals containing chemically and physically adsorbed water. *J. Geophys. Res., 88,* 10635–10644.

Clark R. N. and McCord T. B. (1980) The Galilean satellites — New near-infrared spectral reflectance measurements (0.65–2.5 microns) and a 0.325–5 micron summary. *Icarus, 41,* 323–339.

Clark R. N., Steele A., Brown R. H., and Owensby P. D. (1984) Saturn's satellites — near-infrared spectrophotometry (0.65–2.5 microns) of the leading and trailing sides and compositional implications. *Icarus, 58* 265–281.

Cloutis E. A. (2001) H_2O/OH-associated absorption band depth relationships in mineral reflectance spectra (abstract). In *Lunar and Planetary Science XXXII,* Abstract #1146. Lunar and Planetary Institute, Houston (CD-ROM).

Cloutis E. A. and Burbine T. H. (1999) The spectral properties of troilite/pyrrhotite and implications for the E-asteroids (abstract). In *Lunar and Planetary Science XXX,* Abstract #1875. Lunar and Planetary Institute, Houston (CD-ROM).

Cochran A. L. and Vilas F. (1997) The McDonald Observatory serendipitous UV/blue spectral survey of asteroids. *Icarus, 127,* 121–129.

Cohen B. A. and Coker R. F. (2000) Modeling of liquid water on CM meteorite parent bodies and implications for amino acid racemization. *Icarus, 145,* 369–381.

Consolmagno G. J. and Britt D. T. (2002) Low-density materials and asteroidal macroporosity (abstract). In *Lunar and Planetary Science XXXIII,* Abstract #1701. Lunar and Planetary Institute, Houston (CD-ROM).

Cruikshank D. P., Roush T. L., Bartholomew M. J., Geballe T. R., Pendleton Y. J., White S. M., Bell J. F., Davies J. K., Owen T. C., De Bergh C., Tholen D. J., Bernstein M. P., Brown R. H., Tryka K. A., and Dalle Ore C. M. (1998) The composition of Centaur 5145 Pholus. *Icarus, 135,* 389–407.

Cruikshank D. P., Ore C. M. D., Roush T. L., Geballe T. R., Owen T. C., De Bergh C., Cash M. D., and Hartmann W. K. (2001) Constraints on the composition of Trojan asteroid 624 Hektor. *Icarus, 153,* 348–360.

Cyr K. E. (1998) The distribution of water in the solar nebula: Implications for solar system formation. Ph.D. dissertation, University of Arizona, Tucson.

Cyr K. E., Sears W. D., and Lunine J. I. (1998) Distribution and evolution of water ice in the solar nebula: Implications for solar system body formation. *Icarus, 135,* 537–548.

Dahlgren M. and Lagerkvist C.-I. (1995) A study of Hilda asteroids. I. CCD spectroscopy of Hilda asteroids. *Astron. Astrophys., 302,* 907–914.

Dahlgren M., Lagerkvist C.-I., Fitzsimmons A., Williams I. P., and Gordon M. (1997) A study of Hilda asteroids. II. Compositional implications from optical spectroscopy. *Astron. Astrophys., 323,* 606–619.

Davis D. R., Weidenschilling S. J., Farinella P., Paolicchi P., and Binzel R. P. (1989) Asteroid collisional history: Effects on sizes and spins. In *Asteroids II* (R. P. Binzel et al., eds.), pp. 805–826. Univ. of Arizona, Tucson.

Deer W. A., Howie R. A., and Zussman J. (1962) *Rock-Forming Minerals, Vol. 3: Sheet Silicates.* Longman, London.

Delsemme A. H. (2000) 1999 Kuiper prize lecture. *Icarus, 146,* 313–325.

DuFresne E. R. and Anders E. (1962) On the chemical evolution of the carbonaceous chondrites. *Geochim. Cosmochim. Acta, 26,* 1084–1114.

Dumas C., Owen T., and Barucci M. A. (1998) Near-infrared spectroscopy of low-albedo surfaces of the solar system: Search for the spectral signature of dark material. *Icarus, 133,* 221–232.

Eaton N., Green S. F., McCheyne R. S., Meadows A. J., and Veeder G. J. (1983) Observations of asteroids in the 3- to 4-micron region. *Icarus, 55,* 245–249.

Emery J. P. and Brown R. H. (2001) Near-infrared spectroscopy of Trojan asteroids: Implications for the evolution of the solar system (abstract). In *Lunar and Planetary Science XXXII,* Abstract #1385. Lunar and Planetary Institute, Houston (CD-ROM).

Encrenaz T., Drossart P., Feuchtgruber H., Lellouch E., Bézard B., Fouchet T., and Atreya S. K. (1999) The atmospheric composition and structure of Jupiter and Saturn from ISO observations: A preliminary review. *Planet. Space Sci., 47,* 1225–1242.

Farinella P., Gonczi R., Froeschlé C., and Froeschlé C. (1993) The injection of asteroid fragments into resonances. *Icarus, 101,* 174–187.

Feierberg M. A., Lebofsky L. A., and Larson H. P. (1981) Spectroscopic evidence for aqueous alteration products on the surfaces of low-albedo asteroids. *Geochim. Cosmochim. Acta, 45,* 971–981.

Feierberg M. A., Lebofsky L. A., and Tholen D. (1985) The nature of C-class asteroids from 3-μm spectrophotometry. *Icarus, 63,* 183–191.

Fitzsimmons A., Dahlgren M., Lagerkvist C.-I., Magnusson P., and Williams I. P. (1994) A spectroscopic survey of D-type asteroids. *Astron. Astrophys., 282,* 634–642.

Florczak M., Lazzaro D., Mothé-Diniz T., Angeli C. A., and Betzler A. S. (1999) A spectroscopic study of the Themis family. *Astron. Astrophys. Suppl. Ser., 134,* 463–471.

Fornasier S. and Lazzarin M. (2001) E-type asteroids: Spectroscopic investigation on the 0.5 μm absorption band. *Icarus, 152,* 127–133.

Fornasier S., Lazzarin M., Barbieri C., and Barucci M. A. (1999) Spectroscopic comparison of aqueous altered asteroids with CM2 carbonaceous chondrite meteorites. *Astron. Astrophys. Suppl. Ser., 135,* 65–73.

Gaffey M. J. and Gilbert S. L. (1998) Asteroid 6 Hebe: The probable parent body of the H-type ordinary chondrites and the IIE iron meteorites. *Meteoritics & Planet. Sci., 33,* 1281–1295.

Gaffey M. J., Bell J. F., Brown R. H., Burbine T. H., Piatek J. L., Reed K. L., and Chaky D. A. (1993a) Mineralogical variations within the S-type asteroid class. *Icarus, 106,* 573–602.

Gaffey S. J., McFadden L. A., Nash D., and Pieters C. M. (1993b) Ultraviolet, visible, and near-infrared reflectance spectroscopy: Laboratory spectra of geologic materials. In *Remote Geochemical Analysis: Elemental and Mineralogical Composition* (C. M. Pieters and P. A. J. Englert, eds.), pp. 43–77. Cambridge Univ., Cambridge.

Grimm R. E. and McSween H. Y. (1989) Water and the thermal evolution of carbonaceous chondrite parent bodies. *Icarus, 82,* 244–280.

Grimm R. E. and McSween H. Y. (1993) Heliocentric zoning of the asteroid belt by aluminum-26 heating. *Science, 259,* 653–655.

Haack H., Rasmussen K. L., and Warren P. H. (1990) Effects of regolith/megaregolith insulation on the cooling histories of differentiated asteroids. *J. Geophys. Res., 95,* 5111–5124.

Harris A. W. (2001) The albedos and taxonomy of small asteroids. In *Asteroids 2001: From Piazzi to the Third Millenium,* p. 61. Osservatorio di Palermo, Sicily.

Harris A. W. and Lagerros J. S. V. (2002) Asteroids in the thermal infrared. In *Asteroids III* (W. F. Bottke Jr. et al., eds.), this volume. Univ. of Arizona, Tucson.

Herbert F. and Sonett C. P. (1979) Electromagnetic heating of minor planets in the early solar system. *Icarus, 40,* 484–496.

Hiroi T. and Vilas F. (1996) Discovery and analysis of minor absorption bands in S-asteroid visible reflectance spectra. *Icarus, 119,* 202–208.

Hiroi T., Zolensky M. E., Pieters C. M., and Lipschutz M. E. (1996) Thermal metamorphism of the C, G, B, and F asteroids seen from the 0.7 μm, 3 μm and UV absorption strengths in comparison with carbonaceous chondrites. *Meteoritics & Planet. Sci., 31,* 321–327.

Hiroi T., Zolensky M. E., and Pieters C. M. (2001a) Discovery of the first D-asteroid spectral counterpart: Tagish lake meteorite (abstract). In *Lunar and Planetary Science XXXII,* Abstract #1776. Lunar and Planetary Institute, Houston (CD-ROM).

Hiroi T., Zolensky M. E., and Pieters C. M. (2001b) The Tagish Lake meteorite: A possible sample for a D-type asteroid. *Science, 293,* 2234–2236.

Howell E. S. (1995) Probing asteroid composition using visible and near-infrared spectroscopy. Ph.D. dissertation, University of Arizona, Tucson.

Howell E. S., Britt D. T., Bell J. F., Binzel R. P., and Lebofsky L. A. (1994a) Visible and near-infrared spectral observations of 4179 Toutatis. *Icarus, 111,* 468–474.

Howell E. S., Merényi E., and Lebofsky L. A. (1994b) Classification of asteroid spectra using a neural network. *J. Geophys. Res., 99,* 10847–10865.

Howell E. S., Rivkin A. S., and Vilas F. (2001a) Uneven distribution of aqueously altered minerals on asteroids. In *Asteroids 2001: From Piazzi to the Third Millenium,* p. 62. Osservatorio di Palermo, Sicily.

Howell E. S., Rivkin A. S., Vilas F., and Soderberg A. M. (2001b) Aqueous alteration in low albedo asteroids (abstract). In *Lunar and Planetary Science XXXII,* Abstract #2058. Lunar and Planetary Institute, Houston (CD-ROM).

Hutchison R., Alexander C. M. O., and Barber D. J. (1987) The Semarkona meteorite: First recorded occurrence of smectite in an ordinary chondrite, and its implications. *Geochim. Cosmochim. Acta, 51,* 1875–1882.

Jarosewich E. (1990) Chemical analyses of meteorites: A compilation of stony and iron meteorite analyses. *Meteoritics, 25,* 323–337.

Jarvis K. S., Vilas F., Gaffey M. J., and Fitzsimmons A. (1998) The 0.43-micron absorption feature in reflectance spectra of low-albedo asteroids. *Bull. Am. Astron. Soc., 30,* 1025.

Jarvis K. S., Vilas F., Howell E. S., Kelley M. S., and Cochran A. L. (1999) Testing the prediction of iron alteration minerals on low albedo asteroids. *Bull. Am. Astron. Soc., 31,* 1090.

Jewitt D. C. and Luu J. X. (1990) CCD spectra of asteroids. II – The Trojans as spectral analogs of cometary nuclei. *Astron. J., 100,* 933–944.

Jones T. D. (1989) An infrared reflectance study of water in outer belt asteroids: Clues to composition and origin. Ph.D. dissertation, Univ. of Arizona, Tucson.

Jones T. D., Lebofsky L. A., Lewis J. S., and Marley M. S. (1990)

The composition and origin of the C, P, and D asteroids: Water as a tracer of thermal evolution in the outer belt. *Icarus, 88,* 172–192.

Keller L. P. and Zolensky M. E. (1991) Clay minerals in primitive meteorites and interplanetary dust. II: Smectites and micas. *Clay Min. Soc., 28,* 87.

King T. V. V. and Clark R. N. (1997) The presence of a single absorption feature: What it does and doesn't imply (abstract). In *Lunar and Planetary Science XXVIII,* pp. 727–728. Lunar and Planetary Institute, Houston.

King T. V. V., Clark R. N., Calvin W. M., Sherman D. M., and Brown R. H. (1992) Evidence for ammonium-bearing minerals on Ceres. *Science, 255,* 1551–1553.

Krot A. N., Scott E. R. D., and Zolensky M. E. (1997) Origin of fayalitic olivine rims and lath-shaped matrix olivine in the CV3 chondrite Allende and its dark inclusions. *Meteoritics & Planet. Sci., 32,* 31–49.

Larson H. P., Fink U., Treffers R., and Gautier T. N. (1975) Detection of water vapor on Jupiter. *Astrophys. J. Lett., 197,* L137–L140.

Lazzarin M., Barbieri C., and Barucci M. A. (1995) Visible spectroscopy of dark, primitive asteroids. *Astron. J., 110,* 3058–3072.

Lebofsky L. A. (1978) Asteroid 1 Ceres: Evidence for water of hydration. *Mon. Not. R. Astron. Soc., 182,* 17–21.

Lebofsky L. A. (1980) Infrared reflectance spectra of asteroids: A search for water of hydration. *Astron. J., 85,* 573–585.

Lebofsky L. A. and Spencer J. R. (1989) Radiometry and thermal modeling of asteroids. In *Asteroids II* (R. P. Binzel et al., eds.), pp. 128–147. Univ. of Arizona, Tucson.

Lebofsky L. A., Feierberg M. A., Tokunaga A. T., Larson H. P., and Johnson J. R. (1981) The 1.7- to 4.2-micron spectrum of asteroid 1 Ceres: Evidence for structural water in clay minerals. *Icarus, 48,* 453–459.

Lebofsky L. A., Sykes M. V., Tedesco E. F., Veeder G. J., Matson D. L., Brown R. H., Gradie J. C., Feierberg M. A., and Rudy R. J. (1986) A refined "standard" thermal model for asteroids based on observations of 1 Ceres and 2 Pallas. *Icarus, 68,* 239–251.

Lebofsky L. A., Jones T. D., Owensby P. D., Feierberg M. A., and Consolmagno G. J. (1990) The nature of low albedo asteroids from 3-μm spectrophotometry. *Icarus, 83,* 12–26.

Lewis J. S. and Prinn R. G. (1984) *Planets and Their Atmospheres: Origin and Evolution.* Academic, Orlando. 470 pp.

Magri C., Ostro S. J., Rosema K. D., Thomas M. L., Mitchell D. L., Campbell D. B., Chandler J. F., Shapiro I. I., Giorgini J. D., and Yeomans D. K. (1999) Mainbelt asteroids: Results of Arecibo and Goldstone radar observations of 37 objects during 1980–1995. *Icarus, 140,* 379–407.

McCord T. B., Hansen G. B., Matson D. L., Johnson T. V., Crowley J. K., Fanale F. P., Carlson R. W., Smythe W. D., Martin P. D., Hibbitts C. A., Granahan J. C., and Ocampo A. (1999) Hydrated salt minerals on Europa's surface from the Galileo near-infrared mapping spectrometer (NIMS) investigation. *J. Geophys. Res., 104,* 11827–11852.

Merényi E., Howell E. S., Rivkin A. S., and Lebofsky L. A. (1997) Prediction of water in asteroids from spectral data shortward of 3 μm. *Icarus, 129,* 421–439.

Merline W. J., Close L. M., Dumas C., Shelton J. C., Menard F., Chapman C. R., and Slater D. C. (2000) Discovery of companions to asteroids 762 Pulcova and 90 Antiope by direct imaging. *Bull. Am. Astron. Soc., 32,* 1017.

Metzler K., Bischoff A., and Stoeffler D. (1992) Accretionary dust mantles in CM chondrites — Evidence for solar nebula processes. *Geochim. Cosmochim. Acta, 56,* 2873–2897.

Migliorini F., Manara A., Scaltriti F., Farinella P., Cellino A., and Martino M. D. (1997) Surface properties of (6) Hebe: A possible parent body of ordinary chondrites. *Icarus, 128,* 104–113.

Miyamoto M. and Zolensky M. E. (1994) Infrared diffuse reflectance spectra of carbonaceous chondrites: Amount of hydrous minerals. *Meteoritics, 29,* 849–853.

Morbidelli A., Chambers J., Lunine J. I., Petit J. M., Robert F., Valsecchi G. B., and Cyr K. E. (2000) Source regions and time scales for the delivery of water to Earth. *Meteoritics & Planet. Sci., 35,* 1309–1320.

Mothé-Diniz T., DiMartino M., Bendjoya P., Doressoundiram A., and Migliorini F. (2001) Rotationally resolved spectra of 10 Hygiea and a spectroscopic study of the Hygiea family. *Icarus, 152,* 117–126.

Murchie S. and Erard S. (1996) Spectral properties and heterogeneity of Phobos from measurements by Phobos 2. *Icarus, 123,* 63–86.

Noble S. K., Pieters C. M., Taylor L. A., Morris R. V., Allen C. C., McKay D. S., and Keller L. P. (2001) The optical properties of the finest fraction of lunar soil: Implications for space weathering. *Meteoritics & Planet. Sci., 36,* 31–42.

Ostro S. J., Jurgens R. F., Rosema K. D., Winkler R., Howard D., Rose R., Slade M. A., Yeomans D. K., Campbell D. B., Perillat P., Chandler J. F., Shapiro I. I., Hudson R. S., Palmer P., and de Pater I. (1993) Radar imaging of asteroid 4179 Toutatis. *Bull. Am. Astron. Soc., 25,* 1126.

Pieters C., Taylor L. A., Noble S. K., Keller L. P., Hapke B., Morris R. V., Allen C. C., McKay D. S., and Wentworth S. (2000) Space weathering on airless bodies: Resolving a mystery with lunar samples. *Meteoritics & Planet. Sci., 35,* 1101–1107.

Prinn R. G. and Fegley B. Jr. (1989) Solar nebula chemistry: Origin of planetary, satellite, and cometary volatiles. In *Origin and Evolution of Planetary and Satellite Atmospheres* (S. K. Atreya et al., eds.), pp. 78–136. Univ. of Arizona, Tucson.

Rietmeijer F. J. M. and MacKinnon I. D. R. (1985) Layer silicates in a chondritic porous interplanetary dust particle. *Proc. Lunar Planet. Sci. Conf. 16th,* in *J. Geophys. Res., 90,* D149–D155.

Rietmeijer F. J. M. and Nuth J. A. III (2000) Metastable eutectic equilibrium brought down to Earth. *Eos Trans. AGU, 81,* 409–420.

Rivkin A. S. (1997) Observations of main-belt asteroids in the 3-μm region. Ph.D. dissertation, University of Arizona, Tucson.

Rivkin A. S. and Clark B. E. (2001) Observations of 433 Eros from 1.25–3.35 μm. *Meteoritics & Planet. Sci., 36,* 1729–1729.

Rivkin A. S. and Davies J. K. (2002) Calculated water concentrations on C-class asteroids (abstract). In *Lunar and Planetary Science XXXIII,* Abstract #1414. Lunar and Planetary Institute, Houston (CD-ROM).

Rivkin A. S. and E. S. Howell (2001) E-class asteroids: Troilite, hydrated minerals, both, or neither? *Bull. Am. Astron. Soc., 33,* 1150.

Rivkin A. S., Howell E. S., Britt D. T., Lebofsky L. A., Nolan M. C., and Branston D. D. (1995) 3-μm spectrophotometric survey of M and E-class asteroids. *Icarus, 117,* 90–100.

Rivkin A. S., Clark B. E., Britt D. T., and Lebofsky L. A. (1997a) Infrared spectrophotometry of the NEAR flyby target 253 Mathilde. *Icarus, 127,* 255–257.

Rivkin A. S., Howell E. S., Clark B. E., Lebofsky L. A., and Britt

D. T. (1997b) S-class asteroids — The view from three microns (abstract). In *Lunar and Planetary Science XXVIII,* Abstract #1183. Lunar and Planetary Institute, Houston (CD-ROM).

Rivkin A. S., Trilling D. E., and Lebofsky L. A. (1998) Infrared (1.65–3.5 micrometers) observations of 387 Aquitania. *Bull. Am. Astron. Soc., 30,* 1023.

Rivkin A. S., Lebofsky L. A., Clark B. E., Howell E. S., and Britt D. T. (2000) The nature of M-class asteroids in the 3-μm region. *Icarus, 145,* 351–368.

Rivkin A. S., Davies J. K., Clark B. E., Trilling D. E., and Brown R. H. (2001) Aqueous alteration on S asteroid 6 Hebe? (abstract). In *Lunar and Planetary Science XXXII,* Abstract #1723. Lunar and Planetary Institute, Houston (CD-ROM).

Rivkin A. S., Brown R. H., Trilling D. E., Bell J. F., and Plassmann J. H. (2002) Infrared spectrophotometry of Phobos and Deimos. *Icarus, 156,* 64–75.

Roots C. F., Zolensky M. E., Grady M. M., Clayton R. N., Mayeda T. K., Hildebrand A. R., Brown P. G., and Brook J. L. (2000) Tagish Lake: A special new type 2 carbonaceous chondrite fall (abstract). *Meteoritics & Planet. Sci., 35,* A178.

Rubin A. E. (1996) Mineralogy of meteorite groups. *Meteoritics & Planet. Sci., 32,* 231–247.

Salisbury J. W., D'Aria D. M., and Jarosewich E. (1991a) Mid-infrared (2.5–13.5 μm) reflectance spectra of powdered stony meteorites. *Icarus, 92,* 280–297.

Salisbury J. W., Walter L. S., Vergo N., and D'Aria D. M. (1991b) *Infrared (2.1–25 μm) Spectra of Minerals.* Johns Hopkins Univ., Baltimore.

Sasaki S., Nakamura K., Hamabe Y., Kurahashi E., and Hiroi T. (2001) Production of iron nanoparticles by laser irradiation in a simulation of lunar-like space weathering. *Nature, 410,* 555–557.

Sato K., Miyamoto M., and Zolensky M. E. (1997) Absorption bands near three micrometers in diffuse reflectance spectra of carbonaceous chondrites: Comparison with asteroids. *Meteoritics & Planet. Sci., 32,* 503–507.

Sawyer S. R. (1991) A high resolution CCD spectroscopic survey of low-albedo main belt asteroids. Ph.D. dissertation, University of Texas, Austin.

Saylor J., Zolensky M. E., Bodnar R. J., Le L., and Schwandt C. (2001) Fluid inclusions in carbonaceous chondrites (abstract). In *Lunar and Planetary Science XXXII,* Abstract #1875. Lunar and Planetary Institute, Houston (CD-ROM).

Sears D. W. G. and Dodd R. T. (1988) Overview and classification of meteorites. In *Meteorites and the Early Solar System* (J. F. Kerridge and M. S. Matthews, eds.), pp. 1–31. Univ. of Arizona, Tucson.

Sinton W. M. (1967) On the composition of martian surface materials. *Icarus, 6,* 222–228.

Starukhina L. (2001) Water detection on atmosphereless celestial bodies: Alternative explanations of the observations. *J. Geophys. Res., 106,* 14701–14710.

Takaoka N., Nakamura T., Noguchi T., Tonui E., Gounelle M., Zolensky M. E., Ebisawa N., Osawa T., Okazaki R., Nagao K., and Yoneda S. (2001) Sayama CM2 chondrite: Fresh but heavily altered (abstract). In *Lunar and Planetary Science XXXII,* Abstract #1645. Lunar and Planetary Institute, Houston (CD-ROM).

Thibault C. A., Vilas F., Sawyer S. A., and Fitzsimmons A. (1995) Do low-albedo asteroids show surface mineralogical variations? *Bull. Am. Astron. Soc., 27,* 5.

Thibault C. A., Mickelson M. E., and Vilas F. (1997) Reflectance spectroscopy of phyllosilicates from 0.4 to 1.0 μm. *Bull. Am. Astron. Soc., 29,* 975.

Tholen D. J. and Barucci M. A. (1989) Asteroid taxonomy. In *Asteroids II* (R. P. Binzel et al., eds.), pp. 298–315. Univ. of Arizona, Tucson.

Veverka J., Bell J. F. III, Chapman C., Malin M., McFadden L. A., Murchie S., Robinson M., Thomas P. C., Yeomans D. K., Harch A., Williams B. G., Clark B., Farquhar R. W., Cheng A., and Dunham D. W. (1997) NEAR's flyby of mainbelt asteroid 253 Mathilde. *Bull. Am. Astron. Soc., 29,* 958.

Vilas F. (1994) A cheaper, faster, better way to detect water of hydration on solar system bodies. *Icarus, 111,* 456–467.

Vilas F. and Gaffey M. J. (1989) Phyllosilicate absorption features in main-belt and outer-belt asteroid reflectance spectra. *Science, 246,* 790–792.

Vilas F. and Sykes M. V. (1996) Are low-albedo asteroids thermally metamorphosed? *Icarus, 124,* 483–489.

Vilas F., Hatch E. C., Larson S. M., Sawyer S. R., and Gaffey M. J. (1993) Ferric iron in primitive asteroids — A 0.43-micron absorption feature. *Icarus, 102,* 225–231.

Vilas F., Jarvis K. S., and Gaffey M. J. (1994) Iron alteration minerals in the visible and near-infrared spectra of low-albedo asteroids. *Icarus, 109,* 274–283.

Weidenschilling S. J. and Davis D. R. (2001) Planetary accretion: The first 5 million years (abstract). In *Lunar and Planetary Science XXXII,* Abstract #1894. Lunar and Planetary Institute, Houston (CD-ROM).

Wetherill G. W. (1990) Comparison of analytical and physical modeling of planetesimal accumulation. *Icarus, 88,* 336–354.

Zellner B., Tholen D. J., and Tedesco E. F. (1985) The eight-color asteroid survey: Results for 589 minor planets. *Icarus, 61,* 355–416.

Zolensky M. E. and Keller L. P. (1991) Clay minerals in primitive meteorites and interplanetary dust I. *Clay Mineral Soc., 28,* 184.

Zolensky M. and McSween H. Y. Jr. (1988) Aqueous alteration. In *Meteorites and the Early Solar System* (J. F. Kerridge and M. S. Matthews, eds.), pp. 114–143. Univ. of Arizona, Tucson.

Zolensky M. E., Krot A. N., and Scott E. R. D., eds. (1997) *Workshop on Parent-Body and Nebular Modifications of Chondritic Materials.* LPI Tech. Rpt. 97-02, Part 1, Lunar and Planetary Institute, Houston.

Zolensky M. E., Bodnar R. J., Gibson E. K., Nyquist L. E., Reese Y., Shih C.-Y., and Wiesmann H. (1999) Asteroidal water within fluid inclusion-bearing halite in an H5 chondrite, Monahans (1998) *Science, 285,* 1377–1379.

Zolensky M. E., Bodnar R. J., Schwandt C., and Yang S. V. (2000) Halide minerals in the Monahans (1998) and Zag II chondrite regolith breccias (abstract). In *Lunar and Planetary Science XXXI,* Abstract #1181. Lunar and Planetary Institute, Houston (CD-ROM).

Physical Properties of Near-Earth Objects

Richard P. Binzel
Massachusetts Institute of Technology

Dmitrij F. Lupishko
Institute of Astronomy of Kharkov National University

Mario Di Martino
Astronomico Osservatorio di Torino

Robert J. Whiteley
University of Arizona

Gerhard J. Hahn
DLR Institute of Space Sensor Technology and Planetary Exploration

---

The population of near-Earth objects (NEOs) contains asteroids, comets, and the precursor bodies for meteorites. The challenge for our understanding of NEOs is to reveal the proportions and relationships between these categories of solar-system small bodies and their source(s) of resupply. Even accounting for strong bias factors in the discovery and characterization of higher-albedo objects, NEOs having S-type spectra are proportionally more abundant than within the main asteroid belt as a whole. Thus, an inner asteroid belt origin (where S-type objects dominate) is implied for most NEOs. The identification of a cometary contribution within the NEO population remains one of a case-by-case examination of unusual objects, and the sum of evidence suggests that comets contribute at most only a few percent of the total. With decreasing size and younger surfaces (due to presumably shorter collisional lifetimes for smaller objects), NEOs show a transition in spectral properties toward resembling the most common meteorites, the ordinary chondrites. Ordinary chondritelike objects are no longer rare among the NEOs, and at least qualitatively it is becoming understandable why these objects comprise a high proportion of meteorite falls. Comparisons that can be performed between asteroidal NEOs and their main-belt counterparts suggest that the physical properties (e.g., rotation states, configurations, spectral colors, surface scattering) of NEOs may be representative of main-belt asteroids (MBAs) at similar (but presently unobservable) sizes.

1. INTRODUCTION

Planetary science investigations of asteroids, meteorites, and comets all have a common intersection in the study of near-Earth objects (NEOs), represented schematically in Fig. 1. (Here we define a NEO as an object having a perihelion distance of ≤1.3 AU.) Dynamical calculations (see *Morbidelli et al.,* 2002; *Bottke et al.,* 2002a) show that lifespans for NEOs are typically a few million years, eventually meeting their doom by crashing into the Sun, being ejected from the solar system, or impacting a terrestrial world. With such short lifetimes, NEOs observed today cannot be residual bodies that have remained orbiting among the inner planets since the beginning of the solar system. Instead, the NEO population must have some source of resupply. Understanding the source(s) and mechanism(s) of their resupply is one of the fundamental scientific goals for NEO studies. Key questions include the following: What fraction comes from the asteroid belt? What fraction of the NEOs that do not display a coma or a tail are in fact extinct or dormant comet nuclei? Pinpointing the source regions of NEOs is also a matter of high scientific priority for fully utilizing the wealth of information available from laboratory studies of meteorites (e.g., *Kerridge and Matthews,* 1988). The immediate precursor bodies for meteorites are, by definition of proximity, NEOs objects. Thus, the scientific goal of understanding the source(s) for NEOs is identical to the goal of finding the origin locations for meteorites. A key component in tracing meteorite origins is discovering links between the telescopically measured spectral (compositional) properties of asteroids with those measured in the laboratory for meteorites (see *Burbine et al.,* 2002).

The proximity of NEOs also makes them worlds for which we have substantial practical interest. Those having

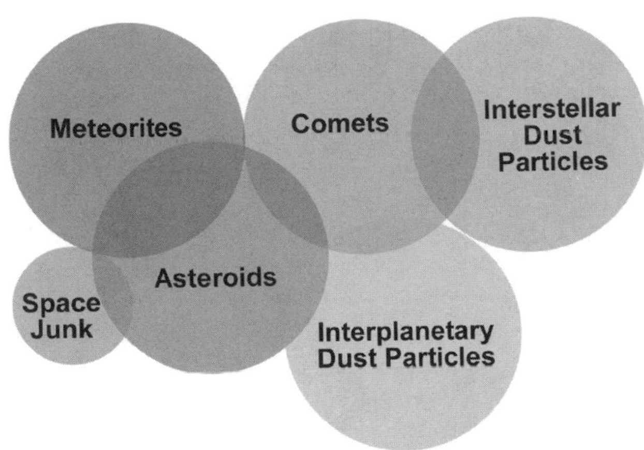

Fig. 1. Cartoon illustration of the many different groups of objects found within near-Earth space. One of the principal objectives for studying NEOs is to understand how these groups may be related. Thus the regions of intersection denote key research areas. As surveys increase their capabilities, human-made space flight hardware ("space junk") is also being increasingly found.

low-inclination and low-eccentricity orbits closest to Earth are among the most accessible spacecraft destinations in the solar system. In terms of the propulsion energy required, more than 20% of the NEOs are known to be more accessible than the Moon for long-duration sample-return missions (*Lewis and Hutson,* 1993). The fact that many NEOs remain well within the inner solar system during their orbits further simplifies thermal-design and power-generation considerations for exploratory spacecraft (*Perozzi et al.,* 2001). The proximity of NEOs also makes them prime targets for radar experiments designed to measure surface properties and achieve image reconstructions. *Ostro et al.* (2002) highlight the spectacular success of this technique and describe results for specific objects. Most renowned of the practical importance of the NEOs is the small, but nonzero, probability of a major impact that could threaten civilization. The hazard issue is addressed in *Morrison et al.* (2002), and the physical properties of NEOs as they pertain to the hazard have been reviewed by *Chapman et al.* (1994) and *Huebner et al.* (2001).

The purpose of this review chapter is to serve as a focal point for what we know about the physical properties of NEOs and how these data serve to illuminate the myriad interrelationships between asteroids, comets, and meteorites. Thus, in a way, we hope this chapter will serve as a "central node" in guiding the reader toward the interconnections that NEOs have to a broad range of planetary-science topics (and chapters within this volume). In particular, we wish to take advantage of the scientific insights that can be achieved by virtue of their proximity: NEOs are the smallest individually observable bodies in our solar system. Thus, these objects, which reside at the crossroads of many different areas of study, are also an end member to the size distribution of measurable planetary worlds. Here we draw upon, build upon, and update previous reviews by *McFadden et al.* (1989) and *Lupishko and Di Martino* (1998).

The terms used to refer to objects in the vicinity of Earth have gone through a rapid convergence as interest in them has increased over the past decade. When speaking broadly of the population, "near-Earth objects" (NEOs) has become the most widely used term, since this inclusive label does not presuppose an origin or nature as an asteroid or a comet. When speaking about objects in the vicinity of the Earth that are presupposed to have an asteroidal origin, the term "near-Earth asteroids" (NEAs) is commonly used. In this chapter we attempt not to make any general suppositions about the origins of these bodies and therefore mostly employ the term "NEO." Objects that appear "asteroidal" (starlike with no apparent coma or tail that would give them the label "comet") dominate the NEO population, with the currently known number having reasonably well-determined orbits approaching 2000 (see *Stokes et al.,* 2002). Only about 50 short-period comets (*Marsden and Williams,* 1999) satisfy the NEO definition of having a perihelion distance ≤1.3 AU.

Asteroidal NEOs are traditionally subdivided into groups based on their orbital characteristics a, q, Q (semimajor axis, perihelion distance, aphelion distance) with respect to Earth's and are called "Amor," "Apollo," and "Aten" asteroids (*Shoemaker et al.,* 1979). Amor objects are defined as bodies residing just outside the orbit of Earth ($a \geq 1$ AU), having $1.017 < q \leq 1.3$ AU. Objects having a semimajor axis > 1 AU and $q \leq 1.017$ AU are known as Apollos. Relatively equal numbers of Amor and Apollo asteroids are currently known; combined they account for ~90% of all currently known NEOs. Atens have orbits substantially inside that of Earth ($a < 1$ AU, $Q > 0.983$ AU), and represent about 8% of the known NEO population. (Short-period comets account for the remaining 2%.) By these definitions, Aten and Apollo objects cross the orbit of Earth while Amor objects do not. However, orbital precession, periodic variations in orbital elements, and planetary perturbations over timescales of centuries are sufficient for objects straddling the boundaries between groupings to change their affiliation. *Milani et al.* (1989) performed an orbital-evolution analysis involving 89 NEOs over a timespan of 200,000 yr. Based on these results, they propose six dynamical classes, named after the best-known and most representative object in each class: Geographos, Toro, Alinda, Kozai, Oljato, and Eros. This classification is indicative of long-term behavior and, of course, differs from the Amor-Apollo-Aten nomenclature, which is based only on the osculating orbital elements. The name "Apohele" (the Hawai'ian transliteration for orbit) has been proposed (*Tholen and Whiteley,* 1998) for one additional group of objects whose orbits reside entirely inside that of Earth ($Q < 0.983$ AU). At present only 1998 DK_{36} (*Tholen and Whiteley,* 1998) has been discovered as a potential member of this class, although this result is controversial due to uncertainties in the values of its orbital elements. *Michel et al.* (2000) refer to these as "inner-Earth objects" (IEOs) and estimate that Atens and IEOs together could constitute 20% of the multikilometer-sized Earth-crossing population.

2. TABULATION OF NEAR-EARTH-OBJECT PHYSICAL PROPERTIES

Over the past decade the growth in measurements of NEO physical properties has increased at a pace nearly commensurate with the increase in their interest and discovery rate. Physical parameters (such as spectroscopic and rotation properties) were known for only a few dozen NEOs at the time of publication of *Asteroids II* (*McFadden et al.,* 1989). An extension of this work is presented by *Chapman et al.* (1994), and a more thorough review of NEO physical properties by *Lupishko and Di Martino* (1998) summarizes results for about 100 objects, where the growth during this time period can largely be credited to the work of Wieslaw Wisniewski (*Wisniewski et al.,* 1997). Since the Lupishko and Di Martino review, a significant amount of new work has pushed the number of NEOs having (at least some) physical characterization up to more than 300 objects (e.g., *Binzel,* 1998, 2001; *Erikson et al.,* 2000; *Hammergren,* 1998; *Pravec et al.,* 2000a; *Rabinowitz,* 1998; *Hicks et al.,* 1998, 2000; *Whiteley and Tholen,* 1999; *Whiteley,* 2001).

Table 1 presents an extensive summary of the currently known physical parameters (derived primarily by spectroscopic and photometric techniques) for asteroidal NEOs. Objects are designated as belonging to the Amor (Am), Apollo (Ap), and Aten (At) groups. Mars-crossing (MC) objects of special interest are also included: 9969 Braille, encountered by the *Deep Space 1* mission in 1999 (*Oberst et al.,* 2001), and (5407) 1992 AX, a likely binary (*Pravec et al.,* 2000b). For most objects, only approximate estimates (guesses) can be made for albedos and diameters. Therefore, analyses and conclusions based on these parameters must be made with considerable caution. Taxonomic classes are from the system defined by *Tholen* (1984) and extended to include the additional designations developed by *Bus* (1999; *Bus and Binzel,* 2002; *Bus et al.,* 2002). When NIR spectral data are available such that the S-class subgroups described by *Gaffey et al.* (1993) are determined, taxonomic designations are given in this system. Rotational periods (in hours) are given if known, along with the range of lightcurve amplitudes represented by these measurements. The final columns present U-B and B-V colors, when available. Physical measurements of NEOs are certainly not limited to those parameters in Table 1, with the most exhaustive additional tabulations available in *Lupishko and Di Martino* (1998). These additional tabulations include information on individual measurements of pole coordinates, senses of rotation and asteroid triaxial shapes, photometric and polarimetric parameters, radiometric albedos and diameters, and radar parameters. More thorough information on some of these latter parameters is presented in *Pravec et al.* (2002), *Harris and Lagerros* (2002), and *Ostro et al.* (2002). Ongoing updates to Table 1, as well as citations for the references to the individual entries, may be found at http://earn.dlr.de/nea/.

3. ANALYSIS

In this analysis of the currently known physical properties of NEOs, we focus on those properties that give the best indication of origin. We particularly focus on the extent to which asteroidal NEOs may be similar to or different from main-belt asteroids in the same size range. Key differences may distinguish the relative importance of asteroid or comet origins for the population. Size dependences in the spectral properties, for example, may also illuminate links for asteroid-meteorite connections.

3.1. Taxonomy of Near-Earth Objects

Figure 2 shows the relative abundance of various taxonomic classes of NEOs, as analyzed from the data in Table 1. Note that there are subtle differences between the asteroid taxonomies derived by *Tholen* (1984) and *Bus* (1999), and these differences affect some of the identifications in Table 1. (Taxonomic designations given are as cited in the published reference.)

Almost all taxonomic classes of main-belt asteroids are represented among classified NEOs, including the P- and D-types most commonly found in the outer asteroid belt, among the Hilda and Trojan asteroids, or possibly among comet nuclei (see *Barucci et al.,* 2002; *Weissman et al.,* 2002). This broad representation of types, including those from distant regions, suggests that the processes delivering objects to the inner solar system are broad in scope (see *Bottke et al.,* 2002b; *Morbidelli et al.,* 2002). A key question we appear to be on the verge of answering is this: How significantly is the delivery of NEOs dominated by processes operating within the inner asteroid belt? S-type asteroids that dominate the inner asteroid belt also dominate the sampled NEO population by a ratio of ~4:1 (Fig. 2). This ratio, however, is subject to selection effects because S-type asteroids have higher albedos than C-types, making their discovery and observation more likely. (In a magnitude-limited survey, their higher reflectivity allows more S-asteroids to be bright enough for detection.) *Luu and Jewitt* (1989) also point out that C-type asteroids fall off in their apparent brightness more rapidly with increasing phase angle than do S-type asteroids (see *Muinonen et al.,* 2002). Since NEOs are typically discovered at larger phase angles, the coupling of this phase-angle effect with the albedo effect can create a strong bias in favor of S-type asteroids. *Luu and Jewitt* (1989) use a Monte Carlo model to estimate this bias factor to be in the range of 5:1 to 6:1.

While bias effects certainly are a major factor in creating the high proportion of S-types observed among NEOs, *Lupishko and Di Martino* (1998) argue that even after bias

TABLE 1. Physical parameters of NEOs (readers utilizing individual entries are reminded to cite the original source for each datum; original source references for each datum listed here, as well as current updates to this table, may be found at http://earn.dlr.de/nea/).

Asteroid Number*	Asteroid Name	Provisional Designation	Group	H (mag)†	Albedo‡	Diameter (km)§	Class¶	Period (hrs)	Amplitude (mag)	U-B	B-V
433	Eros	1898 DQ	Am	11.24	0.21	23.6	S(IV)	5.270	0.03–1.38	0.52	0.90
719	Albert	1911 MT	Am	15.8M	m	2.4		5.80	0.6		
887	Alinda	1918 DB	Am	13.83	0.23	4.2	S	73.97	0.35	0.43	0.84
1036	Ganymed	1924 TD	Am	9.42	0.17	38.5	S(IV)	10.31	0.12–0.40	0.42	0.84
1221	Amor	1932 EA1	Am	17.46	m	1.1					
1566	Icarus	1949 MA	Ap	15.95	0.33	1.3	SU,Q	2.273	0.03–0.18	0.54	0.80
1580	Betulia	1950 KA	Am	14.55	0.17	3.9	C	6.1324	0.13–0.65	0.27	0.66
1620	Geographos	1951 RA	Ap	16.5	0.19	5 × 2 × 1	S	5.2233	0.90–2.00	0.50	0.89
1627	Ivar	1929 SH	Am	13.24	0.26	6.9	S	4.797	0.22–1.15	0.46	0.89
1685	Toro	1948 OA	Ap	13.96	0.31	3	S	10.196	0.55–1.40	0.47	0.88
1862	Apollo	1932 HA	Ap	16.23	0.26	1.4	Q	3.065	0.12–0.70	0.43	0.79
1863	Antinous	1948 EA	Ap	15.81	0.18	1.8	Sq	4.02	0.12	0.37	0.77
1864	Daedalus	1971 FA	Ap	15.02	mh	3.1	Sr	8.57	0.80–1.04	0.50	0.83
1865	Cerberus	1971 UA	Ap	16.97	0.26	1	S	6.810	1.5–2.1	0.40	0.79
1866	Sisyphus	1972 XA	Ap	13.0M	0.14	8.9	S	2.400	0.1	0.45	0.88
1915	Quetzalcoatl	1953 EA	Am	18.97	0.31	0.4	SMU	4.9	0.2	0.43	0.83
1916	Boreas	1953 RA	Am	15.03	mh	3.1	S			0.41	0.85
1917	Cuyo	1968 AA	Am	13.9M	mh	5.2	Sl	2.6905	0.11–0.44		
1943	Anteros	1973 EC	Am	16.01	0.18	1.8	L	2.8695	0.05–0.1	0.45	0.84
1980	Tezcatlipoca	1950 LA	Am	13.95	0.14	6.7	Sl	7.2505	0.47–0.97	0.46	0.96
1981	Midas	1973 EA	Ap	15.18	h	2.2	V	5.220	0.65–0.87	0.48	0.97
2061	Anza	1960 UA	Am	16.56M	m	1.7	TCG	5.75	0.08–0.26	0.35	0.76
2062	Aten	1976 AA	At	17.12	0.20	0.9	Sr	40.77	0.26	0.46	0.93
2063	Bacchus	1977 HB	Ap	17.1M	mh	1.2	Sq	14.904	0.22–0.42		0.84
2100	Ra–Shalom	1978 RA	At	16.07	0.13	2.5	Xc	19.79	0.35–0.41	0.31	0.72
2102	Tantalus	1975 YA	Ap	16.2	m	3.3	Q	2.391	0.07–0.09		
2201	Oljato	1947 XC	Ap	16.86	0.24	2.1	Sq	24	>0.1		
2212	Hephaistos	1978 SB	Ap	13.87M	mh	3.3	SG	20	~0.1	0.41	0.77
2340	Hathor	1976 UA	At	19.2M	mh	5.3	Sq			0.50	0.77
2368	Beltrovata	1977 RA	Am	15.21M	mh	0.5		5.9	0.84	0.52	0.83
2608	Seneca	1978 DA	Am	17.52M	0.16	0.9	S	8	0.35	0.41	0.83
3102	Krok	1981 QA	Am	15.6	m	1.6	S	147.8	1.0	0.52	0.83
3103	Eger	1982 BB	Ap	15.38	0.53	2.5	E	5.709	0.72–1.5		
3122	Florence	1981 ET3	Am	14.20	0.20	2.5	S	2.35812	0.18		
3199	Nefertiti	1982 RA	Am	15.10	0.41	1.8	Sq	3.0207	0.11–0.30	0.38	0.95
3200	Phaethon	1983 TB	Ap	14.32	0.11	5.1	B,F	3.57	0.11–0.26		
3288	Seleucus	1982 DV	Am	15.34	0.17	2.8	S	75	>0.4	0.50	0.82
3352	McAuliffe	1981 CW	Am	15.8	0.18	2.4	S	3.	0.10		
3360		1981 VA	Ap	16.3M	0.07	1.8					
3361	Orpheus	1982 HR	Ap	19.03	m	0.5		3.58	0.32		
3362	Khufu	1984 QA	At	18.27	0.16	0.7					
3551	Verenia	1983 RD	Am	16.75M	0.53	0.9	V	4.93		0.39	
3552	Don Quixote	1983 SA	Am	13.0M	0.02	18.7	D	7	0.5		
3554	Amun	1986 EB	At	15.82M	0.17	2.1	M	2.5300	0.19	0.24	0.71
B3671	Dionysus	1984 KD	Am	16.7	0.16	1.5	Cb	2.705	0.15–0.26		
B3671	Dionysus	1984 KD	Am					27.72			
3691	Bede	1982 FT	Am	14.9M	m	3.6	Xc			0.44	
3752	Camillo	1985 PA	Ap	15.5M	m	2.7		37.846	1.1		
3753	Cruithne	1986 TO	At	15.13	mh	3.3	Q	27.44	0.4–0.95		
3757		1982 XB	Am	18.95	0.34	0.4	S	9.12	0.20	0.51	0.85
3838	Epona	1986 WA	Ap	15.4	m	2.9		4.762	0.04–0.37		
3908	Nyx	1980 PA	Am	17.4M	0.23	0.9	U	4.4257	0.11–0.44	0.44	
3988		1986 LA	Am	18.2M	m	0.8		8			
4015	Wilson–Harri	1979 VA	Ap	15.99	0.05	2	CF	3.556	0.06–0.2		
4055	Magellan	1985 DO2	Am	14.8M	0.24	3	V			0.52	
4179	Toutatis	1989 AC	Ap	15.3	0.13	2.8	S,Sq	129.84	1.2	0.50	0.85
4183	Cuno	1959 LM	Ap	14.4M	mh	4.5	Q,Sq	3.560	0.1–0.84		
4197		1982 TA	Ap	14.88	0.33	1.7	Sq	3.5400	0.28	0.4	0.75
4341	Poseidon	1987 KF	Ap	15.5M	mh	2.5	O	6.262	0.08		
4503	Cleobulus	1989 WM	Am	16.02	m	2.7		3.13	0.22		
4660	Nereus	1982 DB	Ap	18.3	d	1.2	C	15.1	0.6		
4688		1980 WF	Am	19.0M	0.18	0.6	SQ			0.45	0.94
4769	Castalia	1989 PB	Ap	16.9	m	1.4		4.086	0.64–1.0		
4947	Ninkasi	1988 TJ1	Am	18.7M	mh	0.6	Sq				
4953		1990 MU	Ap	14.1M	mh	3.6	S	14.218	0.70		
4954	Eric	1990 SQ	Am	12.6M	mh	9.5	S	12.056	0.57–0.66		
4957	Brucemurray	1990 XJ	Am	15.1M	mh	3.0	S	2.8921	0.10–0.38		
5131		1990 BG	Ap	14.1M	mh	4.7	S				
5143	Heracles	1991 VL	Ap	14.0M	mh	5.0	O	15.8	>0.1		
5324	Lyapunov	1987 SL	Am	15.2M	m	3.1				0.37	0.81
5332		1990 DA	Am	13.9M	mh	5.2	S	5.803	0.35		0.87
5370	Taranis	1986 RA	Am	15.7M	m	2.5			0.02		
5587		1990 SB	Am	13.6M	mh	6.5	Sq	5.052	0.80–1.25		

TABLE 1. (continued).

Asteroid Number*	Name	Provisional Designation	Group	H (mag)†	Albedo‡	Diameter (km)§	Class¶	Period (hrs)	Amplitude (mag)	U-B	B-V
B5407		1992 AX	MC	13.7	mh	5.8	Sk	2.549	0.11		
B5407		1992 AX	MC					13.5	0.35		
5620		1990 OA	Am	17.0M	m	1.4			1.2		
5626		1991 FE	Am	14.7M	mh	3.6	S	2.4860	0.07		
5646		1990 TR	Am	16.05	m	2.1	U	6.25	0.19		
5653	Camarillo	1992 WD5	Am	15.4M	m	2.9		4.8341	0.85		
5660		1974 MA	Ap	15.7M	mh	2.3	Q				
5693		1993 EA	Ap	16.82	mh	1.4	Q		0.13		
5751	Zao	1992 AC	Am	14.93	m	3.5	X	21.7	0.04–0.12	0.29	0.81
5797	Bivoj	1980 AA	Am	19.1M	mh	0.5	S	2.706	0.10–0.17	0.37	0.81
5836		1993 MF	Am	15.03	mh	3.1	S	4.959	0.53–0.76		
5863	Tara	1983 RB	Am	15.5M	m	2.7			>0.02		
6037		1988 EG	Ap	18.7M	m	0.6		4.27	0.2		
6047		1991 TB1	Ap	17.0M	mh	1.2	S				
6053		1993 BW3	Ap	15.23	0.18	3.1	Sq	2.57341	0.45		0.99
6063	Jason	1984 KB	Ap	15.3M	0.16	1.4	S				
6178		1986 DA	Am	15.1M	0.14	2.3	M	3.58	0.10–0.40		
6455		1992 HE	Ap	13.8M	mh	5.4	S				
6489	Golevka	1991 JX	Ap	19.074	0.63	.35 × .25 × .25	Q	6.02640	0.28–1.05		
6491		1991 OA	Am	18.5M	m	0.7		2.69	0.09		0.7
6569		1993 MO	Am	16.2	mh	1.8	Sr	5.9588	0.98		
6611		1993 VW	Ap	16.5M	h	1.2	V				
7025		1993 QA	Ap	18.3M	m	0.8		2.50574	0.32		
7092	Cadmus	1992 LC	Ap	15.4M	d	4.5	C				
7236		1987 PA	Am	18.4M	d	1.1	C				
7335		1989 JA	Ap	17.85	m	0.9					
7336	Saunders	1989 RS1	Am	18.7M	mh	0.6	Sq	6	0.3		
7341		1991 VK	Ap	16.7M	mh	1.4	Sq	4.20960	0.28–0.70		
7358		1995 YA3	Am	14.4M	mh	4.5	Sq	2.75	0.1–0.5		
7474		1992 TC	Am	18.3	m	0.8	X	5.540	0.07		
7480	Norwan	1994 PC	Am	17.45	mh	1.1	S	35.90	0.5		
7482		1994 PC1	Ap	16.8M	mh	1.4	S	2.5999	0.29		
7753		1988 XB	Ap	18.6M	d	1.0	B				
7822		1991 CS	Ap	17.4M	0.25	0.9	S	2.389	0.27–0.32		
7888		1993 UC	Ap	15.3M	mh	3.1	S	2.340	0.10		
7889		1994 LX	Ap	15.3	h	3.1	V	2.741	0.32–0.39		
7977		1977 QQ5	Am	15.4M	mh	2.6	S	7.46	0.56		
8013		1990 KA	Am	17.31	m	1.2		6	0.5		
8034		1992 LR	Am	17.9M	mh	0.8	S	3.638	0.46–0.52	0.47	0.84
8176		1991 WA	Ap	17.1M	mh	1.2	Q	8.3	1.0		
8201		1994 AH2	Ap	16.3	m	2.2	O	23.949	0.3–0.4		
8566		1996 EN	Ap	16.5M	m	1.7	U				
9162		1987 OA	Ap	18.3M	d	1.2	B				
9400		1994 TW1	Am	14.8M	mh	3.4	Sr				
9856		1991 EE	Ap	17.0	0.30	1	S	3.045	0.14		
9969	Braille	1992 KD	MC	15.8M	mh	2.2	Q				
10115		1992 SK	Ap	17.0M	m	1.4		7.320	0.70–1.01		
10165		1995 BL2	Ap	17.1M	m	1.3	L				
10302		1989 ML	Am	19.5	m	0.6	X	15.786	0.6–1.0		
10563	Izhdubar	1993 WD	Ap	17.33	mh	1.2	Q	2.660	0.17		
11066	Sigurd	1992 CC1	Ap	15.00	mh	3.2	S	8.4958	1.02		
11311	Peleus	1993 XN2	Ap	16.5M	mh	1.6	Sq				
11398		1998 YP11	Am	16.27	m	1.9		38.61	0.22		
11405		1999 CV3	Ap	15.0M	m	3.4		5.78	0.25–0.4		
11500		1989 UR	Ap	18.43	mh	0.7	S	73.0	0.46		
12711		1991 BB	Ap	16.04	mh	2.1	Sr	3.48	0.6		
12923		1999 GK4	Ap	16.1M	m	2.1		3.892	0.18		
13651		1997 BR	Ap	17.6M	mh	0.9	S	33.644	1.2		
14402		1991 DB	Am	18.4M	0.16	1.1	B	2.266	0.1		
14827		1986 JK	Ap	18.3M	d	1.2	C				
15817	Lucianotesi	1994 QC	Am	18.6M	m	0.7	X	11.	0.8		
16064		1999 RH27	Am	16.9M	d	2.5	C	178.6	0.6		
16636		1993 QP	Am	17.50	m	1.2		22.05	0.23		
16657		1993 UB	Am	16.9M	mh	1.3	Sr				
16834		1997 WU22	Ap	15.7M	mh	2.3	S	9.348	0.4		
16960		1998 QS52	Ap	14.3	mh	4.3	Sq				
17274		2000 LC16	Am	16.7M	m	1.6		16.495	0.35		
17511		1992 QN	Ap	17.1M	m	1.3	X	5.9902	1.1		
18882		1999 YN4	Am	16.3M	mh	1.7	S				
19356		1997 GH3	Am	17.1M	mh	1.2	S	6.714	0.74		
20086		1994 LW	Am	16.9M	m	1.5		29.1	0.28		
20236		1998 BZ7	Ap	17.6M	mh	1.0	Q	10.17	0.15		
20255		1998 FX2	Am	18.2M	mh	0.7	Sq	6.826	0.22		
20429		1998 YN1	Ap	18.0M	m	0.9		2.72	0.1		
22753		1998 WT	Ap	17.7M	mh	0.9	Q				

TABLE 1. (continued).

Asteroid		Provisional	Group	H (mag)†	Albedo‡	Diameter	Class¶	Period	Amplitude	U-B	B-V
Number*	Name	Designation				(km)§		(hrs)	(mag)		
23548		1994 EF2	Am	17.6M	mh	0.9	Q				
23714		1998 EC3	Am	16.7M	mh	1.4	Q	1.2	0.25		
25143		1998 SF36	Ap	19.2M	0.32	0.36	S(IV)	12.15	1.0		
27002		1998 DV9	Ap	18.2M	mh	0.7	Q				
29075		1950 DA	Ap	17.0M	m	1.4		2.1216	0.2		
B31345		1998 PG	Am	17.64	(0.16)	0.9	Q	2.516	0.11		0.81
B31345		1998 PG	Am					7.003	0.09		
31346		1998 PB1	Am	17.1M	mh	1.2	Q				
33342		1998 WT24	At	17.9M	0.42	0.5	E	3.6977	0.3		
B35107		1991 VH	Ap	16.5	mh	1.4	Sk	2.624	0.08		
B35107		1991 VH	Ap					32.69			
35432		1998 BG9	Am	19.5M	mh	0.4	S				
36017		1999 ND43	Am	19.2M	mh	0.5	Sl	>5	>0.5		
36183		1999 TX16	Am	15.61	m	2.7	Ld	5.611	1.3		
B38071		1999 GU3	Am	19.6M	m	0.4		4.49			
B38071		1999 GU3	Am					9.03d			
		1977 VA	Am	19.0M	m	0.5	XC			0.15	0.7
		1978 CA	Ap	18.0M	h	0.6	M	3.756	0.8	0.14	0.72
		1988 TA	Ap	20.8M	d	0.4	C				
		1989 DA	Ap	18.6M	m	0.7		3.925	0.12		
		1989 UP	Ap	20.5M	m	0.3		6.98	1.16		
		1989 UQ	At	19.0M	d	0.9	B	7.733	0.27		
		1989 VA	At	17.89	mh	0.8	Sq	2.51357	>0.15–0.4		
		1989 VB	Ap	19.82	m	0.4		16,24	>0.32		
		1990 HA	Ap	16.74	m	1.5		8.58	>0.09		
		1990 SA	Am	17.0M	mh	1.2	S				
		1990 UA	Ap	19.64	m	0.4		6.25?	>0.1		
		1990 UP	Am	20.45	m	0.3		20.	0.8		
		1991 AQ	Ap	17.20	mh	1.1	Q,U				
		1991 VA	Ap	26.5M	m	0.02			0.4		
		1991 XB	Am	18.10	mh	0.9	SX				
		1992 BF	At	19.5M	m	0.4	Xc				
		1992 NA	Am	16.5M	d	2.7	C	6.992	0.42		
		1992 UB	Am	16.0M	m	2.1	X				
		1993 BX3	Ap	21.0M	m	0.2		20.463	0.91		
		1993 TQ2	Am	20.0M	mh	0.3	Sa				
		1994 AB1	Am	16.3M	mh	1.7	Sq				
		B1994 AW1	Am	17.5	mh	1.0	Sa	2.519	0.3		
		B1994 AW1	Am					22.40			
		1994 CB	Ap	21.0M	m	0.2		8.676	>0.90		
		1994 GY	Am	17.0M	m	1.4		2.5553	0.06		
		1994 TF2	At	19.3	mh	0.4	S				
		1995 BC2	Am	17.3M	m	1.2	X				
		1995 CR	At	21.5M	mh	0.16	S	2.42			
		1995 EK1	Ap	18.0M	m	0.9		8.444	0.45		
		1995 FJ	Ap	20.5M	m	0.3		9.2	0.3		
		1995 FX	Am	20.0M	m	0.3		5.46	0.2		
		1995 HM	Am	22.5	m	0.11		1.62	2.		
		1995 WL8	Am	18.1M	mh	0.8	Sq				
		1996 BZ3	Am	18.2M	m	0.8	X				
		B1996 FG3	Ap	17.76	m	1.6	X	3.594	0.08		0.71
		B1996 FG3	Ap					16.1	0.25		
		1996 FQ3	Am	21.0M	mh	0.2	Sq				
		1996 JA1	Ap	21.1	0.30	0.2	V	5.227	0.39–0.8		
		1997 AC11	At	21.0M	m	0.2	Xc				
		1997 AQ18	Ap	18.2M	d	1.2	C				
		1997 BQ	Ap	18.0M	mh	0.8	S				
		1997 GL3	Ap	20.0M	h	0.2	V				
		1997 MW1	At	19.2M	m	0.5	X				
		1997 NC1	At	18.0M	d	1.4	B				
		1997 QK1	Am	20.1M	mh	0.3	SQ				
		1997 RT	Am	20.0M	mh	0.3	Q				
		1997 SE5	Am	14.8M	m	3.8	T	9.0583	0.23		
		1997 TT25	Am	19.3M	mh	0.4	Sq				
		1997 UH9	At	18.8M	mh	0.5	Sq	>5	0.15		
		1997 US9	Ap	17.3M	mh	1.2	Q	3.58	0.2		
		1997 VM4	Ap	18.0M	mh	0.8	SQ				
		1998 BB10	Ap	20.4M	mh	0.3	Sq				
		1998 BT13	Ap	26.5M	mh	0.02	Sq				
		1998 FM5	Am	16.0M	mh	2.2	S	6.35	1.0		
		1998 HD14	At	20.9M	mh	0.2	SQ				
		1998 HE3	At	21.8M	mh	0.1	SQ				
		1998 KU2	Am	16.6	d	2.6	F,Cb				
		1998 KY26	Ap	25.5M	d	0.04	CP	0.178	0.30		
		1998 ME3	Am	19.3M	d	0.7	F				
		1998 ML14	Ap	17.6	mh	1.2	Q,S	14.98	0.12		

TABLE 1. (continued).

Asteroid Number* Name	Provisional Designation	Group	H (mag)[†]	Albedo[‡]	Diameter (km)[§]	Class[¶]	Period (hrs)	Amplitude (mag)	U-B	B-V
	1998 MQ	Am	16.6	mh	1.5	S				
	1998 MT24	Ap	14.8M	m	4.0	X	12.07	0.38		
	1998 MW5	Ap	19.2M	mh	0.5	Sq				
	1998 MX5	Am	18.1M	m	0.8	X				
	1998 QA1	Ap	19.1M	d	0.8	C				
	1998 QC1	Ap	19.6M	d	0.7	C				
	1998 QH2	Ap	16.1M	mh	1.9	Q				
	1998 QK28	Ap	19.5M	d	0.7	C				
	1998 QP	Ap	21.5M	m	0.2		5.4	0.1		
	1998 QR15	Am	18.0M	mh	0.9	Sq	2.46	0.1		
	1998 QR52	Ap	18.7M	m	0.6		235.	0.8		
	1998 QV3	Am	20.5M	mh	0.2	Q				
	1998 SG2	Am	19.7M	mh	0.4	Sq				
	B1998 ST27	At	19.5M	m	0.4					
	1998 ST49	Ap	17.7M	mh	0.9	Q				
	1998 TU3	At	14.7M	mh	3.6	Q				
	1998 UT18	Ap	19.1M	d	0.9	G	34	0.8		
	1998 VD31	Ap	19.1	mh	0.5	S				
	1998 VO	Ap	20.4	mh	0.3	S				
	1998 VO33	Ap	16.9	h	1.0	V	8.5	0.24		
	1998 VR	At	18.5	mh	0.6	Sk				
	1998 WB2	Ap	22.8	mh	0.12	S	0.313	0.6		
	1998 WM	Ap	16.8	mh	1.4	Sq				
	1998 WP5	Am	18.4M	mh	0.7	Sl				
	1998 WZ1	Ap	19.9M	mh	0.3	Q				
	1998 WZ6	Ap	17.3	h	0.8	V				
	1998 XA5	Am	18.8M	m	0.6			0.22		
	1998 XS16	Ap	16.46	m	1.7		5.421	1.4		
	1999 CF9	Ap	17.8M	mh	0.9	Q				
	1999 DJ4	Ap	18.5M	mh	0.6	Sq				
	1999 EE5	Am	18.4M	mh	0.7	S				
	1999 FA	Ap	20.7M	mh	0.3	S	10.09	1.2		
	1999 FB	Ap	18.1M	mh	0.8	Q				
	1999 GJ4	Ap	14.97	mh	3.2	Sq	4.956	1.0		
	B1999 HF1	At	14.5M	m	4.3	EMP	2.3191	0.10–0.12		0.72
	B1999 HF1	At					14.02			
	1999 JD6	At	17.2M	m	1.2	K	7.68	1.2		
	1999 JE1	Ap	19.5M	mh	0.4	Sq				
	1999 JM8	Ap	15.15	m	3.3		137.	0.7		
	1999 JO8	Am	17.0M	mh	1.4	S	2.386	0.11		
	1999 JU3	Ap	19.6M	d	0.7	Cg				
	1999 JV3	Ap	19.0M	mh	0.5	S				
	1999 JV6	Ap	19.9M	m	0.4	Xk				
	B1999 KW4	At	16.6M	m	1.6		2.61	0.2		
	B1999 KW4	At					17.5			
	1999 NC43	Ap	16.0M	mh	2.0	Q				
	1999 PJ1	Am	18.0M	m	0.9		6.201	1.1		
	1999 RB32	Am	19.8M	h	0.3	V				
	1999 RQ36	Ap	20.9M	m	0.2		2.146	0.22		
	1999 SE10	m	20.0M	m	0.3	X				
	1999 SF10	Ap	24.0	m	0.06		0.0411	0.58		
	1999 SK10	Ap	19.3M	mh	0.4	Sq				
	1999 SM5	Ap	19.07	m	0.54		6.230	0.77–0.96		
	1999 TA10	Am	17.77	m	1.0		14.	0.1		
	1999 TY2	Ap	23.1	mh	0.08	S	0.121	0.68		0.94
	1999 VM40	Am	14.60	mh	3.8	S	5.185	0.25–0.36		
	1999 VN6	Am	19.5M	d	0.7	C				
	1999 WK13	Am	17.2M	mh	1.1	S				
	1999 XO35	Am	16.8M	mh	1.4	Sq				
	1999 YB	Am	18.5M	mh	0.6	Sq				
	1999 YD	Am	21.1M	mh	0.2	Sk				
	1999 YF3	Am	18.5M	mh	0.6	Sq				
	1999 YG3	Ap	19.1M	mh	0.5	S				
	1999 YK5	At	16.8M	m	1.5	X				
	2000 AC6	At	21.0M	mh	0.2	Q				
	2000 AG6	Ap	25.3M	m	0.03		0.077	0.8		
	2000 AX93	Am	17.7M	mh	0.9	Sq				
	2000 AE205	Am	22.9M	mh	0.08	S				
	2000 AH205	Ap	22.4M	mh	0.1	Sk				
	2000 DO8	Ap	24.8M	m	0.04		0.022	1.39		
	B2000 DP107	Ap	18.2M	m	0.8		2.7755			
	B2000 DP107	Ap					42.23			
	2000 EB14	At	23.0M	m	0.09		1.79	1.7		
	2000 EE14	At	17.1M	mh	1.2	Q				
	2000 ES70	Am	17.1M	mh	1.2	S				
	2000 ET70	At	18.4M	m	0.7	X				

TABLE 1. (continued).

Asteroid Number* Name	Provisional Designation	Group	H (mag)†	Albedo‡	Diameter (km)§	Class¶	Period (hrs)	Amplitude (mag)	U-B	B-V
	2000 EV70	Ap	19.7	mh	0.4	Q				
	2000 EW70	At	21.1	d	0.3	F				
	2000 GK137	Ap	17.4M	m	1.1		4.84	0.27		
	2000 HB24	At	23.3M	m	0.08		0.218	0.24		
	2000 JG5	Ap	18.3M	m	0.8		6.055	1.0		
	2000 JQ66	Am	18.1M	m	0.8		11.11	0.6		
	2000 NM	Ap	15.6M	m	2.6		9.24	0.3–0.5		
	2000 OG8	Am	17.8M	m	1.0		4.07	0.1		
	2000 PH5	At	22.6M	m	0.1		0.2029	0.85		
	2000 QW7	Am	19.8M	m	0.37		long	0.04		
	2000 RD53	Am	20.1M	m	0.33		14.79	0.10		
	2000 SM10	Ap	24.1M	m	0.05		15.	0.2		
	2000 SS164	Am	16.7M	m	1.6		6.894	0.9		
	B2000 UG11	Ap	20.4M	m	0.3					
	B2000 UG11	Ap					0.809 d			
	2000 WH10	Ap	22.5M	m	0.11		0.023	0.25		
	2000 WL107	Am	24.8	m	0.038		0.322	1.2		
	2000 YA	Ap	23.6M	m	0.07		1.33	0.35		
	2001 CB21	Ap	18.5M	m	0.7		3.30	0.19		
	2001 CP36	At	23.7M	m	0.06		10.	0.05		
	2001 OE84	Am	17.8M	m	0.9		0.4865	0.60		
	B2001 SL9	Ap	17.5M	m	1.1		2.40	0.08		
	B2001 SL9	Ap					16.4	0.08		
	2002 BM26	Ap	20.1M	m	0.3		2.7			

* "B" before an asteroid number indicates a possible binary asteroid. For such objects, a second line gives the orbital period (if known) and the lightcurve amplitude contribution of the binary.

† "M" within this column indicates the value is from the Minor Planet Center (http://cfa-www.harvard.edu/cfa/ps/mpc.html).

‡ When albedo is not estimated through physical measurements, an approximation is assigned based on the taxonomic class. These assumed albedos are coded as follows: d for "dark" (0.06), m for "medium" (0.15), mh for "medium high" (0.18), h for "high" (0.30). "m" is assigned in the case of no taxonomic information.

§ When diameter is not directly measured or determined through physical measurements, as is the case for all objects assigned an albedo code, the diameter (D, in km) is estimated from the following relationship (*Fowler and Chillemi,* 1992): $2 \log(D) = 6.247 - 0.4 H - \log(\text{albedo})$.

¶ Taxonomic class. See text in section 2 for the conventions used.

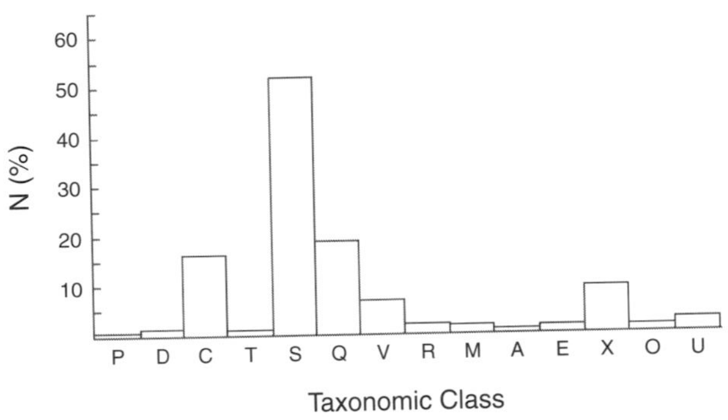

Fig. 2. Histogram of the relative proportions of measured taxonomic properties for more than 300 NEOs listed in Table 1. Almost all taxonomic classes seen among main-belt asteroids are represented within the NEO population. As detailed by *Luu and Jewitt* (1989), strong selection effects favor the discovery and characterization of higher-albedo objects such as S-type (and possibly Q-type) asteroids. Within this histogram, the designation "C" includes both C-types and related subgroups (B, F, G). Those having unusual characteristics that do not fall into any present category, or classes (such as L, K) having <1% representation, are combined within the designation "U."

corrections are accounted for, a clear signature for a dominant contribution from the inner asteroid belt remains. *Benedix et al.* (1992), *Lupishko and Di Martino* (1998), and *Whiteley* (2001) all find that after applying bias-correction factors to the observed NEO population, at any given size there are relatively equal proportions of C- and S-type objects within near-Earth space. However the main belt, in its entirety, is dominated by C-types. [A bias-correction analysis of the main belt performed by *Zellner* (1979) suggests that C-types dominate among all main-belt asteroids by as much as 5:1.] The fact that C-types do not dominate the NEO population (even after strong bias correction) indicates that asteroidal NEOs are not being contributed equally by all regions of the main belt. Thus the inner regions of the asteroid belt, where S-types are most common (*Gradie and Tedesco,* 1982; *Gradie et al.,* 1989) must preferentially contribute to the NEO population. *Benedix et al.* (1992) point out that the region of the 3:1 resonance has roughly equal populations of C- and S-type asteroids in its vicinity, making it a compatible source. Dynamical models (e.g., *Migliorini et al.,* 1998; *Morbidelli and Nesvorný,* 1999; *Vokrouhlický et al.,* 2000; *Bottke et al.,* 2000, 2002a; *Morbidelli et al.,* 2002) certainly support the view of the 3:1 resonance and inner asteroid belt dominating the contributions to the near-Earth population.

General taxonomic and spectral links between the main belt and near-Earth populations have been proposed since the beginning of substantial studies of NEO properties (*McFadden et al.,* 1984, 1985). Unique taxonomic classifications and mineralogic interpretations do show evidence for specific ties to main-belt sources. Most notable among these is the E-type object 3103 Eger, which appears both compositionally and dynamically related to the Hungaria region (high-inclination objects) of the inner asteroid belt (*Gaffey et al.,* 1992). These authors also argue for a connection to the enstatite achondrite meteorites. Basaltic (pyroxene-rich) NEOs having V-type taxonomies and good spectral matches to both the howardite-eucrite-diogenite (HED) classes of meteorites and Vesta were found by *Cruikshank et al.* (1991). The existence of numerous main-belt asteroid fragments apparently excavated from Vesta (*Binzel and Xu,* 1993; *Thomas et al.,* 1997) and the dynamic viability of their delivery into the inner solar system (*Migliorini et al.,* 1997) provides an additional specific link between the main belt and NEOs. Perhaps the objects of most practical interest (from the hazard-assessment and resource-utilization points of view) among the NEOs are the M-types that may be highly metallic in composition (*Tedesco and Gradie,* 1987). The most notable case among NEOs, confirmed as metallic by virtue of its extremely high radar albedo, is (6178) 1986 DA (*Ostro et al.,* 1991). Nevertheless, confirmed M-types and (presumably) highly differentiated, olivine-rich A-types are relatively rare among the NEOs.

3.2. Relationships of Near-Earth Objects to Comets

While taxonomic and mineralogic characterization of NEOs provide confident links to main-belt origins, cometary origins are suggested with substantially less certainty as described in *Weissman et al.* (2002). Most supply models have broadly considered asteroid and comet sources (e.g., *Wetherill,* 1988; *Bottke et al.,* 2002a), and some analyses (e.g., *Rabinowitz,* 1997a,b) suggest that comets may not be required at all as a major contributor to the population. Direct imaging (e.g., *Luu and Jewitt,* 1992) through the discovery and followup process to date has not revealed any other NEO case like that of the dual comet/asteroid citizenship of 4015 Wilson-Harrington (*Fernandez et al.,* 1997). Analysis of images of more than 100 NEOs by *Whiteley* (2001) constrains most of these objects to have production rates 1–2 orders of magnitude lower than weakly active comets such as P/Arend-Rigaux and P/Neujmin 1 (*Campins et al.,* 1987; *Jewitt and Meech,* 1985).

Nevertheless, interesting cases among the NEOs leave the issue open. Cases to be resolved include the meteor-stream association for 3200 Phaethon (*Whipple,* 1983; *Williams and Wu,* 1983; *Cochran and Barker,* 1984; *Fox et al.,* 1985); unusual spectral and possible magnetic signatures from 2201 Oljato (*McFadden et al.,* 1993); and the intermittent cometary properties of 4015 Wilson-Harrington (*Fernandez et al.,* 1997). While the taxonomic classifications (neutrally colored F and CF designations; Table 1) for 3200 Phaethon and 4015 Wilson-Harrington appear consistent with primitive solar-system materials presumed to dominate in comets, the classifications (Sq and SU;Q) for Oljato and Icarus are more like inner main-belt asteroids and do not seem to make "cometary sense." D-type asteroids such as 3552 Don Quixote and 1997 SE5 (*Hicks et al.,* 1998, 2000), however, do add to the list of NEOs having taxonomic characteristics that make them extinct comet candidates.

3.3. Relationships of Near-Earth Objects to Ordinary-Chondrite Meteorites

As described in *Burbine et al.* (2002), measurements of the spectral properties of NEOs have been revealing toward the problem of finding sources for the most common class of meteorites, the ordinary chondrites. *Clark et al.* (2002) outline the considerable debate over whether the most commonly observed asteroids, the S-types, are related to the most common meteorites (see also *Wetherill,* 1985; *Wetherill and Chapman,* 1988). Here we briefly describe and illustrate the role of NEO physical studies toward achieving an understanding that appears to be forging a link between S-type asteroids and ordinary-chondrite meteorites. This link is most likely for a subset of S-type asteroids denoted as S(IV) (*Gaffey et al.,* 1993). Overall, the mineralogy of asteroids across the entire S-class appears to be diverse (see *Gaffey et al.,* 2002).

The proximity of NEOs provides the opportunity for measuring the physical properties of objects in the size range (roughly 10–100 m) of most meteorite precursors. Spectral measurements over a continuity of sizes from meteoroids to main-belt asteroids appears to show a transition between S-type asteroids and ordinary-chondrite meteorites. The tendency toward seeing "ordinary-chondrite-like" spectral properties among S-types at smaller and smaller sizes (measured within the NEO population) has been noted in multifilter color measurements (*Rabinowitz et al.,* 1998; *Whiteley and Tholen,* 1999; *Whiteley,* 2001) and in visible and NIR CCD spectra (*Binzel et al.,* 1996, 1998, 2001). Figure 3 illustrates the trend in spectral properties between S-type asteroids and ordinary-chondrite meteorites revealed by NEO spectral measurements.

Several plausible explanations can be offered for the trend toward ordinary chondritelike spectral properties with decreasing diameter. The first is that spectral variations are due to particle-size effects (*Johnson and Fanale,* 1973), where the decreasing surface gravity results in a different size distribution of regolith and ~1-μm-sized particles on the surface. (These are the particle sizes most responsible for absorption, reflection, and scattering of visible and NIR wavelengths measured by reflectance spectroscopy.) A variety of photometric parameters are indicative of surface particle sizes, as we discuss in section 3.4. However, these parameters show no evidence for a diameter dependence, thereby giving doubt to a surface particle-size explanation for the trend in S-type asteroid spectral properties.

A second explanation is related to the average surface age of smaller bodies (*Binzel et al.,* 1998). Survival lifetimes against catastrophic disruption (see *Davis et al.,* 2002) decrease with decreasing size. Thus, on average, as we examine smaller and smaller objects, we see younger and younger surfaces. If time-dependent surface-alteration processes are effective [e.g., space weathering; see *Clark et al.*

(2002)] the smallest objects will have on average the youngest, freshest, and least-altered surfaces. The finding that smaller S-type NEOs have spectral properties tending increasingly toward those of "fresh" ordinary-chondrite meteorite specimens is fully consistent with the occurrence of a space-weathering process. In our view, the reality of a space-weathering process is strongly supported by the elemental-abundance measurements of Eros made by the *NEAR Shoemaker* spacecraft (*Trombka et al.*, 2000; *McCoy et al.*, 2001; *Cheng*, 2002). These *NEAR Shoemaker* results support the conclusion that Eros, a rather typical S-type NEO, has the same elemental abundance as ordinary-chondrite meteorites, except for a deficiency of S (perhaps explained by solar-wind sputtering). It has become increasingly accepted that the most likely way to reconcile these elemental-abundance results with the mismatch between telescopic spectra of Eros and laboratory spectra of ordinary chondrites is for some space-weathering-like surface alteration process to be active on S-type asteroid surfaces.

There are notable objections to the above idea, however, some of which are described in *Whiteley* (2001). The most significant objection is that S-type asteroids can still be found among very small NEOs, some so small that their collisional lifetimes are 5–10 m.y. or less. There also exist 5-km NEOs that are spectrally very good matches to OC meteorites and that have collisional lifetimes of 0.5–1.0 b.y. If the spectral signatures of SQ-type asteroids are dominated by a strong temporal weathering trend, we should expect to see no (or very few) S-type spectra among the collisionally "youngest" asteroids. There is also some spectral evidence in *Pravec et al.* (2000a) that there are S-types among the monolithic fast-rotating asteroids. This is a significant complication for the space-weathering hypothesis, because such asteroids rotate too quickly to retain a regolith, and thus should be the least-weathered asteroids we can observe. It remains to be seen whether some size-dependent petrological process, or the consideration of more sophisticated surface-age models, would help resolve these contradictions.

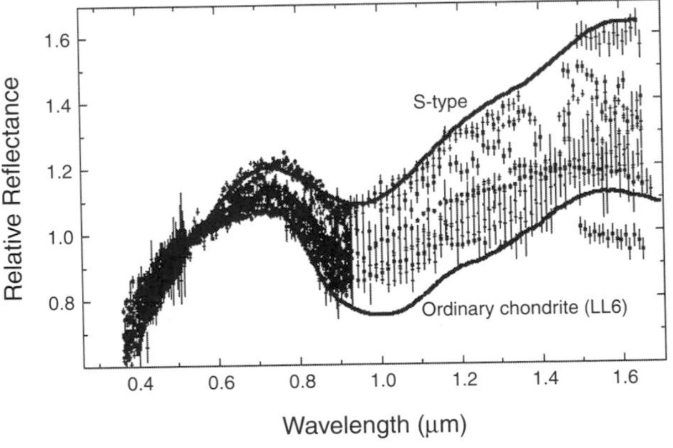

Fig. 3. An apparently continuous distribution of spectral properties is seen between the most commonly observed S-type asteroids and the most common class of meteorites, the ordinary chondrites. One possible explanation is a size-dependent trend where smaller NEOs have (on average) younger and fresher surfaces that have not been subjected to possible space-weathering effects. Thus their spectral properties are most likely to resemble those for meteorites measured in the laboratory. [NEO data from *Binzel et al.* (2001); meteorite data from *Gaffey* (1976).]

Regardless of the exact nature of the relationship between S- and Q-type asteroids, Fig. 4 illustrates that the once-scarce matches between NEOs and ordinary-chondrite meteorites are now increasingly common. As Fig. 2 depicts, ~20% of all observed NEOs have spectral properties placing them in the taxonomic class Q. [Q class asteroids have spectra most similar to laboratory spectra of ordinary-chondrite meteorites (*McFadden et al.*, 1984; *Bus et al.*, 2002).] How do we reconcile that ordinary chondrites, which account for ~80% (by fall statistics) of all meteorites, are derived from objects that seem to account for only 20% of the NEO population? While achieving a rigorous quantitative agreement between these proportions is not yet within our grasp, we can qualitatively conceive a bridge across this disparity. As a first step, we can understand that the higher relative strength of ordinary-chondrite material, compared to more primitive carbonaceous material, will create some amount of overrepresentation of ordinary-chondrite material in our total sample. [In the extreme case of the strongest objects, a vastly greater proportion of iron meteorites populate our collections than their likely abundance in near-Earth space (*Lipschutz et al.*, 1989).] A second qualitative step we can recognize is that the *NEAR Shoemaker* elemental-abundance results for Eros (*Trombka et al.*, 2000; *McCoy et al.*, 2001) suggest the common S-type asteroids (such as Eros) may have ordinary chondritelike compositions. Thus, if ordinary chondrites are in fact derived from S-type (and not just Q-type) asteroids, qualitatively it appears possible to reconcile the high proportion of ordinary-chondrite meteorite falls with the supply of objects in near-Earth space.

3.4. Shapes and Rotations

The shapes and rotation rates of small objects, such as NEOs, arise from a variety of factors. NEOs derived from the asteroid belt are almost certainly second- (or multi-) generation collision fragments from once-larger parent bodies (see *Davis et al.*, 2002). Asteroids in the size range of a few tens of kilometers, or smaller, are not large enough for self-gravity to protect them from catastrophic disruption over the age of the solar system. Just as NEOs are relatively recent (and transient) visitors to the inner solar system, most have shapes and rotations that have been reworked on a timescale short compared with 4.5 b.y. (see *Paolicchi et al.*, 2002). Collision processes have also been active on the cometary component (if any) that contributes to the NEO population, where *Durda and Stern* (2000) calculate that comet nuclei having NEO-like sizes (smaller than 5 km) have also undergone substantial collisional processing since the time of original formation.

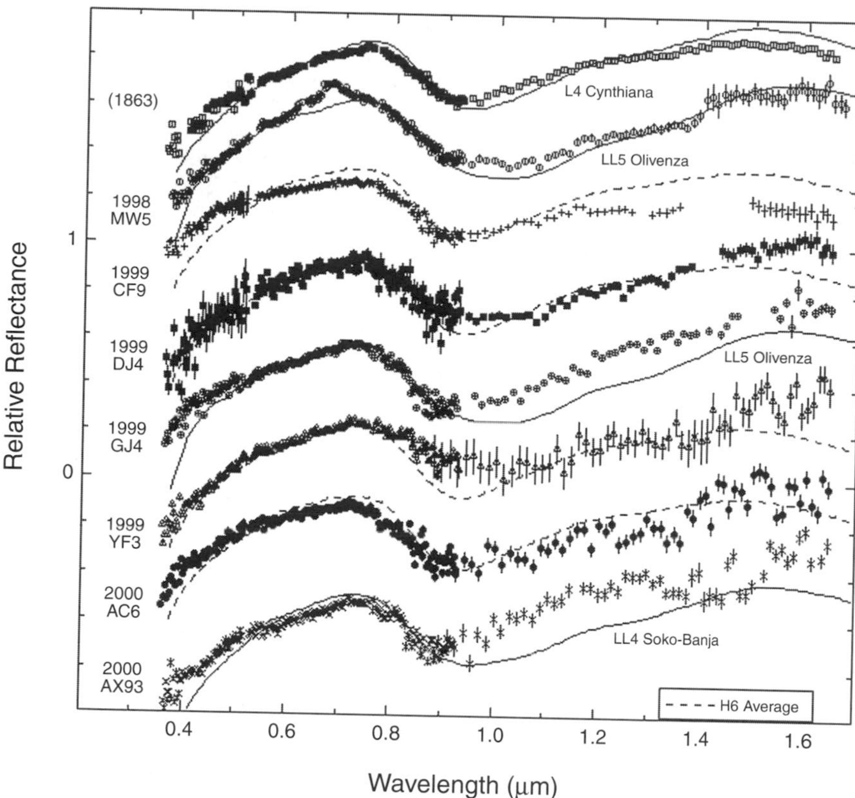

Fig. 4. A decade ago, only one NEO [1862 Apollo (*McFadden et al.,* 1985)] was known to have spectral properties resembling ordinary-chondrite meteorites. At present about 20% of all measured NEOs provide a plausible match to ordinary chondrites, with several examples illustrated here. [NEO data from *Binzel et al.* (2001); meteorite data from *Gaffey* (1976).]

An important result bearing on the collisional (and hence shape and rotation) evolution of NEOs comes from the crater statistics on Eros revealed by the *NEAR Shoemaker* mission (*Veverka et al.,* 2000). These results suggest that at some time since Eros entered near-Earth space, it has been effectively decoupled from the collisional environment of the main belt (*Michel et al.,* 1998). Thus, the shapes and rotations seen for NEOs (with exceptions noted below) may be most strongly determined by the processes occurring at their place of origin. If this is the case, then the shapes and perhaps rotations seen for NEOs that come from the asteroid belt should be representative of what we would observe in the asteroid belt at these small diameters if our observational techniques allowed. Analysis of the YORP (Yarkovsky-O'Keefe-Radzievskii-Paddack) effect by *Rubincam* (2000), however, points out the possibility of anisotropic thermal emission dominating the spin states of kilometer-sized bodies. Thus, the YORP effect might decouple the spin rates for most kilometer-sized asteroids from their initial state at the time of formation. Unfortunately, it is currently below the capabilities of most facilities to measure the detailed physical properties of main-belt asteroids below the size range of 5–10 km.

There is an observational suggestion that asteroidal NEOs are indeed similar in rotation and shape to their comparably sized main-belt counterparts. Using rotational lightcurves to convey information on spin period and approximate shape, Table 2 compares NEOs with two diameter (D) ranges of main-belt asteroids. The first group attempts to provide a comparison for NEOs by using the subset of main-belt asteroids having D < 2 km, approximating (as closely as possible with available data) the size range of NEOs. The second group simply contains the lightcurve characteristics of large (D > 130 km) main-belt asteroids. The sample size for these main-belt groups, and for the NEOs, is more than 100 objects in each case. Because NEOs are typically observed at large phase angles, all data have been reduced to their expected lightcurve amplitudes at 0° solar-phase angle, following the method of *Zappalà et al.* (1990). Table 2 shows that both the reduced-lightcurve amplitudes and the rota-

TABLE 2. Mean values of asteroid amplitudes and rotation rates.

Population	⟨D⟩ (km)	Observed Amplitude (mag)	N	Reduced Amplitude (mag)	Rotation Rate (rev/d)	N
NEOs	2.9 ± 0.5	0.49 ± 0.04	118	0.29	4.80 ± 0.29	119
MBAs (D < 12 km)	6.8 ± 0.3	0.35 ± 0.03	102	0.28	4.34 ± 0.23	100
MBAs (D > 130 km)	186 ± 1	0.22 ± 0.01	100	0.19	2.90 ± 0.12	100

tion rates are statistically indistinguishable between NEOs and the D < 12 km main-belt asteroids. These results give us confidence that the rotation and shape characteristics of asteroidal NEOs are reasonable proxies for similar diameter main-belt asteroids. Among the unusual complexities revealed are a nonprincipal axis rotation for 3288 Seleucus, 4179 Toutatis, 1994 AW1, and 4486 Mithra (see *Pravec et al.*, 2002; *Ostro et al.*, 2002; also *Lupishko and DiMartino,* 1998, and references therein). Super-fast rotators (having periods between 2 and 20 min) have been revealed through a variety of observations (e.g., *Steel et al.*, 1997; *Ostro et al.*, 1999; *Pravec et al.*, 2000a; *Whiteley et al.*, 2002). *Pravec and Harris* (2000) and *Whiteley et al.* (2002) demonstrate that these fast-spinning objects are beyond the rotational breakup limit for aggregates with no tensile strength ("rubble piles") for bulk densities plausible for asteroids.

How well can rotation and shape data distinguish those NEOs that may be of cometary origin? By definition, objects labeled as comets have comae that substantially increase the difficulty of directly measuring the comparable physical properties of their nuclei. Yet those comets that have been measured typically have axial ratios that would produce rotational lightcurves whose amplitude of brightness variation would be in the range of 0.5–1.0 mag (*Hartmann and Tholen,* 1990; *Luu,* 1994; *Nelson et al.* 2001), substantially larger than the 0.29 value estimated for asteroidal NEOs in Table 2. Thus, elongated shapes may provide some suggestion, when combined with dynamical and compositional factors, for discerning NEOs as having a cometary origin. Similarly, *Binzel et al.* (1992) find that slower rotations might also indicate cometary NEOs. However, we emphasize that rotation and shape alone are not sufficient by themselves to conclusively reveal a cometary origin for an individual NEO.

An unresolved question at this time is whether the relatively common occurrence of binary objects among NEOs (*Pravec and Harris,* 2000) is especially intrinsic to the NEO population. Table 1 in *Merline et al.* (2002) lists the detailed properties of the handful of NEOs that (to date) have been revealed to be binary.

NEOs that suffer close encounters with Earth could be distorted into particularly elongated shapes, and these tidal distortions could play a role in forming binaries (*Bottke et al.*, 1996, 1999; *Richardson et al.*, 1998). However, the discovery of binaries within the main-belt population [e.g., 762 Pulcova and 90 Antiope (*Merline et al.*, 2000)] indicates that the process or processes that form them are not unique to the NEO population. These processes are examined in *Paolicchi et al.* (2002).

3.5. Optical Properties and Surface Structure

The small diameters (young age, low surface gravity), proximity, and possibly diverse origins of the NEOs make an understanding of their surface properties a topic of broad interest. A complete and extensive review of these properties is presented by *Lupishko and Di Martino* (1998). Here we present an updated summary.

TABLE 3. Mean optical parameters of S-type NEOs and S-type main-belt asteroids (all wavelength-dependent measurements are with respect to the V band).

Parameter	NEAs	N	MBAs (D > 100 km)	N
Albedo polarimetric	0.183 ± 0.011	9	0.177 ± 0.004	28
Albedo radiometric	0.190 ± 0.014	23	0.166 ± 0.006	27
U-B (mag)	0.445 ± 0.013	30	0.453 ± 0.008	28
B-V (mag)	0.856 ± 0.013	31	0.859 ± 0.006	28
β (mag/deg)	0.029 ± 0.002	9	0.030 ± 0.006	18
P_{min} (%)	0.77 ± 0.04	3	0.75 ± 0.02	28
h (%/deg)	0.098 ± 0.006	9	0.105 ± 0.003	23
α_{inv} (deg)	20.7 ± 0.2	6	20.3 ± 0.2	18

Table 3 compares the surface properties for large main-belt asteroids and NEOs which have S-type asteroid reflectance properties. These parameters include the polarimetric and radiometric albedos, color indices, phase coefficients β_V, and polarimetric parameters such as depth of negative polarization P_{min}, polarization slope h, and inversion angle α_{inv} [for the definition of these parameters see *Dollfus et al.* (1989)]. The table indicates that the smaller S-type NEOs may have higher albedos on average, a result consistent with a limited amount of thermal measurements and modeling of NEOs (see *Harris and Lagerros,* 2002). One explanation for the difference in albedo as a function of diameter (and presumably surface age) could be a space-weathering effect (see *Clark et al.*, 2002). If space weathering involves only a coating of grains [as proposed by *Pieters et al.* (2000)], then only measurements sensitive to spectral colors (and not particle size) would show a diameter-dependent effect. However, we note that because the albedo difference is suggested more strongly in the radiometric data than in the polarimetric data, thermal properties of the surfaces of these smaller bodies (and our success in modeling them) may play a role in creating this effect.

Interestingly, the characterization of the surface properties most sensitive to particle sizes as measured through the parameters β_V, P_{min}, h, and α_{inv} reveals no systematic differences across significant diameter ranges, suggesting that at least the majority of the S-type NEOs have the same surface porosity and roughness at the submicron scale as their larger diameter counterparts in the main belt (*Helfenstein and Veverka,* 1989). A comparison of Hapke parameters between NEOs, main-belt asteroids, and satellites (Table 4) shows similar results: Very few differences appear to be present at microscales across a broad range of diameters. Qualitatively, this may be understood as arising from the fact that numerous forces are at work on micrometer-sized particles. Gravity (and hence diameter dependence) may be relatively inconsequential compared with electrostatic forces and Poynting-Robertson drag (*Lee,* 1996). Thus, the relative presence (or absence) and structure of micrometer-sized particles on the surfaces of asteroids and NEOs may be quite independent of size.

Macroscale (centimeter and larger) differences in surface properties, however, become apparent when comparing small NEOs with large main-belt asteroids. The circular-

TABLE 4. Hapke parameters of NEOs and other small bodies.

Object	Data	Particle Albedo w	Oppostion Surge Width h	Oppostion Surge Amplitude Bo	Asymmetry Parameter g	Microscopic Roughness θ (deg)	Reference
Eros	NEAR	0.44 ±0.044	0.03 ±0.003	1.0 ±0.1	−0.31 ±0.031	28 ±2.8	*Clark et al.* (2000)
Geographos	EB,rad	≥0.22	0.02	1.32 ±0.10	−0.34 ±10	25	*Hudson and Ostro* (1999)
Apollo	EB	0.318 ±0.004	0.034 ±0.007	0.90 ±0.02	−0.32 ±0.01	15 ±1	*Helfenstein and Veverka* (1989)
Toutatis	EB	0.261 ±0.019	0.036 ±0.023	1.20 ±0.32	−0.29 ±0.06	32 ±8	*Hudson and Ostro* (1998)
Castalia N	EB	0.384 ±0.07	—	—	−0.11 ±0.09	46 ±10	*Hudson et al.* (1997)
Castalia S	EB	0.239 ±0.07	—	—	−0.30 ±0.09	25 ±10	*Hudson and Ostro* (1998)
Golevka	EB	0.58 ±0.03	0.0114 ±0.0004	0.758 ±0.014	−0.435 ±0.001	7 ±7	*Mottola et al.* (1997)
Golevka	rad	0.173 ±0.006	0.024 ±0.012	1.03 ±0.45	−0.34 ±0.02	20 ±5	*Hudson et al.* (2000)
Phobos	VK	0.070 ±0.020	0.055 ±0.025	4.0 +6−1	−0.08*	22 ±2	*Simonelli et al.* (1998)
Deimos	VK	0.079 +0.008 −0.006	0.068 +0.082 −0.037	1.65 +0.90 −0.61	−0.29 ±0.03	16 ±5	*Thomas et al.* (1996)
Ida	EB,GL	0.218 +0.024 −0.005	0.020 ±0.005	1.53 ±0.10	−0.33 ±0.01	18 ±2	*Helfenstein et al.* (1996)
Dactyl	GL	0.211 +0.028 −0.010	⟨0.020⟩	⟨1.53⟩	−0.33 ±0.03	23 ±5	*Helfenstein et al.* (1996)
Gaspra	EB,GL	0.360 ±0.07	0.060 ±0.01	1.63 ±0.07	−0.18 ±0.04	29 ±2	*Helfenstein et al.* (1996)

*Effective value for two-term Henyey-Greenstein phase function.

EB = Earth-based (V filter); GL = *Galileo* (GRN filter); VK = Viking (clear filter); NEAR = *NEAR Shoemaker* (0.55 μm); rad = radar observations.

polarization ratio of radar echo power denoted as SC:OC (see *Ostro et al.,* 2002) is diagnostic of surface roughness at scales of the radar wavelength and wave penetration depth. If the SC:OC ratio is very low, the surface should be smooth at scales within an order of magnitude of the radar wavelength (*Ostro,* 1989). The higher mean ratios depicted in Table 5 show that the surfaces of NEOs are much rougher than those of larger-diameter main-belt asteroids at the scale length of decimeters and meters. Asteroid 433 Eros is at present the only NEO for which we have *in situ* images of the surface at centimeter to meter scales, and thus Eros provides some perspective on what these surfaces may be like (*Veverka et al.,* 2001). NEOs on average have rougher surfaces than Eros. However, Eros has an SC:OC value (Table 5) that places it intermediate between NEOs and main-belt asteroids. In addition to their higher mean, the SC:OC ratios of individual NEOs show tremendous diversity and span ~1 order of magnitude, from 0.09 [(6178) 1986 DA, 2.3-km, M-type] to 1.0 (2101 Adonis, 3103 Eger, 1992 QN). Thus among the smallest objects, surfaces range from being highly smooth to incredibly rough. While surface roughness of main-belt asteroids and NEOs are different on average, comparable values of radar albedo (Table 5) imply similar bulk densities and porosities of surface materials.

TABLE 5. Mean radar albedos and circular polarization ratios of NEAs and main-belt asteroids.

Sample	⟨D⟩ km	Radar Albedo	N	SC/OC	N
433 Eros	13 × 13 × 33	0.20 ± 0.01	1	0.22 ± 0.06	1
NEAs, S-type	6.3 ± 2.7	0.16 ± 0.02	15	0.31 ± 0.03	17
MBAs, S-type	136.5 ± 12.2	0.15 ± 0.01	14	0.14 ± 0.02	10
NEAs, all types	1.9 ± 1.8	0.18 ± 0.02	24	0.36 ± 0.04	36
MBAs, all types	179.8 ± 27.3	0.15 ± 0.01	36	0.11 ± 0.01	22

4. CONCLUSIONS AND FUTURE WORK

Achieving an understanding of the population of NEOs provides insights into a broad range of solar-system processes. Progress has been made in recognizing the processes for delivering material to the vicinity of Earth, where dynamical studies and physical measurements show independent and consistent evidence for the inner asteroid belt as a primary source. While the cometary contribution remains uncertain, great progress has been made toward identifying sources for ordinary-chondrite meteorites among the near-Earth population. Key directions for future research include pinpointing more precisely and quantitatively the sources for NEOs. Work also remains to be done for quantitatively reconciling meteorite-fall statistics with the population of objects that intersects the Earth. All evidence points to asteroidal NEOs being representative of similarly sized objects in the main belt. From an exploration perspective, this correlation presents a convenient opportunity to study the diversity of main-belt compositions (such as through sample-return missions) with the comparative ease and convenience of operating in near-Earth space. For these scientific reasons, and for the pragmatic reasons of hazard and resource assessment, NEOs will remain a continuing focus for solar-system small-body research in the decades ahead.

Acknowledgments. R.P.B. acknowledges support for this research from the National Science Foundation, NASA, and The Planetary Society. We thank T. A. Lupishko for help in the original preparation of the data tables.

REFERENCES

Barucci M. A., Cruikshank D. P., Mottola S., and Lazzarin M. (2002) Physical properties of Trojan and Centaur asteroids. In *Asteroids III* (W. F. Bottke Jr. et al., eds.), this volume. Univ. of Arizona, Tucson.

Benedix G. K., McFadden L. A., and Morrow E. M. (1992) Bias correction factors for near-Earth asteroids. In *Asteroids, Comets, Meteors 1991,* pp. 65–68. Lunar and Planetary Institute, Houston.

Binzel R. P. and Xu S. (1993) Chips off of asteroid 4 Vesta: Evidence for the parent body of basaltic achondrite meteorites. *Science, 260,* 186–191.

Binzel R. P., Xu S., Bus S. J., and Bowell E. (1992) Origins for the near-Earth asteroids. *Science, 257,* 779–782.

Binzel R. P., Bus S. J., Burbine T. H., and Sunshine J. M. (1996) Spectral properties of near-Earth asteroids: Evidence for sources of ordinary chondrite meteorites. *Science, 273,* 946–948.

Binzel R. P., Bus S. J., and Burbine T. H. (1998) Size dependence of asteroid spectral properties: SMASS results for near-Earth and main-belt asteroids (abstract). In *Lunar and Planetary Science XXIX,* Abstract #1222. Lunar and Planetary Institute, Houston (CD-ROM).

Binzel R. P., Harris A. W., Bus S. J., and Burbine T. H. (2001) Spectral properties of near-Earth objects: Palomar and IRTF results for 48 objects including spacecraft targets (9969) Braille and (10302) 1989 ML. *Icarus, 151,* 139–149.

Bottke W. F., Nolan M. C., Melosh H. J., Vickery A. M., and Greenberg R. (1996) Origin of the spacewatch small Earth-approaching asteroids. *Icarus, 122,* 406–427.

Bottke W. F., Richardson D. C., Michel P., and Love S. G. (1999) 1620 Geographos and 433 Eros: Shaped by planetary tides? *Astron. J., 117,* 1921–1928.

Bottke W. F., Jedicke R., Morbidelli A., Petit J. M., and Gladman B. (2000) Understanding the distribution of near-Earth asteroids. *Science, 288,* 2190–2194.

Bottke W. F., Morbidelli A., Jedicke R., Petit J. M., Levison H. F., Michel P., and Netcalf T. S. (2002a) Debiased orbital and magnitude distribution of the near-Earth objects. *Icarus, 156,* 399–433.

Bottke W. F. Jr., Vokrouhlický D., Rubincam D. P., and Brož M. (2002b) The effect of Yarkovsky thermal forces on the dynamical evolution of asteroids and meteoroids. In *Asteroids III* (W. F. Bottke Jr. et al., eds.), this volume. Univ. of Arizona, Tucson.

Burbine T. H., McCoy T. J., Meibom A., Gladman B., and Keil K. (2002) Meteoritic parent bodies: Their number and identification. In *Asteroids III* (W. F. Bottke Jr. et al., eds.), this volume. Univ. of Arizona, Tucson.

Bus S. J. (1999) Compositional structure in the asteroid belt: Results of a spectroscopic survey. Ph.D. thesis, Massachusetts Institute of Technology, Cambridge. 367 pp.

Bus S. J. and Binzel R. P. (2002) Phase II of the Small Main-Belt Asteroid Spectroscopic Survey: A feature-based taxonomy. *Icarus, 158,* 146–177.

Bus S. J., Vilas F., and Barucci M. A. (2002) Visible-wavelength spectroscopy of asteroids. In *Asteroids III* (W. F. Bottke Jr. et al., eds.), this volume. Univ. of Arizona, Tucson.

Campins H., A'Hearn M. F., and McFadden L. A. (1987) The bare nucleus of comet P/Neujmin 1. *Astrophys. J., 316,* 847–857.

Chapman C. R., Binzel R. P., and Harris A. W. (1994) Physical properties of near-Earth asteroids: Implications for the hazard issue. In *Hazards Due to Comets and Asteroids* (T. Gehrels, ed.), pp. 537–549. Univ. of Arizona, Tucson.

Cheng A. F. (2002) Near Earth Asteroid Rendezvous: Mission summary. In *Asteroids III* (W. F. Bottke Jr. et al., eds.), this volume. Univ. of Arizona, Tucson.

Clark B. E., Thomas P. C., Veverka J., Helfenstein P., Robinson M. S., and Murchie S. L. (2000) NEAR lightcurves of asteroid 433 Eros. *Icarus, 145,* 641–644.

Clark B. E., Hapke B., Pieters C., and Britt D. (2002) Asteroid space weathering and regolith evolution. In *Asteroids III* (W. F. Bottke Jr. et al., eds.), this volume. Univ. of Arizona, Tucson.

Cochran A. L. and Barker E. S. (1984) Minor planet 1983 TB: A dead comet? *Icarus, 59,* 296–300.

Cruikshank D. P., Tholen D. J., Hartmann W. K., Bell J. F., and Brown R. H. (1991) Three basaltic Earth-approaching asteroids and the source of the basaltic meteorites. *Icarus, 89,* 1–13.

Davis D. R., Durda D. D., Marzari F., Campo Bagatin A., and Gil-Hutton R. (2002) Collisional evolution of small-body populations. In *Asteroids III* (W. F. Bottke Jr. et al., eds.), this volume. Univ. of Arizona, Tucson.

Dollfus A., Wolff M., Geake J. E., Dougherty L. M., and Lupishko D. F. (1989) Photopolarimetry of asteroids. In *Asteroids II* (R. P. Binzel et al., eds.), pp. 594–616. Univ. of Arizona, Tucson.

Durda D. D. and Stern S. A. (2000) Collision rates in the present-day Kuiper belt and centaur regions: Applications to surface activation and modification on comets, Kuiper belt objects, Centaurs, and Pluto-Charon. *Icarus, 145,* 220–229.

Erikson A., Mottola S., Lagerros J. S. V., Lindgren M., Piironen J., Oja T., Hahn G., Lagerkvist C.-I., Harris A. W., Nathues A., and Neukum G. (2000) The near-Earth objects follow-up program. *Icarus, 147,* 487–497.

Fernandez Y. R., McFadden L. A., Lisse C. M., Helin E. F., and Chamberlin A. B. (1997) Analysis of POSS images of comet-asteroid transition object 107P/1949 W1 (Wilson-Harrington). *Icarus, 128,* 114–126.

Fox K., Williams I. P., and Hunt J. (1985) The past and future of 1983 TB and its relationship to the Geminid meteor stream. In *Dynamics of Comets: Their Origin and Evolution* (A. Carusi and G. B. Valsecchi, eds.), pp. 143–148. IAU Colloq. No. 83, Reidel, Dordrecht.

Fowler J. W. and Chillemi J. R. (1992) IRAS asteroid data processing. In *The IRAS Minor Planet Survey* (E. F. Tedesco et al., eds.), pp. 17–43. Phillips Laboratory, Hanscom Air Force Base, Massachusetts.

Gaffey M. J. (1976) Spectral reflectance characteristics of the meteorite classes. *J. Geophys. Res., 81,* 905–920.

Gaffey M. J., Reed K. L., and Kelley M. S. (1992) Relationship of E-type Apollo asteroid 3103 (1982 BB) to the enstatite achondrite meteorites and the Hungaria asteroids. *Icarus, 100,* 95–109.

Gaffey M. J., Bell J. F., Brown R. H., Burbine T. H., Piatek J. L., Reed K. L., and Chaky D. A. (1993) Mineralogical variations within the S-type asteroid class. *Icarus, 106,* 573–602.

Gaffey M. J., Cloutis E. A., Kelley M. S., and Reed K. L. (2002) Mineralogy of asteroids. In *Asteroids III* (W. F. Bottke Jr. et al., eds.), this volume. Univ. of Arizona, Tucson.

Gradie J. C. and Tedesco E. F. (1982) Compositional structure of the asteroid belt. *Science, 216,* 1405–1407.

Gradie J. C., Chapman C. R., and Tedesco E. F. (1989) Distribution of taxonomic classes and the compositional structure of the asteroid belt. In *Asteroids II* (R. P. Binzel et al., eds.), pp. 316–335. Univ. of Arizona, Tucson.

Hammergren M. (1998) The composition of near-Earth objects. Ph.D. Thesis, University of Washington. 110 pp.

Harris A. W. and Lagerros J. S. V. (2002) Asteroids in the thermal infrared. In *Asteroids III* (W. F. Bottke Jr. et al., eds.), this volume. Univ. of Arizona, Tucson.

Hartmann W. K. and Tholen D. J. (1990) Comet nuclei and Trojan asteroids: A new link and possible mechanism for comet splitting. *Icarus, 86,* 448–454.

Helfenstein P. and Veverka J. (1989) Physical characterization of asteroid surfaces from photometric analysis. In *Asteroids II* (R. P. Binzel et al., eds.), pp. 557–593. Univ. of Arizona, Tucson.

Helfenstein P., Veverka J., Thomas P. C., Simonelli D. P., Klaasen K., Johnson T. V. Fanale F., Granahan J., McEwen A. S., Belton M., and Chapman C. (1996) Galileo photometry of asteroid 243 Ida. *Icarus, 120,* 48–65.

Hicks M., Fink U., and Grundy W. (1998) The unusual spectra of 15 near-Earth asteroids and extinct comet candidates. *Icarus, 133,* 69–78.

Hicks M., Buratti B. J., Newburn R. L., and Rabinowitz D. L. (2000) Physical observations of 1996 PW and 1997 SE5: Extinct comets or D-type asteroids? *Icarus, 143,* 354–359.

Hudson R. S. and Ostro S. J. (1998) Photometric properties of asteroid 4179 Toutatis from lightcurves and a radar-derived physical model. *Icarus, 135,* 451–457.

Hudson R. S. and Ostro S. J. (1999) Physical model of asteroid 1620 Geographos from radar and optical data. *Icarus, 140,* 369–378.

Hudson R. S., Ostro S. J., and Harris A. W. (1997) Constraints on spin state and Hapke parameters of asteroid 4769 Castalia using lightcurves and radar-derived shape model. *Icarus, 130,* 165–176.

Hudson R. S. and 27 colleagues (2000) Radar observations and physical model of asteroid 6489 Golevka. *Icarus, 148,* 37–51.

Huebner W. F., Cellino A., Cheng A. F., and Greenberg M. (2001) NEOs: Physical properties. In *International Seminar on Planetary Emergencies XXV Session* (R. Ragaini, ed.), pp. 309–340. World Scientific, Singapore.

Jewitt D. C. and Meech K. J. (1985) Rotation of the nucleus of P/Arend-Rigaux. *Icarus, 64,* 329–335.

Johnson T. V. and Fanale F. P. (1973) Optical properties of carbonaceous chondrites and their relationship to asteroids. *J. Geophys. Res., 78,* 8507–8518.

Kerridge J. F. and Matthews M. L., eds. (1988) *Meteorites and the Early Solar System.* Univ. of Arizona, Tucson. 1269 pp.

Lee P. (1996) Dust levitation on asteroids. *Icarus, 124,* 181–194.

Lewis J. S. and Hutson M. L. (1993) Asteroidal resource opportunities suggested by meteoric data. In *Resources of Near-Earth Space* (J. S. Lewis et al., eds.), pp. 523–542. Univ. of Arizona, Tucson.

Lipschutz M. E., Gaffey M. J., and Pellas P. (1989) Meteoritic parent bodies: Nature, number, size, and relation to present-day asteroids. In *Asteroids II* (R. P. Binzel et al., eds.), pp. 740–777. Univ. of Arizona, Tucson.

Luu J. X. (1994) Comets disguised as asteroids. *Publ. Astron. Soc. Pacific, 106,* 425–435.

Luu J. X. and Jewitt D. C. (1989) On the relative number of C types and S types among near-Earth asteroids. *Astron. J., 98,* 1905–1911.

Luu J. X. and Jewitt D. C. (1992) High resolution surface brightness profiles of near-Earth asteroids. *Icarus, 97,* 276–287.

Lupishko D. F. and Di Martino M. (1998) Physical properties of near-Earth asteroids. *Planet. Space Sci., 46,* 47–74.

Marsden B. G. and Williams G. V. (1999) *Catalogue of Cometary Orbits,* 13th edition. IAU Central Bureau for Astronomical Telegrams and Minor Planet Center. 127 pp.

McCoy T. J., Burbine T. H., McFadden L. A., Starr R. D., Gaffey M. J., Nittler L. R., Evans L. G., Izenberg N., Lucey P., Trombka J. I., Bell J. F. III, Clark B. E., Clark P. E., Squyres S. W., Chapman C. R., Boynton W. V. and Veverka J. (2001) The composition of 433 Eros: A mineralogical-chemical synthesis. *Meteoritics & Planet. Sci., 36,* 1661–1672.

McFadden L. A., Gaffey M. J., and McCord T. B. (1984) Mineralogical-petrological characterization of near-Earth asteroids. *Icarus, 59,* 25–40.

McFadden L. A., Gaffey M. J., and McCord T. B. (1985) Near-Earth asteroids: Possible sources from reflectance spectroscopy. *Science, 229,* 160–163.

McFadden L. A., Tholen D. J., and Veeder G. J. (1989) Physical properties of Aten, Apollo and Amor asteroids. In *Asteroids II* (R. P. Binzel et al., eds.), pp. 442–467. Univ. of Arizona, Tucson.

McFadden L. A., Cochran A. L., Barker E. S., Cruikshank D. P., and Hartmann W. K. (1993) The enigmatic object 2201 Oljato: Is it an asteroid or an evolved comet? *J. Geophys. Res., 98,* 3031–3041.

Merline W. J., Close L. M., Dumas C., Shelton J. C., Menard J., Chapman C. R., and Slater D. C. (2000) Discovery of companions to asteroids 762 Pulcova and 90 Antiope by direct imaging. *Bull. Am. Astron. Soc., 32,* 1017.

Merline W. J., Weidenschilling S. J., Durda D. D., Margot J. L., Pravec P., and Storrs A. D. (2002) Asteroids *do* have satellites. In *Asteroids III* (W. F. Bottke Jr. et al., eds.), this volume. Univ. of Arizona, Tucson.

Michel P., Farinella P., and Froeschlé Ch. (1998) Dynamics of Eros. *Astron. J., 116,* 2023–2031.

Michel P., Zappalà V., Cellino A., and Tanga P. (2000) Estimated abundances of Atens and asteroids on orbits between Earth and Sun. *Icarus, 143,* 421–424.

Migliorini F., Morbidelli A., Zappalà V., Gladman B. J., Bailey M. E., and Cellino A. (1997) Vesta fragments from ν_6 and 3:1 resonances: Implications for V-type near-Earth asteroids and howardite, eucrite and diogenite meteorites. *Meteoritics & Planet. Sci., 32,* 903–916.

Migliorini F., Michel P., Morbidelli A., Nesvorný D., and Zappalà V. (1998) Origin of multikilometer Earth- and Mars-crossing asteroids: A quantitative simulation. *Science, 281,* 2022–2024.

Milani A., Carpino M., Hahn G., and Nobili A. (1989) Dynamics of planet crossing asteroids: Classes of orbital behavior. Project Spaceguard. *Icarus, 78,* 212–269.

Morbidelli A. and Nesvorný D. (1999) Numerous weak resonances drive asteroids toward terrestrial planets orbits. *Icarus, 139,* 295–308.

Morbidelli A., Bottke W. F. Jr., Froeschlé Ch., and Michel P,

(2002) Origin and evolution of near-Earth objects. In *Asteroids III* (W. F. Bottke Jr. et al., eds.), this volume. Univ. of Arizona, Tucson.

Morrison D., Harris A. W., Sommer G., Chapman C. R., and Carusi A. (2002) Dealing with the impact hazard. In *Asteroids III* (W. F. Bottke Jr. et al., eds.), this volume. Univ. of Arizona, Tucson.

Mottola S. and 28 colleagues (1997) Physical model of near-Earth asteroid 6489 Golevka (1991 JX) from optical and infrared observations. *Astron. J., 114,* 1234–1245.

Muinonen K., Piironen J., Shkuratov Yu. G., Ovcharenko A., and Clark B. E. (2002) Asteroid photometric and polarimetric phase effects. In *Asteroids III* (W. F. Bottke Jr. et al., eds.), this volume. Univ. of Arizona, Tucson.

Nelson R. M., Rayman M. D., Varghese P., and Lehman D. H. (2001) The Deep Space One encounter with the comet Borrelly. *Bull Am. Astron. Soc., 33,* 1087.

Oberst J., Mottola S., Di Martino M., Hicks M., Buratti B., Soderblom L., and Thomas N. (2001) A model for rotation and shape of asteroid 9969 Braille from ground-based observations and images obtained during the Deep Space 1 (DS1) flyby. *Icarus, 153,* 16–23.

Ostro S. J. (1989) Radar observations of asteroids. In *Asteroids II* (R. P. Binzel et al., eds.), pp. 192–212. Univ. of Arizona, Tucson.

Ostro S. J., Campbell D. B., Chandler J. F., Hine A. A., Hudson R. S., Rosema K. D., and Shapiro I. I. (1991) Asteroid 1986 DA: Radar evidence for a metallic composition. *Science, 252,* 1399–1404.

Ostro S. J. and 19 colleagues (1999) Radar and optical observations of asteroid 1998 KY26. *Science, 285,* 557–559.

Ostro S. J., Hudson R. S., Benner L. A. M., Giorgini J. D., Magri C., Margot J.-L., and Nolan M. C. (2002) Asteroid radar astronomy. In *Asteroids III* (W. F. Bottke Jr. et al., eds.), this volume. Univ. of Arizona, Tucson.

Paolicchi P., Burns J. A., and Weidenschilling S. J. (2002) Side effects of collisions: Spin rate changes, tumbling rotation states, and binary asteroids. In *Asteroids III* (W. F. Bottke Jr. et al., eds.), this volume. Univ. of Arizona, Tucson.

Perozzi E., Rossi A., and Valsecchi G. B. (2001) Basic targeting for rendezvous and flyby missions to the near-Earth asteroids. *Planet. Space Sci., 49,* 3–22.

Pieters C. A., Taylor L. A., Noble S. K., Keller L. P., Hapke B., Morris R. V., Allen C. C., McKay D. S., and Wentworth S. (2000) Space weathering on airless bodies: Resolving a mystery with lunar samples. *Meteoritics & Planet. Sci., 35,* 1101–1107.

Pravec P. and Harris A. W. (2000) Fast and slow rotation of asteroids. *Icarus, 148,* 12–20.

Pravec P., Hergenrother C., Whiteley R. J. , Šarounová L., Kušnirák P., and Wolf M. (2000a) Fast rotating asteroids 1999 TY_2, 1999 SF_{10}, and 1998 WB_2. *Icarus, 147,* 477–486.

Pravec P., Šarounová L., Rabinowitz D. L., Hicks M. D., Wolf M., Krugly Y. N., Velichko F. P., Shevchenko V. G., Chiorny V. G., Gaftonyuk N. M., and Genevier G. (2000b) Two-period lightcurves of 1996 FG3, 1998 PG, and (5407) 1992 AX: One probable and two possible binary asteroids. *Icarus, 146,* 190–203.

Pravec P., Harris A. W., and Michałowski T. (2002) Asteroid rotations. In *Asteroids III* (W. F. Bottke Jr. et al., eds.), this volume. Univ. of Arizona, Tucson.

Rabinowitz D. L. (1997a) Are main-belt asteroids a sufficient source for the Earth-approaching asteroids? I. Predicted vs observed orbital distributions. *Icarus, 127,* 33–54.

Rabinowitz D. L. (1997b) Are main-belt asteroids a sufficient source for the Earth-approaching asteroids? I. Predicted vs observed size distributions. *Icarus, 130,* 287–295.

Rabinowitz D. L. (1998) Size and orbit dependent trends in the reflectance colors of Earth-approaching asteroids. *Icarus, 134,* 342–346.

Richardson D. C., Bottke W. F. Jr., and Love S. G. (1998) Tidal distortion and disruption of Earth-crossing asteroids. *Icarus, 134,* 47–76.

Rubincam D. P. (2000) Radiative spin-up and spin-down of small asteroids. *Icarus, 148,* 2–11.

Shoemaker E. M., Williams J. G., Helin E. F., and Wolf R. F. (1979) Earth-crossing asteroids: Orbital classes, collision rates with Earth, and origin. In *Asteroids* (T. Gehrels, ed.), pp. 253–282. Univ. of Arizona, Tucson.

Simonelli D. P., Wisz M., Switala A., Adinolfi D., Veverka J., Thomas P. C., and Helfenstein P. (1998) Photometric properties of Phobos surface materials from Viking images. *Icarus, 131,* 52–77.

Steel D. I., McNaught R. H., Garradd G. J., Asher D. J., and Taylor A. D. (1997) Near-Earth asteroid 1995 HM: A highly-elongated monolith rotating under tension? *Planet. Space Sci., 45,* 1091–1098.

Stokes G. H., Evans J. B., and Larson S. M. (2002) Near-Earth asteroid search programs. In *Asteroids III* (W. F. Bottke Jr. et al., eds.), this volume. Univ. of Arizona, Tucson.

Tedesco E. F. and Gradie J. (1987) Discovery of M class objects among the near-Earth asteroid population. *Astron. J., 93,* 738–746.

Tholen D. J. (1984) Asteroid taxonomy from cluster analysis of photometry. Ph.D. thesis, University of Arizona, Tucson.

Tholen D. J. and Whiteley R. J. (1998) Results from NEO searches at small solar elongations. *Bull. Am. Astron. Soc., 30,* 1041.

Thomas P. C., Adinolfi D., Helfenstein P., Simonelli D., and Veverka J. (1996) The surface of Deimos: Contribution of materials and processes to its unique appearance. *Icarus, 123,* 536–556.

Thomas P. C., Binzel R. P., Gaffey M. J., Storrs A. D., Wells E. N., and Zellner B. H. (1997) Impact excavation on asteroid 4 Vesta: Hubble Space Telescope results. *Science, 277,* 1492–1495.

Trombka J., Squyres S., Bruckner J., Boynton W., Reedy R., McCoy T., Gorenstein P., Evans L., Arnold J., Starr R., Nittler L., Murphy M., Mikheeva I., McNutt R., McClanahan T., McCartney E., Goldsten J., Gold R., Floyd S., Clark P., Burbine T., Bhangoo J., Bailey S., and Pataev M. (2000) The elemental composition of asteroid 433 Eros: Results of the NEAR Shoemaker X-ray spectrometer. *Science, 289,* 2101–2105.

Veverka J. and 30 colleagues (2000) NEAR at Eros: Imaging and spectral results. *Science, 289,* 2088–2097.

Veverka J. and 40 colleagues (2001) The landing of the NEAR-Shoemaker spacecraft on asteroid 433 Eros. *Nature, 413,* 390–393.

Vokrouhlický D., Milani A., and Chesley S. R. (2000). Yarkovsky effect on small near-Earth asteroids: Mathematical formulation and examples. *Icarus, 148,* 118–138.

Weissman P. R., Bottke W. F. Jr., and Levison H. F. (2002) Evolution of comets into asteroids. In *Asteroids III* (W. F. Bottke Jr. et al., eds.), this volume. Univ. of Arizona, Tucson.

Wetherill G. W. (1985) Asteroidal sources of ordinary chondrites. *Meteoritics, 20,* 1–22.

Wetherill G. W. (1988) Where do Apollo objects come from? *Icarus, 76,* 1–18.

Wetherill G. W. and Chapman C. R. (1988) Asteroids and meteorites. In *Meteorites and the Early Solar System* (J. F. Kerridge and M. S. Matthews, eds.), pp. 35–67. Univ. of Arizona, Tucson.

Whipple F. L. (1983) *1983 TB and the Geminid Meteors.* IAU Circular No. 3881.

Whiteley R. J. (2001) A compositional and dynamical survey of the near-Earth asteroids. PhD. thesis, University of Hawai'i. 202 pp.

Whiteley R. J. and Tholen D. J. (1999) The UH near-Earth asteroid composition survey: An update. *Bull Am. Astron. Soc., 31,* 945.

Whiteley R. J., Tholen D. J., and Hergenrother C. W. (2002) Lightcurve analysis of four new monolithic fast-rotating asteroids. *Icarus,* in press.

Wisniewski W. Z., Michałowski T. M., Harris A. W., and McMillan R. S. (1997) Photometric observations of 125 asteroids. *Icarus, 126,* 395–449.

Williams I. P. and Wu Z. (1983) The Geminid meteor stream and asteroid 3200 Phaethon. *Mon. Not. R. Astron. Soc., 262,* 231–248.

Zappalà V, Cellino A., Barucci A. M., Fulchignoni M., and Lupishko D. F. (1990) An analysis of the amplitude-phase relationship among asteroids. *Astron. Astrophys., 231,* 548–560.

Zellner B. (1979) Asteroid taxonomy and the distribution of the compositional types. In *Asteroids* (T. Gehrels, ed.), pp. 783–806. Univ. of Arizona, Tucson.

Physical Properties of Trojan and Centaur Asteroids

M. A. Barucci
Observatoire de Paris

D. P. Cruikshank
NASA Ames Research Center

S. Mottola
Deutsches Zentrum für Luft- und Raumfahrt, German Aerospace Center

M. Lazzarin
University of Padova

Trojans and Centaurs are primitive, peculiar objects orbiting in the middle solar system. Both groups characteristically have low albedos and red colors. Physical observations of Trojans reveal featureless reddish spectra, implying surfaces probably rich in complex organic solid materials. The interiors are expected to be rich in H_2O ice and other volatile material. Centaurs have surfaces showing dramatically different spectral reflectances, from neutral to very red. Some spectra are featureless, while others show signatures of water ice, methanol, or other light hydrocarbons. Trojans were formed near Jupiter's orbit, while Centaurs were formed far beyond Jupiter's orbit, but both were formed at low temperatures at which water exists as solid ice.

1. INTRODUCTION

This chapter concerns the physical properties of two populations of minor bodies: the Trojans and the Centaurs, defined primarily on the basis of their specialized orbital characteristics.

The Trojan asteroids that we consider here are members of the population of objects that are coorbital with Jupiter. Mars is also known to have at least four coorbital asteroids, and the outer planets may have similar companions that have not yet been discovered (*Weissman and Levison,* 1997; *Evans and Tabachnik,* 2000), thus expanding the original definition of a Trojan asteroid beyond those that are companions to Jupiter. There are two swarms of jovian Trojans, each consisting of a number of objects that librate about the L_4 and L_5 Lagrangian points in Jupiter's orbit, 60° of heliocentric ecliptic longitude ahead and 60° behind the planet, and possessing sufficient dynamical stability to survive over the age of the solar system (*Marzari and Scholl,* 1998).

Because of their special place in the solar system vis-à-vis Jupiter, the Trojans are of particular interest and importance in our attempts to understand the origin and evolution of Jupiter and its systems of inner (regular) and outer (irregular) satellites. The surface and internal compositions of the Trojans are expected to reflect the materials and conditions in the solar nebula at the time and location of their accretion. The questions that draw us to a study of the physical properties of these objects, using currently available remote-sensing techniques (spectroscopy, radiometry, and photometry), include whether the conditions of formation were uniform for all Trojans, how genesis of the Trojans was related to the formation of Jupiter and the satellites, and how formation of the Trojans relates to small satellites of the other outer planets.

Over the age of the solar system, the gravitational action of Neptune on the inner regions of the Kuiper Disk has caused an erosion of the Kuiper Disk population, ejecting a large fraction of the bodies that formerly occupied near-circular orbits in the range of ~30–45-AU heliocentric distance (*Duncan et al.,* 1995; *Levison and Weissman,* 1999). This process continues at the present time. Most of these bodies have been perturbed to more distant orbits, but some of them are perturbed into orbits that traverse the inner regions of the solar system, giving rise to the short-period comets and to the Centaurs (*Levison and Duncan,* 1997). The Centaur objects occupy orbits in the region of the outer planets, typically crossing the orbits of those planets. They are dynamically unstable, with dynamical lifetimes of ~10^6 10^7 yr before ejection from the planetary region or collision with a planet. The Centaurs are given traditional minor-planet designations, although in terms of their physical properties many of them differ from the main-belt and Trojan asteroids.

Measurements of thermal emission of a few Trojans and the Centaurs indicate that these objects have low surface albedos, in the range of 0.03–0.13 (e.g., *Cruikshank,* 1977; *Fernandez et al.,* 2002). Various lines of observational evidence suggest that much of the surface material is C-rich, in the form of complex organic polymers, polycyclic aromatic hydrocarbons, and other macromolecular compounds (e.g., *Cruikshank and Khare,* 2000; *Khare et al.,* 2001).

Specifically, the occurrence of a vast array of organic solids in the low-albedo nucleus of Comet Halley (e.g., *Kissel and Krueger,* 1987), as well as those in the black carbonaceous meteorites (*Cronin et al.,* 1988; *Cronin and Chang,* 1993; *Ehrenfreund et al.,* 2001), support the link between low-albedo surfaces and the presence of C in many forms. Some combinations of minerals with elemental C of inorganic origin may be able to match the surface reflectance properties of some of the low-albedo objects in the solar system, as discussed below for the case of Trojan asteroid 624 Hektor.

Thus, while the Trojans probably originated in or near their current heliocentric distance, the Centaurs have clearly come to the planetary region from an external reservoir, where they are presumed to have accreted in the distant solar nebula. Because of the low temperature, even at 5 AU heliocentric distance, both groups are thought to consist of primitive material — that is, material that has been relatively little altered chemically since the accretion epoch. The details of, and degree to which, primitive materials differ in composition between two solar nebula regions at ~5 AU and 30–50 AU are matters of compelling interest, thus motivating us to pursue physical studies of all available objects in both classes.

2. TROJANS

As of October 5, 2001, there were 663 known asteroids in the L_4 group and 421 in the L_5 group. From a wide-field observational survey, *Jewitt et al.* (2000) estimate that the population of the L_4 group is ~1.6×10^5 (objects with radii ≥1 km), with a combined mass of ~10^{-4} $M_{\oplus}$. The mean collision velocity in the Trojan clouds is about 5 km s^{-1}, similar to that in the main belt (*Marzari et al.,* 1996). This occurs because the lower Keplerian velocities at the heliocentric distance of the Trojan clouds are compensated by the higher average inclinations of the Trojans. The intrinsic collision probability for the Trojans is about twice that found in the main belt, which, contrary to what was commonly accepted earlier, leads to a picture of a Trojan population where considerable collisional evolution has taken place. This scenario is compatible with the discovery of dynamical families among the Trojans (*Milani,* 1994).

2.1. Rotational Properties

Shape and angular momentum are acquired during the accretion process, and are affected by the subsequent collisional evolution of the bodies. The characterization of these properties can provide important clues to the history of the Trojans. Lightcurve observations represent the basic tool for determining the rotational properties of asteroids, allowing for the determination of the rotation rate and angular momentum direction, as well as an approximation of body shape.

Early work by *Dunlap and Gehrels* (1969) on 624 Hektor, the largest member of the Trojan clouds, reveals a body with a very elongated shape. *Hartmann et al.* (1988) report lightcurve observations for an additional 18 Trojans. Based on these data, the authors suggest that higher than normal lightcurve amplitudes, and therefore considerably elongated shapes, might be characteristic of the Trojans, possibly reflecting a difference in composition and collisional evolution with respect to their counterparts in the main belt. The authors study the distribution of the lightcurve amplitudes of these Trojans by combining them with the available data for the Hilda asteroids in order to enlarge the data sample. The rationale for this choice is based on the consideration that both Trojans and Hildas reside in the so-called outer belt and are comparatively dynamically isolated from the main-belt population. Further, their lower number density and average Keplerian velocity would have protected them from an intense collisional evolution. However, while this argument is certainly true for the Hildas (*Dahlgren,* 1998), it does not apply to the Trojans, thus illustrating the need for separate analysis of the rotation properties of the two populations.

Zappalà et al. (1989) analyze the rotation rates of a group of 29 objects, consisting of Trojans, Hildas, and Cybeles, and find that the distribution can be fitted with a Maxwellian function. *Binzel and Sauter* (1992) report new lightcurve observations of Trojans and, based on a total sample of 31 objects, confirm the presence of high lightcurve amplitudes. The authors, however, find that the amplitudes were significantly larger than in the main belt only for objects larger than ~90 km.

This work focuses mainly on obtaining information on amplitudes for the largest possible number of objects, therefore compromising on coverage. As a result, the determinations of rotation periods are not highly reliable, and no conclusions on the rotation period distribution can be drawn. Another problem related to these initial investigations is that both the Trojan and the main-belt comparison samples used were affected by a considerable observational bias tending to favor the determination of high amplitudes and short periods.

The necessity of determining a sizable, reliable, and unbiased sample of rotation periods and amplitudes has motivated a major observational program to systematically explore the rotational properties of Trojans (S. Mottola et al., personal communication, 2001). In this study, new reliable rotation periods and amplitudes have been measured for 72 Trojans down to an absolute magnitude of H ~10.2, which, combined with the existing dataset, have resulted in determinations for 75 objects, most of which are in the diameter range 70–150 km. Particular care was taken to avoid the observational bias present in previous surveys. Furthermore, a sample of main-belt asteroids in the same size range was observed, in order to provide a suitable, unbiased comparison sample.

2.1.1. Rotation rates. The distribution of the rotational frequencies, f_T, for 75 Trojans is shown in Fig. 1a. The mean value is $f_T = 2.14 \pm 0.12$ rev/d, which compares to $f_M = 2.26 \pm 0.14$ rev/d for the main-belt control sample. The two values fall within 1σ of one another, and therefore the means can be considered statistically indistinguishable.

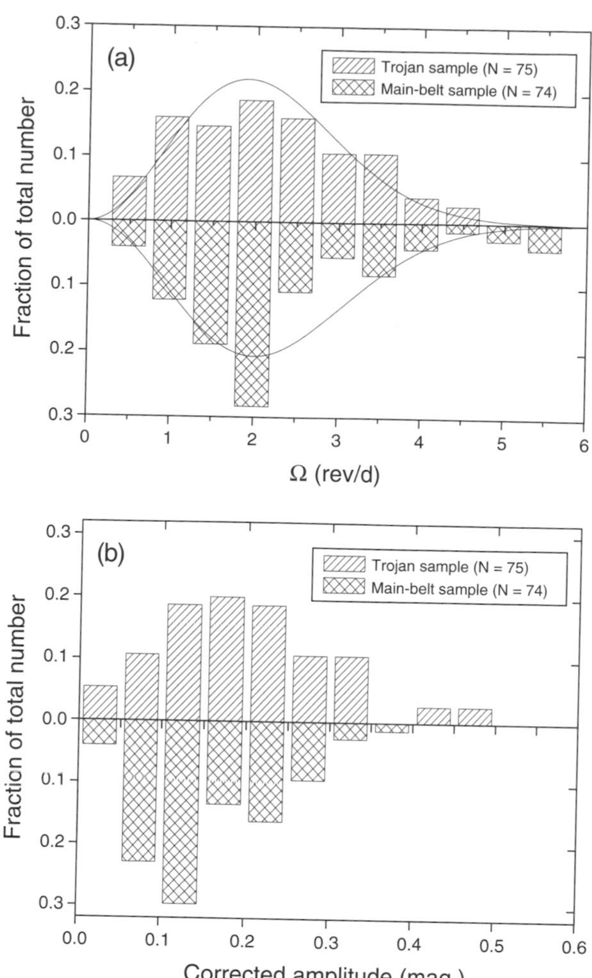

Fig. 1. **(a)** Distribution of the rotation periods for the Trojan and for the main-belt sample. The best-fit Maxwellian for the two distributions are shown as a solid line. **(b)** Distribution of the amplitudes reduced to an aspect of 60° and 0° phase angle. The average amplitude is 0.198 ± 0.011 for the Trojan and 0.155 ± 0.009 for the main-belt sample.

The distributions of the spin rates of asteroids have often been compared to single (or linear combinations of) Maxwellian distributions, which are considered to represent, although with some limitations, the rotational state of a collisionally relaxed population *(Farinella et al.,* 1981; *Binzel et al.,* 1989; *Fulchignoni et al.,* 1995). Figure 1a shows the best-fit Maxwellian distribution as a solid curve for the Trojan and main-belt samples respectively. The Trojan distribution is reasonably approximated by a Maxwellian function. On the other hand, the main-belt sample displays an excess of rotators with a period of ~12 h, and a deficiency of rotators in the range of 8–10 h. A Kolmogorov-Smirnov statistical test confirms these findings: The Trojan sample is found compatible with the Maxwellian distribution, while the hypothesis that the main-belt sample is drawn from the best-fit Maxwellian distribution is formally rejected at the 95% confidence level.

This result suggests that the Trojan population withstood a higher degree of collisional evolution compared to the main belt, which would be consistent with the high values for the intrinsic collisional probability and high average collisional velocity computed for the Trojans.

2.1.2. Lightcurve amplitudes. The amplitude of a lightcurve is an indicator of an asteroid's elongation. The amplitude, however, varies considerably depending on the unknown aspect angle under which the observations are made, with the amplitude being largest with an equatorial aspect and smallest with a polar aspect. When we observe an object at a random epoch, if the spin vectors are distributed isotropically on the celestial sphere, the most probable value for the aspect is 60°. If each object is observed only once, we can directly compare the amplitudes at random epochs. However, if we are confronted with a mixed dataset, where some objects have been observed only once, and some have been observed on several occasions, we have to identify a procedure to select the right amplitudes. *Hartmann et al.* (1988) select the largest available amplitude for their sample, in an attempt to estimate the maximum lightcurve amplitude. The reason for doing so is that the maximum possible amplitude is directly related to the a:b semiaxis ratio by the relation $A_{ma} = 2.5 \log (a{:}b)$ (assuming a triaxial ellipsoid shape).

However, this approach tends to underestimate the maximum amplitudes, and therefore the a:b ratio, for objects that are observed only once. In order to overcome this problem, *Binzel and Sauter* (1992) propose a scheme for reducing the amplitudes to the aspect of 60° in case of multiple observations that we have adopted, in slightly modified form, in the following analysis.

Figure 1b shows the distribution of corrected amplitudes of the Trojan and of the main-belt comparison samples. The Trojans appear to have a larger mean amplitude than the main-belt comparison sample. This difference is significant at the 99% confidence level, with the student's t-test resulting in a probability smaller than 1% that the two distributions do indeed come from the same population, and the observed difference is only the result of a statistical fluctuation. The shapes of the distributions are also statistically different, with the Kolmogorov-Smirnov test rejecting the null hypothesis (the two datasets coming from the same distribution) at the 95% confidence level.

Figure 2 shows the lightcurve amplitude vs. the diameter. From this graph no obvious trend of increasing amplitudes with size is recognizable.

2.1.3. Spin axes. The spin axis orientation has been computed for eight Trojan asteroids (S. Mottola et al., personal communication, 2001), two of which belong to the L_4 cloud and six to the L_5 cloud. The obliquity of the spin appears to be randomly distributed in space, with four of the objects rotating in a prograde and four in a retrograde sense. Even within the limitation of small-numbers statistics, the spin axis distribution appears similar to that of the main belt, which was recently studied by *Erikson* (2000). Clearly, the possibility that the large observed amplitudes

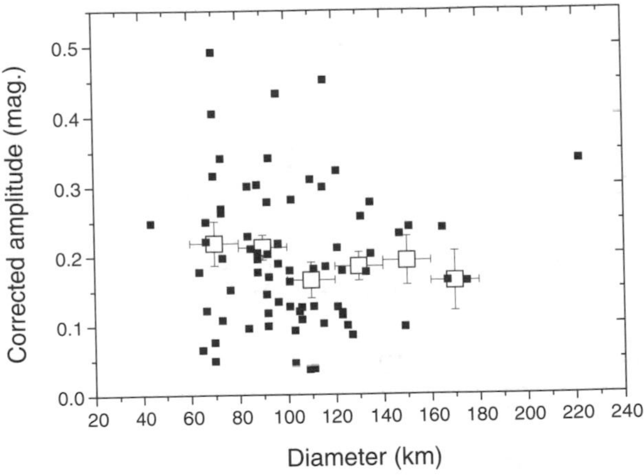

Fig. 2. Amplitude-size distribution of the Trojan sample. Filled squares represent individual asteroids. Empty squares represent the average amplitudes within a 20-km bin. The points at about 45 km and 220 km have not been included in the average.

are due to the alignment of the spin axis toward the ecliptic pole, which would favor observations at equatorial aspect, can be ruled out.

2.2. Taxonomic Classes, Visible and Infrared Spectroscopy

Only a few of the detected Trojan asteroids have an assigned taxonomical class, but almost all of the classified objects are D-type asteroids; only a few belong to the P class. D-type objects have very low albedo (~0.065) and are usually characterized by featureless spectra with neutral to slightly red spectra shortward of 0.55 μm and very red longward of 0.55 μm (*Barucci et al.,* 1987; *Tholen and Barucci,* 1989; *Jewitt and Luu,* 1990). In fact, reflectance spectra in the optical region (*Jewitt and Luu,* 1990; *Fitzsimmons et al.,* 1994; *Lazzarin et al.;* 1995) and in the NIR (*Luu et al.,* 1994; *Dumas et al.,* 1998; *Cruikshank et al.,* 2001) have not yet shown the clear presence of any feature, and are very similar to the spectra of the nuclei of short-period comets and also some Centaurs and KBOs investigated to date.

In particular, the CCD visible spectra of 32 Trojans obtained by *Jewitt and Luu* (1990) have shown slight differences in spectral behavior, with spectral slopes ranging from nearly neutral (S' = 3 ± 1%/10³ Å) to very red (S' = 25 ± 1%/10³ Å), giving a mean value S' = 9.6 ± 4.7%/10³ Å (Fig. 3). The spectra of the 32 Trojans have also been compared with the existing spectra of cometary nuclei in order to study possible relationships between the two groups. The reflectivity gradients of cometary nuclei spectra (mean value S' = 14 ± 5%/10³ Å relative to five cometary nuclei) fall in a range similar to that of the spectra of Trojan asteroids. Some of the 32 Trojans were subsequently observed by others (*Fitzsimmons et al.,* 1994; *Lazzarin et al.,* 1995); the slightly different values of S' that were found might be associated with different regions of the surfaces of these objects, but also could be attributed to different acquisition and analysis of the data.

In the NIR, the spectra of Trojans usually continue to be reddish (*Luu et al.,* 1994). In particular, spectroscopic observations in the I-J-H-K bands obtained by *Dumas et al.* (1998) of four Trojans have not revealed any spectral bands, and they all show reddish spectra reminiscent of the spectra of the nuclei of short-period comets. K-band spectra of five Trojans and L-band (3.0–3.5 μm) spectra of two Trojans were obtained in 1999 by *Emery and Brown* (2001). They show the red slope expected for P and D asteroids and no clear compositional features. Besides the similarity to short-period cometary nuclei, the visible and NIR spectra of the Trojans are similar to the spectra of some KBOs and Centaurs.

Hiroi et al. (2001) have recently report that the reflectance spectrum of the Tagish Lake meteorite is a close match to the spectra of D-type asteroids over the spectral range 0.3–3 μm. They suggest that this meteorite, an ungrouped C2 carbonaceous chondrite that fell in January 2000, originated from a D-type asteroid. While the overall shape and albedo level of the meteorite spectrum do provide a good match, the spectrum at 3 μm shows the presence of hydrous material that is not seen in the spectrum of 624 Hektor, as noted in section 2.3 below.

2.3. The Case of 624 Hektor

The D-type asteroid 624 Hektor is the largest of the jovian Trojan population. It is highly elongated, with maximum and minimum dimensions of 300 and 150 km respectively

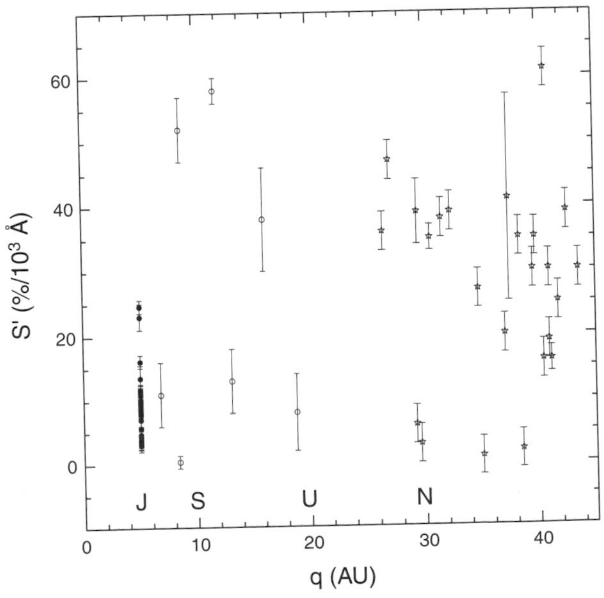

Fig. 3. Visible spectral slope plotted vs. perihelion (q) for Trojans (filled circles), Centaurs (empty circles), and Kuiper objects (stars). The location of the outer planets are marked for reference.

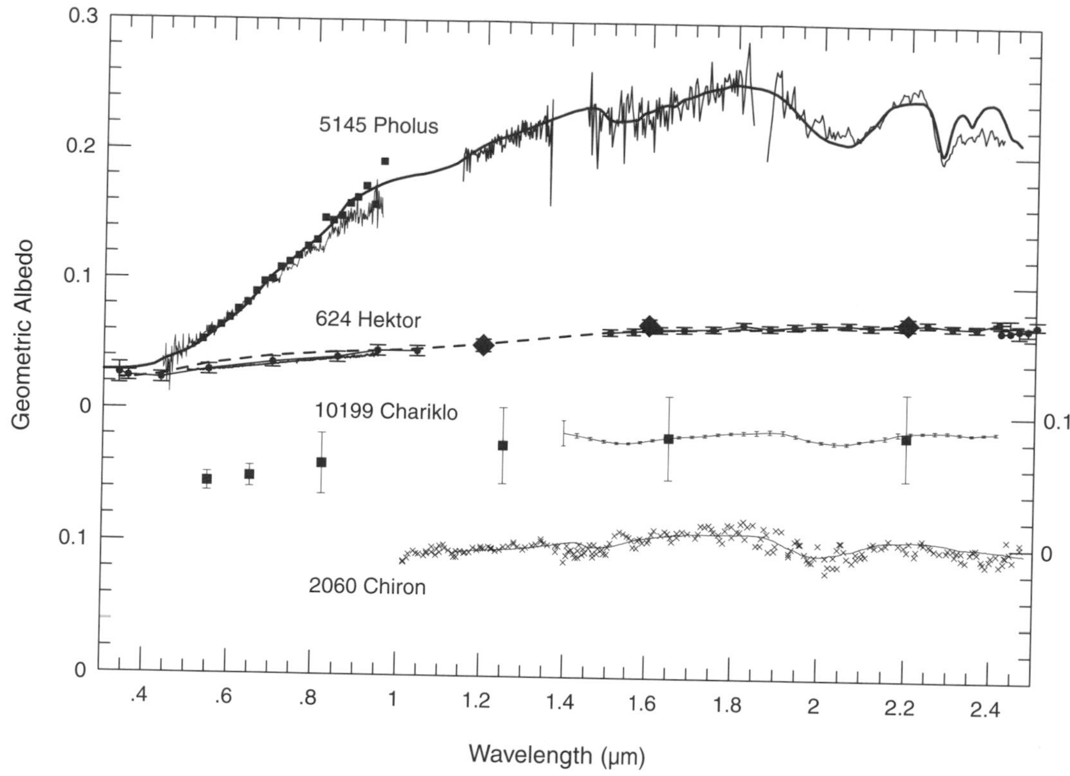

Fig. 4. Reflectance spectra of three Centaurs and Trojan asteroid 624 Hektor. The spectrum of 2060 Chiron is from *Luu et al.* (2000) and shows a model consisting of H_2O ice and olivine. The scaling to geometric albedo is approximate. Data for 10199 Chariklo (originally 1997 CU_{26}) are from *Brown et al.* (1998); the ordinate for the spectrum of Chariklo is given on the right side of the figure. The 5145 Pholus data and model come from *Cruikshank et al.* (1998), while the 624 Hektor data and model come from *Cruikshank et al.* (2001). The ordinate for both Hektor and Pholus is given on the left side of the figure. All three Centaurs show the prominent 2-μm H_2O ice band, with indications of the weaker 1.5-μm band of H_2O ice. There is no spectral evidence for ice on Hektor.

(*Dunlap and Gehrels,* 1969; *Hartmann and Cruikshank,* 1978, 1980). Infrared thermal emission measurements show that Hektor's surface has a low geometric albedo of 0.03 (*Cruikshank,* 1977; *Hartmann and Cruikshank,* 1978), while the spectrum [0.48–0.95 μm (*Vilas et al.,* 1993)] and photometry in the eight-color system shows that it is distinctly red (*Zellner et al.,* 1985). The spectrum of Hektor between 0.3 and 1 μm (*Jewitt and Luu,* 1990) and 1.5 and 2.5 μm (*Cruikshank et al.,* 2001) shows no distinctive or diagnostic absorption bands that would indicate specific mineral, ice, or organic surface components. *Dumas et al.* (1998) recorded segments of the spectrum between 0.9 and 2.4 μm and found no clear absorption bands. The visible spectrum by *Vilas et al.* (1993) and *Lazzarin et al.* (1995) similarly shows no prominent spectral features. While the low albedos and red colors of asteroids and other small bodies in the outer solar system have often been considered suggestive of the presence of macromolecular C-rich surface materials (e.g., *Gradie and Veverka,* 1980; *Cruikshank,* 1987, 1997; *Cruikshank and Khare,* 2000), specific spectral matches have been very hard to establish (e.g., *Luu et al.,* 1994; *Moroz et al.,* 1998).

Cruikshank et al. (2001) extended the spectrum to 3.6 μm. Using the Hapke scattering theory and the complex refractive indexes of several minerals and organic solids, Cruikshank et al. show that the general features of the Hektor spectrum can approximately be matched with a mixture of Mg-rich pyroxene and elemental C (a mixture of graphite and amorphous C). In this model, the pyroxene provides the overall shape of the spectrum (Fig. 4), while the C decreases the albedo to the level observed on Hektor. From the models, the upper limit to the amount of crystalline H_2O ice (30-μm grains) in the surface layer of Hektor is ~3 wt%; larger amounts mixed in the visible surface layer would show an absorption band at 2 μm. The upper limit for serpentine, as a representative of hydrous silicates, is much less stringent, at 40%, based on the shape of the spectral region around 3 μm. Thus, the spectrum at 3 μm does not itself preclude the presence of a few weight percent of volatile material in the surface layer of Hektor. All the models require elemental C, which could be of organic or inorganic origin, to achieve the low geometric albedo that matches Hektor.

The general conclusion of this investigation is that the color and low albedo of Hektor may be attributed to a com-

TABLE 1. Orbital characteristics, diameters, and albedos of Centaurs.

Name	q (UA)	Q (UA)	E	i (°)	Diameter (km)	Albedo
2060 Chiron	8.45	18.77	0.38	6.9	180 ± 10*	0.15 ± 0.05*
					148 ± 8‡	0.17 ± 0.02‡
					190 ± 22†	0.04 ± 0.03†
5145 Pholus	8.66	31.93	0.57	24.7		
7066 Nessus	11.78	37.01	0.52	15.7	≈75	—
8405 Asbolus	6.84	29.01	0.62	17.6	66 ± 4‡	0.12 ± 0.03‡
10199 Chariklo	13.10	18.51	0.17	23.4	302 ± 30*	0.045 ± 0.010*
					273 ± 19§	0.055 ± 0.008§
10370 Hylonome	18.84	31.04	0.24	4.1	≈150	—
1998 SG_{35}	5.83	11.01	0.31	15.6	~35	—
2000 QC_{243}	13.17	19.92	0.20	20.7	≈190	—

**Jewitt and Kalas* (1998).
†*Davies et al.* (1993).
‡*Fernandez et al.* (2002).
§*Altenhoff et al.* (2001).

bination of minerals and material of very low albedo and neutral reflectance, and it therefore does not appear to require an organic component. The degree to which this conclusion extends to other D-type asteroids remains to be investigated.

3. CENTAURS

Centaurs are minor planets having unstable orbits with semimajor axes between those of Jupiter (5.2 AU) and Neptune (30 AU). Their planet-crossing orbits imply a short dynamical lifetime (10^6–10^7 yr) compared to the age of the solar system (*Hahn and Bailey,* 1990; *Asher and Steel,* 1993). The origin of the Centaurs is uncertain, but they are thought to have been ejected from the transneptunian belt by planetary perturbations or mutual collisions (*Duncan et al.,* 1995). *Levison and Duncan* (1997) suggest that these objects could be the source of short-period comets. Later, *Levison et al.* (2001), on the basis of their numerical model, predict that some Centaurs could have originated in the Oort Cloud. Centaurs seem to be located on a boundary between many solar system populations, and they are important for understanding the dynamical evolution of the outer solar system.

To date, 25 such objects have been discovered, following the continuously updated list from the Minor Planet Center (*Marsden,* 2001), and the discoveries continue. Although no formal definition exists, Centaurs have been identified as asteroids at the times of discovery, even though 2060 Chiron was subsequently shown to have cometary activity. *Jewitt and Kalas* (1998) add to the list of Centaurs the comets P/Oterma and P/Schwassmann-Wachmann 1 because their orbits lie inside the orbits of Jupiter and Neptune. Recently, *Marsden* (2001) includes Centaurs in a common list with scattered transneptunian objects, arguing that there are no dynamical reasons to make a distinction, but in this paper we will consider the two populations to be well separated.

Jewitt et al. (1996), on the basis of the number of detections in their ecliptic survey, suggest a population of about 2600 Centaurs with diameters greater than 75 km. From the rate of detections by the Spacewatch automated search program, *Jedicke and Herron* (1997) estimate an upper limit of 2000 Centaurs in the absolute magnitude range H < 10.5. This implies that although the density of the population is transient, the Centaurs can be as numerous as the main-belt asteroids over the same size range. Using a wide-field sky survey, *Sheppard et al.* (2000) estimate that the Centaur population of objects with radius ≥1 km is about 10^7 (if the albedo is 0.04). They predict about 100 Centaurs larger than 50 km in radius, of which only a few have been found. The current total mass of the Centaurs is estimated to be ~10^{-4} $M_{\oplus}$. From their large survey to search for the brightest transneptunians and Centaurs, *Larsen et al.* (2001) found a lower density distribution compared to the Jedicke and Herron survey, but within the error bars.

Only a few objects have been observed at thermal and millimetric wavelengths; thus, only a few have well determined diameters and albedos (Table 1). For others, an indication of the diameter can be obtained from the absolute magnitude, assuming a low albedo value (~0.05).

3.1. The Particular Case of 2060 Chiron

Chiron was discovered by *Kowal* (1978). No other asteroids had been discovered at such large distances (orbital semimajor axis = 13.7 AU), and it was given the name of a Centaur in recognition of its unusual orbit. The orbit was found to be chaotic and subject to strong perturbations by Saturn and Uranus (*Oikawa and Everhart,* 1979; *Scholl,* 1979) and thereby typical of short-period comets (*Hahn and Bailey,* 1990). In 1988, when Chiron was at ~13 AU, *Tholen et al.* (1988) found brightness fluctuations indicating possible cometary behavior of the object; a cometary behavior of Chiron was indicated also by *Bus et al.* (1988, 1989) and *Hartmann et al.* (1990). A coma was detected for the first time by *Meech and Belton* (1989).

Chiron's cometary activity produces an unusual long-term brightness variation. Some investigators found that the cometary activity increases as Chiron approaches perihelion, but the brightness diminishes (*Lazzaro et al.,* 1996). *Luu*

(1993) analyzed photometric data from 1982 to 1992 showing a series of outbursts, with brightness fluctuations as large as 1 mag occurring on timescales of a year. She concluded that long-term (timescale of ~1 yr) and short-term (timescale of days) outbursts are the principal causes of the activity of Chiron, even at large distances such as 13 AU. Brightness variations were also reported by *Bus et al.* (1991), and, in a recent analysis, *Bus et al.* (2001) used archival photographic plates to determine Chiron's brightness between 1969 and 1977. When added to the postdiscovery observations, it is clear that Chiron's brightness does not depend directly on its heliocentric distance; there was an extensive interval of brightening near the most recent aphelion passage.

Lazzaro et al. (1997) investigate changes in Chiron's brightness by performing a photometric survey along the passage of Chiron through perihelion in 1996. The monitoring of Chiron's brightness in 1996 showed that its mean absolute V magnitude varied by ~10% in a few months. *Parker et al.* (1997), using UV data from the Hubble Space Telescope, reveal a 20% magnitude increase in activity in January and April 1996. *Lazzaro et al.* (1997) also found short-period activity of the order of hours during one night. A similar burst of short-period activity was also detected in 1989 by *Luu and Jewitt* (1990) and in 1990 by *Buratti and Dunbar* (1991).

The synodic rotation period of 5.9180 ± 0.0001 h was determined by *Bus et al.* (1989) and is similar to the value determined by *Marcialis and Buratti* (1993) and *Lazzaro et al.* (1997).

Two reflectance spectra in the visible region were obtained in 1996 nearly at the same time as the photometric data by *Lazzaro et al.* (1997), and they show a quite flat, featureless continuum with a little difference in slope: The reddest corresponds to an increase of the brightness of Chiron. The reflectivity gradient of these two spectra (ranging from S' = 2.3 ± 0.1%/10^3 Å to S' = 0.4 ± 0.1%/10^3 Å) is rather different from the mean-reflectance slope of optical spectra of cometary nuclei, which are usually redder in color (S' = 14 ± 5%/10^3 Å) and is more similar to that of C-type asteroids. A negative reflectivity gradient was also found by *Luu* (1993), *Fitzsimmons et al.* (1994), and *Barucci et al.* (1999). This color peculiarity is not unique to Chiron; *Davies et al.* (1998) and *Barucci et al.* (1999) find that the optical reflectivity gradients of the Centaurs are different from one another, indicating a diversity in surface composition probably due to different evolutionary mechanisms. The small variations in the reflectivity gradient S' of Chiron could be due to variations in dust production during episodes of cometary activity. The difference in the reflectivity of Chiron compared to the other Centaurs, which are so far all redder than Chiron, has been attributed (*Luu et al.*, 2000) to cometary activity on Chiron. Such activity would cause a continuous rain of cometary debris on the surface of Chiron, burying a primordial irradiation mantle with unirradiated matter ejected from the interior (*Luu et al.*, 2000).

Chiron is one of the few objects that has shown activity at distances greater than 12 AU, and there has always been a debate on the mechanism responsible for the activity and outbursts of these distant bodies. The sublimation of water ice, which drives the activity in nearby comets (r < 3 AU), would not occur at Chiron's heliocentric distances. More volatile ices such as CO or CO_2 are probably responsible (*Capria et al.*, 2000). *Prialnik et al.* (1995) suggest that the outbursts on Chiron could be due to a combination of energy released by the crystallization of amorphous water ice and the release of trapped CO from a porous matrix of both amorphous and crystalline ice. Gaseous CO was probably found in Chiron's coma in 1995 by *Womack and Stern* (1995) via the molecule's J = 1–0 rotational transition at 115 GHz, but the result still needs to be confirmed. Another molecule that could be in part responsible for the activity on Chiron is CN, found during the outburst of January 29, 1990, by *Bus et al.* (1991). In particular, *Bus et al.* (1991) suggest that CO and CO_2 would drive the activity, and their escape from the surface of Chiron could be responsible of the release of the parent of the CN.

In spite of the earlier featureless spectra in the visible and NIR, Chiron has recently shown (see Fig. 4) the presence of water ice through the 2-μm absorption band (*Foster et al.*, 1999; *Luu et al.*, 2000). The ice present on the surface of Chiron is very likely mixed with dark impurities, a small amount of which is usually enough to mask the spectral band.

3.2. Rotational Periods and Lightcurve Amplitudes

In addition to Chiron, rotational periods have been determined for a few Centaurs:

5145 Pholus has a rotational period of 9.9825 ± 0.004 h determined by *Buie and Bus* (1992) and confirmed with small differences by *Hoffmann et al.* (1993) and *Davies et al.* (1998). The lightcurve seems almost symmetric, with an amplitude of ~0.15 mag. No significant V-R variation is evident with rotation.

8405 Asbolus has been observed photometrically by *Brown and Luu* (1997), who derive a period of 8.87 ± 0.02 h with an amplitude of ~0.34 mag. *Davies et al.* (1998), using new data, argue for a slightly longer period (8.9351 ± 0.0003 h) and larger amplitude (~0.55 mag). *Kern et al.* (2000), on the basis of spectroscopic observations and on the hypothesis that the lightcurve may be dominated by a relatively bright surface spot, suggest the possibility of a rotational period half of the previous value (4.47 h).

10199 Chariklo was observed by *Davies et al.* (1998), who suggest a rotation period close to 24 h or longer and a very small amplitude. *Peixinho et al.* (2001) suggest a possible rotational period of 31.2 h from nine nights of observation, while Alessandrino et al. (personal communication, 2001) propose a very long rotational period of 39 ± 6 d.

1998 SG_{35} was observed by Green (*Davies,* 2000) with the preliminary result that the rotational period seems longer than 12 h and the amplitude larger than 0.04 mag.

1999 UG_5 was observed by *Gutierrez et al.* (2001), who determined a period of ~26.5 h (but other values are possible) and an amplitude of 0.24 ± 0.02 mag.

There are no pole inclinations or shape constraints available for these objects. From the amplitude of the lightcurves we can estimate a lower limit on the axis ratio, assuming an ellipsoidal shape with semiaxes a > b > c. That limit is a:b ≥ $10^{0.4\Delta m}$ if we assume that the lightcurve is influenced only by the shape and not by significant albedo variations. The few observed Centaurs seem to have small amplitudes, except 8405 Asbolus, for which the amplitude is 0.55 mag, corresponding to a:b ≥ 1.66. This elongated shape and its small diameter (~66 km) could give some indication of the collisional evolution of this object.

3.3. Colors

A wide range of broadband colors has been found in the Centaur population. The colors vary from neutral or slightly blue for 2060 Chiron (V-R = 0.37) to very red for 5145 Pholus (V-R = 0.78). 7066 Nessus has a red color very similar to Pholus in the visible range, but its redness is less evident in the IR region. Laboratory experiences suggest that a red color surface can be produced by the energetic processing (exposure to cosmic rays, solar ultraviolet, corona discharge, and/or ion bombardment) of surface ices, organic solids, and even minerals. Such processing forms a coating of dark materials and a very red spectrum (e.g., *Andronico et al.,* 1987; *Thompson et al.,* 1987).

The color distribution of Centaurs, in the B-V vs. V-R plot (Fig. 5), seems to be bimodal, but the sample is still too small. Two groups are evident: one as Chiron with neutral/slightly red colors and the other one with much redder colors, more similar to Pholus. As has been demonstrated by *Doressoundiram et al.* (2001), this dichotomy is further enhanced by the fact that Chiron is an active comet, while no activity has been detected on Pholus, which should have a more pristine surface. In fact, Pholus might possess a thick irradiation mantle that inhibits outgassing of volatiles, while the neutral color on Chiron may be due to outgassing during successive episodes of activity, with a surface most likely dominated by a dust layer created from cometary debris (*Luu et al.,* 2000). Therefore, as *Luu et al.* (2000) and *Doressoundiram et al.* (2001) suggest, two groups may exist: one (like Chiron) in which a reddish crust can be removed or processed by activity or impact and the other one (like Pholus) with older surfaces, probably due to their recent escape from the Kuiper Belt.

No correlation has been found between color variations and heliocentric distance and/or diameter (Fig. 3).

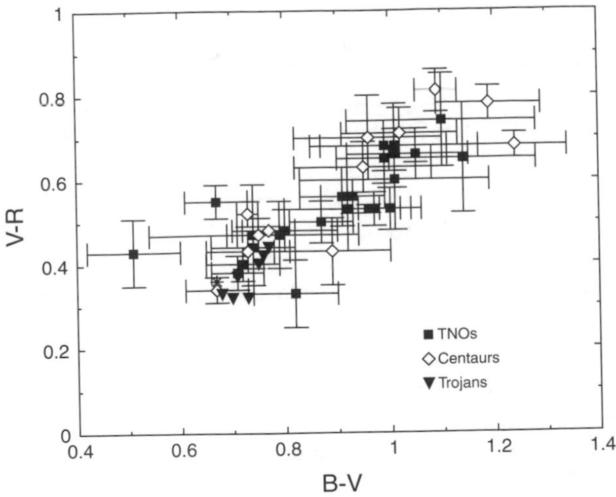

Fig. 5. B-V vs. V-R color of Centaurs (empty diamonds) and Trojans (filled triangle) with comparison of TNOs (filled squares).

3.4. Spectroscopy

As has been shown by their color differences, Centaurs have surfaces showing dramatic spectral differences from neutral to very red. In the visible region, *Barucci et al.* (1999) observe five of them, confirming a great diversity among the reflectances of these objects. They obtain featureless spectra distributed between very flat for 2060 Chiron to very red for Nessus, which appears in this region to be nearly as red as 5145 Pholus, the reddest object known in the solar system so far.

In the IR region, some spectra are featureless, while some others show signature of water ice, methanol, or other light hydrocarbon ices. Very few of these objects have been well studied or rigorously modeled in the visible and NIR (Fig. 4).

5145 Pholus was observed in the visible and NIR by several authors and a synthesis was presented by *Cruikshank et al.* (1998). The NIR spectrum shows a strong absorption band at 2.04 and at 2.27 μm. The large band at 2.04 μm is typical of water ice while the other at 2.27 μm might be due to methanol ice. *Cruikshank et al* (1998) modeled the complete spectrum of Pholus from 0.4 to 2.5 μm with Hapke scattering theory, interpreting the extraordinary red color and the rich NIR features as due to the surface presence of common silicate olivine, a complex organic solid (tholin), water ice, methanol (CH_3OH) (or another light hydrocarbon), and C. They conclude that Pholus is a primitive object hat has not yet been substantially processed by solar heating and probably is a nucleus of a large comet that has never been active.

8405 Asbolus was observed by *Brown* (2000) and *Barucci et al.* (2000). Both spectra show a lack of spectral signatures, in particular no absorptions due to water ice (bands at 1.5 and 2 μm). As in the case of Pholus, *Barucci et al.* (2000) model the complete spectrum of Asbolus from 0.4 to 2.3 μm. Due to the difficulty in interpreting the spectral behavior of these objects, two models are proposed. One model, with a spatial mixture of 50% amorphous C plus 50% kerogen, can reproduce well the shape of the spectrum, but not below 0.5 μm. To investigate the limit of the abundance of water ice, another model, with 91.5% amorphous C, 7.5% tholin, and 1% water ice, was proposed. Depending on the grain size, no more than a few percent of the surface can be covered with pure water ice in order to keep the strong 2-μm band below the level of the noise. *Kern et al.* (2000) obtained a 1–2-μm spectra using the Hubble Space Telescope, revealing a significantly inhomogeneous surface characterized by water ice mixed with unknown low-albedo constituents. The spectrum was initially featureless, but during an interval of 1.7 h, the spectrum developed a strong absorption band at the approximate location of the 1.5-μm H_2O band. *Kern et al.* (2000) explained this with the possibility that an impact has penetrated the object's crust, exposing the underlying ice in the surface region that rotated into view during the observations. *Romon-Martin et al.* (2001) reobserved Asbolus at VLT (ESO, Chile) obtaining five IR spectra covering the full rotational period. The new data show flat spectra with no variation in the rotation and no presence of ice absorption.

10199 Chariklo was observed by *Brown and Koresko* (1998) and *Brown et al.* (1998). The IR spectra show clear evidence of water ice. The spectrum was modeled by *Brown and Koresko* (1998) by distinct surface areas of a dark, neutral surface and 3% water ice. *Brown et al.* (1998) made a model fit assuming an intimate mixture of red-colored material and higher percentage of water ice, probably in an amorphous state.

10370 Hylonome has been observed by Romon-Martin et al. (personal communication, 2001) and on the basis of preliminary results, the spectrum seems featureless.

Unfortunately, these objects are faint and even observations with the largest telescopes (Keck and VLT) do not generally yield good quality spectra. Interpretation is also very difficult because the behavior of the spectra depends on the choice of many parameters. The uncertainty in determining quantitative abundance on the surfaces is large and the derived models are not unique. Furthermore, the number of plausible materials (ices, organic solids, minerals) that can be incorporated into the scattering models is small because so few such materials have had their complex refractive indexes measured in the laboratory.

4. RELATIONS TO OTHER SMALL BODIES

4.1. Relationship of Centaurs and Trojans to Icy Satellites of the Planets

The giant planets have many satellites, the largest of which all show spectroscopic evidence for H_2O and/or other ices on their solid surfaces (the case of Titan is ambiguous). All the regular satellites (low-inclination, near-circular, prograde orbits) of radius >240 km have surface ices, and a number of other smaller satellites also show ice bands in the NIR spectra. Among the irregular satellites (high-inclination, elliptical, retrograde or prograde orbits), Nereid (Neptune) and Phoebe (Saturn) have detectable ice bands (*Brown et al.,* 1999b; *Brown,* 2000; *Owen et al.,* 1999). The irregular satellites, in particular, tend to be low-albedo objects, and their ice bands (when detected) are weak, indicating that the ice is mixed at the granular level with low-albedo minerals, C, organic solids, or some combination of the three. Irregular satellites are also thought to be captured objects that originated elsewhere in the solar nebula. It is possible, perhaps likely, that they were captured after extraction from the Kuiper Disk.

The relationship of the Trojan asteroids to icy satellites is not known. The Trojans have very low albedos, consistent with the C-, P-, and D-type asteroids in the main belt and with many of the small outer satellites of Jupiter and the other giant planets. They represent objects that accreted in the zone around or beyond the orbit of Jupiter, and the heliocentric distance they currently occupy marks the inner boundary of the zone of long-term stability for surface exposures of water ice. There are no reported detections of water or other ices in any Trojan spectra (*Jewitt and Luu,* 1990; *Barucci et al.,* 1994; *Dumas et al.,* 1998), although there may be ices below the surface that are accessible to remote-sensing observation. Neither do we have any measurements of the densities of Trojan asteroids from which the interior composition might be inferred. The work of *Cruikshank et al.* (2001) on 624 Hektor noted above indicates that hydrous minerals might be present in the surface material and remain undetected with the quality of NIR spectroscopic data we now have. Indeed, the spectral feature at 0.7 and 3 μm identified in the spectra of many low-albedo asteroids of the C, G, and F classes in the main belt (*Jones et al.,* 1990; *Vilas et al.,* 1993; *Barucci et al.,* 1998) are associated with hydrous minerals. Hydrous minerals, if present, might indicate the presence of liquid ice that originated through the heating of interior ice in an earlier epoch, or they might indicate the incorporation of materials that were serpentinized by other processes (*Reitmeijer and Nuth,* 2000, 2001).

Hartmann (1980) noted that planetesimals that formed in Jupiter's region of the solar nebula were probably composed of a mixture of roughly equal amounts of water ice and stony material; the amount of organic solids in this mix could be significant. Material condensing to form planetesimals in the region of Jupiter was subjected to temperatures above 40 K (*Lunine et al.,* 2000); if the temperature were as high as 55 K, the volatile molecules N_2, CO, and CH_4 were not efficiently trapped and incorporated into the planetesimals, but refractory organics, NH_3, and compounds of other heavy elements were trapped (*Bar-Nun and Kleinfeld,* 1989). The surface layers of solid bodies at Jupiter's heliocentric distance are much warmer than 55 K in the present epoch; average surface temperatures range from ~89 to 100 K for objects with geometric albedos 0.4 and 0.04 respectively.

The visible surface of an initially ice-rich body in the Trojan population, or a small ice-rich satellite of Jupiter, could be transformed into a low-albedo surface by the impact gardening and selective removal of the volatile component, leaving a buildup of the rocky and organic solid material (*Hartmann,* 1980). The eruption of volatile-rich (primarily water) magmas onto the impacted surface following thermal disturbances would serve to redeposit water on the surfaces, but the long-term effect might produce a net darkening and devolatilization of the surface layer that is observed with optical and IR remote-sensing techniques. As noted above, *Luu et al.* (2000) predict the opposite effect (surface brightening) from cometary activity on Chiron.

Other complex mechanisms might also darken the uppermost surface of an icy-rich body. In particular, the darkening of ices containing simple organic molecules by cold plasma irradiation (*Thompson et al.,* 1987) and ion bombardment (*Andronico et al.,* 1987) has been demonstrated in the laboratory.

4.2. Relations to Kuiper Belt Objects

The transneptunian objects, also called Kuiper Belt objects, represent the new frontier of our solar system. More than 500 objects have been discovered up to the present; these represent a small sample of a much larger population of icy planetesimals orbiting at the outer edge of the solar system. They are expected to be the best-preserved fossils of the protoplanetary disk, as they have been formed at very low temperatures (~40 K) and are believed to be the remnants of solar system formation. Collisions and irradiation have reworked their surfaces, especially in the inner part of the belt, and extensive cratering can be expected to characterize their surfaces (*Durda and Stern,* 2000).

The physical and chemical properties of the Kuiper objects are still poorly known due to their faintness. Spectroscopic data in the visible and the NIR are available for only a few of these objects. The spectra range from neutral to very red with behavior very similar to that of Trojans and Centaurs (Fig. 3). Most of the information on the surface composition comes from studies in the NIR. Few Kuiper objects have been observed in the NIR, and although the spectra show a very low S:N, their surface characteristics show a wide diversity: 1996 TL_{66} (*Luu and Jewitt,* 1998) as well as 2000 EB_{173} (*Brown et al.,* 2000; and *Licandro et al.,* 2001; *Jewitt and Luu,* 2001) have flat featureless spectra similar to that of dirty water ice, while 1996 TO_{66} shows an inhomogeneous surface with the presence of small amounts of water ice mixed with other minor components (*Brown et al.,* 2000). 1993 SC, observed by *Brown et al.* (1997), shows features that may be due to hydrocarbon ices, with a general red behavior suggesting the presence of more complex hydrocarbons, while the spectrum of 1996 SC did not show any similar features when observed by *Jewitt and Luu* (2001) with the same telescope. In contrast, 1999 DE_9 shows solid-state absorption features near 1.4, 1.6, 2.00, and probably at 2.25 μm (*Jewitt and Luu,* 2001). This spectral diversity is confirmed by photometric observations of a larger sample. TNO colors exhibit wide diversity, with quasicontinuum color variation ranging from neutral to very red. *Barucci et al.* (2001) apply for the first time a statistical analysis of B-V, V-R, V-I, and V-J colors of TNO and Centaur populations. The general results show that the sample of objects can be divided into a few groups with spectra ranging from neutral to very red. The quasicontinuum color variation of TNOs can be a consequence of collisions at all scales that resurface the body by fresh debris and of space weathering.

Centaurs appear to have very similar spectral and color characteristics to those of the TNOs (Fig. 3), supporting thc hypothesis that Centaurs are ejected from the Kuiper Belt by planetary scattering; this is the strongest observational argument for a common origin. Centaur colors are more similar to the neutral and to the reddest TNOs, while the Trojans are more similar to the neutral/red group (Fig. 5).

The rotational properties of those few Centaurs and Kuiper Belt objects for which the determinations have been made seem to be similar to those of the Trojans and main-belt asteroids.

5. CENTAURS AND THEIR EVOLUTION FROM THE INTERSTELLAR MEDIUM

Notwithstanding the mechanisms for surface modification noted above, the surface composition of a small (undifferentiated) outer solar system body should to some degree be indicative of bulk composition.

In order to understand the compositions and structures of the Centaurs, we consider the problem from the viewpoint of the condensation of solid material, originally from the interstellar medium, in the regions of the solar nebula beyond Neptune. This is the essence of the problem of the origin and compositions of the short-period comets that arise from the Kuiper Belt. In principle, knowledge gained about the interior and surface compositions of comets can be applied more or less directly to the Centaur objects, bearing in mind the disparity in size between typical comets (1–10 km) and Centaur objects discovered up to the present (~40 km and greater).

Objects in the Kuiper Disk are thought to have condensed from the solar nebula in the zone from ~40–60 AU where they presently reside. The temperature at the time of condensation of these objects may well have been lower than 40 K, thus trapping volatile molecules and heavy noble gases, as well as organic materials that were largely unaltered from their original state in the interstellar medium (*Lunine et al.,* 2000). The degree of processing of interstellar solid material in the outer solar nebula is uncertain. As *Lunine et al.* (2000) have noted, planetesimals formed in a disk that was embedded within the nascent molecular cloud, and it is reasonable to expect a continuum of processing of material between the cloud and the protoplanetary disk. Certainly the water abundance in the nebula was both spatially and temporally variable during the condensation of the planetesimals.

The nuclei of comets, and, by inference, the Kuiper Belt objects and Centaurs, are assemblages of ices, organic solids, and silicate minerals that solidified in a variety of different environments; they are therefore presumed to be in a state of chemical disequilibrium. When they accreted in the outer solar nebula, they incorporated both unaltered material from the nascent molecular cloud and material that had undergone some degree of chemical and thermal pro-

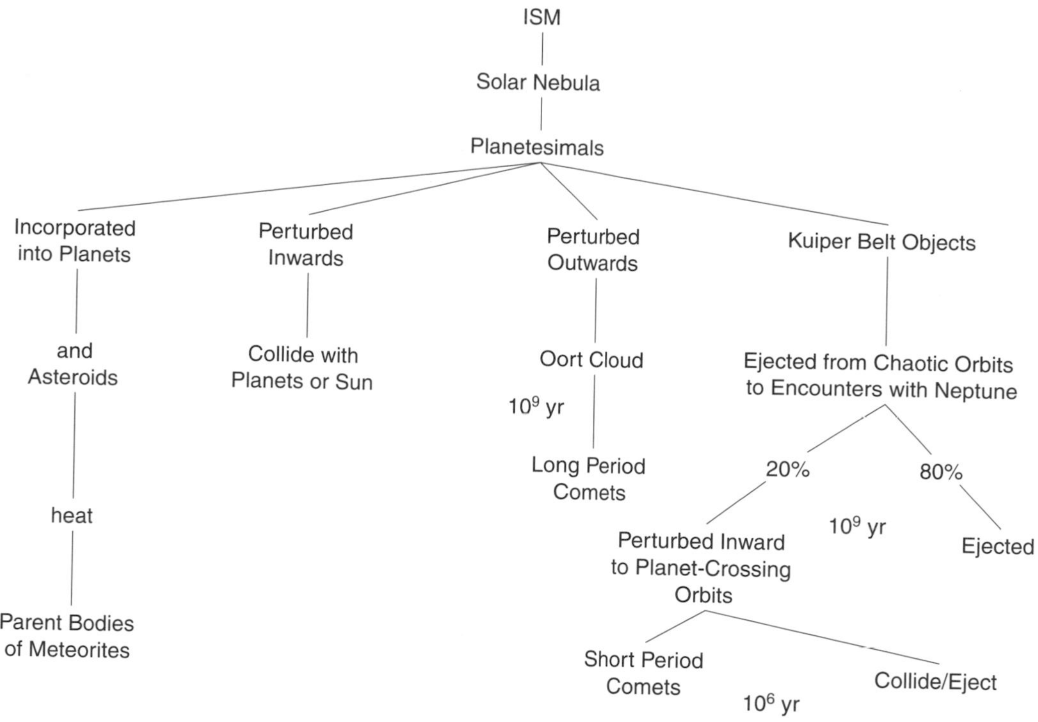

Fig. 6. A scenario for the evolution of the planetesimals in the solar nebula. The approximate dynamical lifetimes in years for some stages in the development are shown. Adapted from *Cruikshank* (1997).

cessing in the solar nebula. On the basis of the ortho/para ratios of H in water, measured in several comets, the accretion temperature of cometary ices is estimated to lie in the range 25–35 K (*Irvine et al.,* 2000). The disequilibrium conditions of formation and accretion of cometary materials appear to be reflected in the variety of (surface) compositions on the small number of Kuiper Belt objects and Centaurs for which such information presently exists (e.g., *Brown et al.,* 1997, 1998, 1999a, 2000; *Brown,* 2000).

How can we integrate these basic ideas into a scenario that embraces the evolution of material from the molecular cloud to the solid bodies we find in the modern solar system? In Fig. 6 we trace the development of material from the interstellar medium into the modern solar system in a very schematic way by following four pathways. One of these pathways, representing the formation and accretion of planetesimals in the inner solar system, leads to asteroids with their wide range of compositions. Another important pathway follows planetesimals in the outer solar nebula to the formation of the Kuiper Disk, from which encounters with Neptune disperse some of these bodies into the planetary region. This extraction from the Kuiper Disk occurs on approximately 10^9-yr timescales, but those objects in planet-crossing orbits (the Centaurs and short-period comets) have dynamical lifetimes of only ~10^6 yr before they are ejected, disintegrate, or collide with the Sun or a planet.

By some other unclear pathway, not shown in the figure, icy or rocky bodies are trapped in orbits around planets to become the irregular satellites that each of the giant planets seems to have. The pathway to the formation of the jovian Trojan asteroids is also unclear, leaving us to wonder if the Trojans are more compositionally similar to the objects in the outer main belt or to the irregular satellites of Jupiter and Saturn. Unfortunately, the information contained in their surface spectral reflectances appears to be very meager and somewhat ambiguous. In any event, the jovian Trojans occupy a transition zone between these populations, and further clarification of their physical properties will eventually lead to a much better understanding of their origin.

6. CONCLUSIONS

Trojans and Centaurs are very interesting populations within the solar system. Trojan asteroids seem to possess the necessary dynamic stability for survival over the age of the solar system (*Levinson et al.,* 1997), and even if the origin of Trojans is still unclear, they surely formed in a region of the solar nebula rich in frozen volatiles. Moreover, it is generally believed that they did not suffer selective induction heating that produced aqueous alteration processes, and probably they still contain frozen water in their interior. The population has undergone a collisional evolution as least as intense as the one that took place in the main belt. This picture is in line with the random orientation of the spin axes of the Trojans and with the distribution of their rotation rates being close to a Maxwellian. The Trojans have on average larger lightcurves amplitudes compared to similarly sized main-belt objects, suggesting that the Trojans have more elongated shapes. Although this feature is not yet understood, possible explanations include a different response to collisions, probably due to their different composition and/or shape exaggeration induced by volatile sublimation in extinct cometary nuclei.

If the Trojans formed near or beyond Jupiter's orbit, their temperatures could be low enough for water to have existed always as solid ice. The hypothesis that Trojans could possess ice-rich interiors similar to the cometary nuclei is not contradicted by any available observations. Unfortunately, physical observations provide only limited clues about the source of Trojans.

The Centaurs are located in unstable orbits with short dynamical lifetimes. They have also suffered many collisions but less intensely than Trojans and probably at lower velocities. The presence of the irradiation mantle on the surface of the Centaurs (*Luu and Jewitt,* 1996) would explain the redness of the spectra of these objects, and their diversity in colors could be probably explained by removal, with different degrees, of the irradiation material exposing primordial ices. However, the slightly neutral spectrum of Chiron distinguishes itself from the behavior of the other Centaurs. *Luu et al.* (2000) believe that the variable spectrum of the object, the revealed presence of water ice and persistent cometary activity would indicate that the surface is not dominated by an irradiation mantle but by a layer of cometary debris, not uniformly distributed on the surface, that was formed by the sublimation of the ices like CO, N_2, etc. On the contrary, an object like Pholus showing a very red color, absorption features probably due to hydrocarbons materials, and the absence of cometary activity is compatible with a surface covered by an irradiation mantle (*Luu et al.,* 2000). They also predict that water ice has to exist on Centaur surfaces and should be ubiquitous among objects originating in the Kuiper Belt; its detectability and possible cometary activity is determined by the water-ice surface coverage.

Outward migration of the planets after their formation and large-scale impacts may have modified the surface composition of objects in the outer solar system. The spectral behavior of Trojans is similar to those of some KBOs, Centaurs, and short-period cometary nuclei. It could be argued that Trojans also have irradiation mantles. However, as opposed to some Centaurs and KBOs, which have shown some spectral features, the lack of any detectable spectral features in the spectra of Trojans could be due to a high level of alteration (dehydrogenation) by solar radiation of the organics present on the surface.

In conclusion, low-albedo objects, in particular Trojans and Centaurs that formed at larger heliocentric distance, have been less thermally processed in comparison with main-belt asteroids and therefore may have better preserved primordial materials. These small bodies can still contain a considerable amount of information about some primordial processes that governed the evolution of the early solar system. The available observations do not allow us at the moment to constraint the evolution of these populations, and for this reason many more NIR and infrared observations are needed to study the physical properties and the composition of these objects. New generation telescopes, and in particular a SIRFT (Space Infrared Telescope Facility) will allow new discoveries and revelations fundamental for the understanding of these small bodies.

REFERENCES

Altenhoff W. J., Menten K. M., and Bertoldi F. (2001) Size determination of the Centaur Chariklo from millimeter-wavelength bolometer observations. *Astron. Astrophys., 366,* L9–L12.

Andronico G., Baratta G. A., Spinella F., and Strazzulla G. (1987) Optical evolution of laboratory-produced organics: Applications to Phoebe, Iapetus, outer belt asteroids, and cometary nuclei. *Astron. Astrophys., 184,* 333–336.

Asher D. J. and Steel D. I. (1993) Orbital evolution of the large outer solar system object 5145 Pholus. *Mon. Not. R. Astron. Soc., 263,* 179–190.

Bar-Nun A. and Kleinfeld I. (1989) On the temperature and gas composition in the region of comet formation. *Icarus, 80,* 243–253.

Barucci M. A., Capria M. T., Coradini A., and Fulchignoni M. (1987) Classification of asteroids using G-mode analysis. *Icarus, 72,* 304–324.

Barucci M. A., Lazzarin M., Owen T., Barbieri C., and Fulchignoni M. (1994) Near-infrared spectroscopy of dark asteroids. *Icarus, 110,* 287–291.

Barucci M. A., Doressoundiram A., Fulchignoni M., Florczak M., Lazzarin M., Angeli C., and Lazzaro D. (1998) Search for aqueously altered materials on asteroids. *Icarus, 132,* 388–396.

Barucci M. A., Lazzarin M., and Tozzi G.P. (1999) Compositional surface variety among the Centaurs. *Astron. J., 117,* 1929–1932.

Barucci M. A., de Bergh C., Cuby J.-G., Le Bras A., Schmitt B., and Romon J. (2000) Infrared spectroscopy of the Centaur 8405 Asbolus: First observations at ESO-VLT. *Astron. Astrophys., 357,* L53–L56.

Barucci M. A., Fulchignoni M., Birlan M., Doressoundiram A., Romon J., and Boehnhardt H. (2001) Analysis of Trans-Neptunian and Centaur colours: Continuous trend or grouping? *Astron. Astrophys., 371,* 1150–1154.

Binzel R. P. and Sauter L. M. (1992) Trojan, Hilda, and Cybele asteroids: New lightcurve observations and analysis. *Icarus, 95,* 222–238.

Binzel R. P., Farinella P., Zappalà V., and Cellino A. (1989) Asteroid rotation rates: Distributions and statistics. In *Asteroids II* (R. P. Binzel et al., eds), pp. 298–315. Univ. of Arizona, Tucson.

Brown M. E. (2000) Near-infrared spectroscopy of Centaurs and irregular satellites. *Astron. J., 119,* 977–983.

Brown M. E. and Koresko C. D. (1998) Detection of water ice on the Centaur 1997 CU26. *Astrophys. J. Lett., 505,* L65–L67.

Brown M. E., Blake G. A., and Kessler J. E. (2000) Near-infrared spectroscopy of the bright Kuiper Belt object 2000 EB173. *Astrophys. J. Lett., 543,* L163–L165.

Brown R. H., Cruikshank D. P., Pendleton Y. J., and Veeder G. (1997) Surface composition of Kuiper Belt object 1993 SC. *Science, 276,* 937–939.

Brown R. H., Cruikshank D. P., Pendleton Y., and Veeder G. J. (1998) Identification of water ice on the Centaur 1997 CU_{26}. *Science, 280,* 1430–1432.

Brown R. H., Cruikshank D. P., and Pendleton Y. J. (1999a) Water ice on Kuiper Belt object 1996 TO66. *Astrophys. J. Lett., 519,* L101–L104.

Brown R. H., Cruikshank D. P., Pendleton Y. J., and Veeder G. J. (1999b) Water ice on Nereid. *Icarus, 139,* 374–378.

Brown W. R. and Luu J. X. (1997) CCD photometry of the Centaur 1995 GO. *Icarus, 126,* 218–224.

Buie M. W. and Bus S. J. (1992) Physical observations of 5145 Pholus. *Icarus, 100,* 288–294.

Buratti B. J. and Dunbar R. S. (1991) Observation of a rapid decrease in the brightness of the coma of 2060 Chiron in 1990 January. *Astrophys. J. Lett., 266,* L47–L49.

Bus S. J., Bowell E., and French L. M. (1988) (2060) Chiron. *IAU Circular 4684.*

Bus S. J., Bowell E., Harris A. W., and Hewitt A. V. (1989) 2060 Chiron: CCD and electronographic photometry. *Icarus, 77,* 223–238.

Bus S. J., A'Hearn M. F., Schleicher D. G., and Bowell E. (1991) Detection of CN emission from 2060 Chiron. *Science, 251,* 774–777.

Bus S. J., A'Hearn M. F., Bowell E., and Stern S. A. (2001) (2060) Chiron: Evidence for activity near aphelion. *Icarus, 150,* 94–103.

Capria M. T., Coradini A., De Sanctis M. C., and Orosei R. (2000) Chiron activity and thermal evolution. *Astron. J., 119,* 3112–3118.

Cronin J. R. and Chang S. (1993) Organic matter in meteorites: Molecular and isotopic analyses of the Murchison meteorite. In *The Chemistry of Life's Origins* (J. M. Greenberg et al., eds.), pp. 209–258. Kluwer, Dordrecht.

Cronin J. R., Pizzarello S., and Cruikshank D. P. (1988) Organic matter in carbonaceous chondrites, planetary satellites, asteroids and comets. In *Meteorites and the Early Solar System* (J. F. Kerridge and M. S. Matthews, eds.), pp. 819–857. Univ. of Arizona, Tucson.

Cruikshank D. P. (1977) Radii and albedos of four Trojan asteroids and Jovian satellites 6 and 7. *Icarus, 30,* 224–230.

Cruikshank D. P. (1987) Dark matter in the solar system. *Adv. Space Res., 7,* 109–120.

Cruikshank D. P. (1997) Organic matter in the solar system: From the meteorites to the Kuiper Belt. In *From Stardust to Planetesimals* (Y. J. Pendleton and A. G. G. M. Tielens, eds.), pp. 315–333. ASP Conference Series 122.

Cruikshank D. P. and Khare B. N. (2000) Planetary surfaces of low albedo: Organic material throughout the solar system. In *Bioastronomy '99 — A New Era in Bioastronomy* (G. A. Lemarchand and K. J. Meech, eds.), pp. 253–261. ASP Conference Series 213.

Cruikshank D. P., Roush T. L., Bartholomew M. J., Geballe T. R., Pendleton Y. J., White S. M., Bell J. F. III, Davies J. K., Owen T. C., de Bergh C., Tholen D. J., Bernstein M. P., Brown R. H., Tryka K. A, and Dalle Ore C. M. (1998) The composition of Centaur 5145 Pholus. *Icarus, 135,* 389–407.

Cruikshank D. P., Dalle Ore C. M., Roush T. L., Geballe T. R., Owen T. C., de Bergh C., Cash M. D., and Hartmann W. K. (2001) Constraints on the composition of Trojan asteroid 624 Hektor. *Icarus, 153,* 348–360.

Dahlgren M. (1998) A study of Hilda asteroids III. Collision velocities and collision frequencies of Hilda asteroids. *Astron. Astrophys., 336,* 1056–1064.

Davies J. K. (2000) Physical characteristics of Trans-Neptunian objects and Centaurs. In *Minor Bodies in the Outer Solar System* (A. Fitzsimmons et al., eds.), pp. 9–24. Springer-Verlag, Berlin.

Davies J., Spencer J., Sykes M., Tholen D., and Green S. (1993) *5145 Pholus.* IAU Circular 5698.

Davies J. K., McBride N., Ellison S. E., Green S. F., and Ballantyne D. (1998) Visible and infrared observations of six Centaurs. *Icarus, 134,* 213–227.

Doressoundiram A., Barucci M. A., Romon J., and Veillet C. (2001) Multicolour photometry of Trans-Neptunian objects. *Icarus, 154,* 277–286.

Dumas C., Owen T., and Barucci M. A. (1998). Near-infrared spectroscopy of low-albedo surfaces of the solar system: Search for the spectral signature of dark material. *Icarus, 133,* 221–232.

Duncan M., Levison H. F., and Budd S. M. (1995) The dynamical structure of the Kuiper Belt. *Astron. J., 110,* 3073–3081.

Dunlap J. L. and Gehrels T. (1969) Minor planets. III. Lightcurves of a Trojan asteroid. *Astron. J., 74,* 797–803.

Durda D. D. and Stern S. A. (2000) Collision rates in the present-day Kuiper Belt and Centaurs regions: Application to surface activation and modification on Comets, Kuiper Belt Objects, Centaurs and Pluto-Charon. *Icarus, 145,* 220–229.

Ehrenfreund P., Glavin D. P., Botta O., Cooper G., and Bada J. L. (2001) Extraterrestrial amino acids in Orgueil and Ivuna: Tracing the parent body of CI type carbonaceous chondrites. *Proc. Natl. Acad. Sci., 98,* 2138–2141.

Emery J. P. and Brown R. H. (2001) Near-infrared spectroscopy of Trojan asteroids: Implications for the evolution of the solar system (abstract). In *Lunar and Planetary Science XXXII,* pp. 1385–1396. Lunar and Planetary Institute, Houston.

Erikson A. (2000) The present distribution of asteroid spin vectors and its relevance to the origin and evolution of main belt asteroids. Ph.D. thesis, Freie Universitaet, Berlin. (DLR Research Report 2000-37, ISSN #1434-8454.)

Evans N. W. and Tabachnik S. A. (2000) Asteroids in the inner solar system II: Observable properties. *Mon. Not. R. Astron. Soc., 319,* 80–94.

Farinella P., Paolicchi P., and Zappalà V. (1981) Analysis of the spin rate distribution of asteroids. *Astron. Astrophys., 104,* 159–165.

Fernandez Y. R., Jewitt D. C., and Shepard S. S. (2002) Thermal properties of Centaurs Asbolus and Chiron. *Astron. J., 123,* 1050–1055.

Fitzsimmons A., Dahlgren M., Lagerkvist C.-I., Magnusson P., and Williams I. P. (1994) A spectroscopic survey of D-type asteroids. *Astron. Astrophys., 282,* 634–642.

Foster M. J., Green S. F., and McBride N. (1999) Detection of water ice on 2060 Chiron. *Icarus, 141,* 408–410.

Fulchignoni M., Barucci M. A., Di Martino M., and Dotto E. (1995) On the evolution of the asteroid spin. *Astron. Astrophys., 299,* 929–932.

Gradie J. and Veverka J. (1980) The composition of the Trojan asteroids. *Nature, 283,* 840–842.

Gutierrez P. J., Ortiz J. L., Alexandrino E., Roos-Serote M., and Doressoundiram A. (2001) Short term variability of Centaur 1999 UG_5. *Astron. Astrophys., 371,* L1–L4.

Hahn G. and Bailey M. E. (1990) Rapid dynamical evolution of giant comet Chiron. *Nature, 348,* 132–136.

Hartmann W. K. (1980) Surface evolution of two-component stone/ice bodies in the Jupiter region. *Icarus, 44,* 441–453.

Hartmann W. K. and Cruikshank D. P. (1978) The nature of Trojan asteroid 624 Hektor. *Icarus, 36,* 353–366.

Hartmann W. K. and Cruikshank D. P. (1980) Hektor, the largest highly elongated asteroid. *Science, 207,* 976–977.

Hartmann W. K., Tholen D. J., Goguen J., Binzel R. P., and Cruikshank D. P. (1988) Trojan and Hilda asteroid lightcurves. I Anomalously elongated shapes among Trojans (and Hildas?).

Icarus, 73, 487–498.

Hartmann W. K., Tholen D. J., Meech K. J., and Cruikshank D. P. (1990) Chiron: Colorimetry and cometary behavior. *Icarus, 83,* 1–15.

Hiroi T., Zolensky M. E., and Peters C. M. (2001) The Tagish Lake meteorite: A possible sample from a D-type asteroid. *Science, 293,* 2234–2236.

Hoffmann M., Fink U., Grundy W. M., and Hicks M. (1993) Photometric and spectroscopic observations of 5145 Pholus. *J. Geophys. Res., 98,* 7403–7407.

Jedicke R. J. and Herron J. D. (1997) Observational constraints on the Centaur population. *Icarus, 127,* 494–507.

Jewitt D. and Kalas P. (1998) Thermal observations of Centaur 1997 CU_{26}. *Astrophys. J. Lett., 499,* L103–L109.

Jewitt D. C. and Luu J. X. (1990) CCD spectra of asteroids. II. The Trojans as spectral analogs of cometary nuclei. *Astron. J., 100,* 933–944.

Jewitt D. C. and Luu J. X. (2001) Colors and spectra of Kuiper Belt objects. *Astron. J., 122,* 2099–2114.

Jewitt D., Luu J., and Chen J. (1996) The Mauna Kea-Cerro-Tololo (MKCT) Kuiper Belt and Centaur survey. *Astron. J., 112,* 1225–1236.

Jewitt D. C., Trujillo C. A., and Luu J. X. (2000) Population and size distribution of small jovian Trojan asteroids. *Astron. J., 120,* 1140–1147.

Jones T. D., Lebofsky L. A., Lewis J. S., and Marley M. S. (1990) The composition and the origin of the C, P, and D asteroids: Water as a tracer of thermal evolution in the outer belt. *Icarus, 88,* 172–192.

Kern S. D., McCarthy D. W., Buie M. W., Brown R. H., Campins H., and Rieke M. (2000) Compositional variation on the surface of Centaur 8405 Asbolus. *Astrophys. J. Lett., 542,* L155–L159.

Khare B. N., Bakes E. L. O., Cruikshank D. P., and McKay C. P. (2001) Solid organic matter on outer solar system bodies. *Adv. Space Res., 27,* 299–307.

Kissel J. and Krueger F. R. (1987) The organic component in dust from comet Halley as measured by the PUMA mass spectrometer on board Vega 1. *Nature, 326,* 755–760.

Kowal C. T. (1978) Surprise in the solar system. *The Sciences, 18,* 12–15.

Irvine W. M., Schloerb F. P., Crovisier J., Fegley B. Jr., and Mumma M. J. (2000) Comets: A link between interstellar and nebular chemistry. In *Planets and Protostars IV* (V. Mannings et al., eds.), pp. 1159–1200. Univ. of Arizona, Tucson.

Larsen J. A., Gleason A. E., Danzl N. M., Descour A. S., McMillan R. S., Gehrels T., Jedicke R., Montani J. L., and Scotti J. V. (2001) The spacewatch wide-area survey for bright Centaurs and Trans-Neptunian objects. *Astron. J., 121,* 562–579.

Lazzarin M., Barbieri C., and Barucci M. A. (1995) Visible spectroscopy of dark, primitive asteroids. *Astron. J., 110,* 3058–3072.

Lazzaro D., Florckzac M. A., Betzler A., Winter O. C., Giuliatti-Winter S. M., Angeli C. A., and Foryta D. W. (1996) 2060 Chiron back to a minimum of brightness. *Planet. Space Sci., 44,* 1547–1550.

Lazzaro D., Florckzac M. A., Angeli C. A., Carvano J. M., Betzler A. S., Casati A. A., Barucci M. A., Doressoundiram A., and Lazzarin M. (1997) Photometric monitoring of 2060 Chiron's brightness at perihelion. *Planet. Space Sci., 45,* 1607–1614

Levison H. and Duncan M. (1997) From the Kuiper Belt to Jupiter-family comets: The spatial distribution of ecliptic comets. *Icarus, 127,* 13–32.

Levison H. F. and Weissman P. R. (1999) The Kuiper Belt. In *Encyclopedia of the Solar System* (P. R. Weissman et al., eds.), pp. 557–583. Academic, San Diego.

Levison H., Shoemaker E. M., and Shoemaker C. S. (1997) The dispersal of the Trojan asteroid swarm. *Nature, 385,* 42–44.

Levison H., Dones L., and Duncan M. (2001) The origin of Halley-type comets: probing in the inner Oort cloud. *Astron. J., 121,* 2253–2267.

Licandro J., Oliva E., and Di Martino M. (2001) *Astron. Astrophys., 373,* 29L–32L.

Lunine J. I., Owen T. C., and Brown R. H. (2000) The outer solar system: Chemical constraints at low temperatures on planet formation. In *Planets and Protostars IV* (V. Mannings et al., eds.), pp. 1055–1080. Univ. of Arizona, Tucson.

Luu J. X. (1993) Cometary activity in distant comets: Chiron. *Publ. Astron. Soc. Pac., 105,* 946–950.

Luu J. X. and Jewitt D. (1990) Cometary activity in 2060 Chiron. *Astron. J., 100,* 913–931.

Luu J. X. and Jewitt D. (1996) Color diversity among the Centaurs and Kuiper Belt objects. *Astron. J., 112,* 2310–2318.

Luu J. X. and Jewitt D. (1998) Optical and infrared reflectance spectrum of Kuiper Belt object 1996 TL_{66}. *Astrophys. J. Lett., 494,* L117–121.

Luu J., Jewitt D., and Cloutis E. (1994) Near-infrared spectroscopy of primitive solar system objects. *Icarus, 109,* 133–144.

Luu J. X., Jewitt D. C., and Trujillo C. (2000) Water ice in 2060 Chiron and its implications for Centaurs and Kuiper Belt objects. *Astrophys. J. Lett., 531,* L151–L154.

Marcialis R. L. and Buratti J. (1993) CCD photometry of 2060 Chiron in 1985 and 1991. *Icarus, 104,* 234–243.

Marsden B. (2001) *List of Centaurs and Scattered-Disk Objects.* http://cfa-www.harvard.edu/iau/lists/Centaurs.html.

Marzari F. and Scholl H. (1998) The growth of Jupiter and Saturn and the capture of Trojans. *Astron. Astrophys., 339,* 278–285.

Marzari F., Scholl H., and Farinella P. (1996) Collision rates and impact velocities in the Trojan asteroid swarms. *Icarus, 119,* 192–201.

Meech K. J. and Belton M. J. S. (1989) (2060) Chiron. *IAU Circular 4770.*

Milani A. (1994) The dynamics of the Trojan asteroids. In *Asteroids Comets Meteors 1993, Proceedings IAU Symposium 160* (A. Milani et al., eds.), pp. 159–174. Kluwer, Dordrecht.

Moroz L. V., Arnold G., Korochantsev A. V., and Wäsch R. (1998) Natural solid bitumens as possible analogs for cometary and asteroid organics. *Icarus, 134,* 253–268.

Oikawa S. and Everhart R. (1979) The past and future orbit of 1977 UB, object Chiron. *Astron. J., 84,* 134–139.

Owen T. C., Cruikshank D. P., Dalle Ore C. M., Geballe T. R., Roush T. L., and de Bergh C. (1999) Detection of water ice on Saturn's satellite Phoebe. *Icarus, 140,* 379–382.

Parker J. M., Stern A., and Festou M. C. (1997) Ultraviolet observations of Chiron with the HST/FOS. Examining a Centaur's gray matter. *Astron. J., 113,* 1899–1908.

Peixinho N., Lacerda P., Ortiz J. L., Doressoundiram A., Roos-Serote M., and Gutierrez P. J. (2001). Photometric study of Centaurs 10199 Chariklo (1997 CU_{26}) and 199 UG_5. *Astron. Astrophys., 371,* 753–759.

Prialnik D., Brosch N., and Ianovici D. (1995) Modelling the activity of 2060 Chiron. *Mon. Not. R. Astron. Soc., 276,* 1148–1154.

Reitmeijer F. J. M. and Nuth J. A. (2000) Metastable eutectic equilibrium brought down to Earth. *Eos Trans. AGU, 81(36),* 409, 414–415.

Reitmeijer F. J. M. and Nuth J. A. (2001) Serpentine by hydrogenation of Fe-rich ferromagnesiosilica PCs in aggregate IDPs (abstract). In *Lunar and Planetary Science XXXII,* Abstract #1219. Lunar and Planetary Institute, Houston (CD-ROM).

Romon-Martin J., Barucci M. A., de Bergh C., and Peixinho N. (2001) Spectroscopy of Centaurs Asbolus and Chiron: Observations over full rotational period. *Bull. Am. Astron. Soc., 33,* 1034.

Sheppard S. S., Jewitt D. C., Trujillo C. A., Brown M. J. I., and Ashley M. C. B. (2000) A wide-field CCD survey for Centaurs and Kuiper Belt Objects. *Astron. J., 120,* 2687–2694.

School H. (1979) History and evolution of Chiron's orbit. *Icarus, 40,* 345–349.

Thompson W. R., Murray B. G. J. P. T., Khare B. N., and Sagan C. (1987) Coloration and darkening of methane clathrate and other ices by charged particle irradiation: Applications to the outer solar system. *J. Geophys. Res., 92,* 14933–14947.

Tholen D. J. and Barucci M. A. (1989) Asteroid taxonomy. In *Asteroids II* (R. P. Binzel et al., eds.), pp. 298–315. Univ. of Arizona, Tucson.

Tholen D. J., Hartmann W. K., and Cruikshank D. P. (1988) *1980 PA and 1985 DO2.* IAU Circular 4770.

Vilas F., Larson S. M., Hatch E. C., and Jarvis K. S. (1993) CCD reflectance spectra of selected asteroids. II. Low-albedo asteroid spectra and data extraction techniques. *Icarus, 105,* 67–78.

Weissman P. R. and Levison H. F. (1997) The population of the trans-neptunian region: The Pluto-Charon environment. In *Pluto and Charon* (S. A. Stern and D. J. Tholen, eds.), pp. 559–604. Univ. of Arizona, Tucson.

Womack M. and Stern S. A. (1995) 2060 Chiron = Comet 95P/Chiron. *IAU Circular 6193.*

Zappalà V., Di Martino M., Cellino A., Farinella P., De Sanctis G., and Ferreri W. (1989) Rotational properties of outer belt asteroids. *Icarus, 82,* 354–368.

Zellner B., Tholen D. J., and Tedesco E. F. (1985) The eight-color asteroid survey: Results for 589 minor planets. *Icarus, 61,* 355–416.

Asteroids *Do* Have Satellites

William J. Merline
Southwest Research Institute

Stuart J. Weidenschilling
Planetary Science Institute

Daniel D. Durda
Southwest Research Institute

Jean-Luc Margot
California Institute of Technology

Petr Pravec
Astronomical Institute of the Academy of Sciences of the Czech Republic

Alex D. Storrs
Towson University

After years of speculation, satellites of asteroids have now been shown definitively to exist. Asteroid satellites are important in at least two ways: (1) They are a natural laboratory in which to study collisions, a ubiquitous and critically important process in the formation and evolution of the asteroids and in shaping much of the solar system, and (2) their presence allows to us to determine the density of the primary asteroid, something which otherwise (except for certain large asteroids that may have measurable gravitational influence on, e.g., Mars) would require a spacecraft flyby, orbital mission, or sample return. Binaries have now been detected in a variety of dynamical populations, including near-Earth, main-belt, outer main-belt, Trojan, and transneptunian regions. Detection of these new systems has been the result of improved observational techniques, including adaptive optics on large telescopes, radar, direct imaging, advanced lightcurve analysis, and spacecraft imaging. Systematics and differences among the observed systems give clues to the formation mechanisms. We describe several processes that may result in binary systems, all of which involve collisions of one type or another, either physical or gravitational. Several mechanisms will likely be required to explain the observations.

1. INTRODUCTION

1.1. Overview

Discovery and study of small satellites of asteroids or double asteroids can yield valuable information about the intrinsic properties of asteroids themselves as well as their history and evolution. Determination of orbits of these moons can provide precise determination of the total (primary + secondary) mass of the system. In the case of a small secondary, the total mass is dominated by the primary. For a binary with a determinable size ratio of components (e.g., double asteroids), an assumption of similar densities can yield individual masses. If the actual sizes of the primary or the pair are also known, then reliable estimates of the primary's bulk density — a fundamental property — can be made. This reveals much about the composition and structure of the primary and will allow us to make comparisons between, for example, asteroid taxonomic types and our inventory of meteorites. In general, uncertainties in the asteroid size will dominate the uncertainty in density. We define satellites to be small secondaries, a double asteroid to be a system with components of similar size, and a binary to be any two-component system, regardless of size ratio.

Similarities and differences among the detected systems reveal important clues about possible formation mechanisms. Systematics are already being seen among the main-belt binaries; many of them are C-like and several are family members. There are several theories to explain the origin of these binary systems, all of them involving disruption of the parent object, either by physical collision or gravitational interaction during a close pass to a planet. It is likely that several of the mechanisms will be required to explain the observations.

The presence of a satellite provides a real-life laboratory to study the outcome of collisions and gravitational interactions. The current population probably reflects a steady-state process of creation and destruction. The nature and prevalence of these systems will therefore help us understand the collisional environment in which they formed and will have further implications for the role of collisions in shaping our solar system. They will also provide clues to the dynamical history and evolution of the asteroids.

A decade ago, binary asteroids were mostly a theoretical curiosity, despite sporadic unconfirmed satellite detections. In 1993, the *Galileo* spacecraft made the first undeniable detection of an asteroid moon with the discovery of Dactyl, a small moon of Ida. Since that time, and particularly in the last year, the number of known binaries has risen dramatically. In the mid to late 1990s, the lightcurves of several near-Earth asteroids (NEAs) revealed a high likelihood of being binary. Previously odd-shaped and lobate NEAs, observed by radar, have given way to signatures revealing that at least six NEAs are binary systems. These lightcurve and radar observations indicate that among the NEAs, the binary frequency may be ~16% (see sections 2.4 and 2.5).

Among the main-belt asteroids, we now know of eight confirmed binary systems, although the overall frequency of these systems is likely to be low, perhaps a few percent (see section 2.2.6). These detections have come about largely because of significant advances in adaptive-optics systems on large telescopes, which can now reduce the blurring of the Earth's atmosphere to compete with the spatial resolution of space-based imaging [which is also, via the Hubble Space Telescope (HST), now contributing valuable observations]. Searches among the Trojans and transneptunian objects (TNOs) have shown that other dynamical populations also harbor binaries.

With new reliable techniques for detection, the scientific community has been rewarded with many examples of systems for study. This has in turn spurred new theoretical thinking and numerical simulations, techniques for which have also improved substantially in recent years.

1.2. History and Inventory of Binary Asteroids

Searches for satellites can be traced back to William Herschel in 1802, soon after the discovery of the first asteroid, (1) Ceres. The first suspicion of an asteroidal satellite goes back to *Andre* (1901), who speculated that the β-Lyrae-like lightcurve of Eros could result from an eclipsing binary system. Of course, we now know definitively that this interpretation is wrong (*Merline et al.,* 2001c), Eros being one of the few asteroids visited directly by spacecraft (cf. *Cheng,* 2002).

The late 1970s saw a flurry of reports of asteroid satellites, inferred from indirect evidence, such as anomalous lightcurves or spurious secondary blinkouts during occultations of stars by asteroids. *Van Flandern et al.* (1979) in *Asteroids* give a complete summary of the evidence at that time. To some, the evidence was highly suggestive that satellites were common. To date, however, none of those suspected binaries has been shown to be real, despite rather intensive study with modern techniques.

In the 1980s, additional lines of evidence were pursued, including asteroids with slow rotation, asteroids with fast rotation, and the existence of doublet craters on, e.g., the Moon or Earth. *Cellino et al.* (1985) studied 10 asteroids that showed anomalous lightcurves, which they compared with predictions from models of equilibrium binaries of varying mass ratios by *Leone et al.* (1984). Model separations and magnitude differences for these putative binaries were given; most of these could have been detected using modern observations, but none have been confirmed as separated binaries, although *Ostro et al.* (2000a), *Merline et al.* (2000b), and *Tanga et al.* (2001) have shown (216) Kleopatra to be a contact binary. In the same decade, radar emerged as a technique capable of enabling study of a small number of (generally nearby) asteroids. In addition, speckle interferometry was used to search for close-in binaries, and the advent of CCD technology allowed for more sensitive and detailed searches. Studies by *Gehrels et al.* (1987), who searched 11 main-belt asteroids using direct CCD imaging and by *Gradie and Flynn* (1988), who searched 17 main-belt asteroids, using a CCD/coronagraphic technique, did not produce any detections. By the end of the decade, previous optimism about the prevalence of satellites had retreated to claims ranging from their being essentially nonexistent (*Gehrels et al.,* 1987) to their being rare (*Weidenschilling et al.,* 1989). *Weidenschilling et al.* (1989) give a summary of the status of the observations and theory at the time of *Asteroids II*.

The tide turned in 1993, when the *Galileo* spacecraft, en route to its orbital tour of the Jupiter system, flew past (243) Ida and serendipitously imaged a small (1.4-km-diameter Dactyl) moon orbiting this 31-km-diameter, S-type asteroid. This discovery spurred new observations and theoretical thinking on the formation and prevalence of asteroid satellites. *Roberts et al.* (1995) performed a search of 57 asteroids, over multiple observing sessions, using speckle interferometry. No companions were found in this survey. A search by *Storrs et al.* (1999a) of 10 asteroids using HST also revealed no binaries. Numerical simulations performed by *Durda* (1996) and *Doressoundiram et al.* (1997) showed that the formation of small satellites may be a fairly common outcome of catastrophic collisions. *Bottke and Melosh* (1996a,b) suggest that a sizable fraction (~15%) of Earth-crossing asteroids may have satellites, based on their simulations and the occurrence of doublet craters on Earth and Venus. Various theoretical studies have been performed on the dynamics and stability of orbits about irregularly shaped asteroids (*Chauvineau and Mignard,* 1990; *Hamilton and Burns,* 1991; *Chauvineau et al.,* 1993; *Scheeres,* 1994).

After the first imaging of an asteroid moon by *Galileo*, several reports of binaries among the NEA population, based on lightcurve shapes, were made by Pravec et al. and Mottola et al., including 1994 AW_1 (*Pravec and Hahn,* 1997), 1991 VH (*Pravec et al.,* 1998a), 3671 Dionysus (*Mottola et al.,* 1997), and 1996 FG_3 (*Pravec et al.,* 1998b). While these systems are likely to be real, they have not been confirmed by direct imaging or radar techniques.

It was not until 1998 that the first definitive and verifiable evidence for an asteroid satellite was acquired from Earth, when 215-km (45) Eugenia was found to have a small moon (13-km Petit Prince) by direct imaging assisted by adaptive optics (AO) (*Merline et al.,* 1999b,c). This discovery was the first result from a dedicated survey with the capability to search for faint companions ($\Delta m \sim 7$ mag) as close as a few tenths of an arcsecond from the primary. This survey detected two more asteroid binaries in 2000: (762) Pulcova (*Merline et al.,* 2000a) and (90) Antiope (*Merline et al.,* 2000a,b). While the moon of Pulcova is small, Antiope is truly a double asteroid, with components of nearly the same size.

After these detections, the first two NEA binaries to be definitively detected by radar were announced: 2000 DP_{107} (*Ostro et al.,* 2000b; *Margot et al.,* 2000) and 2000 UG_{11} (*Nolan et al.,* 2000). In the meantime, Pravec and colleagues have continued to add to the rapidly growing list of suspected binary NEAs from lightcurves.

Starting in 2001, the discovery of binary discoveries really surged. In February, *Brown and Margot* (2001), also using adaptive-optics technology, discovered a moon of (87) Sylvia, a Cybele asteroid beyond the main belt. Soon afterward, *Storrs et al.* (2001a) reported a moon of (107) Camilla, also a Cybele, using HST observations. Four additional radar binaries were announced: 1999 KW_4 (*Benner et al.,* 2001a), 1998 ST_{27} (*Benner et al.,* 2001b), 2002 BM_{26} (*Nolan et al.,* 2002a), and 2002 KK_8 (*Nolan et al.,* 2002b). In addition, *Veillet et al.* (2001, 2002) reported the first binary among transneptunian objects (aside from Pluto/Charon), 1998 WW_{31}, obtained by direct CCD imaging without AO. Six more TNO doubles were reported: 2001 QT_{297} (*Eliot et al.,* 2001); 2001 QW_{322} (*Kavelaars et al.,* 2001); 1999 TC_{36} (*Trujillo and Brown,* 2002); 1998 SM_{165} (*Brown and Trujillo,* 2002); 1997 CQ_{29} (*Noll et al.,* 2002a); and 2000 CF_{105} (*Noll et al.,* 2002b). A small moon was discovered around (22) Kalliope by *Margot and Brown* (2001) and *Merline et al.* (2001a); this was the first M-type asteroid known to have a companion. Later, the first binary Trojan asteroid, (617) Patroclus, was found (*Merline et al.,* 2001b); this asteroid, like Antiope, has components of nearly equal size. *Merline et al.* (2002) then detected a widely spaced binary in the main belt, (3749) Balam, which appears to be the most loosely bound system known. (The list of asteroid satellites in this chapter is complete as of August 2002.)

1.3. Observational Challenges

Direct imaging of possible satellites of asteroids has been hampered by the lack of adequate angular resolution to distinguish objects separated by fractions of an arcsecond and by the lack of sufficient dynamic range of detectors to resolve differences in brightness of many magnitudes. The basic observational problem, detection of a faint object in close proximity to a much brighter one, is common to many areas of astronomy, such as binary and multiple star systems or circumstellar and protostellar disks.

At the inner limit, the smallest separations between the primary asteroid and the companion are determined by orbital instabilities (a few radii of the primary); at the far extreme, they are determined by the Hill stability limit (a few hundred radii of the primary for the main belt). For a 50-km-diameter main-belt asteroid (say at 2.5 AU), observed at opposition, the angular separation at which we might find a satellite spans the range of ~0.05 arcsec to several arcsec. If the satellite has a diameter of 2 km, the brightness difference is 7 mag. Using conventional telescopes, the overlapping point-spread functions of these objects of widely disparate brightness make satellite detection in the near field extraordinarily challenging. The FWHM of the uncorrected point spread function of a large groundbased telescope, under average seeing conditions of 1 arcsec, corresponds to nearly 25 primary radii in the above example. Indeed, both theory and most examples of observed binaries suggest that moons are more likely to be found closer to the primary.

The traditional detection techniques have been deep imaging using multiple short exposures to search the nearfield and the use of "coronagraphic" cameras for the farfield. With modern, low-noise, high-dynamic-range detectors and with the advent of adaptive-optics technology, a groundbased search for and study of, asteroid satellites has been realized.

Radar is a powerful technique for nearby objects because the return signal is proportional to the inverse 4th power of the distance. This has limited study to either very large asteroids at the inner edge of the main belt or to NEAs. Radar has shown tremendous promise and upgrades to the telescopes and electronics have enhanced the range and capabilities. Observations of NEAs, however, have drawbacks because the objects are small and opportunities to observe them may be spaced many years apart. Therefore, it is difficult to make repeat or different observations.

Lightcurve observations generally require the observed system to be nonsynchronous, i.e., having a primary rotation rate different from the orbital rate. In addition, either the system must be eclipsing or the secondary must have an elongated shape. Such a system will show a two-component lightcurve. To be well resolved, both contributions should have an amplitude of at least a few hundredths of a magnitude. The requirements generally restrict efficient observations to close-in binary systems with the secondary's diameter at least approximately one-fifth that of the primary. This technique works best also on NEAs, where these small binaries appear to have a long tidal evolution timescale and therefore can remain nonsynchronous for a long time after formation. These close binaries also lend themselves to having a high probability of eclipse at any given time. This technique suffers from the same problems with NEAs mentioned above: Relatively quick repeat observations over a wide range of viewing geometries are not possible. Thus, in many cases there may be ambiguities in interpretation of the lightcurve signatures.

Direct imaging has been shown to be possible for TNOs because those detected so far have wide separations and large secondary/primary size ratios. So although these objects are far away (~45 AU), loosely bound binaries can be separated with conventional (non-AO) imaging under ideal conditions. HST searches for main-belt binaries have been largely unsuccessful, not because of limitations to instrumentation, but because of the lack of telescope time allocated. HST searches for TNO binaries are now under way and are showing promising results.

2. OBSERVATIONAL TECHNIQUES AND DISCOVERIES

2.1. Searches During Spacecraft Encounters

One of the most effective ways of performing a search for satellites of asteroids is by a flyby or orbital tour with a spacecraft, although this is prohibitively expensive for more than a few objects. Nonetheless, this method produced the first definitive evidence for the existence of asteroid moons. It also allows searches to much smaller sizes than is possible from Earth.

A variety of problems are encountered when searching for satellites in images taken during spacecraft encounters. A major problem is that the images are taken from a rapidly moving platform. This makes quick visual inspection difficult, because one must project the image to a common reference point. If the moon is resolved, as in the case of Dactyl, the problem is more manageable. But it is possible that moons would appear as small, pointlike objects, competing for recognition with stars, cosmic-ray hits, and detector defects. The strategy is normally to take a series of many pictures, in which the detector defects are known and the cosmic rays may be eliminated through lack of persistence. Stars may be eliminated by identification using star catalogs or by common motion. Even with these techniques, however, cosmic-ray hits in a series of images may conspire to cluster in a pattern consistent with the spacecraft motion and an object in a plausible position in three-dimensional space relative to the asteroid. Correlations among all identified point-source candidates on a series of images must be examined.

2.1.1. Discovery of Dactyl. The first image of an asteroid moon was spotted by Ann Harch of the *Galileo* Imaging Team on February 17, 1994, during playback of images from *Galileo*'s encounter with S-type (243) Ida on August 29, 1993. Because of the loss early in the mission of *Galileo*'s high-gain antenna, some data from the Ida encounter were returned months afterward. The first images were returned as "jailbars," or thin strips of a few lines of data separated by gaps. This technique allowed a quick look at the contents of the images to determine which lines contained Ida data. Fortuitously, one of these lines passed through the satellite, as shown in Fig. 1. The presence of the moon was later confirmed by the infrared spectrometer experiment and was announced by *Belton and Carlson* (1994). It was initially dubbed 1993(243)1, as the first satellite of asteroid (243) to be discovered in 1993, and was later given the permanent name Dactyl, after the Dactyli, the children or protectors of Ida.

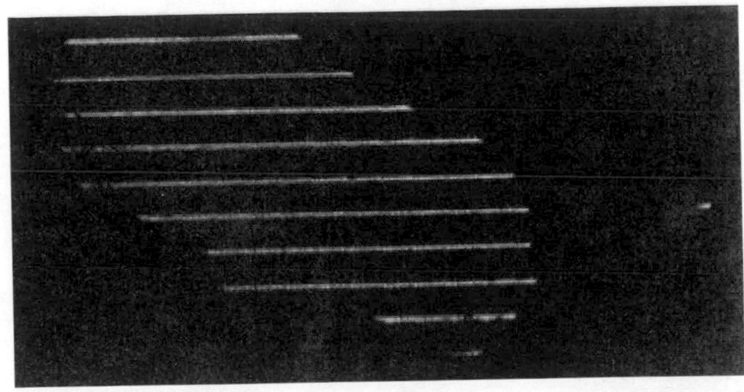

Fig. 1. Discovery image for Dactyl, the first known asteroid satellite (*Belton et al.,* 1996). It was taken by *Galileo* on August 29, 1993, from a range of 10,719 km. The picture has a resolution of ~100 m/pixel. Because of limited downlink, not all images could be returned. Instead, this technique of playing back image strips was used to find the relevant images or portions of images that contained Ida. The resulting "jailbar" image here fortuitously provides the first clue of an extended object, with the expected photometric profile, off the bright limb of Ida.

During the flyby of Ida, 47 images of Dactyl were obtained (*Chapman et al.,* 1995; *Belton et al.,* 1995, 1996). However, because there was no opportunity for feedback to guide an imaging sequence, these pictures were all serendipitous. The spacecraft trajectory was nearly in the plane of the satellite motion, and hence little relative motion was observed, resulting in poorly determined orbital parameters. Followup observations with HST (*Belton et al.,* 1995, 1996) failed to find the satellite, which was not surprising given its separation. But if the object were on a hyperbolic or highly elliptical orbit, there would be some chance of finding it with HST. These additional observations did allow limits to be set on the density of Ida.

Additionally, resolved pictures of Dactyl's surface have allowed for geological interpretation and have provided a glimpse into the possible origin and history of an asteroid moon. The pair is shown in Fig. 2, with a smaller-scale image of Dactyl in Fig. 3. *Chapman et al.* (1996) and *Veverka et al.* (1996b) indicate that the crater size-frequency distribution on both Ida and Dactyl exhibit equilibrium saturation (see also *Chapman,* 2002). Thus, we can estimate only the minimum ages for both objects; the relative age of the two, from crater data alone, is uncertain. Given the observed impactor size-distribution, saturation at the largest craters on Ida, ~10 km, would be expected after about 2 b.y. (*Chapman et al.,* 1996), setting roughly the minimum age of Ida. The largest craters on Dactyl, however, are less than 0.4 km in size, and would saturate in about 30 m.y. Impacts that would create larger craters on Dactyl would instead break

Fig. 2. Full image of Ida and Dactyl, taken from approximately the same range and with the same resolution listed in Fig. 1. The picture is in a green filter. Ida is ~56 km long and Dactyl is roughly spherical with a diameter of ~1.4 km. At this time Dactyl is in the foreground, ~85 km (5.5 R_{Ida}) from Ida's center, and moving at about 6 m s^{-1}. The orbit is prograde with respect to Ida's spin, which itself is retrograde with respect to the ecliptic.

up the object. The mean time between impacts that would destroy Dactyl is estimated by *Davis et al.* (1996) to be, depending on model assumptions, between about 3 m.y. and 240 m.y., the same order as the saturation cratering age. If Dactyl was formed 2 b.y. ago with Ida, via disruptive capture (section 3.3), perhaps during the Koronis-family breakup (*Binzel,* 1988), then it is very unlikely that it would still exist intact, given its short lifetime against collisional breakup. Conceivably, it may have formed from the ejecta of a more recent, large cratering event (section 3.2). Either way, it must have been disrupted and reaccreted several times since

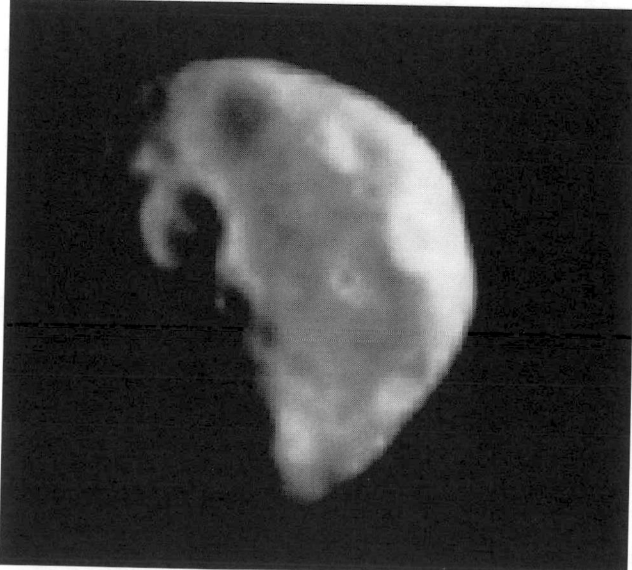

Fig. 3. Highest-resolution picture of Dactyl, at 39 m/pixel, showing shape and surface geology. The topography is dominated by impact craters, without prominent grooves or ridges.

its initial formation, because it is unlikely to have formed only in the last 30 m.y. Additional geological data support the idea of this satellite as a reaccumulated rubble pile. It is roughly spherical, with no obvious evidence of coherent monolithic structure. It displays a softened appearance and likely has a surface regolith (*Veverka et al.,* 1996b).

The spectrum of Dactyl (from *Galileo* imager data, 0.4–1.0 μm) is similar to that of Ida (*Veverka et al.,* 1996a), but with some important differences. Both objects show S-type spectra and have similar albedos. Dactyl, however, shows somewhat less reddening than Ida, possibly indicating less space weathering, which is also consistent with a younger surface age, as expected from the most recent disruption/reaccretion episode (*Chapman,* 1996).

2.1.2. Other searches. Extensive searches were made for additional satellites of Ida in the *Galileo* datasets; no candidates were found that were not consistent with single or multiple cosmic-ray events (*Belton et al.,* 1995, 1996). The searches were made at spacecraft-to-asteroid ranges of 200,000 km (satellite-detection size limit ~800 m), 10,000 km (size limit ~50 m), and 2400 km (encounter; size limit ~10 m).

Cursory searches for satellites were made during the *Galileo* flyby of S-type (951) Gaspra in 1991, with no detections of objects larger than 27 m out to ~10 Gaspra radii (*Belton et al.,* 1992).

The *NEAR Shoemaker* spacecraft made a fast flyby of C-type, inner main-belt asteroid (253) Mathilde in 1997 en route to its orbital encounter with (433) Eros. A well-planned imaging sequence to search for satellites was performed and a thorough search made (*Merline et al.,* 1998; *Veverka et al.,* 1999a). More than 200 images were taken specifically to search for satellites. No unambiguous evidence for satellites larger than 40 m diameter was found within the searchable volume, which was estimated to be ~4% of the Hill sphere. The portion, however, of the Hill sphere searched was an important one, inside roughly 20 radii of Mathilde (almost all of the known main-belt binaries show separations well below 20 primary radii). From approach images, which were less sensitive due to lighting geometry, no satellites larger than 10 km were found in the entire Hill sphere.

The *NEAR Shoemaker* spacecraft continued on to an unplanned flyby of (433) Eros, an S-type NEA, in December 1998 (cf. *Cheng,* 2002). The first critical burn of the main rocket for the rendezvous aborted prematurely, which led to execution of a contingency imaging sequence. This included a search for satellites down about 50 m in the entire Hill sphere (*Merline et al.,* 1999a; *Veverka et al.,* 1999b). About one year later, after engineers had diagnosed the problem and brought the spacecraft slowly back to Eros, the orbital tour of Eros began. During approach to orbit insertion, another, more detailed and thorough search for satellites was made. During this search, both manual and automated searches were performed (*Merline et al.,* 2001c; *Veverka et al.,* 2000). This was the first systematic search for satellites of the entire Hill sphere of an asteroid down to small sizes. The search found no objects at diameter 20 m (95% confidence) and none at 10 m (with 70% confidence).

2.2. Adaptive Optics on Large Groundbased Telescopes

Given the observational challenges just discussed and the number of failed attempts to detect asteroid satellites, it was clear that a new approach was needed. In 1996, Merline and collaborators began to apply a relatively new technology in hopes of achieving high-contrast, high-spatial-resolution imaging on a large number of targets from groundbased telescopes. This new technique, called adaptive optics (AO), ultimately led to the first Earth-based images of satellites.

2.2.1. Method and capabilities. This technique minimizes the distortion in an astronomical image by sensing and correcting, in real time, aberrations due to the Earth's atmosphere, usually by means of a deformable mirror. This new technology can result in diffraction-limited imaging with the largest groundbased telescopes. Compared with conventional direct-imaging techniques, this technique shows a dramatic improvement in the ability to detect asteroid companions. Adaptive optics (1) decreases the light contribution from the primary asteroid at the position of the satellite on the plane of the sky and (2) increases the signal from the secondary asteroid at that position, enhancing the ability to detect, or set limits on the sizes of, satellites. In addition, because IR-imaging cameras are used, no charge bleeding (as for CCDs) occurs in an overexposure of the primary. This effectively gives near-field coronagraphic imaging capability, allowing deep exposures for faint companions.

In adaptive-optics systems, the light from the telescope is processed by a separate optical unit that resides beyond the telescope focal plane. A recollimated beam impinges on a deformable mirror (DM), which has many actuators that can be adjusted rapidly to "correct" the beam back to its undistorted "shape." Light from the DM is then divided, with part (typically near-IR) of it going to the science camera, and part (typically visible) going to a wavefront sensor, which analyzes the deformation of the wavefront and provides correction signals to the DM, forming a closed loop.

Two types of systems are in use. One uses a Shack-Hartmann (SH) wavefront sensor, basically an x-y array of many lenslets in a collimated beam. Each of these lenslets allows sensing of the beam deviation in a different part of the pupil. The other method is curvature-wavefront sensing (CS) (*Roddier,* 1988) in which the wavefront sensor is divided in a radial/sectoral fashion. The illumination pattern of the beam is then sampled rapidly at positions on either side of a focal plane; the differences in illumination are related to the local wavefront curvature. While the Shack-Hartmann systems are more common, the curvature systems can work with fewer elements, at faster speeds, and on fainter objects. CS systems trade the higher-order corrections of an SH system for faster (kHz) sample and correction speeds.

There are many AO systems either in use or under development. Among those that have been used for planetary applications are systems at the Starfire Optical Range (U.S. Air Force), the Mt. Wilson 100", the University of Hawai'i (on 88", UKIRT, and CFHT), the Canada-France-Hawai'i Telescope (CFHT), the Keck, the ESO/Adonis, the Lick, the Palomar, and the Gemini North. Only three of these systems, all located on Mauna Kea, Hawai'i, have resulted in discoveries of asteroid satellites. The 3.6-m CFHT uses a 19-element CS system called PUEO (*Roddier et al.,* 1991; *Rigaut et al.,* 1998). It can reach a limiting magnitude of about V = 14.5 with a resolution of about 0.11 arcsec at H-band. The 10-m Keck uses a 349-element SH system (*Wizinowich et al.,* 2000), allowing compensation to about V = 13 with a resolution of 0.04 arcsec at H-band. The 8.1-m Gemini telescope, with the Hokupa'a 36-element CS system (*Graves et al.,* 1998) of the University of Hawai'i, can reach about V = 17.5, with resolution of ~0.05 arcsec at H-band.

The AO systems must have a reference point source to compute atmospheric turbulence. The systems may either use a natural guide star (NGS) or an artificially generated star (LGS), in which a laser is used to produce a point source in the upper atmosphere. Laser-guide systems have been tested and used largely within military applications. Although there are plans for LGS systems at many astronomical facilities, the progress has been slow and of limited use thus far. Therefore, NGS systems dominate AO systems. For astronomical (fixed-source) applications, a nearby brighter star may be used, provided it is within the isoplanatic patch, which may be about 20 arcsec at 2 μm. But for planetary objects, e.g., main-belt asteroids, their fast motion prohibits use of nearby objects, and one must rely on the object itself as the reference. This presents two limitations: Extended objects will tend to degrade the quality of the compensation, although asteroids are not extended enough to be of concern. In addition, the quality of the AO correction will depend on the brightness of the reference object, so there is a limit to how faint an asteroid can be observed.

Most of the AO systems operate in the near IR, using HgCdTe IR (1–2.5 μm) array detectors as the science camera. Although the ultimate signal-to-noise of the science data is a function of the brightness in the selected IR band, it is the visible light that is used by the wavefront sensor, so the quality of the AO compensation is dependent upon the V magnitude.

The correct wavelength band for observations is adjusted depending on conditions and the telescope. With IR AO observations, there is always a tradeoff between competing effects — the shorter the wavelength, the narrower the PSF for a given telescope. But at shorter wavelengths, the number of cells in the telescope beam that need to be continuously corrected grows beyond the capacity of the AO system — more cells require more AO actuators for compensation. But systems with a large number of actuators means prohibitively high cost, so there is a limit. Of course, the larger the telescope, the larger the number of cells needed to compensate. Therefore, the 10-m Keck usually performs best at K'-band (2.1 μm) and the 3.6-m CFHT at H-band (1.6 μm). Thus, the Strehl ratio (the ratio of peak brightness of acquired image to the peak brightness of a perfectly diffraction-limited point source) increases at longer wavelengths, while the instrumental width also increases. Under good conditions one hopes to achieve about 50% Strehl. On exceptionally good nights, it may be possible to use J-band (1.2 μm) for a narrower PSF.

The future holds great promise for AO, as more telescopes adopt this technology. In addition, the advent of quality LGS systems and the opportunity for systems employing many more actuators, as costs decline and computer speeds increase, means the possibility of visible-light systems and a correspondingly narrower diffraction limit.

Using AO, because the result is a picture of the system on the plane of the sky, we can hope to achieve the same information (and more) about a system as that which can be obtained from visual binary stars, only on a substantially shorter timescale. Basically, all seven dynamic orbital elements required to describe the motion are derivable. These are the elements describing motion along the orbital ellipse — the semimajor axis, the eccentricity, and an indication of orbital phase, such as time of periapse passage or true anomaly — plus the elements describing the orientation in three-dimensional space — e.g., the inclination, the longitude of the ascending node, and the argument of periapse. In addition, because the system mass is unknown (unlike Sun-orbiting objects) we also require determination of the orbital period. From a limited span of observations, say a single orbit or series of a few orbits, there remains a two-fold ambiguity in the orbital pole position (determination of the pole direction is equivalent to determination of the two elements inclination and node). This can be resolved by observing at a different viewing geometry at some later time. The period and orbit size (assuming a circular orbit) are readily obtainable, which immediately yields an estimate of the system (primary + secondary) mass, by Kepler's Third Law. If the secondary is small or if we can independently determine the size ratio (and then make an assumption that the primary and secondary are of the same density), then the primary mass can be estimated. If the primary asteroid size is known, then we can determine the primary's density. Of course, density is clearly one of the most fundamental parameters one hopes to know about any body, and gives direct insight into the composition and structure. Because most of the orbits are small in angular terms (and pixels on a detector), the errors in measurement of positions translate into sizable uncertainties in most of the orbital elements. However, the period can be very accurately determined, and the ultimate uncertainties in density are dominated by uncertainties in the size of the asteroid.

2.2.2. (45) Eugenia. The first binary system discovery using AO was accomplished on November 1, 1998, when a small companion of (45) Eugenia was discovered at the CFHT by *Merline et al.* (1999b,c). The system was tracked for 10 d and again occasionally in the following months and years. It was the first AO system for which the two-fold degeneracy in the orbit pole had been resolved. Further, because of the large brightness difference (about 7 mag), it remains one of the more difficult AO binaries to observe. Figure 4 shows the discovery image of this object, provisionally named S/1998 (45) 1 and later given the permanent name Petit Prince in honor of the prince imperial of France, the only child of Napoleon III and his wife Empress Eugenie (namesake of Eugenia). (The name itself is derived from the popular children's book *Le Petit Prince* by A. Saint-Exupery, whose central character was an asteroid-dwelling Little Prince.) The intention was to keep and solidify the tradition of naming asteroid moons after the children or other derivative of the parent asteroid. Figure 5 shows five epochs of the orbit at the time of discovery. Figure 6 exhibits the tremendous power of modern AO techniques both to resolve the asteroid and to clearly separate a close companion.

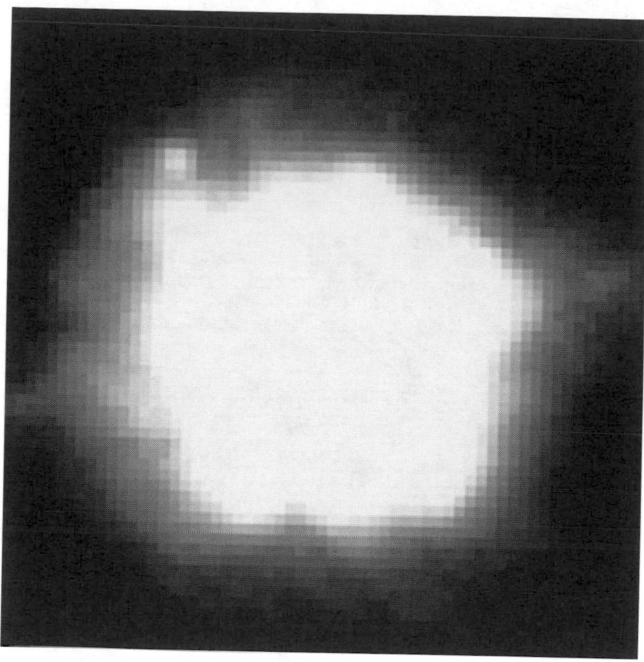

Fig. 4. Discovery image of Petit Prince, moon of (45) Eugenia, taken at the Canada-France-Hawai'i Telescope on November 1, 1998, using the PUEO adaptive-optics system (*Merline et al.*, 1999b). It is the first asteroid moon to be imaged from Earth. The image is an average of 16 images of exposure 15 s. It is taken in H-band (1.65 μm) and has a plate scale of 0.035 arcsec/pixel. The separation of the moon is ~0.75 arcsec from Eugenia and has a brightness ratio of ~7 mag.

The satellite appears to be roughly in the asteroid's equatorial plane and in a prograde orbit (*Merline et al.*, 1999c). A prograde orbit is preferred for a satellite formed from impact-generated orbital debris (*Weidenschilling et al.*, 1989; *Durda and Geissler,* 1996). A retrograde orbit, however, is more stable against perturbing effects of the nonuniform gravitational field of an oblate primary (*Chauvineau et al.*, 1993; *Scheeres,* 1994). An orbit with an opposite sense to the asteroid's orbital motion around the Sun (as it is for Petit Prince — Eugenia's spin is retrograde) is more stable against the effects of solar tides (*Hamilton and Burns,* 1991). Mechanisms for capture of such ejecta into quasi-stable orbits are reviewed by *Scheeres et al.* (2002).

The orbital period was determined to be ~4.7 d for the satellite of this FC-type asteroid and yields a density estimate of ~1.2 g cm^{-3} (*Merline et al.*, 1999c). This result fol-

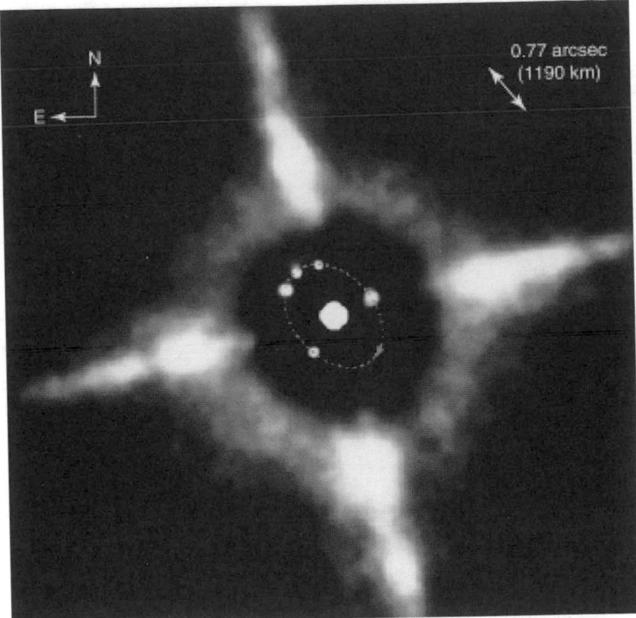

Fig. 5. This infrared image is a composite of five epochs of Eugenia's moon. The moon has a period of 4.7 d, with a nearly circular orbit of ~1190 km (0.77 arcsec). The orbit is tilted ~46° with respect to our line-of-sight. The normal two-fold degeneracy in pole position (i.e., true sense of the moon's orbit) was resolved by observing the system later, when positional differences between the two solutions became apparent. Eugenia is ~215 km in diameter and the moon's diameter is ~13 km. The large "cross" is a common artifact of diffraction from the secondary-mirror support structure. The images are deconvolved, and the brightness of Eugenia has been suppressed to enhance sharpness and clarity.

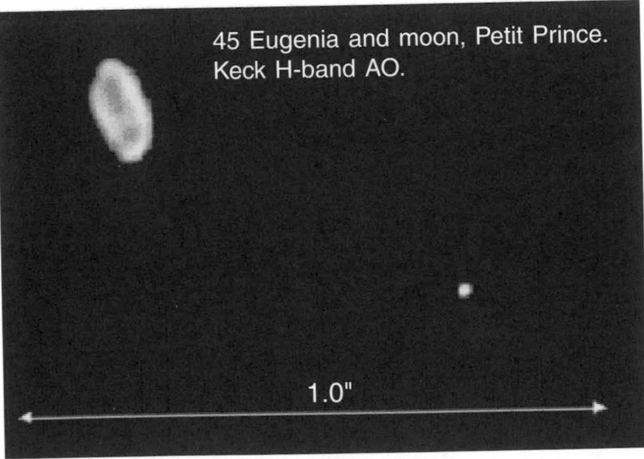

Fig. 6. This deconvolved Keck image in February 2000 shows Petit Prince (*Merline et al.,* 1999c) and a resolved image of the disk of Eugenia (after *Close et al.,* 2000). The pair is well separated enough to get accurate colors or spectra. The unusual elongation of Eugenia's shape was inferred previously from lightcurve amplitudes. Because the lack of detailed fidelity in flux preservation under deconvolution, the brightness variations across the disk are not real. The brightness of the satellite (which is not resolved) has been scaled to appear to have roughly the same "surface brightness" as the primary. The flux ratio of the two objects is about 285.

lowed soon after the surprising announcement that the density to C-type Mathilde was only 1.3 g cm^{-3}, as determined by spacecraft flyby (*Veverka et al.,* 1999a). Such a density requires a significant amount of macroporosity to be consistent with the expected meteorite analog for these objects, namely carbonaceous chondrites (*Britt and Consolmagno,* 2000). Therefore, it is possible that these asteroids are loosely packed rubble piles.

2.2.3. (90) Antiope and (617) Patroclus. The first true double asteroid, (90) Antiope, was discovered in August 2000 by *Merline et al.* (2000a). This main-belt C-type was found to have two nearly equal-sized components of diameter ~85 km, rather than a single object of size 120 km as previously assumed. The orbital period of the pair was found to be ~16.5 h, consistent with the previously observed lightcurve period. Interestingly enough, a lightcurve by *Hansen et al.* (1997) shows a classic eclipsing-binary shape (although they did not make this interpretation), which would be expected to result from equal-sized components, with the orbit viewed edge-on. The derived density for the components of Antiope, assuming they are of the same size and density, is about 1.3 g cm^{-3}, again similar to previous measurements of low-albedo asteroids. Figure 7 shows the components of Antiope as they orbit the common center of mass. Another double, (617) Patroclus, was discovered in September 2001 by *Merline et al.* (2001b). Again, it is a primitive P-type and is the first Trojan to be shown definitively to be binary. Few data were acquired, but it appears that this object also will show a low density.

2.2.4. (762) Pulcova, (87) Sylvia, and (22) Kalliope. Small satellites were also found around two more large, low-albedo asteroids: F-type (762) Pulcova (*Merline et al.,* 2000b) at CFHT and P-type (87) Sylvia (*Brown and Margot,* 2001) at Keck. Sylvia, a Cybele, is the first binary found in the outer main belt. In August/September 2001, a small companion to (22) Kalliope was discovered by *Margot and Brown* (2001) and *Merline et al.* (2001a). This is the first M-type asteroid known to have a companion and gives the hope of getting a reliable density estimate for these controversial objects, which have traditionally been thought to be metallic. Initial estimates put the density near ~2.3 g cm^{-3}. This value is even lower, although not significantly, than the values previously derived for S-types (around 2.5 g cm^{-3}). If so, it clearly indicates that at least Kalliope is not of a

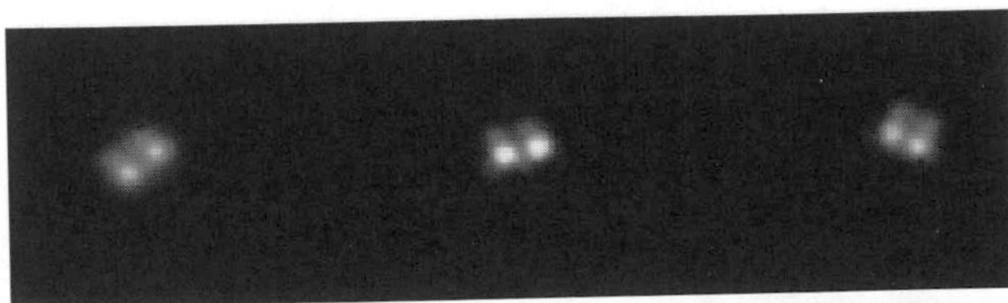

Fig. 7. Double asteroid (90) Antiope as it rotates with a 16.5-h period, soon after its discovery at Keck in August 2000 (*Merline et al.,* 2000a,b). Once thought to be an object ~125 km across, the C-type asteroid Antiope actually has two components, each ~85 km in diameter. The separation is ~170 km.

solid metallic composition. It would also be difficult to imagine an extremely porous rubble pile of metallic composition, because it would imply a macroporosity of more than about 60%. We may be faced with the difficult task of explaining how bodies with metallic spectra and radar reflectivities have rocklike densities.

2.2.5. (3749) Balam. Among the main-belt binaries, this object stands out as an oddity. Discovered at Gemini Observatory in 2002 (*Merline et al.,* 2002), this binary is the most loosely bound system known, even more so than the TNO binaries. The secondary appears to orbit at least 100 (primary) radii from the primary, which itself is rather small (about 7 km in diameter). This is probably the first system known that was formed by "disruptive capture," discussed in section 3.3. Early models of *Durda* (1996) and *Doressoundiram et al.* (1997), as well as the more sophisticated models currently being performed by Durda et al., indicate that such systems (small primaries, with a widely separated secondary) are commonly formed in catastrophic collisions and that a large number of should be found in the main belt.

2.2.6. Systematics. While there appears to be a rash of newly discovered binaries, it turns out that the prevalence of (large) main-belt moons is likely to be low, probably ~2% (*Merline et al.,* 2001d). The largest survey to date, by Merline et al., has sampled more than 300 main-belt asteroids, with five examples of relatively large satellites (few tens of kilometers in diameter). The overall frequency, including small, close-in moons such as Dactyl (currently unobservable from Earth), will undoubtedly rise, but it is unknown by how much. Very small satellites will have a limited lifetime against collisions, although it is possible they may reaccrete. The single known binary among the Trojans, from a sample of about six, hints that the binary frequency may be higher in that population, although it is noted that the collision speeds are comparable to the main belt and the collision frequencies are only higher by about a factor of 2 (*Davis et al.,* 2002).

For those satellites that are found, it would be useful to establish any systematics that may provide clues as to the origin mechanism for the moons. For example, it has been suggested that either slow (from tidal spindown due to a satellite) or fast (from a glancing collision, which might form satellites) rotation might be correlated with the presence of satellites. Family members have been suggested as likely candidates for satellites, because coorbiting pairs may have been created in the family-forming event. The likelihood of moons may even be linked to the taxonomic type or shape of the asteroid.

Most of the observed binaries in the main belt, outer belt, or Trojan region are of primitive type (C, F, P). Are satellites truly more prevalent around these objects, or is there some observational selection effect? Clearly, those asteroids highest in priority for observation are the apparently

TABLE 1. Binary asteroids discovered by adaptive optics or direct imaging techniques.

Object	Type	Taxonomic Classification (Tholen)	Family	Asteroid a (AU)	Primary Rotation Period (h)	Primary Diameter (km)	Discovery Date	Method
(243) Ida	MB	S	Koronis	2.86	4.63	31	Aug. 29, 1993	SC
(45) Eugenia	MB	FC	Eugenia	2.72	5.70	215	Nov. 1, 1998	AO
(762) Pulcova	MB	F		3.16	5.84	137	Feb. 22, 2000	AO
(90) Antiope	MB	C	Themis	3.16	16.50*	85 + 85	Aug. 10, 2000	AO
(87) Sylvia	OB	P		3.49	5.18	261	Feb. 18, 2001	AO
(107) Camilla	OB	C		3.48	4.84	223	Mar. 1, 2001	HST
(22) Kalliope	MB	M		2.91	4.15	181	Aug. 29, 2001	AO
(3749) Balam	MB	S	Flora	2.24		7	Feb. 8, 2002	AO
(617) Patroclus	L5-TROJ	P		5.23		95 + 105	Sep. 22, 2001	AO
1998 WW_{31}	TNO			44.95		150†	Dec. 22, 2000	DI
2001 QT_{297}	TNO			44.80		580‡	Oct. 11, 2001	DI
2001 QW_{322}	TNO			44.22		200§	Aug. 24, 2001	DI
1999 TC_{36}	TNO			39.53		740‡	Dec. 8, 2001	HST
1998 SM_{165}	TNO			47.82	7.98	450‡	Dec. 22, 2001	HST
1997 CQ_{29}	TNO			45.34		300‡	Nov. 17, 2001	HST
2000 CF_{105}	TNO			44.20		170‡	Jan. 12, 2002	HST

*Assuming synchronous rotation.
†Assuming, for both components, albedo ~5.4% and density ~1 g cm^{-3} (*Veillet et al.,* 2002).
‡Values provided by A. W. Harris (personal communication, 2002), assuming albedo 4%.
§Assuming albedo 4% (*Kavelaars et al.,* 2001).

MB = main belt; OB = outer belt; TROJ = Jupiter Trojan; TNO = transneptunian object; SC = spacecraft encounter; AO = adaptive optics; HST = HST direct imaging; DI = direct groundbased imaging.

TABLE 2. Properties of secondaries and derived properties of primaries.

Object	Orbit a (km)	Orbit Period (d)	Orbit Size (a/R_p)	Orbit Sense	Moon Diameter (km)	Size Ratio (D_p/D_s)	Primary Mass ($\times 10^{16}$ kg)	Primary Density (g cm^{-3})	Mass Ratio (M/m)
(243) Ida	108	1.54	7.0	Prograde	1.4	22	4.2	2.6 ± 0.5	11,000
(45) Eugenia	1190	4.69	11.1	Prograde	13	17	610	1.2 ± 0.4	4900
(762) Pulcova	810	4.0	11.6		20	7	260	1.8 ± 0.8	340
(90) Antiope	170	0.69	4.0		85	1.0	41	1.3 ± 0.4	1.0
(87) Sylvia	1370	3.66	10.5		13	20	1500	1.6 ± 0.3	7900
(107) Camilla	~1000		~9		9	25			18,000
(22) Kalliope	1060	3.60	11.7	Prograde	19	10	730	2.3 ± 0.4	870
(3749) Balam	~350	~100	~100		1.5	4.6			95
(617) Patroclus	610	3.41	11.6		95	1.1	87	1.3 ± 0.5	1.3
1998 WW_{31}	22,300	574	300*		120*	1.2	170		1.7
2001 QT_{297}	~20,000		69†			1.4			2.6
2001 QW_{322}	~130,000	~1500§	1300‡		200*	1.0			1.0
1999 TC_{36}	~8000		22†			2.8			21
1998 SM_{165}	~6000		27†			2.4			14
1997 CQ_{29}	~5200		35†			~1?			~1?
2000 CF_{105}	~23,000		270†			1.6			3.9

*Assuming, for both components, albedo ~5.4% and density ~1 g cm^{-3} (*Veillet et al.,* 2002).
†Values provided by A. W. Harris (personal communication, 2002), assuming albedo 4%.
‡Assuming albedo 4% (*Kavelaars et al.,* 2001).
§This period is reasonable, despite the large observed separation, because of a high eccentricity (A. W. Harris, personal communication, 2002).

brighter objects. Among the objects in Merline et al.'s target lists, the S-like and C-like asteroids are about equal in number. (This may mean that the frequency of binaries is more like 4% among the primitive asteroids.) But this is not where the bias ends. To be of equal brightness, a C-like asteroid must be much larger than an S-like, and therefore will have a larger Hill sphere. As such, one can image deeper into the gravitational well of a C-like object than an S-like object of the same apparent brightness, on average. Given that most of the observed companions reside within about 12 primary radii, the companions of C-like objects will be more easily found. Nonetheless, if the frequency of companions were also 4% for the S-like asteroids, some should still have been found. This raises the question as to whether it is more difficult to make satellites around S-types, which may be predominantly fractured-in-place chards, rather than rubble piles (*Britt and Consolmagno,* 2001). If this is true, and because many of the outer-belt and Trojan asteroids are of primitive type, we may ultimately find a higher binary frequency among those populations.

Tables 1 and 2 summarize the properties of known binary systems discovered using adaptive-optics or direct-imaging techniques.

2.3. Discovery by Direct Groundbased Imaging

Despite the difficulty of directly resolving a binary asteroid system from the ground without the assistance of adaptive optics, detections have recently been achieved. By direct imaging with CCDs on large telescopes under exceptional conditions, it has been possible to resolve TNO binaries. *Toth* (1999) discusses some of the issues regarding detectability of these objects. The first of these, 1998 WW_{31}, was discovered by *Veillet et al.* (2001) in December 2000 at CFHT. Followup observations of 1998 WW_{31} from groundbased telescopes and HST, as well as archival searches of previous datasets, indicate that the system has a size ratio of about 1.2, with an eccentric (~0.8) orbit, a semimajor axis near 22,000 km, and a period of ~570 d (*Veillet et al.,* 2002).

Soon afterward, two more TNO binaries were detected in the same way: 2001 QT_{297} (*Eliot et al.,* 2001), showing a separation of 0.6 arcsec at time of discovery and a size ratio of about 1.7; and 2001 QW_{322} (*Kavelaars et al.,* 2001) with a size ratio of ~1.0 and a wide separation of 4 arcsec when discovered. Four additional TNO systems were subsequently discovered using HST (discussed in section 2.6). All of these systems, except one, are classical Kuiper Belt objects, residing at ~45 AU. One system, 1999 TC_{36}, is a Plutino at ~40 AU.

For these objects, AO cannot be used directly because they are too faint, so direct imaging, either from the ground or in ongoing campaigns on the HST, is likely to be the most attractive approach. Because they move slowly past field stars, it is possible to use AO to image these objects during appulses with brighter stars. This technique may improve the overall sensitivity to fainter companions.

The size of the Hill sphere of an object is directly proportional to its distance r from the Sun, but the angular size

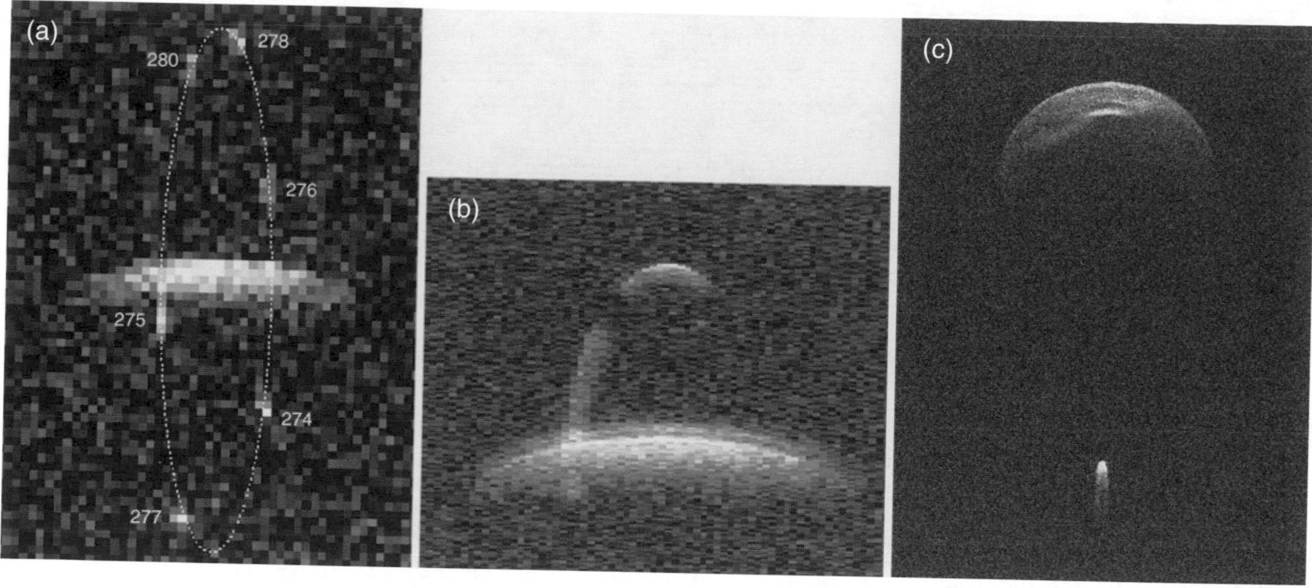

Fig. 8. **(a)** Arecibo delay-Doppler images of binary asteroid 2000 DP_{107} (*Margot et al.,* 2002b) obtained on 2000 DOY 274-280. A dashed line shows the approximate trajectory of the companion on consecutive days. **(b)** Goldstone radar echoes of 1999 KW_4 (S. Ostro et al., personal communication, 2001) accumulated over several hours during its May 2001 close approach to Earth. **(c)** Radar image of 1999 KW_4 obtained at Arecibo on May 27, 2001, with 7.5-m range resolution. Range from the observer increases down and Doppler frequency increases to the right. Dimensions in the cross-range dimension are affected by the primary and secondary spin rates.

of a satellite orbit, as seen from Earth is inversely proportional to the distance from the observer, Δ, which is approximately r. So if satellites reside at the same fraction of their Hill sphere from the primary, there should be no advantage of direct imaging in observing outer solar system objects compared with similar-sized objects in the main belt. Apparently, the main reasons these systems are being found with direct imaging, while those in the main belt are not, is that the secondary to primary size ratios are high, making the secondary easier to detect, while at the same time the satellites are more loosely bound. Additionally, the TNO primaries are rather large, further assisting detection because of the correspondingly larger Hill sphere. Possibly, similar systems are rare in the main belt, and the TNO binaries are formed by a different process.

2.4. Radar Discovery and Characterization of Binary Near-Earth Asteroids

The radar instruments at Goldstone and Arecibo recently provided the first confirmed discoveries of binary asteroids in the near-Earth population (*Margot et al.,* 2002a,b). In the two-year period preceding this writing, six near-Earth objects have been unambiguously identified as binary systems: 2000 DP_{107} (*Ostro et al.,* 2000b; *Margot et al.,* 2000); 2000 UG_{11} (*Nolan et al.,* 2000); 1999 KW_4 (*Benner et al.,* 2001a); 1998 ST_{27} (*Benner et al.,* 2001b); 2002 BM_{26} (*Nolan et al.,* 2002a); and 2002 KK_8 (*Nolan et al.,* 2002b). Previous attempts to detect asteroid satellites with radar date back to the search for a synchronous moon around Pallas (*Showalter et al.,* 1982). S. Ostro (personal communication, 2001) recalls that concrete anticipation for the radar discovery of binary systems arose with the imaging and shape modeling of the strongly bifurcated NEA (4769) Castalia (*Ostro et al.,* 1990; *Hudson and Ostro,* 1994). *Ostro et al.* (2002) provide a thorough description of radar observations of asteroids.

In continuous-wave (CW) datasets, in which echoes resulting from a monochromatic transmission are spectrally analyzed, the diagnostic signature is that of a narrowband spike superposed on a broadband component. The wide-bandwidth echo is distinctive of a rapidly rotating primary object, i.e., with spin periods of order a few hours. The narrowband feature, which does not move at the rate associated with the rotation of the primary, represents power scattered from a smaller and/or slowly spinning secondary. As time goes by, the narrowband echo oscillates between negative and positive frequencies, representing the variations in Doppler shift of a moon revolving about the system's center of mass (COM). The timescale associated with this motion in the small sample of objects studied so far is on the order of a day.

In delay-Doppler images, in which echo power is discriminated as a function of range from the observer and line-of-sight velocity, the signatures of two distinct components are easily observed. Both the primary and secondary are typically resolved in range and Doppler, and their evolution in delay-Doppler space is consistent with the behavior of an orbiting binary pair. Example datasets are shown in Fig. 8.

The observables that can be measured from radar images are (1) visible range extents, which constrain the sizes of each component; (2) Doppler bandwidths, which constrain the spin periods of both the primary and secondary;

TABLE 3. Binary asteroids detected by radar.

Object	a (m)	e	P_{orb} (d)	$(M_1 + M_2)$ (10^9 kg)	R_p (m)	R_s (m)	a/R_p	ρ (g cm^{-3})
2000 DP_{107}	2622 ± 54	0.010 ± 0.005	1.755 ± 0.002	460 ± 50	400 ± 80	150	6.6	1.7 ± 1.1
2000 UG_{11}	337 ± 13	0.09 ± 0.04	0.770 ± 0.003	5.1 ± 0.5	115 ± 30	50	2.9	0.8 ± 0.6
1999 KW_4	2566 ± 24	≤0.03	0.758 ± 0.001	2330 ± 230	600 ± 120	<200	4.3	2.6 ± 1.6
1998 ST_{27}	4000–5000				250–300	<50	13–20	
2002 BM_{26}			<3		300	50		
2002 KK_8					500	100		

Orbital parameters for radar-observed binary NEAs, including semimajor axis in meters, eccentricity, orbital period in days, and inferred total mass. Size and density estimates of the primary are also listed.

(3) range and Doppler separations as a function of time, which characterize the system's total mass and orbital parameters; and (4) reflex motion of the primary about the COM, which constrains the mass ratio of the system. Although the location of the COM is initially uncertain, the process of ephemeris refinement quickly leads to a very precise knowledge of its position in each image frame.

The bulk of the data analysis so far has concentrated on using the range and Doppler separations to fit for the system's total mass and orbital parameters. The model assumes that the orbital motion of the secondary takes place in a plane with an orientation that remains fixed in inertial space during the time of the observations. Such mass estimates, coupled with a detailed knowledge of the component volumes from shape-modeling techniques (*Hudson,* 1993), can lead to precise asteroid density measurements. The density values presented here rely on size estimates from visual inspection of the raw radar images and on the verifiable assumption that most of the system's mass belongs to the primary object.

The current best-fit orbital parameters along with the formal errors of the fit are presented in Table 3. All solutions have χ-squared values of ≃1. The best-fit mass and density estimates are also shown.

The binary systems observed with radar so far share similar characteristics. The primary components all appear roughly spheroidal and have spin periods near the breakup limit. The secondaries have diameters on the order of one-third the diameter of the primary. All radar-observed NEA binaries have satellites orbiting at a distance of a few primary radii. Their orbital periods are on the order of a day. Because the spin periods of the primary are typically a few hours, the systems observed to date cannot be mutually synchronous. The spin periods of the secondaries are indicative of spin-lock configurations, which is consistent with calculations of tidal despinning timescales (*Margot et al.,* 2002b).

The proportion of binary objects among radar-observed NEAs larger than 200 m is ~16% (*Margot et al.,* 2002b). This large proportion requires the formation of binaries to be frequent compared to the ~10-m.y. dynamical lifetime of NEAs. Radar observations show that binary NEAs have spheroidal primaries spinning near the breakup point for strengthless bodies, suggesting that the binaries formed by spinup and fission, probably as a result of tidal disruption during close planetary encounters (section 3.1).

The ability to determine the orientation of the orbital plane using radar depends critically on the plane-of-sky coverage. For 2000 DP_{107}, which had a sky motion of ~40° during radar observations, the orientation of the orbital plane can be constrained to within a 28° cone. In the case of 2000 UG_{11} and 1999 KW_4, with ~60° and ~110° of sky motion respectively, pole solutions are expected to be better constrained. For 2000 DP_{107} and 1999 KW_4, one cannot reject the hypothesis that the orbit is circular, but for 2000 UG_{11} that same hypothesis can be rejected at better than the 1% level.

Reflex motion of the primary is clearly observed in the radar datasets, providing the exciting prospect of measuring the densities of NEA satellites. Improved orbital fits will incorporate the residual motion of the primary with respect to the COM and will include the mass ratio of the system as an additional parameter.

Additional improvements are expected from shape reconstruction techniques (*Hudson,* 1993), in which a series of delay-Doppler images are inverted in a least-squares sense to provide a shape model. Given images with sufficient signal-to-noise ratio and orientation coverage, it is possible to infer shape and spin information for the satellites and to derive solid conclusions regarding spin-orbit resonances. Apart from possibly yielding clues on formation mechanisms, shape models will significantly decrease the uncertainties associated with size/volume estimates, and this will result in considerably lower error bars on the initial density measurements presented here.

The techniques for extracting information about binary systems from the radar data are still very much under active development. At this early stage, it appears that one weakness of the radar method lies in its inability to constrain unambiguously the orientation of the orbital plane, particularly when sky motion is limited. This is an intrinsic limitation of range and line-of-sight velocity measurements obtained without angular leverage. Observations over a range of aspect angles can overcome this ambiguity. The detection of occultations in the radar data or of occultations or eclipses from lightcurve observations can also place tight constraints on the inclination of the orbit. In general, a combination of radar and lightcurve observations will yield the best orbital determinations. The radar data may in turn help the interpretation of lightcurve profiles by distinguishing occultations from eclipses and primary from secondary events. Interesting synergies are therefore expected from the combination of the radar and lightcurve techniques. Because radar shadows are cast in much the same way as their optical counterparts, radar occultations of binary systems will be observed sooner or later, in which case the orientation of the orbital plane would be very tightly constrained.

Radar observations of binary asteroids constitute an emerging field that holds great promise for the future. The information that can be gathered from radar datasets includes determination of bulk properties (e.g., density, rigidity), and of orbital/spin characteristics, of both components. Combined with high-resolution imaging and shape models, these are providing powerful constraints on the formation mechanisms of binary NEAs. The characteristics of eccentricity and spin damping provide insightful clues about the internal structure of asteroids.

2.5. Binary Asteroids Detected by Lightcurves

Serious attempts to reveal the binary nature of some asteroids from their lightcurve features date back to the 1970s (cf. *Cellino et al.,* 1985). A review of the advantages and disadvantages of various methods of extracting such information from asteroid lightcurves is given by *Weidenschilling et al.* (1989). Recent advances in methods for interpretation of lightcurves can be found in *Kaasalainen et al.* (2002). While most techniques have not led to a successful detection of a binary asteroid, one of them, mentioned in the end of section IV.B of *Weidenschilling et al.* (1989), has been recently successful — the detection of nonsynchronous satellites.

Pravec (1995) analyzes a two-period lightcurve of the NEA 1994 AW_1, measured by *Mottola et al.* (1995) and *Pravec et al.* (1995), and interprets the complex lightcurve as being due to occultation/eclipse events in a binary asteroid system combined with a fast rotation of the primary. The results were published also in *Pravec and Hahn* (1997), who present the binary hypothesis as the likely explanation of the 1994 AW_1 lightcurve but also consider the possibility that it might be an asteroid in a complex rotation state. In light of more recent results (see below), the binary status of 1994 AW_1 is quite likely and we consider it to be the first binary asteroid detected by the lightcurve technique. See Table 4 for estimated parameters of this binary system.

The second binary asteroid found from lightcurve observations is 1991 VH (*Pravec et al.,* 1998a). Extensive photometric observations show that the asteroid's lightcurve is doubly periodic and that its long-period component shows occultation-like features; Pravec et al. interpret the data as evidence that 1991 VH is an asynchronous binary system,

TABLE 4. Estimated parameters of binary NEAs, detected by lightcurve.

Object	D_p (km)	D_s/D_p	a/R_p	e	P_{orb} (h)	P_{rot} (h)	A_{rot} (mag)	Taxonomic Class	Orbital Type	References
1994 AW_1	0.9	0.53	4.6	<0.05	22.40	2.5193	0.16		PHA	[1]
1991 VH	1.2	0.40	5.4	0.07	32.69	2.6238	0.11		PHA	[2]
(3671)	0.9	>0.28	5.2		27.72	2.7053	0.16	EM	PHA	[3,12,13]
1996 FG_3	1.4	0.31	3.4	0.05	16.14	3.5942	0.09	C	PHA, VC	[4,5]
(5407)	4.0	≥0.30	(3.4)	(<0.05)	(13.52)	2.5488	0.13	(S)	MC	[4]
1998 PG	0.9	≥0.30	(3.4)		(14.01)	2.5162	0.13	S	Amor	[4]
1999 HF_1	3.5	0.24	4.0		14.02	2.3191	0.13	EMP	Aten, VC	[6]
2000 DP_{107}	0.8	0.38	6.6	0.01	42.2	2.7755	0.22	C	PHA	[7,8,14]
2000 UG_{11}	0.23	≥0.6	3.6	0.12	18.4	(4.44)	0.10	QR	PHA	[13,14]
1999 KW_4	1.2	≥0.3	4.2	≤0.03	17.45	2.765	0.13	Q	PHA, VC	[9,10,13,14]
2001 SL_9	1.0	0.31	3.6		16.40	2.4003	0.09		Apollo	[15]

References: [1] *Pravec and Hahn* (1997); [2] *Pravec et al.* (1998a); [3] *Mottola et al.* (1997); [4] *Pravec et al.* (2000a); [5] *Mottola and Lahulla* (2000); [6] *Pravec et al.* (2002a); [7] *Margot et al.* (2002b); [8] *Pravec et al.* (2000b); [9] *Benner et al.* (2001a); [10] *Pravec and Šarounová* (2001); [11] *Harris and Davies* (1999); [12] P. Pravec et al. (unpublished data, 1997); [13] *Margot et al.* (2002a); [14] P. Pravec et al. (personal communication, 2002); [15] *Pravec et al.* (2001).

The diameter of the primary D_p was estimated from the effective diameter 1.0 km given by *Harris and Davies* (1999) for (3671), and from measured absolute magnitudes assuming the geometric albedo p = 0.06 for 1996 FG_3, and 2000 DP_{107}, and p = 0.16 for the other objects; it was corrected for $D_s/D_p = 0.4$ in cases where only a lower limit on the secondary-to-primary diameter ratio is available. a is the semimajor axis of the mutual orbit, e is its eccentricity, P_{orb} is the orbital period. P_{rot} is the rotation period of the primary, A_{rot} is its amplitude corrected for contribution of the light from the secondary. The values in brackets are derived using the assumptions discussed in *Pravec et al.* (2000a). PHA stands for potentially hazardous asteroid, which is an object approaching closer than 0.05 AU to the Earth's orbit, VC stands for Venus-crosser, MC stands for Mars-crosser. This table has been updated from *Pravec et al.* (2000a). For uncertainties and assumptions made with the estimates, see the original publications. Note that some of these objects are in common with NEAs observed by radar, in Table 3. An updated, combined radar/lightcurve NEA table is maintained at http://www.asu.cas.cz/~asteroid/binneas.htm.

similar to 1994 AW_1. The same or similar observational and analysis techniques have been used to reveal the binary nature of several other objects, shown in Table 4. The general technique has been validated by the radar detection of the binary status of 2000 DP_{107}, for which *Pravec et al.* (2000b) and P. Pravec et al. (personal communication, 2002) observe a two-period lightcurve of the same kind as in the previous cases and estimate parameters of the binary system that are in agreement with results from the radar observations.

This lightcurve technique for detecting binaries has been described in the above-mentioned papers as well as in more recent works by *Pravec et al.* (2000a) and *Mottola and Lahulla* (2000). Briefly, it is based on detecting brightness attenuations caused by mutual occultations or eclipses between components of the binary system superposed on the short-period rotational lightcurve of the primary. An example is shown in Fig. 9. The principles of the technique introduce several selection effects. The technique can reveal the existence of large satellites around asynchronously rotating primaries only under favorable geometric conditions. Another bias is that detection of close binary systems is favored, because observations and their interpretation are easier for systems with shorter orbital periods. Satellites smaller than ~20% of the primary diameter are difficult or impossible to detect unambiguously from lightcurve observations because they produce only small brightness attenuations during occultations or eclipses, less than ~0.04 mag. This may be difficult to separate from other effects, like an evolution of the primary's rotational lightcurve in changing observational geometric conditions. The asynchronous rotation of the primary allows one to resolve the occultation/eclipse events, which occur with a period different from the rotation period of the primary, and therefore one may rule out their possible connection with any peculiar shape feature of the primary. Occultations or eclipses can be observed only when the Earth or Sun, respectively, lie close enough to the mutual orbital plane of the binary system. These selection effects mean that there may be a bias toward binary systems with certain favorable parameters in the sample of known or suspected binary asteroids presented in Table 4. Nevertheless, at least some of the similarities of the characteristics of the binary asteroids cannot be explained by selection effects alone and must be real.

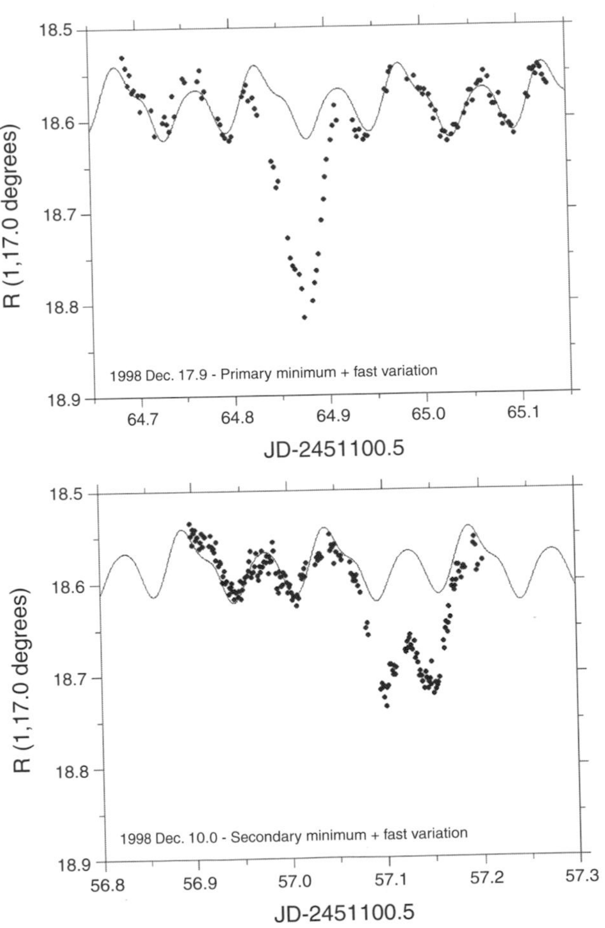

Fig. 9. Observed lightcurves of 1996 FG_3 show the fast-variation, small-amplitude component, caused by the rotation of the primary, with superposed sudden sharp attenuations caused by the eclipse/occultation of the primary by the secondary. The top panel shows the primary minimum, while the bottom panel shows the secondary minimum. The primary rotation component can be seen also during the attenuations. (From *Pravec et al.,* 2000a.)

The similarities of the 13 NEA binary asteroids, known or suspected from lightcurve or radar observations, are:

1. They are small objects with primary diameters 0.7–4.0 km. The lower limit may be due to a bias against detection of small binary systems, because fainter asteroids are normally more difficult to observe. There may exist an upper limit but it is difficult to estimate from the small sample.

2. They all are inner planet-crossers. Most of them approach the orbits of Earth and Venus. This feature may be due, at least partly, to a selection effect, as kilometer-sized asteroids are much easier to observe in near-Earth space than in the main belt. Another possible selection effect is that more observations are being made, in general, of near-Earth objects.

3. All the primaries are fast rotators (periods 2.3–3.6 h), not far below the critical stability spin rate, with low amplitudes (0.1–0.2 mag), suggesting nearly spheroidal shapes (see *Pravec et al.,* 2002b).

4. The secondary-to-primary diameter ratios are almost all in the range of 0.2–0.6. While the lower limit may be just a result of the selection effect mentioned above, it appears that binaries with nearly equal-sized components are rare among kilometer-sized NEAs. The probability that there are twelve objects with the diameter ratios in the range of 0.2–0.6 and one in 0.6–1.0, for a uniform distribution of diameter ratios, is less than 0.2%.

5. Semimajor axes estimates are in the range 3.4–6.6 primary radii. While the upper limit may be due to the selection effect mentioned above, the lower limit (corresponding to orbital periods ~14 h) may be real, and it suggests that very close binary systems are not present (perhaps due to their instabilities).

6. Eccentricities are poorly constrained but appear to be low (less than 0.1).

Pravec et al. (1999) accounted for the bias due to the selection effect related to the geometric observing conditions and estimated, on the basis of the first three known binary NEAs, that the fraction of binaries among NEAs is ≈17% with an uncertainty factor of 2. This is consistent with the estimates from radar data that ~16% of NEAs are binary (*Margot et al.,* 2002b), and the estimates (about 15%) of *Bottke and Melosh* (1996a,b) from models of binary production by tidal disruption (see section 3.1). Based on these studies, we adopt 16% as our working estimate of the NEA binary fraction. We note that ~30% of kilometer-sized asteroids are fast rotators with periods <4 h and that binary NEAs have fast-rotating primaries. Therefore, it may be that roughly half of the fast-rotating NEAs are binary (*Pravec and Harris,* 2000) and that binary asteroids are common among fast-rotating objects on Earth-approaching orbits.

2.6. Hubble Space Telescope (HST) Companion Searches

One of the major projects that *Zellner et al.* (1989) expected to be addressed by HST was the search for asteroid companions. The absence of atmospheric effects on HST images allows diffraction-limited operation over a very large field of view. The spherical aberration of the primary mirror did not stop the execution of an early attempt to survey the asteroid belt (program 4521) as well as an "amateur" program that targeted asteroids thought to have companions, primarily from occultation observations (program 4764). No companions were found but careful restoration of the data was necessary to minimize the effects of the aberration. While aberration did not limit the spatial resolution of the images (the middle two-thirds of the primary was ground correctly), the additional "skirt" of scattered light did limit the dynamic range over which a companion could be detected.

Storrs et al. (1999a) published the data from these two programs. Their reconstruction of the HST images allowed upper limits to be put on the presence of companions to asteroids (9) Metis, (18) Melpomene, (19) Fortuna, (109) Felicitas, (146) Lucina, (216) Kleopatra, (434) Hungaria, (532) Herculina, (624) Hektor, and (674) Rachele. No companions were found to a brightness limit that varied with distance from the primary, as shown in Fig. 10. Barring the companion being in conjunction at the time of observation, Storrs et al. rule out companion objects (suggested by early occultation observations) to asteroids (9) Metis, (18) Melpomene, and (532) Herculina (the brightness and separation of suggested companions are designated by the numbers in Fig. 10).

Program 6559 (*Storrs et al.,* 1998, 1999b) detected no companions to the eight asteroids imaged by HST with the corrected Wide Field Planetary Camera 2 (WFPC2) instrument. Further HST imaging observations are currently under way in program 8583, which is a "snapshot" program designed to fill in gaps in the spacecraft's calendar of observations. The program targets 50 large, main belt asteroids (many of them twice) with the WFPC2 in a manner simi-

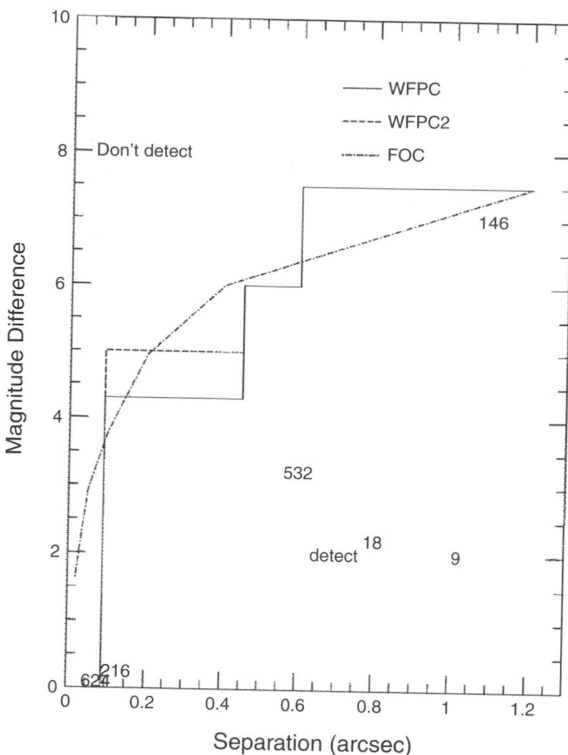

Fig. 10. Brightness difference (in magnitudes) between a primary asteroid and a possible companion as a function of projected distance from the primary asteroid, for well-exposed HST images (after *Storrs et al.,* 1999a). The region below the curves is where companions could be detected. Also shown are the locations of putative binaries (given by asteroid number) previously suspected from occultation or other data.

lar to that used to map (4) Vesta by *Binzel et al.* (1997). This program resulted in the discovery of a companion to (107) Camilla (*Storrs et al.,* 2001a) and confirmed observations of companions to (87) Sylvia (*Storrs et al.,* 2001b) and (45) Eugenia. The companions to (45) Eugenia and (107) Camilla have the same color in the visible range as their primaries. *Storrs et al.* (2001b) report that the companion to (87) Sylvia appears significantly bluer than its primary. The observations of (6) Hebe in this program reveal no companions brighter than 7 mag fainter than the primary, or larger than 8 km in diameter.

Another program for observing main-belt asteroids, that of Zappalà and colleagues, used the HST Fine Guidance Sensor (FGS). The first results of this program confirmed that (216) Kleopatra is a contact binary (*Tanga et al.,* 2001). Two other programs are under way, both of them targeting TNOs; both began to detect binaries in early 2002. A large program by M. Brown has detected two TNO companions: 1999 TC_{36} (*Trujillo and Brown,* 2002) and 1998 SM_{165} (*Brown and Trujillo,* 2002). In a second program, two more binaries have been found: 1997 CQ_{29} (*Noll et al.,* 2002a) and 2000 CF_{105} (*Noll et al.,* 2002b). As in the case of the other known TNO binaries, these objects have a wide separation and relatively large secondaries.

The strengths and weaknesses of HST/WFPC2 observations of asteroids are discussed in *Dotto et al.* (2002). Briefly, WFPC2 observations allow diffraction-limited observation over a large field of view from the vacuum UV to beyond 1-μm wavelength. These high-resolution images can provide information on the shape and mineralogical variegation of the primary as well. Drawbacks include the robotic nature of HST scheduling (ephemerides good to better than 10 arcsec for over a year are necessary to find the asteroid), no sensitivity beyond 1 μm [but see *Dotto et al.* (2002) for a discussion of WF3, which will operate to 1.8 μm], and the difficulty in getting observing time on HST (no immediate follow up of detections). HST observations are complementary to groundbased AO observations because they cover a larger field of view per exposure at a shorter wavelength, but cannot cover the critical near-IR wavelength region.

2.7. Role of Occultations

Described as a technique of searching for asteroid satellites by *Van Flandern et al.* (1979) in *Asteroids*, the method of using stellar occultations suffers from the inability to plan or repeat an experiment, at least reliably. *Reitsema* (1979) has called into question many of the early reports of satellites, indicating that the measurements are susceptible to spurious events. One-time reports of occultations can serve only to alert more rigorous search methods of a potential candidate. In addition, once an asteroid is known to have a moon, systematic networks of observers may be placed as to attempt to see an event from the moon during an occultation of the primary. These observations could greatly constrain our understanding of the sizes and positions of the satellites.

It is important to note, however, that archived occultation records (D. Dunham, personal communication, 2001) have shown that two short events have been recorded accompanying an occultation of Eugenia (diameter 215 km). One was in 1983 (chord equivalent ~9 km) and another in 1994 (chord equivalent ~20 km). Another short event, of chord size 18 km, was recorded in 1997 during an occultation of Sylvia (diameter 271 km). The satellite diameters predicted from AO observations are 13 km for Eugenia and 13 km for Sylvia. It is unlikely that such short chords would have resulted from asteroids of this large size. Therefore, it is possible that these occultations in fact did record satellite events.

3. ORIGIN AND EVOLUTION OF BINARY ASTEROIDS

In *Asteroids II*, *Weidenschilling et al.* (1989) discussed the most promising mechanisms for formation of asteroid binaries. Most of the progress since that time has been observational, but theoretical efforts, especially numerical modeling, have also made advances. With the new examples of actual binary systems to study, there has recently been a renewed interest in theories of formation and in numerical modeling of binary origin. All of the formation mechanisms discussed by Weidenschilling et al. remain viable. Here we revisit these and add others.

3.1. Near-Earth Asteroids: Tidal Encounters

As discussed in sections 2.4 and 2.5, a significant fraction (16%) of NEAs appear to be binary. This is much higher than their apparent abundance in the main belt (although detection is more difficult for the latter), but is consistent with the fraction of recognized doublet craters in impacts on Earth (*Weidenschilling et al.,* 1989). Apparently, some mechanism favors production of binaries among planet-crossers (unless it is possible to get small main-belt binaries to be ejected from the belt intact). A close planetary encounter can subject an asteroid to tidal stresses and torques that may produce a binary. The same process, however, can also disrupt existing binary systems. Because the lifetime of NEAs is relatively short (a few times 10^7 yr) and close encounters are more probable than planetary impacts, this formation/destruction is an equilibrium process. *Bottke and Melosh* (1996a,b) examine the effect of planetary encounters on contact binaries (two components) and conclude that ~15% of Earth-crossers evolve into coorbiting binaries. *Richardson et al.* (1998) and *Bottke et al.* (1999), model the tidal disruption of ellipsoidal rubble-pile asteroids (composed of many small, equal-sized particles) encountering Earth and find that rotational spin-up frequently cause them to undergo mass shedding. In many cases, some of the shed fragments go into orbit around the progenitor, producing binary asteroids. Most of these satellites, however, are much smaller than the primary. Also, the yield of binaries is low; disruption into a string of clumps, as for comet Shoemaker-Levy 9, is more probable than binary formation. The results of these studies suggest that tidal disruption can produce enough binaries to account for the observed population of doublet craters on the terrestrial planets, provided that small asteroids (less than a few kilometers in diameter) are not finely divided gravel piles, but "coarse" structures dominated by a few large chunks. This inference is also consistent with their observed maximum rotation rates (cf. *Paolicchi et al.,* 2002).

3.2. Cratering Ejecta

A cratering event from a subcatastrophic impact on an asteroid produces ejecta with a range of velocities. It is therefore likely that some of the ejecta will have sufficient kinetic energy and angular momentum to go into orbit about the target body. Except in highly oblique impacts, the ejecta leave the crater with a more or less uniform azimuthal distribution as seen in the frame of the target's surface. If the target is rotating, the rotational velocity of the surface at the impact point is added to the ejecta velocity; therefore, more mass will attain orbital velocity in the prograde direction (we assume that the impact is not large enough to make a significant change in the target's rotational state). The problem with this model is how to place the ejecta into stable orbits. If the target is a sphere with a purely radial gravity field, then the ejecta particles have elliptical orbits that would intersect the surface after one revolution. Collisions between fragments, as well as solar perturbations acting on particles with highly eccentric orbits, might prevent immediate reimpact, but these apparently inefficient mechanisms would have to act during the first orbit after the impact. However, many asteroids are significantly nonspherical (triaxial) in shape and usually rotate about their shortest axis. This means that ejecta particles experience a noncentral gravity field, which can significantly alter their orbital parameters on the timescale of a single orbit. Also, a particle launched from a point near the longer equatorial axis may encounter a shorter axis during its first few periapse passages, avoiding impact and prolonging its lifetime. Mutual collisions among fragments during the first few orbits can dampen their eccentricities, yielding orbits that no longer intersect the primary's surface. This material could then accrete into a small satellite. As pointed out by *Weidenschilling et al.* (1989), ejecta velocities must be within the limited range that allows material to go into orbit about the primary without escaping completely. Such orbits have specific angular momentum corresponding to circularized orbits within a distance of about 2 radii from the primary. Unless this distance is outside the synchronous point, any satellite that accreted in this manner would be subject to tidal decay and would eventually collide with the primary. The requirement that the synchronous distance lies within 2 radii implies a spin period of not more than ~6 h. Tidal torque would then cause the satellite to migrate outward; for small secondary/primary mass ratios, the primary's spin would not be slowed significantly. Thus, satellites formed by this mechanism would be small rubble piles in prograde orbits about rapidly rotating primaries. In addition to these criteria listed by Weidenschilling et al., we add the requirement that the primaries be significantly nonspherical.

In a preliminary numerical study to explore the viability of this mechanism for producing small satellites, *Durda and Geissler* (1996) examined the accretion of ejecta particles from three different 10-km-scale craters on Ida. In each case they followed the dynamical evolution of 1000 ejecta particles for 100 h after the cratering impact and searched for "collisions" between orbiting particles, treating each "collision" as an accretion event. That study found that temporary aggregates containing ~0.1% of the ejected debris mass did indeed form while in flight around the primary, but none of these aggregates occupied stable orbits and survived [although the temporary aggregates were primarily on prograde trajectories concentrated near the equatorial plane of Ida, as predicted by *Weidenschilling et al.* (1989)]. The failure of the model to yield small satellites via accretion of ejected cratering debris may not be evidence that this mechanism fails to work or is incredibly inefficient, but instead may be a result of the approximations inherent to the model (the Dactyl-forming impact may also have been larger than modeled). Indeed, several processes that subsequently have been shown to play important roles in placing material into bound orbits (e.g., distortion of the primary's shape, vaporization of some fraction of material, impact angle) were not included in the modeling. Instead, the Durda and Geissler model, which has proved quite successful in explaining the distribution of ejecta on Ida's surface (*Geissler et al.,* 1996), simulated the ejection of crater debris from various locations on Ida by launching particles from a point at a 45° angle to the local surface. The particles were all launched at the same instant at the beginning of the simulations, with no momentum transfer to the asteroid. In reality, excavation flows encompass the entire center-to-rim extent of a crater, the timescale for crater excavation on a low-gravity object can approach a significant fraction of the asteroid rotation period, and translational and rotational momentum is imparted to the primary during the impact (e.g., *Asphaug et al.,* 1996; *Love and Ahrens,* 1997). Thus, a combination of shape/distortion effects and translational/rotational motion during the excavation phase may play an important role in allowing particles to remain in temporary orbit.

This mechanism would operate in the environment of high-velocity impacts in the present main belt. Impacts are also capable of destroying small satellites, which would have shorter lifetimes against disruption than their primaries (although they might reaccrete after such events if the fragments remain in orbit). Thus, we expect the population of such binaries to be in equilibrium between formation and destruction by impacts.

Of the main-belt asteroids known to be binaries, six of eight (22, 45, 87, 107, 243, and 762) have satellites much smaller than their primaries. Assuming equal albedos and densities for both components, the mass ratio is typically $\sim 10^{-3}$. Significantly, all the primaries are rapid rotators; the longest period is 5.84 h for (762) (*Davis,* 2001). Also, they have rather large amplitude lightcurves, with maximum observed amplitudes of at least 0.25 mag. These properties are consistent with the formation of their satellites from impact ejecta. If the direction of an orbit relative to the rotation of the primary is found to be prograde, this would be a strong indication of their origin by this mechanism. The sense of the orbit is known for three of these main-belt binaries. The moons of (243) Ida (*Belton et al.,* 1995, 1996), (45) Eugenia (*Merline et al.,* 1999b,c), and (22) Kalliope orbit in a prograde sense.

3.3. Disruptive Capture

Many asteroids belong to dynamical families that reveal them to be fragments of larger parent bodies that were disrupted by catastrophic collisions. In such a disruptive event, fragments may end up in orbit about each other, as suggested by *Hartmann* (1979). *Weidenschilling et al.* (1989) point out that in a radial-velocity field of fragments escaping from a disrupted primary, geometrical constraints imposed by the finite sizes of fragments would tend to ensure that

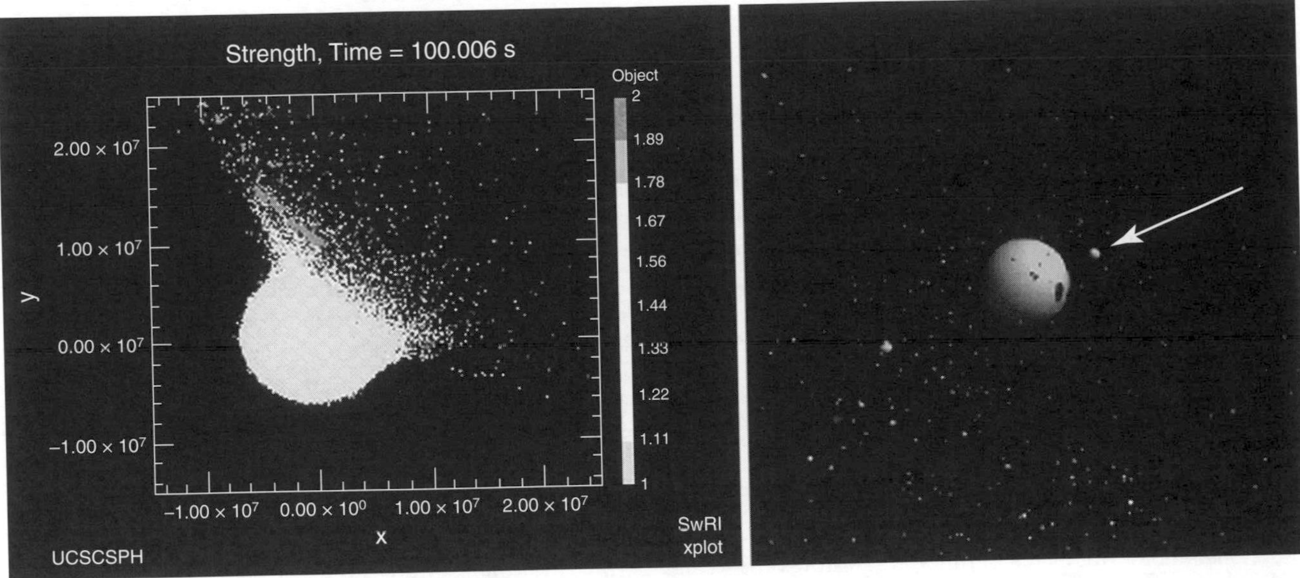

Fig. 11. The "next generation" of numerical models of asteroid satellite formation substantially improve upon past models by (left) conducting detailed three-dimensional smooth-particle hydrodynamics (SPH) models of collisions between asteroids and then (right) following the subsequent dynamics of ejected debris and formation of orbiting satellites (arrow) through fast, state-of-the-art N-body simulations. Shown here is a collision of a 20-km impactor into a 100-km solid basaltic target, as simulated by *Durda et al.* (2001). The ~4-km satellite is captured into an elliptical orbit with a separation of about 6 $R_{primary}$. The final primary diameter is ~75 km.

they would have relative velocities exceeding their mutual escape velocity and in general would not remain gravitationally bound.

This problem was examined in some detail by *Durda* (1996) and by *Doressoundiram et al.* (1997), who simulated disruptions numerically, integrating orbits of fragments in the debris field. They found that the fraction of binaries depends on the magnitude of a random velocity dispersion assumed to be imposed on the general expansion; however, even with no dispersion some binaries were produced, apparently by jostling among fragments. More pairs of fragments in contact were produced than orbiting binaries. The fraction of contact pairs and binaries was small in Durda's models (~0.1%), while the fraction of binaries found by Doressoundiram et al. was ~1%. The limited range of sizes and numbers of particles in the simulations probably limited the binary fraction. Treating larger numbers of smaller fragments would be expected to yield more binaries with smaller satellite:primary mass ratios.

The early, simple numerical models of this mode of satellite formation contained some critical limitations, however. Because the initial conditions simulating the expansion phase following a catastrophic impact were merely treated in a simple empirical fashion, a self-consistent description of the mass-speed distribution of fragments and the direction of fragment ejection was not possible. Variations in these collision outcomes, and therefore in the efficiency of binary-pair formation, with initial conditions, could not be examined in these initial studies. The next generation of numerical models (*Michel et al.*, 2001; *Durda et al.*, 2001) substantially improve upon the limitations of the *Durda* (1996) and *Doressoundiram et al.* (1997) models by conducting detailed three-dimensional smooth-particle hydrodynamics (SPH) models of catastrophic collisions between asteroids (e.g., *Benz and Asphaug,* 1995; *Asphaug et al.*, 1998) and then following the subsequent dynamics of the ejected fragments through fast, state-of-the-art N-body simulations (such as described in *Leinhardt et al.*, 2000).

One of the most important benefits of this scheme over the previous numerical studies is that it includes a rigorous treatment of the impact physics, so that accurate fragment size distributions and velocity fields are established early in the ejection process. Thus, the dependence of satellite formation efficiency with respect to various collision parameters (e.g., speed, impact parameter, impact angle) can be studied in a self-consistent manner. These new models also allow for a far-faster N-body integration scheme with efficient mutual capture and collision detection capabilities. A sample model can be seen in Fig. 11.

Four of the known main-belt binaries (45, 90, 243, and 3749) are members of dynamical families, so this mechanism is plausible (possibly, the fraction of binaries in families is greater than for the general population). There should be no initial preference for rapid rotation of primaries or prograde orbits, but tidal dissipation could cause loss of satellites of slow rotators or in retrograde orbits. We would expect no correlation with the primary's shape, so lightcurves may discriminate between cratering ejecta and disruptive capture.

3.4. Collisional Fission

An impact may shatter an asteroid without disrupting it. As the probability of an exactly central collision is zero, it will also impart angular momentum to the target. If the specific angular momentum exceeds a threshold value, a weak (shattered) self-gravitating body cannot remain single but must fission into a binary, with some of the angular momentum in orbital motion rather than rotation. The angular momentum imparted is proportional to the impact velocity v, while the impact energy scales as v^2. As discussed by *Weidenschilling et al.* (1989), it is difficult to impart enough angular momentum without destroying the target at typical impact velocities (~5 km s^{-1}) in the present belt (although there is a distribution of velocities over a wide range, but at lower impact probabilities). If gravitational binding dominates, then for impacts large enough to impart the critical angular momentum, the ratio of impact energy to binding energy is of order v_{impact}/v_{escape}. For even the largest asteroids, disruption is more likely than rotational fission in the present collisional environment. Conditions were presumably more favorable in the earliest stage of the belt's evolution, before velocities were pumped up; however, only large satellites would have been able to survive its later collisional history. No convincing candidate systems have yet been found in the main belt.

The masses and relatively large separation (~4 radii) of the main-belt double (90) Antiope imaged by *Merline et al.* (2000a,b) mean that this pair has unusually high specific angular momentum. The lightcurve eclipses recorded by *Hansen et al.* (1997) are consistent with the nearly equal-sized components seen in the images. At other times, the lightcurve had a low amplitude consistent with nearly spherical, noneclipsing components (actually, Darwin ellipsoids are an even better match). *Merline et al.* (2000a,b) inferred a density of ~1.3 g cm^{-3}, which suggests that the Antiope components may be "rubble piles" with equilibrium shapes. Such models of equilibrium binaries and the expected lightcurve morphologies were studied by *Leone et al.* (1984). The origin of the Antiope binary is hard to explain. It is a member of the Themis family and so must postdate the disruption of its parent body by a high-velocity impact. Disruptive capture of two equal-mass fragments of such large size in that event is unlikely, and they would have to be converted to rubble piles by later impacts. However, some of the model runs of P. Michel (personal communication, 2001) appear to produce similar-sized components. Collisional fission seems to be the most likely origin for Antiope, but still presents the problem of imparting so much angular momentum in a collision without dispersing the target. Due to the low orbital inclination of the Themis family, collisions between members have a lower mean velocity [~3 km s^{-1} (*Bottke et al.,* 1994)] than between field asteroids (~5 km s^{-1}) but this difference is not very significant. *Weidenschilling et al.* (2001) estimate that the required angular momentum implies an impactor of diameter ~20 km on a 100-km target body, with about 100 times its gravitational binding energy, at the mean encounter velocity. An impact by a larger body at much lower velocity is improbable, even if the Themis family is several billion years old. Low-velocity impacts could have occurred in the immediate aftermath of the disruption of the Themis family's parent body, before Jovian perturbations randomized the nodes and apsides of the fragments. Models by *Dell'Oro et al.* (2002) show an enhancement in the impact probabilities of several orders of magnitude initially after breakup. However, the time available before randomization is short (~10^4 orbital periods), and a collision between two fragments of sufficient size is unlikely. In either scenario, the probability of forming a binary with these properties is only ~10^{-3}, and thus Antiope should be unique in the main belt.

3.5. Primordial Binaries?

Other binaries with components of comparable mass and large separations have been discovered, but at larger heliocentric distances. The Trojan asteroid (617) Patroclus (*Merline et al.,* 2001b) and at least two of the TNO binaries (1998 WW_{31}, 2001 QW_{322}) have size ratios close to one. All have significantly greater separations than Antiope: ~600 km (~12 radii) for Patroclus and 10^4–10^5 km (~10^2–10^3 radii) for the TNOs. In one sense, these properties are not surprising, because detection of smaller and/or closer satellites of such distant objects is impossible with current technology. However, it is unclear how such loosely bound pairs could have formed. If the Patroclus binary formed by a collision, it would have required more extreme parameters (larger impactor and/or lower velocity) than Antiope's formation. The collision rate in the Trojan clouds is somewhat higher than in the main belt (see *Davis et al.,* 2002), while the mean impact velocity is comparable (lower orbital velocity is offset by higher mean inclination). However, a binary of this size would have a collisional lifetime greater than the age of the solar system. It is plausible, therefore, that the Patroclus binary formed by a low-velocity collision before eccentricities and inclinations were pumped up, perhaps before its capture into resonance with Jupiter.

The frequency of transneptunian binaries appears to be ~1%. Their large separations could not have been produced by two-body collisions or tidal evolution. The most plausible origin for such a loosely bound binary seems to be an impact with another body of comparable mass while the two components passed within their mutual Hill radius. The present spatial density in the Kuiper Belt is far too low for three-body encounters; any such events must have occurred when it was more populous and/or dynamically "cold" with low inclinations. Dynamical modeling is needed to determine the efficiency of binary production by this mechanism as a function of population density and orbital parameters. Alternatively, these binaries may represent objects that formed as loosely bound pairs from inherent disk instabilities during accretion (S. A. Stern, personal communication, 2001). Observations of binary TNOs will eventually allow direct determination of their masses and densities, but may also provide a constraint on the formation and early history of the Kuiper Belt.

3.6. Tidal Evolution of Spins and Orbits

Weidenschilling et al. (1989) consider the tidal evolution of orbits of asteroid satellites. Their Fig. 1 shows the timescale for a hypothetical satellite to evolve outward from an orbit initially close to a primary of radius R = 100 km, as a function of the satellite:primary mass ratio. There are now enough data for real binaries to compare this model with observation. Most of the known main-belt binaries have separations a/R ~ 10, and M/m ~ 3 × 10^2–10^4 (Table 2); the inferred tidal evolution timescales are in the range ~10^8–10^9 yr. These values depend on the mechanical properties of the primaries, which are uncertain, but are consistent with collisional production of close binaries and tidal expansion of their orbits to their present distances since the formation of the asteroid belt. All such satellites lie below the line of synchronous stability, with orbits that are still evolving outward (consistent with the observation that their primaries have rotation periods shorter than their orbital periods). The NEAs typically have smaller separations with a/R ~ 5, and smaller M/m ~ 10^1–2 × 10^2. However, they are much smaller than the main-belt binaries, with R ~ 1 km; since the rate of tidal evolution of orbits scales as R^2, they also have timescales ~10^9 yr, consistent with the observation that they have not evolved to a synchronous end state. The binaries with relatively close massive satellites have much shorter evolution times; extrapolating from Weidenschilling et al.'s Fig. 1, (90) Antiope would have reached its tidally locked end state in only a few thousand years, and (617) Patroclus in less than 10^6 yr. However, it can be seen from that figure that Patroclus has too much angular momentum to have evolved by despinning of an initially close binary. This system, and the Kuiper Belt binaries with comparable mass ratios and still larger separations, must have attained their present orbital configurations by a mechanism other than tidal despinning.

The timescale for despinning of a satellite's rotation by tides is generally shorter than that for evolution of its orbit by despinning of the primary. Using the classic formula for the rate of despinning (*Goldreich and Soter,* 1966), the smaller main-belt and NEA satellites have despinning times of ~10^6–10^7 yr, so they would be expected to keep one face toward their primary. The only observational datum for rotation of a main-belt satellite is from the *Galileo* flyby of Ida/Dactyl, which shows that Dactyl had slow rotation, consistent with spin-orbit synchronicity (*Veverka et al.,* 1996b). On the other hand, the known Kuiper Belt binaries have such large separations that their tidal despinning times probably exceed the age of the solar system; they are unlikely to be in synchronous rotation.

Finally, *Harris* (2002) has suggested that the gravitational ejection of a satellite from orbit around an irregularly shaped primary would deplete the rotational energy of the primary, thus slowing substantially the spin of the primary. This ultimately may be shown to be the cause of the anomalously slow rotation of many asteroids, which so far have eluded satisfactory explanation.

3.7. Triple and Multiple Systems

Little work has been done specifically on the formation and stability of triple or multiple asteroid systems. Perhaps the closest analogs are those studies of the stability of satellites around a nonspherical primary (e.g., *Scheeres,* 1994; *Petit et al.,* 1997). Significant progress, however, has been made in the understanding of triple- or multiple-star systems. Many of these results can be applied directly to asteroids for insight into what characteristics might be expected for multiple-asteroid systems. It is generally accepted that the masses would be configured in a hierarchical fashion (cf. *Eggleton and Kiseleva,* 1995). This would involve the superposition of two binary systems: an inner massive object orbited by a satellite and a moon of the satellite (like Sun/Earth/Moon) or a close binary system with a tertiary object in a wide orbit about the central pair. The ratio of the semimajor axes of the two relevant "binaries" must be ~3–4 to be stable (*Harrington,* 1977a,b). For eccentric orbits, the ratio of the periapse of the outer orbit to the inner semimajor axis is the relevant parameter. Eccentric orbits are therefore less stable (*Eggleton and Kiseleva,* 1995; *Kiseleva et al.,* 1994). In addition, the stability depends, in a complicated way, on the mass ratios of the objects (*Black,* 1982). Systems that have the two orbits counter-revolving (retrograde) also display greater stability than if the orbits are both in the same sense (*Harrington,* 1977b). Recent work on evolution of triple systems (*Miller and Hamilton,* 2002) emphasizes the importance of Kozai resonances in stability and indicates a strong preference that the orbits be approximately coplanar. Multiple systems would be formed in successively higher levels of hierarchy as discussed by *Harrington* (1977b).

Unlike triple-star systems, which can form by gravitational capture, (e.g., during the collision of two binary systems), such a formation mechanism would be difficult for asteroids because of the high encounter velocities relative to the orbital speeds (P. Hut, personal communication, 2002). The initial formation of triple/multiple systems were indicated, however, in the early numerical models of *Durda* (1996) and *Doressoundiram et al.* (1997) and are clearly produced by the next-generation models of *Michel et al.* (2001) and *Durda et al.* (2001). These SPH/N-body models of satellite formation show that in addition to producing binary systems with a single satellite in orbit about a primary asteroid, catastrophic disruption events can result (at least initially) in more complex, hierarchical systems with satellites of satellites. The gravitational reaccumulation of clumps of debris in the ejecta field around the largest remnant often leads to Shoemaker-Levy/9-like "strings-of-pearls." Many of these reaccumulating rubble-pile fragments, some of which are gravitationally bound in initially stable orbits around the largest remnant, are themselves surrounded by swarms of smaller orbiting debris. The simulation timescales are too short, thus far, to directly examine the longer-term stability of these hierarchical satellite systems.

4. SUMMARY

The question posed in the title to the Weidenschilling et al. chapter in *Asteroids II*, "Do Asteroids Have Satellites?," has now been answered. Now that we have many examples of binary systems for study, representing diverse collisional and dynamical populations, we may be at the threshold of a revolution in asteroid science. In the *next* decade, we can expect to learn a great deal from the ever-increasing pace of discovery, involving several rapidly improving, complementary techniques, as well as the concomitant numerical modeling and theoretical thinking about how these systems were formed, how they evolve, and what clues they hold to the history of the solar system. These binary systems will provide probes of asteroid interiors and perhaps will eventually allow definitive coupling of asteroid taxonomic type with the meteorite inventory. In fact, they may tell us about asteroid material for which it is unlikely we currently have representation among the meteorites, such as very low-density carbonaceous material that may not survive passage through Earth's atmosphere or primitive material of the outer main-belt, Trojan, or TNO regions. Research in this area will lead to spinoffs in related areas, including improvements to our understanding of the formation of the Earth/Moon or Pluto/Charon systems, dynamics and collisional physics, and assist in the mitigation of impact hazards to Earth.

REFERENCES

André C. (1901) Sur le système formé par la planète double (433) Eros. *Astron. Nachr., 155,* 27–30.

Asphaug E., Moore J. M., Morrison D., Benz W., Nolan M. C., and Sullivan R. J. (1996) Mechanical and geological effects of impact cratering on Ida. *Icarus, 120,* 158–184.

Asphaug E., Ostro S. J., Hudson R. S., Scheeres D. J., and Benz W. (1998) Disruption of kilometre-sized asteroids by energetic collisions. *Nature, 393,* 437–440.

Belton M. and Carlson R. (1994) *1993 (243) 1.* IAU Circular No. 5948.

Belton M. J. S., Veverka J., Thomas P., Helfenstein P., Simonelli D., Chapman C., Davies M. E., Greeley R., Greenberg R., Head J., Murchie S., Klaasen K., Johnson T. V., McEwen A., Morrison D., Neukum G., Fanale F., Anger C., Carr M., and Pilcher C. (1992) *Galileo* encounter with 951 Gaspra: First pictures of an asteroid. *Science, 257,* 1647–1652.

Belton M. J. S., Chapman C. R., Thomas P. C., Davies M. E., Greenberg R., Klaasen K., Byrnes D., D'Amario L., Synnott S., Johnson T. V., McEwen A., Merline W. J., Davis D., Petit J-M., Storrs A., Veverka J., and Zellner B. (1995) Bulk density of asteroid 243 Ida from the orbit of its satellite Dactyl. *Nature, 374,* 785–788.

Belton M. J. S., Mueller B. E. A., D'Amario L., Byrnes D. V., Klaasen K. P., Synnott S., Breneman H., Johnson T. V., Thomas P. C., Veverka J., Harch A. C., Davies M. E., Merline W. J., Chapman C. R., Davis D. R., Denk T., Petit J-M., Greenberg R., Storrs A., and Zellner B. (1996) The discovery and orbit of 1993 (243) 1 Dactyl. *Icarus, 120,* 185–199.

Benner L. A. M., Ostro S. J., Giorgini J. D., Jurgens R. F., Margot J.-L., and Nolan M. C. (2001a) *1999 KW_4.* IAU Circular No. 7632.

Benner L. A. M., Nolan M. C., Ostro S. J., Giorgini J. D., and Margot J.-L. (2001b) *1998 ST_{27}.* IAU Circular No. 7730.

Benz W. and Asphaug E. (1995) Simulations of brittle solids using smooth particle hydrodynamics. *Comput. Phys. Commun., 87,* 253–265.

Binzel R. P. (1998) Collisional evolution in the Eros and Koronis asteroid families: Observational and numerical results. *Icarus, 73,* 303–313.

Binzel R. P., Gaffey M. J., Thomas P. C., Zellner B. H., Storrs A. D., and Wells E. N. (1997) Geologic mapping of Vesta from 1994 Hubble Space Telescope images. *Icarus, 128,* 95–103.

Black D. C. (1982) A simple criterion for determining the dynamical stability of three-body systems. *Astron. J., 87,* 1333–1337.

Bottke W. F. Jr. and Melosh H. J. (1996a) The formation of asteroid satellites and doublet craters by planetary tidal forces. *Nature, 381,* 51–53.

Bottke W. F. Jr. and Melosh H. J. (1996b) The formation of binary asteroids and doublet craters. *Icarus, 124,* 372–391.

Bottke W. F., Nolan M., and Greenberg R. (1994) Velocity distributions among colliding asteroids. *Icarus, 107,* 255–268.

Bottke W. F. Jr., Richardson D. C., Michel P., and Love S. G. (1999) 1620 Geographos and 433 Eros: Shaped by planetary tides? *Astron. J., 117,* 1921–1928.

Britt D. T. and Consolmagno G. J. (2000) The porosity of dark meteorites and the structure of low-albedo asteroids. *Icarus, 146,* 213–219.

Britt D. T. and Consolmagno G. J. (2001) Modeling the structure of high porosity asteroids. *Icarus, 152,* 134–139.

Brown M. E. and Margot J.-L. (2001) *S/2001 (87) 1.* IAU Circular No. 7588.

Brown M. E. and Trujillo C. A. (2002) (26308) *1998 SM_{165}.* IAU Circular No. 7807.

Cellino A., Pannunzio R., Zappalà V., Farinella P., and Paolicchi P. (1985) Do we observe light curves of binary asteroids? *Astron. Astrophys., 144,* 355–362.

Chapman C. R. (1996) S-type asteroids, ordinary chondrites, and space weathering: The evidence from *Galileo*'s fly-bys of Gaspra and Ida. *Meteoritics & Planet Sci., 31,* 699–726.

Chapman C. R. (2002) Cratering on asteroids from *Galileo* and *NEAR Shoemaker.* In *Asteroids III* (W. F. Bottke Jr. et al., eds.), this volume. Univ. of Arizona, Tucson.

Chapman C. R., Veverka J., Thomas P. C., Klaasen K., Belton M. J. S., Harch A., McEwen A., Johnson T. V., Helfenstein P., Davies M. E., Merline W. J., and Denk T. (1995) Discovery and physical properties of Dactyl, a satellite of asteroid 243 Ida. *Nature, 374,* 783–785.

Chapman C. R., Ryan E. V., Merline W. J., Neukum G., Wagner R., Thomas P. C., Veverka J., and Sullivan R. J. (1996) Cratering on Ida. *Icarus, 120,* 77–86.

Chauvineau B. and Mignard F. (1990) Dynamics binary asteroids II. Jovian perturbations. *Icarus, 87,* 377–390.

Chauvineau B., Farinella P., and Mignard F. (1993) Planar orbits about a triaxial body: Applications to asteroidal satellites. *Icarus, 105,* 370–384.

Cheng A. F. (2002) Near-Earth asteroid rendezvous: Mission summary. In *Asteroids III* (W. F. Bottke Jr. et al., eds.), this volume. Univ. of Arizona, Tucson.

Close L. M., Merline W. J., Tholen D. J., Owen T. C., Roddier F., and Dumas C. (2000) Adaptive optics imaging of Pluto-Charon and the discovery of a moon around the asteroid 45 Eugenia: The potential of adaptive optics in planetary astronomy. *Proc. SPIE 4007,* pp. 787–795.

Davis D. R., Chapman C. R., Durda D. D., Farinella P., and Marzari F. (1996) The formation and collisional/dynamical evolution of the Ida/Dactyl system as part of the Koronis family. *Icarus, 120,* 220–230.

Davis D. R., Durda D. D., Marzari F., Campo Bagatin A., and Gil Hutton R. (2002) Collisional evolution of small-body populations. In *Asteroids III* (W. F. Bottke Jr. et al., eds.), this volume. Univ. of Arizona, Tucson.

Davis R. G. (2001) High precision lightcurves for 762 Pulcova. *Minor Planet Bull., 28,* 10–12.

Dell'Oro A., Paolicchi P., Cellino A., and Zappalà V. (2002) Collisional rates within newly formed asteroid families. *Icarus, 156,* 191–201.

Doressoundiram A., Paolicchi P., Verlicchi A., and Cellino A. (1997) The formation of binary asteroids as outcomes of catastrophic collisions. *Planet. Space. Sci., 45,* 757–770.

Dotto E., Barucci M. A., Müller T. G., Storrs A. D., and Tanga P. (2002) Observations from orbiting platforms. In *Asteroids III* (W. F. Bottke Jr. et al., eds.), this volume. Univ. of Arizona, Tucson.

Durda D. D. (1996) The formation of asteroidal satellites in catastrophic collisions. *Icarus, 120,* 212–219.

Durda D. D. and Geissler P. (1996) The formation of asteroidal satellites in large cratering collisions. *Bull. Am. Astron. Soc., 28,* 1101.

Durda D. D., Bottke W. F., Asphaug E., and Richardson D. C. (2001) The formation of asteroid satellites: Numerical simulations using SPH and N-body models. *Bull. Am. Astron. Soc., 33,* 5203.

Eggleton P. and Kiseleva L. (1995) An empirical condition for stability of hierarchical triple systems. *Astrophys. J., 455,* 640–645.

Elliot J. L., Kern S. D., Osip D. J., and Burles S. M. (2001) *2001 QT_{297}*. IAU Circular No. 7733.

Gehrels T., Drummond J. D., and Levenson N. A. (1987) The absence of satellites of asteroids. *Icarus, 70,* 257–263.

Geissler P., Petit J.-M., Durda D. D., Greenberg R., Bottke W., Nolan M., and Moore J. (1996) Erosion and ejecta reaccretion on 243 Ida and its moon. *Icarus, 120,* 140–157.

Goldreich P. and S. Soter (1996) Q in the solar system. *Icarus, 5,* 375–389.

Gradie J. and Flynn L. (1998) A search for satellites and dust belts around asteroids: Negative results (abstract). In *Lunar and Planetary Science XIX,* pp. 405–406. Lunar and Planetary Institute, Houston.

Graves J. E., Northcott M. J., Roddier F. J., Roddier C. A., and Close L. M. (1998) First light for Hokupa'a: 36-element curvature AO system at UH. *Proc. SPIE 3353,* pp. 34–43.

Hamilton D. P. and Burns J. A. (1991) Orbital stability zones about asteroids. *Icarus, 92,* 118–131.

Hansen A. T., Arentoft T., and Lang K. (1997) The rotational period of 90 Antiope. *Minor Planet Bull., 24,* 17.

Harrington R. S. (1997a) Planetary orbits in binary stars. *Astron. J., 82,* 753–756.

Harrington R. S. (1997b) A review of the dynamics of classical triple stars. *Rev. Mex. Astron. Astrofis., 3,* 139–143.

Harris A. W. (2002) On the slow rotation of asteroids. *Icarus, 156,* 184–190.

Harris A. W. and Davies J. K. (1999) Physical characteristics of near-Earth asteroids from thermal infrared spectrophotometry. *Icarus, 142,* 464–475.

Hartmann W. K. (1979) Diverse puzzling asteroids and a possible unified explanation. In *Asteroids* (T. Gehrels, ed.), pp. 466–479. Univ. of Arizona, Tucson.

Hudson R. S. (1993) Three-dimensional reconstruction of asteroids from radar observations. *Remote Sensing Rev., 8,* 195–203.

Hudson R. S. and Ostro S. J. (1994) Shape of asteroid 4769 Castalia (1989 PB) from inversion of radar images. *Science, 263,* 940–943.

Kaasalainen M., Mottola S., and Fulchignoni M. (2002) Asteroid models from disk-integrated data. In *Asteroids III* (W. F. Bottke Jr. et al., eds.), this volume. Univ. of Arizona, Tucson.

Kavelaars J. J., Petit J.-M., Gladman B., and Holman M. (2001) *2001 QW_{322}*. IAU Circular No. 7749.

Kiseleva L., Eggleton P. P., and Orlov V. V. (1994) Instability of close triple systems with coplanar initial doubly circular motion. *Mon. Not. R. Astron. Soc., 270,* 936–946.

Leinhardt Z. M., Richardson D. C., and Quinn T. (2000) Direct N-body simulations of rubble pile collisions. *Icarus, 146,* 133–151.

Leone G., Farinella P., Paolicchi P., and Zappalà V. (1984) Equilibrium models of binary asteroids. *Astron. Astrophys. 140,* 265–272.

Love S. G. and Ahrens T. J. (1997) Origin of asteroid rotation rates in catastrophic impacts. *Nature, 386,* 154–156.

Margot J.-L. and Brown M. E. (2001) *S/2001 (22) 1.* IAU Circular No. 7703.

Margot J.-L., Nolan M. C., Benner L. A. M., Ostro S. J., Jurgens R. F., Slade M. A., Giorgini J. D., and Campbell D. B. (2000) *Satellites of Minor Planets.* IAU Circular No. 7503.

Margot J.-L., Nolan M. C., Benner L. A. M., Ostro S. J., Jurgens R. F., Giorgini J. D., Slade M. A., Howell E. S., and Campbell D. B. (2002a) Radar discovery and characterization of binary near-Earth asteroids (abstract). In *Lunar and Planetary Science XXXIII,* Abstract #1849. Lunar and Planetary Institute, Houston (CD-ROM).

Margot J.-L., Nolan M. C., Benner L. A. M., Ostro S. J., Jurgens R. F., Giorgini J. D., Slade M. A., and Campbell D. B. (2002b) Binary asteroids in the near-Earth object population. *Science, 296,* 1445–1448.

Merline W. J., Chapman C. R., Robinson M., Veverka J., Harch A., Bell J. III, Thomas P., Clark B. E., Joseph J., Carcich B, Murchie S., Cheng A., Izenberg N., McFadden L., and Malin M. (1998) *NEAR*'s encounter with 253 Mathilde: Search for satellites (abstract). In *Lunar and Planetary Science XXIX,* Abstract #1954. Lunar and Planetary Institute, Houston (CD-ROM).

Merline W. J., Chapman C. R., Colwell W. B., Veverka J., Harch A., Bell M., Bell J. III, Thomas P., Clark B. E., Martin P., Murchie S., Cheng A., Domingue D., Izenberg N., Robinson M., McFadden L., Wellnitz D., Malin M., Owen W., and Miller J. (1999a) Search for satellites around asteroid 433 Eros from *NEAR* flyby imaging (abstract). In *Lunar and Planetary Science XXX,* Abstract #2055. Lunar and Planetary Institute, Houston (CD-ROM).

Merline W. J., Close L. M., Dumas C., Chapman C. R., Roddier F., Ménard F., Slater D. C., Duvert G., Shelton C., Morgan T., and Dunham D. W. (1999b) *S/1998 (45) 1.* IAU Circular No. 7129.

Merline W. J., Close L. M., Dumas C., Chapman C. R., Roddier F., Ménard F., Slater D. C., Duvert G., Shelton C., and Morgan T. (1999c) Discovery of a moon orbiting the asteroid 45 Eugenia. *Nature, 401,* 565–568.

Merline W. J., Close L. M., Dumas C., Shelton J. C., Ménard F., Chapman C. R., and Slater D. C. (2000a) *Satellites of Minor Planets.* IAU Circular No. 7503.

Merline W. J., Close L. M, Dumas C., Shelton J. C., Ménard F., Chapman C. R., and Slater D. C. (2000b) Discovery of companions to asteroids 762 Pulcova and 90 Antiope by direct imaging. *Bull. Am. Astron. Soc., 32,* 1309.

Merline W. J., Ménard F., Close L., Dumas C., Chapman C. R., and Slater D. C. (2001a) *S/2001 (22) 1.* IAU Circular No. 7703.

Merline W. J., Close L. M., Siegler N., Potter D., Chapman C. R., Dumas C., Ménard F., Slater D. C., Baker A. C., Edmunds M. G., Mathlin G., Guyon O., and Roth K. (2001b) *S/2001 (617) 1.* IAU Circular No. 7741.

Merline W. J., Tamblyn P., Chapman C. R., Colwell W. B., Gor V., Burl M. C., Bierhaus E. B., and Robinson M. S. (2001c) An automated search for moons during approach of the *NEAR* spacecraft to asteroid Eros. *Proc. 6th International Symposium on Artificial Intelligence, Robotics, and Automation in Space, 2001 (Montreal),* Paper #AM128 (CD-ROM).

Merline W. J., Close L. M., Mènard F., Dumas C., Chapman C. R., Slater D. C. (2001d) Search for asteroidal satellites. *Bull. Am. Astron. Soc., 32,* 1017.

Merline W. J., Close L. M., Siegler N., Dumas C., Chapman C., Rigaut F., Ménard F., Owen W. M., and Slater D. C. (2002) *S/2002 (3749) 1.* IAU Circular No. 7827.

Michel P., Benz W., Tanga P., Richardson D. C. (2001) Collisions and gravitational re-accumulation: Forming asteroid families and satellites. *Science, 294,* 1696–1700.

Miller M. C. and Hamilton D. P. (2002) Four-body effects in globular cluster black hole coalescence. *Astrophys. J., 576,* 894–898.

Mottola S. and Lahulla F. (2000) Mutual eclipse events in asteroidal binary system 1996 FG_3: Observations and a numerical model. *Icarus, 146,* 556–567.

Mottola S., De Angelis G., Di Martino M., Erikson A., Hahn G., and Neukum G. (1995) The near-Earth objects follow-up program: First results. *Icarus, 117,* 62–70.

Mottola S., Hahn G., Pravec P., and Šarounová L. (1997) *S/1997 (3671) 1.* IAU Circular No. 6680.

Nolan M. C., Margot J.-L., Howell E. S., Benner L. A. M., Ostro S. J., Jurgens R. F., Giorgini J. D., and Campbell D. B. (2000) *2000* UG_{11}. IAU Circular No. 7518.

Nolan M. C., Howell E. S., Magri C., Beeney B., Campbell D. B., Benner L. A. M., Ostro S. J., Giorgini J. D., and Margot J-L. (2002a) *2002* BM_{26}. IAU Circular No. 7824.

Nolan M. C., Howell E. S., Ostro S. J., Benner L. A. M., Giorgini J. D., Margot J. L., and Campbell D. B. (2002b) *2002* KK_8. IAU Circular No. 7921.

Noll K., Stephens D., Grundy W., Spencer J., Millis R., Buie M., Cruikshank D., Tegler S., and Romanishin W. (2002a) *1997* CQ_{29}. IAU Circular No. 7824.

Noll K., Stephens D., Grundy W., Spencer J., Millis R., Buie M., Cruikshank D., Tegler S., Romanishin W. (2002b) *2000* CF_{105}. IAU Circular No. 7857.

Ostro S. J., Chandler J. F., Hine A. A., Rosema K. D., Shapiro I. I., and Yeomans D. K. (1990) Radar images of asteroid (1990) 1989 PB. *Science, 248,* 1523–1528.

Ostro S. J., Hudson R. S., Nolan M. C., Margot J-L., Scheeres D. J., Campbell D. B., Margi C., Giorgini J. D., and Yeomans D. K. (2000a) Radar observations of asteroid 216 Kleopatra. *Science, 288,* 836–839.

Ostro S. J., Margot J.-L., Nolan M. C., Benner L. A. M., Jurgens R. F., and Giorgini J. D. (2000b) *2000* DP_{107}. IAU Circular No. 7496.

Ostro S. J., Hudson R. S., Benner L. A. M., Giorgini J. D., Magri C., Margot J.-L., and Nolan M. C. (2002) Asteroid radar astronomy. In *Asteroids III* (W. F. Bottke Jr. et al., eds.), this volume. Univ. of Arizona, Tucson.

Paolicchi P., Burns J. A., and Weidenschilling S. J. (2002) Side effects of collisions: Rotational properties and tumbling and binary asteroids. In *Asteroids III* (W. F. Bottke Jr. et al., eds.), this volume. Univ. of Arizona, Tucson.

Petit J.-M., Durda D. D., Greenberg R., Hurford T. A., and Geissler P. E. (1997) The long-term stability of Dactyl's orbit. *Icarus, 130,* 177–197.

Pravec P. (1995) Rotation studies and orbital distribution of near-Earth asteroids. Ph.D. thesis, Charles University, Prague.

Pravec P. and Hahn G. (1997) Two-period lightcurve of 1994 AW_1: Indication of a binary asteroid? *Icarus, 127,* 431–440.

Pravec P. and Harris A. W. (2000) Fast and slow rotation of asteroids. *Icarus, 148,* 12–20.

Pravec P. and Šarounová L. (2001) *1999* KW_4. IAU Circular No. 7633.

Pravec P., Wolf M., Varady M., and Bárta P. (1995) CCD photometry of 6 near-Earth asteroids. *Earth Moon Planets, 71,* 177–187.

Pravec P., Wolf M., Šarounová L. (1998a) Occultation/eclipse events in binary asteroid 1991 VH. *Icarus, 133,* 79–88.

Pravec P., Šarounová L., and Wolf M. (1998b) *1996* FG_3. IAU Circular No. 7074.

Pravec P., Wolf M., Šarounová L. (1999) How many binaries are there among near-Earth asteroids? In *Evolution and Source Regions of Asteroids and Comets* (J. Svoreò et al., eds.), pp. 159–162. IAU Colloquium No. 173, Astron. Inst. Slovak Acad. Sci., Tatranská Lomnica.

Pravec P., Šarounová L., Rabinowitz D. L., Hicks M. D., Wolf M., Krugly Y. N., Velichko F. P., Shevchenko V. G., Chiorny V. G., Gaftonyuk N. M., and Genevier G. (2000a) Two-period lightcurves of 1996 FG_3, 1998 PG, and (5407) 1992 AX: One probable and two possible binary asteroids. *Icarus, 146,* 190–203.

Pravec P., Kušnirák P., Hicks M., Holliday B., and Warner B. (2000b) *2000* DP_{107}. IAU Circular No. 7504.

Pravec P., Kušnirák P., and Warner B. (2001) *2001* SL_9. IAU Circular No. 7742.

Pravec P., Šarounová L., Hicks M. D., Rabinowitz D. L., Wolf M., Scheirich P., Krugly Y. N. (2002a) Two periods of 1999 HF_1 — another binary NEA candidate. *Icarus, 158,* 276–280.

Pravec P., Harris A. W., and Michałowski T. (2002b) Asteroid rotations. In *Asteroids III* (W. F. Bottke Jr. et al., eds.), this volume. Univ. of Arizona, Tucson.

Reitsema H. J. (1979) The reliability of minor planet satellite observations. *Science, 205,* 185–186.

Richardson D. C., Bottke W., and Love S. G. (1998) Tidal distortion and disruption of Earth-crossing asteroids. *Icarus, 134,* 47–76.

Rigaut F., Salmon D., Arsenault R., Thomas J., Lai O., Rouan D., Véran J. P., Gigan P., Crampton D., Fletcher J. M., Stilburn J., Boyer C., and Jagourel P. (1998) Performance of the Canada-France-Hawaii Telescope adaptive optics bonnette. *Publ. Astron. Soc. Pacific, 110,* 152–164.

Roberts L. C. Jr., McAlister H. A., and Hartkopf W. I. (1995) A speckle interferometric survey for asteroid duplicity. *Astron. J., 110,* 2463–2468.

Roddier F. (1988) Curvature sensing and compensation: A new concept in adaptive optics. *Appl. Optics, 27,* 1223–1225.

Roddier F., Northcott, M. and Graves J. E. (1991) A simple low-order adaptive optics system for near infrared applications. *Publ. Astron. Soc. Pac., 103,* 131–149.

Scheeres D. J. (1994) Dynamics about uniformly rotating tri-axial ellipsoids. *Icarus, 110,* 225–238.

Scheeres D. J., Durda D. D., and Geissler P. E. (2002) The fate of asteroid ejecta. In *Asteroids III* (W. F. Bottke Jr. et al., eds.), this volume. Univ. of Arizona, Tucson.

Showalter M. R., Ostro S. J., Shapiro I. I., and Campbell D. B. (1982) Upper limit on the radar cross section of a Pallas satellite. *Bull. Am. Astron. Soc., 14,* 725.

Storrs A., Wells E., Stern A., and Zellner B. (1998) Surface heterogeneity of asteroids: HST WFPC2 images 1996–1997. *Bull. Am. Astron. Soc., 30,* 1026.

Storrs A., Weiss B., Zellner B., Burleson W., Sichitiu R., Wells E., Kowal C., and Tholen D. (1999a) Imaging observations of asteroids with Hubble Space Telescope. *Icarus, 137,* 260–268.

Storrs A. D., Wells E. N., Zellner B. H., Stern A., and Durda D. (1999b) Imaging observations of asteroids with HST. *Bull. Am. Astron. Soc., 31,* 1089.

Storrs A., Vilas F., Landis R., Wells E., Woods C., Zellner B., and Gaffey M. (2001a) *S/2001 (107) 1.* IAU Circular No. 7599.

Storrs A., Vilas F., Landis R., Wells, E., Woods C., Zellner B., and Gaffey M. (2001b) *S/2001 (87) 1.* IAU Circular No. 7590.

Tanga P., Hestroffer D., Berthier J., Cellino A., Lattanzi M. G., Di Martino M., and Zappalà V. (2001) HST/FGS observations of asteroid (216) Kleopatra. *Icarus, 153,* 451–454.

Toth I. (1999) On the detectability of satellites of small bodies orbiting the Sun in the inner region of the Edgeworth-Kuiper Belt. *Icarus, 141,* 420–425.

Trujillo C. A. and Brown M. E. (2002) *1999* TC_{36}. IAU Circular No. 7787.

Van Flandern T. C., Tedesco E. F. and Binzel R. P. (1979) Satellites of asteroids. In *Asteroids* (T. Gehrels, ed.), pp. 443–465. Univ. of Arizona, Tucson.

Veillet C., Doressoundiram A., Shapiro J., Kavelaars J. J., and Morbidelli A. (2001) *S/2000 (1998* WW_{31}*) 1.* IAU Circular No. 7610.

Veillet C., Parker J. W., Griffin I., Marsden B., Doressoundiram A., Buie M., Tholen D. J., Connelley M., and Holman M. J. (2002) The binary Kuiper-belt object 1998 WW_{31}. *Nature, 416,* 711–713.

Veverka J., Helfenstein P., Lee P., Thomas P., McEwen A., Belton M., Klaasen K., Johnson T. V., Granahan J., Fanale F., Geissler P., and Head J. W. Jr. (1996a) Ida and Dactyl: Spectral reflectance and color variations. *Icarus, 120,* 66–76.

Veverka J., Thomas P. C., Helfenstein P., Lee P., Harch A., Calvo S., Chapman C., Belton M. J. S., Klaasen K., Johnson T. V., and Davies M. (1996b) Dactyl: *Galileo* observations of Ida's satellite. *Icarus, 120,* 200–211.

Veverka J., Thomas P., Harch A., Clark B., Bell J. F. III, Carcich B., Joseph J., Murchie S., Izenberg N., Chapman C., Merline W., Malin M., McFadden L., and Robinson M. (1999a) *NEAR* encounter with asteroid 253 Mathilde: Overview. *Icarus, 140,* 3–16.

Veverka J., Thomas P. C., Bell J. F.III, Bell M., Carcich B., Clark B., Harch A., Joseph J., Martin P., Robinson M., Murchie S., Izenberg N., Hawkins E., Warren J., Farquhar R., Cheng A., Dunham D., Chapman C., Merline W. J., McFadden L., Wellnitz D., Malin M., Owen W. M. Jr., Miller J. K., Williams B. G., and Yeomans D. K. (1999b) Imaging of asteroid 433 Eros during *NEAR*'s flyby reconnaissance. *Science, 285,* 562–564.

Veverka J., Robinson M., Thomas P., Murchie S., Bell J. F. III, Izenberg N., Chapman C., Harch A., Bell M., Carcich B., Cheng A., Clark B., Domingue D., Dunham D., Farquhar R., Gaffey M. J., Hawkins E., Joseph J., Kirk R., Li H., Lucey P., Malin M., Martin P., McFadden L., Merline W. J., Miller J. K., Owen W. M. Jr., Peterson C., Prockter L., Warren J., Wellnitz D., Williams B. G., and Yeomans D. K. (2000) *NEAR* at Eros: Imaging and spectral results. *Science, 289,* 2088–2097.

Weidenschilling S. J., Paolicchi P. and Zappalà V. (1989) Do asteroids have satellites? In *Asteroids II* (R. P. Binzel et al., eds.), pp. 643–658. Univ. of Arizona, Tucson.

Weidenschilling S. J., Marzari F., Davis D. R., and Neese C. (2001) Origin of the double asteroid 90 Antiope: A continuing puzzle (abstract). In *Lunar and Planetary Science XXXII,* Abstract #1890. Lunar and Planetary Institute, Houston (CD-ROM).

Wizinowich P. L., Acton D. S., Lai O., Gathright J., Lupton W., and Stomski P. J. (2000) Performance of the W. M. Keck Observatory natural guide star adaptive optic facility: The first year at the telescope. *Proc. SPIE 4007,* pp. 2–13.

Zellner B., Wells E. N., Chapman C. R., and Cruikshank D. P. (1989) Asteroid observations with the Hubble Space Telescope and the Space Infrared Telescope Facility. In *Asteroids II* (R. P. Binzel et al., eds.), pp. 949–969. Univ. of Arizona, Tucson.

Part III:

In Situ Exploration

Cratering on Asteroids from *Galileo* and *NEAR Shoemaker*

Clark R. Chapman
Southwest Research Institute

Asteroidal craters provide information about not only the geology of asteroids but also the populations of impactors. Significant statistics on crater densities as a function of diameter have been obtained from spacecraft imaging of Gaspra, Ida (and its moon), Mathilde, and Eros. At spatial scales larger than ~100 m, the saturated crater populations on Ida and Eros are similar. Gaspra is undersaturated with such craters, exhibiting a "steep" power-law production function. Mathilde is uniquely dominated by huge craters with diameters similar to the body's radius. The *NEAR Shoemaker* camera sampled craters and boulders down to just a few centimeters in size on Eros. At spatial scales smaller than 10 m, craters are extremely rare on Eros and boulders are several hundred times more prevalent. These data are best understood if small projectiles are unexpectedly rare in the asteroid belt (forming few small craters and fragmenting few boulders); armoring of the surface by boulders, seismic shaking, and levitated dust may also contribute to the unexpected nature of the surface layer of Eros.

1. INTRODUCTION

As airless, solid bodies traveling in interplanetary space, asteroids are cratered by smaller asteroids and meteoroids as is our own Moon. However, since asteroids are small, comparatively distant objects, groundbased telescopic observations had been inadequate to resolve even the shapes of these bodies, let alone details of their surface geology such as impact craters. That changed beginning on October 29, 1991, when the *Galileo* spacecraft obtained images of the 18-km-long, S-type, main-belt asteroid 951 Gaspra from as close as 5300 km (*Veverka et al.,* 1994). On August 28, 1993, Galileo acquired even higher-resolution images of the larger, S-type, main-belt asteroid 243 Ida (*Belton et al.,* 1996) and discovered its moon Dactyl. Since then, *NEAR Shoemaker* has flown past and imaged a larger, C-type, main-belt asteroid, 253 Mathilde (*Veverka et al.,*1999). It then spent a year orbiting and mapping the surface of the S-type, Earth-approaching asteroid 433 Eros (see special *NEAR Shoemaker* issue of *Icarus*, January 2002). A few geological features like impact craters have also been identified in recent years from delay-Doppler radar imaging of several chiefly small, Earth-approaching asteroids (cf. *Benner et al.,* 2001; *Ostro et al.,* 2002), and a large crater has been identified from Hubble Space Telescope imaging of 4 Vesta (*Thomas et al.,* 1997).

Studies of asteroidal cratering elucidate many fundamental issues in planetary science. Asteroids have been geologically inert since formative epochs; ever since, impacts have been the dominant process that has shaped their forms, sizes, and surface geology. Other chapters in this volume focus on the large-scale interasteroidal collisions that catastrophically disrupt them, creating fragments that form families and asteroidal satellites and generally modulate the asteroid size distribution. Such collisions also shape the interior structures of asteroids (e.g., "rubble piles"). Here I focus on the smaller-scale impacts that affect localities but not the bulk geophysical properties of these bodies; necessarily, impact cratering affects the variety of surface morphologies and geological structures generally, as amplified by *Sullivan et al.* (2002).

Asteroid surfaces also serve as witness plates, recording impacts of objects too small to be seen by other techniques. For example, the three fly-by spacecraft imaged craters as small as tens to hundreds of meters in diameter, recording the impacts of bodies meters to tens of meters in size. While the number of such objects in near-Earth space is approximately known from observations of rare, large bolides in the Earth's atmosphere and from the Spacewatch Survey (see *Jedicke et al.,* 2002; *Morbidelli et al.,* 2002), such small objects cannot be detected in the much more distant asteroid belt or Trojan clouds. *NEAR Shoemaker*'s imaging of Eros reveals craters as small as a couple of centimeters (actually, the virtual *absence* of such craters), providing insights about still smaller interplanetary projectiles.

Impact cratering controls the attributes of asteroid surfaces, which will affect potential human operations at asteroids (mining asteroids for resources, attaching devices for purposes of deflection, etc.). Because of the low-gravity fields of asteroids, regolith processes are very different from those well-studied on the Moon, leading to the possibility (actually observed, in the case of Eros) that surface properties on a human scale may differ radically from our lunar-influenced expectations. Associated with regolith evolution, of course, are issues related to the exposure of the optical surfaces of asteroids (and meteorites derived from materials once near the surface) to micrometeorites, solar-wind particles, etc.; these "space-weathering" processes alter the optical properties of asteroids, presenting a challenge to telescopic, remote-sensing studies of asteroids (see *Clark et al.,* 2002).

The holy grail of planetary cratering, of course, is to determine the relative and, especially, absolute ages of cratered surfaces — to apply the principles of "interplanetary correlation of geologic time" (*Shoemaker et al.,* 1963). Actually, absolute chronology is difficult to disentangle from other variables, including the poorly known physical attributes of asteroid surfaces. Nevertheless, there are broad variations in crater densities and morphologies among the four asteroids studied, as well as from place to place on Eros, which permit interpretations that address the fundamental natures of these bodies.

This chapter reviews measurements and interpretations of crater populations on Gaspra, Ida, its moon Dactyl, Mathilde, and Eros. My emphasis is on the general attributes of the crater populations and on the processes that (1) form the craters and (2) degrade/destroy such craters as well as other features on asteroid surfaces. Necessarily, I will also touch on rock/boulder populations on asteroid surfaces, since their production and retention is intimately entwined with cratering.

2. SOME GENERALITIES ABOUT ASTEROIDAL CRATERING

I begin by establishing a qualitative, largely theoretical context in which to place the spacecraft observations of real asteroidal cratering that follow. Many of these precepts have been understood since the original classic papers on asteroid collisions (e.g. *Dohnanyi,* 1971) and asteroid regoliths (e.g., *Housen et al.,* 1979). Other elements of this picture are still evolving, especially as computers have enabled realistic simulations of the processes that fracture asteroids, form asteroid families, and influence dynamical evolution of asteroids.

There are good reasons for expecting that impact cratering on asteroids has, to first order, been similar for the last 3–4 b.y. on nearly all asteroids from near-Earth asteroids (NEAs) out to the Trojans. In general, cratering rates within the main asteroid belt (and in the Trojan clouds) are 2 orders of magnitude greater than impact rates on the Moon. Although the Late Heavy Bombardment (LHB, which is well documented on the Moon at 4 Ga) would be expected to have been a less-important spike in impact rates in the crowded asteroid belt, evidence for the LHB may exist for several meteorite classes whose parent bodies are presumably asteroids (*Bogard,* 1995). However, since the LHB, the numbers of asteroids and their orbital properties have not changed drastically. Any slight, gradual decline in impact rates has been occasionally interrupted by modest spikes in cratering rates following major disruptive collisions. Because of the appreciable eccentricities and inclinations of asteroid orbits, most asteroids and their debris collisionally interact with each other, although the more isolated asteroids between the outer-belt Cybeles and the Trojans are subject to a lower impact flux. While NEAs whose aphelia have been lowered to within the inner edge of the main belt have much lower, lunarlike impact rates, their dynamical lifetimes in such orbits are brief; hence, their surfaces are likely to be dominated by craters formed during their earlier residence in the asteroid belt.

Crater populations on asteroids should roughly mimic the size distribution of impacting projectiles an order of magnitude smaller in size. That is, depending on impact velocity, target size (which determines the importance of gravity), and target strength (especially for smaller impacts and impacts on smaller asteroids, where strength dominates over gravity), a projectile forms a crater roughly 10× its size. For all asteroids, one might simplistically expect that the size distribution of impacting projectiles (and thus the diameter distribution of craters, or the "production function") would be everywhere the same. There is the possibility, however, that the quasi-steady-state size distribution is different within the Trojan clouds, or in the zone between the main belt and the Trojans. Pending understanding of the reasons for differences in bias-corrected size distributions correlated with taxonomic types (*Zellner,* 1979; *Gradie et al.,* 1989), one might expect there to be size-distribution differences between crater populations in the inner and outer main belt. Such differences, however, are probably modest, so the first-order approach is to compare the cratering populations on different asteroids assuming the same production function.

Contrasting with the case of cratering on semi-infinite planetary surfaces, the physics change as craters from larger impacts approach the sizes of finite-sized asteroids. Catastrophic disruption is defined to result when the largest remnant following collision has less than half the mass of the original body. Nevertheless, the outcomes of much-smaller impacts are sensitive to finite body size. First, there is the case where the shockwave penetrates throughout the body and is, perhaps, reflected, fracturing much of the body's interior [see simulations by *Asphaug et al.* (1998, 2002)]. An unfractured original body is forever changed after its first fracturing collision; it responds to subsequent impacts as a fractured, weaker body. Larger impacts may spall off sizable portions of the target body, in addition to creating craters that approach or even exceed the radius of the body itself. Finally, still-larger impacts that would form craters approaching the size of the body instead physically disaggregate the body. Commonly, most fragments do not exceed escape velocity and instead settle back down, or reaccrete, into a so-called "rubble pile." This historic term (see *Chapman,* 1978) has recently been more rigorously defined (see *Richardson et al.,* 2002). Only "supercatastrophic" collisions impart sufficient kinetic energy to launch different portions of a rubble-pile asteroid at greater-than-escape velocity and create an asteroid family, fulfilling the definition of catastrophic disruption (counterintuitively, rubbilization — which weakens an asteroid's material properties — actually strengthens an asteroid in the sense that it takes even larger impacts to disrupt it because of lowered ejecta velocities). Prior to such final disruption, a typical asteroid may undergo many generations of further comminution and rubbilization.

Because of the power-law-like size distribution of asteroids, it is almost inevitable that asteroids will be thoroughly cratered, fractured, and rubblized before there is a chance of a large catastrophic collision creating a family of new, separate asteroids. Even the objects created through such disruption are likely themselves to be rubble piles, according to the simulations of *Michel et al.* (2001). Therefore, it is unlikely that any modern asteroids (except for some very small ones, smaller than some hundreds of meters) are created by catastrophic disruptions as clean, monolithic shards. What theorists once might have considered to be "fresh" surfaces of newly created asteroids must now be considered to be more analogous to "megaregolith" surfaces on the Moon.

The finite sizes of asteroids and their lower gravities result in another major difference in cratering processes compared with the Moon. Velocities of crater ejecta are comparable to or exceed an asteroid's escape velocity. On a high-gravity planet or moon, low-velocity ejecta blocks pile up on the crater rim, higher-velocity ejecta form a continuous ejecta blanket, and still-higher-velocity ejecta form a more extended distribution of secondary craters and rays. Only a tiny fraction of ejecta (comparable to the impactor) actually escape the Moon, some of which become lunar meteorites. On asteroids, in contrast, ejecta are much more widely spread out, resulting in larger but thinner (or absent) continuous ejecta blankets, and many projectiles that would form secondary craters on the Moon instead escape. A few may wind up in usually temporary orbits around the asteroid while most become independent, heliocentrically orbiting bodies. Termed "interplanetary secondaries" by *Hartmann* (1995), these cratering ejecta projectiles mix with the smaller fragments from catastrophic disruptions and become an integral part of the small end of the asteroidal size distribution.

The usual understanding of regoliths is grounded in the descriptions of lunar regolith, laboratory studies of core samples returned by Apollo astronauts, and modeling of lunar regolith processes. Asteroid regolith processes must be very different. In the case of the Moon, there is repetitious in-place churning of the surface materials. While lateral distribution of lunar ejecta occurs on scales wider than the regolith depth, the maintenance of albedo boundaries between the highlands and the maria — visible through binoculars from Earth — proves that regolith formation is essentially a localized, rather than global, process on the Moon. Moreover, with a negligible fraction of regolith escaping the Moon altogether, a fraction of the upper lunar regolith undergoes countless generations of surface exposure intermingled with stirring to intermediate depths. On asteroids, lateral movement of ejecta is much more enhanced, even as the horizon of such smaller bodies is much closer. Any individual impact contributes less ejecta to a nearby locality; instead, ejecta are distributed widely, some globally. Moreover, since a fraction of ejecta escapes, an asteroid is always in net erosion. After only a few generations of redistribution, escape of a particular component becomes likely. Thus, regolith materials do not become "mature" (whether measured by grinding to fine sizes, agglutinization, surface exposure to space weathering, or any other measure of maturity). And, paralleling the erosion of the regolith, considerable new, fresh material is generated from whatever substrate exists at the bottom of the regolith. So regolith material is composed of a higher fraction of "fresh" material than on the Moon, where the regolith more effectively buffers the underlying rock. (In a typical asteroidal rubble pile, however, such "fresh" material may be material that was once near or part of a surficial regolith in a previous-generation rubble pile.)

The theoretical expectation of comparatively thin, immature surficial regoliths on asteroids is sometimes thought of as demonstrated by the differences between lunar regolith samples and the gas- and cosmic-ray-track-rich meteorites known as "regolith breccias." But this is surely a comparison of apples and oranges. We do not have samples of asteroidal regoliths in our meteorite collections. The Earth's atmospheric filter strongly biases our meteorite collections toward the strongest asteroidal rocks. These include breccias formed beneath impact craters, a good fraction of which are composed of materials that spent some time near the asteroid's surface, as revealed by their gas-rich and other attributes. But to view such rocks as simply compacted surface soils is overly simplistic.

3. CRATERING ON GASPRA

Gaspra was the first asteroid revealed to possess craters. Apart from Ida's 1.6-km moonlet Dactyl, Gaspra remains the smallest asteroid to be imaged by spacecraft. The best resolution achieved, 54 m/pixel (covering one side of Gaspra), is twice as coarse as the best resolution achieved on Ida. The opposite side of Gaspra was imaged 15–20× farther away and at a poorer lighting angle, revealing few topographic features.

The most thorough treatment of cratering on Gaspra is by *Chapman et al.* (1996a). *Carr et al.* (1994) independently studied Gaspra cratering in the context of Gaspra's overall geology. *Greenberg et al.* (1994) concentrate on analyzing the largest impacts on Gaspra (transitional to catastrophic fragmentation), including effects on the population of smaller craters. *Stooke and Ford* (2001) reconsider evidence for large impacts on Gaspra.

3.1. Small Crater Population

Gaspra (Fig. 1) has few, if any, large craters, but appears peppered with fresh, small craters. Compared with other asteroids imaged during subsequent years, this attribute of Gaspra's crater population is (so far) unique.

Representative crater counts for Gaspra are shown in Figs. 2 (cumulative) and 3 (R plot), based on counts by two independent research groups (*Chapman et al.,* 1996a). Data of *Carr et al.* (1994) agree. In both studies, the craters are divided into comparatively fresh, bowl-shaped craters and one or more classes of comparatively subdued depressions, generally assumed to be highly degraded impact craters that were originally fresh, although the possibility was expressed by Carr et al. that some subdued craters might be of endogenic origin (e.g., collapse or drainage of regolith into underlying cavities, which might also be related to the formation of the several grooves seen on Gaspra).

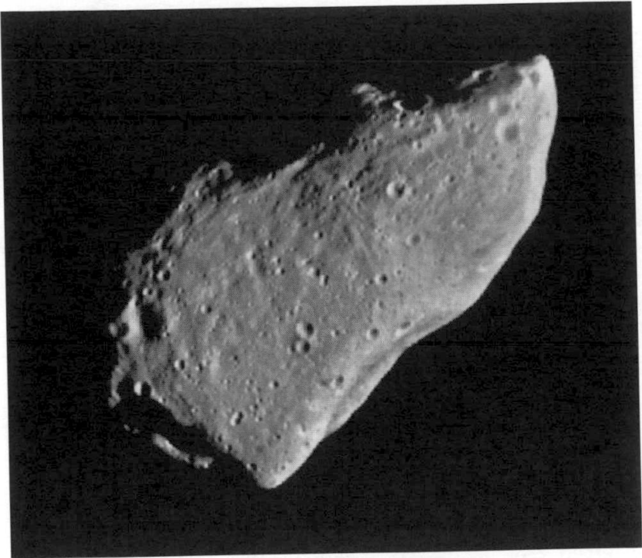

Fig. 1. Highest-resolution image of Gaspra.

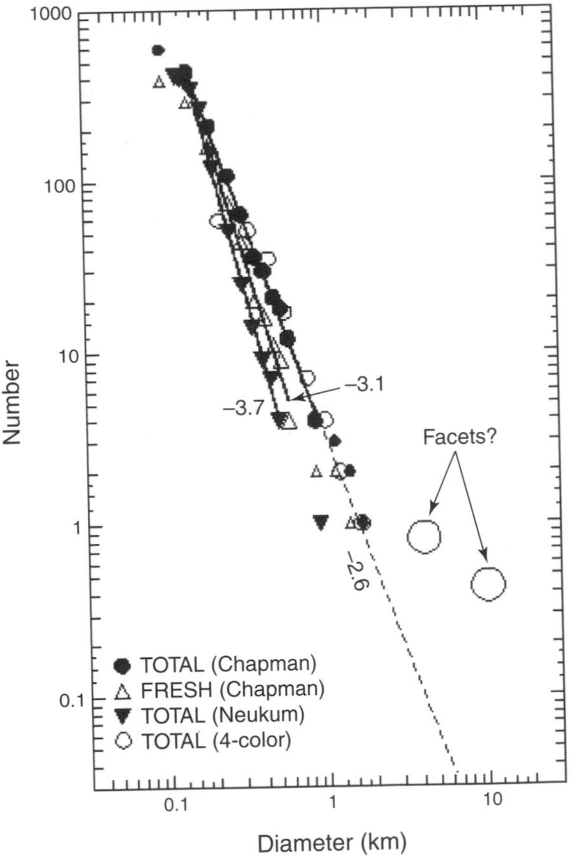

Fig. 2. Cumulative size-frequency relationship for craters on Gaspra. Total crater counts by Neukum's group are similar to Chapman's counts of fresh craters only. Cumulative power-law exponents of least squares fits are indicated; smaller symbols indicate counts based on small counting statistics or incompleteness and are not included in the fits. Counts for possible large impact features called facets are indicated.

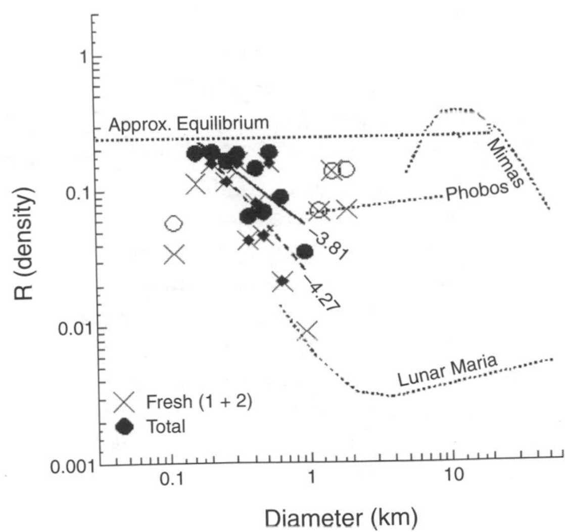

Fig. 3. R plot size-frequency data (spatial densities) for craters on Gaspra. Fresh craters (circles) and all craters (Xs with central dot; includes fresh and degraded craters) are plotted, as well as approximate curves for other small bodies. Differential power-law exponents of least-squares fits are indicated for filled circles and Xs with dots; empty circles and Xs without dots indicate low-quality or incomplete data. The approximate density level for empirical equilibrium is indicated.

Henceforth in this chapter, R plots will be exhibited exclusively, so I explain here several features of R plots (see definition in *Crater Analysis Techniques Working Group,* 1977). Unlike cumulative numbers, which include all craters larger than the size plotted, the data points shown in R plots are counts from within a diameter increment, and thus represent frequencies of craters *near that size* (plotted at the average diameter, D), uncontaminated by data from craters of much larger sizes. R plots differ from standard plots of the differential size distribution (the number within a diameter increment divided by the width of the increment and the surface area counted) in that they are further divided by D^{-3}. This approach has several virtues. First, since typical planetary crater populations follow differential power laws with exponents in the range of –2 to –5, the normalization permits deviations to be measured relative to D^{-3}, which plots as a horizontal line on the R plot; thus, such deviations are easier to see than deviations from steeply sloping trends on the usual log-log differential or cumulative plots. Second, the theoretical curve for saturation equilibrium — in the case of a differential production population with an exponent steeper than –3 — has a slope of –3, which plots as a horizontal line in an R plot (idealized saturation is at unity, while empirically many planetary and satellite crater populations follow horizontal trends near R = 0.2–0.3). Finally, and related to the last point, height in an R plot may be interpreted as spatial density: Points near the top of the plot indicate that craters of those sizes cover the surface, whereas points low on the plot indicate that craters of those sizes are rare and sparsely distributed.

Given Gaspra's dimensions (roughly 18 × 10.5 × 9 km), it could in principle — by analogy with Mathilde — sport prominent, bowl-shaped impact craters up to 5 km or more in size. It does not. The largest fresh crater on the side of Gaspra imaged in the highest resolution frame is a little over 1 km in diameter; one subdued crater is about twice as large. Two craters approximately 3 km in diameter can be recognized in the other, low-resolution images that cover the remainder of Gaspra's surface. [*Stooke and Ford* (2001) claim to see some other large craters.] However, small fresh craters densely pepper Gaspra's surface near the resolution limit. In fact, the steeply sloping trend of Gaspra's fresh crater population reaches empirical saturation near 150-m diameter, just below the size of the smallest craters for which complete counts are available. Since, over the measured size range of 0.16–1.9 km, crater densities are generally well below saturation, the observed distribution must be the "production function," generally uninfluenced by size-dependent, erosive impact processes. Its slope, around –4.3, is similar to that measured on the Moon and Mars for craters formed by similar-sized projectiles. Gaspra may, therefore, have resolved a long-standing debate about whether the steep power-law slope (e.g., as observed on the Moon) might reflect dominance of the lunar small-crater population by secondary craters rather than primaries (cf. *Neukum et al.,* 2001). Gaspra appears to demonstrate that the steep slope is characteristic of the size distribution of small asteroids tens to hundreds of meters in size, a subset of which escape the asteroid belt and dominate cratering on the terrestrial planets.

The more subdued craters, which are most numerous relative to the fresh craters near 0.5-km diameter, may be impact craters that are older than average and have been eroded and degraded by saturation impacts forming craters near and below the resolution limit of the image. Alternatively, they may represent the remnant of an earlier population of impact craters, some of which "show through" whatever process it was that created the comparatively "clean slate" on which the undersaturated, fresh crater population is expressed. *Stooke and Ford* (2001) argue that this older population of preserved large craters is more substantial than reported by *Chapman et al.* (1996a); on the other hand, their counts also agree with those of Chapman et al. in showing that Gaspra is substantially undersaturated by craters ~1 km diameter.

In any case, even the fresh craters appear a bit subdued in profile compared with craters of similar size on other bodies. *Carr et al.* (1994) report depth:diameter = 0.14 on Gaspra, in comparison with values near 0.2 for the Moon, Mars, and Phobos. They suggest possible explanations (e.g., seismic ringing) for why such craters might have formed with initially shallow shapes or might have been subsequently degraded. It is more likely that the craters are shallower because of lower rims due to the wide distribution of crater ejecta on a low-gravity body (*Sullivan et al.,* 1996).

3.2. Facets: Possible Large Craters

From some perspectives in the low-resolution images, Gaspra exhibits a peanutlike shape, as though it is made out of two lumps. However, much thinking about Gaspra has been influenced by the profile it happened to exhibit in the highest-resolution image. As seen in Fig. 1, there are several large structures, seen in profile, that either resemble incipient craters or have planar, facetlike shapes. A portion of the imaged surface is another facet lying roughly in the image plane. According to *Thomas et al.* (1994), one such facet is up to 6 km across, defining a plane to within ±200 m. *Greenberg et al.* (1996) most thoroughly developed the concept that the facets are, in fact, impact scars or spalls (they count as many as eight facets on Gaspra). There is no way to tell, from the images alone, whether this hypothesis is correct, or whether Gaspra's shape (whether of a lumpy peanut or angular/faceted) was created in some other way — e.g., as a result of the catastrophically disruptive collision that originally created Gaspra, from its parent body. For instance, Gaspra might be the remnant "core" of a badly fractured, disrupted body, with its shape determined by spalls and fractures in the final collision. Or it might be a rubble pile, dominated by two large components, each of which has faceted elements to its shape.

Greenberg et al. turn to then-recent hydrocode modeling to support their argument that very large craters could form on a body without disrupting it. (Previously it had been suggested that such large features, relative to body size, would inevitably have disrupted the body.) More recent hydrocode work supports and extends this conclusion. *Chapman et al.* (1996a) object that only a single, latest such impact scar could exist because the topography that defines any earlier ones would have necessarily been obliterated by the concussion and ejecta of the latest large impact. Imaging of Mathilde (see below), which shows several coexisting, very large craters, disproves that objection (cf. *Stooke and Ford,* 2001). On the other hand, Mathilde's large craters look dramatically different from Gaspra's facets, and Mathilde's small crater population differs from Gaspra's, so Mathilde is a poor analog for Gaspra.

If the facets represent impact scars, they would have been formed during a period ~20× as long as that represented by the visible fresh crater population, or 7× as long as the period represented by all of Gaspra's small craters.

3.3. Regolith on Gaspra

Carr et al. (1994) infer, on fairly speculative grounds, that Gaspra has regolith tens of meters deep. Surely there is some regolith, necessary to explain the correlation of slight color variations on Gaspra's surface with topography. *Chapman et al.* (1996a) show that the average depth of regolith, created just from the visible craters [not including facets nor some other possible large craters tabulated by *Stooke and Ford* (2001)], would be <10 m deep; it could be considerably thinner, given the net erosive impact environment applicable to this small asteroid with minimal gravity. This would be the regolith depth, if the "clean slate" on which the small crater population is expressed was created in the disruptive collision that originally formed Gaspra, not counting preexisting megaregolith material originating on the precursor body. On the other hand, if the facets and other possible large craters reflect real impacts, in which case the existing small-crater population would be only the latest cratering generation of many, then the regolith could be much deeper. It is always very difficult to determine a body's internal properties (in the vertical dimension) from surface imaging.

3.4. Age, History, and Nature of Gaspra

The prime hope in cratering studies is the possibility of absolutely dating events in the geological history of solar-system bodies. This expectation is rarely fulfilled with sufficient accuracy. The best absolute chronologic information comes from various isotopic chronometers measured from samples. Failing that, cratering can be a backup, provided (1) the impact rate (and associated crater formation rate, dependent on uncertain scaling relationships) is known, and (2) the craters can be identified and counted unambiguously. The result is, however, only a "crater retention age," which may not be related to the age of the body itself (e.g., when the catastrophic disruption occurred that liberated Gaspra from its parent body).

Carr et al. (1994) attempt to determine the crater-retention age of Gaspra's fresh craters by the following direct approach: They adopt a statistics-of-one estimate of the number of 1-km diameter craters on Gaspra and then extrapolate from the known number of 50-km-diameter main-belt asteroids down to the number of 100-m asteroids (which would make 1-km craters, but whose actual numbers are completely unknown) using several published theoretical models for extrapolation from much-larger bodies. The derived age ranges from 2×10^7 yr to the age of the solar system, which is too unconstrained to be useful.

Chapman et al. (1996a) adopted a more sophisticated approach to estimate the formation rate of 1-km craters on Gaspra, utilizing the known lunar crater size distribution as well as the inherent information from Gaspra imaging about the steeply sloping power law for smaller craters. However, improvement in knowledge of the numbers of 100-m main-belt asteroids was marginal at the time, and the uncertainties are nearly as large as for *Carr et al.* (1994). *Chapman et al.* (1996a) derived a crater-retention age of $\sim 2 \times 10^8$ yr. A self-consistent, but now obsolete and unreliable, estimate of the collisional lifetime for a Gaspra-sized object was $\sim 5 \times 10^8$ yr. Thus Gaspra's crater population was found to be somewhat youthful relative to when the asteroid would likely be catastrophically disrupted, consistent with Gaspra's observed undersaturated crater densities. Since Gaspra's larger precursor body presumably had an even longer expected survivability against catastrophic disruption, the conclusion would be either (1) that Gaspra was formed unexpectedly recently (ca 200 Ma, assuming that the cratering age reflects the age of Gaspra's formation) or (2) that Gaspra has gone through multiple generations of surficial cratering since formation much earlier in solar-system history (in accordance with the interpretation of the facets as large impact craters). The former conclusion, being "unexpected," might be least preferred, except that independent analyses of asteroid families suggest that the Flora family (of which Gaspra may be a part) formed comparatively recently (*Nesvorný et al.*, 2001).

Greenberg et al.'s (1996) interpretation of Gaspra is a rubble pile possibly dominated by a couple of large components that has suffered a lengthy history of subsequent impacts, including those that formed the facets. The large impacts would have, in their model, seismically "jolted" the surface, repeatedly cleansing it of preexisting smaller craters and other topography (but not other facets). Presumably one facet marks the latest such impact, and it created the smoothed surface on which the small, fresh craters have since formed.

A very different interpretation (*Chapman,* 1997) turns to another variable (compositional strength of the body) to explain Gaspra's significant differences from other closely studied asteroids. If Gaspra is a very strong body — for instance, if it is a monolithic, metallic object, like the core of a differentiated parent body — then several attributes that distinguish it from Ida, Mathilde, and Eros might all be explained: (1) Gaspra's apparently nonchondritic, olivine-rich composition may be akin to pallasites, generally regarded as coming from the core-mantle interface of a parent body. (2) Gaspra's undersaturated crater density might reflect the difficulty of forming craters in a solid, strong, metallic surface, rather than youthful age (strength-scaled impacts into solid metal rather than rock would form considerably smaller craters); hence, Gaspra's surface could be very old. (3) Gaspra's angular, faceted shape (if that's the correct characterization) might hint that its composition is fundamentally different from that of other asteroids, which do not share this attribute.

4. CRATERING ON IDA AND DACTYL

Ida is second only to Eros in the resolution and coverage of cratering data obtained, yet — apart from some preliminary reports — only one analysis of Ida's crater statistics has been published (*Chapman et al.,* 1996b). Nevertheless, the discovery of a moon (the first ever found around an asteroid, *Chapman et al.,* 1995), grooves and other interesting geological features, and significant color differences across Ida inspired numerous studies of collisional physics with prime application to Ida (e.g., *Asphaug et al.,* 1996; *Durda,* 1996; *Davis et al.,* 1996; *Geissler et al.,* 1996; *Greenberg et al.,* 1996). In addition, Ida's cratering processes figure centrally in studies of the asteroid's geology (*Sullivan et al.,* 1996) and of its population of ejecta blocks

(*Lee et al.,* 1996). Finally, nearly 30 craters or possible craters were recognized on Dactyl and were analyzed by *Veverka et al.* (1996).

4.1. Crater Population on Ida

Unlike Gaspra, Ida's surface looks very similar to the cratered lunar surface at similar scales (Fig. 4). And, indeed, crater statistics reveal that Ida is sensibly saturated with craters of all sizes below ~1 km diameter, and close to saturation for the largest craters (1–15 km diameter). "Saturation" is here taken to mean empirical saturation with R = ~0.25.

Ida crater counts are presented in the R plot in Fig. 5 (additional counts are displayed in *Chapman et al.,* 1996b). Relative to the R = 0.25 reference line, the largest craters on Ida are relatively numerous. There is a slight depletion for craters a few kilometer in diameter, then a rise back to R = 0.3 near 1 km. The distribution then follows the equilibrium slope down to the smallest craters resolved (diameter 0.1 km). These features are consistent with saturation cratering of Ida by a production population with a shape identical to that observed on the Moon (*Neukum and Ivanov,* 1994) and on Gaspra. Craters <1 km in diameter are certainly in saturation, as reflected by the equilibrium distribution of crater morphologies [high proportion of degraded craters, small but constant fraction of fresh craters; see *Belton et al.* (1994)]. On the other hand, saturation equilibrium is not a unique interpretation for the larger craters, which follow a shallow-sloped production function. They may be in quasiequilibrium, reflecting many generations of cratering, or they may reflect the first generation of cratering approaching saturation for the first time since Ida's surface was created.

There are hints that Ida has a global dichotomy (*Thomas et al.,* 1996), reflected in the crater populations in the sense that one "lobe" of Ida has numerous large craters, while the other has none (except for Vienna Regio, if that is counted as a crater). Perhaps Ida is a rubble pile, dominated by two large components. One lobe is saturated with craters while the other one has been reset (i.e., previous craters destroyed), possibly by the impact that created Vienna Regio. This in-

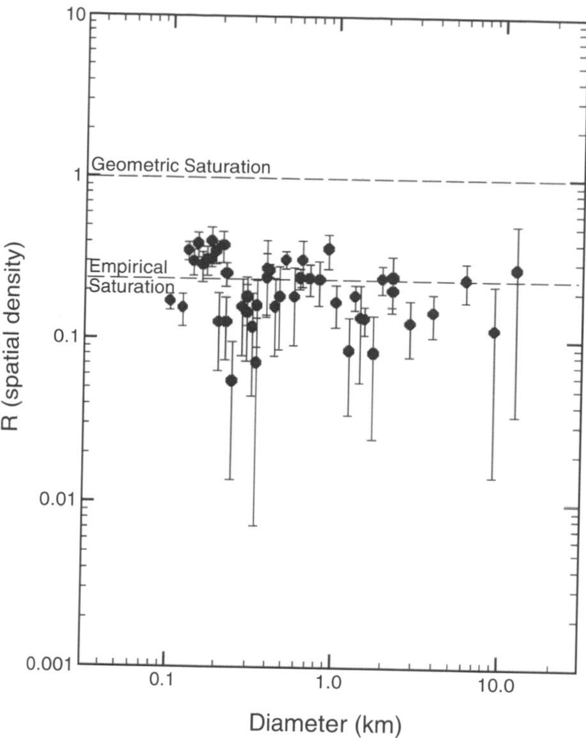

Fig. 5. R plot size-frequency data for craters on Ida. Same data as plotted in Fig. 2 of *Chapman et al.* (1996b). A weighted least-squares fit yields –3.1 for the differential power-law exponent. The data approximate the empirical saturation curve, especially at diameters <1 km.

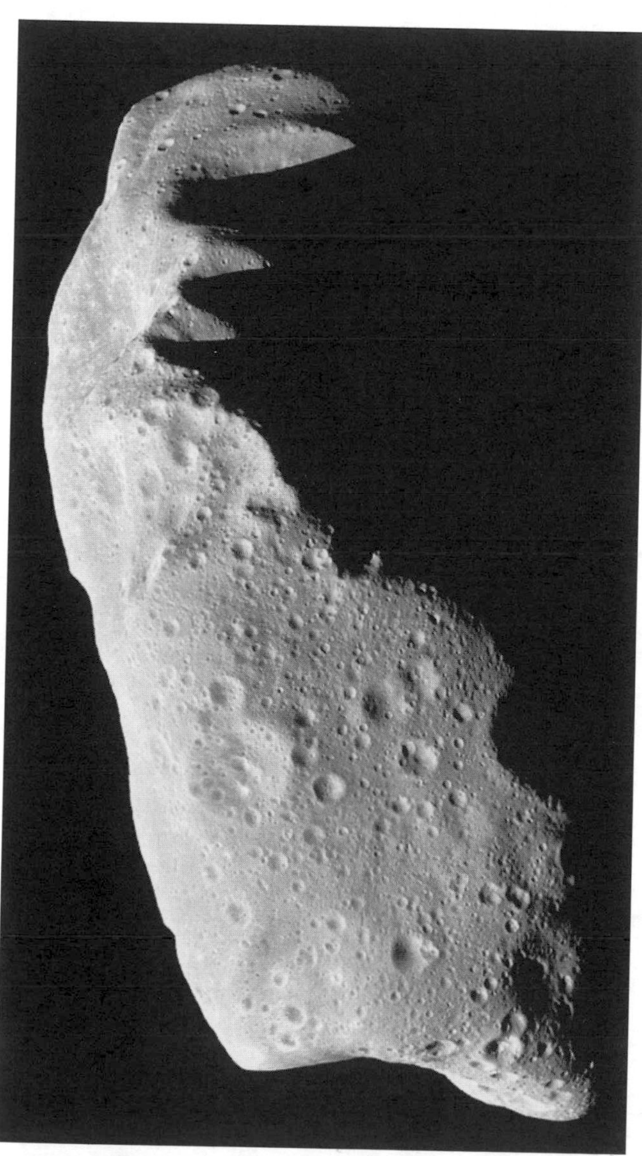

Fig. 4. High-resolution mosaic of Ida.

terpretation is at some variance to the model preferred by *Asphaug et al.* (1996), in which the grooves preferentially located in the cratered lobe were produced by far-field stresswaves created by the formation of Vienna Regio on the other lobe, implying structural coherence between the two apparent components.

4.2. Regolith on Ida

Evidence for regolith on Ida is much more obvious than on Gaspra, partly because of higher-resolution images but also because regolith is more abundant on Ida. The visible population of craters would have produced a global regolith averaging ~150 m in depth. Given that the crater population is plausibly saturated, this is a minimum estimate. *Sullivan et al.* (1996) offer several kinds of photogeological evidence for regolith 50–100 m deep on Ida, but they also note reasons for expecting megaregolith up to several kilometers deep.

Geissler et al. (1996) model the ejecta from Azzurra, a large but rather poorly resolved impact crater on the opposite side of Ida from the well-imaged hemisphere. This crater appears to be morphologically fresh and certainly has high albedo, like the freshest craters on the Moon. Also, Azzurra, like some smaller fresh craters (*Sullivan et al.,* 1996), exhibits spectral reflectance attributes suggesting that it is youthful and has not long been subjected to the space-weathering effects that apparently gradually redden S-type asteroids (*Chapman,* 1996). *Geissler et al.* (1996) find that ejecta from Azzurra (at a few meters per second) are distributed nonuniformly around Ida, tending to match the distribution of bluer-than-average color patches on Ida.

A portion of higher-velocity ejecta blocks, launched at ~10 m/s in Geissler et al.'s model, accumulate in regions on Ida where 17 positive relief features, probable ejecta blocks, are mapped by *Lee et al.* (1996). The blocks range in size from about 50 to 150 m. They are, in all probability, a sampling of the large end of a size distribution that includes innumerable smaller blocks (below the resolution of Ida images, but well expressed on Eros — see below). It is plausible that extant blocks were derived from a recent, large cratering event, since blocks are gradually destroyed by meteoroid impacts over time, in a process analogous to the collisional disruption of small, nearly gravity-free asteroids.

Because of the similarity in appearance of Eros and Ida at scales above the resolution limit for Ida (and in other ways; e.g., both are S-IV types and their sizes are not too dissimilar), it is plausible that *NEAR Shoemaker*'s high-resolution investigations of Eros (see below) would be applicable, in most respects, to Ida.

4.3. Cratering Age and History of Ida

Ida exhibits about 10× the density of craters a few kilometers in size as does Gaspra. Based on the estimate (see above) of Gaspra's crater-retention age at 200 m.y., *Chapman et al.* (1996b) and *Davis et al.* (1996) estimate a cratering age for Ida near 2 b.y. Given that Ida may be saturated with craters, that is a lower limit to the age of Ida as an independent body — there could have been a number of generations of cratering. Contemporaneous models for the age of the Koronis family, of which Ida was a part, suggest family creation ca 2 Ga (see also *Farinella et al.,* 1996); however, approaches to understanding the ages of asteroid families are undergoing revision, particularly as dispersion of family members becomes understood as a result of Yarkovsky forces and minor resonances rather than just the initial ejection velocities, so those earlier family-age estimates must be regarded as obsolete.

Presumably Ida was created as an independent body in the catastrophic disruption of the parent body of the Koronis family. It is probably a rubble pile itself, based on its double-lobed shape, its bulk density of ~2.6 g/cm^3 (compared with ~3.5 for ordinary chondrites), and recent theoretical results on family creation (*Michel et al.,* 2001).

4.4. Cratering on Dactyl

Dactyl is, by far, the smallest asteroid imaged by spacecraft, although the quality of the image is comparable to that of some of the best radar delay-Doppler images of NEAs of similar size. Ida's satellite appears spherical, but is actually best approximated by an ellipsoid with axes 1.6 × 1.4 × 1.2 km. Crater counts are presented by *Chapman et al.* (1996b) and *Veverka et al.* (1996). While nearly 30 possible craters were perceived on the one side of Dactyl that was imaged, the statistical sample of secure craters under good viewing and lighting geometry is restricted to only 9 craters. To first order, Dactyl is saturated with craters 100–300 m in diameter. Some of these craters appear to form a short chain, but the alignment is probably not statistically robust.

A potential problem posed by the existence of Dactyl is that its expected lifetime against collisional disruption (and perhaps erosion) is considerably shorter than the minimal age of Ida, ~2 b.y. (*Davis et al.,* 1996; *Geissler et al.,* 1996). Provided that Dactyl was formed, as seems most plausible, in the same catastrophic disruption of the Koronis family parent body that created Ida, then Dactyl would likely have suffered catastrophic disruption itself after only a few hundred million years (or even sooner, depending on model parameters). The best explanation for Dactyl's retention is that it has indeed undergone several such disruptive impacts, but much of the debris has reaccreted while orbiting Ida. Such a process needs to be quantitatively demonstrated. As shown by Geissler et al., reaccretion of ejecta launched from Dactyl is inefficient; much of it is accumulated on Ida itself. In all probability Dactyl has lost mass and is smaller now than it was when it first entered orbit around Ida.

5. CRATERING ON MATHILDE

The *NEAR Shoemaker* spacecraft was not designed to operate in the main asteroid belt, so the image sequence obtained for this main-belt C-type asteroid during the spacecraft's serendipitous fly-by was unexpectedly and remarkably successful (*Veverka et al.,* 1999). On the other hand,

Fig. 6. *NEAR Shoemaker* image of Mathilde, showing several of its giant craters.

the best resolution was only 0.12 × 0.20 km/pixel (the pixels are rectangular, so resolutions differ in orthogonal directions), about 3× coarser than for Gaspra. While Mathilde is the biggest asteroid imaged by spacecraft, the counting areas for crater statistics were limited by the gaping shadowed areas within the giant craters that characterize this asteroid. Unlike Gaspra and Ida (and, of course, Eros), no images were obtained of the "back side" of this extraordinarily slowly rotating object (Earth-based radar data have been obtained and may eventually reveal something about the unseen side).

Despite these limitations, *NEAR Shoemaker*'s images of Mathilde revolutionized thinking about asteroidal cratering. Defying all expectations, Mathilde not only exhibits one crater exceeding its own radius, it has 4 such craters, not even counting those that may exist on its as-yet-unseen side. This remarkable aspect of Mathilde, which dominates its appearance (Fig. 6), is the subject of several publications, which attempt to explain it: *Cheng and Barnouin-Jha* (1999), *Housen et al.* (1999), *Davis* (1999), *Asphaug and Thomas* (1998), and *Asphaug* (1999). *Chapman et al.* (1999) review the whole population of craters on Mathilde.

5.1. Crater Population on Mathilde

Figure 7 presents an R plot of crater frequencies on Mathilde (*Chapman et al.,* 1999), derived from the total counting area for which the smaller craters have adequate viewing and lighting angles, less than 15% of the total surface area of Mathilde. Craters 0.65–3 km diameter follow the D^{-3} line characteristic of empirical saturation at R ~ 0.25. The wide range of morphologies, from fresh to very subdued, at all sizes confirms the interpretation that Mathilde's smaller craters are in saturation equilibrium, like subkilometer craters on Ida, and sub-150-m craters are presumed to be on Gaspra. Counting statistics are poor for craters in the 3–15-km diameter range, but such craters may fall below the R = 0.25 line, analogous to the depletion of several-kilometer craters on Ida. Remarkably, the seven craters >15 km in diameter define a trend that even exceeds geometrical saturation (R = 1) at 30 km.

The crater densities for small and intermediate-sized craters in Fig. 7 may be biased to higher values since the counting regions are almost wholly exterior to the large craters, whose interiors are generally shadowed or have other unsatisfactory viewing conditions. In those portions of the giant crater interiors that are visible, smaller crater densities are quite low, probably reflecting destruction by downslope mass-wasting.

5.2. Giant Craters, Interior Structure, and Collisional History

Although there have been descriptions of large craters — on the order of radius of the body — on other asteroids and planetary satellites, most are poorly resolved or otherwise indistinct (e.g., the facets on Gaspra). Several of the giant craters on Mathilde are unambiguous concavities: deep, circular, and bowl-shaped. As surprising as their sizes are, their relatively pristine morphologies are astonishing. It was commonly assumed that any impact sufficiently energetic to form such a large crater must have approached the size that

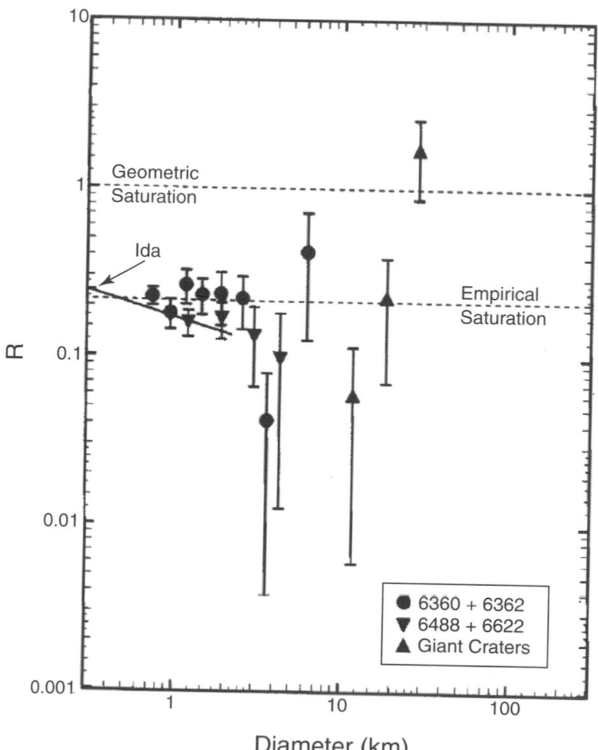

Fig. 7. R plot size-frequency data for craters on Mathilde (legend indicates frame numbers). The data point for the largest craters indicates the extraordinary numbers of giant craters. Smaller craters have similar frequencies to craters of the same size on Ida (line).

would have catastrophically fragmented the target object; whether or not the target actually came close to being disrupted and perhaps reaccreted into a rubble pile, such impacts were expected to have destroyed all preexisting topography by seismic shaking and other obliteration processes (e.g., ejecta blanketing) in the near vicinity of crater rims.

Clearly those suppositions were wrong. Several of the giant craters are proximate to each other, or even overlapping, yet there is little evidence of later ones disrupting preexisting ones. Moreover, there is little evidence of any blanketing of surrounding terrain by low-velocity ejecta from the huge craters. Something must be very different about Mathilde compared with the other asteroids studied. Most researchers who have considered this puzzle conclude that Mathilde's internal constitution is somehow responsible. Indeed, the best estimate of Mathilde's bulk density is 1.3 g/cm^3 (*Veverka et al.,* 1999), or half that of Ida; this estimate is highly uncertain, though, since nearly half of Mathilde's surface that would help define the body's volume is wholly unseen (Veverka et al. point out that most of the uncertainty is on the lower end and that Mathilde's density is very likely to be <1.5 g/cm^3).

Housen et al. (1999) suggest that Mathilde may be composed of weakly bonded porous, crushable materials (consistent with the low density), which are capable of compaction. Laboratory cratering experiments in much media, using a centrifuge to simulate the greater role of gravity in the large-scale impacts on Mathilde, demonstrate that craters form almost solely from compaction, with nearly all ejecta falling back within the cavity rather than being excavated as in cases of cratering in normal geological materials (including sand).

Asphaug and Thomas (1998) and *Asphaug* (1999) have used smooth particle hydrocode (SPH) modeling to offer another option for explaining Mathilde's large craters. Impacts into a rubble pile with high-porosity attenuate and localize shockwave propagation, limiting damage beyond the confines of the crater, and launch most debris at velocities exceeding escape velocity, thus resulting in little blanketing by ejecta. The results are not necessarily contradictory; the hydrocodes lack the resolution to model the fine-scale, porous structure experimented with by *Housen et al.* (1999), and the simulation does not involve inherently crushable material.

While it is not possible to rigorously decide which, if either, of these ideas is correct, there are reasons to prefer the idea of crater formation by crushing and compaction. C-type asteroids, especially those like Mathilde that lack spectral evidence for water, are plausible parent bodies for the least thermally processed meteoritic materials. While nearly all the rocky or dirt-clod meteorites that make it through Earth's atmospheric filter have been strongly selected for strength, the structures of asteroidal interplanetary dust particles (IDPs) suggest that primitive asteroids might not be lithified and may be made of a porous latticework of inherently crushable materials. As described by Housen et al., repetitious impacts serve to compact such material, increasing its density and perhaps even lithifying it in places. The structures modeled by *Asphaug* (1999) (necessarily constrained by computational limitations) appear to be more contrived. After all, we already know that many asteroids — probably including Ida — are likely to be rubble piles with appreciable porosity, yet they exhibit no features, even transitional ones, toward the giant craters on Mathilde. Most likely it is Mathilde's C-type composition, whatever that is actually like, that is responsible for its radically different response to hypervelocity impact cratering.

An important point should be made about Mathilde's structure (cf. *Cheng and Barnouin-Jha,* 1999): The fact that Mathilde has a very low density does not necessarily imply that it is a "rubble pile," in any of the usual senses of that word. While its bulk density is compatible with the proportions of rock and voids that one would get from a rubble-pile structure of a disrupted/reaccreted body made of low-density carbonaceous meteorites, that is hardly a unique model. Primitive objects could have been made, and could remain, of even lower-density materials (including ice) with voids characteristic of the original primordial structure. However, primitive objects existing in the asteroid belt must necessarily have been structurally modified by the pummeling of impacts; whether the rubble-pile gestalt is applicable to such bodies is doubtful.

Standard models of collisional disruption ages and cratering ages (*Davis,* 1999) suggest that Mathilde is ancient and that its surface might even be billions of years old. But such model-dependent calculations may be inapplicable to a body with material properties that respond so oddly to impacts.

6. EROS

Eros was thoroughly studied during a year-long orbital mission by *NEAR Shoemaker* (cf. *Veverka et al.,* 2000) before making several low-altitude flyovers and finally landing on the surface on February 12, 2001 (cf. *Veverka et al.,* 2001). Only preliminary analyses of this vast dataset have been published as of this writing. But attributes of the data are unique — especially imaging at unprecedented resolution down to a couple of centimeters — and even the preliminary results are exceptionally important for understanding asteroidal cratering. A complication, of course, is that Eros is currently in a different impact environment, essentially detached from the asteroid belt, than the other asteroids examined by spacecraft; but, as I will explain, virtually all of the impact cratering expressed on Eros (except at very small spatial scales) must reflect its previous existence in the main asteroid belt.

This review primarily reflects the work of *Chapman et al.* (2002), although it is influenced by other work reported in the January 2002 special issue of *Icarus* by the NEAR Shoemaker Science Team.

6.1. Craters >100 m Diameter

We first consider Eros cratering from the perspective of the early imaging from 200-km altitude, with a resolution

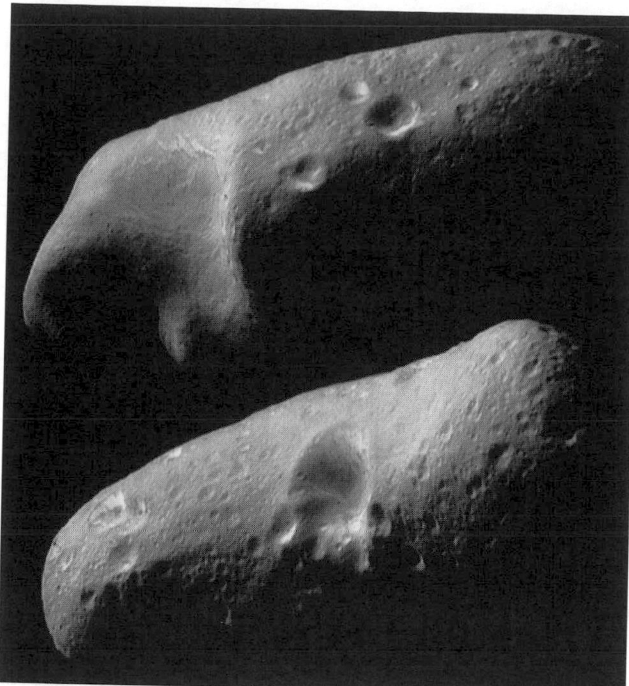

Fig. 8. Images of opposite hemispheres of Eros at moderate resolution. Himeros is near the left end of the upper image, Psyche near the center of the lower image.

of ~25 m/pixel — equivalent to the best resolution obtained on Ida (Fig. 8). The R plot (Fig. 9) shows crater frequencies essentially indistinguishable from those found for Ida. (Crater frequencies are lower by about an order of magnitude in one region, named Shoemaker Regio, which has since been understood to be the interior of an irregular but apparently recent large impact crater. Small craters are also relatively few on the walls of other craters, probably due to mass wasting.) Although quantitative measurements have not yet been made to determine the size-frequency relationship for craters of different morphological classes, the crater populations seem to exhibit a full range of morphologies indicative of saturation equilibrium. Therefore, to first order, Eros looks like Ida and probably for the same reasons.

There are some second-order differences between Eros and Ida at these spatial scales. A global fabric of grooves and ridges is evident on Eros, especially from some lighting geometries, which is not apparent in the more restricted views of Ida. Several features, e.g., the lengthy ridge Rahe Dorsum, clearly have no analog on Ida. These features, and other data, have led NEAR researchers to favor a model of Eros as a heavily fractured "shard" (that is, retaining a roughly coherent rather than disarranged internal structure), implying that Eros may never have been disrupted and reassembled into a rubble pile. Eros also lacks the spatial heterogeneities in color that are important on Ida; perhaps this difference is due to superficial processes (including enhanced space-weathering during Eros' recent residence in Earth-approaching orbits, as well as a possibly enhanced role of levitated dust coating rock surfaces). Conceivably, the stochastic timings of rare large impacts resulted in an unexpectedly recent formation of Ida's Azzurra, making Ida the oddball.

The large, saddle-shaped structure named Himeros, assumed but not proved to be an impact structure, is unique. With a diameter approaching the shortest body dimension of Eros, it could be considered to be analogous to the giant craters on Mathilde. On the other hand, there is only one feature that large on Eros, and it is not so impressive in comparison with Eros' overall dimensions.

6.2. Craters <100 m Diameter and Boulders

Both Eros and Ida appear to have crater populations in saturation equilibrium at crater diameters <1 km (and perhaps at larger sizes, too, as explained in section 4.1). By analogy with the similar-appearing lunar surface, and given a Gaspra- or lunarlike production function, there was every expectation that the equilibrium population would be expressed at all smaller sizes, down to at least the scale of 1–3 m craters, which are near equilibrium densities on the Moon (*Morris et al.,* 1968). As shown in Fig. 10, this is dramatically not the case for smaller craters on Eros.

Instead, the density of craters begins to decline below empirical saturation around 100-m diameter and is only ~0.5% of empirical saturation at 4 m! The images taken during *NEAR Shoemaker*'s landing sequence (Fig. 11) show very few craters centimeters to meters in size, but they would be difficult to recognize in the rocky terrain (some of those that are visible are pits in big rocks) and such low

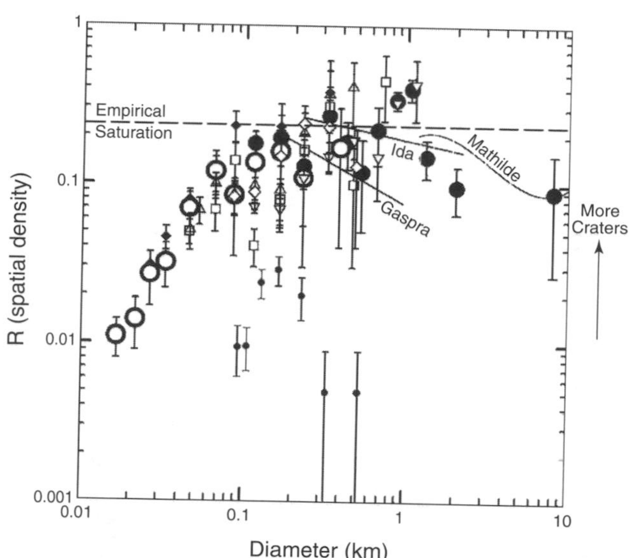

Fig. 9. R plot size-frequency data for craters on Eros >15 m diameter. Various regions sampled are indicated by different symbols. The small, solid dots between 90 and 600 m, which plot at least an order of magnitude lower than most data, are for portions of Shoemaker Regio, a probable recent, large impact crater responsible for distributing many of the larger blocks around Eros. Comparison curves for Gaspra, Ida, and Mathilde are also shown.

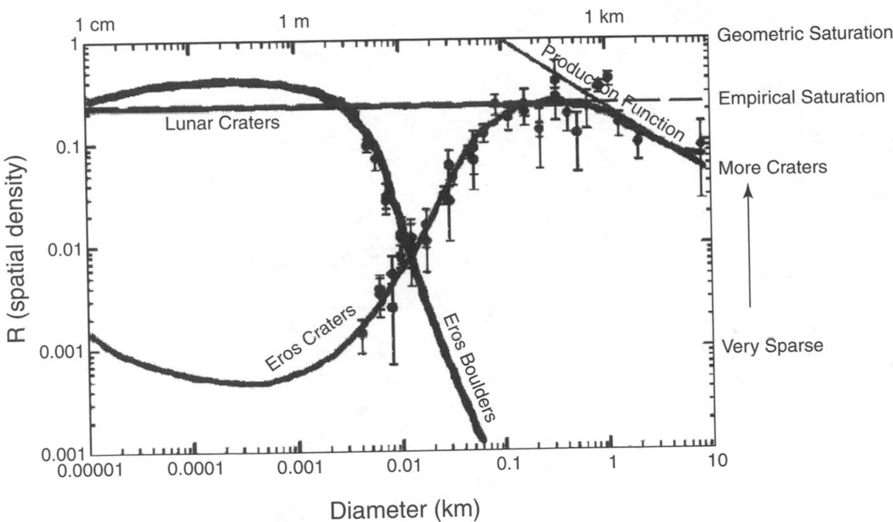

Fig. 10. R plot size-frequency relations for craters and boulders on Eros, from 1 cm to 10 km diameter. A few representative counts are plotted. Solid curves represent, approximately, trends for craters and boulders on Eros, the nominal crater production curve at large diameters, and the lunar maria crater distribution (which follows the production curve at diameters larger than 1 km and then follows the empirical saturation line at smaller sizes). Curves for Eros craters and boulders below 3 m are schematic pending detailed analysis.

spatial densities are difficult to characterize with any statistical significance, because they reflect the presence of only a handful of craters per image.

In a complementary way, boulders increase in spatial density with decreasing size over the same size range that craters decrease (Fig. 10). The largest boulder on Eros is ~150 m, but the boulder population approaches the densities of empirically saturated craters at sizes between 3 and 8 m (boulder density varies regionally much more than does crater density, by factors of tens). While landing-sequence images (taken in the vicinity of the rim of Himeros) are dominated by a rocky and boulder-covered landscape, the rocks are not "piled up" on each other, as would be predicted by an extrapolation to smaller sizes from the very steep power-law size-distribution (differential exponent ~–5) in the size range of 8–30 m.

Fig. 11. One of the last images taken during *NEAR Shoemaker*'s final descent to the surface of Eros. This image shows a region roughly 18 m across; the largest boulders are 2–3 m in size. The surface is essentially covered with rocks and boulders. Solar illumination is from the left.

Despite the shocking predominance of positive relief features (boulders) rather than craters in the higher-resolution images of Eros, there is nothing inconsistent with what might also be true for Ida (if it could be resolved). Indeed *Lee et al.* (1996), from the analysis of 17 blocks on Ida between 50 and 150 m in size, correctly predicted that the largest block on Eros would be "of order 100 m." Lee et al.'s sampling of blocks was from just a modest fraction of Ida's surface and was markedly incomplete; it is entirely consistent with *Thomas et al.*'s (2001) global tabulation of ~100 blocks >50 m in size on Eros. Thus Ida could well have innumerable small blocks, just like Eros. Consistent with theories for the origin of the much rarer blocks on the Moon, large blocks are believed to have been produced by excavation and spallation in association with the formation of large, recent craters. Thomas et al. suggest that many of the larger blocks on Eros might have been excavated by the impact that formed the large (7.5-km) recent crater, Shoemaker Regio. On the other hand, the profusion of smaller, meter-scale blocks on Eros, which are much more numerous than on the Moon, requires explanation.

6.3. Processes Affecting Small-Scale Character of Eros' Surface

High-resolution imaging of Eros has revealed abundant geological features of great interest, including "ponds" and other features. I concentrate here on processes that might explain the unexpected scarcity of small craters and abundance of small boulders on Eros [for more details, cf. *Chapman et al.* (2002)].

Logically, the smaller craters (1) formed at expected frequencies but were subsequently erased (by erosion or blanketing), (2) formed less efficiently due to factors that might inhibit formation of a crater by an impacting projectile, or (3) formed in reduced numbers because of a paucity of small impactors. One would naturally turn to explanation (1) (e.g., ejecta blanketing or seismic shaking from a very large impact), but this explanation fails for several reasons. First, the very strong size dependence of the observed crater depletion does not fit models for crater erasure [typically varying as 1/D, e.g., for infilling of homologously shaped craters (*Chapman,* 1974)]; second, effects due to large impacts would be intermittent and the production function should be prominently reexpressed (unless the major cratering event improbably happened "yesterday"); and, finally, boulders would be buried by blanketing, just as the craters would be. Crater formation might be inhibited by the armoring of the surface by the boulders; this might affect the production of some craters smaller than 10-m diameter, but boulders are not sufficiently pervasive to account for orders-of-magnitude depletion. The most workable explanation is that the impactor population is depleted in projectiles meters in diameter and smaller. While the meteoroid population in near-Earth space is quite well characterized and is not so depleted, the numbers of small asteroids in the main asteroid belt — where, as we will see, Eros acquired most of its cratering — are not known. As *Bell* (2001) has suggested, the Yarkovsky forces that derive meteorites from the asteroid belt deplete the belt of objects of just these sizes. A quantitative model has yet to be done that would demonstrate if such depletion would overcome processes that repopulate the belt with objects of such sizes; this would be needed to explain the apparent orders-of-magnitude depletion. It may be that the Yarkovsky effect must be supplemented by other processes operating at the smallest sizes, including armoring by abundant boulders, seismic shaking, and perhaps small-scale blanketing by electrostatically levitated dust.

Let me turn briefly to the profusion of boulders. These rocks might be produced as ejecta blocks from large craters (perhaps after temporary existence in orbit around Eros for a few of them) and/or by erosive exhumation of blocky megaregolith *in situ*. There is some evidence for alignments and directionalities of boulders that suggests a small fraction may be necessarily formed in the latter way, although the global distributions reported by *Thomas et al.* (2001) favor ejection from large impact craters (chiefly from one crater) for most boulders. A problem is that few boulders exhibit trails (or pits), which were expected to reflect how some of them must necessarily land at very oblique (or vertical) angles in the regolith.

The craggy appearance of many boulders, contrasting with the more rounded/sandblasted appearance of most lunar blocks, is one factor that leads *Bell* (2001) to invoke the Yarkovsky effect. Boulders and rocks have limited lifetimes on the Moon for the same reasons that the ages of meteorites, as measured by cosmic-ray exposure, are limited (cf. *Lee et al.,* 1996). A depletion — for Eros compared with the Moon — of small projectiles relative to the numbers of projectiles that form the large, boulder-producing craters would sustain the larger population of boulders on Eros. In this way, the beauty of the Yarkovsky explanation for the paucity of craters (paucity of small projectiles) simultaneously explains the abundance of boulders. It operates in the asteroid belt but not on the Moon.

6.4. History of Eros

Eros is in a relatively short-lived, temporary Earth-approaching orbit, subject to chaotic dynamics and occasional close encounters with planets. Therefore, its past (and future) history cannot be specified deterministically. Its dynamical evolution as an NEA has been investigated by *Michel et al.* (1998) and more recently by *Bottke* (2001). While *Michel et al.* (1998) find that Eros is in an unusually long-lived orbit for an NEA, perhaps 50–100 m.y., *Bottke*'s (2001) simulations suggest that it has been an NEA for only ~15 m.y. During that time, it may or may not have been collisionally decoupled from the main asteroid belt. For instance, in *Michel et al.*'s (1998) 5-m.y. integrations of 16 Eros clones, 3 have collisional histories comparable to main-belt asteroids (because the clones' aphelia stick out into the main belt for most of the time), while about half the clones have collisional fluxes down by 2 orders of magnitude, similar to the lunar impact rate. In the latter case, Eros may have had a relative hiatus in cratering rates for the past few tenths of a percent of its lifetime; while that is a short fraction of its lifetime, it may be long with respect to timescales for space weathering, thus manifesting observable spectral differences compared with main-belt asteroids that have had no recent cratering hiatus.

Bottke's (2001) analysis shows that Eros was most likely (~50% chance) derived from the ν_6 secular resonance, but may instead have left the main belt via Mars-crossing (30% chance) or 3:1 resonance (20%) routes. In all probability Eros was created by a catastrophic disruption of its parent body ca 2 Ga [very approximately, by analogy with the age of Ida, which has a nearly identical cratering record (*Chapman et al.,* 1996b)] in a collision that may have produced a family of other asteroids. *Zappalà et al.* (1997) propose that Eros and another large NEA, Ganymed, may have been produced at the time of the creation of the Maria family, on the edge of the 3:1 resonance. However, it is more likely that Eros was formed far from a resonance "escape hatch" and slowly drifted for eons under the Yarkovsky force to one of the previously mentioned resonances. During this time, it would have accumulated between ~99% and ~99.99% of its cumulative impact record, depending on its subsequent degree of decoupling from the asteroid belt. Thus its large-scale geology must be that of a main-belt asteroid. However, the possibility that it may have had a cratering hiatus for the last ~10 m.y. or longer, during which it has been subject to solar wind and impact flux more nearly like that of the Moon, implies that its smallest-scale attributes (and even optical properties) might differ from what would be true had it remained in the asteroid belt.

7. CONCLUSIONS

Asteroids, the erstwhile "vermin of the skies," have had a difficult time living down the reputation of "seen one asteroid, seen them all." In yet one more case, asteroidal cratering, we have seen that cratering on different asteroids is more commonly dissimilar than similar. Of the four as-

Fig. 12. Images of Gaspra, Ida, Mathilde, and Eros to the same scale.

teroids whose surfaces have been imaged by spacecraft, only two (Ida and Eros) seem similar, while the other two (Gaspra and Mathilde) are dramatically different from Ida and Eros, and even more dramatically different from each other (Fig. 12). Moreover, the very high-resolution images of Eros have demonstrated that our theoretical expectations of what an asteroid surface might look like close up were badly wrong.

One way to try to understand the differences in cratering traits among these bodies is to ascribe them to dramatic differences in the structural nature of the asteroids. Those bodies with crater populations (100 m to 10 km in diameter) most similar to that on the Moon can be taken to be rocky bodies, structurally similar to the lunar crust. (Differences at high resolution can be ascribed, in part, to the different regolith processes expected for small, low-gravity bodies, perhaps augmented by differences in the small-projectile populations in and outside of the asteroid belt.) The gaping, huge craters on Mathilde may reflect that body's exotic, low-density, crushable composition, perhaps typical of primitive carbonaceous objects. Finally, the faceted shape of Gaspra and its very underabundant population of medium and large craters may exemplify the response of a solid-metallic (e.g., pallasitic) object to impact cratering. The only way to know if these speculations might be correct is to continue to explore the asteroids, close up.

Acknowledgments. The NEAR Shoemaker Project at Johns Hopkins Applied Physics Laboratory supported this research. I have benefited, over the years, from numerous collaborators and discussions with colleagues, both members of the *Galileo* and *NEAR Shoemaker* imaging teams and not.

REFERENCES

Asphaug E. (1999) The large, undisturbed craters of Mathilde: evidence for structural porosity (abstract). In *Lunar and Planetary Science XXXI,* Abstract #1864. Lunar and Planetary Institute, Houston (CD-ROM).

Asphaug E. and Thomas P. C. (1998) Modeling mysterious Mathilde (abstract). In *Lunar and Planetary Science XXX,* Abstract #2028. Lunar and Planetary Institute, Houston (CD-ROM).

Asphaug E., Moore J. M., Morrison D., Benz W., Nolan M. C., and Sullivan R. J. (1996) Mechanical and geological effects of impact cratering on Ida. *Icarus, 120,* 158–184.

Asphaug E., Ostro S. J., Hudson R. S., Scheeres D. J., and Benz W. (1998) Disruption of kilometre-sized asteroids by energetic collisions. *Nature, 393,* 437–440.

Asphaug E., Ryan E. V., and Zuber M. T. (2002) Asteroid interiors. In *Asteroids III* (W. F. Bottke Jr. et al., eds.), this volume. Univ. of Arizona, Tucson.

Bell J. F. (2001) Eros: A comprehensive model (abstract). In *Lunar and Planetary Science XXXII,* Abstract #1964. Lunar and Planetary Institute, Houston (CD-ROM).

Belton M. J. S., Chapman C. R., Veverka J., Klaasen K. P., Harch A., Greeley R., Greenberg R., Head J. W. III, McEwen A., Morrison D., Thomas P. C., Davies M. E., Carr M. H., Neukum G., Fanale F. P., Davis D. R., Anger C., Gierasch P. J., Ingersoll A. P., and Pilcher C. B. (1994) First images of asteroid 243 Ida. *Science, 265,* 1543–1547.

Belton M. J. S., Chapman C. R., Klaasen K. P., Harch A. P., Thomas P. C.,Veverka J., McEwen A. S., and Pappalardo R. T. (1996) Galileo's encounter with 243 Ida: An overview of the imaging experiment. *Icarus, 120,* 1–19.

Benner L. A. M., Nolan M. C., Margot J.-L., Giorgini J. D., Hudson R. S., Jurgens R. F., and Ostro S. J. (2001) Recent radar observations of four near-Earth asteroids (abstract). *Bull. Am. Astron. Soc., 33,* 918.

Bogard D. D. (1995) Impact ages of meteorites: A synthesis. *Meteoritics, 30,* 244–268.

Bottke W. F. (2001) On the orbital and collisional history of (433) Eros (abstract). In *AGU Fall Meeting 2001,* Abstract #P32B-0557.

Carr M. H., Kirk R. L., McEwen A., Veverka J., Thomas P., Head J. W., and Murchie S. (1994) The geology of Gaspra. *Icarus, 107,* 61–71.

Chapman C. R. (1974) Cratering on Mars: I. Cratering and obliteration history. *Icarus, 22,* 272–291.

Chapman C. R. (1978) Asteroid collisions, craters, regoliths, and lifetimes. In *Asteroids: An Exploration Assessment* (D. Morrison and W. C. Wells,eds.) pp. 145–160. NASA Conference Publication 2053.

Chapman C. R. (1996) S-type asteroids, ordinary chondrites, and space weathering: The evidence from Galileo's fly-bys of Gaspra and Ida. *Meteoritics & Planet. Sci., 31,* 699–725.

Chapman C. R. (1997) Gaspra and Ida: Implications of spacecraft reconnaissance for NEO issues. *Ann. N.Y. Acad. Sci., 822,* 227–235.

Chapman C. R. Veverka J., Thomas P. C., Klaasen K., Belton M. J. S., Harch A., McEwen A., Johnson T. V., Helfenstein P., Davies M. E., Merline W. J., and Denk T. (1995) Discovery and physical properties of Dactyl, a satellite of asteroid 243 Ida. *Nature, 374,* 783–785.

Chapman C. R., Veverka J., Belton M. J. S., Neukum G., and Morrison D. (1996a) Cratering on Gaspra. *Icarus, 120,* 231–245.

Chapman C. R., Ryan E. V., Merline W. J., Neukum G., Wagner R., Thomas P. C., Veverka J., and Sullivan R.J. (1996b) Cratering on Ida. *Icarus, 120,* 77–86.

Chapman C. R., Merline W. J., and Thomas P. (1999) Cratering on Mathilde. *Icarus, 140,* 28–33.

Chapman C. R., Merline W. J., Thomas P. C., Joseph J., Cheng A. F., and Izenberg N. (2002) Impact history of Eros: Craters and boulders. *Icarus, 155,* 104–118.

Cheng A. F. and Barnouin-Jha O. S. (1999) Giant craters on Mathilde. *Icarus, 140,* 34–48.

Clark B. E., Hapke B., Pieters C., and Britt D. (2002) Asteroid space weathering and regolith evolution. In *Asteroids III* (W. F. Bottke Jr. et al., eds.), this volume. Univ. of Arizona, Tucson.

Crater Analysis Techniques Working Group (1977) Standard techniques for presentation and analysis of crater size-frequency data. *Icarus, 37,* 467–474.

Davis D. R. (1999) The collisional history of asteroid 253 Mathilde. *Icarus, 140,* 49–52.

Davis D. R., Chapman C. R., Durda D. D., Farinella P., and Marzari F. (1996) The formation and collisional/dynamical evolution of the Ida/Dactyl system as part of the Koronis family. *Icarus, 120,* 220–230.

Dohnanyi J. S. (1971) Fragmentation and distribution of asteroids. In *Physical Studies of Minor Planets* (T. Gehrels, ed.), pp. 263–295. NASA Special Publication 267.

Durda D. D. (1996) The formation of asteroidal satellites in catastrophic collisions. *Icarus, 120,* 212–219.

Farinella P., Davis D. R., and Marzari F. (1996) Asteroid families, old and young. In *Completing the Inventory of the Solar System* (T. W. Rettig and J. M. Hahn, eds.), pp. 45–55. ASP Conference Series 107.

Geissler P., Petit J.-M., Durda D. D., Greenberg R., Bottke W., Nolan M., and Moore J. (1996) Erosion and ejecta reaccretion on 243 Ida and its moon. *Icarus, 120,* 140–157.

Gradie J. C., Chapman C. R. and Tedesco E. F. (1989) Distribution of taxonomic classes and compositional structure of the asteroid belt. In *Asteroids II* (R. P. Binzel et al., eds.), pp. 316–335. Univ. of Arizona, Tucson.

Greenberg R., Nolan M. C., Bottke W. F., Kolvoord R. A., and Veverka J. (1994) Collisional history of Gaspra. *Icarus, 107,* 84–97.

Greenberg R., Bottke W. F., Nolan M., Geissler P., Petit J.–M., Durda D. D., Asphaug E., and Head J. (1996) Collisional and dynamical history of Ida. *Icarus, 120,* 106–118.

Hartmann W. K. (1995) Planetary cratering 1. The question of multiple impactor populations. *Meteoritics, 30,* 451–467.

Housen K. R., Wilkening L. L., Chapman C. R., and Greenberg R. J. (1979) Regolith development and evolution on asteroids and the Moon. In *Asteroids* (T. Gehrels, ed.), pp. 601–627. Univ. of Arizona, Tucson.

Housen K. R., Holsapple K. A., and Voss M. E. (1999) Compaction as the origin of the unusual craters on the asteroid Mathilde. *Nature, 402,* 155–157.

Jedicke R., Larsen J., and Spahr T. (2002) Observational selection effects in asteroid surveys and estimates of asteroid population sizes. In *Asteroids III* (W. F. Bottke Jr. et al., eds.), this volume. Univ. of Arizona, Tucson.

Lee P., Veverka J., Thomas P. C., Helfenstein P., Belton M. J. S., Chapman C. R., Greeley R., Pappalardo R. T., Sullivan R., and Head J. W. III (1996) Ejecta blocks on 243 Ida and on other asteroids. *Icarus, 120,* 87–105.

Michel P., Farinella P., and Froeschlé Ch. (1998) Dynamics of Eros. *Astron. J., 116,* 2023–2031.

Michel P., Benz W., Tanga P., and Richardson D. C. (2001) Collisions and gravitational reaccumulation: Forming asteroid families and satellites. *Science, 294,* 1696–1700.

Morbidelli A., Bottke W. F. Jr., Froeschlé Ch., and Michel P. (2002) Origin and evolution of near-Earth objects. In *Asteroids III* (W. F. Bottke Jr. et al., eds.), this volume. Univ. of Arizona, Tucson.

Morris E. C., Batson R. M., Holt H. E., Rennilson J. J., Shoemaker E. M., and Whitaker E. A. (1968) III. Television observations from Surveyor VI. In *Surveyor VI Mission Report. Part II: Science Results,* pp. 9–45. JPL Technical Report 32-1262.

Nesvorný D., Morbidelli A., Bottke W. F., Vokrouhlický D., and Brož M. (2001) *Do The Asteroid Families Dynamically Evolve?* Available on line at http://www.boulder.swri.edu/~davidn/yarko/yarko.html.

Neukum G. and. Ivanov B. A (1994) Crater size distributions and impact probabilities on Earth from lunar, terrestrial-planet, and asteroid cratering data. In *Hazards Due to Comets and Asteroids* (T. Gehrels, ed.), pp. 359–416. Univ. of Arizona, Tucson.

Neukum G., Ivanov B., and Hartmann W. K. (2001) Cratering records in the inner solar system in relation to the lunar reference system. In *Chronology and Evolution of Mars* (R. Kallenbach et al., eds.), pp. 55–86. Kluwer, Dordrecht.

Ostro S. J., Hudson R. S., Benner L. A. M., Giorgini J. D., Magri C., Margot J.-L., and Nolan M. C. (2002) Asteroid radar astronomy. In *Asteroids III* (W. F. Bottke Jr. et al., eds.), this volume. Univ. of Arizona, Tucson.

Richardson D. C., Leinhardt Z. M., Melosh H. J., Bottke W. F. Jr., and Asphaug E. (2002) Gravitational aggregates: Evidence and evolution. In *Asteroids III* (W. F. Bottke Jr. et al., eds.), this volume. Univ. of Arizona, Tucson.

Shoemaker E. M., Hackman R. J., and Eggleton R. E. (1963) Interplanetary correlation of geologic time. *Adv. Astronaut. Sci., 88,* 70–89.

Stooke P. J. and Ford H. A. (2001) Gaspra: Revised crater counts (abstract). In *Lunar and Planetary Science XXXII,* Abstract #1073. Lunar and Planetary Institute, Houston (CD-ROM).

Sullivan R., Greeley R., Pappalardo R., Asphaug E., Moore J. M., Morrison D., Belton M. J. S., Carr M., Chapman C. R., Geissler P., Greenberg R., Granahan J., Head J. W. III, Kirk R., McEwen A., Lee P., Thomas P. C., and Veverka J. (1996) Geology of 243 Ida. *Icarus, 120,* 119–139.

Sullivan R. J., Thomas P. C., Murchie S. L., and Robinson M. S. (2002) Asteroid geology from *Galileo* and *NEAR Shoemaker* data. In *Asteroids III* (W. F. Bottke Jr. et al., eds.), this volume. Univ. of Arizona, Tucson.

Thomas P. C., Veverka J., Simonelli D., Helfenstein P., Carcich B., Belton M. J. S., Davies M. E., and Chapman C. (1994) The shape of Gaspra. *Icarus, 107,* 23–36.

Thomas P. C., Belton M. J. S., Carcich B., Chapman C. R., Davies M. E., Sullivan R., and Veverka J. (1996) The shape of Ida. *Icarus, 120,* 20–32.

Thomas P. C., Binzel R. P., Gaffey M. J., Storrs A. D., Wells E. N., and Zellner B. H. (1997) Impact excavation on asteroid 4 Vesta: Hubble Space Telescope results. *Science, 277,* 1492–1495.

Thomas P. C., Veverka J., Robinson M. S., and Murchie S. (2001) Shoemaker crater as the source of most ejecta blocks on the asteroid 433 Eros. *Nature, 413,* 394–396.

Veverka J., Belton M., Klaasen K., and Chapman C. (1994) Galileo's encounter with 951 Gaspra: Overview. *Icarus, 107,* 2–17.

Veverka J., Thomas P. C., Helfenstein P., Lee P., Harch A., Calvo S., Chapman C., Belton M. J. S., Klaasen K., Johnson T. V., and Davies M. (1996) Dactyl: Galileo observations of Ida's satellite. *Icarus, 120,* 200–211.

Veverka J., Thomas P., Harch A., Clark B., Bell J. F. III, Carcich B., Joseph J., Murchie S., Izenberg N., Chapman C., Merline W., Malin M., McFadden L., and Robinson M. (1999) NEAR encounter with asteroid 253 Mathilde: Overview. *Icarus, 140,* 3–16.

Veverka J., Robinson M., Thomas P., Murchie S., Bell J. F. III, Izenberg N., Chapman C., Harch A., Bell M., Carcich B.,

Cheng A., Clark B., Domingue D., Dunham D., Farquhar R., Gaffey M. J., Hawkins E., Joseph J., Kirk R., Li H., Lucey P., Malin M., Martin P., McFadden L., Merline W. J., Miller J. K., Owen W. M. Jr., Peterson C., Prockter L., Warren J., Wellnitz D., Williams B. G., and Yeomans D. K. (2000) NEAR at Eros: Imaging and spectral results. *Science, 289,* 2088–2097.

Veverka J., Farquhar R., Robinson M., Thomas P., Murchie S., Harch A., Antreasian P. G., Chesley S. R., Miller J. K., Owen W. M. Jr., Williams B. G., Yeomans D., Dunham D., Heyler G., Holdridge M., Nelson R. L., Whittenburg K. E., Ray J. C., Carcich B., Cheng A., Chapman C., Bell J. F. III, Bell M., Bussey B., Clark B., Domingue D., Gaffey M. J., Hawkins E., Izenberg N., Joseph J., Kirk R., Lucey P., Malin M., McFadden L., Merline W. J., Peterson C., Prockter L., Warren J., and Wellnitz D. (2001) The landing of the NEAR-Shoemaker spacecraft on asteroid 433 Eros. *Nature, 413,* 390–393.

Zappalà V., Cellino A., DiMartino M., Migliorini F., and Paolicchi P. (1997) Maria's family: Physical structure and possible implications for the origin of giant NEAs. *Icarus, 129,* 1–20.

Zellner B. (1979) Asteroid taxonomy and the distribution of the compositional types. In *Asteroids* (T. Gehrels, ed.), pp. 783–806. Univ. of Arizona, Tucson.

Asteroid Geology from *Galileo* and *NEAR Shoemaker* Data

R. J. Sullivan
Cornell University

P. C. Thomas
Cornell University

S. L. Murchie
Applied Physics Laboratory

M. S. Robinson
Northwestern University

The four asteroids visited by spacecraft, Gaspra, Ida, Mathilde, and Eros, have shapes and surface morphologies dominated by the effects of impact cratering. The presence of impact-derived regolith has been inferred on S-class asteroids Gaspra, Ida, and Eros from photometry, morphological features, and subtle color variations related to optical maturation of surface materials. Grooves and ridges suggest the presence of throughgoing structures, implying that interiors are partially competent (e.g., fractured but intact) rock materials, not collections of gravitationally bound fragments reassembled after catastrophic disruption. Mathilde's shape is dominated by impact craters with diameters comparable to the asteroid's radius that formed without disrupting the asteroid, suggesting a less-competent, highly porous interior. The surfaces of all these objects exhibit evidence of downslope motion of loose, fragmental material. There is no unambiguous spectral evidence for kilometer-scale compositional heterogeneity within any of these bodies. S(IV) class asteroids Ida and Eros have some compositional similarities to ordinary chondrites.

1. INTRODUCTION

During the two centuries since Ceres was discovered, advances in our knowledge about asteroids have been made primarily from groundbased observations, as well as from inferences drawn from analyses of meteorites. Only since 1991 have we extended our reach and begun to visit these bodies (vicariously) with spacecraft. Compared with other methods, spacecraft encounters with asteroids provide uniquely detailed information about asteroid shapes, surface features, regolith properties, cratering histories, and clues about asteroid interiors. However, the cost and effort required for obtaining these data are high compared with groundbased methods, and only five asteroids of the several tens of thousands of known bodies have been visited so far. The *Galileo* spacecraft, on its way to Jupiter, flew past the S-type asteroids 951 Gaspra in 1991 and 243 Ida (discovering its satellite Dactyl) in 1993. The *NEAR Shoemaker* spacecraft flew past C-type asteroid 253 Mathilde in 1997 and S-type asteroid 433 Eros in 1998, and began year-long orbital operations at Eros in 2000 that culminated in a successful descent and touchdown on the surface one year later. These asteroids are shown together in a comparative portrait in Fig. 1, and their basic properties are summarized in Table 1.

Modeling of asteroid surface properties and geology has a long history beginning well before our first spacecraft visit (e.g., *Chapman,* 1971, 1978; *Cintala et al.,* 1978, 1979; *Housen et al.,* 1979a,b; *Housen and Wilkening*, 1982; *McKay et al.,* 1989). Spacecraft exploration of asteroids has been guided by expectations and unresolved questions from this and other work, as well as experience with Phobos and Deimos at Mars (e.g., *Pollack et al.,* 1972, 1973; *Noland et al.,* 1973; *Veverka et al.,* 1974; *Goguen et al.,* 1978; *Thomas,* 1979; *Veverka and Thomas,* 1979; *Thomas and Veverka,* 1980; *Avanesov et al.,* 1989, 1991; *Murchie et al.,* 1991; *Asphaug and Benz,* 1994; *Murchie and Erard,* 1996; *Thomas et al.,* 1996a; *Simonelli et al.,* 1998), glimpses of irregular satellites of the outer planets (e.g., *Croft,* 1992; *Stook,* 1993, 1994; *Thomas,* 1999), and especially by our need to pursue questions of asteroid science that cannot be addressed effectively using groundbased methods. To touch upon just a few of these issues, do asteroids have impact-derived regoliths despite their low surface gravity? If so, what is the nature of these regoliths, and how are they different from more familiar impact-derived regolith on the Moon? Are asteroid regoliths subject to optical maturation processes from exposure to the space environment, similar to processes affecting lunar surface materials? If so, how might the analogous asteroid process differ from the Moon

Fig. 1. A composite view of the asteroids visited by *Galileo* and *NEAR Shoemaker*, all shown approximately at the same scale along with the satellites of Mars. Clockwise from upper left are Phobos, Eros, Ida, tiny Dactyl, Mathilde (the largest), and Deimos, with Gaspra in the center of the figure. All portraits have been contrast-enhanced to highlight surface detail and do not represent relative albedos.

because of different composition, impact history, distance from the Sun, and gravity? Are some S-type asteroids the sources for ordinary chondrite meteorites? What is the nature of the impact-cratering process under very low gravity? What are the relative influences of material target strength and gravity on crater geometry at increasingly smaller crater sizes under very low gravitational accelerations? Are there grooves on asteroids, as first seen on Phobos? If so, what do their configurations imply about internal structure and coherence? Are asteroids gravitationally bound rubble piles, are they shards from collisions, or are examples of both configurations abundant? What other internal configurations are possible? Is there evidence for compositional zonation (e.g., from differentiation of larger parent bodies) within aster-

TABLE 1. Basic properties of Gaspra, Ida, Dactyl, Mathilde, and Eros.

	Gaspra	Ida	Dactyl	Mathilde	Eros
Spectral Class	S	S	S	C	S
Semimajor Axis (AU)	2.21	2.86	2.86	2.65	1.46
Mean Radius (km)	6.1 ± 0.4	15.7 ± 0.6	0.7	26.5 ± 1.3	7.311 ± 0.010
Triaxial Ellipsoid Fit	18.2 × 10.5 × 8.9	29.9 × 12.7 × 9.3	1.6 × 1.4 × 1.2	66 × 48 × 46	34.4 × 11.2 × 11.2
Rotational Period	7.04 h	4.63 h	Unknown	17.4 d	5.27 h
Density (g/cm^3)	Unknown	2.6 ± 0.5	Unknown	1.3 ± 0.2	2.67 ± 0.03
Surface Gravity (cm/s^2)	Unknown	0.3–1.1	0.07 if density is 3.5 g/cm^3	~1.0	0.23–0.56
Spacecraft Encounter	*Galileo* flyby 10/29/91	*Galileo* flyby 8/28/93	*Galileo* flyby 8/28/93	*NEAR Shoemaker* flyby 6/27/97	*NEAR Shoemaker* flyby 12/23/98; orbit 2/14/00; landing 2/12/01

oids? How well can asteroid ages be determined from counting craters on their surfaces? What are the densities and internal porosities of asteroids? Direct exploration by spacecraft during the past decade also has presented answers to questions that had not been emphasized previously: Do asteroids have satellites? How does high porosity influence crater size and morphology? Can electrostatic levitation move regolith particles laterally across asteroid surfaces?

The *Galileo* and *NEAR Shoemaker* spacecraft functioned as tools for addressing such questions. *Galileo* carried a comprehensive suite of instruments, some mounted on a scan platform, for its main mission objective at Jupiter. Both asteroids encountered by *Galileo* were S-types, a broadly defined spectral class that is found more commonly in the inner part of the asteroid belt and is thought to be composed of various proportions of pyroxene, olivine, and possibly metallic Fe and Ni (*Gaffey et al.,* 1993a,b). Asteroid 951 Gaspra is a relatively olivine-rich S-type. Asteroid 243 Ida is classified S(IV), a subgroup within the S-class that has been proposed as a potential source for ordinary chondrite meteorites (e.g., *Gaffey et al.,* 1993a,b; *Chapman,* 1996). Ida is part of the Koronis dynamical family of asteroids that probably was created by the collisional fragmentation of a larger parent body (*Binzel,* 1988; *Binzel et al.,* 1993). Most information about Gaspra and Ida was acquired with *Galileo*'s Solid-State Imager (SSI) multispectral camera designed primarily for jovian system science (*Belton et al.,* 1992a). The Near Earth Asteroid Rendezvous (NEAR) mission employed a smaller, simpler spacecraft, *NEAR Shoemaker*, with five body-mounted instruments optimized for asteroid targets (*Hawkins et al.,* 1997; *Warren et al.,* 1997; *Goldsten et al.,* 1997; *Cole et al.,* 1997; *Lohr et al.,* 1997; *Murchie et al.,* 1999). *NEAR Shoemaker*'s flyby with 253 Mathilde represents the only visit so far to a C-type asteroid (the class that dominates the outer asteroid belt). *NEAR Shoemaker*'s primary goal was to explore the largest near-Earth asteroid, S(IV)-type 433 Eros, during extended operations from the first-ever orbit around an asteroid (*Cheng,* 1997). In this chapter we summarize our current knowledge of the geology of Gaspra, Ida (and satellite Dactyl), Mathilde, and Eros as determined from *Galileo* and *NEAR Shoemaker* data. In the following sections we present overviews of the geology of each of these bodies, then conclude with a discussion of their similarities and differences and implications for geological processes acting in an asteroid setting.

2. GASPRA

Our first close-up examination of an asteroid occurred when the *Galileo* spacecraft dipped into the inner asteroid belt and flew past Gaspra on October 29, 1991, while en route to a final gravity-assist encounter with Earth before heading outward to Jupiter. Low transmission rate from the spacecraft allowed only the four highest-resolution color images (162 m/pixel) to be returned initially (*Belton et al.,* 1992b), followed several months later by the rest of the data, including the highest-resolution 54-m/pixel images (*Veverka et al.,* 1994a). These data revealed Gaspra to be an irregular body 18.2 × 10.5 × 8.9 km (*Simonelli et al.,* 1993; *Thomas et al.,* 1994). The 8-km/s flyby speed and Gaspra's relatively slow rotation rate did not allow the entire surface to be observed, but the areas seen best by the spacecraft form several nearly flat facets and broad, shallow concavities several kilometers across, divided by narrower, gentle ridges (*Thomas et al.,* 1994). No blocks or boulders are apparent and the facets and intervening ridges appear relatively smooth, even in the highest-resolution images. At finer scales the surface is marked by craters, and some grooves that occur as linear and pitted depressions 100–200 m wide and up to 2.5 km long (*Veverka et al.,* 1994b). The grooves are much less conspicuous than grooves on Phobos associated with its large crater Stickney (*Thomas et al.,* 1979; *Fujiwara,* 1991; *Asphaug and Melosh,* 1993). Two populations of craters are present that are distinguished both by size and morphology. Many craters <1 km diameter have resolvable, crisply defined rims, and a size-frequency distribution consistent with a steep production function. The second crater population consists of larger, but relatively shallower concavities with more subdued, ambiguous rims (*Belton et al.,* 1992b; *Carr et al.,* 1994; *Greenberg et al.,* 1994; *Chapman et al.,* 1996a). Some workers also consider the broad, slightly concave facets several kilometers across to be the largest members of this second crater population (*Greenberg et al.,* 1994). There is general consensus that the second population probably represents older craters that have been degraded by ejecta blanketing, seismic disturbance, or some other process, perhaps as a result of a single impact event that caused widespread alteration and degradation of surface features. Since that event, which nearly reset the surface age, subsequent impacts have produced smaller craters scattered relatively thinly across the surface according to a steep production function.

The presence of a regolith (loose, fragmental debris, regardless of mode of origin) on so small a body as Gaspra was not predicted with certainty before the *Galileo* encounter (e.g., *Chapman,* 1971, 1976; *Housen et al.,* 1979a,b), although polarimetry work (*Dolfus et al.,* 1977; see also review by *Dolfus and Zellner,* 1979) predicted fine particle mantles covering even very small bodies. Several lines of evidence from *Galileo* images indicate the surface of Gaspra is covered with tens of meters of regolith, and suggest that subtle optical changes occur with increasing exposure age at the surface. The *Galileo* flyby, in combination with groundbased studies, allowed improved photometric analysis of Gaspra across a much wider range of phase angles than from groundbased observations alone. Gaspra's surface photometric behavior is inconsistent with broad expanses of bare rock, and is more similar to other airless, regolith-covered bodies such as the Moon (*Helfenstein et al.,* 1994). Photometric analysis in combination with the shape model derived by *Thomas et al.* (1994) reveals that most of the surface has similar albedo, color, and photometric properties. Albedo contrasts across the surface are 10% or less. Subtle color differences vary across a continuum between two end members, one of which characterizes most terrain in the broad facets. The other color end member is correlated with increased albedo in some cases, and is associated mainly with fresh-appearing craters, especially along ridges. These spectrally distinct craters are bluer and in some cases brighter than materials elsewhere, and have deeper 1-μm absorptions (*Belton et al.,* 1992b; *Helfenstein et al.,* 1994; *Carr et al.,* 1994). These craters are also found, but less commonly, within the more extensive, spectrally typical facets that form broad topographic lows (*Helfenstein et al.,* 1994). *Belton et al.* (1992b) and *Helfenstein et al.* (1994) note similarities to more conspicuous color and age trends observed on the Moon (e.g., *Adams and McCord,* 1971), and interpret the color and brightness differences on Gaspra as due to fresh material exposed along ridges that matures optically with increasing exposure age as it migrates gradually downslope on to the facets. If regolith were capable of migrating downslope to accumulate in the facets, how deep might it be there? *Helfenstein et al.* (1994) point to some examples of fresh, spectrally distinct craters in the facets as evidence for bedrock underlying no more than a few tens of meters of regolith there.

Carr et al. (1994), however, distinguish between the spectral distinctiveness of craters along ridges, and background ridge material itself. These workers find that along ridges, only the fresh-appearing craters (including materials within one radius of a crater rim) are spectrally distinct, and ridge materials further away from fresh crater rims are indistinguishable from average facet material. Measurements by *Carr et al.* (1994) of crater depth:diameter ratios indicate many degraded craters probably have been filled with several tens of meters of debris. Together, these observations suggest a scenario where optically mature (for Gaspra) regolith covers the surface generally everywhere but is thinner along ridges (perhaps "as shallow as 13 m") and is more readily plumbed there by impacts than on the broader facets, where regolith depths are estimated as >50 m (*Carr et al.,* 1994). However, scenarios involving craters excavating through optically matured (for Gaspra) regolith, tens of meters thick, challenge expectations suggested by Gaspra's small surface gravity that regolith might have little chance of being thoroughly gardened through such depths (*Chapman et al.,* 1996a). On such a small body, how could tens of meters of regolith be closely held long enough to be gardened and exposed repeatedly between reburials to optically alter materials throughout such a thick layer? Is reshuffling of regolith from impact-related seismic shaking sufficient to account for the observations?

Pre-encounter navigation uncertainties resulted in a conservative close-approach distance of 1600 km, too distant to allow Gaspra's mass (and thus density) to be determined from slight gravitational deflections of the spacecraft. Speculations about Gaspra's internal structure and history must be guided by other clues. The shape model constructed from the images reveals a subtle constriction around the asteroid's midsection, but the overall shape is not conspicuously bifurcated or suggestive of a debris-draped contact binary (*Thomas et al.,* 1994). Grooves on the surface provide clues about the character of surface materials as well as the strength and continuity of Gaspra's deeper interior. Grooves probably are expressions through regolith of fractures within a more coherent substrate, as first proposed for grooves on Phobos (*Thomas et al.,* 1979). Specifically, pitted grooves probably represent partial drainage of loose surface material into fractures within stronger, deeper materials (*Horstman and Melosh,* 1989). Depths of grooves on Gaspra are consistent with such adjustments under very low gravity only if a very weak surface regolith tens of meters deep is involved (*Veverka et al.,* 1994b). Grooves and minor ridges (distinct from the major, gently convex ridges separating the shape-defining facets, as described above) define two roughly orthogonal patterns, with the more prominent set oriented about 15° from the long axis of the asteroid (*Thomas et al.,* 1994). Parallelism of these groove-and-ridge trends with some of the surface facets, as well as the continuity of this pattern across the asteroid, argue that the interior of Gaspra is coherent enough to propagate and sustain throughgoing structures. Deformation along these structures at various times in the asteroid's history has been expressed at the surface by creation of grooves in its regolith covering. This scenario is consistent with an origin for Gaspra as a single collisional fragment (itself damaged and partly fractured) from a larger precursor body, rather than as a collection of gravitationally bound smaller shards reassembled after a catastrophic disruption (*Thomas et al.,* 1994; *Veverka et al.,* 1994b).

3. IDA AND DACTYL

3.1. Overview

Galileo's second asteroid flyby, of 243 Ida, occurred on August 28, 1993 as the spacecraft was passing through the asteroid belt for the second and final time on its way to Jupiter. The larger available surface area of Ida and Ida's faster rotation rate allowed more comprehensive data to be obtained than at Gaspra. Five 31–38-m/pixel clear-filter images were returned initially (*Belton et al.,* 1994), followed later by other data, including color images and a single 25-m/pixel limb-grazing clear-filter picture (*Belton et al.,* 1996a).

The shape model constructed from the images shows that Ida is crescent shaped, deviating significantly from a best-fit ellipsoid of 29.9 × 12.7 × 9.3 km. On the basis of overall shape, limb roughness, and range of diameters (14.8–55.8 km), Ida remains the most irregular body visited by a spacecraft (*Thomas et al.,* 1996b). Townsend Dorsum, a 40-km ridge spanning nearly 150° of longitude, and a constriction ("waist") on the opposite side of Ida suggest that a near-planar discontinuity exists through the interior of the asteroid (*Thomas et al.,* 1996b). This potential discontinuity divides the two ends of the asteroid into roughly equal areas that have different populations of large impact craters. The first area [including Palisa Regio and Vienna Regio; see *Belton et al.* (1996a) for feature names and nomenclature details] contains no unambiguous craters greater than 6 km across, but the other area (including Pola Regio) contains several depressions 10 km across or larger (*Thomas et al.,* 1996b).

Nine sets of color observations were obtained during approach to Ida, providing complete coverage through a single asteroid rotation of all illuminated areas (*Veverka et al.,* 1996a; *Belton et al.,* 1996a). The SSI 0.89/0.99-μm ratio varies little in the images, implying that the wavelength position of the 1-μm absorption is stable across the surface and that regional variations in the Ca abundance in pyroxenes or in the pyroxene/olivine ratio are minor. The average 0.89/0.99-μm ratio for Ida is 8% lower than for Gaspra, consistent with a slightly longer wavelength location of the 1-μm absorption for Gaspra due to its more olivine-rich composition (*Sullivan et al.,* 1996), anticipated from groundbased work (*Goldader et al.,* 1991) and suggested also by spectral differences in preliminary *Galileo* Near-Infrared Mapping Spectrometer (NIMS) results (*Granahan et al.,* 1994, 1995).

3.2. Crater Morphology and "Space Weathering" of Surface Materials

Morphologically, the surface of Ida is dominated by simple, bowl-shaped impact craters that range from 12-km depressions to much smaller pits a few pixels wide in the highest-resolution images (*Chapman et al.,* 1996b; *Sullivan et al.,* 1996). Craters on Ida display a complete range of degradation states. The most pristine-appearing craters have sharp, circular rims — at least where resolution limitations permit such assessments. Some of these fresh-appearing craters occur within higher albedo spots that have irregular, diffuse boundaries usually extending no more than a crater radius from rims. These bright, morphologically fresh craters are bluer and have deeper 1-μm absorptions than more typical surface materials. The most likely explanation is that impact craters on Ida produce, distribute, or reveal materials that initially are brighter, bluer, and have deeper 1-μm absorptions than most areas on the surface, but these distinctions disappear gradually in the first stage of crater degradation, generally before obvious damage to rims has occurred from subsequent impacts (*Sullivan et al.,* 1996). Although similar trends of albedo and color with increasing exposure age have been observed for crater materials on the Moon (e.g., *Adams and McCord,* 1971; *Hapke et al.,* 1975; *Pieters et al.,* 2000; *Hapke,* 2001), it is unclear whether Ida's spectrally distinct crater materials are fresh, ballistically emplaced ejecta deposits as on the Moon, or zones where a very thin maturing optical surface layer has been seismically disturbed or removed, or some combination. None of these possibilities could be ruled out by numerical simulations (*Asphaug et al.,* 1996).

With increasing exposure age, craters on Ida apparently undergo degradational changes similar to those documented for craters on other rocky, airless bodies, including increased rim irregularities from subsequent impacts, and shallowing of floors from infill of debris. Rim degradation and infill of debris are responsible for most craters on Ida having depth:diameter ratios much lower than the freshest craters. *Sullivan et al.* (1996) measured depth:diameter ratios using the same photoclinometric technique as for Gaspra (*Carr et al.,* 1994, based on *Kirk,* 1987). The freshest craters appear to have depth:diameter ratios of about 0.15, slightly deeper than the 0.14 ratio of the freshest craters on Gaspra, but still not as deep as the 0.20 ratio for fresh craters on the Moon, Mercury, and Mars (*Gault et al.,* 1975; *Malin and Dzurisin,* 1977; *Pike,* 1980a,b, 1988) or even the 0.2 value [±0.04 (P. C. Thomas, personal communication, 2001)] reported for fresh craters on Phobos (*Thomas,* 1978; *Thomas and Veverka,* 1980). *Sullivan et al.* (1996) suggested that lower fresh crater depth:diameter ratios on Ida (and Gaspra) compared with larger bodies might be due to less ejecta being deposited at rims of smaller craters, reducing initial rim heights. A second suggestion was that gravitational packing causing large changes in strength, porosity, and bulk density in the upper 0.6 m of lunar regolith (*Houston et al.,* 1974; *Carrier et al.,* 1992) could, by scaling to Ida's much smaller gravitational acceleration, apply down through much greater depths of tens of meters within any surficial debris layer on Ida. In this scenario, the rim of an expanding transient cavity would expand laterally more easily through substantially weaker near-surface materials compared with how readily its floor would deepen through progressively stronger materials. However, the relatively low fresh crater depth:diameter ratios on Gaspra and Ida also could be a statistical artifact. The crater depth:diameter plots in *Carr et al.* (1994) and *Sullivan et al.* (1996) show that among the many partially degraded, infilled craters, the maximum depth:diameter relationship for each asteroid is a line defined by only a very small number of the freshest craters. In Ida's case, involving 160 craters greater than 8 pixels across, only 4 craters have depth:diameter ratios of 0.15, with a single outlier at 0.19. Difficulties of small numbers are exacerbated for Gaspra, where only a single crater of the 50 measured defines the maximum 0.14 ratio. Is it significant that Ida, with more craters available to measure, approaches a 0.20 value more closely than does Gaspra? Perhaps the freshest craters on both bodies form with depth:diameter ratios closer to 0.20 than observed, but fresh examples of sufficient size to resolve and measure in the images are unavailable. Still, we might expect at least somewhat lower initial depth:diameter ratios if, for smaller crater sizes, rim heights are reduced by wider dispersal of ejecta than on the Moon, or if gravitational loading of surficial debris influences the shapes of transient cavities.

3.3. Regolith: Evidence from Photometry, Blocks, Mass Wasting, and Grooves

The presence of substantial debris on the surface is indicated by photometric modeling, as well as boulders, shallow mass-wasting features, grooves, and infilled craters seen in the images. Photometric modeling of surface properties is consistent with fine particles covering the surface (not exposed bedrock) as was found at Gaspra (*Helfenstein et al.*, 1996). In addition, higher-resolution images obtained at Ida compared with Gaspra allowed more definitive observations of morphological evidence for regolith. About 17 isolated positive-relief features can be resolved on Ida, aided by contrasting shadows where lighting conditions were favorable. These features probably are perched or partly buried large ejecta blocks that represent the largest size-fraction of more abundant, finer-grained ejecta deposits distributed across the surface. Most blocks are found within Mammoth and Lascaux, two large craters constituting about 4% of Ida's surface area that are located on one of the leading rotational edges of the asteroid. Block distribution is consistent with origins as either impact products derived directly from within the largest craters where blocks are found (*Lee et al.*, 1986, 1996), or as blocks ejected at near-escape velocity that later were swept up by the leading rotational edge of the asteroid (*Geissler et al.*, 1996). The distribution of the small number of resolved blocks fits both concepts well. Nonuniform image resolution and illumination across the surface does not allow distinguishing between these two ideas, and both remain viable (*Sullivan et al.*, 1996).

Glancing illumination along an interior wall of the crater Mammoth reveals small, asymmetric crater cavities and other elongated shallow depressions that probably are minor mass-movement scars. For downslope movements to have occurred with such low gravitational accelerations (even if augmented by seismic accelerations from nearby impacts), weakly cohesive debris must be involved, not bedrock (*Sullivan et al.*, 1996). Elsewhere on the asteroid, where high-resolution image illumination is more direct, albedo contrasts forming streaks oriented downslope are seen, suggesting the presence and downward migration of cohesionless surface materials. However, these markings are much more subtle than downslope markings found on Deimos (*Thomas*, 1979; *Thomas and Veverka*, 1980).

Grooves on Ida are up to 4 km long with widths typically ≤100 m and depths estimated as no greater than a few tens of meters. They are most readily identified near and within crater Lascaux and near craters Mammoth and Kartchener, in the Pola Regio end of the asteroid. In Pola Regio grooves are subparallel to the long axis of Ida (*Sullivan et al.*, 1996) and are independent of local slopes calculated by *Thomas et al.* (1996b). Evaluation of directional trends was complicated by illumination, slope, and resolution constraints, but the concentration of grooves in Pola Regio does not appear to be organized into intersecting sets of parallel features as on Phobos (*Thomas et al.*, 1979) or Gaspra (*Veverka et al.*, 1994b). As mentioned previously, the presence of a debris layer is implied if groove morphologies represent drainage of surface debris into underlying fractures within deeper, more coherent materials. Alternative explanations for grooves such as scars from secondary impacts or rolling boulders, as proposed for grooves on Phobos (*Head and Cintala*, 1979; *Wilson and Head*, 1989), seem less likely, based on the much greater widths of grooves compared with the largest observed blocks, and the lack of evidence for blocks located at the ends of grooves.

3.4. Regolith: Thickness and Character

Morphology provides several clues about the thickness of the surface debris layer on Ida. The depth:diameter ratios of five large, degraded craters indicate burial of their original floors by 20–50 m of debris if depth:diameter reduction included complete removal of rims, or as much as 75–110 m of burial if shallowing is due primarily to infill rather than rim erosion (all of these values would be increased slightly if a fresh-crater depth:diameter ratio of 0.2 is assumed, rather than 0.15). Groove depths of a few tens of meters suggest the presence of a surface debris layer at least as thick, and mass movement scars on the walls of Mammoth indicate a potentially mobilized layer of 20–60 m there. These lines of evidence suggest that the upper, potentially mobilized or ballistically emplaced component of Ida's debris layer is typically about 50 m thick but probably varies from place to place across the surface (*Sullivan et al.*, 1996). One factor contributing to this nonuniformity is the influence of asteroid rotation on ejecta trajectories. Numerical experiments predict that ejecta launched near escape velocity that later reaccretes onto Ida causes a net loss of debris from the trailing rotational edges of the asteroid and net accumulation of debris onto the leading rotational edges over time (*Geissler et al.,* 1996). Below any given point on the surface, strength, porosity, and particle size of the debris layer probably change with depth, and with some irregularity according to the stochastic nature of the process creating this material. The uppermost material right at the surface probably consists of a very thin layer that is subject to optical maturation with increasing exposure age, probably related to the deposition of nanophase iron in voids among surface particles (*Pieters et al.,* 2000; *Hapke et al.,* 1975; *Hapke,* 2001; see also *Clark et al.,* 2002b). This optical maturation layer probably is just the very top of a much thicker upper regolith zone of very loosely held material subject to primary and secondary mixing and shifting caused by impacts. The potential for such movement probably declines at greater depths as overburden gradually increases (*Britt and Consolmagno,* 2001), although this material should still be subject to direct excavation and ballistic redistribution by moderate-sized impacts. Through these depths, the debris layer probably contains interleaved but laterally discontinuous thin ejecta sheets from individual impacts, or perhaps sublayers derived from larger, singular impact events that blanketed most or all of the surface of Ida at some point in its history. Deeper down, debris created in place by impact brecciation of target material during larger impacts probably is more dominant, as well as more coherent blocks less likely to have been as heavily fractured or otherwise altered by impacts battering the surface. The transition between debris and more coherent materials at greater depths is gradational, otherwise abrupt changes in strength across a narrow depth interface probably would have been revealed by craters with interior benches or flat floors (*Oberbeck and Quaide,* 1968; *Quaide and Oberbeck,* 1968). Instead, only simple bowl-shaped craters have been observed (*Sullivan et al.,* 1996).

3.5. Internal Structure and Configuration

The range of possible orbits for Ida's satellite Dactyl constrains the bulk density to 2.6 ± 0.5 g/cm^3 (*Belton et al.,* 1995, 1996b; slight refinement by *Petit et al.,* 1997), indicating that porosity is an important contributor to the interior volume of the asteroid but providing few definitive clues about how much porosity might be present. The distribution of porosity through the interior is also not well-constrained. Comparison of the shape model with the spin axis position shows only a 2° offset between the observed spin axis of Ida and its maximum angular momentum vector calculated for a trial assumption of homogeneous density, but this close agreement "does not eliminate [the possibilities of] symmetric variations in density or modest asymmetric variations in porosity or composition" (*Thomas et al.,* 1996b). Such internal heterogeneities were evaluated by Thomas et al. as possible explanations for the nonuniform distribution of grooves and largest craters between Regions 1 and 2, mentioned previously. Grooves are not continuous across the length of the asteroid, but are found almost exclusively in Region 2, the part of the asteroid that also contains all the craters larger than 10 km (with the possible exception of Vienna Regio, an irregular 14-km concavity on the 0°E end of the asteroid in Region 1). Thomas et al. speculate that Region 1 and Region 2 could represent two separate components of different material strengths (weaker material in Region 2 could favor development of larger craters there, as well as grooves) or different resurfacing histories. These authors note, however, that the numerical simulations of *Asphaug et al.* (1996) provide little support for these ideas by predicting that formation of the largest craters on Ida was influenced much more strongly by gravity than by material strength differences. Asphaug et al. examined the question of mechanical coupling between Regions 1 and 2, across the "waist" area and Townsend Dorsum, and found that grooves concentrated in Pola Regio in Region 2 were best explained by an impact at the Vienna Regio site in Region 1 rather than from impacts at Azzurra or Orgnac (both in Region 2); this scenario implies mechanical coupling through the length of the asteroid. However, Vienna Regio, if it is an impact site, appears very degraded compared with the morphologically fresh and presumably more youthful grooves in Pola Regio. Damage to the Pola Regio groove site from the Azzurra impact was predicted to have been insufficient to create grooves there, but modeling the Azzurra impact showed it could have projected sufficient energy to reactivate fractures in Pola Regio created originally by a more ancient impact forming Vienna Regio (*Asphaug et al.,* 1996).

Brighter- and bluer-than-average material is widely distributed across Ida's northern hemisphere, which was not seen at high resolution. This material probably does not reflect compositional heterogeneity within the asteroid, however, because it is spectrally indistinguishable from smaller deposits associated with fresh craters seen elsewhere at higher resolution. Low-velocity ejecta from Azzurra (a bright, spectrally distinct 10-km crater in the northern hemisphere that appears less degraded than other craters of similar size) is the most likely explanation (*Geissler et al.,* 1996).

3.6. Dactyl

Ida's satellite Dactyl was discovered unexpectedly, moving only ~6 m/s 85 km from Ida in images returned after *Galileo*'s encounter (*Chapman et al.,* 1995; *Belton et al.,* 1995, 1996b). Dactyl was found in 47 SSI images, including a five-color view at 110 m/pixel, monochrome views at 89 and 39 m/pixel, and a 24-m/pixel image showing only part of Dactyl's night side, illuminated sufficiently by "Idashine" to contribute to determination of a shape model for this tiny body with an average radius of only 0.7 km (*Veverka et al.,* 1996a). Dactyl's surface is heavily cratered, but most other details are ambiguous. Color data and photometric modeling indicate Dactyl and Ida have approximately similar compositions and surface textures, implying Dactyl's surface is not bare rock (*Helfenstein et al.,* 1996). However, Dactyl has a deeper 1-μm pyroxene-olivine absorption and a steeper violet-green slope compared with the freshest materials on Ida, so Dactyl's surface cannot be fit on a continuum between fresh and "space-weathered" Ida materials

(*Veverka et al.,* 1996b). These differences imply the possibility of slight intrinsic compositional differences between Dactyl and Ida, although within the range of spectral diversity among other Koronis family members (*Binzel et al.,* 1993; *Veverka et al.,* 1996b). *Chapman* (1996) disagrees with some aspects of the analysis of *Veverka et al.* (1996b), maintaining that Dactyl's deeper 1-μm absorption is much more significant than subtle differences in spectral slope shortward of 0.56 μm, and that spectral differences between Ida and Dactyl are due primarily to Dactyl surface materials being less optically mature, on average, than even the freshest materials on Ida.

4. MATHILDE

4.1. Overview

The first asteroid target of the *NEAR Shoemaker* spacecraft was 253 Mathilde, a C-class asteroid encountered June 27, 1997. The flyby occurred at 1.99 AU where sufficient solar power was available only to operate the Multispectral Imager (MSI); all other instruments were turned off. Over 500 images were obtained of about half the surface during a 25-min period, including 500-m/pixel color images in seven bands as well as the highest-resolution 160-m/pixel images of the surface (*Veverka et al.,* 1997). Mathilde's very slow rotation of 17.4 d did not reveal additional illuminated terrain during the brief encounter, so development of a shape model from limb fits and stereogrammetry was complicated by uncertainties accounting for the unobserved terrain. Image-derived minimum- and maximum-volume models were constrained by groundbased light-curve data (*Mottola et al.,* 1995) to yield a best-fit ellipsoid of 66 × 48 × 46 km (*Veverka et al.,* 1997; *Thomas et al.,* 1999).

4.2. Large Craters and Implications for Interior Structure

The observed portion of the surface is dominated by four craters with diameters that exceed the asteroid's average radius of 26.5 ± 1.3 km. These large craters (Karoo, Damodar, Kuznetsk, and Ishikari) appear relatively fresh, with steep interior walls descending from crisp rims, but they do not show any evidence for individual ejecta blankets obscuring intervening terrain or each other, as might be expected for crater geometries more strongly influenced by gravity than material strength (e.g., *Housen et al.,* 1983; *Greenberg et al.,* 1994, 1996; *Asphaug et al.,* 1996; *Nolan et al.,* 2001; *Asphaug et al.,* 2002). The possibility that all four of these large craters were blanketed by ejecta from yet another large, more recent, but so far unseen impact is not a very satisfying notion, because it does not explain how these large craters in close proximity to one another appear to have formed without degrading their neighbors or disrupting the asteroid. If the bulk composition of Mathilde is similar to some classes of carbonaceous chondritic meteorites (*Hiroi et al.,* 1993; *Binzel et al.,* 1996), then high internal porosities of 40–60% are implied by a low bulk density of 1.3 ± 0.3 g/cm^3 determined from the shape model and the mass measured from spacecraft trajectory perturbations (*Yeomans et al.,* 1997). Mathilde's inferred high porosity was suspected to somehow be responsible for the strange morphologies of the largest craters (*Veverka et al.,* 1999; *Chapman et al.,* 1999; *Thomas et al.,* 1999). Since the encounter, numerical hydrocode experiments simulating impacts into porous and nonporous targets much smaller than Mathilde (*Asphaug et al.,* 1998) suggest that high porosity should have a protective effect by localizing impact fragmentation damage in the interior, but partitioning more impact energy into heaving ejecta thus potentially increasing crater size. *Housen et al.* (1999) provide important insights into the Mathilde observations with laboratory experiments involving projectiles impacting into porous targets with underdense packing of grains. Experimental craters produced in these settings form primarily by compaction of target material rather than by fragmentation and excavation, explaining the relatively fresh appearance, absence of obscuring ejecta blankets, and noninterference among the largest craters on Mathilde as a direct consequence of the asteroid's low density and inferred high porosity (*Housen et al.,* 1999). The lack of raised rims around the largest craters (*Thomas et al.,* 1999) is also consistent with this concept. Along with a target's material strength and its surface gravity, target porosity seems to have an important influence on asteroid impact crater geometry in cases where bulk densities below 2–3 g/cm^3 imply underdense packing of grains, at least for most silicate materials (*Housen et al.,* 1999).

4.3. Surface Morphology and Color

Analysis of smaller-scale surface features is restricted by much lower image resolution obtained at Mathilde (160 m/pixel) compared to other bodies. Craters of all degradation states are present, and counts of areas outside the largest crater interiors (which were mostly shadowed, in any case) indicate that crater sizes up to 5 km approach saturation equilibrium, with a size-frequency distribution similar to craters found on Ida (*Chapman et al.,* 1999). No blocks were resolved on the surface, although a few ambiguous positive relief features at the limit of resolution are seen between Kuznetsk and Karoo (*Thomas et al.,* 1999). The interior walls of the largest craters do not show morphological or spectral evidence for layering, but subtle morphologies that may be mass-wasting chutes and slides are found inside Karoo, and a possible slump exists on the floor of the 7-km crater Lublin (*Thomas et al.,* 1999). No grooves are visible at 160 m/pixel in the areas seen by the spacecraft. [With only one example visited, seen with insufficient image resolution, it is too soon to evaluate the prediction of *Thomas and Veverka* (1979) that grooves should also be found on C-class asteroids.] Surface features such as polygonal outlines of some craters and a few subtle markings that have some associated topographic relief provide the only structural insights about the deeper interior (*Thomas et al.,*

1999). Mathilde's dark surface [geometric albedo 0.047 ± 0.005 at 0.55 μm reported in *Clark et al.* (1999), but subsequent MSI recalibration refined the value to 0.043 ± 0.005 (*Murchie et al.,* 2002a)] reveals no spectral or albedo contrasts, in agreement with groundbased results (*Binzel et al.,* 1996; *Rivkin et al.,* 1997).

5. EROS

5.1. Overview

The *NEAR Shoemaker* orbital mission of 433 Eros in 2000–2001 provided global imaging by the Multispectral Imager (MSI) at much higher resolution (~5 m/pixel) than available for any other asteroid (*Veverka et al.,* 2000), as well as coverage of selected regions at submeter/pixel resolutions (*Veverka et al.,* 2001a) and even higher resolutions during landing descent (*Veverka et al.,* 2001b). In addition, most of the asteroid was mapped spectrally at near-infrared and X-ray wavelengths (*Bell et al.,* 2002; *Trombka et al.,* 2000) and γ-ray data were obtained at the landing site (*Evans et al.,* 2001). Because of our detailed knowledge of Eros, its properties are becoming a standard model for our views of S-type asteroids pending similarly detailed studies of other bodies. Applying results from *NEAR Shoemaker* to other small bodies, however, requires taking into account differences in size, asteroid type, and orbit.

Eros is a curved object 34 km in length, with a volume equivalent to that of a sphere 8.4 km in radius (Fig. 1, Table 1). Lightcurve and radar studies (*Zellner,* 1976; *Mitchell et al.,* 1998) conducted before the *NEAR Shoemaker* mission yielded good approximations of the shape and a spin vector within 2° of that obtained by the *NEAR Shoemaker* spacecraft. The shape and volume information in Table 1, from MSI data, are within an average of 20 m of those obtained independently by the NEAR Laser Rangefinder (NLR) (*Zuber et al.,* 2000).

5.2. Compositional and Spectral Properties

Both X-ray data and near-infrared spectra are most consistent with a primitive, largely undifferentiated composition for Eros (*McCoy et al.,* 2001). X-ray measurements probing the uppermost 100 μm of surface material show that abundances of Ca, Al, Mg, Fe, and Si are consistent with compositions either of ordinary chondrites or certain primitive achondrites (*Trombka et al.,* 2000). However, there is a marked depletion of S compared with expectations for these meteorite types. This could be due either to removal of a small fraction of partial melt (which would imply that Eros is primitive achondritic), or to devolatilization of a thin surface layer by impacts (*Trombka et al.,* 2000; *Nittler et al.,* 2001). The positions and relative depths of the 1-μm pyroxene-olivine absorption and the 2-μm pyroxene absorption in the near-infrared spectra have greatest overall similarity to L and possibly LL or H ordinary chondrites (*Veverka et al.,* 2000; *Bell et al.,* 2002; *McFadden et al.,* 2001; *Lucey et al.,* 2002). This is consistent with the early groundbased interpretations of *Pieters et al.* (1976), although uncertainties existed from discrepancies between telescopic data and laboratory meteorite spectra, and other contemporary interpretations were somewhat different (e.g., *Larson et al.,* 1976; *Wisniewski,* 1976). Although groundbased work suggested compositional heterogeneity (*Murchie and Pieters,* 1996), near-infrared spectra from NIS and multispectral measurements from MSI both show that the center of the 1-μm absorption does not vary across the surface, ruling out compositional segregation of surface materials into distinct units hundreds of meters across and larger (e.g., from embedded bodies of partial melt or compositional zones created by differentiation of Eros' parent body) (*Veverka et al.,* 2000; *Bell et al.,* 2002; *Murchie et al.,* 2002b).

MSI results show the northern hemisphere exhibits average reflectances at 30° incidence and 0° emission of 0.136 ± 0.007 at 760 nm and 0.115 ± 0.006 at 950 nm, equivalent to geometric albedos of 0.30 ± 0.02 and 0.26 ± 0.02 respectively (*Murchie et al.,* 2002b; cf. *Clark et al.,* 2002a; *Domingue et al.,* 2002). However, these data also show surprisingly large variations in 760-nm albedo across the asteroid's surface, with bright slope materials typically 1.5× brighter than average surface values (*Murchie et al.,* 2002b; *Bell et al.,* 2002; *Clark et al.,* 2001) (Fig. 2). The 1-μm band depth (which can be affected by compositional variation, particle size, or optical alteration processes) of the brighter materials is several percent lower than average, as indicated from 950/760-nm ratios (Fig. 3). This inverse relationship of increased albedo with reduced 1-μm band depth is characteristic of "space weathering" optical maturation processes first recognized for lunar surface materials (e.g., *Adams and McCord,* 1971; *Hapke et al.,* 1975; *Pieters et al.,* 2000; *Hapke,* 2001; see also *Clark et al.,* 2002b). On Eros, albedo and color variations also correlate with specific geological settings, consistent with optical maturation with increasing exposure age: Brighter-than-average materials with lower 950/760-nm ratios are found almost exclusively on slopes exceeding 25° (*Murchie et al.,* 2002b), where mass wasting should be more likely to reveal freshly exposed materials.

The relationship between Eros' 760-nm albedo and 950/760-nm ratio suggests that space weathering affects the uppermost regoliths of Eros and the Moon differently. For lunar highlands and mare mineral assemblages, *Clementine* observations (*Pieters et al.,* 1994; *Lucey et al.,* 1995, 2000; Fig. 3) have demonstrated correlated variations in albedo and 1-μm band depth with increasing surface exposure age. With increasing Fe content of the starting material, the eventual decrease in depth of the 1-μm absorption is greater, but the eventual decrease in albedo is smaller. For typical mare and highland assemblages, greater "space weathering" causes lunar materials to approach the optical properties of a theoretical end state. In the case of Eros, for a chondritic composition, FeO content must approach that of the lunar maria (*Lucey et al.,* 1995). However, Eros' variation of 950/760-nm ratio with 760-nm albedo is an order of magnitude smaller than for lunar mare materials. Furthermore, inferred evolution in Eros' optical properties from the lower right to the upper left of the data in Fig. 3 implies that the up-

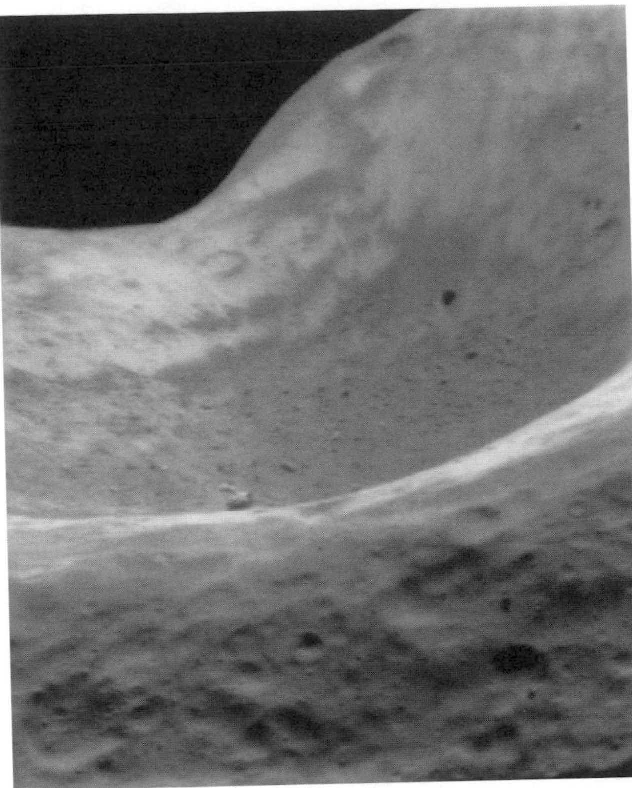

Fig. 2. Bright materials on steep slopes within craters on Eros. **(a)** A view into Psyche, a 5.3-km crater with steep interior slopes exceeding 35° in some places (image width 1.9 km). **(b)** A view into Selene, a smaller crater. In these and many other steep-walled craters brighter-than-average material dominates the upper, steepest slopes. However, the distribution of bright material is not strictly gradient-dependent, as shown by lateral bright and dark zones alternating near the middle parts of some slopes without any clear evidence of correlating/alternating gradient changes. Sharp contacts between dark and bright lobes and streaks oriented downslope are different from diffuse boundaries of downslope streaks on Deimos.

permost layer of Eros' regolith is evolving to an optical end state that is different from that of the lunar maria and highlands. This suggests that compositional differences between fresh materials on the Moon and Eros and/or differences in the microphysical processes of space weathering between these bodies (perhaps due to differences in gravity or position in the solar system) significantly influence the optical outcome of space weathering (*Murchie et al.,* 2002b).

5.3. Craters and Stratigraphy

Impact craters dominate the surface morphology of Eros, as with other asteroids. The three largest depressions, almost certainly of impact origin, are Himeros (about 9 km across), Shoemaker (7.6 km), and Psyche (5.3 km) (proposed names). The shapes of these features, their depth:diameter ratios, and the presence of rims on Psyche and Shoemaker are consistent with gravity-controlled impact excavation (*Thomas et al.,* 2002a). Shadow measurements show craters with diameters up to 2.1 km have a best-fit depth:diameter ratio of 0.134, but some examples within these data have depth:diameter ratios close to 0.2. A subset of smaller craters with diameters less than 0.5 km have a slightly higher best-fit depth:diameter ratio of 0.15, but the deeper (fresher?) examples have depth:diameter ratios close to 0.2. Few craters on Eros appear as pristine as the freshest lunar examples found in *Lunar Orbiter* images. Qualitatively, craters on Eros also appear somewhat less sharp than those on Ida, and do not seem to include examples as sharp as some craters seen on Phobos. Craters on Deimos have been modified by a unique (so far) global covering of ejecta. Comparison of Eros' craters to those on Gaspra is impractical because of low resolution there. Differences between craters on Eros and lunar craters may be due partly to less ejecta retained at rims or to a covering by globally distributed ejecta on Eros. However, these mechanisms fail to explain the differences between craters on Eros and those seen on Ida and Phobos. Further quantitative studies of crater shape are required to elucidate true interobject differences in initial crater morphology.

Craters less than 200 m in diameter are deficient on Eros relative to expectations based on production functions for other airless objects (*Chapman et al.,* 2002). This deficiency is not yet confidently explained. It may result from different impactor populations from those expected, from crater erasure processes not yet understood, or a combination of

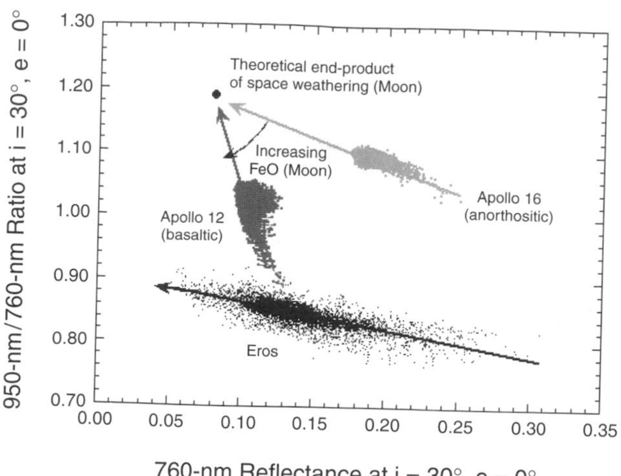

Fig. 3. Comparison of 760-nm reflectances and 950/760-nm color ratios between Eros and lunar mare and highland materials, modified from *Murchie et al.* (2002b). All values are corrected to 30° incidence and 0° phase angle. The lunar 760-nm reflectances and 950/760-nm ratios converge toward a theoretical end state with increasing optical maturity (*Lucey et al.,* 1998, 2000), as shown by the arrows derived from the relationship of *Lucey et al.* (2000). The arrow for Eros is an empirical fit to the data. For the Moon, the slope of the trend formed by a single material as it is space weathered is highly correlated with FeO content. The trend of Eros' 760-nm reflectances and 950/760-nm ratios converges to a different theoretical end product from the lunar materials, showing that the microphysical results of space weathering on Eros are different (*Murchie et al.,* 2002b).

these effects. Crater densities also vary significantly across the surface of the asteroid. Densities of craters 80–300 m in diameter in Himeros and Psyche approach saturation, thus these two areas cannot be placed in stratigraphic order using those crater densities. However, craters 80–300 m in Shoemaker crater are about a factor of 3 reduced in density from those in Psyche and Himeros, implying that Shoemaker is the youngest of the three large craters. The interiors of all these features are subject to downslope transport, which probably affects crater retention and complicates simple crater age comparisons.

5.4. Ejecta and Regolith

Blocks up to 150 m across scattered on the surface were the initial indication from *NEAR Shoemaker* imaging that ejecta was globally distributed across the surface, although a fine particle surface component had long been inferred from polarimetry (*Zellner and Gradie,* 1976) and thermal studies (*Morrison,* 1976; *Lebofsky and Reike,* 1979). Direct inspection at the highest resolutions shows that Eros is covered by regolith (e.g., Fig. 4). While low gravity and escape velocity (less than 17 m/s, geographically dependent) might be expected to allow nearly all ejecta to escape, observations of Phobos and Deimos (*Murchie et al.,* 1991; *Thomas et al.,* 1996a; *Simonelli et al.,* 1998) and numerical modeling of large impacts (e.g., *Asphaug and Melosh,* 1993; *Asphaug et al.,* 1996, 1998; *Love and Ahrens,* 1996; *Nolan et al.,* 1996, 2001) suggest that the majority of ejecta from relatively large impacts could be retained on an object such as Eros.

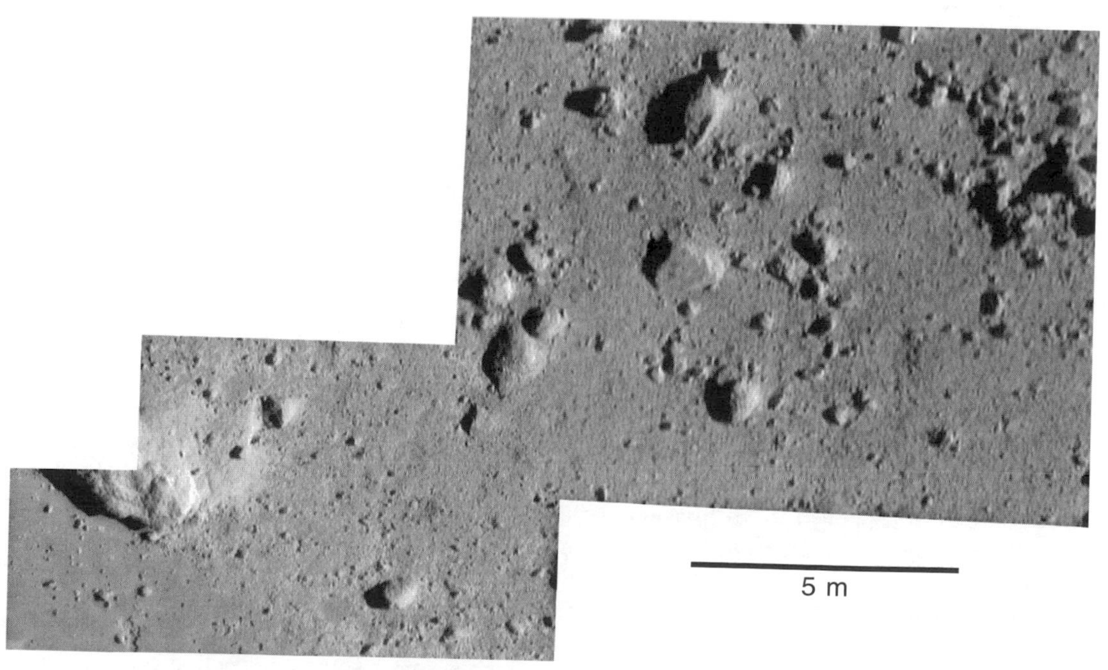

Fig. 4. Our closest view yet of an asteroid: The last three MSI images obtained before touchdown of the *NEAR Shoemaker* spacecraft on Eros. The landing site is located just within the 9-km depression Himeros, inside a block-strewn 100-m crater with a ponded fine particle deposit on its floor visible at lower left.

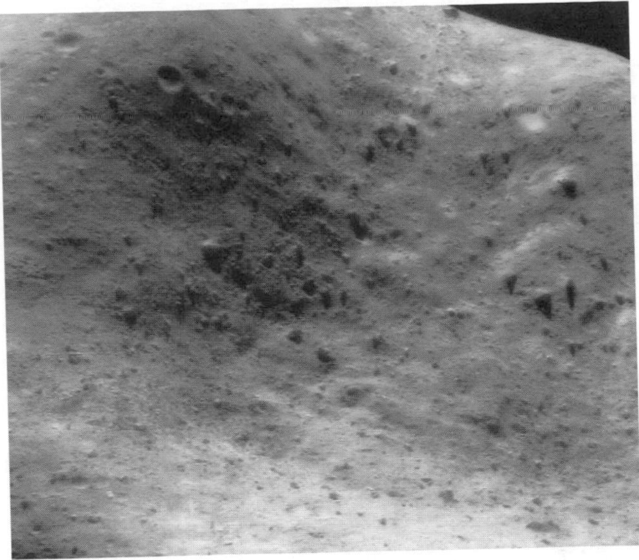

Fig. 5. A view directly into Shoemaker crater on Eros, showing an abundance of large blocks. Assuming block heights average half of block widths, 44% of the volume of all blocks on Eros wider than 15 m are found within this 7.6-km-diameter crater (*Thomas et al.,* 2001). Image width is 4.5 km.

Global mapping of the larger blocks reveals a concentration at low latitudes, and some longitudes that are relatively deficient in blocks (*Thomas et al.,* 2001). There are few blocks near obvious source craters. The single major exception is Shoemaker crater, which contains nearly half the mapped volume of blocks on Eros (Fig. 5). The distribution of other blocks is largely, but not completely, predictable from reimpact of blocks launched from Shoemaker crater (*Thomas et al.,* 2001). The lack of block concentrations consistent with reimpact from Psyche and Selene suggests that ejecta from these other large, older craters is buried or eroded, or did not initially contain abundant large blocks as did Shoemaker's ejecta. Blocks are concentrated near the centers of many craters. Such blocks are almost all in degraded craters, indicating that the blocks originated elsewhere and were trapped rolling down the crater slopes. A few notable exceptions are blocks that remain on 15° slopes in Psyche (Fig. 2a). These slopes would have been impacted nearly normally by Shoemaker ejecta, according to trajectory modeling by *Thomas et al.* (2001). Block size distribution varies over the surface, but the global population of ≥15-m blocks has a cumulative number/size slope (log-log) of about –3.1. Local regions measured at higher resolution show slopes from about –1 to –5. Images of such regions suggest that the greater susceptibility to burial of smaller blocks may have modified an initial size-frequency distribution: In these regions debris movements are apparent and smaller blocks more commonly have tabular exposures than larger blocks (*Veverka et al.,* 2001a).

Blocks have shapes ranging from angular to rounded and clodlike. Varying degrees of degradation are also observed, with aprons of debris surrounding many blocks. There are only a few tracks of bouncing blocks, and none that appear to be the simple result of rolling downhill, although some linear marks on slopes of crater walls may be slide marks from individual boulders. The largest block is consistent with that predicted for a 7.6-km crater on the basis of empirical relations developed from the Moon, Phobos, Deimos, and Ida (*Lee et al.,* 1996). The volume of blocks attributed to Shoemaker crater, ~0.5% of its estimated excavated volume, is well within the range observed on the Moon (*Cintala et al.,* 1982).

Regolith depths of tens of meters are indicated by filled craters and benches (*Barnouin-Jha et al.,* 2001), groove morphology (*Prockter et al.,* 2002), and slump morphology in Himeros. Bounce marks from ejecta blocks locally indicate easily deformed materials several meters deep.

5.5. Regolith Transport and Modification

Regolith has been significantly modified and redistributed by slope processes. Gravity has directed transport of material, and other sorting processes appear to occur. For example, high-albedo features occur preferentially where slopes exceed 25° (*Murchie et al.,* 2002b). These features occur as tapered markings, irregular sheets, and ribbon-shaped areas approximately parallel to the slope contours (*Thomas et al.,* 2002a) (Fig. 2). Unlike downslope motion indicators on other objects such as Deimos, those on Eros tend to have fairly sharp boundaries, possibly indicating relatively little vertical or lateral mixing. Topographic breaks at the margins are absent, with only a few possible exceptions visible in 3–4-m/pixel images. Burial of craters on slopes and in depressions implies minimum depths of materials that have been transported; in some instances, burial of 150-m craters indicates over 20 m of local accumulation (*Barnouin-Jha et al.,* 2001).

The slopes of Himeros exhibit different landforms indicative of mass wasting, including sinuous, dark markings extending 2 km down the eastern slopes (*Thomas et al.,* 2002a). Hummocky topography indicates mass movement of perhaps tens of meters of debris down the northeast side of Himeros as well as slumps tens of meters thick on the crater's western slopes.

5.6. Ponded Deposits

In certain regions of Eros flat deposits with surfaces perpendicular to the local gravitational acceleration fill bottoms of depressions (*Veverka et al.,* 2001b; *Robinson et al.,* 2001). Such morphology is consistent with emplacement with no shear strength, allowing material to act like a fluid and "pond" to an equipotential surface. However, ponded deposits currently support steep-walled grooves, impact craters, and superposed blocks, showing that subsequent to emplacement some compaction or cohesion has occurred to create nonzero shear strength. Estimated volumes of the largest ponds would require less than 10 cm of material to be shed from crater interior walls, indicating that an outside source of material is not required. No analogous deposits have been seen on any solar system object and their origin is not clear. Redistribution of fine material during impact-induced shaking is a simple explanation for the formation of the ponded deposits; however, such a mechanism fails to explain several observations.

Ponded deposits are found preferentially in an equatorial zone — 91% of ponds greater than 30 m in diameter are located within ±30° of the equator (*Robinson et al.,* 2001). Low-latitude regions with the highest concentrations of ponded deposits coincide with areas having low gravity that spend a large fraction of Eros' orbital period near the terminator (due to Eros' 88° obliquity). Many of the ponds are relatively blue (higher 550/760-nm reflectance ratio) and have a deeper infrared band (lower 950/760-nm ratio), consistent with less alteration from the space environment than other materials on Eros (*Robinson et al.,* 2001). Such alteration by micrometeorite impacts produces relatively large aggregates of glass and crystalline material. The long terminator exposure favors creation of photoelectric charge differentials between illuminated and shadowed terrain, capable of lifting and redistributing particles tens of micrometers in size (*Lee,* 1996). *Robinson et al.* (2001) hypothesize that the ponds formed by electrostatic sedimentation processes that preferentially concentrated the finest component of the regolith. This process may be consistent with the observed color properties through one (or both) of two mechanisms. Since electrostatic levitation is an inefficient process the smallest grain size would be most susceptible to transport, possibly the crystalline-rich fraction (relatively blue) of an agglutinate plus crystalline regolith. Alternatively, for mafic minerals, as grain size decreases, color becomes redder at visible wavelengths, but at very fine grain sizes (<50 μm) this trend can be reversed and visible color becomes bluer (*Adams and Felice,* 1967). The extremely fine grain sizes required to explain the pond color (by grain size alone) are consistent with the size range effectively mobilized by electrostatic levitation (*Lee,* 1996). While seismic effects may aid in downslope motion on asteroids (*Houston et al.,* 1973; *Carr et al.,* 1994; *Sullivan et al.,* 1996), the craters containing ponds are notably lacking in indicators of downslope surface motion, evident inside many other craters. Additionally, the peculiar geography of the ponded deposits has no seismic explanation.

5.7. Internal Structure

Wilkison et al. (2002) estimate a bulk porosity for Eros of 21–33% by comparing Eros' density of 2.67 ± 0.03 g/cm^3 (*Veverka et al.,* 2000; *Yeomans et al.,* 2000) with densities of ordinary chondrite meteorites (*Consolmagno and Britt,* 1998; *Flynn et al.,* 1999; *Wilkison and Robinson,* 2000). The total porosity is the sum of the microporosity (porosity on the same scale of grains within handsamples) and macroporosity (larger voids between structural elements within an asteroid). *Wilkison et al.* (2002) estimate a macroporosity on the order of 20% for Eros, suggesting the interior is heavily fractured but probably not an assemblage of reaccumulated collisional debris held together by gravity and particle friction.

Because of the apparently ubiquitous covering of loose material, structural mapping depends largely on locating features in the regolith, such as grooves and ridges, that can be inferred to reflect deeper structures such as faults or fractures (*Prockter et al.,* 2002; *Thomas et al.,* 2002b; *Cheng et al.,* 2002). The 18-km ridge Rahe Dorsum has been interpreted on the basis of morphology as showing compressional failure over most of its length, with possible extension near the ends. There are no features running more than half the length of the asteroid, and only Rahe Dorsum even approaches that distance. There are local groups of nearly parallel features, most prominently in and northwest of Himeros. There are no radial clusters of structures around craters (*Prockter et al.,* 2002).

Clues about "global" internal structural details are best evaluated using large structures, of which there are only a few: Rahe Dorsum; a long ridge system at 25°S, 150°–170°W called Calisto Fossae; and grooves at 50°–60°S, 300°–360°W. Rahe Dorsum is fit by a plane with residuals of up to 500 m and an average absolute residual of 170 m. Over its 7-km length, Calisto Fossae can be fit by a plane with an average of 25-m residuals. Fitting Rahe Dorsum and the Calisto Fossae with a single plane gives a pole little different from the original Rahe Dorsum fit, at 52.9°S, 328.6°W. Thus, Rahe Dorsum and Calisto Fossae are coplanar (*Thomas et al.,* 2002). A plane fit to a facetlike section of the shape model west of Shoemaker has a pole only ~7° from that of the Rahe-Calisto plane, suggesting additional structural control related to this orientation. The grooves at 50°–60°S do not fit the Rahe-Calisto-facet orientation. However, Rahe Dorsum, Calisto Fossae, and the surface facet are of greatly different morphology and interpreted mechanical origins, and probably also are of very different ages. The surface facet is the oldest of the three. Calisto Fossae is heavily cratered and has the appearance of being very degraded (*Prockter et al.,* 2002). Rahe Dorsum has superposed craters, but has a much crisper appearance and cross-cuts Himeros, surface deposits of which are younger than the average Eros surface.

Possible origins for the complex global structural fabric on Eros include compositional layering, preexisting fractures, or a metamorphic fabric. If the surface facet indeed is related to Rahe-Calisto, then the fabric is more pervasive than a single layer or interface. Eros' lack of spectral heterogeneity favors an origin due to fracturing or thermal or pressure-related metamorphosis within Eros' parent body.

Any offset of the center of figure from the center of mass is under 1% of the mean radius, and implies an interior with only modest density variations. *Miller et al.* (2002), using an NLR-based 6th-degree-and-order model, found an offset of (–9.7, 2.4, 32.6 m; x, y, z); *Thomas et al.* (2002a), using the imaging-based shape model, found values of (–25.6, 1.6, 47.4 m). The two measures differ substantially in magnitude, but are similar in direction. The offset is real, and is interpreted here as at least as large as the *Miller et al.* (2002) value. Explanations for this offset, or for the more specific comparison of the gravity model with that predicted from a uniformly dense shape [in essence a Bouger gravity map (*Miller et al.,* 2002)], are nonunique. However, the approximate scale of variations is instructive. Shifting the center of mass by 0.6% of the radius of an otherwise uniform object could be accomplished by changing the density of one end by 30%, for approximately 4.3% of the object's half-length. This is equivalent to having 100–200 m more regolith with 40% porosity in high northern latitudes than in the south. However, the image data do not suggest such a difference. This is a further suggestion that the asymmetry may be related to porosity variations in the main body of Eros. Because the offset is primarily in the z axis and the Rahe-Calisto plane pole is in the x axis, there is not an obvious relationship between the two.

6. DISCUSSION

What have we learned about asteroid geology, in a general sense, from the *Galileo* and *NEAR Shoemaker* missions? Results from these missions have continued to motivate modeling regolith evolution and asteroid interiors (e.g., *Housen,* 1992; *Greenberg et al.,* 1996; *Wilson et al.,* 1999; *Wilkison and Robinson,* 2000; *Britt and Consolmagno,* 2001; *Nolan et al.,* 2001). For the purposes of continued refinement of such models and testing predictions against observations, we currently have closeup images and other spacecraft data from only five asteroids: three S-types, an S-type satellite, and a single C-type. It is of course hazardous to generalize comparisons among such a limited number of bodies to several thousands more objects that have never been observed closely. Nevertheless, the spacecraft visits completed so far were undertaken partly in the hope that such generalizations could be made, with appropriate caution. Our current small collection of closely observed bodies allows recognition of some important similarities and differences among them, and answers some older questions as well as presenting many new ones.

A long-standing controversy has revolved around the unknown source of ordinary chondrite meteorites (reviewed by *Gaffey,* 1993a; *Chapman,* 1996). Proposed links between ordinary chondrites and certain types of S-class asteroids have remained uncertain for decades because of persistent unexplained spectral discrepancies. *Galileo* and *NEAR Shoemaker* have shown that these spectral discrepancies can be explained largely by space weathering of S-type asteroid regoliths. Addressing this controversy was a primary goal of the *NEAR Shoemaker* spacecraft, which found elemental abundances at S(IV)-type Eros to be similar to L and possibly LL or H ordinary chondrites. The issue has not been closed entirely, however; unexplained spectral discrepancies remain between some of the freshest surface materials on Eros seen by NIS and proposed ordinary chondrite analogs (*Clark et al.,* 2001).

Gaspra, Ida, and Eros, the major S-type examples, all appear to have a full range of impact crater degradation states, fresh crater depth:diameter ratios approaching 0.2, and tens of meters of impact-derived debris with an uppermost optically maturing ("space weathering") layer that is probably very thin. Yet closer consideration of each of these major shared characteristics reveals important, sometimes puzzling distinctions.

Differences in cratering histories among these three bodies hint at the diversity of collisional histories probably experienced by many other asteroids. Gaspra's surface shows two populations of impacts, indicating the surface age was nearly reset (by an as-yet-unseen "final" large impact?), while contrasts between Ida's two regions suggest vastly different cratering histories for the largest impacts. Gaspra's spectrally distinct craters appear to be localized along topographic highs while those of Ida are not, and such features are largely absent from Eros. Could this be due somehow to shape differences between Gaspra, which is characterized by ridges separating broader facets, and Ida and Eros, which are crescent-shaped and not strongly defined by facets and intervening prominent ridges?

Although optical maturation of uppermost regolith occurs on Gaspra, Ida, and Eros, magnitudes of optical changes that occur and their geological associations are different among the three bodies. Bright, spectrally distinct, probably freshly exposed materials on Eros are found almost exclusively on slopes steeper than 25°. But on Gaspra and Ida, brighter, spectrally distinct materials have far less albedo contrast and are found only in areas locally disturbed by recent impacts, while such spectrally distinct craters are very rare on Eros. Do these differences reflect fundamental differences in regolith characteristics and/or processes among these bodies? Are resolution differences among data sets partly responsible? (For example, remember that Azzurra on Ida is large, bright, spectrally distinct, and probably relatively young; Azzurra might well have bright interior walls like Psyche on Eros, but resolution during *Galileo*'s Ida flyby was too poor for such details to be recognized.)

One possibility is that the surface of Eros has been affected relatively recently in its history by a large impact, causing a blanket of low-velocity ejecta to return gradually to the surface in a global distribution, filling in and effectively erasing smaller craters while partially burying preexisting blocks, as well as seismically degrading the sharpest rims of other craters (*Veverka et al.,* 2001a). The relative youth of the Shoemaker impact and possible effects on surface features (*Thomas et al.,* 2001) make this event a reasonable candidate. However, if Shoemaker is responsible for the paucity of smaller craters as well as the distribution of the larger blocks across the surface, it seems this would have to have been a very recent event to have obscured almost all other fresh, bright, spectrally distinct craters and their ejecta. Yet, if this is so, why isn't Shoemaker itself bright and spectrally distinct, as Azzurra is on Ida?

Another possibility is that Eros' regolith in the latter part of its history might have been stirred at a different rate and/or by a different population of impactors compared with Gaspra and Ida. These ideas are still under development, with contributions so far by *Barnouin-Jha and Cheng* (2000),

Chapman et al. (2002), *Murchie et al.* (2002b), and *Thomas et al.* (2001), among others. *Chapman et al.* (2002) and *Murchie et al.* (2002b) suggest the spectral homogeneity of Eros compared with Ida and dearth of small craters on Eros compared with the Moon could be due to a dramatic falloff in cratering rates since Eros was injected into its current near-Earth orbit (*Barnouin-Jha and Cheng,* 2000). In this scenario various regolith maturation and redistribution processes would have operated during the latter part of Eros' history with much less disturbance from impacts, resulting in increased preservation of boulders, more extreme and better-preserved redistribution of the finest particles by electrostatic levitation, and space weathering outpacing crater production.

Murchie et al. (2002b) also suggest that currently reduced impact rates at Eros compared with Gaspra and Ida might explain the larger albedo variations and different geological settings of the freshest materials on Eros. The unusually high albedos (relative to 1-μm band depths) of spectrally distinct, probably fresh exposures on steep slopes on Eros are most consistent with initial stages of space weathering (*Hapke,* 2001; *Moroz et al.,* 1996; *Murchie et al.,* 2002b). If bulk compositions as well as processes and products of space weathering are similar on S(IV) asteroids Ida and Eros, then (1) the freshest materials exposed (only) by recent impacts on Ida apparently still would not be as fresh as materials exposed (only) by mass wasting on Eros, and (2) impacts as recent as the bright craters on Ida should be recognizable on Eros if they were present. Both implications are consistent with a reduced impact flux at Eros during the latter part of its history compared with Ida. Under these conditions optical maturation of surface materials would still continue on Eros, but with greater contributions from nonimpact processes such as irradiation and ion bombardment (cf. *Adams et al.,* 1997; *Pieters et al.,* 2000; *Hapke et al.,* 2001). As previously mentioned, albedo boundaries along mass-wasting streaks in Psyche and on other steep slopes on Eros are sharp, even at very high MSI resolution, compared with more diffuse boundaries seen on downslope streaks on Deimos (*Thomas et al.,* 1996a; *Veverka et al.,* 2000; *Murchie et al.,* 2002b). Murchie et al. suggests that a relatively low current impact flux at Eros might restrict vertical and lateral mixing of materials that would otherwise smear these albedo boundaries.

The surface of Mathilde is of course significantly different from the surfaces of the S-class objects we have seen. Mathilde's dark surface presents no spectral or albedo contrasts, at least none that were detectable by the MSI. If some form of optical maturation of materials occurs on C-type asteroids, this process does not produce sufficient albedo or spectral contrast to be detected or to reveal areas of recent disturbance (e.g., fresh craters, ongoing mass wasting).

Grooves are present on all three S-types, and clues from these and other morphological features about interiors, along with density data where available, indicate that each probably is a damaged, heavily fractured collisional fragment, rather than a collection of gravitationally bound smaller shards reassembled after a catastrophic disruption. Mathilde, again, is substantially different, with much higher porosity implied from its density, probably causing the unique morphology of the largest craters that reflect compression of an underdense grain configuration rather than fragmentation and excavation (*Housen et al.,* 1999). Housen et al. speculate that continued compressional effects of such impacts should gradually decrease the overall porosity of the asteroid, implying that highly porous, underdense packing of asteroid components may be a transient condition. Could a condition of underdense packing of components be the initial configuration of an asteroid after massive disruption and reassembly? Could asteroids like Mathilde fluctuate around relatively high porosity values, alternately compacting somewhat for awhile but never achieving enough strength by this process to resist occasional rupturing and return to very low porosities — never achieving the more compact configurations of Ida and Eros (and probably Gaspra) that could have originated, via gradual porosity increases, from even denser internal configurations inherited from each of their precursor parent bodies? Is it possible for an asteroid to evolve from an underdense configuration such as Mathilde into a body with a more coherent interior such as the three S-types visited, or is target compaction too inefficient to ever result in as much coherence as a collisional fragment (albeit battered, damaged, and internally fractured) such as Gaspra, Ida, and Eros? These and other questions concerning asteroid evolution will be addressed most effectively in the future in the context of additional spacecraft visits to asteroids.

REFERENCES

Adams J. B. and Felice A. L. (1967) Spectral reflectance 0.4 to 2.0 microns of silicate rock powders. *J. Geophys. Res., 72,* 5705–5715.

Adams J. B. and McCord T. M. (1971) Alteration of lunar optical properties: Age and composition effects. *Science, 171,* 4829–4836.

Adams J. B., Schultz P., Sugita S., and Goguen J. (1997) Impact flash spectroscopy as a means to characterize planetary surface composition (abstract). In *Lunar and Planetary Science XXVIII,* pp. 3–4. Lunar and Planetary Institute, Houston.

Asphaug E. and Benz W. (1994) The surface and interior of Phobos (abstract). In *Lunar and Planetary Science XXV,* pp. 43–44. Lunar and Planetary Institute, Houston.

Asphaug E. and Melosh H. J. (1993) The Stickney impact of Phobos: A dynamical model. *Icarus, 101,* 144–164.

Asphaug E., Moore J. M., Morrison D., Benz W., Nolan M. C., and Sullivan R. (1996) Mechanical and geological effects of impact cratering on Ida. *Icarus, 120,* 158–184.

Asphaug E., Ostro S. J., Hudson R. S., Scheeres D. J., and Benz W. (1998) Disruption of kilometre-sized asteroids by energetic collisions. *Nature, 393,* 437–440.

Asphaug E., Ryan E. V., and Zuber M. T. (2002) Asteroid interiors. In *Asteroids III* (W. F. Bottke Jr. et al., eds.), this volume. Univ. of Arizona, Tucson.

Avanesov G., Bonev B., Kempe F., Basilevsky A., Boycheva V., Chikov K., Danz M., Dimitrov D., Duxbury T., Gromatikov P., Halmann D., Head J., Heifets V., Kolev V., Kostenko V., Kottsov V., Krasavtsev V., Krasikov V., Krumov A., Kuzmin

A., Losev K., Lumme K., Mishev D., Mohlmann D., Muinonen K., Murav'ev V., Murchie S., Murray B., Neumann W., Paul L., Petkov D., Petuchova I., Possel W., Rebel B., Shkuratov Y., Simeonov S., Smith B., Totev A., Uzonov Y., Fedotov V., Weide G.-G., Zapfe H., Zhukov B., and Ziman Y. (1989) Television observations of Phobos. *Nature, 341,* 585–587.

Avanesov G., Zhukov B., Ziman Y., Kostenko V., Murav'ev V., Fedotov V., Bonev B., Mishev D., Petkov D., Krumov A., Simeonov S., Boycheva V., Uzunov Y., Kempe F., Rebel B., Weide G.-G., Halmann D., Neumann W., Petuchova I., Possel W., Head J., Murchie S., Lumme K., Muinonen K., Peltoniemi J., Duxbury T., Murray B., Herkenhoff K., Fanale F., Irvine W., and Smith B. (1991) Results of TV imaging of Phobos (Experiment VSK-FREGAT). *Planet. Space Sci., 39,* 281–295.

Barnouin-Jha O. S. and Cheng A. (2000) Expected crater distributions on asteroid 433 Eros (abstract). In *Lunar and Planetary Science XXXI,* Abstract #1255. Lunar and Planetary Institute, Houston (CD-ROM).

Barnouin-Jha O. S., Garvin J. B., Cheng A. F., Zuber M., Smith D., Neumann G., Murchie S., Robinson M., and Veverka J. (2001) Preliminary impact crater dimensions on 433 Eros from the NEAR laser range-finder and imager (abstract). In *Lunar and Planetary Science XXXII,* Abstract #1786. Lunar and Planetary Institute, Houston (CD-ROM).

Bell J. F. III, Izenberg N. I., Lucey P. G., Clark B. E., Peterson C., Gaffey M. J., Joseph J., Carcich B., Harch A., Bell M. E., Warren J., Martin P. D., McFadden L. A., Wellnitz D., Murchie S., Winter M., Veverka J., Thomas P., Robinson M. S., Malin M., and Cheng A. (2002) Near-IR reflectance spectroscopy of 433 Eros from the NIS instrument on the NEAR mission. 1. Low phase angle observations. *Icarus, 155,* 119–144.

Belton M. J. S., Klaasen K. P., Clary M. C., Anderson J. L., Anger C. D., Carr M. H., Chapman C. R., Davies M. E., Greeley R., Anderson D., Bolef L. K., Townsend T. E., Greenberg R., Head J. W. III, Neukum G., Pilcher C. B., Veverka J., Gierasch P. J., Fanale F. P., Ingersoll A. P., Masursky H., Morrison D., and Pollack J. B. (1992a) The Galileo Solid-State Imaging experiment. *Space Sci. Rev., 60,* 413–455.

Belton M. J. S., Veverka J., Thomas P., Helfenstein P., Simonelli D., Chapman C., Davies M. E., Greeley R., Greenberg R., Head J., Murchie S., Klaasen K., Johnson T. V., McEwen A., Morrison D., Neukum G., Fanale F., Anger C., Carr M., and Pilcher C. (1992b) Galileo encounter with 951 Gaspra: First pictures of an asteroid. *Science, 257,* 1647–1652.

Belton M. J. S., Chapman C. R., Veverka J., Klaasen K. P., Harch A., Greeley R., Greenberg R., Head J. W. III, McEwen A., Morrison D., Thomas P. C., Davies M. E., Carr M. H., Neukum G., Fanale F. P., Davis D. R., Anger C., Gierasch P. J., Ingersoll A. P., and Pilcher C. B. (1994) First images of 243 Ida. *Science, 265,* 1543–1547.

Belton M. J. S., Chapman C. R., Thomas P. C., Davies M. E., Greenberg R., Klaasen K., Byrnes D., D'Amario L., Synnott S., Johnson T. B., McEwen A., Merline W. J., Davis D. R., Petit J-M., Storrs A., Veverka J., and Zellner B. (1995) Bulk density of asteroid 243 Ida from the orbit of its satellite Dactyl. *Nature, 374,* 785–788.

Belton M. J. S., Chapman C. R., Klaasen K. P., Harch A., Thomas P. C., Veverka J., McEwen A. S., and Pappalardo R. (1996a) Galileo's encounter with 243 Ida: An overview of the imaging experiment. *Icarus, 120,* 1–19.

Belton M. J. S., Mueller B. E. A., D'Amario L. A., Byrnes D. V., Klaasen K. P., Synnott S., Breneman H., Johnson T. V., Thomas P. C., Veverka J., Harch A. P., Davies M. E., Merline W. J., Chapman C. R., Davis D., Denk T., Neukum G., Petit J.-M., Greenberg R., Storrs A., and Zellner B. (1996b) The discovery and orbit of 1993 (243)1 Dactyl. *Icarus, 120,* 185–199.

Binzel R. P. (1988) Collisional evolution in the Eos and Koronis asteroid families: Observational and numerical results. *Icarus, 73,* 303–313.

Binzel R. P., Xu S., and Bus S. J. (1993) Spectral variations within the Koronis family: Possible implications for the surface colors of asteroid 243 Ida. *Icarus, 106,* 608–611.

Binzel R. P., Burbine T. H., and Bus S. J. (1996) Groundbased reconnaissance of asteroid 253 Mathilde: Visible wavelength spectrum and meteorite comparison. *Icarus, 119,* 447–449.

Britt D. T. and Consolmagno G. J. (2001) Modeling the structure of high porosity asteroids. *Icarus, 152,* 134–139.

Carr M. H., Kirk R. L., McEwen A., Veverka J., Thomas P., Head J. W., and Murchie S. (1994) The geology of Gaspra. *Icarus, 107,* 61–71.

Carrier W. D. III, Olhoeft G. R., and Mendell W. (1992) Physical properties of the lunar surface. In *Lunar Sourcebook* (G. Heiken et al., eds.), pp. 475–594. Cambridge Univ., New York.

Chapman C. R. (1971) Surface properties of asteroids. Ph.D. thesis, Massachusetts Institute of Technology, Cambridge. 392 pp.

Chapman C. R. (1976) Asteroids as meteorite parent-bodies: The astronomical perspective. *Geochim. Cosmochim. Acta, 40,* 701–719.

Chapman C. R. (1978) Asteroid collisions, craters, regoliths, and lifetimes. In *Asteroids: An Exploration Assessment* (D. Morrison and W. C. Wells, eds.), pp. 145–160. NASA Conference Publication 2053.

Chapman C. R. (1996) S-type asteroids, ordinary chondrites, and space weathering: The evidence from Galileo's fly-bys of Gaspra and Ida. *Meteoritics & Planet. Sci., 31,* 699–725.

Chapman C. R., Veverka J., Thomas P. C., Klaasen K., Belton M. J. S., Harch A., McEwen A., Johnson T. V., Helfenstein P., Davies M. E., Merline W. J., and Denk T. (1995) Discovery and physical properties of Dactyl, a satellite of asteroid 243 Ida. *Nature, 374,* 783–785.

Chapman C. R., Veverka J., Belton M. J. S., Neukum G., and Morrison D. (1996a) Cratering on Gaspra. *Icarus, 120,* 231–245.

Chapman C. R., Ryan E. V., Merline W. J., Neukum G., Wagner W., Thomas P. C., Veverka J., and Sullivan R. J. (1996b) Cratering on Ida. *Icarus, 120,* 77–86.

Chapman C. R., Merline W. J., and Thomas P. (1999) Cratering on Mathilde. *Icarus, 140,* 28–33.

Chapman C. R., Merline W. J., Thomas P. C., Joseph J., Cheng A. F., and Izenberg N. (2002) Impact history of Eros: Craters and boulders. *Icarus, 155,* 104–118.

Cheng A. F. (1997) Near Earth Asteroid Rendezvous: Mission overview. *Space Sci. Rev., 82,* 3–29.

Cheng A. F., Barnouin-Jha O., Prockter L., Zuber M. T., Neumann G., Smith D. E., Garvin J., Robinson M., Veverka J., and Thomas P. (2002) Small-scale topography of 433 Eros from laser altimetry and imaging. *Icarus, 155,* 51–74.

Cintala M. J., Head J. W., and Veverka J. (1978) Characteristics of the cratering process on small satellites and asteroids. *Proc. Lunar Planet. Sci. Conf. 9th,* pp. 3803–3830.

Cintala M. J., Head J. W., and Wilson L. (1979) The nature and effects of impact cratering on small bodies. In *Asteroids* (T. Gehrels, ed.), pp. 579–600. Univ. of Arizona, Tucson.

Cintala M. J., Garvin J. B., and Wetzel S. J. (1982) The distribution of blocks around a fresh lunar mare crater (abstract). In *Lunar and Planetary Science XIII,* pp. 100–101. Lunar and Planetary Institute, Houston.

Clark B. E., Veverka J., Helfenstein P., Thomas P. C., Bell J. F. III, Harch A., Robinson M. S., Murchie S. L., McFadden L. A., and Chapman C. R. (1999) NEAR photometry of asteroid 253 Mathilde. *Icarus, 140,* 53–65.

Clark B. E., Lucey P. G., Helfenstein P., Bell J. F. III, Peterson C., Veverka J., McConnochie T., Robinson M. S., Bussey B., Murchie S., Izenberg N., and Chapman C. (2001) Space weathering on Eros: Constraints from albedo and spectral measurements of Psyche Crater. *Meteoritics & Planet. Sci., 36,* 1617–1637.

Clark B. E., Helfenstein P., Bell J. F. III, Peterson C., Veverka J., Izenberg N., Domingue D., Wellnitz D., and McFadden L. (2002a) NEAR Infrared Spectrometer photometry of asteroid 433 Eros. *Icarus, 155,* 189–204.

Clark B. E., Hapke B., Pieters C., and Britt D. (2002b) Asteroid space weathering and regolith evolution. In *Asteroids III* (W. F. Bottke Jr. et al., eds.), this volume. Univ. of Arizona, Tucson.

Cole T. D., Boies M. T., El-Dinary A. S., Cheng A., Zuber M. T., and Smith D. E. (1997) The Near-Earth Asteroid Rendezvous laser altimeter. *Space Sci. Rev., 82,* 217–254.

Consolmagno G. J. and Britt D. T. (1998) The density and porosity of meteorites from the Vatican collection. *Meteoritics & Planet. Sci., 33,* 1231–1241.

Croft S. K. (1992) Proteus: Geology, shape, and catastrophic destruction. *Icarus, 99,* 402–419.

Dolfus A. and Zellner B. (1979) Optical polarimetry of asteroids and laboratory samples. In *Asteroids* (T. Gehrels, ed.), pp. 170–183. Univ. of Arizona, Tucson.

Dolfus A., Geake J. E., Mandeville J. C., and Zellner B. (1977) The nature of asteroid surfaces from optical polarimetry. In *Comets, Asteroids, Meteorites — Interrelations, Evolution and Origins* (A. H. Delsemme, ed.), pp. 243–251. Univ. of Toledo, Toledo, Ohio.

Domingue D. L., Robinson M., Carcich B., Joseph J., Thomas P., and Clark B. E. (2002) Disk-integrated photometry of 433 Eros. *Icarus, 155,* 205–219.

Evans L. G., Starr R. D., Brückner J., Reedy R. C., Boynton W. V., Trombka J. I., Goldsten J. O., Masarik J., Nittler R., and McCoy T. J. (2001) Elemental composition from gamma-ray spectroscopy of the NEAR-Shoemaker landing site on 433 Eros. *Meteoritics & Planet. Sci., 36,* 1639–1660.

Flynn G. J., Moore L. B., and Klock W. (1999) Density and porosity of stone meteorites: Implications for the density, porosity, cratering, and collisional disruption of asteroids. *Icarus, 142,* 97–105.

Fujiwara A. (1991) Stickney-forming impact on Phobos: Crater shape and induced stress distribution. *Icarus, 89,* 384–391.

Gaffey M. J., Burbine T. H., and Binzel R. P. (1993a) Asteroid spectroscopy: Progress and perspectives. *Meteoritics, 28,* 161–187.

Gaffey M. J., Bell J. F., Brown R. H., Burbine T. H., Piatek J. L., Reed K. L., and Chaky D. A. (1993b) Mineralogical variations within the S-type asteroid class. *Icarus, 106,* 573–602.

Gault D. E., Guest J. E., Murray J. B., Dzurisin D., and Malin M. C. (1975) Some comparisons of impact craters on Mercury and the Moon. *J. Geophys. Res., 80,* 2444–2460.

Geissler P., Petit J-M., Durda D. D., Greenberg R., Bottke W., Nolan M., and Moore J. (1996) Erosion and ejecta reaccretion on 243 Ida and its moon. *Icarus, 120,* 140–157.

Goldsten J. O., McNutt R. L. Jr., Gold R. E., Gary S. A., Fiore E., Schneider S. E., Hayes J. R., Trombka J. I., Floyd S. R., Boynton W. V., Bailey S., Brückner J., Squyres S. W., Evans L. G., Clark P. E., and Starr R. (1997) The X-Ray/Gamma-Ray Spectrometer on the Near Earth Asteroid Rendezvous mission. *Space Sci. Rev., 82,* 169–216.

Goguen J., Veverka J., and Thomas P. (1978) Phobos: Photometry and origin of dark markings on crater floors. *Geophys. Res. Lett., 5,* 981–984.

Goldader J. D., Tholen D. J., Cruikshank D. P., and Hartmann W. K. (1991) Galileo support observations of asteroid 951 Gaspra. *Astron. J., 102,* 1503–1509.

Granahan J. C., Fanale F. P., Robinson M., Carlson R. W., Kamp L. W., Klaasen K. P., Smythe W. D., Weissman P. R., Belton M., Cook E., Edwards K., McEwen A. S., Soderblom L. A., Carcich B. T., Helfenstein P., Simonelli D., Thomas P., and Veverka J. (1994) A Galileo multi-instrument spectral analysis of 951 Gaspra (abstract). In *Lunar and Planetary Science XXV,* pp. 453–454. Lunar and Planetary Institute, Houston.

Granahan J. C., Fanale F. P., Carlson R., Kamp L. W., Klaasen K. P., Helfenstein P., Thomas P., McEwen A. S., Chapman C., Sunshine J., Belton M. J. S., and the Galileo NIMS and SSI Instrument Teams (1995) Galileo multi-instrument spectral observations of 243 Ida and Dactyl (abstract). In *Lunar and Planetary Science XXVI,* pp. 489–490. Lunar and Planetary Institute, Houston.

Greenberg R., Nolan M. C., Bottke W. F., and Kolvoord R. A. (1994) Collisional history of Gaspra. *Icarus, 107,* 84–97.

Greenberg R., Bottke W. F., Nolan M., Geissler P., Petit J.-M., and Durda D. D. (1996) Collisional and dynamical history of Ida. *Icarus, 120,* 106–118.

Hapke B. (2001) Space weathering from Mercury to the asteroid belt. *J. Geophys. Res., 106,* 10039–10073.

Hapke B., Cassidy W., and Wells E. (1975) Effects of vapor-phase deposition processes on the optical, chemical and magnetic properties of the lunar regolith. *Moon, 13,* 339–354.

Hawkins S. E. III, Darlington E. H., Murchie S. L., Peacock K., Harris T. J., Hersman C. B., Elko M. J., Prendergast D. T., Ballard B. W., Gold R. E., Veverka J., and Robinson M. S. (1997) Multi-Spectral Imager on the Near Earth Asteroid Rendezvous mission. *Space Sci. Rev., 82,* 31–100.

Head J. W. and Cintala M. J. (1979) Grooves on Phobos: Evidence for possible secondary cratering origin (abstract). In *Reports of the Planetary Geology Program, 1978–1979,* pp. 19–21. NASA Technical Memorandum 80339.

Helfenstein P., Veverka J., Thomas P. C., Simonelli D. P., Lee P., Klaasen K., Johnson T. V., Breneman H., Head J. W., Murchie S., Fanale F., Robinson M., Clark B., Granahan J., Garbeil H., McEwen A. S., Kirk R. L., Davies M., Neukum G., Mottola S., Wagner R., Belton M., Chapman C., and Pilcher C. (1994) Galileo photometry of asteroid 951 Gaspra. *Icarus, 107,* 37–60.

Helfenstein P., Veverka J., Thomas P. C., Simonelli D. P., Klaasen K., Johnson T. V., Fanale F., Granahan J., McEwen A. S., Belton M., and Chapman C. (1996) Galileo photometry of asteroid 243 Ida. *Icarus, 120,* 48–65.

Hiroi T., Pieters C. M., Zolensky M. E., and Lipschutz M. E. (1993) Evidence of thermal metamorphism on the C, G, B, and F asteroids. *Science, 261,* 1016–1018.

Horstman K. C. and Melosh H. J. (1989) Drainage pits in cohesionless materials: Implications for the surface of Phobos. *J. Geophys. Res., 94,* 12433–12441.

Housen K. R. (1992) Crater ejecta velocities for impacts on rocky bodies (abstract). In *Lunar and Planetary Science XXIII,* pp. 555–556. Lunar and Planetary Institute, Houston.

Housen K. R. and Wilkening L. L. (1982) Regoliths on small bodies in the Solar System. *Annu. Rev. Earth Planet Sci., 10,* 355–376.

Housen K. R., Wilkening L. L., Chapman C. R., and Greenberg R. (1979a) Asteroidal regoliths. *Icarus, 39,* 317–351.

Housen K. R., Wilkening L. L., Chapman C. R., and Greenberg R. (1979b) Regolith development and evolution on asteroids on the Moon. In *Asteroids* (T. Gehrels, ed.), pp. 601–627. Univ. of Arizona, Tucson.

Housen K. R., Schmidt R. M., and Holsapple K. A. (1983) Crater ejecta scaling laws: Fundamental forms based on dimensional analysis. *J. Geophys. Res., 88,* 2485–2499.

Housen K. R., Holsapple K. A., and Voss M. E. (1999) Compaction as the origin of the unusual craters on the asteroid Mathilde. *Nature, 402,* 155–157.

Houston W. N., Moriwaki Y., and Chang C. S. (1973) Downslope movement of lunar soil and rock caused by meteoroid impact. *Proc. Lunar Sci. Conf. 4th,* pp. 2425–2435.

Houston W. N., Mitchell J. K., and Carrier W. D. III (1974) Lunar soil density and porosity. *Proc. Lunar Sci. Conf. 5th,* pp. 2361–2364.

Kirk R. L. (1987) A fast finite-element algorithm for two-dimensional photoclinometry. Part III of Ph.D. thesis, California Institute of Technology, Pasadena. 258 pp.

Larson H. P., Fink U., Treffers R. R., and Gautier T. N. III (1976) The infrared spectrum of asteroid 433 Eros. *Icarus, 28,* 95–103.

Lebofsky L. A. and Rieke G. H. (1979) Thermal properties of 433 Eros. *Icarus, 40,* 297–308.

Lee P. (1996) Dust levitation on asteroids. *Icarus, 124,* 181–194.

Lee P., Veverka J., Thomas P. C., Helfenstein P., Belton J. J. S., Chapman C. R., Greeley R., Pappalardo R. T., Sullivan R., and Head J. W. III (1996) Ejecta blocks on 243 Ida and on other asteroids. *Icarus, 120,* 87–105.

Lee S. W., Thomas P., and Veverka J. (1986) Phobos, Deimos and the Moon: Size and distribution of ejecta blocks. *Icarus, 68,* 77–86.

Lohr D. A., Zanetti L. J., Anderson B. J., Potemra T. A., Hayes J. R., Gold R. E., Henshaw R. M., Mobley F. F., Holland D. B., Acuña M. H., and Scheifele J. L. (1997) NEAR magnetic field investigation, instrumentation, spacecraft magnetics and data access. *Space Sci. Rev., 82,* 255–281.

Love S. G. and Ahrens T. (1996) Catastrophic impacts on gravity dominated asteroids. *Icarus, 124,* 141–155.

Lucey P., Taylor G., and Malaret E. (1995) Abundance and distribution of iron on the moon. *Science, 268,* 1150–1153.

Lucey P., Blewett D., and Hawke B. R. (1998) Mapping the FeO and TiO_2 content of the lunar surface with multispectral imagery. *J. Geophys. Res., 103,* 3679–3699.

Lucey P., Blewett D., and Jolliff B. (2000) Lunar iron and titanium abundance algorithms based on final processing of Clementine ultraviolet-visible images. *J. Geophys. Res., 105,* 20297–20305.

Lucey P. G., Hinrichs J., Kelly M., Wellnitz D., Izenberg N., Murchie S., Robinson M., Clark B. E., and Bell J. F. III (2002) Detection of temperature-dependent spectral variation on the asteroid Eros and new evidence for the presence of an olivine-rich silicate assemblage. *Icarus, 155,* 181–188.

Malin M. C. and Dzurisin D. (1977) Landform degradation on Mercury, the Moon, and Mars: Evidence from crater depth/diameter relationships. *J. Geophys. Res., 82,* 376–388.

McCoy T. J., Burbine T. H., McFadden L. A., Starr R. D., Gaffey M. J., Nittler L. R., Evans L. G., Izenberg N., Lucey P., Trombka J. I., Bell J. F. III, Clark B. E., Clark P. E., Squyres S. W., Chapman C. R., Boynton W. V., and Veverka J. (2001) The composition of 433 Eros: A mineralogical-chemical synthesis. *Meteoritics & Planet. Sci., 36,* 1661–1672.

McFadden L. A., Wellnitz D. D., Schnaubelt M., Gaffey M. J., Bell J. F. III, Izenberg N., Murchie S., and Chapman C. R. (2001) Mineralogical interpretation of reflectance spectra of Eros from NEAR NIS low phase flyby. *Meteoritics & Planet. Sci., 36,* 1711–1726.

McKay D. S., Swindle T. D., and Greenberg R. (1989) Asteroidal regoliths: What we do not know. In *Asteroids II* (R. P. Binzel et al., eds.), pp. 617–642. Univ. of Arizona, Tucson.

Miller J. K., Konopliv A. S., Antreasian P. G., Bordi J. J., Chesley S., Helfrich C. E., Owen W. M., Scheeres D. J., Wang T. C., Williams B. G., and Yeomans D. K. (2002) Determination of shape, gravity and rotational state of asteroid 433 Eros. *Icarus, 155,* 3–17.

Mitchell D. L., Hudson R. S., Ostro S. J., and Rosema K. D. (1998) Shape of asteroid 433 Eros from inversion of Goldstone radar doppler spectra. *Icarus, 131,* 4–14.

Moroz J., Fisenko A., Semjonova L., Pieters C., and Korotaeva N. (1996) Optical effects of regolith processes on S asteroids as simulated by laser shots on ordinary chondrite and other mafic materials. *Icarus, 122,* 366–382.

Morrison D. (1976) The diameter and thermal inertia of 433 Eros. *Icarus, 28,* 125–132.

Mottola S. W., Sears D., Erikson A., Harris A. W., Young J. W., Hahn G., Dahlgren M., Mueller B. E. A., Owen B., Gil-Hutton R., Licandro J., Barrucci M. A., Angeli C., Neukum G., Lagerkvist C. I., and Lahulla J. F. (1995) The slow rotation of 253 Mathilde. *Planet. Space Sci., 43,* 1609–1613.

Murchie S. and Erard S. (1996) The spectral properties and composition of Phobos from measurements by Phobos 2. *Icarus, 123,* 63–86.

Murchie S. L. and Pieters C. M. (1996) Spectral properties and rotational spectral heterogeneity of 433 Eros. *J. Geophys. Res., 101,* 2201–2214.

Murchie S., Britt D., Head J., Pratt S., Fisher P., Zhukov B., Kuzmin A., Ksanfomality L., Zharkov A., Nikitin G., Fanale F., Blaney D., Robinson M., and Bell J. (1991) Color heterogeneity of the surface of Phobos: Relationships of geologic features and comparison to meteorite analogs. *J. Geophys. Res., 96,* 5925–5945.

Murchie S., Robinson M., Hawkins S. E. III, Harch A., Helfenstein P., Thomas P., Peacock K., Owen W., Heyler G., Murphy P., Darlington E. H., Keeney A., Gold R., Clark B., Izenberg N., Bell J. F. III, Merline W., and Veverka J. (1999) Inflight calibration of the NEAR Multispectral Imager. *Icarus, 140,* 66–91.

Murchie S., Domingue D., Robinson M., Li H., Prockter L., Hawkins S. E., Owen W., and Clark B. (2002a) Inflight calibration of the NEAR Multispectral Imager, 2: Results from Eros approach and orbit. *Icarus, 155,* 229–243.

Murchie S., Robinson M., Clark B., Li H., Thomas P., Joseph J., Bussey B., Domingue D., Veverka J., Izenberg N., and Chapman C. (2002b) Color variations on Eros from NEAR multispectral imaging. *Icarus, 155,* 145–168.

Nittler L. R., Starr R. D., Lim L., McCoy T. J., Burbine T. H., Reedy R. C., Trombka J. I., Gorenstein P., Squyres S. W., Boynton W. V., McClanahan T. P., Bhangoo J. S., Clark P. E., Murphy M. E., and Killen R. (2001) X-ray fluorescence measurements of the surface elemental composition of asteroid 433 Eros. *Meteoritics & Planet. Sci., 36,* 1673–1695.

Nolan M. C., Asphaug E., Melosh H. J., and Greenberg R. (1996) Impact craters on asteroids: Does gravity or strength control their size? *Icarus, 124,* 359–371.

Nolan M. C., Asphaug E., Greenberg R., and Melosh H. J. (2001) Impacts on asteroids: Fragmentation, regolith transport, and disruption. *Icarus, 153,* 1–15.

Noland M., Veverka J., and Pollack J. B. (1973) Mariner 9 polarimetry of Phobos and Deimos. *Icarus, 20,* 490–502.

Oberbeck V. R. and Quaide W. L. (1968) Genetic implications of lunar regolith thickness variations. *Icarus, 9,* 446–465.

Petit J.-M., Durda D. D., Greenberg R., Hurford T. A., and Geissler P. E. (1997) The long-term dynamics of Dactyl's orbit. *Icarus, 130,* 177–197.

Pieters C. M., Gaffey M. J., Chapman C. R., and McCord T. B. (1976) Spectrophotometry (0.33 to 1.07 μm) of 433 Eros and compositional implications. *Icarus, 28,* 105–115.

Pieters C. M., Staid M. I., Fischer E. M., Tompkins S., and He G. (1994) A sharper view of impact craters from Clementine data. *Science, 266,* 1844–1848.

Pieters C. M., Taylor L. A., Noble S. K., Keller L. P., Hapke B., Morris R. V., Allen C. C., McKay D. S., and Wentworth S. (2000) Space weathering on airless bodies: Resolving a mystery with lunar samples. *Meteoritics & Planet. Sci., 35,* 1101–1107.

Pike R. J. (1980a) *Geometric Interpretation of Lunar Craters.* U.S. Geol. Surv. Prof. Paper 1046-C, U.S. Government Printing Office, Washington. 279 pp.

Pike R. J. (1980b) Control of crater morphology by gravity and target type: Mars, Earth, Moon. *Proc. Lunar Planet. Sci. Conf. 11th,* pp. 2159–2189.

Pike R. J. (1988) Geomorphology of impact craters on Mercury. In *Mercury* (F. Villas et al., eds.), pp. 165–273. Univ. of Arizona, Tucson.

Pollack J. B., Veverka J., Noland M., Sagan C., Hartmann W. K., Duxbury T. C., Born G. H., Milton D. J., and Smith B. A. (1972) Mariner 9 television observations of Phobos and Deimos. *Icarus, 17,* 394–407.

Pollack J. B., Veverka J., Noland M., Sagan C., Duxbury T. C., Acton C. H., Born G. H., Hartmann W. K., and Smith B. A. (1973) Mariner 9 television observations of Phobos and Deimos, 2. *J. Geophys. Res., 78,* 4313–4326.

Prockter L., P. Thomas P., Robinson M., Joseph J., Milne A., Bussey B., Veverka J., and Cheng A. (2001) Surface expressions of structural features on Eros. *Icarus, 155,* 75–93.

Quaide W. L. and Oberbeck V. R. (1968) Thickness determinations of the lunar surface layer from lunar impact craters. *J. Geophys. Res., 73,* 5247–5270.

Rivkin A. S., Clark B. E., Britt D. T., and Lebofsky L. A. (1997) Infrared spectrophotometry of the NEAR flyby target 253 Mathilde. *Icarus, 127,* 255–257.

Robinson M. S., Thomas P. C., Veverka J., Murchie S., and Carcich B. T. (2001) Morphology, distribution and origin of ponded deposits on Eros. *Nature, 413,* 396–400.

Simonelli D., Thomas P. C., Carcich B. T., and Veverka J. (1993) The generation and use of numerical shape models for irregular Solar System objects. *Icarus, 103,* 49–61.

Simonelli D. P., Wisz M., Switala A., Adinolfi D., Veverka J., Thomas P. C., and Helfenstein P. (1998) Photometric properties of Phobos surface materials from Viking images. *Icarus, 131,* 52–77.

Stook P. J. (1993) The topography of Epimetheus. *Earth, Moon Planets, 63,* 67–83.

Stook P. J. (1994) The surfaces of Larissa and Proteus. *Earth, Moon Planets, 65,* 31–54.

Sullivan R., Greeley R., Pappalardo R., Asphaug E., Moore J. M., Morrison D., Belton M. J. S., Carr M., Chapman C. R., Geissler P., Greenberg R., Granahan J., Head J. W. III, Kirk R., McEwen A., Lee P., Thomas P. C., and Veverka J. (1996) Geology of 243 Ida. *Icarus, 120,* 119–139.

Thomas P. (1978) The morphology of Phobos and Deimos. Ph.D. thesis, Cornell University, Ithaca, New York. 272 pp.

Thomas P. (1979) Surface features of Phobos and Deimos. *Icarus, 40,* 223–243.

Thomas P. C. (1999) Large craters on small objects: Occurrence, morphology, and effects. *Icarus, 142,* 89–96.

Thomas P. and Veverka J. (1979) Grooves on asteroids: A prediction. *Icarus, 40,* 394–405.

Thomas P. and Veverka J. (1980) Downslope movement of material on Deimos. *Icarus, 42,* 234–250.

Thomas P., Veverka J., Bloom A., and Duxbury T. (1979) Grooves on Phobos: Their distribution, morphology, and possible origin. *J. Geophys. Res., 84,* 8457–8477.

Thomas P. C., Veverka J., Simonelli D., Helfenstein P., Carcich B., Belton M. J. S., Davies M. E., and Chapman C. (1994) The shape of Gaspra. *Icarus, 107,* 23–36.

Thomas P. C., Adinolfi D., Helfenstein P., Simonelli D., and Veverka J. (1996a) The surface of Deimos: Contribution of materials and processes to its unique appearance. *Icarus, 123,* 536–556.

Thomas P. C., Belton M. J. S., Carcich B., Chapman C. R., Davies M. E., Sullivan R., and Veverka J. (1996b) The shape of Ida. *Icarus, 120,* 20–32.

Thomas P. C., Veverka J., Bell J. F. III, Clark B. E., Carcich B., Joseph J., Robinson M., McFadden L. A., Malin M. C., Chapman C. R., Merline W., and Murchie S. (1999) Mathilde: Size, shape, and geology. *Icarus, 140,* 17–27.

Thomas P. C., Robinson M. S., Veverka J., and Murchie S. (2001) Shoemaker crater as a major source of ejecta on asteroid 433 Eros. *Nature, 413,* 394–396.

Thomas P. C., Joseph J., Carcich B., Veverka J., Clark B. E., Bell J. F. III, Byrd A. J., Chomko R., Robinson M., Murchie S., Prockter L., Cheng A., Izenberg N., Malin M., Chapman C., McFadden L. A., Kirk R., Gaffey M., and Lucey P. G. (2002a) Eros: Shape, topography, and slope processes. *Icarus, 155,* 18–37.

Thomas P. C., Prockter L., Robinson M., Joseph J., and Veverka J. (2002b) Global structure of asteroid 433 Eros. *Geophys. Res. Lett., 29,* 46-1 to 46-4.

Trombka J., Squyres S., Brückner J., Boynton W., Reedy R., McCoy T., Gorenstein P., Evans L., Arnold J., Starr R., Nittler L., Murphy M., Mikheeva I., McNutt R., McClanahan T., McCartney E., Goldsten J., Gold R., Floyd S., Clark P., Burbine T., Bhangoo J., Bailey S., and Petaev M. (2000) The elemental composition of asteroid 433 Eros: Results of the NEAR-Shoemaker X-ray spectrometer. *Science, 289,* 2101–2105.

Veverka J. and Thomas P. (1979) Phobos and Deimos: A preview of what asteroids are like? In *Asteroids* (T. Gehrels, ed.), pp. 628–651. Univ. of Arizona, Tucson.

Veverka J., Noland M., Sagan C., Pollack J., Quam L., Tucker R., Eross B., Duxbury T., and Green W. (1974) A Mariner 9 atlas of the moons of Mars. *Icarus, 23,* 206–289.

Veverka J., Belton M., Klaasen K., and Chapman C. (1994a)

Galileo's encounter with 951 Gaspra: Overview. *Icarus, 107,* 2–17.

Veverka J., Thomas P., Simonelli D., Belton M. J. S., Carr M., Chapman C., Davies M. E., Greeley R., Greenberg R., Head J., Klaasen K., Johnson T. V., Morrison D., and Neukum G. (1994b) Discovery of grooves on Gaspra. *Icarus, 107,* 72–83.

Veverka J., Thomas P., Helfenstein P., Lee P., Harch A., Calvo S., Chapman C., Belton M. J. S., Klaasen K., Johnson T. V., and Davies M. (1996a) Dactyl: Galileo observations of Ida's satellite. *Icarus, 120,* 200–211.

Veverka J., Helfenstein P., Lee P., Thomas P., McEwen A., Belton M., Klaasen K., Johnson T. V., Granahan J., Fanale F., Geissler P., and Head J. W. III (1996b) Ida and Dactyl: Spectral reflectance and color variations. *Icarus, 120,* 66–76.

Veverka J., Thomas P., Harch, A., Clark B., Bell J. F. III, Carcich B., Joseph J., Chapman C., Merline W., Robinson M., Malin M., McFadden L. A., Murchie S., Hawkins S. E. III, Farquhar R., Izenberg N., and Cheng A. (1997) NEAR's flyby of 253 Mathilde: Images of a C asteroid. *Science, 278,* 2109–2114.

Veverka J., Thomas P., Harch A., Clark B., Bell J. F. III, Carcich B., Joseph J., Murchie S., Izenberg N., Chapman C., Merline W., Malin M., McFadden L., and Robinson M. (1999) NEAR encounter with asteroid 253 Mathilde: Overview. *Icarus, 140,* 3–16.

Veverka J., Robinson M., Thomas P., Murchie S., Bell J. F. III, Izenberg N., Chapman C., Harch A., Bell M., Carcich B., Cheng A., Clark B., Domingue D., Dunham D., Farquhar R., Gaffey M. J., Hawkins E., Joseph J., Kirk R., Li H., Lucey P., Malin M., Martin P., McFadden L., Merline W. J., Miller J. K., Owen W. M. Jr., Peterson C., Prockter L., Warren J., Wellnitz D., Williams B. G., and Yeomans D. K. (2000) NEAR at Eros: Imaging and spectral results. *Science, 289,* 2088–2097.

Veverka J., Thomas P. C., Robinson M., Murchie S., Chapman C., Bell M., Harch A., Merline W. J., Bell J. F. III, Bussey B., Carcich B., Cheng A., Clark B., Domingue D., Dunham D., Farquhar R., Gaffey M. J., Hawkins E., Izenberg N., Joseph J., Kirk R., Li H., Lucey P., Malin M., McFadden L., Miller J. K., Owen W. M. Jr., Peterson C., Prockter L., Warren J., Wellnitz D., Williams B. G., and Yeomans D. K. (2001a) Imaging of small-scale features on 433 Eros from NEAR: Evidence for a complex regolith. *Science, 292,* 484–488.

Veverka J., Farquhar B., Robinson M., Thomas P., Murchie S., Harch A., Antreasian P. G., Chesley S. R., Miller J. K., Owen W. M. Jr., Williams B. G., Yeomans D., Dunham D., Heyler G., Holdridge M., Nelson R. L., Whittenburg K. E., Ray J. C., Carcich B., Cheng A., Chapman C., Bell J. F. III, Bell M., Bussey B., Clark B., Domingue D., Gaffey M. J., Hawkins E., Izenberg N., Joseph J., Kirk R., Lucey P., Malin M., McFadden L., Merline W. J., Peterson C., Prockter L., Warren J., and Wellnitz D. (2001b) The landing of the NEAR-Shoemaker spacecraft on asteroid 433 Eros. *Nature, 413,* 390–393.

Warren J. W., Peacock K., Darlington E. H., Murchie S. L., Oden S. F., Hayes J. R., Bell J. F. III, Krein S. J., and Mastandrea A. (1997) Near Infrared Spectrometer for the Near Earth Asteroid Rendezvous mission. *Space Sci. Rev., 82,* 101–167.

Wilkison S. L. and Robinson M. S. (2000) Bulk density of ordinary chondrite meteorites and implications for asteroidal internal structure. *Meteoritics & Planet. Sci., 35,* 1203–1213.

Wilkison S. L., Robinson M. S., Thomas P. C., Veverka J., McCoy T. J., Murchie S. L., Prockter L., and Yeomans D. (2002) An estimate of Eros's porosity and implications for internal structure. *Icarus, 155,* 94–103.

Wilson L. and Head J. W. (1989) Dynamics of groove formation on Phobos by ejecta from Stickney (abstract). In *Lunar and Planetary Science XX,* pp. 1211–1212. Lunar and Planetary Institute, Houston.

Wilson L., Keil K., and Love S. J. (1999) The internal structures and densities of asteroids. *Meteoritics & Planet. Sci., 34,* 479–483.

Wisniewski W. Z. (1976) Spectrophotometry and UBVRI photometry of Eros. *Icarus, 28,* 87–90.

Yeomans D. K., Barriot J.-P., Dunham D. W., Farquhar R. W., Giorgini J. D., Helfrich C. E., Konopliv A. S., McAdams J. V., Miller J. K., Owen W. M. Jr., Scheeres D. J., Synnott S. P., and Williams B. G. (1997) Estimating the mass of asteroid 253 Mathilde from tracking data during the NEAR flyby. *Science, 278,* 2106–2109.

Yeomans D. K., Antreasian P. G., Barriot J.-P., Chesley S. R., Dunham D. W., Farquhar R. W., Giorgini J. D., Helfrich C. E., Konopliv A. S., McAdams J. V., Miller J. K., Owen W. M. Jr., Scheeres D. J., Thomas P. C., Veverka J., and Williams B. G. (2000) Radio science results during the NEAR-Shoemaker spacecraft rendezvous with Eros. *Science, 289,* 2085–2088.

Zellner B. (1976) Physical properties of asteroid 433 Eros. *Icarus, 28,* 149–153.

Zellner B. and Gradie J. (1976) Polarization of the reflected light of asteroid 433 Eros. *Icarus, 28,* 117–123.

Zuber M. T., Smith D. E., Cheng A. F., Garvin J. B., Aharonson O., Cole T. D., Dunn P. J., Guo Y., Lemoine F. G., Neumann G. A., Rowlands D. D., and Torrence M. H. (2000) The shape of 433 Eros from the NEAR-Shoemaker Laser Rangefinder. *Science, 289,* 2097–2101.

Near Earth Asteroid Rendezvous: Mission Summary

Andrew F. Cheng
The Johns Hopkins Applied Physics Laboratory

On February 14, 2000, the Near Earth Asteroid Rendezvous spacecraft (*NEAR Shoemaker*) began the first orbital study of an asteroid, the near-Earth object 433 Eros. Almost a year later, on February 12, 2001, *NEAR Shoemaker* completed its mission by landing on the asteroid and acquiring data from its surface. *NEAR Shoemaker*'s intensive study has found an average density of 2.67 ± 0.03, almost uniform within the asteroid. Based upon solar fluorescence X-ray spectra obtained from orbit, the abundance of major rock-forming elements at Eros may be consistent with that of ordinary chondrite meteorites except for a depletion in S. Such a composition would be consistent with spatially resolved, visible and near-infrared (NIR) spectra of the surface. Gamma-ray spectra from the surface show Fe to be depleted from chondritic values, but not K. Eros is not a highly differentiated body, but some degree of partial melting or differentiation cannot be ruled out. No evidence has been found for compositional heterogeneity or an intrinsic magnetic field. The surface is covered by a regolith estimated at tens of meters thick, formed by successive impacts. Some areas have lesser surface age and were apparently more recently disturbed or covered by regolith. A small center of mass offset from the center of figure suggests regionally nonuniform regolith thickness or internal density variation. Blocks have a nonuniform distribution consistent with emplacement of ejecta from the youngest large crater. Some topographic features indicate tectonic deformations. Several regional-scale linear features have related orientations, suggesting a globally consolidated internal structure. Structural control of crater shapes hints that such internal structure is pervasive. From the bulk density, inferred composition, and evidence for global structure, Eros is interpreted to be largely intact but extensively fractured.

1. INTRODUCTION

The Near Earth Asteroid Rendezvous (*NEAR*) spacecraft was launched on February 17, 1996, and performed the first flyby of a C-type main-belt asteroid, 253 Mathilde, on June 27, 1997 (*Veverka et al.,* 1997b; *Yeomans et al.,* 1997b). *NEAR* performed a flyby of Eros on December 23, 1998 (*Yeomans et al.,* 1999; *Veverka et al.,* 1999a), and on February 14, 2000, *NEAR* entered orbit around Eros, becoming the first spacecraft to orbit an asteroid. The spacecraft was then renamed *NEAR Shoemaker* in honor of Eugene M. Shoemaker (1928–1997). After successfully completing its year-long orbital mission, *NEAR Shoemaker* became the first spacecraft to land on an asteroid. On February 12, 2001, it touched down softly on Eros, close to the southeastern edge of the saddle-shaped depression Himeros at 37.2°S by 278.4°W (all names of features on Eros are those proposed by the *NEAR* Science Team to the IAU; see appendix).

NEAR Shoemaker was a three-axis stabilized spacecraft that carried a six-instrument scientific payload. This paper summarizes results obtained by the NEAR Science Team using a multispectral imager (MSI), a near-infrared spectrometer (NIS), a laser rangefinder (NLR), an X-ray spectrometer (XRS), a γ-ray spectrometer (GRS), and a magnetometer (MAG). In addition, *NEAR Shoemaker* included a radio science investigation (RS) using the spacecraft telecommunications system. The scientific background and expected results of these investigations on *NEAR* were summarized in a special issue of the *Journal of Geophysical Research* (*Cheng et al.,* 1997; *Veverka et al.,* 1997a; *Trombka et al.,* 1997; *Zuber et al.,* 1997; *Acuña et al.,* 1997; *Yeomans et al.,* 1997a).

In what follows, a brief overview of the spacecraft, instruments, and mission as flown is presented, after which the science background and original objectives of the mission are summarized. Next, a selective review of asteroid science pertinent to *NEAR* mission objectives is given to provide context. Then a summary of results at Mathilde is given, followed by a summary of Eros results. Some outstanding issues and directions for future work are discussed in the conclusion.

2. *NEAR SHOEMAKER* SPACECRAFT

NEAR was a solar-powered, three-axis-stabilized spacecraft (*Santo et al.,* 1995) with a launch mass, including propellant, of 805 kg and a dry mass of 468 kg. The spacecraft was simple and highly redundant (Fig. 1). It used X-band telemetry to the NASA Deep Space Network. With a 70-m antenna, the data rates from Eros ranged from 17.6 to 26.5 kbps. Two solid-state recorders were accommodated with a combined memory capacity of 1.6 Gb. Spacecraft attitude was determined using a star camera, a fully redundant inertial measurement unit, and redundant digital Sun sensors. The propulsion subsystem was dual mode, using hydrazine as fuel for both monopropellant and bipropellant thrusters. Attitude was controlled by a redundant set of four reaction wheels or by the thruster complement to within

1.7 mrad. *NEAR*'s line-of-sight pointing stability was within 20 µrad over 1 s, and postprocessing attitude knowledge was within 130 µrad.

NEAR contained six scientific instruments, which are detailed in the next section. The MAG was mounted on top of the high-gain antenna feed, where it was exposed to the minimum level of spacecraft-generated magnetic fields. The remaining instruments (MSI, NIS, XRS, GRS, and NLR) were mounted on the aft deck (Fig. 1), all on fixed mounts, and were coaligned to view a common boresight direction. The NIS had a scan mirror that allowed it to look 110° away from the common boresight. Key properties of the mission design permitted the use of this fixed spacecraft geometry. Throughout most of the orbital rendezvous with Eros, the angle between the Sun and the Earth as seen from the spacecraft remained less than ~30°. In addition, the mission aphelion was reached during cruise. Hence, if the solar panels were sufficiently large to sustain *NEAR* at aphelion, there was sufficient power margin at Eros for the spacecraft to pull its solar panels 30° off full illumination to point the HGA at Earth. Moreover, the rendezvous orbit plane was maintained so that the orbit normal pointed approximately at the Sun. In this case, as *NEAR* orbited Eros, it was able to roll around the HGA axis so as to keep the instruments pointed at the asteroid. (The instruments were usually pointed away from the asteroid when the HGA was used to downlink to Earth.) This mode of operation motivated the requirement for onboard data storage. With onboard image compression, *NEAR* could store more than 1000 images and downlink them within 10 h at its maximum data rate of 26.5 kbps.

3. INSTRUMENTS

A brief summary of instrument characteristics is given in this section (Fig. 2). Full descriptions of each instrument's design and science operations can be found in a special issue of *Space Science Reviews* (vol. 82, 1997). Additional information on results of inflight calibrations have been published and are given as the second members of the paired citations in this section.

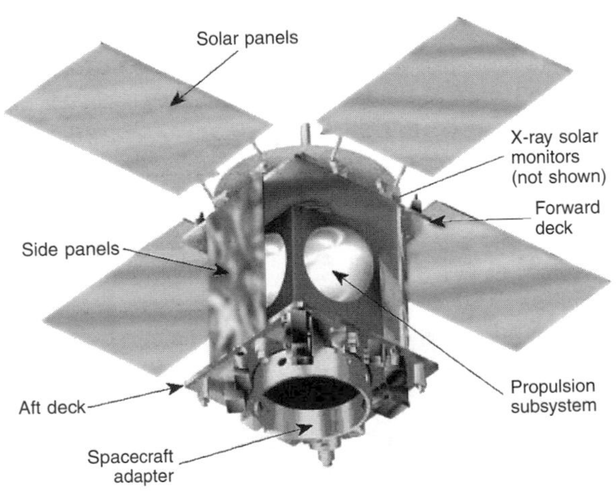

Fig. 1. The *NEAR Shoemaker* spacecraft, showing the aft deck where all instruments except the magnetometer and the X-ray solar monitors were mounted. Side panels have been removed to show the propulsion subsystem.

All data from the *NEAR* mission were downlinked to the NASA Deep Space Network and then forwarded to the Mission Operations Center (MOC) at the Johns Hopkins University's Applied Physics Laboratory (APL) (*Heeres et al.,* 1997). Doppler and ranging data from the spacecraft were analyzed by the *NEAR* navigation team at the Jet Propulsion Laboratory (JPL). The entire spacecraft telemetry stream, including spacecraft and instrument housekeeping data, navigation data, and all science data, was forwarded to the APL MOC together with the radiometric Doppler and range data. The spacecraft telemetry stream was then passed to the Science Data Center at APL, which supported the data analysis and archiving activities. *NEAR* data were released over the Internet at http://near.jhuapl.edu/. All *NEAR* data have been archived in the NASA Planetary Data System. The total archived data volume amounted to approximately 212 GB.

3.1. Multispectral Imager (MSI)

The main goals of the MSI (*Hawkins et al.,* 1997; *Murchie et al.,* 1999, 2002a) were to determine the shape of Eros and map the mineralogy and morphology of features on its surface at high spatial resolution. MSI was a 537 × 244-pixel charge-coupled device camera with five-element radiation-hardened refractive optics. It covered the spectral range from 0.4 to 1.1 µm, and it had an eight-position filter wheel (Table 1). Seven narrow-band filters were chosen to discriminate the major Fe-bearing silicates present (olivine and pyroxene); the eighth, broadband filter provided higher sensitivity for starfield exposures to enhance optical navigation. The camera had an FOV of 2.93° × 2.26° and a pixel resolution of 96 × 162 µrad. It had a maximum framing rate of 1/s with images digitized to 12 bits and a dedicated digital processing unit with an image buffer in addition to both lossless and lossy onboard image compression (lossless compression causes no loss of information, whereas lossy compression yields a more compact dataset with loss of information).

3.2. Near-Infrared Spectrometer (NIS)

The NIS (*Warren et al.,* 1997; *Izenberg et al.,* 2000) measured the spectrum of sunlight reflected from Eros in the NIR range from 0.8 to 2.5 µm to determine the distribution and abundance of surface minerals like olivine and pyroxene. This grating spectrometer dispersed the light from the slit FOV (0.38° × 0.76° in its narrow position and 0.76° × 0.76° in the wide position) across a pair of passively cooled one-dimensional array detectors. A 32-channel Ge array covered the lower wavelengths, with channel centers at 0.82–1.49 µm with a 0.022-µm spacing between channels. A 32-channel In/Ga-As array covered longer wavelengths, with channel centers at 1.37–2.71 µm with a 0.043-µm spacing between channels. Due to the configuration of the optics and

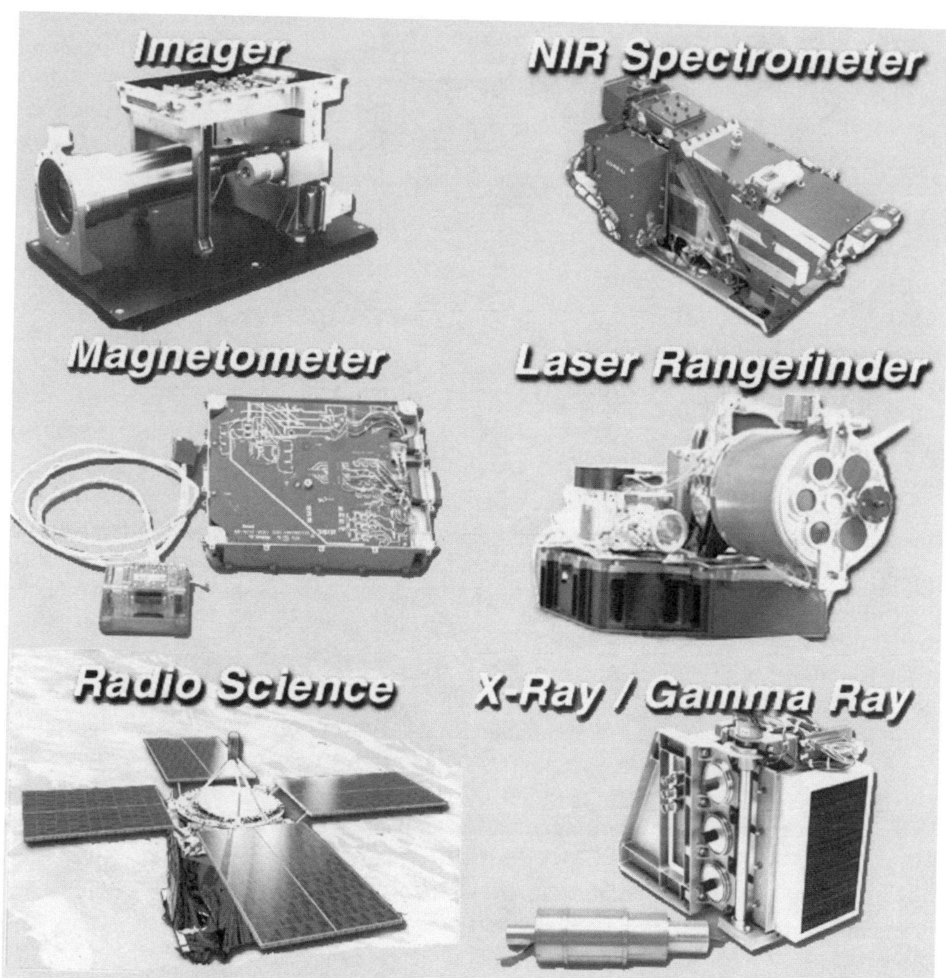

Fig. 2. The *NEAR Shoemaker* instruments and radio science experiment. The high-gain antenna is the circular dish centered amidst the four solar panels.

TABLE 1. NEAR MSI filter positions (*Murchie et al.*, 1999).

Filter Number	Nominal Center (nm)	Nominal Width (nm)	Effective Center (nm)	Effective Width (nm)
2	450	50	462	23
1	550	30	554	24
0	700	200	700	133
3	760	20	755	19
5	900	40	900	33
4	950	40	951	38
6	1000	50	996	44
7	1050	80	1033	51

the sensitivity of this array, useful measurements were acquired over the wavelength range 1.5–2.5 μm. NIS had a scan mirror that enabled it to step across the range from 30° forward of the common boresight to 110° aft, in 0.4° steps. Spectral images were built up by a combination of scan mirror and spacecraft motions. In addition, the NIS had a gold calibration target that viewed at the forward limit of the mirror's scan to provide inflight calibration of instrument stability.

3.3. X-Ray Spectrometer (XRS)

The XRS (*Goldsten et al.*, 1997; *Starr et al.*, 2000) was an X-ray fluorescence spectrometer that measured characteristic X-ray line emissions excited by solar X-rays from major elements in the asteroid's surface. It covered X-rays from 1 to 10 keV using three gas-proportional counters. The balanced, differential filter technique was used to separate the closely spaced Mg, Al, and Si lines below 2 keV. The gas-proportional counters directly resolved higher-energy line emissions from Ca and Fe. A mechanical collimator gave the XRS a 5° field-of-view (FOV), yielding a spatial resolution as fine as 2 km in the low orbits. The XRS also included a solar monitor to measure continuously the incident spectrum of solar X-rays, using both a gas-proportional counter and a high-spectral-resolution Si X-ray detector. The XRS performed inflight calibration using a calibration rod with Fe^{55} sources that could be rotated into or out of the detector FOV.

3.4. Gamma-Ray Spectrometer (GRS)

The GRS (*Goldsten et al.*, 1997; *Evans et al.*, 2000) detected characteristic γ-rays from 0.3 to 10 MeV emitted from the asteroid surface. Some of these emissions were excited by cosmic rays and some arose from natural radioactivity in the asteroid. The GRS used a body-mounted, passively cooled NaI scintillator detector with a bismuth germanate anticoincidence shield that defined a 45° FOV. Abundances of several important elements such as K, Si, O, and Fe were measured.

3.5. *NEAR* Laser Rangefinder (NLR)

The NLR (*Cole et al.,* 1997; *Cheng et al.,* 2000) was a laser altimeter that measured the distance from the spacecraft to the asteroid surface by sending out a short burst of laser light and then recording the time required for the signal to return from the asteroid. It used a Cr-Nd-YAG solid-state laser and a compact reflecting telescope. It sent a small portion of each emitted laser pulse through an optical fiber of known length and into the receiver, providing a continuous inflight calibration of the timing circuit. The ranging data were used to construct a global shape model and a global topographic map of Eros with horizontal resolution of ~400 m. The NLR also measured detailed topographic profiles of surface features on Eros with a best spatial resolution <5 m.

3.6. Magnetometer

The fluxgate magnetometer (*Lohr et al.,* 1997; *Acuña et al.,* 2000) used ring core sensors made of highly magnetically permeable material. It searched for any intrinsic magnetic fields of Eros. The previous *Galileo* flybys of the S-type asteroids Gaspra and Ida yielded evidence that both of these bodies are magnetic, although this evidence is ambiguous (*Kivelson et al.,* 1993). If Eros were as magnetized as typical meteorites, its magnetic field would have been easily measured. Discovery of an intrinsic magnetic field at Eros would have yielded important insights about its thermal and geological history.

3.7. Radio Science

In addition to the six instruments, the coherent X-band transponders (*Yeomans et al.,* 1997a) were used to conduct a radio science investigation by measuring the Doppler shift from the spacecraft's radial velocity component relative to Earth. Accurate measurements of the Doppler shift and the range to Earth as the spacecraft orbited Eros allowed mapping of the asteroid's gravity field. In conjunction with MSI/NIS and NLR data, gravity determinations were combined with global shape and rotation data to constrain the internal density structure of Eros and search for heterogeneity.

4. THE *NEAR* MISSION

The *NEAR* spacecraft was launched on February 17, 1996, by a Delta II 7925 launch vehicle (Fig. 3). The Mathilde encounter occurred on June 27, 1997, one week before the deep space maneuver. The Earth swingby (*Izenberg and Anderson,* 1998) occurred on January 23, 1998, retargeting the *NEAR* spacecraft to Eros. After a main rocket engine abort on December 20, 1998, a flyby of Eros was performed on December 23, 1998, and the Eros rendezvous was rescheduled for February 14, 2000, when orbit insertion occurred. On February 12, 2001, *NEAR* accomplished a soft landing on Eros.

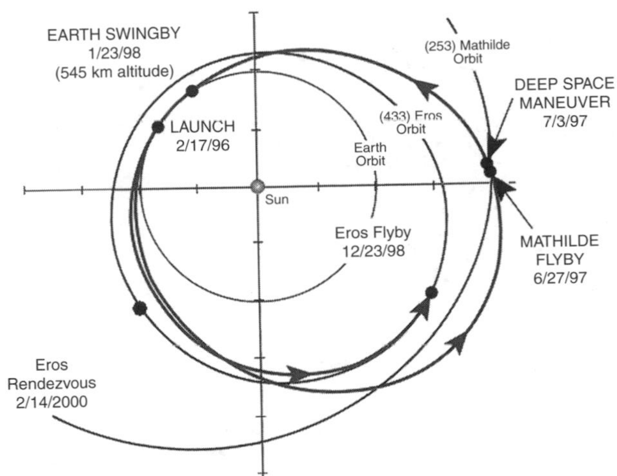

Fig. 3. Heliocentric trajectory of *NEAR Shoemaker,* with vernal equinox to the right, and orbits of Earth, Mathilde, and Eros. After the Eros flyby, *NEAR Shoemaker* followed the orbit of Eros to within the accuracy of this figure.

5. EROS OPERATIONS

Beginning in January 2000, a sequence of small maneuvers decreased the relative velocity between *NEAR* and Eros to only 5 m/s. On February 14, 2000, *NEAR* performed a flyby of Eros on its sunward side at a distance as low as 130 km. In addition to gathering spectra at an optimal illumination geometry, this first pass provided improved estimates of the asteroid's physical parameters (Table 3), such as a mass determination to 1% accuracy, identification of surface landmarks, and an improved estimate of Eros' spin vector. As the spacecraft maneuvered closer to Eros, the mass, moments of inertia, gravity harmonics, spin state, and landmark locations were determined with increasing precision. The year-long prime science phase at Eros began on February 14, 2000. During this phase, *NEAR* operated in a series of orbits that come as close as 3 km to the asteroid's surface, culminating with a soft landing on February 12, 2001. The evolution of low-altitude orbits around Eros was strongly influenced by its irregular gravity field. For some possible orbital conditions, the spacecraft would have crashed into Eros in a few days. Safe orbital operations required close coordination between the science, mission design, navigation, and mission operations teams (*Landshof and Cheng,* 1995).

Figure 4 depicts the *NEAR* spacecraft in a low-altitude, circular orbit around Eros as viewed by an observer on the Sun. This is a convenient reference frame to show *NEAR*'s

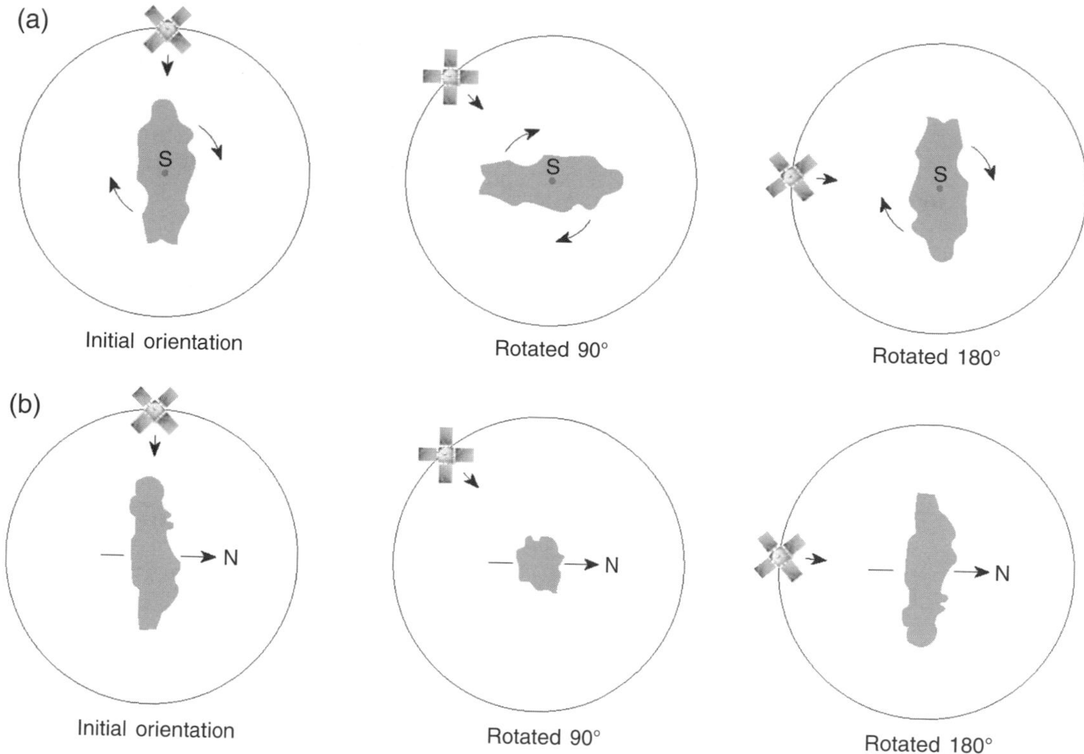

Fig. 4. Examples of orbital geometries at Eros as viewed from the Sun, at an initial rotational phase of Eros (left), after 90° rotation (center), and after 180° rotation (right). The small arrow by the spacecraft shows the instrument boresight. **(a)** Spacecraft is in retrograde equatorial orbit, and the south pole of Eros is marked by S. **(b)** Spacecraft is in polar orbit, and the spin axis of Eros is shown.

orbit because the orbital plane was controlled so that it was always within 30° of a plane perpendicular to the Sun-Eros line. In this orbital geometry, *NEAR*'s fixed solar panels were oriented toward the Sun. The science instruments were pointed at Eros' surface by rolling the spacecraft around its high-gain antenna axis as it orbited the asteroid. Two fundamentally different orbital geometries are shown in Fig. 4. In Fig. 4a, Eros' rotation axis is aligned with the Sun-Eros line. Its south pole points toward the Sun, which means that its northern hemisphere is shadowed. In this case, which applied during roughly the last two months of the orbital phase, the spacecraft orbit was retrograde and approximately equatorial. Earlier in the orbital mission, the asteroid's rotation axis was roughly perpendicular to the Sun-Eros line, and the spacecraft orbit was approximately polar as shown in Fig. 4b. This was the geometry that prevailed roughly from May through August 2000.

Because of the changing orbital geometry, coverage of Eros' surface by the instruments varied throughout the orbital mission. The subsolar point lay in the northern hemisphere early in the mission, crossing to the southern hemisphere on June 26, 2000, and remaining in the south thereafter. All areas on the asteroid were observed many times under diverse illumination conditions and viewing geometries.

The mission profile at Eros is summarized in Fig. 5. To simplify science operations, the orbital mission was divided into distinct phases (*Landshof and Cheng,* 1995). The highest-priority science varied by mission phase, because of the changing orbital geometry. While in orbits at 100 km from the center of Eros or higher, the highest-priority science was global mapping by MSI. In orbits at 50 km or lower, the highest-priority science was compositional measurement by XRS/GRS. A two-week period was allocated to altimetry

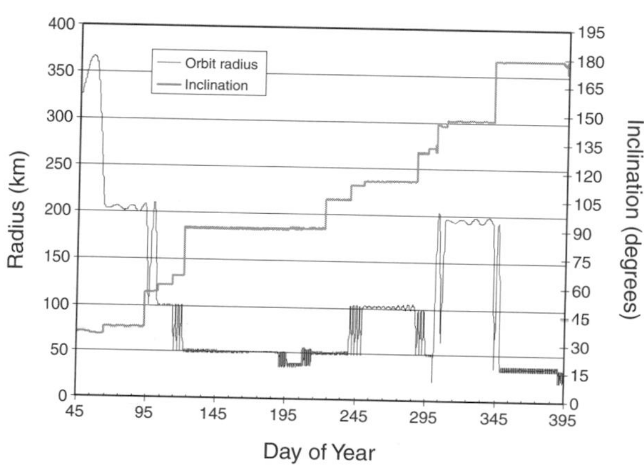

Fig. 5. Eros orbital phase as flown. The thin curve is the orbit radius (left axis) measured from the center vs. day of year 2000. The thick curve is the orbital inclination (right axis), where inclination <90° is prograde and inclination >90° is retrograde. The orbit was retrograde in the second half of the orbital phase. The figure extends from February 14, 2000, through January 29, 2001.

by NLR at the start of the 50-km polar orbits in May 2000. First detection of Eros by NLR occurred at a range of 290 km. *NEAR* spent more than 150 d in orbits at 50 km or less from the center of Eros, plus two additional weeks on the surface acquiring GRS data. In general, all instruments acquired data in all mission phases, except that NIS ceased operating on May 13, 2000, and only GRS was operated on the surface of Eros.

The rendezvous burn sequence was designed to put *NEAR* into an initial slow flyby trajectory past Eros, taking the spacecraft through a zero solar phase angle (Sun-asteroid-spacecraft angle) at a miss distance of 200 km. This flyby enabled NIS to obtain critical science observations. Since the nominal rendezvous orbit plane was kept near the Eros terminator (plane dividing the dayside from the nightside), most of the orbital observations were obtained at large phase angles >60° that were favorable for imaging but unfavorable for IR spectral mapping. The low-phase flyby on February 14, 2000, provided an important opportunity for NIS to obtain infrared spectral maps of the entire illuminated surface (the northern hemisphere) under optimal lighting conditions with minimum shadowing. Since the NIS failed on May 13, 2000, execution of a second low-phase flyby over the southern hemisphere of Eros was canceled. Instead, a low-altitude flyover to within 6 km of the surface was performed on October 26, 2000. The two weeks after the timespan of Fig. 5 were devoted mainly to preparations for the descent to the surface on February 12, 2001.

An optical search for satellites of Eros was performed during the approach to Eros before the initial flyby. No satellites were discovered at Eros (*Veverka et al.,* 2000), with a sensitivity limit of 20 m diameter for an albedo equal to that of Eros.

6. SCIENCE BACKGROUND

The vast majority of known asteroids is found in the main asteroid belt between the orbits of Mars and Jupiter, but those that come within 1.3 AU of the Sun are known as near-Earth asteroids (NEAs). The *NEAR* target 433 Eros is the second largest of the NEAs. The orbits of these dynamically young bodies have evolved on 100-m.y. timescales because of collisions, the Yarkovsky effect (reaction force from anisotropic emission of thermal radiation), and gravitational interactions with planets (*Bottke et al.,* 2000). The present-day orbits of NEAs do not indicate where they formed. NEAs are believed to originate from the main belt, and their orbital evolution can cause them to impact the Sun, collide with a terrestrial planet, or be ejected from the solar system. More than 1500 near-Earth asteroids are known as of October 2001, of which more than 238 have orbits determined well enough to be numbered. The NEAs appear to represent a broad sample of the main-belt population.

Before *NEAR,* knowledge of the nature of asteroids came from three sources: Earth-based remote sensing, data from the *Galileo* spacecraft flybys of the two main-belt asteroids 951 Gaspra and 243 Ida, and laboratory analyses of meteorites. Most meteorites are believed to be collisional fragments of asteroids (a few come from the Moon and from Mars), but they may represent a biased and incomplete sampling of the materials actually found in and on near-Earth asteroids. Firm links between meteorite types and asteroid types have been difficult to establish (*Gaffey et al.,* 1993a). The uncommon HED basaltic achondrite meteorites have been linked by visible and NIR reflectance measurements to the V-type asteroids (*McCord et al.,* 1970; *Binzel and Xu,* 1993).

However, the links between asteroids and meteorites are controversial in several ways. Foremost are the questions of whether and how the most common meteorite types (ordinary chondrites, or OCs) may be linked to the most common asteroid types (S asteroids) in the inner part of the asteroid belt (*Bell et al.,* 1989; *Gaffey et al.,* 1993b). Gaspra and Ida, which were visited by *Galileo,* and the *NEAR* target 433 Eros, are all S asteroids — a diverse class of objects containing the silicate minerals olivine and pyroxene plus an admixture of Fe/Ni metal. Some S asteroids appear to be fragments of bodies that underwent substantial melting and differentiation, whereas others may consist of primitive materials like OCs that underwent little or no melting and that may preserve characteristics of the solid material from which the inner planets accreted. Few asteroids have been found to be clear spectral analogs for OCs, compared to the numbers of asteroids that may represent differentiated cores and crusts (*Bell et al.,* 1989; *Xu et al.,* 1995). Many authors have suggested that a "space-weathering process," due to micrometeorite bombardment and/or radiation processing, may suitably alter spectral properties of optical surfaces on asteroids (*Chapman,* 1996) so that OC materials on asteroids are not recognized spectrally.

The *Galileo* flybys of Gaspra and Ida provided the first high-resolution images of asteroids, revealing complex surfaces covered by craters, fractures, grooves, and subtle color variations (*Belton et al.,* 1992, 1994). *Galileo* also discovered the satellite Dactyl of Ida (*Belton et al.,* 1996), which is a member of the Koronis family. The NIR spectrum of Gaspra indicates a high olivine abundance such that it is inferred to be a fragment of a differentiated body, but Ida and Eros display IR spectra that may be consistent with a silicate mineralogy like that in OCs (*Chapman,* 1996; *Murchie and Pieters,* 1996). The *Galileo* instrument complement did not include any capability to measure elemental composition.

Another fundamental question deals with the collisional history of small bodies in the early solar system, when the terrestrial planets formed and when disruption by violent impacts competed with accretion from gentler collisions. The question is whether asteroids the size of Eros must have been battered into an agglomeration of much smaller components bound mostly by gravity (that is, a rubble pile), or are instead intact collisional fragments from larger parent bodies ("collisional shards"). A collisional shard would be a globally consolidated body with appreciable shear and/or tensile strength, for which self-gravitation would be relatively unimportant. For Gaspra, the faceted shape and the presence of grooves may hint at such a picture (*Thomas et al.,* 1994). However, an alternative picture is that asteroids the size of Eros were thoroughly broken up without being dispersed (or possibly, Eros reaccreted), so they would now be rubble piles (e.g., *Davis et al.,* 1994; *Melosh and Ryan,* 1997). Some small asteroids rotate so rapidly that they must be monolithic (e.g., *Ostro et al.,* 1999), but the vast majority of asteroids larger than 0.2 km have rotation periods shorter than 2 h, suggesting that most of these larger bodies are rubble piles (*Pravec and Harris,* 2001).

The *NEAR* mission made the first comprehensive, spatially resolved measurements of the geology, mineralogy, and elemental composition of an S-type asteroid. Table 2 lists a set of science questions that were originally formulated in 1985 by the Near Earth Asteroid Rendezvous Science Working Group (*NEAR Science Working Group,* 1986). The present *NEAR* mission was designed to address these questions. The table identifies the *NEAR* instruments whose measurements addressed these questions and provides a summary of principal findings. As can be seen in Table 2, the *NEAR* findings exceeded and/or defied expectations in many cases.

7. MATHILDE FLYBY

The *NEAR* encounter with 253 Mathilde is significant because it was the first spacecraft encounter with a C asteroid, a completely different type of object from the S asteroids visited earlier by *Galileo* and explored later by *NEAR*. The dark C asteroids are the most common type in the central portion of the main belt of asteroids. Their carbonaceous composition is inferred from the spectral similarity of C asteroids and the carbonaceous chondrite meteorites. The visible spectrum of Mathilde in particular (*Binzel et al.,* 1996) is similar to the spectra of 1 Ceres and 2 Pallas (the two largest asteroids), but Mathilde has no 3-μm water absorption feature; its NIR spectrum is consistent with heated (to 900°C) carbonaceous chondrites or black chondrites (*Rivkin et al.,* 1997). The nature and origins of the dark, primitive asteroid types (B, C, F, G, P, D, etc.), and their relationships to the comets and the dark objects in the satellite systems of the outer planets are among the most important unresolved issues in solar-system exploration.

Apart from its importance as the first example of the C asteroids to be explored, Mathilde is extremely slow rotating (*Mottola et al.,* 1995). Its 17.4-d period is the third-

TABLE 2. *NEAR* science closure.

Science Question	Instruments	Summary Of Principal Findings
What are the morphological and textural characteristics of the asteroid surface, and how do they compare with those on larger bodies?	MSI, NLR	Tectonic features; pervasive covering of tens-of-meters thick regolith; large spatial variations in density of large craters; dearth of small craters; high boulder densitites; ponded deposits of regolith emplaced in fluidized form.
What is the elemental composition of the asteroid?	XRS, GRS	For areas observed by XRS, elemental composition was consistent with ordinary chondrites, except for sulfur depletion; GRS spectra at landing site suggest depletion in iron from chondritic values, but K chondritic.
What is the mineralogical composition of the asteroid?	NIS, MSI	Mineral composition is consistent with ordinary chondrite or similar material.
Is there evidence of compositional or structural heterogeneity?	MSI, NIS; NLR; RS; XRS	Structural control of crater formation; evidence for faulting; no evidence for compositional heterogeneity.
Is the asteroid a solid fragment of a larger parent body or a rubble pile?	MSI, NIS; NLR; RS; XRS	Eros largely intact but deeply fractured, most likely a fragment of a larger parent body; alignments of linear features suggest through-going internal structures.
Is the asteroid's precursor body(ies) primitive or differentiated?	XRS, GRS; NIS, MSI	Eros precursor was primitive, but some degree of metamorphism or partial melting not ruled out.
Is there evidence of past or present cometary activity?	MSI; NLR	None found.
Is the asteroid related to a meteorite type or types?	XRS, GRS; NIS, MSI	Eros may be similar to ordinary chondrites, but definitive relationship to petrologic group not established.
Does an intrinsic magnetic field exist? What is it like?	MAG	No intrinsic magnetic field found.
Does the asteroid have any satellites, and how might they compare with Eros?	MSI	No satellites found.

TABLE 3. Four asteroids visited by spacecraft.

Asteroid	Dates Visited by Spacecraft	Heliocentric Orbit: a, e, i	Ellipsoidal Diameters	Density (g cm^{-3})	Rotation Period
951 Gaspra	10/29/91 by *Galileo*	a = 2.21 AU, e = 0.173, i = 4.10°	19 km × 12 km × 11 km	?	7.04 h
243 Ida	9/29/93 by *Galileo*	a = 2.86 AU, e = 0.046, i = 1.14°	59.8 km × 25.4 km × 18.6 km	2.6 ± 0.5	4.63 h
253 Mathilde	6/27/97 by *NEAR*	a = 2.65 AU, e = 0.265, i = 6.71°	66 km × 48 km × 46 km	1.3 ± 0.2	17.4 d
433 Eros	12/23/1998; and 2/14/2000 through 2/28/2001 by *NEAR*	a = 1.46 AU, e = 0.223, i = 10.8°	34.4 km × 11.2 km × 11.2 km	2.67 ± 0.03	5.27 h

References: *Belton et al.* (1992, 1996); *Mottola et al.* (1995); *Yeomans* (1995); *Yeomans et al.* (1997b, 2000); *Veverka et al.* (1999b, 2000); *Zuber et al.* (2000).

longest known and is at least an order of magnitude longer than that of typical asteroids. The origin of these very slow rotation states is puzzling (*Harris,* 1994).

Table 3 shows key characteristics of four asteroids visited by spacecraft so far, including the diameters of best-fit ellipsoids, which give a rough indication of size (the shapes are not well described by ellipsoids in any of the four cases). Mathilde is the largest of the four, by far the slowest rotating and the only C asteroid. *NEAR*'s encounter with Mathilde took place under difficult conditions: at an approach phase angle of 140° and a distance of 1.99 AU from the Sun, where the available power from the spacecraft's solar panels was reduced to about 25% of its maximum mission level. The solar panels could not be pointed more than 50° away from the optimal solar direction because of available power, and only the MSI could be operated during the encounter (*Veverka et al.,* 1997b, 1999b). However, spacecraft tracking data for the radio science experiment were obtained for a mass determination (*Yeomans et al.,* 1997b).

The entire imaging sequence was accomplished in about 25 min, during which 534 images were obtained. Closest approach was 1212 km and the flyby speed was 9.93 km/s. The whole illuminated portion of the asteroid was imaged in color at about a 500-m/pixel resolution after closest approach at phase angles near 40°. The best partial views, acquired at higher phase angles, had a resolution of 200–350 m/pixel. At closest approach, the imager was accurately targeted to the center of Mathilde, but happened to view the deeply shadowed interiors of giant craters, which were completely dark (*Veverka et al.,* 1999b).

8. MATHILDE RESULTS

The *NEAR* flyby obtained the first direct mass determination of an asteroid (*Yeomans et al.,* 1997b). The measured mass of 1.03×10^{20} g and estimated volume of 78,000 km^3 imply a density of 1.3 ± 0.2 g cm^{-3}. The volume must be estimated because only one face of Mathilde could be imaged during the 25-min *NEAR* flyby. The inferred density was unexpectedly low, half or less than that of carbonaceous chondrite meteorites that are the closest spectral analogs, and it implies a high porosity of 50% or more (*Veverka et al.,* 1997b, 1999b). No natural satellite of Mathilde was found. The surface of Mathilde is heavily cratered, with at least five giant craters whose diameter is comparable to the 26.5-km mean radius of Mathilde itself. The areal density of smaller craters <3 km diameter is approximately at equilibrium, similar to that of 243 Ida. However, the presence of the five giant craters was a surprise, because impacts of the magnitude required to make such large craters are about the largest that can be survived without disruption of the target (*Greenberg et al.,* 1996). The mere existence of giant craters on Mathilde was not surprising, but the presence of five giant impacts on a single face of the asteroid was remarkable in that the giant craters did not disrupt one another. Finally, Mathilde proved to be highly uniform in both color and albedo. Mathilde was known from groundbased observations to be a C-type asteroid, and therefore dark and spectrally neutral, but the ground observations could not rule out the possibility of spectrally distinct regions. The *NEAR* observations revealed no evidence of any albedo or spectral variations, implying a homogeneous composition. The measured geometric albedo of 0.043 ± 0.005 was consistent with telescopic observations [*Clark et al.* (1999); improved calibration is given by *Murchie et al.* (2002a)].

Mathilde's low density implies a high porosity and is consistent with a rubble-pile structure. This is clearly a significant result for Mathilde's geologic history, but it is unclear whether Mathilde's porosity is microscopic or macroscopic. The nature and distribution of voids within the interior also remain unclear. The porosity of Mathilde may be primordial; that is, Mathilde may have originally accreted as a porous structure and survived as such to the present. Alternatively, the structure may be an agglomerate of larger fragments, subsequently accreted to form Mathilde. For example, Mathilde may have been thoroughly disrupted by impacts but not dispersed. In this case, macroscopic voids would be expected, possibly in addition to microscopic porosity.

The *NEAR* observations provide important clues to the nature of Mathilde. No ejecta blankets, and no ejecta blocks, have been identified on Mathilde, although blocks larger than 300 m are not expected (*Thomas et al.,* 1999); the resolution limit was approximately 500 m for most of the surface that could be imaged. If Mathilde accreted fragments of diverse parent bodies, these must have had remarkably uniform albedos and colors, or else the fragments must be smaller than ~500 m. Evidence for structure has been reported from images; there are polygonal craters, a 20-km-long marking interpreted as a scarp and a sinuous linear feature that may be an exposed layer (*Thomas et al.,* 1999).

The giant craters provide additional clues to Mathilde's history and nature. Mathilde's porosity makes it more difficult to crater and enhances the likelihood of survival of giant impacts (*Chapman et al.,* 1999; *Cheng and Barnouin-Jha,* 1999; *Davis,* 1999). Porosity causes efficient shock damping, and collapse of pore space rapidly dissipates impact energy. Moreover, roughly half of all impacts are more oblique than 45°. An oblique impact generates lower peak pressure and lower peak strain rates than a normal impact that creates the same-sized crater, so oblique impacts are also less likely to disrupt a target. If account is taken of oblique impacts and of Mathilde's porosity, then the probability of making a giant crater is 2.1–2.6× more than the probability of disruption (*Cheng and Barnouin-Jha,* 1999). Recent laboratory experiments on cratering in highly porous targets (*Housen et al.,* 1999) have shown that the crater volume in this case is produced mainly by target compression rather than excavation and that ejecta volumes are reduced. These results may explain not only how Mathilde survived so many giant impacts, but also how a giant crater can be emplaced practically adjacent to another without disrupting it (Fig. 6). Mathilde's high porosity is key to understanding its collisional history, but its structural features, such as the aforementioned scarp, and its polygonal craters indicate that it is not completely strengthless and that at least one of its structural components appears coherent over a few tens of kilometers.

Fig. 6. Asteroids visited by *NEAR*: Mathilde (left) and Eros (right). The images are in the approximate relative scale, but the C asteroid Mathilde is much darker than shown relative to Eros. On Mathilde the giant crater Karoo, diameter 33 km, is marked. On Eros, the 5-km-crater Psyche is at the terminator and is labeled. At the bright limb of Eros, the saddle-shaped depression Himeros is labeled.

9. EROS RESULTS

The 1998 *NEAR* flyby of Eros yielded determinations (*Veverka et al.,* 1999a; *Yeomans et al.,* 1999) of the mass $(7.2 \pm 1.8) \times 10^{15}$ kg and density (2500 ± 800) kg m^{-3}. The size and rotation pole of Eros were found to be consistent with previous groundbased determinations. These data greatly enhanced the Eros orbital mission as actually flown, by facilitating planning and allowing earlier approach at low altitude. For reference, updated values of the mass and density after additional orbital measurements (*Yeomans et al.,* 2000) are $(6.687 \pm 0.003) \times 10^{15}$ kg and (2670 ± 30) kg m^{-3}.

Eros' average density of 2.67 g/cc is less than the average bulk density of ordinary chondrite meteorites as measured in the laboratory (*Wilkison et al.,* 2001), indicating that bulk Eros is significantly porous and/or fractured, but not to the same extent as Mathilde. The interior of Eros is nearly uniform in density, as inferred from its gravity field, which is within a few percent of that which would be expected from a uniform density object of the same shape (*Yeomans et al.,* 2000). There is a small center of mass offset from the center of figure that may be consistent with an underdense regolith layer of up to 100 m depth (*Zuber et al.,* 2000). Surface morphology clearly indicates the presence of a complex regolith (*Veverka et al.,* 2001a), inferred from the dearth of small craters (<100 m), the profusion of blocks and boulders in varying states of burial, and evidence for burial of small craters and mass wasting (*Cheng et al.,* 2002; *Thomas et al.,* 2002). Craters on Eros are typically shallower than craters of the same diameter on the Moon, consistent with an unconsolidated infill depth ranging from tens of meters to >100 m close to the largest impact craters (*Barnouin-Jha et al.,* 2001).

NEAR Shoemaker has shown that Eros is a consolidated body, not an agglomeration of smaller component bodies bound mostly by gravity (*Zuber et al.,* 2000). There are pervasive linear structural features: a variety of ridges, grooves, and chains of pits or craters (*Veverka et al.,* 2000) that display at least regionally coherent alignments over several kilometers across Eros. In a few cases, structural alignments have been demonstrated on global scales [e.g., between the 18-km-long ridge of Fig. 7 and a ridge system on the opposite face of Eros, southwest of Psyche as shown in Fig. 8 (*Thomas et al.,* 2002)]. Further evidence of structural strength is found from crater morphologies; many craters smaller than 1 km appear to be jointed and/or structurally controlled (*Prockter et al.,* 2002), although larger craters (e.g., Psyche in Fig. 6) are bowl-shaped. This finding is also consistent with a typical unconsolidated regolith

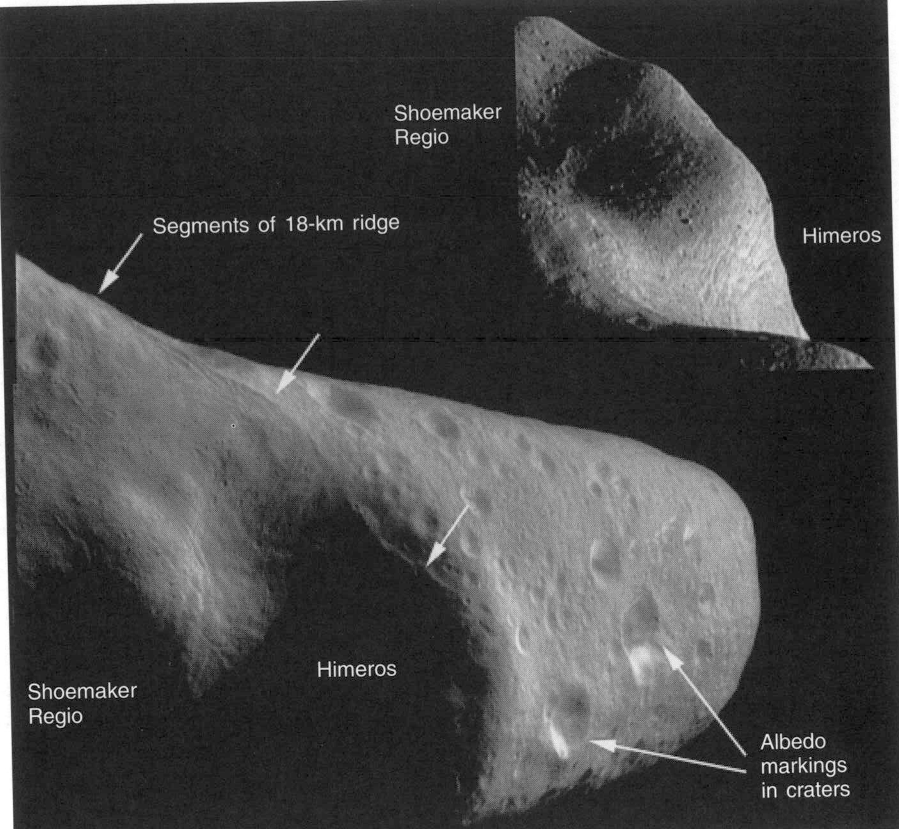

Fig. 7. Eros from 200-km orbit. At lower left, the saddle-shaped depression Himeros and the boulder-rich depression Shoemaker Regio are marked (these regions are in shadow). A portion of the 18-km-long ridge, which continues over the horizon, is marked with short arrows. The density of craters larger than 500 m is much greater outside Himeros and Shoemaker Regio. Systems of crosscutting grooves are seen in Himeros. Many crater walls exhibit bright and dark-albedo markings. The Sun is to the right. At upper right is another view of Himeros and Shoemaker Regio.

depth of <100 m. Additional evidence for a consolidated substrate is found in the presence of steep slopes over a few percent of the surface area that are well above expected angles of repose (*Zuber et al.,* 2000). Taken together, the linear structural features, the tectonic features such as the 18-km-long ridge, the jointed craters, and the indications of internal structural coherence, all suggest that Eros is a collisional fragment from a larger parent body. Nevertheless, throughgoing fractures may exist.

Most of Eros' surface is old and close to equilibrium saturation by craters >200 m (*Veverka et al.,* 2000; *Chapman et al.,* 2002), but some regions appear to be relatively young

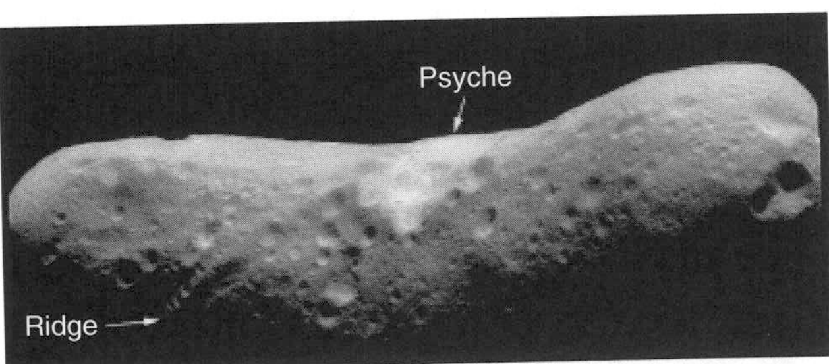

Fig. 8. Western face of Eros from 200-km orbit. Psyche crater is indicated, showing albedo markings on its walls. A prominent ridge, 140 m high, lies in roughly the same plane as the 18-km ridge of Fig. 7. Linear structural features indicate that Eros is a globally consolidated body.

and extensively resurfaced. Blocks and boulders are ubiquitous but are not confined to gravitational lows; most may have originated from an impact that created Shoemaker Regio (*Thomas et al.,* 2001). The density of small craters (<100 m) is markedly depleted on Eros compared to the Moon, but the density of boulders is markedly enhanced, as illustrated in Fig. 9. The surface of Eros is extremely rough and exhibits a fractal structure from scales of a few meters up to hundreds of meters (*Cheng et al.,* 2001, 2002). Examples of downslope motion have been found on Eros, associated with steep slopes in crater walls (*Cheng et al.,* 2002). Bright albedo features are also associated with steep crater walls. The morphological and spectral properties of these features are consistent with a darkened, space-weathered material that slides down crater walls exposing brighter, less-weathered underlying material (*Murchie et al.,* 2002b; *Clark et al.,* 2001).

Another unexpected finding at Eros was the discovery of extremely level, ponded deposits from high-resolution, low-altitude observations toward the end of the mission (*Veverka et al.,* 2001a,b; *Cheng et al.,* 2001; *Robinson et al.,* 2001). An example is shown in Fig. 9. The ponds are characterized by smooth, sharply bounded surfaces and are found in gravitational lows. Processes such as electrostatic levitation and seismic shaking from impacts have been discussed as possible means to mobilize fine particulates on the surface.

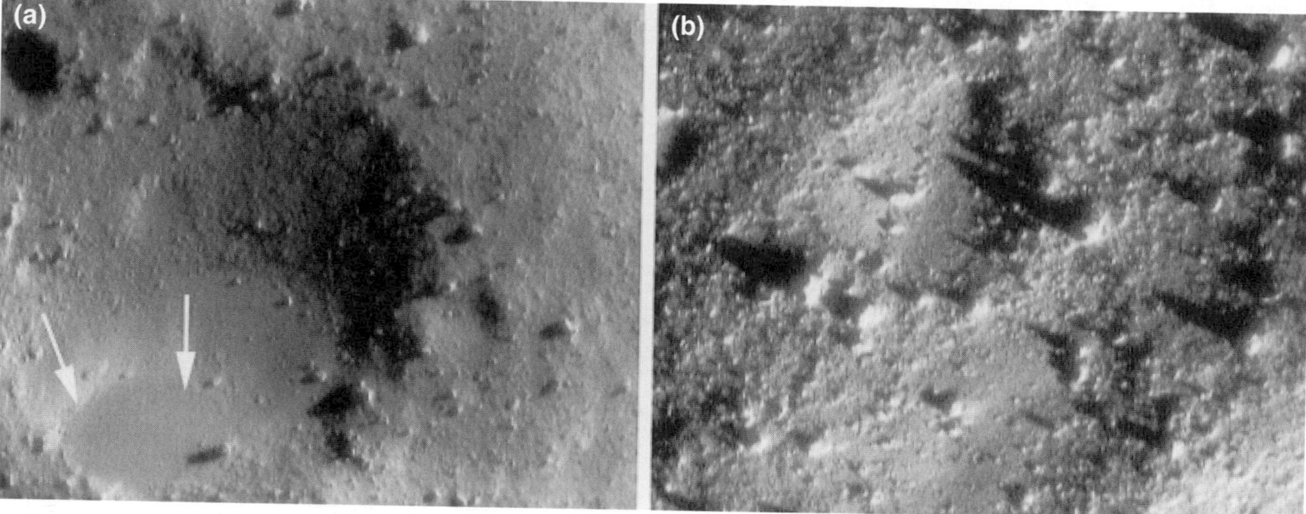

Fig. 9. Eros closeup. **(a)** Arrows indicate a pond within a degraded crater; this frame is about 250 m across. **(b)** Frame is about 150 m across and illustrates the high density of boulders and the paucity of small, fresh craters.

As summarized in Table 2, *NEAR* has shown that Eros is a primitive, intact object, not a rubble pile. The bulk-elemental composition of Eros has been measured independently by XRS (*Trombka et al.,* 2000) and by GRS (*Evans et al.,* 2001). The XRS data from orbit are consistent with an ordinary chondritic composition. The XRS did not find any evidence for spatial heterogeneity in composition. The XRS-measured element ratios Fe/Si, Al/Si, and Mg/Si were consistent with chondritic values, but S/Si was depleted from chondritic values. The XRS is sensitive to element ratios only within the uppermost few tens of microns depth, and it is not clear whether the S depletion is a surface effect due to impact devolatilization or a bulk Eros phenomenon that suggests removal of partial melt at the time Eros was part of its parent body.

The GRS observations had lower signal levels than predicted, and the highest-precision elemental abundance ratios were measured from the landing site on the southeast border of Himeros (*Evans et al.,* 2001), with the instrument at rest within a meter of the surface. Significantly, the landing site is within a pond. The GRS surface data showed the Mg/Si and Si/O ratios and the abundance of K to be consistent with chondritic values, but found Fe/Si and Fe/O to be lower than chondritic. The GRS data are sensitive to a depth of tens of centimeters, so this measurement pertains specifically to a volume on the order of a cubic meter of Eros. It is unclear from the GRS data whether the depletion of Fe results from partial melting or from sorting of the particles in the pond that removed lower-Fe silicates from Fe-Ni metal. Clear evidence from MSI images for depletion of large particles in ponds (*Robinson et al.,* 2001; *Veverka et al.,* 2001a) and the lack of evidence for Fe depletion in XRS data covering more representative areas of the surface support the latter interpretation. However, these analyses are preliminary and their implications are unclear.

The silicate mineralogy of Eros was inferred from visible and NIR spectra to be consistent with ordinary chondrites, using measurements of the 1-μm mafic absorption band center vs. the band area ratio [2-μm pyroxene absorption band area/1-μm mafic band area (*Gaffey et al.,* 1993b)] obtained by NIS and MSI (*Bell et al.,* 2002). The visible and NIR spectra are inconsistent with known primitive achondrites (*Burbine et al.,* 2001). Eros has neither melted nor differentiated fully, but some degree of partial melting or differentiation is possible. The spatially resolved measurements revealed no convincing evidence for compositional variation.

There is no evidence for intrinsic magnetization of Eros (*Acuña et al.,* 2000). No evidence of magnetization at Eros was observed even during the descent to the surface. The absence of magnetization may be consistent with a thermal history in which Eros was never heated to melting.

10. CONCLUSION AND FUTURE PROSPECTS

NEAR substantially increased our knowledge of primitive bodies in the solar system by providing a long, closeup look at the S-type asteroid 433 Eros and the first resolved images of the C-type asteroid 253 Mathilde. *NEAR* was the first mission to a near-Earth asteroid and a C-type asteroid, and it was the first spacecraft to flyby, orbit, and land on a small body. We have learned much from *NEAR,* but much remains to be learned.

At Mathilde, the *NEAR* flyby imaged only one face of the asteroid. The volume and density could only be estimated, but for any plausible volume, the density of Mathilde would be low enough to imply a high porosity. *NEAR* did not obtain any new information on the rotation state or the cause of the slow rotation, nor did it obtain any spatially resolved spectral data. Nevertheless, the *NEAR* flyby of Mathilde provided important confirmation of the existence of large, highly porous main-belt asteroids. *NEAR*'s brief glimpse of Mathilde left us with fascinating questions: Is its composition related to that of carbonaceous chondrites? What is the nature of the porosity? To what extent did internal structures survive within Mathilde?

NEAR gathered much more detailed information at Eros. The low-order gravity harmonics as inferred from the spacecraft orbit agreed well with those inferred from the shape, assuming a constant density (*Yeomans et al.,* 2000), indicating that only small density variations can be present. Initial analyses suggested a small offset of the center of figure from the center of mass, consistent with a lower-density regolith layer up to 100 m thick (*Zuber et al.,* 2000). Further work is needed to assess the significance of this offset.

Images and laser altimetry from *NEAR Shoemaker* have provided strong evidence that Eros has a regolith cover and a consolidated but fractured substrate. Although angles of repose on Eros fall in the expected range for an unconsolidated regolith (*Zuber et al.,* 2000; *Cheng et al.,* 2002), the cohesive strength of bulk Eros is not usefully constrained by *NEAR* observations. The observations of jointed craters and structurally controlled craters imply the presence of a substrate with cohesive strength exceeding the gravitational stress, but the latter is very small on Eros. Even a cohesion of 0.1 MPa, which is easily exceeded by terrestrial clays, would be comparable to ~100 m overburden on Eros. The "consolidated substrate" of Eros may be weak enough to crumble easily in one's hand, or it may be much stronger. The geometric relations of grooves on Eros are in some cases suggestive of fractures in competent rock. The degree to which the interior has been fractured, and the nature of the porosity, also bear upon the efficiency with which seismic energy is transmitted across Eros. This efficiency may depend on the wave amplitude, so that weak seismic waves are transmitted with little damping, whereas stronger waves, which impose stresses sufficient to cause failure or to collapse pore space, may be highly damped.

High-resolution imaging — down to a few centimeters per pixel — revealed a complex and active regolith, with many puzzling features (*Veverka et al.,* 2001a). Why are there so few craters on Eros at small sizes under 100-m diameter, and why are there so many blocks at tens of meters and below? Why does the regolith appear to be more than 100 m deep in some regions, but much less in others? What makes ponds and debris aprons around some large blocks? What produces the albedo contrasts on Eros, which are much greater than on Ida or Gaspra (see, for example, Fig. 10)?

NEAR obtained its principal, spatially resolved NIR dataset from the initial low-phase flyby of Eros on February 14, 2000 (*Bell et al.,* 2002). The NIR spectra were consistent with an ordinary chondrite composition, but did not establish a link to a specific meteorite type. Moreover, why are spectral properties so homogeneous over the observed region in the northern hemisphere? Is the mineral composition in the optically active surface layer itself so uniform?

Fig. 10. Interior of crater Selene, diameter 3.6 km, showing albedo markings. Width of frame is 1.2 km. Streamers of dark material have slid downslope exposing lighter material. Many boulders can be seen (the Sun is to the right).

Even at the much-higher spatial resolutions achieved by the MSI, color variations were extremely subtle, and notably so in comparison with *Galileo* images of Ida and Gaspra (*Veverka et al.,* 2000). On Ida, relatively blue units (with a stronger 1-μm band) could be identified with fresh craters and even associated with ejecta from crater Azzurra (*Geissler et al.,* 1996), but no such phenomena are found on Eros. Is the different orbital history of Eros perhaps responsible for the difference (*Chapman et al.,* 2002)? Eros is presently in a Mars-crossing (but not Earth-crossing) orbit, and numerical simulations suggest that it may evolve into an Earth-crossing orbit within 2 m.y. (*Michel et al.,* 1998). Eros formed in the main belt, where its orbit underwent slow, chaotic diffusion because of numerous weak resonances. When or how Eros evolved onto a planet-crossing orbit cannot now be reconstructed with assurance, although the impact rate on Eros declined drastically once its orbit departed from the main belt (*Michel et al.,* 1998). Does such a cratering history explain the appearance of Eros, with its saturation density of craters >200 m, much lower densities of smaller craters, and extremely homogeneous color properties (*Chapman et al.,* 2002)?

The *NEAR* XRS/GRS experiment, even in its low-altitude orbits, observed both X-ray and γ-ray signal levels that were lower than predicted prelaunch. The X-ray and γ-spectra were puzzling (*McCoy et al.,* 2002; *Nittler et al.,* 2002; *Evans et al.,* 2001) and did not establish a definite meteorite link. Is Eros actually unrelated to any known meteorite type, or is it actually chondritic at depth, below the surface layers that may have been altered and fractionated by unknown weathering processes?

Finally, the lack of magnetization of Eros is a most surprising result. Why are most meteorites — including chondrites — much more strongly magnetized than Eros?

Much more work remains to be done with the wealth of data returned by *NEAR,* but many of the outstanding questions may be resolved only by another mission to Eros, one that might land on the surface, make *in situ* measurements

TABLE 4. Proposed Eros nomenclature.

Number	Name	Latitude	Longitude	Radius* (km)	Diameter (km)	Image
1	Himeros	21.16	282.32	4.99	~10	129267230
2	Psyche	31.60	94.61	3.29	4.85	128771856
3	Jahan	74.16	293.46	5.68	2.06	131586807
4	Mahal	79.36	169.98	5.38	1.16	131586807
5	Zhara	67.14	185.59	5.92	1.03	131585647
6	Fortunio	63.09	221.27	6.40	1.15	131575647
7	Viviane	46.58	236.01	6.85	0.94	131551436
8	Tutanekai	56.36	3.33	6.53	2.10	130420057
9	Narcissus	18.15	7.12	12.60	2.91	131232807
10	Leander	25.61	210.34	9.40	1.39	131530139
11	Valentine	14.64	208.43	10.53	2.21	132023740
12	Cupid	8.29	230.18	9.44	1.84	126711646
13	Don Juan	29.55	356.67	10.09	1.15	127388574
14	Avtandil	–22.53	233.10	8.36	1.16	135754887
15	Lolita	–35.17	197.67	8.58	1.81	137481063
16	Orpheus	25.56	176.74	11.37	1.07	135617746
17	Eurydice	13.49	170.03	13.52	2.24	132074943
18	Aida	7.94	130.50	4.93	1.66	129644551
19	Radames	–5.16	115.09	4.38	1.63	129644551
20	Genji	–19.50	88.62	4.37	1.52	129644551
21	Fujitsubo	–3.75	62.70	5.64	1.70	129644551
22	Hios	–9.41	130.90	4.98	1.26	129644551
23	Gamba	–20.58	54.13	6.41	1.27	129644551
24	Prosperina	–61.01	27.30	6.51	0.82	132640073
25	Selene	–14.18	12.53	13.48	3.64	131112223
26	Leylie	–3.01	23.48	15.87	1.88	131480043
27	Majnoon	3.76	28.80	15.52	2.06	131342359
28	Pao-yu	–73.20	105.58	5.49	0.84	134043704
29	Tai-yu	–47.02	126.06	4.81	1.40	131403667
30	Pygmalion	–1.85	191.09	15.33	1.75	128507342
31	Galatea	–10.17	183.05	15.23	1.38	131694511
32	Don Quixote	–57.70	250.82	6.23	0.88	135340688
33	Dulcinea	–76.13	272.86	6.05	1.40	137502634
34	Valentino	6.76	161.31	12.10	1.66	126723790
35	Abelard	–3.53	12.23	15.18	1.15	131112135
36	Heathcliff	7.35	167.89	15.42	1.09	131343205
37	Catherine	9.13	171.08	16.12	1.12	132207995

*Radius is distance from the Eros center, whereas diameter is the crater diameter.

such as seismic sounding, or perhaps even return samples. A more worthy, important, or fascinating target can hardly be imagined.

APPENDIX: EROS NOMENCLATURE

The Eros nomenclature system used in this chapter was proposed by the *NEAR Shoemaker* science team to the International Astronomical Union but had not been provisionally accepted as of November 2001. In this system, place names on Eros are names of famous lovers from literature, mythology, and history. Table 4 lists names of large craters on Eros. Two names of large features are not included in Table 4 although they have been used in publications. These are Shoemaker Regio [or simply Shoemaker crater (*Thomas et al.,* 2001)], and Rahe Dorsum, the 18-km ridge. Both features can be seen in Fig. 7.

Acknowledgments. I thank Scott Murchie and Peter Thomas for helpful comments. I also thank the many members of the *NEAR* team from the Applied Physics Laboratory, NASA Headquarters, NASA centers, universities, and industry for their hard work and dedication. This work was supported by NASA.

REFERENCES

Acuña M., Russell C., Zanetti L., and Anderson B. (1997) Magnetic field investigation: Science objectives at asteroid 433 Eros and experimental approach. *J. Geophys. Res., 102,* 23751–23759.

Acuña M. H., Anderson B. J., Russell C. T., Wasilewski P., Kletetshka G., Zanetti L., and Omidi N. (2002) NEAR magnetic field observations at 433 Eros: First measurements from the surface of an asteroid. *Icarus,* in press.

Barnouin-Jha O. S., Garvin J. B., Cheng A. F., Zuber M., Smith D., Neumann G., Murchie S., Robinson M., and Veverka J. (2001) Preliminary impact crater dimensions on 433 Eros from the NEAR laser range-finder and imager (abstract). In *Lunar and Planetary Science XXXII,* Abstract #1786. Lunar and Planetary Institute, Houston (CD-ROM).

Bell J., Davis D., Hartmann W., and Gaffey M. (1989) Asteroids: The big picture. In *Asteroids II* (R. P. Binzel et al., eds.), pp. 921–945. Univ. of Arizona, Tucson.

Bell J. F. III, Izenberg N. I., Lucey P. G., Clark B. E., Peterson C., Gaffey M. J., Joseph J., Carcich B., Harch A., Bell M. E., Warren J., Martin P. D., McFadden L. A., Wellnitz D., Murchie S., Winter M., Veverka J., Thomas P., Robinson M. S., Malin M., and Cheng A. (2002) Near-IR reflectance spectroscopy of 433 Eros from the NIS instrument on the NEAR mission: 1. Low phase angle observations. *Icarus, 155,* 119–144.

Belton M. J. S., Veverka J., Thomas P., Helfenstein P., Simonelli D., Chapman C., Davies M. E., Greeley R., Greenberg R., Head J., Murchie S., Klaasen K., Johnson T. V., McEwen A., Morrison D., Neukum G., Fanale F., Anger C., Carr M., and Pilcher C. (1992) Galileo encounter with 951 Gaspra: First pictures of an asteroid. *Science, 257,* 1647–1652.

Belton M. J. S., Chapman C. R., Veverka J., Klaasen K. P., Harch A., Greeley R., Greenberg R., Head J. W. III, McEwen A., Morrison D., Thomas P. C., Davies M. E., Carr M. H., Neukum G., Fanale F. P., Davis D. R., Anger C., Gierasch P. J., Ingersoll A. P., and Pilcher C. B. (1994) First images of 243 Ida. *Science, 265,* 1543–1547.

Belton M. J. S., Mueller B. E. A., D'Amario L. A., Byrnes D. V., Klaasen K. P., Synnott S., Breneman H., Johnson T. V., Thomas P. C., Veverka J., Harch A. P., Davies M. E., Merline W. J., Chapman C. R., Davis D., Denk T., Neukum G., Petit J.-M., Greenberg R., Storrs A., and Zellner B. (1996) The discovery and orbit of 1993 (243)1 Dactyl. *Icarus, 120,* 185–199.

Binzel R. P. and Xu S. (1993) Chips off of asteroid 4 Vesta: Evidence for the parent body of basaltic achondrites. *Science, 260,* 186–191.

Binzel R. P., Burbine T. H., and Bus S. J. (1996) Groundbased reconnaissance of asteroid 253 Mathilde: Visible wavelength spectrum and meteorite comparison. *Icarus, 119,* 447–449.

Bottke W. F., Rubincam D., and Burns J. (2000) Dynamical evolution of main belt asteroids: Numerical simulations incorporating planetary perturbations and Yarkovsky thermal forces. *Icarus, 145,* 301–331.

Burbine T. H., McCoy T. J., Nittler L. R., and Bell J. F. III (2001) Could 433 Eros have a primitive achondritic composition? (abstract). In *Lunar and Planetary Science XXXII,* Abstract #1860. Lunar and Planetary Institute, Houston (CD-ROM).

Chapman C. R. (1996) S-type asteroids, ordinary chondrites, and space weathering: The evidence from Galileo's fly-bys of Gaspra and Ida. *Meteoritics & Planet. Sci., 31,* 699–725.

Chapman C. R., Merline W. J., and Thomas P. (1999) Cratering on Mathilde. *Icarus, 140,* 28–33.

Chapman C. R., Merline W. J., Thomas P. C., Joseph J., Cheng A. F., and Izenberg N. (2002) Impact history of Eros: Craters and boulders. *Icarus, 155,* 104–118.

Cheng A. F. and Barnouin-Jha O. (1999) Giant craters on Mathilde. *Icarus, 140,* 34–48.

Cheng A. F., Santo A., Heeres K., Landshof J., Farquhar F., Gold R., and Lee S. (1997) Near Earth Asteroid Rendezvous: Mission overview. *J. Geophys. Res., 102,* 23695–23708.

Cheng A. F., Cole T., Zuber M., Smith D. E., Guo Y., and Davidson F. (2000) In-flight calibration of the Near Earth Asteroid Rendezvous laser rangefinder. *Icarus, 148,* 572–586.

Cheng A. F., Barnouin-Jha O., Zuber M., Veverka J., Smith D. E., Neumann G., Robinson M., Thomas P., Garvin J., Murchie S., Chapman C., and Prockter L. (2001) Laser altimetry of small-scale features on 433 Eros from NEAR Shoemaker. *Science, 292,* 488–491.

Cheng A. F., Barnouin-Jha O., Prockter L., Zuber M., Neumann G., Smith D. E., Garvin J., Robinson M., Veverka J., and Thomas P. (2002) Small scale topography of 433 Eros from laser altimetry and imaging. *Icarus, 155,* 51–74.

Clark B. E., Veverka J., Helfenstein P., Thomas P. C., Bell J. F. III, Harch A., Robinson M. S., Murchie S. L., McFadden L. A., and Chapman C. R. (1999) NEAR photometry of asteroid 253 Mathilde. *Icarus, 140,* 53–65.

Clark B. E., Lucey P. G., Helfenstein P., Bell J. F. III, Peterson C., Veverka J., McConnochie T., Robinson M. S., Bussey B., Murchie S., Izenberg N., and Chapman C. (2001) Space weathering on Eros: Constraints from albedo and spectral measurements of Psyche Crater. *Meteoritics & Planet. Sci., 36,* 1617–1637.

Cole T. D., Boies M. T., El-Dinary A. S., Cheng A., Zuber M. T., and Smith D. E. (1997) The Near-Earth Asteroid Rendezvous laser altimeter. *Space Sci. Rev., 82,* 217–254.

Davis D. (1999) The collisional history of asteroid 253 Mathilde. *Icarus, 140,* 49–52.

Davis D. R., Ryan E., and Farinella P. (1994) Asteroid collisional evolution: Results from current scaling algorithms. *Planet. Space Sci., 42,* 599–610.

Evans L., Starr R., Trombka J., McClanahan T., Bailey S., Mikheeva I., Bhangoo J., Brueckner J., and Goldsten J. (2000) *Icarus, 148,* 95–117.

Evans L., Starr R., Brückner J., Reedy R., Boynton W., Trombka J., Goldsten J., Masarik J., Nittler L., and McCoy T. (2001) Elemental composition from gamma-ray spectroscopy of the NEAR-Shoemaker landing site on 433 Eros. *Meteoritics & Planet. Sci,* in press.

Gaffey M. J., Burbine T. H., and Binzel R. P. (1993a) Asteroid spectroscopy: Progress and perspectives. *Meteoritics, 28,* 161–187.

Gaffey M. J., Bell J. F., Brown R. H., Burbine T. H., Piatek J. L., Reed K. L., and Chaky D. A. (1993b) Mineralogical variations within the S-type asteroid class. *Icarus, 106,* 573–602.

Geissler P., Petit J-M., Durda D. D., Greenberg R., Bottke W., Nolan M., and Moore J. (1996) Erosion and ejecta reaccretion on 243 Ida and its moon. *Icarus, 120,* 140–157.

Goldsten J. O., McNutt R. L. Jr., Gold R. E., Gary S. A., Fiore E., Schneider S. E., Hayes J. R., Trombka J. I., Floyd S. R., Boynton W. V., Bailey S., Brückner J., Squyres S. W., Evans L. G., Clark P. E., and Starr R. (1997) The X-ray/gamma-ray spectrometer on the Near Earth Asteroid Rendezvous mission. *Space Sci. Rev., 82,* 169–216.

Greenberg R., Bottke W., Nolan M., Geissler P., Petit J.-M., and Durda D. (1996) Collisional and dynamical history of Ida. *Icarus, 120,* 106–118.

Harris A. (1994) Tumbling asteroids. *Icarus, 107,* 209–211.

Hawkins S. E. III, Darlington E. H., Murchie S. L., Peacock K., Harris T. J., Hersman C. B., Elko M. J., Prendergast D. T.,

Ballard B. W., Gold R. E., Veverka J., and Robinson M. S. (1997) Multi-spectral imager on the Near Earth Asteroid Rendezvous mission. *Space Sci. Rev., 82,* 31–100.

Heeres K. J., Holland D., and Cheng A. F. (1997) The NEAR science data center. *Space Sci. Rev., 82,* 283–308.

Housen K. R., Holsapple K. A., and Voss M. E. (1999) Compaction as the origin of the unusual craters on the asteroid Mathilde. *Nature, 402,* 155–157.

Izenberg N. and Anderson B. (1998) NEAR swings by Earth en route to Eros. *Eos Trans. AGU, 79,* 289, 294–295.

Izenberg N. I., Bell J. F. III, Warren J. W., Martin P., Peacock K., Darlington E. H., Heyler G., Murchie S. L., McFadden L., Wellnitz D., Clark B., Joseph J., Carcich B., Harch A., Robinson M., Chapman C., Merline B., and Veverka J. (2000) Inflight calibration of the Near Earth Asteroid Rendezvous mission's Near Infrared Spectrometer I. Initial calibrations. *Icarus, 148,* 550–571.

Kivelson M., Bargatze L., Khurana K., Southwood D., Walker R., and Coleman P. (1993) Magnetic field signatures near Galileo's closest approach to Gaspra. *Science, 261,* 331–334.

Landshof J. and Cheng A. F. (1995) NEAR mission and science operations. *J. Astronaut. Sci., 43,* 477–489.

Lohr D. A., Zanetti L. J., Anderson B. J., Potemra T. A., Hayes J. R., Gold R. E., Henshaw R. M., Mobley F. F., Holland D. B., Acuña M. H., and Scheifele J. L. (1997) NEAR magnetic field investigation, instrumentation, spacecraft magnetics and data access. *Space Sci. Rev., 82,* 255–281.

McCord T., Adams J., and Johnson T. V. (1970) Asteroid Vesta: Spectral reflectivity and compositional implications. *Science, 168,* 1445–1447.

McCoy T. J. et al. (2002) The composition of 433 Eros: A mineralogical-chemical synthesis. *Meteoritics & Planet. Sci., 36,* 1661–1672.

Melosh H. and Ryan E. (1997) Asteroids: Shattered but not dispersed. *Icarus, 129,* 562–564.

Michel P., Farinella P., and Froeschle C. (1998) Dynamics of Eros. *Astron J., 116,* 2023–2031.

Mottola S. W., Sears D., Erikson A., Harris A. W., Young J. W., Hahn G., Dahlgren M., Mueller B. E. A., Owen B., Gil-Hutton R., Licandro J., Barrucci M. A., Angeli C., Neukum G., Lagerkvist C. I., and Lahulla J. F. (1995) The slow rotation of 253 Mathilde. *Planet. Space Sci., 43,* 1609–1613.

Murchie S. and Pieters C. (1996) Spectral properties and rotational spectral heterogeneity of 433 Eros. *J. Geophys. Res., 101,* 2201–2214.

Murchie S., Robinson M., Hawkins S. E. III, Harch A., Helfenstein P., Thomas P., Peacock K., Owen W., Heyler G., Murphy P., Darlington E. H., Keeney A., Gold R., Clark B., Izenberg N., Bell J. F. III, Merline W., and Veverka J. (1999) Inflight calibration of the NEAR multispectral imager. *Icarus, 140,* 66–91.

Murchie S., Domingue D., Robinson M., Li H., Prockter L., Hawkins S. E., Owen W., and Clark B. (2002a) Inflight calibration of the NEAR multispectral imager: 2. Results at Eros. *Icarus, 155,* 229–243.

Murchie S., Robinson M., Clark B., Li H., Thomas P., Joseph J., Bussey B., Domingue D., Veverka J., Izenberg N., and Chapman C. (2002b) Color variations on Eros from NEAR multispectral imaging. *Icarus, 155,* 145–168.

NEAR Science Working Group (1986) *Near Earth Asteroid Rendezvous.* JPL Report 86-7.

Nittler L. R. et al. (2002) X-ray fluorescence measurements of the surface elemental composition of asteroid 433 Eros. *Meteoritics & Planet. Sci., 36,* 1673–1696.

Ostro S., Pravec P., Benner L., Hudson S., Šarounová L., Hicks M., Rabinowitz D., Scotti J., Tholen D., Wolf W., Jurgens R., Thomas M., Giorgini J., Chodas P., Yeomans D., Rose R., Frye R., Rosema K., Winkler R., and Slade M. (1999) Radar and optical observations of asteroid 1998 KY26. *Science, 285,* 557–559.

Pravec P. and Harris A. W. (2001) Asteroid rotations. Presentation at the Asteroids 2001 conference.

Prockter L., Thomas P., Robinson M., Joseph J., Milne A., Bussey B., Veverka J., and Cheng A. (2002) Surface expressions of structural features on Eros. *Icarus, 155,* 75–93.

Rivkin A., Clark B., Britt D., and Lebofsky L. (1997) Infrared spectrophotometry of the NEAR flyby target 253 Mathilde. *Icarus, 127,* 255–257.

Robinson M. S., Thomas P. C., Veverka J., Murchie S., and Carcich B. T. (2001) Morphology, distribution and origin of ponded deposits on Eros. *Nature, 413,* 396–400.

Santo A., Lee S., and Gold R. (1995) NEAR spacecraft and instrumentation. *J. Astronaut. Sci., 43,* 373–398.

Starr R., Clark P., Murphy M., Floyd S., McClanahan T., Nittler L., Trombka J., Evans L., Boynton W., Bailey S., Bhangoo J., Mikheeva I., Brueckner J., Squyres S., McCartney E., Goldsten J., and McNutt R. (2000) Instrument calibrations and data analysis procedures for the NEAR x-ray spectrometer. *Icarus, 147,* 498–519.

Thomas P. C., Veverka J., Simonelli D., Helfenstein P., Carcich B., Belton M. J. S., Davies M. E., and Chapman C. (1994) The shape of Gaspra. *Icarus, 107,* 23–36.

Thomas P. C., Veverka J., Bell J. F. III, Clark B. E., Carcich B., Joseph J., Robinson M., McFadden L. A., Malin M. C., Chapman C. R., Merline W., and Murchie S. (1999) Mathilde: Size, shape, and geology. *Icarus, 140,* 17–27.

Thomas P. C., Robinson M. S., Veverka J., and Murchie S. (2001) Shoemaker crater as a major source of ejecta on asteroid 433 Eros. *Nature, 413,* 394–396.

Thomas P. C., Joseph J., Carcich B., Veverka J., Clark B. E., Bell J. F. III, Byrd A. J., Chomko R., Robinson M., Murchie S., Prockter L., Cheng A., Izenberg N., Malin M., Chapman C., McFadden L. A., Kirk R., Gaffey M., and Lucey P. G. (2002) Eros: Shape, topography, and slope processes. *Icarus, 155,* 18–37.

Trombka J., Floyd S., Boynton W., Bailey S., Brueckner J., Squyres S., Evans L., Clark P., Starr R., Fiore E., Gold R., Goldsten J., and McNutt R. (1997) Compositional mapping with the NEAR x-ray/gamma ray spectrometer. *J. Geophys. Res., 102,* 23729–23750.

Trombka J., Squyres S., Brückner J., Boynton W., Reedy R., McCoy T., Gorenstein P., Evans L., Arnold J., Starr R., Nittler L., Murphy M., Mikheeva I., McNutt R., McClanahan T., McCartney E., Goldsten J., Gold R., Floyd S., Clark P., Burbine T., Bhangoo J., Bailey S., and Petaev M. (2000) The elemental composition of asteroid 433 Eros: Results of the NEAR-Shoemaker x-ray spectrometer. *Science, 289,* 2101–2105.

Veverka J., Bell J. F., Thomas P., Harch A., Murchie S., Hawkins S., Warren J., Darlington E., Peacock K., Chapman C., McFadden L., Malin M., and Robinson M. (1997a) An overview of the NEAR multispectral imager-near infrared spectrometer investigation. *J. Geophys. Res., 102,* 23709–23727.

Veverka J., Thomas P., Harch, A., Clark B., Bell J. F. III, Carcich B., Joseph J., Chapman C., Merline W., Robinson M., Malin

M., McFadden L. A., Murchie S., Hawkins S. E. III, Farquhar R., Izenberg N., and Cheng A. (1997b) NEAR's flyby of 253 Mathilde: Images of a C asteroid. *Science, 278,* 2109–2114.

Veverka J. et al. (1999a) Imaging of asteroid 433 Eros during NEAR's flyby reconnaissance. *Science, 285,* 562–564.

Veverka J., Thomas P., Harch A., Clark B., Bell J. F. III, Carcich B., Joseph J., Murchie S., Izenberg N., Chapman C., Merline W., Malin M., McFadden L., and Robinson M. (1999b) NEAR encounter with asteroid 253 Mathilde: Overview. *Icarus, 140,* 3–16.

Veverka J., Robinson M., Thomas P., Murchie S., Bell J. F. III, Izenberg N., Chapman C., Harch A., Bell M., Carcich B., Cheng A., Clark B., Domingue D., Dunham D., Farquhar R., Gaffey M. J., Hawkins E., Joseph J., Kirk R., Li H., Lucey P., Malin M., Martin P., McFadden L., Merline W. J., Miller J. K., Owen W. M. Jr., Peterson C., Prockter L., Warren J., Wellnitz D., Williams B. G., and Yeomans D. K. (2000) NEAR at Eros: Imaging and spectral results. *Science, 289,* 2088–2097.

Veverka J., Thomas P. C., Robinson M., Murchie S., Chapman C., Bell M., Harch A., Merline W. J., Bell J. F. III, Busey B., Carcich B., Cheng A., Clark B., Domingue D., Dunham D., Farquhar R., Gaffey M. J., Hawkins E., Izenberg N., Joseph J., Kirk R., Li H., Lucey P., Malin M., McFadden L., Miller J. K., Owen W. M. Jr., Peterson C., Prockter L., Warren J., Wellnitz D., Williams B. G., and Yeomans D. K. (2001a) Imaging of small-scale features on 433 Eros from NEAR: Evidence for a complex regolith. *Science, 292,* 484–488.

Veverka J., Farquhar B., Robinson M., Thomas P., Murchie S., Harch A., Antreasian P. G., Chesley S. R., Miller J. K., Owen W. M. Jr., Williams B. G., Yeomans D., Dunham D., Heyler G., Holdridge M., Nelson R. L., Whittenburg K. E., Ray J. C., Carcich B., Cheng A., Chapman C., Bell J. F. III, Bell M., Bussey B., Clark B., Domingue D., Gaffey M. J., Hawkins E., Izenberg N., Joseph J., Kirk R., Lucey P., Malin M., McFadden L., Merline W. J., Peterson C., Prockter L., Warren J., and Wellnitz D. (2001b) The landing of the NEAR-Shoemaker spacecraft on asteroid 433 Eros. *Nature, 413,* 390–393.

Warren J. W., Peacock K., Darlington E. H., Murchie S. L., Oden S. F., Hayes J. R., Bell J. F. III, Krein S. J., and Mastandrea A. (1997) Near infrared spectrometer for the Near Earth Asteroid Rendezvous mission. *Space Sci. Rev., 82,* 101–167.

Wilkison S. L., Robinson M. S., Thomas P. C., Veverka J., McCoy T. J., Murchie S. L., Prockter L., and Yeomans D. (2001) An estimate of Eros's porosity and implications for internal structure. *Icarus,* in press.

Xu S., Binzel R., Burbine T., and Bus S. (1995) Small main belt asteroid spectroscopic survey: Initial results. *Icarus, 115,* 1–35.

Yeomans D. (1995) Asteroid 433 Eros: The target body of the NEAR mission. *J. Astronaut. Sci., 43,* 417–426.

Yeomans D., Konopliv A., and Barriot J. (1997a) NEAR radio science investigations. *J. Geophys. Res., 102,* 23775–23780.

Yeomans D. et al. (1997b) Estimating the mass of asteroid 253 Mathilde from tracking data during the NEAR flyby. *Science, 278,* 2106–2109.

Yeomans D. K., Antreasian P. G., Cheng A., Dunham D. W., Farquhar R. W., Gaskell R. W., Giorgini J. D., Helfrich C. E., Konopliv A. S., McAdams J. V., Miller J. K., Owen W. M. Jr., Thomas P. C., Veverka J., and Williams B. G. (1999) Estimating the mass of asteroid 433 Eros during the NEAR spacecraft flyby. *Science, 285,* 560–562.

Yeomans D. K., Antreasian P. G., Barriot J.-P., Chesley S. R., Dunham D. W., Farquhar R. W., Giorgini J. D., Helfrich C. E., Konopliv A. S., McAdams J. V., Miller J. K., Owen W. M. Jr., Scheeres D. J., Thomas P. C., Veverka J., and Williams B. G. (2000) Radio science results during the NEAR-Shoemaker spacecraft rendezvous with Eros. *Science, 289,* 2085–2088.

Zuber M. T., Smith D. E., Cheng A. F., and Cole T. D. (1997) The NEAR laser ranging investigation. *J. Geophys. Res., 102,* 23761–23773.

Zuber M. T., Smith D. E., Cheng A. F., Garvin J. B., Aharonson O., Cole T. D., Dunn P. J., Guo Y., Lemoine F. G., Neumann G. A., Rowlands D. D., and Torrence M. H. (2000) The shape of 433 Eros from the NEAR-Shoemaker Laser Rangefinder. *Science, 289,* 2097–2101.

Spacecraft Exploration of Asteroids: The 2001 Perspective

Robert Farquhar
Johns Hopkins University

Jun'ichiro Kawaguchi
Institute of Space and Astronautical Science

Christopher T. Russell
University of California at Los Angeles

Gerhard Schwehm
European Space Agency

Joseph Veverka
Cornell University

Donald Yeomans
Jet Propulsion Laboratory

An overview of spacecraft missions to asteroids is presented. Past missions include the *Galileo* flybys of 951 Gaspra and 243 Ida, *NEAR Shoemaker*'s flyby of 253 Mathilde, *Deep Space One*'s flyby of 9969 Braille, and finally *NEAR Shoemaker*'s rendezvous with 433 Eros. Of course, *NEAR Shoemaker*'s yearlong orbital operations at Eros, and subsequent landing on Eros' surface, are the most notable accomplishments thus far, but plans for future asteroid missions are even more ambitious. These plans include a sample-return mission to a near-Earth asteroid, a rendezvous mission to Ceres and Vesta, and flybys of 4979 Otawara and 140 Siwa by the European Space Agency's *Rosetta* spacecraft.

1. INTRODUCTION

In the last ten years, there has been considerable progress in the exploration of asteroids by spacecraft (cf. *Veverka et al.,* 1989). On October 29, 1991, the *Galileo* spacecraft carried out the first every flyby of an asteroid, 951 Gaspra. Less than two years later, on August 28, 1993, *Galileo* encountered a second main-belt asteroid, 243 Ida. Both encounters were accomplished with great technical and scientific success. Especially noteworthy was the discovery of a small natural satellite, Dactyl, in orbit around Ida. Gaspra and Ida are S-type asteroids. The first encounter with a C-type asteroid took place on June 27, 1997, when the *NEAR Shoemaker* spacecraft flew by the main-belt asteroid 253 Mathilde. [On March 14, 2000, the *NEAR* (*Near-Earth Asteroid Rendezvous*) spacecraft was renamed to honor the renowned planetary geologist, Eugene Shoemaker (1928–1997).]

The first spacecraft encounter with a near-Earth asteroid was supposed to occur on August 31, 1994. A Department of Defense spacecraft called "*Clementine*" was scheduled to fly by 1620 Geographos after completing a two-month mission in lunar orbit (*Nozette and Shoemaker,* 1993). Unfortunately, *Clementine* expired shortly before it could be redirected for its intended flyby of Geographos. Nevertheless, the 70-m antenna at Goldstone, California, was used to obtain some spectacular radar images (*Sky & Telescope,* August 1995).

As will be discussed in the next section, the second attempt to investigate a near-Earth asteroid also encountered a few problems. Fortunately, *NEAR Shoemaker* was able to overcome its technical difficulties, and it was eventually inserted into an orbit around the relatively large near-Earth asteroid 433 Eros on February 14, 2000. During its yearlong stay at Eros, *NEAR Shoemaker* obtained a vast quantity of scientific data, including more than 160,000 images.

Future dedicated asteroid missions are likely to rely on advanced spacecraft propulsion to achieve their scientific objectives. Two missions, now in a planning stage, are calling for the use of solar-electric propulsion (SEP). They are Japan's *MUSES-C* sample-return mission to the near-Earth asteroid 1998 SF36, and NASA's *Dawn* mission that is planning to orbit two very large asteroids, Ceres and Vesta. Details of both missions are discussed in this chapter.

2. ASTEROID FLYBYS

Most asteroids to date have been studied during fast flybys at speeds near 10 km/s. Flybys provide a mixed bless-

TABLE 1. Asteroid flybys.

Date	Asteroid	Type	Size (km)	Spacecraft	Closest Approach (km)	Flyby Speed (km/s)	Number of Images	Best Resolution (m/pxl)
October 29, 1991	951 Gaspra	S	18 × 11 × 9	*Galileo*	1600	8	57	54
August 28, 1993	243 Ida	S	60 × 25 × 19	*Galileo*	2391	12.4	96	25
	(Dactyl)	(?)	(1.5)					
June 27, 1997	253 Mathilde	C	66 × 48 × 46	*NEAR*	1212	9.9	330	160
December 23, 1998	433 Eros	S	31 × 13 × 13	*NEAR*	3827	0.9	222	400
July 28, 1999	9969 Braille	?	~2	*Deep Space 1*	28	15.5	1	200
July 11, 2006	4979 Otawara	?	~3	*Rosetta*	2200	10.6	—	—
July 24, 2008	140 Siwa	C	110	*Rosetta*	3500	17.0	—	—

TABLE 2. Physical and orbital parameters of targeted asteroids.

	951 Gaspra	243 Ida	253 Mathilde	9969 Braille	433 Eros	4979 Otawara	140 Siwa
Size (km)	18 × 11 × 9	60 × 25 × 19	66 × 48 × 46	~1 × 2.2	31 × 13 × 13	2.6–4.0	110
Spectral type	S	S	C	V or S	S	V or S	P
Visual albedo	0.23 (0.06)	0.21 (0.03)	0.045 (0.003)		0.25 (0.05)		0.07
Rotation period (h)	7.04	4.633	417	226	5.27	2.7	18.5
Perihelion distance (AU)	1.8	2.7	2.0	1.3	1.1	1.8	2.1
Aphelion distance (AU)	2.6	3.0	3.4	3.4	1.8	2.5	3.3
Orbital period (yr)	3.3	4.8	4.3	3.6	1.8	3.2	4.5
Orbital inclination (degrees)	4.1	1.1	6.7	28.9	10.8	0.9	3.2

ing: They are ideal for obtaining reconnaissance data about asteroids, often on the way to another target, but due to their short duration the information they can capture is limited.

Remote sensing during the flyby is limited by two important factors, distance and the limited duration of the encounter. In terms of distance, past experience suggests that geologic interpretation requires spatial resolution at better than 200 m/pixel. Useful determinations of global characteristics (shape, volume, etc.) require at least 20–50 pixels across the asteroid. The limited duration of flybys leads to two constraints. First, the complement of applicable investigation techniques is limited. To date, X-ray and γ-ray investigations have required integration times that are incompatible with typical flybys. Second, the short duration can limit the completeness of coverage depending on the spin rate and pole orientation of the asteroid. For rapidly rotating Ida (P ~ 4.6 h), *Galileo* was able to see most of the asteroid during its flyby. For slowly rotating Mathilde (P ~ 420 h), *NEAR Shoemaker* saw only a little more than half, leaving the uncertainty in the asteroid's volume 3× larger than in the case of Ida (5% vs. 15%). The December 1998 flyby of Eros by *NEAR Shoemaker* provided an intermediate case. For asteroids in the size range of Gaspra and Mathilde (10–50 km in mean diameter) flybys at no more than 1000–2000 km will produce the best results. Flybys can produce the asteroid mass, which following a reliable estimate of volume, can lead to a determination of the mean density, an important clue to the interior makeup of the asteroid.

One important technical challenge of flybys is the difficulty of pointing accurately at the target during closest approach with high-spatial-resolution cameras that tend to have narrow fields of view. Even following approach, optical navigation downtrack errors remain significant and can amount to several fields of view. Cameras used during past flybys lacked automated tracking capability, a situation that will no doubt continue in the future for flybys carried out as complements to missions with other primary objectives.

To date, five flybys (of six asteroids) have been carried out with varying degrees of success (Table 1). (Physical and orbital parameters of the targeted asteroids are listed in Table 2.) The first was *Galileo*'s flyby of asteroid 951 Gaspra (Fig. 1). *Galileo* showed Gaspra to be a highly irregular, cratered body with principal diameters of 18.2, 10.5, and 8.9 km (average radius = 6.1 ± 0.4 km). Gaspra's irregular shape and the prominence of grooves, linear depressions 100–300 m wide and tens of meters deep, suggested that the asteroid was derived from a larger body by catastrophic collision. Features that appeared to reflect structural grain, including ridges, grooves, and flat surfaces, suggested that Gaspra is a single coherent body.

Analysis of spectral imaging data (0.40–1.10 μm) revealed small but significant color variations over the asteroid's surface. The spectrally most distinct materials on Gaspra were distinguished by a more prominent 1-μm absorption band and tended to be slightly brighter and bluer than average Gaspra. Often such materials are associated

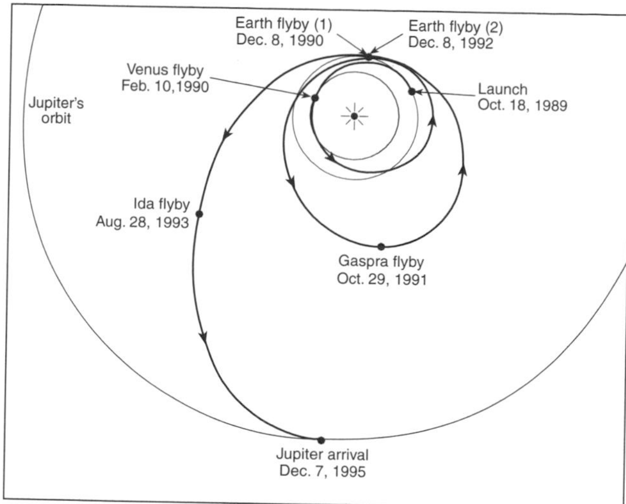

Fig. 1. *Galileo*'s trajectory profile.

with small, fresh-appearing craters along ridges. A strong correlation was found between the infrared/violet color ratio and elevations on Gaspra, a correlation that can be explained in terms of downhill migration of a regolith. No determination of Gaspra's mass was possible.

The biggest surprise during *Galileo*'s second asteroid flyby, that of 243 Ida, was the discovery of a 1.5-km-wide satellite since named Dactyl. From Dactyl's orbit, *Belton et al.* (1996) were able to estimate the mass of Ida. Ida's density (2.6 g/cm^3) turns out to be very similar to that determined for S-type asteroid 433 Eros by *NEAR Shoemaker* (2.67 g/cm^3).

The discovery of Dactyl led to enhanced efforts to search for satellites around Mathilde and Eros. Unfortunately, none were found. However, more than half a dozen satellites of asteroids have been discovered since Dactyl by ground-based observers using optical and radar techniques.

NEAR Shoemaker's encounter with Mathilde is depicted in Fig. 2. This flyby was unusually difficult for a number of reasons. First, the spacecraft approached Mathilde from a phase angle of 140°, which created a severe problem for

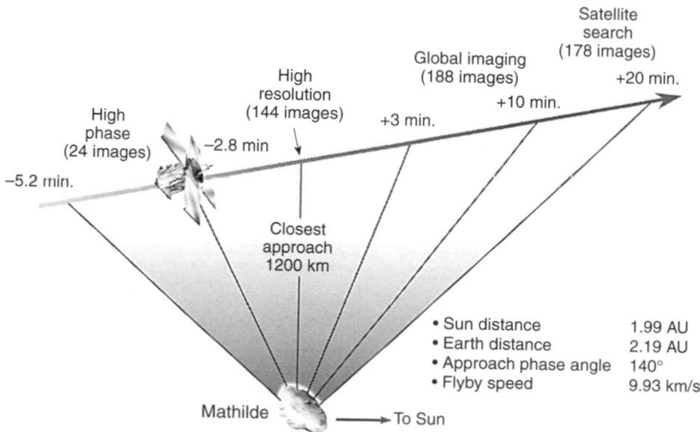

Fig. 2. Mathilde encounter: June 27, 1997.

obtaining optical navigation images. Because *NEAR Shoemaker* was observing Mathilde at a solar elongation angle of only 40°, the asteroid was first detected just 36 h before closest approach as a faint dot almost lost in the Sun's glare. Second, the encounter took place at about 2 AU from the Sun, where the available power from the spacecraft's solar panels was only 25% of its maximum mission level. Furthermore, because the entire spacecraft had to be turned to point the camera at Mathilde, it was necessary to orient the solar panels about 50° away from the optimal solar direction during the encounter, and this reduced the available power by another 36%. Therefore, to conserve power, only one of *NEAR Shoemaker*'s six instruments, the multispectral imager, was on during the encounter. Finally, because *NEAR Shoemaker* did not have a scan platform for the camera, the design of the imaging sequence was more complicated than usual.

Nevertheless, in spite of the aforementioned obstacles, the flyby was flawless. As shown in Fig. 2, the imaging sequence began some 5 min before closest approach, when views of a crescent-illuminated Mathilde were obtained at about 500 m/pixel. The highest-resolution data were obtained at closest approach, when the phase angle was close to 90°. The imaging sequence continued for another 20 min as the spacecraft receded from the asteroid. Although *NEAR Shoemaker* took 534 images, about 200 frames were devoted to a search for satellites.

NEAR Shoemaker's images of Mathilde, a classic C-type asteroid, revealed an irregular and heavily cratered body. Within the roughly 50% of the total area imaged by *NEAR Shoemaker*, there are five craters with diameters between 19 and 35 km. The largest and best-imaged crater is 33 km across and may be 5–6 km deep. The asteroid's surface is very dark (albedo between 0.035 and 0.050) and similar in color to some CM carbonaceous chondrites. No albedo or color variations were detected.

Mathilde's mass was determined by accurately tracking *NEAR Shoemaker* before and after the encounter. Except for a short interval during the close approach period, continuous tracking of the spacecraft was performed from one week before to almost one week after the flyby. The tracking data led to a mass estimate for Mathilde of 1.033 (±0.044) × 10^{20} g (*Yeomans et al.,* 1997). This mass estimate coupled with a volume estimate from the imaging team yielded a bulk density for Mathilde of only 1.3 g/cm^3.

Mathilde's low density indicates that it is probably a "rubble pile," whose interior has been pulverized by a long history of collisions. The existence of such underdense objects has been predicted by several studies. Finally, it is also possible that if C-type asteroids consist of very primitive, unprocessed materials, their low density may in some sense be primordial.

NEAR Shoemaker's second asteroid flyby was unplanned. It was the result of a botched rendezvous maneuver on December 20, 1998. *NEAR Shoemaker*'s control center lost contact with the spacecraft about 37 s after the start of the maneuver. Although communications were restored 27 h later, it was not possible to execute another maneuver before *NEAR Shoemaker* would pass Eros on December 23. With less than 24 h to get ready for the Eros flyby, engineers and scientists worked through the night to update *NEAR Shoemaker*'s observing sequence. Due to uncertainties in the asteroid's position relative to the spacecraft, it would be necessary to image a significant area of the sky to be sure of getting pictures of Eros near closest approach. Unfortunately, the aborted rendezvous burn and ensuing attitude maneuvers had pushed the spacecraft far off its intended course. Instead of the originally planned 1000-km miss distance, *NEAR Shoemaker*'s closest approach to Eros was 3827 km. This meant that the smallest detail resolved by *NEAR Shoemaker*'s camera was about 400 m across. Nevertheless, the first close-up encounter with a near-Earth asteroid yielded 222 images of Eros as well as supporting spectral observations (*Veverka et al.,* 1999).

Fortunately, *NEAR Shoemaker* had a forgiving mission design that had planned for adversity. The design included generous fuel margins and a variety of contingency options. More than any other factor, the resilient mission design was responsible for giving *NEAR Shoemaker* another opportunity to rendezvous with Eros. Although the planned rendezvous date of January 10, 1999, was no longer possible, *NEAR Shoemaker*'s mission planners quickly settled on a strategy that would achieve a rendezvous with Eros in mid-February 2000 (*Dunham et al.,* 2000). The new target date was February 14, Valentine's Day.

The fifth spacecraft flyby of an asteroid was scheduled to take place on July 28, 1999, when the technology spacecraft, *Deep Space 1*, would encounter 9969 Braille. Unfortunately, the science return from this encounter was far less than anticipated. Due to a number of mishaps, only one very distant image was obtained about 15 min after closest approach.

As shown in Table 1, the European Space Agency's *Rosetta* spacecraft is planning to fly by two asteroids, 4979 Otawara in 2006 and 140 Siwa in 2008. Very little is known about Otawara, but Siwa could be an interesting target (*Barucci et al.,* 1998). With a diameter around 110 km, Siwa is larger than any asteroid so far examined by spacecraft. Spectral studies indicate that it is a very black, primitive C-type object that has probably been less altered by collisions than its smaller cousins. It would be particularly interesting to compare Siwa and Mathilde.

Finally, it should be mentioned that the *Stardust* project is considering an encounter with 5535 AnneFrank (diameter ~7 km) on November 2, 2002. However, funding for this potential encounter has not been approved, and the phase angle at encounter is not favorable (~150°).

3. *NEAR SHOEMAKER* AT EROS: ORBITAL OPERATIONS AND A SOFT LANDING

On February 14, 2000, a small propulsive maneuver (ΔV ~10 m/s) was used to place the *NEAR Shoemaker* spacecraft into a 321 × 366-km orbit around Eros. *NEAR Shoemaker*'s orbital history during its first 2.5 months at Eros

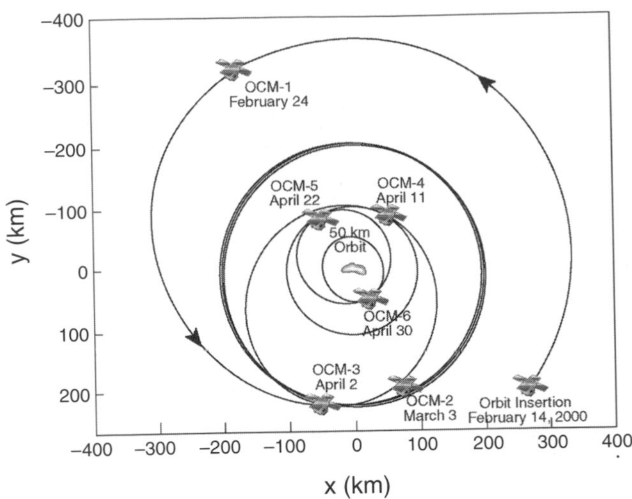

Fig. 3. Early Eros orbit phase (view from Sun).

is shown in Fig. 3. A series of small orbit correction maneuvers (OCMs) gradually brought *NEAR Shoemaker* closer to Eros until it reached its nominal mission orbit of roughly 50 × 50 km on April 30. As the spacecraft descended through these early orbits, physical parameters of Eros such as its mass, gravity field, shape, spin state, and rotation pole orientation were determined with increasing precision (*Yeomans et al.,* 2000; *Zuber et al.,* 2000).

When *NEAR Shoemaker* arrived at Eros in February 2000, Eros' north pole was oriented toward the Sun and its southern hemisphere was dark. About 4 months later, as Eros traveled along its orbit around the Sun, Eros' rotation axis was perpendicular to the Sun-Eros line. *NEAR Shoemaker*'s orbital geometry at this time (June 2000) is shown in the top half of Fig. 4. The *NEAR Shoemaker* spacecraft is shown in a 50-km circular orbit around Eros as viewed

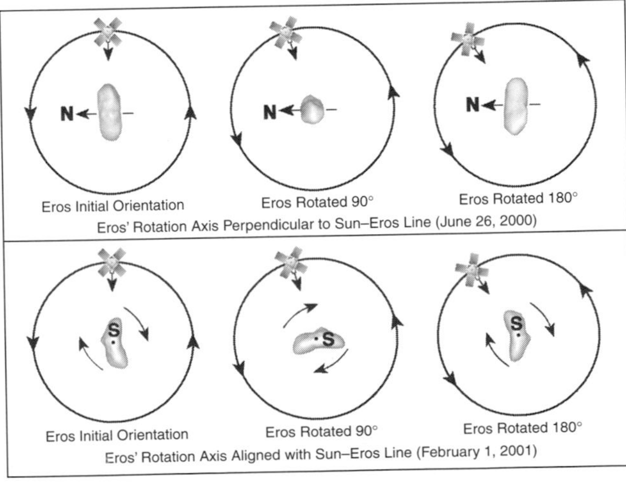

Fig. 4. *NEAR Shoemaker*'s orbital geometry at Eros in June 2000 and February 2001 (view from Sun; orbit size: 50 × 50 km).

by an observer on the Sun. *NEAR Shoemaker*'s orbit and Eros are drawn to scale, but obviously the spacecraft is not. This is a convenient reference frame to show *NEAR Shoemaker*'s orbit because *NEAR Shoemaker*'s orbital plane was controlled so that it was always within 30° of a plane that is normal to the Sun-Eros line. In this configuration, *NEAR Shoemaker*'s fixed solar panels are oriented toward the Sun. The science instruments are pointed at Eros' surface by slowly rolling the spacecraft as it orbits Eros. The orbital geometry in February 2001 is shown in the bottom half of Fig. 4. Here, Eros' south pole is directed at the Sun, and the northern hemisphere is in darkness.

During its initial high-orbit phase, *NEAR Shoemaker* obtained thousands of images of Eros' northern hemisphere at resolutions of about 25 m/pixel (*Veverka et al.,* 2000). Later, when the spacecraft reached its 50 × 50-km orbit, *NEAR Shoemaker*'s camera mapped the surface at scales of 5–10 m. Because *NEAR Shoemaker*'s nominal orbit plane was close to the terminator plane (plane dividing dayside and nightside), most of the images were taken at phase angles near 90°, an ideal geometry for studying the surface because shadows are prominent and reveal details of surface morphology.

In addition to obtaining higher-resolution images, it was necessary to get closer to Eros because the X-ray/γ-ray spectrometer (XGRS) required long observation periods in orbits with radii of 50 km or less. Only the lowest orbits would provide sufficient sensitivity and resolution for the XRGS instrument to measure and map the surface composition of Eros. However, the evolution of low-altitude orbits around Eros is strongly influenced by its irregular gravity field. Orbits exist that are quire unstable, and safe operation in these low-altitude orbits required close coordination between the science, mission design, navigation, and mission operations teams.

During the first low-orbit phase from May 1 to August 26, 2000, *NEAR Shoemaker* spent virtually all its time in a 50 × 50-km orbit. The only exception was a brief 10-day interval in July when it operated in a 35 × 35-km orbit. Because the first operation in 35 × 35-km orbit did not encounter any serious problems, it was decided to go directly to this orbit during the second low-orbit phase (December 2000 to February 2001). This decision was significant because it allowed the XGRS instrument to operate for about 2 months in an orbit that regularly passes by Eros at altitudes under 20 km.

Eros is a small asteroid and hence has only weak gravity. Typically, gravity is 1000× less than on Earth, making it relatively easy for fast-moving ejecta from impacts to escape. While there were previous indications that asteroids even as small as Eros are covered with impact-generated debris, as regolith, some scientists on the *NEAR Shoemaker* project were surprised by how ubiquitous and abundant this impact debris is on the asteroid's surface. Sizeable blocks of ejecta tens of meters across are everywhere and most of the craters are partially filled by finer debris. A global map of all blocks bigger than 30 m was undertaken. This effort produced clear evidence that most of the conspicuous blocks that currently litter the surface of Eros were produced by the impact that formed the most recent large crater on the asteroid: Shoemaker Crater, a scar some 7 km wide. However, the complexity of the regolith raised many questions that could only be answered by getting a closer look at the surface.

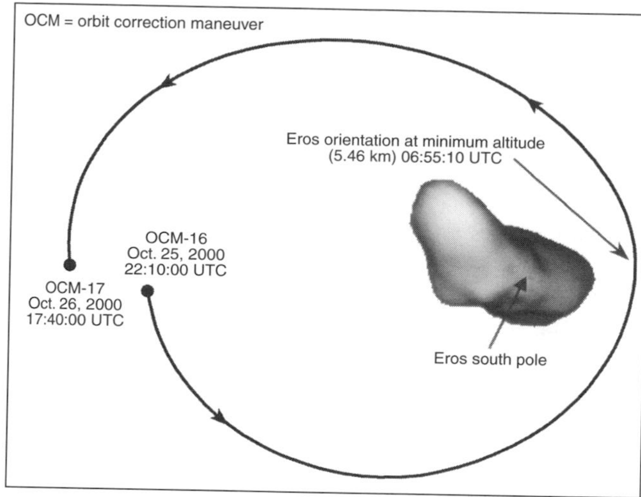

Fig. 5. *NEAR Shoemaker*'s low-altitude flyover of Eros on October 26, 2000 (view from Sun).

The first opportunity for really close images came on October 25, 2000, when *NEAR Shoemaker* swept down to 6.4 km over one of the ends of Eros, allowing the camera to view the surface at a resolution of 1 m/pixel (*Veveka et al.,* 2001a). The minimum altitude for this flyover occurred on October 26 and was roughly 5.5 km (Fig. 5). The success of the October low-altitude flyover led to a second set of low-altitude passes in late January 2001 that ended with a 2.74-km pass on January 28. The January 28 images revealed surface details at resolutions under 0.5 m.

The low-altitude images showed a subdued, gently undulating surface characterized by abundant blocks and conspicuously degraded craters. Many of the degraded craters show evidence of infilling. A novel feature is the occurrence of smooth flat areas (known as "ponds") in the interiors of certain craters. The smoothness of the ponds indicate that there is an efficient process that is able to sort out the finest component of the regolith from the coarser, more blocky portion, and concentrate the fine material into low-lying areas such as crater bottoms.

Even before *NEAR Shoemaker*'s launch, the issue arose as to what should be done with the spacecraft when its primary mission was completed. Should it just be abandoned in its orbit around Eros, or alternatively, could a scientifically useful extended mission be identified?

One adventurous proposal was that *NEAR Shoemaker* should slowly descend to Eros and attempt a landing. During its descent, the spacecraft would keep its high-gain antenna pointed at Earth to transmit images and other science data as quickly as possible. Although the landing idea would definitely attract considerable media attention, several key members of the *NEAR Shoemaker* team were worried that a failure would tarnish the favorable impression of *NEAR Shoemaker*'s earlier successes.

On the other hand, supporters of the landing option argued that this was too good an opportunity to pass up. If everything went according to plan, images of Eros' surface with resolutions 10–20× better than anything obtained earlier would be acquired. Because the images would be returned during the descent, success would not depend on the spacecraft surviving the landing impact.

After listening to all the arguments, both pro and con, NASA decided in favor of a "controlled descent" to Eros' surface. The primary goal of the controlled descent was to obtain high-resolution images. A secondary goal was to achieve a soft landing (i.e., an impact velocity <3 m/s). There was also a slight hope that the spacecraft would survive the landing and transmit a signal from the surface.

In preparation for the descent phase, *NEAR Shoemaker* was placed into a near-circular 36-km orbit on January 28. A deorbit maneuver at about 10:30 a.m. EST on February 12 began *NEAR Shoemaker*'s descent to a targeted region near the edge of the large Himeros depression (Fig. 6). Approximately 3 h and 45 min later, the first of four braking maneuvers was initiated. This maneuver occurred at an altitude of roughly 5 km, and slowed *NEAR Shoemaker*'s rate of descent by about 6 m/s. After three more braking maneuvers, *NEAR Shoemaker* touched down on the surface of Eros at 3:01:51 p.m. EST.

The final 46 min of the descent profile is shown in Fig. 7. Measurements from the laser rangefinder indicated that *NEAR Shoemaker*'s braking maneuvers were very close to

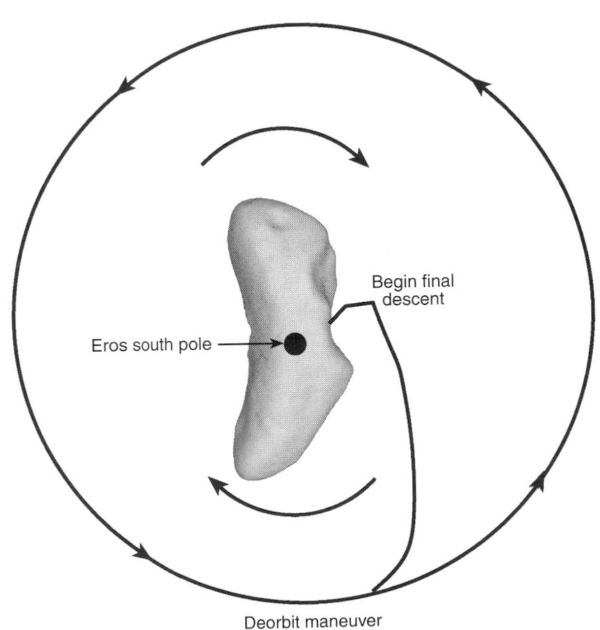

Fig. 6. *NEAR Shoemaker*'s descent from 36-km orbit (view from Sun).

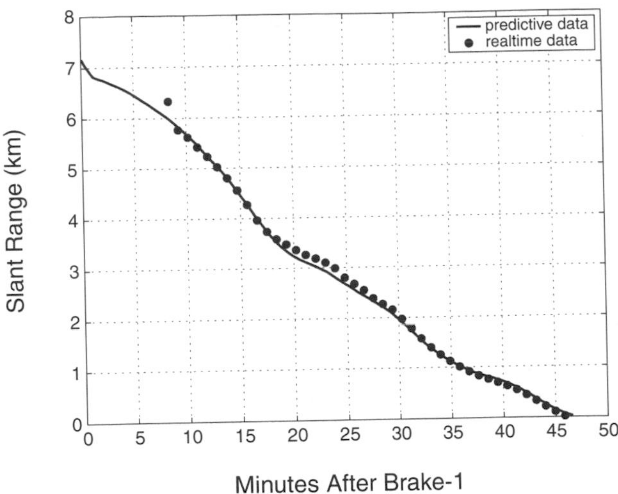

Fig. 7. *NEAR Shoemaker*'s descent profile, slant range vs. time (at start, slant range ~7 km and altitude ~5 km).

Fig. 8. *NEAR Shoemaker* on Eros' surface.

nominal. The nearly perfect maneuver performance resulted in a vertical impact velocity somewhere between 1.5 and 1.8 m/s and a transverse velocity of 0.2–0.3 m/s. This was less than the 2.4-m/s impact velocity for *Viking 1*, and could be the lowest ever. In any case, *NEAR Shoemaker*'s touchdown on Eros was definitely a soft landing and not the "controlled crash" that some had predicted.

During its descent, *NEAR Shoemaker* returned 69 images of Eros, the last one snapped just 125 m above the surface (*Veverka et al.,* 2001b). As the spacecraft reached altitudes below 1 km (resolution ~50 cm), local differences in surface texture became evident. The morphology of the surface remained generally blocky, but smoother, more block-free areas appeared. The final sequence of images showed a clear transition from a blocky surface to a very smooth "pond" area. The last image had a resolution of 1–2 cm.

Although telemetry ceased when the spacecraft hit Eros' surface, carrier lock was maintained, indicating that *NEAR*

TABLE 3. Planetary landers: Historic firsts.

February 3, 1966	Moon	*Luna 9*	USSR
December 15, 1970	Venus	*Venera 7*	USSR
December 2, 1971	Mars	*Mars 3*	USSR
February 12, 2001	Eros	*NEAR*	USA

Shoemaker was still operational. Apparently, the attitude control thrusters tried to keep the high-gain antenna pointed at the Earth, but could not achieve the accuracy needed to support the telemetry link. In any case, the remaining fuel was soon exhausted, and the spacecraft dropped into its final resting on the tips of two solar panels and the bottom edge of the main body (Fig. 8).

Sometime later, *NEAR Shoemaker* automatically switched to its forward low-gain antenna, and a tenuous 10-bps telemetry link was established. Even at this low data rate, it was found that useful surface science could be obtained from two instruments, the γ-ray spectrometer and the magnetometer. The possibility of gathering additional data from the γ-ray spectrometer was especially attractive because of its vastly improved sensitivity on Eros' surface. Therefore, NASA decided to extend *NEAR Shoemaker*'s mission by 14 d. This decision proved to be very wise because the surface data obtained by the γ-ray instrument was far better than all its earlier orbital data (*Sky & Telescope*, May 2001).

NEAR Shoemaker was the first mission in NASA's Discovery Program of low-cost planetary missions. During its 5-year mission, *NEAR Shoemaker* has racked up an impressive list of "firsts": (1) first spacecraft powered by solar cells to operate beyond Mars orbit; (2) first encounter with a C-class asteroid (June 27, 1997); (3) first encounter with a near-Earth asteroid (December 23, 1998); (4) first spacecraft to orbit a small body (February 14, 2000); (5) first spacecraft to land on a small body (February 12, 2001). The final "first" was especially significant because this was the first time that a U.S. spacecraft was the first to land on a solar system body (Table 3).

TABLE 4. *MUSES-C* mission timeline.

	1998 SF36	1989 ML
Launch window	Nov.–Dec. 2002 May 2003	July 2002
Arrival at asteroid	June 2005 (L + 30/25 months)	October 2003 (L + 25 months)
Depart from asteroid	November 2005 (L + 35/30 months)	March 2004 (L + 20 months)
Return to Earth	June 2007 (L + 54/49 months)	June 2006 (L + 47 months)
Sun–s/c distance	0.95–1.20 AU	1.11–1.33 AU
Earth–s/c distance	1.93–2.13 AU	1.86–2.33 AU
Sun–s/c–Earth angle	3.5°–8.8°	8.7°–26.4°
Max. Sun–s/c distance	1.69 AU	1.44 AU
Min. Sun–s/c distance	0.88 AU	1.00 AU
Max. Earth–s/c distance	2.52 AU	2.44 AU

4. *MUSES-C* SAMPLE-RETURN MISSION

Japan's Institute of Space and Astronautical Science (ISAS) and NASA have agreed to cooperate on the first space mission to collect asteroid surface samples and return them to Earth for detailed compositional analyses. Current plans include the launch of the *MUSES-C* spacecraft via the MV launch vehicle from the Kagoshima Space Center, Japan, in November or December 2002 (*Kawaguchi et al.*, 2000). After an Earth swingby in May 2004, the spacecraft will rendezvous with near-Earth asteroid 1998 SF36 in June 2005. After a stay of about 5 months, *MUSES-C* will return to Earth in June 2007 with an asteroid surface sample of a few grams (Table 4). The ISAS Project Manager and Project Scientist are Drs. Jun Kawaguchi and Akira Fujiwara respectively. ISAS is responsible for the mission management, mission design, and operations as well as spacecraft development. NASA participation involves spacecraft tracking by NASA's Deep Space Network (DSN), some navigation support, and participation on the *MUSES-C* science team.

The *MUSES-C* mission is primarily a test flight for four new technologies, including (1) a demonstration of the four ion engines in interplanetary space for up to 18,000 h; (2) the use of optical navigation images to autonomously guide the spacecraft during the asteroid rendezvous; (3) a demonstration of a sample collection device for surface materials retrieval; (4) a demonstration of the sample capsule's ability to effect an Earth atmosphere entry directly from an interplanetary trajectory.

The entire spacecraft is shown in Fig. 9. It is a three-axis stabilized spacecraft whose nominal attitude is pointed to the Sun, so that the solar power can be extracted as much as possible for driving the ion engines. The 1.6-m-diameter high-gain antenna (HGA) is placed atop the spacecraft. The X-band up and down links are baselined with two receivers and one transmitter whose output is boosted by two power amplifiers. Two medium-gain horn antennas (MGA) are mounted on the spacecraft inclined to the HGA radiative direction. One MGA is gimbaled to ensure the downlink communication, while the ion engines are driven toward the prescribed acceleration direction.

While the science objectives are secondary to the technologies being tested, the planned science return is very impressive. The asteroid surface samples will be studied ex-

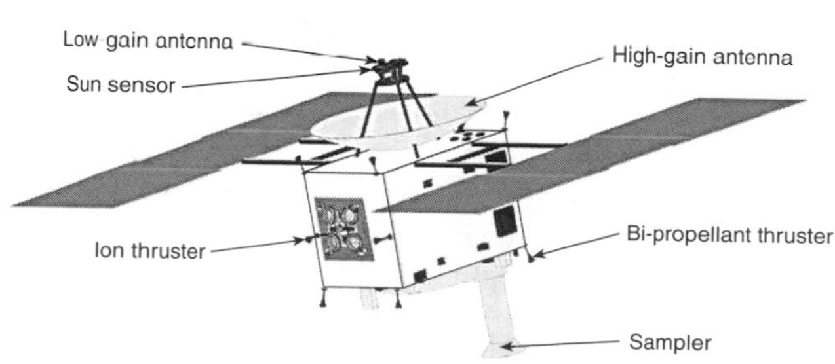

Fig. 9. *MUSES-C* spacecraft.

TABLE 5. Target asteroids for *MUSES-C*.

	1998 SF36	1989 ML
Physical Characteristics		
Magnitude (abs.)	19.1	19.5
Type	S	C, E, or M
Albedo	0.32	0.04–0.4
Diameter (average)	0.36 km	0.25–0.80 km
Major/minor axis ratio	1–2.5 (?)	1.7–2.5
Density	1–3 g/cm^3	1–3 g/cm^3
Rotation period	12 h	~19 h
Temperature at subsolar point	217–445 K	410 K
Orbital Elements		
Semimajor axis (AU)	1.324	1.273
Eccentricity	0.280	0.137
Inclination (degrees)	1.72	4.38
Perihelion (AU)	0.95	1.10
Aphelion (AU)	1.62	1.45
Period (yr)	1.52	1.43

tensively by an international group of scientists in an effort to establish a link between the parent asteroid's spectral type (S-type) and the meteorite type that is compositionally most similar to the asteroid's surface samples. Once this link is forged between an S-type asteroid and the most likely meteorite analog, future ground-based observations could be used to infer the chemical composition of any asteroid that has the spectral classification of asteroid 1998 SF36.

The orbit of the *MUSES-C* target body, 1998 SF36, has a low inclination with respect to the ecliptic plane (1.7°) so that it is one of the more accessible asteroids for a spacecraft rendezvous. Fortunately, this asteroid made a close Earth approach to within 6,000,000 km in late March 2001 and will make an even closer Earth approach to within 2,000,000 km in late June 2004. Optical, infrared, and radar observations were undertaken in March 2001 and the object appears to have a diameter of about 360 m, and an S-type spectra analogous to that of a LL ordinary chondrite. Additional observations, planned for June 2004, will be used to further characterize the shape and spin axis orientation of the target body. Table 5 provides the orbital and physical characteristics for the primary mission target asteroid, 1998 SF36, and the backup target asteroid, (10302) 1989 ML.

As well as the science benefits to be derived from the returned surface samples, the *MUSES-C* spacecraft carries an impressive array of instruments to study the asteroid *in situ*. The *MUSES-C* spacecraft's science instruments and their science objectives are:

AMICA — Asteroid Multiband Imaging Camera (T. Nakamura, Team Leader)

- Characterize the surface morphology and the processes that affect the asteroid's surface
- Determine the global size, shape, and volume of asteroid
- Determine the asteroid's spin rate
- Establish a global map of surface features and colors
- Reveal the history of impacts from other asteroid and comet fragments
- Search for possible asteroid satellites and dust rings
- Determine optical parameters of regolith particles using polarization degree vs. phase curve at large phase angles

LIDAR (T. Mukai, Team Leader)

- Provide accurate shape and mass determinations
- Map the asteroid's surface with a maximum resolution of about 1 m
- Map global surface albedo at a wavelength of 1 μm

X-Ray Spectrometer — XRS (M. Kato, Team Leader)

- Map the major-element composition of the surface as the asteroid rotates under the spacecraft
- Determine the major-element composition at localized areas during asteroid approach phase
- Measure surface composition accurately enough to establish relationship between asteroids and meteorites and identify type of meteorite to which asteroid is linked
- Provide elemental abundance maps to investigate inhomogeneity of regolith
- Characterize surface heterogeneity

Near-IR Spectrometer (M. Abe, Team Leader)

- Map mineralogic composition of asteroid and provide evidence for the rock types present on the surface at scales as small as 20 m
- Together with elemental composition measurements provided by the XRS and color imagery from the camera, the provide link between asteroids and meteorites

In addition to the above four instruments, the spacecraft has a laser rangefinder (LRF), a star tracker, two wide-angle cameras, a surface sampling device, and possibly a surface hopper.

The LRF targets a reflector on the spacecraft sampling horn so that when the horn is deformed upon touching the

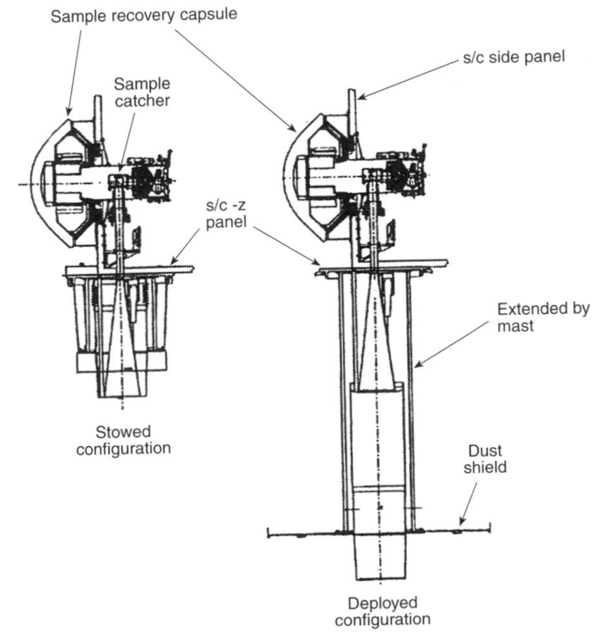

Fig. 10. Capsule and sample collector.

asteroid's surface, the sampling sequence begins. One wide-angle camera will be used for navigation during the spacecraft touchdown sequences while the other will be used as a backup for the star tracker and also for science observations in the terminator region.

The surface sampling will be carried out using one or more pellets fired into the asteroid's surface. The surface ejecta will then be captured when a collecting horn funnels the ejecta to an onboard sample catcher (Fig. 10). This catcher is a canister with separate compartments and once the sampling process is complete, the canister is pushed into an Earth reentry vehicle (40 cm diameter) that eventually makes a direct entry into the Earth's atmosphere. Equipped with a heat shield, the sample return capsule then parachutes down to the recovery site. After an international proposal process, a portion of the few grams of asteroid surface sample will then be made available for scientific study.

One mission option includes a technology test of an asteroid surface hopper (*MINERVA*) equipped with a visible wide-angle imager. The design mass of this surface hopper is only 1 kg and it is equipped with a turntable for orientation and a torque wheel device that provides the necessary hopping motion.

5. THE *DAWN* MISSION: A RENDEZVOUS WITH VESTA AND CERES

Dawn is a Discovery-class mission, approved for development as this volume was in press and scheduled for launch in May 2006. In the *Asteroids II* volume (*Veverka et al.,* 1989) a multiple main-belt asteroid orbiter/flyby was described as the "asteroid gem of the Solar System Exploration Committee (SSEC) program." The objectives of such a mission, as listed in the *SSEC Report* (1983), were to (1) characterize asteroids of various types including determinations of size, shape, rotation, albedo, mass, density, surface morphology, surface composition magnetic field and solar wind interaction; and (2) provide a more detailed study of one or more selected main belt asteroids, emphasizing elemental and mineralogical composition and detailed morphology."

At that time the suggested instruments to be carried on such a multiple asteroid mission and their measurement objectives were:

Imaging	*Size, shape rotation, surface morphology*
X-ray and γ-ray spectrometer	Elemental composition
IR reflectance spectral mapper	Mineralogical composition
Magnetometer	Intrinsic magnetic field; nature of solar wind interaction
Radio science	Mass determination

As conceived in 1983 this mission would rendezvous with a main-belt asteroid and spend 2 months in orbit. It would then fly by several other asteroids, hopefully to rendezvous with a second asteroid.

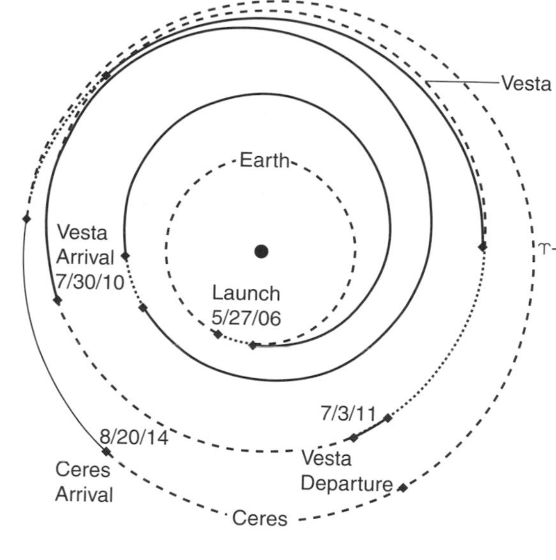

Fig. 11. *Dawn* mission overview and trajectory profile.

The *Dawn* mission was very consciously designed to address the SSEC science objectives for the main belt. An important advance over the intervening years is that it can now use solar electric propulsion, flight qualified on the *Deep Space 1* mission to rendezvous with and orbit for 11 months at each of Vesta and Ceres, the two most massive main-belt asteroids and possibly the only unshattered survivors of the original planetesimals (*Bell,* 1989). A mission overview and trajectory profile for the *Dawn* mission is given in Fig. 11. In the course of its interplanetary trajectory it can be directed to pass arbitrarily close to up to a dozen asteroids. The exact number will be chosen to maximize science return while maintaining simplicity of operations and prudent fuel reserves. The *Dawn* payload does not include an X-ray spectrometer as its γ-ray/neutron spectrometer returns similar information for lower cost and fewer spacecraft resources and because the X-ray measurement becomes less sensitive with increasing distance from the Sun and at times at low solar activity. The *Dawn* payload instead includes a laser altimeter in recognition of the valuable contribution of this instrument to the study of the Moon on the *Clementine* mission, to Mars on the *Mars Global Surveyor* mission, and to Eros on the *NEAR Shoemaker* mission.

Ceres appears to have retained its primordial structure while Vesta has differentiated and formed a core. The relative water content of Ceres and Vesta may have played an important role in allowing such different thermal evolutions. Thus *Dawn* has the added objective of determining the role of water in asteroidal evolution.

An important aspect of the *Dawn* mission will be the involvement of the asteroid and meteorite communities. This will be facilitated through *Dawn*'s participating scientist and data analysis programs over the course of the mission. Open communications and collaborative opportunities with the science team will be a hallmark of the *Dawn* project.

After launch in May 2006, Dawn cruises for four years arriving at Vesta in July 2010. The spacecraft would then leave in July 2011 and arrive at Ceres in August 2014. The *Dawn* science team (A. Coradini, W. C. Feldman, R. Jaumann, A. S. Konopliv, T. B. McCord, L. A. McFadden, H. Y. McSween, S. Mottola, G. Neukum, C. M. Pieters, C. A. Raymond, C. T. Russell, D. E. Smith, M. V. Sykes, B. Williams, and M. T. Zuber) look forward to being able to undertake this mission, which we expect to set new standards for science return by a Discovery-class mission.

REFERENCES

Barucci M. A., Doressoundiram A., Fulchignoni M., Florczak M., Lazzarin M., and Angeli C. (1998) Compositional type characterization of Rosetta asteroid candidates. *Planet. Space Sci., 46,* 75–82.

Bell J. (1989) Asteroids: The big picture. In *Asteroids II* (R. P. Binzel et al., eds.), pp. 921–945. Univ. of Arizona, Tucson.

Belton M. J. S., Chapman C. R., Klaasen K. P., Harch A. P., Thomas P. C., Veverka J., McEwen A. S., and Pappalardo R. T. (1996) Galileo's encounter with 243 Ida: An overview of the imaging experiment. *Icarus, 120,* 1–19.

Dunham D. W., Farquhar R. W., McAdams J. V., Williams B. G., Miller J. K., Helfrich C. L., Antreasian P. G., and Owen W. (2000) Recovery of NEAR's mission to Eros. *Acta Astron., 47,* 503–512.

Kawaguchi J., Uesugi K., and Fujiwara A. (2000) Project status of the MUSES-C sample and return from an asteroid. *IAA Paper 00-11.2.01,* paper presented at the Intl. Astronautical Congress meeting in Rio de Janeiro, Brazil.

Nozette S. and Shoemaker E. M. (1993) Back to the Moon, on to an asteroid: The Clementine mission. *The Planetary Report, 13,* 10–15.

Veverka J., Langevin Y., Farquhar R., and Fulchignoni M. (1989) Spacecraft exploration of asteroids: The 1988 perspective. In *Asteroids II* (R. P. Binzel et al., eds.), pp. 970–993. Univ. of Arizona, Tucson.

Veverka J. and 25 colleagues. (1999) Imaging of asteroid 433 Eros during NEAR's flyby reconnaissance. *Science, 285,* 562–564.

Veverka J. and 32 colleagues (2000) NEAR at Eros: Imaging and spectral results. *Science, 289,* 2088–2097.

Veverka J. and 32 colleagues (2001a) Imaging of small-scale features on 433 Eros from NEAR: Evidence for a complex regolith. *Science, 292,* 484–488.

Veverka J. and 38 colleagues (2001b) NEAR's descent to Eros: Unexpected depositional and erosional features in the regolith. *Nature,* in press.

Yeomans D. K. and 12 colleagues (1997) Estimating the mass of asteroid 253 Mathilde from tracking data during the NEAR flyby. *Science, 278,* 2106–2109.

Yeomans D. K. and 15 colleagues (2000) Radio science results during the NEAR Shoemaker spacecraft rendezvous with Eros. *Science, 289,* 2085–2088.

Zuber M. T., Smith D. E., Cheng A. F., Garvin J. B., Aharonson O., Cole T. D., Dunn P. J., Guo Y., Lemoine F. G., Neumann G. A., Rowlands D. D., and Torrence M. H. (2000) The shape of 433 Eros from the NEAR-Shoemaker laser rangefinder. *Science, 289,* 2097–2101.

Part IV:
Evolutionary Processes

Regular and Chaotic Dynamics in the Mean-Motion Resonances: Implications for the Structure and Evolution of the Asteroid Belt

D. Nesvorný
Southwest Research Institute

S. Ferraz-Mello
Universidade de São Paulo

M. Holman
Harvard-Smithsonian Center for Astrophysics

A. Morbidelli
Observatoire de la Côte d'Azur

This chapter summarizes the achievements over the last decade in understanding the effect of mean-motion resonances on asteroid orbits. The developments from the beginning of the 1990s are many. They range from a complete theoretical description of the secular dynamics in the mean-motion resonances associated with the Kirkwood gaps to the discovery of the three-body resonances and slow chaotic phenomena acting throughout the asteroid belt. Consequences arising from these results have required remodeling the process of asteroid delivery to the Earth-crossing orbits.

1. INTRODUCTION

It is evident from the plethora of studies of mean-motion resonances (MMRs) in the last decade that major advances have been made in this area, and these advances have broad consequences for our understanding of asteroids. An attentive observer of this branch of dynamical astronomy would have noted the following progress.

It was shown that bodies in the main MMRs with Jupiter (3:1, 4:1, 5:2) can reach very high orbital eccentricities (*Ferraz-Mello and Klafke,* 1991; *Klafke et al.,* 1992; *Saha,* 1992). This result suggested that resonant bodies can be transported to Mars-, Earth-, and Venus-crossing orbits and then be efficiently extracted from the resonances due to the larger mass of the two latter planets. The cited works confirmed the pioneering findings of *Wisdom* (1982, 1983, 1985) on the Kirkwood gap at the 3:1 MMR with Jupiter and showed that very high eccentricities are frequent outcomes of the dynamics in MMRs on million-year timescales.

The progress in numerical modeling revealed another surprising alternative: Resonant asteroids can fall into the Sun (*Farinella et al.,* 1993, 1994), as a consequence of their eccentricity approaching unity. The current state of our understanding of the removal of resonant bodies from the 3:1 MMR is that 65–70% are extracted by Earth and Venus, and 25–30% go directly into the Sun (*Gladman et al.,* 1997).

The innovative seminumerical treatment of classical perturbation methods introduced by *Henrard* (1990) later allowed a global and realistic description of the secular dynamics in the main MMRs with Jupiter, taking into account the precession of Jupiter's orbit (*Moons and Morbidelli,* 1995). This supplied convincing evidence that the orbital evolution of simulated bodies toward the planet-crossing orbits is driven by chaotic secular resonances (*Morbidelli and Moons,* 1995). A brief account of this new development is given in section 4.

The 2:1 MMR with Jupiter at 3.27 AU is a unique case, because here the chaotic secular resonances are located at high eccentricities (*Morbidelli and Moons,* 1993). Yet in the early 1990s, only a few asteroids were known to exist on resonant orbits, in contrast to the large nonresonant populations on either side of the 2:1 MMR. The studies devoted to the question of the long-term stability of the resonant orbits were marked by the evolution of the numerical methods (*Wisdom and Holman,* 1991; *Levison and Duncan,* 1994) and the computer speed. Important progress was made when the globally chaotic nature of orbits in the 2:1 MMR was demonstrated (*Ferraz-Mello,* 1994b). Later works (*Morbidelli,* 1996; *Nesvorný and Ferraz-Mello,* 1997; *Ferraz-Mello et al.,* 1998a) showed that this resonance is an intermediate case between the unstable MMRs and stable 3:2 MMR with Jupiter, the latter hosting the Hilda group. The 2:1 MMR possesses a core with dynamical lifetime comparable to the age of the solar system, where currently nearly 30 small asteroids are known (the Zhongguo group). We devote section 5 to this subject.

A major breakthrough, already signaled by the studies of the slow chaos in the 2:1 MMR, was made concerning the phenomenon referred to as "stable chaos." It was previously noted that a large number of asteroids have strongly chaotic orbits yet are stable on long intervals of time (*Milani and Nobili,* 1992; *Milani et al.,* 1997). The main reason for such behavior was revealed by the discovery of the so-called three-body resonances, i.e., MMRs, where the commensurability of orbital periods occurs between an asteroid and two planets [mainly Jupiter and Saturn; (*Murray et al.,* 1998, *Nesvorný and Morbidelli,* 1998)]. The subsequent modeling of these resonances showed that they represent narrow but strongly chaotic layers densely intersecting the asteroid belt (section 6.1).

The slow chaotic evolution observed in numerical simulations in the narrow MMRs in the outer asteroid belt was explored by *Murray and Holman* (1997). This study correctly showed that the mechanism driving the chaotic diffusion lies in the "multiplet structure" of the narrow MMRs (section 6.2).

The study of the dynamics in the inner belt (2.1–2.5 AU) revealed surprising instabilities of asteroid orbits. Numerical simulations showed that many asteroids currently on non-planet-crossing orbits with large eccentricities evolve to Mars-crossing orbits within the next 100 m.y. (*Migliorini et al.,* 1998, *Morbidelli and Nesvorný,* 1999). The responsible resonances in this case are the MMRs with Mars, previously thought unimportant because of the small mass of this planet. The MMRs with Mars were identified as a major source of the large near-Earth asteroids (NEAs; section 6.3). They were also conjectured to disperse the asteroid families in the inner asteroid belt (*Nesvorný et al.,* 2002).

After introducing basic notation (section 2) and showing the global structure of MMRs in the asteroid belt (section 3), we follow the above historical overview. Section 2 is directed toward the reader who wishes to gain some basic insight in the mathematical methods developed and utilized in studies of MMRs. Perspectives are given in the last section.

2. NOTATION AND BASIC TERMINOLOGY

A mean-motion resonance (MMR) occurs when an asteroid has an orbital period commensurate with the orbital period of one or more of the planets. To fix the notation, we define

$$\sigma_{\mathbf{k},k,\mathbf{l},l,\mathbf{m},m} = \sum_{j=1}^{N} k_j\lambda_j + k\lambda + \sum_{j=1}^{N} l_j\varpi_j + l\varpi + \sum_{j=1}^{N} m_j\Omega_j + m\Omega \tag{1}$$

where k, l, m and $\mathbf{k} = (k_1, \ldots, k_N)$, $\mathbf{l} = (l_1, \ldots, l_N)$, $\mathbf{m} = (m_1, \ldots, m_N)$ are integers with zero sum: $k + l + m + \Sigma(k_j + l_j + m_j) = 0$ (because of the invariance of interaction by rotation, only such combinations exist; this and other conditions on integer coefficients required by symmetries of the gravitational interaction are known as D'Alembert rules), and $k \neq 0$ and $\|\mathbf{k}\| \neq 0$ (i.e., MMRs are related to the fast orbital frequencies, unlike the secular resonances). Here λ, ϖ, Ω, λ_j, ϖ_j, Ω_j are orbital angles in the usual notation (index j goes over N planets).

The MMR occurs when $\dot{\sigma}_{\mathbf{k},k,\mathbf{l},l,\mathbf{m},m} = 0$, where $\dot{\sigma}_{\mathbf{k},k,\mathbf{l},l,\mathbf{m},m}$ is the time derivative of $\sigma_{\mathbf{k},k,\mathbf{l},l,\mathbf{m},m}$ given by equation (1). In the case of asteroid motion, the secular frequencies $\dot{\varpi}$, $\dot{\varpi}_j$, $\dot{\Omega}$, $\dot{\Omega}_j$ are small compared to the orbital frequencies $\dot{\lambda}$, $\dot{\lambda}_j$. For this reason, $\dot{\sigma}_{\mathbf{k},k,\mathbf{l},l,\mathbf{m},m} = 0$ is approximately $\Sigma k_j\dot{\lambda}_j + k\dot{\lambda} = 0$, which can be solved for the resonant semimajor axis (a_{res}) in the Keplerian approximation. This means that resonant conditions $\dot{\sigma}_{\mathbf{k},k,\mathbf{l},l,\mathbf{m},m} = 0$ with unique **k**,k but different **l**,l,**m**,m hold at about the same semimajor axis. Thus, each (**k**,k) MMR may be thought to be composed from several resonant terms with different **l**,l,**m**,m (this structure is called the "resonant multiplet").

In practice, there are two important cases: (1) the two-body MMR, when index **k** has only one nonzero integer k_{j_1} where j_1 denotes, in the asteroid belt, either Jupiter or Mars (a few two-body resonances with Saturn and the Earth occur but they are of minor importance), and (2) the three-body MMR, when index **k** has two nonzero integers, k_{j_1} and k_{j_2}, corresponding to two planets (mainly to pairs Jupiter–Saturn and Mars–Jupiter). Without loss of generality we assume that $k_{j_1} > 0$.

The notation of MMRs that we adopt in the following text is that a (**k**,k) two-body MMR with Jupiter ($j_1 = 5$) is k_5J:–k, and with Mars ($j_1 = 4$) is k_4M:–k, where k_4, k_5, and k are integers defining the resonant angle in equation (1). In this notation, 2J:1 is a MMR with Jupiter, where an asteroid has the orbital period of one-half that of Jupiter. Moreover, if it is clear from the text which planet is considered, we drop the letter indicating the planet (2J:1 becomes 2:1 as frequently used in the literature). Concerning the three-body resonances we denote k_5J:k_6S:k for the MMRs with Jupiter and Saturn, and k_4M:k_5J:k for the MMRs with Mars and Jupiter. (This notation attempts to generalize the classical notation, i.e., the minus sign in k_{j_1}:–k for two-body MMRs, to the case of three-body MMRs and MMRs with planets other than Jupiter.)

The equations of motion of an asteroid in the presence of a MMR can be conveniently written in the Hamiltonian formalism. In such formalism, the equations of motion derive from a function of the canonical variables (the Hamiltonian). This formulation is useful because it allows us to use rigorous methods for treating problems where an integrable system is coupled with small perturbations (in the current context, the asteroid's motion about the Sun is perturbed by the planets).

The classical expression for the Hamiltonian of a small body evolving under the gravitational force of the Sun and N planets is

$$\mathcal{H} = -\frac{1}{2a} - \sum_{j=1}^{N} \mu_j \mathcal{R}_j$$
$$\mathcal{R}_j = \frac{1}{\Delta_j} - \frac{\mathbf{r}.\mathbf{r}_j}{r_j^3} \tag{2}$$

where $\Delta_j = |\mathbf{r} - \mathbf{r}_j|$ and $\mathbf{r}.\mathbf{r}_j$ are the heliocentric positions of the small body and planet j respectively, and a is the semimajor axis of the small body. (We adopt units where the product of the gravitational constant and the mass of the Sun is 1.) Here, 1/2a is the heliocentric Keplerian part and $\mu_j\mathcal{R}_j$ is the perturbation exerted by the planet j having mass μ_j.

Appropriate canonical variables of the Hamiltonian equation (2) are the so-called modified Delaunay variables

$$\begin{aligned} &\lambda && \Lambda = L \\ &p = -\varpi && P = L - G \\ &q = -\Omega && Q = G - H \end{aligned} \tag{3}$$

where $L = \sqrt{a}$, $G = L\sqrt{1-e^2}$ and $H = G \cos i$ are the usual Delaunay variables (e and i are the eccentricity and inclination of the small body respectively). The perturbation $\mathcal{R}_j$ can be written as a function of the orbital elements of the small body and planet j [see *Moons* (1993) for such an expression with general validity]. The Hamiltonian equation (2) is then expressed in terms of variables defined in equation (3).

To realistically account for the motion of the planets, the time evolution of the planetary orbital elements, provided by the planetary theory (*Laskar,* 1988; *Bretagnon and Simon,* 1990), must be substituted in $\mathcal{R}_j$. If we denote the planetary orbital elements by index o (as osculating) to distinguish them from the proper orbital elements (the latter being denoted by a_j, e_j, i_j, λ_j, ϖ_j, Ω_j), the general form of the quasi-periodic evolution of the planetary elements is

$$\begin{aligned} a_j^o &= a_j + \sum_v A_v \cos\Psi_v \\ \lambda_j^o &= \lambda_j + \sum_v B_v \cos\Psi_v \\ e_j^o \exp \iota\varpi_j^o &= e_j \exp \iota\varpi_j + \sum_v C_v \cos\Psi_v \\ i_j^o \exp \iota\Omega_j^o &= i_j \exp \iota\Omega_j + \sum_v D_v \cos\Psi_v \end{aligned} \tag{4}$$

where $\Psi_v = \Sigma_j(r_j\lambda_j + s_j\varpi_j + t_j\Omega_j)$, the multiindex v denoting different values of integers r_j, s_j, t_j, and $\iota = \sqrt{-1}$.

By definition, the proper angles evolve linearly with time, with fixed frequencies provided by the planetary theory. We will denote by n_j, g_j, s_j the orbital, perihelion, and node frequencies respectively. In fact, because A_v, B_v, C_v, D_v in equation (4) are small, the new $\mathcal{R}_j$ — functions of the planetary proper elements — are at the first approximation identical to the original functions. There will, however, appear important terms at second and higher orders in the planetary masses generated by the substitution of equation (4) in equation (2). A number of approximations of equation (4) are used in the literature, varying from the planar model with one planet on the circular orbit [$a_j^o = a_j$, $\lambda_j^o = \lambda_j = n_jt$, $e_j = i_j = 0$, where $n_j = \sqrt{(1+\mu_j)/a_j^3}$ and t is time] to more realistic ones.

To summarize, it is understood at this point that although we do not explicitly show such an expression, the Hamiltonian equation (2) is a function of equation (3) and the planets' proper elements (through the substitution in equation (4)). This is an autonomous Hamiltonian of 3(N + 1) degrees of freedom with the general form

$$\mathcal{H} = \mathcal{H}_0 + \mu\mathcal{H}_1 + \mu^2\mathcal{H}_2 + \mathcal{O}(\mu^3) \tag{5}$$

where

$$\mathcal{H}_0 = -\frac{1}{2\Lambda^2} + \sum_{j=1}^{N}\left(n_j\Lambda_j + g_j\Lambda_{g_j} + s_j\Lambda_{s_j}\right) \tag{6}$$

Here, μ denotes the largest of the masses of the involved planets, and Λ_j, Λ_{g_j}, Λ_{s_j} are the canonical momenta conjugate to the proper angles λ_j, ϖ_j, Ω_j.

To first order in μ, the elimination of the fast orbital angles by the Lie-series canonical transformation (see, e.g., *Hori,* 1966) is equivalent to averaging equation (5) over λ_j, j = 1, …, N, maintaining only the terms in which λ, λ_{j_1}, λ_{j_2} appear in resonant combinations $\sigma_{\mathbf{k},k,\mathbf{l},l,\mathbf{m},m}$ [corresponding to the ($\mathbf{k}$,k) MMR in question]. This procedure is straightforward if one utilizes the resonant variables in equation (5)

$$\begin{aligned} &\sigma = \frac{k_{j_1}\lambda_{j_1} + k_{j_2}\lambda_{j_2} + k\lambda + \left(k_{j_1} + k_{j_2} + k\right)p}{k_{j_1} + k_{j_2} + k} && S = P \\ &\nu = -\frac{k_{j_1}\lambda_{j_1} + k_{j_2}\lambda_{j_2} + k\lambda}{k_{j_1} + k_{j_2} + k} && N = \frac{k_{j_1} + k_{j_2} + k}{k}\Lambda + P + Q \\ &\sigma_z = \frac{k_{j_1}\lambda_{j_1} + k_{j_2}\lambda_{j_2} + k\lambda + \left(k_{j_1} + k_{j_2} + k\right)q}{k_{j_1} + k_{j_2} + k} && S_z = Q \\ &\lambda_{j_1} && \tilde{\Lambda}_{j_1} = \Lambda_{j_1} - \frac{k_{j_1}}{k}\Lambda \\ &\lambda_{j_2} && \tilde{\Lambda}_{j_2} = \Lambda_{j_2} - \frac{k_{j_2}}{k}\Lambda \\ &\lambda_j,\ j \neq j_1, j_2 && \Lambda_j,\ j \neq j_1, j_2 \end{aligned} \tag{7}$$

where, in the case of a two-body MMR with the j_1-th planet, $k_{j_2} = 0$. The Hamiltonian equation (2) written in canonical variables defined by equation (7) is then simply averaged over λ_j, j = 1, …, N. The resulting Hamiltonian of the two-body MMR is

$$\begin{aligned} \mathcal{H}^{2BR} = &-\frac{1}{2\Lambda^2} + \frac{k_{j_1}}{k}n_{j_1}\Lambda + \sum_{j=1}^{N}\left(g_j\Lambda_{g_j} + s_j\Lambda_{s_j}\right) + \\ &\mu\bar{\mathcal{H}}_1(S,N,S_z,\sigma,\nu,\sigma_z,\boldsymbol{\varpi},\boldsymbol{\Omega}) + \mathcal{O}(\mu^2) \end{aligned} \tag{8}$$

where

$$\mu\bar{\mathcal{H}}_1 = -\frac{1}{2\pi}\sum_{j=1}^{N}\mu_j\int_0^{2\pi}\mathcal{R}_j\,d\lambda_j \tag{9}$$

and $\boldsymbol{\varpi} = (\varpi_1, \ldots, \varpi_N)$, $\boldsymbol{\Omega} = (\Omega_1, \ldots, \Omega_N)$.

In the case of a three-body MMR, $\bar{\mathcal{H}}_1$ in equation (9) contains only terms dependent on $\varpi = -\sigma - \nu$ and/or $\Omega = -\sigma_z - \nu$, because $\mathcal{R}_j$ depends solely on the variables of the small body and j-th planet. The resonant terms of three-body MMRs appear at second- and higher-order terms in μ. These terms are generated by the substitution of the planetary orbital elements (equation (4)) (the so-called "indirect" contribution) and by iterating the Lie-series transformation elimination of fast orbital angles to higher orders in μ (the so-called "direct" contribution). This procedure was described in *Nesvorný and Morbidelli* (1999). The resonant Hamiltonian of the three-body MMR is

$$\mathcal{H}^{3BR} = -\frac{1}{2\Lambda^2} + \frac{k_{j_1}n_{j_1} + k_{j_2}n_{j_2}}{k}\Lambda + \sum_{j=1}^{N}\left(g_j\Lambda_{g_j} + s_j\Lambda_{s_j}\right) + \mu\bar{\mathcal{H}}_1(S,N,S_z,\sigma + \nu,\sigma_z + \nu,\varpi,\Omega) + \mu^2\bar{\mathcal{H}}_2(S,N,S_z,\sigma,\nu,\sigma_z,\varpi,\Omega) + \mathcal{O}(\mu^3) \tag{10}$$

Note that equations (8) and (10) have 2N + 3 degrees of freedom (i.e., N less than equation (5)).

Thus, we learn an important difference between the two- and three-body MMRs, which is that the magnitudes of the resonant terms are proportional to the planetary mass and to the planetary mass squared respectively. As the planetary masses in our solar system are $<10^{-3}$ (in solar units), this shows that a typical two-body MMR is expected to have a larger effect on an asteroid's orbit than a typical three-body MMR would have. To document this fact, let us consider an Hamiltonian with only one $\sigma_{\mathbf{k},k,\mathbf{l},l,\mathbf{m},m}$ (i.e., fixing $\mathbf{k}$,k,$\mathbf{l}$,l, $\mathbf{m}$,m, and ignoring terms in equations (8) and (10) with other than this multiindex). Such a Hamiltonian accounts for a single isolated multiplet term of the ($\mathbf{k}$,k) MMR. The phase portrait of trajectories of a single multiplet term is basically equivalent to the phase portrait of trajectories of a pendulum. The width in semimajor axis, represented by the maximal extent of the pendulum separatrixes, is

$$\Delta a = 8a_{res}^{3/2}\sqrt{\frac{\beta}{3}} \tag{11}$$

where β is the coefficient in front of the cosine of the specific multiplet term in the Fourier expansion of the resonant parts in equations (8) and (10) (*Murray et al.,* 1998; *Nesvorný and Morbidelli,* 1999). In the case of the two-body MMR, this coefficient is $\propto \mu_{j_1}P(e, e_{j_1}, i, i_{j_1})$, where P is a polynomial in e, e_{j_1}, i, i_{j_1} with the total power of each term being at least $\rho = |k_{j_1} + k|$ (called the resonant order). From equation (11), the width of the two-body resonance is then $\propto \sqrt{\mu_{j_1}}a_{res}^{3/2}P^{1/2}$. For the three-body MMR, $\rho = |k_{j_1} + k_{j_2} + k|$ and its width is $\propto \mu a_{res}^{3/2}P^{1/2}$. Consequently, the two-body MMR is generally larger than the three-body MMR due to the mass factor. On the other hand, there are typically more three-body MMRs than two-body MMRs within a semimajor axis interval, because three integers (k_{j_1}, k_{j_2}, k) allow for a larger number of combinations than two integers (k_{j_1}, k). In some sense, the larger density of three-body MMRs in the orbital space compensates for their smaller widths, so that both two- and three-body MMRs are important for asteroid dynamics.

3. OVERLAPPING MEAN-MOTION RESONANCES AND THE GLOBAL STRUCTURE OF THE ASTEROID BELT

Let us consider the two-body resonances only. Figure 1 shows the structure of resonant trajectories for the 3J:2 (Figs. 1a,b), 3J:1 (Fig. 1c), 5J:2 (Fig. 1d) and 5M:9 (Figs. 1e,f). The model used for the Jupiter's MMRs (Figs. 1a–d) is an approximation of equation (8) assuming coplanar orbits and a circular orbit for Jupiter. In such case the first-order resonant Hamiltonian is

$$\mathcal{H}^{PC} = -\frac{k_5^2}{2k^2}(N - S)^{-2} + n_5(N - S) + \mu_5\bar{\mathcal{H}}_1(S, N, \sigma) \tag{12}$$

It does not depend on S_z, σ_z because of the coplanarity, and does not depend on ν because $\mathcal{R}_5$ in equation (2) is invariant by rotation around an axis perpendicular to the orbital plane. Consequently,

$$N = \sqrt{a}\left[1 - \sqrt{1 - e^2} + \frac{k_5 + k}{k}\right] = \text{const}$$

and trajectories can be obtained as level curves of the Hamiltonian in equation (12).

In Fig. 1, two N = const manifolds are shown for the 3J:2 (first-order MMR, a_{res} = 3.97 AU) corresponding to eccentricities under (Fig. 1a) and above (Fig. 1b) the Jupiter-crossing limit. The Jupiter-crossing limit is $e \simeq 0.31$. Resonant orbits with $e \gtrsim 0.31$ may intersect the orbit of Jupiter. The bold line in Fig. 1b is where collisions take place. There are two equilibria in Fig. 1a: the stable one at $\sigma = 0$ (libration center) and the unstable one at $\sigma = \pi$. The trajectories connected to the unstable equilibrium are called separatrixes. The trajectories enclosing the stable equilibrium are the resonant ones. They are characterized by oscillations of σ about 0 (so-called "libration"). Above the Jupiter-crossing limit (Fig. 1b), the trajectories near the libration center are protected from collisions. This happens due to the so-called "resonant phase-protection mechanism," which guarantees that conjunctions with Jupiter occur when the resonant asteroid is near the perihelion of its orbit, i.e., far from Jupiter.

The 3J:1 MMR (second order) has two libration centers at $\pi/2$ and $3\pi/2$ (Fig. 1c) and occurs closer to the Sun (a_{res} = 2.5 AU), where collisions with Jupiter cannot happen on elliptic heliocentric orbits. The 5J:2 MMR (third order) has, in addition to the libration center at 0, two other centers at $2\pi/3$ and $4\pi/3$ (Fig. 1d). In general, the two-body MMRs interior of the planet's orbit ($|k_{j_1}| > |k|$) have libration cen-

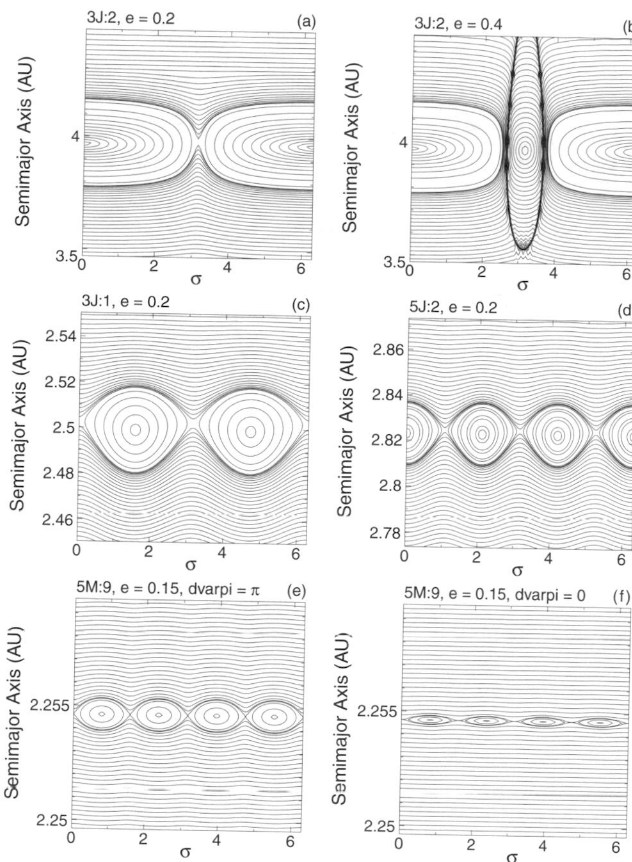

Fig. 1. Trajectories at MMRs. **(a,b)** 3J:2 below **(a)** and above **(b)** the Jupiter-crossing limit; **(c)** 3J:1 MMR; **(d)** 5J:2 MMR; **(e,f)** 5M:9 MMR with $\varpi - \varpi_4 = \pi$ **(e)** and $\varpi - \varpi_4 = 0$ **(f)**. On x-axis, σ is defined by equation (7).

ters at $\sigma_c = 2j_1\pi/\rho$ if ρ is odd and at $\sigma_c = (2j_1 + 1)\pi/\rho$ if ρ is even. The exterior resonances have the libration centers at $\sigma_c = (2j_1 + 1)\pi/\rho$ if $k_{j_1} \neq 1$. Conversely, if $k_{j_1} = 1$, the structure of the exterior resonance is characterized by so-called asymmetric librations where none of the 2ρ libration centers is at $2j_1\pi/\rho$ or $(2j_1 + 1)\pi/\rho$ for most values of N (*Message*, 1958; *Beaugé,* 1994).

An interesting case is that of the two-body MMRs with Mars, due to the large modulation of their widths with the evolution of perihelion longitudes and eccentricities. Figures 1e and 1f show the trajectories near the 5M:9 MMR (a_{res} = 2.253 AU) in the coplanar model with Mars on a fixed elliptic orbit (e_4 = 0.065), assuming two values of perihelion longitudes: $\varpi - \varpi_4 = \pi$ in Fig. 1e and $\varpi - \varpi_4 = 0$ in Fig. 1f. In general, when $\varpi - \varpi_{j_1} = \pi$, the MMRs have their maximum widths. For other phases of the secular angles, the sizes of resonant islands are smaller.

The resonant width (i.e., the width of the libration island) can be conveniently estimated by measuring the distance between separatrixes for σ = 0 (for interior resonances of odd order) or σ = π/ρ (for interior resonances of even order and exterior resonances with $k_j \neq 1$). For the exterior resonances with $k_j = 1$, the separatrix distance is measured at σ_c, which must be determined beforehand as an extreme of equation (12). Iterating the algorithm over levels of N = const, the resonant width Δa can be determined as a function of e.

Figure 2 shows the MMRs in the asteroid belt. It is an extension of a similar result obtained by *Dermott and Murray* (1983), but accounting for all MMRs with $\Delta a > 10^{-4}$ AU. We also plot in this figure the asteroids with magnitudes up to the current level of completeness (*Jedicke and Metcalfe,* 1998). The orbital distribution of these bodies is thus unaffected by observational biases. In the region of the 2J:1, the resonant orbital elements of all known resonant objects, regardless of size, are shown.

To correctly interpret this figure, one should be aware of the now-classical result of *Chirikov* (1979) known as the "resonance overlap criterion." This criterion affirms that whenever two resonances with similar sizes overlap, the corresponding resonant orbits become chaotic. The same criterion was used by *Wisdom* (1980) to show the onset of chaos in the vicinity of a planet because of the overlap of the first-order MMRs with Jupiter. This criterion can be used in the current context to explain why there are so few asteroids located above the threshold in e for the resonance overlap. Moreover, we know from the experience obtained in the last few years with the secular dynamics inside the MMRs that MMRs themselves are usually characterized by chaotic dynamics (see next section). In resonances like the 3J:1 and 5J:2 MMRs (at 2.5 and 2.82 AU), and many others, the chaos causes large-scale instabilities: The resonant bodies evolve to high-e orbits and are removed from resonant orbits by encounters with Mars or Jupiter (*Gladman et al.,* 1997). Such resonances are associated with the Kirkwood gaps in the asteroid distribution. Smaller MMRs, albeit strongly chaotic, do not generate strong instabilities whenever eccentricities are not near a planet-crossing limit. Large first-order MMRs are characterized by either marginal instabilities (like 2J:1 at 3.27 AU) or possess cores in which orbital lifetimes exceed the solar system age (such as the 3J:2 MMR at 3.97 AU and the low-e region in the 4J:3 MMR at 4.29 AU, where 279 Thule is located).

The basic impression one has from Fig. 2 is that the MMRs delimit the orbital space inhabited by the observed asteroids in a and e. Due to chaos and instabilities generated by MMRs, this is quite intuitive. Most of the job in removing the material from the belt was apparently done by several large MMRs with Jupiter (as 3J:1, 5J:2, 7J:3, 9J:4, 7J:4, 5J:3, etc.) and many narrow MMRs with Mars. It is not more than a curiosity that a few MMRs with Saturn and Earth also exist in the inner and central asteroid belts (6S:1 at 2.89 AU, 7S:1 at 2.61 AU, 2E:7 at 2.305 AU, 2E:9 at 2.73 AU, etc.).

We have accounted solely for two-body MMRs in Fig. 2. This happens to be a good approximation when considering the gross structures of the asteroid belt in a and e, but becomes less acceptable when one's objective is to under-

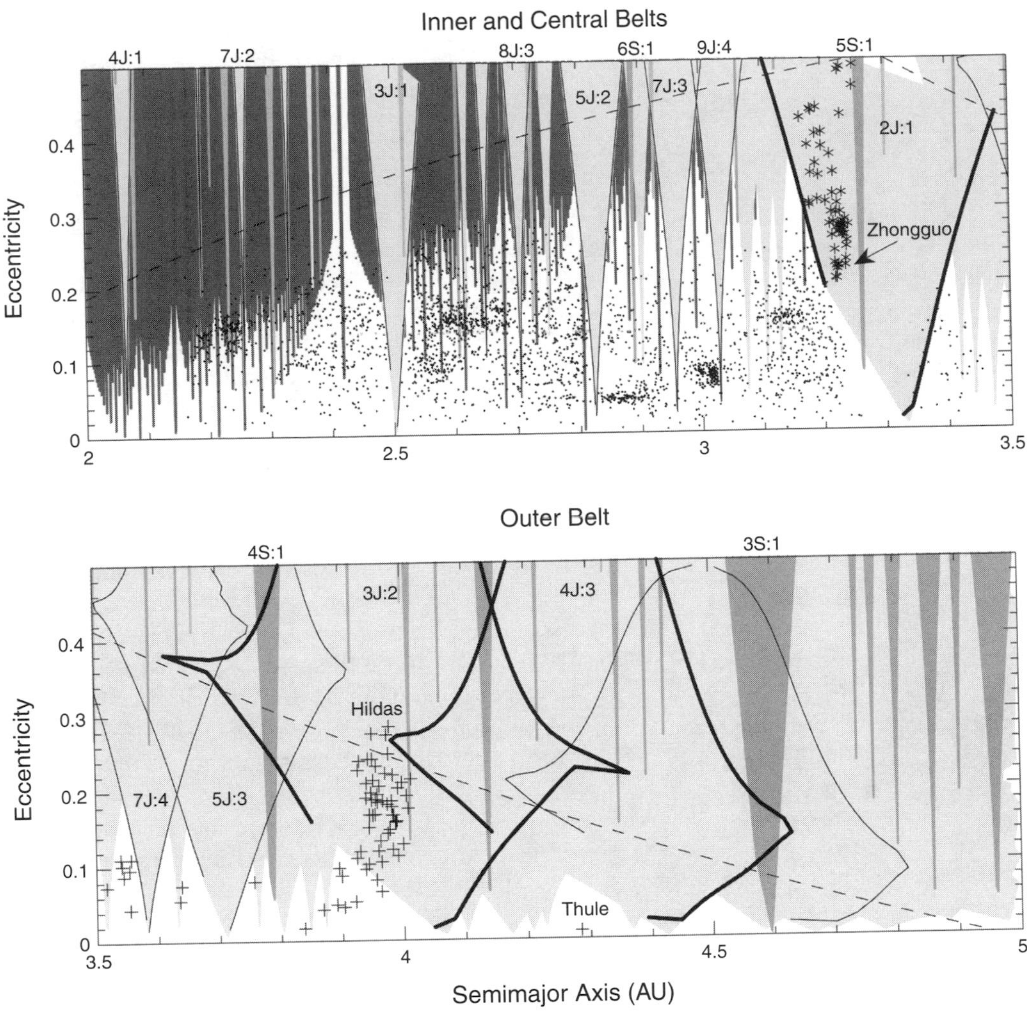

Fig. 2. Global structure of the MMRs in the asteroid belt. There are four different gray shades denoting the regions of resonant motion with planets: light gray for Jupiter MMRs, intermediate for Saturn MMRs, and dark gray for Mars MMRs. Each resonance corresponds to one V-shaped region except the large first-order MMRs with Jupiter, which have particular shapes. Some of the resonances are labeled. For some Jupiter MMRs the projection of separatrixes on the a, e plane is shown by black lines; for 2J:1, 3J:2, and 4J:3, these lines are bold. We also show the proper (dots; *Milani and Knežević,* 1994) or orbital elements (crosses; *Bowell et al.,* 1994) of asteroids with magnitudes up to the completeness level. In case of the group of small asteroids in the 2J:1 MMR (arrow indicates 3789 Zhongguo), the resonant elements are plotted (asterisks; *Roig et al.,* 2002). Other resonant asteroids are the Hilda group in the 3J:2 MMR and 279 Thule in the 4J:3 MMR. The orbits above the dashed lines are planet-crossing.

stand the dynamics of real bodies. The three-body resonances are usually as narrow as two-body resonances with Mars ($<10^{-2}$ AU), but generally being of a low order, they have nonnegligible sizes down to small eccentricities. Moreover, the three-body MMRs are numerous.

4. CHAOTIC SECULAR DYNAMICS IN THE MEAN-MOTION RESONANCES WITH JUPITER: THE KIRKWOOD GAPS

The bodies inside the wide MMRs with Jupiter undergo important secular dynamics, in particular for what concerns the evolution of e (*Wisdom,* 1982; *Yoshikawa,* 1990, 1991; *Ferraz-Mello and Klafke,* 1991; *Morbidelli and Moons,* 1995; *Gladman et al.,* 1997). We briefly review how this can be theoretically justified, restricting for simplicity to the case where the asteroid and planets have coplanar orbits, although elliptic and precessing [see *Morbidelli and Moons* (1993) for a discussion of the inclined case]. We start from the resonant Hamiltonian equation (8), that we rewrite as $\mathcal{H}^{2BR} = \mathcal{H}^{PC} + \mathcal{F}$ where $\mathcal{H}^{PC}$ is given in equation (12), and $\mathcal{F} = \mathcal{H}^{2BR} - \mathcal{H}^{PC}$. The function $\mathcal{F}$ can be considered as a perturbation of $\mathcal{H}^{PC}$, because (for D'Alembert rules) the former is proportional to the planetary eccentricities.

Because $\mathcal{H}^{PC}$ is integrable, we introduce the Arnold action-angle variables in the MMR region where σ librates.

Following *Henrard* (1990), the transformation has the form

$$\psi_\sigma = \frac{2\pi}{T_\sigma}t \qquad J_\sigma = \frac{1}{2\pi}\oint S\,d\sigma$$
$$\psi_v = v - f(\psi_\sigma, J_\sigma, J_v) \qquad J_v = N \tag{13}$$

where the integral is done over a trajectory of S, σ for N = const (Fig. 1), T_σ is a period of the trajectory, and f is a periodic function of ψ_σ, with null average.

We then write $\mathcal{H}^{PC}$ and $\mathcal{F}$ as functions of these variables. By construction, $\mathcal{H}^{PC}$ now depends only on the new action variables J_σ, J_v. By averaging $\mathcal{F}$ over ψ_σ, we obtain an Hamiltonian of the form

$$\mathcal{H}^{SEC} = \sum_j g_j\Lambda_{g_j} + \mathcal{H}^{PC}(J_\sigma,J_v) + \bar{\mathcal{F}}(J_\sigma,J_v,\psi_v,\varpi) \tag{14}$$

where $\varpi = (\varpi_1, ..., \varpi_N)$. Because this Hamiltonian is independent of ψ_σ, the action J_σ is now a constant of motion. Thus, equation (14) describes the secular dynamics inside the MMR, namely the evolution of the eccentricity (through J_v) due to the motion of the perihelia of the asteroid and the planets ($-v$ and ϖ_j respectively).

The mean frequency of the longitude of perihelion can be computed as

$$-\dot{\psi}_v = \frac{\partial\mathcal{H}^{SEC}}{\partial J_v} \tag{15}$$

A first-order perihelion secular resonance occurs when $\dot{\psi}_v + g_j = 0$, where g_j is the frequency of ϖ_j [for numeric values of g_j see, e.g., *Laskar* (1988)].

It often occurs in MMRs with Jupiter that the secular resonances associated with the g_5 and g_6 frequencies are located close to each other. To study the effects of the interaction between these two resonances, we construct a two-resonance model by retaining from equation (14) the har-

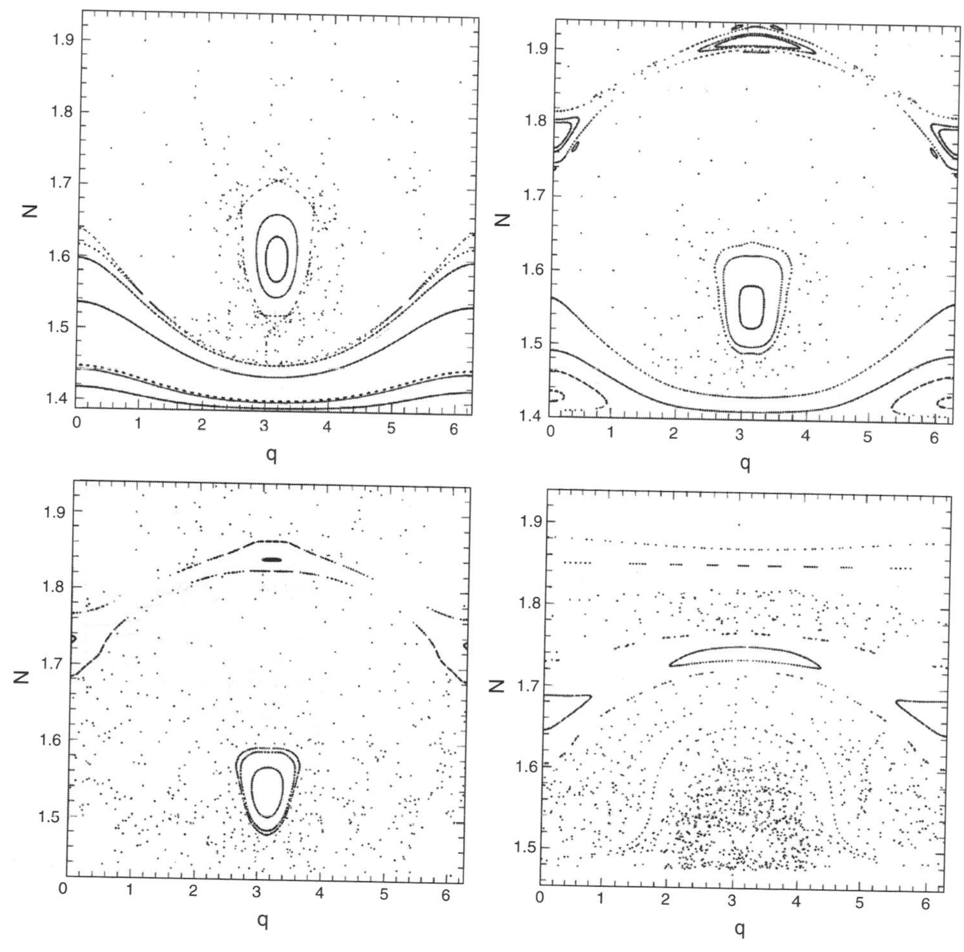

Fig. 3. Sections of the dynamics of equation (16) computed at $\sigma_6 = 0$ for the 3J:1 MMR. The four panels correspond to increasing values of J_σ, the latter being related to the amplitude of oscillation of a. The label q stands for $\varpi - \varpi_5$, while $N = J_v = \sqrt{a}\left(3 - \sqrt{1-e^2}\right)$. At the center of the 3J:1, the values N = 1.4, 1.5, 1.6, 1.7, 1.8, 1.9 correspond to e = 0.2, 0.55, 0.72, 0.84, 0.92, 0.97 respectively. In each panel, the lower limit on the N axis is the value that identifies the separatrix of the MMR. Consequently, all the curves that seem to exit from the bottom border of the panels correspond to trajectories that hit the separatrix of the MMR during their secular evolution, and are therefore expected to be chaotic. From *Moons and Morbidelli* (1995).

monic of $\bar{\mathcal{F}}$ with arguments $\sigma_5 = \psi_v + \varpi_5$ and/or $\sigma_6 = \psi_v + \varpi_6$. Thus, we consider the Hamiltonian

$$\mathcal{H}^{5,6} = g_5\Lambda_{g_5} + g_6\Lambda_{g_6} + \mathcal{H}^{PC}(J_\sigma, J_v) + \bar{\mathcal{F}}^{5,6}(J_\sigma, J_v, \sigma_5, \sigma_6) \tag{16}$$

This is a nonintegrable Hamiltonian that must be studied numerically. The dynamics can be represented on the σ_5, J_v plane, for instance through a section at $\sigma_6 = 0$. The sections depend parametrically on J_σ, roughly corresponding to the libration amplitude of a (A_a).

In the case of the 3:1 MMR with Jupiter (Fig. 3), most of the phase space is covered by a chaotic region, which extends up to e = 1 (top borders of the panels). Only the orbits with small A_a and e (the smooth curves at the bottom of the two top panels of Fig. 3) still have regular dynamics. These orbits, however, periodically reach Mars-crossing eccentricities (e > 0.3); the encounters with Mars give impulsive changes to a and e, which, although generally very small, effectively modify A_a. At large A_a, the chaotic region generated by the overlap of the ν_5 and ν_6 resonances extends to all eccentricities (bottom panels of Fig. 3), and the asteroid can therefore rapidly and chaotically evolve to very large e. This combination of large-scale chaos of the secular dynamics and weak martian encounters explains the behavior of 3:1 resonant asteroids observed in numerical integrations of the full equations of motion (see *Morbidelli et al.,* 2002), and explains the formation of a gap in the asteroid distribution.

The same happens in many other major MMRs with Jupiter that are associated with a Kirkwood gap. *Moons and Morbidelli* (1995) have shown that also the 4J:1, 5J:2, and 7J:3 MMRs are dominated by the chaotic region generated by the overlap of the ν_5 and ν_6 resonances, so that the asteroids in these MMRs can also reach very large e on a timescale of a few million years.

Large evolutions of e in many MMRs are typically caused by the secular resonance ν_5 itself. The structure of the ν_5 resonance can be computed in models neglecting g_6 (*Ferraz-Mello and Klafke,* 1991; *Klafke et al.,* 1992; *Moons and Morbidelli,* 1995). These models show that by the effect of ν_5, the eccentricity suffers large variations, while $\varpi - \varpi_5$ oscillates around 0 or π. In such models, however, only bodies with certain initial orbits can evolve from low to very high eccentricities (*Klafke et al.,* 1992). Other parts of the orbital space are characterized by regular motion bounded at moderate e. This regular motion almost completely vanishes when g_6 is accounted for (as shown above; Fig. 3). Consequently, in a globally chaotic environment, orbits evolve according to the underlying structure of ν_5, and despite their initial location in the orbital space, reach very large e. This is usually accomplished by a series of small transitions because of the interaction of ν_5 and ν_6 and a few large events, when e increases due to the effect of ν_5 (*Morbidelli and Moons,* 1995).

5. TRANSIENT AND STABLE POPULATIONS OF THE 2J:1 AND 3J:2 MEAN-MOTION RESONANCES

In the 2J:1 and 3J:2 MMRs, simple models with Jupiter's orbit fixed also show a high-e regime of motion associated with the ν_5 resonance. In this case, however, the high-e regime is separated from low e by regular orbits, which are robust and persist in the models that account for g_6 (*Henrard and Lemaître,* 1987; *Morbidelli and Moons,* 1993; *Ferraz-Mello,* 1994b; *Michtchenko and Ferraz-Mello,* 1995; *Moons et al.,* 1998). Consequently, unlike in the 3J:1, 4J:1, 5J:2, and 7J:3 MMRs, the secular dynamics do not explain why the observed orbital distribution of asteroids should display a gap at the place of the 2J:1 MMR (the Hecuba gap, see Fig. 2), where only a few tens of small resonant asteroids reside. Conversely, the 3J:2 MMR hosts some 260 resonant asteroids known at the time of writing this text (the Hilda group), with about 30 bodies exceeding 50 km in diameter.

This puzzling difference, known as the "2:1 vs. 3:2 paradox," led some authors to investigate the possibility of opening the Hecuba gap during the primordial stages of the solar system formation by invoking effects that can mutually displace the resonances and amplitudes of asteroids (*Henrard and Lemaître,* 1983). It became clear later that although plausible, such effects are not strictly required to explain the lack of asteroids in the 2J:1 MMR.

An important difference between the 2J:1 and 3J:2 MMRs was noted by *Ferraz-Mello* (1994a,b). He computed the maximum Lyapounov characteristic exponents (LCE) [the maximum LCE measures the rate of divergence of nearby orbits and is an indicator of chaos (*Benettin et al.,* 1976)] of a number of resonant orbits in both MMRs. It turned out that in the 2J:1 MMR, LCE $\sim 10^{-5}$–$10^{-3.5}$ yr^{-1}, while in the 3J:2 MMR, LCE $< 10^{-5.5}$ yr^{-1}. This indicated that orbits in the 2J:1 MMR are chaotic on short timescales (*Morbidelli,* 1996), at variance with the 3J:2 MMR where most trajectories are only weakly chaotic.

The comparative study (*Nesvorný and Ferraz-Mello,* 1997) of the MMRs employing frequency map analysis (*Laskar,* 1988, 1999) provided additional clues. Plate 1 shows how the magnitude of the chaotic evolution varies in the orbital space of the 2J:1 and 3J:2 MMRs. The color coding represents the value of $\log_{10}|\delta\dot{\varpi}|$, where $\delta\dot{\varpi}$ is the relative change of the perihelion frequency per 0.2 m.y. This quantity is a powerful indicator of the rate of chaotic evolution (chaotic diffusion) suffered by orbits in the integration timespan. The results were extrapolated to larger time intervals assuming a chaotic random walk of orbits (see *Ferraz-Mello et al.,* 1998a).

In regions corresponding to the smallest magnitudes of the chaotic diffusion (deep blue), the orbits evolve relatively by less than 10% on 1 G.y. Such orbits should be stable over the age of the solar system. Conversely, in regions where $\log_{10}|\delta\dot{\varpi}| > -2.5$ (red, yellow), the perihelion frequency of a resonant body should typically change by more than 100%

in less than 1 G.y. Such an evolution should be enough to destabilize the orbit. Thus, the orbits with $\log_{10}|\delta\dot{\varpi}| > -2.5$ are expected to be unstable. Despite the large uncertainties involved in the extrapolation from million-year to billion-year timescales, Plate 1 provides a convincing argument suggesting at least 1–2 orders in magnitude shorter lifetimes in the 2J:1 than in the 3J:2 MMR. The lifetimes in the most stable regions of the 2J:1 MMR were estimated to be on the order of 10^9 yr (*Morbidelli,* 1996; *Nesvorný and Ferraz-Mello,* 1997).

Recently, these results has been put on firmer ground by direct simulation of 50 test bodies in the 2J:1 MMR over 4 G.y. (*Roig et al.,* 2002). This simulation showed that the most stable region of the 2J:1 MMR is characterized by marginal instabilities: There is about a 50% probability that a body started at $3.2 < a < 3.3$ AU, $0.2 < e < 0.4$, small i, and $\varpi - \varpi_J = \sigma = 0$ escapes from the resonance in 4 G.y. because of the diffusive chaos. Thus, the lack of a larger asteroid population in the 2J:1 MMR is at least partially due to the slow removal of primordial population in the last 4 G.y. The group of resonant asteroids, now accounting for some 30 small bodies known as the Zhongguo group, is localized in a very small region in the resonance where lifetimes generally exceed 4 G.y.

The origin of the Zhongguo group is unknown. This group has a steep size distribution, which would be expected from the collision injection at the breakup of the parent body of the Themis family (*Morbidelli,* 1996). The large ejection velocities needed for such injection, an offset in e between the Themis family and the Zhongguo group, and possibly incompatible spectral type of (3789) Zhongguo, however, suggest this to be an unlikely origin (*Roig et al.,* 2002). The Zhongguo group seems to be an order of magnitude more depleted than what would be expected from a dynamically eroded population similar to the Hilda group. Assuming Zhongguos to be dynamically primordial bodies in the 2J:1 MMR, their additional depletion could have been achieved by collision processes or during the primordial stages of the solar system formation.

The slow diffusive chaos in the 2J:1 MMR is where the concept of a resonance between an asteroid and two perturbing planets first appeared. As originally pointed out (*Ferraz-Mello,* 1997; *Michtchenko and Ferraz-Mello,* 1997; *Ferraz-Mello et al.,* 1998b) the slow chaos in the 2J:1 MMR is probably generated by commensurabilities between the periods of the "Great Inequality," i.e., the period of $2\lambda_5 - 5\lambda_6$ (≈880 yr), and of $\sigma = 2\lambda_5 - \lambda - \varpi$ (300–500 yr). Symplectic maps of the 2J:1 MMR found the central region less chaotic when the effect of the Great Inequality was switched off. The effect of this resonance was also put into evidence in direct numerical simulations by artificially changing the Great Inequality period to 440 yr. In such situation, which could have occurred when Jupiter and Saturn were slightly closer to each other during the primordial migration phase (*Fernandez and Ip,* 1984), the "beat" between the periods of the Great Inequality and of σ was approximately 1:1, and the instabilities in the 2J:1 MMR were significantly accelerated (inset in Plate 1). The effect of the Great Inequality may be analytically modeled as overlap between the 2J:1 and 5S:1 MMRs (*Morbidelli,* 2002).

6. NUMEROUS NARROW MEAN-MOTION RESONANCES DRIVING THE CHAOTIC DIFFUSION

The narrow MMRs (three-body, high-order two-body, and resonances with Mars) in the asteroid belt are usually not powerful enough to destabilize bodies continuously resupplied into them by the Yarkovsky force (*Bottke et al.,* 2002b) and collisions (*Morbidelli et al.,* 1995; *Zappalà et al.,* 2000). Consequently, the narrow MMRs do not open gaps in the asteroid distribution and are populated by many asteroids at present. In the following three sections, we discuss the three-body MMRs (section 6.1), narrow two-body MMRs (section 6.2), and MMRs with Mars (section 6.3).

6.1. Three-Body Mean-Motion Resonances

The growing evidence that many asteroids move on chaotic orbits (*Milani and Nobili,* 1992; *Milani and Knežević,* 1994; *Holman and Murray,* 1996, *Šidlichovský and Nesvorný,* 1997; *Knežević and Milani,* 2000) has challenged the view of the asteroid belt as an unchanging, fossil remnant of its primordial state. For example, the asteroid 490 Veritas at a = 3.174 AU has a Lyapunov time (defined as an inverse of the LCE) ≈10,000 yr, showing unpredictability of the orbit over $>10^5$-yr intervals. It was moreover noted that e and i of this object chaotically evolve outside the borders of the Veritas family (of which 490 Veritas is the largest fragment) in 50 m.y. Based on this fact, the age of the Veritas family was hypothesized to be of this order (*Milani and Nobili,* 1992).

In our current understanding, 490 Veritas is one among many objects in the asteroid belt residing in the three-body MMRs. Convincing demonstration of this fact was provided by estimates of the LCE for large samples of real and fictitious asteroids (*Nesvorný and Morbidelli,* 1998; *Murray et al.,* 1998; *Morbidelli and Nesvorný,* 1999). Figure 4 shows the profile of the LCE computed for test bodies initially at e = 0.1 and i = 0. The peaks correspond to chaotic regions. Some of the main peaks may be identified with two-body MMRs (3J:1 at 2.5 AU, 5J:2 at 2.82 AU, 7J:3 at 2.96 AU, etc.), while most of the narrow peaks are related to three-body MMRs. Moreover, many two-body MMRs with Mars occur in the inner asteroid belt (2.1–2.5 AU). It has been shown that a large fraction of real asteroids reside in narrow resonances (*Nesvorný and Morbidelli,* 1998). 490 Veritas evolves in the 5J:–2S:–2, the angle $\sigma_{5,-2,-2,0,0,-1,0,0,0}$ having irregular oscillations about 0 correlated with oscillations of semimajor axis about a_{res} = 3.174 AU.

This result raised questions concerning the implications of such a complex chaotic structure of the asteroid belt.

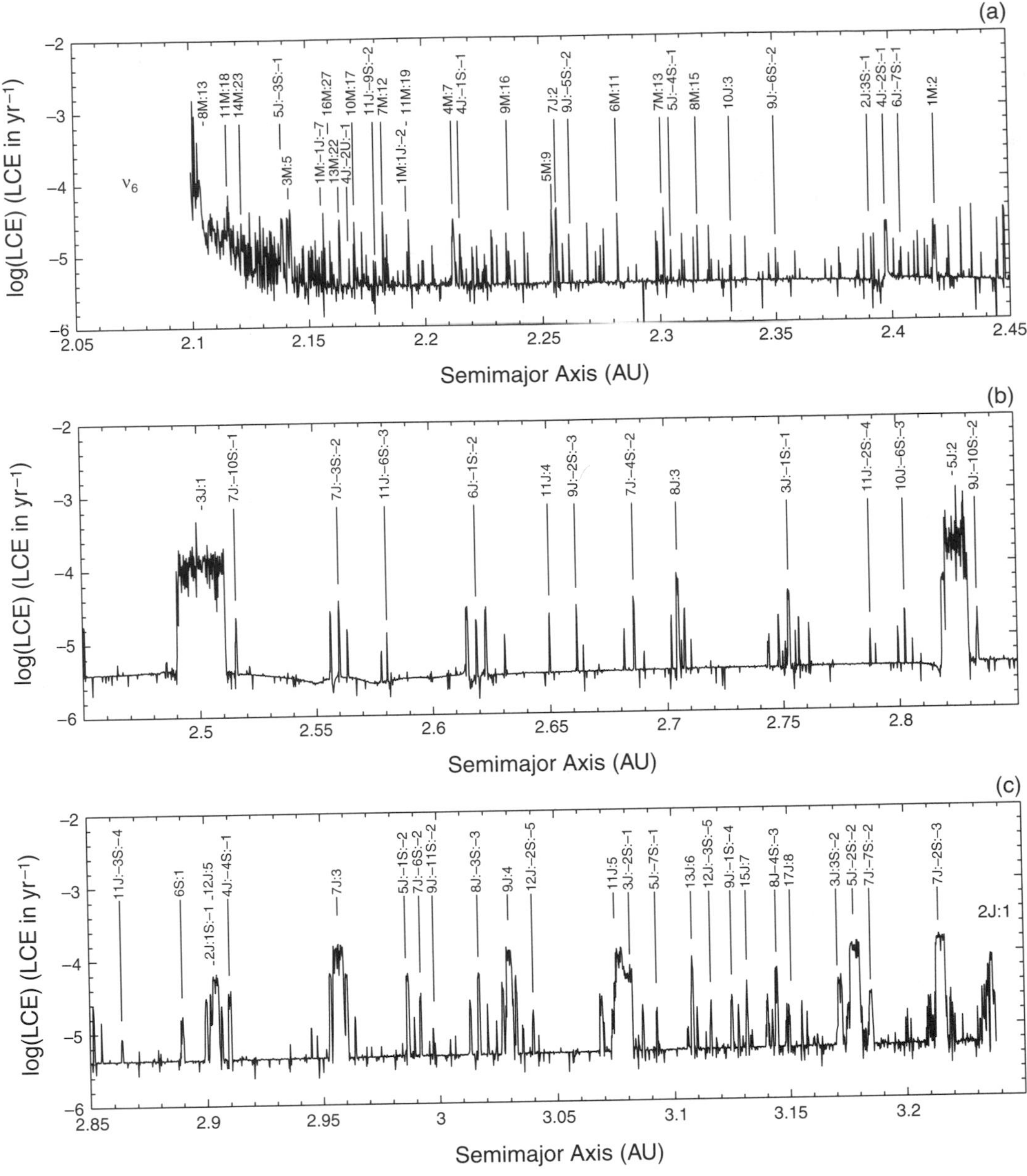

Fig. 4. The estimate of the LCE computed for test bodies initially placed on a regular grid in semimajor axis, with e = 0.1. **(a)** The gravitational perturbations provided by the terrestrial planets were taken into account in addition to the perturbations of four giant planets. **(b,c)** Accounted only for four giant planets. The figure shows many peaks that rise from a background level. The background value of about $10^{-5.3}$ yr^{-1} is dictated by the limited integration timespan (2.2 m.y.); increasing the latter, the background level generally decreases. The peaks reveal the existence of narrow chaotic regions associated with MMRs. The main resonances are indicated above the corresponding peaks. From *Morbidelli and Nesvorný* (1999).

Asteroids were shown to escape from the main belt to Mars- and Jupiter-crossing orbits due to a slow chaotic evolution of eccentricities in narrow resonances. Realistic modeling including the Yarkovsky effect demonstrated that the narrow resonances are an important source of the near-Earth-object population (*Migliorini et al.,* 1998; *Morbidelli and Nesvorný,* 1999; *Bottke et al.,* 2002a). Recently, narrow MMRs were also hypothesized to disperse the asteroid families in e and i, possibly creating their often asymmetric shapes (*Nesvorný et al.,* 2002). If this process is as general as conjectured, many asteroid families could have been created as small groupings and later were dynamically dispersed by chaotic diffusion in narrow MMRs and by the Yarkovsky effect. This mechanism may eventually solve the paradox between hydrocode simulations of breakup events suggesting creation of tightly grouped collision swarms, and the observed, largely dispersed asteroid families.

Theoretical modeling of the three-body MMRs revealed several clues to understanding the behavior of asteroids in numerical simulations. The resonant Hamiltonian of several three-body MMRs with Jupiter and Saturn in the planar model has been computed by *Murray et al.* (1998) and by *Nesvorný and Morbidelli* (1999). Following the notation of the latter work, the Hamiltonian of the $k_5J{:}k_6S{:}k$ MMRs (equation (10)) can be written as a Fourier expansion

$$\bar{\mathcal{H}}_1 = \frac{1}{a_5}\sum_{\nu} \mathcal{P}_{1,\nu}(\alpha_{res})\, e_5^{s_5} e_6^{s_6} e^{s} \cos[l_5\varpi_5 + l_6\varpi_6 - l(\sigma + \nu)]$$

$$\bar{\mathcal{H}}_2 = \frac{1}{a_5}\sum_{\nu} \mathcal{P}_{2,\nu}(\alpha_{res})\, e_5^{s_5} e_6^{s_6} e^{s} \cos[\sigma_{\mathbf{k},k,(l_5,l_6),l,\mathbf{0},0}] \tag{17}$$

where $\alpha_{res} = a_{res}/a_5$, and the multiindex ν denotes different values of integers s_5, s_6, s, l_5, l_6, l. The angle $\sigma_{\mathbf{k},k,(l_5,l_6),l,\mathbf{0},0}$ in equation (17) is given in terms of σ and ν (equation (7)). Denoting

$$\beta = \frac{\mu_5^2}{a_5} \sum_{\substack{\nu \\ l_5 = l_6 = 0}} \mathcal{P}_{2,\nu}\, e^s \tag{18}$$

these coefficients, truncated at two lowest orders in e, are given in Table 1.

Ignoring all but one resonant multiplet term in equation (17) (generically denoted by σ_*), the complete Hamiltonian equation (10) can be reduced to the Hamiltonian of a simple pendulum $\mathcal{H} = \alpha(S - N)^2 + \beta \cos \sigma_*$, N representing a constant parameter (*Nesvorný and Morbidelli,* 1999; *Murray et al.,* 1998). The resonant full width can then be computed from equation (11). Similarly, the period of small amplitude librations (in years) is

$$T = \frac{a_{res}}{|k|}\sqrt{\frac{1}{3\beta}} \tag{19}$$

The position of the libration center σ_c is determined by the signs of α and β. As $\alpha = -3\hat{\lambda}^2/2a_{res}^2$ is always negative, the resonant center is at 0 when $\beta > 0$, and at π when $\beta < 0$. Moreover, according to simple arguments, the Lyapunov time is on the order of the libration period, so that T provides a rough estimate of the LCE (*Benettin and Gallavotti,* 1986; *Holman and Murray,* 1996).

Table 1 shows Δa, T, and σ_c for the most important three-body MMRs with e = 0.1. The values computed from equations (11) and (19) are in good agreement with numerical results (Fig. 4). In particular, 490 Veritas was numerically found to evolve in the 5J:–2S:–2 MMR with 0.004-AU oscillations of a and the libration period of several 10^3 yr, both values being in agreement with the ones in Table 1.

Understanding the chaos and the long-term evolution of asteroids in the three-body MMRs requires to account for more than one multiplet term in equation (17). *Murray et al.* (1998) and *Nesvorný and Morbidelli* (1999) showed that different multiplets of the three-body MMR generally overlap, thus creating a chaotic domain at the position of the MMR. Quantitative estimates of the LCE and diffusion timescales in this domain showed that significant chaotic evolutions of e and i of resonant bodies occur on long timespans (*Murray and Holman,* 1997; *Murray et al.,* 1998). Conversely, the semimajor axis of resonant bodies remains

TABLE 1. Analytic results on the three-body MMRs (from *Nesvorný and Morbidelli,* 1999).

Resonance Designation	Resonant Semimajor Axis a_{res} (AU)	Coefficient β ($\times 10^{-8}$)	Resonance Width (e = 0.1)	Libration Period T (10^3 yr) (e = 0.1)	Libration Center σ_c
4J:–1S:–1	2.2155	$6.05e^2 - 2.22e^4$	0.37	52	0
3J: 1S:–1	2.2994	$3.78e^3 - 1.94e^5$	0.1	217	0
4J:–2S:–1	2.3978	$19.7e - 3.95e^3$	2.4	9.9	0
7J:–2S:–2	2.4479	$46.0e^3 - 53.0e^5$	0.37	33	0
7J:–3S:–2	2.5600	$-4.35e^2 + 5.28e^4$	0.39	36	π
2J: 2S:–1	2.6155	$6.03e^3 + 3.63e^5$	0.15	200	0
6J:–1S:–2	2.6192	$-17.6e^3 + 21.3e^5$	0.25	57	π
4J:–3S:–1	2.6229	$-0.206 - 0.763e^2$	0.9	30	π
7J:–4S:–2	2.6858	$0.255e - 0.356e^3$	0.3	49	0
3J:–1S:–1	2.7527	$-7.99e + 0.360e^3$	1.9	17.8	π
4J:–4S:–1	2.9092	$0.0203e + 0.0691e^3$	0.1	370	0
5J:–1S:–2	2.9864	$20.9e^2 - 21.3e^4$	1.1	19	0
8J:–3S:–3	3.0166	$-10.0e^2 + 24.0e^4$	0.8	19	π
3J:–2S:–1	3.0794	$-3.06 - 15.5e^2$	4.5	9.9	π
6J: 1S:–3	3.1389	$-22.4e^4 + 143e^6$	0.12	132	π
8J:–4S:–3	3.1421	$0.357e - 0.99e^3$	0.5	33	0
3J: 3S:–2	3.1705	$-25.8e^4 + 5.15e^6$	0.13	180	π
5J:–2S:–2	3.1751	$45.6e - 32.3e^3$	5.6	4.3	0
7J:–2S:–3	3.2100	$-115e^2 + 273e^4$	2.8	5.8	π

locked in the interval spanned by the resonant multiplet and although behaving stochastically on short timespans, never directly evolves outside the resonant borders, unless two or more different MMRs overlap (the case of high e in Fig. 2).

6.2. Narrow Two-Body Mean-Motion Resonances with Jupiter

As discussed above, the details of the multiplet structure of each two-body and three-body resonance affect the extent of the chaotic zone associated with the resonance, the typical Lyapunov time for a small body in that resonance, and the timescale on which such a body might chaotically diffuse in the space of orbital elements. Useful analytic models of these effects can be obtained using some simplifying assumptions (*Holman and Murray,* 1996; *Murray and Holman,* 1997; *Murray et al.,* 1998). We show an example of such modeling for a two-body resonance (see also *Morbidelli,* 2002). This modeling can be used with small modifications for the three-body (*Nesvorný and Morbidelli,* 1999) and Mars' MMRs.

To most simply account for the multiplet structure of a two-body (**k**,k) MMR, we restrict ourselves to the planar elliptic three-body model. The resonant Hamiltonian of such model can be written

$$\mathcal{H} = -\frac{1}{2\Lambda^2} + \frac{k_{j_1}}{k} n_{j_1}\Lambda + c_0(\Lambda,P) + \sum_{s=0}^{|k_{j_1}+k|} c_s(\Lambda,P) \times \cos\left[k_{j_1}\lambda_{j_1} + k\lambda + sp - \left(k_{j_1} + k - s\right)\varpi_{j_1}\right] \quad (20)$$

where c_0 is a part of $\mu\bar{\mathcal{H}}_1$ in equation (8) independent of angles. In equation (20), we retained only those terms of the Fourier expansion of equation (8) that are lowest order in the eccentricities of the planet and the small body. For simplicity, we arbitrarily set $\varpi_{j_1} = 0$, and drop it from equation (20).

The canonical variables we use in the following slightly differ from equation (7). We define them by

$$\begin{aligned} \psi &= k_{j_1}\lambda_{j_1} + k\lambda & \Psi &= \frac{1}{k}(\Lambda - \Lambda_0) \\ \phi &= p & I &= P \\ &\lambda_{j_1} & \tilde{\Lambda}_{j_1} &= \Lambda_{j_1} - \frac{k_{j_1}}{k}\Lambda \end{aligned} \quad (21)$$

where $\Lambda_0 = \sqrt{a_{res}}$ is the unperturbed MMR location. Keeping the lowest-order terms in Ψ and I in the Taylor expansion of first three terms of Hamiltonian equation (20), we get

$$\mathcal{H} = \frac{1}{2}\beta\Psi^2 + 2\varepsilon AI + \sum_{s=0}^{|k_{j_1}+k|} c_s(I)\cos(\psi + s\phi) \quad (22)$$

where β and $2\varepsilon A = \partial c_0/\partial I(\Lambda_0,0)$ are the constants associated with the Taylor expansion, and $c_s(I) = c_s(\Lambda_0,I)$.

To apply the resonance overlap criterion to the Hamiltonian equation (22) we must compute the separation and widths of the individual multiplet terms in the resonance. The separation in Ψ between two such terms is given by $\beta\delta\Psi = 2\varepsilon A$, while the width of each multiplet is $\Delta\Psi = 2\sqrt{c_s(I, P_{j_1})/\beta}$. The stochasticity parameter (*Chirikov,* 1979; *Lichtenberg and Lieberman,* 1983) is then

$$K_{eff} = \left(\pi\frac{\Delta\Psi}{\delta\Psi}\right)^2 = \left(\frac{c_{s_0}}{\beta}\right)\left(\frac{\pi\beta}{\varepsilon A}\right)^2 \quad (23)$$

where c_{s_0} is the coefficient of a representative multiplet term (*Murray and Holman,* 1997). The orbits in the MMR will be chaotic if $K_{eff} > K_{crit}$, where K_{crit} ~1. The extent of the chaotic zone in Ψ (and therefore in semimajor axis) is roughly given by the Ψ range of those multiplet terms that overlap. The Lyapunov time in the chaotic zone is given by

$$T_L \approx \frac{\pi}{\varepsilon A} / \ln\left[1 + \frac{K_{eff}}{4} + \sqrt{\frac{K_{eff}}{2} + \left(\frac{K_{eff}}{4}\right)^2}\right] \sim \frac{\pi}{\varepsilon A} \quad (24)$$

which indicates that the Lyapunov time is comparable to the precession period.

If $K_{eff} > K_{crit}$, the diffusion in Ψ and I, can be described by the Fokker-Planck equation (*Lichtenberg and Lieberman,* 1983; *Murray and Holman,* 1997). The associated diffusion coefficients are

$$D_\Psi \approx \frac{K_{eff}^2}{2}\left(\frac{\varepsilon A}{\pi\beta}\right)^2 = \frac{1}{2}\left(\frac{c_{s_0}}{\beta}\right)^2\left(\frac{\pi\beta}{\varepsilon A}\right)^2 = \frac{1}{2}c_{s_0}^2\frac{\pi}{\varepsilon A} \quad (25)$$

and

$$D_I = s_0^2 D_\Psi = \frac{s_0^2}{2}c_{s_0}^2\frac{\pi}{\varepsilon A} \quad (26)$$

In the model of *Murray and Holman* (1997) the resonant bodies execute random walks in Ψ (or semimajor axis) and I (or eccentricity). These random walk are subject to boundary conditions. The permissible values of Ψ lie within the range of values for which the resonant multiplet terms overlap (the chaotic zone). The values of I range from I = 0 (circular orbit) to $I = I_{max}$, where I_{max} corresponds to planet crossing. The bounding values of Ψ and I = 0 can be modeled as reflecting barriers, and the value of $I = I_{max}$ as an absorbing barrier. For a resonant asteroid at I_0, the solution of the Fokker-Planck equation yields typical values of the removal time, T_R, of

$$T_R \sim \frac{I_{max}I_0}{\sqrt{D_I(I_{max})D_I(I_0)}} \quad (27)$$

Figure 5 shows removal times for outer-belt asteroids in a number of two-body MMRs. The solid symbols are the results of numerical integrations; the open symbols are analytic predictions from equation (27). The values of I_0 were determined from short numerical integrations of resonant asteroids, hence the scatter in the predicted removal

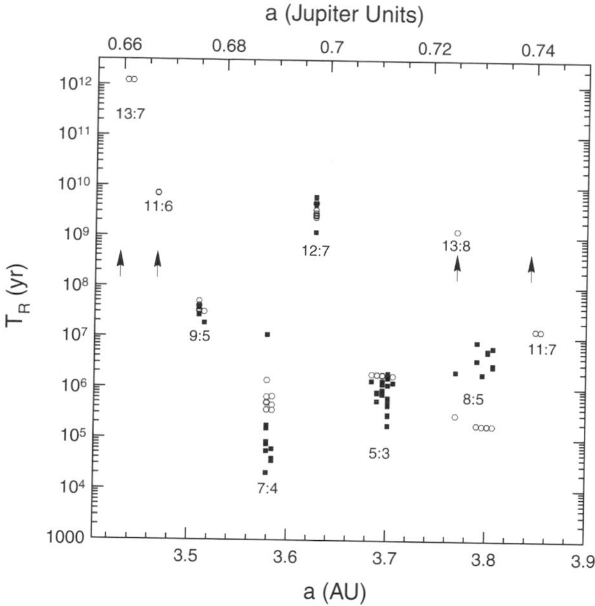

Fig. 5. The removal times for outer-belt asteroids in a number of two-body MMRs with Jupiter. The solid symbols are the results of numerical integrations; the open symbols are analytic predictions from equation (27). The arrows represent lower limits obtained from numerical integration of bodies in MMRs characterized by extremely slow chaotic diffusion. Here, Jupiter units are 5.2 AU. From *Murray and Holman* (1997).

times. In most cases the results of numerical experiments agree with the analytic predictions to an order of magnitude. The results of these calculations indicate that for high-order MMRs in the outer belt, removal times are in some MMRs substantially shorter than the age of the solar system (the case of the 7J:4, 5J:3, and 8J:5 MMRs) and in other resonances substantially longer than the age of the solar system (the 13J:7 and 11J:6 MMRs).

Figure 5 shows that resonances like the 7J:4, 5J:3, and 8J:5 MMR are expected to correspond to gaps in the asteroid distribution. Indeed, no large outer-belt asteroids are observed in them (Fig. 2). The outer asteroid belt, however, seems more depleted than what would be expected from the effect of chaotic resonances. This led some authors to suggest that depletion occurred during the primordial radial migration of the planets (*Holman and Murray,* 1996; *Liou and Malhotra,* 1997). The same scenario can also nicely explain the presence of many asteroids in the 3J:2 MMR, which could have been captured by the resonance accompanying the inward migration of the Jupiter. Alternatively, the outer asteroid belt could have been depleted by the effect of large Jupiter-scattered planetesimals (*Petit et al.,* 1999).

6.3. Mean-Motion Resonances with Mars

It may seem surprising at a first glance that the MMRs with Mars are important in the asteroid belt because Mars is a factor of ≈3000 less massive than Jupiter. Thus, according to the discussion in section 2, the two-body MMRs with Mars should be in general a factor $\sqrt{3000}$ ~55 smaller than the two-body MMRs with Jupiter. In the inner asteroid belt, however, this factor may be compensated for by the small distance between an asteroid and Mars. Indeed, the width of any MMRs near the planet-crossing limit becomes large. According to Fig. 2 (see also Fig. 4), the MMRs with Mars provide a dominant source of chaos in the inner asteroid belt, overwhelming in size and number the MMRs with Jupiter.

The dynamical interaction of asteroids with the MMRs with Mars is complex. As an example, Fig. 5 shows the evolution of a, $\sigma = 9\lambda_4 - 22\lambda + 13\varpi$, and $e^{13/2}$ of 41 Daphne. This asteroid is temporally captured in the 3M:11 MMR when e becomes large during the secular oscillations. The irregular behavior of a and σ witnesses the chaoticity of the orbit, which has a Lyapunov time of $<10^4$ yr (*Šidlichovský,* 1999).

Very little analytical work has been done on the subject of the MMRs with Mars. Figures 1e–f are apparently highly idealized approximations of the real dynamics (Fig. 6). To understand the real dynamics, it is important to account for the rich multiplet structure of the MMRs with Mars and the significant time-modulation of each multiplet's width with varying eccentricities (interesting MMRs with Mars are usually of a very high order).

The chaotic diffusion in the MMRs with Mars provide transfer routes from the asteroid belt to Mars-crossing orbits. This source was hypothesized to be one of the major sources of large NEAs (*Migliorini et al.,* 1998; *Morbidelli and Nesvorný,* 1999; *Bottke et al.,* 2002a). The detailed mechanism of this transfer involves the complex interaction of the Yarkovsky effect and MMRs. In particular, the Yarkovsky effect is thought to be responsible for refilling MMRs by new objects; otherwise the space occupied by the active MMRs would be cleared of the asteroids, and gaps in the asteroid distribution at resonant semimajor axes would be created. Such gaps are not currently observed.

7. PERSPECTIVES

In the last decade, we witnessed fascinating progress from studies of separate MMRs in simplest physical models to the global understanding of the structure and effects of MMRs throughout the asteroid belt. Nevertheless, we feel that many open problems await solution. The least developed is the modeling of narrow resonances, especially the three-body MMRs and the MMRs with Mars. The intricate interaction of MMRs and the Yarkovsky effect is an interesting subject for future studies.

Basically, one would like to know how the probability of capture of a drifting body into narrow MMRs depends on the magnitude and direction of the semimajor axis drift, and whether the phase of secular angles plays any role in the capture process. Ideally, when capture occurs, one would also like to statistically estimate the change of e and i of an asteroid during the time spent on a resonant orbit. Although some hints about these processes may be derived from re-

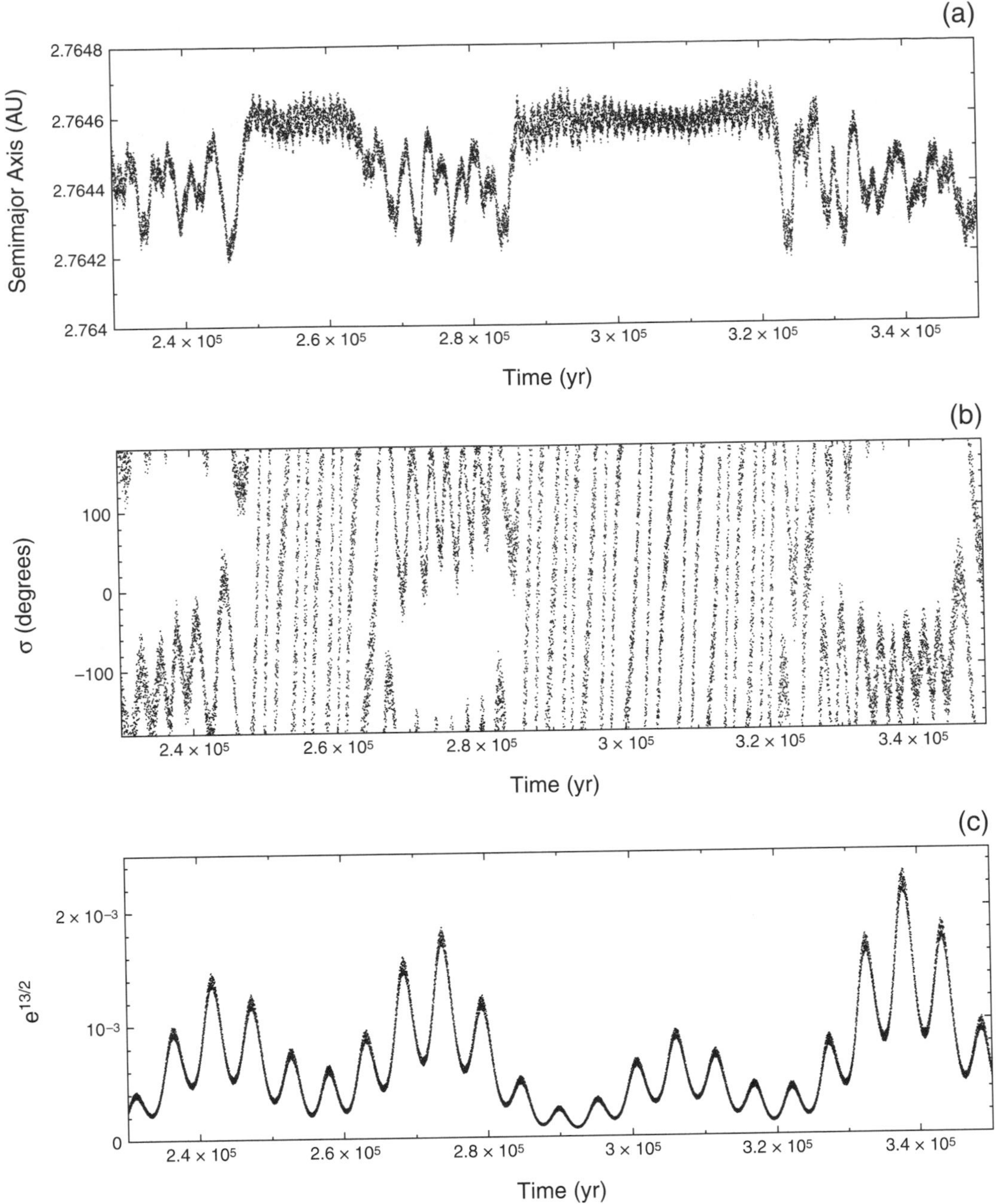

Fig. 6. The evolution of orbital elements of the asteroid 41 Daphne. The panels show **(a)** the semimajor axis, **(b)** $\sigma = 9\lambda_4 - 22\lambda + 13\varpi$, and **(c)** $e^{13/2}$. The evolution of 41 Daphne is characterized by the interaction with the 9M:22 MMR. The width of a MMR is proportional to $e^{\rho/2}$ (see equation (11) and related discussion), where here $\rho = 13$. Consequently, the resonant width seen by the asteroid is largely modulated with the parameter in **(c)**. Resonant behavior occurs only when the eccentricity of 41 Daphne is large. Adapted from *Šidlichovský* (1999).

cent numerical simulations, the general trends in the parameter space are unknown. Yet the interaction of narrow MMRs with the Yarkovsky effect is important for such issues as the origin of NEAs and the dynamical dispersion of the asteroid families. Also remaining are a few puzzles concerning the asteroid population of large MMRs in the asteroid belt, possibly related to the processes involved in the early stages of the solar system formation.

We note that a large number of asteroids exist in the 3J:2 MMR, while both the 2J:1 and 4J:3 MMRs are substantially depleted. Although some depletion of latter resonances can be explained by the chaotic processes acting in the last 4 b.y., additional depopulation likely occurred during the formation of the asteroid belt. Is the resonant structure of the outer asteroid belt a natural outcome of the accretion itself (*Petit et al.,* 1999), or possibly a sign of the primordial planetary migration? We wonder if a unique answer to this question can be given. Apparently, detailed modeling of the primordial phases is required.

Such quantitative modeling should also show whether the current depletion of the outer belt between the 2J:1 and 3J:2 MMRs (3.4–3.9 AU) requires primordial migration (*Holman and Murray,* 1996; *Liou and Malhotra,* 1997). Moreover, what effect would planetary migration have in the primordial main asteroid belt? First attempts to quantify this effect by direct simulations (*Gomes,* 1997; *Levison et al.,* 2001) show that the sweeping ν_6 secular resonance is efficient at deleting the asteroid population. Our hope is that the increasingly detailed information about the current asteroid belt provided by observations, combined with future progress in theoretical modeling, will further reveal signatures that primordial processes imprinted on the orbital structure of the belt, and that have been preserved until now. Understanding the effects of MMRs will certainly be part of that story.

REFERENCES

Beaugé C. (1994) Asymmetric librations in exterior resonances. *Cel. Mech. Dyn. Astron., 60,* 225–248.

Benettin G. and Gallavotti G. (1986) Stability of motions near resonances in quasi-integrable Hamiltonian systems. *J. Stat. Phys., 44,* 293–338.

Benettin G., Galgani L., and Strelcyn J. (1976) Kolmogorov entropy and numerical experiments. *Phys. Rev. A., 14,* 2338–2345.

Bretagnon P. and Simon J.-L. (1990) General theory of the Jupiter-Saturn pair using an iterative method. *Astron. Astrophys., 239,* 387–398.

Bottke W. F., Morbidelli A., Jedicke R., Petit J.-M., Levison H., Michel P., and Metcalfe T. S. (2002a) Debiased orbital and size distributions of the Near Earth Objects. *Icarus,* in press.

Bottke W. F. Jr., Vokrouhlický D., Rubincam D. P., and Brož M. (2002b) The effect of Yarkovsky thermal forces on the dynamical evolution of asteroids and meteoroids. In *Asteroids III* (W. F. Bottke Jr. et al., eds.), this volume. Univ. of Arizona, Tucson.

Bowell E. K., Muinonen K., and Wasserman L. H. (1994) A public-domain asteroid orbit database. In *Asteroids, Comets and Meteors* (A. Milani et al., eds.), pp. 477–481. Kluwer, Dordrecht.

Chirikov B. V. (1979) A universal instability of many-dimensional oscillator systems. *Phys. Rep., 52,* 263–379.

Dermott S. F. and Murray C. D. (1983) Nature of the Kirkwood gaps in the asteroid belt. *Nature, 301,* 201–205.

Farinella P., Froeschlé C., and Gonczi R. (1993) Meteorites from the asteroid 6 Hebe. *Cel. Mech. Dyn. Astron., 56,* 287–305.

Farinella P., Froeschle C., Froeschle C., Gonczi R., Hahn G., Morbidelli A., and Valsecchi G. B. (1994) Asteroids falling onto the Sun. *Nature, 371,* 315–317.

Fernandez J. A. and Ip W.-H. (1984) Some dynamical aspects of the accretion of Uranus and Neptune — The exchange of orbital angular momentum with planetesimals. *Icarus, 58,* 109–120.

Ferraz-Mello S. (1994a) Kirkwood gaps and resonant groups. *IAU Symp. 160,* pp. 175–188.

Ferraz-Mello S. (1994b) The dynamics of the asteroidal 2:1 resonance. *Astron. J., 108,* 2330–2337.

Ferraz-Mello S. (1997) A symplectic mapping approach to the study of the stochasticity in asteroidal resonances. *Cel. Mech. Dyn. Astron., 65,* 421–437.

Ferraz-Mello S. and Klafke J. C. (1991) A model for the study of very-high-eccentricity asteroidal motion. The 3:1 resonance. In *Predictability, Stability and Chaos in N-Body Dynamical Systems* (A. E. Roy, ed.), pp. 177–184. NATO Adv. Stud. Inst. Ser. B Phys. 272, Plenum, New York.

Ferraz-Mello S., Nesvorný D., and Michtchenko T. A. (1998a) Chaos, diffusion, escape and permanence of resonant asteroids in gaps and groups. In *Solar System Formation and Evolution* (D. Lazzaro et al., eds.), pp. 65–82. ASP Conference Series 149.

Ferraz-Mello S., Michtchenko T. A., and Roig F. (1998b) The determinat role of Jupiter's Great Inequality in the depletion of the Hecuba gap. *Astron. J., 116,* 1491–1500.

Gladman B.J., Migliorini F., Morbidelli A., Zappalà V., Michel P., Cellino A., Froeschlé C., Levison H. F., Mailey M., and Duncan M. (1997) Dynamical lifetimes of objects injected into asteroid belt resonances. *Science, 277,* 197–201.

Gomes R. S. (1997) Dynamical effects of planetary migration on the primordial asteroid belt. *Astron.. J., 114,* 396–401.

Henrard J. (1990) A semi-numerical perturbation method for separable Hamiltonian systems. *Cel. Mech. Dyn. Astron., 49,* 43–67.

Henrard J. and Lemaître A. (1983) A mechanism of formation for the Kirkwood gaps. *Icarus, 55,* 482–494.

Henrard J. and Lemaître A. (1987) A perturbative treatment of the 2/1 Jovian resonance. *Icarus, 69,* 266–279.

Henrard J., Watanabe N., and Moons M. (1995) A bridge between secondary and secular resonances inside the Hecuba gap. *Icarus, 115,* 336–346.

Holman M. J. and Murray N. W. (1996) Chaos in high-order mean resonances in the outer asteroid belt. *Astron.. J., 112,* 1278.

Hori G. (1966) Theory of general perturbation with unspecified canonical variable. *Publ. Astron. Soc. Japan, 18,* 287.

Jedicke R. and Metcalfe T.S. (1998) The orbital and absolute magnitude distributions of main belt asteroids. *Icarus, 113,* 245–260.

Klafke J. C., Ferraz-Mello S., and Michtchenko T. (1992) Very-high-eccentricity librations at some higher-order resonances. *IAU Symp. 152,* pp. 153–158.

Knežević Z. and Milani A. (2000) Synthetic proper elements for outer main belt asteroids. *Cel. Mech. Dyn. Astron., 78,* 17.

Laskar J. (1988) Secular evolution of the solar system over 10 million years. *Astron. Astrophys., 198,* 341–362.

Laskar J. (1999) Introduction to frequency map analysis. In *Hamiltonian Systems with Three or More Degrees of Freedom* (C. Simó, ed.), pp. 134–150. Kluwer, Dordrecht.

Lecar M., Franklin F. A., Holman M. J., and Murray N. W. (2001) Chaos in the solar system. In *Annu. Rev. Astron. Astrophys., 39,* 581–631.

Levison H. F. and Duncan M. J. (1994) The long-term dynamical behavior of short-period comets. *Icarus, 108,* 18–36.

Levison H. F., Dones L., Chapman C. R., Stern S. A., Duncan M. J., and Zahnle K. (2001) Could the lunar "Late Heavy Bombardment" have been triggered by the formation of Uranus and Neptune? *Icarus, 151,* 286–306.

Lichtenberg A. J. and Lieberman M. A. (1983) *Regular and Stochastic Motion: Applied Mathematical Sciences, Vol. 38.* Springer, New York.

Liou J. C. and Malhotra R. (1997) Depletion of the outer asteroid belt. *Science, 275,* 375–377.

Message P. J. (1958) Proceedings of the Celestial Mechanics Conference: The search for asymmetric periodic orbits in the restricted problem of three bodies. *Astron. J., 63,* 443.

Michtchenko T. A. and Ferraz-Mello S. (1995) Comparative study of the asteroidal motion in the 3:2 and 2:1 resonances with Jupiter — I. *Astron. Astrophys., 303,* 945–963.

Michtchenko T. A. and Ferraz-Mello S. (1997) Escape of asteroids from the Hecuba gap. *Planet. Space Sci., 45,* 1587–1593.

Migliorini F., Michel P., Morbidelli A., Nesvorný D., and Zappalà V. (1998) Origin of multikilometer Earth and Mars-crossing asteroids: A quantitative simulation. *Science, 281,* 2022–2024.

Milani A. and Knežević Z. (1994) Asteroid proper elements and the dynamical structure of the asteroid main belt. *Icarus, 107,* 219–254.

Milani A. and Nobili A. M. (1992) An example of stable chaos in the solar system. *Nature, 357,* 569–571.

Milani A., Nobili A. M., and Knežević Z. (1997) Stable chaos in the asteroid belt. *Icarus, 125,* 13–31.

Moons M. (1993) On the resonant Hamiltonian in the restricted three-body problem. *Internal Report, 19,* Facultés Universitaires de Namur, Belgique.

Moons M. and Morbidelli A. (1995) Secular resonances in mean-motion commensurabilities. The 4/1, 3/1, 5/2 and 7/3 cases. *Icarus, 114,* 33–50.

Moons M., Morbidelli A., and Migliorini F. (1998) Dynamical structure of the 2/1 commensurability with Jupiter and the origin of the resonant asteroids. *Icarus, 135,* 458–468.

Morbidelli A. (1996) The Kirkwood Gap at the 2/1 commensurability with Jupiter: New numerical results. *Astron. J., 111,* 2453.

Morbidelli A. (2002) *Modern Celestial Mechanics. Aspects of the Solar System Dynamics.* Gordon and Breach, London, in press.

Morbidelli A. and Moons M. (1993) Secular resonances in mean-motion commensurabilities. The 2/1 and 3/2 cases. *Icarus, 102,* 316–332.

Morbidelli A. and Moons M. (1995) Numerical evidences of the chaotic nature of the 3/1 mean-motion commensurability. *Icarus, 115,* 60–65.

Morbidelli A. and Nesvorný D. (1999) Numerous weak resonances drive asteroids toward terrestrial planets orbits. *Icarus, 139,* 295–308.

Morbidelli A., Zappalà V., Moons M., Cellino A., and Gonczi R. (1995) Asteroid families close to mean-motion resonances: Dynamical effects and physical implications. *Icarus, 118,* 132–154.

Morbidelli A., Bottke W. F. Jr., Froeschlé Ch., and Michel P. (2002) Origin and evolution of near-Earth objects. In *Asteroids III* (W. F. Bottke Jr. et al., eds.), this volume. Univ. of Arizona, Tucson.

Murray N. and Holman M. (1997) Diffusive chaos in the outer belt. *Astron. J., 114,* 1246–1259.

Murray N., Holman M., and Potter M. (1998) On the origin of chaos in the solar system. *Astron. J., 116,* 2583–2589.

Nesvorný D. and Ferraz-Mello S. (1997) On the asteroidal population of the first-order resonances. *Icarus, 130,* 247–258.

Nesvorný D. and Morbidelli A. (1998) Three-body mean-motion resonances and the chaotic structure of the asteroid belt. *Astron. J., 116,* 3029–3037.

Nesvorný D. and Morbidelli A. (1999) An analytic model of three-body mean motion resonances. *Cel. Mech. Dyn. Astron., 71,* 243–263.

Nesvorný D., Morbidelli A., and Vokrouhlický D., Bottke W. F., Brož M. (2002) The Flora family: A case of the dynamically dispersed collisional swarm? *Icarus, 157,* 155–172.

Petit J., Morbidelli A., and Valsecchi G. B. (1999) Large scattered planetesimals and the excitation of the small body belts. *Icarus, 141,* 367–387.

Roig F., Nesvorný D., and Ferraz-Mello S. (2002) Asteroids in the 2:1 resonance with Jupiter. Dynamics and size distribution. *Mon. Not. R. Astron. Soc.,* in press.

Saha P. (1992) Simulating the 3:1 Kirkwood gap. *Icarus, 100,* 434–439.

Šidlichovský M. (1999) On stable chaos in the asteroid belt. *Cel. Mech. Dyn. Astron., 73,* 77–86.

Šidlichovský M. and Nesvorný D. (1997) Frequency modified Fourier transform and its applications to asteroids. *Cel. Mech. Dyn. Astron., 65,* 137–148.

Wisdom J. (1980) The resonance overlap criterion and the onset of stochastic behavior in the restricted three-body problem. *Astron.. J., 85,* 1122–1133.

Wisdom J. (1982) The origin of Kirkwood gaps: A mapping for asteroidal motion near the 3/1 commensurability. *Astron. J., 85,* 1122–1133.

Wisdom J. (1983) Chaotic behaviour and the origin of the 3/1 Kirkwood gap. *Icarus, 56,* 51–74.

Wisdom J. (1985) Meteorites may follow a chaotic route to earth. *Nature, 315,* 731–733.

Wisdom J. and Holman M. (1991) Symplectic maps for the N-body problem. *Astron. J., 102,* 1528–1538.

Yoshikawa M. (1990) Motions of asteroids at the Kirkwood gaps. I — On the 3:1 resonance with Jupiter. *Icarus, 87,* 78–102.

Yoshikawa M. (1991) Motions of asteroids at the Kirkwood Gaps. II — On the 5:2, 7:3, and 2:1 resonances with Jupiter. *Icarus, 92,* 94–117.

Zappalà V., Bendjoya P., Cellino A., Di Martino M., Doressoundiram A., Manara A., and Migliorini F. (2000) Fugitives from the Eos family: First spectroscopic confirmation. *Icarus, 145,* 4–11.

The Effect of Yarkovsky Thermal Forces on the Dynamical Evolution of Asteroids and Meteoroids

William F. Bottke Jr.
Southwest Research Institute

David Vokrouhlický
Charles University, Prague

David P. Rubincam
NASA Goddard Space Flight Center

Miroslav Brož
Charles University, Prague

The Yarkovsky effect is a thermal radiation force that causes objects to undergo semimajor axis drift and spinup/spindown as a function of their spin, orbit, and material properties. This mechanism can be used to (1) deliver asteroids (and meteoroids) with diameter D < 20 km from their parent bodies in the main belt to chaotic resonance zones capable of transporting this material to Earth-crossing orbits, (2) disperse asteroid families, with drifting bodies jumping or becoming trapped in mean-motion and secular resonances within the main belt, and (3) modify the rotation rates of asteroids a few kilometers in diameter or smaller enough to possibly explain the excessive number of very fast and very slow rotators among the small asteroids. Accordingly, we suggest that nongravitational forces, which produce small but meaningful effects on asteroid orbits and rotation rates over long timescales, should now be considered as important as collisions and gravitational perturbations to our overall understanding of asteroid evolution.

1. CLASSICAL MODEL OF ASTEROID EVOLUTION

Over the last several decades, it has been assumed that collisions and gravitational forces are the primary mechanisms governing the evolution of asteroids and meteoroids. Using these processes, it is possible to construct an approximate history of how the main-belt and inner solar system asteroid populations have changed over the last several billion years. The main tenets of this model, which we broadly refer to as the "classical" asteroid evolution model, are summarized below.

Asteroids, whose orbits intersect in the main belt, occasionally collide with one another at high velocities (~5 km s^{-1}; *Bottke et al.,* 1994). These events result in cratering and fragmentation, with the collisional physics determining the orbits, spin states, shapes, and internal structures of the surviving bodies. The largest impact events are believed to produce the observed asteroid families. The orbital positions of family members suggest that some ejecta can be launched approximately several 100 m s^{-1} (*Zappalà et al.,* 1996). If true, it is plausible that fragments from asteroid collisions, thrown with just the right trajectory and velocity, can be directly injected into powerful or diffusive resonance zones produced by the gravitational perturbations of the planets (*Farinella et al.,* 1993). Numerical studies have shown that test objects in such resonance regions frequently have their eccentricities pumped up to planet-crossing orbits (e.g., *Wisdom,* 1983). Once on planet-crossing orbits, asteroids have their dynamical evolution dominated by resonances and gravitational close encounters with the planets. Some of these asteroids go on to strike the planets, although most impact the Sun or are ejected from the inner solar system via a close encounter with Jupiter (*Gladman et al.,* 1997). If the object is small, it may also be removed via a catastrophic collision. It is believed that most meteorites and near-Earth asteroids are delivered to the inner solar system (and Earth) by this long chain of events.

Up to now, the classical model (CM) has been useful in helping us interpret asteroid data and broadly understand the evolution of asteroid populations. Nevertheless, some predictions are inconsistent with observations. For example:

CM Prediction 1: Since fresh ejecta is directly injected into chaotic resonances, and the dynamical lifetime of bodies placed in powerful resonances are generally a few million years or less (*Gladman et al.,* 1997), we should expect to see an abundance of meteorites with short cosmic-ray-exposure (CRE) ages (i.e., only a few million years) and a paucity of long-lived meteorites.

Observation 1: Relatively few meteorites have CRE ages less than ~10 m.y. Most stony meteorites have CRE ages between ~10 and 100 m.y., while iron meteorites have CRE ages between ~0.1 and 1.0 G.y. (*Caffee et al.,* 1988; *Marti and Graf,* 1992). In general, CRE ages are comparable to, or longer than, the average dynamical lifetime of Earth-crossing asteroids (~10 m.y.; *Gladman et al.,* 1997; *Migliorini et al.,* 1997; *Morbidelli and Gladman,* 1998).

CM Prediction 2: There are roughly 5000–6000 kilometer-sized asteroids crossing the orbits of the terrestrial planets (*Bottke et al.,* 2001a). These bodies have a wide range of taxonomic types (e.g., *Binzel et al.,* 2001). To keep this population in steady state, disruption events among large, spectrally diverse asteroids must be frequent, particularly since these are the only events capable of injecting kilometer-sized fragments into suitable resonant "escape hatches." Since most of these asteroids come from the inner and central main belt (*Bottke et al.,* 2001a), we should expect these regions to contain numerous asteroid families. Moreover, since the planet-crossing asteroids are "fresh ejecta," they should have a relatively steep size-frequency distribution.

Observation 2: Few asteroid families can be found in the inner and central main belt, while most potential parent asteroids for the kilometer-sized inner solar system asteroids reside in dynamically stable regions far from resonant "escape hatches." Modeling results including these constraints suggest that the direct injection of asteroid fragments into resonances is too inefficient to keep the inner solar system asteroid population in steady state (*Zappalà and Cellino,* 2002). In addition, the size-frequency distribution of kilometer-sized near-Earth objects (NEOs) is fairly shallow (*Bottke et al.,* 2000a).

CM Prediction 3: Studies of asteroid families suggest that many large fragments are ejected from the impact site at high velocities (approximately several 100 m s^{-1}), with the smallest fragments traveling the furthest from the cluster center (*Cellino et al.,* 1999).

Observation 3: The peak velocities of size-velocity distributions derived from numerical hydrocode results are generally much lower than those inferred from the orbital positions of asteroid family members (*Pisani et al.,* 1999). Though it is possible that hydrocodes are inaccurate, they have successfully reproduced results ranging from laboratory impact experiments, where centimeter-sized projectiles are shot into targets, to underground nuclear explosions (e.g., *Benz and Asphaug,* 1999).

CM Prediction 4: Asteroid collisions should produce a wide range of asteroid spin rates. To zeroth order, we would expect the spin rates for large and small asteroids to follow a Maxwellian frequency distribution (e.g., *Binzel et al.,* 1989; *Davis et al.,* 1989).

Observation 4: The distribution of spin rates among observed small asteroids (D < 10 km) contains an excess number of fast rotators and very slow rotators when this data is fit to Maxwellian distribution (*Pravec and Harris,* 2000; *Pravec et al.,* 2002).

We believe there is a connection between these mismatches, and that an important physical mechanism is missing from the classical model, namely how nongravitational forces affect the evolution of asteroids. It is already well known that the dynamical evolution of dust particles can be explained using Poynting-Robertson drag, a radiation effect that causes small objects to spiral inward as they absorb energy and momentum streaming radially outward from the Sun and then reradiate this energy isotropically in their own reference frame (e.g., *Burns et al.,* 1979; *Dermott et al.,* 2002). It is not as well known, however, that a different nongravitational force called the Yarkovsky effect can compel objects between 0.1 m and 20 km to spiral inward or outward at different rates as a function of their spin, orbit, and material properties, or that a variant of this force can also modify the spin rates of asteroids. As we will show in this chapter, this previously known but mostly ignored effect, which can be essentially described as a radiation recoil produced by asymmetrically reradiated thermal energy, has the potential to resolve many of the problems described above. Accordingly, we believe the classical model should now be revised to include nongravitational forces as a third important mechanism, in addition to gravity and collisions, affecting asteroid evolution.

2. INTRODUCTION TO THE YARKOVSKY EFFECT

Ivan Osipovich Yarkovsky (1844–1902), a civil engineer who worked on scientific problems in his spare time, first proposed the effect that now bears his name (*Neiman et al.,* 1965). Writing in a pamphlet around the year 1900, Yarkovsky noted that the diurnal heating of a rotating object in space would cause it to experience a force that, while tiny, could lead to large secular effects in the orbits of small bodies, especially meteoroids and small asteroids (*Öpik,* 1951). Yarkovsky's effect is a radiation force, and is the photonic equivalent of *Whipple*'s (1950) rocket effect.

Yarkovsky's remarkable insight would have been consigned to oblivion had it not been for the brilliant Estonian astronomer Ernst J. Öpik (1893–1985), who read Yarkovsky's pamphlet sometime around 1909. Decades later Öpik, recalling the pamphlet from memory, discussed the possible importance of the Yarkovsky effect for moving meteoroids about the solar system (*Öpik,* 1951). [Curiously, *Öpik*'s (1976) book, which continues the theme of his 1951 paper, makes no mention of Yarkovsky.] Following Öpik and before its current flowering, research on the Yarkovsky-type effect was pursued in Russia by *Radzievskii* (1952, 1954) and *Katasev and Kulikova* (1980); in the United States by *Paddack* (1969, 1973), *Paddack and Rhee* (1975), *Peterson* (1976), *O'Keefe* (1976), *Slabinski* (1977), *Dohnanyi* (1978), and *Burns et al.* (1979); and in Australia by *Olsson-Steel* (1986, 1987). Additional history can be found in *Hartmann et al.* (1999).

2.1. Description of Diurnal Component

The basic idea behind Yarkovsky's diurnal effect is shown in Fig. 1a, which shows a spherical meteoroid in a circular orbit about the Sun. For simplicity, the meteoroid's spin axis is taken to be normal to the orbital plane, so that the Sun always stands on its equator. Insolation heats up the sunward side, with the heat ultimately reradiated into space by the meteoroid (typically in the infrared part of the spectrum,

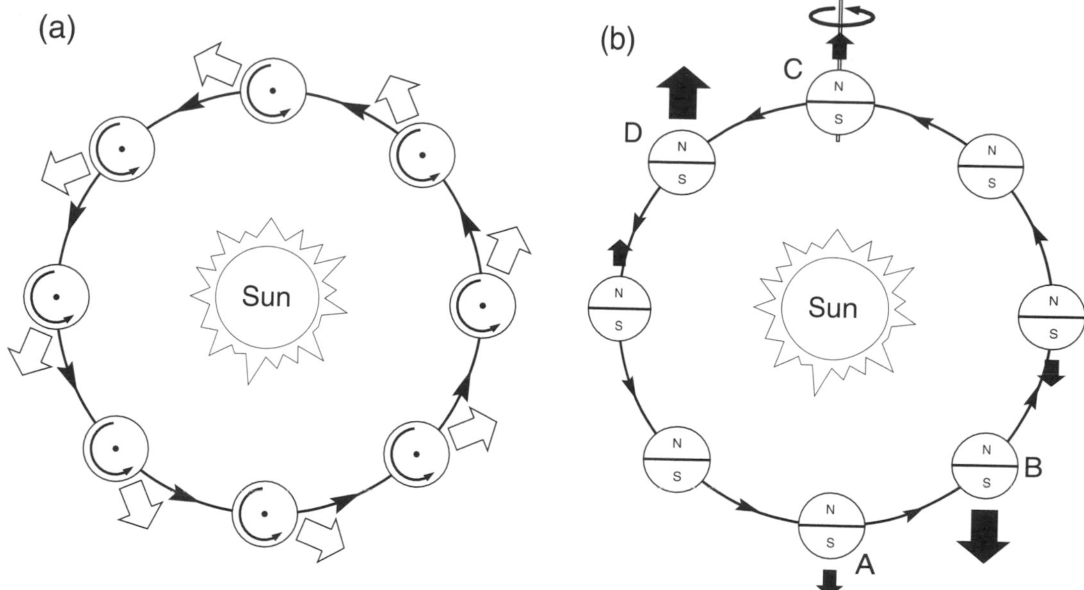

Fig. 1. **(a)** The diurnal Yarkovsky effect, with the asteroid's spin axis perpendicular to the orbital plane. A fraction of the solar insolation is absorbed only to later be radiated away, yielding a net thermal force in the direction of the wide arrows. Since thermal reradiation in this example is concentrated at about 2:00 p.m. on the spinning asteroid, the radiation recoil force is always oriented at about 2:00 a.m. Thus, the along-track component causes the object to spiral outward. Retrograde rotation would cause the orbit to spiral inward. **(b)** The seasonal Yarkovsky effect, with the asteroid's spin axis in the orbital plane. Seasonal heating and cooling of the "northern" and "southern" hemispheres give rise to a thermal force that lies along the spin axis. The strength of the reradiation force varies along the orbit as a result of thermal inertia; even though the maximum sunlight on each hemisphere occurs as A and C, the maximum resultant radiative forces are applied to the body at B and D. The net effect over one revolution always causes the object to spiral inward.

unless the meteoroid is very close to the Sun). An infrared photon carries away momentum when it leaves the meteoroid according to the relation p = E/c, where p is the photon's momentum, E its energy, and c is the speed of light. Because more energy and therefore more momentum departs from the hotter part of the meteoroid than the colder, the meteoroid feels a net kick in the direction away from the hotter part.

If the meteoroid had no thermal inertia, then the temperature distribution would be symmetrical about the subsolar point and the meteoroid would experience a net force radially outward from the Sun. The only consequence of this force would be to weaken the Sun's grip on the meteoroid. However, all bodies have thermal inertia, which causes a delay, so that the hottest part of the meteoroid is its afternoon side rather than the subsolar point. This is similar to the Earth, where the afternoon is typically the warmest time of day. As a result, the force on the meteoroid has not only a component that is radially outward from the Sun, but also has an along-track component.

This along-track component causes a secular increase in the semimajor axis (and, to a lesser degree, eccentricity) for the prograde sense of rotation shown in the figure, so that over time the tiny Yarkovsky force can profoundly change the orbit. The sign of the diurnal Yarkovsky effect depends on the sense of rotation. If the meteoroid shown in Fig. 1a rotated in the retrograde sense, the orbit would shrink instead of expand, while if the rotation axis was in the orbital plane, the diurnal Yarkovsky would be shut off entirely. The magnitude of the diurnal effect also depends on how close a body is to the Sun, the tilt of the body's spin axis with respect to the orbital plane, and the body's physical characteristics (i.e., the size of the body, its shape and thermal properties, and how fast it is rotating). The interplay of these factors means that there is an optimal size for maximizing the diurnal Yarkovsky effect for a given rotation speed and thermal structure. A very large object would have a poor area-to-mass ratio (e.g., the effect is negligible on a large body like the Earth). On the other hand, the smaller the body, the better the area-to-mass ratio, but at some point the radius becomes so small that the thermal wave penetrates all the way across the body, lessening the temperature differences between the night and day sides and weakening the effect (e.g., a slowly rotating dust particle). For rotation periods believed to be typical in the solar system [P ~ 5 h × (D/1 km), where D is the diameter of the body], optimal sizes for the Yarkovsky effect range from centimeters to meters. Objects having zero or infinitely fast rotation rates experience no diurnal Yarkovsky force.

2.2. Description of Seasonal Component

Nearly a century after Yarkovsky wrote his pamphlet, a second Yarkovsky effect emerged. While searching for the cause of the secular decay of the orbit of the LAGEOS satellite, it was realized that there had to be a seasonal effect (*Rubincam,* 1987, 1988, 1990) in addition to Yarkovsky's original diurnal effect. The seasonal effect applies not just to Earth satellites like LAGEOS, but also to objects orbiting the Sun.

The seasonal Yarkovsky effect is illustrated in Fig. 1b. As in Fig. 1a, a spherical meteoroid is assumed to be in a circular orbit about the Sun; but in this case the spin axis lies in the orbital plane. It is the component of force lying along the spin axis that gives rise to the seasonal effect. When the meteoroid is at A (bottom of the figure) the Sun shines most strongly on its northern hemisphere. As with the diurnal effect, there is a delay due to thermal inertia, so that the northern hemisphere is hottest at B. Likewise, the Sun shines most strongly on the southern hemisphere at C but this hemisphere becomes hottest at D. When the along-track force is averaged around the orbit, it turns out to be nonzero. For a body without thermal inertia, however, the along-track force averages to zero when integrated over one revolution about the Sun.

For small orbital eccentricities, the average along-track force is always in opposition to the motion of the meteoroid. Hence in the small eccentricity regime the seasonal force always acts like drag and causes orbital decay; for this reason the seasonal Yarkovsky effect was originally dubbed "thermal drag" (*Rubincam,* 1987). Unlike the diurnal Yarkovsky effect, the seasonal Yarkovsky effect is independent of the sense of rotation of the meteoroid; reversing its spin does not change the effect's sign. Moreover, the relevant timescale for the seasonal effect is the meteoroid's orbital period rather than the usually much quicker rotational period involved in the diurnal effect.

The seasonal effect does depend on the body's proximity to the Sun and on the tilt of the spin axis with respect to the orbit; it vanishes when the spin axis is normal to the orbital plane. Like in the diurnal case, there is an optimum size for maximizing the effect. For basaltic bodies on circular orbits in the inner main belt, 12-m-diameter objects would experience the greatest effects (*Rubincam,* 1998; *Farinella et al.,* 1998). The seasonal Yarkovsky force also affects the other orbital elements in addition to the semimajor axis. For small eccentricities it tends to circularize the orbit, like atmospheric drag does (*Rubincam,* 1995, 1998; *Vokrouhlický and Farinella,* 1999).

3. THEORY OF THE YARKOVSKY EFFECT

The Yarkovsky force computation naturally splits into two parts: (1) determination of the surface temperature distribution, and (2) evaluation of the thermal radiation recoil force (or torque if desired). Mathematically similar derivations of this solution can be found in several modern references (*Rubincam,* 1995, 1998; *Vokrouhlický,* 1998a,b, 1999; *Vokrouhlický and Farinella,* 1999; *Bottke et al.,* 2000a). In this chapter, we follow the formalism of *Vokrouhlický* (2001).

Problem (1) above has already been examined within the context of asteroid radiometry, but the Yarkovsky application requires some special care. For example, thermal inertia of the surface material — often omitted in radiometry — must now be included. On the other hand, the complexity of the heat diffusion problem can be reduced (within reasonable errors) by adopting linearization (e.g., small temperature differences are referred to a suitably chosen mean value). For simple asteroid shapes, this procedure allows us to compute the Yarkovsky force using analytical expressions. Since most applications of the Yarkovsky effect require rapid computations, it is advantageous to sacrifice some precision for speed. More exact solutions, particularly for irregularly shaped bodies and/or inhomogeneous thermal parameters, require more sophisticated (and computationally expensive) treatments (*Vokrouhlický and Farinella,* 1998, 1999; *Spitale and Greenberg,* 2001, 2002).

To compute the surface temperature on a body, we use the heat diffusion equations for energy flows inside the body

$$\nabla \cdot (K\nabla T) = \rho C \frac{\partial T}{\partial t} \quad (1)$$

or across its surface

$$(K\nabla T \cdot \mathbf{n}_\perp) + \varepsilon\sigma T^4 = \alpha\mathcal{E} \quad (2)$$

the latter of which appears as a boundary condition for the temperature (T) determination. Here, K is the thermal conductivity, C is the specific heat at constant pressure, ρ is the material density, ε is the surface thermal emissivity, σ is the Stefan-Botzmann constant, and $\alpha = 1 - A$, with A being the Bond albedo. Equation (2) refers to a surface element with an external normal vector $\mathbf{n}_\perp$, while $\mathcal{E}$ is the flux of solar radiation through this element. Once the insolation function $\mathcal{E}$ for the surface elements is specified (which requires knowledge of the body's shape and its rotation state) and material parameters (K, C, ρ) are known, equations (1) and (2) can be solved numerically. Unfortunately, the complexity of these equations means that orbit perturbation calculations can be computationally expensive. To overcome this, we can make some simplifying assumptions. For example, the nonlinearity of thermal emission on the surface can be dealt with by assuming that the temperature throughout the body is close to an average value T_0 (i.e., $T = T_0 + \Delta T$ with $\delta = (\Delta T/T_0) \ll 1$. If T_0 is constant, equations (1) and (2) may be rewritten for the δ variable, while the fourth-order term in the boundary condition (equation (2)) may be simplified as $T^4 \approx T_0^4\,(1 + 4\delta + \ldots)$.

At this point, we find it useful to scale size and time so that minimum parameters are retained in the mathematical formulation of the problem. For example, dimensional analysis shows that, for a given Fourier term with frequency ν in the decomposition of the insolation function $\mathcal{E}$, the problem involves two fundamental parameters: (1) the penetration depth of the thermal wave $\ell_\nu = \sqrt{K/\rho C\nu}$, and (2) the thermal parameter $\Theta_\nu = \sqrt{K\rho C\nu}/(\varepsilon\sigma T_\star^3)$ (here $T_\star$ is the subsolar temperature defined by $\varepsilon\sigma T_\star^4 = \alpha\mathcal{E}_\star$ with $\mathcal{E}_\star$ being the solar radiation flux at the distance of the body). The thermal parameter Θ_ν is a measure of the relaxation between the absorption and reemission at frequency ν. Thus, as Θ_ν decreases, the difference between the absorption and reemission decreases as well.

Assuming a spherical body rotating about an arbitrary axis, the spectrum of the insolation function consists primarily of the "diurnal line" with rotation frequency ω (and its multiples) and the "seasonal line" with the mean-motion frequency n. Note that orbital eccentricity adds higher multiples of the mean-motion frequency, increasing algebraic complexity, while also weakening the assumption of the linearized approach (temperature changes cannot be represented as small variations around a constant average value). Fortunately, most applications of the Yarkovsky effect involve main belt bodies on low-eccentricity orbits. Assuming a small eccentricity, the solution of the amplitudes of the Fourier representation of δ as a function of the spatial coordinates can be worked out analytically.

Having solved the temperature T, or the linearized quantity δ, we can proceed to compute the recoil force (or torque) due to the thermal radiation (i.e., the Yarkovsky force). Assuming isotropic (Lambert) emission, the corresponding force per unit of mass is given by

$$\mathbf{f} = -\frac{2}{3}\frac{\varepsilon\sigma}{mc}\int_S dS(u,v)T^4\mathbf{n}_\perp \tag{3}$$

where the integration is to be performed over the whole surface parameterized by a system of coordinates u and v (such as the longitude and latitude in the spherical case), m is the mass of the body, and c is the light velocity. The integral in equation (3) may be evaluated numerically, or we may again refer to linearization of the fourth power of the temperature as mentioned above and perform the integration analytically.

Adopting a local coordinate system with the z-axis aligned with the body's spin axis and the xy-axes in its equatorial plane, the linearized solutions suggest a useful classification of two variants of the Yarkovsky force: (1) The out-of-spin components of the Yarkovsky acceleration (f_x, f_y) depend primarily on the rotation frequency ω (with typically unimportant splitting ω ± n due to the orbital motion; *Vokrouhlický,* 1999), while (2) the spin-aligned component of the Yarkovsky acceleration f_z depends only on the mean motion n. The former Yarkovsky-acceleration components are thus called "diurnal," while the latter is called "seasonal" (and they correspond to the qualitative concepts discussed in section 2). It should be noted that splitting the Yarkovsky effect into these two variants is an artifact of the linearized solution. In the more complete formulation, the effects are coupled.

Yarkovsky accelerations primarily change orbital semimajor axis a. Since the perturbation is usually small, we average the variation in a over one revolution. Assuming a spherical body with radius R, and neglecting eccentricity e, the averaged diurnal and seasonal perturbations on da/dt are

$$\left(\frac{da}{dt}\right)_{\text{diurnal}} = -\frac{8\alpha}{9}\frac{\Phi}{n}F_\omega(R',\Theta)\cos\gamma + \mathcal{O}(e) \tag{4}$$

$$\left(\frac{da}{dt}\right)_{\text{seasonal}} = \frac{4\alpha}{9}\frac{\Phi}{n}F_n(R',\Theta)\sin^2\gamma + \mathcal{O}(e) \tag{5}$$

The total da/dt rate is the superposition of the two variants. The albedo-factor α in equation (4) and equation (5) is close to that in equation (2) (*Vokrouhlický and Bottke,* 2001), $\Phi = \pi R^2\mathcal{E}_0/(mc)$ is the usual radiation pressure coefficient, and γ is obliquity of the spin axis. The function $F_\nu(R',\Theta)$ depends on the radius of the body R, scaled by the penetration depth ℓ_ν of the thermal wave ($R' = R/\ell_\nu$) and the thermal parameter Θ_ν, both corresponding to the frequency ν. For the diurnal effect, ν = ω, while for the seasonal effect, ν = n. Note that apart from the different frequency, F is the same in equations (4) and (5). The explicit form of F function may be found in the literature (e.g., *Vokrouhlický*, 1998a, 1999). Here we restrict ourselves to mentioning its dependence on the thermal parameter

$$F_\nu(R',\Theta) = -\frac{\kappa_1(R')\Theta_\nu}{1 + 2\kappa_2(R')\Theta_\nu + \kappa_3(R')\Theta_\nu^2} \tag{6}$$

with κ_1, κ_2, and κ_3 analytic functions of R'. The frequency-index of F reminds us that both the scaling factor ℓ_ν of R and the thermal parameter Θ_ν depend on a given frequency. This parameter is the principle difference between the diurnal and seasonal Yarkovsky effects.

The da/dt rates listed above give us a basic understanding of how the Yarkovsky perturbations depend on a number of parameters:

1. *Obliquity and rotation dependence:* Since the F functions are always negative (i.e., thermal reemission lags behind the absorption) the seasonal Yarkovsky effect always produce a net decrease in a. The seasonal effect is maximum at 90° obliquity and nil at 0° (or 180°) obliquity. On the other hand, the diurnal effect may lead to both a net increase in a (for $\gamma < 90°$) or a net decrease in a (for $\gamma > 90°$). The effect is maximum at 0° (or 180°) obliquity and nil for 90° obliquity. The diurnal Yarkovsky effect becomes negligible in the limit of infinitely fast rotation, since surface temperature variations are smeared along lines of constant latitude, and zero rotation.

2. *Size dependence:* The Yarkovsky effect vanishes for both very small and very large objects. For large objects, (da/dt) ≈ Θ/R', where the ≈1/R dependence arises from the body's cross-section vs. its mass. For small objects, (da/dt) ≈ R'^2/Θ. The maximum drift in a occurs when R' ≈ 1 (i.e., when the body's size is comparable to the penetration depth of the corresponding thermal wave).

3. *Surface-conductivity dependence:* Surface conductivity K is the major thermal material parameter that influences the strength of the Yarkovsky effect. It ranges from very low values for highly porous or regolith-like surfaces

($\approx$0.001 W m^{-1} K^{-1}), to moderate values for bare rocks such as ordinary chondrites or icy objects ($\approx$1 W m^{-1} K^{-1}), up to high values for iron-rich objects like iron meteorites ($\approx$40 W m^{-1} K^{-1}). Variations of K modify ℓ_v and Θ_v. At low conductivities, we expect that Θ will be small and R' large, since the penetration depth of the corresponding thermal wave decays to zero. Thus, (da/dt) $\approx \Theta$ and the Yarkovsky effect disappears. For high conductivities, the thermal parameter diverges and the scaled radius of the body tends to zero, since the penetration depth of the corresponding thermal wave diverges. Thus, (da/dt) $\approx R'^2/\Theta$, yielding very fast decay of the Yarkovsky effect as the body is driven toward thermal equilibrium. Maximum da/dt rates occur when both R' $\approx$ 1 and $\Theta \approx 1$.

4. *Solar-distance dependence:* The Yarkovsky effect decreases with increasing distance to the Sun. In case of the diurnal effect, objects are usually in the high-Θ and high-R' regime, so that (da/dt) $\approx \Phi/(n\Theta)$. From the functional dependence of Φ, n, and Θ on a, we derive (da/dt) $\approx$ a^{-2} (e.g., *Radzievskii,* 1952; *Peterson,* 1976). Thus, the diurnal effect dwindles very fast with increasing distance from the Sun, with very slowly rotating bodies a possible exception. A comparable analysis for the seasonal effect is more involved since F_n cannot be approximated as $\approx 1/\Theta$. An example of this would be 0.1–1-km icy bodies in the Kuiper Belt, whose seasonal da/dt drift rates become much shallower as a function of distance from the Sun. This surprising result occurs because the penetration depth of the seasonal thermal wave ℓ_n increases to ~0.1 km.

4. SEMIMAJOR AXIS MOBILITY OF ASTEROID FRAGMENTS

Using the above equations, *Farinella and Vokrouhlický* (1999) computed the average semimajor axis displacement (Δa) of main-belt meteoroids and asteroids caused by the Yarkovsky effect before undergoing a catastrophic disruption (Fig. 2). The collision lifetime of the objects, τ_{disr}, was assumed to be $\tau_{disr} \simeq 16.8 \sqrt{R}$ m.y., with R being the body's radius in meters. The objects were started with random obliquity orientations (γ), but were also assumed to go through spin axis reorientation events via nondisruptive impacts. The characteristic timescale of these events was assumed to be $\tau_{rot} \simeq 15.0 \sqrt{R}$ m.y. (*Farinella et al.,* 1998). Rotation rates were assumed to be correlated with size through the relation P = 5 R where P is the rotation period in seconds and R the radius in meters. Since surface conductivity K for asteroids is unknown, several different values of K were selected.

We point out several interesting results from Fig. 2. (1) Except for the high-strength iron objects, the maximum expected drift distance from these mean values was on the order of 0.1 AU. (2) Δa becomes smaller for large bodies (down to 0.01 AU at R $\simeq$ 5–10 km). The dependence on surface conductivity, however, becomes much less important. (3) High-conductivity objects (curve 4 in Fig. 2) have maximum mobility for R $\approx$ 10 m, primarily because of the seasonal Yarkovsky effect (*Rubincam,* 1998; *Farinella et al.,* 1998). (4) Characteristic Δa values of $\approx$0.1 AU for smaller asteroids and $\approx$0.01 AU for kilometer-sized asteroids have important dynamical consequences. For instance, 0.1–0.2 AU is a typical distance that a main-belt meteoroid might have to travel to reach a powerful main belt resonance. Similarly, 0.01–0.02 AU is a typical semimajor axis span of asteroid families, whose observed components are dominated by multikilometer bodies. More details about these applications will be given below.

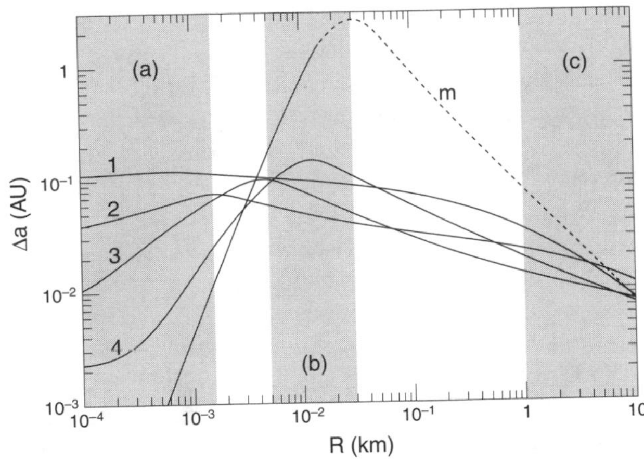

Fig. 2. Mean change of the semimajor axis Δa (in AU) of inner-main-belt asteroids over their collisional lifetimes (see text) vs. their radius R (in kilometers). Both components (diurnal and seasonal) of the Yarkovsky effect are included. Five different values of the surface conductivity K are considered: (1) K = 0.002 W m^{-1} K^{-1}; (2) K = 0.02 W m^{-1} K^{-1}; (3) K = 0.2 W m^{-1} K^{-1}; (4) K = 2 W m^{-1} K^{-1}; and (5) K = 40 W m^{-1} K^{-1} (curve m, for metal-rich bodies). The low-K cases are dominated by the diurnal effect, while for high-K cases the seasonal effect is more important. The dashed strips correspond to three astronomically important classes of bodies: **(a)** preatmospheric meteorite parent bodies (R = 0.1–1.5 m); **(b)** Tunguska-like small NEAs (R = 5–30 m); and **(c)** the largest existing NEAs or the smallest observed family members (R = 1–10 km). Note that Δa depends sensitively on the selected value of K in the **(a)** and **(b)** size ranges, but much less so in range **(c)**.

5. APPLICATIONS OF THE YARKOVSKY EFFECT

5.1. Delivery of Meteoroids from the Main Belt to Earth

The original motivation behind the Yarkovsky effect was related to the transport of small bodies from the main belt to Earth (*Öpik,* 1951; *Radzievskii,* 1952; *Peterson,* 1976). At the time of these papers, it was unclear whether collisional and dynamical processes were efficient enough to explain the overall flux of meteorites reaching Earth, let alone the CRE ages of stony meteorites (e.g., *Wetherill,* 1974). For this reason, these researchers hypothesized that the Yarkovsky effect might deliver meteoroids from the main belt to Earth via a slow decay of their semimajor axes.

The timescales involved with this scenario, however, were too long to be considered practical, particularly when reasonable meteoroid rotation rates were used.

The apparent solution to this meteoroid delivery problem was found in the pioneering works of Williams (see *Wetherill,* 1979) and *Wisdom* (1983), who showed that powerful mean-motion and secular resonances in the inner main belt could potentially move main-belt bodies onto Earth-crossing orbits within relatively short timescales (~1 m.y.). Thus, a plausible scenario for explaining the CRE ages of stony meteorites became the following: (1) collisions in the main belt inject fragments into resonances, (2) the fragments evolve onto Earth-crossing orbits via resonant motion, (3) close encounters remove the objects from resonance, and (4) the objects wander the inner solar system for 10–100 m.y. before striking a terrestrial planet, being ejected from the inner solar system by Jupiter, or experiencing a collisional disruption event. Since Monte Carlo code results verified the main components of this model (e.g., *Wetherill,* 1985), the Yarkovsky effect came to be viewed as an unneeded complication and was summarily dropped from consideration by most dynamicists.

Problems with this scenario began to present themselves in the 1990s as fast workstations and efficient numerical integrations codes began to overtake Monte Carlo codes as the dominant means of tracking the evolution of small bodies in the solar system. The major blows came from *Farinella et al.* (1994), who showed that many resonant objects strike the Sun, and *Gladman et al.* (1997), who showed that bodies escaping the main belt via the 3:1 mean-motion resonance with Jupiter or the ν_6 secular resonance only had a mean dynamical lifetime of ~2 m.y. As described in section 1, these lifetimes are largely discordant with the CRE ages of stony and iron meteorites (e.g., *Morbidelli and Gladman,* 1998).

Farinella et al. (1998), however, recognized that the explanation to the CRE problem might be the Yarkovsky effect, since it could slowly deliver material to powerful resonances inside the main belt. [Note that this scenario had been previously suggested by both *Peterson* (1976) and *Afonso et al.* (1995). Unfortunately, the implications of their work were overlooked, primarily because (1) the CRE age of stony meteorites were consistent with dynamical lifetimes derived from Monte Carlo codes (i.e., *Wetherill,* 1985) and (2) results from more accurate numerical integration codes were not yet in hand (e.g., *Dones et al.,* 1999).] As these bodies drifted toward an escape hatch (typically 0.05–0.15 AU), they would be hit by cosmic rays, which would push their CRE ages into the appropriate range. In addition, because iron meteorites have very different thermal conductivities than stones, their da/dt rates are slow enough to explain their long CRE ages (0.1–1.0 G.y.). Thus, the Yarkovsky effect provides a natural explanation for the paucity of short CRE ages among stony meteorites and the differences in the observed CRE ages of stony and iron meteorites.

The dynamical evolution of main-belt meteoroids can be surprisingly complex. As described in the previous section, the drift rate for meter-sized stones in the main belt is ±(0.01–0.001) AU m.y.$^{-1}$, depending on their spin axis orientation, spin rate, and thermal properties. Numerical integration work by *Bottke et al.* (2000a) and *Brož et al.* (2002) have shown that these da/dt drift rates are fast enough to allow meteoroids to "jump over" most weak resonances, effectively accelerating their drift rate. Most meteoroids will spiral inward or outward until they become trapped in a powerful resonance too chaotic to jump (e.g., the 3:1 or ν_6 resonance). En route, some may become temporarily trapped in weak mean-motion or secular resonances, allowing their e and i values to undergo secular changes while a remains fixed. If a meteoroid's e oscillations reach a high enough amplitude, it may escape the main belt via a close encounter with Mars. Additional complications come from nondisruptive collisions, since they can modify the meteoroid's spin axis orientation and spin rate. Thus, objects drifting via the Yarkovsky effect may well reverse course and speed several times before reaching a powerful resonance.

Although tracking the dynamical evolution of individual meteoroids via the Yarkovsky effect requires careful work, the evolution of large "swarms" of fragments, released by catastrophic break-up events or impacts on large asteroids in the main belt, can be modeled statistically. To this end, the most successful effort so far to combine dynamics, collisions, and the Yarkovsky effect into a meteoroid evolution code has been the work of *Vokrouhlický and Farinella* (2000). In their model, they started with a size distribution of small bodies ejected from a chosen parent asteroid, with each body having its own spin rate and spin axis orientation. Using simplified dynamics, they tracked these bodies across the inner main belt to the 3:1 or ν_6 resonance, assuming that their da/dt drift rates were not influenced by smaller resonances. Collisions were also included, with random impact events producing cascades of new fragments from the disruption of the existing bodies. When the objects reached the 3:1 or ν_6 resonance, Yarkovsky evolution was shut off and the bodies were delivered to Earth via statistical results taken from the numerical simulations of *Morbidelli and Gladman* (1998).

The combination of the two studied phenomena — Yarkovsky drift and collisional dynamics — was found to efficiently supply the 3:1 and ν_6 resonances with small asteroid fragments from nearly all locations in the inner and central main belt. Direct injections, considered in "pre-Yarkovsky" studies (e.g., *Farinella et al.,* 1993), only seem important when a source is close to a resonance. Moreover, the flux of objects to the resonances is, contrary to the direct-injection scenario, spread over hundreds of millions of years, as the collisional cascade creates fast-drifting fragments from larger, slower-drifting progenitors. For example, *Vokrouhlický and Farinella*'s (2000) results indicate that 50–80% of the mass of the initial population of bodies released in the Flora-region are transported to resonances (dominantly the ν_6 resonance) over 0.5 to 1 G.y.

Another important result from this model is that the distribution of accumulated CRE ages in the population of fragments reaching Earth is in reasonable agreement with observations (e.g., *Marti and Graf,* 1992; *Welten et al.,* 1997). In general, the CRE age histograms are found to depend on the age of the last event capable of dominating the local Earth swarm. Relatively old events are likely to generate the background CRE age profiles (like in the case of L chondrites) peaked at 20–50 m.y. for stones and 200–500 m.y. for irons, while comparatively recent and large events may create discrete peaks in the CRE age distributions (such as the 7–8 m.y. prominent peak for the H chondrites). In the latter case, the bulk of the original fragment population may still reside in the main belt and will supply a significant flux of meteorites in the future. Figure 3 shows comparison of the simulated and observed CRE ages for different types of meteorites and different parent asteroids.

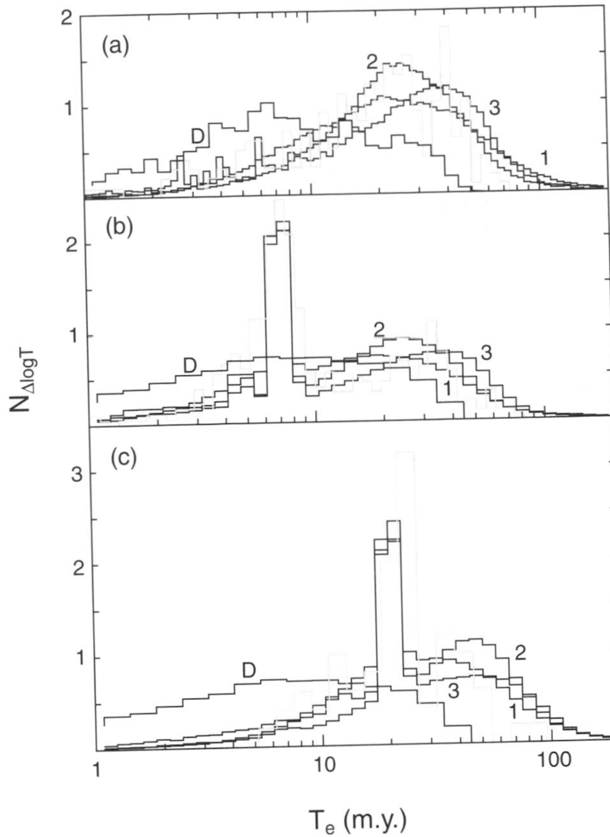

Fig. 3. Comparison of the modeled and observed CRE-age distributions for three different meteorite types (data — gray histograms). We show results of the direct-injection scenario with no Yarkovsky mobility (D histogram) and the model including Yarkovsky mobility of the meteoroids and their precursors (bold full-line histograms). Histograms 1, 2, and 3 refer to thermal conductivity values of 0.0015, 0.1, and 1 W m^{-1} K^{-1} respectively. **(a)** Assumes ejecta from asteroid Flora whose computed CRE ages are compared with the observed distribution for 240 L chondrites. **(b)** Assumes ejecta from asteroid (6) Hebe and the comparison with 444 CRE ages of H chondrites. **(c)** Assumes ejecta from asteroid (4) Vesta, compared to the CRE age data for 64 HED (howardite-eucrite-diogenite) meteorites. In all cases, the intermediate K value appears to provide the best match to the data. Note that the direct-injection scenario would always predict many more short CRE ages than are observed, as well as a shortage of ages between 20 and 50 m.y., which is not observed.

5.2. Escape of Kilometer-sized Asteroids from the Main Belt

Dynamical modeling suggests most Earth- and Mars-crossing asteroids ultimately come from the main belt (e.g., *Bottke et al.,* 2000b, 2002a). The cratering records of the terrestrial planets suggest this joint population, containing 5000–6000 D > 1 km asteroids of various taxonomic types, has been more or less in steady state for the last 3 G.y. (e.g., *Grieve and Shoemaker,* 1994). The primary sources of these bodies are the 3:1 mean-motion resonance with Jupiter, the ν_6 secular resonance, and numerous narrow mean-motion resonances produced by Mars or the combined effects of Jupiter and Saturn (*Wisdom,* 1983; *Morbidelli and Gladman,* 1998; *Migliorini et al.,* 1998; *Morbidelli and Nesvorný,* 1999). The viability of these sources have been checked using sophisticated numerical integration codes that track test asteroids evolving under the combined perturbations of the Sun and planets for ~100 m.y. (*Wisdom and Holman,* 1991; *Levison and Duncan,* 1994).

A possible problem with these simulations, however, is that they do not consider how the test asteroids reach their starting orbits. As described in the introduction, previous work has assumed that asteroids are thrown directly into resonances by main-belt collisions (e.g., *Farinella et al.,* 1993). The combined width of resonances in the inner and central main belt, however, is small enough that collisions alone may be unable to keep them filled with debris (*Farinella and Vokrouhlický,* 1999). Dynamical models suggest a shortage of resonant material could eventually lead to a discernible depletion of inner solar system asteroids (*Migliorini et al.,* 1998; *Michel et al.,* 2000). This problem would also be exacerbated by the fact that most potential parent bodies are located far from resonant escape hatches, and that the disruption of large bodies in the inner main belt should produce observable asteroid families.

For these reasons, *Farinella and Vokrouhlický* (1999) postulated that most main-belt resonances are restocked with D ≤ 20-km asteroids via the Yarkovsky effect. This potential solution could explain the spectral diversity of the inner solar system asteroid population (e.g., *Binzel et al.,* 2001) as well as well as the slope of its size-frequency distribution, which is shallower [N(>D) ∝ $D^{-1.75}$ (*Bottke et al.,* 2000a)] than one might expect if fresh ejecta were being launched directly into resonances [N(>D) ∝ D^{-3} (*Tanga et al.,* 1999)] but, at the same time, only slightly steeper than the population of main-belt asteroids in this size range [N(>D) ∝ D^{-1} (*Durda et al.,* 1998; *Davis et al.,* 2002)].

To investigate this scenario, *Bottke et al.* (2002b) numerically integrated hundreds of test asteroids in the inner (2.1–2.48 AU) and central (2.52–2.8 AU) main belt with and without the Yarkovsky effect. The orbits of the test asteroids were chosen to be a representative sample of the observed population residing near (but not on) Mars-crossing orbits (perihelion q ≥ 1.8 AU). Where possible, these tests duplicated the initial conditions investigated by *Migliorini et al.* (1998) and *Morbidelli and Nesvorný* (1999). All these test asteroids were tracked for at least 100 m.y. using a numerical integration code modified to accommodate Yarkovsky thermal forces (*Levison and Duncan,* 1994; *Bottke et al.,* 2000b; *Brož et al.,* 2002). A wide range of asteroid diameters (0.2 km, 0.4 km, 2 km, 4 km, 10 km) were used. Objects in the inner and central main belt were given S-type and C-type albedos respectively. Thermal conductivities were chosen to be consistent with values expected from regolith-covered asteroids. Random spin axis orientations and size-dependent spin rates were also used (e.g., *Bottke et al.,* 2000a). All these tests were compared to a control case where the Yarkovsky effect was turned off.

Bottke et al. (2002b) found that Yarkovsky-driven objects with D > 2 km reached Mars-crossing orbits at the same rate as the control case, despite the fact that the dynamical evolution of individual bodies in each set could be quite different (Fig. 4). For example, D = 10-km objects, with slow drift rates (e.g., Fig. 2), followed dynamical paths that were more or less analogous to the results of *Migliorini et al.* (1998) and *Morbidelli and Nesvorný* (1999). In this case, secular increases in e were caused predominantly by the bodies interacting with overlapping mean-motion resonances near the main belt periphery. On the other hand, resonant trapping does not appear to be the dominant behavior of D < 2-km objects; their faster drift rates allow many them to jump across many numerous weak resonances as they drift into the Mars-crossing region. In general, small inner-main-belt asteroids do not stop until they reach the wide and powerful 3:1 mean-motion resonance with Jupiter, the ν_6 secular resonance, or the Mars-crossing region itself. *Bottke et al.* (2002b) concluded from these results that the Yarkovsky effect was more efficient at driving subkilometer bodies out of the main belt than multikilometer bodies. The major source regions for subkilometer asteroids in the inner solar system should be powerful resonances like the 3:1 or ν_6 resonances, while an important source for multikilometer bodies would be the numerous tiny resonances scattered throughout the main belt (and possibly secular resonances intersecting asteroid families; see next section). *Bottke et al.* (2002a) estimate that the combined flux of kilometer-sized bodies from these sources is ~220 per million years. This rate is high enough to suggest the Yarkovsky effect, rather than collisional injection, is the dominant mechanism pushing material into resonances. It also suggests that some main-belt asteroid sources may produce more large or small objects than other sources. A considerable amount of work will be needed to fully appreciate all the ramifications of this new asteroid delivery scenario.

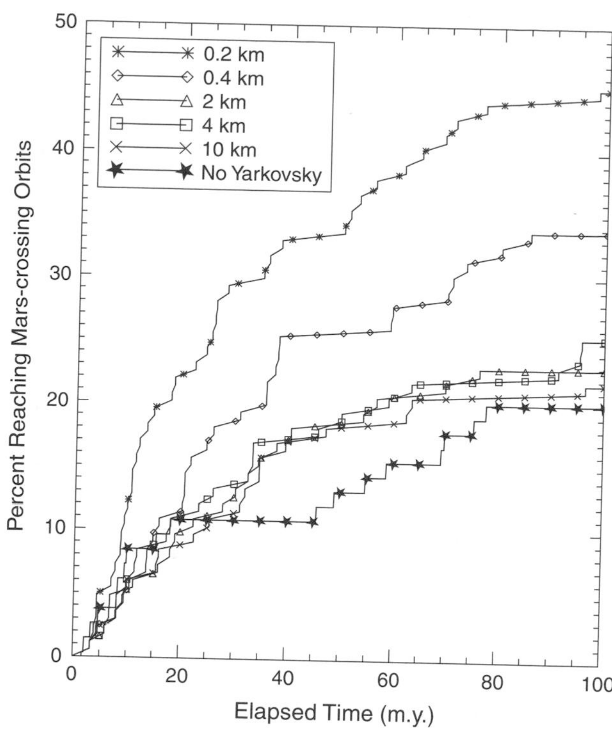

Fig. 4. Evolution of nearly-Mars-crossing bodies under the influence of Yarkovsky thermal forces. The plot shows the fraction of test asteroids, started with perihelion q ≥ 1.8 AU, reaching Mars-crossing orbits after 100 m.y. of integration. The initial conditions of the test asteroids nearly duplicated the initial conditions of *Morbidelli and Nesvorný* (1999). The bottom curve shows the *Morbidelli and Nesvorný* (1999) results. Results indicate that roughly the same fraction of D > 2-km bodies reach Mars-crossing orbits after 100 m.y., with or without Yarkovsky. Asteroids with D < 2 km, however, are much more efficient at escaping the main belt.

5.3. Dynamical Spreading of Asteroid Families

Asteroid families are remnants of large-scale catastrophic collisions. They are usually identified by their orbital elements, which tend to be clustered at similar values (e.g., *Bendjoya et al.,* 2002). By studying asteroid families, we hope to learn more about asteroid impacts, the primary geologic process occurring on asteroids today. Despite extensive work on this topic, however, there are still many issues related to asteroid families that we do not yet understand. We list a few below.

1. *Velocity distributions.* If one assumes that the semimajor axis distribution of family members has been constant since the formation event, it is possible to deduce the original ejection velocities of the fragments (*Zappalà et al.,* 1996). The inferred velocity distributions from this technique, on order of several 100 m s^{-1}, are inconsistent with ejection velocities derived by other means [i.e., laboratory impact experiments and numerical hydrocodes suggest multikilometer bodies typically have ejection velocities on the order of 100 m s^{-1} (*Benz and Asphaug,* 1999)].

2. *Orbital distributions.* Many prominent asteroid families have asymmetric (a, e, i) distribution. For example, Fig. 5 shows the distribution of the Koronis family in proper (a, e). Note that family members with small proper a are far less dispersed in proper e than those with large proper a, while both ends of the family are truncated by powerful

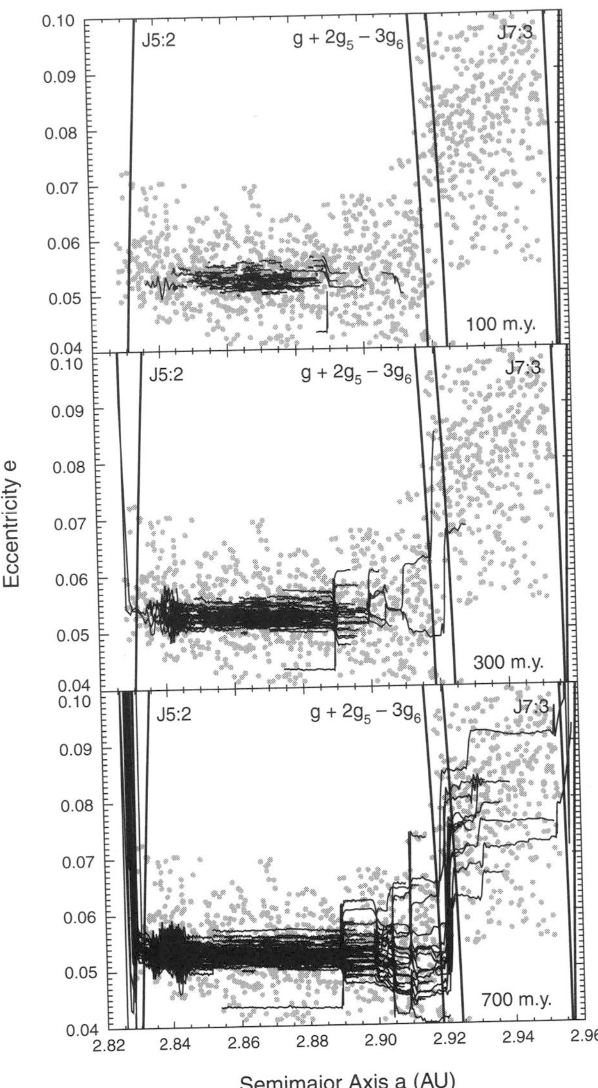

Fig. 5. Evolution of 210 simulated Koronis family members via the Yarkovsky effect (*Bottke et al.*, 2001). The test family members (black lines) were started within ~60 m s^{-1} of (158) Koronis (proper elements a = 2.87 AU, e = 0.045, sin i = 0.038) and were integrated for ~700 m.y., short compared with the estimated age of the family (~2.5 G.y.) but enough to determine evolution trends. The orbital tracks were averaged over a running 10-m.y. window in order to compare them with the proper (a, e) of the Koronis family members (gray dots). Snapshots of the integration tracks, shown at 100 m.y., 300 m.y., and 700 m.y., indicate these bodies interact with several resonances between 2.89 and 2.93 AU, with the secular $g + 2g_5 - 3g_6$ resonance at 2.92 AU being most prominent. These jumps allow the simulated family members to reach the (a, e) positions of many real family members. Fast-drifting bodies are seen to escape the main belt via the 5:2 and 7:3 mean-motion resonances with Jupiter.

mean-motion resonances with Jupiter (i.e., 5:2 on the left, 7:3 on the right). Surprisingly, no family members appear to have crossed either resonance, even though the 5:2 and 7:3 resonances are relatively narrow when compared to the span of the family.

3. *Family members on short-lived orbits.* Some multikilometer members of asteroid families are "on the brink" of entering a resonance [e.g., Koronis family members (*Milani and Farinella,* 1995; *Knežević et al.,* 1997; *Vokrouhlický et al.,* 2001)], are already inside a powerful resonances [e.g., Eos family members (*Zappalà et al.,* 2000)], or are part of the relatively short-lived NEO population [V-type asteroids, which presumably are part of the Vesta family (*Migliorini et al.,* 1997)]. Since most large families are thought to be 1 G.y. old or more (*Marzari et al.,* 1995), it is hard to understand how these family members attained these orbits. Using the classical model, one might assume that secondary fragmentation moved these objects onto their current orbits, but the large size of some of the objects (D > 10 km) makes this scenario improbable.

One way to resolve these issues is to assume that family members, since their formation, have migrated via the Yarkovsky effect. As shown in Fig. 2, an ensemble of D = 5-km asteroids will move inward and outward at mean drift rates of $|da/dt| \sim 2 \times 10^{-5}$ AU m.y.$^{-1}$, while larger asteroids drift more slowly (e.g., D ~20-km asteroids drift at $|da/dt| \sim 6 \times 10^{-6}$ AU m.y.$^{-1}$). Since collisional models suggest that many asteroid families are hundreds of millions of years to billions of years old (*Marzari et al.,* 1995, 1999), the potential drift distances of these objects are large enough to explain the observed dispersions of many asteroid families. Moreover, since Yarkovsky drift is size-dependent, the family members would eventually take on the appearance that they were launched using a size-dependent velocity distribution.

Thus, according to this scenario, the observed asteroid families were created through a multistep process. (1) A large asteroid undergoes a catastrophic disruption and ejects fragments at velocities consistent with those found in laboratory experiments and hydrocode simulations. (2) D < 20-km fragments, whose initial velocity dispersion is smaller than those currently observed among asteroid families, start drifting in semimajor axis under the Yarkovsky effect. D > 20-km fragments, which are less susceptible to the Yarkovsky effect, mainly move in semimajor axis via less efficient processes like collisions and/or close encounters with asteroids like Ceres, Pallas, or Vesta (*Carruba et al.,* 2000; *Nesvorný et al.,* 2002). (3) D < 20-km fragments, whose drift rate is a function of each object's size, spin state, and thermal properties, jump over or become trapped in chaotic mean-motion and secular resonances that change their eccentricity and/or inclination. In some cases, these orbital changes are significant enough that the objects can no longer be easily recognized as family members. (4) Family members that drift far enough may fall into mean-motion or secular resonances capable of pushing them onto planet-crossing orbits. From here, they become members of the Mars-crossing and/or NEO populations.

To check this idea, *Bottke et al.* (2001) tracked the evolution of test asteroids started close to the center of the Koronis family using the symplectic integration code SWIFT-RMVS3, which was modified to accommodate Yarkovsky thermal forces (*Levison and Duncan,* 1994; *Bottke et al.,* 2000b; *Brož et al.,* 2002). Figure 5 shows Yarkovsky forces driving multikilometer asteroids through numerous secular resonances where resonant jumping/trapping events produce noticeable changes in proper e, particularly on the right side of the plot. The most significant jumps are caused by the secular resonance $g + 2g_5 - 3g_6$ at 2.92 AU, which increases e but does not change i. Eventually, objects drifting far enough become trapped in the powerful 5:2 or 7:3 mean-motion resonances, where they are pushed onto planet-crossing orbits and are lost from the main belt.

Overall, these integration results reproduce the (a, e, i) distribution of the Koronis family while also explaining the paucity of family members on the left/right sides of the 5:2 and 7:3 resonances and the short-lived nature of some Koronis family members. The success of this model, together with the previous section's results, make a strong case that the Yarkovsky effect, working in concert with resonances, is the primary mechanism by which $D \lesssim 20$-km asteroids escape the main belt and reach the inner solar system.

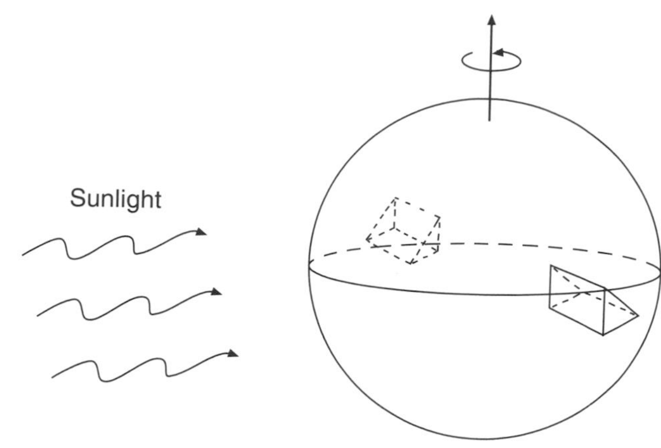

Fig. 6. Spinup of an asymmetrical asteroid. The asteroid is modeled as a sphere with two wedges attached to its equator. The asteroid is considered a blackbody, so it absorbs all sunlight falling upon it and then reemits the energy in the infrared as thermal radiation. Since the kicks produced by photons leaving the wedges are in different directions, a net torque is produced that causes the asteroid to spin up.

5.4. Radiative Spinup/Spindown of Asteroids (YORP Effect)

Besides changing the orbit, Yarkovsky forces can also produce torques that affect the spin rate and spin axis orientation of asteroids and meteoroids. This "sunlight alters spin" mechanism was coined by *Rubincam* (2000) as the Yarkovsky-O'Keefe-Radzievskii-Paddack effect, or YORP for short (*Radzievskii,* 1954; *Paddack,* 1969, 1973; *Paddack and Rhee,* 1975; *O'Keefe,* 1976). YORP comes from two sources: reflection and reemission. *Rubincam* (2000) illustrated its workings using a rotating spherical asteroid with two wedges attached to the equator (Fig. 6). For a Lambertian radiator, the reaction force from photons departing from any given element of area on the sphere will be normal to the surface, such that no torque is produced. Energy reradiated from the wedges, however, can produce a torque because the wedge faces are not coplanar. For the sense of rotation shown in Fig. 6, the wedge-produced YORP torque spins the object up. If the body happened to spin in the opposite sense, the YORP torques would slow it down. Thus, an object must have some "windmill" asymmetry for YORP to work (i.e., it would have no effect on rotating triaxial ellipsoids).

YORP torques can also modify asteroid obliquities, which leads to the concept of the YORP cycle. For the geometry shown in Fig. 6, a fast-spinning asteroid would gradually increase its obliquity as well. When the obliquity becomes large enough, the axial torque changes sign and the object begins to spin down. This can be seen by imagining that the Sun shines down on the object from its north pole rather than the equator; the wedges must spin it the other way. Hence YORP may spin objects up for a while, but when the obliquity becomes large, they slow down and then perhaps tumble until they reestablish principal axis rotation, with the spin axis presumably pointing in a random direction. Then the cycle begins all over again, such that small solid objects probably avoid the "rotational bursting" envisioned by Radzievskii, Paddack, and O'Keefe (i.e., spinning a solid object so fast that it disrupts). Collisions large enough to modify an asteroid's spin axis orientation may also short-circuit a YORP cycle, potentially putting the object into an entirely different rotation state. Thus, YORP is most likely to be important in regimes where the YORP cycle is faster than the spin axis reorientation timescale via collisions (*Rubincam*, 2000; *Vokrouhlický and Čapek,* 2002).

Rubincam (2000) found that YORP is strongly dependent on an asteroid's shape, size, distance from the Sun, and orientation. For example, assuming the Sun remains on the equator, asteroid (951) Gaspra, with R = 6 km and a = 2.21 AU, would in 240 m.y. go from a rotation period P = 12 h to 6 h (and vice versa). We call this value the YORP timescale. If we gave (243) Ida the same R and a values as Gaspra, it would have a YORP timescale half as big, while a body with Phobos' shape would have a YORP timescale of several billion years. Clearly, shapes make a big difference. The YORP timescale is also size-dependent (i.e., it goes as $\approx R^2$), such that smaller sizes spin up much more quickly. If Gaspra was only R = 0.5 km, its YORP timescale would be a few million years. Thus, YORP may be very influential for kilometer-sized and smaller asteroids. YORP is also more effective as you move closer to the Sun. Moving our R = 0.5 km Gaspra to 1 AU allows it to go from P = 12 h to rotational disruption speeds of ~2 h (and vice versa) in ~1 m.y. We caution, however, that YORP-induced obliquity torques may double or possibly triple the above timescales. Moreover, these rates also assume the YORP cycle continues without interruption via collisions, planetary close encounters, etc., and that asteroid thermal properties do not significantly change with size. These real-life complications will be modeled in the future.

If the aforementioned YORP timescales are reasonable values for small asteroids, it is plausible that YORP may spin small gravitational aggregates up so fast that they are forced to "morph" into a new shape and/or undergo mass shedding. Since symmetrical shapes increase the YORP timescale, these shape changes may eventually strand some objects close to the rotational breakup limit. If rotational disruption is common, we hypothesize that YORP may supersede tidal disruption and collisions as the primary means by which binary asteroids are produced.

At this point, we can begin to explore the possible connection between asteroid spin rates and the YORP effect. Observations show that D > 125-km asteroids have rotation rates that follow a Maxwellian-frequency distribution, while 50 < D < 125-km asteroids show a small excess of fast rotators relative to a Maxwellian and D < 50-km asteroids show a clear excess of very fast and slow rotators (*Binzel et al.,* 1989). More recent observations indicate that D < 10-km asteroids have even more pronounced extrema (*Pravec and Harris,* 2000; *Pravec et al.,* 2002). These results suggest that one or more mechanisms are depopulating the center of the spin rate distribution in favor of the extremes, and that these mechanisms may be size-dependent.

The possible mechanisms capable of performing these spin modifications are (1) collisions; (2) tidal spinup/spindown via a close encounter with a planet; (3) tidal evolution between binary asteroids, where spin angular momentum is exchanged for orbital angular momentum (with possible escape); and (4) YORP. The limitations of (1)–(3) are described in *Pravec et al.* (2002) and will not be reviewed here. The advantage of YORP over these other mechanisms is that it can naturally produce both slow and fast rotation rates for small asteroids over relatively short timescales, and that it is a size-dependent effect, helping to explain why the spin rate distributions change with D. The disadvantage of YORP is that it does not appear to be capable of significantly modifying the spin rates of large asteroids by itself. A unified model, which includes these processes and YORP, however, might do a reasonable job at explaining the spin rates of large asteroids like (253) Mathilde. The solution is left to future work.

At the time of this writing, we consider YORP studies to be in their infancy. For example, *Vokrouhlický and Čapek* (2002) recently pointed out that the YORP-evolving rotation state may become temporarily locked in one of the resonances between the precession rate and the proper or forced frequencies of the long-term orbital-plane evolution. These effects may temporarily halt, reverse, or accelerate the YORP influence on the obliquity. Future work on the YORP effect must also take into account thermal relaxation, non-principal-axis rotation, and more refined thermophysics (e.g., *Spitale and Greenberg,* 2001a,b).

6. FUTURE WORK

At the present time, the existence of the Yarkovsky effect is mostly based on inferences (e.g., CRE age distribution of meteorites, origin of large NEAs, and size-dependent dispersion of the asteroid families). To conclusively prove the existence of the Yarkovsky effect, however, it would be useful to directly detect its orbital perturbation on asteroids in a manner consistent with what was done for the LAGEOS artificial satellite (e.g., *Rubincam,* 1987). *Vokrouhlický and Milani* (2000) and *Vokrouhlický et al.* (2000) have suggested that the Yarkovsky perturbations can be computed directly from radar observations of small NEAs like (1566) Icarus, (6489) Golevka, or 1998 KY26 over a period of years. The advantages of radar include precise astrometry (by a factor of 100–1000× better than the usual optical astrometry), information on asteroid physical parameters like shape/surface properties, and its rotation state, all useful for Yarkovsky modeling efforts. At the time of this writing, the necessary plans — including possible preapparition optical observations — are underway. If the modeling work of *Vokrouhlický et al.* (2000) is correct, radar observations during the next close encounters of the most promising candidate asteroids may produce a discernable Yarkovsky "footprint."

The biggest challenge for future Yarkovsky modeling will be combining Yarkovsky accelerations with YORP, particularly since there is a complicated interaction between rotation, orbit precession rates, and spin axis precession rates. For many asteroids, particularly those kilometer-sized and larger, the spin and orbital precession rates are comparable, such that we can expect complicated beatlike phenomena to affect obliquity over relatively short timescales (*Skoglöv,* 1999). Moreover, coupling between the rates may allow the spin axis to be captured into a spin-orbit resonance. All these factors produce complicated feedbacks that (1) can modify asteroid drift rates and rotation rate changes and (2) are difficult to predict without extensive modeling. Since YORP is sensitive to the size, shape, material properties, and asteroid location, this effect will also vary from object to object. Thus, though we have hopefully demonstrated the usefulness of including Yarkovsky forces into the classical asteroid evolution model, there is still much work left to be done.

Acknowledgments. The authors wish to acknowledge helpful reviews by J. Spitale and S. Dermott and support from NASA's Planetary Geology and Geophysics Program. We dedicate this chapter to the memory of John A. O'Keefe and Paolo Farinella, both of whom championed the importance of the Yarkovsky effect during their illustrious and distinguished careers. D. P. Rubincam would like to thank Milton Schach for helping stimulate the thoughts that led to the theoretical development of the seasonal Yarkovsky effect.

REFERENCES

Afonso G. B., Gomes R. S., and Florczak M. A. (1995) Asteroid fragments in Earth-crossing orbits. *Planet. Space Sci., 43,* 787–795.

Benz W. and Asphaug E. (1999) Catastrophic disruptions revisited. *Icarus, 142,* 5–20.

Bendjoya Ph. and Zappalà V. (2002) Asteroid family identification. In *Asteroids III* (W. F. Bottke Jr. et al., eds.), this volume. Univ. of Arizona, Tucson.

Binzel R. P., Farinella P., Zappalà V., and Cellino A. (1989) Asteroid rotation rates — Distributions and statistics. In *Asteroids II* (R. P. Binzel et al., eds.), pp. 416–441. Univ. of Arizona, Tucson.

Binzel R. P., Harris A. W., Bus S. J., and Burbine T. H. (2001) Spectral properties of near-Earth objects: Palomar and IRTF results for 48 objects including spacecraft targets (9969) Braille and (10302) 1989 ML. *Icarus, 151,* 139–149.

Bottke W. F., Nolan M. C., Greenberg R., and Kolvoord R. A. (1994) Velocity distributions among colliding asteroids. *Icarus, 107,* 255–268.

Bottke W. F., Jedicke R., Morbidelli A., Petit J., and Gladman B. (2000a) Understanding the distribution of near-Earth asteroids. *Science, 288,* 2190–2194.

Bottke W. F., Rubincam D. P., and Burns J. A. (2000b) Dynamical evolution of main belt meteoroids: Numerical simulations incorporating planetary perturbations and Yarkovsky thermal forces. *Icarus, 145,* 301–331.

Bottke W. F., Vokrouhlický D., Brož M., Nesvorný D., and Morbidelli A. (2001) Dynamical spreading of asteroid families via the Yarkovsky effect. *Science, 294,* 1693–1696.

Bottke W. F. Jr., Morbidelli A., Jedicke R., Petit J.-M., Levison H. F., Michel P., and Metcalfe T. S. (2002a) Debiased orbital and absolute magnitude distribution of the near-Earth objects. *Icarus, 156,* 399–433.

Bottke W. F., Vokrouhlický D., Brož M., and Morbidelli A. (2002b) Yarkovsky-assisted escape of kilometer-sized asteroids from the main belt. *Icarus,* in press.

Brož M., Vokrouhlický D., and Farinella P. (2002) Interaction of the Yarkovsky-driven orbits of meteoroids and their precursors with the weak resonances in the main belt. *Icarus,* in press.

Burns J. A., Lamy P. L., and Soter S. (1979) Radiation forces on small particles in the solar system. *Icarus, 40,* 1–48.

Caffee M. W., Reedy R. C., Goswami J. N., Hohenberg C. M., and Marti K. (1988) Irradiation records in meteorites. In *Meteorites and the Early Solar System* (J. F. Kerridge and M. S. Matthews, eds.), pp. 205–245. Univ. of Arizona, Tucson.

Carruba V., Burns J. A., Bottke W. F., and Morbidelli A. (2000) Asteroid mobility due to encounters with Ceres, Vesta, Pallas: Monte Carlo codes versus direct numerical integrations. *Bull. Am. Astron. Soc., 32,* 1406.

Cellino A., Michel P., Tanga P., Zappalà V., Paolicchi P., and Dell'Oro A. (1999) The velocity-size relationship for members of asteroid families and implications for the physics of catastrophic collisions. *Icarus, 141,* 79–95.

Davis D. R., Weidenschilling S. J., Farinella P., Paolicchi P., and Binzel R. P. (1989) Asteroid collisional history — Effects on sizes and spins. In *Asteroids II* (R. P. Binzel et al., eds.), pp. 805–826. Univ. of Arizona, Tucson.

Davis D. R., Durda D. D., Marzari F., Campo Bagatin A., and Gil-Hutton R. (2002) Collisional evolution of small body populations. In *Asteroids III* (W. F. Bottke Jr. et al., eds.), this volume. Univ. of Arizona, Tucson.

Dermott S., Durda D. D., Grogan K., and Kehoe T. J. (2002) Asteroidal dust. In *Asteroids III* (W. F. Bottke Jr. et al., eds.), this volume. Univ. of Arizona, Tucson.

Dohnanyi J. S. (1978) Particle dynamics. In *Cosmic Dust* (J. A. M. McDonnell, ed.), pp. 527–605. Wiley-Interscience, New York.

Dones L., Gladman B., Melosh H. J., Tonks W. B., Levison H. F., and Duncan M. (1999) Dynamical lifetimes and final fates of small bodies: Orbit integrations vs Öpik calculations. *Icarus, 142,* 509–524.

Farinella P. and Vokrouhlický D. (1999) Semimajor axis mobility of asteroidal fragments. *Science, 283,* 1507–1510.

Farinella P., Gonczi R., Froeschlé Cl., and Froeschlé Ch. (1993) The injection of asteroid fragments into resonances. *Icarus, 101,* 174–187.

Farinella P., Froeschlé C., Froeschlé C., Gonczi R., Hahn G., Morbidelli A., and Valsecchi G. B. (1994) Asteroids falling onto the Sun. *Nature, 371,* 315–316.

Farinella P., Vokrouhlický D., and Hartmann W. K. (1998) Meteorite delivery via Yarkovsky orbital drift. *Icarus, 132,* 378–387.

Gladman B. J., Migliorini F., Morbidelli A., Zappalà V., Michel P., Cellino A., Froeschlé C., Levison H. F., Bailey M., and Duncan M. (1997) Dynamical lifetimes of objects injected into asteroid belt resonances. *Science, 277,* 197–201.

Grieve R. A. F. and Shoemaker E. M. (1994) The record of past impacts on Earth. In *Hazards Due to Comets and Asteroids* (T. Gehrels and M. S. Matthews, eds.), pp. 417–462. Univ. of Arizona, Tucson.

Hartmann W. K., Farinella P., Vokrouhlický D., Weidenschilling S. J., Morbidelli A., Marzari F., Davis D. R., and Ryan E. (1999) Reviewing the Yarkovsky effect: New light on the delivery of stone and iron meteorites from the asteroid belt. *Meteoritics & Planet. Sci., 34,* 161–167.

Katasev L. A. and Kulikova N. V. (1980) Physical and mathematical modeling of the formation and evolution of meteor streams II. *Astr. Vestnik, 14,* 225–229.

Knežević Z., Milani A., and Farinella P. (1997) The dangerous border of the 5:2 mean motion resonance. *Planet. Space Sci., 45,* 1581–1585.

Levison H. F. and Duncan M. J. (1994) The long-term dynamical behavior of short-period comets. *Icarus, 108,* 18–36.

Marti K. and Graf T. (1992) Cosmic-ray exposure history of ordinary chondrites. *Annu. Rev. Earth Planet. Sci., 20,* 221–243.

Marzari F., Davis D., and Vanzani V. (1995) Collisional evolution of asteroid families. *Icarus, 113,* 168–187.

Marzari F., Farinella P., and Davis D. R. (1999) Origin, aging, and death of asteroid families. *Icarus, 142,* 63–77.

Michel P., Migliorini F., Morbidelli A., and Zappalà V. (2000) The population of Mars-crossers: Classification and dynamical evolution. *Icarus, 145,* 332–347.

Migliorini F., Morbidelli A., Zappalà V., Gladman B. J., Bailey M. E., and Cellino A. (1997) Vesta fragments from ν_6 and 3:1 resonances: Implications for V-type NEAs and HED meteorites. *Meteoritics & Planet. Sci., 32,* 903–916.

Migliorini F., Michel P., Morbidelli A., Nesvorný D., and Zappalà V. (1998) Origin of Earth-crossing asteroids: A quantitative simulation. *Science, 281,* 2022–2024.

Milani A. and Farinella P. (1995) An asteroid on the brink. *Icarus, 115,* 209–212.

Morbidelli A. and Gladman B. (1998) Orbital and temporal distributions of meteorites originating in the asteroid belt. *Meteoritics & Planet. Sci., 33,* 999–1016.

Morbidelli A. and Nesvorný D. (1999) Numerous weak resonances drive asteroids toward terrestrial planets orbits. *Icarus, 139,* 295–308.

Neiman V. B., Romanov E. M., and Chernov V. M. (1965) Ivan Osipovich Yarkovsky. *Earth and Universe, 4,* 63–64 (in Russian). (Translation into English by T. O. Laoghog available.)

Nesvorný D., Morbidelli A., Vokrouhlický D., Bottke W. F., and Brož M. (2002) The Flora family: A case of the dynamically dispersed collisional swarm? *Icarus,* in press.

O'Keefe J. A. (1976) *Tektites and Their Origin.* Elsevier, Amsterdam. 254 pp.

Olsson-Steel D. (1986) The origin of the sporadic meteoroid component. *Mon. Not. R. Astron. Soc., 219,* 47–73.

Olsson-Steel D. (1987) The dispersal of the Geminid meteoroid stream by radiative effects. *Mon. Not. R. Astron. Soc., 226,* 1–17.

Öpik E. J. (1951) Collision probabilities with the planets and the distribution of interplanetary matter. *Proc. R. Irish Acad., 54A,* 165–199.

Öpik E. J. (1976) *Interplanetary Encounters: Close-Range Gravitational Interactions.* Elsevier, New York. 155 pp.

Paddack S. J. (1969) Rotational bursting of small celestial bodies: Effects of radiation pressure. *J. Geophys. Res., 74,* 4379–4381.

Paddack S. J. (1973) Rotational bursting of small particles. Ph.D. thesis, Catholic University,

Paddack S. J. and Rhee J. W. (1975) Rotational bursting of interplanetary dust particles. *Geophys. Res. Lett., 2,* 365–367.

Peterson C. (1976) A source mechanism for meteorites controlled by the Yarkovsky effect. *Icarus, 29,* 91–111.

Pisani E., Dell'Oro A., and Paolicchi P. (1999) Puzzling asteroid families. *Icarus, 142,* 78–88.

Pravec P. and Harris A. W. (2000) Fast and slow rotation of asteroids. *Icarus, 148,* 12–20.

Pravec P., Harris A. W., and Michalowski T. (2002) Spin rates of asteroids. In *Asteroids III* (W. F. Bottke Jr. et al., eds.), this volume. Univ. of Arizona, Tucson.

Radzievskii V. V. (1952) About the influence of the anisotropically reemited solar radiation on the orbits of asteroids and meteoroids. *Astron. Zh., 29,* 162–170.

Radzievskii V. V. (1954) A mechanism for the disintegration of asteroids and meteorites. *Dokl. Akad. Nauk SSSR, 97,* 49–52.

Rubincam D. P. (1987) LAGEOS orbit decay due to infrared radiation from earth. *J. Geophys. Res., 92,* 1287–1294.

Rubincam D. P. (1988) Yarkovsky thermal drag on LAGEOS. *J. Geophys. Res., 93,* 13805–13810.

Rubincam D. P. (1990) Drag on the LAGEOS satellite. *J. Geophys. Res., 95,* 4881–4886.

Rubincam D. P. (1995) Asteroid orbit evolution due to thermal drag. *J. Geophys. Res., 100,* 1585–1594.

Rubincam D. P. (1998) Yarkovsky thermal drag on small asteroids and Mars-Earth delivery. *J. Geophys. Res., 103,* 1725–1732.

Rubincam D. P. (2000) Radiative spin-up and spin-down of small asteroids. *Icarus, 148,* 2–11.

Skoglöv E. (1999) Spin vector evolution for inner solar system asteroids. *Planet. Space Sci., 47,* 11–22.

Slabinski V. J. (1977) Solar radiation torques on meteoroids: Complications for the Yarkovsky effect from spin axis precession. *Bull. Am. Astron. Soc., 9,* 438.

Spitale J. and Greenberg R. (2001) Numerical evaluation of the general Yarkovsky effect: Effects on semimajor axis. *Icarus, 149,* 222–234.

Spitale J. and Greenberg R. (2002) Numerical evaluation of the general Yarkovsky effect: Effects on eccentricity and longitude of periapse. *Icarus, 156,* 211–222.

Tanga P., Cellino A., Michel P., Zappalà V., Paolicchi P., and Dell'Oro A. (1999) On the size distribution of asteroid families: The role of geometry. *Icarus, 141,* 65–78.

Vokrouhlický D. (1998a) Diurnal Yarkovsky effect as a source of mobility of meter-sized asteroidal fragments. I. Linear theory. *Astron. Astrophys., 335,* 1093–1100.

Vokrouhlický D. (1998b) Diurnal Yarkovsky effect as a source of mobility of meter-sized asteroidal fragments. II. Non-sphericity effects. *Astron. Astrophys., 338,* 353–363.

Vokrouhlický D. (1999) A complete linear model for the Yarkovsky thermal force on spherical asteroid fragments. *Astron. Astrophys., 344,* 362–366.

Vokrouhlický D. (2001) Yarkovsky effect: Many-fingered phenomenon in the solar system dynamics. In *The Restless Universe: Applications of Gravitational N-Body Dynamics to Planetary, Stellar and Galactic Systems* (B. A. Steves and A. J. Maciejewski, eds.), pp. 53–78. Institute of Physics, Bristol.

Vokrouhlický D. and Bottke W. F. (2001) The Yarkovsky thermal force on small asteroids and their fragments. Choosing the right albedo. *Astron. Astrophys., 371,* 350–353.

Vokrouhlický D. and Čapek D. (2002) YORP-induced long-term evolution of the spin state of small asteroids and meteoroids. I. Rubincam's approximation. *Icarus,* in press.

Vokrouhlický D. and Farinella P. (1998) The Yarkovsky seasonal effect on asteroidal fragments: A nonlinearized theory for the plane-parallel case. *Astron. J., 116,* 2032–2041.

Vokrouhlický D. and Farinella P. (1999) The Yarkovsky seasonal effect on asteroidal fragments: A nonlinearized theory for spherical bodies. *Astron. J., 118,* 3049–3060.

Vokrouhlický D. and Farinella P. (2000) Efficient delivery of meteorites to the Earth from a wide range of asteroid parent bodies. *Nature, 407,* 606–608.

Vokrouhlický D. and Milani A. (2000) Direct solar radiation pressure on the orbits of small near-Earth asteroids: Observable effects? *Astron. Astrophys., 362,* 746–755.

Vokrouhlický D., Milani A., and Chesley S. R. (2000) Yarkovsky effect on small near-Earth asteroids: Mathematical formulation and examples. *Icarus, 148,* 118–138.

Vokrouhlický D., Brož M., Farinella P., and Knežević Z. (2001) Yarkovsky-driven leakage of Koronis family members. *Icarus, 150,* 78–93.

Welten K. C., Lindner L., van der Borg K., Loeken T., Scherer P., and Schultz L. (1997) Cosmic-ray exposure ages of diogenites and the recent collisional history of the HED parent body/bodies. *Meteoritics & Planet. Sci., 32,* 891–902.

Wetherill G. W. (1974) Solar system sources of meteorites and large meteoroids. *Annu. Rev. Earth Planet. Sci., 2,* 303–331.

Wetherill G. W. (1979) Steady state populations of Apollo-Amor objects. *Icarus, 37,* 96–112.

Wetherill G. W. (1985) Asteroidal source of ordinary chondrites (Meteoritical Society Presidential Address 1984). *Meteoritics, 20,* 1–22.

Whipple F. L. (1950) A comet model. I. The acceleration of Comet Encke. *Astrophys. J., 111,* 375–394.

Wisdom J. (1983) Chaotic behavior and the origin of the 3/1 Kirkwood gap. *Icarus, 56,* 51–74.

Wisdom J. and Holman M. (1991) Symplectic maps for the n-body problem. *Astron. J., 102,* 1528–1538.

Zappalà V. and Cellino A. (2002) A search for the collisional parent bodies of large NEAs. *Icarus,* in press.

Zappalà V., Cellino A., Dell'Oro A., Migliorini F., and Paolicchi P. (1996) Reconstructing the original ejection velocity fields of asteroid families. *Icarus, 124,* 156–180.

Zappalà V., Bendjoya P., Cellino A., Di Martino M., Doressoundiram A., Manara A., and Migliorini F. (2000) Fugitives from the Eos family: First spectroscopic confirmation. *Icarus, 145,* 4–11.

Origin and Evolution of Near-Earth Objects

A. Morbidelli
Observatoire de la Côte d'Azur, Nice, France

W. F. Bottke Jr.
Southwest Research Institute, Boulder, Colorado

Ch. Froeschlé
Observatoire de la Côte d'Azur, Nice, France

P. Michel
Observatoire de la Côte d'Azur, Nice, France

Asteroids and comets on orbits with perihelion distance $q < 1.3$ AU and aphelion distance $Q > 0.983$ AU are usually called near-Earth objects (NEOs). It has long been debated whether the NEOs are mostly of asteroidal or cometary origin. With improved knowledge of resonant dynamics, it is now clear that the asteroid belt is capable of supplying most of the observed NEOs. Particular zones in the main belt provide NEOs via powerful and diffusive resonances. Through the numerical integration of a large number of test asteroids in these zones, the possible evolutionary paths of NEOs have been identified and the statistical properties of NEOs dynamics have been quantified. This work has allowed the construction of a steady-state model of the orbital and magnitude distribution of the NEO population, dependent on parameters that are quantified by calibration with the available observations. The model accounts for the existence of ~1000 NEOs with absolute magnitude $H < 18$ (roughly 1 km in size). These bodies carry a probability of one collision with the Earth every 0.5 m.y. Only 6% of the NEO population should be of Kuiper Belt origin. Finally, it has been generally believed that collisional activity in the main belt, which continuously breaks up large asteroids, injects a large quantity of fresh material into the NEO source regions. In this manner, the NEO population is kept in steady state. The steep size distribution associated with fresh collisonal debris, however, is not observed among the NEO population. This paradox might suggest that Yarkovsky thermal drag, rather than collisional injection, plays the dominant role in delivering material to the NEO source resonances.

1. INTRODUCTION

The discovery of 433 Eros in 1898 established the existence of a population of asteroid-like bodies on orbits intersecting those of the inner planets. It was not until the Apollo program in the 1960s and 1970s, however, that lunar craters were shown to be derived from impacts rather than volcanism. With this evidence in hand, it was finally recognized that the Earth-Moon system has been incessantly bombarded by asteroids and comets over the last 4.5 G.y. In 1980, Alvarez et al. presented convincing arguments that the numerous species extinction at the Cretaceous-Tertiary transition were caused by the impact of a massive asteroid (*Alvarez et al.,* 1980).

These results brought increasing attention to the objects on Earth-crossing orbits and, more generally, to those having perihelion distances $q \leq 1.3$ AU and aphelion distances $Q \geq 0.983$ AU. The latter constitute what is usually called the near-Earth-object population (NEOs). Figure 1 shows the distribution of the known NEOs with respect to their semimajor axis, eccentricity, and inclination. The NEOs are by convention subdivided into Apollos ($a \geq 1.0$ AU; $q \leq 1.0167$ AU), Atens ($a < 1.0$ AU; $Q \geq 0.983$ AU), and Amors ($1.0167 \text{ AU} < q \leq 1.3$ AU).

It is now generally accepted that the NEOs represent a hazard of global catastrophe for human civilization. While the discovery of unknown NEOs is of primary concern, the theoretical understanding of the origin and evolution of NEOs is also of great importance. Together, these efforts can ultimately allow the estimatation of the orbital and size distributions of the NEO population, which in turn makes possible the quantification of the collision hazard and the optimization of NEO search strategies. The purpose of this chapter is to review our current knowledge of these issues. We start in section 2 with a brief historical overview, focused on the many important advances that have contributed to the understanding of the NEO population. In section 3 we discuss how asteroids can escape from the main

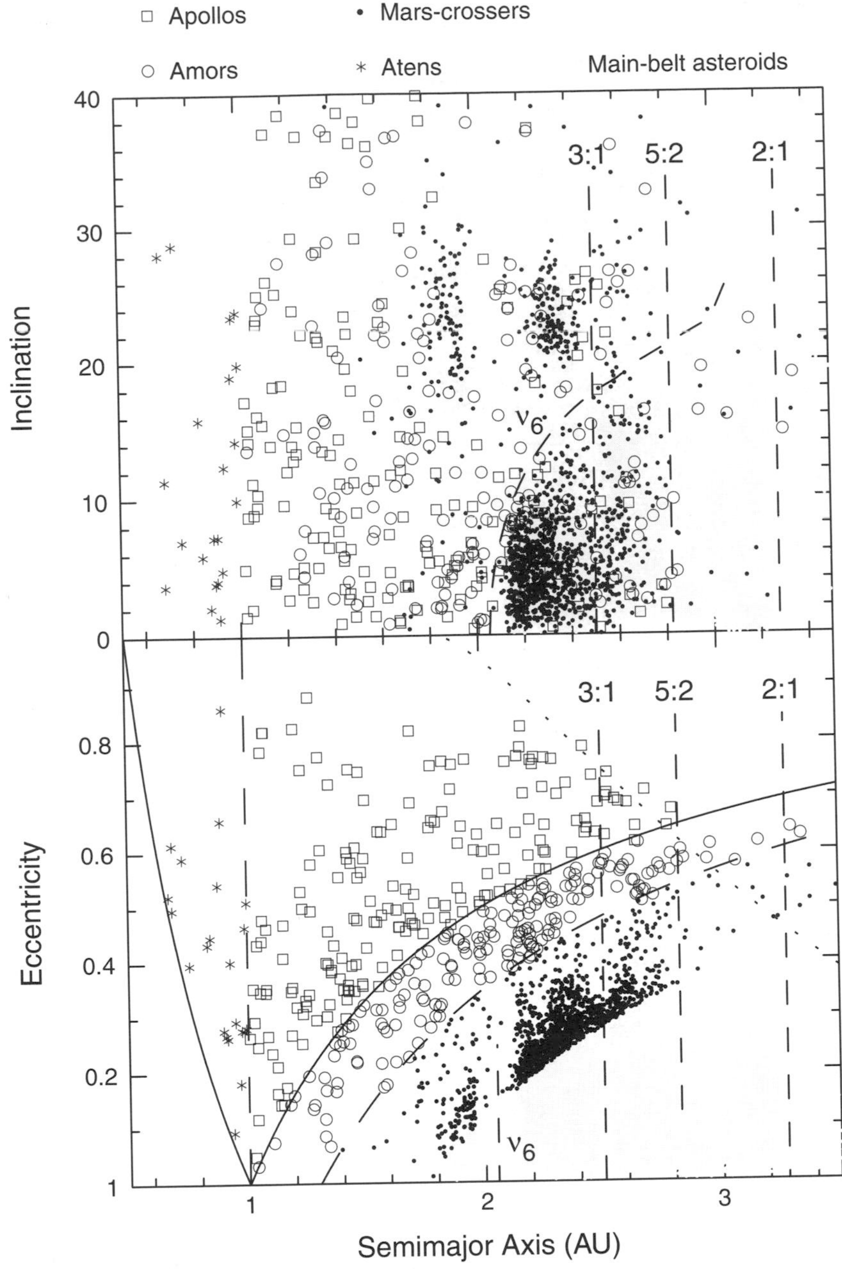

Fig. 1. The distribution of NEOs, Mars-crossers and main-belt asteroids with respect to semimajor axis, eccentricity, and inclination. The first 10,000 main-belt asteroids in are shown in gray. The dots represent the Mars-crossers, according to *Migliorini et al.*'s (1998) definition and identification: bodies with $1.3 < q < 1.8$ that intersect the orbit of Mars within the next 300,000 yr. The Amors, Apollos, and Atens are shown as circles, squares, and asterisks respectively. In boldface, the solid curve bounds the Earth-crossing region; the dashed curve delimits the Amor region at q = 1.3 AU, and the dashed vertical line denotes the boundary between the Aten and Apollo populations. The dotted curve corresponds to Tisserand parameter (1) with respect to Jupiter = 3, at i = 0: Jupiter-family comets reside predominantly beyond the curve T = 3. The locations of the 3:1, 5:2, and 2:1 mean motion resonances with Jupiter are shown by vertical dashed lines. The semimajor axis location of the ν_6 resonance is roughly independent of the eccentricity but is a function of the inclination: It is shown by the dashed curve on the top panel, while in the bottom panel it is represented for i = 0.

belt and become NEOs, but we also mention the cometary contribution to the NEO population. Section 4 will be devoted to a description of the typical evolutions of NEOs. In these two sections, emphasis will be given to NEO dynamical lifetimes and possible end states. Section 5 will discuss how the current knowledge on the origin and evolution of NEOs can be utilized to construct a quantitative model of the NEO population. Section 6 will detail the debiased NEO orbital and magnitude distribution resulting from this modeling effort. It will also discuss the implications of this model for the collision probability of NEOs with the Earth, and the collisional and dynamical mechanisms that continuously supply new bodies to the transportation resonances of the asteroid belt. Finally, we discuss open problems and future perspectives in section 7.

2. HISTORICAL OVERVIEW

The asteroid vs. comet origin of the NEO population has been debated throughout the last 40 years. With calculations based on a theory of Earth encounter probabilities, *Öpik* (1961, 1963) claimed that the Mars-crossing asteroid population was not large enough to keep the known Apollo population in steady state. Based on this result, Öpik concluded that about 80% of the Apollos are of cometary origin. The existence of meteor streams associated with some Apollos seemed to provide some support for this hypothesis. Conversely, *Anders* (1964) proposed that most Apollo objects are small main-belt asteroids that became Earth-crossers as a result of multiple close encounters with Mars. *Anders and Arnold* (1965) concluded that some Apollos with high eccentricities or inclinations (like 1566 Icarus and 2101 Adonis) might be extinct cometary nuclei, whereas the other Apollos (six known at the time!) should have been of asteroidal origin. Possible asteroidal and cometary sources of Apollo and Amor objects were reviewed in *Wetherill* (1976). The author considered several mechanisms (including close encounters with Mars, mean motion resonances, and secular resonances) to produce NEOs from the asteroid belt. Ultimately, he concluded that, although qualitatively acceptable, these mechanisms were unable to supply the required number of NEOs by at least an order of magnitude. Thus, Wetherill concluded that most Apollo objects were cores of comets that had lost their volatile material by repeated evaporation.

The problem at the time was that resonant dynamics was poorly understood and computation speed was extremely limited. Thus, direct numerical integration of asteroid orbits could not be used to determine the evolutionary paths of NEOs. With the underestimatation of the effect of resonant dynamics, NEO modelers were left with collision as the only viable mechanism for moving asteroids from the main belt directly into the NEO region. The typical ejection velocities of asteroid fragments generated in collisions, however, is ~100 m/s, far too small in most cases to achieve planet-crossing orbits (*Wetherill,* 1976).

The first indication that resonances could force main-belt bodies to cross the orbits of the planets came from the Ph.D. thesis work of J. G. Williams. In a diagram reported by *Wetherill* (1979), Williams showed that bodies close to the ν_6 resonance have secular eccentricity oscillations with amplitude exceeding 0.25, and therefore they must periodically cross the orbit of Mars. Shortly afterwards, *Wisdom* (1983, 1985a,b) showed that the 3:1 mean motion resonance has a similar effect: The eccentricity of resonant bodies can have, at irregular time intervals, rapid and large oscillations whose amplitudes exceed 0.3, the threshold value to become a Mars-crosser at the 3:1 location. Following these pioneering works, attention was focused on the 3:1 and ν_6 resonances as primary sources of NEOs from the asteroid belt. The dynamical structure of the 3:1 resonance was further explored by *Yoshikawa* (1989, 1990), *Henrard and Caranicolas* (1990), and *Ferraz-Mello and Klafke* (1991) in the framework of the three-body problem. In a more realistic multiplanet solar system model, numerical integrations by *Farinella et al.* (1993) showed that in the 3:1 resonance the eccentricity evolves much more chaotically than in the three-body problem — a phenomenon later explained by *Moons and Morbidelli* (1995) — and typically it reaches not only Mars-, but also Earth- and Venus-crossing values (see *Moons,* 1997, for a review). Concerning the ν_6 resonance, the first quantitative numerical results were obtained by *Froeschlé and Scholl* (1987), who confirmed the role that this resonance has in increasing asteroid eccentricities to Mars-crossing or Earth-crossing values. A first analytic theory of the dynamics in this resonance was developed by *Yoshikawa* (1987), and later improved by *Morbidelli and Henrard* (1991) and *Morbidelli* (1993) (see *Froeschlé and Morbidelli,* 1994, for a review).

Using these advances, *Wetherill* (1979, 1985, 1987, 1988) developed Monte Carlo models of the orbital evolution of NEOs coming from the ν_6 and 3:1 resonances. In these resonances the dynamics were simulated through simple algorithms designed to mimic the results of analytic theories or direct integrations, while elsewhere reduced to the sole effects of close encounters, calculated using two-body scattering formulae (*Öpik,* 1976). As the Monte Carlo models were refined over time, the cometary origin hypothesis was progressively abandoned as a potential source of NEOs in the inner solar system. Wetherill hypothesized that the ν_6 and 3:1 resonances are continuously resupplied via catastrophic collisions and/or cratering events in the main belt, and that enough material is injected into the resonances to keep the NEO population in steady state. Wetherill's work was later extended by *Rabinowitz* (1997a,b), who predicted the existence of 875 NEOs larger than 1 km, in remarkable agreement with current estimates.

In the 1990s, the availability of cheap and fast workstations allowed the first direct simulations of the dynamical evolution of test particles, initially placed in the NEO region or in the transport resonances, over million-year timescales. Using a Bulirsch-Stoer integrator, a breakthrough result was obtained by *Farinella et al.* (1994), who showed that NEOs with a < 2.5 AU can easily collide with the Sun, which limits their typical dynamical lifetime to a few million years. It became thus rapidly evident that Monte Carlo codes do not adequately treat the inherently chaotic behavior of bodies in NEO space (see *Dones et al.,* 1999, for a discussion). The introduction of a new numerical integration code (*Levison and Duncan,* 1994), which extended a numerical symplectic algorithm proposed by *Wisdom and Holman* (1991), introduced the possibility of numerically integrating a much larger number of particles, to quantify the statistical properties of NEO dynamics. The subsequent studies have contributed to our current understanding of the origin, evolution, and orbital distribution of NEOs, reviewed in the next sections.

3. DYNAMICAL ORIGIN OF NEOs

Asteroids become planet crossers by increasing their orbital eccentricity under the action of a variety of resonant phenomena. It is suitable to separately consider "powerful resonances" and "diffusive resonances." The former can be effectively distinguished from the latter by the existence of associated gaps in the main-belt asteroid distribution. The most notable resonances in the "powerful" class are the ν_6 secular resonance at inner edge of the asteroid belt, and the mean motion resonances with Jupiter 3:1, 5:2, and 2:1 at 2.5, 2.8, and 3.2 AU respectively. Their properties are detailed below. The diffusive resonances are so numerous that they cannot be effectively enumerated. Therefore, we will discuss only their generic dynamical effects. The reader can refer to *Nesvorný et al.* (2002) for a more technical discussion of the dynamical structure of the main asteroid belt.

3.1. ν_6 Resonance

The ν_6 secular resonance occurs when the precession frequency of the asteroid's longitude of perihelion is equal to the sixth secular frequency of the planetary system. The latter can be identified with the mean precession frequency of Saturn's longitude of perihelion, but it is also relevant in the secular oscillation of the eccentricity of Jupiter (see chapter 7 of *Morbidelli,* 2002). As shown in the top panel of Fig. 1, the ν_6 resonance marks the inner edge of the main belt. The effect of the resonance rapidly decays with the distance from the shown curve. To schematize, we divide the ~0.08-AU-wide neighborhood on the righthand side of the curve into a "powerful region" and a "border region," roughly of equal size (about 0.04 AU each).

In the powerful region the resonance causes a regular but large increase of the eccentricity of the asteroids. As a consequence, the asteroids reach Earth- (or Venus-) crossing orbits, and in several cases they collide with the Sun, with their perihelion distance becoming smaller than the solar radius. The median time required to become an Earth-crosser, starting from a quasicircular orbit, is about 0.5 m.y. Accounting also for the subsequent evolution in the NEO region (discussed in section 4), the median lifetime of bodies initially in the ν_6 resonance is 2 m.y., the typical end states being collision with the Sun (80% of the cases) and ejection on hyperbolic orbit (12%) (*Gladman et al.,* 1997). The mean time spent in the NEO region is 6.5 m.y. (*Bottke et al.,* 2002a), and the mean collision probability with Earth, integrated over the lifetime in the Earth-crossing region, is $\sim 10^{-2}$ (*Morbidelli and Gladman,* 1998).

In the border region, the effect of the ν_6 resonance is less powerful, but is still capable of forcing the asteroids to cross the orbit of Mars at the top of the secular oscillation cycle of their eccentricity. To enter the NEO region, these asteroids must evolve under the effect of martian encounters, and the required time increases sharply with the distance from the resonance (*Morbidelli and Gladman,* 1998). The dynamics in this region are complicated by the dense presence of mean motion resonances with Mars, and we will revisit this in section 3.5.

3.2. 3:1 Resonance

The 3:1 mean-motion resonance with Jupiter occurs at ~2.5 AU. Inside the resonance, one can distinguish two regions: a narrow central region where the asteroid eccentricity has regular oscillations that cause them to periodically cross the orbit of Mars, and a larger border region where the evolution of the eccentricity is wildly chaotic and unbounded, so that the bodies can rapidly reach Earth-crossing and even Sun-grazing orbits. Under the effect of martian encounters, bodies in the central region can easily travel to the border region and be rapidly boosted into NEO space (see chapter 11 of *Morbidelli,* 2002). For a population initially uniformly distributed inside the resonance, the median time required to cross the orbit of the Earth is ~1 m.y., the median lifetime is ~2 m.y., and the typical end states are the collision with the Sun (70%) and the ejection on hyperbolic orbit (28%) (*Gladman et al.,* 1997). The mean time spent in the NEO region is 2.2 m.y. (*Bottke et al.,* 2002a), and the mean collision probability with Earth, integrated over the lifetime in the Earth-crossing region, is 2×10^{-3} (*Morbidelli and Gladman,* 1998).

3.3. 5:2 Resonance

The 5:2 mean-motion resonance with Jupiter is located at 2.8 AU. The mechanisms that allow the rapid and chaotic eccentricity evolution in the border region of the 3:1 resonance in this case extend to the entire resonance (*Moons and Morbidelli,* 1995). As a consequence, this resonance is the one that pumps the orbital eccentricities on the shortest timescale. The median time required to reach Earth-crossing orbit is ~0.3 m.y., and the median lifetime is 0.5 m.y. Because the resonance is closer to Jupiter than the previous ones, the ejection on hyperbolic orbit is the most typical end state (92%), while the collision with the Sun accounts only for 8% of the losses (*Gladman et al.,* 1997). The mean time spent in the NEO region is 0.4 m.y. and the mean collision probability with Earth, integrated over the lifetime in the Earth-crossing region, is 2.5×10^{-4}.

3.4. 2:1 Resonance

Despite the fact that this resonance, located at 3.28 AU, is associated with a deep gap in the asteroid distribution, there are no mechanisms capable of destabilizing the resonant asteroid motion on the short timescales typical of the other resonances. In fact, the dynamical structure of the 2:1 resonance is very complicated (*Nesvorný and Ferraz-Mello,* 1997; *Moons et al.,* 1998). At the center of the resonance and at moderate eccentricity, there are large regions where the dynamical lifetime is on the order of the age of the solar system. Some asteroids are presently located in these regions (the so-called Zhongguo group), but it is still not completely understood why their number is so small (see *Nesvorný and Ferraz-Mello,* 1997, for a discussion on a possible cosmogonic mechanism). The regions close to the borders of the resonance are unstable, but several million years are required before an Earth-crossing orbit can be achieved (*Moons et al.,* 1998). Once in NEO space, the dynamical lifetime is only on the order of 0.1 m.y., because the bodies are rapidly ejected by Jupiter onto hyperbolic orbit. The mean collision probability with the Earth, integrated over the lifetime in the Earth-crossing region, is about 5×10^{-5}.

3.5. Diffusive Resonances

In addition to the few wide mean-motion resonances with Jupiter described above, the main belt is densely crossed by hundreds of thin resonances: high-order mean-motion resonances with Jupiter (where the orbital frequencies are in a ratio of large integer numbers), three-body resonances with Jupiter and Saturn (where an integer combination of the orbital frequencies of the asteroid, Jupiter, and Saturn is equal to zero; *Murray et al.,* 1998; *Nesvorný and Morbidelli,* 1998, 1999), and mean motion resonances with Mars (*Morbidelli and Nesvorný,* 1999). Because of these resonances, many — if not most — main-belt asteroids are chaotic (see *Nesvorný et al.,* 2002, for discussion). The effect of this chaoticity is very weak. The mean semimajor axis is bounded within the narrow resonant region; the proper eccentricity and inclination (see *Knežević et al.,* 2002) slowly change with time, in a chaotic diffusion-like process. The time required to reach a planet-crossing orbit (Mars-crossing in the inner belt, Jupiter-crossing in the outer belt) ranges from several 10^7 yr to billions of years, depending on the resonances and starting eccentricity (*Murray and Holman,* 1997).

Integrating real objects in the inner belt (2 < a < 2.5 AU) for 100 m.y., *Morbidelli and Nesvorný* (1999) estimated that chaotic diffusion drives about two asteroids larger than 5 km into the Mars-crossing region every million years. The escape rate is particularly high in the region adjacent to the ν_6 resonance, because of the effect of the latter, but also because of the dense presence of mean motion resonances with Mars. The high rate of diffusion of asteroids from the inner belt can explain the existence of the conspicuous population of Mars-crossers. Following *Migliorini et al.* (1998), we define the latter as the population of bodies with q > 1.3 AU that intersect the orbit of Mars within a secular cycle of their eccentricity oscillation (in practice within the next 300,000 yr). The Mars-crossers are about 4× more numerous than the NEOs of equal absolute magnitude (statistics done on bodies with H < 15, which constitute an almost complete sample in both populations). It was believed in the past that most q > 1.3 AU Mars-crossers were bodies extracted from the main transportation resonances (i.e., 3:1 and ν_6 resonances) by close encounters with Mars. But, as pointed out by Migliorini et al., the eccentricity of bodies in these resonances increases so rapidly to Earth-crossing values that only a few bodies can be extracted by Mars and emplaced in the Mars-crossing region with q > 1.3 AU. This low probability is only partially compensated by the fact that, once bodies enter this Mars-crossing region, their dynamical lifetime becomes about 10× longer. Indeed, work on numerical integrations indicate that if the 3:1 and ν_6 resonances sustained both the Mars-crossing population with q > 1.3 AU and the NEO population, the ratio between these populations would be only 0.25 (i.e., 16× smaller than observed).

Figure 1 shows that the population of Mars-crossers extends up to a ~ 2.8 AU, suggesting that the phenomenon of chaotic diffusion from the main belt extends at least up to this threshold. The (a,i) panel shows that in addition to the main population situated below the ν_6 resonance (called IMC hereafter), there are two groups of Mars-crossers with orbital elements that mimic those of the Hungaria (1.77 < a < 2.06 AU and i > 15°) and Phocaea (2.1 < a < 2.5 AU and i >18°) populations, arguing for the effectiveness of chaotic diffusion as well in these high-inclination regions. To reach Earth-crossing orbit, the Mars-crossers random walk in semimajor axis under the effect of martian encounters until they enter a resonance that is strong enough to further decrease their perihelion distance below 1.3 AU. For the IMC group, the median time required to become an Earth-crosser is ~60 m.y.; about two bodies larger than 5 km become NEOs every million years (*Michel et al.,* 2000b), consistent with the supply rate from the main belt estimated by *Morbidelli and Nesvorný* (1999). The mean time spent in the NEO region is 3.75 m.y. (*Bottke et al.,* 2002a). The median time to reach Earth-crossing orbits from the two groups of high inclined Mars-crossers exceeds 100 m.y. (*Michel et al.,* 2000b).

The paucity of Mars-crossers with a > 2.8 AU is not due to the inefficiency of chaotic diffusion in the outer asteroid belt. It is simply the consequence of the fact that the dynamical lifetime of bodies in the Mars-crossing region drops with increasing semimajor axis toward the Jupiter-crossing limit. In addition, the observational biases for kilometer-sized asteroids are more severe than in the a < 2.8 AU region. In fact, the outer belt is densely crossed by high-order mean-motion resonances with Jupiter and three-body resonances with Jupiter and Saturn, so an important escape rate into the NEO region should be expected. *Bottke et al.* (2002a) have integrated for 100 m.y. nearly 2000 observed main-belt asteroids with 2.8 < a < 3.5 AU and i < 15° and q < 2.6 AU; almost 20% of them entered the NEO region. About 30,000 bodies with H < 18 can be estimated to exist in the region covered by Bottke et al.'s initial conditions. According to Bottke et al. integrations, in a steady-state scenario this population could provide ~600 new H < 18 NEOs per million years, but the mean time that these bodies spend in the NEO region is only ~0.15 m.y.

3.6. Cometary Contribution

Despite the fact that asteroids dominate the NEO population with small semimajor axes, comets are also expected to be important contributors to the overall NEO population. Comets can be subdivided into two groups: those coming from the Kuiper Belt (or, more likely, the scattered disk) and those coming from the Oort Cloud. The first group includes the Jupiter-family comets (JFCs). Their orbital distribution has been well-characterized with numerical integrations by *Levison and Duncan* (1997). Their cometary test bodies, however, remained confined to a > 2.5 AU orbits, probably because terrestrial planet perturbations and nongravitational forces were not included in the simulations.

The population of comets of Oort Cloud origin includes the long periodic and Halley-type groups. To explain the orbital distribution of the observed population, several researchers have postulated that the comets from the Oort Cloud rapidly "fade" away, either becoming inactive or splitting into small components (*Wiegert and Tremaine,* 1999; *Levison et al.,* 2001). Since the number of faded comets to new comets has yet to be determined, calculating the population on NEO orbits is problematic. Despite this, best-guess estimates suggest that impacts from Oort Cloud comets may be responsible for 10–30% of the craters on Earth (see *Weissman et al.,* 2002). However, recent unpublished work (Levison et al., personal communication, 2001) reduces this estimate to only ~1%.

4. EVOLUTION IN NEO SPACE

The dynamics of the bodies in NEO space is strongly influenced by close encounters with the planets. Each encounter provides an impulse velocity to the body's trajectory, causing the semimajor axis to "jump" by a quantity depending on the geometry of the encounter and the mass of the planet. The change in semimajor axis is correlated with the change in eccentricity (and inclination) by the quasiconservation of the so-called Tisserand parameter

$$T = \frac{a_p}{a} + 2\sqrt{\frac{a(1-e^2)}{a_p}}\cos i \qquad (1)$$

relative to the encountered planet with semimajor axis a_p (*Öpik,* 1976). An encounter with Jupiter can easily eject the body from the solar system (a = ∞ or negative), while this is virtually impossible in encounters with the terrestrial planets.

Under the sole effect of close encounters with a unique planet, and neglecting the effects on the inclination, a body would randomly walk on a curve of the (a, e) plane defined by T = constant. These curves are transverse to all mean-motion resonances and to most secular resonances, so that the body can be extracted from a resonance and be transported into another one. Resonances, on the other hand, change the eccentricity and/or the inclination of the bodies, keeping the semimajor axis constant. The real dynamics in the NEO region are therefore the result of a complicated interplay between resonant dynamics and close encounters (see *Michel et al.,* 1996a). A further complication is that encounters with several planets can occur at the same time, thus breaking the Tisserand parameter approximation even in the absence of resonant effects.

As anticipated in the previous section, most bodies that become NEOs with a > 2.5 AU are preferentially transported to the outer solar system or are ejected on hyperbolic orbit. In fact, if the eccentricity is sufficiently large, the NEOs in this region approach the Jupiter-crossing limit where the giant planet can scatter them outward. At that point, their dynamics become similar to that of Jupiter-family comets. A typical example is given by the rightmost evolution in Fig. 2a. The body penetrates the NEO region by increasing its orbital eccentricity inside the 5:2 resonance until e ~ 0.7; the encounters with Jupiter then extract it from the resonance and transport it to a larger semimajor axis. Only bodies extracted from the resonance by Mars or Earth and rapidly transported on a low-eccentricity path to a smaller semimajor axis could escape the scattering action of Jupiter. But this evolution is increasingly unlikely as the initial semimajor axis is set to larger and larger values.

Conversely, the bodies on orbits with a ~ 2.5 AU or smaller do not approach Jupiter even at e ~ 1, so that they end their evolution preferentially by impacting the Sun. Most of the bodies originally in the 3:1 resonance are transported to e ~ 1 without having a chance of being extracted from the resonance. For the bodies originally in the ν_6 resonance, a temporary extraction from the resonance in the Earth-crossing space is more likely (Fig. 2a). Most of the IMCs with 2 < a < 2.5 AU eventually enter the 3:1 or the ν_6 resonance and subsequently behave as resonant particles.

Figures 2b and 2c show evolutionary paths from ν_6 and 3:1, which are unlikely (they occur in ~10% of the cases), but extremely important to understanding the observed NEO orbital distribution. In these cases, encounters with Earth or Venus extract the body from its original resonance and transport it into the region with a < 2 AU (hereafter called the "evolved region"). Once in the evolved region, the dynamical lifetime grows longer (~10 m.y., *Gladman et al.,* 2000) because there are no statistically significant dynamical mechanisms that pump the eccentricity up to Sun-grazing values. To be dynamically eliminated, the bodies in the evolved region must either collide with a terrestrial planet (rare) or be driven back to a > 2 AU, where powerful resonances can push them into the Sun. The enhanced lifetime partially compensates for the difficulty these

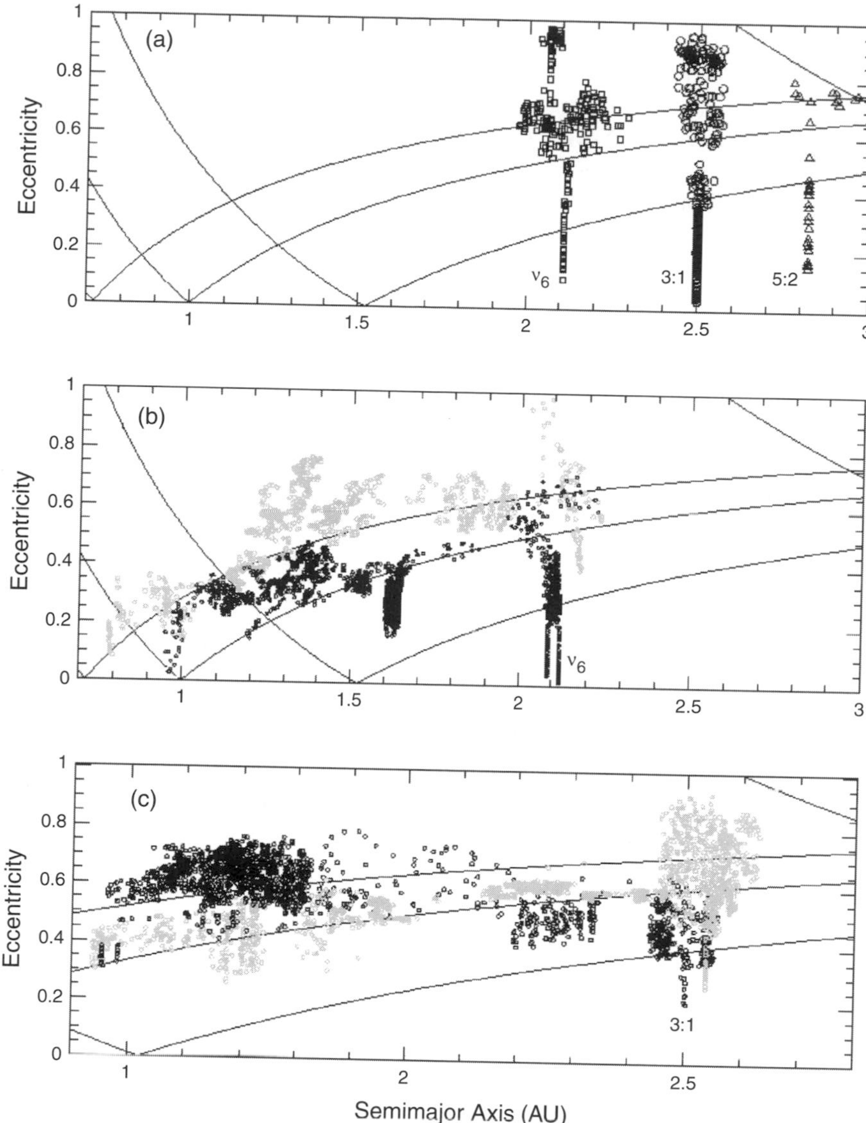

Fig. 2. Orbital evolutions of objects leaving the main asteroid belt. The curves indicate the sets of orbits having aphelion or perihelion at the semimajor axis of one of the planets Venus, Earth, Mars, or Jupiter. **(a)** Common evolution paths from the ν_6, 3:1, and 5:2 resonances. **(b)** A long-lived particle from the ν_6 resonance. The evolution before 20 m.y. is shown in black, and the subsequent one is illustrated in gray. **(c)** A long-lived particle from 3:1; black dots show the evolutions during the first 15 m.y. Adapted from *Gladman et al.* (1997).

objects have in reaching the evolved region. Thus, the latter should host 38% and 70% respectively of all NEOs of 3:1 and ν_6 origin (*Bottke et al.,* 2002a).

The dynamical evolution in the evolved region can be very tortuous, and does not follow any clear curve of constant Tisserand parameter. The main reason is that first-order secular resonances are present and effective in this region, and continuously change the perihelion distance of the body (see *Michel et al.,* 1996b; *Michel and Froeschlé,* 1997; *Michel,* 1997a). In some cases the perihelion distance can be temporarily raised above 1 AU, where, in the absence of encounters with Earth, the body can reside for several million years (notice the density of points at a ~ 1.62, e ~ 0.3 in Fig. 2b). The Kozai resonance (*Michel and Thomas,* 1996; *Gronchi and Milani,* 1999) and the mean-motion resonances with the terrestrial planets (*Milani and Baccili,* 1998) can also provide a protection mechanism against close approaches, thus stabilizing the motion of the bodies temporarily locked into them. Among mean-motion resonance trapping, of importance is the temporary capture in a coorbital state, where the body follows horseshoe and/or tadpole librations about a planet. This phenomenon has been observed in the numerical simulations by *Michel et al.* (1996b), *Michel* (1997b), and *Christou* (2000), and a detailed theory has been developed by *Namouni* (1999). The asteroid (3753) 1986 TO has been shown to currently evolve on a highly inclined horseshoe orbit around Earth (*Wiegert et al.,* 1997).

We point out a few additional features in the evolution of Fig. 2b that we find remarkable. The body passes into the region with a ~ 1 AU and e < 0.1 twice, where the so-called small Earth approachers (SEAs) have been observed (*Rabinowitz et al.,* 1993). This shows that the SEAs could come from the main belt. Also, the evolution penetrates in the region inside Earth's orbit (aphelion distance smaller than 0.983 AU). It is then plausible to conjecture the existence of a population of "interior to the Earth objects" (IEOs), despite the fact that observations have not yet discovered any such bodies (*Michel et al.,* 2000a).

5. QUANTITATIVE MODELING OF THE NEO POPULATION

The observed orbital distribution of NEOs is not representative of the real distribution, because strong biases exist against the discovery of objects on some types of orbits.

Given the pointing history of a NEO survey, the observational bias for a body with a given orbit and absolute magnitude can be computed as the probability of being in the field of view of the survey, with an apparent magnitude brighter than the limit of detection (see *Jedicke et al.,* 2002). Assuming random angular orbital elements of NEOs, the bias is a function B(a, e, i, H), dependent on semimajor axis, eccentricity, inclination, and absolute magnitude H. Each NEO survey has its own bias. Once the bias is known, in principle the real number of objects N can be estimated as

$$N(a, e, i, H) = n(a, e, i, H)/B(a, e, i, H)$$

where n is the number of objects detected by the survey. The problem, however, is resolution; even a coarse binning in the four-dimensional orbital-magnitude space of the bias function and the observed distribution requires the use of about 10,000 cells. The total number of NEOs detected by the most efficient surveys is a few hundreds. Thus n is zero on the vast majority of the cells, and it is equal to 1 in most other cells; cells with n > 1 are very rare. The debiasing of the NEO population is therefore severely affected by small number statistics. To circumvent this problem, one can consider projected distributions of the detected NEOs, such as n(a, e), n(i), and n(H), and try to debias each of them independently, using a mean bias [$\bar{B}(a, e)$, $\bar{B}(i)$, or $\bar{B}(H)$], computed by summation of B(a, e, i, H) over the hidden dimensions (*Stuart,* 2000). However, the study of the dynamics shows that the inclination distribution is strongly dependent on a and e: Dynamically young NEOs with large semimajor axes roughly preserve their main-belt-like inclination, while dynamically old NEOs in the evolved region have a much broader inclination distribution, due to the action of the multitude of resonances that they crossed during their lifetime. As a consequence, this simplified debiasing method can lead to very approximate results.

An alternative way to construct a model of the real distribution of NEOs heavily relies on the dynamics. In fact, from numerical integrations, it is possible to estimate the steady-state orbital distribution of the NEOs coming from each of the main source regions defined in section 3 (see below for the method). In this approach, the key assumption is that the NEO population is currently in steady state. This requires some comments. The analysis of lunar and terrestrial craters suggests that the impact flux on the Earth-Moon system has been more or less constant for the last ~3 G.y. (*Grieve and Shoemaker,* 1994), thus supporting the idea of a gross steady state of the NEO population. This view is challenged, however, by the analysis of the asteroid belt, which shows that the formation of several asteroid families occurred over the age of the solar system (see *Zappalà et al.,* 2002).

The formation of asteroid families may have had two consequences on the time evolution of the NEO population:

1. A large number of asteroids might have been injected into the resonances at the moment of the parent bodies breakup, producing a temporary large increase in the NEO population (*Zappalà et al.,* 1997). Because the median lifetime of resonant particles is ~2 m.y., the NEO population should have returned to a normal "steady-state" value in about 10 m.y. Being short-lived, these temporary spikes in the total number of NEOs should not have left an easily detectable trace in the terrestrial planet crater record. However, evidence that one of these spikes really happened — at least for meteorite-sized NEOs — comes from the discovery of 17 "fossil meteorites" in a 480-m.y.-old limestone (*Schmitz et al.,* 1996, 1997). These discoveries suggest that the influx of meteorites during that 1–2-m.y. period was 10–100× higher than at present. If the general belief that families are several 10^8–10^9 yr old (*Marzari et al.,* 1995) is correct, the current NEO population should not be affected by any of these hypothetical spikes.

2. The rate at which bodies have been continuously supplied to some transporting resonance(s) might have increased after a major family formation event enhanced the population in the resonance vicinity. In this case the NEO population would have evolved to a new and different steady-state situation. Possible evidence for such a transition comes from *Culler et al.* (2000), who claimed that the lunar impactor flux has been higher by a factor of 3.7 ± 1.2 over the last 400 m.y. than over the previous 3 G.y. Even this case, however, would not invalidate the steady-state assumption, essential for constructing a model of the NEO distribution based on statistics of the orbital dynamics. In fact, being short-lived, the NEO population preserves no memory of what happened several 10^8 yr ago; what is important is that the NEO population has been in the same "steady state" during the last several ~10^7 yr, namely on a timescale significantly longer than the mean lifetime in the NEO region.

The recipe to compute the steady-state orbital distribution of the NEOs for a given source is as follows. First, the dynamical evolutions of a statistically significant number of particles, initially placed in the NEO source region(s), are numerically integrated. The particles that enter the NEO region are followed through a network of cells in the (a, e, i) space during their entire dynamical lifetime. The mean time spent in each cell (called residence time hereafter) is computed. The resultant residence time distribution shows where the bodies from the source statistically spend their time in the NEO region. As is well known in statistical mechanics, in a steady-state scenario, the residence time distribution is equal to the relative orbital distribution of the NEOs that originated from the source.

This dynamical approach was first introduced by *Wetherill* (1979) and was later imitated by *Rabinowitz* (1997a). Unfortunately, these works used Monte Carlo simulations, which allowed only a very approximate knowledge of the statistical properties of the dynamics. *Bottke et al.* (2000) have reactualized the approach using modern numerical integrations. They computed the steady-state orbital distributions of the NEOs from three sources: the ν_6 resonance, the 3:1 resonance, and the IMC population. The latter was considered to be representative of the outcome of all diffusive resonances in the main belt up to a = 2.8 AU. The overall NEO orbital distribution was then constructed as a linear combination of these three distributions, thus obtaining a two-parameter model. The NEO magnitude distribution, assumed to be source-independent, was constructed so its shape could be manipulated using an additional parameter. Combining the resulting NEO orbital-magnitude distribution with the observational biases associated with the Spacewatch survey (*Jedicke,* 1996), *Bottke et al.* (2000) obtained a model distribution which could be fit to the orbits and magnitudes of the NEOs discovered or accidentally rediscovered by Spacewatch. To have a better match with the observed population at large a, *Bottke et al.* (2002a) extended the model by also considering the steady-state orbital distributions of the NEOs coming from the outer asteroid belt (a > 2.8 AU) and from the Transneptunian region. The resulting best-fit model nicely matches the distribution of the NEOs observed by Spacewatch, without restriction on the semimajor axis (see Fig. 10 of *Bottke et al.,* 2002a).

Notice that, once the values of the parameters of the model are computed by best-fitting the observations of *one* survey, the steady-state orbital-magnitude distribution of the *entire* NEO population is determined. This distribution is valid also in those regions of the orbital space that have never been sampled by any survey because of extreme observational biases. For instance, Bottke et al. models predict the total number of IEOs, despite the fact that none of these objects has ever been detected. This underlines the power of the dynamical approach for debiasing the NEO population.

6. DEBIASED NEO POPULATION

This section is based strongly on the results of the modeling effort by *Bottke et al.* (2002a). Unless explicitly stated, all numbers reported below are taken from that work.

The total NEO population contains 960 ± 120 objects with absolute magnitude H < 18 (roughly 1 km in diameter) and a < 7.4 AU. The observational completeness of this population as of July 2001 is ~45%. The NEO absolute magnitude distribution is of type $N(<H) = C \times 10^{0.35 \pm 0.02 H}$ in the range 13 < H < 22, implying 24,500 ± 3000 NEOs with H < 22. Assuming that the albedo distribution is not dependent on H, this magnitude distribution implies a power-law cumulative size distribution with exponent –1.75 ± 0.1. This distribution is in perfect agreement with that obtained by *Rabinowitz et al.* (2000), who directly debiased the magnitude distribution observed by the NEAT survey. Also, once scaling laws are taken into account, it is in agreement with the –2 mean exponent of the cumulative crater size distributions observed on all terrestrial planets and the Moon (see *Ivanov et al.,* 2002).

The debiased orbital-magnitude distribution of the NEOs with H < 18 is shown in Fig. 3. For comparison, the figure also reports the distribution of discovered objects, combining the results of all NEO surveys. Most of the undiscovered NEOs have H > 16, e > 0.4, a = 1–3 AU, and i = 5°–40°. Overall, 32 ± 1% of the NEOs are Amors, 62 ± 1% are Apollos, and 6 ± 1% are Atens. Of the NEOs, 49 ± 4% should be in the evolved region (a < 2 AU), where the dynamical lifetime is strongly enhanced. The ratio between the IEO and the NEO populations is about 2%.

With this orbital distribution, and assuming random values for the argument of perihelion and the longitude of node, about 21% of the NEOs turn out to have a minimal orbital intersection distance (MOID) with the Earth smaller than 0.05. The MOID is defined as the minimal distance between the osculating orbits of two objects. NEOs with MOID < 0.05 AU are classified as potentially hazardous objects (PHOs), and the accurate orbital determination of these bodies is considered a top priority.

To estimate the collision probability with the Earth, we use the collision probability model described by *Bottke et al.* (1994). This model, which is an updated version of similar models described by *Öpik* (1951), *Wetherill* (1967), and *Greenberg* (1982), assumes that the values of the mean anomalies of the Earth and the NEOs are random. The gravitational attraction exerted by the Earth is also included. We find that, on average, the Earth collides with an H < 18 NEO once every 0.5 m.y. Figure 4 shows how the collision probability is distributed as a function of the NEO orbital elements or, equivalently, which NEOs carry most of the collisional hazard. For an individual object, the collision probability with the Earth decreases with increasing semimajor axis, eccentricity, and/or inclination (*Morbidelli and Gladman,* 1998). For this reason, the histograms in Fig. 4 are very different from those in Fig. 3: The a distribution peaks at 1 AU instead of 2.2 AU, the e distribution is more symmetric with respect to the e = 0.5 value, and the i distribution is much steeper.

The stated goal of the Spaceguard survey, a NEO survey program proposed in the early 1990s, was to discover 90% of the NEOs with H < 18 (*Morrison,* 1992). However, it would be more appropriate to state the goal in terms of discovering the NEOs carrying 90% of the total collision probability. Figures 3 and 4 show that the two goals are not equivalent. For instance, the Atens, even though they are only 6% of the total NEO population, carry about 20% of the total collision probability; thus, their discovery (of sec-

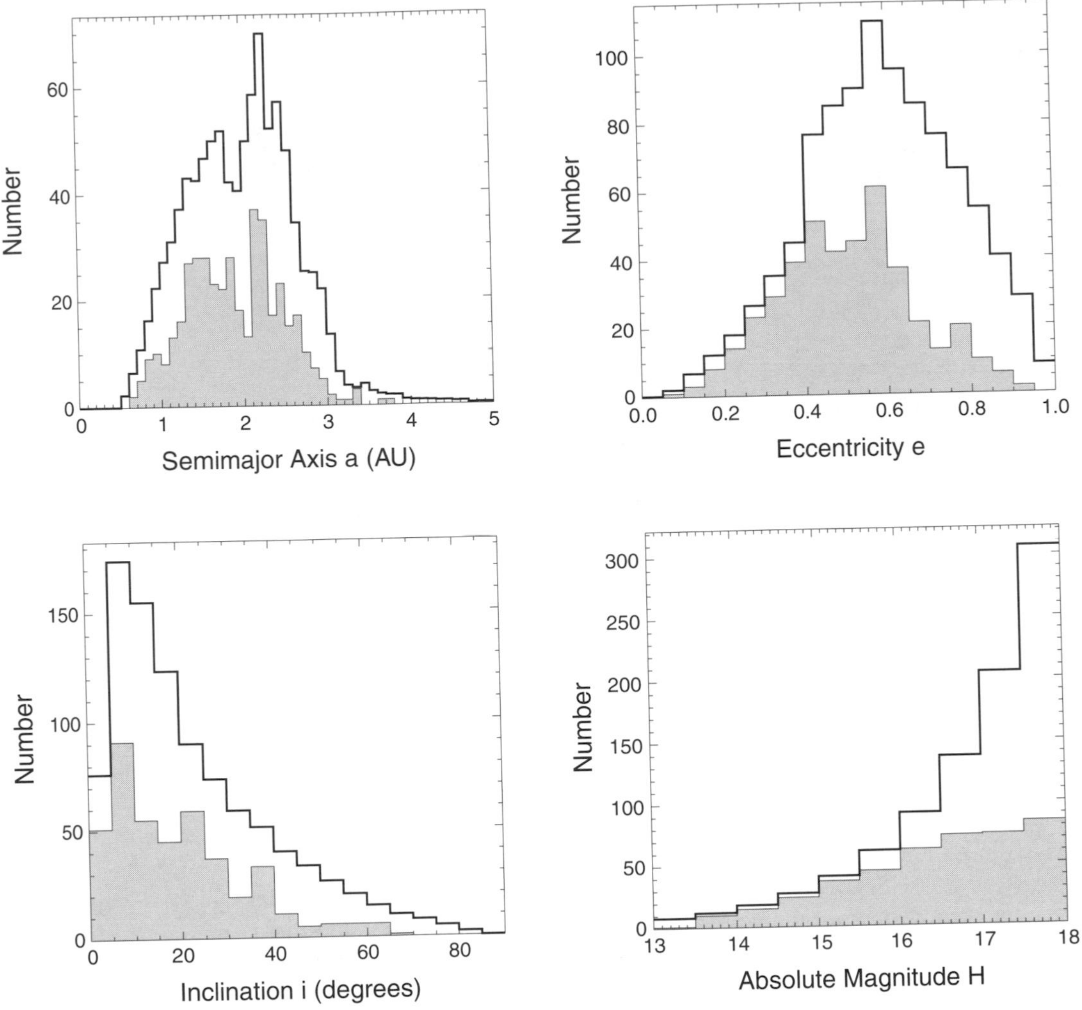

Fig. 3. The steady-state orbital and absolute magnitude distribution of NEOs for H < 18. The predicted NEO distribution (solid line) is normalized to 960 objects. It is compared with the 426 known NEOs for all surveys (shaded histogram). From *Bottke et al.* (2002a).

ondary importance for the original Spaceguard goal) becomes a top priority, when collisional hazards are taken into account.

By applying the same collision probability calculations to the H < 18 NEOs discovered so far, we find that the known objects carry about 47% of the total collisional hazard. Thus, the current completeness of the population computed in terms of collision probability is about the same of that computed in terms of number of objects. This seems to imply that the current surveys discover NEOs more or less evenly with respect to the collision probability with the Earth. In Fig. 4 the shaded histogram shows how the collision probability of the known NEOs is distributed as a function of the orbital elements.

Finally, we discuss the origin of NEOs. The *Bottke et al.* (2002a) model implies that 37 ± 8% of the NEOs with 13 < H < 22 come from the ν_6 resonance, 25 ± 3% from the IMC population, 23 ± 9% from the 3:1 resonance, 8 ± 1% from the outer-belt population, and 6 ± 4% from the Transneptunian region. Thus, the long-debated cometary contribution to the NEO population probably does not exceed 10%. Note, however, that the Bottke et al. model does not account for the contribution of comets of Oort Cloud origin, which is still largely unconstrained for the reasons explained in section 3.6. These comets should be relegated to orbits with a > 2.6 AU and/or orbits with large eccentricities and inclinations.

About 800 bodies with H < 18, 70% of which come from the outer main belt, should escape from the asteroid belt every million years in order to sustain the NEO population. These fluxes may seem huge, but they only imply a mass loss of ~5×10^{16} kg/m.y., or the equivalent of 6% of the total mass of the asteroid belt over 3.5 G.y. (assuming that the flux has been constant over this timescale).

By itself, this flux does not constrain the mechanism (or mechanisms) that resupply the transporting (powerful and diffusing) resonances with new bodies. The shallow size distribution of NEOs, however, suggests that collisional injection probably does not play a dominant role in the delivery of asteroidal material to resonances. If it did, the NEOs, being fragments from catastrophic breakups, would have a much steeper size distribution, at least like that of

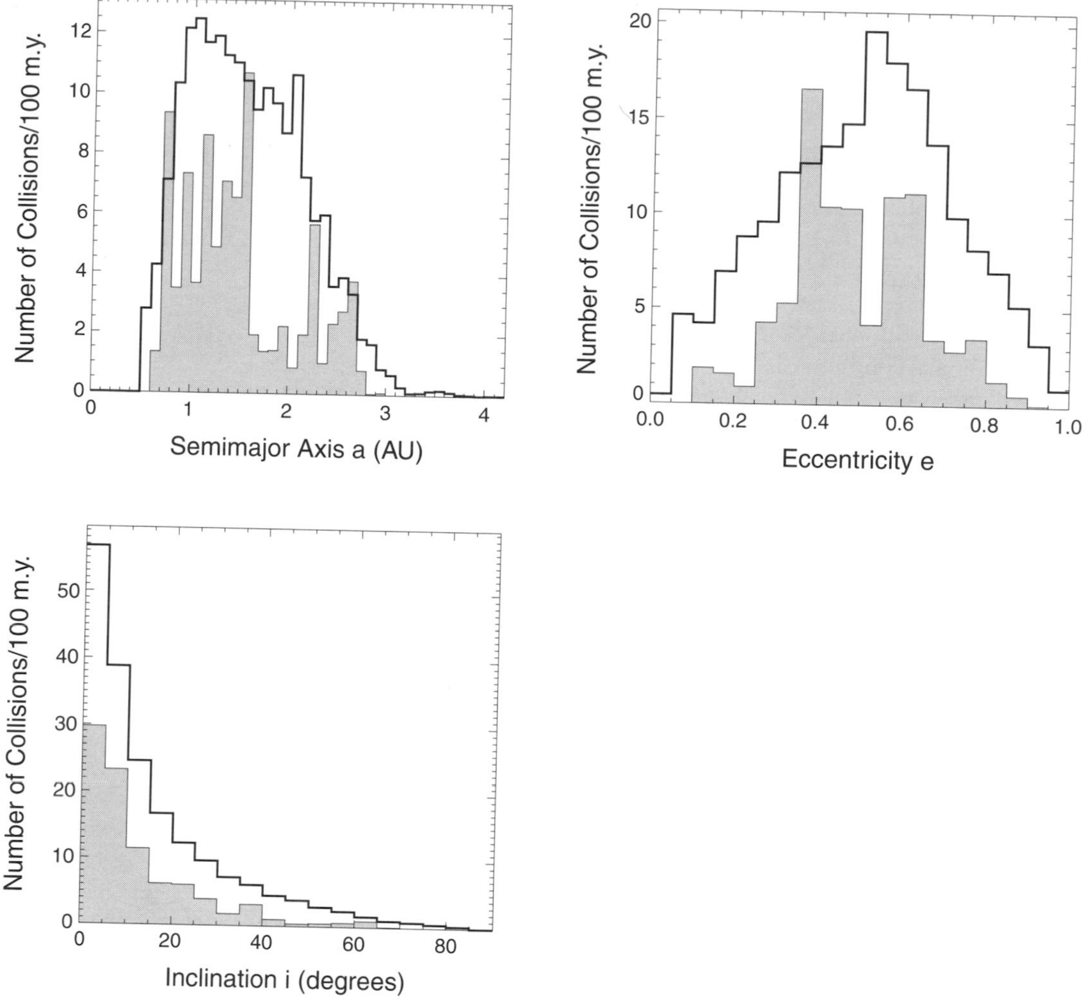

Fig. 4. Number of Earth impacts with H < 18 NEOs as a function of the NEO orbital distribution. The solid histogram shows the expected distribution, deduced from the *Bottke et al.* (2002a) model; the shaded histogram is obtained applying the collision probability calculation to the population of known objects.

the observed asteroid families (*Tanga et al.,* 1999; *Campo Bagatin et al.,* 2000). Recall that the mean lifetime in the NEO region is only a few million years, too short for collisional erosion to significantly reduce the slope of a size distribution dominated by fresh debris (bodies with a diameter of about 170 m have a collisional lifetime >100 m.y.; *Bottke et al.,* 1994). Moreover, it is unclear how collisional injection could explain the relative abundance of multikilometer objects in the NEO population. According to standard collision models, only the largest (and most infrequent) catastrophic disruption events are capable of throwing multikilometer objects into the transporting major resonances (*Menichella et al.,* 1996; *Zappalà and Cellino,* 2002). The NEO size distribution shows some interesting similarities with the main-belt size distribution. At present, a direct estimate of the latter is available only for D > 2 km asteroids (*Jedicke and Metcalfe,* 1998), but its shape has been extrapolated to smaller sizes by *Durda et al.* (1998) using a collisional model. The results suggest the main belt's size distribution for 0.2 < D < 5 km asteroids is NEO-like or possibly shallower. The results of recent surveys (*Ivezić et al.,* 2001; *Yoshida et al.,* 2001) seem to confirm this expectation.

Although more quantitative studies are needed for a definite conclusion, we suspect that this paradox might be solved by the Yarkovsky effect. The latter forces kilometer-sized bodies to drift in semimajor axis by ~10^{-4} AU m.y.$^{-1}$ (*Farinella and Vokrouhlický,* 1999; see also *Bottke et al.,* 2002b), enough to bring into resonance a large — possibly sufficient — number of bodies. The Yarkovsky drift rate is slow enough that the size distribution of fresh collisional debris would have the time to collisionally evolve to a shallower, main-belt-like slope.

7. CONCLUSIONS AND PERSPECTIVES

The massive use of long-term numerical integrations have boosted our understanding of the origin, evolution, and steady-state orbital distribution of NEOs. The "powerful resonances" of the main belt have been shown to transport the asteroids into the NEO region on a timescale of only a few 10^5 yr. On the other hand, we now understand that these resonances eliminate most of these same bodies by forcing them to collide with the Sun. The "diffusive resonances" have been shown to be additional important sources of NEOs. The steady-state orbital distributions of the NEO subpopulations related to the various sources have been computed. This work has allowed the construction of a quantitative model of the debiased orbital and magnitude distribution of the NEO population, calibrated on available observations.

Despite the steady progress, the current understanding of the NEO population is still lacking in several areas. As more and more NEOs are continuously found, the NEO model presented here will certainly need to be refined. Moreover, although it is now clear that the vast majority of NEOs with semimajor axes inside Jupiter's orbit are of asteroidal origin, the contribution of inactive comets coming from the Oort Cloud remains poorly quantified.

The most unclear part of the scenario regarding the origin and evolution of NEOs concerns the mechanisms that resupply the transporting resonances with new asteroids. At the end of the previous section, we discussed several arguments that suggest that the Yarkovsky effect plays the dominant role in moving small asteroids to powerful and diffusive resonances. More definitive conclusions, however, will require a better knowledge of the orbital and size distribution of the main-belt population in the kilometer size range, in order to quantify the availability of material in the neighborhood of the transporting resonances. Our understanding of asteroidal and cometary collisional physics is also limited, and additional work will be needed to determine more accurately the size-frequency and velocity-frequency distributions of asteroid disruption events. Simulations are also required to understand the interaction of the Yarkovsky-drifting bodies with the multitude of diffusive resonances that they encounter. It is possible that small bodies drift too fast to be efficiently captured by these thin resonances. If it turns out that different mechanisms resupply different NEO source regions, it may be inappropriate to assume that the NEO subpopulations coming from the various sources described in this chapter have the same size distribution.

Finally, we need to better understand the changes that occurred over the last 3.8 G.y. (after the so-called late heavy bombardment), in terms of the total number of NEOs and their size or orbital distribution. For this purpose, we need to understand whether the members of asteroid families dominate the rest of the main-belt asteroid population. If they do (as claimed by *Zappalà and Cellino,* 1996), the formation of new families should have significantly changed the main-belt population and consequently altered the feeding rate of the NEO population. If not, the breakup of asteroid families can be considered noise in the history of the NEO population.

We believe that a detailed knowledge of the physical properties of asteroids close to the transporting resonances will eventually allow us to indirectly but accurately deduce the compositional distribution of the NEO population. In turn, this will enable the conversion of NEO absolute magnitudes into diameters, and better quantify the collisional hazard (frequency of collisions as a function of the impact energy) for Earth.

REFERENCES

Alvarez L. W., Alvarez W., Asaro F., and Michel H. V. (1980) Extraterrestrial cause for the Cretaceous Tertiary extinction. *Science, 208,* 1095–1099.

Anders E. (1964) Origin, age and composition of meteorites. *Space Sci. Rev., 3,* 583–574.

Anders E. and Arnold J. R. (1965) Age of craters on Mars. *Science, 149,* 1494–1496.

Bottke W. F., Nolan M. C., Greenberg R., and Kolvoord R. A. (1994) Velocity distributions among colliding asteroids. *Icarus, 107,* 255–268.

Bottke W. F., Jedicke R., Morbidelli A., Petit J. M., and Gladman B. (2000) Understanding the distribution of near-Earth asteroids. *Science, 288,* 2190–2194.

Bottke W. F., Morbidelli A., Jedicke R., Petit J. M., Levison H. F., Michel P., and Metcalfe T. S. (2002a) Debiased orbital and size distribution of the Near Earth Objects. *Icarus,* in press.

Bottke W. F. Jr., Vokrouhlický D., Rubincam D. P., and Brož M. (2002b) Dynamical evolution of asteroids and meteoroids using the Yarkovsky effect. In *Asteroids III* (W. F. Bottke Jr. et al., eds.), this volume. Univ. of Arizona, Tucson.

Campo Bagatin A., Martinez V. J., and Paredes S. (2000) Multinomial fits to the observed main belt asteroid distribution (abstract). In *Bull. Am. Astron. Soc., 32,* 813.

Christou A. (2000) A numerical survey of transient co-orbitals of the terrestrial planets. *Icarus, 144,* 1–20.

Culler T. S., Becker T. A., Muller R. A., and Renne P. R. (2000) Lunar impact history from ^{40}Ar/^{39}Ar dating of glass spherules. *Science, 287,* 1785–1788.

Dones L., Gladman B., Melosh H. J., Tonks W. B., Levison H. F., and Duncan M. (1999) Dynamical lifetimes and final fates of small bodies: Orbit integrations vs Öpik calculations. *Icarus, 142,* 509–524.

Durda D. D., Greenberg R., and Jedicke R. (1998) Collisional models and scaling laws: A new interpretation of the shape of the main-belt asteroid size distribution. *Icarus, 135,* 431–440.

Farinella P. and Vokrouhlický D. (1999) Semimajor axis mobility of asteroidal fragments. *Science, 283,* 1507–1510.

Farinella P., Froeschlé Ch., and Gonczi R. (1993). Meteorites from the asteroid 6 Hebe. *Cel. Mech. Dyn. Astron., 62,* 87–305.

Farinella P., Froeschlé Ch., Gonczi R., Hahn G., Morbidelli A., and Valsecchi G. B. (1994) Asteroids falling onto the Sun. *Nature, 371,* 315–317.

Ferraz-Mello S. and Klafke J. C. (1991) A model for the study of the very high eccentricity asteroidal motion: The 3:1 resonance. In *Predictability, Stability and Chaos in N-Body Dynamical Systems* (A. E. Roy, ed.), pp. 177–187. Plenum, New York.

Froeschlé Ch. and Morbidelli A. (1994) The secular resonances in the solar system. In *Asteroids, Comets, and Meteors, 1993* (A. Milani et al., eds.), pp. 189–204. Kluwer, Boston.

Froeschlé Ch. and Scholl H. (1987) Orbital evolution of asteroids near the secular resonance ν_6. *Astron. Astrophys., 179,* 294–303.

Gladman B., Migliorini F., Morbidelli A., Zappalà V., Michel P., Cellino A., Froeschlé Ch., Levison H., Bailey M., and Duncan

M. (1997) Dynamical lifetimes of objects injected into asteroid belt resonances. *Science, 277,* 197–201.

Gladman B., Michel P., and Froeschlé Ch. (2000) The Near-Earth Object population. *Icarus, 146,* 176–189.

Greenberg R. (1982) Orbital interactions — A new geometrical formalism. *Astron. J., 87,* 184–195.

Grieve R. A. and Shoemaker E. M. (1994) The record of past impacts on Earth. In *Hazards Due to Comets and Asteroids* (T. Gehrels and M. S. Matthews, eds.), pp. 417–462. Univ. of Arizona, Tucson.

Gronchi G. F. and Milani A. (1999) The stable Kozai state for asteroids and comets with arbitrary semimajor axis and inclination. *Astron. Astrophys., 341,* 928–937.

Henrard J. and Caranicolas N. D. (1990) Motion near the 3:1 resonance of the planar elliptic restricted three body problem. *Cel. Mech. Dyn. Astron., 47,* 99–108.

Ivanov B. A., Neukum G., Hartmann W. K., and Bottke W. F. Jr. (2002) Comparison of size-frequency distributions of impact craters and asteroids and the planetary cratering rate. In *Asteroids III* (W. F. Bottke Jr. et al., eds.), this volume. Univ. of Arizona, Tucson.

Ivezić Z. and 31 colleagues (2001) Solar system objects observed in the Sloan digital sky survey commissioning data. *Astron. J., 122,* 2749–2784.

Jedicke R. (1996) Detection of near Earth asteroids based upon their rates of motion. *Astron. J., 111,* 970–983.

Jedicke R. and Metcalfe T. S. (1998) The orbital and absolute magnitude distributions of main belt asteroids. *Icarus, 131,* 245–260.

Jedicke R., Larsen J., and Spahr T. (2002) Observational selection effects in asteroid surveys and estimates of asteroid population sizes. In *Asteroids III* (W. F. Bottke Jr. et al., eds.), this volume. Univ. of Arizona, Tucson.

Knežević Z., Lemaitre A., and Milani A. (2002) Asteroid proper elements determination. In *Asteroids III* (W. F. Bottke Jr. et al., eds.), this volume. Univ. of Arizona, Tucson.

Levison H. F. and Duncan M. J. (1994) The long-term dynamical behavior of short-period comets. *Icarus, 108,* 18–36.

Levison H. F. and Duncan M. J. (1997) From the Kuiper Belt to Jupiter-family comets: The spatial distribution of ecliptic comets. *Icarus, 127,* 13–32.

Levison H. F., Dones L., and Duncan M. J. (2001) The origin of Halley-type comets: Probing the inner Oort Cloud. *Astron. J., 121,* 2253–2267.

Marzari F., Davis D., and Vanzani V. (1995) Collisional evolution of asteroid families. *Icarus, 113,* 168–187.

Menichella M., Paolicchi P., and Farinella P. (1996) The main belt as a source of near-Earth asteroids. *Earth Moon Planets, 72,* 133–149.

Michel P. (1997a) Effects of linear secular resonances in the region of semimajor axes smaller than 2 AU. *Icarus, 129,* 348–366.

Michel P. (1997b) Overlapping of secular resonances in a Venus horseshoe orbit. *Astron. Astrophys., 328,* L5–L8.

Michel P. and Froeschlé Ch. (1997) The location of linear secular resonances for semimajor axes smaller than 2 AU. *Icarus, 128,* 230–240.

Michel P. and Thomas F. (1996) The Kozai resonance for near-Earth asteroids with semimajor axes smaller than 2 AU. *Astron. Astrophys., 307,* 310–318.

Michel P., Froeschlé Ch., and Farinella P. (1996a) Dynamical evolution of NEAs: Close encounters, secular perturbations and resonances. *Earth Moon Planets, 72,* 151–164.

Michel P., Froeschlé Ch., and Farinella P. (1996b) Dynamical evolution of two near-Earth asteroids to be explored by spacecraft: (433) Eros and (4660) Nereus. *Astron. Astrophys., 313,* 993–1007.

Michel P., Zappalà V., Cellino A., and Tanga P. (2000a) Estimated abundance of Atens and asteroids evolving on orbits between Earth and Sun. *Icarus, 143,* 421–424.

Michel P., Migliorini F., Morbidelli A., and Zappalà V. (2000b) The population of Mars-crossers: Classification and dynamical evolution. *Icarus, 145,* 332–347.

Migliorini F., Michel P., Morbidelli A., Nesvorný D., and Zappalà V. (1998) Origin of multikilometer Earth- and Mars-crossing asteroids: A quantitative simulation. *Science, 281,* 2022–2024.

Milani A. and Baccili S. (1998) Dynamics of Earth-crossing asteroids: the protected Toro orbits. *Cel. Mech. Dyn. Astron., 71,* 35–53.

Moons M. (1997) Review of the dynamics in the Kirkwood gaps. *Cel. Mech. Dyn. Astron., 65,* 175–204.

Moons M. and Morbidelli A. (1995) Secular resonances in mean motion commensurabilities: The 4:1, 3:1, 5:2 and 7:3 cases. *Icarus, 114,* 33–50.

Moons M., Morbidelli A., and Migliorini F. (1998) Dynamical structure of the 2:1 commensurability and the origin of the resonant asteroids. *Icarus, 135,* 458–468.

Morbidelli A. (1993) Asteroid secular resonant proper elements. *Icarus, 105,* 48–66.

Morbidelli A. (2002) *Modern Celestial Mechanics: Aspects of Solar System Dynamics.* Gordon and Breach, London, in press.

Morbidelli A. and Gladman B. (1998) Orbital and temporal distribution of meteorites originating in the asteroid belt. *Meteoritics & Planet. Sci., 33,* 999–1016.

Morbidelli A. and Henrard J. (1991) The main secular resonances ν_6, ν_5 and ν_{16} in the asteroid belt. *Cel. Mech. Dyn. Astron., 51,* 169–185.

Morbidelli A. and Nesvorný D. (1999) Numerous weak resonances drive asteroids towards terrestrial planets orbits. *Icarus, 139,* 295–308.

Morrison D. (1992) *The Spaceguard Survey: Report of the NASA International Near Earth Object Detection Workshop.* QB651 N37, Jet Propulsion Laboratory/California Institute of Technology, Pasadena.

Murray N. and Holman M. (1997) Diffusive chaos in the outer asteroid belt. *Astron. J., 114,* 1246–1252.

Murray N., Holman M., and Potter M. (1998) On the origin of chaos in the asteroid belt. *Astron. J., 116,* 2583–2589.

Namouni F. (1999) Secular interactions of coorbiting objects. *Icarus, 137,* 293–314.

Nesvorný D. and Ferraz-Mello S. (1997) On the asteroidal population of the first-order Jovian resonances. *Icarus, 130,* 247–258.

Nesvorný D. and Morbidelli A. (1998) Three-body mean motion resonances and the chaotic structure of the asteroid belt. *Astron. J., 116,* 3029–3026.

Nesvorný D. and Morbidelli A. (1999) An analytic model of three-body mean motion resonances. *Cel. Mech. Dyn. Astron., 71,* 243–261.

Nesvorný D., Ferraz-Mello S., Holman M., and Morbidelli A. (2002) Regular and chaotic dynamics in the mean motion resonances: Implications for the structure and evolution of the asteroid belt. In *Asteroids III* (W. F. Bottke Jr. et al., eds.), this volume. Univ. of Arizona, Tucson.

Öpik E. J. (1951) Collision probabilities with the planets and the distribution of interplanetary matter. *Proc. R. Irish Acad., 54A,* 165–199.

Öpik E. J. (1961) The survival of comets and comet material. *Astron. J., 66,* 381–382.

Öpik E. J. (1963) The stray bodies in the solar system. Part I. Survival of comet nuclei and the asteroids. *Advan. Astron. Astrophys., 2,* 219–262.

Öpik E. J. (1976) *Interplanetary Encounters: Close-Range Gravitational Interactions.* Elsevier, Amsterdam. 155 pp.

Rabinowitz D. L. (1997a) Are main-belt asteroids a sufficient source for the Earth-approaching asteroids? Part I. Predicted vs. observed orbital distributions. *Icarus, 127,* 33–54.

Rabinowitz D. L. (1997b) Are main-belt asteroids a sufficient source for the Earth-approaching asteroids? Part II. Predicted vs. observed size distributions. *Icarus, 130,* 287–295.

Rabinowitz D. L., Gehrels T., Scotti J. V., McMillan R. S., Perry M. L., Wisniewski W., Larson S. M., Howell E. S., and Mueller B. E. (1993) Evidence for a near-Earth asteroid belt. *Nature, 363,* 704–706.

Rabinowitz D., Helin E., Lawrence K., and Pravdo S. (2000) A reduced estimate of the number of kilometre-sized near-Earth asteroids. *Nature, 403,* 165–166.

Schmitz B., Lindstrom M., Asaro F., and Tassinari M. (1996) Geochemistry of meteorite-rich marine limestone strata and fossil meteorites from the lower Ordovician at Kinnekulle, Sweden. *Earth Planet. Sci. Lett., 145,* 31–48.

Schmitz B., Peucker-Ehrenbrink B., Lindstrom M., and Tassinari M. (1997) Accretion rates of meteorites and extraterrestrial dust in the early Ordovician. *Science, 278,* 88–90.

Stuart J. S. (2000) The near-Earth asteroid population from Lincoln near-Earth asteroid research project data (abstract). *Bull. Am. Astron. Soc., 32,* 1603.

Tanga P., Cellino A., Michel P., Zappalà V., Paolicchi P., and Dell'Oro A. (1999) On the size distribution of asteroid families: The role of geometry. *Icarus, 141,* 65–78.

Weissman P. R., Bottke W. F. Jr., and Levison H. F. (2002) Evolution of comets into asteroids. In *Asteroids III* (W. F. Bottke Jr. et al., eds.), this volume. Univ. of Arizona, Tucson.

Wetherill G. W. (1967) Collisions in the asteroid belt. *J. Geophys. Res., 72,* 2429–2444.

Wetherill G. W. (1976) Where do the meteorites come from? A re-evaluation of the Earth-crossing Apollo objects as sources of chondritic meteorites. *Geochim. Cosmochim. Acta, 40,* 1297–1317.

Wetherill G. W. (1979) Steady-state populations of Apollo-Amor objects. *Icarus, 37,* 96–112.

Wetherill G. W. (1985) Asteroidal source of ordinary chondrites. *Meteoritics, 20,* 1–22.

Wetherill G. W. (1987) Dynamic relationship between asteroids, meteorites and Apollo-Amor objects. *Phil. Trans. R. Soc. London, 323,* 323–337.

Wetherill G. W. (1988) Where do the Apollo objects come from? *Icarus, 76,* 1–18.

Wiegert P. A., Innanen K. A., and Mikkola S. (1997) An asteroidal companion to the Earth. *Nature, 387,* 685–685.

Wiegert P. A. and Tremaine S. (1999) The evolution of long-period comets. *Icarus, 137,* 84–121.

Wisdom J. (1983) Chaotic behaviour and origin of the 3:1 Kirkwood gap. *Icarus, 56,* 51–74.

Wisdom J. (1985a) A perturbative treatment of motion near the 3:1 commensurability. *Icarus, 63,* 272–289.

Wisdom J. (1985b) Meteorites may follow a chaotic route to the Earth. *Nature, 315,* 731–733.

Wisdom J. and Holman M. (1991) Symplectic maps for the N-body problem. *Astron. J., 102,* 1528–1534.

Yoshida F., Nakamura T., Fuse T., Komiyama Y., Yagi M., Miyazaki S., Okamura S., Ouchi M., and Miyazaki M. (2001) First Subaru observations of sub-km main belt asteroids. *Natl. Astron. Obs. Report, 99,* 1–11.

Yoshikawa M. (1987) A simple analytical model for the ν_6 resonance. *Cel. Mech., 40,* 233–246.

Yoshikawa M. (1989) A survey on the motion of asteroids in commensurabilities with Jupiter. *Astron. Astrophys., 213,* 436–450.

Yoshikawa M. (1990) On the 3:1 resonance with Jupiter. *Icarus, 87,* 78–102.

Zappalà V. and Cellino A. (1996) Main belt asteroids: Present and future inventory. In *Completing the Inventory of the Solar System,* pp. 29–44. ASP Conference Series 107.

Zappalà V. and Cellino A. (2002) A search for the collisional parent bodies of km-sized NEAs. *Icarus,* in press.

Zappalà V., Cellino A., Gladman B. J., Manley S., and Migliorini F. (1997) Asteroid showers on Earth after family breakup events. *Icarus, 134,* 176–179.

Zappalà V., Cellino A., Dell'Oro A., and Paolicchi P. (2002) Physical properties of asteroid families. In *Asteroids III* (W. F. Bottke Jr. et al., eds.), this volume. Univ. of Arizona, Tucson.

Asteroidal Dust

Stanley F. Dermott
University of Florida

Daniel D. Durda
Southwest Research Institute

Keith Grogan
NASA Goddard Space Flight Center

Thomas J. J. Kehoe
University of Florida

There is good evidence that the high-speed, porous, anhydrous chondritic interplanetary dust particles (IDPs) collected in Earth's stratosphere originated from short-period comets. However, by considering the structure of the solar-system dust bands discovered by IRAS, we are able to show that asteroidal collisions are probably the dominant source of particles in the zodiacal cloud. It follows that a significant and probably the dominant fraction of the IDPs collected in Earth's stratosphere also originated from asteroids. IDPs are the most primitive particles in the inner solar system and represent a class of material quite different from that in our meteorite collections. The structure, mineralogy, and high C content of IDPs dictate that they cannot have originated from the grinding down of known meteorite types. We argue that the asteroidal IDPs were probably formed as a result of prolonged mechanical mixing in the deep regoliths of asteroidal rubble piles in the outer main belt.

1. INTRODUCTION

In our collections on Earth, we have an abundance of meteorite samples from three major sources of extraterrestrial material: the asteroid belt, the Moon, and Mars. Some of the source bodies of these meteorites have experienced major physical, chemical, and mineralogical changes since the time of their formation; hence the resulting meteorites, while providing vital information on the origin and evolution of the parent bodies, give limited information on the nature of the primordial particles out of which these bodies accreted. However, this broad statement does not apply to unequilibriated ordinary chondrites. These primitive meteorites are composed almost entirely of particles that were freely orbiting in the solar nebula and although they show some signs of thermal metamorphism (*McSween et al.,* 1988) and aqueous alteration (*Zolensky and McSween,* 1988), they do provide information on initial conditions.

A fourth class of extraterrestrial material consists of particles recently accreted by Earth from the zodiacal cloud. These are the small, mostly 5–25-μm diameter, interplanetary dust particles (IDPs) collected, gently and mostly unaltered, in Earth's stratosphere and the larger micrometeorites (MMs) and cosmic spherules collected from polar ices and deep-sea floors (*Brownlee,* 1985; *Jessberger et al.,* 2001). Some of the IDPs appear to be the most primitive material in the solar system and, at present, provide our best source of information on the nature of the particles in the preplanetary solar nebula (*Bradley,* 1999). Observations of microcraters on the Long Duration Exposure Facility (LDEF) confirmed that each year Earth accretes 3×10^7 kg of dust particles, a mass influx ~100× greater than the influx associated with the much larger meteorites that have masses between 100 g and 1000 kg. The action of sunlight on these small interplanetary particles causes their orbits to decay into the Sun on timescales of less than 10 m.y. Thus, the dust is not original but must be continuously replenished. The emphasis of this chapter is on the origin of these IDPs and MMs and whether they are asteroidal or cometary. There is now good evidence that interstellar dust particles also penetrate the inner solar system (*Grün et al.,* 1993) and may provide a significant contribution to the dust flux at 1 AU (*Grün et al.,* 1997). In addition, small dust particles that originate in the Kuiper Belt may be transported to the inner solar system (*Flynn,* 1996; *Liou et al.,* 1996). However, the larger Kuiper Belt particles, which have long orbital-decay periods, are likely to be destroyed by collisions or removed by the giant planets before reaching Earth.

In section 2, we give a brief description of our current understanding of the nature of IDPs, emphasizing those features that may separate the putative cometary particles from those that are asteroidal. Despite the enormous advances that have been made in recent years in our understanding of these highly interesting particles, the question of their provenance, whether they are predominantly asteroidal, predominantly cometary, or a useful mixture of both, remains controversial. In section 3, we attempt to answer this key question by looking beyond the particles themselves to the zodiacal cloud as a whole.

Observations by the Infrared Astronomical Satellite (IRAS) in 1983 revealed that the zodiacal cloud is not devoid of large-scale structure (*Low et al.,* 1984). We now know that the Sun is not at the center of symmetry of the cloud (*Kelsall et al.,* 1998; *Dermott et al.,* 1999), that the plane of symmetry is warped (*Dermott et al.,* 2001), and that clouds of dust trail Earth in its orbit that are associated with a circumsolar ring of dust particles in resonant lock with the planet (*Dermott et al.,* 1994a; *Reach et al.,* 1995). Ultimately, after more analysis, these features should contribute significant information on the origin and evolution of the cloud (*Dermott et al.,* 2001). However, for the purposes of this chapter, it is clear that the most direct and important source of information on the origin of the particles in the zodiacal cloud is provided by the solar-system dust bands. Remarkably, these dust bands have now been observed from the ground, without the aid of a telescope, simply by using a wide-angle lens attached to a cooled CCD camera (*Ishiguro et al.,* 1999). In section 3, we show that these dust bands must originate from the disintegration of asteroids (*Low et al.,* 1984; *Dermott et al.,* 1984; *Sykes and Greenberg,* 1986; *Grogan et al.,* 2001), and we use the IRAS observations to estimate, with comparative confidence, the total asteroidal contribution to the zodiacal cloud.

In section 4, we discuss the collisional evolution of the asteroid belt. Earlier discussions on the sources of the particles in the cloud were based on estimates of the dust production rates from comets (*Whipple,* 1967; *Delsemme,* 1976) and asteroids (*Dohnanyi,* 1976). In the case of comets, these calculations may have been strengthened by the discovery of cometary trails (*Sykes and Walker,* 1992), and the analysis of these trails may lead to more accurate estimates of the average cometary input to the cloud (*Lisse et al.,* 1998; *Lisse,* 2001). However, we consider that the uncertainties in these types of calculation, both for the asteroidal and the cometary sources, are still too large to provide definitive results. We support this statement in section 5 by a discussion of the current uncertainties in the asteroidal dust production rate.

2. INTERPLANETARY DUST PARTICLES

Important reviews of the nature of IDPs and MMs have been given by *Brownlee* (1985), *Bradley et al.* (1988), and *Rietmeijer* (1988) and are to be found in the conference proceedings edited by *Zolensky et al.* (1994) and *Gustafson and Hanner* (1996). The most recent summaries, on which the comments in this chapter are largely based, have been given by *Bradley* (1999) and *Jessberger et al.* (2001).

The extraterrestrial origin of IDPs has been confirmed by the presence of solar-wind noble gases (*Rajan et al.,* 1997), solar-flare tracks, and solar-wind-irradiated rims (*Bradley et al.,* 1984). In addition, the nonsolar isotopic abundances of D, H, and N (*Messenger and Walker,* 1996) of some of these particles prove not only that they are extraterrestrial but also that some of these particles existed in the presolar nebula out of which the planets formed. Roughly 15% of the particles are essentially single-mineral grains or a few-mineral grains. However, the most common IDPs (75%) are unequilibriated, fine-grained mixtures of thousands to millions of mineral grains and amorphous components. Their composition is chondritic and similar, within a factor of 2, to the most primitive, C-rich, CI carbonaceous chondrites (*Schramm et al.,* 1989; *Thomas et al.,* 1996). *Jessberger et al.* (2001) have noted that the fact that the composition of these tiny particles is chondritic is remarkable "because most chondritic meteorites do not typically have chondritic elemental composition at the 10-micron size scale. Most meteorites are coarser grained and common 15-micron volumes are single mineral grains." It follows that most IDPs cannot be formed from the simple grinding down of known meteorites, but must result "from the mechanical mixing of large numbers of randomly selected tiny grains." The compositions of MMs are similar to CM- and CR-type carbonaceous chondrites. However, there are some very important differences between carbonaceous chondrites and both MMs (*Maurette et al.,* 1996) and IDPs. In particular, both MMs and IDPs are markedly C-rich. In fact, some IDPs have C contents about 2–3× higher than the most primitive, C-rich, carbonaceous chondrites (*Thomas et al.,* 1994). Moderately volatile elements such as Zn and Ga also show systematic enrichments in IDPs by a factor of 2–4 above CI chondrites (*Flynn et al.,* 1996).

On the basis of mineralogy, as revealed by transmission electron microscopy (TEM), infrared (IR) spectroscopy, and X-ray diffraction analyses, IDPs can be divided into two major groups: chondritic and nonchondritic (*Klöck and Stadermann,* 1994). The nonchondritic group is poorly studied, with only a few very refractory IDPs having received thorough mineralogic and isotopic characterization (*Zolensky,* 1987; *McKeegan,* 1987). The chondritic group is well studied and has been further subdivided into hydrous and anhydrous subgroups. The hydrous chondritic IDPs have been extensively altered by liquid water inside a parent body (*Zolensky and McSween,* 1988; *Zolensky and Barrett,* 1994), and this fact suggests that they are asteroidal. The anhydrous chondritic IDPs have, in general, higher porosities (*Corrigan et al.,* 1997) and consist mainly of olivine and/or pyroxene. In particular, the pyroxene-rich particles are physically, chemically, and isotopically more primitive than any other materials available for laboratory study. Most of these particles are complex admixtures of 0.1–5-μm-diameter single-mineral grains (most commonly enstatite and Fe-Ni sulfides), amorphous material, carbonaceous material, and GEMS [submicrometer spheroidal grains of silicate glass with embedded metal and sulfides (*Bradley,* 1999)]. Some anhydrous chondritic IDPs contain enstatite whisker crystals (*Fraundorf,* 1981) believed to be early nebular condensates. Other chondritic IDPs have very large D/H and

^{15}N-isotopic anomalies (*Messenger and Walker,* 1996), and these provide evidence for the survival of presolar interstellar components. It has even been suggested that these IDPs could actually contain a large quantity of well-preserved aggregates of circumstellar and/or interstellar materials (*Bradley,* 1999).

Reflectance spectra of chondritic IDPs in the visible wavelength range show that whereas most hydrous IDPs have spectra similar to carbonaceous chondrites and C-type asteroids, many of the most porous anhydrous chondritic IDPs have spectra similar to outer-belt P- and D-type asteroids (*Bradley et al.,* 1996). It has therefore been suggested that the highly primitive and porous anhydrous chondritic IDPs probably originate from comets and/or outer-belt (P- and D-type) asteroids. However, a recently fallen unique carbonaceous chondrite (Tagish Lake) with a hydrous mineral assemblage shows an IR reflectance spectra identical to D-class asteroids, which might therefore preclude these asteroids from being the parents of the anhydrous IDPs (*Hiroi et al.,* 2001). The mineralogy of the hydrous chondritic IDPs is roughly similar (but not identical) to that of the CI, CM, and CR carbonaceous chondrites, and these are thus believed to derive from asteroids (*Bradley,* 1999). We note that *Keller et al.* (1993) argue that these well-mixed aggregates could be agglutinates from asteroidal regoliths. *Klöck and Stadermann* (1994) have argued that anhydrous chondritic IDPs, although mineralogically unlike known meteorites, contain forsterite with a unique chemical signature that is also present in primitive chondrites, establishing a possible link between these IDPs and meteorites. Thus, there is some evidence that both hydrated and anhydrous chondritic IDPs could come from asteroids.

We should note that there may be very little difference between an outer-belt P- or D-type asteroid and a cometary nucleus (*Brownlee,* 1996). Nevertheless, the question of the origin of the anhydrous chondritic IDPs is of outstanding importance. We need to know if we already have samples of comets originating in the Kuiper Belt in our collections or whether, in order to analyze a real cometary particle, we must have a sample-return mission. To try to settle this question, *Brownlee et al.* (1995) have used the thermally stepped He-release method developed by *Nier and Schlutter* (1993) to determine the atmospheric entry speed of IDPs. *Jackson and Zook* (1992) claim that there is little overlap between the geocentric encounter velocities of asteroidal and cometary particles and thus that the high-speed, strongly heated particles must be cometary, whereas the low-speed, unheated particles are probably asteroidal. The validity of this entry-speed criterion is discussed in section 5. *Brownlee et al.* (1995) analyzed a set of IDPs with computed entry velocities of 20 km s^{-1}, which they consider to be almost certainly cometary, and found that these particles are all porous anhydrous chondritic IDPs dominated by GEMS. This would appear to be conclusive evidence that the porous anhydrous chondritic IDPs are cometary. However, while accepting the conclusion of Brownlee et al., we argue here that the dominant source of the particles in the zodiacal cloud is probably asteroidal. The cometary particles are certainly of particular interest, because comets may preserve unprocessed material from the presolar molecular cloud that gave birth to our solar system. However, the asteroidal IDPs are equally interesting, partly for the same reason, but also because they appear to be samples of asteroids, probably the more friable and more distant asteroids, that are not well represented in our meteorite collections.

3. SOLAR SYSTEM DUST BANDS

Observations made by IRAS were the first to reveal fine structure in the zodiacal cloud. Figure 1 shows a Fourier-filtered IRAS observation of the zodiacal cloud in the 25-μm infrared wave band, illustrating the smooth low-frequency background cloud and the high-frequency residual structure that we associate with the solar-system dust bands (*Grogan et al.,* 2001). It is important to note here that because the low-frequency component of the dust bands is indistinguishable from the low-frequency background zodiacal cloud, the high-frequency residuals resulting from the

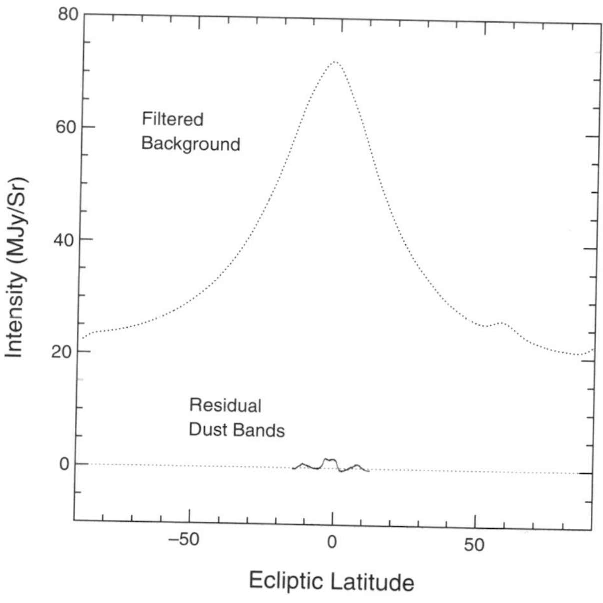

Fig. 1. Fourier-filtered IRAS observation of the zodiacal cloud in the 25-μm infrared wave band, showing the smooth low-frequency background cloud (upper curve) and the high-frequency residual structure (lower curve). This pole-to-pole observation was made at 90° solar elongation (the angle between the telescope pointing direction and the Earth-Sun line) in the direction leading Earth in its orbit when the planet was at an ecliptic longitude of 293°. The high-frequency residuals, which we associate with the solar-system dust bands, can be seen as projecting "shoulders" in the unfiltered IRAS observation near latitudes of ±10° and a central "cap" near 0°. The filtered dust-band profile above merely represents the "tip of the iceberg" in terms of the dust-band material in the zodiacal cloud (*Dermott et al.,* 1994b). The structure around 60° latitude (upper curve) is due to dust in the plane of the galaxy (adapted from *Grogan et al.,* 2001).

filtering process will merely represent the "tip of the iceberg" in terms of the total contribution of dust-band material to the zodiacal cloud (*Dermott et al.,* 1994b). In the unfiltered IRAS observations, the dust bands appear as projecting "shoulders" near latitudes of ±10° and a central "cap" near 0°. We argue that these dust bands are associated with the collisional debris of the *Hirayama* (1918) asteroid families (*Dermott et al.,* 1984; *Sykes and Greenberg,* 1986; *Grogan et al.,* 1997; *Reach et al.,* 1997), the near-ecliptic dust bands with the Themis and Koronis families, and the 10° band with the Eos family; as such they represent a unique observational constraint on the contribution of asteroidal material to the zodiacal cloud.

Our approach to generating a physical model for the various components of the zodiacal cloud, including the dust bands, is essentially a two-step process: (1) Given a postulated source of dust particles, we numerically investigate the orbital evolution of the particles due to Poynting Robertson (P-R) light drag and solar-wind drag (*Burns et al.,* 1979), using equations of motion that also include the effects of radiation pressure and planetary gravitational perturbations (*Dermott et al.,* 1992). (2) Once these dust-particle orbits have been computed for a range of particle sizes, their distribution is visualized in three dimensions via the FORTRAN code SIMUL (*Dermott et al.,* 1988; *Grogan et al.,* 1997), taking into account the thermal and optical properties of the particles and their variation with particle size. Using this tool, the viewing geometry of any telescope can be reproduced exactly, allowing IRAS-type brightness profiles to be created and compared with the observed profiles.

A dust band is a toroidal distribution of asteroidal dust particles with common proper inclinations, common forced inclinations, and common forced longitudes of ascending node (*Dermott et al.,* 1984). The particles' common proper inclination (representing their "intrinsic" inclination) derives from their common source in a given asteroid family, and their common forced inclinations and longitudes of ascending node [reflecting the effect of secular planetary perturbations; for further explanation see, for example, *Murray and Dermott* (1999)] result from the dominant perturbing force of Jupiter on particles located in the outer part of the main asteroid belt. Figure 2 shows some results from our numerical simulations, illustrating the variation of the forced inclination and the forced longitude of ascending node with semimajor axis at the present epoch (Julian Date 2450700.5) for asteroid-family dust particles with diameters of 10, 100, and 200 μm (*Dermott et al.,* 2001). The forced inclination and the forced longitude of ascending node of the asteroid-family particles define the orientation of the mean plane of the dust band associated with the family and are a function of semimajor axis, time, and particle size. The low dispersion of the forced inclinations and longitudes of ascending node of these asteroidal particles in the region of the main asteroid belt (≳2.5 AU), regardless of particle size, is the fundamental reason why dust bands are observed. However, the effects of secular perturbations on particles in highly eccentric orbits, typical of particles that are cometary in origin, prevent such well-defined dust bands from being formed (*Liou et al.,* 1995).

As large dust particles (≳100 μm in diameter) encounter the ν_{16} secular resonance at the inner edge of the asteroid belt (at ~2 AU), the effect of the resonance acts to disperse their forced inclinations and forced longitudes of ascending node. The ν_6 secular resonance (also at ~2 AU) produces analogous behavior in the forced eccentricities and forced longitudes of pericenter of the dust particles. The effects of passage through these secular resonances are far more pronounced on the waves of large (~100-μm-diameter) dust particles compared to the waves of small (~10-μm-diameter) dust particles due to the weaker effects of P-R and solar-wind drag on the large particles (*Wyatt and Whipple,* 1950) and their consequently slower inward evolution toward the Sun. The large particles are thus acted on by the resonances for longer periods of time. The orbital-element distributions of the large asteroidal dust particles therefore lose their characteristic family signatures in the inner region of the main belt and diffuse into the (low-frequency) zodiacal background cloud. We continue to investigate the dynamics of even-larger particles, up to and beyond 500 μm.

In this new large particle regime, we will need to incorporate the effects of particle-particle collisions into our simulations, as the timescales for the particles' orbits to decay into the Sun due to P-R and solar-wind drag become comparable with the particles' collisional lifetimes (*Grün et al.,* 1985; *Wyatt et al.,* 1999). Some particles may therefore not penetrate far into the inner solar system. The nature of the size-frequency distribution of the particles will be a complex function of dust production rates, P-R and solar-wind drag rates, collisional lifetimes and the nature of particle-particle collisions, and will, presumably, vary with heliocentric distance. The situation is further complicated by the fact that even if the size distribution of the debris resulting from an asteroidal collision could be described by a power law (*Dohnanyi,* 1969), the size-frequency power-law index, q, will reflect the characteristics of the parent. The equilibrium size distribution of the collisional cascade originating from a single asteroid has been shown to be a function of the impact strength of that asteroid (*Durda and Dermott,* 1997). Thus, it is possible for the value of q associated with a given family to be different from that of other families and different from the value for the background cloud. In the case of a "rubble pile" (*Davis et al.,* 1989), the value of q associated with the initial disruption may be significantly higher than that associated with the disruption of a solid, coherent asteroid. This provides us with further motivation to relate the dust bands to given parent bodies in the main belt. However, as a first step in answering the fundamental question of the extent to which large and small particles contribute to the dust-band emission, we model the size-frequency distribution as a single power law. We will refine this assumption in the future when we have a better understanding of the role of the complicating factors outlined above.

The forced elements of dust particles obtained from numerical simulations, combined with the proper elements of

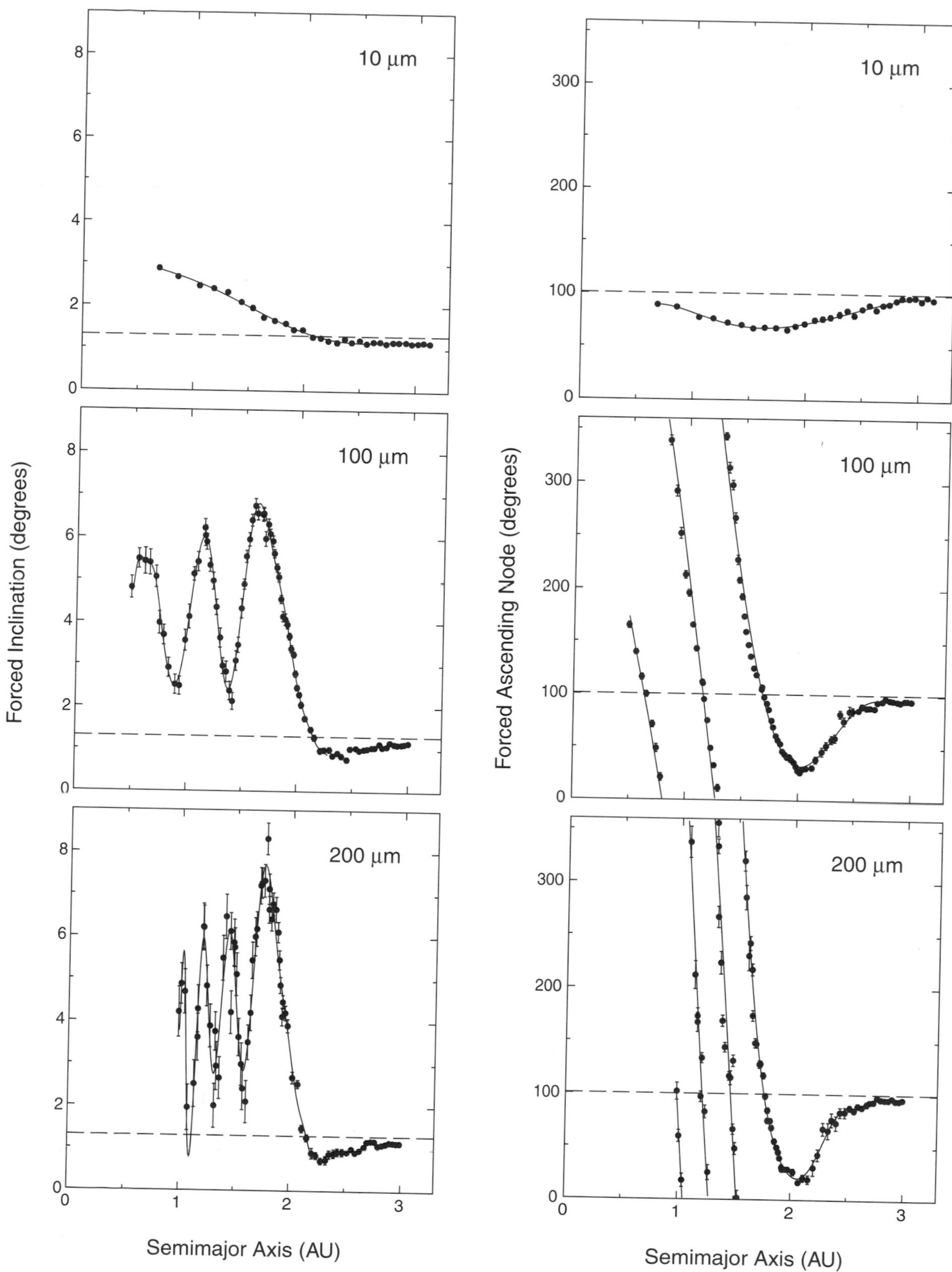

Fig. 2. Variation of the forced inclination (left) and the forced longitude of ascending node (right) with semimajor axis at the present epoch (Julian Date 2450700.5) for asteroidal dust particles composed of astronomical silicate (*Draine and Lee,* 1994) of density 2500 kg m^{-3} with diameters 10, 100, and 200 μm. The dashed lines show the present osculating inclination (left) and osculating longitude of ascending node (right) for Jupiter (*Dermott et al.,* 2001).

the source bodies (see Table 1), are provided as input data to the SIMUL algorithm. For a given range of particle sizes and a given size-frequency power-law index q, the total surface area of material associated with the model bands is adjusted until the amplitudes of the 25-μm model dust bands matches the 25-μm observations; q can then be varied until a single model provides a match in amplitude to the 12-, 25-, and 60-μm observations simultaneously. An estimate for the low-frequency component of the dust bands is obtained by employing the same filter in the modeling process that we use to define the observed dust bands and iterating. Figure 3 shows the best results of our modeling,

TABLE 1. Dust band model parameters: proper element distributions (semimajor axis, a, eccentricity, e, and inclination, I) of the source bodies (mean, σ) and cross-sectional area of material required.

Asteroid Family	$\bar{a}$, σ_a (AU)	$\bar{e}$, σ_e	$\bar{I}$, σ_I (°)	Area (10^9 km^2)
Eos	3.015, 0.012	0.076, 0.009	9.35, 1.5	4.0
Themis	3.148, 0.035	0.155, 0.013	1.43, 0.32	0.35
Koronis	2.876, 0.026	0.047, 0.006	2.11, 0.09	0.35

Dust particles originating from each family are distributed into the inner solar system as far as 2 AU according to a 1/r Poynting-Robertson drag distribution, where r specifies the radial distance from the Sun (*Grogan et al.,* 2001).

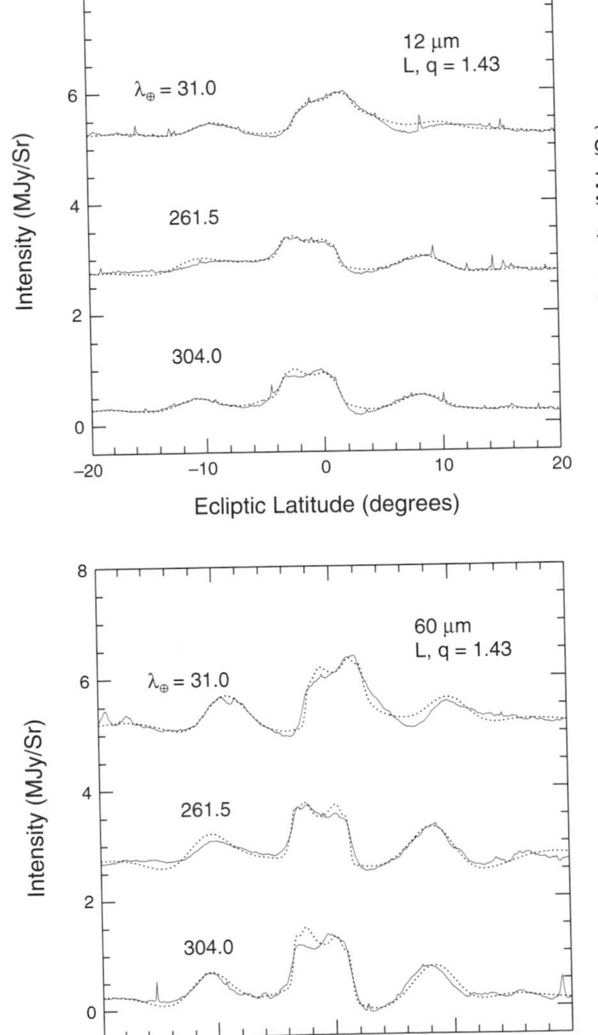

Fig. 3. Filtered IRAS dust band profiles (solid lines) in the 12-, 25-, and 60-μm wave bands are compared with models (dotted lines) having a size-frequency distribution with power-law index q = 1.43, at three different ecliptic longitudes of Earth ($\lambda_\oplus$). All profiles were made at 90° solar-elongation angle in the leading (L) direction of Earth in its orbit. The model dust bands were constructed using particles from the Themis and Koronis families for the central band pair and from the Eos family for the 10° band pair. A dispersion of 1.5° was imposed on the proper inclination of the Eos material in this model and all the material was confined to the asteroid belt (2.0 AU < r < 3.1 AU, where r specifies the radial distance from the Sun). The low value of q (<1.66) implies that large particles (diameters ~10^2 μm) dominate the surface area distribution (adapted from *Grogan et al.,* 2001).

comparing the dust band observations (solid curves) to the dust-band models (dotted curves) in the 12-, 25-, and 60-μm wave bands (*Grogan et al.,* 2001). The model includes particles with diameters in the range 1–100 μm and has a size-frequency power-law index q equal to 1.43. As q < 5/3, large particles dominate the total surface area of this distribution (*Dohnanyi,* 1969). The amplitudes in all wave bands are well matched, and the shapes of the dust-band profiles describe the variation in shape of the observations around the sky very well.

A clear result from our modeling is that a high size-frequency power-law index q, in which small particles dominate (*Dohnanyi,* 1969), fails to account for the observations of the solar-system dust bands. This index has to be reduced to the point where large particles dominate the size distribution, and we place an upper limit of q ≃ 1.4. This is consistent with the cratering record on the LDEF satellite (*Love and Brownlee,* 1993), which suggests a value for q of ~1.15 at Earth and a peak in the particle diameter at ~100–200 μm. Since the Fourier filter preferentially isolates material exterior to the 2 AU secular resonance (in the inner solar system the dust-band material is dispersed into the low-frequency background cloud due to the action of secular resonances), our results are more indicative of the size-frequency index of dust particles in the main asteroid belt.

We also find that models confining the dust-band material to the main asteroid belt (exterior to 2 AU) better match the IRAS observations, as the integrity of the dust bands is lost interior to 2 AU due to the action of secular resonances. In the future, we will incorporate the results from our simulations of large particle dynamics into our models and will populate the inner solar system as well as the main-belt region, allowing the dust bands to disperse naturally into the background cloud. However, we can obtain an estimate for the dust-band contribution to the zodiacal cloud as a whole by simply extending our best fit dust-band models to populate the inner solar system artificially. The distribution of orbits obtained in this manner will not be exactly correct, due to our insufficient treatment of the secular resonances and the effects of jovian mean-motion resonances and gravitational scattering by the terrestrial planets, but it will still be reasonably accurate in terms of the total surface area associated with the dust bands. Figure 4 compares the thermal emission obtained from this raw dust-band model to the corresponding unfiltered profile in the IRAS 25-μm wave band (*Grogan et al.,* 2001). The result is shown for inner solar-system distributions of material corresponding to $1/r^{\gamma}$, where r is the radial distance from the Sun and $\gamma = 1.0$, as expected for a system evolved by P-R drag, or $\gamma = 1.3$, as observed by the *Helios* and *Pioneer 10/11* space probes (*Leinert et al.,* 1983) and predicted in parametric models of the zodiacal cloud, most recently by *Kelsall et al.* (1998). We expect the radial distribution of material in the zodiacal cloud to be a function of the particles' eccentricities (*Dermott et al.,* 1999) and note that the orbital eccentricities of large asteroidal particles are strongly influenced by resonances and gravitational scattering as they evolve into the inner solar system. The effects of particle collisions will also influence the radial distribution, as some collisional debris will be deposited into the cloud, whereas a fraction will be blown out by radiation pressure. The net effect of all these processes has yet to be calculated. For the values of γ assumed above, the dust bands appear to contribute approximately 30% to the total thermal emission. Also shown is the amplitude of the emission from dust-band material confined to the main asteroid belt (exterior to 2 AU). This indicates that ~4% of the in-ecliptic IR emission from the zodiacal cloud is produced by dust-band particles that orbit exterior to 2 AU and also clearly demonstrates the extent to which the dust band contribution is underestimated if it is assumed that the filtered dust-band observations represent the entirety of the dust-band component of the cloud.

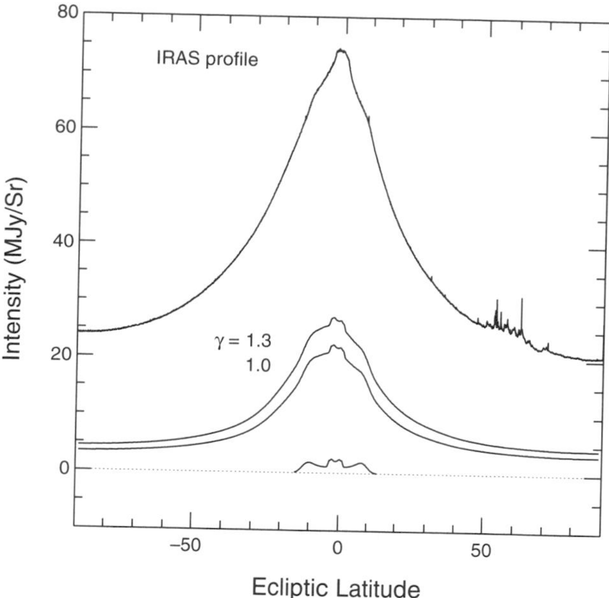

Fig. 4. Contribution of dust-band material to the zodiacal cloud. This figure shows a comparison of the thermal emission obtained from the raw best fit dust-band model to the corresponding unfiltered IRAS profile in the 25-μm wave band. The result is shown for inner-solar-system distributions of material corresponding to $1/r^{\gamma}$, where r is the radial distance from the Sun and $\gamma = 1.0$, as expected for a system evolved by Poynting-Robertson drag, or $\gamma = 1.3$, as observed by the *Helios* and *Pioneer 10/11* space probes (*Leinert et al.,* 1983) and predicted in parametric models of the zodiacal cloud, most recently by *Kelsall et al.* (1998). The dust bands appear to contribute approximately 30% to the total thermal emission. Also shown (bottom profile) is the amplitude of the dust-band material if confined to the main asteroid belt [exterior to 2 AU (*Grogan et al.,* 2001)].

Figure 5 shows the ratio of the area of collisionally evolved material associated with the entire main-belt asteroid population to that associated with all the asteroid families combined, for asteroid diameters greater than 1 km (*Grogan et al.,* 2001). The best-fit lines have a slope corresponding to a size-frequency power-law index q = 1.795 (*Durda and Dermott,* 1997), slightly less than the *Dohnanyi*

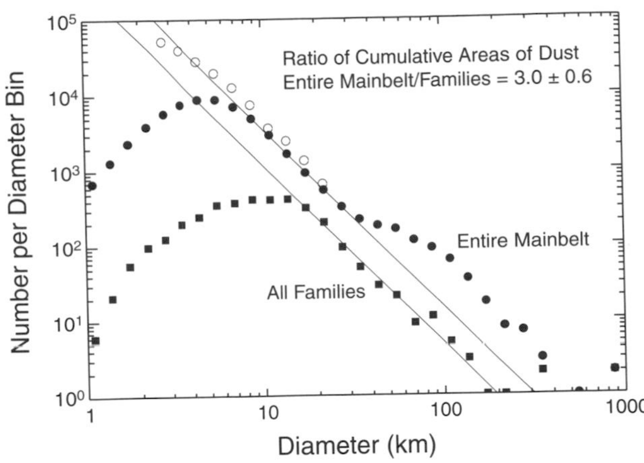

Fig. 5. Diameter-frequency diagram for the main-belt asteroid population, obtained by combining data from the catalogued population and McDonald/Palomar-Leiden surveys (MDS/PLS). Open points represent counts for which the PLS data had to be corrected for incompleteness. These were not included in the least-squares fits to the linear portion of the distribution. The ratio of the area of dust associated with the entire main-belt asteroid population to that of the combined asteroid families alone was calculated to be 3.0 ± 0.6 [*Grogan et al.* (2001); updated from *Durda and Dermott* (1997)].

(1969) value of q = 1.83, which is only expected if the breakup criterion for an asteroid is independent of its size. This diagram can be used to make a rough estimate of the total contribution of main-belt asteroid collisions to the dust in the zodiacal cloud, by extrapolating the observed size distributions of larger asteroids in both populations assuming a collisional equilibrium power-law size-frequency distribution. The result is that the main-belt asteroid population contributes approximately 3× the dust area of the Hirayama families alone, and therefore the total asteroidal contribution to the zodiacal cloud could account for almost the entirety (~90%) of the interplanetary dust complex (*Grogan et al.,* 2001). However, as noted above, employing the same single power-law size-frequency distribution to characterize both the main-belt and family asteroid populations is probably an oversimplification, and while this implies some uncertainty in the precise contribution of asteroidal material to the zodiacal cloud, the asteroidal contribution must still be significant and is probably dominant.

4. COLLISIONAL CASCADE

The size distribution of the collisionally evolved asteroids is given by

$$N(D) = \frac{1}{3(q-1)}\left(\frac{D_0}{D}\right)^{3(q-1)} \tag{1}$$

where N(D) is the number of asteroids with diameter >D, D_0 is a constant, and q is the power-law index (*Dohnanyi,* 1969). If 5/3 < q < 2, then the total area in the cascade population is dominated by contributions from the smallest particles of diameter ~D_{min}, where D_{min} is the lower cutoff of the size distribution, whereas the total volume, and hence also the total mass, of the source population is dominated by the contributions from the larger fragments. If D_{vol} denotes the diameter of the sphere that would contain the total volume of the source material, then observations of the Hirayama families show that $D_0 \simeq 0.83D_{vol}$ (*Dermott et al.,* 1984).

The importance of collisions in determining a particle's evolution depends on its collisional lifetime. For the particles of diameter D_{typ} that constitute most of the zodiacal cloud's cross-sectional area (that is, those particles that are expected to characterize its mid-IR emission), this lifetime can be approximated (in years) by

$$t_{coll}(D_{typ}, r) \simeq \frac{t_{per}}{4\pi\tau_{eff}(r)} \tag{2}$$

where r is the heliocentric radial distance and $\tau_{eff}(r)$ is the zodiacal cloud's effective face-on optical depth, which would be equal to the zodiacal cloud's true face-on optical depth if its particles had unity extinction efficiency (*Artymowicz,* 1997; *Wyatt et al.,* 1999). The orbital period of the particle in years is given by $t_{per} \simeq \sqrt{(a/a_\oplus)^3}$, where a is the semimajor axis of the particle in AU and $a_\oplus$ = 1 AU is the semimajor axis of Earth's orbit.

The combined effect of P-R light drag and solar-wind drag acting on micrometer-sized dust particles results in an evolutionary decrease in both the osculating semimajor axis, a, and eccentricity, e, of the particles' orbits, given by (*Wyatt and Whipple,* 1950)

$$\dot{a}_{PR} = -(\alpha/a)(2 + 3e^2)/(1 - e^2)^{3/2} \tag{3}$$

$$\dot{e}_{PR} = -5(\alpha/a^2)e/2(1 - e^2)^{1/2} \tag{4}$$

where $\alpha = 6.24 \times 10^{-4}\beta(1 + sw)$ AU^2 yr^{-1} and β is the ratio of radiation pressure to the gravitational attraction of the Sun on a particle. The effect of solar-wind drag on a particle is usually approximated as being 30% of the effect of P-R drag (*Gustafson,* 1994); that is, sw is set to an average value of 0.3. The effects of P-R and solar-wind drag therefore cause the orbit of a particle to spiral in toward the Sun. However, it does not affect the orientation of the particle's longitude of pericenter, $\dot{\varpi}_{PR} = 0$. Nor does it change the orientation of the plane of the particle's orbit, $\dot{I}_{PR} = \dot{\Omega}_{PR} = 0$, where I and Ω are the inclination and longitude of ascending node of the orbit respectively. The constant β depends upon the size and other physical properties of a particle: a useful approximation, valid for particles composed of astronomical silicate with diameters D ≳ 1 μm, is given by

$$\beta(D) \simeq 1150/\rho D \tag{5}$$

where ρ is the particle's density measured in units of kg m^{-3} and D is measured in μm.

For a particle with zero eccentricity, equation (3) can be solved trivially to find the time it takes for the particle to spiral in from a heliocentric radial distance r_0 to r_f

$$t_{PR} \simeq 400\left[(r_0/a_\oplus)^2 - (r_f/a_\oplus)^2\right]/\beta(1 + sw) \qquad (6)$$

where r_0 and r_f are specified in AU, and t_{PR} is given in years. For eccentric orbits, equation (4) can be solved in conjunction with equation (3) to find the time taken for the particle's orbital eccentricity to decay from an initial value e_0 to a final value e_f

$$t_{PR} = \frac{2C^2}{5\alpha}\int_{e_f}^{e_0} \frac{e^{3/5}}{(1-e^2)^{3/2}} de \qquad (7)$$

where the constant $C = a_0(1 - e_0^2)/e_0^{4/5}$, with a_0 specified in units of AU and α is defined as above to again obtain t_{PR} in years.

Now consider the fragments created in the breakup of an asteroid at a heliocentric radial distance r. The largest fragments, with $D > D_{crit}$, are broken up by collisions before their orbits have suffered any significant P-R drag evolution, while the smaller fragments, with $D < D_{crit}$, for which the P-R drag evolution is faster, can reach the Sun without a catastrophic collision. By equating the collisional and P-R drag lifetimes, *Wyatt et al.* (1999) estimate that

$$D_{crit} \simeq \frac{0.23}{\rho\tau_{eff}(r)}\sqrt{\frac{a_\oplus}{r}} \qquad (8)$$

where r is measured in units of AU, ρ in kg m^{-3}, and D_{crit} in µm. It is interesting to note here that if the effective normal optical depth, τ_{eff}, were to increase, as expected following the collisional disruption of a large rubble-pile asteroid (see the discussion in section 5), then the value for D_{crit} would be proportionately lower.

The daughter fragments created in the breakup of an "endless" supply of asteroids on orbits with semimajor axis a_s flow toward the Sun due to the effects of P-R and solar wind drag. If we ignore any further disintegrations of the particles that are involved in the flow, then the orbits of all the particles in a given size range will be distributed between the source and the Sun according to (e.g., see *Gor'kavyi et al.,* 1997)

$$N(a) \propto 1/\dot{a}_{PR} \propto a \qquad (9)$$

where N(a)da is the number of orbits with semimajor axes in the range a to a + da and $\dot{a}_{PR}$ is the rate of change of a particle's semimajor axis due to P-R and solar-wind drag given by equation (3). Thus, the spacing of the orbits increases as the particles approach the Sun, and this fact tends to decrease the number density of the particles, defined as the number of particles per unit volume. But given that both the circumferences of the orbits and the vertical extent of the particle distributions also decrease proportionally with decreasing a, it follows that, for particles in near-circular orbits, the number density of these particles, regardless of their size, will increase inversely with heliocentric distance.

However, because the flow rate of a particle is inversely proportional to its diameter, that is, because $\dot{a}_{PR} \propto 1/D$ (equations (3) and (5)), it follows that the size distribution of the particles in the flow region interior to the asteroid belt, must be quite different from that in the source region.

If the collisional processes leading to the size distribution of the large parent bodies, $N_s(D)$, still hold for the production of the P-R drag-affected particles, then the size distribution in the flow region is given by

$$N(D) \propto N_s(D)/\dot{a}_{PR} \propto N_s(D)D \qquad (10)$$

Thus, if $N_s(D)$ is given by equation (1) with q = 1.83 [as expected for a collisionally evolved system that has reached an equilibrium state (*Dohnanyi,* 1969)], then the cross-sectional area of the zodiacal cloud's smaller, P-R drag-affected particles is concentrated in the largest of these small particles, while the cross-sectional area of the particles that are large enough to be unaffected by P-R drag ($D > D_{crit}$) is concentrated in the smallest of these larger particles. The result is that most of the zodiacal cloud's cross-sectional area is expected to be concentrated in particles with $D_{typ} \approx D_{crit}$, justifying the use of equation (2) for the collisional lifetime of these particles.

Observations of the mean polar brightness of the zodiacal cloud at 1 AU can be used to estimate (the results are somewhat model-dependent) that, near Earth, the effective normal optical depth $\tau_{eff} \sim 4 \times 10^{-8}$ (see also equation (11)). If, as we believe, these zodiacal particles originated in the asteroid belt and migrated to 1 AU due to P-R drag, then the zodiacal cloud's volume density should vary ~1/r and, as its column height scales as r, the zodiacal cloud's effective normal optical depth in the asteroid belt should be similar to that at 1 AU. Assuming the zodiacal cloud particles to have a density 2500 kg m^{-3}, the cross-sectional area of material in the asteroid belt should be concentrated in particles with $D_{typ} \sim 10^3$ µm (equation (8)), for which the collisional lifetime and the P-R drag lifetime are both ~10^7 yr (equations (5) and (6)).

However, because their collisional and P-R drag lifetimes are similar, we must expect many of these large particles to be broken up by collisions before they reach the inner solar system, in which case we must expect the cross-sectional area of material at 1 AU to be concentrated in particles smaller than that in the asteroid belt. This is in agreement with the LDEF cratering record that shows the cross-sectional area distribution at 1-AU peaks for particles with D ~ 140 µm (*Love and Brownlee,* 1993) and other evidence (see the review by *Grün et al.,* 1985).

We have shown (see Table 1) that the total cross-sectional area of dust associated with the three major Hirayama families that we need to model the dust bands is 4.7×10^9 km^2, and that 85% of this dust is associated with the Eos family alone. However, the dust in these models is constrained between approximately 2 and 3 AU, whereas the actual dust associated with the Hirayama families must migrate to the inner solar system due to the effects of P-R and solar-wind drag. We therefore estimate that, as $N(a) \propto a$ (equation (9)), the total area of dust associated with these families in the zodiacal cloud is a factor of 9/5 greater. However, we also estimate that the area of dust needed to account for the whole of the zodiacal cloud must be 3× greater than the total contribution from the Hirayama families alone, that is, $A_{cloud} \sim 2.5 \times 10^{10}$ km^2. Note that if this dust is distributed between the Sun and r = 3 AU, then it follows that the effective face-on optical depth

$$\tau_{eff} \simeq A_{cloud}/\pi r^2 \sim 4 \times 10^{-8} \tag{11}$$

in agreement with the observations discussed above.

If A_{cloud} is characterized by N particles with diameter D_{typ}, then

$$A_{cloud} \simeq (\pi/4)ND_{typ}^2 \tag{12}$$

and it follows that the total mass of the cloud, M_{cloud}, is given by

$$M_{cloud} \simeq (2/3)A_{cloud}\rho D_{typ} \tag{13}$$

At 1 AU, $D_{typ} \sim 140$ µm and hence $M_{cloud} \sim 6 \times 10^{15}$ kg. However, in the asteroid belt, D_{typ} is probably closer to 500 µm, and it follows that the total mass of the cloud must be between 6×10^{15} kg and 2×10^{16} kg. This is roughly a factor of 2 less than the estimate made by *Whipple* (1967) of a zodiacal cloud mass between 1.1×10^{16} kg and 4.5×10^{16} kg.

It is worth noting that if all this dust were gathered together into one object with a diameter of D_{equiv}, then

$$D_{equiv} \simeq (4A_{cloud}D_{typ}/\pi)^{1/3} \tag{14}$$

giving $D_{equiv} \sim 25$ km, assuming $D_{typ} \sim 500$ µm. Neglecting the effects of collisions, this dust will reside in the cloud for a period of about 3.0×10^6 yr, the time taken for the orbit of a 500-µm-diameter dust particle composed of astronomical silicate with density 2500 kg m^{-3} to decay into the Sun under the effects of P-R and solar-wind drag from 3 AU (equations (5) and (6)). Any calculation of the dust production rate must therefore be averaged over this period of time.

Alternatively, we may consider the average mass loss rate from the zodiacal cloud due to infall under the effects of P-R and solar-wind drag, which can be expressed as

$$\dot{M}_{cloud} \simeq M_{cloud}/t_{PR} \tag{15}$$

where, in this case, $t_{PR} \sim 3.0 \times 10^6$ yr for $D_{typ} \sim 500$ µm, giving $\dot{M}_{cloud} \sim 6.7 \times 10^9$ kg yr^{-1}. This value compares favorably with the estimated ~8.2×10^9 kg yr^{-1} total Poynting-Robertson mass loss rate inside 1 AU (*Grün et al.,* 1985).

The value determined above for $\dot{M}_{cloud}$ must also be equivalent to the average rate at which material from the zodiacal cloud flows past Earth. According to the results from LDEF (*Love and Brownlee,* 1993), Earth accretes mass from the cloud at the rate of $\dot{M}_{accrete} \sim 3 \times 10^7$ kg yr^{-1}. If $P_{capture}$ is the total probability of Earth capturing a particle from the zodiacal cloud then Earth's mass accretion rate can be expressed as

$$\dot{M}_{accrete} = \dot{M}_{cloud}P_{capture} \tag{16}$$

Rearranging this equation allows $P_{capture}$ to be written directly in terms of the observable quantities $\dot{M}_{accrete}$ and A_{cloud}, giving

$$P_{capture} = 3.6\dot{M}_{accrete}/A_{cloud} \sim 4.3 \times 10^{-3} \tag{17}$$

or roughly 0.4%, where $\dot{M}_{accrete}$ is specified in units of kg yr^{-1} and A_{cloud} in km^2. It is interesting to note that this estimate of $P_{capture}$ does not explicitly depend upon the assumed value of ρD_{typ}.

The total probability of Earth capturing a particle from the zodiacal cloud may also be written as

$$P_{capture} = \dot{P}_{capture}t_{accrete} \tag{18}$$

where $\dot{P}_{capture}$ is Earth's annual rate of capture of particles and $t_{accrete}$ is the timescale, in years, over which these particles are accreted. Particles in the zodiacal cloud are distributed on tori of width 2ae (*Dermott et al.,* 1985; *Wyatt et al.,* 1999) and so this timescale is given by the time taken for a torus of particles to flow past Earth under the effects of P-R and solar-wind drag. The first particles to be accreted by Earth from such a torus will be those that initially have pericenters, $q_0 = a_0(1 - e_0)$ at 1 AU, while the last particles to be accreted will be those that have final apocenters, $Q_f = a_f(1 + e_f)$, at 1 AU. Both the semimajor axis and eccentricity of the particles in this torus will decrease with time due to the effects of P-R and solar-wind drag, and so $t_{accrete}$ is given by equation (7) with e_0 specified such that $q_0 = 1$ AU; e_f is determined such that $Q_f = 1$ AU, consistent with decaying from its initial value, e_0, due to P-R and solar-wind drag; and $C = q_0(1 + e_0)/e_0^{4/5}$. As both $\dot{P}_{capture}$ and $t_{accrete}$ depend on the orbital-element distributions of the particles in each torus, which in turn are determined by the orbital elements of the source body of the particles, this calculation needs to be performed separately for each class of orbit. Results from LDEF (*Love and Brownlee,* 1993) indicate that the mass distribution of particles accreted by Earth reaches a peak for particles near 200-µm diameter (see also *Grün et al.,* 1985), thus this is the value we adopt here for D_{typ} in order to determine a β value for these particles (equation (5)), which we assume to have a density of 2500 kg m^{-3} irrespective of source.

The average values of $\dot{P}_{capture}$ and e_0 determined numerically (*Kortenkamp and Dermott,* 1998a) for 10-µm-diameter dust particles from three different sources are shown in Fig. 6 and listed in Table 2, along with corresponding values for the typical accretion timescale, $t_{accrete}$. Using these values, we calculate that $P_{capture}$ for particles in the zodiacal cloud with typical asteroidal orbits is 4.6×10^{-3}, con-

TABLE 2. Estimates of the total capture probability, $P_{capture}$, and mass accretion rate, $\dot{M}_{accrete}$, by Earth for dust particles derived from three different sources: asteroidal particles, cometary particles previously trapped in jovian mean-motion resonances, and cometary particles not previously trapped in resonance.

Particle Source	$\dot{P}_{capture}$ (10^{-9} yr^{-1})	e_0 (q_0 = 1 AU)	$t_{accrete}$ (10^3 yr)	$P_{capture}$ (× 10^3)	$\dot{M}_{accrete}$ (10^6 kg yr^{-1})
Asteroidal	180	0.05	25	4.6	31
Resonant cometary	34	0.2	96	3.3	22
Nonresonant cometary	4.6	0.6	380	1.7	11

The typical accretion timescales, $t_{accrete}$, were calculated by assuming that the first particles to be accreted from each source had initial pericenters q_0 at 1 AU with eccentricities e_0, and the last particles to be accreted had final apocenters Q_f at 1 AU, while allowing the particles' orbits to decay under the effects of Poynting-Robertson and solar-wind drag. Values for the annual capture rate, $\dot{P}_{capture}$, were obtained from the results of numerical simulations [source data from *Kortenkamp and Dermott* (1998a)].

sistent with the value derived above (equation (17)) from observable quantities. This value for $P_{capture}$ in turn implies a mass-accretion rate by Earth of 3.1×10^7 kg yr^{-1}, in close agreement with the value determined from the LDEF cratering record. Particles released from comets and then trapped into a mean-motion resonance with Jupiter, which then acts to reduce the orbital eccentricities of the particles to near-asteroidal values (*Liou and Zook,* 1996), could also account for the observations. Note that the values shown for the mass-accretion rate, $\dot{M}_{accrete}$, in Table 2 were calculated in each case by making the simplifying assumption that all particles in the zodiacal cloud were produced by the single particle source in question. In reality, dust particles from the three sources shown in Table 2, plus a variety of others, all contribute to the zodiacal cloud in differing proportions.

It is worth noting here that the calculated value of $P_{capture}$ for cometary particles not previously trapped in jovian mean-motion resonances is only a factor of a few less than the observed value. While this implies a mass-accretion rate by Earth that is still probably too small to account for the observed value, even assuming that such cometary particles do comprise the entire zodiacal cloud, previous studies of the capture of dust particles by Earth have suggested a much stronger bias towards accreting asteroidal over cometary dust particles. *Flynn* (1990), for example, came to this conclusion based upon the lower geocentric-encounter velocities, and hence greater effective capture cross-section of Earth due to gravitational focusing, of particles on near-circular, near-ecliptic orbits. *Kortenkamp and Dermott* (1998a), however, did appreciate that the lower geocentric-encounter velocities of particles on these "asteroidal-type" orbits also impinged negatively upon the effective volume swept out by Earth per unit time (see Fig. 6), but were still led to conclude that Earth would have a strong preference for accreting asteroidal rather than cometary particles. We now realize that the latter conclusion may be misleading. It is important here to distinguish between the annual rate of capture of particles from a given torus, while Earth is inside that torus, $\dot{P}_{capture}$, and the total probability of a particle being accreted

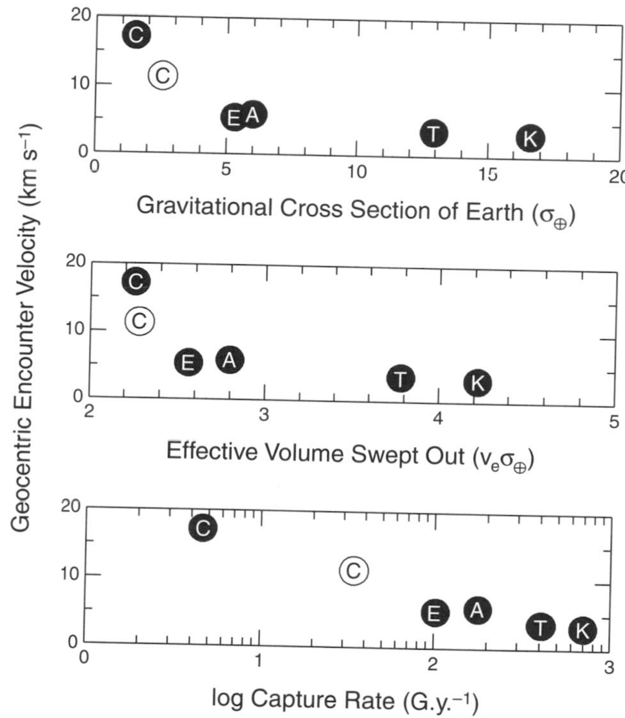

Fig. 6. Mean geocentric encounter velocities (prior to acceleration by Earth) for 10-μm-diameter dust particles from several different source populations are plotted against the mean gravitational capture cross-section of Earth (measured in units of the geometric cross-section of Earth at 100 km altitude, $\sigma_\oplus = 1.32 \times 10^8$ km^2; top), the mean effective volume of each population swept up by Earth (measured in units of $v_e\sigma_\oplus$, where v_e = 11.1 km s^{-1} is the escape velocity of Earth at 100 km altitude; middle), and the log of the mean capture rate (bottom). E, K, and T labels indicate particles from the Eos, Koronis and Themis asteroid families respectively, while A indicates other nonfamily asteroidal particles. Open points labeled C indicate cometary particles that were previously trapped in jovian mean-motion resonances. Solid points labeled C indicate cometary particles that were not previously trapped in jovian mean-motion resonances. Reprinted from *Kortenkamp et al.* (2001) with kind permission from Kluwer Academic/Plenum Publishers.

by Earth, $P_{capture}$. While a bias toward preferentially accreting asteroidal particles certainly exists in the annual capture rates shown in Table 2, and this bias must be included when calculating $P_{capture}$ and hence the average rate of capture of particles from the asteroidal and cometary sources, this asteroidal bias in the annual rates is almost entirely compensated for by the much longer timescale ($t_{accrete}$) over which cometary particles, with their wider tori, are accreted by Earth. Note that there are still significant uncertainties in the values for $P_{capture}$, mainly because the values for $\dot{P}_{capture}$ and e_0 were determined from numerical simulations of 10-μm-diameter dust particles only. These particles are much smaller than those indicated by the peak in the LDEF mass distribution. The orbits of the 10-μm-diameter particles evolve more rapidly under the effects of P-R and solar-wind drag than the orbits of larger particles, and are therefore less prone to the effects of secular resonances, trapping into mean-motion resonances, or gravitational scattering by the terrestrial planets. Larger particles will therefore typically arrive at 1 AU with higher orbital eccentricities and inclinations than the 10-μm-diameter particles and will suffer lower annual rates of capture by Earth (*Kortenkamp and Dermott,* 1998a). These numerical simulations therefore need to be extended to much larger particle sizes.

5. DISCUSSION

Our modeling and analysis of the solar system dust bands indicate that large particles with diameters between 10^2 and 10^3 μm dominate the dust-band structures (*Grogan et al.,* 2001). Numerical investigation of the dynamical behavior of these large particles demonstrates that the action of secular resonances at the inner edge of the main asteroid belt disperses the inclinations and nodes of the particles into the broad-scale zodiacal background cloud, accounting for the natural inner edge of the dust bands just exterior to 2 AU. This leads us to estimate that ~4% of the in-ecliptic infrared emission from the zodiacal cloud is produced by dust-band particles that orbit exterior to 2 AU. However, some of these asteroidal dust particles must migrate to the inner solar system under the effects of P-R light drag and solar-wind drag (*Wyatt and Whipple,* 1950), where they are both warmer and closer to Earth. We therefore conclude that IR emission from the asteroidal dust particles associated with the dust bands alone is likely to be much greater than 4% of the total emission. If these asteroidal particles migrate to the inner solar system without further breakup, we calculate that the contribution is 30%. If the particles are broken up and blown out of the solar system before reaching the inner solar system, then our estimate would, of course, remain closer to 4%. However, it is certainly possible, and perhaps even more likely, that particle breakup leads to an increase in the effective area of the dust, in which case our estimate would be greater than 30%. Whether the effective area of the dust actually increases or decreases is not known and thus 30% is, at present, our best estimate of the contribution of asteroidal dust, from the dust-band particles alone, to the zodiacal cloud. But 30% must be an underestimate of the total asteroidal contribution. From a separate discussion of the ratio of the average rate of dust production in the asteroid belt as a whole to that due to those asteroids in the Eos, Koronis, and Themis families alone, which we estimate to be 3:1, we are led to conclude that asteroidal dust may effectively constitute the entirety of the zodiacal cloud, with a quarter (85% of 30%) of the whole cloud originating from the Eos family alone.

This conclusion is, however, contrary to the long-held belief that comets are the dominant source of particles in the zodiacal cloud (e.g., see *Whipple,* 1967), and is still a matter of considerable debate. *Zook* (2001), for example, estimates that the cometary contribution to the near-Earth flux of particles is ~75%, based on cratering rates from an ensemble of Earth- and lunar-orbiting satellites, in complete contrast to our result. In the argument above we have assumed that the dust production rate in asteroid families is similar to that in the nonfamily asteroidal population. If, however, the dust production rate in asteroid families is significantly greater than that in the nonfamily asteroidal population, as suggested by *Dell'Oro et al.* (2001), then the total asteroid-to-family asteroid dust ratio would be less than 3:1 and our estimate of total asteroidal contribution to the zodiacal cloud would be correspondingly lower.

Weak support for a dominant asteroidal component to the zodiacal cloud is provided by our analysis linking the total cross-sectional area of the cloud needed to account for the observed IR flux, A_{cloud}, to the observed value of the annual mass-accretion rate of particles from the zodiacal cloud by Earth, $\dot{M}_{accrete}$, as measured by LDEF. Particles at 1 AU in near-circular asteroidal orbits (note that this includes particles originating from comets that have had their orbits circularized by resonance with Jupiter) can account for the observed accretion rate, whereas particles originating from comets that have not been affected by resonances give an accretion rate that is probably too small. However, some caution needs to be exercised here as the numerical simulations from which these results were obtained (*Kortenkamp and Dermott,* 1998a) need to be extended to determine the effect of particle size on average rates of capture. There is also roughly a factor of 2 uncertainty in the mass-accretion rates estimated from the LDEF cratering record (*Love and Brownlee,* 1993). In fact, on the basis of elemental abundances in stratospheric aerosols, *Murphy* (2001) argues for an extraterrestrial mass-accretion rate in the lower range of that indicated by LDEF (see also *Cziczo et al.,* 2001), in agreement with previous estimates of the mass-accretion rate obtained by *Hughes* (1978) (using both visual and radio meteor observations and also satellite data) and *Grün et al.* (1985) (based upon *in situ* spacecraft measurements and the lunar microcratering record) that are indeed closer to half the value determined from LDEF. Until the actual mass flux to Earth of extraterrestrial material is more tightly constrained, the predicted fluxes from the various dust sources cannot be usefully employed as a discriminant of the actual source.

More importantly, perhaps, our calculation of the area of dust in the dust bands associated with the three most prominent Hirayama families is consistent with the collisional cascade model of *Dohnanyi* (1969). Figure 7 links the cumulative surface areas associated with the Eos, Koronis, and Themis asteroid families to the observed volumes of the source material in these families for a range of values of the size-frequency power-law index q (*Dermott et al.,* 1984; *Grogan et al.,* 2001). In each case, we see that the area of dust in each family with $D > D_{crit} \sim 500$ μm (the particles largely unaffected by P-R drag) needed to satisfy the dust-band observations can be accounted for if q is close to 1.83, the value predicted by *Dohnanyi* (1969). This is a comparatively robust conclusion, although more work is needed on calculating the area of dust in the transition region between collisional evolution and P-R drag evolution. However, even though we know the orbits of the asteroids in the collisional cascade, and we have an accurate measure of their total volume, we do not know how the strength of these bodies varies over the size range from micrometers to kilometers. Furthermore, we are not entirely confident that a simple power law is an adequate description of the size-frequency distribution. Hence, we do not claim that Dohnanyi-type supply arguments can be employed to predict, with useful precision, the observed quantity of dust from first principles, without the crucial input of the IR dust-band observations. Only consider, for example, that an error in the assumed average value of q of 0.1 would result in an error in the cumulative area of material of $\sim 10^3$ (see Fig. 7). This leads us to state that in the case of comets these supply-type arguments are also not good enough to be decisive, largely because of our ignorance of the numbers and sizes of the comets that may have contributed to the zodiacal cloud over the past 10^7 yr.

There are some outstanding problems with linking the dust bands with the prominent Hirayama families. Firstly, there is a discrepancy between the mean proper inclination of the 10° dust-band model (9.35°; see Table 1) and the

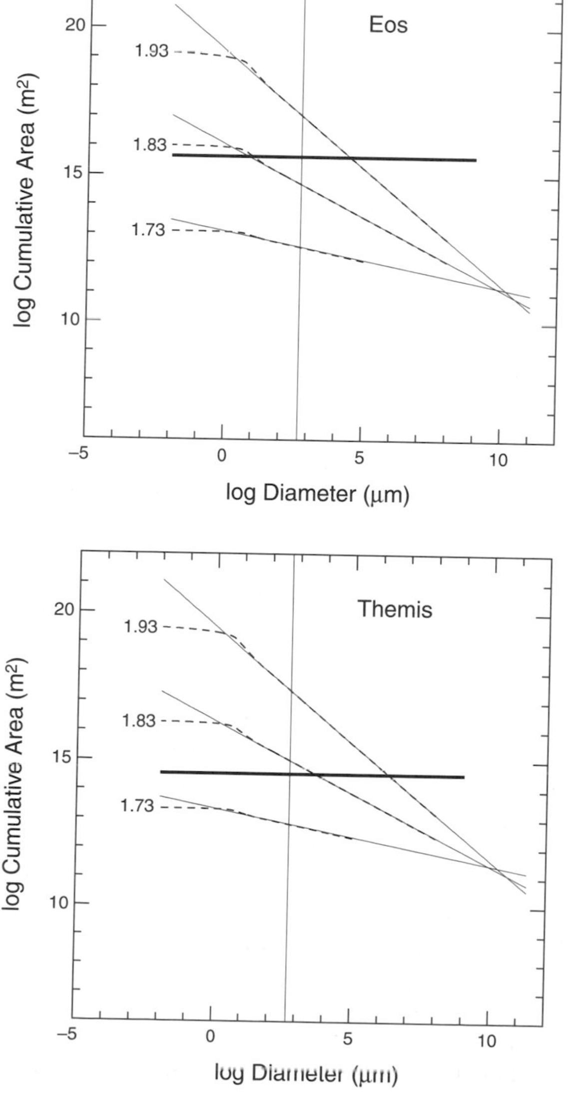

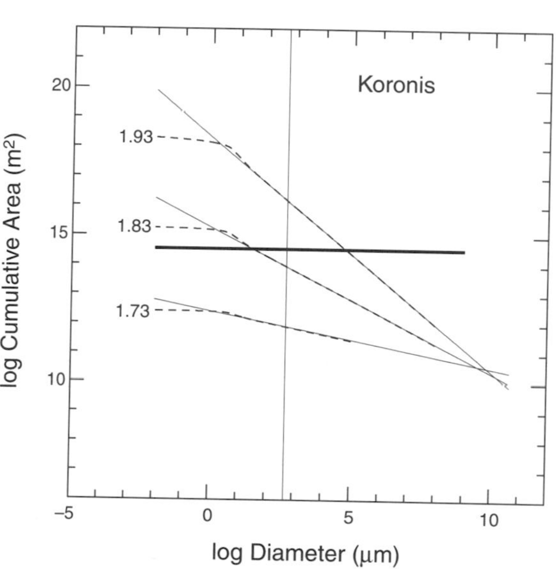

Fig. 7. Cumulative area of dust associated with the Eos, Koronis, and Themis asteroid families [as defined by *Zappalà et al.* (1995)] assuming three different size-frequency distributions for each with power-law indexes q = 1.73, 1.83, and 1.93. The solid lines represent the geometrical area, while the broken lines are the cross-sectional area of emission calculated for spheres of astronomical silicate of density 2500 kg m^{-3}, using Mie theory. In each panel, the cumulative area of dust from each family (Table 1) needed to model the IRAS observations of the dust bands is represented by a heavy, horizontal line. A vertical line corresponding to a diameter of 500 μm is shown for reference. In order to account for the amplitude of the dust bands associated with the three prominent Hirayama families, the average size-frequency power-law index q in each case has to be close to the value of 1.83 predicted by *Dohnanyi* (1969). Adapted from *Grogan et al.* (2001).

mean proper inclination of the Eos asteroid family (10.08°). This suggests that the 10° dust-band material does not trace the orbital-element distribution of the Eos family as a whole, as would be expected from the equilibrium model (*Dermott et al.,* 1984) in which the dust bands represent the continual grinding down of family asteroids. There are two possible explanations for this discrepancy. Forces may have acted on the dust particles in this band, or on the small bodies in the collisional cascade that produced these particles, to reduce their inclinations. The Yarkovsky effect, which preferentially acts upon meter- to kilometer-sized bodies, may be responsible for the migration of the dust particle parent bodies in orbital-element space (see *Bottke et al.,* 2001, 2002). Recent work by *Vokrouhlický* (2001), for example, speculates about the possibility of a jet of small asteroids from the Eos family streaming along the $z_1 = g + s - g_6 - s_6$ secular resonance to lower-inclination orbits due to the dynamic mobility induced by the Yarkovsky effect. This interesting possibility needs to be investigated more fully. More prosaically, we could simply be observing the latest disintegration of a small member of the Eos family, or even an unrelated background asteroid with the appropriate proper inclination. The origin of the large dispersion in proper inclination (1.5°) required to successfully model the 10° dust band (see Table 1), in rough agreement with the 1.4° found by *Sykes* (1990) and the 2° found by *Reach et al.* (1997), remains unclear, although the most likely source of the dispersion is simply the action of the ν_{16} secular resonance near 2 AU. However, this leaves open the question of why a large dispersion is required to model the 10° band, and only the small dispersion of the Koronis and Themis families is required to successfully reproduce the central band observations. One answer may be that the emission associated with the central band is due to relatively recent collisions within these families.

We recognize, of course, that the asteroid belt will contribute dust to the zodiacal cloud one asteroid at a time. Figure 8 shows a simulation of the variation with time of the total cross-sectional area of dust associated with the main belt (*Grogan et al.,* 2001) and describes both the steady decline in the total area of the dust and the inevitable stochastic variations. This numerical approach to describing the collisional evolution of the asteroid belt is detailed by *Durda and Dermott* (1997). The initial main-belt mass is taken to be approximately 3× greater than the present mass (*Durda et al.,* 1998); this population evolves after 4.5 b.y. to resemble the current main belt. The calculation is performed for particles from 100 μm to the largest asteroidal sizes, with an arbitrary fragmentation power-law index q = 1.9. The "spikes" in dust production are due to the breakup of small- to intermediate- (diameter ~40 km) sized asteroids. Therefore, while the volume of the source material in the zodiacal cloud may decay at a fairly constant and well-defined rate, the total area of dust associated with the cloud during that time, given a breakup value of q = 1.9, may fluctuate by an order of magnitude or more. On the surface this appears to be evidence for a catastrophic (non-equilibrium) origin for the dust bands (*Sykes and Greenberg,* 1986), in which the dust bands are produced by the disruption of random main-belt asteroids. However, we note that the grinding down of an asteroid family is also a stochastic process. Spikes in the dust-production rate correspond to the breakup of individual asteroids and can therefore originate from any asteroid within the family. The dust band associated with the family, produced by the fresh injection of material from the most recent fragmentation, may therefore shift in latitude over time.

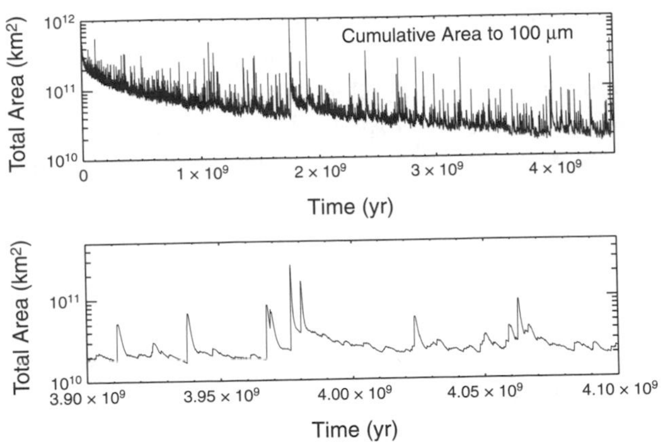

Fig. 8. Variation with time of the total cross-sectional area of dust associated with the breakup of an asteroid that was big enough to supply all the observed collision products of the main belt. This simulation modeled the stochastic breakup of asteroidal fragments down to a diameter of 100 μm. A size-frequency distribution with power-law index q = 1.9 was assumed for the initial breakup of each asteroid and this accounts for the heights of the "spikes" in the plots. The upper panel shows the evolution over the age of the solar system, while the lower panel shows in more detail a 200-m.y. period of evolution centered on 4 b.y. following the start of the simulation [*Grogan et al.* (2001); adapted from *Durda and Dermott* (1997)].

Dust particles generated by the collisional evolution of the asteroid belt eventually spiral into the inner solar system under the effects of P-R light drag and solar-wind drag. Variations in the production rate of these particles, as suggested by the results of the modeling shown in Fig. 8, should therefore result in a variation in the rate of accretion of dust particles by Earth. With certain caveats, it is possible to determine this variation by measuring the concentration of ^{3}He contained in sedimentary layers in Earth's crust, with the ^{3}He acting as a proxy for dust particles of extraterrestrial origin (*Farley,* 2001). Important information on the dust flux at 1 AU should therefore be contained in the geologic record. In fact, exciting data on the variation of the extraterrestrial dust accretion rate over a period of tens of millions of years has recently been published by *Mukhopadhyay et al.* (2001).

Interplanetary dust particles acquire He from the implantation of solar-wind or flare particles in space, and its concentration is therefore strongly surface-correlated. LDEF results (*Love and Brownlee,* 1993) indicate that the cross-sectional area distribution of these particles at 1 AU peaks for those with diameters of ~140 μm. However, *Farley et al.* (1997) calculate that due to the effect of heating during atmospheric entry, volatile components are stripped from large dust particles, and most of the extraterrestrial He (>70%) reaching the sedimentary layers is carried into Earth's atmosphere by particles between about 3 and 35 μm in diameter, with the dominant peak occurring at ~7 μm. The size distribution and total mass of He-bearing particles on the seafloor is therefore distinct from the size distribution and total mass of the parental dust-particle population. *Farley et al.* (1997) estimate that only ~4% of the total surface area and ~0.5% of the total mass of accreted dust particles are delivered to the surface of Earth at temperatures below which He is released (~600°C). Platinum-group elements, on the other hand, have proven to be the most sensitive nonvolatile tracers of extraterrestrial matter in marine sediments (*Peucker-Ehrenbrink,* 2001). Iridium and Os, specifically, can therefore be employed as a proxy for the flux of dust particles across the entire size spectrum and in particular for the large particle population not registered by the He proxy. Furthermore, differences in the dust particle flux estimates based upon these different proxies may indicate variations in the size-frequency distribution of extraterrestrial material accreted by Earth over geologic time (B. Peucker-Ehrenbrink, personal communication, 2001). However, as pointed out by *Farley et al.* (1997), mass-correlated nonvolatile tracers such as Ir are subject to extreme undersampling that may require the use of extraordinarily large sediment samples (many kilograms) in order to accurately infer the dust-particle flux. A comprehensive suite of papers reviewing many aspects of the accretion of extraterrestrial material by Earth has recently been compiled by *Peucker-Ehrenbrink and Schmitz* (2001).

The resolution of the debate between a collisional equilibrium or catastrophic model for the origin of the dust bands depends on whether the disruption of a single asteroid can liberate a large enough cross-sectional area of dust to account for the total emission from any given dust band. In this connection, we note that there is now substantial evidence for the existence of a large population of rubble-pile asteroids in the main belt. For example, asteroids with diameters between about 0.2 and 10 km show a cut-off in rotation period of ~2 h, indicating that they lack tensile strength and are probably "loosely bound, gravity-dominated aggregates" (*Pravec and Harris,* 2000). Larger asteroids (>10 km) may also be rubble piles, but the rotational statistics are insufficient to confirm this possibility (*Pravec and Harris,* 2001). Asteroids smaller than a few hundred meters in diameter typically have spin rates faster than the rotational breakup limit for rubble-pile asteroids, implying that they must have some tensile strength and may possibly be monolithic bodies. It has been argued that these bodies are the collisional fragments of larger rubble-pile asteroids and that the sharp division between the rotational periods of these small bodies and the rubble-pile asteroids indicates that the characteristic size of the largest rubble-pile fragments must be ~0.2 km (refer to *Pravec et al.,* 2002). Recent spacecraft missions and telescopic observations of asteroids have also provided measurements of a number of surprisingly low bulk densities in the 1300 kg m^{-3} range (see *Hilton,* 2002; *Britt et al.,* 2002), indicating bulk porosities of ~50% (*Britt and Consolmagno,* 2001; *Britt et al.,* 2002). This has led some authors to conclude that such asteroids are loosely bound rubble piles containing large voids (e.g., see *Britt et al.,* 2002). The increasing body of evidence suggesting the existence of gravitational aggregates is discussed in more detail in *Richardson et al.* (2002).

If all the dust in the zodiacal cloud is released from the final catastrophic disruption of a single rubble pile of diameter D_{rubble}, then the equivalent depth of the regolith layer, $d_{regolith}$, needed to account for the present observed cross-sectional area is given by

$$d_{regolith} \simeq \frac{2A_{cloud}D_{typ}}{3\pi D^2_{rubble}} \qquad (19)$$

where we have assumed that the regolith forms a uniform, shallow layer on a deep rubble pile. In the case of a 200-km-diameter rubble pile, similar in size to the precursor of the Eos family, the whole of the zodiacal cloud with $A_{cloud} \sim 2.5 \times 10^{10}$ km^2 could have been released from a regolith of depth $d_{regolith} \sim 70$ m, if we assume that $D_{typ} \sim 500$ μm. However, in the case of a rubble pile, D_{typ} is not determined by considerations of the balance between the collisional and P-R drag timescales and D_{typ} could be very much smaller than 500 μm, perhaps by a factor of 10–10^2. The depth of the regolith on an ancient rubble pile is also likely to be very much greater than 70 m, perhaps by a factor as large as 10^2. These considerations suggest that the disruption of those large asteroids that were the precursors of the major asteroid families (*Hirayama,* 1918; *Zappalà and Cellino,* 1993), if these precursor bodies were rubble piles (*Michel et al.,* 2001, *Richardson et al.,* 2002), could have resulted in an increase in the mass of dust in the zodiacal cloud by a factor of 10^2 (equation (13)) and an increase in the cross-sectional area of dust in the zodiacal cloud by a factor of 10^3–10^4 (equation (19)).

If the final stage of dust production is indeed the release of particles from both small and large rubble-pile asteroids, we would expect there to be variations in the dust-production rate in the past, and consequently, variations in Earth's accretion rate of dust particles (*Kortenkamp and Dermott,* 1998b). According to our calculations above, following the collisional disruption of a large rubble pile, Earth's mass-accretion rate of dust particles could have been enhanced by 2 orders of magnitude over the current estimated mass flux of particles measured by LDEF (*Love and Brownlee,* 1993), resulting in mass-accretion rates that may have been as high as several million tonnes per year. The potential effects of such enhanced dust loading of the atmosphere from extraterrestrial sources has previously been considered by, among others, *Muller and MacDonald* (1997), who suggest a possible causal link with glacial cycles. The consequences of a large influx of cometary debris onto Earth is discussed by *Napier* (2001).

The magnitude of any climatic effects caused by atmospheric dust loading depends mostly on the quantity of submicrometer particles deposited above the tropopause (*Toon et al.,* 1994), where they have a residence time on the order of a year (*Murphy,* 2001). Results obtained by LDEF, however, indicate that the mass flux of dust particles accreted by Earth peaks for particles of ~200 μm in diameter (*Love and Brownlee,* 1993), and so the proportion of incoming particles in the submicrometer size range may be relatively small unless the size distribution of particles released from rubble-pile asteroids differs significantly from the usual background flux of particles, as noted above. Nevertheless, particles in the correct size range can also be produced as a result of ablation and fragmentation of incoming dust particles during atmospheric entry (e.g., see *Hunten et al.,* 1980; *Love and Brownlee,* 1991). At present, the fraction of particles in the submicrometer size range resulting from this incoming extraterrestrial dust flux that is retained in the upper atmosphere is not precisely known, although some constraints are provided by measurements of elemental abundances in stratospheric aerosols (e.g., see *Cziczo et al.,* 2001; *Murphy,* 2001). Consequently, there must be a large uncertainty in the magnitude of the climatic effects produced by this enhanced extraterrestrial dust loading. Further, more detailed, modeling of this problem will need to be performed to accurately quantify the anticipated accretion rates of submicrometer particles by Earth during such periods of enhanced dust flux and to ascertain any potential climatic consequences this may have.

To place this atmospheric dust loading in the wider context, *Toon et al.* (1994) estimate that the impact of a large extraterrestrial body can pulverize and hurl into the atmosphere an amount of material equal to roughly $10^2\times$ the mass of the impacting bolide, of which 0.1% will be lofted into the stratosphere as submicrometer-sized particles. The impact event postulated by *Alvarez et al.* (1980) to have occurred at the time of the Cretaceous-Tertiary (K-T) boundary, for example, is estimated to have been caused by a bolide with the equivalent impact energy of 10^8 Mt of TNT and an entry velocity of ~25 km s^{-1}, implying that on the order of 10^{14} kg of submicrometer particles could have been lofted into the stratosphere by this single event alone. The resulting "impact winter" scenario would have been truly devastating for life on Earth and likely played an important role in the K-T mass extinction, although this is a matter of ongoing debate.

Considered as an abbreviated event, the atmospheric loading due to the enhanced accretion of asteroidal dust particles clearly pales in comparison to an event on the scale of the K-T impact. In this regard, the climatic consequences of volcanic eruptions may well serve as a more commensurate parallel for the effects of such dust loading of the atmosphere (e.g., see *Robock,* 2000). In the case of volcanic eruptions, any long-term global effect on climate is largely determined by the quantity of sulfur volatiles injected into the stratosphere where they are converted into sulfuric acid aerosols that have optical properties similar to fine dust particles (*Toon et al.,* 1994) and an e-folding residence time of ~1 yr (*Robock,* 2000). Large volcanic eruptions in recent history are typically estimated to have injected tens of millions of tonnes of chemically and microphysically active aerosol particles into the stratosphere (*Robock,* 2000), although eruptions in the more distant past may have been far more productive in this respect (*Rampino et al.,* 1988). Indeed, examples exist in the geologic record of massive flood basalt eruptions (*Hooper,* 2000) that may have been capable of sustaining, or episodically exceeding, this level of atmospheric aerosol loading over million-year timescales. As a wave of dust released by the destruction of a large rubble pile would take between approximately 10^4 and 10^7 yr to drift inward past Earth under the effect of P-R and solar-wind drag, the cumulative effects of such an eruption perhaps serves as a more appropriate analog for the climatic effects caused by the atmospheric loading of these asteroidal dust particles. It is interesting to note in this context that two of the largest known flood basalt eruptions, the Siberian Traps and the Deccan Traps, appear to coincide with the Permian-Triassic and the K-T mass extinctions, respectively, although no definite causal link has yet been established (*Stothers,* 1993).

Potential impactors large enough to produce globally catastrophic effects on Earth are likely to originate from the complete collisional disruption of one of the larger asteroids and possibly a precursor of one of the major asteroidal families (*Zappalà et al.,* 1998). As discussed above, such a massive collisional event would have certainly also generated an enormous quantity of dust. It follows from this argument that a large impact event on Earth would have also been associated with the influx of a large wave of dust, although the two events may well have been widely separated in time due to the different dynamical mechanisms and timescales involved in their delivery, with a wave of dust associated with the smallest dust particles probably arriving before the impactor. In this scenario, the extinction of life on Earth may have occurred not necessarily because of the effects of the giant impact, although this could have provided the *coup de grâce* (*Hallam,* 2001), but because the production of potential large impactors in the asteroid belt generates a wave of dust that Earth cannot avoid. Most, and in some cases all, of the potential impactors produced by the catastrophic disruption of a large asteroid may have missed Earth, but interaction with the concomitant wave of dust is unavoidable. In this regard, we should not ignore the fact that large increases in the area of dust in the zodiacal cloud between Earth and the Sun could also affect Earth's level of insolation.

We have strong, quantitative evidence that asteroidal dust is a significant component of the zodiacal cloud. The final question is whether this preponderance of asteroidal particles is reflected in our collections of IDPs, MMs, and cosmic spherules that provide a sample of the zodiacal cloud particles at 1 AU. As discussed in section 2, hydrous chondritic IDPs are generally believed to be derived from asteroids because of their chondritic nature, their similarity to C-type asteroids, and the evidence of their aqueous alteration. Their remarkable fine-grained nature at the submicrometer scale, which is more chondritic than the chondritic meteorites (*Jessberger et al.,* 2001), requires that they had to be formed in environments where substantial mechanical mixing occurred. Natural locations for such processes are found in the solar nebula itself or in asteroidal regoliths (*Keller et al.,* 1993), in particular, the surface layer of rubble piles (*Britt and Consolmagno,* 2001; *Britt et al.,* 2002). On the other hand, it has been suggested that the highly porous and primitive composition of anhydrous chondritic IDPs is indicative of a cometary origin.

One of the most promising methods of discriminating between asteroidal and cometary IDPs on dynamical grounds is to determine their precapture orbits from their thermal histories. However, there are some problems with this method of classification. The distinction between the orbital elements of cometary and asteroidal dust particles may not be as sharp as that displayed by the orbits of their parent bodies. Cometary particles that are trapped in mean-motion resonance with Jupiter can have their orbital eccentricities decreased to asteroidal values (*Liou and Zook,* 1996). On the other hand, large asteroidal particles, with diameters in the range ~100–500 µm, move slowly through the inner solar system toward the Sun under the effects of P-R and solar-wind drag and can have their orbital elements increased to the low end of the cometary range due to the action of mean-motion and secular resonances, as well as gravitational scattering by the terrestrial planets. Caution may therefore be in order when classifying particles as either asteroidal or cometary without an appreciation of their orbital histories.

This argument does not apply to those smaller dust particles (with diameters in the range ~5–25 µm) generated directly by interasteroidal collisions in the main belt. These particles traverse the inner solar system more rapidly and arrive at 1 AU in near-circular, near-ecliptic orbits. Thus, the result of *Brownlee et al.* (1995) that the very-high-speed IDPs are dominated by porous anhydrous chondritic IDPs could be strong evidence that these particles are cometary. However, we believe that the other dust particles accreted by Earth must be predominantly asteroidal in origin. Given that these particles are quite different from known meteorites, it follows that the composition of the asteroid belt is largely unexplored and may consist largely of more friable material not well represented by the strong meteorites in our collections. However, it is interesting to note that because 85% of the dust-band material is associated with a single family, Eos, it follows from our calculations that 25% (85% of 30%) of all accreted dust particles could, in fact, be samples of a single K-type asteroid.

Acknowledgments. It is a pleasure to thank M. Zolensky and D. Brownlee for their valuable contributions and also J. Burns for his careful review. This research was supported by NASA ADP Grant NAG5-9280 and NASA PGG Grant NAG5-4531.

REFERENCES

Alvarez L. W., Alvarez W., Asaro F., and Michel H. V. (1980) Extraterrestrial cause for the Cretaceous-Tertiary extinction. *Science, 208,* 1095–1108.

Artymowicz P. (1997) Beta Pictoris: An early solar system? *Annu. Rev. Earth Planet. Sci., 25,* 175–219.

Bottke W. F. Jr., Vokrouhlický D., Brož M., Nesvorný D., and Morbidelli A. (2001) Dynamical spreading of asteroid families by the Yarkovsky effect. *Science, 294,* 1693–1696.

Bottke W. F. Jr., Vokrouhlický D., Rubincam D. P., and Brož M. (2002) The effect of Yarkovsky thermal forces on the dynamical evolution of asteroids and meteoroids. In *Asteroids III* (W. F. Bottke Jr. et al., eds.), this volume. Univ. of Arizona, Tucson.

Bradley J. P. (1999) Interstellar dust — evidence from interplanetary dust particles. In *Formation and Evolution of Solids in Space* (J. M. Greenberg and A. Li, eds.), pp. 485–503. Kluwer, Dordrecht.

Bradley J. P., Brownlee D. E., and Fraundorf P. (1984) Discovery of nuclear tracks in interplanetary dust. *Science, 226,* 1432–1434.

Bradley J. P., Sandford S. A., and Walker R. M. (1988) Interplanetary dust particles. In *Meteorites and the Early Solar System* (J. F. Kerridge and M. S. Matthews, eds.), pp. 861–898. Univ. of Arizona, Tucson.

Bradley J. P., Keller L. P., Brownlee D. E., and Thomas K. L. (1996) Reflectance spectroscopy of interplanetary dust particles. *Meteoritics & Planet. Sci., 31,* 394–402.

Britt D. T. and Consolmagno G. J. (2001) Modeling the structure of high porosity asteroids. *Icarus, 152,* 134–139.

Britt D. T., Yeomans D., Housen K., and Consolmagno G. (2002) Asteroid density, porosity, and structure. In *Asteroids III* (W. F. Bottke Jr. et al., eds.), this volume. Univ. of Arizona, Tucson.

Brownlee D. E. (1985) Cosmic dust: Collection and research. *Annu. Rev. Earth Planet. Sci., 13,* 147–173.

Brownlee D. E. (1996) The elemental composition of interplanetary dust. In *Physics, Chemistry, and Dynamics of Interplanetary Dust* (B. Å. S. Gustafson and M. S. Hanner, eds.), pp. 261–264. ASP Conference Series 104.

Brownlee D. E., Joswiak D. J., Schlutter D. J., Pepin R. O., Bradley J. P., and Love S. G. (1995) Identification of individual cometary IDP's by thermally stepped He release (abstract). In *Lunar and Planetary Science XVI,* pp. 183–184. Lunar and Planetary Institute, Houston.

Burns J. A., Lamy P. L., and Soter S. (1979) Radiation forces on small particles in the solar system. *Icarus, 40,* 1–48.

Corrigan C. M., Zolensky M. E., Dahl J., Long M., Weir J., Sapp C., Burkett P. J. (1997) The porosity and permeability of chondritic meteorites and interplanetary dust particles. *Meteoritics & Planet. Sci., 32,* 509–515.

Cziczo D. J., Thomson D. S., and Murphy D. M. (2001) Ablation, flux, and atmospheric implications of meteors inferred from stratospheric aerosol. *Science, 291,* 1772–1775.

Davis D. R., Weidenschilling S. J., Farinella P., Paolicchi P., and Binzel R. P. (1989) Asteroid collisional history: Effects on sizes and spins. In *Asteroids II* (R. P. Binzel et al., eds.), pp. 805–826. Univ. of Arizona, Tucson.

Dell'Oro A., Paolicchi P., Cellino A., Zappalà V., Tanga P., and Michel P. (2001) The role of families in determining collision probability in the asteroid main belt. *Icarus, 153,* 52–60.

Delsemme A. H. (1976) The production rate of dust by comets. In *Interplanetary Dust and Zodiacal Light* (H. Elsässer and H.

Fechtig, eds.), pp. 314–318. Lecture Notes in Physics Vol. 48, Springer-Verlag, Berlin.

Dermott S. F., Nicholson P. D., Burns J. A., and Houck J. R. (1984) Origin of the solar system dust bands discovered by IRAS. *Nature, 312,* 505–509.

Dermott S. F., Nicholson P. D., Burns J. A., and Houck J. R. (1985) An analysis of IRAS' solar system dust bands. In *Properties and Interactions of Interplanetary Dust* (R. H. Giese and P. Lamy, eds.), pp. 395–410. ASSL Proc. 199, Reidel, Dordrecht.

Dermott S. F., Nicholson P. D., Kim Y., Wolven B., and Tedesco E. F. (1988) The impact of IRAS on asteroidal science. In *Comets to Cosmology* (A. Lawrence, ed.), pp. 3–18. Springer-Verlag, Berlin.

Dermott S. F., Gomes R. S., Durda D. D., Gustafson B. Å. S., Jayaraman S., Xu Y.-L., and Nicholson P. D. (1992) Dynamics of the zodiacal cloud. In *Chaos, Resonance and Collective Dynamical Phenomena in the Solar System* (S. Ferraz-Mello, ed.), pp. 333–347. Kluwer, Dordrecht.

Dermott S. F., Jayaraman S., Xu Y.-L., Gustafson B. Å. S., and Liou J.-C. (1994a) A circumsolar ring of asteroidal dust in resonant lock with the Earth. *Nature, 369,* 719–723.

Dermott S. F., Durda D. D., Gustafson B. Å. S., Jayaraman S., Liou J.-C., and Xu Y.-L. (1994b) Zodiacal dust bands. In *Asteroids, Comets, Meteors 1993* (A. Milani et al., eds.), pp. 127–142. Kluwer, Dordrecht.

Dermott S. F., Grogan K., Holmes E. K., and Kortenkamp S. J. (1999) Dynamical structure of the zodiacal cloud. In *Formation and Evolution of Solids in Space* (J. M. Greenberg and A. Li, eds.), pp. 565–582. Kluwer, Dordrecht.

Dermott S. F., Grogan K., Durda D. D., Jayaraman S., Kehoe T. J. J., Kortenkamp S. J., and Wyatt M. C. (2001) Orbital evolution of interplanetary dust. In *Interplanetary Dust* (E. Grün et al., eds.), pp. 569–639. Springer-Verlag, Berlin.

Dohnanyi J. S. (1969) Collisional model of asteroids and their debris. *J. Geophys. Res., 74,* 2531–2554.

Dohnanyi J. S. (1976) Sources of interplanetary dust: Asteroids. In *Interplanetary Dust and Zodiacal Light* (H. Elsässer and H. Fechtig, eds.), pp. 187–205. Lecture Notes in Physics Vol. 48, Springer-Verlag, Berlin.

Draine B. T. and Lee H. M. (1984) Optical properties of interstellar graphite and silicate grains. *Astrophys. J., 285,* 89–108.

Durda D. D. and Dermott S. F. (1997) The collisional evolution of the asteroid belt and its contribution to the zodiacal cloud. *Icarus, 130,* 140–164.

Durda D. D., Greenberg R., and Jedicke R. (1998) Collisional models and scaling laws: A new interpretation of the shape of the main-belt asteroid size distribution. *Icarus, 135,* 431–440.

Farley K. A. (2001) Extraterrestrial helium in seafloor sediments: Identification, characteristics, and accretion rate over geologic time. In *Accretion of Extraterrestrial Matter Throughout Earth's History* (B. Peucker-Ehrenbrink and B. Schmitz, eds.), pp. 179–204. Kluwer, New York.

Farley K. A., Love S. G., and Patterson D. B. (1997) Atmospheric entry heating and helium retentivity of interplanetary dust particles. *Geochim. Cosmochim. Acta, 61,* 2309–2316.

Flynn G. J. (1990) The near-Earth enhancement of asteroidal over cometary dust. *Proc. Lunar Planet. Sci. Conf. 20th,* pp. 363–371.

Flynn G. J. (1996) Sources of 10 micron interplanetary dust: The contribution from the Kuiper belt. In *Physics, Chemistry, and Dynamics of Interplanetary Dust* (B. Å. S. Gustafson and M. S. Hanner, eds.), pp. 171–175. ASP Conference Series 104.

Flynn G. J., Bajt S., Sutton S. R., Zolensky M. E., Thomas K. L., and Keller L. P. (1996) The abundance pattern of elements having low nebular condensation temperatures in interplanetary dust particles: Evidence for a new chemical type of chondritic material. In *Physics, Chemistry, and Dynamics of Interplanetary Dust* (B. Å. S. Gustafson and M. S. Hanner, eds.), pp. 291–294. ASP Conference Series 104.

Fraundorf P. (1981) Interplanetary dust in the transmission electron microscope: Diverse materials from the early solar system. *Geochim. Cosmochim. Acta, 45,* 915–943.

Gor'kavyi N. N., Ozernoy L. M., Mather J. C., and Taidakova T. (1997) Quasi-stationary states of dust flows under Poynting-Robertson drag: New analytical and numerical solutions. *Astrophys. J., 488,* 268–276.

Grogan K., Dermott S. F., Jayaraman S., and Xu Y.-L. (1997) Origin of the ten degree dust bands. *Planet. Space Sci., 45,* 1657–1665.

Grogan K., Dermott S. F., and Durda D. D. (2001) The size-frequency distribution of the zodiacal cloud: Evidence from the solar system dust bands. *Icarus, 152,* 251–267.

Grün E., Zook H. A., Fechtig H., and Giese R. H. (1985) Collisional balance of the meteoritic complex. *Icarus, 62,* 244–272.

Grün E., Zook H. A., Baguhl M., Balogh A., Bame S. J., Fechtig H., Forsyth R., Hanner M. S., Horanyi M., Kissel J., Lindblad B.-A., Linkert D., Linkert G., Mann I., McDonnell J. A. M., Morfill G. E., Phillips J. L., Polanskey C., Schwehm G., Siddique N., Staubach P., Svestka J., and Taylor A. (1993) Discovery of jovian dust streams and interstellar grains by the ULYSSES spacecraft. *Nature, 362,* 428–430.

Grün E., Staubach P., Baguhl M., Hamilton D. P., Zook H. A., Dermott S. F., Gustafson B. Å. S., Fechtig H., Kissel J., Linkert D., Linkert G., Srama R., Hanner M. S., Polanskey C., Horanyi M., Lindblad B.-A., Mann I., McDonnell J. A. M., Morfill G. E., and Schwehm G. (1997) South-North and radial traverses through the interplanetary dust cloud. *Icarus, 129,* 270–288.

Gustafson B. Å. S. (1994) Physics of zodiacal dust. *Annu. Rev. Earth Planet. Sci., 22,* 553–595.

Gustafson B. Å. S. and Hanner M. S., eds. (1996) *Physics, Chemistry, and Dynamics of Interplanetary Dust.* ASP Conference Series 104. 568 pp.

Hallam A. (2001) Mass extinctions in Phanerozoic time. In *Meteorites: Flux with Time and Impact Effects* (M. M. Grady et al., eds.), pp. 259–274. Geological Society of London Special Publication 140.

Hilton J. L. (2002) Asteroid masses and densities. In *Asteroids III* (W. F. Bottke Jr. et al., eds.), this volume. Univ. of Arizona, Tucson.

Hirayama K. (1918) Groups of asteroids probably of common origin. *Astron. J., 31,* 185–188.

Hiroi T., Zolensky M. E., and Pieters C. M. (2001) The Tagish Lake meteorite: A possible sample from a D-type asteroid. *Science, 293,* 2234–2236.

Hooper P. J. (2000) Flood basalt provinces. In *Encyclopedia of Volcanoes* (H. Sigurdsson et al., eds.), pp. 345–359. Academic, San Diego.

Hughes D. W. (1978) Meteors. In *Cosmic Dust* (J. A. M. McDonnell, ed.), pp. 123–185. Wiley, Chichester.

Hunten D. M., Turco R. P., and Toon O. B. (1980) Smoke and dust particles of meteoric origin in the mesosphere and stratosphere. *J. Atmos. Sci., 37,* 1342–1357.

Ishiguro M., Nakamura R., Fujii Y., Morishige K., Yano H., Yasuda

H., Yokogawa S., and Mukai T. (1999) First detection of visible zodiacal dust bands from ground-based observations. *Astrophys. J., 511,* 432–435.

Jackson A. A. and Zook H. A. (1992) Orbital evolution of dust particles from comets and asteroids. *Icarus, 97,* 70–84.

Jessberger E. K., Stephan T., Rost D., Arndt P., Maetz M., Stadermann F. J., Brownlee D. E., Bradley J. P., and Kurat G. (2001) Properties of interplanetary dust: Information from collected samples. In *Interplanetary Dust* (E. Grün et al., eds.), pp. 253–294. Springer-Verlag, Berlin.

Keller L. P., Thomas K. L., and McKay D. S. (1993) Asteroidal agglutinate particles as a component of anhydrous interplanetary dust. *Meteoritics, 28,* 378–379.

Kelsall T., Weiland J. L., Franz B. A., Reach W. T., Arendt R. G., Dwek E., Freudenreich H. T., Hauser M. G., Moseley S. H., Odegard N. P., Silverberg R. F., and Wright E. L. (1998) The COBE diffuse infrared background experiment search for the cosmic infrared background. II. Model of the interplanetary dust cloud. *Astrophys. J., 508,* 44–73.

Klöck W. and Stadermann F. J. (1994) Mineralogical and chemical relationships of interplanetary dust particles, micrometeorites and meteorites. In *Analysis of Interplanetary Dust* (M. E. Zolensky et al., eds.), pp. 51–87. AIP Conference Proceedings 310.

Kortenkamp S. J. and Dermott S. F. (1998a) Accretion of interplanetary dust particles by the Earth. *Icarus, 135,* 469–495.

Kortenkamp S. J. and Dermott S. F. (1998b) A 100,000-year periodicity in the accretion rate of interplanetary dust. *Science, 280,* 874–876.

Kortenkamp S. J., Dermott S. F., Fogle D., and Grogan K. (2001) Sources and orbital evolution of interplanetary dust accreted by Earth. In *Accretion of Extraterrestrial Matter Throughout Earth's History* (B. Peucker-Ehrenbrink and B. Schmitz, eds.), pp. 13–30. Kluwer, New York.

Leinert C., Röser S., and Buitrago J. (1983) How to maintain the spatial distribution of interplanetary dust. *Astron. Astrophys., 118,* 345–357.

Liou J.-C. and Zook H. A. (1996) Comets as a source of low eccentricity and low inclination interplanetary dust particles. *Icarus, 123,* 491–502.

Liou J.-C., Dermott S. F., and Xu Y.-L. (1995) The contribution of cometary dust to the zodiacal cloud. *Planet. Space Sci., 43,* 717–722.

Liou J.-C., Zook H. A., and Dermott S. F (1996) Kuiper belt dust grains as a source of interplanetary dust particles. *Icarus, 124,* 429–440.

Lisse C. M., A'Hearn M. F., Hauser M. G., Kelsall T., Lien D. J., Moseley S. H., Reach W. T., and Silverberg R. F. (1998) Infrared observations of comets by COBE. *Astrophys. J., 496,* 971–991.

Lisse C. M. (2001) The death of short period comets (abstract). *Bull. Am. Astron. Soc., 33,* 1093.

Love S. G. and Brownlee D. E. (1991) Heating and thermal transformation of micrometeoroids entering the Earth's atmosphere. *Icarus, 89,* 26–43.

Love S. G. and Brownlee D. E. (1993) A direct measurement of the terrestrial mass accretion rate of cosmic dust. *Science, 262,* 550–553.

Low F. J., Beintema D. A., Gautier T. N., Gillet F. C., Beichmann C. A., Neugebauer G., Young E., Aumann H. H., Boggess N., Emerson J. P., Habing H. J., Hauser M. G., Houck J. R., Rowan-Robinson M., Soifer B. T., Walker R. G., and Wesselius P. R. (1984) Infrared cirrus: New components of the extended infrared emission. *Astrophys. J. Lett., 278,* L19–L22.

Maurette M., Engrand C., and Kurat G. (1996) Collection and microanalysis of Antarctic micrometeorites. In *Physics, Chemistry, and Dynamics of Interplanetary Dust* (B. Å. S. Gustafson and M. S. Hanner, eds.), pp. 265–273. ASP Conference Series 104.

McKeegan K. D. (1987) Oxygen isotopes in refractory stratospheric dust particles: Proof of extraterrestrial origin. *Science, 237,* 1468–1471.

McSween H. Y. Jr., Sears D. W. G., and Dodd R. T. (1988) Thermal metamorphism. In *Meteorites and the Early Solar System* (J. F. Kerridge and M. S. Matthews, eds.), pp. 102–113. Univ. of Arizona, Tucson.

Messenger S. and Walker R. M. (1996) Isotopic anomalies in interplanetary dust particles. In *Physics, Chemistry, and Dynamics of Interplanetary Dust* (B. Å. S. Gustafson and M. S. Hanner, eds.), pp. 287–290. ASP Conference Series 104.

Michel P., Benz W., Tanga P., and Richardson D. C. (2001) Collisions and gravitational reaccumulation: Forming asteroid families and satellites. *Science, 294,* 1696–1700.

Mukhopadhyay S., Farley K. A., and Montanari A. (2001) A 35 m.y. record of helium in pelagic limestones from Italy: Implications for interplanetary dust accretion from the early Maastrichtian to the middle Eocene. *Geochim. Cosmochim. Acta., 65,* 653–669.

Muller R. A. and MacDonald G. J. (1997) Glacial cycles and astronomical forcing. *Science, 277,* 215–218.

Murphy D. M. (2001) Extraterrestrial material and stratospheric aerosols. In *Accretion of Extraterrestrial Matter Throughout Earth's History* (B. Peucker-Ehrenbrink and B. Schmitz, eds.), pp. 129–142. Kluwer, New York.

Murray C. D. and Dermott S. F. (1999) *Solar System Dynamics.* Cambridge Univ., Cambridge. 606 pp.

Napier W. M. (2001) The influx of comets and their debris. In *Accretion of Extraterrestrial Matter Throughout Earth's History* (B. Peucker-Ehrenbrink and B. Schmitz, eds.), pp. 51–74. Kluwer, New York.

Nier A. O. and Schlutter D. J. (1993) The thermal history of interplanetary dust particles collected in the Earth's stratosphere. *Meteoritics, 28,* 675–681.

Peucker-Ehrenbrink B. (2001) Iridium and osmium as tracers of extraterrestrial matter in marine sediments. In *Accretion of Extraterrestrial Matter Throughout Earth's History* (B. Peucker-Ehrenbrink and B. Schmitz, eds.), pp. 163–178. Kluwer, New York.

Peucker-Ehrenbrink B. and Schmitz B., eds. (2001) *Accretion of Extraterrestrial Matter Throughout Earth's History.* Kluwer, New York. 492 pp.

Pravec P. and Harris A. W. (2000) Fast and slow rotation of asteroids. *Icarus, 148,* 12–20.

Pravec P. and Harris A. W. (2001) Asteroid rotations (abstract). In *Asteroids 2001: From Piazzi to the Third Millenium,* p. 59. Osservatorio di Palermo, Sicily.

Pravec P., Harris A. W., and Michałowski T. (2002) Asteroid rotations. In *Asteroids III* (W. F. Bottke Jr. et al., eds.), this volume. Univ. of Arizona, Tucson.

Rajan R. S., Brownlee D. E., Tomandl D., Hodge P. W., Farrar H., and Britten R. A. (1977) Detection of ^{4}He in stratospheric particles gives evidence of extraterrestrial origin. *Nature, 267,* 133–134.

Rampino M. R., Self S., and Stothers R. B. (1988) Volcanic winters. *Annu. Rev. Earth. Planet. Sci., 16,* 73–99.

Reach W. T., Franz B. A., and Weiland J. L. (1997) The three-dimensional structure of the zodiacal dust bands. *Icarus, 127,* 461–484.

Reach W. T., Franz B. A., Weiland J. L., Hauser M. G., Kelsall T. N., Wright E. L., Rawley G., Stemwedel S. W., and Spiesman W. J. (1995) Observational confirmation of a circumsolar dust ring by the COBE satellite. *Nature, 374,* 521–523.

Richardson D. C., Leinhardt Z. M., Melosh H. J., Bottke W. F. Jr., and Asphaug E. (2002) Gravitational aggregates: Evidence and evolution. In *Asteroids III* (W. F. Bottke Jr. et al., eds.), this volume. Univ. of Arizona, Tucson.

Rietmeijer F. J. M. (1988) Interplanetary dust particles. In *Reviews in Mineralogy, Vol. 36: Planetary Materials* (J. J. Papike, ed.), pp. 2.1–2.95. Mineralogical Society of America, Washington.

Robock A. (2000) Volcanic eruptions and climate. *Rev. Geophys., 38,* 191–219.

Schramm L. S., Brownlee D. E., and Wheelock M. M. (1989) Major element composition of stratospheric micrometeorites. *Meteoritics, 24,* 99–112.

Stothers R. B. (1993) Flood basalts and extinction events. *Geophys. Res. Lett., 20,* 1399–1402.

Sykes M. V. (1990) Zodiacal dust bands: Their relation to asteroid families. *Icarus, 84,* 267–289.

Sykes M. V. and Greenberg R. (1986) The formation and origin of the IRAS zodiacal dust bands as a consequence of single collisions between asteroids. *Icarus, 65,* 51–69.

Sykes M. V. and Walker R. G. (1992) Cometary dust trails. I. Survey. *Icarus, 95,* 180–210.

Thomas K. L., Keller L. P., Blanford G. E., and McKay D. S. (1994) Quantitative analyses of carbon in anhydrous and hydrated interplanetary dust particles. In *Analysis of Interplanetary Dust* (M. E. Zolensky et al., eds.), pp. 165–172. AIP Conference Proceedings 310.

Thomas K. L., Keller L. P., and McKay D. S. (1996) A comprehensive study of major, minor, and light element abundances in over 100 interplanetary dust particles. In *Physics, Chemistry, and Dynamics of Interplanetary Dust* (B. Å. S. Gustafson and M. S. Hanner, eds.), pp. 283–286. ASP Conference Series 104.

Toon O. B., Zahnle K., Turco R. P., and Covey C. (1994) Environmental perturbations caused by impacts. In *Hazards Due to Comets and Asteroids* (T. Gehrels, ed.), pp. 791–826. Univ. of Arizona, Tucson.

Vokrouhlický D., Brož M., Morbidelli A., Nesvorný D., and Bottke W. F. Jr. (2001) Long-term dynamical diffusion in asteroid families via the Yarkovsky effect (abstract). In *Asteroids 2001: From Piazzi to the Third Millenium,* p. 225. Osservatorio di Palermo, Sicily.

Whipple F. L. (1967) On maintaining the meteoritic complex. In *The Zodiacal Light and the Interplanetary Medium* (J. L. Weinberg, ed.), pp. 409–426. NASA SP-150, Washington, DC.

Wyatt S. P. Jr. and Whipple F. L. (1950) The Poynting-Robertson effect on meteor orbits. *Astrophys. J., 111,* 134–141.

Wyatt M. C., Dermott S. F., Telesco C. M., Fisher R. S., Grogan K., Holmes E. K., and Piña R. K. (1999) How observations of circumstellar disk asymmetries can reveal hidden planets: Pericenter glow and its application to the HR 4796 disk. *Astrophys. J., 527,* 918–944.

Zappalà V. and Cellino A. (1993) Asteroid families. In *Asteroids, Comets, Meteors 1993* (A. Milani et al., eds.), pp. 395–414. Kluwer, Dordrecht.

Zappalà V., Bendjoya P. H., Cellino A., Farinella P., and Froeschlè C. (1995) Asteroid families: Search of a 12,487 asteroid sample using two different clustering techniques. *Icarus, 116,* 291–314.

Zappalà V., Cellino A., Gladman B. J., Manley S., and Migliorini F. (1998) Asteroid showers on Earth after family breakup events. *Icarus, 134,* 176–179.

Zolensky M. E. (1987) Refractory interplanetary dust particles. *Science, 237,* 1466–1468.

Zolensky M. E. and Barrett R. A. (1994) Compositional variations of olivines and pyroxenes in chondritic interplanetary dust particles. *Meteoritics, 29,* 616–620.

Zolensky M. E. and McSween H. Y. Jr. (1988) Aqueous alteration. In *Meteorites and the Early Solar System* (J. F. Kerridge and M. S. Matthews, eds.), pp. 114–143. Univ. of Arizona, Tucson.

Zolensky M. E., Wilson T. L., Rietmeijer F. J. M., and Flynn G. J., eds. (1994) *Analysis of Interplanetary Dust.* AIP Conference Proceedings 310. 371 pp.

Zook H. A. (2001) Spacecraft measurements of the cosmic dust flux. In *Accretion of Extraterrestrial Matter Throughout Earth's History* (B. Peucker-Ehrenbrink and B. Schmitz, eds.), pp. 75–92. Kluwer, New York.

Asteroid Impacts: Laboratory Experiments and Scaling Laws

K. Holsapple
University of Washington

I. Giblin
Planetary Science Institute

K. Housen
The Boeing Company

A. Nakamura
Kobe University

E. Ryan
New Mexico Highlands University

The present states of the small bodies of the solar system are largely an outcome of collisional processes. A rocky main-belt asteroid has endured a multitude of small and large cratering impacts; for example, estimates here show that one starting with a radius of 1 km has been shattered about five times every 10^6 yr and one with a radius of 10 km has been shattered about every 10^7 yr, or perhaps even more frequently in the past when collision rates were higher than they are now. All solar system bodies bear the scars and imprints of those impacts. Much has been learned about these topics since publication of the *Asteroids II* book (*Fujiwara et al.,* 1989). Here we briefly review the previous wisdom, but primarily address new experiments, calculations, and scaling methods.

1. INTRODUCTION

An understanding of collisional processes and collisional evolution is required in order to interpret observations of the solar system. Although laboratory experiments and computer simulations have provided many insights into these processes, our understanding of energetic impacts is still relatively primitive. Each new view of asteroids provided by spacecraft brings new surprises. For example, the substantial regolith on a small body like Gaspra, the huge closely packed craters on Mathilde, and the block-strewn surface of Eros were entirely unexpected. Nevertheless, experiments and modeling, guided by observations of asteroids, will allow us to deduce much about the collision history of these bodies, including crater sizes and frequency, crater morphology, ejecta block distributions, regolith development, and the formation of asteroid families.

Impact processes are very complex and involve extreme ranges of conditions. The initial coupling of energy from a high-speed impactor into another body can occur in microseconds, with the extreme pressures and temperatures sufficient to melt and vaporize the target and projectile material. On large bodies, the latter stages of these processes can continue for hours and involve very low pressures. These conditions range from those encountered in the initial detonation of nuclear bombs down to those involved in the statics of ordinary soil and rock mechanics. The results of a given impact are therefore very difficult to predict, although much progress has been made over the last few decades.

The outcome of a collision depends largely on the ratio of the kinetic energy of the impactor to the mass of the impacted body, a specific energy, commonly denoted as Q. Two threshold values of Q are often defined, although the literature is not consistent in terminology or notation. Impacts with small values of Q form craters, but leave the target body largely intact. Larger values of Q can shatter a body into numerous pieces. The specific energy to shatter, Q^*_S, is defined as the threshold value for which the largest remaining intact piece immediately following a collision has one-half the mass of the original body. We refer to it as the shattering energy. The shattered pieces may reaccumulate or not, depending on their velocity relative to the escape velocity. A higher threshold Q^*_D is the specific energy such that the largest object following reaccumulation is one-half the mass of the original body. This is called the dispersion energy. The term disruption energy is sometimes used in the literature for either of these thresholds, often in a generic way for any significant breaking and/or dispersion.

Numerous unsolved problems remain for collisions in all three regimes. Important questions about cratering include the crater size; the shape; the amount of compaction; and the amount, velocity, and fate of ejected material. For shattering impacts, the distributions of velocity, size, shape, and spin of the pieces, and the ultimate fate of those pieces, are still largely unknown. How many fragments remain in place, how many are lofted and reaccumulate, and what is the structure of that modified body? What conditions determine whether a body is shattered but basically intact, as Eros might be, or completely turned to rubble? Finally, for energy above Q_D^*, when the body is shattered and scattered, significant questions remain about the size distribution of the fragments, the largest piece, and their ultimate fate.

There are three interrelated and complementary approaches to the study of these processes. All three approaches suffer from our lack of detailed knowledge about the materials and structure of solar system bodies and asteroids. In addition, each approach also has its own shortcomings.

The first approach is to conduct laboratory experiments. Laboratory methods use guns to launch a projectile into a target at speeds up to ~7 km/s, or explosive charges to simulate an impact. While experiments allow the study of actual geological materials, their primary shortcoming is the inability to use targets and projectiles of the size of interest. Laboratory experiments are limited to samples of centimeter size, while the asteroids range to many hundreds of kilometers in size. The response of a small target to impact is dominated by its material strength, while that of a large asteroid is dominated by gravitational forces. For the predominantly unidirectional gravity field that governs surface cratering, the lack of sufficient gravity for the experimental small targets can sometimes be overcome by performing experiments at high artificial gravity using a geotechnical centrifuge, but that tool is of little use for the three-dimensional gravity fields dominating catastrophic disruptions. Additional uncertainties are introduced by an inability to perform experiments at velocities of several tens of kilometers per second. Features that are important at low velocities, such as projectile material and shape, probably are not significant at high velocities. As a consequence of these limitations, laboratory experiments can only probe a small part of the parameter space of interest.

The second approach is to use computer calculations based on the underlying physical principles. These methods continue to evolve. Finite-difference, finite-element and, over the last decade, smooth-particle-hydrodynamics (SPH) methods have been particularly useful, giving efficient ways to study impact processes. As computer power grows in leaps and bounds, we can now calculate many more cases in much less time. However, these methods are also limited in important ways, particularly by the relative infancy of material models. While laboratory experiments use actual materials, computer codes must rely on mathematical models of material behavior. The mechanical behavior of geological materials is considerably more difficult to model than the common metals and alloys used in structural applications. This shortcoming is exacerbated by the fact that the processes involve the extreme ranges of conditions mentioned above. The variety of physical material response is enormous, and we have at best a very rough understanding of relevant physical models. One must model the material behavior during shock propagation; crushing of voids in the material; and failure, flow, and fracture. Important features such as compaction, strain softening, nonlinearity, and hysteresis are seldom included. We have only a crude understanding of even the nature of these responses; even when complicated mathematical models are hypothesized and constructed, there is seldom enough data to calibrate them, especially for three-dimensional states. Tests of sensitivities to inputs are rarely made. As a result, code calculations must always be questioned, and even more so when few attempts are made to calibrate them against known experimental results.

The third major approach uses scaling methods. Scaling theories are developed to predict how collisional processes will depend on the parameters of the problem, including the size, impact velocity, gravity, and material type. They are developed from considerations of similarity analysis applied to experimental results, from code calculations themselves, and from observations of asteroids. However, scaling laws are based on assumptions about the importance of various parameters. They always require some tie to experiments or calculations to determine unknown constants and can lead to erroneous conclusions if important parameters are neglected or results are extrapolated into regimes of new physics.

Since the last contributions in *Asteroids II*, it has become clear that there are at least four major issues about the mechanics of small-body impacts that have not been sufficiently addressed. The first is the effect of substantial porosity. Twelve years ago there was conjecture about the existence of rubble-pile asteroids [the term "rubble piles" was first introduced by *Davis et al.* (1977)], but no knowledge of the effects of the implied porosity on the shock processes. There were neither analysis nor code calculations for the cratering and disruption of porous asteroids. Now the scientific community seems to have reached consensus that many and maybe even most large asteroids are reaccumulated rubble piles with possibly large porosity. The discovery of low densities in C-type asteroids such as Mathilde and Eugenia strongly indicate high porosity. For example, Mathilde has a bulk density of about 1.3 g/cm^3, implying a porosity of 50%. Rocky asteroids are estimated to have been shattered many times over the lifetime of the solar system and may reaccumulate into a low-density state. Comets are also thought to be very low-density conglomerates of ice and dirt.

Porosity may be the dominant physical property governing an impact process. The mechanics of impacts into highly porous bodies is substantially different than for low-porosity bodies, due to significant energy losses from the outgoing shock wave as it compacts the target material. Recent observations of the large craters on Mathilde imply substantial differences in basic cratering mechanisms and ejecta existence and fate. While experiments have been conducted in porous materials such as dry soil and sand, the compaction processes in those bodies is much less important than in much more porous bodies. This remains an area of extreme uncertainty, although some progress has been made.

The second issue deals with rate-dependent strength effects. There were also at the time of *Asteroids II* simple theories, but no data, that represented the effective strength of rocky asteroids as size and time dependent. The extrapolation from small-scale experiments to large asteroids is determined primarily by the form of those strength dependences, but the exact nature of those forms was not known. This issue has now been addressed both experimentally and using codes, as is summarized below, but many uncertainties remain.

Thirdly, the discovery of the Kuiper Belt in 1992, together with a general feeling in the community that the distinction between asteroids and comets may only be one of nomenclature, has prompted a number of studies into the impact behavior of icy materials and the interpretation of these data in the context of a potentially large icy planetesimal population in the greater solar system. We will briefly discuss ice experiments in this chapter and provide references for further study. However, the study of ice under impact is far less complete than that of rocky materials and no coherent models have been developed for the scaling of ice impacts to realistic sizes and conditions.

Finally, much remains to be learned about the effects of oblique impacts. To first order the obliquity may be accounted for by a simple reduction of normal component of velocity, but there may be other more subtle effects deserving of further study, particularly for near-grazing impacts.

This chapter is to be considered as an addition to the corresponding chapter in *Asteroids II* (*Fujiwara et al.,* 1989). The previous results are generally not presented again. Here we review some previous wisdom, but primarily address new approaches and results.

Mention of all the research in this broad topic over the last 12 years would produce a reference section alone larger than our allocated space, so we apologize to those whose important work we did not mention.

2. LABORATORY EXPERIMENTS

2.1. Overview

The 12 years since the publication of the *Asteroids II* book have seen many new experimental studies relevant to asteroids, and have produced a valuable database of experimental results (*Fujiwara et al.,* 1989; *Martelli et al.,* 1994), helping us to predict such quantities as the energy required for catastrophic breakup and the post-impact fragment shapes, sizes, and velocities. Target materials used in impact experiments have included rock, glass, clay, sand, loose aggregates, ice and ice-silicate mixtures, and artificial materials such as cement mortar, clay, alumina, and plaster. The impacting projectile has consisted of metals (aluminum, steel, iron), Pyrex, mortar, basalt, nylon, polycarbonate, and ice. Projectile velocities have ranged from a few meters per second to >6 km/s, and most tests have been conducted in evacuated chambers (*Fujiwara et al.,* 1977; *Takagi et al.,* 1984; *Davis and Ryan,* 1990; *Ryan et al.,* 1991; *Nakamura and Fujiwara,* 1991; *Nakamura et al.,* 1992). Some aspects of impact disruption (e.g., fragment velocity) have been studied by multiple researchers, while other aspects (e.g., rotational modes of ejecta) have been measured by only one or two researchers. Extensive use of computerized image processing — essentially impossible in 1989 — has provided significant new data.

This section is intended to provide a comprehensive review of significant experimental studies carried out between 1989 and 2001, to summarize the results, and to briefly discuss their relevance. One must carefully note the wide variations in results in different materials due to the large differences in impact velocities.

2.2. Impact Techniques and Measurements

The methods to launch high-speed projectiles have remained unchanged over several decades, and include powder guns, light-gas guns, and electromagnetic launchers. Velocities range from tens of meters per second to ~7 km/s. Both laboratory and field explosive tests have also been used to study energetic disruptions. Means of measuring the results of such tests have improved in the past few decades, largely due to the availability of fast-framing video cameras and image-processing computer systems. High-speed frame rates between 400 per second (e.g., *Giblin et al.,* 1994a) and 6000 per second (e.g., *Nakamura,* 1993) have been used. Film or video footage is typically digitized for computer analysis, thus enabling researchers to measure the inflight dynamical properties of fragments. Examples include *Nakamura and Fujiwara* (1991) and *Davis and Ryan* (1990). *Giblin et al.* (1994a) used two cameras to measure fragment velocities in three dimensions. *Cintala et al.* (1997) used a strobed laser system to measure cratering ejecta velocities. *Giblin* (1998) discusses methods to recover three-dimensional particle velocity trajectories from filmed records.

2.3. Disruption of Target Materials

As stated above, the kinetic energy per unit target mass Q, a specific energy, is used to measure collision outcomes. A related measure is the kinetic energy per unit volume of target, an energy density, which has the units of stress. The specific energy Q multiplied by the target mass density gives the energy density of the impact. "Impact strength" is sometimes defined as either the specific energy or the energy density needed to produce a largest intact fragment that contains one-half the target mass (see *Fujiwara et al.,* 1977; *Davis and Ryan,* 1990; and *Ryan et al.,* 1991). Here the symbol Q is always used for the specific energy and the energy density is written as ρQ.

Figure 1 shows many experimental results for the ratio of the mass of the largest postimpact fragment to the initial target mass, as a function of the total impact specific energy

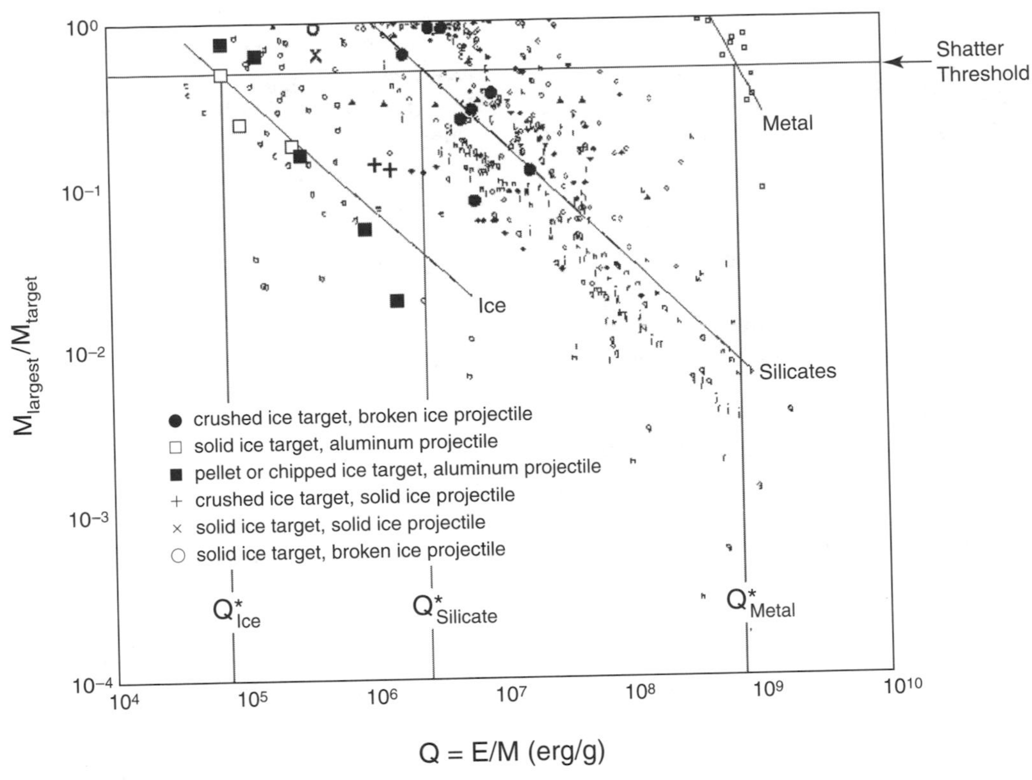

Fig. 1. Summary of disruption (shattering) experiments in various materials. Data from a variety of sources, including *Hartmann* (1969), *Fujiwara et al.* (1977), *Fujiwara and Tsukamoto* (1980, 1981), *Lange and Ahrens* (1981), *Matsui et al.* (1982, 1984), *Kawakami et al.* (1983), *Fujiwara and Asada* (1983), *Takagi et al.* (1984), *Cintala and Hörz* (1984), *Cintala et al.* (1985), *Smrekar et al.* (1985), *Hartmann* (1988), *Davis and Ryan* (1990), *Ryan et al.* (1991), and *Nakamura and Fujiwara* (1991). There is a general grouping by material types, but significant scatter within material types because of differences in impact velocity, temperature, projectile types, target strength, and many other factors.

Q. A value of Q with the ordinate equal to 1/2 as indicated defines the shattering specific energy Q^*_S. Included are results for ice, silicate, and meteoritic metal targets. The data show that the degree of fragmentation is strongly dependent on the target material. In all materials, increasing collisional energy increases the degree of fragmentation.

Experimental studies using rock projectiles to impact solid rock targets reveal that projectile/target density, mass, or strength differences can also have a significant influence on collision outcome, especially for velocities below ~1 km/s (*Matsui et al.*, 1982; *Takagi et al.*, 1984). *Kato et al.* (1992) noted similar effects in ice targets impacted by ice, aluminum, polycarbonate, and basalt projectiles. *Davis and Ryan* (1990), *Ryan et al.* (1991), and *Ryan et al.* (1999) found that there is systematically less collision damage to the target at the same specific energy as the projectile becomes weaker.

A major question about the interpretation of such experiments for asteroids is the possibility of a size- or rate-dependent strength (see *Fugiwara et al.*, 1989), such as from a model based on the growth and coalescence of inherent flaws in natural rock. [Rate dependence gives the same scaling result as size dependence, since for large bodies all impact processes increase in duration with the size of the body, i.e., larger bodies have slower processes. A common trick in the movie industry is to slow down the depiction of small-scale simulations to make them appear to be large-scale.] A method to scale these experiments assuming such a size-dependent strength based on crack growth was given first by the scaling model of *Holsapple and Housen* (1986) as discussed below. The computer simulations by *Ryan and Melosh* (1998) and *Benz and Asphaug* (1999), using the modeling approach of *Melosh et al.* (1992), have since replicated aspects of the scaling, a consequence of their use of a strength model that is rate-dependent.

Of course, those studies do not prove that rocky materials actually have rate or size-dependent strength. The first experimental data on how target size affects collisional outcome was provided by *Housen and Holsapple* (1999a,c) using homogeneous granite targets. Target diameters were varied by a factor of 18 (up to a target diameter of 34.4 cm), and specific energy was kept constant as the size scale was increased. The larger bodies were found to be weaker in impacts than the smaller ones.

Meteoritic targets have also been used in impact studies. A series of high-velocity impact experiments into cooled Gibeon iron-nickel meteorites were performed by *Ryan and Davis* (2001). It was found that at asteroid belt temperatures near 167 K, iron meteorites underwent brittle fracture, and the resultant fragment size distributions had the typical two-slope power law behavior often observed for homogeneous rock targets. The impact strength was determined to be about 500× larger than for basalt. Ejecta velocities were much higher than those observed in rock fragmentation experiments, consistent with predictions of scaling theories in which velocities scale with the square root of the material strength in the strength regime. Ryan and Davis (personal communication, 2001) also conducted impact experiments using iron meteorite targets that were not cooled. For these targets, the crater formed at the impact point had large rim-flaps, and a very "plastic" overall morphology.

2.4. Prefragmented and Shattered Targets

To model fractured asteroids, *Ryan et al.* (1991) constructed samples of previously shattered mortar targets by careful reassembly and weak cementation, and reimpacted them. They found no large differences in impact strength between the preshattered targets and the original targets from which they were constructed, even though the preshattered targets had some porosity and a compressive strength that was less than half that of the original strong mortar target. For some unknown reason, the mean ejecta speeds from these bodies were higher than those measured for the original strong homogeneous targets. The resultant fragment size distributions for the preshattered targets were not significantly different from their homogeneous counterparts, i.e., they were not further fractured upon reimpact. *Giblin et al.* (1994a) tested the effect of void spaces in targets by fabricating a three-section, strong cement mortar body. The sections consisted of a top and bottom spherical cap, with a 1-mm spacer separating these pieces from a middle section. They reported no appreciable difference between the velocity distributions from these targets compared to homogeneous targets of the same composition. They concluded that the 1-mm spacing might not been a large enough to affect shock wave propagation and subsequent target fracture.

Nakamura et al. (1994) also examined the effect of reimpacting previously fractured targets. They performed experiments in which a projectile was shot into a "core" fragment produced from a previous impact event. They found that the outcomes, including largest fragment masses, mass distribution of fragments, and size-velocity distributions, were not changed significantly. However, fine dusts were spewed out with a velocity higher than tens of meters per second within a short time after the impact, probably from the interior of surface cracks.

2.5. Disruption of Porous Targets

Porous targets including gypsum (*Kawakami et al.,* 1991; *Nakamura et al.,* 1992), porous alumina, cement mortar (*Davis and Ryan,* 1990), and sandbags (*Yanagisawa and Itoi,* 1993) have been tested in impact disruption experiments. Exceptionally high impact strengths have been found for such porous target materials. While these porous materials are fragile in terms of tensile or compressive strength, their high porosity readily dissipates the energy delivered in an impact. Thus, unusually large energies are required to damage porous structures.

Ryan et al. (1999) impacted porous and solid ice bodies at a temperature of about −15°C with fractured ice, solid ice, and aluminum projectiles at low velocities, and determined that the impact strength of the porous ice was higher by a factor of about 5 than the impact strength of solid ice targets. The degree of fragmentation also increased with the strength of the projectile. The porous ice targets were as strong as silicates when hit with fractured ice projectiles. Therefore, even though the very porous ice targets had a static material strength well below that for solid ice, they were just as resistant to collision as solid ice. Energy is apparently well dissipated by the void spaces within the target.

Kawakami et al. (1991) constructed an ellipsoidal sample of gypsum with a porosity of 10–30% to model Phobos, and impacted the body to produce a crater equivalent to the Stickney crater. Fracture patterns appeared similar for both low- and high-velocity impacts, which was attributed to the low shock impedance and porosity of the target (i.e., shock wave attenuation occurred). They concluded that target type significantly affects the resulting mode of fragmentation. The impact strength was found to be comparable to that of basaltic bodies, even though the static strength of gypsum is 1–2 orders of magnitude lower than basalt.

To model stone meteorites, *Durda and Flynn* (1999) conducted a series of ~5-km/s experiments using inhomogeneous, porous targets constructed of two materials having different strengths (porphyritic olivine basalt). The largest fragments generated were representative of the bulk composition, while millimeter-sized fragments were composed of isolated olivine crystals. The latter indicates that the target experienced preferential failure along phenocryst-matrix boundaries. They conclude that collisions involving chondritic asteroids may overproduce olivine-rich material in the millimeter size range, and olivine may be underrepresented at smaller sizes in the primary debris.

Nakamura et al. (1992) used gypsum spheres to investigate the fragment velocity distribution of a porous body. The antipodal fragment velocity for the gypsum target was lower than that for a basalt target impacted with a similar energy density. *Yanagisawa and Itoi* (1993) also found that their sand-bag and porous alumina targets produced fragments with significantly lower velocities than a comparable impacts in nonporous basalt. *Love et al.* (1993) found that increasing a target's porosity had the effect of decreasing the speeds of the ejecta for equal collision conditions.

None of those studies give information about gravitational effects for large bodies. Scaling theories predict that lithostatic pressure considerably strengthens a body against disruption, which is why disruption specific energy increases markedly in the gravity regime (see section 3 below). *Housen et al.* (1991) and *Housen* (1993) simulated the effects of gravity by applying external pressure to small, weakly cemented, porous basalt targets. The largest overpressures corresponded to an average lithostatic stress inside a 460-km-diameter body. They used explosives buried at appropriate depths to simulate an impact disruption. They found that as overpressure is increased, more specific energy was required to shatter the body, and that there was a marked increase in the size of the largest fragment, confirming the scaling theory. Also, the specific energies measured for catastrophic disruption of the target body compared well with those estimated from observations of the Themis, Eos, and Koronis asteroid families.

2.6. Cratering and Ejecta in Porous Targets

Compaction mechanisms may dominate the cratering processes of highly porous, pristine bodies such as cometesimals (*Sirono and Greenberg,* 2000) and asteroids such as Mathilde (*Housen et al.,* 1999). *Michikami et al.* (2001) impacted centimeter-sized glass bead targets of various porosities in an evacuated chamber to view and measure ejecta velocities. The ejecta velocities decreased markedly as the porosity increased. For the most porous targets (80% and 60% porosity), measured velocities were more than 2 orders of magnitude below those measured for rocks, and for a porosity of 60% only 2% of the ejecta had a velocity greater than 10 m/s. For the 33-km crater on Mathilde, a velocity of 10 m/s is required for material to escape the crater, so these experiments suggest that almost no visible ejecta would be present. *Love et al.* (1993) also performed cratering experiments in porous glass targets, although the craters were dominated by the spall features of small experiments. *Housen and Holsapple* (1999b) and *Housen et al.* (1999) report impact experiments at 1.9 km/s into a very porous material with a density of 0.9 g/cm^3 and very low crush strength, as a simulant for cratering on Mathilde. The experiments were performed at 500G on a centrifuge in order to reproduce the ejecta ballistics and lithostatic forces involved in large cratering events on Mathilde. Cratering was dominated by compaction, with negligible ejecta. That interpretation was corroborated by post-event computed tomography scans of the targets that clearly showed substantial increases in density below the crater, accounting for the crater volume. Those results implied that the five largest craters on Mathilde would have increased its overall density by about 20%.

More recent cratering experiments at 500G in that same porous material were recently reported by *Housen and Voss* (2001). For porosities of about 50%, only about 10% of the crater mass was ejected; the crater was formed primarily by compaction. Additional experiments at other gravity levels have also been performed to test scaling and gravity-strength transition. The scaling implications of these experiments are under study.

Yamamoto and Nakamura (1997) examined the applicability of the *Housen et al.* (1983) scaling for very-high-velocity ejecta (hundreds of meters per second) from craters generated by oblique impacts into powder glass sphere targets with a porosity of 44%. The laboratory data were estimated to be about an order of magnitude lower than the results extrapolated from the scaling formula. Since the high-velocity ejecta originate from near the impactor the point-source assumption of the scaling theory is not valid, so the discrepancy is not surprising. *Yamamoto* (2001) performed further oblique impact cratering experiments into glass particle targets at various impact angles, and determined the impact angle effects.

The size distribution of the regolith material on an asteroid surface affects the asteroid observational properties such as albedo, spectra, and thermal emission. Ejection processes due to repetitive impacts of micrometeoroids on the surface can alter the size distribution of the regolith. Yamamoto et al. (personal communication, 2001) investigated the size-velocity relation of ejecta from the surface of glass spheres. They mixed three different size glass spheres as the target, and found that the flux of high-velocity ejecta increases as the particle size decreases.

Cintala et al. (1999) analyzed ejecta velocities for craters formed in coarse-grain sand. For impacts between 0.8 and 1.9 km/s, the ejecta velocity distribution was found to be a power law, although the exponent differed significantly from that expected from scaling theory. They suggested this discrepancy might be due to the fact that the sand grains were comparable in size to the impactor.

2.7. Experiments in Ice

Experimental studies in ice are important in the light of studies suggesting that collisions are very important evolutionary processes even in the Kuiper Belt (*Farinella and Davis,* 1996; *Davis and Farinella,* 1997) and the Oort Cloud (*Stern and Weissman,* 2001).

Croft (1982) describes cratering experiments in porous ice in which polyethylene projectiles from 2.3 to 6.3 km/s impacted sifted granular water ice with a density of 0.5 g/cm^3. He reports formation of "a large hemispherical cup whose walls consist of snow compacted to nearly the density of competent ice." Clearly compaction played an important role.

Mizutani et al. (1982) reported experiments in which aluminum and polycarbonate projectiles were fired into silicate and competent –18°C ice targets at 100–1000 m/s. They concluded that the specific energy required for the catastrophic failure of ice is approximately 2 orders of magnitude smaller than that required for basalts, and that the crater morphology in ice at these impact velocities is strongly dependent upon the projectile shape and material properties.

2.8. Fragment Properties and Distributions

An important aspect of impact disruptions and asteroid evolution is the nature of the fragments: sizes, shapes, spins, and numbers. A number of experiments have been performed to measure these quantities.

2.8.1. Fragment sizes. Fragment size distributions following a disruptive impact tend to conform to power laws, although the distribution is often divided into two or three segments with each segment showing a different power-law exponent. The divisions between the segments generally correspond to different fragmentation regimes in the disrupted target. For example, *Di Martino et al.* (1990) impacted alumina cement and basalt targets at 9 km/s, and 94–98% of target mass was recovered after the experiments. They found a knee (a bend in the size distribution) at millimeter sizes, which they attributed to the transition from cratering to bulk fragmentation. This process was studied analytically by *Bashkirov and Vitzayev* (1996). Since 1989 many authors have reported fragment size distribution data that broadly agree with the discussion in *Asteroids II* (see, e.g., *Martelli et al.,* 1993; *Mizutani,* 1993; *Davis,* 1998).

2.8.2. Fragment shapes. Fragment shapes have not generally been reported, although *Giblin et al.* (1998b) goes into some detail. They compare their data to those listed in *Asteroids II* and to the semi-empirical model (SEM) of disruption developed by *Paolicchi et al.* (1989, 1996), and find good agreement. However, in the triaxial ellipsoid model using orthogonal dimensions a, b, c (see, e.g., *La Spina and Paolicchi,* 1996) the values of (b/a) = 0.60 and (c/a) = 0.45 found by those authors differ from their *Asteroids II* counterparts of 0.70 and 0.50. One possible explanation for this difference that the experiments were open-air and there was no secondary fragmentation of ejecta, which would tend to increase c/a and b/a.

In studying the size and shape distribution in a laboratory experiment, it should be borne in mind that many studies suggest that monolithic asteroids are rare, and that the majority of asteroids are rubble piles that have been disrupted and reaccumulated numerous times in their lifetime [see estimates below and *Richardson et al.* (2002)]. We must therefore be cautious extrapolating any data on fragment shapes from the laboratory to real asteroids.

Models of the initial and evolved asteroid size distribution tend to have trouble reproducing the steep slopes of the observed population, an issue addressed by *Tanga et al.* (1999); the paper also includes a review of a geometric approach. They conclude that the steep size distributions may be partly explained by their model, which includes a consideration of the geometry of the ejecta, specifically the convexity of largest fragments.

2.8.3. Velocity distributions. Fragment velocity of disruption experiments has received a large amount of attention since the publication of *Asteroids II*, largely due to the widespread use of image processing and analysis computer systems. The primary conclusions presented in *Asteroids II* still hold: The fastest fragments originate near the impact point, fragments generally do not collide with one another, and core fragments in core-type disruption tend to be traveling at very low velocity. The fast fragments produced near the impact point usually consist of fine dust traveling an order of magnitude faster than the larger ejecta, typically several kilometers per second in hypervelocity impacts (e.g., *Di Martino et al.,* 1990; *Drobyshevski et al.,* 1994), although the impacts into porous targets discussed above have much smaller velocities. The fine fragments may be the result of jetting. This is studied and discussed by *Arakawa and Higa* (1995), who found ejection velocities of fine fragments to be 1.7 to 2.9× the impact velocity in low-velocity (150–690 m/s) collisions of ice spheres, independent of target size and impact velocity. *Arakawa* (1999) continued the program of ice studies by measuring maximum and minimum ejection velocities in oblique impacts, but using relatively large impactors with 13% of the target mass. He confirmed a maximum velocity of ejected fragments as 3× the impact velocity.

Data on ejecta velocities have also been reported for a wide range of other target types: cement mortar (*Davis and Ryan,* 1990), pyrophilite (*Takagi et al.,* 1992), alumina and gypsum (*Nakamura and Fujiwara,* 1991; *Nakamura et al.,* 1992), porous alumina, commercial mortar and sand (*Yanagisawa and Itoi,* 1993), limestone, alumina cement (*Giblin et al.,* 1994a,b) and gabbro (*Polanskey and Ahrens,* 1990). The three-dimensional velocity data have generally been collected using a stereoscopic system. Characteristic ejection velocities are typically between 0.1 and 0.5% of the impactor velocity for high-speed impacts into rocky targets, but much slower for porous targets.

The antipodal velocity has been measured by a number of researchers, since it provides a reasonable and easily measured characteristic velocity for an impact. *Nakamura* (1993) tests the scaling of antipodal velocity with both Q and NDIS (see section 3.1.1 below), finding good agreement with *Fujiwara and Tsukamoto* (1980) and *Davis and Ryan* (1990). *Giblin et al.* (1998a,b), using a contact charge to simulate the impact of a 6.2-km/s projectile, found antipodal velocity to be only one-third of the value predicted by *Fujiwara et al.* (1989), raising questions about the equivalence between a contact charge and an impactor.

Martelli et al. (1993) studied the angular distribution of velocities in several open-air impact experiments and found evidence for collimated jets and other anisotropies in the ejecta field. They discuss these data as a possible mechanism for the formation of asteroid binaries and families.

2.8.4. Velocity-mass correlations. The relations between a fragment's velocity and mass is a key component in models of collisional disruption and asteroid evolution (see, e.g., *Petit and Farinella,* 1993). Results of detailed crater ejecta studies (e.g., *Gault and Heitowit,* 1963; *Vickery,* 1986, 1987) indicated a power-law relationship between fragment velocity and mass in high-velocity cratering of rocklike materials. Laboratory disruption experiments (*Fujiwara and Tsukamoto,* 1980; *Nakamura et al.,* 1992; *Nakamura,* 1993) have suggested that a power law holds to some extent (see Fig. 2).

However, other studies (e.g., *Ryan,* 1992; *Takagi et al.,* 1996; *Giblin,* 1998) have found little correlation between velocity and mass for similar experimental parameters (see Fig. 3). The authors pointed out the lack of complete data as well as the problem of selection effects in the analysis of laboratory fragmentation experiments.

2.8.5. Rotation rate distributions. Very few data on fragment rotation were available at the time of *Asteroids II* publication, but the brief conclusions presented there are still valid. Specifically, the fastest rotators originate near the impact point and, although many fragments have been observed rotating quickly [e.g., 100 rotations/s (*Fujiwara,*

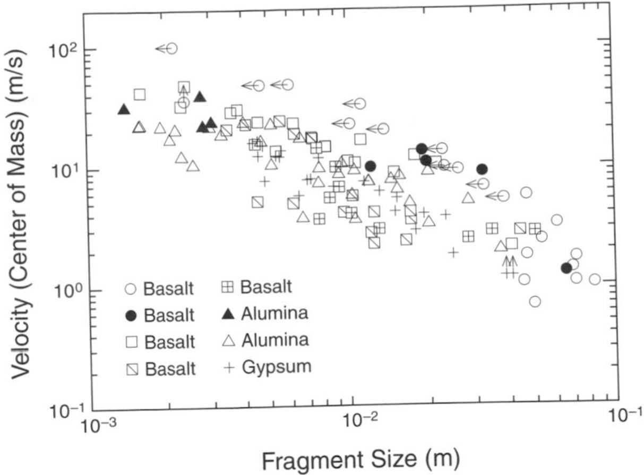

Fig. 2. Collected data on fragment velocities in the center of mass system vs. fragment size (from *Nakamura,* 1993). These data clearly exhibit the slope of –1/6 reported by *Nakamura and Fujiwara* (1991).

1987)], none of these approach the rotational bursting limit for their materials. Generally, large fragments have been found to rotate more slowly than small ones.

The distribution of asteroid rotation rates is often compared to a Maxwellian distribution (e.g., *Harris and Burns,* 1979; *Farinella et al.,* 1981; *Binzel et al.,* 1989; *Fulchignoni et al.,* 1995). Although considered appropriate for a highly evolved asteroid population (*Harris,* 1990; *Farinella et al.,* 1992; *Yanagisawa et al.,* 1991; *Yanagisawa and Hasegawa,* 1999; *Yanagisawa and Hasegawa,* 2000; *Sirono et al.,* 1993; *Kadono,* 1993), the Maxwellian distribution is not expected to give a good fit to "young" fragments from disruption experiments because such fragments have undergone no further collisional processing. Thus experimental data may be used to test the idea that small asteroids are "young," in the sense that most of them may have remained close to their original rotational state. *Giblin et al.* (1998b) report the rotation rate of a total of 811 fragments studied across 8 similar experiments. Data on rotation rate distributions are shown in Fig. 4.

Some asteroids have non-principal-axis rotation ("tumbling") states (see, e.g., *Hudson and Ostro,* 1995). Authors discussing fragment rotation in impact experiments have not generally reported any evidence of tumbling fragments. This may be due to lack of instrumentation (or time for data reduction). However, *Giblin and Farinella* (1997) report data on a number of tumbling fragments from the experiments described in *Giblin et al.* (1994a, 1998b). Their conclusions support a biased distribution of spin vectors that favors principal axis rotation in ejected fragments.

Fujiwara and Tsukamoto (1980) study the issue of rotational bursting and conclude that "generally, no collisions among fragments occurred, but in the exceptional case some spinning fragments were split into smaller ones and collided with other fragments." *Giblin et al.* (1994a) report that they observed several ejected fragments splitting just after ejection from the disrupted target. *Giblin et al.* (1998a) describe a very well studied case from the same experiments of rotational bursting, where a tumbling fragment travels more than five target diameters before breaking into two pieces due to rotational stress. Rotational bursting is most likely in the case of a tumbling prefractured fragment since only in this case will the internal stresses of the body be time-varying. With Monte Carlo simulations those authors show that a fragment ejected from a body can subsequently place

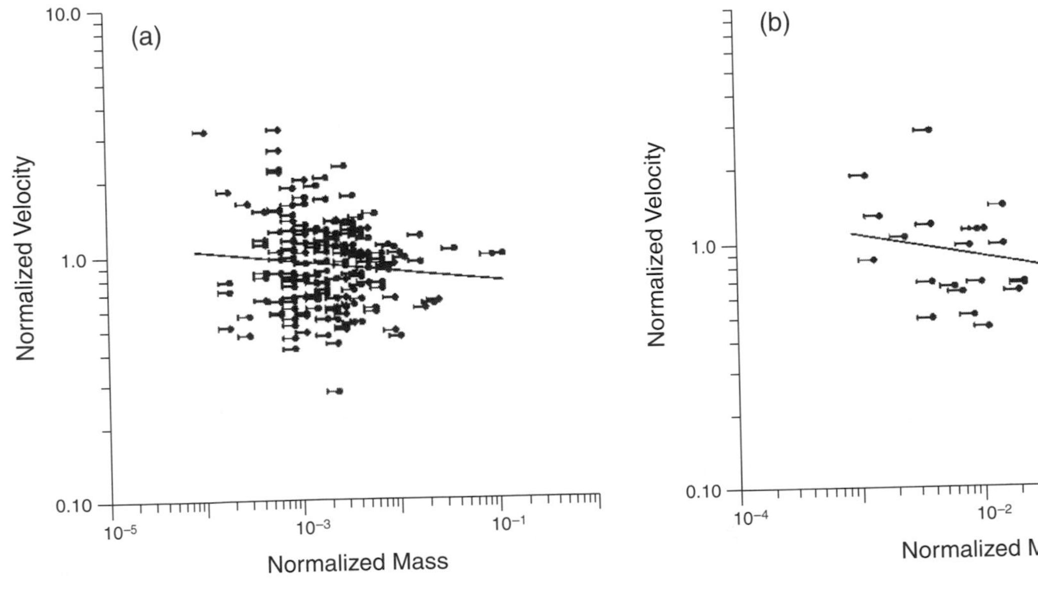

Fig. 3. **(a)** Measured two-dimensional velocity and estimated mass from four similar experiments where alumina cement targets were catastrophically disrupted using a contact charge to simulate an impact at 6.2 km/s. **(b)** Measured three-dimensional velocity vs. mass for a subset of the points in **(a)**. Little correlation is apparent. From *Giblin* (1998).

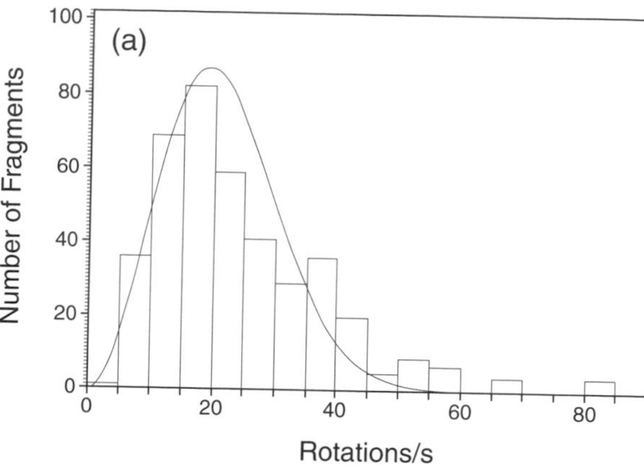

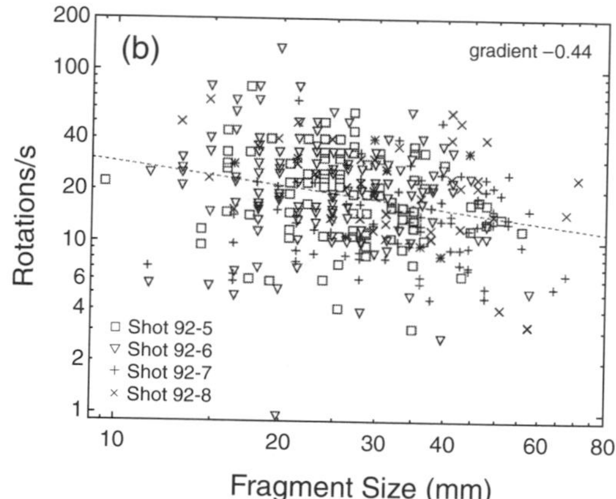

Fig. 4. **(a)** Rotation rate distribution for 386 fragments from catastrophic disruption of alumina cement targets (*Giblin et al.,* 1998b). **(b)** Size and rotation rate for the same 386 fragments showing considerable dispersion around the best-fit line. The deviation from a Maxwellian distribution is due to an overabundance of slow and fast rotators and a correspondingly depleted peak in the distribution, similar to the data for small asteroids.

some of its mass in stable orbit by rotational bursting. This is a plausible mechanism for producing parent-satellite systems such as Ida and Dactyl.

Fujiwara and Tsukamoto (1981), *Nakamura et al.* (1992), and *Nakamura* (1993) discuss the variation of rotation rate with size in ejecta from disruption of rocklike targets, finding a best-fit power-law exponent between –1.5 and –1.0, but note that their data were limited. *Giblin et al.* (1998b) find a significantly shallower slope in their data. Figure 4b shows some of these data; although there is a definite slope, a power-law fit simply does not provide a good description of the data.

2.9. Energy Balances

Energy partitioning is an important consideration in the effects of any impact. The total energy is divided (generally in decreasing order) between heat, comminution of the target, target and ejecta kinetic energy and ejecta rotational energy; and, for sufficiently high-velocity impact, melt and vaporization energies. An important component in the context of modeling and understanding collisional evolution is the fraction of impact energy partitioned into ejecta kinetic energy, since this determines whether or not a disrupted body remains disrupted or reaccumulates.

According to the available results in 1989, this fraction is less than 3% for high-velocity (core-type) catastrophic impacts (*Fujiwara and Tsukamoto,* 1980) and less than 1% for high-velocity cratering impacts into semi-infinite basalt (*Gault and Heitowit,* 1963). The fraction of energy partitioned into ejecta KE is higher for low-velocity impacts; on the order of 10–20% (*Waza et al.,* 1985).

Fujiwara (1987), *Nakamura and Fujiwara* (1991), and *Nakamura et al.* (1992) all reported that the largest proportion of energy goes into heat, comminution of the target, and the relatively high velocity of fine fragments. *Fujiwara* (1987) reports that fragment rotational energy is generally less than 1% of kinetic energy in high-velocity disruption; *Nakamura* (1993) reports between 1% and 10%. A similar study by *Giblin* (1994) reports this ratio to be between 2% and 4% for four alumina cement targets disrupted using a contact charge.

Sugi et al. (1998) impacted water-ice targets using a copper projectile at speeds from 54 to 329 m/s in order to study the degree of impact vaporization. They observed that, below 100 m/s, less than 0.03% of impact energy went into impact vaporization in their experiments, and that this increased to 18–26% between 100 and 180 m/s. *Schultz* (1996) found significant amounts of impact kinetic energy went into impact vaporization in oblique impacts of an Al sphere into competent ice, in experiments where impact speeds were between 4.7 and 5.9 km/s.

3. SCALING THEORIES, ISSUES, AND RESULTS

Issues of scaling are fundamental to our understanding and interpretation of impact processes, and are needed to unravel the meaning of laboratory experiments and code calculations. Various approaches to scaling have been developed over the years. The result of any impact depends on the conditions of the impactor and those of the impacted body (target), perhaps in complicated ways. A successful scaling approach must distill the large number of possibly relevant physical parameters down to the essential few. The goal is to have a rule to predict all impact responses (the effects) from the impactor conditions (the cause).

3.1. Approaches

There are two synergistic approaches to scaling: the theoretical approach using similarity and dimensional analyses;

and numerical computer calculations of a suite of impacts. Each has its strengths and weaknesses. They are reviewed in turn.

3.1.1. Theoretical scaling methods. The impactor has dominant measures of size a, velocity U, and mass density δ that determine its energy, mass, and momentum. It also has an impact angle and many material properties. A common feature of all scaling theories is that one single measure is chosen to represent all properties of the impact initial conditions. In the historic approaches, the impactor kinetic energy was used most of the time, although some questioned whether the measure should be the impactor momentum. Beginning with *Holsapple* (1981, 1983), and in a number of subsequent papers (*Schmidt and Holsapple,* 1982; *Housen et al.,* 1983; *Holsapple,* 1987, 1993; *Holsapple and Schmidt,* 1987), a different measure (the coupling parameter) was introduced. Its existence and its form $C = aU^{\mu}\delta^{\nu}$ was shown to be a consequence of physical and mathematical "point-source solutions" for rapid energy deposition in vanishingly small regions for general materials. Therefore it is a global measure of the impact process for regions away from the immediate details right at the impact site. It generally is intermediate to the energy and momentum measures, depending on the materials. However, it is only as valid as the point-source approximation, and cannot be expected to hold for cratering when, for example, the crater is only a little larger than the impactor, or for phenomena (such as melt and vaporization) that occur only on a scale comparable to the impactor radius. However, it does seem to work surprisingly well over large ranges in velocity and scale. For example, for transient craters in water it correlates to within a few percent experiments with impact velocities ranging over the extreme range of 1 m/s to 6 km/s (see Fig. 2 in *Holsapple,* 1993). Code calculations have showed that it governs impact-generated flows in regions as near as 1–2 impactor radii.

A different measure of the impact, the nondimensional impact stress (NDIS), was introduced by *Takagi et al.* (1984). It was initially based on the maximum pressure generated at the impact point. In a later approach, *Mizutani et al.* (1993) used the stress propagated to the antipodal point of the target. [Strictly speaking, it is a measure of the effects and not the impactor (the cause). Its utility for scaling may rest on an expectation that the transmitted stress is easier to predict than the other effects of the impact.] *Mizutani* (1993) introduced special assumptions about the nature of the stress decay from the source through the body, while recently *Mitani* (2000) used a numerical calculation to estimate the transmitted stress. Thus, the NDIS has evolved from a detailed measure of the impact condition, which is probably not important globally, to a global measure of the response itself.

In all cases of scaling it is also necessary to choose simple measures of the target resistance. Approaches for scaling of cratering or disruption typically use some material strength and/or the gravitational field (determined by body size). Cratering scaling generally uses a constant strength, while disruption scaling has used rate- and/or size-dependent strength. This is done because, for many geological materials, the tensile strength, which is fundamental in disruption, is observed to be more strongly rate dependent than is the shear strength, which is most important in crater formation.

Insofar as the scaling forms, any strength measure with stress units gives the same result. When trying to correlate between results for different materials, some specific measure must be chosen. However, since different strength measures such as compressive, shear, or tensile strength are often in about the same ratios for different materials, the choice is not so important. The choice becomes more uncertain when choosing between a crush strength of a porous material and some other strength measure.

Therefore, there are two measures: one for the impactor and one for the target resistance. Their choice, together with methods of dimensional analysis, lead to definite power-law algebraic forms for scaling. The reader is referred to examples in the literature such as *Housen and Holsapple* (1990), *Mizutani* (1993), and *Holsapple* (1993). These approaches are limited to phenomena where the point-source approximation governs.

3.1.2. Code calculation approaches. Scaling laws and studies of impacts can also come from the outcomes of code calculations. Numerical simulations have become popular: At the last two Lunar and Planetary Science Conferences there were perhaps 30 abstracts by authors using the codes CTH, SOVA, an SPH continuum code, an "n-body" code, or a finite-element code to do either two- or three-dimensional impact calculations. One attractive feature is the ability to investigate effects of specific shape and structure (*Asphaug et al.,* 1996, 1998), and to look at gravitational assemblages (*Richardson et al.,* 1998; *Leinhardt and Richardson,* 2001).

Initial shock processes and energy coupling are calculated in many ways. There are simple equation-of-state models with no thermodynamic coupling (e.g., Murnaghan), simple analytical models for single solid phases (Mie-Gruneisen and Tillotson), complex analytical models including melt and vapor (ANEOS), or complete tabular databases such as the SESAME library. None of these include any kinetic effects, which might be important. Effects of porous crush-up are almost never modeled, since that greatly increases the numerical difficulties. (Calculations with very low density that use the standard forms of the equation of state omit the energy dissipation of a crushable material and do not model porous materials.) Some of these models are only appropriate for special cases, although they are sometimes used in other inappropriate cases.

Strength models include none (hydrodynamic), constant (shear and/or tensile), Mohr-Coloumb shear strength, rate-dependent tensile, and complex damage models. Some have even included a viscous component to model acoustic fluidization. Gravity, either a fixed planar field or a self-gravity central field, may or may not be included.

Obviously these myriad approaches lead to myriad re-sults. The results vary substantially due to the difficulties of correctly modeling the complex geological materials. A recent statement in the literature that "recent exponential increases in computational power have enabled numerical simulations to become the method of choice to investigate these issues (planetary impacts) in greater detail" may be premature, although that may become true in the future.

The theoretical methods make it clear that any scaling outcomes from code calculations are determined primarily by the dominant measures of the impact conditions and the target resistance in the codes. Since the point-source measure is firmly rooted in special mathematical solutions to energy deposition, the codes must reproduce, at least in an approximate manner, point-source results for any phenomena away from the immediate impact region. The point-source measure should fail only for phenomena very near the impact point (for example, the amount of melt and vapor), for very slow impactors, or for very large (compared to the target body) impactors.

Also, since the resistance of the target is one of the primary scaling measures, the choice of the gravity level and the strength model in the code will determine the form of the scaling outcome. Thus, the form of scaling results from code calculations is simply a reflection of the models chosen. The codes cannot determine what physics are important; the creator of the code does that. A good example is the *Benz and Asphaug* (1999) paper, in which the slopes obtained from SPH code calculations for disruption match the slopes given in the scaling theories of *Housen and Holsapple* (1990) and *Holsapple* (1994) in both the strength and gravity regimes. That is because their strength model is based on the same rate-dependent physics that was the basis for the theoretical scaling approach.

The contributions of codes are then, in principle, to calibrate the constants that are unknown from the dimensional analysis approaches, to investigate ranges where the scaling may not hold, to bridge the gap between scaling regimes, and to investigate ranges of material, shape, and structural models. However, there is generally a lack of good material models and data. The material models usually have many parameter choices ("knobs to turn"), and often the magnitude of some primary variable such as some strength is simply "dialed-in" to make the code match some limited experimental data. Furthermore, there is usually no unique way to make that match; other "knobs" might succeed equally well. Other quite different material models might also succeed. Therefore, magnitude calibrations are often specious. Codes are a poor way to determine material properties.

However, having noted the uncertainties, code studies certainly have contributed to the understanding of scaling issues, and provide ways to study phenomena out of the reach of experiments. As they mature and better material models are developed and implemented, they will become even more useful. Important contributions from the codes are included in the sections below.

3.2. Crater Size Scaling Results

3.2.1. Previous results. The current knowledge of crater scaling in rocks, water, and dry sands was summarized in *Schmidt and Housen* (1987) and in *Holsapple* (1993). They give algebraic forms and figures for important aspects of impact cratering events, including volume, diameter, depth, rim heights, timescale, and ejecta velocity. Impact experiments can only be conducted at small scales, but explosive events have been conducted at relatively large scales (megaton nuclear weapons). Therefore much of the scaling is based on an equivalence between impact and explosive events.

Generally, an impact will impart more of its energy and more downward momentum to cratering mechanisms than an explosive on the surface. However, that difference is offset as an explosive is buried; for very deep burial, the explosive has the greater cratering effects. At a depth of burial of about 2× the explosive radius, the tamping effects of soil above the explosive increases both the momentum and energy in the downward directed cratering flow, and the results of the explosion and impact at the same energy and energy density are similar. That has been validated both by many experiments and code calculations (e.g., *Holsapple,* 1980; *Ryan,* 1992).

Terrestrial explosive events in continental sites have generated craters to ~400 m in diameter, while nuclear events in Pacific corals have produced craters in the kilometer range. *Schmidt et al.* (1986) present the explosive cratering database and scaling.

For small impacts (the strength regime) the crater volume is determined only by target strength. Explosive field data show some increase in crater volume per unit energy as the event size increases, a feature attributed to a weakening of the target material with increasing size. However, that increase is only apparent for near-surface explosive energies greater than about 1 t of TNT (4.2×10^9 J), and is not apparent in buried events. Thus, there is little evidence of rate- or size-dependent strength in cratering events, probably because they are dominated by shear flows.

For large events (the gravity regime), the crater volume becomes dependent on the surface gravity and is independent of the target strength. Dry sands, having essentially zero cohesion, are always in the gravity regime. In the gravity regime the crater volume per unit energy decreases with increasing event energy. The point of transition between the strength and gravity regimes has been estimated both by comparing the physical strength of the material to the stresses of gravity and by the actual data from the large explosive events. For near-surface terrestrial events in hard rocks, this transition occurs at about 1 kt of TNT (4.2×10^{12} J, crater diameter ~30 m), while for dry soils it occurs at a few tons of TNT (crater diameter ~10 m) (see *Holsapple,* 1993). These transition crater sizes apply only for the terrestrial gravity field. If the size or rate effects in cratering are minimal, then the transition diameters for other bodies are found by simply dividing these sizes by the magnitude of the gravitational acceleration measured in Earth's gravity (*Holsapple and Schmidt,* 1982).

It should be emphasized that all these crater-scaling results are only for common soils and rocks as found on Earth, and are probably not applicable to highly porous materials. Scaling in materials with high porosity is not yet determined. In the numerous experiments in dry sands, the target material is at or near its "maximum density" or fully packed state, with about 30% porosity. Substantial compaction from that state can only occur by crushing or shearing individual grains, which requires substantial pressures, on the order of kilobars (*Housen and Voss,* 2001). Cratering experiments in dense sands show only a small amount of crushed material remaining in the crater, and no noticeable density increases below the crater. *Housen and Voss* (2001) measured pressure-crush curves for highly porous materials, and the pressure required for 10% compaction of their 53% porous material was a factor of about 30 less than for dense Ottawa sand. Porosity may play only a minor role for cratering in dense sands, but is thought to play a much larger role in a very-low-density material where the material is easily crushed.

3.2.2. Implications of new experiments on cratering. In water ice, new experiments for cratering and ejecta have been performed, as are presented in the experiments section above. However, important questions remain about their relevance and scaling to the large icy solar system bodies. Small craters in ice are typically dominated by surface spall, as is observed also for centimeter-sized cratering in rocks. However, it is known from explosive testing in rocks that spall effects are absent for craters of tens of meters and larger. Thus, small spall-dominated experiments in ice and other brittle materials may involve different physics than do larger cases and, if so, are of limited use for scaling to large sizes.

In addition, many ice experiments have used water ice near its melting point. Crystalline materials are known to have significant rate effects at near-melt temperatures, which are much less important at low temperatures. The materials become more brittle at lower temperature. Thus, there may be dominant rate effects in the small-scale experiments in warm ice (where the rates are very large), which would not be present in large craters (where the timescales are orders of magnitude larger). The exact form of these temperature dependences are not known. For these reasons, the data on cratering and ejecta in ices have not yet been synthesized into an inclusive scaling form, and large gaps in the data exist.

Recently craters have also been studied in highly porous materials (see section 2.6). A major observation is that, as porosity increases, the mechanics of cratering change from one dominated by lateral and upward flow and ejection to one dominated by crushing within the crater. As a consequence, the amount and speed of ejected material decrease substantially as the porosity increases. However, important questions remain about the exact nature of porous materials in asteroids, and whether the materials of the experiments mentioned above are reasonable simulants for low-density asteroids. In any case, the experimental results have not yet been synthesized by a reliable scaling theory. Also, there have been few computer calculations for highly porous targets.

3.2.3. Code calculations of crater scaling. Using the CTH finite-difference code, *O'Keefe and Ahrens* (1993, 1999, 2000) performed many cratering calculations with various impact velocity, size, gravity, and material strengths. The shear strength was modeled by a Mohr-Coloumb plasticity model, with no size or rate effect. The calculations were carried out to relatively late stages. They tabulate constants for the scaling of various crater features. The results compare very well to the theoretical scaling curves mentioned above that have been calibrated by the terrestrial cratering database. They have been carried to the very late stages of crater formation to study complex crater morphologies.

Housen and Holsapple (2000) also used the CTH code to perform cratering calculations, and included a specific porous model for crushing. It was found that the modeling of the crush behavior was difficult but essential. They found that results were very sensitive to the material constants of the crush model and were able to match some, but not all, aspects of well-documented experiments in dry sand.

Rate-dependent models have been available in the codes used in the weapons community since the 1980s. For impact calculations, a rate-dependent strength model was implemented in the SALE finite difference code by *Melosh et al.* (1992). That model was based on an implicit description of the statistics of flaw sizes and growth, an interpretation of the one-dimensional Grady-Kipp model (*Grady and Kipp,* 1980). It used a scalar damage measure; and degraded the material stiffness in both tension and shear to zero as the damage accumulated. It has no plasticity model. It was used to model Stickney Crater on Phobos (*Asphaug and Melosh,* 1993), for disruption calculations (*Ryan and Melosh,* 1998), and to study crater scaling (*Nolan et al.,* 1996). For cratering, *Nolan et al.* (1996) conclude that, for large impacts in basalt, the target material is substantially fractured by the outgoing shock wave, and the crater forms within the fractured region just as it would in an entirely strengthless material.

Those results should be compared to the terrestrial cratering database and the impact scaling theories mentioned above. The equivalent explosive energy for the impacts studied by *Nolan et al.* (1996) (5.3 km/s, 1–120-m-diameter impactors) ranges from a few tons of TNT to ~5 Mt. For tons to a few kilotons of yield, their cratering results can be directly compared to large explosive field events, because in that range the curvature of their target is of little consequence, and the terrestrial events in this range show no effect of gravity. The numerical simulations give a crater volume per unit energy of about 5×10^4 ft^3/t of TNT for the smaller impactor and 2.4×10^5 ft^3/t for the larger impactors. These values are 2 orders of magnitude or more larger than actual large terrestrial explosive events in rocks (*Schmidt et al.,* 1986). (The report by Schmidt et al., which presents scaling curves for explosive cratering, has restricted distribution; it is available to U.S. government agencies and their contractors only. However, the actual explosive cratering data is mostly in the open literature in the form of government agency reports. These include significant explosive craters in dry geologies such as Sailor Hat in rock; Stagecoach and Scooter in alluvium; and the Danny Boy, Sedan, Teapot Ess, JangleU and Johnie Boy nuclear craters, as well as a multitude of smaller field and experimental craters.)

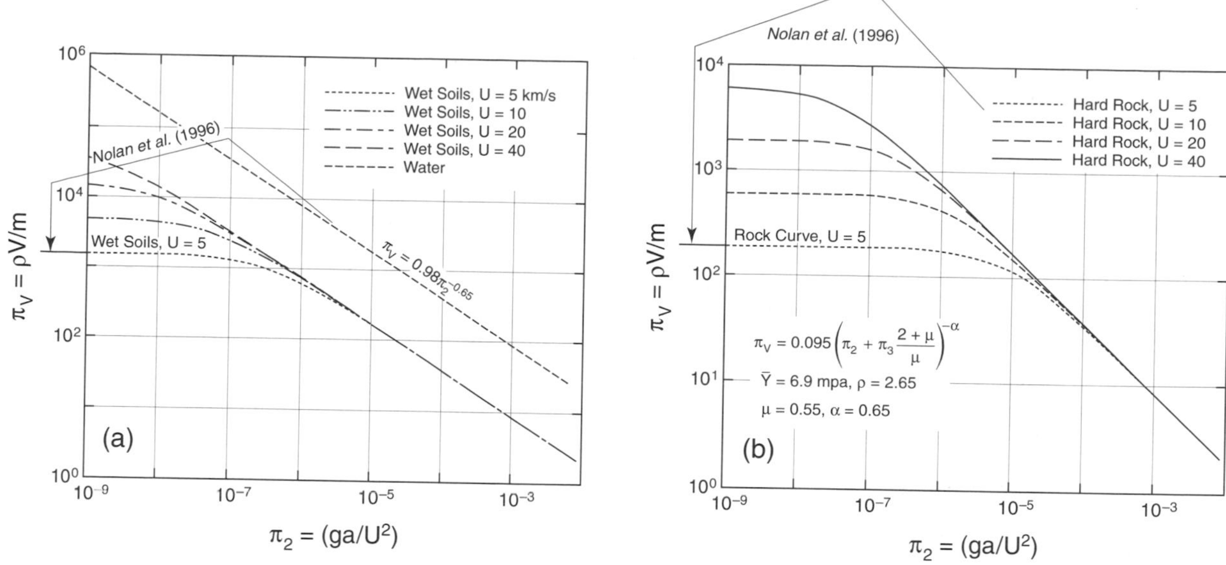

Fig. 5. The cratering results of *Nolan et al.* (1996) compared to existing scaling curves based on terrestrial explosive craters and laboratory impact craters in rocks, soils and water (from *Holsapple,* 1993). The vertical arrows indicate the difference between the code results and the scaling curve in the strength regime. **(a)** The code gravity-regime results fall just about on the cratering curves for water, a decade above the gravity curves for dry sands and soils. **(b)** The calculations are two decades above rock scaling curves.

For example, the 500-t Sailor Hat explosive event in basalt gave a cratering efficiency of 580 ft^3/t (*Vortman,* 1965). This can also be seen in Fig. 5, where the calculated crater sizes are compared to the conventional impact scaling (from *Holsapple,* 1993) in a dimensionless form. On the left at small sizes (Fig. 5a), their result is about an order of magnitude above the value for cratering in wet soils at impacts of 5 km/s (the vertical arrow). At large sizes, in the gravity regime, they are just about on the curve for cratering in water. Figure 5b shows that the numerical results are about a factor of 100 above the curve for the volume of craters in hard rocks in the strength regime and a factor of 10 above them in the gravity regime. It must be emphasized that these theoretical scaling curves are based on the terrestrial crater database, and there are actual explosive craters in this size range to compare to. These results are indeed unexpected.

A different implementation of rate dependence based on an explicit (rather than an implicit) flaw distribution has been developed by *Benz and Asphaug* (1994) and used in a SPH code. They discuss some significant problems of the *Melosh et al.* (1992) numerical implementation of the implicit flaw description. *Asphaug et al.* (1996) also added the newer explicit method to the SALE finite-difference code to study the mechanics of cratering on Ida, and to determine curves of crater scaling in the strength regime. However, their results also give craters about a factor of 100 larger in volume at the same source energy than the terrestrial explosive data as presented in *Schmidt et al.* (1986). These comparisons may illustrate some of the difficulties in the application of code calculations for impacts, and the caution that must be exercised in their interpretation and application.

Some suggestions about possible shortcomings can be given. The primary difference between these recent codes and the earlier ones is the inclusion of the crack growth physics, which are thought to be the primary cause of rate effects. The Grady-Kipp rate models apply to one-dimensional tensile fracture only, and the codes use it to calculate a single scalar damage measure. The correct measure of rate to use is uncertain: that at the shock front, or one based on the pulse duration. Then, when a single crack grows to be as large as (roughly) a calculation zone, the damage is assumed to be complete, indicating a fully failed material. The models then assume there is no remaining rigidity in any tension or shear state, i.e., the material behaves totally as a fluid. All shear stresses in the zone become zero, and unrestricted shear strain and flow is possible. In reality, there should remain substantial shear strength, even along the fracture plane, whenever there is a compression acting on that plane. There should be no loss of strength or stiffness for shear stresses perpendicular to a failure plane. Completely failed material (even with cracks in all directions) should be modeled using the plasticity models of a dry soil, not as a fluid. A limiting behavior of a cohesionless Mohr-Coloumb model such as used for dry sands would seem to be more appropriate. Even at full failure, the cratering results should correspond to the database for dry sands, not that for water. Hence, the fact that the numerical results are a factor of 100 larger than the terrestrial explosion data probably results from incorrect modeling of the shear strength of damaged material.

In addition, cratering flows are dominated by shearing flow, not tensile failures. Shear strength does not appear to have the strong size dependence that tensile fracture does.

The terrestrial cratering data show only a little evidence of decreasing strength with event size. The CTH code results mentioned above using a rate-independent shear strength compare very well to the terrestrial cratering data.

Clearly there is great opportunity for future work here, including the thorough testing of the codes against the terrestrial explosive database and other code models. While the codes were compared to disruption experiments, the constants of the crack distribution models were adjusted to make the best fit. *Melosh et al.* (1992) made comparisons to the *Takagi et al.* (1984) disruption experiments in basalt. They found that the use in the code of published strength and crack size distribution data for basalt gave a poor match to the experimental outcomes, but by adjusting the crack size coefficient down by a factor of about 10^5 (less cracks of a given size, or a stronger material) they were then able to get a good fit to the fragment size distribution of the experiments. However, at a given strain rate, the implied strength of that code model is then a factor of about 2 higher than actual data for similar rock materials. *Ryan and Melosh* (1998) used that same strength model for basalt. Also, for comparison to *Housen et al.* (1991) experiments in a porous grout they again adjusted the Weibull crack distribution parameters of the code until a reasonable match to the disruption experiments was obtained. Then they also made some comparisons to two experiments that were basically cratering events. However, the crater masses they obtained were larger by factors of 3 and 5.5 respectively than the experiments.

Benz and Asphaug (1994) made comparisons of their code results to the disruption experiments by *Nakamura and Fujiwara* (1991) and were able to obtain good results, but again made choices of the fracture parameters to give a best fit. Again, the implied strength at a given strain rate is about 2× published strength data. Later applications of the code with the same material fracture parameters were made by *Asphaug et al.* (1996) for cratering on Ida and by *Nolan et al.* (1996) for cratering on Gaspra-sized bodies. However, comparisons to actual cratering data and scaling curves were not made.

The modeling of the crack growth, shear strengths, and the inclusion of substantial porosity deserve special attention for the code methods. The approaches used in the weapons community, particularly those for impacts into porous ceramics, should be considered. These issues require much further study before code methods can be considered to be entirely reliable.

3.3. Ejecta Scaling

Ejecta scaling laws are important in order to model the evolution of the regolith on small bodies and the production of interplanetary dust particles. *Housen et al.* (1983) present scaling formulas for amounts and velocities of the ejecta from cratering. In the strength regime, the ejecta velocity at a scaled range increases as the square root of the strength. Laboratory experiments typically show velocities sufficient to exceed the escape velocity on Gaspra-sized asteroids. Thus, small, very strong rock asteroids were expected to retain little regolith. However, if the strength decreases considerably with body size, then the laboratory experiments are misleading. In experiments simulating a jointed rock, *Housen* (1992) showed that indeed the ejecta velocity did decrease as the strength decreased, and that a body the size of Gaspra could retain much of its ejecta and develop considerable regoliths.

Impact experiments in porous materials were mentioned in section 2.6. Ejecta velocities are substantially less than in competent nonporous materials or even in dry sands. The tests used materials of different crush strengths, different shear strengths, different porosities, different scaled sizes (G-level), and used different impact velocities. That large range of parameters has so far precluded any identification of the most relevant variables and the synthesis into a comprehensive scaling theory. The one fact that is apparent is that the conventional wisdom on scaling in normal terrestrial materials is not valid.

The images of Mathilde have also raised new questions about the production of ejecta on a porous body. Although Mathilde has several very large craters as wide as the asteroid mean radius, there are no visible ejecta features. This lack has been explained in three very different ways. *Housen et al.* (1999) performed experiments in a material with the same porosity as Mathilde, and concluded that almost no ejecta are produced because the cratering is dominated by (downward) crushing, not (outward and upward) excavation. The experiments by *Michikami et al.* (2001) mentioned above confirm those interpretations, in a different but also highly porous material. On the other hand, *Asphaug et al.* (1998) performed a code calculation for a porous body and deduced that ejecta velocites are greatly enhanced by the porosity, and that all ejecta would escape the asteroid. Finally, *Cheng and Barnouin-Jha* (1999) used conventional crater scaling for dense dry sands to apply to Mathilde and determined that the large crater morphology is consistent with oblique impacts.

There are major differences in these approaches. The use of conventional dry sand crater scaling used by *Cheng and Barnouin-Jha* (1999) is questionable in view of recent experiments in highly porous materials. The fidelity of the sample material constructed by *Housen et al.* (1999) as a Mathilde analog is uncertain. Finally, the code calculation of *Asphaug et al.* (1998) included macroporosity but no microporosity, and therefore did not model the continuum behavior of highly porous materials. What is the most appropriate model for Mathilde? We do not know. No resolution of these major discrepancies has been made, so at present the question of ejecta scaling in porous asteroids is as uncertain as crater size scaling.

3.4. Scaling of Shattering and Dispersion

The scaling of disruptions presented in *Asteroids II* was given in *Fujiwara et al.* (1989, see Fig. 9, p. 259). Since

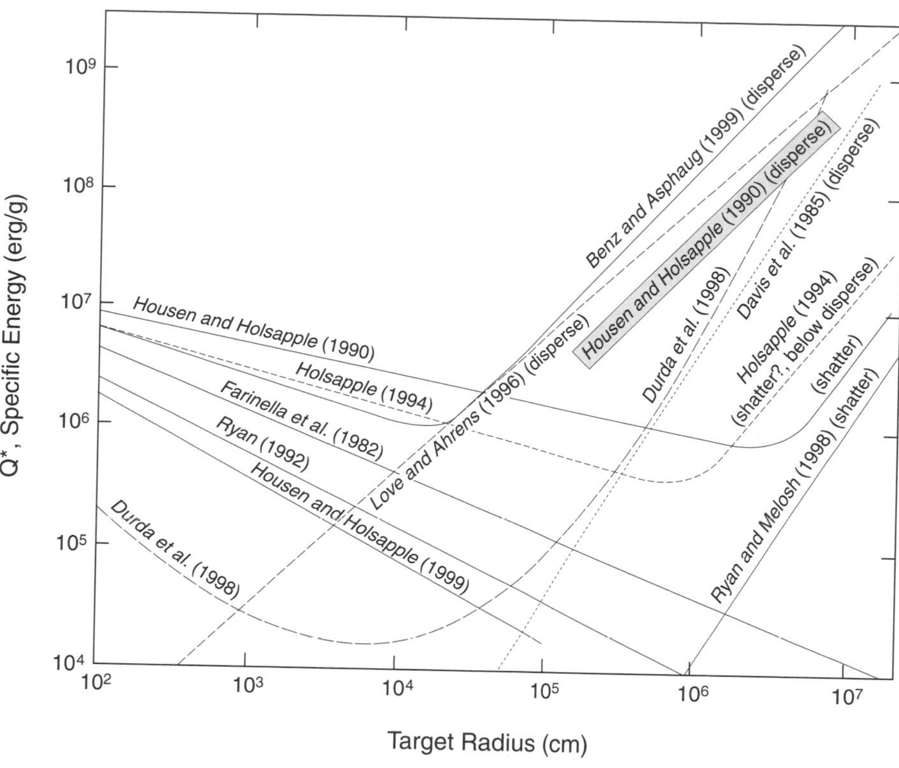

Fig. 6. Specific energy thresholds, some for shattering and some for dispersion, as presented by various authors. For small asteroids (on the left) the specific energy decreases with increasing target size because of a decreasing asteroid effective strength with size. For large asteroids (on the right), the energy increases with increasing target size because of the increasing role of self-gravitation. Note that these results are for rocky bodies only, and not for porous bodies.

that time much has been added. A recent version, Fig. 6, shows the specific energy required for shattering and for dispersion. It is from *Benz and Asphaug* (1999) but with several curves added.

There are three features that are common among the various curves. Each has a negative-slope strength region on the left, a positive-slope gravity region on the right, and a transition size between those two regimes.

For the strength regime, the scaling theory of *Holsapple and Housen* (1986) and *Housen and Holsapple* (1990) and Holsapple (1994) assumes that tensile failures determine the strength, so they use a rate-dependent strength model. The slope of the power-law in the strength regime (*Holsapple,* 1994) is $9\mu/(3 - 2\phi)$ where ϕ is the exponent of a Weibull flaw-size distribution, and μ is the exponent of impact velocity in the point-source coupling parameter measure. (The exponent μ reflects the decrease in coupling efficiency and increasing waste heat with increasing impact velocity. It is determined in the early-time coupling regime, and is primarily a consequence of the high pressure-temperature equation of state. A material with no energy dissipation has $\mu = 2/3$.)

Holsapple (1994) assumes that $\mu = 0.55$ and $\phi = 9$ to give the slope as –0.33. The code calculations of *Benz and Asphaug* (1999) also use $\phi = 9$, and their slope is also about –0.33. *Housen and Holsapple* (1999a,c) postulate that a better value is $\phi = 6$ so the slope is –0.67; furthermore, their curve is fitted to their size-dependent granite experimental results (those data are off this curve to the left). The curve given by calculations by *Ryan and Melosh* (1998) for mortar assumes that $\phi = 6.5$ and their slope is –0.61. Thus, in all cases the slopes obtained are entirely consistent with the scaling prediction based on the assumed value of the parameter ϕ.

For large bodies, in the gravity regime, the scaling theories predict that the slope should be 3μ. If μ were 2/3 (the limiting value when the point source measure is the impactor energy), there would be no decrease of coupled energy with increasing impact velocity, and pure energy scaling would hold. Then the slope in the gravity regime on this plot would be 2. The *Davis et al.* (1985) model assumed such energy scaling, so they have the slope of 2. The value chosen in *Holsapple* (1994), $\mu = 0.55$, was determined by a multitude of results for nonporous materials such as rocks and water (see *Holsapple,* 1993), which gives a slope of 1.65. The *Benz and Asphaug* (1999) curve has essentially that same slope, suggesting that the code calculation reproduces the expected early-time energy coupling. The slightly shallower slope of the *Love and Ahrens* (1996) curve suggests more dissipation in the energy coupling. That might be a consequence of low resolution at the shock front, although *Ryan and Melosh* (1998) got a much steeper slope with a very-low-resolution calculation. The differences also might arise from the equation of states used.

The most striking feature of this plot is the wide discrepancy of the values. However, much of this apparent discrepancy arises simply because some of the curves define the condition for shattering, Q_S^*, while others apply to the condition for dispersal, Q_D^* (see section 1). In the small-body strength regime where there is negligible gravity, these two values of Q are the same, since any shattered body will disperse. There are substantial differences in the predictions in this strength region, although most match the centimeter-sized laboratory data. (The theoretical scaling curves are in fact calibrated to those results. The codes are usually "dialed in" to match those results; see section 3.2.)

In the gravity regime there is a substantial difference between the shattering and the dispersion energy. The *Ryan and Melosh* (1998) curves are for the shattering energy only, while the *Housen and Holsapple* (1990), *Benz and Asphaug* (1999), *Melosh and Ryan* (1997), and *Love and Ahrens* (1996) curves are for the dispersion energy. The *Holsapple* (1994) curve was based on the largest observed craters on various bodies, so it is a lower bound for the dispersion energy, but in principle could be above the shattering energy. The *Housen and Holsapple* (1990) curve was based on velocity scaling, assuming one-half the mass had velocity greater than the escape velocity. *Melosh and Ryan* (1997) also used velocity scaling, but in a different way, to get their dispersion curve.

The estimates of the energy to disperse are about a factor of about 100 above the shattering estimates. *Fujiwara* (1982) and *Davis et al.* (1983) previously noted the large difference "energy gap" between these two thresholds.

Durda et al. (1998) obtained a very different curve by comparing the results of a collisional evolution model with the present asteroid size distribution, and then backing out the scaling law that best matched the present distribution. A more recent paper by *Campo Bagatin and Petit* (2001), with a different model of evolution of asteroids, did not agree with that result.

3.4.1. Uncertainties in disruption scaling. The very wide diversity of curves for either of shattering or dispersion is unsettling. Those estimates have changed a lot in form over the last 20 years. It used to be common to adopt a single value for Q^* at all sizes, and the distinction between shattering and dispersion was overlooked. Then the increasing slope in the gravity regime was introduced for dispersion thresholds, a consequence of requiring energy to launch remnants to escape velocities. The strengthening due to lithostatic pressure was then formulated, so the positive slope applied to shattering also. That was based on the concept that the initial compressive stress due to lithostatic pressure must first be overcome before the stress wave could achieve tensile fracture conditions. Finally, a decrease of required energy in the strength regime was introduced based on rate-dependent strength. That is currently an accepted feature of all present curves. What, if any, evidence is there for any of these effects? Most laboratory experiments cannot address any of them.

In the strength regime, the experiments by *Housen and Holsapple* (1999c) show the size dependence, but only over a diameter range of 18:1. We do not know of any other direct evidence for the size effect. It remains uncertain whether those trends continue to asteroid sizes. Codes reproduce that dependence, but that is simply a consequence of their assumed models.

In the gravity range, there is some indirect evidence for the positive slope. First, the estimates of required energy to form the Themis, Eos, and Koronis families are at least consistent with current estimates for dispersion (*Housen and Holsapple,* 1990), although not definitive by themselves. The use of crater scaling applied to the largest observed craters on relatively dense bodies seems to show an increase of energy with size, and was used as a lower limit by *Holsapple* (1994). The experiments by *Housen et al.* (1991), using external pressure to simulate the gravitational binding, show a definite increase in shattering strength with that external pressure, and were used as an upper limit for the curve by *Holsapple* (1994). The analysis by *Durda et al.* (1998), based on an evolution calculation, has an even larger upward slope in the gravity regime. Finally, the computer calculations also show the upward slope, and here the results are in a regime where the uncertainties of strength modeling discussed above are not important. Therefore, although there is general agreement on the upward slope, there remain substantial differences in estimated magnitudes. This is certainly an area needing further research.

And, finally, all experiments and calculations for impact disruption consider only nonrotating asteroids. If indeed many are rubble-pile bodies, many are rotating at a significant fraction of the maximum allowable rotation for their shape (*Holsapple,* 2001). What effect does that have on required disruption energy? One would certainly think that could have a major effect at lowering the required energy for disruption.

3.4.2. Implications of disruption scaling. There are many important implications of these results, numerous examples of which can be found in the literature. Here only one significant one is considered, that of collisional lifetimes. *Farinella et al.* (1998) give the disruptional lifetime in years for a target of radius R as $t_{disr} = [2.85 \times 10^{-18} R^2 N(r_{disr})]^{-1}$, where $N(r_{disr})$ is the number of asteroids of radius greater than the radius r_{disr} (km) of the impactor required to shatter the target. They use an asteroid number distribution in the main belt as $N(r) = 3.5 \times 10^5 r^{-5/2}$, with asteroid radius r in kilometers [see *Farinella et al.* (1998) for details].

Using the relation for the specific energy to either shatter or disperse (a factor of about 100 different) with an assumed velocity of 5 km/s gives, for any asteroid, estimates of the lifetime between shattering events and between complete dispersion events. The *Housen and Holsapple* (1990) shattering kinetic energy per unit mass in the gravity regime is given in cgs units as $Q^* = (5.8 \times 10^{-8})U^{0.35}R^{1.65}$. Assuming an impact velocity of 5 km/s, and equal densities for the impactor and target, gives the required impactor radius for shattering as

$$r_s(\text{km}) = (2.01 \times 10^{-3})R(\text{km})^{1.55}$$

which gives the time period as

$$t_{shatter}(\text{yr}) = (1.8 \times 10^5)R(\text{km})^{1.875}$$

Taking the *Housen and Holsapple* (1990) dispersion curve as a factor of 100 larger specific energy gives the required impactor radius for dispersion as

$$r_s(\text{km}) = (9.32 \times 10^{-3})R(\text{km})^{1.55}$$

and a lifetime before dispersion as

$$t_{disperse}(\text{yr}) = (8.4 \times 10^6)R(\text{km})^{1.875}$$

Three features of these lifetime estimates are very significant. First is simply the large uncertainty; the time is proportional to the required Q to the power of 0.833, and Q is uncertain to maybe an order of magnitude. The number densities are uncertain to perhaps a factor of 2–3, giving a combined uncertainty factor of about 20 for the lifetime.

The second feature is the large 1.875 exponent on the asteroid radius R. The paper by *Farinella et al.* (1998) has only a power of 0.5, since they assumed a constant dispersion specific energy. The effect of the increasing energy with size in the gravity region is significant. The difference between a 10-km object and a 100-km asteroid is, by the results here, a factor of 80 in both the shattering or dispersion lifetimes: The larger one lasts 80× longer. From these estimates, a 100-km-radius asteroid has a shattering lifetime of ~1 b.y. Its dispersion lifetime is 2 orders of magnitude greater, so it will likely never be impacted by the 20-km-diameter body required to disperse it. Few large asteroids will ever be shattered.

Then consider a 10-km-radius main-belt asteroid, assuming it somehow is an intact rock body. The impactor that will shatter it at 5 km/s has a diameter of ~150 m. Dispersal requires a 700-m-diameter impactor. The lifetime against shattering is about 10^7 yr, while that for dispersal is 50× longer. Therefore asteroids are predicted to average perhaps 50 shattering events with various amounts of disruption before a single dispersion event. Thus a 10-km object cannot remain intact for very long. Various amounts of disruption will occur about every 10 m.y., ranging from just breaking it throughout (like Eros?), to almost complete dispersion where lots of small fragments are ejected but later reaccumulate. A 1-km asteroid is shattered maybe five times every 10^6 yr.

How could any present 1–10-km rocky asteroid not be a rubble-pile? A small coherent asteroid cannot escape shattering for very long. Could a 10-km body be a piece from a larger body, e.g., a 15-km body? Dispersion of that 15-km body requires a factor of 100 in energy over its shattering threshold, so all its pieces before reaccumulation would initially be very small. *Benz and Asphaug* (1999) determine that a dispersal event simply pulverizes the parent body. Thus any large remnant of the 15-km asteroid must itself be an aggregate of smaller pieces. How can any 10-km body ever again become a "rock" with density ~3?

One possibility is indicated from *Housen et al.* (1999), which estimates that the five largest impacts on Mathilde increased its density by 20%. Perhaps a primary effect of impacts into very porous bodies is not to disrupt them at all, but simply to pound them back into relatively dense asteroids, to begin the cycle over again. Perhaps they densify during large impacts by shaking small fragments down into the holes between larger ones (see *Britt et al.,* 2002; *Asphaug et al.,* 2002). Alternatively, other processes could increase the density of asteroids (*Consolmagno and Britt,* 1999). Such processes are a prime area for future research.

The question of the effects of impacts into rubble piles is the primary outstanding question for which we have no answer. Indeed, the present state of knowledge is very tenuous: What we know about scaling for rocky bodies seems to tell us they cannot remain rocky, so the scaling does not apply! As stated in the closing words of a lecture by James Head, "Almost everything is not yet known."

REFERENCES

Arakawa M. (1999) Collisional disruption of ice by high-velocity impact. *Icarus, 142,* 34–45.

Arakawa M. and Higa M. (1995) Measurement of ejection velocities in collisional disruption of ice spheres. *Planet. Space Sci., 44,* 901–908.

Asphaug E. and Melosh H. J. (1993) The Stickney impact of Phobos: A dynamical model. *Icarus, 101,* 144–164.

Asphaug E., Moore J. M., Morrison D., Benz W., Nolan M. C., and Sullivan R. J. (1996) Mechanical and geological effects of impact cratering on Ida. *Icarus, 120,* 158–184.

Asphaug E., Ostro S. J., Hudson D. J., Scheeres D. J., and Benz W. (1998) Disruption of kilometer-sized asteroids by energetic collisions. *Nature, 393,* 437–440.

Asphaug E., Ryan E. V., and Zuber M. T. (2002) Asteroid interiors. In *Asteroids III* (W. F. Bottke Jr. et al., eds.), this volume. Univ. of Arizona, Tucson.

Bashkirov A. G. and Vitzayev A. V. (1996) Statistical mechanics of fragmentation processes in ice and rock bodies. *Planet. Space Sci., 44,* 909–915.

Benz W. and Asphaug E. (1994) Impact simulations with fracture: I. Method and tests. *Icarus, 107,* 98–116.

Benz W. and Asphaug E. (1999) Catastrophic disruptions revisited. *Icarus, 142,* 5–20.

Binzel R. P., Farinella P., Zappalà V., and Cellino A. (1989) Asteroid rotation rates: Distributions and statistics. In *Asteroids II* (R. P. Binzel et al., eds.), pp. 416–441. Univ. of Arizona, Tucson.

Britt D. T., Yeomans D., Housen K., and Consolmagno G. (2002) Asteroid density, porosity, and structure. In *Asteroids III* (W. F. Bottke Jr. et al., eds.), this volume. Univ. of Arizona, Tucson.

Campo Bagatin A. and Petit J.-M. (2001) How many rubble piles are in the asteroid belt? *Icarus, 149,* 198–209.

Cheng A. F. and Barnouin-Jha O. S. (1999) Giant craters on Mathilde. *Icarus, 140,* 34–48.

Cintala M. J. and Hörz F. (1984) Catastrophic rupture experiments: Fragment-size analysis and energy considerations (abstract). In *Lunar and Planetary Science XV,* pp. 158–159. Lunar and

Planetary Institute, Houston.

Cintala M. J., Hörz F., Smrekar S., and Cardenas F. (1985) Impact experiments in H_2O ice, II. Collisional disruption (abstract). In *Lunar and Planetary Science XVI,* pp. 129–130. Lunar and Planetary Institute, Houston.

Cintala M. J., Berthoud L., Hörz F., Petersen R. K., and Jolly G. D. (1997) A method of measuring ejection velocities during experimental impact events (abstract). In *Lunar and Planetary Science XXVIII,* pp. 233–234. Lunar and Planetary Institute, Houston.

Cintala M. J., Berthoud L., and Hörz F. (1999) Ejection-velocity distributions from impacts into coarse-grained sand. *Meteoritics & Planet. Sci., 34,* pp. 605–623.

Consolmagno G. J. and Britt D. T. (1999) Turning meteorites into rock: Constraints on asteroid physical evolution (abstract). In *Lunar and Planetary Science XXX,* Abstract #1137. Lunar and Planetary Institute, Houston (CD-ROM).

Croft S. K. (1982) Impacts in ice and snow: Implications for crater scaling on icy satellites (abstract). In *Lunar and Planetary Science XIII,* pp. 135–136. Lunar and Planetary Institute, Houston.

Davis D. R. (1998) The experimental and theoretical basis for studying collisional disruption in the solar system. In *Impacts on Earth* (D. Benest and C. Froeschlé, eds.), pp. 113–136. Springer-Verlag, Berlin.

Davis D. R. (1998) The experimental and theoretical basis for studying collisional disruption in the Solar System, In *Impacts On Earth,* (Benest and Froeschlé eds.), 113–136. Springer-Verlag.

Davis D. R. and Farinella P. (1997) Collisional evolution of Edgeworth-Kuiper Belt objects. *Icarus, 125,* 50–60.

Davis D. R. and Ryan E. V. (1990) On collisional disruption: Experimental results and scaling laws. *Icarus, 83,* 156–182.

Davis D. R., Chapman C. R., and Greenberg R. (1977) Asteroid fragmentation processes and collisional evolution. In *Reports of the Planetary Geology Program 1976–1977,* pp. 72–73. NASA Tech. Memo X-3511.

Davis D. R., Chapman C. R., Greenberg R., and Weidenschilling S. J. (1983) Asteroid collisions: Effective body strength and efficiency of catastrophic disruption (abstract). In *Lunar and Planetary Science XIV,* pp. 146–147. Lunar and Planetary Institute, Houston.

Davis D. R., Chapman C. R., Greenberg R., and Weidenschilling S. J. (1985) Collisional history of asteroids: Evidence from Vesta and the Hirayama families. *Icarus, 62,* 30–53.

Di Martino M., Martelli G., Smith P. N., and Woodward A. W. (1990) Time evolution and dust production in catastrophic fragmentation by hypervelocity impacts. *Icarus, 83,* 126–132.

Drobyshevski E. M., Rozov S. I., Zhukov B. G., Kurakin R. O., and Sokolov V. M. (1994) Railgun collisional experiments. In *Seventy-Five Years of Hirayama Asteroid Families* (Y. Kozai et al., eds.), p. 231. ASP Conference Series 63.

Durda D. and Flynn G. (1999) Experimental study of the impact disruption of a porous, inhomogemeous target, *Icarus, 142,* 46–55.

Durda D., Greenberg R., and Jedicke R. (1998) Collisional models and scaling laws: A new interpretation of the shape of the main-belt asteroid size distribution. *Icarus, 135,* 431–440.

Farinella P. and Davis D. R. (1996) Short period comets: Primordial bodies or collisional fragments? *Science, 273,* 938–941.

Farinella P., Paolicchi P., and Zappalà V. (1981) Analysis of the spin rate distribution of asteroids. *Astron. Astrophys., 104,* 159–165.

Farinella P., Davis D. R., Paolicchi P., Cellino A., and Zappalà V. (1992) Asteroid collisional evolution: An integrated model for the evolution of asteroid rotation rates. *Astron. Astrophys., 253,* 604–614.

Farinella P., Vokrouhlický D., and Hartmann W. K. (1998) Meteorite delivery via Yarkovsky orbital drift. *Icarus, 132,* 378–387.

Fujiwara A. (1982) Complete fragmentation of the parent bodies of the Themis, Eos, and Koronis families. *Icarus, 52,* 434–443.

Fujiwara A. (1987) Energy partition into translational and rotational motion of fragments in catastrophic disruption by impact: An experiment and asteroid cases. *Icarus, 70,* 536–545.

Fujiwara A. and Asada N. (1983) Impact fracture patterns on Phobos ellipsoids. *Icarus, 56,* 590–602.

Fujiwara A. and Tsukamoto A. (1980) Experimental study on the velocity of fragments in collisional breakup. *Icarus, 44,* 142–153.

Fujiwara A. and Tsukamoto A. (1981) Rotation of fragments in catastrophic impact. *Icarus, 48,* 329–334.

Fujiwara A., Kamimoto G., and Tsukamoto A. (1977) Destruction of basaltic bodies by high-velocity impact. *Icarus, 31,* 277–288.

Fujiwara A., Cerroni P., Davis D. R., Ryan E. V., Di Martino M., Holsapple K., and Housen K. (1989) Experiments and scaling laws for catastrophic collisions. In *Asteroids II* (R. P. Binzel et al., eds.), pp. 240–265. Univ. of Arizona, Tucson.

Fulchignoni M., Barucci M. A., Di Martino M., and Dotto E. (1995) On the evolution of asteroids spin. *Astron. Astrophys., 299,* 929–932.

Gault D. E. and Heitowit E. D. (1963) The partition of energy for hypervelocity impact craters formed in rock. *Proc. 6th Hypervelocity Impact Symposium, Vol. 2,* pp. 419–456. Firestone Rubber Co., Cleveland.

Giblin I. (1994) Development and use of software for the analysis of hypervelocity impact experiments. Ph.D. thesis, Univ. of Sussex, Brighton.

Giblin I. (1998) New data on the velocity-mass relation in catastrophic disruption. *Planet. Space Sci., 46,* 921–928.

Giblin I. and Farinella P. (1997) Tumbling fragments from experiments simulating asteroidal catastrophic disruption. *Icarus, 127,* 424–429.

Giblin I., Martelli G., Smith P. N., Cellino A., Di Martino M., Zappalà V., Farinella P., and Paolicchi P. (1994a) Field fragmentation of macroscopic targets simulating asteroidal catastrophic collisions. *Icarus, 110,* 203–224.

Giblin I., Martelli G., Smith P. N., and Di Martino M. (1994b) Simulation of hypervelocity impacts using a contact charge. *Planet. Space Sci., 42,* 1027–1030.

Giblin I., Petit J. M., and Farinella P. (1998a) Impact ejecta rotational bursting as a mechanism for producing stable Ida-Dactyl systems. *Icarus, 132,* 43–52.

Giblin I., Martelli G., Farinella P., Paolicchi P., Di Martino M., and Smith P. N. (1998b) The properties of fragments from catastrophic disruption events. *Icarus, 134,* 77–112.

Grady D. E. and Kipp M. E. (1980) Continuum modeling of explosive fracture in oil shale. *Intl. J. Rock Mech. Min. Sci., 17,* 147–157.

Harris A. W. (1990) The collisional evolution of the spin of a nonspherical body. *Icarus, 83,* 183–185.

Harris A. W. and Burns J. A. (1979) Asteroid rotation. I. Tabulation and analysis of rates, pole positions and shapes. *Icarus, 40,* 115–144.

Hartmann W. K. (1969) Terrestrial, lunar and interplanetary rock fragmentation. *Icarus, 10,* 201–213.

Hartmann W. K. (1988) Impact strengths and energy partitioning in impacts into finite solid targets (abstract). In *Lunar and Planetary Science XIX,* pp. 451–452. Lunar and Planetary Institute, Houston.

Holsapple K. A. (1980) The equivalent depth of burst for impact cratering (abstract). In *Lunar and Planetary Science XI*, pp. 457–458. Lunar and Planetary Institute, Houston.

Holsapple K. A. (1981) Coupling parameters in cratering (abstract). *Eos Trans. AGU, 62,* 944.

Holsapple K. A. (1983) On the existence and implications of coupling parameters in cratering mechanics (abstract). In *Lunar and Planetary Science XIV,* pp. 319–320. Lunar and Planetary Institute, Houston.

Holsapple K. A. (1987) The scaling of impact phenomenon. *Intl. J. Impact Eng., 5,* 343–355.

Holsapple K. A. (1993) The scaling of impact processes in planetary sciences. *Annu. Rev. Earth Planet. Sci., 21,* 333–373.

Holsapple K. A. (1994) Catastrophic disruptions and cratering of solar system bodies: A review and new results. *Planet. Space Sci., 16,* 1067–1078.

Holsapple K. A. (2001) Equilibrium configurations of solid cohesionless bodies. *Icarus, 154,* 432–448.

Holsapple K. A. and Housen K. R. (1986) Scaling laws for the catastrophic collisions of asteroids. *Mem. Soc. Astron. Ital., 57,* 65–85.

Holsapple K. A. and Schmidt R. M. (1982) On the scaling of crater dimensions. II — Impact processes. *J. Geophys. Res., 87,* 1849–1870.

Holsapple K. A. and Schmidt R. M. (1987) Point-source solutions and coupling parameters in cratering mechanics. *J. Geophys. Res., 92,* 6350–6376.

Housen K. R. (1992) Crater ejecta velocities for impacts on rocky bodies (abstract). In *Lunar and Planetary Science XXIII,* pp. 555–556. Lunar and Planetary Institute, Houston.

Housen K. R. (1993) Simulation of collisional fragmentation with explosives (abstract). In *Lunar and Planetary Science XXIV,* pp. 675–676. Lunar and Planetary Institute, Houston.

Housen K. R. and Holsapple K. A. (1990) On the fragmentation of asteroids and planetary satellites. *Icarus, 84,* 226–253.

Housen K. R. and Holsapple K. A. (1999a) A general scaling law for strength-dominated collisions of rocky asteroids (abstract). In *Lunar and Planetary Science XXX,* Abstract #1095. Lunar and Planetary Institute, Houston (CD-ROM).

Housen K. R. and Holsapple K. A. (1999b) Impact cratering on porous low-density bodies (abstract). In *Lunar and Planetary Science XXX,* Abstract #1228. Lunar and Planetary Institute, Houston (CD-ROM).

Housen K. R. and Holsapple K. A. (1999c) Scale effects in strength-dominated collisions of rocky asteroids. *Icarus, 142,* 21–33.

Housen K. R. and Holsapple K. A. (2000) Numerical simulations of impact cratering in porous materials (abstract). In *Lunar and Planetary Science XXXI,* Abstract #1498. Lunar and Planetary Institute, Houston (CD-ROM).

Housen K. R. and Voss M. E. (2001) Ejecta from impact craters in porous materials. *Lunar and Planetary Science XXXII,* Abstract #1617. Lunar and Planetary Institute, Houston (CD-ROM).

Housen K. R., Schmidt R. M., and Holsapple K. A. (1983) Crater ejection scaling laws: Fundamental forms based on dimensional analysis. *J. Geophys. Res., 88,* 2485–2499.

Housen K. R., Schmidt R. M., and Holsapple K. A. (1991) Laboratory simulations of large scale fragmentation events. *Icarus, 94,* 180–190.

Housen K. R., Holsapple K. A., and Voss M. E. (1999) Compaction as the origin of the unusual craters on the asteroid Mathilde. *Nature, 402,* 155–157.

Hudson R. S. and Ostro S. J. (1995) Shape and non-principal axis spin state of asteroid 4179 Toutatis. *Science, 270,* 84–86.

Kadono T. (1993) Change in spin orientation of satellites by oblique impact. *Icarus, 105,* 580–584.

Kato M., Iijima Y., Okimura Y., Arakawa M., Maeno N., and Mizutani H. (1992) Impact experiments on low temperature H_2O ice. In *Physics and Chemistry of Ice* (N. Maeno and T. Hondoh, eds.), pp. 237–244. Hokkaido Univ., Hokkaido.

Kawakami S., Mizutani H., Takagi Y., and Kato M. (1983) Impact experiments on ice. *J. Geophys. Res., 88,* 5806–5814.

Kawakami S., Kanaori Y., Fujiwara A., Arakawa M., Kato M., Mizutani H., Cerroni P., and Capaccioni F. (1991) An experimental study of impact fracturing of small planetary bodies in the solar system with an application to Phobos. *Astron. Astrophys., 241,* 233–242.

Lange A. and Ahrens T. J. (1981) Impact experiments in low-temperature ice (abstract). In *Lunar and Planetary Science XII,* pp. 592–594. Lunar and Planetary Institute, Houston.

La Spina A. and Paolicchi P. (1996) Catastrophic fragmentation as a stochastic process: Sizes and shapes of fragments. *Planet. Space Sci., 44,* 1563–1578.

Leinhardt A. M. and Richardson D. C. (2001) The effect of the internal configuration of rubble piles on collision outcome (abstract). In *Lunar and Planetary Science XXXII,* Abstract #1400. Lunar and Planetary Institute, Houston (CD-ROM).

Love S. G. and Ahrens T. J. (1996) Catastrophic impacts on gravity dominated asteroids. *Icarus, 124,* 141–155.

Love S. G., Hörz F., and Brownlee D. E. (1993) Target porosity effects in impact cratering and collisional disruption. *Icarus, 105,* 216–224.

Martelli G., Rothwell P., Giblin I., Di Martino M., and Farinella P. (1993) Fragment jets from catastrophic break-up events and the formation of asteroid binaries and families. *Astron. Astrophys., 271,* 315–318.

Martelli G., Ryan E. V., Nakamura A. M., and Giblin I. (1994) Catastrophic disruption experiments: Recent results. *Planet. Space Sci., 42,* 1013–1026.

Martelli G., Smith P. N., Woodward A. J. (1991) Simulation of impacts in a pre-biotic atmosphere. In *Erici international school on the origin of life.*

Matsui T., Waza K., Kani S., and Suzuki S. (1982) Laboratory simulation of planetesimal collisions. *J. Geophys. Res., 87,* 10968–10982.

Matsui T., Waza T., and Kani K. (1984) Destruction of rocks by low velocity impact and its implications for accretion and fragmentation processes of planetesimals. *J. Geophys. Res., 89,* 700–706.

Melosh H. J. and Ryan E. V. (1997) Asteroids: Shattered but not dispersed. *Icarus, 129,* 562.

Melosh H. J., Ryan E. V., and Asphaug E. (1992) Dynamic fragmentation in impacts: Hydrocode simulation of laboratory impacts. *J. Geophys. Res., 97,* 14735–14759.

Michikami T. (2001) Evolution of asteroid regolith layers by cratering. Ph.D. thesis, Univ. of Tokyo, Japan.

Michikami T., Moriguchi K., Abe M., Hasegawa S., and Fujiwara A. (2001) Ejecta velocity distribution for impact cratering experiment on porous target. *Proc. 34th ISAS Lunar Planet. Symp.,* pp. 107–110.

Mitani N. K. (2000) Numerical simulations of shock attenuation in solids induced by hypervelocity impacts. Ph.D. thesis, Univ. of Tokyo, Japan, in press.

Mizutani H. (1993) Scaling law on impact phenomena. In *Primitive Solar Nebula and Origin of Planets* (H. Oya, ed.), pp. 297–317. Terra Scientific, Tokyo.

Mizutani H., Kawakami S., Takagi Y., Kato M., and Kumazawa M. (1982) Low velocity impact experiments on ice and basalt (abstract). In *Lunar and Planetary Science XIII,* pp. 532–533. Lunar and Planetary Institute, Houston.

Mizutani H., Takagi Y., and Kawakami S. (1990) New scaling laws on impact fragmentation. *Icarus, 87,* 307–326.

Nakamura A. M. (1993) *Laboratory Studies on the Velocity of Fragments from Impact Disruptions.* ISAS Report 651, Institute of Space and Astronautical Science, Tokyo.

Nakamura A. and Fujiwara A. (1991) Velocity distribution of fragment formed in a simulated collisional disruption. *Icarus, 92,* 132–146.

Nakamura A., Suguiyama K., and Fujiwara A. (1992) Velocity and spin of fragments from impact disruptions I. An experimental approach to a general law between mass and velocity. *Icarus, 100,* 127–135.

Nakamura A. M., Fujiwara A., Kadono T., and Shirono S. (1994) Fragmentation of intact and pre-impacted basalt targets. In *Seventy-Five Years of Hirayama Asteroid Families* (Y. Kozai et al., eds.), pp. 237–242. ASP Conference Series 63.

Nolan M. C., Asphaug E., Melosh H. J. and Greenberg R. (1996) Impact craters on asteroids: Does gravity or strength control their size? *Icarus, 124,* 359–371.

O'Keefe J. D. and Ahrens T. J. (1993) Planetary cratering mechanics. *J. Geophys. Res., 98,* 17011–17028.

O'Keefe J. D. and Ahrens T. J. (1999) Complex craters: Relationship of stratigraphy and rings to impact conditions. *J. Geophys. Res., 104,* 27091.

O'Keefe J. D. and Ahrens T. J. (2000) Impact bulking, faulting and ejecta (abstract). In *Lunar and Planetary Science XXXI,* Abstract #1429. Lunar and Planetary Institute, Houston (CD-ROM).

Paolicchi P., Cellino A., Farinella P., and Zappalà V. (1989) A semi-empirical model of catastrophic breakup processes. *Icarus, 77,* 187–212.

Paolicchi P., Verlicchi A., and Cellino A. (1996) An improved semi-empirical model of catastrophic impact processes. I. Theory and laboratory experiments. *Icarus, 121,* 126–157.

Petit J-M. and Farinella P. (1993) Modelling the outcomes of high-velocity impacts between solar system bodies. *Cel. Mech. Dyn. Astron., 57,* 1–28.

Polanskey C. A. and Ahrens T. J. (1990) Impact spallation experiments — Fracture patterns and spall velocities. *Icarus, 87,* 140–155.

Richardson D., Bottke W. F. Jr., and Love S. G. (1998) Tidal distortion and disruption of earth crossing asteroids. *Icarus, 134,* 47–76.

Richardson D. C., Leinhardt Z. M., Melosh H. J., Bottke W. F. Jr., and Asphaug E. (2002) Gravitational aggregates: Evidence and evolution. In *Asteroids III* (W. F. Bottke Jr. et al., eds.), this volume. Univ. of Arizona, Tucson.

Ryan E. V. (1992) Catastrophic collisions: Laboratory impact experiments, hydrocode simulations, and the scaling problem. Ph.D. thesis, Univ. of Arizona, Tucson.

Ryan E. V. and Melosh H. J. (1998) Impact fragmentation: From the laboratory to asteroids. *Icarus, 133,* 1–24.

Ryan E. V., Hartmann W. K., and Davis D. R. (1991) Impact experiments 3: Catastrophic fragmentation of aggregate targets and relation to asteroids. *Icarus, 94,* 283–298.

Ryan E. V., Davis D. R., and Giblin I. (1999) A laboratory impact study of simulated Edgeworth-Kuiper Belt objects. *Icarus, 142,* 56–62.

Schmidt R. M. and Holsapple K. A. (1982) Estimates of crater size for large-body impact: Gravity-scaling results. In *Geological Implications of Impacts of Large Asteroids and Comets on Earth* (L. T. Silver and P. H. Schultz, eds.), pp. 93–102. GSA Special Paper 190.

Schmidt R. M. and Housen K. R. (1987) Some recent advances in the scaling of impact and explosive cratering. *Intl. J. Impact Eng., 5,* 543–560.

Schmidt R. M., Housen K. R., and Holsapple K. A. (1986) *Gravity Effects in Cratering.* DNA-TR-86-182, Defense Nuclear Agency, Washington, DC (limited distribution).

Schultz P. H. (1996) Effect of impact angle on vaporization. *J. Geophys. Res., 101,* 21117–21136.

Sirono S. I. and Greenberg J. M. (2000) Do cometesimal collisions lead to bound rubble piles or to aggregates held together by gravity? *Icarus, 145,* 230–238.

Sirono S., Tada M., Nakamura A. M., Kadono T., Rivkin A., and Fujiwara A. (1993) Efficiency of linear and angular momentum transfer in oblique impact. *Planet. Space Sci., 41,* 687–692.

Smrekar S, Cintala M. J., Hörz F., Cardenas F., and Thompson T. D. (1985) Impact experiments in cold rock. II: Collisional fragmentation (abstract). In *Lunar and Planetary Science XVI,* pp. 795–796. Lunar and Planetary Institute, Houston.

Stern S. A. and Weissman P. R. (2001) Rapid collisional evolution of comets during the formation of the Oort cloud. *Nature, 409,* 589–591.

Sugi N., Arakawa M., Kouchi A., and Maeno N. (1998) In-situ mass spectrometric observation of impact vaporization of water-ice at low temeratures. *Geophys. Res. Lett., 25,* 837–840.

Takagi Y., Mizutani H., and Kawakami S. (1984) Impact fragmentation experiments of basalts and pyrophyllites. *Icarus, 59,* 462–477.

Takagi Y., Kato M., and Mizutani H. (1992) Velocity distribution of fragments of catastrophic impacts. In *Asteroids, Comets, Meteors 1991* (A. W. Harris and E. Bowell, eds.), pp. 597–600. Lunar and Planetary Institute, Houston.

Takagi Y., Nakamura A. M., and Fujiwara A. (1996) Fragment velocity dependence on fragment mass of impact fragmentation phenomena (abstract). In *Lunar and Planetary Science XXVII,* pp. 1301–1302. Lunar and Planetary Institute, Houston.

Tanga P., Cellino A., Michel P., and Zappalà V. (1999) On the size distribution of asteroid families: The role of geometry. *Icarus, 141,* 65–78.

Vickery A. M. (1986) Size-velocity distribution of large ejecta fragments. *Icarus, 67,* 224–236.

Vickery A. M. (1987) Variation in ejecta size with ejection velocity. *Geophys. Res. Lett., 14,* 726–729.

Vortman L. J. (1965) *Dimensions of a Crater from a 500-Ton TNT Hemisphere Detonated on Rock.* Sandia Corp. Report SC-RR-76-277.

Waza T., Matsui T., and Kani K. (1985) Laboratory simulation of a planetesimal collision. 2. Ejecta velocity distribution. *J. Geophys. Res., 90,* 1995–2011.

Yamamoto S. (2001) Velocity distribution of powdery ejecta in oblique impacts (abstract). In *Lunar and Planetary Science XXIII,* Abstract #1180. Lunar and Planetary Institute, Houston.

Yamamoto S. and Nakamura A. M. (1997) Velocity measurements of impact ejecta from regolith targets. *Icarus, 128,* 160–170.

Yanagisawa M. and Hasagawa S. (1999) Angular momentum transfer in oblique impacts: Implication for 1989ML. *Earth Planets Space, 51,* 1163–1171.

Yanagisawa M. and Hasagawa S. (2000) Momentum transfer in oblique impacts: Implication for asteroid rotations. *Icarus, 146,* 270–288.

Yanagisawa M. and Itoi T. (1993) Antipodal fragment velocities for porous and weak targets at catastrophic impacts (abstract). In *Lunar and Planetary Science XXIV,* pp. 1551–1552. Lunar and Planetary Institute, Houston.

Yanagisawa M., Eluszkiewicz J., and Ahrens T. J. (1991) Angular momentum transfer in low velocity oblique impacts: Implications for asteroids. *Icarus, 94,* 272–282.

Asteroid Interiors

Erik Asphaug
University of California, Santa Cruz

Eileen V. Ryan
New Mexico Highlands University

Maria T. Zuber
Massachusetts Institute of Technology

Asteroids represent a geophysical frontier of considerable scientific importance, for these building blocks of planets preserve a compositional and mechanical record of the solar system's origin. They also pose a formidable hazard to life on Earth, while offering potentially enormous resources to benefit solar system exploration. This chapter introduces their geologic subtlety through the theme of strength vs. gravity: Asteroids stand apart from rocks and planets because these primary forces (strength short-range and gravity long-range) achieve an intricate mechanical balance that is only partly understood. Recent findings have turned Earth-based intuition on its head. For example, structurally weak asteroids and comets may, through stress dissipation, be the most highly resistant to catastrophic disruption. Perhaps more surprising, the collisional evolution of bodies the size of a city block may be determined by their minuscule self-gravity. While the fundamental geophysical behavior of asteroids continues to elude us, one idea resonates: Instead of competing for dominance, strength and gravity collaborate over a wide spectrum of sizes and shapes to produce some of the richest structures in nature.

1. INTRODUCTION

The gravity of a small asteroid can barely hold on to its rocks, yet the best spacecraft images of ~10-km bodies show fields of boulders, lunarlike craters with dramatic rim deposits, landscapes mantled in mature regolith, and shapes broadly conforming to gravity potentials (see *Sullivan et al.,* 2002). But gravity's influence must decrease with decreasing mass, and asteroids smaller than some diameter are governed by strength. But how small? Smaller than tiny Dactyl? This ~1-km satellite of Ida [the first confirmed asteroid satellite, imaged by *Galileo* en route to Jupiter (*Belton et al.,* 1996)] is a spheroid possessing what appear to be central peak craters. These characteristics indicate gravitational control, yet anything traveling faster than a half meter per second will escape Dactyl's surface. [For a nonrotating asteroid with bulk density $\rho \sim 2$ g cm^{-3}, escape velocity (in meters per second) is about equal to asteroid radius (in kilometers). If you can jump half a meter on Earth, you can leap off a 5-km-diameter asteroid.] Smaller than near-Earth asteroid Castalia? The first imaged asteroid (*Ostro et al.,* 1990), Castalia is believed to consist of two ~800-m lobes resting against one another self-gravitationally. Detailed examination of even smaller objects has just begun (see *Pravec et al.,* 2002; and *Ostro et al.,* 2002), but for reasons we shall see, gravity may control the evolution of asteroids the size of a city block, whose escape velocities are much slower than a page turning.

This is at serious odds with Earth-based intuition. Indeed, most early ideas about asteroids have proven wrong. The recent adoption of modern geophysical ideas (rate- and size-dependent strength, micro- and macroporosity), together with laboratory, numerical, and dynamical advances described below, have led to the overturn of several fundamental concepts regarding asteroid geodynamics and mechanics. Their behavior is certainly not trivial, and it is appropriate to say that the study of asteroids expands our geophysical experience to include low-gravity test beds of processes that remain poorly understood on Earth. Granular flow, impact cratering mechanics, studies of wave attenuation and fault motion in geologic media, and the mechanics of landslides may all be advanced considerably by studying the geophysics of asteroids. Of course, they also deserve our scrutiny for pragmatic reasons. The common kilometer-sized asteroids most hazardous to life on Earth might exist in a no-man's land where neither gravity nor strength predominates, and may be stranger than ever imagined.

2. STRENGTH AND GRAVITY

To understand the collisional evolution of asteroids and comets, researchers in the 1960s began using laboratory impact experiments to determine relations between initial conditions (e.g., speed and angle of impact, structure and composition of target) and outcome (e.g., kinetic energy required for breakup, velocity and rotation of fragments). This database (see *Fujiwara et al.,* 1989; *Martelli et al.,* 1994; *Holsapple et al.,* 2002) guides our understanding of impact physics, although the experiments are limited to small projectiles whose masses can be accelerated to speeds (several kilometers per second) typical of asteroid collisions. The largest laboratory targets are orders of magnitude smaller than any asteroid; Earth's gravity overwhelms self-gravity, to be sure, and also overwhelms delicate strength effects that may be important at large scale. Chemical and nuclear explosions (*Rodionov et al.,* 1972; *Perret et al.,* 1967) provide an analogous experimental database at geologic scales, although an explosion in a half-space is significantly different from impact into a finite, irregular target. Despite these limitations, laboratory-scale impacts and high-yield explosions constitute important benchmarks against which asteroid collisional models must be gauged. But as we shall see, the best benchmarks are the cratered asteroids themselves.

It has long been recognized (*Gault et al.,* 1972; *Fujiwara,* 1980) that impact disruption depends upon target size, so two approaches were taken to extrapolate laboratory data to relevant scales. Hydrodynamical similarity ["scaling" (see *Holsapple et al.,* 2002, and references therein)] is a powerful approach that offers a unique ability to provide meaningful relationships between key parameters, for instance, fragment size as a function of impact energy and momentum, or ejecta velocity as a function of rock strength. The most significant limitation of scaling is that it presumes a homogeneous continuum, i.e., bodies that are either monolithic or finely comminuted. Where competing length scales exist (e.g., size of projectile comparable to or smaller than size of structural components in an aggregate, or crater diameter comparable to target diameter) or where target structure and shape is important (contact binary or coarsely shattered asteroids), the technique may break down. But for well-posed problems (notably half-space cratering into homogeneous media) the technique has a well-established utility. Scaling also has the distinct advantages of providing direct physical intuition (for instance, how crater diameter relates to impact speed) and requiring no major computational effort.

A very different approach is to directly integrate the shock physics relations, together with an equation of state plus the mass, momentum, and energy conservation equations of continuum mechanics. These "hydrocodes" are versatile and on modern computers can be run with adequate resolution and fidelity to model targets with layers, cores, realistic shapes, prefractures, and multiple components. [The term "hydrocode" broadly describes any explicit continuum mechanics code capable of modeling shock physics and the behavior of solid media under high dynamic stress. Originally applied only to extreme conditions where rock strength can be ignored ("hydro"), these codes have evolved to model distinctively nonfluid behaviors such as elasticity and fragmentation.] But hydrocodes are seldom easy to use and the data they produce are not trivial to interpret. As with all tools, hydrocodes are easier to misuse than use. One common culprit is lack of resolution adequate to resolve the impact shock. The shock determines the velocities in the excavation flow, and hence the growth of a gravity-regime crater, or disassembly following catastrophic disruption. Other errors include the use of an inappropriate equation of state, application of a model beyond its designed domain, and misinterpretation of results. Furthermore, although these numerical techniques date back half a century (e.g., *von Neumann and Richtmyer,* 1950), hydrocode modeling is considered by many to be an immature technique. Fortunately, hydrocode outcomes can be tested in detail against laboratory experiments for cratering and catastrophic disruption (*Melosh et al.,* 1992; *Benz and Asphaug,* 1994, 1995; *Ryan and Melosh,* 1998). Thus verified, a code can be applied with some confidence to more complex scenarios and much larger sizes that are typical of asteroids.

While there remains some conflict between scaling vs. numerical modeling (particularly in the half-space cratering regime), the predictions for catastrophic disruption appear to have achieved some convergence, encouraging us to apply laboratory-validated code models to the fundamental issues of gravity, strength, and structure on asteroids. Furthermore, if asteroids are complicated entities with odd shapes, rotations, prefractures, and porosities, numerical approaches may be the only way, short of *in situ* experiments, of understanding their behavior.

2.1. Strong Asteroids

Two decades ago it was widely believed on the basis of scaling from impact experiments (*Gault and Heitowit,* 1963; *Housen et al.,* 1979) that rocky asteroids smaller than ~70 km diameter and icy bodies smaller than ~20 km would lack regolith (cf. *Veverka et al.,* 1986). [Correcting for a typesetting error: A line was skipped on p. 354 of *Satellites,* so that it would appear that the 20-km limit applied to rock, not ice (K. Housen, personal communication, 2001).] Smaller bodies could not hold onto loose material because their craters would form in the strength regime, with nearly all material being accelerated to speeds $>v_{esc}$. [The two classical regimes of impact cratering are the strength regime and the gravity regime. Other regimes can also be defined (see *Housen et al.,* 1983), depending on the dominant force restricting crater growth. Craters forming in the strength regime must break apart rock, in which case the energy of fragmentation (the rock strength) is proportional to the energy of ejection. For craters forming in the gravity regime, ejection velocity scales with v_{esc}.] Nor could asteroids smaller than this size retain primordial regolith, because scaled experiments showed that any surface debris would rapidly erode by impacts down to bedrock. Because regolith is a common component of meteorites (see *Bunch and Rajan,* 1988), regolith breccias were believed to derive

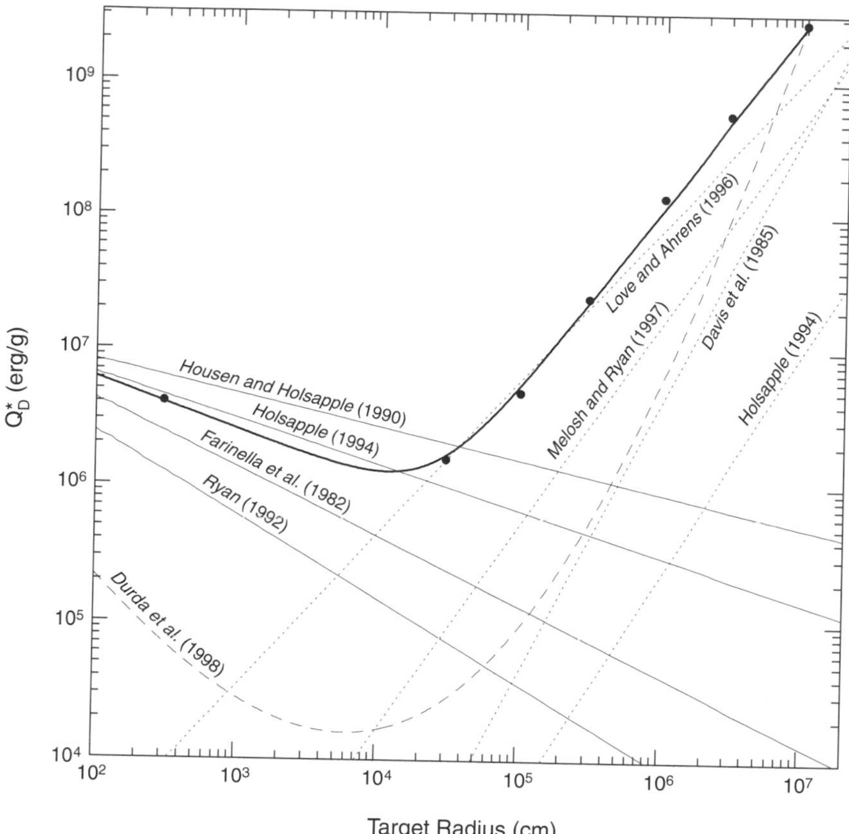

Fig. 1. Values for the catastrophic disruption threshold Q^*_D of asteroids vary widely in the literature. The darker line is the summary of numerical outcomes for basalt spheres by *Benz and Asphaug* (1999). The general trend for all models is that rocks get weaker with size because of size- and/or rate-dependent strength, and then get stronger once self-gravity dominates. For a collisionally evolved population, objects to the right of the minimum (the strength-gravity transition) are likely to be shattered but not dispersed.

from parent bodies hundreds of kilometers across, or from primordial surfaces prior to the onset of erosive collisions (*Chapman,* 1976).

A related early puzzle concerned asteroid families, whose members have orbital elements suggestive of dispersal by catastrophic impact (*Hirayama,* 1923) at speeds of hundreds of meters per second (*Cellino et al.,* 1999). For these family members to have survived such dramatic impact acceleration, it seemed that their parent rock must have been strong [see *Chapman et al.* (1989) for a variety of asteroid family hypotheses]. Strong targets would yield fast, large family members whose strength might thereafter resist catastrophic disruption over the billions of years to the present day. The argument for strong asteroids was not entirely satisfying to those who have seen primitive meteorites fall apart in their hands, but it was largely self-consistent. [High ejection speeds are no longer required by the orbital elements of family members, however. It has recently been discovered that the Yarkovsky effect can perturb small asteroid orbits significantly over time; see *Bottke et al.* (2002).]

2.2. Regolith All the Way Down

For asteroids larger than ~100 km, another concept was introduced in the 1970s: the existence of gravity-controlled rubble piles that consist of nothing but regolith. In developing an asteroid collisional evolution model, *Davis et al.* (1979) defined a threshold specific energy Q^*_D (impact kinetic energy per target mass) required to both shatter mechanical bonds and accelerate half the mass to escaping trajectories. Shattering requires a lower specific energy $Q^*_S < Q^*_D$ to create fragments none larger than half the target mass. For small rocks $Q^*_D \rightarrow Q^*_S$, whereas for large bodies $Q^*_S/Q^*_D \rightarrow 0$. Whenever $Q^*_S \ll Q^*_D$, the probability of a shattering impact becomes far greater than the probability of a dispersal, in which case an asteroid might be expected to evolve into a pile of rubble, unless other effects (such as melting and compaction) were to dominate.

Davis et al. (1979) expressed impact strength as the sum of the shattering strength plus the gravitational binding energy of the target, for example,

$$Q^*_D = Q^*_S + 4/5\ \pi\ \rho\ GR^2 \quad (1)$$

where R is the radius of a spherical target and ρ is its density. Equation (1) is called energy scaling; on a graph of Q^*_D vs. R (see Fig. 1) it plots as a horizontal line ($Q^*_D \approx Q^*_S$ = constant) transitioning at some size to a gravity-regime slope of 2 ($Q^*_D \propto R^2$). The size corresponding to this break in slope is known as the strength-gravity transition for catastrophic disruption. [The strength-gravity transition for catastrophic disruption must be distinguished from the strength-gravity transition for planetary cratering. An object in the gravity regime for disruption (Earth is one) can certainly have strength-controlled craters.] Subsequent analysis has changed the slopes in both regimes (the predicted transition size varies by orders of magnitude from model to model), but the concept was established that beyond some size rubble piles might exist.

Because ρQ^*_S has dimensions of strength and is close to the corresponding static tensile strength of ice and rock in laboratory impact experiments (*Fujiwara et al.,* 1989), tensile strength was used as an easily measured proxy for ρQ^*_S in early disruption theory. Primitive asteroids, comets, and early planetesimals, which are presumably weak, would be easily disrupted. This led to some notable inconsistencies, the most obvious being that primitive bodies smaller than ~100 km would be unlikely to survive over billions of years, in contrast with their abundant population.

The resolution to this dilemma appears to be that impact strength and tensile strength are not simply related, and may even be inversely correlated. Only later did experiments show that loosely bonded aggregates [such as a pile of sintered glass beads (*Love et al.,* 1993)] can survive a projectile that would blast an equal-mass monolithic cylinder to smithereens. It is now believed that some of the most fragile bodies in the solar system — porous aggregates with little or no cohesion — can be highly resistant to catastrophic disruption owing to their ability to dissipate and absorb impact energy. Furthermore, ejection velocities from fragile bodies are correspondingly low, enabling them to hold on to their pieces. Like palm trees that bend in a storm, weak asteroids may survive collisions that would shatter and disperse monolithic solids.

But prior to these and other recent insights to be explored below, a straightforward application of laboratory-derived values for Q^*_S and for the partitioning of impact kinetic energy among fragments (*Fujiwara and Tsukamoto,* 1980) made it fairly obvious that bodies up to ~100 km diameter would be strength-controlled. This fit in well with the idea (*Housen et al.,* 1979; *Veverka et al.,* 1986) that regolith would be thin or absent on such bodies. The ~100-km transition gained further support due to its consistency with two much simpler notions: (1) The transition should occur when central pressure $\sim 2/3\pi G\rho^2 R^2$ equals rock strength; for icy or rocky targets this transition occurs at about 100 km diameter since rock is both stronger and denser. (2) It should occur when gravitational binding energy per volume equals rock strength Y; neglecting constants this yields $R\rho = \sqrt{(Y/G)}$, and R on the order of several hundred kilometers, again whether for ice or rock.

With four distinct ways of viewing asteroid structure converging upon a transition to the gravity regime at ~100-km sizes, the idea seemed safe that all but the largest asteroids were monolithic. Because geochemists predict intense early thermal effects in bodies larger than 100 km (*Scott et al.,* 1989; *McSween et al.,* 2002) the existence of rubble piles remained a conjecture.

2.3. Size-dependent Strength

Impact modelers soon began to appreciate that static strength is sensitive to target size (*Jaeger and Cook,* 1969; *Lawn and Wilshaw,* 1975) because of the greater likelihood of finding large flaws in large volumes. This notion, formalized by engineers who found that it became increasingly difficult to grind coal (for example) down to smaller and smaller sizes, contributes a new power-law size-dependence to asteroid mechanics. Break any brittle solid in two under modest tension, and the number of available flaws decreases by the weakest flaw (plus whatever neighboring flaws are utilized to grow the crack). Continue in this manner to deduce that strength increases monotonically with decreasing size. [The concept was originally demonstrated by measuring the force required to snap smaller and smaller pieces of thread (*Weibull,* 1951)].

If the static failure stress s of a rock decreases with flaw size L as $\sigma \propto 1/\sqrt{L}$ (*Griffith,* 1920), and if the maximum flaw size in a target rock increases in proportion to its radius [as postulated by *Fujiwara* (1980) in his argument for size-dependent strength], then impact strength — if equated with static failure stress — depends on target size R as

$$Q^*_S \propto 1/\sqrt{R} \tag{2}$$

This same relation was derived by *Farinella et al.* (1982) by relating impact energy to the energy required to form new surface area during fracture. When plotted on a graph of Q^*_D vs. R (Fig. 1) this has a slope of –1/2 in the strength regime: According to this relation (and neglecting strength-selection effects of meteorites, which makes them stronger still), a centimeter-sized meteorite is ~1000× stronger than a 10-km asteroid. This has tremendous implications for the survival of asteroids. If strength diminishes this rapidly, self-gravity becomes the dominant binding force at sizes smaller than previously considered.

A problem crops up at this point: The larger asteroid is only weaker somewhere, otherwise every cubic centimeter of a large asteroid would be weaker than every cubic centimeter of a small asteroid of the same material. To avoid such nonsense, concepts of heterogeneity are required. Instead of considering only the single weakest flaw, consider an asteroid of volume $V \sim R^3$ that samples a theoretical continuum riddled with a power-law distribution of flaws. These can be grain boundaries, pores, inclusions, or cracks opened by previous collisions.

Suppose the number of flaws per unit volume that begin to grow at or below a stress σ is described by a simple power law

$$n(\sigma) = k\sigma^m \tag{3}$$

The probability (*Weibull,* 1939) of finding a flaw in a random volume V that will begin to grow at or below σ is then $1\text{-}\exp[-(\sigma/\sigma_{min})^m]$, where

$$\sigma_{min} = (kV)^{-(1/m)} \tag{4}$$

The threshold for failure σ_{min} goes with the –3/m power of radius. For reasons not yet understood, m ~ 6 may be favored by nature (*Housen and Holsapple,* 1999; *Asphaug,*

TABLE 1. Weibull dynamic fracture coefficients for various rocks.

Material	Reference	m	k (cm^{-3})	ln(k)/m
Basalt*	*Melosh et al.* (1992)	9.5	1.0×10^{27}	6.54
Basalt*	*Benz and Asphaug* (1995)	9.0	4.0×10^{29}	7.17
Basalt†	*Lindholm et al.* (1974)	9.5	1.59×10^{30}	7.32
Granite†	*Grady and Lipkin* (1980)	6.2	4.14×10^{17}	6.54
Water Ice*,†,‡	*Benz and Asphaug* (1999)	9.6	1.4×10^{32}	7.71
30% Sand + Water Ice†	*Stewart et al.* (1999)	9.57	1.34×10^{30}	7.25
Concrete†	*Grady and Lipkin* (1980)	5.3	5.27×10^{12}	5.53
Oil Shale†	*Grady and Kipp* (1980)	8.1	1.70×10^{21}	6.04
Limestone†	*Grady and Lipkin* (1980)	57.0	4.26×10^{167}	6.77

* Determined from simulation fits to laboratory data. The two-dimensional axisymmetric simulations of *Melosh et al.* (1992) require stronger fracture parameters than the nonsymmetric three-dimensional simulations of *Benz and Asphaug* (1995) for the same impact experiment.

† Determined experimentally through measurements of tensile strength vs. strain rate.

‡ Earlier published values of m = 8.7, k = 3.2×10^{38} (*Lange and Ahrens,* 1983) were later corrected to similar values [m = 9.57, k = 1.28×10^{32} (*Stewart et al.,* 1999)].

Although k ranges by ~150 orders of magnitude, the situation is not entirely bleak. Homogeneous materials (such as limestone reported below) have large m, and metals are approximated by m → 8. Because the static failure threshold (σ_{min}) is unchanged so long as ln(k)/m remains constant (see text), homogeneous materials of a given strength have very large k. Equation (4) shows how rocks with large m exhibit little size dependence. While dynamic fracture coefficients have yet to be measured for any meteorite, *Housen and Holsapple* (1999) propose that m ~ 6 might generally describe "well-cracked rocks at large scale," and hence might apply to asteroids.

1993), in which case one recovers equation (2) assuming $Q^*_S \propto \sigma_{min}$. But values of m as high as 57 and as low as 2 have been reported for rocks in the literature (*Grady and Kipp,* 1980) (see Table 1) and no values of m or k have been reported for any meteorite.

In some instances m may be determined by direct measurement of the flaws in a geologic specimen. However, this introduces a bias, as flaw sizes approaching the sample size tend to be excluded, and flaws larger than the sample will not exist. More generally m is determined by fitting laboratory experiments to the strength-strain rate relations described below, although these experiments are limited to small specimens and to the small flaws activated by dynamic fragmentation. In most asteroid impact models, fracture parameters for basalt, granite, or ice are assumed; the sensitivity of these models upon m precludes one from having robust faith in blind forward modeling.

2.4. Rate Dependence

The above analysis of σ_{min} provides a static explanation for why Q^*_S should diminish with size. But Q^*_S is further diminished on account of dynamical fracture mechanics: Rock strength is a function of the loading rate $\dot{\varepsilon} = d\varepsilon/dt$, where ε is the mechanical strain. It has long been established that the dynamic failure strength of rocks scales with the ~1/4 to ~1/3 power of $\dot{\varepsilon}$ (*Rinehart,* 1965). Recognizing that $\dot{\varepsilon}$ decreases with the size of the collisional event, *Holsapple and Housen* (1986) explored the implication of low strain rate collisions in a revised set of scaling models.

If $\dot{\varepsilon}$ is approximated as the impact speed divided by the impactor radius, then a 1-km-diameter impactor striking at 5 km/s couples at a strain rate $\dot{\varepsilon} \sim 10\ s^{-1}$. Though undeniably dynamic, this is a far cry from the $\dot{\varepsilon} \sim 10^6\ s^{-1}$ typical of laboratory experiments. Moreover, stress waves broaden and decay with distance, so $\dot{\varepsilon}$ drops with a steep (~4th) power of distance during hypervelocity collisions (*Melosh et al.,* 1992). Strain rates responsible for shattering of ~10-km asteroids can be as small as 10^{-3} to $10^{-5}\ s^{-1}$ (*Asphaug and Melosh,* 1993).

Grady and Kipp (1980) applied Weibull statistics (equation (3)) to derive a dynamic fragmentation model computing fracture stress and fragment size as a function of $\dot{\varepsilon}$. Except for dynamical loads of such intensity that brittle fragmentation is not the mode of failure, cracks grow at a rate c_g that is about half the sound speed (*Lawn and Wilshaw,* 1975). The crack tip is assumed to accelerate instantaneously to this velocity, when in fact it takes some finite time that depends upon the applied stress and upon crack length. This dependence upon crack length (the integral of crack growth velocity) makes any formal treatment of crack tip acceleration difficult in such models. Because cracks relieve stress over finite time, a dynamic equilibrium is established whereby strong flaws become active as needed, when weaker flaws cannot grow fast enough to accommodate the accumulating stress. Therefore the weakest flaws are sufficient to relieve the stress under low strain rates, resulting in large fragments (due to the low density of weak flaws) and a low measured strength. At high strain rates, on the other hand, stronger and more numerous flaws are called into play. This leads to the formation of small fragments (because of the high density of strong flaws) and at high measured strength. The resulting relationships (*Grady and Kipp,* 1980) are

$$\sigma \propto \dot{\varepsilon}^{3/(m+3)} \qquad (5)$$

where exponents of 1/3 and 1/4 (*Rinehart,* 1965) correspond to m = 6 and m = 9 respectively, and

$$L \propto \dot{\varepsilon}^{-m/(m+3)} \qquad (6)$$

so that fragment size L is inversely proportional to strain rate for large m.

These same assumptions are the basis for numerical models of dynamic fracture of brittle solids (*Melosh et al.,* 1992; *Benz and Asphaug,* 1994, 1995), namely that (1) cracks initiate from a Weibull distribution in accordance with the applied tensile stress, (2) the crack tip accelerates instantaneously to half the sound speed, (3) planar cracks of radius a relieve deviatoric and tensile stress in a spherical circumscribed volume; and (4) a volume is damaged (stress-relieved) according to the ratio of cracked volume to total volume. Fully damaged rock, according to these models, is relieved of all stress except compressive pressure, and behaves as a fluid. [A straightforward modification would be to have fully damaged rock behave as a cohesionless Mohr-Coulomb material, where shear stress is still resisted according to the applied normal stress according to the material's angle of internal friction (angle of repose).] Differences in these models depend primarily upon implementation, with *Grady and Kipp* (1980) assuming $\dot{\varepsilon}$ = constant to derive closed integrals, *Melosh et al.* (1992) allowing integration over time-varying strain rate, and *Benz and Asphaug* (1994, 1995) introducing an explicit flaw distribution plus three-dimensional hydrodynamics in addition to time-varying strain rate. [In these numerical models, the total stress tensor is rotated into principal axis coordinates and the maximal value (most tensile) is taken as the stress σ that activates the Weibull flaws.] Without explicit flaws (assigned at random from equation (3) to all discretized volume elements) large rock volumes are modeled as homogeneously weak everywhere, when in fact rock strength is approximately fractal.

Experimental verification of rate dependence in catastrophic collisions was obtained by *Housen and Holsapple* (1999) in a series of experiments involving cylindrical granite targets ranging from 1.9 to 34.4 cm diameter. These targets, identical except for size, were impacted by proportional diameter Al cylinders at uniform speed (~0.6 km/s), with Q^* held constant at ~1.0×10^7 erg/g for each event. The remarkable outcome is shown in Fig. 2. The smallest target is merely chipped by the bullet, while the largest target (everything held constant but size) is catastrophically demolished. *Housen and Holsapple* (1999) found agreement between this granite test data and hydrocode estimates (*Ryan and Melosh,* 1998) for critical specific energy as a function of target size; detailed quantitative comparisons of these specific experiments are in progress.

At low strain rates, not only is strength diminished, but so is ejection velocity. Fragment ejection energy ($1/2v_{ej}^2$) is proportional to the energy of fragmentation (strength), which varies inversely with target size. Thus, the fragments from a large asteroid travel more slowly, not counting the effects of gravity. This greatly reduced ejection velocity is perhaps the most significant aspect of rate-dependent strength for asteroid collisional evolution, since as $v_{ej} \rightarrow v_{esc}$ self-gravity begins to dominate.

Rate-dependent and size-dependent strength, together with the greatly diminished ejection speeds for large/weak targets, combine to greatly increase gravity's influence over catastrophic disruption for small asteroids. By the early 1990s, rate-dependent scaling theories (notably *Holsapple and Housen,* 1986) had pushed the threshold size for catastrophic disruption down to tens of kilometers diameter. This revised threshold (soon to be pushed to even smaller sizes) began to erode confidence in well-established principles, and placed asteroid science at a crossroads. By 1990 there was no longer any uniformly accepted conclusion regarding asteroid internal structure and the existence of regolith. In only a few years, the first detailed views of asteroids would come along, without science having converged upon even the basics of what they are.

3. CONCEPTS

3.1. Structural Nomenclature

Before proceeding, it is useful to introduce various archetypes for asteroid interiors (see *Richardson et al.,* 2002). Simplest is the monolith, which is any rock of low porosity and significant strength. A monolith is a good transmitter of elastic stress. By definition, monoliths tend to be smaller than the nominal strength-gravity size transition. Since this transition is likely to be dependent upon fracture parameters and ρ (at least), any theoretically abrupt transition would be smoothed out over a population. Monoliths formed by impact ejection from a parent body must be stronger than the acceleration stress; kinetic energy densities and angular momentum densities (speeds and rotation rates) should therefore be high.

Monoliths are fractured by impact bombardment, in which case their tensile strength is compromised and may be reduced to zero (shattered). A fractured or shattered monolith might transmit a compressive stress wave fairly well, provided pore space has not been introduced between the major fragments. (Tensile stress, however, is not supported across a fracture.) A rubble pile includes any shattered body whose pieces are furthermore translated and rotated into loose packing (e.g., *Chapman and Davis,* 1975). Stress waves of any sort are poorly transmitted across a rubble pile, although intense shocks may propagate by crushing and vapor expansion. This category also includes primor-

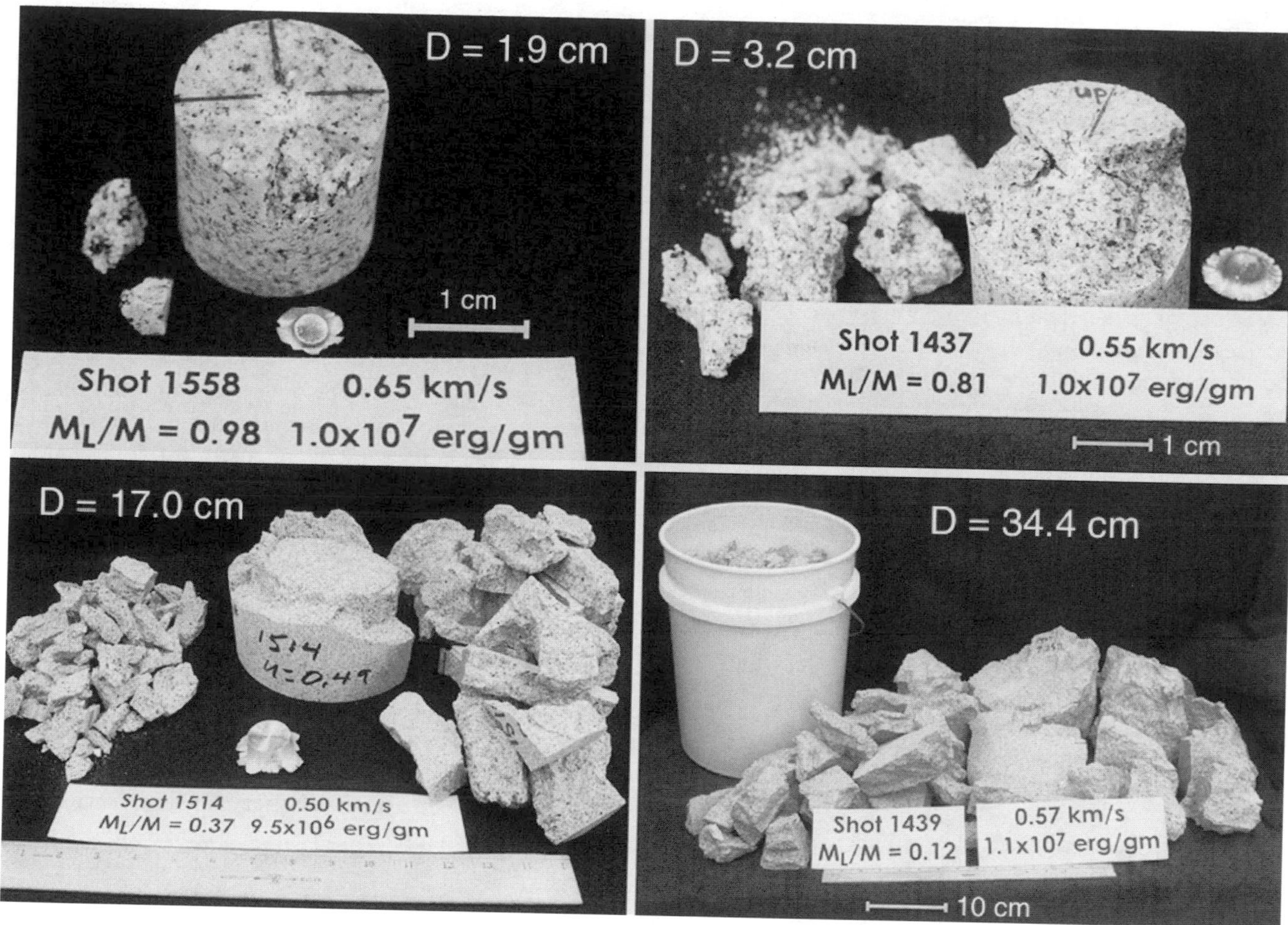

Fig. 2. In a controlled series of impact experiments using a unique ballistics facility at Boeing Aerospace, *Housen and Holsapple* (1999) impacted granite targets ranging from 1.9 cm to 34.4 cm diameter, otherwise identical, with proportional Al bullets fired at ~0.6 km/s, so that $Q^* = 1.0 \times 10^7$ erg/g for each event. The recovered bullet is the "fried egg" next to the first three targets. Because strength is sensitive to both the loading rate and the target size, the mass of the largest remnant drops by a factor >8 between the smallest and largest impacts. Analysis of the fragment size distributions leads to a Weibull exponent m ~ 6 for this material, in agreement with thin-section examinations of the target rock's flaw structure. This change of outcome happens across little more than an order or magnitude in size; imagine the outcome for an ~10-km granite cylinder floating in space. Because rocks get weaker with the ~3/m power of size (see text), asteroids as small as ~1 km are now believed to be gravity-dominated entities.

dial rubble piles, those objects that accreted as uncompacted cumulates. These objects (e.g., *Whipple,* 1949; *Weissman,* 1986) may have formed heterogeneously from grains into clusters, clusters into larger aggregates, and so on hierarchically into comets and asteroids (e.g., *Weidenschilling and Cuzzi,* 1993).

3.2. Onion Shells and Cosmic Sediment

Many asteroids trace their lineage to differentiated protoplanets, and this remains the conceptual model for family-related asteroids (*Hirayama,* 1923; see also *Zappalà et al.,* 2002). Meteoriticists have strong evidence for disrupted parent bodies; the H and L chondrites can be convincingly reassembled (*Keil et al.,* 1994) as "onion shells" of increasing metamorphic grade with depth, and HED achondrites appear to be samples of the crust and upper mantle of asteroid Vesta (*Drake,* 1979; *Binzel and Xu,* 1993; *Asphaug,* 1997; *Keil,* 2002; *Burbine et al.,* 2002). Vesta is the only known survivor of the original "onions" (basalt or related partial melts on the outside, and presumably olivine and then Fe on the inside), so the remainder of the original differentiated asteroids — dozens of parent bodies judging by the number of distinct classes of iron meteorites — have been catastrophically disrupted (see *McSween et al.,* 2002).

In contrast to differentiated asteroids and their monolithic fragments are the rubble piles, which either accreted as cumulates but never metamorphosed, or which reaccreted in the aftermath of a collision when $Q^*_S < Q < Q^*_D$. Indeed, disruption and reaccumulation of planetesimals is likely to have been a common and ongoing process in the early solar system. In a terrestrial context we would call such an object sedimentary, of local or transported origin — this may be a more appropriate paradigm for asteroids since the igneous "onions" appear to be mostly gone. Indeed, observation and modeling no longer support the idea of simple parentage for most surviving asteroids. For example, M-type spectra are no longer thought to be indicative of metallic composition (see *Rivkin et al.,* 2002). S-type asteroids, apparently primitive in composition (*Trombka et al.,* 2000), appear in some cases (e.g., Gaspra) to derive from much larger parent bodies that one might expect to have undergone extensive metamorphism. Strangest of all is the quandary that olivine-rich materials, representing mantle rocks from the dozens of disrupted original companions to Vesta, is unaccounted for in the meteorite collection (see *Burbine et al.,* 2002), despite the fact that mantle rock should represent at least half of each disrupted parent body's mass.

3.3. Meteorites

Samples of asteroid interiors exist on Earth. However, meteorites remain for the most part "samples without geologic context" (*McSween,* 1999), and any connections between meteorite class and asteroid taxonomy are tentative. To further obscure the asteroid-meteorite relationship, meteorites are small, highly selected specimens whose mechanical (and perhaps compositional) properties are seldom representative of asteroidal parents, as the latter are certainly weaker and more porous (*Flynn et al.,* 1999).

Selection effects prevail when a meteorite is blasted from a parent body. Because strong rock fragments are ejected at the highest speed, strong meteoroids are most likely to leave their parent body at sufficient speed to encounter Earth at 1 AU. Fragments must furthermore survive a long and convoluted journey through space, including the threat of catastrophic disruption by smaller meteoroids (*Greenberg and Nolan,* 1989; *Burbine et al.,* 1996) and must finally survive passage through Earth's atmosphere and then residence upon our planet. They must also wind up in a meteorite collection, although the Antarctic search plus direct followup of fireballs has removed much of the bias against finding extraterrestrial stones that may look like terrestrial rocks.

Only one meteorite parent body to date has been visited with a sample return mission: A comparison of lunar meteorites with Apollo samples (*Warren,* 1994) reveals that lunar meteorites are on the average much stronger than what the astronauts gathered and brought home. Martian meteorites are also highly selected, almost all of them being relatively young basalts, whose high strength and wave speed may have facilitated their acceleration to escape velocity during a cratering event (*Head and Melosh,* 2000).

3.4. Cohesion

A volatile-rich aggregate is more cohesive than a dry aggregate because of the facilitation of mechanical bonding, either directly (e.g., van der Waals forces) or indirectly during episodes of sublimation and frost deposition (*Bridges et al.,* 1996). Some propose the existence of volatile reservoirs in the deep interiors of primitive asteroids, with their mantles depleted through sublimation (*Fanale and Salvail,* 1990) or impact (see *Rivkin et al.,* 2002). For gravity as low as on a typical asteroid, frost or other fragile bonds can be critical to long-term survival during impact or tidal events. Comet Shoemaker-Levy 9, for example, could never have disrupted during its 1992 tidal passage near Jupiter (a nearly parabolic encounter with periapse of 1.3 jovian radii) had the tensile strength across the comet exceeded ~1000 dyn/cm^2 (*Sekanina et al.,* 1994), weaker than snow. *Asphaug and Benz* (1994b, 1996) calculated a maximum tensile strength of only ~30 dyn/cm^2 for Shoemaker-Levy 9 in order for it to fragment from a coherent body into ~20 pieces, and therefore proposed a rubble-pile structure for this comet and gravitational clumping (as opposed to fragmentation) as the cause of its "string of pearls" postperiapse structure. It matters a great deal whether a body is truly strengthless or only extraordinarily weak: Volatile inventory is critical, and poorly known.

3.5. Fast and Slow Rotation

Recent discussion regarding monolithic asteroids has centered around a simple but profound observational result by *Pravec and Harris* (2000; see also *Pravec et al.,* 2002). None of the ~1000 asteroids larger than 150 m with reliably measured spin periods rotates fast enough to require global cohesion. Specifically, for reasonable density estimates none rotates faster than $\omega_0^2 = 4\pi G\rho/3$, where ω_o is the frequency at which material on a sphere's equator becomes orbital. The corresponding minimum rotation period is $P_{crit} = 3.3\ \text{hrs}/\sqrt{\rho}$, with ρ in g/cm^3. The most dense common asteroids (S type) appear to have $\rho \sim 2.7$ g/cm^3 (*Belton et al.,* 1996; *Yeomans et al.,* 2000); thus an S-type gravitational aggregate can spin no faster than $P_{crit} \sim 2.0$ hrs [see *Holsapple* (2002) for a more detailed analysis of rotational breakup]. From the Pravec-Harris asymptote at 2.2 hrs one might infer a common maximum density $\rho \sim 2.3$ g/cm^3, but as few asteroids are spherical the corresponding density is probably greater: Equatorial speeds are faster on an elongated body and the mass distribution is noncentral.

The easiest interpretation is that nearly all asteroids larger than ~150 m lack cohesion. There are other plausible interpretations. Larger asteroids might be internally coherent but possess thick regolith, shedding mass into orbit whenever random spinups [or spinup by the Yarkovsky effect (*Rubincam,* 2000)] cause them to transgress the 2.2-hr rotation period, and then transferring angular momentum to this orbiting material. In this case the ~150-m transition may be the minimum size required for a body to retain regolith, and may have less to do with asteroid internal structure.

Another possibility is angular momentum drain (*Dobrovolskis and Burns,* 1984). Angular momentum is lost whenever prograde ejecta (that launched in the direction of an asteroid's rotation) preferentially escapes while retrograde ejecta remains bound. This is a good explanation for the relatively slow rotations of ~50–100-km-diameter asteroids, but for this process to apply to ~150-m asteroids the ejecta mass-velocity distribution must have a significant component slower than $v_{esc} \sim 5$–10 cm/s. Such slow ejection speeds are typical of gravity regime cratering; for craters on ~150-m bodies to be governed by gravity, the target must be strengthless.

A final possibility is that small asteroids are simply collision fragments with more angular momentum per unit mass than the larger remnants. In that case one would anticipate a smooth transition from fast, small rotators to large, slow rotators, whereas the observed transition appears to be abrupt. Furthermore this does not explain the absence of anomalous fast-rotating asteroids larger than 150 m. In summary, the rotation rate data appear to require that asteroids larger than 150 m are either strengthless or else mantled in deep regolith; perhaps future rotation rate surveys around this transition size will constrain our explanations.

An even bigger mystery is that only one asteroid smaller that 150 m diameter (out of ~25 total) with measured rotation rate rotates slower than this limit. [These statistics change on a monthly basis and are by the time of reading out of date; see *Pravec et al.* (2002)]. While no asteroid larger than ~150 m shows evidence for global cohesion, almost all asteroids smaller than this must be cohesive. One might suppose that every larger asteroid is a gravitational aggregate of smaller pieces, whereas every smaller asteroid is a fast-rotating collisional shard. But the term "monolith" for these smallest asteroids is misleading. Consider a spherical object of uniform density ρ rotating with a frequency ω; the mean stress across its equator is $\sim R^2\rho\omega^2$. For the well-studied fast rotator 1998 KY26 (*Ostro et al.,* 1999), self-gravity is not capable of holding it together; however, its ~11-min period and ~30-m diameter requires only a tensile strength of ~300 dyn/cm^2 (presuming $\rho \sim 1.3$ g/cm^3 for this C-type), orders of magnitude weaker than the tensile strength of snow.

3.6. Regolith and Structural Porosity

Besides the rotationally induced regolith loss just considered, a number of fundamental processes can take place in the components and regoliths of shattered and fractured monoliths and rubble piles.

3.6.1. Grain sorting. *Horstman and Melosh* (1989) proposed that regolith might drain into the interior of the martian satellite Phobos in response to block motion during the collision responsible for its large crater Stickney. Over time this process would expose large crustal blocks to the surface as impact-comminuted materials work their way down (*Asphaug and Melosh,* 1993). *Sears and Akridge* (1998) proposed that grains might become compositionally sorted in a regolith rendered dynamic by volatile outflow or meteorite gardening, resulting in iron-silicate fractionation on an unmelted parent body.

Indeed, an asteroid's regolith or a rubble pile's entire mass might be expected to undergo dynamical granular processes (reviewed by *Jaeger et al.,* 1996) on a variety of timescales in response to a variety of perturbations such as impact excavations, vibrations, tides, differential expansion and electrostatic repulsion (*Lee et al.,* 1996). *Asphaug et al.* (2001a) proposed that the extraordinarily high spatial densities of blocks on Eros (*Chapman,* 2001) represent the migration of large blocks through size-sorting dynamics towards the surface. Because size-sorting (e.g. the Brazil-nut effect) may work best in low gravity (*Jiongming et al.,* 1998), it is even possible that the shapes of rubble-piles might be governed by the very largest blocks working their way out towards the lowest gravitational potential.

A different view of particulate dynamics in rubble piles leads to the stacked block model of *Britt and Consolmagno* (2001), whereby regolith drains inwards only until small grains clog the gaps between large blocks. Thereafter a mantle of fine material settles over an interior of large blocks and voids: a body with fairly high bulk density near the surface, and significant interior porosity. These and other ideas regarding size-sorted asteroid structure will remain conjecture until the sciences of granular dynamics and of asteroid geology make significant advances.

3.6.2. Microporosity vs. macroporosity. One must distinguish between macroscopic and microscopic porosity in aggregate materials. An asteroid consisting of quintillions of tiny grains might exhibit considerable cohesion and perhaps support a fractal-like porosity. The total energy of contact bonds divided by the total mass of a granular asteroid (its overall cohesional strength) is inversely proportional to grain diameter, so that a coarse aggregate is weaker than a fine one. On the other hand a highly porous, finely comminuted body can be crushable: cratering on microporous asteroids might be a strange event involving compaction (*Housen et al.,* 1999) rather than ejection.

A coarse rubble pile by contrast would have far fewer contact surfaces distributed over the same total mass, and would therefore behave much differently. *Asphaug et al.,* (1998) used a coarse rubble pile as a starting condition for impact studies, and found that the impact shock wave gets trapped in the impacted components, with few pathways of transmission to neighboring components. Whether an asteroid is macroporous or microporous, stress wave transmission is hindered due to the great attenuation of poorly consolidated rock, making the survival of porous asteroids during impact more likely, as demonstrated by the experiments of *Ryan et al.* (1991) and *Love et al.* (1993) and as illustrated below.

3.7. Overburden Pressures

Small asteroids have correspondingly low internal pressures. Because our geophysical intuition of monoliths and rubble piles is based upon our familiarity with terrestrial landforms, it may be helpful to consider the equivalent depth z_{eq} of a planar stack of the same material, under Earth gravity, for which overburden pressure $\rho g z_{eq}$ is equal to the central pressure in an asteroid of radius R. Solving, this gives

$$z_{eq}(R) = \tfrac{2}{3}\pi G\rho^2 R^2/\rho g = 1.4 \times 10^{-10} \rho R^2 \qquad (7)$$

in cgs units, where $g = 980$ cm/s^2. For Phobos $z_{eq} \sim 3$ m with a central pressure of about 2/3 bar. For Mathilde, $z_{eq} \sim 13$ m. For Vesta, $z_{eq} \sim 3$ km. One might anticipate analogous conditions within rock and soil masses at depths on Earth corresponding to equal applied pressure. It is food for thought that the 150-m-diameter transition observed by *Pravec and Harris* (2000) corresponds to a z_{eq} of less than a millimeter, and to a central pressure of less than 100 dyn/cm^2 — one-millionth the tensile strength of rock. It is therefore easy to appreciate the continued resistance to ideas of gravity dominance at such small scales.

The importance for asteroid geology is considerable, as already discussed. For spacecraft exploration these details are also critical: The maximum shear stress supportable by asteroid regolith will be miniscule, in proportion to this vanishingly small normal stress. The behavior of materials at exploration landing sites is likely to be strange compared with the operational test beds here on Earth.

4. COLLISIONAL EVOLUTION

The physical geology of asteroids is largely the aftermath of several global-scale collisions plus a fusillade of smaller impacts; accretion itself is nothing but collisions of a gentler sort. Tides and surface process play a secondary role, at least in the present solar system. As is now believed to be the case for planet formation (*Wetherill,* 1985), accretion and cratering of asteroids may be dominated in terms of energy, mass, and angular momentum contribution by the largest events. For planets, the largest event might trigger core formation (*Tonks and Melosh,* 1992); for asteroids it may determine the interior structure of the body and hence its response to future collisions. If rubble-pile asteroids are resistant to further disruption (*Love et al.,* 1993; *Asphaug,* 1998) then the first noncatastrophic ($Q < Q^*_D$) impact exceeding Q^*_S might therefore determine an asteroid's long-term survival.

As discussed by *Richardson et al.* (2002), the 150-m threshold discovered by *Harris* (1996) and *Pravec and Harris* (2000) had been deduced by hydrocode simulations of asteroid collisions. *Love and Ahrens* (1996), *Melosh and Ryan* (1997), and *Benz and Asphaug* (1999) applied a variety of code and analytical techniques to explicitly derive the values of Q^*_D for asteroids as a function of size. The most detailed of these studies (*Benz and Asphaug,* 1999) applied a laboratory-derived size- and rate-dependent explicit fragmentation model (*Benz and Asphaug,* 1995) together with a postimpact search for the largest surviving clump, whether bound by self-gravity or strength. All these approaches derive a strength-gravity transition, for initially coherent spherical bodies of rock or ice, at a few hundred meters diameter.

Benz and Asphaug (1999) further found an abrupt transition in the structure of the largest remnant after a disruption. For targets smaller than ~1 km (whether rock or ice) suffering catastrophic disruption, the largest surviving remnant consists of a single monolith in their models. For targets larger than ~1 km, on the other hand, the largest surviving remnant is a gravitationally bound aggregate of pieces all much smaller than the target. This is taken as further evidence that the ~150-m Pravec-Harris transition is due to structure, for it appears to hold true for target materials of dissimilar density, thermodynamical behavior, and flaw distribution. Then again, these hydrocode outcomes are for near-threshold disruption of spherical monoliths, whereas actual asteroids have less-pristine original states and more intricate collisional histories. In particular, a detailed analysis of the impact evolution of primordial rubble piles (or a population evolving through myriad hypervelocity collisions) has yet to be conducted. In an initial exploration (see below), *Asphaug et al.* (1998) find that simple variations in target structure (a crack down the middle, for instance) yields dramatically different disruption outcomes. [For low-impact velocities (~10 m/s) typical of planetesimals in the "cold disk" before planet formation, *Benz* (2000) and *Leinhardt et al.* (2000) find that rubble piles are easily dispersed. This suggests a problematic bottleneck at the initial stage of planetary accretion.]

Given that small impacts are more common than large ones, an event of $Q^*_S < Q < Q^*_D$ is expected before catastrophic dispersal ($Q > Q^*_D$). This means that shattering is expected before catastrophic disruption. The end state of collisional evolution of asteroids larger than a few hundred meters is, by this logic, a gravitational aggregate. Those primitive bodies that are rubble piles from the start (e.g., *Weissman,* 1986) probably follow a different disruption scaling than the monolithic rocks simulated by *Ryan and Melosh* (1998) and *Benz and Asphaug* (1999), with shattering energy Q^*_S actually greater for a rubble pile with modest cohesion than for a monolithic rock (*Ryan et al.,* 1991; *Love et al.,* 1993). It remains unclear how the energy for dispersal, Q^*_D, will change for rubble piles, but it appears this, too, will be greater than for monolithic rock. The two current models for impact into porous asteroids differ fundamentally in this regard (*Asphaug and Thomas,* 1999; *Housen et al.,* 1999) (see below), yet both conclude that aggregate bodies resist disruption and dispersal in the hypervelocity regime.

While there is much to be done in understanding the behavior of shattered monoliths and rubble piles, it would appear that such aggregates are the natural end state of asteroids larger than ~1 km.

5. SIMULATING THE GIANT CRATERS

Asteroids suffering catastrophic disruption have lost at least half their volume and bear little sign of their previous incarnation. Those suffering giant craters, by contrast, still exhibit their original shape minus a huge divot or two, and may tell us how their interiors responded. It is hardly a coincidence that most asteroids imaged to date exhibit craters with diameters comparable to their own mean radius; this simply reflects that asteroids are capable of surviving enormous blows with relative impunity, perhaps owing to their unconsolidated interior structure.

Giant craters probe the brink of catastrophic disruption and thereby elucidate impacts at geologic scales — a process masked by gravity on Earth. With three-dimensional hydrocodes including rate-dependent strength running on powerful computers, one can model the details of their formation. Indeed, because one must resolve both the target and the impactor, large craters are actually easier to model than small ones for a finite target.

Most importantly, impact models can now be tested against targets of asteroidal composition and asteroidal scale. Rather than relying on blind extrapolation from laboratory experiments, we can match the outcomes of detailed impact models against existing giant craters on asteroids. This begins with an asteroid shape model transformed into a simulation grid [for modeling specifics, see *Asphaug et al.* (1996)]. For a giant crater, one must then reconstitute the shape so as to fill in the volume deficit, hopefully leaving a good approximation of the preimpact target. One then makes initial guesses as to interior structure and composition, as these are to be tested. For the impactor, one can assume nominal incidence angle and impact velocity and adopt some value [perhaps that predicted by gravity scaling (*Housen et al.,* 1983)] for the mass. Craters larger than about ~1 km diameter appear to form in the gravity regime for ~10-km monolithic asteroids (*Asphaug et al.,* 1996), so this is frequently a good starting assumption. One can then iterate upon interior geology, and if required, upon the mass and velocity of the impactor, until one arrives at a satisfactory fit to observables: distribution or absence of crater ejecta, global or regional fractures associated with the event, boulder distributions, etc.

Because high-fidelity modeling remains computationally expensive, and data analysis is not yet automated for this kind of inversion, only rough iterations have been performed to date for a handful of asteroids, as presented here.

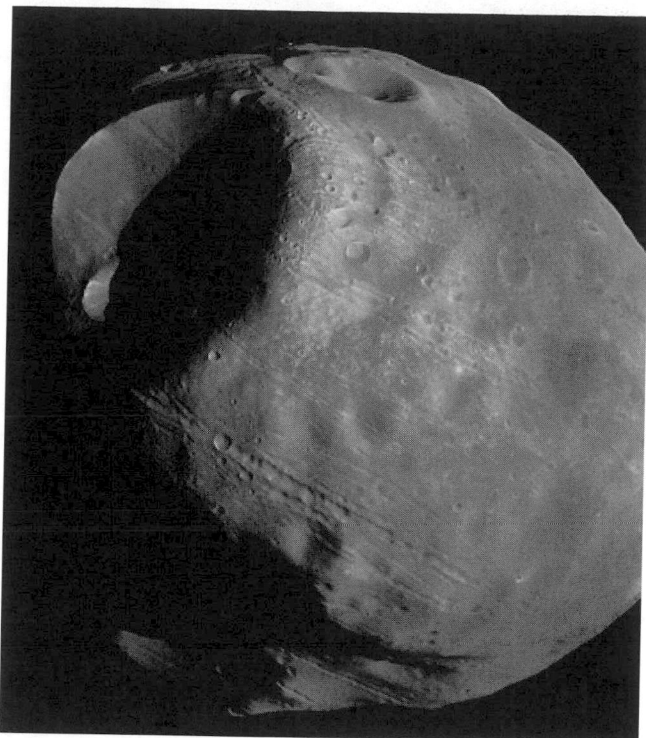

Fig. 3. A photomosaic of Phobos by the *Viking* Mars orbiter (see *Thomas et al.,* 1979). The large crater Stickney is visible at the upper left, and a number of grooves are prominent. Groove geometry has been shown (*Fujiwara,* 1991) to correlate with impact stress. Phobos is approximately a triaxial ellipsoid with dimensions 19 × 21 × 27 km; Stickney's diameter is ~10 km. Phobos is mysterious for its subsynchronous unstable orbit about Mars, its low density, and the fact that Stickney did not disrupt a satellite inside the Roche limit.

5.1. Phobos

Phobos, the ~22-km-diameter innermost satellite of Mars, was the first small planetary body observed at high resolution (*Veverka and Thomas,* 1979). Its ~10-km crater Stickney was a major curiosity until giant craters were found to be the norm. Indeed, the smaller martian moon Deimos has an even larger crater in proportion to its size (*Thomas,* 1998), although it took decades for this to be acknowledged. Its prevalent fracture grooves (Fig. 3), probably correlated with Stickney (*Thomas et al.,* 1979; *Fujiwara,* 1991), provide an ideal application for strength models.

Using a two-dimensional hydrocode in axisymmetry (*Melosh et al.,* 1992), *Asphaug and Melosh* (1993) attempted to model Stickney as a strength-regime event; i.e., they used *Housen et al.* (1983) to predict the impactor mass required to yield an equal-sized crater in a geologic half-space, and applied this same impactor to Phobos. Not knowing any parameters other than shape and bulk density [~1.95 g/cm^3 was assumed from the recently concluded *Phobos 2* flyby (*Avanesov et al.,* 1989)], they adopted Weibull fracture constants and equations of state for basalt and for water ice, hoping to span the range of possibilities. They first discovered that a strength-scaled impactor (730 m diameter at 6 km/s, for the size-dependent strength of 0.1 kbar) destroyed Phobos (exceeded Q^*_S) and launched most of the impacted hemisphere into orbit, turning Phobos inside-out. This was in stark contrast to the satellite's state of preservation. Using instead a gravity-scaled impactor [230 m diameter at 6 km/s, from *Housen et al.* (1983)], they found that this fragmented more than enough bedrock (whether for icy or rocky targets) to allow an ~10-km crater to evolve into shock-fragmented material. This greatly diminishes the effect of strength control over crater excavation (see *Nolan et al.,* 1996).

Equally significant, the flow field behind the shock in their model had particle velocities consistent with gravity scaling's predictions, with mean flow of a few meters per second throughout the rubble, and an excavation timescale of an hour. *Asphaug and Melosh* (1993) concluded that Stickney was a gravity-regime event, and showed that huge craters on asteroids as small as a few kilometers in diameter should be gravitationally determined. At slow excavation speeds, however, dynamic grain friction begins to play the role envisioned by *Housen et al.* (1983). This is not yet included in any model. Neglecting friction and the influence of Mars, *Asphaug and Melosh* (1993) found that Phobos should be covered by hundreds of meters of Stickney ejecta,

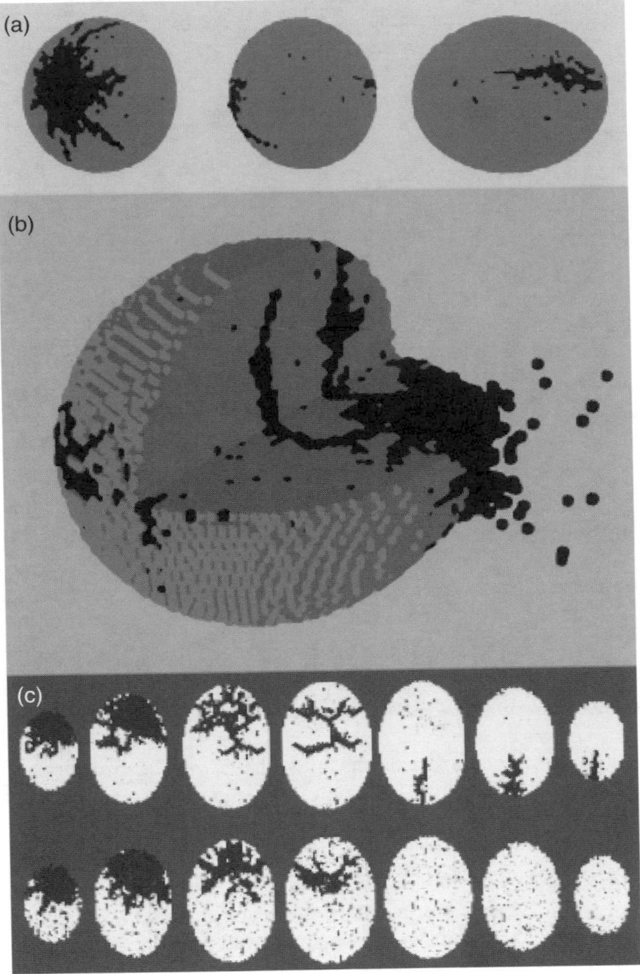

Fig. 4. **(a)** Fracture damage (black) on the surface of an ellipsoidal Phobos viewed during half a rotation, one minute after the Stickney-forming event (*Asphaug and Benz,* 1994a). The initial SPH hydrocode target has a flaw distribution and other material properties (other than density) derived from basalt (used in the absence of better alternatives), substituting a reference density of 1.95 g/cm^3. The impactor is a 230-m-diameter sphere of the same material impacting at 6 km/s. The interior, homogeneous except for the explicit flaw distribution (see text), allows for the transmission of coherent stresses to the backside of the target, leading to focusing and antipodal rupture broadly consistent with observations. **(b)** Cross section through the final Phobos target 12 s after impact. Fast ejecta is escaping, but the crater bowl has only begun to open. An astronaut witnessing the event could read this entire chapter in the meantime. **(c)** Parallel slices 2 km thick through the postimpact targets for two SPH models of Phobos. The top slices show the competent "monolith" with Weibull flaws [**(a)** and **(b)** above], with cracks throughout the cratered hemisphere and at the antipode. The bottom slices show an object of the same bulk density (1.95 g/cm^3) suffering the same impactor, but whose porosity is 30% with material specific density 2.7 g/cm^3. The flecks of black in the bottom slices are interstitial voids (random particles that were removed), not damaged rock. The porous target is a poor transmitter of impact stress; collisional effects are confined to the impacted hemisphere — a crater and little more.

in agreement with observation (*Thomas et al.,* 2000; *Horstman and Melosh,* 1989). Stickney's formation in the gravity regime was dramatically unlike the millisecond-scale laboratory experiments upon which asteroid evolution models had thus far been derived.

Fujiwara (1991) pioneered an important structural geological technique, correlating fracture geometry on small bodies (Phobos) with the responsible impact, thereby constraining elastic mechanical properties of the target rock. More detailed fracture modeling became possible with three-dimensional strength hydrocodes at high resolution, resulting in a kind of asteroid seismology. *Asphaug and Benz* (1994a) performed the same calculations as *Asphaug and Melosh* (1993), but using three-dimensional smooth particle hydrodynamics (SPH), explicit flaws, and an ellipsoidal target (see Fig. 4) so that the actual fracture pattern could be discerned in the model outcome. The modeled Phobos, being a low-density (ρ = 1.95 g/cm^3) elastic solid, transmits stress energy well, and fractures open up radial to the crater and also around the antipode (Fig. 4a). Figure 4b shows ejecta evolution 12 s after impact in this gravity-regime event. One difference between two-dimensional and three-dimensional hydrocode models is that the crater in three dimensions (for the identical gravity-regime projectile) is smaller than in two dimensions. The three-dimensional results show the gravity-regime impactor fragmenting barely enough rock to allow the flow field to open up unencumbered by strength. The discrepancy arises because axisymmetric models cannot relieve stress along radial cracks, resulting in a larger crater bowl. With strength-regime impactors too big and gravity-regime impactors too small, Stickney on Phobos may be intermediate between the classic scaling regimes.

A strongly heterogeneous Phobos was found to be incompatible with groove formation. By removing 30% of the SPH particles at random from a basalt (ρ = 2.7 g/cm^3) target, *Asphaug and Benz* (1994a) constructed a macroscopically porous, highly scattering target of the appropriate bulk density. The top frame of Fig. 4c shows seven "potato chip" slices, 2 km thick, through the final target interior of Fig. 4a. The bottom frame shows the same slices through the 30% porous body (the flecks of black are the interstitial voids). As porous targets are poor transmitters of impact stress, fragmentation is confined to the impacted hemisphere. Phobos cannot have had this kind of highly scattering interior if fracture stresses propagated coherently to the antipode.

With sufficient numerical resolution and time, one might tune the interior model of Phobos to yield the best possible agreement between crack morphologies, crater diameter, ejecta placement, and interior structure. That is the promise of crater profiling of asteroid interiors. At present we arrive at some preliminary conclusions: For either model, the velocities subsequent to impact are too low throughout Phobos to greatly reduce bulk density. If Phobos was not highly porous before Stickney (so that distal cracks could form), and if Stickney did not introduce significant porosity, then the low density of Phobos is compositional or microstructural in nature.

5.2. Ida

By the time *Galileo* encountered asteroid Ida (*Belton et al.,* 1996), model resolution had improved to the point that image-derived shapes could be used in impact simulations. Figure 5 shows an SPH model of Ida based upon the shape derived by *Thomas et al.* (1996): a highly irregular body with mean radius 15.7 km. As with Gaspra (*Carr et al.,* 1994) and Eros (*Prockter et al.,* 2000), this S-type asteroid exhibits expressions of internal shattering in the form of linear furrows almost certainly related to large collisions. Using this Ida model, *Asphaug et al.* (1996) attempted to recreate a few major craters including the ~12-km-diameter Vienna Regio (Fig. 5a). Figure 5b plots particle velocity through a slice of the target 9, 10, and 12 s after a gravity-scaled 330-m-diameter projectile strikes at 3.55 km/s [$\overline{v_{impact}}$ at Ida (*Bottke et al.,* 1994)]. Gravity scaling assumes constant gravity (in this case, the local effective gravity at the impact site) when in fact gravity varies by a factor of ~4 across the asteroid (*Thomas et al.,* 1996), and by a factor of ~2 over the region of this particular collision. Wave interference is complex in this irregular target, and potentially important focusing might occur if the asteroid is a homogeneous transmitter of compressive stress. The simulation (again using fracture and elastic constants for laboratory basalt, pending better alternatives) shows the opening of fractures in the same narrow opposite end of Ida (Pola Regio) where grooves are found. A heterogeneous or porous interior would have dissipated or scattered these stresses (*Asphaug and Benz,* 1994a), so this model outcome lends support to the hypothesis that the deep interior of Ida is competent and homogeneous, in agreement with its presumably low porosity [from its satellite-derived density of 2.6 ± 0.5 g/cm³ (*Belton et al.,* 1996)]. As with fractured Phobos, mechanical competence is needed only under compression. A well-connected interior is required, but tensile strength is not.

5.3. Eros

The *NEAR Shoemaker* mission to asteroid Eros, completed in 2001, represented a quantum leap for asteroid geology but left deep puzzles about the asteroid's interior. The mission was instrumented to determine asteroid composition, the only interior probe being the radio science experiment, which tracked gravitational influence on orbit

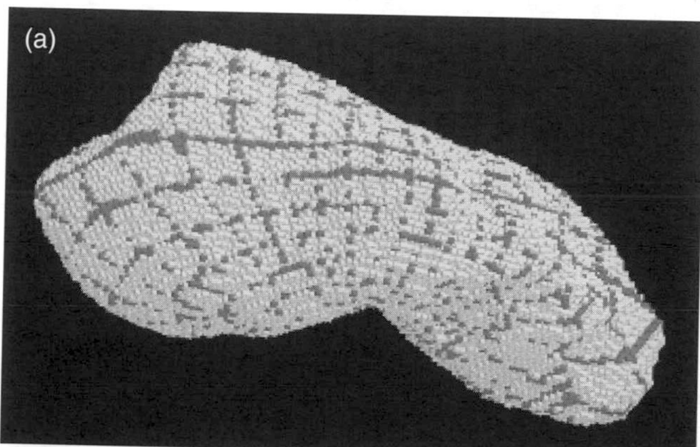

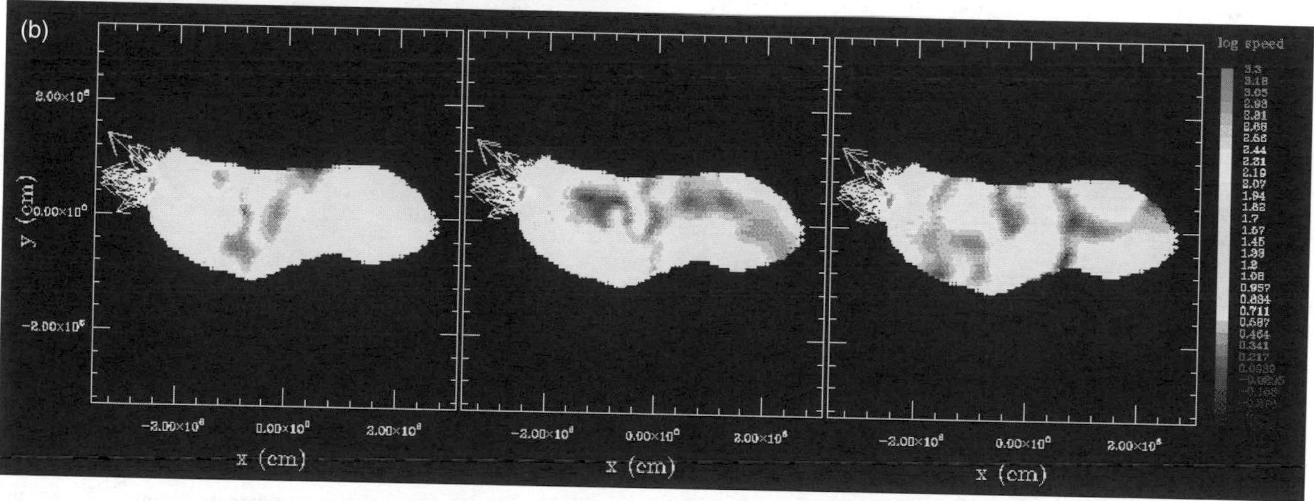

Fig. 5. **(a)** Smooth particle hydrodynamics (SPH) model of asteroid Ida, using the shape derived by *Thomas et al.* (1996) from *Galileo* imaging. This model was used by *Asphaug et al.* (1996) to recreate three major craters on Ida and a suite of smaller craters. Prior to the impact, Vienna Regio (~12-km crater on the upper left) was filled in and then impacted by a gravity-scaled impactor in an attempt to simulate the key aspects of its formation. Particles close to 10° grid lines are color coded in this view. **(b)** Particle velocity is shaded within a slice of the Ida target at t = 8, 9 and 12 s after the 330-m-diameter impactor strikes at 3.55 km/s to create Vienna Regio. Reflections from the complex shape lead to complex waveforms and stress wave interferences as can be seen in the dark zero-velocity nodes. This model resulted in distal, isolated fracture on the far narrow end where grooves are observed, and lent support to the hypothesis that the deep interior of Ida is a good transmitter of compressive stress.

(*Yeomans et al.,* 2000). This enabled determination of density and broad-frequency (approximately kilometer-scale) variations in density; of the latter it found none. Details of this mission are recounted in *Farquhar et al.* (2002); here we consider the implications of these results for asteroid interior structures. In composition and shape, and roughly in size, Eros is akin to Ida, also an S-type with bulk density ~2.7 g/cm^3. This is somewhat lower than the mean density of ordinary chondrites [the closest compositional analog according to *Trombka et al.* (2000)], so Eros is probably as nominally porous as any heavily fractured rock mass. Given its battered appearance (*Prockter et al.,* 2002), Eros' density is not a mystery.

Despite this evidence for disruption, a number of observations exhibit structural competence: the twisted planform of Eros, its clustered regions of high slopes (*Zuber et al.,* 2000), its long, continuous grooves (*Prockter et al.,* 2002), its polygonal craters (indicative of fault structure, just as at Meteor Crater in Arizona), and its subdued or missing crater rims (particularly at the low-gravity ends of the asteroid). The latter suggest crater formation in the strength-regime, where ejecta velocities are higher than v_{esc}. But geological competence is different from dynamical competence, and this has led to a healthy debate about structural nomenclature [perhaps now resolved; see *Richardson et al.* (2002)] and asteroid mechanics.

Nearly all agree the saddle-shaped depression Himeros is an impact crater. Crater fracture models (*Asphaug et al.,* 1996) and seismic profiles at Meteor Crater (*Ackerman et al.,* 1975) show that impact craters have major fractures extending about one crater radius beyond the crater in all directions. For Himeros, fractures would extend through the asteroid's narrow waist. If so, the asteroid is disconnected, possibly evidenced by the complex fault structure Rahe Dorsum, which strikes through Himeros. Dynamicists might then call Eros a "rubble pile" since it comes apart into pieces under modest tension (rotations or tides). To impact modelers, it might fall into that category as well, because a binary object can, like an assemblage of smaller rubble, absorb considerable collisions by trapping impact energy inside the impacted lobe (*Asphaug et al.,* 1998). To reconcile dynamical ideas, that the body might simply float apart if gravity were turned off, with geological ideas, that Eros exhibits considerable strength-controlled structure, we call this S asteroid a shattered monolith, and apply the same term to Ida based upon the modeling just presented.

Despite its irregular shape (see Plate 1) actual slopes on Eros, as on Ida, are moderate. Variations in the gravity vector (including centrifugal force) are extreme, largely canceling extreme variations in topography. One can quantify slopes by plotting a histogram of cumulative area steeper than a given angle (see *Zuber et al.,* 2000), although such histograms depend upon the measurement baseline: Geologic (fractal) surfaces become jagged at small scale. At 100 m baseline, ~2% of Eros is steeper than can be maintained by a pile of talus (see Table 2), and ~3–4% of the slopes are steeper than can be maintained by sand.

TABLE 2. Approximate angles of repose for common materials.

Material	Angle of Repose
Glass beads	~20°
Common unconsolidated materials (e.g., sand)	~33°
Steepest value for highly angular, poorly sorted rocks	~40°–50°
Water-rich soils	up to 90°

Is this evidence for or against structural control? One can argue both ways from the dataset. In order to have some slopes steeper than the presumed angle of repose, Eros must possess some cohesion at ~100-m scales. Conversely, the total area steeper than repose is less than a few percent, and highly localized. Plate 1, derived from the *NEAR Shoemaker* laser rangefinder (NLR) experiment (*Zuber et al.,* 2000), shows these local slopes plotted on a shape model of the asteroid. Almost all slopes exceeding ~30° lie inside the rims of craters, which on Eros (as on the Moon and Earth and other bodies) have slopes consistent with the angle of repose of unconsolidated rock.

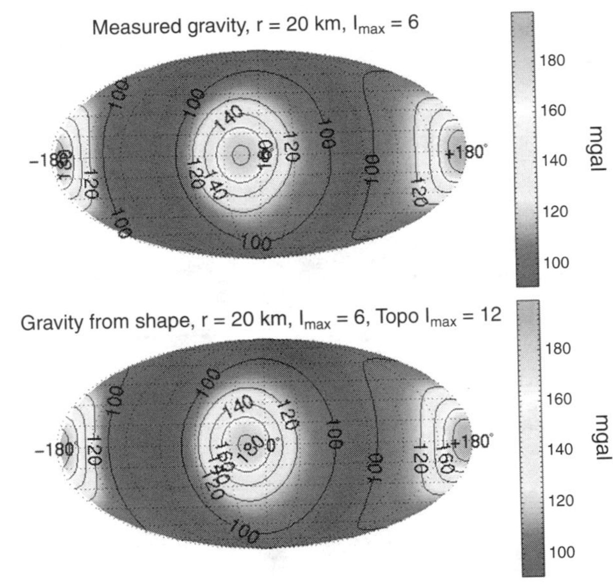

Fig. 6. The *NEAR Shoemaker* gravity experiment (*Yeomans et al.,* 2000) detected no sign of heterogeneity within Eros, as evidenced by these almost indistinguishable gravity maps. The upper figure shows gravity as determined by the radio science experiment, and the lower shows gravity computed from the NLR shape model (*Zuber et al.,* 2000) assuming uniform density. Gravity is only resolved to low degree and order by spacecraft Doppler tracking, corresponding to a resolution at scales of 1 km or larger. Gravity is also more sensitive to exterior than interior components. If Eros is of heterogeneous density, it must be well-mixed at kilometer scales.

The only interior probe of Eros was the radio science experiment that Doppler-tracked the spacecraft position and velocity (*Yeomans et al.,* 2000) and yielded a gravity map of the asteroid's interior. Because orbital speeds were only a few meters per second, intrinsic limits to Doppler accuracy resulted in coarse (approximately kilometer-scale) determination of interior density distribution. Moreover, most of the power in gravity variation is found in the asteroid's exterior, so that interior structure is not well characterized other than bulk density. Figure 6 shows the gravity field of Eros as measured by the spacecraft, plotted at 20 km radius from the center of mass, vs. the gravity field that is deduced from the NLR shape model, assuming a uniform bulk density. The variations are extremely slight, and can be accounted for by tens of meters of regolith thickness variations overlying a uniform substrate, or by a density gradient across the asteroid of 0.006 g cm^{-3} km^{-1}.

Another aspect of Eros is its highly fractured state (*Prockter et al.,* 2002; see also *Sullivan et al.,* 2002), which has been used to support the notion of geologic competence. This dataset also argues both ways. On the one hand it shows that the asteroid can support global fault structures. Conversely, in a dynamical sense, those global fault structures are probably what disconnect the asteroid. The distinction is critical, for as we shall see, an asteroid that is broken behaves very differently during impact than one that is not.

5.4. Mathilde

En route to Eros, *NEAR Shoemaker* encountered the primitive C-type asteroid 253 Mathilde and retrieved high-quality images despite limited solar power and challenging illumination (*Veverka et al.,* 1997) (see Fig. 7a). Mathilde is the first visited of those dark, primitive asteroids that dominate the main belt; it may be typical of unequilibrated or carbonaceous asteroids and perhaps representative of early planetesimals. Orbiting the Sun at a = 2.6 AU, Mathilde measures 66 × 48 × 46 km (best-fit ellipsoid). Its mass, determined by deflection of the spacecraft, is ~1.0×10^{20} g, yielding a bulk density of ρ ~ 1.3 g/cm^3, a low density that would be consistent with an icy composition except that groundbased astronomy detects no water on the surface (*Rivkin et al.,* 1997). Mathilde is a typically red C-type asteroid whose spectrum is best matched by primitive carbonaceous meteorites with densities over twice as great, and this has led to the idea of >50% porosity. Mathilde's rotation period is among the longest measured, P = 17.4 d, which is difficult to reconcile with its incredible cratering history: five to seven giant craters, at least four larger in diameter than the asteroid's mean radius, none exhibiting clear signs of structural degradation or overprinting by subsequent collisions, and no evidence for the impact fragmentation that is widespread on Phobos and Eros (*Thomas et al.,* 1999). Regarding the spin period, it is hard to believe (e.g., *Davis,* 1999) that the momentum contribution from all these impacts just happened to cancel out.

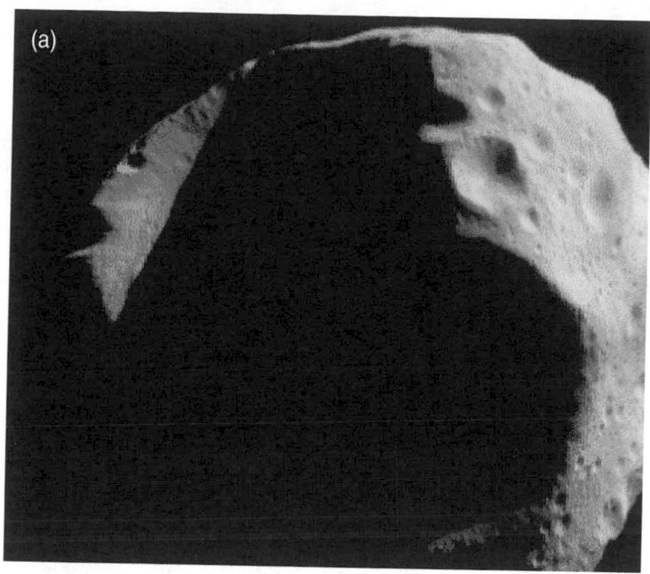

(b)

Nonporous Mathilde Impact:
1.2-km-diameter, 1.3-g/cm3 impactor,
at 5 km/s

Three-dimensional
Surface

Impact
Fractures

Fig. 7. **(a)** The second image mosaic of asteroid Mathilde (~66 × 48 × 46 km), the only C-type asteroid imaged to date, taken just prior to closest approach (*Veverka et al.,* 1997). Two vast craters span the foreground; there are ~5–7 craters of comparable size (*Thomas et al.,* 1999). Equally baffling are the observations that Mathilde's craters lack raised rims or any other evidence of ejecta deposition, and that Mathilde at ~5 km/pixel shows no sign of fracture grooves, impact disruption, or even rim-disturbance of preexisting craters. Each giant crater left no apparent trace of its formation other than an enormous cavity; we cannot tell which was the last to form. **(b)** A failed attempt using SPH to produce ~33-km-diameter Karoo Crater on Mathilde by introducing a 1.2-km-diameter impactor at 5 km/s (*Asphaug and Thomas,* 1999). The result is a sufficiently large fractured region, but also (as revealed in this side view) widespread impact fragmentation, and bulk ejecta flow velocities slow enough to be gravity-controlled. The result is a seriously fragmented asteroid covered in gravity-regime ejecta deposits — the opposite of what is observed.

Following *NEAR Shoemaker*'s determination of Mathilde's low density, a half dozen primitive asteroids have been found to have similar densities, computed from orbital periods of their newly discovered companion satellites (see *Merline et al.,* 2002). Once density is known, structure and composition are inseparable, since molecular weight then depends on the distribution of voids. Along with previous determinations of low density for the martian moons and Comets Halley and Shoemaker-Levy 9 (*Sagdeev et al.,* 1987; *Asphaug and Benz,* 1994b), these low C-type asteroid densities support a consensus that primitive small bodies are highly porous — rubble piles, primordial aggregates, or volatile-depleted residues. Whatever the origin of a body's porosity, it is significant to impact mechanics (*Trucano and Grady,* 1995) and the topic must be revisited.

The absence of ejecta deposits around a crater can signify formation in the strength regime. Despite earlier modeling to the contrary for a much smaller low-density body, Phobos (*Asphaug and Melosh,* 1993), a strength-regime crater model was attempted for Mathilde by *Asphaug and Thomas* (1998) and the result is shown in Fig. 7b. As in Fig. 4c for Phobos, a size-dependent strength was computed for Mathilde's volume, the largest crater Karoo was filled in, and a strength-scaled impactor (1.2 km diameter) was introduced at 5 km/s into the 1.3-g/cm^3 continuum. The result can be seen as a fractured region larger than Karoo, and also as widespread impact fragmentation all around the asteroid. Nevertheless, flow velocities within the resultant crater are slow enough to be gravity controlled, as they were for Stickney. The result is a globally fragmented asteroid covered in gravity-regime ejecta deposits — entirely the opposite of observation. A bold modeler might remedy this by making Mathilde stronger than basalt, but primitive meteorites are weak, often falling apart in one's hands.

5.4.1. Survival of the weakest. An alternative has emerged from these failed attempts: Perhaps Mathilde is too weak to be disrupted by collisions. This scenario (*Asphaug,* 1999) is borrowed from armament lore, and has roots in the experimental literature (*Love et al.,* 1993). *Davis* (1999), commenting upon Mathilde, recounts the use of porous cactus ribs to dissipate the energy of cannon balls colliding into the walls of Sonoran forts. Perhaps a similar heterogeneity or porosity has enabled Mathilde to survive each giant cratering event without any noticeable disturbance to its preexisting morphology.

The effects of structure and prefragmentation on collisional evolution was examined by *Asphaug et al.* (1998), following the work on asteroid porosity by *Asphaug and Benz* (1994a) and motivated by the experimental work of *Ryan et al.* (1991) and *Love et al.* (1993). For these simulations, shown in Plate 2a, the approximately kilometer-sized near-Earth asteroid Castalia (*Ostro et al.,* 1990) was rendered in a variety of ways: a monolith, a contact binary, and a rubble-pile. Each target was impacted by the same projectile (16 m diameter, 2.7 g/cm^3, 5 km/s) whose kinetic energy happened to equal the yield of one Hiroshima bomb. (This work was published the same week as the release of the Hollywood blockbuster *Deep Impact,* in which brave astronauts use nuclear devices to blow up a rogue comet, so the comparison was unfortunate.) This modeling showed what laboratory studies had hinted at: that monolithic asteroids are actually easier to disperse than rubble piles or gravitational aggregates. The contact binary (not shown) trapped impact energy in the impacted lobe, reflecting it at the discontinuity and preventing any severe perturbation of the distal lobe.

The same modeling technique was used to assemble Mathilde out of basalt spheres (Plate 2b) ranging in diameter from 0.5 to 3 km, just touching, with bulk porosity ~0.5 and bulk density ~1.3 g/cm^3, this time using the material density of basalt (2.7 g/cm^3). Resolving shock waves in each sphere requires a resolution of ~1000 particles per component, requiring supercomputer modification of the code. This places approximately six particles across each component's radius. Because numerical shock waves cannot be treated with fewer than three zones (*von Neumann and Richtmyer,* 1950), that is probably a safe minimum for such simulations. The contact portion of touching spheres is replaced with damaged rock of slightly lower material density (1.7 g/cm^3) and can be seen as red flecks in the bottom of the center figure. Damaged rock cannot support tensile or shear stress, so the modeled Mathilde is just like a contact binary, only with thousands of components. The size of the component spheres is established by the resolution requirement of modeling each individual sphere by ~1000 particles, not by any guiding philosophy of rubble pile structure.

A shock wave can propagate freely through a rubble pile so long as it melts, vaporizes, or collapses pores in the rock it encounters. But shocks attenuate rapidly even in competent rock (*Rodionov et al.,* 1972), and even the most powerful elastic wave has difficulty propagating in an aggregate. Stress energy is trapped, and as a result the impact energy is confined to a local region (the shattered region colored red in the central figure of Plate 2b), so this receives all the energy that would have otherwise been transmitted to distant regions.

Ejection velocities (left) in the damaged region (middle) are therefore greater due to this energy confinement. Indeed, the nonescaping fraction (right) equals all the damaged rock. The conclusion is that structural porosity can lead to stalled shocks, and hence crater ejecta in a weak asteroid, just as in a strength-controlled asteroid, might never return to the asteroid. Furthermore, because almost no stress energy propagates beyond the crater bowl, preexisting craters on Mathilde would not feel the occurrence of subsequent large impacts, and would preserve their original forms, as observed in the *NEAR Shoemaker* images.

5.4.2. Compaction cratering. An alternative model for cratering on Mathilde has very different implications for asteroid structure and evolution. *Housen et al.* (1999) propose that Mathilde is so underdense that craters form by crushing rather than ejection. In their model no ejecta leaves the crater — just opposite the previous scenario in which

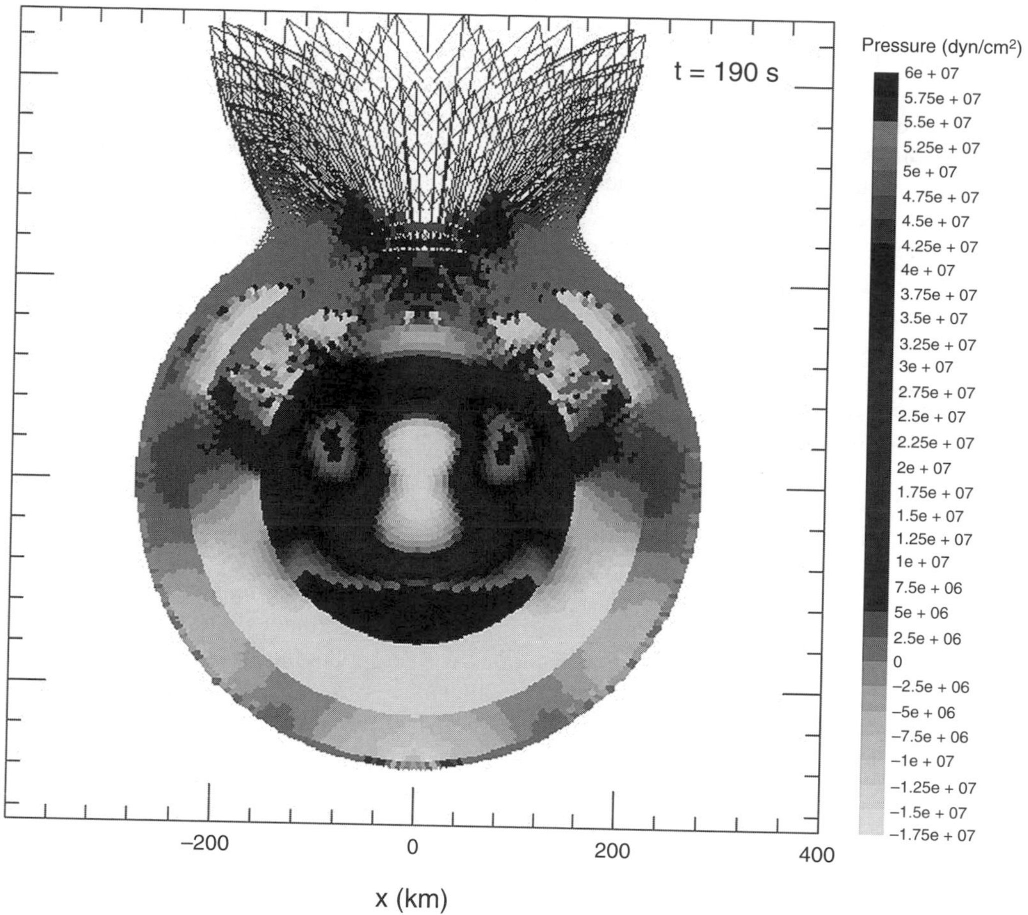

Fig. 8. An impact into the differentiated ~530-km-diameter asteroid Vesta using the two-dimensional hydrocode of *Melosh et al.* (1992). In this calculation a 34-km-diameter impactor strikes at 8 km/s in an attempt to reproduce the hemispheric crater on Vesta and the V-type asteroids (*Asphaug,* 1997; *Keil,* 2002). It turns out to be relatively easy to launch crustal meteorites from parent bodies at high speed, whereas excavation of interior Fe requires truly gargantuan events. Interestingly, much of the stress energy (shown here three minutes after impact as the primary wave reflects from the core/mantle boundary) is trapped in the asteroid's core, reverberating for the remainder of the calculation.

all ejecta leaves the crater. Under compaction cratering, an asteroid crushes up to higher density over time, leaving mass concentrations at each crater floor. Compaction cratering, if it occurs, would have facilitated planetary growth, as it allows primitive asteroids to accrete material with ease. That is also its problem, however, given Mathilde's near-absence of spin. In a regime of perfectly inelastic collisions (see *Agnor et al.,* 1999) accreting bodies almost always begin to spin with periods of hours, not tens of days.

6. LARGER ASTEROIDS

Little has been said in this chapter about larger asteroids, those that have differentiated into cores, mantles, and crusts. For these, meteoriticists would like impact modelers to help them remove their mantles so as to expose their iron cores and deliver iron meteorites to Earth, while at the same time getting rid of mantle material not evident in the meteorite collection (see *Burbine et al.,* 2002). This has proven to be a difficult prospect.

While large asteroid interiors are treated in other chapters in this book (*McSween et al.,* 2002; *Keil,* 2002), one aspect is worth mentioning in passing — that cores can trap impact stress energy in a manner reminiscent of how contact binaries trap energy in their impacted lobe. Figure 8 shows an impact simulation where asteroid Vesta (*Asphaug,* 1997) is modeled using the two-dimensional hydrocode of *Melosh et al.* (1992) to resolve a basaltic crust, a denser mantle, and an iron core. The impact (an attempt to recreate the hemispheric crater on Vesta and the V-type asteroids) transmits a shock deep into the asteroid, where it reverberates throughout the course of the calculation.

7. SEISMIC AND RADAR IMAGING OF ASTEROID INTERIORS

Geophysical exploration tools that are commonly deployed on Earth can be flown on spacecraft to image asteroid interiors directly. Many are designed for field deployment in remote areas and are compact and lightweight.

Radio reflection tomography (ground-penetrating radar or GPR) is commonly used to image sinkholes, pipes, burial grounds, etc.; similar imaging is in principle obtainable on asteroids, although the instruments would probably have to be deployed from orbit, reducing resolution and posing challenges with regard to data inversion and echo noise. The European Space Agency's *Rosetta* spacecraft will use transmission radio tomography (*Kofman et al.,* 1998) to investigate the interior of Comet Wirtanen during its 2011–2016 rendezvous as the comet approaches perihelion; in this mode of exploration radio energy is transmitted from a lander and received by the orbiter. Another popular field-deployable tool, magnetotelluric imaging, probes subsurface geology using the fact that the magnetic to electric-field ratio (impedance) is constant at given frequency for constant resistivity. On Earth this technique takes advantage of natural fluctuations in the background magnetic field, whereas on an asteroid field generation would require the deployment of antennae across the surface.

Certain asteroid materials (clays, rocks flecked with metal) may be opaque to electromagnetic energy and might be better explored by other means. *Deep Impact,* a NASA Discovery mission, will pioneer the technique of kinetically blasting holes in small bodies [*Belton and A'Hearn* (1999); see *Farquhar et al.* (2002) for an overview of recent and planned small-body missions]. In July 2005, *Deep Impact* will slam ~350 kg of copper into Comet Tempel 1 at ~10 km/s for the purpose of investigating outer crust and mantle physical and compositional properties.

A more refined technique, more complex to deploy, is seismic imaging. This has taught us the detailed structure of Earth, and more recently has been refined for small-scale imaging (e.g., *Wu and Yang,* 1997). Seismic imaging may prove to be an ideal complement to electromagnetic imaging of asteroid interiors, since its spatial resolution can be comparable but its means of data acquisition and inversion are entirely distinct. In some cases where seismic imaging is challenging (highly attenuative porous bodies such as comets) radio imaging may be optimal, and perhaps vice versa.

Seismic imaging requires surface probes. In one scenario instrumented penetrators detect signals broadcast from cratering grenades. Blasts produce white noise and are not optimal for tomographic data inversion (compared with drills or thumpers), but they are convenient, reliable, and cheap. They also produce small-scale cratering experiments as a bonus, enabling the imaging and spectroscopy of shallow layers. Alternatively, and at much lower cost, an armada of ballistic penetrators could be deployed to strike an asteroid in the manner that the pieces of Comet Shoemaker-Levy 9 struck Jupiter, hitting it one after another, with each embedded penetrator acquiring seismic reverberations from successive impacts at diverse locations.

One can deploy, at the NASA Discovery level, a dual-wavelength radar tomography mission pursuing multiple rendezvous with a variety of near-Earth objects (*Asphaug et al.,* 2001b). This population is fairly representative of the asteroid population at large, and may include dormant comet nuclei (see *Morbidelli et al.,* 2002; *Weissman et al.,* 2002). Anchored seismology landers would be too costly, given the surface material uncertainties. Spring-fired grenades (a precursor to seismic studies) would blast several small (~5-m) craters at selected sites on each asteroid or comet nucleus. Ejecta ballistics and crater formation filmed from orbit at high time and spatial resolution would facilitate the development of comet- and asteroid-analog simulation chambers at impact research laboratories. Together with long-term imaging of ejecta orbital evolution, this would greatly reduce uncertainties and design parameters for future landed spacecraft.

Until we proceed with direct geophysical exploration of asteroids and comets, our understanding of their interiors shall remain a matter of educated speculation. Science aside, there are practical reasons to learn more in the short term. To divert or disrupt a potentially hazardous near-Earth asteroid, one must understand its internal structure, as porous or discontinuous asteroids can absorb or divert disruptive energy. Composite targets appear to be able to sacrifice small regions near an impact or explosion without perturbing disconnected regions. Monoliths, rubble piles, and porous ice-dust mixtures might each require a different mode of diversion, disruption, or resource exploitation (*Huebner and Greenberg,* 2000). Except for the mean densities of a dozen asteroids and the broadly resolved homogeneous mass distribution of Eros, we know none of the basic bulk constitutive properties of any asteroid. Obviously the modeling presented here is fraught with guesswork. Lacking such knowledge, a standoff blast (*Ahrens and Harris,* 1992) might fail to impart the expected momentum or might disrupt a body without diverting it, sending a cluster of fragments toward Earth. Nonnuclear scenarios (*Melosh et al.,* 1994) are greatly preferred, but these also require detailed awareness of asteroid geology and composition for the purposes of anchoring, momentum loading, and resource extraction.

8. CONCLUSIONS

Asteroids include about a million objects between ~100 m and ~100 km across. Many have satellites (see *Merline et al.,* 2002), and they have long served as test particles for understanding the detailed evolution of planetary dynamics (see *Gladman et al.,* 1997). Their geophysics is a complex interplay between long- and short-range forces (self-gravity and mechanical cohesion), making Earth-based intuition a fickle guide, yet until we peer inside an asteroid directly we can only strive to interpret their exterior geology.

We presently believe that most asteroids larger than ~1 km are gravitational aggregates. Highly tentative is our conclusion that Gaspra, Ida, Eros (all S-types), and perhaps Phobos are shattered monoliths sufficiently competent to transmit compressive stress. As for Mathilde and other low-density primitive bodies, detailed explanations diverge, but they appear to require a mechanically uncoupled interior, either structural or microstructural. Their porosity may be primordial or the aftermath of collisional evolution; it helps them withstand the gargantuan collisions that have been the norm in the evolving solar system.

There is much to learn. Our present state of knowledge comes as no surprise, for complex characteristics are expected of objects at the boundary between primary forces of nature.

REFERENCES

Agnor C. B., Canup R. M., and Levison H. F. (1999) On the character and consequences of large impacts in the late stage of terrestrial planet formation. *Icarus, 142,* 219–237.

Ahrens T. J. and Harris A. W. (1992) Deflection and fragmentation of near-earth asteroids. *Nature, 360,* 429–433.

Asphaug E. (1993) Dynamic fragmentation in the solar system. Ph.D. thesis, Univ. of Arizona, Tucson.

Asphaug E. (1997) Impact origin of the Vesta family. *Meteoritics & Planet. Sci., 32,* 965–980.

Asphaug E. (1999) Survival of the weakest. *Nature, 402,* 127–128.

Asphaug E. and Benz W. (1994a) The surface and interior of Phobos (abstract). In *Lunar and Planetary Science XXV,* pp. 43–44. Lunar and Planetary Institute, Houston.

Asphaug E. and Benz W. (1994b) Density of Comet Shoemaker-Levy-9 deduced by modelling breakup of the parent rubble pile. *Nature, 370,* 120–124.

Asphaug E. and Benz W. (1996) Size, density, and structure of Comet Shoemaker-Levy 9 inferred from the physics of tidal breakup. *Icarus, 121,* 225–248.

Asphaug E. and Melosh H. J. (1993) The Stickney impact of Phobos: A dynamical model. *Icarus, 101,* 144–164.

Asphaug E. and Thomas P. C. (1999) Modeling mysterious Mathilde (abstract). In *Lunar and Planetary Science XXX,* Abstract #2028. Lunar and Planetary Institute, Houston (CD-ROM).

Asphaug E., Moore J. M., Morrison D., Benz W., Nolan M. C., and Sullivan R. J. (1996) Mechanical and geological effects of impact cratering on Ida. *Icarus, 120,* 158–184.

Asphaug E., Ostro S. J., Hudson R. S., Scheeres D. J., and Benz W. (1998) Disruption of kilometre-sized asteroids by energetic collisions. *Nature, 393,* 437–440.

Asphaug E., King P. J., Swift M. R., and Merrifield M. R. (2001a) Brazil nuts on Eros: Size-sorting of asteroid regolith (abstract). In *Lunar and Planetary Science XXXII,* Abstract #1708. Lunar and Planetary Institute, Houston (CD-ROM).

Asphaug E., Belton M. J. S., and Kakuda R. Y. (2001b) Geophysical exploration of asteroids: The Deep Interior Mission concept (abstract). In *Lunar and Planetary Science XXXII,* Abstract #1867. Lunar and Planetary Institute, Houston (CD-ROM).

Avanesov G.A. and 39 colleagues (1989) Television observations of Phobos. *Nature, 341,* 585–587.

Belton M. J. S. and A'Hearn M. F. (1999) Deep sub-surface exploration of cometary nuclei. *Adv. Space Res., 24,* 1175–1183.

Belton M. J. S. and 20 colleagues (1996) The discovery and orbit of 1993 (243)1 Dactyl. *Icarus, 120,* 185–199.

Benz W. (2000) Low velocity collisions and the growth of planetesimals. *Space Sci. Rev., 92,* 279–294.

Benz W. and Asphaug E. (1994) Impact simulations with fracture: I. Method and tests. *Icarus, 107,* 98–116.

Benz W. and Asphaug E. (1995) Simulations of brittle solids using smooth particle hydrodynamics. *Comput. Phys. Commun., 87,* 253–265.

Benz W. and Asphaug E. (1999) Catastrophic disruptions revisited. *Icarus, 142,* 5–20.

Binzel R. P. and Xu S. (1993) Chips off of asteroid 4 Vesta — Evidence for the parent body of basaltic achondrite meteorites. *Science, 260,* 186–191.

Bottke W. F., Nolan M. C., Greenberg R., and Kolvoord R. A. (1994) Collisional lifetimes and impact statistics of near-Earth asteroids. In *Hazards Due to Comets and Asteroids* (T. Gehrels et al., eds.), pp. 337–357. Univ. of Arizona, Tucson.

Bottke W. F. Jr., Vokrouhlický D., Rubincam D. P., and Brož M. (2002) The effect of Yarkovsky thermal forces on the dynamical evolution of asteroids and meteoroids. In *Asteroids III* (W. F. Bottke Jr. et al., eds.), this volume. Univ. of Arizona, Tucson.

Bridges F. G., Supulver K. D., Lin D. N. C., Knight R., and Zafra M. (1996) Energy loss and sticking mechanisms in particle aggregation in planetesimal formation. *Icarus, 123,* 422–435.

Britt D. T. and Consolmagno G. J. (2001) Modeling the structure of high porosity asteroids. *Icarus, 152,* 134–139.

Bunch T. E. and Rajan R. S. (1988) Meteorite regolith breccias. In *Meteorites and the Early Solar System* (J. F. Kerridge and M. S. Matthews, eds.), pp. 144–164. Univ. of Arizona, Tucson.

Burbine T. H., Meibom A., and Binzel R. P. (1996) Mantle material in the main belt: Battered to bits? *Meteoritics & Planet. Sci., 31,* 607–620.

Burbine T. H., McCoy T. J., Meibom A., Gladman B., and Keil K. (2002) Meteoritic parent bodies: Their number and identification. In *Asteroids III* (W. F. Bottke Jr. et al., eds.), this volume. Univ. of Arizona, Tucson.

Carr M. H., Kirk R. L., McEwen A., Veverka J., Thomas P., Head J. W., and Murchie S. (1994) The geology of Gaspra. *Icarus, 107,* 61.

Cellino A., Michel P., Tanga P., Zappalà V., Paolicchi P., and Dell'Oro A. (1999) The velocity-size relationship for members of asteroid families and implications for the physics of catastrophic collisions. *Icarus, 141,* 79–95.

Chapman C. R. (1976) Asteroids as meteorite parent bodies: The astronomical perspective. *Geochim. Cosmochim. Acta, 40,* 701–719.

Chapman C. R. (2001) Eros at very high resolution: Meteoritical implications (abstract). *Meteoritics & Planet. Sci., 36,* A39.

Chapman C. R. and Davis D. R. (1975) Asteroid collisional evolution — Evidence for a much larger early population. *Science, 190,* 553.

Chapman C. R., Paolicchi P., Zappala V., Binzel R. P., and Bell J. F. (1989) Asteroid families: physical properties and evolution. In *Asteroids II* (R. P. Binzel et al., eds.), pp. 386–415. Univ. of Arizona, Tucson.

Davis D. R. (1999) The collisional history of asteroid 253 Mathilde. *Icarus, 140,* 49–52.

Davis D. R., Chapman C. R., Greenberg R., Weidenschilling S. J., and Harris A. W. (1979) Collisional Evolution of asteroids: Populations, rotations and velocities. In *Asteroids* (T. Gehrels, ed.), pp. 528–557. Univ. of Arizona, Tucson.

Davis D. R., Chapman C. R., Weidenschilling S. J., and Greenberg R. (1985) Collisional history of asteroids: Evidence from Vesta and Hirayama families. *Icarus, 62,* 30–53.

Dobrovolskis A. R. and Burns J. A. (1984) Angular momentum drain — A mechanism for despinning asteroids. *Icarus, 57,* 464–476.

Drake M. J. (1979) Geochemical evolution of the eucrite parent body: Possible nature and evolution of asteroid 4 Vesta? In *Asteroids* (T. Gehrels, ed.), pp. 765–782. Univ. of Arizona, Tucson.

Durda D. D., Greenberg R., and Jedicke R. (1998) Collisional models and scaling laws: A New interpretation of the shape of the main-belt asteroid size distribution. *Icarus, 135,* 431–440.

Fanale F. P. and Salvail J. R. (1990) Evolution of the water regime of Phobos. *Icarus, 88,* 380–395.

Farinella P., Paolicchi P, and Zappalà V. (1982) The asteroids as outcomes of catastrophic collisions. *Icarus, 52,* 409–433.

Farquhar R., Kawaguchi J., Russell C., Schwehm G., Veverka J., and Yeomans D. (2002) Spacecraft exploration of asteroids: The 2001 perspective. In *Asteroids III* (W. F. Bottke Jr. et al., eds.), this volume. Univ. of Arizona, Tucson.

Flynn G. J., Moore L. B., and Klöck W. (1999) Density and porosity of stone meteorites: Implications for the density, porosity, cratering, and collisional disruption of asteroids. *Icarus, 142,* 97–105.

Fujiwara A. (1980) On the mechanism of catastrophic destruction of minor planets by high-velocity impact. *Icarus, 41,* 356–364.

Fujiwara A. (1991) Stickney-forming impact on Phobos: Crater shape and induced stress distribution. *Icarus, 89,* 384–391.

Fujiwara A. and Tsukamoto A. (1980) Experimental study on the velocity of fragments in collisional breakup. *Icarus, 44,* 142–153.

Fujiwara A., Cerroni P., Davis D. R., Ryan E. V., Di Martino M., Holsapple K., and Housen K. (1989) Experiments and scaling laws for catastrophic collisions. In *Asteroids II* (R. P. Binzel et al., eds.), pp. 240–265. Univ. of Arizona, Tucson.

Gault D. E. and Heitowit E. D. (1963) The partition of energy for hypervelocity impact craters formed in rock. *Proc. 6th Hypervelocity Impact Symposium, Vol. 2,* pp. 419–456.

Gault D. E., Horz F., and Hartung J. B. (1972) Effects of microcratering on the lunar surface. *Proc. Lunar Sci. Conf. 3rd,* pp. 2713–2734.

Gladman B. J., Migliorini F., Morbidelli A., Zappalà V., Michel P., Cellino A., Froeschlé C., Levison H. F., Bailey M., and Duncan M. (1997) Dynamical lifetimes of objects injected into asteroid belt resonances. *Science, 277,* 197–201.

Grady D. E. and Kipp M. E. (1980) Continuum modeling of explosive fracture in oil shale. *Intl. J. Rock Mech. Min. Sci. Geomech. Abstr., 17,* 147–157.

Grady D. E. and Lipkin J. (1980) Criteria for impulsive rock fracture. *Geophys. Res. Lett., 7,* 255–258.

Greenberg R. and Nolan M. C. (1989) Delivery of asteroids and meteorites to the inner solar system. In *Asteroids II* (R. P. Binzel et al., eds.), pp. 778–804. Univ. of Arizona, Tucson.

Griffith A. A. (1920) The phenomena of rupture and flow in solids. *Philos. Trans. R. Soc. London, A221,* 163–198.

Harris A. W. (1996) The rotation rates of very small asteroids: Evidence for "rubble pile" structure (abstract). In *Lunar and Planetary Science XXVII,* p. 493. Lunar and Planetary Institute, Houston.

Head J. N. and Melosh H. J. (2000) Launch velocity distribution of the martian clan meteorites (abstract). In *Lunar and Planetary Science XXXI,* Abstract #1937. Lunar and Planetary Institute, Houston (CD-ROM).

Hirayama K. (1923) Families of asteroids. *Ann. Tokyo Astron. Obs., 11,* 55.

Holsapple K. A. (1994) Catastrophic disruptions and cratering of solar system bodies: A review and new results. *Planet. Space Sci., 42,* 1067–1078.

Holsapple K. A. (2002) Equilibrium configurations of solid cohesionless bodies. *Icarus,* in press.

Holsapple K. A. and Housen K. R. (1986) Scaling laws for the catastrophic collisions of asteroids. *Mem. Soc. Astron. Ital., 57,* 65–85.

Holsapple K., Giblin I., Housen K., Nakamura A., and Ryan E. (2002) Asteroid impacts: Laboratory experiments and scaling laws. In *Asteroids III* (W. F. Bottke Jr. et al., eds.), this volume. Univ. of Arizona, Tucson.

Housen K. R. and Holsapple K. A. (1990) On the fragmentation of asteroids and planetary satellites. *Icarus, 84,* 226–253.

Housen K. R. and Holsapple K. A. (1999) Scale effects in stength-dominated collisions of rocky asteroids. *Icarus,* in press.

Housen K. R., Wilkening L. L., Chapman C. R., and Greenberg R. (1979) Asteroidal regoliths. *Icarus, 39,* 317–351.

Housen K. R., Schmidt R. M., and Holsapple K. A. (1983) Crater ejecta scaling laws: Fundamental forms based on dimensional analysis. *J. Geophys. Res., 88,* 2485–2499.

Housen K. R., Holsapple K. A., and Voss M. E. (1999) Compaction as the origin of the unusual craters on the asteroid Mathilde. *Nature, 402,* 155–157.

Horstman K. C. and Melosh H. J. (1989) Drainage pits in cohesionless materials: Implications for the surface of Phobos. *J. Geophys. Res., 94,* 12433–12441.

Huebner W. F. and Greenberg J. M. (2000) Needs for determining material strengths and bulk properties of NEOs. *Planet. Space Sci., 48,* 797–799.

Jaeger J. C. and Cook N. G. W. (1969) *Fundamentals of Rock Mechanics.* Chapman and Hall, London. 515 pp.

Jaeger H. M., Nagel S. R., and Behringer R. P. (1996) The physics of granular materials. *Phys. Today, 49,* 32–39.

Jiongming S., Binglu S., and Bin W. (1998) Dynamics of size segregation of granular materials by shaking. *Nuovo Cimento, 20,* 1443.

Keil K. (2002) Geological history of asteroid 4 Vesta: The "smallest terrestrial planet." In *Asteroids III* (W. F. Bottke Jr. et al., eds.), this volume. Univ. of Arizona, Tucson.

Keil K., Haack H., and Scott E. R. D. (1994) Catastrophic fragmentation of asteroids: Evidence from meteorites. *Planet. Space Sci., 42,* 1109–1122.

Kofman W. and 20 colleagues (1998) Comet nucleus sounding experiment by radiowave transmission. *Adv. Space Res., 21,* 1589–1598.

Lange M. A. and Ahrens T. J (1983) The dynamic tensile strength of ice and ice-silicate mixtures. *J. Geophys. Res., 88,* 1197–1208.

Lawn B. R. and Wilshaw T. R. (1975) *Fracture of Brittle Solids.* Cambridge Univ., Cambridge. 160 pp.

Lee P. C. (1996) Dust levitation on asteroids. *Icarus, 124,* 181–194.

Leinhardt Z. M., Richardson D. C., and Quinn T. (2000) Direct N-body simulations of rubble pile collisions. *Icarus, 146,* 133–151.

Lindholm U. S., Yeakley L. M., and Nagy A. (1974) The dynamic strength and fracture properties of Dressler basalt. *Intl. J. Rock Mech. Min. Sci. Geomech. Abstr., 11,* 181–191.

Love S. G. and Ahrens T. J. (1996) Catastrophic impacts on gravity dominated asteroids. *Icarus, 124,* 141–155.

Love S. G., Hörz F., and Brownlee D. E. (1993) Target porosity effects in impact cratering and collisional disruption. *Icarus, 105,* 216–224.

Martelli G., Ryan E. V., Nakamura A. M., and Giblin I. (1994) Catastrophic disruption experiments: Recent results. *Planet. Space Sci., 42,* 1013–1026.

McSween H. Y. (1999) *Meteorites and Their Parent Planets,* 2nd edition. Cambridge Univ., New York. 322 pp.

McSween H. Y. Jr., Ghosh A., Grimm R. E., Wilson L., and Young E. D. (2002) Thermal evolution models of asteroids. In *Asteroids III* (W. F. Bottke Jr. et al., eds.), this volume. Univ. of Arizona, Tucson.

Melosh H. J. and Ryan E. V. (1997) Asteroids: Shattered but not dispersed. *Icarus, 129,* 562–564.

Melosh H. J., Ryan E. V., and Asphaug E. (1992) Dynamic fragmentation in impacts: Hydrocode simulation of laboratory impacts. *J. Geophys. Res., 97,* 14735–14759.

Melosh H. J., Nemchinov I. V., and Zetzer Yu. I. (1994) Non-nuclear strategies for deflecting comets and asteroids. In *Hazards Due to Comets and Asteroids* (T. Gehrels, ed.), pp. 1111–1132. Univ. of Arizona, Tucson.

Merline W. J., Weidenschilling S. J., Durda D. D., Margot J.-L., Pravec P., and Storrs A. D. (2002) Asteroids *do* have satellites. In *Asteroids III* (W. F. Bottke Jr. et al., eds.), this volume. Univ. of Arizona, Tucson.

Morbidelli A., Bottke W. F. Jr., Froeschlé Ch., and Michel P. (2002) Origin and evolution of near-Earth objects. In *Asteroids III* (W. F. Bottke Jr. et al., eds.), this volume. Univ. of Arizona, Tucson.

Nolan M. C., Asphaug E., Melosh H. J., and Greenberg R. (1996) Impact craters on asteroids: Does gravity or strength control their size? *Icarus, 124,* 359–371.

Ostro S. J., Chandler J. F., Hine A. A., Rosema K. D., Shapiro I. I., and Yeomans D. K. (1990) Radar images of asteroid 1989 PB. *Science, 248,* 1523–1528.

Ostro S. J., Pravec P., Benner L. A. M., Hudson R. S., Sarounova L., Hicks M. D., Rabinowitz D. L., Scotti J. V., Tholen D. J., Wolf M., Jurgens R. F., Thomas M. L., Giorgini J. D., Chodas P. W., Yeomans D. K., Rose R., Frye R., Rosema K. D., Winkler R., and Slade M. A. (1999) Radar and optical observations of asteroid 1998 KY26. *Science, 285,* 557–559.

Ostro S. J., Hudson R. S., Benner L. A. M., Giorgini J. D., Magri C., Margot J.-L., and Nolan M. C. (2002) Asteroid radar astronomy. In *Asteroids III* (W. F. Bottke Jr. et al., eds.), this volume. Univ. of Arizona, Tucson.

Perret W. R., Rutter R. L., Millsap F. K., Thornbrough A. D., and Hansen G. H. (1967) *Free-Field Particle Motion from a Nuclear Explosion in Salt, Part I, Project Dribble, Salmon Event.* Sandia Rept. VUF-3012, Sandia Laboratory.

Pravec P. and Harris A. W. (2000) Fast and slow rotation of asteroids. *Icarus, 148,* 12–20.

Pravec P., Harris A. W., and Michałowski T. (2002) Asteroid rotations. In *Asteroids III* (W. F. Bottke Jr. et al., eds.), this volume. Univ. of Arizona, Tucson.

Prockter L., Thomas P., Robinson M., Joseph J., Milne A., Bussey B., Veverka J., and Cheng A. (2002) Surface expressions of structural features on Eros. *Icarus, 155,* 75–93.

Richardson D. C., Leinhardt Z. M., Melosh H. J., Bottke W. F. Jr., and Asphaug E. (2002) Gravitational aggregates: Evidence and evolution. In *Asteroids III* (W. F. Bottke Jr. et al., eds.), this volume. Univ. of Arizona, Tucson.

Rinehart J. S. (1965) Dynamic fracture strength of rocks. *Proc. 7th Symposium on Rock Mechanics, Vol. 1,* pp. 205–208.

Rivkin A. S., Clark B. E., Britt D. T., and Lebofsky L. A. (1997) Infrared spectrophotometry of the NEAR flyby target 253 Mathilde. *Icarus, 127,* 255–257.

Rivkin A. S., Howell E. S., Vilas F., and Lebofsky L. A. (2002) Hydrated minerals on asteroids: The astronomical record. In *Asteroids III* (W. F. Bottke Jr. et al., eds.), this volume. Univ. of Arizona, Tucson.

Rodionov V. N., Adushkin V. V., Kostyuchenko V. N., Nikolaevskii V. N., Romashov A. N., Sadovskii M. A., and Tsvetkov V. M. (1972) *Mechanical Effect of an Underground Explosion.* U.S. Atomic Energy Commission, Los Alamos.

Rubincam D. P. (2000) Radiative spin-up and spin-down of small asteroids. *Icarus, 148,* 2–11.

Ryan E. V. (1992) Catastrophic collisions: Laboratory impact experiments, hydrocode simulations, and the scaling problem. Ph.D. thesis, Univ. of Arizona, Tucson.

Ryan E. V. and Melosh H. J. (1998) Impact fragmentation: From the laboratory to asteroids. *Icarus, 133,* 1–24.

Ryan E.V., Hartmann W. K., and Davis D. R. (1991) Impact experiments III — Catastrophic fragmentation of aggregate targets and relation to asteroids. *Icarus, 94,* 283–298.

Sagdeev R. Z., Elaysberg P. E., and Moroz V. I. (1987) An estimate of the mass and density of the Comet Halley nucleus. *Pisma Astron. Zh., 13,* 621–629.

Scott E. R. D., Taylor G. J., Newsom H. E., Herbert F., and Zolensky M. (1989) Chemical, thermal and impact processing of asteroids. In *Asteroids II* (R. P. Binzel et al., eds.), pp. 701–739. Univ. of Arizona, Tucson.

Sears D. W. G. and Akridge D. G. (1998) Nebular or parent body alteration of chondritic material: Neither or both? *Meteoritics & Planet. Sci., 33,* 1157–1167.

Sekanina Z., Chodas P. W., and Yeomans D. K. (1994) Tidal disruption and the appearance of periodic comet Shoemaker-Levy 9. *Astron. Astrophys., 289,* 607–636.

Stewart S. T., Ahrens T. J., and Lange M. A. (1999) Correction to the dynamic tensile strength of ice and ice-silicate mixtures (Lange and Ahrens 1983) (abstract). In *Lunar and Planetary Science XXX,* Abstract #2037. Lunar and Planetary Institute, Houston (CD-ROM).

Sullivan R. J., Thomas P. C., Murchie S. L., and Robinson M. S. (2002) Asteroid geology from Galileo and NEAR Shoemaker data. In *Asteroids III* (W. F. Bottke Jr. et al., eds.), this volume. Univ. of Arizona, Tucson.

Thomas P. C. (1998) Ejecta emplacement on the martian satellites. *Icarus, 131,* 78–106.

Thomas P., Veverka J., Bloom A., and Duxbury T. (1979) Grooves on Phobos: Their distribution, morphology, and possible origin. *J. Geophys. Res., 84,* 8457–8477.

Thomas P. C., Belton M. J. S., Carcich B., Chapman C. R., Davies M. E., Sullivan R., and Veverka J. (1996) The shape of Ida. *Icarus, 120,* 20–32.

Thomas P. C. and 11 colleagues (1999) Mathilde: Size, shape, and geology. *Icarus, 140,* 17–27.

Thomas P. C., Veverka J., Sullivan R., Simonelli D. P., Malin M. C., Caplinger M., Hartmann W. K., and James P. B. (2000) Phobos: Regolith and ejecta blocks investigated with Mars Orbiter Camera images. *J. Geophys. Res., 105,* 15091–15106.

Thomas P. C., Veverka J., Robinson M. S., and Murchie S. (2001) Shoemaker crater as the source of most ejecta blocks on the asteroid 433 Eros. *Nature, 413,* 394–396.

Tonks W. B. and Melosh H. J. (1992) Core formation by giant impacts. *Icarus, 100,* 326–346.

Trombka, J. I. and 23 colleagues (2000) The elemental composition of asteroid 433 Eros: Results of the NEAR Shoemaker X-ray spectrometer. *Science, 289,* 2101–2105.

Trucano G. T. and Grady D. E. (1995) Impact shock and penetration fragmentation in porous media. *J. Impact Eng., 17,* 861–872.

Veverka J. and Thomas P. (1979) Phobos and Deimos — A preview of what asteroids are like? In *Asteroids* (T. Gehrels, ed.), pp. 628–651. Univ. of Arizona, Tucson.

Veverka J., Thomas P., Johnson T. V., Matson D., and Housen K. (1986) The physical characteristics of satellite surfaces. In *Satellites* (J. A. Burns and M. S. Matthews, eds.), pp. 342–402. Univ. of Arizona, Tucson.

Veverka J. and 16 colleagues (1997) NEAR's flyby of 253 Mathilde: Images of a C asteroid. *Science, 278,* 2109–2112.

von Neumann J. and Richtmyer R. D. (1950) A method for the numerical calculation of hydrodynamic shocks. *J. Appl. Phys., 21,* 232–237.

Warren P. H. (1994) Lunar and martian meteorite delivery services. *Icarus, 111,* 338–363.

Weibull W. A. (1939) A statistical theory of the strength of materials (translation). *Ingvetensk. Akad. Handl. (Stockholm), 151,* 5–45.

Weibull W. A. (1951) A statistical distribution function of wide applicability. *J. Appl. Mech., 18,* 293–297.

Weidenschilling S. J. and Cuzzi J. N. (1993) Formation of planetesimals in the solar nebula. In *Protostars and Planets III* (E. H. Levy and J. I. Lunine, eds.), pp. 1031–1060. Univ. of Arizona, Tucson.

Weissman P. R. (1986) Are cometary nuclei primordial rubble piles? *Nature, 320,* 242–244.

Weissman P. R., Bottke W. F. Jr., and Levison H. F. (2002) Evolution of comets into asteroids. In *Asteroids III* (W. F. Bottke Jr. et al., eds.), this volume. Univ. of Arizona, Tucson.

Wetherill G. W. (1985) Occurrence of giant impacts during the growth of the terrestrial planets. *Science, 228,* 877–879.

Whipple F. L. (1949) Comets, meteors and the interplanetary complex. *Astron. J., 54,* 179–180.

Wu R. and Yang F. (1997) Seismic imaging in wavelet domain: Decomposition and compression of imaging operator. In *Wavelet Applications in Signal and Image Processing* (V. A. Aldroubi et al., eds.), pp. 148–162. Proc. SPIE Vol. 3169.

Yeomans D. K. and 15 colleagues (2000) Radio science results during the NEAR Shoemaker spacecraft rendezvous with Eros. *Science, 289,* 2085–2088.

Zappalà V., Cellino A., Dell'Oro A., and Paolicchi P. (2002) Physical and dynamical properties of asteroid families. In *Asteroids III* (W. F. Bottke Jr. et al., eds.), this volume. Univ. of Arizona, Tucson.

Zuber M. T. and 11 colleagues (2000) The shape of 433 Eros from the NEAR Shoemaker Laser Rangefinder. *Science, 289,* 2097–2101.

Asteroid Density, Porosity, and Structure

D. T. Britt
University of Tennessee

D. Yeomans
NASA Jet Propulsion Laboratory

K. Housen
Boeing Aerospace

G. Consolmagno
Vatican Observatory

New data from observations of asteroid mutual perturbation events, observations of asteroid satellites, and spacecraft encounters have revolutionized our understanding of asteroid bulk density. Most asteroids appear to have bulk densities that are well below the grain density of their likely meteorite analogs. This indicates that many asteroids have significant porosity. High porosity attenuates shock propagation, strongly affecting the nature of cratering and greatly lengthening the collisional lifetimes of porous asteroids. Analysis of density trends suggests that asteroids are divided into three general groups: (1) asteroids that are essentially solid objects, (2) asteroids with macroporosities ~20% that are probably heavily fractured, and (3) asteroids with macroporosities >30% that are loosely consolidated "rubble pile" structures.

1. OVERVIEW: THE DENSITY AND POROSITY OF ASTEROIDS

Data from a number of sources indicate that many asteroids have significant porosity. In some cases, porosities are large enough to affect asteroid internal structure, gravity field, impact dynamics, and collisional lifetimes. Porosity can also affect a range of asteroid physical properties including thermal diffusivity, seismic velocity, cosmic-ray exposure, and dielectric permeability. The thermal and seismic effects can in turn affect asteroid internal evolution, metamorphism, shock dissipation, and elastic properties, which can determine whether colliding asteroids accrete or disrupt. The study of asteroid bulk density, along with supporting studies of meteorite porosity and physical properties, is just beginning. These data allow for new views of asteroid belt structure and evolution. This chapter will review some of the basic data accumulated on asteroid bulk density and meteorite grain densities. Using these data we will estimate the porosities of asteroids, outline the implications of high porosity on internal structure, examine how porous asteroids respond to impacts and shock, and interpret these data to characterize the distribution of porous objects in the asteroid belt.

1.1. Terms and Definitions

Density, expressed as mass per unit volume, is a fundamental physical property of matter. For common rock-forming minerals, the crystal structures, lattice volumes, and elemental compositions are well defined, so the densities of geologic materials common in asteroids are similarly well defined. Examples of these densities include 2.2–2.6 g/cm^3 for clays, 3.2–4.37 g/cm^3 for the mafic silicates pyroxene and olivine, and 7.3–7.7 g/cm^3 for Ni-Fe. These densities refer to a grain density, which is the mass of an object divided by the volume occupied only by mineral grains. This is the average density of the solid portions of a rock. The density value returned by spacecraft measurements is bulk density, which is the mass of an object divided by its volume (including the volume of its pore spaces). The ratio between grain and bulk density is the porosity, the percentage of the bulk volume of a rock that is occupied by empty space. Porosity can be a major component of asteroid volume, and some porosity is found in most meteorites. The type of porosity measured in meteorites is microporosity, which is the meteorite's fractures, voids, and pores on the scale of tens of micrometers. Microporosity is subject to strong selection effects since these features cannot be so large or thoroughgoing as to prevent the meteorite from surviving transport to Earth. Microporosity can represent both voids and pores that have survived from the earliest formation of these aggregates as well as post-lithification impact-induced fractures. Both types have been reported in scanning electron microscopy (SEM) studies of a limited number of meteorite thin sections, but at least in ordinary chondrites it appears that impact fractures dominate (*Flynn et al.,* 1999; *Consolmagno et al.,* 1999). Large-scale voids and fractures on asteroids are called macroporosity and are probably produced by the impact history of the asteroid. These are features that are large on the scale of meteorites and are the zones of struc-

TABLE 1. Asteroid bulk density measurements: A compilation of current data on asteroid mass, volume, and bulk density.

Asteroid	Mass (10^{-10} $M_\odot$)	Mass (10^{19} kg)	Diameter (km)	Bulk Density (g/cm^3)	References
1 Ceres (G)	4.762 ± 0.015	94.7	948.8 ± 11.2	2.12 ± 0.04	*Standish* (2001), *Drummond et al.* (1998)
	4.70 ± 0.04	93.5			*Michalak* (2000)
	4.39 ± 0.04	87.3			*Hilton* (1999)
	4.70	93.5			*Standish* (1998)
	4.759 ± 0.023	94.7			*Viateau and Rapaport* (1998)
	5.0 ± 0.2	99			*Viateau and Rapaport* (1995)
		103			*Hilton et al.* (1996)
	4.74 ± 0.3	94			*Goffin* (1991)
	5.0 ± 0.2	99			*Standish and Hellings* (1989)
	5.21 ± 0.3	103			*Landgraf* (1988)
	5.9 ± 0.3	117			*Schubart and Matson* (1979)
			848.4 ± 19.7		*Tedesco et al.* (1992)
			941.4 ± 34		*Millis et al.* (1987)
2 Pallas (B)	1.078 ± 0.038	21.4	532.6 ± 6	2.71 ± 0.11	*Standish* (2001), *Dunham et al.* (1990)
	1.17 ± 0.03	23.3			*Goffin* (2001)
	1.21 ± 0.26	24.1			*Michalak* (2000)
	1.59 ± 0.05	31.6			*Hilton* (1999)
	1.00	21.4			*Standish* (1998)
	1.4 ± 0.2	28			*Standish and Hellings* (1989)
	1.08 ± 0.22	21	538		*Schubart and Matson* (1979)
			498.1 ± 18.8		*Tedesco et al.* (1992)
			538 ± 12		*Wasserman et al.* (1979)
			524.4 ± 15.2		*Drummond and Cocke* (1989)
4 Vesta (V)	1.341 ± 0.015	26.7	529 ± 10	3.44 ± 0.12	*Standish* (2001), *Thomas et al.* (1997)
	1.36 ± 0.05	27.0			*Michalak* (2000)
	1.69 ± 0.11	34			*Hilton* (1999)
	1.30	25.9			*Standish* (1998)
	1.33	26			*Goffin* (1991)
	1.5 ± 0.3	30			*Standish and Hellings* (1989)
	1.38 ± 0.12	27	525		*Schubart and Matson* (1979)
	1.17 ± 0.10	23			*Hertz* (1966)
			468.3 ± 26.7		*Tedesco et al.* (1992)
10 Hygeia (C)	0.49 ± 0.21	10	407.1 ± 6.8	2.76 ± 1.2	*Goffin* (1991), *Tedesco et al.* (1992)
	0.47 ± 0.23	9			*Scholl et al.* (1987)
11 Parthenope (S)	0.0258 ± 0.001	0.513	153.3 ± 3.1	2.72 ± 0.12	*Viateau and Rapaport* (1997), *Tedesco et al.* (1992)
15 Eunomia (S)	0.042 ± 0.011	0.84	255.3 ± 15.0	0.96 ± 0.3	*Hilton* (1997), *Tedesco et al.* (1992)
16 Psyche (M)	0.087 ± 0.026	1.73	253.2 ± 4.0	2.0 ± 0.6	*Viateau* (2000), *Tedesco et al.* (1992)
20 Massalia (S)	0.0264 ± 0.0041	0.525	145.5 ± 9.3	3.26 ± 0.6	*Bange* (1998), *Tedesco et al.* (1992)
22 Kalliope (M)				2.5 ± 0.3	*Margot and Brown* (2001)
24 Themis (C)	0.289 ± 0.126	5.75			*Lopez Garcia et al.* (1997)
45 Eugenia (F)	0.030	0.60	214.6 ± 4.2	$1.2^{+0.6}_{-0.2}$	*Merline et al.* (1999)
87 Sylvia (P)	0.076	1.51 ± 0.15	260.94 ± 13.3	1.62 ± 0.3	J.-L. Margot (personal communication, 2001), *Tedesco et al.* (1992)
90 Antiope (C)			120.07 ± 4.0	1.3	*Weidenshilling et al.* (2001), *Tedesco et al.* (1992)
121 Hermione (C)	0.047 ± 0.008	0.93	209.0 ± 4.7	1.96 ± 0.34	*Viateau* (2000), *Tedesco et al.* (1992)
243 Ida (S)	0.00021	0.0042 ± 0.0006	31.4	2.6 ± 0.5	*Belton et al.* (1995)
253 Mathilde (C)	0.00052	0.0103 ± 0.0004	53.02	1.3 ± 0.2	*Yeomans et al.* (1997)
433 Eros (S)	0.0000336	0.00067 ± 0.00003	17.36 ± 1.2	2.67 ± 0.03	*Yeomans et al.* (2000)
704 Interamnia (F)	0.37 ± 0.17	7 ± 3	316.6 ± 5.2	4.4 ± 2.1	*Landgraf* (1992), *Tedesco et al.* (1992)
762 Pulcova (FC)			137.09 ± 3.2	1.8 ± 0.8	*Merline et al.* (2000), *Tedesco et al.* (1992)
804 Hispania (PC)	0.05 ± 0.04	0.995 ± 0.796	157.3 ± 5.3	4.9 ± 3.9	*Landgraf* (1992), *Tedesco et al.* (1992)
1999 KW4		$2.16 \pm 0.43 \times 10^{-7}$	1.2 ± 0.12	2.39 ± 0.9	J.-L. Margot (personal communication, 2001)
2000 DP107 (C)		$4.34^{+1.91}_{-0.56} \times 10^{-8}$	0.8 ± 0.15	$1.62^{+1.2}_{-0.9}$	J.-L. Margot (personal communication, 2001)
2000 UG11		$9.35^{+1.87}_{-3.74} \times 10^{-10}$	0.23 ± 0.03	$1.47^{+0.6}_{-1.3}$	J.-L. Margot (personal communication, 2001)

Current best estimates of mass and volume are shown in the top line of each asteroid listing along with the bulk density determination. For completeness, previous mass and volume determinations are also included.

tural weakness that break apart during impacts to form what become meteorites. Macroporosoity defines the internal structure of an asteroid. Those with low macroporosity are solid, coherent objects, while high macroporosities values indicate loosely consolidated objects that may be collections of rubble held together by gravity (see *Richardson et al.,* 2002).

1.2. Current Measurements of Asteroid Bulk Density

Spacecraft missions and advances in asteroid optical and radar observations have revolutionized our knowledge of asteroid bulk density. Shown in Table 1 and Fig. 1 is a summary of published mass and volume measurements. The methods of mass and volume determination are discussed in section 2.0, but a glance at Table 1 shows that before the 1990s bulk-density measurements were limited to a handful of the largest asteroids. In the past 10 years, the accuracy and breadth of these measurements has exploded and produced our first picture of the density structure of the asteroid belt. The largest three asteroids, Ceres, Pallas, and Vesta, have been studied for decades and have well-constrained values. These objects make up most of the mass of the asteroid belt. As shown in Fig. 1 in comparison with meteorite grain densities, these density values seem to make mineralogical sense. Because common geologic materials can vary by almost a factor of 4 in their grain density, asteroid bulk-density measurements need to be interpreted in terms of the object's mineralogy. The differentiated V-type asteroid 4 Vesta has a bulk density consistent with basaltic meteorites overlying an olivine mantle and metal-rich core. The primitive C-type asteroid 1 Ceres has a bulk density similar to primitive CI meteorites (for definitions of meteorite types see *McSween,* 1999). However, the smallest of these three asteroids is an order of magnitude more massive than the next well-characterized asteroids and these less-massive asteroids exhibit some intriguing trends. In general, S-type asteroids appear to have higher bulk densities than C-type asteroids, but the range in both groups is large. The M-type asteroid 16 Psyche, which is interpreted to have a mineralogy analogous to Fe-Ni meteorites, shows a bulk density in the range of hydrated clays. This indicates either very high porosity or a misidentification of the mineralogy. In the case of 16 Psyche, in addition to spectra and albedo consistent with metal, radar-albedo data strongly indicate a largely metallic surface.

2. THE DETERMINATION OF ASTEROID MASSES, VOLUMES, AND BULK DENSITIES

Though the number of asteroid density measurements has begun to increase rapidly in the last few years, still only a tiny fraction of the known asteroids have usable density measurements. A short history of the efforts to determine the masses of asteroids has been provided by *Hilton* (2002). Asteroid masses have been reliably determined from asteroid-asteroid or asteroid-spacecraft perturbations. That is, the mass

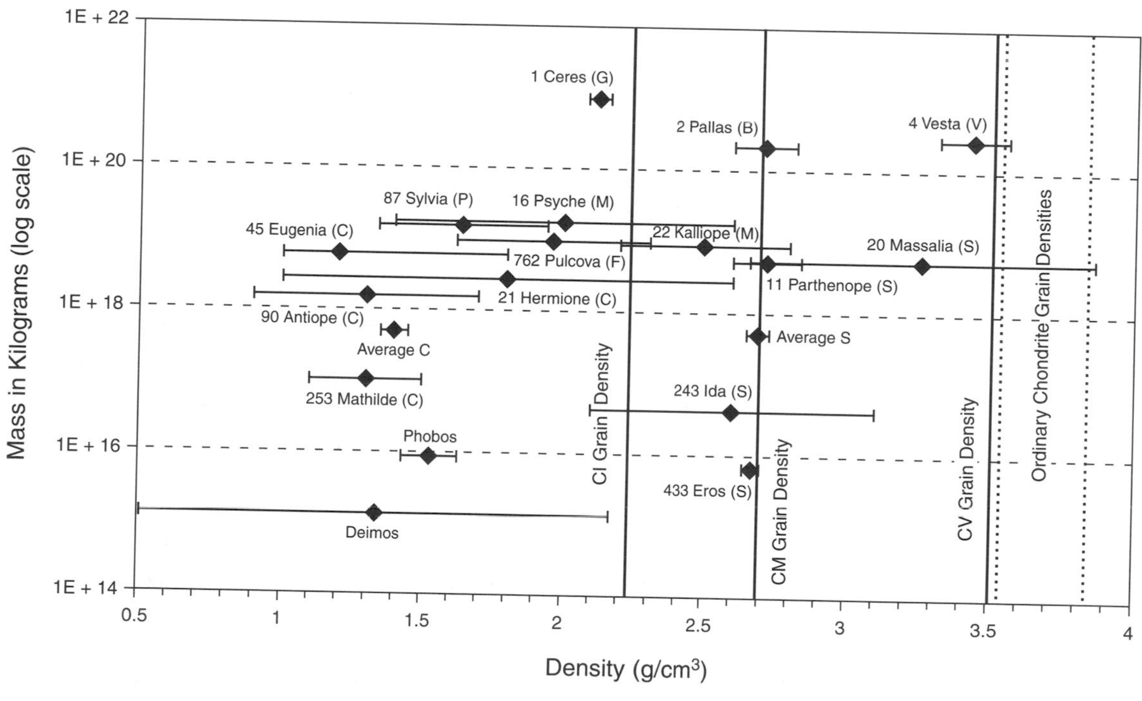

Fig. 1. Bulk densities of measured asteroids with the grain densities of common meteorites for comparison. Also included in the plot are the asteroidlike moons of Mars, Phobos and Deimos, as well as estimates for the average C- and S-type asteroids (*Standish,* 2001). Several asteroids in Table 1 with large error bars have been left off the plot for clarity.

of an asteroid is determined by observing the trajectory deviations of another asteroid, planet, or spacecraft after one or more close encounters. An asteroid's mass can also be determined by observing the motion of a satellite in orbit about the primary body since by Kepler's third law, the mass of the primary body can be determined if the orbital period and semimajor axis of the satellite are known. By far the most accurate method for determining an asteroid's mass, and even its mass distribution, is to track the motion of a spacecraft in orbit about it.

2.1. Asteroid Mass Determination Techniques

2.1.1. Asteroid mass determinations using their perturbations on neighboring spacecraft. The heliocentric change in velocity of an asteroid or spacecraft after a close asteroid approach is directly proportional to the mass of the perturbing asteroid and inversely proportional to both the close approach distance and relative velocity of the two bodies. For a perturbed spacecraft, the line-of-sight component of this velocity change is determined by observing the change in the Doppler tracking data during a close encounter. The close approach distance and relative velocity are determined from a spacecraft orbital solution that includes not only the spacecraft Doppler and range data but also the optical images of the asteroid before, during, and after the close flyby (see *Yeomans et al.,* 1997; *Yeomans et al.,* 1999). In practice, this solution provides not only the asteroid mass, close approach distance, and relative velocity but also several hundred other parameters associated with the solar radiation pressure, maneuvers and stochastic accelerations affecting the spacecraft, the surface landmark locations and spin characteristics associated with the asteroid, and the corrections required to the ephemerides of the spacecraft and asteroid. As a result of the Doppler and range tracking of the spacecraft and the optical asteroid landmark locations gathered by *NEAR Shoemaker* while in orbit about the asteroid Eros, this asteroid's gravity field, or mass distribution, was computed along with the total mass value (*Yeomans et al.,* 2000).

2.1.2. Mass determinations using the motions of their natural satellites. As of this writing, some nine asteroids have been identified as having satellites, and most of these objects were observed well enough that mass and density determinations were possible. Using *Galileo* imaging data of 243 Ida and its satellite Dactyl in August 1993, *Belton et al.* (1995) estimated the mass, volume, and bulk density for Ida. Groundbased adaptive optics at the Canada-France-Hawai'i Telescope (CFHT) were used to discover and observe satellites about asteroids 45 Eugenia (*Merline et al.,* 1999) and the transneptunian object 1998 WW31 (*Veillet et al.,* 2001), while Keck II adaptive optic images were used to discover and observe satellites about asteroids 87 Sylvia (J.-L. Margot, personal communication, 2001), 90 Antiope, and 762 Pulcova (*Merline et al.,* 2000). Goldstone and Arecibo radar imaging observations were used to identify and observe satellites about asteroids 1999 KW4 (*Benner et al.,* 2001), 2000 DP107 (*Margot et al.,* 2001), and 2000 UG11 (*Nolan et al.,* 2001). With the exception of 1998 WW31, at least preliminary values for the densities of all the double asteroids have been announced.

2.1.3. Mass determination using their perturbations on Mars. *Standish and Hellings* (1989) used the perturbations of asteroids on the motion of Mars to directly determine masses for the three largest asteroids. For perturbations with periods of 10 years or less, only Ceres, Pallas, and Vesta produce perturbative amplitudes of more than 50 m on the motion of Mars. The Mars observational dataset includes optical data back to the introduction of the impersonal micrometer in 1911 as well as more recent radio, lunar laser ranging, and radar data. However, it was primarily the *Viking* lander spacecraft data that allowed Mars range measurements to about 7 m (1976–1980) and 12 m (1980–1981) — well below the perturbative effects of these three minor planets. With additional observations of Mars, including the ranging and Doppler data from the *Mars Pathfinder* and *Mars Global Surveyor* spacecraft, this analysis has been updated recently (*Standish,* 2001) with the results being presented in Table 1. Although other, smaller asteroids had nonnegligible effects upon the orbit of Mars, a direct solution for their individual masses was not feasible because their perturbative effects were not substantially larger than the observational accuracy. However, Standish found it necessary to model these perturbative effects in the Mars ephemeris development effort. A mass was computed for each of a few hundred of the largest asteroids by using its estimated diameter and assuming a particular bulk density based upon its spectral class. By accumulating the perturbations of each spectral class together, it was possible to solve for the mean bulk density for the spectral class as a whole. For a recent solution completed after the release of JPL Development Ephemeris number 405 (DE405), the mean bulk densities were computed to be 1.4 (±0.05), 2.69 (±0.04), and 4.7 (±0.5) for the C-, S-, and M-asteroid spectral types respectively (E. M. Standish, personal communication, 2001).

2.1.4. Mass determinations from asteroid-asteroid interactions. The largest number of asteroid masses has been computed using the observed motions of asteroids that have interacted with other asteroids either one or more times. The more recent results are given in Table 1. Relatively precise mass determinations have been computed in this fashion for the three largest asteroids, 1 Ceres, 2 Pallas, and 4 Vesta (see Table 1). *Hilton* (1999) concluded that attempts to compute the mass of Ceres or Pallas by itself would lead to errors since the mass determination for either one was dependent upon the other's assumed mass. To avoid what he considered a computational degeneracy, he simultaneously solved for the masses of Ceres, Pallas, and Vesta. However, the masses he obtained for Ceres and Pallas are discordant with other solutions, including recent ones by *Michalak* (2000), *Goffin* (2001), and *Standish* (2001), where simultaneous mass solutions were also determined for more than one asteroid. *Goffin* (2001) pointed out that his results for the mass of Pallas were relatively insensitive to the assumed mass of Ceres. While the masses of Vesta determined by *Standish* (2001), *Michalak* (2000), *Goffin* (1991), *Schubart and Matson* (1979), and even the very first reliable asteroid mass determination by *Hertz* (1966) are all consistent, the result by *Hilton* (1999) is again discordant with the others. In Table 1, we have selected the Ceres, Pallas, and Vesta masses determined by *Standish* (2001) for the given bulk-density determinations.

2.2. Volume Estimates

Asteroid volume estimates are most often made using an effective radius determined by the Infrared Astronomical Satellite (IRAS) Minor Planet Survey (*Tedesco et al.,* 1992). Because of corrections and refinements to the data-reduction process, care must be taken to use the 1992 publication rather than the earlier results presented by *Tedesco* (1989). For a select few asteroids, their shapes and hence their volumes can be determined by occultation techniques. For the bulk density of 2 Pallas displayed in Table 1, an occultation derived volume has been employed (*Dunham et al.,* 1990). Volume estimates based upon Hubble Space Telescope and groundbased adaptive optics techniques have been used for 1 Ceres (*Drummond et al.,* 1998), 4 Vesta (*Thomas et al.,* 1997), 45 Eugenia (*Merline et al.,* 1999), and 87 Sylvia (J.-L. Margot, personal communication, 2001). Radar images of double asteroids 1999 KW4, 2000 DP107, and 2000 UG11 have been taken and J.-L. Margot (personal communication, 2001) has provided preliminary estimates for their masses, radii, and bulk densities. Spacecraft imaging data were used to determine the shapes and volumes for 243 Ida (*Belton et al.,* 1995) and 253 Mathilde (*Veverka et al.,* 1997). For asteroid Ida, complete (but distant) imaging coverage from the *Galileo* spacecraft allowed a volume determination to the 12% level. Because of Mathilde's 17.4-d rotation period, complete imaging coverage of the surface was not achieved during the *NEAR Shoemaker* spacecraft flyby, allowing only a 15% volume determination. On the other hand, the volume for 433 Eros was determined to the few percent accuracy level using the imaging and laser rangefinder data onboard the orbiting *NEAR Shoemaker* spacecraft (*Veverka et al.,* 2000; *Yeomans et al.,* 2000; *Zuber et al.,* 2000). In Table 1, the effective diameters given for Ceres, Pallas, and Vesta were determined from the volume estimates computed using their major and minor figure axes.

2.3. Bulk Density Determinations

In Table 1, the best mass and volume determinations together with their respective uncertainties have been employed to determine the given bulk densities and their associated uncertainties. For asteroids for which multiple estimates are provided for either the mass or volume, the best estimate listed in each case has been used to compute the given bulk density. The mass estimate for 20 Massalia was taken from *Bange* (1998) assuming the mass of Vesta is that given by *Standish* (2001). A few data gaps appear in Table 1 because some values were not available in the published literature. For example, only the bulk densities for 90 Antiope and 762 Pulcova have been published, so the corresponding mass estimates are not available and we could not locate a reliable effective diameter for 24 Themis.

3. CONSTRAINTS ON ASTEROIDS FROM METEORITE DATA

3.1. Selection Effects and Asteroid Analogs

To what degree can we use meteorites measured in our laboratory to put constraints on the density and structure of asteroids? There is little doubt that meteorites do sample the asteroid belt. There is somewhat less certainty that the meteorites are a complete or representative sample of the belt (cf. *Sears,* 1998; *Meibom and Clark,* 1999). Selection effects in ejection from the parent asteroid, transportation to Earth, and atmospheric entry certainly play a major role in the population of meteorites reaching our collections.

The overwhelming majority of meteorites (some 75% of those seen to fall, ~85% of those collected from dry deserts, and >90% of those found in Antarctica) are stony meteorites of the class known as "ordinary chondrites" (*Grady,* 2000). A comparison of observed fireball rates with meteorite recoveries in desert areas (*Bland et al.,* 1998) indicates that the total number of objects hitting the Earth's atmosphere is within a factor of 2, at most, of the number of similar-sized objects hitting the ground. This still allows the possibility that the vast majority of weak meteorites, like the most friable carbonaceous types (e.g., CIs) could be broken up and lost in the atmosphere. But such meteorites are so rare in our collections that even if 99% of them were lost in Earth's atmosphere, they would still only have an abundance in space comparable to ordinary chondrites. That would be consistent with the fireball statistics. Even in that event, however, a significant fraction if not the absolute majority of meteorites falling today would still be ordinary chondrites. Thus, at a minimum, at least half the meteoroids arriving at Earth today are ordinary chondrites. (It is important to note that samples of the most abundant types of carbonaceous chondrites, the CV and CO classes, are essentially as dense and as strong as ordinary chondrites; passage through the atmosphere should not strongly discriminate against them.)

Another kind of selection effect is temporal. We know with some certainty what sort of meteorites have hit Earth in historic times and we know, with a bit more uncertainty, that similar distributions of meteorites have accumulated in Antarctica over the past 10,000 to 100,000 yr. Attempts to "mine" meteorites from beds of various geologic ages (or from the Moon) have proved difficult so far, and in any event, the populations do not show any significant changes in the kinds of materials recovered. One cannot rule out the possibility that the flux of meteorites hitting Earth today represents a chance sampling from a small number of sources. Different types of meteorites may have been prevalent on Earth 1 b.y. ago. More to the point, different types of meteorites, unsampled by anything hitting Earth now, may be prevalent in the asteroid belt today. Indeed, there is a range of spectra among the S-type asteroids that show relative abundances of pyroxene and olivine with no direct analogs in our collections (*Gaffey et al.,* 1993). Given that possibility, how reasonable is it to use meteorite physical characteristics to set limits on asteroidal structure?

There are, however, three arguments to suggest that meteorites in our collections today are a reasonable place to start to predict the density and structure of material making up the asteroids. First, it is noteworthy that the spectra even of the S types not sampled in our collections remain essentially mixtures of these same minerals, pyroxene and olivine, albeit in differing proportions with differing FeO contents. Furthermore, the abundances of the rock-forming elements seen in solar spectra are not significantly different from the abundances seen in ordinary chondrite meteorites. There are only so many ways that these elements can be formed into minerals, and it is unlikely that the vast majority of material in the asteroid belt should be anything but these expected minerals (olivine, pyroxene, and plagioclase). Finally, there are physical limits to the porosity state in which coherent rocks can be found; the same evidence that suggests a body like asteroid Eros can maintain geological coherence also places limits on the microstructure of the rock that transmits that coherence. In any event, it is clearly better to start with data of material we know comes from the asteroid belt than to speculate, without data, about what may or may not be present there.

3.2. Measurement of Meteorite Density and Porosity

Measurement of meteorite density is hampered by two concerns: the risk of contamination and the (*a priori* unknown) effect of microporosity. The traditional methods for measuring the density of rocks (and other irregularly shaped objects) are based on the method of Archimedes: immersing the sample in a fluid and measuring either the sample's displacement of a measured volume of fluid or the relative weight of the sample in the fluid compared to its weight in air. But for meteorites, the problem of contamination is serious: Water can alter or remove indigenous salts and other soluble minerals in the meteorite and promote the oxidation of metallic Fe often present in tiny grains that are especially vulnerable to weathering. Organic fluids such as CCl_4, which have also been used, carry with them their own possible organic contaminants.

Even without these contamination worries, however, a second issue is the question of porosity. One must assume these samples could have significant porosity; indeed, a goal of our measurements is to determine that porosity, which is the ratio of the bulk volume (given by the outer shape of the sample) to the grain volume (that of the minerals within the rock, not including voids, microcracks, and other flaws in the rock fabric). But any liquid may enter into the pore spaces to some unknown extent; thus the volume measured will be neither the bulk volume nor the grain volume.

To get around these issues, gases are commonly used as the Archimedean fluid in grain-density measurements. In a typical pycnometer, the sample is placed in a chamber of known volume and flushed with He at a known pressure. A second chamber, also of known volume, contains helium at a higher (known) pressure. The two chambers are connected and pressures equalized; the final pressure indicates how much of the given chamber volumes is taken up by the grain volume of the sample. Helium is known to be nonreactive, and at least in ordinary chondrites it can be expected to penetrate quickly into all pore spaces and voids. (An examination of meteorite thin sections shows that most porosity comes as a dense network of microcracks, generally <10 μm wide. Helium would only have to diffuse a distance of <100 μm from these cracks, through fine-grained material, to reach any voids not connected by these cracks.)

Consolmagno and Britt (1998) adapted an unusually large pycnometer to measure large (up to several kilograms) meteorite specimens. To measure bulk densities, they developed a method using 40-μm glass spheres as the Archimedean fluid; the spheres are nonreactive, are easy to see and brush off the samples after measuring, and clearly do not penetrate into the cracks and pores. From these data, porosities could be measured directly on dozens of meteorites. This work has been supplemented by new data from *Flynn et al.* (1999) and *Wilkison and Robinson* (2000), who measured additional grain and bulk densities respectively.

3.3. Results of These Measurements: Porosity and Density of the Meteorite Classes

Along with their data analysis, *Consolmagno et al.* (1998) incorporated a detailed literature search of meteorite densities and porosities (most notably a large database collected by *Terho et al.,* 1993) into the study of general trends in ordinary chondrites. In all, more than 160 meteorite porosities and nearly 400 density measurements were analyzed. They concluded that weathering of meteorite falls reduces most ordinary chondrite porosities as a function of exposure time in the terrestrial atmosphere.

Consolmagno et al. (1998) noted that one could assume a grain density for each meteorite chemical type (H, L, or LL) since the grain density should be only a function of composition, which is specified by the type classification. They suggested that bulk density is not significantly changed due to mild weathering, since the filling of pore spaces by weathering product will not change the outer shape of the meteorite and the weathering-related mass flux (mostly terrestrial O reacting with meteoritic Fe) will amount to only a few percent of the total meteorite mass. Thus, they conclude that a model preweathering porosity can be estimated for each meteorite by comparing measured bulk densities against theoretical grain densities. When applied to the ordinary chondrites for which bulk densities have been measured (presently numbering ~300), they find the distribution of model porosities shows no dependence on meteorite type or metamorphic grade. All meteorite porosities so estimated fall into a bell-shaped distribution centered on 11% porosity with a 1σ spread of ±5%. This distribution is shown in Fig. 2. The average model grain density, bulk density, and model porosity for each subtype of ordinary chondrite are shown in Table 2.

Consolmagno et al. (1999) explain this distribution by noting the visible evidence of porosity in meteorite thin sections. Virtually all the porosity of an ordinary chondrite can be attributed to a network of cracks, roughly 10 μm wide, that have been emplaced onto the meteorite fabric (presumably by impacts) after the meteorite was accreted and lithified. In find meteorites, these cracks are filled with weathering product (the microprobe detects Fe and Cl, but no Ni) and in some older falls, these cracks can be seen to be half-filled with weathering product, extending out from Fe-Ni grains; in the freshest falls, the cracks are empty. This is in line with the hypothesis that weathering in ordinary chondrites primarily fills porosity with oxides (cf. *Bland et al.,* 1998) and the presence of the weathering product also confirms that these cracks are indeed present in the meteorite and not an artifact of the thin sectioning.

The situation for C-type meteorites is more complex. These meteorites arrive on Earth already quite oxidized, and so it is more difficult to determine the extent to which terrestrial weathering has filled pore spaces. These meteorites are also much rarer than ordinary chondrites, and tend to come in much smaller samples, thus making the collection of their density and porosity data much more difficult. Only a handful of these meteorites have had their porosities and densities reported (cf. *Britt and Consolmagno,* 2000). The range of porosities is much higher — from zero to over 30% for various samples of the CI, Orgueil, and from near-zero (Vigarano) to 25% (Allende) for two different CVs.

The bulk densities of C-type meteorites, however, fit into two ranges: CI and CM meteorites, characterized by significant water content and low Fe, have bulk densities around 2 g/cm³, while most dry (CO and CV) or more Fe-rich (CR) carbonaceous meteorites have densities between 2.9 and 3.5 g/cm³, essentially indistinguishable from the bulk densities of ordinary chondrites.

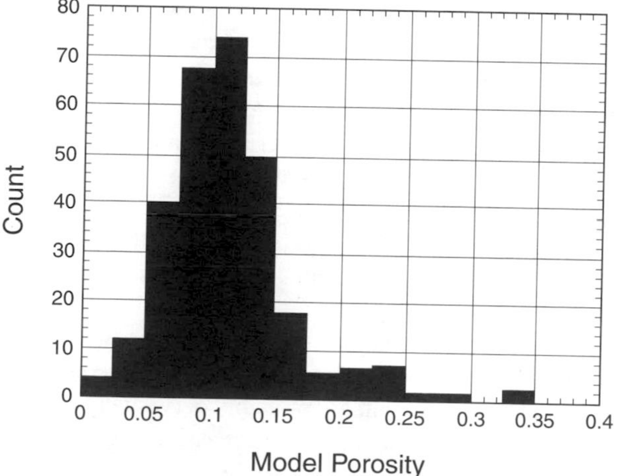

Fig. 2. Ordinary chondrite model porosities.

TABLE 2. Average densities for meteorite types.

Meteorite Type	Grain Density	Average Bulk Density	Average Porosity
H ordinary chondrites	3.84	3.40	11.5%
L ordinary chondrites	3.75	3.34	10.8%
LL ordinary chondrites	3.56	3.19	10.4%
Achondrites	3.20	2.97	7.0%
CI carbonaceous chondrites	2.27	2.12	11%
CM carbonaceous chondrites	2.71	2.21	12%
CR carbonaceous chondrites	3.11	3.15	6%
CV carbonaceous chondrites	3.51	3.10	11%
CO carbonaceous chondrites	3.69	3.11	16%

4. ASTEROID POROSITY

Given knowledge of meteorite porosity and bulk density, and the growing database of asteroid bulk-density measurements, we can start to estimate the porosity of asteroids and assess their internal structure. The starting points are the three pieces of data we have about an individual asteroid: (1) the bulk density, (2) the reflectance spectrum that provides a meteoritic compositional analog, and (3) thc grain density and average porosity of the analog meteorite. Since the meteorite grain density gives the density of the asteroid if it was a completely solid object, the deviation of the asteroid's bulk density from this provides an estimate for the bulk porosity of the asteroid. This is essentially scaling the raw bulk-density data by the asteroid's meteorite analog. The assumption that the pore space is empty establishes a lower bound on the porosity of the asteroid. Since empty space has no density itself, this is the minimum amount of low-density volume needed to reconcile the observed bulk density with the grain density of its mineralogy. If the pore spaces were filled with some material such as water-ice, the amount of pore space required to explain the low bulk density of the object increases dramatically. In the case of low-density objects like 243 Mathilde or 90 Antiope with bulk densities of 1.3 g/cm³, their bulk densities are not significantly higher than water-ice (at 0.97 g/cm³) and would require unrealistically huge (i.e., >75 vol%) components of water-ice to account for their low densities given the grain density of their meteorite analogs (*Britt and Consolmagno,* 2000).

The estimated bulk porosities of the asteroids in Fig. 1 are shown in Fig. 3. This figure highlights the emerging structure of the asteroid belt. First, low-porosity objects are relatively rare. Aside from the three largest asteroids, the only other asteroid with an estimated porosity of <15% is 20 Massalia. Most asteroids have significant porosity. There is a cluster of objects around 30% porosity and a scattering of objects on the high end all the way out to 16 Psyche and 22 Kalliope with total porosities of more than 70%.

These data reflect the bulk porosity, which is the sum of all the asteroid's porosity, including the microporosity found

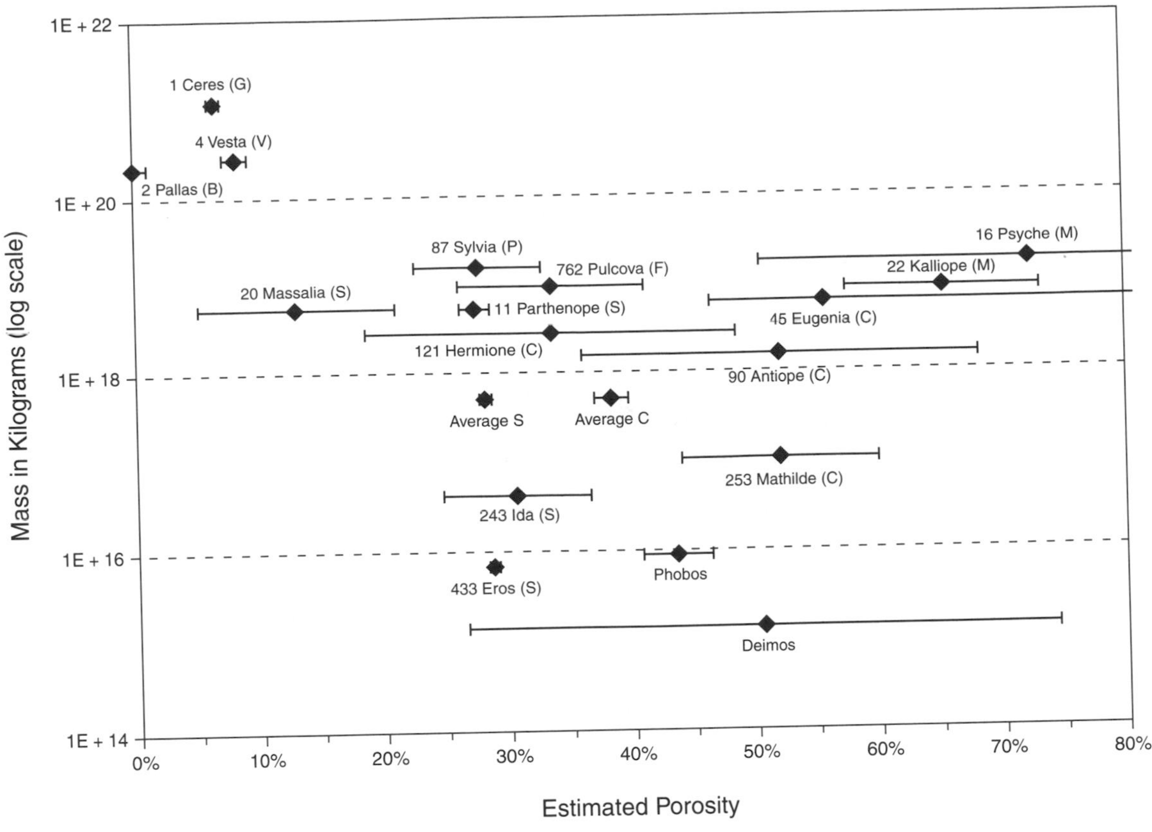

Fig. 3. Estimated porosity of measured asteroids. The bulk density of the asteroid is scaled by the grain density of its best meteoritic spectral analog to provide an estimate of the asteroid's total porosity. This includes the small-scale microporosity common in meteorites and the large-scale macroporosity that can affect the asteroid's structure.

in meteorites as well as the large-scale macroporosity. Microporosity probably does not seriously affect the parent asteroid's internal strength since this level of small-scale porosity survives in meteorites that have survived ejection, transportation, and delivery to Earth. For insight into the asteroid's internal structure, we need to extract an estimate of the asteroid's macroporosity from its bulk porosity. Subtracting the average meteorite analog microporosity from the bulk porosity of an asteroid provides a rough estimate of the asteroid's large-scale macroporosity. Note that this estimate requires two assumptions: first, that we know the asteroid's meteorite analog and, second, that the meteorites delivered to Earth are a representative sample of that material.

Shown in Fig. 4 are the estimated macroporosities for these asteroids. These data appear to divide into three rough groups. The first group includes the large asteroids 1 Ceres, 2 Pallas, and 4 Vesta, as well as the somewhat smaller 20 Massalia. Their bulk densities are very close to the grain densities of their analog meteorites, indicating essentially zero macroporosity. These asteroids are probably strong, coherent objects that have not been disrupted throughout solar-system history. It is interesting that all three asteroids with diameters >500 km fall in the zero-macroporosity group and only one other smaller asteroid has low macroporosity. This probably indicates a very strong size selection process within the asteroid belt for survival as a coherent object and perhaps the relative rarity of coherent smaller asteroids.

The second group includes the S asteroids 433 Eros and 243 Ida as well as 762 Pulcova and 121 Hermione. These asteroids have between 15% and 25% macroporosity, indicating that they have been extensively fractured. Spacecraft images of both 433 Eros and 243 Ida show numerous morphological indications of pervasive fracturing (*Belton et al.*, 1995; *Veverka et al.*, 1997; *Veverka et al.*, 2000). However, this fracturing was probably not extensive enough to disrupt the object and asteroids with less than approximately 25% macroporosity probably have some measure of coherent strength. In terrestrial geology, well-sorted sedimentary rocks can have up to 30% porosity and still be coherent. However, porosities larger than ~30% usually indicate loose rubble or soils.

The third group are those with greater than 30% macroporosity. These objects, some of which have more empty space than solid material, are probably pervasively fractured and may have been disrupted and reassembled by mutual gravity. Asteroid 16 Psyche is likely the most porous object observed so far. Its reflectance spectra and radar albedo strongly indicate a metallic surface composition. Assuming

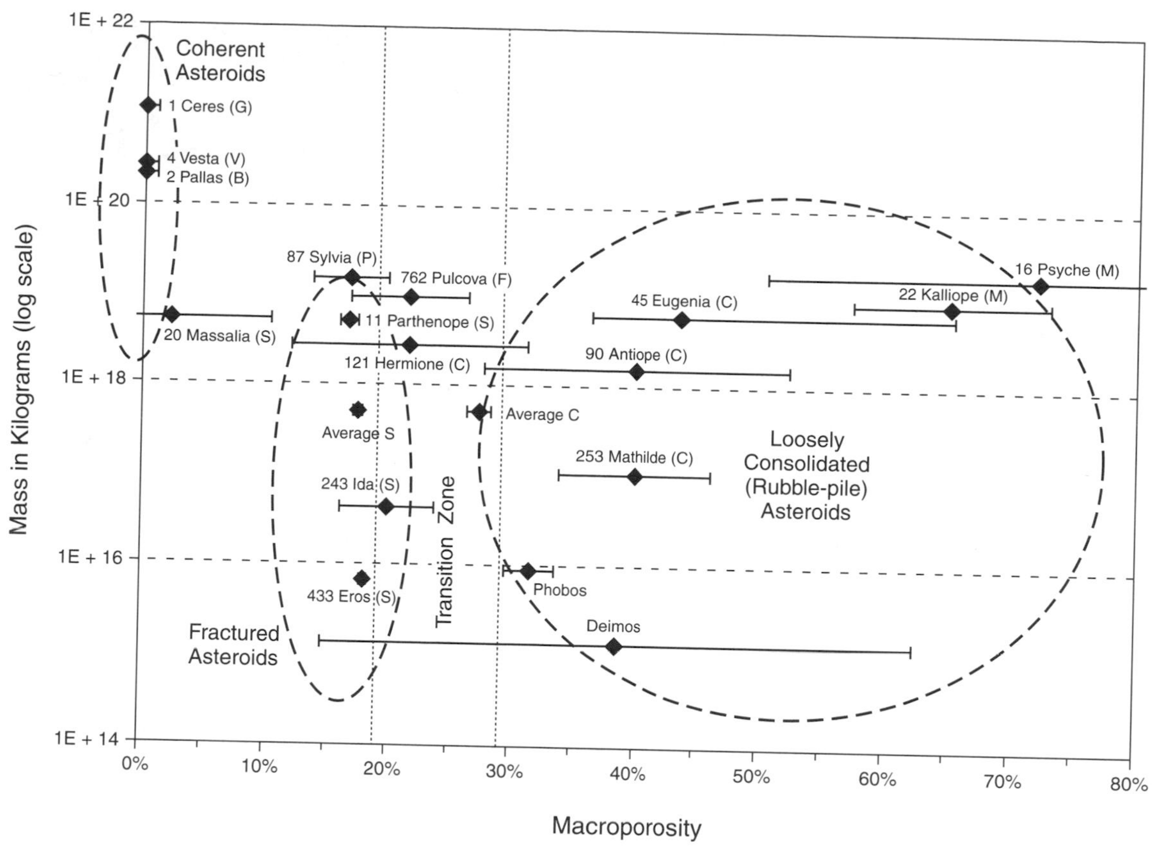

Fig. 4. Asteroid macroporosity estimated by subtracting the average porosity of asteroid's meteorite analogue from the bulk porosity shown in Fig. 3. Since microporosity probably does not seriously effect the structural integrity of asteroids, this is a direct estimate of the large-scale fractures and voids that determine the asteroid's internal structure.

a iron meteorite grain density of 7.4 g/cm^3, this would require a bulk porosity of 75%. Iron meteorites have essentially zero microporosity, so the bulk porosity in this case likely equals the macroporosity suggesting a pervasively disrupted object that has been loosely reassembled and held together by mutual gravitation.

It is also interesting that each of these three groups contain representatives of a range of asteroid spectral types. The group coherent asteroids contains of two primitive spectral types, the C-type 1 Ceres and the B-type 2 Pallas that have mineralogies of low-temperature condensates, organics, and hydrated minerals. But this low-porosity group also contains the differentiated V-type 4 Vesta with a basaltic crust and the high-temperature, mafic silicate-rich S-type 20 Massalia, both of which have strikingly different mineralogies from the first pair.

In general, the dark, primitive asteroids seem to be more prevalent in the "rubble-pile" group, while the high-albedo, higher-temperature S-type asteroids seem to be more common in the "fractured" group. The average S-type bulk density determined by E. M. Standish (personal communication, 2001) falls in the fractured group while the average C-type bulk density is closer to the rubble-pile group. This may indicate that primitive asteroids are fundamentally weaker and more likely to be loosely consolidated, while the metamorphic and igneous asteroids tend to be stronger and more likely to be coherent. However, there are glaring exceptions in each group, such as the low-porosity P-type 87 Sylvia. Fundamentally, what these data tell us is that the collisional disruption history of the asteroid belt is complex and probably stochastic, that mineralogy affects but does not control macroporosity, and that most asteroids have and retain significant porosity.

5. ASTEROID POROSITY AND IMPACT PHYSICS

The response of an asteroid to a collision event is determined by the manner in which stress waves generated by the impact are propagated and attenuated throughout the body. Stress-wave propagation is, in turn, governed by the mechanical properties and shape of an asteroid and its internal structure, i.e., the size and spatial distribution of fractures and pore spaces. Similarly, the occurrence of an impact can drastically alter the internal structure of an asteroid, through the creation of additional fractures and large-scale pore spaces or the reduction of porosity through compaction. It is important to understand these coupled processes because they are major drivers in the collisional evolution of asteroids.

This section considers the interplay between the porosity of an asteroid and the outcomes of collision events. This is a complex and relatively immature field, with important and interesting problems yet to be studied. This section highlights some of the important issues for collisions involving porous bodies.

5.1. Effect of Porosity on Collisions

The principal effect of target porosity on collisional outcomes is to attenuate the stress wave generated in an impact. As a strong shock propagates though a porous target, energy is expended (in PdV work) to collapse the pore spaces, which contributes to irreversible heating of the target. In addition, pore spaces, depending on their size scale, disperse or scatter the shockwave front. Both of these mechanisms serve to make porous materials much better attenuators of shock pressure than nonporous materials.

The dissipative nature of porous materials has been observed in experimental simulations of collisional fragmentation (*Holsapple et al.,* 2002). For example, *Stewart and Ahrens* (1999) performed experiments with plaster targets having a porosity of 65%. The tensile strength of the plaster was about a factor of 10 below that of competent basalt. Even so, the targets behaved as if they were quite strong and the projectile energy per unit target mass for catastrophic disruption was very close to that measured for basalt. Apparently, the shock attenuation in the porous target was large enough to offset the low tensile strength. This, along with similar results for other materials (e.g., *Nakamura et al.,* 1992; *Love et al.,* 1993) suggest that weak porous asteroids could be quite resistant to collisional disruption. This is discussed in more detail below.

Efficient shock dissipation is also evident in craters formed in porous materials. For example, the large craters on the porous asteroid Mathilde are packed closely together with little evidence of shock-induced disturbance of adjacent craters (*Veverka et al.,* 1997; *Chapman et al.,* 1999). This effect has been reproduced in laboratory experiments in which impact craters were formed in highly porous materials (*Housen et al.,* 1999). The experiments were performed on a centrifuge in order to satisfy the similarity requirements needed to replicate the formation of large craters on asteroids. Figure 5 shows an example of two craters formed in a material with ~60% porosity. The experiments demonstrated that craters form in close proximity (rims nearly touching), with no damaging effects on earlier craters. Evidently, shocks damp out extremely quickly in the presence of high porosity.

Numerical simulations of collisions have provided another interesting insight into the propagation of shocks in porous asteroids. *Asphaug et al.* (1998) reported on smoothed particle hydrodynamic (SPH) simulations of an impact into a model asteroid represented by a collection of nonporous boulders. The size scale of the pores was comparable to the size of the impactor. They too found that the shock was effectively dissipated due to the energy used in pore collapse. In addition, scattering of the shock due to the large-scale heterogeneity of the target caused the impact energy to be deposited in a localized volume, resulting in high velocities of ejected debris. This could have significant implications for disruption and ejecta retention on rubble pile asteroids composed of fragments that are comparable to, or larger than, an impacting body.

Fig. 5. Two craters formed sequentially in a target having initial porosity of 60%. Images of the first crater (furthest from camera) before and after the second impact showed no noticeable damage caused by the second impact, even though the crater rims were nearly touching. The porous material efficiently damps the shock pressure.

5.2. Effect of Collisions on Asteroid Porosity

Because fractures and porosity can have such a dramatic effect on the collisional response of asteroids, it is important to also consider how collisions affect the internal structure of asteroids. Numerous laboratory and numerical studies have addressed this problem (*Holsapple et al.,* 2002; *Asphaug et al.,* 2002).

It is generally accepted that impacts below the threshold energy for catastrophic disruption leave a body highly fractured, but mostly intact. Eros is probably an example of such a fractured shard (*Britt and Consolmagno,* 2000; *Veverka et al.,* 2000; *Wilkison and Robinson,* 2000). Impact-generated fractures provide for modest increases in porosity. Much larger increases are probably attained in the energetic collisions that shatter a body and temporarily disperse the fragments, which then reaccumulate into a rubble pile. Using terrestrial soils and rocky debris as an analogy, it is reasonable to expect catastrophic collisions to produce rubble pile structures having macroporosity in the range of 30–50% (e.g., *Lambe and Whitman,* 1969; *Britt and Consolmagno,* 2000).

Interestingly, an analogy to rubble-pile formation can be found in the literature of explosive field cratering. When the depth of burial of an explosive charge is less than critical in value, a crater is formed by excavation and ejection of material in a manner similar to the formation of most impact craters. On the other hand, explosives buried slightly deeper than the critical depth launch debris nearly vertically. These ejected fragments do not "escape" the crater, but instead fall back to the surface with little horizontal translation. The jumbled material may bulk considerably, leaving a mound instead of a crater.

Figure 6 shows the result of a deep explosion in competent basalt. Ejected rock fragments rose to a height of some 30 m and reaccumulated into a rubble pile with a diameter of 46 m and a height of 8 m. The bulk porosity increased from an initial value of ~8% to a final value of ~30%. A trench dug through the mound showed fragments ranging from very small sizes up to more than 1 m, with significant pore space over a corresponding range in size scale. The largest blocks in the mound had dimensions comparable to the spacing of fractures in the preshot material, suggesting that block size was controlled by the spacing of *in situ* fractures. The same mechanism probably applies to rubble-pile formation of asteroids. That is, the size distribution of fragments that end up in a rubble pile from a catastrophic collision will depend on the varied fracture systems generated in previous noncatastrophic impacts.

In contrast to the above mechanism, an asteroid that already has high porosity, either as a remnant of accretion processes or from rubble-pile formation, can experience a net reduction in bulk porosity during collisions. Shocks in highly porous asteroids act as compaction waves closing pore spaces as they damp out in the body (*Menikoff,* 2001).

In granular geological materials, compaction occurs in two ways. First, at low pressures, grains can slide past one another into a more compact configuration. This occurs when the pressure is sufficient to cause elastic deformation in the grains, overcome frictional forces at grain boundaries, or break the intergranular bonds in cemented aggregates. This simple geometrical rearrangement is possible as long as the bulk density is below a threshold value, which is referred to in the soil-mechanics community as the maximum relative density. (In soil mechanics, the term "maximum relative density" refers to the highest density that can naturally be attained without fracturing the constituent grains. Tapping, vibration, and pluviation are typically used to attain the maximum-density state for soils.) Therefore, as long as the constituent grains or rubble fragments of an asteroid are not in their maximum-density configuration, a reduction in porosity can occur at relatively low pressures (the magnitude of the pressure depends on the shape and the mechanical and frictional properties of the grains). This provides an opportunity for subcatastrophic collisions to reduce the bulk porosity of a rubble pile by seismically shaking the fragments into a more compact configuration.

Fig. 6. Side view of the rubble pile produced in a buried nuclear explosion (codename SULKY) in competent basalt. The mound is approximately 46 m in diameter and 8 m high.

The second mechanism that reduces porosity in granular materials occurs when the applied pressure is large enough to fracture the grains, thereby allowing additional movement and better packing. For silicate soils, grain fracture initiates when the pressure exceeds ~10 MPa, although significant compaction requires more than ~100 MPa (*Hagerty et al.,* 1993; *Yamamuro et al.,* 1996). During impacts into dry sand at a few kilometers per second, these pressures are reached only in a localized region near the impact point. As a result, the volume of compacted material produced by impacts in dry sand is very small compared to the crater volume. This is true even for sands that have been intentionally fluffed up to increase their porosity. Viewing a rocky rubble-pile asteroid as a collection of large sand grains, one might conclude that subcatastrophic collisions would not be very efficient at compacting the rubble by this mechanism.

However, it is important to note that while silicate grains require high pressures for fracture, the "grains" in an asteroid rubble pile could range from typical mineral grains up to fragments or aggregates of considerable size. The pressure required to fracture large fragments is probably much smaller than 10 MPa. For example, chunks of rock tend to weaken with increasing size, roughly as $L^{-1/2}$, where L is the linear size of the rock sample (*Lundborg,* 1967). Moreover, in a C-type rubble-pile asteroid, the fragments might be very weak, with strengths comparable to those of carbonaceous meteorites. As such, the crushing strength of the "grains" might be small enough that shock compaction could be quite efficient.

An example of shock compaction was reported by *Housen et al.* (1999) and *Housen and Voss* (2001). They performed impact-cratering experiments in silicate materials and varied the porosity from 35% to 70%. Compaction, as opposed to excavation, was found to be the dominant mechanism of crater formation when the target porosity was greater than ~40%. When the porosity exceeded ~55%, only a small part of the material originally within the crater was excavated and emplaced into an external ejecta deposit. That is, most of the crater volume was formed by permanent compression of the target material. A consequence of such compaction events is that the target volume is reduced, thereby increasing its bulk density. Applying this result to Mathilde, the five largest craters on Mathilde would have increased its bulk density by ~20%. This value would be even larger if additional big craters are discovered on the as-yet-unobserved hemisphere. Hence, if Mathilde's large crater population is typical of initially porous asteroids, compaction cratering could be an important mechanism for reduction of porosity.

It should be emphasized that porosity may not be the most important property that determines whether collisions will compact an asteroid. For example, it is likely that two asteroids with the same porosity, one metallic and one carbonaceous, would respond differently in collisions. A more relevant property might be a measure of the compaction strength, since this determines the volume of material crushed by the shock. One of the outstanding problems in studies of asteroid collisions is to determine the fundamental mechanical and physical properties that determine collisional outcomes.

5.3. Collisional Lifetimes of Porous Asteroids

The collisional lifetime of an asteroid is determined by the size distribution and flux of impacting bodies, the relative velocity of a collision, and the threshold size of the impactor capable of shattering or dispersing the body. The size distribution, flux, and relative velocity of objects in the main belt have been determined with reasonable certainty (e.g., *Farinella and Davis,* 1994; *Bottke et al.,* 1994). On the other hand, the impactor conditions required to shatter or disperse an asteroid are much more uncertain (*Holsapple et al.,* 2002). Numerical simulations can provide some guidance, but the material models used in hydrocodes are not yet sufficiently robust to predict collisional outcomes with certainty. This is especially true when considering porous asteroids; none of the numerical models in use at present have incorporated equations of state or constitutive models for porous materials. This is partially due to the fact that porous asteroids have only recently become of interest, but can also be attributed to the fact that the mechanical behavior of porous materials is not fully understood.

Laboratory experiments can provide some insights. However, it is very difficult to perform collision experiments under conditions that are directly applicable to collisions of kilometer-scale asteroids. By necessity, these studies must be conducted at small scales, where the effects of gravity are negligible compared to the cohesive strength of the target. As a result, the conditions required to shatter a small model asteroid are quite different from those required for a large object (*Holsapple,* 1993; *Housen and Holsapple,* 1999). While the scaling issues are understood to some degree for brittle nonporous materials (like rock), highly porous targets are completely different. Fundamental questions remain about which material properties are important and how porous asteroids fail during collisions. With these caveats in mind, we consider the current state of knowledge concerning catastrophic collisions of porous asteroids and how their collisional lifetimes might differ from those of nonporous bodies.

Asphaug et al. (1998) used their SPH code to model an impact into nonporous and porous models of Castalia. The impactor they considered was able to fracture the nonporous model significantly, but only ~10% of the mass of fragments were able to escape. An identical impact into a weak porous model produced more localized damage, but accelerated nearly half of the excavated fragments to escape velocity. This suggests porous weak asteroids could have shorter collisional lifetimes than their nonporous counterparts. But it is important to note that these simulations considered porosity only at scales comparable to the impactor size. Different results might be obtained for a model target consisting of porous boulders.

A somewhat different result is indicated by laboratory experiments. *Ryan et al.* (1999) impacted porous and nonporous ice targets with various projectiles and found that, even though the porous targets were mechanically weaker than their solid counterparts, the porous targets required about the same kinetic energy per unit target mass, Q^*, for shattering. Evidently, the shock dissipation in the porous targets was enough to offset their lower strength. *Stewart and Ahrens* (1999) reached a similar conclusion for targets made of plaster. *Love et al.* (1993) impacted sintered glass bead targets having various porosities and strengths. Based on a small number of experiments, *Love et al.* presented the following tentative relationship for Q^*

$$Q^* \propto S^{0.45} (1 - n)^{-3.6} \qquad (1)$$

where S is the strength and n is the bulk porosity of the target. This is in qualitative agreement with the other experimental results in that low target strength tends to offset high porosity. For a constant target strength, this relation implies that Q^* for an asteroid with 60% porosity would be ~8× larger than for a body with 30% porosity.

The dependence of collisional lifetime on bulk porosity can be estimated as follows: By definition, $Q^* = 0.5a^{*3}\delta U^2/(\rho R^3)$, where a^* is the threshold impactor radius required for shattering, ρ is the bulk density of the asteroid, δ is the impactor density, U is the impact velocity, and R is the asteroid radius. For a given impactor density, impact velocity, and asteroid radius, $Q^* \propto a^{*3}/\rho$. Additionally, the bulk density is related to the porosity n and the grain density ρ_g by $\rho = (1 - n)\rho_g$. Substituting these expressions into equation (1) shows that $a^* \propto (1 - n)^{-0.87}$. *Farinella and Davis* (1994) estimate that the flux of impacting bodies in the main belt is proportional to the –2.5 power of the impactor radius. The collisional lifetime against shattering (which is inversely proportional to the flux) is therefore proportional to $(1 - n)^{-2.2}$. Using this relation, an asteroid with 60% porosity survives 3.4× longer than an asteroid of the same size that has 30% porosity.

We note that any estimates of Q^* for porous asteroids are highly uncertain at present. It is not clear what strength measure should be used in relationships like equation (1). Equally uncertain is the range of event sizes over which target strength is important. In nonporous materials, a ratio of lithostatic pressure ρgR (ρ = target density, g = gravitational acceleration, R = asteroid radius) to the size-dependent tensile strength S is typically used to delineate the regimes where gravity or strength determine the outcome of a collision. Large or small values of ρgR/S indicate collisions in a gravity-dominated or strength-dominated regime respectively. Consider a similar approach for porous targets. There is reason to believe, based on as-yet-unpublished results by one of the authors (K.H.), that crushing strength provides a good correlation of cratering data for impacts into porous materials. In this case, a strength/gravity transition would be determined by the ratio $\rho gR/S_c$, where S_c is the crushing strength of the target material. Normally, a large value of this ratio would indicate a collision is gravity-dominated. However, by definition, a crushable material under the condition of $\rho gR/S_c \gg 1$ would self-compact until reaching a state near $\rho gR/S_c = 1$. On this basis, a gravity-dominated regime may never exist for collisions involving highly porous asteroids. Such bodies might always be restricted to a regime where strength is either dominant or, at best, comparable to lithostatic pressure. These questions must be studied further using both laboratory and numerical methods before reliable estimates of collisional lifetimes can be made.

5.4. Structure of Porous Asteroids

A large proportion of asteroids appear to have significant macroporosity, and in some cases the level of macroporosity suggests that the object has almost as much (or more) empty space within its volume as solid material. How can highly porous asteroids maintain so much empty macropore space? Spacecraft imaging of asteroid surfaces show that they are covered with a loose particulate regolith with particle sizes that are much smaller than the limit of resolution of the images (*Belton et al.*, 1995; *Veverka et al.*, 2000). In the case of the highest-resolution images of an asteroid surface, provided by the *NEAR Shoemaker* landing on 433 Eros, the finest surface material was again smaller than the limit of imaging resolution, suggesting dust-sized particles. But 433 Eros has an estimated macroporosity of 18%. Why doesn't this fine material filter into the interior cracks and voids and eliminate the object's macroporosity over time? As discussed in section 4, asteroid 253 Mathilde has an estimated macroporosity of 40%. Why doesn't the particulate material on the surface work its way into an object that is almost as much empty space as solid material? Can an asteroid maintain high internal porosity for what appears to be millions of years while meteoroid impacts on its surface grind down material to finer particle sizes, while shaking should cause that material to settle into pore spaces?

There are probably three major factors acting to create and preserve high levels of macroporosity in asteroids. First is the original structure of disrupted asteroids. When a previously coherent asteroid is catastrophically disrupted by a high-velocity impact, the physics of impacts favor a strong size sorting of the fragments that reaccrete. The largest disrupted pieces will have lowest relative velocities imparted by the collision and the largest relative gravitational attraction, so they will tend to be the ones to reaccrete first, forming an irregular core for the asteroid (*Melosh and Ryan*, 1997; *Wilson et al.*, 1999). Smaller size fractions would have progressively larger relative velocities, have smaller gravitational attractions, and take progressively longer to reaccrete. The tendency would be to size sort the rubble-pile asteroid with the largest (and most irregularly shaped) pieces in the center and progressively finer sizes building up in size-sorted layers. This size-sorted structure would create large void spaces in the interior of the asteroid while segregating the finer particle sizes on the outside layers of an asteroid.

The second factor is that in addition to the finest size fractions accreting last, the energy and processes that generate fine particles over time are impact-related regolith processes. These processes crush and pulverize asteroidal material, but necessarily operate only on the asteroid's surface. What is needed to fill the internal void space is a mechanism to move the fine material from the surface to the interior.

The third factor is the forces on those fine particles. Using a simple model of gravitational and frictional forces *Britt and Consolmagno* (2001) point out that while gravitational forces on particles increase linearly with the distance from the center of an object, the frictional forces preventing a piece of rubble from falling downward follow an inverse square law, increasing with the square of the depth into the body. While frictional force is weakest near the surface, it increases rapidly with depth. The small particles of the surface regolith tend to have the smallest gravitational force and a higher surface area to volume ratio than the larger particles; thus, they have proportionally the largest ratio of frictional to gravitational force. The smallest particles at any level are thus the ones most likely to be held up by friction. Calculations based on this model show that friction rapidly increases with depth and tends to quickly dominate gravitational force. For example, in a 40-km-diameter asteroid, frictional forces on a centimeter-sized particle would be 2 orders of magnitude larger than gravitational forces at only 10 m depth. By 500 m the difference is 4 orders of magnitude. The low gravitational acceleration on asteroids would also make it difficult for any particle to move a significant distance during the jolting and jostling that results from impact events. For example, for a 40-km-diameter asteroid, the gravitational acceleration at the surface is only 1.3 cm/s^2. For brief periods where interparticle contact is reduced or eliminated, this low acceleration would not allow much movement.

Another effect of friction may be to provide shattered bodies with what is in effect some level of coherent strength. Since friction will resist the displacement of the particles within a shattered object, pushing already-broken pieces apart would require larger forces than would be required by just gravitational scaling.

The implication is that for reaccreted rubble piles, the fine material generated and/or accreted at the surface tends to stay in the regolith zone. The major voids in the asteroid's deep interior that may date from the original reaccretion of the object remain unfilled. The major source of macroporosity reduction would probably come, as discussed earlier in this section, from the collapse of macroporosity due to major impacts. However, this is a very localized effect because of the rapid dissipation of the shock energy in the porous medium. Asteroid 253 Mathilde is an excellent example of an object with very high macroporosity that has survived repeated major impacts.

6. CONCLUSIONS AND FUTURE DIRECTIONS FOR ASTEROID POROSITY RESEARCH

6.1. Conclusions

Research into asteroid porosity and structure is still in its earliest stages. Shown in Fig. 1 are only 17 asteroids, some with substantial error bars. However, the rate of new asteroid bulk density measurements has increased remarkably over the last 10 years, and new measurements are being announced at every major planetary science meeting. This is an exciting time to see the density structure of the asteroid belt start to take shape. What we can interpret from these early and very limited data is extremely interesting:

1. Observations of asteroid mutual perturbation events, observations of asteroidal satellites (both optically and with radar), and spacecraft missions to asteroids have revolutionized our knowledge of asteroid bulk densities.

2. Meteorite studies have started to characterize the microporosity of meteorite groups. Using these data along with the grain density of meteorites, we can estimate bulk porosity and macroporosity of asteroids.

3. Most asteroids tend to have significant bulk porosity and significant macroporosity.

4. The three asteroids with diameters >500 km are coherent objects without macroporosity. Other than these three, low-macroporosity asteroids are rare.

5. A group of asteroids, including 433 Eros, have macroporosities of ~20%. These moderate-macroporosity asteroids have probably been significantly fractured, but not collisionally disrupted. Alternately, if they were disrupted and reassembled, they must have experienced significant compression by subsequent impacts.

6. A group of asteroids, including 253 Mathilde, have macroporosities >30% and in some cases >50%. These objects are probably loosely consolidated piles of collisional rubble whose internal structure is highly porous.

7. C-type asteroids tend to have greater macroporosities than S-type asteroids, suggesting that S-type asteroids are somewhat more resistant to catastrophic disruption, but both types are represented in the different groups.

8. Asteroids can have substantial interior macroporosity while retaining a fine loose surface regolith. Fine particle size fractions tend to be accreted and/or generated at the surface of an asteroid, but frictional and not gravitational forces dominate the movement of small particles, preventing infilling fractures and voids within the object.

9. High macroporosity creates a different and unusual impact regime in asteroids. Effects include rapid attenuation of shock, low-ejecta velocities, crater formation by compaction rather than ejection of material, and significantly enhanced survival against impacts that would otherwise disrupt a coherent object of the same size.

6.2. Future Directions

The little that is known about asteroid porosity whets the appetite for more knowledge and hints at exciting areas of research. First and foremost is the need for continuing observations of asteroids, both for mutual perturbation events and discovering satellites. Most of the mutual event and satellite data come from relatively large asteroids (i.e., masses $>10^{18}$ kg) and the figures of this paper are sparsely populated with asteroids less than that mass. Observations of asteroids a few hundred meters in diameter are only now becoming available, albeit with large error bars. Also needed are more porosity measurements of meteorites. Ordinary chondrites have been well characterized, but the surface has only been scratched on the other meteorite groups and a number of major questions remain to be answered.

The impact physics of porous asteroids is perhaps the most significant frontier. Porous targets appear to dominate the asteroid belt, but very little is known about how the nature of those targets affects the processes and outcomes of impacts. What we do know is that porosity does have a major effect on these processes and outcomes as illustrated by the images of 253 Mathilde that show at least five impact craters that should be large enough to disrupt a coherent object. With more measurements of asteroids and meteorites, and a greater understanding of the physics of impacts into porous targets, our understanding of the internal structure of asteroids will greatly expand. Already with our limited data we can see: an asteroid belt that contains large, coherent asteroids coexisting with somewhat smaller, shattered rubble piles; asteroids of the same spectral type and therefore similar mineralogy with factors of 3 differences in their bulk densities; and asteroids with more empty space than solid material. It is clear that the asteroid belt is a very interesting place, full of surprises and intellectual adventures.

REFERENCES

Asphaug E., Ostro S. J., Hudson R. S., Scheeres D. J., and Benz W. (1998) Disruption of kilometre-sized asteroids by energetic collisions. *Nature, 393,* 437–440.

Asphaug E., Ryan E. V., and Zuber M. T. (2002) Asteroid interiors. In *Asteroids III* (W. F. Bottke Jr. et al., eds.), this volume. Univ. of Arizona, Tucson.

Bange J. (1998) An estimation of the mass of asteroid 20-Massalia derived from the Hipparcos minor planets data. *Astron. Astrophys., 340,* L1–L4.

Belton M. J. S., Chapman C. R., Thomas P. C., Davies M. E., Greenberg R., Klassen K., Byrnes D., D'Amario L., Synnott S., Johnson T. V., McEwen A., Merline M. J., Davis D. R., Petit J.-M., Storrs A., Veverka J., and Zellner B. (1995) Bulk density of asteroid 243 Ida from the orbit of its satellite Dactyl. *Nature, 374,* 785–788.

Benner L. A. M., Ostro S. J., Giorgini J. D., Jurgens R. F., Margot J. L., and Nolan M. C. (2001) *1999 KW4; N Aql 2001.* IAU Circular 7632.

Bland P. A., Sexton A. S., Jull A. J. T., Bevan A. W. R., Berry F. J., Thornley D. M., Astin T. R., Britt D. T., and Pillinger C. T. (1998) Climate and rock weathering: A study of terrestrial age dated ordinary chondritic meteorites from hot desert regions. *Geochim. Cosmochim. Acta, 62,* 3169–3184.

Bottke W. F., Nolan M. C., Kolvoor R. A., and Greenberg R. (1994) Velocity distribution among colliding asteroids. *Icarus, 107,* 255–268.

Britt D. T. and Consolmagno G. J. (2000) The porosity of dark meteorites and the structure of low-albedo asteroids. *Icarus, 146,* 213–219.

Britt D. T. and Consolmagno G. J. (2001) Modeling the structure of high-porosity asteroids. *Icarus, 152,* 134–139.

Chapman C. R., Merline W. J., and Thomas P. (1999) Cratering on Mathilde. *Icarus, 140,* 28–33.

Consolmagno G. J. and Britt D. T. (1998) The density and porosity of meteorites from the Vatican collection. *Meteoritics & Planet. Sci., 33,* 1231–1242.

Consolmagno G. J., Britt D. T., and Stoll C. P. (1998) The porosities of ordinary chondrites: Models and interpretation. *Meteoritics & Planet. Sci., 33,* 1221–1230.

Consolmagno G. J., Bland P. A., and Strait M. M. (1999) Weathering and porosity: A preliminary SEM study of weathered meteorites (abstract). In *Lunar and Planetary Science XXX,* Abstract #1158. Lunar and Planetary Institute, Houston (CD-ROM).

Drummond J. D. and Cocke W. J. (1989) Triaxial ellipsoid dimensions and rotational pole of 2 Pallas from two stellar occultations. *Icarus, 78,* 323–329.

Drummond J. D., Fugate R. Q., Christou J. C., and Hege E. K. (1998) Full adaptive optics images of asteroids Ceres and Vesta; rotational poles and triaxial ellipsoid dimensions. *Icarus, 132,* 80–99.

Dunham D. W. and 45 colleagues (1990) The size and shape of (2) Pallas from the 1983 occultation of 1 Vulpeculae. *Astron. J., 99,* 1636–1662.

Farinella P. and Davis D. R. (1994) Will the real size distribution please step forward? (abstract). In *Lunar and Planetary Science XXV,* pp. 365–366. Lunar and Planetary Institute, Houston.

Flynn G. J., Moore L. B., and Klock W. (1999) Density and porosity of stone meteorites: Implications for the density, porosity, cratering, and collisional disruption of asteroids. *Icarus, 142,* 97–105.

Gaffey M. J., Bell J. F., Brown R. H., Burbine T. H., Piatek J. L., Reed K., and Chaky D. A. (1993) Mineralogical variations within the S-type asteroid class. *Icarus, 106,* 573–602.

Grady M. M. (2000) *Catalogue of Meteorites,* 5th edition. Cambridge Univ., Cambridge. 696 pp.

Goffin E. (1991) The orbit of 203 Pompeja and the mass of Ceres. *Astron. Astrophys., 249,* 563–568.

Goffin E. (2001) New determination of the mass of Pallas. *Astron. Astrophys., 365,* 627–630.

Hagerty M. M., Hite D. R., Ulrich C. R., and Hagerty D. J. (1993) One-dimensional high pressure compression of granular media. *J. Geotech. Eng., 119,* 1–18.

Hertz H. (1966) *T Pyx; 1966e; The Mass of Vesta; Notice Re. Precise Positions of Comets.* IAU Circular 1983 (see also *Science, 160,* 299–300).

Hilton J. L. (1997) The mass of the asteroid 15 Eunomia from observations of 1313 Berna and 1284 Latvia. *Astron. J., 114,* 402–408.

Hilton J. L. (1999) US Naval Observatory ephemerides of the largest asteroids. *Astron. J., 117,* 1077–1086.

Hilton J. L. (2002) Asteroid masses and densities. In *Asteroids III* (W. F. Bottke Jr. et al., eds.), this volume. Univ. of Arizona, Tucson.

Hilton J. L., Seidelmann P. K., and Middour J. (1996) Prospects for determining asteroid masses. *Astron. J., 112,* 2319–2329.

Holsapple K. A. (1993) The scaling of impact processes in planetary sciences. *Annu. Rev. Earth Planet. Sci., 21,* 333–373.

Holsapple K., Giblin I., Housen K., Nakamura A., and Ryan E. (2002) Asteroid impacts: Laboratory experiments and scaling laws. In *Asteroids III* (W. F. Bottke Jr. et al., eds.), this volume. Univ. of Arizona, Tucson.

Housen K. R. and Holsapple K. A. (1999) Scale effects in strength-dominated collisions of rocky asteroids. *Icarus, 142,* 21–33.

Housen K. R. and Voss M. E. (2001) Ejecta from impact craters in porous materials (abstract). In *Lunar and Planetary Science XXXII,* Abstract #1617. Lunar and Planetary Institute, Houston (CD-ROM).

Housen K. R., Holsapple K. A., and Voss M. E. (1999) Compaction as the origin of the unusual craters on asteroid Mathilde. *Nature, 402,* 155–157.

Lambe T. W. and Whitman R. V. (1969) *Soil Mechanics.* Wiley and Sons, New York. 553 pp.

Landgraf W. (1988) The mass of Ceres. *Astron. Astrophys., 191,* 161–166.

Landgraf W. (1992) A determination of the mass of (704) Interamnia from observations of (993) Moultona. In *Chaos, Resonance, and Collective Dynamical Phenomena in the Solar System* (S. Ferraz-Mello, ed.), pp. 179–182. Kluwer, Dordrecht.

Lopez Garcia A., Medvedev Yu. D., and Morano Fernandez J. A. (1997) Using close encounters of minor planets for the improvement of their masses. In *Dynamics and Astrometry of Natural and Artificial Celestial Bodies* (I. M. Wytrzyszczak et al., eds.), pp. 199–204. Kluwer, Dordrecht.

Love S. J., Hörz F., and Brownlee D. E. (1993) Target porosity effects in impact cratering and collisional disruption. *Icarus, 105,* 216–224.

Lundborg N. (1967) The strength-size relation of granite. *Intl. J. Rock Mech. Min. Sci., 4,* 269–272.

Margot J. L. and Brown M. E. (2001) Discovery and characterization of binary asteroids 22 Kalliope and 87 Sylvia. *Bull. Am. Astron. Soc., 33,* 1133.

Margot J. L., Nolan M. C., Benner L. A., Ostro S. J., Jurgens R. F., Giorgini J. D., Slade M. A., and Campbell D. B. (2001) Radar observations of binary asteroid 2000 DP107 (abstract). In *Lunar and Planetary Science XXXII,* Abstract #1754. Lunar and Planetary Institute, Houston (CD-ROM).

McSween H. Y. (1999) *Meteorites and Their Parent Planets.* Cambridge Univ., New York. 310 pp.

Meibom A. and Clark B. E. (1999) Evidence for the insignificance of ordinary chondritic material in the asteroid belt. *Meteoritics & Planet. Sci., 34,* 7–24.

Melosh H. J. and Ryan E. V. (1997) Asteroids: Shattered but not dispersed. *Icarus, 129,* 562–564.

Menikoff R. (2001) Meso-scale simulations of compaction waves in granular beds. *23rd Intl. Symp. Shock Waves,* Paper #5016. Fort Worth, Texas.

Merline W. J., Close L. M., Dumas C., Chapman C. R., Roddler F., Menard F., Slater D. C., Duvert G., Shelton C., and Morgan T. (1999) Discovery of a moon orbiting the asteroid 45 Eugenia. *Nature, 401,* 565–568.

Merline W. J., Close L. M., Dumas C., Shelton J. C., Menard F., Chapman C. R., and Slater D. C. (2000) Discovery of companions to asteroids 762 Pulcova and 90 Antiope by direct imaging. *Bull. Am. Astron. Soc., 32,* 1017.

Michalak G. (2000) Determination of asteroid masses I. (1) Ceres, (2) Pallas and (4) Vesta. *Astron. Astrophys., 360,* 363–374.

Millis R. L. and 41 colleagues (1987) The size, shape, density, and albedo of Ceres from its occultation of BD+8o471. *Icarus,*

72, 507–518.

Nakamura A., Suguiyama K., and Fujiwara A. (1992) Velocity and spin of fragments from impact disruptions. I. An experimental approach to a general law between mass and velocity. *Icarus, 100,* 127–135.

Nolan M. C., Margot J.-L., Howell E. S., Benner L. A. M., Ostro S. J., Jurgens R. F., Giorgini J. D., and Campbell D. B. (2001) Radar observations of near-Earth asteroids 2000 UG11 and 2000 UK11 (abstract). In *Lunar and Planetary Science XXXII,* Abstract #2055. Lunar and Planetary Institute, Houston (CD-ROM).

Richardson D. C., Leinhardt Z. M., Melosh H. J., Bottke W. F. Jr., and Asphaug E. (2002) Gravitational aggregates: Evidence and evolution. In *Asteroids III* (W. F. Bottke Jr. et al., eds.), this volume. Univ. of Arizona, Tucson.

Ryan E. V., Davis D. R., and Giblin I. (1999) A laboratory impact study of simulated Edgeworth-Kuiper belt objects. *Icarus, 142,* 56–62.

Sears D. W. G. (1998) The case for rarity of chondrules and calcium-aluminum-rich inclusions in the early solar system and some implications for astrophysical models. *Astrophys. J., 498,* 773–778.

Scholl H., Schmadel L.D., and Roeser S. (1987) The mass of the asteroid (10) Hygiea derived from observations of (829) Academia. *Astron. Astrophys., 179,* 311–316.

Schubart J. and Matson D. L. (1979) Masses and densities of asteroids. In *Asteroids* (T. Gehrels, ed.), pp. 84–97. Univ. of Arizona, Tucson.

Standish E. M. (1998) JPL Planetary and Lunar Ephemerides, DE405/LE405. *JPL Interoffice Memorandum 312.F-98-048,* August 26, 1998.

Standish E. M. (2001) *JPL Interoffice Memorandum 312.F-01-006,* April 11, 2001.

Standish E. M. and Hellings R.W. (1989) A determination of the masses of Ceres, Pallas, and Vesta from their perturbations on the orbit of Mars. *Icarus, 80,* 326–333.

Stewart S. and Ahrens T. J. (1999) Porosity effects on impact processes in solar system materials (abstract). In *Lunar and Planetary Science XXXI,* Abstract #2020. Lunar and Planetary Institute, Houston (CD-ROM).

Tedesco E. F. (1989) Asteroid magnitudes, UBV colors, and IRAS albedos and diameters. In *Asteroids II* (R. P. Binzel et al., eds.), pp. 1090–1150. Univ. of Arizona, Tucson.

Tedesco E. F., Veeder G. J., Fowler J. W., and Chillemi J. R. (1992) *The IRAS Minor Planet Survey.* Phillips Laboratory Report PL-TR-92-2049.

Terho M., Pesonen L. J., Kukkonen I. T., and Bukovansk M. (1993) The petrophysical classification of meteorites. *Studia Geoph. Geod., 37,* 65–82.

Thomas P. C., Binzel R. P., Gaffey M. J., Storrs A. D., Wells E. N., and Zellner B. H. (1997) Impact excavation on asteroid 4 Vesta: Hubble Space Telescope results. *Science, 277,* 1492–1495.

Veillet C., Doressoundiram A., and Shapiro J. (2001) *S/2000 (1998 WW31) 1.* IAU Circular 7610.

Veverka J., Thomas P., Harch A., Clark B., Bell J. F. III, Carcich B., Joseph J., Chapman C., Merline W., Robinson M., Malin M., McFadden L. A., Murchie S., Hawkins S. E. III, Farquhar R., Izenberg N., and Cheng A. (1997) NEAR's flyby of 253 Mathilde: Images of a C asteroid. *Science, 278,* 2109–2114.

Veverka J., Robinson M., Thomas P., Murchie S., Bell J. F. III, Izenberg N., Chapman C., Harch A., Bell M., Carcich B., Cheng A., Clark B., Domingue D., Dunham D., Farquhar R., Gaffey M. J., Hawkins E., Joseph J., Kirk R., Li H., Lucey P., Malin M., Martin P., McFadden L., Merline W. J., Miller J. K., Owen W. M. Jr., Peterson C., Prockter L., Warren J., Wellnitz D., Williams B. G., and Yeomans D. K. (2000) NEAR at Eros: Imaging and spectral results. *Science, 289,* 2088–2097.

Viateau B. (2000) Mass and density of asteroids (16) Psyche and (121) Hermione. *Astron. Astrophys., 354,* 725–731.

Viateau B. and Rapaport M. (1995) The orbit of (2) Pallas. *Astron. Astrophys. Suppl. Ser., 111,* 305–310.

Viateau B. and Rapaport M. (1997) The Bordeaux meridian observations of asteroids. First determination of the mass of (11) Parthenope. *Astron. Astrophys., 320,* 652–658.

Viateau B. and Rapaport M. (1998) The mass of (1) Ceres from its gravitational perturbations on the orbits of 9 asteroids. *Astron. Astrophys., 334,* 729–735.

Wasserman L. H., Millis R. L., Franz O. G., Bowell E., White N. M., Giclas H. L., Martin L. J., Elliot J. L., Dunham E., Mink D., Baron R., Honeycutt R. K., Henden A. A., Kephart J. E., A'Hearn M. F., Reitsema H. J., Radick R., and Taylor G. E. (1979) The diameter of Pallas from its occultation of SAO 85009. *Astron. J., 84,* 259–268.

Weidenshilling S. J., Marzari F., Davis D. R., and Neese C. (2001) Origin of the double asteroid 90 Antiope: A continuing puzzle (abstract). In *Lunar and Planetary Science XXXII,* Abstract #1890. Lunar and Planetary Institute, Houston (CD-ROM).

Wilkison S. L. and Robinson M. S. (2000) Bulk density of ordinary chondrite meteorites and implications for asteroidal internal structure. *Meteoritics & Planet. Sci., 35,* 1203–1213.

Wilson L., Keil K., and Love S. J. (1999) The internal structures and densities of asteroids. *Meteoritics & Planet. Sci., 34,* 479–483.

Yamamuro J. A., Bopp P. A., and Lade P. V. (1996) One-dimensional compression of sands at high pressures. *J. Geotech. Eng., 121,* 147–155.

Yeomans D. K., Barriot J.-P., Dunham D. W., Farquhar R. W., Giorgini J. D., Helfrich C. E., Konopliv A. S., McAdams J. V., Miller J. K., Owen W. M. Jr., Scheeres D. J., Synnott S. P., and Williams B. G. (1997) Estimating the mass of asteroid 253 Mathilde from tracking data during the NEAR flyby. *Science, 278,* 2106–2109.

Yeomans D. K., Antreasian P. G., Cheng A., Dunham D. W., Farquhar R. W., Gaskell R. W., Giorgini J. D., Helfrich C. E., Konopliv A. S., McAdams J. V., Miller J. K., Owen W. M. Jr., Thomas P. C., Veverka J., and Williams B. G. (1999) Estimating the mass of asteroid Eros during the NEAR spacecraft flyby. *Science, 285,* 560–561.

Yeomans D. K., Antreasian P. G., Barriot J.-P., Chesley S. R., Dunham D. W., Farquhar R. W., Giorgini J. D., Helfrich C. E., Konopliv A. S., McAdams J. V., Miller J. K., Owen W. M. Jr., Scheeres D. J., Thomas P. C., Veverka J., and Williams B. G. (2000) Radio science results during the NEAR Shoemaker spacecraft rendezvous with Eros. *Science, 289,* 2085–2088.

Zuber M. T., Smith D. E., Cheng A. F., Garvin J. B., Aharonson O., Cole T. D., Dunn P. J., Guo Y., Lemoine F. G., Neumann G. A., Rowlands D. D., and Torrence M. H. (2000). The shape of 433 Eros for the NEAR Shoemaker laser rangefinder. *Science, 289,* 2097–2101.

Gravitational Aggregates: Evidence and Evolution

D. C. Richardson
University of Maryland at College Park

Z. M. Leinhardt
University of Maryland at College Park

H. J. Melosh
University of Arizona

W. F. Bottke Jr.
Southwest Research Institute at Boulder

E. Asphaug
University of California at Santa Cruz

It has been suggested that asteroids between ~100 m and ~100 km in size may be gravitational aggregates of loosely consolidated material. Recently evidence has been mounting to support this idea. Comet breakups, crater chains, doublet craters, giant craters, grooves, asteroid spins, underdense asteroids, asteroid satellites, and unusual asteroid shapes can all be explained to some degree by the properties of aggregate structures. Moreover, laboratory and numerical experiments indicate that solid asteroids, if they existed, would likely be completely shattered by now, owing to their violent collisional histories. Once shattered, a body becomes increasingly difficult to disrupt since it can efficiently absorb energy deposited at the impact site. In this chapter we present the evidence for gravitational aggregates and discuss the implications of their existence for asteroid evolution. We also propose a new classification scheme for asteroid structure parameterized by fundamental quantities such as strength and porosity.

1. DEFINITIONS

Over 50 years ago, *Jeffreys* (1947) and *Öpik* (1950) introduced the idea that some asteroids and comets (collectively, planetesimals) may not be solid objects governed solely by material strength. While investigating the distortion and disruption effects of planetary tides, both came to the conclusion that tidal forces in the solar system could not significantly affect solid objects. Nearly three decades later, *Chapman* (1978) used the term "rubble pile" to describe a gravitationally bound collection of boulders, arguing that high-speed collisions between main-belt asteroids would cause extensive fracturing and erosion, turning such asteroids into rubble. Later, *Weissman* (1986) suggested comets may be "primordial rubble piles" in order to explain the comparatively high frequency of comet-splitting events. Currently, "rubble pile" is used in the planetary science community to describe a variety of configurations that range from theoretical constructs like piles of marbles to more realistic speculations on planetesimal interiors, and some confusion has arisen as to precise definitions.

In an effort to standardize terms, we propose a new scheme for describing the range of possible asteroid (and, to a certain extent, comet) configurations. We have selected parameters that convey a sense of how a given planetesimal will react to the most important geological processes affecting such bodies today, namely short-term stresses like collisions (and possibly manmade explosions), and long-term stresses such as tidal forces and rotational spinup. (Here "short" and "long" refer to the interval over which the load is applied divided by the time it takes a seismic wave to travel the diameter of the object.) The definitions also provide some idea of the origin or collisional history of the object. It is important to emphasize that the classification is meant as a bridge between theory and observation, not a tool for mapping observed asteroids. Some parameters may be directly measurable (from ground or flyby observation, or even from *in situ* seismic mapping in the future) but others perhaps can only be inferred from theory. The aim is to provide a useful starting point from which both observers and theorists can make firm predictions.

In the proposed scheme, the term "rubble pile" becomes more specific, referring to a particular class of broken-up objects. We assign, at a minimum, two parameters that uniquely locate the object on a diagram that ranges from zero consolidation (rubble pile) to perfect consolidation (monolith). In addition, we augment the classification by the size of the object's largest component, if applicable, and treat regolith, contact binaries, and other such compound structures as special cases. We omit differentiated objects (such as Vesta) entirely. The classification is not exhaustive, nor is it meant to be: The hope is that future study will add detail to this simple scheme, without sacrificing the fundamentals.

The two principal parameters in our classification scheme are porosity

$$\text{porosity} = 1 - \frac{\text{sum of component volumes}}{\text{bulk volume}} = \frac{\text{sum of void space}}{\text{bulk volume}} \quad (1)$$

and relative tensile strength

$$\text{RTS} = \frac{\text{tensile strength of object}}{\text{mean tensile strength of components}} \quad (2)$$

where "tensile strength" is the maximum stress (force per unit area) — applied in a way that tends to cause separation across the plane on which it acts — that a body can withstand before fracture or rupture occurs. These parameters are dimensionless and vary from 0 to 1, both desirable properties. We are deliberately vague in our notion of "component" since the aim is to measure and compare bulk quantities, not detailed internal properties, although evidently some assumption as to the nature of the components is present in the definition of RTS. Similarly, we do not distinguish between micro- and macroporosity, since a macroporous asteroid (with voids between components) could be made up of microporous components (with voids inside the components), adding an undesirable level of complication. Such a distinction, however, may be important for understanding detailed collisional physics: Both configurations damp impact energy, but in a different way, the former via excavation and ejection, the latter through compaction and crushing (cf. sections 2.2 and 2.3). These details are beyond the scope of our simple classification but can always be added later. Finally, we recognize that various factors can contribute to overall strength, including shear strength (the internal resistance of a body to tangential stress, which typically includes a frictional component) and even geometrical locking between pieces, but again for our purposes we choose not to distinguish between these cases. Instead, using just porosity and RTS as defined, we can roughly predict how asteroids and comets will react to the primary geological processes affecting these bodies today, which is our primary goal.

To these parameters we adopt a secondary measure proposed by *Campo Bagatin and Petit* (2001), the mass fraction of the largest component

$$\text{MF} = \frac{\text{mass of largest component}}{\text{total mass of object}} \quad (3)$$

With this definition we can state that bodies with $\text{MF} \geq 0.5$ are more properly termed compound objects, such as regolith overlying a large consolidated component, or a contact binary with one lobe thoroughly damaged. Detail regarding the size distribution of the components can always be added, such as a single power-law exponent (if appropriate), or even a ratio of higher-order moments of the size distribution (e.g., sum of component volumes divided by sum of component surface areas, normalized by the mean diameter of the object). This level of detail may be more appropriate for theorists discussing a numerical model (e.g., monodispersive vs. size distribution) or for observers who have achieved a comprehensive characterization of the particle sizes in the surface layer of the object in question, for example. It may be better in some cases to simply distinguish between "fine" and "coarse" particles.

Figure 1 shows two realizations of the RTS vs. porosity parameter space in this scheme, with the plot on the left showing proposed structural classifications for distinct regions and the plot on the right showing how objects in these regions will react to stress. We deliberately omit quantitative values or sharp divisions, since many details regarding these structures are unknown at this time. In simple terms, objects with low RTS are more susceptible to tidal disruption than objects with high RTS, and objects with low porosity are less efficient at absorbing impact energy than objects with high porosity. We elaborate on these points in the following brief qualitative descriptions of the regions.

Monolithic (bottom right): These objects are essentially unaffected by long-term stresses like planetary tidal forces. For highly energetic events such as collisions or explosions, the compressive wave easily reaches the farside of the object, reflecting as a tensive wave that can produce damage and spalls.

Fractured (bottom middle): These objects may have a significant number of faults and/or joints and may lack enough tensile strength to resist disruption by long-term stresses. Our definition implies that the original structure of the object is still largely in place (hence the low porosity). However, the definition is open to more complicated histories, for example, a previously reassembled object may have been tamped down into a less-porous structure, resembling a simple fractured body.

Shattered (bottom left): These objects are related to fractured bodies, except that they are even more susceptible to disruption from long-term stresses like tidal forces or rotational spinup. A body in this category could be characterized as an object cleaved into a few large components or one whose structure is dominated by joints and cracks. Like the fractured body, the original structure of the object is mostly in place. The RTS of a shattered object may be larger than zero, since jagged or interlocking edges and friction may allow the object to resist some stresses. The reaction of this body to short-term stresses like collisions is different from a simple monolith. Spalls are damped and the tensile wave is suppressed, such that this object is more difficult to disrupt. Small craters formed on single components of the body may act like they were formed on a monolith.

Shattered with rotated components (lower middle): An object in this category has had its original pieces displaced

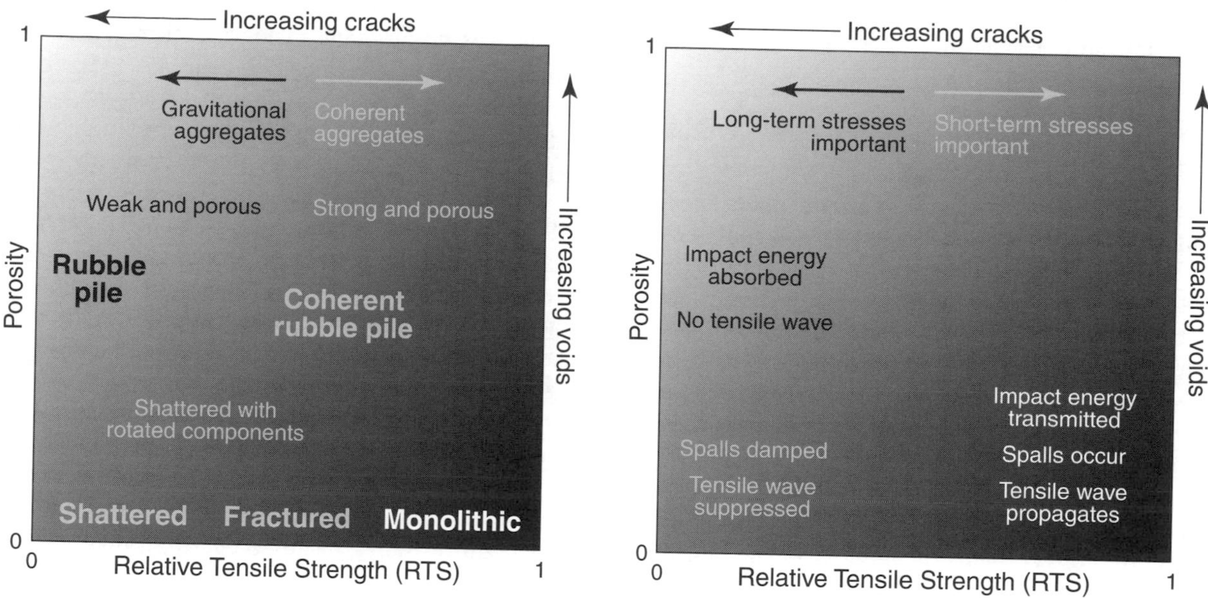

Fig. 1. The RTS-porosity parameter space. The plot on the left assigns labels to distinct regions on the basis of internal structure. The plot on the right describes how objects correspondingly react to stress in these regions. The divisions are deliberately vague and qualitative since the quantitative details are poorly known.

and reoriented somewhat, although much of the original structure may remain. It may contain void spaces, or possibly some cracks may be filled with regolith, although these features do not dominate its overall structure.

Rubble pile (middle left): This structure is literally a pile of rubble, with the organization that you might expect from a bunch of rocks dumped from a truck. A body that has been completely shattered and reassembled may fit into this category. The "bowling-ball," "marble," or "sand" piles used in numerical models (as simplified constructs of real rubble piles) also fit in this category. Typically a real rubble pile has moderate porosity because of the disorganization, but note that the porosity of a group of equal-size spheres can range from 0.26 (hexagonal or cubic closest packing) to 0.48 (simple cubic packing), and yet both configurations are highly ordered.

We deliberately omit any notion of particle size from this definition. Formally, "rubble" is breccia (angular rock fragments) without any cement or glue, distinguished from a shattered body by its rotated components (*Jackson,* 1997). A researcher may have a certain notion for the fragment sizes in this context, though it is not a formal part of the definition. Instead, it is more appropriate to qualify the sizes as needed, using formal definitions of particle types (i.e., silt, sand, pebble, boulder) if desired. With this in mind, for our classification, "rubble" can be anything from tiny grains of clay to boulders tens of kilometers across or even larger, unless specified otherwise.

The reaction of rubble piles to short-term stresses like collisions is absorption of impact energy via compression, with little to no tensile wave developed in the structure. Thus, impact energy is muffled and often goes into heat. Conversely, long-term tidal stresses can completely pull a rubble pile apart.

Coherent rubble pile (middle): This group is for rubble piles whose components have somehow become attached or cemented to one another. This would give the objects some RTS.

Weak and porous/strong and porous (top): These objects have low to high tensile strength with high porosity, something akin to dust bunnies, pumice, or aerogel. Comet nuclei are often described as weak and porous (e.g., *Bockelée-Morvan et al.,* 2001).

What is missing so far in this categorization is an overall term that describes bodies made up of multiple components, i.e., bodies that are not pure monoliths. We propose "aggregate" for this purpose, without being too specific about the detailed configuration. This can be refined to "gravitational aggregate" for objects with low to zero RTS, conveying the notion of gravity keeping a collection of particles together, or "coherent aggregate" for objects with moderate to high RTS, something similar to a conglomerate but with no explanation about the "glue," except that it is not gravity (see Fig. 1).

In the remainder of this chapter we will outline the evidence for why we think aggregates, and gravitational aggregates in particular, exist among the small-body population in the solar system and what the implications are for the origin and evolution of these bodies. The observational evidence is presented first, with detailed theoretical considerations and discussion of origin and evolution deferred to the later sections. We will endeavor to employ the new classification scheme as consistently as possible throughout the chapter with the hope that use of the new terms in their proper context will help solidify the definitions.

2. EVIDENCE

Recently the possibility that a large percentage of ~0.1–100-km-sized bodies in the solar system, including asteroids, are fragile gravitational aggregates has gained greater acceptance. The reason for this is the mounting evidence from observations, experiments, and simulations that such configurations are (or should be) common. In what follows we provide an overview of this evidence.

2.1. Observations

Observational evidence for gravitational aggregates comes primarily from direct optical imaging or radar. Targets for optical imaging include both the bodies themselves and the geologic record of their past existence. Some of the earliest evidence actually comes from meteorites, mesosiderites in particular, which are mixtures of mantle basalt and core iron. The presence of both mantle and core material in a single meteorite suggests gravitational reaccumulation of the parent body following catastrophic disruption (*Scott et al.,* 2001). We must turn to other evidence, however, to determine whether such bodies still exist today.

2.1.1. Comet breakup. Comet breakups, such as that of Comet D/Shoemaker-Levy 9 (SL9) at Jupiter, may offer some insight into present-day asteroid aggregates. In July 1992, SL9 passed very close to Jupiter (*Chodas and Yeomans,* 1996), well inside the tidal breakup (Roche) limit for unconsolidated water ice (*Asphaug and Benz,* 1996). *Scotti and Melosh* (1993) estimated the tidal stress on the inferred parent body and found it to be so small (~10^{-4} bar) that the comet was likely an incoherent aggregate of fragments before breakup. *Asphaug and Benz* (1994) used an N-body code to model the gravitational forces between the constituent fragments and found that the tidal encounter elongated the rubble pile until it suffered "clump instabilities" that formed a fragment train reminiscent of SL9. Tidal breakup simulations of rubble pile asteroids show similar behavior (section 3.2).

Other comet breakups have been observed, the bulk of which were apparently caused by spontaneous nucleus splitting (*Weissman,* 1982). The most recent example is the breakup of Comet C/1999 S4 (LINEAR) (*Weaver et al.,* 2001). Some other tidal disruptions are known, however: P/Brooks 2 broke into at least eight fragments when it approached within 2 jovian radii of Jupiter in 1886 (*Sekanina and Yeomans,* 1985); various Sun grazers may also have been broken up by tides (*Weissman,* 1980; *Sekanina,* 2000).

2.1.2. Crater chains. Crater chains, or catenae, are linear configurations of up to several dozen equally spaced, similarly sized impact craters spread out over tens of kilometers. They are distinguished from endogenic features by the well-defined rims on the regular and circular craters, and the presence of chains in nonvolcanic regions (*Wichman and Wood,* 1995; *McKinnon and Schenk,* 1995). Some crater chains are known to have formed from the ejecta of a single impact event owing to their radial alignment with respect to a nearby large crater. But there are about a dozen crater chains on Ganymede and Callisto (*Schenk et al.,* 1996), and even one or two on our own Moon (*Melosh and Whitaker,* 1994; *Wichman and Wood,* 1995), that have no obvious source crater. [*Schenk et al.* (1996) found several potential crater chains on the saturnian satellites Dione, Rhea, Enceladus, and Triton, but they are all either severely degraded or in historically active regions. *Asphaug and Benz* (1996) do not think SL9-type disruption can occur near Saturn because of the planet's low density.]

Melosh and Schenk (1993) and *Bottke et al.* (1997) have suggested that these catenae are impact signatures of SL9-like fragment trains. In this scenario, a tidally disrupted body strikes one of the planet's moons on the outbound orbit, several hours to a few days after disruption. This is long enough for distinct clumps to form through instability and yet short enough for the fragment train to form a recognizable crater chain. The spacing of the craters is a function of the somewhat random orientation of the train with respect to the impact surface. The absence of any correlation between the inferred parent body mass and the number of craters in the catena supports the idea that the fragments reaccumulated via gravitational instability just prior to impact (*Asphaug and Benz,* 1996; *Schenk et al.,* 1996). *Bottke et al.* (1997) find that Earth-crossing asteroids could be pulled apart by tides if the bodies are sufficiently weak, and therefore may account for the presence of one or two crater chains on the Moon since the late heavy bombardment. Suggestions of crater chains formed on Earth by tidal disruption of an asteroid by the Moon (e.g., *Rampino and Volk,* 1996; *Ocampo and Pope,* 1996; *Spray et al.,* 1998) are discounted by *Bottke et al.* (1997) because in their model they find that, owing to the Moon's smaller size and density, for every crater chain on Earth there should be dozens of fresh crater chains on the Moon.

2.1.3. Doublet craters. Roughly 10% of the largest (mean diameter $D \geq 20$ km) impact structures on the Earth and Venus, and 2% on Mars, are doublet craters, i.e., well-separated pairs of similarly sized craters that formed simultaneously (*Bottke and Melosh,* 1996a,b). The craters are too separated (for their size) to have been formed by tidal disruption or aerodynamic breakup of an asteroid just prior to impact since these forces do not give the components sufficient tangential separation during the short interval before collision (*Melosh and Stansberry,* 1991). Also, shallow impact angles can be ruled out by the lack of crater elongation. *Bottke and Melosh* (1996a,b) argue that the impactor must be a binary with well-separated components before the final impact encounter. They showed that tidal disruption of a gravitational aggregate (modeled as a simple two-component contact binary) by a terrestrial planet could result in detached components that evolve to larger separation via repeated distant encounters with terrestrial planets, or through mutual tidal interactions. Combining their code with a Monte Carlo code based on the work of *Chauvineau et al.* (1995), they found that ~15% of kilometer-sized Earth- and Venus-crossing asteroids and ~5% of solely Mars-crossing asteroids evolve into well-separated binaries (the Mars fraction is smaller due to the relative inefficiency of its cross section at causing tidal disruption). They further showed that a steady-state population of binaries from tidal fission and disruption events could account for the present fraction of doublets on Venus, Earth, and Mars, and could imply a sizable percentage of doublets on the Moon (~10%) and Mercury (~5%) as well (the actual number has not been determined due to the difficulty in distinguishing doublets from the saturated small-crater population on these bodies).

2.1.4. Spins. Recent measurements of asteroid spin periods from lightcurve analysis have placed interesting constraints on asteroid properties. *Pravec and Harris* (2000) analyzed data for 750 main-belt, near-Earth, and Mars-crossing asteroids. They found that the smallest asteroids (mean diameter D between 0.2 and 10 km, inferred from the mean visual magnitude assuming an average albedo consistent with the asteroid classes studied) have a nearly bimodal distribution of fast and slow rotators, unlike larger asteroids, which have a more Maxwellian distribution of spins. The small, fast rotators typically have small lightcurve amplitudes, indicating a tendency toward spherical shapes. Most importantly, there is a sharp cutoff at 2.2 h: No asteroid larger than 200 m has been observed spinning faster than this limit, which corresponds roughly to the critical breakup period for a strengthless body of bulk density ~3 g cm^{-3} (e.g., *Weidenschilling,* 1981). Since *a priori* there is no reason why a strong body would be precluded from spinning faster than this limit, the authors conclude that few (if any) asteroids larger than 200 m have tensile strength. There are, however, some very small asteroids (D < 200 m) with spin periods as fast as 2.5 min (*Pravec et al.,* 2002); these objects must have some tensile strength, though they need not be monoliths. Interestingly, sizes of a few hundred meters are thought to lie at the transition between the strength and gravity regimes of crater formation (see section 2.3).

2.1.5. Underdense asteroids. Another surprising recent observation is the apparent underdensity of some C-class asteroids. Such asteroids, found primarily in the inner part of the main belt (*Britt and Consolmagno,* 2000), are thought by virtue of their similar spectral and albedo characteristics to be the sources of carbonaceous chondrites (mean density ~2.6 g cm^{-3}; *Flynn et al.,* 1999; *Burbine et al.,* 2002). However, the *NEAR Shoemaker* spacecraft, on its way to 433 Eros, passed close enough to C-class asteroid 253 Mathilde (dimensions 66 × 48 × 44 km) to obtain a detailed shape model of the visible portion along with a mass estimate of the body, which together imply a bulk density of 1.3 ± 0.2 g cm^{-3} (*Yeomans et al.,* 1997; *Veverka et al.,* 1997, 1999; *Thomas et al.,* 1999; *Cheng et al.,* 2002). If Mathilde consists mostly of chondritic material, then the effective porosity is ~40%. As discussed in section 2.3, porosities of 20–40% can result if a body is completely shattered and reassembled, creating a gravitational aggregate. The underdensity is not believed to be due to embedded water ice because no hydrated materials were detected during the flyby (*Veverka et al.,* 1999).

There have been several other recent asteroid density measurements, some owing to advances in groundbased observation (cf. *Merline et al.,* 2002). *Merline et al.* (1999) reported on the discovery of a 13-km satellite (dubbed "Petit Prince") orbiting main-belt asteroid 45 Eugenia, a C- or F-class asteroid with an estimated mean diameter of ~215 km. The 4.7-d orbit of the satellite implies a bulk density for Eugenia between 1.2 and 1.8 g cm^{-3}, depending on the shape of the asteroid. In subsequent observations *Merline et al.* (2000) found a companion to 762 Pulcova, implying a bulk density of 1.8 g cm^{-3}. Lightcurve measurements have also revealed binaries with sufficient precision to estimate bulk densities: *Pravec et al.* (1998) find a bulk density of 1.7 g cm^{-3} for the primary of the Apollo asteroid system 1991 VH by inferring a companion from mutual eclipse events; *Pravec and Hahn* (1997) also estimate a bulk density of 1.7 g cm^{-3} for 1994 AW1, although this value is less certain; and *Mottola and Lahulla* (2000) find a bulk density of 1.4 g cm^{-3} for C-class asteroid 1996 FG3. Again, such low densities imply high porosities for these asteroids.

It is also worth noting that the bulk densities of the martian satellites Phobos and Deimos, thought to be captured C/P/D-class asteroids, are below 2 g cm^{-3} (*Hartmann,* 1990; *Kieffer et al.,* 1992; *Smith et al.,* 1995; *Murchie and Erard,* 1996). Some S-class asteroid densities have been measured by spacecraft: 2.6 ± 0.5 g cm^{-3} for 243 Ida (*Belton et al.,* 1995), thanks to the discovery of its satellite Dactyl by the *Galileo* spacecraft, and ~2.7 g cm^{-3} for 433 Eros (*Yeomans et al.,* 2000), by virtue of the *NEAR Shoemaker* orbiter (*Cheng et al.,* 2002). The higher densities imply lower porosities for these asteroids (although some of their components do contain Fe), but the values are not inconsistent with fractured or shattered configurations of low strength.

2.1.6. Giant craters. About 50% of small objects imaged to date have "giant" craters with diameters on the order of the radius of the object (*Thomas,* 1999). The most intriguing example is 253 Mathilde (section 2.1.5), which has at least four craters in this category (*Veverka et al.,* 1999; *Chapman et al.,* 1999). Mathilde also apparently lacks ejecta blankets and other global signatures of the large impacts such as fracturing and erasure of older craters. These findings support both the compaction and excavation models for crater formation on Mathilde (sections 2.2 and 2.3). Note that Mathilde is unusual in at least one other respect: It is currently the third slowest rotator known, with a spin period of 17.4 d, and is also in a tumbling state (*Mottola et al.,* 1995).

Other examples of giant craters on small bodies imaged by spacecraft include one ~23-km crater and five ~10-km craters on S-class asteroid 243 Ida (dimensions 60 × 26 × 18 km; *Belton et al.,* 1994, 1995; *Thomas et al.,* 1996); as many as eight ~4-km or larger craters on 951 Gaspra (dimensions 18 × 11 × 9 km; *Belton,* 1994; *Greenberg et al.,* 1994); the 11-km Stickney Crater on the martian moon Phobos (dimensions 27 × 22 × 19 km; *Asphaug and Melosh,* 1993; *Murchie and Erard,* 1996); and possibly the 10-km concavity on the other martian moon, Deimos (mean radius ~6.2 km; *Thomas et al.,* 1996).

Numerical hydrocode models of asteroid collisions (section 2.3) suggest that craters this large relative to the body size could have formed only in weak or fragmented targets capable of absorbing the collision energy close to the impact site. A solid monolithic body would have been completely disrupted, erasing any sign of a crater (cf. Fig. 1). High-resolution spacecraft observations of some of these bodies support this conclusion. Crater saturation on the surface and the irregular shape of Ida suggest an internal structure that is at least megaregolith or possibly large blocks covered by rubble (*Chapman et al.,* 1996a; *Greenberg et al.,* 1996; *Sullivan et al.,* 1996). Similarly, Gaspra is probably covered with megaregolith but its lumpy structure is also consistent with a blocky interior (*Greenberg et al.,* 1994; *Chapman et al.,* 1996b).

2.1.7. Grooves. Another feature indicating that the asteroids we know best are at least partially fractured is the apparently universal occurrence of linear grooves on their surfaces. First discovered in Viking images of Mars' satellite Phobos (*Veverka and Duxbury,* 1977), grooves are mostly rimless, sometimes beaded linear depressions that have been observed on every asteroid for which we have high-resolution images: Gaspra (*Veverka et al.,* 1994), Ida (*Belton et al.,* 1994), and most recently Eros (*Veverka et al.,* 2001). These grooves are currently believed to form where loose, incohesive regolith drains into underlying gaping fissures (*Thomas et al.,* 1979). The grooves' width and the spacing of beads along them are proportional to the depth of the regolith (*Horstman and Melosh,* 1989). Their lengths indicate the lateral continuity of the fissures that underlie them. The fissures may not have been initially formed by impacts, but they probably open every time a large impact jostles the interior of the asteroid, so the grooves may postdate the fissures themselves. The internal volume created by the fissures is at least as large as the volume deficit of the grooves themselves. Physical experiments suggest that large voids may develop beneath blocks trapped in the narrow fissures, so the internal void space could be larger. The presence of grooves on an asteroid thus suggests that its interior is coherent but fractured, and so its tensile strength is reduced from that of a pristine body and that there must be voids within the asteroid, increasing its porosity.

2.1.8. Unusual shapes and binaries. Delay-Doppler radar imaging of near-Earth asteroids (*Ostro,* 1993; *Ostro et al.,* 2002) has revealed some unusual asteroid shapes: 4769 Castalia (*Ostro et al.,* 1990; *Hudson and Ostro,* 1994), 4179 Toutatis (*Ostro et al.,* 1999), 2963 Bacchus (*Benner et al.,* 1999), 12 Victoria (*Mitchell et al.,* 1995), and 216 Kleopatra (the "dogbone" asteroid; *Ostro et al.,* 2000) have distinct lobes, while 1620 Geographos (*Ostro et al.,* 1995) — the most highly elongated body known in the solar system (aspect ratio ~2.8:1) — has a shape reminiscent of a porpoise. The *NEAR Shoemaker* mission revealed 433 Eros to be somewhat kidney-bean shaped, with a large saddle depression in the middle (*Zuber et al.,* 2000). In some cases, like Geographos and Eros, the spin rate is so high the effective gravity at the extreme ends is close to zero. In contrast, Toutatis is a slow, tumbling rotator, like Mathilde. Castalia and Bacchus look very much like contact binaries. Finally, Kleopatra, an M-class asteroid, has a very high density at the surface (3.5 g cm^{-3} as inferred from the radar reflectivity), consistent with its metallic composition, but shows no slopes in excess of the angle of repose for typical grains (~34°–37°). This, together with the overall shape, led *Ostro et al.* (2000) to speculate that Kleopatra is a metallic rubble pile. [Recently *Viateau* (2000) found a bulk density of 1.8 ± 0.8 g cm^{-3} for M-class asteroid 16 Psyche from astrometric measurements. However, the moderately high radar albedo implies a surface bulk density of $2.8^{+0.5}_{-0.6}$ g cm^{-3} (C. Magri, personal communication, 2001). These values are marginally consistent, but may suggest size sorting in the surface layers (cf. *Britt and Consolmagno,* 2001)]. In much earlier work based on lightcurves, *Weidenschilling* (1980) suggested that Kleopatra is a contact or close binary and that Trojan asteroid Hektor is a close binary (*Cruikshank et al.,* 2000, suggest Hektor may be a contact binary).

Simulations of "mild" asteroid tidal disruption, for which mass loss is minimal but significant reshaping still occurs, have given rise to distinctive shapes like Geographos and Eros (section 3.3). Contact binaries and dumbbell/dogbone shapes are a natural byproduct of tidal fission (section 3.2) or damping low-speed collisions (section 3.4). These results are all facilitated by a loose aggregate structure among the progenitors.

The near-spherical shapes of many fast rotators inferred from low-amplitude lightcurves (*Pravec and Harris,* 2000) must also be considered unusual since solid bodies smaller than a few hundred kilometers cannot be rounded by gravity. A more fluid body on the other hand, i.e., a gravitational aggregate, can relax to a spherical shape during the reaccumulation process following collisional or tidal breakup. However, to maintain a spherical shape with fast spin, friction and particle size effects are needed to prevent rotational flattening.

Radar, high-resolution optical imaging, and lightcurve observations are also revealing for the first time the presence of asteroid satellites (about 10 known so far), with an extrapolated frequency of a few percent among the main-belt population and ~10–20% among near-Earth asteroids (*Pravec et al.,* 2002). Interestingly, *Pravec and Harris* (2000) find that a significant fraction of the observed (near-Earth) population of fast rotators are binaries. Binaries provide the best possible measure of the primary mass (strictly, the sum of the masses), and together with shape estimates these lead to bulk density determinations, revealing in many cases some surprisingly low values (section 2.1.5).

2.2. Experiments

Naturally the best way to understand asteroids, aggregate asteroids in particular, is to study them *in situ*. Recently researchers have carried out successful flybys of several asteroids (Ida, Gaspra, Mathilde) and even managed a landing on one (Eros). In 2005, the *Deep Impact* spacecraft will fire a 350-kg projectile into Comet P9 Temple 1. Taking this one step further, an ideal but expensive mission would include detailed seismic measurements from multiple points on the surface of an asteroid to deduce the interior structure. However, it is also possible to gain insight into asteroid properties through carefully conceived experiments in laboratories on Earth, despite practical restrictions on impactor sizes and speeds. For example, experimentalists can investigate the effect of porosity on collision outcome, or even simulate the gravity regime with a centrifuge.

Typically, experimentalists find that asteroids with low strength and/or high porosity can damp impact energy more effectively than their stronger, less porous counterparts, a conclusion shared by numerical modelers (section 2.3). For example, *Love et al.* (1993) conducted hypervelocity experiments using targets of variable porosity and found that the more porous targets had deeper craters, lower ejecta velocities, and less distal damage. They concluded that porosity is an effective damper of impact energy and could be an important factor in calculating the lifetimes of small solar system objects. *Ryan et al.* (1991) dropped objects made of shattered concrete fragments held together by glue and found these resisted impact damage a factor of 2–3× more efficiently than the original concrete. They attributed this characteristic to energy dissipation.

Two convenient measures of impact outcomes are the critical specific shattering energy, Q_S^*, and dispersing energy, Q_D^* (*Durda et al.,* 1998; also see *Davis et al.,* 1979, for an earlier convention). Q_S^* is the projectile kinetic energy per unit target mass required to shatter the target so that the largest intact fragment contains 50% of the target mass. Q_D^* is the specific energy required to shatter and permanently disperse the target against gravity so that the largest (possibly reaccumulated) fragment has 50% of the target mass. Evidently Q_S^* is all that can be measured directly in the laboratory, since strength dominates over gravity for target sizes up to a few hundred meters. Generally Q_S^* decreases with target diameter due to the increasing likelihood of finding larger flaws in bigger bodies (*Housen and Holsapple,* 1990; *Holsapple,* 1994). Note that a body with RTS = 0 (e.g., a previously shattered body, or a rubble pile) by definition also has $Q_S^* = 0$. In the gravity regime, Q_S^* is negligible and Q_D^* dominates, increasing in value with target size as the gravity well deepens. The point where Q_S^* is comparable to Q_D^* is the transition from the strength to the gravity regime.

Housen et al. (1991) used a pressure chamber to expose a target of low strength to variable amounts of overpressure in order to simulate the stresses inside a self-gravitating asteroid, in other words, to estimate Q_D^*. Due to practical constraints, they could not conduct actual impacts with this configuration; instead they used buried explosive charges to simulate projectile impacts. They found that as the overpressure increased, the amount of material retained by the target also increased. By relating the target diameter to the amount of overpressure, they derived a Q_D^* scaling law consistent with the specific energies estimated for several asteroid families. However, the use of an explosion in place of an impact, and the 1/r scaling of pressure (instead of $1/r^2$ for gravity), introduces some uncertainty in the result, but the experiment does provide an upper bound on Q_D^* (*Holsapple,* 1994).

Another way of simulating the gravity regime on large bodies is to use a centrifuge. *Housen et al.* (1999) did this in an experiment designed to study the unusual cratering on Mathilde (section 2.1.6). A gas gun mounted on the centrifuge arm was used to impact a crushable porous target at 1.9 km s^{-1}. At 500 g the small craters formed by centimeter-sized projectiles have the same gravitational effects as the giant craters on Mathilde. They found that in this regime the impacts compacted the target material and produced almost no ejecta inside or outside the crater. They concluded that if compaction is responsible for the craters on Mathilde and other similar asteroids (no ejecta blankets were detected on Mathilde, although they would have been at the limit of resolution, and neighboring impacts appear to have produced little distal damage; cf. section 2.1.6), the characteristics of the craters are governed by neither strength nor gravity but rather the crushing material property of the target itself. For significant compaction to occur in silicates, densities must be below 2 g cm^{-3}. If compaction is the relevant mechanism on Mathilde, the five impacts that created the giant craters increased its bulk density by about 20%. This suggests that over time a porous body could be processed into a denser object. This mechanism also suggests that porous, compactable bodies are not efficient producers of meteorites, since most of the material is retained. This is in contrast to numerical studies that suggest excavation is highly efficient in porous bodies (section 2.3).

2.3. Simulations

The final piece of evidence concerning the existence of aggregate asteroids comes from direct numerical simulations, which in many respects are simply theoretical experiments. The two most common approaches use hydrodynamic codes, where state variables such as pressure and temperature are followed explicitly, and particle codes, where only gravity and collisions are considered. The former are better suited for modeling short-term stresses such as hypervelocity impacts (where fracturing, crushing, and melting are important) while the latter are optimal for studying long-term stresses such as tidal disruption or very low speed collisions (where the evolution takes place over many dynamical times). In this section only those experiments designed to compare the reaction of solid and aggregate bodies to short-term stresses will be discussed. Simulations that start solely from the premise of aggregate bodies are discussed in section 3.

Since laboratory experiments are limited to very small impactor sizes and only moderate velocities, numerical simulations are used to extrapolate over the many orders of magnitude in energy represented by typical collisions between small bodies in the solar system. Prior to the availability of simulations, analytic scaling laws relating Q^*_S or Q^*_D were used (cf. section 2.2). In the strength regime, the function $Q^*_S(D)$ (D is the mean target diameter) depends on the adopted strength model (i.e., equation of state, distribution of flaws, etc.). In the gravity regime, $Q^*_D(D)$ in principle has a fixed power-law slope, since the fragmentation is dominated by gravity and not the strength of the constituent pieces. However, analytic scaling laws often disagree with numerical simulations on the details of these functions, largely because the former rely on simplifications needed for practical considerations (*Benz and Asphaug,* 1994; see Fig. 5 of *Holsapple et al.,* 2002, for a graphical comparison of analytical, experimental, and simulation results — estimates of the transition from the strength regime to the gravity regime vary from D ~100 m to ~10 km). Numerical simulations differ among each other as well due to simplifying assumptions of the geometry, the range of parameters tested, differences in the numerical resolution, or, in the strength regime, the adopted strength model. As computing power improves the severity of these issues will be reduced.

Perhaps the most detailed numerical model to date, spanning both the strength and gravity regimes, is the three-dimensional smoothed particle hydrodynamics (SPH) model of *Benz and Asphaug* (1999). Because their code incorporates a brittle fragmentation model, it has been calibrated with laboratory experiments, adding an element of robustness not found in previous models, although it still requires extrapolation along a power law for the size dependence (*Asphaug et al.,* 2002). They find the transition from the strength to the gravity regime occurs around D ~ 300 m, making these the weakest bodies in the solar system. *Love and Ahrens* (1996) found a similar value using a simpler SPH code for the gravity regime and extrapolating to the *Holsapple* (1994) scaling curve for the strength regime. *Melosh and Ryan* (1997) found a somewhat larger (but still small) value of D = 800 m using a Lagrangian hydrocode with a material strength model included. These results imply that most asteroids of a few hundred meters in size or larger (up to a few hundred kilometers in size, where gravity can begin to crush the interiors) may be gravitational aggregates, since the probability of a dispersing impact is far less than the probability of a shattering impact for these bodies. This provides a natural explanation for the sharp cutoff seen in asteroid spin rates at about this size (section 2.1.4). Objects below this size are presumably young intact fragments of recent collisions that have so far escaped destruction.

Benz and Asphaug (1999) show that the inefficiency of momentum transfer in collisions contributes to the steep Q^*_D power-law slope in their simulations, a finding shared by *Ryan and Melosh* (1998). Both groups also find that, for fixed collisional energy, larger projectiles are more efficient at transferring momentum, i.e., they give rise to higher fragment speeds. This has important implications for planetesimal growth (section 3.4). *Benz and Asphaug* (1999) also note that Q^*_D can be a factor of 10 larger for a head-on collision compared to a glancing collision.

In related work, *Asphaug et al.* (1998) compared the hypervelocity-impact response of a coherent body with that of a moderately fragmented (but not strengthless) porous aggregate. For realism, both bodies were shaped like 4769 Castalia (cf. section 2.1.8). They found that fractures and voids in the porous body damp the propagation of the impact shock wave, confining the deposition of energy (kinetic or thermal) to a small volume near the impact site. This can result in excavation of a large crater and ejection of material at the impact site, perhaps beyond the surface escape speed, so there is no ejecta blanket and little evidence of the impact elsewhere on the body. Such a scenario may explain the giant craters on Mathilde (section 2.1.6; *Asphaug et al.,* 2002). In a solid body, the shock wave propagates freely so there is less energy deposition at the impact site and the collision may lead to extensive fracturing and distal damage. The authors conclude that the first impact to significantly fragment an asteroid will determine much of its subsequent collisional evolution. For compound bodies, such as a contact binary or a solid body with a deep regolith mantle, impact energy may be confined to one component, i.e., one lobe in the case of a contact binary, assuming there is an impedance barrier of rubble between the two lobes, or to the regolith in the second example. Evidently a detailed knowledge of the internal properties of an asteroid is required to fully predict, for instance, its response to nuclear explosions for the purpose of hazard mitigation, or even for understanding small-body evolution.

Finally, numerical simulations of hypervelocity impacts have revealed that reaccumulation of gravitationally bound material following a catastrophic collision can create objects with moderate porosities, between 20% and 40% (*Wilson et al.,* 1999; cf. "shattered with rotated components," Fig. 1). Since this high porosity impedes energy transmission through the body, subsequent impacts may not greatly alter the porosity. Moreover, although small-scale collisions may generate new regolith on the surface, it may be that friction prevents the small particles from filtering into the void spaces (*Britt and Consolmagno,* 2001; *Britt et al.,* 2002). Perhaps moderate-scale collisions provide the only mechanism for lowering the porosity, by gradually collapsing void spaces without ejecting much material.

3. EVOLUTION

Having established various lines of evidence that asteroids with low tensile strength and a range of porosity may exist in the solar system, we now consider the origin and evolution of these bodies.

3.1. Origin

Numerical simulations have shown that solid bodies can be completely shattered by hypervelocity (≥1 km s^{-1}) impacts without being dispersed (section 2.3). Moreover, the larger a body is, beyond ~100 m or so, the more likely it is to be shattered before ever being dispersed. This is because the critical dispersal energy is a steep function of the target diameter and the size distribution of impactors is a power law with large negative exponent (*Davis et al.,* 1989). In addition, the first shattering, nondispersive impact creates a body with even greater resistance to disruption. This then is a natural explanation for the origin of gravitational aggregates in the solar system, at least among the asteroid population: The long collisional history of such bodies suggests they suffered a shattering impact in their past. This does not preclude the possibility that asteroids (and comets; cf. *Weidenschilling,* 1994, 1997) were actually born with a very loosely consolidated structure. For example, planetesimal growth via gravitational instabilities could lead naturally to aggregate structure (*Ward,* 2000). But many asteroids, particularly those ~100 km or smaller, most likely have suffered at least one shattering impact during their lifetime (*Davis et al.,* 1989; *Bottke et al.,* 1994).

Of related interest is the origin of binary asteroids (section 2.1.8). Although such systems may be long lived, it is unlikely they have survived for the age of the solar system. [*Chapman et al.* (1996a) suggest Ida/Dactyl may be a few billion years old, but *Davis et al.* (1996) note that Dactyl, due to its small size, may have suffered several completely disruptive impacts while orbiting Ida.] Instead binaries probably formed more recently. Gravitational capture of a satellite can be ruled out since it requires a special mechanism such as an improbable interaction with a third body to extract energy from the otherwise hyperbolic encounter orbit. Collisions would occur more frequently than such chance encounters. Tidal disruption could work for bodies that approach terrestrial planets (cf. section 2.1.3), but cannot explain the main-belt binaries [even if there were terrestrial planets in the main-belt region in the past (cf. *Chambers and Wetherill,* 2001), they were ejected long ago]. Retention of ejecta in orbit following a moderately dispersive collision is a possible way to create tiny satellites (e.g., Ida/Dactyl, Eugenia/Petit Prince; cf. section 3.4) since this involves a many-body ($N > 3$) interaction in which escaping ejecta carries away energy, leaving some material in orbit. This mechanism cannot produce orbiting components of near-equal size, however; these must always fall back to produce at best a contact binary since there is insufficient escaping mass to alter the essentially two-body interaction of the components. The most promising explanation for binaries of any mass ratio is that they were mutually captured following a highly dispersive impact into a much larger body (*Durda,* 1996; *Greenberg et al.,* 1996; *Michel et al.,* 2001). This suggests that binary asteroids may be reaccumulated gravitational aggregates as a result of the catastrophic impacts that formed them. So far the low bulk densities measured for binary asteroids supports this conclusion (section 2.1.5).

3.2. Tidal Breakup

The observed breakup of comets (section 2.1.1) has led to speculation about whether dense terrestrial planets could play a significant role in the evolution of asteroids via tidal encounters. The familiar Roche criterion for tidal breakup (cf. *Chandrasekhar,* 1969) implies, for example, that a body with a bulk density of 2 g cm^{-3} and no tensile strength can be disrupted if it passes within 3.4 $R_\oplus$ of Earth's center, where $R_\oplus$ is Earth's radius. The Roche limit in general applies only to synchronous rotating liquid satellites in circular orbit. *Sridhar and Tremaine* (1992) extended Roche's treatment to nonrotating viscous bodies undergoing parabolic encounters and found the disruption limit to be about two-thirds smaller than the Roche limit. *Asphaug and Benz* (1994) confirmed this value in their simulations of the SL9 breakup. Later, *Asphaug and Benz* (1996) generalized their results to obtain a scale-invariant description of tidal encounters, finding that the same tidal outcome will result for the same ratio of periapse distance to Roche limit (for parabolic encounters with nonrotating spherical bodies). In earlier work on the tidal disruption of planetesimals by terrestrial planets, *Boss et al.* (1991) used an SPH model to show that Earth's tidal forces can cause planetesimal mass shedding, sometimes even resulting in SL9-like outcomes. More recently, *Bottke and Melosh* (1996a,b) invoked tidal fission of rotating contact binaries by terrestrial planets to explain the doublet crater population (section 2.1.3).

In an attempt to explore the possible outcomes of tidal encounters between strengthless asteroids and the Earth more fully, *Richardson et al.* (1998) performed hundreds of simulations using a particle-based N-body code. Starting with kilometer-sized spheroidal and ellipsoidal progenitors of 2 g cm^{-3} bulk density each made up of 247 identical 3.6 g cm^{-3} self-gravitating spheres, they investigated the encounter outcome as a function of periapse, speed at infinity, spin period, spin axis orientation, and phase angle at periapse. All encounters were hyperbolic and covered a range of speeds representative of the near-Earth asteroid population. A hard-sphere model with dissipation was used for treating particle collisions. Generally the outcome was found to be only weakly sensitive to the amount of dissipation, so long as there was at least some dissipation (otherwise reaccumulation resulted in particle swarms rather than aggregates). Encounter outcomes were parameterized chiefly by the mass retained, orbiting, or escaping the largest remnant. Figure 2 shows some examples of post-tidal-encounter configurations (not previously published).

Richardson et al. (1998) found that the closer and slower the encounter, i.e., the more time spent within the tidal limit, the more violent the outcome, ranging from SL9-type disruptions (with reaccumulation into fragment trains) at one extreme to mild distortion and/or spinup at the other. The spin period and orientation also strongly affected the outcome, with fast prograde rotation assisting breakup and fast retrograde rotation resisting it. An elongated asteroid was found to be much easier to disrupt than a spherical one if the orientation (phase) of the ellipsoid was favorable at periapse (long axis rotating toward the Earth). *Richardson et al.* (1998) showed that their results can account for the presence of one or two crater chains on the Moon and none on the Earth (section 2.1.2; also see *Bottke et al.,* 1997).

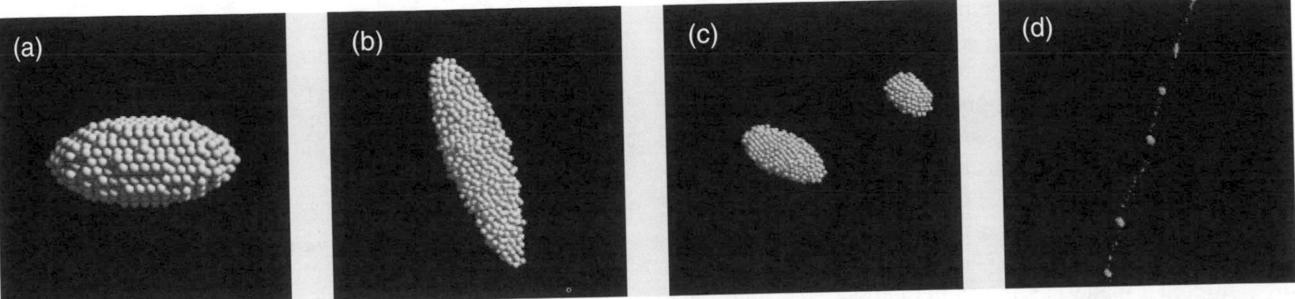

Fig. 2. Snapshots of simulated gravitational aggregate tidal encounters with Earth at different encounter speeds (close approach distance $q = 1.4\ R_\oplus$): **(a)** 4 × 2 km, 2 g cm^{-1} ellipsoidal progenitor with 993 identical spheres prior to encounter; **(b)** mild disruption at encounter speed $v_\infty = 15$ km s^{-1}, stretching the progenitor to 6 km length and stripping away a few particles (not shown); **(c)** moderate disruption at $v_\infty = 12$ km s^{-1} resulting in a stable binary of mass ratio 0.36 and eccentricity 0.16; **(d)** SL9-like disruption at $v_\infty = 6$ km s^{-1} yielding 21 large fragments (not all shown).

They found that ~10% of the ejecta ends up in orbit around the largest remnant, suggesting binaries could be created by tidal disruptions. They also found that the irregular shapes of some near-Earth asteroids could be explained by tidal encounters (section 3.3). This includes double-lobed bodies such as Castalia, Bacchus, Toutatis, and Kleopatra, which may be formed by the gentle mutual reaccretion of similar-sized bodies with moderate angular momentum following a tidal breakup or a low-speed collision. Tidal disruption could also explain the enhancement of the local population of small (≤50 m) bodies (*Bottke et al.,* 1998).

3.3. Spinup and Reshaping

Even though a tidal encounter may be too weak for significant mass loss to occur, the progenitor can still suffer substantial changes to its spin or, for bodies with low tensile strength, its shape. *Solem and Hills* (1996) used a strengthless aggregate model consisting of spherical frictionless boulders to demonstrate that shapes even more elongated than Geographos (section 2.1.8) could be achieved through tidal stretching. *Bottke et al.* (1999) were able to match details of the unusual porpoise-like shape of Geographos using their model with a mild low-mass-loss tidal encounter, arguing that the orbital parameters of the Earth-crossing asteroid are favorable for past tidal encounters with the Earth or Venus. They also suggested that Eros could be shaped by planetary tides. More generally, for mild interactions, *Richardson et al.* (1998) showed that the post-encounter spin and ellipticity of the progenitor are correlated with the preencounter values: A prograde progenitor spins up a bit and becomes more elongated, a retrograde progenitor spins down and becomes more spherical. *Scheeres et al.* (2000) used a combination of analytical theory and numerical simulations to show that spin changes (including tumbling for an asteroid like Toutatis) can occur even for relatively large encounter distances, although this work does not require a low-strength progenitor. They also showed that spin states of comparable-sized bodies can be dramatically altered during the brief interval of their evolution following fragmentation from a larger body. This kind of interaction may explain the origin of the slow-rotator population among small asteroids: An escaping fragment could rob its partner of rotational energy in order to escape the system (*Harris,* 2001). Finally, nongravitational thermal forces may also play a role in changing an asteroid's spin (*Bottke et al.,* 2002).

It should be noted that comprehensive numerical studies of allowed shapes and spin states of gravitational aggregates, including a determination of the damping time to principal axis rotation, have not yet been performed. Analytical models that approximate asteroids as fluid Maclaurin spheroids or Jacobi ellipsoids (cf. *Binney and Tremaine,* 1987) have been available for some time (*Weidenschilling,* 1981; *Farinella et al.,* 1981; both studies demonstrated that asteroid spin and shape are diagnostic of strength and porosity), but it remains to compare these in detail with numerical models, even those restricted to simple spheres with dissipation and surface friction without normal resting forces. More sophisticated models with irregular particle shapes and rolling and sliding modes are still a long way off.

3.4. Impacts

Hydrocode models of hypervelocity impacts into aggregate asteroids were discussed in section 2.3. It was shown that porous or damaged structure damps impact energy very efficiently, shielding the rest of the asteroid from damage. A solid body, on the other hand, may be completely shattered by a high-speed impact and yet retain most of its material if it has enough mass. At even higher impact energies, dispersal will occur, but the ejecta can reaccumulate into smaller aggregate bodies. *Durda* (1996) proposed this mechanism to explain the origin of the Ida/Dactyl system (cf. section 3.1), which is a member of the Koronis family. (Asteroid families share similar physical and dynamical characteristics and are thought to have originated from catastrophic disruptions of a single larger body; see the chapters on asteroid families in this volume for a com-

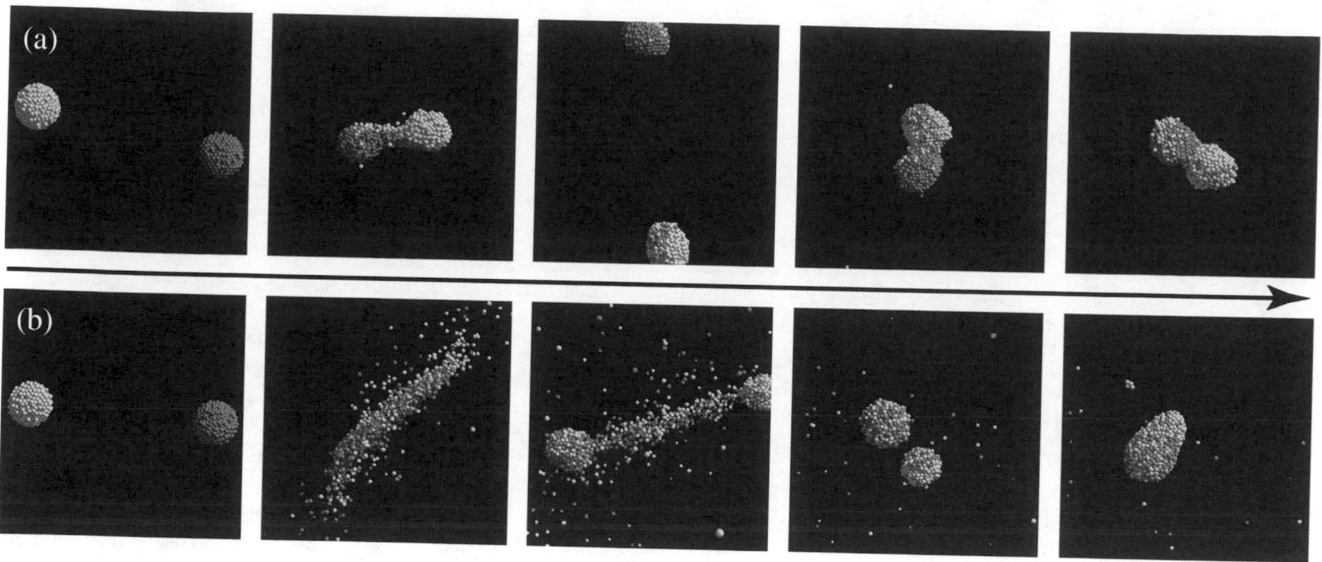

Fig. 3. Simulations of collisions between kilometer-sized gravitational aggregates: **(a)** impact angle $\phi \sim 60°$, impact speed $v_c \sim v_e$ (v_e is the escape speed from one body, ~1 m s^{-1} in this case), resulting in a contact binary; **(b)** $\phi \sim 17°$, $v_c \sim 2.5\ v_e$, opposite 6-h spins, giving rise to a teardrop-shaped aggregate with a tiny reaccumulated satellite (upper left). See *Leinhardt et al.* (2000) for more details.

prehensive review.) *Campo Bagatin and Petit* (2001) used a semianalytical approach to argue that gravitational aggregates in the size range of 10–100 km should be common in the asteroid belt. Recently *Michel et al.* (2001) combined an SPH fragmentation code with an N-body gravity code to show that hypervelocity impacts can form systems that resemble asteroid families and that during this process many binaries can form. In their simulations they found that family-forming impacts always completely shattered the target, yet postimpact reaccumulation gave rise to a mass spectrum and velocity distribution that were a good match for present-day families (allowing for erosional and dynamical evolution). They concluded that large family members must be gravitational aggregates consisting of reaccumulated fragments. This result resolves the long-standing dilemma that family members, previously believed to be solid fragments, have ejection speeds too high for the fragments to have survived the family-forming impact intact.

It is also of interest to consider the response of gravitational aggregates to low-speed collisions (≲100 m s^{-1}). Low encounter speeds are appropriate during the planetesimal growth phase, when relative speeds are on the order of the surface escape speed for the most gravitationally dominant body, e.g., ~1 m s^{-1} for a body of 1 km radius and 2 g cm^{-3} bulk density. Low collision speeds may also be relevant during the reaccumulation phase following a catastrophic impact. Using an SPH code that incorporates material strength and brittle fracture, *Benz* (2000) found that low-speed collisions between equal-sized bodies (both weak and strong) are surprisingly more efficient at fragmentation and dispersal than high-speed collisions for equal projectile kinetic energy per unit target mass. The reason is that momentum transfer is more efficient in collisions involving bodies of comparable size. *Benz* (2000) suggests that this relative fragility may hamper planetesimal growth in the 1-m to 1-km size range.

Leinhardt et al. (2000) found similar results using an N-body particle code (see *Richardson et al.*, 2000, for code details). In this study, kilometer-sized strengthless gravitational aggregates comprised of 955 identical spherical particles were collided at speeds less than 10 m s^{-1} (see Fig. 3 for some snapshots). Various spin configurations were sampled, with a range of impact angles and impact speeds for each. The impactors were of equal size except in one model where a mass ratio of 10:1 was used. They found that higher impact speeds (kinetic energy) yielded smaller remnant sizes and greater mixing, while larger impact angles (angular momentum) generated fast-spinning elongated or double-lobed remnants. Critical disruption was found to occur at roughly 4× the gravitational binding energy for these bodies. When initial spins were oppositely oriented, triaxial remnant shapes sometimes formed. The results were sensitive to the value of the dissipation parameter, with greater dissipation yielding larger remnant masses and more efficient clumping overall. The results were not sensitive to the numerical resolution: Simulations with 5× as many particles gave results similar to the low-resolution cases for the same impact parameters. The largest mass accumulation found orbiting the remnant in these simulations was ~2% of the total mass (fragments orbiting bodies smaller than the largest remnant were not searched for). Collisions between objects of equal size were found to result in net accretion 35% of the time (for a nominal value of the dissipation parameter), indicating that these bodies are relatively fragile. Again this may imply there is a bottleneck for planetesimal growth at small sizes, unless the growth mode is predominantly through collisions between bodies of large mass ratio. Although planetesimals of different mass typically have higher collision speeds owing to size-dependent gas drag effects (*Weidenschilling and Cuzzi,* 1993), the critical disruption energy is also higher for these cases, so accretion may still be favored (*Leinhardt and Richardson,* 2002).

4. CONCLUSIONS

Many, if not most, kilometer-sized asteroids (and comets) in the solar system may be gravitational aggregates of low tensile strength. Growth of planetesimals via collective instabilities may have led to aggregate structure, but even if asteroids were solid at one time in the past, repeated impacts over their lifetimes have likely shattered them by now. Once broken up, an asteroid can absorb impacts more efficiently, improving its survival chances. Apart from these theoretical and experimental considerations, the existence of gravitational aggregates is supported by observations: The cometary breakups that have been directly witnessed indicate very weak structures for these primordial bodies; crater chains on the Moon are consistent with tidal breakup of weak asteroids by the Earth; the doublet crater population on Earth, Venus, and Mars can be understood in terms of asteroid binary formation via tidal fission; the sharp asteroid spin-rate cutoff suggests asteroids larger than a few hundred meters cannot support tension; low bulk densities measured for several asteroids imply high porosities that can be explained by the reassembly of fragments; the presence of giant craters on several asteroids indicates efficient damping or compaction at the surface, consistent with damaged, rubblized, or highly porous and compactable structures; grooves imaged on asteroids suggest regolith drainage between fractured blocks; the unusual shapes of some asteroids can be explained by tidal stripping or shaping; and recent observations of binary asteroids suggest they may have formed from reaccumulated fragments following catastrophic disruptions of larger bodies.

Until direct seismic measurements of asteroids are made, or an extremely fortuitous observable collision or tidal breakup occurs, evidence for gravitational aggregates will remain largely circumstantial. This means the onus is on experimentalists and theorists to improve their models of the mechanical and dynamical properties of aggregate configurations. As the weight of evidence increases, even the most stubborn of skeptics will be convinced that gravitational aggregates play a crucial role in our understanding of the solar system.

Acknowledgments. The authors thank A. Campo Bagatin and D. Davis for their critical reviews of the manuscript. We also wish to thank the many contributors to section 1 in the wake of the Asteroids 2001 meeting. We hope we got it right!

REFERENCES

Asphaug E. and Benz W. (1994) Density of Comet Shoemaker-Levy 9 deduced by modelling breakup of the parent "rubble pile." *Nature, 370,* 120–124.

Asphaug E. and Benz W. (1996) Size, density, and structure of Comet Shoemaker-Levy 9 inferred from the physics of tidal breakup. *Icarus, 121,* 225–248.

Asphaug E. and Melosh H. J. (1993) The Stickney impact of Phobos: A dynamical model. *Icarus, 101,* 144–164.

Asphaug E., Ostro S. J., Hudson R. S., Scheeres D. J., and Benz W. (1998) Disruption of kilometre-sized asteroids by energetic collisions. *Nature, 393,* 437–440.

Asphaug E., Ryan E., and Zuber M. (2002) Asteroid interiors. In *Asteroids III* (W. F. Bottke Jr. et al., eds.), this volume. Univ. of Arizona, Tucson.

Belton M. J. S. (1994) Galileo encounter with 951 Gaspra: First pictures of an asteroid. *Science, 257,* 1647–1652.

Belton M. J. S. and 19 colleagues (1994) First images of asteroid 243 Ida. *Science, 265,* 1543–1547.

Belton M. J. S. and 12 colleagues (1995) Bulk density of asteroid 243 Ida from the orbit of its satellite Dactyl. *Nature, 374,* 785–788.

Benner L. A. M. and 12 colleagues (1999) Radar observations of asteroid 2063 Bacchus. *Icarus, 139,* 309–327.

Benz W. (2000) Low velocity collisions and the growth of planetesimals. *Space Sci. Rev., 92,* 279–294.

Benz W. and Asphaug E. (1994) Impact simulations with fracture. I. Method and tests. *Icarus, 107,* 98–117.

Benz W. and Asphaug E. (1999) Catastrophic disruptions revisited. *Icarus, 142,* 5–20.

Binney J. and Tremaine S. (1987) *Galactic Dynamics.* Princeton Univ., Princeton, New Jersey. 733 pp.

Bockelée-Morvan D. and 11 colleagues (2001) Outgassing behavior and composition of Comet C/1999 S4 (LINEAR) during its disruption. *Science, 292,* 1339–1343.

Boss A. P., Cameron A. G. W., and Benz W. (1991) Tidal disruption of inviscid planetesimals. *Icarus, 92,* 165–178.

Bottke W. F. Jr. and Melosh H. J. (1996a) Formation of asteroid satellites and doublet craters by planetary tidal forces. *Nature, 381,* 51–53.

Bottke W. F. Jr. and Melosh H. J. (1996b) Binary asteroids and the formation of doublet craters. *Icarus, 124,* 372–391.

Bottke W. F. Jr., Richardson D. C., and Love S. G. (1997) Can tidal disruption of asteroids make crater chains on the Earth and Moon? *Icarus, 126,* 470–474.

Bottke W. F. Jr., Richardson D. C., and Love S. G. (1998) Production of Tunguska-sized bodies by Earth's tidal forces. *Planet. Space Sci., 46,* 311–322.

Bottke W. F. Jr., Nolan M. C., Greenberg R., and Kolvoord R. A. (1994) Velocity distributions among colliding asteroids. *Icarus, 107,* 255–268.

Bottke W. F. Jr., Richardson D. C., Michel P., and Love S. G. (1999) 1620 Geographos and 433 Eros: Shaped by planetary tides? *Astron. J., 117,* 1921–1928.

Bottke W. F. Jr., Vokrouhlický D., Rubincam D. P., and Brož M. (2002) The effect of Yarkovsky thermal forces on the dynamical evolution of asteroids and meteoroids. In *Asteroids III* (W. F. Bottke Jr. et al., eds.), this volume. Univ. of Arizona, Tucson.

Britt D. T. and Consolmagno G. J. (2000) The porosity of dark meteorites and the structure of low-albedo asteroids. *Icarus, 146,* 213–219.

Britt D. T. and Consolmagno G. J. (2001) Modeling the structure of high porosity asteroids. *Icarus, 152,* 134–139.

Britt D. T., Yeomans D. K., Housen K. R., and Consolmagno G. J. (2002) Asteroid densities, porosities, and structures. In *Aster-*

oids III (W. F. Bottke Jr. et al., eds.), this volume. Univ. of Arizona, Tucson.

Burbine T. H., McCoy T. J., Meibom A., Gladman B., and Keil K. (2002) Meteoritic parent bodies: Their number and identification. In *Asteroids III* (W. F. Bottke Jr. et al., eds.), this volume. Univ. of Arizona, Tucson.

Campo Bagatin A. and Petit J.-M. (2001) How many rubble piles are in the asteroid belt? *Icarus, 149,* 198–209.

Chambers J. E. and Wetherill G. W. (2001) Planets in the asteroid belt. *Meteoritics & Planet. Sci., 36,* 381–399.

Chandrasekhar S. (1969) *Ellipsoidal Figures of Equilibrium.* Yale Univ., New Haven, Connecticut. 252 pp.

Chapman C. R. (1978) Asteroid collisions, craters, regolith, and lifetimes. In *Asteroids: An Exploration Assessment* (D. Morrison and W. C. Wells, eds.), pp. 145–160. NASA Conf. Publ. 2053.

Chapman C. R., Merline W. J., and Thomas P. (1999) Cratering on Mathilde. *Icarus, 140,* 28–32.

Chapman C. R., Ryan E. V., Merline W. J., Neukum G., Wagner R., Thomas P. C., Veverka J., and Sullivan R. J. (1996a) Cratering on Ida. *Icarus, 120,* 77–86.

Chapman C. R., Veverka J., Belton M. J. S., Neukum G., and Morrison D. (1996b) Cratering on Gaspra. *Icarus, 120,* 231–245.

Chauvineau B., Farinella P., and Harris A. W. (1995) The evolution of Earth-approaching binary asteroids: A Monte Carlo dynamical model. *Icarus, 115,* 36–46.

Cheng A. F. (2002) Near Earth Asteroid Rendezvous: Mission summary. In *Asteriods III* (W. F. Bottke Jr. et al., eds.), this volume. Univ. of Arizona, Tucson.

Chodas P. W. and Yeomans D. K. (1996) The orbital motion and impact circumstances of Comet Shoemaker-Levy 9. In *The Collision of Comet Shoemaker-Levy 9 and Jupiter* (K. S. Noll et al., eds.), pp. 1–30. Proc. IAU Colloq. 156, Cambridge Univ., Cambridge.

Cruikshank D. P., Dalle Ore C. M., Geballe T. R., Roush T. L., Owen T. C., Cash M., de Bergh C., and Hartmann W. K. (2000) Trojan asteroid 624 Hektor: Constraints on surface composition. *Bull. Am. Astron. Soc., 32,* 1027.

Davis D. R., Chapman C. R., Greenberg R., Weidenschilling S. J., and Harris A. W. (1979) Collisional evolution of asteroids: Populations, rotations, and velocities. In *Asteroids* (T. Gehrels, ed.), pp. 528–557. Univ. of Arizona, Tucson.

Davis D. R., Weidenschilling S. J., Farinella P., Paolicchi P., and Binzel R. P. (1989) Asteroid collisional history: Effects on sizes and spins. In *Asteroids II* (R. P. Binzel et al., eds.), pp. 805–826. Univ. of Arizona, Tucson.

Davis D. R., Chapman C. R., Durda D. D., Farinella P., and Marzari F. (1996) The formation and collisional/dynamical evolution of the Ida/Dactyl system as part of the Koronis family. *Icarus, 120,* 220–230.

Durda D. D. (1996) The formation of asteroidal satellites in catastrophic collisions. *Icarus, 120,* 212–219.

Durda D. D., Greenberg R., and Jedicke R. (1998) Collisional models and scaling laws: A new interpretation of the shape of the main-belt asteroid size distribution. *Icarus, 135,* 431–440.

Farinella P., Paolicchi P., Tedesco E. F., and Zappalà V. (1981) Triaxial equilibrium ellipsoids among the asteroids? *Icarus, 46,* 114–123.

Flynn G. J., Moore L. B., and Klöck W. (1999) Density and porosity of stone meteorites: Implications for the density, porosity, cratering, and collisional disruption of asteroids. *Icarus, 142,* 97–105.

Greenberg R., Nolan M. C., Bottke W. F. Jr., Kolvoord R. A., and Veverka J. (1994) Collisional history of Gaspra. *Icarus, 107,* 84–97.

Greenberg R., Bottke W. F. Jr., Nolan M., Geissler P., Petit J., Durda D., Asphaug E., and Head J. (1996) Collisional and dynamical history of Ida. *Icarus, 120,* 106–118.

Harris A. W. (2001) On the slow rotation of asteroids. *Icarus,* in press.

Hartmann W. K. (1990) Additional evidence about an early intense flux of C asteroids and the origin of Phobos. *Icarus, 87,* 236–240.

Holsapple K. A. (1994) Catastrophic disruptions and cratering of Solar System bodies: A review and new results. *Planet. Space Sci., 42,* 1067–1078.

Holsapple K., Giblin I., Housen K., Nakamura A., and Ryan E. (2002) Asteroid impacts: Laboratory experiments and scaling laws. In *Asteriods III* (W. F. Bottke Jr. et al., eds.), this volume. Univ. of Arizona, Tucson.

Horstman K. C. and Melosh H. J. (1989) Drainage pits in cohesionless materials: Implications for the surface of Phobos. *J. Geophys. Res., 94,* 12433–12441.

Housen K. R. and Holsapple K. A. (1990) On the fragmentation of asteroids and planetary satellites. *Icarus, 84,* 226–253.

Housen K. R., Holsapple K. A., and Voss M. E. (1999) Compaction as the origin of the unusual craters on the asteroid Mathilde. *Nature, 402,* 155–157.

Housen K. R., Schmidt R. M., and Holsapple K. A. (1991) Laboratory simulations of large scale fragmentation events. *Icarus, 94,* 180–190.

Hudson R. S. and Ostro S. J. (1994) Shape of asteroid 4769 Castalia (1989 PB) from inversion of radar images. *Science, 263,* 940–943.

Jackson J. A., ed. (1997) *Glossary of Geology,* 4th edition. American Geological Institute, Alexandria, Virginia. 800 pp.

Jeffreys H. (1947) The relation of cohesion to Roche's limit. *Mon. Not. R. Astron. Soc., 107,* 260–272.

Kieffer H. H., Jakosky B. M., and Snyder C. W. (1992) The planet Mars: From antiquity to the present. In *Mars* (H. H. Kieffer et al., eds.), pp. 1–33. Univ. of Arizona, Tucson.

Leinhardt Z. M., Richardson D. C., and Quinn T. (2000) Direct N-body simulations of rubble pile collision. *Icarus, 146,* 133–151.

Leinhardt Z. M. and Richardson D. C. (2002) N-body simulations of planetesimal evolution: Effect of varying impactor mass ratio. *Icarus,* in press.

Love S. G. and Ahrens T. J. (1996) Catastrophic impacts on gravity dominated asteroids. *Icarus, 124,* 141–151.

Love S. G., Horz F., and Brownlee D. E. (1993) Target porosity effects on impact cratering, and collisional disruption. *Icarus, 105,* 215–224.

McKinnon W. B. and Schenk P. M. (1995) Estimates of comet fragment masses from impact crater chains. *Geophys. Res. Lett., 22,* 1829–1832.

Melosh H. J. and Ryan E. V. (1997) Asteroids: Shattered but not dispersed. *Icarus, 129,* 562–564.

Melosh H. J. and Schenk P. (1993) Split comets and the origin of crater chains on Ganymede and Callisto. *Nature, 365,* 731–733.

Melosh H. J. and Stansberry J. A. (1991) Doublet craters and the tidal disruption of binary asteroids. *Icarus, 94,* 171–179.

Melosh H. J. and Whitaker E. A. (1994) Lunar crater chains. *Nature, 369,* 713–714.

Merline W. J., Close L. M., Dumas C., Chapman C. R., Roddier F., Menard F., Slater D. C., Duvert G., Shelton C., and Morgan T. (1999) Discovery of a moon orbiting the asteroid 45 Eugenia. *Nature, 401,* 565–568.

Merline W. J., Close L. M., Dumas C., Shelton J. C., Menard F., Chapman C. R., and Slater D. C. (2000) Discovery of companions to asteroids 762 Pulcova and 90 Antiope by direct imaging. *Bull. Am. Astron. Soc., 32,* 1017.

Merline W. J., Weidenschilling S. J., Durda D. D., Margot J. L., Pravec P., and Storrs A. D. (2002) Asteroids *do* have satellites. In *Asteroids III* (W. F. Bottke Jr. et al., eds.), this volume. Univ. of Arizona, Tucson.

Michel P., Benz W., Tanga P., and Richardson D. C. (2001) Collisions and gravitational reaccumulation: Forming asteroid families and satellites. *Science, 294,* 1696–1700.

Mitchell D. L., Ostro S. J., Rosema K. D., Hudson R. S., Campbell D. B., Chandler J. F., and Shapiro I. I. (1995) Radar observations of asteroids 7 Iris, 9 Metis, 12 Victoria, 216 Kleopatra, and 645 Zelinda. *Icarus, 118,* 105–131.

Mottola S. and Lahulla F. (2000) Mutual eclipse events in asteroidal binary system 1996 FG3: Observations and a numerical model. *Icarus, 146,* 556–567.

Mottola S. and 15 colleagues (1995) The slow rotation of 253 Mathilde. *Planet. Space Sci., 43,* 1609–1913.

Murchie S. and Erard S. (1996) Spectral properties and heterogeneity of Phobos from measurements by Phobos 2. *Icarus, 123,* 63–86.

Ocampo A. C. and Pope K. O. (1996) Shuttle imaging radar (SIR-C) images reveal multiple impact craters at Aorounga, northern Chad (abstract). In *Lunar and Planetary Science XXVII,* pp. 977–978. Lunar and Planetary Institute, Houston.

Öpik E. J. (1950) Roche's limit: Rings of Saturn. *Irish Astron. J., 1,* 25–26.

Ostro S. J. (1993) Planetary radar astronomy. *Rev. Mod. Phys., 65,* 1235–1279.

Ostro S. J., Chandler J. F., Hine A. A., Rosema K. D., Shapiro I. I., and Yeomans D. K. (1990) Radar images of asteroid 1989 PB. *Science, 248,* 1523–1528.

Ostro S. J. and 11 colleagues (1995) Extreme elongation of asteroid 1620 Geographos from radar images. *Nature, 375,* 474–477.

Ostro S. J. and 15 colleagues (1999) Asteroid 4179 Toutatis: 1996 radar observations. *Icarus, 137,* 122–139.

Ostro S. J., Hudson R. S., Nolan M. C., Margot J., Scheeres D. J., Campbell D. B., Magri C., Giorgini J. D., and Yeomans D. K. (2000) Radar observations of asteroid 216 Kleopatra. *Science, 288,* 836–839.

Ostro S. J., Hudson R. S., Benner L. A. M., Giorgini J. D., Magri C., Margot J.-L., and Nolan M. C. (2002) Asteroid radar astronomy. In *Asteroids III* (W. F. Bottke Jr. et al., eds.), this volume. Univ. of Arizona, Tucson.

Pravec P. and Hahn G. (1997) Two-period lightcurve of 1994 AW1: Indication of a binary asteroid? *Icarus, 127,* 431–440.

Pravec P. and Harris A. W. (2000) Fast and slow rotation of asteroids. *Icarus, 148,* 12–20.

Pravec P., Wolf M., and Šarounová L. (1998) Occultation/eclipse events on binary asteroid 1991 VH. *Icarus, 133,* 79–88.

Pravec P., Harris A. W., and Michałowski T. (2002) Asteroid rotations. In *Asteroids III* (W. F. Bottke Jr. et al., eds.), this volume. Univ. of Arizona, Tucson.

Rampino M. R. and Volk T. (1996) Multiple impact event in the Paleozoic: Collision with a string of comets or asteroids? *Geophys. Res. Lett., 23,* 49–52.

Richardson D. C., Bottke W. F. Jr., and Love S. G. (1998) Tidal distortion and disruption of Earth crossing asteroids. *Icarus, 134,* 47–76.

Richardson D. C., Quinn T., Stadel J., and Lake G. (2000) Direct large-scale N-body simulations of planetesimal dynamics. *Icarus, 143,* 45–59.

Ryan E. V. and Melosh H. J. (1998) Impact fragmentation: From the laboratory to asteroids. *Icarus, 133,* 1–24.

Ryan E. V., Hartmann W. K., and Davis D. R. (1991) Impact experiments. 3. Catastrophic fragmentation of aggregate targets and relation to asteroids. *Icarus, 94,* 283–298.

Scheeres D. J., Ostro S. J., Werner R. A., Asphaug E., and Hudson R. S. (2000) Effects of gravitational interactions on asteroid spin states. *Icarus, 147,* 106–118.

Schenk P. M., Asphaug E., McKinnon W. B., Melosh H. J., and Weissman P. R. (1996) Cometary nuclei and tidal disruption: The geological record of crater chains on Callisto and Ganymede. *Icarus, 121,* 249–274.

Scott E. R. D., Haack H., and Love S. G. (2001) Formation of mesosiderites by fragmentation and reaccretion of a large differentiated asteroid. *Meteoritics & Planet. Sci., 36,* 869–891.

Scotti J. V. and Melosh H. J. (1993) Estimate of the size of comet Shoemaker-Levy 9 from a tidal breakup model. *Nature, 365,* 733–735.

Sekanina Z. (2000) Secondary fragmentation of the Solar and Heliospheric Observatory sungrazing comets at very large heliocentric distance. *Astrophys. J. Lett., 542,* L147–L150.

Sekanina Z. and Yeomans D. K. (1985) Orbital motion, nucleus precession, and splitting of periodic Comet Brooks 2. *Astron. J., 90,* 2335–2352.

Smith D. E., Lemoine F. G., and Zuber M. T. (1995) Simultaneous estimation of the masses of Mars, Phobos, and Deimos from spacecraft distant encounters. *Geophys. Res. Lett., 22,* 2171–2174.

Solem J. C. and Hills J. G. (1996) Shaping of Earth-crossing asteroids by tidal forces. *Astron. J., 111,* 1382–1387.

Spray J. G., Kelley S. P., and Rowley D. B. (1998) Evidence for a late Triassic multiple impact event on Earth. *Nature, 392,* 171–173.

Sridhar S. and Tremaine S. (1992) Tidal disruption of viscous bodies. *Icarus, 95,* 86–99.

Sullivan R. and 17 colleagues (1996) Geology of 243 Ida. *Icarus, 120,* 119–139.

Thomas P. C. (1999) Large craters on small objects: Occurrence, morphology, and effects. *Icarus, 142,* 89–96.

Thomas P. C., Veverka J., Bloom A., and Duxbury T. (1979) Grooves on Phobos: Their distribution, morphology, and possible origin. *J. Geophys. Res., 84,* 8457–8477.

Thomas P. C., Adinolfi D., Helfenstein P., Simonelli D., and Veverka J. (1996) The surface of Deimos: Contribution of materials and processes to its unique appearance. *Icarus, 123,* 536–556.

Thomas P. C. and 11 colleagues (1999) Mathilde: Size, shape, and geology. *Icarus, 140,* 17–27.

Veverka J. and T. C. Duxbury (1977) Viking observations of Phobos and Deimos: Preliminary results. *J. Geophys. Res., 82,* 4213–4223.

Veverka J., Thomas P., Simonelli D., Belton M. J. S., Carr M., Chapman C., Davies M. E., Greeley R., Greenberg R., and Head J. (1994) Discovery of grooves on Gaspra. *Icarus, 107,* 72–83.

Veverka J. and 16 colleagues (1997) NEAR's flyby of 253 Mathilde: Images of a C asteroid. *Science, 278,* 2109–2114.

Veverka J. and 13 colleagues (1999) NEAR encounter with asteroid 253 Mathilde: Overview. *Icarus, 140,* 3–16.

Veverka J. and 32 colleagues (2001) Imaging of small-scale features on 433 Eros from NEAR: Evidence for a complex regolith. *Science, 292,* 484–491.

Viateau B. (2000) Mass and density of asteroids (16) Psyche and (121) Hermione. *Astron. Astrophys., 354,* 725–731.

Ward W. R. (2000) On planetesimal formation: The role of collective particle behavior. In *Origin of the Earth and Moon* (R. M. Canup and K. Righter, eds.), pp. 75–84. Univ. of Arizona, Tucson.

Weaver H. A. and 20 colleagues (2001) HST and VLT investigations of the fragments of Comet C/1999 S4 (LINEAR). *Science, 292,* 1329–1333.

Weidenschilling S. J. (1980) Hektor: Nature and origin of a binary asteroid. *Icarus, 44,* 807–809.

Weidenschilling S. J. (1981) How fast can an asteroid spin? *Icarus, 46,* 124–136.

Weidenschilling S. J. (1994) Origin of cometary nuclei as "rubble piles." *Nature, 368,* 721–723.

Weidenschilling S. J. (1997) The origin of comets in the solar nebula: A unified model. *Icarus, 127,* 290–306.

Weidenschilling S. J. and Cuzzi J. N. (1993) Formation of planetesimals in the solar nebula. *Protostars and Planets III* (E. Levy and J. I. Lunine, eds.), pp. 1031–1060. Univ. of Arizona, Tucson.

Weissman P. R. (1980) Physical loss of long-period comets. *Astron. Astrophys., 85,* 191–196.

Weissman P. R. (1982) Terrestrial impact rates for long- and short-period comets. In *Geological Implications of Impacts of Large Asteroids and Comets on Earth* (L. T. Silver and P. H. Schultz, eds.), pp. 15–24. GSA Special Paper 190.

Weissman P. R. (1986) Are cometary nuclei primordial rubble piles? *Nature, 320,* 242–244.

Wichman R. W. and Wood C. A. (1995) The Davy crater chain: Implications for tidal disruption in the Earth-Moon system and elsewhere. *Geophys. Res. Lett., 22,* 583–586.

Wilson L., Keil K., and Love S. J. (1999) The internal structures and densities of asteroids. *Meteoritics & Planet. Sci., 34,* 479–483.

Yeomans D. K. and 12 colleagues (1997) Estimating the mass of asteroid 253 Mathilde from tracking data during the NEAR flyby. *Science, 278,* 2106–2109.

Yeomans D. K. and 15 colleagues (2000) Radio science results during the NEAR-Shoemaker spacecraft rendezvous with Eros. *Science, 289,* 2085–2088.

Zuber M. T. and 11 colleagues (2000) The shape of 433 Eros from the NEAR-Shoemaker laser rangefinder. *Science, 289,* 2097–2101.

Side Effects of Collisions: Spin Rate Changes, Tumbling Rotation States, and Binary Asteroids

P. Paolicchi
University of Pisa

J. A. Burns
Cornell University

S. J. Weidenschilling
Planetary Science Institute

Energetic collisions in the asteroid belt produce several observable effects, altering their size distribution and rotational properties. Moreover, interesting objects or systems may be created: dynamical families, or binaries, and asteroids in unusual rotation states (fast rotators or tumbling bodies). While the evolution of the size distribution and properties of families as well as binaries are discussed in separate chapters, the other relevant effects of collisions will be dealt with here. During collisions, both the projectile and the target bring their own spin and orbital angular momenta to the system. In the collision the fragments may acquire a "breakup spin" connected to the breakup process. The "breakup spin" is related to the ejection velocity and size of the fragment; fast and small ejecta can achieve rapid spin rates. Usually, freshly created fragments leaving a collisional site do not spin around their axis of maximum momentum of inertia; thus, tumbling (observed also in laboratory experiments) is "natural." The presence of dissipative processes damps the wobbling, with a timescale that is usually short compared to the age of the solar system. Observations of main-belt asteroids suggest a lower limit of ~2 h to their spin periods. Very fast rotators, which must be monolithic bodies held together by internal strength, are present only among the smallest observed asteroids (all of which are NEAs, owing to obvious selection effects). Several asteroids have very long rotational periods, which remains puzzling. Theoretical arguments give preliminary and qualitative explanations for the observed properties. Asteroidal binary systems are an expected outcome from catastrophic collisions, through the effect of the mutual gravitational interaction or as a consequence of later bursting fission.

1. INTRODUCTION

Collisions have shaped many of the observable properties of the asteroids. The most interesting and relevant cases are those in which catastrophic disruption or high-energy cratering processes create a very large number of sizable fragments from a parent system (the two bodies that impact each other are typically referred to as the smaller projectile and the larger target). Throughout the history of the solar system, essentially all asteroids (see *Davis et al.,* 2002) have been influenced by such processes.

These kinds of events have been widely analyzed through laboratory experiments (*Holsapple et al.,* 2002) that have unfortunately involved bodies smaller by many orders of magnitude than corresponding asteroidal systems. Analysis of possible observable consequences of such processes among asteroids suggests various peculiar objects and systems, as well as the overall properties. In particular, it is usually assumed that the so-called dynamical families (*Zappalà et al.,* 2002) are typical outcomes of catastrophic breakup.

Despite the difficulties connected with scaling from the decimeter-sized targets used in experiments to the bodies involved in astronomical events (*Housen and Holsapple,* 1990, 1991; *Holsapple,* 1993, 1994), the outcomes of laboratory experiments and astronomical observations may be interpreted in terms of a unique, qualitatively consistent description of the process (*Paolicchi et al.,* 1996; *Doressoundiram et al.,* 1997; *Zappalà et al.,* 1996; *Cellino et al.,* 1999; *Tanga et al.,* 1999). The basic features of this description (discussed below) provide a sort of a "minimal standard model," useful, at least as a logical framework, for the topics discussed here.

Nevertheless, some recent indications from theoretical studies cast serious doubt on the validity of the standard model in the size range of asteroids. Several numerical computations have been performed, using hydrodynamical codes adapted to deal with processes typical of solid bodies, such as plastic processes, creation and growth of fractures, and breakup (see *Asphaug et al.,* 2002). These simulations have proved accurate enough to reproduce many of the results of laboratory experiments (*Melosh et al.,* 1992; *Ryan and Melosh,* 1998). The extension of the same models to cases involving astronomical systems leads to controversial results only partially consistent with observations (*Benz and Asphaug,* 1999; *Love and Ahrens,* 1996). The possibility of reconciling the results of "hydrocodes" and the whole body of observational data [in particular those concerning families, as discussed by *Pisani et al.* (1999)] is one of the main challenges facing asteroidal science. Perhaps a different interpretation of the data obtained from astronomical observation has to be given. For example, families could be outcomes of peculiar collisional processes. This might be seen in the preliminary computations by *Michel et al.* (2001), which reproduce some families as resulting from impacts with high specific energy, moderate relative velocity, and high transfer of angular momentum. In this case, most of the observable features of asteroids should be due to the post-event gravitational reaccumulation, rather than properties of individual fragments, and the physical meaning of the "minimal standard model" would be seriously limited. Alternatively, perhaps the existing hydrocodes are not yet sophisticated enough to reproduce impact processes involving bodies endowed with a complex internal structure, such as the asteroids certainly are, with that structure itself produced by collisional history. That is, the result of a collision could depend on several unknown parameters: internal texture of target, shape, impact point and direction, etc. Some numerical simulations (*Asphaug et al.,* 1998) give a qualitative indication in this sense. In this latter case, the "minimal standard model" should be a tool allowing a qualitative and empirical — but general — description of outcomes. Here we decided to keep to the standard model for a qualitative description of the processes, guessing that several features will remain valid even if, in the future, a consistent paradigm and a related general consensus on it becomes available.

2. "STANDARD MODEL" OF COLLISIONS

The basic features of the standard model of collisions, derived from the analysis of laboratory experiments and of astronomical observations (mainly involving dynamical families) are as follows.

1. The largest remnant(s) of a catastrophic process originate far from the impact point. Several experiments (*Giblin et al.,* 1994b, 1998a) suggest the formation of an antipodal cap as the largest remnant, surrounded by a few other large fragments; according to other experimental findings (*Fujiwara et al.,* 1989), the largest remnant is a central core. Experiments performed at low impact velocities generate a conelike fragmentation, with a set of large fragments, again formed not too close to the impact point, and elongated approximately along the impact direction. The numerical models support these ideas, even if they show, when applied to astronomical sizes, that fragmentation might be rather fine throughout the entire target. Since fragments close to the impact point are ejected at higher velocities, the possible gravitational reaccumulation should be more effective for fragments formed at distances farther from the impact point.

2. Some of the original mass is dispersed into very fine debris, which cannot be recovered in laboratory experiments and cannot be observed at all as independent astronomical bodies. Formation of debris is the rule in the region around the impact point, but small fragments can come out also from different regions of the target, even from the region eventually dominated by large fragments: These last, in fact, allow only an approximate reconstruction of the parent body, as if it were a jigsaw puzzle. Some interstitial (missing) mass is usually lost from between the large fragments.

3. The fragments have different shapes. If we define the shape of an irregular fragment in terms of an equivalent ellipsoid [for a discussion see *Verlicchi et al.* (1994); *La Spina and Paolicchi* (1996)], the shape distributions from experiments usually peak at ratios among the c:b:a axes, close to 0.5:0.7:1 (*Giblin et al.,* 1994a,b, 1998a). A similar shape distribution can be found to hold for the asteroids (*Capaccioni et al.,* 1984). Nevertheless, this shape similarity does not necessarily require that the asteroids are fragments created in some impact-disruption process; in fact, their shapes are also consistent with a rubble-pile model for bodies with internal stresses (*Holsapple,* 2001) that originated directly from the collision or resulted from later reaccumulation. Note that because of frictional and other effects, it is not required for a rubble pile to assume an equilibrium shape, such as the Maclaurin or Jacobi ellipsoids, typical of a "fluid" rotating body.

4. The fastest fragments are usually small (*Martelli et al.,* 1994). The largest remnant and the other big fragments (if present) usually leave at moderate velocities. In general, the ejection velocities depend strongly on the degree of fragmentation. The correct scaling of velocities to processes involving asteroids remains a major open problem, as thoroughly discussed in *Zappalà et al.* (2002).

5. Fragments always rotate. The initial total angular momentum of the parent bodies, including both spin and orbital contributions, is partially distributed to invisible debris. Thus the angular momentum may not appear to be "conserved" when we compare the rotational properties of the final observable bodies with the initial conditions. Of course, individual fragments may develop spins even when the total angular momentum before and after the breakup is zero, as shown by laboratory experiments in which the initial impacting bodies are not rotating and the collision is central. This "breakup spin" is obviously imparted in different directions (*Giblin et al.,* 1994a,b) with a geometry that can be easily understood if we assume that the kinematical properties of fragments may be explained in terms of a "breakup velocity field" for which the ejection velocities are larger close to the impact point and decrease regularly from the impact to the antipodal direction. The gradient of the velocity field results in rotation of the fragments. If one assumes also [as in the semiempirical model by *Paolicchi et al.* (1996)] that the sizes of the fragments are somehow connected to the gradient of the velocity field, then the observed correlation between sizes and velocities, and a similar one between sizes and spins (smaller bodies

rotate faster), comes out naturally (*Giblin et al.,* 1994b; *Martelli et al.,* 1994).

No physical mechanism exists such that the original direction of the spin of a fragment should be correlated with its principal axes. According to laboratory impact experiments, the axis of rotation is not usually close to the c axis, corresponding to the maximum moment of inertia (*Fujiwara and Tsukamoto,* 1981), even if the consequent tumbling is not experimentally found as often as one might expect (*Giblin and Farinella,* 1997). This behavior may be understood in terms of semiempirical modeling (A. Verlicchi, personal communication, 1998) and is consistent with results from hydrocodes, applied to fragmentation of the kilometer-sized asteroid 4769 Castalia (*Asphaug and Scheeres,* 1999). Fragments that are not initially rotating about their axes of maximum moment of inertia will eventually evolve to that state by internal dissipation of energy while conserving angular momentum. Before that happens, they will tumble, and experience time-varying internal stresses that may cause delayed breakup into smaller fragments (*Giblin et al.,* 1998b). This secondary splitting could result in formation of bound pairs, i.e., binary asteroids.

Several aspects of the standard model can be understood with the aid of Fig. 1, based on the semiempirical model by *Paolicchi et al.* (1996) and adapted from a similar figure by *D'Abramo et al.* (1999). In the figure we represent the original starting points of the fragments produced in a catastrophic process. The vertical line passes through the impact point (at the top) and the center of the parent body, while the abscissa represents the distance of a fragment from this line. Thus the parent body (assumed to be spherical) is projected into a semicircle.

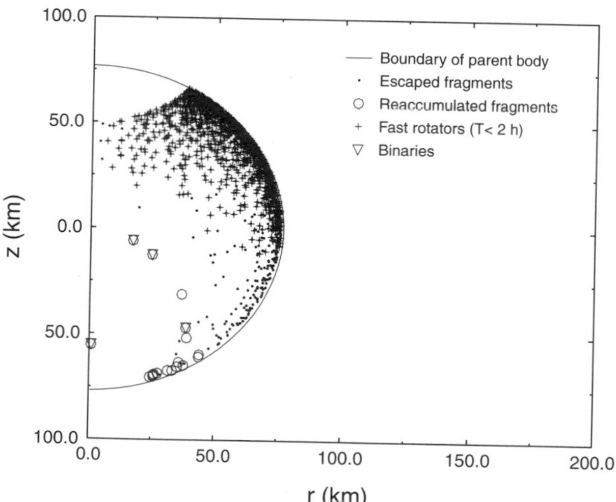

Fig. 1. Representation of the original position within the parent body (projected onto a semicircle, the vertical diameter corresponding to the one passing through the impact point, while the abscissa represents the distance from this line) of fragments with different fates (reaccumulation, formation of a binary, free escape, free escape with a period smaller than 2 h). The empty region on the top corresponds to the region very close to the impact point, at which very small fragments are formed; these are not dealt with in the model. Data adapted from *D'Abramo et al.* (1999).

Different symbols indicate the fragments that are later reaccumulated by mutual self-gravity (or give rise to binary systems; see the discussion in section 5), according to the results of an N-body integration (see *Doressoundiram et al.,* 1997), and those that definitely escape from the system (some of them rapidly spinning). Note that usually the fragments that undergo reaccumulation are larger than those that freely escape. For this reason the former ones are less "dense" in the figure. The sizes of the various regions, from which are derived reaccumulated bodies, escaping fragments, and so on, depend on the model, i.e., the impact energy and strength and size of the target. They may be larger or smaller, but the qualitative pattern remains the same.

3. ROTATIONAL PROPERTIES

A theoretical understanding of the rotational outcomes of catastrophic collisions is far from complete, in spite of recent valuable efforts (*Love and Ahrens,* 1997; *Asphaug and Scheeres,* 1999). On the other hand, our ideas on rotational properties, derived from impact experiments, and accepted by the "minimal standard model," cannot be readily incorporated into a general model of collisional evolution of asteroids. In contrast to what happens in laboratory experiments, the rotational properties of the fragments are not always dominated by the "breakup spin." The contributions of the original spin angular momenta of target and projectile and of the transferred orbital angular momentum are dominant for large and intermediate-sized bodies.

There is no satisfactory model of the collisional evolution of asteroids that includes spin properties. After *Davis et al.*'s (1989) description of the state of the art, few attempts have been made to further refine the model (*Farinella et al.,* 1992). The problem is difficult mainly because several critical parameters are not known. Other noncollisional processes, such as radiative torques (*Rubincam,* 2000) or close encounters with other asteroids or planetary bodies (*Scheeres et al.,* 2000; *Richardson et al.,* 1998), may also play a role in the evolution of spin rates. The creation of a robust theoretical scenario for the evolution of rotational properties of asteroids is a rather ambitious task facing future researchers.

Critical uncertainties include the scaling of catastrophic processes to astronomical sizes, and the precise distribution of angular momentum among the fragments (experiments and empirical arguments suggest a faster rotation for smaller bodies, but the steepness of the rotation vs. size relation is not well known). Moreover, and primarily, we do not know what fraction of the original angular momentum of the parent system (arising out from spins of both projectile and target and from the orbital angular momentum) is imparted to the macroscopic fragments. According to some evidence from experiments, a significant fraction of this angular momentum is dispersed by the cloud of debris, which is formed close to the impact point; nevertheless, as previously mentioned, the astronomical case is different and has been only preliminarily analyzed by means of numerical simulations (*Love and Ahrens,* 1997; *Asphaug and Scheeres,* 1999). If one assumes the simple conservation of angular momentum, unrealistically rapid spins would result for most asteroids, with the exception of the few largest bodies.

The observed rotational properties of asteroids are discussed in *Pravec et al.* (2002) (see also Fig. 2 below). Here we only sketch those features whose interpretation involves collisional processes.

1. For asteroids larger than ~40 km diameter, the shortest spin periods are about 4 h. The observed distribution is consistent with a Maxwellian distribution (*Binzel et al.,* 1989), and the absence of shorter periods may be due to the small number of bodies in this size range. However, this limit also corresponds to the fastest rotation allowed for weak, self-gravitating bodies in hydrostatic equilibrium, for plausible densities of 2–3 g/cm^3 (*Weidenschilling,* 1981). Any addition of angular momentum results in slower rotation, by formation of a Jacobi ellipsoid or binary. Thus, the observed limit is also consistent with predictions of collisional evolution models (*Davis et al.,* 1989) that most large asteroids have been shattered.

2. In the intermediate size range (~10–100 km), the mean rotation period is slightly longer than for smaller or larger bodies. This effect has been explained as the result of "angular momentum drain" (*Dobrovolskis and Burns,* 1984; see also *Davis et al.,* 1989), the preferential escape of ejecta in the prograde direction following cratering impacts. A similar mechanism, with similar consequences, has been shown to work for larger catastrophic collisions; this has been called "angular momentum splash" (*Cellino et al.,* 1990).

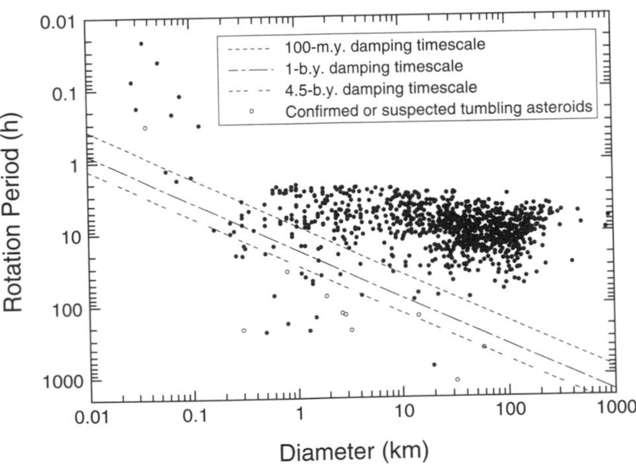

Fig. 2. Spin rate plotted vs. diameter for all (~1000) asteroids with high-quality spins as of March 31, 2001. Open circles identify those asteroids interpreted to be tumbling. Lines of constant damping timescale t_d for wobble (equation (6)) are shown. Figure courtesy of Alan W. Harris, JPL.

3. Among very small asteroids, those less than a few hundred meters in diameter (all NEAs due to observational selection), a high proportion are fast rotators, with periods of <2 h [the shortest known is only a few minutes (*Pravec and Harris,* 2000)]. These cannot be bound by gravity, and their interiors must be in tension; i.e., they are monolithic bodies. Their internal stresses can be estimated to be $\rho r^2\omega^2$, generally in the range ~10^5–10^6 dynes cm^{-2}, which can be sustained by unfractured meteoritic material. Their rapid spins are probably acquired during breakup of parent bodies disrupted by catastrophic collisions, although radiative torques may also be a factor.

4. Asteroids smaller than a few tens of kilometers but larger than ~150 m (main-belt objects and most NEAs) show a lower limit to the observed spin period of about 2.1 h. It seems unlikely that this limit corresponds to a tensile bursting strength of competent bodies; the stresses would be only ~10^2–10^3 dynes cm^{-2}. However, the observed limit corresponds essentially to the surface orbit period for a sphere of plausible density (2.5 g cm^{-3}), i.e., the rate at which loose material would not be bound to the surface by gravity at the equator of a spherical body. It is likely that bodies in this size range are not monolithic, but not so completely shattered as to have equilibrium shapes; they may consist of a core or a few large components covered by a layer of regolith. Such a body may have a naturally regulated rotational "speed limit" imposed by the loss of material from its equatorial region.

An elementary and rough argument may be useful for exploring this process and understand its potential consequences. Let us assume that a homogeneous spherical body of radius r rotates at a rate $\omega_0[= (4\pi G\rho/3)^{1/2}]$, for which the equatorial centrifugal force exactly equals gravitational attraction. The mean specific angular momentum of the body is

$$L_0 = 2/5r^2\omega_0 = (2/5)(GMr)^{1/2} \tag{1}$$

Suppose that the body gains an increment of angular momentum that tends to increase its spin rate. Some material at the equator is no longer bound to the surface and is removed. It carries off the local specific angular momentum at the equator, which is 5/2× the mean value. This loss of mass must continue until either the equatorial rotation speed decreases to the surface orbital velocity or the supply of loose regolith material at the equator is depleted. If the body is initially spherical, loss of material from the equatorial region would also alter the mass distribution so that its rotation is no longer about the axis of maximum moment of inertia. The asteroid would tumble until dissipation produced a spin state about a new principal axis with larger moment of inertia and a lower angular velocity.

In general, angular momentum would be added by an impact that would also tend to remove mass from the target. More gradual spinup might also occur due to radiative torques (*Rubincam,* 2000). Material shed by these processes would accrete into a satellite. In the aftermath of a disruption of a parent body, a rapidly rotating fragment may also experience rotational bursting (*Giblin et al.,* 1998b). Any of these processes may produce a binary system. The angular momentum is divided between the spins of the components and the orbital term. The latter may be dominant, especially for a tidally synchronized system with a mass ratio not too close to unity. The orbital specific angular momentum is expressed in terms of the separation R by

$$L_f = x(1 - x)(GMR)^{1/2} \quad (2)$$

where x is the fraction of the total mass in the smaller component and M the total mass. The final orbital frequency, which is equal to the rotation rate of tidally synchronized components, is given by

$$\Omega = (GM/R^3)^{1/2} \quad (3)$$

By equating L_0 and L_f, one obtains the final value of Ω

$$\Omega_f = \omega_0 \left[\frac{5x(1-x)}{2} \right]^3 \quad (4)$$

We see that $\Omega_f < \omega_0$ for all values of x.

5. A few very slow rotators are observed, among bodies not too large. These rotations cannot be original: It is almost impossible to obtain a quasizero rotational rate as the result of several processes transferring angular momentum in a stochastic way (randomly oriented vectors behave as in a random walk process; it is difficult to go back close to the origin). The above-discussed fission processes may have something to do with their origin [as qualitatively suggested by *Farinella et al.* (1982), and recently pointed out by A. W. Harris at the Asteroids 2001 Conference], even if, according to the discussion of *Pravec and Harris* (2000), it is not easy to explain in this way the slow rotation of asteroids such as 253 Mathilde (*Mottola et al.,* 1995; see also *Cuk and Burns,* 2001).

4. TUMBLING ASTEROIDS

Almost all asteroids with measured lightcurves are in pure spin; that is, they rotate at a constant rate around a direction fixed in space and in the body's reference frame (*Pravec et al.,* 2002). This condition requires that the body's angular momentum **L** and angular velocity ω (each measured relative to an inertial reference system at the body's center of mass) are parallel to one another. This will only occur if **L** and ω lie along one of three principal inertia axes of the body, whether it be the maximum, minimum, or intermediate axis. In the most general rotation state **L** and ω are not aligned with one another or with the body's principal axes. Then we will say that the body "tumbles" or "wobbles" as ω moves. In the case of axially symmetric bodies, various authors describe this motion as Chandler wobble or free precession or nutation.

When an asteroid is in pure rotation about its ith axis, its rotational kinetic energy is $T = I_i\omega_i^2/2$, or, correspondingly, $T = L^2/2I_i$, where I_i is the principal moment of inertia about the ith axis. As long as only internal forces act, **L** is constant regardless of how the rotation state evolves. Thus the body's kinetic energy is an absolute minimum (vs. maximum) for rotation about the principal axis of maximum (vs. minimum) inertia. Hence the observed pure spin of typical asteroids can be most simply understood as indicating that those objects have damped down into their minimum rotational energy well. Alternatively, one may think of tumbling bodies as being in a "heated" or excited energy state. This minimum-energy condition is remarkable since asteroids continually suffer mutual collisions that usually misalign the body's principal axes from **L**'s direction.

Considering the mutual collision between a small (projectile) asteroid with a much larger (target) asteroid, angular momentum is transferred from the orbital motion of the projectile to the spin of the target. Moreover, impacts often cause some mass and angular momentum to escape the target. Each of these effects has the capacity to knock **L** out of the parallelism it has when the asteroid rotates purely about one of the body's principal axes, here presumably the axis of maximum inertia. Since this bombardment is surely occurring, but apparently is not effective in producing a "heated" rotation state, some process must be draining rotational energy from the asteroid just as fast as collisions can impart it. To give a feel for the ease of misalignment, we mention that, for typical collisions between typical asteroids, a "bullet" of relative mass ~10^{-5}R, where R is the asteroid target's radius in kilometers, carries roughly the same angular momentum as the target originally possessed in its spin. We caution, however, that some fraction, perhaps a large fraction, of the angular momentum will be removed in the collisional ejecta.

Processes other than collisions may also cause tumbling. When an asteroid passes closely by a planet, that planet will gravitationally tug on the asteroid's irregular shape (*Richardson et al.,* 1998; *Black et al.,* 1999). Since flyby times are usually a few hours, they are comparable to spin periods, meaning that asteroids present roughly fixed orientations during flybys so that the transmitted torque has a constant sign. Similar physics allows impact ejecta escaping an asteroid to exert gravitational torques on that body, and to significantly alter its spin state (*Scheeres et al.,* 2000). Radiation forces, whether the direct solar flux (*Paddack,* 1969) or the (Yarkovsky) recoil from reemitted thermal radiation (*Rubincam,* 2000), may generate torques that induce nutation. Mass redistribution — for example, produced as craters are gouged out or as ice sublimates — will usually reorient the body's principal axes and will commonly produce wobble (*Peale and Lissauer,* 1989).

Laboratory experiments (e.g., *Giblin et al.,* 1998a), albeit at relatively tiny scales, show how impacts generate wobble as mentioned above. Moreover, *Giblin and Farinella* (1997) concluded that some tens of percent, as determined both from lab experiments and estimated in a simple

TABLE 1. Tumbling asteroids.

Asteroid	Class*	a (AU)	e	D (km)	$2\pi/\omega$ (h)[†]	References[‡]
253 Mathilde	MB	2.64	0.27	58	417	*Mottola et al.* (1995)
288 Glauke	MB	2.75	0.21	32	1200	*Harris et al.* (1999); *Ostro et al.* (2001)
1689 Floris-Jan	MB	2.45	0.21	14.3	145	*Harris* (1994); *Pych* (1999)
3288 Seleucus	A	2.03	0.46	1.9	75	*Harris* (1994); *Harris et al.* (1999)
3691 Bede	MX	1.77	0.29	3.3	227	*Pravec et al.* (1998)
4179 Toutatis	A	2.52	0.63	2.7	130	*Spencer et al.* (1995)
4486 Mithra	A	2.20	0.66	2.4	days	*Ostro et al.* (2000)
13651 (1997BR)	A	1.34	0.31	0.8	33.6	*Pravec et al.* (1998)
1999 GU3	A	2.08	0.51	0.3	216	*Pravec et al.* (2000)
1999 JM8	A	2.72	0.65	2.9	137	P. Pravec, personal communication (2001)[§]
2000 WL107	A	2.90	0.65	0.037	0.322	P. Pravec, personal communication (2001)[§]

*MB = main belt; MX = Mars-crosser; A = Apollo-Amor.
†Rotation periods are often poorly determined because of the complex lightcurve.
‡Exhaustive references are not given, but rather just enough to allow an introduction to the data.
§Unpublished lightcurve data (available on line at http://sunkl.asu.cas.cz/~ppravec/neo.html).

model, will display visible tumbling initially. In a numerical simulation of breakup, *Asphaug and Scheeres* (1999) notice that all disrupted fragments leave in a state of complex rotation, with widely varying spin rates and very little memory of the parental target's spin.

The rotational states of real asteroids can be ascertained from an analysis of their brightnesses, usually observed at unevenly spaced times (*Burns and Tedesco,* 1979). Fourier techniques are used to produce a lightcurve and extract the rotation period from these brightness measurements. Lightcurves that exhibit two periods indicate a precessing object (e.g., *Kaasalainen,* 2001) or perhaps an eclipsing pair (e.g., *Pravec and Hahn,* 1997), while complex (nonrepeatable) lightcurves suggest tumbling. Radar returns (e.g., *Ostro,* 1993, *Hudson et al.,* 1997) may more directly indicate complex rotations and shapes.

The rotation rates for all objects with clear determinations are plotted against size in Fig. 2, similar to diagrams in *Harris* (1994) and *Pravec and Harris* (2000). Of the nearly 1000 asteroids with well-resolved rotation periods, just 11 are known to wobble, with 4179 Toutatis the first asteroid to be identified as a complex rotator (*Spencer et al.,* 1995). Since that time, an increasing number of tumblers (*Harris,* 1994; *Pravec et al.,* 1998, 2000; A. W. Harris, personal communication, 2001) have been discovered (see Table 1).

Most of the tumblers known today are small and/or slowly rotating, as seen in the lower left corner of Fig. 2. A sizable fraction of them (7 out of the 11) are Earth-crossing objects and another is a Mars-crossing object, but it is uncertain whether this result occurs because most small objects with measured lightcurves are predominately Earth crossers or because planet-crossing events themselves may engender the wobble (*Black et al.,* 1999). Among the very fastest-spinning asteroids, only 2000 WL107, a 20-m-radius NEA turning in ~20 min, is tumbling (P. Pravec, personal communication concerning unpublished observations by himself, C. Hergenrother, and S. Mottola, 2001).

For an oblate body with moments of inertia (C, A, A), the wobble timescale is $\tau_\omega = (P/\Delta)(C/A)$, where P is the spin period and $\Delta = (C - A)/C$ (*Burns and Tedesco,* 1979). If such a motion can be detected, the relative difference in the moments of inertia is determined, and this constrains density inhomogeneities in the asteroid's interior. Unfortunately, for the more likely case when the body's inertia-ellipsoid is triaxial, the dynamical motion becomes much more complex and often indecipherable (cf. *Kaasalainen,* 2001).

The mechanism that dissipates rotational energy, and thereby damps wobble, can be simply stated, but it is much more difficult to compute in a general circumstance. Internal friction (or anelasticity) occurs in all real materials: Due to processes occuring at the atomic or grain level, energy is lost whenever materials undergo cyclic stress-strain (recall the warmth produced when a paper clip is rapidly flexed). Each element of a tumbling body is subject to variable stress and strain as that element's acceleration changes due to the rotation axis moving through the body. Thus, energy is lost, and the only available source to supply it is the rotational energy. Of course, once the body moves into pure spin, the internal stresses no longer vary and energy loss ceases.

One can estimate the timescale for damping from dimensional analysis alone. It is

$$\tau_d = K_1 \mu Q/(\rho R^2 \omega^3) \tag{5}$$

where μ, Q, ρ, and ω are the object's rigidity, anelasticity (or quality) factor, mass density, and rotation rate; K_1 is a nondimensional scaling coefficient. Since Q is defined to be E/dE, where dE is the energy lost over a cycle with E the maximum stored strain energy, it is a nondimensional quantity; thus it does not arise from the dimensional analysis mentioned above and should be strictly incorporated in

K_1. The functional dependence given in equation (5) has been derived by numerous researchers in many contexts using several different approaches. For example, to argue why so few asteroids wobble and to estimate the likely nutation angles for mutually colliding minor planets, *Burns and Safronov* (1973) computed the strain energy stored in bending stresses and the polar bulge. *Yoder and Ward* (1979), in investigating Venus's spin state, approached the problem somewhat similarly. *Peale* (1973) derived nutational damping times for planets and satellites applying a Hamiltonian formulation. *Purcell* (1979), to study the alignment of interstellar dust, attacked the problem with classical rotational dynamics and conventional elasticity theory. *Lambeck* (1980) calculated the alignment time for the Earth's Chandler wobble by considering the Earth as a Maxwellian material and using the Liouville equations of rotational dynamics.

A recent series of papers (e.g., *Efroimsky and Lazarin,* 2000; *Efroimsky,* 2001; hereafter referred to as *EL*) has sharply criticized these earlier works as missing important dynamics, failing to satisfy boundary conditions perfectly and using unphysical choices for the parameters in equation (5). Nevertheless, all the previous solutions come to similar conclusions and have the same functional dependence, differing only in K_1's magnitude. Using an approach similar to Purcell's, *EL* derive a comparable expression but maintain that the actual damping time is 2–6 orders of magnitude shorter than earlier researchers had estimated. *Burns and Safronov* (1973) assert that K_1 will depend on the asteroid's shape; *EL* concur and also point out that τ_d may change depending on the magnitude of nutation. Because of its relevance to many dynamics problems for small solar system bodies, this problem is being reconsidered by several groups. Most maintain that *EL*'s claims are overstated; accordingly, for the purposes of further discussion, we will select the numerical values chosen by *Harris* (1994). *Sharma et al.* (2001) compare damping times calculated by various authors and discuss the reasons for the slight differences in expressions, before deriving their own result.

The damping time may be alternatively expressed as

$$\tau_d = K_2P^3/D^2 \quad (6)$$

where τ_d is given in billions of years, P is the rotation period in hours, D is the asteroid's diameter in kilometers and $K_2 = (1/17)^3$ to within an order of magnitude (*Harris,* 1994, *Pravec et al.,* 2002). To obtain equation (6) from (5), *Harris* (1994) chose $\mu Q = 5 \times 10^{12}$ and $\rho = 2.5$, each in cgs units; *Burns and Safronov* (1973) had assumed a much stronger and harder material.

Various damping times, corresponding to different (P, D) combinations in equation (6), are shown on Fig. 2. We see that, according to this criterion, almost all asteroids with well-known spin periods damp in times much less than the solar system's age. In contrast, all the identified tumblers — except for the smallest, fastest one — have damping times that are comparable to or longer than the age of the solar system. Hence, by this standard, regardless of when these wobblers last received a significant angular momentum kick, they should be still wobbling, just as they were found to be. Note also that, whereas some apparently nontumbling objects lie below the 4.5-Ga line, in many cases their rotations are not well enough determined as yet to rule out tumbling (*Harris,* 1994).

Of all the wobblers, only the fast-spinning 2000 WL107 has a damping time much less than the solar system's age. In fact, on Fig. 2 its damping time is ~5 m.y., a time comparable to, but shorter than, the estimated collisional age [5–20 m.y. (*Davis et al.,* 2002)] of such a minor planet in the main belt where this object presumably originated. Thus it is no surprise that this NEA is wobbling. On the other hand, if the much shorter scale of *EL* were chosen, the damping time would be a mere 5×10^4 yr or so, making it highly improbable to discover this object tumbling.

5. BINARIES

There is now ample observational evidence for the existence of binary asteroids. A full discussion of observations and possible mechanisms for their origins is given in *Merline et al.* (2002). Here we mention only those aspects of binary origins relevant to disruptive collisions.

Hartmann (1979) suggests that, in the collisional disruption of a parent body, individual fragments might have sufficiently low relative velocities to be gravitationally bound; if their relative angular momentum were large enough, they would remain in orbit about each other. *Weidenschilling et al.* (1989) point out that due to purely geometrical effects, adjacent fragments moving fast enough to escape from the parent body would also tend to have relative velocities large enough to escape from each other. That argument depends on the assumption of a smoothly varying radial velocity field during the disruption, however. *Durda* (1996) performs numerical simulations of disruptive events and shows that a small dispersion of radial velocities increases the fraction of binaries among the fragments (as well as reaccumulation of pairs in contact). Even with no imposed velocity dispersion, some (<1%) bound pairs were produced, apparently by nonradial motions and nonuniformities due to jostling of fragments during the early stage of disruption. Similar results, based on the implementation of a N-body algorithm into the semiempirical model by *Paolicchi et al.* (1996), are obtained by *Doressoundiram et al.* (1997) and *Paolicchi et al.* (1999). According to them, binary systems are formed within the overall process of partial reaccumulation of fragments, due to self-gravity. As discussed by *D'Abramo et al.* (1999) (see Fig. 1), the reaccumulation involves a localized region from which the fragments originated; the size of this region depends on the amount of total reaccumulation. The binary systems are formed close to the boundaries of this region. According to these computations, binary systems are not very numerous.

The limited number of binaries in both models is to some degree due to the limited range of sizes in the simulations;

if more numerous, smaller fragments were considered, there would be a higher probability of bound pairs having very unequal masses. A system such as Ida-Dactyl fits properly within this scenario.

On the other hand, if the ejection velocities of the fragments were larger (as suggested by different considerations; see the discussion in *Zappalà et al.,* 2002), little self-gravity-driven reaccumulation would take place, and the formation of binaries through the above-discussed process might be hindered [note that a possible channel for the formation of binaries may be through the formation of the so-called "jets" observed in some experiments (*Martelli et al.,* 1993)]. Nevertheless, in the case of high ejection velocities, many bodies rotating faster than 2 h are presumably created. Since such bodies are absent, they (if formed) are subject to fission immediately after formation or later, triggered perhaps by a further minor collisional event. Thus a widespread formation of binary systems follows. In this case, the mass ratio between the components should depend on unknown initial conditions or parameters with no preferential value (*Paolicchi et al.,* 1999).

The potential binary systems that develop in a catastrophic collision are represented in Fig. 1 (adapted from *D'Abramo et al.,* 1999). Among the fragments reaccumulated by self-gravitation, a few give rise to binary systems. Moreover, numerous fragments that freely escape rotate faster than 2 h and are thus subject to a later bursting fission, eventually ending up as binary systems. The size of the various regions (reaccumulation, escape, fast rotation) depends on the model; nevertheless, when we have a small amount of reaccumulation (and thus few "reaccumulated" binaries), we usually have many fast rotators (and thus many "bursting" binaries) and vice versa.

6. OPEN PROBLEMS AND CONCLUSIONS

In spite of some progress made in the last decade, we are presently unable to draw a general and satisfactory picture of what happens when two asteroids collide. Due to these difficulties, the analysis of collisional side effects presents several open problems.

We can identify a few points:

1. Most asteroids larger than a few hundred meters have a composite structure. Due to this structure, the bodies cannot rotate too quickly; otherwise, they lose regolith or fission. Also, the presence of very slow rotators can possibly be explained in terms of similar processes. Only small asteroids may have a monolithic structure, thus being allowed to rotate at essentially any rate, limited only by internal tensile strength.

2. The slowing down observed on average among medium-large asteroids may be explained in terms of selective reaccumulation processes.

3. As a consequence of collisional processes, asteroids (or fragments formed by the breakup of the parent body) can naturally acquire complex rotational properties (such as tumbling); on the other hand, the observation of such properties is usually possible only for a relatively short time after the causative event.

4. As a result of catastrophic disruption, binary systems may be formed; a few different channels may lead to this outcome. The uncertainties of the physics affect the efficiency of different channels and thus the expected detections. Nevertheless, several binaries are probably always formed among the numerous fragments, although their relative proportion is low.

A more quantitative analysis of the above-presented features (and the related development of a general model for the collisional evolution of asteroids) will require understanding a few critical points. Among them the most relevant open problem is how large bodies react to energetic impacts, i.e., how strong or weak they are and how large are the ejection velocities of the fragments. Another question is of paramount importance for the problems we have discussed: How effective is the transfer of angular momentum during the impact, or, in other terms, what percentage of angular momentum is carried away unnoticed with the debris?

The above-discussed properties (and the related open problems) are relevant also for other aspects of asteroidal science. For example, the rotational properties influence the diurnal Yarkovsky effect; thus the collisional evolution of rotational properties plays to strengthen/weaken this effect, which in turn is relevant for the overall evolution of the belt and for the observed properties of asteroid families.

Acknowledgments. We are grateful to A. Cellino, A. Dell'Oro, and Ishan Sharma for useful discussions. We are also grateful to the referee Petr Pravec. P.P. was supported by a special grant from the University of Pisa. J.A.B. and S.J.W. received funding from NASA's PG&G program.

REFERENCES

Asphaug E. and Scheeres D. J. (1999) Deconstructing Castalia: Evaluating a postimpact state. *Icarus, 139,* 383–386.

Asphaug E., Ostro S. J., Hudson R. S., Scheeres D. J., and Benz W. (1998) Disruption of kilometer-sized asteroids by energetic collisions. *Nature, 393,* 437–440.

Asphaug E., Ryan E. V., and Zuber M. T. (2002) Asteroid interiors. In *Asteroids III* (W. F. Bottke Jr. et al., eds.), this volume. Univ. of Arizona, Tucson.

Benz W. and Asphaug E. (1999) Catastrophic disruptions revisited. *Icarus, 142,* 5–20.

Binzel R. P., Farinella P., Zappalà V., and Cellino A. (1989) Asteroid rotation rates: Distributions and statistics. In *Asteroids II* (R. P. Binzel et al., eds.), pp. 416–441. Univ. of Arizona, Tucson.

Black G. J., Nicholson P. D., Bottke W. F., and Burns J. A. (1999) On a possible rotation state of (433) Eros. *Icarus, 140,* 239–242.

Burns J. A. and Safronov V. S. (1973) Asteroid nutation angles. *Mon. Not. R. Astron. Soc., 165,* 403–411.

Burns J. A. and Tedesco E. F. (1979) Asteroid lightcurves: Results for rotations and shapes. In *Asteroids* (T. Gehrels, ed.), pp. 494–527. Univ. of Arizona, Tucson.

Capaccioni F., Cerroni P., Coradini M., Farinella P., Flamini E.,

Martelli G., Paolicchi P., Smith P. N., and Zappalà V. (1984) Shapes of asteroids compared with fragments from hypervelocity impact experiments. *Nature, 308,* 832–834.

Cellino A., Zappalà V., Davis D. R., Farinella P., and Paolicchi P. (1990) Asteroid collisional evolution. I. Angular momentum splash — Despinning asteroids through catastrophic collisions. *Icarus, 87,* 391–402.

Cellino A., Michel P., Tanga P., Zappalà V., Paolicchi P., and Dell'Oro A. (1999) The velocity-size relationship for members of asteroid families and implications for the physics of catastrophic collisions. *Icarus, 141,* 79–95.

Cuk M. and Burns J. A. (2001) Tidal slowdown of Mathilde's spin (abstract). In *Asteroids 2001 Conference,* p. 234. Osservatorio di Palermo, Sicily.

D'Abramo G., Dell'Oro A., and Paolicchi P. (1999) Gravitational effects after the impact disruption of a minor planet: Geometrical properties and criteria for the reaccumulation. *Planet. Space Sci., 47,* 975–986.

Davis D. R., Weidenschilling S. J., Farinella P., and Paolicchi P. (1989) Asteroid collisional history: Effects on sizes and spins. In *Asteroids II* (R. P. Binzel, et al., eds.), pp. 806–825. Univ. of Arizona, Tucson.

Davis D. R., Durda D. D., Marzari F., Campo Bagatin A., and Gil-Hutton R. (2002) Collisional evolution of small-body populations. In *Asteroids III* (W. F. Bottke Jr. et al., eds.), this volume. Univ. of Arizona, Tucson.

Dobrovolskis A. R. and Burns J. A. (1984) Angular momentum drain: A mechanism for despinning asteroids. *Icarus, 57,* 464–476.

Doressoundiram A., Paolicchi P., Verlicchi A., and Cellino A. (1997) Formation of binary asteroids as outcomes of catastrophic collisions. *Planet. Space Sci., 45,* 757–770.

Durda D. D. (1996) The formation of asteroidal satellites in catastrophic collisions. *Icarus, 120,* 212–219.

Efroimsky M. (2001) Relaxation of wobbling asteroids and comets — theoretical problems, perspectives on experimental observations. *Planet. Space Sci., 49,* 937–955.

Efroimsky M. and Lazarin A. (2000) Inelastic dissipation in wobbling asteroids and comets. *Mon. Not. R. Astron. Soc., 311,* 269–278.

Farinella P., Paolicchi P., and Zappalà V. (1982) The asteroids as outcomes of catastrophic collisions. *Icarus, 52,* 409–433.

Farinella P., Davis D. R., Paolicchi P., Cellino A., and Zappalà V. (1992) Asteroid collisional evolution — an integrated model for the evolution of asteroid rotation rates. *Astron. Astrophys., 253,* 604–614.

Fujiwara A. and Tsukamoto A. (1981) Rotation of fragments in catastrophic impact. *Icarus, 48,* 329–334.

Fujiwara A., Cerroni P., Davis D., Ryan E., Di Martino M., Holsapple K., and Housen K. (1989) Experiments and scaling laws for catastrophic collisions. In *Asteroids II* (R. P. Binzel et al., eds.), pp. 240–265. Univ. of Arizona, Tucson.

Giblin I. and Farinella P. (1997) Tumbling fragments from experiments simulating asteroidal catastrophic disruption. *Icarus, 127,* 424–430.

Giblin I., Martelli G., Smith P. N., and Di Martino M. (1994a) Simulation of hypervelocity impacts using a contact charge. *Planet. Space Sci., 42,* 1027–1030.

Giblin I., Martelli G., Smith P. N., Cellino A., Di Martino M., Zappalà V., Farinella P., and Paolicchi P. (1994b) Field fragmentation of macroscopic targets simulating asteroidal catastrophic collisions. *Icarus, 110,* 203–224.

Giblin I., Martelli G., Farinella P., Paolicchi P., Di Martino M., and Smith P. N. (1998a) The Properties of fragments from catastrophic disruption events. *Icarus, 134,* 77–112.

Giblin I., Petit J. M., and Farinella P. (1998b) Impact ejecta rotational bursting as a mechanism for producing stable Ida-Dactyl systems. *Icarus, 132,* 43–52.

Harris A. W. (1994) Tumbling asteroids. *Icarus, 107,* 209–211.

Harris A. W., Young J. W., Bowell E., and Tholen D. J. (1999) Asteroid lightcurve observations from 1981–1983. *Icarus, 142,* 173–201.

Hartmann W. K. (1979) Diverse puzzling asteroids and a possible unified explanation. In *Asteroids* (T. Gehrels, ed.), pp. 466–479. Univ. of Arizona, Tucson.

Holsapple K. A. (1993) The scaling of impact processes in planetary sciences. *Annu. Rev. Earth Planet. Sci., 21,* 333–373.

Holsapple K. A. (1994) Catastrophic disruption and cratering of solar system bodies: A review and new results. *Planet. Space Sci., 42,* 1067–1078.

Holsapple K. A. (2001) Shape limits for rock-pile asteroids (abstract). In *Asteroids 2001 Conference,* p. 73. Osservatorio di Palermo, Sicily.

Holsapple K., Giblin I., Housen K., Nakamura A., and Ryan E. (2002) Asteroid impacts: Laboratory experiments and scaling laws. In *Asteroids III* (W. F. Bottke Jr. et al., eds.), this volume. Univ. of Arizona, Tucson.

Housen K. R. and Holsapple K. A. (1990) On the fragmentation of asteroids and planetary satellites. *Icarus, 84,* 226–253.

Housen K. R. and Holsapple K. A. (1991) Laboratory simulation of large-scale fragmentation events. *Icarus, 94,* 180–190.

Hudson R. S., Ostro S. J., and Harris A. W. (1997) Constraints on spin state and Hapke parameters of asteroid 4769 Castalia using lightcurves and a radar-derived shape model. *Icarus, 130,* 165–176.

Kaasalainen M. (2001) Interpretation of lightcurves of precessing asteroids. *Astron. Astrophys., 376,* 302–309.

Lambeck K. (1980) *The Earth's Variable Rotation: Geophysical Causes and Consequences.* Cambridge Univ., Cambridge. 449 pp.

La Spina A. and Paolicchi P., (1996) Catastrophic fragmentation as a stochastic process: Sizes and shapes of fragments. *Planet. Space Sci., 44,* 1563–1578.

Love S. G. and Ahrens T. J. (1996) Catastrophic impacts on gravity-dominated asteroids. *Icarus, 124,* 141–155.

Love S. G. and Ahrens T. J. (1997) Origin of asteroid rotation rates in catastrophic impacts. *Nature, 386,* 154–156.

Martelli G., Rothwell P., Giblin I., Smith P. N., Di Martino M., and Farinella P. (1993) Fragment jets from catastrophic breakup events and the formation of asteroid binaries and families. *Astron. Astrophys., 271,* 315–318.

Martelli G., Ryan E. V., Nakamura A. M., and Giblin I. (1994) Catastrophic disruption experiments: Recent results. *Planet. Space Sci., 42,* 1013–1026.

Melosh H. J., Ryan E. V., and Asphaug E. (1992) Dynamic fragmentation in impacts — Hydrocode simulation of laboratory impacts. *J. Geophys. Res., 97,* 14735–14759.

Merline W. J., Weidenschilling S. J., Durda D. D., Margot J.-L., Pravec P., and Storrs A. D. (2002) Asteroids *do* have satellites. In *Asteroids III* (W. F. Bottke Jr. et al., eds.), this volume. Univ. of Arizona, Tucson.

Michel P., Benz W., Tanga P., and Richardson D. C. (2001) New simulations of collisions between asteroids in the gravity regime: Comparison with the properties of some observed asteroid families. In *Asteroids 2001 Conference,* p. 92. Osservatorio di Palermo, Sicily.

Mottola S. and 15 colleagues (1995) The slow rotation of 253 Mathilde. *Planet. Space Sci., 42,* 1609–1612.

Ostro S. J. (1993) Planetary radar astronomy. *Rev. Mod. Phys., 65,* 1235–1279.

Ostro S. J., Hudson R. S., Benner L. A. M., Nolan M. C., Margot J.-L., Magri C., Giorgini J. D., Jurgens R. F., Rose R., and Yeomans D. K. (2000) Radar observations of asteroid 4486 Mithra. *Bull. Am. Astron. Soc., 32,* 1003.

Ostro S. J., Nolan M. C., Margot J.-L., Magri C., Harris A. W., and Giorgini J. D. (2001) Radar observations of asteroid 288 Glauke. *Icarus, 152,* 201–204.

Paddack S. J. (1969) Rotational bursting of small celestial bodies: Effects of radiation pressure. *J. Geophys. Res., 74,* 4379–4381.

Paolicchi P., Verlicchi A., and Cellino A. (1996) An improved semi-empirical model of catastrophic impact processes. I: Theory and laboratory experiments. *Icarus, 121,* 126–157.

Paolicchi P., Dell'Oro A., Verlicchi A., Doressoundiram A., Cellino A., and Michel P. (1999) Are binary asteroids formed always during catastrophic processes? (abstract). In *ACM Conference,* p. 8.

Peale S. J. (1973) Rotation of solid bodies in the solar system. *Rev. Geophys. Space Phys., 11,* 763–793.

Peale S. J. and Lissauer J. J. (1989) Rotation of Halley's comet. *Icarus, 79,* 396–430.

Pisani E., Dell'Oro A., and Paolicchi P. (1999) Puzzling asteroid families. *Icarus, 142,* 78–88.

Pravec P. and Hahn G. (1997) Two-period lightcurve of 1994 AW1: Indication of a binary asteroid? *Icarus, 127,* 431–440.

Pravec P. and Harris A. W. (2000) Fast and slow rotation of asteroids. *Icarus, 148,* 12–20.

Pravec P., Wolf M., and Šarounová L. (1998) Lightcurves of 26 near-Earth asteroids. *Icarus, 136,* 124–153.

Pravec P. and 11 colleagues (2000) Slowly rotating asteroid 1999 GU3. *Icarus, 148,* 589–593.

Pravec P., Harris A. W., and Michałowski T. (2002) Asteroid rotations. In *Asteroids III* (W. F. Bottke Jr. et al., eds.), this volume. Univ. of Arizona, Tucson.

Purcell E. M. (1979) Suprathermal rotation of interstellar grains. *Astrophys. J., 231,* 404–416.

Pych W. (1999) Short period oscillations in the light curve of the asteroid 1689 Floris-Jan. *Astron. Astrophys., 343,* L75–L77.

Richardson D. C., Bottke W. B., and Love S. G. (1998) Tidal distortion and disruption of Earth-crossing asteroids. *Icarus, 134,* 47–76.

Rubincam D. P. (2000) Radiative spin-up and spin-down of small asteroids. *Icarus, 148,* 2–11.

Ryan E. V. and Melosh H. J. (1998) Impact fragmentation: From the laboratory to asteroids. *Icarus, 133,* 1–24.

Scheeres D. J., Ostro S. J., Werner R. A., Asphaug E., and Hudson R. S. (2000) Effects of gravitational interactions on asteroid spin states. *Icarus, 147,* 106–118.

Sharma I., Burns J. A., and Hui C. Y. (2001) Nutational damping times in solids of revolution. *Bull. Am. Astron. Soc., 33,* 1114.

Spencer J. R. and 48 colleagues (1995) The lightcurve of 4179 Toutatis: Evidence for complex rotation. *Icarus, 117,* 71–89.

Tanga P., Cellino A., Michel P., Zappalà V., Paolicchi P., and Dell'Oro A. (1999) On the size distribution of asteroid families: The role of geometry. *Icarus, 141,* 65–78.

Verlicchi A., La Spina A., Paolicchi P., and Cellino A. (1994) The interpretation of laboratory experiments in the framework of an improved semi-empirical model. *Planet. Space Sci., 42,* 1031–1041.

Weidenschilling S. J. (1981) How fast can an asteroid spin? *Icarus, 46,* 124–126.

Weidenschilling S. J., Paolicchi P., and Zappalà V. (1989) Do asteroids have satellites? In *Asteroids II* (R. P. Binzel et al., eds.), pp. 643–658. Univ. of Arizona, Tucson.

Yoder C. F. and Ward W. R. (1979) Does Venus wobble? *Astrophys. J. Lett., 233,* L33–L37.

Zappalà V., Cellino A., Dell'Oro A., Migliorini F., and Paolicchi P. (1996) Reconstructing the original ejection velocity fields of asteroid families. *Icarus, 124,* 156–180.

Zappalà V., Cellino A., Dell'Oro A., and Paolicchi P. (2002) Physical and dynamical properties of asteroid families. In *Asteroids III* (W. F. Bottke Jr. et al., eds.), this volume. Univ. of Arizona, Tucson.

The Fate of Asteroid Ejecta

D. J. Scheeres
The University of Michigan

D. D. Durda
Southwest Research Institute

P. E. Geissler
University of Arizona

The distribution of regolith on asteroid surfaces has only recently been measured directly by *in situ* observations from spacecraft. To the surprise of many researchers, most of the classical predictions for the distribution of asteroid impact ejecta have not rung true, with regoliths appearing to be geologically active at small scales on asteroid surfaces. This indicates that significant insight into geological processes on asteroids may be inferred by detailed studies of the distribution of impact ejecta on asteroids. This chapter has been written to support these future investigations, by trying to identify and clarify all the important elements for such a study, to point to the recent history of such studies, and to indicate the current gaps in our understanding. The chapter begins with a discussion of the initial conditions of ejecta fields generated from impacts on the asteroid surface. Then the relevant physical laws and forces affecting asteroid ejecta, in orbit and on the surface, are reviewed and the basic dynamical equations of motion for ejecta are stated. Some general results and constraints on the solutions to these equations are given, and a classification scheme for ejecta trajectories is given. Finally, recent studies of asteroid ejecta are reviewed, showing the application of these techniques to asteroid science.

1. INTRODUCTION

Whether the debris ejected from impacts on asteroids escapes or reimpacts has important implications for the erosion of asteroids, retention and distribution of regolith, dispersal or reaccretion of fragments after catastrophic disruptions, and the formation of temporary satellites and permanent moons. Asteroids present complex dynamical environments because of their low gravitational accelerations, nonspherical shapes, complex geological makeup, and diverse rotation states. Additionally, critical parameters related to the flux and size distribution of impactors and the resulting initial ejecta fields are only poorly known. Thus the physics and dynamics of regolith processes are complicated and not fully understood. Finally, physical observations of asteroids are only now approaching the resolution necessary to seriously constrain and delineate between competing theories of the asteroid environment, making the study of asteroid ejecta a timely endeavor.

Within the last decade we have obtained closeup pictures of asteroids Gaspra, Ida and Dactyl, Mathilde, and Eros to supplement earlier images of the martian moons. Radar and telescopic observations have revealed the shapes and rotation states of many more objects. Morphological indications of regolith on these asteroids include blocks, landslides, buried craters, and color variations. Observational tests of dynamical theories include nonuniform regolith, ejecta block distributions and asymmetric crater ejecta blankets, rays, and strings of secondaries. The importance of erasure mechanisms, such as seismic shaking (*Greenberg et al.,* 1994) and electromagnetic forces (*Lee,* 1996), that compete with dynamical effects to shape the surfaces of asteroids has recently been emphasized by detailed studies of Eros by *NEAR Shoemaker*.

Rapid advances are expected in our understanding of impact cratering on diverse objects through *in situ* experiments such as NASA's Deep Impact mission to Comet Tempel 1, numerical simulation of ejecta trajectories that employ realistic shape and gravity models and consider third-body and nongravitational forces, and geological evidence from analysis of spacecraft data. The fundamental motivation for the study of asteroid regoliths arises from the meteoritics community and the interpretations of (primordial) asteroid regoliths as observed in the meteorite database. Future motivation for ejecta studies will include the necessity of providing a complete mechanical understanding of the asteroid environment. Ultimately, a detailed understanding of ejecta dynamics will be crucial to characterize the safety of the orbital environment about asteroids for rendezvous missions, landed operations on asteroid surfaces, and other close-proximity operations. A key issue of concern for any surface operation on an asteroid, such as sampling, will be the trajectories of ejecta disturbed and lofted into orbit during routine operations, as disturbed regolith may reimpact on the surface with speeds on the order of the surface escape speed long after being dislodged (*Scheeres and Asphaug,* 1998).

The study of asteroid ejecta is intimately tied to the study of impacts on asteroids [see the reviews by *Asphaug et al.* (2002) and *Holsapple et al.* (2002)], transient and long-term orbital dynamics close to asteroids, and the study of natural and artificial asteroid satellites (*Merline et al.,* 2002). Additionally, the dynamics of asteroid ejecta have many similarities to the dynamics of cometary ejecta. In general, methods devised for the study of each area will have application to both areas.

Out of necessity, this chapter brings together several diverse areas of asteroid and dynamical science. Ideally, this chapter will serve as a starting point for future investigations into the dynamics of ejecta from the surfaces of small bodies. As such, we have injected many topics into the chapter, in some instances without detailed descriptions of the background theory or its development. Thus, to supplement the work described herein, we suggest that the following textbooks be referenced: *Melosh* (1989) for an introduction to the basic principles and physics of impacts, *Murray and Dermott* (1999) for an introduction to the orbital dynamics of natural bodies, and *Szebehely* (1967) for an introduction to advanced orbital dynamics theory.

2. A BRIEF HISTORY

The study of asteroid ejecta has been considered in the books *Asteroids* and *Asteroids II,* distributed among chapters on asteroid regoliths (*Cintala et al.,* 1979; *Housen et al.,* 1979b; *Veverka and Thomas,* 1979; *McKay et al.,* 1989) and asteroid satellites (*van Flandern et al.,* 1979; *Weidenschilling et al.,* 1989). Yet the study of ejecta and satellite dynamics about asteroids was only fully validated with the discovery of Dactyl in orbit about Ida (*Belton et al.,* 1996). In recent years the relevance of this topic has continued to grow, with the rapid rate at which asteroid satellites have been discovered (see the chapter by *Merline et al.,* 2002) and recent realizations from the *NEAR Shoemaker* mission that the small-scale structure of asteroid surfaces are not well understood (*Veverka et al.,* 2001; *Cheng et al.,* 2001).

Initial studies of asteroid ejecta, and orbital dynamics about asteroids, assumed that their dynamical environment was analogous to, and directly scalable from, planetary satellite dynamics (*van Flandern et al.,* 1979; *Weidenschilling et al.,* 1989). However, *Weidenschilling et al.* (1989) already noted that the potential for complex dynamics close to asteroids existed and proposed that further studies on this topic be done. Prior to this, *Dobrovolskis and Burns* (1980) had already performed detailed ejecta trajectory analysis for the asteroids Phobos and Deimos, noting that ejecta trajectories were strongly influenced by the rotation state and gravity field. For these bodies, however, the Mars tidal force is so strong that there is no direct analogy between those works and the evolution of ejecta about asteroids.

Detailed dynamical studies of orbital motion about asteroids has blossomed since the publication of *Asteroids II.* Initial studies focused on the stability of binary asteroids. *Chauvineau et al.* (1990a,b, 1991) investigated the stability of binary asteroid systems relative to the solar tide, jovian perturbations, and collisions in a series of papers, based on the earlier studies of the dynamics of the "Hill problem" (*Hénon,* 1969). *Hamilton and Burns* (1991a,b) investigated the limits of stable motion about an asteroid, with a special interest given to the safety of the planned *Galileo* flybys of asteroids Gaspra and Ida. These papers, taken together, provide a clear picture of the stability of asteroid binaries, and place constraints on the stability of asteroid ejecta that move relatively far from the asteroid. Such studies are still continuing, and additional progress in understanding the dynamics of trajectories far from an asteroid have been made (*Richter and Keller,* 1995; *Hamilton and Krivov,* 1997).

Chauvineau et al. (1993) and *Scheeres* (1994) initiated the study of dynamics in the near-asteroid environment, studying the motion of particles and ejecta close to rotating ellipsoids. These early studies showed that the near-asteroid orbital environment was fundamentally different from the environment found in the vicinity of a planet or larger satellite. Studies along these lines have continued with the detailed analysis of specific asteroid shapes (*Geissler et al.,* 1996; *Petit et al.,* 1997; *Scheeres et al.,* 1996, 1998a, 2000a) and the theoretical analysis of motion in generalized models of asteroid gravity fields (*Scheeres,* 1999). Now this area of study has its first precision set of data with the results of the *NEAR Shoemaker* mission to asteroid 433 Eros (*Yeomans et al.,* 2000; *Miller et al.,* 2001), the fruits of which are already being published (*Thomas et al.,* 2001, *Robinson et al.,* 2001).

3. EJECTA GENERATION

The study of asteroid regolith mechanics and the dynamics of impact ejecta fields must first concern itself with the mechanics of impact ejecta generation. *Weidenschilling et al.* (1989) noted that there are tight constraints on ejecta speed before all ejecta immediately escape from the asteroid and into heliocentric space. Indeed, early estimates on regolith depth (or lack thereof) on smaller asteroids predicted little, if any, retained regolith. This view of asteroid surfaces has changed with recent observations of asteroids from spacecraft and radar. In the following we review some basic results on the ejecta fields resulting from impact events, with an emphasis on the implications of these models for the initial conditions of an ejecta fragment field.

3.1. Mathematical Models and Scaling-Law Predictions

One approach to the understanding of generation and redistribution of regoliths on small bodies is through theoretical modeling of impacts. These models have been guided by observations of crater ejecta and regolith on the Moon and by the results of numerous laboratory impact experiments. Detailed models of regolith emplacement and evolution on small bodies, incorporating quantitative treatments of cratering rates and ejecta thickness, were considered by *Housen et al.* (1979a,b) and *Housen* (1981). Those models predicted that for the smallest asteroidal bodies (D < 10 km), nearly all impact ejecta escapes and such objects should have only thin (on the order of 1 mm) coatings of comminuted debris. With increasing asteroid size, more ejecta is retained and regoliths are predicted to be thicker, on the order of hundreds of meters for asteroids ~100 km in diameter and larger. In *Veverka et al.* (1986) the minimum diameter for regolith retention was estimated to be 20 km for icy bodies and 70 km for rocky bodies. Now that the surfaces of several asteroids have been imaged at resolutions allowing small-scale surface features to be examined in detail, regolith thickness and spatial distribution may be more directly estimated as a function of asteroid size, shape, and rotation state.

The volume of ejecta material and the mass of the largest fragments excavated from an impact crater on an asteroid may be assumed to scale with the crater size, although it must be recognized that factors including target surface gravity, porosity, layering and structure in the target, and impact angle may all play a role in complicating the predictions of simple scaling relations. The simplest estimates for ejecta volume may be made by scaling from the apparent diameters of craters (e.g., *Lee et al.,* 1996). Photoclinometry applied to fresh craters on Ida (*Sullivan et al.,* 1996) indicates that a diameter D to depth h ratio D:h is ~6:5. Craters on Eros show similar diameter:depth ratios (*Veverka et al.,* 2000). Assuming that craters are spherical segments with depth h ~ D/6.5 and diameter D, their volume is $V \sim 0.06\ D^3$.

With ejecta volumes estimated in this fashion, the total volume of material ejected from craters larger than 0.5 km diameter on Ida amounts to ~500 km^3, which would amount to a regolith layer ~130 m thick, if retained and distributed evenly over the ~3800-km^2 surface area of Ida. If Ida responds to impacts as a strong, competent object, then ejecta would have escaped the surface to space and such an estimate of regolith thickness is not valid. However, if craters of this size formed in the gravity regime, preexisting regolith might be present, in which case the estimate is a lower limit to regolith depth. As pointed out in *Hartmann* (1978), "Regolith begets regolith."

From observations of blocks on Ida and a review of previous work of blocks on the rims of lunar craters, *Moore* (1971) and *Lee et al.* (1996) derive a general relationship between the largest ejecta block size, L, and crater diameter, D, for craters in rocky targets, $L \sim 0.25\ D^{0.7}$, where L and D are in meters. This ejected material ranges in size from the largest ejecta blocks down to dust-sized particles, with a cumulative mass distribution expressed as $N(>m) = Cm^{-b}$, where b commonly ranges between 0.8 and 0.9, C is a normalizing constant, and m is the cumulative mass fraction. A fraction of this material is jetted from the impact site at high speed or spalled from the near-surface interference zone of the growing crater at speeds up to half the impact speed of the projectile (*Melosh,* 1989, p. 73). With typical main-belt impact speeds of ~5 km/s, most of this spalled material immediately escapes the target asteroid. Excavation flow speeds are much lower, however, with even the highest speeds between one-sixth and one-tenth the impact speed, so that some portion of the crater ejecta may be retained on the surface of the asteroid.

3.2. Laboratory Experiments

The results of laboratory-scale impact experiments can provide a useful guide in understanding the generation of regoliths on asteroids and estimating the amount of debris retained on their surfaces or ejected to escape. Indeed, mathematical models as described above rely in part on data from these experiments. There is extensive literature on laboratory experiments of impactors, reviewed by *Holsapple et al.* (2002). From these laboratory experiments essential, basic relations for ejecta volumes, speeds, ejecta field orientations, and asymmetries have been established. Fundamental results from these areas can be found in *Gault et al.* (1963) and are reviewed by *Fujiwara et al.* (1989).

Recent evidence suggests that C- and F-type asteroids have remarkably low densities (in the range of 1.2–1.8 g cm^3) and that high porosity probably plays a significant role in limiting the excavation of debris from, and the disruption of terrain surrounding, craters on these bodies. One possible explanation for this is given in *Housen et al.* (1999), where they used a centrifuge and impacts into porous, highly crushable silicate materials to experimentally simulate crater formation and ejecta deposition on low-density, porous asteroids. Their results show that the ratio of ejecta mass (material deposited outside the crater rim) to crater mass (crater volume multiplied by initial target density) steadily decreases as the target porosity increases. This is due to ejecta speeds being very low in porous materials, and the fact that much of the crater volume is formed by compression of the target material as opposed to excavation. Large craters on porous asteroids should exhibit only minor ejecta deposits; centrifuge experiments at 250 g in material with porosity 70%, the conditions of the formation of largest impact crater on Mathilde (Karoo, 33 km diameter), indicates that only 10% of the crater mass will be ejected outside the crater. (For a competing explanation of the Mathilde craters found using numerical experimentation, see the next section.)

Finally, recent laboratory work has shown mineral-specific comminution. This hints that some mineral components isolated as chondrules or phenocrysts may be ejected at different speeds, which might be of interest in segregating mineralogically distinct portions of regolith (*Hörz et al.,* 1985; *Durda and Flynn,* 1999).

3.3. Numerical Experiments

Hydrocode experiments incorporate gravity and what is known about the fracture mechanics of rock into numerical simulations of the contact, compression, and excavation stages of an impact. Hydrocode simulations can be used to model impacts at scales far too large to be directly accessible through laboratory experimentation (*Benz and Asphaug,* 1999).

Using such hydrocode techniques, *Asphaug et al.* (1996) modeled the formation of craters on Ida ranging in size from 60 m to 8 km, and compared the ranges of ballistic ejecta from these craters with the area of seismically disturbed regolith surrounding them. The results confirm that ballistically emplaced ejecta deposits around small craters on Ida, and by inference, on asteroids of similar size and composition, should be diffuse and widespread, so that bright halos around small craters could be due to seismic disturbance of surrounding regolith rather than continuous ejecta blankets. Larger-scale impacts create a "megaregolith"-like zone of intense fracturing within a depth approaching one crater diameter below large craters and appear to be able to deposit a significant amount of debris in irregular blankets around them.

Numerical experiments also give insight into the role that high porosity may have in modifying the resultant impact ejecta field. In contrast to the laboratory results described above, *Asphaug* (2000) shows that high porosity can lead to high ejecta speeds, which also matches the observed lack of impact blocks on the Mathilde surface. This is due to energy confinement in the immediate crater fracture zone, due to the inability of the porous asteroid material to efficiently transmit the impact energy though the entire body. In *Asphaug* (2000) it was shown that the crater Karoo on Mathilde would have launched almost all its ejecta faster than 30 m/s, sufficient for escape, if Mathilde was 50% porous.

3.4. Strength and Gravity Regimes

Two of the most important questions regarding regolith generation and ejecta escape are (1) how much ejecta is created during any given impact, and (2) at what speeds the ejecta are launched. Calculation of crater volumes and the velocities of the ejecta expelled relies on scaling the outcomes of laboratory and field experiments to vastly different sizes and gravitational accelerations. The theoretical basis for such scaling is dimensional analysis, introduced by *Holsapple and Schmidt* (1982) and *Housen et al.* (1983) (see review by *Holsapple,* 1993). This approach groups variables into dimensionless ratios to reduce the complexity of arbitrary expressions relating impactor and target properties to crater volumes or ejecta speeds. In cases where the number of variables is one greater than the number of dimensions (mass, length, time), the expression can be reduced to a simple power-law relationship between the variables.

By this analysis, the volume of a crater produced by an impactor of a given mass scales neither with the energy nor the momentum of the impact, but varies according to a power law with an exponent between 1 (momentum) and 2 (energy). For strengthless targets, or in the case of large impacts in the gravity regime, the volume V of a crater created by an impactor of diameter D and speed U is given by (*Holsapple and Schmidt,* 1982)

$$V(D) = Ag^{-\alpha} U^{2\alpha} D^{3-\alpha} \tag{1}$$

where A is a constant that includes the densities of the projectile and the target [see *Geissler et al.* (1996) for a specific application], g is the gravitational acceleration, and α is a scaling exponent that depends on the target properties but falls in the range of 3/7 to 3/4. Similar arguments (*Housen et al.,* 1983) suggest that ejecta follow a power-law speed distribution with an exponent $e_v = 6\alpha/(3 - \alpha)$. When the target has substantial cohesive strength, or in the limit of small impactors or small target sizes, the volume of the crater scales linearly with the mass of the impactor. This is the strength regime, exemplified by shooting at boulders with a rifle. Ejecta velocities tend to be higher than those of gravity-dominated craters but the volume of material excavated is much lower. As the size of the impact increases, even cohesive materials gradually transition to the gravity regime due to the scale- and strain-rate-dependence of material strength. The transition between strength- and gravity-dominated impacts depends on the strength and gravity of the target, but for asteroids that are a few tens of kilometers in diameter the transition is expected to occur at crater diameters in the range of 10 to 1000 m. It may seem surprising that the relatively weak gravitational grasp of an asteroid could control the formation of kilometer-sized craters. The transition to gravity-controlled cratering is aided by the presence of regolith and rubble left by earlier impacts, and by fragmentation of even strong targets by shock waves that precede the crater excavation (*Asphaug and Melosh,* 1993; *Nolan et al.,* 1996).

3.5. Impact Statistics and Asteroidal Erosion

Erosion of asteroids competes with regolith generation/retention and can sometimes yield important constraints on the ages of small gravitationally bound objects (e.g., *Geissler et al.,* 1996). A knowledge of impactor size distributions, impact collision probabilities, and impact speeds is needed in order to estimate ejecta generation and escape rates and timescales for the creation and destruction of ejecta blocks. These quantities are in general poorly known and constitute the greatest uncertainties in erosion rate calculations.

The size-frequency distribution is well constrained by telescopic observations only for the largest asteroids (e.g., *Van Houten et al.,* 1970). Smaller size ranges must be estimated by (1) extrapolation of power laws for the observed asteroids, (2) collisional models predicting the production of small fragments, and (3) observations of the size distribution of the craters produced by these small impactors on asteroid and planetary surfaces. Simple calculations assuming that the impact efficiency is independent of target size yield a differential power law index of –3.5 (*Dohnanyi,* 1969), i.e., the relation between the number of fragments n and their radius r should follow the power law dn(r) ~ $r^{-3.5}$. However, *Galileo* observations of small craters on Gaspra and Ida suggest a much steeper size distribution for small impactors in the main belt. The differential power-law index for small impactors on these objects (<175 m diameter) is estimated to be near –4.0 (*Belton et al.,* 1992; *Chapman et al.,* 1996a,b). For such steep size distributions (indexes >4), infinite mass is found in the smallest fragments. Thus some minimum projectile size must exist, below which a steep size distribution no longer applies. This cut-off size threshold is poorly constrained by observations, and determines whether asteroid erosion is dominated by big bites taken during large impacts, or if asteroid surfaces are mainly sandblasted by particles that are centimeter-scale or smaller. An upper limit to this minimum impactor size is ~1 m, the size of a projectile that would produce craters in the diameter range of 10 to 100 m (the resolution limit of the *Galileo* observations).

The rate and efficiency of impacts depends upon the size, location, and orbit of the target asteroid. The intrinsic collision probability and the distribution of impact velocities for any specific target can be calculated by integrating these quantities over the population of asteroids on intersecting trajectories. Many estimates of intrinsic collision probabilities and impact velocities have been made using both direct numerical integration (*Marzari et al.,* 1996, 1997; *Dahlgren,* 1998) and statistical methods (*Wetherill,* 1967; *Greenberg,* 1982; *Farinella and Davis,* 1992; *Bottke et al.,* 1994; *Vedder,* 1996; *Dell'Oro and Paolicchi,* 1998). For main-belt asteroids, collision probabilities are typically on the order of 10^{-18} km^{-2} yr^{-1}, and impact velocity distributions are broad, non-Gaussian, and often contain spikes. For the purposes of evaluating impact efficiency, a value between the mean impact velocity and the root mean square impact velocity should be used, depending on the target crater scaling exponent α (equation (1)).

As an example we calculate the rate of production of 10-m-scale ejecta blocks on Eros. *Merline et al.* (1999, 2001) completed a search for satellites around Eros during the *NEAR Shoemaker* flyby and rendezvous and, at a 70% confidence level, found no objects near the asteroid with a diameter larger than 10 m, and a 95% confidence level for diameters greater than 20 m. Estimates of the rate of production from impacts on Eros and knowledge of their dynamical lifetimes could thus be compared with the observed lack of such objects to constrain in an iterative fashion the present impact rate on Eros. *Bottke et al.* (1995) give an intrinsic collision probability for near-Earth asteroids (NEAs) hitting other NEAs of ~15×10^{-18} km^{-2} yr^{-1} at a most probable relative speed of ~18 km s^{-1}. Multiplying this by the cross-sectional area of Eros and by the number of impactors capable of making 200-m and larger craters [the minimum crater size capable of producing 10-m ejecta blocks, according to the ejecta-block scaling law of *Lee et al.* (1996)] on Eros yields the block production rate. A sphere with the same 1106-km^2 surface area of Eros would have a radius of r = 9.4 km, so r^2 = 88 km^2 (the factor of π in the cross-section is not required since it is included in the instrinsic collision probability). Estimates based on strength regime cratering in soil (*Holsapple,* 1993) indicate that projectiles roughly 2–10 m in diameter are capable of producing the requisite craters. The NEA population may contain ~2×10^9 objects of this size (*Neukum et al.,* 2001; *Rabinowitz et al.,* 2000; *Ivanov et al.,* 2002); herein lies the greatest uncertainty in such calculations. Multiplying, $P_i \times A \times N = 15 \times 10^{-18}$ km^{-2} yr$^{-1} \times 88$ km$^2 \times 2 \times 10^9 = 2.7 \times 10^{-6}$ yr^{-1}, or a mean time between impacts of ~370,000 yr. Ejecta blocks launched from the surface of Eros may not be dynamically stable for such long times, so *Merline et al.*'s (1999, 2001) result of not finding any orbiting boulder larger than 10 m does not, unfortunately, constrain the time of formation of the last 200-m-scale crater on Eros.

The fraction of ejecta that escapes from an asteroid during a given impact depends on the target size and strength and the size of the impact. For both the strength and gravity regimes, the mass eroded per mass of impactor is independent of the size of the impactor (*Geissler et al.,* 1996). Impacts into strong targets impart ejecta with speeds much greater than the escape velocity of a typical asteroid. Although only a small fraction of ejecta reaches escape speed during gravity-dominated cratering, erosion of a gravitationally bound rubble or sand pile is much more efficient than that of a coherent object of similar size. Because craters created in soft targets are much larger than corresponding strength-regime craters, the total volume of ejecta that escapes in this case can be much greater than the volume of material excavated by a similar impact into a strong target. For example, *Geissler et al.* (1996) found that the mass eroded (ejected and escaped) from a soft Dactyl (made of sand) per unit mass of impactor should be at least 36× greater than that eroded from a Dactyl made of solid rock.

3.6. Surface Launch Conditions

For studying the subsequent motion of ejecta the most crucial item is its initial position and velocity relative to the asteroid surface. Assume that the crater is measured from a nominal vector $\mathbf{r}_0$ on the asteroid surface, that the asteroid surface normal vector at that point is $\hat{\mathbf{n}}_z$, and that there are two orthogonal unit vectors $\hat{\mathbf{n}}_x$ and $\hat{\mathbf{n}}_y$ tangent to the asteroid surface. The location of a single ejecta, as measured from the asteroid center of mass, is

$$\mathbf{r} = \mathbf{r}_0 + \delta\mathbf{r} \tag{2}$$

$$\delta\mathbf{r} = x\hat{\mathbf{n}}_x + y\hat{\mathbf{n}}_y + z\hat{\mathbf{n}}_z \tag{3}$$

where $|\delta\mathbf{r}| \ll |\mathbf{r}|$ in general. The ejecta velocity relative to the crater site is then specified as

$$\mathbf{V}_e = V_e(\delta\mathbf{r})\left[\cos\beta\hat{\mathbf{n}} + \sin\beta\cos\lambda\hat{\mathbf{n}}_x + \sin\beta\sin\lambda\hat{\mathbf{n}}_y\right] \tag{4}$$

where the angle β and λ define the direction of the velocity vector relative to the crater normal and the ejecta speed V_e will depend on its position within the crater. Nominal assumptions are that $\beta = 45°$, and that $\lambda \in [0,360]°$.

For some dynamical computations the ejecta velocity must be transformed into an inertially oriented frame. Then the asteroid rotational velocity vector $\mathbf{\Omega}$ must be introduced

$$\mathbf{V}_I = \mathbf{V}_e + \mathbf{\Omega} \times \mathbf{r} \tag{5}$$

Note that $\mathbf{\Omega}$ is not necessarily constant and can have a significant time variation for bodies in nonuniform rotation.

4. EJECTA DYNAMICS

Once the ejecta has left the asteroid surface, it becomes subject to one of the more strongly perturbed environments that can be found in the solar system. Any serious study of asteroid ejecta must start with dynamical models that capture the main elements of these perturbations, since each of them can skew the global nature of ejecta dynamics into significantly different evolutions than would be expected from the simple application of two-body orbital dynamics.

4.1. General Force Models

For the general study of dynamics about asteroids one must, at the onset, determine which force perturbations will be significant for the system. Due to the wide variety of shapes, sizes, densities, rotation states, and orbits found for asteroids, this determination must usually repeated for each new asteroid.

4.1.1. Gravity fields. Several approaches to the modeling of asteroid gravity fields are available. In general, the most accurate formulations for a gravity field are spherical harmonic expansions where the gravity coefficients are measured from spacecraft radiometric tracking (*Yeomans et al.,* 2000; *Miller et al.,* 2001). Despite the high accuracy of these fields, they are in general inapplicable to the study of ejecta motions that arise from the surface of an asteroid due to the divergence of the expansion within the circumscribing sphere surrounding the asteroid (the circumscribing sphere is the sphere of minimum radius, centered at the asteroid center of mass, that encloses the asteroid). A modification to this technique using ellipsoidal harmonics is available (*Garmier and Barriot,* 2001) that decreases the region of divergence to within the circumscribing ellipsoid that fits about the body. Even this, however, does not completely eliminate the problem, as there will still be significant regions of divergence when close to or within this circumscribing ellipsoid (*Garmier et al.,* 2002).

To overcome this, recourse is usually made to the known closed-form gravitational potentials, a class that includes the sphere, the general ellipsoid (*Danby,* 1992), the tetrahedron (*Werner,* 1994), and a general polygonal shape (*Werner and Scheeres,* 1997). The main restriction to these potentials is that the mass density is assumed to be constant, or at the very least is constrained to follow a very specific mathematical variation (which in general may not be physical). The main approaches to gravity field modeling have used collections of point mass gravity potentials, collections of tetrahedron gravity potentials (making up a single, polyhedral shape), and the use of the simple ellipsoidal shape model.

The point mass (or mascon) approach consists of taking a defined shape model of the asteroid and populating its interior by a distribution of point masses, properly scaled to yield the correct total mass. This approach can lead to regions of poor gravity field computation on the surface and is inefficient if a high resolution is desired (*Werner and Scheeres,* 1997). The polyhedron approach takes a polygon shape model and computes the gravity potential (and its attendant partials) directly from this model. Even though the individual computations needed to compute the polyhedron gravity field are more involved than those used for the mascon approach, the overall efficiency of a polyhedral gravity field computation is often better than a mascon approach, since the mascon approach must sum over the entire volume of the body while the polyhedron approach must only sum over the surface elements of the body. The polyhedron approach also has the advantage of giving a direct indication of whether the point is inside or outside the asteroid. The ellipsoid model is useful for situations when a precision model of a gravity field is not needed. Its advantages are that it is relatively simple to code, has no singularities (such as are found for collections of point masses), and can be specified based on light-curve analysis alone.

In the following, the gravitational force potential is specified as $U(\mathbf{r})$ where $\mathbf{r}$ is the position vector relative to the asteroid-fixed frame. The gravitational attraction acting on a particle is $\partial U/\partial \mathbf{r}$. The potential U is often split into the main contribution of the monopole (μ/r) plus the perturbation contribution (R) as

$$U(\mathbf{r}) = \frac{\mu}{|\mathbf{r}|} + R(\mathbf{r}) \tag{6}$$

However, on an asteroid surface the perturbation contribution can often compete with the main contribution, and thus this form is only used for notational convenience.

When relatively far from an asteroid, MacCullagh's formula can be used to approximate the gravity field of a general asteroid with a reasonable degree of accuracy (*Danby,* 1992)

$$R(\mathbf{r}) = \frac{\mu R_o^2}{r^3}\left[C_{20}\left(1 - \frac{3}{2}\cos^2\delta\right) + 3C_{22}\cos^2\delta\cos 2\lambda \right] \tag{7}$$

where C_{20} and C_{22} are gravity coefficients, δ is the declination, and λ is the body-fixed longitude. This formulation assumes that the coordinate system is aligned with the inertial axes, with the maximum inertia axis along z and the minimum axis along x. In many cases, consideration of just this contribution to the gravity field of the asteroid can adequately capture the major departures of orbit dynamics from the simple Keplerian case. It must be noted that equation (7) cannot be used on the surface of the asteroid or within the circumscribing sphere, as it will give nonphysical values of potential and acceleration.

4.1.2. Rotation state. There are two classes of rotational motion that must be considered. The first, simplest, and most common is asteroid rotation about its maximum axis of inertia. It is well known that this is a stable rotational end state for an asteroid when dissipation of energy is taken into account, as it provides the minimum energy rotational state for a given value of angular momentum (*Burns and Safronov,* 1973). To completely specify the rotational dynamics for a uniformly rotating asteroid requires the rotational velocity vector and a phase angle for the asteroid. The rotational angular momentum of the asteroid will be subject to solar torques and nongravitational effects, and can be altered by impact events or planetary flybys. Still, over long periods of time it is acceptable to treat such a rotation state as a constant.

More interesting, but rarer, are cases where the asteroid has a nonuniform rotation state, usually corresponding closely to the general solution to Euler's equations for a torque-free rotating body. Examples of such bodies include the asteroids Toutatis, Mathilde, and Alinda. Most, if not all, of the asteroids observed to have a nonuniform rotation state are slow rotators, which makes physical sense as the time to relax to uniform rotation scales with the generalized rotation period cubed. Thus, a body such as Toutatis has a predicted relaxation time longer than the age of the solar system (*Harris,* 1994). In modeling ejecta dynamics about a nonuniform rotator the most efficient modeling approach is to use the classical solution for rotational dynamics in a torque-free environment (*MacMillan,* 1960); a summary of such an application is given in *Scheeres et al.* (1998a).

In the following we specify the asteroid angular velocity vector as $\mathbf{\Omega}$. For the case of a uniformly rotating asteroid, $\mathbf{\Omega}$ is constant in both the asteroid-fixed frame and in an inertially oriented frame since it is aligned with the asteroid's total rotational angular momentum vector. When modeling an asteroid with a nonuniform rotation state, the vector $\mathbf{\Omega}$ is no longer aligned with the asteroid's rotational angular momentum vector, but has a precession and nutation relative to this vector. If we model the asteroid nonuniform rotation using the solution for torque-free motion, the angular velocity vector $\mathbf{\Omega}$ is a periodic function of time in the asteroid-fixed frame, i.e., an observer sitting on the asteroid tracing out the path of this rotational velocity vector would see that it repeats itself exactly after a characteristic period (which is a function of the body's moments of inertia, rotational energy, and rotational angular momentum). The implications of this are discussed in greater detail in *Scheeres et al.* (1998a).

4.1.3. Solar effects. When far from the asteroid a particle must contend with strong perturbations from the solar gravity and radiation pressure. For precision computation, detailed models of the ejecta shape and interaction with solar radiation could be developed if desired. This level of detail is not always necessary for understanding the basic effect of the solar radiation pressure on the ejecta.

The solar gravity and radiation pressure forces are derived from a force potential written as

$$V_S = \mu_S \frac{1-\beta}{|\mathbf{d}+\mathbf{r}|} \tag{8}$$

where μ_S is the gravitation parameter of the sun, $\mathbf{d}$ is the asteroid position vector from the Sun, $\mathbf{r}$ is the ejecta position vector from the asteroid, and β is the ratio of solar radiation pressure force to solar gravity force acting on the ejecta. A reasonable assumption for ejecta motion relative to the asteroid is $|\mathbf{r}| \ll |\mathbf{d}|$, leading to the simplified potential

$$V_S = \frac{\mu_S(1-\beta)}{d}\left[\begin{array}{c} 1-\frac{1}{d}\hat{\mathbf{d}}\times\mathbf{r}- \\ \frac{1}{2d^2}\left\{\mathbf{r}\times\mathbf{r}-3\left(\hat{\mathbf{d}}\times\mathbf{r}\right)\right\}^2 \end{array}\right]+\ldots \tag{9}$$

4.1.4. Other nongravitational forces. For specific purposes, other nongravitational forces may also be modeled. This is especially true for the modeling of comets, where there is significant gas pressure that emanates from the nucleus surface (*Weeks,* 1995; *Scheeres et al.,* 1998b). There has been speculation in the past on an outgassing environment for asteroids as well, but evidence for this has not been detected to date, and the perturbations that would result from such outgassing would be very small. Another nongravitational force that has been recently considered in many contexts is the Yarkovsky effect (*Bottke et al.,* 2000, 2002; *Rubincam,* 2000), which essentially consists of a thermal imbalance on a body. Since ejecta will have definite shapes and rotations, the Yarkovsky effect may be able to modify a particle's orbital dynamics if it falls into a long-term stable orbit. There have been no studies performed on the application of this force to ejecta dynamics to date, however.

4.2. Equations of Motion

The measured force parameters and models define the dynamical problem of motion in the vicinity of the asteroid. Depending on the force parameters, the character of motion in these equations will take on a variety of forms. Specifically, for smaller asteroids the regions where solar and gravity field perturbations are important can coincide, leading to very complicated dynamics. For larger asteroids these regions of influence do not coincide, making it possible to distinguish between a far-field regime dominated by solar effects and a close-field regime dominated by asteroid gravity and rotation. Of course, a single ejecta trajectory can transition between these regimes as it passes from apoapsis to periapsis and back again.

The relevant equations of motion in an inertially oriented frame for the ejecta relative to the asteroid can be stated as (*Scheeres et al.,* 2001)

$$\ddot{\mathbf{r}}_I = \frac{\partial V(\mathbf{r}_I)}{\partial \mathbf{r}_I} \tag{10}$$

$$V(\mathbf{r}) = \frac{\mu}{|\mathbf{r}|} + R(\mathbf{r}) + \frac{\mu_S\beta}{d^2}\hat{\mathbf{d}}\times\mathbf{r} - \frac{1}{2}\frac{\mu_S}{d^3}\left[\mathbf{r}\times\mathbf{r}-3\left(\hat{\mathbf{d}}\times\mathbf{r}\right)^2\right] \tag{11}$$

where $d = |\mathbf{d}|$, $\hat{\mathbf{d}} = \mathbf{d}/d$, and $\mathbf{r}_I$ denotes that the position vector is referenced to an inertial, nonrotating frame. These equations are entirely general and only incorporate a few assumptions (noted above). Note that it is necessary to have a solution for the motion of the asteroid relative to the Sun; however, it is generally sufficient to use a Keplerian orbit for the computation of d and $\hat{\mathbf{d}}$. Exceptions occur when the asteroid has a close encounter with a planet, but this would also require the addition of the tidal effect of that planet on the motion of a particle and on the rotation state of the asteroid, situations we do not directly discuss here (*Chauvineau and Mignard,* 1990b; *Scheeres et al.,* 2000b).

4.2.1. Perturbation formulation. In the course of analyzing motion about an asteroid, it is often convenient to use the constants of motion of the two-body problem in order to characterize the strength and effect of the perturbations acting on the ejecta. The classical orbital elements can be defined as the semimajor axis, a, the eccentricity, e, the inclination, i, the longitude of the ascending node, Ω, and the argument of periapsis, ω. Frequently, the true or mean anomaly of the orbit, f or M respectively, are used to replace the classical sixth orbit element of the time of periapsis passage. The variation of these constants due to orbital perturbations are generally specified using the Lagrange planetary equations with a perturbation function. An extended discussion of these equations can be found in *Brouwer and Clemence* (1961). For our system, the general force perturbation potential can be given as $V(\mathbf{r}) - \mu/|\mathbf{r}|$.

4.2.2. Asteroid-fixed frame. For the analysis of ejecta motion close to the asteroid surface it is more convenient to shift the equations into an asteroid-fixed frame. In doing so we must allow for the fact that the asteroid is rotating with an angular velocity vector $\boldsymbol{\Omega}$ with respect to inertial space, so the equations of motion relative to the asteroid have the form

$$\ddot{\mathbf{r}} + \dot{\Omega} \times \mathbf{r} + 2\Omega \times \dot{\mathbf{r}} + \Omega \times \Omega \times \mathbf{r} = \frac{\partial V(\mathbf{r})}{\partial \mathbf{r}} \quad (12)$$

If the asteroid is uniformly rotating, $\dot{\Omega} = 0, \boldsymbol{\Omega}$ is constant, and the equations simplify. On the other hand, if the asteroid is in nonprincipal axis rotation, then the vectors $\boldsymbol{\Omega}$ and $\dot{\Omega}$ are time periodic.

These equations of motion have no integrals of motion in general. However, for motion close to an asteroid we can often disregard the solar perturbation terms. Then, if the asteroid is in uniform rotation, equation (12) is time-invariant and a Jacobi integral exists

$$J = \frac{1}{2}\dot{\mathbf{r}} \times \dot{\mathbf{r}} - \frac{1}{2}(\Omega \times \mathbf{r}) \times (\Omega \times \mathbf{r}) - \frac{\mu}{r} - R \quad (13)$$

This integral is often helpful in constraining and understanding the limits on motion near the asteroid surface. In application, this integral can be used just as the Jacobi integral is used in the restricted three-body problem (*Szebehely,* 1967) and was used extensively for the analysis of Phobos (*Dobrovolskis and Burns,* 1980).

5. THE DYNAMICAL FATE OF EJECTA

The equations of motion reviewed above can lead to extremely complicated motion that cannot in general be solved analytically. However, there are many insights to be had from the study of these equations, both numerically using the full models and analytically using suitably simplified models. In all such investigations it is important to remember the guiding dynamical questions: What is the dynamical evolution of an impact ejecta field, what fraction of an ejecta fragment field will escape, what fraction will reimpact, and what fraction will be captured in a transient orbit?

5.1. Stability of Synchronous Motion

First, a special note must be given on the stability of synchronous (1:1) motion about an asteroid. In the past, many authors have made a tacit assumption that synchronous orbits about an asteroid would follow the basic pattern found for geosynchronous orbits (such as described in *Kaula,* 2000, p. 54) with two stable, synchronous orbits and two unstable, synchronous orbits. Application of these assumptions lead to predictions for the stability of orbital motion about asteroids and to the tidal evolution of asteroid satellites that are not valid. In the following, "stable" means that a trajectory close to the synchronous orbit will remain close to it for arbitrarily long periods of time, while "unstable" means that a trajectory close to the synchronous orbit will rapidly leave its vicinity.

In *Scheeres* (1994) it is shown that synchronous orbits about asteroids are unstable in general. For an asteroid whose shape is spheroidal, or for Earth, we find four synchronous orbits, two of which are hyperbolically unstable and two of which are linearly stable. An analogy can be made with the restricted three-body problem, where for a small mass ratio we find three synchronous orbits that are hyperbolically unstable and two orbits (the so-called equilateral points) that are linearly stable. Now, as is well known, in the restricted three-body problem these stable equilateral points become unstable if the mass ratio between the primaries is increased to greater than ~0.1 (cf. *Szebehely,* 1967). A similar phenomenon occurs in the asteroid problem, where the stable synchronous orbits become unstable if the body's shape is sufficiently elliptic (more precisely, this involves both the asteroid's rotation rate and its ellipticity). Furthermore, the instability timescale of these orbits is on the order of the rotation period of the asteroid, and hence operate very quickly. Thus, particles placed near a 1:1 resonance with a rotating asteroid will in general either impact with or escape from the asteroid, usually in a matter of hours or days at most.

5.2. Final Outcomes for Ejecta

We can delineate several distinct final outcomes for ejecta trajectories whose initial conditions lie beneath or on the surface of an asteroid. Using a Keplerian dynamics model applied to a spherical asteroid there are three dis-

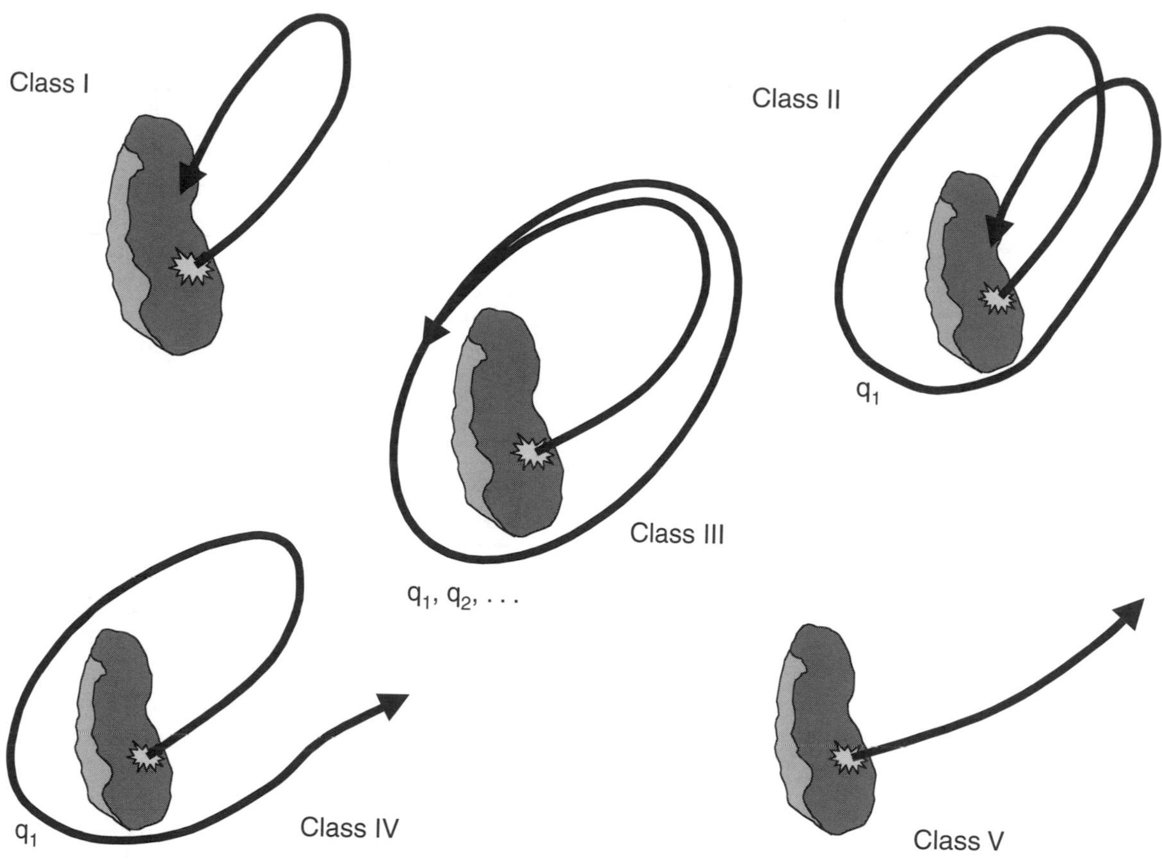

Fig. 1. The five classes of ejecta fate.

tinct classes of motion. First, if the orbital energy is negative then the ejecta will reimpact as periapsis is initially on or beneath the asteroid surface. Second, if the energy is zero or positive the ejecta will escape. Third, a subset of the second class of escaping ejecta may eventually reimpact on the asteroid after an extended period of time in orbit about the Sun. We will discount this third class, however, as it is practically indistinguishable from other impacts.

As additional perturbations are considered, the possible classes of motion expand. It is useful to use periapsis passage relative to the asteroid to delimit between different classes of motion. At launch the ejecta are starting from an initial radius $r_0 \geq q_0$, since in general the initial periapsis (q_0) lies beneath the body's surface. In the absence of perturbations the next periapsis passage q_1 will either equal q_0, and thus will be an impact, or will never occur, indicating escape. When force perturbations are incorporated, or even if nonspherical shapes are allowed, it becomes possible for multiple periapsis passages to occur. We denote these as a series q_i; i = 0, 1, 2, …. Associated with each periapsis passage is the periapsis vector, $\mathbf{q}_i$, representing the periapsis location in the asteroid-fixed space. If we denote the set of points that constitute the asteroid body as $\mathcal{B}$, then if $\mathbf{q}_i \in \mathcal{B}$ the sequence stops and an impact has occurred. Conversely, given a periapsis passage q_i, if q_{i+1} does not ensue, then the ejecta has escaped. Finally, if the sequence never terminates ($i \to \infty$), then the ejecta is in a stable orbit about the asteroid.

Based on this understanding, we tender the following classifications (see Fig. 1): *Class I* — Immediate reimpact: Ejecta reimpacts with the surface prior to first periapsis passage. *Class II* — Eventual reimpact: Ejecta does not reimpact at the first periapsis passage, but eventually reimpacts in the future. *Class III* — Stable motion: Ejecta is placed into a long-term stable orbit about the asteroid. *Class IV* — Eventual escape: Ejecta has at least one periapsis passage by the asteroid before it escapes. *Class V* — Immediate escape: Ejecta escapes from the asteroid prior to its first periapsis passage.

Classes I and V are clear carryovers from the nonperturbed case, and we expect most ejecta to fall into these two catagories. The fraction of ejecta that fall into classes II, III, and IV is an important consideration for understanding the formation of asteroid regolith and asteroid binaries. Given a specific system it is relatively easy to find regions of ejecta initial conditions that fall into classes II and IV. In terms of celestial mechanics and astrodynamics, it would appear to be very difficult to place a particle into class III

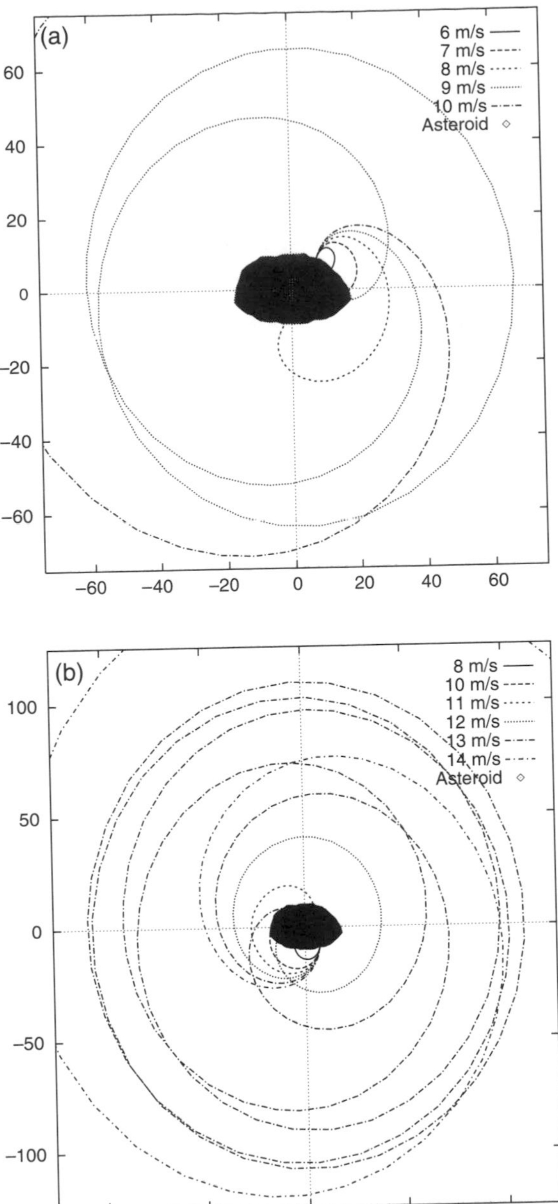

Fig. 2. Effects of location and ejection speed on ejecta trajectories launched from a uniformly rotating asteroid shape. **(a)** Launch from the leading edge of an asteroid; because of the asteroid's rotation, the local escape speed is lower. **(b)** Launch from the trailing edge of an asteroid; because of the asteroid's rotation, the local escape speed is higher.

due to the nature of these dynamical systems; however, there are other physical forces that can cause a particle to transition into a stable orbital motion, and these will be reviewed below. Figure 2 shows the trajectory evolution of a number of different ejecta particles launched off an asteroid at different speeds. We note that changes in initial speed or location on the asteroid can have dramatic consequences for the final state of the ejecta.

5.3. Impact and Escape Conditions

By applying analytical theories to the motion of ejecta it is possible in many instances to immediately determine if an individual ejecta particle will fall into class I or V, based only on its initial conditions. Using such determinations can greatly decrease the amount of computational effort needed to evaluate the outcome of a high-resolution impact cratering event. Also, such methods can directly compute the fraction of an ejecta fragment field that falls within these classes, and hence provides an indication of the fraction that may reside in classes II–IV. In *Scheeres et al.* (1996, 1998a) a number of analytical results directly pertaining to the computation of reimpact conditions and escape conditions are given. Specific results developed in these papers include the computation of guaranteed reimpact speed and guaranteed escape speed as a function of location on an asteroid. The guaranteed escape speed is the speed at which an ejecta, launched normal to the surface, will have sufficient energy to escape the asteroid. The guaranteed return speed is the maximum speed an ejecta can have while still being energetically trapped by the zero-velocity curves surrounding the asteroid (see section 5.6). These methods have also been applied to the Eros dataset, which is definitive since all the force model parameters have been measured (*Yeomans et al.,* 2000; *Miller et al.,* 2001). For Eros the escape speeds range from 3.3 to 17.3 m/s over its surface. This large variation is due to the combined shape/gravity field variation and the rapid rotation rate of the asteroid. At the other end of the spectrum, the guaranteed return speeds computed over the surface of Eros range from 1 to 5 m/s, but are always less than the escape speed at any particular point. Many of these ideas can be developed in additional detail, and can provide sharper conditions on the fraction of an ejecta fragment field that will immediately escape, or that will redistribute itself on the asteroid.

5.4. Transient Classes

Of particular interest are classes II–IV, as they define the space where interesting things can happen to an ejecta fragment field. If we draw a "spectrum" of outcomes, class III will lie at the intersection between classes II and IV, as it is the limit of these cases. Thus, one of the fundamental questions concerning the fate of asteroid ejecta is how a particle can be placed into one of these transient classes, and what its subsequent evolution will be. Additionally, if a fraction of an ejecta fragment field falls into class II or IV for an extended period of time, there is a higher probability that additional perturbations or impacts may push it into class III, creating a binary asteroid.

5.4.1. Problem of initial capture. The basic dynamical problem is how to generate class II and IV ejecta, and subsequently transition these into class III ejecta. The real problem is not whether such trajectories exist, as we can firmly establish the existence of trajectories that fall into class III.

Specifically, families of unstable periodic orbits and equilibrium points (in the asteroid-fixed frame) exist close to the asteroid surface. Each of these orbits has a stable manifold that asymptotically approach these special solutions. It can be shown that many of the stable manifolds of these objects intersect (or emanate from) the surface of the asteroid, and hence provide exact initial conditions that lead to orbital capture (*Scheeres et al.,* 1996). The problem with these solutions, of course, is that they are unstable and the set of initial conditions that leads to capture is vanishingly small. Thus, the real question is whether there are any significant regions of initial conditions that lead to long-term, trapped orbits about an asteroid. Again, the answer here appears to be yes, but the proof is not as direct, and the full extent of initial conditions that actually lead to such capture has yet to be fully explored.

5.4.2. Methods of analysis. There are several different approaches to determining if a particle falls into one of the transient classes. The first is direct numerical simulation of discretized elements of the ejecta field. This approach, used by *Geissler et al.* (1996) in studying the evolution of ejecta about Ida, provides definite results, subject to modeling assumptions used in setting up the computations, and allows for the use of a full perturbation model. It is limited by the finite number of ejecta that can be propagated and due to the discrete nature of each propagation. Indeed, in an actual ejecta field we expect a near continuum flow of particles, which should in general lead to higher probabilities for capture into transient dynamical situations.

Analytically motivated approaches can give greater insight into the evolution of larger numbers of particles and can model the ejecta field as a continuous flow in some situations, at the cost of lost precision in the computed trajectories. In *Scheeres and Marzari* (2000) an averaging approach is used that provides analytical solutions to ejecta evolution following ejection from the surface of a small body. Their approach only incorporated solar radiation pressure perturbations, but could be generalized to include other effects. With this approach it is possible to rapidly compute the evolution of ejecta fields, which allows for more precise estimates on the fraction of an ejecta fragment field that is injected into a transient class, potentially allowing for direct computation of probabilities of different outcomes from a given ejecta field.

Application of advanced understandings of dynamical systems could also be used to evaluate the likely outcomes for an ejecta fragment field. As mentioned earlier, the space around an asteroid is filled with periodic orbits, both stable and unstable, each of which have manifolds that can influence the dynamical flow of an ejecta fragment field. Recent advances in the application of dynamical systems theory to spacecraft trajectory design (*Koon et al.,* 2000) could also be brought to bear on the evolution of asteroid ejecta, and provide qualitative descriptions of ejecta field flow that may allow for specific quantitative predictions in some cases. Initial approaches to this sort of application have been developed (*Scheeres et al.,* 1996, 1998a, 2000a) and, at the least, can be used to establish the existence of transient orbits of extremely long duration.

5.4.3. Mechanisms for capture into stable orbits. There are many mechanisms that have been hypothesized that lead to ejecta becoming captured into stable orbits. We will provide a very brief summary of these approaches. Issues of long-term stability and the lifetime of such orbits are not considered, but are discussed in more detail in *Merline et al.* (2002).

Direct initial condition generation. In this scenario a parent asteroid is subject to an intense impact event, which shatters and disperses the original body into many fragments, all imparted with a range of speeds. In general, the smaller particles have higher speeds and the larger have slower speeds. Given such a random distribution of particle positions and speeds it is probable that some of the fragments will be placed into mutually bound orbits as they escape (indeed this postdisruption environment can even influence their motion during the short period when the asteroid disperses), leading to primitive binary systems. Such bound orbits will have large eccentricities in general, but assuming long-term stability against impact and escape, energy dissipation (i.e., tidal effects with energy dissipation) can cause the orbits to circularize over time, leading to the types of stable binaries now being found. *Hartmann* (1979) first suggested this scenario, which has been investigated analytically by *Weidenschilling et al.* (1989), and more recently has been simulated by *Durda* (1996), *Dorresoundiram et al.* (1997), and *Michel et al.* (2001). They have found that small numbers of bound asteroid pairs do result, but these studies have not addressed the long-term stability and evolution of these pairs.

Mass shedding in tidal flybys. Additional mechanisms not involving impacts have also been suggested to increase the rotation rate of asteroids to the point of mass shedding. *Richardson et al.* (1998) and *Bottke et al.* (1999) numerically simulated the tidal disruption of asteroids modeled as "rubble piles" (see *Richardson et al.,* 2002) composed of numerous equal-sized spherical components encountering Earth, and found that rotational spinup frequently induces debris to be cast off the primary bodies. In many cases, the shed fragments were found to go into initially bound orbits around the progenitor. *Bottke and Melosh* (1996a,b) and *Richardson et al.* (1998) have shown that tidal disruption can create enough satellites in the NEA population to explain the statistics of doublet craters seen on the terrestrial planets.

Mutually impacting ejecta. A postimpact ejecta field will have a distribution of particle sizes and speeds resembling, in some aspects, a continuous field distribution. Thus, it is likely that mutual impacts between elements of the ejecta field will ensue, and that these slow-velocity impacts will mutually alter the trajectory of the particles, in some cases leading to capture orbits. *Weidenschilling et al.* (1989) considered this mechanism in an analytical argument, and concluded that satellites formed from reaccreted ejecta are expected to be small and found in prograde orbits. *Durda and Geissler* (1996) simulated impact ejecta fields to search for such impacts, but did not find any that evolved into stable trajectories. Their approach used 1000 ejecta particles, which may be too few to reliably find such outcomes. This approach becomes more likely to yield stable trajectories when portions of the ejecta fields are captured into transient orbits that may not reimpact for many orbits (classes II–IV), since there will be a higher probability of mutual impacts and repeated impacts that could yield stable trajectories. Such a long-term analysis has not been performed to date, however.

Rotational bursting. A novel idea for injection of ejecta into stable orbits was posited in *Giblin et al.* (1998), based on observations of laboratory impact events. In this scenario, specific particles in the ejecta field have large rotational velocities and are placed in tension. In some situations, these rotating fragments have been observed to spontaneously "burst," or disassemble into smaller fragments, shortly after ejection from the laboratory target. This situation, if found in nature, creates a situation such as found in the paragraph above on direct initial condition generation, and can plausibly lead to particles placed directly into stable orbits.

External force perturbation. Underlying many of the above mechanisms, and indeed a mechanism in itself, is the effect of force perturbations on the trajectory. As mentioned earlier, the ejecta are subject to an extremely perturbed force environment, first from the asteroid gravity field, and second from the solar radiation pressure and tidal perturbations. Any of these can place a particle into an orbit that persists for some time about the asteroid. Specifically, in *Scheeres et al.* (1998a) a particle orbit perturbed only by the asteroid gravity field is described that has a "hang time" of over 100 d. In *Fulle* (1997) and *Scheeres and Marzari* (2000) the effect of solar radiation pressure on a ejecta particle is shown to capture regions of initial ejecta conditions into bound orbits that do not reimpact for hundreds of days in some cases. A study that combines solar and asteroid gravity effects has not yet been performed, but may provide mechanisms that could extend the lifetime of such transient orbits to multiple asteroid years. At these timescales it becomes possible for small perturbation forces to influence the orbits, potentially leading to stable orbits.

5.4.4. Long-term lifetime and evolution of captured orbits. Once in orbit about an asteroid, a particle is subject to a variety of perturbation forces that can cause orbital evolution over long timespans. These effects include the asteroid gravity field and tidal effects (*Petit et al.,* 1997), the solar tide (*Chauvineau and Mignard,* 1990a; *Hamilton and Burns,* 1991a), solar radiation pressure (*Hamilton and Burns,* 1991b; *Richter and Keller,* 1995), four-body effects (*Chauvineau and Mignard,* 1990b), and disruption by impacts (*Davis et al.,* 1996). The majority of these analyses have only considered these perturbations in isolation. From such studies, it is clear that long-term, stable orbits can exist about asteroids, in some cases with minimal orbital evolution. Perhaps the most interesting results, related to the lifetime and evolution of an asteroidal satellite, are found in *Davis et al.* (1996) and based on the work by *Geissler et al.* (1996), where it is posited that Ida and Dactyl may actually be in an equilibrium state, exchanging mass between the bodies, driven by impacts and ejecta field evolution on each body. As the statistics on asteroid binaries is improved, with increasing numbers of detections, a firmer context for such studies can be established and, most likely, real distinctions between different classes of binary systems will be found.

5.5. Reimpact Dynamics

An unanswered question involves the dynamics of a particle after it reimpacts on the asteroid surface. A distinction should be made between high-energy secondary impacts that may occur in the fractions of a second after a primary impact (due to ricochets) and low-energy impacts that may occur immediately or months after the primary impact with speeds less than or equal to surface escape speed. For the second type of reimpacts, it can be hypothesized that collision with the surface may not be disruptive nor completely inelastic, so that some amount of rebound energy will exist. If true, there are significant implications for the modeling of reimpact ejecta. This is especially interesting in light of the recent returns from the *NEAR Shoemaker* mission, which found that the asteroid surface at high resolution was dominated by ejecta blocks, with a paucity of craters, which raises a host of scientific questions on the nature of the Eros surface at centimeter scales (discussed at the end of this chapter). The existence of transient dynamical behavior of a reimpacting ejecta block has implications for the extent of downslope motion a particle will experience, and hence the degree of ponding at lows in the potential that will occur prior to a particle settling on the surface.

This issue has been studied in an engineering application in the context of the settling time of a navigational aid deployed on the surface of an asteroid (*Sawai et al.,* 2001). In that study it was found that, even for relatively low coefficients of restitution on the order of 0.1, settling times of 10–20 min were common. This is ample time for a particle to migrate toward potential lows on an asteroid's surface.

5.6. Surface Forces and Dynamics

Finally, we must consider the environment that is felt by an impact ejecta once it comes to rest on the surface of an asteroid. The mechanical forces felt on the surface can be reduced to surface normal and transverse frictional forces acting on a particle. These are, in turn, defined by the asteroid's gravity field, surface, and rotation state. Recent interest in other forces acting on the asteroid surface have been revived by the unexpected morphology of the Eros surface. Indeed, electromagnetic forces operating on small dust particles are being considered to explain some of the dust ponding seen on Eros (*Lee,* 1996; *Robinson et al.,* 2001). In addition to these are occasional impulsive forces that may jolt asteroid regolith, due to impacts of other asteroids on the asteroid surface (*Greenberg et al.,* 1994).

The total acceleration that a particle feels when at rest on the surface of a rotating asteroid is

$$\mathbf{N} = \dot{\mathbf{\Omega}} \times \mathbf{r} + \mathbf{\Omega} \times \mathbf{\Omega} \times \mathbf{r} - \frac{\partial V}{\partial \mathbf{r}} \tag{14}$$

If the local surface normal is $\hat{\mathbf{n}}_z$, then the surface force is split into a normal and tangential component

$$N_z = \hat{\mathbf{n}}_z \cdot \mathbf{N} \tag{15}$$

$$\mathbf{N}_t = \mathbf{N} - N_z \hat{\mathbf{n}}_z \tag{16}$$

and the local slope of the system is defined as

$$\phi = \arctan\left[\frac{|\mathbf{N}_t|}{N_z}\right] \tag{17}$$

The surface slope can be related to the coefficient of friction on the surface, μ, as $\mu \geq \tan \phi$ (*Greenwood,* 1988). The rotational dynamics of the body can take a significant role in modifying the surface environment, and may change the stability and structure of motion on the surface. For bodies in complex rotation the slopes and surface forces are time-periodic, and could potentially add sufficient "shaking" (physically realized by slowly varying slopes at each point on the surface) to cause the surface to relax, reducing the potential energy stored in local slopes. Any asteroid subject to nonuniform rotation following a large impact or planetary flyby will have these time-periodic forces acting on its surface, which could play a role in smoothing a surface after an impact. This is distinguished from seismic shaking, where the asteroid frequently feels small seismic events due to the flux of impactors striking the asteroid (*Greenberg et al.,* 1994). While the magnitude of shaking expected from impactors should be larger than from nonuniform rotation, the nonuniform rotation will act continuously on the asteroid over the time it takes for it to relax into uniform rotation. Estimates of this effect for Toutatis are given in *Scheeres et al.* (1998a).

Finally, it should be noted that if an asteroid would actually describe a figure of equilibrium (*Weidenschilling,* 1981), then the surface slope would be identically zero over the entire body. In fact, deviations of surface slope from zero indicate deviations from a figure of equilibrium. Slope distributions of asteroids have been measured from spacecraft observations and from radar imaging of asteroids. Some bodies measured in this way, such as Toutatis and Kleopatra, have uniformly low slopes that, at the least, could be indicators of their rotational and impact past (*Scheeres et al.,* 1998a; *Ostro et al.,* 2000).

A second parameter of interest for the surface environment is the effective potential, defined by the combined gravitational potential and rotational potential terms. For a uniformly rotating body this is just the Jacobi integral discussed earlier. This gives a direct measure of the available energy that can be converted to kinetic energy (and hence easily dissipated) based on the location of a particle in the asteroid frame (*Thomas,* 1993). The effective potential energy function of an asteroid is

$$C(\mathbf{r}) = -\frac{1}{2}(\Omega \times \mathbf{r}) \times (\Omega \times \mathbf{r}) - U(\mathbf{r}) \tag{18}$$

Using this, the dynamical height of the asteroid surface can be computed, a relative measure from a locally defined average gravity (*Thomas,* 1993).

On the surface of a uniformly rotating asteroid, this same effective potential energy can also be related to the minimum amount of energy a particle requires before it can escape from the asteroid [the guaranteed reimpact speed in *Scheeres et al.* (1996)]. Specifically, the value of $C(\mathbf{r})$ at the synchronous orbits (C_R) defines the zero-velocity surface that surrounds and encloses the asteroid in three-dimensional space. The effective potential energy evaluated at this synchronous orbit defines the minimum energy that a particle must have before it becomes possible to escape from the asteroid; i.e., a particle with Jacobi constant greater than this value could, theoretically, escape from the asteroid following a purely ballistic trajectory. If a particle has an energy less than this, and is within the zero-velocity curve, then it is impossible for it to leave the vicinity of the asteroid. This surface has also been referred to as the Roche lobe, and was studied in the particular case of Phobos (*Dobrovolskis and Burns,* 1980), and more recently has been computed for Eros (*Yeomans et al.,* 2000; *Miller et al.,* 2001). Phobos was found to "fill" this minimum energy surface, meaning that particles on its surface were prone to escape that body when given sufficient speeds. Conversely, Eros lies entirely within this energy surface, although 56% of that asteroid's surface lies within 1 km of this energy surface, the closest point lying only 90 m from the energy surface. Figure 3 shows the computed Eros Roche lobe projected into the Eros equatorial plane.

6. CURRENT DATA CONSTRAINTS ON EJECTA EVOLUTION

The interpretation and analysis of asteroid regolith and the dynamics of impact ejecta will ultimately be constrained by *in situ* observations of asteroids. Historically, the analysis of Phobos and Deimos from spacecraft images has allowed for a rather complete understanding of the regolith and dynamical environment of these bodies to be developed. Several obvious indications of regolith on the martian moons Phobos and Deimos were noted in *Viking Orbiter* pictures (e.g., *Thomas,* 1979; *Thomas and Veverka,* 1980; *Lee et al.,* 1986) and recently scrutinized with new images from *Mars Global Surveyor* (*Thomas et al.,* 2000). The dynamics of ejecta lofted from the martian moons are complicated by tidal forces from nearby Mars as well as the rapid rotation

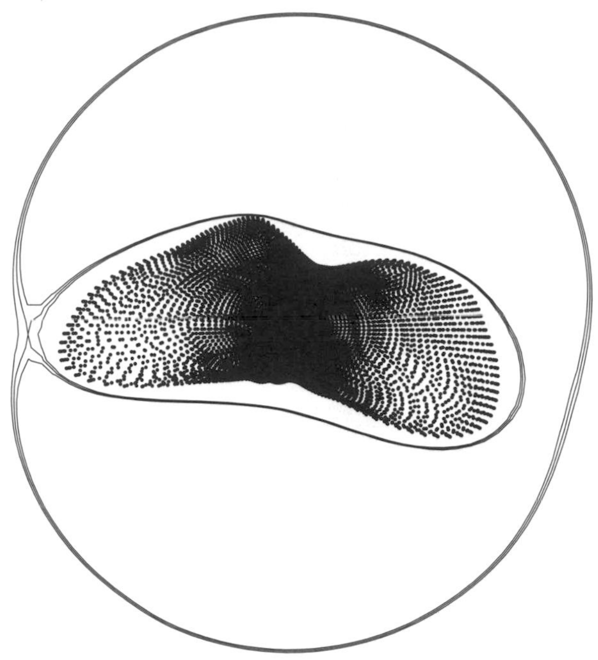

Fig. 3. Eros Roche lobe computed from *NEAR Shoemaker* data.

and irregular shapes of these satellites (*Dobrovolskis and Burns,* 1980; *Davis et al.,* 1981). Changes in the moons' semimajor axes due to orbital evolution should have produced gradual variations in the effects of tides over time. Using the asymmetric ejecta deposit from the giant crater Stickney, *Thomas* (1998) deduced that the impact occurred while Phobos was slightly farther from Mars than it is at present. The thick mantle of regolith on Deimos, presumably derived from a giant impact near the satellite's south pole, has been subjected to significant postdepositional redistribution, perhaps due to seismic shaking.

The morphology of surface features on Gaspra and Ida (and Ida's satellite Dactyl), imaged at moderate to high resolution by the *Galileo* spacecraft, indicate the existence of impact ejecta retained on their surfaces (e.g., *Belton et al.,* 1992; *Sullivan et al.,* 1996; *Geissler et al.,* 1996). Morphological indications of regolith include (1) numerous isolated positive relief features, which appear to be ejecta blocks, the largest size fraction of the regolith; (2) chutes and albedo streaks oriented down local slopes, interpreted as mass-wasting scars in regolith; (3) grooves, which may be the surface expression of deep-seated fractures partially filled by regolith; and (4) color/albedo variations associated with slopes and apparently fresh impact craters consistent with regolith maturity variations. Applications of basic models of impact ejecta fields and their dynamics have been able to explain the observed regolith features on these bodies (see Fig. 4). While the explanation of Dactyl has been a challenge, several reasonable ideas on its formation and evolution do exist.

The asteroids imaged by the *NEAR Shoemaker* spacecraft, Mathilde and Eros, have not fit as well with the expected theory. Several of the observations made were, in essence, totally unexpected. For Mathilde this includes the size and extent of its craters, along with a lack of observed depositional features (*Veverka et al.,* 1997). Theories of impact physics that describe this situation have been posited that are consistent with its low measured density (*Housen et al.,* 1999; *Davis,* 1999; *Asphaug,* 2000). Still, the ability of that asteroid to survive intact is surprising, as is its extremely slow rotation rate. Planned radar observations of Mathilde will hopefully provide additional insight into this primitive body.

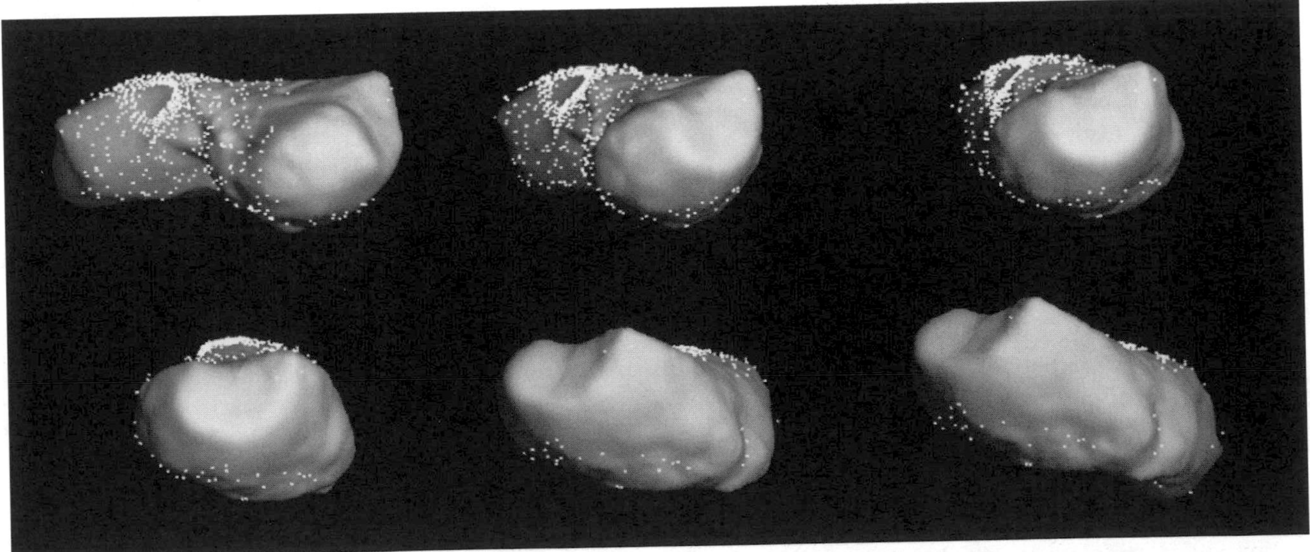

Fig. 4. Theoretical landing locations of ejecta launched from the giant crater Azzurra on Ida. This ejecta distribution, calculated considering Ida's irregular shape and rapid rotation, provides a close match to the bright, relatively blue spectral unit found on the asteroid. The distinct color of Azzurra's ejecta deposits suggests that relatively fresh, unweathered materials were excavated by the impact (see *Geissler et al.,* 1996, for a more detailed description).

For Eros, detailed mapping of the largest ejecta blocks coupled with calculations of trajectories from candidate source craters led *Thomas et al.* (2001) to suggest the Shoemaker Crater as the source of most of the ejecta blocks on Eros. Moreover, the lack of large blocks associated with the other giant impacts on Eros confirms that these house-sized boulders are rapidly destroyed or buried. While some aspects of the distribution of ejecta blocks have been explained, the observed lack of cratering at high resolution is an outstanding puzzle. *NEAR Shoemaker* imaging of Eros (*Veverka et al.,* 2000, 2001) revealed a surprising lack of small craters, and closeup imaging showed pools of fine sediment in topographic lows with no dust deposits on top of the boulders. This suggests that the regolith has been shaken and stirred since deposition. Seismic disturbances due to distant impacts could cause vertical redistribution of the regolith and perhaps erase the smaller craters (*Greenberg et al.,* 1994). Alternatively, electromagnetic forces may act on sunlit dust particles, causing them to rise and subsequently fall, leading to ponding of small particles on the asteroid surface (*Robinson et al.,* 2001). Unknown is whether the surface morphology of Eros implies a dearth of small projectiles in the main belt, or is caused by postdepositional modification during Eros' recent history as a NEA.

The radar asteroid dataset has also expanded greatly in recent years. There are now many radar-derived asteroid shapes in existence, from which many insights can be extracted (see *Ostro et al.,* 2002). Each asteroid measured to date with this technique has some unique features that defy a uniform classification. Even without direct measurements of an asteroid's mass or density, it is still possible to place constraints on its surface environment. First, all asteroids measured by radar to date have exhibited no strong density inhomogeneities, which would be observable by a mismatch between the observed center of rotation and the shape center of the body. Some asteroids have been observed to have uniformly low slopes on their surfaces, indicating that they may be covered with regolith, particularly asteroids Toutatis and Kleopatra (*Scheeres et al.,* 1998a; *Ostro et al.,* 2000). Other asteroids clearly have regions of exposed, monolithic material; in particular, Golevka has regions of high slope on its surface (up to 60°) consistent with bare rock (*Hudson et al.,* 2000). The recent observation of binary asteroids with this technique make it possible to determine the mass of these objects in addition to their size, shape, and rotation state (*Margot et al.,* 2001). Once these observations are properly reduced it should be possible to test a range of specific theories of satellite formation and evaluate the fate of asteroidal ejecta in these regimes.

7. FUTURE OBSERVATIONS

Future observations should answer some of the outstanding questions concerning ejecta capture and regolith retention on asteroids. Substantial uncertainties about the population of impactors that collide with asteroids and the process of ejecta generation remain. The generalizations that we are tempted to draw from a handful of S-type objects may not apply to asteroids elsewhere or asteroids of different types, i.e., craters may look very different on metallic objects and extinct comets. Closeup observations of main-belt asteroids will ultimately decide whether the dearth of small craters on Eros is due to the lack of small projectiles in the main belt or due to postimpact surface processes. The observed depth, or lack, of regolith on smaller asteroids should directly constrain the strength-to-gravity transition for impactors. Direct experimentation, such as that planned for the Deep Impact mission (*A'Hearn et al.,* 1999), will shed light on low-gravity cratering mechanics. Finally, continued development and advancement of the mathematical tools, methods, and simulations need to understand regolith and ejecta will enable a better understanding of this particular aspect of asteroid science.

REFERENCES

A'Hearn M. F. and the Deep Impact Project Team (1999) Deep impact (abstract). *Bull. Am. Astron. Soc., 31,* 1114.

Asphaug E. (2000) The large, undisturbed craters of Mathilde: Evidence for structural porosity (abstract). In *Lunar and Planetary Science XXXI,* Abstract #1864. Lunar and Planetary Institute, Houston (CD-ROM).

Asphaug E. and Melosh H. J. (1993) The Stickney impact of Phobos — A dynamical model. *Icarus, 101,* 144–164.

Asphaug E., Moore J. M., Morrison D., Benz W., Nolan M. C., and Sullivan R. J. (1996) Mechanical and geological effects of impact cratering on Ida. *Icarus, 120,* 158–184.

Asphaug E., Ryan E. V., and Zuber M. T. (2002) Asteroid interiors. In *Asteroids III* (W. F. Bottke Jr. et al., eds.), this volume. Univ. of Arizona, Tucson.

Belton M. J. S., Veverka J., Thomas P., Helfenstein P., Simonelli D., Chapman C., Davies M. E., Greeley R., Greenberg R., and Head J. (1992) Galileo encounter with 951 Gaspra — First pictures of an asteroid. *Science, 257,* 1647–1652.

Belton M. J. S., Mueller B. E. A., D'Amario L. A., Byrnes D. V., Klaasen K. P., Synnott S., Breneman H., Johnson T. V., Thomas P. C., Veverka J., Harch A. P., Davies M. E., Merline W. J., Chapman C. R., Davis D., Denk T., Neukum G., Petit J.-M., Greenberg R., Storrs A., and Zellner B. (1996) The discovery and orbit of 1993 (243) Dactyl. *Icarus, 120,* 185–199.

Benz W. and Asphaug E. (1999) Catastrophic disruptions revisited. *Icarus, 142,* 5–20.

Bottke W. F. Jr. and Melosh H. J. (1996a) The formation of asteroid satellites and doublet craters by planetary tidal forces. *Nature, 381,* 51–53.

Bottke W. F. Jr. and Melosh H. J. (1996b) The formation of binary asteroids and doublet craters. *Icarus, 124,* 372–391.

Bottke W., Nolan M., Greenberg R., and Kolvoord R. (1994) Velocity distributions among colliding asteroids. *Icarus, 107,* 255–268.

Bottke W. F., Nolan M. C., Greenberg R., and Kolvoord R. A. (1995) Collisional lifetimes and impact statistics of near-Earth asteroids. In *Hazards Due to Comets and Asteroids* (T. Gehrels and M. S. Matthews, eds.), pp. 337–357. Univ. of Arizona, Tucson.

Bottke W. F. Jr., Richardson D. C., Michel P., and Love S. G. (1999) 1620 Geographos and 433 Eros: Shaped by planetary tides? *Astron. J., 117,* 1921–1928.

Bottke W. F., Rubincam, D. P., and Burns J. A. (2000) Dynamical evolution of main belt meteoroids: Numerical simulations incorporating planetary perturbations and Yarkovsky thermal forces. *Icarus, 145,* 301–331.

Bottke W. F. Jr., Vokrouhlický D., Rubincam D. P., and Brož M. (2002) The effect of Yarkovsky thermal forces on the dynamical evolution of asteroids and meteoroids. In *Asteroids III* (W. F. Bottke Jr. et al., eds.), this volume. Univ. of Arizona, Tucson.

Brouwer D. and Clemence G. M. (1961) *Methods of Celestial Mechanics*. Academic, New York. 598 pp.

Burns J. A. and Safronov V. S. (1973) Asteroid nutation angles. *Mon. Not. R. Astron. Soc., 165,* 403–411.

Chapman C. R., Veverka, J., Belton M. J. S., Neukum G., and Morrison D. (1996a) Cratering on Gaspra. *Icarus, 120,* 231–245.

Chapman C. R., Ryan E. V., Merline W. J., Neukum G., Wagner R., Thomas P. C., Veverka J., and Sullivan R. J. (1996b) Cratering on Ida. *Icarus, 120,* 77–86.

Chauvineau B. and Mignard F. (1990a) Dynamics of binary asteroids. I. Hill's Case. *Icarus, 83,* 360–381.

Chauvineau B. and Mignard F. (1990b) Dynamics of binary asteroids. I. Jovian perturbations. *Icarus, 87,* 377–390.

Chauvineau B., Mignard F., and Farinella P. (1991) The lifetime of binary asteroids vs. gravitational encounters and collisions. *Icarus, 94,* 299–310.

Chauvineau B., Farinella P., and Mignard F. (1993) Planar orbits about a triaxial body: Applications to asteroidal satellites. *Icarus, 105,* 370–384.

Cheng A. F., Barnouin J. O., Zuber M. T., Veverka J., Smith D. E., Neumann G. A., Robinson M., Thomas P., Garvin J. B., Murchie S., Chapman C., and Prockter L. (2001) Laser altimetry of small-scale features on 433 Eros from NEAR-Shoemaker. *Science, 292,* 488–491.

Cintala M. J., Head J. W., and Wilson L. (1979) The nature and effects of impact cratering on small bodies. In *Asteroids* (T. Gehrels, ed.), pp. 579–600. Univ. of Arizona, Tucson.

Dahlgren M. (1998) A study of Hilda asteroids. III. Collision velocities and collision frequencies of Hilda asteroids. *Astron. Astrophys., 336,* 1056–1064.

Danby J. M. A. (1992) *Fundamentals of Celestial Mechanics,* 2nd edition. Willmann-Bell, Richmond. 483 pp.

Davis D. R. (1999) The collisional history of asteroid 253 Mathilde. *Icarus, 140,* 49–52.

Davis D. R., Housen K. R., and Greenberg R. (1981) The unusual dynamical environment of Phobos and Deimos. *Icarus 47,* 220–233.

Davis D. R., Chapman C. R., Durda D. D., Farinella P., and Marzari F. (1996) The formation and collisional/dynamical evolution of the Ida/Dactyl system as part of the Koronis family. *Icarus, 120,* 220–230.

Dell'Oro A. and Paolicchi P. (1998) Statistical properties of encounters among asteroids: A new, general purpose, formalism. *Icarus, 136,* 328–339.

Dobrovolskis A. R. and Burns J. A. (1980) Life near the Roche limit: Behavior of ejecta from satellites close to planets. *Icarus, 42,* 422–441.

Dohnanyi J. (1969) Collisional model of asteroids and their debris. *J. Geophys. Res., 74,* 2531–2554.

Doressoundiram A., Paolicchi P., Verlicchi A., and Cellino A. (1997) The formation of binary asteroids as outcomes of catastrophic collisions. *Planet. Space. Sci., 45,* 757–770.

Durda D. D. (1996) The formation of asteroidal satellites in catastrophic collisions. *Icarus, 120,* 212–219.

Durda D. D. and Flynn G. J (1999) Experimental study of the impact disruption of a porous, inhomogeneous target. *Icarus, 142,* 46–55.

Durda D. D. and Geissler P. E. (1996) The formation of asteroidal satellites in large cratering collisions. *Bull. Am. Astron. Soc., 28,* 1101.

Farinella P. and Davis D. R. (1992) Collision rates and impact velocities in the main asteroid belt. *Icarus, 97,* 111–123.

Fujiwara A., Cerroni P., Davis D., Ryan E., Di Martino M., Holsapple K., and Housen K. (1989) Experiments and scaling laws for catastrophic collisions. In *Asteroids II* (R. P. Binzel et al., eds.), pp. 240–265. Univ. of Arizona, Tucson.

Fulle M. (1997) Injection of large grains into orbits around comet nuclei. *Astron. Astrophys., 325,* 1237–1248.

Garmier R. and Barriot J.-P. (2001) Ellipsoidal harmonic expansion of the gravitational potential: Theory and application. *Cel. Mech. Dyn. Astron., 79,* 235–275.

Garmier R., Barriot J.-P., Konopliv A. S., and Yeomans D. K. (2002) Modeling of the Eros gravity field as an ellipsoidal harmonic expansion from the NEAR Doppler tracking data. *Geophys. Res. Lett., 29,* in press.

Gault D. E., Shoemaker E. M., and Moore H. J. (1963) *Spray Ejected from the Lunar Surface by Meteoroid Impact.* NASA TND-1767, Washington, DC.

Geissler P., Petit J.-M., Durda D. D., Greenberg R., Bottke W., Nolan M., and Moore J. (1996) Erosion and ejecta reaccretion on 243 Ida and its moon. *Icarus, 120,* 140–157.

Giblin I., Petit J.-M., and Farinella P. (1998) Impact ejecta rotational bursting as a mechanism for producing stable Ida-Dactyl systems. *Icarus, 132,* 43–52.

Greenberg R. (1982) Orbital interactions — A new geometrical formalism. *Astron. J., 87,* 184–195.

Greenberg R., Nolan M. C., Bottke W. F., Kolvoord R. A., and Veverka J. (1994) Collisional history of Gaspra. *Icarus, 107,* 84.

Greenwood D. T. (1988) *Principles of Dynamics.* Prentice Hall. 552 pp.

Hamilton D. P. and Burns J. A. (1991a) Orbital stability zones about asteroids. *Icarus, 92,* 118–131.

Hamilton D. P. and Burns J. A. (1991b) Orbital stability zones about asteroids. II. The destabilizing effects of eccentric orbits and of solar radiation. *Icarus, 96,* 43–64.

Hamilton D. P. and Krivov A. V. (1997) Dynamics of distant moons of asteroids. *Icarus, 128,* 241–249.

Harris A. W. (1994) Tumbling asteroids. *Icarus, 107,* 209–211.

Hartmann W. K. (1978) Planet formation: Mechanism of early growth. *Icarus, 33,* 50–61.

Hartmann W. K. (1979) Diverse puzzling asteroids and a possible unified explanation. In *Asteroids* (T. Gehrels, ed.), pp. 466–479. Univ. of Arizona, Tucson.

Hénon M. (1969) Numerical exploration of the restricted problem. V. *Astron. Astrophys., 1,* 223–238.

Holsapple K. (1993) The scaling of impact processes in planetary sciences. *Annu. Rev. Earth Planet. Sci., 21,* 333–373.

Holsapple K. and Schmidt R. (1982) On the scaling of crater dimensions. 2. Impact processes. *J. Geophys. Res., 87,* 1849–1870.

Holsapple K., Giblin I., Housen K., Nakamura A., and Ryan E. (2002) Asteroid impacts: Laboratory experiments and scaling laws. In *Asteroids III* (W. F. Bottke Jr. et al., eds.), this volume. Univ. of Arizona, Tucson.

Hörz F., Cintala M. J., Olds S., See T. H., and Cardenas F. (1985) Experimental regolith evolution: Differential comminution of plagioclase, pyroxene and olivine (abstract). In *Lunar and Planetary Science XVI,* pp. 362–363. Lunar and Planetary Institute, Houston.

Housen K. R. (1981) The stochastic evolution of asteroidal regoliths and the origin of brecciated and gas-rich meteorites. Ph.D. thesis, Univ. of Arizona, Tucson.

Housen K. R., Wilkening L. L., Chapman C. R., and Greenberg R. J. (1979a) Asteroidal regoliths. *Icarus, 39,* 317–351.

Housen K. R., Wilkening L. L., Chapman C. R., and Greenberg R. J. (1979b) Regolith development and evolution on asteroids and the Moon. In *Asteroids* (T. Gehrels, ed.), pp. 601–627. Univ. of Arizona, Tucson.

Housen K. R., Schmidt R. M., and Holsapple K. A. (1983) Crater ejecta scaling laws: Fundamental forms based on dimensional analysis. *J. Geophys. Res., 88,* 2485–2499.

Housen K. R., Holsapple K. A., and Voss M. E. (1999) Compaction as the origin of the unusual craters on the asteroid Mathilde. *Nature, 402,* 155–157.

Hudson R. S. and 26 colleagues (2000) Radar observations and physical model of asteroid 6489 Golevka. *Icarus, 148,* 37–51.

Ivanov B. A., Neukum G., Bottke W., and Hartmann W. K. (2002) The comparison of size-frequency distributions of impact craters and asteroids and the planetary cratering rate. In *Asteroids III* (W. F. Bottke Jr. et al., eds.), this volume. Univ. of Arizona, Tucson.

Kaula W. M. (2000) *Theory of Satellite Geodesy: Applications of Satellites to Geodesy.* Dover, Mineola, New York. 124 pp.

Koon W. S., Lo M., Marsden J. E., and Ross S. D. (2000) Heteroclinic connections between periodic orbits. *Chaos, 10,* 427–469.

Lee P. (1996) Dust levitation on asteroids. *Icarus, 124,* 181–194.

Lee S. W., Thomas P., and Ververka J. (1986) Phobos, Deimos, and the moon — Size and distribution of crater ejecta blocks. *Icarus, 68,* 77–86.

Lee P., Veverka J., Thomas P. C., Helfenstein P., Belton M. J. S., Chapman C. R., Greeley R., Pappalardo R. T., Sullivan R., and Head J. W. (1996) Ejecta blocks on 243 Ida and on other asteroids. *Icarus, 120,* 87–105.

MacMillan W. D. (1960) *Dynamics of Rigid Bodies.* Dover, New York. 478 pp.

Margot J.-L., Nolan M. C., Benner L. A. M., Ostro S. J., Brown M. E., Howell E. S., Jurgens R. F., Giorgini J. D., Slade M. A., Campbell D. B., De Pater I., and Roe H. (2001) Discovery and characterization of three binary asteroids, 2000 DP107, 2000 UG11, and 87 Sylvia (abstract). In *Asteroids 2001,* Palermo, Italy, June 2000.

Marzari F., Scholl H., and Farinella P. (1996) Collision rates and impact velocities in the Trojan asteroid swarms. *Icarus, 119,* 192–201.

Marzari F., Farinella P., Davis D. R., Scholl H., and Bagatin A. C. (1997) Collisional evolution of Trojan asteroids. *Icarus, 125,* 39–49.

McKay D. S., Swindle T. D., and Greenberg R. (1989) Asteroidal regoliths: What we do not know. In *Asteroids II* (R. P. Binzel et al., eds.), pp. 617–642. Univ. of Arizona, Tucson.

Melosh H. J. (1989) *Impact Cratering: A Geologic Process.* Oxford Univ., New York. 245 pp.

Merline W. J., Chapman C. R., Colwell W. B., Veverka J., Harch A., Bell M., Bell J., Thomas P., Clark B. E., Martin P., Murchie S., Cheng A., Domingue D., Izenberg N., Robinson M., McFadden L., Wellnitz D., Malin M., Owen W., and Miller J. (1999) Search for satellites around asteroid 433 Eros from NEAR flyby imaging (abstract). In *Lunar and Planetary Science XXX,* Abstract #2055. Lunar and Planetary Institute, Houston (CD-ROM).

Merline W. J., Tamblyn P., Chapman C. R., Colwell W. B., Gor V., Burl M. C., Bierhaus E. B., and Robinson M. S. (2001) An autonomous search for moons during the approach of the NEAR spacecraft to asteroid Eros. In *Proc. 6th Intl. Symp. on Artificial Intelligence, Robotics, and Automation in Space,* Paper #AM128. i-SAIRAS, Montreal, Canada (CD-ROM).

Merline W. J., Weidenschilling S. J., Durda D. D., Margot J. L., Pravec P., and Storrs A. D. (2002) Asteroids *do* have satellites. In *Asteroids III* (W. F. Bottke Jr. et al., eds.), this volume. Univ. of Arizona, Tucson.

Michel P., Benz W., Tanga P., and Richardson D. C. (2001) New simulations of collisions between asteroids in the gravity regime: Comparison with the properties of some observed asteroid families. In *Asteroids 2001,* Palermo, Italy, June 2000.

Miller J. K., Konopliv A. S., Antreasian P. G., Bordi J. J., Chesley S., Helfrich C. E., Owen W. M., Scheeres D. J., Wang T. C., Williams B. G., and Yeomans D. K. (2002) Determination of shape, gravity and rotational state of asteroid 433 Eros. *Icarus, 155,* 3–17.

Moore H. J. (1971) Large blocks around lunar craters. In *Analysis of Apollo 10 Photography and Visual Observations,* pp. 26–27. NASA SP-232, Washington, DC.

Murray C. D. and Dermott S. F. (1999) *Solar System Dynamics,* Cambridge Univ., Cambridge. 592 pp.

Neukum G., Ivanov B. A., and Hartman W. K. (2001) Cratering records in the inner soalr system in relation to the lunar reference system. In *Composition and Origin of Cometary Materials,* pp. 1–32. Kluwer, Dordrecht.

Nolan M. C., Asphaug E., Melosh H. J., and Greenberg R. (1996) Impact craters on asteroids: Does gravity or strength control their size? *Icarus, 124,* 359–371.

Ostro S. J., Hudson R. S., Nolan M. C., Margot J.-L., Scheeres D. J., Campbell D. B., Magri C., Giorgini J. D., and Yeomans D. K. (2000) Radar observations of asteroid 216 Kleopatra. *Science, 288,* 836–839.

Ostro S. J., Hudson R. S., Benner L. A. M., Giorgini J. D., Magri C., Margot J.-L., and Nolan M. C. (2002) Asteroid radar astronomy. In *Asteroids III* (W. F. Bottke Jr. et al., eds.) this volume, Univ. of Arizona, Tucson.

Petit J.-M., Durda D. D., Greenberg R., Hurford T. A., and Geissler P. E. (1997) The long-term dynamics of Dactyl' orbit. *Icarus, 130,* 177–197.

Rabinowitz D. L., Helin E., Lawrence K., and Pravdo S. (2000) A reduced estimate of the number of kilometer-sized near-Earth asteroids. *Nature, 403,* 156–165.

Richardson D. C., Bottke W. F. Jr., and Love S. G. (1998) Tidal distortion and disruption of Earth-crossing asteroids. *Icarus, 134,* 47–76.

Richardson D. C., Leinhardt Z. M., Melosh H. J., Bottke W. F. Jr., and Asphaug E. (2002) Gravitational aggregates: Evidence and evolution. In *Asteroids III* (W. F. Bottke Jr. et al., eds.), this volume. Univ. of Arizona, Tucson.

Richter K. and Keller H. U. (1995) On the stability of dust particle orbits around cometary nuclei. *Icarus, 114,* 355–371.

Robinson M. S., Thomas P. C., Veverka J., Murchie S., and Carcich B. (2001) The nature of ponded deposits on Eros. *Nature, 413,* 396–400.

Rubincam D. (2000) Radiative spin-up and spin-down of small

asteroids. *Icarus, 148,* 2–11.

Sawai S., Kawaguchi J., Scheeres D. J., Yoshizawa N., and Ogawara M. (2001) Development of a target marker for landing on asteroids. *J. Spacecraft Rockets, 38,* 601–608.

Scheeres D. J. (1994) Dynamics about uniformly rotating tri-axial ellipsoids. Applications to asteroids. *Icarus, 110,* 225–238.

Scheeres D. J. (1999) The effect of C_{22} on orbit energy and angular momentum. *Cel. Mech. Dyn. Astron., 73,* 339–348.

Scheeres D. J. and Asphaug E. I. (1998) Debris and sample transport about asteroids. In *Space, 1998* (R. G. Galloway and S. Lokaj, eds.), pp. 340–346. American Society of Civil Engineering, Reston, Virginia.

Scheeres D. J. and Marzari F. (2000) Temporary orbital capture of ejecta from comets and asteroids. *Astron. Astrophys., 356,* 747–756.

Scheeres D. J., Ostro S. J., Hudson R. S., and Werner R. A. (1996) Orbits close to asteroid 4769 Castalia. *Icarus, 121,* 67–87.

Scheeres D. J., Ostro S. J., Hudson R. S., DeJong E. M., and Suzuki S. (1998a) Dynamics of orbits close to asteroid 4179 Toutatis. *Icarus, 132,* 53–79.

Scheeres D. J., Marzari F., Tomasella L., and Vanzani V. (1998b) ROSETTA mission: Satellite orbits around a cometary nucleus. *Planet. Space Sci., 46,* 649–671.

Scheeres D. J., Williams B. G., and Miller J. K. (2000a) Evaluation of the dynamic environment of an asteroid: Applications to 433 Eros. *J. Guidance Control Dyn., 23,* 466–475.

Scheeres D. J., Ostro S. J., Werner R. A., Asphaug E., and Hudson R. S. (2000b) Effects of gravitational interactions on asteroid spin states. *Icarus, 147,* 106–118.

Scheeres D. J., Yeomans D. K., and Miller J. K. (2001) The orbital dynamics environment of 433 Eros. AAS Paper 01-373, presented at the *2001 Astrodynamics Specialist Meeting,* July 2001, Quebec City, Canada.

Sullivan R., Greeley R., Pappalardo R., Asphaug E., Moore J. M., Morrison D., Belton M. J. S., Carr M., Chapman C. R., Geissler P., Greenberg R., Granahan J., Head J. W., Kirk R., McEwen A., Lee P., Thomas P. C., and Veverka J. (1996) Geology of 243 Ida. *Icarus, 120,* 119–139.

Szebehely V. (1967) *Theory of Orbits, the Restricted Problem of Three Bodies.* Academic, New York. 668 pp.

Thomas P. (1979) Surface features of Phobos and Deimos. *Icarus, 40,* 223–243.

Thomas P. C. (1993) Gravity, tides, and topography on small satellites and asteroids: Application to surface features of the martian satellites. *Icarus, 105,* 326–344.

Thomas P. C. (1998) Ejecta emplacement on the martian satellites. *Icarus, 131,* 78–106.

Thomas P. and Veverka J. (1980) Downslope movement of material on Deimos. *Icarus, 42,* 234–250.

Thomas P. C., Veverka J., Sullivan R., Simonelli D. P., Malin M. C., Caplinger M., Hartmann W. K., and James P. B. (2000) Phobos: Regolith and ejecta blocks investigated with Mars Orbiter Camera images. *J. Geophys. Res., 105,* 15091–15106.

Thomas P. C., Veverka J., Robinson M. S., and Murchie S. (2001) Shoemaker crater as the source of most ejecta blocks on the asteroid 433 Eros. *Nature, 413,* 394–396.

van Flandern T. C., Tedesco E. F., and Binzel R. P. (1979) Satellites of asteroids. In *Asteroids* (T. Gehrels, ed.), pp. 443–465. Univ. of Arizona, Tucson.

Van Houten C. J., van Houten-Groeneveld I., Herget P., and Gehrels T. (1970) The Palomar-Leiden survey of faint minor planets. *Astron. Astrophys. Suppl. Ser., 2,* 339.

Vedder J. D. (1996) Asteroid collisions: Estimating their probabilities and velocity distributions. *Icarus, 123,* 436–449.

Veverka J. and Thomas P. (1979) Phobos and Deimos: A preview of what asteroids are like? In *Asteroids* (T. Gehrels, ed.), pp. 628–654. Univ. of Arizona, Tucson.

Veverka J., Thomas P., Johnson T. V., Matson D., and Housen K. (1986) The physical characteristics of satellite surfaces. In Satellites (J. A. Burns and M. S. Matthews, eds.), pp. 342–402. Univ. of Arizona, Tucson.

Veverka J., Thomas P., Harch A., Clark B., Bell J. F., Carcich B., Joseph J., Chapman C., Merline W., Robinson M., Malin M., McFadden L. A., Murchie S., Hawkins S. E., Farquhar R., Izenberg N., and Cheng A. (1997) NEAR's flyby of 253 Mathilde: Images of a C asteroid. *Science, 278,* 2109–2114.

Veverka J., Robinson M., Thomas P., Murchie S., Bell J. F., Izenberg N., Chapman C., Harch A., Bell M., Carcich B., Cheng A., Clark B., Domingue D., Dunham D., Farquhar R., Gaffey M. J., Hawkins E., Joseph J., Kirk R., Li H., Lucey P., Malin M., Martin P., McFadden L., Merline W. J., Miller J. K., Owen W. M., Peterson C., Prockter L., Warren J., Wellnitz D., Williams B. G., and Yeomans D. K. (2000) NEAR at Eros: Imaging and spectral results. *Science, 289,* 2088–2097.

Veverka J., Thomas P. C., Robinson M., Murchie S., Chapman C., Bell M., Harch A., Merline W. J., Bell J. F., Bussey B., Carcich B., Cheng A., Clark B., Domingue D., Dunham D., Farquhar R., Gaffey M. J., Hawkins E., Izenberg N., Joseph J., Kirk R., Li H., Lucey P., Malin M., McFadden L., Miller J. K., Owen W. M., Peterson C., Prockter L., Warren J., Wellnitz D., Williams B. G., and Yeomans D. K. (2001) Imaging of small-scale features on 433 Eros from NEAR: Evidence for a complex regolith. *Science, 292,* 484–488.

Weeks C. J. (1995) The effect of comet outgassing and dust emission on the navigation of an orbiting spacecraft. *J. Astronaut. Sci., 43,* 327–343.

Weidenschilling S. J. (1981) How fast can an asteroid spin? *Icarus, 46,* 124–126.

Weidenschilling S. J., Paolicchi P., and Zappalà V. (1989) Do asteroids have satellites? In *Asteroids II* (R. P. Binzel et al., eds.), pp. 643–658. Univ. of Arizona, Tucson.

Werner R. A. (1994) The gravitational potential of a homogeneous polyhedron or don't cut corners. *Cel. Mech. Dyn. Astron., 59,* 253–278.

Werner R. A. and Scheeres D. J. (1997) Exterior gravitation of a polyhedron derived and compared with harmonic and mascon gravitation representations of asteroid 4769 Castalia. *Cel. Mech. Dyn. Astron., 65,* 313–344.

Wetherill G. W. (1967) Collisions in the asteroid belt. *J. Geophys. Res., 72,* 2429–2444.

Yeomans D. K., Antreasian P. G., Barriot J.-P., Chesley S. R., Dunham D. W., Farquhar R. W., Giorgini J. D., Helfrich C. L., Konopliv A. S., McAdams J. V., Miller J. K., Owen W. M. Jr., Scheeres D. J., Thomas P. C., Veverka J., and Williams B. G. (2000) Radio science results during the NEAR-Shoemaker spacecraft rendezvous with Eros. *Science, 289,* 2085–2088.

Collisional Evolution of Small-Body Populations

Donald R. Davis
Planetary Science Institute

Daniel D. Durda
Southwest Research Institute

Francesco Marzari
Università di Padova

Adriano Campo Bagatin
Universidad de Alicante

Ricardo Gil-Hutton
Félix Aguilar Observatory

Asteroid collisional evolution studies are aimed at understanding how collisions have shaped observed features of the asteroid population in order to further our understanding of the formation and evolution of our solar system. We review progress in developing more realistic collisional scaling laws, the effects of relaxing over-simplifying assumptions used in earlier collisional evolution studies, and the implications of including observables, such as collisionally produced families, on constraining the collisional history of main-belt asteroids. Also, collisional studies are extended to include Jupiter Trojans and the Hilda population. Results from collisional evolution models strongly suggest that the mass of main-belt asteroids was only modestly larger, by up to a factor of 5 or so, at the time that the present collisionally erosive environment was established, presumably early in solar system history. Major problems remain in identifying the appropriate scaling algorithm for determining the threshold for catastrophic disruption as well as understanding the resulting size and velocity distribution of fragments. Dynamical effects need to be combined with collisional simulations in order to understand the structure of the small asteroid size distribution.

1. INTRODUCTION

The importance of collisions in shaping the present asteroid belt has been recognized for nearly 50 years, following the pioneering work of *Piotrowski* (1953) on the frequency of asteroidal collisions. Since that time, increasingly sophisticated tools have been developed for tracing the collisional history of asteroid populations and understanding the role of collisions in producing observed features of various small-body populations. Here we review the present state of understanding of asteroid collisional evolution, concentrating on progress since the *Asteroids II* volume.

Observations relevant to collisional processes are described in section 2, including recent work on extending the asteroid size distribution to subkilometer sizes. Determining the size distribution is a key constraint for collisional models and is a major discriminant among the various scaling laws proposed for describing collisional outcomes. In section 3 we summarize work on scaling laws that describes the outcome of collisions involving bodies of various sizes hitting at speeds of multikilometers per second. Understanding collisional outcomes for asteroid-sized bodies is fundamental to understanding asteroid collisional history; however, a consensus has not yet developed among workers as to the appropriate methodology to go from laboratory-scale experiments to giant collisions involving bodies to hundreds of kilometers in size. Gravity becomes significant in the asteroid size range between subkilometer-sized bodies and the ~850-km-diameter Ceres and thus adds to the complexity of understanding collisional outcomes. The effects of size-dependent collisional outcomes as well as boundary conditions on the overall population distribution are active topics of research. As discussed in section 3, the existence of a small size cutoff in a collisionally interacting population can produce a wavelike structure, a feature not recognized in earlier work on collisional evolution.

Collisional evolution of Trojans and Hildas has been studied in depth in recent years, as reviewed in section 3. These populations are particularly interesting to study because they are dynamically isolated from much of the main asteroid belt, although they do collisionally interact with each other. Additionally, these populations are dynamically stable, but collisions can eject fragments into unstable orbits that are then lost from the system. Comparing the collisional evolution of these populations with each other provides insight to features common to the collisional evolution of small-body populations. Another collisionally evolved population discovered recently is the Kuiper Belt, a vast population of remnant planetesimals orbiting beyond Neptune. Kuiper Belt objects are thought to share many features with main-belt asteroids, including a significant role for collisions in shaping the present Kuiper Belt. However, this chapter will be restricted to the role of collisions on asteroids in the inner solar system, and the interested reader is referred to *Farinella et al.* (2000) for a comprehensive review of the role of collisions in the Kuiper Belt.

Finally, section 5 gives our perspective on the major outstanding problems in the field of asteroid collisional studies. We expect that by the time of the publication of *Asteroids IV,* most of these will be solved and an equally challenging set of questions will be posed for the next decade.

2. OBSERVATIONAL CONSTRAINTS

The litmus test for models of the collisional history of asteroids is how well they match the observed characteristics of the present asteroid belt, i.e., those features of the belt that are the product of 4.5 b.y. of collisional history. So, what are the collisionally produced observables in the asteroid belt and other small body populations? First is the size-frequency distribution; Fig. 1 shows the inferred size-frequency distribution for (a) main-belt asteroids, (b) Trojans, and (c) Hildas. [Throughout this chapter we use the cumulative diameter distribution, i.e., the number of bodies larger than diameter D, to describe the size distribution of populations. This relationship is expressed mathematically as $N(>D) = K D^{-b}$ where b is the cumulative diameter population index. We also use the incremental size distribution for clarity in some cases; however, the population index is numerically the same in these two formulations. An excellent discussion of the various power laws used to describe small-body populations is given by *Colwell* (1994).] It is widely believed that the largest asteroids (although where "largest" begins is a debated question) are primordial objects whose sizes have not been significantly altered by collisions; only the small size end of the population is collisionally evolved.

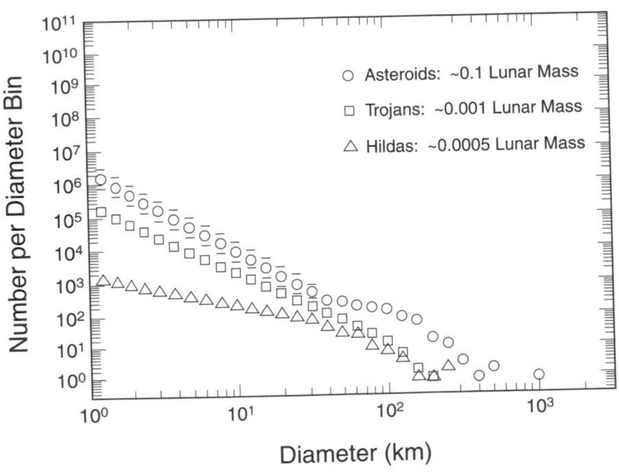

Fig. 1. Comparison of the size-frequency distribution for main-belt asteroids (MBA), Trojans, and Hildas to show the relative abundance as a function of size in the three populations. Note that while the Trojan population is numerically 10–25% of the main-belt population at sizes ≤100 km diameter, the total mass of the Trojans is only about 1% of the MBA. Most of the mass for these populations is in the largest bodies. Each diameter bin spans a factor of 2 in mass.

The so-called "bump" in the main-belt asteroid size distribution centered at about 100 km diameter has been explained in three different ways: (1) *Anders* (1965) proposed that this was the signature of collisional modification of an initial Gaussian-like planetesimal size distribution. (2) *Davis et al.* (1979, 1984) found that this feature marked the transition from the gravity strength regime to the strength dominant size regime and hence was a product of collisional physics. (3) *Durda et al.* (1998) argued that this feature was a secondary "bump" produced by the wave from the strength-gravity transition that occurs at much smaller sizes (see section 3).

While there is structure in the small asteroid size distribution, a power-law fit to the cumulative absolute magnitude distribution of the Palomar Leiden Survey (PLS) (*Kresak,* 1977; *Van Houten et al.,* 1970) gives a slope of 1.95 over the size range larger than a few kilometers (D ≥ 2–5 km). Figure 2 shows various recent estimates of the small end of the main-belt size small distribution (see also *Jedicke et al.,* 2002).

It is worth remarking that several lines of evidence point to the existence of a variable index in the size distribution of the small asteroids. For main-belt asteroids smaller than about 20 km, the PLS (*Van Houten et al.,* 1970) found variable indexes. Also, *Cellino et al.* (1991) analyzed IRAS data for different zones of the asteroid belt and different size ranges (but always exceeding a few tens of kilometers, to avoid discovery biases) and found indexes ranging from ~–2 to ~–4. Similar variability occurs for objects with orbits crossing those of the inner planets, as suggested by the lunar and martian cratering record (*Grün et al.,* 1985; *Strom et al.,* 1992), the *in situ* dust detection experiments carried out on spacecraft (*Grün et al.,* 1985; *Love and Brownlee,* 1993), and telescopic searches for small Earth-approaching asteroids (*Ceplecha,* 1992; *Rabinowitz,* 1993). Analyses of the cratering record on the surface of 951 Gaspra, imaged in 1991 by the *Galileo* probe, have indicated that a cumulative size distribution index of –2.7 ± 0.5 over the size range 0.4–1.5 km (*Belton et al.,* 1992), while for Ida, the index varies from ~–3.3 to –3.5 for crater diameters ≥1 km (*Belton et al.,* 1994). The craters were produced by projectiles (~20 to 200 m across) that have cratered the surfaces of Gaspra and Ida.

A powerful constraint on the asteroid collisional history is the existence today of the basaltic crust of Vesta, which, if it is the ultimate source of the eucrite meteorites as is widely believed, dates to the earliest era of the solar system some 4.54 b.y. ago. Any collisional model must preserve this 25–40-km-thick crust (*Gaffey et al.,* 1993) during the collisional bombardment over solar system history (*Davis et al.,* 1984). Recent Hubble Space Telescope (HST) observations

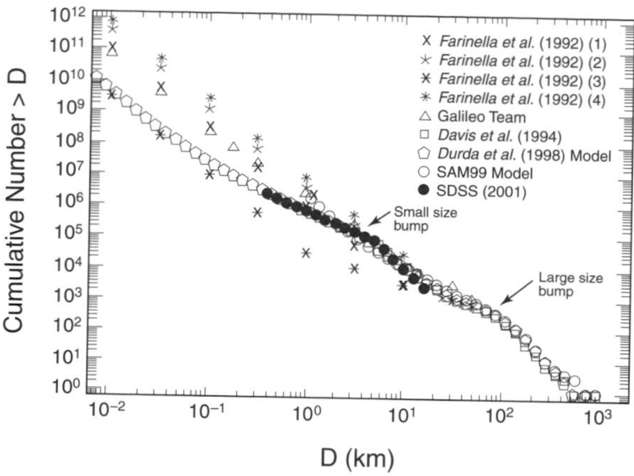

Fig. 2. Nine estimates of the main-belt asteroid size distribution. The four models from *Farinella et al.* (1992b) are based on assuming N(>50 km) = 700, a cumulative diameter slope of –1.0 between 50 km > D > D_{tr} and the *Dohnanyi* (1969) slope (cumulative diameter) slope of –2.5 for D < D_{tr}. The values of D_{tr} are taken to be 10 km, 25 km, 1 km, and 40 km, respectively. The model labeled "Galileo" is the adopted asteroid population for calculating the projectile flux onto Gaspra. They used the PLS slope (–1.95) for D > 0.175 km and the Dohnanyi slope for D < 0.175. The *Davis et al.* (1994) model derives from extrapolating the known population (assumed to be complete) at D = 44 km using the PLS slope and the *Cellino et al.* (1991) "corrected random" albedo model down to D = 5.5 km. The *Durda et al.* (1998) model is the fit to the asteroid size distribution determined by *Jedicke and Metcalfe* (1998), who used Spacewatch data to estimate the small size main-belt distribution. The SAM99 model is the Statistical Asteroid Model of *Tedesco et al.* (2002), who combined the observed size distribution of the main asteroid families with that of the background population into a single unified population estimate. The small size distribution is obtained by extrapolating the observed large size trends. Finally, the SDSS 2001 model is the fits to the Sloan Digital Sky Survey data by *Ivezić et al.* (2001), who find that a broken power law well represents their data. For 5 km < D < 40 km, the slope is –3.0, while for 0.4 km < D < 5 km, a –1.3 slope is a good fit to their data.

of Vesta revealed the existence of a ~450-km-diameter basin thought to be formed by the impact of a ~40-km-diameter projectile (*Marzari et al.,* 1996; *Asphaug,* 1997). This impact, presumably the largest since Vesta's crust formed, provides a very specific constraint on Vesta's collisional history. Yet this event, which apparently excavated through the crust to some degree, was not energetic enough to shatter Vesta and destroy its basaltic crust. The fundamental reason that the large impact did not destroy Vesta's crust is that it occurs in the gravity — rather than the strength — regime, as shown by detailed numerical simulations (*Asphaug,* 1997).

At the same time, collisions are argued to have produced the parent bodies of iron meteorites, widely interpreted to be the metallic cores of disrupted differentiated asteroids. These metallic cores were identified with the M-class asteroids; however, this interpretation has been challenged by the discovery of water of hydration on many M asteroids, leading to the proposed "W" (or wet) taxonomic class (*Rivkin et al.,* 2000). However, the M-class asteroid 16 Psyche appears to be the collisionally exposed core of a parent body that was virtually identical to Vesta. How is it possible to disrupt the Psyche parent body and remove all traces of this giant collision (there is no family associated with Psyche today) except for Psyche itself, while still preserving the crust of Vesta (*Chapman,* 1985)? This problem was investigated by *Davis et al.* (1999), who concluded that it was very difficult, but not impossible, based on collisional modeling, to create Psyche and preserve Vesta's crust, but other explanations for Psyche should be sought. Furthermore, recent work suggests a mean density for Psyche of 1.8 ± 0.6 g/cc (*Viateau,* 2000), which is inconceivably low for the Fe core of a differentiated body. The more general problem for collisional effects on asteroid materials is how to disrupt the dozens of parent bodies needed to explain the iron meteorite collection while eliminating from the asteroid belt most of the volumetrically more abundant dunite mantles of these bodies. Collisional grinding of this mantle material has been suggested, but such a process would need to be extremely efficient [see the section on the great dunite shortage in *Bell et al.* (1989); see also *Burbine et al.* (1996)].

Asteroid families are another observable that is the product of collisional evolution. There are approximately 25 reliable families and over 60 statistically significant clusters identified in asteroid proper elements [see *Zappalà et al.* (2002) and *Bendjoya and Zappalà* (2002) for detailed discussions of asteroid families].

Another constraint on the asteroid small size population comes from the cosmic-ray-exposure (CRE) ages of stony meteorites, which are measured to be around 20 m.y. The CRE age measures the time that the meteorite was in meter-sized bodies (or near the surface of a larger body); however, the collisional lifetime of meter-sized objects is calculated to be around 1 m.y. (*O'Brien and Greenberg,* 1999). Any successful asteroid collisional history must be consistent with the observed CRE ages.

Collisions have modified asteroid rotation rates over solar system history, so information about the collisional history of the belt is embedded in the spin rates of asteroids of all sizes. However, as discussed by *Davis et al.* (1989) and *Farinella et al.* (1992a), the uncertainties in modeling how collisions alter rotation rates are so large that any definitive work on this topic is yet to be done. However, recent progress has been made on understanding fragment rotation rates from catastrophic disruption events using numerical hydrocodes by *Love and Ahrens* (1997) and *Asphaug and Scheeres* (1999). These investigations provided deeper insights into fragmental spin rates as a function of fragment size as well as providing data on critical parameters needed to model the collisional evolution of asteroid spin rates.

In summary, any successful collisional evolution scenario for main-belt asteroids must satisfy the observable constraints imposed by the size distribution, the number of families, and the existence of an old basalt crust on Vesta; the CRE ages of strong meteorites; and the cratering projectile flux onto asteroids. Other observables, such as aster-

TABLE 1. Intrinsic probability of collision and mean impact velocity for different populations of minor bodies.

Asteroid Populations	Intrinsic Probability (10^{-18} yr^{-1} km^{-2})	Impact Velocity (km s^{-1})	Reference
Main Belt (MB)			
MB-MB	2.85 ± 0.66	5.81 ± 1.88	*Farinella and Davis* (1992)
MB-MB	3.97	—	*Yoshikawa and Nakamura* (1994)
MB-MB	2.86	5.3	*Bottke et al.* (1994)
MB-MB	4.38	4.22	*Vedder* (1996, 1998)
MB-SPC	1.51	10.86	*Gil-Hutton* (2000)
MB-UN	1.08	14.0	*Gil-Hutton and Brunini* (1999)
Hildas (Hil)			
Hil-Hil	2.31 ± 0.10	3.09 ± 1.47	*Dahlgren* (1998)
Hil-Hil	1.93	3.36 ± 1.52	*Dell'Oro et al.* (2001)
Hil-Tro	0.24 ± 0.06	4.59 ± 1.71	*Dahlgren* (1998)
Hil-Tro	0.27	4.49 ± 1.68	*Dell'Oro et al.* (2001)
Hil-MB	0.62 ± 0.04	4.78 ± 1.78	*Dahlgren* (1998)
Hil-MB	0.66	4.80 ± 1.77	*Dell'Oro et al.* (2001)
Hil-UN	6.5	10.98	*Gil-Hutton and Brunini* (1999)
Trojans (Tro)			
L_4-L_4	6.46 ± 0.09	4.90 ± 0.07	*Marzari et al.* (1996)
L_4-L_4	7.79 ± 0.67	4.66	*Dell'Oro et al.* (1998)
L_5-L_5	5.30 ± 0.10	4.89 ± 0.10	*Marzari et al.* (1996)
L_5-L_5	6.68 ± 0.18	4.51	*Dell'Oro et al.* (1998)
L_4-SPC	0.33	6.78 ± 2.71	*Dell'Oro et al.* (2001)
L_5-SPC	0.35	6.67 ± 2.59	*Dell'Oro et al.* (2001)
Tro-UN	0.50	8.19	*Gil-Hutton and Brunini* (1999)

For each population the values calculated by various authors with the relative errors, when available, are reported. The more recent values are usually derived from a larger sample of orbits and are probably more accurate. Collision rates with external populations are given for (a) short-period comets (SPC), and (b) Uranus/Neptune scattered planetesimals (UN).

oid rotation rates, dust production, etc., are altered by, or are the product of, collisional processes. However, our understanding of how collisions affect them is not sufficiently mature at this time to allow imposition of additional constraints.

3. SCALING LAWS, BOUNDARY CONDITIONS, AND COLLISIONAL EVOLUTION

3.1. Basic Collisional Parameters: Collision Rates and Impact Speeds

The frequency of collisions and the collisional impact speeds are fundamental quantities for collisional evolution studies. While the basic theory for calculating collision rates and speeds was developed by *Öpik* (1951) and *Wetherill* (1967) and applied by various workers in subsequent years, there has been recent work to refine such calculations and to extend them to both Trojans and Hildas. The intrinsic collision probability, P_i, gives the collision rate per unit cross-section area of target and projectile per unit time, while the actual number of collisions onto a target of radius R_t by projectiles of radius R_p within a time (ΔT) is then given by

$$N_{coll} = \langle P_i \rangle (R_t + R_p)^2 V\ N_p \Delta T$$

where V is the mean impact speed and N_p is the spatial density of projectiles.

For each population several estimates of $\langle P_i \rangle$ and V are given in Table 1. Different authors improved the Öpik/Wetherill analytical formulation with corrections related to specific features of the orbital distribution of the population under study or to statistical mechanics (*Greenberg,* 1982; *Namiki and Binzel,* 1991; *Farinella and Davis,* 1992; *Bottke and Greenberg,* 1993; *Bottke et al.,* 1994; *Vedder,* 1996, 1998; *Dell'Oro and Paolicchi,* 1998; *Dell'Oro et al.,* 1998, 2001). These methods also allow fast computations of updated values of $\langle P_i \rangle$ and V whenever the discovery of new objects significantly enriches the known population. Alternatively, different authors have resorted to a direct numerical approach based on the integration of the orbits of the asteroids over a sufficiently long timespan. The derived distribution of close encounters and mutual speeds recorded during the integration can be extrapolated to infer the collision probability and characteristic impact speed (*Yoshikawa and Nakamura,* 1994; *Marzari et al.,* 1996; *Dahlgren,* 1998).

It is interesting to note that P_i and V for each of the three populations treated here are rather similar, differing by at most a factor of 4 in P_i and a little more than a factor of 2 in V. The collision rates between populations, though (e.g., Hildas with main-belt asteroids or Hildas with Trojans), can be an order of magnitude smaller than the intrapopulation collision rates. However, the collision speeds for the interpopulation collisions are similar to those within a population.

While various authors have attempted to infer the primordial population of asteroids by integrating backward in time from the present belt, the inherent instability of the collisional problem prevents such results from converging. Indeed, the earliest result of *Dohnanyi* (1969), which showed that under specific circumstances the asteroid population was collisionally relaxed and independent of the starting population, meant that it would be impossible to uniquely infer the initial asteroid population by going back in time from the present belt. All recent models of asteroid collisional evolution have assumed various hypothetical initial populations and propagated these forward in time using a variety of scaling laws, and have compared the terminal model population with the present observed belt.

3.2. Scaling Laws and Collisional Evolution

The modeling of the collisional evolution of populations of small bodies is generally performed in two main stages: (1) the simulation of the outcomes of single collisions, and (2) the integration of the equations of evolution, assuming given boundary conditions. A set of simplifying assumptions is usually made in these kind of models: Bodies are considered to be spherical and homogeneous, and they move and collide within a finite volume around the Sun with relative impact speeds determined by the orbits of the target and projectile. Simulations are performed by numerically integrating a given set of first-order, nonlinear, differential equations, taking into account the rate of variation of the population of asteroids of any given mass due to the number of fragments produced or destroyed by collisions, and to the removal of bodies by nongravitational effects (e.g., *Campo Bagatin et al.,* 1994a).

A useful mathematical analysis for the collisional evolution of the asteroid belt was introduced some 30 years ago by *Dohnanyi* (1969, 1971) and independently by *Hellyer* (1970, 1971). Three crucial assumptions of these models are: (1) the interacting bodies are modeled as spheres of equal density, and their collisional cross-section is simply proportional to the squared sum of the radii; (2) all the collisional response parameters are size-independent, implying that fragmentation occurs for a fixed projectile-to-target mass ratio and no self-gravitational effect is taken into account; (3) the population has an upper cutoff in mass (a largest body, Ceres in the asteroid belt) but no lower cutoff. These workers analytically found a stationary solution to the governing differential equations for collisional evolution. To trace the detailed time history of collisional evolution and to include more realistic collisional physics requires numerical models. The first such models led to new insights regarding asteroid collisional history (*Chapman and Davis,* 1975; *Davis et al.,* 1979) and numerical models have been used by nearly all workers subsequently.

The most important result from Dohnanyi's work is that as the collisional process gives rise to a cascade of fragments shifting mass toward smaller and smaller sizes, a simple power-law equilibrium size distribution is approached, which has a cumulative diameter slope of –2.5. Other authors have confirmed this result, both analytically and numerically (e.g., *Paolicchi,* 1994; *Williams and Wetherill,* 1994; *Tanaka et al.,* 1996). In the real world, however, the basic assumptions of Dohnanyi's theory are not fulfilled. It was realized early on that gravitational binding, a mass dependent process, plays a dominant role in holding large asteroids together, hence the assumption that self-similar collisional outcomes could not hold in the gravity regime (*Davis et al.,* 1979). Below we separately explore progress made in recent years in understanding the consequences of relaxing the above assumptions 2 (collisional parameters are size-independent) and 3 (no lower cutoff in the size distribution). Relaxing assumption 1 (on collisional cross-sections) has only a minor effect on the overall collisional evolution.

Recent collisional models replace Dohnanyi assumption 2 with a scaling law describing how the collisional outcomes vary with target and impactor size. In its most general form, a scaling law is simply an algorithm for extrapolating the outcomes of collisions for which we have direct, physical experimental experience to the outcomes of collisional events at size scales much larger or smaller than can be practically treated under laboratory conditions. One important component of scaling laws is determining the critical impact specific energy, Q^*_D, the energy per unit target mass delivered by the projectile required for catastrophic disruption of the target, i.e., such that the largest resulting object has a mass one-half that of the original body. Notice that this object may be formed by partial reaccumulation of fragments due to self-gravity. Sometimes, as in laboratory experiments, the shattering impact specific energy is given in terms of the energy per unit target mass required for the catastrophic shattering of the target, Q^*_S (regardless of reaccumulation of fragments), such that the largest fragment produced has a mass of one-half that of the original target. The two definitions only differ in the regime in which the effect of gravity is much larger than that of solid-state forces.

A number of experimental and analytical studies have focused on quantifying just how Q^*_D scales with target size in the strength-dominated and gravity-dominated regimes (see *Durda et al.,* 1998, for a recent review) and at what sizes the transition from strength-scaling to gravity-scaling takes place. Dimensional analyses (*Farinella et al.,* 1982;

Housen and Holsapple, 1990), impact experiments (*Holsapple,* 1993; *Housen and Holsapple,* 1999), and hydrocode studies (*Ryan,* 1992; *Benz and Asphaug,* 1999) show that Q^*_D for strength-dominated targets scales as roughly $D^{-0.24}$ to $D^{-0.61}$, where D is the target diameter. On the other hand, determining the size dependence of Q^*_D in the gravity-dominated regime has proven to be more difficult, particularly due to the lack of direct experimental experience in the disruption of asteroid-sized targets. Numerous analytical arguments and hydrocode studies (e.g., *Davis et al.,* 1985; *Housen and Holsapple,* 1990; *Love and Ahrens,* 1996; *Melosh and Ryan,* 1997; *Benz and Asphaug,* 1999) yield size-dependent Q^*_D that scale as $D^{1.13}$ to D^2. Recent hydrocode studies (*Benz and Asphaug,* 1999) and numerical collisional models (*Durda et al.,* 1998) indicate that the transition from the strength-dominated to the gravity-dominated regime occurs at target diameters of about 100–300 m, while many earlier studies generally place the transition in the ~1–10-kilometer size range. Observations of asteroid spin rates (*Pravec and Harris,* 2000) confirm these theoretical findings, showing a lack of objects rotating with periods less than 2.2 h among asteroids with absolute magnitude $H < 22$, evidence that asteroids larger than a few hundred meters across are indeed mostly gravity-dominated aggregates with negligible tensile strength ("rubble piles"; see *Richardson et al.,* 2002).

Durda (1993) and *Durda and Dermott* (1997) examined the influence of Q^*_D on the shape of an evolving size distribution by showing that the power-law slope index of a population in collisional equilibrium is a function of the size dependence of Q^*_D. In the case of pure energy scaling (size-independent Q^*_D), the slope index of the size distribution is –2.5, as *Dohnanyi* (1969) showed. When Q^*_D decreases with increasing target size, as is the case in the strength-scaling regime, the slope index of the equilibrium size distribution is steeper than Dohnanyi's; instead, when Q^*_D increases, as in the gravity-scaling regime, the resulting slope index is shallower than –2.5.

Given the wide range of Q^*_D proposed by different scaling laws, the question arises as to which one best describes the real world? Experiments involving asteroid-sized bodies are unfortunately as yet impossible to arrange, so the next best test is to compare model results using different scaling laws with asteroid collisional observables (section 2). *Davis et al.* (1994) tested several proposed scaling laws and found that strain-rate scaling of *Housen et al.* (1991) and simple energy scaling best satisfied all constraints. Further work testing scaling laws was carried out by *Durda et al.* (1998), who presented results from numerical experiments illustrating in a more systematic fashion the sensitivity of an evolved size distribution on the shape of the strength-scaling law. A linear relationship between $\log Q^*_D$ and $\log D$ yields a power-law size distribution; if there is nonlinearity in this relationship (due to a transition from strength- to gravity-scaling, for instance), then structure is introduced into the evolving size distribution due to a nonlinearity in collision lifetimes of the colliding objects.

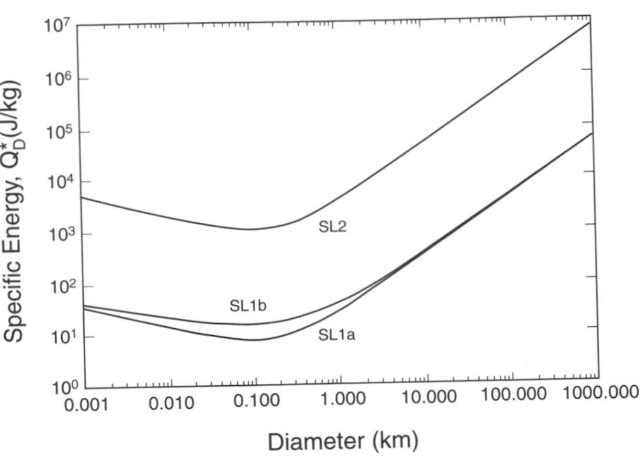

Fig. 3. Three hypothetical scaling laws used to explore the behavior of evolved size distributions.

Figure 3 shows three hypothetical scaling laws used to explore the behavior of evolved size distributions (*Durda et al.,* 1998). For scaling law SL1a, at small sizes we assume a strength-scaling law with a $D^{-0.4}$ size dependence. At large sizes a gravity-scaling relationship with a $D^{1.1}$ size dependence is similar to the predictions of hydrocode models. Scaling law SL1b is similar to SL1a with the exception of a more gradual transition between the strength and gravity-scaling regimes. Scaling law SL2 is identical in shape to SL1a, but is 100× stronger everywhere.

Figure 4 shows the resulting evolved size distributions for colliding populations with the properties of disruption specific energy governed by scaling laws SL1a, SL1b, and SL2. In all three cases, the transition from strength- to gravity-scaling at D = 100 m results in a "bump" in the size distribution for objects just larger than this size. This bump results from the increased disruption lifetimes of objects in this size range over what they would have been had gravity-scaling not taken over to strengthen larger objects. Objects in this bump represent an excess supply of projectiles that can disrupt targets roughly an order of magnitude larger than themselves, resulting in a wavelike perturbation to the size distribution at larger sizes, as described in further detail below. The abruptness of the transition from strength- to gravity-scaling can significantly influence the shape of the resulting size distribution, with more gradual transitions (SL1b vs. SL1a) resulting in smaller amplitude waves. Greater strengths at all sizes (SL2 vs. SL1a) result in higher frequency waves, since the critical projectile size for target disruption is closer to the size of the targets themselves, and the mechanism driving the wave occurs over a smaller range of sizes.

These results demonstrate that evolved size distributions generated by collisional models depend strongly (and understandably) upon the shape of the size-strength scaling law. With this in mind, *Durda et al.* (1998) adopted a least-squares approach and adjusted the strength law for asteroi-

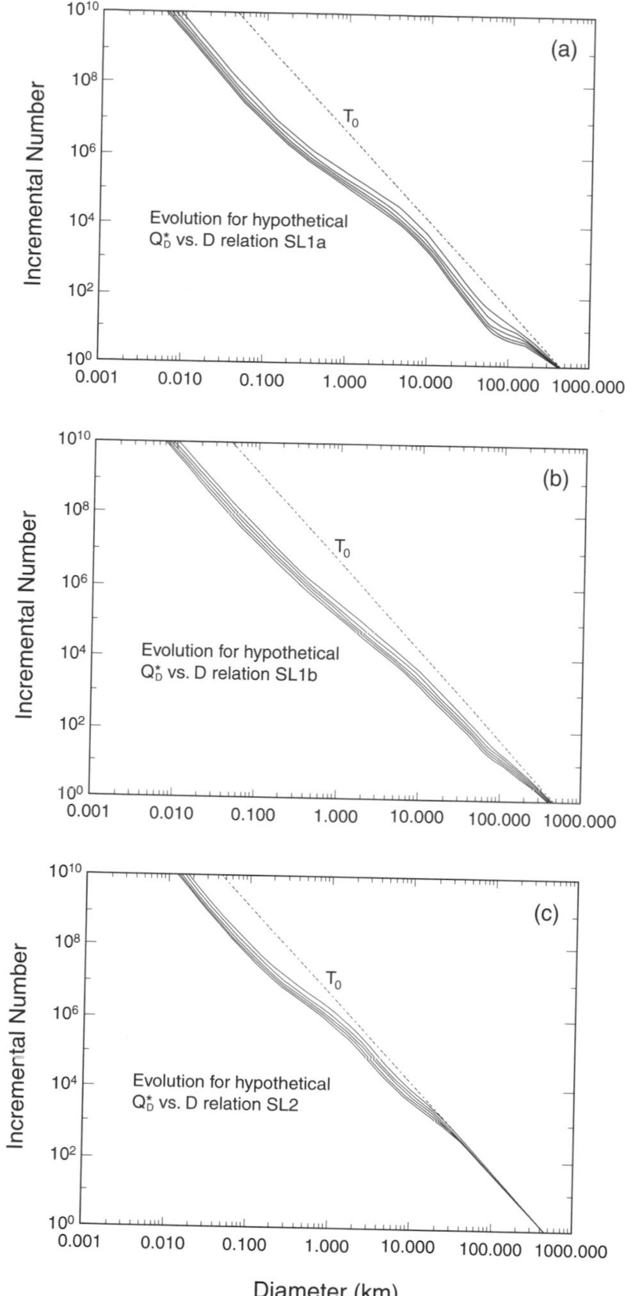

Fig. 4. Resulting evolved size distributions for colliding populations with strength properties governed by scaling laws: **(a)** SL1a, **(b)** SL1b, and **(c)** SL2. Solid lines give the population at 1, 2, 3, 4, and 4.5 b.y.

dal bodies to obtain a best fit to the actual asteroid size distribution determined from the cataloged asteroids and Spacewatch data. Their derived strength scaling law, with a strength minimum at ~150 m diameter as independently suggested by hydrocode models (e.g., *Benz and Asphaug,* 1999) and observed rotation rates (e.g., *Pravec and Harris,* 2000), naturally gives rise to an evolved main-belt asteroid size distribution in good agreement with the two-bump structure observed in the actual size distribution of main-belt asteroids. *Durda et al.* (1998) interpret the bump in the size distribution between ~3 and 30 km (see Fig. 2) as a primary bump due to the transition from strength scaling to gravity scaling for asteroids larger than ~150 m; the well-known bump at ~50–200 km is a secondary feature resulting from a wavelike structure induced in the size distribution by the ~3–30-km primary bump.

3.3. Boundary Conditions and Collisional Evolution

Even more severe departures from simple power-law distributions result from relaxation of Dohnanyi assumption 3 (no lower cutoff in the size distribution). When a small-size cutoff in the initial distribution is imposed, a remarkable feature appearing in the simulations is that the final distribution is not a fixed-exponent power-law, but rather displays a wavy pattern of variable wavelength and amplitude, superimposed on an "average" power law of slope close to Dohnanyi's canonical value.

The wavy structure is produced as follows. We divide the asteroid mass range into a series of discrete mass bins, so the smallest one in the distribution has particles removed only in two ways: collisions within the bin itself or by its particles hitting larger targets. Thus there is only a small number of bodies removed from this bin. The next larger bin, however, has bodies removed not only by the above processes, but also due to impacts by projectiles from the smallest bin. This leads to an enhanced removal rate from this bin, and its population is more rapidly depleted than is that of the smallest bin. As we proceed to ever larger sizes, the depletion rate increases relative to that of smaller bins due to the ever-increasing number of projectiles per target body. Thus the population is increasingly depleted, and develops a steeper slope than Dohnanyi's equilibrium value, up to the size bin for which the smallest available projectile can shatter its members. Beyond this size, larger target bins no longer have an enhanced depletion rate, but rather have a somewhat reduced removal rate due to the rapid decrease in the number of available projectiles with increasing size. This leads to a flattening of the size distribution and to a slope lower than the equilibrium one — thus a wavelike pattern arises. This pattern propagates itself to larger and larger sizes due to the rise and fall of the relative number of projectiles capable of shattering larger bodies. It is interesting to note that this shift propagates in such a way that the final distribution at sizes between 1 and 100 km is substantially changed when the cutoff is changed. Figure 5, for instance, shows the effect of shifting the cutoff size from 1 cm to 1 mm. Thus, the behavior of very small particles can in principle affect the size distribution of the observable asteroids in a very significant way. Only when averaged over several orders of magnitude in size does the overall size distribution still approximately match the canonical Dohnanyi equilibrium slope.

Is such a small-particle cutoff actually present in the actual asteroid population? The answer to this question is

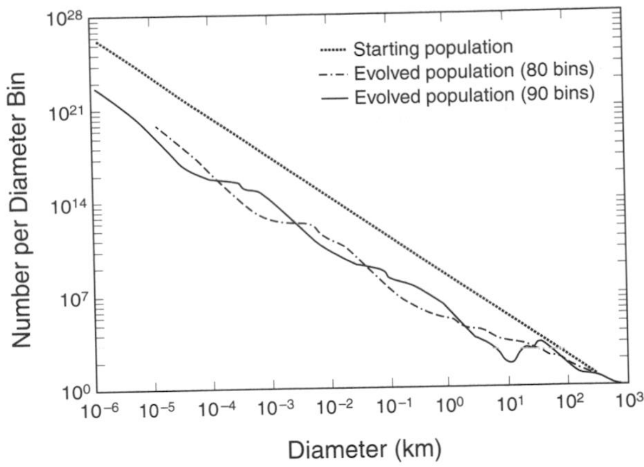

Fig. 5. Propagation of the wave effect with different small-size cutoffs (solid line: 1 cm; dashed line: 1 mm).

not obvious, but many nongravitational forces do indeed act on interplanetary matter and are efficient at different sizes: the solar wind, Poynting-Robertson drag, and the Yarkovsky effect. There is strong observational evidence that most of the micrometeoroid mass is concentrated at particles sizes around 100 μm, and that the corresponding mass distribution flattens considerably for smaller sizes (*Grün et al.,* 1985; *Love and Brownlee,* 1993). Thus the real low-mass cutoff of the asteroid population probably lies at sizes of micrometers or smaller. Solar radiation pressure on particles of size ~1 μm (*Burns et al.,* 1979) converts them into hyperbolic β meteoroids, which, according to *Grün et al.* (1985), provide the main loss mechanism for the meteoritic complex. *Durda* (1993) and *Durda and Dermott* (1997) imposed a small-mass cutoff in their collisional model that empirically matched the *Grün et al.* (1985) data and found it too gradual to induce significant wave structure in the evolved population. Models that attempt to simulate the removal of small particles from the colliding population through radiation forces (*Campo Bagatin et al.,* 1994a) do show a noticeable wave structure only when Poynting Robertson drag is artificially enhanced; the Yarkovsky effect — much more effective than the Poynting-Robertson — might instead be responsible for such a pattern. Clearly, more work is needed in this area to explicitly treat within collisional models the various nongravitational forces that act to remove small objects from the asteroid population and to determine how strong and abrupt a small-mass cutoff must be in order for wavelike features to appear in the evolved size distribution.

3.4. Model Parameters and Collisional Evolution

Besides the scaling laws for Q_D^* or Q_S^*, which affect the disruption/fragmentation threshold and the number of fragments produced in any collision, fragmentation models depend on a number of poorly known critical parameters that govern the mass distribution and the reaccumulation of fragments after their formation (see *Campo Bagatin et al.,* 1994b). Numerical models generally calculate the fragmental size distribution, assumed to be a power law, from Q_S^* and the collisional energy. The inelasticity parameter, f_{KE}, which determines what fraction of the collisional kinetic energy goes into the fragments, is used to calculate the ejecta fraction that escapes the bodies' gravity. Its value depends on the composition, internal structure, and size of thc bodies, and it is often taken between 1% and 10%. On the other hand, f_{KE} is implicitly embedded in Q_D^* in that a large Q_D^* implies a small f_{KE}, while a small Q_D^* goes with a larger f_{KE} (*Campo Bagatin et al.,* 2001).

The fragment mass-speed relation is critical to calculating collisional outcomes and relationship of the form $V(m) = Cm^{-r}$ (*Nakamura et al.,* 1992; *Giblin,* 1998), with r ranging from 0 (no mass-velocity dependence) to 1/6, as found by *Giblin* (1998). The effects of this dependence have been studied by *Petit and Farinella* (1993) and *Campo Bagatin et al.* (1994b). Even a shallow dependence of ejection speed on fragment mass makes a significant difference in the amount of mass that can be reaccumulated by objects in the range 1 to 100 km (*Campo Bagatin et al.,* 2001). Also, the size at which a significant ratio of aggregate rubble-pile asteroids is present seems to depend strongly on the reaccumulation model. For example, assuming a mass-velocity relationship with r = 1/6 provides some 10% rubble-pile asteroids for 2-km-sized bodies, while r = 0 prevents reaccumulated objects for sizes below some 10 km (Fig. 6). It is thus clear that further efforts are necessary to understand the dependence of ejection speeds of fragments on their masses. On the other hand, recent work using hydrocodes to model collisional fragmentation (*Benz and Asphaug,* 1999; *Michel et al.,* 2001; see also section 5) suggests that essentially all asteroids from ~1 to 100 km diameter are reaccumulated rubble piles.

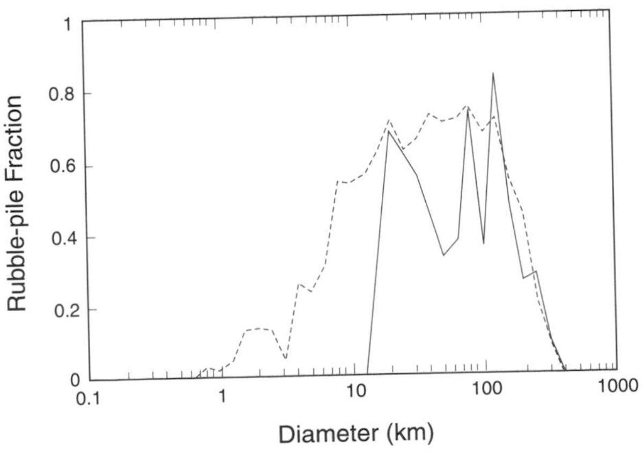

Fig. 6. Fraction of rubble piles in the asteroid belt as a function of size, according to different reaccumulation models (solid line: mass-velocity dependence with r = 1/6; dashed line: no mass-velocity relation, i.e., r = 0), and keeping $f_{KE} = 0.01$ (see text).

Recent models of asteroid collisional evolution consider two different kinds of objects (*Campo Bagatin et al.,* 2001): monoliths (high f_{KE}), and rubble-pile reaccumulated objects (low f_{KE}), usually called rubble-piles, or aggregate asteroids. The number of aggregate asteroids present in the main asteroid belt can be estimated through collisional modeling to range from 30% to 100% in the range 10 to 200 km, depending on the adopted scaling law, the exponent r, and the choice of other collisional parameters, especially the inelasticity parameter, f_{KE}.

A characteristic pattern of collisional evolution outcomes is that the final size distribution of bodies appears to be fairly insensitive to the initial conditions, provided most of the mass is in bodies that can be collisionally disrupted at impact speeds characterizing the population. For any population characterized by an impact speed, V, there is a largest-sized body that can be collisionally disrupted even if hit by an equally large body. As long as most of the population mass is in bodies smaller than this critical size, the final distribution is essentially independent of the starting distribution. Even for the extreme initial distribution without bodies smaller than 1 km, the subkilometer size range is populated by fragments within a few million years, and the stationary distribution is attained.

3.5. Asteroid Observables and Collisional Evolution

The previous three sections tested asteroid collisional evolution models against the observed size distribution of asteroids, but as noted in section 2, there are other collisionally produced observables:

Families. *Marzari et al.* (1999) used the ~85 recognized families and statistically significant grouping recognized in the present belt as a constraint on the overall asteroid collisional history. Their work showed that a small mass initial belt best reproduced the observed number and types of families produced by disruption of parent bodies larger than 100 km diameter.

Vesta's crust. Preservation of the basaltic crust of Vesta requires that the mass of the asteroid belt was at most several times the mass of the present belt at the time that the present dynamical environment was established. This statement assumes a power-law size distribution for the early asteroid population; it would be possible for the early asteroid belt to contain significantly more mass than it does today if such mass were in a few massive bodies that were removed from the belt by dynamical, rather than collisional, processes. Work by *Davis et al.* (1984) and *Davis et al.* (1994) showed the range of initial asteroid populations that were consistent with the preservation of Vesta's crust.

Cosmic-ray exposure ages of meteorites and the Gaspra/Ida cratering record. Reconciling these constraints with the overall asteroid size distribution remains a challenging problem (*O'Brien and Greenberg,* 2001). Collisional models that include Poynting-Robertson drag and the Yarkovsky effect, along with hydrocode-derived scaling laws, offer promise for understanding the asteroid size distribution from the largest bodies down to subdecameter-sized asteroids.

4. COLLISIONAL EVOLUTION OF TROJANS AND HILDAS

4.1. Trojans

Jovian Trojan asteroids have a peculiar dynamical behavior: They librate around the Lagrangian equilibrium points L_4 and L_5 located at the maxima of the "pseudopotential" defined within the frame of the restricted three-body problem (*Greenberg and Davis,* 1978). The resonant link with Jupiter confines them in a limited volume of space and, notwithstanding that they have an average high inclination on the ecliptic, their collisional activity is comparable to that of main-belt asteroids. Collisions for Trojans are a relevant evolutionary mechanism not only because they grind down the primordial size distribution, but also because they significantly alter the parameters of the librational motion. The velocity imparted to fragments in a collision generate new orbital parameters that, for librating bodies, may even drive them out of the resonance. The present Trojan population has almost surely lost memory of the primordial distribution of libration amplitude and proper eccentricity.

The first step to quantitatively model the collisional evolution of Trojans is to derive precise estimates of the typical collisional frequencies and impact velocities for the two swarms. Since the librational motion sets kinematical constraints to the values of the orbital angles respective to those of Jupiter, the standard statistical approaches based on *Wetherill*'s (1967) theory, even in more recent formulations (*Bottke and Greenberg,* 1993; *Vedder,* 1996, 1998), cannot be applied. Critical assumption of these models is that the perihelion argument and longitude of nodes of the bodies precess regularly, while this condition is violated for Trojans. More sophisticated analytical models like that of *Dell'Oro et al.* (1998) or, alternatively, a direct numerical approach (*Marzari et al.,* 1996) have been employed to determine the collisional parameters (Table 1).

Armed with improved values of the impact parameters, *Marzari et al.* (1997) numerically simulated the collisional evolution of Trojan asteroids over the age of the solar system. They found that the slope of a primordial planetesimal-like initial population truncated below 10 km in diameter and with a steep slope (–5.5 incremental) gives a good match to the present population (Fig. 7a). An initial distribution of this kind would have been collisionally very active since it is far from Dohnanyi's equilibrium slope and, as a consequence, we would expect a large number of asteroidal families to be produced over solar system history. Some of these families have been identified, even with a sample of only 174 Trojan orbits by *Milani* (1993, 1994), and by *Beaugé and Roig* (2001), based on a set of 533 Trojans. Another important aspect of the collisional evolution is the large flux of breakup fragments out of the Trojan swarms. As pointed

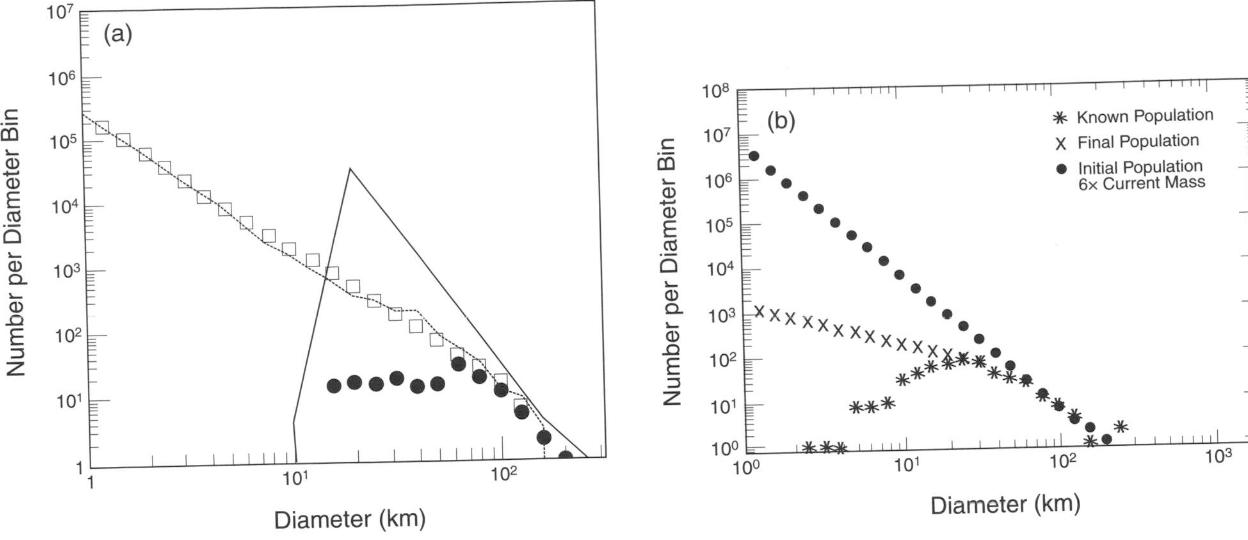

Fig. 7. **(a)** Numerical simulation of the collisional history of Trojans. The continuous line is the starting population, the dotted line the evolved population, the open squares show the inferred population by *Shoemaker et al.* (1989), and the filled circles the known distribution of Trojans in 1997. The initial population is computed as a power law with a –5.5 slope from 250 to 20 km in diameter, then it has a rapid decline to 0 between 20 and 10 km. It is a planetesimal-like initial population where the accretion process produced bodies as large as 250 km from 10- to 20-km-sized primordial planetesimals. After 4.5 b.y. the evolved population reproduces well the present population. **(b)** Numerical simulation of the collisional evolution of Hildas where the collisional process with a planetesimal population scattered from the Uranus-Neptune region was taken into account. The filled circles show one possible initial population with 6× the current mass, the crosses the evolved population, and the stars the known distribution of Hildas in 2000. The initial population is computed as a power law with a –3 slope (incremental) and it is one case among other possibilities.

out by *Marzari et al.* (1995), the collisional disruption of a Trojan asteroid can inject some fragments into unstable orbits that end up in Jupiter-crossing chaotic cometary orbits. A quantitative estimate of the actual flux based on the numerical simulation of *Marzari et al.* (1997) suggests that the Trojan swarms could supply approximately 10% of the short-period comet and Centaurs populations. This dynamical connection between Trojans and short-period comets is reinforced by the spectrophotometric similarity with D-type asteroids, which are dominant among Trojans, and inactive cometary nuclei (*Hartmann et al.,* 1987; *Shoemaker et al.,* 1989).

The primordial Trojan population found by *Marzari et al.* (1997) showed only a possible lower limit to the real primordial population due to neglecting the depleting effect of dynamics on the population. As shown by *Levison et al.* (1997), the stability region populated by present Trojans in the phase space is surrounded by isochrone curves of limited lifetime. The primordial population of Trojans had then two independent sink mechanisms: collisions and long-term dynamical diffusion. Collisions grind down larger bodies to small fragments; dynamical diffusion ejects them out of the swarm into chaotic orbits. However, these mechanisms also work in synergy: Collisions move the bodies in the phase space, possibly refilling regions cleared up by instability, and at the same time, bodies on unstable orbits take part in the collisional evolution over a limited timespan before escaping. Further work is needed to understand the long-term effects on the size and orbital distribution of the Trojan population.

4.2. Hildas

The Hildas are objects that reside in the 3:2 mean-motion resonance with Jupiter at ~4.0 AU. These asteroids are relatively isolated from physical interactions with objects of the main belt, and are weakly coupled to other outer-belt asteroids such as Cybeles or Trojans. This is the reason their collisional activity is not very intense: The intrinsic collision probability for Hilda asteroids is lower than for main-belt Cybele and Trojan asteroids by about a factor of 2.2 and 3, respectively (*Dahlgren,* 1998; *Dell'Oro et al.,* 2001).

The low collisional activity is the result of two effects. First, the 3:2 mean motion resonance with Jupiter has a 0.1-AU-wide dynamically stable zone, surrounded by a strongly chaotic boundary with very short characteristic diffusion times (*Ferraz-Mello et al.,* 1998). Therefore, if any asteroid is extracted from the resonance, it is quickly ejected far away from the stable zone, no longer participating in the collisional evolution of the population. Then, if a fragment produced by a collision reaches a relative velocity enough to escape from the resonance, a depletion of the size distribution occurs (*Gil-Hutton and Brunini,* 2000). Assuming that the relative velocity of a fragment is small with respect to the orbital velocity of the parent body V_c, the semimajor axis change is

$$\Delta a = 2a\frac{\Delta V_T}{V_c}$$

which is a zeroth-order form of Gauss' equation, where a is the semimajor axis and ΔV_T is the component of the relative velocity along the direction of the motion. Using a = 3.96 AU, and Δa ~ 0.05 AU, and assuming that the relative velocity of the fragment with respect to the parent body is equally partitioned between the radial, tangential, and normal components, the ejection velocity needed to escape from the center of the resonance is ΔV > 0.163 km/s. Thus, small fragments produced by collisions closer to the boundary of the resonance or having large relative velocities can easily escape, depleting the Hildas' size distribution and producing a permanent loss of projectiles.

Second, the initial mass for the Hildas was not large. It is possible to obtain a rough value for it using the model of planetary nebulae mass distribution proposed by *Weidenschilling* (1977) to calculate the ratio between the total mass of the main-belt and Hildas region. Another possibility is to assume that the number of objects currently observed is proportional to the initial mass, and find the ratio between the number of objects larger than a certain radius that are observed in both populations. These methods give similar results: The initial mass for the Hildas was ~22× less than the initial mass in the main belt [between 1.5×10^{-3} and 3.6×10^{-5} $M_\oplus$ (*Gil-Hutton and Brunini,* 2000)].

As a consequence of these effects, it is expected that there has been only a modest collisional attrition to the primordial size distribution. Nevertheless, the observed size distribution shows a change in the slope at a radius of ~12 km, which was traditionally explained as an observational bias due to the faintness of small objects in the outer belt. However, it could be the result of a collisional process: *Fernández and Ip* (1983) and *Brunini and Fernández* (1999) found that the accretion process of Uranus and Neptune was extremely inefficient and a large number of bodies was scattered into the inner solar system in a period not longer than few times 10^7 yr (see section 4.3). Using this result, *Gil-Hutton and Brunini* (2000) found that collisions between Hildas and scattered planetesimals from the Uranus-Neptune zone could produce a large number of fragments that could easily escape from the resonance. Thus, the current size distribution of Hildas could be a result of an intense collisional process, which preferentially depleted the small end of the population, leaving the Hildas without enough objects to have further collisional evolution. As a consequence of the fast depletion and low collisional activity after the planetesimal bombardment, it is possible that the population of currently observed Hildas is complete or nearly complete at sizes ≥20 km in diameter (Fig. 7b), and the slope change in the size distribution could be the result of this collisional process (*Gil-Hutton and Brunini,* 2000). This result is contrary to the usual assumption that the outer belt observed populations are far from complete, but it is observationally testable: Any future survey of Hildas asteroids should not modify the currently observed size distribution to a substantial degree.

4.3. Interrelations with Other Populations

The existence of planetesimal populations is an integral part of planetary formation by accretion. Once the protoplanets grew to masses several tenths the current ones, they started to scatter unaccreted bodies well beyond their accretion zones. As a possible byproduct of this process, large numbers of bodies from these populations diffused inward, reaching the inner solar system, and a certain fraction of them have eventually interacted with asteroids in the main belt.

Collisional interactions between the asteroid belt and scattered planetesimals from the Uranus and Neptune zone were first considered by *Brunini and Gil-Hutton* (1998). The possible presence of a nonnegligible amount of H and He in the outer planetary nebula (*Pollack et al.,* 1996) suggests that the formation timescales of Uranus and Neptune could not have been much longer than the timescale of dissipation of the gaseous component of the nebula, perhaps not longer than a few 10^7 yr (*Brunini and Fernández,* 1999). A significant number of scattered planetesimals reached the region of the inner planets (*Fernández and Ip,* 1983), some of which collided with asteroids in the main belt. *Gil-Hutton and Brunini* (1999) studied the collisional process between these populations and found that the planetesimal bombardment was extremely efficient, removing mass of the primordial asteroid belt in a very short time, allowing very massive initial populations. In spite of the fact that the collisional process was very intense (a total mass infall of 15 $M_\oplus$ was scattered to the inner solar system from the Uranus-Neptune zone), it lasted only 1–3 × 10^7 yr, so Vesta had a good chance of retaining its basaltic crust largely intact in this scenario.

The interaction between the asteroid belt and the short-period comet population was considered by *Gil-Hutton* (2000). This author obtained values for the intrinsic collision probabilities and collision velocities of 159 short-period comets (SPC) colliding with the main belt using the algorithm developed by *Bottke et al.* (1994; see Table 1). The intrinsic collision probability is about one-half that of main-belt collisions, but the mean impact speed is significantly higher. However, SPC are not a serious threat for asteroids due to their small population.

5. FUTURE CHALLENGES

Asteroid collision studies made sizeable strides in the past decade, particularly with the increased volume of data on the asteroid size distribution and a better understanding of fragmentation physics. However, there is still a wide discrepancy in scaling-law prediction for the collisional energy needed to disrupt bodies. As numerical models become more sophisticated and computers ever faster, new insights are expected on this fundamental aspect of collisional studies.

Recent results have been presented based on the marriage of smooth particle hydrocode calculations of the fragmentation process with an N-body integrator to follow the trajectories of fragments once the material interactions have ceased (*Michel et al.,* 2001; *Durda et al.,* 2001). *Michel et al.* (2001) have successfully reproduced the observed size distribution for asteroid families with such an approach and, at the same time, have introduced a new paradigm for collisional outcomes. In contrast to the earlier work, which argued that the size distribution from disruptive collisions was controlled by shattering the material bonds of the body, the new model suggests that gravitational reaccumulation is the fundamental mechanism for forming individual bodies during disruptive collisions. Hence, the velocity field established by the collision would determine the number and size of fragments, not the propagation of cracks that fracture material bonds. In the *Michel et al.* (2001) model, the collision is sufficiently energetic to pulverize the material of the body to very small sizes, but many of these fragments gravitationally reaccumulate to form the fragments that make up asteroid families. Further work is certainly needed in this area, but this approach suggests a very different physical basis for understanding the outcomes of disruptive collisions.

Virtually all asteroid collisional studies point to a small mass initial belt, i.e., the total mass of the asteroid belt was only a few times larger than the present belt at the time the present disruptive dynamical environment was established. What is needed now is to link collisional models with accretion models for the formation of asteroids to produce a seamless, self-consistent scenario for the origin and evolution of asteroids. Work by *Wetherill* (1992) and *Chambers and Wetherill* (2001) suggests that massive bodies accreted in the primordial asteroid belt. These bodies acted to gravitationally stir up the population, thus terminating accretion among small bodies. The massive bodies were then removed by mutual gravitational scattering into resonances in the asteroid belt that formed when Jupiter grew to be a massive body. The residual mass from this process is what we see as the asteroid belt today.

Modeling the collisional evolution of asteroids populations is still a very complex problem, due to the fact that a number of poorly known free parameters are embedded in the modeling of the involved physical phenomena. More observational data and more refined modeling techniques are needed in order to get a comprehensive and self-consistent picture of fragmentation and collisional evolution of small bodies populations. Specific challenges in this field for the next decade include (1) achieving a better understanding, and accurate modeling, of the collisional response of rubble-pile asteroids to energetic impacts; (2) developing a numerical model that combines collisional processes and dynamics, e.g., the collisional evolution of asteroids including the Poynting-Robertson and Yarkovsky effects, the major dynamical resonances and, possibly, even three-body resonances; and (3) modeling the collisional evolution of asteroid rotations.

The relatively isolated Hildas and Trojans regions are natural laboratories to compare their collisional evolution with that of main-belt asteroids. As in the case for main-belt asteroids, one needs better knowledge about the size distribution of these populations down to kilometer-sized bodies. Also, combining dynamical depletion mechanism with collisional models will yield deeper insights into the evolution of these populations. The next decade promises to show major progress in deciphering the collisional history of small body populations.

REFERENCES

Anders E. (1965) Fragmentation history of asteroids. *Icarus, 4,* 399–408.

Asphaug E. (1997) Impact origin of the Vesta family. *Meteoritics & Planet. Sci., 32,* 965–980.

Asphaug E. and Scheeres D. J. (1999) Deconstructing Castalia: Evaluating a post-impact state. *Icarus, 139,* 383–386.

Beaugé C. and Roig F. (2001) A semi-analytical model for the motion of the Trojan asteroids: Proper elements and families. *Icarus, 153,* 391–415.

Bell J. F., Davis D. R., Hartmann W. K., and Gaffey M. J. (1989) Asteroids: The big picture. In *Asteroids II* (R. P. Binzel et al., eds.), pp. 921–945. Univ. of Arizona, Tucson.

Belton M. J. S., Veverka J., Thomas P., Helfenstein P., Simonelli D., Chapman C., Davies M. E., Greeley R., Greenberg R., Head J., Murchie S., Klaasen K., Johnson T. V., McEwen A., Morrison D., Neukum G., Fanale F., Anger C., Carr M., and Pilcher C. (1992) Galileo encounter with 951 Gaspra: First pictures of an asteroid. *Science, 257,* 1647–1652.

Belton M. J. S., Chapman C. R., Veverka J., Klaasen K. P., Harch A., Greeley R., Greenberg R., Head J. W. III, McEwen A., Morrison D., Thomas P. C., Davies M. E., Carr M. H., Neukum G., Fanale F. P., Davis D. R., Anger C., Gierasch P. J., Ingersoll A. P., and Pilcher C. B. (1994) First images of asteroid 243 Ida. *Science, 265,* 1543–1547.

Bendjoya Ph. and Zappalà V. (2002) Asteroid family identification. In *Asteroids III* (W. F. Bottke Jr. et al., eds.), this volume. Univ. of Arizona, Tucson.

Benz W. and Asphaug E. (1999) Catastrophic disruptions revisited. *Icarus, 142,* 5–20.

Bottke W. F. and Greenberg R. (1993) Asteroidal collision probabilities. *Geophys. Res. Lett., 20,* 879–881.

Bottke W. F., Nolan M. C., Greenberg R., and Kolvoord R. A. (1994) Velocity distribution among colliding asteroids. *Icarus, 107,* 255–268.

Brunini A. and Fernández J. A. (1999) Numerical simulations of the accretion of Uranus and Neptune. *Planet. Space. Sci., 47,* 591–605.

Brunini A. and Gil-Hutton R. (1998) The cometary bombardment on the primitive asteroid belt. *Planet. Space Sci., 46,* 997–1001.

Burbine T. H., Binzel R. P., Bus S. J., and Sunshine J. M. (1996) Mantle material in the main belt: Battered to bits? *Meteoritics & Planet. Sci., 31,* 607–620.

Burns J. A., Lamy P. L., and Soter S. (1979) Radiation forces on small particles of the solar system. *Icarus, 40,* 1–48.

Campo Bagatin A., Cellino A., Davis D. R., Farinella P., and Paolicchi P. (1994a) Wavy size distributions for collisional systems with a small-size cutoff. *Planet. Space Sci., 42,* 1079–1092.

Campo Bagatin A., Farinella P., and Petit J-M. (1994b) Fragment ejection velocities and the collisional evolution of asteroids. *Planet. Space Sci., 42,* 1099–1107.

Campo Bagatin A., Petit J-M., and Farinella P. (2001) How many rubble piles in the asteroid belt? *Icarus, 149,* 198–209.

Cellino A., Zappalà V., and Farinella P. (1991) The size distribution of main-belt asteroids from IRAS data. *Mon. Not. R. Astron. Soc., 253,* 561–574.

Ceplecha Z. (1992) Influx of interplanetary bodies onto Earth. *Astron. Astrophys., 263,* 361–366.

Chambers J. E. and Wetherill G. W. (2001) Planets in the asteroid belt. *Meteoritics & Planet. Sci., 36,* 381–399.

Chapman C. R. (1985) Implications of the inferred composition of asteroids for their collisional evolution. *J. Ital. Astron. Soc., 57,* 103–112.

Chapman C. R. and Davis D. R. (1975) Asteroid collisional evolution: Evidence for a much larger early population. *Science, 190,* 553–556.

Colwell J. E. (1994) Power-law confusion: You say incremental, I say differential (abstract). In *Lunar and Planetary Science XXIV,* pp. 325–326. Lunar and Planetary Institute, Houston.

Dahlgren M. (1998) A study of Hilda asteroids. III. Collisional velocities and collision frequencies of Hilda asteroids. *Astron. Astrophys., 336,* 1056–1064.

Davis D. R., Chapman C. R., Greenberg R., Weidenschilling S. J., and Harris A. W. (1979) Collisional evolution of asteroids: Populations, rotations, and velocities. In *Asteroids* (T. Gehrels, ed.), pp. 528–537. Univ. of Arizona, Tucson.

Davis D. R., Chapman C. R., Weidenschilling S. J., and Greenberg R. (1984) Asteroid collisional evolution studies (abstract). In *Lunar and Planetary Science XV,* pp. 192–193. Lunar and Planetary Institute, Houston.

Davis D. R., Chapman C. R., Weidenschilling S. J., and Greenberg R. (1985) Collisional history of asteroids: Evidence from Vesta and the Hirayama families. *Icarus, 62,* 30–53.

Davis D. R., Weidenschilling S. J., Farinella P., Paolicchi P., and Binzel R. P. (1989) Asteroid collisional history: Effects on sizes and spins. In *Asteroids II* (R. P. Binzel et al., eds.), pp. 805–826. Univ. of Arizona, Tucson.

Davis D. R., Ryan E. V., and Farinella P. (1994) Asteroid collisional evolution: Results from current scaling algorithms. *Planet. Space. Sci., 42,* 599–610.

Davis D. R., Farinella P., and Marzari F. (1999) The missing Psyche family: Collisionally eroded or never formed? *Icarus, 137,* 140–151.

Dell'Oro A. and Paolicchi P. (1998) Statistical properties of encounters among asteroids: A new, general purpose, formalism. *Icarus, 136,* 328–339.

Dell'Oro A., Marzari F., Paolicchi P., Dotto E., and Vanzani V. (1998) Trojan collision probability: A statistical approach. *Astron. Astrophys., 339,* 272–277.

Dell'Oro A., Marzari F., Paolicchi P., and Vanzani V. (2001) Updated collisional probabilities of minor body populations *Astron. Astrophys., 366,* 1053–1060.

Dohnanyi J. W. (1969) Collisional models of asteroids and their debris. *J. Geophys. Res., 74,* 2531–2554.

Dohnanyi J. W. (1971) Fragmentation and distribution of asteroids. In *Physical Studies of Minor Planets* (T. Gehrels, ed.), pp. 263–295. NASA SP-267, Washington, DC.

Durda D. D. (1993) The collisional evolution of the asteroid belt and its contribution to the zodiacal cloud. Ph.D. thesis, Univ. of Florida, Gainesville. 129 pp.

Durda D. D. and Dermott S. F. (1997) The collisional evolution of the asteroid belt and its contribution to the Zodiacal Cloud. *Icarus, 130,* 140–164.

Durda D. D., Greenberg R., and Jedicke R. (1998) Collisional models and scaling laws: A new interpretation of the shape of the main-belt asteroid size distribution. *Icarus, 135,* 431–440.

Durda D. D., Bottke W. F., Asphaug E., Richardson D. C., Leinhardt Z. M., Merline W. J., and Flynn K. W. (2001) Numerical models of the formation of asteroids satellites (abstract). In *Asteroids 2001: From Piazzi to the 3rd Millennium,* pp. 88–89. Palermo, Italy.

Farinella P. and Davis D. R. (1992) Collision rates and impact velocities in the main asteroid belt. *Icarus, 97,* 111–123.

Farinella P., Paolicchi P., and Zappalà V. (1982) The asteroids as outcomes of catastrophic collisions. *Icarus, 52,* 409–433.

Farinella P., Davis D. R., Paolicchi P., Cellino A., and Zappalà V. (1992a) Asteroid collisional evolution: An integrated model for the evolution of asteroid rotation rates. *Astron. Astrophys., 253,* 604–614.

Farinella P., Davis D. R., Cellino A., and Zappalà V. (1992b) The collisional lifetime of asteroid 951 Gaspra. *Astron. Astrophys., 257,* 329–330.

Farinella P., Davis D. R., and Stern A. (2000) Formation and collisional evolution of the Edgeworth-Kuiper Belt. In *Protostars and Planets IV* (V. Mannings et al., eds.), pp. 1255–1282. Univ. of Arizona, Tucson.

Fernández J. A. and Ip W.-H. (1983) On the time evolution of the cometary influx in the region of the terrestrial planets. *Icarus, 54,* 377–387.

Ferraz-Mello S., Nesvorný D., and Michtchenko T. A. (1998) Chaos, diffusion, escape and permanence of resonant asteroids in gaps and groups. In *Solar System Formation and Evolution* (D. Lazzaro et al., eds.), pp. 65–82. ASP Conference Series, 149.

Gaffey M. J., Bell J. F., Brown R. H., Burbine T. H., Piatek J. L., Reed K. L., and Chaky D. A. (1993) Mineralogical variations within the S-type asteroid class. *Icarus, 106,* 573–602.

Giblin I. (1998) New data on the velocity-mass relation in catastrophic disruption. *Planet. Space. Sci., 46,* 921–928.

Gil-Hutton R. (2000) Collisional rates and impact velocities of short-period comets with interplanetary particles. *Planet. Space Sci., 48,* 545–551.

Gil-Hutton R. and Brunini A. (1999) Collisional evolution of the early asteroid belt. *Planet. Space Sci., 47,* 331–338.

Gil-Hutton R. and Brunini A. (2000) Collisional evolution of the outer asteroid belt. *Icarus, 145,* 382–390.

Greenberg R. (1982) Orbital interactions — A new geometrical formalism. *Astron. J., 87,* 184–195.

Greenberg R. and Davis D. R. (1978) Stability at potential maxima: The L4 and L5 points of the restricted three-body problem. *Am. J. Phys., 36,* 1068–1070.

Grün E., Zook H. A., Fechtig H., and Giese R. H. (1985) Collisional balance of the meteoritic complex. *Icarus, 62,* 244–272.

Hartmann W. K., Tholen D. J., and Cruikshank D. P. (1987) The relationship of active comets, "extinct" comets, and dark asteroids. *Icarus, 69,* 33–50.

Hellyer B. (1970) The fragmentation of the asteroids. *Mon. Not. R. Astron. Soc., 148,* 383–390.

Hellyer B. (1971) The fragmentation of the asteroids — II. Numerical calculations. *Mon. Not. R. Astron. Soc., 154,* 279–291.

Holsapple K. (1993) The scaling of impact processes in planetary sciences. *Annu. Rev. Earth Planet. Sci., 22,* 333–374.

Housen K. and Holsapple K. (1990) On the fragmentation of asteroids and planetary satellites. *Icarus, 84,* 226–253.

Housen K. R. and Holsapple K. A. (1999) Scale effects in strength-dominated collisions of rocky asteroids. *Icarus, 142,* 21–33.

Housen K., Schmidt R. M., and Holsapple K. A. (1991) Laboratory simulations of large-scale fragmentation events. *Icarus, 184,* 180–190.

Ivezić Z. and 31 colleagues (2001) Solar system objects observed in the Sloan Digital Sky Survey commissioned data. *Astron. J., 122,* 2749–2784.

Jedicke R. and Metcalfe T. S. (1998) The orbital and absolute magnitude distributions of main belt asteroids. *Icarus, 131,* 245–260.

Jedicke R., Larsen J., and Spahr T. (2002) Observational selection effects in asteroid surveys and estimates of asteroid population sizes. In *Asteroids III* (W. F. Bottke Jr. et al., eds.), this volume. Univ. of Arizona, Tucson.

Kresák L. (1977) Mass content and mass distribution in the asteroid system. *Bull. Astron. Inst. Czech., 28,* 65–82.

Levison H., Shoemaker E. M., and Shoemaker C. S. (1997) The dispersal of the Trojan asteroid swarm. *Nature, 385,* 42–44.

Love S. G. and Ahrens T. J. (1996) Catastrophic impacts in the gravity regime (abstract). In *Lunar and Planetary Science XXVII,* pp. 777–778. Lunar and Planetary Institute, Houston.

Love S. G. and Ahrens T. J. (1997) Origin of asteroid rotation rates in catastrophic impacts. *Nature, 386,* 154–156.

Love S. G. and Brownlee D. E. (1993) A direct measurement of the terrestrial mass accretion rate of cosmic dust. *Science, 262,* 550–553.

Marzari F., Farinella P., and Vanzani V. (1995) Are Trojan collisional families a source for short-period comets? *Astron. Astrophys., 299,* 267–276.

Marzari F., Scholl H., and Farinella P. (1996) Collision rates and impact velocities in the Trojan asteroid swarms. *Icarus, 19,* 192–201.

Marzari F., Farinella P., Davis D. R., Scholl H., and Campo Bagatin A. (1997) Collisional evolution of Trojan asteroids. *Icarus, 125,* 39–49.

Marzari F., Farinella P., and Davis D. R. (1999) Origin, aging, and death of asteroid families. *Icarus, 142,* 63–77.

Melosh H. J. and Ryan E. V. (1997) Asteroids: Shattered but not dispersed. *Icarus, 129,* 562–564.

Michel P., Benz W., Tanga P., and Richardson D. C. (2001) Collisions and gravitational re-accumulation: A recipe for forming asteroid families and satellite. *Science, 294,* 1696–1700.

Milani A. (1993) The Trojan asteroid belt: Proper elements, stability, chaos and families. *Cel. Mech. Dyn. Astron., 57,* 59–94.

Milani A. (1994) The dynamics of the Trojan asteroids. In *Asteroids, Comets, Meteors 1993* (A. Milani et al., eds.), p. 159. Proc. 160th Inst. Astron. Union, Kluwer, Dordrecht.

Nakamura A., Suguiyma K., and Fujiwara A. (1992) Velocity and spin of fragments from impact disruptions. I. An experimental approach to a general law between mass and velocity. *Icarus, 100,* 127–135.

Namiki J. and Binzel R. (1991) 951 Gaspra: A pre-Galileo estimate of its surface evolution. *Geophys. Res. Lett., 18,* 1155–1158.

O'Brien D. P. and Greenberg R. (1999) Asteroidal collisional evolution: Three contradictory lines of evidence (abstract). In *Asteroids, Comets, Meteors,* 02.07. Cornell, Ithaca.

O'Brien D. P. and Greenberg R. (2001) The collisional evolution of really tiny asteroids: Implications for their size distribution and lifetimes (abstract). In *Asteroids 2001: From Piazzi to the 3rd Millennium,* pp. 101–102. Palermo, Italy.

Öpik E. J. (1951) Collision probabilities with the planets and the distribution of interplanetary matter. *Proc. R. Irish Acad., 54,* 165.

Paolicchi P. (1994) Rushing to equilibrium: A simple model for the collisional evolution of asteroids. *Planet. Space Sci., 42,* 1093–1097.

Petit J-M. and Farinella P. (1993) Modeling the outcomes of high velocity impacts between small solar system bodies. *Cel. Mech., 57,* 1–28.

Piotrowski S. (1953) The collisions of asteroids. *Acta Astron., A5,* 115.

Pollack J. B., Hubickyj O., Bodenheimer P., Lissauer J. J., Podolak M., and Greenzweig Y. (1996) Formation of the giant planets by concurrent accretion of solids and gas. *Icarus, 124,* 62–85.

Pravec P. and Harris A. W. (2000) Fast and slow rotation of asteroids. *Icarus, 148,* 12–20.

Rabinowitz D. L. (1993) The size distribution of the Earth-approaching asteroids. *Astrophys. J., 407,* 412–427.

Richardson D. C., Leinhardt Z. M., Melosh H. J., Bottke W. F. Jr., and Asphaug E. (2002) Gravitational aggregates: Evidence and evolution. In *Asteroids III* (W. F. Bottke Jr. et al., eds.), this volume. Univ. of Arizona, Tucson.

Rivkin A. S., Howell E. S., Lebofsky L. A., Clark B. E., and Britt D. T. (2000) The nature of M-class asteroids from 3-micron observations. *Icarus, 145,* 351–368.

Ryan E. V. (1992) Catastrophic collisions: Laboratory impact experiments, hydrocode simulations, and the scaling problem. Ph.D. thesis, Univ. of Arizona, Tucson. 228 pp.

Shoemaker E. M., Shoemaker C. S., and Wolfe R. F. (1989) Trojan asteroids — Populations, dynamical structure and origin of the L4 and L5 swarms. In *Asteroids II* (R. P. Binzel et al., eds.), pp. 487–523. Univ. of Arizona, Tucson.

Strom R.G., Croft S. K., and Barlow N. G. (1992) The Martian impact cratering record. In *Mars* (H. H. Kieffer et al., eds.), pp. 383–423. Univ. of Arizona, Tucson.

Tanaka H., Inaba S., and Nakazawa K. (1996) Steady-state size distribution for the self-similar collision cascade. *Icarus, 123,* 450–455.

Tedesco E. F., Cellino A., and Zappalà E. (2002) The statistical asteroid model. I: The main-belt population for diameters greater than 1 km. *Astron. J.,* in press.

Van Houten C. J., Van Houten-Groeneveld I., Herget P., and Gehrels T. (1970) The Palomar-Leiden Survey of faint minor planets. *Astron. Astrophys. Suppl., 2,* 339–448.

Vedder J. D. (1996) Asteroid collisions: Estimating their probabilities and velocity distributions. *Icarus, 123,* 436–449.

Vedder J. D. (1998) Main belt asteroid collision probabilities and impact velocities. *Icarus, 131,* 283–290.

Viateau B. (2000) Mass and density of asteroids (16) Psyche and (121) Hermione. *Astron. Astrophys., 354,* 725–731.

Weidenschilling S. J. (1977) The distribution of mass in the planetary system and the solar nebula. *Astrophys. Space Sci., 51,* 152–158.

Wetherill G. W. (1967) Collisions in the asteroid belt. *J. Geophys. Res., 72,* 2429–2444.

Wetherill G. W. (1992) An alternative model for the formation of the asteroids. *Icarus, 100,* 307–325.

Williams D. R. and Wetherill G. W. (1994) Size distribution of collisionally evolved asteroid populations: Analytical solution for self-similar collision cascades. *Icarus, 107,* 117–128.

Yoshikawa M. and Nakamura T. (1994) Near-misses in orbital motion of asteroids. *Icarus, 108,* 298–308.

Zappalà V., Cellino A., Dell'Oro A., and Paolicchi P. (2002) Physical and dynamical properties of asteroid families. In *Asteroids III* (W. F. Bottke Jr. et al., eds.), this volume. Univ. of Arizona, Tucson.

Thermal Evolution Models of Asteroids

Harry Y. McSween Jr.
University of Tennessee

Amitabha Ghosh
University of Tennessee

Robert E. Grimm
Blackhawk Geoservices

Lionel Wilson
Lancaster University

Edward D. Young
University of California at Los Angeles

Thermal evolution models for asteroids that experienced metamorphism (ordinary chondrites), aqueous alteration (carbonaceous chondrites), and melting and differentiation (HED achondrites) are compared. These models, based on decay of ^{26}Al, can be used to study a variety of asteroidal processes such as the insulating effect of regolith, the buffering effect of ice and fluid flow, and the complications arising from redistribution of heat sources during differentiation. Thermal models can also account for an apparent relationship between peak temperature and heliocentric distance of asteroids in the main belt. Thermal evolution models using other heat sources (electromagnetic induction, collisions) are poorly constrained at this point and have been used primarily for simple plausibility calculations.

1. INTRODUCTION

Many asteroids and the meteorites derived from them have been heated, as manifested in metamorphism, aqueous alteration, melting, and differentiation. Almost half a century ago, Harold Urey recognized that decay of long-lived radioactive isotopes (K, U, Th), the primary heating mechanism for planets, was not an effective heat source for asteroids, because the timescale for energy release is long compared to that for conductive heat loss from small bodies. *Urey* (1955) suggested decay of the short-lived radionuclide ^{26}Al and performed a back-of-the-envelope calculation of the heat produced — a precursor to the first asteroid thermal evolution model. During the next several decades, thermal models were used as plausibility tests for various proposed heat sources. More recently, thermal models have been used to describe quantitatively the geologic evolution of asteroids, thereby linking their formation to measurable parameters in meteorites.

The case for ^{26}Al heating of asteroids has become increasingly robust. Live ^{26}Al in the early solar system was widespread (*MacPherson et al.,* 1995; *Huss et al.,* 2001), and its decay product has been found in most classes of chondrites (*Lee et al.,* 1976; *Russell et al.,* 1996; *Kita et al.,* 2000) and several achondrites (*Srinivasan et al.,* 1999; *Nyquist et al.,* 2001). Reasons why evidence for ^{26}Al might be obscured in other achondrites have been given (*LaTourrette and Wasserburg,* 1997; *Ghosh and McSween,* 1998). This heat source appears capable of explaining the full range of temperature excursions of asteroids within the main belt (*Grimm and McSween,* 1993). Although nebular heterogeneity of ^{26}Al has been suggested (*Ireland and Fegley,* 2000), the consistency of ^{26}Al/^{27}Al ratios in calcium-aluminum-rich inclusions (CAIs) and in chondrules, regardless of chondrite class, implies broad nebular homogeneity and indicates that differences in initial ratios reflect formation time (*Huss et al.,* 2001).

A competing hypothesis that asteroids were heated by electromagnetic induction (*Sonnett et al.,* 1968) is based on resistance to flow of electric currents induced by outflows from the young Sun. However, studies of T-Tauri stars have found that solar winds are focused at high latitudes, avoiding the nebular disk where planetesimals form (*Edwards et al.,* 1987), and mass losses, the rates of which govern magnetic fields, have been revised downward significantly (*DeCampli,* 1981). Induction models thus hinge on the choice of reasonable parameters where, as noted by *Wood and Pellas* (1991), most parameters are unconstrained. Nevertheless, several recent thermal models (*Herbert,* 1989; *Shimazu and Terawawa,* 1995) suggest that electromagnetic induction heating could melt asteroids, so in the absence of other information this heat source cannot be ruled out.

Numerous authors (e.g., *Mittlefehldt,* 1979; *Wasson et al.,* 1987; *Rubin,* 1995) have appealed to impact heating to explain metamorphism and melting in meteorite parent bodies. However, this process, by itself, cannot account for global thermal effects in meteorite parent bodies. The global temperature rise from near-disruptive collisions is no more than a few degrees, even for high-porosity asteroids with greater impact strength (*Keil et al.,* 1997). This stems from the fact that collisional energy is proportional to gravitational potential energy, which is negligible in bodies of asteroidal dimensions (*Melosh,* 1990). The high relative abundance of chondrites heated to high temperatures argues that asteroid metamorphism was a global process, unlike the low proportion of metamorphic target rocks in impact craters. However, a correlation between metamorphic grade and shock stage in chondrites (*Rubin,* 1995) may support collisional heating. Although partial melting of phases with low melting points or low shock impedance has been suggested to have produced some achondrites and iron meteorites, shock experiments and studies of impacted materials demonstrate that impact produces either total melts or localized incomplete melts on a microscopic scale that cannot segregate into pools of substantial size (*Keil et al.,* 1997).

The heat transfer equation is the basis for most model calculations (for a detailed discussion, see *Ghosh and McSween,* 1998). Three methods exist for its numerical solution: the classical series solution, the finite difference method, and the finite element method, with the latter being most accurate. By necessity, asteroid thermal models must make assumptions that address uncertainties in initial conditions (e.g, asteroid temperature at the beginning of the simulation), boundary conditions (e.g., nebular ambient temperature, asteroid emissivity), and model parameters (e.g., specific heat capacity, thermal diffusivity, presence of regolith, voids, or ice). Initial temperatures are usually constrained from nebular models (e.g., *Wood and Morfill,* 1988), and many thermal models assume asteroid accretion was instantaneous. Boundary conditions are implemented in two ways: The Dirichlet boundary condition forces the asteroid surface temperature to that of the ambient nebula, and the radiation boundary condition calculates a heat flux depending on temperature difference between the asteroid surface and the nebula. Although the radiation boundary condition is numerically unstable, it is probably more realistic. Model parameters are constrained, to the extent possible, using meteorite and asteroid data (e.g., peak temperatures, cooling rates, closure ages, ^{26}Al contents, asteroid sizes). Published asteroid thermal evolution models are briefly summarized in Table 1.

2. ORDINARY CHONDRITE ASTEROIDS AND THE EFFECT OF A REGOLITH

Construction of thermal models for the parent asteroids of ordinary chondrites (Oc) is relatively straightforward, because heat movement through these asteroids is domi-

TABLE 1. Chronological summary of published asteroid thermal evolution models.

Reference	Model
Urey (1955)	First feasibility calculation of ^{26}Al as an asteroid heat source
Sonnett et al. (1968)	First proposal for electromagnetic induction heating of asteroids
Herndon and Herndon (1977)	Feasibility study of ^{26}Al as an asteroid heat source
Fujii et al. (1979)	Comparison of internal and external heating models for asteroids
Minster and Allegré (1979)	^{26}Al heating model for the H-chondrite parent body
Wood (1979)	Model to reproduce metallographic cooling rates of iron meteorites
Miyamoto et al. (1981)	^{26}Al heating model to constrain sizes of Oc parent bodies using cooling rates, isotopic closure ages, and fall statistics
Yomogida and Matsui (1984)	^{26}Al heating model for small, unsintered asteroids
Grimm (1985)	Model of asteroid metamorphism with fragmentation and reassembly
Grimm and McSween (1989)	^{26}Al heating model of ice-bearing planetesimals, to account for aqueous alteration in Cc
Herbert (1989)	Model of electromagenetic induction heating that causes melting
Haack et al. (1990)	Thermal model of a differentiated asteroid based on decay of long-lived radionuclides
Miyamoto (1991)	^{26}Al heating model to account for aqueous alteration in Cc asteroids
Grimm and McSween (1993)	Explanation of inferred thermal stratification of the asteroid belt based on heliocentric accretion and ^{26}Al heating
Shimazu and Terasawa (1995)	Model of electromagnetic induction heating
Bennett and McSween (1996)	Updated ^{26}Al heating model for Oc asteroids, using revised chronology and thermophysical properties
Akridge et al. (1998)	Model for ^{26}Al heating of Oc asteroid (6 Hebe) with a megaregolith
Ghosh and McSween (1998)	^{26}Al heating model of HED parent body 4 Vesta
Wilson et al. (1999)	Overpressure and explosion resulting from heating Cc asteroids
Young et al. (1999)	^{26}Al heating model of Cc asteroids with fluid flow, to explain O-isotopic fractionations
Cohen and Coker (2000)	Short- and long-lived radionuclide heating model of Cc parent bodies used to study racemization of amino acids
Wilson and Keil (2000)	Thermal effects of magma migration in 4 Vesta
Ghosh et al. (2001)	Effect of incremental accretion on inferred thermal distribution of asteroids in the main belt

nated by conduction (only minor fluids were present and rock fabrics indicate no solid-state convection occurred) and rigorous model constraints are provided by meteorite data. Ordinary chondrite metamorphism occurred at temperatures ranging up to ~1175 K (*McSween et al.,* 1988), i.e., below the melting point for a eutectic mixture of metal and sulfide. Peak temperatures for highly metamorphosed (type 6) chondrites are estimated from geothermometry based on pyroxene compositions (*Olsen and Bunch,* 1984) and on crystallographic ordering in plagioclase (*Nakamura and Motomura,* 1999), and those for the least-metamorphosed (type 3) chondrites are based on thermoluminescence sensitivity (*Sears et al.,* 1980). Meteorite cooling rates are determined from measurements of the temperatures and times at which specific radiogenic isotope systems ceased to equilibrate and fission tracks ceased to anneal. The derived chondrite cooling curves (*Pellas and Storzer,* 1981) show that heating commenced at the time of asteroid accretion (consistent with ^{26}Al decay as the heat source), cooling was rapid (in a small body), and chondrites at higher metamorphic grades cooled more slowly than less-metamorphosed chondrites (implying that the asteroid interior was hotter than the near-surface regions). The thermal structure of such a body resembles an onion, with each successive layer representing a limited interval of temperature corresponding to a particular metamorphic grade.

The thermal model of *Miyamoto et al.* (1981), which incorporated ^{26}Al heating and an extensive set of thermophysical data from Oc, described a 100-m.y.-long thermal evolution of several asteroids with onion-shell stratigraphy. This thermal model was updated (*Bennett and McSween,* 1996) by incorporating refined thermophysical properties of chondrites and a shortened thermal history of 60 m.y. based on Pb-Pb isotope chronology (*Göpel et al.,* 1994). The revised H-chondrite asteroid model (the L-chondrite model is similar) is illustrated in Fig. 1. The initial chondritic $^{26}Al/^{27}Al$ ratio requires an interval of ~2 m.y. between the formation of CAIs (the earliest formed nebular materials) and asteroid accretion, in conformity with constraints on the timing of asteroid formation from radiogenic isotope systematics (*Lugmair and Shukolyukov,* 2001). Higher metamorphic grades in the asteroid interior reach peak temperatures later than low-grade chondrites that were closer to the surface. The bulk of the asteroid is composed of highly metamorphosed type 6 chondrites, with only thin veneers of less-metamorphosed material. A test of this model is that

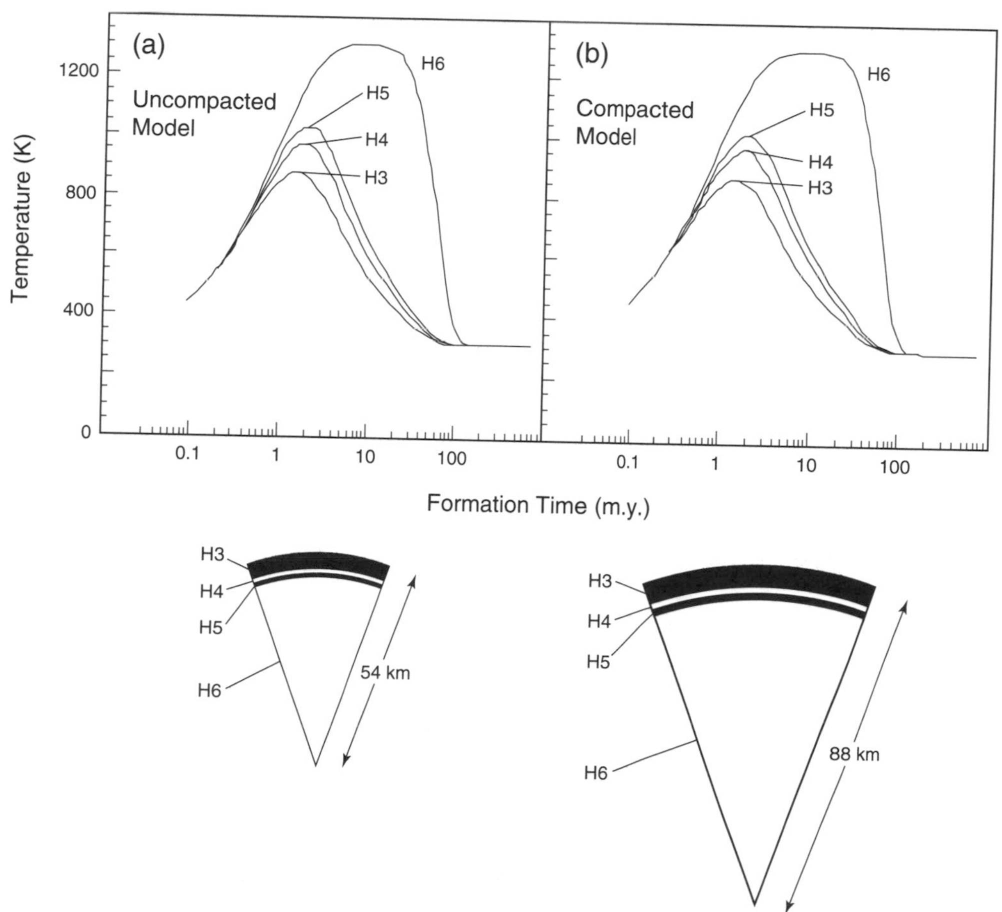

Fig. 1. Time-temperature curves plotted at various depths in the **(a)** uncompacted and **(b)** compacted Oc (H-chondrite) parent bodies of *Bennett and McSween* (1996). Sketches illustrating the corresponding volume proportions of petrologic types are plotted below for each case.

it approximately reproduces the cooling histories of H4, H5, and H6 chondrites (*Bennett and McSween,* 1996). The calculated radius of the H-chondrite parent body (88 km) is similar to the measured radius of asteroid 6 Hebe (~93 km), thought to be the probable source of H chondrites (*Gaffey and Gilbert,* 1998).

Particulate materials have much lower thermal conductivity than consolidated rock, and their effects on thermal models are appreciable. *Wood* (1979) and *Yomogida and Matsui* (1984) considered asteroids to be composed originally of powder that became sintered as temperatures rose during the calculations. *Bennett and McSween* (1996) used measured thermophysical data for high-porosity chondritic breccias to model uncompacted asteroids, and *Akridge et al.* (1998) and *Ghosh and McSween* (2001) modeled asteroids having particulate regoliths of varying thickness. The insulation afforded by even 120 m of regolith (the thickness threshold for insulation has not yet been established) results in a nearly isothermal asteroid interior with a large thermal gradient in the unconsolidated regolith. In effect, this increases the proportion of highly metamorphosed chondrite and moves the metamorphic boundaries (the onion shells) closer to the asteroid surface. Another consequence is that chondritic asteroids must be smaller, to preclude protracted thermal histories and melting. For example, the uncompacted H-chondrite parent body of *Bennett and McSween* (1996) has a radius of only 54 km, relative to the compacted model of 88 km (Fig. 1). Based on their thermal calculations, *Yomogida and Matsui* (1984) even suggested that each metamorphic grade of ordinary chondrite might have been derived from a different, small body. However, H chondrites of different metamorphic grade share the same (8 Ma) cosmic-ray exposure age, implying that they were parts of the same asteroid when launched by impact.

Metallographic cooling rates, determined from measured Ni diffusion profiles in taenite, in some Oc regolith breccias show extreme variations of as much as 1000 K/m.y. (*Williams et al.,* 1999). These cooling rates correspond to burial depths spanning the interval from the asteroid surface to ~100 km (the approximate asteroid radius) and are independent of metamorphic grade. It is inconceivable that an asteroid could survive an impact that sampled its center. The existence of breccias that sample such a depth interval implies that the parent body was disrupted and gravitationally reassembled, producing a rubble-pile structure (*Taylor et al.,* 1987). *Grimm* (1985) reasoned that asteroids shattered during accretion would reaccrete promptly (on the free-fall timescale) and therefore metamorphic grades would be set by initial position within the body but cooling rates would be determined by position following reassembly.

To facilitate calculation, thermal models for Oc parent bodies have generally assumed that asteroid accretion was instantaneous. This approximation can introduce errors, since it ignores the period during which ^{26}Al was most potent as a heat source. *Wood* (1979) and *Yomogida and Matsui* (1984) followed the progressive thermal evolution of small bodies of accreted dust that sintered into rock at a specific temperature. *Ghosh and McSween* (2000) devised a thermal model for the H-chondrite parent body that accreted incrementally, based on a constant growth rate. Peak temperatures in instantaneous accretion models must be reached, by definition, after accretion is complete. However, model runs with long duration of accretion (>2 m.y. from the time accretion starts) can reach peak temperature in the asteroid center while accretion is happening (Fig. 2a).

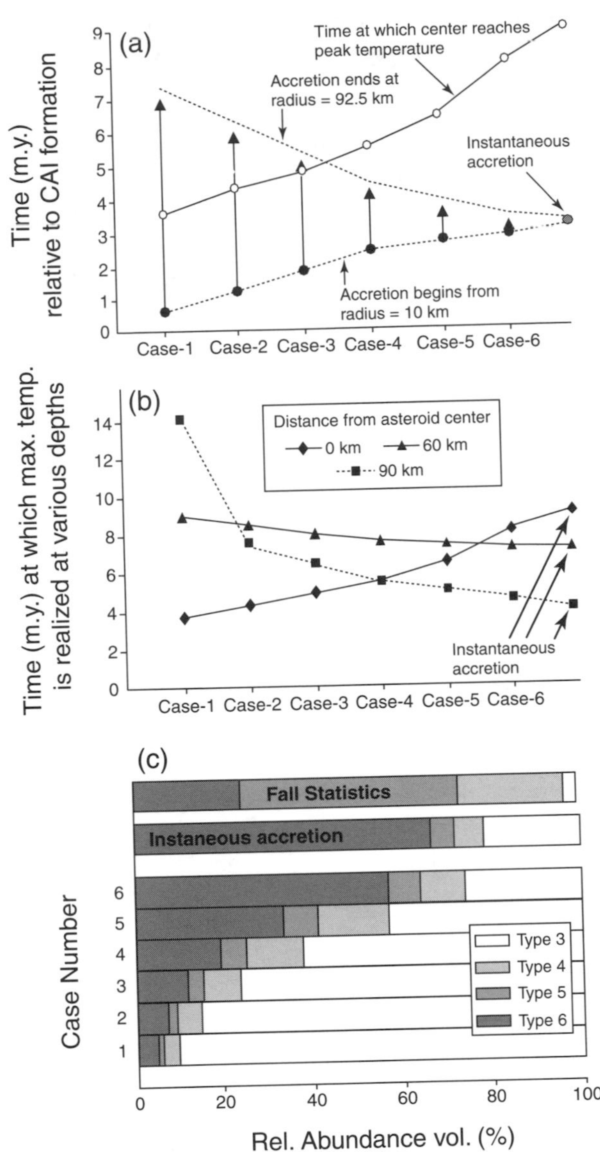

Fig. 2. **(a)** Timelines for the thermal evolution of asteroid 6 Hebe are shown for six cases with different accretion times and durations. Arrows represent periods of asteroid growth. The time (relative to CAI formation) at which peak temperature is attained at the asteroid center is indicated. Note that in cases 1, 2, and 3, peak temperature at the asteroid center is attained before accretion ends. **(b)** Time (relative to CAI formation) at which peak temperature is attained at various distances from the asteroid center. **(c)** Volume proportions of petrologic types obtained in cases 1–6 compared with results for an instantaneous accretion model. After *Ghosh and McSween* (2000).

Instantaneous accretion models also consistently underestimate the time at which the peak temperature is realized, because they fail to account for heating during accretion. The time at which the peak temperature is achieved decreases from the center to the surface of the asteroid in instantaneous accretion models. However, for incremental accretion with long duration, the opposite relationship is observed (Fig. 2b). Finally, the volumetric proportions of metamorphic grades may differ considerably. Instantaneous accretion models overestimate the amount of highly metamorphosed chondrite in the asteroid interior (Fig. 1), but neither incremental or instantaneous accretion models are able to match the observed chondrite fall statistics (Fig. 2c).

3. CARBONACEOUS CHONDRITE ASTEROIDS AND EFFECTS OF WATER ICE AND FLUID FLOW

Aqueous alteration is characteristic of many carbonaceous chondrites (Cc). Alteration produced secondary minerals that either contain water or hydroxyls (phyllosilicates) or formed by precipitation from hydrous fluids (carbonates and sulfates) (*Zolensky et al.,* 1989). Petrographic (*Brearley,* 1997) and kinetic (*Prinn and Fegley,* 1987) arguments support the assumption that melting of H_2O-rich ice incorporated into Cc parent bodies caused the aqueous alteration, although some hydrous alteration has been suggested to have occurred prior to asteroid accretion (*Metzler et al.,* 1992). The presence of free water profoundly influenced the thermal and chemical evolution of Cc parent bodies.

Oxygen-isotopic partitioning in CM and CI chondrites indicates that temperatures within many Cc parent bodies were within ~50° of the melting temperature of water ice during aqueous alteration (*Clayton and Mayeda,* 1984, 1999, *Leshin et al.,* 1997; *Young et al.,* 1999). *Grimm and McSween* (1989) first suggested that the large fusion heat of ice, the high heat capacity of water, and the ability of circulating water to enhance heat loss all may have contributed to thermal buffering of primordial heat sources in Cc parent objects. This fundamental difference in Cc and Oc initial composition led to low-temperature aqueous alteration instead of high-temperature metamorphic recrystallization.

Detailed modeling of hydration reactions — which liberate large amounts of heat — has been more difficult, as the cooling effect of endothermic melting of water ice is insufficient to negate the larger exothermic enthalpies of hydration. Where reaction rates are rapid compared to rates of thermal dissipation, temperatures of hundreds of degrees in excess of the constraints imposed by O-isotopic data would have resulted throughout large portions of Cc parent bodies (*Grimm and McSween,* 1989; *Cohen and Coker,* 2000). Low temperatures associated with aqueous alteration therefore imply either slow hydration-reaction rates or dissipation of heat by mechanisms more efficient than conduction. Reaction times must effectively exceed the conductive cooling time of the body for the former to hold.

Large temperature excursions could have been mitigated by hydrothermal convection. Hydration reactions as fast as 10^4 yr can be hydrothermally buffered with permeabilities comparable to those of fractured crystalline rocks and unconsolidated sands. Such permeabilities are comparable to the upper limit suggested previously by *Grimm and McSween* (1989), but are still far smaller than the maximum permeabilities of basaltic lavas. Hydrothermal convection is likely to have been important for parent bodies larger than several tens of kilometers in diameter. Flowing water in Cc parent bodies is supported by apparent water/rock volume ratios approaching or exceeding unity from oxygen-isotopic data (*Leshin et al.,* 1997; *Clayton and Mayeda,* 1999). The convective model of *Grimm and McSween* (1989) produced uniformly low temperatures and pervasive alteration throughout the asteroid interior, or allowed alteration within a surficial regolith when water was introduced from below.

Young et al. (1999) reinterpreted the O-isotopic data in terms of progress of moving reaction fronts caused by flow of water. They reasoned that the trend of Cc O isotopes upward (toward higher $^{18}O/^{16}O$) along a mass fractionation line could best be explained by progressive partial reequilibration of aqueous fluid as it flowed down a thermal gradient. A monotonic thermal gradient is obtained in the presence of fluid flow by allowing "exhalation" of water under internal gas pressure. Isotopic exchange in both silicates and carbonates is tied to the kinetics of aqueous alteration in the exhalation model, which in turn depends on thermal history. The model successfully explains patterns of variation in Cc O-isotopic ratios and is consistent with the hypothesis that different Cc classes are samples of various horizons within asteroid precursors that had similar geological histories.

A Cc thermal model for a body thought to be too small for convection of water (radius = 9 km) is shown in Fig. 3. The model is based on the approach of *Young et al.* (1999) and uses a chondritic concentration of Al with an initial $^{26}Al/^{27}Al$ of 1×10^{-5} (corresponding to accretion at 1.6 m.y. after CAIs). The results are summarized using two-dimensional time vs. radius plots (the solutions are spherically symmetrical and thus one-dimensional in space). The Cc parent body was considered to be composed initially of forsterite olivine and water ice in these calculations. Forsterite was converted to secondary hydrous minerals (represented by talc) and carbonate minerals (represented by magnesite). Progress of the hydration and carbonation reactions was driven by the amount of CO_2 in the fluid rather than by temperature alone. It is envisaged that CO_2 would have come from oxidation of C within the parent body and/or from the ice itself.

Important features of the thermal evolution of small icy bodies like that in Fig. 3 are the short time span associated with geological evolution (<1 m.y.) and the presence of protracted temperature gradients that permit coexistence of metamorphosed rocks deep in the interior and aqueously altered rocks toward the surface (Fig. 3). The rocks exposed to intensive aqueous alteration are spatially removed from

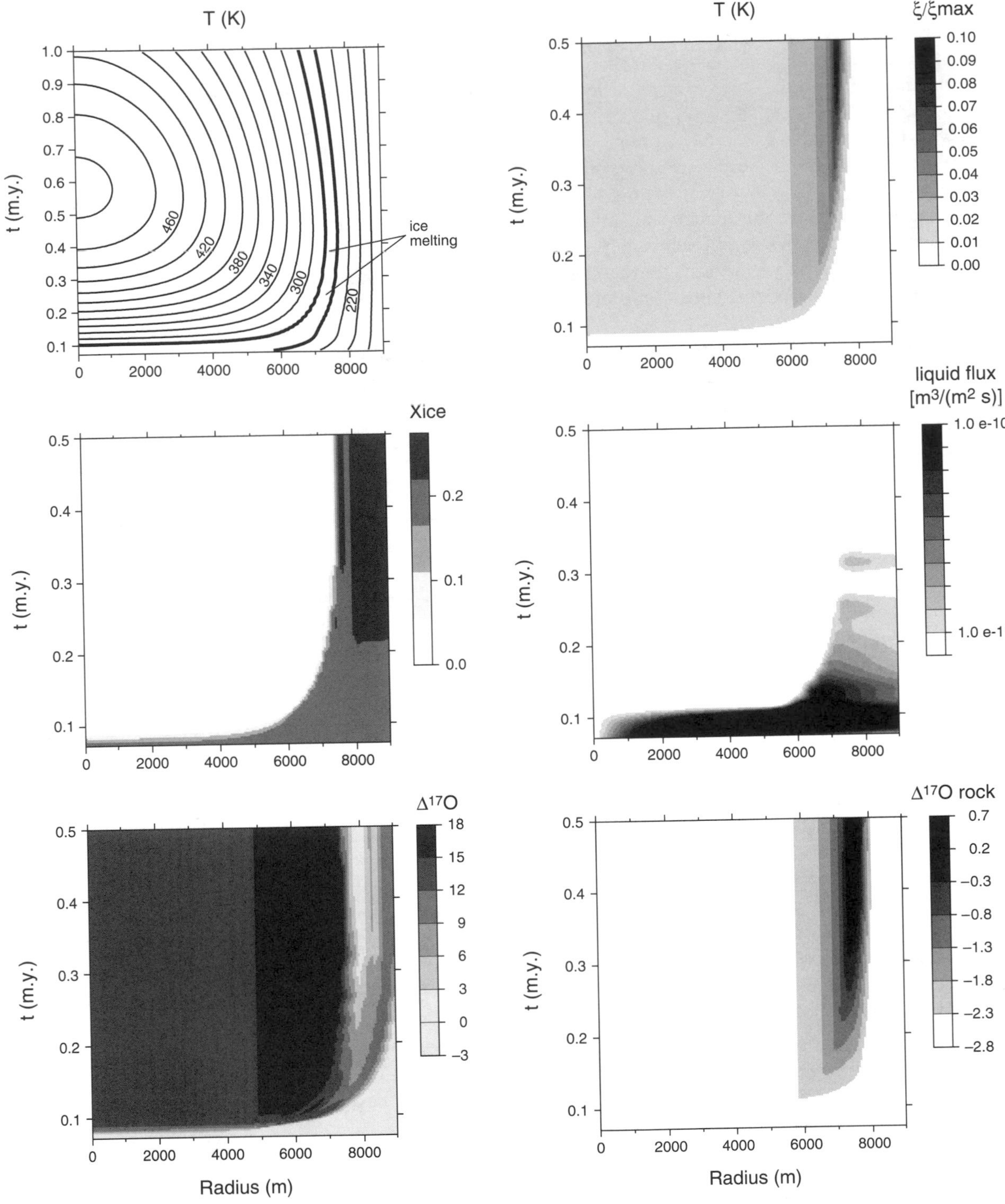

Fig. 3. Plots of time vs. radius for a small Cc parent body (radius = 9 km). Upper left shows the temperature history up to 1 m.y.; all other panels show history up to 0.5 m.y. Upper right panel shows progress of the model hydration and carbonation reaction relative to the maximum progress in mol. units. Middle left shows distribution of water ice in vol. fraction with time and radial position. Middle right shows flux of liquid water as a function of time and position (note the episodic nature of the flux at radial positions beyond ~7 km). Lower left shows changes in $\Delta^{17}O$ in liquid water with time and position. Lower right shows evolution of rock $\Delta^{17}O$ with time and position. Note that the zone of maximum mineralogical alteration coincides with the zone of maximum shift in $\Delta^{17}O$. Initial conditions for the model (*Young et al.,* 1999) included 0.2 vol. fraction water ice, 0.1 vol. fraction empty pore space, surface temperature of 180 K (a simple approximation to radiation to space), bulk ice mol. fraction CO_2 of 0.2, rock $\delta^{18}O$ and $\delta^{17}O$ values of –3.6 and –4.6 respectively, and water $\delta^{18}O$ and $\delta^{17}O$ values of 35.0 and 34.0 respectively (corresponding to a water $\Delta^{17}O$ of 15.8). Ice $\Delta^{17}O$ values substantially greater than those of rock are consistent with other studies (e.g., *Clayton and Mayeda,* 1999).

those subjected to thermal metamorphism (Fig. 3). The suggestion (*Brearley,* 1999) that matrixes of some largely anhydrous Ccs (CVs) could be dehydrated equivalents of intensively altered Ccs (CMs and CIs) may not be consistent with these models.

An analogous model for a large Cc asteroid, e.g., having the size of 1 Ceres, must invoke a smaller initial $^{26}Al/^{27}Al$ of 6.8×10^{-7} (accreting 4.4 m.y. after CAIs) in order to avoid driving peak temperatures well above the maximum recorded in Cc. Important features of large-body models are the long time span prior to aqueous alteration (>5 m.y. after accretion), lack of temperature gradients in the interior where aqueous alteration can occur, and absence of aqueous alteration where temperatures are sufficient for metamorphism. In addition, in the absence of convection, large bodies with heat production sufficient for metamorphism displace and expel water too rapidly for fluid-rock reaction to occur using realistic reaction rates.

High vapor pressures associated with ice melting may also have profoundly affected the geological evolution of Cc asteroids. Consideration of vapor permeabilities appropriate for chondrites and vapor pressures within icy planetesimals suggests that Cc parent bodies may have fractured and vented gases (*Grimm and McSween,* 1989) and could have exploded due to vapor overpressures once water ice began to melt (*Wilson et al.,* 1999; *Cohen and Coker,* 2000). Observations that Cc clasts are common in other meteorite groups (*Zolensky et al.,* 1996) and that highly altered Ccs are brecciated (*Wilson et al.,* 1999) may suggest that explosive disaggregation was an integral part of the evolution of Cc parent bodies.

Although different in fundamental ways, the Cc thermal models of *Young et al.* (1999) and *Cohen and Coker* (2000), as well as the regolith alteration model of *Grimm and McSween* (1989), suggest that low-temperature aqueous alteration was restricted to relatively narrow horizons within the asteroids. The depth of the alteration zone and the timescale for alteration depend upon the size of the body and the rate of heat production. Icy bodies with radii ≤50 km would have experienced aqueous alteration and metamorphism within ~1 m.y. of accretion. Aqueous alteration on much larger bodies would have been delayed by ~5 m.y. or more relative to the time of accretion. The model of *Young et al.* (1999) suggests that a single small body could have produced both metamorphosed and aqueously altered Cc rocks. The same may not be true of larger bodies. In the absence of convection of water, rapid heating of larger bodies to metamorphic temperatures drives water outward with such speed that no aqueous alteration can occur. Recent suggestions that aqueous alteration in Ccs occurred over intervals on the order of 8 m.y. (e.g., *Hutcheon et al.,* 1999) may be consistent with diachronous aqueous alteration within large parent bodies with radii of hundreds of kilometers.

While there has been considerable progress in thermal modeling of Cc parent bodies, there is still no self-consistent model that incorporates reaction heat, isotopic exchange, and fluid flow. Hydrothermally convective interiors are consistent with gross isotopic water/rock ratios, relatively uniform compositions of Cc, and heat loss (*Grimm and McSween,* 1989), but recirculating water may not satisfy isotopic constraints. The "exhalation" model precisely matches the isotopic constraints (*Young et al.,* 1999), but as presently formulated may not produce sufficient alteration, nor is it likely to be able to extract heat without very slow reaction kinetics. Better knowledge of the rates of hydration and carbonation reactions at low temperatures would be useful for judging the relative importance of convection (recirculation) vs. exhalation (single-pass flow) in the evolution of Cc parent bodies.

4. DIFFERENTIATED ASTEROID 4 VESTA AND THE EFFECT OF REDISTRIBUTING HEAT SOURCES

The eucrites and closely related diogenites and howardites (collectively called HED achondrites) are basalts, pyroxenites, and regolith breccias thought to have been extracted from asteroid 4 Vesta (*Consolmagmo and Drake,* 1977; *Binzel and Xu,* 1993; *Farinella et al.,* 1993; *Drake,* 2001). Unlike models of chondrite parent bodies, thermal calculations for achondrite parent bodies require incorporation of complexities introduced by melting and differentiation. *Ghosh and McSween* (1998) modeled the thermal history of Vesta from instantaneous accretion to cooling, using decay of short-lived radionuclides (primarily ^{26}Al, although ^{60}Fe was included) as heat sources. Achondrites and iron meteorites demonstrate that many other differentiated asteroids existed, and a thermal model for a differentiated body based on long-lived radionuclide decay has also been formulated (*Haack et al.,* 1990).

Although Vesta's radius is known (*Thomas et al.,* 1997), the mass of Vesta as determined by its gravitational effect on a nearby asteroid has considerable uncertainity (*Standish and Hellings,* 1989), which introduces a corresponding uncertainty in bulk density. This, in turn, makes it impossible to reliably estimate the size of the core or the asteroid's metal content. *Ghosh and McSween* (1998) preferred H chondrite as the starting composition, which has a metal content similar to Vesta estimates by *Dreibus et al.* (1997). Initial compositions of L and LL chondrites produce slightly higher temperatures for the same parameter set, due to increases in the relative amounts of ^{26}Al. Bulk compositions of H, L, and LL chondrites yield core radii of 123, 108, and 90 km respectively.

Jones (1984) estimated the mantle composition of the HED asteroid based on olivine-melt partition coefficients for Sc, Mg, and Si. He concluded that the undifferentiated mantle could be approximated by a mixture of 25% eucrite and 75% olivine. In the absence of a better model, *Ghosh and McSween* (1998) assumed the crust composition to be eucrite and the depleted mantle composition to be pure olivine. The degree of partial melting of Vesta's mantle was assumed to be 25% based on experimental studies of eucrites (*Stolper,* 1977; *Grove and Bartels,* 1992; *Jurewicz et al.,* 1995). A competing model based primarily on trace-element abundances suggests a much larger degree of melting, producing a magma ocean (*Righter and Drake,* 1997).

The mechanisms that lead to sulfide or silicate melt segregation in asteroids, and thus the formation of cores and crusts, are poorly understood. There exist two schools of thought about the degree of melting required for separation of metal-sulfide liquids from a silicate matrix: one requiring extensive melting (*Stevenson,* 1990; *Taylor,* 1992), and the other limited melting (*Larimer,* 1995). Neither approach takes into account the rate of melt generation. In addition to physical properties of the melt and enclosing rock, the rate of melt migration depends on how fast melting takes place, which in turn depends upon the rate of heat generation by ^{26}Al. When the eutectic temperature of the Fe-FeS system is reached at a particular depth, a melt of eutectic composition is generated. Separation of the metal-sulfide liquid promotes further melting, because the residue has a higher Al content than the melt plus residue. Thus, migration of metal-sulfide liquid results in a positive feedback mechanism: The greater the amount of metal-sulfide melt drained away, the greater will be the melting of the residue, and hence the amount of melt generated will increase. *Ghosh and McSween* (1998) reasoned that if melt migration were somehow triggered, thermal considerations point to rapid core separation.

The timeframe of crust formation on Vesta is difficult to constrain. In regions of the upper mantle where upward movement and decompression of rocks during solid-state convection allow partial melting, melt segregation occurs initially by percolation along grain boundaries. Deformation of the matrix allows melt to be concentrated (*Richter and McKenzie,* 1984; *Barcilon and Lovera,* 1989). However, the region in which melt is concentrated itself rises buoyantly by deforming the surrounding rocks (*Marsh,* 1989). In both cases the timescale is controlled by the viscosity of the matrix, the size of the concentration zone, and the gravitational acceleration (and is therefore slower in an asteroid than on Earth). However, at some stage in the upward segregation process, the rheological response of the surrounding rocks changes from plastic to elastic and a liquid-filled fracture, i.e., a dike, forms (*Sleep,* 1988). The propagation speed of the dike is controlled by the viscosity of the fluid rather than the viscosity of the enclosing rocks, and the melt rise speed is therefore likely to increase by many orders of magnitude. As soon as dikes dominate the process, transfer of melt to shallow depths or to the surface is essentially instantaneous (*Wilson and Keil,* 1996). For convenience in coding, *Ghosh and McSween* (1998) assumed temperature "windows" for both metal-sulfide and silicate melting, and assumed instantaneous formation of core and crust.

Ghosh and McSween (1998) divided the evolution of Vesta into three stages (Fig. 4): (1) radiogenic heating of a homogeneous asteroid until core separation, (2) subsequent heating of the mantle until crust formation, and (3) subsequent heating and cooling of the differentiated asteroid. Two end members, which assumed that either all or no melt erupted, were evaluated since it is not known what proportion of the silicate magma generated at depth eventually erupts. The model places instantaneous accretion of Vesta at 2.9 m.y. after CAI formation. Core formation occurs at 4.6 m.y., and crust formation at 6.6 m.y. The model ages compare favorably with constraints on the timing of core and crust formation from ^{182}Hf-^{182}W (*Lee and Halliday,* 1997), ^{26}Al-^{26}Mg (*Srinivasan et al.,* 1999), and ^{53}Mn-^{53}Cr (*Lugmair and Shukolyukov,* 2001) isotope systematics in HED meteorites. This model illustrates the thermal effect of redistributing ^{26}Al during differentiation. After core formation, the core contains no ^{26}Al and its abundance of ^{60}Fe is too low (*Shukolyukov and Lugmair,* 1996) to contribute significant heat. Thus, the heat engine in the core is shut off, whereas the temperature in the overlying mantle increases (Fig. 4b). This gives rise to a reverse thermal gradient where temperature decreases with increasing depth. In terms of cooling history, this means that not only is heat loss from the core inhibited, but some heat in fact flows into the core by thermal diffusion from the overlying mantle. This reverse gradient persists for ~100 m.y., and is responsible for minimizing heat loss from Vesta's interior during this time interval. Interestingly, this phenomenon is not observed in planets, where core formation takes place long after ^{26}Al decay, but should be observed in small planetesimals that underwent metal-sulfide melting and segregation at a time when ^{26}Al was still potent. A similar reversed thermal gradient is observed in one model end member (Fig. 4c) after ^{26}Al is sequestered in the crust, causing the crust to attain higher temperatures than the underlying mantle.

This study may provide answers to several longstanding problems with the hypothesis of heating by ^{26}Al. The rarity of excess ^{26}Mg, the decay product of ^{26}Al, in eucrites can be explained because the timing of volcanism is such that the ^{26}Al concentration would commonly fall below detectable limits. Excess ^{26}Mg has since been detected in several eucrites (*Srinivasan et al.,* 1999; *Nyquist et al.,* 2001). Chronologic data suggest a time interval of ~100 m.y. between the formation of noncumulate and cumulate eucrites (*Tera et al.,* 1997). Since ^{26}Al is not potent beyond a few million years after the solar system formed, the long time interval was thought to be problematic (*Wood and Pellas,* 1991). A combination of factors — the reverse thermal gradients in the core and crust after metal segregation and crust formation, respectively, and the low thermal diffusivity — produced a prolonged cooling history for Vesta. Figures 4c,d show that temperatures in the mantle stay hot enough after 100 m.y. to prevent geochemical closure in cumulate eucrites.

Ghosh and McSween (1998) suggested the possibility that chondritic percursor rocks, present in the outer layer of Vesta before development of a crust, may still exist. The radiation boundary condition ensures that the temperature in near-surface layers remains low. The thickness of the unaltered carapace decreases with increasing degrees of

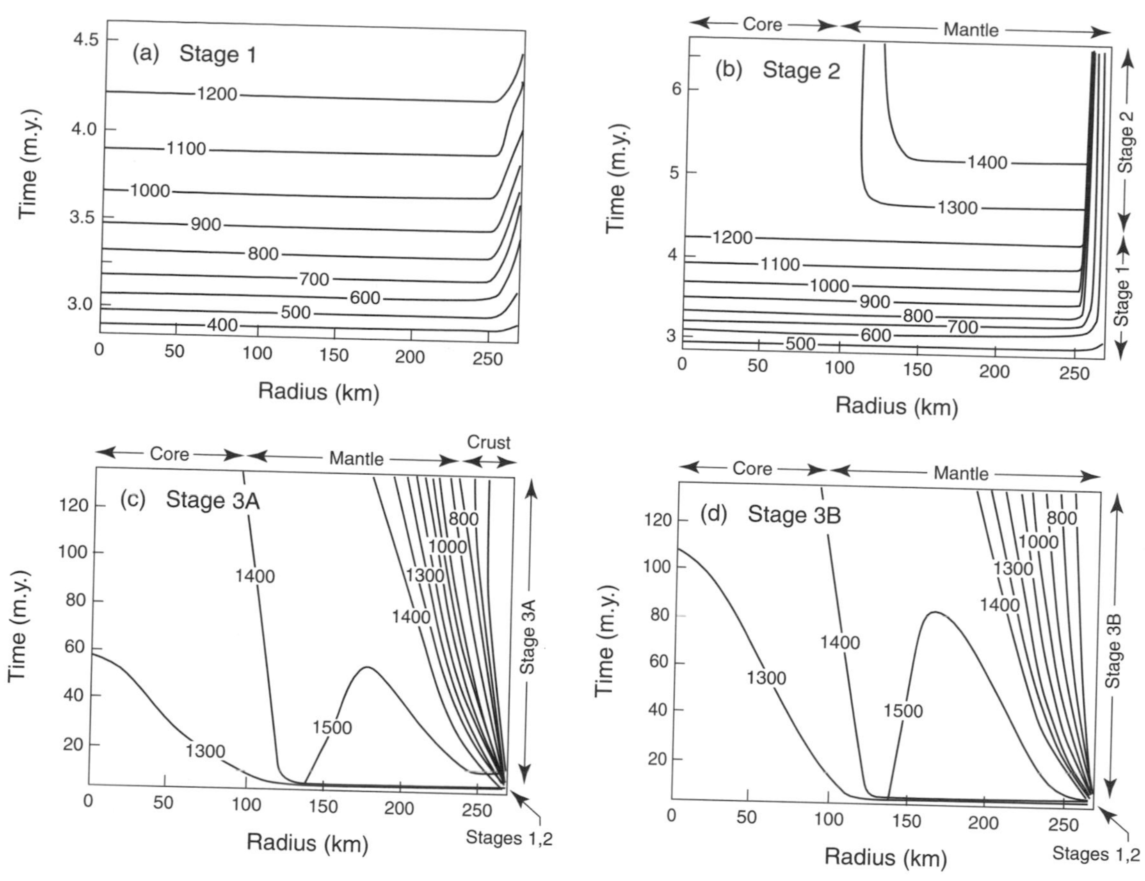

Fig. 4. Temperature contours for 4 Vesta, on plots of time elapsed since CAI formation and radial distance from the asteroid center, after *Ghosh and McSween* (1998). **(a)** Stage 1 is the interval from accretion to core separation. **(b)** In stage 2, core formation has redistributed ^{26}Al, causing heat generation in the core to stop. Mantle temperatures continue to rise, causing silicate melting for production of the crust. Comparison of stages 3A and 3B illustrates the difference in heat transfer between a configuration **(c)** where the entire melt generated is extruded onto the surface and **(d)** where the melt entirely solidifies as plutons.

melting and, for 25% partial melting, the outer 10 km of the asteroid never achieves melting temperatures, although parts of the layer are metamorphosed. However, eruptions of silicate melt, or intrusions of dikes or sills at shallow depth, must cause local metamorphism (*Yamaguchi et al.,* 1997; *Wilson and Keil,* 2000). Further work is needed to establish whether all the unmelted carapace will be destroyed by igneous crust formation or by increased melting in the mantle as in the magma ocean scenario (*Righter and Drake,* 1997). As in chondrite parent bodies, small impacts are not capable of widespread melting (*Melosh,* 1990). Large impacts can cause some melting, but the effect is restricted to the hemisphere that is impacted, leaving the other hemisphere unaltered or at most slightly metamorphosed (*Williams and Wetherill,* 1993).

5. THERMAL STRUCTURE OF THE ASTEROID BELT

The heliocentric distribution of asteroid spectral types (*Gradie and Tedesco,* 1982) has been interpreted to indicate high peak temperatures appropriate for melting or metamorphism for bodies closer to the Sun, with mildly heated or unaltered bodies at greater distances (*Bell et al.,* 1989). This pattern persists, despite some subsequent dynamical stirring of asteroid orbits and ejection of bodies from the main belt. *Grimm and McSween* (1993) devised a quantitative model to explain this radial thermal structure. Because accretion time increases with heliocentric distance (*Wetherill,* 1980), objects that accreted at greater distances had smaller proportions of live ^{26}Al available to drive heating.

The results, expressed as contours of peak temperature on a plot of asteroid size vs. semimajor axis (the latter is equivalent to accretion time relative to CAI formation), are shown in Fig. 5. In this diagram, bodies inward of 2.7 AU are anhydrous (90% rock, 10% voids), whereas those farther from the Sun contain ice (60% rock, 30% ice,10% voids). The vertical bar at 2.7 AU marks the approximate distance for the transition from melted or metamorphosed asteroids to those that experienced aqueous alteration, and the bar at 3.4 AU denotes the transition to unaltered asteroids in which ice was never melted. The accretion times at the top of Fig. 5 produce appropriate peak temperature

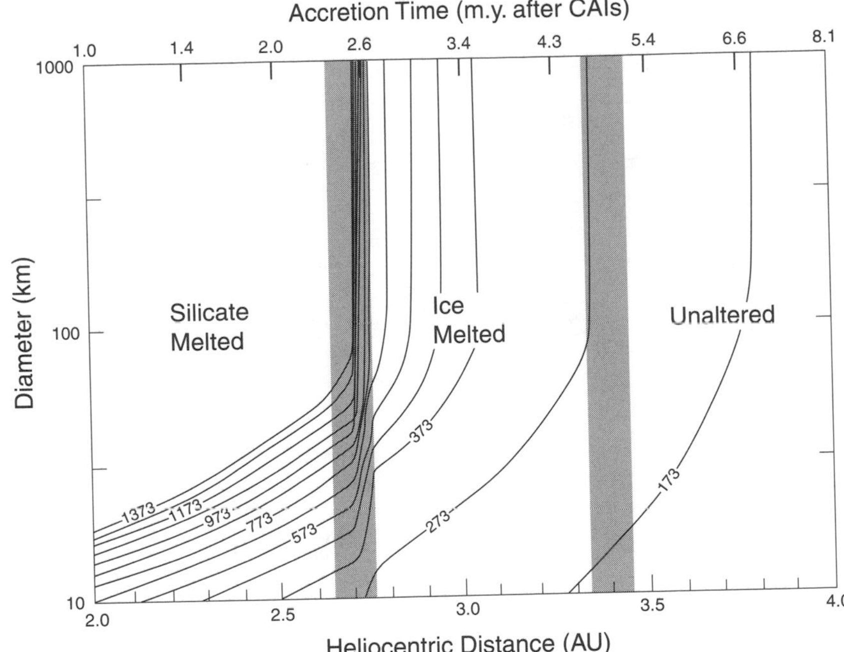

Fig. 5. Contours of peak temperature in asteroids as functions of size and semimajor axis (or accretion time, relative to CAI formation). Shaded bands mark major divisions in the asteroid belt based on interpretation of spectra. Modified from *Grimm and McSween* (1993).

contours (1375 K for silicate melting, 273 K for ice melting) for ~100-km-diameter bodies at these heliocentric distances. The accretion times in Fig. 5 have been slightly increased from that published by *Grimm and McSween* (1993), to correct a coding error in the fusion heat of water.

Ghosh et al. (2001) formulated a more complex model that incorporates incremental rather than instantaneous accretion. The multizone accretion model (*Weidenschilling et al.,* 1997) allows accretion to begin simultaneously (as 0.5-km planetesimals) throughout the belt, but growth rates still vary with swarm density and semimajor axis. Although accretion in the inner asteroid belt is faster than in the outer belt, the difference in accretion rates by itself is not sufficient to produce thermal stratification in a model of ^{26}Al heating. The buffering effect of ice in the outer belt lowers peak temperatures for bodies in this region. Other factors that may contribute to the thermal stratification are differences in accretion temperature between the inner and outer belt, and the accretion of planetesimals that are unsintered and hence capable of achieving higher temperatures for smaller asteroid sizes. Bodies that are too small to sustain metamorphic temperatures comprise most of the mass of the multizone accretion code. Thus, unmetamorphosed small bodies dominate the inner belt. The thermal distribution can be made to conform approximately with the observed distribution of asteroids if these small bodies are destroyed by mutual collisions (*Davis et al.,* 1989).

These calculations demonstrate that heliocentric thermal zoning of the asteroid belt can be achieved by ^{26}Al heating with realistic accretion scenarios. This Sun-centered pattern might also be consistent with solar electromagnetic induction heating, but that mechanism is not sufficiently constrained to allow a similar computation. Neither heating mechanism, by itself, provides an obvious explanation for why Vesta is differentiated while Ceres, at double its size, is not. The thermal histories of individual asteroids must reflect complex interactions between their sizes, accretional timescales, physical states, and chemical compositions.

6. CONCLUSIONS AND FUTURE WORK

Thermal evolution models using ^{26}Al as a heat source have been used to address a spectrum of problems, including metamorphism of Oc parent bodies, aqueous alteration of Cc parent bodies, and melting of differentiated asteroids. Models based on ^{26}Al heating and either instantaneous accretion varying with heliocentric distance or stochastic, incremental accretion appear to be broadly consistent with the thermal stratification of the asteroid belt inferred from the taxonomic distribution of asteroids.

However, ^{26}Al heating requires a longer time interval (~2 m.y.) for accretion to match asteroid peak temperatures than is allowed by most nebular accretion models. This may imply that metamorphism and melting occurred in smaller bodies than currently envisioned, or that ^{26}Al was heterogeneously distributed so that its overall abundance was less than the canonical value. Although electrical induction heating of asteroids is plausible, the hypothesis is difficult to test quantitatively because it hinges on the choice of parameters that are largely unconstrained. Collisional heating appears to be insufficient to account for global thermal metamorphism or significant partial melting in bodies of asteroidal size.

The most straightfoward asteroid thermal models are for metamorphosed Oc parent bodies. An added complexity in these models is the presence of a regolith during heating, which effectively insulates the asteroid interior and profoundly affects its thermal evolution. Carbonaceous chondrite parent bodies originally contained ice, the melting of which acts as a thermal buffer to limit temperature excursions. Fluid flow was also apparently important in controlling heat loss, but still must be fully reconciled with stable-isotopic data. Thermal models for asteroids that experience partial melting and differentiation are more complex, because heat sources migrate within the body during the simulation. These models also have many unconstrained parameters, including the extent of melting and the depth range of melt emplacement.

Many parameters in existing thermal models need revision or refinement. Better theoretical estimates of the initial temperatures of originally accreted materials, as well as a way to anchor timescales in nebular models to CAI formation, are required. Improved constraints from meteorites are also needed. For example, additional measurements of specific heat capacity and diffusivity, as well as accurate peak temperatures from geothermometry and more precise ages from high-resolution chronometers, would improve chondrite thermal models. Overprinted shock effects must be disentangled from meteorite cooling rates. Spacecraft missions to asteroids will hopefully provide data on regolith thicknesses, thermal properties, and ages. Thermal models for differentiated bodies require better constraints on the relative timing of core separation and mantle melting, and existing models do not yet adequately account for heat loss by convection. Also, it is critical to tie whole-asteroid thermal models to magma migration models.

The essence of thermal evolution models is knowing what can be simplified without sacrificing accuracy. The most common simplifying assumption in existing models is that accretion happened instantaneously. However, preliminary attempts to account for the heat budget during asteroid growth show that the rate of accretion can profoundly affect thermal evolution. Incorporation of realistic, incremental accretion scenarios for both chondritic and achondritic asteroids would be a major step forward in thermal modeling.

REFERENCES

Akridge G., Benoit P. H., and Sears D. W. G. (1998) Regolith and megaregolith formation of H-chondrites: Thermal constraints on the parent body. *Icarus, 132,* 185–195.

Barcilon V. and Lovera O. (1989) Solitary waves in magma dynamics. *J. Fluid Mech., 204,* 121–133.

Bell J. F., Davis D. R., Hartmann W. K., and Gaffey M. J. (1989) Asteroids: The big picture. In *Asteroids II* (R. P. Binzel, T. Gehrels, and M. S. Matthews, eds.), pp. 921–945. Univ. of Arizona, Tucson.

Bennett M. E. and McSween H. Y. Jr. (1996) Revised model calculations for the thermal histories of ordinary chondrite parent bodies. *Meteoritics & Planet. Sci., 31,* 783–792.

Binzel R. P. and Xu S. (1993) Chips off of asteroid 4 Vesta: Evidence for the parent body of basaltic achondrite meteorites. *Science, 260,* 186–191.

Brearley A. J. (1997) Disordered biopyriboles, amphibole, and talc in the Allende meteorite: Products of nebular or parent body aqueous alteration? *Science, 276,* 1103–1105.

Brearley A. J. (1999) Origin of graphitic carbon and pentlandite in matrix olivines in the Allende meteorite. *Science, 285,* 1380–1381.

Clayton R. N. and Mayeda T. K. (1984) The oxygen isotope record in Murchison and other carbonaceous chondrites. *Earth Planet. Sci. Lett., 67,* 151–161.

Clayton R. N. and Mayeda T. K. (1999) Oxygen isotope studies of carbonaceous chondrites. *Geochim. Cosmochim. Acta, 63,* 2089–2104.

Cohen B. A. and Coker R. F. (2000) Modeling of liquid water on CM meteorite parent bodies and implications for amino acid racemization. *Icarus, 145,* 369–381.

Consolmagmo G. and Drake M. J. (1977) Compositional evolution of the eucrite parent body: Evidence from rare earth elements. *Geochim. Cosmochim. Acta, 41,* 1271–1282.

Davis D. R., Weidenschilling S. J., Stuart J., Farinella P., Paolicchi P., and Binzel R. P. (1989) Asteroid collisional history — Effects on sizes and spins. In *Asteroids II* (R. P. Binzel, T. Gehrels, and M. S. Matthews, eds.), pp. 805–826. Univ. of Arizona, Tucson.

DeCampli W. M. (1981) T Tauri winds. *Astrophys. J., 244,* 124–146.

Drake M. J. (2001) The eucrite/Vesta story. *Meteoritics & Planet. Sci., 36,* 501–513.

Dreibus G., Brückner J., and Wänke H. (1997) On the core mass of asteroid 4 Vesta. *Meteoritics & Planet. Sci., 32,* A36.

Edwards S., Cabrit D., Strom S. E., Heyer I., Strom K. M., and Anderson E. (1987) Forbidden line and Hα profiles in T Tauri star spectra: A probe of anisotropic mass outflows and circumstellar disks. *Astrophys. J., 321,* 473–495.

Farinella P., Gonczi R., Froeschlé Ch., and Froeschlé C. (1993) The injection of asteroid fragments into resonances. *Icarus, 101,* 174–187.

Fujii N., Miyamoto M., and Ito K. (1979) The role of external heating and thermal metamorphism of chondritic parent body. *Planet. Sci., 1,* 84.

Gaffey M. J. and Gilbert S. L. (1998) Asteroid 6 Hebe: The probable parent body of the H-type ordinary chondrites and the IIE iron meteorites. *Meteoritics & Planet. Sci., 33,* 1281–1295.

Ghosh A. and McSween H. Y. Jr. (1998) A thermal model for the differentiation of asteroid 4 Vesta, based on radiogenic heating. *Icarus, 134,* 187–206.

Ghosh A. and McSween H. Y. Jr. (2000) The effect of incremental accretion on the thermal modeling of asteroid 6 Hebe (abstract). *Meteoritics & Planet. Sci., 35,* A59.

Ghosh A., Weidenschilling S. J., and McSween H. Y. Jr. (2001) Thermal consequences of the multizone accretion code on the structure of the asteroid belt (abstract). In *Lunar and Planetary Science XXXII,* abstract #1760. Lunar and Planetary Institute, Houston (CD-ROM).

Göpel C., Manhes G., and Allegré C. J. (1994) U-Pb systematics of phosphates from equilibrated ordinary chondrites. *Earth Planet. Sci. Lett., 121,* 153–171.

Gradie J. C. and Tedesco E. F. (1982) Compositional structure of the asteroid belt. *Science, 216,* 1405–1407.

Grimm R. E. (1985) Penecontemporaneous metamorphism, fragmentation, and reassembly of ordinary chondrite parent bodies. *J. Geophys. Res., 90,* 2022–2028.

Grimm R. E. and McSween H. Y. Jr. (1989) Water and the thermal evolution of carbonaceous chondrite parent bodies. *Icarus, 82,* 244–280.

Grimm R. E. and McSween H. Y. Jr. (1993) Heliocentric zoning of the asteroid belt by alumimum-26 heating. *Science, 259,* 653–655.

Grove T. L. and Bartels K. S. (1992) The relation between diogenite cumulates and eucrite magmas. *Proc. Lunar Planet. Sci., Vol. 22,* pp. 437–445.

Haack H., Rasmussen K. L., and Warren P. H. (1990) Effects of regolith/megaregolith insulation on the cooling histories of differentiated asteroids. *J. Geophys. Res., 95,* 5111–5124.

Herbert F. (1989) Primoridal electrical induction heating of asteroids. *Icarus, 78,* 402–410.

Herndon J. M. and Herndon M. A. (1977) Aluminum-26 as a planetoid heat source in the early solar system. *Meteoritics, 12,* 459–465.

Huss G. R., MacPherson G. J., Wasserburg G. J., Russell S. S., and Srinivasan G. (2001) Aluminum-26 in calcium-aluminum-rich inclusions and chondrules from unequilibrated ordinary chondrites. *Meteoritics & Planet. Sci., 36,* 975–997.

Hutcheon I. D., Weisberg M. K., Phinney D. L., Zolensky M. E., Prinz M., and Ivanov A. V. (1999) Radiogenic ^{53}Cr in Kaidun carbonates: Evidence for very early aqueous activity (abstract). In *Lunar and Planetary Science XXX,* abstract #1722. Lunar and Planetary Institute, Houston (CD-ROM).

Ireland T. R. and Fegley B. Jr. (2000) The solar system's earliest chemistry: Systematics of refractory inclusions. *Intl. Geol. Rev., 42,* 865–894.

Jones J. H. (1984) The composition of the mantle of the eucrite parent body and the origin of eucrites. *Geochim. Cosmochim. Acta, 48,* 641–648.

Jurewicz A. J. G., Mittlefehldt D. W., and Jones J. H. (1995) Experimental partial melting of the St. Severin (LL) and Lost City (H) chondrites. *Geochim. Cosmochim. Acta, 59,* 391–408.

Keil K., Stöffler D., Love S. G., and Scott E. R. D. (1997) Constraints on the role of impact heating and melting in asteroids. *Meteoritics & Planet. Sci., 32,* 349–363.

Kita N. T., Nagahara H., Togashi S., and Morishita Y. (2000) A short duration of chondrule formation in the solar nebula: Evidence from ^{26}Al in Semarkona ferromagnesian chondrules. *Geochim. Cosmochim. Acta, 64,* 3913–3922.

Larimer J. W. (1995) Core formation in asteroid-sized bodies. *Meteoritics, 30,* 552.

LaTourette T. and Wasserburg G. J. (1997) Mg diffusion in anorthite: Implications for the formation of early solar system planetesimals. *Earth Planet. Sci. Lett., 158,* 91–108.

Lee D.-C. and Halliday A. N. (1997) Core formation on Mars and differentiated asteroids. *Nature, 388,* 854–857.

Lee T., Papanastassiou D. A., and Wasserburg G. J. (1976) Demonstration of ^{26}Mg excess in Allende and evidence for ^{26}Al. *Geophys. Res. Lett., 3,* 41–44.

Leshin L. A., Rubin A. E., and McKeegan K. D. (1997) The oxygen isotopic composition of olivine and pyroxene from CI chondrites. *Geochim. Cosmochim. Acta, 61,* 835–845.

Lugmair G. W. and Shukolyukov A. (2001) Early solar system events and timescales. *Meteoritics & Planet. Sci., 36,* 1017–1026.

MacPherson G. J., Davis A. M., and Zinner E. K. (1995) The distribution of aluminum-26 in the early solar system — a reappraisal. *Meteoritics & Planet. Sci., 30,* 365–386.

Marsh B. D. (1989) Magma chambers. *Annu. Rev. Earth Planet. Sci.,17,* 439–474.

McSween H. Y. Jr., Sears D. W. G., and Dodd R. T. (1988) Thermal metamorphism. In *Meteorites and the Early Solar System* (J. F. Kerridge and M. S. Matthews, eds.), pp. 102–113. Univ. of Arizona, Tucson.

Melosh H. J. (1990) Giant impacts and the thermal state of the early Earth. In *Origin of the Earth* (H. E. Newsom and J. H. Jones, eds.), pp. 69–84. Oxford Univ., New York.

Metzler K., Bishoff A., and Stöffler D. (1992) Accretionary dust mantles in CM chondrites: Evidence for solar nebula processes. *Geochim. Cosmochim. Acta, 56,* 2873–2897.

Minster J. F. and Allegré C. J. (1979) ^{87}Rb-^{87}Sr chronology of H chondrites: Constraint and speculations on the early evolution of their parent body. *Earth Planet. Sci. Lett., 42,* 333–347.

Mittlefehldt D. W. (1979) The nature of asteroid differentiation processes: Implications for primordial heat sources. *Proc. Lunar Planet. Sci. Conf. 10th,* pp. 1975–1993.

Miyamoto M. (1991) Thermal metamorphism of CI and CM carbonaceous chondrites: A internal heating model. *Meteoritics, 26,* 111–115.

Miyamoto M., Fujii N., and Takeda H. (1981) Ordinary chondrite parent body: An internal heating model. *Proc. Lunar Planet. Sci. 12B,* pp. 1145–1152.

Nakamuta Y. and Motomura Y. (1999) Sodic plagioclase thermometry of type 6 ordinary chondrites: Implications for the thermal histories of parent bodies. *Meteoritics & Planet. Sci., 34,* 763–772.

Nyquist L. E., Reese Y., Wiesmann H., Shih C.-Y., and Takeda H. (2001) Live ^{53}Mn and ^{26}Al in an unique cumulate eucrite with very calcic feldspar (An-98). *Meteoritics & Planet. Sci., 36,* A151–A152.

Olsen E. J. and Bunch T. E. (1984) Equilibration temperatures of the ordinary chondrites: A new evaluation. *Geochim. Cosmochim. Acta, 48,* 1363–1365.

Pellas P. and Storzer D. (1981) ^{244}Pu fission track thermometry and its application to stony meteorites. *Proc. R. Soc. Lond., A374,* 253–270.

Prinn R. G. and Fegley B. J. (1987) The atmospheres of Venus, Earth and Mars: A critical review. *Annu. Rev. Earth Planet. Sci., 15,* 171–212.

Richter F. M. and McKenzie D. (1984) Dynamical models for melt segregation from a deformable matrix. *J. Geol., 92,* 729–740.

Righter K. and Drake M. J. (1997) A magma ocean on Vesta: Core formation and petrogenesis of eucrites and diogenites. *Meteoritics & Planet. Sci., 32,* 929–944.

Rubin A. E. (1995) Petrologic evidence for collisional heating of chondritic asteroids. *Icarus, 113,* 156–167.

Russell S. S., Srinivasan G., Huss G. R., Wasserburg G. J., and MacPherson G. J. (1996) Evidence for widespread ^{26}Al in the solar nebula and constraints for nebula time scales. *Science, 273,* 757–762.

Sears D. W., Grossman J. N., Melcher C. L., Ross L. M., and Mills A. A. (1980) Measuring metamorphic history of unequilibrated ordinary chondrites. *Nature, 287,* 791–795.

Shimazu H. and Terasawa T. (1995) Electromagnetic induction heating of meteorite parent bodies by the primordial solar wind. *J. Geophys. Res., 100,* 16923–16930.

Shukolyukov A. and Lugmair G. W. (1996) ^{60}Fe-^{60}Ni systematics in the eucrite Caldera. *Meteoritics, 31,* A129.

Sleep N. H. (1988) Tapping of melt by veins and dikes. *J. Geophys. Res., 93,* 10255–10272.

Sonnett C. P., Colburn D. S., and Schwartz K. (1968) Electrical heating of meteorite parent bodies and planets by dynamo induction from a premain sequence T Tauri "solar wind." *Nature, 219,* 924–926.

Srinivasan G., Goswami J. N., and Bhandari N. (1999) ^{26}Al in eucrite Piplia Kalan: Plausible heat source and formation chronology. *Science, 284,* 1348–1350.

Standish E. M. and Hellings R. W. (1989) A determination of the masses of Ceres, Pallas and Vesta from their perturbation upon the orbit of Mars. *Icarus, 80,* 326–333.

Stevenson D. J. (1990) Models of the Earth's core. *Science, 214,* 611–619.

Stolper E. M. (1977) Experimental petrology of eucritic meteorites. *Geochim. Cosmochim. Acta, 41,* 587–611.

Taylor G. J. (1992) Core formation in asteroids. *J. Geophys. Res., 97,* 14717–14726.

Taylor G. J., Maggiore P., Scott E. R. D., Rubin A. E., and Keil K. (1987) Original structures, and fragmentation and reassembly histories of asteroids: Evidence from meteorites. *Icarus, 69,* 1–13.

Tera F., Carlsson R. W., and Boctor N. Z. (1997) Radiometric ages of basaltic achondrites and their relation to the early history of the Solar System. *Geochim. Cosmochim. Acta, 61,* 1713–1731.

Thomas P. C., Binzel R. P., Gaffey M. J., Storrs A. D., Wells E. N., and Zellner B. H. (1997) Impact excavation of 4 Vesta: Hubble Space Telescope results. *Science, 277,* 1492–1495.

Urey H. (1955) The cosmic abundances of potassium, uranium, and thorium and the heat balances of the Earth, the Moon and Mars. *Proc. Natl. Acad. Sci. U.S., 41,* 127–144.

Wasson J. T., Rubin A. E., and Benz W. (1987) Heating of primitive, asteroid-size bodies by large impacts. *Meteoritics, 22,* 525–526.

Weidenschilling S. J., Spaute D., Davis D. R., Mazari F., and Ohtsuki K. (1997) Accretional evolution of a planetesimal swarm. *Icarus, 128,* 429–455.

Wetherill G. W. (1980) Formation of terrestrial planets. *Annu. Rev. Astron. Astrophys., 18,* 77–213.

Williams C. V., Keil K., Taylor G. J., and Scott E. R. D. (1999) Cooling rates of equilibrated clasts in ordinary chondrite regolith breccias: Implications for parent body histories. *Chem. Erde, 59,* 287–305.

Williams D. R. and Wetherill G. W. (1993) Equilibrium models of mass distribution and collisional lifetime of asteroids (abstract). In *Lunar and Planetary Science XXIV,* pp. 1523–1524. Lunar and Planetary Institute, Houston.

Wilson L. and Keil K. (1996) Volcanic eruptions and intrusions on the asteroid 4 Vesta. *J. Geophys. Res., 101,* 18927–18940.

Wilson L. and Keil K. (2000) Crust development on differentiated asteroids (abstract). In *Lunar and Planetary Science XXXI,* abstract #1576. Lunar and Planetary Institute, Houston (CD-ROM).

Wilson L., Keil K., Browning L. B., Krot A. N., and Bourcher W. (1999) Early aqueous alteration, explosive disruption, and reprocessing of asteroids. *Meteoritics & Planet. Sci., 34,* 541–557.

Wood J. A. (1979) Review of the metallographic cooling rates of meteorites and a new model for the planetesimals in which they formed. In *Asteroids* (T. Gehrels, ed.), pp. 849–891. Univ. of Arizona, Tucson.

Wood J. A. and Morfill G. (1988) A review of solar nebular models. In *Meteorites and the Early Solar System* (J. F. Kerridge and M. S. Matthews, eds.), pp. 329–347. Univ. of Arizona, Tucson.

Wood J. A. and Pellas P. (1991) What heated the parent meteorite planets? In *The Sun in Time* (C. P. Sonnett and M. S. Giampapa, eds.), pp. 740–760. Univ. of Arizona, Tucson.

Yamaguchi A., Taylor G. J., and Keil K. (1997) Metamorphic history of the eucritic crust of 4 Vesta. *J. Geophys. Res., 102,* 13381–13286.

Yomogida K. and Matsui T. (1984) Multiple parent bodies of ordinary chondrites. *Earth Planet. Sci. Lett., 68,* 34–42.

Young E. D., Ash R. D., England P., and Rumble D. III (1999) Fluid flow in chondrite parent bodies: Deciphering the compositions of planetesimals. *Science, 286,* 1331–1335.

Zolensky M. E., Bourcier W. L., and Gooding J. L. (1989) Aqueous alteration on the hydrous asteroids: Results of EQ3/6 computer simulations. *Icarus, 78,* 411–425.

Zolensky M. E., Weisberg M. K., Buchanan P. C., and Mittlefehldt D. W. (1996) Mineralogy of carbonaceous chondrite clasts in HED achondrites and the Moon. *Meteoritics & Planet. Sci., 31,* 518–537.

Geological History of Asteroid 4 Vesta: The "Smallest Terrestrial Planet"

Klaus Keil
University of Hawai'i

The asteroid 4 Vesta is the only known differentiated asteroid with an intact internal structure, probably consisting of a metal core, an ultramafic mantle, and a basaltic crust. Considerable evidence suggests that the HED meteorites are impact ejecta from Vesta, and detailed studies of these meteorites in terrestrial laboratories, combined with ever more sophisticated remote sensing studies of the asteroid, have resulted in a good understanding of the geological evolution of this fascinating object. Extensive mineralogical, petrological, geochemical, isotopic, and chronological data suggest that heating, melting, and formation of a metal core, a mantle, and a basaltic crust took place in the first few million years of solar system history. It is likely that many more Vesta-like asteroids formed at the dawn of the solar system but were destroyed by impact, with the iron meteorites being remnants of their cores. Such differentiated objects may have played an important role in the accretion and formation of the terrestrial planets, and it is therefore highly desirable to explore by spacecraft this world that can be viewed as the smallest of the terrestrial planets.

1. INTRODUCTION

The world's collections contain meteorites from at least three sources: asteroids (tens of thousands), Earth's Moon (~26), and Mars (~26). Incredible progress has been made in recent years in the study of asteroidal meteorites, stimulated by the discoveries of thousands of new specimens in Antarctica and hot deserts. Among these many thousands of meteorites are many new types and rare individuals that represent asteroidal parent bodies heretofore unrepresented in our collections. Detailed studies of asteroidal meteorites have shown that, based on their mineralogical, chemical, and isotopic properties, an astonishing ~135 different asteroids are represented (*Meibom and Clark,* 1999). Although nearly all asteroids of which we have samples have been affected by postaccretionary heating to some degree (*Keil,* 2000), many (108 of the 135) were actually partially or completely melted and differentiated (*Meibom and Clark,* 1999).

One of the most fascinating asteroids is 4 Vesta. Remote sensing shows that it is a differentiated, nearly intact object with a basaltic crust (e.g., *McCord et al.,* 1970) and ultramafic mantle rocks (pyroxenite; olivine-bearing) exposed in a huge impact crater (e.g., *Binzel et al.,* 1997; *Gaffey,* 1997; *Thomas et al.,* 1997a). Modeling (e.g., *Righter and Drake,* 1997; *Dreibus et al.,* 1997) and density estimates (e.g., *Thomas et al.,* 1997b) further suggest that Vesta has a metal core and thus the asteroid is a differentiated object with crust, mantle, and core, analogous in its structure to the terrestrial planets, albeit much smaller in size. Vesta can therefore be thought of as the smallest of the terrestrial planets. Fortuitously, we are not restricted to remote sensing data for understanding the geological history of Vesta. The howardite-eucrite-diogenite meteorites (or HEDs; *Takeda et al.,* 1983), a large suite of differentiated basalts (eucrites), pyroxenites (diogenites), and breccia mixtures principally of these two rock types (howardites) (e.g., *Mason,* 1962; *Takeda et al.,* 1976; *Takeda,* 1979; *Mittlefehldt et al.,* 1998, and references therein) are, in all likelihood, impact-produced fragments of Vesta (e.g., *McCord et al.,* 1970; *Consolmagno and Drake,* 1977; *Drake,* 1979, 2001; *Binzel and Xu,* 1993; *Gaffey,* 1997). Their detailed study in terrestrial laboratories, combined with the remote sensing data and numerous modeling studies, have contributed to a reasonably good understanding of the complex history of this unique world.

Because much of the geological history of Vesta (i.e., heating, melting, fractionation, extrusion, and solidification of the basaltic crust) took place in the first 10 m.y. of solar system history (e.g., *Lugmair and Shukolyukov,* 1998; *Srinivasan et al.,* 1999; *Nyquist et al.,* 2001), it is of great interest for understanding the earliest differentiation of solar system bodies at the dawn of the solar system. The asteroid is also of interest because differentiated objects of this type are thought to have been the embryos that may have played an important role in the accretion and formation of the larger terrestrial planets (e.g., *Taylor and Norman,* 1990; *Carlson and Lugmair,* 2000).

In the present paper, I summarize the most important properties of Vesta, based on remote sensing data and the study of HEDs. I also use these data, supported by modeling studies, to outline Vesta's early geological history.

2. ORBIT, SIZE, SHAPE, MASS, AND DENSITY OF 4 VESTA

Asteroid 4 Vesta was discovered by H. W. Olbers in Bremen, Germany, on March 29, 1807 (*Pilcher,* 1979). It orbits the Sun at a mean heliocentric distance of a = 2.36 AU, has a proper eccentricity of e = 0.097, and a proper sine of inclination of sin i = 0.112 (*Williams,* 1989). Based on Hubble Space Telescope (HST) images, *Thomas et al.* (1997b) concluded that the asteroid is a triaxial ellipsoid of radii 289, 280, and 229 km, all ±5 km. This compares well with the earlier data of *McCarthy et al.* (1994) of 278 ± 12, 261 ± 9, and 232 ± 5 km (see also *Drummond et al.,* 1998). Its mean radius is 258 ± 12 km and its volume is ~7.19 ± 0.87 × 10^7 km^3 (*Thomas et al.,* 1997b). The mass of the asteroid has been estimated by *Schubart and Matson* (1979) to be 1.38 ± 0.12 × 10^{-10} $M_\odot$ or 2.75 ± 0.24 × 10^{20} kg, and by *Standish and Hellings* (1989) as 1.5 ± 0.3 × 10^{-10} $M_\odot$ or 2.99 ± 0.60 × 10^{20} kg [note that the recent mass estimate, based on perturbations of Vesta on 26 selected asteroids of 1.36 ± 0.05 × 10^{-10} $M_\odot$ by *Michalak* (2000), is very close to that of *Schubart and Matson* (1979)]. Using the *Thomas et al.* (1997b) volume and the *Schubart and Matson* (1979) mass yields a mean density of 3800 ± 600 kg/m^3, whcreas the *Standish and Hellings* (1989) mass yields a density of 4100 ± 950 kg/m^3 [*Thomas et al.* (1997a) give 3500 ± 400 and 3900 ± 800 respectively].

3. THE HED-VESTA-VESTOID CONNECTION

The proposal that the HED meteorites are impact ejecta from Vesta has important implications for understanding the geological history of the asteroid. Detailed studies of the HEDs in terrestrial laboratories with sophisticated analytical techniques have provided a wealth of ground truth data that supplement the "global field geology observations" made possible by ever-improving remote sensing techniques. The link between the HED meteorites and Vesta was originally based on the similarities in the compositions of the HEDs and the surface mineralogy of Vesta, as determined by spectroscopy. More recently, dynamical considerations that connect the so-called "vestoids" of the Vesta family to Vesta and to the 3:1 and ν_6 escape hatch resonances have added strong evidence in favor of this proposition (*Binzel and Xu,* 1993).

McCord et al. (1970) were the first to determine, through spectroscopy in the visible and near-infrared using ground-based telescopes, that the surface of Vesta exhibits absorption features typical for low-Ca pyroxene and that it is similar in composition to certain basaltic achondrites (i.e., eucrites) (Fig. 1). Since then, many Earth-based measurements in the visible and infrared (e.g., *Larson and Fink,* 1975; *McFadden et al.,* 1977; *Feierberg et al.,* 1980; *Gaffey,* 1983, 1997, and references therein; *Cochran and Vilas,* 1998) have confirmed the *McCord et al.* (1970) findings. For example, *Feierberg et al.* (1980) determined that the surface of Vesta consists of a mixture of pyroxene ($Fs_{50\pm5}$) and plagioclase, with a pyroxene/plagioclase ratio of 1.5–2.0, consistent with Vesta's surface being covered by a mixture of eucrites and howardites, suggesting that Vesta is an intact, differentiated object with a basaltic crust. Furthermore, *Cochran and Vilas* (1998) observed a spectral absorption feature centered at 5065 Å in spectra across the complete surface of Vesta that is indicative of the presence of relatively Ca-rich augite. Recent Earth-based (*Gaffey,* 1997) and HST images (*Binzel et al.,* 1997) of Vesta as it rotates have confirmed earlier observations of geological diversity across the asteroid (e.g., *Gaffey,* 1983) and revealed a heterogeneous surface, consistent with a composition similar to that of the HED meteorites. Specifically, *Gaffey* (1997) noted that Vesta appears to have an old, age-darkened surface akin to the howardites and polymict eucrites, with fresher rocks akin to diogenites and olivine-bearing material exposed in impact craters. *Binzel et al.* (1997) note that the eastern hemisphere of Vesta is dominated by what appear to be impact-excavated plutonic rocks made of Mg-rich and Ca-poor pyroxene, akin to diogenites, with some units containing a substantial amount of olivine. The western hemisphere, on the other hand, is dominated by Fe-rich and relatively Ca-rich pyroxene, consistent with eucrites.

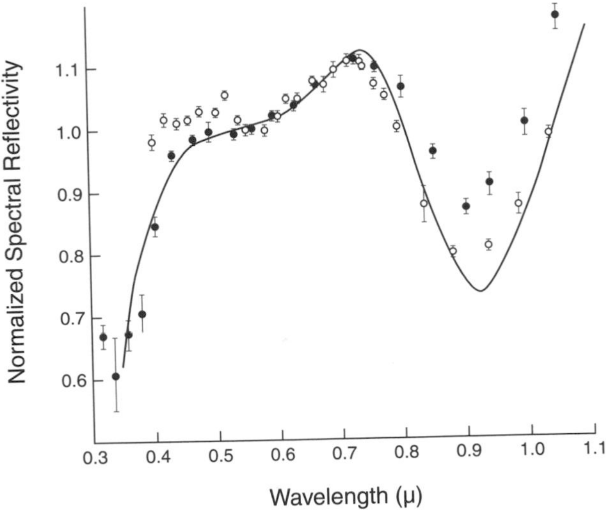

Fig. 1. Laboratory measurements of the spectral reflectivity of the Nuevo Laredo eucrite (solid line), compared to telescope data points from Vesta, suggesting that Vesta is covered by eucritelike material. Solid circles, data obtained at Cerro Tololo, Chile, December 1969; open circles, data obtained at Mount Wilson, California, October 1968. The standard deviation for each average value is shown as an error bar. Reprinted with permission from *McCord et al.* (1970). Copyright 1970, American Association for the Advancement of Science.

One argument against the origin of the HEDs from Vesta and the vestoids has been that space weathering will make basaltic asteroids look like S asteroids. Hence, the sharp, uniquely defined basaltic reflection spectra of Vesta and the vestoids must be due to recent (<10 Ma) impact resurfacing and covering of their surfaces by fresh basalt debris (*Wasson and Chapman,* 1996; *Wasson et al.,* 1996). These authors suggested that many S asteroids might be space-weathered basaltic chunks from the differentiation of the many iron meteorite parent bodies and that the parent body of the HEDs is hiding among the S asteroids. However, *Gaffey* (1983) concluded that the maturation of the Vesta surface (e.g., reddening of the IR curve) is ~100–1000× lower than that of the surface of the Moon, and *Hiroi et al.* (1994) suggested that the good matches between the reflectance spectra and brightness of Vesta and the HED meteorites indicate that Vesta's surface is free from heavy space weathering. In addition, *Yamada et al.* (1999) have recently shown that olivine may be the principal mineral altered by space-weathering processes. Since Vesta's surface is dominated by pyroxene rather than olivine as the mafic mineral, it is much less susceptible to space weathering than, for example, ordinary chondrites would be. Thus, Vesta's surface does not have to be "made fresh" in order for it to provide the exceptionally well-defined spectral matches to the HEDs

Dynamical difficulties to get impact debris off Vesta and into the far-away 3:1 jovian and ν_6 secular resonances have also been cited as evidence against an origin of the HEDs from Vesta, as this would require implausibly high launch velocities (e.g., *Wasson and Wetherill,* 1979; *Wetherill,* 1987). However, the discovery by *Binzel and Xu* (1993) of main-belt asteroids <10 km in diameter with surface compositions similar to that of Vesta (basalt) that constitute a small Vesta family located between Vesta and the major dynamical resonances are powerful evidence in favor of the HED-Vesta link. Specifically, these authors found 12 of the Vesta family members and 8 other small main-belt asteroids, collectively referred to as vestoids, to have spectra like typical basaltic achondrites, and 14 of these are similar to Vesta and have the spectral characteristics of the eucrites (i.e., a strong absorption band near 9000 Å). The remaining six are interpreted as diogenitic in composition and were designated as J asteroids (mnemonic for the Johnstown diogenite). Currently, more than 40 vestoids <10 km in diameter with Vesta-like spectral characteristics are known that fall inside 2.5 AU, the Kirkwood gap defined by the 3:1 jovian resonance (*Binzel et al.,* 1999). Note that *Binzel and Xu* (1993) also suggested that the then-known three near-Earth V-type asteroids with diameters of 1–4 km (*Cruikshank et al.,* 1991) were also derived from Vesta, a view that is not shared by *Cruikshank et al.* (1991) but is advocated by *Migliorini et al.* (1997) for the newly discovered ones (now totaling seven) as well. Recently, *Vilas et al.* (2000) confirmed the *Binzel and Xu* (1993) results and found that of the 13 asteroids dynamically linked to Vesta that they studied, 9 show deep 9000 Å absorption bands, suggesting a pyroxene composition consistent with rocks of the HED clan. In addition, six of these asteroids also show an absorption feature centered at 5065 Å, consistent with the occurrence of a high-Ca pyroxene, as previously shown for Vesta (*Cochran and Vilas,* 1998). This supports the notion that the vestoids are indeed impact ejecta from Vesta, but the fact that not all objects show this feature suggests that the vestoids are derived from different layers of the parent body [or from a different parent body altogether (*Vilas et al.,* 2000)]. *Burbine et al.* (2001) also concluded, based on new measurements of 20 vestoids, that they hail from Vesta and that there is a clear connection to the HEDs. It should be noted that dynamical modeling studies show that the Vesta family asteroids can, in fact, be launched off Vesta by large impacts (e.g., *Zappalà et al.,* 1995; *Marzari et al.,* 1996; *Asphaug,* 1997).

The recent Hubble Space Telescope discovery near the south pole of Vesta of a large impact crater ~460 km in diameter and ~13 km deep with a central uplift and bounding rim, as well as of several smaller depressions ~160 km and ~150 km in diameter (*Thomas et al.,* 1997a) (Plate 7), is further evidence in favor of the origin of the vestoids and HEDs from Vesta. The dimensions of the huge crater are consistent with excavation deep into the high-Ca pyroxene-rich crust or olivine-rich upper mantle, or both. The volume of this cratering event of ~1 vol% of Vesta (~10^6 km^3) is more than sufficient to account for the volume of the vestoids, which amount to only a few percent of this volume.

Cosmic-ray-exposure ages (CREAs) are also consistent with the HED-Vesta-vestoids connection. *Eugster and Michel* (1995), for example, found two major peaks at 21 ± 4 and 38 ± 8 Ma and three minor peaks at 6 ± 1, 12 ± 2, and 73 ± 3 Ma in the CREAs of the HEDs. This suggests that the immediate parent bodies of the HEDs (the vestoids) should contain all three meteorite types. *Welton et al.* (1997) determined the CREAs of diogenites and found that 10 ages cluster at 21–25 and 4 at 35–42 Ma (while all 20 studied range from 6 to 50 Ma). The two peaks coincide with the two major peaks for the HEDs. Since the surface compositions of some of the vestoids are dominated by only one type of HED meteorite (*Binzel and Xu,* 1993; *Xu et al.,* 1995) but all HED CREAs peak in two major peaks, *Welton et al.* (1997) favor Vesta rather than the vestoids as the source for the HEDs, although the recent work of *Burbine et al.* (2001) has shown a great diversity of objects among the vestoids.

Could the HED meteorites have originated from a main-belt asteroid other than Vesta and the vestoids? The only other large main-belt asteroid with a basaltic surface that has no known dynamical link to Vesta, the vestoids, any family, or any nearby large asteroid is 1459 Magnya (*Lazzaro et al.,* 2000). These authors argue that Magnya is not a fragment of Vesta, as it would require an ejection velocity of ~5 km/s. Rather, they suggest that this 30-km-diameter object is a fragment of a large, differentiated asteroid that was disrupted long ago and must have had a remarkably thick (at least 30 km) basaltic crust. Since the delivery efficiency of fragments to Earth from Magnya at 3.15 AU is much less than that from Vesta and the vestoids, *Lazzaro et al.* (2000) do not consider Magnya as a likely alternative source to Vesta for the HED meteorites.

Are all eucrites from the same parent body? They presumably are, except for a recently recovered, fascinating, highly metamorphosed rock originally classified as a noncumulate eucrite, Northwest Africa 011, with FeO/MnO ratios of pyroxenes considerably higher (~65) than those of normal ordinary eucrites (≤40), has been described by *Yamaguchi* (2001). More recent work, including O-isotopic measurements, suggests that this is a new type of basaltic achondrite that originated on a different parent body from the HEDs, but one that was probably similar in size and history to Vesta (*Yamaguchi et al.,* 2001b, 2002).

Finally, it has been suggested that the HED meteorites and certain other meteorite groups, because of their mineralogical and, particularly, O-isotopic similarities, come from one and the same parent body, although there are conflicting statements in the literature. For example, *Mittlefehldt* (1980), *Clayton and Mayeda* (1996), and *Mittlefehldt et al.* (1998) suggested that the HEDs, mesosiderites, main-group pallasites, and IIIAB irons may have originated on the same parent body, although they do not propose that this is Vesta [note that *Mittlefehldt* (1990) and *Rubin and Mittlefehldt* (1993) argued that mesosiderites and HEDs come from different parent bodies]. Since Vesta has a nearly intact basaltic crust that formed in the first few million years of solar system history (e.g., *Shukolyukov and Lugmair,* 1993a,b; *Lugmair and Shukolyukov,* 1998; *Srinivasan et al.,*1999; *Nyquist et al.,* 2001), the asteroid clearly did not go through a catastrophic breakup and reassembly episode since formation of the crust (*Gaffey,* 1983). Such a breakup would be required to excavate the extremely slowly cooled mesosiderites, main-group pallasites, and IIIAB irons from great depths. For that reason, *Cruikshank et al.* (1991) suggested that this parent body could not have been Vesta, although the identical O-isotopic compositions of mesosiderites, main-group pallasites, IIIAB irons, and HEDs are consistent with formation on the same body. Instead, they suggested that the HEDs come from the three near-Earth V-type asteroids they discovered, and that these are fragments of some other, totally disrupted, differentiated asteroid. In view of the recent evidence that the HED meteorites originated on Vesta and the vestoids, it is evident that the mesosiderites, main-group pallasites, and IIIAB irons, in spite of their similar O-isotopic compositions to those of the HEDs, did not form on Vesta but must have originated on a different parent body(ies), a view now shared by D. Mittlefehldt (personal communication, 2001). However, they clearly formed from a related O-isotopic reservoir.

4. MINERALOGY, PETROLOGY, CHEMISTRY, AND PETROGENESIS OF THE HED METEORITES

The identical O-isotopic compositions of the HEDs, their similarities in mineralogy and composition [e.g., the Fe/Mn ratios of pyroxenes (*Papike,* 1998)], the occurrence of polymict breccias among them [e.g., howardites that contain fragments of eucrites and diogenites, polymict eucrites that contain fragments of diogenitic material, and polymict diogenites that contain fragments of eucrites (e.g., *Delaney et al.,* 1984)] and the existence of rocks intermediate between diogenites and cumulate eucrites (*Takeda and Mori,* 1985) are strong evidence that all these meteorites come from the same parent body.

The classification of the HED meteorites and their most important mineralogical characteristics are summarized in Table 1 and their textures are shown in Plate 8. Rocks are listed arranged in order of decreasing depth of origin on the parent body as proposed in the layered crust model of *Takeda* (1979, 1997).

Diogenites are coarse-grained, usually brecciated, cumulates (Plate 8a) from a fractionally crystallizing magma and consist on average of (in vol%) 92.2 orthopyroxene, 4.2 olivine, 1.2 clinopyroxene, 0.9 chromite, 0.4 plagioclase, 0.1 metallic Fe,Ni, 0.6 troilite, and 0.4 silica phase (average of 21 diogenites) (*Bowman et al.,* 1997). Diogenites are thought to be the most deep-seated known lithology from the HED parent body and, in the layered crust model of *Takeda* (1997), for example, are thought to be overlain successively by the cumulate eucrites, ordinary eucrites, unequilibrated eucrites, and howardites. One would therefore expect that cooling rates of the diogenites should be the slowest of all HEDs, but this is not the case. Cooling rates estimated from Fe^{2+}-Mg ordering in orthopyroxene (*Zema et al.,* 1997) are ~50°C/ka for Johnstown and ~800°C/ka for Roda, at closure temperatures between 311° ± 29°C and 408° ± 10°C respectively. These rates are much faster than those for the cumulate and noncumulate eucrites, which are thought to have overlain the diogenites (see below). *Zema et al.* (1997) suggested that their diogenite cooling rates are not the result of cooling deep within Vesta, but are due to cooling at different (and shallower) burial depths in the ejected fragments. Note that H. Takeda (personal communication, 2001) pointed out that "Fe^{2+}-Mg ordering in the orthopyroxene structure is easily modified or reset by subsequent minor events and it cannot be used as an estimate of the cooling rate of the original rocks."

Cumulate eucrites are coarse-grained and mostly unbrecciated gabbros (Plate 8b). To estimate the depth of origin (and thus the thickness of the basaltic crust) of Vesta, *Miyamoto and Takeda* (1994) computed the cooling rate of the cumulate eucrite Moore County. Their estimate is based on the Ca compositional gradients and the widths of augite lamellae in pyroxene, and assumes a parent body 250 km in radius. They found that Moore County originally cooled at 0.16°C/ka, which corresponds to a burial depth of about 8 km, assuming a rocklike thermal diffusivity. After this, a sudden increase in temperature is recorded, probably as the result of excavation due to an impact, from which Moore County cooled down at 350°C/ka. This corresponds to a burial depth of ~100 m assuming rocklike or ~10 m assuming regolithlike thermal diffusivity. The original burial depth of 8 km suggests that the basaltic crust of Vesta was at least that thick, and this is in agreement with the largest eucritic vestoid being ≤10 km in diameter (*Binzel and Xu,* 1993).

Noncumulate (basaltic) eucrites formed originally as quickly cooled and hence unequilibrated surface lava flows (Plate 8d) (i.e., the unequilibrated eucrites; also referred to as unmetamorphosed or least-metamorphosed eucrites) at cooling rates, based on their textures, of ~0.001°–100°C/h (*Walker et al.,* 1978). Their pyroxenes (pigeonite; mg# ~70–20) are compositionally zoned and have exsolution lamellae on the TEM scale. However, most unequilibrated euc-

TABLE 1. Classification and principal properties of HED meteorites (modified after *Takeda*, 1997).

Diogenites: Orthopyroxenites (usually brecciated), with ~84–~100 vol% orthopyroxene [$(Mg,Fe)SiO_3$] with mg# ~85–67*.

Eucrites: Pyroxene-plagioclase basalts.

Cumulate eucrites of the Binda and Moore County types: Coarse-grained gabbros, often not brecciated. Orthopyroxene inverted from low-Ca clinopyroxene (mg# ~67–58); orthopyroxene inverted from pigeonite (mg# ~57–45); thick high-Ca pyroxene exsolution lamellae.

Noncumulate (basaltic) eucrites

Ordinary ("equilibrated; metamorphosed") eucrites: Monomict-brecciated or unbrecciated basalts, with homogeneous host pigeonite (a low-Ca clinopyroxene; mg# ~42–30) and fine high-Ca pyroxene exsolution lamellae. Includes the main-group eucrites (Juvinas type) and the Stannern and Nuevo Laredo types.

Unequilibrated ("unmetamorphosed; least metamorphosed") eucrites (Pasamonte type): Surface lavas that experienced minor metamorphism. Pigeonites (mg# ~70–20) with compositional zoning and TEM scale exsolution.

Polymict eucrites: Polymict breccias of various types of eucrites with <10 vol% orthopyroxene.

Howardites: Polymict breccias of eucritic and diogenitic materials with >10 vol% orthopyroxene.

*mg# = Mg × 100/(Mg + Fe) atomic %.

rites subsequently experienced extensive thermal metamorphism to form the metamorphosed eucrites (e.g., *Takeda and Graham,* 1991; *Yamaguchi et al.,* 1996, 1997).

The metamorphosed (equilibrated), noncumulate eucrites (Plate 8c) are collectively referred to as the ordinary eucrites and include the main-group eucrites (Juvinas type) and the Stannern and Nuevo Laredo types. They are monomict-brecciated or unbrecciated, metamorphosed basalts, with homogeneous host pigeonite and fine high-Ca pyroxene exsolution lamellae. The widths of their exsolved pyroxene lamellae allows estimation of the cooling rates of the rocks and hence their burial depths. *Miyamoto et al.* (2001), for example, calculated a subsolidus cooling rate of 20°C/ka down to 550°C for the highly metamorphosed nonbrecciated eucrite Ibitira. Depending upon assumptions for thermal diffusivity of the overlaying material, this corresponds to a burial depth of ~550 m in solid rock, ~30 m in regolith with 50% porosity, and ~90 m in a compacted regolith.

Polymict eucrites are polymict breccias consisting mostly of eucritic material with <10 vol% orthopyroxene (diogenitic component) (Plate 8e). These rocks were originally recognized in the collections from Antarctica (e.g., *Delaney et al.,* 1984).

Howardites are polymict regolith breccias consisting mostly of eucritic and diogenitic components, with >10 vol% orthopyroxene. They also contain some olivine, suggesting yet another source rock, as well as carbonaceous chondrite xenoliths. Their matrixes are fine-grained, and into these are embedded rock and mineral clasts, including a variety of impact-melt rocks and breccias (e.g., *Fredriksson and Keil,* 1963; *Metzler et al.,* 1995; *Pun et al.,* 1998) (Plate 8f). The clastic matrix contains solar-wind-implanted gases, indicating that its constituents were once exposed at the very top surface of the parent body (e.g., *Suess et al.,* 1964).

A number of models have been proposed to explain the petrogenesis of the HED meteorites (i.e., the origin of the eucrites and diogenites). *Mason* (1968) was the first to discuss their origin by fractional crystallization, and a recent summary of this and other models of eucrite petrogenesis is given by *Takeda* (1997). *Stolper* (1977), for example, carried out melting experiments on basaltic eucrites and concluded that they formed by crystallization of primary partial melts. These melts are thought to have formed by low degrees [~4–15 vol% (*Consolmagno and Drake,* 1977)] of partial melting in the temperature interval of ~1150°–1190°C, possibly from a chondritic (CM) source material (*Jurewicz et al.,* 1993). Cumulate eucrites formed from a fractionally crystallizing melt of mafic composition, and *Stolper* (1977) pointed out that these rocks are too Fe-rich to have crystallized from the melts from which the basaltic eucrites crystallized. However, *Treiman* (1997) suggested that they could have formed as mixtures of cumulus minerals and trapped melt. *Stolper* (1977) further suggested that the diogenites crystallized from melts of essentially orthopyroxene composition, and *Warren* (1997) proposed that they formed as early cumulates from a large magma system, probably a global magma ocean. On the other hand, *Shearer et al.* (1997) suggested that diogenites are cumulates from ~10–20% fractional crystallization of specific basaltic melts.

While the *Stolper* (1977) model of eucrite petrogenesis requires low degrees of partial melting, a variety of models have been proposed that visualize the origin of these rocks as the products of fractional crystallization of residual melts that formed by higher degrees of partial melting, such as from a magma ocean on the HED parent body (*Ikeda and Takeda,* 1985; *Taylor et al.,* 1993; *Ruzicka et al.,* 1997; *Warren,* 1997; *Takeda,* 1997). The magma ocean concept has also been championed by *Righter and Drake* (1997) and *Drake* (2001) who, based on the abundances of moderately siderophile elements (Ni, Co, Mo, W, P) in the HED mantle, visualize melting of the entire HED parent body, including formation of a metal core.

5. GEOLOGICAL HISTORY OF 4 VESTA

5.1. Accretion, Melting, Differentiation, Core Formation, and Heat Source

One of the most fascinating developments in the chronology of the HEDs is the recognition that accretion, melting, and fractionation of their parent body, Vesta, and extrusion to, and crystallization on, its surface of eucritic basalts, occurred in the first few million years of solar system history (e.g., *Srinivasan et al.,* 1999; *Nyquist et al.,* 2001; *Carlson and Lugmair,* 2000, and references therein). This knowledge stems from much progress that has been made in recent years in the determination of precise absolute U-Pb ages of meteorites based on the still-extant radionuclide ^{235}U, and the measurement of the decay products of now-extinct radionuclides (e.g., ^{53}Mn with a half-life of 3.7 m.y. decaying into ^{53}Cr) that allow an astonishingly precise timescale for these processes to be established. For example, the oldest-recorded Pb-Pb age of calcium-aluminum-rich inclusions (CAIs; high-temperature condensates and, hence, thought to be the most ancient solar system materials) from the Allende CV3 chondrite is 4566 ± 8 Ma (*Chen and Wasserburg,* 1981) and 4566 ± 2 Ma (*Allègre et al.,* 1995), and the Pb model age for the differentiated angrite Angra dos Reis is 4557.8 ± 0.5 Ma (see references in *Carlson and Lugmair,* 2000). Taking into account the somewhat uncertain radial heterogeneity in ^{53}Mn in the formation regions, these ages can be used to translate the ^{53}Mn-^{53}Cr formation intervals of eucrites into an absolute age for the differentiation of Vesta. Based on measurements of the basaltic eucrite Chervony Kut, for example, an age of 4563.6 ± 0.9 Ma is derived, only a few million years younger than the formation of CAIs (*Lugmair and Shukolyukov,* 1998). Furthermore, the great antiquity of the eucrites and hence the evidence for melting and differentiation of Vesta on a timescale of a few million years is further supported by the detection of the decay products of other extinct radionuclides such as ^{26}Mg from the decay of ^{26}Al (half-life 0.73 m.y.) (*Srinivasan et al.,* 1999; *Nyquist et al.,* 2001) and ^{60}Ni from the decay of ^{60}Fe (half-life 1.5 m.y.) (see references in *Carlson and Lugmair,* 2000). Finally, it should be noted that the "most pristine," unmetamorphosed surface lava flow (i.e., unequilibrated eucrite), which is represented by clasts in the polymict eucrite Y 75011, also yields very old ages, albeit with larger error bars. These clasts are coarse-grained, mesostasis-rich basalts with subophitic texture and extensive Mg-Fe zoning in their pyroxenes. The Rb-Sr ages of the largest clast are 4.60 ± 0.05 and 4.50 ± 0.05 Ga, depending upon the assumed value of the ^{87}Rb decay constant, and the internal Sm-Nd isochron age for clasts and matrix is 4.55 ± 0.14 Ga (*Nyquist et al.,* 1986).

Cumulate eucrites, which crystallized and cooled at depth, have younger ages of <4480 Ma, due to thermal processing at depth over a longer time span than the quickly-cooled basaltic eucrites (sec references in *Carlson and Lugmair,* 2000). Modeling of the thermal history of Vesta by *Ghosh and McSween* (1998) suggests that heating by ^{26}Al would keep the mantle hot for ~100 m.y., consistent with the younger ages of cumulate eucrites. Also, many eucrites (and howardites) experienced extensive impact processing and reheating that resulted in resetting of some chronometers (e.g., *Nyquist et al.,* 1997) and, for example, partial or complete resetting of ^{39}Ar-^{40}Ar ages in the relatively narrow time interval of 3.4–4.1 G.y. ago (the "cataclysmic" early bombardment) (e.g., *Bogard,* 1995). However, a few rare metamorphosed noncumulate eucrites such as Ibitira and EET 90020 have relatively old ^{39}Ar-^{40}Ar ages of 4.485 and 4.49 Ga respectively, and thus escaped the widespread cataclysmic impact resetting (*Bogard and Garrison,* 1995; *Yamaguchi et al.,* 2001a).

There is convincing geochemical evidence that Vesta experienced a high degree of (or possibly complete) melting that resulted in the formation of a metal core. For example, the depletion in moderately siderophile incompatible elements (e.g., Ni, Co, Mo, W, P) relative to nonsiderophile incompatible elements in HED meteorites suggests metal segregation and hence core formation (e.g., *Hewins and Newsom,* 1988, and references therein; *Righter and Drake,* 1997). However, estimates of the amount of metal in Vesta vary widely between 0 and 50 wt% (see references in *Ruzicka et al.,* 1997). For example, *Ruzicka et al.* (1997) estimated the mass of the core by mass balance from the density of Vesta and the density of the silicate fraction to be between ~0 and 25 wt%, with the best estimate being ~5 wt%. They also suggested that the core is <130 km in radius, the olivine-rich mantle is ~65–220 km thick, the lower crustal diogenite unit is ~12–43 km thick, and the upper crustal eucrite unit is ~23–42 km thick. *Dreibus et al.* (1997) estimated the mass of the core from their calculated composition of the bulk silicate portion of Vesta (assuming CI abundances for Fe and Ni) to be 21.7 wt%. They also calculated the density of the mantle to be 3400 kg/m^3 and, with a core density of 7900 kg/m^3, calculated the bulk density of Vesta to be 3800 kg/m^3, in good agreement with the astronomically determined values (see above). With a radius of 263 km and a core mass of 21.7 wt%, they calculated a core radius of 123 km.

The time of formation of the core was estimated by *Lee and Halliday* (1997). This estimate is based on the measurement in two eucrites (ALHA 78132, Juvinas) of excess ^{182}W from the decay of short-lived ^{182}Hf (half-life 9 m.y.). Since Hf fractionates into silicates and W into metal, the W-isotopic composition is a function of the timing of the Hf/W fractionation and hence the silicate-metal segregation (i.e., core formation). They concluded that accretion, differentiation, and core formation on Vesta took place in the first 5–15 m.y. of solar system history (note that these two meteorites may not be the best samples for this work, as they are extensively brecciated and metamorphosed; H. Takeda, personal communication, 2001).

What was the heat source for the melting of Vesta and the parent bodies of the many other differentiated asteroids of which we have samples in our collections? In recent years, discussions have largely focused on electrical conduction heating by the T-Tauri solar wind from the pre-main-sequence Sun (e.g., *Sonett et al.,* 1970) and the decay of now extinct ^{26}Al, as first proposed by *Urey* (1955). The discovery of excess ^{26}Mg from the decay of ^{26}Al in CAIs from ordinary, carbonaceous, and enstatite chondrites; in chondrules from carbonaceous and ordinary chondrites (see references in *MacPherson et al.,* 1995; *Huss et al.,* 2001); and in plagioclase from the equilibrated noncumulate eucrite Piplia Kalan (*Srinivasan et al.,* 1999) and the crystalline eucrite A 881394 (*Nyquist et al.,* 2001) suggests that ^{26}Al was widespread in the early solar system and most likely was the heat source for heating and melting of Vesta. It should be noted that the discovery of excess ^{26}Mg in these eucrites also suggests that they formed only 4–5 m.y. after CAIs, lending support to the notion that Vesta melted, fractionated, and differentiated, and that basalts extruded and crystallized, within the first few million years of solar system history. Thermal models based on ^{26}Al decay are broadly consistent with this constraint, producing Vesta's core and crust within 7 m.y. of CAI formation (*Ghosh and McSween,* 1998).

5.2. Volcanism and Metamorphism on 4 Vesta

As noted above, accretion, partial to perhaps total melting, fractionation, and differentiation of Vesta and extrusion and crystallization of basaltic partial melts all took place in the first few million years of solar system history. Thus, unlike on Earth, tectonically controlled magmatism, which requires long times of melting and fractionation to develop, did not occur on Vesta. Instead, the basaltic crust formed in a few million years.

Although the world's collections of meteorites contain samples from some 108 differentiated asteroidal parent bodies (*Meibom and Clark,* 1999), the only abundant basalts are the eucrites. Clearly, many other differentiated asteroids must have produced basaltic partial melts, so the question is, where are the basalts that should have crystallized from these partial melts? *Wilson and Keil* (1991) proposed that if early basaltic partial melts contained a few hundred parts per million of volatiles, these melts would have been ejected from parent bodies ≤100 km in radius as small pyroclastic droplets by explosive volcanism as these melts migrated to the asteroidal surfaces and encountered the vacuum of space. The small pyroclasts, whose sizes have been estimated on the basis of pyroclastic droplet sizes in the lunar regolith to range between ~30 and 4000 μm, would be accelerated into space and lost by spiraling into the Sun (*Wilson and Keil,* 1996a). As is indicated in Fig. 2, basaltic partial melts on the much bigger Vesta (~258 km in radius) would have had to contain ~3.8% of volatiles to be ejected with velocities that would exceed the escape velocity of ~390 m/s. Such high magma volatile contents are rare among basaltic magmas on Earth and are particularly unlikely to have formed on the "dry" Vesta (e.g., *Mittlefehldt,* 1987; *Grady et al.,* 1997), and thus basaltic melts were retained on Vesta and formed its basaltic crust.

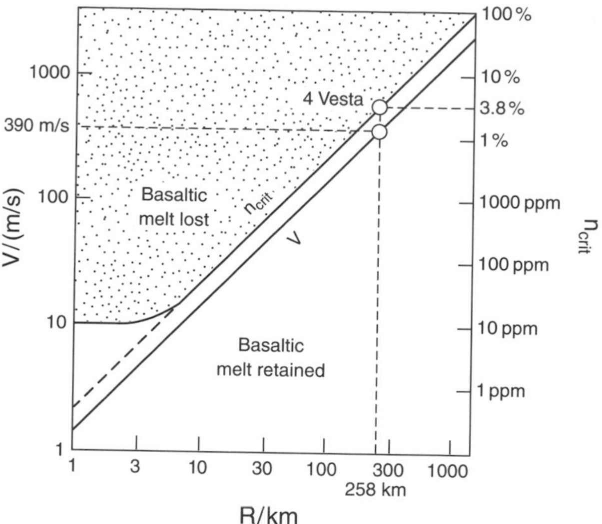

Fig. 2. Values of the escape speed, V, for asteroids of radii, R, assuming a density of 3500 kg/m^3. Also shown as a function of R are the critical values of magma gas content, n_{crit}, at which the eruption velocity, U, equals the escape speed, V. For the approximate radius of 258 km for Vesta, an unrealistically high magma gas content n_{crit} of >3.8 wt% would be required to overcome the escape velocity of ~390 m/s to eject into space basaltic pyroclast/gas mixtures. Thus, Vesta is sufficiently large to retain basalts on its surface. Modified from *Wilson and Keil* (1991). Reprinted from *Earth and Planetary Science Letters,* copyright 1991, with permission of Elsevier Science Publishers.

Cooling rates of cumulate eucrites (*Miyamoto and Takeda,* 1994) and the sizes of the largest vestoids of eucritic composition (*Binzel and Xu,* 1993) suggest that the basaltic crust of Vesta is at least 10 km thick. What were the physical processes that formed this crust? *Wilson and Keil* (1996b), for example, used basic principles of volcanic fluid dynamics to model the nature of the basaltic intrusions and extrusions on a body with the physical properties of Vesta and compared these modeling studies to relevant properties of the HEDs. They concluded that dikes carrying basaltic magma upward from zones of partial melting in the mantle of Vesta may have had vertical extents of 1–30 km and widths of ~10 mm to 4 m. Typical surface lava flows might have had lengths of a few kilometers to several tens of kilometers, widths of a few hundred meters to several kilometers, and thicknesses of 5–20 m. These modeled thicknesses are in excellent agreement with those calculated for unequilibrated eucrites based on comparisons of plagioclase crystal sizes in eucrites and in experimental charges of eucrite composition of known cooling rates (*Walker et al.,* 1978). Eruptions from dikes that reached the surface are estimated to have lasted 8–60 h and, per meter of horizontal surface fissure outcrop, may have had effusion rates of ~0.05 to >3 m^3 s^{-1}, similar to those of current basaltic eruptions on Earth. The erupted lava volumes for individual eruptions may have ranged from negligibly small to up to

3 km^3. Assuming that a 20-km-thick crust formed on Vesta in ~1 m.y., then the time interval between successive lava flow emplacements is estimated to be ~1000 yr. Intrusions with vertical extents of <10 km and widths of ~1 m could have formed at very shallow depth in the forming basaltic crust as cooling dikes stalled before they erupted completely to the surface. Furthermore, intrusions with lateral extents of ~1–30 km, thicknesses of 10 mm to 3 m, and volumes of 10^4 m to 3 km^3 could have formed at the base of the crust because magmas in these locations would have been neutrally buoyant as the density structure of Vesta evolved.

Pyroclastic rocks that formed by explosive eruptions of volatile-bearing magmas are abundant on Earth and Moon (e.g., *Wilson and Head,* 1981, 1983). Unless magmas on Vesta were totally free of volatiles, which is not likely (e.g., *Grady et al.,* 1997), pyroclastic material should therefore occur in HEDs. However, no such pyroclasts were found during detailed microscopic studies of numerous polished thin sections of howardites (A. Yamaguchi, personal communication, 1995). Modeling studies by *Wilson and Keil* (1997) indicate that the physics of pyroclastic eruptions on the relatively small Vesta are sufficiently different from those of Earth and Moon to explain the lack of products of pyroclastic eruptions in HEDs. Specifically, lava fountains on Vesta would have been extremely optically dense, and hence pyroclastic melt/gas mixtures would tend not to cool in flight but land as liquid and form lava lakes that would feed lava flows indistinguishable from those that formed by extrusion. For example, for a typical magma discharge rate from a fissure of 0.3 m^3/s/m and a released gas content of 300 ppm, the gas-droplet eruption speed would be 33.9 m/s; the maximum droplet range would be 4.4 km; and the thickness of the outer layer within which pyroclastic droplets would cool before landing would be 40 m. This corresponds to a fractional area of only 0.9% within which pyroclasts would cool, and an area of 99.1% within which pyroclasts would land as hot, molten droplets, thus readily explaining the lack of pyroclastic materials within HEDs.

Although most eucrites originally formed as quickly cooled surface lava flows (the unequilibrated eucrites), most were subsequently metamorphosed (the equilibrated eucrites) at >800°C for periods of about 1 m.y. (e.g., *Yamaguchi et al.,* 1997). As most eucrites have experienced this metamorphism, it has been suggested that Vesta experienced global crustal metamorphism (e.g., *Yamaguchi et al.,* 1996). Eucrite metamorphism was recognized early and led *Takeda and Graham* (1991) to establish a metamorphic scale for eucrites, based on compositional and textural relationships of their pyroxenes [it is important to remember that this scale concerns the nature of the pyroxenes of eucrites and thus is different from the petrologic type scale of chondritic meteorites of *Van Schmus and Wood* (1967)].

Various heat sources and physical settings have been proposed for the origin of the global metamorphism. *Takeda and Graham* (1991), for example, suggested that impact heating during cratering events was an important process for thermal metamorphism, but *Keil et al.* (1997) argued that on relatively small asteroids impact into cold rock is not an efficient heat source for global thermal events. *Takeda et al.* (1997) later proposed that, after formation of an early scarflike crust that experienced more heating from the magma beneath it, additional heating may have been supplied by impacts into the hot crust. *Sears et al.* (1997) also suggested that metamorphism was triggered by impacts, but they are alone in proposing that metamorphism took place relatively late in the history of Vesta, at ~3.9 Ga at the end of the cataclysmic bombardment and at the time the Ar-Ar ages were reset. They further argued that the different degrees of metamorphism of eucrites resulted from burial under regolith ejecta blankets of highly variable thicknesses (from nonexistent to 2 km), rather than from different initial burial depths.

Clearly impacts have played a role in reheating eucrites, but this process alone cannot be responsible for the global thermal metamorphism experienced by most noncumulate eucrites. *Metzler et al.* (1995), for example, argued that the textures of HEDs reflect complex postigneous histories dominated by multistage thermal and impact metamorphism. Specifically, they recognized six evolutionary phases: (I) crystallization of primary magmas and formation of unequilibrated, noncumulate eucrites; (II) slow subsolidus cooling or reheating during which pyroxene equilibrated; (III) and (V) periods of impact brecciation during which rocks were brecciated *in situ* or, in the case of the polymict HED breccias, mixed with various other rock types; (IV) and (VI) breccias suffered annealing and recrystallization due to thermal metamorphism. They noted that thermal events that caused recrystallization and equilibration of HED lithologies were active prior to, during, and after the formation of impact breccias, indicating that thermal input by impact might be responsible for some thermal overprinting.

However, the relatively old ^{39}Ar-^{40}Ar ages of 4.485 and 4.49 Ga of the highly-metamorphosed noncumulate eucrites Ibitira and EET 90020 respectively indicate that these rocks escaped the widespread cataclysmic impact resetting in the time interval of 3.4–4.1 G.y. ago (*Bogard and Garrison,* 1995; *Yamaguchi et al.,* 2001a), suggesting that impact alone cannot have been the heat source for eucrite metamorphism and that an internal heat source is responsible for global metamorphism of noncumulate eucrites. Note, however, that *Takeda et al.* (1997) argue that impacts into a hot crust could be responsible for global metamorphism. *Warren* (1997), for example, argued that the thermal metamorphism resulted from baking of earlier eucrite lava flows by superjacent flow units. *Yamaguchi et al.* (1996, 1997), on the other hand, modeled the development of the eucritic crust of Vesta by serial volcanism. They argued that the eruption of lava flows and subsequent burial of earlier by later lava flows in the growing basaltic crust of Vesta in the first few million years of Vesta's history resulted in metamorphism of deeply buried flows from heat from within the asteroid as a result of a steep thermal gradient in the growing crust of Vesta.

6. WHAT NEXT?

Among all the asteroids, it appears Vesta is the only one that has preserved nearly intact its original differentiated internal structure with a metal core, an ultramafic mantle, and a basaltic crust. In its structure and composition, it can be viewed as the smallest terrestrial planet, and its further exploration, particularly by spacecraft such as the now approved Dawn Discovery Mission (*Farquhar et al.,* 2002), is highly desirable and anticipated. This mission will add much to our understanding of the accretion, heating, melting, fractionation, and differentiation of small asteroid-sized objects that may have played an important role in the accretion and formation of the terrestrial planets.

Acknowledgments. I thank Peter Thomas for providing Plate 7, Tim Fagan and Akira Yamaguchi for assistance in preparing the photomicrographs in Plate 8, and Rick Binzel, Hap McSween, Hiroshi Takeda, Lionel Wilson, and Akira Yamaguchi for thorough and helpful reviews of the manuscript. This work was supported in part by NASA Grants NAG 5-4212 and NAG 5-11591 (K. Keil, P.I.). This is Hawai'i Institute of Geophysics and Planetology Publication No. 1224 and School of Ocean and Earth Science and Technology Publication No. 5998.

REFERENCES

Allègre C. J., Manhés G., and Göpel C. (1995) The age of the Earth. *Geochim. Cosmochim. Acta, 59,* 1445–1456.

Asphaug E. (1997) Impact origin of the Vesta family. *Meteoritics & Planet. Sci., 32,* 965–980.

Binzel R. P. and Xu S. (1993) Chips off of asteroid 4 Vesta: Evidence for the parent body of basaltic achondrite meteorites. *Science, 260,* 186–191.

Binzel R. P., Gaffey M. J., Thomas P. C., Zellner B. H., Storrs A. D., and Wells E. N. (1997) Geologic mapping of Vesta from 1994 Hubble Space Telescope images. *Icarus, 128,* 95–103.

Binzel R. P., Bus S. J., and Burbine T. H. (1999) The orbital distribution of Vesta-like asteroids (abstract). In *Lunar and Planetary Science XXX,* Abstract #1214. Lunar and Planetary Institute, Houston (CD-ROM).

Bogard D. D. (1995) Impact ages of meteorites: A synthesis. *Meteoritics, 30,* 244–268.

Bogard D. D. and Garrison D. H. (1995) ^{39}Ar-^{40}Ar age of the Ibitira eucrite and constraints on the time of pyroxene equilibration. *Geochim. Cosmochim. Acta, 59,* 4317–4322.

Bowman L. E., Spilde M. N., and Papike J. J. (1997) Automated energy dispersive spectrometer modal analysis applied to diogenites. *Meteoritics & Planet. Sci., 32,* 869–875.

Burbine T. H., Buchanan P. C., Binzel R. P., Bus S. J., Hiroi T., Hinrichs J. L., Meibom A., and McCoy T. J. (2001) Vesta, Vestoids and the HEDs: Relationships and the origin of spectral differences. *Meteoritics & Planet. Sci., 36,* 761–782.

Carlson R. W. and Lugmair G. W. (2000) Timescales of planetesimal formation and differentiation based on extinct and extant radioisotopes. In *Origin of the Earth and Moon* (R. M. Canup and K. Righter, eds.), pp. 25–44. Univ. of Arizona, Tucson.

Chen J. H. and Wasserburg G. J. (1981) The isotopic composition of uranium and lead in Allende inclusions and meteorite phosphates. *Earth Planet. Sci. Lett., 52,* 1–15.

Clayton R. N. and Mayeda T. K. (1996) Oxygen-isotope studies of achondrites. *Geochim. Cosmochim. Acta, 60,* 1999–2018.

Cochran A. L. and Vilas F. (1998) The changing spectrum of Vesta: Rotationally resolved spectroscopy on surface. *Icarus, 134,* 207–212.

Consolmagno G. J. and Drake M. J. (1977) Composition and evolution of the eucrite parent body: Evidence from rare earth elements. *Geochim. Cosmochim. Acta, 41,* 1271–1282.

Cruikshank D., Tholen D. J., Hartmann W. K., Bell J. F., and Brown R. H. (1991) Three basaltic Earth-approaching asteroids and the source of the basaltic meteorites. *Icarus, 89,* 1–13.

Delaney J. S., Prinz M., and Takeda H. (1984) The polymict eucrites. *Proc. Lunar Planet. Sci. Conf. 15th,* in *J. Geophys. Res., 89,* C252–C288.

Drake M. J. (1979) Geochemical evolution of the eucrite parent body: Possible nature and evolution of asteroid 4 Vesta. In *Asteroids* (T. Gehrels, ed.), pp. 765–782. Univ. of Arizona, Tucson.

Drake M. J. (2001) The eucrite/Vesta story. *Meteoritics & Planet. Sci., 36,* 501–513.

Dreibus G., Brückner J., and Wänke H. (1997) On the core mass of the asteroid Vesta (abstract). In *Meteoritics & Planet. Sci., 32,* A36.

Drummond J. D., Fugate R. Q., Christo J. C., and Hege E. K. (1998) Full adaptive optics images of asteroids Ceres and Vesta; rotational poles and triaxial ellipsoid dimensions. *Icarus, 132,* 80–99.

Eugster O. and Michel T. (1995) Common asteroid break-up events of eucrites, diogenites, and howardites and cosmic-ray production rates for noble gases in achondrites. *Geochim. Cosmochim. Acta, 59,* 177–200.

Farquhar R., Kawaguchi J., Russell C., Schwehm G., Veverka J., and Yeomans D. (2002) Spacecraft exploration of asteroids: The 2001 perspective. In *Asteroids III* (W. F. Bottke Jr. et al., eds.), this volume. Univ. of Arizona, Tucson.

Feierberg M. A., Larson H. P., Fink U., and Smith H. A. (1980) Spectroscopic evidence for two achondrite parent bodies: Asteroid 349 Dembowska and 4 Vesta. *Geochim. Cosmochim. Acta, 44,* 513–524.

Fredriksson K. and Keil K. (1963) The light-dark structure in the Pantar and Kapoeta stone meteorites. *Geochim. Cosmochim. Acta, 27,* 717–739.

Gaffey M. J. (1983) The asteroid 4 Vesta: Rotational spectral variations, surface material heterogeneity, and implications for origins of basaltic achondrites (abstract). In *Lunar and Planetary Science XIV,* pp. 231–232. Lunar and Planetary Institute, Houston.

Gaffey M. J. (1997) Surface lithologic heterogeneity of asteroid 4 Vesta. *Icarus, 127,* 130–157.

Ghosh A. and McSween H. Y. Jr. (1998) A thermal model for the differentiation of asteroid 4 Vesta, based on radiogenic heating. *Icarus, 134,* 187–206.

Grady M. M., Wright I. P., and Pillinger C. T. (1997) Carbon in howardite, eucrite and diogenite basaltic achondrites. *Meteoritics & Planet. Sci., 32,* 863–868.

Hewins R. H. and Newsom H. E. (1988) Igneous activity in the early solar system. In *Meteorites and the Early Solar System* (J. F. Kerridge and M. S. Matthews, eds.), pp. 73–101. Univ. of Arizona, Tucson.

Hiroi T., Pieters C. M., and Takeda H. (1994) Grain size of the surface regolith of asteroid 4 Vesta estimated from its reflectance spectrum in comparison with HED meteorites. *Meteor-*

itics, 29, 394–396.

Huss G. R., MacPherson G. J., Wasserburg G. J., Russell S. S., and Srinivasan G. (2001) Aluminum-26 in calcium-aluminum-rich inclusions and chondrules from unequilibrated ordinary chondrites. *Meteoritics & Planet. Sci., 36,* 975–997.

Ikeda Y. and Takeda H. (1985) A model for the origin of basaltic achondrites based on the Yamato 7308 howardite. *Proc. Lunar Planet. Sci. Conf. 15th,* in *J. Geophys. Res., 90,* C649–C663.

Jurewicz A. J. G., Mittlefehldt D. W., and Jones J. H. (1993) Experimental partial melting of the Allende (CV) and Murchison (CM) chondrites and the origin of asteroidal basalts. *Geochim. Cosmochim. Acta, 57,* 2123–2139.

Keil K. (2000) Thermal alteration of asteroids: Evidence from meteorites. *Planet. Space Sci., 48,* 887–903.

Keil K., Stöffler D., Love S. G., and Scott E. R. D. (1997) Constraints on the role of impact heating and melting in asteroids. *Meteoritics & Planet. Sci., 32,* 349–363.

Larson H. and Fink U. (1975) Infrared spectral observations of asteroid 4 Vesta. *Icarus, 26,* 420–427.

Lazzaro D., Michtchenko T., Carvano J. M., Binzel R. P., Bus S. J., Burbine T. H., Mothé-Diniz T., Florczak M., Angeli C. A., and Harris A. W. (2000) Discovery of a basaltic asteroid in the outer main belt. *Science, 288,* 2033–2035.

Lee D. C. and Halliday A. N. (1997) Core formation on Mars and differentiated asteroids. *Nature, 388,* 854–857.

Lugmair G. W. and Shukolyukov A. (1998) Early solar system timescales according to ^{53}Mn-^{53}Cr systematics. *Geochim. Cosmochim. Acta, 62,* 2863–2886.

MacPherson G. J., Davis A. M., and Zinner E. K. (1995) The distribution of aluminum-26 in the early solar system — A reappraisal. *Meteoritics, 30,* 365–386.

Marzari F., Cellino A., Davis D. R., Farinella P., Zappalà V., and Vanzani V. (1996) Origin and evolution of the Vesta asteroid family. *Astron. Astrophys., 316,* 248–262.

Mason B. (1962) *Meteorites.* Wiley, New York. 274 pp.

Mason B. (1968) Composition of stony meteorites. In *Extraterrestrial Matter* (C. A. Randall Jr., ed.), pp. 2–24. Northern Illinois Univ., DeKalb.

McCarthy D. W. Jr., Freeman J. D., and Drummond J. D. (1994) High resolution images of Vesta at 1.65 μm. *Icarus, 108,* 285–297.

McCord T. B., Adams J. B., and Johnson T. V. (1970) Asteroid Vesta: Spectral reflectivity and compositional implications. *Science, 168,* 1445–1447.

McFadden L. A., McCord T. B., and Pieters C. (1977) Vesta: The first pyroxene band from new spectroscopic measurements. *Icarus, 31,* 439–446.

Meibom A. and Clark B. E. (1999) Evidence for the insignificance of ordinary chondritic material in the asteroid belt. *Meteoritics & Planet. Sci., 34,* 7–24.

Metzler K., Bobe K.-D., Palme H., Spettel B., and Stöffler D. (1995) Thermal and impact metamorphism of the HED parent body. *Planet. Space Sci., 43,* 499–525.

Michalak G. (2000) Determination of asteroid masses I. (1) Ceres, (2) Pallas and (4) Vesta. *Astron. Astrophys., 360,* 363–374.

Migliorini F., Morbidelli A., Zappalà V., Gladman B. J., Bailey M. E., and Cellino A. (1997) Vesta fragments from ν_6 and 3:1 resonances: Implications for V-type near-Earth asteroids and howardite, eucrite and diogenite meteorites. *Meteoritics & Planet. Sci., 32,* 903–916.

Mittlefehldt D. W. (1980) The composition of mesosiderite olivine clasts and implications for the origin of pallasites. *Earth Planet. Sci. Lett., 51,* 29–40.

Mittlefehldt D. W. (1987) Volatile degassing of basaltic achondrite parent bodies: Evidence from alkali elements and phosphorous. *Geochim. Cosmochim. Acta, 51,* 267–278.

Mittlefehldt D. W. (1990) Petrogenesis of mesosiderites: I. Origin of mafic lithologies and comparison with basaltic achondrites. *Geochim. Cosmochim. Acta, 54,* 1165–1173.

Mittlefehldt D. W., McCoy T. J., Goodrich C. A., and Kracher A. (1998) Non-chondritic meteorites from asteroidal bodies. In *Planetary Materials* (J. J. Papike, ed.), pp. 4-1 to 4-195. *Reviews in Mineralogy, Vol. 36,* Mineralogical Society of America.

Miyamoto M. and Takeda H. (1994) Evidence for excavation of deep crustal material of a Vesta-like body from Ca compositional gradients in pyroxene. *Earth Planet. Sci. Lett., 122,* 343–349.

Miyamoto M., Mikouchi T., and Kaneda K. (2001) Thermal history of the Ibitira noncumulate eucrite as inferred from pyroxene exsolution lamellae: Evidence for reheating and rapid cooling. *Meteoritics & Planet. Sci., 36,* 231–237.

Nyquist L. E., Takeda H., Bansal B. M., Shih C.-Y., Wiesmann H., and Wooden J. L. (1986) Rb-Sr and Sm-Nd internal isochron ages of a subophitic basalt clast and a matrix sample from the Y75011 eucrite. *J. Geophys. Res., 91,* 8137–8150.

Nyquist L. E., Bogard D., Takeda H., Bansal B., Wiesmann H., and Shih C.-Y. (1997) Crystallization, recrystallization, and impact-metamorphic ages of eucrites Y792510 and Y791186. *Geochim. Cosmochim. Acta, 61,* 2119–2138.

Nyquist L. E., Reese Y., Wiesmann H., Shih C.-Y., and Takeda H. (2001) Dating eucrite formation and metamorphism (abstract). *Antarctic Meteorites, XXVI,* pp. 113–115. National Institute of Polar Research, Tokyo.

Papike J. J. (1998) Comparative planetary mineralogy: Chemistry of melt-derived pyroxene, feldspar, and olivine. In *Planetary Materials* (J. J. Papike, ed.), pp. 7-1 to 7-11. *Reviews in Mineralogy, Vol. 36,* Mineralogical Society of America.

Pilcher F. (1979) Circumstances of minor planet discovery. In *Asteroids* (T. Gehrels, ed.), pp. 1130–1154. Univ. of Arizona, Tucson.

Pun A., Keil K., Taylor G. J., and Wieler R. (1998) The Kapoeta howardite: Implications for the regolith evolution of the howardite-eucrite-diogenite parent body. *Meteoritics & Planet. Sci., 33,* 835–851.

Righter K. and Drake M. J. (1997) A magma ocean on Vesta: Core formation and petrogenesis of eucrites and diogenites. *Meteoritics & Planet. Sci., 32,* 929–944.

Rubin A. E. and Mittlefehldt D. W. (1993) Evolutionary history of the mesosiderite asteroid: A chronologic and petrologic synthesis. *Icarus, 101,* 201–212.

Ruzicka A., Snyder G. A., and Taylor L. A. (1997) Vesta as the howardite, eucrite and diogenite parent body: Implications for the size of a core and for large-scale differentiation. *Meteoritics & Planet. Sci., 32,* 825–840.

Schubart J. and Matson D. L. (1979) Masses and densities of asteroids. In *Asteroids* (T. Gehrels, ed.), pp. 84–97. Univ. of Arizona, Tucson.

Sears D. W. G., Symes S. J. K., Batchelor J. D., Akridge D. G., and Benoit P. H. (1997) The metamorphic history of eucrites and eucrite-related meteorites and the case for late metamorphism. *Meteoritics & Planet. Sci., 32,* 917–927.

Shearer C. K., Fowler G. W., and Papike J. J. (1997) Petrogenetic models for magmatism on the eucrite parent body: Evidence from orthopyroxene in diogenites. *Meteoritics & Planet. Sci., 32,* 877–889.

Shukolyukov A. and Lugmair G. W. (1993a) Live iron-60 in the

early solar system. *Science, 259,* 1138–1142.
Shukolyukov A. and Lugmair G. W. (1993b) ^{60}Fe in eucrites. *Earth Planet. Sci. Lett., 119,* 159–166.
Sonett C. P., Colburn D. S., Schwartz K., and Keil K. (1970) The melting of asteroidal-sized parent bodies by unipolar dynamo induction from a primordial T Tauri sun. *Astrophys. Space Sci., 7,* 446–488.
Srinivasan G., Goswami J. N., and Bhandari N. (1999) ^{26}Al in eucrite Piplia Kalan: Plausible heat source and formation chronology. *Science, 284,* 1348–1350.
Standish E. M. Jr. and Hellings R. W. (1989) A determination of the masses of Ceres, Pallas, and Vesta from their perturbations upon the orbit of Mars. *Icarus, 80,* 326–333.
Stolper E. (1977) Experimental petrology of eucritic meteorites. *Geochim. Cosmochim. Acta, 41,* 587–681.
Suess H. E., Wänke H., and Wlotzka F. (1964) On the origin of gas-rich meteorites. *Geochim. Cosmochim. Acta, 28,* 595–607.
Takeda H. (1979) A layered-crust model of a howardite parent body. *Icarus, 40,* 455–470.
Takeda H. (1997) Mineralogical records of early planetary processes on the howardite, eucrite, diogenite parent body with reference to Vesta. *Meteoritics & Planet. Sci., 32,* 841–853.
Takeda H. and Graham A. L. (1991) Degree of equilibration of eucritic pyroxenes and thermal metamorphism of the earliest planetary crust. *Meteoritics, 26,* 129–134.
Takeda H. and Mori H. (1985) The diogenite-eucrite links and the crystallization history of a crust of their parent body. *Proc. Lunar Planet. Sci. Conf. 15th,* in *J. Geophys. Res., 90,* C626–C648.
Takeda H., Miyamoto M., Ishii T., and Reid A. M. (1976) Characterization of crust formation on a parent body of achondrites and the moon by pyroxene crystallography and chemistry. *Proc. Lunar Sci. Conf. 7th,* pp. 3535–3548.
Takeda H., Mori H., Delaney J. S., Prinz M., Harlow G. E., and Ishii T. (1983) Mineralogical comparison of Antarctic and non-Antarctic HED (howardites-eucrites-diogenites) achondrites. *Mem. Natl. Inst. Polar Res., Spec. Issue No. 30,* 181–205.
Takeda H., Ishii T., Arai T., and Miyamoto M. (1997) Mineralogy of the Asuka 87 and 88 eucrites and crustal evolution of the HED parent body. *Antarct. Meteorite Res., 10,* 401–413.
Taylor G. J., Keil K., McCoy T., Haack H., and Scott E. R. D. (1993) Asteroid differentiation: Pyroclastic volcanism to magma oceans. *Meteoritics, 28,* 34–52.
Taylor S. R. and Norman M. D. (1990) Accretion of differentiated planetesimals to the Earth. In *Origin of the Earth* (H. E. Newsom and J. H. Jones, eds.), pp. 29–43. Oxford Univ., New York.
Thomas P. C., Binzel R. P., Gaffey M. J., Storrs A. D., Wells E. N., and Zellner B. H. (1997a) Impact excavation on asteroid 4 Vesta: Hubble Space Telescope results. *Science, 277,* 1492–1495.
Thomas P. C., Binzel R. P., Gaffey M. J., Zellner B. H., Storrs A. D., and Wells E. N. (1997b) Vesta: Spin pole, size and shape from HST images. *Icarus, 128,* 88–94.
Treiman A. H. (1997) The parent magmas of the cumulate eucrites: A mass balance approach. *Meteorit. & Planet Sci., 32,* 217–230.
Urey H. C. (1955) The cosmic abundances of potassium, uranium and thorium and the heat balance of the Earth, the Moon, and Mars. *Proc. Natl. Acad. Sci. U.S., 41,* 127.
Van Schmus W. R. and Wood J. A. (1967) A chemical-petrologic classification for the chondritic meteorites. *Geochim. Cosmochim Acta, 31,* 747–765.
Vilas F., Cochran A. L., and Jarvis K. S. (2000) Vesta and the Vestoids: A new rock group? *Icarus, 147,* 119–128.
Walker D., Powell M. A., Lofgren G. E., and Hays J. F. (1978) Dynamic crystallization of a eucrite basalt. *Proc. Lunar Planet. Sci. Conf. 9th,* pp. 1369–1391.
Warren P. H. (1997) Magnesium oxide-iron oxide mass balance constraints and a more detailed model for the relationship between eucrites and diogenites. *Meteoritics & Planet. Sci., 32,* 945–963.
Wasson J. T. and Chapman C. R. (1996) Space-weathering of basalt-covered asteroids: Vesta an unlikely source of the HED meteorites (abstract). In *Workshop on Evolution of Igneous Asteroids: Focus on Vesta and the HED Meteorites,* pp. 38–39. LPI Tech. Rpt. 96-02, Part 1, Lunar and Planetary Institute, Houston.
Wasson J. T. and Wetherill G. W. (1979) Dynamical, chemical and isotopic evidence regarding the formation locations of asteroids and meteorites. In *Asteroids* (T. Gehrels, ed.), pp. 926–974. Univ. of Arizona, Tucson.
Wasson J. T., Chapman C. R., Grogan K., and Dermott S. F. (1996) Possible formation of the Vesta-family asteroids and the main IRAS dust band by an oblique impact on Vesta (abstract). In *Lunar and Planetary Science XXVII,* pp. 1387–1388. Lunar and Planetary Institute, Houston.
Welton K. C., Lindner L., van der Borg K., Loeken T., Scherer P., and Schultz L. (1997) Cosmic-ray exposure ages of diogenites and the recent collisional history of the howardite, eucrite and diogenite parent body/bodies. *Meteoritics & Planet. Sci., 32,* 891–902.
Wetherill G. W. (1987) Dynamical relations between asteroids, meteorites and Apollo-Amor objects. *Phil. Trans. R. Soc. London, A323,* 323–337.
Williams J. G. (1989) Asteroid family identifications and proper elements. In *Asteroids II* (R. P. Binzel et al., eds.), pp. 1034–1072. Univ. of Arizona, Tucson.
Wilson L. and Head J. W. (1981) Ascent and eruption of basaltic magma an the Earth and Moon. *J. Geophys. Res., 86,* 2971–3001.
Wilson L. and Head J. W. (1983) A comparison of volcanic eruption on Earth, Moon, Io and Venus. *Nature, 302,* 663–669.
Wilson L. and Keil K. (1991) Consequences of explosive eruptions on small Solar System bodies: The case of the missing basalts on the aubrite parent body. *Earth Planet. Sci. Lett., 104,* 505–512.
Wilson L. and Keil K. (1996a) Clast sizes from ejecta from explosive eruptions on asteroids: Implications for the fate of the basaltic products of differentiation. *Earth Planet. Sci. Lett., 140,* 191–200.
Wilson L. and Keil K. (1996b) Volcanic eruptions and intrusions on the asteroid 4 Vesta. *J. Geophys. Res., 101,* 18927–18940.
Wilson L. and Keil K. (1997) The fate of pyroclasts produced in explosive eruptions on the asteroid 4 Vesta. *Meteoritics & Planet. Sci., 32,* 813–823.
Xu S., Binzel R. P., and Bus S. J. (1995) Small main-belt asteroid spectroscopic survey — Initial results. *Icarus, 115,* 1–35.
Yamada M., Sasaki S., Nagahara H., Fujiwara A., Hasegawa S., Yano H., Hiroi T., Obashi H., and Ohtake H. (1999) Simulation of space weathering of planet-forming materials: Nanosecond pulse laser irradiation and proton implantation on olivine and pyroxene samples. *Earth Planets Space, 51,* 1255–1265.
Yamaguchi A. (2001) Mineralogical study of a highly metamorphosed eucrite, Northwest Africa 011 (abstract). In *Lunar and*

Planetary Science XXXII, Abstract #1578. Lunar and Planetary Institute, Houston (CD-ROM).

Yamaguchi A., Taylor G. J., and Keil K. (1996) Global crustal metamorphism of the eucrite parent body. *Icarus, 124,* 97–112.

Yamaguchi A., Taylor G. J., and Keil K. (1997) Metamorphic history of the eucritic crust of 4 Vesta. *J. Geophys. Res., 102,* 13381–13386.

Yamaguchi A., Taylor G. J., Keil K., Floss C., Crozaz G., Nyquist L. E., Bogard D. D., Garrison D., Reese Y., Wiesman H., and Shih C-Y. (2001a) Post-crystallization reheating and partial melting of eucrite EET90020 by impact into the hot crust of asteroid 4 Vesta ~4.50 Ga ago. *Geochim. Cosmochim. Acta, 65,* 3577–3599.

Yamaguchi A., Misawa K., Haramura H., Kojima H., Clayton R. N., Mayeda T. K., and Ebihara M. (2001b) Northwest Africa 011, a new basaltic meteorite (abstract). *Meteoritics & Planet. Sci., 36,* A228.

Yamaguchi A., Clayton R. N., Mayeda T. K., Ebihara M., Oura Y., Miura Y. N., Haramura H., Misawa K., Kojima H., and Nagao K. (2002) A new source of basaltic meteorites inferred from Northwest Africa 011. *Science, 296,* 334–336.

Zappalà V., Bandjoya Ph., Cellino A., Farinella P., and Froeschlé C. (1995) Asteroid families: Search of a 12,487-asteroid sample using two different clustering techniques. *Icarus, 116,* 291–314.

Zema M., Domeneghetti M. C., Molin G. M., and Tazzoli V. (1997) Cooling rates of diogenites: A study of Fe^{2+}-Mg ordering in orthopyroxene by single-crystal x-ray diffration. *Meteoritics & Planet. Sci., 32,* 855–862.

Asteroid Space Weathering and Regolith Evolution

Beth Ellen Clark
Cornell University

Bruce Hapke
University of Pittsburgh

Carlé Pieters
Brown University

Daniel Britt
University of Tennessee

Over time, exposure of airless bodies to the space environment results in optical changes to their surfaces. These optical changes are functions of the porosity, grain size distribution, and composition of the surface, and they depend on the relative rates of surface modification processes. Collectively, surface modification processes (such as impacts, solar wind ion implantation, sputtering, and micrometeorite bombardment) and their resulting optical effects have come to be known as "space weathering." Studies of lunar rocks and soils are the most important foundation we have on which to build an understanding of space weathering on asteroids. We cannot directly measure asteroid surfaces in a laboratory environment; therefore, we describe the lunar case, and compare it with the evidence for asteroids. In this chapter we review the evidence for space weathering on asteroids, including spectroscopy of optical effects, microscopy of physical effects, simulations of processes, lunar soils, meteorite breccias, spacecraft observations, and theoretical modeling. An understanding of space weathering is important to all remote-sensing studies of asteroid surfaces.

1. INTRODUCTION

We use the term space weathering to mean any surface modification process (or processes) that may tend to change the apparent traits (optical properties, physical structure, or chemical or mineralogical properties) of the immediate, remotely sensed surface of an airless body from analogous traits of the body's inherent bulk material. Studies of lunar soils and rocks brought back by the Apollo astronauts have provided important constraints on space weathering on the Moon (*Pieters et al.,* 2000; *Hapke,* 2001). These constraints have been used to account for the effects of space weathering in order to reveal underlying geologic/mineralogic patterns (*Lucey et al.,* 1995, 2000). However, space weathering effects on asteroids are not well understood because we do not have soils taken directly from the surface of an asteroid. We must therefore utilize indirect means to study space weathering on these bodies. A variety of evidence suggests that asteroid surfaces are characterized by complex particle size distributions (*Hörz and Cintala,* 1997), impact shock effects (*Keil et al.,* 1992), and mineral specific responses to impacts, solar wind, and cosmic-ray bombardment (*Sasaki et al.,* 2001; *Moroz et al.,* 1996; *Clark and Johnson,* 1996).

There is a variety of evidence that asteroids, except for the smallest ones, possess regoliths (of unknown depth). Asteroid regoliths were thought to be dominated by grains of a coarser average size than lunar regolith because asteroids are smaller (have weaker gravitational fields) and probably retain less impact ejecta (*Housen and Wilkening,* 1982; *Dollfus et al.,* 1989). Also, impact velocities are probably lower, so agglutination is less effective (*McKay et al.,* 1989). Polarization studies indicate surface grain sizes in the range of 30–300 μm (*Dollfus et al.,* 1989). Observations of 951 Gaspra and 243 Ida by the *Galileo* spacecraft revealed surfaces with degraded crater morphology and evidence of retained crater ejecta, suggesting that regolith formation and evolution processes were at work (*Sullivan et al.,* 1996; *Lee et al.,* 1996; *Carr et al.,* 1994; *Geissler et al.,* 1996). In addition, *NEAR Shoemaker* images of 433 Eros reveal infilled craters, distributed boulders, and abundant slump features. Taken together, these are strong evidence for a substantial regolith layer (*Veverka et al.,* 2001).

In this chapter we review the literature for constraints on asteroid space weathering. We discuss spectroscopic studies of optical effects, microscopic studies of physical effects, laboratory simulations of processes, lunar and meteorite evidence, pertinent spacecraft observations of asteroid surfaces, and theoretical modeling. This topic is important to asteroid researchers because space weathering affects the interpretation of all spectroscopic (compositional) observations of asteroids.

2. CHANGING PERSPECTIVE

One of the first to suggest that the space environment darkens material exposed on the lunar surface was *Gold* (1955). His prediction was amply confirmed when the Apollo samples of regolith were found to be darker, with redder continuum spectra, and more subdued absorption bands than pulverized lunar rocks of similar composition. It was postulated, based on experiments in which lunar rocks were vitrified in a N atmosphere, that the darkening was caused by impact-melted glass in the regolith. By an unfortunate quirk of nature, ferric impurities in glasses melted in N mimicked the spectral effects of space weathering and seemed to provide a ready explanation for them. Consequently, the idea that space weathering is caused by impact vitrification rapidly became widely accepted in the planetary science community, in spite of the evidence that vacuum-melted glasses are not dark. A suggestion by *Hapke et al.* (1975) that the spectral effects are due to vapor condensates was ignored, largely because there seemed to be no evidence for them in the lunar samples.

One of the first asteroids to be measured spectroscopically was Vesta, whose spectrum turned out to be strongly similar to those of particulate basaltic achondrites (*Bobrovnikoff,* 1929; *McCord et al.,* 1970). This seemed to provide a direct link between a group of meteorites and a probable parent body, and gave rise to the hope that other groups of meteorites and asteroids could be similarly linked. The strong resemblance between the spectra of Vesta and the achondrites also implied that space weathering did not operate in the asteroid belt. This was consistent with the paradigm that the optical changes were caused by agglutinate formation, because the velocities with which micrometeorites hit the surfaces of asteroids are generally thought to be too low for appreciable melting to occur (*Chapman and Salisbury,* 1973; *Matson et al.,* 1977; *McKay et al.,* 1989).

However, as time went on, the attempt to connect groups of asteroids and meteorites spectrally met with only limited success. In particular, no clear parent bodies were found for the most abundant class of meteorites, the ordinary chondrites, and no class of meteorites could be matched with one of the largest groups of asteroids, the S-class asteroids. Thus, the question arose as to whether the S asteroids might, in fact, be bodies of the same composition as ordinary chondrites, but with surfaces that have been altered by some kind of space weathering, either the same or different from the process that operates on the lunar surface. This gave rise to considerable discussion and sometimes heated debate (cf. *Matson et al.,* 1977; *Chapman,* 1995, 1996; *Bell,* 1995a,b, 1997, 1998; *Britt et al.,* 1992).

Meanwhile, optical, magnetic, electron spin resonance, and X-ray photoelectron spectroscopic evidence that reduced Fe is present in abundance on or near the surfaces of most lunar regolith particles continued to accumulate (see *Hapke,* 2001, and *Taylor et al.,* 2001a, for detailed descriptions). In addition, *Pieters et al.* (1993) showed that the spectral effects of lunar space weathering are concentrated in the smaller regolith particles (rather than agglutinates), and that the optical effects are caused by a surface, rather than a volume, phenomenon. Finally, Keller and his associates (*Keller and McKay,* 1993, 1997; *Keller et al.,* 1998, 1999; *Pieters et al.,* 2000), using improved transmission electron technology, documented that the regolith particles of a mature lunar soil are almost universally covered with thin, vapor-deposited coatings that contain abundant grains of submicroscopic metallic Fe, as predicted by the vapor deposition hypothesis.

There is no generally accepted term for the fine-grained Fe particles in the surface coatings. *Pieters et al.* (2000) used nanophase reduced Fe, $npFe^0$, which emphasized the scale of particles produced and their reduced nature. However, a significant amount of the optically active Fe in lunar soil is not nanometer-sized, but is in the >100-nm size range (*Hapke,* 2001). In addition, for at least 25 years the more general term "submicroscopic metallic iron" (SMFe) has been used (*Hapke et al.,* 1975).

Vapor condensates can be generated by two processes: solar wind sputtering and micrometeorite impact vaporization. Because of the lower velocities of the impactors in the asteroid belt, the second process is probably less effective there. However, the first process must operate, although at a reduced rate. *Hapke* (2000, 2001) has shown that the addition of as little as 0.025% SMFe can alter the spectrum of a powdered ordinary chondrite so that it strongly resembles that of an S asteroid. This may be compared to the 0.5% Fe required to convert the spectrum of a pulverized lunar rock into a regolith spectrum. Hence, the relevant question no longer is whether or not space weathering occurs on asteroids, but rather why some bodies appear pristine, while others have spectra that are altered.

3. ASTEROID-METEORITE CONNECTION

Studies of the asteroid-meteorite compositional connection concentrate on the evidence from spectral reflectance properties of materials in the wavelength region of ~0.35–3.2 μm. In this range, particulate minerals at the surfaces of airless bodies impart absorption bands on reflected light spectra that are diagnostic of the surface composition. Common parametric measurements of spectral reflectance include the albedo (total amount of light reflected), the depth of absorption bands (relative to the continuum), and the slope of the continuum, generally measured as rise over run from 0.7 to 1.5 μm (e.g., *McCord et al.,* 1981; *Clark and Roush,* 1984; *Pieters,* 1986). Telescopic and laboratory spectral work over the last 30 years (cf. *McCord et al.,* 1970; *Chapman and Salisbury,* 1973; *Gaffey et al.,* 1993; *Pieters and McFadden,* 1994) have established links between many asteroid spectral types and meteorite groups. Statistically, however, asteroid and meteorite spectra show consistent offsets in spectral parameters, possibly as a result of optical differences between bulk meteorites measured in the laboratory and asteroid surfaces measured remotely. For example, *Britt et al.* (1992) and *Burbine* (1991) have com-

TABLE 1. Space weathering and asteroid types.

Asteroid Type	Meteorite Analogs	Mineralogy	Alteration Effects
S	H, L, LL chondrites; stony irons, IAB irons, lodranites; winonites, siderophiles, ureilites	Olivine, pyroxene, metal	Albedo: up to 50% decrease Red slope: weak to moderate increase Band suppression: 50% decrease
V	Basaltic achondrites	Pyroxene, feldspar	Band suppression: 20%
A	Brachinites, pallasites	Olivine	Red slope: moderate increase Band suppression: 20% decrease?
Q	H, L, LL chondrites	Olivine, pyroxene, metal	Albedo: 20–30% decrease Band suppression: 20–30% decrease
Lunar	Lunar meteorites	Basalt, feldspar	Albedo: 50–75% decrease Red slope: moderate to strong increase Band suppression: 50–75% decrease

pared spectral parameters for asteroid and meteorite surveys and found significant mismatches even in the linked populations. Although space weathering may occur on all asteroids, many types lack the high albedo and strong spectral band contrasts that make weathering effects easily detectable. Asteroid types for which there is additional evidence of space weathering relative to their meteorite analogs include the S, V, A, and Q types. Shown in Table 1 are the probable meteorite analogs, estimated mineralogy, and inferred weathering effects. Note that the range in continuum slope change is 20–40% on average, although some lunar samples show strong increases of up to 300%. To illustrate weathering-related spectral alteration, Fig. 1 shows two examples of asteroid spectra with their likely meteorite analogs.

The Q asteroids, which have ordinary chondrites as their direct meteorite analog, show evidence of band suppression and albedo reduction. The olivine-rich A class shows a red continuum slope and modest band suppression consistent with weathering of olivine (*Sasaki et al.,* 2001). Within the V class, the largest V asteroid, 4 Vesta, shows subdued absorption features relative to many of the much smaller "vestoids" as well as the V-type meteorite analog, the basaltic achondrite meteorites (*Binzel and Xu,* 1993; *Hiroi et al.,* 1994, 1995). This suggests that some physical process or minor weathering has occurred on Vesta's surface, although Vesta may have been resurfaced by a major impact in its recent past (*Binzel and Xu,* 1993; *Pieters and Binzel,* 1994; *Hiroi et al.,* 2000). Figure 1 compares Vesta with a eucrite meteorite, thought to be an excellent match (*McCord et al.,* 1970). While the band centers and shapes are similar, Vesta shows a reduction in band depth indicative of space weathering. The strongest evidence for space weathering is for the S-type asteroids and the ordinary chondrites (OCs). The spectral band center data show that OCs have similar mineralogy to the S asteroids, but weathering processes appear to produce differences in band depth, red slope, and albedo.

Although the spectral data indicate that space weathering might occur on asteroids, some of the spectral trends are difficult to interpret and differ from lunarlike space weathering. Within the S-type asteroids there is a trend toward reduced 1.0-μm band depth with increasing diameter shown in Fig. 2 (*Gaffey et al.,* 1993). If larger-diameter asteroids have longer lifetimes and therefore older surfaces, this indicates an attenuation of band depth with increasing surface age.

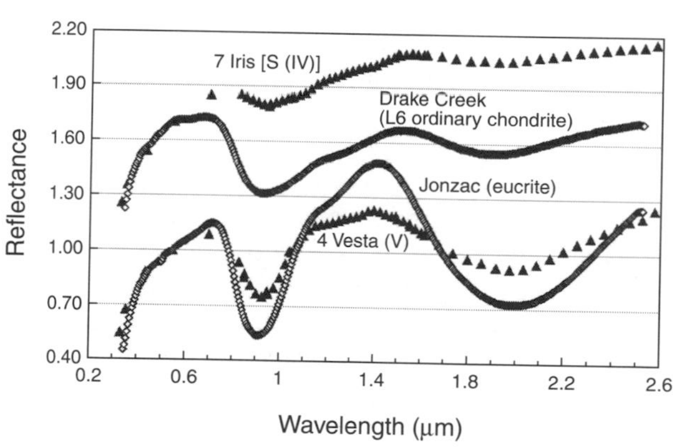

Fig. 1. Two examples of reflectance spectra of asteroids paired with their meteorite analogs. One plausible explanation for the spectral mismatches is that space weathering processes affect the surfaces of the asteroids, altering them from their original spectral properties. Asteroids are displayed as closed triangles and meteorites as open diamonds. The top pair is offset vertically by 0.7 for clarity and show the S (IV) asteroid 7 Iris with the L6 ordinary chondrite Drake Creek. Although both objects have approximately the same mineralogy as shown by their band minima, the spectra of 7 Iris is reddened by 30% relative to Drake Creek. Also, Iris's major absorption bands are significantly reduced. The pair of 4 Vesta and the eucrite Jonzac also show similar mineralogy, but the band depth of 4 Vesta is reduced by 40% relative to its analog meteorite.

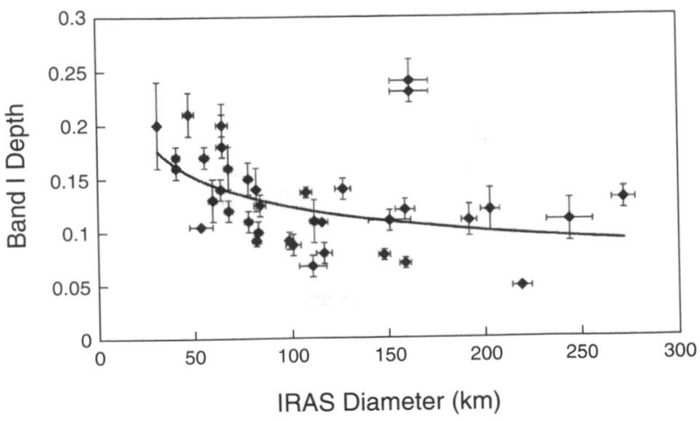

Fig. 2. For the population of S-type asteroids the depth of the 1.0-μm band (from *Gaffey et al.,* 1993) is plotted against estimated diameter (from *Tedesco et al.,* 1989). The trend toward shallower 1.0-μm bands with increasing size could show the effects of weathering increasing with surface age.

This tracks well with the idea that weathering reduces band depth over time. However, the other two spectral components of this process from the lunar case, albedo and spectral slope, tell a different story. There is essentially no trend with size for either slope or albedo in the S-type asteroids.

This lack of correlation may be explained by two factors. First, diameter is probably a fair proxy for surface age; however, for weathering purposes, what is critical is the age of the last global resurfacing event, which is unknown for any asteroid and probably random for this population. Numerical simulations indicate that gravitational retention of ejecta is less important on asteroids several kilometers in diameter, and these are not represented in Fig. 2. Second, the S-type asteroids are represented by a wide range of mineralogies that vary from almost pure olivine to almost exclusively pyroxene with wide ranges of mixtures between these extremes. Differing mineral chemistry probably responds differently to space weathering. However, restricting the analysis to mineralogically similar subgroups does not clear up this problem. For example, within the OC-like S(IV) subclass there is a strong trend toward reduced spectral slope with increasing asteroid size, which suggests that red slope is anticorrelated with surface age proxy in this critical subclass (*Gaffey et al.,* 1993). Also, for this subclass, increasing band depth is correlated with increasing red slope, which is the opposite of what would be expected with lunarlike weathering. On the other hand, there is a weak trend within the S(IV)s of reduced albedo with increasing red slope that follows the lunar trends.

Another line of evidence comes from telescopic observations of small (<20-km-diameter) S- and Q-type asteroids (*Binzel et al.,* 1996). In this distribution the 1-μm band depth forms a continuum between the relatively shallow band depths for S types to the relatively deep and therefore fresh Q types. The implication is that we are seeing a weathering continuum between fresh and weathered objects that corresponds to the surface age of the asteroid. This is probably true, but there is no way to determine surface age short of a sample return. These seemingly contradictory trends suggest that space weathering is occurring on S-type asteroids, but the details and the weathering products are different from the lunar case. It is perhaps not surprising that different mineralogies and different energy inputs produce different results.

4. SPACE WEATHERING: THE LUNAR EXAMPLE

Studies of the Moon have provided excellent materials for laboratory investigation of the physical and optical effects of lunar space weathering processes. Important reviews are available in *Hapke* (2001) and *Pieters et al.* (2000). In this section we briefly summarize our present knowledge, and we conclude with a description of the current model for lunarlike space weathering.

4.1. Surface Modification Processes

Processes known to affect the surface of the Moon include, but are not limited to, interplanetary dust and micrometeorite bombardment; electromagnetic radiation; solar wind ion implantation and sputtering; cosmic-ray bombardment; and meteoroid, asteroid, and comet impacts. One important result of these processes is the formation of reduced SMFe particles on grain surfaces, and this SMFe seems to dominate the optical effects. The SMFe also causes a characteristic electron spin resonance signal whose strength is denoted by I_s. A measure of the abundance of SMFe normalized to the abundance of FeO (I_s/FeO) is used as a good indicator of cumulative surface exposure, with "mature" soils having values greater than 60 (*Morris,* 1976, 1977, 1978, 1980).

4.2. Effects Observed with Lunar Samples

The physical effects of space-weathering processes on lunar soils have been studied carefully in the laboratory (e.g., *Heiken et al.,* 1991, Chapter 7). For example, particle size distributions of Apollo 17 lunar soils are shown in Fig. 3. Individual soils do not exhibit a history of continuous exposure, but can be described as a random product of reworking by micrometeorites and mixing with other soils. With time, finer particles are often combined by agglutination into larger particles. A mature soil reaches a "steady state" in which the process of pulverization is countered by agglutination and replenishment of coarser particles (*McKay et al.,* 1974). The mass fraction of particles <25 μm is typically on the order of 25% or less.

On the Moon, the optical effects of weathering in the space environment accumulate over time such that fresh subsurface materials "mature" at the surface. With exposure to the space environment, lunar soils generally darken and diagnostic absorption bands weaken. A mature lunar soil has a characteristic red-sloped continuum (reflectance increasing toward longer wavelengths). Representative bidirectional reflectance spectra of several Apollo 16 lunar samples (I = 30°, e = 30°) are shown in Fig. 4. Most Apollo 16 rocks are feldspathic breccias and contain small amounts of low-Ca pyroxene. Because of their high plagioclase content, some can be quite bright. The mature soils of the region, however, are all relatively uniform: significantly darker especially at shorter wavelengths and with very little spectral contrast. Regolith breccias are typically darker than other more "pristine" rocks, but only rarely exhibit properties comparable to natural lunar soils.

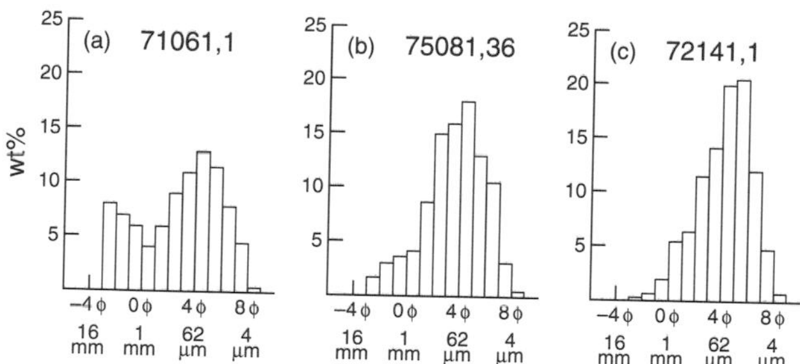

Fig. 3. Distribution of particle sizes in three Apollo 17 lunar soils in terms of mass (or volume %) of each size fraction (after *McKay et al.,* 1974): **(a)** An immature soil with a bimodal population of coarse and fine grains. **(b)** A submature soil. **(c)** A mature well-developed soil. The volumetric mean grain size of most well-developed lunar soils range from 45 to 75 μm.

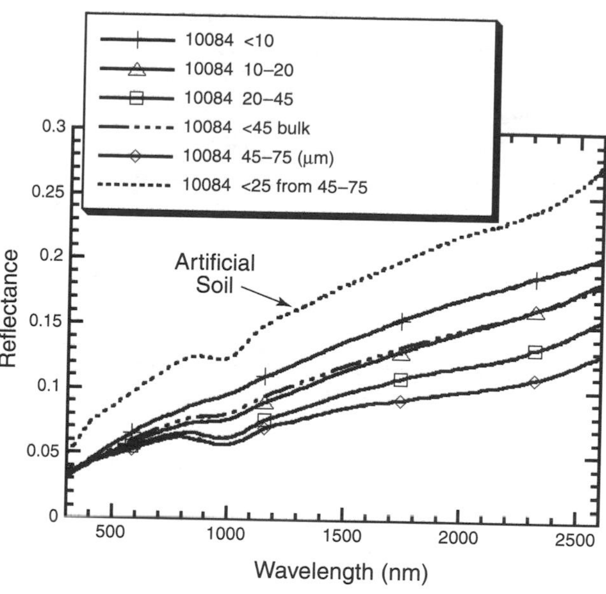

Fig. 5. Bidirectional reflectance spectra for size separates from a mature soil from Apollo 11 (10084) (data from *Taylor et al.,* 2001a; *Pieters et al.,* 1993). The bulk soil is indicated with a dot-dashed line and wet-sieved particle size separates are shown as solid lines. Note the bulk soil spectrum is dominated by the finest fractions rather than by the volumetrically dominant fractions. Note also that all size fractions of these natural soils converge at shorter wavelengths. An artificial fine fraction was prepared by grinding a subsample of the agglutinate-rich 45–75 fraction to <25 μm and is shown as a dotted line. The legend gives the size fraction in micrometers.

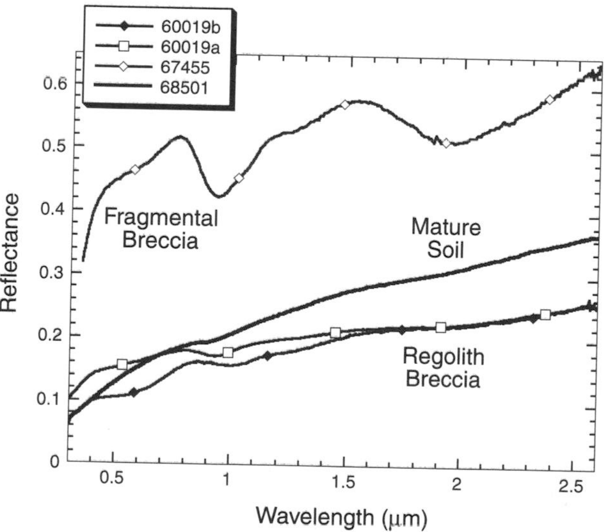

Fig. 4. Bidirectional reflectance spectra of typical Apollo 16 lunar samples (I = 30°, e = 0°). Mature soils such as 68501 have relatively weak features and exhibit a distinctive continuum with reflectance increasing toward longer wavelengths. Spectra for two different parts of regolith breccia 60019 are shown. These spectra from the same rock are quite different, but neither sample is similar to mature regolith from the region. In contrast, a typical noritic breccia (67455) from the site is bright and exhibits prominent absorption bands from low-Ca pyroxene.

The initial study of the optical properties of lunar soils was carried out in the 1970s by several research groups (J. B. Adams, T. B. McCord, B. Hapke, and D. Nash). When laboratory facilities later became available to accurately measure the bidirectional reflectance properties of small (30 mg) subsamples of lunar soil, additional properties of lunar soils became apparent. Specifically, the finest fractions were shown to dominate the optical properties of bulk soils even though the greater mass is in the larger particles (*Pieters et al.,* 1993). Shown in Fig. 5 are spectra of the bulk soil and size separates for Apollo 11 mature soil 10084. Even though the more abundant (by weight) size fractions greater than 20 μm exhibit relatively strong absorption bands, the bulk soil is clearly most similar to the <20-μm fractions. Furthermore, the optical properties of natural lunar soils cannot be reproduced artificially. All size fractions are agglutinate rich, but when the 45–75-μm fraction was ground to <25 μm it is unnaturally bright at all wavelengths. The optical properties of this lunar artificial <25-μm soil separate are totally inconsistent with those of the natural <25-μm fraction. The anomalous properties of the manually ground soil sample was an important indication that the weathering products of natural lunar soils are surface correlated (*Pieters et al.,* 1993).

A detailed coordinated analysis of the compositional and optical properties of a suite of selected lunar soils has recently been carried out for mare soils (*Taylor et al.,* 2001a) and is in progress for highland soils (*Taylor et al.,* 2001b). Several critical insights have come from this modern study: (1) A regular variation of composition with grain size is well documented and is dominated by concentration of feldspathic components in the finer fractions. (2) The amount of agglutinitic glass increases with decreasing grain size, as does the amount of nanophase-reduced Fe, but the nanophase Fe increases at a *much* greater rate. (3) The surfaces of individual grains were shown to contain compositions clearly foreign to the grain itself (Fe-rich rims on plagioclase; Al-rich rims on pyroxene), indicating mobility of atoms during exposure to the space environment. The latter two observations by this consortium are additional indications that the products of space weathering are surface correlated. These observations of where space-weathering products occur have now been beautifully documented with transmission electron microscopy (TEM) images of individual soil grains carried out by *Keller et al.* (1999) and *Pieters et al.* (2000). TEM images of anorthosite grains from a mature lunar soil shown in Fig. 6 exhibit abundant nanophase-reduced Fe on the rims. Since anorthosite is almost Fe-free, the origin of the observed SMFe must be from other sources. Such images document the surface deposition of nanophase Fe from solar wind sputtering and/or vapor fractionation that occurs during micrometeorite impacts. Both processes release atoms from one grain to be deposited on another.

Bidirectional reflectance spectra of the finest fraction (<10 μm) of lunar soils from the consortium study (*Noble et al.,* 2001) illustrate the optical effects of nanophase-reduced Fe accumulation on the surface of soil grains (Fig. 7). The finest fraction is well suited to document the effects of space weathering since the path length through these particles is small but the surface area/volume large. The optical effects of nanophase-reduced Fe are nonlinear (*Pieters et al.,* 2000; *Noble et al.,* 2001; *Hapke,* 2001): Small amounts redden the visible portion of the spectrum with little effect on longer wavelengths, whereas larger amounts dominate the spectrum, producing an almost linear continuum. This is readily seen in the spectra of Fig. 7. The highland soils have the least nanophase-reduced Fe and the mare soils the most, a direct result of the availability of Fe in the soil.

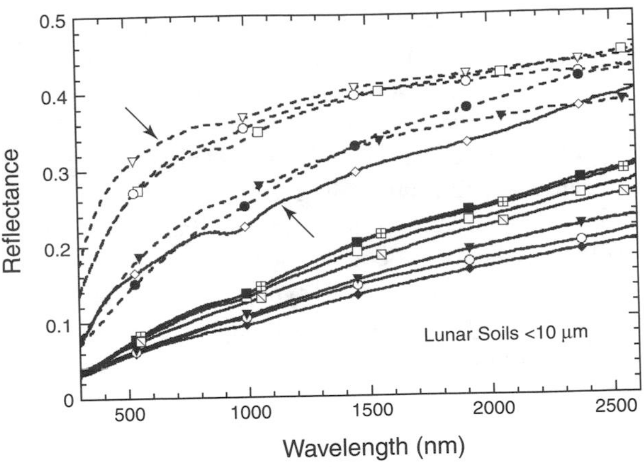

Fig. 7. Bidirectional reflectance spectra of the finest fraction (<10 μm) of lunar soils (after *Noble et al.,* 2001) illustrating the optical effects of SMFe accumulation on the surface of soil grains. Highland soils are indicated with dashed lines and mare soils with solid lines. The most immature soil of each (I_s/FeO < 15) is indicated with an arrow. The finest fraction is well suited to document the effects of space weathering since the path length through these particles is small but the surface area/volume is large. The highland soils have the least SMFe and the mare soils the most, a direct result from the availability of iron in the soil.

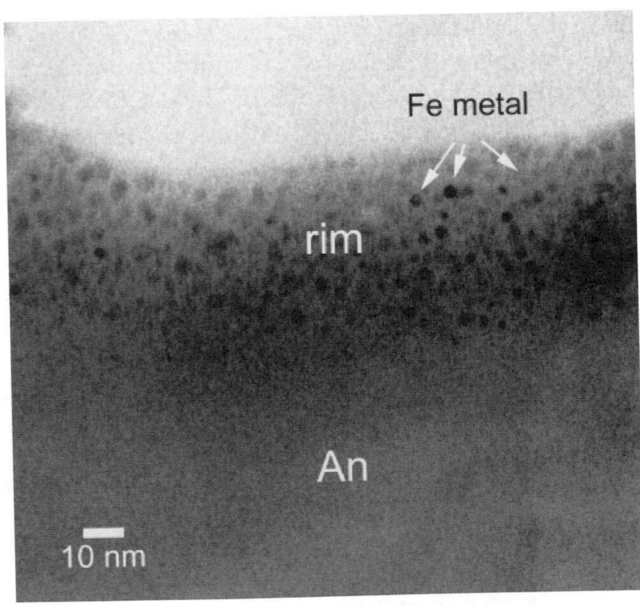

Fig. 6. TEM image of an anorthosite (An) grain from a mature lunar soil that exhibits a rim of Fe metal particles (SMFe) (from *Keller et al.,* 1999; *Pieters et al.,* 2000).

4.3. Lunar Model

It is important to realize that we have only the lunar studies to serve as a foundation on which to build a model for space weathering of the asteroids. Because we cannot directly describe asteroid space weathering, we instead describe the lunar case and compare it with the evidence for asteroids (*McKay et al.,* 1989). Currently the model for space weathering on the Moon that is most consistent with all the observations is that micrometeorite and solar wind bombardment produce a vapor from target particulate materials that, upon redeposition at the surface, is chemically reduced such that ferrous Fe previously existing in silicates condenses out as metallic Fe in submicroscopic spheres embedded in the mineral coatings on individual grains (*Hapke,* 2000). The formation and accumulation of the SMFe results in changes of the chemical, mineral, optical, and physical properties of the immediate surface and all subsequent generation of products (agglutinates). Although this process probably dominates the optical maturation of the lunar regolith, it does not explain all the differences observed between mature and immature soils. In addition, the combination of the processes of agglutinate formation, particle shock and comminution, crystal damage due to cosmic rays, and SMFe deposition are probably important. Lunar surface alteration can take millions of years and is an ongoing process since major impacts excavate and distribute fresh material.

5. METEORITE EVIDENCE

Much of what we know about the geochemistry of asteroids comes from the "ground truth" provided by meteorites. Samples from several meteorite groups show evidence of surface exposure and carry at least some information on the effects of regolith processes. The "solar-wind-implanted gas-rich" meteorites contain grains that have low-energy solar wind particles implanted into their surfaces. For this implantation to occur, the grain had to be directly exposed to the solar wind and thus had to be on the surface of an atmosphereless body. The gas-rich meteorites are typically breccias that contain a mixture of fine-grained gas-rich regolith soil and larger rock fragments that have been fused together by grain-boundary melting. This mixture probably was once a loose regolith that was relithified by impact shock and heating. Gas-rich ordinary chondrites are characterized by a light-dark structure (*Heymann,* 1967) where the gas-rich fine grained material is much darker than the rock fragments. This structure is shown in Fig. 8 in the gas-rich H6 ordinary chondrite Dwaleni. Spectrally the gas-rich portions of these meteorites can be up to a factor of 2 darker than the adjacent rock fragments and the spectral absorption bands are similarly suppressed. This suggests that some of the effects of asteroidal regolith processes in OCs are to darken the surface material and suppress spectral bands. Similar effects are seen in the black OCs, which show darkening and band suppression from the effects of shock (*Britt and Pieters,* 1994). These meteorites are not solar-wind gas-rich, but probably are highly shocked material from the lower levels of large impact craters (*Keil et al.,* 1992).

Other meteorite groups that show evidence of regolith-related alteration include enstatite chondrites and eucrites. The enstatite chondrite Abee (E4) is brecciated and shock-blackened (*Rubin et al.,* 1997). Its reflectance is about a factor of 2 lower than other enstatite chondrites. The eucrite Juvinas has shock-blackened areas and shows reduced reflectance and moderately subdued absorption bands (*Gaffey,* 1976). Solar-wind-implanted gases are also found in many carbonaceous chondrites, particularly CI and CM chondrites (*Keil,* 1982). However, there are essentially no spectral differences between gas-rich and gas-poor carbonaceous chondrites. These meteorites may not show any optical effects from regolith action because the abundance of opaque oxides and carbon compounds that characterize these meteorites normally produce a low albedo and suppressed absorption features.

Fig. 8. The gas-rich ordinary chondrite Dwaleni (H6) shows the effects of exposure to the space environment. The dark, fine-grained material in this meteorite was probably once part of the regolith soil of its parent asteroid and preserves some of the darkening and spectral band reduction caused by space weathering.

Although gas-rich OCs show strong evidence of once being part of the surface layer of asteroids, and of darkening and band attenuation, there is no evidence of the characteristic red spectral slope of the S-type asteroids (*Britt and Pieters,* 1994). Shown in Fig. 9 is a plot of albedo vs. red slope for ordinary chondrites, S-type asteroids, and lunar soils. In lunar terms, all S-type asteroids are immature as shown in Fig. 9. This relative immaturity is probably the result of a combination of several factors: (1) The lower average impact velocities in the asteroid belt (~5 km/s vs. ~15 km/s on the Moon) produce much less melting, vaporization, and alteration. It may take much longer for an asteroidal surface to mature because of these lower energies, and because of the lower gravity. (2) The fundamental mineralogical differences between the feldspar normative Moon and the olivine normative S-type asteroids produce somewhat different alteration products during micrometeorite impact (*Keil,* 1982). (3) The much smaller sizes and lower gravity regimes of asteroids produce much more global distributions of fresh ejecta from major impacts. Although the range of red slopes in OCs slightly overlap the range in S-type asteroids, there is essentially no red slope in OCs and gas-rich OCs are no redder than other OCs.

Why aren't gas-rich ordinary chondrites spectrally red? There are several factors working against the preservation of a strong red slope signal in gas-rich OCs. First, the formation of meteorites strong enough to survive excavation, transport to Earth, atmospheric entry, and landing requires

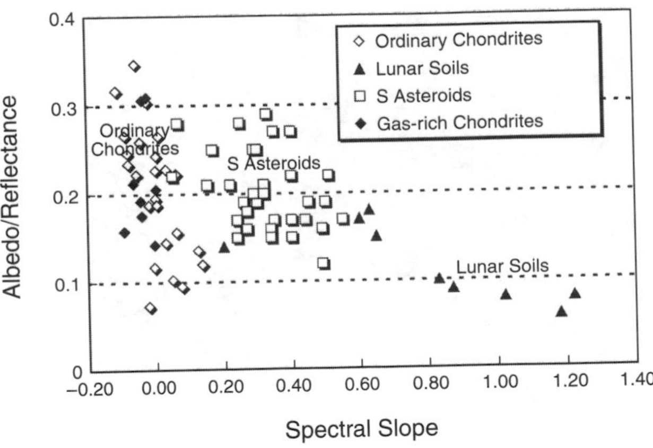

Fig. 9. The reflectance of ordinary chondrite meteorites, S-type asteroids, and lunar soils compared to their red continuum slope. The ordinary chondrites show essentially no red slope with slope increasing through the S-type asteroids to an extreme in the mature lunar soils. Note that increasing slope is correlated with decreasing reflectance. S-type asteroids are moderately red-sloped, but would be considered immature in lunar terms. Ordinary chondrites exhibit the full range of reflectance change, but show no increase in red slope.

the relithification of the regolith soil by grain-boundary melting. This kind of melting would strongly affect any SMFe metallic surface layers, which would remelt long before the silicates, and tend to form larger, less optically active particles. Second, gas-rich ordinary chondrites are intimate mixtures of surface and nonsurface materials (Fig. 2). Typically, the regolith-derived grains are much less than 50% of the volume of a gas-rich meteorite. Any red slope signal in these mixtures would be severely diluted by the stronger spectral signature of the fresh material. Finally, the red slope in S-type asteroids is fundamentally weaker than the lunar signal. It is easier to dilute the S-type red spectral slope than it is to affect the lunar red spectral slope because of the relative strengths of the signatures.

Fanale et al. (1992) compared S-type asteroid albedos and OC meteorite reflectances and showed that there was significant overlap in values, indicating a possible genetic relationship between the two populations. However, in that study Fanale et al. did not take into account the fact that OCs are measured at a viewing geometry that differs from that used for the calculation of asteroid albedos. Asteroid albedo values are measurements of the geometric albedo — defined as the ratio of the brightness of a body at 0° phase angle to the brightness of a perfect Lambert disk of the same radius and at the same distance as the body, but illuminated and observed perpendicularly (*Hapke,* 1993). Laboratory reflectance quantities are generally bidirectional reflectance factors, defined as the ratio of the brightness of a sample at 30° incidence and 0° emission, to the brightness of a Lambert surface illuminated identically (e.g., *Pieters,* 1983). At any given wavelength, the value of the geometric albedo (0° phase) is therefore systematically higher than the value of the bidirectional reflectance of a laboratory sample (30° phase). According to the photometric studies of *Shkuratov et al.* (1999) and *Clark et al.* (2002), the difference can be up to a factor of 2.5 due to significant opposition surge at 0° phase. Thus, taking viewing geometry into account, the results of *Fanale et al.* (1992) indicate that meteorites are actually up to 2.5× brighter than asteroids on average. In other words, S-type asteroids are generally much darker than OC meteorites. Thus, if OCs come from S types, this could be evidence that optical darkening (space weathering) occurs on asteroids.

6. LABORATORY SIMULATIONS

A large number of experiments have been carried out in the laboratory in attempts to simulate possible lunar or asteroidal space-weathering processes. These include the following: vitrification to simulate impact melting, ion irradiation to simulate solar wind bombardment, evaporation and subsequent condensation to simulate impact vaporization, and pulsed laser irradiation to simulate micrometeorite bombardment. These experiments are discussed briefly here, and they are described in detail and critically analyzed in *Hapke* (2001).

6.1. Vitrification

The lunar regolith contains abundant glass, mostly as part of agglutinate particles. In fact, some confusion exists in the literature because quench glass is often not distinguished from amorphous agglutinates. *Conel and Nash* (1970) and *Adams and McCord* (1971) melted lunar rocks in a nitrogen atmosphere and produced a material that had a low albedo and a reddened spectrum with subdued absorption bands. They initially proposed that lunar space weathering is due to simple impact vitrification. This was later modified to attribute the effects to accumulation of agglutinates. However, when *Hapke et al.* (1975) and *Wells and Hapke* (1977) melted lunar rocks in vacuum the resulting glass had a high albedo with a spectrum that was unreddened and had strong, broad absorption bands. A similar study by *Bell et al.* (1976) documented the effects of f_{O_2} and composition on the optical properties of quench glass. Quench glasses are quite different from agglutinitic glass and appear to have little effect on the properties of most lunar soils. *Hapke et al.* (1975) also suggested that the early N-melted quench glasses were oxidized and that their optical properties were caused by ferric Fe, which is not present in the lunar regolith. Similar objections apply to later vitrification experiments reported by *Clark et al.* (1992) and *Cloutis and Gaffey* (1993).

6.2. Ion Irradiation

Bombardment of silicate rock powders by H and He ions of kilovolt energies have been carried out by several groups, including Wehner and his colleagues (*Rosenberg and Wehner,* 1964; *KenKnight et al.,* 1967); *Hapke* (1966, 1968,

1973); and *Dukes et al.* (1999). Irradiation of a smooth surface has no effect on the optical properties. However, irradiation of a powder results in darkening, reddening, and decreased depth of absorption bands caused by the accumulation of absorbing sputter-deposited coatings on the grains. *Dukes et al.* (1999) showed that He^+ irradiation is more effective than H^+ in surface reduction of Fe in olivine.

6.3. Evaporation-Condensation

Hapke et al. (1975) and *Cassidy and Hapke* (1975) evaporated lunar and terrestrial silicates in an electron beam furnace and condensed some of the vapor onto fused silica microscope slides. The condensates had a transmission spectrum that decreased monotonically with increasing wavelength, with no discrete absorption bands. The vapor deposits were also found to exhibit the characteristic lunar ferromagnetic resonance and electron spin resonance exhibited by metallic Fe particles tens of nanometers in size (*Morris,* 1976). *Hapke et al.* (1975) showed that the condensation process is inherently reducing and results in the production of the Fe particles. They also showed that such particles can account for the optical properties of the coatings. They proposed that lunar space weathering was caused by coatings containing SMFe generated by both solar wind sputtering and impact vaporization in the lunar regolith.

6.4. Laser Shots

Hapke et al. (1975) irradiated a lunar glass with a pulsed laser. The resulting condensate was brownish. *Moroz et al.* (1996) and *Yamada et al.* (1999) bombarded silicate powders with laser pulses and found that the powders darkened and reddened. *Moroz et al.* (1996) interpreted their results as caused by melting; however, it is likely that the darkening is also due to condensates of the vapor generated by the laser shots. This was corroborated by the results of *Sasaki et al.* (2001), who used nanopulse laser shots to simulate space weathering, and found direct evidence for SMFe-bearing condensates on grain surfaces resulting from the laser evaporation and condensation. These experiments indicate that olivine is more easily weathered than pyroxene (*Yamada et al.,* 1999; *Hiroi and Sasaki,* 2002).

7. MODELING OPTICAL EFFECTS

The effects of the SMFe in the vapor condensates on the optical properties of a regolith can be modeled analytically using the reflectance spectroscopy formalism in *Hapke* (1993). The model is described in detail in *Hapke* (2001) and will only be summarized here.

The bidirectional reflectance of a medium of isotropically scattering particles can be described to a good approximation by

$$r(i, e, g) = (w/4\pi)[\cos i/(\cos i + \cos e)]H(w, i)H(w, e)$$

where w = single scattering albedo, i = angle of incidence, e = angle of emergence, $H(w, x) = (1 + 2x)/[1 + 2x(1 - w)^{1/2}]$.

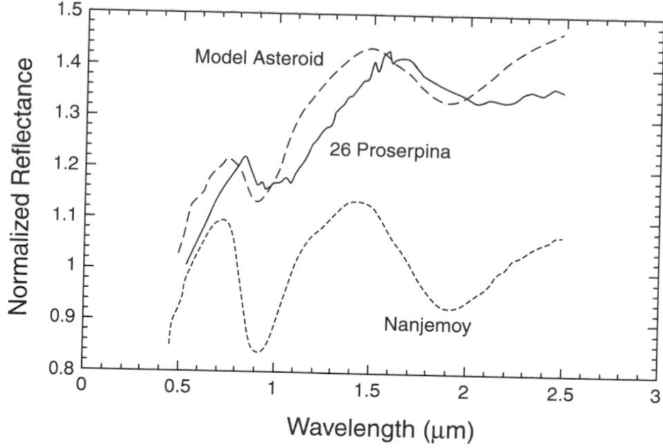

Fig. 10. Plot showing the effect of adding a 0.025% SMFe to a pulverized ordinary chondrite, Nanjemoy. In this example, n_h = 1.70, and the complex refractive index of Fe measured by *Johnson and Christy* (1974) was used. The resulting normalized spectrum is compared with that of a typical S asteroid, 26 Proserpina (shown offset for clarity). Lunar regolith typically contains about 0.5% SMFe. Thus, only about one-twentieth as much space weathering apparently occurs in the asteroid belt as on the lunar surface. Note that adding the Fe does not change the wavelengths of the band centers, only their depths.

The equivalent slab model expression for the single scattering albedo is

$$w = S_e + (1 - S_e)(1 - S_i)e^{-\alpha D}/(1 - S_i e^{-\alpha D})$$

where S_e = reflection coefficient for light incident on the particle surface from outside, S_i = reflection coefficient for light incident on the particle surface from inside, α = particle absorption coefficient, and D = mean particle size.

To model the effects of space weathering, α is increased by adding to it α_{Fe}, the absorption coefficient of a suspension of Fe particles. If the particles are all much smaller than the wavelength, α_{Fe} can be calculated from the Maxwell-Garnett effective medium theory

$$\alpha_{Fe} = 36\pi z f \rho_h/\lambda\rho_{Fe}$$

where f = mass fraction of SMFe averaged over the whole medium, ρ_h = density of the host silicate medium, ρ_{Fe} = density of Fe particles, λ = wavelength, and

$$z = n_h^3 n_{Fe} k_{Fe}/[(n_{Fe}^2 - k_{Fe}^2 + 2n_h^2)^2 + (2n_{Fe}k_{FE})^2]$$

where n and k are the real and imaginary parts respectively of the refractive index and the subscripts h and Fe refer to the host medium and Fe respectively. Figure 10 shows an example of an application of this model.

8. SPACECRAFT OBSERVATIONS

Several asteroids have been visited by spacecraft. In particular, we have images and a few spectra of the main-belt asteroids 951 Gaspra and 243 Ida obtained by *Galileo* (see *Sullivan et al.,* 2002). In addition, the *NEAR Shoemaker* spacecraft obtained images of 253 Mathilde (*Veverka et al.,*

1999) and then orbited the near-Earth asteroid 433 Eros for one year (see *Cheng,* 2002), mapping the surface and obtaining images (*Veverka et al.,* 2000), 0.8–2.4-μm spectra (*Bell et al.,* 2002), X-ray and γ-ray spectra (*Trombka et al.,* 2000), and laser altimetry (*Zuber et al.,* 2000). In this section we briefly describe some relevant observations for each asteroid, paying special attention to the Eros results because of the high quantity and quality of data obtained on that asteroid.

8.1. 951 Gaspra

Gaspra holds the distinction of being the first main-belt asteroid to be visited by spacecraft. Gaspra is an S-type asteroid, and has been linked with olivine-rich meteorites. *Galileo* flew by Gaspra in 1991, and multispectral images of the surface revealed minor spectral variations associated with morphologically fresh craters on ridges. Specifically, craters on ridges tend to be bluer in color than nearby surrounding terrain. Although the level of color heterogeneity is extremely low (less than 5%), the association of color with craters strengthened the interpretation that these craters excavated relatively fresh subsurface materials (*Clark,* 1993; *Chapman,* 1996).

8.2. 243 Ida

Ida is also an S-type asteroid, and has been linked with pyroxene-rich meteorites such as OCs. On Ida, regolith processes such as crater formation and ejecta emplacement are linked with color variations (*Sullivan et al.,* 1996; *Geissler et al.,* 1996; *Lee et al.,* 1996; *Chapman,* 1996). These color variations were attributed to the exposure of fresh subsurface materials that contrasted with the surrounding surface (*Chapman,* 1996). Color variations on Ida mimic the sense of the color differences due to optical maturity observed on the Moon; however, the magnitude of the variations on Ida is much smaller.

8.3. 253 Mathilde

Mathilde is a C-type asteroid, and has been generally linked with carbonaceous chondrites. *NEAR Shoemaker* images show that Mathilde is uniformly dark, with no fresh exposures (albedo or color variations) near craters or in crater walls (*Veverka et al.,* 1999; *Clark et al.,* 1999). The lack of color and albedo variations on Mathilde's surface argues that surface processes on this dark asteroid do not result in optical alteration. This is not surprising because Mathilde is composed of dark C-rich (opaque) particles, whereas the dominant weathering processes cause the most observable changes to bright transparent particles.

8.4. 433 Eros

Eros is an S-type asteroid that has been linked with L and LL OCs (*McCoy et al.,* 2001). Spectral differences between dark ambient surface and fresher exposures of subsurface materials on Eros mimic the optical alteration expected from space weathering. Psyche, the largest fresh crater on Eros (5.3 km diameter) shows the clearest evidence of space-weathering effects on Eros. Combined imaging and spectrometer observations of Psyche crater reveal distinctive brightness patterns consistent with downslope motion of dark regolith material overlying a substrate of brighter material (Fig. 11) (*Veverka et al.,* 2000; *Murchie et al.,* 2002; *Clark et al.,* 2001). At spatial scales of 620 m, Psyche crater wall materials exhibit albedo contrasts of 30–40% at 0.95 μm, with associated spectral variations at a much lower level of 4–8% (Fig. 12). At spatial scales of 50 m the albedo contrasts are as high as 80% with associ-

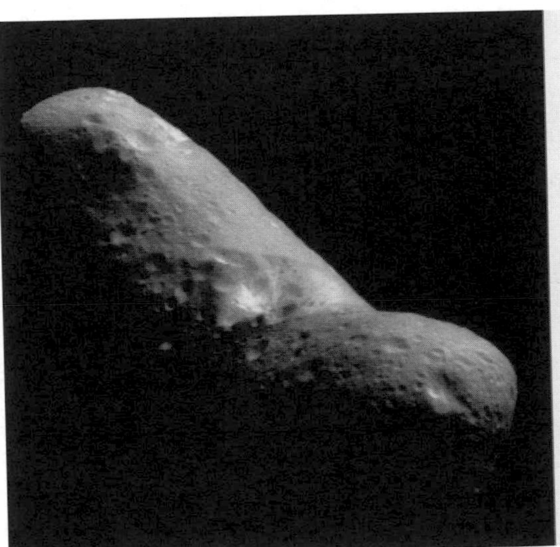

Fig. 11. (left) A NEAR image of Psyche crater on Eros. (right) A zoom image of an area with nearby exposures of bright and dark materials in the walls of Psyche crater (from *Clark et al.,* 2001).

ated spectral contrasts of less than 10% (*Murchie et al.,* 2002). It is unusual to observe such stark albedo contrasts with so little associated color differences. Scattering model and lunar analogy investigations into several possible causes for these albedo and spectral trends reveal that: the contrasts are not consistent with a cause due solely to variations in grain size, olivine, pyroxene, troilite, or lunarlike optical maturation (denoted SMFe) (Fig. 13). A grain size change sufficient to explain the albedo contrasts would result in strong color variations that are not observed. Olivine and pyroxene variations cannot reproduce the albedo contrasts, and would produce strong band-correlated variations that are not observed. Troilite, or dark neutral spectral variations (a proxy for shock-related darkening), would produce bluer color contrasts than are observed, and a simple lunarlike optical maturation effect would produce strong reddening that is not observed. The actual albedo contrasts and associated spectral variation trends are most consistent with a combination of enhanced dark spectrally neutral components and lunarlike optical maturation (*Clark et al.,* 2001). These results suggest that space-weathering processes may explain the differences between bright and dark materials on Eros. However, there are significant spectral differences between Eros' proposed analog meteorites and Eros' freshest exposures of subsurface bright materials. After accounting for all differences in the measurement units of reflectance comparisons, the bright materials on Eros have reflectance values at 0.95 μm consistent with meteorites, but spectral continua that are much redder than meteorites from 1.5 to 2.4 μm. Average Eros surface materials are 30–40% darker than meteorites (*Clark et al.,* 2001).

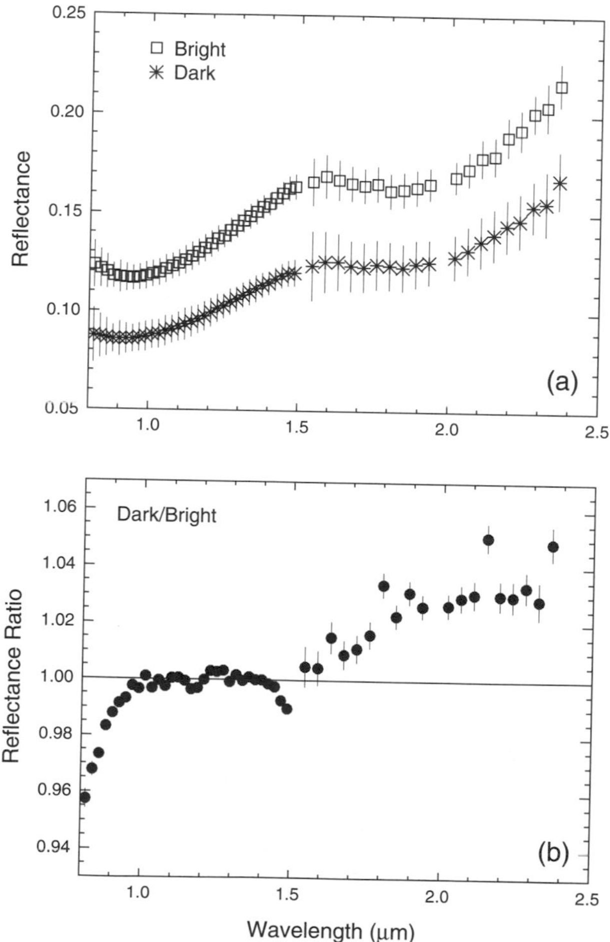

Fig. 12. **(a)** Spectra for dark and bright regions in Psyche crater. **(b)** A normalized spectral ratio of dark/bright materials. This ratio brings out the differences in spectral properties between the dark and bright materials. In general, dark materials tend to be redder from 1.5 to 2.4 μm, and slightly broader in 1-μm band depth from 0.8 to 1.0 μm (from *Clark et al.,* 2001).

The association between color units on Eros is entirely different from associations observed on Gaspra and Ida (*Sullivan et al.,* 1996). On Eros, albedo contrasts are high, color contrasts are low, and bright materials are correlated with steep crater walls (*Thomas et al.,* 2002; *Murchie et al.,* 2002). On Gaspra and Ida albedo contrasts are low, color contrasts are high, and color units appear to be correlated with ejecta emplacement. These geomorphologic associations are important constraints on space weathering of asteroid surfaces. It is possible that Eros does not have bright rimmed craters like Gaspra and Ida have because it is in a different weathering environment in the solar system. Eros is in near-Earth space, whereas Gaspra and Ida are in the main belt. This argues for different rates of the competing processes of surface maturation and impact cratering between near-Earth orbits and the asteroid main belt (for more detail see *Chapman et al.,* 2002).

9. SUMMARY AND FUTURE WORK

9.1. Summary

A description of space weathering on asteroids that is most consistent with the available evidence is as follows. (1) Some lunarlike optical maturation occurs and its strength is dependent on the composition of the target material; however, it is not as effective on asteroids as it is on the Moon. The main process is probably deposition of condensates bearing SMFe on grain surfaces from vaporization of target material by solar wind sputtering and micrometeorite bombardment. (2) Some spectrally neutral darkening occurs and is probably related to the processes of shock, solar wind gas implantation, and contamination by carbonaceous material.

Space-weathering effects on asteroids strongly depend on the composition of the target. Some asteroid types show very little evidence of optical alteration (C types), while other types show strong evidence of optical alteration (S types). These trends indicate that objects composed of dark, relatively opaque components exhibit minimal space-weathering effects while objects composed of bright, relatively transparent components exhibit maximal space-weathering effects. Availability of Fe in target minerals influences the abundance of SMFe. Experiments indicate that olivine is more easily weathered than pyroxene, perhaps explaining

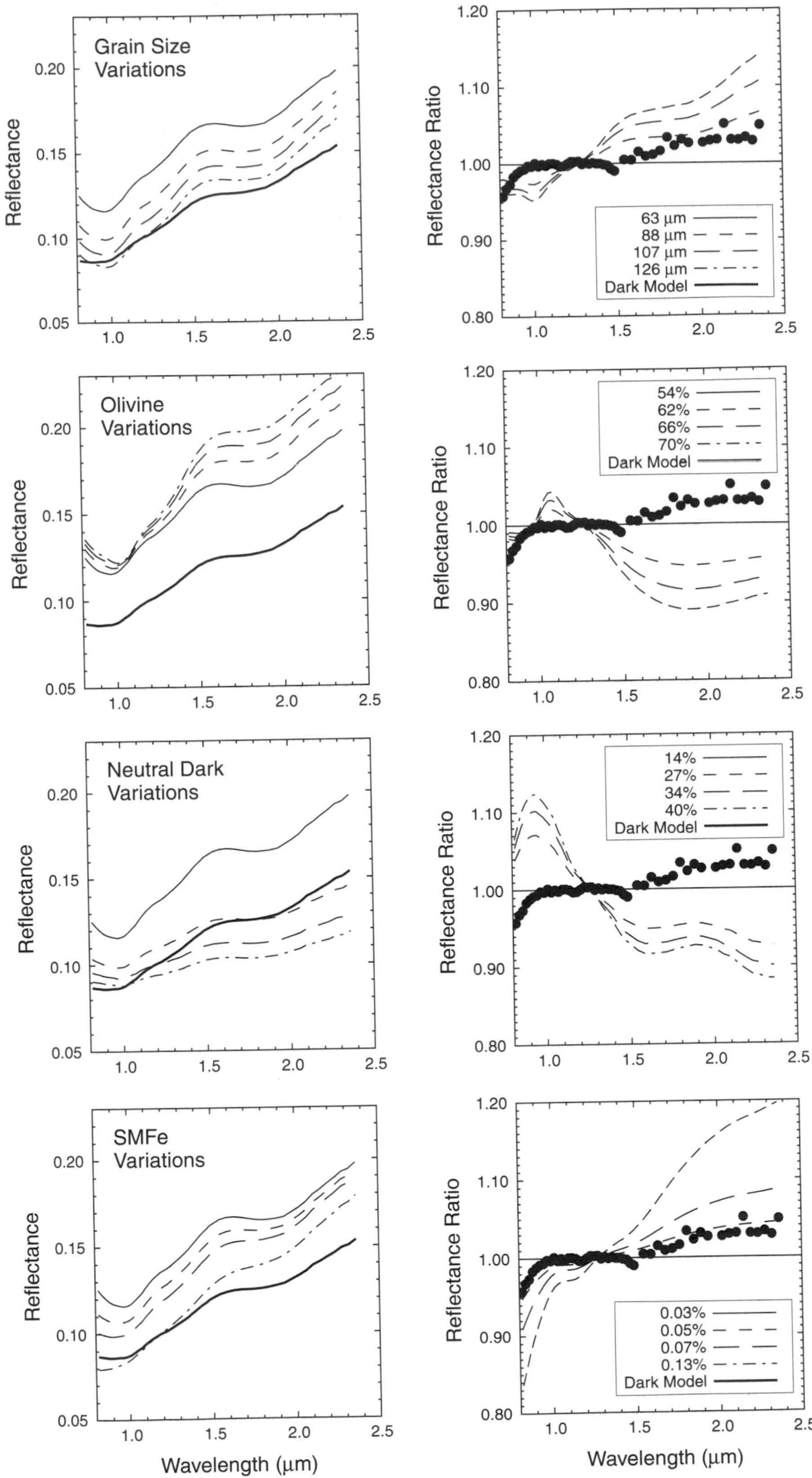

Fig. 13. (left) Each panel shows the bright material spectrum (solid thin line) and the dark material spectrum (solid thick line) compared with variations in the indicated model parameter. (right) In terms of normalized reflectance ratio, each panel compares the observed color trends (solid black dots) to expected color variations given the parameter variations in the left panels. Most parameter variations (except olivine abundance) can produce the observed albedo changes; however, the reflectance ratios indicate that the expected associated color variations are too strong to match the observations. A good explanation for the observed trends would be a combination of the SMFe effects with enhanced neutral dark component (from *Clark et al.,* 2001).

some of the variations in the degree of weathering observed within an asteroid class.

A prediction of the model is that a pristine sample of asteroid regolith in which space weathering has occurred should possess a weak ESR ferromagnetic resonance.

9.2. Remaining Unresolved Issues and Problems

Soil samples from several different asteroid spectral types are needed to verify the compositional dependencies of space-weathering effects. There is no quantitative understanding of the relative rates of the space-weathering processes and their optical effects. In particular, it is not clear why the color and albedo trends due to space weathering on 951 Gaspra and 243 Ida differ from those on 433 Eros.

The values of the complex spectral refractive index of Fe measured by various workers vary greatly (by more than a factor of 2), probably because of surface oxidation effects. Accurate values appropriate to the space environment are badly needed for reliable modeling.

REFERENCES

Adams J. and McCord T. (1971) Optical properties of mineral separates, glass, and anorthositic fragments from Apollo mare samples. *Proc. Lunar Sci. Conf. 2nd,* pp. 2183–2195.

Bell J. F. (1995a) Do ordinary chondrites come from S-class asteroids? Invited talk, AAS Div. for Planetary Sciences, annual meeting, Kona, Hawai'i.

Bell J. F. (1995b) Q-class asteroids and ordinary chondrites (abstract). In *Lunar and Planetary Science XXVI,* pp. 93–94. Lunar and Planetary Institute, Houston.

Bell J. F. (1997) Geochemical space weathering effects on asteroids: Implications for spacecraft instrumentation (abstract). In *Lunar and Planetary Science XXVIII,* pp. 83–84. Lunar and Planetary Institute, Houston.

Bell J. F. (1998) The Vesta asteroid family: Fact or fiction (abstract). In *Lunar and Planetary Science XXIX,* abstract #1851. Lunar and Planetary Institute, Houston (CD-ROM).

Bell J. F. III, Izenberg N. I., Lucey P. G., Clark B. E., Peterson C., Gaffey M. J., Joseph J., Carcich B., Harch A., Bell M. E., Warren J., Martin P. D., McFadden L. A., Wellnitz D., Murchie S., Winter M., Veverka J., Thomas P., Robinson M. S., Malin M., and Cheng A. (2002) Near-IR reflectance spectroscopy of 433 Eros from the NIS instrument on the NEAR mission. 1. Low phase angle observations. *Icarus, 155,* 119–144.

Bell P. M., Mao H. K., and Weeks R. A. (1976) Optical spectra and electron paramagnetic resonance of lunar and synthetic glasses: A study of the effects of controlled atmosphere, composition, and temperature. *Proc. Lunar Sci. Conf. 7th,* pp. 2543–2559.

Binzel R. P. and Xu S. (1993) Chips off asteroid 4 Vesta: Evidence for the parent body of basaltic achondrite meteorites. *Science, 260,* 186–191.

Binzel R. P., Bus S. J., Burbine T. H., and Sunshine J. M. (1996) Spectral properties of near-Earth asteroids: Evidence for sources of ordinary chondrite meteorites. *Science, 273,* 946–948.

Bobrovnikoff N. T. (1929) The spectra of minor planets. *Lick Observatory Bull., 14 (407),* 18–27.

Britt D. and Pieters C. M. (1994) Darkening in black and gas-rich ordinary chondrite meteorites: The spectral effects of opaque morphology and distribution. *Geochim. Cosmochim. Acta., 58,* 3905–3919.

Britt D., Tholen D. J., Bell J. F., and Pieters C. M. (1992) Comparison of asteroid and meteorite spectra: Classification by principal components analysis. *Icarus, 99,* 153–166.

Burbine T. (1991) Principal component analysis of asteroid and meteorite spectra from 0.3 to 2.5 microns. Master's thesis, Univ. of Pittsburgh, Pittsburgh, Pennsylvania.

Carr M., Kirk R., McEwan A., Veverka J., Thomas P., Head J., and Murchie S. (1994) The geology of Gaspra. *Icarus, 107,* 61–71.

Cassidy W. and Hapke B. (1975) Effects of darkening processes on surfaces of airless bodies. *Icarus, 25,* 371–383.

Chapman C. (1995) Do ordinary chondrites come from S-class asteroids? Invited talk, AAS Div. for Planetary Sciences, annual meeting, Kona, Hawai'i.

Chapman C. (1996) S-type asteroids, ordinary chondrites and space weathering: The evidence from Galileo's fly-bys of Gaspra and Ida. *Meteoritics & Planet. Sci., 31,* 699–725.

Chapman C. and Salisbury J. (1973) Comparisons of meteorite and asteroid spectral reflectivities. *Icarus, 19,* 507–522.

Chapman C. R., Merline W. J., Thomas P. C., Joseph J., Cheng A. F., and Izenberg N. (2002) Impact history of Eros: Craters and boulders. *Icarus, 155,* 104–118.

Cheng A. F. (2002) Near Earth Asteroid Rendezvous: Mission summary. In *Asteroids III* (W. F. Bottke Jr. et al., eds.), this volume. Univ. of Arizona, Tucson.

Clark B.E. (1993) Spectral reflectance studies and optical surface alteration in the search for links between meteorites and asteroids. Ph.D. dissertation, Univ. of Hawai'i.

Clark B. E. and Johnson R. E. (1996) Interplanetary weathering: Surface erosion in outer space. *Eos Trans. AGU, 77,* 141–145.

Clark B. E., Fanale F., and Salisbury J. (1992) Meteorite-asteroid spectral comparison: The effects of comminution, melting and recrystallization. *Icarus, 97,* 288–297.

Clark B. E., Veverka J., Helfenstein P., Thomas P., Bell J. F. III, Harch A., Robinson M., Murchie S., McFadden L., and Chapman C. (1999) NEAR photometry of asteroid 253 Mathilde. *Icarus, 140,* 53–65.

Clark B. E., Lucey P., Helfenstein P., Bell J. F. III, Peterson C., Veverka J., McConnochie T., Robinson M., Bussey B., Murchie S., Izenberg N., and Chapman C. (2001) Space weathering on Eros: Constraints from albedo and spectral measurements of Psyche crater. *Meteoritics & Planet. Sci., 36,* 1617–1637.

Clark B. E., Helfenstein P., Bell J. F. III, Peterson C., Veverka J., Izenberg N. I., Domingue D., Wellnitz D., and McFadden L. (2002) NEAR infrared spectrometer photometry of asteroid 433 Eros. *Icarus, 155,* 189–204.

Clark R. N. and Roush T. L. (1984) Reflectance spectroscopy: Quantitative analysis techniques for remote sensing applications. *J. Geophys. Res., 89,* 6329–6340.

Cloutis E. and Gaffey M. (1993) Lunar regolith analogues: Spectral reflectance properties of compositional variations. *Icarus, 102,* 203–224.

Conel J. and Nash D. (1970) Spectral reflectance and albedo of Apollo 11 lunar samples: Effects of irradiation and vitrification and comparison with telescopic observations. *Proc. Apollo 11 Lunar Sci. Conf.,* pp. 2013–2024.

Dollfus A., Wolff M., Geake J., Lupishko D., and Dougherty L. (1989) Photopolarimetry of asteroids. In *Asteroids II* (R. P. Binzel et al., eds.), pp. 594–616. Univ. of Arizona, Tucson.

Dukes C. A., Baragiola R. A., and McFadden L. A. (1999) Surface modification of olivine by H^+ and He^+ bombardment. *J.*

Geophys. Res., 104, 1865–1872.

Fanale F., Clark B. E., and Bell J. F. (1992) A spectral analysis of ordinary chondrites, S-type asteroids and their component minerals: Genetic implications. *J. Geophys. Res., 97,* 20863–20874.

Gaffey M. J. (1976) Spectral reflectance characteristics of the meteorite classes. *J. Geophys. Res., 81,* 905–920.

Gaffey M. J., Bell J. F., Brown R. H., Burbine T. H., Piatek J. L., Reed K. L., and Chaky D. A. (1993) Mineralogical variations within the S-type asteroid class. *Icarus, 106,* 573–602.

Geissler P., Petit J. M., Durda D., Greenberg R., Bottke W., Nolan M., and Moore J. (1996) Erosion and ejecta reaccretion on 243 Ida and its moon. *Icarus, 120,* 140–157.

Gold T. (1955) The lunar surface. *Mon. Not. R. Astron. Soc., 115,* 585–604.

Hapke B. (1966) Optical properties of the moon's surface. In *The Nature of the Lunar Surface* (W. Hess et al., eds.), pp. 141–154. Johns Hopkins, Baltimore.

Hapke B. (1968) The composition of the lunar surface inferred from optical properties. *Science, 159,* 76–79.

Hapke B. (1973) Darkening of silicate rocks by solar wind sputtering. *Moon, 7,* 342–355.

Hapke B. (1993) *Theory of Reflectance and Emittance Spectroscopy.* Cambridge Univ., New York.

Hapke B. (2000) How to turn OC's into S's: Space weathering in the asteroid belt (abstract). In *Lunar and Planetary Science XXXI,* abstract #1087. Lunar and Planetary Institute, Houston (CD-ROM).

Hapke B. (2001) Space weathering from Mercury to the asteroid belt. *J. Geophys. Res., 106,* 10039–10073.

Hapke B., Cassidy W., and Wells E. (1975) Effects of vapor-phase deposition processes on the optical, chemical and magnetic properties of the lunar regolith. *Moon, 13,* 339–354.

Heiken G. H., Vaniman D. T., and French B. M., eds. (1991) *Lunar Sourcebook: A User's Guide to the Moon.* Cambridge Univ., New York. 736 pp.

Heymann D. (1967) On the origin of hypersthene chondrites: Ages and shock effects of black chondrites. *Icarus, 6,* 189–221.

Hiroi T. and Sasaki S. (2002) Importance of space weathering simulation products in compositional modeling of asteroids: 349 Dembowska and 446 Aeternitas as examples. *Meteoritics & Planet. Sci., 36,* 1587–1596.

Hiroi T., Pieters C., and Takeda H. (1994) Grain size of the surface regolith of asteroid 4 Vesta estimated from its reflectance spectrum in comparison with HED meteorites. *Meteoritics, 29,* 394–396.

Hiroi T., Binzel R., Sunshine J., Pieters C., and Takeda H. (1995) Grain sizes and mineral compositions of surface regoliths of Vesta-like asteroids. *Icarus, 115,* 374–386.

Hiroi T., Pieters C. M., Vilas F., Sasaki S., Hamabe Y., and Kurahashi E. (2000) Possible space weathering of Vesta and Vestoids evident in their reflectance spectra (abstract). *Eos Trans. AGU, 81,* WP1-1-102.

Hörz F. and Cintala M. (1997) Impact experiments related to the evolution of planetary regoliths. *Meteoritics & Planet. Sci., 32,* 179–209.

Housen K. R. and Wilkening L. L. (1982) Regoliths on small bodies in the solar system. *Annu. Rev. Earth Planet. Sci., 10,* 355–376.

Johnson P. and Christy R. (1974) Optical constants of transition metals: Ti, V, Cr, Mn, Fe, Co, Ni and Pd. *Phys. Rev., B9,* 5056–5070.

Keil K. (1982) Composition and origin of chondritic breccias. In *Workshop on Lunar Breccias and Soils and Their Meteoritic Analogs* (G. J. Taylor and L. L. Wilkening, eds.), pp. 65–83. LPI Tech. Rpt. 82-02, Lunar and Planetary Institute, Houston.

Keil K., Bell J. F., and Britt D. T. (1992) Reflection spectra of shocked ordinary chondrites and relations to C-type and K-type asteroids. *Icarus, 98,* 43–53.

Keller L. P. and McKay D. S. (1993) Discovery of vapor deposits in the lunar regolith. *Science, 261,* 1305–1307.

Keller L. P. and McKay D. S. (1997) The nature and origin of rims on lunar soil grains. *Geochim. Cosmochim. Acta., 61,* 2331–2341.

Keller L. P., Wentworth S. J., and McKay D. S. (1998) Surface-correlated nanophase iron metal in lunar soils: Petrography and space weathering effects. In *Workshop on New Views of the Moon: Integrated Remotely Sensed, Geophysical, and Sample Datasets,* pp. 44–45. LPI Contribution No. 958, Lunar and Planetary Institute, Houston.

Keller L. P., Wentworth S. J., McKay D. S., Taylor L. A., Pieters C. M., and Morris R. V. (1999) Space weathering in the fine size fractions of lunar soils: Mare/highland differences. In *Workshop on New Views of the Moon II: Understanding the Moon Through the Integration of Diverse Datasets,* pp. 32–33. LPI Contribution No. 980, Lunar and Planetary Institute, Houston.

KenKnight C., Rosenberg D., and Wehner G. (1967) Parameters of the optical properties of the lunar surface powder in relation to solar wind bombardment. *J. Geophys. Res., 72,* 3105–3130.

Lee P., Veverka J., Thomas P., Helfenstein P., Belton M. J. S., Chapman C., Greenberg R., Pappalardo R., Sullivan R., and Head J. (1996) Ejecta blocks on 243 Ida and on other asteroids. *Icarus, 120,* 87–105.

Lucey P. G., Taylor G., and Malaret E. (1995) Abundance and distribution of iron on the Moon. *Science, 268,* 1150–1153.

Lucey P. G., Blewett D. T., Taylor G. J., and Hawke B. R. (2000) Imaging of lunar surface maturity. *J. Geophys. Res., 105,* 20377–20386.

Matson D., Johnson T., and Veeder G. (1977) Soil maturity and planetary regoliths: The Moon, Mercury and the asteroids. *Proc. Lunar Sci. Conf. 8th,* pp. 1001–1011.

McCord T. B., Adams J., and Johnson T. (1970) Asteroid Vesta: Spectral reflectivity and compositional implications. *Science, 168,* 1445–1447.

McCord T. B., Clark R. N., Hawke B. R., McFadden L. A., Owensby P. D., Pieters C. M., and Adams J. B. (1981) Moon: Near infrared spectral reflectance, a first good look. *J. Geophys. Res., 86,* 10883–10892.

McCoy T. J., Burbine T. H., McFadden L. A., Starr R. D., Gaffey M. J., Nittler L. R., Evans L. G., Izenberg N., Lucey P., Trombka J., Bell J. F. III, Clark B. E., Clark P. E., Squyres S. W., Chapman C. R., Boynton W. V., and Veverka J. (2001) The composition of 433 Eros: A mineralogical-chemical synthesis. *Meteoritics & Planet. Sci., 36,* 1661–1672.

McKay D. S., Fruland R. M., and Heiken G. H. (1974) Grain size and evolution of lunar soils. *Proc. Lunar Sci. Conf. 5th,* pp. 887–906.

McKay D., Swindle T., and Greenberg R. (1989) Asteroidal regoliths: What we do not know. In *Asteroids II* (R. P. Binzel et al., eds.), pp. 617–642. Univ. of Arizona, Tucson.

Moroz L., Fisenko A., Semjonova L., Pieters C., and Korotaeva N. (1996) Optical effects of regolith processes on S-asteroids as simulated by laser shots on ordinary chondrite and other mafic materials. *Icarus, 122,* 366–382.

Morris R. V. (1976) Surface exposure indices of lunar soils: A comparative FMR study. *Proc. Lunar Sci. Conf. 7th,* pp. 315–335.

Morris R. V. (1977) Origin and evolution of the grain-size dependence of the concentration of fine-grained metal in lunar soils: The maturation of lunar soils to a steady-state stage. *Proc. Lunar Sci. Conf. 8th,* pp. 3719–3747.

Morris R. V. (1978) The surface exposure (maturity) of lunar soils: Some concepts and I_s/FeO compilation. *Proc. Lunar Planet. Sci. Conf. 9th,* pp. 2287–2297.

Morris R. V. (1980) Origins and size distribution of metallic iron particles in the lunar regolith. *Proc. Lunar Planet. Sci. Conf. 11th,* pp. 1697–1712.

Murchie S., Robinson M., Clark B., Li H., Thomas P., Joseph J., Bussey B., Domingue D., Veverka J., Izenberg N., and Chapman C. (2002) Color variations on Eros from NEAR multispectral imaging. *Icarus, 155,* 145–168.

Noble S. K., Pieters C. M., Taylor L. A., Morris R. V., Allen C. C., McKay D. S., and Keller L. P. (2001) Optical properties of the finest fraction of lunar soils: Implications for space weathering environments. *Meteoritics & Planet. Sci., 36,* 31–42.

Pieters C. M. (1983) Strength of mineral absorption features in the transmitted component of near-infrared light: First results from RELAB. *J. Geophys. Res., 88,* 9534–9544.

Pieters C. M. (1986) Composition of the lunar highland crust from near-infrared spectroscopy. *Rev. Geophys. Space Phys., 24,* 557–578.

Pieters C. M. and Binzel R. P. (1994) Young Vesta (regolith)? (abstract). In *Lunar and Planetary Science XXV,* pp. 1083–1084. Lunar and Planetary Institute, Houston.

Pieters C. M. and McFadden L. (1994) Meteorite and asteroid reflectance spectroscopy: Clues to early solar system processes. *Annu. Rev. Earth Planet. Sci., 22,* 457–497.

Pieters C. M., Fischer E., Rode O., and Basu A. (1993) Optical effects of space weathering: The role of the finest fraction. *J. Geophys. Res., 98,* 20817–20824.

Pieters C. M., Taylor L., Noble S., Keller L., Hapke B., Morris R., Allen C., McKay D., and Wentworth S. (2000) Space weathering on airless bodies: Resolving a mystery with lunar samples. *Meteoritics & Planet. Sci., 35,* 1101–1107.

Rosenberg D. and Wehner G. (1964) Darkening of powdered basalt by simulated solar wind. *J. Geophys. Res., 69,* 3307–3308.

Rubin A. E., Scott E. R. D., and Keil K. (1997) Shock metamorphism of enstatite chondrites. *Geochim. Cosmochim. Acta, 61,* 847–858.

Sasaki S., Nakamura K., Hamabe Y., Kurahashi E., and Hiroi T. (2001) Production of iron nanoparticles by laser irradiation in a simulation of lunar-like space weathering. *Nature, 410,* 555–557.

Shkuratov Y., Kreslavsky M., Ovcharenko A., Stankevich D., Zubko E., Pieters C., and Arnold G. (1999) Opposition effect from Clementine data and mechanisms of backscatter. *Icarus, 141,* 132–155.

Sullivan R., Greeley R., Pappalardo R., Asphaug E., Moore J., Morrison D., Belton M. J. S., Carr M., Chapman C., Geissler P., Greenberg R., Granahan J., Head J., Kirk R., McEwan A., Lee P., Thomas P., and Veverka J. (1996) Geology of 243 Ida. *Icarus, 120,* 119–139.

Sullivan R. J., Thomas P. C., Murchie S. L., and Robinson M. S. (2002) Asteroid geology from *Galileo* and *NEAR Shoemaker* data. In *Asteroids III* (W. F. Bottke Jr. et al., eds.), this volume. Univ. of Arizona, Tucson.

Taylor L. A., Pieters C. M., Morris R. V., Keller L., and McKay D. (2001a) Lunar mare soils: Space weathering and the major effects of surface-correlated nanophase Fe. *J. Geophys. Res., 106,* 27985–28000.

Taylor L. A., Cahill J. T., Patchen A., Pieters C. M., Morris R. V., Keller L., and McKay D. (2001b) Mineralogical and chemical characterization of lunar highland regolith: Lessons learned from mare soils (abstract). In *Lunar and Planetary Science XXXII,* abstract #2196. Lunar and Planetary Institute, Houston (CD-ROM).

Tedesco E. F., Matson D., and Veeder G. (1989) Classification of IRAS asteroids. In *Asteroids II* (R. P. Binzel et al., eds.), pp. 290–297. Univ. of Arizona, Tucson.

Thomas P. C., Joseph J., Carcich B., Veverka J., Clark B. E., Bell J. F. III, Byrd A. W., Chomko R., Robinson M., Murchie S., Prockter L., Cheng A., Izenberg N., Malin M., Chapman C., McFadden L. A., Kirk R., Gaffey M., and Lucey P. G. (2002) Eros: Shape, topography, and slope processes. *Icarus, 155,* 18–37.

Trombka J., Squyres S., Bruckner J., Boynton W., Reedy R., McCoy T., Gorenstein P., Evans L., Arnold J., Starr R., Nittler L., Murphy M., Mikheeva I., McNutt R., McClanahan T., McCartney E., Goldsten J., Gold R., Floyd S., Clark P., Burbine T., Bhangoo J., Bailey S., and Pataev M. (2000) The elemental composition of asteroid 433 Eros: Results of the NEAR Shoemaker X-ray spectrometer. *Science, 289,* 2101–2105.

Veverka J., Thomas P., Harch A., Clark B. E., Bell J. F. III, Carcich B., Joseph J., Murchie S., Izenberg N., Chapman C., Merline W., Malin M., McFadden L., and Robinson M. (1999) NEAR encounter with asteroid 253 Mathilde: Overview. *Icarus, 140,* 3–16.

Veverka J., Robinson M., Thomas P., Murchie S., Bell J. F. III, Izenberg N., Chapman C., Harch A., Bell M., Carcich B., Cheng A., Clark B. E., Domingue D., Dunham D., Farquhar R., Gaffey M., Hawkins E., Joseph J., Kirk R., Li H., Lucey P., Malin M., Martin P., McFadden L., Merline W., Miller J., Owen W., Peterson C., Proktor L., Warren J., Wellnitz D., Williams B., and Yeomans D. (2000) NEAR at Eros: Imaging and spectral results. *Science, 289,* 2088–2097.

Veverka J., Thomas P. C., Robinson M., Murchie S., Chapman C., Bell M., Harch A., Merline W. J., Bell J. F. III, Bussey B., Carcich B., Cheng A., Clark B. E., Domingue D., Dunham D., Farquhar R., Gaffey M. J., Hawkins E., Izenberg N., Joseph J., Kirk R., Li H., Lucey P., Malin M., McFadden L., Miller J. K., Owen W., Peterson C., Procktor L., Warren J., Wellnitz D., Williams B., and Yeomans D. K. (2001) Imaging of small-scale features on 433 Eros from NEAR: Evidence for a complex regolith. *Science, 292,* 484–488.

Wells E. and Hapke B. (1977) Lunar soil: Iron and titanium bands in the glass fraction. *Science, 195,* 977–979.

Yamada M., Sasaki S., Nagahara H., Fujiwara A., Hasegawa S., Yano H., Hiroi T., Ohashi H., and Otake H. (1999) Simulation of space weathering of planet-forming materials: Nanosecond pulse laser irradiation and proton implantation on olivine and pyroxene samples. *Earth Planets Space, 51,* 1255–1265.

Zuber M., Smith D., Cheng A., Garvin J., Aharonson O., Cole T., Dunn P., Guo Y., Lemoine F., Neumann G., Rowlands D., and Torrence M. (2000) The shape of 433 Eros from the NEAR Shoemaker Laser Rangefinder. *Science, 289,* 2097–2102.

Part V:

History and Interrelations with Other Solar System Bodies

The Determination of Asteroid Proper Elements

Z. Knežević
Astronomical Observatory, Belgrade

A. Lemaître
Facultes Universitaires Notre-Dame de la Paix, Namur

A. Milani
University of Pisa

Following a brief historical introduction, we first demonstrate that proper elements are quasi-integrals of motion and show how they are used to classify asteroids into families and study the long-term dynamics of asteroids. Then, we give a complete overview of the analytical, semi-analytical, and synthetic theories for the determination of proper elements of asteroids, with a special emphasis on the comparative advantages/disadvantages of the methods, and on the accuracy and availability of the computed proper elements. We also discuss special techniques applied in some particular cases (mean motion and secular resonant bodies). Finally, we draw conclusions and suggest directions for future work.

1. INTRODUCTION

The computation of asteroid proper elements is certainly one of the fields of asteroid research that has undergone the most remarkable development in the last decade. The accuracy and efficiency of the methods introduced in this period improved dramatically. Thus we were able to solve many problems that puzzled researchers in previous times. We could also recognize and investigate an entire spectrum of new problems, from novel classes of dynamical behavior to different phenomena that were previously either completely unknown or impossible to investigate with available tools.

The history, definition, and applications of proper elements are described in great detail in a number of reviews (e.g., *Valsecchi et al.,* 1989; *Shoemaker et al.,* 1989; *Lemaître,* 1993; *Knežević and Milani,* 1994; *Knežević,* 1994). However, for the sake of completeness, these topics are tackled in the following sections.

1.1. Historical Overview

A classical definition states that proper elements are quasi-integrals of motion, and that they are therefore nearly constant in time. Alternatively, one can say that they are true integrals, but of a conveniently simplified dynamical system. In any case, proper elements are obtained as a result of the elimination of short and long periodic perturbations from their instantaneous, osculating counterparts, and thus represent a kind of "average" characteristics of motion.

A concept of proper elements has been introduced by *Hirayama* (1918) in his celebrated paper in which he announced the discovery of asteroid families. Even if not using the technical term "proper," he employed Lagrange's classical linear theory of asteroid secular perturbations to demonstrate that certain asteroids tend to cluster around special values of the orbital elements, which very closely correspond to the constants of integration of the solutions of the equations of their motion, i.e., to a sort of averaged characteristic of their motion over very long timespans. In his later papers *Hirayama* (1923, 1928) explicitly computed just the proper elements (proper semimajor axis, proper eccentricity, and proper inclination) and used them for the classification of asteroids into families.

The next important contribution came from *Brouwer* (1951). He computed asteroid proper elements again using a linear theory of secular perturbations, but in combination with an improved theory of motion of the perturbing planets (*Brouwer and Van Woerkom,* 1950). By including more accurate values of planetary masses, and the effect of the "great inequality" of Jupiter and Saturn, he was able to get a more realistic value for the precession rate of the perihelion of Saturn.

Williams (1969) developed a semianalytic theory of asteroid secular perturbations that does not make use of a truncated development of the perturbing function, and which is therefore applicable to asteroids with arbitrary eccentricity and inclination. Williams' proper eccentricity and proper inclination are defined as values acquired when the argument of perihelion $\omega = 0$ (thus corresponding to the minimum of eccentricity and the maximum of inclination over a cycle of ω). The theory is linearized in planetary masses, eccentricities, and inclinations, so that the proper elements computed by means of this theory (*Williams,* 1979, 1989), even if much better than the previously available ones, were still of limited accuracy.

Another achievement to be mentioned is that by *Kozai* (1979), who used his theory of secular perturbations for high-inclination asteroids (*Kozai,* 1962) to define a set of proper parameters to identify the families. The selected parameters were semimajor axis, z component of the angular momentum (integral of motion in a first-order theory, with perturbing planets moving on circular, planar orbits), and the minimum value of inclination over the cycle of the argument of perihelion (corresponding to $\omega = \pi/2$).

Finally, we refer readers to the work by *Schubart* (1982, 1991), *Bien and Schubart* (1987), and *Schubart and Bien* (1987), who pioneered attempts to determine the proper parameters for resonant groups, i.e., for Hildas and Trojans. Since the usual averaging methods do not apply in this case, they adopted slightly different definitions of the proper parameters, the most important difference being the substitution of a representative value measuring the libration of the critical argument instead of the usual proper semimajor axis.

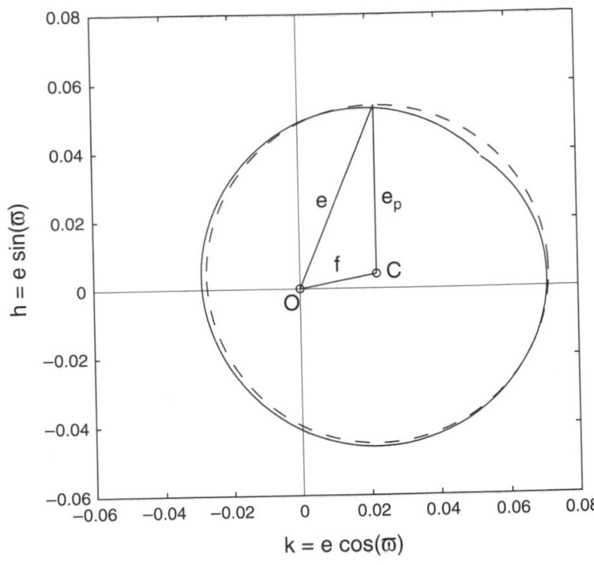

Fig. 1. The orbit of the numbered asteroid 27633 over 20,000 yr projected in the (k,h) plane (full line). Data have been digitally smoothed to remove short periodic perturbations. Point C represents average value of the forced term, f is forced eccentricity, and the dashed circle of radius e_p represents the best-fitting epicycle. The value of the eccentricity e, obtained as the length of the vectorial sum of the forced and the free terms, is an approximation of the current value.

1.2. Proper Elements/Parameters

The notion of proper elements is based on the linear theory of secular perturbations, which dates back to Lagrange. Linear theory neglects the short periodic perturbations, containing anomalies in the arguments; this results in a constant semimajor axis that becomes the first proper element a_p. The long-term evolution of the other variables is obtained by approximating the "secular" equations of motion with a system of linear differential equations. Because of the linearity assumption, the solutions can be represented in the planes $(k,h) = (e \cos \varpi, e \sin \varpi)$ as the sum of "proper modes," one for each planet, plus one for the asteroid. Thus the solution can be represented by epicyclic motion: For the asteroid, the sum of the contributions from the planets represents the "forced" term, while the additional circular motion is the so-called "free oscillation" and its amplitude is the proper eccentricity e_p. The same applies to the plane $(q,p) = (\sin i \cos \Omega, \sin i \sin \Omega)$, with amplitude of the free term given by the (sine of) proper inclination $\sin i_p$. Figure 1 shows the output of a numerical integration of an asteroid's orbit for 20,000 yr plotted in the (k,h) plane, and an epicyclic model fitting to the data. As is apparent from the figure, the approximation of the linear secular perturbation theory is good enough for a timespan of the order of the period of circulation for the longitude of perihelion ϖ. However, even over such a timespan the linear theory is only an approximation, and over a much longer timespan (e.g., millions of years) it would be a rather poor approximation in most cases.

Proper elements can also be obtained from the output of a numerical integration of the full equations of motion: The simplest method is to take averages of the actionlike variables a, e, i, over times much longer than the periods of circulation of the corresponding angular variables. However, this method provides proper elements of low reliability: The dynamical state can change for unstable orbits and in such cases the simple average wipes out this essential information. Thus, if the goal is to compute proper elements stable to 1% of their value or better, over timespans of millions of years, it is necessary to use much more complicated theories. This is in essence saying that the approximate dynamical system, rather than a simple linear one, is by itself complex, even if it admits integrals that are used as proper elements.

Whatever the type of theory, on the other hand, if it is to be accurate enough to represent the dynamics in the framework of a realistic model, its full-detail description requires delving into a very long list of often cumbersome technicalities. For this reason, in the present paper we only give a qualitative description of the computational procedures, and then proceed to discuss the quality of the results.

Several different sets of proper parameters have been introduced over time, but the most common set, usually referred to as "proper elements," includes proper semimajor axis a_p, proper eccentricity e_p, (sine of) proper inclination (sin) i_p, proper longitude of perihelion ϖ_p, and proper longitude of node Ω_p, the latter two angles being accompanied by their precession rates (fundamental frequencies g and s respectively).

The analytical theories and the previously mentioned theory by *Williams* (1969) use a different definition of proper eccentricity and inclination. Other authors introduced completely different parameters to replace the standard proper elements. Still, the common feature of all these parameters is their supposed near constancy in time (or more precisely, stability over very long timespans), and one can say that in this sense the term "proper" is practically a synonym for "invariable."

As explained by *Lemaître* (1993), even if quite different in terms of the proper parameters and the ways to compute them, all the theories follow several basic steps, which can be summarized as (1) modeling of the asteroid motion (N-body, restricted three-body) and distinguishing the fast and slow angles (that is, separating the perturbations depending on the mean longitudes from the rest); (2) removal of the short periodic perturbations (analytical or numerical averaging, online filtering) and computation of the mean elements; (3) splitting of the resulting Hamiltonian into two parts, the integrable (secular) part, and the perturbation (the long-periodic part depending on slow angles) — in the case of synthetic theories this step corresponds to the removal of the forced terms; (4) removal of the long-periodic terms and computation of the proper values — at this stage, analytic and semianalytic theories resort to averaging over the slow angles and iterative procedures to compute the inverse map of canonical transformation, while in the case of synthetic theories this phase includes the Fourier analysis and extraction of the principal harmonics from the time series of mean elements (freed from the forced terms); (5) output of the proper values, possibly accompanied with their errors (standard deviations, maximum excursions), fundamental frequencies, quality, and resonant codes, chaotic behavior indicators, and other information.

1.3. Applications

For a long time proper elements have been used for the sole purpose of identifying the asteroid families. Even today this is one of their most common applications (*Zappalà et al.,* 1990, 1995). Since family identification is the subject of another chapter in this book (*Bendjoya and Zappalà,* 2002), we are not going to discuss it here in further detail.

On the other hand, the proper elements are computed for very different asteroids, from NEAs (*Gronchi and Milani,* 2001) to Trojans (*Milani,* 1994; *Beaugé and Roig,* 2001). They are used to describe very different dynamical phenomena, like stable chaos (*Milani et al.,* 1997), resonant dynamics (*Morbidelli,* 1993), or nongravitational effects (*Farinella and Vokrouhlický,* 1999). The list of problem-solving applications of proper elements is long, ranging from the determination of locations of secular resonances in the asteroid belt (*Milani and Knežević,* 1994) and in a wider planetary region (*Knežević et al.,* 1991, *Michel and Froeschlé,* 1997), through physical and dynamical studies of asteroid families (*Morbidelli et al.,* 1995) and the determination of their age (*Milani and Farinella,* 1994; *Knežević,* 1999), to such "exotic" uses as the study of an optimum strategy for the orbit maintenance of a low lunar polar orbiter (*Knežević and Milani,* 1998), to mention but a few.

2. THEORIES

2.1. Analytical Theory

The main problem in the manufacturing of fully analytical theories of proper elements is in the complete *degeneracy* of the unperturbed dynamics, the two-body problem. Degeneracy means that some of the fundamental frequencies are zero, and indeed in the two-body approximation both the perihelia and the nodes do not precess at all. In the Hamiltonian formalism, this is expressed by the statement that the unperturbed Hamiltonian function $H_0 = H_0(L)$ is a function only of one variable L, which in turn is a function of the semimajor axis. The perturbed problem with Hamiltonian $H_0(L) + \mu H_1(\ell, g, L, G)$ (the small parameter μ representing the ratio of the mass of the planets to the mass of the Sun) can be handled with different perturbative approaches, but they all have in common the use of a solution of the unperturbed two-body problem to be substituted into the perturbing function H_1. Thus they also have in common the problem that the angle variable ℓ, the mean anomaly conjugate to L, can be eliminated, but the angles g, conjugate to the other action variables not appearing in H_0, cannot be removed by averaging.

In practice, this implies that the procedure to compute proper elements must always be decomposed into two computational steps: the transformation from osculating orbital elements to mean elements, free from the short periodic perturbations (with arguments containing the fast variable ℓ), and the transformation of the mean elements into proper elements. A fully analytical theory performs both steps by means of the computation of functions for which analytical expressions, in practice truncations of some series, are available. Note that it is possible, sometimes even convenient, to mix two different methods: As an example, *Lemaître and Morbidelli* (1994) use mean elements computed analytically as a starting point for their semianalytic computation of proper elements.

Different perturbation techniques can be used, with the Lie series technique being the most convenient for theories pushed to higher order (and therefore based upon series with many terms). In essence, analytical perturbation techniques exploit the approximation of the perturbing function H_1 by means of a finite sum of terms, each with a simple expression of the form

$$\mu\, b(L)\, e^h\, e'^k\, i^j\, i'^m \cos(p\ell + q\ell' + \delta)$$

where h, k, j, m, p, and q are integers; the primed elements refer to some perturbing planet; b is a known function; and δ is some combination (with integer coefficients) of the angles included in g, the perihelia and nodes of the asteroid and the planet. The truncation of the series is mostly based on the degree in the small parameters eccentricity and inclination (of both the asteroid and the planets), although truncation for large values of the integers p and q is also possible. Thus we can describe the degree of completeness of a theory by means of the order (in the small parameter μ) and of the degree (in the eccentricities and inclinations). From this arises the main limitation of the analytical method: The accuracy, and stability with time, of the proper elements decreases as the asteroid eccentricity and inclination increase. There is a boundary between the region where the analytical proper elements are most suitable and the region where more computationally intensive methods, such as the semianalytic ones, need to be used; *Knežević et al.* (1995) have mapped this boundary (see section 2.4). The simple analytical form of the terms implies that it is possible to perform both derivatives and integrals analytically; thus the corresponding operations can be applied to the series term by term. An analytical theory can be expressed by means of derivatives, integrals, and arithmetic operations on these series; thus they can be explicitly computed by means of a finite, although large, number of elementary operations. In practice, the current analytical theories use series with several tens of thousands of terms.

The series used in the current theories are essentially based upon the expansions computed by *Yuasa* (1973) and corrected and completed by *Knežević* (1989, 1993). These are complete to degree 4 in eccentricities and inclinations; only a few special terms of degree 6 have been added later (*Milani and Knežević,* 1994). *Yuasa* (1973) defined an algorithm enabling the computation of proper elements with a theory containing the main terms of order 2 in μ and complete to degree 4 in eccentricity and inclination, but one essential step was missing. *Milani and Knežević* (1990) found that, at orders >1 in μ, the formulas of perturbation theory explicitly provide a map between proper and mean elements, in the opposite direction from the one we deal with in practice. Thus the computation of proper elements from mean elements requires the solution of an inverse function problem, and this is possible only by an iterative procedure. *Milani and Knežević* (1999) later applied the same argument to the computation of mean elements from osculating ones.

There are two reasons to use analytical proper elements, even when other methods (for example, synthetic ones) are available for the same objects.

First, the large number of terms and the iterative procedure notwithstanding, even with complications introduced in later versions (*Milani and Knežević,* 1992, 1994), the computational complexity of the fully analytical computation of proper elements is not important (with respect to the computing power available today). To recompute from scratch the proper elements for more than 67,000 asteroids requires currently less than 2 CPU hours on a standard workstation. Thus the only limit to the size of proper element catalogs, when computed analytically, is the size of the catalogs of osculating elements. Taking into account that asteroids observed only at a single opposition generally have osculating elements affected by uncertainties larger than the inaccuracies in the computation of proper elements, the AstDys information system (http://hamilton.dm.unipi.it/astdys/) provides analytical proper elements for all numbered and multiopposition asteroids. This catalog is updated every month with new discoveries, new numberings, and new recoveries providing multiopposition orbits. Since the asteroid families are statistical entities, the advantage of such a large set of proper elements may largely compensate for an accuracy that is not as good as that for the synthetic proper elements.

The second main advantage of the analytical proper elements is that the computational algorithm automatically provides information on all the resonances. Indeed, the iterative procedures used in the computation can be divergent where a *small divisor,* resulting from the attempt to integrate one of the trigonometric terms with very slowly varying arguments, occurs. On the other hand this is, by definition, a resonance. Difficulties in the computation of mean elements (from the osculating ones) are due to mean-motion resonances, involving the anomalies of the asteroid and a planet. Difficulties in the convergence of the iterations for the computation of proper elements from mean elements indicate secular resonances, i.e., very small frequencies resulting from combinations of the frequencies g and s of the perihelia and nodes with the corresponding frequencies for the perturbing planets. Thus the analytical proper elements come with quality codes indicating not only lower accuracy due to failure of convergence, but also the resonance responsible for the problem. By the same theories it is possible to construct maps of secular resonances (*Knežević et al.,* 1991, *Milani and Knežević,* 1994), including combinations of up to four secular frequencies (and in some special cases six or eight frequencies). In Fig. 2 we give an example of such a map with the most important two- and four-frequency secular resonances occurring in the main asteroid belt.

2.2. Semianalytical Theory

The semianalytic calculation of proper elements was initiated by *Williams* (1969) and then revisited by *Lemaître and Morbidelli* (1994). It is a classical perturbation method, where the two averaging processes (the first one on the mean longitudes and the second on the pericenters and nodes) are performed numerically. This avoids the expansion in the eccentricity (e) and inclination (i) of the asteroid, and makes the method particularly suitable for about 20% of the asteroids with large values for these elements.

The latest version of the theory is written in a Hamiltonian formalism and computed up to the second order in the perturbing masses (presently only Jupiter and Saturn) and up to the first degree in the eccentricities (e') and inclinations (i') of the perturbers.

Elimination of the short-periodic terms is performed numerically by the calculation of the double integrals (over the two mean longitudes). After this averaging, the semimajor axis is constant and represents the first proper element. To compute the averaged Hamiltonian, a Fourier series of the slow angles is used, with the coefficients evaluated on a three-dimensional grid (in a, e, and i) and stored; a triple linear interpolation is used each time the Hamiltonian and its derivatives have to be evaluated.

The averaged Hamiltonian is split in two parts, based on the smallness of the parameters e' and i'; the integrable problem corresponds to circular and planar motion of the perturbing planets, while the perturbation part gathers all the first-order contributions in e' and i'. The dynamics of the integrable problem has already been analyzed by *Kozai* (1962), and it reveals different behaviors for low and high

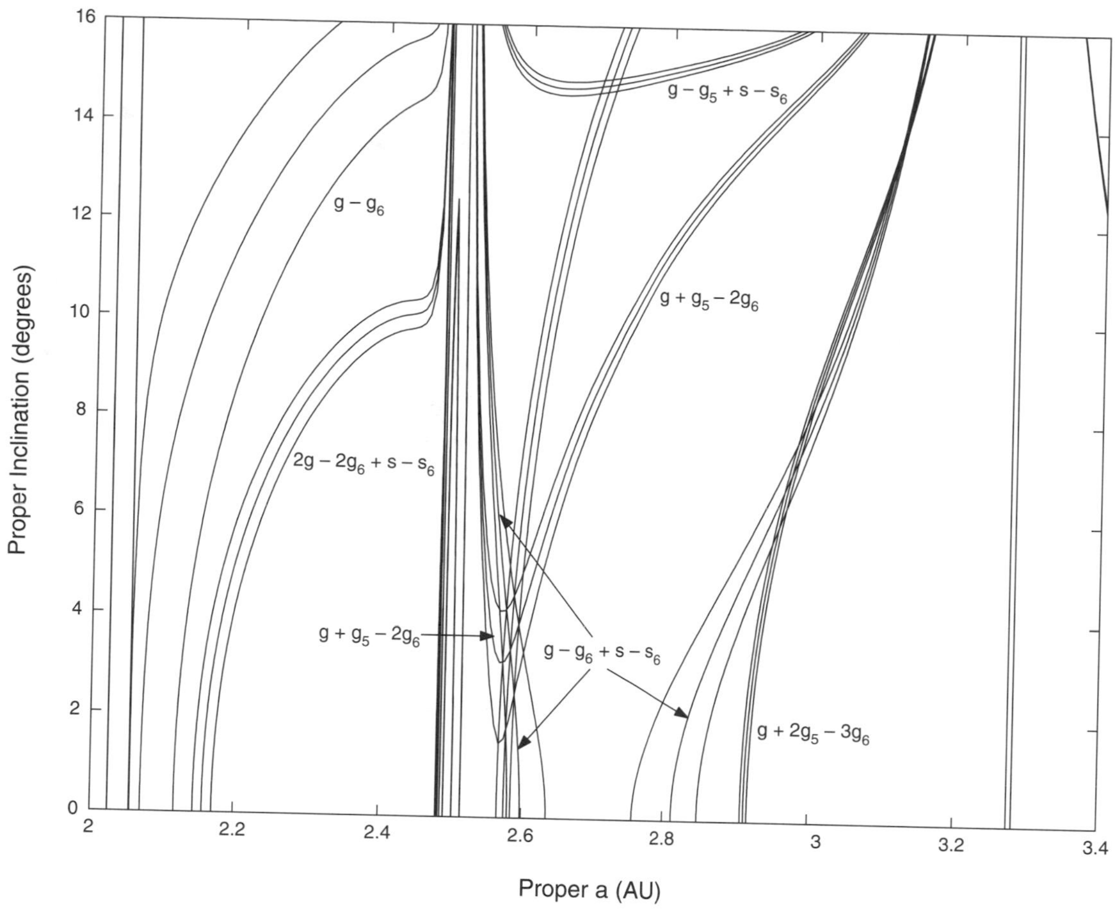

Fig. 2. Map of a sample of secular resonances in the asteroid main belt, involving up to six fundamental frequencies. The locations of the resonances are computed for $e_p = 0.1$ and are marked with three lines corresponding to the zero value of the corresponding divisor and to ±0.5 arcsec/yr (with the exception of $g - g_6$ for which we plot wider boundaries corresponding to ±2 arcsec/yr). g and s are the frequencies of the asteroid perihelion and node, while g_5, g_6, and s_6 are the corresponding values of Jupiter and Saturn.

inclinations; in the latter case, a critical curve separates the phase space into two regions, corresponding to librations (about 90° or 270°) and to circulations of the argument of the pericenter ω of the asteroid.

The removal of the long periodic terms is done by using the action angle variables and is based on the semianalytic method developed by *Henrard* (1990). The resulting Hamiltonian K, after the second averaging process, depends only on two proper actions, J and Z, which are both constant. The result is a proper orbit of area proportional to J, located in a plane identified by the value of Z. Each point on the proper orbit is characterized by a value of the phase ψ, which can be considered as a proper argument of perihelion. From the values of ψ, J, and Z, the corresponding values of e, i, and ω can be calculated. Any point can be chosen as a "representative" of this orbit: For example, ψ = 0°, 180° corresponds to the minimum value of the eccentricity and maximum of the inclination along the proper orbit. On the other hand, ψ = 90°, 270° corresponds to the maximum of the eccentricity and the minimum of the inclination; this option allows the definition of proper elements even for ω librators, since the libration center is either 90° or 270°.

The proper orbit is also characterized by the two basic frequencies, g and s, calculated as the partial derivatives of the Hamiltonian K with respect to the actions J and Z. The set a, g, and s is also a set of "proper elements" and is independent of the choice of the representative point on the proper orbit.

The existence of two high inclined groups of asteroids was clearly shown by this method: a Pallas family at about 35° inclination (*Lemaître and Morbidelli,* 1994), and a Hungaria family at 2 AU (*Lemaître,* 1994).

The precision is limited by the first-order development in e' and i' and could not be easily improved. However, this semianalytic procedure has some advantages, such as the fact that it allows the calculation not only of the proper frequencies, but also of their derivatives with respect to the action variables; to our knowledge, this last possibility has not been explored in specific applications.

The last calculation of a complete set of proper elements by the semianalytic method was done in 1994, but only for the asteroids for which the analytic method proved not to be applicable. Catalogs of semianalytic elements for approximately 2500 asteroids are available from http://www.fundp.ac.be/sciences/math/equadif.html.

2.3. Synthetic Theory

The latest contribution to the field of asteroid proper-element determination is the synthetic theory by *Knežević and Milani* (2000). This is, in fact, a set of purely numerical procedures by means of which one can derive classical proper elements (a_p, e_p, sin i_p, ϖ_p, Ω_p) and fundamental frequencies (g, s). The theory employs the approach used by *Carpino et al.* (1987) for the major planets, and consists of the following steps: First, one numerically integrates the asteroid orbits for a sufficient timespan, together with the orbits of perturbing planets included in the model (the indirect effects of the planets not included in the dynamical model are accounted for by applying the so-called baricentric correction to the initial conditions). The short periodic perturbations are removed by means of an online filtering of the osculating elements, performed in the course of the integration itself; simultaneously, the maximum Lyapunov characteristic exponents are derived from the variational equations to monitor the chaotic behaviors. The forced oscillations are then removed from the output of the integration and the resulting time series spectrally resolved under the constraints set by the d'Alembert rules to extract principal harmonics (proper values) together with the associated fundamental frequencies and error estimates (standard deviations and maximum excursions).

Because it is purely numerical, this theory is comparatively simple in principle. It furnishes results of a superior accuracy with respect to the analytical theory and, more importantly, provides a straightforward way to estimate errors of all the proper values and for each asteroid included in the computation. We used a simple running box method, computing proper values over a number of shorter time intervals and then deriving the corresponding rms and maximum deviation values. The length and shift (i.e., number) of boxes has been selected in such a way to best fit the dynamics, thus in a different way for the inner and the outer belt, for shorter and longer integration timespans, etc. The theory applies equally well to the asteroids of any eccentricity and inclination, with the obvious exception of the planet-crossing objects; since it is based on the averaging principle, it does not work for the resonant asteroids either.

As an example, in Fig. 3 the number frequency distributions of error estimates are given for the synthetic proper eccentricities and inclinations of numbered asteroids. These results are on average better by more than a factor of 3 than the typical accuracy of the proper elements derived from the analytical theory (*Milani and Knežević,* 1994).

The numerical integration of tens of thousands of orbits is, on the other hand, a time-consuming procedure, so that the synthetic theory is ostensibly less efficient in terms of CPU time and more demanding in terms of necessary computing power. This in turn leads to other potential drawbacks, such as getting results supposedly valid over very long timespans from the integrations over comparatively short intervals; such short integrations might affect the determination of maximum LCEs for the objects subject to a slow chaos, etc.

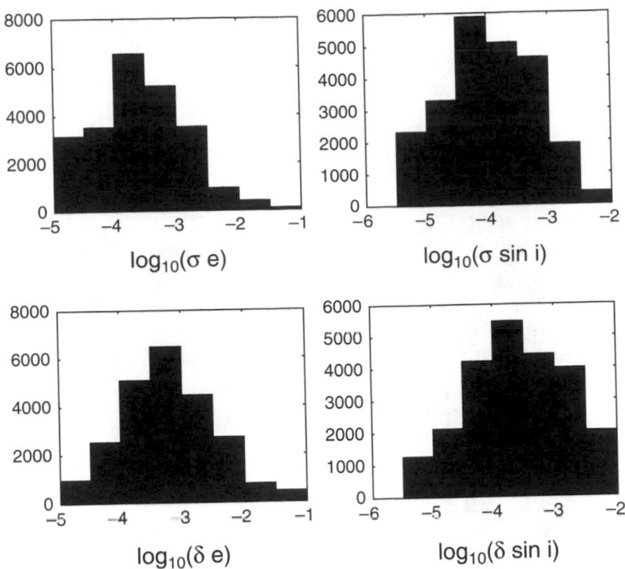

Fig. 3. Number frequency distributions of standard deviations (top) and maximum excursions (bottom) of the proper eccentricity (left) and proper sine of inclination (right). About 80% of the asteroids have standard deviation of proper eccentricity below 0.001, while for the sine of proper inclination this percentage rises to more than 90%. Median rms is 0.00024 in proper eccentricity and 0.00011 in sine of proper inclination. Further improvement is still possible, depending on the available computing power.

These drawbacks, however, are more than amply compensated for by the accuracy achieved and by the other advantages mentioned above. The synthetic theory is at the moment certainly the best theory available on the market.

The synthetic proper elements are available from the AstDys service at http://hamilton.dm.unipi.it/astdys. There are three catalogs available for downloading; these catalogs contain the proper elements themselves, their standard deviations, and their maximum excursions (plus some additional information). The files are updated on a monthly basis for the numbered asteroids between 2.0 and 4.0 AU, with the exception of the planet-approaching asteroids; the catalogs as of July 2001 contain proper elements for 24,910 asteroids.

2.4. Comparison of Theories in Terms of Accuracy and Efficiency

The only direct comparison between different theories to compute asteroid proper elements has been made by *Knežević et al.* (1995). They integrated the orbits of five selected asteroids for 4.5 m.y., simultaneously computing the proper elements and their instabilities by means of the analytical (*Milani and Knežević,* 1994) and semianalytical (*Lemaître and Morbidelli,* 1994) theories. Comparing the outcomes (see Fig. 4 for an example) they concluded that analytic proper elements are more accurate for the asteroids up to about 15° of inclination, while the semianalytic ones are better above about 17°. In the intermediate transition region both methods have roughly the same stability.

There is no direct comparison on a case-by-case basis for the synthetic theory with respect to any of the other theories. Thus, at the moment we can resort only to the general estimate mentioned above of an improvement by a factor of more than 3, on average, of the synthetic proper elements with respect to the results coming from the analytical theory (section 2.3).

As for the efficiency in the computation, the analytical theory is by far the fastest and the synthetic theory is the most time consuming, with the semianalytical theories falling somewhere in between these two. In practice, therefore, when choosing between the theories to be applied under a given circumstance, one always has to find an optimum tradeoff between the conflicting demands of efficiency and accuracy.

3. SPECIAL CASES

3.1. Resonant Proper Elements

The removal of the short- and long-periodic terms becomes problematic in the neighborhood of a resonance — a mean-motion resonance in the first average procedure, a secular resonance in the second one. If one of the frequencies is close to zero, then the usual averaging methods have to be adapted to this particular situation. The nonresonant harmonics can be eliminated but the critical angle is still present in the averaged problem.

Of course the first reaction is to treat the problem purely numerically, integrating the orbits over millions of years and using a frequency analysis to calculate the proper frequencies. This method seems very suitable, because the real cases of resonant asteroids are not so numerous and the time required by the integration is still reasonable. The Trojans, well-known 1:1 jovian resonant asteroids, or the Hilda group were integrated following this idea (section 3.2).

However, the simulations concerning fictitious fragments of asteroids, for example, in order to statistically investigate the efficiency of a particular resonance in delivering

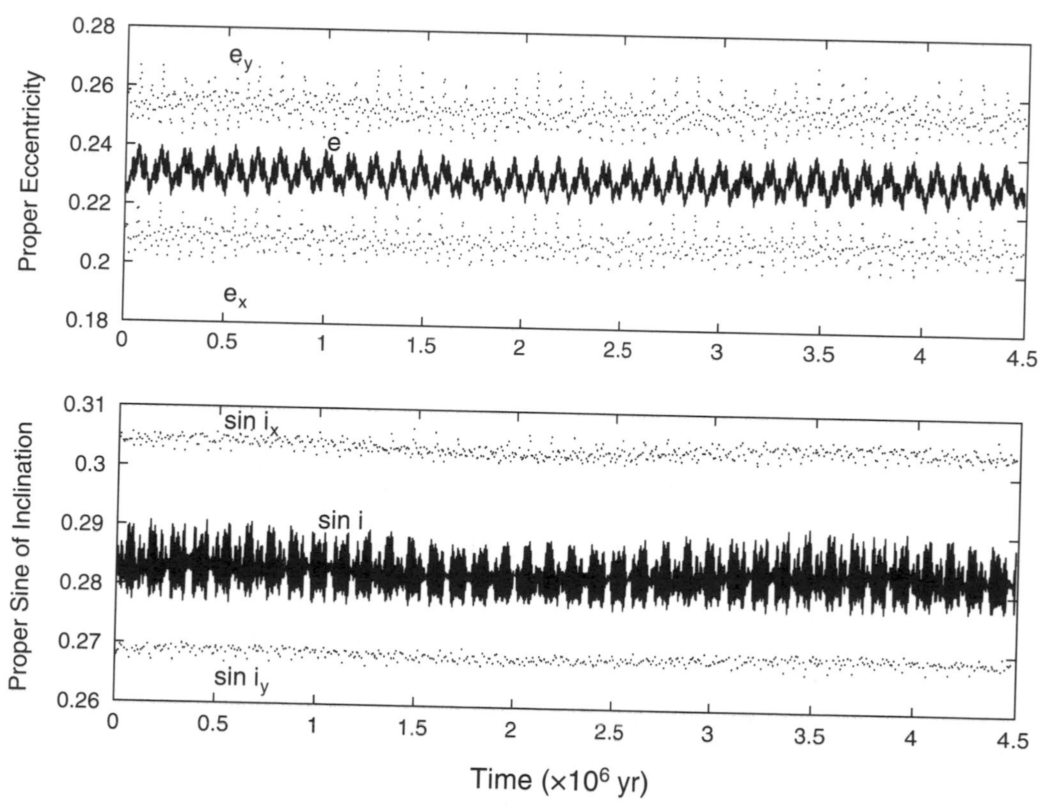

Fig. 4. Comparison of the proper eccentricities (top) and proper inclinations (bottom) for asteroid 387 Aquitania. Proper values determined by means of the semianalytic theory are defined as the values occurring for $\omega = 0°$ (e_x, i_x) and for $\omega = 90°$ (e_y, i_y), thus in the former case corresponding to the minimum of eccentricity and maximum of inclination over a cycle of the argument of perihelion, and in the latter case corresponding to the maximum of e and minimum of i. Proper values obtained by means of the analytic theory are in the middle (e, sin i). For this particular asteroid, having comparatively high orbital inclination, the unremoved instabilities of proper eccentricity as obtained from the two theories are about the same, while for the proper inclination the semianalytical theory already gives better results.

meteorites to the Earth, require the calculation of thousands of orbits. This encourages the development of specific resonant seminumerical methods, allowing a picture of the phase space to be drawn and the computation of "resonant proper elements" in a much faster way.

Let us emphasize that these resonant proper elements are not the same as in the nonresonant cases; they can be defined as any identifier of the trajectories obtained in the model after the resonant averaging process. Their definition is arbitrary (e.g., the values of the eccentricity or of the inclination when the critical angle equals some specific value, for example, zero) and cannot be directly connected to the classical nonresonant proper eccentricity and inclination. Thus it is not possible to mix resonant and nonresonant proper elements for purposes such as family classification.

In this approach, let us mention the calculation of secular resonant proper elements by *Morbidelli* (1993), who applies the same semianalytic method as *Lemaître and Morbidelli* (1994) but adapted for the resonant case. After the elimination of the short-periodic terms, instead of eliminating all secular perturbation terms in the second step, a critical angle ($g - g_6$, for example) is still present in the integrable part of the Hamiltonian. Typical cases such as 582 Olympia, 945 Barcelona, or 739 Mandeville have been treated by this method and show a good qualitative agreement with the numerical integrations. However, no catalogs or listings of secular resonant real asteroids with corresponding proper elements or identifiers are available; anyone interested in this calculation can contact the author of the method, A. Morbidelli (morby@obs-nice.fr).

3.2. Trojans and Hildas

The Trojan asteroids are a special case, because they are locked in the 1:1 resonance with Jupiter. Because this is such a strong resonance, librations with very large amplitude are possible: The critical argument $\lambda - \lambda'$ can have small librations around 60° and also oscillate between ≈30° and ≈100°. Moreover, the inclination can be very large, up to 33°. As a result, the development of an analytical theory is very difficult, and indeed it has been achieved only within a simplified three-body model (*Érdi,* 1988). A semianalytical theory is conceivable, but also difficult to compute within a realistic model.

The synthetic approach has been pioneered by Bien and Schubart (see *Bien and Schubart,* 1987; *Schubart and Bien,* 1987; and references therein). They use as proper elements a proper eccentricity and proper inclination defined essentially in the same way as in the synthetic theories for main-belt asteroids and an amplitude of libration of $\lambda - \lambda'$. *Milani* (1993, 1994) used a synthetic approach adapted to *Érdi*'s (1988) theory, and computed proper elements for 188 numbered, multiopposition, and well-observed single-opposition Trojans. Although the results suggested the possible existence of some Trojan families, the number of asteroids in the catalog was too small for a firm conclusion. *Milani* (1994) also mapped the secular resonances involving the node by fitting the results of the synthetic computations of the frequencies, especially s, with a polynomial in the action variables.

More recently there has been interesting semianalytic work by *Beaugé and Roig* (2001), who produced a catalog (available from the authors upon request) of proper elements for 533 Trojans, also proposing a number of potential dynamical families. However, these elements are of lower accuracy than those produced by means of a synthetic approach. As of July 2001 there are 679 Trojans with good quality orbits (either numbered or multiopposition), thus the computation of a new catalog of accurate proper elements should allow statistically significant conclusions about the existence of Trojan families.

In regard to Hildas, even if important results on their dynamics have been obtained in recent years (e.g., *Ferraz-Mello et al.,* 1998), very little work has been done about their proper elements. The computation of "three characteristic parameters" of orbits of Hilda-type asteroids initiated by *Schubart* (1982) has been later extended to newly discovered objects by the same author (*Schubart,* 1991). Even at present the only source of data for this group of bodies can be found on the Web site maintained by J. Schubart (http://www.rzuser.uni-heidelberg.de/~s24/hilda.htm).

4. CONCLUSIONS AND FUTURE WORK

Activity in the field of determination of asteroid proper elements has been very intense in the previous decade, and one can safely claim that more work has been done and more improvement achieved in this period than in all the preceding time.

The proper elements are now computable for essentially all asteroids for which there are accurate osculating orbits; thus we currently have proper elements determined for about 70,000 objects. The three principal theories intended for production of large catalogs of proper elements for the main-belt objects, plus a number of specially adapted ones for specific groups of objects, have been developed and successfully applied during this period.

The accuracy, long-term stability, and reliability of the computed proper elements have all remarkably improved (by orders of magnitude). This enabled a lot of important problems to be tackled and provided explanations for many interesting properties of individual asteroids and of their families, which, in turn, decisively contributed to a better understanding of the dynamical and collisional evolution of the asteroid belt as a whole.

Apart from providing a reliable basis for the identification of families, the studies related to the determination of proper elements led to the development of complete theories of asteroid motion, revealing at the same time many novel dynamical features and concepts (nonlinear secular resonances, stable chaos, dynamical mass loss, etc.)

All the main theories to compute asteroid proper elements have now reached a high level of "maturity," being developed almost to their corresponding theoretical limits

(*Milani and Knežević,* 1994; see also section 2.2), while on the other hand being amply and thoroughly tested through years of application. Therefore, it is plausible to conjecture that in the near future we can expect more improvement in terms of their practical application than in the theories themselves.

There are two principal directions in which the development can reasonably take place: improvement of the accuracy and reliability of determination of proper elements (e.g., by extending the timespan over which the averaging is done in the synthetic theory), and computation of proper elements for an ever-increasing number of asteroids with different dynamical properties (for the purpose of enlarging and completing the database); in this latter class one can include computation of proper elements for objects such as Centaurs and TNOs for which this has not yet been done.

Acknowledgments. We thank referees F. Namouni and C. Murray for useful suggestions, and A. Morbidelli for providing input on the resonant case.

REFERENCES

Beaugé C. and Roig F. (2001) A semi-analytical model for the motion of the Trojan asteroids: Proper elements and families. *Icarus, 153,* 391–415.

Bendjoya Ph. and Zappalà V. (2002) Asteroid family identification. In *Asteroids III* (W. F. Bottke Jr. et al., eds.), this volume. Univ. of Arizona, Tucson.

Bien R. and Schubart J. (1987) Three characteristic parameters for the Trojan group of asteroids. *Astron. Astrophys., 175,* 292–298.

Brouwer D. (1951) Secular variations of the orbital elements of minor planets. *Astron. J., 56,* 9–32.

Brouwer D. and Van Woerkom A. J. J. (1950) The secular variations of the orbital elements of the principal planets. *Astron. Papers Am. Ephem. Naut. Almanac, 13,* 81–107.

Carpino M., Milani A., and Nobili A. M. (1987) Long-term numerical integrations and synthetic theories for the motion of the outer planets. *Astron. Astrophys., 181,* 182–194.

Érdi B. (1988) Long periodic perturbations of Trojan asteroids. *Cel. Mech. Dyn. Astron., 43,* 303–308.

Farinella P. and Vokrouhlický D. (1999) Semimajor axis mobility of asteroidal fragments. *Science, 283,* 1507–1510.

Ferraz-Mello S., Michtchenko T. A., Nesvorný D., Roig F., and Simula A. (1998) The depletion of the Hecuba gap vs. the long-lasting Hilda group. *Planet. Space Sci., 46,* 1425–1432.

Gronchi G. and Milani A. (2001) Proper elements for Earth-crossing asteroids. *Icarus, 152,* 58–69.

Henrard J. (1990) A semi-numerical perturbation method for separable Hamiltonian systems. *Cel. Mech. Dyn. Astron., 49,* 43–67.

Hirayama K. (1918) Groups of asteroids probably of common origin. *Astron. J., 31,* 185–188.

Hirayama K. (1923) Families of asteroids. *Japan J. Astron. Geophys., 1,* 55–93.

Hirayama K. (1928) Families of asteroids, second paper. *Japan J. Astron. Geophys., 5,* 137–162.

Knežević Z. (1989) Asteroid long-periodic perturbations: The second order Hamiltonian. *Cel. Mech. Dyn. Astron., 46,* 147–158.

Knežević Z. (1993) Minor planet short-periodic perturbations: The indirect part of the disturbing function. *Cel. Mech. Dyn. Astron., 55,* 387–404.

Knežević Z. (1994) Asteroid proper elements: Past and present. In *Seventy-Five Years of Hirayama Asteroid Families: The Role of Collisions in the Solar System History* (Y. Kozai et al., eds.), pp. 129–139. ASP Conference Series 63.

Knežević Z. (1999) Veritas family age revisited. In *Evolution and Source Regions of Asteroids and Comets* (J. Svoreň et al., eds.), pp. 153–158. Astron. Inst. Slovak Acad. Sci. Tatranská Lomnica.

Knežević Z. and Milani A. (1994) Asteroid proper elements: The big picture. In *Asteroids, Comets, Meteors 1993* (A. Milani et al., eds.), pp. 143–158. Kluwer, Dordrecht.

Knežević Z. and Milani A. (1998) Orbit maintenance of a lunar polar orbiter. *Planet. Space. Sci., 46,* 1605–1611.

Knežević Z. and Milani A. (2000) Synthetic proper elements for outer main belt asteroids. *Cel. Mech. Dyn. Astron., 78,* 17–46.

Knežević Z., Milani A., Farinella P., Froeschlé Ch., and Froeschlé C. (1991) Secular resonances from 2 to 50 AU. *Icarus, 93,* 316–330.

Knežević Z., Froeschlé Ch., Lemaître A., Milani A., and Morbidelli A. (1995) Comparison between two theories of asteroid proper elements. *Astron. Astrophys., 293,* 605–612.

Kozai Y. (1962) Secular perturbations of asteroids with high inclination and eccentricity. *Astron. J., 67,* 591–598.

Kozai Y. (1979) The dynamical evolution of the Hirayama family. In *Asteroids* (T. Gehrels, ed.), pp. 334–358. Univ. of Arizona, Tucson.

Lemaître A. (1993) Proper elements: What are they? *Cel. Mech. Dyn. Astron., 56,* 103–119.

Lemaître A. (1994) Hungaria: A potential new family. In *Seventy-Five Years of Hirayama Asteroid Families: The Role of Collisions in the Solar System History* (Y. Kozai et al., eds.), pp. 140–145. ASP Conference Series 63.

Lemaître A. and Morbidelli A. (1994) Proper elements for highly inclined asteroidal orbits. *Cel. Mech. Dyn. Astron., 60,* 29–56.

Michel P. and Froeschlé Ch. (1997) The location of secular resonances for semimajor axes smaller than 2 AU. *Icarus, 128,* 230–240.

Milani A. (1993) The Trojan asteroid belt: Proper elements, chaos, stability and families. *Cel. Mech. Dyn. Astron., 57,* 59–94.

Milani A. (1994) The dynamics of Trojan asteroids. In *Asteroids, Comets, Meteors 1993* (A. Milani et al., eds.), pp. 159–174. Kluwer, Dordrecht.

Milani A. and Farinella P. (1994) The age of the Veritas asteroid family deduced by chaotic chronology. *Nature, 370,* 40–42.

Milani A. and Knežević Z. (1990) Secular perturbation theory and computation of asteroid proper elements. *Cel. Mech. Dyn. Astron., 49,* 347–411.

Milani A. and Knežević Z. (1992) Asteroid proper elements and secular resonances. *Icarus, 98,* 211–232.

Milani A. and Knežević Z. (1994) Asteroid proper elements and the dynamical structure of the asteroid main belt. *Icarus, 107,* 219–254.

Milani A. and Knežević Z. (1999) Asteroid mean elements: Higher order and iterative theories. *Cel. Mech. Dyn. Astron., 71,* 55–78.

Milani A., Nobili A. M., and Knežević Z. (1997) Stable chaos in the asteroid belt. *Icarus, 125,* 13–31.

Morbidelli A. (1993) Asteroid secular resonant proper elements. *Icarus, 105,* 48–66.

Morbidelli A., Zappalà V., Moons M., Cellino A., and Gonczi R. (1995) Asteroid families close to mean motion resonances:

Dynamical effects and physical implications. *Icarus, 118,* 132–154.

Schubart J. (1982) Three characteristic parameters of orbits of Hilda-type asteroids. *Astron. Astrophys., 114,* 200–204.

Schubart J. (1991) Additional results on orbits of Hilda-type asteroids. *Astron. Astrophys., 241,* 297–302.

Schubart J. and Bien R. (1987) Trojan asteroids: Relations between dynamical parameters. *Astron. Astrophys., 175,* 299–302.

Shoemaker E. M., Shoemaker C. S., and Wolfe R. F. (1989) Trojan asteroids: Population, dynamical structure and origin of the L4 and L5 swarms. In *Asteroids II* (R. P. Binzel et al., eds.), pp. 487–523. Univ. of Arizona, Tucson.

Valsecchi G. B., Carusi A., Knežević Z., Kresàk L., and Williams J. G. (1989) Identification of asteroid dynamical families. In *Asteroids II* (R. P. Binzel et al., eds.), pp. 368–385. Univ. of Arizona, Tucson.

Williams J. G. (1969) Secular perturbations in the solar system. Ph.D. thesis, Univ. of California, Los Angeles. 270 pp.

Williams J. G. (1979) Proper orbital elements and family memberships of the asteroids. In *Asteroids* (T. Gehrels, ed.), pp. 1040–1063. Univ. of Arizona, Tucson.

Williams J. G. (1989) Asteroid family identifications and proper elements. In *Asteroids II* (R. P. Binzel et al., eds.), pp. 1034–1072. Univ. of Arizona, Tucson.

Yuasa M. (1973) Theory of secular perturbations of asteroids including terms of higher order and higher degree. *Publ. Astron. Soc. Japan, 25,* 399–445.

Zappalà V., Cellino A., Farinella P., and Knežević Z. (1990) Asteroid families I: Identification by hierarchical clustering and reliability assessment. *Astron. J., 100,* 2030–2046.

Zappalà V., Bendjoya Ph., Cellino A., Farinella P., and Froeschlé C. (1995) Asteroid families: Search of a 12,487-asteroid sample using two different clustering techniques. *Icarus, 116,* 291–314.

Asteroid Family Identification

Ph. Bendjoya
University of Nice

V. Zappalà
Astronomical Observatory of Torino

Asteroid families have long been known to exist, although only recently has the availability of new reliable statistical techniques made it possible to identify a number of very "robust" groupings. These results have laid the foundation for modern physical studies of families, thought to be the direct result of energetic collisional events. A short summary of the current state of affairs in the field of family identification is given, including a list of the most reliable families currently known. Some likely future developments are also discussed.

1. INTRODUCTION

The term "asteroid families" is historically linked to the name of the Japanese researcher Kiyotsugu Hirayama, who was the first to use the concept of orbital proper elements to identify groupings of asteroids characterized by nearly identical orbits (*Hirayama,* 1918, 1928, 1933). In interpreting these results, Hirayama made the hypothesis that such a proximity could not be due to chance and proposed a common origin for the members of these groupings. His interpretation, still accepted today and strengthened by theoretical and observational results described elsewhere in this book (*Zappalà et al.,* 2002; *Cellino et al.,* 2002), was that members of a family are the fragments produced by the disruption of a common parent body resulting from a catastrophic collision. In his classic analysis, Hirayama was able to identify a set of five families still known today as the Hirayama families: Eos, Themis, Koronis, Flora, and Maria.

Several years passed before the problem of asteroid family identification was again taken into consideration by *Brouwer* (1951), who substantially confirmed the Hirayama families, although he divided Flora into four different groups. Moreover, 20 new additional groups were added to the family list. Since then, and before 1990, several authors have proposed their own catalogs of families (*Arnold,* 1969; *Lindblad and Southworth,* 1971; *Carusi and Massaro,* 1978; *Williams,* 1979; *Kozai,* 1979; *Williams,* 1989). No uniform agreement has been achieved among the results of different analyses, and the discrepancies were generally very large. Authors using slightly different databases (in terms of numbers of objects) of asteroidal proper elements found numbers of families ranging from 15 (*Carusi and Massaro,* 1978) to 117 (*Williams,* 1989). Different methods of classification were used by different authors. Most of them were based on a visual analysis of the data, only in some cases complemented by statistical tests to evaluate the level of significance of the different groupings. Some kind of subjectivity was certainly present. Moreover, in some cases, the results of previous investigations were used to calibrate new identification methods. According to the original papers published in the literature, *Brouwer* (1951) used a fairly subjective criterion to subdivide the Flora family delineated by Hirayama. *Arnold* (1969) assumed that the asteroids are dispersed in the proper-element space in a Poisson distribution. *Lindblad and Southworth* (1971) calibrated their method in such a way as to find good agreement with Brouwer's results. *Carusi and Massaro* (1978) adjusted their method in order to again find the classical Hirayama families. *Williams* (1979, 1989) developed a technique based on a visual inspection of the data, complemented by an *a posteriori* statistical test. *Kozai* (1979) explicitly recognized the significant degree of arbitrariness present in his family identification technique. Of course, the quoted discrepancies shed serious doubts regarding the reliability of the proposed families, apart from the classical ones already found by Hirayama. The situation in 1989 is effectively summarized in *Valsecchi et al.* (1989).

The disagreement on the number of existing families and their associated memberships had not completely prevented any kind of physical analysis of the proposed groupings (*Zappalà et al.,* 1984), but there was widespread agreement that only the classical, very prominent Hirayama families could be taken seriously into consideration for detailed physical studies, whereas many of the other groupings proposed by different authors were much more questionable. *Chapman et al.* (1989) pointed out also a number of cosmochemical inconsistencies among several of these smaller families, including objects belonging to taxonomic classes hardly compatible in terms of likely thermal histories.

2. SITUATION AFTER 1990

Beginning in 1990, the situation in the field of family identification began to improve significantly. The main reason has been the development of more objective identification procedures. In addition, the availability of increasingly larger datasets of accurate proper elements has allowed the researchers to analyze increasingly larger samples of the whole asteroid population, giving them better possibilities to reliably identify the presence of dense, compact clusters.

Proper elements are crucially important for family identification procedures, due to the fact that the orbital elements of the asteroids are subject to variations over different timescales, as a consequence of planetary perturbations. Assuming that a family is formed after the disruption of parent body, one can expect that the osculating (instantaneous) orbital elements of the fragments are very similar just after the family's formation, but then start to rapidly diverge as an effect of perturbations. If there were not any way to identify some conserved feature in the orbital evolution of the objects, one could expect not to be able to infer from the osculating elements of different objects at a given epoch strong clues about a possible common origin from a collisional event. Fortunately, the orbital evolution of the objects, at least in the case of "regular" orbits not affected by resonant conditions, is not totally unpredictable in practice over reasonably long timescales. In particular, some degree of orbital characterization describing the long-term behavior can be discerned by the proper elements. These are quasi-constant parameters representing a kind of average of the osculating elements over fairly long timescales, in which short-period and long-period fluctuations have been removed. A more exhaustive and rigorous treatment of asteroid proper elements can be found in *Knežević et al.* (2002). For family identification purposes, the important fact is that the osculating elements of family members vary as a function of time, but the proper elements do not; therefore, it is possible to analyze the orbital similarities of the objects by examining their proper elements, rather than their osculating elements.

The techniques of proper-element computations have improved over time, since the epoch of the first pioneering studies of Hirayama. In particular, the refinement of the dynamical theories and computational techniques has steadily produced increasingly larger datasets of asteroid proper elements.

On the other hand, the choice of the proper-element database was probably not the main source of the discrepancies in the family lists obtained by different authors before 1990, as mentioned in the previous section. In particular, *Zappalà et al.* (1992) show that by applying the new Hierarchical Clustering Method (HCM) (see below) to the dataset of proper elements used by *Williams* (1989), they can identify 26 families having more than 5 members. By applying the same HCM method to a different proper-element database containing a comparable (though not equal) number of objects, computed by *Milani and Knežević* (1990) using a different dynamical theory, the number of resulting families turns out to be 20. The number of families found in both experiments is 16. This is a good agreement, whereas the number of resulting families found by *Williams* (1989) by applying his own identification method to his proper-element database (the same just mentioned above) is 117. These results, summarized in Table 1, show that the choice of the proper elements was not the critical issue in explaining the observed discrepancies in family searches. Most of the problems were thus due to the adopted identification methods. In this respect, the situation changed quickly starting in 1990.

TABLE 1. Comparison between the numbers of families identified by applying the HCM method (see text) to different proper element databases, according to *Zappalà et al.* (1992).

	N ≥ 5	N ≥ 10
Williams	26	15
Milani-Knežević	20	14
Intersection	16	14

The first two lines indicate the numbers of families having numbers of objects (N) larger than 5 or larger than 10 found by using the Williams and the Milani-Knežević proper elements respectively. The third line indicates the number of common families found in both cases, being formed mostly or exclusively by objects present in both databases. The numbers shown indicate a good agreement, mainly when considering larger (more reliable) families.

2.1. Improved Identification Methods

To quantify the similarity of two orbits, a metric in the space of the proper elements (phase space) a', e', and i' (proper semimajor axis, eccentricity, and inclination respectively) must be defined. *Zappalà et al.* (1990) develops different choices for a metric in the phase space. All of them give quantifications of the "distance" between two orbits, in terms of quantities having all the dimensions of a velocity. The reason is that the well-known Gaussian equations define a relation between the change in orbital elements produced by a sudden velocity change and the components of this change. Therefore, differences in a', e', and i' can be measured in terms of velocity changes needed to produce these differences. This approach is very suitable for family-determination purposes, since the physical origin of a family is thought to be the breakup of its parent body. The different fragments would be expected to have small ejection velocities compared with the orbital velocity of their parent (this being one necessary condition for applying the Gaussian formulas). Hence, computing the mutual distances of the objects in the proper-element space allows for the detection of groups of asteroids that may be genetically related. Moreover, the knowledge of the velocity distribution within a family gives important constraints for any further kinematic modeling of these events (see *Zappalà et al.*, 2002).

A crucial issue when dealing with clustering identification in the proper-element space is the ability to evaluate the statistical significance of the detected groupings against chance fluctuations. This can be done by analyzing random distributions of simulated objects generated according to the constraint of reproducing separately the overall a, e, and i distributions of the real asteroids under consideration (quasirandom populations). The techniques for generating such quasirandom distributions were first developed by *Zappalà et al.* (1990). Using these techniques, no underlying model of statistical distribution of the objects (Gaussian or Poissonian) is needed, and the overall properties of the local distributions of the asteroids in different regions of the main belt are reproduced. By performing the clustering analysis on both the original dataset and several realizations of the quasirandom populations, it becomes possible to evaluate and quantify the probability that any given cluster is due to chance fluctuations. In particular, detected groupings having negligible probabilities of being due to chance are the obvious candidates to be considered as real, collisionally produced families.

Automated techniques managed by the computer present several advantages with respect to techniques based mostly on visual and subjective inspection of the data: (1) Because they are implemented in the form of computer algorithms, they are much less prone to subjectivity and more easily reproducible. (2) Alternative choices of the metric can be easily made and immediately used to produce alternative results. In this way, it is possible and easy to test the sensitivity of the obtained classifications upon the choice of the metric. (3) Tests of the robustness of the results as a function of the nominal uncertainties of the data can be easily performed. (4) An overall quantitative parameterization of the overall robustness and statistical significance of the identified groupings can be "objectively" defined.

The following section recalls the most important identification techniques developed by different authors starting in 1990. Most of them have been found to produce results in very good agreement, producing an important database of reliable families to be investigated from the point of view of physical characterization by means of different techniques (described in other chapters in this book).

2.2. Hierarchical Clustering Method (HCM)

This method is based on the classical tree construction for hierarchical classification purposes. *Zappalà et al.* (1990) modifies, adapts, and optimizes this method for the identification of significant groupings in the three-dimensional proper-element space. The tree is obtained by agglomerating, at each step of the procedure, the two nearest objects of the considered sample into a single object and reiterating the process until only one object survives. At each step, the distance d(i ∪ j, k) between an agglomerated i ∪ j object and the generic object k is defined to be equal to the minimum between d(i, k) and d(j, k). The results are given in the form of stalactite diagrams like that shown in Fig. 1. By comparing the stalactites obtained from the real dataset and those obtained from quasirandom simulated populations, it is then possible to cut the hierarchy to derive clusters that are statistically significant. The only crucial choice that must be taken by the human operator is the final assessment of the critical distance level, below which all the existing groupings are assumed to be real families. As mentioned above, this level corresponds to the deepest level reached by a set of simulations of quasirandom populations of fictitious objects. In this context, the choice to be made is not the critical distance level itself, but the minimum number of objects required to exist at the deepest levels of the quasirandom populations. In the successive implementations of this technique (*Zappalà et al.,* 1990, 1995; *Zappalà and Cellino,* 1994), the choice of the critical distance levels has generally been made according to restrictive criteria, leading in some cases to likely underestimation of the family memberships (*Migliorini et al.,* 1995).

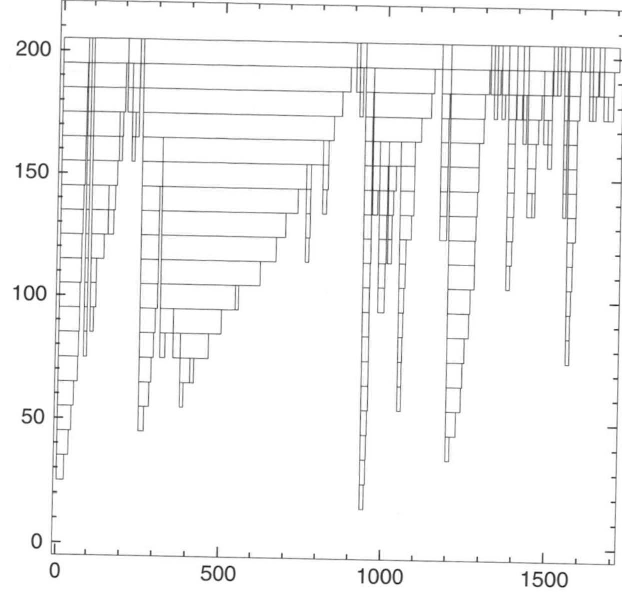

Fig. 1. Example of a typical stalactite diagram, referring to the asteroids belonging to the intermediate region of the main belt ("zone 4" according to *Zappalà et al.,* 1995). The different stalactite "branches" correspond to different identified families.

2.3. Wavelet Analysis Method (WAM)

This method is fully described *by Bendjoya et al.* (1991, 1993). It is a density-evaluation method based on the use of a particular function called a "wavelet" having a characteristic size defined according to the adopted metric. The use of the wavelet transform allows to detect local overdensities of points belonging to a given N-dimensional space (in this case the a', e', and i' space) at different scales. By superimposing a grid in the phase space, it is possible to compute for each node of the grid a wavelet coefficient. This coefficient is the sum of the contribution of each data point weighted by the zero mean wavelet function. The higher the value of this coefficient, the denser a grouping in the vicinity of the node. The closer to zero the coefficient, the more uniform the local distribution. By using the same technique on pseudorandom distributions, one can derive a quantified criterion to retain significant coefficients characterizing a given set of asteroid proper elements. These significant coefficients allow one to reconstruct the underlying clusters and to assign a level of detection significance against chance.

2.4. D Criterion

This method was first developed by *Lindblad and Southworth* (1971) and was later applied to increasingly larger datasets in its most recent applications by *Lindblad* (1992, 1994). It is based on the so-called D criterion, first developed to identify meteor streams (*Southwork and Hawkins,* 1963). The D criterion coupled with the neighbor-linking technique is basically another cluster-analysis method based on a different definition of the metric in the proper-element space. Also in this case, the analysis is carried out by means of a computer algorithm. The only human input is the choice of the cut-off distance used to reject spurious groupings.

3. PRESENT STATE OF THE ART

The HCM, WAM, and D-criterion techniques have been repeatedly applied in recent years to increasingly larger proper-element datasets.

The first applications of the methods were performed on a dataset of about 4000 proper elements (*Zappalà et al.,* 1990; *Bendjoya et al.,* 1991; *Linblad,* 1994). These data were taken from a version of the proper-element dataset computed by *Milani and Knežević* (1990). The results from the three techniques were found to be generally in excellent agreement, both in terms of number of identified families (about 15 having a high statistical reliability) and in terms of derived memberships.

The next step was a study devoted to test the performances of HCM and WAM in the case of a number of fictitious situations, in which artificial families were generated by using a suitable breakup model (*Bendjoya et al.,* 1993). The family members being known *a priori*, it was possible to test the effectiveness of HCM and WAM under different conditions including different compactness of the simulated families and different, more or less dense, background populations. The ability of the methods to separate overlapping and anisotropic families was also tested. The results confirmed the overall reliability of HCM and WAM, and their mutual consistency leading to an increased confidence in the results previously obtained in the case of real asteroids. The tests also suggested some possible improvements in the implementation of the two techniques, and this was taken into account in the next application of the methods to a new, larger dataset of asteroid proper elements, including about 6000 asteroids with a refined proper-element computation theory (*Milani and Knežević,* 1992). Again, the results were found to be in good agreement (*Zappalà and Cellino,* 1994; *Bendjoya,* 1993).

The most recent application of HCM and WAM was performed on a dataset of more than 12,000 proper elements (*Zappalà et al.,* 1995). Even if the difficulty of separating different groupings mutually overlapping was found to grow with the number of available data, this analysis was able to again give results in a very good mutual agreement. Most of the most-recent physical analyses of asteroid families have been based on the family lists published in the above paper.

Also in this latest implementation of the HCM and WAM techniques, robustness coefficients were derived for each grouping in order to assess the stability against modifications of several parameters, including choice of metric, noise in the proper elements, and statistical detection threshold. Table 2 gives a summary of the most-robust families found by both identification methods. These families can be considered the most important candidates for physical studies, and most of them have already been analyzed (see *Zappalà et al.,* 2002) and observed extensively, mainly by means of spectroscopic techniques (*Bus et al.,* 1996; *Cellino et al.,* 2002).

TABLE 2. List of the most reliable families found by at least one of the HCM and WAM methods (see text) in the *Zappalà et al.* (1995) classification.

Family Identification		Members	
HCM	WAM	HCM	WAM
8 Flora	43 Ariadne	604	575
44 Nysa	135 Hertha	381	374
4 Vesta	4 Vesta	231	242
163 Erigone	163 Erigone	45	49
1 Ceres	93 Minerva	89	88
170 Maria	170 Maria	77	83
668 Dora	668 Dora	77	79
145 Adeona	145 Adeona	63	67
808 Merxia	808 Merxia	26	29
569 Misa	569 Misa	25	27
410 Chloris	410 Chloris	21	27
1644 Rafita	1644 Rafita	21	23
1128 Astrid	1128 Astrid	10	11
24 Themis	24 Themis	550	517
221 Eos	221 Eos	477	482
158 Koronis	158 Koronis	325	299
137 Meliboea	137 Meliboea	13	16
845 Naema	845 Naema	6	7
20 Massalia	20 Massalia	49	45
15 Eunomia	15 Eunomia	439	303
110 Lydia	110 Lydia	26	50
128 Nemesis	58 Concordia	20	38
1639 Bower	342 Endymion	10	15
10 Hygiea	10 Hygiea	103	175
490 Veritas	92 Undina	22	36
293 Brasilia	293 Brasilia	10	18

The columns list the family identifications according to the two methods and the numbers of members found by each method. Only families having an intersection (numbers of objects in common) of 75% (upper block) or 50% (lower block) are listed. These families represent therefore the most prominent and reliable groupings currently identified in the main belt.

In addition to the families listed in Table 2, many less-robust groupings were also found and identified as "clumps," "tribes," or "marginal groupings" (see *Zappalà et al.,* 1995). Even in the case of these less-robust groupings, some good agreement could be found between the two techniques, but discrepancies also were found to become important. Of course, the real reliability of these groupings must wait for future, likely imminent, investigations. The interested reader will find the entire set of family members posted on the Web at http://www.obs-azur.fr/cerga_bdd.html.

The investigations mentioned above were based on datasets of proper elements in which high-inclination and high-eccentricity objects [having sin(i) and/or e > 0.3] were not included. The high-e, high-i region of the proper-element space has been investigated in detail by other working teams. In particular, the identification of a family associated with the asteroid Pallas was made by *Lemaître and Morbidelli* (1994) and later confirmed by *Bus* (1999, see below). Also, the possible existence of Hungaria (*Lemaître,* 1994) and Hansa (*Hergenrother et al.,* 1996) families has been proposed. The presence of these groupings has been proposed based on analyses of a set of proper elements computed by *Lemaître and Morbidelli* (1994).

Family searches have not been limited to the main-belt population, but have been performed also in the Trojan clouds by *Milani* (1993). The existence of families among Trojans raises interesting questions about a possible relation with the formation of short-period comets (*Marzari et al.,* 1995).

We should also mention an another important contribution in this field by *Migliorini et al.* (1995), who made a statistical analysis of published families in order to derive the likely number of random interlopers as a function of member sizes. The problem of interlopers is important since one of the ultimate goals of asteroid family studies is to reconstruct the original collisional events from which families were created. An interloper is an asteroid having nothing to do with the real family, but sharing by chance the same orbital proper elements. An estimate of the number of interlopers as a function of the diameter is fundamental in determining the correct barycenter and size distribution of the families. This investigation showed that in most cases the family identification criteria were not too liberal, leading to modest numbers of predictable interlopers. At the same time, it was also shown that in several cases (like Flora), the nominal membership numbers were likely too conservative, and the nominal member lists in these cases are expected to severely underestimate the real family membership.

4. FUTURE WORK

After the most recent application of the HCM and WAM techniques, the number of asteroids for which reliable proper elements have been computed has been steadily increasing. Currently (July 2001), the dataset of proper elements includes about 24,000 numbered asteroids and some thousands of unnumbered objects.

Therefore, it seems that it is time to propose a new family classification, to make it possible to confirm previous results and assess the real family status of several marginal groupings identified in previous searches. Some technical problem will certainly arise due to the large size of the available database. In particular, it is easy to expect that the mutual overlapping of different families will make it hard to separate the memberships of single objects in the overlapping regions. This problem has already been encountered by previous investigations, mainly in the inner belt, where the asteroid inventory is complete down to small sizes. A good example is given by the mutually overlapping Mildred and Polana families (formerly known as the "Nysa clan" following the nomenclature proposed by *Farinella et al.,* 1992), as shown by *Cellino et al.* (2001). This example suggests that in future family classifications, a major support to purely statistical data will come from observational data, mainly spectroscopy and spectrophotometry. In other words, we will be able to separate the memberships of families overlapping in the space of the orbital proper elements only when additional information on the reflectance properties of the objects is available. *Bus* (1999) has been able to show how a hybrid classification method, in which purely spectral information is added to proper-element data, can be developed. Such techniques will likely play an important role in the near future (see also *Cellino et al.,* 2002).

Family members have been found to account for an increasingly larger fraction of the considered asteroid samples in the analyses based on applications of HCM and WAM from 1990 to 1995. Apart from complications due to modifications of the identification algorithms (increasingly more conservative for increasing sizes of the proper-element databases), this generally means that the fraction of family members tends to increase with increasing faintness (decreasing size) of the objects. This is an important and still-controversial issue, covered in more details in *Zappalà et al.* (2002). It will be very interesting to check whether the above-mentioned trend is still apparent when trying to classify families using updated and larger databases. According to *Zappalà and Cellino* (1996), families' members might dominate the whole asteroid population at sizes <10 km. If this can be proved to be true, there are relevant implications for our overall understanding of the asteroid population.

Nearly 100 years after the first discoveries by Hirayama, asteroid families are still a very important branch of modern asteroid science. Being able to reliably identify them is the first necessary prerequisite for any further physical characterization study.

REFERENCES

Arnold L. R. (1969) Asteroid families and jet streams. *Astron. J., 74,* 1235–1242.

Bendjoya Ph. (1993) A classification of 6479 asteroid into families by means of the wavelet clustering method. *Astron. Astrophys. Suppl., 102,* 25–55.

Bendjoya Ph., Slézak E., and Froeschlé C. (1991) The wavelet transform: A new tool for asteroid family determination. *Astron. Astrophys., 272,* 651–670.

Bendjoya Ph., Cellino A., Froeschlé C., and Zappalà V. (1993) Asteroid dynamical family: A reliability test for new identification methods. *Astron. Astrophys., 272,* 651–670.

Brouwer D. (1951) Secular variations of the orbital elements of the minor planets. *Astron. J., 56,* 9–32.

Bus S. J. (1999) Compositional structure in the asteroid belt: Results of a spectroscopic survey. Ph.D. dissertation, Massachusetts Institute of Technology, Cambridge.

Bus S. J., Binzel R. P., and Burbine T. H. (1996) Asteroid families: Myth or reality? *Am. Astron. Soc., 28,* 1097.

Carusi A. and Massaro E. (1978) Statistic and mapping of asteroid concentrations in the proper element space *Astron. Astrophys. Suppl., 34,* 81–90.

Cellino A., Zappalà V., Doressoundiram A., Di Martino M., Bendjoya Ph., Dotto E., and Migliorini F. (2001) The puzzling case of the Nysa Polana family. *Icarus, 153,* 225–237.

Cellino A., Bus S. J., Doressoundiram A., and Lazzaro D. (2002) Spectroscopic properties of asteroid families. In *Asteroids III* (W. F. Bottke Jr. et al., eds.), this volume. Univ. of Arizona, Tucson.

Chapman C. R., Paolicchi P., Zappalà V., Binzel R. P., and Bell J. F. (1989) Asteroid families: Physical properties and evolution. In *Asteroids II* (R. P. Binzel et al., eds.), pp. 386–415. Univ. of Arizona, Tucson.

Farinella P., Davis D. R., Cellino A., and Zappalà V. (1991) From asteroid clusters to families: A proposal for a new nomenclature. In *Asteroids, Comets, Meteors* (A. W. Harris and E. Bowell, eds.), pp. 363–366. Lunar and Planetary Institute, Houston.

Hergenrother C. W., Larson S. M., and Spahr T. B. (1996) The Hansa family: A new high inclination asteroid family. In *Bull. Am. Astron. Soc., 28,* 1097.

Hirayama K. (1918) Groups of asteroids probably of common origin. *Astron. J., 31,* 185–188.

Hirayama K. (1928) Families of asteroids. *Jap. J. Astron. Geophys., 5,* 137–162.

Hirayama K. (1933) Present state of the families of asteroids. *Proc. Imp. Acad. Japan, 9,* 482–485.

Knežević Z., Lemaître A., and Milani A. (2002) The determination of asteroid proper elements. In *Asteroids III* (W. F. Bottke Jr. et al., eds.), this volume. Univ. of Arizona, Tucson.

Kozai Y. (1979) The dynamical evolution of the Hirayama families. In *Asteroids* (T. Gehrels, ed.), pp. 197–210. Univ. of Arizona, Tucson.

Lemaître A. (1994) Hungaria: A potential family. In *Seventy-Five Years of Hirayama Asteroid Families* (Y. Kozai et al., eds.), pp. 140–145. ASP Conference Series 63.

Lemaître A. and Morbidelli A. (1994) Calculation of proper elements for high inclined asteroidal orbits. *Cel. Mech. Dyn. Astron., 60,* 29–56.

Lindblad B. A. (1992) A computer search for asteroid families. In *Asteroids, Comets, Meteors* (A. W. Harris and E. Bowell, eds), pp. 363–366. Lunar and Planetary Institute, Houston.

Lindblad B. A. (1994) A study of asteroid dynamical families. In *Seventy-Five Years of Hirayama Asteroid Families* (Y. Kozai et al., eds.), pp. 62–75. ASP Conference Series 63.

Lindblad B. A. and Southworth R. B. (1971) In *Physical Studies of Minor Planets* (T. Gehrels, ed.), pp. 338–352. NASA SP 267, Washington, DC.

Marzari F., Farinella P., and Vanzani V. (1995) Are Trojan collisional families a source for short-period comets? *Astron. Astrophys., 299,* 267.

Migliorini F., Zappalà V., Vio R., and Cellino A. (1995) Interlopers within asteroid families. *Icarus, 118,* 271–291.

Milani A. (1993) The Trojan asteroid belt: Proper elements, stability, chaos and families. *Cel. Mech., 57,* 59–94.

Milani A. and Knežević Z. (1990) Secular perturbation theory and computation of asteroid proper elements. *Cel. Mech., 49,* 316–330.

Milani A. and Knežević Z. (1992) Asteroid proper elements and secular resonances. *Icarus, 98,* 211–232.

Southworth R. and Hawkins G. (1963) Statistics of meteor streams. *Smithson. Contrib. Astrophys., 7,* 261–285.

Valsecchi G., Carusi A., Knežević Z., Kresak L., and Williams J. G. (1989) Identification of asteroid dynamical families. In *Asteroids II* (R. P. Binzel et al., eds.), pp. 368–385. Univ. of Arizona, Tucson.

Williams J. G. (1979) Proper orbital elements and family memberships of the asteroids. In *Asteroids* (T. Gehrels, ed.), pp. 1040–1063. Univ. of Arizona, Tucson.

Williams J. G. (1989) Asteroid family identification and proper elements. In *Asteroids II* (R. P. Binzel et al., eds.), pp. 1034–1072. Univ. of Arizona, Tucson.

Zappalà V. and Cellino A. (1994) Asteroid families. In *Asteroids, Comets, Meteors 1993* (A. Milani et al., eds.), pp. 395–414. Kluwer, Dordrecht.

Zappalà V. and Cellino A. (1996) Main belt asteroids: Present and future inventory. In *Completing the Inventory of the Solar System* (T. W. Rettig and J. M. Hahn, eds.), pp. 29–44. ASP Conference Series 107.

Zappalà V., Farinella P., Knežević Z., and Paolicchi P. (1984) Collisional origin of the asteroid families: Mass and velocity distributions. *Icarus, 59,* 261–285.

Zappalà V., Cellino A., Farinella P., and Knežević Z. (1990) Asteroid families. I. Identification by hierarchical clustering and reliability assessment. *Astron. J., 100,* 2030–2046.

Zappalà V., Cellino A., and Farinella P. (1992) A comparison between families obtained from different proper elements. In *Asteroids, Comets, Meteors 1991* (A. W. Harris and E. Bowell, eds.), pp. 675–678. Lunar and Planetary Institute, Houston.

Zappalà V., Bendjoya Ph., Cellino A., Farinella P., and Froeschlé C. (1995) Asteroid families: Search of a 12487 asteroid sample using two different clustering techniques. *Icarus, 116,* 291–314.

Zappalà V., Cellino A., Dell'Oro A., and Paolicchi P. (2002) Physical and dynamical properties of asteroid families. In *Asteroids III* (W. F. Bottke Jr. et al., eds.), this volume. Univ. of Arizona, Tucson.

Physical and Dynamical Properties of Asteroid Families

V. Zappalà, A. Cellino, A. Dell'Oro
Istituto Nazionale di Astrofisica,
Osservatorio Astronomico di Torino

P. Paolicchi
University of Pisa

The availability of a number of statistically reliable asteroid families and the independent confirmation of their likely collisional origin from dedicated spectroscopic campaigns has been a major breakthrough, making it possible to develop detailed studies of the physical properties of these groupings. Having been produced in energetic collisional events, families are an invaluable source of information on the physics governing these phenomena. In particular, they provide information about the size distribution of the fragments, and on the overall properties of the original ejection velocity fields. Important results have been obtained during the last 10 years on these subjects, with important implications for the general understanding of the collisional history of the asteroid main belt, and the origin of near-Earth asteroids. Some important problems have been raised from these studies and are currently debated. In particular, it has been difficult so far to reconcile the inferred properties of family-forming events with current understanding of the physics of catastrophic collisional breakup. Moreover, the contribution of families to the overall asteroid inventory, mainly at small sizes, is currently controversial. Recent investigations are also aimed at understanding which kind of dynamical evolution might have affected family members since the time of their formation. In addition to potential consequences on the interpretation of current data, there is some speculative possibility of obtaining some estimate of the ages of these groupings. Physical characterization of families will likely represent a prerequisite for further advancement in understanding the properties and history of the asteroid population.

1. INTRODUCTION

The improvements of the techniques of family identification, leading to the establishment of a database of about 20 statistically reliable families recognized during the last decade (see *Bendjoya and Zappalà,* 2002), and the subsequent confirmation of the cosmochemical self-consistency of these groupings from dedicated spectroscopic campaigns (see *Cellino et al.,* 2002) has introduced new exciting perspectives for the physical studies of asteroids.

Since the early times of the first family discoveries by Hirayama at the beginning of the past century, these groupings have been interpreted as the observable outcomes of energetic collisional events, leading to complete disruption of a number of original parent bodies, and to dispersion of the collisional fragments. In this respect, asteroid families are nothing but nice examples of the collisional phenomena that have marked the history of the solar system, although generally at much smaller energy regimes with respect to major events involving large planetary bodies (origin of the Earth's Moon, tilt of the spin axis of Uranus, etc.).

In the framework of asteroid studies, families are crucially important in many respects. In general terms, they represent a constraint for any attempt at modeling the overall collisional history of the asteroid belt. In other words, the models should be able to justify the number of currently observed families, and to estimate the typical times needed to progressively erode freshly formed families in such a way to make them no longer identifiable at later epochs. The most recent investigations of this particular aspect have been published by *Marzari et al.* (1999).

Another essential aspect concerns the information that can be obtained about the physics of the breakup events from which families originated. Understanding how asteroids react to mutual collisions typically characterized by relative velocities around 5–6 km/s (*Farinella and Davis,* 1992; *Bottke et al.,* 1994) is an essential piece of information for any attempt at modeling their overall history, as well as their internal properties. Apart from the purely theoretical issues, understanding the internal structure of asteroids is a necessary prerequisite to developing any reasonable strategy of mitigation of the near-Earth asteroid (NEA) impact hazard.

In the past, most of our understanding of the way asteroids behave when they are collisionally disrupted came from laboratory experiments, based both on hypervelocity impacts and explosive charge techniques [for reviews of this subject, see *Fujiwara et al.* (1989) or *Martelli et al.* (1994)]. The major source of uncertainty, in this respect, was the scaling problem. In other words, it was not clear how to extrapolate the results of laboratory tests involving centimeter-sized targets, to the astronomical situation of kilometer-sized bodies, including relevant gravitational effects (see *Holsapple et al.,* 2002). In this context, the availability of a number of reliable families placed at the disposal of theoreticians a number of breakup experiments that nature performed for us, directly in a range of masses and energies absolutely beyond current and future laboratory capabilities.

As a consequence, a great effort has been made in recent years to derive from the observed properties of family members clues about the physics and the kinematics of the events from which families originated. These studies have been able to obtain relevant results, whose interpretation is still somewhat debated, as we shall see in the following sections.

We should still mention the problem that the presence of random interlopers represents for physical studies of families. The likely numbers of objects sharing purely by chance the same orbital proper elements of "true" family members have been predicted on the basis of pure statistics by *Migliorini et al.* (1995) as a function of size. At the same time, spectroscopic observing campaigns have been able to identify in several cases a number of likely interlopers, as described in *Cellino et al.* (2002). Interlopers are more "harmful" when they are relatively big, since they can affect the overall behavior of the family size distributions, and the reconstruction of the original ejection velocity fields of the fragments (see below). However, we can now be confident that nearly all the biggest interlopers have been identified in most cases. Thus, the results described in the following section are not expected to be appreciably affected by the presence of these undesired guests.

2. AVAILABLE DATA AND COMMON INTERPRETATIONS

Physical studies of dynamical families are based on analyses of essentially three pieces of information: (1) the coordinates of the family members in the space of the proper elements a', e', and i' (proper semimajor axis, eccentricity, and inclination respectively); (2) the size distributions of family members; and (3) spectroscopic and spectrophotometric properties of the members.

In addition to the items listed above, we should mention that some photometric data are also available. The major problem for this kind of investigation is certainly the faintness of most family members, apart from the generally few largest members of the different groupings. A few lightcurves are not sufficient for any relevant statistical inference. For this reason, no definitive indications have yet come from photometry. The only exception is a tentative clue of a likely young age of the Koronis family, based on the observed distribution of lightcurve periods and amplitudes (*Binzel,* 1988). This investigation found also evidence that in the case of the Eos family the distribution of spin periods seems to suggest collisional relaxation and a likely old age.

Another important source of information has come from direct *in situ* explorations by means of space probes. In particular, the Ida-Dactyl binary system observed by the *Galileo* probe is a nominal member of the Koronis family. Moreover, 951 Gaspra, also observed by *Galileo*, might be a member of the Flora family, although it has not been listed as such in the most recent family classification by *Zappalà et al.* (1995). It is known that for the Flora family the nominal membership has been likely underestimated due to the severe statistical criteria used for identification (*Migliorini et al.,* 1995). The important results of the *Galileo* observations of these two objects, including analyses of the size distributions of impact craters on their surfaces (consequences of their collisional histories), are extensively discussed in *Chapman* (2002) and *Sullivan et al.* (2002).

The family spectroscopic data (item 3 in the above list) deserve a separate analysis, and are the subject of *Cellino et al.* (2002). In what follows, we then focus our attention on the properties mentioned in items 1 and 2 of the above list.

2.1. Ejection Velocity Fields

The structure of the families in the space of proper elements are used to derive information on the ejection velocities of the fragments in family-forming events. This is possible because we can interpret the differences in orbital elements in terms of differences in ejection velocity from the original parent body. The conversion from velocities to orbital elements or vice versa is given by the well-known Gaussian formulas, which can be written as follows, under the assumption (verified for families) that the ejection velocities are much smaller than the orbital velocity of the parent body

$$\begin{cases} \delta a/a = \dfrac{2}{na(1-e^2)^{1/2}}\left[(1+e\cos f)V_T + (e\sin f)V_R\right] \\[2ex] \delta e = \dfrac{(1-e^2)^{1/2}}{na}\left[\dfrac{e+2\cos f+e\cos^2 f}{1+e\cos f}V_T + (\sin f)V_R\right] \\[2ex] \delta I = \dfrac{(1-e^2)^{1/2}}{na}\dfrac{\cos(\omega+f)}{1+e\cos f}V_W \end{cases} \qquad (1)$$

where na is the mean orbital velocity, and V_T, V_R, and V_W are the components of the ejection velocity vector along the direction of the motion, radial, and normal to the orbital plane respectively. The parameters f and ω are the true anomaly and the argument of perihelion of the parent body at the epoch of its disruption. These angles are *a priori* unknown, and this fact has long prevented any attempt at reconstructing the original ejection velocity fields from the presently observed locations of the members in the proper elements space.

By means of extensive numerical simulations, however, *Zappalà et al.* (1996) showed that very reasonable reconstructions of the velocity fields in family-forming events are possible when the real velocity fields were not completely random, but were characterized by a wide variety of possible field structures, including spherical, ellipsoidal, conical (cratering), and more complicated combinations of the above fields, in which some symmetry property is (even weakly) satisfied. The basic idea was to use some adimensional parameters, given by some combinations of the distributions of the velocity components of the family members, to estimate the most likely values of the unknown f and ω angles. Of course, in practical cases the reconstruction can be attempted only when a sufficiently large number of members is available, because of statistical reliability. In many situations a reconstruction of the fields was found to be possible, as in the cases of Vesta, Dora, Merxia (*Zappalà et al.,* 1996), and Maria (*Zappalà et al.,* 1997). The reconstructed fields turn out to be generally symmetric and similar, in terms of general structure, to what is observed in laboratory experiments involving much smaller targets. These results could be obtained in spite of the difficulty related to the fact that families are identified in the space of the orbital proper elements, whereas the Gaussian formulas describe the expected behavior in the osculating elements space. According to *Bendjoya et al.* (1993) the overall structure of the ejection velocity field is conserved in the transformation from osculating to proper elements. The major effect of the transformation was found to be a parallel displacement of the whole set of family members, leading to conservation of the overall velocity field structure. Of course, one can expect that over long timescales the stability of the proper elements can change, making a reconstruction of the velocity fields harder to obtain. On the other hand, one can also expect that over long timescales families tend in any case to disappear due to progressive collisional erosion (*Marzari et al.,* 1995, 1999).

An indication that aging of the proper elements is likely present in several cases is given by the fact that the resulting values of the computed f and ω angles turn out not to be homogeneously distributed within their natural range of variation. This is related to the well-known fact that most families appear to be more elongated in proper eccentricity and inclination with respect to semimajor axis (*Zappalà et al.,* 1984).

Another problem is that the computed ejection velocities of family members resulting from their differences in orbital proper elements turn out to be considerably large (see section 3.3).

2.2. Size Distributions and Size-Velocity Relationship

The study of the fragment size distributions is another major field of investigation. Asteroid sizes are generally not directly measured, but they are derived from knowledge of absolute magnitude and albedo. Families have been found to be characterized by very homogeneous albedo distributions according to the available radiometric data, and this fact has also been used to derive an estimate of the limit of completeness of the current asteroid inventory in terms of apparent magnitude (*Zappalà and Cellino,* 1996).

The size frequencies of all the major families have long been known to exhibit a power-law trend, characterized by steep slopes, much larger than the typical slopes of samples of nonfamily asteroids in different regions of the main belt (*Cellino et al.,* 1991; *Klačka,* 1992).

For a long time such behavior has been essentially misunderstood. General models based on an assumed power-law trend coupled with simple requirements of conservation of the total mass of the parent body led *Petit and Farinella* (1993) to make predictions that were able to fit satisfactorily only a minor fraction of the whole set of known family size distributions. In most cases, the slopes of the observed power laws were much steeper than the predictions. A possible explanation was developed by *Tanga et al.* (1999), based on a very simple concept first proposed by *Paolicchi et al.* (1996). The idea was that the fragmentation processes must satisfy other constraints in addition to conservation of the parent body's mass. In particular, the fragments are formed in a finite volume, and cannot mutually overlap. A simple model taking into account these geometric constraints was developed by *Tanga et al.* (1999), and was found to give excellent fits of the observed size distributions of a large number of families (see Fig. 1). As a byproduct of this technique, some estimates of the original parent body size and of the actual ratio m_{LR}/m_{PB} between the mass of the largest remnant and that of the parent body were also obtained. More recently, the likely role of geometric constraints in fragmentation phenomena has been also recognized by *Campo Bagatin and Petit* (2001), although the authors did not find any real proof that the modeled geometric effects are in any way related to the actual physics of collisions.

These above-mentioned results have important and debated implications concerning the overall inventory of the asteroid population, since they suggest that family members might dominate the asteroid population at small sizes, at least if we can believe in the reliability of extrapolations of the observed size frequencies to small sizes (around 1 km). This will be discussed more thoroughly in the following section.

Another recent result that must be mentioned concerns the relationship found between size and ejection velocity of the fragments in family-forming events. This problem has been analyzed by *Cellino et al.* (1999). The result was that a relation exists between the sizes of the fragments and their maximum possible ejection velocities. In particular, families generally fit the predictions of a model that assumes that the maximum amount of kinetic energy achievable by fragments in family-forming events is a constant. As a consequence, the maximum possible ejection velocity as a function of size follows a power-law trend, and is represented by a straight line in a log-log plot of size vs. velocity, as shown in Fig. 2. In particular, the following relation holds under the assumptions mentioned above, and is found to fit satisfactorily the observational data for most families

$$\log\left(\frac{d}{D}\right) = -\frac{2}{3}\log V - K'$$

where d and D are the diameters of the ith fragment and of

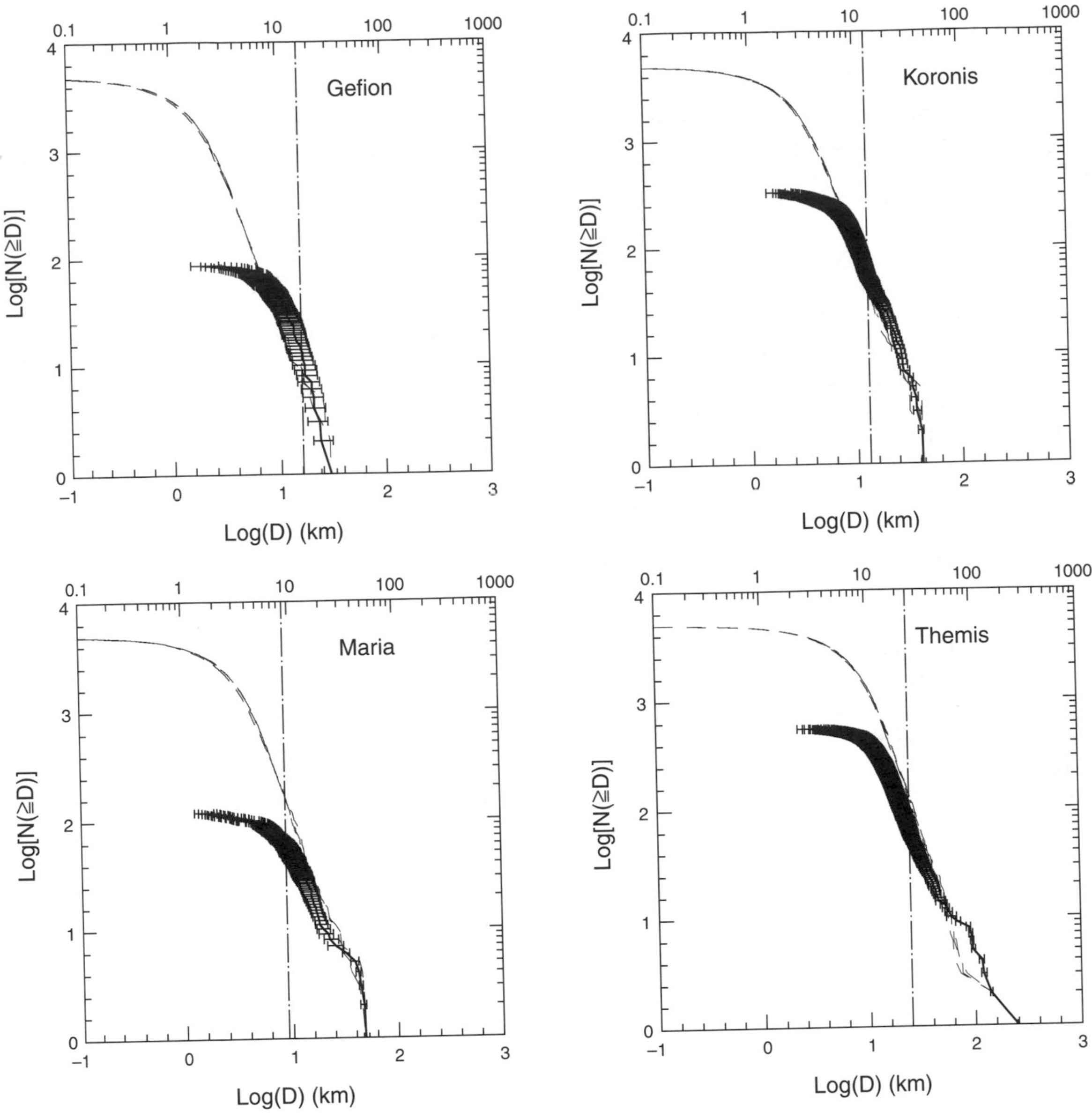

Fig. 1. The size distributions of four prominent families identified in the asteroid main belt are shown, with corresponding error bars due to the uncertainties in the sizes of the single family members. The lines represent fits of the size distributions obtained by *Cellino et al.* (1999) using their model. Three different curves (practically identical) are shown for each family, corresponding to different choices of some random seeds determining the results of different simulations. The simulations show a very good agreement with the data. Also shown are the limits of completeness for each family, due to the fact that the available inventory is not complete at small sizes due to the corresponding faintness of the smallest objects.

the parent body respectively, and V is the maximum possible value for the ejection velocity of the fragment of size d. The K' parameter is given by

$$K' = -\frac{1}{3} \log (2Af_{KE}E/M)$$

where E/M is the specific energy of the impact (defined as the ratio between the kinetic energy of the projectile and the mass of the impacted body), f_{KE} is the so-called anelasticity parameter, defined as the fraction of E that is converted into kinetic energy of the fragments, and A is an *a priori* unknown parameter, describing the maximum fraction of the total kinetic energy of the set of fragments that can be delivered to fragments of any size.

Apart from the most technical issues, an important result of the analysis has been to recognize that not only a fixed slope around (–2/3) is found to actually fit the behavior of most families, but also the value of the intercept K' turns out to be a linear function of log m_{LR}/M (m_{LR} being the mass of the largest family member, and M the mass of the parent body). This prediction is respected by most fami-

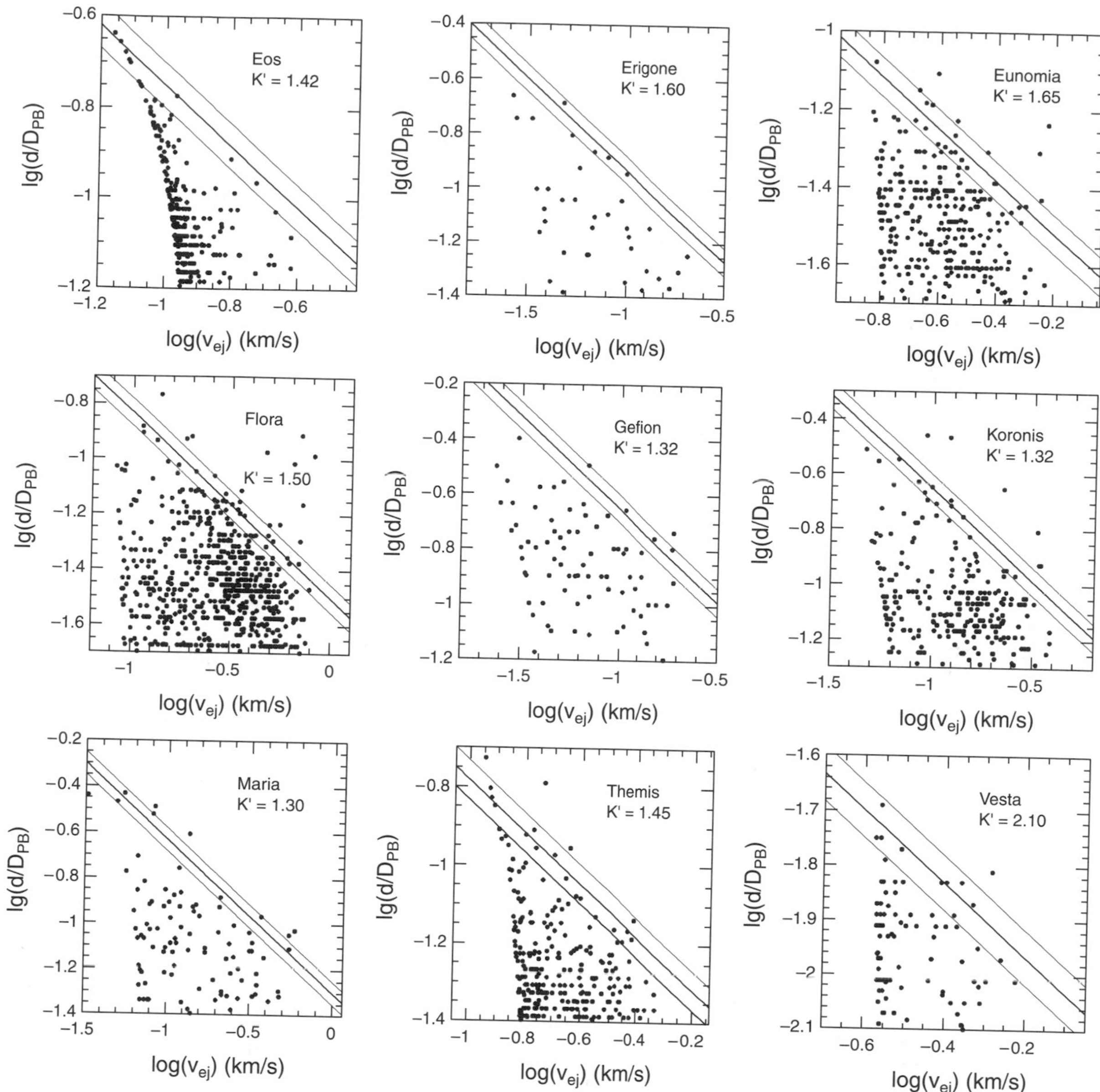

Fig. 2. Size-ejection velocity plots for nine of the most prominent families in the asteroid main belt. The straight lines shown in each plot have a fixed slope equal to –2/3, and correspond to a computed best-fit value, plus or minus its uncertainty. The value K' of the intercept corresponding to the best-fit line are also written in each plot. From *Cellino et al.* (1999).

lies for which $m_{LR}/M < 0.8$. This fact, if fully confirmed, has some important, although not yet fully explored, implications for our understanding of the physics of catastrophic breakup processes. In particular, it would suggest that the currently observed ejection velocity fields were determined mostly by the impact energies (directly related to the m_{LR}/m_{PB} ratio) and that the original size-velocity trend has been scarcely affected by subsequent dynamical and/or physical evolution, being still visible today. Such a conclusion may be still premature, however, and what can be said at this stage is only that the *Cellino et al.* (1999) size-velocity model seems to be in good agreement with available data for a large number of known families, and also with some laboratory experiments.

3. IMPLICATIONS AND OPEN PROBLEMS

The observational facts mentioned above are generally self-consistent, and suggest some kind of coherent scenario. In particular, it turns out that families were formed by events that produced ejection velocity fields having "reasonable" structures (when we compare them with the outcomes of laboratory experiments), although the resulting velocities were generally high. The fragmentation of the parent bodies produced huge swarms of fragments, with size frequencies characterized by steep power-law trends, still recognizable today. However, several facts must be taken into account before accepting literally all the above conclusions. Moreover, the exact implications of the above-mentioned scenario must be adequately stressed and discussed. We touch here on a number of very delicate points affecting our overall understanding of the properties and history of the asteroid population.

In what follows, we give (in no special order) a list of open problems and general implications of the results coming from analyses of the family data at our disposal.

3.1. Dynamical and Physical Aging of the Families

It is certain that, over time, asteroid families undergo an evolution modifying their originally observable features. This evolution is expected to affect both the physical properties of the members and their location in the proper elements space.

In particular, the orbital proper elements a', e', and i' cannot be assumed to be really constant over time (see also *Knežević et al.,* 2002). They are expected to vary slowly over timescales on the order of 10–100 m.y. An estimation of the range of fluctuation of the proper elements was given by *Milani and Knežević* (1992, 1994). As a consequence, the inferred kinematical properties of families (including a reconstruction of their original ejection velocity fields) should slowly vary as a function of the time elapsed since family formation. This does not prevent us from identifying today a number of around 20 "robust" families in the space of proper elements, and to derive information on the original collisional events from which they originated. However, it is reasonable to expect that the apparent kinematical properties of families can be somewhat "aged" with respect to the original configurations at the epoch of their formation.

Not all the proper elements behave in the same way. The semimajor axis seems to be the most stable parameter according to available dynamical theories (see *Knežević et al.,* 2002). Dynamical aging of the proper elements affects mainly eccentricities and inclinations. This may influence the determination of the unknown f and ω angles in the Gaussian equations when the techniques of reconstruction of the ejection velocity fields are applied. As shown by the simulations performed by *Zappalà et al.* (1996), this effect can produce a spurious clustering of the inferred values of f around 90°. In turn, this can affect the derived ejection velocities, since an enhancement of the dispersion of the eccentricities and the inclinations produces a spurious increase of the radial and normal ejection velocity components, leaving the transversal component essentially unchanged (as can be easily seen from equation (1)).

Another possible source of variation of the proper elements is the so-called Yarkovsky effect (*Vokrouhlický and Farinella,* 1998; *Farinella and Vokrouhlický,* 1999; *Vokrouhlický,* 1999; *Bottke et al.,* 2000; *Bottke et al.,* 2002). Unlike the dynamical aging processes mentioned above, the Yarkovsky effect produces changes of the orbital semimajor axis, and is size dependent. Several parameters, whose values are currently uncertain, determine the effectiveness of the Yarkovsky effect, and the range of sizes for which it is really important. The effect depends on the thermal properties of the surface, on the orientation of the spin axis, and on the obliquity angle (the angle between the spin axis and the normal to the orbital plane). In its "diurnal" variant, the Yarkovsky effect can cause the orbital semimajor axis to drift either inward or outward, depending on the sense of rotation of the body. In addition, there is a "seasonal" variant, due to the periodic preferential heating of different hemispheres during the orbital motion. The net effect of this variant is a systematic decrease of the orbital semimajor axis.

An estimate of the expected variation of the semimajor axis due to the Yarkovsky effect has been recently computed by *Spitale and Greenberg* (2001). The effectiveness of this kind of nongravitational force is stronger for smaller objects, and should be negligible for bodies larger than kilometer-sized (the exact value depending on many poorly known parameters). Moreover, another kind of relevant long-term dynamical effect that is expected to affect the orbital semimajor axes of all asteroids, including family members, is that of mutual close encounters with the massive Ceres, Pallas, and Vesta objects (*Carruba et al.,* 2000).

In summary, dynamical aging can increase the dispersion of the eccentricities and inclinations of family members, while the Yarkovsky effect and close encounters can also increase the dispersion of the semimajor axes. Dynamical aging does not depend on the size of the body, whereas the Yarkovsky effect is certainly negligible for objects larger than, e.g., 10–20 km.

The interplay of dynamical aging and the Yarkovsky effect can be important, because during a Yarkovsky-driven drift in semimajor axis the objects can be trapped into some of the many dynamically unstable regions crossing the asteroid main belt (*Nesvorný and Morbidelli,* 1998; *Morbidelli and Nesvorný,* 1999), with the effect of being eventually removed from the family. Of course, the timescales of dynamical and Yarkovsky-driven evolutions must be compared with typical collisional lifetimes for bodies of comparable sizes. We touch here on some key problems that will be likely the subject of extensive analyses in the near future.

In fact, collisions independently produce a "physical" aging of the families. Collisional lifetimes of main-belt asteroids are size-dependent, and range between 10^7 and 10^9 yr. The exact estimates are model-dependent, since they are derived from estimates of both the impact strength of the objects and the number of existing projectiles capable of disrupting an object of a given size. For a 100-km asteroid, the average collisional lifetime should be on the order of 10^9 yr. As a consequence, and according to *Marzari et al.* (1995), the smallest family members tend to be collisionally disrupted over shorter timescales with respect to the bigger members. As a consequence, collisions not only modify the size distribution of families, but also produce an erosion in the space of the orbital proper elements, leading to progressive disappearance of the families as recognizable groupings (*Marzari et al.,* 1999).

Moreover, the possibility of secondary events involving first-generation family members must also be taken into account, since this may potentially produce very complicated structures in the space of proper elements, as possibly in the cases of the so-called "clans," such as the Flora family (*Farinella et al.,* 1992).

All the facts mentioned above must be seriously taken into account when interpreting the data at our disposal. Moreover, we observe today families that likely have different ages, and show different stages of evolution. Understanding the complicated interplay of the different evolutionary tracks will certainly help in understanding the properties of families, and will possibly help in determining their current ages, with important implications for the studies of the collisional evolution of the asteroid belt.

3.2. Size Distributions and Asteroid Inventory

In many cases the slopes of the power-law distributions fitting the size frequencies of asteroid families turn out to be very steep. According to *Tanga et al.* (1999), the resulting exponents of the cumulative size distributions reach often values beyond –3. In these cases, it is easy to show that integrations of the size distributions would lead to infinite reconstructed masses for the corresponding parent bodies. For this reason, it is clear that the size distributions must change slope somewhere at small sizes. A difficult problem is thus determining the exact value of this transition size, and whether this is generally constant or varies for different families. Some general indications come from different kinds of evidence.

First, some families in the inner region of the main belt are complete down to sizes on the order of 5–6 km. This means that any change in slope of the size distributions must occur at smaller sizes for these particular families. Of course, this does not imply that the same is necessarily true for all the families in the main belt. Second, *Campo Bagatin and Petit* (2001) explored some generalizations of the *Tanga et al.* (1999) model, and found analytical reasons to predict that the value of the power-law exponent should asymptotically converge toward a fixed value, found to be around –2.8, for any model based on a similar treatment of the geometric constraints affecting the production of fragments. The authors stressed that the above result can be a pure artifact of the assumed geometric model, and there is not any proof that the model is really representative of the actual physics of the phenomena of catastrophic disruption. If proven to be adequate to represent the real world, the above result would in any case prevent the possibility of obtaining distributions diverging to infinity in terms of reconstructed mass.

Another very important consideration is that further collisional evolution experienced by the freshly formed family members should lead to changes of the size frequencies. According to *Marzari et al.* (1995), the family size distributions should be expected to converge progressively, starting from the low-size end of the distribution, toward the equilibrium value expected for a collisionally relaxed population, being equal to –2.5 according to classical results by *Dohnanyi* (1969, 1971). The convergence should take place over timescales depending on the size, the smallest members being expected to run into relaxation in shorter times (*Marzari et al.,* 1995). The problem in this scenario can be that the Dohnanyi-predicted value of the slope was model-dependent (the outcomes of the collisions were assumed to be independent of size), and it seems to be unconfirmed by general analyses of the overall size distribution of main-belt asteroids (*Cellino et al.,* 1991; *Jedicke and Metcalfe,* 1998). In particular, it turns out that nonfamily asteroids, for instance, exhibit a size frequency much shallower than the predicted Dohnanyi slope.

Apart from causing some uncertainty about the exact value of the slope to which families should be expected to converge, the above fact should also be taken into account when another controversial consequence of the steepness of the family size distributions is discussed. The subject here is that of the possible predominance of family members in the general asteroid inventory at small sizes. According to *Zappalà and Cellino* (1996), family members increasingly dominate the asteroid population going toward smaller sizes. This conclusion is a consequence of both the steep slopes of the family size distributions and the apparently shallow slopes of the nonfamily population. The situation is controversial also because family members should be expected to become dominant at values of size (~1 km) for which the current asteroid inventory is severely incomplete. Many authors tend to dislike the idea that 99% of the main-belt asteroids larger than 1 km should be family members (*Zappalà and Cellino,* 1996). On the other hand, avoiding this type of conclusion is not trivial, if we can believe in extrapolations (even moderate in some cases) of the observed family size frequencies. Allowing for some important change of slope in the size frequencies of families somewhere between 1 and 5 km is reasonable. However, even in this case, families could hardly become so shallow as to not give a predominant contribution to the main-belt asteroid inventory at small sizes.

Only observations will definitely clarify the real situation. Models of the general asteroid inventory based on the dominance of families down to sizes of 1 km are currently being developed. Some definite predictions on the number and apparent magnitudes of the asteroids that should be found in any given sky field can be made, and observations will soon confirm or rule out these models (see, e.g., *Tedesco and Désert,* 1999).

3.3. Ejection Velocities and Comparison with Hydrocodes

The results mentioned in sections 2.1 and 2.2 concerning the derivation of the original ejection velocity fields of families, and the discovery of a defined size-velocity relationship, can certainly in principle be affected by subsequent evolutionary effects experienced by family members (section 3.1). However, it seems difficult to reconcile the expectation of very important evolutionary changes with the

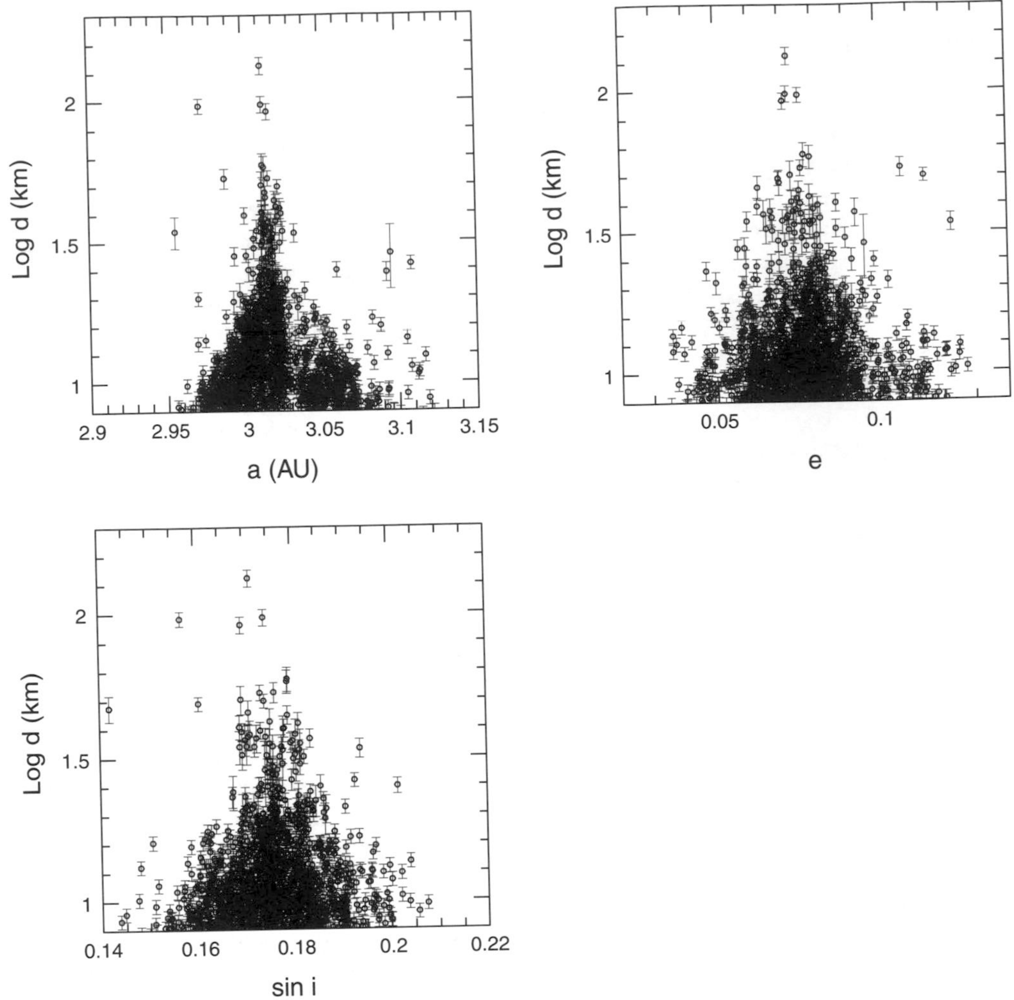

Fig. 3. Proper semimajor axis, eccentricity, and sinus of inclination plotted as a function of the logarithm of diameter for the Eos family (1645 members identified). The data refer to a new, yet-unpublished analysis for the identification of asteroid families based on a dataset of more than 30,000 asteroid proper elements computed by Milani, and available at the AstDys Web site (http://hamilton.dm.unipi.it/cgi-bin/astdys/astibo). These plots show the general existence of a well-defined trend, possibly related to the size-ejection velocity relationship. The same trend is recognizable in a large number of currently identified families, in spite of the existence of dynamical and physical mechanisms that are expected to progressively modify the original locations of the family members in the space of the proper elements.

conservation of regular patterns such those that are observed for the velocity field structures and the size-velocity relations of the most robust families. Figure 3 shows a typical example concerning the Eos family. The existence of defined, "triangular" trends in the diameter vs. proper semimajor axis, eccentricity, and inclination plots can be easily recognized in spite of any level of noise possibly affecting the proper elements.

In particular, we should note that in principle a size-dependent spreading in semimajor axis can be interpreted in terms of the Yarkovsky effect. What seems more difficult to explain, however, is the size-dependent spreading in eccentricity and inclination. For this, effects like Yarkovsky can be only indirectly effective, by moving objects into narrow resonant zones. The resulting distribution of the objects in the proper elements space should be due to a complicated interplay of the residence times in the resonances experienced by bodies of different sizes, and the relative rate of variation of the proper elements under the influence of the size-dependent Yarkovsky effect and the size-independent resonant behavior. It seems fairly difficult to expect that the final result of the evolution should be such a regular behavior like the one shown in Fig. 3.

Although conclusions are certainly premature at the present stage, it seems that if the current interpretation of the data at our disposal is not grossly wrong, we should conclude that either the evolutionary effects are weaker than currently believed, or the families themselves are relatively young, and still poorly affected by the above effects. Of course, this conclusion would have important consequences for our understanding of the collisional evolution of the asteroid belt (see *Davis et al.,* 2002).

From the point of view of the physics of collisions, the ejection velocities that are derived from application of the Gaussian formulas constitute another debated subject. Generally speaking, the problem is that according to the most sophisticated hydrocode models of catastrophic breakup processes (*Love and Ahrens,* 1996; *Nolan et al.,* 1996; *Asphaug et al.,* 1998), the impact energies that are apparently needed to disperse the family members with the observed speeds should have been sufficiently high as to completely pulverize the family parent bodies.

In particular, the reconstructed velocity fields of families indicate that ejection velocities well beyond 100 m/s are not rare for the smallest fragments. As a comparison, typical speed values in laboratory experiments are 5–10× smaller, velocities of 100 m/s being only typical of pulverized fragments produced very close to the impact point (*Martelli et al.,* 1994). This is a well-known problem. When trying to reproduce the reconstructed ejection velocity field of the Maria family using the *Paolicchi et al.* (1996) semi-empirical model (SEM) of catastrophic disruption events, which is generally able to give a good fit to laboratory experiments, *Zappalà et al.* (1997) found that SEM gives maximum velocities a factor of 5 too low to reproduce the family data.

According to hydrocode results, thc observed velocities are not possible unless we assume that the observed family members are essentially reaccumulated rubble piles. In this scenario, all asteroids larger than some hundreds of meters should be rubble piles. *Pisani et al.* (1999), however, made some quantitative tests, and concluded that reaccumulation cannot be sufficiently efficient to justify the observed size distributions of several known families. In particular, important phenomena of reaccumulation were found to be possible for the largest remnant alone, but certainly not for a large number of objects. Moreover, preliminary results of the *in situ* exploration of the near-Earth asteroid 433 Eros, a 31 × 13 × 13-km object, thought to be a collisional fragment from the disruption of a larger parent body, indicate that it is very unlikely that this asteroid can be a rubble-pile [for the meaning of terms like "rubble pile," see *Cheng* (2002) and *Richardson et al.* (2002)].

Of course, a possible dynamically induced spreading of families in the proper elements space might artificially increase the apparent ejection velocities of family members. Taking this into account one could conclude that the real ejection velocities at the epoch of formation were lower, as required by hydrocode results. However, there is not yet any conclusive evidence of a dominating role played by post-formation dynamical evolution, as mentioned above. It is true that when observations do not fit theory, scientists tend to believe that observations are likely wrong. But at this stage we think that the self-consistency of much observational data is very notable, and would tend to indicate that something can be wrong in current theoretical models of collisional processes. Of course, this is another field in which new observational and theoretical activities are needed to achieve some reliable interpretation of what is observed.

3.4. Formation of Binaries

Strictly related to the modeling of family formation is the problem of the possible formation of binaries. This subject is currently very hot, after the recent discoveries of several binary systems among both near-Earth and main-belt asteroids (see *Merline et al.,* 2002). Interestingly enough, the first convincing discovery of an asteroid satellite came from the *Galileo* flyby of the Koronis family member 243 Ida. More recently, the asteroid 90 Antiope, a member of the Themis family, has also been found to be an outstanding example of a binary system with components of comparable size (*Merline et al.,* 2000; *Weidenschilling et al.,* 2001). It is clear that the formation of binaries in collisional processes is an interesting field of research for which we can expect to have important results in the near future. This subject is discussed extensively in *Paolicchi et al.* (2002).

3.5. Collision Probabilities

The formation of big asteroid families is not only an interesting phenomenon per se, but it also has a number of important consequences. *Dell'Oro et al.* (2001) have found that a long-term increase of the collision probabilities in the region of the main belt swept by the orbital motion of the newly formed family members must be expected. Again, this is due to the huge numbers of kilometer-sized members that are apparently produced by these events. When a large family is produced, its members will cross the orbits of the majority of the main-belt population, apart from objects located far enough in terms of semimajor axis and eccentricity (mostly the low-eccentricity asteroids orbiting close to the inner and outer edges of the belt). In this respect, it is not unlikely that major collisional events like those that produced the big Eos and Themis families can trigger the subsequent collisional regime over substantial timescales. This is just another example of family-related mechanisms that should be taken into account by modern models of the overall collisional evolution of the asteroid belt (see *Davis et al.,* 2002).

Another fact that should be mentioned is that according to recent investigations (*Dell'Oro et al.,* 2002), the very early times after formation of any given family are characterized by a larger impact probability among members of the same family, with respect to the typical values controlling the steady collisional regime in the main belt. Although these epochs of enhanced interfamily collisions were found to be short, and were on average characterized by low impact velocities, there is still the possibility that the freshly formed surfaces of family members can have been marked in a significant way, with consequences on the cratering record observed by *in situ* exploration by space probes (243 Ida and 951 Gaspra were observed by *Galileo*). We should also note that, according to *Weidenschilling et al.* (2001), the existence of binary asteroids like 90 Antiope (a member of the Themis family), characterized by components of comparable size, might also be explained by the occurrence of relatively mild interfamily collisions likely occurring during the very early history of a newly born family.

4. FAMILIES AS POSSIBLE SOURCES OF NEAR-EARTH ASTEROIDS

Families have been found to also play an important role in the production of near-Earth asteroids and meteorites. The first indications were found by *Morbidelli et al.* (1995), who convincingly showed that a number of well-known families (such as Themis) are sharply cut by the edges of important mean-motion resonances with Jupiter. This qualitatively showed two important things: First, the formation of families in several cases led to the immediate injection of substantial numbers of sizeable fragments into resonant orbits. Many of them (mostly in the case of resonances located in the inner belt, like the 3:1 mean-motion resonance or the ν_6 secular resonance; a smaller fraction in the case of resonances located in the outer belt) achieved near-Earth-like orbits. This process was quantitatively analyzed by means of numerical integrations of the orbits by *Gladman et al.* (1997). The results showed that the transfer times are extremely short for resonances like 3:1 and 5:2, on the order of a couple of million years. Based on these results and on a preliminary assessment of the numbers of objects actually injected by several families at the epochs of their formation, *Zappalà et al.* (1998) found that in the past several episodes of paroxysmal near-Earth-asteroid production, with consequent enhancements in the collision rate with the terrestrial planets, took place, with durations on the order of several million years in most cases. This also indicates that the inventory of near-Earth objects is not steady, and can be strongly influenced by the occurrence of important collisional events in the main belt.

Second, family-forming events left groupings of objects that are very close to the borders of several important resonances. This fact is important, because it can be expected that the normal collisional evolution experienced by these objects can likely produce a steady rate of production of small fragments directly injected into the most efficient dynamical routes to the inner solar system (see also *Zappalà and Cellino,* 2001). A good example is given by the Maria family, whose reconstructed ejection velocity field, sharply cut by the 3:1 resonance with Jupiter, is shown in Fig. 4 (*Zappalà et al.,* 1997).

An important result obtained by *Morbidelli et al.* (1995) was the discovery that several objects are currently located into the 9:4 mean-motion resonance with Jupiter. This relatively thin resonance cuts across the big Eos family, and was shown by *Gladman et al.* (1997) to give a modest but not fully negligible contribution to the near-Earth population. The reason is that the dynamical evolution inside the 9:4 resonance is relatively slow, with eccentricity being pumped up over timescales sufficiently long to allow Mars to capture some objects before they become dominated by Jupiter (with subsequent ejection from the solar system). Most of the 9:4 resonant objects cannot be recognized as

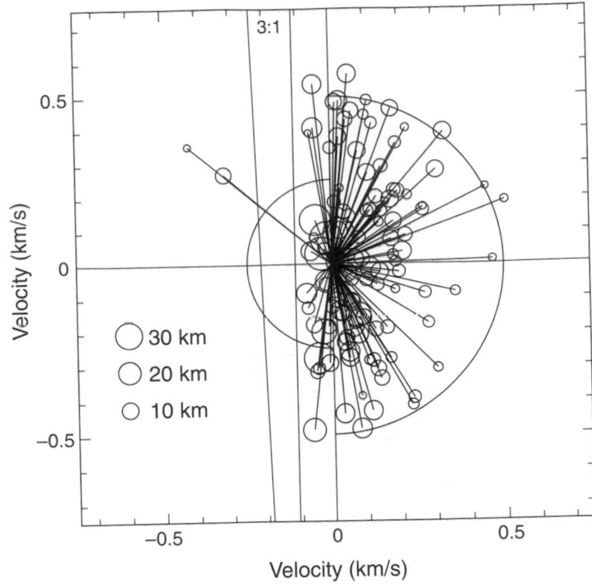

Fig. 4. Reconstruction of the ejection velocity field of the fragments produced in the disruption of the parent body of the Maria family. The location of the edges of the 3:1 mean-motion resonance with Jupiter are also indicated. The family turns out to be sharply cut by the border of the resonance. From *Zappalà et al.* (1997).

Eos family members on the basis of their orbital elements, but it was straightforward to assume that these objects might be original members of this family, observed during the early stages of a dynamical evolution that will decouple them from the asteroid main belt. This conjecture was fully confirmed by subsequent observations by *Zappalà et al.* (2000) (see also *Cellino et al.,* 2002).

Another fact that should be mentioned when speaking about the relation between near-Earth objects and families is the following: If direct collisional injection into some of the dynamically unstable regions in the main belt (such as the 3:1 resonance with Jupiter, the ν_6 secular resonance, and the region of Mars-crossing) is assumed to be the most effective mechanism for supplying kilometer-sized NEAs, some constraints must be respected. In particular, direct injection cannot be accepted if it can be shown that it necessarily predicts the formation of too many observable families. The reason is simply that the number of known families in the main belt is limited. This is a strong constraint for any mechanism of steady NEA supply, mainly for sizeable objects having diameters beyond 1 or 2 km. According to *Zappalà et al.* (2002), it does not seem trivial to explain the currently observed number of NEAs larger than 2 km only by means of steady collisional injection into chaotic regions in the main belt. Many candidate parent bodies exist (the most effective ones have been listed in Table 1), but the ranges of "permitted" impact energies that satisfy the nonfamily formation constraint for most of them are fairly narrow.

TABLE 1. A list of the most effective NEA candidate parent bodies found by *Zappalà et al.* (2001).

ID Number	Proper Elements			Transfer Region	Diameter (km)	Taxonomic Class	
	a' (AU)	e'	sin i'			*Tholen* (1994)	*Bus* (1999)
304	2.404	0.122	0.257	ν_6	68	C	XC
313	2.376	0.236	0.201	MC	96	C	—
495	2.487	0.118	0.045	3:1	39	—	—
512	2.190	0.199	0.142	MC	23	S	S
753	2.329	0.214	0.153	MC	24	S	L
877	2.487	0.142	0.059	3:1	38	F	—
930	2.431	0.121	0.265	ν_6	36	—	Ch
1080	2.420	0.243	0.086	MC	23	F	—
1715	2.400	0.249	0.181	MC	23	—	X
2143	2.281	0.194	0.130	MC	19	—	—

5. FUTURE DEVELOPMENTS

From the facts discussed in the previous sections, it is easy to see that the physical properties of asteroid families play an important role in our current understanding of the collisional and dynamical evolution of the asteroid population. The interpretation of the data at our disposal is not straightforward, and a coordinated observational and theoretical effort will be needed in order to confirm or rule out a number of conclusions that are suggested by the data that are currently available. A determination of the times needed to collisionally erode and dynamically disperse a newly formed family will certainly be a primary field of investigation in the near future. At the same time, trying to understand how families can achieve their steep size distributions, and how long they can keep them unchanged in the range of sizes covered by current datasets, will be another major field of research. If families evolve rapidly, and those that are currently observed can be shown to be really young, this will deeply influence current ideas of the collisional evolution of the main belt, and will provide new important constraints in addition to those currently recognized, like the preservation of the basaltic crust of 4 Vesta; the existence of metallic meteorites; the scarcity of taxonomic types directly interpreted in terms of mantle material from differentiated parent bodies; the observed cratering records of Ida, Gaspra, Mathilde, and Eros, etc. In general, it will be possible to obtain a better and more reliable estimate of the asteroid inventory at small sizes. Two hundred years after the discovery of the first asteroids, we are still far from fully understanding these interesting objects. As the direct outcomes of mutual collisions, families certainly represent a very important piece of the asteroid puzzle.

REFERENCES

Asphaug E., Ostro S. J., Hudson R. S., Scheeres D. J., and Benz W. (1998) Disruption of kilometer-sized asteroids by energetic collisions. *Nature, 393,* 437–440.

Bendjoya Ph. and Zappalà V. (2002) Asteroid family identification. In *Asteroids III* (W. F. Bottke Jr. et al., eds.), this volume. Univ. of Arizona, Tucson.

Bendjoya Ph., Cellino A., Froeschlé C., and Zappalà V. (1993) Asteroid dynamical family: A reliability test for new identification methods. *Astron. Astrophys., 272,* 651–670.

Binzel R. P. (1988) Collisional evolution in the Eos and Koronis asteroid families — Observational and numerical results. *Icarus, 73,* 303–313.

Bottke W. F., Nolan M. C., Greenberg R., and Kolvoord R. A. (1994) Velocity distributions among colliding asteroids. *Icarus, 107,* 255–268.

Bottke W. F., Rubincam D. P., and Burns J. A. (2000) Dynamical evolution of main belt asteroids: Numerical simulations incorporating planetary perturbations and Yarkovsky thermal forces. *Icarus, 145,* 301–330.

Bottke W. F. Jr., Vokrouhlický D., Rubincan D. P., and Brož M. (2002) The effect of Yarkovsky thermal forces on the dynamical evolution of asteroids and meteoroids. In *Asteroids III* (W. F. Bottke Jr. et al., eds.), this volume. Univ. of Arizona, Tucson.

Bus S. J. (1999) Compositional structure in the asteroid belt: Results of a spectroscopic survey. Ph.D. thesis, Massachusetts Institute of Technology, Cambridge. 367 pp.

Campo Bagatin A. and Petit J. M. (2001) Effects of the geometric constraints on the size distributions of debris in asteroidal fragmentation. *Icarus, 149,* 210–221.

Carruba V., Burns J. A., Bottke W. F., and Morbidelli A. (2000) Asteroid mobility due to encounters with Ceres, Vesta, Pallas: Monte Carlo codes versus direct numerical integrations (abstract). *Bull. Am. Astron. Soc., 32,* 1019.

Cellino A., Zappalà V., and Farinella P. (1991) The size distribution of asteroids from IRAS data. *Mon. Not. R. Astron. Soc., 253,* 561–574.

Cellino A., Michel P., Tanga P., Zappalà V., Paolicchi P., and Dell'Oro A. (1999) The velocity-size relationship for members of asteroid families and implications for the physics of catastrophic collisions. *Icarus, 141,* 79–95.

Cellino A., Bus S. J., Doressoundiram A., and Lazzaro D. (2002) Spectroscopic properties of asteroid families. In *Asteroids III* (W. F. Bottke Jr. et al., eds.), this volume. Univ. of Arizona, Tucson.

Chapman C. R. (2002) Cratering on asteroids from Galileo and NEAR Shoemaker. In *Asteroids III* (W. F. Bottke Jr. et al., eds.), this volume. Univ. of Arizona, Tucson.

Cheng A. F. (2002) Near Earth asteroid rendezvous: Mission summary. In *Asteroids III* (W. F. Bottke Jr. et al., eds.), this volume. Univ. of Arizona, Tucson.

Davis D. R., Durda D. D., Marzari F., Campo Bagatin A., and Gil-Hutton R. (2002) Collisional evolution of small-body populations. In *Asteroids III* (W. F. Bottke Jr. et al., eds.), this volume. Univ. of Arizona, Tucson.

Dell'Oro A., Paolicchi P., Cellino A., Zappalà V., Tanga P., and Michel P. (2001) The role of families in determining collision probability in the asteroid main belt. *Icarus, 153,* 52–60.

Dell'Oro A., Paolicchi P., Cellino A., Zappalà V., Tanga P., and Michel P. (2002) Primordial collisional evolution of the asteroid families. *Icarus,* in press.

Dohnanyi J. S. (1969) Collisional model of asteroids and their debris. *J. Geophys. Res., 74,* 2531–2554.

Dohnanyi J. S. (1971) Fragmentation and distribution of asteroids. In *Physical Studies of Minor Planets* (T. Gehrels, ed.), pp. 263–295. NASA SP-267.

Farinella P. and Davis D. R. (1992) Collision rates and impact velocities in the main asteroid belt. *Icarus, 97,* 111–123.

Farinella P., Davis D. R., Cellino A., and Zappalà V. (1992) From asteroid clusters to families: A proposal for a new nomenclature. In *Asteroids, Comets, Meteors 1991* (A. W. Harris and E. Bowell, eds.), pp. 165–166. Lunar and Planetary Institute, Houston.

Farinella P. and Vokrouhlický D. (1999) Semimajor axis mobility of asteroidal fragments. *Science, 283,* 1507–1511.

Fujiwara A., Cerroni P., Davis D. R., Ryan E. V., Di Martino M., Holsapple K., and Housen K. (1989) Experiments and scaling laws on catastrophic collisions. In *Asteroids II* (R. P. Binzel et al., eds.), pp. 240–268. Univ. of Arizona, Tucson.

Gladman B. J., Migliorini F., Zappalà V., Michel P., Cellino A., Froeschlé Ch., Levison H., Bailey M., and Duncan M. (1997) Dynamical lifetimes of objects injected into asteroid belt resonances. *Science, 277,* 197–201.

Holsapple K., Giblin I., Housen K., Nakamura A., and Ryan E. (2002) Asteroid impacts: Laboratory experiments and scaling laws. In *Asteroids III* (W. F. Bottke Jr. et al., eds.), this volume. Univ. of Arizona, Tucson.

Jedicke R. and Metcalfe T. S. (1998) The orbital and absolute magnitude distributions of main belt asteroids. *Icarus, 131,* 245–260.

Klačka J. (1992) Mass distribution in the asteroid belt. *Earth Moon Planets, 56,* 47–52.

Knežević Z., Lemaître A., and Milani A. (2002) The determination of asteroid proper elements. In *Asteroids III* (W. F. Bottke Jr. et al., eds.), this volume. Univ. of Arizona, Tucson.

Love S. G. and Ahrens T. J. (1996) Catastrophic impacts on gravity dominated asteroids. *Icarus, 124,* 141–155.

Martelli G., Ryan E. V., Nakamura A. M., and Giblin I. (1994) Catastrophic disruption experiments: Recent results. *Planet. Space Sci., 42,* 1013–1026.

Marzari F., Davis D. R., and Vanzani V. (1995) Collisional evolution of asteroid families. *Icarus, 113,* 168–187.

Marzari F., Farinella P., and Davis D. R. (1999) Origin, aging and death of asteroid families. *Icarus 142,* 63–77.

Merline W. J., Close L. M., Dumas C., Shelton J. C., Menard F., Chapman C. R., and Slater D. C. (2000) Discovery of companions to asteroids 762 Pulcova and 90 Antiope by direct imaging (abstract). *Bull. Am. Astron. Soc., 32,* 1017.

Merline W. J. et al. (2002) Asteroid satellites. In *Asteroids III* (W. F. Bottke Jr. et al., eds.), this volume. Univ. of Arizona, Tucson.

Migliorini F., Zappalà V., Vio R., and Cellino A. (1995) Interlopers within asteroid families. *Icarus, 118,* 271–291.

Milani A. and Knežević Z. (1992) Asteroid proper elements and secular resonances. *Icarus, 98,* 211–232.

Milani A. and Knežević Z. (1994) Asteroid proper elements and the dynamical structure of the asteroid main belt. *Icarus, 107,* 219–254.

Morbidelli A., Zappalà V., Moons M., Cellino A., and Gonczi R. (1995) Asteroid families close to mean motion resonances: dynamical effects and physical implication. *Icarus, 118,* 132–154.

Morbidelli, A. and Nesvorný D. (1999) Numerous weak resonances drive asteroids toward terrestrial planets orbits. *Icarus, 139,* 295–308.

Nesvorný D. and Morbidelli A. (1998) Three-body mean motion resonances and the chaotic structure of the asteroid belt. *Astron. J., 116,* 3029–3037.

Nolan M. C., Asphaug E., Melosh H. J., and Greenberg R. (1996) Impact craters on asteroids: Does gravity or strength control their size? *Icarus, 124,* 359–371.

Paolicchi P., Verlicchi A., and Cellino A. (1996) An improved semi-empirical model of catastrophic impact processes. *Icarus, 121,* 126–157.

Paolicchi P., Burns J. A., and Weidenschilling S. J. (2002) Side effects of collisions: Spin rate changes, tumbling rotation states, and binary asteroids. In *Asteroids III* (W. F. Bottke Jr. et al., eds.), this volume. Univ. of Arizona, Tucson.

Petit J. M. and Farinella P. (1993) Modelling the outcomes of high-velocity impacts between solar system bodies. *Cel. Mech. Dyn. Astron., 57,* 1–28.

Pisani E., Dell'Oro A., and Paolicchi P. (1999) Puzzling asteroid families. *Icarus, 142,* 78–88.

Richardson D. C., Leinhardt Z. M., Melosh H. J., Bottke W. F. Jr., and Asphaug E. (2002) Gravitational aggregates: Evidence and evolution. In *Asteroids III* (W. F. Bottke Jr. et al., eds.), this volume. Univ. of Arizona, Tucson.

Spitale J. N. and Greenberg R. (2001) Numerical evaluation of the general Yarkovsky effect: Effects on semimajor axis. *Icarus, 149,* 222–234.

Sullivan R., Thomas P., Murchie S., and Robinson M. (2002) Asteroid geology from *Galileo* and *NEAR Shoemaker* data. In *Asteroids III* (W. F. Bottke Jr. et al., eds.), this volume. Univ. of Arizona, Tucson.

Tanga P., Cellino A., Michel P., Zappalà V., Paolicchi P., and Dell'Oro A. (1999) On the size distribution of asteroid families: The role of geometry. *Icarus, 141,* 65–78.

Tedesco E. F. and Désert F. X. (1999) The ISO faint asteroid survey (abstract). *Bull. Am. Astron. Soc., 31,* 1091.

Tholen D. J. (1984) Asteroid taxonomy from cluster analysis of photometry. Ph.D. thesis, Univ. of Arizona, Tucson. 150 pp.

Vokrouhlický D. (1999) A complete linear model for the Yarkovsky thermal force on spherical asteroid fragments. *Astron. Astrophys., 344,* 362–366.

Vokrouhlický D. and Farinella P. (1998) The Yarkovsky seasonal effect on asteroidal fragments: A nonlinearized theory for the plane-parallel case. *Astron. J., 116,* 2032–2041.

Weidenchilling S. J., Marzari F., Davis D. R., and Neese C. (2001) Origin of the double asteroid 90 Antiope: A continuing puzzle (abstract). In *Lunar and Planetary Science Conference XXXII,* Abstract #1890. Lunar and Planetary Institute, Houston (CD-ROM).

Zappalà V. and Cellino A. (1996) Main belt asteroids: Present and future inventory. In *Completing the Inventory of the Solar System* (T. W. Rettig and J. M. Hahn, eds.), pp. 29–44. ASP Conference Series 107.

Zappalà V. and Cellino A. (2001). Formation of asteroid families and delivery of NEO showers from the main belt. In *Collisional Processes in the Solar System* (M. Marov and H. Rickman, eds.), pp. 303–321. Kluwer, Dordrecht.

Zappalà V., Farinella P., Knežević Z., and Paolicchi P. (1984) Collisional origin of the asteroid families: Mass and velocity distributions. *Icarus, 59,* 261–285.

Zappalà V., Bendjoya Ph., Cellino A., Farinella P., and Froeschlé C. (1995) Asteroid families: Search of a 12487 asteroid sample using two different clustering techniques. *Icarus, 116,* 291–314.

Zappalà V., Cellino A., Dell'Oro D., Migliorini F., and Paolicchi P. (1996) Reconstructing the original ejection velocity fields of asteroid families. *Icarus, 124,* 156–180.

Zappalà V., Cellino A., Di Martino M., Migliorini F., and Paolicchi P. (1997) Maria's family: Physical structure and possible implications for the origin of giant NEAs. *Icarus, 129,* 1–20.

Zappalà V., Cellino A., Gladman B. J., Manley S., and Migliorini F. (1998) Asteroid showers on Earth after family break-up events. *Icarus, 134,* 176–179.

Zappalà V., Bendjoya P., Cellino A., Di Martino M., Doressoundiram A., Manara A., and Migliorini F. (2000) Fugitives from the Eos family: First spectroscopic confirmation. *Icarus, 145,* 4–11.

Zappalà V., Cellino A., and Dell'Oro A. (2002) A search for the collisional parent bodies of large NEAs. *Icarus,* in press.

Spectroscopic Properties of Asteroid Families

A. Cellino
Istituto Nazionale di Astrofisica, Osservatorio Astronomico di Torino

S. J. Bus
University of Hawai'i

A. Doressoundiram
Observatoire de Paris–Meudon

D. Lazzaro
Observatório Nacional, Rio de Janeiro

Asteroid families have been the target of several dedicated campaigns of spectroscopic observations during the last 10 years. Preliminary studies were mainly devoted to obtain a confirmation of the cosmochemical reliability of groupings identified by purely statistical analyses of the distributions of objects in the space of the orbital proper elements. These early attempts led to some spectacular confirmations of the common collisional origin of some families, like that associated with 4 Vesta. Subsequently, spectroscopic investigations started to be mostly focused on the issue of characterizing the overall mineralogical compositions of different families, at the same time looking for possible evidence of thermal differentiation of the original parent bodies and for possible evidence of ongoing space-weathering processes. Spectroscopy has also proven to be crucially needed to identify likely interlopers that can seriously affect the derived size distributions of families and the reconstruction of the original fields of ejection velocity of the fragments. At the same time, spectroscopic properties have been recognized as an invaluable tool to assess the real memberships of families mutually overlapping in the space of proper elements. Moreover, spectroscopic surveys have in general been found to be an excellent complement to conventional family searches purely based on the identification of concentrations of objects in the proper-element space. A challenging unsolved problem comes from the fact that several families have been found to exhibit peculiar spectroscopic properties. This opens up new exciting possibilities for future developments in the interpretation of asteroid taxonomic classes.

1. INTRODUCTION

In the early 1990s, the availability of large datasets of asteroid proper elements (*Milani and Knežević,* 1990, 1992) and the simultaneous development of new statistical techniques for the identification of asteroid families (*Zappalà et al.,* 1990; *Bendjoya et al.,* 1991; *Bendjoya,* 1993; *Zappalà et al.,* 1994, 1995) promised to open new perspectives for the physical studies of these groupings, thought to be the direct outcomes of collisional events that occurred in the asteroid belt. Previously, extensive physical studies of families had been prevented or seriously slowed down by strong discrepancies in the family lists proposed by different authors [see the review chapter by *Valsecchi et al.* (1989) in the *Asteroids II* book], and by cosmochemical inconsistencies in some of the proposed family memberships (*Chapman et al.,* 1989).

In that uncertain situation, the first important goal to be pursued was to find some convincing evidence that the most recently identified families could really be considered as "true" collisionally originated groups, and not simply statistical flukes. In this respect, it was immediately clear that some definite answer could come from spectroscopy. Spectroscopy was able to find, through the observations performed by *Binzel and Xu* (1993), the first spectacular confirmation of the collisional origin of a family, the one associated with the large asteroid 4 Vesta (see Fig. 1). In that case, a spectroscopic check was particularly suitable, due to the fact that 4 Vesta was a unique (at that time) case of an object belonging to the V taxonomic class, characterized by spectroscopic properties similar to those of basaltic achondrites. Subsequently, the activities of spectroscopic monitoring of families have been steadily increasing. From the point of view of spectroscopy, families provide a unique opportunity to obtain information on the inner layers of their parent bodies. As a consequence, intense observational activity has been devoted to family members in order to characterize their plausible mineralogical composition. At the same time, spectroscopy has been used as a very important tool for identifying random interlopers within families. These objects share by chance the same orbital properties as "true" family members, but they were not produced by the disruption of the family's parent body. Interlopers can be identified when they are characterized by spectroscopic properties

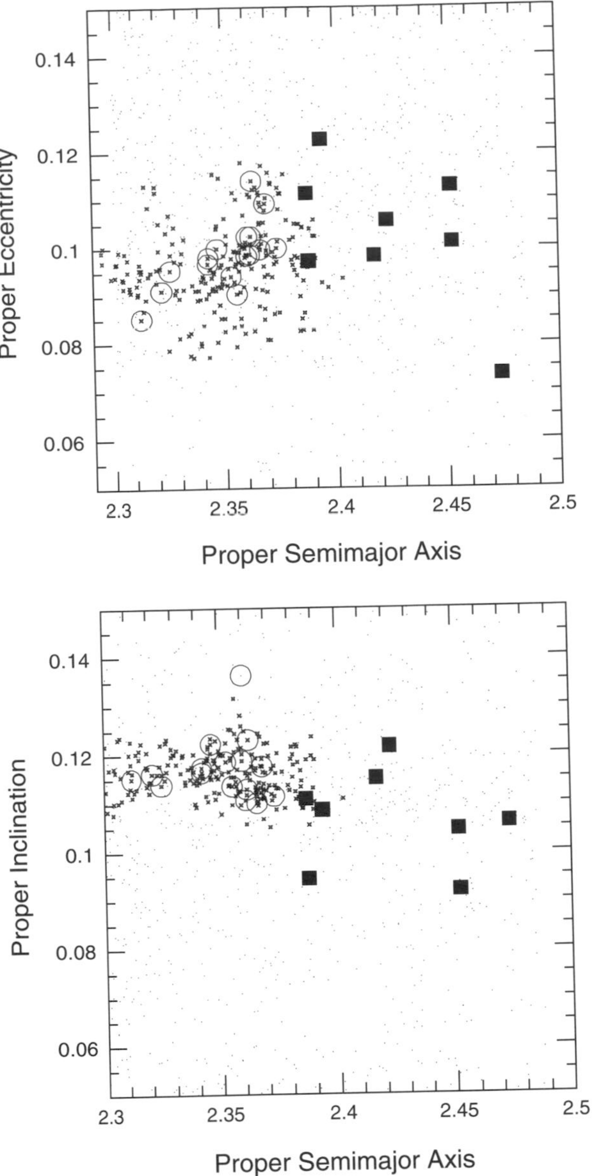

Fig. 1. Plot of the asteroids located around the family of Vesta in the space of the proper elements. Crosses indicate family members. Open circles identify the basaltic objects found by *Binzel and Xu* (1993). Filled squares indicate the basaltic objects not belonging to the nominal family identified by the same authors in the region between the family and the border of the 3:1 mean-motion resonance with Jupiter.

incompatible with those of their family. For instance, an S-type object belonging to a family of C-type members should be considered as a very likely interloper. The presence of interlopers and their plausible numbers can be predicted in purely statistical terms (*Migliorini et al.,* 1995), but it is clear that direct observations can quantify not only how many nominal members are actual interlopers, but they can also identify which ones. In turn, this is very useful for investigations concerning the physical properties of the families (size distributions, ejection velocity fields, etc.) described in *Zappalà et al.* (2002).

Spectroscopic properties can also be used to enlarge the nominal membership of some families, in cases in which some peculiar spectral features can be found to characterize their members. In this way, objects sharing the same feature, but located beyond the nominal family borders, can be added to the list of plausible members. A good example of this was given, again, by the quoted spectroscopic observations of the Vesta family by *Binzel and Xu* (1993). In particular, these authors discovered a number of genetically related objects (belonging to the V taxonomic class) well beyond the nominal borders of the family, all along the region separating the family from the 3:1 mean-motion resonance with Jupiter (Fig. 1). In this way, it was convincingly shown for the first time that large ejection velocities of the fragments, of the order of several hundred meters per second, are possible in family-forming events.

Since then, several families have been extensively observed, including the most important groupings identified by statistical investigations, like Koronis (*Binzel et al.,* 1993), Eos (*Doressoundiram et al.,* 1998a; *Zappalà et al.,* 2000), Eunomia (*Lazzaro et al.,* 1999), Veritas (*Di Martino et al.,* 1997), Hoffmeister (*Migliorini et al.,* 1996), Flora (*Florczac et al.,* 1998), Maria (*Zappalà et al.,* 1997), Nysa/Polana (*Doressoundiram et al.,* 1998b; *Cellino et al.,* 2001), and Hygiea (*Mothé-Diniz et al.,* 2001). In addition to the above specific campaigns, a major observational contribution has also been produced by the SMASS (*Xu et al.,* 1995) and SMASSII (*Bus,* 1999) surveys. The infrared JHK spectrophotometry performed by *Veeder et al.* (1995) for members of the Eos, Koronis, and Maria families also deserves a mention in this context.

The general results of these investigations, as well as a number of problems they were expected to solve, are briefly summarized in the following sections.

2. STATE OF THE ART

A first point deserves mention here. Spectroscopy of asteroids has profited in recent years from improvements in instrumental performance because of the availability of larger telescopes and increasingly better CCDs. Wide spectral windows have been sampled, with an extension of the traditional wavelength range of the classical UBVRI photometry toward longer wavelengths in the near-IR (see *Bus et al.,* 2002). However, this does not mean that, on average, the wavelength coverage in single observing campaigns has been steadily increasing. In most cases, spectra have been obtained for asteroid families in the interval from 5000 Å up to 9,500–10,000 Å. This means that, with respect to the older spectrophotometric investigations based on UBV filters, something has been generally lost at the ultraviolet end of the obtained spectra, whereas the longest wavelengths covered have been rarely beyond 1 µm. This is all because of the limitations of the instruments. Although this did not prevent the observers from obtaining important results, as we shall see, it is important to stress that in several cases family investigations suffered from a less than ideal coverage of the spectra, mainly around the region of the silicate absorption bands around 1 μm and beyond, or shortward of 5000 Å, an important region for objects belonging to the "primitive" taxonomic classes (C, B, F, and G).

In spite of the above limitations, spectroscopy of families has produced a wealth of useful data in recent years. A summary of the current situation is given in Table 1, listing for each family (ordered according to the proper semimajor axis of the lowest-numbered member) the number of members observed spectroscopically, the identified interlopers (by purely spectroscopic means; this does not include interlopers already known on the basis of taxonomic classification), the general spectral properties characterizing the members, and possible analogs among the near-Earth asteroid (NEA) population and/or the meteorite classes. Some additional notes are also included when appropriate, mainly dealing with possible indications of evidence of space-weathering phenomena and/or the inferred properties of the original parent bodies. All these data have been obtained by several dedicated observing campaigns (see section 1), and the main references are also listed in Table 1. Taxonomic information is also given, and is very relevant to the subject of this chapter. For this reason, we have summarized the main taxonomic classes identified by the classical Tholen classification (*Tholen,* 1984) and the most recent SMASSII-based classification (*Bus,* 1999) in Table 2, together with a (forcedly synthetic) description of the main features characterizing each class. Note also that the W taxonomic class mentioned in Table 1 corresponds to M-type objects exhibiting hydration features around 3 μm (*Rivkin et al.,* 2000).

Spectroscopy of family members has been carried out by many authors with the purpose of finding some answers to a number of important questions (listed in no particular order): (1) Did some families originate from a differentiated parent body? (2) Can some peculiar spectroscopic features be identified for different families, allowing us to improve the member lists beyond those derived from the nominal statistical procedures? (3) If the answer to the previous question is "yes," can we use these spectroscopic features to disentangle the memberships of different, mutually overlapping families? (4) Can we identify families by purely spectroscopic means? (5) Is there any evidence of some ongoing space-weathering processes affecting the members of different families? (6) Can some families be identified as likely sources of NEAs, based on comparisons of spectroscopic properties and closeness to some dynamical transfer region? (7) Can we draw conclusions about the mixing of different taxonomic types in the main belt as a consequence of collisional events?

Currently, we cannot say that we have found definitive answers to all the above questions, but a large body of evidence has been collected, allowing us to draw at least some clear indications of the likely solutions.

2.1. Spectral Homogeneity

A first important result of the spectroscopic campaigns carried out so far is that, apart from the peculiar case of Vesta, no convincing evidence of any other family originating from a differentiated parent body has been found. A possible exception might be a tentative association of the Bellona and Phaeo groupings, whose members belong to the S and X taxonomic classes respectively. This possible association has been proposed by *Bus* (1999), but it is not firmly established, being at most a tentative conjecture according to the same author. Another tentative candidate might be the Lydia family, but in this case the evidence is also quite weak.

As a general rule, spectroscopic properties turn out to be fairly homogeneous among members of the same family. This fact can look disappointing from the point of view of the observers (finding evidence of members coming from a metallic core, an olivine-rich mantle, and a basaltic crust might be *a priori* much more exciting!), and certainly it has important implications that will be extensively discussed in the next section. On the other hand, spectral homogeneity of the families has been found to be useful for some applications. In particular, it is possible to characterize in a reliable way the average reflectance properties and surface albedos of different families. This allowed, for instance, *Zappalà and Cellino* (1996) to derive an estimate of the limit of completeness of the asteroid inventory in the main belt, based on the behavior of the distributions of apparent magnitudes for families at different heliocentric distances. These estimates have been later confirmed by independent studies (*Jedicke and Metcalfe,* 1998).

However, the above findings do not mean that all the members of any given family are essentially equal in terms of spectral reflectance. Some limited variations are certainly present, and are in some cases beyond the instrumental uncertainties. Some examples of the typical ranges of spectroscopic properties (from very homogeneous to mildly dispersed) observed in different families are shown in Fig. 2.

Apart from the cases of macroscopic differences related to the presence of random interlopers [the identified C-type members of the Eunomia family being a good example; see *Lazzaro et al.* (1999)], in some cases the differences can be qualitatively ascribed to mild variations in the mineralogical properties of the parent body, and/or to possible phenomena of space weathering. No convincing quantitative analysis has yet been performed regarding this particular subject. In particular, the occurrence of space weathering has so far been demonstrated mainly in the case of asteroids belonging to the S taxonomic class (see *Chapman,* 2002), but little is known about possible space-weathering processes affecting different taxonomic classes. In this respect, data from families like Eos (*Doressoundiram et al.,* 1998a; *Zappalà et al.,* 2000) might indicate that space weathering also affects at least the K class.

2.2. Spectroscopy as a Tool for Family Identification

The generally homogeneous spectral properties of families can be usefully applied to extend family memberships beyond the nominal lists found by purely statistical techniques of family identification. In this respect, a major advance has been the extensive SMASSII spectroscopic sur-

TABLE 1. Summary of spectroscopic studies of families.

Family	Number of Observed Spectra*	Identified Interlopers	Spectral Characteristics	Meteorites and/or NEA Analogs	Notes†	References
Flora‡	47	298, 2093, 3533, 3875, 4278	S type Max. ~750 nm Range of slopes	L3, L4, L5, LL	Homogeneous PB. Evidence of SW	*Florczak et al.* (1998)
Vesta‡	20		V type	HED V-type NEAs	Differentiated PB with basaltic surface. Cratering event.	*Binzel and Xu* (1993)
Nysa/Mildred‡,§	11	Nysa? (E type) Hertha? (W type)	S type		Likely young family	*Cellino et al.* (2001)
Polana‡,§	11	Nysa? (E type) Hertha? (W type) 3881, 3997, 4797	F type	Metamorphic CI/CM F-type NEA?	Possible links with Nysa and/or 135 Hertha	*Cellino et al.* (2001)
Henan‡	16	3430	L type		Uncertain boundaries	*Bus* (1999)
Maria‡	12	4167?	S type Similar spectral slope and 1-μm band Max. ~770 nm	Ordinary chondrites.	Possible source of big NEAs, like 433 Eros and 1036 Ganymede.	*Zappalà et al.* (1997)
Eunomia‡	44 + 10	85, 141, 546	S type Max. ~750 nm Slope spanning a continuous but limited range		Some degree of differentiation of PB?	*Lazzaro et al.* (1999), *Bus* (1999)
Weringia§	4		S type		Very dispersed family	*Bus* (1999)
Eugenia‡	12	2715	Spectra span the boundary between C and X type			*Bus* (1999)
Chloris‡	9		C type Steep UV drop-off Absorption band at 0.7 μm			*Bus* (1999)
Lydia‡	10		Spectra span the boundary between C and X type		Differentiated PB?	*Bus* (1999)
Liberatrix‡	9	6704	C type Steep UV drop-off		Uncertain boundaries	*Bus* (1999)
Merxia‡	10	1987 UF5	S type			*Bus* (1999)
Watsonia§	8	1659	L type	CO3/CV3 affinities	Spinel-bearing PB. Dispersed family. Old?	*Burbine et al.* (1992) *Bus* (1999)
Coelestina‡	8	127	S type Decrease of spectral slope for smaller objects		Uncertain boundaries possible SW	*Bus* (1999)
Thisbe§	5	2730	B type			*Bus* (1999)
Pallas‡	16		B or C type		One or more cratering events	*Bus* (1999)
Phaeo‡	4		X type		Moderate robustness	*Bus* (1999)
Bellona§	14	322, 1427, 1730 2879, 5467	S type		Uncertain boundaries Might include Phaeo	*Bus* (1999)
Astrid‡	5		C type Moderate spectral slopes			*Bus* (1999)
Agnia‡	17		S type Decrease of spectral slope for smaller objects		Size-dependent resurfacing or SW	*Bus* (1999)
Gefion‡	36		S type		Interlopers known from taxonomy/radiometry: 1, 255, 374	*Bus* (1999)

TABLE 1. (continued).

Family	Number of Observed Spectra*	Identified Interlopers	Spectral Characteristics	Meteorites and/or NEA Analogs	Notes†	References
Hoffmeister‡	9		C or F type Small concave absorption feature around 0.9 μm		Carbon-rich PB	*Migliorini et al.* (1996) *Bus* (1999)
Dora‡	33	7081 6907?, 9970?	C type Absorption band at 0.7 μm			*Bus* (1999)
Koronis‡	8		S type Range of 1-μm band depth and center.		Modest degree of defferentiation of PB	*Binzel et al.* (1993)
Eos‡	45 + 7	1910, 4455	K type Max ~800–850nm. Range of 1-μm band depth and center. Range of slopes	CO/CV First evidence of a dynamical route through the 9:4 resonance	Modest degree of internal differentiation within PB. Possible SW	*Doressoundiram et al.* (1998a) *Zappalà et al.* (2000)
Themis‡	36	461, 1171	C type Mostly with absorption band at 0.7 μm		Slight heterogeneity of PB	*Florczak et al.* (1998)
Hygiea‡	11	100, 1109, 1209, 1599	C (10 Hygiea) and B type		Possible inhomogeneities in PB	*Mothé-Diniz et al.* (2001)
Veritas‡	8		C, P, and D type Wide range of spectral slopes		Stratified PB and/or SW. Presence of hydrated members. Dynamical instabilities	*Di Martino et al.* (1997)

*From the studies referenced in the last column; for more details on each family refer to *Zappalà et al.* (2002).
†PB and SW indicate parent body and space weathering respectively.
‡Family identified by means of proper-element analysis.
§Family identified mainly by means of spectroscopy.

TABLE 2. Summary of asteroid taxonomic classes.

Tholen Class	Bus Class	Albedo	Spectral Features
A	A	Moderate	Very steep red slope shortward of 0.75 μm; moderately deep absorption feature longward of 0.75 μm.
B, C, F, G	B, C, C_b, C_h, C_g, C_{hg}	Low	Linear, generally featureless spectra. Differences in UV absorption features and presence/absence of narrow absorption feature near 0.7 μm.
D	D	Low	Relatively featureless spectrum with very steep red slope.
E, M, P	X, X_c, X_e, X_k	From low (P) to very high (E)	Generally featureless spectrum with reddish slope; differences in subtle absorption features and/or spectral curvature and/or peak relative reflectance.
Q	Q	Moderate	Reddish slope shortward of 0.7 μm; deep, rounded absorption feature longward of 0.75 μm.
R	R	Moderate	Moderate reddish slope downward of 0.7 μm; deep absorption longward of 0.75 μm.
S	S, S_a, S_k, S_l, S_q, S_r	Moderate	Moderately steep reddish slope downward of 0.7 μm; moderate to steep absorption longward of 0.75 μm; peak of reflectance at 0.73 μm. Bus subgroups intermediate between S and A, K, L, Q, R classes.
T	T	Low	Moderately reddish shortward of 0.75 μm; flat afterward.
V	V	Moderate	Reddish shortward of 0.7 μm; extremely deep absorption longward of 0.75 μm.
—	K	Moderate	Moderately steep red slope shortward of 0.75 μm; smoothly angled maximum and flat to blueish longward of 0.75 μm, with little or no curvature.
—	L, L_d	Moderate	Very steep red slope shortward of 0.75 μm; flat longward of 0.75 μm; differences in peak level.
—	O	—	Peculiar trend, known so far only for asteroid 3628.

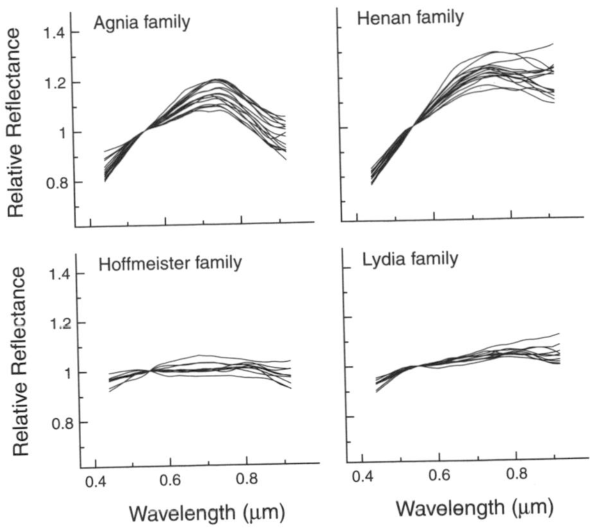

Fig. 2. Available spectra for the Agnia, Henan, Hofmeister, and Lydia families (from *Bus,* 1999). While in some cases families appear to be very homogeneous in terms of spectral properties, in some other cases there is some range of heterogeneity (like in the case of the Agnia family plotted here) that might be explained in terms of possible space-weathering processes and/or in terms of possible mild variations in the mineralogical properties of the parent body.

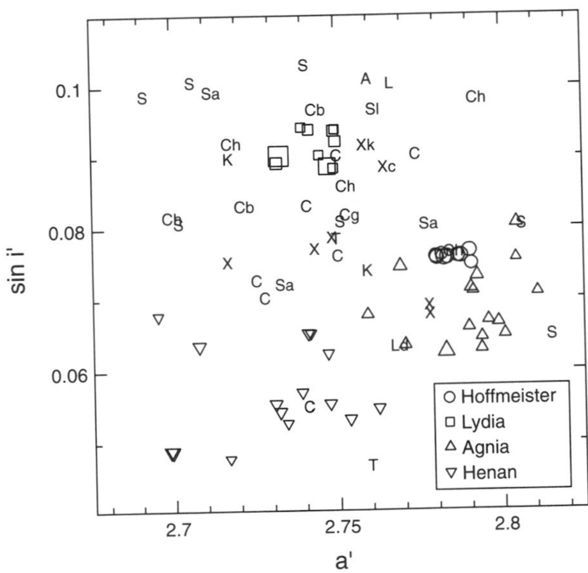

Fig. 3. Plot of the location of the spectroscopically observed members of the Hoffmeister, Lydia, Henan, and Agnia families in the proper inclination vs. semimajor axis plane (from *Bus,* 1999). Different symbols are used for each family, while nonfamily, background objects located in the same region of the sin i'–a' plane are also indicated using their corresponding taxonomic type. The sizes of the family symbols are proportional to the corresponding diameters of the objects. It can be seen that in this region of the belt several families are found to mutually overlap, and spectroscopy is an essential tool for discriminating among the members of different groupings.

vey carried out by *Bus* (1999). In particular, spectroscopic features have made it possible to disentangle the memberships of a number of families identified by means of purely statistical techniques, and found to mutually overlap in the region of the main belt between 2.7 and 2.8 AU in heliocentric distance. For instance, the location in the proper inclination vs. semimajor axis of the spectroscopically observed members of the Lydia, Liberatrix, Henan, and Hoffmeister families (spectra shown in Fig. 2) are plotted in Fig. 3. It is easy to see that, apart from the very compact Hoffmeister grouping, the other families shown are noticeably spread, and it is not surprising that the classical techniques of family identification gave uncertain results, due to the difficulty in separating mutually overlapping groupings (we should take into account, however, that family identification is made by also considering a third dimension, given by the location of the objects in proper eccentricity).

Of course, specroscopy is a powerful tool when the overlapping families are distinct in terms of reflectance spectra. Luckily enough, this happens in several cases of interest. A very nice example, in this respect, is the big Nysa family, which has recently been found to be composed by two mutually overlapping families (named after the asteroids Mildred and Polana) whose members belong to the S and F taxonomic class respectively (*Cellino et al.,* 2001).

In cases like those mentioned above, spectroscopy is decisive for enabling any further analysis of the physical properties of these families, since this would be impossible without reliable indications about the real memberships of the groupings. Even in cases in which the membership is already well established, however, spectroscopy is very important to identify random family interlopers. This task is more important than would appear at first glance. The reason is that interlopers can strongly affect the apparent physical properties of families. Figure 4 shows a good example of this effect. The figure shows the relation between the sizes of the nominal members of the Dora family and the corresponding values of the proper elements (semimajor axis, eccentricity, and inclination) of the same objects. This type of plot is very important, since the differences in proper elements can be interpreted in terms of differences in ejection velocities (see *Zappalà et al.,* 2002). In particular, Fig. 4 shows a typical "triangular" trend in the three plots, indicating that smaller fragments were ejected at higher speeds in the original disruption of the family's parent body. A few objects do not follow the general trend in the plots shown in Fig. 4 (indicated by full circles). *A priori,* this might indicate that the size-velocity relation does not actually exist, or it is fairly weak. However, the discrepant objects were observed by *Bus* (1999) in the SMASSII spectroscopic survey. One of them (a K-type asteroid) turned out to be certainly an interloper. The other three objects were found to be separated by the core of the family in terms of spectroscopic properties. Though not being certain interlopers, they must be considered suspect. In particular, two of them do not belong to the same subset of the big C taxonomic complex (Ch), which characterizes the core of the Dora family (*Bus,* 1999). This

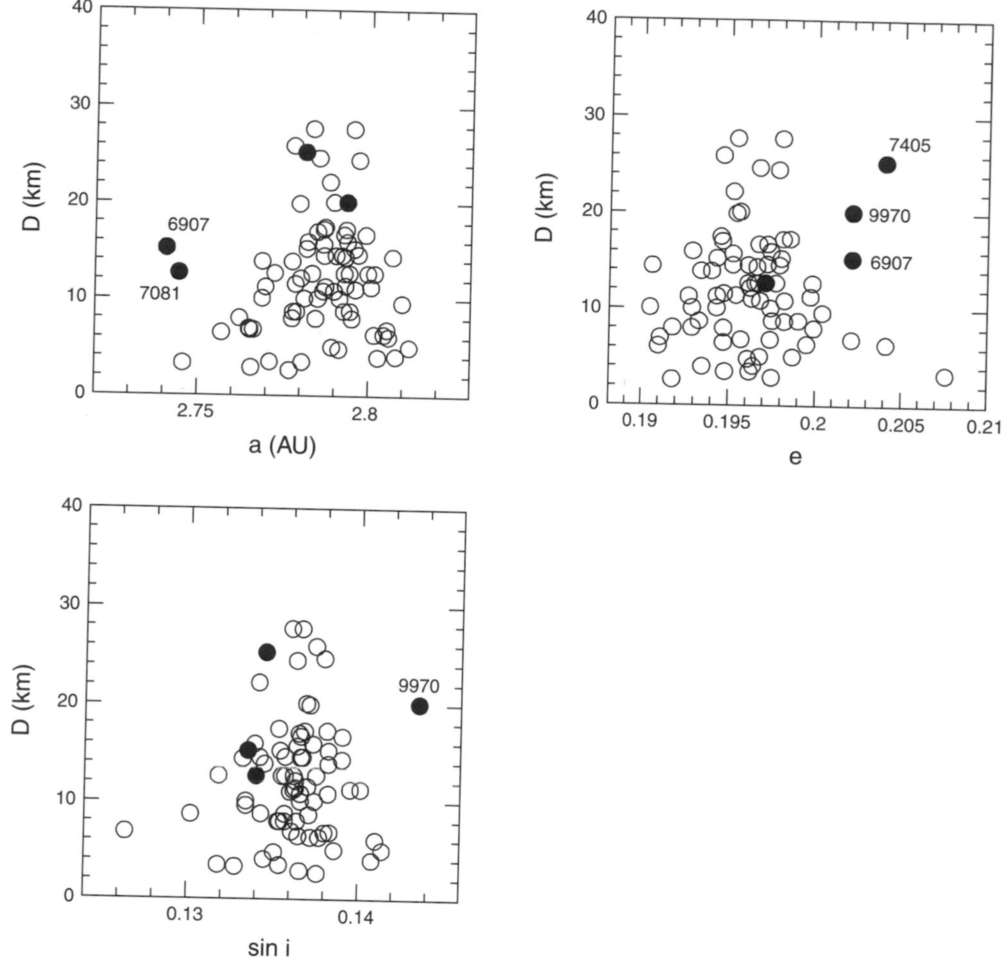

Fig. 4. Plots of the diameter vs. proper semimajor axis, eccentricity, and inclination for the Dora family. Four objects, indicated by their identification numbers and full circles, do not fit the general trend exhibited by the other family members. These objects were observed by *Bus* (1999). One (7081) is surely an interloper. The other three are separated by the core of the family in terms of spectroscopic properties. Though not being certain interlopers, they must be considered suspect.

example shows that the spectroscopic data are a very important element for any physical analysis of the identified families. At the same time, we also have here a potential method for identifying some lists of likely candidate interlopers for different families, to be confirmed by future spectroscopic observations.

The first serious attempt to identify new families by purely spectroscopic means has been made by *Bus* (1999). In particular, he found a number of groupings including objects that would suggest a common collisional origin primarily on the basis of spectroscopic properties. This does not mean that the location in the proper-element space is not taken into account. Instead, a hybrid metric was defined, in which the distance between the objects is defined according to both the distance in the proper-element space and the spectral similarity (quantitatively defined). In this way, it was possible to identify some groupings not previously considered by family searches based on proper elements alone. These new candidate families (listed in Table 1) are associated with the asteroids Thisbe, Watsonia, Weringia, and Bellona. Of course, more detailed physical analyses will be useful to confirm or reject these identifications.

As a general comment, it should be mentioned that spectroscopy can be effective for identifying families formed a long time ago, and subsequently eroded and dispersed by collisional and dynamical evolution. We certainly know that the families that we see today are only the most recent and/or populous groupings among those that necessarily have been produced during the history of the solar system (*Marzari et al.,* 1999). The older, "ghost" families no longer can be found by looking at pure concentrations in the proper-element space. On the other hand, spectroscopic properties can be the only possible diagnostics of the existence of such groupings. Some examples are given by the proposed Watsonia family (*Bus,* 1999), a possible L-type family associated with the asteroids 387 Aquitania and 980 Anacostia as

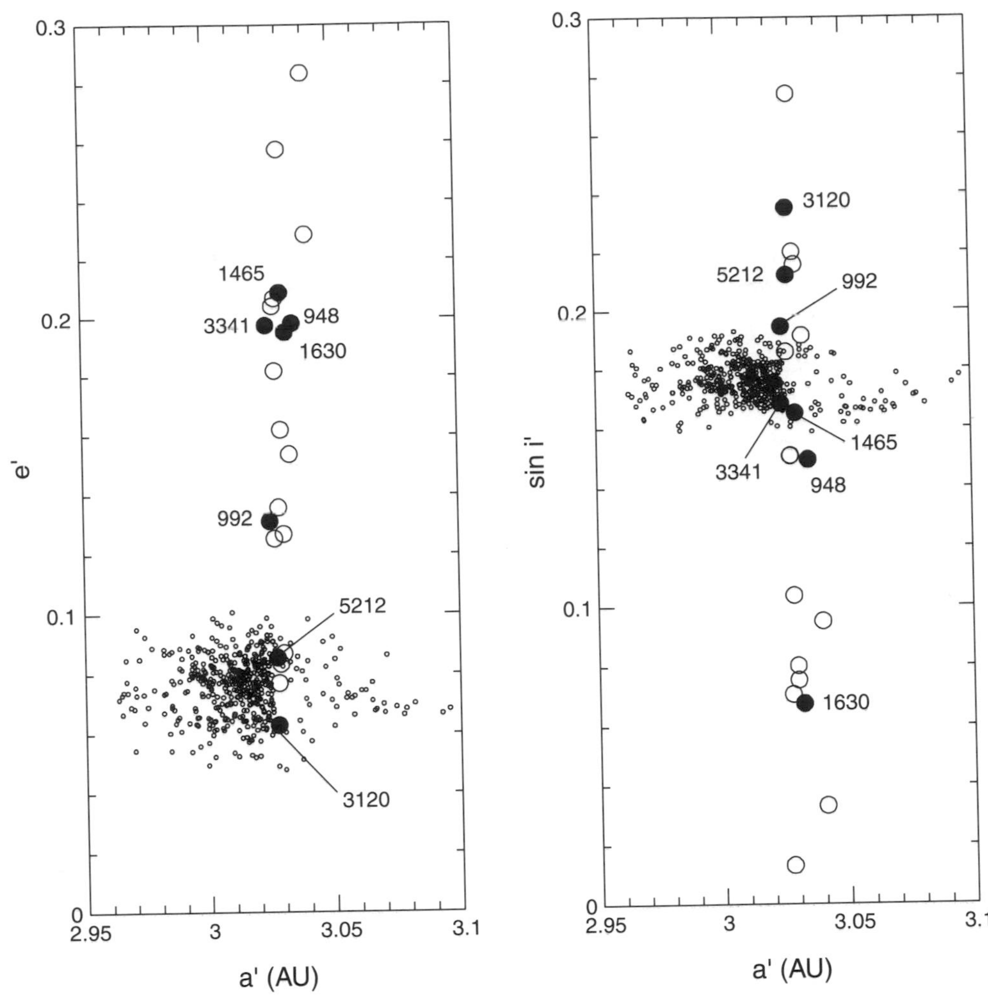

Fig. 5. Location in the proper eccentricity–proper semimajor axis (left) and proper inclination–proper semimajor axis (right) planes of the Eos family, and a group of asteroids currently located into the 9:4 mean-motion resonance with Jupiter. Several of the resonant objects were spectroscopically observed by *Zappalà et al.* (2000) (full circles and numbers) and were found to share the same characteristic spectral features of common Eos family members. This means that the resonant objects were originally Eos members, and are currently observed during the first stages of a dynamical evolution that would lead most of them out of the solar system. A fraction of these objects, however, can be captured by Mars and achieve typical near-Earth orbits.

first suggested by *Burbine et al.* (1992) and an anomalous concentrations of F-type asteroids around, but well beyond the borders of, the Polana family (*Cellino et al.,* 2001).

2.3. Spectroscopic Features

Having been produced by the disruption of "normal" asteroids in the main belt, families are not expected *a priori* to exhibit any exceptional spectroscopic feature. Spectroscopic campaigns have been devoted to observe families discovered by analyses of the clusterings of objects in the space of the proper elements (see *Bendjoya and Zappalà,* 2002) without any *a priori* bias in favor of exotic spectroscopic properties. However, actual observations have shown that, surprisingly, rare features are found fairly often. Apart from the quoted example of the basaltic Vesta family, three other outstanding cases are those of the Eos, Polana, and Henan families. The latter was found by *Bus* (1999) to belong to his newly proposed L taxonomic class. As already mentioned above, the Polana family is a major subcluster of the Nysa clan (*Zappalà et al.,* 1995), and has been found to constitute a single, distinct family composed of objects belonging to the fairly rare F taxonomic class (*Cellino et al.,* 2001).

The case of the Eos family, one of the most populous families in the current main belt, is even more interesting. The members of this family belong to the (elsewhere) rare K taxonomic class, characterized by a well-defined spectroscopic behavior, somewhat intermediate between C- and S-type asteroids (*Gaffey et al.,* 1993). This fact led *Zappalà et al.* (2000) to identify some asteroids presently located into the 9:4 mean-motion resonance with Jupiter as likely Eos members, observed well beyond the family's borders, during the first phases of a dynamical evolution that will remove these asteroids from the main belt (Fig. 5). Some of these objects might be perturbed by Mars and forced to achieve typical near-Earth orbits (*Zappalà et al.*, 2000; *Gladman et al.*, 1997). This is a first example of a positive answer to the question about a possible role of family-forming events in contributing to the flux of NEAs and meteorites. In this respect, we should also quote the results of the spectroscopic analysis of the Maria family performed by *Zappalà et al.* (1997), showing that this family can be considered as a likely source of ordinary chondrite-like material, and a possible parent of the "giant" near-Earth asteroids 433 Eros and 1036 Ganymed. This latter hypothesis has been more recently questioned by *Zappalà and Cellino* (2001) because of the difficulty in reconciling the very short lifetimes of objects injected into the 3:1 mean-motion resonance, with the idea that Eros and/or Ganymede were directly injected into the 3:1 mean-motion resonance during the Maria family-forming event.

3. DISCUSSION

Spectroscopy of families has so far been able to supply a great deal of useful information, as discussed in the previous section. From the point of view of the most general concepts we have learned about the asteroid population, a couple of facts should probably be emphasized at this stage. First, the lack of any clear evidence of differentiated parent bodies is interesting. The occurrence in the past of extensive phenomena of melting and thermal differentiation among asteroids has been inferred basically by three kinds of observational evidence: the basaltic crust of 4 Vesta, the existence of iron and stony-iron meteorites, and the existence of the M-type asteroids, generally interpreted as the metal-rich cores of differentiated parent bodies. These facts have always constituted a delicate constraint for our understanding of the likely origin and subsequent collisional evolution of the asteroid belt. If iron meteorites come from M-type asteroids, and the latter are really the cores of large differentiated parent bodies, the apparent lack of objects having spectroscopic properties that should be expected for fragments originating in the mantle (like the fairly rar olivine-rich A-type asteroids) is a long-debated problem (*Burbine et al.*, 1996). If collisional evolution was sufficiently intense to pulverize the supposedly weak mantle-originating bodies, how can we explain the presence of the intact basaltic crust of Vesta? Moreover, the disruption of the very large bodies that should have been the parents of some large M-type objects observed today (like 16 Psyche) should have produced families that should be still observable (*Davis et al.*, 1999).

More generally, the melting mechanisms are not so obvious for objects as small as asteroids. If radiogenic heating is the natural process to explain the differentiation of a large body like Vesta, how can we explain that 1 Ceres, which is twice as large, apparently has a primitive composition and likely never experienced important heating episodes? These are long-debated problems that are not yet solved. Some plausible mechanisms are mentioned in *McSween et al.* (2002). The lack of any convincing evidence of differentiated family parent bodies, as well as recent discoveries of hydration features in a large fraction of M-type asteroids, could be the first steps toward a new kind of understanding of the likely thermal history of asteroidal bodies. Of course, the existence of iron meteorites indicates that in any case several sources of metallic material must have been produced during the history of the asteroid belt.

The second important fact that we have learned from family spectroscopy has already been noted in the previous section, and is the surprisingly high number of cases in which families are characterized by unusual spectroscopic features. This is true not only in literal terms, but also from the point of view of the general distribution of the taxonomic classes in the main belt. A major example is given by the Eos family. The fact that this family accounts for the vast majority of the K-type asteroids (non-Eos K-type objects being very rare) is puzzling. The existence of the Eos family should imply that a unique prototype of a K-type asteroid existed in the outer belt, in a region where the vast majority of objects exhibits very different spectroscopic properties. This unique parent body was also fairly large [around 220 km according to *Tanga et al.* (1999)] and suffered a collision able to produce this huge family of K-type fragments. All this looks like a sequence of unlikely coincidences. Another example is given by the quoted case of the Polana family in the inner belt, characterized by an F-type taxonomic classification, generally interpreted as altered primitive bodies (*Bell et al.*, 1989), but located in a region in which primitive bodies are not very abundant. The newly proposed L-type class is another example, since most of the identified objects belonging to this class are interpreted as family members (*Bus*, 1999).

In other cases, some families do not belong to unusual taxonomic classes, but they exhibit some spectral reflectance features that are somewhat unusual within their classes, and can be used to characterize their members. Some examples are given by the Maria (*Zappalà et al.*, 1997), Chloris, and Lydia (*Bus*, 1999) families (see also Table 1).

An explanation of the observational evidence is not straightforward, but some simple ideas can already be proposed. One possibility is that we observe relatively young objects, since they were more recently produced. Their spectroscopic features might be due to a shorter exposure to space-weathering phenomena. An alternative explanation might be that we are possibly observing features that are more directly due to thermal and/or physical metamorphism produced by the energetic collisional events responsible of the formations of families. There is not any real proof that this might be true, apart from some preliminary analyses by *Rubin* (1995), and some further work is needed in order to analyze in more detail the problem.

Another relevant fact appears to be reasonably established according to present evidence. Family-forming events were likely able to eject fragments at high speeds, as indicated by the quoted *Binzel and Xu* (1993) observations of the Vesta family (Fig. 1). According to recent results (*Cellino et al.*, 1999), the smallest fragments in family-forming events can achieve the maximum velocities produced in these events. This means that families, and more generally collisional events, have likely been responsible (mainly at small sizes) for a significant mixing of different taxonomic types in the main belt, partly destroying a possible more regular gradient in composition inherited by the original protoplanetary disk (*Cellino,* 2000). Other mechanisms possibly affecting the composition gradient in the asteroid main belt (scattering events produced by massive planetary embryos during the first few million years of the solar system's history) are also discussed in *Petit et al.* (2002).

4. FUTURE DEVELOPMENTS

Family spectroscopy is a vigorous field of research, and its importance is not going to decrease in the near future. The next steps will probably be an extension of the wavelength coverage of the surveys, in order to analyze more satisfactorily the full range of important spectral features, both at short wavelengths, and, probably more important, up to 2 or 3 μm. This will allow the observers to obtain data much more directly diagnostic of the true mineralogical assemblages present on the surfaces of the objects, including very important clues on the presence or absence of hydrated materials (*Jones et al.,* 1990; *Rivkin et al.,* 1995, 2000). An important development will also be an extension of the observations to smaller family members. Observations of these objects can potentially be used to derive possible inferences about the role of space-weathering phenomena related to the different collisional lifetimes of objects of different sizes. It is known, for instance, that S-type NEAs exhibit spectra much more similar to those of ordinary chondrites, with respect to the bigger S-type objects in the main belt (see *Binzel et al.,* 2002). A comparison with family members in the same size ranges might confirm that small main-belt objects surviving over longer timescales exhibit the same "weathered" surface properties of bigger objects of the same taxonomic class. At the same time, observations will likely be extended also to the families that were found by statistical analyses, but with a lesser degree of statistical confidence. Moreover, new examples of families identified by mostly spectroscopic means will be possible. In all these cases, we can expect some important discoveries in the future. In particular, it will be extremely important to see whether the frequency of "strange," "peculiar" spectroscopic features will increase when new families will be analyzed. If this will be the case, we will have probably to modify some of the general ideas we have on the meaning of taxonomic classes, and the importance of space weathering and collisionally induced thermal metamorphism effects.

REFERENCES

Bell J. F., Davis D. R., Hartmann W. K., and Gaffey M. J. (1989) Asteroids: The big picture. In *Asteroids II* (R. P. Binzel et al., eds.), pp. 921–945. Univ. of Arizona, Tucson.

Bendjoya Ph. (1993) A classification of 6479 asteroid into families by means of the wavelet clustering method. *Astron. Astrophys. Suppl., 102,* 25–55.

Bendjoya Ph. and Zappalà V. (2002) Asteroid family identification. In *Asteroids III* (W. F. Bottke Jr. et al., eds.), this volume. Univ. of Arizona, Tucson.

Bendjoya Ph., Slézak E., and Froeschlé C. (1991) The wavelet transform: A new tool for asteroid family determination. *Astron. Astrophys., 272,* 651–670.

Binzel R. P. and Xu S. (1993) Chips off asteroid 4 Vesta: Evidence for the parent body of basaltic achondrite meteorites. *Science, 260,* 186–191.

Binzel R. P., Xu S., and Bus S. J. (1993) Spectral variations within the Koronis family: Possible implications for the surface colors of asteroid 243 Ida. *Icarus, 106,* 608–611.

Binzel R. P., Lupishko D. F., Di Martino M., Whitely R. J., and Hahn G. J. (2002) Physical properties of near-Earth objects. In *Asteroids III* (W. F. Bottke Jr. et al., eds.), this volume. Univ. of Arizona, Tucson.

Burbine T. H., Gaffey M. J., and Bell J. F. (1992) S-asteroids 387 Aquitania and 980 Anacostia: Possible fragments of the breakup of a spinel-bearing parent body with CO3/CV3 affinities. *Meteoritics, 27,* 424–434.

Burbine T. H., Meibom A., and Binzel R. P. (1996) Mantle material in the main belt: Battered to bits? *Meteoritics & Planet. Sci., 31,* 607–620.

Bus S. J. (1999) Compositional structure in the asteroid belt: Results of a spectroscopic survey. Ph.D. thesis, Massachusetts Institute of Technology.

Bus S. J., Vilas F., and Barucci M. A. (2002) Visible-wavelength spectroscopy of asteroids. In *Asteroids III* (W. F. Bottke Jr. et al., eds.), this volume. Univ. of Arizona, Tucson.

Cellino A. (2000) Minor bodies: Spectral gradients and relationships with meteorites. *Space Sci. Rev., 92,* 397–412.

Cellino A., Michel P., Tanga P., Zappalà V., Paolicchi P., and Dell'Oro A. (1999) The velocity-size relationship for members of asteroid families and implications for the physics of catastrophic collisions. *Icarus, 141,* 79–95.

Cellino A., Zappalà V., Doressoundiram A., Di Martino M., Bendjoya Ph., Dotto E., and Migliorini F. (2001) The puzzling case of the Nysa-Polana family. *Icarus, 152,* 225–237.

Chapman C. R. (2002) Cratering on asteroids from *Galileo* and *NEAR Shoemaker*. In *Asteroids III* (W. F. Bottke Jr. et al., eds.), this volume. Univ. of Arizona, Tucson.

Chapman C. R., Paolicchi P., Zappalà V., Binzel R. P., and Bell J. F. (1989) Asteroid families: Physical properties and evolution. In *Asteroids II* (R. P. Binzel et al., eds.), pp. 386–415. Univ. of Arizona, Tucson.

Davis D. R., Farinella P., and Marzari F. (1999) The missing Psyche family: Collisionally eroded or never formed? *Icarus, 137,* 140–151.

Di Martino M., Migliorini F., Zappalà V., Manara A., and Barbieri C. (1997) Veritas asteroid family: Remarkable spectral differences inside a primitive parent body. *Icarus, 127,* 112–120.

Doressoundiram A., Barucci M. A., and Fulchignoni M. (1998a) Eos family: A spectroscopic study. *Icarus, 131,* 15–31.

Doressoundiram A., Cellino A., Di Martino M., Migliorini F., and Zappalà V. (1998b) The puzzling case of the Nysa-Polana family finally solved? *Bull. Am. Astron. Soc., 30,* 505.

Florczac M., Barucci M. A., Doressoundiram A., Lazzaro D., Angeli C. A., and Dotto E. (1998) A visible spectroscopic survey of the Flora clan. *Icarus, 133,* 233–246.

Gaffey M. J., Burbine T. H., and Binzel R. P. (1993) Asteroid spectroscopy: Progress and perspectives. *Meteoritics, 28,* 161–187.

Gladman B. J., Migliorini F., Zappalà V., Michel P., Cellino A., Froeschlé Ch., Levison H., Bailey M., and Duncan M. (1997) Dynamical lifetimes of objects injected into asteroid belt resonances. *Science, 277,* 197–201.

Jedicke R. and Metcalfe T. S. (1998) The orbital and absolute magnitude distributions of main belt asteroids. *Icarus, 131,* 245–260.

Jones T. D., Lebofsky L. A., Lewis J. S., and Marley M. S. (1990) The composition and origin of the C, P and D asteroids: Water as a tracer of thermal evolution in the outer belt. *Icarus, 88,* 172–192.

Lazzaro D., Mothé-Diniz T., Carvano J. M., Angeli C., Betzler A. S., Florczac M., Cellino A., Di Martino M., Doressoundiram A., Barucci M. A., Dotto E., and Bendjoya P. (1999) The Eunomia family: A visible spectroscopic survey. *Icarus, 142,* 445–453.

Marzari F., Farinella P., and Davis D. R. (1999) Origin, aging and death of asteroid families. *Icarus, 142,* 63–77.

McSween H. Y. Jr., Ghosh A., Grimm R. E., Wilson L., and Young E. D. (2002) Thermal evolution models of asteroids. In *Asteroids III* (W. F. Bottke Jr. et al., eds.), this volume. Univ. of Arizona, Tucson.

Migliorini F., Zappalà V., Vio R., and Cellino A. (1995) Interlopers within asteroid families. *Icarus, 118,* 271–291.

Migliorini F., Manara A., Di Martino M., and Farinella P. (1996) The Hoffmeister family: Inferences from physical data. *Astron. Astrophys., 310,* 681–685.

Milani A. and Knežević Z. (1990) Secular perturbation theory and computation of asteroid proper elements. *Cel. Mech., 49,* 316–330.

Milani A. and Knežević Z. (1992) Asteroid proper elements and secular resonances. *Icarus, 98,* 211–232.

Mothé-Diniz T., Di Martino M., Bendjoya P., Doressoundiram A., and Migliorini F. (2001) Rotationally resolved spectra of 10 Hygiea and a spectroscopic study of the Hygiea family. *Icarus, 152,* 117–126.

Petit J-M., Chambers J., Franklin F., and Nagasawa M. (2002) Primordial excitation and depletion of the main belt. In *Asteroids III* (W. F. Bottke Jr. et al., eds.), this volume. Univ. of Arizona, Tucson.

Rivkin A. S., Howell E. S., Britt D. T., Lebofsky L. A., Nolan M. C., and Branston D. D. (1995) Three-micron spectrometric survey of M- and E-class asteroids. *Icarus, 117,* 90–100.

Rivkin A. S., Howell E. S., Lebofsky L. A., Clark B. E., and Britt D. T. (2000) The nature of M-class asteroids from 3-μm observations. *Icarus, 145,* 351–368.

Rubin A. E. (1995) Petrologic evidence of collisional heating of chondritic asteroids. *Icarus, 113,* 156–167.

Tanga P., Cellino A., Michel P., Zappalà V., Paolicchi P., and Dell'Oro A. (1999) On the size distribution of asteroid families: The role of geometry. *Icarus, 141,* 65–78.

Tholen D. J. (1984) Asteroid taxonomy from cluster analysis of photometry. Ph.D. thesis, Univ. of Arizona, Tucson. 150 pp.

Veeder G. J., Matson D. L., Owensby P. D., Gradie J. C., Bell J. F., and Tedesco E. F. (1995) Eos, Koronis and Maria family asteroids: Infrared (JHK) photometry. *Icarus, 114,* 186–196.

Valsecchi G., Carusi A., Knežević Z., Kresák Ľ., and Williams J. G. (1989) Identification of asteroid dynamical families. In *Asteroids II* (R. P. Binzel et al., eds.), pp. 368–385. Univ. of Arizona, Tucson.

Xu S., Binzel R. P., Burbine T. H., and Bus S. J. (1995) Small main-belt asteroid spectroscopic survey: Initial results. *Icarus, 115,* 1–35.

Zappalà V. and Cellino A. (1996) Main belt asteroids: Present and future inventory. In *Completing the Inventory of the Solar System* (T. W. Rettig and J. M. Hahn, eds.), pp. 29–44. ASP Conference Series 107.

Zappalà V. and Cellino A. (2001). Formation of asteroid families and delivery of NEO showers from the main belt. In *Collisional Processes in the Solar System* (M. Marov and H. Rickman, eds.), pp. 303–321. Kluwer, Dordrecht.

Zappalà V., Cellino A., Farinella P., and Knežević Z. (1990) Asteroid families. I. Identification by hierarchical clustering and reliability assessment. *Astron. J., 100,* 2030–2046.

Zappalà V., Cellino A., Farinella P., and Milani A. (1994) Asteroid families II: Extension to unnumbered multiopposition asteroids. *Astron. J., 107,* 772–801.

Zappalà V., Bendjoya Ph., Cellino A., Farinella P., and Froeschlé C. (1995) Asteroid families: Search of a 12487 asteroid sample using two different clustering techniques. *Icarus, 116,* 291–314.

Zappalà V., Cellino A., Di Martino M., Migliorini F., and Paolicchi P. (1997) Maria's family: Physical structure and possible implications for the origin of giant NEAs. *Icarus, 129,* 1–20.

Zappalà V., Bendjoya P., Cellino A., Di Martino M., Doressoundiram A., Manara A., and Migliorini F. (2000) Fugitives from Eos family: First spectroscopic confirmation. *Icarus, 145,* 4–11.

Zappalà V., Cellino A., Dell'Oro A., and Paolicchi P. (2002) Physical and dynamical properties of asteroid families. In *Asteroids III* (W. F. Bottke Jr. et al., eds.), this volume. Univ. of Arizona, Tucson.

Asteroid Meteoroid Streams

Tadeusz J. Jopek
Obserwatorium Astronomiczne Uniwersytet im. Adama Mickiewicza

Giovanni B. Valsecchi
Istituto di Astrofisica Spaziale

Claude Froeschlé
Observatoire de la Côte d'Azur

We briefly review the methods for classifying meteoroids into streams and their association with parent bodies. Most streams have comets as parent bodies, while relatively little work has been done on streams of asteroidal origin, which also seem harder to identify. We discuss how the use of the geocentric variables introduced for stream identification by *Valsecchi et al.* (1999) can help in the identification of asteroidal meteoroid streams.

1. INTRODUCTION

Meteoroid streams consist of small solid particles released from active comets and, perhaps, from some near-Earth asteroids (NEAs). The ejection from comets takes place when they are sufficiently close to the Sun so that the volatiles on the surface of their nuclei can sublimate, carrying off particles. This process was first quantitatively described by *Whipple* (1950, 1951) for comets, but we still lack a satisfactory model for how an asteroid could produce and sustain a meteoroid stream. We observe a meteoroid when it penetrates the atmosphere and ablates, producing a meteor observable with various techniques (radar, photographic, TV), depending on its brightness.

The reason that meteoroids are often found to belong to streams, with orbits similar to that of the parent body, is that they are ejected at very low velocity relative to the parent, up to tens of meters per second in most cases, compared to the tens of kilometers per second of the heliocentric orbital speed. Thus the resulting orbital dispersion is very small immediately after the release. This dispersion, however, is bound to grow because the dynamical evolution of meteoroid stream orbits is chaotic because of the Earth-crossing condition. All streams will first increase in size, both in configuration and in orbital element space, and will eventually become totally dispersed into the background.

The problem of identifying meteoroid streams is similar to that of the identification of asteroid families. The difference between the orbits of potential family/stream members has to be measured by a suitably defined distance function, clusters have to be found among asteroid/meteor orbits, and then the statistical significance of the groupings obtained in this way has to be assessed.

If the asteroid/meteoroid, after having separated from the parent, were orbiting the Sun in the absence of any perturbation, it would continue to do so forever, and its orbital elements would bear a perennial memory of its origin. This is, of course, not the case, which leads to an important difference between the problems of asteroid families and of meteoroid streams. Main-belt asteroid orbits are not very chaotic, which allows the calculation of "proper elements," that is, quasi-integrals of motion that are stable over millions of years to possibly billions of years. This is not possible for meteoroid orbits, on the other hand, since they are strongly chaotic because they cross the orbit of at least one planet, Earth, and often cross the orbits of other planets.

However, *Gronchi and Milani* (2001) have recently introduced planet-crossing orbital proper elements for the secular problem in which the perturbations come from a number of planets in circular and coplanar orbits. These proper elements are quasi-integrals of the motion only as long as close encounters with the planets do not take place. Therefore, while the timescales over which asteroid families remain recognizable are of the order of the age of the solar system, meteoroid streams can only be recognized over much shorter timescales, typically on the order of 10^3–10^4 yr. With the steady growth of the number of known NEAs, a satisfactory solution to the problem of asteroidal meteoroid stream identification would open a further "window" to study this class of asteroids, as has happened for comets through the physical study of the associated meteors.

2. DATA SAMPLES

The public domain meteor data available at the IAU Meteor Data Center (MDC) were collected and tested for internal consistency by *Lindblad* (1987, 1991, 1999) and *Lindblad and Steel* (1994). The data at the MDC include the geocentric and heliocentric parameters of about 62,000 radar meteors and about 6000 meteors observed using optical techniques (photographic and TV). An extensive review of meteor data, including plots illustrating distributions of the orbital elements, was published by *Steel* (1996).

The quality of radar meteor data is poor in comparison with photographic data, but the precision of photographic orbits, where q and e are determined to ~1 ppt and angles to ~0.01°, is inferior to asteroidal orbits. Among photographic meteors, the dataset of highest quality, the Harvard dataset, includes 139 small-camera meteors (*Whipple,* 1954), 413 Super Schmidt meteors (*Jacchia and Whipple,* 1961), and 313 Super Schmidt meteors (*Hawkins and Southworth,* 1958, 1961), for a total of 865 precise orbits.

3. STREAM IDENTIFICATION

3.1. Distance Functions

A meteoroid stream may be defined in terms of geocentric and/or heliocentric parameters, but in any definition this measure of meteoroid orbital similarity is crucial. *Southworth and Hawkins* (1963) first formulated such a quantitative measure by drawing an analogy with a five-dimensional orthogonal coordinate system and considering each orbital heliocentric element as a coordinate. In that space, a meteoroid orbit is represented by a point, and the distance between two points is a measure of the degree of similarity between two meteoroid orbits. These authors developed all the components necessary for a computer cluster analysis:

1. Distance function D. In terms of the osculating orbital elements q, e, i, ω, Ω, for two meteoroids k and l, the D criterion for orbital similarity is defined by *Southworth and Hawkins* (1963) (D_{SH}) as

$$D_{SH}^2 = (e_k - e_l)^2 + (q_k - q_l)^2 + \left(2 \sin \frac{I_{kl}}{2}\right)^2 + \left(\frac{e_k + e_l}{2}\right)^2 \left(2 \sin \frac{\pi_{kl}}{2}\right)^2 \qquad (1)$$

where I_{kl} is the angle between the orbital planes and π_{kl} is the difference between the longitudes of perihelion measured from the common node of the orbits

$$\left(2 \sin \frac{I_{kl}}{2}\right)^2 = \left(2 \sin \frac{i_k - i_l}{2}\right)^2 + \sin i_k \times \sin i_l \left(2 \sin \frac{\Omega_k - \Omega_l}{2}\right)^2$$

$$\pi_{kl} = \omega_k - \omega_l \pm 2 \arcsin \left(\cos \frac{i_k + i_l}{2} \times \sin \frac{\Omega_k - \Omega_l}{2} \sec \frac{I_{kl}}{2} \right)$$

where the minus sign applies when $|\Omega_k - \Omega_l| > 180°$.

2. Rule for calculating the threshold value D_c for orbital similarity

$$D_c = 0.2\left(\frac{360}{N}\right)^{1/4} \qquad (2)$$

where N is the meteor data sample size.

3. A stream-searching algorithm. Together with the D criterion and D_c, this leads to the definition of a meteoroid stream.

Since that time several similar D functions, all based on the orbital elements, have been proposed. In particular, *Drummond* (1981) and *Jopek* (1993) introduced modified functions, in which the relative weights of the different elements were varied with respect to those of D_{SH}.

The suitability of the above mentioned D functions was established empirically on known streams or by comparison of the results obtained by two different D functions. Many authors, starting from *Southworth and Hawkins* (1963), have proven the utility of D functions in computer stream classifications. However, all D functions based on the osculating orbital elements have a critical shortcoming: The elements q, e, i, ω, Ω, are dynamical invariants only in the two-body problem. Meteoroids are subject to other perturbations, and their orbital elements can change significantly on timescales of about 10^4 yr because of secular perturbations (*Hamid and Whipple,* 1963; *Kozai,* 1962; *Babadzhanov and Obrubov,* 1980; *Froeschlé et al.,* 1993).

Recently a new approach to the problem of measuring meteoroid orbital similarity was introduced. Based on geocentric variables, *Valsecchi et al.* (1999) proposed the following D function for two meteoroids k and l

$$D_N^2 = (U_k - U_l)^2 + w_1(\cos \theta_k - \cos \theta_l)^2 + \Delta\xi^2 \qquad (3)$$

where

$$\Delta\xi^2 = \min(w_2\Delta\phi_I^2 + w_3\Delta\lambda_I^2,\ w_2\Delta\phi_{II}^2 + w_3\Delta\lambda_{II}^2)$$

$$\Delta\phi_I = 2 \sin \frac{\phi_k - \phi_l}{2}$$

$$\Delta\phi_{II} = 2 \sin \frac{180° + \phi_k - \phi_l}{2}$$

$$\Delta\lambda_I = 2 \sin \frac{\lambda_k - \lambda_l}{2}$$

$$\Delta\lambda_{II} = 2 \sin \frac{180° + \lambda_k - \lambda_l}{2}$$

and w_1, w_2, w_3 are weighting factors that can be set to 1 or can be determined on the basis of information from modeling the background or stream dispersion. The quantities U, θ, ϕ come from Öpik's theory of close encounters, described in *Öpik* (1976) and *Carusi et al.* (1990), and λ is the ecliptic longitude of the meteoroid at the time of the meteor apparition.

To define U, θ, and ϕ, let us assume that Earth moves on an unperturbed circular orbit of radius equal to 1, lying on the ecliptic, that the constant of gravity and the mass of the Sun are equal to 1, and that the heliocentric velocity of Earth is exactly $V_\oplus = 1$ instead of $V_\oplus = \sqrt{1 + M_\oplus}$. With these conventions the magnitude of the unperturbed geocentric velocity of the meteoroid $\vec{U}$ when crossing the Earth's orbit is

$$U = \sqrt{3 - \frac{1}{a} - 2\sqrt{a(1 - e^2)}\cos i} \qquad (4)$$

In a reference frame centered on Earth, with the z-axis perpendicular to the ecliptic, the y-axis in the direction of Earth's velocity, and the x-axis pointing away from the Sun, $\vec{U}$ has components

$$\begin{pmatrix} U_x \\ U_y \\ U_z \end{pmatrix} = \begin{pmatrix} U \sin\theta \sin\theta \\ U \cos\theta \\ U \sin\theta \cos\theta \end{pmatrix}$$

where θ is the angle between $\vec{U}$ and the y-axis (the direction of motion of Earth) and ϕ is the angle between the y-z plane and the plane containing $\vec{U}$ and the y-axis. For encounters at the ascending node, $-90° < \phi < 90°$, and for encounters at the descending node, $90° < \phi < 270°$. Expressions can be given to pass directly from any of the three sets (U, θ, ϕ), (U_x, U_y, U_z), (a, e, i) to any other (*Carusi et al.,* 1990; *Valsecchi et al.,* 1999).

In the case of meteoroids, U, θ, and ϕ can be computed from U_x, U_y, and U_z, which in turn can be calculated from the observed quantities that characterize a meteor, the geocentric velocity V_G, and the equatorial coordinates of the meteor radiant α_G and δ_G. In fact, we have

$$U = \frac{V_G}{29.8} \qquad (5)$$

$$\begin{pmatrix} U_x \\ U_y \\ U_z \end{pmatrix} = \hat{\mathbf{r}}(\lambda) \times \hat{\mathbf{p}}(\varepsilon) \times U \times \begin{pmatrix} -\cos\delta_G \cos\alpha_G \\ -\cos\delta_G \sin\alpha_G \\ -\sin\delta_G \end{pmatrix} \qquad (6)$$

where V_G is in km/s, $\hat{\mathbf{p}}(\varepsilon)$, $\hat{\mathbf{r}}(\lambda)$ are rotational matrixes around the x- and z-axis respectively, and the angle ε is the obliquity of the ecliptic plane to the plane of the celestial equator.

This set of geocentric parameters has an important property: U and θ are invariants of the problem in which the proper elements by *Gronchi and Milani* (2001) are defined. In fact, in the absence of close planetary encounters, the most important secular perturbation to take into account is the one related to the cycle of ω (*Kozai,* 1962). Assuming that all the perturbing planets are on circular and coplanar orbits, and furthermore assuming that the small body is not near any mean-motion or secular resonance, this secular perturbation leaves invariant the z component of the orbital angular momentum

$$L_z = \sqrt{a(1 - e^2)}\cos i \qquad (7)$$

and also, because of the assumed lack of close encounters, the specific orbital energy

$$E = -\frac{1}{2a}$$

But then

$$U = \sqrt{3 - \frac{1}{a} - 2\sqrt{a(1 - e^2)}\cos i} = \sqrt{3 - 2(L_z - E)}$$

is constant. However, if a and U are conserved, so is cos θ, since

$$\cos\theta = \frac{1 - U^2 - 1/a}{2U} \qquad (8)$$

(*Carusi et al.,* 1990). Thus, the point representing a meteoroid orbit in the U-cos θ plane does not change with time under the action of the Kozai perturbation even if both e and i undergo very large variations, such as in the case of the Quadrantids and δ Aquarids streams (*Hamid and Whipple,* 1963; *Babadzhanov and Obrubov,* 1993; *Valsecchi et al.,* 1999).

3.2. Meteoroid Association into Streams

Two meteoroids k and l are considered associated if the value of $D_{k,l}$ calculated by a distance function, does not exceed a certain threshold D_c. The choice of the threshold D_c is therefore the crucial point of any association procedure; its value can be estimated by equation (2) or the slight modification given by *Lindblad* (1971b). Both formulae are very easy to apply, but they are only approximations. According to *Jopek* (1993), the application of these formulae may not always be justified because of differing precision and statistical distributions in meteor data. Therefore, *Jopek and Froeschlé* (1997) developed a numerical procedure to obtain values of D_c for meteoroid association thresholds corresponding to the probability for the chance occurrence of a stream with a given number of members to occur in a random data sample.

A meteoroid stream may be considered as a concentration of points around some center representing the mean orbit of the stream. A number of different ways to define the concentration have been used:

1. A stream consists of the points inside a small hyperparallelepiped in the space of either the orbital elements (*Nilsson,* 1964) or the geocentric quantities V_G, α_G, δ_G, T, where T is the time of the meteor fall (*Kramer and Shestaka,* 1983).

2. A stream consists of the points inside a hypersphere of radius D_c, given *a priori*, centered at the mean orbit in the space of orbital elements, with the distance from the center evaluated with D (*Southworth and Hawkins,* 1963).

3. *Sekanina* (1970, 1976) also defined a meteoroid stream using D_{SH}, but instead of requiring that stream members be inside a hypersphere of radius D_c, he proposed an iterative procedure to determine the mean stream orbit. Consequently, a meteoroid stream is the set of points inside a hypersphere of radius D_c, but as one iterates, the center of the sphere moves until it converges to the final mean orbit.

4. *Southworth and Hawkins* (1963) used a hierarchical clustering based on the linking of the nearest neighbor. Starting from any point of the data sample, and accepting the link only if the distance is below a certain value D_c, this method can be applied either in the space of orbital elements using D_{SH} (*Southworth and Hawkins,* 1963; *Lindblad,* 1971a,b), or in the space of U, θ, φ, λ, using D_N (*Jopek et al.,* 1999).

The last definition has the advantage of not requiring any *a priori* orbital information on the meteoroid stream, while the others are useful for finding members of known streams that may exist in a given meteor sample.

The very first computer search by *Southworth and Hawkins* (1963) applied the second and fourth definitions to a sample of 359 meteoroid orbits. Since then, many authors have searched meteoroid orbits for streams, making use of one of the definitions given above in most cases. The number of identified streams has ranged from 7 to 275, with streams including from 7.5% to 83% of the searched orbital samples. This wide spectrum of results looks suspicious. The differing sizes of the meteor samples are one of the obvious reasons for the different numbers of streams detected. Other factors relate to observational selection effects, e.g., the location of the stations on Earth's surface and the timespan over which the data of a specific sample were collected.

At any rate, there are still a number of open problems: Which cluster analysis method is the best for a given meteor sample? What is the optimal way to find the threshold for meteoroid orbital similarity? What parameters should one use for this purpose? At the moment we do not have satisfactory answers to these questions, and the situation appears in some way similar to that of asteroid families in the 1980s (*Carusi and Valsecchi,* 1982; *Valsecchi et al.,* 1989), when only a small number of major families could be trusted and all the smaller ones looked questionable. Likewise, at present the reliability of most minor meteoroid streams is rather poor.

4. IDENTIFICATION OF PARENT BODIES

The fact that some comets are the parent bodies of meteoroid streams was first noticed in the nineteenth century. When meteoroid orbits began to be determined photographically, the origin of several meteoroid streams was established. The first hyphotheses about an asteroidal origin for some meteoroids can be found in works by Whipple and Hoffmeister. These hypotheses were not considered much more than speculations at the time they were put forward. [Nevertheless, Hoffmeister was always in favor of the asteroidal nature of some meteoroid streams (*Lindblad,* 1986).]

More thorough studies could take place as the number of NEA discoveries increased significantly. To our knowledge, *Sekanina* (1973, 1976) and *Drummond* (1981, 1982) were the first to make computer searches for meteoroid parent bodies among the known comets and NEAs. A notable result of these studies was the possibility that a single parent body (comet or asteroid) could give origin to more than one shower, as was confirmed by the search done by *Olson-Steel* (1988) among the southern hemisphere radio data. Furthermore, various studies (*Drummond,* 1991, 2000; *Obrubov,* 1991; *Shestaka,* 1994; *Babadzhanov,* 2001) have pointed to the existence of dynamical groups, called complexes, that include comets, asteroids, and meteoroid streams. A natural consequence of the establishment of these complexes is the hypothesis that their members originate from a single large parent body. However, many of these results seem to be uncertain, since some authors have given conflicting lists of members. The picture is further complicated since a number of NEAs are considered to be the exinct cometary nuclei, sometimes because of the presumed association with a meteoroid stream (see *Weissman et al.,* 1989).

In our opinion, these difficulties are similar to those described by *Valsecchi et al.* (1989) in the case of asteroid families, where disagreements between classifications were due to various causes, including the poor quality of proper elements and, in many cases, the absence of strict statistical criteria to assess the significance of the groupings. Improvements in both these areas over the last decade have led to a much more satisfactory situation at present (see *Bendjoya and Zappalà,* 2002). However, in the case of asteroidal meteoroid streams, we are still far behind.

Attempts to identify minor stream parent bodies among NEAs have been made with several methods using various meteor data. We prefer to discuss here the streams identified among the 865 most precise photographic meteors, in order to minimize the effects of generally low precision in meteoroid orbital data. The first search in this dataset of photographic meteors was performed by *Lindblad* (1971a) and was recently followed by another using D_N (*Jopek et al.,* 1999). The two searches give essentially equivalent results, except for some low-inclination streams, and in the following we discuss the streams found by *Jopek et al.* (1999). Table 1 reports the relevant data, i.e., the mean orbital and geocentric data for the streams, and the parent body, if known, of the streams with three or more members

TABLE 1. Stream orbital and geocentric data.

Stream Name	q	e	i	ω	Ω	V_G	θ	φ	Parent
Leonids	0.98	0.92	162	173	235	71	170	168	55P/Tempel-Tuttle
ε Geminids	0.81	0.96	174	231	203	70	166	258	C/1964N1 Ikeya*
Orionids	0.57	0.97	165	83	29	66	155	288	1P/Halley
Perseids	0.95	0.95	113	150	139	59	139	164	109P/Swift-Tuttle
σ Hydrids	0.24	0.98	126	122	78	58	137	295	
Lyrids	0.92	0.99	80	214	32	47	119	198	C/1861G1 Thatcher
Monocerotids	0.18	1.00	37	129	80	43	112	286	D/1917F1 Mellish
Quadrantids	0.98	0.68	72	171	282	41	116	176	96P/Machholz 1
S. δ Aquarids	0.08	0.97	27	151	308	41	118	278	96P/Machholz 1
N. δ Aquarids	0.10	0.95	21	328	142	38	117	262	96P/Machholz 1
Geminids	0.14	0.90	24	324	261	35	117	258	(3200) Phaethon
α Virginids (S)	0.32	0.87	7	118	198	31	103	275	
Taurids (N)	0.32	0.85	3	299	213	30	104	267	2P/Encke
Taurids (S)	0.34	0.82	6	118	27	28	104	275	2P/Encke
χ Orionids (N)	0.38	0.83	3	291	265	28	101	267	
α Virginids (N)	0.41	0.83	8	288	34	28	99	263	
κ Cygnids	0.98	0.76	39	202	147	25	88	194	
χ Orionids (S)	0.51	0.79	5	96	77	25	93	276	
α Capricornids (N)	0.58	0.78	6	268	134	23	89	262	45P/Honda-Mrkos-Pajdušáková
ε Piscids (N)	0.58	0.76	5	268	190	22	89	263	
ε Piscids (S)	0.61	0.73	4	85	5	21	88	276	
α Capricornids (S)	0.63	0.62	4	89	329	18	89	276	45P/Honda-Mrkos-Pajdušáková
α Pegasids	0.97	0.68	7	200	230	11	42	226	

* *Olsson-Steel* (1987) found that Comet C/1987B1 Nashikawa-Takamizawa-Tago is as good a candidate to be the parent of the ε Geminids as is C/1964N1 Ikeya (*Drummond,* 1981).

found by *Jopek et al.* (1999) using D_N among 865 precise photographic meteoroid orbits. The orbital elements i, ω, and Ω are given for B1950.0. All angles are in degrees, q is in AU, and V_G is in km/s. For streams identified as possessing both a northern and a southern branch, the data are given separately for each branch. The streams are listed in order of decreasing V_G, and the last column gives the presumed stream parent, taken from the literature.

There are at least two immediate conclusions that can be drawn from an inspection of Table 1: (1) In almost all cases in which a generally accepted association with a parent body has been established, the parent body is an active comet. To date, the only significant exception is the Geminid stream, whose parent is (3200) Phaeton. (2) Parent bodies have been found for almost all the streams with $V_G > 37$ km/s, and for very few others.

These two points are linked to each other because the geocentric velocities of comets are generally higher than those of asteroids. To better appreciate the situation, we can look at the distributions of meteoroids, comets, and asteroids in the U-cos θ plane as shown in Figs. 1, 2, and 3. In the U-cos θ plane, meteoroid orbits are not found everywhere below the parabolic limit curve, but rather in specific areas. The density of points varies significantly in the various regions of the plot, providing clues about their parent bodies. In fact, parent bodies might be found in the regions where the density of meteoroid orbits is high.

Looking at the distributions of comets and asteroids, onc notices that the vast majority of these two classes occupy almost disjoint regions of the U-cos θ plane, with essentially very little degree of overlap. Very few asteroids can be found above the line a = 3 AU, consistent with the fact that asteroids leaving the main belt on chaotic routes associated with the 3:1 and especially the 5:2 mean-motion resonances with Jupiter are quickly removed by encounters with

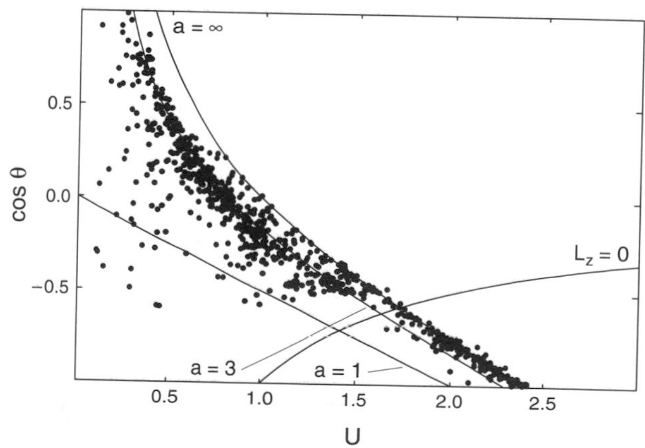

Fig. 1. The positions of 865 precisely measured photographic meteoroid orbits in the U-cos θ plane. Four continuous lines denote various conditions. The two curves going from cos θ = 1 to cos θ = −1 represent heliocentric parabolic orbits (upper curve) and orbits with a = 3 AU, the straight line on the lower left represents orbits with a = 1, and the remaining curve shows the condition $L_z = 0$.

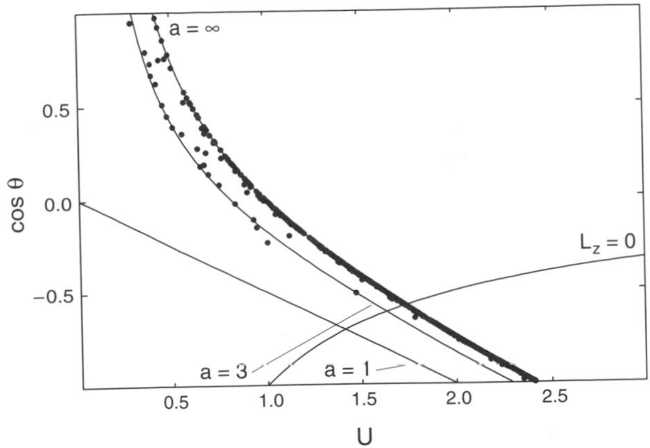

Fig. 2. The positions of comets in the U-cos θ plane. The various lines have the same meaning as in Fig. 1.

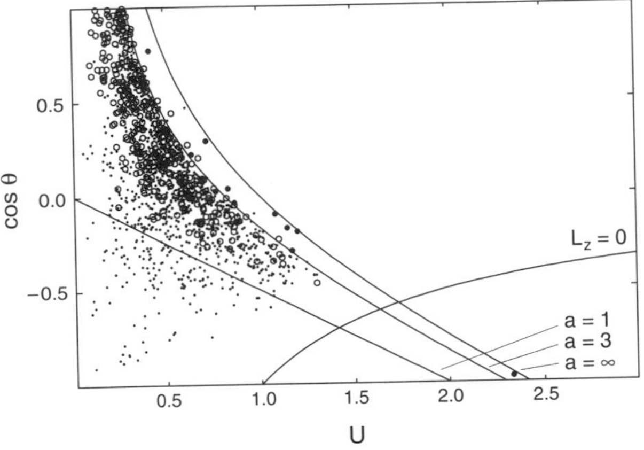

Fig. 3. The positions of Apollos, Atens, Amors, and other asteroids in unusual orbits in the U-cos θ plane. The various lines have the same meaning as in Fig. 1. Apollos and Atens are represented with small dots and are recognizable by their position with respect to the a = 1 line, Amors are represented by empty circles, and the asteroids on unusual orbits are represented by large full dots.

that planet as soon as their aphelion distances become large enough, as found in numerical integrations (*Farinella et al.,* 1994; *Valsecchi et al.,* 1995; *Gladman et al.,* 2000).

The opposite is true for comets, whose population lies almost entirely above the a = 3 AU line in a rather narrow band confined on the other side by the parabolic limit. In fact, not many Jupiter-family comets reach semimajor axes below the 2:1 mean-motion resonance with Jupiter at 3.27 AU, and only some of these can become Earth-crossing (*Fernàndez,* 1984). It must be noted that not all asteroidal and cometary orbits can be plotted on the U-cos θ plane, but only those that have values of E and L_z such that

$$T = \frac{1}{a} + 2\sqrt{a(1 - e^2)}\,\cos i \;\leq\; 3$$

$$\left|\frac{1 - U^2 - 1/a}{2U}\right| \;\leq\; 1$$

where T is the Tisserand parameter with respect to Earth. These are necessary conditions for a small body orbit to become Earth-crossing; whether it actually can become Earth-crossing depends on the value of the Kozai integral (*Kozai,* 1962), the third proper element of *Gronchi and Milani*'s (2001) theory.

Comparing the distribution of meteoroid orbits with those of asteroids and comets, it appears clear that a sizable fraction of meteoroids, those below the a = 3 AU line, must be of asteroidal origin, because there are less than a handful of comets below that line (one of them being Comet 2P/Encke). The meteoroids in this asteroidal region, however, are found far from the few comets here.

The preceding discussion has been based on the positions of individual orbits in the U-cos θ plane, and the condition a = 3 AU has turned out to be a reasonably good discriminator between asteroidal and cometary orbits in that diagram. One may then ask, why not simply adopt the same dividing line (at a = 3 AU) in orbital-element space when analyzing meteoroid orbits? The answer has to do with the observable quantities from which the orbital elements of meteoroids are deduced, and in particular with the strong nonlinearities existing in the relationships between the two sets of quantities.

As evidenced by equation (8), the semimajor axis of a meteoroid orbit can be determined by simply measuring the magnitude of U and the angle θ. However, for a given U, we have cos θ proportional to −1/a. Therefore, a small error in the measurement of U and/or θ has a large effect on a, as shown by the varying distances among the three lines a = 1 AU, a = 3 AU and a = ∞ in Figs. 1, 2, and 3. In particular, close to the line a = ∞, and especially for negative values of cos θ, an error of, say, 3 km/s can make the difference between a hyperbolic orbit and an orbit with a < 3 AU. The fraction of meteoroid orbits whose velocity is measured to better than this accuracy is very small in nonphotographic data samples. This suggests that in order to make a meaningful search for asteroidal meteoroid streams, only good-quality photographic meteors should be used. Unfortunately, the best such unbiased data is the 865-meteor Harvard dataset discussed above, collected almost half a century ago.

5. DISCUSSION AND CONCLUSIONS

The problem of the identification of asteroidal meteoroid streams seems to be still largely unresolved. While the comparison of Figs. 1 and 3 shows that a good fraction of the photographic meteors is in the same region occupied by asteroids, Table 1 shows that the majority of low-velocity streams, those for which an asteroidal origin is more likely, still lack an identified parent body. This situation is of course unsatisfactory, but unfortunately it does not seem likely that it will change in the near future. Although it is true that the study of the dynamics of objects in Earth-crossing orbits has recently made substantial progress, both on the numerical and on the theoretical side, and we can expect an improvement in the dynamical criteria for stream identification, the obstacle seems to be the availability of high-precision meteoroid orbits.

As mentioned, the only dataset of photographic meteors with the quality needed for meaningful work on this subject is composed of only 865 orbits, and was produced a long time ago. Because of the limitations of the then-available technology, only bright meteors could be recorded, thus limiting the number of slow meteoroid orbits in the sample, which are precisely the most interesting for us since the meteor magnitude is a very steep function of V_G (*Hughes and Williams,* 2000). We therefore need a new campaign aimed at obtaining high-precision meteoroid orbits in substantial numbers, and extending over several years in order to sample the vicinity of the orbit of Earth at all longitudes. The number of orbits, as well as their precision, must be considerably higher than those in the Harvard dataset if we want to have the resolution needed to distinguish among the ever-growing number of NEAs. These goals should be achieved using precise optical and radar techniques from not only the northern hemisphere but from the southern hemisphere as well, since NEAs approaching Earth do so from all directions.

Acknowledgments. We thank referees D. Steel and J. Drummond for their constructive criticism, and especially J. Drummond for his help in streamlining the paper. T.J.J. and G.B.V. worked on this subject during extended stays at the Observatoire de la Côte d'Azur, whose hospitality is gratefully acknowledged.

REFERENCES

Babadzhanov P. B. (2001) Search for meteor showers associated with near-Earth asteroids. I. Taurid complex. *Astron. Astrophys., 373,* 329–335.

Babadzhanov P. B. and Obrubov Yu. V. (1980) Evolution of orbits and intersection conditions with the Earth of the Geminid and Quadrantid meteor showers. In *Solid Particles in the Solar System* (I. Halliday and B. A. McIntosh, eds.), pp. 157–162. Proc. IAU Symp. 90, Kluwer, Dordrecht.

Babadzhanov P. B. and Obrubov Yu. V. (1993) Dynamics and relationships between interplanetary bodies. I. Comet Machholz and its meteor showers. In *Meteoroids and their Parent Bodies* (J. Štohl and I. P. Williams, eds.), pp. 49–52. Slovakian Academy of Sciences, Bratislava.

Bendjoya Ph. and Zappalà V. (2002) Asteroid family identification. In *Asteroids III* (W. F. Bottke Jr. et al., eds.), this volume. Univ. of Arizona, Tucson.

Carusi A. and Valsecchi G. B. (1982) On asteroid classifications in families. *Astron. Astrophys., 115,* 327–335.

Carusi A., Valsecchi G. B., and Greenberg R. (1990) Planetary close encounters — Geometry of approach and post-encounter orbital parameters. *Cel. Mech. Dyn. Astron., 49,* 111–131.

Drummond J. D. (1981) A test of comet and meteor shower associations. *Icarus, 45,* 545–553.

Drummond J. D. (1982) Theoretical meteor radiants of Apollo, Amor, and Aten asteroids. *Icarus, 49,* 149–153.

Drummond J. D. (1991) Earth-approaching asteroid streams. *Icarus, 89,* 14–25.

Drummond J. D. (2000) The D discriminant and near-Earth asteroid streams. *Icarus, 146,* 453–475.

Farinella P., Froeschlé Ch., Froeschlé Cl., Gonczi R., Hahn G., Morbidelli A., and Valsecchi G. B. (1994) Asteroids falling onto the Sun. *Nature, 371,* 314–317.

Fernàndez J. A. (1984) The distribution of the perihelion distances of short-period comets. *Astron. Astrophys., 135,* 129–134.

Froeschlé Cl., Gonczi R., and Rickman H. (1993) New results on the connection between comet P/Machholz and the Quadrantid meteor stream: Poynting-Robertson drag and chaotic motion. In *Meteoroids and their Parent Bodies* (J. Štohl and I. P. Williams, eds.), pp. 269–272. Slovakian Academy of Sciences, Bratislava.

Gladman B., Michel P., and Froeschlé Ch. (2000) The near-Earth object population. *Icarus, 146,* 176–189.

Gronchi G. F. and Milani A. (2001) Proper elements for Earth crossing asteroids. *Icarus, 152,* 58–69.

Gustafson B. A. S. (1989) Comet ejection and dynamics of nonspherical dust particles and meteoroids. *Astrophys. J., 337,* 945–949.

Hamid S. E. and Whipple F. L. (1963) Common origin between the Quadrantids and the δ Aquarids streams. *Astron. J., 68,* 537.

Hawkins G. S. and Southworth R. B. (1958) *Smithson. Contrib. Astrophys., 2,* 349–364.

Hawkins G. S. and Southworth R. B. (1961) *Smithson. Contrib. Astrophys., 4,* 85–95.

Hughes D. W. and Williams I. P. (2000) The velocity distributions of periodic comets and stream meteoroids. *Mon. Not. R. Astron. Soc., 315,* 629–634.

Jacchia L. G. and Whipple F. L. (1961) Precision orbits of 413 photographic meteors. *Smithson. Contrib. Astrophys., 4,* 97–129.

Jopek T. J. (1993) Remarks on the meteor orbital similarity D-criterions. *Icarus, 106,* 603–607.

Jopek T. J. and Froeschlé Cl. (1997) A stream search among 502 TV meteor orbits. An objective approach. *Astron. Astrophys., 320,* 631–641.

Jopek T. J., Valsecchi G. B., and Froeschlé Cl. (1999) Meteoroid streams identification: A new approach. II Application to 865 photographic meteor orbits. *Mon. Not. R. Astron. Soc., 304,* 751–758.

Kozai Y. (1962) Secular perturbations of asteroids with high inclination and eccentricity. *Astron. J., 67,* 591.

Kramer E. N. and Shestaka I. S. (1983) *Meteornaja material v atmosferie ziemli I okolosolnetchnom kosmitcheskom prostranstvie.* Nauka, Moskva. 136 pp.

Lindblad B. A. (1971a) A computerized stream search among 865 precise photographic meteor orbits. *Smithson. Contrib. Astrophys., 12,* 1–12.

Lindblad B. A. (1971b) A computerized stream search among 2401 photographic meteor orbits. *Smithson. Contrib. Astrophys., 12,* 14–24.

Lindblad B. A. (1986) Physics and orbits of meteoroids. In *The Evolution of the Small Bodies of the Solar System* (M. Fulchignoni and Ľ. Kresák, eds.), p. 229. Proc. of the Enrico Fermi Summer School, North-Holland, Amsterdam.

Lindblad B. A. (1987) The IAU Meteor Data Center in Lund. In *Interplanetary Matter* (Z. Ceplecha and P. Pecina, eds.), pp. 201–204. Proc. of 10th European Regional Astronomy Meeting of the IAU, Praha, Czechoslovakia.

Lindblad B. A. (1991) The IAU Meteor Data Center in Lund. In *Origin and Evolution of Interplanetary Dust* (A. C. Levasseur-Regourd and H. Hasegawa, eds.), pp. 311–314. Proc. IAU Colloq. 126, Kluwer, Dordrecht.

Lindblad B. A. (1999) A survey of meteoroid orbits obtained by two-station video-observations. In *Meteoroids 1998* (W. J. Baggaley and V. Porubcan, eds.), pp. 274–282. Astr. Inst. Slovak Acad. Sci., Bratislava.

Lindblad B. A. and Steel D. I. (1994) Meteoroid orbits available from the IAU Meteor Data Center. In *Asteroids, Comets, Meteors 1993* (A. Milani et al., eds.), pp. 497–501. Proc. IAU Symp. 160, Kluwer, Dordrecht.

Nilsson C. S. (1964) A southern hemisphere radio survey of meteor streams. *Austral. J. Phys., 17,* 205.

Obrubov Y. V. (1991) Complexes of minor bodies in the solar system. *Astron. Zh., 68,* 1063–1073.

Olsson-Steel D. (1987) Comet Nishikawa-Takamizawa-Tago (1987c) and the Epsilon Geminid meteor shower. *Mon. Not. R. Astron. Soc., 228,* 23P–28P.

Olsson-Steel D. (1988) Identification of meteoroid streams from Apollo asteroids in the Adelaide radar orbit surveys. *Icarus, 75,* 64–96.

Öpik E. J. (1976) *Interplanetary Encounters: Close-Range Gravitational Interactions.* Elsevier, New York. 155 pp.

Sekanina Z. (1970) Statistical model of meteor streams. I — Analysis of the model. *Icarus, 13,* 459–474.

Sekanina Z. (1973) Statistical model of meteor streams. III — Stream search among 19303 radio meteors. *Icarus, 18,* 253–284.

Sekanina Z. (1976) Statistical model of meteor streams. IV — A study of radio streams from the synoptic year. *Icarus, 27,* 265–321.

Shestaka I. S. (1994) Asteroidno-meteornyje kompleksy. *Astron. Vestnik, 28,* 85–03.

Southworth R. B. and Hawkins G. S. (1963) Statistics of meteor streams. *Smithson. Contrib. Astrophys., 7,* 261–285.

Steel D. I. (1996) Meteoroid orbits. *Space Sci. Rev., 78,* 507–553.

Valsecchi G. B., Carusi A., Knežević Z., Kresák Ľ., and Williams J. G. (1989) Identification of asteroid dynamical families. In *Asteroids II* (R. P. Binzel et al., eds.), pp. 368–385. Univ. of Arizona, Tucson.

Valsecchi G. B., Morbidelli A., Gonczi R., Farinella P., Froeschlé Ch., and Froeschlé Cl. (1995) The dynamics of objects in orbits resembling that of P/Encke. *Icarus, 118,* 169–180.

Valsecchi G. B., Jopek T. J., and Froeschlé Cl. (1999) Meteoroid streams identification: A new approach. I. Theory. *Mon. Not. R. Astron. Soc., 304,* 743–750.

Weissman P. R., A'Hearn M. F., and McFadden L. A. (1989) Evolution of comets into asteroids. In *Asteroids II* (R. P. Binzel et al., eds.), pp. 743–750. Univ. of Arizona, Tucson.

Whipple F. L. (1950) A comet model. I. The acceleration of Comet Encke. *Astrophys. J., 111,* 375–394.

Whipple F. L. (1951) A comet model. II. Physical relations for comets and meteors. *Astrophys. J., 113,* 464.

Whipple F. L. (1954) Photographic meteor orbits and their distribution in space. *Astron. J., 59,* 201–217.

Meteoritic Parent Bodies: Their Number and Identification

Thomas H. Burbine
Smithsonian Institution

Timothy J. McCoy
Smithsonian Institution

Anders Meibom
Stanford University

Brett Gladman
Observatoire de la Côte d'Azur

Klaus Keil
University of Hawai'i

Extensive collection efforts in Antarctica and the Sahara in the past 10 years have greatly increased the number of known meteorites. Groupings of meteorites according to petrologic, mineralogical, bulk- chemical, and isotopic properties suggest the existence of 100–150 distinct parent bodies. Dynamical studies imply that most meteorites have their source bodies in the main belt and not among the near-Earth asteroids. Spectral observations of asteroids are currently the primary way of determining asteroid mineralogies. Linkages between ordinary chondrites and S asteroids, CM chondrites and C-type asteroids, the HEDs and 4 Vesta, and iron meteorites, enstatite chondrites, and M asteroids are discussed. However, it is difficult to conclusively link most asteroids with particular meteorite groups due to the number of asteroids with similar spectral properties and the uncertainties in the optical, chemical, and physical properties of the asteroid regolith.

1. INTRODUCTION

Since the publication of *Asteroids II*, our knowledge concerning the composition, orbital dynamics, and geologic histories of meteorites and asteroids has increased substantially. These advances in meteoritics have been primarily due to the thousands of new meteorites discovered in Antarctica (e.g., *Zolensky,* 1998) and the Sahara (e.g., *Bischoff,* 2001) coupled with more precise and sensitive analytical equipment to characterize them. The near-exponential increase in computational power and improvements in integration algorithms have allowed for vastly improved studies of the transport of meteorites from the asteroid belt. Our knowledge of the diversity of asteroid mineralogies has increased considerably from the use of charge-coupled device (CCD) detectors to obtain visible and near-infrared reflectance spectra of smaller and smaller objects.

With these advances, it may be possible to determine parent bodies for particular meteorites. Geochemical and petrological evidence (e.g., *Lipschutz et al.,* 1989) implies that almost all meteorites originated on subplanetary-sized objects that formed ca. 4.56 Ga. The exceptions are more than 50 unpaired meteorites that appear to originate from the Moon and Mars. Meteorites from other sources such as Mercury (*Love and Keil,* 1995), Venus (*Melosh and Tonks,* 1993), comets (e.g., *Campins and Swindle,* 1998), and even Earth (*Melosh and Tonks,* 1993) also appear possible (*Gladman et al.,* 1996). Interplanetary dust particles (IDPs) (e.g., *Rietmeijer,* 1998; *Dermott et al.,* 2002) sample comets and also asteroids. Meteorites from planets in other stellar systems appear to be very unlikely (*Melosh,* 2001).

These linkages between asteroids and meteorites are important for a variety of reasons. Scientifically, they allow for an understanding of compositional and thermal gradients in the solar nebula by allowing the orbital locations of objects with different chemical and isotopic compositions to be pinpointed. Economically, many Fe-rich asteroids could be important resources for elements relatively rare on Earth's surface (*Kargel,* 1994). For the preservation of the human race, an asteroid on a collision course with Earth may have to be diverted or destroyed, and it would be vital to know the object's composition when formulating scenarios for keeping the approaching body from impacting Earth.

The goal of this chapter is to determine the number and identity of meteoritic parent and/or source bodies. A parent body is the body from which the meteorite acquired its current chemical and mineralogical characteristics. A source body is a fragment of the parent body by which the meteorite was completely shielded from cosmic rays for most of

TABLE 1. Meteorite groups and their compositional characteristics.

Groups	Composition*	$\Delta^{17}O$†	Compositional Linkages with Other Groups
Carbonaceous Chondrites			
CI	phy, mag	0	
CM	phy, toch, ol,	–3	
CO	ol, px, CAIs, met	–4	
CR	phy, px, ol, met	–1.5	
CH	px, met, ol,	–1.5	
CV	ol, px, CAIs	–4	
CK	ol, CAIs	–4	
Enstatite Chondrites			
EH	enst, met, sul, plag, ±ol	0	
EL	enst, met, sul, plag	0	
Ordinary Chondrites			
H	ol, px, met, plag, sul	0.73	IIE (*Olsen et al.,* 1994)
L	ol, px, plag, met, sul	1.07	
LL	ol, px, plag, met, sul	1.26	
R Chondrites			
R	ol, px, plag, sul	2.7	
Primitive Achondrites			
Acapulcoites	px, ol, plag, met, sul	–1.04	lodranites (*Nagahara and Ozawa,* 1986)
Lodranites	px, ol, met, ±plag, ±sul	–1.18	acapulcoites
Winonaites	ol, px, plag, met	–0.50	IAB, IIICD (*Bild,* 1977)
Differentiated Achondrites			
Angrites	TiO_2-rich aug, ol, plag	–0.15	
Aubrites	enst, sul	0.02	
Brachinites	ol, cpx, ±plag	–0.26	
Diogenites	opx	–0.27	eucrites, howardites (*Mason,* 1962)
Eucrites	pig, plag	–0.24	howardites, diogenites
Howardites	eucritic-diogenitic breccia	–0.26	eucrites, diogenites
Ureilites	ol, px, graph	–1.20	
Stony-Irons			
Mesosiderites	basalt-met breccia	–0.24	
Main group pallasites	ol, met	–0.28	
Irons			
IAB	met, sul, ol-px,-plag incl	–0.48	IIICD, winonaites
IC	met, sul		
IIAB	met, sul, schreib		
IIC	met, sul		
IID	met, sul		
IIE	met, sul, ol-px-plag incl	0.59	H chondrites
IIF	met, sul		
IIIAB	met, sul	–0.21	
IIICD	met, sul, ol-px-plag incl	–0.43	IAB, winonaites
IIIE	met, sul		
IIIF	met, sul		
IVA	met, SiO_2-px incl	1.17	
IVB	met		

*Minerals or components are listed in decreasing order of average abundance. Abbreviations: ol = olivine, px = pyroxene, opx = orthopyroxene, pig = pigeonite, enst = enstatite, aug = augite, cpx = clinopyroxene, plag = plagioclase, mag = magnetite, met = metallic iron, sul = sulfides, phy = phyllosilicates, toch = tochilinite, graph = graphite, CAIs = Ca-Al-rich refractory inclusions, schreib = schreibersite, incl = inclusions, ± = may be present.

† Average values (e.g., *Clayton et al.,* 1991; *Clayton and Mayeda,* 1996, 1999) for $\Delta^{17}O$, where $\Delta^{17}O = \delta^{17}O - 0.52 \times \delta^{18}O$.

the solar system's history and from which the meteorite was recently liberated at a time measured by its recent cosmic-ray-exposure age.

We first detail the number of postulated distinct parent bodies. This is followed by a short discussion on the dynamical issues for delivering meteorites to Earth. We then discuss the telescopic and spacecraft data used to identify meteoritic parent bodies and some proposed asteroid-meteorite linkages. We conclude the chapter with a discussion of the work that needs to be done to better identify meteoritic parent bodies.

2. THE DIVERSITY OF METEORITES AND THEIR PARENT BODIES

Meteorites provide the most tangible evidence of the chemical and physical makeup of asteroids, the processes by which asteroids were formed and modified during the history of the solar system, and the number of distinct asteroidal bodies for which we currently have samples. Classification of the more than 22,500 known meteorites (*Grady,* 2000) provides a means for grouping samples formed from common constituents and processes and, more importantly for the purposes of this discussion, from a likely common parent body.

The bulk composition, mineralogy, and petrology of a meteorite are functions of the original bulk composition of its parent body and the amount of heating and melting it has experienced. The precursor assemblages ranged from highly reduced to highly oxidized material. Meteorites can be broken into two types: those that experienced heating but not melting (chondrites) and those that experienced melting and differentiation (achondrites, primitive achondrites, stony-irons, irons). A more detailed discussion of chondritic meteorites can be found in *Brearley and Jones* (1998) and a more complete discussion of differentiated meteorites can be found in *Mittlefehldt et al.* (1998).

Chondrites are the most primitive material in the solar system. Mild thermal metamorphism and aqueous alteration has not destroyed the nebular components they contain or erased the record of their formation in the solar nebula ca. 4.56 Ga. On the other hand, differentiated meteorites have experienced significant degrees of melting, leading to an erasing of most of the evidence of their chondritic precursors.

2.1. Classification

Meteorites that are similar in terms of petrologic, mineralogical, bulk-chemical, and isotopic properties are separated into groups (Table 1). Particularly important parameters for the classification of silicate-bearing meteorites include refractory lithophile ("oxygen-loving") elements (e.g., Ca, Al, Ti), the FeO concentration in olivine ($(Fe,Mg)_2SiO_4$) and orthopyroxene ($(Fe,Mg)_2Si_2O_6$), and the whole-rock O-isotopic composition (Figs. 1 and 2). Iron meteorites are classified according to siderophile ("iron-loving") element (Ga, Ge, Ir, Ni) concentrations. It is noteworthy that the proper-

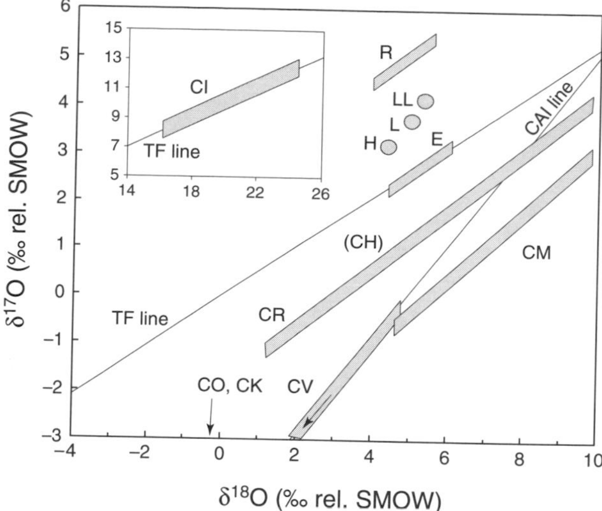

Fig. 1. Oxygen-isotopic compositions relative to standard mean ocean water (SMOW) for the chondritic meteorite groups. The terrestrial fractionation (TF) line and the refractory inclusion (CAI) mixing line are also plotted. CI chondrites are plotted in the inset in the upper left part of the diagram. The CO and CK chondrites plot off the diagram along the CAI line. The label for the CH chondrites is in parentheses because the CH chondrites occupy only part of the CR chondrite region. The region labeled E contains both the EH and EL chondrites. Figure revised from *Meibom and Clark* (1999).

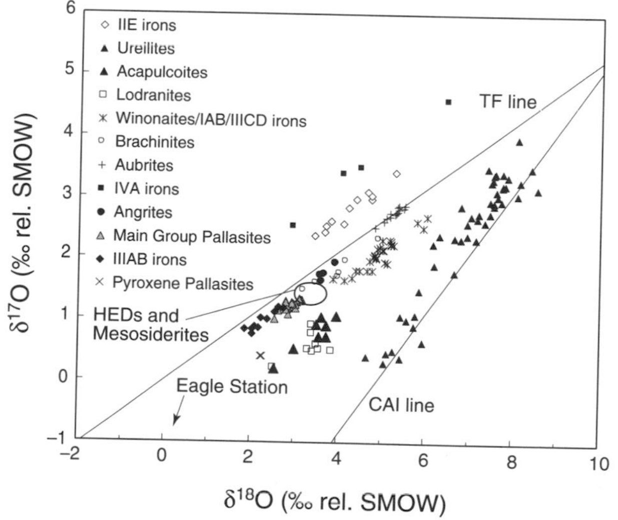

Fig. 2. Oxygen-isotopic compositions relative to standard mean ocean water (SMOW) for the differentiated meteorite groups plus the members of the pyroxene-pallasite grouplet. The terrestrial fractionation (TF) line and the refractory inclusion (CAI) mixing line are also plotted. Members of the Eagle Station pallasite grouplet ($\delta^{18}O \approx -2.80‰$, $\delta^{17}O \approx -6.15‰$) plot off the diagram. The standard deviations for $\delta^{18}O$ and $\delta^{17}O$ are ~0.1‰ for individual points. Figure revised from *Meibom and Clark* (1999).

TABLE 2. Meteorite groups and their postulated parent or source bodies.

Group	Fall Percentage (%)*	Postulated Parent or Source Bodies†
L	38.0	S(IV) asteroids (*Gaffey et al.,* 1993)
H	34.1	6 Hebe [S(IV)] (*Gaffey and Gilbert,* 1998)
LL	7.9	S(IV) asteroids (*Gaffey et al.,* 1993)
Irons	4.2	M asteroids (*Cloutis et al.,* 1990; *Magri et al.,* 1999)
Eucrites	2.7	4 Vesta (V) (*Consolmagno and Drake,* 1977; *Drake,* 2001)
Howardites	2.1	4 Vesta (V) (*Consolmagno and Drake,* 1977; *Drake,* 2001)
CM	1.7	19 Fortuna (G, Ch) (*Burbine,* 1998)
Diogenites	1.2	4 Vesta (V) (*Consolmagno and Drake,* 1977; *Drake,* 2001)
Aubrites	1.0	3103 Eger (E) (*Gaffey et al.,* 1992)
EH	0.8	M asteroids (*Gaffey and McCord,* 1978)
EL	0.7	M asteroids (*Gaffey and McCord,* 1978)
Mesosiderites	0.7	M asteroids (*Gaffey et al.,* 1993)
CV	0.6	K asteroids (*Bell,* 1988)
CI	0.5	C asteroids (*Gaffey and McCord,* 1978)
CO	0.5	221 Eos (K) (*Bell,* 1988)
Pallasites	0.5	A asteroids (*Cruikshank and Hartmann,* 1984; *Lucey et al.,* 1998)
Ureilites	0.5	S asteroids (*Gaffey et al.,* 1993)
"Martian"	0.4	Mars (*McSween,* 1994)
CR	0.3	C asteroids (*Hiroi et al.,* 1996)
CK	0.3	C asteroids (*Gaffey and McCord,* 1978)
Acapulcoites	0.1	S asteroids (*McCoy et al.,* 2000)
Angrites	0.1	S asteroids (*Burbine et al.,* 2001a)
Lodranites	0.1	S asteroids (*Gaffey et al.,* 1993; *McCoy et al.,* 2000)
R	0.1	A or S asteroids
Winonaites	0.1	S asteroids (*Gaffey et al.,* 1993)
(Tagish Lake)‡	0.1	D asteroids (*Hiroi et al.,* 2001)
Brachinites	Only finds	A asteroids (*Cruikshank and Hartmann,* 1984; *Sunshine et al.,* 1998)
CH	Only finds	C or M asteroids
"Lunar"	Only finds	Moon (*Warren,* 1994)

*Fall percentages are calculated from the 942 classified falls that are listed in *Grady* (2000), *Grossman* (2000), and *Grossman and Zipfel* (2001).

†Asteroid classes are a combination of those of *Tholen* (1984), *Gaffey et al.* (1993), and *Bus* (1999).

‡Tagish Lake is a newly discovered type of carbonaceous chondrite (*Brown et al.,* 2000) and is listed in the table because of its spectral similarity to D asteroids.

ties most useful for classifying meteorites are not readily measured by Earth-based remote-sensing techniques.

Oxygen-isotopic data (Figs. 1 and 2) are very useful for distinguishing different meteorite groups because samples from the same parent body will cluster together or tend to fall on a mass-fractionation line with a slope of 0.52 (e.g., *Clayton,* 1993). The exceptions are the CK, CO, and CV chondrites, which fall on a line with a slope of ~1 due to the mixing of two distinct nebular components (*Clayton,* 1993). We note that although a common O-isotopic composition is often taken as evidence of a common parent-body origin, it only requires formation in a similar region and does not preclude separate parent bodies.

In general, meteorite groups contain five or more members. Grouplets contain two to four members. Meteorites that do not fit in any well-defined group are termed anomalous or ungrouped. Each group is generally thought to represent a distinct parent body, although some cases exist where multiple meteorite groups apparently originated from a common body (Table 1). It is more difficult to interpret the origin of anomalous or ungrouped meteorites, but many of these are thought to be the only samples of their respective parent bodies in our meteorite collections.

2.2. Chondrites

Currently, 13 groups (Table 1) of chondritic meteorites have been defined, largely based on refractory lithophile element abundances and O-isotopic compositions. Chondrites, particularly ordinary chondrites (~80% of all falls), dominate the current flux of meteorite landings (Table 2). The classification of ordinary and enstatite chondrites has changed little in the past 10 years, with ordinary chondrites divided into the classic H, L, and LL groups, and enstatite (FeO-poor pyroxene) chondrites divided into the EH and EL groups on the basis of total Fe content.

In contrast, several new groups of carbonaceous chondrites have been defined. In addition to the traditional CI,

CM, CO, and CV groups, researchers now recognize the CK group of thermally metamorphosed carbonaceous chondrites and the metal-rich CH and CR groups. Also recently recognized are the highly oxidized R chondrites (e.g., *Schulze et al.,* 1994). Chondritic groups are subdivided according to petrologic type (1–6), with 1 being the most aqueously altered, 3.0 being the least altered, and 6 being the most thermally altered. Some CI1 and CM2 chondrites have also undergone some late-stage thermal metamorphism where heating to 500°–700°C (Lipschutz *et al.,* 1999) has dehydrated their phyllosilicates.

In addition to the 13 well-defined groups, ~14 chondritic grouplets or unique meteorites (*Meibom and Clark,* 1999; *Brown et al.,* 2000; *Weisberg et al.,* 2001) have been recognized. Thus, the full range of chondritic meteorites probably requires at least 27 distinct parent bodies.

2.3. Differentiated Meteorites

In contrast to chondrites, the differentiated meteorites (Table 1) represent a much larger number of meteorite parent bodies. The differentiated meteorites range from those that experienced only limited differentiation (primitive achondrites) to those (achondrites, stony-irons, irons) that were produced by extensive melting, melt migration, and fractional crystallization. Although classification of differentiated meteorites is relatively straightforward, igneous differentiation by its very nature can produce radically different lithologies of a common parent body.

2.3.1. Primitive achondrites. Among the primitive achondrites, the acapulcoites and lodranites appear to have originated on a single parent body, as evidenced by the transitional nature of some members of these groups. The winonaites, which tend to be similar in mineralogy and chemistry to acapulcoites and lodranites, appear to sample another parent body based on distinct O-isotopic compositions.

2.3.2. Achondrites. Among the fully differentiated achondrites, the basaltic angrites (containing a TiO_2-rich augite), the ultrareduced, pyroxenitic aubrites, the olivine-dominated brachinites, and the C-rich ureilites each require a unique parent body. The basaltic eucrites and orthopyroxenitic diogenites appear to sample a common parent body, as evidenced by the occurrence of polymict breccias known as howardites, which contain both eucritic and diogenitic material. These meteorites are referred to as the HEDs. However, the recent discovery (*Yamaguchi et al.,* 2002) of a eucrite with an O-isotopic value very different (near the CR region) from the HEDs argues for the formation of another HED-like body in the belt.

2.3.3. Stony-iron meteorites. Stony-iron meteorites include the pallasites and mesosiderites. All pallasites are composed primarily of centimeter-sized olivine grains imbedded in metallic Fe. They differ in terms of O-isotopic composition, olivine composition, and pyroxene abundance, and these features have been used to delineate the main group, the Eagle Station grouplet, and the pyroxene-pallasite grouplet, each of which requires a separate parent body. Pallasites are generally thought to be fragments of the core-mantle boundary.

Mesosiderites are breccias composed of HED-like clasts of basaltic to orthopyroxenitic material mixed with metallic clasts. They likely formed by impact mixing of a core fragment on the surface of a basaltic asteroid. Their O-isotopic compositions are indistinguishable from the HEDs. Whether they originated on the same parent body as the HEDs or sample yet another basaltic asteroid remains uncertain.

2.3.4. Iron meteorites. The number of types of differentiated meteorites is roughly doubled by 13 groups of iron meteorites defined by siderophile-element (Ga, Ge, Ir, Ni) compositions. The large differences in the most volatile siderophile elements (Ga and Ge) between groups suggest that each iron group formed in a separate parent body. Evidence that these irons existed in cores include fractional crystallization trends suggestive of prolonged cooling in 10 of the groups (excluding IAB, IIE, and IIICD) and the subsolidus exsolution of Ni-rich and Ni-poor phases (the familiar Widmanstätten pattern) requiring cooling of 1–100 K/m.y. at low temperatures.

The IAB, IIE, and IIICD irons all contain abundant silicate inclusions and do not display well-developed fractional crystallization trends. The IAB irons and primitive achondritic winonaites appear to sample a common parent body and also have been linked with the IIICD irons (e.g., *Mittlefehldt et al.,* 1998). H chondrites and silicate inclusions in IIE irons share similar bulk and mineral compositions, textures, and O-isotopic compositions (*Olsen et al.,* 1994; *Casanova et al.,* 1995) and appear to be related.

The largest number of distinct parent bodies is probably sampled by the ungrouped irons, which numbered 95 in *Grady* (2000) and account for ~10% of known irons. *Wasson* (1995) argues that the ungrouped irons required ~70 distinct parent bodies. However, some of these ungrouped irons are probably related to the 13 major groups, perhaps representing an extreme composition produced by fractional crystallization from a poorly represented parent body. We suggest that these ungrouped irons more likely represent ~50 distinct parent bodies.

The large number of ungrouped irons appears to be due to the strength of metallic Fe and its resistance to terrestrial weathering compared to silicates. These two factors allow for more parent bodies rich in metallic Fe to be sampled in our meteorite collections. Irons have cosmic-ray exposure ages ranging from hundreds of millions to a few billion years (e.g., *Voshage and Feldman,* 1979) while stony meteorites tend to have much shorter exposure ages of <100 m.y. (e.g., *Marti and Graf,* 1992; *Scherer and Schultz,* 2000). (Cosmic-ray exposure ages record the time an object has spent as a meter-sized (or less) body in space or within a few meters of the surface.) The greater physical strength of irons allows them to survive as meter-scale objects in space much longer than silicate bodies. The relative resistance of irons to terrestrial weathering increases the probability that iron meteorites that fall to Earth will survive to the present day.

2.4. Number of Parent Bodies

In total, our meteorite collections could represent as few as ~100 distinct asteroidal parent bodies (~27 chondritic, ~2 primitive achondritic, ~6 differentiated achondritic, ~4 stony-iron, ~10 iron groups, ~50 ungrouped irons) or perhaps as many as 150, if assumed relationships between ungrouped iron meteorites prove untrue. This number is evolving because of thousands of new meteorites collected each year, primarily from Antarctica and the Sahara, resulting in a continuous supply of new samples that expand our perception of the diversity of the asteroid belt.

It is interesting to note the enormous disparity between the 100–150 meteorite parent bodies sampled on Earth and the approximately 1,000,000 asteroids (e.g., *Ivezić et al.,* 2001) in the main belt with diameters greater than 1 km. Although we cannot expect to have sampled the vast number of asteroids in their entirety, two explanations for this discrepancy are worth noting. First, meteorite researchers focus on "parent bodies," the primordial asteroids as they existed in the first tens of millions of years of solar system history. In contrast, asteroid researchers study fragments produced by 4.5 b.y. of impact and fragmentation. A single "parent body" may have produced tens, hundreds, or thousands of current asteroids. Secondly, although we commonly refer to a single parent body (e.g., the H-chondrite parent body), we have no direct evidence that rules out multiple asteroids composed of essentially identical material. Thus, our meteorite collections may well sample a large percentage of the types of materials present in the asteroid belt, despite the apparent mismatch.

We also do not understand how biased our meteorite collection is compared to the asteroid belt as a whole. Many types of carbonaceous chondritic material may be too weak (e.g., *Sears,* 1998) to be able to make it through the atmosphere to Earth's surface as meteorites and may only be sampled as IDPs. It is also unclear, as discussed in the next section, how well the dynamical mechanisms that deliver meteorites to Earth sample the asteroid belt.

3. ASTEROIDS TO EARTH: THE DYNAMIC CONNECTION

Since the review of meteorite transport in *Asteroids II* (*Greenberg and Nolan,* 1989), advances in computer speed and numerical algorithms have produced several surprising revelations. *Wetherill* (1985) showed that the expected flux from continual interasteroid collisions in the main belt could be sufficient to push the necessary mass of ejected meteorite-sized bodies into the main orbital resonances. These resonances would then be responsible for increasing the orbital eccentricity to planet-crossing values, allowing some fraction of the material to impact Earth. However, modern computer power has allowed *Farinella et al.* (1994) to show that the ν_6 resonance (Fig. 3) could rapidly raise orbital eccentricities to 1, leading to collision with the Sun. *Gladman et al.* (1997) show this result to be generic for all main orbital resonances in the inner belt out to the 3:1 resonance

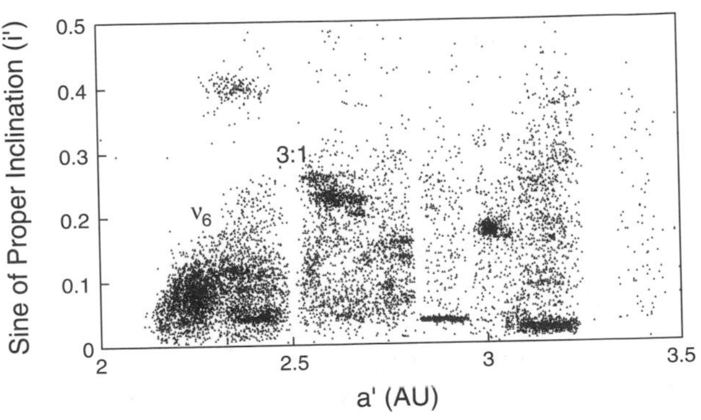

Fig. 3. Proper semimajor axis (a') (AU) vs. sine of proper inclination (i') for the first ~10,000 numbered main-belt asteroids. The 3:1 and ν_6 resonances are labeled.

at 2.5 AU (Fig. 3). Outside 2.5 AU, most emerging meteoroids fall under the control of Jupiter and are ejected from the solar system, strongly reducing the fractional contribution from the outer belt to the meteorite flux at Earth.

Dynamical studies of meteorite delivery are most easily constrained by the cosmic-ray-exposure ages measured in falls. *Morbidelli and Gladman* (1998) extend the previous dynamical studies by calculating the entry geometries, speeds, and delivery times from the main inner-belt resonances and show that that the distribution of radiants and entry speeds is in good agreement with that determined by the fireball camera networks. However, the delivery timescales calculated were typically 2–4 m.y., which is 3–10× shorter than that recorded in most ordinary chondrites. This disagreement is serious and robust; the dynamical calculations are now direct approximation-free N-body integrations, and similar studies on the transport of lunar and martian meteorites (*Gladman et al.,* 1996) match the spectrum of transfer ages to Earth from these two source bodies extremely well. Therefore, the inescapable conclusion is that the "classic" scenario of meteorite delivery is incorrect: Meteorites (especially ordinary chondrites) are not delivered to Earth by being liberated by large collisions in the main belt and directly injected into the belt's orbital resonances for transport to the inner solar system. So how do meteorites arrive at Earth? Where are their source bodies, and can we learn about their parent bodies from dynamical studies? One possibility is that most source bodies are near-Earth asteroids [NEAs; see *Morbidelli et al.* (2002)] distributed throughout terrestrial space. However, it is unlikely that the meteoroid mass flux off NEAs (which are in a less-collisional environment) can rival that of the entire main belt, and *Morbidelli and Gladman* (1998) showed that the orbital distribution of fireballs does not match an NEA source. Therefore, the major chondrite groups almost certainly have their source bodies in the main belt.

A main-belt source for meteorites (especially ordinary chondrites) requires that the bulk of their cosmic-ray exposure occurs after collisional liberation of the meteoroids but before they reach the main resonant "escape hatches." The most-developed models for doing this use the Yarkovsky effect, which causes a slow drift in semimajor axis due to the anisotropic emittance of thermal radiation of a meteoroid (see *Bottke et al.,* 2002). In this context, meteoroid orbits spiral in or out by a distance determined by their collisional lifetimes, producing a sampling zone around the resonances, which evacuate them from the belt. Note that this process allows the ejection velocities during collisional events that liberate meteorites to be effectively zero (in contrast to the classic picture that required meteoroids to be thrown out at hundreds of meters per second to reach the resonances). This allows even small (kilometer-scale) asteroids to be source bodies (*Vokrouhlický and Farinella,* 2000).

The number of source bodies for the various meteorite classes is a problem perhaps best illustrated by using the HEDs as an example. If the widely accepted link between HEDs and Vesta is correct, given that Vesta is almost as far as possible from any orbital resonance (being midway between the ν_6 and 3:1), it would seem that identifying "actual" source bodies is a hopeless quest because all main-belt bodies can be sampled. But perhaps Vesta is not the source body for the HEDs and is rather just the parent body of the "vestoids," which are closer to the resonances and easier to sample. However, distinct peaks in the cosmic-ray-exposure age distribution among the howardites, eucrites, and diogenites (*Welten et al.,* 1997) argue for major impacts on one body (4 Vesta) ejecting all three types of material.

The contribution to the meteoroid flux from big and small parent bodies depends on the size-frequency distribution of objects in the main belt down to subkilometer sizes, the physics of collisional disruption, and the relative importance of slow transport mechanisms like the Yarkovsky effect or orbital diffusion. If the most sophisticated size-frequency distribution models (e.g., *Durda et al.,* 1998) are correct, *Bottke et al.* (2000) found that the flux of meteorite-sized ejecta produced by the largest asteroids (with the largest collisional cross sections) should dominate the flux produced by the smaller asteroids (which lose nearly all their ejecta due to low escape velocities). Hence, many of the chondrites and HEDs falling on Earth today may ultimately be derived from a few large source objects (e.g., 4 Vesta and 6 Hebe), despite that fact that nearly all main-belt asteroids can potentially provide meteorite samples to Earth. If true, the peaks seen in histograms of cosmic-ray-exposure ages for many meteorite groups would represent individual impact events on large asteroids, while the background continuum would represent meteorites produced by numerous smaller impacts occurring across the main belt. If not, then the peaks are due to the large relative mass contribution from a recent major disruption of a "medium-sized" asteroid that was well situated for efficient delivery.

4. DETERMINING METEORITE PARENT BODIES

From analyzing a meteorite in the laboratory, we can learn a variety of details on its composition, the history of its parent body, and its passage through space. However, until samples are returned to Earth, all compositional measurements of asteroids will need to be done through observations using telescopes and spacecrafts.

4.1. Telescopic Data

Compositional data on asteroids are usually derived from the analysis of sunlight reflected from their surfaces. Many minerals (e.g., olivine, pyroxene) have diagnostic absorption bands in the visible and NIR. Spectral surveys (e.g., *Zellner et al.,* 1985; *Bus,* 1999) are primarily done in the visible (~0.4–1.1 μm) due to the peaking of the illuminating solar flux in the visible and the relative transparency of the atmosphere at these wavelengths. More than 2000 asteroids have been observed in the visible. CCD detectors now allow for objects as small as a few hundred meters in near-Earth orbit and a few kilometers in the main belt to be observed by Earth-based telescopes.

Asteroids are generally grouped into classes based on their visible spectra (~0.4 to ~0.9–1.1 μm) and visual albedo (when available). The most widely used taxonomy (*Tholen,* 1984) classifies objects observed in the eight-color asteroid survey (ECAS) (*Zellner et al.,* 1985). *Bus* (1999) develops an expanded taxonomy with many more classes and subclasses to represent the diversity of spectral properties found in his CCD spectra of more than 1300 objects. However, NIR observations (e.g., *Gaffey et al.,* 1993; *Rivkin et al.,* 1995, 2000) of a few hundred objects show that most asteroid classes contain a variety of surface assemblages.

Reflectance spectra of an asteroid can be directly compared to spectral data obtained on tens-of-milligram-sized to gram-sized samples of meteorites. A few hundred meteorites have had their spectra measured. *Gaffey* (1976) shows that different meteorite types tend to have distinctive spectra from 0.3 to 2.5 μm. Besides mineralogy, the shapes and depths of absorption bands and the spectral slope are a function of many other surface parameters including particle size (e.g., *Johnson and Fanale,* 1973) and temperature (e.g., *Singer and Roush,* 1985; *Hinrichs et al.,* 1999). It has also been proposed (e.g., *Chapman,* 1996; *Sasaki et al.,* 2001; *Hapke,* 2001) that the optical properties of an asteroidal surface will change over time due to processes such as micrometeorite impacts and sputtering due to solar wind.

4.2. Spacecraft Data

The best way to identify the parent body of a meteorite is through a spacecraft mission. Meteorites from the Moon have been identified (e.g., *Warren,* 1994) from lunar samples retrieved by Apollo astronauts, since these samples have distinctive petrologic and chemical (e.g., bulk Mn:Fe) properties. The most conclusive evidence that meteorites originate from Mars is the finding (e.g., *Bogard and Johnson,* 1983) that the measured abundances and isotopic ratios of trapped noble gases in glasses in these meteorites are similar to those measured for the martian atmosphere by the *Viking* landers. Other arguments for linking these meteorites to Mars can be found in *McSween* (1994).

Spacecraft have visited a number of S-type asteroids (e.g., 243 Ida, 433 Eros, 951 Gaspra) and one C-type asteroid (253 Mathilde). Spacecraft are able to obtain data that is difficult to impossible to obtain from Earth, including high-resolution images, reflectance spectra of different lithologic units, bulk densities, magnetic field measurements, and bulk-elemental compositions. For determining meteoritic parent bodies, bulk-elemental compositions are probably the most important pieces of information that can be obtained because the data can be directly compared to meteorite bulk compositions (e.g., *Jarosewich,* 1990). Only one previous mission (*NEAR Shoemaker*) has determined elemental ratios (*Nittler et al.,* 2001; *Evans et al.,* 2001) of the surface of an asteroid (433 Eros) using measurements of discrete-line X-ray and γ-ray emissions.

5. METEORITES AND POSSIBLE PARENT BODIES

The following sections will discuss a number of postulated asteroid-meteorite linkages for some of the most common meteorites to fall on Earth. A more complete list of postulated meteorite parent bodies are in Table 2.

5.1. Ordinary Chondrites and S Asteroids

Ordinary chondrites are composed (e.g., *Gomes and Keil,* 1980; *Brearley and Jones,* 1998) of abundant chondrules containing olivine, pyroxene, plagioclase feldspar, and glass with lesser amounts of an olivine-rich matrix, metal, and sulfides. Although grouped under the heading "ordinary chondrites," they actually comprise three separate chemical groups (H, L, and LL) and a range of petrologic types. They range substantially in mineralogy (ol:px ratio of ~54:46 in H chondrites to ~66:34 in LL), metal concentration (~18 vol% in H chondrites to ~4% in LL), mineral composition (particularly the FeO concentrations in olivine and pyroxene), and degree of thermal metamorphism (from virtually none in type 3 to extensive in type 6).

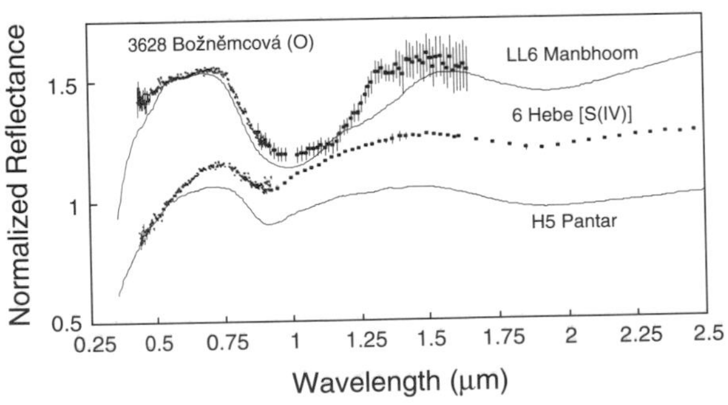

Fig. 4. Normalized reflectance vs. wavelength (μm) for S-type 6 Hebe vs. H5 chondrite Pantar and O-type 3628 Božněmcová vs. LL6 chondrite Manbhoom. All meteorite spectra are from *Gaffey* (1976). All spectra are normalized to unity at 0.55 μm. The asteroid spectra are offset by 0.5 in reflectance. Visible asteroid data (points) are from *Bus* (1999). Near-infrared data (dark squares) for Hebe are from *Bell et al.* (1988) and Božněmcová are from *Burbine* (2000) and *Binzel et al.* (2001).

Spectrally, ordinary chondrites have features due to olivine and pyroxene (Fig. 4). LL chondrites, because of their higher olivine contents, have more distinctive olivine bands while H chondrites have more distinctive pyroxene bands. Even though ordinary chondrites contain significant amounts of metallic Fe, their spectra are not appreciably reddened (reflectances tending to increase with increasing wavelength) as found in other metallic Fe-rich meteorites.

S asteroids are the most abundant type of asteroid observed in the inner main belt. S asteroids have long been considered possible parent bodies of the ordinary chondrites, because a large number of these objects have spectral features due to both olivine and pyroxene. However, S asteroids are spectrally redder than ordinary chondrites (Fig. 4) and tend to have weaker absorption bands. It has long been unclear (e.g., *Wetherill and Chapman,* 1988) if these spectral difference are due to a compositional difference or an alteration process that can redden ordinary chondrite material. Some researchers (e.g., *Chapman,* 1996) argue that some percentage of the S asteroids have ordinary chondrite compositions. Others (*Bell et al.,* 1989) believe that the ordinary chondrite parent bodies are found among small (diameters <10 km) objects and not among the S asteroids, which are believed to be primarily differentiated or partially differentiated bodies.

One asteroid (3628 Božněmcová) was originally announced (*Binzel et al.,* 1993) as the first main-belt object with a visible spectrum similar to ordinary chondrites. However, NIR spectra (*Burbine,* 2000; *Binzel et al.,* 2001) of Božněmcová (Fig. 4) show an unusual bowl-shaped 1-μm feature unlike any measured meteorite spectrum from ~1.2 to 1.5 μm.

Because of the diversity of assemblages found in the S-asteroid population, *Gaffey et al.* (1993) devised a classification system based on analyses of NIR spectral data (*Bell et al.,* 1988). Using the ratio of the areas of Band II (2-μm feature) to Band I (1-μm feature) and the Band I centers, the S asteroids were broken into seven subtypes, S(I) through S(VII). S(I) objects have surfaces mineralogies dominated by olivine, and S(VII) objects have surfaces dominated by pyroxene. The pyroxene abundance tends to increase with increasing number of the subtype.

Only a subset (~25%) of the measured S asteroids, designated as S(IV) objects, fall within the region defined by the ordinary chondrites. However, a few other meteorite types will also plot within this region. Primitive achondrites such as lodranites and acapulcoites tend to overlap the most pyroxene-rich part of the ordinary chondrite area (*Burbine et al.,* 2001b). Some ureilites have compositions (e.g., *Mittlefehldt et al.,* 1998) consistent with falling in this region.

S(IV) asteroids include many of the largest asteroids in the belt including 3 Juno, 6 Hebe, 7 Iris, and 11 Parthenope. Hebe is an often-discussed candidate due to its location near the 3:1 and ν_6 resonances; if ejecta speed distribution favors production from large bodies, then Hebe could be a major contributor to the terrestrial meteorite flux (*Farinella et al.,*

1993). Mineralogies derived from visible and NIR spectra of Hebe (*Gaffey and Gilbert,* 1998) appear consistent with H chondrites.

However, Hebe is spectrally redder than measured ordinary chondrites (Fig. 4). *Gaffey and Gilbert* (1998) argue that the reddening on Hebe is due to "red" metallic Fe similar to that found in iron meteorites. Assuming a very coarse metallic Fe component, they propose that a surface mixture of ~60% H-chondrite material and ~40% metallic Fe would duplicate Hebe's spectral characteristics. They argue that the existence of H6 chondrite Portales Valley, which contains numerous metallic veins with a distinctive Widmanstätten pattern (*Kring et al.,* 1999), and H-chondrite silicate inclusions in IIE irons are evidence that the H-chondrite parent body contains numerous metallic regions. Other researchers (e.g., *Sasaki et al.,* 2001; *Hapke,* 2001) propose that the spectral reddening is due to some type of surface alteration processes (e.g., micrometeorite impacts or solar wind sputtering) believed to produce vapor-deposited coatings of nanophase Fe to redden the spectra. Another chapter (*Clark et al.,* 2002) discusses "space weathering" mechanisms on asteroids.

However, the best spectral matches to ordinary chondrites are in the near-Earth population (*Binzel et al.,* 2002). For example, 1862 Apollo has a visible spectrum (*McFadden et al.,* 1985) similar to an LL chondrite and was classified as a Q type by *Tholen* (1984). In their spectral survey of NEAs, *Binzel et al.* (2001) discovered that one-third of their observed objects resembled ordinary chondrites.

Binzel et al. (2001) also found that there is a continuum of spectral properties between the S asteroids and the ordinary chondrites among the Q- and S-type NEAs in the visible and NIR. This spectral continuum may be related to surface gravity and/or surface age, because the NEAs are much smaller and should have much younger surfaces (on average) than main-belt objects. These observations argue that only "fresher" asteroidal surfaces would resemble ordinary chondrites.

It was hoped that many of these questions on the composition of S asteroids would be answered by the *NEAR Shoemaker* mission to S(IV) asteroid 433 Eros. The average olivine-to-pyroxene composition derived from band area ratios (*McFadden et al.,* 2001) and elemental ratios (Mg:Si, Fe:Si, Al:Si, and Ca:Si) derived from X-ray measurements (*Nittler et al.,* 2001) of Eros are consistent (*McCoy et al.,* 2001) with ordinary chondrites. However, the S:Si ratio derived from X-ray data (*Nittler et al.,* 2001) and the Fe:O and Fe:Si ratios derived from γ-ray data (*Evans et al.,* 2001) are significantly depleted relative to ordinary chondrites. *McCoy et al.* (2001) found that the best meteoritic analogs to Eros are an ordinary chondrite, whose surface chemistry has been altered by the depletion of metallic Fe and sulfides, or a primitive achondrite, derived from a precursor assemblage of the same mineralogy as one of the ordinary chondrite groups. They note that the biggest obstacle in identifying a meteoritic analog for Eros is the lack of understanding (e.g., *McKay et al.,* 1989) of the nature of the chemical and physical processes that affect asteroid regoliths. Returned samples are needed to answer these questions concerning the properties of asteroidal regoliths.

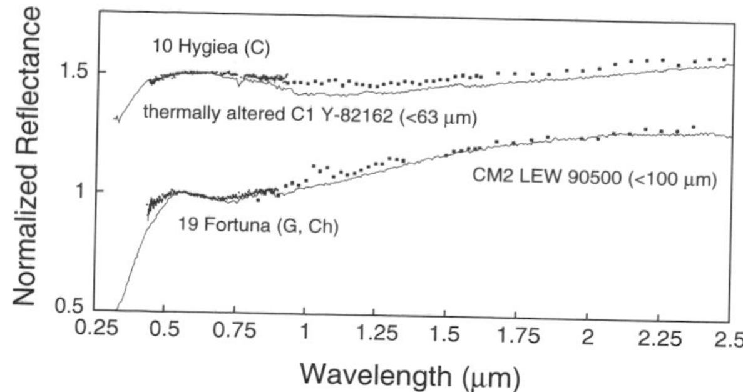

Fig. 5. Normalized reflectance vs. wavelength (μm) for G-type (also classified as a Ch-type) 19 Fortuna vs. CM chondrite LEW 90500 (*Burbine,* 1998) and C-type 10 Hygiea vs. thermally altered CI chondrite Y-82162 (*Hiroi et al.,* 1993). All spectra are normalized to unity at 0.55 μm. The asteroid spectra are offset by 0.5 in reflectance. Visible asteroid data (points) are from *Bus* (1999). Near-infrared asteroid data (dark squares) are from *Bell et al.* (1988).

5.2. CM Chondrites and C-type Asteroids

CM chondrites are typically composed of small chondrules set in an aqueously altered matrix (e.g., *Buseck and Hua,* 1993; *Brearley and Jones,* 1998) of Fe^{3+}-bearing phyllosilicates (e.g., cronstedtite, greenalite) and tochilinite (alternating layers of a sulfide and a hydroxide). The chondrules themselves tend to be FeO-poor, while the phyllosilicates tend to be FeO-rich. In some cases, alteration has been considerably more extensive, leading to aqueous alteration of the chondrules and the replacement of tochilinite but preserving the chondritic structure (*Zolensky et al.,* 1997). CM chondrites typically contain 6–12 wt% water (*Jarosewich,* 1990).

Spectrally, CM chondrites (Fig. 5) have low visible albedos, (~0.04), relatively strong UV features, a number of weak features between 0.6 and 0.9 μm, and relatively featureless spectra from 0.9 to 2.5 μm (*Johnson and Fanale,* 1973; *Gaffey,* 1976; *Vilas and Gaffey,* 1989). CM chondrites have 3-μm features with average band strengths of ~45% (*Jones,* 1988).

The strongest (~5% band strength) of the weak features between 0.6 and 0.9 μm is a band centered at ~0.7 μm, attributed to an absorption due to ferric Fe (Fe^{3+}) in the phyllosilicates. This 0.7-μm band has been found in a number of CM chondrite spectra (*Vilas and Gaffey,* 1989; *Burbine,* 1998), but not in CI- or CR-chondrite spectra.

CM chondrites have been generally linked to the C-type asteroids because they also have low albedos and relatively featureless spectra (Fig. 5) from low-resolution photometric data. *Burbine* (1998) identified two asteroids (13 Egeria and 19 Fortuna) as possible CM-chondrite parent bodies. These asteroids have a 0.7-μm band, similar spectral slopes to CM chondrites out to 2.5 μm, and relatively strong 3-μm features (strengths of 40 ± 15% for Egeria and 25 ± 6% for

Fortuna) (*Jones et al.,* 1990). Egeria and Fortuna are classified as G asteroids by *Tholen* (1984) but as Ch objects by *Bus* (1999) on the basis of having a 0.7-μm band in his higher-resolution spectra.

However, *Bus* (1999) finds that almost half of the C-type asteroids observed in the main belt have this 0.7-μm feature. What makes Egeria and Fortuna likely candidates as CM-chondrite parent bodies is that they are the two largest objects (diameters >200 km) with this feature, and both are located relatively near the 3:1 resonance at ~2.5 AU. For samples to survive passage to Earth, the parent body or bodies of the CM chondrites would have to be relatively near a meteorite-supplying resonance due to their very fragile nature (*Scherer and Schultz,* 2000). CM chondrites have relatively low cosmic-ray-exposure ages (<7 m.y.) (e.g., *Eugster et al.,* 1998) with approximately half the CMs having ages <1 m.y., consistent with the expectation that they could not survive long in space.

Hiroi et al. (1993, 1996) measured the spectra of a number of CI and CM chondrites that have undergone late-stage thermal metamorphism. These meteorites and laboratory-heated CM chondrite material tend to have UV features that are weaker than "typical" carbonaceous chondrites and no ~0.7-μm band, which disappears at temperatures of ~400°C. Heating also tends to weaken the strength of the 3-μm feature.

Hiroi et al. (1993, 1996) find that the weaker UV and 3-μm features were similar to those found for a number of large C-type asteroids such as 10 Hygiea (Fig. 5) and 511 Davida. They suggest that these bodies might have been heated sufficiently to destroy some of their aqueous alteration products. The rarity of thermally altered carbonaceous chondrite material in our meteorite collections could be due to these bodies being located too far from meteorite-supplying resonances.

5.3. HEDs and 4 Vesta

The V-type asteroid 4 Vesta, the vestoids, and Vesta's relationship to the basaltic achondrites (HEDs) are discussed in detail in *Keil* (2002). The one question that will be discussed here is whether other bodies like Vesta have existed in the belt.

According to arguments put forth by *Consolmagno and Drake* (1977), Vesta appears to be the parent body of the HEDs because it is the only large (few-hundred-kilometer) body surviving with an intact "basaltic" crust. Spectrally, Vesta is most similar to a howardite (*Hiroi et al.,* 1994), consistent with a surface mixture of eucritic and diogenitic material. The much smaller (~10-km-sized) vestoids (e.g., *Binzel and Xu,* 1993) have been found in the Vesta family and between Vesta and the 3:1 and the ν_6 resonances, consistent with derivation from Vesta.

However, the grouped and ungrouped irons imply the formation and later disruption of at least 50 differentiated parent bodies. Recent observational and meteoritical evidence now unequivocally supports the existence of at least one other "Vesta." An object (1459 Magnya) with a "Vesta-like" spectrum (*Lazzaro et al.,* 2000) has been identified at 3.15 AU, too far to be easily related to Vesta at 2.36 AU. Magnya appears spectrally indistinguishable from inner-main-belt vestoids. The discovery (*Yamaguchi et al.,* 2002) of the eucrite with a very different O-isotopic value from the HEDs also confirms the formation of other "Vesta-like" bodies. Vesta appears to be the parent body of most HEDs and vestoids, but not all of them.

5.4. Iron Meteorites, Enstatite Chondrites, and M Asteroids

Iron meteorites are composed (e.g., *Buchwald,* 1975; *Mittlefehldt et al.,* 1998). primarily of metallic Fe with usually 5–20 wt% Ni. The variations in bulk Ni content for most meteorite groups is a few weight percent except for the IAB and IICD irons, which have variations over 15 wt%. Accessory phases include troilite (FeS), schreibersite ($(Fe,Ni)_3P$), and graphite. IAB, IIICD, and IIE irons contain abundant inclusions of olivine, pyroxene, and plagioclase feldspar, while IVA irons contain abundant inclusions of trydymite (SiO_2) and pyroxene. The Widmanstätten pattern found in most irons is an oriented intergrowth of body-centered cubic α-Fe,Ni (<6 wt% Ni) (kamacite) and high-Ni (30–50 wt%) regions composed of a variety of phases.

Spectrally, iron meteorites (e.g., *Cloutis et al.,* 1990) have relatively featureless spectra (Fig. 6) with red spectral slopes and moderate albedos (~0.10–0.30). The reflectance properties of metal are functions of composition and grain size/surface roughness of the sample. Contrary to the conclusion of *Gaffey* (1976), *Cloutis et al.* (1990) find no simple correlation between Ni abundance and spectral redness in measured metallic Fe samples. This finding argues that it may not be possible to differentiate between different iron meteorite groups by their spectral properties. *Cloutis et al.* (1990) also found that the spectra of iron meteorites become redder with increasing grain size.

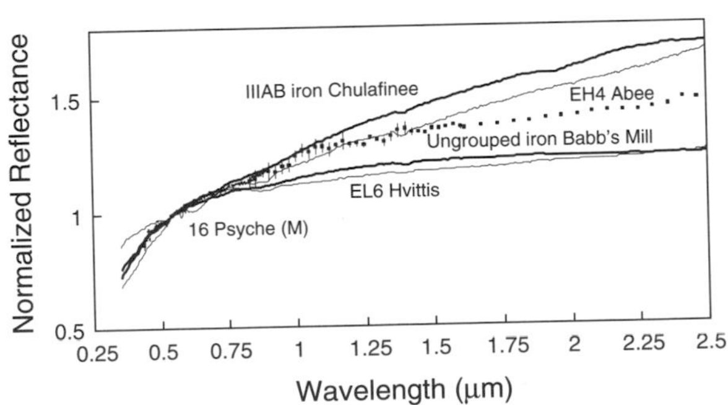

Fig. 6. Normalized reflectance vs. wavelength (μm) for M-type 16 Psyche vs. iron and enstatite chondrite meteorites (lines). All meteorite spectra are from *Gaffey* (1976). In order of increasing reflectance at 2.0 μm, the meteorites are EL6 chondrite Hvittis, ungrouped iron Babb's Mill, EH4 chondrite Abee, and IIIAB iron Chulafinee. All spectra are normalized to unity at 0.55 μm. Visible asteroid data (points) are from *Bus* (1999). Near-infrared asteroid data (dark squares) are from *Bell et al.* (1988).

Enstatite chondrites also have relatively featureless spectra (Fig. 6) with red spectral slopes and moderate albedos (*Gaffey,* 1976). Besides enstatite, they are composed of FeO-poor chondrules, metal, and a range of sulfides. Like aubrites, they are extremely reduced, with virtually all iron occurring in the metallic form (~13–28 vol%) (*Keil,* 1968). They are divided into two groups (EH and EL) and exhibit a range of metamorphic types (3–6). These meteorites have no distinctive absorption features from 0.3 to 2.5 μm because of the absence of Fe^{2+} in their silicates.

M asteroids have moderate visual albedos (~0.10 to ~0.30), relatively featureless spectra, and red spectral slopes. *Tholen* (1989) identifies approximately 40 M asteroids. Because metallic Fe has the same reflectance characteristics in this wavelength region (Fig. 6), the M asteroids have been historically identified as the disrupted cores of differentiated objects that have had their silicates removed. However, enstatite chondrites have similar reflectance characteristics (Fig. 6) to metallic Fe and have also been proposed (e.g., *Gaffey and McCord,* 1978) as meteoritic analogs to the M asteroids. Complicating possible interpretations, *Rivkin et al.* (2000) find that more than one-third of observed M asteroids have 3-μm absorption features, implying hydrated silicates on the surfaces of the objects with weight percents of a few tenths of a percent. Further discussion on whether these absorption features on M asteroids really indicate hydrated assemblages can be found in *Rivkin et al.* (2002) and *Gaffey et al.* (2002).

Radar observations have been used to estimate the metal contents of asteroids. Radar albedos are functions of the near-surface bulk density, which is related to both the solid-rock density and the surface porosity of the object. M asteroids (e.g., 16 Psyche, 216 Kleopatra) tend to have higher radar albedos than C or S asteroids (*Magri et al.,* 1999). M asteroids 6178 1986 DA (*Ostro et al.,* 1991) and 216 Kleopatra (*Ostro et al.,* 2000) have the highest observed radar albedos (~0.6–0.7), implying surfaces of metallic Fe with "lunarlike" porosities (35–55%) or solid enstatite chondritic material with little to no porosity. Without knowing the porosity of the surface, it is impossible to conclusively differentiate between these two types of assemblages. However, recent groundbased measurements of bulk densities, determined from astrometric observations, for M asteroids 16 Psyche (*Viateau,* 2000; *Britt et al.,* 2002) and 22 Kalliope (*Margot and Brown,* 2001) are ~2 g/cm^3. Their bulk densities are much lower than expected and may imply that these objects are not metallic in composition. If these objects are exposed Fe cores, they must be extremely porous with substantially more internal empty space than solid material (*Britt et al.,* 2002).

6. CONCLUSIONS AND FUTURE WORK

In our meteorite collections, there is evidence for 100–150 asteroidal parent bodies. The "true" number is dependent on how well we can discern the chemical, mineralogical, petrologic, and isotopic characteristics of asteroidal-sized parent bodies from gram- to kilogram-sized meteorite samples.

It is very difficult to "conclusively" identify the parent bodies of even the most well-studied meteorite groups with current telescopic and spacecraft data. Asteroids and meteorites with similar spectral and mineralogical characteristics can be identified, and theoretical models for delivering fragments to Earth from these objects can be formulated. However, most postulated parent bodies are not unique spectrally so it is very difficult to rule out all other possible asteroids as parent bodies. Likely parent bodies can be easily identified, but it is very difficult to identify the "true" parent body in most cases. The obvious exception is 4 Vesta, which appears to be the parent body of almost all HEDs.

Making this problem even harder is our limited knowledge of the chemical and physical properties of asteroid regoliths. We do not understand the significance of spectral and chemical differences between meteorites measured in the laboratory and asteroids observed from Earth and by spacecraft.

How can we better answer these questions in the next decade?

1. Samples need to be returned to Earth from the upper (top few millimeters) and lower (tens of centimeters) surface layers of a variety of asteroid types to understand the optical, chemical, and physical properties of asteroid regoliths. *MUSES-C,* a Japanese sample return mission (*Zolensky,* 2000), is currently being prepared for launch to a near-Earth object and it is hoped that it will return ~1 g of material back to Earth.

2. Spacecraft orbiters and landers should be sent to taxonomic types (e.g., A, E, M, V) not yet visited by spacecrafts. These missions should focus on determining chemical compositions of the asteroids.

3. Near-infrared spectra from ~0.9 to 3.5 μm are needed to complement visible spectral surveys. Only with this extended wavelength coverage can we more accurately determine asteroid mineralogies.

4. Reflectance spectra of rare meteorite types need to be obtained.

5. More research needs to be done on the petrology of ungrouped irons to understand possible relationships with other meteorite groups.

6. Processes by which small meteoroids reach the resonances need to be explored in a model that considers all the meteorite types in a self-consistent scenario. In particular, how can fragile carbonaceous chondrites reach the resonances in sufficient numbers to provide the observed flux?

REFERENCES

Bell J. F. (1988) A probable asteroidal parent body for the CV or CO chondrites (abstract). *Meteoritics, 23,* 256–257.

Bell J. F., Owensby P. D., Hawke B R., and Gaffey M. J. (1988) The 52-color asteroid survey: Final results and interpretation (abstract). In *Lunar and Planetary Science XIX,* pp. 57–58. Lunar and Planetary Institute, Houston.

Bell J. F., Davis D. R., Hartmann W. K., and Gaffey M. J. (1989) Asteroids: The big picture. In *Asteroids II* (R. P. Binzel et al., eds.), pp. 921–945. Univ. of Arizona, Tucson.

Bild R. W. (1977) Silicate inclusions in group IAB irons and a relation to the anomalous stones Winona and Mt. Morris (Wis.). *Geochim. Cosmochim. Acta, 41,* 1439–1456.

Binzel R. P. and Xu S. (1993) Chips off of asteroid 4 Vesta: Evidence for the parent body of basaltic achondrite meteorites. *Science, 260,* 186–191.

Binzel R. P., Xu S., Bus S. J., Skrutskie M. F., Meyer M. R., Knezek P., and Barker E. S. (1993) Discovery of a main-belt asteroid resembling ordinary chondrite meteorites. *Science, 262,* 1541–1543.

Binzel R. P., Harris A. W., Bus S. J., and Burbine T. H. (2001) Spectral properties of near-Earth objects: Palomar and IRTF results for 48 objects including spacecraft targets (9969) Braille and (10302) 1989 ML. *Icarus, 151,* 139–149.

Binzel R. P., Lupishko D. F., Di Martino M., Whitely R. J., and Hahn G. J. (2002) Physical properties of near-Earth objects. In *Asteroids III* (W. F. Bottke Jr. et al., eds.), this volume. Univ. of Arizona, Tucson.

Bischoff A. (2001) Meteorite classification and the definition of new chondrite classes as a result of successful meteorite search in hot and cold deserts. *Planet. Space Sci., 49,* 769–776.

Bogard D. D. and Johnson P. (1983) Martian gases in an Antarctic meteorite? *Science, 221,* 651–654.

Bottke W. F., Rubincam D. P., and Burns J. A. (2000) Dynamical evolution of main belt meteoroids: Numerical simulations incorporating planetary perturbations and Yarkovsky thermal forces. *Icarus, 145,* 301–331.

Bottke W. F. Jr., Vokrouhlický D., Rubincam D. P., and Brož M. (2002) The effect of Yarkovsky thermal forces on the dynamical evolution of asteroids and meteoroids. In *Asteroids III* (W. F. Bottke Jr. et al., eds.), this volume. Univ. of Arizona, Tucson.

Brearley A. J. and Jones R. H. (1998) Chondritic meteorites. In *Reviews in Mineralogy, Vol. 36: Planetary Materials* (J. J. Papike, ed.), pp. 3-1 to 3-398. Mineralogical Society of America, Washington.

Britt D. T., Yeomans D., Housen K., and Consolmagno G. (2002) Asteroid density, porosity, and structure. In *Asteroids III* (W. F. Bottke Jr. et al., eds.), this volume. Univ. of Arizona, Tucson.

Brown P. G., Hildebrand A. R., Zolensky M. E., Grady M., Clayton R. N., Mayeda T. K., Tagliaferri E., Spalding R., MacRae N. D., Hoffman E. L., Mittlefehldt D. W., Wacker J. F., Bird J. A., Campbell M. D., Carpenter R., Gingerich H., Glatiotis M., Greiner E., Mazur M. J., McCausland P. J., Plotkin H., and Mazur T. R. (2000) The fall, recovery, orbit, and composition of the Tagish Lake meteorite: A new type of carbonaceous chondrite. *Science, 290,* 320–325.

Buchwald V. F. (1975) *Handbook of Iron Meteorites. Their History, Distribution, Composition and Structure.* Univ. of California, Berkeley. 1418 pp.

Burbine T. H. (1998) Could G-class asteroids be the parent bodies of the CM chondrites? *Meteoritics & Planet. Sci., 33,* 253–258.

Burbine T. H. Jr. (2000) Forging asteroid-meteorite relationships through reflectance spectroscopy. Ph.D. thesis, Massachusetts Institute of Technology, Cambridge. 303 pp.

Burbine T. H., McCoy T. J., and Binzel R. P. (2001a) Spectra of angrites and possible parent bodies (abstract). In *Lunar and Planetary Science XXXII,* Abstract #1857. Lunar and Planetary Institute, Houston (CD-ROM).

Burbine T. H., McCoy T. J., Nittler L. R., and Bell J. F. III (2001b) Could 433 Eros have a primitive achondritic composition? (abstract). In *Lunar and Planetary Science XXXII,* Abstract #1860. Lunar and Planetary Institute, Houston (CD-ROM).

Bus S. J. (1999) Compositional structure in the asteroid belt: Results of a spectroscopic survey. Ph.D. thesis, Massachusetts Institute of Technology, Cambridge. 367 pp.

Buseck P. R. and Hua X. (1993) Matrices of carbonaceous chondrite meteorites. *Annu. Rev. Earth Planet. Sci., 21,* 255–305.

Campins H. and Swindle T. D. (1998) Expected characteristics of cometary meteorites. *Meteoritics & Planet. Sci., 33,* 1201–1212.

Casanova I., Graf T., and Marti K. (1995) Discovery of an unmelted H-chondrite inclusion in an iron meteorite. *Science, 268,* 540–542.

Chapman C. R. (1996) S-type asteroids, ordinary chondrites, and space weathering: The evidence from Galileo's fly-bys of Gaspra and Ida. *Meteoritics & Planet. Sci., 31,* 699–725.

Clark B. E., Hapke B., Pieters C., and Britt D. (2002) Asteroid space weathering and regolith evolution. In *Asteroids III* (W. F. Bottke Jr. et al., eds.), this volume. Univ. of Arizona, Tucson.

Clayton R. N. (1993) Oxygen isotopes in meteorites. *Annu. Rev. Earth Planet. Sci., 21,* 115–149.

Clayton R. N. and Mayeda T. K. (1996) Oxygen isotope studies of achondrites. *Geochim. Cosmochim. Acta, 60,* 1999–2017.

Clayton R. N. and Mayeda T. K. (1999) Oxygen isotope studies of carbonaceous chondrites. *Geochim. Cosmochim. Acta, 63,* 2089–2104.

Clayton R. N., Mayeda T. K., Goswami J. N., and Olsen E. J. (1991) Oxygen isotope studies of ordinary chondrites. *Geochim. Cosmochim. Acta, 55,* 2317–2337.

Cloutis E. A., Gaffey M. J., Smith D. G. W., and Lambert R. St. J. (1990) Reflectance spectra of "featureless" materials and the surface mineralogies of M- and E-class asteroids. *J. Geophys. Res., 95,* 281–293.

Consolmagno G. J. and Drake M. J. (1977) Composition and evolution of the eucrite parent body: Evidence from rare earth elements. *Geochim. Cosmochim. Acta, 41,* 1271–1282.

Cruikshank D. P. and Hartmann W. K. (1984) The meteorite-asteroid connection: Two olivine-rich asteroids. *Science, 223,* 281–283.

Dermott S. F., Durda D. D., Grogan K., and Kehoe T. J. J. (2002) Asteroidal dust. In *Asteroids III* (W. F. Bottke Jr. et al., eds.), this volume. Univ. of Arizona, Tucson.

Drake M. J. (2001) Presidential address: The eucrite/Vesta story. *Meteoritics & Planet. Sci., 36,* 501–513.

Durda D. D., Greenberg R., and Jedicke R. (1998) Collisional models and scaling laws: A new interpretation of the shape of the main-belt asteroid size distribution. *Icarus, 135,* 431–440.

Eugster O., Eberhardt P., Thalmann Ch., and Weigel A. (1998) Neon-E in CM-2 chondrite LEW90500 and collisional history of CM-2 chondrites, Maralinga, and other CK chondrites. *Geochim. Cosmochim. Acta, 62,* 2573–2582.

Evans L. G., Starr R. D., Brückner J., Reedy R. C., Boynton W. V., Trombka J. I., Goldsten J. O., Masarik J., Nittler L. R., and McCoy T. J. (2001) Elemental composition from gamma-ray spectroscopy of the NEAR-Shoemaker landing site on 433 Eros. *Meteoritics & Planet. Sci., 36,* 1639–1660.

Farinella P., Gonczi R., Froeschlé Ch., and Froeschlé C. (1993) The injection of asteroid fragments into resonances. *Icarus, 101,* 174–187.

Farinella P., Froeschlé C., Froeschlé C., Gonczi R., Hahn G., Morbidelli A., and Valsecchi G. B. (1994) Asteroids falling into the Sun. *Nature, 371,* 314–317.

Gaffey M. J. (1976) Spectral reflectance characteristics of the

meteorite classes. *J. Geophys. Res., 81,* 905–920.
Gaffey M. J. and Gilbert S. L. (1998) Asteroid 6 Hebe: The probable parent body of the H-type ordinary chondrites and the IIE iron meteorites. *Meteoritics & Planet. Sci., 33,* 1281–1296.
Gaffey M. J. and McCord T. B. (1978) Asteroid surface materials: Mineralogical characterizations from reflectance spectra. *Space Sci. Rev., 21,* 555–628.
Gaffey M. J., Reed K. L., and Kelley M. S. (1992) Relationship of E-type Apollo asteroid 3103 (1982 BB) to the enstatite achondrite meteorites and the Hungaria asteroids. *Icarus, 100,* 95–109.
Gaffey M. J., Bell J. F., Brown R. H., Burbine T. H., Piatek J. L., Reed K. L., and Chaky D. A. (1993) Mineralogical variations within the S-type asteroid class. *Icarus, 106,* 573–602.
Gaffey M. J., Cloutis E. A., Kelley M. S., and Reed K. L. (2002) Mineralogy of asteroids. In *Asteroids III* (W. F. Bottke Jr. et al., eds.), this volume. Univ. of Arizona, Tucson.
Gladman B., Burns J. A., Duncan M., Lee P., and Levison H. (1996) The exchange of impact ejecta between terrestrial planets. *Science, 271,* 1387–1392.
Gladman B., Migliorini F., Morbidelli A., Zappalà V., Michel P., Cellino A., Froeschlé Ch., Levison H., Bailey M., and Duncan M. (1997) Dynamical lifetimes of objects injected into asteroid belt resonances. *Science, 277,* 197–201.
Gomes C.B. and Keil K. (1980) *Brazilian Stone Meteorites.* Univ. of New Mexico, Albuquerque. 162 pp.
Grady M. M. (2000) *Catalogue of Meteorites.* Cambridge Univ., Cambridge. 689 pp.
Greenberg R. and Nolan M. C. (1989) Delivery of asteroids and meteorites to the inner solar system. In *Asteroids II* (R. P. Binzel et al., eds.), pp. 778–804. Univ. of Arizona, Tucson.
Grossman J. N. (2000) The Meteoritical Bulletin, No. 84, 2000 August. *Meteoritics & Planet. Sci., 35,* A199–A225.
Grossman J. N. and Zipfel J. (2001) The Meteoritical Bulletin, No. 85, 2001 September. *Meteoritics & Planet. Sci., 36,* A293–A322.
Hapke B. W. (2001) Space weathering from Mercury to the asteroid belt. *J. Geophys. Res., 106,* 10039–10073.
Hinrichs J. L., Lucey P. G., Robinson M. S., Meibom A., and Krot A. N. (1999) Implications of temperature-dependent near-IR spectral properties of common minerals and meteorites for remote sensing of asteroids. *Geophys. Res. Lett., 26,* 1661–1664.
Hiroi T., Pieters C. M., Zolensky M. E., and Lipschutz M. E. (1993) Evidence of thermal metamorphism on the C, G, B, and F asteroids. *Science, 261,* 1016–1018.
Hiroi T., Pieters C. M., and Takeda H. (1994) Grain size of the surface regolith of asteroid 4 Vesta estimated from its reflectance spectrum in comparison with HED meteorites. *Meteoritics, 29,* 394–396.
Hiroi T., Zolensky M. E., Pieters C. M., and Lipschutz M. E. (1996) Thermal metamorphism of the C, G, B, and F asteroids seen from the 0.7 μm, 3 μm, and UV absorption strengths in comparison with carbonaceous chondrites. *Meteoritics & Planet. Sci., 31,* 321–27.
Hiroi T., Zolensky M. E., and Pieters C. M. (2001) The Tagish Lake meteorite: A possible sample from a D-type asteroid. *Science, 293,* 2234–2236.
Ivezić Ž., Tabachnik S., Rafikov R., Lupton R. H., Quinn T., Hammergren M., Eyer L., Chu J., Armstrong J. C., Fan X., Finlator K., Geballe T. R., Gunn J. E., Hennessy G. S., Knapp G. R., Leggett S. K., Munn J. A., Pier J. R., Rockosi C. M., Schneider D. P., Strauss M. A., Yanny B., Brinkmann J., Csabai I., Hindsley R. B., Kent S., Lamb D. Q., Margon B., McKay T. A., Smith J. A., Waddel P., and York D. G. (for the SDSS Collaboration) (2001) Solar system objects observed in the Sloan Digital Sky Survey commissioning data. *Astron. J., 122,* 2749–2784.
Jarosewich E. (1990) Chemical analyses of meteorites: A compilation of stony and iron meteorite analyses. *Meteoritics, 25,* 323–337.
Johnson T. V. and Fanale F. P. (1973) Optical properties of carbonaceous chondrites and their relationship to asteroids. *J. Geophys. Res., 78,* 8507–8518.
Jones T. D. (1988) An infrared reflectance study of water in outer belt asteroids: Clues to composition and origin. Ph.D. thesis, Univ. of Arizona, Tucson. 281 pp.
Jones T. D., Lebofsky L. A., Lewis J. S., and Marley M. S. (1990) The composition and origin of the C, P, and D asteroids: Water as a tracer of thermal evolution in the outer belt. *Icarus, 88,* 172–192.
Kargel J. S. (1994) Metalliferous asteroids as potential sources of precious metals. *J. Geophys. Res., 99,* 21129–21141.
Keil K. (1968) Mineralogical and chemical relationships among enstatite chondrites. *J. Geophys. Res., 73,* 6945–6976.
Keil K. (2002) Geological history of asteroid 4 Vesta: The "smallest terrestrial planet." In *Asteroids III* (W. F. Bottke Jr. et al., eds.), this volume. Univ. of Arizona, Tucson.
Kring D. A., Hill D. H., Gleason J. D., Britt D. T., Consolmagno G. J., Farmer M., Wilson S., and Haag R. (1999) Portales Valley: A meteoritic sample of the brecciated and metal-veined floor of an impact crater on an H-chondrite asteroid. *Meteoritics & Planet. Sci., 34,* 663–669.
Lazzaro D., Michtchenko T., Carvano J. M., Binzel R. P., Bus S. J., Burbine T. H., Mothé-Diniz T., Florczak M., Angeli C. A., and Harris A. W. (2000) Discovery of a basaltic asteroid in the outer main belt. *Science, 288,* 2033–2035.
Lipschutz M. E., Gaffey M. J., and Pellas P. (1989) Meteoritic parent bodies: Nature, number, size, and relation to present-day asteroids. In *Asteroids II* (R. P. Binzel et al., eds.), pp. 740–777. Univ. of Arizona, Tucson.
Lipschutz M. E., Zolensky M. E., and Bell M. S. (1999) New petrographic and trace element data on thermally metamorphosed carbonaceous chondrites. *Antarct. Meteorite Res., 12,* 57–80.
Love S. G. and Keil K. (1995) Recognizing mercurian meteorites. *Meteoritics & Planet. Sci., 30,* 269–278.
Lucey P. G., Keil K., and Whitely R. (1998) The influence of temperature on the spectra of the A-asteroids and implications for their silicate chemistry. *J. Geophys. Res., 103,* 5865–5871.
Magri C., Ostro S. J., Rosema K. D., Thomas M. L., Mitchell D. L., Campbell D. B., Chandler J. F., Shapiro I. I., Giorgini J. D., and Yeomans D. K. (1999) Mainbelt asteroids: Results of Arecibo and Goldstone radar observations of 37 objects during 1980–1985. *Icarus, 140,* 379–407.
Margot J. L. and Brown M. E. (2001) Discovery and characterization of binary asteroids 22 Kalliope and 87 Sylvia (abstract). *Bull. Am. Astron. Soc., 33,* 1133.
Marti K. and Graf T. (1992) Cosmic-ray exposure history of ordinary chondrites. *Annu. Rev. Earth Planet. Sci., 20,* 221–243.
Mason B. (1962) *Meteorites.* Wiley and Sons, New York. 274 pp.
McCoy T. J., Nittler L. R., Burbine T. H., Trombka J. I., Clark P. E., and Murphy M. E. (2000) Anatomy of a partially differentiated asteroid: A "NEAR"-sighted view of acapulcoites and lodranites. *Icarus, 148,* 29–36.

McCoy T. J., Burbine T. H., McFadden L., Starr R. D., Gaffey M. J., Nittler L. R., Evans L. G., Izenberg N., Lucey P., Trombka J. I., Bell J. F. III, Clark B. E., Clark P. E., Squyres S. W., Chapman C. R., Boynton W. V., and Veverka J. (2001) The composition of 433 Eros: A mineralogical-chemical synthesis. *Meteoritics & Planet. Sci., 36,* 1661–1672.

McFadden L. A., Gaffey M. J., and McCord T. B. (1985) Near-Earth asteroids: Possible sources from reflectance spectroscopy. *Science, 229,* 160–163.

McFadden L. A., Wellnitz D. D., Schnaubelt M. W., Gaffey M. J., Bell J. F. III, Izenberg N., and Chapman C. R. (2001) Mineralogical interpretation of reflectance spectra of Eros from NEAR NIS low phase flyby. *Meteoritics & Planet. Sci., 36,* 1711–1726.

McKay D. S., Swindle T. D., and Greenberg R. (1989) Asteroidal regoliths: What we do not know. In *Asteroids II* (R. P. Binzel et al., eds.), pp. 617–642. Univ. of Arizona, Tucson.

McSween H. Y. Jr. (1994) What we have learned from Mars from SNC meteorites. *Meteoritics, 29,* 757–779.

Meibom A. and Clark B. E. (1999) Evidence for the insignificance of ordinary chondritic material in the asteroid belt. *Meteoritics & Planet. Sci., 34,* 7–24.

Melosh H. J. (2001) Exchange of meteorites between stellar systems (abstract). *Meteoritics & Planet. Sci., 36,* A130–131.

Melosh H. J. and Tonks W. B. (1993) Swapping rocks: Ejection and exchange of surface material among the terrestrial planets (abstract). *Meteoritics & Planet. Sci., 28,* 398.

Mittlefehldt D. W., McCoy T. J., Goodrich C. A., and Kracher A. (1998) Non-chondritic meteorites from asteroidal bodies. In *Reviews in Mineralogy, Vol. 36: Planetary Materials* (J. J. Papike, ed.), pp. 4-1 to 4-195. Mineralogical Society of America, Washington.

Morbidelli A. and Gladman B. (1998) Orbital and temporal distributions of meteorites originating in the asteroid belt. *Meteoritics & Planet. Sci., 33,* 999–1016.

Morbidelli A., Bottke W. F. Jr., Froeschlé Ch., and Michel P. (2002) Origin and evolution of near-Earth objects. In *Asteroids III* (W. F. Bottke Jr. et al., eds.), this volume. Univ. of Arizona, Tucson.

Nagahara H. and Ozawa K. (1986) Petrology of Yamato-791493, "lodranite": Melting, crystallization, cooling history and relationship to other meteorites. *Mem. NIPR Spec. Issue, 41,* 181–205.

Nittler L. R., Starr R. D., Lim L., Burbine T. H., Reedy R. C., Trombka J. I., Gorenstein P., Squyres S. W., Boynton W. V., McClanahan T. P., Bhangoo J. S., Clark P. E., Murphy M. E., and Killen R. (2001) X-ray fluorescence measurements of the surface elemental composition of asteroid 433 Eros. *Meteoritics & Planet. Sci., 36,* 1673–1695.

Olsen E., Davis A., Clarke R. S. Jr., Schultz L., Weber H. W., Clayton R., Mayeda T., Jarosewich E., Sylvester P., Grossman L., Wang M-S., Lipschutz M. E., Steele I. M., and Schwade J. (1994) Watson: A new link in the IIE iron chain. *Meteoritics, 29,* 200–213.

Ostro S. J., Campbell D. B., Chandler J. F., Hine A. A., Hudson R. S., Rosema K. D., and Shapiro I. I. (1991) Asteroid 1986 DA: Radar evidence for a metallic composition. *Science, 252,* 1399–1404.

Ostro S. J., Hudson R. S., Nolan M. C., Margot J. L., Scheeres D. J., Campbell D. B., Magri C., Giorgini J. D., and Yeomans D. K. (2000) Radar observations of asteroid 216 Kleopatra. *Science, 288,* 836–839.

Rietmeijer F. J. M. (1998) Interplanetary dust particles. In *Reviews in Mineralogy, Vol. 36: Planetary Materials* (J. J. Papike, ed.), pp. 2-1 to 2-195. Mineralogical Society of America, Washington.

Rivkin A. S., Howell E. S., Britt D. T., Lebofsky L. A., Nolan M. C., and Branston D. D. (1995) 3-μm spectrophotometric survey of M- and E-class asteroids. *Icarus, 117,* 90–100.

Rivkin A. S., Lebofsky L. A., Clark B. E., Howell E. S., and Britt D. T. (2000) The nature of M-class asteroids from 3-μm observations. *Icarus, 145,* 351–368.

Rivkin A. S., Howell E. S., Vilas F., and Lebofsky L. A. (2002) Hydrated minerals on asteroids: The astronomical record. In *Asteroids III* (W. F. Bottke Jr. et al., eds.), this volume. Univ. of Arizona, Tucson.

Sasaki S., Nakamura K., Hamabe Y., Kurahashi E., and Hiroi T. (2001) Production of iron nanoparticles by laser irradiation in a simulation of lunar-like space weathering. *Nature, 410,* 555–557.

Scherer P. and Schultz L. (2000) Noble gas record, collisional history, and pairing of CV, CO, CK, and other carbonaceous chondrites. *Meteoritics & Planet. Sci., 35,* 145–153.

Schulze H., Bischoff A., Palme H., Spettel B., Dreibus G., and Otto J. (1994) Mineralogy and chemistry of Rumuruti: The first meteorite fall of the new R chondrite group. *Meteoritics, 29,* 275–286.

Sears D. W. G. (1998) The case for rarity of chondrules and CAI in the early solar system and some implications for astrophysical models. *Astrophys. J., 498,* 773–778.

Singer R. B. and Roush T. L. (1985) Effects of temperature on remotely sensed mineral absorption features. *J. Geophys. Res., 90,* 12434–12444.

Sunshine J. M., Binzel R. P., Burbine T. H., and Bus S. J. (1998) Is asteroid 289 Nenetta compositionally analogous to the Brachinite meteorites? (abstract). In *Lunar and Planetary Science XXIX,* Abstract #1430. Lunar and Planetary Institute, Houston (CD-ROM).

Tholen D. J. (1984) Asteroid taxonomy from cluster analysis of photometry. Ph.D. thesis, Univ. of Arizona, Tucson. 150 pp.

Tholen D. J. (1989) Asteroid taxonomic classifications. In *Asteroids II* (R. P. Binzel et al., eds.), pp. 1139–1150. Univ. of Arizona, Tucson.

Viateau B. (2000) Mass and density of asteroids (16) Psyche and (121) Hermione. *Astron. Astrophys., 354,* 725–731.

Vilas F. and Gaffey M. J. (1989) Phyllosilicate absorption features in main-belt and outer-belt asteroid reflectance spectra. *Science, 246,* 790–792.

Vokrouhlický D. and Farinella P. (2000) Efficient delivery of meteorites to the Earth from a wide range of asteroid parent bodies. *Nature, 407,* 606–608.

Voshage H. and Feldman H. (1979) Investigations on cosmic-ray produced nuclides in iron meteorites, 3. Exposure ages, meteoroid sizes and sample depths determined by mass spectrometric analyses of potassium and rare gases. *Earth Planet. Sci. Lett., 45,* 293–308.

Warren P. W. (1994) Lunar and martian meteorite delivery services. *Icarus, 111,* 338–363.

Wasson J. T. (1995) Sampling the asteroid belt: How biases make it difficult to establish meteorite-asteroid connections (abstract). *Meteoritics, 30,* 595.

Weisberg M. K., Prinz M., Clayton R. N., Mayeda T. K., Sugiura N., Zashu S., and Ebihara M. (2001) A new metal-rich chondrite grouplet. *Meteoritics & Planet. Sci., 36,* 401–418.

Welten K. C., Lindner L., Van Der Borg K., Loeken T., Scherer

P., and Schultz L. (1997) Cosmic-ray exposure ages of diogenites and the recent collisional history of the howardite, eucrite and diogenite parent body/bodies. *Meteoritics & Planet. Sci., 32,* 891–902.

Wetherill G. (1985) Asteroidal sources of ordinary chondrites. *Meteoritics, 20,* 1–22.

Wetherill G. W. and Chapman C. R. (1988) Asteroids and meteorites. In *Meteorites and the Early Solar System* (J. F. Kerridge and M. S. Matthews, eds.), pp. 35–67. Univ. of Arizona, Tucson.

Yamaguchi A., Clayton R. N., Mayeda T. K., Ebihara M., Oura Y., Miura Y. N., Haramura H., Misawa K., Kojima H., and Nagao K. (2002) A new source of basaltic meteorites inferred from Northwest Africa 011. *Science, 296,* 334–336.

Zellner B., Tholen D. J., and Tedesco E. F. (1985) The eight-color asteroid survey: Results for 589 minor planets. *Icarus, 61,* 355–416.

Zolensky M. E. (1998) The flux of meteorites to Antarctica. In *Meteorites: Flux with Time and Impact Effects* (M. M. Grady et al., eds.), pp. 93–104. Geological Society Special Publication 140, Geological Society, London.

Zolensky M. E. (2000) The MUSES-C asteroid sample return mission (abstract). *Meteoritics & Planet. Sci., 35,* A178.

Zolensky M. E., Mittlefehldt D. W., Lipschutz M. E., Wang M.-S., Clayton R. N., Mayeda T. K., Grady M. M., Pillinger C., and Barber D. (1997) CM chondrites exhibit the complete petrologic range from type 2 to 1. *Geochim. Cosmochim. Acta, 61,* 5099–5115.

Evolution of Comets Into Asteroids

Paul R. Weissman
Jet Propulsion Laboratory

William F. Bottke Jr.
Southwest Research Institute

Harold F. Levison
Southwest Research Institute

In the past decade, new observations and new research tools have afforded us a better understanding of the interrelationships between comets and asteroids. The extensive automated surveys for near-Earth objects (NEOs) have serendipitously discovered many objects in comet-like orbits. Dynamical simulation codes have improved to the point where we can simulate the detailed orbital behavior of large swarms of test particles as they move out from various storage reservoirs, such as the Oort cloud and the Kuiper belt. Physical studies of both comets and asteroids have given us a far better understanding of the nature of these bodies and have identified possible discriminators to be used in comparing them. As a result, we can now identify likely dormant or extinct comet candidates among the asteroid population. It appears that $\sim 6 \pm 4\%$ (or perhaps more) of the NEO population is derived from Jupiter-family comets. Also, it is highly likely that many asteroids in eccentric orbits with large semimajor axes and large inclinations are derived from the Oort cloud. However, we must also recognize that some small fraction of the Oort cloud population is likely to consist of asteroidal bodies ejected there during the clearing of the planetary zones in the early solar system. Additional physical and dynamical studies are required to continue improving our knowledge of the interrelationships between comets and asteroids and to help identify likely extinct comet candidates.

1. INTRODUCTION

Solar system astronomers have long speculated on the possible existence of extinct or dormant comets, objects that appear asteroidal but in truth are icy objects that had their origins as comets. *Kresák* (1979) suggested a number of extinct comet candidates based primarily on dynamical criteria. *Degewij and Tedesco* (1982) compared physical studies of comets and asteroids and concluded that no extinct cometary nucleus had yet been found. *Weissman et al.* (1989) suggested a list of cometary candidates among the Apollo and Amor asteroids, as well as some outer solar system objects such as Chiron (later discovered to be active) and Hidalgo. *Wetherill* (1988, 1991) investigated the dynamical evolution of Jupiter-family comets in the inner solar system and concluded that as many as half of the Apollo-Amor objects could be of cometary origin.

The problem is not a simple one. The chaotic nature of the dynamical evolution of objects in planet-crossing orbits, as well as nongravitational accelerations on comets caused by outgassing, make it impossible to track orbits accurately backward (or forward) in time more than a few decades or centuries. The obscuring effects of comae make it difficult to apply the same physical observation techniques to comets that have been so successful with asteroids, when the comets are relatively close to the Earth. [Observations of cometary nuclei at large heliocentric distances when they are presumably inactive have had considerable success (e.g., *Lowry,* 2001), but are necessarily difficult to perform and interpret.] Finally, our understanding of comets and cometary processes is still in a primitive state, and has yet to benefit from the intense scrutiny that the asteroid-orbiting *NEAR* mission has afforded us of asteroid 433 Eros, a typical Earth-approaching, S-type asteroid. The first comet rendezvous mission, *Rosetta*, is scheduled to be launched in January 2003 and will arrive at periodic comet 46P/Wirtanen in November 2011.

In the past decade, several lines of research have provided new data and new tools with which to pursue these questions. First is the extensive automated surveys for near-Earth objects (NEOs) that have serendipitously discovered many objects throughout the planetary system in unusual, sometimes cometlike orbits. Second is the advances in numerical simulation of solar system dynamics made possible by advanced numerical codes, in particular symplectic integrators, and by the availability of low-cost computers that can be dedicated to running these codes for weeks or even months. Third is the substantial increase in physical observations of comets and asteroids that permit us a better understanding of the interrelationships between these two very important classes of objects.

In this chapter we will discuss how these new tools have made it possible to argue for, at least statistically, the existence of extinct or dormant comets among the observed asteroid population, in particular among the NEOs. Our approach is considerably different from that of *Weissman et al.* (1989, hereafter *Paper 1*) in *Asteroids II*, and we refer the reader to that chapter for more detailed discussions of physical observations of comets and extinct/dormant comet candidates. Section 2 provides definitions of the classes of bodies we are discussing. Section 3 presents a discussion of the possible physical end states of cometary nuclei. Section 4 discusses cometary dynamics and section 5 reviews physical observations of comets and extinct comet candidates. Section 6 examines a number of asteroidal objects that have been identified as likely extinct comet candidates. Finally, section 7 provides additional discussion and a summary of our results.

2. COMETARY DEFINITIONS

Comets and asteroids represent different parts of a continuous spectrum of planetesimals that were formed and processed in the primordial solar nebula. Two parts of that "stratigraphic" record, the asteroid belt between Mars and Jupiter and the Kuiper belt beyond Neptune, have been preserved *in situ*, at or close to their formation locations. However, an important part of the record, the comets in and just beyond the giant planets zone, have been scattered out of that region by gravitational interactions with those planets. (Many of those scattered comets are preserved in the Oort cloud.) As a result, the critical transition zone between rocky and icy objects has likely been lost from the "stratigraphic" record.

In general, the distinction between comets and asteroids has been based on the existence of a substantial fraction of ices within the object, thus providing the volatiles necessary for the object to develop a coma. Traditionally, the boundary between the formation zones for comets and asteroids has been taken to be the orbit of Jupiter at 5.2 AU, although the icy nature of several of the Galilean satellites suggests that the boundary might be placed somewhat closer to the Sun, perhaps at ~4 AU.

At present, the choice between a cometary vs. an asteroidal designation for a newly discovered object is based on the presence or absence of a visible coma. This has occasionally led to objects being classified as both types, in particular when a coma has been discovered long after the object had been cataloged as an asteroid. The most notable case is the Centaur asteroid 2060 Chiron, also known as comet 95P/Chiron. In addition, this scheme has led to Kuiper belt objects being numbered as asteroids, although their location in the solar system argues strongly for a significant icy fraction in their compositions. Since no formal definitions of comets and asteroids exist, we will use the following, slightly modified definitions from *Paper 1*.

1. *Comet (or cometary nucleus)*: a body formed in the outer planets region (or beyond) containing a significant fraction of volatiles in the form of ices and capable of developing a coma if its orbit brings it close enough to the Sun.

2. *Active comet*: a comet nucleus losing volatiles and dust in a detectable coma.

3. *Inactive comet*: a comet nucleus that is active during part of its orbit, but presently is in a part of the orbit where volatile loss is negligible and there is no detectable coma.

4. *Dormant comet*: a comet nucleus that, although once active, has lost the ability to generate a detectable coma in any part of its present orbit. A dormant comet perturbed to a smaller perihelion distance might be reactivated. Or an impact might remove an overlying nonvolatile crust and expose fresh icy materials to sublimation, reactivating the comet.

5. *Extinct comet*: a comet nucleus that has lost its ices or has its ices so permanently buried under a nonvolatile crust that it is incapable of generating a coma.

6. *Asteroid*: an interplanetary body formed without significant ice content, and thus incapable of displaying cometary activity.

7. *Near-Earth object*: an asteroid or comet in an orbit with perihelion distance of less than 1.3 AU.

The informed reader will recognize exceptions and shades of gray that do not fit these definitions. For example, Kuiper belt objects, Trojan asteroids, and some outer main-belt asteroids all likely have significant ice content, which would make them comets under our definition, but are officially cataloged (and often referred to) as asteroids. Also, our criteria themselves are subjective and open to interpretation. As detector technology has improved, our ability to recognize comae around distant objects has similarly improved. In addition, the distinction between dormant and extinct comets may be very subtle. If an extinct comet could somehow be split open (for example, by a catastrophic collision), exposing the ices within, would it not return to being an active comet? We do not mean for the definitions to be perfect, but only intend them to serve as a guide in our discussion below. Note that in that discussion we use the term "extinct comet" quite frequently, meaning either an extinct or dormant comet.

3. PHYSICAL END STATES OF COMETARY NUCLEI

Comets are transient bodies, losing some fraction of their mass, perhaps 0.01–0.1%, on each return. Dynamical studies of the evolution of the orbits of long-period comets have shown that the $1/a_0$ (inverse original semimajor axis) distribution for those orbits can only be explained if there is some physical mechanism that destroys comets or renders them unobservable (*Oort,* 1950; *Whipple,* 1962; *Weissman,* 1979; *Bailey,* 1984; *Wiegert and Tremaine,* 1999). This process is often referred to as "fading," although the actual physical mechanism or mechanisms are not identified.

In fact, compilations of observations of three Halley-type comets (1P/Halley, 109P/Swift-Tuttle, and 55P/Tempel Tuttle) show that these comets remained remarkably constant in maximum brightness over periods of ~600–2200 yr (*Stephenson et al.,* 1985; *Yau et al.,* 1994; *Yeomans et al.,* 1996). These periods covered 30, 17, and 19 returns for Halley, Swift-Tuttle, and Tempel-Tuttle respectively. Similarly, *Kamél* (1991) showed that 2P/Encke has not faded over 155 yr, encompassing ~47 returns. Thus, if comets do fade, they likely do so very slowly, or episodically.

This result is in contrast to that of *Kresák* (1987), who noted that several short-period comets had been missed on prediscovery apparitions when the comets were close to Earth and should have achieved naked-eye brightness. Kresák proposed that these comets went through irregular dormant phases in which their activity dropped to near-zero. However, *Stephenson et al.* (1985), *Yau et al.* (1994), and *Yeomans et al.* (1996) showed that the three Halley-type comets noted above were not discovered by naked-eye observers until they reached a visual magnitude of 3.5–4.0. Examination of Kresák's list of the expected brightness ranges for 14 potentially dormant comets shows that all of them could have been fainter than V = 3.5–4.0, and only four might have been brighter than those values. Although the total magnitude of each comet might have exceeded the threshold for naked-eye detection of stars, V ≈ 6, the diffuse nature of comae in fact make them very difficult to detect unless they are significantly brighter than that threshold.

Many cases of comets that disappeared can likely be attributed to poorly determined orbits. For example, comet 107P/Wilson-Harrington, discovered in 1949 and observed for only 6 days, was lost for 30 yr before it was accidentally recovered in 1979 (see section 6). But in other cases, well-observed comets seem to have simply disappeared. These include comets 3D/Biela (observed 1772–1852, see below), 5D/Brorsen (1846–1879), 11D/Tempel-Swift (1869–1908), 18D/Perrine-Mrkos (1896–1968), 20D/Westphal (1852–1913), 25D/Neujmin 2 (1916–1926), and 34D/Gale (1927–1938). Note that the designation "D" refers to "defunct" comets. Several of these comets displayed unusual brightness changes on their last apparition (outbursts and/or fading), perhaps indicative of their imminent demise (*Kronk,* 1984).

Three physical mechanisms have been proposed to explain the disappearance of comets: (1) random disruption, (2) loss of all volatiles, and (3) formation of a nonvolatile crust or mantle on the nucleus surface. Unfortunately, none of these mechanisms are well understood or quantified. Nevertheless, some physical loss mechanism(s) is needed to match the energy distribution of the orbits of the known long-period comets. *Weissman* (1979) achieved this by assuming that comets split destructively at a rate of ~12% per perihelion passage, but that only 85% of comets were capable of splitting. *Wiegert and Tremaine* (1999) proposed that (1) the fraction of comets remaining visible after m apparitions is proportional to $m^{-0.6 \pm 0.1}$ or (2) ~95% of comets live for only approximately six returns and the remainder last indefinitely.

Random disruption of comets has been observed on many occasions. The classic example is comet 3D/Biela, a Jupiter-family comet with a period of 6.6 yr that was observed in 1772, 1805, 1826, and 1832. The comet was observed to be splitting during its 1846 apparition and returned as a double comet in 1852. It was never observed again. More recently, comet LINEAR, D/1999 S4, was observed to completely disrupt as it passed through perihelion in July 2000 (*Weaver et al.,* 2001) (see Fig. 1).

Weissman (1980) compiled records of observations of disrupted or split comets and showed that 10% of dynamically new comets from the Oort cloud split, vs. 4% for returning long-period comets, and only 1% for short-period comets (per orbit; see also *Sekanina,* 1982). The splitting events did not show any correlation with perihelion distance, distance from the ecliptic plane, or time of perihelion passage. The statistics suggest that the tendency of cometary nuclei to split may reflect some intrinsic property, such that comets that are likely to split do so early on, and those that are unlikely to split can survive for hundreds or even thousands of returns.

Note that splitting events do not always lead to total disruption of the nucleus. For example, comet 73P/Schwassmann-Wachmann 3 has been observed to shed fragments on at least three perihelion passages, yet still returns every 5.3 yr. Also, since splitting appears to eventually lead to destruction of the cometary nucleus, it cannot explain dormant or extinct nuclei that might be observed as asteroidal objects.

Loss of all volatiles is a slow acting process. *Weissman* (1980) estimated lifetimes of ~600, 4500, and 4 × 10^5 returns for 1-km-radius water-ice spheres with surface albedo = 0 and density of 0.6 g cm^{-3}, for long-period comets with perihelia of 1, 2, and 3 AU respectively. (Lifetimes for icy spheres in short-period comet orbits with comparable perihelia would be similar though somewhat shorter because of their less eccentric orbits.) However, it is not clear that a comet could sublimate all its volatiles, leaving a coherent, nonvolatile remnant nucleus. Evolving gases would tend to

Fig. 1. Hubble Space Telescope image of comet LINEAR, D/1999 S4, on August 5, 2000, showing fragments of the disintegrating nucleus. This long-period comet disrupted close to perihelion at 0.765 AU from the Sun on July 26, 2000 (*Weaver et al.,* 2001).

carry away much of the nonvolatile matter, and any residual material would form an insulating layer that would prevent insolation from penetrating to the volatiles at depth. Thus, total loss of volatiles appears to be an unlikely end state, or one that would not leave a substantial extinct nucleus.

Formation of a nonvolatile crust or mantle on the nucleus surface, first proposed by *Whipple* (1950), is the one process that would presumably lead to asteroidlike objects. There are two ways that such crusts may form. First, irradiation of comets by galactic cosmic rays and solar protons will sputter away volatiles and transform organic molecules to more refractory forms during the comets' long storage in the Oort cloud or Kuiper belt (*Johnson et al.,* 1987; *Moore et al.,* 1983; see also *Weissman and Stern,* 1997, and references therein). This irradiated crust would extend ~1/ρ m below the nucleus surface, where ρ is the density of the cometary materials in g cm^{-3}. An interesting and unsolved problem is how such a crust is removed when the comet reenters the planetary region on it first perihelion passage, allowing it to become active.

A second method for forming a cometary surface crust is through the lag deposit of nonvolatile grains, left behind or launched on suborbital trajectories that do not achieve escape velocity, as water and more volatile ices sublimate from the nucleus surface during perihelion passage. It is not clear how such large nonvolatile grains, which presumably are too heavy to be lifted off the nucleus by evolving gases, might form. However, if they did, thermal models have shown that a layer only a few to perhaps 10 cm thick would be sufficient to insulate the underlying ices from sublimation (*Brin and Mendis,* 1979; *Fanale and Salvail,* 1984; *Prialnik and Bar-Nun,* 1988).

The existence of crusts, or more specifically, inactive areas on cometary nuclei, was shown dramatically by spacecraft images of comet Halley in 1986 (Fig. 2) (*Keller et al.,* 1986) and again by images of the nucleus of comet Borrelly in 2001 (*Soderblom et al.,* 2002). Dust jets emanated from distinct and relatively small regions on the nucleus surface. From the Halley images, it was estimated that only 10% of the visible surface, or ~20–30% of the sunlit surface, was active. This agreed well with thermal modeling of the water production from Halley, which showed that only 30% of the sunlit surface needed to be active to match the total water production rate (*Weissman,* 1987). At the time of this writing, the Borrelly images are still undergoing analysis, but the apparently active regions are again fairly small. Although one cannot say for certain that the apparently inactive areas in Fig. 2 are not emitting gas (dust emission can likely be ruled out as it would be visible), the Halley and Borrelly images appear to support the idea that comets can develop crusts and thus evolve to dormant or extinct objects.

Further support comes from estimates of the active fraction of other cometary nuclei. *A'Hearn* (1988) showed that many Jupiter-family comets have active fractions of only a few percent. For comets 49P/Arend-Rigaux and 28P/

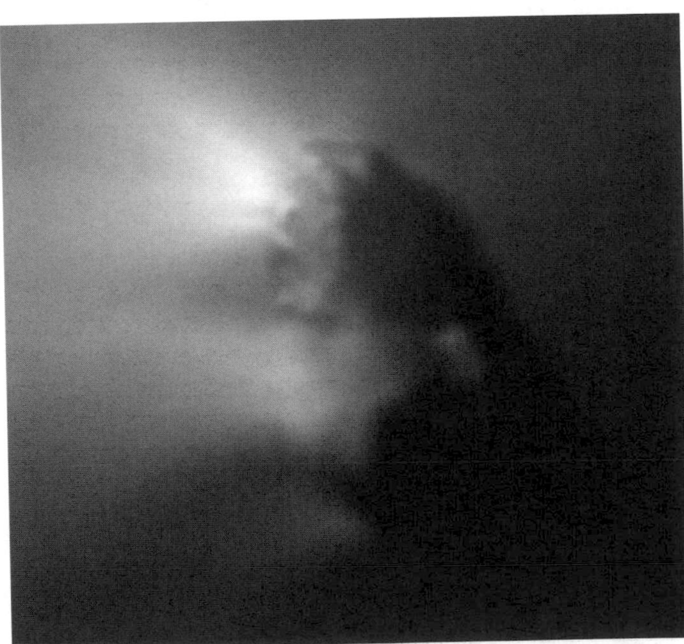

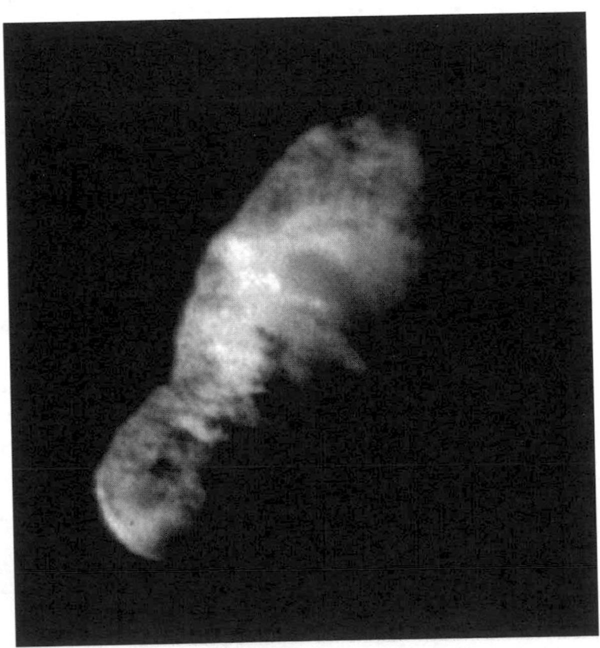

Fig. 2. Composite image of the nucleus of comet Halley (left), taken by the *Giotto* spacecraft on March 14, 1986, and the nucleus of comet Borrelly (right) taken by the *Deep Space 1* spacecraft on September 22, 2001. The Sun is at the left in both images. The *Giotto* spacecraft approached the nucleus from the darkside at a phase angle of ~107°. Several distinct jets of gas and dust are seen emanating from the dayside of the nucleus. The entire Halley nucleus shape is silhouetted against the bright dust coma. The nucleus is ~16 × 8 × 8 km in diameter. The Borrelly image, taken at a phase angle of ~52°, shows a dark, irregular nucleus ~8 × 4 × 4 km with a highly irregular surface topography and active jets (not visible in this version of the image) originating near the center of the nucleus. Halley image: Copyright H. U. Keller, Max-Planck-Institut für Aeronomie. Borrelly image from *Soderblom et al.* (2002).

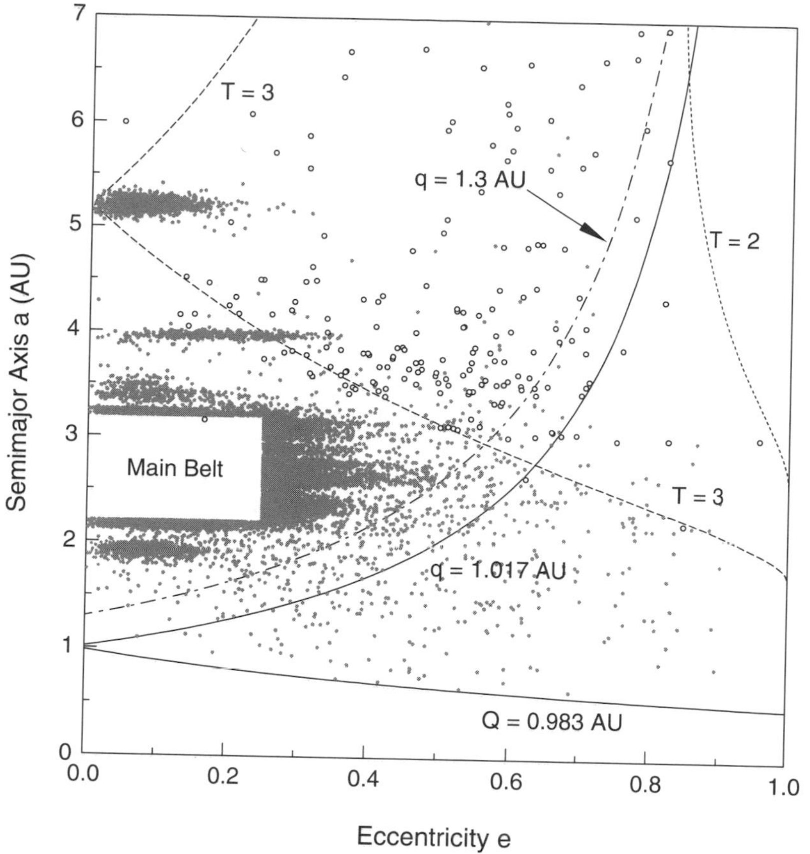

Fig. 3. Location of all known comets and asteroids (H brighter than 18) on an (a, e) scatter diagram of the inner solar system. Comets are shown by open circles, asteroids by filled circles. Solid lines show the boundaries of the region where objects are Earth-crossing (q < 1.017 AU and Q > 0.983 AU). The dot-dash line is for q = 1.3 AU, which defines the Amor-type (Earth-approaching) objects. Dashed lines show T = 3, the traditional boundary between cometary and asteroidal orbits; the dotted line shows T = 2, the boundary between ecliptic and nearly isotropic comets. (Data taken from the Minor Planet Center orbit files for comets and asteroids, March 8, 2001.)

Neujmin 1, both of which are trapped in dynamical resonances that have prolonged their lifetimes in the active comets region (*Marsden,* 1970), the active fractions are estimated as only 0.08% and 0.1% respectively. This suggests that comets may slowly age and evolve toward total inactivity, either by developing nonvolatile crusts on their surfaces or by some other as-yet-unrecognized mechanism. Thus, crust formation does appear to provide a mechanism for evolving cometary nuclei to dormant or extinct states.

4. COMETARY DYNAMICS: EVOLUTION TO ASTEROIDAL ORBITS

Comets have traditionally been divided into two major dynamical groups: long-period (LP) comets with orbital periods >200 yr, and short-period (SP) comets with periods <200 yr. Long-period comets typically have random orbital inclinations while short-period comets typically have inclinations relatively close (within ~35°) to the ecliptic plane. In the past decade it has become common to divide the SP comets into two subgroups: Jupiter-family comets (JFC), with orbital periods <20 yr and a median inclination of ~11°, and Halley-type comets (HTC), with periods of 20–200 yr and a median inclination of ~45°.

A more formal dynamical definition of the difference between the JFC and HTC comets was proposed by *Carusi and Valsecchi* (1987) based on the Tisserand parameter. The Tisserand parameter is an approximation to the Jacobi constant, which is an integral of the motion in the circular restricted three-body problem. It was originally devised to recognize returning periodic comets that may have been perturbed by Jupiter and is given by

$$T = a_J/a + 2\sqrt{(a/a_J)(1 - e^2)}\cos(i) \tag{1}$$

where a_J is Jupiter's semimajor axis, and a, e, and i are the comet's semimajor axis, eccentricity, and inclination respectively. T is also a measure of the relative velocity between a comet and Jupiter during close encounters

$$v_{rel} = v_J\sqrt{3 - T} \tag{2}$$

where v_J is Jupiter's circular orbit velocity about the Sun. Objects with T close to but smaller than 3 have very slow, and thus very strong, encounters with Jupiter. Objects with T > 3 are not Jupiter-crossing.

Jupiter-family comets have T > 2 while long-period and Halley-type comets have T < 2. *Levison* (1996) suggested a new nomenclature to recognize this new classification scheme. He proposes that comets with T > 2 be known as ecliptic comets, while comets with T < 2 be known as nearly isotropic comets. We will use this nomenclature in the discussions that follow.

The different dynamical regimes occupied by comets and asteroids are seen in Fig. 3, which is a scatter diagram in semimajor axis and eccentricity of the known asteroids brighter than H = 18 (corresponding to ~1-km-diameter objects) and all known ecliptic comets. The various lines in the diagram depict lines of constant perihelion or aphelion distance for q = 1.017 and 1.3 AU, Q = 0.983 AU, and lines of constant T = 2 and 3. Objects in the center of the main belt have been removed for clarity. The orbital data is taken from the Minor Planet Center file of orbits for March 8, 2001 (cfa-ftp.harvard.edu/pub/MPCORB/MPCORB.DAT and COMET.DAT). Asteroids are plotted as filled circles in Fig. 3; comets are open circles.

The effect of the T = 3 barrier is clearly seen. It is difficult for ecliptic comets to dynamically detach themselves from Jupiter and evolve to orbits with T > 3. The only real exception is comet 2P/Encke (a = 2.217 AU, e = 0.847, i = 11.8°, T = 3.026), which will be discussed below. In total, 11 JFC comets have orbits with T > 3; most are Jupiter-approaching (with aphelia very close to or crossing Jupiter's eccentric orbit) to the extent that they are still under the gravitational influence of Jupiter. Also, Fig. 3 is slightly misleading as it attempts to plot a three-dimensional function, T(a, e, i) on a two-dimensional (a, e) plane.

Asteroids, for the most part, have T > 3; the Jupiter Trojans are of course a notable exception. Historically, the failure to find asteroidal-appearing bodies with T < 3 was used as an argument against the existence of extinct cometary nuclei. However, in the mid-1980s search programs began to find these asteroids; many of these can be seen in Fig. 3. As we will see below, it is now possible to argue on statistical grounds which of these objects have a high probability of being derived from ecliptic comets.

Comets on planet-crossing orbits are transient members of the solar system. Close and/or distant encounters with the giant planets, in particular Jupiter, limit their mean dynamical lifetimes to ~0.4–0.6 m.y. (*Weissman,* 1979; *Levison and Duncan,* 1997). Thus, they must be continually resupplied from long-lived dynamical reservoirs. The different inclination distributions of the ecliptic and nearly isotropic comets reflect their different source reservoirs. Nearly isotropic comets (LP and HTC) are believed to originate from the nearly spherical Oort cloud, the vast cloud of some 10^{12}–10^{13} comets surrounding the solar system at distances of 10^3–10^5 AU (*Oort,* 1950; for a review see *Weissman,* 1996a). Ecliptic comets are fed into the planetary system from the highly flattened Kuiper belt of some 10^9–10^{10} comets beyond the orbit of Neptune, extending from ~35 to perhaps several hundred AU (for reviews see *Weissman and Levison,* 1997; *Malhotra et al.,* 2000). A third reservoir that may supply ecliptic comets to the planetary region is the scattered disk (*Duncan and Levison,* 1997), icy planetesimals scattered out of the Uranus-Neptune zone and inner Kuiper belt to large semimajor axes, several hundred to ~1000 AU (though with perihelia still close to Neptune's orbit), but not large enough for them to be captured into the Oort cloud.

Much of the research into cometary dynamics in recent years has had the goal of understanding the structure and evolution of these mostly invisible reservoirs. These reservoirs were formed at the same time as the planets. Indeed, they can be viewed as the remnants of planet formation. We discuss the evolution of comets from each of these reservoirs to asteroidal orbits below.

4.1. Ecliptic Comets

The observed ecliptic comets have a very flattened inclination distribution; the Jupiter-family comets have a median inclination of only 11°. In the last 15 yr, research attempting to explain this inclination distribution has been extremely active. Indeed, attempts to understand these comets led to one of the most important discoveries in planetary science in the twentieth century: the discovery of the Kuiper belt.

Historically, the ecliptic comets were thought to originate from nearly isotropic comets that had been captured into short-period orbits by gravitational encounters with the planets (*Newton,* 1891, 1893; *Everhart,* 1977). *Fernández* (1980) argued that this process is too inefficient and suggested that a belt of distant icy planetesimals beyond Neptune could better serve as the source of most of the ecliptic comets; such a belt was suggested by *Kuiper* (1951) and later investigated by *Whipple* (1964). *Duncan et al.* (1988) greatly strengthened this argument by performing dynamical simulations that showed that a cometary source beyond Neptune with a low initial inclination distribution (which they named the Kuiper belt) was far more consistent with the observed orbits of the JFC comets than the randomly distributed inclinations of comets in the Oort cloud (see also *Quinn et al.,* 1990). The first Kuiper belt object, 1992 QB_1 (now numbered asteroid 15760), was discovered in 1992 (*Jewitt and Luu,* 1993). Since that time, more than 560 objects have been discovered in orbits beyond Neptune.

A second potential source of ecliptic comets is the scattered disk, predicted by *Duncan and Levison* (1997). Although the scattered disk is related to the Kuiper belt, it is dynamically distinct, consisting of objects in highly eccentric and inclined orbits that are typically Neptune-crossing or approaching. Approximately 50 scattered disk objects have now been found. However, once an object leaves the Kuiper belt or the scattered disk, its dynamical evolution is indistinguishable. Thus, for the remainder of this discussion, we will not distinguish between these two source regions.

Levison and Duncan (1997, hereafter *LD97*) have presented the most comprehensive simulations to date of the dynamical evolution of objects from the Kuiper belt. They found that as objects evolve inward from the Kuiper belt or scattered disk, they tend to be under the dynamical control of only one planet at any time, and the Tisserand parameter with respect to that planet is always close to 3. The planet scatters the comets randomly inward and outward. However, the comets can reach maximum eccentricities on the order of only ~0.25 due to the large Tisserand parameter. Thus, the planet can only hand the comets off to the planet directly interior or exterior to it. However, once the comets come under the gravitational influence of Jupiter, they can have much larger eccentricities, even up to 1.0, at which point they can be ejected from the solar system.

LD97 estimated the ratio of active to extinct ecliptic comets by comparing the orbital inclination distribution of simulated JFCs to that of the observed JFCs. *LD97* estimated the physical lifetimes of JFCs, restricting the analysis to JFCs with perihelion distances <2.5 AU and absolute cometary magnitudes, H_{10}, brighter than 9, which they argued are observationally complete. (The cometary absolute magnitude, H_{10}, is the total magnitude the active comet would have at 1 AU from the Sun and 1 AU from Earth, assuming that cometary brightness varies with heliocentric distance, r, as r^{-4}. Note that this is different from the traditional H magnitude system for asteroids.)

As described above, the observed JFCs have a very flattened inclination distribution with a median inclination of only 11°. However, *LD97* showed that the inclinations were even lower when objects were first injected into this population and subsequently increased as a function of time (see also *Levison and Duncan,* 1994). This can be seen in Fig. 4. The heavy solid curve in Fig. 4 shows the cumulative inclination distribution of the known JFCs, while the top dotted curve shows the cumulative inclination distribution of *LD97*'s simulated comets when they were first injected into the visible JFC region, q < 2.5 AU. As can be seen in the figure, the median inclination of the model is significantly smaller than the observations. A Kolmogorov-Smirnov test shows that the probability that the two distributions are derived from the same parent distribution is only 7×10^{-3}.

The bottom dotted curve in Fig. 4 shows the inclination distribution that *LD97* predicted if the JFCs are in steady-state dynamical equilibrium and if they remain active throughout their time in the JFC region. In this case the median inclination is significantly larger than is observed. This implies that we are not seeing all the comets in the JFC population, and that the dynamically older comets remain undetected or unrecognized as comets. Alternatively, some physical mechanism removes the comets before they reach dynamical maturity.

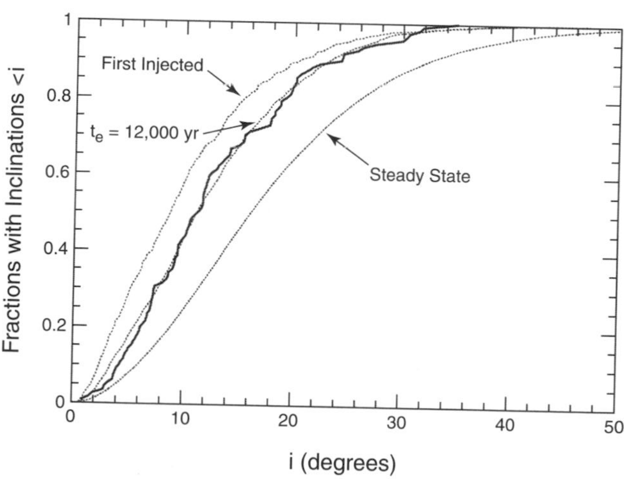

Fig. 4. The cumulative inclination distribution of Jupiter-family comets with perihelion distances less than 2.5 AU. The bold solid curve is the observed distribution taken from *Marsden and Williams'* (1999) catalog. The top dotted curve shows the cumulative inclination distribution of *LD97*'s simulated comets when they were first injected into the JFC region with q < 2.5 AU. The bottom dotted curve shows the inclination distribution that *LD97* predicts if the JFCs are in steady-state equilibrium and if they remain active throughout their time in the JFC region. The middle dotted curve presents the distribution of *LD97*'s best-fit model with t_e = 12,000 yr.

Indeed, it is possible to match the model to the observations by assuming a simple model for the physical aging of comets. *LD97* modeled the effects of physical aging by assuming that all comets remain active for t_e years, after which they are permanently extinct. They found good agreement to the observed inclination distribution of JFCs for values of t_e between 3000 and 25,000 yr. The most likely value is 12,000 yr, or about 1600 perihelion passages, assuming a typical orbital period of ~7.5 yr. The middle dotted curve in Fig. 4 shows the predicted inclination distribution for t_e = 12,000 yr. The match to the observations is excellent. A Kolmogorov-Smirnov test shows that the probability that the two distributions are derived from the same parent distribution is >0.9.

The fact that *LD97* needed to invoke a physical lifetime for comets that is significantly smaller than their dynamical lifetime (2.7×10^5 yr, although much of that time is spent at perihelia >2.5 AU) suggests that there may be a significant number of extinct comets in orbits similar to the JFCs. *LD97* estimated the ratio of extinct to active JFCs with perihelion distance <2.5 AU. For values of t_e between 3000 and 25,000 yr, this ratio is between 7 and 2. If one assumes t_e = 12,000 yr, then the ratio is 3.5. Based on data from *Fernández et al.* (1992), *LD97* estimated that there are 108 active JFCs with q < 2.5 AU and H_{10} < 9 in the solar system. Thus, there may be ~380 extinct JFCs with q < 2.5 AU and H_{10} < 9.

There are several caveats that must be understood before interpreting this result. First, the above estimates assume that all comets become extinct rather than disintegrating. As discussed in the previous section, splitting or disruption may be a common end state for ecliptic comets. Thus, we must assume that the above estimate is an upper limit to the population of extinct JFCs.

Also, *LD97*'s integrations fail to produce objects with orbits similar to that of comet 2P/Encke. P/Encke is a bright, low-inclination comet whose orbit is distinguished by an aphelion distance of only 4.1 AU, well inside the orbit of Jupiter. Although similar to JFCs in all other respects, 2P/Encke is dynamically decoupled from Jupiter. *LD97* most likely did not produce objects on orbits similar to this comet because their integrations did not include the terrestrial planets or the effects of nongravitational forces (forces from outgassing on the dayside of the rotating cometary nucleus). Therefore, the above estimate of the extinct comets does not include this population.

Harris and Bailey (1998) performed dynamical integrations that included the terrestrial planets and nongravitational forces and showed that the capture rate into decoupled, Encke-like orbits was substantially increased. They find that nongravitational forces, in particular, substantially enhance the capture rate into decoupled orbits, if the nongravitational forces act in a uniform manner throughout the comet's dynamical evolution. However, the latter may not be the case as there is evidence for apparent precession of cometary rotation pole orientations and for variations in the nongravitational accelerations (see, e.g., *Sekanina,* 1991). Also, *Weissman* (1979) showed that uniformly acting nongravitational forces lead to predictions for the long-period comet population that are not, in fact, observed.

Finally, the estimate of ~380 extinct JFCs refers to objects that at one time had absolute magnitudes brighter than H_{10} = 9. Ideally, we would like this estimate to be with respect to objects larger than some known size. Relationships that equate H_{10} with nucleus size have been proposed by *Weissman* (1990) and *Bailey and Stagg* (1988) but others have found such relationships poorly founded and unreliable (*Zahnle et al.,* 1998; *Levison et al.,* 2000).

A new method for estimating the number of extinct JFCs is now possible based on studies of the origin of the NEO population. *Bottke et al.* (2002) modeled the orbital and absolute magnitude distributions of the NEO population, using 138 discoveries or accidental rediscoveries from a single source, the Spacewatch survey (*Scotti,* 1996), and then corrected for observational selection effects. The details of this modeling effort are reviewed by *Morbidelli et al.* (2002); we will summarize them here.

Bottke et al. (2002) numerically integrated thousands of test particles from five source regions believed to have provided most NEOs to the terrestrial planets region. These source regions are (1) the ν_6 secular resonance at ~2.1 AU; (2) the Mars-crossing asteroid population adjacent to the main belt; (3) the 3:1 mean-motion resonance with Jupiter at 2.5 AU; (4) the outer main-belt population between 2.8 and 3.5 AU; and (5) the Jupiter-family comets. The dynamical simulation of the JFCs uses the integrations of *LD97*.

By comparing the orbital element distributions of hypothetical NEOs originating from these five regions to the observed NEO orbital distributions derived from the Spacewatch survey, *Bottke et al.* (2002) estimated the relative importance of each NEO source region. Although there is considerable overlap between the NEOs produced by the individual source regions, particularly for NEOs with semimajor axes <2.8 AU, the JFCs produced NEOs that were largely distinct from the other source regions, with the exception of the outer main belt asteroid source region. JFC-derived NEOs had semimajor axes >2.6 AU and eccentricities >0.55.

Bottke et al. found that ~6 ± 4% of NEOs with a < 7.4 AU, e < 1.0, i < 90°, and 13 < H < 22 are extinct JFCs. (H in this case refers to the standard absolute magnitude system for asteroids.) The total number of NEOs with H < 18 was found to be ~960 ± 120. These results imply that there are ~61 ± 50 extinct JFCs with H < 18 among the NEO population.

If we assume that an extinct comet has an albedo of 0.04, like those measured for cometary nuclei (*Fernández,* 1999), we obtain a diameter of 1.7 km for H = 18 (*Bowell et al.,* 1989). *Shoemaker and Wolfe* (1982) estimated that the cumulative cometary nucleus size distribution has the form $N(>D) \propto D^{-2.2}$, although *Lowry* (2001) more recently found a shallower slope of –1.6; *Weissman and Lowry* (2002) extended the latter work and found a similar slope of –1.6. In contrast, *Fernández et al.* (1999) found a slope of –2.65, although the sample they fit is quite small, only nine comets. Thus, the number of 1-km-diameter extinct JFCs among the NEOs (q < 1.3 AU) is likely ~140–200 objects. *Bottke et al.* (2002), using their estimated NEO size distribution of $N(>D) \propto D^{-1.75}$, calculated ~150 1-km-diameter extinct JFCs among the NEOs. If one assumes a uniform perihelion distribution vs. heliocentric distance, these numbers compare favorably with the *LD97* estimate of ~380 extinct JFCs with q < 2.5 AU (although we do not know if we are talking about the same sized objects).

Bottke et al. (2002) also used their model to estimate the probability that any individual known NEO is an extinct comet, based on its semimajor axis, eccentricity, and inclination. Taking the list of asteroids from the November 2000 database of T. Bowell (http://asteroid.lowell.edu), Bottke et al. found 57 NEOs with a P_{JFC} > 10% chance of having originally been an ecliptic comet. These objects are listed in Table 1, in order of decreasing P_{JFC}. In addition to the orbital elements a, e, and i for each object, Table 1 lists the Tisserand parameter for that orbit, and the probability that the object comes from each of the five source regions used by Bottke et al. (The last column, which refers to the object's spectral type and/or albedo, will be discussed in the next section.) Note that as one reads down the table, the most probable source region shifts from the JFC population to the outer main belt, and then to asteroid regions at successively smaller semimajor axes. Most of the objects in Table 1 have 2 < T < 3. A plot of the (a,e) positions of the 57 objects listed in Table 1 is shown in Fig. 5.

There are several caveats that must be noted with regard to the work of *Bottke et al.* (2002). First, the results quoted are for Bottke et al.'s best-fit model, and the solution may not be unique. Second, Bottke et al. do not include the nearly isotropic comets as a potential source of NEOs. Third, the integrations of the orbits of the JFCs do not include the terrestrial planets (which were included for all of the asteroid source-region integrations) or nongravitational forces. The inclusion of the terrestrial planets and nongravitational forces would likely have resulted in more comets being captured to smaller semimajor axes, in particular to orbits dynamically detached from Jupiter, and this would likely have increased the fraction of NEOs that may have originated from the JFC population.

4.2. Nearly Isotropic Comets

Nearly isotropic comets (NICs) are believed to originate from the Oort cloud. Monte Carlo simulations by *Weissman* (1979) showed that the typical long-period comet from the Oort cloud makes an average of 5 perihelion passages

TABLE 1. Extinct comet candidates among the NEO population, in order of P_{JFC} probability (*Bottke et al.*, 2002).

Number	Name	a (AU)	e	i (degrees)	T	P_{JFC}	P_{OB}	$P_{3:1}$	P_{MC}	P_{ν_6}	Type
3552	Don Quixote	4.232	0.714	30.816	2.314	1.000	0.000	0.000	0.000	0.000	D, A = 0.045
	1997 SE5	3.727	0.667	2.609	2.656	1.000	0.000	0.000	0.000	0.000	D, T
	1982 YA	3.657	0.700	35.270	2.401	0.971	0.022	0.006	0.000	0.000	
	1984 QY1	2.939	0.914	17.732	2.351	0.961	0.037	0.000	0.002	0.000	
	2000 PG3	2.825	0.859	20.454	2.549	0.929	0.025	0.037	0.002	0.007	A = 0.021
	2000 EB107	3.032	0.585	25.283	2.836	0.904	0.036	0.058	0.002	0.000	D
	2000 KE41	3.000	0.865	50.450	2.220	0.842	0.000	0.021	0.000	0.137	
	1999 LT1	2.976	0.658	42.608	2.587	0.738	0.037	0.187	0.007	0.031	C, F
	1994 LW	3.167	0.619	22.999	2.771	0.709	0.232	0.038	0.010	0.011	
	1998 GL10	3.183	0.668	8.673	2.786	0.677	0.316	0.003	0.003	0.001	
	1998 FR11	2.797	0.711	6.597	2.885	0.653	0.220	0.062	0.052	0.014	
	2000 DN1	2.884	0.669	7.769	2.901	0.645	0.299	0.023	0.029	0.005	
	1995 SA15	2.753	0.739	0.971	2.870	0.599	0.332	0.028	0.024	0.017	
	1998 SH2	2.710	0.722	2.484	2.918	0.599	0.332	0.028	0.024	0.017	
	1998 MX5	2.918	0.611	9.707	2.952	0.578	0.398	0.010	0.011	0.004	Sq, X
	1999 LD30	2.901	0.606	8.729	2.968	0.578	0.398	0.010	0.011	0.004	
	1998 HN3	3.132	0.614	9.215	2.870	0.543	0.448	0.005	0.003	0.001	
14827	1986 JK	2.800	0.680	2.139	2.933	0.534	0.438	0.011	0.013	0.004	C
	1998 SY14	2.850	0.665	3.517	2.929	0.534	0.438	0.011	0.013	0.004	
	1999 RD32	2.630	0.777	6.681	2.868	0.534	0.312	0.093	0.041	0.020	
	1999 DB2	2.999	0.620	11.608	2.902	0.424	0.547	0.014	0.012	0.003	D
	2000 LF6	2.911	0.611	14.826	2.932	0.424	0.547	0.014	0.012	0.003	D
	1992 UB	3.070	0.582	15.945	2.896	0.412	0.560	0.013	0.012	0.003	S
	1998 SE35	3.005	0.594	14.817	2.914	0.401	0.573	0.010	0.012	0.003	
	1999 VX15	3.010	0.599	12.337	2.919	0.401	0.573	0.010	0.012	0.003	
	2000 QS7	2.701	0.665	3.202	3.001	0.373	0.517	0.052	0.043	0.015	
	1998 KO3	2.622	0.773	54.642	2.506	0.354	0.025	0.457	0.137	0.028	
	1983 LC	2.686	0.716	1.528	2.940	0.349	0.479	0.086	0.049	0.037	
	1997 VM4	2.622	0.812	14.137	2.788	0.290	0.360	0.230	0.084	0.034	Q
	1998 VD31	2.652	0.803	10.234	2.800	0.290	0.360	0.230	0.084	0.034	S
	1985 WA	2.831	0.607	9.803	2.993	0.287	0.623	0.027	0.058	0.005	
	1995 QN3	3.304	0.644	14.793	2.754	0.280	0.714	0.003	0.002	0.000	
	1995 DV1	2.802	0.650	3.512	2.970	0.218	0.737	0.016	0.025	0.004	
5370	Taranis	3.342	0.632	19.027	2.731	0.205	0.782	0.006	0.006	0.001	
	1998 US18	2.623	0.680	9.661	3.010	0.195	0.465	0.190	0.115	0.035	
5324	Lyapunov	2.959	0.615	19.495	2.880	0.190	0.790	0.008	0.009	0.003	
	1999 HA2	2.789	0.700	15.085	2.875	0.163	0.417	0.169	0.184	0.067	
	1997 EN23	3.261	0.634	6.966	2.811	0.157	0.841	0.001	0.000	0.001	
	1999 SE10	3.210	0.621	6.897	2.843	0.157	0.841	0.001	0.000	0.001	X
	2000 PF5	3.237	0.642	6.156	2.810	0.157	0.841	0.001	0.000	0.001	
	2000 QN130	2.902	0.573	2.564	3.016	0.156	0.836	0.004	0.002	0.001	
	1997 YM3	3.242	0.673	4.014	2.770	0.155	0.809	0.017	0.005	0.013	
6178	1986 DA	2.811	0.586	4.307	3.039	0.152	0.812	0.015	0.019	0.002	M, A = 0.14
16064	1999 RH27	2.885	0.577	4.396	3.016	0.152	0.812	0.015	0.019	0.002	C
	1994 AB1	2.850	0.590	4.523	3.017	0.152	0.812	0.015	0.019	0.002	Sq
	1999 GT6	2.830	0.578	4.277	3.039	0.152	0.812	0.015	0.019	0.002	
	1997 UZ10	2.868	0.618	12.763	2.953	0.148	0.735	0.038	0.064	0.015	
	1999 AF4	2.828	0.618	12.571	2.971	0.148	0.735	0.038	0.064	0.015	C
	1991 XB	2.942	0.590	16.305	2.934	0.139	0.842	0.009	0.007	0.002	S
	2000 HD74	2.922	0.594	49.373	2.566	0.138	0.480	0.244	0.104	0.033	
	2000 GV127	2.823	0.622	17.936	2.941	0.120	0.737	0.045	0.084	0.014	
	1998 ST4	2.820	0.597	9.292	3.011	0.114	0.828	0.012	0.043	0.003	
	1999 RU2	2.807	0.560	5.449	3.065	0.114	0.828	0.012	0.043	0.003	
	1999 VQ11	2.810	0.595	7.940	3.022	0.114	0.828	0.012	0.043	0.003	
	1997 QK1	2.794	0.642	2.886	2.985	0.109	0.776	0.056	0.043	0.017	SQ
	2000 GC147	2.735	0.601	2.278	3.061	0.109	0.776	0.056	0.043	0.017	
	1983 VA	2.608	0.694	16.261	2.974	0.100	0.183	0.401	0.208	0.108	A = 0.07

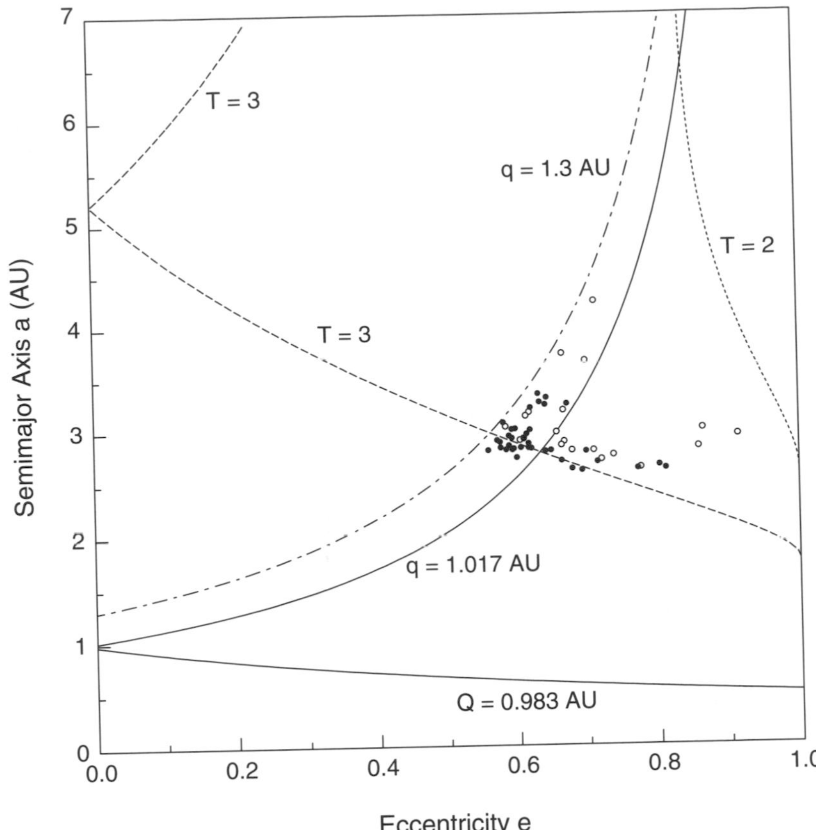

Fig. 5. Location on the (a, e) diagram of the 57 extinct comet candidates with $P_{JFC} > 10\%$ from *Bottke et al.* (2002). Twenty candidates with $P_{JFC} > 50\%$ are shown by open circles; 37 candidates with $50\% > P_{JFC} > 10\%$ are shown by filled circles. Labeled curves are the same as in Fig. 3.

through the planetary region (for perihelia <4 AU) with a mean time of 0.6 m.y. between the first and last passage. Approximately 65% of LP comets are dynamically ejected to interstellar space (primarily by Jupiter), 27% are lost to random disruption (or some other physical mechanism), and the remainder are lost to a variety of processes including loss of all volatiles or perturbation onto a Sun-impacting orbit. The mean semimajor axis of an observed "dynamically new" LP comet (a comet passing through the planetary region for the first time) is ~22,000 AU (*Marsden et al.,* 1978).

Oort cloud comets have been stored at great distances from the Sun for most of the history of the solar system. However, it is most likely that the Oort cloud comets were formed in the giant planets region (*Kuiper,* 1951). That is, the comets were formed as icy planetesimals in nearly circular orbits between and beyond the giant planets and are the probable remnants of planetary formation (*Duncan et al.,* 1987; *Fernández,* 1997). As the giant planets grew, the comets were gravitationally scattered to larger semimajor axes. When the semimajor axes had grown sufficiently, to several thousand AU, perturbations by the tidal field of the galactic disk and by random passing stars scattered the perihelia of the cometary orbits, raising many of them away from the planetary region. Comets from the Oort cloud are currently being fed back into the planetary region by these same galactic and stellar perturbations.

Dynamical modeling of the Oort cloud (*Duncan et al.,* 1987; *Dones et al.,* 2000) suggests that there are two dynamical regimes within the cloud. The outer, dynamically active cloud, as first described by *Oort* (1950), is approximately spherical and comets there are fed directly into the planetary system by the perturbations of the galactic tide and random passing stars. These are the long-period comets that we see. The outer Oort cloud reaches from ~15,000 to 100,000 AU (in semimajor axis), though as noted above, most observed LP comets have semimajor axes of ~20,000–30,000 AU. Cometary orbits in the outer cloud have been randomized by stellar and galactic perturbations, with random inclinations and a mean eccentricity of ~0.7 (*Hills,* 1981).

The inner Oort cloud stretches from ~3000 to 15,000 AU in semimajor axis (*Duncan et al.,* 1987). The inner cloud is dynamically less active because, being closer to the Sun, it does not experience the tidal effect of the galactic disk and random passing stars. An exception to this occurs if a passing star penetrates the Oort cloud (*Hills,* 1981, *Weissman,* 1996b), in which case large numbers of comets may be perturbed into the planetary region, resulting in a "cometary shower." The best current evidence suggests that we are not presently in a cometary shower (*Fernández and Ip,* 1991; *Weissman,* 1993). Because it is not subject to the same external perturbations as the outer cloud, cometary orbits in the inner Oort cloud have not been completely randomized and the inner cloud is believed to be flattened, with inclinations relatively close, within ~45° of the ecliptic plane.

Weissman (1996a) estimated a population of 10^{12} comets in the outer Oort cloud brighter than $H_{10} = 11$ by comparing dynamical models with the observed flux of LP comets through the planetary system. *Dones et al.* (2000) found that the populations of the inner and outer clouds were roughly equal. However, that model assumes that the Oort cloud formed in a galactic environment similar to the one that the solar system is in today. The inner Oort cloud could contain a significantly larger population if the Sun formed in a star cluster or molecular cloud, which seems likely (*Fernández,* 1997; *Eggers,* 1999; *Fernández and Brunini,* 2000). Also, *Levison et al.* (2001) argued that a massive inner Oort cloud must exist in order to explain the observed Halley-type comet population. Thus, the population of the inner Oort cloud is very uncertain but is thought to be at least a few times 10^{12} comets.

Galactic tides and stellar perturbations change the perihelion distances of Oort cloud comets, feeding them into (and out of) the planetary region. As a comet comes through the planetary system for the first time, it receives a random kick in orbital energy that is, on average (if it is Saturn- or Jupiter-crossing), much larger than its binding energy. Thus, depending on the sign of the kick, the comet is either ejected from the solar system or captured onto an orbit with a smaller semimajor axis (and removed from the realm where galactic and stellar perturbations are important). This random walk in orbital energy continues until the comet reaches some dynamical or physical end state.

There are two dynamical paths that an Oort cloud comet can take in order to become a visible comet. Comets from the outer Oort cloud can be perturbed from orbits with perihelia beyond the orbit of Neptune to perihelia inside the orbit of Jupiter. This involves a substantial decrease in the angular momentum of the orbit. *Hills* (1981; also see *Duncan et al.,* 1987) showed that the rate of change in a comet's perihelion distance due to galactic tides is proportional to a^2, where a is the orbit semimajor axis. Thus, only orbits sufficiently distant from the Sun feel the strong tidal effects of galactic and stellar perturbations, and can achieve this direct perturbation into the visible region. This is why the outer Oort cloud is dynamically active and the inner cloud is less so.

If a comet is not perturbed to a perihelion distance inside Jupiter's orbit, it will still undergo a random walk in energy due to perturbations by the giant planets, but likely without ever becoming visible. We note that some NICs have been discovered with perihelia beyond Jupiter; these are always dynamically new comets, which are known to be anomalously bright on their first perihelion passage [see *Weissman* (1996a) for a discussion of possible reasons]. This inability of comets to evolve into the inner-planets region unless they are thrown there directly has been called the "Jupiter barrier" and has been examined by *Weissman* (1985), *Wetherill* (1994), and others.

However, comets from both the inner and outer clouds can follow a second path into the visible region. As noted above, comets that are thrown into orbits with perihelia among the giant planets undergo an orbital evolution that can lead to some of them being captured in intermediate-period, Halley-type orbits. Because this evolution occurs relatively far from the Sun, beyond the water ice sublimation region, the comets may not be subjected to the same insolation-related physical loss mechanisms as those with perihelia inside the orbit of Jupiter. This dynamical path was first proposed by *Everhart* (1977) as a possible source of Jupiter-family comets.

As with the ecliptic comets, a diagnostic for the origin of the HTCs is their inclination distribution (*Levison et al.,* 2001; hereafter *LDD01*). The cumulative inclination distributions of LP comets (dotted line), HTCs (dashed line), and JFCs (solid line), taken from *Marsden and Williams'* (1999) catalog, is shown in Fig. 6. While the distribution of the LP comets is isotropic, the HTCs are mainly prograde with a median inclination of only 45°, and the JFCs are all prograde with a median inclination of only 11°.

The most complete work on the subject of the HTC inclination distribution was performed by *LDD01*. They integrated the orbits of 27,700 test particles initially entering the planetary system from the Oort cloud. They found that an isotropically distributed Oort cloud does not reproduce the observed orbital element distribution of the HTCs. In order to match the observations, the initial inclination distribution of the progenitors of the HTCs must be similar to the observed HTC inclination distribution. *LDD01* was able to match the observations with an Oort cloud that consisted of an isotropic outer cloud and a massive disklike inner

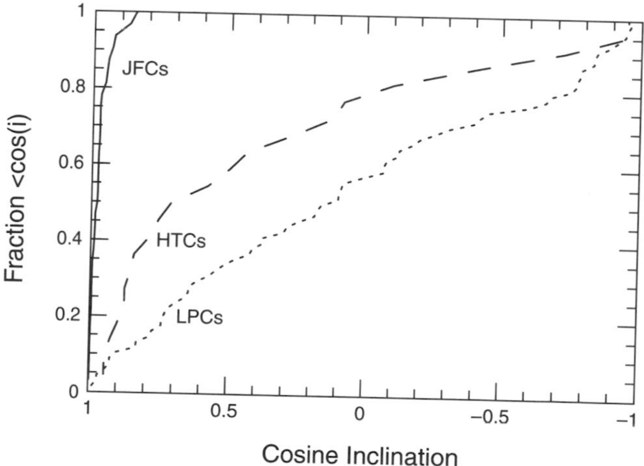

Fig. 6. The cumulative inclination distribution of observed comets with perihelia within 1.3 AU of the Sun taken from the catalog of *Marsden and Williams* (1999). The solid, dashed, and dotted curves are the JFCs, HTCs, and LPCs, respectively. We only include LPCs with a "quality code" of 1A or 1B. A random distribution would be a diagonal line from the lower left to the upper right in this figure.

TABLE 2. Asteroids in cometlike orbits with Tisserand parameter < 2.

Number	Name	a (AU)	e	i (degrees)	H	T	q (AU)	P (yr)	Spectral Type
5335	Damocles	11.818	0.8667	62.047	13.3	1.145	1.575	40.63	D
15504	1998 RG33	9.433	0.7726	34.936	12.1	1.953	2.145	28.97	
20461	1999 LD31	24.429	0.9025	160.213	13.8	−1.543	2.381	120.74	
	1996 PW	305.934	0.9917	29.782	14.0	1.731	2.547	5351.10	D
	1997 MD10	26.865	0.9427	59.047	16.0	0.974	1.540	139.25	
	1998 WU24	15.207	0.9069	42.564	15.0	1.403	1.415	59.30	
	1999 LE31	8.155	0.4718	151.879	12.4	−1.309	4.307	23.29	A = 0.031
	1999 XS35	18.044	0.9474	19.473	17.2	1.412	0.949	76.65	
	2000 AB229	52.163	0.9561	68.715	14.0	0.773	2.290	376.74	
	2000 DG8	10.773	0.7930	129.426	13.1	−0.630	2.230	35.36	A = 0.027
	2000 HE46	24.191	0.9029	158.340	14.6	−1.507	2.349	118.98	A = 0.023

cloud. Their best-fit model consisted of an inner Oort cloud with a median inclination of only 20°.

However, *LDD01*'s model predicts too many observed HTCs as compared to the observed number of dynamically new long-period comets. This inconsistency may be related to the fading described in section 3 for the LP comets, though *LDD01*'s model requires a more severe fading law than has been suggested previously. Thus, the origin of the Halley-type comets remains problematic.

If physical-loss mechanisms do produce extinct LP and HTC nuclei, rather than simply destroying them, it may be possible to find such objects among the asteroid population. These objects should have Tisserand parameters, T < 2, and semimajor axes outside of the radius of Jupiter's orbit. They also are likely to be in high-inclination orbits.

It is also important to be able to show that these objects cannot have an asteroidal origin. *Bottke et al.*'s (2002) integrations show that when objects leaving the asteroid belt cross the orbit of Jupiter, they have T > 2. Further, *LD97*'s integrations show that objects with T > 2 are unlikely to evolve onto orbits similar to visible HTCs or LP comets.

Table 2 lists 11 asteroids from the Minor Planet Center database (July 27, 2001) with semimajor axes >8 AU and eccentricities >0.4. Dynamically, it is most probable that these objects are all derived from NIC-type orbits and not from the asteroid belt. Note the large number of high-inclination objects among the sample. The median inclination of these objects is 62°, which is close enough to that of the observed HTCs, given the small number of objects in both samples, to suggest that they may be the same.

Given that we believe that the objects in Table 2 came from the Oort cloud, can we conclude that they are extinct comets? Unfortunately not. *Weissman and Levison* (1997) examined the orbit of 1996 PW and showed that it was almost certainly derived from the Oort cloud. But they found that the probability of it being an extinct cometary nucleus was roughly equal to the probability of it being an asteroid ejected during the clearing of the planetary zones in the early solar system. *Hicks et al.* (2000) found D-type colors for 1996 PW, consistent with it being either an extinct nucleus or an outer main-belt asteroid. *Weissman and Levison* (1997) proposed that ~0.8–2.3% of the Oort cloud population is likely asteroidal, material ejected from the outer primordial asteroid belt and from dynamical resonances during the late stages of planetary-zone clearing. These objects would most likely be indistinguishable from extinct comets, using Earth-based observations, except perhaps by their spectral types (see section 5). Thus we are unable to determine with certainty if an asteroidal-appearing object in a NIC-like orbit is an extinct comet or an asteroid from the Oort cloud.

5. PHYSICAL STUDIES OF COMETARY NUCLEI: COMPARISON WITH ASTEROIDS

The same techniques that can be used to study and classify asteroids, e.g., BVRI photometry, reflection spectroscopy, thermal IR observations, etc., have now begun to be applied to cometary nuclei. Observations of "bare" cometary nuclei are necessarily difficult because they must be performed when the comets are far from the Sun and apparently inactive, and thus relatively faint when observed from Earth. For example, a bare nucleus with a radius of 2 km and an albedo of 0.04 would be V magnitude ~20.0, 21.5, and 22.6 at 3, 4, and 5 AU, respectively, at opposition. Also, it is difficult to rule out the presence of unresolved coma in such studies.

Nevertheless, a consistent picture of cometary nuclei has begun to emerge. Nuclei are typically elongated bodies, a few kilometers in radius, with surface albedos of 0.04 ± 0.02 (see *Meech,* 2002, and references therein; also *Fernández,* 1999). Their visible spectra are typically featureless and range from gray to very red objects, similar to C- or D-type asteroids (*Luu,* 1993; see also Fig. 2 in *Paper 1*). The distribution of rotation periods tends to be similar but somewhat slower than main-belt asteroids; there are no recognized fast rotators among the cometary nuclei. The latter is suggestive of low tensile strengths for the cometary nuclei. Bulk densities for cometary nuclei have only been estimated indirectly, based either on the effect of nongravitational forces on the orbit of comet 1P/Halley: 0.2–1.3 g cm^{-3} (*Rickman,* 1986; *Peale,* 1989) or on the tidal disruption and reassembly of the fragments of comet D/Shoemaker-Levy 9: 0.6–1.1 g cm^{-3} (*Asphaug and Benz,* 1994, 1996). Preferred models of nuclei suggest they are "primordial rubble piles" or "fluffy aggregates" (*Weissman,* 1986; *Donn and Hughes,* 1986) having received relatively little physical processing since they accreted in the solar nebula 4.5 G.y. ago. The spacecraft images of comets Halley and Borrelly, shown in Fig. 2, are consistent with this description. The topography of the nucleus surfaces appears to be chaotic and rough at all scales.

Although none of these properties provide absolute discriminators between comets and asteroids, we can test *Bottke et al.*'s (2002) sample of extinct cometary candidates, listed in Table 1, for consistency with a cometary origin. The final column of Table 1 lists the identified spectral types (and/or measured albedos) of 20 of the candidate comets, drawn from the literature and from personal communications with M. Hicks, R. Whiteley, S. Bus, R. Binzel, and Y. Fernández. We find that those objects with $P_{JFC} > 0.5$ are dominated by primitive spectral types; six out of seven observed objects are spectral type C, D, F, or low albedo (see below). On the other hand, only 5 out of 13 observed objects with $P_{JFC} < 0.5$ are primitive types. Thus, Bottke et al.'s dynamical argument for many of these objects being extinct comets is independently substantiated by the spectral data. We emphasize that although this is not a proof that these objects are cometary, it is consistent with them having a cometary origin.

Similarly, *Fernández et al.* (2001) compared measured albedos for 14 cometary nuclei and 10 NEOs with Tisserand parameters <3, and 34 NEOs with T > 3. They showed that all of the comets and 9 out of 10 of the NEOs with T < 3 had low albedos, ≤0.07, while most of the NEOs with T > 3 had albedos >0.15. Again, this is consistent with the T < 3 objects having a cometary origin.

In Table 2 we list the spectral types or measured albedos for 5 of the 11 Oort cloud candidates. All of these are either primitive spectral types or low albedo, consistent with a cometary origin. So once again we have a consistent picture of candidate objects with primitive, possibly cometary, surface types.

Clearly, additional physical studies are required of the other objects listed in Tables 1 and 2 to gain a more complete picture of these extinct comet candidates. Although it may not be possible to conclusively identify these bodies as cometary using only Earth-based observations, such studies will help to identify the most probable candidates, which might then become the targets of spacecraft missions that will study them in far greater detail.

6. OTHER COMETARY CANDIDATES

Many asteroids have been suggested as extinct comet candidates over the past two decades, based on their physical or dynamical characteristics. In this section we will examine some of these candidates in greater detail. To begin with, we consider the list of nine "probable" extinct comet candidates in *Paper 1*, listed in Table 3. In light of current data, how many of these would we still consider to be good candidates?

To begin with, 2060 Chiron, the first Centaur asteroid discovered, has since been found to display cometary activity (*Meech and Belton,* 1989) and is clearly a comet, likely derived from the Kuiper belt or scattered disk. However, *Bottke et al.* (2002) find that only one of five NEOs in Table 3, 3552 Don Quixote, likely has a cometary origin. The other NEOs appear to be better explained, at least dynamically, as having come from the asteroid belt. One of these, 2201 Oljato, is spectral type Sq, suggesting a surface composition similar to ordinary chondrites. Although Oljato has been a popular candidate for an extinct cometary nucleus in the past (see *Paper 1*; also *McFadden et al.,* 1993), both the dynamical and spectral data appear to rule it out.

Asteroid 944 Hidalgo, with T = 2.067, is in a dynamical regime that would argue very strongly for it being a cometary body, and equally likely for it being derived from the Kuiper belt, Oort cloud, or scattered disk. Asteroids 5201 Ferraz-Mello and 1984 BC are in dynamical regimes where they are possibly extinct JFCs, but without additional physical studies, it is difficult to say for sure.

TABLE 3. Probable extinct comet candidates from *Weissman et al.* (1989).

Number	Name	a (AU)	e	i (degrees)	T	Spectral Type	Likely Source Region (*Bottke et al.,* 2002)
944	Hidalgo	5.752	0.661	42.557	2.067	D	
2060	Chiron	13.602	0.380	6.940	3.352	B	
2101	Adonis	1.874	0.765	1.349	3.550		ν_6 43%; OB 32%
2201	Oljato	2.171	0.713	2.517	3.302	Sq	ν_6 55%; MC 27%
2212	Hephaistos	2.166	0.834	11.769	3.100		ν_6 44%; MC 38%
3200	Phaethon	1.271	0.890	22.166	4.512	BF	ν_6 52%; MC 24%; 3:1 24%
3552	Don Quixote	4.232	0.714	30.816	2.314	D	JFC 100%
5201	Ferraz-Mello	3.173	0.518	4.053	2.973		
	1984 BC	3.493	0.534	21.364	2.780		

Asteroid 3200 Phaethon is an interesting case because it has been identified with the Geminid meteor shower (*Whipple,* 1983; *Williams and Wu,* 1993), one of the major regular showers each year. Our expectation is that meteor showers originate from comets; the Geminids is the only major shower definitively identified with an asteroid. Could 3200 Phaethon be an extinct comet nucleus? Its spectral type, BF, is a primitive class, more likely to originate from the middle or outer asteroid belt than from the ν_6 resonance, as suggested by *Bottke et al.* (2002). However, its albedo has been estimated as 0.11–0.17, more typical of a non-primitive asteroid (*Veeder et al.,* 1984; *Green et al.,* 1985). Also, a source of F-type asteroids has been identified close to the 3:1 resonance in the Polana family in the main belt (*Doressoundiram et al.,* 1998). As noted in section 4, the dynamical integrations used by *Bottke et al.* (2002) to simulate the JFC evolution did not include the terrestrial planets or nongravitational forces, and this probably precluded them from producing objects in orbits like comet 2P/Encke or even perhaps asteroid 3200 Phaethon. Searches for cometary activity in 3200 Phaethon have all been negative (*Cochran and Barker,* 1984; *Chamberlin et al.,* 1996). So we must conclude that the origin of this object remains an open question.

Two interesting objects are comets 107P/Wilson-Harrington and 133P/Elst-Pizarro, which are also cataloged as asteroids 4015 and 7968 respectively. In each case, these objects were briefly observed to display narrow, taillike structures, which have been interpreted by some as cometary outbursts (*Fernández et al.,* 1997; *Boehnhardt et al.,* 1998). *Lien* (1998) has alternatively suggested that the structures were dust trails resulting from impact events. Both objects have Tisserand parameters greater than 3 (3.084 for P/Wilson-Harrington and 3.184 for P/Elst-Pizarro). In fact, these are the highest values of T for any objects classified as comets, with the exception of 95P/Chiron (which is not Jupiter-crossing). The orbit of Elst-Pizarro is particularly unusual as it is a typical low-eccentricity, low-inclination, main-belt asteroid orbit ($a = 3.158$ AU, $e = 0.165$, $i = 1.38°$). This orbit places it in the Themis collisional family. It is highly improbable that an extinct comet would evolve to such an orbit. A collision with other Themis family debris seems a more plausible explanation for 7968's transient activity.

In the case of Wilson-Harrington, the orbit is more typical of a JFC ($a = 2.643$ AU, $e = 0.621$, $i = 2.78°$) but *Bottke et al.* (2002) assigned it only a 4% probability of having originated from the JFCs. They found the most probable source to be the outer main belt (65%), although that result may again change if more complete integrations of the JFCs (including nongravitational forces) were to be included in their work. Wilson-Harrington's spectral type is CF, consistent with a cometary (or an outer main belt) origin, though as noted above, a source of F-type asteroids has also been identified close to the 3:1 resonance in the main belt (*Doressoundiram et al.,* 1998). The case of Wilson-Harrington is problematic because its activity was observed on only one night, on photographic plates taken for the original Palomar Sky Survey in 1949, and the object is trailed on both images obtained (see *Fernández et al.,* 1997); no activity was detected on plates taken three nights later. Subsequent searches for cometary activity in 4015 Wilson-Harrington have all been negative (e.g., *Chamberlin et al.,* 1996).

Some researchers have suggested possible links between weak meteoroid streams and Earth-crossing asteroids (*Drummond* 1982). For example, *Kostolanský* (1998) suggests possible stream identifications with 2102 Tantalus, 5496 1973 NA, and 1996 BT. However, *Bottke et al.* (2002) did not identify any of these as cometary candidates. Also, the possibility that these are extinct comets is generally not supported by taxonomic identifications, where they are available. Either the meteoroid stream identifications are coincidence or perhaps asteroids can be a source of meteoroid streams.

Yeomans (1991) searched for nongravitational motion in the orbits of 12 near-Earth asteroids and reported that he detected such evidence for 1566 Icarus and 1862 Apollo. However, he later showed (*Yeomans,* 1992) that this was because of the failure to include relativistic terms in the motion of these objects. Since Icarus is spectral type Sq and Apollo is type Q, it would seem unlikely that these could be extinct comets.

7. DISCUSSION

It now appears highly likely that there are a significant number of dormant or extinct cometary bodies among the asteroid population. We now have the tools and the understanding to begin identifying the most likely cometary candidates and to follow up with telescopic physical studies that will help to confirm or refute those identifications. In addition, we have the promise of possible low-cost spacecraft missions that can study these objects *in situ* in far greater detail.

The work of *Bottke et al.* (2002) allows us to identify likely cometary candidates among the NEO population. The most likely candidates, those with $P_{JFC} > 50\%$, all have Tisserand parameters < 3. It is thus likely that additional asteroids with $T < 3$ but with $q > 1.3$ are also dormant or extinct comets. Potential improvements to the work of Bottke et al. will further refine our ability to identify cometary candidates through their dynamical properties. These improvements include better integrations of the Jupiter-family comet population where the terrestrial planets and nongravitational forces are included, improved integrations of the nearly isotropic comet population, including in particular the Halley-type comets, and improved statistics for the unbiased distribution of NEO orbits, based on increased discovery statistics and data from some of the other ongoing NEO search programs.

Additional evidence for extinct comets comes from the discovery of objects in unusual orbits with large semimajor axes, eccentricities, and/or inclinations, and with $T < 2$.

It is far more likely that these objects are derived from the Oort cloud and from objects in Halley-type orbits than from the main asteroid belt. However, we must also remember that some small fraction of the Oort cloud is actually composed of asteroidal bodies ejected during the planetary-zone clearing phase in the early solar system. Thus, these objects with $T < 2$ may still be asteroidal in nature rather than cometary. Improved dynamical simulations of the ejection of these asteroids to the Oort cloud would be valuable in helping to understand this process.

The other key component is physical studies of the candidate comets, starting with those listed in Tables 1 and 2. The most important measurements are spectral identifications and albedo determinations, but rotation light curves, shape modeling, and other measurements will also provide useful clues to the true nature of the objects. Also, the same measurements need to be made of the known cometary nuclei so as to better determine the distributions of physical parameters for the progenitor population. Potential differences in these distributions, such as a lack of fast rotators among comets, will provide useful clues in determining the probability that a particular object has its origin as a comet or an asteroid. Radar studies may help to identify cometary surface types (*Harmon et al.,* 1999), which appear to be low density and rough at scales from meters to kilometers.

It is also conceivable that progress in instrument technology as well as the increasing availability of very large aperture telescopes will allow us to identify weak cometary activity among the candidate comets, observations that so far have been unsuccessful.

Ultimately, some of the greatest gains will come from spacecraft missions, which will permit us to study active cometary nuclei and perhaps some extinct ones at close range. The recent success of *Deep Space 1* in imaging the nucleus of P/Borrelly has been very valuable. Similarly valuable results can be expected from the *Stardust* mission, now en route to 81P/Wild 2, and the *CONTOUR* and *Deep Impact* missions, to be launched in 2002 and 2004 respectively. However, the greatest results will come from ESA's *Rosetta* rendezvous and lander mission, to be launched in 2003, which will permit an in-depth study of the nucleus of 46P/Wirtanen as it travels around the Sun. Since comets appear to be a very diverse population, additional comet rendezvous missions are highly desirable in order to compare results and provide criteria and clues for the detection of extinct nuclei among the asteroid population.

Our understanding of both comets and asteroids has increased remarkably since the *Asteroids II* book and conference 14 years ago. At the same time we can only guess at how much further we will need to go to understand the complex interrelationships between these primordial bodies. There are many questions yet to be answered: How do comets physically evolve to dormant or extinct objects? What fraction of comets reach this end state? Can comets become truly extinct or only temporarily dormant? What fraction of the Oort cloud population is asteroidal bodies? Can we find physical criteria that will definitively identify extinct cometary nuclei among the asteroid population, without the necessity for actually visiting them with spacecraft? It will be most interesting to see how much progress has been made in answering these questions at the Asteroids IV meeting.

Note added in proof: Comet 11D/Tempel-Swift, one of the lost comets listed in section 3, was recovered by the LINEAR automated NEO survey in September 2001 (*Hergenrother et al.,* 2001). The comet was active at recovery, though faint, near its perihelion of 1.58 AU. The comet had not been seen since 1908.

Acknowledgments. We thank M. Hicks, R. Whiteley, S. Bus, and R. Binzel for the use of their unpublished asteroid spectra and Y. Fernández for his comet and asteroid albedos. We thank D. Yeomans, A. Chamberlin, and especially Y. Fernández for useful comments on an earlier draft of this paper. This work was performed in part at the Jet Propulsion Laboratory and was funded in part by the NASA Origins and Planetary Geology and Geophysics Programs.

REFERENCES

A'Hearn M. (1988) Observations of cometary nuclei. *Annu. Rev. Earth Planet. Sci., 16,* 273–293.

Asphaug E. and Benz W. (1994) Density of comet Shoemaker-Levy-9 deduced by modeling breakup of the parent rubble pile. *Nature, 370,* 120–121.

Asphaug E. and Benz W. (1996) Size, density, and structure of comet Shoemaker-Levy 9 inferred from the physics of tidal breakup. *Icarus, 121,* 225–248.

Bailey M. E. (1984) The steady-state 1/a distribution and the problem of cometary fading. *Mon. Not. R. Astron. Soc., 211,* 347–368.

Bailey M. E. and Stagg C. R. (1988) Cratering constraints on the inner Oort Cloud: Steady-state models. *Mon. Not. R. Astron. Soc., 235,* 1–35.

Boehnhardt H., Sekanina Z., Fiedler A., Rauer H., Schulz R., and Tozzi G. (1998) Impact-induced activity of the asteroid-comet P/1996 N2 Elst-Pizarro: Yes or No? *Highlights Astron., 11A,* 233–237.

Brin G. D., and Mendis D. A. (1979) Dust release and mantle development in comets. *Astrophys. J., 229,* 402–408.

Bottke W. F., Morbidelli A., Jedicke R., Petit J.-M., Levison H., Michel P., and Metcalfe T. S. (2002) Debiased orbital and absolute magnitude distributions of near-Earth objects. *Icarus, 156,* 399–433.

Bowell E., Hapke B., Domingue D., Lumme K., Peltoniemi J., and Harris A. W. (1989) Application of photometric models to asteroids. In *Asteroids II* (R. P. Binzel et al., eds.), pp. 524–556. Univ. of Arizona, Tucson.

Carusi A. and Valsecchi G. (1987) Dynamical evolution of short-period comets. In *Interplanetary Matter* (Z. Ceplecha and P. Pecina, eds.), pp. 21–28. Czech. Acad. Sci., Ondrejov.

Chamberlin A. B., McFadden L., Schulz R., Schleicher D. G., and Bus S. J. (1996) 4015 Wilson-Harrington, 2201 Oljato, and 3200 Phaethon: Search for CN emission. *Icarus, 119,* 173–181.

Cochran A. L. and Barker E. S. (1984) Minor planet 1983 TB: A dead comet? *Icarus, 59,* 296–300.

Degewij J. and Tedesco E. F. (1982) Do comets evolve into asteroids? Evidence from physical studies. In *Comets* (L. L. Wilkening, ed.), pp. 665–695. Univ. of Arizona, Tucson.

Dones L., Levison H., Duncan M., and Weissman P. (2000) Formation of the Oort cloud revisited (abstract). *Bull. Am. Astron. Soc., 32,* 1060.

Donn B. and Hughes D. (1986) A fractal model of a cometary nucleus formed by random accretion. In *20th ESLAB Symposium on the Exploration of Halley's Comet* (B. Battrick et al., eds.), pp. 523–524. ESA SP-250.

Doressoundiram A., Cellino A., di Martino M., Migliorini F., and Zappalà V. (1998) The puzzling case of the Nysa-Polana family finally solved? (abstract). *Bull. Am. Astron. Soc., 30,* 505.

Drummond J. D. (1982) Theoretical meteor radiants of Apollo, Amor, and Aten asteroids. *Icarus, 49,* 143–153.

Duncan M. J. and Levison H. F. (1997) A scattered comet disk and the origin of Jupiter family comets. *Science, 276,* 1670–1672.

Duncan M., Quinn T., and Tremaine S. (1987) The formation and extent of the solar system comet cloud. *Astron. J., 94,* 1330–1338.

Duncan M., Quinn T., and Tremaine S. (1988) The origin of short-period comets. *Astrophys. J. Lett., 328,* L69–L73.

Eggers S. (1999) Cometary dynamics during the formation of the solar system. Ph.D. thesis, Max-Planck-Institüt für Aeronomie.

Everhart E. (1977) The evolution of comet orbits as perturbed by Uranus and Neptune. In *Comets, Asteroids, Meteorites: Interrelations, Evolution and Origins* (A. H. Delsemme, ed.), pp. 99–104. Univ. of Toledo, Toledo.

Fanale F. P. and Salvail J. R. (1984) An idealized short period comet model: Surface insolation, H_2O flux, dust flux and mantle development. *Icarus, 60,* 476–511.

Fernández J. A. (1980) On the existence of a comet belt beyond Neptune. *Mon. Not. R. Astron. Soc., 192,* 481–491.

Fernández J. A. (1997) The formation of the Oort cloud and the primitive galactic environment. *Icarus, 129,* 106–119.

Fernández J. A. and Brunini A. (2000) The buildup of a tightly bound comet cloud around an early Sun immersed in a dense galactic environment: Numerical experiments. *Icarus, 145,* 580–590.

Fernández J. A. and Ip W.-H. (1991) Statistical and evolutionary aspects of cometary orbits. In *Comets in the Post-Halley Era* (R. L. Newburn Jr. et al., eds.), pp. 487–536. Kluwer, Dordrecht.

Fernández J. A., Rickman H., and Kamél L. (1992) The population size and distribution of perihelion distances of the Jupiter family. In *Periodic Comets* (J. A. Fernández and H. Rickman, eds.), pp. 143–157. Univ. Republica, Montevideo.

Fernández J. A., Tancredi G., Rickman H., and Licandro J. (1999) The population, magnitudes, and sizes of Jupiter family comets. *Astron. Astrophys., 352,* 327–340.

Fernández Y. R. (1999) Physical properties of cometary nuclei. Ph.D. thesis, Univ. of Maryland.

Fernández Y. R., McFadden L. A., Lisse C. M., Helin E. F., and Chamberlin A. B. (1997) Analysis of POSS images of comet-asteroid 107P/1949 W1 Wilson Harrington. *Icarus, 128,* 114–126.

Fernández Y. R., Jewitt D. C., and Shephard S. S. (2001) Low albedos among extinct comet candidates. *Astrophys. J. Lett., 553,* L197–L200.

Green S. F., Meadows A. J., and Davies J. K. (1985) Infrared observations of the extinct cometary candidate minor planet (3200) 1983 TB. *Mon. Not. R. Astron. Soc., 214,* 29–36.

Harmon J. K., Campbell D. B., Ostro S. J., and Nolan M. C. (1999) Radar observations of comets. *Planet. Space Sci., 47,* 1409–1422.

Harris N. W. and Bailey M. E. (1998) Dynamical evolution of cometary asteroids. *Mon. Not. R. Astron. Soc., 297,* 1227–1236.

Hergenrother C., Muraoka K., Nakano S., and Marsden B. G. (2002) *Comet 11D (Tempel-Swift) = P/2001 X3 (LINEAR).* IAU Circular 7779.

Hicks M. D., Buratti B. J., Newburn R. L., and Rabinowitz D. L. (2000) Physical observations of 1996 PW and 1997 SE5: Extinct comets or D-type asteroids? *Icarus, 143,* 354–359.

Hills J. G. (1981) Comet showers and the steady-state infall of comets from the Oort cloud. *Astron. J., 86,* 1730–1740.

Jewitt D. and Luu J. (1993) Discovery of the candidate Kuiper belt object 1992 QB_1. *Nature, 362,* 730–732.

Johnson R. E., Cooper J. F., Lanzerotti L. J., and Strazzula G. (1987) Radiation formation of a non-volatile comet crust. *Astron. Astrophys., 187,* 889–892.

Kamél L. (1991) The evolution of P/Encke's light curve — No secular fading, a vanishing perihelion asymmetry. *Icarus, 93,* 226–245.

Keller H. U., Arpigny C., Barbieri C., Bonnett R. M., Cazes S., Coradini M., Cosmovici C. B., Curdt W., Delamere W. A., Huebner W. F., Hughes D. W., Jamar C., Kramm R., Malaise D., Reitsema H., Schmidt H. U., Schmidt K., Schmidt W. K. H., Seige P., Whipple F. L., and Wilhelm K. (1986) Observations by the Halley multicolour camera. In *20th ESLAB Symposium on the Exploration of Halley's Comet* (B. Battrick et al., eds.), pp. 347–350. ESA SP-250.

Kostolanský E. (1998) On asteroidal meteoroid streams detection. *Contrib. Astron. Obs. Skalnaté Pleso, 28,* 22–30.

Kresák L. (1979) Dynamical interrelations among comets and asteroids. In *Asteroids* (T. Gehrels, ed.), pp. 289–309. Univ. of Arizona, Tucson.

Kresák L. (1987) Dormant phases in the aging of periodic comets. *Astron. Astrophys., 187,* 906–908.

Kronk G. W. (1984) *Comets, A Descriptive Catalog.* Enslow, Hillside, New Jersey. 331 pp.

Kuiper G. P. (1951) On the origin of the solar system. In *Astrophysics* (J. A. Hynek, ed.), pp. 357–424. McGraw-Hill, New York.

Levison H. F. (1996) Comet taxonomy. In *Completing the Inventory of the Solar System* (T. W. Rettig and J. M. Hahn, eds.), pp. 173–191. ASP Conference Series 107.

Levison H. F. and Duncan M. J. (1994) The long-term dynamical behavior of short-period comets. *Icarus, 108,* 18–36.

Levison H. F. and Duncan M. J. (1997) From the Kuiper belt to Jupiter-family comets: The spatial distribution of ecliptic comets. *Icarus, 127,* 13–32.

Levison H. F., Duncan M. J., Zahnle K., Holman M., and Dones L. (2000) Planetary impact rates from ecliptic comets. *Icarus, 143,* 415–420.

Levison H. F., Dones L., and Duncan M. J. (2001) The origin of Halley-type comets: Probing the inner Oort cloud. *Astron. J., 121,* 2253–2267.

Lien D. (1998) Asteroid debris trails: Evidence for recent collisions in the asteroid belt (abstract). *Bull. Am. Astron. Soc., 30,* 1035.

Lowry S. C. (2001) Physical properties of cometary nuclei and studies of distant activity. Ph.D. thesis, Queens Univ., Belfast.

Luu J. X. (1993) Spectral diversity among the nuclei of comets. *Icarus, 104,* 138–148.

Malhotra R., Duncan M. J., and Levison H. F. (2000) Dynamics of the Kuiper Belt. In *Protostars and Planets IV* (V. Mannings et al., eds.), pp. 1231–1254. Univ. of Arizona, Tucson.

Marsden B. G. (1970) On the relationship between comets and minor planets. *Astron. J., 75,* 206–217.

Marsden B. G. and Williams G. V. (1999) *Catalogue of Cometary Orbits 1999.* Minor Planet Center. 127 pp.

Marsden B. G., Sekanina Z., and Everhart E. (1978) New osculating orbits for 110 comets and the analysis of the original orbits of 200 comets. *Astron. J., 83,* 64–71.

McFadden L. A., Cochran A. L., Barker E. S., Cruikshank D. P., and Hartmann W. K. (1993) The enigmatic object 2201 Oljato — Is it an asteroid or an evolved comet? *J. Geophys. Res., 98,* 3031–3041.

Meech K. J. (2002) Physical properties of comets. In *Asteroids, Comets, Meteors 1996* (A.-C. Levasseur-Regourd, ed.). Kluwer, Dordrecht, in press.

Meech K. J. and Belton M. J. S. (1989) *2060 Chiron.* IAU Circular 4770.

Morbidelli A., Bottke W. F. Jr., Froeschlé Ch., and Michel P. (2002) Origin and evolution of near-Earth objects. In *Asteroids III* (W. F. Bottke Jr. et al., eds.), this volume. Univ. of Arizona, Tucson.

Moore M. H., Donn B., Khanna R., and A'Hearn M. F. (1983) Studies of proton-irradiated cometary-type ice mixtures. *Icarus, 54,* 388–405.

Newton H. A. (1891) Capture of comets by planets. *Astron. J., 11,* 73–75.

Newton H. A. (1893) On the capture of comets by planets, especially their capture by Jupiter. *Mem. Natl. Acad. Sci., 6,* 7–23.

Oort J. H. (1950) The structure of the cloud of comets surrounding the solar system and a hypothesis concerning its origin. *Bull. Astron. Inst. Neth., 11,* 91–110.

Peale S. J. (1989) On the density of Halley's comet. *Icarus, 82,* 36–49.

Prialnik D. and Bar-Nun A. (1988) The formation of a permanent dust mantle and its effect on cometary activity. *Icarus, 74,* 272–283.

Quinn T., Tremaine S., and Duncan M. (1990) Planetary perturbations and the origin of short-period comets. *Astrophys. J., 355,* 667–679.

Rickman H. (1986) Masses and densities of comets Halley and Kopff. In *The Comet Nucleus Sample Return Mission* (O. Melita, ed.), pp. 195–205. ESA SP-249.

Sekanina Z. (1982) The problem of split comets in review. In *Comets* (L. L. Wilkening, ed.), pp. 251–287. Univ. of Arizona, Tucson.

Sekanina Z. (1991) Outgassing asymmetry of periodic Comet Encke. II — Apparitions 1868–1918 and a study of the nucleus evolution. *Astron. J., 96,* 1455–1475.

Scotti J. V. (1996) Near-Earth object surveying in the late 20th century. In *Completing the Inventory of the Solar System* (T. W. Rettig and J. M. Hahn, eds.), pp. 107–113. ASP Conference Series 107.

Shoemaker E. M. and Wolfe R. F. (1982) Cratering time scales for the Galilean satellites. In *Satellites of Jupiter* (D. Morrison, ed.), pp. 277–339. Univ. of Arizona, Tucson.

Soderblom L. A., Becker T. L., Bennett G., Boice D. C., Britt D. T., Brown R. H., Buratti B. J., Isbell C., Geise B., Hare T., Hicks M. D., Howington-Kraus E., Kirk R. L., Lee M., Nelson R. M., Oberst J., Owen T. C., Rayman M. D., Sandel B. R., Stern S. A., Thomas N., and Yelle R. V. (2002) Observations of comet 19P/Borrelly by the Miniature Integrated Camera and Spectrometer aboard Deep Space 1. *Science, 296,* 1087–1091.

Stephenson F. R., Yau K. K. C., and Hunger H. (1985) Records of Halley's comet on Babylonian tablets. *Nature, 314,* 587–592.

Veeder G. J., Kowal C., and Matson D. L. (1984) The Earth-crossing asteroid 1983 TB (abstract). In *Lunar and Planetary Science XV,* pp. 878–879. Lunar and Planetary Institute, Houston.

Weaver H. A., Sekanina Z., Toth I., Delahodde C. E., Hainaut O. R., Lamy P. L., Bauer J. M., A'Hearn M. F., Arpigny C., Combi M. R., Davies J. K., Feldman P. D., Festou M. C., Hook R., Jorda L., Keesey M. S. W. Lisse C. M., Marsden B. G., Meech K. J., Tozzi G. P., and West R. (2001) HST and VLT investigations of the fragments of comet C/1999 S4 (LINEAR). *Science, 292,* 1329–1334.

Weissman P. R. (1979) Physical and dynamical evolution of long-period comets. In *Dynamics of the Solar System* (R. L. Duncombe, ed.), pp. 277–282. Reidel, Dordrecht.

Weissman P. R. (1980) Physical loss of long-period comets. *Astron. Astrophys., 85,* 191–196.

Weissman P. R. (1985) Dynamical evolution of the Oort cloud. In *Dynamics of Comets: Their Origin and Evolution* (A. Carusi and G. D. Valsecchi, eds.), pp. 87–96. Reidel, Dordrecht.

Weissman P. R. (1986) Are cometary nuclei primordial rubble piles? *Nature, 320,* 242–244.

Weissman P. R. (1987) Post-perihelion brightening of Halley's Comet: Spring time for Halley. *Astron. Astrophys., 187,* 873–878.

Weissman P. R. (1990) The cometary impactor flux at the Earth. In *Global Catastrophes in Earth History* (V. L. Sharpton and P. D. Ward, eds.), pp. 171–180. Geol. Soc. Am. Spec. Paper 247.

Weissman P. R. (1993) No, we are not in a cometary shower (abstract). *Bull. Am. Astron. Soc., 25,* 1063.

Weissman P. R. (1996a) The Oort cloud. In *Completing the Inventory of the Solar System* (T. W. Rettig and J. M. Hahn, eds.), pp. 265–288. ASP Conference Series 107.

Weissman P. R. (1996b) Star passages through the Oort cloud. *Earth Moon Planets, 72,* 25–30.

Weissman P. R. and Levison H. F. (1997) Origin and evolution of the unusual object 1996 PW: Asteroids from the Oort cloud? *Astrophys. J. Lett., 488,* L133–136.

Weissman P. R. and Lowry S. C. (2002) The size distribution of cometary nuclei. *Science,* in press.

Weissman P. R. and Stern S. A. (1997) Physical processing of cometary nuclei. In *Workshop on Analysis of Returned Comet Nucleus Samples* (S. Chang, ed.), pp. 119–166. NASA Conference Publication 10152.

Weissman P. R., A'Hearn M. F., McFadden L. A., and Rickman H. (1989) Evolution of comets into asteroids. In *Asteroids II* (R. P. Binzel et al., eds.), pp. 880–920. Univ. of Arizona, Tucson.

Wetherill G. W. (1994) Possible consequences of absence of Jupiters in planetary systems. *Astrophys. Space Sci., 212,* 23–32.

Wetherill G. W. (1988) Where do the Apollo objects come from? *Icarus, 76,* 1–18.

Wetherill G. W. (1991) End products of cometary evolution: Cometary origin of Earth-crossing bodies of asteroidal appearance. In *Comets in the Post-Halley Era* (R. L. Newburn Jr. et al., eds.), pp. 537–556. Kluwer, Dordrecht.

Whipple F. L. (1950) A comet model. I. The acceleration of comet Encke. *Astrophys. J., 111,* 375–394.

Whipple F. L. (1962) On the distribution of sernimajor axes among comet orbits. *Astron. J., 67,* 1–9.

Whipple F. L. (1964) Evidence for a comet belt beyond Neptune. *Proc. Natl. Acad. Sci. U.S., 51,* 711–718.

Whipple F. L. (1983) *1983 TB and the Geminid Meteors.* IAU Circular 3881.

Wiegert P. and Tremaine S. (1999) The evolution of long-period comets. *Icarus, 137,* 84–121.

Williams I. P. and Wu Z. (1993) The Geminid meteor stream and asteroid 3200 Phaethon. *Mon. Not. R. Astron. Soc., 262,* 231–248.

Yau K., Yeomans D., and Weissman P. (1994) The past and future motion of comet Swift-Tuttle. *Mon. Not. R. Astron. Soc., 266,* 305–316.

Yeomans D. K. (1991) A comet among the near-Earth asteroids? *Astron. J., 101,* 1920–1928.

Yeomans D. K. (1992) Erratum: A comet among the near-Earth asteroids? *Astron. J., 104,* 1266.

Yeomans D. K., Yau K., and Weissman P. R. (1996) The upcoming approach of periodic comet Tempel-Tuttle and the Leonid meteors. *Icarus, 124,* 407–413.

Zahnle K., Dones L., and Levison H. F. (1998) Cratering rates on the Galilean satellites. *Icarus, 136,* 202–222.

Chronology of Asteroid Accretion and Differentiation

A. Shukolyukov
University of California, San Diego

G .W. Lugmair
Max-Planck-Institute for Chemistry, Mainz, and University of California, San Diego

The study of short-lived, now extinct radionuclides offers constraints on the duration of planetesimal formation and on other processes and events, which took place in the early solar system. The relative chronometers based on the decay of ^{53}Mn to ^{53}Cr ($T_{1/2}$ = 3.7 m.y.) and of ^{26}Al to ^{26}Mg ($T_{1/2}$ = 0.73 m.y.) provide an adequate time resolution of at least 1 m.y. A calibration of these relative chronometers with an absolute Pb-Pb chronometer permits converting relative ages into absolute ages. Using reasonable assumptions, an absolute timescale for events in the early solar system can be constructed. Based on the studies of these isotope systems in various meteorites, we estimated the formation time of the first high-temperature condensates (Ca-Al-rich meteorite inclusions) in the solar system, inferred constraints on the duration of planetesimal accretion, calculated the time of planetary melting, mantle, and core formation, and constrained the timing of igneous processes and thermal metamorphism within planetesimals.

1. INTRODUCTION

The main topic of this chapter is the discussion of the timing of asteroid accretion, differentiation, and interior thermal processes. These issues are closely related to a more general subject: the reconstruction of a timescale for various processes and events in the early solar system. We will not discuss theoretical chronological constraints obtained with the use of dynamical models (e.g., *Weidenschilling,* 2000). Instead, we will focus on the constraints for asteroid formation and evolution that are provided by radioactive clocks — the isotopic chronometers based on the radioactive decay of various nuclides.

It has been known for several decades that the age of the solar system is approximately 4.55–4.57 b.y. (e.g., *Patterson,* 1956; *Tatsumoto et al.,* 1976; *Chen and Tilton,* 1976). At present, however, it is possible to determine ages of solar system materials with much higher precision. Newly developed isotopic chronometers permit a time resolution approaching 1 m.y. or even better. The development of techniques to define high-resolution timescales is crucial for understanding the processes that occurred during the very early stages of solar system evolution.

The results that we will discuss below are based on the experimental study of meteorites — the fragments of asteroids that are transported from the asteroid belt to the Earth due to asteroid collisions in space. The wide variety of meteorite classes found on Earth implies a broad diversity of asteroid types. We do not know how many asteroid parent bodies we actually sample with the meteorites in our collections. Nevertheless, this number must be large, and, according to some estimates, may be as high as >135 (*Meibom and Clark,* 1999).

The objective here is to briefly describe some recent chronological results obtained in laboratory meteorite studies and to relate them to the main question we are addressing: How old are the asteroids?

2. ISOTOPIC CHRONOMETERS

The isotope chronometers based on the radioactive decay of the long-lived radionuclides are well known and have been in use for several decades. They include Rb-Sr, Sm-Nd, U-Pb, Re-Os, K-Ar, and some other chronometers. The half-lives of the parent nuclides are billions to tens of billions of years, e.g.,

$$^{87}\text{Rb} \rightarrow {}^{87}\text{Sr}\ (T_{1/2} \approx 4.9 \times 10^{10}\ \text{yr})$$

$$^{147}\text{Sm} \rightarrow {}^{143}\text{Nd}\ (T_{1/2} \approx 1.1 \times 10^{11}\ \text{yr})$$

$$^{238}\text{U} \rightarrow {}^{206}\text{Pb}\ (T_{1/2} \approx 4.5 \times 10^{9}\ \text{yr})$$

$$^{235}\text{U} \rightarrow {}^{207}\text{Pb}\ (T_{1/2} \approx 0.7 \times 10^{9}\ \text{yr})$$

The principles and use of these isotopic systems in geochronology and cosmochronology are described in detail elsewhere (e.g., see *Dickin,* 1995) and will not be repeated here.

The important requirements for dating early events in the solar system are that these chronometers should provide an adequate time resolution and should be insensitive to late secondary events (such as a late bombardment of an asteroid's surface). At the present state of technology only the U-Pb chronometer can provide an adequate time resolution of ~1 m.y. (e.g., *Lugmair and Galer,* 1992; *Göpel et al.,* 1994). However, the drawback of this system is that in many cases it was disturbed by natural phenomena such as shock or reheating, which can compromise the age information obtained with this method (e.g., *Carlson and Lugmair,* 2000). The Sm-Nd-isotopic system appears to be more robust, although the uncertainties in the Sm-Nd ages are usually not less than ~25 m.y. and, thus, a high-resolution timescale cannot be constructed with this chronometer.

The other type of chronometers is based on relatively short-lived nuclides. There is a series of short-lived radioactive nuclides with lifetimes on the order of a million to a hundred million years. These lifetimes are long enough to survive the interval between their production and the formation of solids in the solar system, but are still comparatively short, so that these nuclides are practically fully decayed and now extinct in the solar system. The presence of short-lived radionuclides in the early solar system may be due either to nucleosynthetic production within stars and subsequent rapid transport to and injection into the nascent solar system (e.g., *Cameron et al.,* 1997; *Boss and Vanhala,* 2000) or, possibly, to local production by high-energy particles from the Sun (e.g., *Clayton and Jin,* 1995; *Shu et al.,* 1997). The study of these now-extinct radionuclides in meteorites has offered reliable constraints on the time and duration of the formation of solid bodies in the early solar system. The former presence in meteorites of more than ten short-lived radionuclides has been discovered so far (see the recent review by *Podosek and Nichols,* 1997). Several examples are

$$^{41}Ca \rightarrow {}^{41}K \ (T_{1/2} = 0.10 \text{ m.y.})$$

$$^{26}Al \rightarrow {}^{26}Mg \ (T_{1/2} = 0.73 \text{ m.y.})$$

$$^{10}Be \rightarrow {}^{10}B \ (T_{1/2} = 1.5 \text{ m.y.})$$

$$^{60}Fe \rightarrow {}^{60}Ni \ (T_{1/2} = 1.5 \text{ m.y.})$$

$$^{53}Mn \rightarrow {}^{53}Cr \ (T_{1/2} = 3.7 \text{ m.y.})$$

$$^{129}I \rightarrow {}^{129}Xe \ (T_{1/2} = 17 \text{ m.y.})$$

$$^{146}Sm \rightarrow {}^{142}Nd \ (T_{1/2} = 103 \text{ m.y.})$$

The importance of these radionuclides is that due to their short half-lives, they could serve as very sensitive chronometers for events in the early solar system and as tracers for early planetary evolution. Together with longer-lived radionuclides the abundances of these short-lived species provide constraints on nucleosynthetic timescales and models of nucleosynthesis. The short-lived nuclide ^{26}Al is of special importance. It is considered as a viable heat source for early heating of planetesimals and, in some cases, for their differentiation (see *McSween et al.,* 2002).

The former presence of an extinct radionuclide is revealed in the form of isotopic anomalies in the daughter element. Obviously, a daughter element should have at least two isotopes so that the anomaly can be observed in principle. In addition, since the abundances of these radionuclides are usually small, it is often necessary to study objects with a high parent-to-daughter element ratio. In this case, even a small addition of the radiogenic daughter nuclide will result in a measurable shift in the isotopic composition of the daughter nuclide.

One of these radionuclides, ^{53}Mn, with a half-life of 3.7 m.y., is especially suitable for high-resolution chronological studies of the first ~20 m.y. of solar system history. The former presence of ^{53}Mn is indicated by excesses relative to the terrestrial "standard" value of the daughter nucleus, ^{53}Cr. One of the advantages of the ^{53}Mn-^{53}Cr chronometer is that Mn and Cr are abundant elements in solar system materials and that the ^{53}Mn-^{53}Cr system, therefore, is potentially useful for dating a wide variety of ancient objects. Another advantage is the presence in many objects of a Cr-spinel phase which has a very low Mn/Cr ratio. This permits high precision measurements of initial $^{53}Cr/^{52}Cr$ ratios for the time of isotopic closure of the Mn-Cr system. After the pioneering finding of the vestiges of ^{53}Mn in Allende refractory inclusions (*Birck and Allègre,* 1985), excesses of ^{53}Cr were detected in various solar system objects: carbonaceous chondrites, enstatite chondrites, ordinary chondrites, pallasites, iron meteorites, angrites, eucrites, mesosiderites, SNC meteorites, and primitive achondrites (see references in *Lugmair and Shukolyukov,* 1998). In many cases, the measured relative $^{53}Cr/^{52}Cr$ excesses in mineral phases of these meteorites are correlated with their respective Mn/Cr ratios, which indicates *in situ* decay of ^{53}Mn. From the slope of the correlation line of measured $^{53}Cr/^{52}Cr$ vs. $^{55}Mn/^{52}Cr$ data pairs a relative abundance of ^{53}Mn (that is the $^{53}Mn/^{55}Mn$ ratio) at the time of isotopic closure can be obtained. The relative ages of two meteorites, 1 and 2, are then calculated from their $^{53}Mn/^{55}Mn$ ratios

$$\Delta T_{1-2} = 1/\lambda \ \ln[(^{53}Mn/^{55}Mn)_1/(^{53}Mn/^{55}Mn)_2] \qquad (1)$$

where λ is the decay constant of ^{53}Mn.

Using this and other short-lived radionuclides, we can date processes that occurred 4.56 b.y. ago with a resolution of 1 m.y. or better. However, these chronometers have a disadvantage: Because the parent nuclides are extinct, only relative ages can be obtained. Thus we need to map these relative ages onto an absolute timescale.

3. MAPPING OF THE RELATIVE CHRONOMETERS ONTO AN ABSOLUTE TIMESCALE

Figure 1 shows an example of such a procedure. The graph illustrates our results for the Mn-Cr-isotopic system in a differentiated meteorite, the angrite LEW 86010 (*Lugmair and Shukolyukov,* 1998). This meteorite is an ideal sample for calibration purposes. The angrites are early equilibrated planetary differentiates that cooled fast, do not show any signs of later disturbance, and have absolute ages that are precisely known. The Pb-Pb age of LEW 86010 is 4557.8 ± 0.5 Ma (*Lugmair and Galer,* 1992). In addition, in LEW

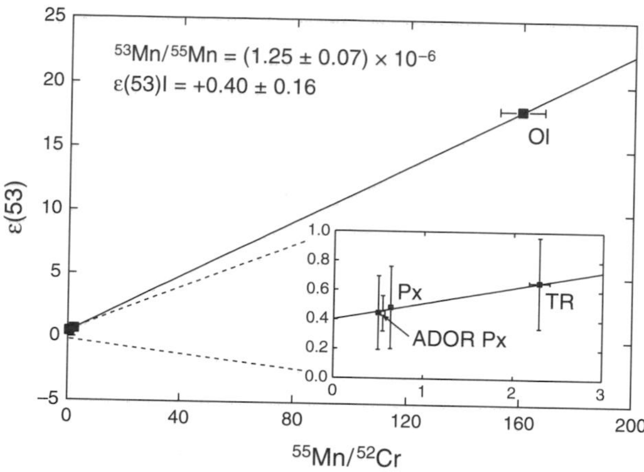

Fig. 1. Manganese-53–chromium-53 systematics in the angrite LEW 86010 (LEW). Here and in the following figures the y-axis represents the deviations of the measured $^{53}Cr/^{52}Cr$ ratios relative to the standard terrestrial value. These deviations are expressed in ε units (1 part in 10^4). TR = total rock, Px = pyroxene, Ol = olivine. The slope of the best-fit line yields the $^{53}Mn/^{55}Mn$ ratio at the time of isotopic closure. The precise absolute age of LEW (*Lugmair and Galer,* 1992) and the obtained $^{53}Mn/^{55}Mn$ ratio allow the relative Mn-Cr chronometer to be mapped onto an absolute timescale. The insert shows enlarged the results for the mineral phases with small Mn/Cr ratios. The data point for a pyroxene separate from the angrite Angra dos Reis (ADOR Px) is shown for comparison. It is totally consistent with the LEW data that reflects a contemporaneous formation of these two angrites (*Lugmair and Galer,* 1992). ε(53)I is the initial $^{53}Cr/^{52}Cr$ ratio at the time of isotopic closure.

86010 the range of the $^{55}Mn/^{52}Cr$ ratios is rather large — from ~0.5 in pyroxenes up to ~160 in olivines — allowing for the precise determination of the $^{53}Mn/^{55}Mn$ ratio at the time of isotopic closure. The y-axis in Fig. 1 represents the relative abundance of ^{53}Cr ($^{53}Cr/^{52}Cr$) with respect to the terrestrial Cr composition, expressed in ε units (1ε is 1 part in 10^4, or 0.01%). The x-axis is the $^{55}Mn/^{52}Cr$ ratio. The $^{53}Cr/^{52}Cr$ ratios in the mineral phases are correlated with the respective $^{55}Mn/^{52}Cr$ ratios. The slope of this line yields the relative abundance of extinct ^{53}Mn at the time this meteorite formed. The fact that the angrites cooled fast (*Störzer and Pellas,* 1977) suggests that both the U-Pb and the Mn-Cr isotope systems closed approximately at the same time or, more correctly, within the timespan provided by the time resolution of our isotopic chronometers. Thus, the obtained value of $(1.25 \pm 0.07) \times 10^{-6}$ represents the $^{53}Mn/^{55}Mn$ ratio at 4557.8 Ma. This data pair now make it possible to map the relative ^{53}Mn-^{53}Cr chronometer, that is the relative Mn-Cr ages obtained for other meteorite samples, onto an absolute timescale.

A necessary condition for this method to yield valid absolute ages is that both the material of the absolute time marker and that of the meteorite to be dated come from an isotopically uniform reservoir. The isotopic studies of various solar system bodies have shown that the solar nebula has been well mixed: The isotopic compositions of elements are essentially the same in all studied objects. There are a few exceptions, but these anomalies are confined to a limited number of objects and do not play a significant role on a large planetary scale. The isotopic homogeneity of the solar system material is a result of the prolonged and thorough mixing of numerous nucleosynthetic components of various origins that were part of the solar nebula. However, the short-lived radionuclides were injected into or formed within the solar nebula just prior to or during solar system formation. During the course of a few 10^5 to 10^6 yr, this material may not have been distributed homogeneously within the solar system.

We have shown recently that the relative ^{53}Cr abundance in the Earth-Moon system, the martian meteorites, and the "asteroid belt" meteorites is a function of their present heliocentric distance (*Lugmair and Shukolyukov,* 1998). This led us to suggest that the observed radial gradient in the relative ^{53}Cr abundance may be due to an original radially heterogeneous ^{53}Mn distribution. The alternative explanation is an early, volatility-driven, Mn/Cr fractionation in the nebula with an originally homogeneous ^{53}Mn distribution (e.g., see *Cassen and Woolum,* 1997). The discussion of the details of these scenarios (*Lugmair and Shukolyukov,* 1998; *Shukolyukov and Lugmair,* 2000a), however, is beyond the scope of this chapter. What is important in the present context is that no detectable variations of the $^{53}Cr/^{52}Cr$ ratios were found among the studied bulk asteroid belt bodies (*Shukolyukov and Lugmair,* 2000a). This, therefore, implies that the original ^{53}Mn abundance was essentially the same among these objects. Thus, the ^{53}Mn-^{53}Cr system can be used as a chronometer, at least for samples from the asteroid belt.

4. ABSOLUTE TIMESCALES FOR EVENTS IN THE EARLY SOLAR SYSTEM

4.1. Refractory Inclusions, Chondrules, and Planetesimals

Among all early solar system objects dated in the past, Ca-Al-rich inclusions (CAIs) provided the oldest high-precision absolute ages. These refractory inclusions (see *MacPherson et al.,* 1988, for a review on CAI mineralogy and models of formation) are considered the first condensates of matter in the solar system. An average absolute Pb-Pb age of several CAIs from the Allende chondrite is 4566 ± 2 Ma (*Göpel et al.,* 1991). This age is often taken to be the best estimate for the solar system age. However, one cannot exclude the possibility that this CAI age reflects processes of alteration or late isotopic reequilibration rather than the true formation age. If so, this value would only be a lower limit for the true time of CAI formation.

Additional constraints for the solar system age can be derived from the ^{53}Mn-^{53}Cr system. Among planetary samples, the terrestrial $^{53}Cr/^{52}Cr$ value is the lowest known so far. The lower $^{53}Cr/^{52}Cr$ ratios (by ~1ε) found in CAIs from carbonaceous chondrites (*Birck and Allègre,*1985; *Nyquist et al.,* 2001) most likely do not represent bulk solar system material. Indeed, CAIs contain not only the vestiges of some ^{53}Mn decay but also exhibit pronounced anomalies in other isotopes of Cr (*Papanastassiou,* 1986; *Rotaru et al.,* 1992; *Podosek et al.,* 1997). This indicates that the measured ^{53}Cr/^{52}Cr ratios in CAIs reflect a complex superposition of the ^{53}Mn decay process and mixing of at least two Cr components (possibly of presolar origin) with anomalous isotopic composition. We believe, therefore, that the chronological meaning of the initial ^{53}Cr/^{52}Cr ratios in CAIs is rather tenuous. Thus, we use the terrestrial ^{53}Cr/^{52}Cr ratio, defined as 0ε, as the upper limit for the solar system initial value (SSI). Using the Mn-Cr data obtained for the HED (howardite-eucrite-diogenite) parent body (see Fig. 2 and the next section for details), with an initial ^{53}Cr/^{52}Cr ratio of +0.25ε and a chondritic ^{55}Mn/^{52}Cr ratio of 0.76, we estimated a minimum solar system age of ~4568 Ma (*Lugmair and Shukolyukov,* 1998). This value is still in agreement with the above mentioned CAI Pb-Pb age of 4566 ± 2 Ma. Using a similar approach and SSI = –0.42ε, as derived from an extrapolation of the previously discussed ^{53}Cr/^{52}Cr gradient to zero heliocentric distance, one can obtain an estimate for an upper limit of the solar system age of ~4571 Ma (*Lugmair and Shukolyukov,* 1998, 2001). Thus, these estimates for the age of the solar system provide a rather narrow range of only about 3 m.y. It has to be noted that estimates of the "solar system age" based on short-lived radioactive nuclides require that these nuclei were either injected into or produced within the nebula itself within a timespan after T_0, which is short relative to their respective half-lives.

A recent study of the Mn-Cr isotopic system in carbonates from the Kaidun carbonaceous chondrite (*Hutcheon et al.,* 1999) provides further constraints. These results indicate that the ^{53}Mn/^{55}Mn ratio at the time when these carbonates formed was ~9.4 × 10^{-6}. Using this value and the angrite LEW 86010 as a time marker yields a time for this event of ~4569 Ma. Clearly, carbonate formation in a meteorite parent body can only occur after the formation of the first solids. Hence, a time of ~4569 Ma can only be a lower limit for the solar system age, making ~4571 Ma a more reasonable age estimate. This also strongly suggests that the lower Pb-Pb age obtained for CAIs most likely is a result of later alteration.

Additional constraints can be obtained from the ^{26}Al-^{26}Mg chronometer. The majority of CAIs are characterized by a "canonical" initial ^{26}Al/^{27}Al ratio of ~5 × 10^{-5} (see *MacPherson et al.,* 1995, for a review). This sharp clustering implies that the inclusions formed within an interval of less than 1 m.y. The comparison of this "canonical" value with the ^{26}Al data from feldspars in the H4 chondrite Ste. Marguerite [(2.0 ± 0.6) × 10^{-7} (*Zinner and Göpel,* 1992)] yields a time difference between the isotopic closure of the ^{26}Al-^{26}Mg system in Ste. Marguerite and CAIs of 5.6 m.y. Using the Pb-Pb age of phosphates, 4562.7 ± 0.6 Ma (*Göpel et al.,* 1994) gives an absolute age for CAIs of ~4568 Ma. If, however, the feldspars in Ste. Marguerite predate the

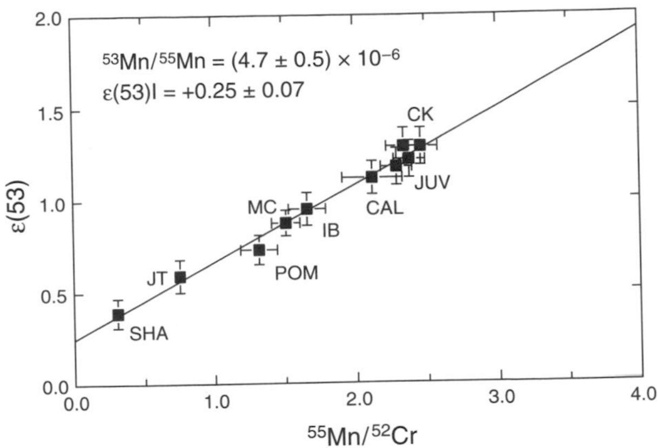

Fig. 2. Manganese-53–chromium-53 systematics in the HED parent body. The measured ^{53}Cr/^{52}Cr ratios for bulk meteorites are plotted vs. their respective ^{55}Mn/^{52}Cr ratios (data from *Lugmair and Shukolyukov,* 1998). The samples include two diogenites [Johnstown (JT) and Shalka (SHA)] and six noncumulate eucrites [Chervony Kut (CK), Juvinas (JUV), Caldera (CAL), Ibitira (IB), Pomozdino (POM), and the cumulate eucrite Moore County (MC)]. The data points form a well-defined isochron whose slope corresponds to the ^{53}Mn/^{55}Mn ratio of (4.7 ± 0.5) × 10^{-6}, which dates the global differentiation of the HED parent body at 4564.8 ± 0.9 Ma (see text).

phosphates by 3–4 m.y., as suggested by their I-Xe age of ~4566 Ma (*Brazzle et al.,* 1999; *Gilmour,* 2000) and the recently obtained metamorphic ^{53}Mn-^{53}Cr age of 4565 ± 0.7 Ma (*Polnau and Lugmair,* 2001), then the CAI age would be close to 4571 Ma.

In summary, a time of ~4571 Ma is our current "best estimate" for the solar system age. This age estimate of ~4571 Ma would then mark the time when the first high-temperature condensates (CAIs) were starting to form. Considering the recently measured variation of ^{26}Al/^{27}Al between petrographically distinct components within the same Allende CAI (*Hsu et al.,* 2000), it appears that CAI formation may have persisted for several 10^5 yr. The relative ^{53}Mn and ^{26}Al abundances at that time were ~14 × 10^{-6} and ~5 × 10^{-5} respectively.

Nyquist et al. (2001) have studied the Mn-Cr-isotopic system in individual bulk chondrules from the unequilibrated primitive ordinary chondrites Chainpur and Bishunpur. Chondrules are individual submillimeter- to millimeter-sized particles whose mineralogy and internal texture testify to crystallization from a melt. They are an abundant constituent of primitive (undifferentiated) meteorites. The Mn-Cr-isotopic systematics from Chainpur and Bishunpur chondrules suggest a ^{53}Mn/^{55}Mn ratio of ~9.5 × 10^{-6}. If this value corresponds to the time when these chondrules were formed, this time would be ~11 m.y. prior to the crystallization of LEW 86010, or ~4569 Ma. This value is approximately the same as the time of carbonate formation in Kaidun but ~2 m.y. lower than our preferred estimate for the solar system age of 4571 Ma. Thus, we suggest that the main episode of primary chondrule formation must have occurred within the first ~2 m.y. of solar system history (*Lugmair and Shukolyukov,* 2001).

In several recent studies of ^{26}Mg excesses in Al-rich chondrules (*Russell et al.,* 1996), Mg-rich chondrules (*Kita et al.,* 1998), and ferromagnesian chondrules (*McKeegan et al.,* 2000), it was shown that the inferred ^{26}Al/^{27}Al ratios at the time of their formation (or last isotopic equilibration) were >3–7 × 10^{-6}. The comparison of these values with the "canonical" CAI value of 5 × 10^{-5} indicates that the timespan between CAI formation and formation of these chondrules was not more than ~2 m.y. Thus, both the Mn-Cr and the Al-Mg isotopic systems are consistent in that primary chondrule formation occurred mainly within the first ~2 m.y. of solar system history. This process would have been largely concluded ~4569 Ma. This suggests that most chondrules with inferred ^{26}Al/^{27}Al ratios significantly lower than 3–7 × 10^{-6} and CAIs with ^{26}Al/^{27}Al ratios much lower (by a factor of >~2) than the canonical value of ~5 × 10^{-5} have most likely been metamorphosed or even totally recrystallized (i.e., no clear indication of metamorphism) in the deeper layers of early generations of planetesimals. [*Marhas et al.* (2000), however, prefer a different interpretation. They argue that the absence of a correlation between mineralogical evidence of alteration in carbonaceous chondrite chondrules and their ^{26}Al content suggests either extremely heterogeneous original distribution of ^{26}Al in the solar nebula or an extended timescale of chondrule formation.]

The formation of the first planetesimals may be required to have occurred at the very early stages of solar system evolution. This is because CAIs, which were formed at ~4571 Ma, had to be stored within small planetesimals (several hundred meters to kilometers in size?) in order to prevent their rapid loss into the Sun by gas drag (*Weidenschilling,* 1977). Thus, it appears to be likely that formation of CAIs and small planetesimals took place contemporaneously during several 10^5 yr of CAI formation. The small planetesimals were later disrupted by already larger objects, which caused further mixing of different types of CAIs. At any rate, these small objects were the original carriers of CAIs, which were responsible for later distribution of CAIs among chondrule-bearing planetesimals. Within the following ~2 m.y., large numbers of planetesimals of several tens of kilometers in size had formed. Their interior started to heat up and melt through the decay of ^{26}Al (*Lugmair and Shukolyukov,* 2001; *McSween et al.,* 2002). It is envisioned that chondrule formation may have occurred during this phase by the release of the interior melt as finely dispersed droplets into the surrounding space, caused by collisional disruption of these planetesimals (*Lugmair and Shukolyukov,* 2001). Repeated collisional destruction and reaccretion of these planetesimals was followed by the formation of larger planetary objects several hundred kilometers in size. In many of these larger planetesimals, there was sufficient residual heat and still-extant ^{26}Al for melt to form in their interior. Within ~3 m.y., the molten interior started to chemically differentiate to form a stratified mantle and, most likely, a core. By ~4565 Ma this process came to a conclusion (at least on some asteroids; see below). In the following sections, we will discuss several examples of Mn-Cr dating of differentiation events and other thermal processes occurring within asteroids.

4.2. Asteroid Vesta

An example of a differentiated asteroid is Vesta. A genetic link between the family of HED (howardite-eucrite-diogenite) meteorites and the asteroid Vesta was suggested in the 1970s (*McCord et al.,* 1970). This suggestion was based on the fact that the reflectance spectrum of Vesta matches that of the HED meteorites. However, this idea has long been a subject of debate because of the dynamical difficulties of transporting the HED meteorites to Earth. The finding of more than twenty small asteroids having Vesta-like spectral properties indicates that they are distributed from Vesta to the 3:1 resonance and, thus, demonstrate a dynamically viable route for samples from Vesta to Earth (*Binzel and Xu,* 1993). Most students of Vesta now believe that this asteroid is the parent body (PB) of the HED meteorites [for a review on the history of Vesta, see *Keil et al.* (2002)].

Some of the noncumulate eucrites — abundant constituents among HED meteorites — reveal the former presence of the short-lived radionuclides ^{60}Fe ($T_{1/2}$ = 1.5 m.y.) (*Shukolyukov and Lugmair,* 1993) and ^{26}Al ($T_{1/2}$ = 0.73 m.y.) (*Srinivasan et al.,* 1999). This indicates the antiquity of these meteorites and implies that the differentiation processes within Vesta occurred very early.

Figure 2 illustrates the results of our study of the Mn-Cr-isotopic system in various constituents of the HED PB — noncumulate and cumulate eucrites and diogenites (*Lugmair and Shukolyukov,* 1998). We plotted the measured ^{53}Cr/^{52}Cr ratios for bulk meteorites vs. their respective ^{55}Mn/^{52}Cr ratios. The data points form a well-defined correlation line. Since this line represents a bulk meteorite isochron, no information on the time of crystallization or cooling of individual meteorites can be derived from this diagram. Instead, the slope of the line corresponds to a ^{53}Mn/^{55}Mn ratio of (4.7 ± 0.5) × 10^{-6} at the time of the last Mn/Cr fractionation in the HED mantle. Since any resolvable scatter of the data points from the line does not exist, the source reservoirs of all these meteorites can be assumed to have formed contemporaneously. The Mn-Cr systems of the bulk samples of these meteorites must have remained closed since their formation. From the ^{53}Mn/^{55}Mn ratio for the HED PB of (4.7 ± 0.5) × 10^{-6} and that of LEW 86010 (see above) a relative time for the HED PB mantle fractionation is 7.1 ± 0.8 m.y. prior to angrite crystallization, yielding an absolute time of 4564.8 ± 0.9 Ma. Thus, the time of global mantle fractionation on Vesta postdates the beginning of the solar system by only ~6 m.y. This demonstrates that planetary differentiation processes must have occurred very early in solar system history. The results of thermal modeling of Vesta (*Ghosh and McSween,* 1998) are consistent with this finding.

The ages of individual basaltic achondrites from Vesta (*Lugmair and Shukolyukov,* 1998) suggest that basaltic volcanism occurred within only a few million years of global differentiation. Figure 3 illustrates several examples of internal ^{53}Mn-^{53}Cr isochrons for noncumulate eucrites. The moderately brecciated eucrite Chervony Kut (CK) reveals the highest ^{53}Mn abundance. The slope of the best-fit line yields the ^{53}Mn/^{55}Mn ratio of (3.7 ± 0.4) × 10^{-6} at the time of isotopic closure. The difference between this value and that in LEW 86010 corresponds to a time difference of 5.8 ± 0.8 m.y., yielding an absolute age of 4563.6 ± 0.9 Ma. Thus, CK formed almost contemporaneously with the global differentiation of Vesta. The isochron for the eucrite Juvinas (JUV) is shown schematically. Its slope yields the ^{53}Mn/^{55}Mn ratio of (3.0 ± 0.5) × 10^{-6}. With a Mn-Cr age of 4562.5 ± 1.0 Ma, JUV is slightly younger than CK. In contrast to CK and JUV, the mineral fractions from the eucrite Caldera (CAL) have totally equilibrated ^{53}Cr/^{52}Cr ratios (Fig. 3) (*Wadhwa and Lugmair,* 1996). This indicates that ^{53}Mn had practically fully decayed by the time the ^{53}Mn-^{53}Cr system closed in this meteorite. The upper limit of its ^{53}Mn/^{55}Mn ratio of 1.2 × 10^{-7} implies that the age of CAL is ≤4545 Ma. Although this young age may correspond to a "cooling age," it is more likely the result of impact melting that reequilibrated the Cr isotopes. Similar examples for reequilibrated Cr isotopes were observed in the mineral fractions of other noncumulate eucrites such as Pomozdino

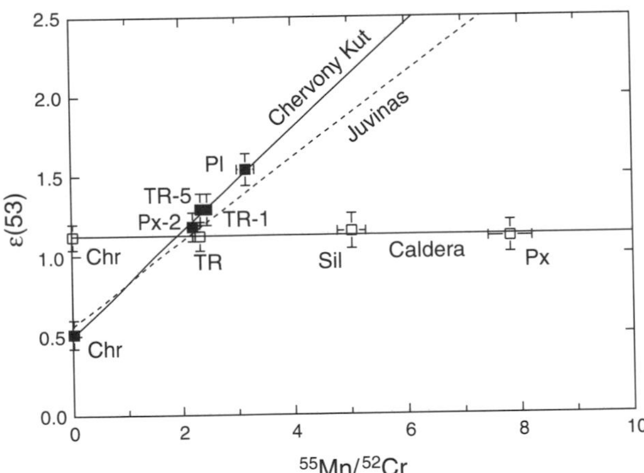

Fig. 3. Manganese-53–chromium-53 systematics for several noncumulate eucrites. Chr = chromite, TR = total rock, Px = pyroxene, Sil = silicates, Pl = plagioclase. The oldest is Chervony Kut (filled squares), clearly showing the presence of live ^{53}Mn at the time of crystallization. The slope of the best-fit line yields the ^{53}Mn/^{55}Mn ratio of (3.7 ± 0.4) × 10^{-6} at the time of isotopic closure, corresponding to an absolute age of 4563.6 ± 0.9 Ma. The isochron for the eucrite Juvinas is shown schematically (dashed line). A ^{53}Mn/^{55}Mn ratio of (3.0 ± 0.5) × 10^{-6} for Juvinas yields an age of 4562.5 ± 1.0 Ma. In contrast, Caldera (open symbols) had isotopically equilibrated at a time when ^{53}Mn was no longer extant. See text for more details. Data from *Lugmair and Shukolyukov* (1998) and *Wadhwa and Lugmair* (1996).

(T ≤ 4554 Ma) and EET 87520 (T ≤ 4549 Ma). The diogenites Shalka and Johnstown and the cumulate eucrite Moore County (not shown here) also show a flat ^{53}Cr/^{52}Cr-isotopic pattern. However, this most likely reflects slow cooling in the deeper zones of the HED PB (*Lugmair and Shukolyukov,* 1998). While the basaltic eucrites are believed to originate from basaltic deposits in the upper layers of the crust, the diogenites and cumulate eucrites are derived from deeper layers.

4.3. Other Differentiated and Undifferentiated Asteroids

Another example for early igneous activity on asteroids is the meteorite Brachina. Brachinites are a unique group of primitive achondrites, which, although near-chondritic in their major-element bulk composition, are distinctly igneous textured with an olivine-rich mineralogy (e.g., *Nehru et al.,* 1983). The study of the Mn-Cr-isotopic system in this unique meteorite has shown that at the time of isotopic closure, the ^{53}Mn/^{55}Mn ratio was (3.8 ± 0.4) × 10^{-6} (*Wadhwa et al.,* 1998). This value translates into an absolute age of 4563.7 ± 0.9 Ma and most likely corresponds to the time of extensive thermal metamorphism subsequent to Brachina's formation. Within 2σ errors, this age is similar to that of CK and to the time of global differentiation of the HED PB (4564.8 ± 0.9 Ma). A result of the earliest phases of igneous activity in the asteroid belt, Brachina may be among the earliest generations of achondrites formed in the solar system.

In contrast, the Mn-Cr ages of another type of differentiated meteorites, the pallasites, are much younger. These meteorites consist mainly of metal and olivine and are probably formed at the boundary between the silicate mantle and the metal core of large planetesimals. The Mn-Cr-isotopic system in two pallasites, Omolon (OM) (*Shukolyukov and Lugmair,* 1997) and Eagle Station (ES) (*Shukolyukov and Lugmair,* 2001), is illustrated in Fig. 4. The slope of the best-fit line for ES yields a ^{53}Mn/^{55}Mn ratio of (1.17 ± 0.09) × 10^{-6} at the time of isotopic closure. This value corresponds to an absolute age of 4557.5 ± 0.6 Ma. The best-fit line for OM results in a ^{53}Mn/^{55}Mn ratio of (1.29 ± 0.19) × 10^{-6}, yielding an age of 4558.0 ± 1.0 Ma. This age is indistinguishable from that of ES. The correlation line for ES, although of similar slope to that of OM, passes below the OM best-fit line and the other isochrons for the meteorites with chondritic Mn/Cr ratio (*Shukolyukov and Lugmair,* 2001). This indicates that the ES precursor material had a lower than chondritic Mn/Cr ratio, possibly similar to CV3-type carbonaceous material (*Shukolyukov and Lugmair,* 2000b). This finding, as well as the presence in ES of an anomalous presolar component with an excess of ^{54}Cr, an anomalous O-isotopic composition (*Clayton and Mayeda,* 1996), and the enrichment in refractory siderophile trace elements (*Scott,* 1977), indicates that this pallasite formed within a parent body that is different from that of OM. The younger Mn-Cr ages of the pallasites (~4558 Ma)

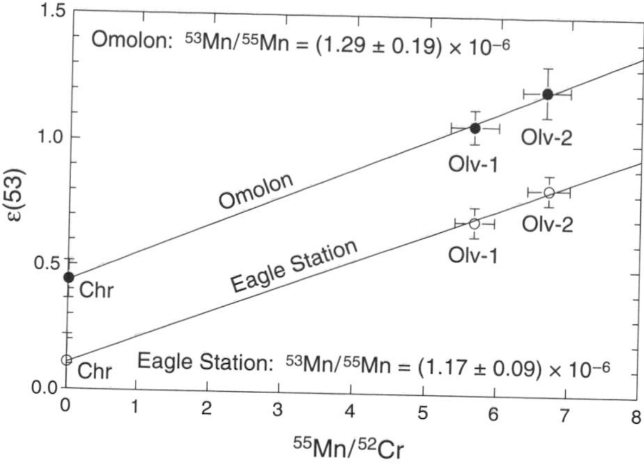

Fig. 4. Manganese-53–chromium-53 systematics in the pallasites Eagle Station and Omolon. Chr = chromite, Ol = olivine. The $^{53}Mn/^{55}Mn$ ratios at the time of isotopic closure are indistinguishable within the uncertainties and yield an age of ~4558 Ma for both meteorites. However, Eagle Station and Omolon originate from different parent bodies (see text). The relatively young Mn-Cr age of the pallasites corresponds to the time when either the interiors of their parent bodies cooled or the parent bodies were disrupted, which allowed rapid cooling of the exposed interior.

reflect the time when either the interiors of large chemically stratified, differentiated planetesimals cooled below the isotopic closure temperature or the planetesimals were disrupted to allow rapid cooling of the exposed interior.

Because of either their small size or relatively late time of formation, some of the planetary objects did not accumulate enough energy (^{26}Al and gravitational energy) for melting but experienced only different degrees of thermal metamorphism. The parent bodies of the sampled chondrites may be some examples of this. The Mn-Cr-isotopic system in the primitive H4 chondrite Ste. Marguerite closed at 4565.0 ± 0.7 Ma (*Polnau and Lugmair,* 2001). The EH4 chondrites Indarch and Abee cooled essentially at the same time (4564–4566 Ma) (*Shukolyukov and Lugmair,* 1999). These results are in good agreement with the recent investigations with the ^{129}I-^{129}Xe chronometer (*Brazzle et al.,* 1999; *Gilmour,* 2000). However, many other chondrites of higher metamorphic grades show metamorphic ages of up to more than 10 m.y. younger. For example, the U-Pb age of the L5–6 chondrite Barwell is ~4538 Ma (*Göpel et al.,* 1994). Most likely this is a result of slow cooling in the deeper zones of the parent asteroid.

5. SUMMARY

We have constructed an absolute timescale for the events in the early solar system based mostly on the data obtained with the Mn-Cr-isotopic chronometer. Some of the results are summarized in Fig. 5. First solids in the early solar system (CAIs) formed ~4571 Ma. Both the Mn-Cr and the Al-Mg timescales are anchored to this time. The formation of small planetesimals (100 m to km in size) occurred almost simultaneously. An early and swift formation of small planetesimals is necessary in order to prevent the rapid loss of CAIs into the Sun. Larger planetesimals (tens of kilometers) accreted during the next few million years. Both the Mn-Cr and the Al-Mg systems are consistent in that primary chondrule formation occurred mostly within the

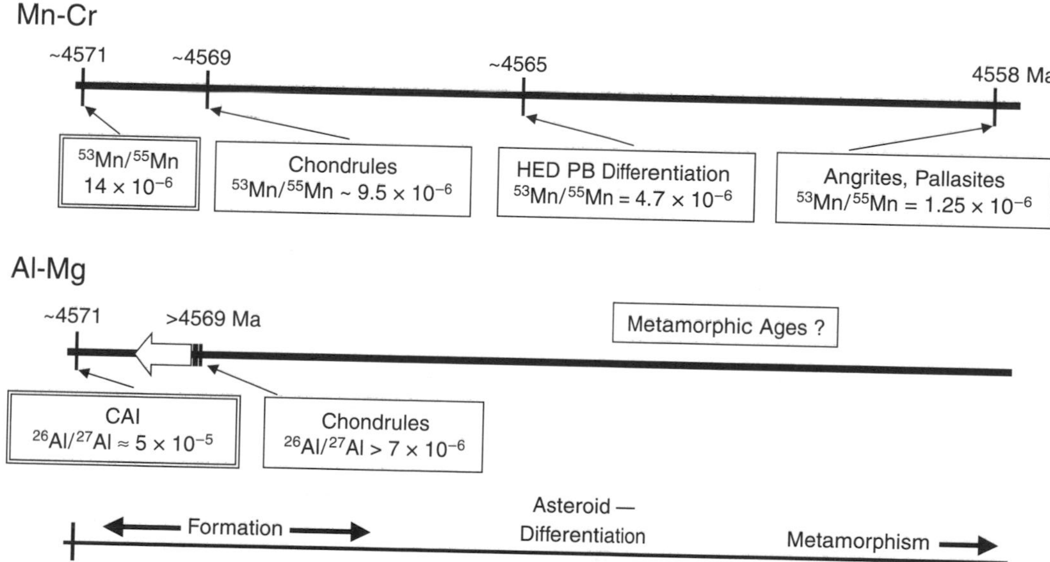

Fig. 5. Manganese-chromium and Al-Mg early solar system timelines. Both the Mn-Cr and Al-Mg system are anchored to the time of 4571 Ma, when most refractory meteorite inclusions (CAI) are believed to have formed. Small planetesimals were formed almost simultaneously to preserve the CAIs. The interiors of many intermediate-sized planetesimals were melted through the decay of ^{26}Al. Within a timespan of about 2 m.y. most primary chondrules appear to have been formed. Larger planetesimals accreted after a few million years. Chemical differentiation, formation of a stratified mantle and a core in these larger bodies came to an end at ~4565 Ma (Vesta). Metamorphic alteration of asteroidal material due to latent heat and collisions persisted for many millions of years. See text for details.

first ~2 m.y. of solar system history and would have been largely concluded ~4569 Ma. It appears that this time was an era of intense thermal processing in inner solar system material due to decay of ^{26}Al. Later collisional disruption of these larger planetesimals distributed chondrules to their ultimate meteorite parent bodies. On yet larger planetesimals (several 100 km) chemical differentiation, formation of a stratified mantle and a core came to a conclusion a few million years later at ~4565 Ma (Vesta). At about the same time (4564–4566 Ma), the outer zones of some chondrite PBs, such as those of the H chondrites and the enstatite chondrites, cooled below temperatures where isotopic closure occurs. The younger ages of pallasites (~4558 Ma) mark the time when either the interiors of their PBs cooled or the PBs were disrupted to allow rapid cooling of the exposed interior. The chondrites of higher metamorphic grade show metamorphic ages up to more than 10 m.y. younger, which is most likely a result of slow cooling due to a deep burial in their PBs. Clearly, accretion, accumulation of heat, and chemical differentiation on larger asteroids occurred on a relatively short timescale, on the order of only ~6 m.y.

Acknowledgments. We are grateful to A. Ghosh and M. Chaussidon for constructive reviews. This work was supported in part by grants from NASA.

REFERENCES

Birck J.-L. and Allègre C. J. (1985) Evidence for the presence of ^{53}Mn in the early solar system. *Geophys. Res. Lett., 12,* 745–748.

Binzel R. P. and Xu S. (1993) Chips off of asteroid 4 Vesta: Evidence for the parent body of basaltic achondrite meteorites. *Science, 260,* 186–191.

Boss A. P. and Vanhala H. A. T. (2000) Triggering protostellar collapse, injection, and disk formation. *Space Sci. Rev., 92,* 13–22.

Brazzle R. H., Pravdivtseva O. V., Meshik A. P., and Hohenberg C. M. (1999) Verification and interpretation of the I-Xe chronometer. *Geochim. Cosmochim. Acta, 63,* 739–760.

Cameron A. G. W., Vanhala H., and Hoflich P. (1997) Triggering presolar cloud collapse and injecting material into the presolar nebula. In *Astrophysical Implications of Laboratory Study of Presolar Materials* (T. J. Bernatowicz and E. K. Zinner, eds.), pp. 649–664. American Institute of Physics, New York.

Carlson R. W. and Lugmair G. W. (2000) Timescales of planetesimal formation and differentiation based on extinct radioisotopes. In *Origin of the Earth and Moon* (R. M. Canup and K. Righter, eds.), pp. 25–44. Univ. of Arizona, Tucson.

Cassen P. and Woolum D. S. (1997) Nebular fractionation and Mn-Cr systematics (abstract). In *Lunar and Planetary Science XXVIII,* pp. 211–212. Lunar and Planetary Institute, Houston.

Chen J. H. and Tilton G. R. (1976) Isotopic lead investigations on the Allende carbonaceous chondrite. *Geochim. Cosmochim. Acta, 40,* 635–643.

Clayton D. D. and Jin L. (1995) A new interpretation of ^{26}Al in meteoritic inclusions. *Astrophys. J. Lett., 451,* L87–L91.

Clayton R. N. and Mayeda T. K. (1996) Oxygen isotope studies of achondrites. *Geochim. Cosmochim. Acta, 60,* 1999–2017.

Dickin A. P. (1995) *Radiogenic Isotope Geology.* Cambridge Univ., Cambridge. 452 pp.

Ghosh A. and McSween H. Y. Jr. (1998) A thermal model for the differentiation of asteroid 4 Vesta, based on radiogenic heating. *Icarus, 134,* 187–206.

Gilmour D. J. (2000) The extinct radionuclide timescale of the early solar system. *Space Sci. Rev., 92,* 123–132.

Göpel C., Manhés G., and Allègre C. J. (1991) Constraints on the time of accretion and thermal evolution of chondrite parent bodies by precise U-Pb dating of phosphates (abstract). *Meteoritics, 26,* 338.

Göpel C., Manhés G., and Allègre C. J. (1994) U-Pb systematics of phosphates from equilibrated ordinary chondrites. *Earth Planet. Sci. Lett., 121,* 153–171.

Hsu W., Huss G. R., and Wasserburg G. J. (2000) Compound CAI and multi-stage formation: Evidence of ^{26}Al abundances from the Allende 5241 CAI (abstract). In *Lunar and Planetary Science XXXI,* Abstract #1734. Lunar and Planetary Institute, Houston (CD-ROM).

Hutcheon I. D., Weisberg M. K., Phinney D. L., Zolensky M. E., Prinz M., and Ivanov A. V. (1999) Radiogenic ^{53}Cr in Kaidun carbonates: Evidence for very early aqueous activity (abstract). In *Lunar and Planetary Science XXX,* Abstract #1722. Lunar and Planetary Institute, Houston (CD-ROM).

Keil K. (2002) Geological history of asteroid 4 Vesta: The "smallest terrestrial planet." In *Asteroids III* (W. F. Bottke Jr. et al., eds.), this volume. Univ. of Arizona, Tucson.

Kita N. T., Nagahara H., Togashi S., and Morishita Y. (1998) New evidence of aluminum-26 from ferrous-oxide-rich chondrule in Semarkona (LL3.0) (abstract). *Meteoritics & Planet. Sci., 33,* A83–A84.

Lugmair G. W and Galer S. J. G. (1992) Age and isotopic relationships among the angrites Lewis Cliff 86010 and Angra dos Reis. *Geochim. Cosmochim. Acta, 56,* 1673–1694.

Lugmair G. W. and Shukolyukov A. (1998) Early solar system timescales according to ^{53}Mn-^{53}Cr systematics. *Geochim. Cosmochim Acta, 62,* 2863–2886.

Lugmair G. W. and Shukolyukov A. (2001) Early solar system events and timescales. *Meteoritics & Planet. Sci., 36,* 1017–1026.

MacPherson G. J., Wark D. A., and Armstrong J. T. (1988) Primitive material surviving in chondrites: Refractory inclusions. In *Meteorites and the Early Solar System* (J. F. Kerridge and M. S. Matthews, eds.), pp. 436–461. Univ. of Arizona, Tucson.

MacPherson G. J., Davis A. M., and Zinner E. K. (1995) The distribution of aluminum-26 in the early solar system: A reappraisal. *Meteoritics, 30,* 365–386.

Marhas K. K., Hutcheon I. D., Krot A. N., Goswami J. N., and Komatsu M. (2000) Aluminum-26 in carbonaceous chondrite chondrules (abstract). *Meteoritics & Planet. Sci., 35,* A102.

McCord T. B., Adams J. B., and Johnson T. V. (1970) Asteroid Vesta: Spectral reflectivity and compositional implications. *Science, 168,* 1445–1447.

McKeegan K. D., Greenwood J. P., Leshin L. A., and Cosarinsky M. (2000) Abundance of ^{26}Al in ferromagnesian chondrules of unequilibrated ordinary chondrites (abstract). In *Lunar and Planetary Science XXXI,* Abstract #2009. Lunar and Planetary Institute, Houston (CD-ROM).

McSween H. Y. Jr., Ghosh A., Grimm R. E., Wilson L., and Young E. D. (2002) Thermal evolution models of asteroids. In *Asteroids III* (W. F. Bottke Jr. et al., eds.), this volume. Univ. of Arizona, Tucson.

Meibom A. and Clark B. E. (1999) Evidence for the insignificance of ordinary chondritic material in the asteroid belt. *Meteoritics & Planet. Sci., 34,* 7–24.

Nehru C. E., Prinz M., Delaney J. S., Dreibus G., Palme H., Spettel B., and Wänke H. (1983) Brachina: A new type of meteorite, not a chassignite. *Proc. Lunar Planet. Sci. Conf. 14th*, in *J. Geophys. Res., 88,* B237–B244.

Nyquist L. E., Lindstrom D., Mittlefehldt D., Shih C.-Y., Wiesmann H., Wentworth S., and Martinez R. (2001) Manganese-chromium formation intervals for chondrules from the Bishunpur and Chainpur meteorites. *Meteoritics & Planet. Sci., 36,* 911–938.

Papanastassiou D. A. (1986) Cr isotopic anomalies in the Allende meteorite. *Astrophys J. Lett., 308,* L27–L30.

Patterson C. C. (1956) Age of meteorites and the Earth. *Geochim. Cosmochim. Acta, 10,* 230–237.

Podosek F. A. and Nichols R. H. Jr. (1997) Short-lived radionuclides in the solar nebula. In *Astrophysical Implications of the Laboratory Study of Presolar Materials* (T. J. Bernatowicz and E. K. Zinner, eds.), pp. 617–647. American Institute of Physics, New York.

Podosek F. A., Ott U., Brannon J. C., Neal C. R., Bernatowicz T. J., Swan P., and Mahan S. E. (1997) Thoroughly anomalous chromium in Orgueil. *Meteoritics & Planet. Sci., 32,* 617–627.

Polnau E. and Lugmair G. W. (2001) Mn-Cr isotope systematics in the two ordinary chondrites Richardton (H5) and Ste. Marguerite (H4) (abstract). In *Lunar and Planetary Science XXXII,* Abstract #1527. Lunar and Planetary Institute, Houston (CD-ROM).

Rotaru M., Birck J.-L., and Allègre C. J. (1992) Clues to early solar system history from chromium isotopes in carbonaceous chondrites. *Nature, 358,* 465–470.

Russell S. S., Srinivasan G., Huss G. R., Wasserburg G. J., and MacPherson G. J. (1996) Evidence for widespread ^{26}Al in the solar nebula and constraints for nebula time scales. *Science, 273,* 757–762.

Scott E. R. D. (1977) Geochemical relationships between some pallasites and iron meteorites. *Mineral. Mag., 41,* 265–272.

Shu F. H., Shang H., Glassgold E., and Lee T. (1997) X-rays and fluctuating X-wind from protostars. *Science, 277,* 1475–1479.

Shukolyukov A. and Lugmair G. W. (1993) ^{60}Fe in eucrites. *Earth Planet. Sci. Lett., 119,* 159–166.

Shukolyukov A. and Lugmair G. W. (1997) The ^{53}Mn-^{53}Cr isotope system in the Omolon pallasite and the half-life of ^{187}Re (abstract). In *Lunar and Planetary Science XXVIII,* pp. 1315–1316. Lunar and Planetary Institute, Houston.

Shukolyukov A. and Lugmair G. W. (1999) The ^{53}Mn-^{53}Cr isotope systematics of the enstatite chondrites (abstract). In *Lunar and Planetary Science XXX,* Abstract #1093. Lunar and Planetary Institute, Houston (CD-ROM).

Shukolyukov A. and Lugmair G. W. (2000a) On the ^{53}Mn heterogeneity in the early solar system. *Space Sci. Rev., 92,* 225–236.

Shukolyukov and Lugmair G. W. (2000b) Cr isotope anomalies in the carbonaceous chondrites Allende and Orgueil and a potential connection between 54-Cr and oxygen isotopes (abstract). *Meteoritics & Planet. Sci., 35,* A146.

Shukolyukov A. and Lugmair G. W. (2001) Cr isotope systematics in the pallasite Eagle Station: Chronology and evidence for a genetic link to carbonaceous chondrites (abstract). In *Lunar and Planetary Science XXXII,* Abstract #1365. Lunar and Planetary Institute, Houston (CD-ROM).

Srinivasan G., Goswami J. N., and Bhandari N. (1999) ^{26}Al in eucrite Piplia Kalan: Plausible heat source and formation chronology. *Science, 284,* 1348–1350.

Störzer D. and Pellas P. (1977) Angra dos Reis: Plutonium distri bution and cooling history. *Earth Planet. Sci. Lett., 35,* 285–293.

Tatsumoto M., Unruh D. M., and Desborough G. A. (1976) U-Th-Pb and Rb-Sr systematics of Allende and U-Th-Pb systematics of Orgueil. *Geochim. Cosmochim. Acta, 40,* 617–634.

Wadhwa M. and Lugmair G. W. (1996) Age of the eucrite "Caldera" from convergence of long-lived and short-lived chronometers. *Geochim. Cosmochim. Acta, 60,* 4889–4893.

Wadhwa M., Shukolyukov A., and Lugmair G. W. (1998) ^{53}Mn-^{53}Cr systematics in Brachina: A record of one of the earliest phases of igneous activity on an asteroid (abstract). In *Lunar and Planetary Science XXIX,* Abstract #1480. Lunar and Planetary Institute, Houston (CD-ROM).

Weidenschilling S. J. (1977) Aerodynamics of solid bodies in the solar nebula. *Mon. Not. R. Astron. Soc., 180,* 57–70.

Weidenschilling S. J. (2000) Formation of planetesimals and accretion of the terrestrial planets. *Space Sci. Rev., 92,* 295–310.

Zinner E. and Göpel C. (1992) Evidence for ^{26}Al in feldspars from the H4 chondrite Ste. Marguerite (abstract). *Meteoritics, 27,* A311–A312.

Meteorite Evidence for the Accretion and Collisional Evolution of Asteroids

Edward R. D. Scott
University of Hawai'i

Meteorites contain a record of impacts during all stages of asteroid origin and evolution: the formation and accretion of chondritic particles; the alteration, metamorphism and melting of asteroids; and the erosion and disruption of asteroids by hypervelocity impacts. A review of meteorite classification shows that numerous meteorites are not readily classified because they do not fit simple models for asteroid formation and evolution that assume impacts were only important during the final stage of asteroid evolution and because of inadequate understanding of asteroidal impacts. Chronological, textural, and thermal constraints allow us to identify meteorite impact breccias that formed during accretion (e.g., Kaidun), when asteroids were partly molten (e.g., mesosiderites), and during the subsequent disruption of asteroids (e.g., L chondrites). Studies of chondrites including Kaidun suggest that chondrules accreted with similar-sized fragments of preexisting bodies that formed at greater heliocentric distances. In the inner solar system, chondrules appear to have been crucial for initiating accretion. Without chondrules and rock fragments, dry dust failed to accrete to the nebular midplane because of nebular turbulence and spiraled into the protosun. The tiny mass of asteroids may be partly due to inefficient chondrule formation beyond 2 AU or less-efficient delivery of chondrules from near the protosun.

1. INTRODUCTION

Meteorites and returned asteroid samples are keys to understanding the chemical, mineralogical, and physical properties of near-Earth and main-belt asteroids. We know a lot about the chemical, mineralogical, and physical properties of meteorites but are still very ignorant about the corresponding properties of their parent asteroids and possible matches between meteorite classes and asteroid types (*Burbine et al.,* 2002). To make progress in understanding the formation and evolution of asteroids, the origin of meteorites, and the formation of chondritic components we urgently need to know more about the composition and structure of asteroids. Our poor understanding of how impacts modified asteroids and meteorites is a major barrier in achieving these goals. Impacts controlled the formation of asteroids and their destruction and were a major factor in their geological evolution (e.g., *Scott et al.,* 1989). Understanding the role of impacts is critical if we wish to learn how diverse types of chondritic materials accreted into asteroids, how and when 99.9% of the solid material in the asteroid belt was lost, and how impacts changed the chemical, mineralogical, and physical properties of chondritic and differentiated asteroids.

A simple model for the geological evolution of chondritic and differentiated asteroids has four stages. Chondrules, metal grains, and other components were formed, probably in the solar nebula. These accreted together to form parent asteroids. Some of these bodies were subsequently melted, forming metallic cores and olivine-rich mantles; others were metamorphosed or aqueously altered. After the asteroids had cooled, they were eroded by impacts for 4.5 G.y. or disrupted. However, this simple model is inadequate as detailed petrologic studies and radiometric ages suggest that the four stages overlapped in time. Chondrules, for example, appear to have accreted with fragments of preexisting asteroids (e.g., *Scott et al.,* 1996). Some chondrites and eucrites are breccias of materials excavated by impacts from different depths that were metamorphosed after assembly (e.g., *Yamaguchi et al.,* 1997). In addition, few chondrites and differentiated meteorites experienced simple cooling histories after peak temperatures were reached, probably because of disruption by impacts (*Scott and Keil,* 1999).

Here I review the meteorite evidence for impact processing, both during and after accretion, in the light of studies of terrestrial, lunar, and laboratory impacts and impact modeling. I first review the variety of meteorites, emphasizing those features and rocks that do not fit simple models for the accretion and geological evolution of asteroids.

2. METEORITE TYPES AND PARENT BODIES

Meteorites are divided into two classes: chondrites and differentiated or nonchondritic meteorites (*Brearley and Jones,* 1998; *Mittlefehldt et al.,* 1998). Chondrites are aggregates of silicate and metallic particles and are remarkably similar in composition to the condensable material in the Sun. Except for H, He, C, N, O, and the inert gases, elemental ratios are very largely within a factor of 2 of those in the solar photosphere (*Wasson and Kallemeyn,* 1988). Differentiated or nonchondritic meteorites are igneous rocks and differ considerably from the Sun in composition. Three types are distinguished: irons, achondrites, and stony irons, most of which are thought to be derived respectively from the core, mantle and crust, and core-mantle boundary of melted asteroids. Differentiated meteorites have elemental ratios that may be up to 10–1000× different from solar values.

Largely as a result of meteorite discoveries in Antarctica and various deserts around the world, the number of known meteorites has increased in the past 20 years from 3000 to around 20,000, allowing for pairing of fragments (*Bischoff,* 2001). Detailed studies of these meteorites suggest they come from 100–150 different parent asteroids (*Burbine et al.,* 2002). However, a surprising number cannot be readily classified as chondritic or differentiated. This suggests that our ideas about the formation of meteorites and the evolution of their parent asteroids may require modification.

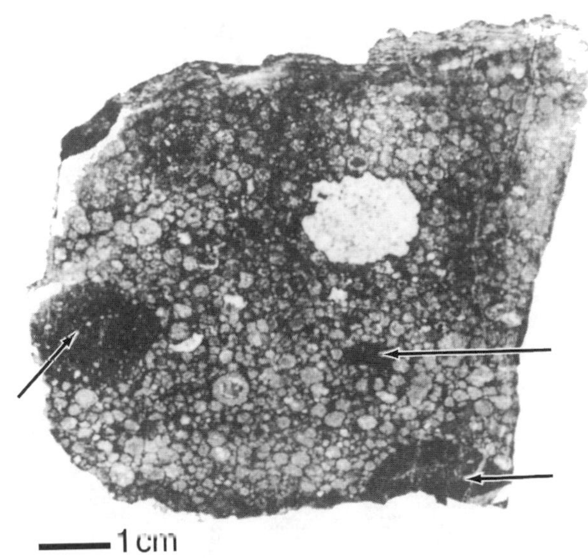

Fig. 1. Polished slab of the Leoville CV3 chondrite showing numerous chondrules, one large and several small, white Ca-Al-rich inclusions (CAIs), and three dark inclusions, which are marked by arrows. Chondrules and CAIs formed in the solar nebula but dark inclusions are probably fragments of preexisting bodies that accreted with chondrules and CAIs. (Photo courtesy of Alfred Kracher and Klaus Keil.)

2.1. Chondrites

2.1.1. Components. Chondrites are mixtures of diverse types of materials that formed in different parts of the solar nebula under very different thermal conditions. Chondrules are rounded particles, typically millimeter-sized, that formed from molten silicate droplets (*Rubin,* 2000). Various kinds of chondrules, associated metal grains, and the rarer Ca-Al-rich inclusions (CAIs) formed at high temperatures (1200°–2000°C) in the solar nebula (*Ireland and Fegley,* 2000). Before these components accreted they were commonly coated or mixed with volatile-rich, fine-grained matrix material, which contains micrometer- and submicrometer-sized grains of circumstellar grains and clearly experienced less thermal processing (*Ott,* 2001).

Chondrites also contain angular rock fragments. Many are clearly pieces of chondrules and CAIs, some are foreign fragments of known meteorite types (*Lipschutz et al.,* 1989), but most are matrix-rich chondritic fragments called dark inclusions (see Fig. 1) (*Kracher et al.,* 1985; *Johnson et al.,* 1990; *Endress et al.,* 1994). Most dark inclusions are heavily altered, have sizes comparable to those of chondrules and CAIs, may contain smaller chondrules than their host chondrites, and some are rimmed by matrix material (e.g., Fig. 1 of *Scott et al.,* 1996). Their I-Xe ages suggest they were altered before their host chondrites (*Hohenberg et al.,* 2001). Thus some dark inclusions appear to be fragments of chondritic bodies that may have spiraled toward the protosun and then accreted with chondrules to form new kinds of chondrites.

2.1.2. Formation of chondrules and CAIs: How and where? Most authors infer that chondrules and CAIs formed in the solar nebula by some kind of localized heating mechanism such as nebular lightning, magnetic reconnection flares, or gas dynamic shock wave (*Rubin,* 2000; *Jones et al.,* 2000). Some authors invoke prior formation of chondrule-free planetesimals. *Weidenschilling et al.* (1998), for example, suggest that planetesimals in resonance with Jupiter attained high eccentricities, generating bow shocks in the nebula gas that melted dust, forming chondrules. A few authors have argued that chondrules formed by impacts of molten planetesimals (e.g., *Sanders,* 1996; *Lugmair and Shukolyukov,* 2001) or impacts into asteroidal regoliths (e.g., *Symes et al.,* 1998). However, formation of chondrules by impacts between solid or molten asteroids appears inconsistent with many properties of chondrules and chondrites (*Taylor et al.,* 1983; *Taylor and Scott,* 2001).

It is commonly thought that CAIs and chondrules formed in the asteroid belt and were quickly accreted into asteroids (e.g., *Alexander et al.,* 2001). However, new models suggest that they may have formed close to the protosun and were then distributed across the solar nebula by a bipolar jet flow (*Liffman and Brown,* 1996; *Shu et al.,* 2001; *Ireland and Fegley,* 2000; *Krot et al.,* 2001). If this model is correct, the conclusion derived from inferred ^{26}Al concentrations that CAIs predate chondrules by a million years or more (*Wadhwa and Russell,* 2000) may not be valid, as ^{26}Al may have been produced by intense particle irradiation near the young Sun.

2.1.3. Chondrite classes, groups, and origins. Three main classes of chondrites, enstatite, ordinary, and carbonaceous, are subdivided into 12 groups. One group, R chondrites, and several ungrouped chondrites do not fit into these three classes (Fig. 2). Within each chondrite group, with the possible exception of CR and CH chondrites, the proportions of chondrules and other chondritic components and the bulk composition of each chondrite are surprisingly uniform. As for all meteorite groups, it is plausible that the members of each group come from a single body, but multi-

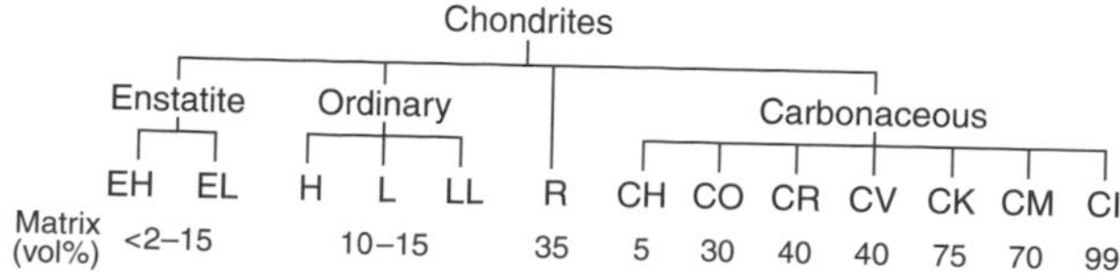

Fig. 2. Classification and matrix concentrations in 13 chondrite groups, 12 of which belong to the enstatite, ordinary, and carbonaceous classes. R chondrites and ungrouped chondrites, which are not shown, do not fit into these three classes. Chondrite groups are arranged in approximate order of increasing matrix concentration (data from *Scott et al.,* 1996). Properties of meteorites and asteroids suggest that this sequence corresponds very approximately to increasing formation distance from the Sun (*Rubin and Wasson,* 1995). Almost all the other mineralogical, chemical, and isotopic properties of these groups do not correlate with the matrix concentration, suggesting that chondritic asteroids did not form simply from a disk of solid materials with compositions that varied smoothly with heliocentric distance.

ple bodies cannot be excluded. The ungrouped chondrites probably come from another 13 or more bodies (*Meibom and Clark,* 1999). The rate at which new types of carbonaceous chondrites are being discovered and the strange properties of recently discovered groups suggest that the known chondrites may represent only a minor fraction of the chondritic asteroids and the precursor material of differentiated asteroids.

Oxygen-isotopic analyses show that chondrules in each class are quite distinct and must have been derived from at least three separate reservoirs (*Clayton,* 1993). Chemical and mineralogical studies confirm that chondrules in H, L, and LL chondrites are not fundamentally different and might be derived from a single reservoir by size sorting. A similar relationship holds for EH and EL chondrites, but the properties of carbonaceous chondrites are much more diverse and appear to require at least four separate sources. (Carbonaceous chondrites have little in common except high abundances of refractory elements.) Thus the parent asteroids of the chondrites probably formed from at least six separate reservoirs of chondrules with remarkably little mixing between the reservoirs.

2.1.4. Chondritic asteroids. *Sears* (1998) and *Akridge et al.* (1998) argued that chondrites are biased samples derived from the outer few kilometers of asteroids where chondrules formed by impact and that we have not sampled the asteroids' chondrule-free dusty interiors. However, concentrations of chondrules and matrix material are rather uniform in each group (Fig. 2) and we certainly have meteorite samples from the mantles and cores of differentiated asteroids (see below). Since ordinary chondrite parent bodies were baked at high temperatures, any loosely consolidated dirt should have been converted to rock capable of surviving the journey to Earth. In view of other difficulties with impact origins for chondrules (see above) and the likelihood that our samples come from ~24 or more chondritic asteroids, it seems unwise to conclude that the interiors of chondritic asteroids are devoid of chondrules. In the absence of returned samples from asteroids, we will assume that we have sampled the interiors of chondritic asteroids.

Possible links between chondrite groups and asteroid types are discussed by *Burbine et al.* (2002). Enstatite chondrites, which may be derived from two or more M asteroids, probably formed closer to the Sun than other chondrites (*Wasson,* 1988; *Rubin and Wasson,* 1995; *Lugmair and Shukolyukov,* 1998, 1999). The E-type asteroids, which appear to be the source of the closely related aubrites (enstatite achondrites), are also found at the inner edge of the belt (*Bell et al.,* 1989). Ordinary chondrites are probably derived from three or more S(IV)-type asteroids, fragments of which may be classed as Q-type asteroids among near-Earth asteroids. CI, CM, CR, and CK chondrites are probably derived from C-type asteroids, which are concentrated further out in the asteroid belt (2.5–3.5 AU). CV and CO chondrites may come from K-type asteroids, which are largely from the Eos family at 3.0 AU (*Bell et al.,* 1989). The most porous and possibly very primitive chondrite, Tagish Lake, is a CM-like chondrite that is very rich in C (~5 wt%), H_2O, and presolar grains (*Brown et al.,* 2000). It contains fewer chondrules and CAIs than CM chondrites, and is thought to be derived from one of the D-type asteroids (*Hiroi et al.,* 2001), which lie beyond 2.9 AU and probably dominate the asteroid population beyond 4 AU (*Jones et al.,* 1990).

Properties of chondrites and asteroids suggest there were some systematic variations in the properties of accreting planetesimals that survive today in the asteroid belt despite some mixing due to subsequent orbital evolution of the asteroids and their impact debris. At the inner edge of the asteroid belt, chondritic asteroids tend to be chondrule-rich and matrix-poor, like E and O chondrites (Fig. 2), and volatile concentrations in the matrix material are relatively low. In the outer parts of the belt, chondritic asteroids tend to have much higher proportions of matrix to chondrules and the matrix material is richer in water and other volatiles.

The degree of alteration and metamorphism experienced by the parent bodies of chondrites also varied across the belt (*Keil,* 2000). At the inner edge, peak temperatures were generally higher. Chondrites in the EH, EL, H, L, and LL groups were all heated in the temperature range ~400°–950°C. The most deeply buried probably experienced the highest temperatures (type 6), whereas the least-metamorphosed (type 3) were closer to the surface. Enstatite chondrites experienced little if any aqueous alteration prior to metamorphism, and ordinary chondrites experienced only mild alteration in the most-fine-grained portions. The exceptional CH chondrites resemble ordinary chondrites in that they contain very little water and matrix material but experienced lower peak temperatures (<300°C). They may have formed closer to the Sun than other chondrites.

Further out in the asteroid belt, the carbonaceous parent bodies experienced aqueous alteration at temperatures of ~0°–300°C (CM, CR, and CI bodies). In bodies that were heated to 300°–500°C, volatiles were lost and aqueous alteration merged into fluid-assisted metamorphism (CO and CV). The most metamorphosed carbonaceous chondrites are found in CK chondrites (types 4–6), which are almost completely dry.

2.2. Differentiated Meteorites

Differentiated meteorites formed in asteroids that melted shortly after they accreted (*Wadhwa and Russell,* 2000; *Taylor et al.,* 1993), and are commonly divided into achondrites, irons, and two kinds of stony irons, pallasites and mesosiderites (*Shearer et al.,* 1998). Conventional views suggest that differentiated meteorites should consist of easily characterized rocks from the core, mantle, or crust of asteroids.

The largest class of achondrites are the HED meteorites (~80% of achondrite falls): the howardites, eucrites, and diogenites, which are very probably derived from Vesta (*Keil,* 2002). Another Vesta-like body is required for the basaltic meteorite, Northwest Africa 011 (*Yamaguchi et al.,* 2001), conceivably the parent asteroid of 1459 Magnya (see *Burbine et al.,* 2002). Studies of the HED meteorites and remote sensing indicate that Vesta differentiated to form a core, mantle, and crust. It experienced a high degree of melting, possibly complete melting, probably due to ^{26}Al decay. Thus Vesta can be considered as the smallest terrestrial planet (*Keil,* 2002). However, it may be incorrect to consider Vesta as the archetypal differentiated asteroid as most may have lost their basaltic magma because of explosive volcanism (*Keil,* 2000).

Excluding meteorites from Mars and the Moon, there are four major groups of achondrites: ureilites, aubrites, brachinites, and angrites (*Mittlefehldt et al.,* 1998). These meteorites appear to be derived from four parent asteroids with very different igneous and impact histories from Vesta. Ureilites are largely unbrecciated, coarse-grained, igneous rocks made of olivine and pyroxene. Aubrites are brecciated rocks largely composed of coarse-grained enstatite that formed from a body distinct from the parent bodies of the EH and EL chondrites (*Keil,* 1989). Brachinites are unbrecciated, igneous rocks containing 80–95 vol% olivine and 0–10 vol% plagioclase, and angrites are unbrecciated, plagioclase-poor, basaltlike igneous rocks. None of these groups of achondrites is a close match for the products predicted by simple models of asteroid differentiation, suggesting that igneous processes in asteroids were much more complex than commonly appreciated (*Mittlefehldt et al.,* 1998).

Brachinites may be derived from A-type asteroids, aubrites from E-type asteroids, and ureilites and angrites from S asteroids (*Burbine et al.,* 2002).

2.3. Other Types of Meteorites

Textbooks imply that chondrites and differentiated meteorites are so different that even a simple-minded robot should be able to distinguish these two basic types of asteroidal material! However, our museums contain many strange meteorites that required detailed laboratory studies before they could be assigned to one of these two classes, and others that still have controversial classifications. Most of the difficulties have arisen because of disagreements about the role of impacts in the formation of these meteorites.

2.3.1. Impact melts. Meteorites identified as chondritic impact melts are mixtures of unmelted chondritic material and melt veins with chondritic composition and widths of less than a few centimeters; e.g., Rose City and Chico (*Bogard et al.,* 1995). However, two rocks that appear to be melted chondrites are quite different. Patuxent Range 91501 is an 8.5-kg igneous-textured rock containing a few volume percent of metal-troilite nodules (Fig. 3). Without detailed bulk chemical and O-isotopic analyses and mineral compositions, we would not know that it is probably an impact melt from the L-chondrite body (*Mittlefehldt and Lindstrom,* 2001). The second, Abee, was for many years the best-studied EH4 chondrite, but subsequent studies showed that it is actually an impact melt and that its chondrule-shaped features are merely ghosts from the original structure (*Rubin and Scott,* 1997). Its origin is still poorly understood.

2.3.2. Metal-chondrite breccias. The best-studied meteorite breccias that consist of metal veins or regions and chondritic fragments are Netschaevo (classed as a IIE iron) and Portales Valley [classified as an H6 chondrite (*Grady,* 2000)]. Both must have been deeply buried when they cooled as their metal portions developed Widmanstätten patterns (Fig. 4). Impacts created the chondritic fragments but it is not known how molten metal was formed and how

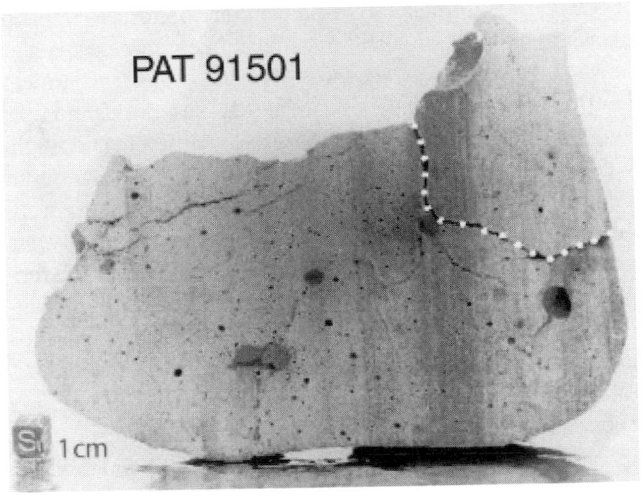

Fig. 3. Polished face of the PAT 91501 meteorite showing a uniform fine-grained rock with black vesicles under 3 mm in size and rounded or elongated, gray nodules of metal and troilite, which are up to a centimeter in size. Chemical and O-isotopic analysis of the silicates suggest that it is an impact melt from the L-chondrite parent body (*Mittlefehldt and Lindstrom,* 2001). However, unlike nearly all other impact-melted meteorites, it contains very few unmelted, chondritic grains, and none of these grains are shocked. A cometary projectile may have been responsible for the high degree of melting and vaporization. (Vertical lines are saw marks. Photo courtesy of D. Mittlefehldt and NASA Johnson Space Center.)

Fig. 4. Slice of the Portales Valley meteorite showing dark angular clasts of what appears to be an H6 chondrite separated by light-colored veins of metallic Fe,Ni, which have been etched to reveal a Widmanstätten pattern of oriented kamacite plates. An impact created the angular clasts but it is not known how the molten metal was formed, mixed with chondritic clasts, and then cooled slowly to form the Widmanstätten pattern. (Photo courtesy of Marvin Killgore.)

it was mixed with chondritic material. Both impact heating and internal heating have been proposed (see *Rubin et al.,* 2001a).

2.3.3. Metal-rich chondrites. Most chondrites contain <15 vol% metallic Fe,Ni, but two newly discovered types contain much more metal. CH chondrites contain 20 vol% metallic Fe,Ni, while two CB chondrites, QUE 94411 and HaH 237, have ~70 vol% metallic Fe,Ni; one of these was initially classified as an iron meteorite (*Weisberg et al.,* 2001). Metal particles in these two chondrites are chondrule-sized and appear to have similar condensation histories to the chondrules (*Krot et al.,* 2001). The chondrites also contain small amounts of hydrated matrix (*Greshake et al.,* 2002) and might be derived from hydrated M-class asteroids (*Rivken et al.,* 1995). Because they are so different from other chondrites in their composition and texture, some authors have suggested that metal-rich chondrites might be impact products, not true chondrites (*Wasson and Kallemeyn,* 1990). Their bulk Fe/Mg ratios are 1.7–9× above the solar ratio and their volatile elements are depleted by factors of up to 30×, placing them well outside the envelope of chemical compositions of other chondrites.

Three related meteorites, Bencubbin, Weatherford, and Gujba, all contain ~60 vol% metallic Fe,Ni and may also be metal-rich chondrites. However, they lack CAIs and their classification is less certain (*Rubin et al.,* 2001b). (Bencubbin was once classified as a mesosiderite.)

2.3.4. Partly differentiated meteorites. Acapulcoites and lodranites appear to represent the transition between chondrites and achondrites. Acapulcoites contain relict chondrules and have chondritic compositions but were heated up to 1000°C so that millimeter- to centimeter-sized metallic veins formed. They were not hot enough to lose much metal or basalt. Lodranites were heated to higher temperatures and are depleted in plagioclase and troilite, which are both concentrated in early silicate and metallic melts. Most authors envisage that internal heating was responsible, but impact heating has also been invoked (*Rubin et al.,* 2001c). *McCoy et al.* (2000) infer that the parent asteroid of these rocks would have internal chemical and mineralogical heterogeneities on scales of >200 m. The asteroid Eros lacks heterogeneities on this scale but minor igneous differentiation cannot be excluded as impact mixing may have homogenized its surface (*McCoy et al.,* 2001).

2.3.5. Metal-rich achondrites. Two meteorites, Mt. Egerton and Itqiy, appear to be igneous rocks made of ~20 vol% of metallic Fe,Ni intergrown with coarse-grained enstatite crystals (*Mittlefehldt et al.,* 1998; *Patzer et al.,* 2001). Since molten silicate and metallic Fe,Ni should readily separate in internally heated asteroids, they may have formed by impact. However, neither contains relict chondrules or clasts and their bulk composition is far from chondritic. Five other metal-bearing enstatite meteorites have been identified as impact melts of enstatite chondrite parentage by *McCoy et al.* (1995) and *Burbine et al.* (2000). Other workers favor an origin from indigenous melts from the aubrite parent body.

2.3.6. Achondritic breccias containing chondritic material. Most meteorite breccias contain only a tiny fraction of material that could have come from a separate parent body (e.g., *Lipschutz et al.,* 1989). However, the aubrite, Cumberland Falls, contains abundant chondritic clasts that were probably derived from a single projectile.

2.3.7. Chondritic breccias containing achondritic clasts. Some chondrites contain achondritic clasts that are obviously impact melted material. But a few clasts in type 3 ordinary chondrites appear to be fragments of differentiated asteroids and may have been acquired during accretion (see *Mittlefehldt et al.,* 1998, pp. 168–171).

2.3.8. Other oddball meteorites. A few meteorites, e.g., Enon, Bocaiuva, and Northwest Africa 176, are composed of ~50 vol% chondritic silicate inclusions in metallic Fe,Ni (*Liu et al.,* 2001). In these meteorites, the silicate inclusions do not resemble rock fragments and the meteorites might be strongly metamorphosed metal-rich chondrites or some kind of impact melt product.

The abundance of meteorites that lie outside the envelope of standard chondrites and differentiated meteorites suggests that our schemes for classifying meteorites are too simple-minded and that geological processes on asteroids, especially impacts, were more complex than we commonly appreciate. In addition, these meteorites emphasize the limitations of remote analyses of asteroids and the need for careful characterization of returned samples.

3. METEORITE EVIDENCE FOR ASTEROIDAL IMPACTS

Information about asteroidal impacts has been culled from three types of meteoritic evidence: mineral and rock textures, radiometric ages, and other constraints on the thermal histories of minerals. The degree of shock in minerals and clasts and the extent and nature of brecciation in meteorites provide direct clues to the impact history of meteorites and their constituents (e.g., *Scott et al.,* 1989; *Stöffler et al.,* 1991; *Bischoff and Stöffler,* 1992). However, shock and impact deformation effects are commonly heterogeneous and notoriously difficult to interpret. For example, several studies of the martian meteorite ALH 84001 have failed to agree on whether it suffered one major impact that formed a large complex crater (*Scott et al.,* 1998) or numerous impacts and one thermal metamorphic event (*Treiman,* 1998).

Radiometric ages of meteorites combined with other kinds of chronological constraints such as cosmic-ray-exposure ages can provide invaluable constraints on the nature of impact events (*Bogard,* 1995). However, radiometric chronometers are reset in impacts by postshock heating, which depends on shock pressure and the crater setting. Thus any proposed relationship between a radiometric age and a specific impact may be controversial. For example, *Turner et al.* (1997) argued that their 3.9-G.y. ^{40}Ar-^{39}Ar age for ALH 84001 was a metamorphic age and that the plagioclase glass and other shock metamorphic features formed during an impact 13 m.y. ago that removed the rock from Mars and exposed it to cosmic rays without significant reheating. However, constraints from the magnetization record suggest that the rock was not shocked or reheated significantly when it left Mars, unlike other martian meteorites, and that the 3.9-Ga event was the major impact that damaged the rock (*Weiss et al.,* 2000).

The ~3.9-G.y. ^{40}Ar-^{39}Ar ages and thermal histories of mesosiderites, which are essentially unshocked metal-silicate breccias, were initially interpreted to result from a near-catastrophic impact that reheated the breccias at the time of heavy lunar bombardment (*Bogard et al.,* 1990). However, additional studies of the thermal histories of metallic and silicate minerals suggested that mesosiderites were probably cooling slowly deep inside a large parent body at that time. The major impact that created the metal-silicate breccias occurred much earlier, probably at ~4.45 Ga (see *Scott et al.,* 2001).

These studies show that understanding the impact histories of meteorites may require several independent thermal constraints in addition to radiometric ages. Below I summarize how meteorite evidence can illuminate our understanding of various impact phenomena during the accretion and evolution of asteroids. The final section describes in more detail how impact studies have been used to infer the impact history of specific meteorite parent bodies.

3.1. Accretion of Asteroids

For accretion to occur, impact velocities must be less than about twice the escape velocity, which for a 100-km-radius asteroid is ~140 m/s. At these velocities, silicates are broken but not shocked or heated sufficiently to reset radiometric clocks. Thus accretional impacts can only be dated indirectly.

Models for accretion in the asteroid belt generally envisage micrometer- or submicrometer-sized dust grains sticking together and settling to the midplane of the nebula to form a dense layer of dust, which because of gravitational instabilities or some other mechanism was converted into a disk of kilometer-sized planetesimals. Several arguments suggest that asteroids may not have accreted this way. First, as discussed earlier, many chondritic asteroids accreted largely from chondrules, not dust. In many cases, the fine-grained dust accreted as rims on chondrules or other large particles. In addition, chondrules appear to have accreted with fragments of preexisting bodies that were large enough to undergo alteration and possibly also melting. Second, dry, solid silicate grains would probably not adhere to each other (*Wood,* 1996), and nebular turbulence may have prevented loose grain aggregates from settling to the midplane to form a gravitationally unstable layer (*Cuzzi et al.,* 1996, 2001). Third, as discussed below, the extraordinarily diverse menagerie of chondrite groups appear to require more complex accretion models.

3.1.1. Chondrite compositions. Although the matrix concentration in chondrite groups appears to be correlated with formation location (Fig. 2), it is not correlated with the concentration of refractory or moderately volatile elements or the bulk O-isotopic composition. Thus chondritic planetesimals did not simply form from a disk of solids with compositions that varied smoothly with radial distance. The bulk composition of a chondrite group or chondritic asteroid must have depended on two or more nebular parameters, e.g., the time and place of accretion (e.g., *Clayton,* 1993).

Since chondrules accreted from at least six separate reservoirs, the accretion process must have ensured that each asteroid acquired chondrules from a single reservoir. Whether chondrules formed near the protosun or in the asteroid belt, the time for an asteroid to accrete from a single reservoir of chondrules must have been much less than the time required to generate a new reservoir or chondrules at that location. In addition, early-formed asteroids had to be isolated from late-formed reservoirs of accreting chondrules. Thus chondrule formation and accretion may have been sporadic processes, ensuring that the composition of accreting material did not vary smoothly with time or radial distance.

3.1.2. Role of chondrules. Chondrites with the lowest concentrations of chondrules, CI, CM, and Tagish Lake, all have heavily hydrated matrix material. Not one of the undifferentiated asteroids that we appear to have sampled delivers chondrites that are composed largely of dry matrix material with few chondrules. Even without chondrules, such materials should be recognizable as chondritic from their near-solar bulk compositions. It is therefore probable that planetesimals in the inner part of the asteroid belt and in the terrestrial planet region did not form without chondrules and similar-sized particles. Either chondrule formation was remarkably efficient at these locations or fine dust failed to accrete without coating chondrules.

Cuzzi et al. (1996, 2001) infer that chondrules played a key role in triggering the accretion because chondrules and particles with similar aerodynamic stopping distances were concentrated at transient locations between turbulent eddies. Some of these accumulations settled to the midplane, drifted inward, and accreted into planetesimals. Dust and fluffy grain aggregates were carried along by the gas into the protosun. Episodic addition of chondrules to the nebula and planetesimal accretion may have allowed chondrules from different reservoirs to accrete without significant cross contamination.

The apparent tendency for the matrix concentration to increase with increasing inferred formation distance suggests that in the cooler parts of the solar nebula, fine-grained solids accreted without the assistance of chondrules. Ice crystals and sticky organics may have allowed grain aggregates to grow large enough and dense enough to decouple from the turbulent nebular gas and descend to the midplane to accrete. In the inner solar system, temperatures were probably too high for ice and organics to coat grains so that only chondrules or particles with comparable aerodynamic stopping distances acccreted.

3.1.3. Kaidun chondritic breccia. The Kaidun chondrite (Fig. 5) is unique as it consists almost entirely of millimeter- and submillimeter-sized fragments of various types of chondritic material: CI, CM, CR, EH, and EL chondrites, material similar to the Tagish Lake chondrite, and unidentified material (*Zolensky et al.,* 1996; *Zolensky and Ivanov,* 2001). Some Kaidun clasts resemble the dark clasts in CR chondrites (*Zolensky et al.,* 1996), which accreted with CR chondrules. In addition, many clasts are unshocked and probably formed very early as the carbonates in all lithologies appear to have formed <10^6 yr after CAIs (*Hutcheon et al.,* 1999). It seems probable that the clasts impacted at relatively low velocities and that the parent body of Kaidun accreted from fragmental debris from preexisting asteroids.

Fig. 5. Backscattered electron image of a thin section of the unique Kaidun chondrite, which is mostly composed of millimeter- and submillimeter-sized fragments from many other chondritic bodies. Some identified fragments include impact melt (a), enstatite chondrite (j), CI chondrite (c), CM chondrite (d), and Tagish Lake-type material (e). Kaidun, which has the highest proportion of rock fragments to chondrules of any chondrite, may be an accretionary breccia that formed by turbulent accretion of fragments in the nebula when few chondrules were available. (Image supplied by Michael Zolensky.)

The Kaidun breccia might usefully be considered as a chondrule-poor chondrite that is exceptionally rich in chondritic clasts. If there was a pause in chondrule formation or deposition at some nebular location, the abundance of tiny asteroidal fragments suspended in turbulent gas may have built up until rock chips, rather than chondrules, triggered planetesimal accretion.

Asteroidal fragments may also have played a role in transporting water across the asteroid belt during accretion. Experiments and theory suggest that slow kinetics prevented the formation of hydrous silicates in the nebula and that nebular temperatures in the asteroid belt probably exceeded the condensation temperature of water ice [~180 K (*Fegley,* 2000)]. Water may have accreted into asteroids in the form of particles of ice that drifted inward from beyond the nebular snowline (*Cyr et al.,* 1998), or fragments of outer planetesimals that contained hydrous silicates.

Many details of asteroid accretion remain obscure but the turbulent accretion model of *Cuzzi et al.* (2001) seems to offer a good framework for understanding an early stage in the accretion of components in chondrites. Further studies of chondrule and fragment sizes in Kaidun and other chondrites are needed to test the ideas discussed above.

3.2. Mass Loss from the Asteroid Belt

Models for the solar nebula based on the existing bodies suggest that ~99.9% of the solid material in the asteroid belt, 1 $M_\oplus$ or more, failed to accrete into asteroids or was removed soon after. At Mars' location, ~95% of the solids may have been lost. If Jupiter formed rapidly in ~100 yr from gravitational instabilities, it is possible that Jupiter prevented planetary embryos from accreting in the belt while allowing the terrestrial planetary embryos to develop (*Kortenkamp and Wetherill,* 2000). But if Jupiter formed in 10^7 yr from a solid core that subsequently accreted nebular gas, planetary embryos may have formed in the asteroid belt. In this case, the mass of the asteroid belt may have been removed largely by gravitational perturbations or collisional fragmentation followed by nongravitational removal of asteroid chips by, for example, gas drag, or by some combination of all three processes (*Wetherill,* 1989). The survival of Vesta's crust argues against collisional fragmentation as the major process. Most authors have therefore investigated gravitational perturbations by the embryos themselves and the giant planets to clear the asteroid belt (*Chambers and Wetherill,* 2001; *Petit et al.,*

2002). However, it is not clear whether lunar-sized bodies could have been removed from the asteroid belt without leaving a more traumatic collisional record in the meteorites and surviving asteroids.

If chondrules were as critical for asteroid formation as we infer, the efficiency of chondrule formation and accretion may have been important factors affecting the mass of the asteroids. Chondrule production, or chondrule deposition from bipolar flows, may have decreased with increasing distance from the protosun. If a large mass deficiency existed in the asteroid belt prior to the accretion of planetesimals, few asteroids would have been able to accrete, according to *Wetherill* (1989). However, accreting solids may have been heterogeneously distributed in the belt. For example, turbulent accretion of chondrules and particles with similar aerodynamic stopping distances may have focused solids to specific locations where planetesimals formed.

3.3. Impact Heating

The effectiveness of impact heating of asteroids has been a controversial issue in meteorite studies. Various authors have argued that impacts caused extensive metamorphism or melting on the parent bodies of almost every kind of meteorite (e.g., *Takeda,* 1993; *Rubin,* 1995). However, studies of terrestrial craters, shock experiments, and theoretical considerations suggest that impacts were probably not a significant source of heat for metamorphism and melting for asteroids under a few hundred kilometers in diameter (*Keil et al.,* 1997). The current mean impact velocity of asteroids, ~5 km/s (*Bottke et al.,* 1994), is scarcely above the threshold impact velocity required to generate shock pressures high enough to completely melt materials. Much of the impact melt that is generated on asteroids is ejected with velocities that exceed the asteroid's escape velocity. The volume of melt may increase for very porous asteroids, but this does not alter the basic conclusion that mean temperatures of asteroids increase by <100°C even in the largest impacts (*Keil et al.,* 1997).

3.4. Regolith Formation

For nearly all groups of chondrites and achondrites, we have samples that contain solar-wind gases and solar-flare tracks due to exposure in the top millimeter of their parent-body regoliths (*Bunch and Rajan,* 1988; *McKay et al.,* 1989). These regolith breccias provide our only tangible samples of grains that have been on the top surfaces of asteroids. However, the proportion of grains exposed to solar flares, which is lower than in typical lunar regolith breccias, is too low to allow definitive inferences to be drawn about the optical properties of asteroidal surfaces. Possibly because of efficient mixing of irradiated and unirradiated grains, there are only very minor chemical and mineralogical differences between regolith breccias and unirradiated fragmental breccias or, in the case of chondritic bodies, unbrecciated samples. Regolith breccias contain only a percent or so of foreign clasts, mostly carbonaceous chondrite material as most projectile material is lost during impacts on asteroids.

Achondritic and many chondritic regolith breccias appear to have been lithified by localized shock melting. We lack iron and stony-iron meteorite breccias with solar-wind gases, possibly because regolith on their parent bodies could not be converted to coherent rocks by localized shock melting of minerals. Carbonaceous chondrite regolith breccias tend to have the lowest concentrations of track-rich grains and solar-wind gases and may have been lithified by alteration. Shocked carbonaceous chondrites containing hydrous minerals are not known, perhaps because of devolatilization on shock release.

3.5. Fragmentation and Reaccretion

For seven groups of meteorites, there is textural and cooling rate evidence suggesting that their parent bodies were fragmented and then largely reaccreted so that material from diverse depths of the target asteroid were mixed together. These seven bodies are the parent bodies of the Shallowater enstatite achondrite, mesosiderites, H chondrites, L chondrites, ureilites, IVA irons, and the parent body of IAB irons and winonaites (*Keil et al.,* 1994; *Benedix et al.,* 2000; *Scott et al.,* 2001). For all bodies except the H- and L-chondrite bodies, the targets appear to have been hot enough prior to impact to contain at least minor amounts of melt. This does not necessarily imply that few cold meteorite parent bodies were fragmented and reaccreted as it is easier to recognize mixtures of materials from different depths when the targets contain some melt.

Low-resolution, smoothed-particle hydrodynamics computer simulations for 50–300-km-diameter targets suggest that for undifferentiated and fully differentiated target asteroids, near-catastrophic impacts can mix significant proportions of materials from all depths (*Love and Ahrens,* 1996; *Scott et al.,* 2001). However, these studies have done little more than prove the concept and more detailed modeling of near-catastrophic impacts is needed.

Existing studies suggest that impacts do not gradually strip mantles from the cores of differentiated asteroids. Cores and mantles of differentiated asteroids were probably mixed together before disruption by catastrophic impacts. Thus the rarity of olivine-rich, metal-free asteroids may reflect impact scrambling rather than near-total fragmentation of differentiated asteroids smaller than Vesta (*Burbine et al.,* 1996). Some S asteroids may be differentiated asteroids that were disguised by impact scrambling (*Scott et al.,* 2001).

3.6. Case Histories

3.6.1. Ordinary chondrites. Despite many constraints from mineral cooling rates, ^{40}Ar-^{39}Ar ages, and studies of regolith and fragmental breccias, there are disagreements about the nature of the parent bodies of ordinary chondrites.

The parent bodies of the H and L chondrites probably suffered one or more major impacts that may have mixed materials from all depths (*Keil et al.,* 1994). Some ordinary chondrites are breccias consisting of the most metamorphosed and presumably most deeply buried type 6 material mixed with the least-metamorphosed type 3 material. A few such breccias experienced temperatures above 500°C after assembly suggesting that at least one major impact occurred >4.4 G.y. ago when the parent asteroid was cooling but still hot. The wide range of cooling rates in metal grains in some regolith breccias combined with impact and thermal modeling suggest that the H and L parent bodies were broken up and reassembled between 4.4 G.y. and exposure to cosmic rays <100 m.y. ago (*Taylor et al.,* 1987). However, the thermal modeling depends critically on assumptions about the porosity and thermal conductivity of the ordinary chondrite bodies (see *McSween et al.,* 2002). Layers of regolith <1 km thick can drastically reduce the rate of cooling of 100-km-radius asteroids and enhance the thermal gradients near the surface of the asteroid (*Haack et al.,* 1990). Thus *Akridge et al.* (1998) argue that H chondrites are derived from depths of <10 km from a 100-km radius asteroid and that most of the interior of the parent body has not been sampled, contrary to *Bennett and McSween* (1996). Further thermal modeling to fit the constraints imposed by ^{40}Ar-^{39}Ar ages would be useful.

One candidate parent body of H chondrites, 6 Hebe, is an atypical S(IV) type as it shows rotational spectral variations. *Gaffey and Gilbert* (1998) suggest that these are due to impact-formed melt sheets on Hebe or residues of metal-rich projectiles, and that these supply meteorites such as Portales Valley (Fig. 4) and the IIE iron, Netschaevo. However, the link between IIE and H chondrites is not well established (*Bogard et al.,* 2000), Portales Valley probably formed at depth as it cooled slowly, and known asteroids are too small to generate impact-formed melt sheets (*Keil et al.,* 1997). Studies of complementary metal-poor samples would help us to understand the formation of metal-chondrite breccias and possible links with H chondrites but such samples have not yet been identified.

The high proportion of heavily shocked L chondrites with ^{40}Ar-^{39}Ar ages around 500 m.y. shows that the L-chondrite parent body suffered a major impact at this time. *Haack et al.* (1996) argue that the L chondrites are derived from a kilometer-sized rubble pile or fragment ejected from near the impact point of a catastrophic dispersion event. This event may have been responsible for the deposition of numerous ordinary chondrites in 450-m.y.-old limestone in Sweden when the meteorite flux on Earth may have been 10–100× higher for ~30 m.y. (*Schmitz et al.,* 1997). *Nesvorný et al.* (2002) suggest that the parent body of L chondrites and the Flora asteroid family may have been the same body.

3.6.2. Mesosiderites. These differentiated meteorites are all breccias composed of roughly equal proportions of metallic Fe,Ni and silicate. The silicate portion consists of fragments of rocks and silicate minerals in a fine-grained fragmental or igneous matrix (Fig. 6). The rock and mineral fragments are pieces of basalts, gabbros, and pyroxenites with minor amounts of dunite. Thus their constituents could have been derived from all levels of an igneously stratified asteroid that was similar but not identical to the parent body of the howardites, eucrites, and diogenites, presumably Vesta (*Keil,* 2002). Geochemical evidence suggests that the metal was molten prior to the impact that mixed metal and silicate. The major puzzle for understanding mesosiderites has been the small amount of olivine, which should have been abundant in differentiated asteroids. One model suggests that the olivine is confined to metal-poor regions that are poorly sampled and that the ingredients in mesosiderites are derived from a single body that was scrambled by a large impact (*Scott et al.,* 2001). An alternative idea is that the mesosiderite body formed when molten metal from the core of a projectile was mixed with crust from the mesosiderite parent body in a 1-km/s impact (see *Rubin and Mittlefehldt,* 1993). However, impacts at this velocity would probably have been very rare at the time of metal-silicate mixing, which has been inferred to be 100–150 m.y. after the solar system formed. At the current mean impact speed of 5 km/s, projectile material is almost entirely lost (*Love and Ahrens,* 1996).

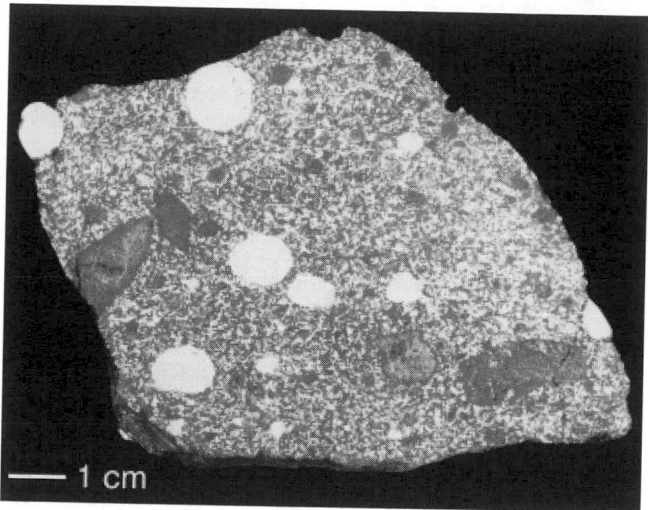

Fig. 6. Polished slice of the Barea mesosiderite showing a mixture of centimeter-sized spherules of metallic Fe,Ni (light) and angular fragments of basaltic rock (dark) embedded in an intimate metal-silicate matrix. This breccia formed when molten metal from the core of an asteroid was mixed with mantle and crustal rocks. The impact was probably a near-catastrophic event that scrambled metal and silicate from a single asteroid that broke up and reaccreted (*Scott et al.,* 2001). The alternative explanation, that the molten metal and rock were derived from separate asteroids that collided at low velocity during accretion (*Rubin and Mittlefehldt,* 1993), appears less plausible as radiometric ages suggest that mixing occurred 100–150 m.y. after the solar system formed, long after asteroids accreted.

The large M-type asteroid, 16 Psyche, has been suggested as a scrambled parent body for mesosiderites as it lacks a dynamical family that might have formed if Psyche were an exposed core from a 500-km-diameter asteroid (*Davis et al.*, 1999). However, an estimate of the density of Psyche by *Viateau* (2000) is too low to be consistent with this model. Field work would be useful to test these ideas on mesosiderite origins and to develop better models for large impacts.

3.6.3. Ureilite achondrites. Ureilites, which probably formed at depth in a partly melted asteroid, consist largely of coarse-grained olivine and pyroxene grains with minor interstitial metal and graphite. If these phases were in equilibrium, ureilites must have formed at depth in a body that was at least 100 km in radius (*Mittlefehldt et al.*, 1998). One of the most remarkable features of ureilites is that they all cooled rapidly from 1250° to <650°C at ~10°C/h (see *Goodrich et al.*, 2001, 2002). To cool in a few days, the hot rock must have been broken into meter-sized pieces. Fragmentation of impact debris into such tiny pieces may have been aided by formation of CO-CO_2 gas in grain boundaries on pressure release (*Keil et al.*, 1994).

A large number of the ureilites contain shock-formed diamonds and other shock metamorphic features. The high proportion of shocked ureilites and their unusual thermal history suggest that they are derived from near the point of impact. Since ureilite regolith breccias are made from similar material, it seems likely that ureilites are derived from one member of a family of asteroids that accreted a few days or more after the catastrophic breakup of the first ureilite body. Since the catastrophic impact probably occurred 4.5 G.y. ago, the asteroid family may no longer be recognizable.

4. IMPLICATIONS

The meteorite record contains a wealth of information about the origin and evolution of asteroids and the critical role of impacts at all stages. Meteorite breccias and features can be identified that reflect impacts during accretion; during melting, metamorphism, and alteration of asteroids; and during the erosion and disruption of cold asteroids. However, many meteorites do not fit comfortably into our simple models for understanding asteroid accretion and evolution. This probably results from the complexity of geological processes on asteroids and our inadequate understanding of what impacts do to hot and cold asteroids during and after accretion.

New meteorites and better constraints on the chronology, impact, and thermal history of chondrites and differentiated meteorites have greatly helped to advance our understanding of asteroid formation and evolution since the review by *Scott et al.* (1989). But they have also emphasized the urgent need for fieldwork on main-belt and near-Earth asteroids and laboratory study of returned samples. Without these we cannot hope to understand fully how meteorites, asteroids, and planets formed; how asteroids and planets interact; and how the harmful effects of such interactions can be minimized.

Acknowledgments. I thank D. W. Mittlefehldt, M. Killgore, M. Zolensky, and A. Kracher for supplying meteorite photos; G. J. Taylor, A. N. Krot, B. A. Cohen, C. A. Goodrich, K. Keil, and other colleagues for valuable discussions and comments; and H. Y. McSween and W. Bottke for helpful reviews. This work was supported in part by NASA Grant NAG 5-4212 (K. Keil, P.I.). This is Hawai'i Institute of Geophysics and Planetology Publication No. 1231 and School of Ocean and Earth Science and Technology Publication No. 6025.

REFERENCES

Akridge G., Benoit P. H., and Sears D. W. G. (1998) Regolith and megaregolith formation of H-chondrites: Thermal constraints on the parent body. *Icarus, 132,* 185–195.

Alexander C. M. O'D., Boss A. P., and Carlson R. W. (2001) The early evolution of the inner solar system: A meteoritic perspective. *Science, 293,* 64–68.

Bell J. F., Davis D. R., Hartmann W. K., and Gaffey M. J. (1989) Asteroids: The big picture. In *Asteroids II* (R. P. Binzel et al., eds.), pp. 921–945. Univ. of Arizona, Tucson.

Benedix G. K., McCoy T. J., Keil K., and Love S. G. (2000) A petrologic study of the IAB iron meteorites: Constraints on the formation of the IAB-winonaite parent body. *Meteoritics & Planet. Sci., 35,* 1127–1141.

Bennett M. E. and McSween H. Y. (1996) Revised model calculations for the thermal histories of ordinary chondrite parent bodies. *Meteoritics & Planet. Sci., 31,* 783–792.

Bischoff A. (2001) Meteorite classification and the definition of new chondrite classes as a result of successful meteorite search in hot and cold deserts. *Planet. Space Sci., 49,* 769–776.

Bischoff A. and Stöffler D. (1992) Shock metamorphism as a fundamental process in the evolution of planetary bodies: Information from meteorites. *Eur J. Mineral., 4,* 707–755.

Bogard D. D. (1995) Impact ages of meteorites: A synthesis. *Meteoritics & Planet. Sci., 30,* 244–268.

Bogard D. D., Garrison D. H., Jordan J. L., and Mittlefehldt D. W. (1990) ^{39}Ar-^{40}Ar ages and thermal history of mesosiderites: Evidence for major parent body disruption <4 Ga ago. *Geochim. Cosmochim. Acta, 54,* 2549–2564.

Bogard D. D., Garrison D. H., Norman M., Scott E. R. D., and Keil K. (1995) ^{39}Ar-^{40}Ar age and petrology of Chico: Large-scale impact melting on the L chondrite parent body. *Geochim. Cosmochim. Acta, 59,* 1383–1399.

Bogard D. D., Garrison D. H., and McCoy T. J. (2000) Chronology and petrology of silicates from IIE iron meteorites: Evidence of a complex parent body evolution. *Geochim. Cosmochim. Acta, 64,* 2133–2154.

Bottke W. F., Nolan M. C., Greenberg R., and Kolvoord R. A. (1994) Velocity distribution among colliding asteroids. *Icarus, 107,* 255–268.

Brearley A. J. and Jones R. H. (1998) Chondritic meteorites. In *Reviews in Mineralogy, Vol. 36, Planetary Materials* (J. J. Papike, ed.), pp. 3-1 to 3-398. Mineralogical Society of America, Washington.

Brown P. G. and 21 colleagues (2000) The fall, recovery, orbit and composition of the Tagish Lake meteorite: A new type of carbonaceous chondrite. *Science, 290,* 320–325.

Bunch T. E. and Rajan R. S. (1988) Meteorite regolithic breccias. In *Meteorites and the Early Solar System* (J. F. Kerridge and M. S. Matthews, eds), pp. 144–164. Univ. of Arizona, Tucson.

Burbine T. H., Meibom A., and Binzel R. P. (1996) Mantle material in the main belt: Battered to bits? *Meteoritics & Planet. Sci., 31,* 607–620.

Burbine T. H., McCoy T. J., and Dickinson T. L. (2000) Origin of plagioclase-"enriched", igneous, enstatite meteorites (abstract). *Meteoritics & Planet. Sci., 35,* A36.

Burbine T. H., McCoy T. J., Meibom A., Gladman B., and Keil K. (2002) Meteoritic parent bodies: Their number and identification. In *Asteroids III* (W. F. Bottke Jr. et al., eds.), this volume. Univ. of Arizona, Tucson.

Chambers J. E. and Wetherill G. W. (2001) Planets in the asteroid belt. *Meteoritics & Planet. Sci., 36,* 381–399.

Clayton R. N. (1993) Oxygen isotopes in meteorites. *Annu. Rev. Earth Planet. Sci., 21,* 115–149.

Cuzzi J. N., Dobrovolskis A. R., and Hogan R. C. (1996) Turbulence, chondrules, and planetesimals. In *Chondrules and the Protoplanetary Disk* (R. H. Hewins et al., eds.), pp. 35–43. Cambridge Univ., New York.

Cuzzi J. N., Hogan R. C., Paque J. M., and Dobrovolskis A. R. (2001) Size selective concentration of chondrules and other small particles in protoplanetary nebula turbulence. *Astrophys. J., 546,* 496–508.

Cyr K. E., Sears W. D., and Lunine J. I. (1998) Distribution and evolution of water ice in the solar nebula: Implications for solar system body formation. *Icarus, 135,* 537–548.

Davis D. R., Farinella P. and Marzari F. (1999) The missing Psyche family: Collisionally eroded or never formed? *Icarus, 137,* 140–151.

Endress M., Keil K., Bischoff A., Spettel B., Clayton R. N., and Mayeda T. K. (1994) Origin of dark clasts in the Acfer 059/El Djouf 001 CR2 chondrite. *Meteoritics, 29,* 26–40.

Fegley B. Jr. (2000) Kinetics of gas-grain reactions in the solar nebula. *Space Sci. Rev., 92,* 177–200.

Gaffey M. J. and Gilbert S. L. (1998) Asteroid 6 Hebe: The probable parent body of the H-type ordinary chondrites and the IIE irons. *Meteoritics & Planet. Sci., 33,* 1281–1295.

Goodrich C. A., Fioretti A. M., Tribaudino M., and Molin G. (2001) Primary trapped melt inclusions in olivine in the olivine-augite-orthopyroxene ureilite Hughes 009. *Geochim. Cosmochim. Acta, 65,* 621–652.

Goodrich C. A., Krot A. N., Scott E. R. D., Taylor G. J., Fioretti A. M., and Keil K. (2002) Formation and evolution of the ureilite parent body and its offspring (abstract). In *Lunar and Planetary Science XXXIII,* Abstract #1379. Lunar and Planetary Institute, Houston (CD-ROM).

Grady M. M. (2000) *Catalogue of Meteorites.* Cambridge Univ., Cambridge. 689 pp.

Greshake A., Krot A. N., Meibom A., Weisberg M. K., Zolensky M. E., and Keil K. (2002) Heavily hydrated lithic clasts in CH chondrites and the related metal-rich chondrites Queen Alexandra Range 94411 and Hammadah al Hamra 237. *Meteoritics & Planet. Sci., 37,* 281–293.

Haack H., Rasmussen K. L., and Warren P. H. (1990) Effects of regolith/megaregolith on the cooling histories of differentiated asteroids. *J. Geophys. Res., 95,* 5111–5124.

Haack H., Farinella P., Scott E. R. D., and Keil K. (1996) Meteoritic, asteroidal, and theoretical constraints on the 500 Ma disruption of the L chondrite parent body. *Icarus, 119,* 182–191.

Hiroi T., Zolensky M. E., and Pieters C. M. (2001) The Tagish Lake meteorite: A possible sample from a D-type asteroid. *Science, 293,* 2234–2236.

Hohenburg C. M., Meshik A. P., Pravdivtseva O. V., and Krot A. N. (2001) I-Xe dating: Dark inclusions from Allende CV3 (abstract). *Meteoritics & Planet. Sci., 36,* A83.

Hutcheon I. D., Weisberg M. K., Phinney D. L., Zolensky M. E., Prinz M., and Ivanov A. V. (1999) Radiogenic ^{53}Cr in Kaidun carbonates: Evidence for very early aqueous activity (abstract). In *Lunar and Planetary Science XXX,* Abstract #1722. Lunar and Planetary Institute, Houston (CD-ROM).

Ireland T. R. and Fegley B. (2000) The solar system's earliest chemistry: Systematics of refractory inclusions. *Intl. Geol. Rev., 42,* 865–894.

Johnson C. A., Prinz M., Weisberg M. K., Clayton R. N., and Mayeda T. K. (1990) Dark inclusions in Allende, Leoville, and Vigarano: Evidence for nebular oxidation of CV3 constituents. *Geochim. Cosmoshim. Acta, 54,* 819–830.

Jones R. H., Lee T., Connolly H. C., Love S. G., and Shang H. (2000) Formation of chondrules and CAIs: Theory and observation. In *Protostars and Planets IV* (V. Mannings et al., eds.), pp. 927–962. Univ. of Arizona, Tucson.

Jones T. D., Lebovsky L. A., Lewis J. S., and Marley M. S. (1990) The composition and origin of the C, P, and D asteroids: Water as a tracer of thermal evolution in the outer belt. *Icarus, 88,* 172–192.

Keil K. (1989) Enstatite meteorites and their parent bodies. *Meteoritics, 24,* 195–208.

Keil K. (2000) Thermal alteration of asteroids: Evidence from meteorites. *Planet. Space Sci., 48,* 887–903.

Keil K. (2002) Geological history of asteroid 4 Vesta: The "smallest terrestrial planet". In *Asteroids III* (W. F. Bottke Jr. et al., eds.), this volume. Univ. of Arizona, Tucson.

Keil K., Haack H., and Scott E. R. D. (1994) Catastrophic fragmentation of asteroids: Evidence from meteorites. *Planet. Space Sci., 42,* 1109–1122.

Keil K., Stöffler D., Love S. G., and Scott E. R. D. (1997) Constraints on the role of impact heating and melting in asteroids. *Meteoritics & Planet. Sci., 32,* 349–363.

Kortenkamp S. J. and Wetherill G. W. (2000) Formation of the asteroid belt (abstract). In *Lunar and Planetary Science XXXI,* Abstract #1813. Lunar and Planetary Institute, Houston (CD-ROM).

Kracher A., Keil K., Kallemeyn G. W., Wasson J. T., Clayton R. N., and Huss G. I. (1985) The Leoville accretionary breccia. *Proc. Lunar Planet. Sci. Conf. 16th,* in *J. Geophys. Res., 90,* D123–D135.

Krot A. N., Meibom A., Russell S. S., Alexander C. M. O'D., Jeffries T. E., and Keil K. (2001) A new astrophysical setting for chondrule formation. *Science, 291,* 1776–1779.

Liffman K. and Brown M. J. I. (1996) The protostellar jet model of chondrule formation. In *Chondrules and the Protoplanetary Disk* (R. H. Hewins et al., eds), pp. 285–302. Cambridge Univ., New York.

Lipschutz M. E., Gaffey M. J., and Pellas P. (1989) Meteoritic parent bodies: Nature, number, size and relationship to present-day asteroids. In *Asteroids II* (R. P. Binzel et al., eds.), pp. 740–777. Univ. of Arizona, Tucson.

Liu M., Scott E. R. D., Keil K., Wasson J. T., Clayton R. N., Mayeda T., Eugster O., Crozaz G., and Floss C. (2001) Northwest Africa 176: A unique iron meteorite with silicate inclusions related to Bocaiuva (abstract). In *Lunar and Planetary Science XXXII,* Abstract #2152. Lunar and Planetary Institute,

Houston (CD-ROM).

Love S. G. and Ahrens T. J. (1996) Catastrophic impacts on gravity-dominated asteroids. *Icarus, 124,* 141–155.

Lugmair G. W. and Shukolyukov A. (1998) Early solar system timescales according to ^{53}Mn-^{53}Cr systematics. *Geochim. Cosmochim. Acta, 62,* 2863–2886.

Lugmair G. W. and Shukolyukov A. (1999) The ^{53}Mn-^{53}Cr systematics of the enstatite chondrites (abstract). In *Lunar and Planetary Science XXX,* Abstract #1093. Lunar and Planetary Instute, Houston (CD-ROM).

Lugmair G. W. and Shukolyukov A. (2001) Early solar system events and timescales. *Meteoritics & Planet. Sci., 36,* 1017–1026.

McCoy T. J. and 9 colleagues (1995) Origin and history of impact-melt rocks of enstatite chondrite parentage. *Geochim. Cosmochim. Acta, 59,* 161–175.

McCoy T. J., Nittler L. R., Burbine T. H., Trombka J. I., Clark P. E., and Murphy M. E. (2000) Anatomy of a partially differentiated asteroid: A "NEAR"-sighted view of acapulcoites and lodranites. *Icarus, 148,* 29–36.

McCoy T. J., Burbine T. H., McFadden L., Starr R. D., Gaffey M. J., Nittler L. R., Evans L. G., Izenberg N., Lucey P., Trombka J. I., Bell J. F. III, Clark B. E., Clark P. E., Squyres S. W., Chapman C. R., Boynton W. V. and Veverka J. (2001) The composition of 433 Eros: A mineralogical-chemical synthesis. *Meteoritics & Planet. Sci., 36,* 1661–1672.

McKay D. S., Swindle T. D., and Greenberg R. (1989) Asteroidal regoliths: What we do not know. In *Asteroids II* (R. P. Binzel et al., eds.), pp. 617–642. Univ. of Arizona, Tucson.

McSween H. Y. Jr., Ghosh A., Grimm R. E., Wilson L., and Young E. D. (2002) Thermal evolution models of asteroids. In *Asteroids III* (W. F. Bottke Jr. et al., eds.), this volume. Univ. of Arizona, Tucson.

Meibom A. and Clark B. E. (1999) Evidence for the insignificance of ordinary chondrite material in the asteroid belt. *Meteoritics & Planet. Sci., 34,* 7–24.

Mittlefehldt D. W. and Lindstrom M. M. (2001) Petrology and geochemistry of the Patuxtent Range 91501, a clast-poor impact melt from the L-chondrite parent body and Lewis Cliff 88663, an L7 chondrite. *Meteoritics & Planet. Sci., 36,* 439–457.

Mittlefehldt D. W., McCoy T. J., Goodrich C. A., and Kracher A. (1998) Non-chondritic meteorites from asteroidal bodies. In *Reviews in Mineralogy, Vol. 36, Planetary Materials* (J. J. Papike, ed.), pp. 4-1 to 4-195. Mineralogical Society of America, Washington.

Nesvorný D., Morbidelli A., Vokrouhlický, Bottke W. F., and Brož M. (2002) The Flora family: A case of the dynamically dispersed collisional swarm? *Icarus, 157,* 155–172.

Ott U. (2001) Presolar grains in meteorites: An overview and some implications. *Planet. Space Sci., 49,* 763–767.

Patzer A., Hill D. H., and Boynton W. V. (2001) Itqiy: A metal-rich enstatite meteorite with achondritic texture. *Meteoritics & Planet. Sci., 36,* 1495–1505.

Petit J.-M., Chambers J., Franklin F., and Nagasawa M. (2002) Primordial excitation and depletion of the main belt. In *Asteroids III* (W. F. Bottke Jr. et al., eds.), this volume. Univ. of Arizona, Tucson.

Rivken A. S., Howell E. S., Britt D. T., Lebovsky L. A., Nolan M. C., and Branston D. D. (1995) 3-μm spectrophotometric survey of M- and E-class asteroids. *Icarus, 117,* 90–100.

Rubin A. E. (1995) Petrologic evidence for collisional heating of chondritic asteroids. *Icarus, 113,* 156–167.

Rubin A. E. (2000) Petrogic, geochemical and experimental constraints on models of chondrule formation. *Earth Sci. Rev., 50,* 3–27.

Rubin A. E. and Mittlefehldt D. W. (1993) Evolutionary history of the mesosiderite asteroid: A chronologic and petrologic synthesis. *Icarus, 101,* 201–212.

Rubin A. E. and Scott E. R. D. (1997) Abee and related EH chondrite impact-melt breccias. *Geochim. Cosmochim. Acta, 61,* 425–435.

Rubin A. E. and Wasson J. T. (1995) Variations of chondrite properties with heliocentric distance (abstract). *Meteoritics & Planet. Sci., 30,* 569.

Rubin A. E., Ulff-Møller F., Wasson J. T., and Carlson W. D. (2001a) The Portales Valley meteorite breccia: Evidence for impact-induced melting and metamorphism of an ordinary chondrite. *Geochim. Cosmochim. Acta, 65,* 323–342.

Rubin A. E., Kallemeyn G. W., Wasson J. T., Clayton R. N., Mayeda T. K., Grady M. M., and Verchovsky A. B. (2001b) Gujba: A new Bencubbin-like meteorite fall from Nigeria (abstract). In *Lunar and Planetary Science XXXII,* Abstract #1779. Lunar and Planetary Institute, Houston (CD-ROM).

Rubin A. E., Wasson J. T., Kallemeyn G. W., Clayton R. N., and Mayeda T. K. (2001c) Ungrouped iron NWA 468: An impact-melt product with low-Ca clinopyroxene-bearing silicate inclusions (abstract). *Meteoritics & Planet. Sci., 36,* A178–179.

Sanders I. (1996) A chondrule-forming scenario involving molten planetesimals. In *Chondrules and the Protoplanetary Disk* (R. H. Hewins et al., eds.), pp. 327–334. Cambridge Univ., New York.

Schmitz B., Peucker-Ehrenbrink B., Lindström M., and Tassinari M. (1997) Accretion rates of meteorites and cosmic dust in the early Ordovician. *Science, 278,* 88–90.

Scott E. R. D. and Keil K. (1999) Metamorphism of meteorites: Clues to the geology of asteroids (abstract). *Meteoritics & Planet. Sci., 34,* A103–A104.

Scott E. R. D., Taylor G. J., Newsom H., Herbert F., Zolensky M., and Kerridge J. F. (1989) Chemical, thermal and impact processing of asteroids. In *Asteroids II* (R. P. Binzel et al., eds.), pp. 701–739. Univ. of Arizona, Tucson.

Scott E. R. D., Love S. G., and Krot A. N. (1996) Formation of chondrules and chondrites in the protoplanetary nebula. In *Chondrules and the Protoplanetary Disk* (R. H. Hewins et al., eds.), pp. 87–96. Cambridge Univ., New York.

Scott E. R. D., Krot A. N., and Yamaguchi A. (1998) Carbonates in fractures in martian meteorite Allan Hills 84001: Petrologic evidence for impact origin. *Meteoritics & Planet. Sci., 33,* 709–719.

Scott E. R.D., Haack H., and Love S. G. (2001) Formation of mesosiderites by fragmentation and reaccretion of a large differentiated asteroid. *Meteoritics & Planet. Sci., 36,* 869–881.

Sears D. W. G. (1998) The case for rarity of chondrules and calcium-aluminum-rich inclusions in the early solar system and some implications for astrophysical models. *Astrophys. J., 498,* 773–778.

Shearer C. K., Papike J. J., and Rietmeijer F. J. M. (1998) The planetary sample suite and environments of origin. In *Reviews in Mineralogy, Vol. 36, Planetary Materials* (J. J. Papike, ed.), pp. 1-1 to 1-28. Mineralogical Society of America, Washington.

Shu F. H., Shang H., Gounelle M., and Glassgold A. E. (2001) The origin of chondrules and refractory inclusions in chondritic meteorites. *Astrophys. J., 548,* 1029–1050.

Stöffler D., Keil K., and Scott E. R. D. (1991) Shock metamorphism of ordinary chondrites. *Geochim. Cosmochim. Acta, 55,* 3845–3867.

Symes S. J. K., Sears D. W. G., Akridge D. G., Huang S., and Benoit P. (1998) The crystalline lunar spherules: Their formation and implications for the origin of meteoritic chondrules. *Meteoritics & Planet. Sci., 33,* 13–29.

Takeda H. (1993) Chemical differentiation during collision and accretion of meteorite parent bodies. In *Primitive Solar Nebula and Origin of Planets* (H. Oya, ed.), pp. 375–394. Terra Scientific, Tokyo.

Taylor G. J. and Scott E. R. D. (2001) Molten planetesimals in the early solar system (abstract). *Meteoritics & Planet. Sci., 36,* A204.

Taylor G. J., Scott E. R. D., and Keil K. (1983) Cosmic setting for chondrule formation. In *Chondrules and Their Origin* (E. A. King, ed.), pp. 262–278. Lunar and Planetary Institute, Houston.

Taylor G. J., Maggiore P., Scott E. R. D., Rubin A. E., and Keil K. (1987) Original structures and fragmentation and reassembly histories of asteroids: Evidence from meteorites. *Icarus, 69,* 1–13.

Taylor G. J., Keil K., McCoy T., Haack H., and Scott E. R. D. (1993) Asteroid differentiation: Pyroclastic volcanism to magma oceans. *Meteoritics, 28,* 34–52.

Treiman A. H. (1998) The history of Allan Hills 84001 revised: Multiple shock events. *Meteoritics & Planet. Sci., 33,* 753–764.

Turner G., Knott S. F., Ash R. D., and Gilmour J. D. (1997) Ar-Ar chronology of the Martian meteorite ALH84001: Evidence for the timing of the early bombardment of Mars. *Geochim. Cosmochim. Acta, 61,* 3835–3850.

Viateau B. (2000) Mass and density of asteroids (16) Psyche and (121) Hermione. *Astron. Astrophys., 354,* 725–731.

Wadhwa M. and Russell S. S. (2000) Timescales of accretion and differentiation in the early solar system. In *Protostars and Planets IV* (V. Mannings et al., eds.), pp. 995–1018. Univ. of Arizona, Tucson.

Wasson J. T. (1988) The building stones of planets. In *Mercury* (F. Vilas et al., eds.), pp. 622–650. Univ. of Arizona, Tucson.

Wasson J. T. and Kallemeyn G. W. (1988) Compositions of chondrites. *Phil. Trans. R. Soc. Lond., A325,* 535–544.

Wasson J. T. and Kallemeyn G. W. (1990) Allan Hills 85085: A subchondritic meteorite of mixed nebular and regolithic heritage. *Earth Planet. Sci. Lett., 101,* 148–161.

Weidenschilling S. J., Marzari F., and Hood L. L. (1998) The origin of chondrules at Jovian resonances. *Science, 279,* 681–684.

Weisberg M. K., Prinz M., Clayton R. N., Mayeda T. K., Sugiura N., Zashu S., and Ebihara M. (2001) A new metal-rich chondrite grouplet. *Meteoritics & Planet. Sci., 36,* 401–418.

Weiss B. P., Kirschvink J. L., Baudenbacher F. J., Vali H., Peters N. T., Macdonald F. A., and Wikswo J. P. (2000) A low temperature transfer of ALH84001 from Mars to Earth. *Science, 290,* 791–795.

Wetherill G. W. (1989) Origin of the asteroid belt. In *Asteroids II* (R. P. Binzel et al., eds.), pp. 661–680. Univ. of Arizona, Tucson.

Wood J. A. (1996) Unresolved issues in the formation of chondrules and chondrites. In *Chondrules and the Protoplanetary Disk* (R. H. Hewins et al., eds.), pp. 55–69. Cambridge Univ., New York.

Yamaguchi A., Taylor G. J., and Keil K. (1997) Metamorphic history of the eucritic crust of 4 Vesta. *J. Geophys. Res., 102,* 13381–13386.

Yamaguchi A., Misawa K., Haramura H., Kojima H., Clayton R. N., Mayeda T. K., and Ebihara M. (2001) Northwest Africa 011, a new basaltic meteorite (abstract). *Meteoritics & Planet. Sci., 36,* A228.

Zolensky M. E. and Ivanov A. V. (2001) Kaidun: A smorgasbord of new asteroid samples (abstract). *Meteoritics & Planet. Sci., 36,* A233.

Zolensky M. E., Ivanov A. V., Yang S. V., Mittlefehldt D. W., and Ohsumi K. (1996) The Kaidun meteorite: Mineralogy of an unusual CM1 lithology. *Meteoritics & Planet. Sci., 31,* 484–493.

Primordial Excitation and Depletion of the Main Belt

Jean-Marc Petit
Centre National de la Recherche Scientifique and Observatoire de la Côte d'Azur (now at Observatoire de Besançon)

John Chambers
NASA Ames Research Center

Fred Franklin
Harvard-Smithsonian Center for Astrophysics

Makiko Nagasawa
Tokyo Institute of Technology

The asteroid belt is characterized by a large dynamical excitation, an important mass depletion compared to its original density, and a radial mixing of bodies with different physical and chemical properties. Two mechanisms have been proposed to explain this structure: the sweeping of secular resonances through the belt, and the scattering action by embryos later ejected from the solar system by planetary perturbations. The resonance sweeping is due to the dissipation of the gas-drag-inducing primordial nebula. Resonance sweeping and gas drag are responsible for mass depletion and orbital migration. Embryos are naturally created in the inner solar system during planet formation, producing a strong dynamical excitation. Although both approaches are fairly good at structuring the asteroid belt and depleting it, it is likely that both mechanisms occurred at one time or another. More work is needed to assess the combined effect of the nebula and the excitation by embryos.

1. INTRODUCTION

The different populations of small bodies (asteroid belt, Edgeworth-Kuiper Belt, irregular satellites, and comets) provide strong constraints about the early evolution of the solar system. The size and chemical and orbital distributions of small bodies, by their large number, carry statistically significant information, while the properties of the planets may be altered by unlikely stochastic events.

In this chapter, we review the main processes that tend to reproduce the observed characteristics of the asteroid belt. The Edgeworth-Kuiper Belt is also of great statistical significance, but the small amount of accurate data prevents discrimination among the models that already exist, and makes any refinement difficult. As for the planets, their detailed dynamical characteristics are not considered here but models must still account for the chemical and isotopic observations.

We now present the most important characteristics of the asteroid belt. In discussing these, we primarily consider asteroids with diameters larger than 50 km for the following reasons. During the 4.5 G.y. of existence of the solar system, the asteroids have evolved greatly through high-velocity collisions. Collision velocities are typically a few kilometers per second, very often resulting in the complete shattering and disruption of the colliding bodies. Therefore, most of the asteroids one sees today are not primordial, but fragments of larger asteroids destroyed in a collision. Only the largest asteroids retain characteristics that relate to the formation of the asteroid belt and that were not drastically changed by the later evolution. Asteroids with diameters D > 50 km have a collisional lifetime of the order of the age of the solar system or longer (*Geissler et al.,* 1996). Most of these are primordial asteroids, i.e., they were already present in the belt at the end of its excitation and mass depletion, after the terrestrial planets were completely formed. The few large bodies that were destroyed generally yielded at most one large fragment (larger than 50 km), with mostly unchanged dynamical characteristics, and a swarm of smaller fragments (*Tanga et al.,* 1999). In addition, it is very likely that all the asteroids larger than 50 km have been discovered; the completeness limit is likely to be about 25 km (*Jedicke and Metcalfe,* 1998). So our statistics are not contaminated by observational biases. From Fig. 1a, one can naturally distinguish three zones: the inner belt, with semimajor axis a < 2.5 AU (3:1 mean motion resonance with Jupiter); the central belt, 2.5 < a < 3.28 AU (2:1 resonance); and the outer belt, beyond 3.28 AU. In the outer belt, all asteroids beyond 3.8 AU are in mean-motion resonances with Jupiter. The most striking aspects that one would like to explain with a unitary model are the following:

1. *Strong dynamical excitation.* The median eccentricity e_m and inclination i_m respectively are 0.15 and 6° in the inner belt, 0.14 and 10.7° in the central belt, and 0.1 and

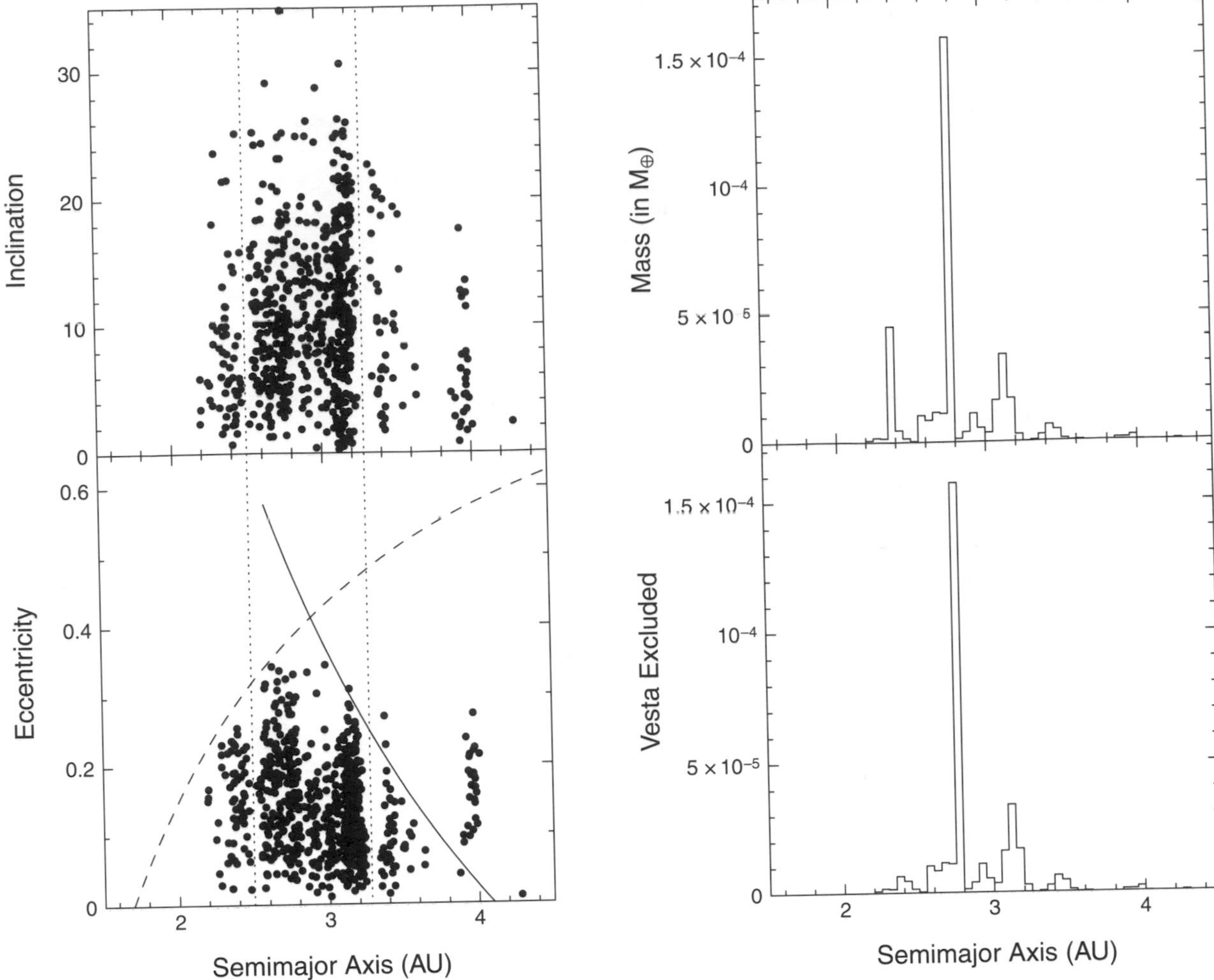

Fig. 1. **(a)** Osculating inclination (top) and eccentricity (bottom) vs. semimajor axis for the asteroid belt for bodies larger than 50 km in diameter (solid line: aphelion distance of 4.1 AU; dashed line: perihelion distance of 1.7 AU). **(b)** Mass distribution of asteroids vs. semimajor axis for all asteroids larger than 50 km (top), and excluding Vesta (bottom). The dotted lines give the boundaries of the inner belt (right), central belt (middle), and outer belt (right). From *Petit et al.* (2001).

12.1° in the outer belt. In the outer belt, e_m, with the exception of the bodies in the 3:2 and 4:3 resonances, is lower than in the other belts because of the instabilities due to Jupiter that tend to deplete the region above the solid line in Fig. 1a. The eccentricity e and inclination i of the asteroids are much larger than those of the planets in the solar system (with the exception of Mercury and the special case of Pluto), and much larger than would allow accretion. Actually, the absence of bodies at low e (<0.05) and i (<2°), which would be stable over the age of the solar system (*Duncan,* 1994; *Holman and Murray,* 1996), indicates that the primordial excitation in the outer belt has been even larger than in the rest of the asteroid belt, as confirmed by the larger median value of inclination in the former. In the central belt, modeling the existence of (2) Pallas on an orbit with e = 0.23 and i = 34.8° has only recently been successful (*Petit et al.,* 2001).

2. *Large mass depletion.* The present total mass of the asteroid belt is estimated to be on the order of $5 \times 10^{-4} M_\oplus$ ($M_\oplus$ = Earth mass), namely 10^3–$10^4\times$ smaller than its primordial mass (*Lecar and Franklin,* 1973). The accretion of the largest asteroids on a timescale comparable with the meteoritic solidification age also implies the primordial existence of at least 100× more material than at present (*Wetherill,* 1989). This shows that the important mass deficiency of the asteroid belt is not due to the presence of a gap in the primordial disk, but is the result of some process that occurred after the formation of the asteroids. From Fig. 1b, one sees that the mass deficiency is larger in the inner and outer belts than in the central belt.

3. *Radial mixing of asteroid types.* The optical properties of the asteroids depend roughly on their distance from the Sun: S-types dominate the inner belt, C-types are the most abundant in the central belt, and P-types dominate in

the outer belt (with exception of the Trojan population, which is mainly D-type). This radial compositional zoning likely reflects the temperature gradient of the primitive nebula. However, the boundaries between compositional zones are not sharp: Asteroids of different types are mixed over scales of 1 AU (*Gradie and Tedesco,* 1982). Such mixing is not easily explained by the turbulence of the primitive nebula or by the radial decay of pristine bodies due to gas drag (*Ruzmaikina et al.,* 1989).

The gravitational perturbations arising from all the planets in their current orbits, even acting over times on the order of a billion years, are unable to account for either the range in, or average values of, the e and i of the asteroids, nor for the striking mass depletion of the belt (*Duncan et al.,* 1989). Intense primordial collisional activities have been invoked to explain the large mass deficiencies of the asteroid belt (*Chapman and Davis,* 1975). Because large bodies cannot be collisionally destroyed, the collisions could have reduced the total mass to its present value only if the original number of large bodies was basically the same as the present one and the original size distribution was very steep (*Davis et al.,* 1979; *Wetherill,* 1989). However, the survival of the fragile basaltic crust of Vesta is an important argument against a primordial collisional activity significantly more important than the present one (*Davis et al.,* 1994). Therefore it seems that the large mass deficiency of the belt also requires dynamical mechanisms capable of displacing most of the material into unstable regions.

Several mechanisms have been proposed over the last 25 years to explain the dynamical structure of the asteroid belt, with much renewed interest over the last decade. They can be classified in two main categories: those that invoke the sweeping of the belt by resonances of various types, most especially secular resonances, and those that invoke the scattering action of embryos that have later been ejected from the solar system by planetary perturbations. Up to now, these two approaches seem to have ignored each other completely, the first one mostly assuming the dominant dynamical effect of the gaseous primordial nebula, the other one relying on the gravitational perturbations of embryos in a gas-free environment.

The idea that secular resonances could have swept through the asteroid belt as a result of the dissipation of the primordial nebula has been first proposed by *Ward et al.* (1976). *Heppenheimer* (1980) and *Ward* (1981) developed a planar analytic linear model that showed that the current eccentricities of the asteroids could be explained by the passage of two secular resonances through the asteroid belt. Much work has been done during the last decade on the effect of uniform dissipation of the primordial nebula, culminating with the work of *Lecar and Franklin* (1997) and *Franklin and Lecar* (2000). More recently, some authors (*Nagasawa et al.,* 2000) have started investigating a nonuniform dissipation of the nebula to circumvent some limitations of the previous approach. All these works are reviewed in section 2.

The idea that large Jupiter scattered planetesimals (LJSPs, also called large embryos hereafter) of mass comparable to that of Earth could have dynamically heated the asteroid belt was first proposed by *Safronov* (1979). The existence of large embryos as leftovers from planetary formation is predicted by all the current theories (*Fernandez and Ip,* 1996). The tilts of the spin axes of Saturn, Uranus, and Neptune support the idea of collisions with massive bodies (about 1 $M_\oplus$). The most recent work on this hypothesis concluded that it could not solve the problem (*Petit et al.,* 1999; and section 3.1).

Introducing gravitational perturbation from massive bodies, *Wetherill* (1992) alternatively proposed an endogenic source of excitation, i.e., that the asteroid belt was originally a massive dynamically cold system, which contained about 200 sublunar- to martian-sized embryos among its population. These embryos excited each other by mutual gravitational interactions until they came close to Jupiter and were ejected from the solar system. At the end of this phase, only a small fraction of the original asteroids survived in the stable regions of the belt, on eccentric and inclined orbits. Due to the availability of fast computers and efficient integration algorithms, many simulations have been recently performed on this topic using direct N-body integrations (*Wetherill and Chambers,* 1997; *Chambers and Wetherill,* 2001). Extending the disk of embryos into the terrestrial planet region, *Chambers and Wetherill* (1998) have shown that this approach is actually one of the most promising for the formation of the terrestrial planets. *Petit et al.* (2001) used these models to study the excitation and depletion of the asteroid belt (section 3 summarizes all these findings).

In section 4, we identify what, in our minds, are the areas that need further investigation and we try to list the particular questions that should be addressed in the coming years.

2. SWEEPING SECULAR RESONANCES

This section focuses on the effects of secular resonances, specifically ones that are not fixed in phase space. The term secular resonance applies to commensurabilities between the apsidal or nodal frequencies of an (asteroid's) orbit and one of the similar eigenfrequencies that describe the complete solar system. An important example in the present solar system is one that defines the inner edge of the main belt at a = 2.0 AU (*Knežević et al.,* 1991), which is a narrow region where both the nodal and apsidal frequencies of an orbit (as driven principally by Jupiter) resonate with one of the secular terms (owing to Saturn) in Jupiter's orbit itself.

2.1. What Is It, Why Is It of Interest?

We refer the reader to *Brower and Clemence* (1961) and *Nagasawa et al.* (2000) for a full derivation of the equations of secular resonance. Here we only give a hint as to what a secular resonance is, and how it is influenced.

Let e_j, i_j, ϖ_j, and Ω_j be the eccentricity, inclination, longitude of perihelion, and ascending node of body j. We introduce the canonical variables

$$h_j = e_j \sin \varpi_j$$

$$k_j = e_j \cos \varpi_j$$

$$p_j = \tan i_j \sin \Omega_j$$

$$q_j = \tan i_j \cos \Omega_j$$

Calling R_j the perturbing function for body j, i.e., the variation from a pure Keplerian two-body Hamiltonian, the time evolutions of these variables are

$$\frac{dh_j}{dt} = \frac{1}{n_j a_j^2} \frac{\partial R_j}{\partial k_j}$$

$$\frac{dk_j}{dt} = -\frac{1}{n_j a_j^2} \frac{\partial R_j}{\partial h_j}$$

$$\frac{dp_j}{dt} = \frac{1}{n_j a_j^2} \frac{\partial R_j}{\partial q_j}$$

$$\frac{dq_j}{dt} = -\frac{1}{n_j a_j^2} \frac{\partial R_j}{\partial p_j}$$

In a system with an asteroid, Jupiter, Saturn, and the Sun, assuming e_{ast}, $i_{ast} \ll 1$, the solutions are

$$h_{ast} = \nu \sin(gt + \beta) + \sum_{j=1,2} \frac{\nu_j}{g - g_j} \sin(g_j t + \beta_j)$$

$$k_{ast} = \nu \cos(gt + \beta) + \sum_{j=1,2} \frac{\nu_j}{g - g_j} \cos(g_j t + \beta_j)$$

$$p_{ast} = \mu \sin(ft + \gamma) + \sum_{j=1,2} \frac{\nu_j}{f - f_j} \sin(f_j t + \gamma_j)$$

$$q_{ast} = \mu \cos(ft + \gamma) + \sum_{j=1,2} \frac{\nu_j}{f - f_j} \cos(f_j t + \gamma_j)$$

where f and g are the frequencies of the free oscillations due to gravitational perturbations, while f_j and g_j are those of the forced oscillations due to Jupiter and Saturn. In the real solar system, they correspond to the eigenfrequency of the whole system. Whenever one of the denominators vanishes, we have what we call a secular resonance. The presence of a massive nebula influences all these frequencies, and may create resonances at various locations in the solar system.

In working out the details of a different secular resonance problem, *Ward et al.* (1976) suggested that a decaying solar nebula might force these resonances to sweep through the asteroid belt. It was also clear from that paper's mathematical development that both e and i could in principle be markedly elevated during a secular resonance passage that was slower than characteristic apsidal/nodal periods. *Heppenheimer* (1980) worked out a practical example of e-pumping in the planar nebula case and *Ward* (1981) treated the broader theoretical topic of secular resonance sweeping. These early efforts led to a fair amount of interest in scanning secular resonances as a naturally occurring means of sculpturing the asteroid belt for many reasons: (1) The effects of secular resonances on the motions of bodies in evolving solar system models can be carried out in a systematic, highly quantitative and partly analytic way. (2) The existence and eventual decay of a solar nebula (which would automatically move these resonances) had already been, and continued to be, established on the basis of very different yet compelling theoretical and observational grounds. (3) The required nebula density was eminently reasonable. To be more precise, the two important secular resonances, labeled ν_5 and ν_6, that arise from the principal terms defining the apsidal motions of Jupiter and Saturn currently lie near semimajor axes, a, of 0.65 and 2.0 AU. These resonances need to be initially located at a > 4 AU to be able to sweep through the asteroid belt and affect it. The minimum mass (*Hayashi,* 1981) solar nebula model, with volume density given by

$$\rho = 1.4 \times 10^{-9} \left(\frac{r}{1\ \mathrm{AU}}\right)^{-11/4} \exp\left[-\left(\frac{z}{0.047(r/1\ \mathrm{AU})^{5/4}}\right)^2\right] \mathrm{g/cm^3}$$

where r < 36 AU and (r, z) are the cylindrical coordinates centered at the Sun, has precisely this property (see Fig. 4 of *Nagasawa et al.,* 2000). (4) Estimates of the characteristic timescales required by secular resonance sweeping to generate the observed asteroid distribution from an initial field of low e, i values yield times from (a few) × 10^4 to about 10^5 yr for a uniformly dispersing nebula. This is a plausible value for the e-folding decay time of a T-Tauri-like nebula (*Kenyon and Hartmann,* 1995). Thus the nebula might remain quiescent for 10^6–10^7 yr before it starts to decay in this shorter interval. For nonuniform depletion the decay time can be a few × 10^5 yr and longer (section 2.3). (5) Moving secular resonances potentially can account for features other than the e and i distributions. They can explain both the extreme depopulation in the outer belt and at least contribute mightily to the very considerable mass loss from the inner and central belts. Coupled with the aerodynamic gas drag also characteristic of the nebula, secular resonances can help to diversify the remaining asteroid population so that various taxonomic types can be found at similar semimajor axes.

2.2. Uniform Nebula Depletion

The first nebula models (*Heppenheimer,* 1980) to investigate secular resonance sweeping were planar, continuous ones with a radial density profile of $r^{-3/2}$ and a time decay, independent of r, governed only by e^{-t/t_0}. Nonetheless, encouraging results followed: e values were pumped to likely levels, easily reaching 0.3 during nebula dissipation times in the range $10^4 < t_0 < 10^5$ yr. During these times both ν_5 and ν_6 would traverse the entire belt, commencing near Jupiter and coming to rest, as the nebula density fell to zero, at their current locations. Passage of two resonances would not only elevate asteroidal e, but also aid in generating a broader range of e values. Should only one secular resonance sweep through the belt, then some other mechanism (e.g., particle close encounters) may be needed to produce a dispersion in e and/or in i, as we shall discuss later. But note (*Ward et al.,* 1976) that the amplitude of e and i depend on phasing of the relevant apsidal/nodal angles. Thus two bodies at the same semimajor axis, feeling the same secular resonance at the same time, may acquire very different e and i.

Although these timescales and amplitudes are very satisfactory, a corollary of *Ward*'s (1981) discussion made it clear that the excitation of the i would not be successful if approached in such a simple way. All disks with densities proportional to $r^{-3/2}$, whether or not a gap exists around the orbit of Jupiter (and/or Saturn), will increase the nodal regression of an asteroid's orbit. Since nodes in the belt regress more rapidly than Jupiter's (or more precisely, more rapidly that the value characteristic of the ν_{16} resonance) in the absence of a nebula, the needed equality for resonance will never occur. The failure of the $r^{-3/2}$ uniformly decaying nebula to pump the i has been duly confirmed numerically by *Nagasawa et al.* (2000).

It is important to recognize that all calculations above have employed the $r^{-3/2}$ density law characteristic of the Hayashi model and supported by *Weidenschilling* (1977). But this choice of exponent does not agree with the extensive studies by *Cameron* (1995), who favors an r^{-1} relation. *Bell et al.* (1997) propose even shallower gradients, between r^{-1} and $r^{-0.5}$, in the asteroid belt. *Lemaître and Dubru* (1991), assuming a nebula with constant vertical density, examined three examples of the density law, r^{-s}, with s = 1, 3/2, and 2 (and uniform time decay), and found that s = 1, Cameron's preferred value, allows ν_{16} to sweep through the entire asteroid belt. However, *Nagasawa et al.* (2000), adopting a more realistic Gaussian vertical profile, found that no nodal resonance would sweep through the entire belt for positive values of s. At present we do not know whether this claim is very general and if a more detailed probing is required. Still, there are grounds for a pessimistic assessment so that, as we shall discuss in section 2.3, "nonuniform" models of nebula dissipation may be the most likely — and possibly the only — way to elevate i throughout the entire belt.

We turn now to the question of mass loss from the belt. The reason for the marked emptiness of the outer belt has bothered dynamicists for several decades (*Liou and Malhotra,* 1997, and references therein). *Ida and Lin* (1996) examined the effects of drag combined with mean-motion resonance with Jupiter, and found that this could explain the extreme depletion of the outer belt, provided a minimum mass nebula disapears in 10^{6-7} yr. This prompted *Lecar and Franklin* (1997) to consider drag and secular resonance inasmuch as both are integral properties of any nebula. One needed to evaluate the resulting behavior when the eccentricity reducing drag interacted with the e-enhancing resonances. Such a combined process would promote mass loss via a reduction in the semimajor axes of the asteroids. This paper, still using a two-dimensional Hayashi nebula, showed that e-pumping by secular resonance would empty the outer belt, whether drag was important or not, provided only that nebula removal times were greater than about 10^4 yr, a fact that attests to the rapidity of e-pumping and the efficiency of gravitational scattering of high-e bodies by Jupiter. For bodies in the inner belt, it pointed out that drag would operate on the eccentric orbits of objects several tens of kilometers in radius and spiral them into the inner solar system if the removal time, t_0, were greater than about 60,000 yr.

An informal question asked at a past Gordon Conference, wondering if dynamical means could explain why meteorites, apparently originating from the same part of the belt, could have different physical properties, prompted a return to this general problem (*Franklin and Lecar,* 2000) in order to obtain a more complete statistical sampling. Some results from that paper, again relying on the two-dimensional minimum mass model that included Jupiter and Saturn moving in eccentric, precessing orbits, are contained in Tables 1 and 2. Table 1 provides the fate in four categories of some 8000 randomly chosen outer belt bodies, lying in four size ranges, as a function of four timescales stretching from 30,000 to 240,000 yr. This survey reconfirms that the combined action of secular resonance sweeping and drag readily depopulates the outer belt, but that the "end" for any body depends upon its radius: Drag quickly shoots small objects into the inner belt and the inner solar system, but larger ones, being less affected by drag, are more likely to encounter Jupiter because of their augmented eccentricities. However, some material is transferred from the outer belt into the inner and central belts in most of the examined cases.

Turning now to the inner and central belts, because the two shorter timescales (cases A and B in Table 1) failed to produce more than a token mass reduction in the inner belt, Table 2 concentrates on case C — hence a dispersal time, T_n, of 120,000 yr — for four different asteroidal radii. Mass loss now becomes considerable, potentially reaching 80–90% for all four radii. Increasing T_n by a factor of 2 would elevate these percentages to >95%. Beyond Tables 1 and 2, this paper makes the following additional comments on the

TABLE 1. Fate of bodies of various radii, r, with initial semimajor axes, in the range 3.22 < a < 3.85.

Case	Close Encounter with Jupiter	Remaining in Outer Belt	Driven to Inner Belt	a < 2.1 AU
		5 < r < 10 km		
A	4.6	2.4	54.4	38.6
B	6.4	1.2	42.2	50.2
C	3.6	1.0	12.8	82.6
D	1.8	0.4	0	97.8
		10 < r < 20 km		
A	17.2	2.0	25.2	55.6
B	22.7	1.6	23.9	51.8
C	40.6	1.4	7.0	49.9
D	27.0	0.4	1.2	71.4
		20 < r < 50 km		
A	48.2	3.4	4.0	44.4
B	65.8	1.6	3.4	29.3
C	73.9	1.6	0.7	23.7
D	80.5	0.5	0.4	18.5
		50 < r < 100 km		
A	91.1	1.4	1.8	5.6
B	90.0	0.9	1.0	7.2
C	95.9	1.3	0.2	2.6
D	97.9	0.6	0.1	1.4

Entries are percentages drawn from a total of about 8000 orbits, evenly divided between the four ranges in r. Labels on columns 2–5 refer to (1) close encounter with Jupiter, (2) and (3) capture into the outer or inner asteroid belts, and (4) passage into the sunward side of the secular resonance a, where a ≃ 2.1 AU. Cases A–D correspond to nebula (e-folding) dispersal times of 2,500, 5,000, 10,000, and 20,000 P_J. P_J is the jovian orbital period, 11.862 yr. A close encounter with Jupiter means a jovian approach within 10 planetary radii. From *Franklin and Lecar* (2000).

TABLE 2. Fate of bodies of various radii, r, with initial semimajor axes, in the range 2.18 < a < 3.12 AU.

a < 2.1 AU	Inner Belt with e < 0.4	Mars Crosser	Close Encounter with Jupiter
	r = 60 km		
84.2	7.2	5.8	2.8
	r = 120 km		
57.4	21.6	17.2	3.8
	r = 240 km		
49.2	20.8	18.0	12.0
	r > 1000 km		
0	24.0	27.8	48.2

Nebula dispersal time is 10,000 P_J. Entries are percentages based on a total of 2000 orbits, evenly divided between the four ranges in r. Labels on columns 1–4 refer to objects that have (1) passed into the sunward side of the belt; (2) continued in stable orbits in the central or inner belts; (3) also in one of these belts, but with eccentricities that force a crossing of Mars' orbit; and (4) approached Jupiter to within 10 radii and hence no longer move in minor-planet-like orbits. Nearly half the bodies that cross Mars' orbit will cross Earth's as well. Of the 224 bodies in the last two "r" categories that remain in the inner belt, 13% have proper e in the range $0 < e_p < 0.10$; 26% from $0.10 < e_p < 0.20$; 38% from $0.20 < e_p < 0.30$, and 23% with $0.30 < e_p < 0.40$. From *Franklin and Lecar* (2000).

inner belt for T_n = 120,000 yr: (1) Removal of bodies does become increasingly rare when their radii are greater than 800 km, but (2) it is virtually certain for radii less than about 50 km and (3) this latter limit approximately scales exponentially with T_n. As is the case for collisional erosion, depletion by gas drag seems to require a steep size distribution. We should also note that it is quite possible for asteroids several hundreds of kilometers in radius [e.g., (4) Vesta] once to have orbited at distances much larger than ones where they are found today.

Finally we return in Fig. 2 to the query raised earlier, whether material from various regions of the belt might be collected at essentially the same semimajor axis. In the heuristic numerical example presented, a body at a_o = 3.6 AU [0.69, a(J) = 1.0] and another at 2.9 AU both come to rest at a = 2.73 AU, near the center of the inner belt. The outer body "feels" secular resonance at t = 15,000 P_J; its e was increased to 0.35 and semimajor axis consequently reduced by drag. The nebula density has fallen from its initial value by a factor of about 10 by the time (t = 25,000 P_J) a secular resonance has encountered the second object. It is much less affected by the drag and therefore spirals inward by a lesser amount. *Franklin and Lecar* (2000) furnish additional examples.

2.3. Nonuniform Nebula Depletion

Although the details of nebula depletion are not clear even in the solar system, it is plausible that the dependence of the decay on distance from the central star is not uniform. Several mechanisms have been proposed for the nebula depletion: tidal interaction with a protoplanet (*Lin and Papaloizou,* 1993), photoevaporation by ultraviolet radiation (*Shu et al.,* 1993; *Hollenbach et al.,* 2000), and magnetorotational instability (*Balbus and Hawley,* 1991; *Hawley et al.,* 1995; *Papaloizou and Lin,* 1995). Any nebula depletion mechanism in the inner solar system may differ from that in the outer one. When Jupiter starts gas accretion, a gap would open along its orbit. The gap can cover the entire inner solar system and extend to ~15 AU in about 10^6 yr (*Takeuchi et al.,* 1996). Photoevaporation tends to deplete the outer region (exterior to 10 AU), where the solar gravitational potential is reduced. Magnetorotational instability tends to develop in the outer regions (outside 20 AU), since the ionization fraction is high there (*Sano et al.,* 2000).

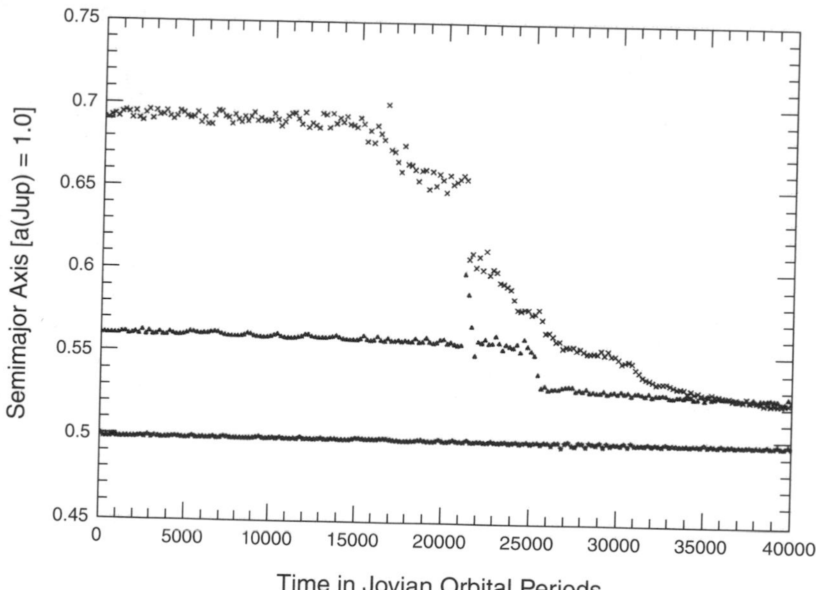

Fig. 2. An example showing the ability of scanning secular resonances, by pumping eccentricities and by acting in the presence of drag, to deposit bodies initally at very different semimajor axes, finally at nearly the same one. Crosses correspond to an object with r = 70 km, triangles with r = 400 km, and stars with one too large (r = 5000 km) to be affected by drag. Nebula's e-folding decay time is 10,000 jovian periods. *Franklin and Lecar* (2000) discuss additional cases.

To study the effects of nonuniform nebula depletion, *Nagasawa et al.* (2000) considered two simple models of nebula depletion, with edges that migrate either inwardly or outwardly. If the jovian gap formation process is followed by photoevaporation or magnetorotational instability, the nebula depletion is similar to the outwardly migrating edge model. If photoevaporation or magnetorotational instability is followed by jovian gap formation, the nebula depletion follows the inwardly migrating edge model. As explained later, only the timing of the depletion throughout the system is important, not other details. Hence, one can learn much from the study of these simple models.

The effects of sweeping of secular resonances are quite different between these two models. As seen in section 2.2, resonances ν_5 and ν_6 can sweep through the belt for a large variety of nebula depletion models. But for ν_{16} to sweep through the belt, the eigenfrequency associated to that resonance needs to regress faster than today. The effect of the nebula is exactly to speed up the regression of the nodes. But it has been shown that the precession frequency of the nodes is mostly influenced by the nebula in the vicinity of the point considered (*Ward,* 1981; *Lemaître and Dubru,* 1991). Hence, as explained in section 2.2, in the inwardly migrating edge model, the node of Jupiter slows down to its current value while the asteroids still regress much faster, so no resonance can occur. On the contrary, in the outwardly migrating edge model, the asteroid nodes slow to their current rate, while Jupiter's node is regressing faster. Hence a resonance may occur. *Nagasawa et al.* (2000), using a surface density profile proportionanl to $r^{-3/2}$ and Gaussian vertical profile, showed that the inclination could be pumped up when the edge migrates from 5 AU to 10 AU, while the nebula is already depleted in the asteroid belt. They even found an effect of the ν_{15} resonance, because the midplane of their nebula coincided with the ecliptic plane rather than with the invariant plane of the solar system.

More recently, we have studied the case of nebula depletion caused by Jupiter opening a gap, essentially corresponding to the outwardly migrating edge model. We take the Hayashi minimum mass nebula to begin with. The gap first opens interior to Jupiter's orbit according to $\rho(r) = 0$ for $r_{in}(t) < r < 5.5$ AU, the nebula density being unchanged elsewhere. Next, the gap opens outward: $\rho(r) = 0$ for $0 < r < r_{out}(t)$, where r_{out} starts from 5.5 AU and increases. In the model, we include Jupiter and Saturn with their current masses, eccentricities of about 0.048 and 0.054, and inclination with respect to the invariant plane of 0.0063 and 0.016 radians respectively. Contrary to *Nagasawa et al.* (2000), we use a nebula whose midplane coincides with the invariant plane of Jupiter and Saturn. However, the nebula midplane is fixed, and the nebula does not respond to the gravitational perturbations of the giant planets. Therefore the ν_{15} still exists, although it has a much lower effect than in *Nagasawa et al.* (2000).

As in the outwardly migrating edge model, this gap formation model excites both e and i all over the asteroid belt and avoids the strong gas drag that leads to rapid inward migration of asteroids with pumped-up e and i. Figure 3 shows the excitation magnitudes of eccentricity Δe (Fig. 3a) and inclination Δ tan i (Fig. 3b) due to a single sweeping of the secular resonance, analytically calculated as in *Nagasawa et al.* (2000), and the location of ν_{15} and ν_{16} with time. The excitation magnitudes are inversely proportional to the square root of edge migration velocity, $\dot{r}_{edge} = -\dot{r}_{in} = \dot{r}_{out}$ (*Ward et al.,* 1976, *Nagasawa et al.,* 2000). All four resonances pass through the asteroid belt, exciting the eccentricities and inclinations. Note that the strong resonances sweep the asteroid belt when the outer nebula is depleted. The second sweeping of the ν_5 and ν_{15} resonances while the edge is between 5 AU and 10 AU (between 5×10^5 and 10×10^5 yr) is mainly responsible for the total pumping up of e and i. The e and i of asteroids are pumped up to the

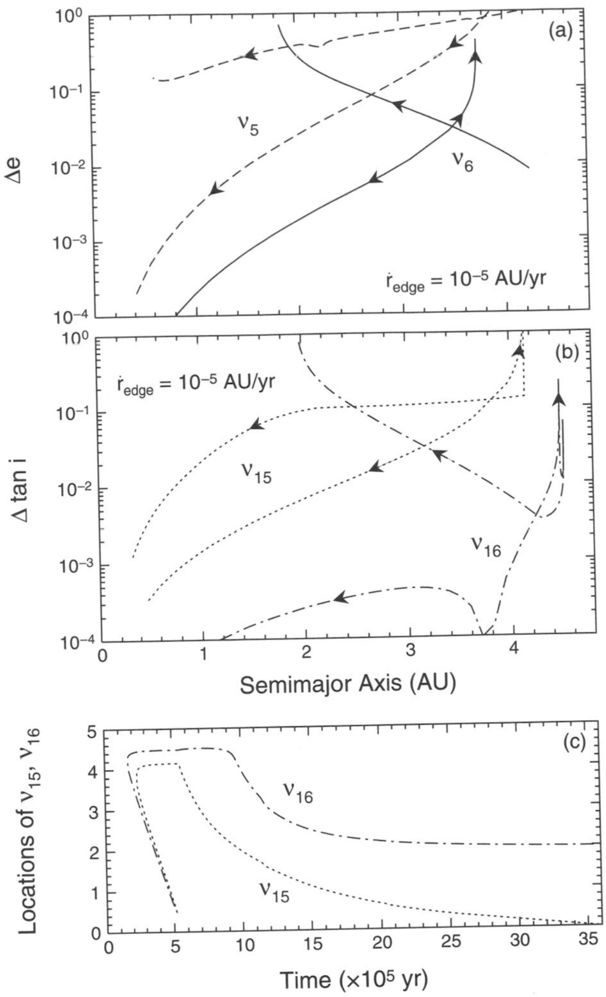

Fig. 3. **(a)** Changes in eccentricity due to a single sweep of the ν_5 (dashed lines) and ν_6 (solid lines) resonances in the jovian gap formation model. The arrows show the directions of resonances migration. **(b)** Changes in inclination due to the ν_{15} (dotted lines) and ν_{16} (dash-dotted lines) resonances. **(c)** Locations of secular resonances ν_{15} and ν_{16} vs. time of disk depletion. The paths of ν_5 and ν_6 are almost the same as those of ν_{15} and ν_{16} respectively. Here the edge velocity $\dot{r}_{edge} = 10^{-5}$ AU/yr. As the inner edge moves, the resonances migrate in two directions until the inner disk is gone (about 5×10^5 yr). After that, the resonances migrate from outside to inside as the outer edge migrates.

observed value in the entire asteroid belt if the nebula is depleted (i.e., the edge moved by 5 AU) on a timescale longer than $\tau = 5 \times 10^5$ yr.

The existence and amplitude of the ν_{15} resonance depends on the inclination between the nebula midplane and the invariant plane of the system. More work is needed to infer the effect of the interaction between the planets and the nebula gas.

In the case of the outwardly migrating edge model and the jovian gap formation model, the secular resonances pump up e and i of the asteroids after nebula gas has already been depleted in that region. Therefore, the orbital excitation of asteroids does not lead to rapid inward migration due to strong gas drag even if the nebula depletion time is as long as that inferred from observed T-Tauri stars (10^6–10^7 yr). The resultant eccentricities and inclinations remain therefore consistent with those observed.

3. GRAVITATIONAL PERTURBATION FROM EMBRYOS

This section focuses on the gravitational effect that embryos in the inner solar system would have on the asteroid belt. We consider two cases, one where large embryos are within a few tenths of an astronomical unit of Jupiter, and thus greatly perturbed by the giant planet, and another one where the embryos are spread over the asteroid belt and the terrestrial planet region.

3.1. Large Jupiter-Scattered Planetesimals

A number of works have quantitatively explored the effects of LJSPs on the asteroid belt in the hope of explaining its present structure and providing a constraint on the number, mass, and lifetime of these embryos. *Davis et al.* (1979), using a gaslike model, estimated that five 1-$M_\oplus$ LJSPs, crossing the belt during 3 m.y., could provide the required eccentricities. *Ip* (1987) was the first to attempt to follow the dynamical evolution of LJSPs and asteroids. For this purpose, he used a Monte Carlo code based on the method of *Arnold* (1965) to account for the statistical effects of close encounters, but neglecting distant perturbations and resonant phenomena, as well as the mutual encounters among the LJSPs. He found that the LJSPs have dynamical lifetimes too short to be efficient asteroidal perturbers; only if Jupiter were smaller, i.e., one-tenth of its present mass for ~10 m.y., could the LJSPs' action be more important. In particular, 50 LJSPs, each of mass 3×10^{27} g, could excite the asteroid belt in this approximation. *Wetherill* (1989), also using Monte Carlo simulations but including the mutual encounters among LJSPs, contradicted the conclusions of *Ip* (1987). He found that, even in the case of a full-sized Jupiter, a LJSP may decrease its eccentricity due to an encounter with another LJSP, isolating itself in the asteroid belt, safe from Jupiter encounters. The isolated body would then have a sufficiently long lifetime to excite the asteroid belt by gravitational scattering. In this case, only 10% of the asteroids would survive more than 700 m.y., and they would show a similar distribution of e and i to that observed. Moreover, the gravitational scattering would also provide an important mixing in the radial distribution of asteroidal compositional types.

Using the modern tools of celestial mechanics, *Petit et al.* (1999) revisited this question. They first studied the dynamical evolution of large embryos initially distributed regularly in the range a ∈ [4.0, 4.8] AU, with low eccentricity (0.01) and inclination (<1°) moving under the gravitational perturbation of Jupiter in its current orbit, and with its current mass. Only 15% of the embryos survived longer than 1 m.y. in low-eccentricity stable orbits, never encountering Jupiter. In a second set, they let the mass of Jupiter grow linearly with time from 5 to 15 $M_⊕$ over 10 m.y. During this time, only 2% of the embryos were ejected. At this point, Jupiter should accrete the surrounding gas very rapidly and reach its present mass of more than 300 $M_⊕$ in a very short time (*Pollack et al.,* 1996). After this event, the following evolution would be statistically equivalent to that of the previous case.

Building upon these results, *Petit et al.* (1999) determined the dynamical excitation and mass loss of the asteroid belt by a Monte Carlo method and direct integration of test particles. In this model, the asteroid belt shows very little excitation between 2 and 3 AU, and a rather large excitation beyond 3 AU. The excitation exhibits a marked gradient, which is not seen in the actual belt. A very important point is the very low inclination excitation, barely reaching 3°, when an equipartition of energy would imply an inclination on the order of 10° in the outer belt. Finally, the mass depletion is very small, only a few percent. This is because the embryos are ejected by Jupiter very quickly, and do not get a chance to cross the inner belt. In the case of a growing Jupiter, the results are similar, since most of the embryos do not move until Jupiter reaches its current mass.

The only way to solve this problem is to have scattered embryos in long-lived orbits in the main belt. Due to mutual gravitational interactions, some LJSPs could decrease their eccentricity, isolating themselves in the main asteroid belt, and therefore avoiding close encounters with Jupiter. Petit et al. performed integrations of a system composed of the present Jupiter and five embryos of 1 $M_⊕$. They found only 2% of the cases producing an embryo decoupled from Jupiter that (1) crossed the entire asteroid belt, (2) did not affect the region of the inner planets, and (3) was in a region dynamically unstable on a moderately long timescale, leaving the embryo time to excite the asteroids but eventually allowing it to escape. The dynamical excitation of the asteroids was then very large, and the radial displacement also quite large, reaching more than 1 AU. The depletion of test particles was complete after about 50 m.y.

Even though these last results are encouraging, they are very unlikely, and it is not clear how the unstability of the embryo and its ejection would affect the inner solar system and the terrestrial planets.

3.2. Endogenic Dynamical Excitation and Mass Depletion

We now consider the case where the planetary embryos formed in the asteroid belt region. It is widely believed that the early stages of the formation of the terrestrial planets were characterized by runaway growth, in which large solid bodies grew rapidly while small bodies grew more slowly or were themselves accreted (*Wetherill and Stewart,* 1989; *Kokubo and Ida,* 2000). After 10^5–10^6 yr, when runaway growth ceased, much of the solid material in the terrestrial-planet region had accreted into "planetary embryos" with masses comparable to the Moon or Mars (*Wetherill and Stewart,* 1993; *Weidenschilling et al.,* 1997). In the currently-favored "core formation" model for the formation of Jupiter and Saturn (*Pollack et al.,* 1996; *Ikoma et al.,* 2000), runaway growth also occurred in the region between 5 and 10 AU from the Sun, producing embryos of 2–15 $M_⊕$ (*Lissauer,* 1987; *Inaba and Wetherill,* 2001). The largest of these objects were sufficiently massive to accrete large atmospheres directly from the gas component of the Sun's protoplanetary nebula.

The formation of planetary embryos in both the inner and outer solar system makes it plausible that embryos also formed in the region now occupied by the main asteroid belt. Numerical simulations (*Wetherill and Stewart,* 1993; *Weidenschilling and Davis,* 2001) suggest that in the absence of external perturbations, embryos would have formed in the belt within ~10^6 yr. If this was the case, the absence of planetary embryos or fully formed planets in the modern belt can be viewed as an additional aspect of the great depletion of solid material from the belt that has occurred since the time when the nebula existed.

More interestingly, the existence of planetary embryos in the belt actually provides a mechanism for producing the mass depletion. *Wetherill* (1992) first proposed and examined this mechanism, which involves a combination of gravitational perturbations by embryos and orbital resonances. The main belt currently contains a number of powerful mean-motion and secular resonances associated with the giant planets. An asteroid in some of these resonances develops an unstable orbit on timescales of ~10^6 yr (*Gladman et al.,* 1997). As a result, the asteroid's orbital eccentricity undergoes large changes until the asteroid either falls into the Sun or passes close to Jupiter and is gravitationally scattered out of the solar system. Today these resonances are associated with the Kirkwood gaps in the belt, while the regions between the resonances contain large numbers of asteroids with apparently stable orbits. At the time when the planets were accreting, asteroids and embryos in the resonances would have been removed rapidly. However, more asteroids and embryos would have been scattered into the resonances by close encounters with other embryos. Thus a continuous stream of material was perturbed into the resonances and removed from the asteroid belt.

Wetherill (1992) found that a combination of resonances and embryo perturbations was sufficient to remove all embryos from the belt in half of the cases he studied. More recent N-body integrations suggest that the belt would be cleared of embryos in two-thirds of the cases (*Chambers and Wetherill,* 2001). The same mechanism would also have removed many asteroids from the belt. In fact, small bodies were lost preferentially since "dynamical friction" with the larger embryos tended to increase the eccentricities of the former more than the latter. Objects with large eccentricities were more likely to be removed since the resonances occupy a larger fraction of the belt for eccentric orbits, and these orbits are also more likely to be planet-crossing.

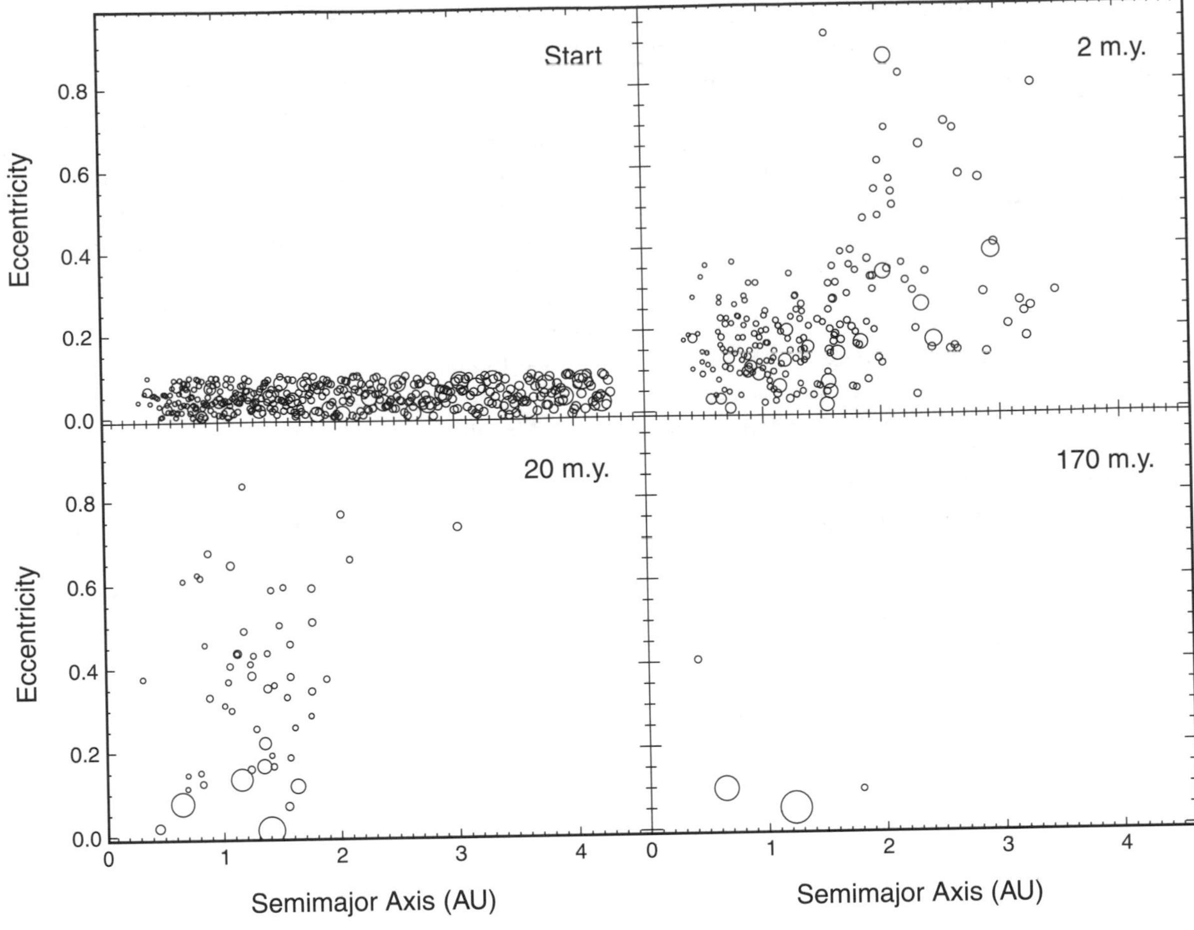

Fig. 4. Orbital evolution of 385 planetary embryos in the terrestrial planet and asteroid regions. Symbol radius is proportional to radius of the embryo. Jupiter and Saturn are included starting on orbits with an eccentricity of 0.1.

It is probable that ~99% of the original population of asteroids was lost (*Petit et al.,* 2001), so that the observed asteroids are a small group of lucky survivors from a much larger primordial population. Even these objects will have been scattered a number of times by close encounters with embryos. As a result, the observed asteroids typically have large e and i, and their orbital distribution essentially fills the volume of phase space that has been stable over the age of the solar system. In addition, if an asteroid's spectral class reflects the original location in the nebula at which it formed, gravitational scattering by embryos explains why the different asteroid classes are partially mixed in terms of heliocentric distance. *Petit et al.* (2001) found a typical radial displacement of ~0.5 AU. In addition, close encounters with embryos in the belt may have tidally stripped the mantles from many small differentiated bodies, leaving intact metal-rich cores that survive today as M-type asteroids (*Asphaug et al.,* 2000).

Figure 4 shows the embryo-clearing mechanism in action in an N-body integration. The simulation begins with a disk containing 385 embryos spread across the inner solar system. In addition, fully formed Jupiter and Saturn are present, starting at their current heliocentric distances. Within 1 m.y., the resonances increase the eccentricities of embryos in the main belt and material starts to be removed. After 20 m.y., hardly any embryos remain in the belt. Conversely, many embryos remain in the terrestrial region, where powerful resonances are absent, and eventually these accrete to form a system of four terrestrial planets. Such rapid clearing of material from the main belt may help explain why the primordial crust of (4) Vesta has avoided disruption due to large impacts since it was formed in the first few miillion years of the solar system.

One conceptual difficulty with the embryo-clearing mechanism is how the last embryo was removed from the asteroid belt. Once this body had scattered all the other embryos into unstable resonances it could have become trapped on a stable orbit with no mechanism to remove it from the belt. In fact, the last embryo probably collided with or was scattered into a resonance by an embryo in the terrestrial-planet region (*Chambers and Wetherill,* 2001). Terrestrial embryos would also have cleared many other embryos and asteroids from the inner belt, and this places strong constraints on the masses and orbital evolution of embryos in the region now occupied by Mars (*Petit et al.,* 2001).

The clearing of mass from the belt affected the orbits of the giant planets, since these bodies ultimately provided the changes in energy and angular momentum needed to eject asteroids and embryos from the belt or cause them to hit the Sun. In response, Jupiter's orbit migrated inward, while the orbits of both Jupiter and Saturn became more circular (*Chambers and Wetherill,* 2001). Hence, Jupiter originally lay further from the Sun than today, and the giant planets had more eccentric orbits in the past (for this reason the giants initially had e = 0.1 in the simulation shown in Fig. 4). The strength of the main-belt resonances depends sensitively on the eccentricities of the giant planets: Doubling e of Jupiter and Saturn increases the flux of material from the belt by a factor of ~10 (*Chambers and Wetherill,* 2001). Hence the formation times and early eccentricities of the giant planets determined when most of the primordial mass was removed from the asteroid belt. This in turn affects how much asteroidal material was accreted by the terrestrial planets that were forming at the same time (*Morbidelli et al.,* 2000).

In summary, the formation of planetary embryos in the asteroid belt provides a plausible and efficient mechanism for depleting much of the mass that originally existed in this region of the protoplanetary nebula. It is likely that this process would leave behind a population of asteroid-sized bodies with orbital and taxanomic distributions similar to the observed ones.

4. CONCLUDING REMARKS AND FUTURE WORK

In summary, we note that the belt could have been dynamically excited and cleared by a combination of the process discussed in section 3.2 and resonance sweeping, provided that embryos formed quickly before resonance sweeping had removed much material. Both mechanisms do a good job at exciting the eccentricities and inclinations of the asteroids, and removing most of the mass. The sweeping of secular resonance has trouble with the radial mixing since it would tend to produce a fairly homogeneous mixture instead of the overlapping zoning seen today. The gravitational perturbation by planetary embryos may not have occurred if it was the case that Jupiter formed before runaway growth took place in the solar system (*Kortenkamp and Wetherill,* 2000) as a result, for example, of disk instability (*Boss,* 2000). As we have implied earlier, the case for secular resonance sweeping does have some unfinished business. Models based on nonuniform nebula depletion clearly excite both e and i, but do not lead to a major decrease in the main belt's population. On the other hand, although uniform depletion can excite e and depopulate the belt, a clearer demonstration is needed as to the range of conditions under which it can excite the i. Perturbation by embryos provides a mechanism to deliver water (and other volatile elements) to Earth, but not necessarily to the other terrestrial planets, by letting an embryo from the outer belt collide with the forming Earth (*Morbidelli et al.,* 2000).

What future developments would be worthwhile? It seems a fair comment that processes naturally arising during the early stages of the solar system may account for many of the properties of the asteroid distribution. But it is also fair to remark that studies to date have proceeded in a rather piecemeal fashion. Astronomers are now approaching the stage when a more comprehensive, integrated discussion is desirable. Such a study must combine the previous mechanisms and include considerations such as (1) gravitational perturbation from planetary embryos with a collective response of the disk of asteroids (as in planetary rings, or in the Kuiper Belt; *Ward and Hahn,* 1998) and in the presence of gas drag — this would probably not directly affect the embryos, but certainly the asteroids; (2) possible ranges of giant planet migration; and (3) limits imposed by timescales set by giant and terrestrial planet formation.

Another specific question of interest is how, in the sweeping resonance and the endogenic excitation mechanisms, the 2:1 mean motion resonance with Jupiter can be such a dividing line between the low density of the central and inner belts, and the much lower density of the outer belt. *Nesvorný and Morbidelli* (1998) have shown that chaotic regions in the outer belt are far more abundant than previously noted, thanks primarily to weak three-body mean-motion resonances. What is not presently known is whether the chaos is sufficient to lead to widespread instability, consequently emptying the region in times less than the solar system's age. If chaos does so prevail, then the paucity of objects in the outer belt (but only in the outer belt) can no longer be used as clear evidence favoring any other ejection mechanism.

We hope that this review will help motivate further interpretive studies of the asteroid belt and so keep pace with the rapid advance of observational knowledge, and at the same time promote additional work on the Edgeworth-Kuiper Belt (*Hahn and Malhotra,* 1999; *Petit et al.,* 1999) to determine which of those processes that have been applied to the asteroid belt may also prove fruitful there.

REFERENCES

Arnold J. R. (1965) The origin of meteorites as small bodies. III. General considerations. *Astrophys. J., 141,* 1548–1556.

Asphaug E., Bottke W. F. Jr., Morbidelli A., and Petit J-M. (2000) Tides versus collisions in the primordial main belt (abstract). *Bull. Am. Astron. Soc., 32,* 1100.

Balbus S. A. and Hawley J. F. (1991) A powerful local shear instability in weakly magnetized disks. I – Linear analysis. II – Nonlinear evolution. *Astrophys. J., 376,* 214–233.

Bell K. R., Cassen P. M., Klahr H. H., and Henning Th. (1997) The structure and appearance of protostellar accretion disks: Limits on disk flaring. *Astrophys. J., 486,* 372–387.

Boss A. P. (2000) Possible rapid gas giant planet formation in the solar nebula and other protoplanetary disks. *Astron. J. Lett., 536,* L101–L104.

Cameron A. G. W. (1995) The first ten million years of the solar nebula. *Meteoritics, 30,* 133–161.

Chambers J. E. and Wetherill G. W. (1998) Making the terrestrial planets: N-body integrations of planetary embryos in three dimensions. *Icarus, 136,* 304–327.

Chambers J. E. and Wetherill G. W. (2001) Planets in the asteroid belt. *Meteoritics & Planet. Sci., 36,* 381–400.

Chapman C. R. and Davis D. R. (1975) Asteroid collisional evolution: Evidence for a much larger early population. *Science, 190,* 553–556.

Davis D. R., Chapman C. R., Greenberg R., Weidenschilling S. J., and Harris A. W. (1979) Collisional evolution of asteroids, populations, rotations and velocities. In *Asteroids* (T. Gehrels, ed.), pp. 528–557. Univ. of Arizona, Tucson.

Davis D. R., Ryan E. V., and Farinella P. (1994) Asteroid collisional evolution: Results from current scaling algorithms. *Planet. Space Sci., 42,* 599–610.

Duncan M. J. (1994) Orbital stability and the structure of the solar system. In *Circumstellar Dust Disks and Planet Formation* (R. Ferlet and A. Vidal-Madjar, eds.), pp. 245–255. Editions Frontières.

Duncan M. J., Quinn T., and Tremaine S. (1989) The long-term evolution of orbits in the solar system: A mapping approach. *Icarus, 82,* 402–418.

Fernandez J. A. and Ip W. H. (1996) Orbital expansion and resonant trapping during the late accretion stages of the outer planets. *Planet. Space Sci., 44,* 431–443.

Franklin F. A. and Lecar M. (2000) On the transport of bodies within and from the asteroid belt. *Meteoritics & Planet. Sci., 35,* 331–340.

Geissler P., Petit J. M., Durda D., Greenberg R., Bottke W., and Nolan M. (1996) Erosion and ejecta redistribution on 243 Ida and its moon. *Icarus, 120,* 140–157.

Gladman B. J., Migliorini F., Morbidelli A., Zappalà V., Michel P., Cellion A., Froeschlé Ch., Levison F., Bailey M., and Duncan M. (1997) Dynamical lifetimes of objects injected into asteroid belt resonances. *Science, 277,* 197–201.

Gradie J. C. and Tedesco E. F. (1982) Compositional structure of the asteroid belt. *Science, 216,* 1405–1407.

Hahn J. and Malhotra R. (1999) Orbital evolution of planets embedded in a planetesimal disk. *Astron. J., 117,* 3041–3053.

Hawley J. F., Gammie C. F., and Balbus S. A. (1995) Local three-dimensional magnetohydrodynamic simulations of accretion disks. *Astrophys. J., 440,* 742–763.

Hayashi C. (1981) Structure of the solar nebula, growth and decay of magnetic fields and effects of magnetic and turbulent viscosities on the nebula. *Prog. Theor. Phys. Suppl., 70,* 35–53.

Heppenheimer T. A. (1980) Secular resonances and the origin of the eccentricities of Mars and the asteroids. *Icarus, 41,* 76–88.

Hollenbach D. J., Yorke H. W., and Johnstone D. (2000) Disk dispersal around young stars. In *Protostars and Planets IV* (V. Mannings et al., eds.), pp. 401–428. Univ. of Arizona, Tucson.

Holman M. J. and Murray N. W. (1996) Chaos in high-order mean motion resonances in the outer asteroid belt. *Astron. J., 112,* 1278–1293.

Ida S. and Lin D. N. C. (1996) Long term gas drag effects on the structure of the asteroid belt and its implications for the solar nebula. *Astron. J., 112,* 1239–1246.

Ikoma M., Nakazawa K., and Emori H. (2000) Formation of giant planets: Dependences on core accretion rate and grain opacity. *Astrophys. J., 537,* 1013–1025.

Inaba S. and Wetherill G. W. (2001) Formation of Jupiter: Core accretion model with fragmentation (abstract). In *Lunar and Planetary Science XXXII,* Abstract #1384. Lunar and Planetary Institute, Houston (CD-ROM).

Ip W. H. (1987) Gravitational stirring of the asteroid belt by Jupiter zone bodies. *Gerl. Beitr. Geophys., 96,* 44–51.

Jedicke R. and Metcalfe T. S. (1998) The orbital and absolute magnitude distributions of main belt asteroids. *Icarus, 131,* 245–260.

Kenyon S. and Hartmann L. (1995) Pre-main sequence evolution in the Taurus-Auriga molecular cloud. *Astrophys. J. Supp. Series, 101,* 117–171.

Knežević Z., Milani A., Farinella P., Froeschlé Ch., and Froeschlé C. (1991) Secular resonances from 2 to 50 AU. *Icarus, 93,* 316–330.

Kokubo E. and Ida S. (2000) Formation of protoplanets from planetesimals in the solar nebula. *Icarus, 143,* 15–27.

Kortenkamp S. J. and Wetherill G. W. (2000) Terrestrial planet and asteroid formation in the presence of giant planets. *Icarus, 143,* 60–73.

Lecar M. and Franklin F. A. (1973) On the original distribution of the asteroids. I. *Icarus, 20,* 422–436.

Lecar M. and Franklin F. A. (1997) The solar nebula, secular resonanccs, gas drag, and the asteroid belt. *Icarus, 129,* 134–146.

Lemaître A. and Dubru P. (1991) Secular resonances in the primitive solar nebula. *Cel. Mech. Dyn. Astron., 52,* 57–78.

Lin D. N. C. and Papaloizou J. C. B. (1993) On the tidal interaction between protostellar disks and companions. In *Protostars and Planets III* (E. H. Levy and J. I. Lunine, eds.), pp. 749–835. Univ. of Arizona, Tucson.

Liou J. C. and Malhotra R. (1997) Depletion of the outer asteroid belt. *Science, 275,* 375–377.

Lissauer J. (1987) Timescales for planetary accretion and the structure of the protoplanetary disk. *Icarus, 69,* 249–265.

Morbidelli A., Chambers J., Lunine J. I., Petit J-M., Robert F., Valsecchi G. B., and Cyr K. E. (2000) Source regions and time scales for the delivery of water to Earth. *Meteoritics & Planet. Sci., 35,* 1309–1320.

Nagasawa M., Tanaka H., and Ida S. (2000) Orbital evolution of asteroids during depletion of the solar nebula. *Astron. J., 119,* 1480–1497.

Nesvorný D. and Morbidelli A. (1998) Three-body mean motion resonances and the chaotic structure of the asteroid belt. *Astron. J., 116,* 3029–3037.

Papaloizou J. C. B. and Lin D. N. C. (1995) Theory of accretion disks I: Angular momentum transport processes. *Annu. Rev. Astron. Astrophys., 33,* 505–540.

Petit J-M., Morbidelli A., and Valsecchi G. (1999) Large scattered planetesimals and the excitation of the small body belts. *Icarus, 141,* 367–387.

Petit J-M., Morbidelli A., and Chambers J. E. (2001) The primordial excitation and clearing of the asteroid belt. *Icarus, 153,* 338–347.

Pollack J. B., Hubickyj O., Bodenheimer P., Lissauer J. J., Podolak M., and Greenzweig Y. (1996) Formation of the giant planets by concurrent accretion of solids and gas. *Icarus, 124,* 62–85.

Ruzmaikina T. V., Safronov V. S., and Weidenschilling S. J. (1989) Radial mixing of material in the asteroidal zone. In *Asteroids II* (R. P. Binzel et al, eds.), pp. 681–700. Univ. of Arizona, Tucson.

Safronov V. S. (1979) On the origin of asteroids. In *Asteroids* (T. Gehrels, ed.), pp. 975–993. Univ. of Arizona, Tucson.

Sano T., Miyama S. M., Umebayashi T., and Nakano T. (2000) Magnetorotational instability in protoplanetary disks. II. Ionization state and unstable regions. *Astrophys. J., 543,* 486–501.

Shu F. H., Johnstone D., and Hollenbach D. (1993) Photoevaporation of the solar nebula and the formation of the giant planets. *Icarus, 106,* 92–101.

Takeuchi T., Miyama S. M., and Lin D. N. C. (1996) Gap formation in protoplanetary disks. *Astrophys. J., 460,* 832–847.

Tanga P., Cellino A., Michel P., Zappalà V., Paolicchi P., and Dell'Oro A. (1999) On the size distribution of asteroid families: The role of geometry. *Icarus, 141,* 65–78.

Ward W. R. (1981) Solar nebula dispersal and the stability of the planetary system. I. Scanning secular resonance theory. *Icarus, 47,* 234–264.

Ward W. R. and Hahn J. M. (1998) Dynamics of the trans-Neptune region: Apsidal waves in the Kuiper belt. *Astron. J., 116,* 489–498.

Ward W. R., Colombo G., and Franklin F. A. (1976) Secular resonance, solar spin down, and the orbit of Mercury. *Icarus, 28,* 441–452.

Weidenschilling S. J. (1977) The distribution of mass in the planetary system and solar nebula. *Astrophys. Space Sci., 51,* 153–158.

Weidenschilling S. J. and Davis D. R. (2001) Planetary accretion: The first 5 million years (abstract). In *Lunar and Planetary Science XXXII,* Abstract #1894. Lunar and Planetary Institute, Houston (CD-ROM).

Weidenschilling S. J., Spaute D., Davis D. R., Marzari F., and Ohtsuki K. (1997) Accretional evolution of a planetesimal swarm. 2: The terrestrial zone. *Icarus, 128,* 429–455.

Wetherill G. W. (1989) Origin of the asteroid belt. In *Asteroids II* (R. P. Binzel et al., eds.), pp. 661–680. Univ. of Arizona, Tucson.

Wetherill G. W. (1992) An alternative model for the formation of the asteroids. *Icarus, 100,* 307–325.

Wetherill G. W. and Chambers J. E. (1997) Numerical integration study of primordial clearing of the asteroid belt (abstract). In *Lunar and Planetary Science XXVIII,* p. 1547. Lunar and Planetary Institute, Houston.

Wetherill G. W. and Stewart G. R. (1989) Accumulation of a swarm of small planetesimals. *Icarus, 77,* 350–357.

Wetherill G. W. and Stewart G. R. (1993) Formation of planetary embryos: Effects of fragmentation, low relative velocity, and independent variation of eccentricity and inclination. *Icarus, 106,* 190–209.

Origin and Evolution of Trojan Asteroids

F. Marzari
University of Padova, Italy

H. Scholl
Observatoire de Nice, France

C. Murray
University of London, England

C. Lagerkvist
Uppsala Astronomical Observatory, Sweden

The regions around the L_4 and L_5 Lagrangian points of Jupiter are populated by two large swarms of asteroids called the Trojans. They may be as numerous as the main-belt asteroids and their dynamics is peculiar, involving a 1:1 resonance with Jupiter. Their origin probably dates back to the formation of Jupiter: the Trojan precursors were planetesimals orbiting close to the growing planet. Different mechanisms, including the mass growth of Jupiter, collisional diffusion, and gas drag friction, contributed to the capture of planetesimals in stable Trojan orbits before the final dispersal. The subsequent evolution of Trojan asteroids is the outcome of the joint action of different physical processes involving dynamical diffusion and excitation and collisional evolution. As a result, the present population is possibly different in both orbital and size distribution from the primordial one. No other significant population of Trojan asteroids have been found so far around other planets, apart from six Trojans of Mars, whose origin and evolution are probably very different from the Trojans of Jupiter.

1. INTRODUCTION

As of May 2001, about 1000 asteroids had been classified as Jupiter Trojans (http://cfa-www.harvard.edu/cfa/ps/lists/JupiterTrojans.html), some of which had only been observed for a few nights and some that had no measured absolute magnitude. The preceding cloud, L_4, consists of 618 known members; there are 375 objects in the trailing cloud, L_5. Out of these 1000, only 426 are numbered asteroids with reliable orbits; 284 are at L_4 and 142 at L_5, according to Bowell's list (ftp.lowell.edu). From spectroscopic surveys it appears that most Trojans belong to the D taxonomic type, while only a few are classified as P and C type. All these objects have low albedos (average around 0.065) and share spectral similarities with short period comets, Centaurs, and transneptunian objects. In addition to Jupiter Trojans, there are five Trojans orbiting in the trailing cloud of Mars. Apart from Jupiter and Mars, observational searches have so far failed to detect Trojan asteroids of any other planet and it is still uncertain if this is due to an intrinsic instability of the tadpole orbits of those planets or to detection difficulties. Putative Trojans of Saturn, Uranus, and Neptune would in fact be very faint, while a possible population of Trojans of Earth or Venus would occupy a large projected area in the sky and would be unfavorably placed with respect to the Sun. A few interesting Trojan configurations have been found among the satellites of Saturn, and they may originate from the collisional disruption and subsequent reaccumulation of larger primordial bodies.

A basic understanding of why asteroids can cluster in the orbit of Jupiter was developed more than a century before the first Trojan asteroid was discovered. In 1772, Joseph-Louis Lagrange demonstrated the existence of five equilibrium points in the restricted three-body problem where an object of negligible mass orbits under the gravitational effect of two larger masses (*Lagrange,* 1772). Three of these points, L_1, L_2, and L_3 lie on the line joining the two masses and are unstable to small perturbations. Each of the remaining two points, L_4 and L_5, lies at the apex of an equilateral triangle with base equal to the separation of the two masses (see Fig. 1); stable motion is possible around them. Successive attempts to understand the dynamical properties of the three-body problem have provided a rich mathematical vein that continues to be mined to this day. Recent summaries of the literature are contained in *Szebehely* (1967) and *Marchal* (1990). There are also particularly relevant chapters in *Brown and Shook* (1993) and *Murray and Dermott* (1999). A series of papers have also been published by *Érdi* in *Celestial Mechanics* (1978, 1979, 1981, 1984, 1988) on the dynamics of Trojan asteroids. He has derived a second-order solution to the three-dimensional motion of Trojans within the framework of the elliptical restricted three-body problem. His perturbative approach leads to analytical expressions for the secular evolution of

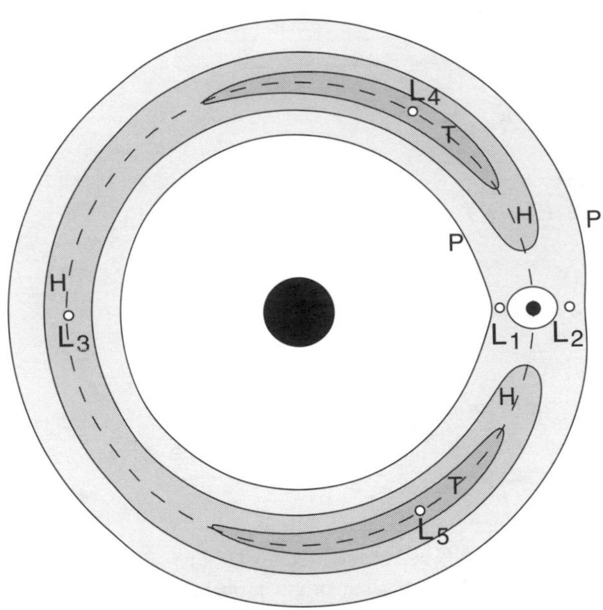

Fig. 1. The location of the five Lagrange equilibrium points in the circular-restricted three-body problem. The primary and secondary masses (the Sun and planet in our examples) are denoted by the large and small filled circles respectively. The selected zero-velocity curves (see text) are closely related to the types of orbits that can occur in the system. The letters P, H, and T denote passing, horseshoe, and tadpole orbits respectively. Note that the two masses and each of the L_4 and L_5 points form an equilateral triangle.

the perihelion and eccentricity and for the node as a function of the libration amplitude and inclination.

An important cosmogonical and dynamical question concerning Trojan orbits is their long-term stability. The first attempt to solve the problem analytically was made by *Rabe* (1967) who derived a region in the eccentricity-libration amplitude space for indefinite stability in the frame of the planar-circular-restricted three-body problem. More recently, *Georgilli and Skokos* (1997) used the same model but with a higher-order perturbative expansion and a different method to determine stability. They were able to mathematically proof the stability in only a small fraction of the region found by *Rabe* (1967). Unfortunately, the stability area found by *Georgilli and Skokos* (1997) does not compare with the Jupiter Trojan population; in fact, it comprises only a few of the observed Trojans. The same authors admit that their criterion is probably too restrictive.

When comparing the results obtained within the restricted three-body problem to the Jupiter Trojan populations, there are additional potential sources of instability that have to be taken into account. These are related to the gravitational influence of the other planets. *Levison et al.* (1997) resorted to a numerical approach to study the stability of Jupiter Trojans within a full N-body model that included all the outer planets. They found that there is a region for Jupiter Trojans where the asteroids may orbit in a stable fashion over the age of the solar system, and that this region, as expected, is significantly wider than that defined by *Georgilli and Skokos* (1997). An immediate question arising from the *Levison et al.* (1997) results arises: What causes the instability outside this region? Is the destabilizing mechanism intrinsic in the three-body problem, or are the external perturbations of the other planets responsible for this instability? Secular resonances cross the region in the phase space where Trojans may orbit, and even the 5:2 near-resonance between Jupiter and Saturn (the "great inequality") may cause Trojan orbits to become unstable. Eventually, a synergy among all these mechanisms may explain the presently observed structure of the Jupiter Trojan clouds.

The same reasoning concerning the stability should be applied to the study of the evolution of tadpole orbits for the other giant planets Saturn, Uranus, and Neptune. Of course, the orbital frequencies of the Trojan orbits and of the perturbing forces change from planet to planet, rendering the stability analysis an independent task for each planet. It is even possible that the strong mutual perturbations among the outer planets prevent Saturn, Uranus, and Neptune from having stable Trojans over the age of the solar system. The few observational surveys have failed so far to detect any Trojans. The reason might be purely dynamical, but we cannot exclude the possibility that Trojans exist but are too dark and distant to be readily detected. The possibility of terrestrial planet Trojans is a different story. While for the outer planets, the stability issue can be investigated over the age of the solar system, in the case of tadpole orbits for the Earth, Venus, and Mars, we can only perform numerical surveys limited in time, since to maintain the accuracy in the integration of the equation of motion a short timestep has to be used. The present studies on the dynamics of tadpole orbits for Earth, Venus, and Mars are in fact limited to a timespan of 100 m.y. Under this condition, we can only have a glimpse at their long-term dynamical behavior, and we cannot give a definitive answer to whether, for example, the six Mars Trojans are primordial, sharing a common origin with Mars, or were captured in more recent times. Even Earth and Venus may well possess stable Trojan-type orbits, but, for the same reason, we cannot asses their stability over the age of the solar system. Moreover, the possible detection of these Trojans would be very difficult because of unfavorable observational conditions.

Another conundrum in the Trojan saga is represented by the high average inclination of Jupiter Trojans. The most accepted theory on their origin assumes that they were planetesimals orbiting near Jupiter that were then captured in the final phase of the planetary formation process. If this were the case, their present orbits should lie in the plane of the proto-nebula disk and close to the orbital plane of Jupiter. Which dynamical mechanism acted after their capture as Trojans to drive them into high-inclination orbits? Alternative scenarios include the raising of inclinations by secular resonances (*Yoder,* 1979; *Marzari and Scholl,* 2000) or by close encounters with lunar-sized bodies similar to a mechanism proposed by *Petit et al.* (1999) for main-belt asteroids. It is even possible that they were stirred up prior to capture by proto-Jupiter (*Kokubo and Ida,* 2000).

In this chapter we summarize the present state of the research on Trojans. After an overview of the known orbital and size distributions of Jupiter Trojans, we analyze the general dynamical properties of tadpole orbits in the context of the three-body model. We then explore models of the origin of Trojans and investigate how they may have subsequently evolved to their present configuration through interactions with other planets of the solar system.

2. PRESENT POPULATION OF JUPITER TROJANS

Dedicated surveys for Trojans have been made by *van Houten et al.* (1970a) with the Palomar Schmidt telescope during several apparitions: March 1971 (L_5), September 1973 (L_4), and October 1977 (L_5). The Trojans discovered during these runs were designated T-1, T-2, and T-3. Also, the Palomar Leiden Survey of Asteroids (*van Houten et al.,* 1970b) was close to the preceding Lagrangian point. *Lagerkvist et al.* (2000) used the ESO Schmidt Telescope for a dedicated survey of L_4 with observations during the oppositions in 1996, 1997, and 1998. *Jewitt et al.* (2000) made a campaign to estimate the number of small Trojans in L_4 (radii <20 km) with the University of Hawai'i Mauna Kea Observatory 2.2-m telescope.

2.1. Asymmetry Between L_4 and L_5?

The first question Trojan surveys could help to answer is whether there is any difference between the leading and trailing Trojan populations. From a theoretical point of view, there is no known difference between the dynamics of the L_4 and L_5 points for Jupiter. However, it has been shown (*Peale,* 1993; *Murray,* 1994; *Marzari and Scholl,* 1998a) that in the presence of gas drag, the orbits of small Trojan asteroids around L_5 are more stable than those around L_4. On the contrary, planetary migration seems to destabilize more easily the orbits of Jupiter Trojans around L_5 (*Gomes,* 1998). A possible difference between the two populations in size distribution or in orbital distribution would confirm either of the hypotheses: that the capture of Trojans occurred in presence of gas drag or that a significant planetary migration took place in the early phases of the solar system. Lagerkvist et al. (in preparation, 2002) applied a simple statistical test, the Kruskal-Wallis test (*Siegel and Castellan,* 1988), to compare the orbital elements of the two Trojan populations. They compared the distributions of the mean distance (a), eccentricity (e), and inclination (i) between the L_4 and L_5 populations, discriminating between numbered and unnumbered (more recently discovered and hence presumably smaller) Trojans. The elements a and e showed no such differences between L_4 and L_5, or between numbered and unnumbered objects. However, they found the inclination to be different between the two clouds in the sense that L_5 contained more high-inclination orbits. The reason for this is not clearly understood, but may be due to observational bias. An investigation of the discovery circumstances might reveal the true nature of this difference. There is, however, no difference between the inclinations of the numbered and unnumbered Trojans in the L_5 cloud. For absolute magnitudes, they found no difference between the two Trojan clouds. Even the Spacewatch data, once bias-corrected, show no significant discrepancies between the magnitude distributions in the two swarms (see *Jedicke et al.,* 2002). The two populations have therefore been treated together in the following discussion of the size distribution even if this is focused on L_4, the most-studied Lagrangian point. In recent years, only a few new Trojans have been found to be brighter than absolute magnitude H = 9.5, and the population seems to be more or less complete down to this limit (H = 9.5 corresponds to a radius of 43 km for an assumed geometric albedo of 0.04; H = 10 corresponds to 34 km). Presently there are 60 Trojans with H ≤ 10.0, of which as many as 18 have been found during the last five years.

Of interest for the stability of the Trojan orbits are the proper elements. These have been calculated numerically for the L_4 members known to date by Karlsson (personal communication, 2001), following the method given in *Schubart and Bien* (1987). There is a large spread in the libration amplitudes ranging from 0.6° to 88.7°, with a mean around 32.7° for the L_4 cloud. The Kruskal-Wallis test gave no significant differences (Lagerkvist et al., in preparation, 2002) of the libration amplitudes of the numbered and unnumbered Trojans in L_4. Even in the proper elements computed by *Milani* (1993) with a slightly different method and with a smaller sample of Trojan orbits (174), there was no significant difference between the distribution of the libration amplitude in L_4 compared to that in L_5. Recently, *Beaugè and Roig* (2001) have developed a semianalytical method to estimate proper elements and they have applied it to a sample of 533 Trojans, observing only minor differences between the L_4 and L_5 populations. From the proper elements they proceeded to identify asteroid families as in *Milani* (1993) and found a possible asymmetry between the two swarms: The L_4 region shows more families than L_5 and they are even more robust. Is this an indication of a larger number of potential projectiles in L_4 for catastrophic disruption events, or is it the orbital distribution to be really asymmetric in the two swarms leading to a different collisional rate? We need additional data to answer to this question.

2.2. Size Distribution

The total number of Trojans down to an absolute magnitude of H = 13.0 was estimated by *van Houten et al.* (1970a) to be about 700. From the same observational data, *Shoemaker et al.* (1989) deduced that there were 995 L_4 Trojans, a factor of 1.4 larger than that claimed by *van Houten et al.* (1970a). *Lagerkvist et al.* (2000; in preparation, 2002), from a large observational sample, after correcting for incompleteness down to the limiting magnitude, suggested that

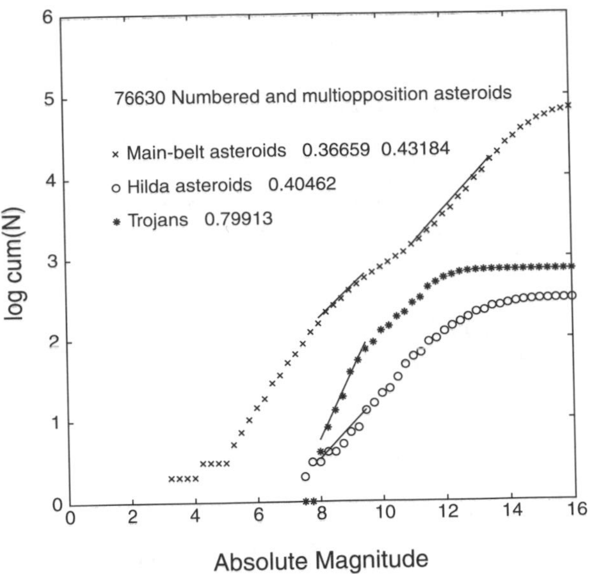

Fig. 2. Cumulative numbers of main-belt asteroids, Hildas, and Trojans plotted on a logarithmic scale vs. absolute magnitude. For each population, the slope of the magnitude distribution is given for selected magnitude ranges. At the high-magnitude end, Trojan asteroids have the steepest slope.

there were 1100 L_4 Trojans in the same cloud. *Jewitt et al.* (2000) got as many as 3300 L_4 Trojans down to same sizes. The large discrepancies between some of these estimates might be due to some extent to the statistical limitations in the detection capabilities of the surveys. Bias correction for the selection effects in the observed populations is a difficult task, in particular for Trojans. Their librational motion can generate nonuniform density patterns once projected onto the sky. Figure 1 shows, for example, that the distribution of Trojans around L_4 is by no means the same in the direction of Jupiter as in the opposite direction and also that latitude plays an important role.

Figure 2 gives the cumulative absolute magnitude distributions (logarithmic) of the main belt asteroids, the Hildas (asteroids captured in the 3:2 mean-motion resonance with Jupiter) and the Trojans (known objects). Only data from numbered and multiopposition objects have been used. For the Trojans, this means that somewhat fewer objects are available, but the data should still be complete down to H = 9.5 for all three categories. The slope of the Trojan distribution at small magnitudes (large size end) is about twice that of main-belt asteroids and Hildas of similar size. Assuming a mean geometric albedo of 0.04 from IRAS data (*Tedesco,* 1989) the differential size distribution is adequately fitted by a power law with index q = 5.5 ± 0.9 for radii larger than 30–45 km (*Jewitt et al.,* 2000). For smaller Trojans, both *Shoemaker et al.* (1989) and *Jewitt et al.* (2000), with a more detailed survey, found a slope around 0.4 in magnitude that corresponds to an index of q = 3.0 ± 0.3. *Jewitt et al.* (2000) predicts, however, a cumulative number of small Trojans larger than that of *Shoemaker et al.* (1989), due to a significantly different scaling of the power laws. What is the interpretation of the different slopes at large and small sizes for Trojans? A possible answer is that large objects represent a primordial planetesimal population captured during the formation of Jupiter, while smaller Trojans represent fragments produced during the subsequent collisional evolution of the population (*Shoemaker et al.,* 1989; *Marzari et al.,* 1997). This interpretation is reinforced by the fact that the index q of the differential size distribution is close to the Dohnanyi's value of q = 3.5 (*Dohnanyi,* 1969), typical of a collisionally relaxed population. Moreover, from the larger mean light-curve amplitude of large-sized Trojans compared to their low albedo main-belt counterparts, *Binzel and Sauter* (1992) argued that large Trojans might have retained their primordial aspherical forms.

If future, more complete surveys of the Trojan swarms confirm the slopes and size distribution derived by *Jewitt et al.* (2000), then Trojan asteroids would outnumber main-belt asteroids. From the debiased observational data of the SDSS survey (*Ivezić et al.,* 2001), there would be about 5.3×10^5 main-belt asteroids with a diameter larger than 1 km, while Trojans, thanks to their steeper slope at the small size end, would number around 1.28×10^6 according to *Jewitt et al.* (2000).

3. DYNAMICS OF THE THREE-BODY PROBLEM

One of the great triumphs of celestial mechanics was the prediction by Lagrange in the eighteenth century that points of stable equilibrium could exist in the orbit of Jupiter and that objects might one day be found there. However, it was only in the twentieth century that the first Trojan asteroid was discovered, thereby validating Lagrange's prediction. Many of the world's greatest mathematicians have studied the "three-body problem," i.e., the motion of a small object moving under the gravitational effect of a planet and the Sun. In the simplest version of the three-body problem, Jupiter and the Sun move in circular orbits around their common center of mass while the asteroid is treated as a test particle perturbed by the two masses but without, in turn, affecting their motion. This is a reasonable approximation because Jupiter's eccentricity is small ($e_J \approx 0.048$) and even the largest asteroid, (1) Ceres, has a mass that is $<10^{-6}$ that of Jupiter. Although the angular momentum and energy in this idealized system are not conserved quantities, the circular restricted problem has a constant of the motion, the Jacobi constant C, that is a function of the position and velocity of the asteroid (e.g., see *Murray and Dermott,* 1999). Although it is not possible to find a practical, analytical solution for the resulting motion of the asteroid, the existence of the Jacobi constant implies that it is sometimes possible to identify regions from which the asteroid is excluded. This is achieved by plotting the so-called "zero-velocity curves" for the asteroid. Figure 1 shows three sets of such curves for a mass ratio $\mu = m_J/(m_S + m_J)$ of 10^{-2}, or approximately 10× larger than the actual value for the Sun-Jupiter system. The three shaded areas denote excluded regions for the zero-velocity curve associated with the corresponding value of C.

It is clear from Fig. 1 that the locations of the Lagrangian equilibrium points (small unfilled circles) are related to the critical points of zero-velocity curves. In appropriate units for small values of μ, the value of the Jacobi constant at L_4 and L_5 is a minimum ($C_{L_{4,5}} = 3 - \mu$) while it is a maximum at L_1 ($C_{L_1} = 3 + 3^{4/3}\mu^{2/3} \approx C_{L_2}$) with an intermediate value at L_3 ($C_{L_3} = 3 + \mu$). Therefore, an asteroid with $C < C_{L_{4,5}}$ will have no zero-velocity curve and hence, on this basis, there are no regions from which it is excluded. That does not imply that its motion is unbounded or that it will escape from the system; it just means that the Jacobi constant cannot be used to place bounds on its motion.

The fact that Jupiter and the Sun maintain a fixed separation in the circular problem means that it is customary to consider the motion of the asteroid in a frame rotating with the (constant) angular velocity or mean motion, n, of Jupiter. One consequence of choosing such a reference frame is that the motion of the asteroid can appear to be quite complicated. However, it is important to remember that the asteroid's motion is dominated by the Sun and that in the nonrotating frame, it is following a near-constant Keplerian elliptical path around the Sun. The effect of Jupiter is to perturb that ellipse and introduce variations in the orbital elements of the asteroid. We can see this and gain a useful insight into the dynamics of Trojan asteroids by investigating the asteroid's motion in the vicinity of the L_4 point.

In addition to identifying the triangular equilibrium points, Lagrange also investigated their linear stability. This involved carrying out an analysis of how an object (the asteroid) would respond if it underwent a small displacement from the equilibrium point. It can be shown (e.g., see *Murray and Dermott,* 1999) that provided the displacement is small and the condition

$$\mu < \mu_{\text{crit}} = \left(27 - \sqrt{621}\right)/54 \approx 0.0385$$

is satisfied, the asteroid moves in a stable path with two distinct components. The first is a long-period, elongated librational motion around L_4 with a period

$$T_1 = T_J / \sqrt{(27/4)\mu}$$

where T_J is the orbital period of the secondary mass (Jupiter) around the central mass (the Sun). Taking $\mu \approx 10^{-3}$ and $T_J \approx 12$ yr gives a librational period of ~146 yr for Jupiter Trojans. Superimposed on this is a short-period motion of period

$$T_2 = T_J / \sqrt{1 - (27/8)\mu}$$

This is usually referred to as the epicyclic motion of the asteroid and is simply its Keplerian motion viewed in the rotating reference frame. Note that $T_2 \approx T_J$ for small μ. Furthermore, as the eccentricity of the test particle (asteroid) tends to zero, so the amplitude of the epicyclic motion is suppressed.

It can be shown that for small amplitude librations around L_4 or L_5, the ratio of the axes for the elongated librational ellipse is $\sqrt{3\mu}$ while that for the epicyclic motion ellipse is 1/2. Figure 3a shows the two components of the motion for a sample case, while Fig. 3b shows the equivalent resulting path in the rotating frame. In this case, there is a zero-velocity curve, albeit a very narrow one, that the particle just touches at the cusps of its motion (see the interior part of the trajectory in Fig. 3b). Again, although the path is complicated, this is simply a consequence of viewing the trajectory in the rotating frame. Note that, except for Pluto

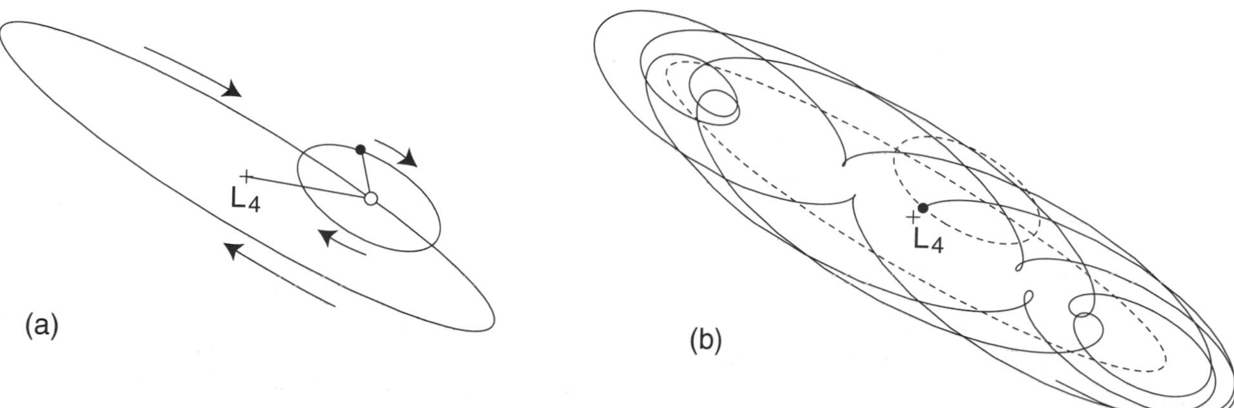

Fig. 3. The typical motion of an asteroid in a small amplitude tadpole orbit around L_4. **(a)** The motion has two separate contributions: a long-period motion (elongated ellipse) about the equilibrium point combined with a short-period motion (small ellipse) arising from the Keplerian motion of the asteroid. **(b)** The resulting path appears complicated but the dynamical mechanism is straightforward. Adapted from *Murray and Dermott* (1999).

and its moon Charon, the condition for linear stability, $\mu < \mu_{crit}$ is satisfied by every planet-Sun and satellite-planet pair in the solar system, even though the vast majority show no signs of maintaining Trojan objects.

It is tempting to think of the curves shown in Fig. 1 as defining an asteroid's path in the rotating frame, despite the fact that these are zero-velocity curves rather than trajectories. In fact, as suggested above, for Trojan asteroids on near-circular orbits, there is a close correspondence between actual paths in the rotating frame and their associated zero-velocity curves [see *Murray and Dermott* (1999) for a more complete description; note that for larger eccentricities the correspondence breaks down and the paths can be significantly different as shown by *Namouni* (1999)]. More importantly, the relationship allows us to see how the orbits in the vicinity of L_4 and L_5 change as the initial displacement from the equilibrium points increases. The symmetry of the librational ellipse around the equilibrium point is lost and it develops an elongated "tail." These are referred to as tadpole orbits and the two examples in Fig. 1 are labeled with a T. As the amplitude of libration increases, the tail of the tadpole extends toward the L_3 point. For an object with $C_{L_3} < C < C_{L_1}$, a type of motion that encompasses L_3, L_4, and L_5 is possible. This is the so-called horseshoe orbit; an example is labeled H in Fig. 1. Although no Jupiter Trojans are currently known to exist in horseshoes, the saturnian satellites Janus and Epimetheus are known to be involved in a variation of this configuration. By increasing the Jacobi constant and hence the libration amplitude further, the horseshoe configuration can be lost and the object can circulate either interior or exterior to the orbit of the secondary mass. These can be referred to as passing orbits and are labeled P in Fig. 1.

Although many of the above results have been derived in the context of the circular-restricted problem, they can be extended to the case where the planet moves in an elliptical rather than a circular path. For example, *Danby* (1964) investigated the linear stability of L_4 and L_5 in the elliptical problem. Analytical and numerical studies show that the phenomena of tadpole and horseshoe orbits carry over into the elliptical regime and form part of a more general class of orbits that demonstrate what is called coorbital motion with important applications to planetary rings and studies of the formation of planets and satellites. In all these cases, the key quantity is the semimajor axis of the asteroid relative to that of the planet. If we define the coorbital width of a planet as the region within which tadpole or horseshoe motion is possible, this region has a half-width $\Delta = [m_p/3\ (m_p + m_S)]^{1/3}\ a_p$, where a_p is the semimajor axis of the planet and m_p is its mass. In fact, this is nothing more than the approximate distance of the L_1 or L_2 points from the planet (e.g., see *Murray and Dermott,* 1999).

The early analytical work on the dynamics of coorbital motion concentrated on the case where the eccentricity, e, and inclination, I, were small. This was because it was relatively easy to tackle analytically using expansions in terms of small quantities. However, *Namouni* (1999) showed analytically that previously unknown types of coorbital motion were possible in the case of relatively large values of e and I. He predicted that in addition to the T, H, and P orbits, one could also expect to find, among others, RS-T (retrograde satellite-tadpole), RS-H (retrograde satellite-horseshoe), and T-RS-T (tadpole-retrograde satellite-tadpole) orbits. Indeed, additional work by *Namouni et al.* (1999) identified asteroids in such orbits associated with Earth and Venus. If variable a denotes the semimajor axis of the asteroid and a_p is that of the planet, we can then define the relative semimajor axis $a_r = (a - a_p)/a_p$. Similarly, we can define the relative longitude $\lambda_r = \lambda - \lambda_p$ (where λ denotes mean longitude) to help identify librational motion. Figure 4 is taken from *Namouni et al.* (1999) and shows plots of the time evolution of a_r and λ_r for three asteroids. Numerical integrations in a planetary system including all planets indicate the possibility that (3753) Cruithne and (3362) Khufu are or have been involved in coorbital interactions with Earth, while 1989 VA will be a coorbital of Venus. Figure 4 clarifies and explains the orbit of (3753) Cruithne (e = 0.515, I = 19.8°) and shows that it will become a retrograde satellite of Earth within the next 10^4 yr. Although one might have expected such orbits not to survive the effect of perturbations from other planets, what was observed in the numerical integrations was a switching from one type of orbit to another. For example, 2×10^4 yr ago (3753) Cruithne was in a large eccentricity T orbit around Earth's L_4 point, before evolving into a T-RS-T orbit, a P orbit, and a H-RS orbit (its current state). In contrast, (3362) Khufu is currently in a P orbit but prior to that had been a retrograde satellite of Earth. *Christou* (2000a) confirmed and extended these results. There is no inherent reason why such orbits cannot exist among asteroid themselves. Indeed, using numerical integrations over 2×10^6 yr, *Christou* (2000b) found four asteroids temporarily trapped in coorbital configurations with (1) Ceres and (4) Vesta. As with all chaotic orbits, it is important to note that many numerical integrations of "clone" asteroids have had to be carried out before definite statements can be made regarding the exact nature and timings of the orbital evolution.

4. ORIGIN AND EVOLUTION

4.1. Jupiter Trojans

A natural question about Jupiter Trojans arises: Did they form where they are now? If this were the case, they would be the remnant of the planetesimal population that populated the feeding zone of the proto-Jupiter embryo, and their composition would give important clues on the interior of Jupiter. If the Trojan precursors were indeed planetesimals orbiting near Jupiter, they must have been trapped before the planet reached its final mass, since a fully formed Jupiter would have cleared up the region around its orbit on a very short timescale. There are various ideas on how planetesimals could have been captured as Trojans in the early phases of Jupiter growth, based on different physical pro-

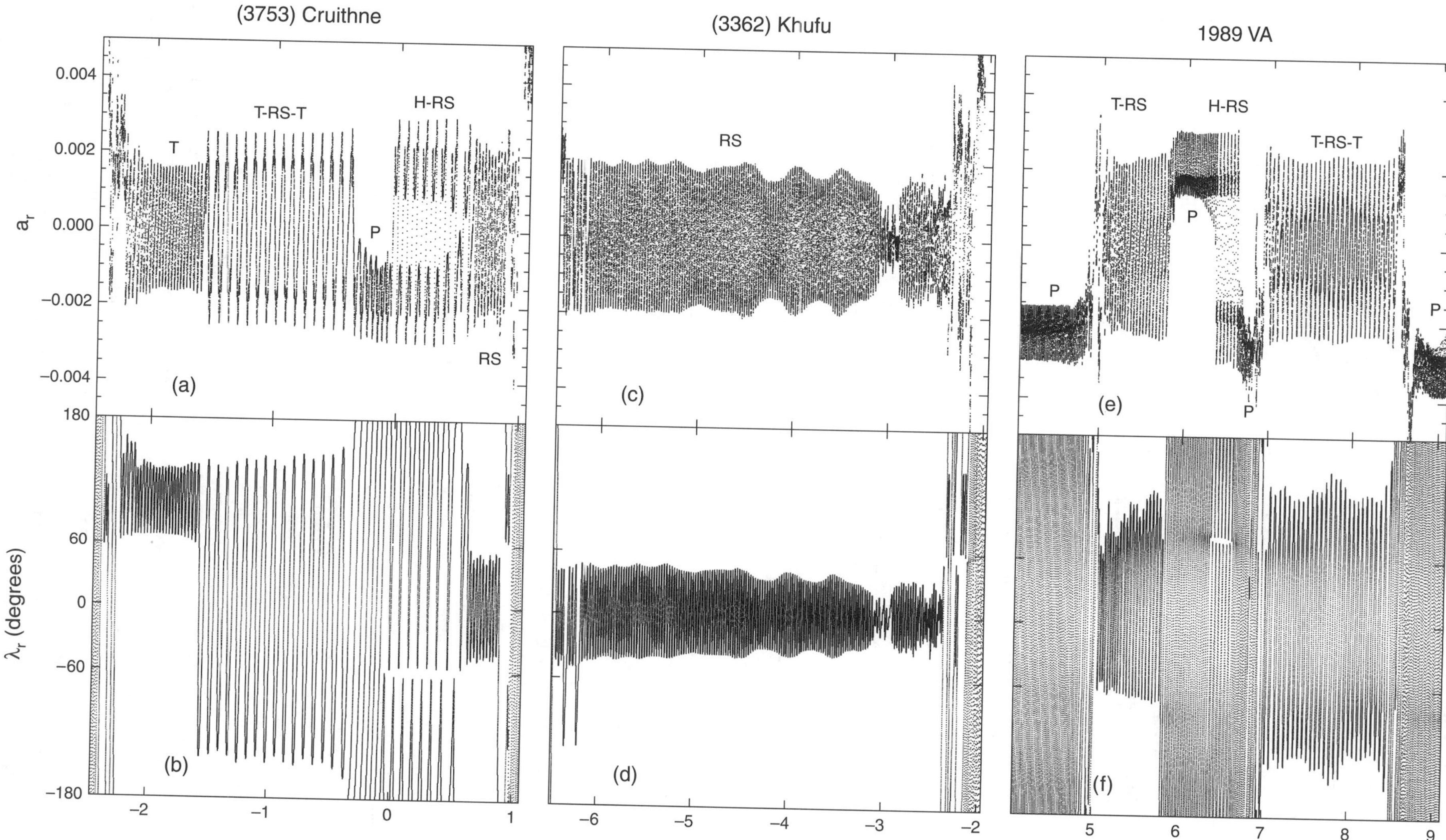

Fig. 4. Plots of relative semimajor axis a_r and relative longitude λ_r (see text) as a function of time for the asteroids (3753) Cruithne, (3362) Khufu, and 1989 VA. The letters in the a_r plots indicate the type of coorbital motion that the asteroid is exhibiting. (3753) Cruithne and (3362) Khufu have coorbital interactions with Earth, while 1989 VA has interactions with Venus (courtesy of *Namouni et al.,* 1999).

cesses. *Shoemaker et al.* (1989) proposed that mutual collisions between planetesimals populating the region around Jupiter orbit might have injected collisional fragments into Trojan orbits. *Yoder* (1979), *Peale* (1993), and *Kary and Lissauer* (1995) showed that nebular gas drag could have caused the drift of small planetesimals into the resonance gap, where they could have grown by mutual collisions to their present size. The mass growth of Jupiter is also an efficient mechanism for trapping planetesimals into stable Trojan orbits, as shown by *Marzari and Scholl* (1998a,b). These mechanisms are by no means mutually exclusive, and each of them may have contributed in a synergistic manner to create the observed population.

There are also alternative ideas on the origin of Jupiter Trojans not related to early stages of the planet's formation. *Rabe* (1972) suggested that Trojans might be comets captured throughout the history of the solar system, while again *Rabe* (1954) and *Yoder* (1979) proposed that Trojans might be fragments of jovian satellites that leaked through the interior Lagrangian point. It seems implausible that these two mechanisms could have contributed significantly to the large populations of L_4 and L_5 Trojans. The origin of Trojan asteroids therefore appears to be more consistent with an early trapping of planetesimals during the formation of Jupiter.

Considerable quantitative work has been done to model the process by which planets like Jupiter may have formed first by accumulation of planetesimals and then by gas infall. *Pollack et al.* (1996) showed that Jupiter and Saturn could have formed on a timescale from 1 to 10 m.y., assuming an initial surface density of planetesimals $\sigma = 10$ g cm^{-2} in the region of the solar nebula where Jupiter formed. This value is about 4× that of the minimum-mass solar nebula proposed by *Hayashi et al.* (1985). According to their numerical simulations, in the final phase of its growth the planet increases its mass from a few tens of Earth mass to its present mass on a very short timescale, on the order of some 10^5 yr.

In this scenario, the capture of planetesimals into Trojan-type orbits by the mass growth of the planet described by *Marzari and Scholl* (1998a,b), and later revisited by *Fleming and Hamilton* (2000), is the most efficient mechanism for explaining the origin of Trojan asteroids. Unaccreted planetesimals within the feeding zone of the growing planet can be captured by the fast expansion of the libration regions that encompass the L_4 and L_5 Lagrangian points, where the expansion is caused by the increasing gravity field of the planet. Figure 5 illustrates the outcome of a numerical integration (*Marzari and Scholl,* 1998a) in which a planetesimal initially on a horseshoe orbit transits to a tadpole orbit when the mass of Jupiter grows at an exponential rate with a characteristic time of 10^5 yr from 10 $M_\oplus$ to its present mass. A similar behavior is also observed for planetesimals orbiting near the planet and not necessarily in a horseshoe orbit. Hence, during the formation of Jupiter, a large number of planetesimals were presumably captured around the planet's Lagrangian points and the symmetrical nature of the tadpole solutions should have led to approximately equal size populations around L_4 and L_5. The symmetry of the situation is, however, broken for small planetesimals if we consider the effects of the nebular gas drag. *Peale* (1993) showed by numerical simulation that the orbits around L_5 are more stable than those around L_4 for bodies smaller than ~100 m in diameter when Jupiter is on an eccentric orbit. *Marzari and Scholl* (1998a) found a similar effect on the trapping probability and explained this phenomenon by introducing in the equation for the Trojan motion derived by *Érdi* (1978, 1979, 1981) a term simulating the perturbation due to gas drag. This primordial asymmetry between the small size end of the two swarms might have been slowly erased by the subsequent collisional evolution of Trojan asteroids (*Marzari et al.,* 1997) so that at present it would be nearly undetectable.

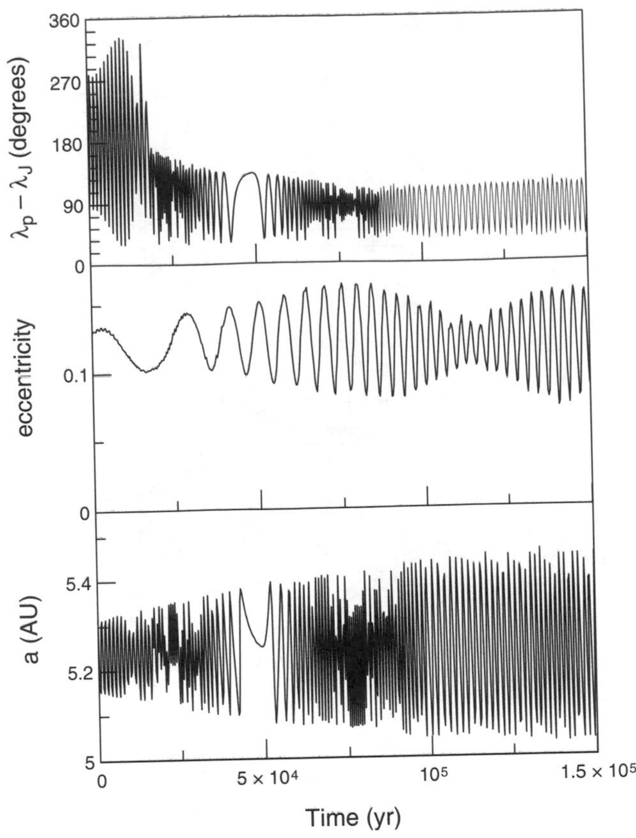

Fig. 5. Trapping of a planetesimal into a Trojan-type orbit by the mass growth of Jupiter. The planetesimal, originally in a horseshoe orbit with the protoplanet, starts to librate around L_4 as the mass of the planet grows. After the capture, the libration amplitude of the critical argument of the Trojan resonance $\lambda_p - \lambda_J$ (top plot), with λ_p and λ_J mean longitudes of the planetesimal and of Jupiter respectively, is further reduced by the increase in mass of the planet. The oscillation of the semimajor axis (bottom plot) widens as D decreases, while the average eccentricity is practically unchanged (middle plot).

Once trapped as a Trojan, the libration amplitude of a planetesimal continues to shrink under the influence of the growing Jupiter, as can be seen in Fig. 5. *Marzari and Scholl* (1998b) have shown numerically how Jupiter's growth from a mass of about 10 $M_\oplus$ to its present mass would cause a reduction of the libration amplitude of a Trojan orbit to ~40% of its original size. This behavior, independent of the capture scenario, is relevant since it leads to a substantial stabilization of Trojan orbits when the planet gains mass. *Fleming and Hamilton* (2000) confirmed the results of *Marzari and Scholl* (1998b) and gave an analytical interpretation of this mechanism on the basis of a simplified Hamiltonian approach that describes the basic librational motion of a body in a 1:1 resonance. They derived a relation between the mass M_J, semimajor axis a_j of the planet, and the libration amplitude A of a Trojan orbit, defined as the difference between the maximum and minimum values of the critical argument, when the mass of the planet M_J changes sufficiently slowly ("adiabatically")

$$\frac{A_f}{A_i} = \left(\frac{M_{Jf}}{M_{Ji}}\right)^{-1/4} \left(\frac{a_{Jf}}{a_{Ji}}\right)^{-1/4} \tag{1}$$

The subscripts i and f indicate the initial and final value of the variables. When i corresponds to the stage with a proto-Jupiter of 10 $M_\oplus$ and j to the final stage with Jupiter fully formed, this equation predicts a reduction of ~42% between A_f and A_i in agreement with the numerical results. According to the above equation, radial migration of the planet toward the Sun caused by tidal interactions with the gas and planetesimal disk and, to a lesser extent, by the mass-accretion process, would not significantly affect the stability of Trojan orbits. A relative change of 20% in the semimajor axis of Jupiter (i.e., ~1 AU) would imply that the libration amplitude A is increased by only ~4%. However, Jupiter and Saturn during their migration may have crossed temporary orbital configurations, where the combination of gravitational perturbations can destabilize Trojan-type orbits of either planet as argued by *Gomes* (1998). In particular, mean-motion resonances between the two planets are very effective and can cause a strong instability of tadpole orbits (*Michtchenko and Ferraz-Mello,* 2001).

The capture efficiency of the mass growth mechanism is high: Between 40% and 50% of the planetesimals populating a ring extending 0.4 AU around Jupiter's orbit are trapped as new Trojans during the formation of Jupiter (*Marzari and Scholl,* 1998a). The total mass of Trojan precursors, assuming a surface density of planetesimals at Jupiter's orbit of $\sigma = 10$ g cm^{-2} as in *Pollack et al.* (1996), should be about 5 $M_\oplus$, much larger than the estimated present mass of ~10^{-4} $M_\oplus$ (*Jewitt et al.,* 2000). However, a simple estimate based on the value of σ does not take into account the density decrease in the planet feeding zone by accretion onto the core and by gravitational scattering of neighboring planetesimals by the protoplanet. This might have opened a gap around the protoplanet orbit in spite of the supply of new planetesimals due to (1) the expansion of the feeding zone caused by the planet growth, (2) gas drag orbital decay of small planetesimals into the feeding zone, and (3) collisional injection of planetesimals close to the borders of the feeding zone. A further significant reduction of the Trojan population occurred throughout the history of the solar system because of collisional evolution (*Marzari et al.,* 1997; *Davis et al.,* 2002) and dynamical outflow (*Levison et al.,* 1997). The latter loss mechanism might have been particularly effective if most of the Trojans captured by the mass growth mechanism had libration amplitudes extending beyond the stability zone.

This seems to be the case of the Trojan precursors, most of which are trapped by the mass growth mechanism in tadpole orbits with large libration amplitudes, in spite of postcapture shrinkage due to mass growth. A process that can reconcile the present population with a primordial population of large librators is, again, the collisional evolution. With a Monte Carlo method, *Marzari and Scholl* (1998b) showed that collisions, possibly with other local planetesimals before they were scattered by Jupiter, can alter significantly the distribution of the libration amplitudes by injecting initially large librators into more stable, small librating orbits. All the large librators would subsequently become unstable and would be ejected from the Trojan swarms, eventually becoming short-period comets (*Marzari et al.,* 1995, 1997; *Levison et al.,* 1997).

The effects of Jupiter's mass growth on the eccentricities and inclinations of the Trojan precursors are essentially negligible (*Marzari and Scholl,* 1998a,b; *Fleming and Hamilton,* 2000). Hence, the present orbital distribution of the two swarms should reflect the original distribution of the planetesimals in the proto-Jupiter accretion zone. Large eccentricities are expected, since a massive protoplanet would stir up the random velocities of the neighboring planetesimals. However, it is difficult to explain the high orbital inclinations observed among Jupiter Trojans. Three mechanisms have been devised that might possibly have excited high inclinations. The first is related to the capture process: Planetesimals with large eccentricity and trapped in a Kozai resonance with the protoplanet can be captured by the mass growth mechanism while they are in the high-inclination stage (*Marzari and Scholl,* 1998b). This dynamical path can explain inclinations up to 10° assuming reasonable values of orbital eccentricities for the planetesimals. The second mèchanism is described in *Marzari and Scholl* (2000) and is based on the synergy between the ν_{16} secular resonance and collisions. The resonance would have excited Trojans with a libration amplitude of ~60° to high-inclination orbits, as suggested for the first time by *Yoder* (1979). Subsequently, collisions would have injected a fraction of these bodies into low-libration orbits. Recently, *Marzari and Scholl* (2000) have demonstrated that ν_{16} not only covers a significant portion of the phase space populated by Jupiter Trojans, larger than previously expected, but that it also extends its effects to initially low-inclination orbits. Figure 6 shows the evolution of the inclination, the critical angle of the Trojan resonance $\lambda_T - \lambda_J$, and the critical argument of the ν_{16} secular resonance of an L_4 Trojan orbit with a starting inclination of 4°. After frequent resonance crossings, the asteroid reaches a final inclination of ~30°. The third hypothesis is suggested by the work of *Petit et al.* (1999) on the excitation of the asteroid belt by large

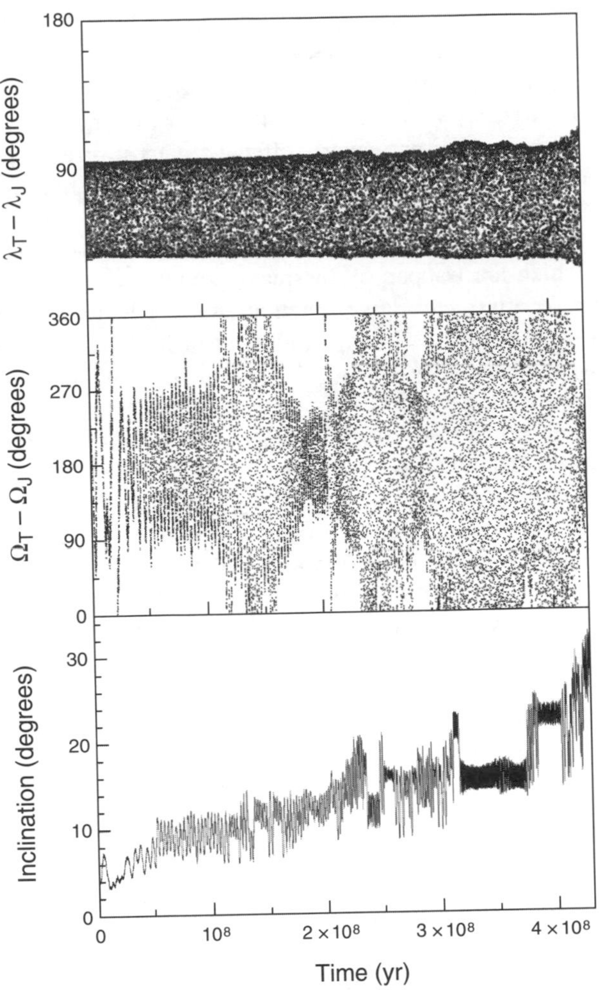

Fig. 6. Evolution of a Trojan orbit trapped in the ν_{16} secular resonance. The initial inclination of the body is ~4° and the libration amplitude D is ~60°. The body has the inclination slowly pumped up by the resonant perturbations and it ends after 400 m.y. in a high-inclination orbit (~30°).

Uranus-Neptune scattered planetesimals. If large embryos, locally grown or coming from outside regions, were trapped in tadpole orbits in the final stage of the planet formation, they would have induced large inclinations by gravitationally scattering smaller Trojans, and, at the same time, would have possibly reduced the large populations trapped during the mass growth of the planet. The pitfall of this model is that it requires two almost-equal embryos to be trapped (one at L_4 and one at L_5), surviving on tadpole orbits for approximately the same time to produce equally stirred populations.

4.2. Saturn Trojans

The stages envisaged for the growth of Jupiter were presumably reproduced during the formation of Saturn, and, as a consequence, the planet should also have trapped local planetesimals as Trojans. However, even high-resolution surveys (*Chen et al.,* 1997) have so far failed to detect any Saturn Trojan. A critical aspect of the Trojan-type orbits of Saturn is that they are mostly unstable. Early numerical experiments by *Everhart* (1973) suggested lifetimes for Saturn Trojans longer than 1 m.y. Subsequent numerical integrations by *Holman and Wisdom* (1993) over timescales of 20 m.y. found only small stability areas in the phase space characterized by large libration amplitudes. A more recent work by *de la Barre et al.* (1996) showed that on longer timespans (400 m.y.), only a few orbits remain stable, and that these are in a peculiar ω-librating state. The cause of instability was attributed by many authors to the close 5:2 resonance with Jupiter known as the "great inequality" (*Innanen and Mikkola,* 1989; *Mikkola and Innanen,* 1992; *de la Barre et al.,* 1996). By investigating the behavior of the critical argument of the most-relevant secular resonances, *Marzari and Scholl* (2000) found that the mixed secular resonance $2\tilde{\omega}_S - \tilde{\omega}_J - \tilde{\omega}_T$, where $\tilde{\omega}_S$, $\tilde{\omega}_J$, and $\tilde{\omega}_T$ are the longitudes of perihelion of Saturn, Jupiter, and

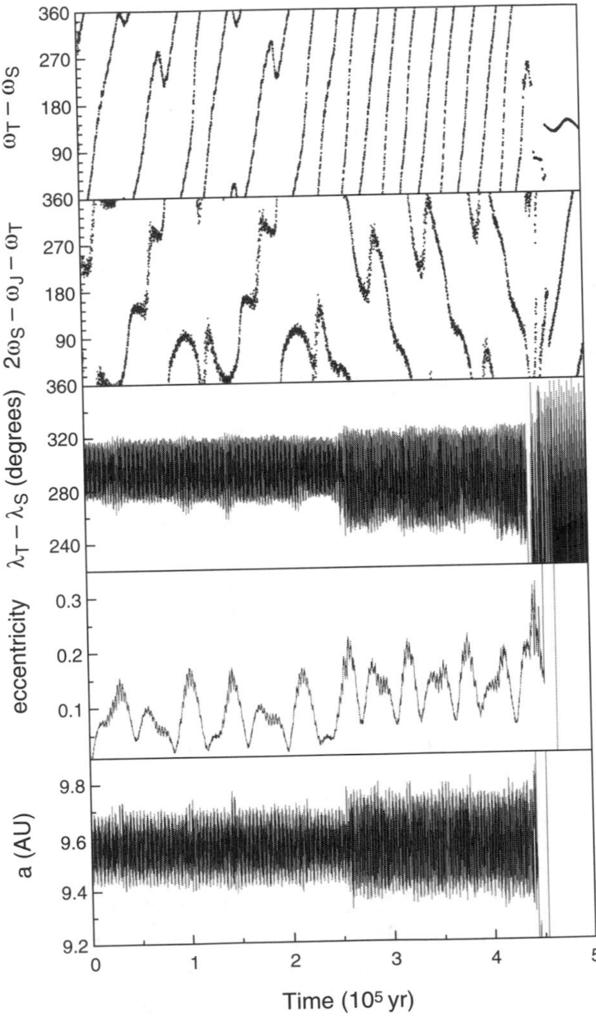

Fig. 7. Instability of a Saturn Trojan orbit caused by the crossing of the mixed secular resonance $2\tilde{\omega}_S - \tilde{\omega}_J - \tilde{\omega}_T$. The crossing occurs at about 2.5×10^5 yr and it is marked by the reversion of the circulation trend of the critical argument and by a sudden jump in both the semimajor axis oscillation amplitude and libration amplitude D.

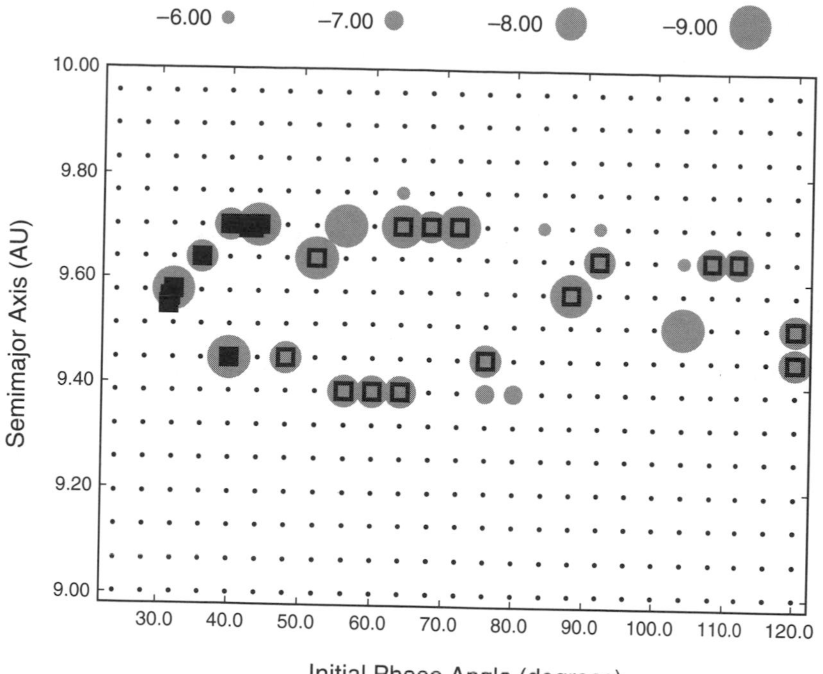

Fig. 8. Stability niches for Saturn Trojans adapted from *Melita and Brunini* (2001). In the plane of the initial semimajor axis and phase angle $\lambda_T - \lambda_S$, the stable niches are shown as filled circles with radius proportional to the diffusion coefficient D that is an indicator of the decimal logarithm of the inverse of the dynamical lifetime of the orbit. Solid squares indicate bodies whose orbits have been integrated with starting conditions within the niche and that are stable over 1 b.y. Blank squares are orbits with stability timescales less than 1 b.y.

the Trojan respectively, also has a dominant role in destabilizing Saturn Trojans. Figure 7 shows the evolution of a test Saturn Trojan belonging to the L_5 swarm. After 2.5 × 10^5 yr, the body crosses the mixed secular resonance and becomes unstable until it is ejected by a close encounter with the planet. The resonance crossing is clearly marked by the change of the circulation direction and a corresponding step in the libration amplitude, semimajor axis oscillation amplitude, and eccentricity. The ν_6 secular resonance seems also to be involved in the escape of possible Saturn Trojans from the 1:1 resonance. The most appealing scenario of the Saturn Trojan evolution is that, after capture by the mass growth mechanism, the low-amplitude librators ($A < 50°$), if any, escaped on a short timescale due to the perturbing effects of the 5:2 near-resonance. On longer timescales, the mixed secular resonance and, eventually, the ν_6 secular resonance depleted the regions at high-libration amplitudes, leaving only a few bodies on the small stability niches found in the phase space by *Melita and Brunini* (2001) from an analysis of diffusion maps of the proper elements (Fig. 8). Their results are limited to orbits initially close to the orbital plane of Jupiter and should be extended to higher inclinations. There might, however, be an upper limit to the possible inclination of Saturn Trojans, according to *Teger* (1999): Beyond 30° no orbit would be stable. However, the work by Teger is based on a few short-term numerical integrations and needs to be confirmed by further simulations.

4.3. Other Trojans

More uncertain is the possibility that planetesimals were trapped as Trojans during the growth of both Uranus and Neptune. According to *Holman and Wisdom* (1993), the two planets have stable regions around both the equilateral points where Trojans could have been captured, but their study was limited in time (20 m.y.). *Weissman and Levison* (1997) performed a numerical exploration of the Trojan orbits for Neptune over 4 b.y., the age of the solar system, and found a small stability region at libration amplitudes lower than 60° and proper eccentricities smaller than 0.05. Their study is, however, limited to 70 test bodies and the timestep of the numerical integration was 1 yr, long compared to the orbital period of Jupiter. Additional tests are needed to assess the possible existence of long-living Trojan orbits at the Lagrangian points of Uranus and Neptune. Even the formation process of the two smaller ice giants is not well understood at present. It has been proposed that the proto-Uranus and proto-Neptune formed via the core-accretion model on a timescale of ~10^7 yr (*Bryden et al.,* 2000). Close encounters between leftover planetesimals and the core would have led to orbital drift and migration of the two planets into outer orbits. In this scenario, the capture of local planetesimals as Trojans may have occurred in the final phase of growth of the two planets, but it is crucial to understand whether Trojan orbits would have been stable during the subsequent phase of orbital migration. *Gomes* (1998) and *Fleming and Hamilton* (2000) have shown that a low or moderate migration would not remove the Trojan populations of the two planets.

According to a recent theory by *Thommes et al.* (1999), Uranus and Neptune would not have formed in their present location but in the region between Jupiter and Saturn at the same time as the two giant planets. When Jupiter and, maybe, Saturn accreted the nebular gas, proto-Uranus and proto-Neptune would have been gravitationally scattered outward. The subsequent evolution caused by gravitational interactions with the planetesimals of the outer disk of the protosolar nebula would have pushed them to their present orbits. In this context, primordial Trojans could not have survived the phase of scattering by Jupiter.

Surprisingly enough, the only known Trojan asteroids in the solar systems not related to Jupiter are martian Trojans. (5261) Eureka, 1998 VF31, 2001 DH47, 2001 FG24, and 2001 FR127 orbit around the L_5 Lagrangian point of Mars, while 1997 U37 orbits around L_4. The stability of the orbits of martian Trojans has been investigated by *Mikkola and Innanen* (1994) and *Tabachnik and Evans* (1999, 2000). Via numerical integrations, they showed that there are stable zones located at inclinations between approximately 15° and 40° with respect to Mars' orbit and at libration amplitudes between 0° and 45°. According to *Mikkola and Innanen* (1994) and *Tabachnik and Evans* (2000), the destabilization of low-inclination martian Trojans is mainly due to a secular resonance with the perihelion of Jupiter $3\Omega_T + 2\tilde{\omega}_T - \tilde{\omega}_J$, with the subscripts T and J referring to Trojan and Jupiter respectively, while additional instability for inclinations around 30° is caused by the secular resonance $\tilde{\omega}_T - \tilde{\omega}_J$. Unfortunately, the numerical investigations performed so far only cover timescales up to some 10^7 yr, since a short timestep must be used in the integration to follow accurately the orbits of the terrestrial planets. This timespan is too short to allow even speculations on the stability of martian Trojans over the age of the solar system. Certainly, if they are primordial, they were not trapped by the mass growth mechanism, since the mass-accretion rate of Mars was very slow. It is possible that they were injected into Trojan orbits by collisions in the highly violent environment that accompanied the final stage of planetary formation. A giant impact on Mars might have created a ring of fragments surrounding the orbit of the planet, from which a few were captured as Trojans. Alternatively, they may have been trapped in temporary tadpole orbits in relatively recent times, and it is possible they will become unstable within timescales of some million years. Even for Earth and Venus, *Tabachnik and Evans* (2000) found rich families of stable tadpole orbits. Trojan orbits of Venus with mean eccentricity ~0.03 and inclinations lower than 16° survive for about 25 m.y., while in the case of Earth there appear to be two extended bands of stability over 50 m.y.: for inclinations lower than 16°, as for Venus, and for inclinations between 24° and 34°. Mercury seems to be unable to retain any Trojan orbit for more than 100 m.y.

Many of the resonant structures we find in our solar system are shared in miniature by the Saturn system. Among the satellites, there are two examples of Trojan configurations discovered during the Pioneer 11 and Voyager 1 and 2 flybys to Saturn: Helene shares the orbit of Dione 60° ahead of its larger companion, while Telesto and Calypso are the leading and trailing Trojans of Tethys respectively. What is the origin of these Trojan satellites? It is believed (*Greenberg,* 1984; *Pollack and Consolmagno,* 1984) that during the early history of Saturn the inner satellites were involved in repeated collisional disruption and reaccretion. After the breakup of a satellite, a ring of debris is formed around the orbit of the largest remnant. Most of the debris reaccumulates onto it on timescales of a few thousand years (*Burns et al.,* 1984; *Stevenson et al.,* 1986; *Marzari et al.,* 1998), but some large fragments may be trapped in Trojan orbits and protected from collisions with the larger coorbiting body.

5. OPEN PROBLEMS AND PERSPECTIVES

Future observational surveys should concentrate on the regions around the Lagrangian points of Saturn, Uranus, and Neptune, and of Earth and Venus. Dynamical models do not exclude the existence of Trojan swarms for these planets. Detection or evidence of absence of Trojans around Uranus and Neptune may contribute to the still-open problem of planet formation and the Trojan capture process. A collisional origin seems to be the most appealing interpretation of the martian Trojans, although a temporary capture event may be another possible explanation.

Detection of smaller objects in the Jupiter Trojan clouds is necessary to lower the limit of completeness, which is at present around 70–80 km in diameter. It would then be possible to confirm the existence of families in the Trojan clouds (*Milani,* 1993; *Beaugè and Roig,* 2001) and to estimate their size distribution, giving stringent constraints on the collisional evolution of the entire Trojan population. The morphology of the families may also shed new light on the physical properties of the Trojans. Comparing kilometer-sized Trojans at L_4 and L_5 might reveal the asymmetry predicted theoretically in presence of gas drag during capture or caused by planetary migration.

The origin of the comparatively high inclination of the Trojans is another open problem presumably related to the cosmogony of the planetary system. The small libration amplitudes of Trojans are an enigma in the context of the capture hypothesis by a growing Jupiter. Further investigation on the effects of collisions among Trojans during Jupiter growth might help to understand this problem.

The stability of Trojans around all planets and satellites over the age of the solar system is not yet fully explored. Jupiter Trojans are the most intensively studied population, with integrations covering at least 10^9 yr. Similar long-term simulations covering a wide range of orbital parameters are still missing for the other outer planets. Furthermore, additional progress on our understanding of the long-term stability for the tadpole orbits of the terrestrial planets is needed to complete our general view of the Trojan behavior.

REFERENCES

Beaugè C. and Roig F. (2001) A semi-analytical model for the motion of the Trojan asteroids: Proper elements and families. *Icarus, 153,* 391–415.

Binzel R. P. and Sauter L. M. (1992) Trojan, Hilda, and Cybele asteroids: New lightcurve observations and analysis. *Icarus, 95,* 222–238.

Brown E. W. and Shook C. A. (1993) *Planetary Theory.* Cambridge University, Cambridge. 302 pp.

Bryden G., Lin D. N. C., and Ida S. (2000) Protoplanetary formation: I. Neptune. *Astrophys. J., 544,* 481–495.

Burns J. A., Showalter M. R., and Morfill G. E. (1984) The ethereal rings of Jupiter and Saturn. In *Planetary Rings* (R. Greenberg and A. Brahic, eds.), pp. 200–272. Univ. of Arizona, Tucson.

Chen J., Jewitt D., Trujillo C., and Luu J. X. (1997) Mauna Kea Trojan survey and statistical studies of L4 Trojans. *Bull. Am. Astron. Soc., 29,* 1020–1021.

Christou A. A. (2000a) A numerical survey of transient coorbitals of the terrestrial planets. *Icarus, 144,* 1–20.

Christou A. A. (2000b) Coorbital objects in the main asteroid belt. *Astron. Astrophys., 356,* L71–L74.

Danby J. M. A. (1964) Stability of the triangular points in the elliptic restricted problem of three bodies. *Astron. J., 69,* 165–172.

David D. R., Durda D. D., Marzari F., Campo Bagatin A., and Gil-Hutton R. (2002) Collisional evolution of small-body populations. In *Asteroids III* (W. F. Bottke Jr. et al., eds.), this volume. Univ. of Arizona, Tucson.

de La Barre C. M., Kaula, W. M., and Varadi F. (1996) A study of orbits near Saturn's triangular Lagrangian points. *Icarus, 121,* 88–113.

Dohnanyi J. W. (1969) Collisional models of asteroids and their debris. *J. Geophys. Res., 74,* 2531–2554.

Érdi B. (1978) The three-dimensional motion of Trojan asteroids. *Cel. Mech., 18,* 141–161.

Érdi B. (1979) The motion of the perihelion of Trojan asteroids. *Cel. Mech., 20,* 59–67.

Érdi B. (1981) The perturbations of the orbital elements of Trojan asteroids. *Cel. Mech., 24,* 377–390.

Érdi B. (1984) Critical inclination of Trojan asteroids. *Cel. Mech., 34,* 435–441.

Érdi B. (1988) Long periodic perturbations of Trojan asteroids. *Cel. Mech., 43,* 303–308.

Everhart E. (1973) Horseshoe and Trojans orbits associated with Jupiter and Saturn. *Astron. J., 78,* 316–328.

Fleming H. J. and Hamilton D. P. (2000) On the origin of the Trojan asteroids: effects of Jupiter's mass accretion and radial migration. *Icarus, 148,* 479–493.

Georgilli A. and Skokos C. (1997) On the stability of Trojan asteroids. *Astron. Astrophys., 317,* 254–261.

Gomes R. S. (1998) Dynamical effects of planetary migration on primordial Trojan-type asteroids. *Astron. J., 116,* 2590–2597.

Greenberg R. (1984) Orbital resonances among Saturn's satellites. In *Saturn* (T. Gehrels and M. S. Matthews, eds.), pp. 593–608. Univ. of Arizona, Tucson.

Hayashi C., Nakazawa K., and Nakagawa Y. (1985) Formation of the solar system. In *Protostars and Planets II* (D. C. Black and M. S. Matthews, eds.), pp. 1100–1154. Univ. of Arizona, Tucson.

Holman M. J. and Wisdom J. (1993) Dynamical stability in the outer solar system and the delivery of short period comets. *Astron. J., 105,* 1987–1999.

Innanen K. A. and Mikkola S. (1989) Studies on solar system dynamics: I. The stability of saturnian Trojans. *Astron. J., 97,* 900–908.

Ivezić Ž. and 31 colleagues (2001) Solar system objects observed in the Sloan digital sky survey commissioning data. *Astron. J., 122,* 2749–2784.

Jedicke R., Larsen J., and Spahr T. (2002) Observational selection effects in asteroid surveys and estimates of asteroid population sizes. In *Asteroids III* (W. F. Bottke Jr. et al., eds.), this volume. Univ. of Arizona, Tucson.

Jewitt D. C., Trujillo C. A., and Luu J. X. (2000) Population and size distribution of small jovian Trojan asteroids. *Astron. J., 120,* 1140–1147.

Kary D. M. and Lissauer J. J. (1995) Nebular gas drag and planetary accretion: II. Planet on an eccentric orbit. *Icarus, 117,* 1–24.

Kokubo E. and Ida S. (2000) Formation of protoplanets from planetesimals in the solar nebula. *Icarus, 143,* 15–27.

Lagerkvist C.-I., Mottola S., Carsenty U., Hahn G., Doppler A., and Gnädig A. (2000) The Uppsala-DLR Trojan survey of the preceding Lagrangian cloud. In *Minor Bodies in the Outer Solar System* (A. Fitzsimmons et al., eds.), pp. 179–185. Springer-Verlag, Berlin.

Lagrange J.-L. (1772) Essai sur le Problème des Trois Corps. *Prix de l'Académie Royale des Sciences de Paris, tome IX*; reprinted in (1873) *Oeuvres de Lagrange, Tome sixième* (J.-A. Serret, ed.), pp. 229–331. Gauthiers-Villars, Paris.

Levison H., Shoemaker E. M., and Shoemaker C. S. (1997) Dynamical evolution of Jupiter's Trojan asteroids. *Nature, 385,* 42–44.

Marchal C. (1990) *The Three-Body Problem.* Elsevier, Amsterdam. 576 pp.

Marzari F., Farinella P., and Vanzani V. (1995) Are Trojan collisional families a source for short-period comets? *Astron. Astrophys., 299,* 267.

Marzari F., Farinella P., Davis D. R., Scholl H., and Campo Bagatin A. (1997) Collisional evolution of Trojan asteroids. *Icarus, 125,* 39–49.

Marzari F. and Scholl H. (1998a) Capture of Trojans by a growing proto-Jupiter. *Icarus, 131,* 41–51.

Marzari F. and Scholl H. (1998b) The growth of Jupiter and Saturn and the capture of Trojans. *Astron. Astrophys., 339,* 278–285.

Marzari F., Dotto E., Davis D. R., Weidenschilling S. J., and Vanzani V. (1998) Modelling the disruption and reaccumulation of Miranda. *Astron. Astrophys., 333,* 1082–1091.

Marzari F. and Scholl H. (2000) The role of secular resonances in the history of Trojans. *Icarus, 146,* 232–239.

Melita M. D. and Brunini A. (2001) A possible long-lived asteroid population at the equilateral Lagrangian points of Saturn. *Mon. Not. R. Astron. Soc., 322,* L17–L21.

Michtchenko T. A. and Ferraz-Mello S. (2001) Resonant structure of the outer solar system in the neighbourhood of the planets. *Astron. J., 122,* 474–481.

Mikkola S. and K. Innanen (1994) On the stability of martian Trojans. *Astron. J., 107,* 1879–1884.

Mikkola S. and Innanen K. (1992) A numerical exploration of the evolution of Trojan-type asteroidal orbits. *Astron. J., 104,* 1641–1649.

Milani A. (1993) The Trojan asteroid belt: Proper elements, stability, chaos, and families. *Cel. Mech. Dyn. Astron., 57,* 59–94.

Murray C. D. (1994) Dynamical effects of drag in the circular restricted three-body problem: 1. Location and stability of the Lagrangian equilibrium points. *Icarus, 112,* 465–484.

Murray C. D. and Dermott S. F. (1999) *Solar System Dynamics.* Cambridge Univ., Cambridge. 592 pp.

Namouni F. (1999) Secular interactions of coorbiting objects. *Icarus, 137,* 293–314.

Namouni F., Christou A. A., and Murray C. D. (1999) Coorbital dynamics at large eccentricity and inclination. *Phys. Rev. Lett., 83,* 2506–2509.

Peale S. J. (1993) The effect of the nebula on the Trojan precursors. *Icarus, 106,* 308–322.

Petit J.-M., Morbidelli A., and Valsecchi G. (1999) Large scattered planetesimals and the excitation of small body belts. *Icarus, 141,* 367–387.

Pollack J. B. and Consolmagno G. (1984) Origin and evolution of the Saturn system. In *Saturn* (T. Gehrels and M. S. Matthews, eds.), pp. 811–866. Univ. of Arizona, Tucson.

Pollack J. B., Hubickyj O., Bodenheimer P., Lissauer J. J., Podolak M., and Greenzweig Y. (1996) Formation of the giant planets by concurrent accretion of solids and gas. *Icarus, 124,* 62–85.

Rabe E. (1954) The Trojans as escaped satellites of Jupiter. *Astron. J., 59,* 433.

Rabe E. (1967) Third-order stability of the long-period Trojan librations. *Astron. J., 72,* 10–17.

Rabe E. (1972) Orbital characteristics of comets passing through the 1:1 commensurability with Jupiter. In *The Motion, Evolution of Orbits, and Origin of Comets* (G. A. Chebotarev et al., eds.), p. 55. IAU Symp. 45, Reidel, Dordrecht.

Schubart J. and Bien R. (1987) Trojan asteroids: Relations between dynamical parameters. *Astron. Astrophys., 175,* 299–302.

Shoemaker E. M., Shoemaker C. S., and Wolfe R. F. (1989) Trojan asteroids: Populations, dynamical structure and origin of the L4 and L5 swarms. In *Asteroids II* (R. P. Binzel et al., eds.), pp. 487–523. Univ. of Arizona, Tucson.

Siegel S. and Castellan N. J. (1988) *Nonparametric Statistics for the Behavioral Sciences.* McGraw-Hill Higher Education, New York. 422 pp.

Stevenson D. J., Harris A. W., and Lunine J. I. (1986) Origins of satellites. In *Satellites* (M. A. Burns and M. S. Matthews, eds.), pp. 39–88. Univ. of Arizona, Tucson.

Szebehely V. (1967) *The Theory of Orbits.* Academic, New York. 662 pp.

Tabachnik K. S. and Evans N. W. (1999) Cartography for martian Trojans. *Astrophys. J., 517,* L63–L66.

Tabachnik K. S. and Evans N. W. (2000) Asteroids in the inner solar system: I. Existence. *Mon. Not. R. Astron. Soc., 319,* 63–79.

Tedesco E. F. (1989) Asteroids magnitudes, UBV colors, and IRAS albedos and diameters. In *Asteroids II* (R. P. Binzel et al., eds.), p. 1090–1138. Univ. of Arizona, Tucson.

Teger F. (1999) On the stability of the L4 and L5 triangular Lagrangian points of Saturn. *Bull. Am. Astron. Soc., 31,* 1107.

Thommes E. W., Duncan M. J., and Levison H. F. (1999) The formation of Uranus and Neptune in the Jupiter-Saturn region of the solar system. *Nature, 402,* 635–638.

van Houten C. J., van Houten-Groeneveld I., and Gehrels T. (1970a) Minor planets and related objects V: The density of Trojans near the preceding Lagrangian point. *Astron. J., 75,* 659–662.

van Houten C. J., van Houten-Groeneveld I., Herget P., and Gehrels T. (1970b) The Palomar-Leiden survey of faint minor planets. *Astron. Astrophys. Suppl. Ser., 2,* 339–448.

Yoder C. F. (1979) Notes on the origin of the Trojan asteroids. *Icarus, 40,* 341–344.

Weissman P. R. and Levison H. F. (1997) The population of the trans-neptunian region: The Pluto-Charon environment. In *Pluto and Charon* (S. A. Stern and D. J. Tholen, eds.), p. 559. Univ. of Arizona, Tucson.

Dealing with the Impact Hazard

David Morrison
NASA Astrobiology Institute

Alan W. Harris
NASA Jet Propulsion Laboratory

Geoffrey Sommer
RAND Corporation

Clark R. Chapman
Southwest Research Institute

Andrea Carusi
Istituto di Astrofisica Spaziale, Roma

The small fraction of asteroids with Earth-crossing or Earth-approaching orbits is of special interest to us because many will eventually impact our planet. The time-averaged impact flux as a function of projectile energy can be derived from lunar-cratering statistics, although we have little information on the possible variability of this flux over time. Alternatively, we can use current observations of near-Earth asteroids (NEAs) to derive the size distribution and flux of impactors. The effects of impacts of various energies can be modeled, using data from historic impacts (such as the K/T impactor 65 m.y. ago) and the observed 1994 bombardment of Jupiter by fragments of Comet Shoemaker-Levy 9. Such models confirm that the terrestrial biosphere is highly vulnerable to severe perturbation from impacts, so that even such a small event as the K/T impact (by a projectile 10–15 km in diameter) can lead to a mass extinction. Combining the impact flux with estimates of environmental and ecological effects reveals that the greatest contemporary hazard is associated with impactors near 1,000,000 MT energy. The current impact hazard is significant relative to other natural hazards, and arguments can be developed to illuminate a variety of public-policy issues. These include the relative risk of different impact scenarios and the associated costs and probabilities of success of countermeasures. It is generally agreed that the first step is to survey and catalog the larger NEAs. To that end, we review the status of the Spaceguard Survey, which has already discovered more than half of the NEAs larger than 1-km diameter, out of a total population estimated to be between 1000 and 1200. We compare the efficiency of survey approaches and consider the challenges of international coordination and the problems and opportunities associated with communicating the results with the press and the public. It is also important to reflect on how the impact hazard might be dealt with by both national governments and international decision-making bodies and to anticipate ways of mitigating the danger if a NEA were located on an apparent Earth-impact trajectory. As the most extreme known example of a natural risk with low probability but severe global consequences, the NEA impact hazard calls for the most careful consideration and planning.

1. INTRODUCTION

Among asteroids, those populations that have the potential for impacting Earth have a special status. We generally refer to them as near-Earth asteroids (NEAs), a category that includes many objects (with perihelion out to q = 1.3 AU) not currently on threatening orbits (*Chapman et al.,* 1994; *Rabinowitz et al.,* 1994; *Shoemaker et al.,* 1994). Closer to home are the Earth-crossing asteroids (ECAs) and their subgroup, the potentially hazardous asteroids (PHAs). In this chapter we will generally use the broader term, NEAs, or even near-Earth objects (NEOs), a term that embraces comets as well. Because of their unstable, planet-approaching orbits, NEAs have impacted the surfaces of the inner planets, including Earth, influencing both geological and biological evolution. Since there is reason to expect further impacts in the future, NEAs are a topic with profound political and societal overtones. The impact hazard represents the intersection of asteroid science with public welfare and governmental policy. As Carl Sagan frequently pointed out (e.g., *Sagan,* 1994), the long-term future of human civilization is linked to our ability to understand and ultimately to control the impact environment of our planet.

It is only during the past two decades that scientists have become aware of the scope of the asteroid impact hazard. This topic was broadly reviewed in 1993, leading to publication of a 1000-page book, *Hazards Due to Comets and Asteroids* (*Gehrels,* 1994), that remains the primary reference in this field. With surprising speed, this concern has been communicated to governments and the public (e.g., *Morrison et al.,* 1994). Due to the advocacy of NEA researchers (with timely publicity from the collision of Comet Shoemaker-Levy 9 with Jupiter and two feature movies), policy makers and their constituents have become aware that impacts are possible. It is less clear, however, that decision makers are convinced that any major action needs to be taken to deal with the impact hazard. The advocacy role of the science community is pivotal, because the abstract nature of the low-probability threat diminishes the likelihood of a response by either policy makers or their constituents. In this chapter we discuss both the "facts" of the impact hazard and the associated issues of public perception and governmental response.

Much of the material in this chapter is associated with estimating the frequency of impacts and evaluating their consequences, particularly for Earth's biosphere. In the abstract, the hazard lends itself to such statistical analysis. However, from a policy perspective we do not need precise estimates of either the frequency of impacts or their consequences. We recognize that the actual risk is not statistical; if there is any sizable NEA on a collision course with Earth, it can be found and the impact predicted decades (or more) in advance. If and when this happens, our attention will focus on that particular object and the circumstances of its predicted impact.

At first, there was considerable skepticism toward proposals for a comprehensive survey to identify any potential impactor decades in advance. Perhaps influenced by their experience with antimissile concepts, many members of the U.S. and Russian defense communities proposed various schemes for shooting down incoming asteroids with only a few days, or even a few hours, of warning [e.g., papers from a Los Alamos workshop collected by *Canavan et al.* (1993)]. However, there is no warning system in place or likely to be built that would focus on such a short-term threat. Almost any asteroid that is on an impact trajectory will repeatedly pass close to Earth on previous orbits, with multiple opportunities for detection. An optical survey system has negligible probability of finding an object on its final plunge to Earth, relative to discovery on some previous close pass. The Spaceguard Survey, discussed in detail later in this chapter, is just such a comprehensive optical search, with nearly continuous coverage of the space around Earth to distances of ~10^8 km. Already, we have found and calculated accurate orbits for more than half of the thousand-odd NEAs larger than 1 km. None of these poses any impact threat on the timescale of a human lifetime. On the other hand, it is still impossible to say anything about the orbits of the undiscovered ones. This Spaceguard Survey approach also has limited use against long-period comets. Fortunately, these comets constitute a rather small fraction of the total impact threat, and we generally omit them from consideration in this chapter.

While it is highly improbable that a large (diameter >1 km) NEA will hit Earth within our lifetimes, such an event is entirely possible. In the absence of specific information, such a catastrophe is equally likely at any time, including next year. Society needs to be prepared to deal with this eventuality. In the meantime, however, the search for possible impactors will inevitably lead to false positives, NEAs that appear for some time to be a real threat. We need to consider the effect of such reports on society. As we discuss in the final sections of this chapter, impact-hazard studies can be considered an applied science — that is, science applied to the tangible needs of society. In determining an optimum or even advisable hazard-mitigation strategy, the reaction of society to scientific information should be considered. The NEO community has a social responsibility to ensure that its message is not just heard but comprehended by society at large. Since the hazard knows no national boundaries, it also behooves us to seek solutions that recognize the international constituency with a stake in impact prediction and prevention.

2. IMPACT HISTORY OF EARTH AND THE MOON

Earth and the Moon have shared a history of impact by asteroids and comets. The flux was high during accretionary epochs, but the course of general decline during the subsequent 0.5 b.y. is uncertain. A relatively short (~100 m.y.) period of high flux ended about 3.85 b.y. ago, forming a number of lunar basins (probably accompanied by smaller craters). Since this so-called Late Heavy Bombardment, the impact rate has been much lower and fairly constant, to within factors of a few (at least averaged over intervals of 10^8 yr). The impactors are, by definition, near-Earth objects (NEOs), which are currently dominated by asteroids derived from the main belt. The types of bodies responsible for impacts in the last ~3.5 b.y. have probably not changed appreciably (*Bottke et al.,* 2002). New analyses of lunar samples suggest variations in flux by up to a factor of 4 (*Culler et al.,* 2000), basically a decrease followed by an increase in the last few hundred million years. There is no reliable evidence concerning shorter, sharper spikes in the flux due to hypothesized comet showers or major asteroidal breakups, except that they can have made at most a modest contribution to the cumulative impacts.

Modern recognition that Earth is impacted by asteroids and comets is barely half a century old (cf. *Baldwin,* 1949); even Meteor Crater (Arizona) wasn't definitively proven to be of impact origin until the work of *Shoemaker et al.* (1960). The fortunate fact that the atmosphere protects us from impacting bodies smaller than a few tens of meters in diameter (except for the rare iron meteorites) has the consequence that we have almost no direct experience with cosmic impacts. The idea is still developing that impacts have dramatically shaped our planet's climate and the evolution of life. Whatever will eventually be found to be the comparative dominance of impacts vs. other causes of climate change (e.g., exploding volcanos, ice ages, snowball Earth instabilities), impacts necessarily make progressively less contribution to the forces shaping our environment on the short timescales relevant to humans. This is because impacts are instantaneous events rather than continuously operating processes, and the largest of those likely to have occurred during recorded human history are much smaller than the largest events expected over tens or hundreds of million years.

Impacts large enough to affect dramatically the fragile climate and ecosystems of the whole planet can cause prominent changes in the geologic record of fossilizable species of life. Such mass extinctions, first recognized two centuries ago, are increasingly being explained as a consequence of impacts. Possible connections between impacts and extinctions had been suggested earlier (e.g., *Urey,* 1973), but the first solid evidence came from the work of *Alvarez et al.* (1980), who proposed that the dinosaur-killing K/T mass extinction was due to an impact, as inferred from the chemical signature of extraterrestrial material in the boundary layer at the end of the Cretaceous. This hypothesis was widely discussed and debated, but generally accepted only after identification of the Chicxulub crater (still among the largest identified on Earth) as the "smoking gun" (*Swisher et al.,* 1992; *Sharpton et al.,* 1992). Evidence is increasing that the very largest mass extinction (the Permian/Triassic) was also due to an impact (*Becker et al.,* 2001). Nevertheless, the hypothesis of *Raup* (1990, 1991) that mass extinctions of all magnitudes are predominantly due to impacts has been resisted by many workers for at least two reasons. First, some paleontologists remain skeptical of what seems to astronomers as the inevitability of impacts and their enormous consequences (*Chapman,* 2002); many still consider impact to be an explanation of last resort. Second, the signature Ir-enriched layer found worldwide at the K/T boundary is generally lacking at other extinction boundaries. It is possible that Ir-enriched material in other large impactors was mainly ejected back into space or their composition was less enriched in Pt-group metals than the K/T impactor.

On the timescale of a human lifetime, the 1908 impact in the Tunguska River region of Siberia stands out as the most remarkable. Estimated (from distant barographic records) as having an 10–15 MT TNT equivalent energy when it exploded perhaps 8 km above the ground, the impactor affected an unusually remote part of the world; the first expedition to study Tunguska did not arrive until 1927. The early Russian investigators, failing to find meteoritic fragments at the site, suggested that the culprit was a small comet. This assumption became the accepted wisdom, although *Sekanina* (1983) made an early case for an asteroidal impactor. Not until the 1990s did numerical modeling of the entry physics clearly indicate that a comet (composed of low-density, friable material) with this kinetic energy could not penetrate into the troposphere, while a rocky (asteroidal) projectile could (*Chyba et al.,* 1993; *Hills and Goda,* 1993; *Chyba,* 1993). Note that by "asteroidal" we mean composed of solid materials of either stony or carbonaceous composition. While we prefer the traditional energy for Tunguska of 10–15 MT, this value is quite uncertain, and modeling by *Boslough and Crawford* (1997) suggests the energy could have been as low as 3 MT.

A simple qualitative argument for the stony (asteroidal) nature of the Tunguska impactor has been given by *Zahnle* (1996). Consider the alternative. If a low-density icy (cometary) projectile in this energy range penetrates into the lower atmosphere, then a stronger rocky object will make it to the surface and produce a crater. Unless there are very few rocky objects in the small NEA population, we should see many of these fresh kilometer-sized craters. But they are clearly not evident on Earth; indeed most of the small young craters (like Meteor Crater) are associated with rare metallic impactors. Thus the common-sense observation that there are few small craters supports the numerical models that indicate that the Tunguska projectile was a rocky NEA and not a comet. If the energy is as low as 3 MT, this qualitative argument is even stronger. Although one still sees occasional statements in the literature that the Tunguska impactor was cometary, the asteroidal hypothesis is generally accepted (e.g., *Vasilyev,* 1998; *Sekanina,* 1998; *Farinella et al.,* 2001).

Impacts or impact scars that most people can personally witness or study are unrepresentative of the sort of impact that could, with very low probability, present a significant hazard to modern life. We have no direct experience with the kind of impact, by an object hundreds of meters or larger, that could cause a regional or global catastrophe. However, there are current data from orbital surveillance systems on impact frequencies for objects a few meters in diameter, helping us to tie down the low-energy end of the population distribution. Although coverage is incomplete and there are substantial uncertainties in the luminous efficiency calibrations, *Tagliaferri et al.* (1994) estimate, from 136 reported atmospheric entries between 1975 and 1992, that the annual maximum impactor has an energy of roughly 10 kT, similar in energy to the Hiroshima atom bomb but about a factor of a 1000 less than Tunguska. We have used data since 1992 in U.S. Air Force press releases to derive similar fluxes. The single largest impact reported in the past 25 yr, which is several tens of kilotons (*McCord et al.,* 1995), is also consistent with these estimates. Thus we know that collisions with Earth continue for small impactors, although we have little direct experience with the large events that dominate the overall impact risk.

3. PHYSICAL AND BIOLOGICAL EFFECTS OF IMPACTS

The most revolutionary insight of the *Alvarez et al.* (1980) paper was that even small impacts (on a geological or astronomical scale) could severely damage the fragile terrestrial ecosystem. The K/T impactor had a mass a billion times less than that of Earth, yet the ensuing extinction fundamentally redirected the course of biological evolution. In the two decades since this discovery, considerable work has been done to understand the mechanisms of mass extinction and to evaluate the ways environmental stress might be affected by the energy of the impact. In this section, we summarize the discussion by *Toon et al.* (1997) of the environmental perturbations due to impacts.

The energy of the K/T impact is estimated at 10^8 MT from the size of the Chicxulub crater, and a consistent value of the size of the impactor (10–15-km diameter) is derived from the observed extraterrestrial component in the boundary layer. Immediate effects of the impact included blast and the generation of tsunami (since the impact occurred in a shallow sea). However, the primary agents of global stress appear to have been a short-lived firestorm from atmospheric heating of reentering ejecta followed by a persistent (months to years) blackout due to particulates suspended in the stratosphere. Other possibly important effects could have included chemical changes in the oceans and atmosphere and large climatic oscillations following the impact.

Toon et al. (1997) review all these environmental effects and their dependence on impact energy. Their chief goal was to provide relatively simple prescriptions for evaluating the importance of impacting objects over a range of energies and compositions. Since mass extinction events such as the K/T impact are rare (intervals of tens to hundreds of million years), we are especially interested in downscaling to determine the thresholds for damage on timescales more relevant to human history (cf. *Toon et al.,* 1994, and *Covey et al.,* 1994, for earlier discussions).

The threshold for atmospheric penetration of impacts, required for the blast effects to reach the ground, is at a few megatons (*Chyba et al.,* 1993; *Hills and Goda,* 1993; *Chyba,* 1993). Below this energy, the atmosphere protects us against all but the rare metallic projectiles. For impacts above this threshold, the primary effects of both airbursts and ground impacts are local blasts and earthquakes, together with the setting of local fires. The Tunguska explosion of an NEA ~60 m in diameter provides a relatively small example. Such impacts cause little harm if they enter over the oceans. However, at sizes of hundreds of meters, oceanic impacts dominate the hazard calculations as a result of impact-induced tsunami (*Hills and Goda,* 1993; *Hills et al.,* 1994; *Hills and Mader,* 1997; *Toon et al.,* 1997; *Crawford,* 1998; *Ward and Asphaug,* 2000). Tsunami waves provide a relatively efficient way to carry the impact energy to large distances. This fact, coupled with concentration of human habitation near the shore, makes tsunami important for energies of tens of thousands of megatons or greater (NEA diameters of hundreds of meters). While there is considerable uncertainty in both the height of the open-ocean wave and the run-up as it reaches the shore, *Toon et al.* (1997) conclude that large tsunami, occurring with average frequency of tens of thousands of years, contribute much more to the hazard than do terrestrial impacts in the same energy range.

The global environmental stress from the K/T event was dominated by a prompt firestorm followed by longer-lasting dust loading of the atmosphere. There is direct evidence in the boundary clay for the soot produced by burning a large fraction of the terrestrial biomass. In addition, analogous effects seen following the impacts of Comet Shoemaker-Levy 9 with Jupiter in July 1994 have been extensively modeled (*Boslough et al.,* 1994; *Zahnle and MacLow,* 1994, 1995). A global firestorm can be ignited by hot debris falling back into the atmosphere on ballistic trajectories from the ejecta plume, as first suggested by *Melosh et al.* (1990). Most of the energy is deposited in the mesosphere (where meteors shine), with radiative heating of the lower atmosphere and surface. Toon et al. conclude that while this mechanism was important in the K/T event, where it was the probable direct cause of the extinction of large land animals such as the dinosaurs, it does not produce surface temperatures high enough for ignition at impact energies below 10^7 MT.

Global darkness from the absorption of sunlight by ejected dust was the prime agent of the K/T extinction as suggested by *Alvarez et al.* (1980). *Toon et al.* (1997), drawing in part on a large literature dealing with volcanic dust and the stratospheric soot from nuclear war, calculated the effects of dust loading on atmospheric circulation under various scenarios. General circulation models (GCMs) permitted them to follow the postimpact development of the suspended dust and calculate the resulting surface temperatures. As we might expect, the results depend in significant ways on the target material (land or ocean) and the season of impact, but less on exact geographic location, since the dust cloud quickly expands to global scales. Since these effects extend down to impacts as small as 10^5 MT, they dominate in determining the threshold for global disaster, defined by *Chapman and Morrison* (1994) as an environmental catastrophe capable of killing 25% of the world's population, primarily from the agricultural losses of an "impact winter." *Toon et al.* (1997) conclude that the energy range between 10^5 and 10^6 MT is transitional between regional and global effects, with a mean value for the threshold of global catastrophe near 10^6 MT energy, corresponding to an NEA diameter of ~2 km.

Pope (2002) has recently questioned the assumptions made by *Toon et al.* (1997) (and others previously) concerning the quantity of submicrometer dust injected into the stratosphere. This fine dust has not been measured directly in the K/T boundary layer, and inferring its quantity indirectly introduces substantial uncertainty into the estimate of the threshold energy for a global catastrophe. This uncertainty could be as great as a factor of ±10 in energy, corresponding to a diameter for an asteroid of 1–4 km.

4. CONTEMPORARY HAZARD OF NEAR-EARTH-ASTEROID IMPACTS

The hazard associated with NEA impacts — that is, the probability for an individual of premature death as a consequence of impact — depends on the frequency of occurrence as well as the destructive effects. Quantitative estimates of this risk were presented in the *NASA Spaceguard Survey Report* (*Morrison,* 1992) and amplified by *Chapman and Morrison* (1994) and *Morrison et al.* (1994). Such estimates are substantially uncertain due to lack of precision (and possible time dependence) in the impact flux as a function of projectile energy, possibly wide variability in the environmental effects that depend on properties of the impactor and target, and especially ignorance of the societal response to disasters of a magnitude never experienced. However, we can estimate the order of magnitude of the risk and its approximate dependence on impact energy, as a guide to possible societal responses and efforts at mitigation.

For these order-of-magnitude estimates, it is sufficient to approximate the frequency distribution of the impactor flux by a simple, time-invariant power law not very different from the form proposed by *Shoemaker* (1983). In the subsequent section we illustrate such a curve (Fig. 3) and examine this assumption in light of current NEA observations, but these considerations do not change the qualitative conclusions of the hazard analysis. This power-law distribution, which is close to the expected mass distribution of an NEA population resulting from fragmentation, is essentially catastrophic in form, in that the rare large impacts dominate the cumulative effects of more frequent smaller impacts. Coupled with the nonlinear response of the environment to impact perturbations discussed above, we conclude (as do *Chapman and Morrison,* 1994) that the greatest risk is associated with large impacts, those that create a global ecological catastrophe.

The threshold impact of ~10^6 MT from *Toon et al.* (1997) is expected to take place roughly twice per 1 m.y. *Chapman and Morrison* (1994) define the threshold as an event that would kill 25% of Earth's population — far less than an "extinction level event" but large enough to rank as the worst catastrophe in human history. Adding the lesser casualties from smaller but more frequent impacts, we estimate very roughly that an average individual on Earth today runs a risk of death from an impact on the order of 1 in 1,000,000 each year. For comparison, this is about the same level of risk associated with one roundtrip commercial air flight per year. Depending on where a person lives, this impact risk may be either higher or lower than the risk of more familiar natural disasters such as earthquakes or flooding.

Although impacts below the threshold for global catastrophe are much more frequent, the total hazard from such impacts is less. At energies between 10^4 and 10^6 MT, the dominant risk is from tsunami created by deep ocean impacts, as discussed in the previous section. From 10 to 10^4 MT, the blast effects for land impacts dominate. The average risk level from tsunami is roughly an order of magnitude lower than that of the threshold global catastrophe, and that of smaller (blast-dominated) land impacts is down another order of magnitude. However, such estimates depend critically on assumptions about human population distribution, warning time, and societal response (e.g., discussions by *Canavan,* 1993, 1994, 1997).

The primary results from this hazard analysis are qualitative but robust. First, in spite of the substantial uncertainties, we have shown that the statistical impact risk is on the same order of magnitude as other more familiar hazards that are taken seriously by both individuals and governments. This conclusion motivates the quest to find out whether there is or is not an object presently on a trajectory that will hit Earth. Second, we find that the larger impacts are more hazardous than the sum of all the smaller ones; we thus focus our attention on asteroids 1 km or larger.

Unlike more familiar hazards, the impact risk is primarily from extremely rare events — literally unprecedented in human history. Although there is a chance on the order of 1 in 1,000,000 that an individual will die in any one year from an impact, it is not the case that 1 out of 1,000,000 people dies each year from an impact. The impact hazard represents the extreme case of a calamity of low probability but high consequences — consequences that include the possible end of civilization as we know it. It is logical to concentrate first on mitigating the risk from global catastrophes, both because they dominate the risk and because catastrophes of this scale threaten global society, making them qualitatively different from any other known hazard. Later, it may be desirable to extend mitigation efforts to smaller impacts that, while they do not threaten society as a whole, are much more likely to happen within our lifetimes. This will be a political decision, of course, related to the value of mitigating the impact hazard relative to other natural hazards (such as earthquakes) that pose comparable local or regional threats.

5. THE SPACEGUARD SURVEY

The first formal proposal for a survey of potentially threatening NEOs was made by the U.S. Congress in 1990. At the request of the U.S. House of Representatives, NASA appointed a study group chaired by David Morrison that met during 1991 to evaluate the impact hazard and propose ways to dramatically increase the detection rate of Earth-crossing objects. That group proposed an international "Spaceguard Survey" to be carried out by groundbased optical telescopes equipped with state-of-the-art widefield detectors and automated search capability (*Morrison,* 1992). The term "Spaceguard" was borrowed (with permission) from Arthur C. Clarke, who had used it in his novel *Rendezvous with Rama* to describe a radar warning system designed to protect Earth from impacts.

In 1994 the U.S. Congress asked the NASA Administrator to submit a plan for locating all NEOs greater than 1 km in diameter (roughly the lower limit to the threshold for global catastrophe). The resulting NASA study chaired by Gene Shoemaker articulated the "Spaceguard goal" of discovering and cataloging at least 90% of all NEAs larger than 1 km in diameter within 10 yr (*Shoemaker,* 1995). A strong rationale was presented that the NEAs with $D > 1$ km are the most dangerous and deserve the highest priority for detection, as discussed in the previous section (the probable range is 1–4 km diameter). However, the 1-km objects specified in the goal can also be thought of as a convenient metric, since an optical survey does not distinguish between small nearby objects and large distant objects in the telescope field of view. While the largest (brightest) objects are the easiest to discover, at no point has anyone suggested "throwing the little ones back," as in fishing. The Spaceguard goal is a metric for assessing progress, not an endpoint after which surveying should cease. As we approach the present goal (which is likely to be reached near 2008, assuming continuing incremental improvements in search systems), it might be well to switch to a new metric (smaller reference diameter for completeness), as has been suggested (for example) in the recommendations of the UK NEO Task Force (*Atkinson et al.,* 2000).

In order to design an optimum search system, it is sensible to simulate discovery efficiency as a function of sky area covered, limiting magnitude, and various other parameters. This was done by Muinonen and Bowell as a part of the *Spaceguard Survey Report* (see also *Bowell and Muinonen,* 1994) and has been extended both for evaluating survey efficiency and for bias-correcting survey discoveries to estimate asteroid populations (*Muinonen,* 1998; *Jedicke et al.,* 2002). *Harris* (1998, 2001) has published the most thorough discussion of such a survey simulation, showing that it is generally better to sacrifice depth of coverage (limiting magnitude) in favor of sky coverage in order to maximize the discovery rate. One gains breadth of coverage inversely proportional to integration time, but one gains depth of coverage only proportional to the square root of integration time. For example, by cutting integration time by one-fourth, 4× the area can be searched to half the depth (in units of intensity). This strategy is of course limited by cycle time (to move the telescope and process the image) and ultimately by the finite area of sky available.

Currently operating surveys cover most of the visible sky each month, with the exception of the southern sky below about –40° declination, so to a good approximation our evaluation can assume "all-sky" coverage. In evaluating completeness of an all-sky survey vs. depth of coverage (limiting magnitude), *Harris* (1998) notes that the two magnitude parameters, the threshold limiting visual magnitude of detection (m_{lim}) of the putative survey system and the absolute magnitude (H) of the asteroid, are nearly 100% correlated with respect to detection probability. This simply means that a survey system with a limiting magnitude $m_{lim} = 20$ will achieve the same completeness of absolute magnitude H = 20 objects as a system with $m_{lim} = 19$ will achieve of H = 19 objects. This simple equivalence allows one to simulate detection efficiency using ($m_{lim} - H$) as the independent variable. A single calculation can thus be done to estimate fraction completeness as a function of ($m_{lim} - H$) and then scale the result for specific values of m_{lim} and/or H.

Figure 1 shows plots of completeness vs. diameter for a 10-yr, all-sky survey with the same "universal completeness curve" scaled for specific values of survey-limiting magnitude. These curves have been modified from those presented by *Harris* (1998) to represent integral completeness (fraction of objects larger than a given size, rather than fraction at a given size), and in units of diameter D rather than absolute magnitude H. An integral-completeness curve depends on a population model, where we have assumed N(>D) proportional to $D^{-2.5}$. The conversion from H to D requires an albedo model, where we have assumed a 50/50 mix of light and dark asteroids (albedos 0.20 and 0.05 respectively). From these plots it appears that achieving the Spaceguard goal should require a 10-yr, all-sky survey to $m_{lim} = 20.2$. This is about half a magnitude fainter than the detection limit of current surveys (*Stokes et al.,* 2002), which are near visual magnitude 19.5–19.7.

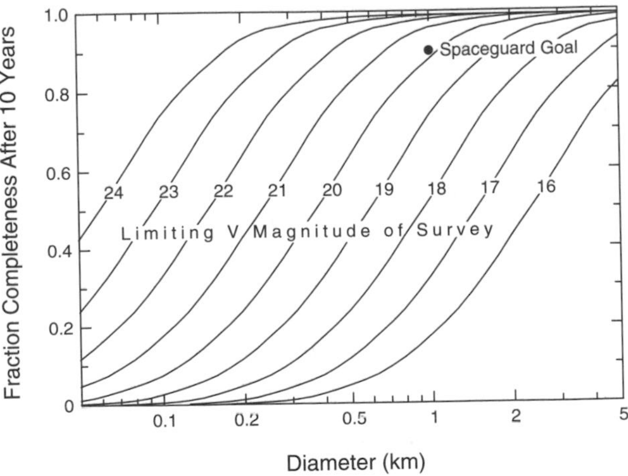

Fig. 1. Completeness vs. diameter for a 10-yr, all-sky survey of NEAs; the model assumptions are described in the text. These plots suggest that achieving the Spaceguard goal of 90% completeness for NEAs larger than 1 km requires a 10-yr survey to limiting visual magnitude of ~20.2.

6. POPULATION AND SURVEY COMPLETENESS

To assess current progress toward achieving the Spaceguard goal, we begin with the number of presently discovered NEAs down to H = 18. As of January 28, 2002, using the discoveries and H magnitudes as listed by the Minor Planet Center, that number is N(H < 18) = 587 (Fig. 2). The total of NEAs of all sizes on this date was 1743. During 2001, more than 100 NEAs were discovered brighter than H = 18. The discoveries have been dominated by the LINEAR system of two telescopes (*Stokes et al.,* 2000), which have accounted for more than 75% of all discoveries since 1999. Because of the steady improvement in detection efficiency, the discovery rate of NEAs brighter than H = 18 has not yet shown the dropoff that would be expected as we pass the halfway mark in meeting the Spaceguard goal.

It is a more difficult task to determine the total population of NEAs in order to obtain a measure of the completeness of the present survey. Figure 3 is a plot of several estimates of the cumulative population N(<H) vs. absolute magnitude (H), using three general methods described below. Also included in this figure is the present discovered population, which obviously forms a lower bound for the

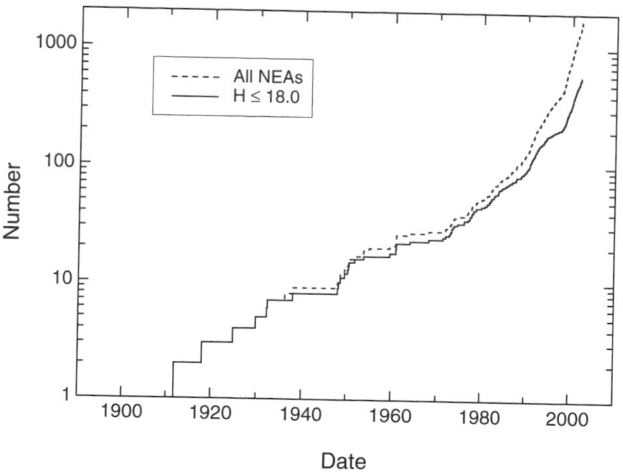

Fig. 2. Discovery history showing cumulative number for all NEAs and for those NEAs with absolute magnitude H = 18 or brighter (corresponding approximately to 1 km or larger diameter). The totals as of the end of January 2002 were 587 brighter than H = 18, and 1743 total.

total population. Below the bottom of the plot is a scale of estimated NEA diameter, adjusted on the assumption that H = 18.0 is equivalent to D = 1.0 km, and representing typical impact velocities. The top and righthand scales will be discussed in the next section.

D'Abramo et al. (2001) have estimated the total population by comparing the rate of new detections of NEAs by LINEAR with the rate of redetections of known objects in the same sample interval. The total population is estimated as the ratio of all detections (new and already known) to redetections times the previously known population. D'Abramo et al. find N(H < 18) = 855 ± 101, which because of various biases should be regarded as a lower bound.

Rabinowitz et al. (2000) and *Stuart* (2001) have utilized a different method to estimate the NEA population from discovered objects. They simulate the expected fraction to be discovered, in a manner similar to the method used by *Harris* (1998) for survey analysis, but using the actual sky coverage achieved by the survey in question. They normalize the numbers actually detected by the fraction of an assumed population that are detected by the simulation to derive a population estimate. *Rabinowitz et al.* (2000) use Spacewatch and NEAT discoveries to estimate N(H < 18) = 689 ± 184 and 708 ± 161 respectively. Both of these estimates are based on very small-number statistics, only ~100 of the more than 1000 objects presently known. *Stuart* (2001) has applied similar methods to the much larger sample of LINEAR discoveries, obtaining an estimate of N(H < 18) = 1227 (+150, –50). The method used by these investigators has biases that can work in either direction, so it is hard to know if the estimates are likely to be too high or too low. Finally, *Bottke et al.* (2000, 2002) have estimated a value of 960 ± 120 NEAs brighter than H = 18, based on a sample of 138 NEAs discovered or rediscovered by Spacewatch and using a debiased orbital-element distribution for the NEO population.

All the population estimates above are derived from the currently known NEAs. In contrast, it is possible to estimate the average population over the past several billion years from lunar-crater statistics (*Opik,* 1960; *Shoemaker,* 1983). Properly, this population will include both comets and asteroids, but several lines of evidence indicate that asteroids dominate the current impacts, as they may have done in the past. This is the approach taken recently by *Werner et al.* (2002), who use a relative size-frequency distribution (SFD) derived from the lunar mare crater SFD, scaled according to theoretical calculations of projectile:crater diameter ratio vs. crater size (e.g., *Melosh,* 1989). Even without a detailed knowledge of the dynamics of the impacting population, this curve can be moved vertically to be tangent to the discovered population in the size range where it is presumably very near complete (e.g., for NEAs larger than a few kilometers). Figure 3 shows the results from *Werner et al.* (2002) assuming, as is customary, that H = 18 is equivalent to D = 1 km, which corresponds to an average albedo of 0.11. In addition, we have shown in Fig. 3 a curve corresponding to an assumed albedo 0.25. Werner et al. suggest that the albedo of smaller NEAs may be higher than those of larger NEAs such that the crater curve, transformed onto a scale of H, might be a smooth blend

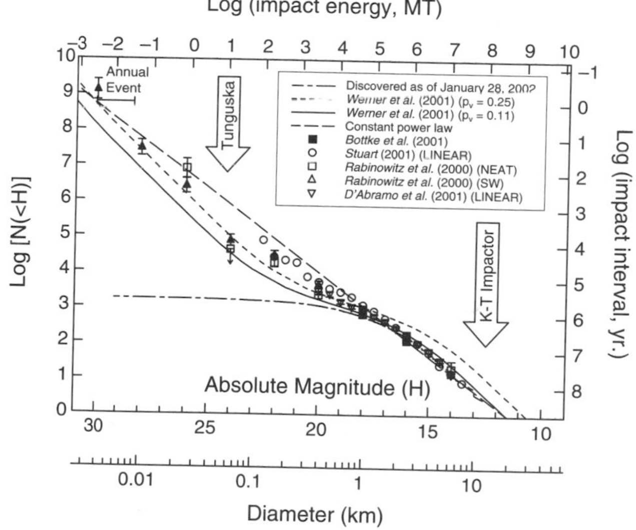

Fig. 3. Cumulative population of NEAs vs. absolute magnitude, with equivalent scales for diameter in kilometers, impact energy in megatons, and expected impact interval in years. See text for detailed discussion of scale equivalence factors and uncertainties of the various quantities plotted. The current observed rate of atmospheric impacts is plotted at the upper left, and the energies of the Tunguska and K/T (end-Cretaceous) impacts are also indicated. The straight line is a simple power law that approximates the data.

from the dashed curve on the left to the solid one on the right side of the plot.

In addition to the various estimates discussed above, we have included in Fig. 3 a simple power law consistent with most estimates from current NEAs.

These estimates, using three different methods and a variety of datasets, span a rather wide range in values (from 689 to 1227 plus uncertainties) for the total NEA population to H = 18. The lowest values appear to be inconsistent with the current total numbers and the continuing high rate of discovery. A conservative lower bound is probably about 800. Considering all these estimates, it appears that a reasonable summary value to use in evaluating survey completeness is N(H < 18) = 1000 ± 200. We will return to discussion of Fig. 3 in the next section.

How are we doing with respect to the Spaceguard goal of 90% discovery, if the total population brighter than H = 18 is 1000 ± 200? Consider the three cases of total population of NEAs (H < 18) of 800, 1000, and 1200, as of the beginning of 2002 when 587 had been discovered. Based on a model of completion vs. time of survey [e.g., Fig. 3 of *Harris* (1998) or Fig. 4 of *Harris* (2001)], we reach the following approximate conclusions. For a total population of 800, we are already 73% complete and have to find only 133 more to reach the 90% goal. This requires a current discovery rate of only 3 per lunation to meet the goal by 2008. For our nominal population of 1000, there are 313 more that must be discovered, and the required rate is 9 per lunation. For a total population of 1200, there are 493 more that must be discovered, and we need to be finding 15 per lunation to be 90% complete in 2008. Since the discovery rate in 2000/2001 has averaged about 10 per lunation, we are on target for the nominal population of 1000, but the real answer to how the survey is progressing depends sensitively on the total population.

It is perhaps illuminating to compare these conclusions with our previous discussion of Fig. 1. If the total population is 1200 (near our upper limit from Fig. 3), then the conclusion from Fig. 1 that achieving the Spaceguard goal requires an all-sky survey to limiting visual magnitude m_{lim} = 20.2 is consistent with the performance of the current survey at its limiting magnitude near 19.6, which yields a discovery rate (10 per lunation) lower than that required for 90% completeness by 2008 (which is 15 per lunation). If the population is ~1000, then there is a modest discrepancy between the model limiting magnitude and the current system performance. If the population is as low as 800, then this discrepancy becomes implausibly large. This contradiction, together with the fact that discoveries continue at a rapid pace even above the 600 mark, suggest that total population of NEAs to H = 18 is probably between 1000 and 1200. From a historical perspective, the most important recent change has been the drop in estimated total NEA population (H < 18) from 1500–2000 to approximately 1000 — a revision that has brought achievement of the Spaceguard goal within reach even though most of the survey telescopes are in the 1-m class and do not reach a limiting visual magnitude of 20.

The current telescopes in the Spaceguard Survey are not necessarily an optimum design, but they are doing the job. If we wish to augment the capability of the system, the primary requirement is to reach fainter magnitudes without giving up sky coverage. This could be achieved with larger apertures; today's survey telescopes are almost all in the 1-m class, which is very small by current astronomical standards. It would also be desirable to have at least one telescope in the southern hemisphere, since currently ~20% of the sky is not being covered. However, we note that while a southern telescope is desirable, it is not an absolute requirement. A NEA that is missed one year because it is too far south will likely be picked up on a subsequent pass. This gap (or "blind spot") in the south is not qualitatively different from, for example, the gap in coverage caused by the monsoon weather that typically closes down observatories in Arizona and New Mexico during the summer months. The primary effect of these gaps is simply to slow completion of the survey. Fortunately, a southern hemisphere survey telescope at a good site could go a long way toward filling both gaps.

Telescopes in space could also be used to augment the survey, but most of the systems that have been proposed are not likely to be cost effective compared with groundbased observatories. The cost-effectiveness would be greatly improved, of course, if the NEA survey activity were incorporated as a secondary goal into spacecraft being launched for other purposes. There is no intrinsic advantage of Earth-orbiting observatories, other than continuously clear sky (in fact, some orbiting telescopes actually have lower duty cycles than groundbased telescopes at good sites). Telescopes looking from interior to Earth's orbit have an advantage in discovering asteroids that spend most of their time inside Earth's orbit, but we already know that there are relatively few of these. Any given survey system should be judged on its merits, of course, and there is no reason that a mix of spacebased and groundbased instruments could not contribute to NEA surveys.

Recent experience with the output of the NEA survey programs has led to more sophisticated treatments of impact probability (e.g., *Milani and Valsecchi,* 1999; *Chodas and Yeomans,* 1999; *Milani et al.,* 2000a,b, 2002; *Chesley et al.,* 2002). In particular, researchers have explored the dynamics of complex encounters. NEAs that might pose a threat usually pass close to Earth on previous orbits. On sufficiently close passes, Earth's gravitational field substantially alters the NEA orbit. Typically, only a very few specific possibilities for resonant return will lead to an impact or even another close pass, at least on a scale of decades. "Keyholes" leading to a possible future impact may be found on the error ellipse of the NEA on the first encounter's target plane (passing through Earth and normal to the asteroid orbit). These keyholes will initially be only a tiny fraction of the target error ellipse. There may be several keyholes corresponding to different resonances, and thus to possible impacts on different future dates. The estimate of risk then depends on the probability that the actual trajectory will take the NEA through one of the keyholes. The rest of that specific target plane is safe, corresponding to the NEA being scattered back into the general population with an impact probability that is not substantially greater than that of typical newly discovered objects.

From a hazard perspective, the goal (for those error ellipses that are noninclusive of Earth) is to assure that the NEA does not pass through a keyhole. Followup observations or archival "precovery" may shrink the error ellipse away from the keyholes in the same manner that they can shrink an Earth-inclusive ellipse off Earth. In cases of a short warning where a short arc of observation cannot eliminate the possibility of keyhole passage, it may be necessary to make negative observations along the virtual trajectories leading through the keyholes.

Of course, a specific NEA may impact Earth without ever passing through a keyhole and experiencing a resonant return. Certain NEAs can impact in a nonresonant return (a return at the opposite node). There are also pathological cases of anomalously large keyholes resulting from "interrupted returns." These are all treated in *Milani et al.* (2002). In the case of one of the best-observed NEAs, 1950 DA (with a 50-yr arc and radar range-Doppler data), the uncertainty in evaluating a possible impact nearly a millennium in the future is actually dominated by the unknown influence of the Yarkovsky effect, which in turn depends on the albedo distribution, surface shape, and spin dynamics (*Giorgini et al.,* 2002)

One way to look at the Spaceguard Survey is as an effort to find each NEA and declare it safe. "Safe" can be defined as having a target plane error ellipse that is well clear of Earth and of keyholes leading to Earth. For this purpose, it is not necessary to calculate most orbits with extreme precision, although there may be good scientific reasons for doing so (such as identifying future targets for radar imaging). So far, nearly all the NEAs discovered by Spaceguard have been declared safe, and the very few exceptions [such as 1950 DA (*Giorgini et al.,* 2002)] will remain under surveillance until we can sound the "all clear."

7. CURRENT NEAR-EARTH-ASTEROID IMPACT FREQUENCY

While the long-term historic impact frequency can be estimated from lunar cratering, we must turn to the current NEA population to estimate current impact rates, since there is no assurance that such rates have remained constant over time. To relate energy to size of object, we need the average impact velocity, and to relate numbers of bodies to impact frequency requires some rather complicated considerations of the orbital distribution of the NEA population. One way to arrive at both of these quantities is to calculate the circumstances of close approaches to Earth of the presently known population. Following approaches described in *Harris* (1998, 2001), we took orbit elements and H magnitudes of all objects known as of July 3, 2001, from the ASTORB data file maintained by Bowell (ftp.lowell.edu/pub/elgb/astorb.dat). We selected only objects with $H < 18.0$ and perihelion $q < 1.0$ AU, that is Earth-crossing asteroids (ECAs) with estimated $D > 1$ km according to the usual assumed albedo of 0.11. The number of asteroids satisfying these criteria was 244, compared to a total of 488 NEAs ($q < 1.3$ AU) in the same size range. We then calculated all approaches to Earth closer than 0.1 AU for a full century, from 2002 to 2102, choosing the dates to avoid the bias of discovery apparitions, using unperturbed osculating elements to provide a statistically valid measure of close approach frequency.

We found 273 close approaches (2.73/yr) by a total of 100 different NEAs. For each close approach, we tabulated the encounter velocity so we could compute the mean squared impact velocity of the flux to relate that to energy. The radius of Earth is 4.25×10^{-5} AU, so the frequency of actual impacts, not including the effect of gravitational focusing, should be $(4.25 \times 10^{-4})^2$ times less than the frequency of passes to within 0.1 AU. Since we kept track of encounter velocity for each event, we could correct for the effects of gravitational focusing, which enhanced the impact frequency by a factor of 1.66. Combining all these factors yields an impact frequency from the presently known population of 8.2×10^{-7} yr^{-1}. The "per object" impact frequency for NEAs is thus $8.2 \times 10^{-7}/488 = 1.68 \times 10^{-9}$ yr^{-1}. The impact frequency implied for the ECA population is twice as high (but there are half as many of them). In Fig. 3 we have taken this per-object impact frequency to relate the expected impact interval, righthand scale, to the population N(>D), lefthand scale.

An additional result that comes out of the close-encounter calculation above is the RMS velocity to relate mass of impactor to energy. We have weighted the individual encounter velocities to account for the higher probability of slower impacts due to gravitational focusing and also added in the contribution of Earth's gravity to obtain an RMS impact velocity of 20.2 km/s, in excellent agreement with a recent calculation by Bottke (personal communication, 2001), in which he also includes a bias correction for discovery selection effects. If we assume a mean density of asteroids of 2.5 g/cm^3, the mass of a 1-km-diameter asteroid is 1.3×10^{12} kg and its kinetic energy at 20 km/s is 2.6×10^{20} J. In Fig. 3, the impact energy (top) and diameter (bottom) are thus scaled so that the impact energy is 6.25×10^4 MT for a 1-km-diameter NEA.

Steel (1998) carries out a similar derivation based on a sample of the 116 ECAs with $H < 18.5$ that were known in June 1996. He finds an RMS impact speed of 19 km/s and an average terrestrial probability of impact of 5×10^{-9} yr^{-1}, about twice as large as the results from our work described above. Other earlier derivations also exist, going back to the work of *Shoemaker* (1983) and even *Opik* (1958, 1960), all giving generally consistent results.

To provide "ground truth" in terms of contemporary impact rates for relatively small NEAs, we plot in Fig. 3 the estimated energy, 10 kT (with total uncertainty a factor of 10), of the largest event seen annually in Earth's atmosphere, based on space surveillance data [*Tagliaferri et al.* (1994) and USAF press releases since 1992]. This value is plotted as a horizontal bar in the upper left, representing the uncertainty in the average largest annual event. Note that the straight-line, power-law approximation to the NEA population in Fig. 3 falls within a factor of 2 or so of almost all of the population and impact-frequency estimates.

We can compare the expected impact frequencies in Fig. 3 with the two well-known benchmarks, the Tunguska event (at ~10 MT) and the K/T impact (at ~10^8 MT). The range of uncertainties spans about a factor of 10 in energy for each, and of course we do not know the frequency for these singular events (except that it is not a great deal less than 100 yr for Tunguska or 65 m.y. for K/T). From the plot, the expected frequency for Tunguska-class events ranges from about twice in a millennium for the power law to about once in 10,000 yr for the lunar curves. Only if Tunguska was as small as 3–5 MT in energy does the expected frequency approach the century timescale even for the power-law size-frequency distribution. The K/T event frequency from the plot is generally consistent with an average interval on the order of 100 m.y.

While most of the data are approximately consistent with a power law, the lunar-derived NEO population curve of *Werner et al.* (2002) shows an obvious departure, usually interpreted as a shortage of small (diameter less than a few hundred meters) impactors, although it might also suggest an early excess of large asteroids or comets not currently represented in the NEA flux. Interpreted in the usual way, however, the lunar curve indicates that the frequency of Tunguska-size impactors is roughly one per 10,000 yr, more than an order of magnitude below the usually quoted frequency of such impacts, and a surprising result given that we experienced such an event within the last century. We don't know where the problem lies, but we suggest that the NEA population derived by Werner et al. from the lunar cratering statistics warrants consideration of alternative interpretations of the data.

We conclude this section with a few words about uncertainties. The relationship between population and impact frequency, as well as that between diameter and impact energy, is probably quite well determined. However, since the observed NEA population curve is derived mainly from sky brightness of asteroids, the biggest uncertainty is the relation between absolute magnitude and diameter. The assumed mean albedo could be wrong by as much as about a factor of ±2 (range from 0.05 to 0.22), translating to a factor of ±1.4 in diameter or ±4 in mass (energy). For the slope of the population line, this horizontal uncertainty is equal to a vertical (population at a given size) uncertainty of more than a factor of ±2. Note that the factor of 2 in each direction amounts to choosing mean albedos close to the high- and low-albedo peaks in the observed bimodal distribution and ignoring the other fraction. Thus the total uncertainty, observation plus conversion from magnitude to diameter, leads to an uncertainty of perhaps a factor of 3 in our knowledge of the population vs. diameter. This is especially true at the very small end of the distribution, where the observational basis is thin and we have almost no direct measure of albedos.

Finally, it is important to reiterate that the entire population of NEOs, or even of NEAs, does not contribute equally to the impact hazard. The NEOs include comets with very long periods. Obviously the individual impact risk for a comet is inversely proportional to its period. This is also true of NEAs in cometlike orbits, such as the recently discovered Earth-crosser 2001 OG18, which has a period of 50 yr. Several large NEAs, including Eros, are currently in safe orbits and will not pose any threat for at least a million years. From the perspective of risk, fewer than half of the NEAs are important. Estimates of risk (and metrics of the performance of the Spaceguard Survey) should properly give higher weight to the subset of NEAs that are in the most risky orbits [as is done, for example, in the file of possibly hazardous asteroids (PHAs) maintained by the Minor Planet Center]. A corollary is that once a newly discovered NEA is shown not to belong to one of these hazardous subsets of the population, it can safely be ignored. Several such nondangerous NEAs have been discovered and lost, but their orbits were determined with sufficient accuracy to ensure that they were not a threat.

8. SOCIETAL CONTEXT FOR NEAR-EARTH-OBJECT SURVEYS

While NEO research embodies classic scientific objectives, studies of impact hazards constitute an applied science that may be judged by different criteria. In determining an NEO hazard-mitigation strategy, we must consider the reaction of society. Such considerations are familiar to specialists in other fields of natural hazard, such as meteorology (with respect to storm forecasts) and seismology. NEO hazard specialists have the added difficulty of explaining a science that is arcane (orbital dynamics) and beyond personal experience (no impact disaster within recorded history). As the NEO community has begun to realize, it has a social responsibility to ensure that its message is not just heard but comprehended by society at large. The adoption of the Torino Impact Scale (*Binzel,* 1997, 2000) was a notable first step toward public communication, although the unique aspects of NEO detection and warning (particularly the evolution of uncertainty) continue to cause communications difficulties (*Chapman,* 2000).

Once it is accepted that the impact hazard is a social and not just a scientific problem, it is a short step to allow that considerations of maximum social benefit may well constrain the scope and form of scientific investigation. That is, while the scientifically optimum level of uncertainty is zero, the socially optimum level is nonzero. It is neither possible nor affordable to remove risk and uncertainty entirely. This is not just a trite benefit-cost argument. Rather, scientific information can have marginal disutility. As an example, many might argue that society incurs a net cost for the science of nuclear physics, since nuclear proliferation is facilitated thereby. Nuclear test-ban treaties rest upon a presumption of the disutility of the scientific and technical information derived from the tests. The inescapable conclusion is that if, despite its best intentions, the NEO community levies a perceived cost to society through mishandled or garbled communication, then society may well act to remove that cost by choosing not to support NEO surveys and related work.

To date, international NEO survey programs have been conducted and coordinated by an eclectic mix of state and nongovernmental organizations operating within the scientific paradigm of openness. The emphasis has been on generating discoveries. Some of these discoveries have indicated (initially at least) a nonzero possibility of a future impact, raising the issue of whether to issue a warning. In some cases (e.g., 1997 XF11, 2000 BF19) individuals or organizations have made public warnings that were widely reported by the press, only to be quickly withdrawn when additional data or more refined calculations became available.

When and under what circumstances should public warnings be made? The trigger threshold for a "confirmed warning" is a key parameter for both NEO scientists and those (primarily science journalists) who make decisions about what information to disseminate to the public. The International Astronomical Union (IAU) NEO Working Group adopts as a threshold the "prediction of impacts with probability larger than one in a million (10^{-6}) in the near future (less than 100 yr)." This accords with the Torino Scale threshold for a 1-km object to achieve Level 1 (that is, to rise out of the background risk). The Torino Scale "raises the bar" for smaller objects — for example, a 100-m object requires a 1-in-10,000 (10^{-4}) collision probability to reach Level 1. All these guidelines are informal, and the IAU leaves any decision about public release to the discoverer of the threat. In practice, each case that has received wide publicity (1997 XF11, 1999 AN10, 2000 BF19, 2000 SG344, 1950 DA, 2002 NT7) has had its own unique nature, demonstrating both that guidelines must be flexible and that it is impossible to control the behavior of either astronomers or the media by fiat from above. Thus the recent historical average of approximately one warning (or rumor thereof) per year may continue.

Although much thought has been applied to modeling the discovery rate of survey programs (as discussed above), no researcher has attempted to model warning rates, yet that question is of paramount interest to policy makers. There have been no confirmed warnings to date that have survived for more than a few days (not counting 1950 DA, with a low-probability impact possibility nearly a millennium in the future), so when the first such occurs, society is in uncharted territory. Of course, most confirmed warnings will become false alarms when new data are acquired, and it can be expected that major astronomical facilities can be quickly turned to NEO followup given sufficient priority to do so. It is an interesting situation: Since most warnings will be false alarms, it would seem to make sense to raise the warning threshold, yet doing so might result in less effort to make new observations and thereby prolong the perception of a potential threat. The reason given for most announcements has been to stimulate additional observations, with warning to the public a secondary issue.

As discussed previously, the Spaceguard Survey will shortly experience diminishing returns in its primary goal of discovering NEAs larger than 1 km, a natural consequence of population sampling without replacement. On the other hand, discoveries of smaller NEAs will continue in proportion to sky coverage and aperture, as their population has been barely sampled. There will be a natural incentive to shift the survey goal down in size to the region of highest return. There is already a "policy hook" for extending the search to smaller objects, in the form of the Council of Europe's Parliamentary Assembly Declaration 1080 "on the detection of asteroids and comets potentially dangerous to humankind." This document called for establishment of an "inventory of NEOs as complete as possible with an emphasis on objects larger than 0.5 km in size." More recently, the U.K. NEO Task Force (*Atkinson et al.,* 2000) called for a "new 3 meter-class survey telescope for surveying substantially smaller objects than those now systematically observed by other telescopes." In the United States, the National Research Council has recommended the construction of a 6–8-m large-aperture Synoptic Survey Telescope (LSST), with one goal: to "catalog 90% of the NEOs larger than 300 m" (*National Research Council,* 2001). Note that the smaller telescopes now used for followup will be unable to keep up with the newer, larger-aperture survey telescopes in terms of limiting magnitude. This situation is recognized by the planners of the LSST, who expect that this telescope will carry out all its own followup observations, sufficient to yield a good orbit for each NEA discovered (S. Wolff, personal communication, 2002).

9. MITIGATION PROGRAMS AND INSTITUTIONS

Mitigation of the NEA impact hazard can take three forms. The preferred but most technically challenging option is to deflect the threatening NEA, changing its orbit so that it will miss Earth (*Ahrens and Harris,* 1992, 1994; *Simonenko et al.,* 1994; *Melosh et al.,* 1994; *Morrison and Teller,* 1994; *Weissman,* 1994; *Asphaug et al.,* 1998; *Benz and Asphaug,* 1999). Alternatively, we can predict an impact and prepare the planet (or at least the target area) to survive the event as well as possible. Least desirable but perhaps most likely, we can follow the example of other natural hazards such as earthquakes and severe storms, focusing not on prevention but on dealing with the aftermath of an impact (*Garshnek et al.,* 2000). In practice, these options are complementary, depending primarily on the length of advance warning of the threat. Note that a short lead time for an NEA is extremely unlikely — we can expect either decades of warning or none at all. It is actually quite difficult to envision a system that provides only short warnings; a detection system that will find an incoming NEO 10 days before impact is also likely to find any impactor decades before it strikes. There are also legal implications of decisions to mitigate or not, as introduced by *Gerrard* (1997) and *Seamone* (2002).

Consider the option of interception and deflection, which would require new and expensive defensive systems. Should such systems be developed now? From the standpoint of an allocator of society's resources, an uncertain threat calls for adaptive policies, delaying potentially costly action but informing later decision by investing in uncertainty-reduction measures. In the context of the NEO impact hazard, this means avoiding the costs of standing organizational structures and capital expenditures until a threat materializes, while continuing modest support for surveys and inexpensive studies of mitigation options. Measures to gather information about the hazard (such as space missions to NEOs) could also be supported, especially if they can be justified on other scientific grounds.

In an organizational sense, planning for adaptivity entails establishing a chain of responsibility prior to the materialization of an emergency — that is, a shadow institution. It is illustrative to examine a hypothetical NEO emergency organizational plan, using the U.S. government as an example. We need an office in the executive branch of government, and the National Science and Technology Council (part of the White House Office of Science and Technology Policy, or OSTP) seems to have natural purview over the impact hazard. NSTC has five multiagency committees, each of which pertains to an aspect of the impact hazard: Environment and Natural Resources; International Science, Engineering, and Technology; National Security; Science; and Technology. The formal assignment of the NEO impact hazard to an NSTC committee would be accomplished by the drafting of a Presidential Decision Directive (PDD/NSTC) or Presidential Review Directive (PRD/NSTC). Little else would need to be done until there is confirmed warning of a threat. In all likelihood the draft PDD has remained in the desk drawer of an NSTC staffer. The President's signature would be sought only upon confirmation of the warning. In the unlikely event that lead time is short, the issue will be moved out of the NSTC to the National Security Council (NSC). In that case, the PDD can be issued as a joint NSC/NSTC document, for which there is precedent in the National Space Policy of September 19, 1996 (PDD/NSTC 8 and PDD/NSC 49). Undoubtedly similar procedures exist in other countries, and it may be that multinational organizations (including the United Nations) would also wish to develop contingency plans.

The principles of adaptive planning in the face of uncertainty fundamentally affect mitigation investment decisions. Civil-defense measures have the advantage that improvements can be gained due to synergism with more mundane natural hazards. To date, very little attention has been given to the demands that would be placed on governmental and private disaster-response systems by even a small (Tunguska-class) impact in a populated region (*Garshnek et al.,* 2000; *Chapman,* 2000). However, consider the more challenging question of interception to deflect a threatening NEO and the expenses of a standing force of anti-NEO launchers. It has been the position of most NEO researchers that these expenses are best deferred, since in all likelihood there will be sufficient warning time before impact to develop an interception system from scratch. To this, the advocates of interception systems reply that "in all likelihood" does not mean "always," and then play the trump card of the cometary threat. An example of a defense architecture oriented toward this most challenging case is found in *Gold* (1999).

The diverse nature of the NEO population (particularly with regard to mechanical strength and composition) has been used as an argument to defer investment in interception capability until a specific target object has been identified (e.g., *Morrison et al.,* 1994; *Harris et al.,* 1994; *Sagan,* 1992, 1994; *Sagan and Ostro,* 1994). On the other hand, in a survey regime characterized by many false warnings, value can be gained from a system that has uncertain effect, to the degree that it reassures the population and prevents panic. For example, the U.S. rushed Patriot antiaircraft missile batteries to Israel during the Desert Storm conflict with Iraq as a defense against ballistic missiles, despite the fact that they were not designed to intercept missiles. The Patriots proved militarily ineffective but politically very useful.

An adaptive planning approach could also accommodate the short warning scenario associated with long-period comets, requiring that a relatively low-cost generic interception system be built, tested, then shelved. In the event of emergency, the system would enter surge production, with industrial capacity commandeered from other programs. In this manner, there would be a tailored response to the threat, and operational flexibility would be enhanced. Salvoes could be launched, and in many cases, shoot-look-shoot would be possible.

It is facile but probably misleading to focus on a scenario where an NEO progresses in a step function from zero threat to Earth impactor. The threat that stays a threat will experience an overall rise in impact probability, as the error ellipse shrinks while Earth stays within it. Many more threats than not, however, will suddenly see their impact probability go to zero as the error ellipse shrinks to exclude Earth or shrinks to exclude dangerous keyholes for the case of a resonant return. This feature of the evolution of impactor uncertainty will encourage those who wish to defer commitment to interception or who just want to keep the public purse closed. The net effect is that the system reaction time will need to be much shorter than the warning time from the point of confirmed threat. This already-challenging situation will only be worsened by failure to examine scenarios and develop appropriate contingency plans. To date the NEO community has not made much effort to pursue such options or enter into dialogue with government organs that deal with security issues.

Many of these issues are discussed by *Parks et al.* (1994), who conclude that societies will not sustain indefinitely a defense against an infrequent and unpredictable threat. Governments often respond quickly to a crisis but are less well suited to remaining prepared for extended periods. But these conclusions reflect a history in which the less-frequent threats are generally of less consequence than those encountered more often. In contrast, the greatest NEA impact hazard is from the very rare large impacts. Put simply, each reader of this chapter has a greater chance of dying within the next month from a globally catastrophic impact than from any of the smaller more frequent impacts. It remains to be seen how governments and other institutions of society will respond to this unique problem.

10. INTERNATIONAL COOPERATION

While several governments have officially recognized the impact hazard and the value of surveying for NEAs, the bulk of the financial and material support for such surveys has come from the United States. With the support of the U.S. Congress, NASA has taken the leading role in organizing and funding the Spaceguard Survey, while the U.S. Air Force has supplied equipment and key personnel. Current NEA discoveries are dominated by a few groups in the U.S., notably the LINEAR survey using two telescopes in New Mexico (*Stokes et al.,* 2000, 2002). Other major programs are NEAT (a JPL-USAF partnership), Spacewatch (University of Arizona), LONEOS (Lowell Observatory), and the Catalina Survey (University of Arizona). However, discovery alone is not sufficient. To avoid losing newly discovered NEAs because of imprecise orbits, discovery must be confirmed by numerous followup observations, suitably distributed in time and aimed at collecting the data necessary to compute accurate orbits. These efforts are widely international in scope and most depend on volunteer labor. As the number and size of survey telescopes increases, this followup becomes increasingly difficult.

The coordination of these efforts is informal and voluntary. When an object is discovered, its position in the sky is computed and this measurement is transmitted to the Minor Planet Center (Cambridge, Massachusetts), which acts as a clearinghouse on behalf of the International Astronomical Union (IAU). If the observations pertain to an NEO and can be verified by additional data, they are posted on a special Web page, the NEO Confirmation Page (http://cfa-www.harvard.edu/cfa/ps/NEO/ToConfirm.html), where observers can get preliminary ephemerides. All these data are available to observers around the world, and followup observations are performed by more than 80 centers. However, the MPC has neither the capability nor the purpose of coordinating observations at a higher level. Consequently, there is overlap and inefficiency in both the original surveys and the astrometric followup. In this sense, there is no single Spaceguard Survey, but rather a group of individual teams who cooperate or compete (usually both) based on their own capabilities and priorities.

More specific coordination is the objective of the Spaceguard Central Node of The Spaceguard Foundation (SGF). This professional international organization, created in 1996, began its coordinating efforts in 1999 (http://spaceguard.ias.rm.cnr.it). There are at least two types of searches that are coordinated by the SGF: the "real" followup, consisting of repeated observations of known objects and aiming at an improved knowledge of their orbits, and the archival searches intended to identify prediscoveries. In both cases, a better coordination is extremely helpful and is obtained by providing support to the interested people and teams. Additional support comes from the NEO Dynamic Site at the University of Pisa (http://newton.dm.unipi.it/neodys), the NASA NEO Program Office of JPL in Pasadena, California (http://neo.jpl.nasa.gov), and the Asteroid Observing Service of the Lowell Observatory in Flagstaff, Arizona (http://asteroid.lowell.edu).

Apart from the interest demonstrated by the Office of Outer Space Affairs of the United Nations, which organized a meeting on this subject in 1995 at the UN headquarters in New York, the first international, nonscientific organization to examine the issue of the impact hazard was the Council of Europe, in 1996. In its Resolution 1080, "on the detection of asteroids and comets potentially dangerous to humankind," the council underlined the relevance of the impact threat and invited all member states (about 30) and the European Space Agency (ESA) to take steps in supporting and funding an international program of NEO studies. Unfortunately, no European nation has taken any initiative in this matter until recently.

In 1999 the United Nations Committee for the Peaceful Uses of Outer Space (COPUOS) organized its third conference in Vienna (UNISPACE III). The impact hazard was discussed in an *ad hoc* working group, which produced a document asking for a deep involvement of all nations, and of the UN in particular, in NEO research. This document has moved the assembly (composed of representatives of the UN member states) to include a mention of the necessity of an internationally coordinated effort on this subject in the Declaration of Vienna, the final document of the conference approved by the UN General Assembly later that year.

In addition to the U.S., two more nations have taken initiatives concerning NEOs: Japan and the United Kingdom. The Japanese government has funded an observational facility in Bisei that expects to be operational in 2002 (*Isobe,* 1999). This center is intended to perform simultaneously studies on space debris and NEOs. The most visible recent action has come from the United Kingdom, where the government nominated a Task Force on NEOs in 1999, chaired by Harry Atkinson. The main purpose of this group was to investigate the current status of the NEO issue and to suggest to the British government actions to be taken to cope with it. The report of the NEO Task Force (*Atkinson et al.,* 2000) contains 14 recommendations to the government. The position of the British government is that any initiative in this sense must be agreed on and funded internationally. The British government has been the only European government as of this writing to make a public commitment to some kind of action.

As discussed in the previous section, we can expect greater governmental interest when the entire NEO impact problem is addressed, not just the astronomical search for NEAs. Scientific investigations alone are not sufficient, nor completely appropriate, for the solution of the problem. Other, nonscientific organizations must be involved, especially the civil-defense organizations in different countries. Of particular interest is the decision in January 2002 that the Organization for the Economic Cooperation and Development (OECD) be directly involved in the elaboration of an action plan, because of the authoritative influence that this organization has on scientific matters of global interest (through its Global Science Forum).

11. CONCLUSIONS

It is now evident that the impact problem has several peculiar characteristics that make it an issue of great scientific and social importance. Scientifically, the core of the problem is to understand the dynamical and physical processes that lead to impacts and that may affect the development and evolution of life on Earth. Socially, we need to understand the effects of impacts (and warnings of impacts) on human society. This double character of the problem is one of the reasons for its confused status at a political level and is the most urgent aspect that needs to be solved on an international basis.

NEO scientists, primarily astronomers, have succeeded in alerting the world to the existence of an impact hazard, and they have been very successful in undertaking the Spaceguard Survey, focused (so far) on the threat of global disaster from collision with a NEA of diameter >1 km. It is fortunate that the greatest danger, in both quantitative and qualitative terms, arises from the larger NEAs, which are the most readily discovered by astronomical techniques. Our success is the product of synergy between the technology (widefield CCD detectors and powerful computational tools) and a recognition of a societal desire to discover and track any projectiles large enough to threaten the survival of human civilization. Currently we have discovered more than half of the NEAs larger than 1 km, and perhaps more than 80% of those larger than 2 km. As a result, the present risk of being struck without warning by a large NEA has been cut by more than half. However, we should note that achieving the second half of the survey will be much more challenging than the first, and that there is still little formal coordination of observing teams (and followup) into a coherent international survey.

It is possible, however, that we will be a victim of our own success. As current surveys become more complete at sizes of 1 km or greater, the rate of discovery of 1-km NEAs will drop. There is as yet no consensus as to the next critical target. Do we want to find all NEAs down to 500 m, or 300 m, roughly the sizes at which the tsunami danger is greatest? Or should we ultimately anticipate a survey and warning system that is complete down to 50 m, so that we can with confidence predict the "next Tunguska"? How much is society willing to invest to reach these goals?

It is in this context that the NEO science community needs to engage in active dialogue with other professionals with greater experience in disaster mitigation and national security. We need to consider the societal context of NEO searches and approaches to mitigation. If the resources required for NEO studies increase, these other considerations will play a critical role in determining what priority will be placed on protecting our planet from cosmic impacts.

Acknowledgments. The work of A.W.H. at the Jet Propulsion Laboratory, California Institute of Technology, was done under contract from NASA. C.R.C. and G.S. acknowledge partial support from the NEO Program Office at JPL. We thank referees D. Steel and D. Yeomans, editors R. Binzel and W. Bottke, and colleagues E. Bowell, K. Zahnle, and the other Alan W. Harris (of DLR) for useful comments and suggestions.

REFERENCES

Ahrens T. J. and Harris A. W. (1992) Deflection and fragmentation of near-Earth asteroids. *Nature, 360,* 429–433.

Ahrens T. J. and Harris A. W. (1994) Deflection and fragmentation of near-Earth asteroids. In *Hazards Due to Comets and Asteroids* (T. Gehrels, ed.), pp. 897–928. Univ. of Arizona, Tucson.

Alvarez L. W., Alvarez F., Asaro, and Michel H. V. (1980) Extra-terrestrial cause for the Cretaceous-Tertiary extinction. *Science, 208,* 1095–1108.

Asphaug E., Ostro S. J., Hudson R. S., Scheeres D. J., and Benz W. (1998) Disruption of kilometre-sized asteroids by energetic collisions. *Nature, 393,* 437–440.

Atkinson H., Tickell C., and Williams D. (2000) *Report of the Task Force on Potentially Hazardous Near Earth Objects.* British National Space Center, London (available on line at http://www.nearthearthobject.co.uk).

Baldwin R. B. (1949) *The Face of the Moon.* Univ. of Chicago, Chicago. 239 pp.

Becker L., Poreda R. J., Hunt A. G., Bunch T. E., and Rampino M. (2001) An extraterrestrial impact at the Permian-Triassic boundary? *Science, 291,* 1530.

Benz W. and Asphaug E. (1999) Catastrophic disruptions revisited. *Icarus, 142,* 5–20.

Binzel R. P. (1997) A near-Earth object hazard index. *Ann. N.Y. Acad. Sci., 822,* 545.

Binzel R. P. (2000) The Torino Impact Hazard Scale. *Planet. Space Sci., 48,* 297–303.

Boslough M. B. E. and Crawford D. A. (1997) Shoemaker-Levy 9 and plume-forming collisions on the Earth. *Ann. N.Y. Acad. Sci., 822,* 236–282.

Boslough M., Crawford D., Robinson A., and Trucano T. (1994) Mass and penetration depth of Shoemaker-Levy 9 fragments from time-resolved photometry. *Geophys. Res. Lett., 21,* 1555–1558.

Bottke W. F., Jedicke R., Morbidelli A., Petit J.-M., and Gladman B. (2000) Understanding the distribution of near-Earth asteroids. *Science, 288,* 2190–2194.

Bottke W. F., Morbidelli A., Jedicke R., Petit J.-M., Levison H. H., Michel P., and Metcalfe T. S. (2002) Debiased orbital and absolute magnitude distribution of the near-Earth objects. *Icarus, 156,* 399–433.

Bowell E. and Muinonen K. (1994) Earth-crossing asteroids and comets: Groundbased search strategies. In *Hazards Due to Comets and Asteroids* (T. Gehrels, ed.), pp. 49–198. Univ. of Arizona, Tucson.

Canavan G. H. (1993) Value of space defenses. In *Proceedings of the Near-Earth-Object Interception Workshop* (G. Canavan et al., eds.), pp. 261–274. Los Alamos Publication LANL 12476-C.

Canavan G. H. (1994) Cost and benefit of near-Earth object detection and interception. In *Hazards Due to Comets and Asteroids* (T. Gehrels, ed.), pp. 1157–1190. Univ. of Arizona, Tucson.

Canavan G. H. (1997) Optimal detection of near-Earth object threats. *Ann. N.Y. Acad. Sci., 822,* 539.

Canavan G. H., Solem J., and Rather D. G., eds. (1993) *Proceedings of the Near-Earth-Object Interception Workshop.* Los Alamos Publication LANL 12476-C.

Chapman C. R. (2000) The asteroid/comet impact hazard: Homo sapiens as dinosaur? In *Prediction: Science, Decision Making, and the Future of Nature* (D. Sarewitz et al., eds.), pp. 107–134. Island Press, Washington, DC.

Chapman C. R. (2002) Impact lethality and risks in today's world: Lessons for interpreting Earth history. In *Catastrophic Events and Mass Extinctions: Impacts and Beyond* (C. Koeberl and K. G. MacLeod, eds.), pp. 7–19. Geol. Soc. Am. Spec. Paper 356.

Chapman C. R. and Morrison D. (1994) Impacts on the Earth by asteroids and comets: Assessing the hazard. *Nature, 367,* 33–39.

Chapman C. R., Harris A. W., and Binzel R. (1994) Physical properties of near-Earth asteroids: Implications for the hazard issue. In *Hazards Due to Comets and Asteroids* (T. Gehrels, ed.), pp. 537–550. Univ. of Arizona, Tucson.

Chesley S. R., Chodas P. W., Milani A., Valsecchi G. B., and Yeomans D. K. (2002) Quantifying the risk posed by potential Earth impacts. *Icarus,* in press.

Chodas P. and Yeomans D. (1999) Predicting close approaches and estimating impact probabilities for Near-Earth Objects. AAS/AIAA Astrodynamics Specialist Conference, AAS 99-462, Girdwood, Alaska, 16–19 August 1999.

Chyba C. F. (1993) Explosions of small Spacewatch objects in the Earth's atmosphere. *Nature, 363,* 701–702.

Chyba C. F., Thomas P. J. and Zahnle K. J. (1993) The 1908 Tunguska explosion: Atmospheric disruption of a stony asteroid. *Nature, 361,* 40–44.

Covey C., Thompson S. L., Weissman P. R., and McCracken M. C. (1994) Global climatic effects of atmospheric dust from an asteroid or comet impact on Earth. *Global and Planetary Change, 9,* 263–273.

Crawford D. A. (1998) Modeling asteroid impact and tsunami. *Sci. Tsunami Hazards, 16,* 21–30.

Culler T. S., Becker T. A., Muller R. A., and Renne P. R. (2000) Lunar impact history from (40)Ar/(39)Ar dating of glass spherules. *Science, 287,* 1785–1788.

D'Abramo G., Harris A. W. (JPL), Boattini A., Werner S. C., Harris A. W. (DLR), and. Valsecchi G. B (2001) A simple probabilistic model to estimate the population of near Earth asteroids. *Icarus, 153,* 214–217.

Farinella P., Foschini L., Froeschle C., Gonczi R., Jopek T. J., Longo G., and Michel P. (2001) Probable asteroidal origin of the Tunguska cosmic body. *Astron. Astrophys., 377,* 1081–1097.

Garshnek V., Morrison D., and Burkle F. M. (2000) The mitigation, management, and survivability of asteroid/comet impact with the Earth. *Space Policy, 16,* 213–222.

Gehrels T., ed. (1994) *Hazards Due to Comets and Asteroids.* Univ. of Arizona, Tucson. 1300 pp.

Gerrard M. B. (1997) Asteroids and comets: U.S. and international law and the lowest-probability, highest consequence risk. *N.Y. Univ. Environ. Law J., 6:1.*

Giorgini J. D., Ostro S. J., Benner L. A. M., Chodas P. W., Chesley S. R., Hudson R. S., Nolan M. C., Klemola A. R., Standish E. M., Jurgens R. F., Rose R., Chamberlin A. B., Yeomans D. K., and Margot J.-L. (2002) Asteroid 1950 DA's encounter with Earth: Physical limits of collision probability prediction. *Science, 296,* 132–136.

Gold R. E. (1999) *SHIELD, a Comprehensive Earth Protection System.* Report of the NASA Institute for Advanced Studies.

Harris A. W. (1998) Evaluation of ground-based optical surveys for near-Earth asteroids. *Planet. Space Sci., 46,* 283–290.

Harris A. W. (2001) Near-Earth Asteroid Surveys. In *Collisional Processes in the Solar System* (M. Y. Marov and H. Rickman, eds.), pp. 323–332. Kluwer, Dordrecht.

Harris A. W., Canavan G. H., Sagan C., and Ostro S. J. (1994) The deflection dilemma: Use versus misuse of technologies for avoiding interplanetary collision hazards. In *Hazards Due to Comets and Asteroids* (T. Gehrels, ed.), pp. 1145–1156. Univ. of Arizona, Tucson.

Hills J. G. and Goda M. P. (1993) The fragmentation of small asteroids in the atmosphere. *Astron. J., 105,* 1114–1144.

Hills J. G. and Mader C. L. (1997) Tsunami produced by the impacts of small asteroids. *Ann. N.Y. Acad. Sci., 822,* 381.

Hills J. G., Nemchinov I. V., Popov S. P., and Teterev A. V. (1994) Tsunami generated by small asteroid impacts. In *Hazards Due to Comets and Asteroids* (T. Gehrels, ed.), pp. 779–790. Univ. of Arizona, Tucson.

Isobe I. (1999) Japanese 0.5 m and 1.0 m telescopes to detect space debris and near-Earth asteroids. *Adv. Space Res., 23,* 33–35.

Jedicke R., Larsen J., and Spahr T. (2002) Observational selection effects in asteroid surveys and estimates of asteroid population sizes. In *Asteroids III* (W. F. Bottke Jr. et al., eds.), this volume. Univ. of Arizona, Tucson.

McCord T. B., Morris J., and Persing D. (1995) Detection of a meteoroid entry into the Earth's atmosphere on February 1, 1994. *J. Geophys. Res., 100,* 3245–3250.

Melosh H. J. (1989) *Impact Cratering: A Geologic Process.* Oxford Univ., New York (see also http://www.lpl.arizona.edu/tekton/crater.html).

Melosh H. J., Schneider N. M., Zahnle K., and Latham D. (1990) Ignition of global wildfires at the Cretaceous/Tertiary boundary. *Nature, 343,* 251–254.

Melosh H. J., Nemchinov I. V., and Zetzer Yu. I. (1994) Non-nuclear strategies for deflecting comets and asteroids. In *Hazards Due to Comets and Asteroids* (T. Gehrels, ed.), pp. 1111–1134. Univ. of Arizona, Tucson.

Milani A. and Valsecchi G. (1999) The asteroid identification problem — II. Target plane confidence boundaries. *Icarus, 140,* 408–423.

Milani A., Chesley S. R., Boattini A., and Valsecchi G. B. (2000a) Virtual impactors: Search and destroy. *Icarus, 144,* 12–24.

Milani A., Chesley S. R., and Valsecchi G. B. (2000b) Asteroid close encounters with Earth: Risk assessment. *Planet. Space Sci., 48,* 945–954.

Milani A., Chesley S. R., Chodas P. W., and Valsecchi G. B. (2002) Asteroid close approaches: Analysis and potential impact detection. In *Asteroids III* (W. F. Bottke Jr. et al., eds.), this volume. Univ. of Arizona, Tucson.

Morrison D., ed. (1992) *The Spaceguard Survey: Report of the NASA International Near-Earth-Object Detection Workshop.* NASA, Washington, DC (available on line at http://impact.arc.nasa.gov).

Morrison D. and Teller E. (1994) The impact hazard: Issues for the future. In *Hazards Due to Comets and Asteroids* (T. Gehrels, ed.), pp. 1135–1144. Univ. of Arizona, Tucson.

Morrison D., Chapman C. R., and Slovic P. (1994) The impact hazard. In *Hazards Due to Comets and Asteroids* (T. Gehrels, ed.), pp. 59–92. Univ. of Arizona, Tucson.

Muinonen K. (1998) Discovery and follow-up simulations for small Earth crossing asteroids. *Planet. Space Sci., 46,* 291–297.

National Research Council (2001) *Astronomy and Astrophysics in the New Millennium.* National Academy Press, Washington, DC.

Opik E. J. (1958) On the catastrophic effect of collisions with celestial bodies. *Irish Astron. J., 5,* 36.

Opik E. J. (1960) The lunar surface as an impact counter. *Mon. Not. R. Astron. Soc., 120,* 404.

Park R. L., Garver L. B., and Dawson T. (1994) The lesson of Grand Forks: Can defense against asteroids be sustained? In *Hazards Due to Comets and Asteroids* (T. Gehrels, ed.), pp. 1225–1232. Univ. of Arizona, Tucson.

Pope K. O. (2002) Impact dust not the cause of the Cretaceous-Tertiary mass extinction. *Geology, 30,* 99–102.

Rabinowitz D. L., Bowell E., Shoemaker E. M., and Muinonen L. (1994) The population of Earth-crossing asteroids. In *Hazards Due to Comets and Asteroids* (T. Gehrels, ed.), pp. 285–312. Univ. of Arizona, Tucson.

Rabinowitz D., Helin E., Lawrence K., and Pravdo S. (2000) A reduced estimate of the number of kilometre-sized near-Earth asteroids. *Nature, 403,* 165–166.

Raup D. M. (1990) Impact as a general cause of extinction: A feasability test. In *Global Catastrophes in Earth History* (V. L. Sharpton and P. D. Ward, ed.), pp. 27–32. Geol. Soc. Am. Spec. Paper 247.

Raup D. M. (1991) *Extinction — Bad Genes or Bad Luck.* Norton, New York. 210 pp.

Sagan C. (1992) Between enemies. *Bull. Atom. Scientists, 48,* 24.

Sagan C. (1994) The Marsh of Camarina. In *Pale Blue Dot: A Vision of the Human Future in Space.* Random House, New York. 429 pp.

Sagan C. and Ostro S. (1994) Dangers of asteroid deflection. *Nature, 369,* 501.

Seamone E. R. (2002) When wishing on a star just won't do: The legal basis for international mitigation of asteroid impacts and similar transboundary disasters. *Iowa Law Rev., 87,* 1091–1139.

Sekanina Z. (1983) The Tunguska event: No cometary signature in evidence. *Astron. J., 88,* 1382–1414.

Sekanina Z. (1998) Evidence for asteroidal origin of the Tunguska object. *Planet. Space Sci., 46,* 191–204.

Sharpton V. L., Dalrymple G. B., Marin L. E., Ryder G., Schuraytz B. C., and Urrutia-Fucugauchi J. (1992) New links between the Chixculub impact structure and the Cretaceous/Tertiary boundary. *Nature, 359,* 819–821.

Shoemaker E. M. (1983) Asteroid and comet bombardment of the Earth. *Annu. Rev. Earth Planet. Sci., 11,* 461–494.

Shoemaker E. M., ed. (1995) *Report of the Near-Earth Objects Survey Working Group.* NASA, Washington DC. (Available on line at http://impact.arc.nasa.gov.)

Shoemaker E. M., Chao E. C. T., and Madsen B. M. (1960) First natural occurrence of coesite from Meteor Crater, Arizona. *Science, 132,* 220–222.

Shoemaker E. M., Weissman P. R., and Shoemaker C. S. (1994) The flux of periodic comets near the Earth. In *Hazards Due to Comets and Asteroids* (T. Gehrels, ed.), pp. 313–336. Univ. of Arizona, Tucson.

Simonenko V. A., Nogin V. N., Petrov D. V., Shubin O. N., and Solem J. C. (1994) Defending the Earth against impacts from large comets and asteroids. In *Hazards Due to Comets and Asteroids* (T. Gehrels, ed.), pp. 929–954. Univ. of Arizona, Tucson.

Steel D. I. (1998) Distributions and moments of asteroid and comet impact speeds upon the Earth and Mars. *Planet. Space Sci., 46,* 473–478.

Stokes G. H., Evans J. B., Viggh H. E. M., Shelly F. C., and Pearce E. C. (2000) Lincoln Near-Earth Asteroid Program (LINEAR). *Icarus, 148,* 21–28.

Stokes G. H., Evans J. B., and Larson S. M. (2002) Near-Earth asteroid search programs. In *Asteroids III* (W. F. Bottke Jr. et al., eds.), this volume. Univ. of Arizona, Tucson.

Stuart J. S. (2001) A near-Earth asteroid population estimate from the LINEAR Survey. *Science, 294,* 1691–1693.

Swisher C. C. III, Grajales-Nishimura J. M., Montanari A., Margolis S. V., Claeys P., Alvarez W., Renne P., Cedillo-Pardo E., Maurasse F. J.-M., Curtis G. H., Smit J., and McWilliams M. O. (1992) Coeval $^{40}Ar/^{39}Ar$ ages of 65.0 million years ago from Chixculub crater melt rock and Cretaceous-Tertiary boundary tektites. *Science, 257,* 954–958.

Tagliaferri E., Spalding R., Jacobs C., Worden S. P, and Erlich A. (1994) Detection of meteoroid impacts by optical sensors in Earth orbit. In *Hazards Due to Comets and Asteroids* (T. Gehrels, ed.), pp. 199–220. Univ. of Arizona, Tucson.

Toon O. B., Zahnle K., Turco R. P., and Covey C. (1994) Environmental perturbations caused by impacts. In *Hazards Due to Comets and Asteroids* (T. Gehrels, ed.), pp. 791–826. Univ. of Arizona, Tucson.

Toon O. B., Zahnle K., Morrison D., Turco R. P., and Covey C. (1997) Environmental pertubations caused by the impacts of asteroids and comets. *Rev. Geophys., 35,* 41–78.

Urey H. C. (1973) Cometary collisions and geological periods. *Nature, 242,* 32–33.

Vasilyev N. V. (1998) The Tunguska meteorite problem today. *Planet. Space Sci., 46,* 129–150.

Ward S. N. and Asphaug E. (2000) Asteroid impact tsunami: A probabilistic hazard assessment. *Icarus, 145,* 64–78.

Weissman P. R. (1994) The comet and asteroid impact hazard in perspective. In *Hazards Due to Comets and Asteroids* (T. Gehrels, ed.), pp. 1191–1212. Univ. of Arizona, Tucson.

Werner S. C., Harris A. W., Neukum G., and Ivanov B. A. (2002) The near-Earth asteroid size-frequency distribution: A snapshot of the lunar impactor size-frequency distribution. *Icarus, 156,* 287–290.

Zahnle K. (1996) Leaving no stone unburned. *Nature, 383,* 674–675.

Zahnle K. and MacLow M.-M. (1994) The collision of Jupiter and Comet Shoemaker-Levy 9. *Icarus, 108,* 1–17.

Zahnle K. and MacLow M.-M. (1995) A simple model for the light curve generated by a Shoemaker-Levy 9 impact. *J. Geophys. Res., 100,* 16885–16894.

Part VI:

Databases

Asteroid Data Archiving in the Planetary Data System

C. L. Neese
Planetary Science Institute

E. J. Grayzeck
University of Maryland

M. V. Sykes
University of Arizona

The archiving of observational data in asteroid science is an essential component contributing to the productivity of the field. The usefulness of present-day research efforts requires the context provided by data from past missions and groundbased observational programs, and new knowledge allows new interpretations of old data. The rapidly expanding volume of new data and results requires the full use of modern information technologies for the ongoing preservation and documentation of asteroid data for the future. The Planetary Data System, created by NASA for the archiving of planetary mission data, serves many of these functions for the benefit of the asteroid science community. This chapter outlines the role of archiving in asteroid science and how individual researchers can contribute by archiving their basic and derived observational data.

1. IMPORTANCE OF DATA ARCHIVING IN ASTEROID SCIENCE

Science is an endeavor that advances as we build on our observations and theories, testing our understanding against the accumulating body of new and old data. Archiving those data and making them easily accessible to the science community is an essential element in our science enterprise. The more complete our holdings are and the easier it is to identify needed data, the better.

In the year 2001, the number of asteroids for which orbits were known exceeded 146,000. The number of asteroids for which we have more than orbital elements and an estimate of their absolute magnitude is small, but inexorably growing. Virtually all asteroid data taken over time have unique information that is waiting to be extracted. Some of this is coarse, such as a rotational period from a lightcurve or principal mineral components from a spectrum. Some is more subtle, such as the effects of surface scattering properties on that lightcurve or spectrum. What we observe is often subject to rotational phase, sub-Earth latitude, heliocentric distance, illumination and exitance angles, and many other factors. It is almost impossible to precisely reproduce all the conditions of a given asteroid observation — something will be different. These differences may give rise to new insights.

Data can give rise to science long after being collected. The shape model derived from spacecraft data may be used to reinterpret old groundbased photometry. Rotational pole orientations determined from Hubble Space Telescope or lightcurve studies may be used to reinterpret old radiometric observations. Whether old data are useful in the future depends upon their adequate characterization. Often deficits can be addressed by contacting the person who collected the data or was otherwise involved in the experiment. However, life or death can intervene, making that information unavailable. If a data archive is to have long-term value, the data themselves are not enough — the context of the data must be documented while the means or people needed to fill any holes in that description are still available.

Short-term access is also critical. Unavailable data are not useful to anyone. Historically, data have been distributed to the science community through its literature — print media. Space and expense have generally been a factor limiting the information presented to that of immediate interest. The possibility of future usefulness in a different context, addressing different scientific issues, has been at best a secondary consideration. The advent of the capability for electronic distribution of large volumes of data and associated information has removed these constraints. As a consequence, modern archiving offers an enhancement of science productivity unknown in the past.

2. ROLE OF THE PLANETARY DATA SYSTEM IN ASTEROID DATA ARCHIVING

The NASA Planetary Data System (PDS) was created to archive and distribute NASA solar system mission data. It also preserves other data needed by the planetary research community, including groundbased observations needed to place data from missions in context and comparative data on many small bodies for use in planning possible future missions (*Grayzeck et al.,* 1996).

The PDS is divided into a number of discipline nodes, each responsible for archiving data that fall within a general subject area. The underlying reason for this organizational design is the conscious involvement of research scientists in both archiving and peer review to best ensure the long-term viability and usefulness of the data.

Data related to asteroids, comets, and interplanetary dust are handled by the Small Bodies Discipline Node (PDSSBN). From the outset, the PDSSBN has recognized that both short- and long-term usefulness of spacecraft data in this area requires the context provided by groundbased telescope and laboratory data. Furthermore, mission design and planning depends upon groundbased telescope and laboratory data. This presents special challenges.

For an asteroid mission, such as *NEAR Shoemaker*, detailed information is required about the instruments, their calibration, data architecture, spacecraft position and pointing, and many other factors. Groundbased data on asteroids are acquired by numerous investigators using different instruments on different telescopes in different regions of the world. As with spacecraft data, sufficient information must be gathered about groundbased instruments, their detectors, calibration, filter profiles, observing conditions, and many other factors. The diversity of data sources in this case greatly amplifies the problem of ensuring long-term and uniform viability and usability in the PDS archive. To mitigate this, the PDSSBN investigates and utilizes modern information technologies in the acquisition, evaluation, and distribution of data.

3. COMMUNITY INPUT TO PLANETARY DATA SYSTEM ARCHIVING

The success of the PDS depends upon the community contributing its data and supporting documentation, providing advice and feedback to the PDS, and serving on its peer-review panels. For the PDS to be something more than a data mortuary, it populates its discipline nodes with scientists having a knowledge of and research interest in the data being archived. This ensures a common basis of understanding with the provider community and the expertise to evaluate data completeness and description. It also allows for community support in accessing and understanding the data archived.

In addition, the SBN has an advisory panel of scientists who participate in data reviews, and attend the annual node meeting where data ingestion priorities are set annually. The node meetings are usually organized to coincide with international gatherings such as the Division of Planetary Science meetings. At the 2000 DPS meeting, the PDS had a hands-on demonstration in which SBN personnel participated. One of the features of that session was a public Web-based survey to assess the usefulness of PDS and the characteristics of our users. (These surveys are repeated every five years to assess the direction of the overall PDS and report back to the general science advisers.) The results of that survey are that over 75% of respondents use PDS, 70% are full-time planetary scientists, and 80% receive NASA Code S funding. A majority get data from a node; this total is more than the number who get data from a mass distribution. Future improvements requested are more Internet distribution, better Web interfaces, and finer granularity in searching algorithms. DVD distribution and custom volumes were relatively unpopular as future directions.

Finally, community participation in peer reviews is critical. All data formally incorporated into the PDS archives must first undergo a peer review. The purpose of the review is to determine that submitted data are complete, correct, and suitable for archiving. External reviewers are invited who are knowledgeable in the type of data being reviewed, but not directly connected with the project that produced it. The review panel includes these external peer reviewers as well as PDS representatives who review the formatting and structure of the datasets. Non-PDS users are often included on these panels so that they can become familiar with the archive and give feedback both regarding the data and the tools supplied for data access.

4. ARCHIVING DATA IN THE PLANETARY DATA SYSTEM SMALL BODIES NODE

This section applies primarily to individual researchers with asteroid datasets that they would like to submit for archiving in the PDS. The PDSSBN ordinarily prefers to archive published data, although in special cases a particularly interesting but unpublished dataset can be archived. Note that the purpose of the PDSSBN asteroid archive is the long-term preservation of data to insure its continued availability and usefulness after the original researchers are no longer available to interpret it.

Submission of asteroid data for archiving is a public service to the future of the asteroid research community. Additional more immediate benefits include improved data availability and cross-fertilization for present-day researchers and the value of groundbased asteroid data for planning future NASA missions. PDS-archived datasets can also be cited as a publication by any subsequent research making use of the data. See the PDSSBN Web site (address below) for current guidelines for citing PDS datasets in publications.

To arrange to submit asteroid data to the PDSSBN, contact the Asteroid Subnode archive scientist (currently Carol Neese at the Planetary Science Institute). You will be asked to supply electronic copies of the data files themselves as well as all ancillary information, including observational circumstances, calibration information, definition of all terms used, other needed documentation, and the citation of publications in which the data have been published. Oftentimes for small datasets, all or most of the needed information can be compiled from the published papers. In other cases you will be asked to supply additional information to support the data.

The submitted datasets undergo external peer review on an annual basis, usually in the spring. Reviewers will check the datasets for completeness, intelligibility, interpretability, and compatibility with the archive goals of the PDS. They will recommend any needed changes to the dataset to insure the quality of the archive, and if needed, you will be asked to implement changes to the dataset before final archiving.

5. PLANETARY DATA SYSTEM ASTEROID DATA ARCHIVE ON THE WEB

The PDSSBN Asteroid Data Archive is available at pdssbn.astro.umd.edu. There you can find access to the currently archived datasets, including asteroid spectrophotometric catalogs, lightcurves, polarimetry, radar data, and many other basic datasets. In addition to the basic data are many derived datasets, including asteroid taxonomic classifications, spin vectors, rotational periods, diameters, and others. There are also some historical datasets including discovery circumstances for the numbered asteroids.

At the same site are software archives making available SBN-developed tools for preparation, manipulation, and interpretation of archived datasets. Also available here are a SBN interactive database that allows interactive access to the data, and a small bodies cross-identification database for both comets and asteroids.

6. OTHER ASTEROID DATA ARCHIVES

An essential asteroid data archive is provided by the Minor Planet Center (MPC) at cfa-www.harvard.edu/cfa/ps/mpc.html. The MPC, well known to all asteroid researchers, collects astrometric and photometric asteroid data from observers, computes orbits, and disseminates both data and current orbits on a short enough timescale to permit immediate follow-up observations of the objects by observers worldwide. The scope of the MPC includes all small solar system bodies, not just asteroids. The fast turnaround provided by the MPC, unattainable by the PDS with its annual data review cycle geared primarily to spacecraft and ground-based physical observations, is necessary for the efficient collection and dissemination of the very dynamic asteroid orbital data.

Other important asteroid archives accessible on the Web include the European Asteroid Research Node (EARN) at www.astro.uu.se/planet/earn, the Near Earth Object Dynamics Site (NeoDys) at newton.dm.unipi.it/cgi-bin/neodys/neoibo, and the Asteroids Dynamic Site at hamilton.dm.unipi.it/cgi-bin/astdys/astibo. EARN is a cooperative archive with contributors from many European institutions doing active asteroid research. Its data archive includes asteroid photometric lightcurves, proper elements, spin vectors, family memberships, rotation periods, and physical properties of NEOs. NeoDys and AstDys provide an interactive relational database for data and information on near-Earth asteroids and main-belt asteroids respectively, including astrometry, physical parameters, orbits, proper elements, and radar observations.

Many individual researchers maintain distribution sites for their data on the Web. These sites can be very valuable for current access to asteroid data, but their explosive proliferation makes it impractical to cite them individually in this chapter.

REFERENCES

Grayzeck E. J. Jr., A'Hearn M. F., Raugh A. C., Sykes M. V., Davis D. R., and Tholen D. (1996) Services of the Small Bodies Node of the Planetary Data System. *Planet. Space Sci., 44,* 47–54.

Glossary

a — semimajor axis of an orbit.

A — bolometric Bond or spherical albedo. A = pq, where p is the bolometric geometric albedo and q the phase integral.

Å — angstrom = 10^{-8} cm.

ablation — removal of material by attrition, e.g., by passage through the atmosphere.

absolute magnitude — the magnitude of an asteroid at zero phase angle and at unit heliocentric and geocentric distances.

accretion — the process by which matter assembles to form larger bodies such as stars, planets, and satellites.

accumulation — *see* accretion.

achondrite — meteorite of nonsolar composition, also known as differentiated stony meteorite.

A class — a rare asteroid taxonomic classification denoted by moderately high albedos and extremely reddish spectra shortward of 0.7 μm. A very strong near-infrared absorption feature centered around 1.05 μm is interpreted as being due to olivine.

active comet — a comet nucleus losing volatiles and dust in a detectable coma. *See also* comet, inactive comet, dormant comet, and extinct comet.

adaptive optics — a technique for removing temporally and spatially varying optical wavefront aberrations (seeing effects) in real time in order to obtain images with resolutions close to the theoretical diffraction limits of telescopes.

agglutinate — small objects consisting of glass and fragments of minerals or rocks, all welded together into an aggregate; they are produced by micrometeorite impact into fine-grained unconsolidated regolith.

albedo, Bond — fraction of the total incident light reflected by a spherical body. Bolometric Bond albedo refers to reflectivity over all wavelengths.

albedo, geometric — ratio of planet brightness at zero phase angle to the brightness of a perfectly diffusing disk with the same position and apparent size as the planet. Bolometric geometric albedo refers to reflectivity over all wavelengths.

albedo, hemispherical — fraction of incident light scattered by a surface as a function of angle of incidence.

albedo, normal — the brightness of a surface at zero phase angle relative to a perpendicularly illuminated, perfectly diffusing (Lambert) surface at the same distance as the surface.

albedo, physical — *see* albedo, geometric.

albedo, single particle scattering — the fraction of incident light scattered by a particle.

albedo, spherical — *see* albedo, Bond.

aliasing (radar) — overlapping of radar echoes at different frequencies or at different time delays.

α — *see* phase angle.

alteration — a change in mineralogy due to chemical reactions between solids and fluids at subsolidus temperatures.

Amor asteroids — asteroids having perihelion distance 1.017 AU < q ≤ 1.3 AU.

anorthosite — an igneous rock made up almost entirely of plagioclase feldspar.

antenna gain — ratio of an antenna's sensitivity in the direction toward which it is pointed to its average sensitivity in all directions.

aphelion — Q, distance of greatest heliocentric separation for a body in an eccentric orbit.

Apohele asteroids — proposed name for a class of objects whose orbits are entirely inside the Earth's (aphelion distance Q < 0.983 AU).

Apollo asteroids — asteroids having semimajor axis a ≥ 1.0 AU, and perihelion distance q ≤ 1.017 AU.

arcsec — arcseconds, or second of arc, equal to 1/3600 degree.

aspect — angle between the rotation axis of the body and the radius vector to Earth.

asteroid — one of a number of objects ranging in size from subkilometer to about 1000 km, many of which lie between the orbits of Mars and Jupiter. These bodies formed without significant ice content, and thus are incapable of displaying cometary activity.

asteroid belt — a region of space lying between Mars and Jupiter, where a large fraction of the asteroids are found.

asteroid pole — expression used to indicate the coordinates in the celestial sphere of the direction corresponding to that of the spin axis of an asteroid.

Aten asteroids — asteroids having semimajor axis a < 1.0 AU, and aphelion distance Q > 0.983 AU.

AU — astronomical unit. The mean distance of Earth from the Sun, equal to 1.496×10^{13} cm.

aubrite — alternate name for enstatite achondrite. A differentiated stony meteorite consisting predominantly of enstatite with very low Fe content, highly reduced, possibly related to enstatite chondrites.

B(1,0) — old-style B-band absolute magnitude, now superseded by H. H ≈ B(1,0)–1.0 mag.

backscatter — scattering of radiation (or particles) through angles greater than 90° with respect to the original direction of motion.

barycenter — the center of mass of a system.

basalt — a dark, fine-grained, mafic igneous rock composed primarily of plagioclase and pyroxene.

basaltic achondrite — collective name for eucrites and howardites that superficially resemble terrestrial basalts or their fragmentation products.

B class — a subclass of the C asteroids, distinguished by higher albedos than the average C type.

binary — a system composed of two components that are physically separated but gravitationally bound to one another.

blackbody — an idealized body that absorbs all radiation of all wavelengths incident on it. The radiation emitted by a blackbody is a function of temperature only. Because it is a perfect absorber, it is also a perfect emitter.
bolometric — including radiation over all wavelengths.
breccia — a clastic rock composed of angular, broken rock fragments that are embedded into a finer-grained matrix. *See also* genomict, monomict, and polymict.
brecciation — breakage of a rock into smaller fragments.
brightness temperature — the temperature that a blackbody would have order to emit radiation of the observed intensity at a given wavelength. This quantity is particularly useful when the Rayleigh-Jeans approximation is valid (as it often is in radio astronomy) because in this approximation it is directly proportional to the specific intensity. It is useful whenever there is reason to believe that it corresponds to a physical temperature; in other cases it merely indicates the radiation's intensity at a given wavelength.
bulk density — the mass of an object divided by its volume including the volume of the pore spaces.
c — speed of light in a vacuum = 2.998×10^{10} cm s^{-1}.
CAI — calcium-aluminum-rich inclusions. Inclusions rich in these elements are abundant in CV and, to a lesser degree, CM chondrites.
carbonaceous chondrite — a chondritic meteorite, generally containing more than about 0.2 wt% C. Most such chondrites are highly oxidized and have nearly solar composition for all but the most volatile elements. It is the most primitive (least processed) type of meteorite.
cataclastic — a type of structure produced in a rock by severe deformation resulting in fracturing and rotation of mineral grains.
catastrophic disruption — term applied to collisional breakup when the mass of the largest postimpact fragment is ≤50% of the original target mass. Note that the largest postimpact fragment may be formed by partial reaccumulation of fragments due to self-gravity. This term only differs from catastrophic shattering in a regime where the effect of gravity is much larger than that of solid state forces.
catastrophic shattering — term applied to collisional breakup when the largest fragment produced has a mass of one-half the original target, regardless of the reaccumulation of fragments due to self-gravity. This term is frequently used when describing the results of laboratory experiments.
CCD — charge-coupled device. An electronic detector with a quantum efficiency (efficiency with which an incident photon is likely to be detected) as high as 90%.
C class — a very common asteroid type in the outer part of the main belt, typically having flat spectra longward of 0.4 µm and presumably similar in surface composition to some carbonaceous chondrites. The relative strength of a UV absorption feature may be correlated with the presence of water of hydration. B, F, and G are subclasses of the C class.
celestial equator — the projection of Earth's equator onto the celestial sphere.
Centaurs — icy objects that have orbits among the giant planets and are thought to be Kuiper Belt objects whose orbits have been perturbed by Neptune. These objects have orbits entirely exterior to Jupiter. Some classify them as having Tisserand parameter $T > 3$ and $a > a_{jupiter}$. These objects are a subclass of ecliptic comets.
chaotic orbit — unpredictable orbit characterized by at least one Lyapunov characteristic exponent strictly positive.
chassignite — a very rare type of achondrite (one known, Chassigny) consisting of olivine with minor amounts of pyroxene, plagioclase, chromite, and sulfide.
chondrite — originally defined as a meteorite that contained chondrules; now also implies a chemical composition similar to that of the Sun, for all but the most volatile elements.
chondrite, equilibrated — chondrite that has closely approached or reached internal equilibrium, presumably as a result of thermal metamorphism, so that individual grains of the same mineral have similar compositions.
chondrules — approximately spherical assemblages, characteristic of most chondrites, that existed independently prior to incorporation in the meteorite and that also show evidence for partial or complete melting.
CHON particles — light-element particles rich in carbon, hydrogen, oxygen, and nitrogen that are thought to exist in comets.
circular polarization ratio — ratio of echo power received in the same sense of circular polarization as transmitted (the same circular sense, or SC) as opposed to that received in the opposite (opposite circular, or OC) sense.
clast — a rock fragment produced by mechanical weathering of a larger rock and included in another rock.
clathrate — a structure formed by the systematic inclusion of certain molecules in cavities within a crystal lattice.
clinopyroxene — a mineral of the pyroxene group that crystallizes in the monoclinic system.
color index — the difference in magnitudes between any two spectral regions. Color index is always defined as the short-wavelength magnitude minus the long-wavelength magnitude. In the UBV system, the color index for an AO star is defined as B–V = U–B = 0; it is negative for hotter stars and positive for cooler ones.
column density — number of atoms or molecules per cm^2 in the line of sight.
coma — the usually spherical region of diffuse gas, typically 150,000 km in diameter, which surrounds the nucleus of a comet. Together, the coma and the nucleus form the comet's head.
comet — a body formed in the outer planets region (or beyond) containing a significant fraction of volatiles in the form of ices, and capable of developing a coma if its orbit brings it close enough to the Sun.
comet nucleus — the solid part of a comet, typically a few kilometers (up to tens of kilometers) in diameter, consisting of a mixture of ices and solid silicate and carbon-

aceous grains.

commensurate orbits — a term applied to two bodies orbiting around a common barycenter when the period of one is an integral multiple of that of the other.

communition — the reduction of a rock to progressively smaller particles by weathering, impacts, erosion, etc.

condensation — transformation from the gaseous to a solid or liquid phase. In the context of this book, it is generally taken to refer to the formation of solid grains from nebular gas.

conjunction — *see* elongation.

cosmic-ray exposure age — the period of time during which a meteorite was exposed to cosmic radiation, commonly the time between its final reduction in size by impact and its arrival on Earth. More generally, it is the time spent within a few meters of the space environment. Nuclear reactions between the radiation and nuclides in the meteorite produce new nuclides or associated phenomena such as tracks, whose abundances can be used to estimate the exposure age.

cosmic rays — atomic nuclei (mostly protons) that are observed to strike Earth's atmosphere with exceedingly high energies.

crust — the outermost, highly differentiated, solid layer of a planet or satellite, mostly consisting of crystalline rock or ice.

Cybele asteroids — asteroids located beyond the main belt with semimajor axes between 3.3 and 3.6 AU. The Cybele asteroid group is bracketed by the 2:1 and 5:3 mean-motion resonances with Jupiter.

D class — an asteroid type that is rare in the main belt, but becomes increasingly dominant beyond the 2:1 jovian resonance. Their spectra are neutral to slightly reddish shortward of 0.5 μm, very red longward of 0.55 μm, and for some objects the spectrum tends to flatten longward of 0.95 μm. Coloring may be due to kerogen-like materials.

depolarization — randomization of polarization by multiple scattering or single scattering off wavelength-scale structure.

differentiation (in a planet) — a process whereby the primordial substances are separated. Generally metal sinks to the center to form a core, displacing the lighter silicates that form the crust plus mantle.

Doppler shift — difference between the frequencies of the radar echo and the transmission, caused by the relative velocity of the target with respect to the radar.

dormant comet — a comet nucleus that, although once active, has lost the ability to generate a detectable coma in any part of its present orbit. A dormant comet perturbed to a smaller perihelion distance may be reactivated, or an impact might remove an overlying nonvolatile crust and expose fresh icy materials to sublimation, reactivating the comet. *See also* comet, active comet, inactive comet, and extinct comet.

dunite — an ultramafic rock composed of at least 90% olivine.

dynamical family — a statistically significant cluster of asteroids in proper element space. *See also* family.

e — eccentricity of an elliptical orbit. The amount by which the orbit deviates from circularity: e = c/a, where c is the distance from the center to a focus and a is the semimajor axis.

ECAs — Earth-crossing asteroids. *See* ECOs.

ECCs — Earth-crossing comets. *See* ECOs.

echo bandwidth — dispersion in Doppler frequency of an echo, that is, the width of the echo power spectrum.

E class — a rare asteroid type with featureless 0.3- to 1.1-μm spectra (identical to M and P classes) but distinguished by high albedos. Surface composition may be similar to enstatite achondrites.

ecliptic — plane of Earth's orbit.

ecliptic comets — comets with Tisserand parameter $T > 2$. Most members of this population have small inclinations. Subclasses of this population include the Jupiter-family comets, the Encke-type comets, and the Centaurs.

ECOs — Earth-crossing objects. Asteroids and comets with perihelion distances $q \leq 1.017$ AU and aphelion distances $Q \geq 0.983$. Subcategories include the Apollo and Aten objects.

ejecta — materials ejected from a crater either by the action of volcanism or a meteoroid impact.

ejecta blanket — the deposit surrounding an impact crater composed of material ejected from the crater during its formation.

elongation — the angle planet-Earth-Sun. Eastern elongations appear east of the Sun in the evening; western elongations, west of the Sun in the morning. An elongation of 0° is called conjunction, one of 180° is called opposition, and one of 90° is called quadrature.

emissivity — ratio of the radiation emitted by a body to that emitted by a blackbody at the same temperature.

Encke-type comets — comets with Tisserand parameter $T > 3$ and $a < a_{jupiter}$. This combination of T and a implies that the orbits of these objects are entirely interior to Jupiter. The most prominent member of this class is Comet 2P Encke. These objects are a subclass of ecliptic comets.

endogenic — originating within a planetary or planetesimal object.

endothermic reaction — a reaction that consumes energy.

enstatite chondrite — collective name for the EH and EL classes of chondritic meteorite, highly reduced chondrites with Mg/Si around 0.83.

ephemeris — (pl., ephemerides) a list of computed positions occupied by a celestial body over successive intervals of time.

equilibrium condensation model — a model for the chemical composition of the planets in which solids are hypothesized to have condensed from an initially hot nebula of solar composition, which cools slowly enough so that chemical equilibrium is maintained, and in which accretion takes place rapidly enough so that the solids may be characterized as being due to condensates at a particular temperature, which decreases with increasing distance of the planet from the Sun.

escape velocity — the velocity required to escape entirely from the gravitational field of an object; also the minimum impact velocity for any body arriving from a very great distance.

eucrite — a class of achondritic meteorite consisting of Ca-pyroxene and plagioclase.

eV — electron volt = 1.602×10^{-12} ergs.

extinct comet — a comet nucleus that has lost its ices or has its ices so permanently buried under a nonvolatile crust that it is incapable of generating a coma. *See also* comet, active comet, inactive comet, and dormant comet.

exogenic — originating externally to a planetary or planetesimal object.

exposure age — *see* cosmic-ray exposure age.

fall — a meteorite that was seen to fall. Such meteorites are usually recovered soon after the fall and are relatively free of terrestrial contamination and weathering effects.

family — a statistically significant cluster of asteroids in proper-element space that may share a common origin, perhaps by the collisional disruption of a larger parent body.

fast-rotating thermal model — a simplistic thermal model for asteroids and other airless bodies that assumes the ideal situation of a rapidly rotating and/or rocky spherical body with the Sun and Earth in the equatorial plane. Thus, the body has a temperature distribution that is isothermal in longitude, i.e., the thermal flux is a function only of equatorial distance (latitude).

F class — a subclass of the C asteroids, distinguished by a weak to nonexistent UV absorption feature.

feldspars — common aluminous silicate minerals in meteorites and other rocks. Plagioclase feldspars are members of a solid solution series that varies continuously from Na-rich to Ca-rich compositions.

find — a meteorite that was not seen to fall but was found and subsequently recognized.

fireball — *see* meteor.

Fourier analysis — the analysis of a periodic function into its simple harmonic components.

fractured asteroid — an asteroid with a significant number of faults and/or joints that may lack enough tensile strength to resist disruption by long-term stresses such as tidal forces or rotational spinup. The original structure of the object is still largely in place so the macroporosity is low. The reaction of such a body to short-term stresses like collisions is between that of a monolith and a shattered asteroid. See chapter by Richardson et al.

g — local gravitational acceleration.

G — V-band slope parameter in the H,G magnitude system. $G \approx 0$ pertains to steep phase curves, such as those of low-albedo asteroids; $G \approx 1$ to shallow phase curves, such as those of icy satellites.

gardening — reworking and overturning of a regolith, principally by micrometeoroid bombardment.

gas-retention ages — the age of a meteorite as calculated from the abundance of gaseous daughter products.

G class — a subclass of the C asteroids, distinguished by a strong UV absorption feature.

genomict — a breccia in which the components originated in distinct but genetically closely related rocks.

geocentric — Earth-centered.

grain density — the mass of an object divided by the volume occupied only by mineral grains. The grain densities of many rock-forming minerals in asteroids are known because their crystal structures, lattice volumes, and elemental compositions are well defined. See chapter by Britt et al.

granite — an igneous rock associated primarily with Earth's continental crust, composed chiefly of quartz and alkali feldspar.

gravitational aggregate — a collection of asteroid or cometary components held together by self-gravity rather than physical strength. Such bodies are susceptible to disruption by long-term stresses such as tidal forces or rotational spinup. See chapter by Richardson et al.

gravitational constant, G — the constant of proportionality in the attraction between two unit masses a unit distance apart. $G = 6.668 \times 10^{-8}$ dyn cm^2 g^{-2}.

G.y. — gigayear = 10^9 yr.

H — absolute magnitude in the H,G magnitude system. H pertains to the V band unless subscripted otherwise (e.g., H_B). It is the time-averaged magnitude of an asteroid, calculated at zero phase angle and unit heliocentric and geocentric distances.

half-life — the time required for half of the radioactive atoms in a sample to disintegrate.

H(α) — reduced V-band magnitude of an asteroid at phase angle α as calculated using the H,G magnitude system: $H(\alpha) = H - 2.5\log[(1 - G)\Phi_1(\alpha) + G\Phi_2(\alpha)$, where H is the absolute magnitude, G is the slope parameter, and Φ_1 and Φ_2 are specified phase functions.

heavy bombardment — the period of time, beginning during planetary formation, when the cratering rate was high throughout at least the inner solar system. The duration of this bombardment is unknown, though some claim it may have lasted until about 3.8 billion years ago. *See also* late heavy bombardment.

heliocentric — Sun-centered.

H-function — (Chandrasekhar H-function) the solution to an integral equation that appears in radiative transfer problems. The H-function describes the results of multiple-scattering in an atmosphere or surface, and depends on the angle of incident or emitted radiation, and on the single-scattering albedo.

H,G magnitude system — a means of calculating the magnitudes of asteroids adopted by the IAU in 1985.

Hilda asteroids — asteroids at low eccentricities located near or inside the 3:2 mean-motion resonance with Jupiter at 4.0 AU.

Hill sphere — the approximately spherical region within which a planet, rather than the Sun, dominates the motion of particles.

Hirayama family — same as "family," although it may refer specifically to one of the clusters first noted by K. Hirayama in the early twentieth century.

horseshoe orbit — the motion of an orbiting particle that alternately nearly overtakes another body and then slows down so as to be nearly overtaken by the other body. In a reference frame rotating with the orbit of the other body's orbital motion, the particle follows a horseshoe-shaped path.

howardite — polymict brecciated achondrite consisting predominantly of lithic units similar to eucrites and diogenites, although more extreme compositions are also found.

HST — Hubble Space Telescope.

Hungaria asteroids — asteroids in the innermost portion of the main belt with semimajor axes near 1.9 AU, low eccentricities, and inclinations between ~16° and 30°. The Hungaria asteroid group lies near the edges of the ν_5 and ν_{16} secular resonances and is bracketed by the 4:1 and 5:1 mean-motion resonances with Jupiter.

hydrated mineral — broadly speaking, any mineral containing water or hydroxyl (OH), either loosely bound or as part of its structural formula. These minerals are found in meteorites (most commonly in carbonaceous chondrites of types CI and CM) and there is evidence for them on many different asteroid types based on their reflectance spectra.

hydration — term indicating the changes in mineralogical properties due to the exposure of materials to chemical reactions with water.

i — inclination of an orbit. The angle between an asteroid's orbit and the plane of the ecliptic (or between a satellite's orbit and the planet's equatorial plane).

IAU — International Astronomical Union.

IAUC — *International Astronomical Union Circulars.*

IDP — interplanetary dust particle, also known as a micrometeoroid or, after entry into the Earth's atmosphere, a micrometeorite.

IEOs — inner Earth objects (also referred to as objects inside Earth's orbit). *See* Apohele asteroids.

igneous — a term used to describe the melting and subsequent solidification of a rock.

impact melt — target material that was melted by the heat generated by an impact.

impact strength (S_0) — energy density (specific energy times target density ρ) required to produce a barely catastrophic outcome.

inclusions — aggregates of mineral grains that existed independently prior to incorporation in the meteorite. *See* CAI.

infrared (IR) — that part of the electromagnetic spectrum that lies beyond the red, having wavelengths from about 7500 Å to a few millimeters (about 10^{11}–10^{14} Hz). Infrared radiation can be produced by atomic transitions, or by vibrational (near-IR) and rotational (far-IR) transitions in molecules. Planetary thermal emissions peak in the infrared.

interstellar grains — small solid particles (including silicates) that exist in interstellar space; some may have become incorporated into comets and meteorite parent bodies and preserved (i.e., not melted or vaporized through the formative and later periods of solar system history).

IRAS — Infrared Astronomical Satellite.

iron meteorite — a meteorite composed primarily of metallic Fe and Ni and thought to represent material from the core of a differentiated parent body.

isochemical — without change in bulk chemical composition.

isomer — one of a number of molecules that all have the same elemental composition but differ from each other in structure.

isothermal latitude model — *see* fast rotating thermal model.

isotope — any of two or more forms of the same elements, whose atoms all have the same number of protons but different numbers of neutrons.

Jacobi ellipsoid — a triaxial figure assumed by a rapidly rotating body of low strength if its specific angular momentum exceeds a critical value. Its shape is determined by self-gravity and centrifugal force, and depends only on the body's density and rotation rate.

Jupiter-family comets — comets with Tisserand parameter $2 < T < 3$ that are on Jupiter-crossing orbits and are dynamically dominated by that planet. These objects are a subclass of ecliptic comets.

Jy — Jansky; 1 Jy = 10^{-26} W m^{-2} Hz^{-1}.

k — Boltzmann constant = 1.38×10^{-16} erg deg^{-1}; alternately, = 8.62×10^{-5} eV deg^{-1}.

K — degrees Kelvin, 0 K = –273 C.

kamacite — Fe,Ni alloy of 7 wt% Ni or less with the body-centered-cubic structure. It occurs as large plates or single crystals in iron meteorites, abundant grains in chondrites, and rare grains in most achondrites.

K class — a taxonomic classification distinguished by spectra having a moderately steep UV absorption feature shortward of 0.75 μm, while having little or no near-IR absorption feature.

Kepler's laws — 1. Each planetary orbit is an ellipse with the Sun at one focus. 2. Equal areas are swept out in equal times (law of areas). 3. The square of the period is proportional to the cube of the distance (harmonic law). Newton's generalized formula for the third law is $P^2 = 4\pi^2 a^3/[G(m_1 + m_2)]$.

Kepler velocity — the orbital velocity of a gravitationally bound object around the central object, i.e., the velocity that leads to a centrifugal force exactly balancing the gravitational attraction between the two objects.

kerogen — insoluble macromolecular organic matter, operationally defined as the organic residue left after acid demineralization of a rock.

Kirkwood gaps — regions in the asteroid zone that have been swept clear of asteroids by the perturbing effects of Jupiter. They were named for the American astronomer Daniel Kirkwood, who first noted them in 1866.

Klystron — vacuum-tube amplifier used in planetary radar transmitters.

km — kilometer = 10^5 cm.

KREEP — lunar basaltic material rich in radioactive elements (K for potassium, REE for rare earth elements, P for phosphorus).

K/T event — the major break in the history of life on Earth (a mass extinction) that occurred 65 m.y. ago, between the Cretaceous and Tertiary periods, apparently due to the impact of an asteroidal object.

Kuiper Belt — a vast population of small bodies orbiting the sun beyond Neptune. Observations suggest these transneptunian objects (TNOs) are mostly confined within a thick band around the ecliptic. It is believed that this population (or the scattered disk) may be the source of the ecliptic comets.

L_4 — one of the five Lagrangian points.

$L_\odot$ — solar luminosity = 3.826×10^{33} erg s^{-1}.

Lagrangian orbit — an orbit in which a particle oscillating about one of the stable Lagrangian equilibrium points defined by the restricted three-body problem moves.

Lagrangian points — the five equilibrium points in the restricted three-body problem. Two of the Lagrange points (L_4 and L_5) are located at the vertexes of equilateral triangles formed by the two primaries (e.g., Sun and Jupiter) and are stable; the other three are unstable and lie on the line connecting the two primaries. *See* Trojans.

Lambert's law — a simple scattering law according to which the intensity of scattered light is independent of the emission angle. An ideal Lambert surface scatters light uniformly in all directions (i.e., a diffuse scatterer).

Late Heavy Bombardment — a period of time from about 4.2 to 3.8 G.y. ago when most of the basins and other craters were formed on the Moon (and possibly the terrestrial planets). *See also* heavy bombardment.

lava — molten rock erupted onto the surface of a planet.

L class — a taxonomic classification distinguished by spectra having a very steep UV absorption feature shortward of 0.75 μm, while having no near-IR absorption feature.

libration — a small oscillation around an equilibrium configuration, such as the angular change in the face that a synchronously rotating satellite presents toward the focus of its orbit.

lightcurve — brightness values plotted as a function of time. Note that this plot does not necessarily have to show variability.

lightcurve amplitude — peak-to-peak value in magnitudes of a lightcurve showing variability.

limb — the edge of the apparent disk of a celestial body, as of the Sun, the Moon, a planet, or a satellite.

lithology — the physical character of a rock.

lithophile — one of the geochemical classes of elements. Lithophile elements are those that tend to concentrate in the silicate phase, e.g., Si, Mg, Ca, Al, Na, K, and rare-earth elements.

lithosphere — the stiff upper layer of a planetary body, including the crust and part of the upper mantle, lying above the weaker asthenosphere.

lithostatic pressure — pressure due to the weight of overlying rock.

Lommel-Seeliger surface — a surface with large-scale roughness where shadowing effects are important.

Lyapunov characteristic exponents (LCE) — numbers that indicate how fast nearby orbits diverge and thus the degree of unpredictability of such orbits.

$M_\oplus$ — mass of Earth = 5.976×10^{27} g.

$M_\odot$ — mass of Sun = 1.989×10^{33} g.

macroporosity — large-scale voids and fractures on asteroids and comets. Macroporosity generally defines the internal structure of the asteroid or comet. See chapters by Britt et al. and Richardson et al.

mafic — term used to describe a silicate mineral whose cations are predominantly Mg and/or Fe. It is also used for rocks made up principally of such minerals.

magma — mobile or fluid rock material, lava, generalized to refer to any material that behaves like silicate magma in Earth.

magnitude — an arbitrary number, measured on a logarithmic scale, used to indicate the brightness of an object. If l_i is the brightness of star i, and m_i its magnitude, then $m_1 - m_2 = 2.5 \log (l_2/l_1)$. Two stars differing by 5 magnitude differ in luminosity by a factor of 100. One magnitude difference is the fifth root of 100, or a factor of about 2.512. The bright objects have negative magnitudes. The star Vega (α Lyrae) is defined to be magnitude zero in the UBV system.

main-belt asteroids — asteroids that occupy the main asteroid belt between Mars and Jupiter, sometimes limited specifically to the most populous parts of the belt, from 2.2 to 3.3 AU from the Sun.

mantle — the interior zone of a planet or satellite below the crust and above the core, which is divided into the upper mantle and the lower mantle with a transition zone in between.

mare — (pl., maria) an area on the Moon or Mars that appears darker and smoother than its surroundings. Lunar maria are scattered basaltic flows.

Mars crosser — object on an orbit that crosses the orbit of Mars. Because secular perturbations cause Mars' eccentricity to undergo substantial variations over million-year timescales, the orbital parameters defining this population fluctuate over time.

matrix — the fine-grained material that occupies the space in a rock, such as a meteorite, between the larger, well-characterized components such as chondrules, inclusions, etc.

M class — a fairly common asteroid type in the main belt with featureless 0.3- to 1.1-μm spectra (identical to E and P classes) but distinguished by moderate albedos. Presumed to have metallic (Ni-Fe) compositions, but with varying metal contents.

mean motion — usually denoted by μ or ν. Average daily motion for an orbiting body = $2\pi/P$ radians d^{-1} or $360/P°d^{-1}$, where P is the orbital period. P may be derived from the semimajor axis using Kepler's third law.

mean-motion resonance — a situation in which the ratio of the mean motions of two bodies (or the reciprocals of the revolution periods) is a simple fraction. For example, an asteroid with a semimajor axis of 2.5 AU has an orbital period that is one-third that of Jupiter and is said to be in a 3:1 resonance. The gravitational influence for the resulting repeated alignments tends to be large and such resonances with Jupiter tend to result in depletions or concentrations of asteroids. See chapter by Nesvorný et al.

megaregolith — regolith structure throughout the asteroid.

mesosiderite — class of stony-iron meteorite consisting of subequal proportions of silicate material (related to eucrites and diogenites) and Fe-Ni metal.

metamorphic rock — any rock produced by the physical and chemical alteration (without melting) of another rock that has been subjected to high temperature and pressure.

metamorphism — solid-state modification of a rock, e.g., recrystallization, caused by elevated temperature (and possibly pressure).

meteor — the light phenomenon produced by a meteoroid experiencing frictional heating when entering a planetary atmosphere; also used for the glowing meteoroid itself. If particularly large, it is described as a fireball.

meteorite — a natural object of extraterrestrial origin that survives passage through the atmosphere.

meteoroid — a natural small (subkilometer) object in an independent orbit in the solar system.

meteor shower — many meteors appearing to radiate from a common point in the sky caused by the collision of Earth with a swarm of meteoritic particles.

micrometeorite — a small extraterrestrial particle that has survived entry into Earth's atmosphere. The actual size is not rigorously constrained but is operationally defined by the collection procedure because small particles are more abundant than large ones. In practice, the micrometeorites being studied in the laboratory after collection in the stratosphere are rarely as large as 50 μm.

microporosity — fractures, voids, and pores on the scale of tens of micrometers. Microporosity is typically the type of porosity measured on meteorites. See chapters by Britt et al. and Richardson et al.

microwave — an electromagnetic wave (in the radio region just beyond the infrared) with a wavelength from about 1 mm to 30 cm (about 10^9–10^{11} Hz).

Mie theory — a theory of light scattering by small spherical particles.

minor planet — *see* asteroid.

minor satellite — a satellite of an asteroid.

MOID — minimum orbital intersection distance, defined as the closest possible approach distance between the osculating orbits of two objects, provided there are no protective resonances in action.

moment of inertia — the product of the mass of a body and the square of its radius of gyration.

monoclinic — a crystal system characterized by either a single twofold symmetry axis, a single symmetry plane, or a combination of the two.

monomict — a breccia in which all components originated in the same type of rock.

monolith — an asteroid with few internal cracks or faults that is essentially unaffected by long-term stresses. The compressive wave from highly energetic collisions or explosions easily reaches the farside of such an asteroid, reflecting as a tensive wave that can produce damage and spalls. See chapter by Richardson et al.

Monte Carlo technique — a computational procedure in which random numbers are used to approximate the solution to otherwise intractable mathematical or physical problems.

MPC — *Minor Planet Circulars.*

ms — millisecond = 10^{-3} s.

μm — micrometer; 1 μm = 10^{-4} cm.

m.y. — 10^6 yr.

nakhlite — a rare type of achondritic meteorite consisting of calcic pyroxene (augite) and olivine.

near-infrared (near-IR) — expression normally used to indicate the range of wavelengths in the electromagnetic spectrum between 1 and 5 μm.

nearly isotropic comets — comets with Tisserand parameter $T < 2$ that are derived mainly from the Oort cloud. These objects have nearly isotropic inclinations.

NEAs — near-Earth asteroids. *See* NEOs.

NECs — near-Earth comets. *See* NEOs.

NEOs — near-Earth objects. Asteroids and comets with perihelion distances $q \leq 1.3$ AU and aphelion distances $Q \geq 0.983$. Subcategories include the Apollo, Amor, and Aten objects.

noble gases — the gases He, Ar, Kr, Ne, Xe, and Rn, which rarely undergo chemical reactions; also known as inert gases and rare gases.

nodes — the points at which a planet's orbit crosses the plane of the ecliptic. The longitude of the ascending node is one of the six orbital elements and measures the angle between the ascending node and vernal equinox, measured in the plane of the ecliptic.

ν_5 — the secular resonance at which the apsidal motion of an asteroid is equal to the fifth secular apsidal frequency. Across the asteroid belt this resonance occurs at a proper inclination near 30° and tends to limit the asteroidal proper inclinations to smaller values.

ν_6 — the secular resonance at which the apsidal motion of an asteroid is equal to the sixth secular apsidal frequency. This resonance occurs at a proper inclination that rapidly increases from near 0° just inside the inner edge of the main asteroid belt to a broad maximum around 20° and therefore separates the main belt from the Hungaria and Phocaea regions.

ν_{16} — the secular resonance at which the nodal motion of an asteroid is equal to the sixth secular nodal frequency. This resonance remains near heliocentric distance 2.0 AU even for high proper inclination and divides the Hungaria and Phocaea regions.

obliquity — the angle between a planet's axis of rotation and the pole of its orbit.

observational bias — *see* selection effect. May also refer to the quantification of the selection effect (e.g., the probability of measuring a particular value for a parameter during an observational campaign and analysis). See chapter by Jedicke et al.

occultation — the cutoff of light or radiation from a celestial body (or spacecraft) due to its passage behind another body.

Ockham's razor — *Entia non sunt multiplicanda praeter necessitatem* ("Things should not be multiplied beyond necessity"). A doctrine formulated by William of Ockham in the fourteenth century. Any hypothesis should be shorn of all unnecessary assumptions; if two hypotheses fit equally well, the one that makes the fewest assumptions should be chosen.

O class — a rare taxonomic classification distinguished by spectra having a moderately steep UV absorption feature from 0.54 to 0.7 μm, while also having a deep near-IR absorption feature longward of 0.75 μm.

olivine — the most abundant mineral in chondritic meteorites, $(Mg,Fe)_2SiO_4$.

ω — argument of perihelion or periapse for a planet or satellite. Angular distance (measured in the plane of a body's orbit) in the direction of motion from the ascending node to the perihelion point.

ϖ — longitude of perihelion or periapse for a planet or satellite = $\Omega + \omega$.

Ω — longitude of ascending node. The angle between some line in the reference plane (usually the direction to the vernal equinox) and the point where the body crosses the reference plane moving south to north.

Oort Cloud — a spherical cloud of comets extending out to heliocentric distances larger than 100,000 AU that surrounds the solar system. The existence of this cloud was deduced by J. H. Oort in his empirical study of the orbits of long-period comets. Comets in this shell can be sufficiently perturbed by the gravitational effects of the galaxy and by bodies external to the solar system that a fraction of them acquire orbits that take them within the orbits of Jupiter and Saturn.

opacity — a loosely defined term referring to the ability of a medium to extinguish radiation of any given wavelength. In various applications, opacity has been used to mean (1) optical thickness divided by physical thickness; (2) optical or radio thickness; or (3) mass extinction coefficient.

opposition — *see* elongation.

opposition effect — an enhancement in the brightness of an object when observed at phase angles ≤7°, in excess of that predicted by a linear extrapolation of the brightness vs. phase relation from larger phase angles.

orbital elements — six quantities that fully describe an orbit; along with time, they specify the position of an orbiting body along its path. A typical set of orbital elements are (1) semimajor axis a; (2) eccentricity e; (3) inclination i; (4) longitude of the ascending node Ω; (5) argument of periapse ω; and (6) the time of passage through periapse, T. An alternative set, as in the *Ephemerides of Minor Planets,* gives the mean anomaly M at an epoch t. These are related by $M = \mu(t - T)$ where μ is the mean motion.

ordinary chondrite — collective name for the most common variety of chondritic meteorite, subdivided into H, L, and LL groups on the basis of Fe content and distributions.

orthopyroxene — a mineral of the pyroxene group that crystallizes in the orthorhombic form.

orthorhombic — a crystal system characterized by three mutually perpendicular two-fold symmetry axes.

osculating orbit — the path that an asteroid would follow if it were subject only to the inverse-square attraction of the Sun or other central body. In practice, secondary bodies such as Jupiter produce perturbations. Osculating orbital elements are thus subject to variations over time.

oxidation — the process of adding oxygen to, or removing hydrogen from, an element (or of increasing the element's valence, i.e., oxidation state).

p — geometric albedo. The ratio of the brightness of an asteroid to that of a perfectly scattering screen of the same cross-sectional area and in the same place, both being illuminated and viewed normally.

p_H — V-band geometric albedo calculated using the H, G magnitude system.

p_v — the geometric albedo with the V filter of the UBV system.

paired falls — meteorite specimens originally recovered some distance apart and hence given separate names, but later recognized as fragments of a single parent mass on the basis of classification, cosmic-ray or gas-retention age, texture, or other diagnostic features.

pallasite — class of stony-iron meteorites in which the Fe-Ni metal forms a continuous framework enclosing nodules of the silicate olivine.

parent body — the object on or in which a given meteorite or class of meteorites was located prior to ejection as approximately meter-sized objects. Also used to refer to the precursor body of an asteroid family.

P class — a fairly common asteroid type in the outer main belt with a heliocentric distribution that peaks near the 3:2 jovian resonance. Their spectra are featureless from 0.3 to 1.1 μm (identical to E and M classes), but the class is distinguishable by low albedos.

perihelion — q, distance of least heliocentric separation for a body in an eccentric orbit.

permanent designation — the numbers and names, beginning with 1 Ceres, given to asteroids for which orbits are accurately determined. The numbers are supplied by the Minor Planet Center after orbits have been fitted to observations at several different oppositions. Names are proposed by discoverers to the Minor Planet Names committee of IAU Commission 20 and cannot be adopted until at least two months after an asteroid is numbered. If the discoverer is deceased there is a waiting period of six months, and names are usually proposed by the orbit identifier or by the discoverer's former colleagues.

perturbation — the gravitational effect of one object on the orbit of another, if this effect is very small.

phase angle — α, the solar phase angle; the angle subtended at the center of the planet by the vector's direction to the Sun and observer.

phase angle bisector — the direction, or position in the sky, that is the mean between the geocentric position and the heliocentric position of the asteroid. If one were to bisect the angle formed by the lines to the Sun and Earth, from the asteroid, the resultant line would be in the direction of the phase-angle bisector.

phase curve — a plot showing the brightness (reduced to common heliocentric and observer distances) of a planet, satellite, or asteroid vs. phase angle.

phase integral — the relationship between the geometric and Bond albedos of a body: A = pq, where A is the bolometric Bond albedo, p is the bolometric geometric albedo, and q is the phase integral.

PHO — potentially hazardous object; near-Earth objects whose minimum orbital intersection distance (MOID) with the Earth is 0.05 AU or less and whose absolute magnitude H is 22.0 or brighter.

Phocaea asteroids — asteroids in the inner main belt that are surrounded by the ν_5, ν_6, ν_{16} secular resonances and the 3:1 mean motion resonance with Jupiter. These objects have semimajor axes between 2.2 and 2.5 AU, low eccentricities, and inclinations between 20° and 28°.

photometry — the measurement of light intensities.

phyllosilicate — one of a family of silicate minerals characterized by a structure that consists of sheets or layers, invariably hydrated.

pixel — picture element. Electronic images (like those obtained by modern CCDs) are composed of pixels arranged in rows and columns.

plagioclase — a mineral group, formula (Na,Ca)Al(Si,Al)Si_2O_8; a solid solution series from $NaAlSi_3O_8$ (albite) to $CaAl_2Si_2O_8$ (anorthite), triclinic. It is one of the most common rock-forming minerals.

planetesimal — small rocky or icy body formed from the primordial solar nebula, perhaps having sizes ranging up to 10 km, out of which all larger solar system members are presumed to have accumulated.

planetocentric — centered on a planet. A satellite is in a planetocentric (as opposed to heliocentric) orbit. A planetocentric coordinate system is subtended at the planet's center.

plasma — the completely ionized gas, the so-called fourth state of matter in which the temperature is too high for atoms, as such, to exist and that consists of free electrons and free atomic nuclei.

polarization — the action or process of affecting radiation, especially light, such that the vibrations assume some definite form. Light that has encountered an index of refraction boundary will have different reflection coefficients depending on the orientation of the electric vector. Polarization is defined as negative if the light reflected from a boundary is greater in the plane given by the scattering plane (source-boundary-observer) than in the perpendicular plane. If the light intensity is the same in both perpendicular and parallel directions, the light is unpolarized, and if it is greater in the perpendicular direction, the polarization is called positive.

polymict — a breccia in which the components originated in two or more rocks of differing compositions.

porosity — the percentage of the bulk volume of a rock that is occupied by empty space. It is also a measure of the ratio between grain and bulk densities. See chapter by Britt et al.

Poynting-Robertson effect — an effect of radiation on a small particle orbiting the Sun that causes it to spiral slowly toward the Sun. It occurs because the orbiting particle absorbs energy and momentum streaming radially outward from the Sun, but reradiates this energy isotropically in its own frame of reference.

ppm — parts per million, generally by weight. This type of measurement is also often given by the term μg/g.

precession — a slow, periodic conical motion of the rotation axis of a spinning body.

preliminary designation — the system for designating asteroids upon discovery and before their orbits are well-enough determined that they can be given a permanent number and name. The designations are supplied by the Minor Planet Center and consist of the year of discovery, an uppercase letter to indicate the half month in that year (A = Jan. 1–15, B = Jan. 16–31, etc., Y = Dec. 16–31, I being omitted), a second uppercase letter in sequence, and when this sequence of 25 has been completed it is repeated and followed by a sequential number (written as a subscript, if possible). Before 1925 the system consisted simply of a sequential pair of letters (and initially only a single letter), and the subsequent interpolation of additional objects was troublesome. The modern system is therefore now also used for unnumbered pre-1925 objects, except that the initial 1 in the year is replaced by A.

primitive — in planetary science and meteoritics, a type of object or rock that is little changed chemically since its formation, hence representative of the conditions in the solar nebula at the time of formation of the solar system.

prograde motion — motion in the same direction as the prevailing direction of motion. As viewed from the north, prograde motion is counterclockwise, or west to east.

proper elements — quasi-integrals of motion that are nearly constant in time. They are obtained by removing the short- and long-term perturbations from osculating elements. See chapter by Knežević et al.

protoplanet — a precursor body from which a planet develops.

PSF — point spread function. The function describing how the image from a point source (like a single star) is degraded in the focal plane of a telescope as a consequence of seeing and nonideal optics. *See* seeing.

pyroxenes — a group of common rock-forming silicates that have ratios of metal oxides (MgO, FeO, or CaO) to SiO_2 of 1:1. These are called metasilicates. Pure members of this group are $MgSiO_3$ (enstatite) and $FeSiO_3$ (ferrosilite). Pure $CaSiO_3$ does not crystallize with the pyroxene structure. Calcium does substitute for up to 50% of the Mg and Fe in the pyroxene structure.

q — perihelion distance, q = a(l – e).

q — phase integral. The brightness of an asteroid summed over all phase angles.

Q — aphelion distance, Q = a(l + e).

Q — specific energy, defined as the ratio of the kinetic energy of the impactor to the mass of the impacted body. See chapters by Holsapple et al. and Davis et al.

Q_D^* — critical impact specific energy, defined as the energy per unit target mass delivered by the projectile required for the catastrophic disruption of the target. This term is also known as the dispersion energy. See chapters by Holsapple et al. and Davis et al.

Q_S^* — shattering impact specific energy, defined as the energy per unit target mass delivered by the projectile required for the catastrophic shattering of the target. This term is also known as the shattering energy. See chapters by Holsapple et al. and Davis et al.

Q class — a rare asteroid classification denoted by moderate albedos and spectra with a strong absorption feature short-ward of 0.7 μm and a modest absorption feature centered near 1 μm. The spectra are interpreted as being similar to ordinary chondrites. This asteroid type has been identified for 1862 Apollo and other near-Earth asteroids.

quadrature — *see* elongation.

$R_\odot$ — solar radius = 6.96×10^{10} cm.

radar albedo — ratio of a target's radar cross section in a specified polarization to its projected area; hence a measure of the target's radar reflectivity.

radar cross section — most common measure of a target's scattering efficiency; equal to the projected area of that perfect metal that would give the same echo power as the target if observed at the target's location.

radioactive age dating — the technique of determining the ages of rocks, meteorites, or other specimens by the amount of decay of certain radioactive elements contained therein.

rare earth element (REE) — one of the elements with atomic numbers from 57 to 71, inclusive, in the lanthanide series of the periodic table.

Rayleigh-Jeans Law — an approximation of Planck's blackbody formula valid at long wavelengths ($h\nu \ll kT$): $B_\nu(T) \approx 2kT\nu^2/c^2$. It is often used in radio astronomy; the brightness temperature of a source observed with a radio telescope is the temperature required to fit the observed specific intensity with the Rayleigh-Jeans formula.

R class — a rare asteroid classification exemplified by 349 Dembowska and denoted by moderately high albedos and spectra with a strong absorption feature shortward of 0.7 μm and a fairly strong absorption feature centered near 1 μm.

refractory — term describing the high-temperature stability of an element or phase. The opposite of volatile.

regolith — the layer of fragmental incoherent rocky debris that nearly everywhere forms the surface terrain; it is produced by meteoritic impact on the surfaces of the planets, satellites, or asteroids.

regolith breccia — fragmental breccias containing some identifiable regolith component such as solar-wind gas.

resonance — the enhanced response of any oscillating system to an external stimulus that has the same driving frequency as the natural frequency of the system; higher-order resonances occur when these frequencies are commensurable.

restricted three-body problem — two bodies assumed to be point masses and called primaries revolve around their center of mass under the influence of their mutual attraction. The problem is to determine the motion of a third body attracted by the previous two but not influencing their motion.

retrograde motion — the opposite of prograde motion.

rubble pile — gravitational aggregate with the organizational structure one might expect if you dumped rocks from a truck (literally, a pile of rubble). The reaction of such a body to short-term stresses like collisions is absorption of impact energy via compaction, with little to no tensile wave developed in the structure. See chapter by Richardson et al.

rms — root mean square. The square root of the mean square value of a set of numbers.

Roche limit — the minimum distance at which a fluid satellite influenced by its own gravitation and that of a central mass can be in mechanical equilibrium. For a satellite of zero tensile strength and the same mean density as its primary, in a circular orbit around its primary, this critical distance is 2.42× the radius of the primary.

saturation equilibrium — the ease whereby a surface has accumulated so many craters (of a particular size) that subsequent craters tend to destroy (by overlapping and other processes) roughly equal numbers of preexisting craters.

scattered disk — a population of objects associated with the Kuiper Belt that reside on large, eccentric, inclined orbits with perihelion distances near q = 35 AU. These objects can be perturbed by Neptune onto planet-crossing orbits, making them a potential source of ecliptic comets.

scattered law (radar) — function giving the dependence of a surface element's radar cross section on viewing angle.

Schmidt telescope — a type of reflecting telescope (more accurately, a large camera) in which the coma produced by a spherical concave mirror is compensated for by a thin correcting lens placed at the opening of the telescope tube. The Palomar 122-cm Schmidt telescope has a usable field of 6°.

S class — a very common asteroid class in the inner main belt with moderate albedos and reddish spectra shortward of 0.7 μm and moderate to nonexistent absorption features in the near-IR. Mineralogically, the S-class objects appear to be olivine-pyroxene mixtures with a full range of mixing ratios (pyroxene-rich and olivine-poor to olivine-rich and pyroxene-poor). Their meteoritical interpretation is uncertain, but S-class objects having relatively equal olivine and pyroxene abundances may be similar to ordinary chondrites.

secular perturbations — averaged perturbations experienced by planets and asteroids when the effects that depend upon the actual positions of the objects in their orbits are eliminated. The classical procedure, due originally to Lagrange, expresses the long-term variations of the eccentricities and perihelion longitudes in terms of constant "proper" eccentricities, one corresponding to each object, and "proper" perihelion longitudes that are linear functions of the time. An analogous process gives the long-term variations of the inclinations and nodal longitudes in terms of constant "proper" inclinations and "proper" nodal longitudes that are linear functions of the time. In this scheme there are, by definition, no secular changes in the semimajor axis. See chapter by Knežević et al.

secular resonance — a situation in which the rate of the precession of the proper longitudes of the nodes or perihelion of an asteroid's orbit (called nodal and apsidal frequencies) is equal to one of the nodal or apsidal frequencies associated with the mutual secular perturbations of the major planets. Over a long period of time an asteroid near a secular resonance can experience large perturbations. For orbits of low (proper) eccentricity and inclination, the principal secular resonances, largely arising from the mutual perturbations of Jupiter and Saturn, occur at a heliocentric distance of about 2.0 AU. See chapter by Nesvorný et al.

seeing — general term describing the overall quality of the sky for astronomical observations. The effect of seeing is that of blurring astronomical images, and to make it impossible to obtain images corresponding to the theoretical resolving power of the telescope.

selection effect — a modification of the distribution of any observational parameter from its actual distribution as a consequence of some aspect of the observational or analysis strategy. See chapter by Jedicke et al.

shattered asteroid — an asteroid with even more faults/joints/cracks than a fractured asteroid, and therefore more susceptible to disruption from long-term stresses such as tidal forces or rotational spinup. The original structure is still mostly in place (though rotation of some fragments may have increased the macroporosity) and friction or interlocking between jagged fragments may provide some tensile strength. The reaction to short-term stresses like collisions is different from a monolith. Spalls are damped and the tensile wave is suppressed, such that this object is more difficult to disrupt. Small craters formed on single components of the body may act like they were formed on a monolith. See chapter by Richardson et al.

shergottite — a rare type of meteorite, consisting of pyroxene (pigeonite) and maskelynite.

shock wave (or shock front) — discontinuity in temperature and pressure propagating in a solid, liquid, or gas with supersonic velocity, caused by impact or explosion.

sidereal period — the time it takes for a planet or satellite to make one complete rotation or revolution relative to the stars.

siderophile — one of the geochemical classes of elements. Siderophile elements are those that tend to go into the metal phase, e.g., Ni, Co, Au, As, Ge, Ga, Ir, Os, Re.

silicate — any of a wide range of rocks and minerals composed in part of silica (Si and O).

SMASS — Small Main-belt Asteroid Spectroscopic Survey.

SNC meteorites — a group of uncommon but apparently genetically related meteorite types that are highly differentiated (the shergottites, nahklites, and Chassigny). They may originate from Mars.

solar nebula — the gas-dust disk that surrounded the protosun. Mass of the solar nebula is usually assumed to be in the range from 0.02 to 0.05 $M_{\odot}$. The term protoplanetary cloud is also sometimes used as a synonym for the solar nebula.

solar wind — the energetic charged particles that flow radially outward from the solar corona, carrying mass and angular momentum away from the Sun.

solidification age — the most common age determined by radioactive dating techniques; the time since the rock or mineral grain being tested solidified from the molten state, thus isolating itself from further chemical changes.

space weathering — any process that changes the spectroscopic properties of airless bodies. Processes that are believed to produce space weathering include impacts, solar wind ion implantation, sputtering, and micrometeorite bombardment. See chapter by Clark et al.

specific energy — kinetic energy per unit mass.

speckle interferometry — technique used to obtain astronomical images with resolution close to the theoretical diffraction limit of a telescope, based on the idea of recording images during very short exposure times in order to "freeze" the effect of the continually changing seeing.

standard thermal model (STM) — a simple thermal model for asteroids and other airless bodies that assumes the ideal situation of a nonrotating spherical body in instantaneous equilibrium with insolation. It also assumes that the subsolar and sub-Earth points on the body coincide. The thermal emission is a function only of subsolar distance.

steradian — the solid angle that, having its vertex in the center of a sphere, cuts out an area of the surface of the sphere equal to that of a square with sides of length equal to the radius of the sphere. A complete sphere contains 4π steradians.

Stokes parameters — four parameters to describe fully a beam of polarized light. They involve the maximum and minimum intensity, the ellipticity, and the direction of polarization.

stony-iron meteorite — a fairly rare kind of differentiated meteorite, composed of a mixture of silicates with metallic Fe-Ni, thought to have originated near the core-mantle boundary of a differentiated parent body.

sub-Earth point — point on the surface of a body from which Earth is seen at the zenith.

subsolar point — point on the surface of a body from which the Sun is seen at the zenith.

superclasses — groupings of asteroid taxonomic classes into large categories such as primitive, igneous, and metamorphic.

surface of section — subspace of the phase space of a dynamical system or, more generally, the set of the successive in intersections of a trajectory with this subspace.

synchronous rotation — rotation of a body so that it always keeps the same face toward another object; the situation where the periods of rotation and revolution of an orbiting body are equal.

synodic period — the period of revolution of one body about another with respect to Earth (synodic period)$^{-1}$ = ± (sidereal period)$^{-1}$ ∓ (Earth's period)$^{-1}$.

synodic rotation period — apparent rotation period of a target that is moving relative to the observer, to be distinguished from the "sidereal" rotation period measured with respect to the fixed stars.

taxonomic classifications — a system for categorizing similar observed properties of asteroids, such as color or spectral properties and albedo. Letters such as A, B, C, D, E, F, G, M, P, Q, S, T, and V are used to denote currently recognized distinct classes.

T class — a class of low albedo asteroids having spectra with a moderate absorption feature shortward of 0.85 μm and generally flat in the near-IR.

terrestrial age — the period of time since the fall of a meteorite.

thermal conductivity — the proportionality constant that gives the amount of heat conducted through a unit cross section in unit time under the influence of unit heat gradient, cal cm^{-2}s^{-1}k^{-1}.

thermal emission — the emission of electromagnetic radiation from a body due to its temperature and emissivity.

thermal emission spectrum — the thermal emission from a body measured as a function of wavelength.

thermal inertia — a material parameter that indicates the rate at which a body's temperature responds to changing heat input. It is proportional to the square root of the product of thermal conductivity and volume heat capacity.

thermophysical model — a thermal model for asteroids and other airless bodies that lies between the standard thermal and isothermal latitude models. It takes into account the thermophysical properties of the body and may also include spin axis and direction as well as shape.

Thule asteroids — asteroids with low eccentricities located near or inside the 4:3 mean motion resonance with Jupiter at 4.3 AU. The only known member of this class is 279 Thule.

time delay — time between transmission of a radar signal and reception of the echo.

Tisserand invariant/parameter — a pseudoconstant of the motion in the restricted three-body problem based on the Jacobi integral. When defined with respect to Jupiter, it can be used to identify various classes of comets.

$$T = a_j/a + 2\sqrt{(a/a_j)(1 - e^2)}\cos i$$

where a, e, and i are the semimajor axis, eccentricity, and inclination (to the plane of Jupiter's orbit), respectively, of the comet's orbit, and a_j is Jupiter's semimajor axis.

Titius-Bode law — a mnemonic device discovered by Titius in 1776 and advanced by Bode in 1772, used for remembering the distances of the planets from the Sun. Take the series 0, 3, 6, 12, …; add 4 to each member of the series, and divide by 10. The resulting sequence 0.4, 0.7, 1.0, 1.6, … gives the approximate distance from the Sun (in AU) of Mercury, Venus, Earth, Mars, …, out to Uranus. The law fails for Neptune and beyond. Its value at 2.8 spurred a search for a "missing" planet between Mars and Jupiter.

TNOs — transneptunian objects. This population includes the Kuiper Belt and scattered-disk objects.

Trojans — asteroids that occur in orbits librating around two of the Lagrangian points. The most prominent populations are the namely the ones preceding and following Jupiter in its orbit, equidistant from the Sun and Jupiter. See chapter by Marzari et al.

T-Tauri stars — young, late-type stars that are precursors to solar-mass stars characterized by emission line spectra, infrared excesses, and irregular variability. The prototype for this class of stars is T Tauri.

T-Tauri wind — outflow from a T-Tauri star.

UBV system — system of stellar magnitudes devised by Johnson and Morgan at the Yerkes Observatory that consists of measuring an object's apparent magnitude through three color filters: ultraviolet (U) at 3600 Å; blue (B) at 4200 Å; and "visual" (V) in the green-yellow spectral region at 5400 Å. It is defined so that, for AO stars, B–V = U–B = 0; it is negative for hotter stars and positive for cooler stars. Filters at other wavelengths are also used and indicated with letters R, I, H, J, K, L, M, etc.

U class — asteroids that have unusual visible wavelength spectral properties that do not resemble any defined taxonomic class (perhaps due to large measurement uncertainties) are often denoted by "U" for "unclassifiable."

ultramafic — an igneous rock consisting predominantly of mafic silicate minerals.

ultraviolet (UV) — that part of the electromagnetic spectrum that lies at wavelengths shortward of about 3500 Å. UV absorption features in asteroid spectra result from charge transfer mechanisms. X-rays and γ-rays occur at wavelengths shortward of 300 Å.

Universal Time (UT) — the local mean time of the prime meridian. It is the same as Greenwich mean time, counted from 0 h beginning at Greenwich mean midnight.

ureilite — class of C-rich achondritic meteorite in which the silicates consist of olivine and pigeonite.

Väisälä orbits — orbits computed on the assumption that the heliocentric radial velocity is zero, i.e., the object is taken to be at perihelion or aphelion (or in a circular orbit as a special case). Series of Väisälä orbits with different eccentricities can be derived very simply from only two observations, and they are often useful in identifying further observations.

V(α) — reduced magnitude. The observed magnitude of an asteroid at phase angle a converted to unit heliocentric and geocentric distances. In the V band $V(\alpha) = V_{obs}\alpha - 5\log r\Delta$, when r and Δ are, respectively, the heliocentric and geocentric distances in AU.

V class — a rare asteroid classification exemplified by 4 Vesta. Spectra are very red shortward of 0.5 µm, moderately red from 0.5 to 0.7 µm, and show a strong near-IR absorption feature centered around 0.95 µm. Surface composition may be similar to basaltic achondrites.

V magnitude — the magnitude (on the astronomical brightness scale) for an object as measured at a wavelength of 0.55 µm.

vernal equinox — the intersection of ecliptic and celestial equator where the Sun is moving from south to north.

visible wavelength — generally refers to the wavelength range where modern "visible" CCD detectors are sensitive, typically 0.4 to 1.0 µm.

volatile — an element that condenses from a gas or evaporates from a solid at a relatively low temperature.

X class — an object whose visible wavelength spectral properties are consistent with it residing in either the E, M, or P classes. Generally, either albedo measurements or the identification of specific spectral features are required to refine the classification.

xenolith — fragment in a rock or meteorite foreign to its host.

Yarkovsky effect — the effect of rotation on the trajectory of a small particle orbiting the Sun (compare Poynting-Robertson effect). If the orbiting body is rotating, there is a temperature variation over its surface, and it will reradiate the incident radiation anisotropically. See chapter by Bottke et al.

YORP effect — The Yarkovsky-O'Keefe-Radzievskii-Paddack, or YORP effect, is a variant of the Yarkovsky effect. Thermal reemission on and radiation reflection off irregularly shaped bodies can result in a torque that may secularly change both the body's rotation rate and the orientation of its spin axis. See chapter by Bottke et al.

zodiac — a belt around the sky that is 18° wide and centered on the ecliptic, within which are found the Moon and planets.

zodiacal light — a faint glow that extends away from the Sun in the ecliptic plane of the sky, visible to the naked eye in the western sky shortly after sunset or in the eastern sky shortly before sunrise. Its spectrum indicates it to be sunlight scattered by interplanetary dust. The zodiacal light contributes about a third of the total light in the sky on a moonless night.

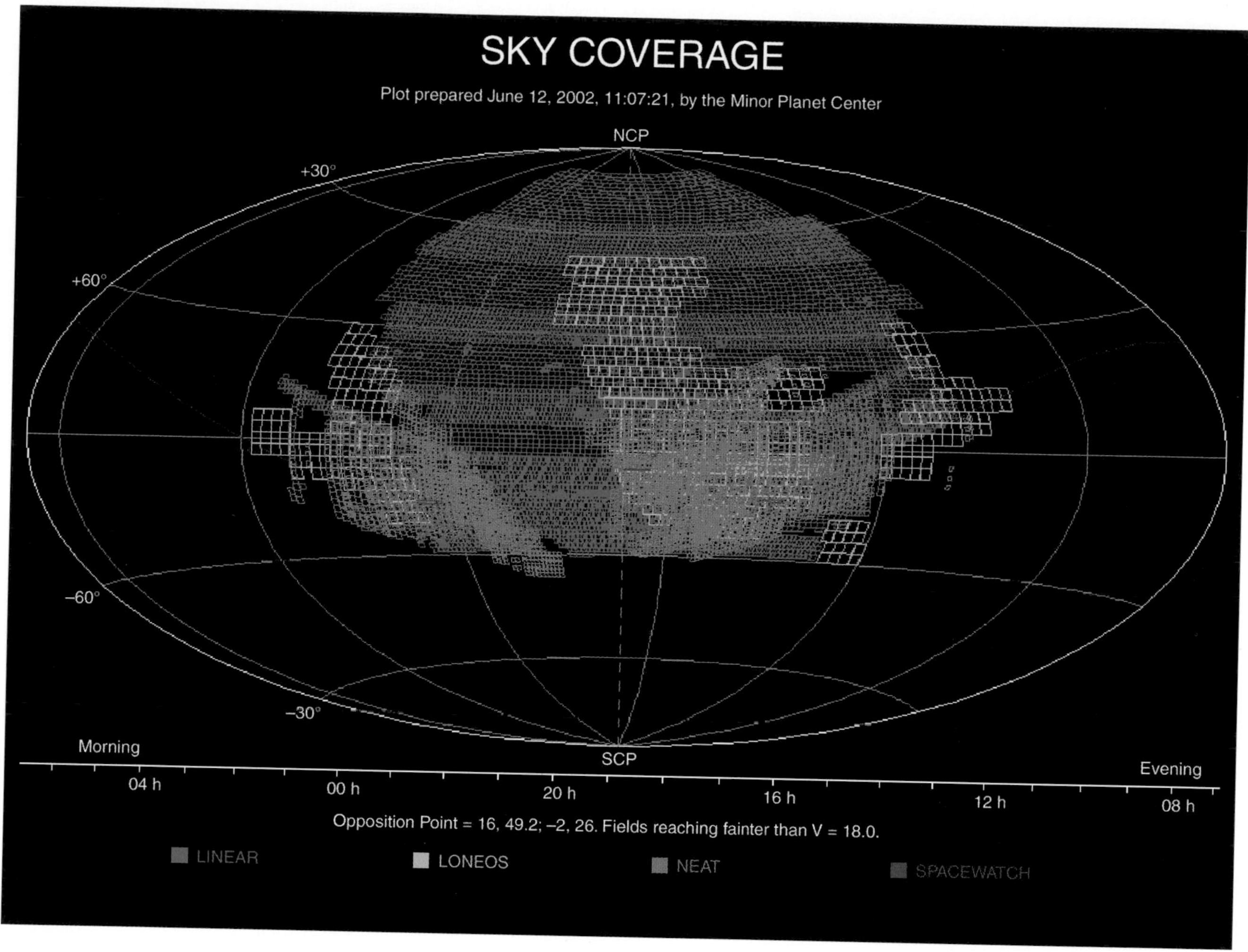

Plate 1. Plot from MPC Web page showing sky coverage for NEAT, LONEOS, LINEAR, and Spacewatch for the first two weeks during a dark period (*Minor Planet Center,* 2001).

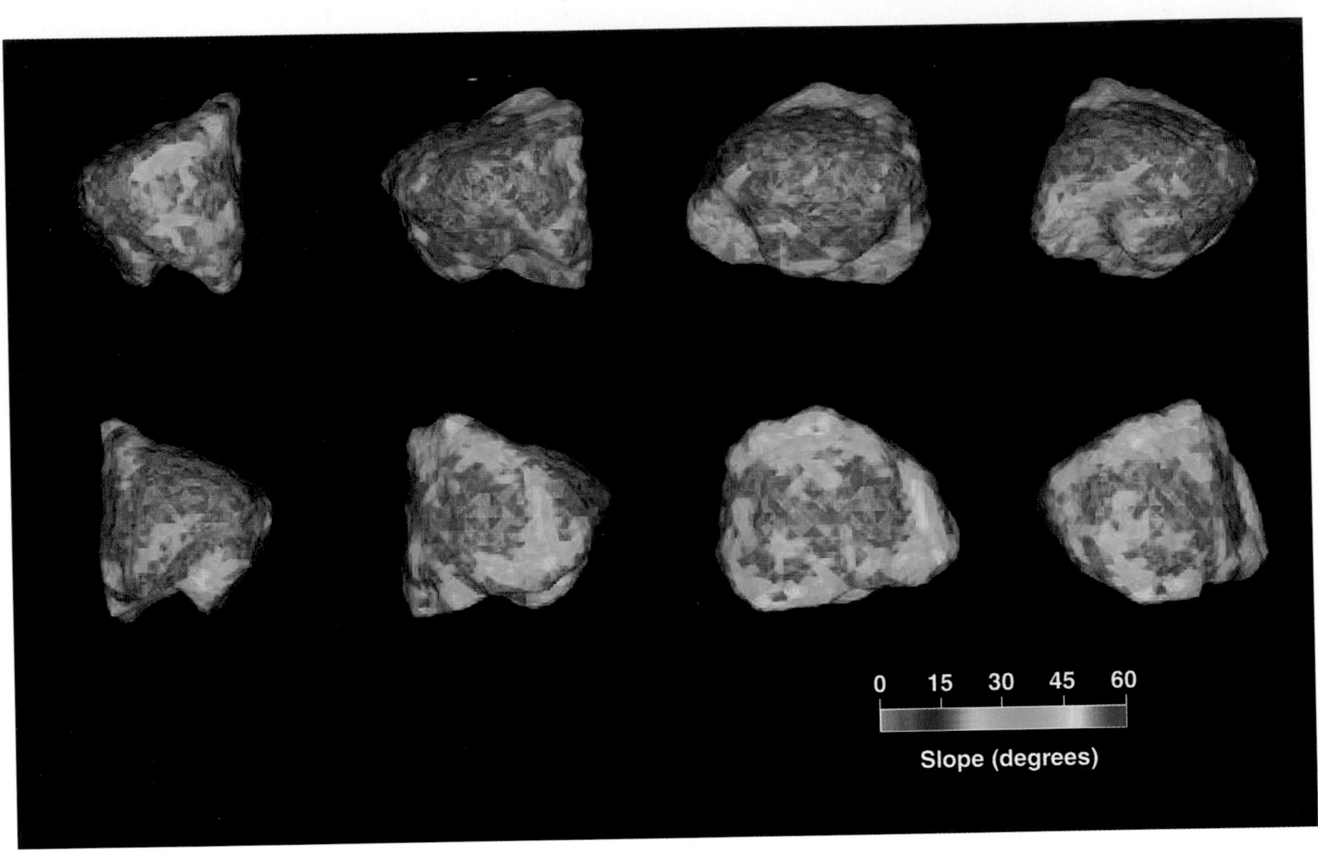

Plate 2. Rotational sequence of the radar-derived shape model of 6489 Golevka (*Hudson et al.,* 2000a), color coded for gravitational slope calculated under the assumption of uniform density. Surfaces with slopes as large as 40° exceed angles of repose for plausible geologic materials and are likely to be exposed, solid rock. The asteroid's equivalent diameter is ~530 m.

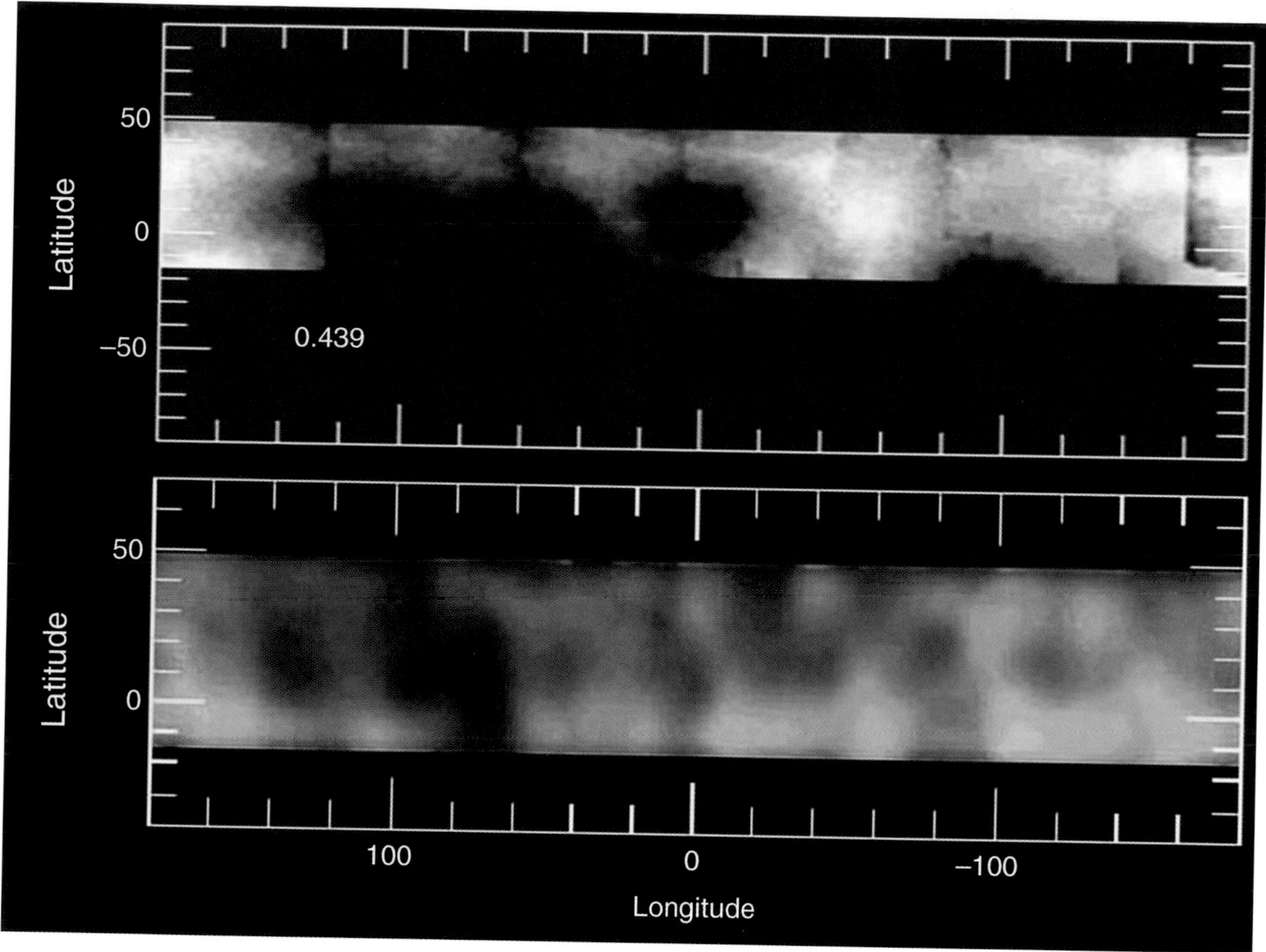

Plate 3. The albedo map of 4 Vesta (top) as imaged in the HST filter at 0.439 μm. The obtained false color mineralogical map of 4 Vesta (bottom) is from *Binzel et al.* (1997).

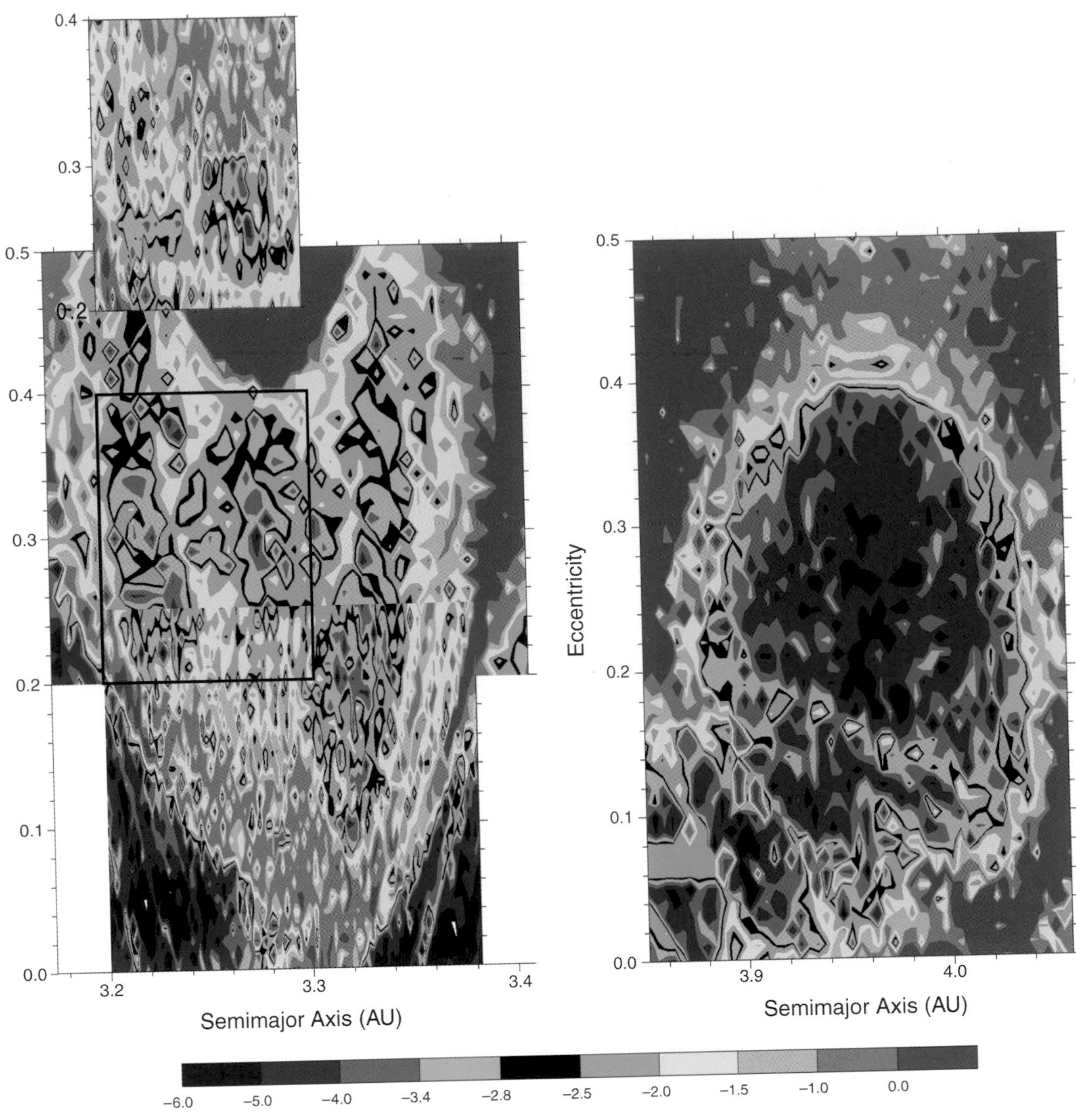

Plate 4. The charts of chaotic diffusion in the 2J:1 (left) and 3J:2 (right) MMRs. The plotted parameter is $\log_{10}|\delta\dot{\varpi}|$. The color coding is such that the starting positions of trajectories characterized by slow chaotic diffusion are shown in blue, while fast-diffusing trajectories are shown in red/yellow. The high-eccentricity orbits in both resonances ($e > 0.4$) are instable due to the overlap of the secular resonances (*Morbidelli and Moons,* 1993). The low-eccentricity orbits in the 2J:1 MMR ($e < 0.2$) are instable due to the so-called secondary resonances, commensurabilities between the periods of σ and ϖ (*Henrard and Lemaître,* 1987; *Henrard et al.,* 1995; *Lecar et al.,* 2001). The orbits in the central region of the size of the rectangle suffer from commensurabilities with the Great Inequality period. The inset on the top left shows how the instabilities in the central region of the 2J:1 MMR are enhanced when the period of $2\lambda_5 - 5\lambda_6$ is set to be 440 yr, a value possibly taken during the primordial phase of the planetary migration. Adapted from *Nesvorný and Ferraz-Mello* (1997).

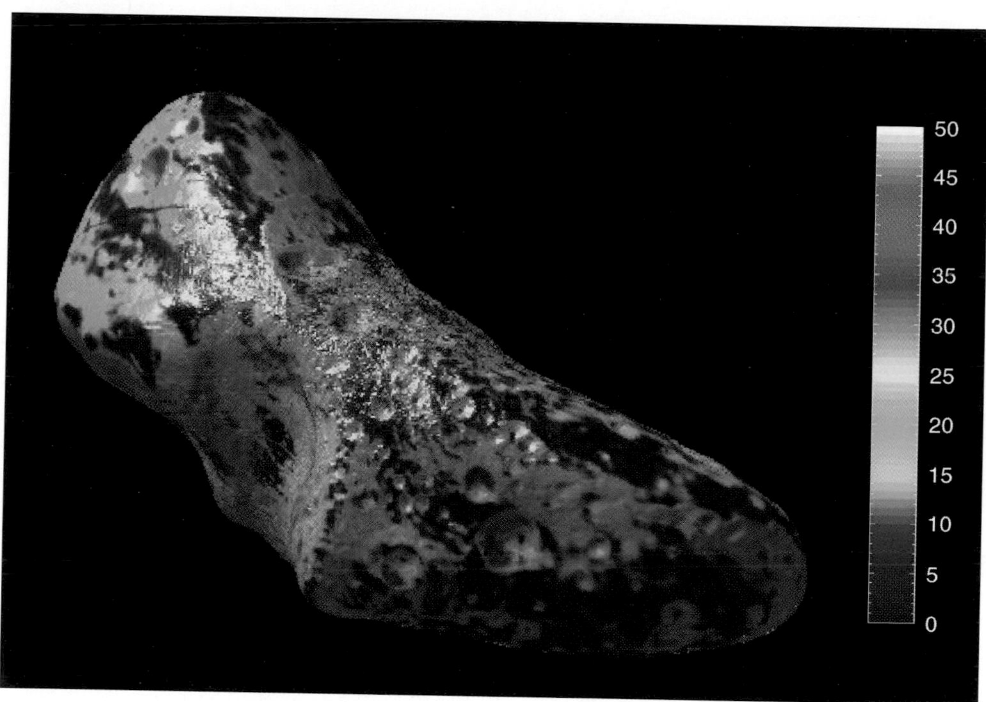

Plate 5. Surface slopes plotted onto the NLR shape model for Eros (*Zuber et al.*, 2000). The broad concavity in the left foreground is the impact structure Himeros, and to its left in profile is Shoemaker Regio, believed to be a recent impact structure (*Thomas et al.*, 2001) responsible for many of the blocks littering the surface of Eros. With only a few exceptions, the steepest slopes (>30°) are located inside the rims of craters. One expects such slopes in angular impact breccias, and such slopes are nominal for simple craters.

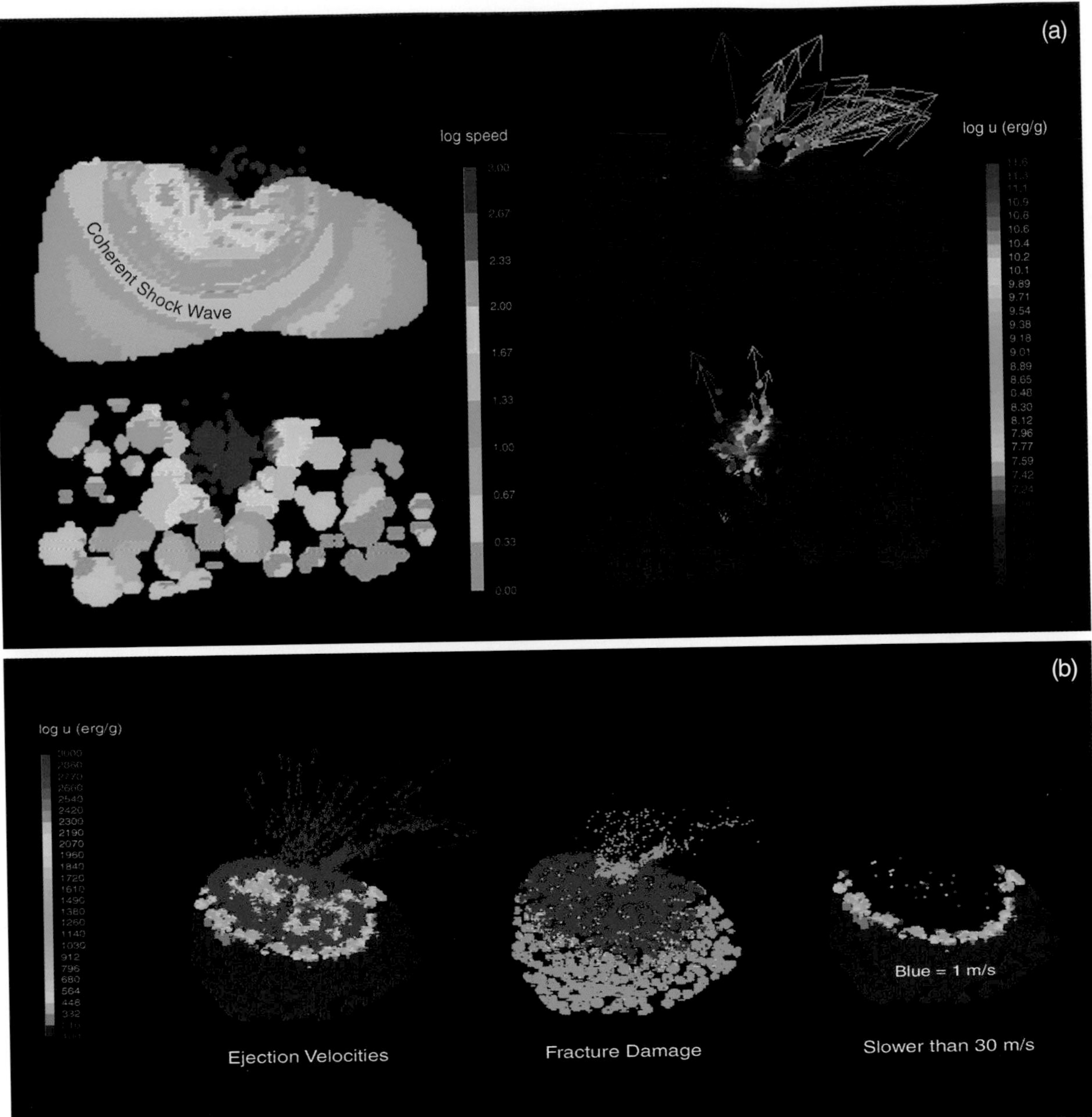

Plate 6. **(a)** One possibility for Mathilde's giant craters is that the asteroid is too weak to be disrupted by collisions. This was initially explored by *Asphaug et al.* (1998) in their models exploring how interior structure affects its collisional evolution. In the figure an ~16-m-diameter, 2.7-g/cm³, 5-km/s projectile strikes between the ~800-m-diameter lobes of asteroid Castalia, modeled here as a monolithic solid (top) and a porous aggregate (bottom), each with the same bulk density (1.3 g/cm³), but the 50% porous target having twice the material density. This is like the *Asphaug and Benz* (1994a) Phobos model (Fig. 4c, bottom), only with particles removed in an organized manner to leave behind contacting spheres. The figures on the left show final particle velocity ("log speed" in centimeters per second) for an ~100-m-thick slice through each target, while the figures on the right plot energy ("log u" in ergs per gram). The porous target does not transmit energy past the cratered region; note the abrupt transition to low velocity. The energy plot (right) also shows velocity vectors: Note the scattering of ejecta in the porous body, much of which has worked its way inside the asteroid. In the end, about half as much of the rubble pile asteroid is accelerated to escape velocity, as of the monolith. **(b)** In this supercomputer simulation (half million particles resolution) Mathilde is modeled as basalt spheres ranging in diameter from 0.5 to 3 km, just touching, with a total porosity of ~0.5 and a bulk density of 1.3 g/cm³. (The contact portion of touching spheres is replaced with damaged rock, as can be seen in the red flecks in the bottom of the center figure.) The shock rapidly attenuates until transmission depends upon mechanical connectedness, which is poor. As a result impact energy is confined to a local region, so that the shattered region (red in central figure) receives all the energy that would have been transmitted globally in a competent body (e.g., Fig. 7b). Ejection velocities (left) in the damaged region (middle) are consequently greater due to this confined impact energy. Indeed, the nonescaping fraction (right) is equal to all the damaged rock: a crater with no ejecta deposits and no external damage.

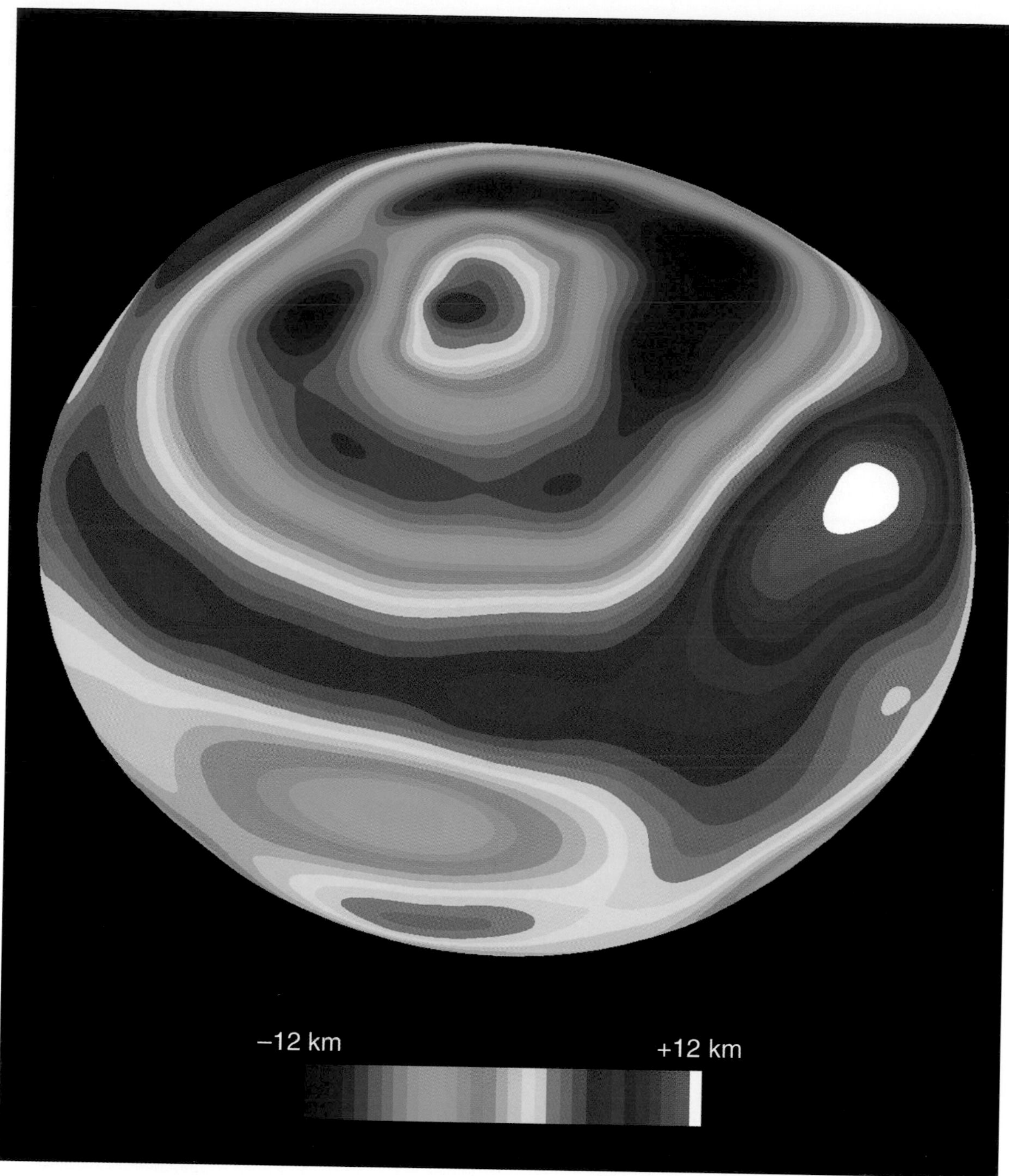

Plate 7. Elevation map of Vesta, based on Hubble Space Telescope images (*Thomas et al.,* 1997a). The large impact crater ~460 km in diameter and ~13 km deep with a pronounced central uplift and bounding rim is visible in the south pole region near the top of the image. Courtesy of Peter C. Thomas.

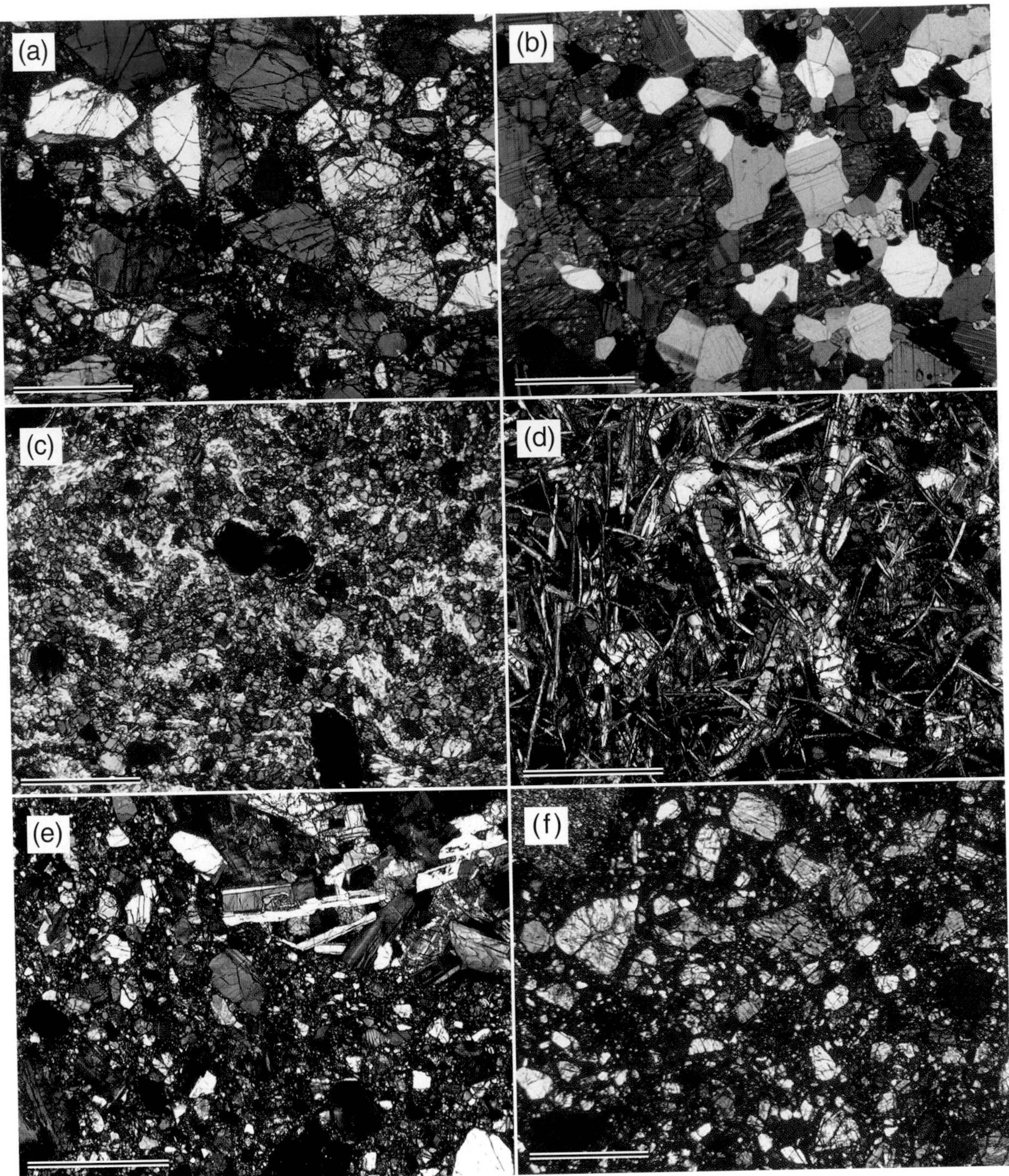

Plate 8. Thin section photomicrographs in transmitted light with crossed polars of typical HED meteorite types (all scale bars are 1 mm). **(a)** Diogenite Johnstown, UNM section #1033, illustrating the highly brecciated nature of the rock that consists essentially entirely of orthopyroxene. **(b)** Cumulate eucrite Serra de Mage, UNM section #217. The rock consists of large crystals of plagioclase (lighter material, with straight twin lamellae) and mostly orthopyroxene with complex augite exsolution lamellae (darker material, with irregular, sometimes wormlike exsolution lamellae). **(c)** Metamorphosed (equilibrated) eucrite Ibitira, UH section #273, showing a recrystallized texture, with plagioclase (white) and pyroxene (colored). The round, dark areas in the center, bottom, and left of image are vesicles. **(d)** Unequilibrated eucrite Pasamonte, showing the typical basaltic texture of plagioclase (light) and pyroxene (colored). **(e)** Polymict eucrite Y-74450-10, consisting of basaltic rock (upper right) and mineral fragments. **(f)** Howardite Kapoeta, section from the K. Keil collection at UH, consisting of mineral fragments of highly variable grain sizes.

Index

Kaiser Wilhelm II (1859–1941) ruled imperial Germany from his accession in 1888 to his enforced abdication in 1918 at the end of the First World War. This book, based on a wealth of previously unpublished archival material, provides the most detailed account ever written of the first half of his reign. Following on from John Röhl's definitive and highly acclaimed *Young Wilhelm: The Kaiser's Early Life, 1859–1888* (1998), the volume demonstrates the monarch's dynastic arrogance and the wounding abuse he showered on his own people as, step by step, he built up his personal power. His thirst for glory, his overweening nationalism and militarism and his passion for the navy provided the impetus for a breathtaking long-term goal: the transformation of the German Reich into the foremost power in the world. Urgent warnings from all sides both about the revival of a semi-absolute Personal Monarchy on the threshold of the twentieth century and about the challenge his goal of 'world power' implied for the existing world powers, Great Britain, France and Russia, were brushed aside by the impetuous young ruler with his faithful military retinue and blindly devoted court favourites. Soon the predicted consequences – constitutional crisis at home and diplomatic isolation abroad – began to make their alarming appearance.

JOHN C. G. RÖHL is Research Professor of German History at the University of Sussex. He is the author of a number of acclaimed works on Kaiser Wilhelm II and the Second Reich. His book *The Kaiser and his Court: Wilhelm II and the Government of Germany* (1994) won the Wolfson History Prize for 1994. His biography of the Kaiser was recently awarded the Gissings Prize.

The Kaiser in 1891

WILHELM II

THE KAISER'S PERSONAL MONARCHY, 1888–1900

JOHN C. G. RÖHL

TRANSLATED BY

SHEILA DE BELLAIGUE

PUBLISHED BY THE PRESS SYNDICATE OF THE UNIVERSITY OF CAMBRIDGE
The Pitt Building, Trumpington Street, Cambridge, United Kingdom

CAMBRIDGE UNIVERSITY PRESS
The Edinburgh Building, Cambridge, CB2 2RU, UK
40 West 20th Street, New York, NY 10011–4211, USA
477 Williamstown Road, Port Melbourne, VIC 3207, Australia
Ruiz de Alarcón 13, 28014 Madrid, Spain
Dock House, The Waterfront, Cape Town 8001, South Africa

http://www.cambridge.org

Originally published in German as *Wilhelm II: Der Aufbau der Persönlichen Monarchie* by C. H. Beck 2001 and © John C. G. Röhl 2001
First published in English as *Wilhelm II: The Kaiser's Personal Monarchy, 1888–1900* by Cambridge University Press, 2004
English translation © John C. G. Röhl 2004

Printed in the United Kingdom at the University Press, Cambridge

Typeface Walbaum 12/14 pt. *System* LaTeX 2ε [TB]

A catalogue record for this book is available from the British Library

Library of Congress Cataloguing in Publication data
Röhl, John C. G.
[Wilhelm II. English]
Wilhelm II: the Kaiser's personal monarchy, 1888–1900 / John C. G. Röhl; translated by Sheila de Bellaigue.
p. cm.
Includes bibliographical references and index.
ISBN 0 521 81920 2
1. William, II, German Emperor, 1859–1941. 2. Germany – Kings and rulers – Biography.
I. Title.
DD229.R6412813 2004
943.08′4′092 – dc22 2003065264
[B]

ISBN 0 521 81920 2 hardback

For
ROSEMARIE

Contents

Illustrations

Preface to the English edition

After many decades of neglect, Kaiser Wilhelm II is at last coming to be recognised internationally as the powerful and pernicious ruler that he actually was,[1] a kind of missing link, as it were, between Bismarck and Hitler, thereby helping to render not only the vagaries of his own long reign but also Germany's twisted road to the Third Reich historically a little more understandable. Biographies, monographs and dissertations about him are now appearing thick and fast, including the first-ever study of his life to be written in French,[2] some penned by the very historians who have hitherto downplayed his significance by emphasising the supreme influence of anonymous structures and social forces. Wolfgang J. Mommsen, one of the leading German authorities on the history of the Wilhelmine era, who only recently proclaimed that those historians who placed Wilhelm 'into the centre of things' were simply 'bad historians',[3] has himself now published a biography highlighting the Kaiser's central importance, beginning with a quotation from a letter of 1906 from Max Weber to Friedrich Naumann which could well serve as the leitmotiv for the present volume: 'The degree of contempt heaped upon us as a nation abroad (in Italy, America, everywhere!) – and quite *rightly* so! That is the decisive thing – *because* we are willing to put up with *this* regime of *this* man, has now become a factor of first-rate "world political" significance for us.'[4]

True, Mommsen presents his study as a mild corrective to my own interpretation of Wilhelm's role, arguing that the 'Prusso-German power elites' were sometimes able to manipulate the Kaiser and should therefore be saddled with greater blame than I have allowed, but the observant reader will soon establish that the disagreement between us, at least for the period here at issue, is more apparent than real. For one thing, it is a gross exaggeration to suggest, as Mommsen's title *War der Kaiser an allem schuld?* seems to do, that I hold the

Kaiser 'alone' responsible for all of imperial Germany's ills. The build-up of his personal power was gradual, and then after a few years of – always dysfunctional – dominance it disintegrated into polycratic confusion. For another, as the following pages will demonstrate, I heap sharp criticism on the 'Prusso-German power elites', though for rather different reasons from Mommsen. Whereas he sees them as at times successfully controlling the monarch in their own interest, I both emphasise the blindness of those at court who assisted the young monarch in his bid for personal power and bewail the political and moral weakness of those in the Wilhelmstrasse and elsewhere who tried and failed to stop him. Mommsen appears to me to overestimate the cohesion of the 'Prusso-German power elites', who were in fact, politically speaking, at each other's throats and allowed themselves all too readily to be controlled by the Kaiser and his court through what Norbert Elias has termed the 'kingship mechanism' – the monarch's ability to choose which of several rival cliques and conflicting courses he preferred at any given moment. The evidence for this interpretation is overwhelming. It is clearly set out in this volume and will, I am confident, prove persuasive in the end. But even now such differences are trivial when compared with the times, not so long ago, when I stood virtually alone and most historians, Wolfgang Mommsen included, banished Kaiser Wilhelm II to the shadows as a political irrelevancy and an embarrassing buffoon.

Gratifying as the international acceptance of the important if catastrophic role played by Wilhelm II in the history of Germany and the wider world may be, it was by no means axiomatic that a work as voluminous as this should be translated into English. Making my research accessible to English-speaking readers became possible only as a result of the generous financial support I received from the Arts and Humanities Research Board and the Stiftungsfonds Deutsche Bank, together with the award in 1994 of the Wolfson Prize and in 2002 of the Gissings Prize, and I should like to thank Dr Ulrich Cartellieri of the Deutsche Bank and Mr Sean Breslin of Gissings in particular for their extraordinary generosity.

The almost superhuman task of faithfully translating the many hundreds of pages that follow, replete as they are with at times rather idiosyncratic extracts from the letters and diaries of Europe's royalty, aristocracy and political and military leadership of more than a century ago, was undertaken by Sheila de Bellaigue, formerly Registrar of the Royal Archives at Windsor Castle and hence well acquainted with the personalities and issues of that vanished world. Not only did she fulfil the daunting task of translating my German text with characteristic professionalism and selfless commitment, however; she also checked and rechecked many of the quotations from the extensive correspondence held in the Royal Archives to ensure their accuracy. Puzzling archaic phrases, obscure aristocratic titles and arcane terms of courtly ritual were discussed at length between us in the hope of arriving at the closest equivalent in current English usage. I cannot praise or thank Sheila de Bellaigue enough for her uncomplaining dedication during what must at times have seemed a Sisyphean sentence. The footnotes were expertly transformed into English by Anna Menge. The entire text has been copy-edited with admirable virtuosity, sensitivity and devotion to the English language by Karen Anderson Howes. The responsibility for any errors that may remain naturally rests entirely with me.

I have seized the opportunity presented by this happy collaboration on the English edition to review the entire text, to shorten the account where it appeared overloaded with detail and to streamline the argument wherever it seemed unnecessarily serpentine, without, I hope, losing the immediacy and authenticity which it has been my aim throughout to achieve. If nothing else, the plentiful and often previously unknown archival evidence gathered together in this volume should stand as the solid foundation upon which future informed debate on this question of world historical significance, which has, for whatever reason, been obscured for far too long, will be able to build.

John Röhl
Sussex, June 2003

Preface to the German edition

I began the first volume of this biography of Wilhelm II, which dealt with the future Kaiser's childhood and youth up to his accession, with Heraclitus's saying that the soul of another person is a distant continent that cannot be visited or explored.[1] In this second volume on the first half of his long and much-debated reign I am returning to a country I first visited more than forty years ago, some of whose inhabitants seem more familiar to me than my own contemporaries. Both for my first book, *Germany without Bismarck*[2] and during the many years I spent compiling the three-volume edition of the correspondence of Philipp Eulenburg,[3] Wilhelm II's best friend, I studied the events and personalities of the transitional epoch between Bismarck and Bülow and examined the almost incalculable wealth of letters, diaries and notes written by the Kaiser, the imperial family and the statesmen and diplomats of the time, both in Germany and beyond, which has survived in public and private archives.

During my archival research for these early studies it became incontrovertibly clear to me, contrary to the opinion of professional German historians, that Wilhelm II was a key figure in modern German history as it followed its fatal course from Bismarck to Hitler. His narcissistic coldness, his dynastic arrogance, the aggressive attitude he adopted towards the federal constitution of the Reich, parliamentarism, political Catholicism, social democracy, the Jewish minority and the democratic aspirations which were beginning to emerge everywhere, even among the Conservatives, led him from the very outset to use the most wounding terms of abuse against the majority of his own people, thereby very rapidly undermining the stability of the Hohenzollern throne. His thirst for glory, his overweening nationalism, his pronounced militaristic tendencies and his passion for the navy provided the impetus for a breathtakingly ambitious, long-term goal: the transformation of imperial Germany into the foremost great power – *the* European world power – on earth, an endeavour which represented so mortal a threat to the established and hitherto far from harmonious world powers, Russia, France and Great Britain, that it caused them to unite against the common danger.

Although these perceptions gained acceptance in the wake of the great controversies over the work of Fritz Fischer in the 1960s and 1970s as far as Wilhelmine Germany in general was concerned, the decisive role which Wilhelm II played in introducing this illusory and self-destructive *Weltmachtpolitik* and naval policy was not recognised, and was even – despite the overwhelming weight of documentary evidence – vehemently denied. The last Kaiser, so the argument ran, had been no more than a *Schattenkaiser* – a shadow Kaiser – too impulsive to be able to intervene effectively in politics, too ridiculous to be taken seriously by historians.[4] The practical result of this neglect, however, was a virtual exoneration of the Kaiser, for in the absence of critical research the ample documentary evidence which could have thrown light on his true character and his disastrous political activity remained undiscovered. Diametrically opposed judgements on Wilhelm II were left abruptly juxtaposed in a limbo of unexplained contradiction – always a sign that the research essential to the furtherance of consensus has not been carried out.[5] For many years I was almost alone in my views on the true nature of the monarchical constitution which Bismarck had created – even at the moment of his dismissal in 1890 the founder of the Reich proudly described the system of Personal Monarchy, in which the King and Kaiser actually ruled, as one of his greatest achievements – and on the overriding decision-making power of Wilhelm II.[6] It is true that, especially in the English-speaking world, several other studies based on empirical archival research had reached similar conclusions.[7] But the general revision of the historical picture of this era that should have followed on from these findings was still lacking.[8]

In this volume I therefore return to the land I know so well, the Prusso-German Kaiserreich of the 1890s. With the benefit of richer sources than were available to me on my first expedition, and from a higher vantage-point – that of the imperial court rather than of the bureaucracy in the Wilhelmstrasse – I have set out to describe in detail the way in which Kaiser Wilhelm II's personal power grew within the monarchical and militaristic structures provided by the Bismarckian constitution and to point to the consequences of such a regime. Clearly, what is in question here is much more than the history of personalities; indeed we are dealing with some of the most fundamental issues of history in general. For there is no doubt that German policy would have taken a different course if the power struggle between the Kaiser and the leading German statesmen, which forms the main theme of this biography, had turned out differently. This book is therefore concerned not only with Kaiser Wilhelm II but also with the question of who controlled the powerful machinery of government in the German Empire at the heart of Europe, and how the guidelines of domestic and foreign policy were decided upon. It is concerned with the conditions necessary for the survival of the monarchical form of government in the twentieth century. And it is concerned with the question of the continuity, or inevitability, of German history in the course it followed from the foundation of the Second Reich by Bismarck to the Third Reich of Adolf Hitler.

For the gradual assumption of power by Wilhelm II in the period from 1889 to 1896, as this volume very clearly demonstrates, was anything but inevitable or predestined. Rather, it was the result of innumerable crises and machinations at the Hohenzollern court, in the officer corps and in the administrative offices of the Wilhelmstrasse, and it had to be more or less forcibly carried through against the growing democratic spirit of the time. From the earliest days urgent warnings came not only from members of parliament and journalists but also from the Reich chancellors, Prussian ministers, Reich secretaries of state, privy councillors, diplomats, generals, federal princes, foreign monarchs and ambassadors and even members of the Kaiser's own family. They all cautioned against the reintroduction, a hundred years after the French Revolution, of a semi-absolute Personal Monarchy in the federalist Reich, in an age of industrialised and urbanised mass society with unrestrained freedom of thought and expression. Everywhere, and with increasing alarm, comparisons were made with the fall of the Stuarts in England in the seventeenth century and of the Bourbons in France in the eighteenth century. The end of the Hohenzollern monarchy, the encirclement of the Reich and a war that would destroy everything were widely predicted. Not long after Wilhelm's accession there was already widespread speculation as to whether an All-Highest Sovereign who made such anachronistic and megalomaniac claims to power could still be in his right mind, and a few years later, as the last chapter shows, serious consideration was being given to the idea of a forced abdication. But in vain. In the end all his opponents, numerous, prominent and influential though they were, proved helpless against the elemental will to power of their young ruler with his faithful military retinue (the Flügeladjutanten), cunning court officials and blindly devoted favourites. From mid-1896 the dramatic power struggle behind the scenes was as good as over, with only the two Berlin parliaments, the Reichstag and the Prussian Landtag, still providing some constraint on Wilhelm II's decision-making power, but only in domestic and armaments policy. In accordance with the Bismarckian system of Personal Monarchy their opinion was not even sought where foreign and military policy were concerned. It was not long before the anticipated consequences began to show themselves.

In view of the unresolved state of research and of the controversy still surrounding the fundamental questions involved, I have again opted for a narrative style in which the very expressive letters and diary entries of the main protagonists are quoted extensively. In this way the reader can get to know the archival sources which have led me to my own view of the growing power and the pernicious influence of Wilhelm II. From these it should become plain that the interpretation offered here of the dramatic developments behind the glittering scenes of imperial Germany is not mine alone, but was also very much that of the people directly involved. Through the documents on which this biography draws heavily, long-forgotten expressions of opinion by Wilhelm II's contemporaries, kept secret at the time for good reason, are brought back to life, enabling us a hundred years later to enter into the thought patterns, convictions, passions, intrigues and human weaknesses of that generation. They should help to deepen our understanding of the vanished civilisation of the last days of the Hohenzollern throne, with a Kaiser who either could not or would not understand what was required of monarchy if it were to survive in a modern pluralist society.

Another reason for adopting a closely source-based narrative form is that several printed editions of letters and diaries which were published after the trauma of 1918 – Article 227 of the Treaty of Versailles stipulated that Wilhelm, who had taken refuge in Holland, was to be handed over to an international tribunal and tried as a war criminal – can be shown to have been corrupted by omissions from and distortions of the authentic text calculated to gloss over the truth. As a result of this falsification a highly inaccurate historical picture, above all of Kaiser Wilhelm II and the manner in which he ruled, has been perpetuated until this day. It can be corrected only by returning to the original sources. This is particularly true of the diaries of the ambitious, arch-conservative and always well-informed General Count Alfred von Waldersee, which were published in 1922–3 in a scandalously expurgated version.[9] The original text of the diaries, with his scathing comments on Wilhelm II, has been fully utilised for the first time in this biography; the disparities with the printed version are indicated in the notes.

A further important source, but one to which little attention has been paid in Germany, provides striking evidence of the extent of Kaiser Wilhelm II's power, and that is the correspondence of his British relations which is preserved in the Royal Archives at Windsor Castle. In drawing on these papers and in giving them their due place in the narrative I hope to have been able to throw more light on the personality of the last German Emperor, and above all to have contributed to our understanding of the international implications of his restless and scheming political activity. I thank Her Majesty Queen Elizabeth II for again allowing me to make use of the Royal Archives for this part of my biography.

As I stated in the preface to the first volume, I began the archival research for this book more than twenty years ago, after I had finished editing Eulenburg's correspondence. The work has required many visits to a variety of archives, and in this I have been greatly helped by generous grants from the British Academy, the Robert Bosch Foundation and the Robert Bosch Jubilee Foundation 1986 in the Stifterverband für die Deutsche Wissenschaft. Once again I thank the archivists of the then Central State Archives of the German Democratic Republic in Merseburg and Potsdam (both now in Berlin) for their help, above all in assisting me to consult the holdings in the Brandenburg–Prussian House Archive and the records of the Kaiser's Civil Cabinet. The Bundesarchiv in Koblenz, the Bundesarchiv-Militärarchiv in Freiburg and the Politisches Archiv of the German Foreign Office, then in Bonn and now also back in Berlin, also hold important collections – papers of major statesmen and military figures, diplomatic reports and the secret reserve collection with the Kaiser's outrageous marginal comments, which had to be locked away on the orders of successive Reich chancellors and foreign secretaries – all of which I was able to consult for this biography, for which I wish to express my thanks. I am most particularly grateful to HRH Landgrave Moritz of Hesse and Prince Rainer of Hesse for permission to work for several months in the Hesse Family Archives in Schloss Fasanerie near Fulda and to make unrestricted use of the extensive papers of Wilhelm II's remarkably progressive mother Victoria, the Empress Frederick, for this volume as for the previous one. My thanks are also due to HH Prince Ferdinand von Bismarck and the Director of the Otto von Bismarck Foundation, Dr Michael Epkenhans, for making available the secret file at Friedrichsruh on the purchase of Wilhelm II's embarrassing letters to Miss Love. And last but not least, the archivists of the Royal Archives, particularly Sheila de Bellaigue, have earned my thanks for their enduring support for my work, both in making material available to me and through their expert knowledge and wise advice, and have saved me from many an error.

In addition, many private individuals have given me access to papers in their possession relevant to this biography. In this connection my particular thanks go to the Earl of Lonsdale, at Lowther, Cumbria; Ramona Countess von Oeynhausen-Sierstorpff and Peter Count von Wedel, at Bad Driburg; Gustav Count von Wedel, at Frankfurt am Main; Hermann Count Stolberg-Wernigerode, at Frankfurt am Main; Karl-Wilhelm Freiherr von Plettenberg, at Essen; Herr and Frau Joachim von Natzmer at Munich; Frau Margot Leo-Hoffmann at Freiburg; Frau Ruth von Santen at Wennigsen, Lower Saxony; Adrian Freiherr von Holzing-Berstett, at Bollschweil, Breisgau; and Freifrau von Seyfried-Marschall at Oberkirch, Baden. My thanks for information and material of great value to this book are also due to Professor Dr Margarete Jarchow, Dr Anastasia Hackett, Dr Ragnhild Fiebig-von Hase, Hagen Count Lambsdorff, Dr Thomas Otte, Dr Roderick McLean, Dr Annika Mombauer, Dr Matthew

Seligmann, Dr Matthew Stibbe, Dr Christopher Duggan, Dr Jost Rebentisch, Miss Frances Dimond, Mr Arthur Addington, Frau Liesbeth Ruitenberg, Frau Dr Friedhild den Toom-Jacobi and Drs Th. L. J. Verroen.

I was able to begin writing this second volume of the biography in ideal conditions in January 1994 at the Institute for Advanced Study in Princeton, and to continue during the academic year of 1997–8 at the National Humanities Center in North Carolina. Stays at the Moses Mendelssohn Centre for European Jewish Studies in Potsdam and at the Bellagio Study and Conference Center of the Rockefeller Foundation on Lake Como contributed considerably to my progress on the manuscript. My own university, the University of Sussex, also generously helped further the work by allowing me leave of absence from teaching and administrative duties on several occasions. During the last stage of the preparation of the book, thanks to a generous research grant from the Arts and Humanities Research Board, I had the good fortune to be able to obtain scholarly assistance, without which this study would have taken a great deal longer to complete. The help which I received from Annika Mombauer and above all Pauline von Hellermann, both in carrying out research on particular aspects of the biography and in the stylistic revision of my drafts, was outstanding. Dr Holger Afflerbach, who visited the University of Sussex in the spring of 2001 as the holder of a Feodor Lynen scholarship, assisted me in the concluding phases of the work with his valuable advice and specialist knowledge. My pupils, and above all my old friends Wilhelm Deist, Hartmut Pogge von Strandmann, Bernd and Britta Sösemann and Manfred Count von Roon encouraged me at moments when the work seemed too much to cope with. I thank my editors at the C. H. Beck Verlag in Munich, Dr Stefan von der Lahr and Peter Schünemann, as well as Frau Dr Jäger, for their accurate and sensitive work in editing the final manuscript. But above all I reserve my warmest thanks for my children Stephanie, Nicky and Christoph and my wife Rosemarie von Berg Röhl, to whom this volume is dedicated, for their patient interest in this lifelong project.

John Röhl
Sussex, June 2001

CHAPTER ONE

The accession

IN PLACE OF A CORONATION

In any other monarchy the death of two Emperors within three months would no doubt have been followed by the coronation of their successor in a solemn and magnificent ceremony designed to demonstrate the stability of the throne. In the complicated German Reich constructed by Bismarck, with its delicately counterbalanced forces, such a ceremony would have been impossible. The coronation of Wilhelm II as Kaiser was ruled out by the federal structure of the Reich, with its four kingdoms (Prussia, Bavaria, Saxony and Württemberg), six grand duchies (Baden, Hesse-Darmstadt, Mecklenburg-Schwerin, Mecklenburg-Strelitz, Oldenburg and Saxe-Weimar-Eisenach), five duchies (Anhalt, Brunswick, Saxe-Altenburg, Saxe-Coburg-Gotha and Saxe-Meiningen), seven principalities (Lippe, Schaumburg-Lippe, Reuss of the senior and junior lines, Schwarzburg-Rudolstadt, Schwarzburg-Sonderhausen and Waldeck), three free Hanseatic cities (Hamburg, Bremen and Lübeck) and the imperial provinces of Alsace-Lorraine. Nor would it have struck the right note in the new Reich for Wilhelm to follow the example of his grandfather Wilhelm I in October 1861, by crowning himself king of Prussia far away in Königsberg, and this idea was not seriously considered by anyone. Instead, following the disruption which this 'year of the three Kaisers' had brought with it, other methods were sought to stabilise the Hohenzollern throne and to legitimise the rule of the 29-year-old former Crown Prince as German kaiser and king of Prussia.

As early as 15 June 1888, immediately after the death of his father, Wilhelm II issued two proclamations which attracted universal attention as indications of what the new regime might bring. In an 'Army Order' of the day, his 'first word' to 'His Army', Wilhelm emphasised the special relationship which his 'glorious ancestors' had always had with the military. 'In the army', its new supreme commander declared, 'the firm, inviolable bond with the commander-in-chief is the inheritance which passes from father to son, from generation to generation, and for that reason I point to my grandfather, who stands before you all in your mind's eye, the picture of the glorious, revered Commander, and no finer or more stirring picture can be imagined – and to my dear Father, who already as crown prince won himself a place of honour in the annals of the army – and to a long series of renowned ancestors, whose names shine brightly in history and whose hearts beat warmly for the army. And so we belong together, I and the army; we were born for one another and it will be our endeavour to hold together in an indissoluble union, through calm or storm, according to God's will.' The army, he continued, would now take the oath of loyalty and obedience to him, and he on his side would pledge 'always to be mindful that the eyes of my forefathers look down upon me from the next world and that I shall one day have to answer to them for the glory and the honour of the army!'[1]

At the same time Wilhelm – this time in his capacity as German Kaiser – issued a proclamation entitled 'To the Navy!' in which he spoke in similar tones of his particularly close relationship to the fleet. Only the previous year his grandfather had expressed 'in the warmest terms his lively satisfaction and appreciation with regard to the development of the navy under his glorious rule', and his beloved father had also taken 'such great pleasure and such a lively interest in the growth and progress of the navy'. He wished the navy to know that he too, the new Kaiser, had been held by bonds of 'lively and warm interest in the navy, since my earliest youth, in complete accord with my dear brother, Prince Heinrich of Prussia'. Thus he could 'declare with full confidence that we shall stand firm and fast together in good and in bad times, in storm as in sunshine, always mindful of the German Fatherland and always ready to give our life's blood for the honour of the German flag'.[2] This was the first time that a Prussian monarch had addressed the navy as well as the army on his accession, and the warm words of the young Kaiser filled his naval officers, as Admiral Gustav Freiherr von Senden-Bibran later recalled in his memoirs, 'with great joy and raised our hopes for the future'. Even in these circles, however, no one could have suspected – as Senden added – 'that Wilhelm II would see it as his task to raise up not only the army but also the navy, with all the means at his command, with his thorough knowledge of it, his great devotion to it, and his whole personality, and to set it beside the army on as near as possible an equal footing'.[3]

It was not until three days later that Wilhelm, reverting this time to his capacity as king of Prussia, issued a proclamation to the Prussian people. In it he invoked dynastic tradition and Christianity in order to weave a special bond between ruler and subjects. 'Called to the throne of my Fathers,' he announced, 'I enter upon my reign with my eyes raised to the King of all Kings, and I have made a pledge to God, following the example of my Fathers, to be a just and merciful ruler of my people, to foster piety and the fear of God, to defend peace, to further the welfare of this country, to be a helper to the poor and oppressed and a true guardian of justice. In asking God for strength to fulfil these Royal duties which His will lays upon me, I rely upon the trust in the Prussian people which I derive from looking back on our history. In good and evil times the people of Prussia have remained ever loyal to their King; upon this loyalty, which has proved an indissoluble bond to My Fathers in all times of difficulty and danger, I also depend, in the knowledge that I return it with all my heart, as the loyal Prince of a loyal People, both equally strong in their devotion to their common Fatherland. This knowledge of the mutual love which binds Myself and My People gives me confidence that God will grant me strength and wisdom to discharge the duties of my Royal office for the good of the Fatherland.'[4]

All three proclamations were received with much sympathy, although the conspicuous singling out of the army drew criticism from the left-wing liberals and in Western foreign countries.[5] The 'Army Order' was greeted with enthusiasm in the Prussian officer corps above all. Helmuth von Moltke, the future Chief of the General Staff, who was serving as personal adjutant to his uncle the old Field Marshal at this time, wrote delightedly to his Swedish wife: 'You will have read in the newspapers the young Kaiser's splendid Proclamation to the army. – There is a decidedly new wind blowing in everything. The young Kaiser is in constant activity, he has spent the whole day conferring, issuing orders and signing documents. We received the first Cabinet Order with the signature Imperator Rex the night before last [15 June]' in the General Staff. The younger Moltke was convinced that Wilhelm had composed all three proclamations himself; there was 'no other hand involved', he claimed; the 'young Master' had 'rejected all suggestions and seen to the matter himself'.[6] It is more probable, however, that Wilhelm had sought Count von Waldersee's advice on the composition of all three proclamations,[7] and that Prince Bismarck and his son had also seen and approved the texts in advance; had it not been so, it would certainly have been remarked upon later in the numerous embittered writings of the Bismarcks about their dismissal. At any rate Waldersee, the elder Moltke's deputy, who in a few weeks would himself be promoted to chief of the General Staff, was more than satisfied when he wrote in his diary on 19 June: 'The . . . proclamations of the Kaiser "To the Army and Navy" and "To My People" have made an excellent impression; in the army there is general jubilation over the new sovereign. There is a widespread feeling that we have successfully overcome a serious illness and are now moving forward to good times. The uncertainty, the mistrust, the dissatisfaction were having a dreadfully oppressive effect on everyone's minds. The gentlemen of the Progressive Party and their Jewish hangers-on have certainly been dealt a hard blow.'[8]

On 24 June 1888 the new Kaiser and Kaiserin moved out of the small Marmorpalais in Potsdam into the Stadtschloss in Berlin. As their carriage drove through the Brandenburg Gate, accompanied by two outriders and with detachments of the aristocratic Gardes-du-Corps regiment preceding and following the carriage, Unter den Linden was as crammed with onlookers as it had been three months earlier for the funeral of the old Kaiser.[9] The next day the grandiose ceremony of the opening of the Reichstag took place; it was even more effective than the proclamations of 15 and 18 June as a spectacular prelude to the new era. The impetus for it came from a suggestion made by Grand Duke Friedrich I of Baden, who, because of the double change of sovereign, wished to rally the German federal princes demonstratively around the young Kaiser.[10] As there was no precedent for a ritual of this kind the ceremony had to be improvised behind the scenes, which also necessitated – not surprisingly, with such a large number of exalted participants – making allowances for numerous special requests and soothing a series of wounded vanities. The first difficulty arose from the fact that the Kingdom of Bavaria, which as the biggest of the federated states could claim the first place after Prussia in the ceremony, was formally under the rule of the hopelessly deranged King Otto; but the Prince Regent, Prince Luitpold, was of inferior rank to King Albert of Saxony. The problem was resolved by the King of Saxony yielding to Prince Regent Luitpold 'the precedence to which he was unquestionably entitled' for the duration of the ceremony.[11] A comparable problem arose with Württemberg, whose highly eccentric King Karl I had been for some years completely under the influence of two American favourites named Woodcock and Jackson and had not the least inclination to travel to Berlin. He appointed as his representative his nephew, the heir presumptive, Prince Wilhelm of Württemberg, to whom the Grand Duke of Baden – Württemberg being a kingdom – yielded precedence.[12] Thus, with the single exception of Grand Duke Friedrich Wilhelm of Mecklenburg-Strelitz, all of the federal princes (or their representatives) were present at the magnificent ceremony of the opening of the Reichstag in the White Hall of the imperial palace – a fact that made a strong impression on all observers. The Austro-Hungarian ambassador Count Imre Széchényi reported to Vienna: 'This great act of homage paid by Germany's princes to the youthful Kaiser Wilhelm II acquires increased significance because no invitations of any kind had been sent out for this purpose; the homage arose from a completely spontaneous decision by the German princes, a fact that understandably gave the greatest pleasure in influential circles here.'[13]

Until the last moment, however, organisational blunders threatened to upset the ceremony. When the two Moltkes, the Field Marshal and his nephew, returned to Berlin from Ratzeburg on the evening of 24 June, they discovered that the old Chief of the General Staff, victor of three wars of unification, was not mentioned anywhere in the programme for the opening ceremony. The 88-year-old was, as his nephew remarked, 'deeply hurt and declared at first that he wished to leave at once' and to hand in his resignation. The following morning he sent a letter to the Flügeladjutant (aide-de-camp) on duty, in which he announced that 'since he, as the most senior Field Marshal, Chancellor of the Order of the Black Eagle etc. might well have expected to find a place in His Majesty's suite, namely directly behind the Reich Chancellor, but is not in fact mentioned at all in the programme, and finds it incompatible with his military dignity to appear as a member of parliament, he would request that His Majesty be informed that he found himself obliged to absent himself from the ceremony'. The younger Moltke attributed the blame for this omission entirely to court toadies who, he suspected, wished to be done with the old man and attach themselves to the new stars. But the Kaiser, beside himself with anger, sent a Flügeladjutant to Bismarck immediately before the ceremony to ask whether it would be permissible to have the old Field Marshal appear together with the federal princes, to which Bismarck assented. Moltke, however, who was still in the palace, declared 'no, he did not belong there', and chose for himself the place behind the crown insignia, directly in front of the Kaiser.[14]

The impressive festivities of 25 June 1888 in the White Hall of the Stadtschloss in Berlin reminded many of the proclamation of the Kaiser in the hall of mirrors at Versailles nineteen years earlier, which had been immortalised – as this ceremony was to be – in a celebrated group portrait by Anton von Werner.[15] The ceremony began at 12 noon with a service in the chapel of the imperial palace. The text of the court chaplain's sermon was 'By God's grace I am what I am' and had been selected by Wilhelm himself.[16] After the service the members of the Reichstag, almost four hundred in number – only the eleven Social Democrats had refused to take part in the ceremony – the members of the government of the Reich and of the Prussian state government, and finally the court, assembled in the White Hall, which had been decorated under the direction of Anton von Werner. 'The throne richly draped with yellow velvet, the red velvet dais for the Kaiserin, the seats right and left of the throne for the princes' – everything made a magnificent impression, Moltke recorded.[17]

Particularly impressive was the entrance of a company of the Palace Guard, marching with shouldered arms, the officers with drawn swords. Bismarck, clad in his cuirassier's uniform, led in the members of the Bundesrat 'like a flock of lambs'. When they were all assembled he announced to the Kaiser that the members of the court could now enter. 'First pages in black breeches with mourning ribbons at the knee, then the insignia of the Reich', followed by Field Marshal Count von Moltke with his marshal's baton supported on his hip, then the Kaiser with the Prince Regent of Bavaria on his left and the King of Saxony on his right, followed by the remaining German princes, twenty in number, all in the great red velvet mantles of the Supreme Order of the Black Eagle, the highest Prussian honour.[18] Behind the Kaiser the new, six-year-old Crown Prince took his place with the young Kaiserin, pregnant with Prince Oskar; next to them sat the Hereditary Princess Charlotte of Saxe-Meiningen, the Kaiser's eldest sister, and a handful of other ladies, all still dressed in black. Facing the Kaiser were ranged the highest court officials, the Reich Chancellor, his son Herbert and the other secretaries of state of the Reich, together with the Prussian ministers of state and the remaining members of the Bundesrat. The people's representatives, the elected members of the German Reichstag, the opening of which was the ostensible reason for the ceremony, stood in the background and melted into the obscurity of insignificance. The speech from the throne, which the young Kaiser was thus to address both to all the federal princes and to the elected members of parliament, and which was seen as a kind of programme for the coming reign, was received 'with breathless attention'.[19]

On his entrance into the White Hall the Kaiser struck the younger Moltke as 'extraordinarily dignified and deeply serious'. The manner in which Wilhelm 'walked with a firm step up to the throne dais and greeted the assembly with a solemn inclination of the head' had been 'truly majestic'. Once complete silence was established, the Reich Chancellor handed him the text of the speech from the throne. Wilhelm grasped the manuscript, set the helmet on his head 'with a resolute gesture' and threw his mantle back, 'his head held high, allowing his glance to roam over the silently waiting assembly'. Although his hand did not shake – Moltke watched closely for this – his voice was at first 'muffled and unclear. The sentences came out jerkily and with difficulty, and in spite of the dead silence he could hardly be understood. But gradually the voice rose, the delivery became fluent, and when he came to the words: I am resolved to maintain peace with everyone, insofar as this depends on me, he emphasised the word *me* so loudly and clearly that it flashed through all his audience like an electric spark, it was so charged with meaning, the full consciousness of the ruler's power, while at the same time there resounded in it the warning: but woe to him who dares to tread too close to me; an extraordinary strength and confidence lay in that one word, so that everyone broke out spontaneously into loud and enthusiastic applause. He spoke the last sentences of the speech in a fine, penetrating voice, every trace of self-consciousness had vanished, and he stood there, strong and proud, the powerful, self-confident ruler of a mighty empire.'[20]

1. The young Kaiser surrounded by the German princes, Bismarck and the ministers at the opening of the Reichstag on 25 June 1888

After a few introductory sentences expressing his grief at the loss of his grandfather and father Kaiser Wilhelm II's speech declared: 'I have called you together, gentlemen, to announce to the German people, in your presence, that as kaiser and king I am resolved to follow the same path by which my late, revered grandfather won the trust of his allies, the love of the German people and the goodwill of other nations. Whether I too shall succeed in this lies with the Almighty; I will strive towards it with earnest endeavour. The greatest tasks of the German kaiser lie in safeguarding the military and political position of the Reich with regard to foreign relations, and, in internal matters, in watching over the execution of the laws of the Reich. The chief of these laws is the constitution of the Reich; to guard and defend it, in all the rights which it vouchsafes to both legislative bodies of the nation and to every German citizen, but also in those vouchsafed to the kaiser and to each of the federated states and their rulers, is among the foremost rights and duties of the kaiser. In the legislation of the Reich, according to the constitution, it is my duty to act more in my capacity as king of Prussia than as German Kaiser; but in both roles it will be my endeavour to continue the work of the legislation of the Reich in the same way as my late, revered grandfather began it. In particular I adopt in its entirety the declaration issued by him on 17 November 1881 [concerning social policy], and I shall continue to work in the spirit of this declaration, to ensure that the legislation of the Reich strives further to give the working population the protection that it is able, in accordance with the principles of Christian morality, to provide for the weak and the oppressed in the struggle for existence. I hope that in this way we shall succeed in bringing closer the elimination of unhealthy social differences, and I am confident that in my care for the internal well-being of the nation I shall receive the unanimous support of all true adherents of the Reich and of the federated governments, support undivided by party differences. At the same time I consider it imperative to maintain our national and social development on the path of the law, and to take firm action against all aspirations which have the aim and effect of undermining national order. With regard to foreign policy I am resolved to maintain peace with everyone, insofar as this depends on me. My love for the German army and my relationship with it will never lead me into the temptation of depriving my country of the benefits of peace unless war becomes a necessity forced upon us by an attack on the Reich or on its federated states. Our army shall safeguard peace, and if in spite of it our peace is breached, it shall be ready to fight for it with honour. This it will be able to do with God's help, thanks to the strength which it has been given by the recent legislation on national defence, passed unanimously by you. To use this strength in wars of aggression is far from my intent. Germany needs neither new military glory nor any conquests, having finally won itself the right to exist as a single and independent nation. Our alliance with Austria-Hungary is publicly known; I hold fast to it with German loyalty, not only because it has been entered into, but because I see in this defensive bond one of the basic elements of the balance of power in Europe, as well as a legacy of German history, the sense of which is shared today by public opinion throughout the German people, and which accords with traditional European international law as it stood, undisputed, until 1866. Similar historical links and similar national needs of the present bind us to Italy. Both countries desire to hold fast to the blessings of peace . . . Our existing agreements with Austria-Hungary and Italy allow me, to my satisfaction, to nurture my personal friendship with the Tsar of Russia and the peaceful relations which have existed for a hundred years with the Russian Empire, as is consistent both with my own feelings and with the interests of Germany. Both in the conscientious preservation of peace and in caring for our army I place myself willingly in the service of the Fatherland, and I rejoice in the traditional relations with foreign powers through which my endeavours towards the former goal will be furthered. Trusting in God and in the valour of our people I am confident that it will be granted to us for the foreseeable future to preserve and to consolidate what was won in battle under the leadership of my two predecessors on the throne, who now rest in God.'[21]

The speech, the beginning of which was received with complete silence, was interrupted several times by cheers from all sides. When it was over the Kaiser handed the text back to Bismarck and held out his hand to him. The Chancellor, deeply moved, kissed the hand of his Kaiser and King. Bavarian Minister-President Freiherr von Lutz called for three cheers for the Kaiser, with which the assembled company joined in; then the monarch and the remaining ruling princes left the hall in the same order as that in which they had entered.[22]

The ceremony with the carefully considered speech from the throne made an excellent impression in all circles in Germany and almost everywhere abroad.[23] Wilhelm himself could look back on this favourable beginning to his reign with satisfaction. 'The first days left nothing to be desired in the way of impressive moments', he wrote delightedly to his grandmother Augusta, 'and the gift of God which was bestowed on me in the gathering of the German princes by common consent must surely be the most magnificent gift that He has given to a young sovereign! May he continue to stand as graciously by my side! Uncle Fritz [of Baden] and Uncle Albert [of Saxony] above all showed themselves quite outstandingly excellent and kind in their longstanding loyalty and friendship; may Heaven reward them both for it. They can both count on me.'[24] With similar enthusiasm Hildegard von Spitzemberg wrote: 'The most magnificent thing of all was the opening of the Reichstag with the German princes!' It was an act 'which must fill every German heart with joy and pride and which vis-à-vis foreign countries is the equivalent of winning a war . . . It filled the eyes of all with tears, who saw the young ruler standing in his place of great responsibility, surrounded by all the German princes, Moltke behind him, the Iron Chancellor before him, bending over his hand, the young Kaiserin with child, beside her the little Crown Prince!'[25]

Only in the Progressive camp, in London and in the British royal family were critical voices heard. The Empress-Mother Victoria – she now took the title Kaiserin Friedrich or the Empress Frederick – pronounced the whole 'pageant and pomp' of the opening of the Reichstag 'very silly and absurd and out of place'. She was indignant that her daughter Charlotte had attended the ceremony in spite of being in deep mourning. In the eyes of the widowed Kaiserin the true meaning of the opening ceremony lay in Bismarck's wish to demonstrate to the world his pleasure at the beginning of a new era which was incomparably more congenial to him than the short reign of Friedrich III. She was bitter in her condemnation of the initiative taken by the Grand Duke of Baden, who had taken a fancy to the role of protector of the Reich and could not see how much he was playing into the hands of Bismarck.[26] 'The opening of the Reichstag with this completely unnecessary pomp hurts me *deeply*!', she wrote in her diary. 'Calling together of all German Princes – a *demonstration* in favour of the present system! . . . It was Prince Bismarck who wanted this farce! And Fritz v. Baden whose idea it was for the Princes to come here. All this will go to Wilhelm's head even more! Anton v. Werner has been brought here to decorate the White Hall as if for a *celebration*! How *cruel* and unfeeling, with the grave only just sealed! What absence of dignity and decency! – if the German Reich needs such an artificial *mise-en-scène* and hocus-pocus to prove to the world that it is not breaking up, then I am *sorry* for it! Quiet, solemn silence, and *mourning* . . . would have been more dignified and fitting than this childish haste to create a spectacle.'[27]

Two days later the ritual was virtually repeated in the same place, only this time without the participation of the federal princes, when Wilhelm opened the Prussian parliament in his capacity as king of Prussia, with a speech from the throne in which he took the oath on the Prussian constitution. In this speech, which had been communicated to the Prussian ministers and approved by them that morning at the first Crown Council at which Wilhelm presided, he again emphasised his wish to carry on quietly pursuing the same policies as his grandfather. 'I have no intention', he declared, 'of shaking the confidence of our people in the permanence of our legislative system by attempting to broaden the rights of the Crown. The legal position as regards my rights, so long as it is not called into question, is quite sufficient to secure the measure of monarchical intervention which Prussia requires in accordance with her historical development, her present structure, her position in the Reich and the feelings and usages of her own people.' Wilhelm closed his speech by taking to heart Frederick the Great's principle that the king is the first servant of the state.[28] On his return to Wilhelmstrasse from the ceremony Bismarck threw his cap on to the table and, beaming with pleasure, exclaimed: 'Im Sattel hew ick en jetzt!' ('I've got him in the saddle now!')[29] He did not ask himself whither the proud rider would now ride on his high horse.

THE FATEFUL ALLURE OF CHARISMATIC KAISERDOM

The newfangled ceremony of the opening of the Reichstag by the Kaiser, with the court, the federal princes and the elected members of the Reichstag all playing their part, illustrated more clearly than any other ritual could have done the complicated constitutional structure of Bismarck's Reich. The competing forces of authoritarian and popular, centralist and federalist elements were held together in a state of unsteady equilibrium by a delicately balanced mechanism. Any shift of power towards one side or the other – whether towards the Prussian hegemonic state, the centralised Reich authorities, the intermediate states led by the Kingdoms of Bavaria, Saxony and Württemberg or the democratically elected Reichstag – would have imperilled the whole edifice of the Reich. Nevertheless such shifts were inevitably happening as a result of population growth, internal migration, industrialisation and urbanisation. With every year of unification the German people grew closer together, in spite of all confessional and particularistic differences, in such a way that the initially feared separatist aspirations in Bavaria, Saxony and Württemberg gradually lost ground. Nonetheless the fear of the disintegration of the Reich (possibly in the course of a civil war with foreign intervention) continued to haunt the highest authorities in Berlin right up to the turn of the century, as we shall see. On the other hand these same powerful social and intellectual fluctuations were strengthening democratic tendencies throughout the land.

To create a bulwark against the dangerous growth of parliamentary and democratic power in the Reich Bismarck was obliged to bring the so-called state-supporting monarchical principle increasingly to the fore, even though this involved an uncomfortable paradox for him: for reasons of *state* he needed to propound the romantic fiction of a genuinely ruling monarch. As his son Herbert expressed it on 5 October 1888, the 'state-supporting Prussian principle' of the *genuinely* ruling monarch must be defended by all possible means; everything depended on preventing 'the modern parliamentary system without a personally ruling monarch' from establishing itself in the Kaiserreich – on ensuring that the Prussian–German monarch did not degenerate into an 'automatic signing-machine', as for example in England or Belgium.[30] The German Kaiser and King of Prussia must continue to wear 'the most powerful crown on the entire planet'.[31] This 'monarchical principle' was elaborated into a political ideology through which Prussia–Germany increasingly detached itself from the parliamentary monarchies and republics in the north, west and south of Europe. In the autumn of his chancellorship Bismarck let loose upon the world a dubious form of monarchical myth which would cast a widespread and lasting spell and would ultimately undermine his own position.

How this 'state-supporting Prussian principle' of 'personal monarchy' was projected on to the young Kaiser Wilhelm II in 1888 can be seen from the speech which the Free Conservative member of the Reichstag, Count Hugo Sholto von Douglas, made in the small provincial town of Aschersleben on 4 October. As the full text of this sensational speech was printed three days later in the official *Norddeutsche Allgemeine Zeitung* and then published as a semi-official pamphlet entitled 'What we can expect from our Kaiser', it is hardly surprising to discover that Douglas was not its author, but that it had been drafted in the Wilhelmstrasse by Franz von Rottenburg, the head of Bismarck's Reich Chancellery. The speech was originally to have been made by District Magistrate Dietz, as the Empress Frederick discovered, and it was only when he refused to act as Bismarck's mouthpiece that the speech was passed on to Douglas, 'who was servile enough to make it. *C'est ainsi que l'on fait l'Histoire!*' was the Dowager Empress's biting comment on this episode.[32] Douglas's action may seem more comprehensible, however, in the light of the fact that he had been created a count by Wilhelm not long beforehand, on 20 August 1888.[33]

The speech declared the dual aims of the Kaiser to be the strengthening of the power of the German Reich vis-à-vis the outside world and the establishment of a strong, genuinely ruling monarchy within the country. 'In the memorable speeches from the throne with which he opened the Reichstag and the Landtag in June, Kaiser Wilhelm proclaimed his loyalty to the legacy of his great predecessors', Douglas declared. 'And this legacy consists in the strengthening and consolidation of the German Reich on the one hand and the strengthening and consolidation of the monarchical principle on the other . . . What Kaiser Wilhelm I won for Germany with his victorious sword Kaiser Wilhelm II will and shall hold fast for Germany. No less will he and shall he hold fast to all that Kaiser Wilhelm did to instil the idea and the consciousness of a strong monarchy into the hearts of his people. There was a time, not long ago, when people mocked the influence of the sovereign and sought to consign him to the role of a purely representative figurehead, as is still the aim today of the democratic and democratically inclined parties. The achievements of Kaiser Wilhelm I showed the whole world what great things a powerful ruler can accomplish, and if there were ever any doubt about it, everyone who loves his Fatherland today has the deep conviction that Prussia has not only become great through the Hohenzollerns, but that the whole future of Germany and Prussia, indeed of the peace of the world, is inseparably bound up with the House of Hohenzollern. That is the great legacy that has come down to our illustrious Kaiser from his predecessors, and he would be no Hohenzollern if he did not see it as his foremost task to build further upon the foundations laid by his fathers, to Germany's, to Prussia's, honour and blessing.'[34]

Undeniably, the personal characteristics of the Kaiser were a problematic aspect of this daring monarchical concept, for in a hereditary monarchy the continuity of suitably statesmanlike talent from generation to generation cannot be guaranteed. But on the very grounds of this inner logic, in drafting the speech Rottenburg was obliged to emphasise how outstandingly well qualified Kaiser Wilhelm II was for the mighty Prussian–German throne. That in so doing he had to set the truth on its head in every single particular, against his better judgement, was quite a different matter.

It was therefore no accident that the speech began with a heroic description of the quite exceptional preparation Wilhelm had undergone for the heavy responsibilities of office.[35] Never before had a Prussian prince 'seen so much of everyday life with his own eyes', Douglas declared. With 'perceptive wisdom' Kaiser Friedrich III had broken through the 'traditional boundaries of royal education' by sending his son to school in Kassel, and had thus brought him 'into contact with circles of everyday life'. 'It was a bold experiment to set the heir to the Prussian throne on the school bench at the gymnasium in Kassel and to give him the task of learning what his fellow pupils had to learn'; yet thanks to the 'tenacious will-power' of the prince the experiment had been a 'brilliant' success, 'so that the practical impressions gained during his time at the gymnasium' could 'continue to be felt and to develop further in his academic life in Bonn'. Still today, according to Douglas's idealised account, Wilhelm's fellow students at Bonn spoke enthusiastically of 'his lively, truly comradely character and of his unrivalled talents as a forceful swordsman'. 'Our young Kaiser has had the benefit of a thorough, scientific education; he has learnt to work with persistence and perseverance, and he has succeeded most happily in uniting the impressions of practical life with the conclusions drawn from his scientific studies.' This successful education in combination with his 'exceptional natural gifts' was 'undoubtedly responsible for the excellent judgement and the quick, felicitous perception which distinguish our present Kaiser, to the astonishment of all those to whom it falls to have dealings with him'. After Kassel and Bonn, as was well known, a 'strict military training' had ensued. 'With the purposeful energy characteristic of him, and with the warm interest in the army innate in the House of Hohenzollern . . . Prince Wilhelm became a smart and enthusiastic soldier' who 'learnt to obey . . . in order to be able to command in days to come'. In spite of all his passion for the military life, however, the Kaiser never allowed his interest in the civil affairs of the nation to recede. 'Already as prince, while he was receiving instruction from the highest officials of our national administration about the work of the civil authorities, of the general provincial administration, of the church and educational authorities, about the economic and social functions of the state, about the fiscal, financial and budgetary principles on which the prosperity of the Prussian state so heavily depends, he surprised everyone with his quick understanding, his healthy, practical intelligence, his lively interest in all these things, characterised by a desire to get to the root of the matter on every subject. So it has remained after his accession', Count Douglas assured his audience.

This 'all-embracing education' of the Prince had brought out a straightforward and yet regal disposition in Wilhelm. Thanks to it the young sovereign had developed 'a restless instinct for activity and a genuine love of work', together with the conviction 'that the duties of a sovereign were also service, a service which must be carried out faithfully and without respite, and to which all personal considerations had to be subordinated'. The magic of the Hohenzollerns, which was so characteristic of all three German Kaisers, showed itself also in Wilhelm II 'in the most natural simplicity and geniality, heart-conquering benevolence, combined with the highest degree of majesty'. Kaiser Wilhelm II knew 'no inactivity, no respite. No hour is too early for him, no discomfort or exertion too great, if it is a matter of fulfilling his duties as ruler. Good health, steeled by energetic physical exercise and temperance, and an admirable steadiness, achieved by iron self-discipline, in all his conduct and habits of life, enable him to bring an efficacy and dedication to his royal calling which, recalling so much our great King Friedrich II [Frederick the Great] and continuing the ideal example of his illustrious grandfather, must indeed stand as a model for all classes of our people.' In all his indefatigable work for land and people the Kaiser always maintained an 'imperturbable calm', Douglas claimed. 'Never does he become impatient, never hasty or agitated.' Moreover, by 'generous planning of his day', in spite of heavy military commitments the Kaiser 'always found enough time to settle government business conscientiously in every sphere, with the same love, the same interest and the same lively intelligence'. It should also be noticed that the Kaiser 'allowed no one to speak to him, or listened to what he had to say, about matters which that person's office did not entitle him to raise with the Kaiser', and it was precisely this which provided 'the guarantee which could not be valued too highly, that the Kaiser is the determined enemy of all coteries and camarillas, and that he never listens to whisperings from unauthorised sources . . . He is likewise characterised by absolute imperviousness to personal flattery and sycophancy and a sure eye for the true evaluation of those who approach him, a resolute rejection of all importunate ambition, as well as, on the other hand, a rare receptiveness to an open, honest and true word, which the Kaiser knows how to appreciate even if it does not correspond to his personal views. These are qualities of the young Kaiser which not only do his character the greatest honour, but are also of incalculable worth for the country.' Furthermore he showed a 'truly royal gratitude' and a 'reverent appreciation' of the services of his leading advisers.

With regard to internal politics the Kaiser stood on the lofty ground of the 'monarchical principle', above all parties. He wanted no 'party government'; his vision was 'far too broad and impartial' for that. Of his government he demanded 'objective action . . . and he has repeatedly declared that his life belongs to the whole people regardless of religious belief, ethnic origin or party political affiliation. The Kaiser knows only one yardstick in this connection: real love for the Fatherland and the throne. This is the only yardstick which fits the monarchical principle. He alone remains and will remain when the ephemeral catch-phrases of the party political activity of the day have long since faded away.' Certainly Wilhelm had a strong 'interest in all humanitarian and welfare work', asserted the speaker. 'Wherever it is a question of alleviating want and creating institutions that are designed to improve the lot of our oppressed fellow creatures, there is no more willing and ready hand than his.' At the same time Wilhelm II had the ability to look calmly and dispassionately at the practical consequences of a new idea. 'That unfortunate romantic impulse which, particularly in the highest echelons of human society, can so easily become a reef on which practical judgement and creative power are shipwrecked, is quite foreign to our Kaiser and finds no place in the conduct of his life.'

The deeper source of all these magnificent qualities, according to the speech, is 'to be found in the quiet assurance which is the fruit of a genuine trust in God grounded in personal experience'. This 'high moral seriousness of the Kaiser, resting on a deep inner religious conviction', nevertheless did not prevent him from 'joyfully looking real life straight in the eye, with a healthy, clear vision, and nothing is more foreign to our young Kaiser than a gloomy, morbid view of life. Anyone who has ever had the opportunity of meeting him will have gained the impression that in him he has a healthy, manly, cheerful nature that places all its strength at the service of land and people with joyous energy, but is absolutely inaccessible to any gloomy, sluggish and melancholy quietism.' 'All these traits are united in an overall picture of such fresh, promising colours, that our people are indeed justified in looking up with the fullest confidence to him who wears the crown.' No sovereign understood his times as Wilhelm did, no king would turn this understanding into deeds better than he, and the people would 'thank him with ever-increasing loyalty'. Thus the monarchy, which under Kaiser Wilhelm I had already attained hitherto unknown glory, and which in the last twenty years had 'sunk its roots deeper than ever into the heart of our people', would in future be still more firmly anchored. All Germans, regardless of which party orientation they favoured, felt assured that the monarchy was 'the guardian of our freedom, our culture, our national welfare'. And 'our present young Kaiser' was the man 'not only to uphold the blessings of a strong and free monarchy for our Fatherland, but to develop them to an even richer and finer degree'.[36]

Even if Bismarck thought of the 'monarchical principle' as a politically necessary fiction which would safeguard his own leadership as 'first counsellor of the crown' under the new Kaiser as before, it was nevertheless clear that other forces were also at work seeking to turn the myth of Personal Monarchy into reality. The first among these was of course Kaiser Wilhelm II himself, who was impatient to bring about the actual beginning of his personal rule, as we shall see. But other influential people also spoke up for a strengthening of monarchical power on constitutional and conservative socio-political grounds. They too argued in favour of building up the 'monarchical principle' under the dynamic young Kaiser as the 'guardian of the concept of the Reich' and as a bastion against the ever-rising democratic tide. Thus for instance the Saarland industrialist Carl Freiherr von Stumm-Halberg expressed the conviction that, in view of the growing democratic threat, 'it was absolutely out of the question for our kings to avoid any longer intervening personally in the fight against the disintegrating tendencies'.[37] Nonetheless even the most ardent royalist could see that there were also great dangers in endowing the Hohenzollern monarchy with such charisma. Bernhard von Bülow, the future Reich Chancellor, spoke of 'Untergehen [downfall]' if the experiment failed,[38] and Philipp Eulenburg, the Kaiser's best friend, captured the desperate 'all-or-nothing' quality of this crisis strategy in the words: 'No, under such circumstances the only outcome can be to give one's support *to the Kaiser sans phrase*. If we do not work to consider *Him* the personification of Germany – even if his character makes it hard work for us! – we shall lose *everything*.'[39]

THE MATERIAL COST OF PERSONAL MONARCHY

The renewed emphasis on the monarchical principle found its first audible echo when shortly after the accession of Wilhelm II a lively disagreement broke out both in government circles and among the public over the financial circumstances of the imperial court. Everyone accepted that the relatively modest means which had been at the disposal of the Prussian–German crown under Wilhelm I would not be sufficient under the dynamic new sovereign. Everyone also knew that while Wilhelm I, whose old-style Prussian frugality was proverbial, thought of himself primarily as king of Prussia and considered the office of kaiser almost as a tiresome honorary position, his grandson had completely different ideas of his tasks as German kaiser – ideas that were shared by many Germans of all classes.

There was no reluctance to talk about the sums involved. According to conversations with 'one of the highest court officials here' (it was probably Count Stolberg), the Austrian chargé d'affaires Arthur von und zu Eissenstein-Lhotta was able to report to Vienna on 28 October 1888, the old Kaiser Wilhelm I had left a private fortune of 24 million marks in securities in addition to landed property to the value of 3 million marks. Out of the securities his widow the Kaiserin Augusta had received 3 million marks at his death, and his daughter Luise, Grand Duchess of Baden, had received 1.7 million marks, while the landed properties went to Prince Heinrich, the brother of the present Kaiser. The remainder of the estate of Wilhelm I had been turned over to the Prussian crown entail fund (Kronfideikommiss). With the knowledge and approval of Wilhelm II, Kaiser Friedrich III had taken the sum of 9 million marks from Wilhelm I's estate; of that, at Friedrich's death, his widow Victoria had inherited 1 million marks and each of her four daughters had received 2 million. Both Dowager Empresses – Augusta and Victoria – also received a widow's pension of 600,000 marks a year, in addition to which the costs of Kaiserin Augusta's household were defrayed from Wilhelm II's privy purse, but not those of the Kaiser's mother. She, however, could draw on 200,000 marks a year from her private means. Kaiser Wilhelm II himself, as the chargé d'affaires was informed, had at his disposal 'an annual income of 20 million marks, of which 13 ½ million was civil list and the remainder was revenue from the Prussian Kronfideikommiss'.[40] These latter sums, however, were far from adequate for the manner in which Wilhelm intended to conduct his reign. The grandiose, rather heavy and overladen interior redecoration of the rooms of the Berlin Schloss had alone cost twice as much as the renovation of Charlottenburg which Friedrich and Victoria had undertaken.[41] A new imperial yacht was ordered for four and a half million marks.[42] The reorganisation and 'expansion of the household in general' – soon to be swelled by 'two men accustomed to the sea, a royal cook and a valet', as well as two brushers, five porters, one cellarman and two court quartermasters[43] – which arose from the restless lifestyle of the young Kaiser, led to demands soon after his accession for an increase both of the Prussian Crown Endowment (or civil list) and of the 'Allerhöchster Dispositionsfonds', the sum allocated to the monarch for the fulfilment of his duties in his capacity as German Kaiser.

Oberhofmarschall Eduard von Liebenau, the Senior Marshal of the Court, persuaded the Kaiser to ask the Prussian parliament for an additional 6 million marks a year – he also pressed for a royal train to be built – while House Minister von Wedell favoured an increase of the civil list of 3 million marks and was of the opinion that such a request would 'go through without a hitch' in the Prussian parliament.[44]

At first these proposed increases came up against determined resistance from Bismarck, who feared that a parliamentary debate on the cost of the monarchy might endanger the monarchy itself. He warned that even in the Prussian parliament, where thanks to the three-class franchise at least there were no Social Democrats, the deliberations should last for no more than five minutes and no debate should be allowed, for 'any discussion harms the prestige of the crown'; if the increase were rejected the ministry would even have to resign. Only the Conservatives could be relied upon to support an increased civil list, and even among them there were several who would agree only 'with trembling hearts', the Chancellor opined. Among the Free Conservatives there were many who were 'already uncertain', and support for the measure from the National Liberals was probably quite out of the question. That the Kaiser 'had heavy expenses because of the 2 dowager empresses, who each received 900,000 [sic] marks a year and because of his 5 children, was quite clear', he admitted; an exceptional request based on this fact would therefore be only fair. 'But a permanent increase, when all the world knew that Kaiser Wilhelm I had made savings of 27 million in 25 years, would cause much bad blood.' So it was on this very question, which foreshadowed the coming power struggle between crown and government, that the first direct confrontation between Kaiser and Chancellor threatened to arise. On 2 December 1888 General Count Alfred von Waldersee, the successor to the old Field Marshal Count von Moltke as chief of the General Staff, recorded: 'the affair is causing . . . the Chancellor serious anxiety; he must . . . soon come to a decision . . . He will be in a difficult position.'[45] In the end a compromise solution was reached, after Wilhelm II had held personal discussions with National Liberal and Free Conservative Party leaders (in the first instance with Bennigsen and Douglas).[46]

When the proposal came up for discussion in the Prussian parliament the Free Conservative Party leader Christoph von Tiedemann spoke in favour of unanimous acceptance. 'You may depend upon it that the whole German people is behind you on this question', he declared to the house in ringing tones. He pointed out that since the creation of the Reich the representational duties of the Prussian crown, particularly with regard to the South German and the non-German states, had increased enormously. Addressing the question of whether the Reich as a whole was not 'under an obligation to provide an endowment for the imperial crown', he argued 'that the duty to provide the crown with an endowment must be regarded as a point of honour for Prussia'. The only opposition came from the ranks of the Progressive Party: Rudolf von Virchow abstained from voting, and Eugen Richter spoke out against a permanent increase of the annuity from the Kronfideikommiss by 3.5 million marks as an unjustified measure, although he recognised the necessity of a one-off extra payment. Both in committee and in plenary session Richter pointed out that the old Kaiser had been able to save up millions, and maintained that the additional duties of the Kaiser's role had already been provided for by the increase of the crown endowment in 1868. In the end, however, the proposal was accepted with a large majority. The Prussian parliament approved an increase of the civil list by 3.5 million marks a year, raising it from 12.2 million to 15.7 million per annum.[47]

Not only this parliamentary success but also other indications suggest that Bismarck's fear of a general debate on the purpose of the monarchy may have been exaggerated.[48] Since the creation of the Reich seventeen years earlier the imperial role of the Hohenzollerns had taken root even in the south and west of the Reich and even in sections of the working class.[49] The leading role which the dynamic young monarch intended to play as German Kaiser was in perfect accord with a widespread desire for it in the people and in the Reichstag.[50] After the sense of stagnation produced by an old Kaiser followed by a sick one, a powerful new impetus was in the air. Thus, for example, in Frankfurt am Main the construction of a new Reich Post Office was approved by an All-Highest Cabinet Order of 26 January 1889 with the proviso that 'lodgings for His Majesty's occasional use' must be incorporated into it; the new apartments cost no less than 2 million marks.[51] At the same time the Reichstag took the decision that a palace should be built for the Kaiser in Strassburg. The popularity of the imperial concept was so great that some saw in it a threat to the federal character of the Reich: the Prince Regent of Bavaria expressed the fear that if Wilhelm II received an additional civil list as *Kaiser*, the civil lists of the other German monarchs – at least at their existing levels – might be called into question. The Bavarian envoy and member of the Bundesrat in Berlin, Hugo Count von Lerchenfeld-Köfering, acknowledged that the Kaiser had onerous public duties, that these would increase greatly under the energetic new monarch and that therefore some remuneration from the Reich's resources was indeed required. He nevertheless argued for the retention of the ersatz solution represented by the 'Allerhöchster Dispositionsfonds', and rejected a 'civil list as such' for the Reich.[52] Bismarck also took a decisive stance against an imperial civil list at first. In August 1888 he declared that he had 'never yet heard the present Kaiser make any mention of it, nor did he expect him to. He had opposed or diverted similar suggestions in the time of Kaiser Wilhelm I *a limine*, and he would continue to do this if they should be made to him.'[53]

In spite of these federalist misgivings the pressure for an imperial civil list continued to grow. As the Austrian chargé d'affaires Eissenstein was able to report in November 1888, among the members of the Reichstag there was a firm determination to introduce soon 'a motion for the creation of a civil list for the German Kaiser'; it was to be set at a level of 5 million marks per annum. The semi-official newspapers had also been suggesting for some time 'that the German Kaiser, to whom in this capacity a great deal of heavy expenditure had accrued, could not make do with the income of a king of Prussia, which was in any case modest. The German Reich therefore had the obligation to do away with this unsatisfactory state of affairs, and must place at the disposal of the Kaiser a certain sum every year to cover the many expenses which have to be incurred in the interest of the nation.' As the government-supporting newspapers had emphasised, the old Kaiser Wilhelm had been 'reluctant, out of delicacy, to accept that the Reich should vote a civil list to him, since His Imperial Majesty, as first German kaiser, chose to see the bestowal of the imperial crown, for Himself, as no more than an honorary position, to which no pecuniary advantages of any kind were attached'. The old Kaiser had always insisted, however, that his successors could not be expected to defray the expenses which accrued to the king of Prussia in his role as German Kaiser out of their Prussian income alone. There was therefore no doubt, Eissenstein reported, 'the Reichstag would accept with acclaim a resolution to this effect to which it had itself given birth'.[54]

In early 1889 the Reichstag voted in favour of a large increase in the funds made available annually to the Kaiser, even though the Prussian crown endowment had already taken the duties of the imperial crown expressly into account. In deference to the character of the Reich as a monarchical alliance the name 'imperial civil list' was avoided, but the so-called Allerhöchster Dispositionsfonds, introduced for this purpose in 1874, was increased tenfold, that is to say from 300,000 to 3 million marks. Thus the total amount of money approved by the two parliaments which Wilhelm II received rose immediately after his accession, as Liebenau had demanded, by 6.2 million marks, from 12,500,000 to 18,700,000 marks per annum.[55]

THE 'WILHELMINE ERA' BEGINS

A new era – the 'Wilhelmine epoch', as it was to be named – had undoubtedly dawned. For anyone who had had the opportunity to observe the Kaiser's proclamations, the opening ceremonies of the Reichstag and of the Prussian parliament in the White Hall of the Berlin Schloss, the officially published flattering speeches and loyal addresses,[56] the public debates on the increase of the crown endowments in Prussia and in the Reich, the countless inaugural visits to German and foreign courts or the newly devised celebration of the Kaiser's birthday, it was obvious that the accession of the 29-year-old Kaiser was an event of no ordinary significance, but a crucial turning-point in what was as yet the short history of Bismarck's Reich.[57] The combination of the inherited power of the Hohenzollern crown with the youthful energy and strength of will which Kaiser Wilhelm II radiated represented, both in the direct political realm and in the social and cultural spheres beyond, an almost invincible – if as yet still latent – new force.[58]

Shortly after his accession, pamphlets and newspaper articles about the Kaiser appeared in numerous cities of the Reich, echoing – partly from ignorance, partly from calculation – the whitewashing and wishful thinking that characterised Count von Douglas's speech. In September 1888 the *Schwäbische Merkur* was already applauding the fact that the German people had now acquired 'what we need at the helm of the Reich: a *man of character*'. The certainty that Wilhelm II's regime 'would not lag behind that of his grandfather in honour and constancy quickly won all hearts for him', the Stuttgart newspaper commented. On all sides there was 'a firm confidence that Germany's destiny [was] on the right path'.[59] In the summer of 1889 the anonymous publication *Wallende Nebel und Sonnenschein*, which even in the opinion of the Wilhelmstrasse amounted to a 'glorification of the reigning Kaiser', caused a great stir throughout the country.[60] The *Badische Presse*, a Karlsuhe newspaper, rejoiced that the 'youthful Reich now [has] the good fortune to acquire a Kaiser who is capable of being his own chancellor'. Hitherto, since Kaiser Wilhelm I had been 'already too advanced in years' and 'his noble son Kaiser Friedrich already too ill', the 'beneficial influence' of an 'active, strong monarchy' had not reached its full potential. 'But Kaiser Wilhelm II in his youthfulness and energy is, for the next phase of the development of the Reich, . . . the right man in the right place.'[61] Even abroad there was great admiration for him at first. In the summer of 1889 a commentary on the German sovereign which was astonishingly favourable for a French publication appeared under the title of *L'Empereur Guillaume II et la première année de son règne*.[62] The writer of this work, Edouard Simon, was admittedly German by birth and had taken French nationality, but the renowned French historian Ernest Lavisse, writing on the German Kaiser in the *Figaro*, remarked no less admiringly: 'To reign [régner] is not sufficient for Wilhelm II: he wishes to rule [gouverner] like his forefathers, who put Prussia together piece by piece with their own royal hands.'[63] An Italian newspaper spoke of the great 'love . . . which [is] shown by the German people towards the Hohenzollern family and in particular towards the illustrious head of this dynasty'.[64] The American friend of the

Kaiser's youth, Poultney Bigelow, unburdened himself of a flattering character sketch of the Kaiser in a Polish newspaper.[65] The Englishman Harold Frederic compiled a 'glorification' of the early life of Wilhelm II, which was well received everywhere except in Bad Homburg and Windsor.[66] But behind the scenes of the apparatus of power, even in these very early days of the Wilhelmine era, grave anxieties about the new state of affairs and the new ruler were perceptible and would indeed never again disappear.

Although the 73-year-old founder of the Reich was still firmly in the saddle, when Wilhelm II came to the throne on 15 June 1888 the balance of power between Chancellor and Kaiser had shifted noticeably and with lasting effect. Scarcely three weeks later the British ambassador Sir Edward Malet summed up the fundamental transformation which had taken place with Wilhelm II's accession, even though there had been no formal constitutional change, and gave a perceptive warning of the fragility of Bismarck's position: 'The scene has changed: a young Prince has come to the throne. The power of Prince Bismarck is dependent upon the will of the Kaiser & . . . whereas in all the latter years of the Emperor William's reign we had only to consider the intentions of Prince Bismarck, we should now in a considerable measure turn our attention to the Emperor, who has just succeeded . . . If we were dealing with a country in which the foreign policy was guided by the Government & not by the Sovereign this would be a matter of small moment, but that is not the case . . . It is therefore of special importance that the Emperor should be on our side. His sentiments will count as a strong factor in the policy which may be adopted towards us. The Chancellor's position is no longer as strong as it was during the reign of the Emperor William, & he will have to humour the young Sovereign in order to maintain his own power.'[67]

The solemn proclamations and theatrical ceremonies of the first days contrasted sharply with the hectic, unconventional style which Wilhelm rapidly brought to his conduct of affairs. There was little sign of grieving for his dead father; on the contrary, observers spoke of a state of intoxication which lasted the whole summer.[68] Wilhelm's mother expressed the opinion to Queen Victoria and also to Kaiserin Augusta that nothing could have been more destructive for the Kaiser's character than his early accession. 'I cannot imagine anything *worse* for his disposition than becoming Kaiser!, or than rising to this position so utterly unripe and so ignorant of the world, of *other* countries & other *sets* than the one he has unfortunately lived & moved in & identified himself with against our will! His good qualities have *not* been developed & the bad ones have been *forced* as in a hot bed', she warned in April 1889.[69] Not only Wilhelm and his Kaiserin but also the entire court and the whole new style of life were 'so far from what I think dignified, "tactvoll" – & refined!'[70]

Most observers were inclined to attribute the 'furious' pace which Wilhelm had kept up in Berlin since mid-June 1888 to a personality which had not yet found its equilibrium, and they found some consolation in the fact that Bismarck was still in office to stave off the dangers of such an intemperate style of conducting business. Only three weeks after the succession the astute Austro-Hungarian ambassador Széchényi reported apprehensively to Vienna: 'When Kaiser Friedrich III came to the throne and, although his reign had hardly begun, made one change and innovation after another in rapid succession with a morbid urge, one could understand this, taking into account the weaknesses of human nature, in a ruler who had to live with the premonition, if not the certainty, that his time for action might be of short duration and passing rapidly by; but if a strong young monarch who probably has a long life ahead of him, at a time when the mortal remains of his father and predecessor are scarcely cold in the grave, begins organising, dismissing, pensioning off with almost greater haste and speed, this is of a nature to arouse serious misgivings.' In such circumstances it was anything but surprising that people in Berlin were 'anxious . . . , looked towards the future not without concern, and that many in the higher posts in the government and in the army are beginning to feel unsafe in their skins'.[71] Similarly Baroness Spitzemberg recorded in her diary in mid-August: 'Yes, the young Kaiser is having a fine time chasing his people around, especially the soldiers, and the wind that is blowing in Berlin at present might seem more dangerous than a military campaign to many! The senile pace of the old Kaiser is followed directly by that of a boisterous young man thirsting for action – it gives one a certain pleasure, and the army needed sorting out. But if our Chancellor, as the last of the heroes of the great days, were not standing behind the young daredevil, one might occasionally become alarmed by the excessive zeal that rushes in all too keenly.'[72] Even Waldersee, who remained one of Wilhelm II's closest advisers after his accession and who was one of those who profited most from the favour of the young ruler, was worried about the Kaiser's restlessness.[73] On 26 August 1888 he recorded in his diary: 'the Kaiser is incredibly, not to say even immoderately, active, and indeed rather too much so in the military sphere. There is hardly any time left for audiences and so it is very difficult to approach the Kaiser.'[74]

Széchényi interpreted as a characteristic 'signature of the new era' the new-style celebration of the Kaiser's birthday which was introduced in January 1889. He reported thus to Vienna on it: 'Whereas under Kaiser Wilhelm I the congratulations were received only in the private apartments of the Palace, and of the foreign diplomats only the ambassadors took part, this time there is to be a great *church ceremony*, which the congratulating parties have to attend and to which the entire male Diplomatic Corps is invited.'[75] The old ambassador also found it strange that the new Kaiser expressed his 'All-Highest thanks' for the numerous congratulations offered to him on the occasion of his birthday in an order addressed to the Reich Chancellor, which he commanded to be published.[76] 'To the astonishment of all' Prince Bismarck had appeared at the birthday ceremony in the Schloss chapel, and when Széchényi remarked to him that this was the first time that he had seen him at a ceremony at court the Reich Chancellor replied that this was 'the first birthday of the present Kaiser and if I did not come my young master might think that I meant to leave him in the lurch'. The ambassador considered this significant episode as 'proof' that Bismarck 'is avoiding everything which might displease the young Kaiser'.[77]

Another novelty introduced by Wilhelm was the New Year meeting of the commanding generals at the Berlin Schloss – an institution which was to last until January 1914. Quite new – and in itself refreshing and modern – was also the tendency of the new ruler to venture into wider, informal social circles and thereby to throw the traditional court etiquette into confusion. Széchényi reported with surprise and mild irony in March 1889 on a 'hitherto non-existent circumstance that causes much talk here, and not without mockery' – a 'beer party' in the house of the Bavarian envoy Lerchenfeld, who like the Kaiser and Herbert Bismarck was a member of the Bonn student fraternity Borussia, at which the Kaiser was able to meet about twenty-eight gentlemen, principally 'Bonn Borussians' and 'a selection of the Bavarians who are living here while in the service of the Reich', in an informal gathering, and at which 'by way of reinforcement for the inner man . . . sausages with sauerkraut, augmented with cold meats and the host's own beer, ordered from his home, [was] served'. The Kaiser, the ambassador was able to report, had left this all-male party towards midnight, but the remaining 'more or less aged youths . . . continued to amuse themselves until around three in the morning carousing, making speeches and conversing together in the spirit of student brethren'. No one had drawn the Kaiser's attention to the precedent he was setting by his participation in this 'beer party', Széchényi commented with disapproval. 'Kaiser Wilhelm I accepted invitations from the diplomatic corps only if they came from ambassadors', he asserted, 'but if this happened to one of them, naturally all the others competed for the same honour.' After the evening entertainment at Lerchenfeld's house it was therefore to be feared 'that the other accredited envoys, particularly those who are also members of the Bundesrat, would conclude from the above-mentioned "beer party" that they were justified in claiming a similar favour'. So the Chief Marshal of the Kaiser's household had had to make it known through the press that the Kaiser 'is accustomed to visit only those private houses to which He Himself announces His intention of going. If however His Majesty wishes to continue as he has done before in this respect, certain contradictions or inconsistencies might nevertheless be impossible to avoid altogether', Széchényi feared. For instance the Kaiser had recently accepted an invitation to dine with his Flügeladjutant Gustav von Kessel and had attended the baptism of the newborn child of his old regimental comrade Oskar von Chelius in Potsdam. The ambassador warned: 'Although one sees a noble and pleasing trait of character in the fact that Kaiser Wilhelm, now that he has come to the throne, has not ceased to maintain the youthful friendships he struck up when he was still Prince Wilhelm, yet one fears that if this is not handled with more moderate and systematic restraint the Imperial Master could be placed in the undesirable position either of seeing the devaluation of the great favour bestowed by his personal support, or of neglecting or offending people who might have a longer-standing and better claim to it.'[78]

The 'beer party' at Lerchenfeld's house indeed proved no exception. Only a few days later a 'parliamentary dinner at Prince Bismarck's house caused a great stir', as Count August zu Eulenburg remarked, and immediately after that the Kaiser attended a large party at the British Embassy.[79] The dinner at Bismarck's house was, according to Malet's report to London, the first time that a Prussian sovereign had honoured members of parliament with his presence, and the impression of a new relationship between the monarch and the parliamentarians had been further strengthened by the Kaiser's declaration that it was his intention to get to know the representatives of the people better than had been customary hitherto for a head of state.[80] Széchényi also pointed out the unprecedented nature of this event in his report to Vienna: the attention attracted by Wilhelm's presence had been all the greater 'as it was the *first time* that the ruling Kaiser [had taken part] in a parliamentary banquet'. 'His Majesty had all the members of parliament present, who included some Poles, presented to him, and after the dinner he circulated among all the Reich Chancellor's guests in the friendliest manner.'[81]

In a strikingly perceptive confidential report, which was transmitted to all ambassadors of the Habsburg monarchy, the Austrian ambassador summed up his views of the political, social and mass psychological significance of the new reign in these words: 'If the youthful Kaiser Wilhelm II holds His head somewhat higher than He should considering His age and the natural absence of a successful past, this is in truth not to be wondered at, if one considers how and in what circumstances He came to the throne, and what the personal satisfactions were that have already fallen to His lot in such a short time. The young ruler ascends the throne almost without a transition, when scarcely half a year earlier, by all human reckoning, this was only a very distant prospect. He wishes that all the sovereigns of the German Empire should gather around him for this ceremony; it is done, and takes place before one's very eyes like a formal act of homage. He wishes that all commanders of the German army corps should take part in the New Year congratulations, something which was never demanded by Kaiser Wilhelm I; they appear in full strength, and among them two royal princes: Prince Georg of Saxony and Prince Leopold of Bavaria. If the Kaiser drives out, in the courtyard of the Schloss around the waiting royal carriages the crowds are scarcely smaller than those which gathered in honour of the "venerable Kaiser-Hero" in front of the Imperial Palace and the so-called historic window; and what is more, in the streets during his drive the All-Highest is often applauded, which even with his grandfather generally happened only on special occasions.' Furthermore, the entourage of the young ruler 'not only owes its present existence to the favour of its Imperial Master but also counts on Him for its future', with the consequence that it 'always guards against expressing a view contrary to His Majesty's intentions . . . so that in many cases the youthful and hasty initiative of the sovereign does not find the counterweight that would be desirable'. Although those closest to the monarch could still exercise some influence over him, they could not necessarily be relied upon to put this to good effect. 'So it is possible that, just as Kaiser Wilhelm I was sometimes taken advantage of during the last years of his reign in consequence of his old age, and Kaiser Friedrich was taken advantage of in consequence of his incurable physical illness, so Kaiser Wilhelm II might be taken advantage of because of his youth', the ambassador warned. The Kaiser's new manner of conducting his reign had met with very differing reactions among the population, observed Széchényi in his remarkable *tour d'horizon*. 'The most dissatisfied are probably all the middle-aged people in both civilian and military life who had pinned all their hopes on the reign of Kaiser Friedrich, and who now, as things have turned out, see themselves to a certain extent passed over, as well as all free-thinking and Progressive Party groups which hoped to see their political ideals realised by the deceased Kaiser. Dissatisfaction is also felt by all those who had a role to play during the era of Kaiser Wilhelm I and who now find themselves relegated to complete obscurity, although their day lasted quite unusually long, beyond any human reckoning. Those who are really content are only the young people in military, civil or court service, who consider not only the future, but already also the present, as belonging to them. And yet it cannot be denied that in spite of everything Kaiser Wilhelm II is much more popular, especially among the public at large and in the broader social classes, than one would think judging by the given circumstances and the way people talk. Yes, youth is indeed that powerful talisman which is capable of finding a softer side in even the most embittered character.'[82]

CHAPTER TWO

The first state visits

IMMEDIATELY after his accession to the throne, Wilhelm was keen to show himself off in the great capitals of Europe, but there were strict limits to where he could go. Paris was quite out of the question, not only because of the fresh memory of the defeat of 1870–1 and the still festering wound of Alsace and Lorraine, but also because the young sovereign epitomised 'Christian–German' racial ideals, Prussian militarism and authoritarian monarchy which were anathema on the Seine. But even Brussels, which was briefly considered for a state visit in 1888, was speedily ruled out as a no-go area for a German Kaiser. As Wilhelm recalled many years later, the German embassy in the Belgian capital warned when the idea of a royal visit was mooted that Brussels must be regarded as a suburb of Paris: it was strongly French ethnically and in sentiment, and one would in addition have to reckon with the influx of whole trainloads of Parisians of all sorts from the south. 'The Brussels mob is the just about the worst in the whole of Europe', Wilhelm wrote in 1904, remembering the warnings of the embassy. 'Miners and other workers would stream in in their thousands from the surrounding countryside and from Paris etc., in short such a huge rabble would gather there that the very poor police force and the weak, wholly inadequate army would be unable to master it. On top of that, all the anarchists of London, Paris and Brussels itself would be present in great numbers, so that the Belgian government could not guarantee the safety of my person. Moreover, one could never be sure that several thousand bribed Camelots [du Roi] – whether Parisians or from within the country itself – would not at any moment provoke demonstrations with shrieks, whistles etc., leading to the most unpleasant situations and the most painful consequences.' Both Prince Bismarck and the Belgian government concurred with this advice, the Kaiser recalled, and hence the proposed visit was cancelled. The ugly scenes that he had had to contend with when embarking some months later at Ostend were quite bad enough and proved that a visit to Belgium should be considered only in return for 'quite massive concessions by the King [such as] an alliance or agreement with us'.[1] In eastern Europe the situation was of course utterly different.

THE JOURNEY TO PETERHOF

As he had announced in his speech from the throne on 25 June 1888, it was Wilhelm's intention to build up a particularly close relationship between Germany and its ally Austria-Hungary and at the same time to maintain the personal friendship between the German and Russian imperial houses. The idea, which Wilhelm had vigorously championed on his two visits to Russia in 1884 and 1886, of the three Emperors uniting to form a bastion against democracy and socialism came distinctly to the fore again now. The day before the speech Széchényi had been able to ascertain that the Kaiser's address would 'make a show of the greatest devotion to peace and that Kaiser Wilhelm would refer to the Austro-German alliance in the warmest terms'.[2] Shortly before the ceremony there had been an exchange of letters between Kaiser Franz Joseph and the young German Kaiser which would have long-term repercussions on the foreign policy of the Reich. One gains the impression from this correspondence that the Habsburg monarch, with whom Wilhelm had hunted so often and with such enjoyment in his youth, had now inherited the position which for Wilhelm had hitherto been held by his grandfather: that of the revered older emperor as role-model. In an initial telegram after the death of Friedrich III, Franz Joseph had assured the young Wilhelm that he could count on his loyal friendship in all circumstances,[3] and in a letter written in Budapest on 17 June 1888 and brought to Berlin by his brother Paul, the Habsburg monarch emphasised 'how sincerely I feel for you in everything which Providence has meted out to you before and at your accession to the throne'. Franz Joseph expressed his firm conviction 'that you – true to your conservative principles and to the tradition inherited from your unforgettable grandfather – will do your utmost to lead your people into happiness and well-being with a firm and calm hand, and to maintain the blessings of peace for them. It is, first and foremost, in the immutability of our friendship, in the strength of the ties which bind our empires together, that the pledge of Europe's peaceful future lies. May both our endeavours be directed above all to achieve and maintain this goal: you may be as sure that you will always find me at your side on this path as I am sure of my trust in your inviolable friendship.'[4]

In response to this letter Wilhelm II sent the Austrian Emperor on 23 June the text of his forthcoming speech from the throne and added, using a draft prepared by Herbert Bismarck: 'My dear, valued Friend. Your affectionate, kind letter in which you express your sympathy for me in such a warm and friendly manner has moved me most deeply; and I thank you most sincerely for your kind words, and particularly for the fact that you have such complete and justifiable confidence in my firm attachment to the traditions inherited from my dear Grandfather. I take possession unreservedly of this inheritance, in domestic as in foreign policy, and notably as regards the immutability of our friendship and the strength of the ties which bind our empires. With joyful gratitude I receive your assurance of this and return it with all my heart. As I have the good fortune of having been known to you from my earliest youth, there is no longer any need for me to assure you of these feelings of mine, to which, as you will see from my enclosed speech for the opening of the Reichstag, I am giving public utterance. The more strongly my late Grandfather was convinced of the necessity of warding off Russian attacks in partnership with you, the more earnestly he strove to prevent them from taking place. I follow him in the endeavour to do everything to bring the blessings of peace to our empires and to preserve them there, and to this end to nurture our relations with Emperor Alexander. The friendship that binds me to you stands so firm that it needs no outward action. But I believe I shall be acting in the interest of our mutual aims if I openly refute all the calumnies concerning our love of peace circulating in Russia as in France, by paying a visit to the Emperor of Russia in the near future. I intend to combine an inspection of my fleet with a recreational sea voyage at the end of July, which I shall conclude by calling on the Tsar at Peterhof. I have no doubt that you and I will be satisfied with the result of this visit. I hope that you will allow me to pay you a visit in the course of the autumn or in the late summer. For I should be very happy if you would agree that we should uphold the tradition of my grandfather also with regard to meeting each other in person where possible every year, even if perhaps not always at Gastein itself, in order to keep up our old friendship by exchanging ideas in person . . . In constant loyalty and devotion I remain your true Friend and Brother Wilhelm.'[5] This hand-written letter was brought to Franz Joseph by Waldersee, who had been commissioned by Wilhelm to convey the formal notification of his accession to the throne to the Austrian Emperor.[6]

In a letter to his grandmother Kaiserin Augusta, Wilhelm likewise emphasised the importance of a German–Russian understanding for the future of Austria as well as for European peace. 'With the Chancellor's full agreement and approval I shall have a meeting with the Tsar near Kronstadt, in order to make the peace of Europe even more secure', Wilhelm informed her. 'But I am making the journey principally for the sake of our dear ally the Emperor of Austria. I am hopeful that if we are able to exchange ideas I may succeed in persuading the Tsar to adopt a calm and sensible attitude towards the various ups and downs and problems with Austria. So that there might then be a calmer time in which diplomatic ways could be found of settling justified claims on both sides. Of course much depends on the extent to which the Emperor [Alexander] still holds the real power in his hands.'[7] Just before his departure for Kiel Wilhelm also informed his 'friend', Crown Prince Rudolf of Austria, 'in all frankness' of the 'threefold aim' of his coming journey to visit the Tsar. In a letter of 12 July he explained that he wanted to set the economic and financial relations between Germany and Russia in order, to settle the dispute between Austria and Russia over Bulgaria and to put an end to the dangerous situation on the Russian–Austrian frontier.[8]

It was in a distinctly cooler, indeed almost frosty, atmosphere that the news of his travelling plans reached his English grandmother. If Queen Victoria and Lord Salisbury had already been taken aback by the absence of mention of Britain in the speech from the throne, and by the indications of a rapprochement between Germany and Russia,[9] the Queen made no secret of her annoyance when Wilhelm's travelling plans were reported in the press. 'There are many rumours of your going & paying visits to Sovereigns', she wrote to her grandson on 3 July. 'I hope at least you will let some months pass before anything of this kind takes place, as it is not 3 weeks yet since your dear beloved Papa was taken & we are all still in such deep mourning for him.'[10] In his reply Wilhelm emphatically rejected the Queen's intervention and in so doing clearly revealed the foreign and domestic policy aspects of his 'Kaiser-ideology': he laid strong emphasis on the importance of his personal relations with Franz Joseph and the Tsar in the battle against parliamentarism, democracy and socialism. 'At the end of this month I shall inspect the fleet & take a trip in the Baltic; where I hope to meet the Emperor of Russia', he wrote, 'which will be of good effect for the peace of Europe & for the rest & quiet of my allies. I would have gone later if possible; but state interest goes before personal feelings, & the fate which sometimes hangs over nations does not wait till the etiquette of court mournings has been fulfilled. And as I am quite d'accord with Prince Bismarck, I hope & trust that much good will come of the proposed meeting; as I deem it necessary that monarchs should meet often & confer together to look out for dangers which threaten the monarchical principle from democratical & republican parties in all parts of the world. It is far better that we Emperors keep firm together with Italy, than that two of them should go pitching into one another without any earthly reason except for a few miserable villages more or less; which would only be arranging & preparing the way for anarchists at home & abroad.' The curious signature, 'Willy I. R.', on this first letter as Kaiser to his English grandmother betrayed how closely personal and family ties were here entangled with high politics, a state of affairs which, as we shall see, was to lead to countless conflicts.[11]

In his unpublished memoirs Herbert von Bismarck claims that the idea of a visit to the Tsar's court had been deliberately conceived as an act of revenge against the British royal family, and that it had been in Wilhelm's mind even before the death of Friedrich III, when Queen Victoria was visiting Berlin.[12] 'The anti-English urge, once again strongly prevalent, the wish to "astonish" at the very start of his reign, had allowed the plan of the Baltic journey to ripen as early as May – shortly after the departure of the Queen, and after Grand Duke Vladimir had come to Berlin at the end of April: after the funeral [of Kaiser Friedrich III on 18 June] it was divulged to Grand Duke Vladimir and was accepted with alacrity on the Russian side.'[13] Herbert Bismarck's claim is given some credence by the letter from Wilhelm to Crown Prince Rudolf of 12 July mentioned earlier, in which he speaks of the necessity of putting an end, in St Petersburg, to the machinations of France and above all of Britain against him. He would have to 'destroy, in St Petersburg, the last manifestations of French intrigues, which have been augmented since my accession by the intrigues of England . . . The latter, which under certain circumstances could increase the strength and effectiveness of the former, I even consider as the more dangerous', the Kaiser declared.[14]

Wilhelm II did not in fact wait until the end of the month but left Berlin as early as the evening of 13 July, to sail in the *Hohenzollern* from Kiel via Stettin and Danzig to Kronstadt. Only a week after the death of his father he had issued an All-Highest Order cancelling the planned fleet manoeuvres and commanding the navy to hold itself in readiness for his further orders.[15] His mother wrote in her diary on 13 July: '*How* unseemly this journey seems to me, when not even 4 weeks have gone by, I really cannot say! It hurts me unspeakably – and *must* cause *painful surprise* abroad! – As if one could not wait to show oneself, to amuse oneself, to go out, to be fêted, to enjoy the outward honours of the new position. It disgusts me! Of course he is also pushed into it so as to turn his head completely!'[16] In this opinion the Dowager Empress was not alone. The Austrian ambassador reported to Vienna that everyone was wondering 'not without misgivings whether it was really necessary to announce this visit as early as the first days of the reign, and also to carry it out so soon, after scarcely a month had passed??'[17]

Certainly the Kaiser's first visit to his Baltic neighbours did not show much evidence of mourning. On his arrival in Kiel the admirals and commanders on parade at the station were delighted to see him descending from the train dressed in the new rear-admiral's uniform which he had himself designed: 'Nothing like this has ever happened before!', exclaimed of one of the commanders in high delight.[18] In the harbour the fleet was dressed overall. A salute of 330 guns was fired for the Kaiser. A naval review followed, to the accompaniment of music and cheering.[19] On the eighteen-day tour to St Petersburg, Stockholm and Copenhagen the imperial yacht, under the command of Prince Heinrich, was escorted by a squadron of ten larger warships – four iron-clads, four training ships and two corvettes – and over a dozen torpedo boats.[20] With the Kaiser and his brother travelled Secretary of State Count von Bismarck, the serving Adjutant-General Adolf von Wittich and the Kaiser's aide-de-camp, Flügeladjutant Gustav von Kessel, House Minister Maximilian Freiherr von Lyncker, physician-in-ordinary Dr Rudolf Leuthold, the diplomat Alfred von Kiderlen-Wächter as representative of the Foreign Office and Privy Court Councillor Willisch as head of the cipher office, together with Prince Heinrich's adjutant, Albert Freiherr von Seckendorff. The convoy sailed at a deliberately leisurely pace so as not to arrive at St Petersburg before the afternoon of 19 July. The atmosphere on board the *Hohenzollern* was exuberant and unconstrained. 'H.M. is in high spirits', reported Herbert Bismarck to his father on 14 July, 'and we are all inhaling the invigorating sea air with delight. The review this morning was a magnificent spectacle, the 10 big ships of our squadron sailed by in perfect order, and the two torpedo divisions (14 little ships) sped by twice at top speed in a shower of spray, overtaking us.'[21] Even when Wilhelm and a few of the other passengers became seasick he remained in the best of moods and declared his intention of undertaking a similar sea-voyage every year.[22]

Despite all the high spirits the diplomats and court officials knew only too well what risks lay hidden in a meeting between two imperial courts. In an audience with the Kaiser the Foreign Secretary urged him 'not to *initiate* any discussion on politics with the Tsar and to pay special attention to his comfort while riding etc.'[23] The first misunderstanding was nevertheless not long in coming. The German naval attaché in St Petersburg, Freiherr von Plessen, came on board at 4 a.m. with the Russian pilots and brought the programme of festivities for the coming days, together with the news that the Tsar would leave Peterhof at half past one in the afternoon. That was interpreted to mean that the Tsar, in his steamer *Alexandrie*, would sail to meet the *Hohenzollern*. At the same time the German squadron was informed of the order in which the ten ships should sail in and drop anchor before Kronstadt: the training ships should moor in front of the iron-clads. Wilhelm II, as Kiderlen-Wächter reported, was 'very angry' at this, as he wanted 'to show off his battleships first'. Seckendorff finally succeeded in calming him with the argument 'that by coming along behind the battleships he would make a far greater impression!!! Next he wanted absolutely to sail past the Tsar as admiral at the head of his squadron, could also not be induced to put on Russian uniform, but continued to wear his own newly designed naval uniform! Despite all the peeping through the telescope, no Tsar came into sight and we hove to and anchored.' Not for another two hours, when the Tsarevich

Nicholas came on board, was it discovered that the Russian court – the Tsar in his yacht, the Tsarina Maria Feodorovna on the landing stage – had been waiting for the *Hohenzollern* in the harbour since half past one. The Kaiser, who had meanwhile hurriedly donned a Russian uniform, was therefore two and a half hours late in setting foot on the Tsar's yacht, which was filled 'with grand dukes of every age and every kind, ministers, and some of the highest officers of the court'. Alexander III's greeting was warm; the 'remaining public had become definitely nervous on account of the wait! We were asked by everyone about our delay', Kiderlen reported to Berlin. 'The two Emperors sat by themselves for a long time. Ours spoke every now and then; one could see how the other was searching for a subject.'[24] Among the German naval officers it was widely felt that the 'cold' reception was intended as a deliberate snub to the German monarch. It was 'very noticeable', wrote one of them, 'that the Tsar did not sail to meet our Kaiser, that only the son, and not the father too, had come on board the Hohenzollern to greet him'.[25]

In spite of this faltering start the six-day programme passed off fairly well, at least outwardly. The political discussions centred on the private audience which Herbert Bismarck had with Tsar Alexander on 22 July, when the German Foreign Secretary, with staggering frankness, talked at length about the most intimate relationships in the German imperial family. As Herbert recorded, the Tsar had the strange idea that the position of German Kaiser was not automatically linked to the crown of Prussia, so that at the death of Kaiser Friedrich III it would have been possible for another German monarch to put forward a claim to the imperial crown. Herbert was able to convince the Tsar not only that the German kaisership was tied hereditarily to the Prussian royal house under the constitution of the Reich, but also that, of the four kings in Germany, only the Prussian monarch was a king by virtue of indigenous sovereignty, as the Kings of Bavaria, Saxony and Württemberg had only received their crowns from Napoleon in the disastrous year of 1806. But in any case, the Secretary of State continued, the princes of the German Reich had long since recognised that 'the firm cohesion of the different parts of the Reich' provided 'the only sure bastion against radical levelling tendencies and threats to the rights of the monarchical crown'. For that very reason the princes had decided on the 'magnificent demonstration that they made by their unanimous appearance at the opening of the Reichstag', as a result of which 'the edifice of the Reich stands more firmly held together now than ever'.

Alexander III then turned his attention to the relationship between Wilhelm II and his mother, whereupon the Foreign Secretary replied with the greatest frankness that the estrangement between mother and son continued beneath the surface, and that the widowed Kaiserin was alone to blame for it, for she had 'never shown . . . what is [understood] in normal life as motherly feelings' towards the present Kaiser. The Dowager Empress thought of herself 'entirely as an Englishwoman' and until the death of her husband she had seen her life's goal in 'ruling the Germany she disliked for as long as possible, as an Englishwoman'. Not satisfied with that, the Chancellor's son then gave a cynical description of the conduct of the various members of the imperial family in the Battenberg affair; he spoke of the interference of the British royal family in it and 'of the systematic efforts of Kaiserin Victoria to prejudice Kaiser Friedrich against our present Master. This went back for many a year and unfortunately had not always been unsuccessful', he added. Many years before, Herbert continued, Wilhelm II himself had once said to him, when he 'was speaking of his fruitless efforts with his exalted Mother: "I see that everything I can do will be in vain: we stand on different ground, my mother will always remain an Englishwoman and I am a Prussian; so how can we ever be in harmony with one another?"'

As if even that were not enough, the younger Bismarck then spoke 'about the role of the English doctors, about the nursing, about the reasons behind the glossing over of the facts, about the conduct of business, as well as about the behaviour of the English–Progressive camarilla' during the last Kaiser's death throes. He further informed the Tsar that 'the Prince of Wales [had] pestered our Kaiser with tiresome advice' and had 'expressed all kinds of wishes regarding Cumberland through his wife and in person'. 'In the English family and its closest offshoots' there was 'a kind of cult of the family principle', Herbert Bismarck explained to the Tsar, 'and Queen Victoria is considered as a kind of absolute head of all members of the Coburg tribe and its offshoots'. Moreover it was Britain's political aim to 'create discord and conflict on the continent and to goad the other great powers for England's benefit'. For precisely this reason the Prince of Wales had 'blackened' the name of the present Kaiser to him, the Tsar, the previous autumn, 'in order to create mutual distrust, and the Queen had written to Kaiser Wilhelm 3 days before our departure to advise him strongly against the journey here. His Majesty, however, has had enough of this uninvited tutelage and has sent the Queen such a plain answer that it is scarcely likely that it will happen again.'[26] Even if one were to take the view that the disclosure of such intimate family relationships to a foreign monarch is acceptable in the national interest, the lack of restraint with which Herbert Bismarck spoke of the Tsar's brother-in-law in this conversation is still astounding – for the Tsarina Maria Feodorovna was of course the sister of Alexandra Princess of Wales.

Otherwise the visit to Russia consisted of ceremonies, banquets and reviews. Wilhelm and Heinrich laid wreaths on the grave of the assassinated Tsar Alexander II.[27] The bachelors Herbert Bismarck and Kiderlen-Wächter, who spun out their nights with visits to bars or to ladies, sometimes did not get to bed until 6 or

7 a.m.[28] On the second day the entire party drove out to Krasnoe Selo, the Tsar's summer residence a few kilometres south-west of St Petersburg, where they all spent the night. Everywhere there were troops lining the roads and cheering, not three times as in the West, but continuously for as long as any member of the suite was still in sight. 'One became quite dizzy from all this Asiatic howling', complained Kiderlen. After a lengthy ride the whole party gathered round a tent where 'a great tattoo, prayers, firing of cannon, etc.' ensued. On the next morning there was a parade which, as Kiderlen reported, was 'led by the Russian Tsar mounted on a small, horribly fat, horse. Our Kaiser mounted on his big elegant chestnut subsequently led his regiment in parade, to the universal admiration of the Russians.' The parade, at which 'in the Russian fashion there were all sorts of extra turns', went off in fine style. 'The white Tsar was obviously in a good humour that reached its height when our sovereign at lunch in the marquee replied in Russian to the toast given by the Tsar in French. This made a really tremendous impression on all the Russians, all the more so because it was quite unexpected. Our sovereign said his little piece quite smartly and without stumbling', Kiderlen reported.[29]

According to the original programme, after the Russian mass, the luncheon and a state banquet, the German party was to re-embark on the evening of Sunday 22 July and sail that night for Stockholm. But by Wilhelm's wish the visit was prolonged for two days, as he wanted to take part in a large cavalry exercise – the horses, specially broken in for the Kaiser, had been transported from Berlin – and above all to show off one of his iron-clads to the Tsar. The cavalry exercise was proposed to the Kaiser by Grand Duke Vladimir and the Tsarevich Nicholas and was at first accepted with delight; but then Wilhelm began to have doubts, as he wondered whether the good impression created by the visit might not be endangered by too long a stay. In addition he was concerned that the invitation had not come from the Tsar himself, who was known to be no lover of riding. These problems were initially solved by the Tsar extending the invitation in person, and at the same time announcing that in any case he did not wish to accompany Wilhelm and Vladimir to the camp. 'On Sunday the Tsar, whose mood was improving all the time, suddenly said that he would go too. Since, however, the Tsar does not like to get up as early as would have been necessary if they were to leave for the camp in the early morning, everybody left that evening for Krasnoe, where they stayed the night.' Kaiser Wilhelm, as Kiderlen reported, was far from pleased with this change of plan. 'He had particularly hoped to have a really good gallop round on his horse and to take the jumps etc., as he did in Potsdam when he last inspected the Hussars.' He had been irritated by the appointment of the 'very ancient' General Glinka to be in honorary attendance on him – this eighty-year-old had been chosen simply because he was the only holder of the ribbon of the Order of the Red Eagle! – and had been looking forward 'enormously' to 'chasing the old man *à la* Albedyll over the jumps'. It was probably quite fortunate that this did not happen, the diplomat considered, and that instead 'our sovereign had to watch the manoeuvres in a neat and proper way from a marquee in the Tsar's company', for 'it would hardly have made a good impression if an old Russian general had broken his neck'.[30]

On the last evening there was a family dinner for Wilhelm and his brother with their Russian relations. The imperial occasion reached its high point with a battle with wet towels between the Tsar and his son on one side and Wilhelm and Heinrich on the other, which in Kiderlen's opinion 'certainly struck the typical domestic tone of the Peterhof court of the Autocrat of all the Russias'. On the Tuesday morning before the departure both Emperors, each with his retinue, visited the smartly turned-out German flagship *Baden*.[31] Among the Russian visitors were the Grand Duke Vladimir with his wife Marie of Mecklenburg-Schwerin, together with Grand Duke Sergei and his Hessian wife Ella, sister-in-law of Prince Heinrich and Wilhelm's first love.[32] 'Nothing was spared us', commented Kiderlen after the visit was over. 'Upstairs and downstairs – everything had to be seen, engine-room, galley, men's quarters etc. Then gunnery drill, marines' drill with blank cartridges etc.'[33] Finally there was lunch on board the *Hohenzollern*, in bright sunshine on a glassy sea. 'We took our leave with the greatest cordiality', accompanied by 'a continuous thunder of cannon-fire', Herbert Bismarck wrote to his brother-in-law Kuno Count Rantzau. 'There was no further talk of politics at all, our master has kept quite quiet, and this is probably what has put the Tsar so much at ease.'[34] The mood of the German Kaiser was no less gratified. 'Whether it was the feeling as we steamed away of being free from ceremonial obligations, whether it was a good lunch combined with pleasure that the visit had gone off so well that induced a peculiarly festive mood', reflected Kiderlen in his report to Holstein, 'H.M. was in the most boisterous spirits and . . . gave us one in the tummy, only this time quite literally.' On the journey from Kronstadt to Stockholm on 25 July Kiderlen summed up his general impression of the St Petersburg visit with the comment that the Russians had never been 'so conciliatory, so humble, so compliant, and in everything one could clearly see their efforts and desires to get peace and quiet, and to be freed from the nightmare caused by a worsening in their relations with us; hence also the fearful avoidance of all "ticklish" questions'.[35]

If Kaiser Wilhelm II's first official visit to St Petersburg can count as a success at all, it is only in the sense that it did not turn out to be a disaster, as many had feared.[36] Both sides registered with relief that during the entire visit political conversations had been successfully avoided. The First Secretary at the German Embassy in St Petersburg, Count Friedrich von Pourtalès, regarded it as a success 'that we have shown that we did not *want* anything from this visit, but at the same time that we are not disposed to comply with certain demands that they would like to make of us'. His one hope was that the two monarchs had gained a favourable impression of one another, Pourtalès wrote to Holstein. 'I believe that our young Kaiser with his frank guileless character found favour in Tsar Alexander's eyes and by his mere presence dispelled much of the artificially created mistrust of him.'[37] Pourtalès and also the ambassador, General Lothar von Schweinitz, were seriously mistaken, however, if they thought that Wilhelm had won back the Tsar's earlier confidence in him.[38] In the imperial family and in the Russian upper class in general there was great indignation at Wilhelm's lack of piety and Herbert Bismarck's spiteful enmity towards the late Kaiser Friedrich and his widow. After the visit the Tsar commented to the future Foreign Minister Count Vladimir Lamsdorff that the young Kaiser was a 'rascally young fop, who throws his weight around, thinks too much of himself and fancies that others worship him'.[39] As the British ambassador in St Petersburg, Sir Robert Morier, ascertained, the family of Alexander III, whose life was a totally domestic one, were '*outrés* by this hideous persecution under the sanction of the son of the memory of the father'. He reported to London on the 'profound effect of disgust produced by the order to put off mourning during the Emperor's stay here and the giving the signal for every sort of feasting & jollity'. The behaviour of Herbert Bismarck had been particularly offensive: 'as usual beastly drunk', in a loud voice which could be heard from the other side of the street, he had welcomed Friedrich's early death as a deliverance. His theme was 'all that wd. have befallen Germany, & how the flood gates of democracy & Jews & God knows what more besides wd. have been opened if the reign of the Emperor Frederick & the Empress Victoria had been prolonged. But, he added, *un bon petit cancer nous a sauvé!*' This story was 'quite authentic, quite in character and has left behind the profoundest sense of loathing', Morier wrote.[40]

The Kaiser's physician also considered that the outcome of the Russian visit, in spite of all the professed friendliness, had been unsatisfactory, and that the young Kaiser was aware of this. In a conversation with a diplomat at the Austro-Hungarian embassy Leuthold acknowledged that 'the reception accorded to Kaiser Wilhelm in St Petersburg had certainly been very warm outwardly, and in particular the personal relations between the All-Highest Persons had left nothing to be desired; nevertheless the German gentlemen had received the impression that all these manifestations lacked spontaneity and that "it had been more a matter of acting than of feeling". The free, unaffected tone which Wilhelm had adopted in everything seemed to please the Russians, yet it never came to the point of real inner warmth of feeling, everyone was extremely polite, but remained fundamentally cold and distant.' When the German ships finally sailed away, Leuthold continued, 'the Russian gentlemen were as pleased that everything had gone off in the best possible way in accordance with the programme, as that the visit had come to an end'. The doctor had been able to establish that the Kaiser had come away with the same impression.[41]

When Gustav Freiherr von Senden-Bibran, who had taken part in the voyage as commander of the *Bayern*, looked back over the visit it was, likewise, the 'frosty' reserve of the Russians that struck him most. At both higher and lower levels he had the distinct feeling of a 'cold reception'. During the entire visit it had been clear 'that the Russian naval officers were not anxious to make any closer contact with our officer corps'. The ships' commanders were invited to meet them, certainly, but even these meetings had had a 'stiff quality', and none of the other German officers was invited, which was 'quite unusual' on a foreign visit. He had noticed on numerous occasions, Senden wrote, that the Russian officers 'did not *want* to speak German', but 'always began the conversation in French'. In the Russian officers' cabins there had been many pictures of Napoleon hanging next to portraits of the Tsar. He had not seen a single German flag flying. All in all he had the impression 'that the cold reception was instigated from on high'. There had also been much talk in St Petersburg of a party sympathetic to France which was gaining influence under Alexander III.[42]

Of the ambitious objectives which Wilhelm had announced in his letters to the Emperor Franz Joseph, Crown Prince Rudolf, Kaiserin Augusta and Queen Victoria, not one was realised. Before the journey Waldersee had taken the view that the antagonism between Russia and Austria was much too strong to be resolved by anything short of a European war. In a letter to the future Minister of War General von Verdy du Vernois, a friend of his, the acting Chief of the General Staff predicted: 'There are indications that the Kaiser's journey to Russia will change nothing in the political sphere; the difference between Austria's and Russia's interests is so great, and on both sides the inclination to yield in any way at all is so slight, that a clash is inevitable; all that we shall achieve is to delay it a little, which is Bismarck's aim; you will probably agree with me that it would be better for us to bring it on. For the moment the Russians still seem to be short of money, but they will find it after we have helped to raise the exchange.'[43] When the visit to the Tsar was over Waldersee drove the point home with his laconic verdict: however favourable an impression Kaiser Wilhelm might have made in St Petersburg, 'that does not alter the situation as a whole'.[44]

THE STATE VISIT TO THE SCANDINAVIAN COURTS

The four-day visit to Sweden was an undoubted success, which – not least because the visit to Peterhof had proved disappointing – left a lasting impression on Wilhelm II. He named his fifth son, who was born at this time, after King

Oscar II of Sweden and Norway.[45] On their arrival in Stockholm on 26 July the beauty of the city and the warmth of their reception delighted the Kaiser and his suite.[46] The King sailed out to meet the Kaiser's yacht, and on the way into the harbour 'the people were standing in large crowds on the shore; they greeted our Kaiser in the heartiest fashion, both from the land and from steamers . . . Wherever one looked there were . . . waving handkerchiefs and friendly faces', Senden commented. 'It was quite a different matter from Russia.' As the Queen of Sweden was in Norway, the Kaiser led in the Swedish Crown Princess Viktoria, his cousin from Baden, to dinner.[47] Looking back on the visit Herbert Bismarck recorded with relish: 'After the dinner in the enchantingly situated Drottningholm Castle, from which the return journey on Lake Mälaren, gilded by the sinking sun, was fabulously beautiful, we continued with a *souper* on the Hohenzollern, with cannon-fire and fireworks, and this after the harbour had already rocked on its foundations with the din of salutes from our big iron-clads at midday, in confirmation of the birth of the 5th Prince. The Swedes were extremely warm-hearted; the monarch was so moved that huge tears rolled down his cheeks; and he was immediately asked to be godfather. Stockholm is magnificently positioned: probably the only royal city in which the iron-clads can tie up directly at the quay of the imposing castle.'[48]

At the end of August, when Wilhelm II looked back on his Baltic cruise, it was the visit to Stockholm which brought back the happiest memories. He wrote to his friend Philipp Eulenburg, the author of the 'Skalden Songs' evoking the Viking past, with rapturous enthusiasm for the Nordic world: 'What an eventful summer it has been! But what marvellous moments it brought me! What would I have given if you could have been in Stockholm when we sailed in! How proud I felt when we sailed in round Sandhammer and then passed through the magical girdle of the skerry reefs with our mighty squadron. Impressive as my "dragons" looked as they sailed on the open sea, now in the serpentine narrows through which they made their way, they seemed truly gigantic. And then came the moment when we rounded the last point, the splendid city lay before us and the first German squadron sent its brazen greeting to the palace of the Nordic King, as it steamed in before the eyes of hundreds of thousands of people. Yes, that was indeed a moment to remember! I felt as if I could see our old Norse friends in the clouds, watching the procession of "dragons" with smiles of pleasure, and rejoicing that a race worthy of them was coming to its ancient ancestral land with warriors a thousand in number! – The whole of Stockholm seemed to me like a dream, for I was transported to the days of Erik, Hokan and Fritjof and hence almost forgot about living people, and yet not quite, for I have never in all my life seen such a quantity of beautiful girls and women together as in Stockholm. And how fresh and rosy, gay and friendly they all looked. I thought to myself that if they were Valkyries, it would be a real delight to be a hero in Valhalla. I picked up flowers thrown by the most beautiful of them, to the gratification of the King, and put them in my buttonhole; the King pointed this out to each lovely benefactor with a wave, which gave great pleasure. When we mounted the stairs of the telephone exchange – unfortunately the only thing we mounted in it – the most beautiful telephonists were standing on the landings and scattering flowers. The lower floor was dressed in tulle montant, the upper floor in very low-cut satin, and furthermore a lovely redhead with a dazzling complexion! Which I particularly admire! What more could one ask? Answer: darkness and a bed! – That was beloved Sweden . . . The King of Sweden made me à la suite of his Navy, just as he wears the uniform of ours, so now I am partly Swedish too, and I am getting closer to the ancient land of the Jarls and Vikings.'[49] King Oscar knew well how to exploit the Germanic racial cult to secure for his country the goodwill of the new ruler of the powerful neighbouring empire.[50] At the state banquet in Stockholm he proposed a cordial toast in German to the 'brother peoples of Germanic origin, both to those who live in Germany and to those who are settled in the North'.[51] Unfortunately in his words of thanks the Kaiser spoke 'very fast', so that his speech failed to make 'as much of an impression as the King's, with his dignified way of speaking'.[52]

The danger that lay in this blend of Nordic romanticism with the Kaiser's naval passion was not something to which German diplomats were as yet particularly alive. Kiderlen merely scoffed at the Kaiser's disappointment that 'the two heaviest armoured ships had . . . to be left behind at Oskar-Friederiksborg because their draught was too great to allow them to enter the harbour. And oh! the pity of it!', he exclaimed sarcastically, 'at Copenhagen we will have to leave *three* behind!' The diplomat was irritated that in Stockholm the Kaiser insisted on going through the 'whole programme' again on the iron-clad *Baden*. He mocked Wilhelm's obvious delight in the joint salute fired by the German and Swedish ships for the birth of his fifth son, making 'an infernal noise' which echoed round the cliffs surrounding the harbour. The bibulous Swabian *bon vivant* was more inclined to appreciate the warm hospitality of the Swedish court, whose 'sole endeavour was to amuse us to the utmost from morning till night'.[53]

After four days the journey continued to Copenhagen, where the imperial squadron arrived on 30 July, although for a stay of only twelve hours, which in view of the shortage of accommodation in the Tivoli and of the strongly anti-German mood of the Danish people seemed only advisable.[54] Senden-Bibran recalled: 'The decoration of the city was minimal and the crowds also very small. We saw no German flags on private buildings. We heard not a single cheer' as the two monarchs drove through the streets.[55] After the journey was over Wilhelm's physician commented that the reception in Copenhagen had been 'very cool and reserved . . . In the streets of Copenhagen the Prussian national anthem and the *Wacht am Rhein* were, it is true, sung as the Kaiser drove by, but this was done by the numerous Germans who are there at the moment, and the Danes were not at all happy at these demonstrations.'[56]

The longstanding distrust which existed between the Danish and the Prussian courts, and which was further inflamed by letters from Queen Victoria,[57] had become acute again as a result of the Danish royal family's championship of the rights of Ernst August, Duke of Cumberland, whose father had been forced to surrender the Kingdom of Hanover to Prussia in 1866. Thyra – the youngest sister of the Tsarina Maria Feodorovna, of Alexandra Princess of Wales and of the King of Greece – was married to the Guelph pretender. Just as the Prince and Princess of Wales had spoken up for the cause of their Hanoverian brother-in-law in Berlin, and the Tsarina likewise in St Petersburg, the mother of the three influential Danish princesses, Queen Luise, born a princess of Hesse-Cassel in 1817, now took up the cudgels in Copenhagen for the interests of the House of Cumberland. As the Kaiser wrote to his grandmother Augusta on his return to Potsdam: 'The Empress [of Russia] was more charming than ever, but as we were leaving she regaled me with a few Cumberland complaints and wishes, of which I naturally could do no more than take note . . . The Queen [of Denmark], whose intelligent and incredibly youthful features betray an inclination for mischief-making, regaled me with the whole Cumberlandiana again, in the most charming way while we were at dinner, which I handled as coolly as in Russia. Later she approached Count H. Bismarck as well; he gave her a fairly plain answer in mine and the Chancellor's names.'[58]

On the evening of 31 July 1888 the *Hohenzollern* arrived back at Kiel, and the party travelled on that night by special train to Friedrichsruh, the elderly Reich Chancellor's residence east of Hamburg. Out of consideration for the Chancellor, Wilhelm II had originally wished to spend another night on the yacht and to travel to Friedrichsruh on 1 August, but Herbert Bismarck explained to him that his father would sleep less well in the expectation of the Kaiser's arrival than if he knew him to be safely under his roof. To this Wilhelm replied: 'Very well, my only concern is to disturb your father as little as possible; so please telegraph to him my express order that he should go to bed at the customary time and should not receive or wait up for me.' When Herbert smilingly answered 'As Your Majesty commands', the Kaiser 'raised his finger threateningly and said "You don't think that your Father will act directly against my imperial command?" – But joking aside, H.M. simply wants to go from the station straight to bed', the Foreign Secretary added reassuringly.[59]

With this Kaiser Wilhelm II's first official visits to the neighbouring courts to the east and north came to an end. In a letter to his grandmother Augusta of 9 August 1888 he commented on how the journey had gone and expressed decided optimism about its foreseeable consequences. 'I did not discuss politics with the Emperor of Russia, as he did not speak to me about them and I had nothing in particular to say', he declared. 'That had been agreed with the Reich Chancellor. On the other hand Count H. v. Bismarck had a long audience with the Tsar in which the latter was extremely open and friendly towards the Count and spoke to him for a long time. For the first time not a word about Austria or Bulgaria passed his lips, which may be considered as a good sign . . . The visible success in Russia is that, according to news just received, a Russian division destined for the Austrian frontier has received the order to return to its quarters. In Stockholm we were shown the highest degree of warmth and friendliness and we felt quite at home. The King was also extremely anxious to accentuate as sharply as possible their firm dependence on Germany; and the French influence which was formerly dominant has almost completely disappeared. Copenhagen did not lag behind the other countries in any way. The family was as friendly as the others and the King was particularly charming . . . So I can confidently declare, as the general result of the journey, that the French influence, formerly predominant everywhere in the North, has suffered a palpable and perhaps fatal blow, everywhere where it previously held sway.'[60]

THE VISIT TO THE SOUTH GERMAN COURTS, VIENNA AND ROME

Scarcely had Wilhelm returned from his Baltic cruise when preparations had to be made for his first official visits to Saxony, Baden, Württemberg, Bavaria, Vienna and Rome, which he had announced immediately after his accession. The Chancellor was decidedly in favour of the visits, and when Wilhelm hinted in mid-August that he might prefer to leave out the South German courts, Bismarck let him know how important he considered the Kaiser's visits to the capitals in question to be. As Kuno Rantzau informed Under-Secretary of State Count Maximilian von Berchem on 16 August, the Kaiser had 'told [the Chancellor] during the All-Highest stay at Friedrichsruh that His Majesty had the intention of going to see the King of Saxony from Berlin, and of visiting the King of Württemberg and the Prince Regent of Bavaria in Friedrichshafen and Munich, from Baden-Baden, where the Kaiser wished to go for the birthday of Kaiserin Augusta, on his way to Vienna. From your letter the Reich Chancellor perceived that these intended visits to the German courts had become uncertain again. He would ask you to put it to His Majesty, when the occasion arises, that he would regret it if the visits did not take place, seeing that the German princes had come to Berlin for the opening of the Reichstag and, by their presence, had rendered considerable service to the Kaiser in reinforcing his position. The character of these gentlemen being what it is, the good relations brought about by their co-operation could be preserved only through careful nurture of personal links with them, which in certain circumstances are more valuable than political links. After His Majesty had visited the monarchs of Russia, Sweden and Denmark it would be all the more remarked upon if the All-Highest did not visit the German princes, who had not only been the first to visit the Kaiser, but had also thereby rendered His Majesty an important political service.'[61]

On 25 September 1888 the Kaiser left Berlin again, to go hunting at Detmold. At the end of September he travelled from there via Frankfurt (where Herbert Bismarck joined him) and Stuttgart to Mainau Island on Lake Constance, to celebrate Kaiserin Augusta's birthday. On 1 October he arrived in Munich for a two-day stay, travelling on to Vienna where he arrived on the morning of 3 October.[62] From Mürzsteg, where he hoped to hunt with the Emperor Franz Joseph as in the old days, he wrote to his grandmother Augusta with the news that the weather was appalling and the hunting had therefore been cancelled, but that otherwise the journey was going extremely well, as 'the friendly feelings of the people and of the ruling houses had tried to outdo each other in the warmth of the reception' that they gave him.[63] Seasoned diplomats in the Kaiser's suite judged the Viennese reception more soberly. Ludwig von Raschdau reported that in comparison with Stuttgart and Munich the welcome in Vienna had been 'a *succès d'estime*, to put it mildly'. There had been 'more curiosity than jubilation', and 'nor were there many decorations. The people in the Hofburg, on the other hand were very friendly, and relations with the imperial family seemed to me affectionate.' Franz Joseph, however, had been put out by the deliberate refusal of Wilhelm II to confer a decoration on Minister-President Taaffe – the Bismarcks and the Kaiser considered Taaffe the 'ruination of his country', guilty of undermining the Habsburg monarchy by his pro-Czech policy in Bohemia.[64] As in St Petersburg, Herbert Bismarck made himself unpleasantly conspicuous by his tactless behaviour.[65] Differences between Wilhelm II and Crown Prince Rudolf, which will be discussed later, together with the ceaseless rain which spoiled the days in the mountains, contributed further to turning the German Emperor's first visit to Austria into a failure. 'What pleases me least about the result of the Kaiser's journey', wrote Waldersee, 'is the fact that our relations with Austria have become not better, but worse. On both sides mistrust is taking root again . . . I am afraid that Herbert Bismarck, who cannot bear the Austrians and always stirs the Kaiser up against Austria, is partly to blame here.'[66]

The public reception in Rome on 11 October was decidedly more grandiose and demonstrative.[67] The young Italian state took every imaginable trouble to impress its illustrious guest from the powerful empire to which it was allied – Wilhelm II was the first reigning monarch to pay a state visit to the united Italy – and his huge retinue. Workmen spent weeks removing unsightly buildings from the neighbourhood of the Palazzo Quirinale; the route from the station was adorned with the banners of all the Italian cities, and the entrance to the Via Nazionale with a gigantic arch of papier mâché painted with brightly coloured knights and horses. Before the Quirinal itself stood a triumphal arch of wood and cardboard. The crowds shouted 'Viva la Germania' and other such greetings. In order to demonstrate Italy's martial strength to its mighty ally, on 13 October a colossal military review was held, at which 32,500 infantry and 4,650 cavalry troops with 108 pieces of field artillery marched past the Kaiser. (To bring this mass of humanity to Rome the Italians 'borrowed' 1,000 railway carriages from Austria-Hungary, which they afterwards somehow forgot to return.) A state banquet on the Capitol, with 3,000 guests, ended with a toast by King Umberto, to which the Kaiser replied in Italian. Three days later in Naples Wilhelm witnessed the launch of the latest Italian iron-clad and an impressive naval review with no fewer than forty-seven warships.[68]

There is no doubt that outwardly the Kaiser's visit was a great success. The German Cardinal Prince Gustav zu Hohenlohe-Schillingsfürst wrote delightedly that the 'great Kaiser days' had filled 'the whole population with incredible enthusiasm and myself likewise. It was altogether wonderful, and I thanked God heartily for allowing me to live long enough to witness this magnificent historic moment, through which the two nations will become ever more closely united, which for both of them and for the whole world is a great blessing.'[69] As Raschdau reported to Berlin, the scene beneath the balcony of the royal palace in Rome on 11 October was 'le comble de l'enthousiasme'. 'On such ceremonial occasions', he commented to Holstein, 'the Kaiser looks very serious, almost hard, and in my opinion this suits the occasion. The soldiers are especially impressed by this imperial attitude. But in general this manner is effective in those countries where the monarchy adopts a frivolous attitude. In society, however, the Kaiser displays his winning charm of manner.'[70] Herbert Bismarck likewise remarked that 'our Master, who is as charming as he could conceivably be, is playing his part quite superbly'.[71] With satisfaction he reported back to Wilhelmstrasse that 'everything goes splendidly, H.M. makes a first-rate impression everywhere, and for his part is thoroughly content'.[72]

Behind the scenes, however, Wilhelm II and his retinue showed another side of themselves. As in Russia and Austria, in Italy also many people were repelled by the malicious way in which the young Kaiser spoke of his parents and of Queen Victoria.[73] His 'somewhat imperious and brusque' manner displeased the Italians, with the exception of the military. He made no attempt to conceal his contempt for parliaments and civilians and scarcely spoke to the Lord Mayor of Rome or the presidents of the Senate and the Chamber of Deputies.[74] He made

2. Kaiser Wilhelm II and Prince Heinrich of Prussia with the Italian royal family in Rome, October 1888

a disagreeable impression by his tactlessness, above all towards the diminutive Italian Crown Prince. The British chargé d'affaires in Rome, Kennedy, sent a secret dispatch to Lord Salisbury which the latter immediately forwarded to the Queen, remarking that it showed 'that the Emperor William's rude ways are beginning to attract notice'.[75] The Kaiser's military suite took the liberty of inspecting all barracks, munitions stores, port installations and ministries. Herbert Bismarck, who according to Holstein 'gave way once or twice to his habitual outbursts of rage',[76] spoke openly to the Italian Prime Minister Francesco Crispi about the coming war to be waged by both countries against France, and Wilhelm himself told the King that they would soon meet again in Paris after the victory over France. 'After that you will return to Italy with your three hundred thousand victorious men and then you will be able to kick the Chambers out of the door', the Kaiser assured the shocked Umberto.[77]

If, in spite of all this, his first visit to the leaders of the Italian state can be considered a success on the whole, Wilhelm's audience with Pope Leo XIII on 12 October was a complete disaster, for which, however, the chief blame lay less with the Kaiser himself than with Herbert Bismarck. The embittered relationship between the pope, who in 1870 had lost all territorial power and was a virtual prisoner in the Vatican, and the anti-clerical Italian state demanded the highest degree of diplomatic sensitivity, and that was exactly what the German visitors lacked.[78] According to Holstein, whose account was based on a note dictated by Wilhelm, in his private conversation with the Kaiser the Pope had called for the 'restoration of temporal power' and to this end demanded that Germany should conclude an alliance with Russia, Austria and Spain against Italy and France. If Germany did this the Vatican would ensure that the 15 million Catholics in Germany would remain loyal to the Kaiser and the Reich; if not he would be forced to regard the German Empire as a hostile power. To this the Kaiser had made the 'spirited reply' that if Rome were no longer the capital of Italy the Italian monarchy would not be able to survive. But the radical republic which would then follow would throw the Pope and all the cardinals out of the Vatican. As to the loyalty of the German or at least of the Prussian Catholics, he, Wilhelm, could see to that himself.[79] The private audience was to have lasted half an hour, by previous agreement between the Cardinal Secretary of State Rampolla, the prelate Johannes de Montel and the Prussian envoy Kurd von Schlözer, but it was forcibly interrupted by Herbert Bismarck, who sent Prince Heinrich into the audience chamber prematurely, with the brusque declaration that 'a Prussian prince does not wait around in an antechamber; either immediately or not at all'.[80] The Pope was a 'hypocritical Lenten preacher' who had 'very greatly bored us', Herbert Bismarck commented afterwards. Wilhelm was 'very disappointed with him', the Chancellor's son added.[81]

What had really happened in the Vatican Waldersee discovered later from conversations with several of the men who had taken part in the trip to Rome. 'The programme had been arranged', he recounted, 'so that the Kaiser should lunch with Schlözer [in the Palazzo Capranica, the residence of the Prussian envoy] and should go from there, that is from German soil, to the Vatican, and not in Italian royal carriages but in his own.' The carriage and horses had been brought from Berlin expressly for this purpose. 'After the visit had gone on for a precise length of time Prince Heinrich was then to appear and to be taken in to see the Pope too. Now at Schlözer's house there was a great deal of drinking – the Kaiser is always very moderate – and Herbert as well as Prince Heinrich had drunk more than was good for them, and had also used the time after the Kaiser's departure for further drinking. They arrived in the Vatican in a very obstreperous mood, were politely received and asked to wait in an antechamber as the Kaiser was still in conversation with the Pope. This did not please Herbert; in the most uncouth manner possible he declared that a Prussian prince could not wait, and had then held forth very rudely in German – which naturally various people in the papal household understood very well – about the behaviour of courtiers, guards etc. Finally he had transgressed so far as to use violence to gain entry: he simply pushed aside a senior court official in front of the door. Thereupon Prince Heinrich entered and unfortunately arrived at a moment when the conversation between the Kaiser and the Pope had just begun to get warm. The Pope broke off completely at this, and we have probably done ourselves a great deal of harm. The Kaiser deeply deplored the incident. The unbelievably tactless Herbert bragged that evening at the royal court about the way he had behaved, and made lamentable jokes about popish mummery.' In the Vatican people had been 'violently angry' at this incident, Waldersee noted, and had 'decided in the first flush of rage to take up the Kulturkampf again with full force'. After negotiations with Herbert Bismarck they had then 'withdrawn the declaration of war', but he, Waldersee, was convinced 'that the war will still break out again now. The people in the Vatican are too conceited to forgive us this day, they will set everything in motion to do us harm. We shall soon see from our Centre Party how things stand. That the Pope virtually used France as a threat is also very characteristic.'[82] In the following days Waldersee heard from various sides and on good authority that the Kaiser's visit to the Pope, 'but particularly the forceful irruption into the papal apartments by Prince Heinrich, will have a disastrous influence on the mood of the Catholic world'. In such circles people were 'beside themselves' at the 'violent entrance by Prince Heinrich into the Pope's apartments' and were determined to have a question asked in the Reichstag about the incident. Leo XIII had 'complained bitterly about Prince Heinrich's forcing his way in' while he had been in the process of 'discussing with the Kaiser a subject which would have been agreeable to him: the fight against social democracy'. At any rate with this 'uncivil treatment . . . of the Holy Father' the Catholics had been handed 'an excellent means of stirring up agitation', the Chief of the General Staff commented disapprovingly.[83] Months after the incident the Vatican was still 'indignant' and had the lowest opinion of Germany's leadership. Bismarck was old, 'Herbert drinks too much, and the Kaiser is too busy and has too many demands on him' was the verdict of one cardinal.[84]

WILHELM II'S 'PASSION FOR TRAVEL'

Soon the Kaiser's 'passion for travel'[85] became a problem of general political concern. Queen Victoria observed at the beginning of November 'that the young Emp's unfeeling & indecent conduct in rushing abt so early uninvited to other Courts, his marked slights & rudeness to England & his relatives is much remarked upon'.[86] On 6 November she wrote to her daughter: 'It is beginning to be much talked of here; & every where abroad, W's indecently early & hurried Journies have shocked people & while of course one wishes not to quarrel *if possible* with the German Govt. there is a *limit* to that.'[87] The Grand Duke of Hesse expressed anxiety that Wilhelm's 'glorious triumphal progress' could go to his head, as he was inclined to take the ovations he received as demonstrations of admiration for his own person.[88] Only with difficulty and with Liebenau's help could the Bismarcks prevent Wilhelm from returning to Vienna on his way back to Germany from Italy.[89] When suggestions were advanced for a visit to Spain and a cruise to Athens and Constantinople while they were still in Mürzsteg, Herbert Bismarck tried in vain to convince the Kaiser that he would weaken the impact of the visits he had already made by adding more visits. He wrote to his father: 'I would regard those journeys as politically impractical and ill-advised for the simple reason that they would inevitably reduce the impression made by the great state visits of this year. Multiplication weakens. Furthermore there is nothing for us to gain in the Orient; the grandeur of an imperial visit is too much to fritter away on those shabby regions. The sense of the mighty weight of an imperial progress is not yet sufficiently developed [in the Kaiser], the inclination towards recreational cruises with Henry the Navigator [Prince Heinrich, the Kaiser's brother] prevails too much as yet.'[90] The Kaiser's journeys to Athens and Constantinople took place nevertheless, as we shall see.

The new Emperor's excessive love of travelling had thus proved to be a double-edged sword within the first few months of his reign. His first official visits to St Petersburg, Stockholm, Copenhagen, Vienna and Rome may indeed have served to show the world that, following the aged Kaiser Wilhelm I and the sick Kaiser Friedrich III, the empire in the centre of Europe was now ruled by a dynamic young monarch with whom everyone would have to reckon; but not every court or government found such a demonstration of strength particularly edifying. There was no question of a visit to republican Paris – that was perfectly comprehensible; but the fact that in the first round of visits his closest relatives in London had been passed over attracted lively comment and was widely interpreted as a deliberate snub to the mighty British nation. Not only in the British royal family, but also in German liberal circles, Wilhelm's restless drive to travel so soon after his father's death was seen as a gross impiety which cast a harsh light on his adulation-seeking character and his political ambitions. Although reasons of foreign policy were advanced to justify the journeys, it was nevertheless clear to everyone that the real reasons were of a personal, emotionally motivated nature: that Wilhelm wanted to show himself, to be paid court to, to 'astonish' – and also that he lacked inner repose.[91] In the capital cities he visited, highly complex political circumstances prevailed, of which he understood very little, and for which the alcoholic Herbert Bismarck and the insomniac old Prince-Chancellor could prepare him only very inadequately. The intrusive German attempt to strengthen monarchical forces and to undermine the position of those dynasties and statesmen who championed constitutional forms of government by ostentatiously cold-shouldering them made an offensive impression and was often counterproductive. Thus these first visits, with the exception of that to Stockholm, on balance brought more damage than profit. The damage was, however, small in comparison to the truly incredible conflict with the British royal family which Wilhelm stirred up soon after his accession. The background to it lay in the disastrous relationship between him and his mother.

CHAPTER THREE

The Kaiser and his mother

THE EMPRESS FREDERICK AND KAISER WILHELM II

THE mental and physical state of the Dowager Empress after the death of her husband was pitiful. Furthermore, the new circumstances in which she found herself were very hard to bear. For the first weeks she wept and lamented incessantly;[1] she had frequent attacks of 'rheumatism' and 'neuralgia';[2] she shut herself away completely with her three younger daughters and Frau von Stockmar, and as far as possible avoided any contact with the world around her.[3] There were moments when she harboured thoughts of suicide.[4] For three months she had stood at the centre of events as German empress (and sometimes even beforehand, as crown princess); now she was avoided, isolated, attacked in the pro-government press. On 4 July the ministers of state – '*not* the Reich Chancellor, of course' – and the aides-de-camp came to offer their condolences and take their leave from her formally. 'Apart from Friedberg, I think the only ones who spoke from the heart or had even a glimmer of sympathy for me were – at the most – Bronsart or Lucius!', the Empress commented.[5] The isolation which she had at first sought soon weighed heavily on her. 'I have to get accustomed to be a person who is *not* considered or remembered by the present Régime, & I find it rather hard!', she complained in September.[6] She saw her eldest son only rarely. He visited his grandmother every day at Babelsberg, but to her, his mother, he came as good as never. '*This* House does *not exist!* William never comes; & I am taken no notice of!'[7] As ever, she had no illusions about her son's loveless, cold, self-centred character. '*Affection* people cannot give – who do not know the feeling & have it not in their heart!', she wrote to her mother on 21 August. 'He will get through the world a great deal easier than his beloved Papa or I, but he will never attach people to him or have half the love that Fritz had, & I really do not think he wants it, or would care for it or understand it!'[8] Comparing Wilhelm with his brother, she commented that both had equally repugnant political opinions, but Heinrich had 'certainly more feeling, & is less spoilt, & is less selfish & rücksichtslos [inconsiderate]!'[9] Wilhelm's coldness made itself felt in both minor and major matters.[10] She was hurt that Wilhelm did not name his fifth son after her but after the King of Sweden, since the fourth son, August Wilhelm, had been named in honour of Kaiserin Augusta.[11] Even the sixth son, Joachim, did not receive the name Victor at baptism, as both the Dowager Empress and her mother had hoped. 'I think they *might* have called *one* of the 6 boys *Victor* after me! Particularly as they know how much I wished it', the younger Victoria wrote to *her* daughter Victoria. It was not until the birth of Wilhelm's daughter Viktoria Luise that her wish was fulfilled.[12]

Victoria was now completely excluded from political affairs, never asked for her opinion, avoided by officialdom, and thus no longer had the slightest influence on politics.[13] At the court too she was relegated to an inferior position which she found very hard to bear. Of the three Empresses she now held the lowest place, third after Kaiserin Augusta and Kaiserin Auguste Viktoria ('Dona'), as she commented in August 1888. 'You know how *very* indifferent rank and etiquette, honours etc. are to me', she wrote to the Queen. 'But yet I am often shocked at the want of courtesy and considerate behaviour I meet with. I am *quite* ready to give way to the Empress Augusta on account of her age and her being my mother-in-law, but to have to knock under to my own daughter in law besides, makes it rather trying & almost ludicrous sometimes.'[14] As her 'widow's pension' amounted to only 600,000 marks, from which she had to pay all salaries and allowances, she reduced her household to a minimum: she retained only one marshal of the court (Hofmarschall Hugo Freiherr von Reischach), one chief master of the court (Oberhofmeister Götz Graf von Seckendorff), one senior lady in waiting (Palastdame Hedwig Countess von Brühl) and two other ladies in waiting (Hofdamen). She was forced to limit her expenditure to a third of what it had been while she was Crown Princess. Kaiserin Augusta was much richer and had a very much larger household.[15] Sorrowfully Victoria compared her widowhood with earlier, happier years. 'Our House here is really like a convent, nothing but Ladies, & only 2 gentlemen, sometimes only *one*, and at other times (though rarely!) *none*! Formerly it used to be principally gentlemen, and so many coming and going – people wanting to speak to one every minute, carriages driving up & away etc . . . Now *all* is *silent*. I am only *too* thankful to be quiet at this moment . . . But the contrast is terrible. The Bees seem to have deserted this hive and flown to the next one, & one is left in the empty one alone!'[16] For his part, Wilhelm complained bitterly about his mother and was still claiming at the end of 1890 that 'he had given her a generous allowance and various castles and had put her generally in the position of being able to do or not do whatever she wanted. But there was no question of thanks. Recently she had even threatened him and prophesied that his autocratic conduct must lead to disaster.'[17]

The contrast with the new lifestyle of her son could hardly have been more flagrant. Wilhelm's grandiose ceremonies, the innumerable court festivities and parades so soon after his father's death, his visits to St Petersburg, Stockholm and Copenhagen, his journeys to Vienna and Rome and the unmistakable cold-shouldering of London[18] – all this the widowed Empress took as a personal insult. 'Since our terrible Loss', she wrote in September 1888, '*not 2* days have been devoted to mourning or to quiet! . . . It has been *one whirl*, of visits, receptions, *dinners*, journeys, parades, manoeuvres, shooting & entertaining. Of course it *jars* on my feelings.'[19] She looked back on the early days of her son's reign with nothing but scorn for his achievements. 'What has William done in those 3 months. Certainly nothing lasting or useful! He has rushed & torn about all over the place fr. morning till night; visited half the courts in Europe, made 2 or 3 most illadvised speeches, allowed that abominable Pamphlet to be published about his Father's illness . . . Some new excitement must be provided for every day!!! Receptions, Dinners, Parades etc . . . To me it seems utterly childish, when I think of how his beloved Father read, & wrote & worked, & tried to help on each good & useful thing.'[20] Two years later the Dowager Empress was still complaining of his astonishing behaviour: Wilhelm was doing 'the very oddest things, the element of show off, – noise, and "sensation", dramatic effects etc. is very preponderant and in these serious times seems to me *very* youthful.'[21]

THE 'CRUSADE' AGAINST THE CROWN PRINCE'S PARTY

In addition to the deplorable personal relations between mother and son, in the first months after Wilhelm's accession the public attacks on the Empress Frederick and the late Kaiser's political intimates took on an increasingly brutal form. The Empress wrote in disgust to her mother that the Bismarcks and their 'pupil' Wilhelm were now trying 'to *crush* & annihilate *all* that is in any way *liberal*, or *independent* or cosmopolitan' in Germany.[22] She found the attacks on her in the pro-government press particularly despicable – she cited above all the semi-official *Norddeutsche Allgemeine*, the right-wing conservative *Kreuzzeitung* and the Free Conservative *Post* – but she was completely powerless to defend herself against them.[23] 'The language of the semi-official press is insolent and disgraceful', she commented in her diary only a few days after the death of her husband. 'I am no longer astonished at *any lies* or impertinence. The most impudent gang in the world, without principles or conscience, is now in power!'[24] 'I feel *utterly without any protection* whatsoever!', she wrote to her mother at the end of July. '*Any*body fr. Pce. Bismark [sic] to a Dr. or Professor can accuse, blacken or calumniate me, excite the public against me! William or Henry would *never* stop it or interfere & it would be the easiest thing in the world to make them believe the *greatest falsehoods* & say such attacks were patriotic, necessary for German interests, the authority of the Crown or some such rubbish!!'[25]

Almost anything that her son ordered, did or said was repugnant to her.[26] She was 'half dead with despair' and '*mad* with sorrow, anger and agitation' when she heard that 'at William's wish and command' an autopsy had been carried out on the body of her husband.[27] When she heard of the 'abominable' decision of Bergmann, Gerhardt and other German doctors to publish an official pamphlet about the illness of Kaiser Friedrich III, she implored Wilhelm 'verbally, in writing and by telegram, most earnestly not to allow this', as it would be unbearably painful for her, but 'all in vain, – he had already given his consent . . . He has no tact or delicacy.'[28] She was naturally horrified, too, when in August 1888 Wilhelm sent a telegram to Heinrich von Treitschke expressing his praise and thanks for a pamphlet in which the historian derided Friedrich III and his principles, and portrayed his reign as one of the most regrettable episodes in German history.[29]

A series of draconian measures were taken in the early stages of his reign by Wilhelm and the Bismarcks against members of the so-called Crown Prince's party. The Hamburg diplomat and constitutional lawyer Heinrich Geffcken was arrested for high treason because he published extracts from the Franco-Prussian war diaries of the then Crown Prince Friedrich Wilhelm in the *Deutsche Rundschau*. When the authorities had no alternative but to discharge him, as a broken man, after ninety-nine days of imprisonment, the Jewish Minister of Justice Heinrich von Friedberg – the 'only remaining friend and prop' of the Empress Frederick in government circles – was made a scapegoat and dismissed, while the lawyer who had drawn up the indictment against Geffcken, Ober-Reichsanwalt Tessendorff, was awarded the cross of a commander of the Hohenzollern royal family order by Wilhelm II in his birthday honours.[30] Shortly after Geffcken's arrest the German translation of Sir Morell Mackenzie's book refuting the Bergmann pamphlet was confiscated (the government prosecutor released it again soon afterwards as unobjectionable), while Mackenzie's arch-enemy Sir Felix Semon was honoured by Wilhelm with the Order of the Red Eagle.[31]

Then the *Kieler Zeitung* was confiscated on the personal orders of Wilhelm II because it had published extracts (which had long since been public knowledge) from the Crown Prince's diary – this time from the 1866 campaign. As if that were not enough, Franz von Roggenbach's house was searched in his absence by the secret police – at Geffcken's arrest letters and writings by Roggenbach had been found, which were said to have contained 'crude outbursts' against the young Kaiser,[32] the widowed Bogumilla von Stockmar was cross-questioned in her home by an examining magistrate of the supreme court of the Reich, and the British ambassador in St Petersburg, Sir Robert Morier, was accused in the semi-official *Kölnische Zeitung* of having passed on German military secrets to the French in the Franco-Prussian war.[33] There was no respite: as the Empress commented in one of her stream of letters of lament, when the left-wing liberal *Volkszeitung* was also confiscated in March 1889, 'The crusade goes on against all & everything that we approved of & stood by!'[34]

'These things are allowed & sanctioned by William against his Father's most trusted & oldest friends!!!', Victoria wrote bitterly to her mother.[35] She was afraid that the Reich was well on the way to becoming 'a sort of military Paraguay'.[36] Many people were already talking of a 'reign of terror' or of 'despotism' which might perhaps be customary in Russia, but which could in no way be permitted in Germany in the nineteenth century. 'Russian conditions hold sway here, the police-state flourishes, it makes me feel faint!!', the dead Kaiser's widow recorded in her diary. 'Such nightmares are possible now. There is no longer any right, decency, respect, piety or justice', she lamented; and the Grand Duke of Hesse and his daughter Princess Irène of Prussia took a very similar view.[37] But Wilhelm, according to his mother, considered it all 'very grand'.[38] He regarded any public mention of his father or his mother as 'an offence to *him*!'[39] He behaved 'unspeakably', 'abominably', 'heartlessly', 'harshly and slightingly' and acted with 'brutal ruthlessness' like a 'tyrant', complained his despairing mother.[40] 'Fritz saw it all, & it *broke* his *heart*, and I am sure the mental worry and distress predisposed him to this disease.'[41]

Writing in November 1888 the Empress Frederick poured scorn on what she considered the reactionary German system of government, and particularly on the role which her son played in it. She told the Queen that Wilhelm was '*so green* and *so suspicious* and prejudiced!!' He read 'only the papers prepared for him, does not understand or care for all the difficult and intricate questions of internal government, & is *utterly ignorant* of *social*, industrial, agricultural, commercial, financial questions etc . . . *only* occupied with military things, with a little smattering of foreign affairs, & *constantly* being *fétéd* [sic], – travelling about having dinners, receptions &c. . . . It is *sad* indeed for me as a mother, but it is *not* surprising. The Clique supported by the Emp: William and the Empss Augusta (who both meant no harm & thought they were right) have brought this about and we *could not* prevented [sic].'[42] In a last letter from Germany before she left for England in mid-November 1888 with her three unmarried daughters, the Empress gave vent to her feelings about her son's superficial character and reactionary views. He was 'considered the right & real representant of his Grandfather's views – & Pce Bismark's [sic] policy, – & is *much* elated at this, – *much flattery* is poured upon him, so he never doubts that all he does & thinks is perfect, and there is no counterbalance or moderating influence in his wife that I can see! She quite approves the present system, gives it her full support & is very happy.' Wilhelm would never replace 'the antediluvian & autocratic ideas of most of the Hohenzollerns' with a 'wider, more *humane*, liberal, tolerant & moderate spirit', she foretold. He was too obstinate, and the people who could influence him in the right way were unknown to him and had no opportunity of approaching him. 'His *whole* mode of thought is so completely different that he would never *read* or understand or study anything which *could* open his eyes. He has never travelled and he has *not one* eminent man as a friend.' His entourage consisted of men like Gustav von Kessel and Maximilian von Lyncker, the former false and dangerous, the latter narrow-minded, violent and always advocating the strongest measures, and their influence on him was pernicious. The young Kaiser was like 'a child that pulls off a flies [sic] legs or wings & does not think the fly minds it, or that it matters'. Even Prince Bismarck had recently expressed surprise at Wilhelm's unstable personality, the Empress Frederick informed her mother. It appeared that the Chancellor had remarked to the ambassador Count Münster that 'the Emperor is like a balloon, if one did not hold him fast on a string, he would go no one knows whither'.[43]

As for the question of who was the driving force behind the new regime – the Bismarcks or the young Kaiser – and in particular who was responsible for the terrorisation of her free-thinking supporters, the Dowager Empress was in no doubt as to the answer. Writing bitterly to her mother and her brother Albert Edward ('Bertie'), her constant refrain was that Wilhelm was now wholly in the hands of 'Bismarck and his gang', and that he was being exploited by them as a tool.[44] Her son was surrounded by those whom she and her husband had fought against for thirty years.[45] Bismarck's wish was for 'his head to be thoroughly turned, his vanity & pride to be still greater than they are already, and *then* he will of course dash into *anything* they may propose'.[46] Her three eldest children, Wilhelm, Charlotte and Heinrich, were almost completely unaware of how they were being used. 'By nature they *do not understand* politics, nor do they *care* about them, they only join the general cry of the circles in which they move, & support William with all the roughness & violence of his disposition.'[47] 'I close my *eyes* & *ears* to the official world and find it the only way not to feel the profoundest irritation with W.', she wrote. 'I am only too ready to make *all* allowances for him, when I think of the deplorable friends he has had, & of all the nonsense with wh. his head has been *so* systematically stuffed!!'[48] There was no point in getting irritated with this or that action of Wilhelm's, she commented stoically, for once one began one would never know when to stop.[49] 'I try never to think of William, so as not to make myself ill with bitterness, indignation and anger', she noted in her diary at Christmas 1888. Nevertheless her bitterness repeatedly broke through, as at the end of December 1888, when she wrote of the Bismarcks and Wilhelm: 'What else can one expect from such a rough, vile person [as Herbert Bismarck] and from such an *infamous* system as that of our Government – with an inexperienced, rather rough and heartless young man at its helm, who is *completely* in the hands of these people – blind and deaf to *all truth*, stuffed with prejudices, indoctrinated with wrong opinions, full of *himself* and his dignity – and of the new power which has gone to his head.'[50]

The Empress Frederick believed that she had detected the political goal behind the hate campaign against her and her political friends orchestrated by the Bismarcks. 'It is now a struggle for Prince Bismarck's power to shake off all obligations and fetters which might be a *gêne* – to those belong Fritz's memory and my person! I must be run down and annihilated, as I am a relic of that fabric of hopes and plans he wishes to destroy once and for all! He fears that William might some day fall under my influence and therefore this must be prevented in time by making me out a danger to the state and an enemy to the Government.' In this struggle for power Bismarck '*could* not have a better tool than William. He has carefully had him prepared by his *own* son Herbert for two years. *All* other voices and views are excluded.'[51] So she was sharing 'the fate of *all* our friends, all the *best*, most experienced & enlightened patriots who wd. have been our support & help. War to the knife is waged against them *all*, with the most unheard-of & unjustifiable means!', she wrote at the beginning of November.[52] At a time when Queen Victoria was having a statue of Friedrich III placed at Windsor and a bust of him at Balmoral, where she had seen him for the last time,[53] his widow in Potsdam was coming to the conclusion that the Bismarcks' goal was the systematic obliteration of all memory of the liberal hopes of the short reign of Friedrich III, so as to create the impression among the people that Wilhelm II had followed directly on from Wilhelm I.[54] She had already observed in early July 1888 that it was the aim of the gentlemen now in power in Berlin 'to wipe out all trace of Fritz's reign, as of an *interlude* without importance . . . William the IInd succeeds William the 1st – in a *perfect* continuity of system, aims and tradition! Frederick III *would* have had to be *submitted* to, but was happily removed by Providence before he had time to set *his* mark and *his* stamp on the German Empire; the sooner *he* is forgotten the better, therefore the sooner his widow disappears the better *also*.'[55] In government circles it was considered a blessing that Friedrich III had not lived longer, for as they saw it he would have destroyed the Reich with his liberal principles. 'These are the *sentiments* & this is the *Language* which has been held during 30 years, but especially during the *2 last*, in *Government*, *Court*, Society & *Berlin* military circles, with which our 3 eldest children have been embued.'[56]

If the Bismarcks used the glorification of Kaiser Wilhelm I chiefly as a calculated tactic in the battle against left-wing liberalism, for Wilhelm II a feeling of almost mystical reverence for his grandfather still played a part, the Empress Frederick observed. The young Wilhelm had tried all his life 'either consciously or unconsciously to place his Grandfather on a pinnacle, in *contrast* to his father. That it should pain, annoy & irritate *me*, you cannot wonder. Certainly I am the last to wish that the Emperor William [I] should [not] have his due. But neither in *character*, intelligence nor education was he to be *compared* to his son! He was a thorough Tyrant & autocrat, & could never understand his task as a sovereign from the same elevated & modern point of view as beloved Fritz did; William has not learnt to appreciate or understand his Father as he should, nor to feel that *he*, W., cannot do better than try to reign in *his* spirit! He considers himself the representant of his *Grandfather's* ideas; & this is very unfortunate indeed! He is of course encouraged in this by everyone around him. Either he will find out his mistake too late or not at all.'[57] It is not difficult to guess the effect of such reports from Germany on the royal family in Britain.

FROM POTSDAM TO KRONBERG

In the first weeks after Wilhelm's accession the relationship between mother and son was put under further strain by the question of accommodation for the widowed Empress, and almost inevitably the discussions over this led to conflict between the new Kaiser and the British royal family. For the grief-stricken widow it was hard to leave Friedrichskron – as the Neues Palais in Potsdam was now called – where her husband had been born and had died, and where she had spent thirty happy years with him.[58] Her eldest brother, the future King Edward VII, who had come to Potsdam with his Danish wife Alexandra to attend the funeral of their late brother-in-law in the Friedenskirche on 18 June 1888, persuaded her that Wilhelm would let her keep the palace if she would only ask him for it directly.[59] 'If you put it to him personally he could not refuse his own Mother & widow of his beloved Father!', he assured her.[60] When this hope was disappointed, the Prince of Wales urged that his sister should at least be given the nearby palace of Sans Souci as her residence, not least because he hoped that if Victoria could continue to live at Potsdam she might in time gain some influence over her son. 'That you have bitter feelings about people & things is indeed not to be wondered at after what you have told me & all I have heard', he wrote to her at the end of June after his return to Windsor. 'But I would beg you dearest Sister to endeavour to dispel those feelings as much as you possibly can . . . Above all things endeavour to have influence over W. This is I know no easy task, but do not despair, do not give up the hope eventually that the example & views of his ever to be regretted Father may yet be an example to him in the difficult task which he has to fulfil. Let no estrangement exist between you both, & remember he is his father's eldest son & your first born.'[61] The Prince entreated her never to quarrel with Wilhelm, whatever happened.[62] 'There will I know be endless difficulties with W.', he wrote to her, 'but you must not be disheartened, dearest Vicky, & try to surmount them. Above all if possible try & have some influence with him – so that he may not be entirely at the mercy of those in whose political opinions you cannot share.'[63] The Empress Frederick was well aware of how illusory such hopes were,[64] but on the other hand she was 'also very keen' to be given Sans Souci as her residence; she thought 'it would make it easier for me to leave Friedrichskron, and it would be so close to the Friedenskirche', where not only her husband but also her two favourite sons, Sigismund and Waldemar, were buried.[65] But this wish was likewise rejected by Wilhelm, and so too was her request to be allocated at least a few rooms in Charlottenburg. She noted bitterly in her diary for 28 June: 'I would have liked Sans Souci – it was curtly and rudely refused! Fritz had wished that I should be given rooms in Charlottenburg – this is also *refused* to me. I am to leave Friedrichskron and only the Villa Liegnitz is to be reserved for me.'[66] This small house, built by Friedrich Wilhelm III for his second wife Auguste Countess Harrach, in which first Vicky's daughter Charlotte and then Prince Heinrich had lived, would at least serve as an overnight lodging for her on her visits to the Friedenskirche and to the school in Bornstedt, she reported to Windsor.[67]

Although she received the Crown Prince's Palace Unter den Linden as her Berlin residence, a stay of any length in Potsdam was made impossible for her by these decisions. Again the Empress Frederick felt she knew what calculations lay behind them. 'They would like me to go right away, for fear I might gain some influence over William, remind the public of Fritz, keep the love for him – the longing for him – alive, and possibly retain some connections with the Liberal Party and give them support', she conjectured.[68] 'The Court wants me away from here, the whole of Wilhelm's entourage would like to be rid of me! They find me tiresome and uncomfortable! I have been asked whether I would not like to go to Italy for the winter?'[69]

Given such circumstances, why did she remain in Germany? She gave a partial answer to this question in a letter to her mother of August 1890: 'It is out of love – to my own darling's memory & to his country – to my girls – & so many charitable institutions that need me, – & kind friends that would be miserable if I went, – that I remain in Germany *at all* – after the treatment I have received!', she explained. It would be 'foolish and wrong – & undignified – to allow oneself to be frightened away or bullied out of the country – for wh. one has toiled as best one could for 30 years; – still it is gall and wormwood to me to have to remain at Berlin under the present circumstances.'[70] Whether in addition to this other considerations of a personal or material kind influenced her decision need not detain us here.[71]

Even before her husband's death, in fact, Victoria had been planning to buy a property well away from Berlin, although for financial reasons it was nevertheless to be in Germany, and if possible in Prussia. Towards the end of June two houses were offered to her through the agency of Roggenbach and Stosch, one of which, in the Rhine-Main area which she had always favoured, seemed particularly attractive.[72] After turning down houses near Gotha and Dresden, as well as Schloss Weikersheim in the Hohenlohe domains, she finally chose the Villa Reiss in Kronberg, which she enlarged to create Schloss Friedrichshof.[73] The prospect of building herself a home of her own in the Taunus, where she could live independently of the Berlin court, consoled her to a certain degree for having to give up Friedrichskron.[74] When she subsequently received a legacy of 5 million French francs from the Duchesse de Galliéra at the beginning of 1889 the building of the new castle, which she directed from Bad Homburg, became a source of happiness for her.[75] Nevertheless she found it unbearably difficult to leave the old rooms in Potsdam. 'Now everything, everything is over!', she wrote in her diary on 1 October 1888. 'Never again will I set foot in this dear house! . . . The rooms, the walls will *not* be able to speak to those who move in here after us!' She could not take this in, she recorded, but only write it down in floods of tears. Now she had no home any more, for the new one in the Taunus would not be ready for two years.[76]

KAISER FRIEDRICH III'S PAPERS

The tense relationship between the young Kaiser and his mother was put under still further pressure by the question of the papers left by Kaiser Friedrich III which had been sent to Windsor in several boxes before 15 June 1888, as described in the first volume of this biography.[77] The volume ends with an account of Wilhelm's action immediately after the death of his father: he had the Neues Palais surrounded by Hussars to prevent documents from being smuggled out. On his instructions General von Winterfeldt and Gustav von Kessel rifled through every desk in the palace in search of secret papers. They were looking in particular for the book containing the secret ciphers of the Foreign Office which every reigning monarch and every crown prince carried about with him, and when this could not be found Wilhelm told his mother that he was convinced that Dr Morell Mackenzie had stolen the cipher-book on orders from a foreign power – that is, from Britain. All the servants were interrogated in the House Ministry and depositions taken down. Then it was said that the Empress herself had removed the cipher-book; hints to that effect even appeared in the press. She was said to have sent abroad not only the cipher-book – which was discovered in late October, sealed, in the Crown Prince's Palace! – but also the secret state papers of the deceased Kaiser. Even in the semi-official press 'W[illiam], Bism[arck] and accomplices' (as the Empress Frederick called them in her diary) spread the word that she, Victoria, had 'purloined the papers of Kaiser Friedrich and sent them abroad to *publish* them there'.[78] The papers which had been deposited in Windsor by Friedrich himself or by his express wish were without exception private papers, she assured her mother. 'He wished *my private* correspondence & *his* to be safe fr. the ruthless hands of people who were traitors to him & to me! – Never never give them up *please*, even if they *try* to make you!' The Queen could deny having any papers of her late son-in-law, her daughter urged, for no one could know what Friedrich had sent to Windsor, if anything, or indeed where he had secreted his papers at all. In any case he had torn up and burned heaps of his papers in Charlottenburg, in Potsdam, in Baveno and San Remo, and even before their last journey to England.[79]

Not long afterwards, however, Friedberg and the new House Minister Wilhelm von Wedell succeeded in gradually allaying the Empress Frederick's mistrust. Wedell visited her on 10 July and persuaded her, in a 'kind and polite' manner, to hand over to him the keys to her husband's desk, tables and cupboard in the Crown Prince's Palace, which had hitherto been sealed, having promised to show her all the papers before they were deposited in the family archives.[80] On 18 July she went to Berlin for the first time since the death of her husband. In the presence of Friedberg, Wedell and Count von Unruh from the House Ministry the writing tables and the iron safe cupboard were unsealed and the papers examined. The Empress handed over to the House Ministry all documents which were not of a private character: she was allowed to keep every paper she wished to have. It was a 'very painful' task for her, but one that also helped restore her confidence.[81]

Four days previously Friedberg had come to her with the urgent advice that she should have all the boxes which the late Kaiser had deposited in Windsor sent back to Berlin, as Wilhelm II, Bismarck and the ministers of state were expressing the view that it was she who had removed Friedrich III's papers, and that they were German state property. Friedberg promised to examine the papers with Wedell in her presence, so that she would be in a position to prove that they did in fact consist purely of private papers. Once she had convinced Friedberg and Wedell of this, they would be able to speak authoritatively in her support to Bismarck and the other ministers. On the strength of this advice the

Empress Frederick wrote confidentially to her mother on 12 July 1888 asking her after all to return all the boxes stored at Windsor to Potsdam. The boxes should be sent as unobtrusively as possible with the luggage of her sister Helena ('Lenchen'), who was coming to Potsdam soon.[82]

The boxes, guarded by a footman named Schobert, arrived safely at Friedrichskron as early as 19 July, with Lenchen.[83] Friedberg went out to Potsdam the very next day and satisfied himself that the papers were indeed the private property of the Empress Frederick and that the House Ministry could not stake any claim to them.[84] On the following day Friedberg returned with Wedell, and the House Minister now ascertained for himself, as the widowed Empress recorded, 'that in *every* box a note in beloved Fritz's hand lay on top – "*All the papers in this box* are the property of the Crown Princess my wife" . . . These were all his *private papers* & now are *mine*; but nevertheless I made a selection with the gentlemen & had a quantity sent to the Family Archives to be deposited there.' As they were her private property these documents could be examined only with her express permission, she insisted.[85] After this 'obliging' co-operation from the Empress Frederick Friedberg and Wedell were in fact able to fend off Herbert Bismarck's demand that the Foreign Office should also participate in the sorting of the papers.[86] To her mother Victoria justified her decision to hand over some of the records with the explanation that 'Whatever there is wh. I think Fritz would not mind their having, or wh. would be of *use*, I will give up to them, retaining all else for ever more! I shall then put them *in a safe place where no one* can get at them!' Depositing the papers in Windsor had however been their salvation, she wrote gratefully to the Queen, for at the time immediately after the death of her husband the palace was searched by 'foolish & false people' who had been ready 'like Harpies to seize everything and anything they could lay hands on'. Now she was alone with Lenchen and could arrange the papers undisturbed.[87] But in this task she missed both her faithful old adviser Stockmar and her former Marshal of the Court Karl von Normann, who had wanted to help her sort the papers but had suddenly died on the way to join her.[88]

The two sisters arranged the papers left by Kaiser Friedrich III in three groups. The first consisted of letters which were to be returned to their authors. Thus the letters of Friedrich's Weimar grandparents were returned to Weimar, and similarly Kaiserin Augusta, Grand Duchess Luise of Baden, King Leopold of the Belgians, Friedrich's tutor Pasteur Godet, General Albrecht von Stosch, General Albert von Mischke, Eduard von Liebenau, Albert Freiherr von Seckendorff, Dr Georg Hinzpeter, Dr Heinrich Geffcken and Pastor Persius received their letters. The documents which were to be kept securely, including Queen Victoria's and Prince Albert's letters to their daughter and son-in-law, were put back into the cases, which were sent back to Windsor temporarily via the British embassy on 26 July, until the Empress could build a safe hiding place for them in her new castle in the Taunus.[89] The third group was destroyed. 'Arranged the remaining correspondence with Lenchen', Victoria noted in her diary for 22 July 1888. 'All from my brothers and sisters – *30* years' worth – *burnt*, with a heavy heart – likewise those from many friends and acquaintances.'[90] At the same time she made arrangements for her own letters to other people to be destroyed, including, for instance, her entire correspondence with Stockmar.[91]

KAISER FRIEDRICH'S WAR DIARY

It was in fact Victoria's intention, as her enemies feared, to write a biography of her husband. In September 1888 she asked her mother to have extracts made from the letters she had written to her, which would provide source material for such a work, since she, unlike her husband, had kept no diary. 'Some day the world *shall* have a true picture of him and all he suffered', she vowed in a letter of 14 September, 'but *now* it is *much* too soon.'[92] A week later she, like everyone else, was startled by the publication of extracts from the war diary of her husband in the *Deutsche Rundschau*. She at once feared that the public might think her responsible for the publication.[93] Shortly thereafter she recorded that 'great excitement' had been caused both on the left and on the right by the article in the *Rundschau*: 'Delight among all well-meaning & non-partisan people, great anger in the Marmor Palais, at Court etc. The publication contains fine and true material, but also incautious remarks, and it is altogether a great and inexplicable indiscretion! How it happened I do not know! . . . The Secret [diary] I have never seen – it is in the Archives! But I am abused again as always and at every opportunity.'[94] Many people thought that *she* had arranged the publication to get her revenge on Bismarck. 'Of course, this is all a mischievous lie! in order to excite his Party, William, &c . . . against me.'[95] In the following days it turned out that the Empress had mistakenly handed over her husband's Franco-Prussian war diary, which he himself had sent on a roundabout journey from San Remo first to the British embassy in Berlin and from there to Windsor,[96] to the House Ministry on 21 July when she had allowed the 'non-private' papers of her husband to be deposited in the House Archives by Friedberg and Wedell; she had taken the sealed diaries for a collection of military writings.[97]

No one was more infuriated by the publication of the war diary than Wilhelm II.[98] As his mother reported to Windsor on 24 September, 'The Marmor Palais & Berlin are in a state of fury & excitement about the publication of Fritz's "Tagebuch". It does *not suit* the "powers that be" at all of course. William was in a rage & called it "high treason" and theft of State papers! Of course this is nonsense!' The only explanation she could think of, she wrote, was that her husband had several lithographic copies of the diary made, which he had given to his more intimate friends.[99] Although it was thoroughly well meant, she commented, the publication of the diaries had only poisoned the atmosphere in Berlin still more, deepened the gulf between her and Wilhelm and inflamed 'evil passions' in the government and in Bismarck's party; 'their violence is untold!' she wrote on 26 September. 'This publication utterly infuriates them. *Where* it was got from, *what it is*, I do not know! I possess nothing of the kind in my papers, and yet every word is true, and the facts are correct, – the writing seems to me Fritz's own, they are his words and opinions, but I never saw them put together in this form! An *outburst* of delight in the public has been followed by an outburst of fury by William, who bitterly criticized his Papa to me, & said "*how could* he write such imprudent things down . . ." I only thought to myself – how deeply is William to be pitied for so little understanding his Father.'[100] Bismarck's anger over the publication was easy to understand, commented Victoria, for did not the diary entries make clear the important role which the Crown Prince had played in the foundation of the German Reich, which was a revelation to the German public, schooled as it was to believe that Wilhelm I and Bismarck alone had founded the Reich? The liberal ideas that her husband had confided to his diary were those which Bismarck and his people had fought against and tried to discredit for decades. 'You can imagine *how indignant* I feel at the tone in which the Government and Bismarck papers *dare* to speak of Fritz & of his Diary. It *is not* "überarbeitet" [reworked], there is *not* a word that is not his *very own* and in his own dear writing. *Of course* it ought *not* to have been published without my permission and *not now*. It was done in a good intention, & the public are delighted! The facts *long known* to me, and which now *leak out*, are of course *odious* to the Govt. and Bismark [sic] Party, – & the *opinions* wh. Fritz *so modestly* & simply puts forth are of course "gall and wormwood" to them, – as they are the very principles they have been treading & trampling down, & holding up to opprobrium for *20* years, *calumniating* and persecuting each individual who dared to uphold them! *Now* this Party B. try to cast doubt, contempt & ridicule on Fritz's word & on his character, wh. makes me feel quite *savage*! . . . I *want* the Tagebuch *back*! I am so afraid William and Bismark [sic] will order it to be burnt, & it is such a *valuable* and precious *record* of the *real* truth of things, that *if* they do *that*, I do *not* see *how* I can *ever* be on a footing of peace with them again! . . . I have *not* published this Diary, nor had anything to do with it! I fear it was Dr Geffcken who did it.' She predicted that he and all her friends would now be persecuted 'à la "Arnim"'.[101]

Even as Victoria wrote these lines the hunt was already on for Heinrich Geffcken, who (as we have already seen) was indeed responsible for the publication of the diary extracts. When the police discovered that he was staying on the island of Heligoland (at that time still British) Wilhelm II ordered police searches of all passenger steamers from Heligoland. At his command a gunboat was ordered to Heligoland in case Geffcken should try to reach the German coast by sailing boat.[102] The next day Geffcken was arrested in Hamburg. 'What an everlasting *disgrace*!', the Empress Frederick commented in her diary. 'Bism. is behaving *unspeakably*!'[103] She wrote to her mother: 'The way Bism. has behaved, and *how* the matter has been treated, is simply *disgraceful*, *much*, *much* worse than the indiscretion and the want of tact in publishing the diary! . . . These arbitrary acts of high handed despotism seem to go down with the people of Berlin in the most extraordinary way! The "Party" are of course exultant & triumphant. "Brutality" in every shape and form is what they admire, practise and preach!'[104]

THE RUPTURE OF RELATIONS BETWEEN MOTHER AND SON

The official discrediting of the dead liberal Kaiser which accompanied the Geffcken affair – the government went so far as to publish a report to the crown by Bismarck which contained insulting passages concerning Friedrich III – brought the widowed Empress's feelings towards Bismarck and her own son to new heights of bitterness in the autumn of 1888. In her letters to England she deplored the 'terrorisation' by the Bismarcks and Wilhelm of the now leaderless free-thinking forces in Germany, in language of mounting vehemence. '*All* must be done to raise William on a Pinnacle, because he is Bismark's [sic] *Pedestal*, wh. Fritz would never have stooped to be! So Fritz must be diminished in the eyes of the nation, and *I* must be calumniated, accused, vilified . . . Independent people are *silent*, cowed into holding their tongues! The whole machinery of the press is in Bismark's hands – in Berlin alone, the Government employés, "Beamten", are 33,000 people; all of these have *no* other opinion than what *he orders* them to have! *Caprice*, *tyranny* and despotism are *rampant*. It is very sad indeed . . . William allows his Father and me to be insulted & attacked, and *sanctions* it! . . . *Fritz* of Baden, Louise & the Empress Augusta are on Bism: side. Fritz of Baden especially has completely changed in politics & sails with William. It is his *interest* to do so . . . Seeing my sons side with our enemies makes me guess what Caesar felt when Brutus stabbed him.'[105] '*What am* I to think and feel', she asked despairingly on 12 October 1888, 'when I see my *own* son approve of and *encourage* the *insults* to his father's memory & his mother's reputation! He is either too lazy & careless, or he *does not* understand, or he intends to break the 5th Commandment [to honour one's parents], or he is so blunt of perception & so blind – in his prejudices – that he does *not* understand *how* disgraceful is the part he *has played*, is playing, or is *made* to play!! He has had a long and careful training & preparation in the Bism: atmosphere – so that his sense of right and wrong, of *gratitude*, *chivalry*, respect, affection for his Parents, – and *pity* for those who are so stricken has been thoroughly destroyed! It well nigh broke Fritz's heart, it may have been the cause of his illness – when he saw how his sons were having their minds wharped [sic] and their judgement & opinions perverted. They were young, easily caught – and their Grand Parents contributed *largely* to this result!'[106]

Victoria's position became so intolerable that she began to suspect that it was the Bismarcks' intention to force her to leave Germany. She would not do that, she assured her mother, although she was completely defenceless against attacks and calumnies. To appeal to Wilhelm in her hour of need would be useless. 'William does not read letters – if they are unpleasant to him he tosses them on one side! He does not *see* or *feel* what is an insult or injury to his Parents, – & does not think it worth his while to trouble about it; – to get on easily and undisturbed with the Chancellor, to do exactly what he pleases – with as little bother as possible is all he cares about . . . His Mama is a consideration he never dreams of remembering! As Pce B. & Herb: know this very well, – they become more and more daring, as they know & feel that against my darling Fritz, against *me* and *mine*, they may say, write, print, do – *what* they like with *complete* impunity! I have *no* one here to defend or advise me!!'[107]

As we have seen, in October she discovered that on Bismarck's orders the police had drawn up a list of all those who had been friendly with the late Kaiser and herself, and that he intended to have the houses of all these friends searched. As a few letters from Roggenbach containing critical remarks about the Bismarcks and Wilhelm II had been found among Geffcken's papers, Roggenbach's house in Schopfheim was searched, his writing table broken open and his papers ransacked. 'These things are allowed and sanctioned by William against his Father's most trusted and oldest friends!!!', the Empress commented indignantly to her mother.[108] Nevertheless Geffcken – meanwhile reduced to physical and mental breakdown – had to be released and the proceedings against him dropped. The cup was now full to overflowing. Shortly before she left Friedrichskron the Empress Frederick wrote her son a letter expressing her profound concern over his attitude towards the memory of his dead father. She could not have made her meaning plainer. 'If you had not enough feeling of filial affection and reverence towards your beloved Father to make you prevent the publication of that passage (in the "Immediat-Bericht" of Prince Bismark [sic] to yourself relative to your Father's Diary) – I am sorry for *you*. It was a passage as unnecessary as it was unjustifiable and a *direct* insult to your Father's memory – and a *distortion* of the bright historical Picture he left behind him. I repeat, that I regret it for yourself, as it has injured you in the eyes of all impartial and rightminded people, destroying the respect which should be felt towards a ruler. But *one* thing I would beg you to remember, – that even though you do not feel any obligation of respect towards *us your Parents*, it is still binding on you as Emperor & King to *uphold* & cultivate *such* in your People. *Truth* & Justice, and a reverence for that which should inspire loyalty – are the surest foundation for the state, which your own hands should be the last to destroy. It is *not* for our sakes, but for your *own*, that I bid you remember this; & remain with deepest concern Your sorrowing Mother Victoria.'[109]

Even before the Empress could send this letter to Wilhelm, violent arguments broke out between mother and son which led to the formal breaking-off of relations between them. On 5 November, on the strength of a piece of tittle-tattle, Wilhelm accused his mother of having arranged to meet Prince Alexander of Battenberg at the railway station in Frankfurt. At Bornstedt the next day her daughter Viktoria ('Moretta') approached her imperial brother with the wish to be allowed at long last to marry Battenberg. Her request was brusquely rejected by Wilhelm, who gave no reasons. 'W. behaved like a tyrant', the Empress recorded despondently in her diary. His 'unheard of tyranny', his 'brutal lack of consideration', his 'want of piety' had 'deeply wounded' her.[110] On 7 November she wrote him a second, bitterly angry letter, in which she broke off all relations with him. 'You have, with perfectly heartless disregard for your sister's feelings, in direct opposition to the will of your Parents, & for no urgent reason of State policy or expediency, *refused* to allow her marriage with Pce. Alexander (Prince of Battenberg), a Prince whose bravery compelled the admiration, & whose illtreatment enlisted the sympathies of the greater part of Europe, because you choose to say that such a marriage would be a mesalliance. You have *repeatedly* proved beyond doubt, that your Father's dying wishes have absolutely no weight with you. To *me*, they are *sacred*. You have shown that the happiness of your sister is a matter of perfect indifference to you. – It is the only thing I have left in the world to care for, & watch over, for my son who *should* be my natural support and protector seems to find his chief delight in doing all he can think of to insult & to wound me. – It is *best* therefore we should understand each other. I cannot prevent my son from insulting his Father's memory if it is his good pleasure to do so, but such conduct coming from a son must necessarily inspire me with the deepest indignation and horror. I therefore intend for the future to hold no communication with you, beyond what is absolutely necessary. – One word of warning. – You are now young, well and prosperous, and arrogant & overbearing in the pride of your newly acquired power, you have not experience enough to know that a selfish disregard of the wishes & feelings of others *always* brings sooner or later a heavy punishment on the selfish person himself, & want of filial piety is never left unavenged. Yr. sorrowing Mother V.' In a postscript the Empress added: 'It is needless to say that it gives me great pain to have to write all this to my own dear child. I *had hoped* that it could have been possible to live in *Harmony*, but so long as your ear is given to those who make it their business to sow discord between us, I fear there is *no* hope of it. Later in Life you will perhaps learn to understand me better, and realize that I *had your interest* at heart *after all*. – I expect no answer to either of my letters.'[111]

THE INTERVENTION OF QUEEN VICTORIA

The Empress Frederick saw clearly the dangers for Anglo-German relations to which her ever-worsening relations with her son were giving rise, but it did not occur to her to moderate the tone of her letters on that account. 'Of course – it must be our endeavour, that the relations of England and Germany should *not* suffer in spite of Pce Bismark's [sic] wickedness and William's folly', she wrote to Queen Victoria on 2 November 1888. 'You & dear Bertie & I – & your Ministers will do all that is possible – to keep everything on the best footing, but still I hope that this state of things is *not* ignored in England, & that the sorrows and sufferings of your daughter are known, as well as their sources and reasons.'[112] Her own despairing letters, together with the verbal accounts of the Prince and Princess of Wales of their experiences in Germany, ensured that the British royal family and above all Queen Victoria herself received a distinctly unfavourable impression of developments in Berlin, and especially of Wilhelm's behaviour. This effect was reinforced by numerous other letters from family members.[113] As early as 27 June 1888 the Queen had a long conversation with Lord Salisbury about 'this all engrossing misfortune of poor darling Fritz's death, which is such an untold tragedy, – of the symptoms in William's Opening Speech, of a leaning towards Russia, & there having been no mention of England, – of Pce Bismarck's violent language, when talking to Bertie, which showed how untrue & heartless he is, after all he seemed to promise me, & after poor Fritz had placed Vicky's hand in his, as if to recommend her to him! It is incredible & disgraceful.'[114] In the first days of July she wrote bitterly to one of her granddaughters: 'It is too dreadful to us all to think of Willy & Bismarck & Dona being the supreme head of all now! Two so unfit & one so wicked. He [Bismarck] spoke so shamefully about dear Aunt V[icky] to Uncle Bertie & A[unti]e Alix!'[115] And in a letter to Princess Maria Anna of Prussia on 30 July 1888 the Queen wrote: 'The loss of our dear Fritz is quite irreplaceable and a great misfortune for Germany and Europe, and it is fearful for my poor child, . . . whose whole existence is quite ruined! William seems to be enjoying himself a great deal – and it is only 6 weeks since the death of his beloved father!!'[116]

The Queen was particularly incensed, as was seen in the previous chapter, by Wilhelm's decision to visit the courts of St Petersburg, Stockholm, Copenhagen, Vienna and Rome immediately after the death of his father. The remonstrances she addressed to her grandson on 3 July merely gave offence and aroused further defiance, and the cold attempt at justification, cited earlier, which Wilhelm wrote to her on 6 July did nothing to lessen the Queen's annoyance.[117] In her letters to relatives near and far she roundly and repeatedly condemned the impious, pleasure-seeking conduct of her grandson. On 17 July, while Wilhelm was steaming towards St Petersburg, she complained in a letter to her eldest son: 'William has started his Journey far too soon – & it is certainly rather significant that he shd. pay his 1st visit to any Sovereign to the Russian Tsar – being so far more nearly related to us.'[118] A week later she wrote again: 'How sickening it is to see Willy not 2 months after his beloved & noble Father's death going to Banquets & Reviews. It is very indecent & very unfeeling! Why does he have to go to Copenhagen? It is so unnecessary & I am sure (& hope) your Parents in Law will not receive him with open arms tho' they must be civil.'[119] On 25 July she assured her eldest daughter that she entered into her feelings: 'If I do not express my more than regret & surprise at Willy's Visit to St. Petersburg & at the Reviews & Parades & at the visits to Stockholm & Copenhagen – all so foolish & unnecessary, not to speak of the gt. impropriety of doing so so soon, it is because I do not wish to pain you still more by alluding to it – or to add fuel to the irrita[tio]n & deep pain it must cause you. All I can say is – that he is v[er]y young & foolish, – has certainly not the right sort of feeling & is not well surrounded. It is much remarked on here.'[120]

Urged by the Prince of Wales to intervene in the argument over a residence at Potsdam for her daughter, on 3 July Queen Victoria wrote to Wilhelm: 'I am naturally very much occupied with poor dear Mama's future home. She feels probably a certain awkwardness amounting to pain to ask for anything, where so lately all was her own, but as Uncle Bertie told me you were only too anxious to do what she wished in that respect. Would you not (if you are not going to live there yourself) offer her to stay – at any rate for the present let her have Friedrichskron – or else Sans Souci? Uncle Bertie told me you had mentioned the Villa Liegnitz but that is far too small & would not do I think for *your* mother who is the first after you – & who is the 1st Princess after Aunt Alix in Great Britain. An Empress could not well live in a little Villa where Charlotte & afterwards Henry lived. Mama does *not* know I am writing to you on this subject nor has she ever mentioned it to me; but after talking it over with Uncle Bertie he advised me to write direct to you.'[121]

3. The Empress Frederick and Queen Victoria in mourning for Kaiser Friedrich III, 1889

Given all that we know of the personality of Wilhelm II, and notably of his fear of being treated like a child by his mother's family, it is no surprise that his reaction to his grandmother's letter – in which she also reproved him for not replying to her last letter – was one of offended vanity. In the reply he wrote in his own hand on 6 July 1888 he was certainly at pains to stress his co-operative attitude towards his mother, but he remained unbending on the matter in hand. 'In regard to what you say about Mama I am doing my uttermost to fullfill her desires. Today I had a long talk with her relating to her wishes concerning her future home – somewhere on the Rhine – & when I found out that the capital she wanted was not free because it is meant as a legacy for us children after her death, I at once renounced for my part once for all, & promised to bring my sisters to do the same! With respect to her living at Friedrichskron or Villa Liegnitz Uncle Bertie seems not to have been informed in detail. I have proposed to Mama to stay at "Friedrichskron" for this year, & to help her find a house of her own. On the other hand, *Mama herself* named the Villa, as one of the houses to be left to her by Papa when her "Witthum" [the arrangements for her widowhood] was beeing [sic] drawn up, besides the Schloss at Homburg, at Wiesbaden & the Palace at Berlin were also named by her as houses she most preferred to live in; so they were entered in the will accordingly. Sans-Souci is the only castle left to me in case of strangers coming to us, as all the other palaces, – except the small Stadtschloss at Potsdam, town –, are in the hands of Grandmama. She told me – Mama – that she only wanted a "pied à terre" here and that was why she wished to have the Villa which she adores.'[122] But the conflict was only seemingly resolved by this exchange of letters. Wilhelm was infuriated by what he regarded as the unacceptable interference of his grandmother and uncle, a state of affairs that Herbert Bismarck skilfully exploited,[123] and on both sides of the North Sea anger and mistrust took on increasingly serious forms.

Another and more far-reaching effect of the conflict between mother and son became apparent at the very outset of the new reign. Just as Waldersee was sent to Vienna and Schlieffen to Bucharest and Sofia to convey the formal announcement of the accession of Wilhelm II, the Kaiser sent Adjutant-General Hugo von Winterfeldt together with his aide-de-camp Albano von Jacobi to London as his emissaries. Unlike Waldersee, who had been received in the Hofburg with particular warmth,[124] Winterfeldt, who was regarded by the Queen, on the strength of her daughter's letters,[125] as Bismarck's spy and an odious enemy of her deceased son-in-law, was deliberately badly treated at Windsor.[126] After his return to Berlin Winterfeldt reported to Herbert Bismarck on his frosty reception: 'Her Majesty addressed a few words of condolence to me on the decease of His Late Majesty Kaiser Friedrich, greeted Lieutenant von Jacobi whom Her Majesty knew from former days, announced Her intention of having a letter for His Majesty the Kaiser delivered to me and took leave of my companion and myself . . . Later I learned from a letter from Sir Henry Ponsonby that Her Majesty had changed Her mind and would not be giving me a letter. Consequently my companion and I left London the following morning.'[127] When Herbert laid this report before him Wilhelm wrote a furious marginal note on it, which Bismarck had to have placed under restriction: 'The reception seems to confirm the bad manners and discourtesy shown several times to German officers at the English Court. Our treatment of the British in the future ought to be modified accordingly.'[128] On 4 July the young Kaiser spoke to the British military attaché Colonel Leopold Swaine, with whom he had been on friendly terms for several years, and who wrote the same day in great alarm to the Queen's Private Secretary: 'The Emperor is much hurt. I gather from my interview with him after the parade today that he feels he is treated as a grandson and not as German Emperor. I don't think he will resent it this time, but I am very anxious even on that head, for there are many advisers here who, feeling as he does, are ready to recommend it. No man is striving harder than Malet [the British ambassador] to bring about and foster a good understanding between the two countries, and it is literally cutting away the ground from under his feet if all he does is undermined by our court. I know you are doing all you know to throw oil on the troubled waters, and you will see . . . how necessary it is to continue to do so at every opportunity. I am quite upset by this unfortunate turn matters have taken and am longing to get away from here.'[129]

Queen Victoria, however, was not at all inclined to regret her treatment of Winterfeldt. When Ponsonby sent her Swaine's letter her response was emphatic: 'The Queen is *extremely* glad to hear that Gl. Winterfeldt says he was recd. coldly tho' civilly for such was her *intention*. He was a traitor to his beloved Master & never mentioned *his* name even, or a word of regret, & spoke of the *pleasure* wh. he experienced at being chosen to announce *his* new Master's accession!! *Could* the Queen, devoted as she is to the dear memory of the beloved & noble Emperor Frederic to whom & to her daughter Gl. Winterfeldt has behaved so treacherously receive him otherwise – & she *hopes* & *asks* Sir Henry to *tell Col. Swaine* what she has just *said*. – Ld. Salisbury *does not* think Col. S. at all a wise man. – He certainly does not see *thru' people*.'[130]

Despairingly the British ambassador in Berlin, Sir Edward Malet, drew attention to the dangers posed by this feud in the Anglo-German royal family for the relations between the two empires, particularly since from the accession of Wilhelm II the monarchical element in the German Reich had gained so much in importance vis-à-vis the power of the state as embodied by Bismarck. As things now stood the most personal feelings of the Kaiser would inevitably play a part in shaping German policy towards Britain. As Swaine had done, so too Malet warned that Britain must therefore avoid everything which might antagonize the immature and oversensitive young Kaiser. 'Nothing is more calculated to change his feelings with regard to England than the suspicion that all his efforts to show friendliness towards us in the reception which he has given to English officers, & other minor matters . . . have been unavailing to correct the false impression which has got abroad regarding his [anti-English] views. He is a high mettled Prince, & he has given his Royal word that we are mistaken. The fear is that a continued show of want of confidence in him at this especial period of his life may produce a revulsion of the feelings which have hitherto weighed the balance, that he may renounce the effort to prove himself friendly to England . . . & cease his friendliness at the same time.'[131] In the fearful storm that was to break out between Germany and Britain in the autumn of 1888, however, such words of statesmanlike wisdom were barely audible.

CHAPTER FOUR

An ominous family quarrel

THE KAISER'S SPEECH AT FRANKFURT AN DER ODER

On 16 August 1888, the anniversary of the battle of Thionville and Mars la Tour, Kaiser Wilhelm II made a speech in Frankfurt an der Oder in honour of the victorious commander, Prince Friedrich Karl of Prussia. The speech was seen, and was indeed intended, as a deliberate insult to his uncle Albert Edward (Bertie), the Prince of Wales. 'There are people', declared the Kaiser, 'who have the impudence to maintain that my father wanted to give up what he, together with the late Prince [Friedrich Karl], won on the field of battle. But we knew him too well to be able to tolerate such an insult to his memory, even for a single instant. He held the view, as we do, that nothing of what was gained in those great days can be surrendered. I believe that every one of us . . . in the army knows that there can be only one opinion on this matter, namely, that we would rather see the whole of our eighteen army corps and our forty-two million inhabitants perish on the battlefield than surrender a single stone of what my father and Prince Friedrich Karl gained.'[1] After the speech the Kaiser turned to Field Marshall von Blumenthal and said loudly: 'I hope my uncle the Prince of Wales will understand that!!'[2]

This furious outburst reflected the general feeling of annoyance and offended pride which the attempts of his British relations to interfere in his life aroused in the Kaiser. But in particular they were Wilhelm's response to remarks which the Prince of Wales was said to have made eight weeks earlier, when he and his wife had come to Berlin to attend the funeral of his brother-in-law Friedrich and to support his sister. Inevitably, given the deep mistrust on both sides, it is well-nigh impossible to establish clearly what the heir to the British throne actually did say. According to the Empress Frederick, her brother had criticised the German administration in Alsace-Lorraine in a conversation with Herbert Bismarck and had also expressed the hope that the Kingdom of Hanover, annexed by Prussia in 1866, would one day be returned to Ernst August, Duke of Cumberland. Count Bismarck passed this on to Wilhelm, which the Empress considered 'very nasty' of him, whereupon Wilhelm publicly attacked his uncle in his speech at Frankfurt.[3] In the pro-government press she herself was described as being responsible for the Prince of Wales's remarks, which she vehemently denied. As she complained to her mother, the Bismarcks wanted to create the impression that '*I* instigated Bertie & Alix, wh. is most absurd, as I really hardly know what they *did* say!' It was 'rather unfortunate that anything *was* said, as the Bismarks [sic] use it as a weapon against *me*, not only have they represented it so to William and caused him to make that foolish speech at Frankfort a.d.O., but they also spread it through the Norddeutsche and Cölnische Zeitung to injure me, and it is then largely believed. I am utterly innocent of all this . . . It is rather silly, to talk of my intriguing for "Danish aspirations", as Fritz & I always did what we could for Schleswig Holstein aspirations and *not* Danish ones . . . One is really ashamed of such rubbish, but it *all profits* the Bismarks & William in the eyes of a widespread class in Germany. Their superior "Patriotism" is aired again on this occasion, and *distrust* sown against me, and doubt cast on Fritz's intentions. It is an abominable game, & apparently always succeeds with a certain set!'[4] The Prince of Wales gave a rather different account of his conversation in Berlin. He had in fact believed, he said, that Friedrich III had considered handing back Alsace-Lorraine in order to put an end to the hostility between France and Germany. In his conversation with Herbert Bismarck in June 1888, however, he had only asked whether this was true. When Herbert replied that it was nothing but a rumour without substance, he had let the matter drop. As to the return of northern Schleswig to Denmark and the restitution to the Duke of Cumberland of the King of Hanover's private property, confiscated in 1866, the Prince of Wales maintained in a letter of April 1889 to his brother-in-law Prince Christian of Schleswig-Holstein that he had mentioned these matters only in passing. He presumed that Herbert Bismarck, in repeating this conversation to his father and to Wilhelm II, had turned this *question* into a concrete *suggestion*, which the Chancellor had then shamelessly used against his sister, the widowed Empress. Herbert had further misused this incident to prejudice Wilhelm against Britain and in favour of Russia. In view of this distortion of events by the Bismarcks it was not surprising that the young Kaiser had been so enraged.[5]

For his part, the Reich Chancellor was anxious to put the best complexion on the Frankfurt speech with regard to Russia. He instructed the Wilhelmstrasse to make clear, in confidential conversations with the Russian Foreign Minister and the Russian grand dukes, that in his speech the Kaiser 'had intended to repudiate the continual insinuations of his uncle, the Prince of Wales, and that one would be on quite the wrong track to see a threat against France in it. One would find His Majesty's remarks quite understandable if one knew that the Prince and, notably, the Princess of Wales, had made all kinds of demands for concessions to the Guelphs, and that in the days of Kaiser Friedrich, partly with and partly without Battenberg interference, the hope had been aroused and nourished that Germany might somehow be inclined to hand back the hard-won national bulwark of Alsace to the French as an offensive base for them. His Majesty's irritation at these demands and insinuations was certainly understandable and had given rise to the vehemence of his language, but if this outburst of displeasure had had any particular target, it was an English one, and more especially the Anglo-French one to which the Prince of Wales had belonged.'[6]

In Herbert Bismarck's retrospective account, written in 1891, the conversation with the Prince and Princess of Wales is shown in yet another light, and the responsibility for Wilhelm's overreaction is ascribed to tale-bearing by Princess Charlotte and intrigues on the part of the Russian Grand Duke Vladimir. During the funeral of his father, the Kaiser had been treated by his uncle 'not enough as . . . Kaiser, but too much as . . . nephew, and still as a green "bad boy"', according to this account. The Prince of Wales's claim that 'the Emperor Frederick would have handed back the Reich's territory etc.' was reported to the Kaiser 'with malicious pleasure' by his sister Charlotte of Saxe-Meiningen and 'with skilful political calculation' by Grand Duke Vladimir, writes Herbert, with the result that Wilhelm's 'anti-English urge' was again brought strongly to the fore. The direct incentive for Wilhelm's speech against the Prince of Wales in Frankfurt probably came from 'reminiscences of Peterhof and renewed provocation from the Meiningen Princess', the Chancellor's son concluded. In any case he himself had certainly not played the role later ascribed to him as the 'scapegoat' in this affair. The Prince of Wales's remarks in June had not occurred in conversation with him, Herbert, but when the Prince and Princess had met the Reich Chancellor. 'At that time they both received the Chancellor and embarrassed him with various indiscreet questions *& demands* concerning Cumberland, Brunswick, Guelphic funds, so that H[is] H[ighness] was much put out when he returned from the audience. The next morning Malet sent me a memorandum of several pages drawn up by the Pr[ince] of W[ales] about the previous day's audience, with the request that I lay it before H.H., so that he could ratify and if necessary correct it. I sent for Malet at once, and asked him to take the document back *brevi manu* [there and then] and in confidence', the Foreign Secretary recalled. 'It could not be regarded as an official document, as Malet had received no instructions from Salisbury, and I should be rightly criticised by the Chancellor and would have been put in the most awkward position vis-à-vis H.M. as well as H.H. if by officially accepting it from the ambassador I had obliged my Government to explain itself and bind itself in writing quite unilaterally and on important internal matters to the Pr. of W., who had no official standing. Malet became red-faced and embarrassed, agreed that I was quite right, excused himself by saying that the Prince was staying with him and would not leave him in peace, so he had given in, although he had foreseen my reaction. He concluded: "I wish the Royalties had not come at all, they have now been staying for a week with me but I feel sure that their presence has only done harm." Later [Lord] Rosebery told me that the Prince of W. had been so "hurt", because that memorandum was the first "State Paper" he had written, and he had been very proud of it. But Rosebery also agreed that I had been perfectly right.'[7]

It cannot be denied that Herbert Bismarck's detailed description of the incident has a more plausible ring to it than the vague version given by the Prince of Wales. Herbert's account is supported by the information which he gave Tsar Alexander in St Petersburg on 22 July 1888 (more than three weeks before the Frankfurt speech), and which he set down in writing a few days later. At his audience Herbert told the Tsar 'that the Prince of Wales had importuned our Kaiser with tiresome suggestions which were refused'. This, as Herbert had learned from the British ambassador, had annoyed the Prince. The younger Bismarck went on to tell the Tsar 'that the Prince of Wales had expressed all kinds of wishes with regard to the Cumberlands, both through his wife and direct. For reasons of state he had to be answered with a *fin de non recevoir* [rejection], which had evidently awakened in the Prince the painful feeling that he had gone too far.' Then it had been discovered in Berlin that the Prince of Wales had tried to 'blacken' Prince Wilhelm in the eyes of Alexander III in the autumn of 1887, which had naturally strained relations between Kaiser Wilhelm II and his English uncle. Finally the Kaiser had been greatly angered by a letter from his grandmother in which she had strongly advised him not to go to Russia – advice which he considered gratuitous and patronising.[8] The records of the German Foreign Office confirm Herbert Bismarck's claim that Grand Duke Vladimir, the Tsar's brother, who was married to Duchess Marie of Mecklenburg-Schwerin, had contributed to Wilhelm's ill-feeling towards his English uncle. In a conversation with a member of the German embassy in St Petersburg in September 1888 Vladimir stated that in Berlin the Prince of Wales had spoken to him 'in all seriousness of the possibility of a surrender of Alsace-Lorraine by Germany'.[9]

But regardless of whether the Prince of Wales made suggestions or merely asked questions in a meeting with Bismarck, whether Kaiser Wilhelm considered his grandmother's letter as patronising, or whether Grand Duke Vladimir and Charlotte Meiningen increased Wilhelm's ill-feeling by their tale-bearing, there was no doubt that public abuse of the heir to the British throne in a speech by the Kaiser eight weeks later amounted to an insult which was bound to have lasting repercussions. It would be tempting to describe the Frankfurt speech by the young Kaiser as an absolute low point in his relationship with his British family, if the next confrontation, which followed hard on its heels, had not cast an even more singular light on Wilhelm's feelings for them. This arose from his decision, to which historians have paid very little attention hitherto and which a psychologist would be hard put to explain, to ban his Uncle Bertie from Vienna throughout the several days of his own first official visit to the Austrian capital in October 1888.

THE VIENNA INCIDENT

At the beginning of September 1888 a group of British officers, including Count 'Eddie' Gleichen, son of Prince Victor of Hohenlohe-Langenburg and Laura Seymour, visited Potsdam and Berlin to attend army manoeuvres. Wihelm received the officers with great friendliness. He laid stress on the importance of such meetings between officers in helping to maintain good relations between the two countries. He treated Count Gleichen as a prince, spoke German to him and used the familiar 'du' form of address, thus emphasising the fact that they were related.[10] There were other signs, too, of Wilhelm's wish to restore good relations with Great Britain. For a brief period he visited his mother several times to discuss the wish of the young Crown Prince of Greece, Constantine, to marry Princess Sophie of Prussia, Wilhelm's sister; Constantine's father was the brother of the Princess of Wales, the Tsarina and the Duchess of Cumberland.[11] Also around this time Wilhelm expressed the wish to have a railway carriage like that of the Prince of Wales, and sent an official from the Marshal of the Court's office to Gmunden to inspect the Prince's carriage.[12] For one deceptive moment the mutual mistrust and annoyance of the summer seemed to have fled and the relations between mother, uncle and son appeared at last to be improving.

On 2 September the Prince of Wales, who was staying at Bad Homburg on his way to Austria, wrote the Kaiser a warm letter thanking him for his friendly reception of the British officers and ending with the hope that he would be seeing Wilhelm in Vienna shortly. He never received a reply. When he arrived in Vienna on 10 September and discussed the programme for his stay in Austria-Hungary with Kaiser Franz Joseph and Crown Prince Rudolf, the Prince of Wales learnt that Wilhelm would be arriving in Vienna on 3 October. Although he was scheduled to be in Hungary at that time, the Prince felt that he should return to Vienna, as his absence from the Austrian capital during his nephew's visit might be misinterpreted. He even decided to send a messenger to Berlin to

4. The Prince of Wales in Marienbad around 1890

fetch the uniform of his Prussian Hussar regiment to wear at his meeting with Wilhelm. On 11 September, however, he discovered that the German ambassador in Vienna, Prince Heinrich VII Reuss, had informed the Austro-Hungarian Foreign Minister Count Kálnoky, the Obersthofmeister (principal Grand Master of the Court) Prince Constantine zu Hohenlohe-Schillingsfürst and the British ambassador in Vienna, Sir Augustus Paget, 'that it would not be agreeable to the Emperor William to meet his uncle at Vienna – in fact that it might mar the success of his visit were they to be there simultaneously'. This official communication through the German ambassador, of which the Prince of Wales learnt with the 'utmost surprise' and with 'great pain', was soon common knowledge throughout Vienna and led to serious consequences. The Prince's equerry, General Arthur Ellis, commented: 'I have never seen the P. of W. so upset about anything and he is racking his brains in vain to discover the cause.' The Prince found it particularly regrettable that the Kaiser should not have written to him directly, if he had some personal grudge against him, his uncle, instead of choosing the official route through the ambassador. In order to avoid placing Emperor Franz Joseph in an even more difficult position, Ellis wrote, the Prince of Wales would go to Romania during Wilhelm II's stay in Vienna, but Colonel Swaine should find out as soon as possible what was the reason behind this quite extraordinary insult. The Prince could only imagine 'that this unfortunate mal attendu [sic] must be based on some unaccountable misunderstanding to clear up which no time must be lost'.[13]

Although he was otherwise on such friendly terms with the Kaiser, the British military attaché was unable to obtain any clear explanation of Wilhelm's motives. His request for a personal audience was refused by Wilhelm with the comment that Swaine could speak to him before the military review. But when the colonel tried to approach him there he found the Kaiser surrounded by generals, so that it was impossible for him to bring up the delicate subject of what had happened in Vienna. He decided instead to write to the Kaiser about it, enclosing General Ellis's letter. On 19 September the Kaiser had several opportunities of exchanging a word with Swaine, but he acted as if he had not seen him and, although Swaine was standing very close to the Kaiser, he was not even given the chance of greeting him. On 20 September the Kaiser left Berlin for Munich, Vienna and Rome without having addressed a single word to Swaine. The colonel, who had placed so much confidence in Wilhelm, felt that to remain in Berlin in such circumstances would be intolerable. 'With all the kindness and almost attachment I feel for the young Emperor I cannot allow or countenance any such treatment of Your Royal Highness', he told the Prince of Wales, 'and my position here would certainly lead to a complete estrangement between The Emperor and myself, which would not be to the advantage of the service and would destroy the object I have had in view and to which I have devoted these last six years.'[14] At the end of September Colonel Swaine handed in his resignation.[15]

Swaine's letter, together with General Ellis's, were sent to Queen Victoria and copied to the Prime Minister Lord Salisbury.[16] The Queen wrote indignantly to her son Arthur, Duke of Connaught, who was married to Princess Louise Margaret of Prussia, about 'the most *outrageous* behaviour of Willie the Gt. (& I fear "the bad hearted") towards Bertie'. Wilhelm, she said, had actually informed his uncle officially through Reuss, Kálnoky and Paget that it would not be agreeable to him to meet him in Vienna, 'Bertie having graciously written

3 weeks before, how much pleasure it wd. give him to meet him there!!! Bertie was naturally shaken & pained beyond measure. However to give Wm. every opportunity of getting out of it, Bertie had Col: Swaine written to to ask to see Wm. & taken him Bertie's or rather Genl. Ellis's letter (dictated by B[ertie], & an excellent letter) to Wm. asking what this meant & what was the cause of this. Wm. while saying he *wd.* see him on Parade avoided Col: Swaine in every way & then went away! Bertie is, as you will easily imagine, *furious*, as well as *deeply pained* & says if he does not receive an apology he "will never speak to him again". Col: Swaine is so shocked that he says if W. does not answer B. he will resign & leave Berlin where he cannot remain after such an insult. To treat the Pce of W. – the oldest son of one of the gtst Sovereigns in the World [sic], & his own kind Uncle in such a manner is one of the greatest insults ever comitted! B. has had to change his arrangements & doubtless this shameful affair will soon come out. People will be furious here & everywhere & I am sure Louischen will be horrified.'[17]

The disastrous affair continued to spread. Soon the story reached the gutter press.[18] Crown Prince Rudolf, who had invited the Prince of Wales to hunt with him in Siebenbürgen and wrote to his wife Stephanie that he would 'only invite Wilhelm along . . . so as to hasten him out of this world through an elegant hunting adventure', stoked up the fire in Vienna by making indiscreet remarks.[19] In this dangerous situation the British ambassador in Vienna, Paget, took the unusual step of warning the Prince of Wales not to say anything about Wilhelm II, for in Vienna, as everywhere, there were malicious gossips at both high and low levels who would only too gladly misrepresent and pass on *every* remark of the Prince about his nephew. He knew for a certainty that this affair itself arose from just such tittle-tattle – people had passed on to Kaiser Wilhelm remarks that the Prince of Wales was said to have made.[20]

The question arose of how the ambassador was to behave on meeting the Kaiser. Lord Salisbury telegraphed from Nice, where he was holidaying, instructing Paget to absent himself, if he could, from any ceremony at which Wilhelm II would be present, and to avoid him as far as possible without causing a scandal.[21] That, however, was a contradiction in terms, as Paget pointed out in his reply, for if he wished to avoid a scandal he would have to behave like every other ambassador; any departure from the norm would be noticed and unfavourably commented upon.[22] The Prince of Wales and Ellis asked the ambassador, in his audience with the Kaiser, to signal to Wilhelm that his uncle was eager for a reconciliation, and to ask him directly whether he had any message to convey to the Prince. Unlike Paget, who feared that this would merely invite a rejection from the Kaiser and thus aggravate the situation still further, they were confident that it would bring about a clarification of the whole mysterious affair.[23]

But in a second telegram from Salisbury Paget received instructions not to raise the matter on any account unless the Kaiser himself spoke of it. Although in the event the audience lasted for more than ten minutes, the ambassador had no opportunity to carry out the Prince of Wales's request, as the Kaiser spoke only of indifferent matters.[24] Wilhelm did not even enquire after the Queen's health! Later Queen Victoria and the Prince of Wales heard through Crown Prince Rudolf that Wilhelm had even intended, originally, to turn his back on the ambassador, and had only been dissuaded from doing so by Franz Joseph and Rudolf. When Rudolf – at the request of the Prince of Wales – asked Wilhelm whether he had any message for his uncle, whom Rudolf would be seeing shortly in Siebenbürgen, Wilhelm answered in the negative and merely said that if his uncle wrote him a very kind letter he, Wilhelm, '*might perhaps answer it!!*'[25]

The crisis became more acute still when on 26 September 1888 the German Foreign Secretary informed Sir Edward Malet in Berlin that the following year – possibly in May – Kaiser Wilhelm II wished to pay a visit to Britain![26] Not only the British royal family but the British government too found it unthinkable that such an invitation should be issued until the Kaiser had made a full apology for his behaviour to the Prince of Wales. At this point at the latest the feelings of both royal families stood in direct opposition to the political interests of the two nations.

It soon transpired that Count Paul von Hatzfeldt, the German ambassador in London, was not only in the dark about the Vienna incident but was positively terrified of bringing up the subject with the Bismarcks. On 13 October Hatzfeldt read Salisbury a lengthy memorandum from the Reich Chancellor in which Bismarck attempted to justify the enforced absence of the Prince of Wales during the Kaiser's visit to Vienna on foreign policy grounds: the Tsar, he claimed, would have been annoyed by a meeting between the heir to the British throne and the German and Austrian Emperors, and this would have been dangerous for Germany, as Britain was not yet prepared, by way of compensation for the mistrust the meeting would arouse in Russia, to conclude a genuine alliance with Germany. Bismarck's memorandum did then touch on the personal aspects of the Vienna incident, if only tentatively and indirectly. Kaiser Wilhelm II, Salisbury gathered from the Chancellor's statement, had three things with which to reproach the Prince of Wales. First, Albert Edward had told a Russian grand duke, who had repeated it to the Kaiser, that if Emperor Friedrich III had lived longer he would have offered concessions on the Alsace and North Schleswig questions and recognised the claims of the Duke of Cumberland; second, the Prince of Wales and his wife, in a conversation with Bismarck, had put forward Cumberland's interests and thereby abused the respect the Chancellor had to show to the Princess, and had subsequently tried to force a written record of this conversation on Prince Bismarck; and, third, the Prince of Wales had treated Wilhelm as his nephew and not as an emperor, even though the latter, although still young, had been of age for some time. Bismarck's memorandum, according to Salisbury, had said nothing about 'the request that the Prince would leave Vienna on personal grounds; or of the Emperor William's announcement that he would not go to Vienna while the Prince of Wales was there; or of the failure to notice Col: Swaine's letter; or of the omission to ask after the Queen in speaking to Sir A. Paget'. This was clearly the first time Hatzfeldt had heard about these matters. Salisbury warned him strongly against encouraging a visit to Britain by the Kaiser in these circumstances, for any such suggestion would be quite unacceptable in the present situation. The Prime Minister nevertheless insisted that the personal conflict between Kaiser Wilhelm and the Prince of Wales must not be allowed to affect the general policy of the two nations towards each other. This view, Hatzfeldt assured Salisbury, was entirely shared by Prince Bismarck.[27]

On the following evening Salisbury and Hatzfeldt met again. The Prime Minister was disturbed to discover, as he reported to the Queen, that the ambassador had not communicated with Berlin at all on the matter of the Vienna incidents, out of fear of Bismarck, and in particular he had failed to warn his government 'that so long as those incidents were unexplained it was impossible for Your Majesty to receive the Emperor. He was simply afraid to do so.' Salisbury again urged the ambassador to send some kind of warning to Berlin, since if the Kaiser proposed himself and the Queen were obliged to turn him away, the gulf between the two nations would only become deeper. Salisbury spoke of the 'terror of the Ambassador' and concluded from his behaviour that Hatzfeldt's post was at risk. The situation made him very uneasy, for 'if nobody dares tell Prince Bismarck the truth, there is no knowing what he might do'.[28]

These conversations, together with further information about Wilhelm's behaviour which she received through the Princess of Wales, convinced Queen Victoria that her grandson was no longer quite responsible for his actions. Anxious as she was to maintain good relations between Britain and Germany, she nonetheless doubted that this would be possible as long as Wilhelm was on the throne and the Reich was in the hands of the two Bismarcks. She found the reasons given by the Chancellor for the Kaiser's impossible behaviour 'simply absurd'. How could Bismarck claim that the Tsar would have been annoyed by a meeting between Wilhelm and his uncle, when the Prince of Wales was the Tsar's brother-in-law, she asked in astonishment. It was not yet clear what Albert Edward had said about Alsace-Lorraine. He was certainly on very intimate terms with all the Russian grand dukes, so that it was not unthinkable that he had made a remark about Friedrich III's willingness to make concessions. What he had said in his conversations with both Bismarcks about Brunswick and Cumberland was well known to the Prime Minister, the Queen declared. She continued indignantly: 'As regarding the Prince's not treating his nephew as Emperor this is really too *vulgar* and too absurd as well as untrue almost *to be believed.* We have always been very intimate with our grandson and nephew, and to pretend that he is to be treated *in private* as well as in public as "His Imperial Majesty" is *perfect madness*! He has been treated just as we should have treated his beloved Father & even Grandfather, and as the Queen *herself* was always treated by her dear uncle King Leopold. *If* he has *such* notions he better *never* come *here*. The Queen will not swallow this affront.' As for what Crown Prince Rudolf had told the Prince of Wales, namely that Wilhelm had intended to turn his back on the British ambassador and had been deterred from doing so only at the urgent instances of Rudolf and Franz Joseph, and that Wilhelm had told Rudolf that he would 'perhaps' be willing to answer a letter from his uncle, but only if it were a particularly friendly letter, Queen Victoria's verdict was that 'all this shows a very unhealthy and unnatural state of mind & he *must* be made to feel that his Grandmother and Uncle will not stand such insolence. The Prince of Wales must *not* submit to such treatment. As regards the political relations of the 2 Governments, the Queen quite agrees that that should not be affected (if possible) by these miserable personal quarrels, but the Queen much *fears* that, with such a hot headed, conceited and wrongheaded young man, devoid of all feeling, this may at *any* moment become *impossible*.'[29]

In spite of all efforts to keep the political repercussions of the affair within bounds, the crisis continued to deepen. Since Bismarck had cited as one of the reasons for Wilhelm's behaviour the lack of proper respect with which Queen Victoria had corresponded with her grandson, copies of her letter to Wilhelm of 3 July and of his answer of 6 July were made and sent to Vienna. All who read the correspondence – Sir Augustus Paget, Count Kálnoky, Kaiser Franz Joseph – were struck by the inexplicable discrepancy between the irreproachable wording of both letters and the way in which they were represented subsequently.[30] On 14 November Queen Victoria wrote to Kaiserin Augusta to complain 'about our grandson Wilhelm', who 'has behaved in such an unkind, inconsiderate and peculiar way towards his Uncle Bertie, who has always been so very kind, and in general scarcely treats his British relations as relations at all'. It would be a very good thing, she advised, if Augusta would draw the young monarch's attention to the consequences of such conduct, which could lead to a cooling of relations between the two countries. Wilhelm evidently did not know how to behave, did not keep good company and did not listen to good advice.[31] The ninety-year-old Kaiserin retorted that she could intervene only if Wilhelm asked for her advice. But he was 'altogether less amenable to advice' than the Queen supposed. Augusta expressed the opinion that Wilhelm's ill feeling towards Britain went back for more than a year, for he had felt 'hurt' when he returned to Germany after the Queen's Jubilee. Moreover the Battenberg marriage project had played a very damaging role.[32]

The Prince of Wales, like his mother, could see no other explanation for Wilhelm's 'outrageous' behaviour than that it was a sign of insanity.[33] He even commented that this explanation was more advantageous for Wilhelm than the only other possible one – that he knew what he was doing! After his return from Austria he wrote to his sister: 'Nothing that your son William does surprises me – my only doubts are whether he is sane! Perhaps it is more charitable to think that he is insane.' Be that as it may, 'till he makes some apology for the gross insults he has heaped upon me in a foreign country I must naturally cease having any further acquaintance with him'.[34]

Although a political crisis was avoided for the time being, Wilhelm II's behaviour had undoubtedly done great long-term harm. On 17 October 1888, four months after Wilhelm's accession, the Princess of Wales wrote to her son, the future King George V, of the atonement that would one day have to be sought for this insult. Wilhelm was an 'ass' whose behaviour, particularly to his mother, was getting worse every day. Now he had also been 'personally most *frightfully rude* & *impertinent* towards Papa' and had 'actually refused to meet him at Vienna!! He is perfectly infuriated against *England* that beast . . . Oh he is mad & a conceited ass – who also says that Papa & Grandmama don't treat him with proper respect as the *Emperor* of *all & mighty Germany*! But my hope is that *pride* will have a *fall* some day!! – Won't we rejoice then.'[35]

THE ROLE PLAYED BY THE BISMARCKS AND KAISER FRANZ JOSEPH

The only other explanation which suggested itself in London for Wilhelm II's extraordinary behaviour was that the young Kaiser had been manipulated by the Bismarcks. For the Prince of Wales there was no question that Wilhelm had been 'aided & abetted by that scoundrel Herbert Bismarck, whose ingratitude for all the kindness I have shown him for so many years knows no bounds'.[36] Queen Victoria's third son, the Duke of Connaught, also thought that Wilhelm's 'most extraordinary & unbecoming' behaviour must be the result of an anti-British and anti-constitutional intrigue by the Bismarcks. 'My own impression is that *someone* has intentionally made mischief between William & Bertie with a view to prevent their meeting. I should not be surprised if the Bismarks [sic] had something to say to it as they appear to be so afraid of *any* English influence being brought to bear on William. They are clearly afraid of any constitutional ideas being put into Willie's head & they think that any of his family might interfere with the dictatorial gov't they wish him to exercise & that Bertie in particular may wish to give William good advice.'[37]

What was the role of the two Bismarcks in this puzzling incident? The account written by Herbert Bismarck after his dismissal is once again of particular interest. As we have seen, he denied all responsibility and claimed that the cause of the incident was to be found at the Austrian court. He thought it quite possible that Wilhelm II had not even read the closing passage of the Prince of Wales's letter, in which the latter expressed his pleasure at their forthcoming meeting in Vienna. Furthermore, 'with the growing number of hurried journeys and because of the manoeuvres . . . many things [had been] neglected, many official papers not even opened for weeks. At any rate the Pr. of Wales received no answer, and this, combined with the gossip about the genesis of H.M.'s Frankfurt speech, certainly put him into a bad temper. When the Archduke Carl Ludwig came to Potsdam at the end of August or the beginning of September H.M. spoke very disparagingly to him about the Pr. of Wales, so that the Archduke was left with the impression that a meeting between the two of them in Vienna would be very undesirable, indeed painful, for the Kaiser: this impression was all the more justified because H.M. had made no mention of the Prince of Wales's letter nor of his publicly announced presence in Austria, so that Kaiser Franz Joseph also thought he was acting in accordance with H.M.'s wishes if he sent the Pr. of W. away. He took the initiative and had the British ambassador informed through Kalnoky that he regretted that because of the large German suite he had no room in the Burg, and wished to offer the Prince hunting in Hungary for the period in question. The Pr. of W. did not take this hint, or did not want to take it, and sent a message to say that he would go to a hotel during the Kaiser's visit, but would wish to be invited to all festivities. Kaiser Franz Joseph, with his back to the wall, now made himself quite clear and told the ambassador that it would not be possible for the Pr. of W. to participate in the meeting of the Kaisers. Whether Franz Joseph also gave as a reason for this the fact that the presence of his uncle would not suit H.M. I have not been able to discover, but I think it not unlikely. The result was that the Pr. of W. was deeply offended, departed in great anger and immediately wrote furious letters to the Queen and the Empress Frederick to the effect that "his imperial nephew had had him thrown out of Vienna". The Queen wrote indignantly to various people (the Duchess of Edinburgh among others) "that Willie has grossly insulted Uncle Bertie", using all possible epithets about her grandson. As those on the British side did not want to push the whole responsibility for these "insults" onto H.M. I was chosen as scapegoat, and his "unheard-of behaviour" was credited to me.' But he, Herbert Bismarck, had been on holiday in Ostend when the incident occurred in Vienna on 11 September. 'Nor was there anything in our records about it, and it was not until several weeks later that I was in a position to reconstruct the episode. In the subsequent discussions we were at first faced with the difficulty that we did not want to expose Kaiser Franz Joseph and therefore always had to refer to Paget.'[38]

Herbert Bismarck was indeed away from Berlin at the critical moment. On 31 August he travelled to Ostend and he did not return to Friedrichsruh until 15 September. On the following day, as he noted in his diary, he attended a 'dinner at the Schloss with Archduke Albrecht, Grand Duke Nicholas etc.', after which he had time only for a 'hurried audience with H.M.'.[39] On 17 September Herbert left Berlin again for five days' hunting on the Livonian estates of the Russian ambassador Count Paul Shuvalov. He did not return to Berlin until 24 September and the next morning he and his father had an audience at the Marmorpalais before the Kaiser left for Detmold to go shooting. On 26 September Herbert Bismarck travelled to Frankfurt am Main, as has already been described, from where he journeyed on, 'united with H.M.', to Stuttgart, Munich and Vienna.[40]

Despite Herbert's claim that the records of the Auswärtiges Amt contained nothing about the incident, documents preserved there do in fact show precisely how the Prince of Wales came to be given this insulting brush-off. After the departure of the Foreign Secretary for Ostend on 31 August a report arrived from the military attaché in Vienna, Adolf von Deines, stating that the heir to the British throne would be arriving there on 10 September and would accompany the Austrian Emperor to the manoeuvres in Croatia the following day.[41] To this Kaiser Wilhelm II responded with an acerbic note written on the envelope containing the report: 'It would be advisable to draw the Kaiser [Franz Joseph]'s attention to the Prince's taste for intrigue. Inform him confidentially of the claim about Papa surrendering Reich territory and warn him against telling the Prince militarily important things about troop deployment or frontier defence. As such things could easily become known in Russia via Copenhagen, through the sisters-in-law.'[42]

When this note reached Friedrichsruh Prince Bismarck, through his son-in-law Count Kuno Rantzau, gave orders for the Kaiser's suspicions about the Prince of Wales to be communicated immediately to Vienna. The Chancellor was anxious to keep the relationship between Germany and Great Britain on a distant footing, and saw an opportunity to sow discord between Wilhelm and his British relations by exaggerating the importance of the Kaiser's scribbled comment. The Wilhelmstrasse was instructed to inform the German ambassador in Vienna, Reuss, and the military attaché, Deines, that the Kaiser had 'expressly commanded that attention should be drawn in Vienna to the advisability of observing caution towards the Prince of Wales with regard to military matters, as he is unreliable by nature, and to judge by his general attitude and inclination he would not be deterred even by goodwill for us and Austria from misusing secret information. His Majesty had grounds for being ill-disposed towards his uncle because he had spread all kinds of untrue claims about Kaiser Friedrich being disposed to make concessions to France. His Highness [Prince Bismarck] did not himself know the details of these matters, but he knew from elsewhere that our French, Danish and Guelphic conquests had been regarded with disfavour at the English court . . . His Majesty had expressly ordered that the Kaiser of Austria should be asked in some discreet way not to consider the Prince of Wales as a safe recipient of secret information, since information given to him would immediately go to Petersburg and Copenhagen.'[43] On 6 September 1888 the Under-Secretary of State at the Foreign Office, Count Maximilian von Berchem, accordingly drew up instructions for Prince Reuss, emphasising the 'express wish' of Kaiser Wilhelm that he should draw Kaiser Franz Joseph's attention to the unreliability of the Prince of Wales, particularly with regard to secret military matters. By way of explanation for this unusual step, Reuss was informed that Kaiser Wilhelm was annoyed with his uncle 'on the one hand because of the continued insinuations made by the Prince and his wife, aimed at concessions in favour of the Guelphs, and also because of the fact that in the days of Kaiser Friedrich, partly through the agency of the Prince of Wales, hopes had been raised and nourished suggesting that there was an inclination on the part of the late Emperor to hand back the national bulwark of Alsace and German Lorraine, won with heavy sacrifices, to the French as an offensive bastion for them'. These matters had also given rise to the well-known speech by Wilhelm II in Frankfurt an der Oder, which had been directed not against France, but rather against 'that English, and indeed *French*–English party to which the Prince of Wales belongs and which regards our political power in Europe with jealousy and mistrust'.[44]

It is not surprising that when Wilhelm II's 'express wish' was officially communicated to him, Franz Joseph gained the impression that the German Kaiser bore his British uncle a deep grudge and did not wish to meet him during his visit to Vienna. The Austrian Emperor decided to accompany the Prince of Wales as far as Gödöllö after the manoeuvres in Croatia, and to take his leave of him there, because – as Reuss reported to Berlin on 13 September – 'it did not please him to have the Prince remaining here as his guest during the visit of our All-Gracious Master'. But the Prince of Wales, who 'always likes to be present wherever exceptional events are taking place', embarrassed his host by stating that he wanted to come to Vienna after all, to 'shake the hand' of his nephew. When the Grand Master of the Court of Vienna, Prince Constantin zu Hohenlohe-Schillingsfürst, discussed the ticklish situation that had now arisen with the German ambassador, Reuss strongly supported Hohenlohe 'in his intention . . . to prevent the Prince from reappearing at court here', since he assumed 'that His Majesty our All-Gracious Master would likewise prefer not to meet other princely personages here, and that this assumption would be particularly applicable with regard to His Royal Highness the Prince of Wales'. Significantly for the subsequent course of events, Reuss also spoke directly to the British ambassador, Sir Augustus Paget, and urged him too to dissuade the Prince of Wales from returning to Vienna. According to his own report the German ambassador gave two reasons for this plea: first, the Prince of Wales would 'only embarrass the court *here*' by such a visit, and second 'His Majesty [Wilhelm II] would probably not' expect 'to find the Prince here'. Paget's first attempt to keep the Prince of Wales away from Vienna was unsuccessful, however, as Albert Edward refused to understand these objections. He declared that he would stay at a hotel so as not to be a burden to the Viennese court; he considered it 'useful for the public impression and for the relations between the two families "because of all the stories of the last few months" to have a friendly meeting with our Kaiser here'. The Prince's deliberate unwillingness to understand the situation only increased the embarrassment of Kaiser Franz Joseph. Reuss reported that as the Habsburg monarch 'was aware, through the secret communication that I was instructed to make to Count Kálnoky about the Prince of Wales, that the confidence which our All-Gracious Master places in His Uncle is not exactly of the highest order, the Kaiser [Franz Joseph] seized the opportunity to talk to Sir A. Paget after the court dinner the day before yesterday, and to ask him to tell the Prince that He would prefer it if His Royal Highness did not come here on the occasion of the Kaiser's visit in October, because He had reason to believe that our All-Gracious Master would prefer not to meet any other foreign Princes here'. This 'imperial message', which he received through the British ambassador, 'greatly surprised' the Prince of Wales, as Reuss reported. 'He has sensed that the court here [in Vienna] is of the opinion that his presence would not please our All-Gracious Master, and it seems that this realisation, which has at last dawned upon him, has upset him very much.' The ambassador would very much have welcomed it, he wrote, 'if the Prince's staying away could have been achieved without employing the harsh method used by Kaiser Franz Joseph'. But Kálnoky had assured him that the problem could not be solved in any other way, after the Prince 'absolutely refused to take the hints given him in the beginning'. For the Austrian Emperor, Kálnoky had explained, it had been chiefly a question of ensuring that Kaiser Wilhelm 'did not have any meetings here which could have been unwelcome to Him, and He therefore had no alternative but to speak as he had done to the British Ambassador'.[45] This report makes it clear beyond doubt that the root cause of the Vienna incident lay in Wilhelm II's grudge against his uncle; that Franz Joseph found himself obliged to act purely on account of the secret information conveyed to him officially several times; and that the explanation of the incident which attributed the initiative for putting off the Prince of Wales entirely to Franz Joseph was not in accordance with the facts.[46] It is quite plain that in his conversations with the British ambassador, the Austrian Foreign Minister and Obersthofmeister Prince Hohenlohe, Prince Reuss strongly and repeatedly stressed that a meeting between Wilhelm II and his uncle in Vienna would be unwelcome. In so doing Reuss was acting on the express official instructions of Prince Bismarck, who was evidently keen to bar the way to the 'French–British' influence of the Prince of Wales on the young Kaiser. Although Wilhelm II had never directly expressed the wish not to see the Prince in Vienna, the origin of the Vienna incident is nevertheless to be seen in his 'grudge' against his uncle, which had been noticeable not only in his Frankfurt speech and in his remarks to the Archduke Carl Ludwig, but also and above all in an angry marginal note whose contents had been conveyed officially, on the personal orders of the Reich Chancellor, through the German ambassador and the Austrian Foreign Minister to Kaiser Franz Joseph. Franz Joseph did *not* act 'entirely on his own initiative', but because he believed that he had to take account of the deep antipathy of Wilhelm II towards the Prince of Wales, to which his attention had been drawn repeatedly. As early as 16 September 1888 Herbert Bismarck wrote gleefully to his father, after his 'hurried' audience, that 'H.M. was wild with delight at Kaiser Franz Joseph turning away the Prince of Wales, who wanted to join in on the Vienna visit. *He bears him a grudge*.'[47] The Habsburg Emperor saw far more clearly than the Bismarcks the damaging political risks which lay in the unprecedented slight to the Prince of Wales. After an audience of an hour and a half with Franz Joseph on 4 October 1888, Herbert Bismarck commented laconically in his diary on the elderly monarch's attitude: 'Prince of Wales, unnecessary worry about repercussion on England'.[48]

The Vienna episode was especially damaging because Germany's international position was anything but secure. As the Empress Frederick commented in a letter to Windsor on 20 April 1889, 'all *serious*, *importan*t and well informed people' were concerned about the internal and external situation of the Reich. In his European policy, Bismarck had made the 'fatal mistake' of weakening Austria so much that it had become almost useless as an ally. As soon as the Russian regiments stationed in Poland were ready, Russia would 'attack Austria to *a certainty*', even if the Tsar were against it. The French were not quite ready yet either, it was true, but their new infantry rifle would be introduced in the course of the next twelve months, while '*ours* in Germany will *not*. If the Russians attack Austria and we are forced to help them, the French will *not* be able to resist the opportunity of falling upon us! We should then . . . face both *French & Russians*!! How awful that would be!! . . . Perhaps all this *need* not come to pass, but we seem to me drifting in this direction.' Suddenly Bismarck, alarmed by the threatening situation both on the continent and overseas, was 'anxious now for England's friendship'.[49] However, the most serious obstacle to a good understanding between the German Reich and the great maritime empire in the West consisted precisely in the tense relationship between the two royal families.

REPERCUSSIONS OF THE VIENNA INCIDENT

In February 1889 Wilhelm II came back to the idea that he had first expressed in the previous September: he wished to visit Britain early in the coming summer. But the enormity of what had happened in Vienna was anything but forgotten, and at the British court the Kaiser's wish was greeted, as before, with vehement objections. 'William must *not* come *this* year', Queen Victoria declared to her son, when she was informed of her grandson's wish on 7 February. 'You cd not meet him & I cld *not* after all he has said & done. He wd not meet with a very cordial reception by anyone I am sure.'[50] The Prince of Wales fully acknowledged the political advantages of the visit, but he too refused resolutely to receive Wilhelm until he had expressed at least a word of regret for the Vienna incident. 'As regards William coming here next summer – of course it would be the proper & right thing that he should having visited the chief countries of Europe excepting England. But it would be impossible for me to meet him till he makes "l'amende honorable" & in some way or other expresses his regret to have treated me with such uncalled for discourtesy when I was the guest of a Foreign Sovereign. My great wish is that complete harmony should always exist betwixt England & Germany & if we remain friends & allies we can do more to keep the world at peace than any other two nations, but if one's nearest Relations abuse one – & are rude to one – it is impossible to "pocket it" without a murmur. The result is coldness & later enmity, & all this is so unnecessary & deplorable.'[51] Once again, therefore, emotionally motivated monarchical politics and national expediency stood in direct opposition to one another, an opposition dangerous for the prestige of the monarchy in both countries.

At the end of February Queen Victoria set out her views more fully in a letter to the Prime Minister, explaining that she still hoped that Wilhelm would postpone his visit 'till another year, as it is too soon after all that has passed and will be very disagreeable to The Queen', but if he should insist on coming this year, he would have to apologise to the Prince of Wales first.[52] Salisbury, on the other hand, put forward reasons of international politics which made a reconciliation between the two royal families and a warmer relationship between Great Britain and Germany highly desirable. From a telegram just received from the Kaiser it was clear, he told the Queen on 9 March 1889, 'that he has a very strong desire to be received back into Your Majesty's favour. Probably his relations with Russia are less satisfactory than they were. But it is also probable that he has now thoroughly awakened from the temporary intoxication of last summer. It is Your Majesty's interest to make his penitential return as easy to him as possible.' The Prime Minister also hinted at the disastrous consequences for the standing of the British monarchy which would arise from a rebuff to the German Kaiser, while the German ambassador, Hatzfeldt, who was well liked in London, feared that he might even lose his post if the visit did not take place.[53] In the end the Queen, who had meanwhile gone to Biarritz, reluctantly agreed that the 'rude boy' Wilhelm should visit her in the Isle of Wight in his yacht – but not in London – and that the visit should take place at the end of July following the Kaiser's Scandinavian cruise.

Herbert Bismarck later recalled: 'When I took the telegram with the invitation to my audience [on 6 March] H.M. became almost ecstatic – like a child on Christmas Eve; he could think of nothing else, immediately composed a much too affectionate telegram to the Queen [it read: "Quite overjoyed to be allowed to come to that dear old Home at Osborne"], and wanted to telegraph joyfully and gratefully to Salisbury at once. I succeeded with the greatest difficulty in stopping the latter; but I was not able to bring down the tone of the other telegram, which was so little in accordance with preceding events, to the right level. H.M. countered all objections with the words: "it is the very intimate family relationship speaking". That is something which I had never detected before!', Herbert commented sarcastically, before continuing his account: 'Hardly had I left the palace when H.M. telegraphed en clair, at least not to Salisbury, but to Hatzfeldt, instructing him to thank the British Prime Minister most warmly. On this day', so it appeared to the Foreign Secretary, 'began the enslavement of H.M. to England.'[54] The Kaiser's delight at the invitation was all the greater, in Herbert Bismarck's opinion, because Wilhelm believed that 'the prospective brilliant British reception' would enable him to 'stand up to the widow Absalom [the Empress Frederick] and her family agitations'.[55]

Wilhelm had now been invited to Osborne House, but the Queen warned that the Prince of Wales would not be present at the meeting of the two sovereigns in the Isle of Wight unless a modus vivendi between him and his nephew had been found by then; if not, his place at the meeting would have to be taken by her second son Alfred, Duke of Edinburgh.[56] Various suggestions for mediation between the Prince and Kaiser Wilhelm foundered on the strong sense of insult which the former continued to harbour. Albert Edward declined the proposal brought to him from the Kaiser by Swaine that he should first write a letter to Wilhelm, to which the latter promised to send a conciliatory reply. Similarly the Prince refused to accept Hatzfeldt's suggestion that he should simply let the unpleasant episode rest until both men had the opportunity to express their mutual regret over the misunderstanding when they met. For his part, the heir to the throne asked his brother-in-law Prince Christian of Schleswig-Holstein to mediate for him during his forthcoming visit to Berlin. Christian, however, was not known for his skill as a negotiator,[57] and the Prince of Wales's Private Secretary, Sir Francis Knollys, took the view that no one was in a better position to settle the family quarrel than Queen Victoria herself.[58]

With the agreement of the Chancellor and Wilhelm and much to the surprise of the British government and the German ambassador, Herbert Bismarck turned up in London at the end of March. His intention was to bring about an improvement in Anglo-German relations, in view of the danger of a military confrontation on the continent, or of a French invasion of Italy, and also of the growing American influence in the Pacific. According to Waldersee it was the German Foreign Secretary's aim to achieve 'a firm union between us and England against Russia'. Herbert had explicit instructions from Wilhelm II to 'emphasise his pleasure at the kind invitation' while he was in London.[59] In two conversations with Salisbury he stressed that it was important 'that England & Germany should appear friendly before the world' and pointed to the significance of the Kaiser's forthcoming visit in this regard. He went so far as to draw the Prime Minister's attention to the danger to the British monarchy of openly opposing public opinion as embodied in Parliament.[60] The Kaiser, he informed Salisbury, wished to come 'quite quietly' to Osborne for two or three days at the end of July. He proposed to spend two or three more days visiting dockyards, weapons arsenals and other military installations in the south of England. The Foreign Secretary spoke in a markedly friendly manner of the Kaiser's 'great affection & veneration' for the Queen and described the death of Kaiser Friedrich III as 'an appalling calamity'.[61] Even so, the Vienna incident continued to cast a dark shadow over Anglo-German relations.

In a letter to Prince Christian of Schleswig-Holstein of 3 April 1889 the Prince of Wales gave a detailed account of the incident and asked him to use his influence with Kaiser Wilhelm during his forthcoming visit to Berlin to try to induce the latter to express his regret over the affair in some form or other. 'If he does not, I shall be obliged to absent myself during his visit which it is needless to say would have a most deplorable effect. I have always been on the terms of the greatest friendship & intimacy with every member of his family, & with many of them before he was born!, but the close relationship between him & me would render an estrangement between us a matter of serious importance!'[62]

Prince Christian had an initial conversation with Wilhelm II on 8 April 1889. Earlier the same morning the Kaiser had conferred with Herbert Bismarck about 'Christian & Wales'.[63] Although Wilhelm claimed several times that he had never expressed the wish not to see his uncle in Vienna, Christian nevertheless detected 'a certain indisposition towards the Prince of Wales'. Above all, the Kaiser remained unbending on the crucial matter of a written explanation. He declared with some vehemence, which Christian nevertheless did not quite understand, that he could not write to the Prince of Wales 'as this was not a simple affair between uncle and nephew, but between Emperor and Prince of Wales'.[64] On the following day Christian asked the British ambassador Malet to telegraph the Prince of Wales to say that the Kaiser certainly wished to settle the quarrel but was not willing to take the first step in writing. Instead he had suggested that the Austrian Foreign Minister Kálnoky should be asked, through Reuss, how he had gained the – ostensibly false – impression that the Kaiser had not wanted to see the Prince.[65] Malet suspected that this suggestion would only cause more trouble, and therefore asked the Prince of Wales to send a cipher telegram expressing his satisfaction with the Kaiser's explanation that Kálnoky must have been given the wrong impression. This would enable the unedifying quarrel to be finally laid to rest, relieving Anglo-German relations of the strain it had caused.[66] The Prince again expressed astonishment at his nephew's behaviour and commented that if the whole affair had been a misunderstanding the Kaiser could surely write a few lines signalling his regret.[67] Salisbury also believed that the means of putting an end to the whole tiresome story now lay within reach: if the Kaiser would write a letter to Prince Christian repeating his declaration that he had never expressed the wish not to see the Prince of Wales, there would be no further reason for the quarrel, and Wilhelm and Albert Edward would be able to meet in the Isle of Wight.[68] In a letter to the Prince of Wales, the Prime Minster went even further and said that such a declaration would purge the personal injury to His Royal Highness even if it were *not* in accordance with reality. The truthfulness of the Kaiser's declaration was not a matter for the British government; it could happily be left to Reuss and his imperial master to sort out between them.[69]

On 11 April Christian saw the Kaiser for the second time and again received a verbal assurance from him that neither he nor the Foreign Office had ever said that he did not wish to meet the Prince of Wales in Vienna. But again the Kaiser 'most explicitly' declared that he could not write to the Prince to this effect. The reason for this, he said, was that he did not know whether Prince Bismarck would agree to his doing so. At any rate Wilhelm was prepared this time to *dictate* to Prince Christian a telegram to be sent to the Prince of Wales, giving the assurance that 'the Emperor has not ever expressed any such wish;

consequently the affair was a misunderstanding'. On his own initiative Christian added that Wilhelm undoubtedly wanted to see the Prince of Wales and had expressed the hope that he would not absent himself during the Kaiser's visit to England.[70] But this attempt at reconciliation also failed because of the Kaiser's refusal, incomprehensible to all on the British side, to set down in writing the verbal declaration he had made to Christian.[71] The Prince of Wales, for his part, refused to be satisfied either by his nephew's oral declaration or by a dictated telegram, and demanded that his letter of 3 April 1889 to Christian should now be shown to the Kaiser. 'If he really wishes to be friends with me again the request I ask of him is very easy to grant', the Prince insisted.[72] 'I *must* have an expression of regret on paper; no verbal message will suffice.'[73] But the Kaiser maintained his stand: he could go no further.[74] A disaster with momentous consequences seemed inevitable.

Unwilling to aggravate what was already an extremely critical situation, Christian refused to show the Prince of Wales's letter to the Kaiser. He pointed out that 'the Emperor is still too young in his position to feel quite sure of himself and always to do the right thing. He is constantly apprehensive of doing something incompatible with his dignity, and he is especially haunted by the thought that his older relatives might treat him as "nephew" and not as "Emperor". Under these circumstances I am firmly convinced that, even if I had given the letter to the Emperor, he would not have complied with the wish of the Prince of Wales. Possibly the situation might have become still more acute and an irreparable rupture would have been caused.' Christian also warned against any attempt to unearth the real reasons for the Vienna incident. 'To do this', he cautioned anxiously, 'some diplomatic steps would have to be taken the consequences of which are beyond my calculations and which might be attended by the most serious dangers.'[75]

Malet tried despairingly to avert a public scandal, which would have been damaging not only to Anglo-German relations but also to the image of the crown in both countries. There were two theories in circulation, he reported, about the Kaiser's refusal to set down his declaration in writing. The first claimed that 'the Kaiser' had indeed expressed the wish not to see the Prince of Wales in Vienna, but that the Kaiser in question had been Franz Joseph, not Wilhelm II. According to this theory Wilhelm did not want to embarrass the Austrian Kaiser, whom he greatly respected, by making a written declaration. The second theory retailed by Malet was that a high-ranking German official (in all probability the ambassador Prince Reuss), in a zealous attempt to prevent a meeting between Wilhelm and Albert Edward in Vienna at all costs, had spoken out on behalf of the Kaiser, but without authorisation. Although Wilhelm was very anxious to be on friendly terms with his uncle again, he was held back from making a written declaration by the feeling, in itself honourable, that he could not disavow this high-ranking official. Both theories allowed the Kaiser's desire for reconciliation with his uncle to appear genuine; the reasons for his refusal to apologise in writing, according to these interpretations, lay elsewhere.[76] Queen Victoria's Private Secretary, General Sir Henry Ponsonby, after reading the numerous reports from Berlin, came to the conclusion that a mysterious game was being played out. Like Queen Victoria he suspected that the Bismarcks were behind the Vienna incident. 'They probably originated the whole difficulty at Vienna and when they found matters had gone too far and that it was necessary to have a reconciliation before the Emperor came to England, Count Herbert was sent.'[77]

But even this perceptive guesswork did not help resolve the problem. The Prince of Wales wrote bluntly to the Prime Minister: 'In asking the Emperor to write me a few lines instead of sending me a vague, verbal message, I do not think that my demand is excessive but should the Emperor refuse to do so, I shall look upon it as a clear proof that the Emperor does not wish to make it up with me, and I shall ask the Queen's permission to absent myself when the Emperor comes over to England.'[78] To his brother-in-law he stated flatly that he had a duty to protect his own dignity. If Wilhelm did not write a letter, he would stay away during the visit of the 'German Emperor and his Fleet' and would also take care 'that every one knows the reason'.[79] He sent Christian a note for the latter to show to the Kaiser, saying: 'Please let William *clearly* understand that unless I do receive a few lines from him as I have asked for in my previous letters to you, I shall be unable to meet him when he comes to England this summer.'[80] Queen Victoria fully shared her son's view that he must insist on receiving 'a few kind words' in writing from Wilhelm.[81]

The Prince of Wales's eldest sister, Wilhelm's mother, was of the same opinion. She considered it absolutely essential that the Kaiser apologise in writing for the Vienna incident. Vicky saw Wilhelm's behaviour as an expression not of malevolence but of embarrassment. 'It is *only laziness* & indifference a little awkward shyness wh. makes William reluctant to write, as he *must* feel that he has behaved very ill, has made a great *bévue* & been very rude! He now does not like to own it, and as he is singularly "*gauche*" & "*ungeschickt*" – does not know how to get out of it.' She nevertheless found it incomprehensible that Wilhelm had not long since written a few simple lines to his uncle.[82]

Prince Christian's attempts at mediation had thus succeeded in worsening, rather than improving, relations between Wilhelm and his British family.[83] Queen Victoria wondered whether Christian had been the most suitable go-between, in view of his close family relationship to Kaiserin Auguste Viktoria, and suggested another German son-in-law, Grand Duke Ludwig of Hesse, instead.[84] But as Christian explained to the Prince of Wales, the Kaiser had been most reluctant to enter into any kind of discussion of the affair with him. 'He was always in a hurry. My interview the other day only lasted a few moments, and when we left the room together we stumbled on Herbert B., the last person I wished to see. W. told him to speak to me on the subject, which I much disliked . . . I showed him [Wilhelm] your memo upon which he again said he could not write.' Christian nevertheless had the impression that 'if he were left to himself his better nature would prevail'.[85] He blamed Wilhelm's attitude on the harmful influence of Herbert Bismarck, who had behaved so 'boorishly and roughly' that Christian 'nearly lost all patience' with him.[86]

This unpleasant affair was still completely unresolved in mid-May 1889. The Prince of Wales, in a solemn memorandum on the subject, spoke of the Kaiser's 'ominous silence' and stated that he, the heir to the British throne, could only interpret Wilhelm's conduct 'as an admission that the affront was intended and that the Emperor William declines any offer of reconciliation'.[87] Relations between Berlin and London were so tense that the Kaiser's visit seemed in jeopardy and Hatzfeldt's position again looked insecure. Prince Ernst zu Hohenlohe-Langenburg, who had recently taken up a post at the German embassy in London, and who as a close relative of the British royal family had an insider's view of the entangled situation, wrote to his father on 11 May to say that 'disagreements' had 'yet again' arisen over the planned visit by the Kaiser. 'The way in which the Kaiser and H. Bismarck have made a point of ignoring the Prince of Wales, whom they have treated, from what I gather, with the utmost lack of consideration, will not improve matters, as relations between the Queen and the Prince are very good. Hatzfeldt is said to have behaved very skilfully in the affair and to have shown great tact in overcoming all the difficulties which Berlin's rudeness and suspicion caused him. If the visit should now be thwarted by new thunderbolts from the Berlin sky, he would have no alternative but to take his leave.'[88]

On the same day Salisbury – acting on Herbert Bismarck's hint – wrote to the Queen warning her again of the serious political consequences, above all for the standing of the crown, which must inevitably result from an estrangement between Britain and Germany caused by personal disagreements in the two royal families. He expressed the anxious hope 'that no decision may be arrived at, which will, in effect or in appearance, give ground to the world to think that a rupture or estrangement with so powerful & important a nation as Germany has arisen out of discussions of a personal or family character. Such an impression would produce a formidable effect on public opinion: which would be regrettable at a time when so many politicians seem ready to rush into extreme opinions. Lord Salisbury dreads the *internal* effect of a rupture with Germany arising out of non-political differences, much more than the general effect upon European politics – though that of course would be prejudicial.'[89] Other ministers and advisers of the Queen likewise recognised the danger for the monarchy that would arise if republican agitators seized upon the royal family quarrel to further their cause among the lower classes.[90]

At this dangerous juncture Sir Francis Knollys and Sir Henry Ponsonby, private secretaries to the Prince of Wales and the Queen respectively, reverted to the suggestion that only Queen Victoria had the sureness of touch and the authority required to bring about a reconciliation. 'I have *long* thought that the only person to take it in hand was the Queen, and no one else . . . She would do it much better than all the Christians, Hatzfeldts & Malets in the world', wrote Knollys. 'I don't say she would succeed, but the best chance of success would be for her to write a carefully composed letter to the Emperor, and for her to suggest at the end of it that he should consult Prince Bismarck on the subject. This however I doubt Her being able to make up Her mind to propose.'[91] On 13 May Ponsonby advised the Queen that she should write a friendly letter to Kaiser Wilhelm II expressing her regret at the conflict between him and the Prince of Wales; Wilhelm would then have no alternative but to reply expressing his own regret at the misunderstandings.[92]

Ponsonby submitted to the Queen a draft letter to the Kaiser drawn up by Lord Salisbury. She made several corrections to it in her own hand, on the grounds that the wording proposed by the Prime Minister simply accepted the Kaiser's denial 'and makes us swallow all the lies'. She also crossed out a passage which could have given the impression that Wilhelm's visit was awaited with pleasure in Britain. To Salisbury's horror she insisted on adding a passage expressing the hope that the Kaiser would order an official investigation into the Vienna incident.[93] After much heart-searching the Prince of Wales, who still felt deeply injured, agreed to the draft as altered by the Queen. He let it be known that he would refuse to dine with the Prime Minister on the Queen's birthday if Salisbury raised any objections to his mother's amendments.[94] Evidently the royal family had reached the limits of its willingness to compromise. But at least its advisers at court were convinced that the Queen's letter would have the desired result and put an end to the quarrel. As Ponsonby remarked on 23 May: 'If the Germans don't like the letter they may put in their pipes & smoke it. Then the fault wd. be theirs. But of course they will answer and I believe all will come right. Very possibly they won't "enquire" – because an enquiry wd. lead to awkward revelations. But we can't make them enquire.'[95]

After the customary exchange of affectionate telegrams on the Queen's birthday on 24 May 1889,[96] the moment arrived to launch the much-discussed correspondence. The final version of the crucial passage in the Queen's letter read: 'My dear Willie, . . . I was very glad to hear from Uncle Christian that you entirely deny the assertion of your having expressed to the Emperor of Austria a wish not to meet Uncle Bertie at Vienna. Lord Salisbury reports to me that you have also instructed Count Hatzfeldt to repeat to him that you never expressed any such wish and I will tell this to your Uncle who will be much pleased to hear it. I cannot understand how the mistake could have arisen which might really have led to very serious consequences, & I hope you will enquire into the circumstances.'[97] The Prince of Wales sent a copy of the letter to his sister in the Taunus, commenting that he thought it too mild and that he feared Wilhelm might consider that it put an end to the matter. 'We must hope', he wrote, 'that in turn he will see how unjust he has been to an Uncle who has always been a kind friend & affect[ionat]e relation to him. I will however do him the justice to say that I think if he *had* been left alone without the pernicious advice of the B[ismarck]s he would have sent me a line, telegram or conciliatory message before now.'[98]

The hopes of the British court and government for a prompt end to this smouldering conflict proved premature, for the Kaiser's rude reply – probably composed with the collaboration of Herbert Bismarck, and at any rate approved by him on 28 May and sent to London by dispatch rider[99] – only made matters worse.[100] 'With respect to the Vienna mistake I have had the matter enquired into immediately when I heard of it from Uncle Christian. I even had Count Kalnoky asked. The result of the enquiry is that the whole affair is absolutely invented, there not having been an atom of a cause to be found. The whole thing is purely a fixed idea which originated either in Uncle Bertie's own imagination or in somebody else's, who put it into his head. I am very glad to hear that this affair has at last come to an end. If you wish for any detailed information Count Hatzfeldt is in possession of all the documents which have reference to the case & can explain the facts to you at any time.'[101]

The Queen sent Wilhelm's letter to her son with the remark that the Kaiser's answer 'will and *must* annoy you very much. It is incredible!!' 'The affair is becoming more and more disagreeable', she commented despairingly. It would now be necessary to recall Sir Augustus Paget from Vienna and carefully compare both versions of the incident. It was clear that several people had told 'great *untruths*'. 'I will believe, at least I *hope* so, that William did not give the message – & that that horrid Herbert & Prince Reuss did it. I shall of course have to answer William . . .'[102] At Marlborough House, the Prince of Wales's residence, the Emperor's answer to the carefully formulated attempt at reconciliation was greeted with incredulity. The Kaiser was trying to dismiss the entire affair as nonsense, Knollys wrote indignantly. 'He simply pooh poohs the whole matter.' This confirmed Knollys's belief that it would have been better not to have been so 'milk and watery' with the Kaiser. It was of course possible that Wilhelm had really had nothing to do with the Vienna episode, and that either the Bismarcks or Crown Prince Rudolf were behind it. But the Kaiser's assertion that the Prince of Wales had simply dreamt up the whole affair was 'a most extraordinary and offensive one and increases the affront which has already been put upon the Prince of Wales'. As the latter had now decided not to forgo his visit to Cowes after all, a meeting with the Kaiser in the Isle of Wight was inevitable, but in the existing circumstances any contact between them must be limited to the purely outward, ceremonial minimum. Naturally the Queen would have to receive her grandson, Knollys opined, 'as otherwise from the Platform and in the Press a howl of abuse will emanate from all the Radical and Socialistic elements in the country', but 'if the Prince were to *receive* the Emperor it would look as if he were *welcoming* him', and he could not be expected to do that.[103] The anger felt at Windsor against Wilhelm was scarcely assuaged by news from the Empress Frederick that her son had been boasting that he could 'do what he liked with his Grandmama'.[104]

Opinion at the German embassy in London was no less critical than at the British court. As the Hereditary Prince of Hohenlohe reported to his father on 9 June, the planned visit by the Kaiser was still 'a precarious matter'. The Prince of Wales, he said, was 'piqued, chiefly because last year he was pushed aside on account of the Kaiser's visit to Vienna, and now this is stubbornly denied, and it is made to appear as if it were an *idée fixe* of his, although in fact there are official documents which show that the greatest zeal was used to make sure that he was out of the way on that occasion. I have read the relevant correspondence and must say that judging from it the Prince has every right to be piqued. It is possible that Reuss will be disavowed so as to bring about a reconciliation. Hatzfeldt is doing the best he can and making use of every possible diplomatic ruse. In the end politics are nothing but a vile game of intrigue, and as we are not having much success at the moment with our brutality, which is given the fine names of sincerity and frankness, and are telling clumsy lies, our position is not exactly a happy one.'[105]

In London there were feverish attempts to formulate yet another letter to the Kaiser which would persuade him to offer some explanation for his insulting behaviour in a manner acceptable to the Prince of Wales. In a first draft of a reply from the Queen to Wilhelm, Salisbury suggested that she should declare herself 'very satisfied' with her grandson's letter, with the single exception of one sentence. 'I wish you had not suggested that the story about your not having wished to see your Uncle at Vienna might possibly be "a fixed idea originating in his own imagination". The proofs that the idea did not originate in *his* imagination are in writing, for my Ambassador Sir Augustus Paget sent home reports at the time showing that this story was communicated to him by people who should have been better informed and was communicated by him to the Prince of Wales. I do not refer to this for the purpose of reviving a matter which has now been happily set at rest but to remove the notion in your mind that the Prince of Wales had any prejudice or any unkind thought towards you.'[106]

For the second time Queen Victoria replaced the title 'Prince of Wales' with 'your Uncle'; but above all she insisted that the German ambassador, Prince Reuss, should be named as the man who had conveyed the Kaiser's wish to Paget, not least so as to avoid creating the impression that she wished to lay the blame for the 'misunderstanding' on Kaiser Franz Joseph or (the now deceased) Crown Prince Rudolf.[107] The draft composed at Marlborough House by the Prince of Wales with the help of Prince Christian of Schleswig-Holstein and Sir Francis Knollys was more direct. The heir to the throne and his advisers put forward a text strongly emphasising the offence caused to him and thus decidedly less conciliatory than the Prime Minister's draft.[108]

On 7 June, however, Salisbury wrote to Ponsonby expressing his 'most distinct disapproval' both of the alteration of his own draft and of the Prince of Wales's proposed wording. Any express mention of Prince Reuss by Queen Victoria would cause 'endless trouble', he warned. He pointed out that it would provoke two possible reactions, both of which would be harmful to British interests. The most probable result would be a flat, insulting denial by the Kaiser which would make further contact with him impossible; alternatively, a large-scale inquiry into the affair, including German complaints against the British side, might actually be launched in Berlin, which would leave the two Kaisers, Wilhelm and Franz Joseph, with a 'rancorous feeling' at a critical moment when a good understanding between the three empires was of the greatest importance. The Prime Minister objected categorically to the naming of Reuss, since that would immediately give the correspondence, which had hitherto been 'personal', a 'political' nature. He strongly advised against replying to the Kaiser's letter at all; but if an answer must be sent, it should be his original draft. What the Queen might later say privately to her grandson was her own affair.[109]

Salisbury's resolute opposition, on grounds of national interest, put an end to the matter, since the Queen could not carry out any action for which her Prime Minister was not prepared to take responsibility,[110] but his attitude was deeply resented in the royal family and generally at court. Prince Christian spoke of it as a 'deplorable end', and Knollys described it as the 'miserable termination' of an affair which had put his master in a 'very awkward' situation. The general verdict of the court was that in adopting a conciliatory stance Salisbury had sacrificed the Prince of Wales on the altar of political expediency.[111] But it was precisely this willingness to allow reasons of state, put forward by a constitutional government, to prevail over personal wishes in matters of politics, that marked the essential difference between the British parliamentary monarchy on the one hand and the 'personal monarchy' of Prussia–Germany on the other. No one was to feel this difference more keenly in the coming months than the Bismarcks themselves.

THE ADMIRAL'S UNIFORM

Meanwhile Salisbury not only maintained his conciliatory stance – under pressure from Hatzfeldt it was decided not to send any reply to the Kaiser's letter[112] – but also pulled off a coup which, at a stroke, transformed Wilhelm's attitude to Britain – at least temporarily – into rapturous gratitude. During Herbert Bismarck's visit to London in March he had raised with Leopold Swaine, the recently recalled military attaché in Berlin, the question of whether the Kaiser might be made colonel of a British regiment, since – as the German Foreign Secretary put it – 'if he receives any English uniform he will be beside himself with joy . . . He sets far more store by new uniforms than even his Uncle Wales.'[113] Now, in June, Salisbury informed the Kaiser confidentially through the British ambassador in Berlin that Queen Victoria intended to create him a British admiral before his forthcoming visit to the Isle of Wight. The Queen would be glad to be given a German regiment in return; the First Dragoon Regiment was selected, as the only cavalry regiment in which Kaiser Friedrich III had served.[114] Finally, to flatter the vanity of the increasingly touchy Prince Bismarck, the British Prime Minister sent word that the Queen had consented to confer her portrait on the Chancellor in honour of the Kaiser's visit.[115]

Wilhelm II was indeed beside himself with delight at being created a British admiral, and all anglophobia seemed forgotten. On 14 June 1889, the eve of the first anniversary of his accession, he wrote to Sir Edward Malet: 'You can not imagine with what joy I read the welcome news, which you so kindly sent me this evening. I shall immediately set to work in order to select a Regiment fit for the great honour of having Her Majesty as Honorary Colonel. But the last sentence of your letter fairly overwhelmed me! What a surprise and an agreeable one too! I am indeed deeply grateful for the intention of Her Majesty to make me a British Admiral! Fancy wearing the same uniform as St. Vincent and Nelson; it is enough to make one quite giddy. I feel something like Macbeth must have felt, when he was suddenly received by the witches with the cry "All Hail who art Thane of Glamis and of Cawdor too!" I shall of course gladly accept the kindness Her Majesty so graciously has preferred with all my heart. Of course I shall not breathe a word to anybody.' Wilhelm added jokingly that in quoting Macbeth he naturally did not mean to liken the ambassador to a witch; on the contrary he regarded him as a good fairy.[116]

5. Oil portrait by Rudolf Wimmer of Kaiser Wilhelm II in the uniform of a British admiral of the fleet on the terrace of Osborne House, 1889

The bestowal of this new uniform, according to Herbert Bismarck, had the effect of intensifying 'to the utmost' the 'enslavement to England' which he had already observed in Wilhelm in March.[117] The Kaiser was 'like a child' in his delight at being created a British admiral, Philipp Eulenburg commented during the Scandinavian cruise.[118] From India the Duke of Connaught wrote that the Kaiser was overjoyed at his appointment and that the political effect in Germany would certainly be excellent.[119] The full significance of the admiral's uniform for Wilhelm himself was apparent from his words to Herbert Bismarck when the latter joined him at Wilhelmshaven after his summer cruise. He declared that 'his appointment as an English admiral was a great event; apart from the flattering honour for him (through which field marshal's rank in the British armed forces had been conferred on him, a distinction that he did not even hold at home) he would now have the right, as admiral of the fleet, to have a say in English naval affairs and to give the Queen his expert advice. I looked up in surprise, but H.M. was perfectly serious in what he said . . . During the visit to England General Wittich repeated several times the remark he had made the previous year, "things are going the same way as with [King] Ludwig II [of Bavaria]"', the German Foreign Secretary recalled.[120]

With this skilful move Salisbury had swept away the last hindrance to Wilhelm II's visit to England. As early as 15 June the First Secretary at the German embassy in London, Count Casimir von Leyden, sent the Queen's Private Secretary the list which Wilhelm had personally drawn up of his suite for the journey.[121] Through the London embassy the Queen asked the Kaiser to come to the Isle of Wight on 2 August, as delegates from the British Parliament had decided to attend the naval review planned for that day. Again Wilhelm was delighted. 'I am very much gratified by this mark of kindness shown to us by the Representatives of the whole British nation; which shows the world, that the Country fully concur & sympathise with their illustrious Sovereign in tightening of the bonds of friendship between our two families & countries', he wrote to the Queen on 23 June 1889. He referred to the unpleasant Vienna episode only in a single, unobjectionable sentence: 'At the same time I am happy to see that you regard the Vienna affair as concluded in which I heartily concur & I shall be happy to meet the [sic] Uncle Bertie in Osborne.'[122]

THE KAISER'S VISIT TO ENGLAND

When Kaiser Wilhelm, his brother Heinrich and his suite arrived in the *Hohenzollern* off the Isle of Wight on the evening of 2 August 1889, the Prince of Wales, accompanied by his two sons Albert Victor ('Eddy') and George (the future King George V), together with his brothers-in-law Prince Christian of Schleswig-Holstein and Prince Henry of Battenberg, sailed out to meet the imperial yacht and brought the German guests back to Osborne House, where they were received by Queen Victoria with her daughters Helena, Louise and Beatrice and her daughter-in-law Alexandra, the ladies and gentlemen of the royal household and Lord Salisbury. The Kaiser, who his grandmother thought had grown 'very large & puffed in the face',[123] was wearing, with evident delight, his new admiral's uniform. Here in Osborne, as Herbert Bismarck contemptuously remarked, he became 'the complete anglomaniac'.[124] He greeted the Queen warmly and kissed her 'very affectionately' on both cheeks. He led the Queen in to the formal dinner for forty guests, held in a marquee in the garden – all the gentlemen were in uniform, with the band of the marines playing softly in the background – and, as on every evening, the Queen sat between Wilhelm and her other Prussian grandson, Heinrich.[125] Among the political guests there were, on the German side, Count Herbert von Bismarck, Count Paul von Hatzfeldt, Generalleutnant Wilhelm von Hahnke, Geheimrat Hermann von Lucanus, Oberhofmarschall Eduard von Liebenau, Generalleutnant Adolf von Wittich and Kapitän zur See Gustav Freiherr von Senden-Bibran.[126] The Queen did her best to conceal her disgust at 'that horrid Herbert B. . . . & that traitor [Flügeladjutant Gustav von] Kessell [sic]'.[127] Outwardly, there was no sign of the ill-feeling of the past weeks and months.

Over the following days, too, the full programme for the Kaiser's visit proceeded perfectly satisfactorily both for him and for his British hosts. Wilhelm described the result of his conversations with Lord Salisbury as 'eminently reassuring' and commented that a 'thorough consonance of views' had prevailed.[128] Accompanied by General von Hahnke and the full complement of princes and princesses, the Kaiser watched the British naval review on 5 August, while Queen Victoria reviewed the German ships, which played *God Save the Queen* one after the other as she sailed past. In the evening, before dinner, Wilhelm introduced to his grandmother, with a 'very pretty speech', a delegation of 'her' First Dragoon Regiment of which she had become colonel-in-chief.[129] On the morning of 7 August Wilhelm travelled to Aldershot with his Uncle Bertie and his brother Heinrich to attend a Field Day.[130] On his return he commented happily to Herbert Bismarck: 'Things are going better and better, tomorrow The Queen has even invited me to her breakfast in the tent, that is a rare honour.' The Foreign Secretary retorted that it was in fact an honour for the Queen 'that Your Majesty should give yourself the trouble to do so', but the Kaiser answered 'irritably': 'You do not understand, . . . you do not know what a secluded life my Grandmother always lives and how much she dislikes receiving in the morning.' For Herbert this remark was one more proof 'that in his English family relationships H.M. had not yet grown up; he was still completely under the influence of his earlier visits to the Isle of Wight when he was treated as a child and as a youngster as his mother thought fit'.[131]

On 8 August, the last day of the visit, while the two monarchs breakfasted in the tent, all the German sailors and marines came ashore to be inspected by the Queen. Wilhelm led the march-past in person. The Queen commented in her journal on how fine the men looked and how well they marched, although 'in that peculiar Prussian way, throwing up their legs'. She conferred the Order of the Garter on Prince Heinrich, which greatly delighted both him and his imperial brother. For his part, the Kaiser bestowed the Order of the Black Eagle on Prince George, and presented the Queen with a bust of himself by Reinhold Begas.[132] Herbert Bismarck and the German officers present saw this demonstration of family affection rather differently. As Herbert commented in his memoirs: 'The worst part was the march-past, led by H.M. himself, of 1200 sailors who had come ashore, on the lawn in front of the tent in which the Queen sat in her armchair waving. H.M. gave the commands and lined them up like a lieutenant in a barracks square, and led the detachment, with drawn sword, at goosestep past the tent. Our generals turned away angrily, and murmured the words "undignified farce".'[133]

On the British side, and for Wilhelm II too, there was relief that the visit had passed off so successfully.[134] 'William came up to see me, & we talked generally of things', wrote the Queen of her grandson's departure. 'He was all the time very amiable, & seemed delighted with his visit.'[135] Salisbury congratulated the Queen on the great success of the visit and above all 'upon the admirable effect produced upon the mind of the Emperor and those who were with him by your Majesty's cordial and most gracious welcome'. He hoped that with time the Empress Frederick would also have reason to be thankful that her son's visit to England had led to happier and more peaceful family relationships.[136]

On his arrival back at Wilhelmshaven the Kaiser sent the Queen an affectionate telegram of thanks: 'I reiterate from the depth of my heart the thanks for your unbounded love and kindness to me, Henry, and all of us, which will never be forgotten, especially for the commission as Admiral of the Fleet.'[137] A few days later in Berlin the Kaiser, the Prince Regent of Brunswick and a number of gentlemen from the British embassy attended a banquet in honour of Queen Victoria's appointment as colonel-in-chief of the First Dragoon Regiment, at which all the speakers proposed toasts to the long and successful co-operation of the German and British armed forces and to the friendship of the two nations.[138] On 17 August, in a letter from Bayreuth, Wilhelm expressed his confidence that, with British maritime power making common cause with the German army and navy, world peace would be safeguarded, or if it should come to war, that both countries would fight shoulder to shoulder. 'May I be allowed to reiterate my most fervent thanks for all the kindness which you lavished on Henry & me, and which I scarcely know how to repay?', he wrote. 'We have felt so comfortable & quite at home at Osborne through all the pains you gave your self to arrange everything for me . . . But personally I beg to be allowed once more to express my warmest thanks for the quite unexampled honour, which you conferred on me with the commission as "Admiral of the Fleet". It really gave me such an immense pleasure, that I now am able to feel & take interest in your fleet as if it were my own; & with keenest sympathy shall I watch every phase of its further development.' Here too he proclaimed it as his most earnest wish that in a future war the German fleet might fight alongside the English, and the 'Pomeranian Grenadier' shoulder to shoulder with the 'Red Coat'.[139] When the Channel Squadron of the British navy sailed in to Kiel at the beginning of October Wilhelm not only went there to attend a large naval dinner, but also made arrangements – as he proudly announced to the Queen – for Admiral Baird and his officers to attend a reception held by the Empress Frederick in Berlin.[140]

The astonishing, almost alarming effect on the Kaiser of the admiral's commission became most clearly apparent in October 1889 during his cruise to Athens. For days on end Wilhelm hoped that the British Mediterranean squadron would come to meet him, 'but no one came'; early each morning he went on deck and surveyed the horizon, telling the ship's commander to hold the Union Jack ready for hoisting if British warships came in sight; at midday he did not eat with the officers but had his meal brought up to him – increasingly tense with expectation – in the chartroom on deck. Eventually, when the British ships hove into view in the Bay of Phaleron he had the Union Jack – which is the badge of rank of the British admiral of the fleet when flown at topmast – hoisted beside the imperial standard. When the Kaiser sailed into the harbour of Piraeus in the *Hohenzollern* the imperial standard was lowered on the great German warship, 'but the English admiral of the fleet's badge was left flying on the order of the Kaiser. A German warship with an English admiral's flag!', exclaimed the ship's commander. The passing ships of all nations were compelled to salute the admiral of the fleet's flag with nineteen guns.[141] The Kaiser, dressed in his admiral of the fleet's uniform, spent several hours inspecting the ships of the British Mediterranean fleet off Athens, and reported to the Queen by telegraph on the excellent state of both equipment and crews.[142] In a 'brilliant' speech in honour of the Queen and the Royal Navy he pointed out that he had felt a bond with the British fleet since his earliest childhood. 'From the time when he was a "little urchin" he accompanied his parents to Osborne and was taken over to see the Royal Dockyard at Portsmouth. It was then, said His Imperial Majesty, that his uncle the Duke of Edinburgh took him by the hand, and gave him his first instruction in nautical matters. Since then his naval education had, he hoped, steadily progressed, till it culminated in the much valued honour which the Queen had recently bestowed upon him. The Emperor said further [so the British envoy in Athens reported] that the German Navy was beholden to that of England for instruction and example, and that the best officers derived the best part of the knowledge from their association with and experience of English ships.'[143]

A remarkable change had come about thanks to an invitation, a uniform and a few friendly gestures. As Lord Salisbury – who had no doubt reflected more than once in the previous months on John Erichsen's warning that Wilhelm would never be a normal person and would be subject to periodic fits of rage[144] – was able to assure the Queen in the autumn of 1889, the attitude of the Kaiser to his grandmother and to Britain was now very satisfactory. 'He is a changed man, from what he was twelve months ago.'[145] 'His present mood is all that could be desired & shows a great improvement in tone & mental grasp.'[146] At the German embassy in London too there was jubilation at the predominantly German-friendly tone of the British press, which since the Kaiser's visit had praised 'Germany, the Kaiser and the Reich Chancellor to the skies'.[147]

The transformation of the Kaiser's personal stance towards Britain had immediately discernible consequences for the international political scene in Europe. Following a visit to Berlin by Kaiser Franz Joseph, the Austrian Foreign Minister, in mid-August 1889, expressed his pleasure at the Anglo-German understanding, which would also enable the Danube monarchy to act more confidently in eastern and southern Europe. If a war should break out, he would now be able to reckon on British participation on the side of Austria and Germany. During his stay in Berlin, Kálnoky reported, he had furthermore been able to establish that the close relationship which had prevailed between the St Petersburg and Berlin courts in the days of Wilhelm I no longer existed, so that the ambivalence which had formerly characterised German policy towards Russia and Turkey would henceforth be replaced by a more clear-cut attitude.[148] The Shah of Persia, in Budapest at the end of his great European tour, also spoke to the British diplomat Sir Arthur Nicolson of his pleasure at the successful visit of Kaiser Wilhelm II to England: it was necessary for all the great powers to hold together, he declared, in order to keep Russia under control.[149]

Less obvious, but nevertheless perceptible under the surface, was the effect of the Anglo-German royal family reconciliation on the position of the Bismarcks, not only because their leanings were towards friendship with Russia and scepticism towards Austria, but above all because in recent months they had taken their attacks on Wilhelm's mother and uncle to new heights. Herbert Bismarck had already sensed the danger on 30 July 1889 in Wilhelmshaven before the departure of the imperial party for England; but with characteristic cynicism he tried to make light of the Kaiser's changing mood, writing to his brother-in-law Rantzau: 'For the time being joy at the British admiral's uniform has the upper hand; but if the Tsar in his turn produces a similar token the scales will tilt back to the other side again.'[150] All too soon his eyes were to be opened to the seriousness of the new situation. Officially he expressed his satisfaction that the visit to England had brought an end to the 'ill-natured stories' that 'evil-disposed persons' had spread about to create 'bad feelings between the families',[151] but he feared that in private conversations the Queen might have stirred up her grandson 'against the name of Bismarck' and also against Russia, for he was in no doubt that 'H.M. had become decidedly more distant after the journey.'[152]

Wilhelm's mother could not but greet the improvement in Anglo German relations resulting from the Kaiser's visit to Osborne as a natural and necessary development that she herself had long sought. It had after all been the chief goal of her husband, in the interests of world peace and progress within Germany, to bring about the closest possible relationship between the two countries. Nonetheless, 'the part wh. only appears so *ironically* sad to me is that the *very persons* should suddenly profess this creed *now*, that did *all* in their power to *thwart* this, who denounced *me* as a traitor to Germany & *abused* Fritz in the most unjust & ungrateful way for remaining true to his Principles. These same people now make all these demonstrations of friendship!! I see *no* more urgent reason for it now, than *last year*, or than *always*, but this fitful capricious policy has long shocked me!'[153]

CHAPTER FIVE

The young Kaiser: a sketch drawn from life

In the months after his accession several attempts to throw light on the mercurial personality and as yet unfathomable mindset of the dynamic new ruler were published. At the same time as Count Douglas made the glowing eulogy of 4 October 1888 discussed in chapter 1,[1] the Kaiser's former tutor Hinzpeter, now living in retirement in Bielefeld, published a pamphlet entitled *Kaiser Wilhelm II. A Sketch Drawn from Life*, which was rapidly reprinted in dozens of editions, and which began, with the brazenness characteristic of its author: 'In all Germany the question is now being asked, with nervous anxiety, warm interest or at least with lively curiosity: What is the young Kaiser like?'[2] In astonishingly frank terms, Hinzpeter described the difficulties he had faced with Wilhelm's education. Quoting almost word for word from his anxious correspondence with the Prince's parents in the 1870s, he spoke of the 'curiously strongly developed individuality' of his pupil, the 'curiously crystalline composition' of his nature, the 'curious capacity of this mind, so unwavering in its course, to seize what pleases it everywhere', and the striking resistance which the 'very girlish boy, whose delicacy was increased to the point of weakness by a very troublesome clumsiness of the left arm', had shown towards any attempt to force his inner nature 'into a particular form'. The 'Doctor' described Wilhelm's 'muddle-headedness in thinking and condescending attitude in feeling', his 'dangerously low powers of concentration' which he, Hinzpeter, had tried to combat with 'the utmost rigour', but without any lasting success, for 'the inner being of the growing Prince always eluded ... even this pressure, sometimes immense ... ; it developed steadily according to his own nature, touched, modified, directed by outward influences, but never fundamentally changed or displaced'. Hinzpeter openly admitted his 'bitter feelings of disappointment' when it became clear that in spite of his efforts '[Wilhelm's] essential being remained unchanged.' He recorded with satisfaction, however, the strong, always easily aroused German national feeling and the 'eminently Prussian' nature of the Kaiser, and in a breathtaking (but thoroughly characteristic) sideswipe at his pupil's English mother he wrote of the 'discomfort of the hen who hatches a duck's egg': the 'pair of eagles', the Crown Prince and Princess, had had no right, Hinzpeter opined, to 'blame the egret who chooses his own trajectory'. If the 'young Prince bubbling over with words and show' had now become a 'steady, reserved, dignified ruler', this 'almost tropically fast ripening' could only be attributed to the profound emotional shocks which Wilhelm had suffered during the last months before his accession, his tutor claimed.[3]

The previous chapters have already shown the disturbing extent to which the political pronouncements and actions of the new Kaiser were determined by uncontrolled emotions. It is now time to take a closer look behind the façade of the Berlin Schloss and to echo Hinzpeter's question: what sort of a person was the young Kaiser? The aim will be to use his own writings, together with contemporary sources from his closest entourage, to investigate the pattern of the Kaiser's working day, the picture he had of his rights and duties as emperor and king, how he viewed the other reigning princes of Europe and how he stood in relation to the various parties and social classes in the German Reich in this initial phase of his reign. The next chapter will seek to throw light on the thinking of the Kaiser with regard to foreign affairs.

THE KAISER AND AFFAIRS OF STATE

In his first letters after coming to the throne Wilhelm self-righteously stressed the great burden of work which had fallen upon him as a result of the double crisis of the previous months, during which countless matters of state had not been attended to. In a letter to the Kaiserin Augusta on 11 July 1888 he claimed that he had been 'not at all prepared for the avalanche of work which engulfed me'. 'I would not have thought it possible for there to be such a stagnation in affairs of state as there was! In the last month under poor dear Papa things had come *de facto* to a complete standstill. All that had to be worked through now! Documents requiring signature piled up like sand on the seashore, and these now have to be gradually dealt with. For instance at Albedyll's audiences, which almost always last for 2 hours, I always have to sign 200 or more documents. Principally patents which date from last summer in Gastein! . . . Today I put in a good day's work with 250 signatures, and am really longing for a moment of rest.'[4] A week earlier he had used similar arguments to justify to Queen Victoria his failure to answer letters. 'The complete stagnation which had set in during the second half of Papa's time, left such an enormous amount of work to be done, especially unnumbered heaps of unsigned orders, Papers, Patents etc., so that for the first 3 weeks I had to work nearly night & day to get rid of all these things', he wrote to her.[5]

On the other hand the experiences of statesmen, diplomats, court officials and military officers in dealing with him immediately after the accession bear unmistakable witness to the very opposite of a regular, dutiful, 'dignified' way of life and work. During his first state visit to Russia and Sweden in July 1888 none of his travelling companions succeeded in persuading him to do any work. 'Nobody here has any luck with "reports", which H.M. avoids as far as possible', Kiderlen-Wächter observed; Wilhelm 'only wants to amuse himself'.[6] Kiderlen, the Foreign Office representative in the imperial suite, then reported to Holstein how Herbert Bismarck had tried in vain for days 'to get hold of H.M. to report to him' until he had at last succeeded, half an hour before lunch on 18 July, in reading out and handing over to Wilhelm the Reich Chancellor's instructions for his meeting with the Tsar. But the Kaiser, Kiderlen continued acidly, 'seems to regard all this as very insignificant compared with the question of how he will bring in his fleet'.[7] In his own memoirs Herbert Bismarck recorded that his father, at the request of the Kaiser, had drawn up a memorandum on the attitude which Wilhelm should adopt at the Russian court during his visit, and had given it to Herbert. 'General Wittich, to whom I gave it to read on the journey, was full of admiration for it, said H.M. should read it twice every day, instead of carrying out tedious naval manoeuvres, and was quite stupefied when I remarked that H.M. had not even looked at it yet. Eventually, just before Kronstadt, H.M. read it through quickly on the fore-deck.'[8] During the visit to Russia, Kiderlen went on to report, the Kaiser had only once taken an interest in political affairs. As a 'small point characteristic of our new sovereign', he told Holstein, Wilhelm 'could hardly be persuaded to attend to business' until he spent the night in Krasnoe Selo for the second time. But then he had suddenly had the military plenipotentiary von Villaume hauled out of bed at half past midnight 'and questioned him for a whole hour about all sorts of things!!!'[9]

On the journey from St Petersburg to Stockholm, Kiderlen again expressed surprise at the attitude of 'our new sovereign' to his official duties. The diplomat repeatedly complained of the difficulty everyone had in persuading the Kaiser to sign papers or grant audiences. 'To conduct business with W. II is, however, in many respects not easy', he wrote to Holstein. 'It is so difficult to get hold of him for that purpose. Wittich has carried around with him during the entire voyage the new drill regulations, which are to be submitted in draft form to a commission if approved by H.M. But H.M. could not be got hold of to approve them. Moreover Wittich reckons that there will be much more travelling, for apart from Vienna and Rome . . . H.M. will probably accept a number of hunting invitations; Wi[ttich] says he still lacks the ability to discriminate as to which should be accepted. People are already asking what will happen when he gets home, where a whole pile of things await him. Everything is being staked on the energy of Herr von Lucanus.'[10]

During the Baltic cruise the diplomats and military officers accompanying the new Kaiser had the opportunity of observing other personality traits which were scarcely in accord with the public image that was being propagated of him. Thus for instance Kiderlen-Wächter reported disapprovingly in a letter to Holstein on the argument which took place between the Kaiser and his brother Heinrich when the latter was given a regiment by the Tsar. When Wilhelm heard of the award he remonstrated with his brother, saying that this was 'quite out of the ordinary and could not be done, one could only be a colonel-in-chief when one was a colonel, but he, Prince H., was only a lieutenant-commander, that is to say major. Prince H. has now established that there were no majors or lieut.-colonels in the Russian Guards, and therefore the next highest rank after captain was colonel. He had this next highest rank and therefore was a Russian colonel; *quod erat demonstrandum!!!*'[11]

As to the 'high moral seriousness of the Kaiser' over which the Douglas speech and countless other publications enthused, the imperial travelling companions on board the *Hohenzollern* had another tale to tell. As early as the third day of the cruise Kiderlen wrote to Holstein: 'We always take our meals at H.M.'s table; up to now the chief subjects of conversation have been shitting, vomiting, pissing, fucking; pardon me for hurting your ear with these harsh words, but I cannot choose any others if I am to give you a true picture.'[12] The Kaiser's personal physician likewise had ample opportunity to reflect on the immaturity of his exalted patient. He blamed Wilhelm's excessively long service in Potsdam for the uncouth demeanour of the imperial entourage, which he considered positively dangerous. 'He [Wilhelm] knows too little of the world and still judges everything from the standpoint of the former colonel in the Hussars', Leuthold commented in a conversation with the Austrian embassy counsellor Eissenstein. 'I am very much afraid', the physician continued, 'that my gracious master will have many a disappointment and many a sad experience to face before his judgement becomes mature and unprejudiced. In his own circle the Kaiser has not yet become accustomed to his new position', he said. 'So for instance on board ship, particularly in the evening, there were wild goings-on, and it was very difficult to strike the right note again the next day. Many of the gentlemen had already been spoilt by the atmosphere of intimacy which so often takes over, and consequently they allow themselves various liberties, which then displeases the Kaiser. All this will no doubt change with time, but for now these incidents hold dangerous germs of unseemly conduct which could easily lead to storms and crises.'[13] Waldersee, who learned of it through Holstein, was likewise disturbed by the 'sort of table-talk that is going on' on the imperial yacht. He had hoped, he remarked, that accession to the throne would have brought about a change in this respect, and blamed Herbert Bismarck's influence for the unimperial tone on board.[14] After conversations with many of those who travelled in the yacht he noted in his diary at the end of August 1888: 'Unfortunately the tone did not improve on the cruise, chiefly of course because of the presence of [Herbert] Bismarck. The Kaiser is certainly inclined to adopt a rather light tone, but nevertheless he sees quite clearly that it is tactless for B. to permit himself to take the same tone towards him which was already scarcely acceptable when Wilhelm was still a Prince.'[15]

In court and government circles people wondered nervously whether such lapses of conduct on the part of the Kaiser were only temporary problems of adjustment, or whether they would prove to be a lasting characteristic. Only a few days after Count Douglas's speech in October 1888, Waldersee, who at this time was still cautiously optimistic, remarked that the Kaiser's entourage continued to be concerned about his restless and unfocused lifestyle. 'The people who have to work with him have been complaining for some time now that it is difficult to get him to grant audiences – he likes finding excuses and puts things off until the last moment', the General noted. 'If one considers what he has taken on and how occupied his time is as a result, it is certainly understandable that there is little time left for audiences; but on the other hand the work must be done and the other activities ought therefore to be limited. I hope that when the [foreign] journeys are over the Kaiser's whole way of life will become more calm and regular, for dealing with affairs in a rush is to be avoided at all costs. The inevitable and deplorable result will be that the Kaiser puts himself too much into the hands of the people to whom he gives audience.' Waldersee urged the introduction of an agreed schedule to bring more regularity and stability into the exercise of imperial authority. 'If the Kaiser wishes to conduct business like his grandfather – and that is what he would like – first the week and then the day must be precisely planned; without a firm plan it certainly cannot be done. There are some people who have misgivings in this regard; I am not among them. The Kaiser is indeed unusually lively, interests himself in countless matters which in the end always take up some of his time, and also aims not to give up his own pleasures entirely. But on the other hand the core of his nature is so good that he will certainly find the right way. The first period of his reign, with so many journeys, is not at all a good basis on which to form a clear judgement. Nor should it be forgotten that the Kaiser can work extraordinarily fast. He has a quick understanding, an excellent memory and makes swift decisions. That makes up for lost time.'[16] In spite of his readiness to enumerate Wilhelm's good qualities and to blame the Kaiser's initial failures on the particular circumstances of his accession, Waldersee saw clearly the danger that his new position and the intoxication of success might go to Wilhelm's head. It was to be feared, he warned on 13 October 1888, 'that the great successes which the young lord has achieved in his short reign, and the fact that he is cheered wherever he goes, and that all destructive elements already fear him . . . could give him false impressions of his own worth. The danger definitely exists that he considers himself more important than he is and that he is becoming less open to good advice. But the successes will not continue in the same way and there will inevitably be setbacks. He is rising too quickly.'[17]

Months after the accession there were still complaints on all sides that Wilhelm took plenty of interest in military matters but none in domestic affairs. 'It is said that Kaiser Wilhelm II practically never reads a political report', Count Széchényi reported to Vienna in January 1889, 'while he always examines the military ones closely himself. Thus as regards the former he will only listen to what he is told in audience, so that naturally a great deal depends on the way the matter is explained to him.'[18] In the imperial family it was even said that letters that were unpleasant to Wilhelm were thrown unread into a corner.[19] On the other hand he read the newspaper articles which were laid before him with passionate interest. As early as October 1888 Prince Bismarck had to reject, as both hopeless and dubious, a demand from Wilhelm for an amendment of the law which would restrict press freedom: Wilhelm would simply have to develop 'the necessary thick skin towards the press' and not allow himself to be so upset by newspaper criticism. The Chancellor likewise rejected his son's suggestion that the Under-Secretary of State in the Prussian State Ministry, rather than the Literary Bureau as hitherto, should be responsible for selecting the newspaper extracts to be laid before the Kaiser. Prince Bismarck thought this unworkable because no official would be able, in making the selection, to find 'the middle way between the dull and the dubious'. The extracts for the Kaiser needed to be 'piquant', and that would always be dangerous as long as the monarch allowed himself to be influenced and upset by newspaper articles.[20]

In the summer of 1889, a year after the accession, Herbert Bismarck summarised for his father the chief characteristics of the Kaiser's method of working. In a letter to his brother-in-law Rantzau he described it thus: 'In his politics H.M. acts hurriedly and in fits and starts, and that is because up to now he lacks a solid grounding based on study and reflection. H.M. does not like reading long documents (Wittich told me that whenever they run to more than 4 pages he regularly writes "oral report" on them without reading them); at least, whenever there are arguments and differences of opinion: he prefers narrative reports, and best of all newspaper articles. Ph. Eulenburg has also repeatedly observed that whenever papers are sent in to him he likes reading the newspaper extracts first, and usually puts written documents on one side to begin with. Thus it is that H.M. forms his views on the basis of *oral* information and discussions, occasionally influenced by press opinions. I have repeatedly spoken to Papa of my experience that one gets much further with H.M., and can do business with him more easily, by speaking rather than writing to him. So I often request an audience even if I only have things to read out to him, for I am sure that H.M. always raises one question or another on which it is then easy to reach agreement *in conversation. Written* reports, as I said, are read only in passing, and if they do not square with H.M.'s preconceived opinion they provoke contradiction and obstinacy and have rude marginal comments added. In conversation H.M. is always very polite and obliging. His grandfather was exactly the same, in fact.'[21] It was all very well for Herbert to speak, for he had almost unlimited access to the sovereign; apart from military officers and naval commanders, the other ministers and secretaries of state did not get near him. Even Holstein, who at this time still admitted to having 'a certain liking for him', recorded in November 1889 the 'complaints from everyone that H.M. dodges political reports'. Meanwhile Wilhelm would read thirty to forty newspaper clippings at one sitting, making marginal comments on them. He was 'a curious personality'.[22]

The complaints about the Kaiser's lack of order, both outward and inward, grew ever louder. At the time of the Douglas speech, which praised his 'imperturbable calm', and of the publication of Hinzpeter's pamphlet, which drew attention to Wilhelm's 'steady, reserved, dignified' character, in government circles there was widespread criticism of his 'passion for travel'.[23] In Potsdam, because of his constant absences, he was nicknamed 'Wilhelm the External'.[24] After his first Scandinavian cruise in the summer of 1889 Philipp Eulenburg deplored the Kaiser's hyperactivity, commenting: 'The health of the Kaiser was excellent – his restlessness immeasurable. His changeable appearance unfortunately indicates a certain nervousness of disposition which worries Leuthold, but then Leuthold is a pessimist of the worst kind.'[25]

From 21 October until 12 November 1889 a middle-class outsider, the commander of the battleship *Kaiser*, Vice-Admiral Paul Hoffmann, had the opportunity of observing Wilhelm II closely during the Mediterranean cruise from Genoa to Athens and Constantinople. He too was struck by the obvious reluctance of the monarch to grant audiences. On numerous occasions the Kaiser had said to the Chief of the Military Cabinet, who was accompanying him, 'at sea no audiences are held'.[26] During the entire voyage Wilhelm only once granted audiences, and that was in Athens; 'otherwise he refuses everything', Hoffmann commented in amazement. When the Flügeladjutant Freiherr von Senden-Bibran, on their arrival in Athens, asked permission to make a report on behalf of the Foreign Office, the Kaiser turned to Hoffmann and said: 'I surely cannot be expected to listen to reports when I have my first opportunity to see Greece.'[27]

If the Kaiser refused to hold audiences at sea, he nevertheless spent several hours every day in childish games. Hoffmann commented with bewilderment on the 'great tenacity' with which Wilhelm II played the quoits game 'Bleiglatt' on deck. 'This game is played to the point of exhaustion and I have absolutely no taste for it. All the other participants became heartily tired of it eventually, but the Kaiser is indefatigable.'[28] For light reading Wilhelm had brought with him the recently published lectures of his former German teacher Carl Werder on Schiller's *Wallenstein*,[29] but there was no question of serious work, the Vice-Admiral concluded, having observed the Kaiser for more than two weeks. This is how he described the course of a typical day on board: 'In the afternoon "Bleiglatt" was played again on the aft deck until dark. After luncheon the Kaiser sat down to play piquet. On extremely rare occasions the Kaiser occupies himself for an hour, when he reads for pleasure, e.g. lectures on Schiller's Wallenstein. But I have never yet known him to "work" on his own, that is, see to official business by himself. As a rule he demands social entertainment, distraction of some kind, and whatever kind it is he joins in with great vigour, more persistently than all other participants.'[30]

In November 1889 Waldersee noted in his diary for the first time a decline in the popularity of the Kaiser. Suddenly, he reported, there were complaints on all sides that Bismarck was being too lenient with the Kaiser. Only the Reich Chancellor had the authority to raise objections to two of Wilhelm's extravagant projects – the new *Hohenzollern* for four and a half million marks and the imperial residence in Frankfurt for two million. The initial popularity of the Kaiser, which rested on the fact that apart from the extreme left every party wanted to win his support, had already passed its climax and would soon 'go backwards', in Waldersee's view. 'It soared to the heights far too quickly and a fall is therefore natural and not at all harmful; the Kaiser must go through hard times in order to reach the heights to which he is entitled by virtue of his talents and his many excellent qualities. Very gradually a certain disappointment is setting in; the many journeys, the restless activity, the innumerable and varied interests naturally lead to a lack of thoroughness.'[31]

For the first time, now that he had gained a deeper insight, the Chief of the General Staff criticised Wilhelm's personality and the manner in which he exercised his rule, although he interpreted the latter as a consequence of the Bismarckian system. 'The Cabinet chiefs complain that they have difficulty in obtaining audiences and when they do, they are too short and hurried. The ministers have the feeling that the Kaiser ought to have thorough discussions of their departmental affairs with each of them from time to time, but he almost never does so. The Minister of War has regular audiences, and I myself have an adequate number, but Count Bismarck has far too many.'[32] Waldersee urged strongly that more 'regularity in the Kaiser's conduct of business' must be introduced. 'It is frequently asserted that he goes away too much and that this inevitably detracts from the thoroughness with which business is conducted. Unfortunately there is some truth in this', he admitted, 'although one must not forget that the Kaiser has a very quick understanding, and therefore works more quickly than many other people. Nevertheless I too wish that he would concentrate a little more. He shows plenty of interest in the matters which I bring before him, and no doubt likewise in military affairs in general, most particularly naval affairs. But his audiences on civil business cannot be very thorough; although he hears more from the Foreign Office, he hears very little indeed from the other ministries. I think, however, that all this will sort itself out. This year [1890] will bring many a serious internal problem in the country, and consequently the necessity of paying more attention to these matters; I hope that 1890 will prove a serious but excellent school for the Kaiser.'[33]

Waldersee failed to see that his own intimate relationship with Wilhelm was inconsistent with the well-regulated, responsible conduct of affairs of state. As with Wilhelm's friendship with Philipp Eulenburg, Waldersee's close contact with the Kaiser was the living contradiction of the claim made in Count Douglas's speech and elsewhere that the Kaiser was inaccessible to personal flattery and sycophancy and hostile to all types of coterie and camarilla. Like Bismarck, Waldersee – and later Eulenburg – was to experience in person how matters really stood with Wilhelm's willingness to receive an 'open, honest and true word' and with the 'truly royal gratitude' of the Kaiser.

THE KAISER AND THE 'MONARCHICAL PRINCIPLE'

It was certainly one of the foremost aims of Wilhelm II, as Douglas pointed out in his speech, to secure the 'monarchical principle' in Prussia and Germany, but his conception of this task differed in several essential respects from that of Bismarck. Whereas for the founder of the Reich the constitutional principle of Personal Monarchy was a legal fiction with which the authoritarian state could defend itself against the floodtide of parliamentarism and democracy, Wilhelm took the theory literally and saw it as the legitimisation of his personal power, indeed more than that, as the obligation laid upon him by heaven to defend monarchy by Divine Right. Nothing enraged him more than the suspicion that the Reich or state government, the Reichstag, the political parties or the press might be trying to interfere with his prerogatives as Kaiser, king of Prussia and supreme war lord. 'I am accustomed to being obeyed, . . . I do not enter into discussions . . . The word of a Kaiser should not be twisted or quibbled with', he insisted, even before Bismarck's dismissal.[34] In his eyes the rights of the Prussian crown inherited from absolutist times were now augmented by the dignity and immense power of the new kaiserdom, which he saw – and here his concept of the constitution differed sharply from Bismarck's interpretation – as the continuation of the medieval emperorship with its claims to universal domination. This was perhaps most clearly expressed in his memorable comment to his mother shortly after Bismarck's death: 'For ever & for ever, there is only *one real Emperor* in the world, & that is the *German*, regardless of his Person & qualities, but by *right* of a *thousand years tradition*. And his Chancellor has to *obey*!'[35]

From the beginning of his reign until the end of his life Wilhelm II regarded the battle for the defence of the monarchical principle against democracy as an international conflict between good and evil, the conduct of which devolved first and foremost on him as German Kaiser. In 1893 he declared to Pope Leo XIII: 'We monarchs represent Divine Right and conservative politics. Republicanism, and with it radicalism, on the other hand, is based on regicide, the abolition of God, and aims to overthrow all existing order . . . The people [of France] cannot find calm and stability because they have beheaded their King whom God set over them, dishonoured the Church and mocked the deity.'[36] Even in the First World War, when millions had died, he declared to the incredulous American envoy that only emperors and kings like himself, that is the Tsar and George V, had the right to decide over peace and war; mere republics like the United States and France could not have a say in such questions.[37]

With every year – with every electoral victory of social democracy, with every new sign of the democratisation in the Catholic Centre Party, with the success of the United States of America in achieving world power status – the Kaiser should have recognised that he was fighting a losing battle in defending his monarchical ideology. Moreover, the principle of Personal Monarchy which he upheld was increasingly called into question, not only from below and in the Western world abroad, but also from within – by his weaker and less autocratically inclined kingly 'colleagues', and Wilhelm reacted to these tendencies with a violence probably unprecedented in the history of the ruling houses of Europe. In the summer of 1895 he wrote to Crown Prince Gustav of Sweden and Norway, urging that his father Oscar II pull himself together and act with greater severity if he did not want to go down in world history as a traitor to the monarchical principle. 'The monarchical principle itself would suffer most seriously', he warned, if King Oscar did not take decisive action against the freedom movement in Norway. The monarchical principle had 'already fallen into disrepute through the Kings of Portugal, Serbia and Greece. May your father be preserved from being counted among such colleagues. His duty as monarch and king is to set his personal feelings aside and to do his duty, which requires of him that he impose respect and obedience to the royal authority in his lands.'[38]

The most remarkable source from which insight can be gained into the idiosyncratic thinking of Kaiser Wilhelm consists of his marginal notes on diplomatic reports. Countless such annotations, in the early days often written on small mourning envelopes, have remained unknown until today, as the Bismarcks had them secreted in a metal trunk in the Foreign Office: they were too explosive even for internal circulation within the Wilhelmstrasse offices.[39] Even while Wilhelm was Crown Prince, deputising for his father, the Reich Chancellor had to have documents bearing his marginalia locked away.[40] In the first months of the new reign Bismarck gave orders for several dozen foreign dispatches to be confiscated because of excessively crude marginal comments by Wilhelm. Even the founder of the Reich, it seems, did not have the courage to advise the young Kaiser to abstain from this disastrous habit. Instead a clerk was given the task of copying out the dispatches without the marginalia, as the original text was needed for the work of the Foreign Office.[41] Bismarck's successor General von Caprivi at first continued Bismarck's practice, but in the latter half of his time in office he allowed more and more documents to be circulated with Wilhelm's marginal comments. Under Reich Chancellor Prince Hohenlohe, and even more under Bernhard von Bülow, no one thought it possible or necessary any more to conceal Wilhelm's marginalia. This creeping moral decline in public life exactly corresponds to the rise of the Kaiser's power vis-à-vis the state and Reich administration in the first twelve years of Wilhelm II's reign. But it later led to a kind of optical illusion, when the records of the Foreign Office were published in more than forty volumes in the 1920s. Unaware of the confiscated marginalia the observant reader could not help gaining the impression from the published records that Wilhelm had only gradually acquired the taste for decking out the official reports of his diplomats with scathing annotations. In actual fact, as we have seen, the mischief had begun even before his accession. The early marginalia, which today are preserved in a reserve collection in the archive of the Foreign Office, throw a harsh light not only on Wilhelm II's strange ideology but also on his contemptuous opinion of other monarchs at the very beginning of his reign.

Wilhelm's dizzying, quasi-absolutist conception of his role as Kaiser expressed itself (as in a mirror image that shows one's own face in reverse, but nevertheless truthfully) in his derisive contempt for other rulers who sought or were obliged to maintain constitutional forms. He especially despised his Coburg relation Dom Pedro, who had abdicated as emperor of Brazil. Since then the country had been in a dreadful state, he thundered, and all 'because the cowardly monarch abandoned the post which God had entrusted to him!'[42] Contempt was mixed with indignation when he heard that Wilhelm, Hereditary Grand Duke of Luxemburg, who as a member of the Nassau family was descended from William of Orange, was to be received into the Catholic Church. The Kaiser described this conversion as 'disgraceful' and 'frightful!' 'The epigones are utterly unworthy of the great Prince of Orange', he pronounced, and declared that 'now the duty [to be the bulwark of Protestantism] passes to my House as next in line'.[43]

With disgust Wilhelm observed developments in Romania, which were strikingly portrayed in the colourful reports, carefully tailored for him, written by the German envoy Bernhard von Bülow, the future Foreign Secretary and Reich Chancellor. The Kaiser reproached King Carol, who was descended from the related (Catholic) House of Hohenzollern-Sigmaringen, with working 'valiantly' towards the overthrow of monarchical order in his country through his constitutional style of government. This 'short-sighted' monarch did not even notice that people were trying to get rid of him; he did not listen to Berlin, which had already tried 'who knows for how long' to open his eyes to the danger of a pro-parliamentary policy. Thanks to the 'limpness' and blindness of the King, Romania was going the best way to become a second Poland, which would be simply 'gobbled up' by Russia. 'Polish conditions' already ruled in Romania, Wilhelm claimed in February 1889, and he annotated a report with a quotation from *Wallenstein*: 'You are blind with your seeing eyes!' As we have seen, he had just been reading Carl Werder's lectures on Schiller. The Romanian monarchy would 'be destroyed by its muddle-headed idealism!!', prophesied the young German ruler. 'Incredible!' was his verdict on the appointment, reported by Bülow, of a liberal government under Lascar Catargis in February 1889: the new Prime Minister, he declared, was a 'cur!', he and his foreign minister were 'milksops!', the Cabinet was a 'fine rabble!' that resembled 'a Richter-Windthorst-Rickert-Virchow ministry'. The 'blessing of parliamentarism combined with constitutional kingship' would be anarchy, Wilhelm predicted in November 1889. Carol's willingness to make concessions – Wilhelm mockingly named him 'Carol the Great'[44] – he could only understand as a wrong-headed calculation that he should give the 'billy-goat . . . a few cabbages so that he does not devour the whole garden'. 'Landgrave, Landgrave, be firm', on the other hand, would have been the only wise motto for the King. When Bülow reported the comment of a Romanian politician that the King ought to cut himself out of politics, Wilhelm was shocked and remarked, 'The public prosecutor ought to be called in here!'[45]

Of the Romanian Queen Elisabeth, born a Princess of Wied, who wrote poems under the pen-name of Carmen Sylva, Wilhelm had as disparaging an opinion as of the King himself. 'Obstinacy and vanity on the part of the monarch and poetic, humanistic fantasising and attacks verging on nymphomania on the part of his consort are hard to combat, but give one little confidence in a level-headed reign and future', he commented in an annotation of March 1889.[46] The Queen's constitutional ideas – she was said to have expressed the wish to be queen of a republic – the Kaiser considered 'high treason and treason to the country'; in his view Elisabeth belonged in a 'padded cell!'[47] The King really ought to put her away 'in a hydrotherapic institution', he remarked; she was after all nothing but 'a crazy bluestocking!'[48] In any case she ought to keep her fingers out of politics, for 'petticoats don't belong in politics, especially when they have blue stockings'.[49] 'Wherever there is mischief: chercher la femme!', he scribbled in the margin of one report from Bucharest.[50] In view of the progressive parliamentarisation of Romania, Wilhelm was only too glad not to have any closer connection with this country. Commenting on a complaint by a Romanian nationalist that Romania, because of its German ruling house, had sunk to the level of a German province, the Kaiser retorted: 'Thank God no, what would I want with such brutes?! You are all frog-arses in Germany's eyes!'[51] It would do nothing but good if the Russian influence in Romania became ever stronger. 'In any case the ruble goes on doing its work at every level.'[52]

Wilhelm also poured scorn and mockery on Bulgaria, where his cousin Ferdinand of Saxe-Coburg and Gotha had been elected king in July 1887 by the national assembly. On the back of a mourning envelope, which Bismarck locked away at Friedrichsruh, the young Kaiser wrote: 'Petit Ferdinand holds out only because the powers cannot agree on anyone else, and only so long as mama [born a princess of Bourbon-Orléans] throws money around. A few clauses of the old Treaty of Berl[in] are energetically defended because they seem to be appropriate at the moment; and they are apparently to be considered sacred and unassailable.'[53]

Wilhelm II took an equally avid interest in the dramatic events in Serbia, where King Milan, the friend of his youth, divorced his wife in October 1888 and then renounced the throne on 6 March 1889. At first the Kaiser did not believe Milan's intentions were serious. 'If someone runs about with a pistol in his pocket and assures everyone he is going to kill himself', he observed in November, 'then one can wager 100 to 1 that he certainly will not do it! And likewise in this case!'[54] But as the rumours of the forthcoming abdication grew stronger Wilhelm responded by suggesting ways of persuading Milan to stay, in order to maintain Austrian influence in the Balkan state. In December 1888, when Prince Reuss reported from Vienna that the Russians might offer the Serbian king, whom they hated, a pension to help him decide to give up the throne, Wilhelm declared: 'As far as I know King Milan's character and personality, the idea of the "Russ. Pension" seems perfectly probable and credible. Could he not be helped with an Austrian pension? Austria's influence in Serbia must not be lost, and must be upheld there by all means. Otherwise the Austrians will later have a threat to their flank!' Patiently, Bismarck had to point out the absurdity of this idea with the comment: 'But if Milan *has* abdicated then nothing more can be gained by a subvention. Austr. would have to bribe him to stay!'[55] A few days later Wilhelm suggested that as he was 'very close to the King and intimate' with him, he could himself write a 'persuasive and encouraging letter'.[56] This suggestion was accepted by the Chancellor, who drafted a letter from the Kaiser to the Serbian King designed to persuade the latter to give up the idea of abdication. When Reuss thereupon sent word to Berlin that Milan was insisting that he had no alternative but to abdicate, 'because he feared that he would certainly go mad if he remained in his present situation', a distinct change took place in Wilhelm's attitude. Several times in his marginal comments he recommended the dispatch of his friend to the cold-water treatment institute in Bad Godesberg, and wrote: 'Cold water on the head and Friedrichshall [hot salt] waters everywhere else will at least help!' At the beginning of March, when the news of Milan's imminent abdication reached the German capital, Wilhelm commented spitefully: 'Better a nasty end than prolonging the agony!'[57] And in response to a further report, on the oriental journey on which Milan embarked after his abdication, the Kaiser observed that the ex-King seemed 'to have been very busy f . . . ing [sic]'.[58]

The fascination with the sexual life of other monarchs and members of their families apparent in such comments was a characteristic of Wilhelm's marginalia in general, from which interesting conclusions may be drawn about his own psychosexual constitution. He once joked that he would 'very gladly' visit the Sultan again, 'if I could go into the harem'.[59] Of Crown Prince Ferdinand of Romania, who for a time contemplated marrying Wilhelm's youngest sister, the Kaiser wrote on an official diplomatic report: 'He can't do anything but f . . . [sic], and he already takes care of that here anyway.'[60] When news arrived to the effect that King Carol was concerned about Ferdinand's delicate health, His Majesty noted in the margin: 'He survived very well in Potsdam and the way he bedded the girls was a delight!'[61] The marriage of the former ruling Prince of Bulgaria, Prince Alexander of Battenberg, to the actress Johanna Loisinger in February 1889, as might be expected, drew equally disgraceful comments from the Kaiser. He described the Prince as a 'cur!' and accused the bride of imprudent behaviour 'of a horizontal nature'. On Prince Alexander's changing his name to Count Hartenau after his marriage, the Kaiser commented contemptuously, 'Once a Polack always a Polack.'[62] He perhaps sank lowest with his jibes in July 1889 when Schweinitz reported from St Petersburg that the Russian Princes Ivan and Gabriel Constantinovich would have to look for Orthodox brides in the Balkans one day. Kaiser Wilhelm II saw fit to record, on the ambassador's official report, his opinion that the Princes would 'probably look after themselves elsewhere meanwhile!'[63] At this time the Grand Duke Constantine's sons were respectively three and two years old.

Such comments were not at all exceptional. When Schweinitz reported that the Tsarevich Nicholas had 'gained a broader knowledge of the fair sex' on his educational tours in Egypt, India and Japan, the Kaiser observed: 'That would be a blessing, for up to now he has refused to go near any woman.'[64] Commenting on a dispatch from Count Casimir Leyden in Cairo according to which the Russian heir and the Greek Crown Prince George had given diamond brooches to the little girls who offered them flowers on their Nile cruise, the All-Highest wrote: 'As these girls are usually naked, it might be interesting to know where they put the brooches?'[65] Learning from a further report from St Petersburg of a feud between the 'bigoted' Grand Duke Sergei and an archimandrite of the Russian Orthodox Church, Wilhelm II gave his explanation of the incident: 'That is because the worthy old man discovered that Sergei was buggering his handsome young domestic chaplain. He transferred the latter at once. This enraged the pious prince so much that he contrived to have the old man transferred!! I have spoken out before now of my fear that Grand Duke Sergei would bring about the downfall and destruction of his family. It almost looks as if that were so.'[66] He accused the (originally middle-class) wife of the aristocratic ambassador of the French Republic in St Petersburg, Count Gustave Lannes de Montebello, of being 'also available for adjutant generals and grand dukes', and moreover of having 'travelled successfully on the Isle of Lesbos'.[67] When the German embassy reported on the great social success of Countess Montebello especially in grand-ducal circles of St Petersburg, the Kaiser observed: 'So it is as I said before, they will all have had their turn.'[68]

Other ruling princes served simply as targets for imperial mockery. Thus for instance Wilhelm noted on a report from Athens, in which attention was drawn to the organisational talents of the King of Greece, a scion of the House of Denmark, that the King was perhaps better fitted to be a court marshal 'than to be Rex'.[69] On a British newspaper article praising the constitutional methods of the Greek King, Wilhelm wrote angrily 'balderdash!', 'rubbish!' and 'utter nonsense!'[70] When it was reported to him that French diplomats feared that the engagement of his (Wilhelm's) sister to the heir to the Greek throne could turn Greece away from France and towards Germany, Wilhelm saw it as merely a sign of incipient madness on the part of the French and wrote one of his favourite sayings, 'Quos deus perdere vult! etc.' in the margin of the report. Nevertheless he hoped that this French attitude would bring about a beneficial effect on the Danish royal house. 'I trust this conduct will deal a hearty blow to the Francophile inclinations probably still latent in Copenhagen', he commented.[71]

When the Empress Frederick told her mother of the primitive conditions in which her daughter Sophie had brought her first child into the world in Athens, she begged her not to repeat any of it to Wilhelm, for 'He always abuses the King & the Greeks – & it *comes round* to them again, & makes things difficult & uncomfortable, he is *so* rash imprudent injudicious & thoughtless.'[72] After Sophie's conversion to the Greek Orthodox Church[73] the relationship of the German Kaiser to the Greek royal house, which had long been tense, became increasingly hostile. The King of Greece had 'perhaps been coached by his Mama in Copenhagen', Wilhelm wrote scornfully on a report from Athens.[74] 'These petticoats should just keep their fingers out of things', he commented angrily on receiving a report in 1897 on the influence of Danish, Russian and British relations in Athens.[75] Echoing a comment by the King of Saxony, Wilhelm wrote that the King of Greece 'is and remains the "Greek gamin" as King Albert always calls him. And rules over a band of robbers.'[76] 'How can one expect courtesy from such a worthless and ill-mannered man as the "Greek gamin"!!', he asked.[77] The King's sons were 'louts without any education'.[78] In further marginal comments the Kaiser characterised the Greeks as 'miserable scoundrels' and the Athenians as 'unreliable blackguards'.[79] On a report from the German envoy in Athens in which it was stated that neither the Greek court nor the Greek people had taken much notice of the Kaiser's birthday, Wilhelm commented that the coolness of the Greek King and Queen was 'very indicative', for 'we do not love each other'. The Greeks were 'piqued', he considered, 'because I have not arranged for them to have Constantinople'.[80] When a report arrived from Lisbon announcing that on 27 January the Portuguese King Carlos had not sent anyone to the German embassy to bring his congratulations, Wilhelm II denounced this 'colleague' too, as a 'bad-mannered boor like the Hellene'.[81]

The gathering of monarchs in Athens on the occasion of the marriage of his sister Sophie on 27 October 1889 confirmed the predominantly negative opinions of the Kaiser about his fellow rulers. When the Kaiser arrived in Piraeus in the *Hohenzollern* there was no welcoming party of Greek ships, the reception was disorganised, nothing had been prepared and neither the King nor the Queen of Greece was present. Wilhelm would not even have credited the King with the ability of a court marshal now, as he gave vent publicly to 'his displeasure at the prevailing confusion'.[82] From now on the Greeks were nothing but a 'wretched, pretentious people' in his eyes.[83] The commander of the battleship *Kaiser*, who attended the banquet at Athens, gave a vivid description of the royalties assembled for the wedding: the bride's mother, the Empress Frederick, looked much younger again dressed in a pale grey gown; her brother the Prince of Wales was as usual the picture of health; his Danish wife Alexandra looked 'almost as young as her two daughters, the son Eddy [Albert Victor] rather foppish with his swan-neck, the second [George] very sensible'. The Tsarevich Nicholas, on the other hand, gave the impression of a 'juvenile Muscovite youth' and was 'as brash as could be in his manner'; the King and Queen of Denmark seemed more like a 'fusty bourgeois councillor and his wife'. At the ceremony known as the 'baisemain' next day, at which Princess Sophie appeared in modern Greek costume, the American and French envoys – 'no doubt as republicans!' – refused to kiss the bride's hand.[84]

On 31 October the German squadron, consisting of the *Kaiser*, *Hohenzollern*, *Deutschland*, *Preussen*, *Friedrich der Grosse*, *Irene*, *Pfeil*, *Wacht*, the Lloyd's steamer *Danzig* which had been hired for the very large suite and the flagship of Admiral Deinhard, the *Leipzig*, sailed onwards to Constantinople, where three large steamers packed with Germans came out to meet them, greeting them 'with continuous cheering, singing of patriotic songs and waving of handkerchiefs'.[85] For the reception by Sultan Abdul Hamid II, Kaiser Wilhelm II put on his red Hussar uniform, 'doubtless', Vice-Admiral Hoffmann assumed, 'because the green sash of the order looks very Turkish against it'.[86] Here, for German eyes, an exotic, almost sinfully opulent monarchical world revealed itself. The pavilion in which the Emperor and Empress were accommodated in Yildiz was 'excessively rich' and furnished with every possible luxury.[87] 'The dinner was very elaborate, the table was a magnificently arranged, we ate off nothing but gold. The centre-pieces on the table were all of gold and bronze, likewise spoons, knives and forks . . . The extravagance which is shown off in the Sultan's palace disgusts one because one knows how wretched things are in the country. The Sultan is in his middle forties', Hoffmann observed; 'he is thin, has very sharp, Jewish features and gives the impression of having some nervous disorder.'[88] As a present for the Sultan the Kaiser had brought a large rococo clock decorated with figures costing 1,800 marks and two nine-branched candelabra worth 1,100 marks,[89] but these costly gifts were far outdone by the Sultan's presents to his guest. On the Kaiser's departure Abdul Hamid presented him with 'an extraordinarily costly sabre' and the Kaiserin with gifts including a piece of jewellery which alone was valued at £10,000.[90] In general the members of the imperial entourage were repelled by the 'senseless luxury', the 'court flim-flam, the humbug about orders, the miserable business of etiquette' in Constantinople.[91] When they embarked on 6 November Hoffmann observed that even 'our exalted master and mistress' were glad to be back on board, particularly as the Kaiserin claimed to have found bed-bugs in her rooms.[92] In political terms Wilhelm II's visit to Constantinople, which was viewed with suspicion in Russia, brought few advantages. On taking leave of the Sultan the Kaiser expressed the hope that he would look on 'his (the Kaiser's) policies from now on no longer with mistrust', to which Abdul Hamid replied that people were 'unfortunately attempting to arouse his suspicion of

German policies and that this had even been attempted again in the course of his exalted guest's visit'.[93]

Thus the impressions that Kaiser Wilhelm II formed of his royal 'colleagues' in Athens and Constantinople were not of a kind to reinforce his belief in the power of the monarchical system to withstand the struggle against the democratic current of the times. Nor was it exactly strengthened by the secret information about the death of King Luiz of Portugal which arrived during the wedding festivities in Athens: when the doctors began the autopsy on the 'corpse', they heard the apparently dead king begging them not to hurt him so much![94] The Kaiser's meeting with Shah Nassr ed-Din of Persia, the 'King of Kings', in Berlin in June 1889 was equally ill calculated to sweep away any lingering inner doubts. On his Mediterranean cruise in November the Kaiser was still telling 'all kinds of tales' about the Shah, and said to Hoffmann: 'He was a queer customer. When my wife showed him our children he said "Sont tous de vous Madame?"'[95] A certain surprise speaks out of the letter in which Wilhelm reported to the Reich Chancellor on his conversations with the Persian ruler. 'The Shah *wants*, without a doubt, to be amiable. He makes what is for him a very considerable effort to show his admiration and satisfaction.' Nevertheless he was clearly upset by the treatment he had received in Russia, and above all by having been bundled off to Warsaw, where he had been forced to spend thirteen days. 'Varsovie pas belle ville', the Shah had complained to the Kaiser. At dinner on 11 June he had questioned Wilhelm closely about the state of the Italian army and navy and had wanted to know, when the 'March for the Entry into Paris' was played to him, 'whether we would march in there again, to which I replied yes, if we have to'. Thereupon the Shah had whispered softly to him 'Français méchants, très méchants', and when the Kaiser admitted 'that it was not very easy to live with them', the Shah responded: 'Ce sont des Saltimbanques.'[96] Wilhelm subsequently told the British ambassador that the Tsar had threatened the Shah with an invading force of 100,000 men if he made concessions to the British; the Persian monarch had twice asked him, the German Kaiser, for help in such an event.[97]

The Kaiser could scarcely contain his contempt for his British – and admittedly very odd – cousin Albert Victor, the elder son of the Prince of Wales and presumptive future king of England, when he was granted the title of Duke of Clarence in 1890. In a lengthy marginal note, in which he went so far as to allude to current rumours about the homosexuality of the newly created Duke, the Kaiser wrote: 'Well! I must say! He can't speak German, nor French; he didn't know whether Munich was a river or a town and had no idea who Fred. the Great was! And furthermore he was under serious suspicion last autumn. And that is called being superbly prepared [for his calling as a monarch]!!! In the end it is of no consequence in England as he is not expected to have anything to say anyway, nor will he.'[98] When the news of the sudden death of the Duke reached him in January 1892 the Kaiser did not even break off his shooting expedition in Bückeburg, which was taken very much amiss by the family.[99] Colonel Leopold Swaine, who returned to Berlin as military attaché in early 1892 after an absence of three years, was shocked by Wilhelm's coldness and especially by his refusal to attend the service held in Berlin in honour of his dead cousin. 'The extraordinary behaviour – almost heartlessness – of the Emperor on the occasion of the death of the Duke of Clarence caused surprise and regret even amongst the immediate surroundings of the Court, but none of them had the courage openly to express their opinion', the military attaché wrote in disgust.[100]

As for the German federal princes, Kaiser Wilhelm II expected them to maintain a 'national' demeanour, an expectation in which he found himself not infrequently disappointed. This was particularly true with the Catholic Bavarian royal house, which demonstrated a strongly particularist attitude. When the Württemberg envoy in Munich, Soden, found himself being attacked in Bavaria because of military agreements which had been reached between Württemberg and Prussia, Wilhelm II wrote on the report from Munich: 'Then Soden should do his duty and tell the stuck-up Wittelsbachs that what the King of Württemberg does is nobody else's business!'[101] The Kaiser regarded the close relationship between Grand Duke Ludwig IV of Hesse-Darmstadt and the British and Russian royal families with suspicion. Ludwig had been married to Princess Alice of Great Britain, the Empress Frederick's sister, who had died in 1878, and his daughter Ella was married to the younger brother of Tsar Alexander III. When Schweinitz reported early in 1889 that the Grand Duke was to prolong his stay in St Petersburg until after the Tsar's birthday, Wilhelm wrote scornfully on the report: 'I'll be damned a million times! He'll be given the title of a Russian grand duke next!'[102] To the amazement of many, Wilhelm II's behaviour was quite different in the case of another German prince with British connections. When his great-uncle Duke Ernst II of Saxe-Coburg and Gotha died in August 1893, the Kaiser was said to have 'knelt, or as some say lain, before the coffin, bathed in tears', as Waldersee reported, although he was unable to take the imperial grief seriously, as Duke Ernst had 'never been really close to the Kaiser', and it could 'not have escaped the latter's attention how little esteem the Duke had enjoyed in the world'. In Waldersee's eyes there could indeed be 'scarcely anyone who does a throne so little honour as this Coburger'. The late Duke, he maintained, had been 'a man of a very low moral calibre', who was 'a thoroughly bad character, false, lying, a braggart and an intriguer'. The successor to the dukedom was Alfred, Duke of Edinburgh, Queen Victoria's second son, which was at first considered undesirable, as a British prince and admiral seemed unlikely to make a satisfactory German ruling prince. But the mood changed when the new Duke 'chased away' some of his predecessor's favourites and mistresses.[103]

The Kaiser was particularly enraged by King Karl of Württemberg's habit of spending the winter enjoying the delights of life on the French Riviera with his homosexual American favourites. On a report from Stuttgart he recommended that 'the person in question be advised that it is not fitting for ruling princes of the Reich to amuse themselves in Cannes in international company, while German citizens have no rights or protection in France'. Wilhelm had to admit, however, that 'unfortunately' other German princely personages were just as disinclined to allow the tense political situation to deter them from spending every winter in Cannes or Nice.[104] After King Karl's death in October 1891 Waldersee warned the Kaiser against the particularist inclinations of his successor, Wilhelm II of Württemberg, whose anti-Prussian attitude was reinforced by the influence of his Hofmarschall Plato, 'a very nasty fellow', and of his scheming wife. 'The Kaiser, however, considered that he had such a hold over the Prince [Wilhelm] that he would not be able to step out of line.'[105] In the end there were only two federal princes for whom Wilhelm II retained a certain respect: King Albert of Saxony and Grand Duke Friedrich of Baden, but even their influence on the young Kaiser was limited and temporary, as we shall see.

There was, however, one instance in which the Kaiser's closest confidants thought they could still detect an uncharacteristic youthful diffidence on his part, which indicated that he 'still did not [have] a sufficiently strong sense of being sovereign and kaiser in dealing with his older relatives'. 'Time must change that', Herbert Bismarck observed.[106] This remark related to the uneasy impression which the Kaiser made in his conversations with Philipp Eulenburg on 2 January 1889 about the 'strange Guelphic ambitions' of Prince Albrecht of Prussia, an uncle of the Kaiser, who had been acting as regent in Brunswick since 1885. Albrecht was married to Princess Marie of Saxe-Altenburg, whose aunt was the mother of Duke Ernst August of Cumberland, the pretender to the throne of Hanover, Brunswick and Lüneburg, hence Wilhelm's suspicions. Yet the Kaiser, so Eulenburg reported to Herbert Bismarck, was unusually restrained in his condemnation of his uncle's 'Guelphic ambitions'. 'I have so much admiration for him – he has so much admiration for me', Wilhelm had declared. Eulenburg gained the impression from this conversation 'that it is painful for His Majesty to play preceptor to his uncle'.[107] Later Wilhelm confided to his best friend that he was deeply concerned for Prince Albrecht's 'character and frame of mind in general'. 'The various cures which my uncle undertakes in Dresden are officially intended to strengthen his nerves. In fact matters are much more serious: by taking colossal quantities of snuff he has overstrained the nerves in his head and nose so much that the effects of the overstrain are already spreading dangerously towards the brain. In short, nicotine poisoning of the brain through snuff is looming in the background; this could lead to a cold-water treatment institution or a lunatic asylum.'[108]

Nothing demonstrated more clearly the human and political fragility of the traditional monarchical world which Kaiser Wilhelm II represented and was determined to uphold against the tide of changing times than the death of his contemporary and erstwhile friend, Crown Prince Rudolf of Austria, in Mayerling on 29 January 1889. With good reason Herbert Bismarck reported to Friedrichsruh that the Kaiser was 'very shocked' by this event and could 'even now scarcely believe that it was suicide'.[109] The Austrian ambassador in Berlin reported that after hearing the news of the death the Kaiser had spent a considerable time with him, 'and I can assure you that he did not put on any conventional demonstration of sorrow, but showed a profound sense of shock and a sincere, heartfelt sympathy. The Kaiser could not keep still, moving about constantly, sometimes standing, sometimes sitting, and everything he said revealed his inner agitation.' Wilhelm expressed the wish, 'only natural' for him although it remained unfulfilled, of going to Vienna in person for the funeral, Széchényi reported. Again, after the memorial service in Berlin, Wilhelm and the Kaiserin lingered for some time at the Austrian embassy and demonstrated 'the heartfelt sympathy and deep distress with which Their Majesties were both filled'. The Kaiserin had been exceedingly upset by the news, Széchényi reported, and had broken down in tears.[110] Through Rudolf's friend Prince Philipp of Coburg, however, Wilhelm heard, as he wrote to Queen Victoria, 'that lunacy was looming in the background & that the monomania of suicide had done its silent but sure work on the overexcitable brain'.[111] When Wilhelm was about to visit the grave of 'the unfortunate Rudolf' for the first time in September 1890 Auguste Viktoria wrote feelingly: 'A friend of your youth in the same position in life, and *how* different, thank God, the course of your lives! One can easily see there what a difference it makes whether someone has built on the right ground or not!!'[112]

In the judgement of many contemporaries, above all those who knew nothing about the torments of his childhood, Kaiser Wilhelm II's blatant arrogance was a consequence of the apparently carefree life he had led hitherto. Thus for instance Vice-Admiral Hoffmann commented in 1889 while on board ship with the Kaiser: 'Great self-esteem is an essential characteristic of the Kaiser . . . With a young, very gifted person, who suddenly becomes Kaiser, who has never had any unpleasant experiences, for whom everything is made palatable, it can hardly turn out otherwise. The great eloquence which he possesses to a rare degree is also part of it. Yesterday he told us that neither the Kaiser of Austria nor the King of Italy was able to say a single word of their own accord; they read every after-dinner speech etc. from a paper.'[113] Whatever the truth of that may be, it is quite clear that his strange imperial ideology and his extraordinary self-satisfaction led him to despise weaker and less imperious fellow monarchs to an extent which is probably unequalled in the history of the royal houses of Europe.

Surveying the ruling dynasties at the time of Kaiser Wilhelm II's accession, however, one is forced to admit that his sense of superiority, viewed objectively, was not altogether unfounded. With the possible exception of the British royal family, among monarchs and their heirs there was no prince who came anywhere near Wilhelm in eloquence, energy or breadth of interests, no charismatic power-figure to match him. Wherever he looked he saw suicide, abdication, imminent (or actual) madness, marital breakdown, sexual excess, conceited narrow-mindedness, petty bourgeois obtuseness, sheer incapacity and poisonous scheming. In comparison with his 'colleagues' Wilhelm could indeed be seen, in spite of all his failings, as 'the most important man in Europe', as a British peer was later to describe him,[114] and likewise in 1910, after the death of King Edward VII, a South German baron was able to say of Wilhelm II: 'Now that his Uncle-King is dead, he is more than ever the decisive voice in the world. He and Roosevelt will be the masters of history from now on.'[115] At any rate his experiences with his fellow monarchs can only have reinforced his conviction that Prussia–Germany under the Hohenzollerns had been chosen by God and world history to be the bulwark of the 'monarchical principle'. Doubts as to whether his dizzyingly ambitious monarchical programme was at all practicable under such circumstances clearly did not enter his mind in these early years.

THE KAISER AND GERMAN SOCIETY

Internally the Kaiser's monarchical ideology expressed itself on the one hand in his glorification of the armed services, the Protestant Church and the aristocracy as the natural pillars of the throne, and on the other in his contempt for civilians and above all for Left Liberal and Social Democratic parliamentarians, Catholics and Jews. In August 1888 he took over the patronage of the Protestant Order of St John of Jerusalem and gave a speech in Sonnenburg in which he alluded to the 'great tasks' which lay before him 'in the sphere of the inner development of my people'. Elaborating on this theme, he declared: 'In order to raise up the people, to strengthen and develop their moral and religious feeling, I need the support of the noblest of them, of my noblemen . . . I hope with all my heart that I may succeed . . . in bringing about the elevation of the spirit of religion and Christian discipline and morality in the people, and thus in achieving the high aims which I have set myself as ideals.'[116] Not only Wilhelm's mother and her left liberal Progressive Party friends were incensed at this speech; even aristocratic diplomats found that 'to address that band of frauds the Order of St John of Jerusalem as "the noblest of the nobility" . . . was going a little too far'.[117]

Civilians, above all democratically minded ones, were third-class citizens for Wilhelm. Not long after his accession he declared to the Italian ambassador Count Launay: 'After the French, the people I hate most are diplomats and deputies.'[118] He made it abundantly clear what sort of behaviour he expected of his diplomats and officials in a crude comment in the margin of a submission in which Prince Hohenlohe had argued for the title of governor to be conferred on the most senior officials in all German colonies. 'As far as I am concerned the gentlemen can stick a blue button in their navels and a peacock feather up their arses too! So long as they pay increased attention to their duty to match their increased rank.'[119] Some of his remarks about civil servants were simply insulting. He described Burghard Freiherr von Cramm, the veteran delegate to the Bundesrat from Brunswick, as 'this evil little toad!'[120] Referring to one of his diplomats, who presumably had a conspicuously large nose, Wilhelm wrote on a report submitted to him by the Reich Chancellor: 'With this gentleman's facial characteristics he only needs to put on a fez and anyone would take him for a Cretan fellow at the very least, if not indeed for an Armenian!'[121] He spoke of members of parliament and other elected politicians with deep contempt. When he visited the King of Italy in Rome in October 1888 he remarked: 'At home [in Germany] if one sees a black coat and asks "Who is that?", if the answer is a deputy, one turns one's back on him.'[122] When he was informed from Madrid that the American President Cleveland did not wish to call on the King of Spain for 'democratic' reasons, Wilhelm II wrote: 'Once a bumpkin always a bumpkin!'[123] On a diplomatic report from St Petersburg about the French ambassador there, Count Gustave Lannes de Montebello, the Kaiser wrote that one could not expect decent behaviour 'from a republican'.[124] When Wilhelm and his entourage were compared by a French newspaper to Friedrich Wilhelm I and his tobacco club, the Kaiser noted in the margin: '*He* would not have sat down with civilians.'[125] But on reading another report from France, according to which the French army command was financially independent of parliament and had raised army units without budgetary authorisation, Wilhelm commented: 'That is exactly right . . . I shall do the same if the Reichstag does not consent.' He gave orders for the 'excellent' article to be 'thoroughly aired and discussed in the press'.[126]

The Progressive movement, which advocated the parliamentarisation of the Prusso-German military monarchy on British lines, and which had been particularly closely associated with his father and mother, was singled out for Kaiser Wilhelm II's special contempt. On 18 October 1888, Friedrich III's birthday, an article in memory of the late Kaiser appeared in the *Berliner Zeitung*, which

Wilhelm II passed on to Bismarck as, in his words, 'proof of the attitude bordering on lèse-majesté of the democratic Press'.[127] A few days later, in a speech to a deputation from the city of Berlin, which had come to greet him on his return from Rome and to present him with the deed of gift of the fountain designed for the Schloss by Reinhold Begas, the Kaiser declared that he had encountered warm sympathy for the German Reich among the foreign rulers and people he had visited, but that he had also brought back a 'most painful' memory from his journey. 'While I have invested all My health and strength in creating bonds of friendship in order to ensure the peace and welfare of our Fatherland, and thus also of its capital city, the daily newspapers of My capital city and My residence have dragged the affairs of My family before the public and discussed them in a way which a private citizen would never have accepted. This does not only painfully affect Me, but it has also aroused My anger. Above all I must insist that the continual invoking of My Father against My person should finally cease. It offends Me very deeply as a son and is unfitting in the highest degree. It is My expectation that, if I choose Berlin as My principal residence – and as a Berliner I am always drawn back here – others will abstain from making the intimate relationships of My family a subject for discussion in the press.'[128] Thereupon the Kaiser dismissed the deputation without shaking hands with the Chief Burgomaster Max von Forckenbeck, who was a Progressive, and without allowing the members of the deputation to be presented to him.

There was general astonishment and bewilderment at this speech.[129] The Austrian chargé d'affaires reported to Vienna: 'Various gentleman of the court, high-ranking officers and government officials expressed to me the fear that the incident in question would make a highly unfavourable impression on the broad mass of the population, and would be, so to speak, grist to the mill of the Progressive Party. It is considered that the occasion was ill chosen for the Kaiser to express his otherwise well-justified indignation against the activities of a certain party, just at the moment when the city is preparing to offer a lasting tribute to the Kaiser in grateful remembrance of His Majesty's journey, and it is thought that the imperial words were misdirected, as one cannot hold the Chief Burgomaster of Berlin responsible for the behaviour of a section of the national press.'[130] The British chargé d'affaires, Beauclerk, was likewise of the opinion that Wilhelm's speech would cost the government camp many votes in the forthcoming elections. No one really knew, he commented, which newspaper reports the Kaiser meant to condemn – whether it was the article about the Battenberg marriage project, Sir Morell Mackenzie's published defence of himself or Friedrich III's war diary that had aroused the Kaiser's fury. Everyone thought that to castigate the Berlin deputation for newspaper articles which they had neither inspired nor could control had been extremely inappropriate and unwise. 'In fact the speech can only be regarded as one more of the regrettable incidents which seem to be becoming of almost daily occurrence in this city . . . The speech forms the one subject of conversation in all circles here. It is remarked that His Majesty showed scant courtesy to the Members of the Town Council who, though nearly all prominent Liberals, have little or nothing to do with the Control of the Press; and the words addressed to them are generally deplored and condemned.'[131]

Friedrich von Holstein was also concerned about the effect of the speech, and thought that its timing was 'extraordinary', just when the City of Berlin had spent several thousand marks on the fountain to please the Kaiser. He could only imagine that the Kaiser's 'remarkable address' had been a consequence of Herbert Bismarck's campaign against the Progressive newspapers, 'in which Kaiser Friedrich III's reign and personality are always contrasted favourably with the present'.[132] Holstein was undoubtedly right in this assumption. When the malicious rumour was spread in the Left Liberal and Catholic press that the Kaiser's anger had been directed not at the Progressive press, but at the pro-government Kartell parties and especially at Count Douglas's speech, the Bismarcks felt compelled to persuade the Kaiser to issue a categorical statement. 'It seems to me', Herbert Bismarck wrote to his father, 'that after this systematic, deliberate, uniform pack of lies has been circulated by the radical press of the whole world we should make . . . an authoritative declaration that: "H.M. meant the Progressive press, and in particular the exploitation of the diary [of Friedrich III]." Shall I put this to H.M. in audience? I am afraid that otherwise the false rumour will take root.'[133] With the Chancellor's approval Herbert 'extracted' an appropriate declaration against the left-liberal newspapers from the Kaiser in an audience on 3 November.[134]

Almost as strong as Wilhelm II's hatred for the Progressives was his antipathy at this time towards the Catholic Centre Party led by Ludwig Windthorst. On a report from Dresden according to which the death of both Counts Platen had removed the principal supporters in Saxony of the Guelphic dynasty, the Kaiser wrote in January 1889: 'Good . . . May Windhorst [sic] follow soon!'[135] Another Centre Party leader he dubbed a 'blockhead' who had 'ridiculous ideas' and talked 'rubbish'.[136] He called the participants in the German Catholic assembly in Munich 'insolent hounds' against whom one must 'at all costs give cover' to the pro-Prussian Bavarian government of Freiherr von Lutz. 'Give Lutz strong support against the blacks [i.e. the clericals] and assure him from here of my warmest approval and interest', he ordered. On the Kaiser's personal instructions the Bavarian Minister-President was given the Order of the Black Eagle in the summer of 1889.[137] In countless sarcastic marginal notes the Kaiser also mocked the Pope and Vatican policy. Very soon after acceding to the throne he called Pope Leo XIII a 'sheep!' and an 'old idiot!', and exclaimed: 'The old man must have gone mad!'[138] Ten years later the imperial tone showed no sign of mellowing. The Pope was an 'old donkey' and 'an incorrigible old bletherer', who was probably 'cracked', Wilhelm wrote on the reports of his envoys to the Holy See.[139] Of the papal hopes for a restoration of temporal power with Germany's help Wilhelm wrote: 'Might it already have got very hot in Rome? In that case an icy blast would be a good thing . . . We'll be damned if we do any such thing!'[140]

As far as the Kaiser's attitude to his half-million Jewish subjects is concerned, the evidence shows that in the period immediately following his accession Wilhelm's anti-Semitism remained as vehement as before. Waldersee's diary bears witness to the fact that during his visits to the Marmorpalais and then to the Berlin Schloss 'the Jewish question' was often the subject of 'lively conversation'.[141] In September 1888 the Chief of the General Staff recorded that the Kaiser 'in truth [could] not bear Jews', and that the monarch had also 'often stated' this.[142] The following month Waldersee noted indignantly in a lengthy diary entry that people were now trying 'to depict the Kaiser as the Jews' friend, or at least as no enemy of the Jews'; that alone was enough to show how influential the Jews had become. When Wilhelm II came to the throne, Waldersee continued, the Jews had been very worried about their future because they did not know how the new Kaiser would act towards them; but now, as no measures had been taken against them, they were full of confidence and had 'gone on the offensive; they know that they have many friends among the National Liberals and Free Conservatives and they are trying to ally themselves with them so as to attack the Conservatives, in whom they see their only enemies'. Waldersee was convinced, however, that Kaiser Wilhelm II's basic anti-Semitic attitude had not changed in the slightest. 'In my view the Kaiser's antipathy towards Jewry, at least towards the arrogance it shows so plainly, towards the sucking-dry of Christians' as well as towards the excessive influence which the Jews exercised, 'is so deeply rooted that not even Bismarck can do anything about it'. Of course, Waldersee reflected, 'whether the Kaiser is strong enough to keep up a perpetual front against the Jews is another question; no doubt we have already sunk too far into this affliction to be able to find a legal [sic!] way out of it again'.[143] Towards the end of 1891, when the anti-Semitic movement was growing alarmingly and spreading into ever broader circles, Waldersee, who must have known precisely what the situation was, again noted: 'The Kaiser is a decided enemy of the Jews and has often said so, right up until very recently, and has called Bleichröder a dog and a scoundrel often enough, but he is not in a position to see through what is going on because he has many dealings with people who are of Jewish origin or who are completely dependent on Jews.'[144]

The marginal comments made by Wilhelm II and locked away in the Foreign Office safe by Bismarck leave the accuracy of Waldersee's assessment of Wilhelm in no doubt. When in the spring of 1889 the Kaiser read a report in a British newspaper of a conspiracy against the Tsar, in which Jews were said to have been implicated, he annotated the report with 'Those dogs, of course!!'[145] A few weeks later he read of the appointment of Nathaniel Mayer Rothschild, on whom Queen Victoria had already conferred a peerage in 1885, as lord lieutenant of Buckinghamshire. He wrote derisively in the margin: 'I'll be damned!! Incredible! Perhaps we should consider appointing Bleichröder governor of Pomerania now!?'[146] In November 1889, when Schlözer reported to Berlin that the papal nuncio in Vienna, Cardinal Luigi Galimberti, 'had been staying, in the company of a beautiful woman of Semitic origin, with whom he seemed to have a very close relationship', at the Adriatic resort of Abbazia, Wilhelm commented: 'The lucky fellow! It is better that he sees the Semitic beauty only in the evening!'[147] The contrast with the attitude of his uncle, the Prince of Wales, was stark. Prince Ernst zu Hohenlohe-Langenburg, who was attached to the German embassy in London, wrote in the summer of 1889 that the heir to the British throne often took him with him to visit Jewish families, 'who give very good dinners in magnificent houses. They play an important part in society here. It is at their houses that one sees the most elegant people and the most beautiful women and enjoys oneself the most. As the Prince is very much in favour of them, I have no reason to be fastidious.'[148] When Wilhelm II read a newspaper account in June 1889 of the Prince of Wales's visit to Auteuil, during which his uncle had accepted an invitation from Baron Gustave de Rothschild to go out for a drive in his carriage, he wrote on the report: 'Good Heavens! My son sitting next to Bleichröder's coachman would be a very odd sight in the Thiergarten.'[149] If a diplomat criticised a foreign envoy in an official dispatch, the Kaiser might well note in the margin: 'He was a Jew too!'[150] When Schweinitz reported in 1891 that a Mr White had arrived in St Petersburg to negotiate with Pobedonostzew on behalf of Baron Hirsch over the emigration of Russian Jews to Argentina, Wilhelm II commented 'Fancy that!', adding that Hirsch was after all 'the friend of "Ich Dien"' – that is, of the Prince of Wales. But he added regretfully: 'If only we could send ours [i.e. Germany's Jews] there too.'[151]

Kaiser Wilhelm II's animosity towards Jews, recorded in such marginalia and also in Waldersee's diary, was anything but peripheral: rather, it formed a key element of his thinking, even if for tactical reasons it had to be hidden from public view. His anti-Semitic attitude lay at the root of many of his actions and utterances and it explains much that contemporary and later observers found baffling. We have seen how for most people the Kaiser's hostile speech to the Berlin city deputation at the presentation of the Schloss fountain on 27 October 1888 seemed utterly incomprehensible. Waldersee and his ultra-conservative party friends knew, however, exactly what was going on in Wilhelm's mind: the deputation had consisted principally of 'Progressive Jewish gentlemen', whose intention it had been 'to get themselves on a good footing with the Kaiser in the eyes of the world, and if possible to reclaim him for their own'. Wilhelm's first concern had been to defend himself against this by taking harsh action.[152] We shall examine the demonic role that his hatred of Jews was to play in the Bismarck crisis in more detail later.

The Kaiser's anti-Semitic attitude almost inevitably brought him close to the Wagnerian circle out of which, even in these early days, a dangerous counterforce to the pragmatic statecraft embodied by Bismarck was growing. Immediately after his accession Wilhelm II wrote a letter to Cosima Wagner which led her to hope that the Kaiser might agree to become the patron of the Bayreuth festival. She sent a fulsome reply on 23 August: 'Your Majesty's graciousness towards me is too great to allow me to feel any apprehension that I aroused Your Majesty's displeasure with my request. Nevertheless I beg Your Majesty most humbly to forgive me that I . . . took the liberty of laying this supplication at Your Majesty's feet . . . Your Majesty's most gracious words to me filled my soul like a divine summons, and with hot tears, in the solitude in which my feeling for myself and for the work is most profound, I praised and thanked God that the protection which we need is coming to us from the most exalted height. Now I could understand why I had suffered, and far more powerful than any suffering was the feeling of hope which penetrated my heart with Your Majesty's words . . . Whatever decision Your Majesty reaches, . . . all will be sacred to me, and with feelings that I could never put into words and which turn to God as my most fervent prayer, I beg the All-Highest Lord for grace to be allowed to subscribe myself Your Majesty's most thankful, humblest servant Cosima Wagner.'[153]

This letter, which recalls Philipp Eulenburg's style, initially did not fail to achieve its aim. As Head of the Civil Cabinet Hermann von Lucanus informed the Reich Chancellor on 18 September, 'His Majesty . . . [was] not disinclined to grant this request, particularly as the All-Highest has reason to believe that His Royal Highness the Regent of Bavaria does not take a very close interest in the festival.'[154] The Kaiser's inclination to accept the patronage was encouraged by Philipp Eulenburg, who had been responsible for establishing the connection between Wilhelm and Bayreuth through their visit to see *Parsifal* together in the summer of 1886.[155] In the press too there was support for the Kaiser's taking over the patronage of Bayreuth 'in the interests of national art': his predilection for the works of Wagner was described as characteristic of 'the Kaiser's disposition and taste', and the imperial patronage was presented as a fait accompli.[156] Bismarck, however, was firmly opposed to the idea, expressing grave doubts about it in a submission to the Kaiser on 20 September. For one thing, he wrote, he feared that the monarch would find himself being called upon increasingly for financial support for the festival. Moreover he believed that 'closer personal contacts between Your [Majesty] and the festival would give new nourishment to the mistrust of the Bavarians, which we have with difficulty but successfully overcome, with regard to the maintenance of their position vis-à-vis the kaiserdom, and would furnish the Ultramontane and Jesuit party with plausible excuses for stirring up feeling against us in Bavaria. Precisely because Bayreuth belongs to the Old Brandenburg lands, and because no one in Bavaria has forgotten that it was only after lengthy and difficult negotiations that Kaiser Wilhelm I granted the cession of Bayreuth, and because the similarity of religious denomination in itself already inclines the Franconian provinces to look northwards, precisely for these reasons Bayreuth is a sensitive point for Bavarian rivalries and fears.' He was therefore convinced 'that the fulfilment of the wish of Frau Cosima Wagner would be a favour which would not fail to be financially exploited at the expense of Your Majesty, and that a patronage held by Your [Majesty] in Bayreuth would harm the internal political situation of the German Reich in no small degree'. As in the case of the Stoecker meeting in the autumn of 1887, Bismarck again warned the Kaiser 'in the All-Highest interest to decline connections with private associations, whether in the form of patronage or in any other form, as it is very difficult for the sovereign to defend himself from the financial and even political exploitation which inevitably arises from such connections'.[157] This submission was laid before the Kaiser on his return from Italy by Herbert Bismarck, but for lack of time the Kaiser could look at it 'only fleetingly', and he therefore took no immediate decision.[158] On 31 October 1888, the Reich Chancellor was able to have a personal audience with the monarch to discuss this awkward affair. As Herbert informed the Head of the Civil Cabinet afterwards, Wilhelm had decided 'to decline the patronage'.[159] Prince Regent Luitpold of Bavaria took over the patronage a few months later.

Harmony between Kaiser and Chancellor was more easily achieved when it came to freemasonry, since Wilhelm showed he fully understood the necessity of maintaining the connection between the monarchy and the freemasons, even though – unlike his father and grandfather – he refused to join the movement himself. As Wilhelm's brother Heinrich had no wish to become a member either, the Kaiser suggested that Prince Friedrich Leopold of Prussia – who was to marry the Kaiserin's younger sister in June 1889 – should be appointed honorary patron of the three Old Prussian lodges, for it was 'an old tradition' that 'a royal prince' should be a freemason.[160] Bismarck was 'in complete agreement with H.M.'s intention to nominate Prince Leopold to the freemasons . . .; he said that freemasonry, like Judaism, was an element that had considerable influence and should therefore not be alienated from the monarchy'.[161] The arch-conservative Waldersee naturally had less understanding for this statesmanlike gesture, and when a few years later Wilhelm 'created' a special order for Prince Leopold in his role as a high dignitary of the freemasons and even allowed his brother-in-law to wear this order 'at non-masonic occasions', the General poured scorn on this symbolic concession to the Liberals and remarked that he was prepared to believe that the Kaiser 'is even beginning to flirt with the idea of becoming a freemason himself'.[162]

In the domain of foreign politics, which both sides – the thirty-year-old monarch and the 74-year-old Chancellor – regarded as their sphere of activity par excellence, it was far harder for them to agree. Since the 'Supreme War Lord' also saw himself as an 'officer' and looked on military questions as a field in which he alone, by virtue of his powers of command and without any civilian 'interference', should give the orders, the risk of conflict breaking out over foreign policy or military matters was particularly high. What were Wilhelm's views in this sphere, and how far did they coincide or conflict with the policies pursued by the Bismarcks?

CHAPTER SIX

First steps in foreign affairs

WILHELM II had shown a lively interest in foreign affairs even before his accession. From the very beginning of his reign he took to annotating diplomatic and military reports with angry, contemptuous and often crude comments describing statesmen and diplomats of foreign powers as 'scoundrels', 'blackguards', 'villains', 'swine', 'hounds', 'filthy hounds', 'insolent dogs', 'curs', 'apes', 'dolts', 'asses', 'toads', 'milksops' and other such epithets.[1] Whatever other conclusions may be drawn from marginalia of this kind, they are clear evidence of the passionate intensity with which the Kaiser followed the course of international affairs. But can one detect in these spontaneous and sporadic utterances any coherent elements which might indicate, if not a systematic foreign policy, at least a pattern of thought where foreign affairs were concerned? To answer this question we shall need to take a closer look at Wilhelm's attitude towards the major European powers during the period immediately following his accession.

THE 'APPROACHING BATTLE WITH FRANCE AND RUSSIA'

The Kaiser's initial hopes of reviving the eastern Three Kaiser Alliance against France, Great Britain and the parliamentary systems prevalent in western Europe were dashed within a matter of weeks. The reason for this was the cool, if not positively frosty conduct of the Russian imperial family, together with the continuing Russian and French rearmament which seemed to indicate that a general war was in the offing. The aspirations of friendship with which Wilhelm had set out on his state visit to St Petersburg were replaced by feelings of frustration and offence. He did not set much store by the influence of Alexander III, whom he considered to be under the thumb of his Danish wife and his francophile advisers: the Tsar, he commented, clearly had 'not the faintest idea' about the belligerent intentions of the Russian military establishment.[2] The Kaiser's relations with Nicholas, the heir to the Russian throne, remained cool and distant. When the latter planned to stop for an hour and a half in Berlin early in the morning of 13 November 1888 on his way to Darmstadt he asked to be left alone and to be received by the Kaiser only on his return journey.[3] The subsequent visit of the 'strikingly small, . . . very ill-at-ease and gauche' Tsarevich to Berlin on 21–2 November was not a success. 'People were very polite to him here', Waldersee noted, and the Kaiser even went so far as to be 'very warm', but Nicholas 'seemed to me to remain extraordinarily cold', and his entourage likewise avoided anything 'which could in any way be connected with politics'.[4] The Kaiser's disillusionment was so profound that at the official reception for his thirtieth birthday on 27 January 1889 he shook hands demonstratively with the Austrian, Italian and British ambassadors but not with the Russian ambassador Count Shuvalov. 'I am convinced that it was on purpose', the anti-Russian Chief of the General Staff noted with satisfaction, 'for the Kaiser is very displeased with the conduct of the Russian imperial family.'[5] In March 1889 Waldersee concluded that as far as Russia was concerned the Kaiser was 'without any illusions and quite certain that we have no friends there any more'.[6]

Wilhelm II now kept a very close eye on the Russian troop movements and the mobilisation of the Russian Empire by land and sea. Before his accession he had noted on a military attaché's report on Russian operational plans that it was clear 'that a very precise system is being followed in order to ensure at all events – while taking advantage of our loyally standing by and consenting – an initial victory, and to disrupt their opponent's deployment of his forces . . . So there is twice as much hurry for our artillery!!'[7] After his failed attempt to ingratiate himself at Peterhof in July 1888 he returned to this earlier attitude. In early August 1888 a dispatch from Count Yorck von Wartenburg reached Berlin from St Petersburg, reporting that Russian non-commissioned officers who were prepared to remain at the front until July 1890 were to receive a bonus of 150 rubles. This information prompted an excited marginal note by the Kaiser: 'It would be of interest to check to see whether within the period of time specified here (until 1890) the big ships in the Black Sea are fitted out and made ready to sail. For if that were the case, an advance on Stamboul possibly through Bulgaria would not be unthinkable.'[8] In the ensuing months Wilhelm wrote note after note on diplomatic dispatches from St Petersburg and Paris. 'Yet another proof of the calm and uninterrupted progress of what are obviously systematic preparations for a war! All this persistence would be worthy of a better cause!'[9] Or, similarly: 'The view I have often expressed, that the French war preparations are tacitly but indisputably linked to those of Russia, is further borne out by this report! Avis au lecteur! So we must get to work strengthening our army at home, especially the artillery!'[10] When in late 1888 he heard of Russian troop movements towards the west he concluded that they marked the last stage of a programme that had begun several years earlier and declared: 'So now the deployment is completed and – – –. What comes next we shall no doubt find out!'[11] In November 1888 when news arrived from Warsaw of the transfer of the 19th Russian Division from the Caucasus to the Austro-Hungarian border, Wilhelm again inferred that a Russian attack was in the offing and wrote: 'If it has already come to the point when Ural Cossacks, Kirghizes etc. are brought over from the steppes, that is deployment in preparation for a later attack in the fullest sense of the word!'[12]

Confronted with such russophobic marginalia, the Reich Chancellor felt obliged to reveal the secrets of his foreign policy in a memorandum for Wilhelm dated 19 August 1888. Bismarck asked the monarch to burn his letter 'after due perusal', as it referred to matters and questions which as a rule he did not think it expedient to commit to paper. In particular, the Chancellor's letter mentioned the secret Reinsurance Treaty concluded between Germany and Russia in 1887 for a period of three years, although he conceded that this agreement clashed 'with our Austro-Italian obligations'. Alluding to the marginalia in which Wilhelm warned of an approaching attack by Russia on Turkey, the Chancellor argued that in view of confidential remarks made repeatedly by high-ranking Russian statesmen during the treaty negotiations, he was of the opinion that, although both the Russian army and the Russian Black Sea fleet would indeed be 'ready' by 1890, this fact 'was important only as an indication of preparedness, not of intended action'. On the other hand he did not doubt Russia's intention 'to make a play for Constantinople and once the Black Sea fleet is ready, that is to say in the early 1890s, choose the moment for action according to the dictates of the situation in Europe. In my most humble estimation', the Chancellor continued, revealing the guiding principle of his foreign policy, 'it is not part of *our* policy to hinder Russia in the fulfilment of her plans for Constantinople, but to leave this entirely to other powers, if they think it in their interest; our interest in the Bosphorus question is not worth the major war on two fronts that a clash with Russia would entail; on the contrary, if Russia becomes involved there, she will become less dangerous for us because she will withdraw from our frontier and because there will be a provocative tension in her relations with the Mediterranean powers, particularly with England and in the long run also with France.' If Russia eventually secured 'control over the opening and closing of the Bosphorus by occupying a fortified position, she will be able to use her expansionist power against Persia and India. That will make it impossible for England to maintain her present fiction that she is an indifferent spectator, and we can wait and see how the constellation of the remaining powers develops, since a Russian attack on Constantinople does not in itself constitute a casus foederis between Austria and ourselves.'[13]

This machiavellian, indeed almost diabolical peace strategy was beyond the comprehension of both Kaiser Wilhelm II and his mentor, the newly appointed Chief of the General Staff Count von Waldersee. Both dismissed Bismarck's hope for a Russian war against Turkey in which Germany would not intervene as a 'delusion'.[14] Instead, they considered that a war in the near future, between Germany, Austria-Hungary and Italy on the one hand and Russia and France on the other, was not only inevitable but also – at the right moment – thoroughly desirable. For Wilhelm and Waldersee, both Russia and France were irreconcilable enemies of the German Reich and had to be defeated in good time before they had the chance to destroy Germany. As his notorious speech of 16 August 1888 at Frankfurt an der Oder had already made plain,[15] in addition to his mounting russophobia a constant factor in Wilhelm II's view of the world was his pronounced hatred for republican France.[16] When for example in November 1888 the French General de Miribel made a speech expressing confidence that the sons would win through where their fathers had won through before them, Wilhelm wrote, just as he had done as Prince, that 'Lutetia' – Paris – would have to be 'destroyed'. 'Just as the fathers came over our border as prisoners, the sons will have to do so too! Nomen omen! Caeterum [sic] censeo Lutetiam etc.'[17] The following month, when Schweinitz, the German ambassador in St Petersburg, reported to Berlin that the Russian army had ordered 1.5 million rifles, but that it would take at least five more years for them to be made, Wilhelm noted on the report that this news at any rate gave Germany the 'opportunity to smash the Gauls'.[18]

In this bellicose attitude the young Kaiser was powerfully seconded by Waldersee. The Chief of the General Staff convinced the monarch that the Russians, in systematic collusion with the French, were preparing to attack Germany and Austria-Hungary soon. Russia was aiming to be 'ready by the end of the year', even if it did not make an immediate strike then, Waldersee claimed in September 1888.[19] During an audience in March 1889 the two discussed in detail German prospects in a war with France, and Waldersee stated his view that in war it was 'absolutely immaterial who had been right or wrong at the outset. In the long run only the victor was right.' When the Kaiser expressed the opinion that 'we must also be prepared to fight the great battle alone', the Chief of the General Staff added that in fact 'all alliances suffer from great weaknesses, and I hoped that we would be victorious even if we were alone; Germany was still in the ascendant, and I had the greatest faith in her future. The Kaiser agreed with me and spoke with complete confidence.'[20] Waldersee admitted in April 1889, however, that in fact the German army still needed 'a year and a half of peace . . . in order to be ready', as 'the French [were] still a year to eighteen months ahead'. 'So we must not act too boldly', he warned.[21] Nevertheless the great war could not be put off for too long, in his view, since Austria would be 'encircled' as a result of Russian efforts to undermine the countries in the Russian sphere of influence south and east of the Danube monarchy, and would thus become increasingly incapable of taking action. This development was putting Austria 'into a worse and worse situation . . . and unfortunately the same is true for us'.[22] Reading these documents one might easily believe they were reports from the years immediately preceding the First World War.

The latent opposition between Kaiser and Chancellor in their assessments of the international political and military situation created a potential time bomb of which the Chief of the General Staff was fully aware. In the autumn of 1888 he noted in his diary: 'The Chancellor does not entirely trust the Kaiser as regards either internal or external political affairs and lives in a very salutary state of anxiety that the Kaiser may occasionally come out with his own ideas; meanwhile I myself am always regarded with a certain disquiet.'[23] Nevertheless Waldersee availed himself of every opportunity to 'contest the Chancellor's view that our opponents are gradually becoming weaker and weaker'.[24] When reports on improvements in the French army reached Berlin in October 1888, Waldersee saw them as renewed proof of his 'long-held view that we have missed our moment. In the Foreign Office the light is slowly dawning about this. They are very much afraid there that I could become troublesome for the Chancellor by explaining the true state of affairs. I now intend', the General observed, 'to remain quite quiet until the end of the year; but then I shall conscientiously report what I believe to the Kaiser.'[25] For Waldersee, as he recorded as early as 1 November, it was 'an incontrovertible fact that the Chancellor has run aground with his policy; all his art has not helped him to prevent the Franco-Russian alliance or to get the Russians engaged in the Orient. For years things have been patched up and put off and opportune moments have been missed. Our adversaries are not getting worse as the Chancellor claims, but better. Only a particularly lucky chance, like discord in France or Russia, can preserve us from a war which we shall have to undertake under not very favourable circumstances.'[26]

The first direct clash between Wilhelm and Waldersee on the one hand and the two Bismarcks on the other came in November 1888. After he had submitted a report to the monarch on French rearmament measures and had been received for a two-hour audience on the progress of Russian rearmament, the Chief of the General Staff commented on 3 November that the Kaiser was now convinced that the general situation 'indicated a war in the course of the next year. In the face of the immense power of Russia and France the assistance of Austria and Italy should not be overestimated. We agreed', Waldersee noted after the audience, that 'the Russian measures had reached a degree which [was] no longer tolerable.' 'Unfortunately it turns out that I was right in my predictions and also that I was not wrong when I claimed that Bismarck had run himself thoroughly aground with his policy.' Waldersee rejected the Kaiser's hope for Turkish participation in a war against Russia on the grounds that the Sultan would take part in a war only if he were compelled by force to do so, and that Russia had more opportunity to use force in that regard than Germany or any other power. In the course of the discussion, Waldersee recorded, he had floated his favourite idea 'that given the seriousness of the situation we had no alternative but to bring the Poles on to our side by working towards the restoration of Poland'. The Kaiser and his Chief of the General Staff were in complete agreement as to the necessity of increasing the number of troops on Germany's eastern borders and strengthening the artillery.[27] It is no surprise that after this audience the Chancellor accused the Chief of the General Staff of stirring the Kaiser up against Russia. Waldersee, however, protested: '*I* did not excite the Kaiser's feelings at all, but found him already in this state; indeed he became even more excited after Herbert's audience which followed mine, and after which the Kaiser said to me: Schweinitz is at last beginning to think that the situation is serious.'[28]

In the following days and weeks Wilhelm II was seriously preoccupied with the thought of an approaching European war. While hunting at Königswusterhausen on 10 November he called Waldersee to see him several times 'to talk about our political situation and, in relation to that, about our armaments question'. The same theme then became the subject of a discussion lasting several hours that evening at the Berlin Schloss, in which the Foreign Secretary also took part. Waldersee described this memorable conversation as follows: 'The Kaiser was very grave and is beginning to be alarmed; he thinks that Russia and France have [agreed] to go to war soon, and he is now confirmed in this belief by the outline I gave him recently of the progress of the Russian arms preparations and by the news that the Russians are intending to take out a loan of a billion marks, while in France a similar sum is being sought for army purposes. He wishes to increase our peacetime artillery establishment substantially, and to expedite the production of the new infantry rifle.' The three men then discussed the likely attitude of Great Britain towards a European war, upon which Herbert Bismarck put forward the view that the British would be 'very foolish if they did not side with us in the great war because, if we are defeated, they would be hopelessly overpowered by France and Russia. The Kaiser quite rightly remarked', Waldersee commented, that the English were too 'short-sighted' for that. 'They wanted above all to make money and a war on the continent was the best means of doing so. What might follow was too far off for them; moreover they were scarcely in a position at the moment to achieve anything significant with their navy or army, nor did they show any inclination to prepare themselves for anything more. In addition there is Ireland, which is causing them a great deal of trouble.' The Chief of the General Staff expressed the opinion that in all probability the war would not begin on both fronts simultaneously; so it would be 'the business of the diplomats . . . to tell us how much time [we] had before we needed to reckon with the other adversary as well', for that would determine 'the deployment of our resources'. Finally the General gave 'further information on the probable intentions of the Russians' in the coming war. Summing up, he commented: 'the whole conversation was very lively and I cannot deny that [Herbert] Bismarck was very quick-witted in his remarks. The Kaiser mostly listened, but from time to time he cut in with forceful and clear opinions. The conclusion was that I reckoned our requirement with regard to increasing the peacetime artillery establishment at approximately 100 batteries; the Kaiser accepted this and ordered [Herbert] Bismarck [to inform] his father' of this decision.[29] Not surprisingly, as we learn from Waldersee's diary, the Reich Chancellor was 'very put out'[30] by the Kaiser's demand for an increase in the artillery, and no doubt still more by the manner in which this order had come about, although his subsequent invitation to the Chief of the General Staff to a confidential discussion at Friedrichsruh, in order to talk over matters which he was 'unwilling to commit to paper', was politeness itself.[31]

The attitude which Wilhelm II and Waldersee took towards a European war was, however, far less confident at this time than it had been six months earlier, for they did not rate the actual state of preparedness of the German and Austrian armies very highly in comparison to that of the French and Russian forces. At the end of November 1888 the Chief of the General Staff frankly stated his fatalistic views. Among the people in the know in Berlin, there was no one who believed that peace could be maintained until the earliest possible moment for the termination of the treaty of alliance with Austria in 1891, he said. It was therefore important 'that we should be on a good footing with Austria'. Above all Germany should 'not make Kaiser Franz Joseph, who is utterly loyal by nature, suspicious'. At the same time it was necessary for Germany 'to continue to work diligently at rearming our infantry, and if possible the artillery too, so that we shall be in good shape in early 1890. But if the circumstances are such that we must be prepared for a war next summer [1889], I suggest that it would be better to start the war at once; for at this moment it suits neither the French nor the Russians.'[32]

Towards the end of 1888, Waldersee recorded after speaking to the Kaiser that Wilhelm regarded 'the general political situation . . . as extremely grave'.[33] The General summarised his own view, which undoubtedly coincided with that of the Kaiser, in pessimistic terms: 'France and Russia are preparing to go to war together against us. We have watched this coming for years, we have seen how the clouds have gathered increasingly, but we have found no way of dispersing them, nor do we dare take the great decision to forestall them. On the other hand we do nothing to bind our friends more closely to us, and have even [made] our trustiest friend, Austria, suspicious by various measures we have taken. The Chancellor is assured from time to time that we can never win the Russians back, but he does not believe it and persists in the illusion that they might make an attack on Turkey. Unfortunately his agents scarcely dare to tell him the plain truth. I cannot deny that I do not at all like the general situation. As things are now, I have strongly urged that war should be avoided; a year ago, or best of all early this year, we would have had good prospects if we had taken action; now it is too late. The Russians have carried out considerable troop redeployments towards the west; they have also strengthened their field artillery; France has introduced a new rifle and has definitely overtaken us in this field. We must not have a war in 89 in any circumstances; we may even have to avoid it in 90. But I have not yet given up hope that our rifles will be ready by [1 April 1890]. Whether we shall succeed in postponing war for so long, however, seems to me questionable, as our weaknesses are very well known abroad.'[34]

At the beginning of 1889, Waldersee repeated that he was doing all he could to speed up the arming of the infantry so as to be 'ready to strike' in 1890, but the Reich Chancellor was opposed to this intensified rearmament 'because he is suspicious and thinks that the Kaiser could strike out whenever he considers the army ready'.[35] Philipp Eulenburg also concluded, after a visit to the Foreign Office, that Bismarck was afraid that the Kaiser might 'strike out' as soon as the army was 'prepared', and that he was therefore '*in favour of* proceeding slowly in the formation of the artillery etc.'. Eulenburg found that this view was not shared by Holstein and other officials in the Foreign Office, however; instead, they harboured 'a strong desire for very rapid preparation of the army', for 'only with *immediate* preparation would we be ready to face the situation of the moment'.[36] To his satisfaction Holstein had been able to establish from the Kaiser's marginal comments 'that the mistrust of Kaiser Wilhelm towards Russia is growing steadily'.[37] Eulenburg's interlocutors in the Wilhelmstrasse also considered 'any suspicion of Austria on His Majesty's part' very dangerous; they thought the alliance with the Danube monarchy '*absolutely essential for us*'. This was an allusion to Herbert Bismarck's anti-Austrian influence, and they deplored the fact that the 'bad treatment Herbert once received from a pair of Viennese countesses' still rankled with him.[38]

KAISER WILHELM II AND AUSTRIA-HUNGARY

If the Kaiser and Waldersee shared the same view of the French and Russian threat, they did not at first see eye to eye over Germany's ally Austria-Hungary.

Waldersee deplored the Kaiser's mistrust of the Danube monarchy, which he attributed to Herbert Bismarck's influence and to the 'immoral' policy of the Reich Chancellor in favouring Russia at the expense of Austria.[39] He set out to open Wilhelm's eyes and to bring home to him the military necessity of strengthening the bond between Germany and Austria. 'I hope the Kaiser will take firm action once he understands the situation', Waldersee remarked in January 1889.[40] Although Wilhelm veered between the Bismarcks and Waldersee over this question it is clear that here too there was increasing potential for a future explosion.

Waldersee's attempts to support Austria and undermine the Bismarcks' pro-Russian policy can be clearly traced in his diary. On 1 November he noted: 'My impression that the Kaiser has been prejudiced against Austria has been confirmed. I am convinced that this is chiefly the work of Count [Herbert] Bismarck . . . It is a misfortune that the Bismarcks, father and son, have such an attachment to Russia and however badly things go they always come back to it. I am now quite determined to speak up resolutely for our alliance with Austria; we certainly cannot stand alone or we should undoubtedly be defeated; we need allies, as France and Russia are collaborating against us, and I do not know on whom we can count if not on Austria; certainly not on Italy. I would understand their policy if it were possible to believe that we could gain any advantage by the collapse of Austria; but this is not the case and it must be in our interest not only to uphold but also to strengthen her.'[41] Not long afterwards the Chief of the General Staff sent Bismarck a report in which he tried 'to give a better impression of the Austrian army', and drew attention to Austria's military measures in Galicia, which revealed the 'clearly discernible intention' of the Austrians to take an offensive stance in the forthcoming war.[42] But even Waldersee was indignant at the conciliatory policy followed by the government in Vienna towards the Czechs: in that way Austria would become 'first a federal state and then fall apart completely'. Bismarck's unequivocal warning to Austria that a federal state 'is considerably less acceptable to us as an ally' met with Waldersee's approval.[43]

Waldersee's subversive influence in favour of Austria and against Bismarck's policy is reflected almost verbatim in the Kaiser's marginalia. Wilhelm greeted as 'very pleasing' the news sent by Adolf von Deines, the military attaché in Vienna, that Austria wished to be as strong an ally of the German army as possible.[44] When reports arrived of the 'excellent morale' in the Austrian army, the Kaiser again welcomed this as a sign that the 'idea of *offensive* action, which is otherwise so foreign to the Austrian character', was now spreading to all parts of the army. 'If that is the case', he commented, 'there are very strong grounds indeed to hope that the Austrian army will have some healthy surprises in store for the Russians in any clash with them!'[45] Like Waldersee, the Kaiser believed that the 'logical consequence' of the Vienna government's pro-Czech policy would be 'a coronation in Prague!'[46] A report by the German Consul-General in Budapest, Count Anton Monts, which contrasted the pleasure-seeking indolence of the German-Austrian aristocracy with the industriousness of the 'small, but energetic and politically talented race' of the Hungarians, who formed a bulwark against the 'Slavic deluge', earned a delighted comment by Wilhelm II: 'A quite outstandingly well-written report! So the "transfer of the centre of gravity" to Budapest will still go ahead thanks to the greater insight and cleverness of the Hungarians!'[47] The Kaiser also spoke out strongly for the maintenance of Austrian influence in Serbia.[48] After the abdication of King Milan, which led to a marked increase of Russian influence in Serbia at Austria's expense, the Kaiser remarked while sailing between Corfu and the Albanian coast: 'One gets a real craving for annexation here; it is just as well that we shall soon be cooling off again in our cold North.'[49]

In the spring of 1889 Waldersee observed with disquiet that Wilhelm II seemed once more to have come under the pro-Russian, anti-Austrian influence of the Bismarcks. During a long walk he took with the Chief of the General Staff, the Kaiser remarked critically that Austria had refused to agree to a partition of Turkey: this 'foolishness' on the part of the Austrians, he believed, arose from their 'expansionist ideas', which were aimed at creating a huge Danubian empire stretching down to the Black Sea and across to Bessarabia and Odessa. Waldersee thought he recognised an 'old idea' of Bismarck's in these remarks. The Chancellor wished for the 'gradual downfall' of Austria, because he still believed he could come to an agreement with Russia. Waldersee, however, considered this idea 'quite wrong' and 'very sad'. 'I have nothing against helping to break up Austria, if it is necessary in our interest, but I must first be convinced that we should really gain some advantage and benefit from it; as things are at the moment Russia would be happy to join us if it were against Austria; but it would not last long and we would find ourselves alone, without allies, and would be overpowered by Russia and France.' Waldersee was astonished to learn from Wilhelm that he had never discussed Austria's supposed 'plans for the future' directly with Kaiser Franz Joseph himself.[50]

BRITAIN, THE UNITED STATES AND COLONIAL POLICY

Since a great war on two fronts on the continent was expected shortly, a firm understanding with Britain, which must at least remain neutral in the conflict, was urgently needed. In order to ensure this neutrality there could be no question of Germany pursuing an expansionist colonial policy for the time being. On this point Wilhelm II and Waldersee were at first in complete agreement with the Bismarcks. The Chief of the General Staff considered the 'whole colonial policy' to be 'nonsense' and sharply condemned the colonial enthusiasts, who wanted to 'take energetic action and impress people by a show of strength' overseas.[51] His influence on the Kaiser is unmistakable here too. 'I told the Kaiser plainly that I considered that our African colonies could never amount to anything, and that I was even convinced that in 30 years all Europeans would be driven off African soil', he declared perceptively in January 1889, adding that 'this might be a slight exaggeration; I should have said: "wherever the Mohammedans rule". But this is still my view and it can be seen that their power is constantly spreading. In the South, however, it may take longer for the native population to become aware of its strength. The days of colonies are altogether over. My only wish now is that we should avoid becoming more deeply embroiled.'[52] In early 1889 Waldersee recorded anxiously that German policy in East Africa had ventured further than was wise.[53] He was afraid, in particular, that Germany might easily 'get into a squabble . . . with England' in East Africa.[54]

Still more dangerous, in the eyes of the Chief of the General Staff, was the deployment of German troops in Samoa, for this Pacific adventure could even lead to differences with the United States. 'An entanglement with America is just what we need!', Waldersee exclaimed despairingly.[55] In March 1889 he recorded that light had at last dawned in the Foreign Office, which had recognised 'that our colonial policy has been extremely foolish'.[56] He strongly criticised those naval officers and diplomats who favoured 'energetic action even to the point of fighting America', for Germany could not conduct a war with the United States; it would be 'the greatest misfortune . . . which could happen to us'. He implored the Kaiser to avoid any aggressive measures over Samoa because the consequences were unforeseeable and 'we should have nothing on our minds now except preparing ourselves for the forthcoming struggle with France and Russia'. 'I found the Kaiser calm, and he had evidently given much thought to the question; he promised me that serious complications would be avoided, and also thought that the Chancellor was taking a calmer view of the question.'[57]

The colonial question had become acute because the situation on the island of Zanzibar, where both Germany and Britain had substantial interests, had been threatening to spiral out of control since the autumn of 1888. A British officer reported that the German East Africa Company, supported by a German squadron, had already executed several natives and an uprising against the Sultan of Zanzibar was imminent. 'Zanzibar itself is very unpleasant, full of Germans who drink all day, rail at the Sultan & scowl at the English', he informed London.[58] This critical view was fully borne out by German reports. In September 1889 Prince Ernst zu Hohenlohe wrote to his father, the President of the

German Colonial Society, from the German embassy in London: 'I have recently heard (from a German) that our representatives in Africa are behaving quite incredibly and are ruining our standing both with the natives and with the Europeans by their behaviour. [Carl] Peters is said to have acted like a madman on his arrival in Africa, [Hermann von] Wissmann is said to have been completely drunk when he arrived in Cairo . . . and to have scandalised everyone by his behaviour there. The officers and officials in Africa are reported to be making the natives everywhere rebellious by their immoral and brutal conduct. All this and more was told me, as I said, not by an Englishman but by a German, who is close to the Foreign Office and is very much concerned with colonial policy.'[59]

In this tense situation Kaiser Wilhelm at first showed an uncharacteristic level-headedness, to the relief of the Bismarcks and Waldersee. After an audience on 16 September 1888 Herbert Bismarck was able to report to his father: 'With regard to East Africa H.M. was as cool and sensible as could be, agreeing entirely with your opinion and needing no persuasion. H.M. said of his own accord that in East Africa we ought to do everything *through* the Sultan, to make use of his Mohammedan authority as a wire-puller without appearing on the scene ourselves, and on no account to allow ourselves to become involved in enterprises in the interior – and even on the coast only on condition that the Sultan's standing is not compromised. [The Kaiser spoke] of the East Africa Company with contemptuous disparagement and called Peters a misguided fantasist. So when Berchem delivers your P[ro] M[emoria] . . . to H.M. the day after tomorrow it will receive the fullest All-Highest approval.'[60] Even when the Kaiser manifested 'a little colonial chauvinism' in the summer of 1889, attributed by Herbert Bismarck to the influence of the 'transoceanically minded' Bennigsen, the Foreign Secretary did not take it seriously, since the Kaiser was in complete agreement with his principle of 'harmony with England on matters of colonial policy'.[61] The younger Bismarck believed he had all the more reason to feel reassured since even senior naval officers were speaking to the Kaiser in support of his dispassionate view of German interests in Zanzibar.[62]

It seems not entirely unlikely that the restraint which the Kaiser showed in his attitude to colonial politics in the first eighteen months of his reign was partly rooted in his half-English identity. It is certainly difficult, given the constant oscillation between hatred and love in his attitude towards his mother's country, to detect any consistent line. But Wilhelm's extensive comments on the activities of British military forces in northern East Africa during this period bear witness to a personal interest in their fate which is absent from his marginal comments on other international events. When, for instance, Ambassador Count Hatzfeldt reported in November 1888 on a government statement in the House of Commons announcing that an expeditionary force of ten officers and three hundred men, drawn from Egyptian and Scottish regiments, under the command of General Francis Grenfell, had left Cairo and was marching through the desert to the Sudan in order to suppress the Mahdi's Islamic rebellion, the Kaiser criticised both the size and the composition of the force as inadequate. 'They are always making the same mistake! A totally inadequate detachment is sent off with great show – with at least 1 General. It is supposed to act purely defensively but nevertheless it later finds itself forced to undertake offensive action, for which it is of no use. The combination with black regiments seems risky to me. To judge from their performance up to now they are not worth a great deal. And add to that a march through the desert!'[63] Similarly, two weeks later, it is clear from the Kaiser's detailed comments on Lord Salisbury's optimistic assurances that the 5,000 British soldiers assembled in Suakin on the Red Sea would be perfectly adequate to defend the port against 2,000 rebels how anxious he was for a British victory. 'M[eo] V[oto] the detachment in Suakin is *not* sufficient', he observed. 'Even a simple defence of the most defensive nature requires occasional offensive action. They have only a very imprecise idea of the strength of the enemy. The reconnaissances have failed because the mounted detachments were driven back by the besieging forces; according to private reports they even ran away. The enemy is well armed and has artillery which has been dug into trenches for weeks, and which is said to fire with great precision. If one reckons on 2000 men to cover occupation and guard duties, sick, wounded &c, that leaves a figure of 2,500–2,000 [sic] infantry and about 500–700 cavalry for the sortie. This would suffice only to drive the enemy out of, or destroy, the trenches. And then, if successful, to go back, and to have to do the same thing again 14 days later. But in the face of an enemy such as the one besieging Suakin the only thing that is of any use is an energetic offensive. Suakin can be protected only by capturing Sinkrat and Tokar and by occupying and fortifying them. For that at least one division of 12000 infantry; 1 cavalry regiment of 1000 horse and 2 batteries of mounted cannon and 4 of rapid fire cannon are required. If the news of Emin's fall is true, the Mahdi or Osman Digna has sufficient reserves of victorious troops to hand to make the Englishmen's task very difficult.'[64] Even though Wilhelm did not send these words of advice to Queen Victoria, in many respects they anticipate the two highly controversial imperial memoranda which he sent to London at the turn of the century, to help the British to victory in their war against the Boers.[65] In the event the Anglo-Egyptian army under Grenfell defeated the rebels in August 1889 even without the benefit of the Kaiser's advice.

As has already been shown, the anglophile side of the Kaiser came to the fore suddenly and much more strongly than ever before after his visit to Osborne and his appointment as a British admiral. After his dismissal, Herbert Bismarck recalled that the British government, having agreed with Germany on a joint approach to the forthcoming Samoa Conference, had proposed the exchange of Heligoland for parts of German South-West Africa. 'My letter about this was submitted by H.H. [the Chancellor] to the Kaiser, who was absolutely delighted, and wanted to accept *immediately*. We had difficulty', wrote Herbert, 'in making it clear to H.M. that it was not yet time for this.' Instead, if the Kaiser set such store by Heligoland it would be better not to show any enthusiasm, but to wait for a situation to arise 'in which England needed us, and this would happen soon enough'. The Kaiser yielded to 'these rational arguments, even if very reluctantly'. At the request of the Foreign Secretary, who feared that the acquisition of Heligoland might be 'made known through the confidential conversations which the Kaiser liked to have with naval officers . . . and as a result only made more expensive for us', the Kaiser ordered the exchange plan to be locked away. But as soon as Herbert had left for his summer holiday in June 1889 the Kaiser asked Under-Secretary of State Count Berchem to take the matter up, so that he could sign the treaty while he was at Osborne. It was only with great difficulty that Herbert succeeded in making it clear to the Kaiser that if *Germany* suggested the cession of Heligoland at this moment, Britain would either put forward an unacceptable counterclaim or reject the whole arrangement, 'because Salisbury would not want to expose himself to opposition accusations of giving up English territory for the sake of a family visit to *Osborne*'. The idea of signing the pact at Osborne was a particularly unfortunate one, in Herbert's eyes, and moreover it would be impossible to achieve an acceptable basis for negotiations so quickly. The Kaiser gave way with bad grace.[66]

At the end of 1889, however, the first indications of imperial enthusiasm for colonialism began to appear. In December Wilhelm ordered the Foreign Secretary to inform Hermann von Wissmann by telegraph 'that in recognition of his achievements hitherto, and for the accomplishment of his duties, I am giving him a battery of 6 naval guns with ammunition (including landing carriages)'.[67] If Herbert Bismarck was disturbed by this evidence of the influence of the navy on the Kaiser, he was horrified to discover, a few weeks later, that Wilhelm was suddenly 'fiercely against England in the colonial question, especially Zanzibar, which He wishes to take'. He was 'embittered against England because of East Africa', and 'inclined to violent action'.[68] Once again it was the Empress Frederick who recognised the mechanisms that were at work here. Prince Bismarck, she wrote, had 'never seriously thought' of acquiring colonies or fighting for them, but he had stoked up enthusiasm for colonies because he thought he could 'use it for electioneering purposes, & that flourishing the patriotic flag, & blowing the national trumpet' would enable him to put pressure on the Reichstag. But he had been caught in his own trap, for meanwhile 'not only the Chauvinist

Party but William have taken it *quite* "au sérieu" [sic], – & wish it followed up. The Chancellor does not *dare* to say that it would be wiser to drop all such undertakings for the present, & while the state of European Peace is so uncertain, but I have no doubt he *thinks* it.'[69]

THE KAISER'S 'PASSION FOR THE NAVY'

Still more significant for the future than the Kaiser's growing enthusiasm for colonies was, of course, his enthusiasm for the navy, which even in these early days struck all observers as strange. In political circles in Berlin it was either mocked as a personal fad or criticised as a dangerous threat to the necessary supremacy of the army. Herbert Bismarck described it contemptuously as 'hydrophilia',[70] and the young Alfred von Kiderlen-Wächter, reporting on the Baltic cruise in July 1888, commented mockingly that Wilhelm was manifesting 'a colossal interest in naval matters'. He was constantly reading naval signals with a telescope. A naval battle had been arranged 'to the accompaniment of much thunder from the guns', followed by a great sailing exercise on another day. Then 'we one and all shot with a revolver-cannon at a barrel thrown overboard as a target, which H.M. to his great joy also hit'.[71] Even senior naval officers considered some aspects of the imperial passion for the navy questionable. Thus for instance the future Chief of the Naval Cabinet, Gustav Freiherr von Senden-Bibran, criticised the fact that throughout his state visit to Russia, apart from the uniform of his Russian regiment, the Kaiser had worn only the new admiral's uniform which he himself had designed; this would cause bad feelings in the army, he predicted.[72] A year later Vice-Admiral Paul Hoffmann commented that on their voyage from Genoa to Athens the Kaiser aspired 'to act completely as an admiral, to give direct orders, to receive direct reports as commanding officer of the squadron, and in addition to dress and behave as a seaman'. His enthusiasm for the navy made itself felt everywhere.[73] 'His interest in these things takes precedence over everything else',[74] the Vice-Admiral recorded; and the Kaiser even found it embarrassing to be seasick.[75] However gratified Hoffmann was by the Kaiser's naval interests, he clearly recognised the dangers that lay in the superficiality of his knowledge, combined with his articulacy and with the inhibitions that his exalted status called forth in others. On several occasions the monarch had started lively discussions on naval matters and marine tactics, and it was pleasing to note 'how clearly and calmly the Kaiser thinks, and how superbly gifted he is intellectually. The fault lies only in his dilettantism; he thinks he can give an opinion on anything, even if his knowledge is only superficial. If an opposing, clear and well-founded view is expressed to him, he disputes it, but then agrees with the opposing view and recognises its force. The only difficulty is being able to defend one's opinion straight away and skilfully to the Kaiser. If his opponent is not practised in this, which of course unfortunately often happens, then the Kaiser is even more convinced of his own opinion and very self-satisfied at having confounded the experts.'[76] Waldersee, as Chief of the General Staff, was shortly to have very similar experiences during the army manoeuvres.

Wilhelm was deeply imbued with the idea that his grandfather had, in a sense, entrusted him with the task of building up the navy to be 'a sister-force to the army, of equal value'. Many years later, in exile in Holland, the Kaiser could still vividly remember the 'solemn moment' during the laying of the foundation stone of the Kaiser Wilhelm Canal at Kiel in June 1887, when the old Kaiser surprised him with the words: 'Heinrich has just told me that your interest in and understanding of the navy is so great that the navy wishes that the bond which already links you with it should be outwardly visible as well. After the splendid impressions left by today I am especially glad to fulfil this wish, which you yourself no doubt also cherish in your heart. Therefore I now place you à la suite of the I. Marine Battalion.'[77] 'I was completely overcome with joy and surprise, and speechless', the exiled Kaiser recalled. 'It was the fulfilment of a fervent wish! Heinrich embraced me at once and pressed me to him so tightly that I could scarcely breathe. The senior naval officers who were present for our departure . . . shook my hand warmly after I had kissed the Kaiser's hand in respectful thanks.' On the train journey from Kiel to Potsdam, however, conflict arose with the Prussian Minister of War as representative of the interests of the army. Admittedly Chief of the Military Cabinet Emil von Albedyll and most of the other generals who were in the compartment congratulated the young Prince 'most warmly' on the new honour and Albedyll went so far as to say, 'You deserved it, for I have already heard from naval circles how surprisingly at home you are in maritime affairs.' But Prussian Minister of War General Paul Bronsart von Schellendorf went on reading a document 'without taking any notice of the event. When I announced the news to him, full of happy excitement [the Kaiser continued] he looked at me over his pince-nez and said with cold condescension: "I see! – Well, you certainly should take an interest in such things", and resumed his reading. Burning with anger I answered him: "I beg your pardon, Excellency. The navy is the equal sister-force of the army! It represents the honour of the German flag abroad, and it is high time and quite proper that the army should also take an interest in such things. In the regiments to which I have the honour to belong I have already awakened this interest", and abruptly left the compartment. When we arrived at Potsdam His Excellency von Albedyll came up to me and while strongly disapproving the conduct of His Exc. von Bronsart . . . he expressed the latter's regret over the affair, upon which I thanked the General and added: "When it is for me to speak, things will be different."' As the Kaiser explained, Bronsart, like Caprivi, belonged 'to the group of older Prussian generals, who lacked any understanding of questions of naval prestige and of a correspondingly strong navy. They looked on it as an appendage to the army, and considered money spent on the navy as wrongfully taken away from the army.'[78] Shortly after Wilhelm II's accession, both Bronsart as Prussian minister of war and Caprivi as chief of the Admiralty were obliged to hand in their resignations.

In his aim of building up the navy into 'the equal sister-force of the army' Kaiser Wilhelm received warm support from his brother Heinrich, who was at this time in command of the torpedo boat division, and was soon to be promoted to admiral and fleet commander.[79] After the bill on naval organisation ordered by Wilhelm had been passed by the Reichstag early in 1889, giving the navy a similar administrative structure to the army (more will be said about this in the next chapter), Heinrich sent an enthusiastic reply to a telegram from his brother: 'I shall never forget your words and shall inform my new officer corps of them tomorrow, as on a day on which I shall hoist our proud flag to your and the Fatherland's glory. Only in such a spirit shall we be able to be named in the same breath as your glorious army.'[80]

In the army, on the other hand, the Kaiser's obsession with the navy aroused grave misgivings. Only a few weeks after the accession Waldersee, who on this point at least was in complete agreement with his rival War Minister Paul von Bronsart, was strongly criticising the 'passion' of the new Kaiser for the sea. The navy must realise, Waldersee warned, that it 'is only one part of our armed forces, and by no means the most important, though certainly the most costly'. He castigated the naval officers and others in the imperial entourage who fuelled this dangerous passion in Wilhelm.[81] At the naval exercises in September 1888 Waldersee was shocked by the evident strength of 'the Kaiser's passion for the navy'. Wilhelm was 'incredibly enthusiastic' and spent the whole day on deck, except for meals. This 'very pronounced' obsession of the Kaiser with the navy was 'stronger than is good for us' and was being exploited improperly by the naval authorities. 'The Kaiser is now determined to make considerable demands for money for the navy so as to build bigger ships again, which Caprivi did not think right', Waldersee complained on 9 October 1888.[82] Only a month later the Chief of the General Staff expressed the hope that the Kaiser would give up his plans for a large naval building programme for the sake of urgently needed reinforcement of the army. 'My wish', he confessed, 'is for the Kaiser to drop his huge demand for the navy – I think it is 100 million; he will be very reluctant to do so, but I hope that he will agree nevertheless, if he acknowledges that we shall have a war soon. Iron-clads take three years to build, but the war will not be kept waiting that long.'[83]

The ample evidence of Wilhelm II's enthusiasm for naval affairs at this initial stage of his reign should not, of course, mislead us into thinking that we are already dealing with plans to build a battlefleet against Britain. The idea that the new German Reich with its minimal naval strength might aspire within a few years to dispute the maritime supremacy of Great Britain seemed so fantastic, so far removed from reality, to both the supporters and the opponents of the Kaiser's naval passion that it was not even thought of as a remote possibility. Indeed, Wilhelm clearly regarded his love for the navy as an expression of his English heritage and declared to one of his Flügeladjutanten in Doorn: '[M]y very particular passion for the navy . . . arose not least from the English blood I inherited from my mother.'[84] And Bülow, as Reich chancellor, once had to point out to him that it would not do, as he proposed, to describe the expensive and highly risky German battleship building programme in public as the expression of his personal predilection and of his youthful experiences in Portsmouth and Plymouth![85]

In these first years Wilhelm II's naval plans were, on the contrary, rooted in the wishful notion that the Royal Navy, the German fleet and the Prusso-German army would together safeguard world peace and that if a war should nevertheless prove inevitable they would carry off the victory jointly. In this expectation the Kaiser positively urged an expansion of the British navy and deplored the fact that the British were 'hardly in a position at the moment to achieve anything significant with either navy or army', nor did they seem to show any desire 'to prepare themselves for more'.[86] When the Westminster Parliament approved the funding of seventy new warships in early 1889 Wilhelm expressed his delight.[87] After his appointment as admiral of the fleet and his successful visit to England in the summer of 1889 Wilhelm's wish for Anglo-German military co-operation by land and by sea seemed to be on the verge of fulfilment. In August that year he wrote to Queen Victoria saying that he knew that 'the British ironclads coupled with mine & my army are the strongest guarantees of peace; which Heaven may help us preserve! Should however the Will of Providence lay the heavy burden on us of fighting for our homes & destinies, then may the British fleet be seen forging ahead side by side with the German, and the "Red Coat" marching to Victory with the "Pommeranian Grenadier"!'[88] In the same spirit he annotated an English newspaper report in which much was made of the independence of the British navy and of the fact that it did not exist to serve the interests of either Germany or other countries, with the comment: 'Short-sighted! History will teach them to know better.'[89] This Anglo-German dream reached its peak in a memorable Christmas letter to Queen Victoria of 22 December 1889, in which the Kaiser strongly urged doubling the strength of the British Mediterranean fleet. 'My journey to Athens and Constantinople was immensely interesting', he wrote.

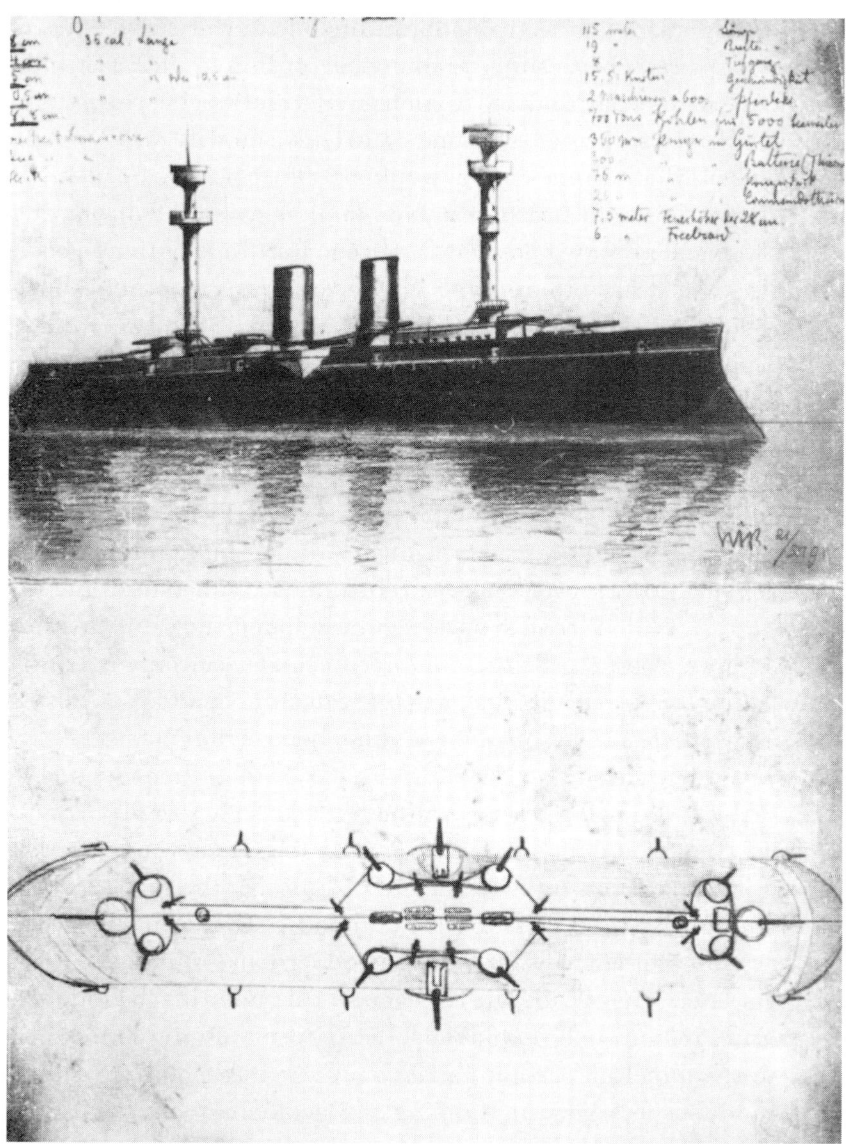

6. A ship of the line designed by Kaiser Wilhelm II in May 1891

'I was very much pleased & interested in seeing the ships of the Mediterranean Squadron. It is in most able & excellent hands; except Adm. Hornby I hardly ever met [?] a more distinguished gentleman, finer seaman, & better political head than S[ir] Anth. Hoskins; he is what the sailors say a "splendid fellow". I only wish – this as Adm. of the Fleet – that I saw a *dozen* 1 class battleships under his command instead of *5*! Admiral Hoskins' Squadron is to[o] small to fullfill [sic] all the duties incumbent upon it, especially in case of war, I among many other points only take the liberty of submitting one to you. France has now in commission *9 first class* battleships at Toulon; according to their last full speed trial this fleet took one day to reach Toulon from Villefranche. Now, should the French, – this boite à surprise for Europe – suddenly run wild & fall foul of Italy, for instance, their Mediterranean fleet would be able to pounce upon any Italian town or division of their fleet in 1 or 2 days. Well, if, as Lord Salisbury told me at Osborne, England never would allow France to hurt Italy without good cause, Adm. Hoskins would have to do something or other to help them. But what? What can he do with 5 ships against *9*? These 9, which will be followed in 24 hours by the whole of the first reserve of Toulon? Adm. Hoskins must be reinforced, as soon as is deemed expedient; a man in whose hands the responsibility for the British prestige in the Mediterranean is placed must be able to muster such a commanding number of battleships, that neither the simplest gamin in Paris, Rome or Constantinople can be in doubt for one second who will be the victor when it comes to fighting. I can assure you, dearest Grandmama, this question has weighed heavily on me ever since my return! I have culled information from all sources in the South & East & have regularly found the same answer: "the French look down upon the British Mediterranean Squadron with disdain, & are sure of doing away with it in short time after the opening of hostilities!" Fancy! What would Lord Nelson say! I sent a scheme I drew up some time ago through L[ord] Ch. Beresford to Ld. Salisbury, a copy of a scheme as it is worked out for my Navy. It shows the British Navy & French Navy told off in Squadrons for war; I believe it would interest you to see it. – But I must be[g] your pardon for taking up your precious time; when this "Admiral of the Fleet" gets hold of me, then I am for a while unable to shake him off.'[90] One wonders how Queen Victoria reacted to this advice from her Prusso-German grandson!

Only years later, when Wilhelm II began to carry out his 'boundless' naval plans, did critics recognise the explosive force, in terms of international politics, which lay in the imperial 'passion for the navy'. At the beginning of his reign only the Grand Duchess Augusta Caroline of Mecklenburg, born in England in 1822, sent out a warning after a visit by the restless young Kaiser to Strelitz, that Wilhelm was 'a passionate *Sailor*, more even than his brother Henry is, his aim being to out-do *all* other *Fleets*'. The old Grand Duchess underlined the word 'all' three times.[91]

CHAPTER SEVEN

The pillars of imperial power

As soon as he came to the throne Wilhelm embarked upon a thorough and – in the view of many observers – overhasty reorganisation of his court. The appointments he made were his own strictly personal choices, and they ranged from officials of the household and court, adjutant-generals and Flügeladjutanten (aides-de-camp), the Secret Cabinets for Civil, Military and Naval Affairs, to the highest posts in the army and in the civil administration of the state. They included appointments of men who were little suited to their posts; they led to conflicts, both among the new court officials themselves and between them and the military suite – conflicts which were exacerbated by the power struggles between the Bismarcks and their adversaries in the army and within ultra-conservative circles. Nevertheless, through this reorganisation of the court Wilhelm II created for himself a compliant instrument of power which enabled him to carry out a successful campaign against Bismarck and then, over the course of the next ten years, gradually undermine and finally control the power of the state. As early as 1896 the influential Privy Councillor Friedrich von Holstein described the imperial entourage as one of the three factors – the others being the Kaiser himself and the government – which had to be reckoned with in political decision-making.[1] In contrast to the chancellors and ministers, a surprisingly large number of these men appointed at the beginning of the reign retained their posts for decades, some until death and others until the fall of the monarchy in November 1918. The court, which consisted of over 2,000 officials and military officers, thus served literally as the Kaiser's own power base.[2]

THE MINISTRY OF THE ROYAL HOUSEHOLD

The Household Ministry was situated at 73 Wilhelmstrasse, close to the Reich chancellor's palace and the Foreign Office, and it administered the business affairs, estates and Royal Archive of the Hohenzollern family and all its branches. It also included the Office of Heralds, which dealt with matters of noble rank and privilege, and consequently played an important role in the upper echelons of German society. It nevertheless stood outside the constitution, in the personal domain of the reigning monarch. The minister of the Royal Household was thus the only minister in Prussia who was not a member of the Ministry of State.

The House Minister at the time of the double succession in 1888, Otto Count zu Stolberg-Wernigerode, was not only a member of one of the most illustrious aristocratic families in Europe, but had also made a brilliant career in government service, having been Oberpräsident of Hanover, a member of the Reichstag, ambassador at Vienna, vice-chancellor and vice-president of the Ministry of State under Bismarck; since 1884 he had served as principal chamberlain, and from 1885 also as minister of the Royal Household (as successor to Count von Schleinitz).[3] At the grand opening ceremony of the Reichstag on 25 June 1888, Stolberg stood very near the Kaiser, just beside the imperial crown. It therefore came as a painful shock to court society when Stolberg was dismissed as house minister shortly after the accession. Wilhelm II had simply invited him to hand in his resignation. Stolberg, as the Kaiser announced to the Prussian ministers in his first Crown Council, had missed the opportunity to buy up for 30,000 marks the entire edition of the controversial memoirs of the Hofrat Louis Schneider, who had spent many years as reader and private librarian to Friedrich Wilhelm IV and Wilhelm I, and had thus damaged the reputation of the old Kaiser.[4] It was true that Stolberg had in any case intended, as was generally known, to ask to be relieved of his post in the course of the next few months. 'But the fact that although he wanted to go, he has been pushed out now, gave him the greatest offence', Count Imre Széchényi reported. The change at the Ministry of the Royal Household caused all the more head-shaking because the successor named by Wilhelm II, the ultra-conservative former President of the Reichstag, Wilhelm von Wedell-Piesdorf, could, as Széchényi commented acidly, 'offer no special qualifications for the post except that he is a friend of Count Waldersee and likewise belongs to the party of the Protestant zealots'. No one would ever have thought, 'even as the remotest possibility', of Wedell as a candidate for this post, Széchényi maintained.[5] He could hardly have suspected that the new House Minister would remain in office until 1907. Even Waldersee criticised the unnecessary haste with which Wilhelm had made the change. Stolberg – who much against his will remained in his post as principal chamberlain until

1893 – was 'undoubtedly the most respected man in our society and the most distinguished member of the Protestant aristocracy', the General commented. 'With his position, fortune and family connections he is a personality the Kaiser must reckon with.'[6] Wilhelm, however, treated the matter lightly; he wrote to his grandmother Augusta concealing the true facts of the case and saying that Stolberg had 'asked, in a most firm and decided manner, to be relieved of his post, as he could no longer manage it because he had to look after his own domains'. The Kaiser added that he had known Wedell well for a long time, and he was 'in every respect perfectly fitted for the position'.[7] When Prince Stolberg died in November 1896, Waldersee paid tribute to his popularity, distinction and tact, but added: 'He no longer liked the way things were in Berlin, and in this he was not alone . . . His relations with the Kaiser have always remained good outwardly, but at heart they were cool.'[8]

Stolberg's dismissal attracted all the more attention because numerous other court officials were retained and even promoted at the accession. Physician-General Dr Rudolf Leuthold, who had been on terms 'almost of friendship' with Wilhelm for years, was appointed his personal physician.[9] Others were taken over from his father's household and court establishment, as for instance were Hugo Count Radolinski, who (now as Prince Radolin) was appointed Lord High Steward, Maximilian Freiherr von Lyncker, the Hausmarschall (Marshal of the Household), and Count August zu Eulenburg, the Senior Master of Ceremonies. They were later to exercise an unparalleled influence at the court in Berlin. Countess Therese von Brockdorff was granted the title 'Excellency' and became Oberhofmeisterin (Senior Mistress of the Court) to Kaiserin Auguste Viktoria. The Kaiser's mother wrote of her that she was certainly 'devoted to her mistress, but not my idea of a Lady!', adding that she was very narrow-minded, spoke 'such common German', and altogether had 'the manners and ideas of a "Bourgeoise"'. In general, she claimed, the people now at the new court were 'very second rate'.[10] The two ladies in waiting, Mathilde Countess von Keller and Claire von Gersdorff likewise remained in the Kaiserin's close entourage. Her erstwhile Chamberlain, the zealously Protestant Ernst Freiherr von Mirbach, was promoted to be her Senior Master of the Court and Chief of her Cabinet.[11]

THE 'MILITARY RETINUE'

With equally unseemly haste (and cunning dissimulation of his feelings, as in Stolberg's case) Wilhelm brought about the dismissal of the elderly generals who had set the tone for decades at his grandfather's court, and who had remained in office for the time being because of the difficult circumstances of his father's accession.[12] Adjutant-Generals Karl Friedrich Count von der Goltz, Heinrich

August Count von Lehndorff and Anton Prince Radziwill, who belonged to 'the closest entourage of Kaiser Wilhelm I, and for whom the old monarch had feelings of genuine friendship', as Széchényi recorded in his report of 4 July 1888, found themselves 'pensioned off without warning, and numerous other measures of a similar kind may well be forthcoming soon for the senior generals'. The ambassador's comment on this brusque new style was telling: 'If all this had happened, not in the very first days of the new reign, but only quietly and gradually after several months, no one would have been surprised, given the advanced age of those affected; but as it was, the impression made on the circles to whom the latter belong was a very painful one.'[13]

The elderly gentlemen were replaced by younger career officers whom Wilhelm had got to know during his service at Potsdam. The 'dashing' General Max von Versen, who had been in command of the Hussar Guards regiment when Wilhelm was colonel, was appointed adjutant-general, while General Adolf von Wittich, who had been giving him lectures on military history since the beginning of 1888, was promoted to adjutant-general in waiting. The period during which the Kaiser had two adjutant-generals in his suite did not last long, however, for in March 1889 Versen was appointed general in command of the III Army Corps in Metz and was hence discharged from his appointment at court.[14] Even Versen's friend and defender Waldersee had to admit with hindsight that the Adjutant-General had exercised a considerable, but in his eyes 'questionable' influence on the young monarch: he was 'not the man to give the Prince the right advice, after the problems began', Waldersee considered. It was true that Versen had 'made a bold horseman' of Prince Wilhelm, and had 'really taken a great deal of trouble' in so doing. But Waldersee considered that this extraordinarily energetic, hard-working and physically tough, but only moderately intelligent and increasingly vain man had been partly responsible for the rough tone of the court. In addition it was he who was chiefly to blame for the arming of the cavalry with lances, and in general for the military deployment of the cavalry, 'about which the Kaiser [had] the unhealthiest ideas'. Above all Waldersee reproached the Adjutant-General with having whipped up Wilhelm 'against the entire entourage of his grandfather and especially against Albedyll'. At the time when his father's cancer became known, Wilhelm had thought 'very highly' of Versen and had discussed numerous 'future plans' with him, in the course of which 'much confused nonsense' had been talked. In the first months of his reign Wilhelm had 'singled him out a great deal' and 'showered him with favours', but soon his star began to wane. Versen had become 'an uncomfortable presence' for the Kaiser 'by constantly putting himself forward'; the Kaiser had suddenly withdrawn his confidence from the General and since then had avoided being alone with him. Although Versen's influence decreased sharply as a result, and although it became increasingly apparent that he was suffering from a brain disorder, he continued to be regarded as an important factor even after his transfer to Metz, and was thus numbered among the most hated people in the Kaiser's circle.[15] The Empress Frederick considered him one of her worst enemies and thought him 'half mad', but others too had nothing but contempt for him and mocked Versen with the punning nickname of 'das Versehen' (the mistake).[16] The British military attaché Swaine, who spoke of Versen's 'domination', later recognised what inner court circles had long known, that the General was 'nothing but a vulgar flatterer'.[17] Both Versen and Waldersee were convinced that the transfer of the Adjutant-General to Metz was the result of an intrigue, blaming in particular the 'ambitious' Flügeladjutant Adolf von Bülow for his downfall.[18] The second Adjutant-General, Adolf von Wittich, remained in office until the summer of 1892, in spite of countless disagreements with his imperial master.[19]

Wilhelm's personal adjutants, Moritz Ferdinand Freiherr von Bissing and Major Kurt Wolf von Pfuel, were both promoted to the position of Flügeladjutant.[20] According to Colonel Swaine, Bissing was the only man in the entire imperial entourage who had the courage to talk openly and frankly to Wilhelm.[21] Friedrich von Holstein likewise considered him the only reliable military officer in the Kaiser's entourage and later, after Bissing had been transferred elsewhere, tried to bring him back to court as Wilhelm's 'bodyguard', although Philipp Eulenburg was of the opinion that Bissing's '*excessive irascibility*' made him seem, 'even to His Majesty, not quite suited for the position'.[22] Among Wilhelm II's Flügeladjutanten in waiting in these early days were also Lieutenant-Colonel von Lippe, Colonel Carl Count von Wedel and Majors Gustav von Kessel, Adolf von Bülow (his former military tutor), Cölestin von Zitzewitz and Friedrich von Scholl. Bissing, Bülow, Wedel and Kessel were repeatedly demonised by the Empress Frederick as 'dreadful', 'selfish', 'ambitious', '*dangerous*' and 'very pernicious', while Zitzewitz (who was married to an Englishwoman) and Scholl she considered merely as 'second rate, inferior, common & vulgar'. The tone and manners of these men were 'far from what they should be at Court & about W.', she maintained; and their inclination towards 'abject flattery' was sheer poison for him.[23] Kessel in particular was an unscrupulous opportunist in her eyes, who had purchased his position with Wilhelm by treachery towards his dying sovereign. 'Of course W. is beholden to him & looks upon his conduct as a proof of K's devotion to his person; *we* were *sacrificed*... My blood boils with indignation the same as it did *then* when I think of it all', she declared in the summer of 1891.[24] When Kessel was promoted to colonel a few months later but not put in command of a regiment, the widowed Empress rightly assumed that this meant that he 'is going to be kept *on* in the Household and then become "General Adjutant" wh. is a post of influence & confidence!

Considering the *past* this fact speaks *volumes* to *me*, and they are *not* consolatory or encouraging.'[25]

Some of these officers remained Wilhelm II's closest companions and advisers until the end of the monarchy. All officers on the new Kaiser's staff, including Waldersee, belonged from December 1888 onwards to the 'Military Retinue of Kaiser Wilhelm II', while those who had served with his grandfather and father were designated as belonging to the retinue of Kaiser Wilhelm I or of Kaiser Friedrich III. The old name of 'Maison Militaire' was abandoned.[26] In its place came the 'Headquarters of His Majesty the Kaiser and King', established on 7 July 1888, and comprising all the adjutant-generals, generals à la suite and Flügeladjutanten. As commandant of the 'Kaiser's Headquarters', Adolf von Wittich was responsible for the division of duties of the 'Military Retinue'.[27]

The adjutant-generals and Flügeladjutanten gave the Wilhelmine court its particular character, and hence the atmosphere in which Wilhelm II took his decisions. Even under the old Kaiser, a Master of Ceremonies had declared proudly that nothing showed the predominantly military character of the Prusso-German court more clearly than the fact that 'in Prussia, more than at many other courts, the King's Flügel-Adjutanten are at the same time his gentlemen in waiting, and the ceremonial of the Prussian court in many ways reveals the military organisation beneath it'. It was, he maintained, 'the Prussian national character which is also clearly reflected in the life of the court'.[28]

The reverse of the medal, as far as this military accent was concerned, was the roughness of tone and the fondness for obscene stories which it often encouraged. The Chief of the Military Cabinet later admitted that he had suggested to the Kaiser early in 1889 that he should appoint Carl Wedel to his entourage, because 'its tone left much to be desired at that time. The Kaiser', Hahnke recalled, 'accepted this suggestion with enthusiasm, because he himself had been keenly aware that not everything was quite as it should be.'[29] However, observers continued to be struck by 'the bad tone' of the Kaiser's entourage. 'Scholl distinguished himself particularly in that respect', Waldersee commented some years later.[30]

One of Wilhelm II's Flügeladjutanten has left behind a precise account of the daily routine at the Schloss which shows how omnipresent the Military Retinue was. There were always two Flügeladjutanten in attendance on the Kaiser. They drove or rode out with him, they advised him on the questions on which he had to take decisions, they answered enquiries addressed to him from the outside world, they announced the people who were granted audience or whom the Kaiser had summoned, they assisted him in arranging his daily programme and they kept the 'All-Highest diary'. Normally an adjutant was on duty for two days, followed by two free days, so that each of them spent about fourteen days a month with the monarch. While on duty they lived at the Schloss, where they had a spacious study and two bedrooms beneath the Sternensaal at their disposal. In Berlin the two Flügeladjutanten usually began their work at about 8.30 a.m. when they accompanied the Kaiser and Kaiserin for an hour's ride in the Tiergarten. After returning to the Schloss the Kaiser gave audience to hear reports (Immediatvorträge) until just before 'second breakfast' (luncheon). At one o'clock (and then again at 8 p.m.) the Flügeladjutanten sat with the Kaiserin's ladies in waiting and the Hofmarschall at the imperial table, at which the Kaiser led the conversation, while the Kaiserin made no effort to disguise her preference for taking meals without a retinue, so as to have 'more of her beloved husband' to herself. About twice a week guests were also invited to luncheon. After the meal the Kaiser smoked a cigar with his two Flügeladjutanten in his study or in the adjutants' room. If the Kaiser and Kaiserin were driving out together in the afternoon, the Flügeladjutanten remained behind in the Schloss, but otherwise one of them drove out with the Kaiser. From 5 p.m. until dinner the Kaiser generally withdrew to his own apartments. After dinner Their Majesties both went into the library with their retinue, 'and one often remained sitting there for a very long time, sometimes without the Kaiser entering into any interesting conversation at all', the Adjutant recalled.[31]

The Flügeladjutanten may have had little independent influence on the Kaiser in the very early part of his reign. They were too new to court service, Wilhelm's personality was too strong and his self-assurance too great. Waldersee, who knew all the members of the military suite personally and kept a close eye on them, commented in August 1888: 'It is quite clear that he [Wilhelm] is ruled by no one in his entourages . . . The Flügeladjutanten have no influence.'[32] Little had changed by the summer of 1891, when Wilhelm's sister Viktoria (Moretta) observed that his entourage was too afraid of him to give him any advice.[33] Waldersee was dismayed to discover, shortly after his dismissal as chief of the General Staff, that of all the high-ranking army officers whose duty it was to advise the Kaiser – Chief of the General Staff Count Schlieffen, Minister of War von Kaltenborn-Stachau, Adjutant-General von Wittich and Chief of the Military Cabinet von Hahnke – 'not a single one' had 'exchanged a word with the Kaiser' during the autumn manoeuvres of 1891.[34] But within a few years it was widely recognised that the Military Retinue had become a powerful factor in German politics; in parliamentary and journalistic circles 'adjutant politics' and 'adjutant intrigues' became catch-phrases, and Bismarck, in discussing with a Bavarian newspaper editor the question of whether the Kaiser could be deposed, went so far as to sound the warning that ousting Wilhelm II 'from the midst of his generals' would 'not be as easy as the removal of King Ludwig II of Bavaria'.[35] Some of the adjutant-generals and Flügeladjutanten even developed a mystical, Knights-of-the-Holy-Grail relationship to their monarch, and for that reason alone would not have dreamt of contradicting, let alone ousting him. When in 1893 the Kaiser appointed as Flügeladjutanten not only Colonel Hans von Arnim, the son of a Potsdam architect, but also the musical and effete Kuno Count von Moltke, the latter's close friend Philipp Count zu Eulenburg wrote ecstatically: 'Your Majesty will become increasingly aware of what a pearl Your Majesty has acquired in this Adjutant – and I am filled with a pleasant, comforting feeling to know that *he* of all people is with my dearly beloved Kaiser.'[36] The world-famous equestrian Max Freiherr von Holzing-Berstett, who was appointed Flügeladjutant to the Kaiser around the turn of the century, openly admitted to his mother: 'I love him',[37] adding later, 'I rave about him like a silly girl, or rather, like a man who appreciates power and importance.'[38] When another officer, Heinrich Prince von Schönburg-Waldenburg, was addressed by the Kaiser during manoeuvres on 14 August 1900 with the question: 'Will you be my Flügeladjutant?', he was too stunned to reply, but expressed his gratitude 'wordlessly' by bowing 'to kiss the hand that is always so good to me'.[39] Gustav von Neumann-Cosel, as Flügeladjutant in attendance, kissed the Kaiser's hand at every possible opportunity. But even he found service at court so stressful that on returning to his bachelor apartments he 'first swore three times very loudly and crudely across the room and then retired to bed for 24 hours'.[40]

LIEBENAU AND THE MARSHALS OF THE COURT

If the Flügeladjutanten had no noticeable influence on the Kaiser in the early days, the same could not be said of his Oberhof- und Hausmarschall (Senior Marshal of the Court and the Household), Eduard von Liebenau and his assistants, the Hofmarschälle. In his short time as Crown Prince Wilhelm had expressed the intention of 'breaking the power of the marshals of the court'; to this end he had planned with Waldersee to appoint a strong-willed general as commandant of his Imperial Headquarters.[41] In reality, however, during the first months of the reign of Wilhelm II the influence of the marshals grew markedly – and indeed chiefly because in August 1888 the Kaiser appointed Liebenau, hitherto his Hofmarschall, to the post of senior marshal of the court and the household, conferring on him the title of 'Excellency'. Liebenau was assisted by the Senior Cupbearer Karl Count von Pückler, who had served as lieutenant in the elite Gardes-du-Corps regiment and was a nephew of the Senior Master of Ceremonies under Wilhelm I. The immaculately dressed Pückler, whom Wilhelm II nicknamed 'Suleiman the Magnificent', was described by a middle-class naval officer as 'every inch the courtier, an elegant air, obliging cordiality mingled with slight condescension'.[42] Under Liebenau and Pückler the 'power of the marshals' grew steadily. In October 1888 Waldersee grumbled that their 'influence [has] grown greater than ever'.[43] 'It is quite extraordinary how this Kaiser allows the influence of these gentlemen to increase, and everywhere one hears complaints of their arrogance.'[44]

In Waldersee's eyes, Liebenau's appointment was incomprehensible except as an expression of gratitude for past services for, apart from the two Bismarcks and Carl Wedel, the Oberhofmarschall had no friends at court and, although he was economical and efficient as an administrator, in every other respect he was thoroughly unsuited to his post. He was 'so vain and puffed-up, so uncouth and at the same time so domineering, that he will give great offence and will perhaps come to grief one day', Waldersee observed in August 1888.[45] A few months later the Chief of the General Staff remarked that Liebenau's influence had grown greater still. 'It is really very much to be deplored that the Kaiser has made Herr von Liebenau Oberhofmarschall', he wrote in his diary. 'There is nothing distinguished about him; even his outward appearance is far from agreeable; and now he suffers from a kind of megalomania which brings him into conflict with everyone. At the court there is not *one* gentleman who does not complain about him; his subordinates are unhappy because of his rough tone and his lack of consideration. Everywhere he tries to exceed the authority of his office . . . He has been on a bad footing with the Kaiserin and the ladies for a long time; this alone speaks against him, for the Kaiserin is an excellent woman, of a most noble and distinguished nature, a model for all women.'[46] It was clear to the Chief of the General Staff that the Reich Chancellor and his son were promoting Liebenau's position in order to put a stop to the pious-orthodox influence of the Kaiserin and her close entourage.[47] Not long afterwards Waldersee began to suspect that Liebenau was even aiming '*to bring about discord or coolness between the Kaiser and the Kaiserin* by every means at his disposal'. At any rate Liebenau was trying 'to be the sole influence' on the Kaiser, and was 'jealous of anyone else who approaches the Kaiser; so it is particularly unpleasant for him when Their Majesties are together, because he feels that the Kaiserin's excellent influence is making itself felt'.[48] Liebenau had long since become the sworn enemy of Freiherr von Mirbach, who as the Kaiserin's Hofmarschall acted as the hub of the strict religious faction at court.[49]

The Oberhofmarschall was indeed making more and more enemies with his tactless behaviour.[50] In November 1888, when arranging a visit to Breslau for the Kaiser, he managed to exclude both the new Chief of the Military Cabinet, General Wilhelm von Hahnke, and the new Chief of the Civil Cabinet, Hermann von Lucanus, and failed to invite the Commanding General in Silesia, Oktavio von Boehn, either to shoot or to dine with the Kaiser. 'That is a crass violation of all traditions, which would have been absolutely impossible under Kaiser Wilhelm

I', was Waldersee's verdict on the incident. Liebenau was 'on the way to losing the Kaiser the support of the army. Boehn wants to hand in his resignation as a result and he is quite right; the act does him credit. Hahnke remonstrated, but the damage had already been done.'[51]

It was not long before Liebenau's nerves and those of other senior court officials gave way. They had overtaxed themselves 'in a quite ridiculous way' in their office work, Senior Master of Ceremonies August Count zu Eulenburg wrote in March 1889. Liebenau had had to go to Wiesbaden for a fortnight 'for his nerves to recuperate' – 'oh if only he would never come back!', sighed Waldersee – and Pückler had 'completely collapsed for months'.[52] The newspapers speculated about Liebenau's retirement and his replacement by August Eulenburg, but Waldersee knew better. It was certainly almost everyone's wish that Liebenau should retire, and one could not deny that he was 'a very regrettable element in the Kaiser's entourage', but he, Waldersee, knew 'for certain that the Kaiser will continue to hold on to him for the time being'.[53] And that was indeed the decisive factor. In spite of many a fiasco and much hostility Liebenau managed to cling on until May 1890, which was after all two months longer than the Reich Chancellor and his son the Foreign Secretary.

THE NEW CABINET CHIEFS

In the political sphere the changes in the two Cabinet posts were probably even more significant than the new appointments to the court and household and to the Kaiser's Military Retinue. A change at the top of the Secret Civil Cabinet, which was responsible for liaison with the non-military authorities of the Reich and Prussia, was certainly overdue. The 71-year-old Karl Freiherr von Wilmowski had been chief of the Civil Cabinet since 1861; his political role had been further enhanced in recent years because in addition to his customary work he had the responsibility of reporting to the monarch on matters concerning the Prussian Ministry of State. Lately, however, he had several times asked to be relieved of his post because of a serious eye disorder. It was therefore perfectly understandable on humanitarian grounds that permission was given for Wilmowski's retirement on 1 July 1888.[54] As Wilhelm informed his grandmother on 11 July, 'poor Wilmowski . . . had become so completely blind that he had to be operated on and he looked so incredibly pleased when I reluctantly gave my consent to his resignation, which had been refused several times, that he must really have wanted it very badly'. His successor was to be the former Under-Secretary in the Prussian Ministry of Ecclesiastical Affairs, Dr Hermann von Lucanus, who, as Wilhelm added, is 'qualified in every respect to fill the post'.[55] As the son of an apothecary, Lucanus had risen in the administration purely on

7. Hermann von Lucanus, Head of the Kaiser's Secret Civil Cabinet 1888–1908

his own merits. He had been ennobled only weeks earlier, in the last days of the reign of Friedrich III, and thus became a very isolated example at court of the tradition of the intelligent, profoundly loyal and dedicated middle-class Prussian government official. That of course did not please everyone in court society. Count Waldersee complained that Lucanus had no 'independent opinions', and danced 'entirely to Bismarck's tune'.[56] Years later, when he was awarded the Order of the Black Eagle, he was still being mocked in aristocratic circles as 'the apothecary'. When Privy Councillor von Holstein heard that Count Philipp zu Eulenburg used the phrase 'most obedient' in his letters to Lucanus, he declared that he was shocked: it was 'enough to turn the heads of all the apothecaries in the Prussian monarchy'.[57] Aristocratic court officials and army officers complained that the 'characterless' Lucanus 'complied with the Kaiser's wishes in everything', in order to remain in his well-paid post.[58] Nonetheless Lucanus led the Secret Civil Cabinet until his death in the summer of 1908, and became in time a key figure in Wilhelm II's reign.[59] Simultaneously with the replacement of Wilmowski by Lucanus, at the Kaiser's request Privy Councillor Miessner took over the running of the imperial private chancellery as secret correspondence secretary. Some time later he was also entrusted with the administration of Wilhelm II's Privy Purse.[60] The Correspondence Secretary, as we have already seen in the first volume, found himself immediately faced with a number of highly delicate matters to settle.[61] Like Lucanus he too remained in office until his death in 1909.

8. General Wilhelm von Hahnke, Head of the Kaiser's Military Cabinet 1888–1901

The Secret Military Cabinet, which was responsible for all matters of army personnel, also received a new chief directly after Wilhelm II's accession. General Emil von Albedyll imagined himself quite secure in the Kaiser's favour, although privately Wilhelm had decided to have no more to do with him since their conflict over the Union Club.[62] It was therefore to the General's considerable surprise that he found himself dismissed in the first few days of the new reign and replaced by General Wilhelm von Hahnke, who was appointed adjutant-general at the same time.[63] Like Versen and Wittich, the monarch had known Hahnke, a man of gigantic height, from his soldiering days in Potsdam: he had been Wilhelm's brigade commander when the young Prince was serving in the First Regiment of Foot Guards, and he was related by marriage to Wilhelm's erstwhile military tutor and current Flügeladjutant Adolf von Bülow.[64] He made an excellent impression on Queen Victoria, who met him in August 1889 during Wilhelm's visit to the Isle of Wight: she pronounced him 'a very pleasing man, with a straightforward manner', who was 'the principal person about William'.[65] Waldersee's verdict on Hahnke, on the other hand, was critical in the extreme: he described him as 'a man below the level of mediocrity' and 'a notoriously narrow-minded person' who was not capable of fulfilling the duties of his post properly.[66] When Waldersee asked a Commanding General who had come to Berlin in the autumn of 1890 whether he intended to call on the Chief of the Military Cabinet, the General replied: 'Why? He knows nothing about anything.'[67] And in

1891, shortly before his own downfall, the Chief of the General Staff said of the 'small-minded' Hahnke that he had not given up hope 'that the Kaiser will come to realise what kind of an adviser he has; the feeling against the man is very much on the increase'.[68] Only once, in the spring of 1891, did Hahnke proffer his resignation; 'the Kaiser, who will certainly not find a more compliant man again, naturally declined it', noted his rival Waldersee in his diary.[69] Hahnke served until 1901 as an influential chief of the Military Cabinet. His son married the daughter of Waldersee's successor as chief of the General Staff, Count Schlieffen.

THE NEW NAVAL CABINET

Shortly after coming to the throne Wilhelm II proposed fundamental changes to the organisation of the imperial navy. Instead of a single body, the Admiralty, which had hitherto held sole responsibility for all naval affairs, the highest level of naval administration was to be divided into three authorities, analogous to the command structure of the army. The Reich Navy Office would be responsible for ship-building, and would thus have the duty of making the case for the naval establishment in the Reichstag, as the Prussian minister of war did for the army budget. The High Command would be responsible for naval operations and the strategic planning, echoing the General Staff's role for the army. Finally, a newly established Naval Cabinet, like the Military Cabinet, would have charge of personnel matters.[70] The Kaiser's aim was obvious: he was not prepared 'to tolerate an intermediate authority with powers such as the Chief of the Admiralty had possessed between himself and his favourite creation'.[71] A major administrative department of the Reich, which was directly answerable to the Reich chancellor, was broken up so that the Kaiser could intervene more easily in naval affairs.[72] The highly respected Chief of the Admiralty, General Leo von Caprivi, warned that although the army was 'so firmly grounded . . . that the interference of the Kaiser [could] not cause any serious damage', matters were 'quite different' in the navy.[73] The fact that on 5 July 1888 the young Kaiser was already ordering extensive changes to naval uniform only confirmed Caprivi in his conviction that he could not continue in office under the new sovereign.[74] To the delight of the naval officer corps it was Vice-Admiral Count Alexander XVI von Monts de Mazin, rather than yet another general, who was placed in charge of the Admiralty, at least until the reorganisation of the administrative structure of the navy could come into effect.[75]

In November 1888 Kaiser Wilhelm appointed a naval officer as Flügeladjutant, for only the second time in the history of the Prussian court. The first had been Albert Freiherr von Seckendorff, who had meanwhile been promoted to the post of Hofmarschall to Prince Heinrich.[76] Wilhelm had got to know the then

9. Admiral Gustav Freiherr von Senden-Bibran, Head of the Kaiser's Naval Cabinet 1889–1906

41-year-old Captain Gustav Freiherr von Senden-Bibran during his state visits to St Petersburg, Stockholm and Copenhagen and also at the naval exercises in September 1888, when Senden had been commandant of the iron-clad *Bayern*; he was also warmly recommended to the Kaiser by Prince Heinrich and Seckendorff. When Senden received the imperial order appointing him Flügeladjutant on 13 November he became 'fearfully agitated', for 'the post as Flügeladjutant [and] the life of a courtier which went with it' seemed to him 'an absolutely horrifying idea'. 'It did not suit me at all', he reminisced in his memoirs, 'to move in exalted circles, be cautious in expressing opinions . . . or make stimulating drawing-room conversation.' Nevertheless he travelled to Potsdam the next day in his new Flügeladjutant's uniform to present himself to the Kaiser. In his agitation he was somewhat reassured to find, in the small office in the Marmorpalais allocated to the Flügeladjutanten, officers of aristocratic birth such as Moritz Freiherr von Bissing, whom he and his brothers had known since childhood. The Kaiser received Senden in Hussar's uniform and said that he was glad 'that he would now always have a naval officer about him'.[77] Commandant of the Imperial Headquarters General von Wittich instructed him in the duties of the Flügeladjutanten. Over the next few days he called personally on all the princes of the royal house, and wrote to all adjutant-generals, generals à la suite and Flügeladjutanten, to introduce himself as Flügeladjutant in attendance. 'This is an old custom which is intended to demonstrate the solidarity between the officers attached to the person of the Kaiser', he explained. He was received cordially everywhere, even by the most senior naval authorities in Berlin, as the man who 'was to become the mouthpiece of the Kaiser'.[78]

Although the reorganisation of the naval authorities desired by the Kaiser was still being debated, it was already settled that Senden would in future administer naval personnel questions in addition to his personal service with the Kaiser. The acting Chief of the Admiralty, Count Monts, who like Caprivi was an opponent of the reorganisation, resigned himself to the new order, not least because Senden had been put in charge of personnel matters.[79] Prior to the establishment of a Naval Cabinet Senden-Bibran was detailed to serve in the Military Cabinet under Hahnke, in order to learn from the latter and from the two departmental heads, General à la suite von Brauchitsch and Colonel von Oidtman, how army personnel questions and disciplinary matters were dealt with.[80] Then, before the year was out, a Naval Cabinet analogous to the Military Cabinet was set up with Senden as chief, with responsibility for all naval personnel questions.

With the appointment of Gustav Freiherr von Senden-Bibran as Flügeladjutant and head of the Naval Cabinet the imperial court acquired a figure who ruthlessly stoked up Wilhelm II's passion for the navy and who, after Bismarck's dismissal, looked down with the greatest contempt on the statesmen who bore the responsibility for the policies of the Reich.[81] German diplomats regarded him as 'a monstrosity', who gave them 'terrible . . . headaches'.[82] Count Münster, the elderly German ambassador in Paris, commented after a meeting with Senden that it was 'a positive disgrace' to have to listen to so much nonsense and megalomania.[83] Yet all attempts to get rid of Senden from the court misfired. Until his retirement in 1906, as Privy Councillor von Holstein was forced to acknowledge through gritted teeth, he remained one of the most influential men in the Kaiser's entourage.[84]

THE GENERAL STAFF AND THE MINISTRY OF WAR

In August 1888 the old Field Marshal Count von Moltke stepped down from his post as chief of the Great General Staff and was succeeded by Wilhelm's fatherly friend Count Alfred von Waldersee, whose influence now became so apparent that he was regarded everywhere – even abroad – as the future Reich chancellor.[85] Immediately after Wilhelm II's accession Waldersee noted in his diary that he had reason to be 'almost too proud', for he was 'definitely greatly in favour with the Kaiser, which the whole world knows, and plenty of people run after me in consequence. The Kaiser thinks a great deal of me, likes to hear my opinion on many things and in fact has friendly feelings towards me.'[86] After his appointment as chief of the General Staff Waldersee reflected with some self-satisfaction: 'I am well aware that the Kaiser has placed great confidence in me and that a tremendous responsibility may fall on my shoulders. If war breaks out, my position is the most important not only in our army, but also in the country, even in the whole world. On the achievements of our army depend the future of Germany and thereby of the whole European states system . . . How wonderful are God's ways! I have one of the most brilliant careers behind me and the whole world looks to me, for as things are the chief of the General Staff of the German Army is in fact the most respected of all military officers.'[87]

Since Wilhelm II listened to Waldersee's advice on all military decisions and frequently disregarded the established division of responsibilities, the power of the General Staff within the army increased so fast and to such an extent that Minister of War General Paul Bronsart von Schellendorf felt obliged to hand in his resignation at the end of 1888. The Kaiser positively welcomed this battle between Waldersee and Bronsart, particularly as the latter had aroused his anger earlier that year with his contemptuous attitude towards the navy.[88] In December Wilhelm declared that he was glad 'to have direct fights with the Minister of War and others, which enabled Him to sharpen His teeth; yesterday He had already had Bronsart in tears!'[89] As Waldersee confided to his friend General Julius von Verdy du Vernois, who was to be appointed Bronsart's successor as minister of war at Waldersee's suggestion, Bronsart had lost his 'balance' since the death of Wilhelm I; he had become nervy and allowed 'trifles' to annoy him. 'From the very beginning he did not know how to get on with the present Kaiser, and there is an uncomfortable feeling on both sides. Now Bronsart has got it into his head that the Kaiser has more confidence in me than in him, which he thinks is an intolerable situation, and of course in this he is just seeing things. The position is that the young Kaiser is very independent, does not always observe divisions of responsibility strictly enough and sometimes expresses His will to the Minister of War in a very decided manner. So the position of a minister is naturally different from before, and all the more so because the Kaiser wishes to give the chief of the General Staff more powers.' If, as Waldersee had recommended, the Kaiser should appoint Verdy minister of war, it would be conditional on Verdy declaring himself in favour of a 'further demarcation of powers between the Ministry and the General Staff', Waldersee wrote. He added with machiavellian disingenuousness: 'I hope that you will not make any difficulties, and I am sure that we shall agree. I feel that this is a great moment and in this way you can begin to render great service to the army and the Fatherland.'[90] At this Verdy himself suggested that on his appointment 'a few of the powers of the minister of war' could be taken away from him.[91] A few days later Waldersee had to inform Verdy that Bronsart had withdrawn his resignation for the time being, at Bismarck's urgent request, and that the Minister of War had received strong support from the officers in his ministry in his battle against the increase in the power of the General Staff. Nevertheless the Kaiser would triumph, Waldersee assured him. 'I believe that the Kaiser has the intention of making you the successor', he wrote to Verdy on 2 December 1888. 'He brought the matter up himself today. It could be that he encounters opposition from the Chancellor; but as things stand at the moment it will not be strong. The Chancellor is no longer what he used to be; he knows the strength of his sovereign's will and just now he has other plans for which he badly needs to win concessions, so he will be careful not to play too high a card here.'[92] The observation was accurate: after Verdy's appointment as minister of war in April 1889 Waldersee was able to note with satisfaction: 'Another sign that the Chancellor is very ready to give in to the Kaiser is that he raised absolutely no objection to Verdy's appointment.' Indeed, Bismarck had even put a brave face on it and told 'all the world that he was very pleased that the Kaiser [had] chosen Verdy'.[93] Only after his dismissal did Bismarck show his resentment at the appointment of a long-standing personal opponent to be his colleague as a minister, when he wrote: 'Verdy treated me as an enemy from 1870 in France (demi-gods) and because he falsely blamed me for his transfer to Königsberg; also because he is of a progressive–liberal turn of mind.'[94]

Since the Kaiser exercised unlimited 'power of command' in the military sphere as supreme war lord, and moreover (after hearing the advice of the Chief of the Military Cabinet) took decisions on all matters of army personnel entirely on his own, the Reich chancellor and Prussian minister-president had absolutely no say in the choice of the minister of war. And yet, because the minister of war was a member of the Prussian Ministry of State, and also because of the anomaly that he had to present the army budget in the Reichstag since there could be no Reich minister of war in the federated German Reich, he had to work closely with the chancellor. The new distribution of functions between the Ministry of War and the General Staff therefore represented – like the introduction of a Naval Cabinet and a Naval High Command – both a reduction of the authority of the state that was more than symbolic, and an increase in the powers of the crown. But even after the concessions made by Verdy to the General Staff in the spring of 1889 the 'all-highest, all-gracious Ministry of War' was still too powerful in the eyes of Wilhelm II. As the Kaiser's brother-in-law, himself a commanding general, commented ten years after Verdy's appointment, Wilhelm was convinced that 'the omnipotence of the Ministry of War must be broken; there is only one way of doing this: greater decentralisation'.[95]

The Kaiser, of course, made full use of his right of appointment to select the commanding generals and to fill the remaining high-ranking posts in the army with his personal choices. As late as 1904 we find an officer, commenting on the appointment of one of the Flügeladjutanten as 'commander of the most illustrious brigade in the army,' writing: 'It is quite clear that the Kaiser's wish is absolutely decisive in such nominations.'[96] What such appointments meant for the social life of a province is vividly illustrated by a letter from the Commanding General in Breslau, Bernhard, Hereditary Prince of Saxe-Meiningen. 'In accordance with the longstanding tradition everyone looks to the commanding general, how he conducts himself here and there, how he receives people and where he goes, and where he does not. The nobility and society instinctively take the lead from the commanding general, which is at present even more the case than hitherto.'[97] At the beginning of each year and in times of crisis the Kaiser summoned the commanding generals to Berlin – this too was an innovation brought in by him. They were expected to act as a body, like an Areopagus hovering over the state, to reinforce his authority and legitimise his decisions. Thus it was that they were summoned to Berlin on 18 March 1890, when Bismarck was forced to hand in his resignation.[98]

'THE ENTOURAGE OF THE YOUNG SOVEREIGN'

The radical reorganisation of the structure and personnel of the court and of the high command of the country's defence forces which took place immediately after the accession of Wilhelm II was undoubtedly the Kaiser's own work. Waldersee was quite right in recording on 23 June 1888 that the Kaiser had 'selected the personalities and the adjutants himself'.[99] Consistent with this was his observation four months later that Wilhelm was 'undoubtedly the master . . . where his entourages are concerned',[100] for the power of appointment and dismissal is the key to power itself. All too soon the Hohenzollern court acquired an atmosphere of mawkish sycophancy which repelled many observers. 'God knows', one of the Flügeladjutanten exclaimed in disgust at the sight of the 'swarm of courtiers', 'one cannot help laughing, but one often feels sick at this wretched craning upwards of eyes and ears!'[101] The military suite, however, showed no more character than the civilian courtiers. As the British military attaché Leopold Swaine remarked in 1892 on being reappointed to Berlin after an absence of three years: 'What we suffer under here is a hornets nest of fulsome flatterers whose sole interest is to remain in favour in order to draw their salleries [sic] and obtain decorations from all arriving Royalties . . . The Emperor is either no judge of character or else he prefers having silly people about him who can't, or dare not, give him contrary advice to his own inclinations.' Swaine preferred the latter explanation.[102]

The men whom Wilhelm had gathered around himself were mostly young and inexperienced. Many of them were professional soldiers from minor aristocratic families, while a few, like Lucanus and Müller, were of middle-class stock. Apart from the holders of the senior honorary posts at court, the higher aristocracy and great landowners were not represented in the new imperial entourage. The court lacked the kind of older, more experienced and prudent person who could have dissuaded the young, immature, impulsive monarch from his notorious 'suddennesses'. The Hungarian *grand seigneur* Count Széchényi reported from Berlin on 29 December 1888 that 'the entourage of the young sovereign' was 'a subject of serious concern to most right-thinking and well-disposed people' and certainly left much to be desired 'in that the Kaiser has no one about him with sufficient experience of the world and knowledge of what is customary, traditional and seemly to make up for what His Majesty himself, being so young, naturally lacks in that regard'. The Kaiser's court consisted of 'nothing but insignificant young people', he wrote, 'who have been so to speak catapulted out of the limited circumstances of the small princely court at the Stadtschloss in Potsdam on to the broad, high plateau of the Imperial Court'. Even General von Wittich, the senior Adjutant General, and General von Hahnke, the Chief of the Military Cabinet, were 'undoubtedly two outstanding experts in their field and in that respect a very fortunate choice; but since they have hitherto lived only within and for their profession' they were 'lacking in any knowledge or experience of the world'. The only person who had had the opportunity to acquire such knowledge, namely Count Herbert Bismarck, either did not possess 'the necessary tact' to make appropriate use of his experience, or lacked influence, perhaps 'because he does not yet feel safe enough in the saddle to use [his influence] without endangering it'. The Reich Chancellor, who could have played the most effective role in this respect, was staying 'well out of the line of fire' at Friedrichsruh, following his ploy of not intervening in insignificant questions in order to preserve the full weight of his influence for matters of great political import. 'The state of affairs here' in the new Reich, the ambassador reflected, 'is in general so very unsettled and still only at the formative stage, especially as it has not evolved step by step and gradually, but has arisen suddenly and developed by leaps and bounds. It is a state which can well be compared with that of fermenting, bubbling young wine. The finer the juice of the vine, the more it works and the better and stronger the wine will then become, when all impure and foreign elements have at last settled. – May it be so!'[103]

At the beginning of 1889 the perspicacious Széchényi returned to this central theme and emphasised how much Wilhelm II dominated his entourage. 'When one . . . considers', he wrote, 'that the entourage of the young sovereign is of the kind that I tried to describe in my . . . secret report . . . of the 29th of last month, and that it not only owes its present existence to the favour of its Imperial Master but also counts on Him for its future, it is perfectly comprehensible that it always guards against expressing a view contrary to His Majesty's intentions, and only ventures such a view when He expressly asks for it, so that in many cases the youthful and hasty initiative of the sovereign does not find the counterweight that would be desirable. At the same time this close entourage nonetheless retains, at least in those cases where it is consulted, sufficient room for manoeuvre to be able to exercise a not inconsiderable influence, although it can by no means be taken for granted that the best use is always made of this.'[104] In the Bismarck crisis that began to emerge soon after the accession, the court that Wilhelm had built around himself formed a bastion of imperial power against which the Reich Chancellor, even with the support of Liebenau, could not in the end prevail.

CHAPTER EIGHT

The domination of the Bismarcks

THE KAISER AND THE BISMARCKS

IN the first months of his reign Wilhelm II's relationship with the Bismarcks was nothing if not harmonious. In October 1888, when it was suggested to him that he ought to develop 'his own ideas as Kaiser', he exclaimed: 'It is too stupid that people simply refuse to understand that the young and the old generations can work excellently together. There is no such thing as the Kaiser and the Chancellor having separate ideas; the troublemakers always forget that I worked in the Foreign Office for $2^1/_2$ years.'[1] Also forgotten, apparently, were the Bismarcks' grim premonitions of the previous year[2] and even the conflicts of the previous few weeks, for while Friedrich III was still on the throne Herbert Bismarck had had to seek a vote of confidence from the then Crown Prince Wilhelm, so badly had the latter treated him.[3] Now, however, after the accession, the Foreign Secretary reported with gratification to Friedrichsruh that the young Kaiser was 'very affable, just as always'.[4] 'Our new sovereign is very attentive, and at the same time calm and objective; he is excellent to work with.'[5] 'Our Kaiser does his job extremely well', the Chancellor's son wrote to Holstein on 15 October 1888 from Rome. 'He pleases and impresses everywhere he goes.'[6] Two days later, in a further letter to the influential Privy Councillor, he commented: 'It is very easy to get along with H.M., at all events easier than with any other sovereign with whom I have hitherto been acquainted.' Herbert added that Wilhelm had been 'quite exceptionally nice' to him on the journey to Austria and Italy; he had 'often come into my room in the evening with his brother to recover from the day's exertions with the help of beer, cigars, and jokes. Everything goes splendidly, H.M. makes a first-rate impression everywhere, and for his part is thoroughly content.'[7]

No doubt some of this optimism was only for show, for the Bismarcks were of course fully aware of the instability of their position, dependent as it was entirely on the 'confidence' of the dynamic, impatient and unpredictable young monarch. That is evident from their carefully calculated conduct. In August 1888 the old Reich Chancellor refused to travel to Hamburg with Wilhelm, 'because he feared becoming a burden to the Kaiser, and putting H.M. in the shade through the ovations that he receives'. In order to 'get out of everyone's way' Prince Bismarck, who had been living at Friedrichsruh since 12 July 1888, took the decision to withdraw even further to his Pomeranian property of Varzin. His daughter Marie and her husband Kuno Count von Rantzau tried to talk him out of this idea 'because of the cold and the great distance from Berlin', but the Chancellor remained at first 'quite firm, and said that he had things to do at Varzin, which to my knowledge is not true, and that Mama wanted to go there'. Rantzau decided to co-opt Bismarck's doctor Schweninger as an ally against the move to Pomerania, for quite apart from the danger that it would scarcely be possible for government business to be carried on during parliamentary sessions if he were at Varzin, to stay so far away from Berlin would be inadvisable on health grounds as well.[8] Although Bismarck did remain at Friedrichsruh for the time being – Herbert also considered a move to Varzin would be 'disastrous' and set out to 'frustrate' his parents' decision[9] – the Chancellor's arguments nevertheless show how precarious he considered his position to be as far as the Kaiser was concerned. This is also demonstrated by the overcautious, courtier-like attitude which the Chancellor adopted towards the young Kaiser during his absence from Berlin. 'Don't forget at the next audience to thank His Majesty for his greetings and for allowing me to remain here longer for my health in spite of the Reichstag', he telegraphed in cipher from Friedrichsruh to his son in Berlin on 7 December 1888.[10]

Wilhelm for his part found Bismarck's absence extremely agreeable. At the end of December 1888, when Herbert Bismarck handed him a letter from the Chancellor, the Kaiser read it with a smile and remarked that it was a very good thing that the Chancellor had 'remained in the country until now', both because it suited Prince Bismarck better and because it obliged him, the Kaiser, to deal directly with the ministers.[11] In June 1889, when the Chancellor carried out his original plan of going to Varzin,[12] Wilhelm expressed pleasure at this decision. The Kaiser was 'in the happiest of moods', Herbert reported to his father, and had said 'it was a real blessing that Schweninger had at last succeeded in persuading you to go to Varzin, and he hoped that you would have a complete rest there'.[13] Apart from a few short trips to Berlin to greet visiting foreign monarchs, Bismarck was not to return to the capital until 24 January 1890.

Given the international significance of this development, it is not surprising to find that the latent threat to the power of the Bismarck family under the new Kaiser was a central theme in the reports written by German and foreign diplomats at the time. As early as October 1888 Count Hatzfeldt, the German ambassador in London, intimated to the British Prime Minister 'that the young Emperor was very difficult to manage, that Prince Bismarck was in great perplexity, and his temper had consequently become more than usually unbearable'.[14] The Austro-Hungarian ambassador recorded in February 1889 that the old Prince was avoiding 'everything which might displease the young Kaiser.' As evidence of this Széchényi cited the fact that Bismarck, 'to the astonishment of all', had attended the service in the Schloss chapel to celebrate the Kaiser's birthday on 27 January 1889. When Széchényi expressed his surprise at this, the Prince replied that if he had not come, his 'young sovereign' might have thought that he was failing to show due respect. As the ambassador aptly observed, it was widely felt in Berlin 'that Prince Bismarck is no longer the only driving-wheel of the state, and that as such he often chooses to keep still, accommodating himself to the new circumstances, whether in order to avoid wearing himself out unnecessarily, or to pave the way for his son'. The undeniable result of this restraint, however, was that other political figures in Berlin 'generally have the tendency to follow whichever paths they think will lead to good, secure positions in the new era'.[15]

Only a few days later Széchényi was commenting on the numerous 'very significant' assertions in the official and pro-government press to the effect that 'the relationship between the Reich Chancellor and Kaiser Wilhelm II continued to be an untroubled one'. As mouthpiece for this announcement Bismarck had chosen the *Hannover'sche Courier*, but the article which appeared in that newspaper entitled 'Kaiser and Chancellor' had been reprinted in the *Norddeutsche Allgemeine Zeitung*, and again in the *Post*, with the additional assurance that the article had received 'full approval in the highest and most authoritative political circles'. The conclusion which one must draw from these 'resounding and persistent trumpet blasts', Széchényi opined, was 'that Prince Bismarck still possessed the confidence of his sovereign, and that his position was therefore in no way compromised or indeed weakened, but that he had ceased to be all things in all respects, partly from the prudent, self-imposed moderation of his influence, partly because of the drive towards independent action which is natural both to the age and to the character of the young sovereign'.[16] Others were less diplomatic in their language. The Empress Frederick, disgusted at the old Chancellor's behaviour, wrote on 20 April 1889 that he 'toadies William as he never did his grandfather or his father'.[17]

10. Count Herbert von Bismarck, German Foreign Secretary 1886–1890

For months on end the Reich Chancellor deliberately remained away from the capital, returning only for a few short visits, and these during the time when Herbert Bismarck was abroad.[18] The actual conduct of government business therefore rested on the shoulders of his son Herbert, who however kept in close touch with his father through his brother-in-law Kuno Rantzau and the Chief of the Reich Chancellery, Franz von Rottenburg. A vivid description of the outward appearance of this little-known key figure in German politics has reached us from the pen of Heinrich von Eckardt, a Baltic baron. The Chancellor's son, he wrote, was 'tall, broad-shouldered and heavily built'; he had 'a rather reddened face with a very determined expression, large eyes with a steady, rather overpowering gaze' and wore his hair 'not smooth but curly', so that he looked as if he had run his hand through his hair. He spoke very fast, 'without pronouncing the words very clearly'. In all, the Foreign Secretary gave the impression of a 'very determined, rather rough, perhaps even brutal personality', but he did not allow himself to be governed by emotions, but demanded 'manly German conduct' from his colleagues. With him one had to be 'able to work very hard; also to be lively and able to drink beer, smoke and talk frankly'.[19]

In the interval between Wilhelm's accession on 15 June 1888 and the beginning of the fateful year of 1890, Herbert Bismarck had 112 private audiences or other opportunities to discuss political questions face to face with the Kaiser; *in addition* to this he spent no fewer than seventy-four days travelling with the

Kaiser. Over the same period the Reich Chancellor saw the monarch – apart from at a few banquets for foreign sovereigns – only about ten times altogether, and Herbert was also present at four of his private audiences during these eighteen months.[20] Undoubtedly, therefore, Herbert Bismarck's relationship with the young Kaiser formed the cornerstone of the whole system of government. As Széchényi reported to Vienna in January 1889, it was 'the people in daily, close contact who gain most influence on His Majesty, and these are Count Herbert von Bismarck and Herr von Liebenau'.[21] For every appointment, every dismissal, every measure of domestic, foreign or colonial policy, Herbert had to 'extract' the consent of the Kaiser, but he succeeded in doing so. As the Empress Frederick observed in April 1889: 'Herb: Bismarks [sic] influence is supreme.'[22]

That the Foreign Secretary was doing his utmost to foster his relationship with Wilhelm can be seen in the – cynically calculating – obsequious tone of his letters, which dripped with expressions like 'most condescending', 'respectfully', 'all-gracious' and 'all-humblest'.[23] He advised Rantzau to remember the Kaiser's penchant for gossip when he wrote his reports from Munich.[24] Waldersee's verdict on this system and on Herbert Bismarck's servile attitude was contemptuous: 'This pathetic adviser throws himself at the Kaiser in an almost laughable way; people rightly see it as proof of the Bismarck system's weakness; it thinks itself very shaky if it cannot constantly work on the Kaiser. Young people are already laughing at Count Bismarck when they see him in society or at dinners etc. laying siege to the Kaiser and watching everyone who speaks to him. The Kaiser takes a very firm attitude towards everyone . . . and easily overcomes any opposition.' Under no circumstances, maintained Waldersee, could one say that the Kaiser had any respect for Herbert Bismarck.[25]

The fact that in a system like this decisions were sometimes based more on a desire to maintain an individual's position of power than on objective considerations led to criticism from many sides. Széchényi, for instance, found it incomprehensible that the Berlin government openly launched a strong attack on the British ambassador in St Petersburg, although 'they need England's support for their policy both on peace and on colonial affairs'. He likewise commented incredulously on the arrest of Geffcken for the publication of extracts from the war diary of the late Kaiser Friedrich, for this affair was undeniably 'a dangerous gamble with the prestige of the throne and the reputation of the courts'.[26] According to Széchényi, in well-informed circles in Berlin erroneous decisions such as these were attributed not to the Reich Chancellor, but to the relationship between Herbert Bismarck and the Kaiser. 'If one asks who was the driving force behind these matters, the answer is usually Count Herbert Bismarck, or even His Majesty the Kaiser, but as for the Reich Chancellor himself having played any part in them, that is denied outright.'[27]

The 39-year-old Herbert Bismarck was vulnerable not only because of the huge burden of work which he had to carry and his fear of rivals in the imperial favour, but also because of his unconventional lifestyle. He was still unmarried at this time: his father, by threatening suicide, had forced him to abandon his marriage in 1881 to the divorced Princess Elisabeth Carolath, née Hatzfeldt-Trachenberg, who was related to the Loë and Schleinitz families.[28] Herbert's diary shows that he regularly went out drinking into the early hours of the morning at large receptions or in smaller groups; occasionally he would have to dress hurriedly after only three or four hours' sleep because the Kaiser was at his door early in the morning with one of his Flügeladjutanten.[29] As secretary of state at the Foreign Office he frequently had to hold receptions for well over 100 guests, but other evenings were of a less formal order. The prudish Protestant Waldersee recorded indignantly in February 1889 that in Berlin 'much scandal is caused by Count Bismarck's parties. He invites ladies now too, the tone is very vulgar from the start and becomes worse by the hour as the drink takes its effect; by the end – yesterday morning the end was between 7 and 8 o'clock – the whole thing was said to have resembled a party in a brothel.' After this particular 'dance' the Kaiser called at Herbert's residence at 9.45 a.m. with Major von Pfuel.[30] Waldersee identified the female members of Herbert Bismarck's circle as the Dutch-born Frau Alide von Schrader, Countess Sibylle von Bismarck (née Arnim-Kröchlendorff, wife of his brother Bill Bismarck), Princess Maria Radziwill (née Countess Branicki in Paris in 1863, daughter-in-law of Prince Anton Radziwill) and Hereditary Princess Dorothee von Fürstenberg (née Talleyrand-Périgord in 1862, daughter of the Duc de Sagan). As foreign diplomats were also present at the parties, the whole world knew about these scandalous proceedings, the Chief of the General Staff commented, adding a note of warning: 'It cannot go on much longer like this.'[31] At the beginning of June 1889 Prince Heinrich and Waldersee were at one in agreeing that 'Count Bismarck is a real misfortune for the Kaiser.' On all his journeys, whether to St Petersburg, Stockholm, Copenhagen, Stuttgart, Munich, Vienna or Rome, Herbert Bismarck had drawn attention to himself by his 'uncouth, brutal and tactless nature'.[32] 'Wherever he has been abroad, and unfortunately he has been to many places, . . . people are horrified by his manners, by the rough tone of his conversation and by innumerable tactless remarks. It is deeply to be deplored that he has accompanied the Kaiser so often; the Kaiser has done everything well abroad, while Count Bismarck has undone much of the good he had done.'[33]

The Foreign Secretary was criticised not only for his marked weakness for drink[34] and his uninhibited bachelor life. His general manner and his misanthropic contempt for his colleagues and for subordinate officials made him positively hated. 'That he is really loathed by his subordinates for his inconsiderate, brutal behaviour is certainly not a good sign', Waldersee noted in autumn 1889. 'The diplomatic profession has unfortunately declined so much that there are no men of character to be found in it any more; some of them unwillingly resign themselves and keep quiet; others try to survive by toadying and flattery. This is true of those abroad as well as of the officials here at home. Abroad as well as among the diplomats here he [Herbert Bismarck] is really detested . . . He has inherited his father's worst characteristics, but unfortunately none of the good ones.'[35]

Whether Waldersee's harsh criticism of his arch-rival was justified or not, it is undeniable that for the time being the Bismarck family maintained tight control on the power of the state through this system. In February 1889 Waldersee complained that the position of the Chancellor was 'more powerful than ever'; no minister, no head of a Reich Office, no diplomat dared to express an independent opinion or make an independent judgement. 'The ministers are completely his creatures, he is in command of the Ministry of State and tolerates no opposition; all diplomatic agents report only what they think he likes to hear. Everyone pays court to him, and not only to him but to his family as well. It is truly disgusting to see how lamentably servile people are to his sons.' Although the Chancellor sensed that he had come up against 'a self-willed Kaiser', his handling of the monarch was 'masterful'. 'He flatters him to an unbelievable degree, sometimes gives in to him in lesser things, acting as if he were making a great sacrifice; he shows himself ostentatiously before the world as the humble servant of his sovereign, whom he nevertheless *de facto* controls. He makes his son go to see the Kaiser as often as he possibly can, and the latter's strength lies in all kinds of stories and jokes with which he keeps the Kaiser amused. Unfortunately the Kaiser has no idea that he is being led, and is equally unaware that he is constantly lied to in the most shameless manner.'[36]

Yet in the course of 1889 the situation gradually changed in two respects. First, not least because of the ever-growing and ever more conspicuous independence of the Kaiser, criticism of the Bismarckian style of government became increasingly vocal in political circles in Berlin. Second, the old Chancellor's powers were visibly on the wane. Both developments were closely watched and recorded by Waldersee. Until the summer of 1889, he wrote, the admirers of the Reich Chancellor were 'still very numerous'. They remained convinced that it was best 'to let the old, experienced Chancellor do as he thought fit; he would be sure to see that everything turned out well'. The entire diplomatic corps, Waldersee complained, was in any case well schooled in the habit of 'blindly obeying the orders they received without expecting any explanations'.[37] But since the autumn of 1889 a different mood had begun to prevail. In government circles there was an increasing recognition of the 'web of lies' spun by Bismarck, from

11. Before the storm – Kaiser Wilhelm II with Bismarck at Friedrichsruh in autumn 1888

which ministers and close colleagues had greatly suffered.[38] There was much ill-feeling among them: they all complained that they 'do not know what the Chancellor wants and that he changes his mind every instant'.[39]

As to the health of the Reich Chancellor, in April 1889, immediately after Bismarck's seventy-fourth birthday, Waldersee commented on how much the Chancellor had aged recently. 'He is far less decisive, less efficient and much feebler.'[40] Six months later, after visiting Friedrichsruh on 16 October, Waldersee noted that, although Bismarck had been complaining about his state of health for years, this time he had emphasised that 'his powers were visibly failing, his walks were becoming shorter and shorter, his capacity and zest for work were greatly diminished'.[41] A year after Wilhelm II's accession Herbert noted in his diary: 'Papa overtired, depressed', and added that he had put on weight again and weighed 200 pounds.[42] At the beginning of 1890 other visitors to Friedrichsruh reported that the Chancellor had become very old and weak and sometimes broke down in tears. As Waldersee wrote: 'He is witnessing the collapse of what he has built up and he no longer has the strength for bold action. His great days were over with the death of Kaiser Wilhelm I. Since then . . . , i.e. in the time of the present Kaiser, his power [has] gradually declined and he says quite openly that the Kaiser is a self-willed sovereign with whom it is difficult to work; he sees that he will escape from his control and will not take his son as his successor; at the same time his powers are failing, he has lost his resolve and has become discouraged. But it could not be otherwise; such a web of lies as he and his creatures have spun round the Kaiser cannot last forever. I still believe strongly', Waldersee affirmed, 'that divine providence will in the long run cause good to triumph and evil to be defeated, and I am therefore firmly convinced that Bismarck will come to an inglorious end and that his son will depart, unlamented by anyone.'[43]

At the beginning of the decisive year of 1890 Waldersee's diary provides a revealing analysis of the Bismarckian system of government in its last throes. 'The Chancellor wishes to control everything', the General commented, 'but no longer has the strength to do so. He is Foreign Minister and interferes in each of the Reich departments without considering the views of their heads; he is the Prussian Minister-President and Minister of Trade and treats the individual ministers as mere subordinates; he feels free to interfere in the most brazen fashion in their areas of responsibility. What is more, he sits at Friedrichsruh and is therefore difficult to get hold of. Not a single minister nor a single head of a Reich department dares to contradict him. They all complain about the lack of instructions or about uncertainties in the decisions taken, and especially about the Chancellor's lies, and in times like the present, when the Reichstag and the Landtag are assembled, they are often placed in the most difficult position.'[44] Waldersee was by no means alone in thinking that this state of affairs was intolerable and degrading for the Kaiser.

PHILIPP EULENBURG, 'THE KAISER'S BEST FRIEND'

One of the first signs that the Bismarckian ascendancy was under threat from the determination of the passionate and strong-willed young Kaiser to go his own way appeared very early, and not by coincidence, in the form of a dispute over an appointment. Wilhelm's deep-rooted friendship with Philipp Count zu Eulenburg, who had been Prussian secretary of legation in Munich since 1881, had begun in May 1886 and had been reinforced by the enthralling reports Eulenburg had written on the dramatic death of King Ludwig II of Bavaria. Wilhelm had taken a cure lasting several weeks at Bad Reichenhall in the summer of that year, when he and Eulenburg had travelled together to the Bavarian

12. Count (later Prince) Philipp zu Eulenburg-Hertefeld, the Kaiser's best friend

royal castles and to Bayreuth; they had also gone on rowing expeditions together (with the young fisherman Jakob Ernst) on the Starnberg lake, all of which had served to strengthen the bond between Wilhelm and the diplomat, poet, novelist, singer and composer, who was some twelve years his senior. Their common interest in spiritualist seances with mediums and secret writing provided further stimulus to the friendship between the two men.[45] Soon it became clear to all those in the know that Kaiser Wilhelm II 'loves Philipp Eulenburg more than any other living person'.[46] And Eulenburg loved the Kaiser no less. In the reign of Wilhelm II he was destined to play the powerful but dangerous role of the imperial protégé, the favourite. It was a relationship which threatened the Bismarcks' monopoly of influence at its weakest point.

In the first few weeks after the accession Eulenburg was still uncertain of how the Kaiser's feelings for him would be affected by the new circumstances. His letters to the monarch continued to be warm, almost affectionate. 'I need not say how painful it is to me that our journey to Bayreuth now also has to be given up', he wrote on 5 July 1888, scarcely three weeks after his 'dearly beloved' friend had succeeded to the 'most powerful crown on this earth'. 'Your Majesty will understand how deeply it affects me to have to remain so far away, after everything Your Majesty has gone through . . . How proud it made me to picture the princes gathered around Your Majesty – around my King!' Eulenburg did not miss the opportunity of adding a detailed description of the strange goings-on at the court of King Karl I of Württemberg, knowing that it would appeal to the Kaiser's taste for such tales.[47]

At the end of July Eulenburg received a sign that Wilhelm's friendship for him had not changed. In an allusion to one of the Nordic 'Skaldic Songs' composed by Eulenburg with the biographical title 'How they became friends', the Kaiser telegraphed to him from Sweden to say that he was writing 'from the land of Skaldic Songs, of beautiful women and of the Sea King. From the land of the Nordic Urwala, from the skerries, my greeting rings out to my friend the Skald.' As the Kaiser signed the telegram with the name 'Hokan', identifying himself with the powerful hero-figure in Eulenburg's Skaldic Song, Eulenburg replied with a new poem, 'King Hokan' which included the telling lines:

Auf loderndem Drachen, im rauschenden Meer
Versank König Helge mit Schild und Speer . . .
'Jetzt gilt unser Ruf einem neuen Mann!
Herr Hokan, der junge, in strahlender Wehr,
Der führt uns wohl sieghaft weit über das Meer!'

[On his flame-spewing dragon, in the stormy sea drear
Sank downwards King Helge with shield and with spear . . .
'Now sounds out our summons to a new man!
Lord Hokan, the youthful, in bright armour arrayed
Will lead us victorious o'er seas far and wide!'][48]

After his return from Scandinavia Wilhelm wrote again stressing how glad he would have been to have Eulenburg with him in Stockholm. He gave a rapturous description of the great parade which he had held on 1 September in Berlin in the presence of the King of Sweden, exclaiming: 'What a feeling it is to call these troops *mine*!' Finally he expressed his pleasure that he and Eulenburg were soon to meet in Munich, where they would be able to talk 'undisturbed'.[49] This letter threw Eulenburg into a 'frenzy of joy'. He had 'the greatest difficulty' in controlling himself, 'so as not to write four pages of thanks!' The Kaiser would know well, he declared, 'what feelings stirred me at the sight of this beloved handwriting!' He was 'delighted with the soaring spirit of imagination which dwells in Your Majesty's breast'. 'That is the element which provides the counterweight to the energetic realism which Your Majesty's high calling demands.' If the Kaiser were able to come to Eulenburg's home in Munich, he would sing him a new ballad over a glass of Swedish punch from a Viking recipe.[50]

The Kaiser's meeting with Eulenburg in Munich in early October brought another unmistakable demonstration of friendship. As Eulenburg reported afterwards to his mother, he had '*never* doubted the Kaiser's sentiments towards me – but I was surprised by the *demonstrative* manner in which he favoured and honoured me'. Wilhelm had obviously wished 'to show others that he is fond of me – and how touching that is! A young man who does not flinch at the mightiest crown on the earth, who holds to his convictions with simplicity and truth, cannot but have a fine, distinguished nature. The mediocre man loses his head – not he!' When he saw Eulenburg again on 1 October at the train station in Munich, Wilhelm told him that he needed to speak to him again that evening. Eulenburg therefore waited until the royal family dinner at the Residenz was over and visited the Kaiser in his room at 11.30 p.m. 'He kept me with him until 1 o'clock, tête à tête – quite his old self – talking about everything which concerns us.'[51] The next morning at 10 o'clock, when Eulenburg arrived at the Residenz, the Kaiser had already sent for him. He gave him '*personally*', as Herbert Bismarck immediately reported to his father, 'the Hohenzollern Order . . . , although he had received the Order of the Red Eagle Class IV only in April'.[52] The other men at the legation received no decorations at all. The Kaiser also singled out Eulenburg's Swedish wife 'very conspicuously' and invited her and all her six children to Potsdam. 'Finally the Kaiser gave the Bavarians the following demonstration of his esteem for me', Eulenburg told his parents. 'On his departure, while he was standing at the carriage window, with the whole Bavarian royal family, all the ministers etc. assembled in front of him after he had taken leave of them, he beckoned me to the carriage last. I had to climb on to the footboard and he spoke to me quietly for a few minutes – again expressing the wish to come and hunt at Liebenberg in the autumn. This demonstration was intended to show Munich that he values me – and that this should be noted for the future.'[53]

It was not until 15 October that Eulenburg felt able to thank the Kaiser for these marks of friendship, for 'if I had tried earlier, I should have become sentimental, and that is as repugnant to Your Majesty as to myself! But now I write these lines with a cool head, although with a burning heart – my thanks for the expression of true friendship which every word and every look proved to me! The most powerful crown on this earth has no power over Your Majesty's heart and true mind – and he who wears such a crown unwaveringly is a noble man indeed! If it were possible, my love for my Kaiser would be many thousand times stronger now!' The impression made in Munich by Wilhelm's visit had been 'very striking and beneficial', Eulenburg reported; it had produced 'an overflow of German national enthusiasm' which would provoke 'the particularist gentlemen' into 'hoisting the blue and white colours of Bavaria again for some time to come'. He, Eulenburg, could see through these aspirations better than anyone else, he claimed. 'My experiences in Bavaria lead me to this conclusion: enthusiasm for the Kaiser must be instilled into the Germans *even more*, for only *a few months ago* there was still a possibility that the whole of southern Germany might be thrown off balance by Battenberg nonsense!'[54] With ardent assurances of friendship and political hints such as these, targeted at the Kaiser's innermost wishes, Eulenburg struck home with deadly accuracy. The longer-term significance of this friendship can scarcely be overestimated; but it also concealed an immediate danger. During the reign of Friedrich III, Bismarck had appointed his son-in-law, Count Rantzau, Prussian envoy in Munich, thus making him Eulenburg's superior, although Wilhelm – Crown Prince at the time – had strongly urged the appointment of Eulenburg to this influential post. With the new reign it was inevitable that this issue should again raise its head. During the great ceremony of the opening of the Reichstag at the Berlin Schloss on 25 June 1888 Wilhelm spoke to Bavarian Minister-President Freiherr von Lutz of his intention of appointing Eulenburg envoy in Munich, adding: 'Unfortunately one cannot always do as one wishes – but what has not happened can still be made to happen!' Both Lutz and Bavarian Foreign Minister Christoph Freiherr von Crailsheim had the impression, as Eulenburg sensed, 'that the Kaiser means business'. Herbert Bismarck, on the other hand, was intent on sending Eulenburg as far away as possible, to a distant post abroad as secretary at an embassy.[55]

A serious argument broke out between Wilhelm II and the Foreign Secretary at the end of September 1888 on Mainau Island. The Kaiser declared that 'he would like to make Philipp Eulenburg an envoy as soon as possible, preferably in Munich'. Herbert Bismarck at once recognised the danger concealed in this arrangement and wrote to his father: 'From the seriousness with which H.M. spoke of Eulenburg's promotion I conclude that he is absolutely set on it and will be angry if he is opposed. Liebenau merely confirms what I have known for a long time, that H.M. loves Ph. Eulenburg more than any other living person. We shall therefore have to reckon with him, and I am in favour of making this concession to the Kaiser, who is otherwise so tractable, sensible and easy to win over. When I told H.M. that Eulenburg ought first to go to an embassy in order get away from the one-sided view from Munich, he argued that this was not necessary, as he knew him well and he was an intelligent person. H.M. brushed aside the question of seniority . . . and said that Eulenburg was more senior in years (which is true) and had lost time, which should not be allowed to count against him. When I objected that there was no available vacancy [for Rantzau] he retorted that one could easily be made – he seemed quite inclined to get rid of [Alfred von] Bülow-Bern or Wesdehlen [the Prussian envoy in Stuttgart] immediately. I then said that Rantzau had only just been appointed; to which H.M. replied that that was exactly why he did not know the form in Munich as well as Eulenburg and would be better off in another post, as soon as there was a vacancy.'[56]

In the course of their conversation in the Munich Residenz on the night of 1 October the Kaiser told Eulenburg bluntly that he intended to appoint him envoy in Munich, but that '*for the moment* there were a few problems to be overcome'. Eulenburg pointed out 'that the sudden and forcible removal of Rantzau would *inevitably* put [him, Eulenburg] in a very difficult position in relation to Bismarck and Herbert'; it would therefore be preferable to find a 'peaceful way' of achieving the goal. When Eulenburg met Herbert Bismarck the next morning the latter was 'tight-lipped and would not talk freely'. He then offered Eulenburg the position of counsellor at the Constantinople embassy, which Eulenburg turned down on financial grounds: he would have been glad to go to Constantinople had he been a bachelor, but with a wife and six children he could barely afford to live even in Munich. Eulenburg then frankly declared that he would not misuse the influence which he had over the Kaiser to cause trouble to the Chancellor and his family. 'I shall not *force* Rantzau out of Munich – but I cannot set myself against the specific wish of my Kaiser.' Herbert should show him the way 'out of this dilemma'. At these words, Eulenburg reported to his parents, the Foreign Secretary 'almost fell on my neck', and said: 'I am impressed by the Kaiser's brazenness, talking to me so calmly of getting rid of Rantzau and my sister. That is how a King should be!' On the other hand the Kaiser's wish put him in an awkward position where his brother-in-law was concerned. 'Of course I *have to* bear this business in mind and I shall do so', Herbert promised. 'Perhaps there will be a rearrangement soon and the Kaiser will put you into another post for a short time, from which you can then move to Munich. Oldenburg is free at the moment, but it is so frightful that you cannot go there. You are also much more valuable in Munich, which is what the Kaiser wishes. If only Wesdehlen would clear off!'[57]

Writing to his father from Vienna, the Foreign Secretary explained the whole dilemma and strongly supported the appointment of Eulenburg as an envoy – but not in Munich for the present. If only to secure Rantzau's post in Munich for a few years, he thought it advisable to lose no time in finding a posting elsewhere for Eulenburg. The Prussian legation to the minor states of Oldenburg, Brunswick and Lippe was vacant and could perhaps provide a temporary solution. What was more important, however, was to avoid provoking Wilhelm II by opposing his wishes in this matter. 'As he has such a particular liking for Ph. Eulenburg,

I believe there is little that would give him more pleasure than making him an envoy. Oldenburg would do very well, and no doubt Stuttgart too. I have my doubts about Bern because of all the tricky Socialist and legal questions.' Herbert therefore advocated Eulenburg's promotion as envoy and asked his father to decide whether this would be better achieved by the enforced retirement of Count Wesdehlen from Stuttgart or by appointing Eulenburg to the vacant post in Oldenburg, Brunswick and Lippe.[58]

The reaction of the Reich Chancellor to this suggestion speaks volumes, and throws into sharp focus the difference between the old statecraft which he represented and what was now becoming the personal style of rule practised by Wilhelm II. Prince Bismarck commented that if important government appointments were filled according to the wishes of the Kaiser, and government ministers did not dare take a stand against the monarch's candidates because they did not want to provoke him, the ministers would be reduced to the status of royal cabinet counsellors. Furthermore, as secretary of state it was well beyond Herbert Bismarck's lawful competence to negotiate with the Kaiser on these matters, for according to the Deputisation Law of March 1878 only the Reich Chancellor was authorised to submit proposals to the Kaiser on appointments in the foreign service. 'They cannot be lawfully settled without my *personal* participation, therefore not until *after* His Majesty has returned home.' This was a question of absolutely fundamental significance, and he, Bismarck, could not 'fly in the face of the experience in these matters which I have accumulated over forty years'. He declared that he could not 'abandon, in the last act of my drama, the clear conscience which I have insisted on maintaining towards the *country* in hard battles fought with my old master, despite all the love I bore him'.[59]

As far as Eulenburg's character and abilities were concerned, the Reich Chancellor thought him wholly unsuited for the difficult post of envoy in Munich. He was certainly 'amiable, but hitherto lacking in political judgement'. He allowed himself to be influenced by 'carping gossip', had 'no political insight, makes no distinction between what is important and what is not, listens to gossip and slander and can thereby cause much damage'. He could have the insignificant post in Oldenburg with pleasure, the Chancellor wrote, but Munich was out of the question: he could not 'pull it off'. To transfer Rantzau or to sack Wesdehlen or another envoy merely to create a vacancy for Eulenburg would be dishonourable and would also be very harmful to the diplomatic service, for such 'acts of violence', through which the better and more reliable elements in the foreign service were deliberately 'sacrificed for *untried* and undeserving favourites', created insecurity and acted as a deterrent. If the Kaiser really loved Philipp Eulenburg more than any other human being he should be at court, not in a legation. In circumlocutory but unmistakable terms the Chancellor hinted that he knew things about Eulenburg which he could not commit to paper but would keep until he could speak privately to Herbert.[60]

Eulenburg was at first relieved that he had been able to ward off the looming conflict with the Bismarck family by his frank conversation with Herbert Bismarck in Munich. He was aware, however, that the Kaiser had not given up his wish but only postponed it, and sounded a note of warning to the Foreign Secretary: 'You know the power of his will, as I do!'[61] Writing to the Kaiser on 15 October to report his conversation with Herbert, Eulenburg conveyed the impression that Rantzau's transfer to another legation in order to free Munich for him was almost a settled arrangement, although one which could not take effect for some months. 'Knowing Your Majesty's will with regard to the post of envoy, he [Herbert] wriggled like an eel and spoke of future rearrangements. For Rantzau, Brussels, The Hague or the Vatican would be as good as Munich', Eulenburg maintained. At the Vatican there was little work in summer, so that Rantzau's service at Friedrichsruh would scarcely be affected. 'I took leave of Herbert on good terms – but it was not easy! I had to be devilishly careful what I said.'[62]

On 31 October Wilhelm II surprised his friend with the news that he had appointed him envoy to Oldenburg, Brunswick and the two Lippe states.[63] Although he may have been disappointed by this development, Eulenburg must also have been relieved, both because of the financial advantages of the promotion and because, for the time being, it removed the danger of conflict with the Bismarcks. He knew, moreover, that his transfer to the lowlands of north-western Germany would not last long. So his reply to the Kaiser was effusive in its thanks: 'I am so overjoyed and grateful – but also so completely overcome at receiving such an unexpected honour from Your Majesty, that I scarcely know what to say. Your Majesty is indescribably good to me and has shown a truly fatherly concern for me. For through this appointment I shall be spared any conflict with my superiors – everything will take its natural course. I cannot get over the surprise! . . . I have never known such a day in my life – and I owe it to the Kaiser, my truest friend.'[64] Since he had expected to be left at Munich Eulenburg could only imagine that the Oldenburg appointment was attributable to the Bismarcks, and he wrote also to Herbert, expressing his 'boundless' thanks.[65] Bismarck's son-in-law Rantzau, who was thus able to remain in his post as envoy in Munich for the present, was relieved at Eulenburg's transfer both for this and for other reasons. 'I wish Eulenburg well', he wrote to Herbert, 'but I am heartily glad that he can now practise his spiritualist mumbo-jumbo in a different setting.'[66] In reality, however, the Kaiser had by no means abandoned his intention of appointing his friend in Rantzau's place. He detested Rantzau and said to Eulenburg: 'If he [Rantzau] had not made the arrangement during my father's reign, you would be in Munich now! But I have declared *definitely* that I wish to have you there!'[67] So the first storm clouds were brewing on the horizon.

It has to be said that Eulenburg continued skilfully to exploit both his advantages and Wilhelm's interests and weaknesses to court favour with the young Kaiser. On 16 December 1888, before his departure for Oldenburg, he had a lengthy private audience with the Kaiser during which the two friends discussed not only the political situation in Munich and the question of imperial patronage for the Bayreuth festival, but also various spiritualist experiences.[68] Eulenburg believed strongly in the continued existence of the 'individuality' of human beings, 'in a purified, continually self-purifying form' after death. 'Related feelings, earthly relationships of many kinds between humans, live on in wonderfully transfigured form; for what is highest, purest and best on this earth is not the ideal, but the actual seed-corn for the form of our existence in the hereafter.'[69] He wrote ballads and Viking fables for the Kaiser and sang or read them to him. 'How wonderful was the last evening which I was permitted to spend with Your Majesty', he wrote in his first letter to the Kaiser from Oldenburg. 'My thoughts are still filled with it and with Your Majesty's understanding for my old-fashioned feelings. How could I ever have dreamed that my Kaiser would be the *only* one who really and truly understands these feelings!!'[70] Though he warned the Kaiser against spiritualists who would try to force themselves on his attention, and urged him to rely only on *his* mediation in these 'delicate and exciting matters', Eulenburg at the same time kept the Kaiser's interest in the spirit world alive and zealously collected secret writings and reports of seances with the intention of reading them to the monarch when they next met.[71] 'H.M. spoils me terribly', he wrote in February 1889, adding that it was difficult to 'keep lurking envy within bounds!'[72]

In the autumn of 1888 Eulenburg invited Wilhelm to hunt on his parents' estate of Liebenberg in the Uckermark, less than an hour's journey north of Berlin. While the Kaiser was still in Italy Eulenburg wrote to him: 'If only Your Majesty were here with us in the old Mark again. One sleeps better beneath fir trees than pines, and since Signora Lucretia Borgia left the Vatican, nothing much happens there any more. Her lovely eyes would have opened wide at the sight of the German Kaiser with his black Trakhener stallion! I know such marvellous stories about Lucretia – what a pity that I cannot tell them! It gives me the greatest pleasure to think that I shall be able to sing a few ballads to Your Majesty again, and the prospect of a day's hunting in Liebenberg makes me quite *overjoyed*!!'[73] On 19 November 1888 news arrived at Liebenberg from the Flügeladjutant Gustav von Kessel – a cousin of Eulenburg – that the Kaiser wished to hunt there from 25 to 27 November. 'I am quite beside myself with joy and happiness and do not know how to express my gratitude!', Eulenburg assured his imperial friend. 'What a great occasion! What an honour for my beloved old home!' He promised the Kaiser to arrange 'a sort of singing contest' with his musical friends Kuno Count von Moltke and Oskar von Chelius. If his cousin Eberhard Count Dohna could also come, he would make 'a tolerable Lady Venus'. In addition, he wanted to invite Herbert Bismarck, Gustav von Kessel, Richard Count Dohna, August Count zu Eulenburg and two more cousins – Heinrich von Keszycki and Walther Freiherr von Esebeck – to the imperial hunt at Liebenberg. 'The thought of having my beloved Kaiser staying comfortably here makes me quite mad with joy! . . . How good Your Majesty always is to me – but *how* I love my Kaiser too!!'[74] Because of the visit of the Duke of Aosta and other obligations, the imperial hunt at Liebenberg had to be postponed until the beginning of January 1889. But in the following years Wilhelm paid regular visits there in October or November.[75]

Immediately after Wilhelm's first visit to Schloss Liebenberg, the observant Austro-Hungarian ambassador reported to Vienna: 'This is probably the moment to tell Your Excellency of a man who, although not part of the Kaiser's official entourage, nevertheless plays a significant role about the person of the monarch, which could become even more significant with time. This is a certain Count Philipp Eulenburg, who is already quite often described as "the Kaiser's best friend". He . . . is neither a contemporary of the ruling Kaiser, for he is already in his forty-second year, nor is he a member of the military profession. His Majesty first became closely acquainted with him when, as Prince Wilhelm, he went to Reichenhall to take a cure of several months because of his ear complaint. The Count in question, who is very pleasant company, artistically talented, writes prose and poetry and makes music most delightfully, was then Secretary of Legation in Munich and had been instructed to go to Reichenhall from time to time to put himself at the disposal of the young Prince. He succeeded in winning the latter's approval and affection to such an extent that from that time onwards he remained in constant and growing favour with the future heir to the throne. When Kaiser Wilhelm II came to the throne one of his first acts was to appoint the said Secretary of Legation to be envoy in Oldenburg, over the heads of all the counsellors. However, the new Head of Mission is to be found far less at his post than in Berlin, where he spends almost all his time in the royal palace, attending Their Majesties' meals and generally spending the evening in the bosom of the Imperial Family. Only a short time ago Kaiser Wilhelm returned from a hunting trip of several days, for which he had betaken himself to Liebenberg, which is half an hour by train from here, . . . less on account of the hunting than for the undisturbed company, which he enjoys so much, of the son of the house. At present the threads by which the Kaiser feels himself bound to Count Philipp Eulenburg seem to be rather of an aesthetic nature, spun from literature and music; but whether they are destined to grow into serious political bonds only the future will reveal. However that may be, the said Count is certainly a personality on whom one should keep a close eye.'[76]

A few days after the first imperial hunt at Liebenberg, Wilhelm and Eulenburg met again in Bückeburg, for Eulenburg was also accredited to both Lippe courts. There the Kaiser introduced him to his tutor Hinzpeter, who had come over from Bielefeld, with the words: 'My bosom friend Philipp Eulenburg – and the only one I have.' Presciently, Eulenburg commented: 'Such remarks please me and make me happy – but if repeated they will make *many* people envious of me.'[77] He soon had an opportunity, however, to make himself useful to the monarch in yet another sphere. Wilhelm entrusted him with a secret mission to investigate the private circumstances of Prince Adolf of Schaumburg-Lippe, the youngest son of the ruling Prince, born the same year as Wilhelm himself. After the final failure of the Battenberg marriage project Wilhelm was looking for a suitable husband for his sister Moretta, not least in order to pre-empt his mother's active efforts in this direction.[78] When Eulenburg's favourable secret report on Prince Adolf's financial position and private life arrived at the Berlin Schloss at the end of February, Wilhelm thanked his friend 'with all [his] heart' for his communication, which he burned at once, and told him that it was 'exactly what I need, and invaluable. I am delighted to see that in you I have found the right person for such services, and that you take the right view of your position as my friend; please continue in this way.'[79] Eulenburg's joy at these words of thanks knew no bounds. 'It is a glorious feeling', he wrote, 'to receive words of appreciation from the one to whom one owes everything. It makes me indescribably happy to feel how Your Majesty has taken the spirit of my friendship. Until my last breath I shall always be the same!'[80] Eulenburg was virtually predestined to play a central role, with Waldersee, in the Bismarck crisis which began in the spring of 1889.

MISS LOVE – THE END OF THE AFFAIR

There was, apart from his closest friend's promotion, a second personal matter in the life of the young Kaiser which urgently needed to be dealt with but which was no less fraught with danger for the Bismarck family. Whereas ambassador Prince Reuss in Vienna was able, with Waldersee's help, to buy the silence of Ella Sommssich and of Anna Homolatsch, who had borne Wilhelm a daughter,[81] and whereas Geheimrat Miessner, the monarch's Privy Correspondence Secretary, took on the messy case of the emotionally disturbed Countess Elisabeth von Wedel,[82] it was left to Herbert Bismarck to confront the Kaiser with the embarrassing news that another of his mistresses, Emilie Klopp, or Miss Love as she called herself, had approached him with a demand for money in return for the six letters and a signed photograph she claimed to have received from Wilhelm. Readers of the first volume of this biography will recall how the Prince had first met this Parisienne cocotte while on manoeuvres in Alsace and had then, in 1887, summoned her to Potsdam, where she lived for many months in a furnished apartment in the Russian quarter, though without ever receiving payment for her – evidently rather exotic – services.[83]

Upon his accession to the throne, Wilhelm turned his back on the fiercely proud Miss Love. Dangerously disaffected, she left Potsdam in October 1888 to return to her home outside Strassburg, determined to exact what she felt was her due, which she valued at around 20,000 marks. In November 1888 Emilie contacted another former lover of hers, Bismarck's younger son Bill, and then, when that approach appeared to be leading nowhere, in April 1889 she wrote to Waldersee. As was shown in the first volume, Herbert and Bill Bismarck were more acutely aware than either the Reich Chancellor or the Chief of the General Staff of the dangers inherent in this situation, both in regard to the damaging international scandal that could ensue if Emilie Love's demands were not met, and also with respect to the hypersensitive young monarch's continued confidence in anyone who caused him to lose face through familiarity with his peculiar sexual proclivities as revealed by the letters. Despite his sons' forebodings, however, Prince Bismarck insisted that Herbert must confront the Kaiser with the problem and act only with his express consent.

The Foreign Secretary's dilemma deepened immeasurably when, in an audience on 28 November 1888, the Kaiser, while eventually admitting to intercourse with Emilie during the manoeuvres in Alsace, flatly denied all further contact with her, just as he was to do when Waldersee offered to mediate in the affair some months later. There were no letters or photographs, Wilhelm insisted, the entire Potsdam story was pure fabrication, and therefore no measures of any kind were to be undertaken. Miss Love's threats to publish the Kaiser's acutely embarrassing letters in the French press could not, on the other hand, be ignored. It was without Wilhelm's prior knowledge, therefore, that Bill Bismarck met Emilie Love in a hotel in Frankfurt on 1 May 1889 and handed over an envelope containing 25,000 marks sent to him by his brother, in exchange for four – not six – handwritten letters and the signed photograph.

How the Foreign Secretary resolved the tricky question of submitting the letters – he kept the photograph – to the Kaiser is not clear from the evidence. His younger brother's suggestion that Herbert should send them in a sealed package and claim that he had acquired the documents without reading them testifies by virtue of its implausibility to the delicacy of the task. Herbert Bismarck's diary for 9 May 1889 records cryptically only that on that day he discussed the issue of the 'burned letter' with Wilhelm.[84] But are we mistaken in sensing that this was the moment the Kaiser came to feel the dominant power of the Bismarcks, which he would have to confront in the event of any disagreement? Did he now feel himself compromised in their eyes and begin to turn against them? Many years later, in conversation with Philipp Eulenburg, Wilhelm II identified this early period of his reign as the point at which he had broken with Herbert Bismarck. The issue which would divide the Kaiser from Herbert 'for all time to come', Eulenburg informed Geheimrat Friedrich von Holstein in 1894 after his conversation with Wilhelm, was the younger Bismarck's attempts to '*separate* the imperial couple inwardly – and outwardly from one another and work towards a regime of mistresses'[85] – presumably a reference, however distorted by self-righteousness and moral indignation, to the Miss Love imbroglio. What is clear is that from May 1889 onwards the Foreign Secretary was received less frequently in audience, even if we take into account that from mid-June to mid-November the monarch was almost constantly on his travels.[86] What is also clear is that, from now on, a new tone is noticeable in the Kaiser's dealings with the Bismarck family, not just in personal matters, but in all issues of home and foreign policy. The great crisis which was to culminate in the Iron Chancellor's dismissal ten months later was about to begin.

CHAPTER NINE

The Bismarck crisis begins

WALDERSEE AS THE KAISER'S CONFIDANT

While the domination of the Bismarcks was increasingly viewed as oppressive by both their opponents and their underlings in the Wilhelmstrasse, at the same time the not exactly underdeveloped self-confidence of 'our courageous and energetic Kaiser' – as Waldersee put it on 3 June 1889 – was growing stronger by the day.[1] The enthusiastic ovations that greeted the young monarch wherever he went – even in Bayreuth, Alsace, the Catholic Rhineland and Guelphic Hanover – visibly increased his self-assurance.[2] Even in South-West Germany the press welcomed the more prominent role he had adopted and enthusiastically demanded that the young Kaiser take personal control of governmental power.[3] After a ceremonial presentation of colours in Potsdam in May 1889 the Chief of the General Staff praised Wilhelm's 'outstanding' oratorical gifts and commented: 'He has developed very quickly in this respect and is already a confident speaker. All his speeches have a fresh, soldierly and highly patriotic note, always combined with pious respect for his predecessors.'[4] Nine months after his accession there were increasing signs everywhere that for the Kaiser the time of waiting was now over. In his dealings with the Reich Chancellor and his son there was an unmistakable new tone which showed clearly that in future he was determined to decide for himself the essential guidelines of both foreign and domestic policy. Not only in questions of high politics but also in difficult personal and family matters[5] Wilhelm II relied on the advice of his father-figure, Waldersee, whom he had appointed Chief of the General Staff in November 1888. On 26 January 1889 Széchényi reported that it was widely remarked 'that the young Kaiser is beginning to like Count Waldersee better than Prince Bismarck'.[6] Ten months later Waldersee was able to note with satisfaction that 'a better relationship' with the Kaiser was 'scarcely imaginable'. 'But I am very cautious, I do not run after him or bother him unnecessarily, and I am convinced that he respects me, which is definitely not the case e.g. with Herbert Bismarck.'[7] Although he was not an uncritical admirer of the young monarch,[8] nor did the latter by any means always follow his advice,[9] there is no doubt that during the first two years of the Wilhelmine era Waldersee exercised a greater and more constant influence on Wilhelm II than any other person in his entourage, and that this influence extended into every area of politics. Despite the harsh criticism that he meted out to others who overstepped the boundaries of their authority, at no time did Waldersee apply the same criterion to himself. On the contrary, he prided himself on being the only person in the imperial entourage, with the exception of the Kaiserin, who could speak openly and frankly to the monarch on any subject. In February 1889 he noted in his diary: 'Lucanus is nothing but a creature of Bismarck, Liebenau has given himself over to them entirely, neither of the two adjutant-generals has been able to create any kind of position for himself; of the Flügeladjutanten, Bissing was the only one who could still put in a word, and unfortunately he has now gone, so the Kaiserin and I are the only ones left who can get anywhere with the Kaiser.' Waldersee confidently predicted that as far as the Bismarckians at court were concerned, 'the Kaiser's eyes will one day be opened and then this band of villains and wretches will be thrown out'.[10]

The General had frequent opportunities to speak to the Kaiser in private, for in addition to the official audiences granted to him as chief of the General Staff, Wilhelm regularly called at his house to take him off for walks and rides[11] and invited him to numerous court events.[12] If he fell ill the Kaiser sent his personal physician to make sure that he had the best possible medical care. Waldersee often accompanied Wilhelm on his regular morning rides as if he were a Flügeladjutant.[13] Even on Bismarck's birthday on 1 April 1889, when the Kaiser appeared at the Reich Chancellor's palace to offer his congratulations, he made a point of taking his Chief of General Staff with him.[14]

At the beginning of March 1889 Waldersee took advantage of a formal audience, when he was alone with Wilhelm and could therefore talk 'completely openly and freely', to spur the Kaiser on towards greater independence. He proposed that, since diplomats simply echoed everything Bismarck said, the Kaiser should grant the military attachés, who were 'independent characters' and 'spoke their minds fearlessly', the right to report directly to the General Staff and thereby to the Kaiser. In general it would be a good thing for the monarch to hear a variety of opinions, Waldersee asserted, for, 'if everyone he listened to was in agreement, that would be a kind of conspiracy against which he would be powerless. He would be able to rule with more security if he kept his advisers

13. General Count Alfred von Waldersee

separate. The *divide et impera* principle would work here too', Waldersee urged, in effect advocating the future system of 'personal rule'. The Kaiser 'laughed heartily and nodded his head meaningfully', the Chief of the General Staff noted in his diary.[15] It may have been this conversation to which Herbert Bismarck was alluding when he recorded that Waldersee had told the Kaiser that Frederick the Great would never have become 'the Great' if he had been confronted with such a powerful chancellor at his accession and had kept him on.[16]

From the spring of 1889 Waldersee's criticism of the personal power of the Bismarck family became increasingly harsh; at the same time he denounced what he saw as their dishonest practices and corruption. He complained incessantly that lying had now become a 'routine practice' for the Bismarcks.[17] In April 1889, when Bismarck had the Prussian parliament dissolved prematurely despite the opposition of the ministers, the Chief of the General Staff saw this as nothing but a corrupt action by which the Chancellor sought to prevent the passing of the new income tax law announced by the Kaiser in his speech from the throne. Prince Bismarck, the General alleged, suspected that he would have to 'pay much more under the new system of assessment than before, and at the same time to allow an inspection of his fortune, which is probably colossal. Both would be very unwelcome to him, as since he has become rich, he has also become greedy.' It was thus purely for personal advantage that Bismarck had had the assembly closed, Waldersee claimed.[18]

But in spite of all his criticisms it would be quite wrong to categorise Waldersee as an advocate of Bismarck's dismissal at this early stage, for he was far too keenly aware of the catastrophic consequences which a premature *belle sortie* by the founder of the Reich would entail for the future position of the Kaiser. Indeed, the Chief of the General Staff occasionally went so far as to toy with the idea of harnessing the Reich Chancellor's unrivalled prestige to the furtherance of two great projects: abroad, the preparation of a war against Russia and France; at home, to anti-democratic revision of the Reich constitution. The shrewd General recognised that the Reich Chancellor's international reputation would work to his advantage in his preparations for what was, in his eyes, an inevitable war. On 15 April 1889, alluding to the now 74-year-old Bismarck, he declared: 'We must use his great skill and his great position throughout the world . . . to maintain peace for a time, but we must also recognise that as soon as we have finished arming it will be our duty to bring about the battle, of which our opponents are hoping to determine the timing, ourselves. Until then we need the Chancellor, but not when matters become serious, and if necessary we must even go against him; but I am convinced that he will then be glad to withdraw voluntarily, since he would scarcely be equal to such great stress.'[19] Waldersee then pressed for an internal *coup d'état* with the aim of abolishing universal manhood suffrage, while the right-leaning 'Kartell' majority still existed in the Reichstag. Bismarck's declared intention of getting rid of run-off elections[20] was not radical enough, in the General's eyes. 'I am of the opinion that if we want to do something, we must do it thoroughly. The universal suffrage that we have is proving to be increasingly dangerous; since we now have a good Reichstag, [we must] make the attempt to get another electoral law passed. It can be done now, but later only with great difficulty, perhaps only by force.'[21] In general, he commented in the spring of 1889, 'the whole party system' was in a shambles, 'and no one can foresee how it will develop'.[22] Only the recognition that pursuing such a policy of force would greatly strengthen the position of the Bismarcks held the reactionary Chief of the General Staff back.

THE FIRST CONFLICTS

Bismarck may have succeeded in temporarily circumventing Wilhelm II's stubborn insistence on Philipp Eulenburg's appointment as envoy in Munich by adopting the compromise solution of the Oldenburg-Brunswick-Lippe post in October 1888,[23] and he had put a good face on the choice, made entirely without reference to him, of General von Verdy as minister of war.[24] But although these potential conflicts were avoided, in the spring of 1889 a long tug-of-war began between the Chancellor on the one hand and the Kaiser and Kaiserin with Waldersee on the other, over the position of the 'Christian–Social' (in other words anti-Semitic) agitator Adolf Stoecker as court preacher. This was the first warning signal marking the beginning of a long-drawn-out crisis.

In response to the official press attacks on Stoecker instigated by Bismarck – Waldersee described them as a 'witch-hunt by the Wilhelmstrasse' – the deeply anti-Semitic Chief of the General Staff asserted, as he had in previous years, that Stoecker was fighting 'for the Christian faith against the Jews and their countless Christian camp-followers', who he was convinced reached 'into the highest circles'; ultimately, he argued, Stoecker was in fact fighting 'for the crown, against which heavy storm clouds are gathering'. Waldersee succeeded in persuading the Kaiser, who had at first agreed to disciplinary proceedings against Stoecker,[25] that the latter's fall would signify a 'triumph for the Jews and Progressives' and that the 'evil man' Bismarck wanted 'at all costs to deliver up a sacrifice to the Jews and to inflict a heavy blow on the Conservative Party'.[26] This anti-Jewish influence directed against Bismarck by Waldersee was not without effect, for the Kaiser decided to leave Stoecker in his post as court preacher, on condition that he abstained from political agitation. Waldersee commented jubilantly: 'The Chancellor has once again failed to enforce his will.' Bismarck's defeat in this question, he said, was 'yet another proof that the Chancellor is losing his touch'. Moreover it would not last long, the Chief of the General Staff predicted confidently, and 'Stöcker will be asked to take up his political activity again.'[27] In the course of the next chapters we shall have further occasion to observe the hitherto-unsuspected role played by anti-Semitism in the crisis which began in the spring of 1889 and ended a year later with the dismissal of both Bismarcks.

At the beginning of April 1889 Waldersee's diary recorded a further instance in which the Kaiser acted 'completely in accordance with his own feelings and not at the behest of the Chancellor'. He had 'simply but very decisively refused' the Chancellor's wish for an order to be conferred on the Russian Consul-General in Hamburg, Count Artur Pavlovich Cassini.[28] The lengthy marginal comment in which the monarch rejected the Reich Chancellor's suggestion survives in the confidential records of the Foreign Office. 'In the staff records of the Hussar Guards Regiment', Wilhelm recalled, 'from the time when I was in command of the Regiment, there are various *reports from Kusserow* [the Prussian envoy in Hamburg] about Cassini. He is described as a *very nasty piece of work*, who enjoys the worst possible reputation in Hamburg and is seen in society only on sufferance, because he is the Russian Consul. He has a mad passion for gambling and took a lot of money off an officer of my Regiment. What is more he lives in sin with a lady of rank, and that in itself causes such anger in Hamburg that decent people will have nothing to do with him. I know him personally, and I have rarely seen anyone looking more like a consummate rogue and crook. I consider the whole panegyric on Cassini in the report [from Bismarck] to be either a gross exaggeration or a piece of collusion with the Russian Embassy.'[29] (Several months later Wilhelm II called Cassini a 'swine of the first order!' in a note on an official report.[30]) The Chancellor was 'furious' at this rebuff, but would have to resign himself to it, Waldersee recorded with glee.[31] Cassini, the 'very nasty piece of work', was eventually promoted to Russian ambassador in Washington and was to represent his country at the Algeciras Conference in 1906, at which Germany was made painfully aware of its international isolation.

In mid-May 1889 the General took the opportunity of a train journey to Potsdam to warn the Kaiser about 'the Chancellor's Russian leanings', to which Wilhelm II responded: 'I have recently told the Chancellor exactly what I think about that.' The Kaiser had spoken out 'very bluntly against Russia' on this occasion.[32] Both Bismarcks again repudiated the accusation so often directed at them of being 'russophile', and adopted a much more hard-headed, realistic attitude, which the Foreign Secretary, after a long morning walk with his father on 22 July 1889, summed up tersely with the comment that one must 'forge links with Russia as long as there is still the smallest connecting link', since the Triple Alliance with the disintegrating Austria-Hungary and the increasingly republican Italy did not offer any reliable security.[33] Waldersee took no heed, but continued his anti-Russian and anti-Bismarck 'mischief-making', as Herbert Bismarck called it.[34]

In May Otto von Bismarck invited the Chief of the General Staff to visit him and urged him to write to Colmar Freiherr von der Goltz, the Prussian colonel acting as a military adviser in Turkey, ordering him not to allow the Bosphorus to be fortified and to keep the Turkish army weak, on the grounds that Russia would then be more easily tempted to capture the defenceless Constantinople.[35] Waldersee tried unsuccessfully to convince the Chancellor that the Russian military measures of recent years had been aimed exclusively at a war in the West: it was therefore necessary to strengthen, not weaken, the Turkish army. The situation was rendered even more critical by the fact that – as Waldersee was able to establish – 'the Kaiser entirely shares my opinion' and was demanding that he, the Chief of the General Staff, send Goltz instructions in accordance with *his* views. 'It is a very unhealthy state of affairs that the Kaiser and the Chancellor are pursuing different policies', he observed.[36] On 20 May Waldersee took advantage of a military review in Tempelhof to bring up this subject again with the Kaiser and noted afterwards that 'the Kaiser holds fast to his point of view and is even inclined to recall Ambassador Radowitz from Constantinople; his wife is Russian, sister of Nelidow's [the Russian ambassador in Constantinople] wife and he has long feared that Nelidow hears more about our intentions than is desirable.'[37] The damaging effects on the diplomatic corps of such fundamental differences of opinion between Kaiser and Chancellor were all too soon apparent. On 2 June Waldersee commented in his diary: 'Radowitz, who came to see me recently, is beginning to see which way the wind is blowing and says what the Kaiser wants, and claims to have told the Chancellor that we must persuade Turkey to make itself strong. The Kaiser does not entirely trust him, however, and is sure that his wife, with her tendency to dabble in politics, and as a Russian and the sister of Nelidow's wife, is a danger for us.'[38] Not long afterwards Holstein too observed that 'the Kaiser harbours a mistrust towards Radowitz which makes his political future look risky'.[39]

Waldersee was also directly involved in a clash between Kaiser and Chancellor over the future role of the Reich Railway Office. At an audience attended jointly by the Chief of the General Staff and the newly appointed Minister of War von Verdy on 4 May 1889 the Kaiser sided unambiguously with the generals and against the Reich Chancellor and Prussian Minister of Public Works Albert von Maybach. The two army chiefs demanded an expansion of the powers of the small Reich Railway Office, which they intended to put in the charge of a senior army officer. Bismarck categorically refused to accept this take-over of one of the Reich Offices by the army and argued that it would be tantamount to centralisation, which would upset the delicate federal balance in the Reich. He stood up for Maybach, about whom the Kaiser had written an insulting marginal comment accusing him of lacking energy. 'We have been negotiating for a long time', Waldersee observed, 'but now the struggle is getting lively.'[40] Bismarck and Maybach had to give way at first because – as the General gleefully noted – 'the Kaiser was strongly on my side'. At the end of November 1889 the matter came to the fore again, however, and played a not insignificant role in the last phase of the Bismarck ascendancy, as we shall see.[41]

From now on the conflicts between Kaiser and Chancellor multiplied, and the subjects on which they disagreed became increasingly serious. In May 1889 the miners in the Westphalian coalfields went on strike; the strike spread rapidly and caused much bloodshed in some areas. Large numbers of troops were sent into the mining districts, but they were not used because Wilhelm, as Herbert Bismarck claimed, 'abruptly intervened, from Kiel, spurred on by Hinzpeter's bragging'. On 12 May the Kaiser burst unannounced into a meeting of the Prussian Ministry of State and declared that he 'did not share the views [of the Reich Chancellor]. The employers and shareholders must give way, the workers were His subjects whom He had to look after, and if the industrial millionaires did not comply with His wishes, He would withdraw His troops, and if the villas of the rich owners and directors were set on fire and their gardens trampled, they would soon have to eat humble pie.' Bismarck's objection that 'the owners were subjects too, and had a claim on the protection of the sovereign', was ignored by the Kaiser, as Herbert recalled: 'If no coal is produced', Wilhelm went on to assert, 'our navy will be defenceless; we shall be unable to mobilise the army, because the lack of coal prevents deployment by railway; and we shall be in such a precarious position that if I were Russia I should declare war at once.' He then departed 'in considerable agitation'.[42] Soon after this dramatic intervention Wilhelm, entirely on his own initiative, received deputations from both employees and employers.[43] Whatever enraged the Chancellor delighted Waldersee, who noted in his diary of 19 May that the rapid ending of the miners' strike had been achieved 'principally by the intervention of the Kaiser, who had received deputations from both parties and had addressed them very earnestly. If this success should prove lasting, it would be of the greatest possible significance for the Kaiser. He acted entirely on his own account, without, and partly against, the advice of the Chancellor and the Minister of the Interior; it is a victory for healthy common sense over bureaucracy, and is very likely to win the Kaiser enormous support.'[44] When the Chancellor submitted a report on 25 May informing the monarch of the resolution of the Prussian Ministry of State asking him to abstain from further such audiences, the Kaiser scribbled in the margin that the Ministry had 'never been asked for its opinion by me'.[45] Bismarck was still able to remark, half-mockingly, that 'the young Master' had 'Friedrich Wilhelm I's conception of his powers', and that it was 'very necessary' to 'protect him from overhasty actions in this regard'.[46] But barely six months later it was precisely this question which was to herald the last stage of the Bismarck crisis.

Conflicts arose over lesser matters too, now that the wills of the two men at the summit of power were pitted against each other. On the occasion of the Kaiser's visit to Oldenburg in the spring of 1889 Eulenburg, the newly appointed Prussian envoy there, had applied for an unusually large number of high-ranking orders to be distributed among the small population of the Grand Duchy (a quarter of a million). Eulenburg was formally reprimanded for this by the Chancellor, and a copy of the Chancellor's order containing the reprimand was laid before the Kaiser. Not long afterwards, while Wilhelm was visiting Brunswick and summoned Eulenburg – who was also accredited there – to an audience to discuss the distribution of orders, the latter replied that he should no doubt restrict himself to the most essential ones, since the Chancellor was of the opinion that too many orders had been conferred the previous month in Oldenburg. At once the 'All-Highest displeasure' made itself felt. 'Very angrily' the Kaiser asked how Eulenburg knew of Bismarck's opinion, and was 'furious when I told Him that it had been conveyed to me through an order'.[47] Eulenburg could not have guessed that the Reich Chancellor's order had been copied to the Kaiser only *after* it had been dispatched, and that the latter had covered the copy with irate marginal comments.[48] The Kaiser, Eulenburg recounted, had 'struck the table with his fist and shouted furiously, who rules now, I or Bismarck!'[49] Not long afterwards Eulenburg returned from the imperial deer hunt in East Prussia and reported – as Herbert Bismarck recalled after his dismissal – that 'H.M. was irritated by a submission [from the Chancellor] advising him against repeating his visit to Rome this year. H.M. said sharply that He did not wish to be treated like a child all the time, only He knew what He should do etc.'[50]

As if that were not more than enough, in the same month differences arose between Germany and Switzerland which led to further serious arguments between Bismarck and Wilhelm II. The Chancellor's aim – such at least was Waldersee's suspicion – was nothing less than to bring Germany and Russia together in a joint invasion of Switzerland. According to Waldersee, Bismarck clearly wished to use the arrest and 'brutal' treatment of the German police agent Wohlgemuth in Switzerland to bring about a 'great coup', in which he hoped 'to win Russia over to a move against Switzerland, and the moment is favourable in so far as the anarchists in Zürich' had been caught 'in the process of making bombs' which were destined for Russia. Bismarck was even contemplating 'war against Switzerland and a partition of the country', claimed the Chief of the General Staff, who was vehemently opposed to the Chancellor on this matter too. He expressed concern at 'the atmosphere of hostility' towards Germany that had arisen in Switzerland as a result of Bismarck's policy. The Chief of the General Staff even feared that in the event of Germany going to war against

France the sympathies of the Swiss would be so much on the side of the French that their neutrality would be called into question.[51] He drew up a memorandum on Swiss neutrality which he hoped would prevent 'overhasty steps', and urged this view on the Kaiser.[52]

Wilhelm II had at first given his approval to the action against Switzerland, but without fully realising its implications.[53] Waldersee claimed that he had been 'badly served', especially by Herbert Bismarck, who 'showed great lack of skill in the matter'. 'The father knew very well how poor the support offered by the son had been, but did not want to disavow him and therefore probably became more deeply involved in the matter than he wanted.' Bismarck's policy gave rise to increasing protests, especially in South Germany and in Progressive circles, whereupon the Grand Duke of Baden – the German ruling princes had assembled in Stuttgart on 25 and 26 June 1889 for the twenty-fifth anniversary of the accession of King Karl of Württemberg – felt obliged to intervene with the Kaiser. At Holstein's instigation, Waldersee also had a discussion with the Kaiser in which he drew his attention to 'the utterly foolish idea of a war with Switzerland'.[54]

The result of this pressure on the Kaiser was soon apparent. Even before he left for Stuttgart he had instructed Under-Secretary of State Berchem to write to the Reich Chancellor 'that he hoped we would not get too heated about it and would succeed in avoiding a conflict'. After his return from Stuttgart the Kaiser had 'expressed the same idea in more detail and more definitely through Berchem to H.H. [the Reich Chancellor] and positively *refused* to sign the Cabinet Order for the control of passports'.[55] Bismarck complained bitterly that he had 'already noticed that contrary influences are at work on His Majesty the Kaiser along the lines of the policies advocated by His Royal Highness the Grand Duke' and for the first time since Wilhelm's accession he threatened to resign.[56] He was 'very annoyed' when conciliatory remarks made by the Kaiser to a Swiss general appeared in the newspapers and were greeted with approval throughout Germany.[57] For the Kaiser the effect of the ill-thought-out and highly unpopular policy of the Bismarcks in the Wohlgemuth affair was to undermine still further his confidence in them. Waldersee summed it up with the words: 'The whole affair did nothing . . . to enhance Bismarck's reputation; it was a completely failed enterprise on his part, and it began to shake the Kaiser's confidence in the superiority of his statecraft.'[58] And that was not all: the Wohlgemuth case had also produced the anti-Bismarck coalition between Waldersee, Holstein, the Grand Duke of Baden and his representative in Berlin, Adolf Freiherr Marschall von Bieberstein. Together with Philipp Eulenburg, these were to be the Kaiser's most influential advisers in the crisis leading to the Bismarcks' dismissal.[59]

THE 'DECISIVE TURNING-POINT'

When he looked back in later years on the Bismarck crisis, Waldersee picked out the fierce conflict which broke out in mid-June 1889 over the question of the Russian conversion as the 'decisive turning-point' in Wilhelm II's relations with the two Bismarcks; from that point, he claimed, the Kaiser had 'broken with the Bismarcks, father and son, in his heart'.[60] What had happened? What was so important that it led to an irreparable breach in the critical relationship between the Kaiser and the Chancellor?

Within only a few days of making his absurd anti-Semitic accusations against Bismarck – that the Chancellor wanted 'at all costs to deliver up a sacrifice to the Jews'[61] – in the Stoecker affair at the beginning of 1889, Waldersee repeated the same sentiments with regard to Bismarck's financial policy towards Russia. At this time the Chief of the General Staff was watching, with growing anxiety, the increasing strength of Russian finances, which had resulted in German capital flowing to Russia again. For Waldersee there was never any question of who was responsible for what he considered an extremely undesirable development. On 25 March he wrote: 'The House of Rothschild seems to have helped considerably, and so Bleichröder and his gang were of course also involved.'[62] Gerson Bleichröder, the Reich Chancellor's banker, was accused by Waldersee not only of strengthening Russian military power but also of simultaneously attempting to undermine the financial stability of Austria-Hungary. 'Lately that rogue Bleichröder, who is helping the Russians to convert their loans, has been working with all his might to damage Austrian credit, and has been sending bleak reports of the situation there', he claimed. Reuss, the German ambassador in Vienna, was 'convinced that the Jew is exaggerating greatly, and this is also my opinion'.[63]

In June 1889 this question, on which, as Waldersee stressed, he took 'exactly the same point of view as the Kaiser', came to the fore once more. During a shooting display for the Shah of Persia at Tegel on 11 June 1889 the Kaiser called the Chief of the General Staff over and said: 'There is fine news from Russia again. According to Yorck's latest report it now seems that a conversion of 250 million rubles is to be arranged in Berlin. This is going too far; I will not stand for it. What is your opinion?' When Waldersee asked for time to reflect, the Kaiser called out to him: 'Here comes the Minister of War, talk to him and then you can both give me your advice together.' After the display Waldersee and Verdy rode up to Wilhelm and advised him to have the conversion stopped by Bismarck. The Kaiser had wanted to give orders for a press campaign through Rudolf Lindau, the press officer of the Foreign Office, but the two generals doubted that this would be effective; the Chancellor himself must intervene with all his authority, they urged. 'If the Jews know that he is strongly against it they will keep their hands off, but not otherwise', they declared.[64] Upon returning from Tegel the Kaiser sent his Flügeladjutant Carl Wedel to the Foreign Office to summon the Foreign Secretary to the Schloss immediately. Herbert Bismarck found the Kaiser 'in a very excitable state', he recalled later, as those about him had been whipping up his feelings against the Chancellor's alleged 'russophilia'. He had already sent for Lindau. 'I drove into the Schloss where Lindau was standing in the ante-room; I was received, and H.M. at once gave orders that the new Russian loan must be condemned in the press, He did not wish any more German money to go to Russia in exchange for Russian paper, as Russia would only use it to pay for its war preparations. "One of my senior army officers has drawn my attention to this danger today, and something must be done absolutely at once."' The Foreign Secretary tried to explain to the Kaiser that 'that was not how matters lay, it was merely a question of a conversion of earlier Russian loans, and thus the best opportunity for German stockholders to take cash and to get *rid* of Russian papers . . . The French would take the Russian papers which were got rid of here. The business would be done in *Paris*. H.M. seemed not to understand', Herbert recorded in some perplexity, 'for He insisted that articles must be written in our press against this Russian financial operation, and He had summoned Lindau to give him instructions accordingly.' The Foreign Secretary insisted: 'If I have not succeeded in making the situation clear to Your Majesty, I beg Your Majesty to request a report from the Finance Ministry. Official articles cannot be written to that effect without reference to H.H. [the Reich Chancellor], because they would influence overall policy substantially; and in any case I consider them pointless in the present case.' At this the Kaiser directed Herbert to write to his father and urge the importance of the press campaign on him. He also spoke 'sharply' to Lindau, who was 'quite dumbfounded at his arrogant tone'. Not content with that, the Kaiser ordered the Reich Secretary for the Interior, Heinrich von Boetticher, to write to the Reich Chancellor; he also gave instructions to his Flügeladjutant, Kurt Wolf von Pfuel, to write to Lucanus. He sent another adjutant, Friedrich von Scholl, to the Prussian Finance Minister's deputy, Meinecke, with orders that the latter should instruct the Supervisory Board of the Stock Exchange, by imperial order, to 'put a stop to the Russian loan issue'. Meinecke, taken by surprise during a meal, replied that it was not the Finance Ministry but the Prussian Ministry of Trade that was responsible for the matter, and then sought Herbert Bismarck's help, 'in a state of consternation'.[65]

The Kaiser's frenetic activity reached its peak when he himself wrote a letter to the Reich Chancellor on 12 June 1889, demanding, in stridently imperious terms, the prohibition of the Russian conversion. 'Yesterday in the course of the day I heard by chance that Bleichröder and his associates intend in the very near future to bring a very large quantity (250 Million) of Russian rubles back here and to accommodate them in Germany. I am most strongly against this. That the Russians arm themselves and mobilise has become a matter of indifference to us; that they insult and scorn us, likewise; but I cannot allow German money to flow into Russian hands so that they can all the more surely attack [and] destroy us. I therefore request Your Highness . . . to give orders for a vigorous campaign to be waged against the Russian securities.'[66]

Waldersee, who was kept informed by Wilhelm himself of the progress of this extraordinary interlude, also made detailed notes on it which provide an interesting supplement to Herbert Bismarck's recollections. 'After [Herbert] Bismarck had admitted that the news of the conversions was correct, there were fierce arguments', he wrote. 'The Kaiser was especially vehement in attacking Bleichröder, of whom he said, among other things, that he frequented the Chancellor's house. With great determination he expressed his intention of preventing the conversion. Lindau was given a direct order to influence the newspapers in this sense. Of course [Herbert] Bismarck took this direct attack very much amiss.'[67] This was how Waldersee recorded the episode in his diary. Later, in a retrospective note on the 'very fierce' discussion between Herbert and Wilhelm, the General added that the monarch had railed against Bleichröder 'in strong language', to which the Foreign Secretary had 'angrily' retorted that he had absolutely no connection with the banker and the Kaiser had countered: 'I know that, but it makes no difference at all to me, for he is a constant visitor at your father's house.' Herbert also reacted very badly to the fact that the Kaiser gave Lindau direct orders to have articles against the conversion published in the press. Even in the eyes of the Chief of the General Staff this was an unwarranted interference by the Kaiser in the domain of the Foreign Secretary and therefore also in that of the Reich Chancellor himself.[68]

The Bismarcks, furthermore, found the role played by the Flügeladjutanten in the quarrel deeply insulting. They considered that such 'orders by adjutant' issued by the Kaiser were 'quite inadmissible' and could not be accepted by a minister or civil servant. For the Reich Chancellor this point of political principle was even more important than the conversion itself, and he told his son through Rantzau that 'in political matters it was not permissible to act upon an order conveyed through the Flügeladjutant on duty, for then the position of the Foreign Minister would become quite untenable, and one ought not to let the sovereign get away with that'.[69] Herbert pointed out that the Kaiser would have to be told this in person, otherwise he would not notice the 'stab' aimed at him.[70]

The younger Bismarck had no doubt that 'the affair had been exploited to stir up trouble', but he could not at first find out 'who had informed H.M. about it'.[71] Later he discovered that, among others, the King of Saxony had told the Kaiser that it was a scandal that Berlin bankers were supplying the Russians with cheap money which would only be used for military purposes against Germany. Others, Herbert reported to Varzin, had expressed the view that 'one had to be rough with the Russians'; then they always 'knuckled under'. With arguments such as these, 'troublemakers' had persuaded the Kaiser that 'Papa's policy was too pro-Russian'.[72] Among the Flügeladjutanten too the general opinion was that 'the pernicious Bleichröder–Rottenburg coalition must be broken up *à tout prix*'.[73] But not surprisingly, and justifiably, Herbert Bismarck's suspicions rested above all on Waldersee and Verdy. At the state banquet for the Shah he asked Verdy point blank: 'What gave you the idea of stirring up H.M. to involve himself in things which are none of your business? You are interfering with foreign policy.' He warned Verdy 'in future to refrain from making comments that have nothing to do with the responsibilities of the minister of war'. But he was well aware that the generals would persist in their subversive work.[74] And indeed the counter-measures which Waldersee advised the Kaiser to take to thwart 'all the Chancellor's clever ruses' are symptomatic of the course of the long crisis which was to end nine months later with the dismissal of both Bismarcks. The General recommended him to discuss the Russian conversion directly with Prussian Finance Minister Adolf von Scholz. This 'greatly displeased the Chancellor', who arranged for Scholz to go on leave before the monarch could speak to him. On 30 June 1889, at the beginning of the first Scandinavian cruise, Waldersee allied himself with Philipp Eulenburg – both believed that 'a major conflict was possibly in the offing' – and together they urged the Kaiser to have a daily report on the state of the Russian loan on the Berlin stock exchange telegraphed to him.[75]

In the last days of June and in early July 1889 this question actually threatened to bring about the resignation of the Reich Chancellor and with him the entire Prussian Ministry of State. Bismarck's colleagues – Rottenburg, Boetticher, Berchem, Magdeburg – had, as Holstein warned Eulenburg in a letter of 3 July, acted only on the instructions of the Chancellor; 'so that if His Majesty is rude to one of them and he is forced to resign', Bismarck would resign too. 'But I think', Holstein continued, 'that the Ministry [of State] will also ally itself to the Chancellor. Maybach is happy to take the opportunity to do so whenever he has the Chancellor nearby, and moreover the other ministers are very annoyed that the Kaiser only deals with them through Lucanus, like Friedrich Wilhelm III in the days of Cabinet politics. In short, there would be a fearful scene.' The forthcoming Reichstag elections, which in any case threatened to go very badly, would be catastrophic and the Kaiser would 'be obliged to tackle an internal conflict without Bismarck . . . with France and Russia looking on'. The Kaiser must realise 'that it really is dangerous for there to be an *open* breach between him and the Chancellor now, just because He wants to prevent a Russian financial operation. In that way war could actually break out sooner than we would like.' If Wilhelm did not want to bring about a break with Bismarck, he ought not to 'go too far in his rudeness, so as not to give anyone . . . an excuse to hand in his resignation'.[76] Writing to Herbert Bismarck, Holstein expressed the hope that the Kaiser would give in. The Chancellor's reply to the Kaiser's letter of 12 June had certainly looked 'rather like a resigning matter', but Wilhelm would want to avoid a chancellor crisis 'if for no other reason than that he would have to give up his travelling plans in the event of a change of ministry. Hopefully we shall be spared the mournful sight of Prince Bismarck disappearing from the scene on account of Bleichröder.'[77]

While Holstein warned against further 'rudeness' on the part of the Kaiser, Herbert Bismarck put pressure on his father to make some concession to the monarch over the Russian loans question. 'H.M. has got his teeth very firmly into this affair, and He is not entirely wrong in saying that the Russians would not have been able to pull it off and thereby save themselves about 20 millions in annual interest unless the big Berlin bankers had agreed to go along with it', he wrote to Rantzau on 27 June.[78] On 4 July 1889 he wrote again urging that his father should 'make some allowance' for the Kaiser's '*present* disposition', 'while at the same time guiding him into better ways'. He warned of the efforts of those who were trying to stir up antagonism at court towards the Chancellor, 'first and foremost Waldersee and the orthodox cathedral preachers with their hangers-on', who wished to bring about a rupture between Bismarck and the Kaiser. Their tactic with the Chancellor was 'to make him lose his patience and annoy him into leaving'; they expected that Bismarck would eventually be driven into saying: 'If H.M. treats me like this and interferes so much, I would rather go, I am too old and too tired to allow myself to be tormented.' Herbert Bismarck, however, was determined not to let the 'unprincipled troublemakers, whom H.M. does not see through', win the day, and therefore argued strongly in favour of making concessions to the Kaiser. 'I would consider a great war a lesser misfortune than if Papa resigned', he wrote to Rantzau, 'for *if he does,* war will come *at once*, and with unskilled, new diplomatic leadership we shall have to be prepared for the worst: that would be to court disaster. H.M. holds very strongly to Papa, when all is said and done. But if Papa allows himself to be annoyed and wants to leave on his own account, the Fatherland will be ruined.'[79] Herbert went on to criticise Gerson Bleichröder bitterly, complaining that he had 'long wished for Papa to get rid of this dangerous Jew as a banker, he is too unscrupulous a liar, and Papa has had more annoyance and trouble because of him than he realises: so long as this money-grubbing Semite can earn a few millions he does not care what becomes of Papa and our Fatherland'.[80] Only recently the Kaiser had been told (as was unfortunately true, Herbert commented) that 'Bleichröder was spreading the lie around that the Reich Chancellor was not against the conversion [Prince Bismarck wrote in the margin beside this: 'That is quite right'], and through years of unscrupulous lying at Papa's expense Bleichröder had built himself up such a position at the stock exchange that the other members all now believed that his transactions had the blessing of the government, *otherwise he would certainly not make them.*'[81]

At the beginning of July 1889 the Reich Chancellor was indeed on the point of giving up. He wrote despondently to the Prussian envoy in Karlsruhe, Karl von Eisendecher, 'I can only conduct policy as I understand it, and in any case I am heartily tired of it all . . . My strength is not sufficient to cope with contrary influences in counselling His Majesty.'[82] Through Rantzau, he replied to his son with the comment that 'if the sovereign believes he can conduct political affairs more skilfully on his own, then let him do so'; it was his, Prince Bismarck's, duty to avoid arousing ill-feeling on the part of Russia 'so long as we are not prepared', 'if only because we have no way of knowing what might erupt in France . . . If the sovereign did not wish to carry on with him, there was nothing to be done about it; as to that, he was far more inclined than you not to give a damn about it.'[83] Herbert's answer to this letter speaks volumes. He wrote to Rantzau on 8 July: 'The reason why I do give a damn about Papa's relationship with H.M. is that we would face great danger and perhaps go to the devil if H.M., with his present immaturity and lack of experience, set out to reign without Papa with *homines novi*. With our old sovereign it was quite different: he had a strong sense of duty, no desire to seek his own pleasure, long experience and, last but not least, he had not only gone through heavy misfortunes in his childhood and in 1848, but had ruled himself into such an impasse in 1862 that he wanted to abdicate. He *knew* what he had in Papa, and would never have let him go, even in the most serious disagreement. But our present sovereign is not like that, he has only known the sunny side of life *et il ne doute de rien*. I was therefore thinking only of *the country* when I wrote recently that one ought to accommodate oneself a little to H.M.'s disposition from time to time.'[84]

If this highly explosive difference of opinion was eventually overcome without the major crisis anticipated by some, it was purely thanks to the Reich Chancellor's unusual willingness to make concessions. On 4 July Bismarck sent telegraphic instructions to the Foreign Office to advise the German public – as the Kaiser had demanded – through a notice in the *Norddeutsche Allgemeine Zeitung*, to cash in the old Russian securities, and not to take up the converted stock.[85] When the news of the Chancellor's intervention reached the *Hohenzollern*, it was greeted with 'the greatest delight'; Kiderlen-Wächter was instructed by the Kaiser to express 'the All-Highest thanks for the highly satisfactory news' to the Chancellor.[86] Eulenburg reported on 17 July to Berlin that the Russian conversion had certainly been 'of lively concern' to the Kaiser in the first days of the Scandinavian cruise, but since the affair had been 'brought to a happy conclusion, there is no more dancing to political tunes'. The Kaiser's travelling companions were 'all taking care not to disturb the restful atmosphere', and for his part the Kaiser was so absorbed by the impressions made on him by the journey itself that he had more or less lost interest in the conversion.[87] But we should not be misled by this outcome into underestimating the significance of the incident itself. As Waldersee pointed out retrospectively, even if the Kaiser had little desire 'to spoil the cruise for himself by getting angry with the Bismarck family', he had nevertheless 'by no means forgotten the whole affair'.[88] Quite the contrary: from this point onwards he had in fact 'done no more than to put on an act with the Chancellor'.[89]

THE CONFLICT BETWEEN BISMARCK AND WALDERSEE

In the summer of 1889 the Reich Chancellor decided to make a frontal attack on Waldersee, whom he considered increasingly as a threat, by means of a press campaign against the 'military undercurrents' which were pushing the Kaiser towards war with Russia.[90] The General, who proudly admitted to having repeatedly put forward his anti-Russian views to the Kaiser 'in person and in writing', commented defiantly: 'As these views are not those of the Chancellor, and as the great man discovered that I was in correspondence with the Kaiser, war had to be waged against me once again, and this time it was with great force.'[91] He asserted, however, that 'for everyone capable of reading it was unmistakable that the thrust of the attacks was directed at the Kaiser, and one could see with total clarity the outrageous effrontery of the Chancellor'. He, Waldersee, had held back 'in order not to cause any difficulties for the Kaiser, who was still obliged to keep the Chancellor for a little longer'; but he had also had the reassuring feeling of knowing that the Kaiser was 'absolutely on my side'.[92] The solidarity between Wilhelm and Waldersee was again demonstrated when on 20 June 1889 the Kaiser wrote on a copy of an article inspired by Bismarck in the *Berliner Tageblatt*, against 'military interference in politics', a furious marginal note: 'Please find out where this shameless concoction comes from and who the author is. And give the paper a thorough dressing-down at the same time. I think it comes from Bleichröder's financial circles?? and his associates?? I shall want an answer on 28th or 29th.'[93]

During the Scandinavian cruise of 1889 – when the Kaiser and his Chief of General Staff spent almost four weeks at close quarters in the cramped cabins of the old *Hohenzollern*, while the Bismarcks were far away in Varzin or Berlin – the public feud between Bismarck and Waldersee took on increasingly dangerous forms. From the *Hohenzollern* Wilhelm issued a rebuke to the editor of the semi-official *Norddeutsche Allgemeine Zeitung* for an article which was patently aimed against Waldersee. With his 'cunning bordering on genius' (as Philipp Eulenburg put it), Waldersee took advantage of the imperial rebuke to send an almost simultaneous telegram of protest to the *Hamburger Nachrichten*, which had published similar attacks on him.[94] The telegram did not fail to attract attention. Writing of the explosive atmosphere on board at this time, Eulenburg recorded that Waldersee was intending to exploit the situation still further by writing an open letter, and could not be dissuaded by either Eulenburg or Kiderlen. It was plain to Eulenburg that 'the rupture between the Chancellor and Waldersee would inevitably become an *open* one as a result', and he tried another tack. 'In desperation I dropped hints to the Kaiser, and on the last morning in the "Hohenzollern" Waldersee told me that the Kaiser had spoken to him so kindly and reassuringly about his relations with the Chancellor that he would now give up the idea of the letter.' Eulenburg nevertheless had a gloomy feeling that such peaceful outcomes were only of a transitory nature. 'This unfortunate antagonism will go on and on doing damage both in the Fatherland and abroad.' Moreover the differences of opinion between Herbert Bismarck and Waldersee, 'in spite of their shared opposition to the power of the stock exchange', were even more intense than those between Waldersee and the Chancellor. 'The antagonism lies in the insuperable antipathy of their whole beings towards each other', Eulenburg observed with alarm. The Kaiser was certainly '*convinced* of the necessity' of keeping the Chancellor; but he would never drop Waldersee, with whom he had a very deep and close relationship.[95]

This incident led to a definitive break between the influential anti-Semitic Chief of the General Staff and the Bismarcks. In his hatred for the Reich Chancellor, the General strayed into the realms of hallucination – one can scarcely call it otherwise – in persuading himself that Prince Bismarck and his closest colleagues were completely 'Jew-ridden' and that therefore every aspect of the foreign and domestic policy of the German Reich was controlled by Jews. For the Chief of the General Staff there was no longer any question that the Reich Chancellor and his closest counsellors were in league with the Jewish bankers. 'It is quite clear that filthy money-interests now play an important role, and quite probably the only role', he claimed. Bismarck had got himself 'so involved with Bleichröder', Waldersee alleged, that it was obvious that 'he was doing a lot of financial business with him'. The Chancellor knew that his banker would lose 12 million marks if the Russian conversion did not go through; that was the only way in which 'the Chancellor's fury' at the Kaiser's intervention could be interpreted. The head of the Reich Chancellery, Franz von Rottenburg, was very intimate with the banker Schwabach, and the 'otherwise so honest and hitherto completely upright' Boetticher had also got caught 'in the Bleichröder net': Bleichröder had saved Boetticher's father-in-law Berg from bankruptcy, and 'the cur does not let anyone go so easily, once he has him in his clutches'.[96] 'Wherever one looks', Waldersee wrote in December 1889, 'one finds Bleichröder's traces. When people said to me a year ago: "The Chancellor is completely Jew-ridden", I laughed or said that it was an enormous exaggeration. After what I have been through since then – e.g. Russian conversions – I have been converted. It is truly terrible: the Jew Bleichröder has a great influence on our foreign and domestic policies.'[97] In the eyes of the Chief of the General Staff the 'pro-Russian' policy adopted by Bismarck could be explained only by this Jewish influence. On 7 December 1889 he noted in his diary: 'The more I am able to look into the situation, the more firmly I am convinced and the more anxious I am that we are gradually becoming more and more Jew-ridden. The Chancellor has thrown himself completely into the arms of Herr Bleichröder; we are dabbling in friendly relations with Russia so as to raise Russian funds and to make their conversions easier for them. It is frightful but unfortunately true.'[98]

RELATIONS WITH RUSSIA AND THE VISIT OF TSAR ALEXANDER

It was not only the Kaiser's anti-Semitic prejudices which Waldersee managed to exploit so skilfully against Bismarck, but also Wilhelm's antipathy towards the Russian imperial family. Since early in 1889 Wilhelm had become increasingly annoyed with the 'boorishness' of the Tsar, who had still given no indication of when he intended to make a return visit to Berlin.[99] 'The other great sovereigns had announced their arrival months beforehand and the Tsar could at least have sent word of whether he intended to come at all, even if he wanted to keep the date a secret', he complained.[100] In June Waldersee welcomed the news that the Tsar had again 'given considerable offence' to the Kaiser by toasting the Prince of Montenegro as Russia's 'only true and honest friend'.[101] In October he noted with satisfaction, when Alexander III at last announced that he would visit Berlin, that 'both Kaiser and Kaiserin strongly resent his discourtesy in delaying so long over the return visit. The Kaiser is quite convinced that nothing more can be done with Russia through friendly means; he says: I have made up my mind, I shall not speak to the Chancellor about it any more, for he has his own views and we shall never agree.'[102]

The sudden improvement in relations between Wilhelm and the British court which had resulted from his August visit to Osborne and his appointment as admiral of the fleet also brought about a noticeable cooling of relations between the Kaiser and Bismarck.[103] Although on the one hand warmer relations between Germany and Britain were very much in accordance with the Reich Chancellor's thinking – 'the mere prospect of England's co-operating with us in the Mediterranean in case of war' contributed materially 'to the reinforcement of peace', according to Ernst Hohenlohe[104] – on the other hand the Kaiser's growing mistrust of Russia presented a growing danger to the Bismarcks' pro-Russian policy. Herbert Bismarck believed that there was a hidden connection between these two developments, for he attributed Wilhelm's suspicion of Russia to 'gossip against the Tsar at Osborne'. He presumed that it was above all 'highly spiced versions of remarks that the Tsar had made about H.M. and which had become known through the Princess of Wales' which had influenced the Kaiser in turning against Russia. There was no other explanation, he asserted, for 'the animosity . . . which H.M. felt towards the Tsar at the beginning of the autumn of 1889', for there were no tangible political reasons for it.[105] In this the Foreign Secretary was seriously underestimating the effects on the Kaiser of the anti-Russian sentiment fomented by the Chief of the General Staff.

Waldersee watched the Russian war preparations and troop movements with an eagle eye, although at this time he did not believe that there was any immediate intention on the part of the Russians to go to war. In his judgement Russia's armaments programme would not be completed until the spring of 1891, and even then Russia would not necessarily have any desire 'to start a war; but she wishes to be in a position to be able to call the tune in case of any complication that might arise, or to be able to intervene at a moment that suited her in any war between us and France. That would be the most inconvenient situation in which we could possibly find ourselves.'[106] Waldersee nevertheless continued to regard Russia with the deepest suspicion. The Russians 'are arming in great haste and on a huge scale and are gradually adopting a position that is scarcely tolerable for us', he observed.[107] In direct contrast to the pro-Russian policy of the Chancellor the Chief of the General Staff warned the Kaiser 'that the war for which we are preparing ourselves will be the mightiest that has ever been waged and that everything is at stake'. During a joint audience with Minister of War von Verdy in October 1889 Waldersee did his best 'to leave the Kaiser in no doubt as to the seriousness of the situation. He is after all the principal person concerned.'[108]

Whether it was English gossip or Waldersee's promptings, the Kaiser at first refused to receive the Tsar in Berlin for a full state visit. As Herbert Bismarck later recalled: 'Although King Umberto had been received in May and Kaiser Franz Joseph in August with maximum military honours, the whole garrison lining the route, mounted escorts, gun salutes etc., H.M. could not be persuaded to do the same for the Tsar in October; I had to call in Wittich and Hahnke to help, so that from the point of view of personal safety there would at least be a full military cordon from the station for the Tsar; at my request Hahnke then went a little beyond the imperial authorisation, because he realised that any difference in the reception from what was done for the other great sovereigns would be politically undesirable and there was nothing to justify it.'[109]

The next obstacle that the Reich Chancellor and the Foreign Secretary had to tackle with regard to the Tsar's visit was the Kaiser's selfish attitude to hunting. They had difficulty in making the young monarch understand that on this occasion he must be 'simply a refined and charming host', who must make allowances for his illustrious Russian guest's 'slow and hesitant' shooting and take care 'that the Tsar shot more than he did'. The Kaiser must not fail to invite the three high-ranking members of the Tsar's suite, Voronzov, Richter and Cherevin; it would be far better to give up the idea of the hunt altogether than to limit the Tsar to fruitless deer-stalking trips and not to invite his three companions. The Kaiser himself could, after the Tsar had left, 'shoot away to his heart's content. *This* is a *political* act, to make a pleasant impression on the Russian guests by helping them get a good bag', the responsible statesmen urged.[110]

The omens were therefore not particularly good for Alexander III's visit to Berlin from 11 to 13 October 1889. The reception of the Russian guests by the population was 'icy'.[111] No fewer than six times the Tsar complained of Waldersee's 'warlike' influence on the Kaiser.[112] He also expressed anxiety about Wilhelm II's forthcoming visit to Constantinople: he could not help taking it amiss, as it looked 'as if we wanted to turn the Turks against him'.[113] But Wilhelm suddenly swung round 'to the opposite extreme', Herbert Bismarck recalled. At a luncheon for the Alexander Regiment 'H.M. proposed a rousing toast to the old brotherhood in arms, with particular emphasis on Borodino and Sebastopol, in the anti-French sense, and ordered it to be published in full in the press, in spite of all arguments against this, which were based on the fact that casting aspersions on France was rash with regard to the Tsar at present, and as a guest he ought to be treated with consideration.'[114] If the Kaiser greatly overstepped the mark with this speech, even in Herbert's eyes, it was the russophobe Waldersee who found the most to regret in the imperial faux-pas. 'Things have gone somewhat wrong with our Kaiser', he lamented; 'he drank to the memory of the heroes of Borodino, Arcis sur Aube, Sebastopol and Plevna and to the Russian army. I do not think that this will please the Russians, the Turks, the French or the Austrians.'[115] In general, the Kaiser had positively besieged the Tsar with his attentions, so that someone in Alexander's retinue had commented that in Berlin 'everyone grovelled before us'.[116] In the Russian imperial family it was firmly believed that on this occasion Wilhelm had suggested to the Tsar the division of 'the whole of Europe' between Germany and Russia, whereupon Alexander – who could not help wondering about the Kaiser's mental state – immediately retorted: 'Stop whirling about like a dervish, Willy, just look at yourself in the mirror!'[117] After taking leave of the Tsar in his saloon carriage at the Lehrter station Wilhelm hurried up to Herbert Bismarck with the words: 'Everything has gone famously, the Tsar has invited me to Krasnoe and we have given each other our cipher codebooks.'[118] The Kaiser took this as a guarantee that Germany could count on peace at least for the coming year; it was 'a fine success', he said, which he attributed entirely to his own charm.[119] Later Waldersee learned that during a lively conversation with the Tsar in the saloon carriage Wilhelm had 'announced his intention of visiting Russia next year'; but 'no invitation from Tsar Alexander was forthcoming'. When Wilhelm informed the Reich Chancellor of this, in the belief that he had carried off a skilful coup, Bismarck was 'very embarrassed and dissatisfied', as Waldersee observed, for he too was of the opinion that one ought 'not to run after the Russians in any circumstances'.[120] Soon there were reports of 'bad-tempered remarks' by the Tsar about the Kaiser's proposed visit, which had been 'forced on him'.[121] As for the exchange of personal cipher codebooks, Herbert Bismarck commented: 'The A. cipher codebook for H.M. was soon ready and he carried it about everywhere, even in shooting costume. When Villaume received the W. cipher codebook to give to the Tsar, he reported that the latter expressed his thanks and put it in his desk.' In the eyes of the Tsar, such cipher codebooks were for adjutants to carry; 'a sovereign [could] not outwardly brand himself the adjutant of another'.[122]

The failed attempt to court the friendship of the House of Romanov reached its culmination in a painful epilogue. 'Certain incidents' which took place while the heir to the Russian throne, the Tsarevich Nicholas, was taking part in the imperial hunt at Springe am Deister in December 1889 embarrassed the Chief of the General Staff so much that he did not have the courage to mention the affair in his diary until several years later, and then only in veiled terms. The mysterious events in Springe, 'which I hesitate to set down on paper', Waldersee wrote after Nicholas's accession in December 1894, had undoubtedly 'made a lasting impression on the then heir to the throne'.[123]

In the weeks following the Tsar's visit Wilhelm's feelings towards Russia remained 'changeable, but still tolerable'. Although Herbert Bismarck recorded that he made many spiteful, mistrustful comments about Russia and the Tsar, 'full of the antipathy he still felt', he continued to have great faith in 'his own irresistible charm, . . . to the influence of which . . . the Tsar would again yield. To give this charm as much scope as possible to do its work, H.M. was even thinking of extending his announced visit, and inviting himself for 14 days' autumn shooting at the little hunting lodge at Spala, for then – by H.M.'s reckoning – the Tsar would be completely and lastingly under his spell.'[124] The relationship between the Kaiser and the Tsar was seen in quite another light by Bismarck's arch-rival Waldersee and yet confirmed in its essential elements. Waldersee complained that the Reich Chancellor, using a different tactic, had begun to flatter the Kaiser and not without success. Bismarck had assured Wilhelm 'that it is entirely to his, that is the Kaiser's, credit if Tsar Alexander now has quite different views about us, and then he tells the Kaiser quite shamelessly that he has considerable diplomatic talents. Unfortunately the Kaiser swallows this only too easily. Today the *Norddeutsche* sings the Kaiser's praises in a positively disgusting way as a great and wise sovereign, on the occasion of his visit to Constantinople. The Chancellor is unfortunately the cleverer and more cunning of the two.'[125]

In spite of the repeated requests of the Tsar, it did not occur to Wilhelm to give up his journey to Athens and Constantinople. He and the Kaiserin travelled to Italy, visiting the King and Queen in Monza on 19 October 1889 and, as we have already seen, continued their journey by sea from Genoa to Athens, where the Kaiser's sister Sophie was married to Crown Prince Constantine on 27 October. While at Monza on 20 October he wrote a lengthy report to Kaiser Franz Joseph on his meeting with Alexander III, which shows how he himself assessed the Tsar's visit.[126] He claimed that 'the Berlin public behaved very well, and to my astonishment greeted the Tsar much more warmly than I expected and than was described. As to the visit itself, it was more or less as follows: (Prince Bismarck and I compared our impressions afterwards and talked them over). The Tsar came to Berlin with a heart full of heavy forebodings and anxious cares. He had again been hard pressed last year, and the 2 months in Fredensborg had not been left unexploited by the illustrious members of the fair sex assembled there. For instance, from the questions which his tormented heart prompted him to ask the Chancellor, whom he had summoned to an audience, it became clear that he had been fooled into believing that I had allied myself with you, Umberto and the Queen of England – and now the Sultan was to be brought in too – in order to make a united and unexpected attack on the Tsar in the near future, to destroy his empire and to annihilate him and his family!! The Prince, in his calm, clear and conciliatory way, managed to contradict every bit of this nonsense so masterfully, and also to give an oversight of European politics in general, that afterwards the Tsar said to me in high delight: "Ah je suis tout a fait soulagé maintenant, et la conversation du Prince de Bismarck m'a dissipé toute[s] mes craintes, ce qui me laisse entierement satisfait." He had also personally been told many lies about me personally to make him as *soupçonneux* as possible, but all that was swept away with a magic wand. He was cheerful, contented, felt at home and was so loyal at the luncheon with the Alexander Regiment that he gave the toast in German and drank toasts with almost all his lieutenants.

Woronzoff told me that that was the first time for 25 years that he had heard the Tsar speak German in public. He set off on his journey home in the best of spirits and invited me to Krasnoe Selo for his big manoeuvres next year, an outcome which came as a complete but very pleasant surprise to the Chancellor. At any rate for the time being we are sure of peace for another year, and I hope for longer, if it is God's will. This was how things went, which I venture to tell you about. With many sincere greetings to the Kaiserin, I remain ever Your sincere Friend and Cousin Wilhelm.'[127]

In order to neutralise what he saw as the dangerous pro-Russian effect of the Tsar's visit, Waldersee held discussions with the Austrian military attaché von Steininger and the chargé d'affaires von Eissenstein and proposed a meeting between Wilhelm II and Kaiser Franz Joseph.[128] The latter made use of his letter of thanks for Wilhelm's long report to express the hope that on the return journey from Constantinople, 'somewhere on the way through Austria', he would have an opportunity to embrace the German Kaiser.[129] The Chief of the General Staff was nervous, however, that this might not be as warm an encounter as he hoped, since under Bismarck's influence the Kaiser was again speaking of Austria 'in a dismissive manner'.[130] The meeting took place on 14 November in Innsbruck and passed off fairly well, not least because the German military attaché, Major Adolf von Deines, went to meet the Kaiser at Bozen and had the opportunity to talk Wilhelm II out of Herbert Bismarck's disparaging views on Austria. The atmosphere was further 'improved' – in the anti-Russian, pro-Austrian sense – by a secret report which Waldersee sent to the Kaiser in Venice and which Wilhelm burned[131] during his second visit to Monza, where he had insisted on going in spite of strong opposition from Herbert Bismarck.[132] After the Kaiser's return to Berlin on 15 November Waldersee was relieved to discover that Wilhelm had not spent much time with Herbert during the journey to the eastern Mediterranean. 'While they were at sea – that is for more than 14 days – they were apart; in Athens and Constant[inople] the days were so fully occupied that the Kaiser could speak to him only infrequently and briefly. Of course the worthy Count again drew attention to himself by his rough and tactless behaviour', the General remarked.[133] Nonetheless Waldersee was still apprehensive, even after the meeting between Wilhelm and Franz Joseph, 'that we are once again faced with a cooling-off of relations with Austria. What we have so often seen before is happening again: the Chancellor is retreating into his old love of Russia, but Russia and France are continuing to rearm with all their might, and we are getting into a worse and worse situation, all the more so because things do not look good in Italy and Austria and each is full of sensitivities towards the other.'[134] After a conversation with Ambassador Count Münster on 8 December 1889 the General expressed the fear 'that we are going the best way about falling out with Austria without making sure of Russia's friendship'.[135]

Bismarck's lack of concern over Russian and French rearmament remained quite incomprehensible to Waldersee. When the Reichstag reassembled at the end of October during the Kaiser's absence, Bismarck sent instructions through Rottenburg to Minister of War von Verdy to postpone the planned creation of the new army corps until the following autumn, because France would take such redeployments amiss, and 'the most serious complications' could arise. Waldersee was furious at the Reich Chancellor's interference in military affairs and wrote: 'It is outrageous that he should express such wishes at this moment, and Verdy has sent him word that it is too late, that all the preparations are fully under way, the Kaiser having approved of the deployment &c. I am extremely anxious to know what will come of this; the Kaiser's being so far away may be a problem. But it is clear how afraid of war the Chancellor is. If our redeployments do not please the French, let them have it out with us, we are ready.'[136] Bismarck's attitude was all the more puzzling to Waldersee since everyone was aware that 'the new Reichstag which is to be elected . . . will be bad'; it would therefore have been 'much better to put forward even greater demands, while they can still be obtained'.[137] After making enquiries into the reasons for Bismarck's attitude and being able to find no convincing explanation, Waldersee once again suspected the influence of Jewish profiteers. 'It can only be Bleichröder who is behind it again' – as a 'well-informed person' had confirmed to him.[138]

At the beginning of December 1889 Waldersee noted with concern that 'thanks to the skill of the Chancellor' Kaiser Wilhelm II had 'without doubt gone over to the Russian camp altogether'.[139] As Waldersee saw it, Bismarck's foreign policy consisted in trying 'to get on a good footing with Russia through flattery and solicitousness'; but this policy destabilised the whole Triple Alliance, for it was making the army and leading circles in Austria feel unsafe and suspicious; the old differences between Austria and Italy were coming to the fore again, and France was making great efforts towards reconciliation with Italy. But the Chancellor could only continue with his policy by practising a massive confidence trick on the Kaiser. It was 'really infuriating' to see, Waldersee commented, 'how the Kaiser is deceived by the Chancellor; he only gives him reports which are written in accordance with his policy'. In any case Bismarck's diplomats reported nothing but what Bismarck ordered, and no newspaper articles which revealed the true atmosphere in Russia were allowed to be laid before the Kaiser. 'Yet in spite of all these despicable measures the Chancellor is not at all sure of the Kaiser', and was therefore feeling 'extremely uncomfortable', Waldersee observed at the beginning of 1890.[140]

The Chief of the General Staff was still confident, however, that 'from one day to the next' the Kaiser would 'begin to see clearly, in spite of all Bismarck's attempts to deceive him'.[141] On 23 December 1889 he recorded, after a conversation with the Kaiser in the Neues Palais, that Wilhelm had again become more sceptical about Russia. 'I am very glad to see that the Kaiser realises that the Russians are unsteady people, and that he is much less drawn along in their wake than he was 6 weeks ago. It was inevitable that this would happen.'[142] The Chief of the General Staff was also delighted to note, during a journey with Wilhelm to Dessau, that the Kaiser had 'recognised the restoration of Poland as a necessity'.[143]

WALDERSEE'S FALL FROM GRACE

It is plain from all these conflicts that throughout the critical year of 1889 Waldersee was the Kaiser's closest confidant not only in the military sphere but also in foreign and domestic policy and political appointments. Indeed his role was almost that of a rival chancellor. But in December came the first signs – the reasons for it are not clear – of a certain disillusionment, which already indicated that the Chief of the General Staff would not after all become Bismarck's successor, as most observers expected, but would himself fall from grace soon after Bismarck's downfall. As the course of German and European history would certainly have been different if Waldersee had become Reich Chancellor in 1890, the slight ill-feeling which can be seen to have arisen between Kaiser and General from the end of 1889 – in early December Wilhelm called him, half-jokingly, a 'poisoner'[144] – is of no small significance.

The cooling of relations between Wilhelm and Waldersee became noticeable above all in the rejection of the renewed attempt by the Chief of the General Staff and the Minister of War to set up a kind of 'military railway dictatorship', as Herbert Bismarck described it.[145] The Chancellor wrote no fewer than three papers setting out his objections to an increased military influence over the Reich Railway Office, and stressing the necessity of securing the strategic interest of the army through the *Prussian* railway administration.[146] As Bismarck was more concerned with the constitutional aspects of this disagreement, while the Minister of War's preoccupation was with practical matters, an agreement was reached in a discussion between Herbert Bismarck and Verdy on 10 December.[147] It was indicative that relations between the Kaiser and Waldersee were no longer so close that – as Herbert observed – Verdy was not unwilling to distance himself from his 'rather sinister protector Waldersee', against whom he was 'desperately seeking a counterweight'.[148] Neverthless there was still deep suspicion on both sides: from the Reich Chancellor's memoranda Verdy gained the impression 'that he intends to make me responsible for the result of the elections if these amount to a vote of no confidence in his domestic policies',[149] while for their part the Bismarcks suspected that the minister in charge of the army wanted to be able to say, if there were a war with France and Germany were defeated, 'Il vivrait, s'il m'avait suivi!' ('He would still be alive if he had followed me!') They were afraid that Verdy's aim was to collect written evidence of Bismarck's refusal, so that he would be able to claim later that it was not he but the Chancellor who had rejected the army's proposals, who was responsible for the defeat.[150] However that may be, the decisive factor was that in this intense conflict of political and military interests the Kaiser finally came down against the army's plans. On 10 December the Chief of the General Staff was still writing hopefully in his diary: 'I am anxious to know what attitude the Kaiser will adopt in the matter.'[151] Herbert Bismarck heard from one of the Flügeladjutanten on duty that Waldersee had been stirring up feeling against the Chancellor at luncheon in the Neues Palais, whereupon the Kaiser had complained 'in strong terms' about Prince Bismarck, who was always causing him problems.[152] But the next day the monarch had already adopted a different tone, and Waldersee was disappointed to realise that Wilhelm was inclined to agree with Bismarck on the railway question. 'I am very worried about his giving way like this', the General lamented for the first time since Wilhelm II's accession.[153] Finally, during a lengthy audience on the journey to Springe, Herbert Bismarck succeeded in winning over the Kaiser to the Chancellor's point of view, and in convincing him of the Minister of War's 'incredible ignorance of the constitution &c'.[154] The monarch had complained of the 'obstinacy and the purely civilian attitude' of Railways Minister Maybach, which would have seriously harmed the country's defence capabilities, but he had quickly seen the force of the constitutional objections put forward by the Reich Chancellor to the 'Waldersee–Verdy plan' and had expressed his pleasure that Bismarck was not opposed to the reform itself. Finally he said to Herbert: 'Good, then let us give up the Reich for the moment, and take up the matter in Prussia first. All I want is to improve our railway network quickly.' The Kaiser had been 'very calm and in an excellent mood', the Foreign Secretary reported to Friedrichsruh. 'I sat alone with him for an hour and three-quarters and discussed all kinds of important things that would take too long to write down and are partly too confidential as well.'[155]

The cooling of relations with Waldersee corresponded with an improvement – albeit only temporary – in Wilhelm II's relationship with the Bismarcks. When Herbert Bismarck called on the monarch on New Year's Eve he was 'very warmly' received. Wilhelm wanted 'detailed news about every member of the family' and added that the Reich Chancellor should 'of course stay at Friedrichsruh' and not put himself at the mercy of the Reichstag again before its dissolution. 'You were too precious for that', Herbert reported to his father, and went on: 'H.M. was simply as nice and considerate as he could be and sent warm greetings to you and Mama.'[156] A few days later, when Kaiserin Augusta died of influenza, the Kaiser sent the Reich Chancellor '*his command* not to come on any account'; he ought not to leave his wife alone and could catch something in Berlin.[157] Meanwhile, during a conversation with the Chief of the General Staff, Holstein observed that Waldersee was 'very nervous of the Kaiser'. Suddenly the General seemed to Holstein 'not a clear thinker after all, nor a consistent, logical character, more a blabbermouth than a worker. I would not be surprised', he commented in a letter to Eulenburg, 'if our sharp-witted Kaiser soon had no more use for him.'[158] Eulenburg thought he knew his 'most beloved Kaiser'[159] better, and cautioned Holstein: 'Do not lose patience with Waldersee! You are one of the few who have a conciliatory influence when there are differences between him and the Chancellor. If he becomes a burden to you the balance will shift and we shall be closer to a collision, the consequences of which *must* be *very* harmful to the relationship between the Kaiser and the Chancellor. For Waldersee is still in high favour. He will remain so for longer than you perhaps suppose. The Kaiser loves certain people *with* their weaknesses. Why should not Waldersee be one of these?'[160] Nevertheless, in the final phase of the Bismarck crisis which began in the autumn of 1889, the arch-conservative Chief of the General Staff no longer stood in the limelight. At this decisive stage the leading roles were played by Eulenburg and Holstein, seconded by the Kaiser's tutor Dr Hinzpeter, who suddenly re-emerged from obscurity at Bielefeld, Grand Duke Friedrich I of Baden and the Badenese envoy in Berlin, Adolf Marschall Freiherr von Bieberstein. All five were far more firm in supporting the Kartell of 'middle parties'. It was they, and not Waldersee, who were to join with Wilhelm in determining the policies of the post-Bismarckian era.

CHAPTER TEN

The Kaiser, the Chancellor and the Kartell

MANY observers, remembering Wilhelm's participation in the Stoecker meeting at Waldersee's house in November 1887, assumed that after his accession, and spurred on by the Kaiserin, Mirbach and Waldersee, he would maintain his alliance with the zealously Lutheran, reactionary wing of the German Conservative Party around the dismissed Interior Minister von Puttkamer, the editor of the *Kreuzzeitung* Freiherr von Hammerstein, the right-wing leader of the Conservatives Rauchhaupt and Adolf Stoecker himself, against the party's moderate wing and the liberal 'middle parties', that is to say the Free Conservatives and the National Liberals. And indeed in the autumn of 1889 a young diplomat at the British embassy was still bewailing the fact that 'poor' Friedrich III had omitted to sack Puttkamer, Stoecker and Waldersee the very day he came to the throne, for since then the 'political bigotry' of this right-wing 'rival government' at court had been gaining more and more influence over the Kaiser. This was fatal, for 'the men that are *very fast* destroying him [Bismarck] are our 10 times worse foes, the representatives of that tendency which ruined Fred[erick] W[illia]m II and VI'.[1] This contemporary interpretation of Bismarck's dismissal as the result of an extreme-right conspiracy has maintained its hold on historical research until the present day. It has much to be said for it, for Wilhelm's spiritual affinity with these ultra-conservative and anti-Semitic circles in the early days is beyond doubt. But it is equally clear that in his public utterances at the beginning of his reign the Kaiser supported the 'Kartell' of the 'middle parties' against the ultra-conservatives and it was this very fact that brought him into sharp conflict with Bismarck in the winter of 1889–90. Let us try to unravel this paradox, which from mid-1889 onwards forms the kernel of the confrontation between the Kaiser and the Chancellor over domestic policy.

THE 'KARTELL KAISER'

In the initial phase of his reign Wilhelm II had no quarrel with Bismarck from a party-political point of view, at least outwardly. On several occasions he spoke out openly in support of the Kartell coalition which Bismarck had brought together in 1887 between the German Conservatives, the Free Conservative Reich party and the National Liberals. In so doing he was in effect showing preference for the latter party and administering a rebuff to the extreme orthodox, anti-Semitic right wing of the Conservative Party, which would rather have allied itself with the Catholic Centre Party than with the anti-clerical Liberals. Within weeks of Wilhelm's accession, when the leader of the Conservative group in the Prussian parliament, Wilhelm von Rauchhaupt, called for a campaign against the National Liberals in the forthcoming Prussian elections, the Kaiser did not conceal his displeasure. Herbert Bismarck reported to the ministers on 4 August 1888 that 'H.M. wished to call Rauchhaupt to order directly' and had personally suggested that, as a counter-demonstration against the right wing of the Conservatives, the National Liberal Party leader Rudolf von Bennigsen should be appointed Oberpräsident of the province of Hanover.[2] The next day the Kaiser informed Herbert that 'he had seen to it that Rauchhaupt was given a sharp dressing-down, so that he would probably keep quiet now'.[3] When Bennigsen accepted the post in Hanover – Bismarck had sent him word that this was the Kaiser's personal wish – Herbert commented that the move would 'make an excellent impression everywhere'.[4]

It was thus perfectly logical that in Count Douglas's election speech of October 1888, which as we have seen was written by Rottenburg on Bismarck's orders,[5] the Kaiser's pro-Kartell attitude should be strongly emphasised. In a striking passage in the speech the Free Conservative deputy declared that the young Kaiser could quite rightly be called the 'Kartell Kaiser', and not in the narrow, party political meaning of the word, but in the broadest, most patriotic sense. Since the Kartell parties were those which 'associate themselves with the Kaiser's endeavours, since they gladly and joyfully declare their support for kaiserdom and the monarchy, without any ulterior motives or subsidiary aims, they can also say with good reason of themselves: "We stand on the ground on which our Kaiser stands."' The appointment of Bennigsen, he continued, which was 'in accordance with the *personal wish of our Kaiser* . . . is above all a proof that in his reign the Kaiser is determined to call on the support of all those who are at one with him on the fundamental questions, regardless of their particular party allegiances'. As long as the Catholic Centre and the left-liberal Progressive Party refused to adopt the same position on the fundamental questions affecting the life of the nation, but persisted in 'carping, constantly negative opposition on principle',[6] the Kartell parties would remain the only ones on which the Kaiser's government could and must rely for support.

It was also completely in line with the Bismarck's thinking that Douglas should emphatically reject the idea that the Kaiser supported Adolf Stoecker's views on religious policy or indeed his anti-Semitism.[7] 'As you know', he pointed out in his speech, 'a meeting which was held at the house of the present Chief of the General Staff, Count Waldersee, and at which the then Prince Wilhelm was present, was exploited to bring the Prince under suspicion in the public mind and to identify him with the party political aspirations of strict church circles, in particular with those of Court Preacher Stoecker.' All such interpretations of the Kaiser's position, however, were based on 'sheer distortion of the truth', Douglas asserted. He must emphasise that, on the contrary, 'the links which Kaiser Wilhelm had with Court Preacher Stoecker were only very temporary, and were limited purely to genuinely Christian and hence genuinely humane endeavours with regard to giving practical assistance to the lower classes in their need . . . Apart from this there has been no connection with Court Preacher Stoecker, and least of all does our Kaiser subscribe to the extreme political and confessional party opinions which are customarily linked with the name of this deputy. On this matter the position is absolutely, unambiguously clear. And if attempts have been made even to link the Kaiser with the anti-Semitic movement, this is yet more effrontery, which I can refute with the greatest certainty.'[8]

The appointment of Bennigsen as Oberpräsident, the Douglas speech, Wilhelm's visit to Friedrichsruh on 29 October 1888, which was intended as 'a demonstration of opposition to all extremes',[9] and finally Stoecker's endangered position as court preacher, which from March 1889 he seemed increasingly likely to lose,[10] gradually led to a change in the public perception of Wilhelm II's party political inclinations. As the Austrian chargé d'affaires, Eissenstein, noted in a perceptive report to Vienna, at Wilhelm's accession it was still widely thought that he subscribed to 'extreme-conservative Christian views'; the Stoecker party had not failed to exploit this belief, which undeniably had some basis in utterances made by Wilhelm as prince, in their own interest. The 'reactionary tendencies ascribed to the young Kaiser, however, have not been confirmed subsequently', Eissenstein commented, and as a result 'the hopes of the Stoecker party [have been] greatly reduced'. The best evidence that Wilhelm had no intention of 'throwing himself into the arms of the ultra-conservatives' was – according to Eissenstein – Bennigsen's appointment.[11]

Outwardly, therefore, in party political terms the elderly founder of the Reich and the youthful Hohenzollern Kaiser were in perfect harmony, supported by the moderate Conservative–National Liberal Kartell coalition. But was it all an empty pretence? In spite of public assurances and demonstrations, many in the inner circles of the ruling elite still had their doubts as to whether the ideas officially described by the Wilhelmstrasse as representing the Kaiser's thinking really corresponded to his innermost convictions. Thus for instance no less a person than the Kaiserin herself wrote to her husband after reading the Douglas speech: 'I think much of it is good but some of what he [writes] about your political and religious opinions is a pity, because it is not quite right.'[12] Similarly, the ultra-conservative Waldersee was of the opinion that, even if Wilhelm were obliged to give public support to the Kartell for tactical reasons, in his heart he continued to sympathise with the 'strict religious' tendency in the Conservative Party around Stoecker and Hammerstein, and had only reluctantly accepted the alliance with the progressively inclined National Liberals. In September 1888 he recorded in his diary: 'The Chancellor is still afraid that the Kaiser is actually on the side of the *Kreuzzeitung* in his heart of hearts . . . The Kaiser is [however] much too clever to identify himself with one party, even if he feels otherwise.'[13] Waldersee was able to note with satisfaction in November of that year that the 'Kaiser's ideas' were still 'as sound as a bell'.[14]

Precisely because there was always a risk that the Kaiser could switch sides and join his right-wing conservative enemies, Bismarck could not but regard the latter as his most dangerous rivals. On 23 October Waldersee wrote that the Chancellor, whose secret aim was to ensure the succession for his son Herbert, needed to fight 'the Conservatives above all'. 'The only rivals he fears are people from the Conservative Party, especially on account of the Kaiser's opinions, and he is now trying to discredit both the party and the individuals whom he considers worthy of notice.'[15] To eliminate the danger that arose from Wilhelm's inner inclination towards the ultra-conservatives, the Bismarcks – according to Waldersee – were intent on splitting the Conservative Party and drawing political and moral odium on to its right wing. With the help of their agents at court – Waldersee picked out Liebenau and Lucanus above all – the Bismarcks were even trying to neutralise the influence of the pious Kaiserin on the Kaiser. In October 1888 he recorded in his diary: 'The Chancellor wants to break up the Conservative Party; he is now using all his press . . . to attack the *Kreuzzeitung* and cast suspicion on it in a most despicable way. They have no scruples in accusing it of hostility to the Reich and would like to make the public at large think that it is in league with the real enemies of the Reich, that is to say Ultramontanes, Progressives and Socialists . . . The Kaiser is being drawn into this and claims that he is not a firm supporter of any extreme party, and thus not of the strict church party either . . . And now Bismarck is flirting with the National Liberals', Waldersee continued in disgust. 'But he is in danger of being seriously misled and then he will find out that many among the National Liberals are of a strong left-wing tendency, and that . . . the result will be to strengthen the Progressives. Naturally every possible means is being tried to influence the Kaiser and persuade him that the Conservatives . . . have no support in the country; the person most involved in this is of course Herbert Bismarck, and then Liebenau, these two being the people who see him most here; then Lucanus, who has no independent opinions but dances entirely to Bismarck's tune. But I do not think that they will have complete success. The Kaiser rightly takes the view – and I have often encouraged him in it – that he must base himself on the Kartell parties, and consequently that the machinery of government must work towards keeping these parties together, and not try to split off the right wing, as Bismarck is now doing.'[16]

The political agitation stirred up by the ultra-conservatives, so dangerous for Bismarck, reached a peak in January 1889 when Hammerstein published an article in the *Kreuzzeitung* condemning government policy – and in particular the publication of the charges against Geffcken – as un-Prussian and harmful to the monarchy. In tones that were quite obviously aimed at winning over the Kaiser, the newspaper editor wrote: 'The monarchical sentiments of Old Prussian patriots cannot but have been deeply wounded by recent events. Highly distressing as it then was to observe, after the Waldersee meeting, that nearly all German newspapers – with the single exception of those of the conservative right – had no compunction in conspiring with the entire foreign Jewish–liberal press to give the world the sad example of shameful criticism of, and malicious allegations against, members of the Imperial Family, without even a word from the official organs to call them to order; . . . deeply as all men of a truly monarchical way of thinking have deplored the tactless attempt by Count Douglas, Konstantin Rössler and others to draw the person of our present Imperial Master into the election campaign and party conflicts; and finally, profoundly displeased as every patriot has been by the shameless utterances of the Progressive press concerning the persons of the Kaisers Friedrich and Wilhelm II in connection with the publication of the Crown Prince's diary – the Prussians, whose motto is "With God for King and Fatherland!", are now confronted with the publication of the indictment [against Geffcken], with all its accompanying circumstances, and are filled with sadness and anxiety . . . The principle of authority not majority, the basis of the Christian state, is beginning to falter. But this is the soil in which Prussia's greatness is rooted, the Prussia in which His Majesty is King by the Grace of God.'[17] For the first time since the Manteuffel era the *Kreuzzeitung* was impounded on account of this article, and its editor, Hammerstein, was arrested.[18] The Kartell block, the parliamentary base of the government in Prussia and in the Reich, was already beginning to develop rifts. 'The Kartell is now a year old, but still has no firm framework; on the contrary, it seems to be coming apart', Waldersee noted in March 1889.[19]

THE CHANCELLOR AND CATHOLIC GERMANY

If the situation in Berlin was already complicated enough, the party-political confusion was greatly increased by the federal structure of the Reich. Bavaria, the second-largest federal state, was governed, to a certain extent against the will of its Catholic majority and also against the personal convictions of Prince Regent Luitpold, by a National Liberal ministry led by Freiherr Johann von Lutz. His government was regarded as a guarantor of the Reich principle and of Prussian hegemony, against Bavarian particularist aspirations. In political circles everyone knew that Lutz was safe in his position as long as he was openly supported by Berlin; but if a more Centre-friendly policy, or a wish to form a 'black–blue' clerical–conservative coalition in the Reichstag, were signalled from the capital, the Prince Regent would be only too glad to replace the Protestant Lutz by some aristocratic Catholic Centre leader such as Georg Arbogast Freiherr von Franckenstein. Many in Berlin feared, however, that such a development in Bavaria might lead to a break-up of the Reich, for, as they argued, if a Bavarian Catholic and particularist came to power in Munich, the leaders in Austria and the Vatican (and perhaps even in France and Russia) would be tempted to sabotage Bismarck's Reich and seek to return the central European states system to its pre-1866 status quo. The result might easily be – and this was the nightmare which haunted many statesmen in Berlin – a civil war with foreign intervention.

Even the orthodox Protestant Waldersee was convinced of the necessity of strengthening the position of the Liberal Prime Minister of Bavaria. 'The Catholic Church is busier than ever', he wrote in his diary. 'It has got itself on to a good footing with the French government and is now trying to do the same with Russia. In Bavaria it is making great efforts to overthrow Lutz's ministry. It seems that the Holy Father is joining the Franco-Russian alliance and I grieve to see our situation worsening all the time.'[20] In the summer of 1889 the Chief of the General Staff called on Minister-President Lutz in Munich to convey greetings to him from Wilhelm II, and to assure him of the latter's support in his struggle against his numerous enemies in South German aristocratic circles and in the Centre Party. He persuaded the Kaiser to confer the Order of the Black Eagle, the highest Prussian order, on Lutz. 'The first time he was unwilling', the Chief of the General Staff recalled; 'but the second time he was more accommodating. I was very glad that Lutz received the Order a few days after my visit.'[21]

Since mid-1889 Lutz's government had been under threat as a result of an insignificant question of religious policy which was to acquire a disproportionately important place in the conflict between the Kaiser and the Chancellor: the recall of the Redemptorists, a small religious order which, together with the Jesuits, had been banned from the whole of Germany during the *Kulturkampf.* When the question of allowing the return of the Redemptorists became acute in June 1889, Herbert Bismarck reminded his father that he had rejected this eighteen months earlier after the Bundesrat had established that the Redemptorists were still affiliated to the Jesuits. The Foreign Secretary went on to warn that it would be senseless to use this measure to gain the co-operation of the ultramontane Catholics: they would immediately demand that the Jesuits be allowed to return next. 'We shall never win over the intransigents in any case', he advised his father. 'They would only be satisfied if *all* Germans became Catholics and the Reich were split up into a mass of small powerless states with a constitution *à la* Paraguay.'[22] Accordingly the Reich Chancellor wrote to Lutz on 6 August 1889 refusing permission for the return of the order.[23]

The Kaiser's two friends, Waldersee and Eulenburg, were also decidedly of the opinion that the Redemptorists should on no account be allowed to return, for this would at once be seen as signalling support for the Centre, it would cause the fall of Lutz's pro-Reich government in Bavaria, and the moderate Conservative–National Liberal Kartell in the Reichstag would be broken up. Indeed, in their eyes even the existence of the Triple Alliance would be threatened, as such a change in German domestic policy would alienate anti-clerical Italy. Remarkably enough, these two influential advisers of Wilhelm II suspected that the Bismarcks, of all people, intended to adopt a pro-Centre course with the aim of provoking both internal and external unrest, which they hoped would increase the strength of their own position. The Kaiser must at all costs be warned of this devilish plot. 'The Kaiser, thank God, is very steady now', Waldersee wrote in July 1889 during the Scandinavian cruise, 'and here on the "Hohenzollern" he shall not be influenced in the other direction.' But he added presciently: 'I hope different influences will not be brought to bear later, for there will certainly be attempts.'[24] Just as a little stream can split a cliff when it freezes, this question, insignificant in itself, developed enormous explosive potential and threatened to break up the powerful German Reich and its alliance system.

IMPERIAL SUPPORT FOR THE KARTELL

In autumn 1889 the crisis came to a dramatic head. On his return from a walking holiday Friedrich von Holstein, the so-called Grey Eminence in the Foreign Office, heard that Bismarck's son-in-law Rantzau, as Prussian envoy in Munich, had sent a report directly to the Reich Chancellor at Friedrichsruh, in which he stated that Bishop of Fulda Georg Kopp had claimed in a letter to Lutz that the recall of the Redemptorists was now wished for 'in Berlin'.

The Chancellor had forwarded the report to Holstein, who was directed to lay it before the Kaiser, after consultation with Prussian Minister for Ecclesiastical Affairs Gustav von Gossler. The Kaiser, Holstein informed Eulenburg with relief on 28 September, had declared himself 'categorically' opposed to the recall of the order.[25] Encouraged by this imperial declaration of confidence in Lutz and in the Kartell combination in the Reichstag, Holstein devised a tactical coup through which the Kaiser could achieve three important aims: by making a public declaration in favour of the Kartell, Wilhelm would secure a majority for his government in the next Reichstag; he would unite moderate Conservative, nationalist and liberal public opinion behind him; and he would thereby deprive Bismarck of the chance of using a serious internal crisis to set himself up as a saviour in the hour of need, to the detriment of the monarch. Holstein's exposition of his plan to Eulenburg was shrewd and subtle. 'The elections will be bad, i.e. the moderate parties will suffer heavy losses, if the Kaiser himself does not in some way express his sympathy for the moderate groups. He would be doing himself a great service, for if his *first* Reichstag elections are won by the opposition, that is a vote of no confidence by the people in *him*, or at least it will be seen as such. The Chancellor does not need to worry about that. I have even come across some supporters of the Chancellor who think that bad elections are good for the Chancellor, although otherwise they take a pessimistic view of the regime. So the Kaiser is doing *himself* a good turn if he does something for the elections. It is nonsense when thoughtless people tell him that it does not matter for him whether there is a majority or a minority.' If the Reichstag elections due at the beginning of 1890 did indeed go badly, Bismarck would try 'to govern with the Centre and the Conservatives', Holstein predicted. But this would mean the appointment of a Centre Party deputy like Karl Freiherr von Huene as minister in Berlin, and the replacement of Lutz by someone like Franckenstein in Munich. 'That will dislocate the whole structure of the Reich, because Bavaria will assume the role destined for it by the curia . . . as the *Catholic* hegemonial power in Germany. If that happens, whenever we want to obtain a war loan from parliament we shall first have to negotiate with the Pope, and buy his agreement.' Holstein asked Eulenburg whether he agreed in principle on the usefulness of a declaration in favour of the Kartell by the Kaiser; they could discuss the *modus operandi* later.[26]

Holstein himself was surprised by the lightning speed with which his suggestion took effect. Two days after Holstein's letter to Eulenburg, Herbert Bismarck returned to the Foreign Office from an audience with the Kaiser with instructions to publish a solemn imperial declaration in favour of the Kartell in the *Reichsanzeiger*, the official government gazette. The Kaiser, as Holstein informed Eulenburg, had stipulated the contents of the declaration with the words: 'His

Majesty is indignant at the misuse made of his name, in that several newspapers dared to claim that he was against the Kartell. On the contrary, &c . . .' Holstein presumed that Eulenburg had been responsible for the imperial order. 'Did *you* have something to do with it, you little rogue?', he asked. 'That would be a commendable achievement.' Holstein was curious to know how the Reich Chancellor, to whom the declaration would have to be shown before publication, would react. But in taking this action the Kaiser had undoubtedly shown 'a high degree of political *sangfroid*... such as many much older people do not possess'.[27]

So on 2 October 1889 the *Reichsanzeiger* published the following sensational declaration: 'His Majesty the Kaiser and King has taken notice of the contents of the *Kreuzzeitung* of the 26th of the month and strongly disapproves of the political opinions expressed therein and of the attacks on other political groups. His Majesty does not allow any party to claim the distinction of possessing the imperial ear. But the Kaiser sees the understanding and mutual restraint shown to each other by the parties supporting the state as a genuinely useful benefit to our parliamentary life, and has unambiguously expressed All-Highest disapproval of the attacks and insinuations directed against it by the *Kreuzzeitung*. His Majesty sees in the Kartell a political formation which corresponds to the principles of His government, and cannot see how the manner in which the *Kreuzzeitung* attacks it can be reconciled with the respect due to the All-Highest Person and to our constitutional institutions.'[28] The Kaiser was 'delighted with the effect of his anti-Kreuzzeitung declaration' and 'wished to see as many supportive articles as possible', Herbert Bismarck subsequently reported to Friedrichsruh. In complete accordance with the line taken by Waldersee, Holstein and Eulenburg, the monarch declared to his astonished Foreign Secretary that 'the Kartell is as essential for peace at home as the Triple Alliance is for peace abroad'.[29]

Eulenburg spent 8 October 1889 in Berlin, where for several hours he held a 'serious discussion' with Holstein and the German ambassador at the Quirinal, Eberhard Count zu Solms-Sonnenwalde, on the importance of the survival of the Kartell for domestic policy and of the Triple Alliance for foreign policy. That same evening Holstein recorded the main points of agreement in an aide-mémoire. 'No concessions (e.g. Redemptorists) through which Lutz could be brought down. An ultramontane ministry in Bavaria *must* be avoided. *No manifestation of any kind* through which advances might be made to the Catholics at the expense of our relations with the *Italian government*. The latter must on no account be made mistrustful or allowed to think that our relations with the Pope are more important to us than those with Italy. Even if the Pope, instead of being Franckenstein's ally, were imbued with the best possible intentions towards us and decided to give expression to these intentions, his *power* would not be nearly significant enough to replace the loss of strength which we should suffer if Italy ceased to trust us and turned away from us. There are plenty of enticements for her to do so. – The Centre will never become a *German* party. It remains a foreign element even if it *sometimes* deigns to support the government; this support remains dependent on the will of the Pope, behind whom hide Ledochowski [the Polish cardinal], Windthorst and the Jesuits. For the imperial government it will be easier, in the case of real need, to come to an agreement with *any German* party *at all*, than with the two international parties: the Centre and Social Democracy.'[30]

As a reaction against the apparently pro-Centre shift in Bismarck's policy, in the autumn of 1889 the strict church, ultra-conservative Waldersee surprisingly also went over to the Kartell camp. He was just as intent as before on using religion, as Stoecker urged, as a weapon against the dangers of democracy and socialism, which he regarded as the 'most important of all questions', but he was deeply suspicious of the Catholic Church and therefore resolutely opposed to any rapprochement between Bismarck and the Centre.[31] On 4 November he noted in his diary: 'There is general uncertainty as to what the clever Chancellor is planning; there is a feeling that an evolution is in the offing, but no one knows in what direction. I still fear that he wants to get involved with the ultramontanes. The Redemptorist question is being pursued very hard in Bavaria and people hope it will bring down Lutz; if the Chancellor decided to give in over this I should consider it highly deplorable.'[32] In reaction to Bismarck's swing towards the Centre, Waldersee not only made contact publicly with the Free Conservatives, who embodied the principle of the Kartell, but even 'became slightly more friendly' towards the National Liberals, as he said on 21 October 1889. The Chief of the General Staff knew very well that this change of direction on his part would be a 'considerable blow' for Bismarck.[33] Waldersee's action strengthened his ties with Holstein, who had repeatedly urged him since July 1889 to speak out 'for the Kartell, the sooner the better'.[34] From then on the two men, whose personalities and views could not have been more contradictory, met regularly for lengthy discussions.[35] The circle of Bismarck's opponents was beginning to close around him.

At the beginning of October 1889 the Kaiser's secret clique of advisers thought they had received confirmation of their worst fears: that the Chancellor was engineering a 'black–blue' clerical–conservative party block behind Wilhelm's back. During the Kaiser's three-week absence in Italy, Greece and Turkey, Eulenburg (who, it should be remembered, was now no longer accredited to Munich but to Oldenburg, Brunswick and Lippe) heard that the categorical 'expression of his will' opposing the recall of the Redemptorists made by the Kaiser at the end of September had not 'reached the ears of the Bavarian ministry' but had been 'held up at the Prussian legation'. On 5 October Eulenburg informed Wilhelm that instead of conveying his, the Kaiser's, clear decision to the Bavarian government, Bismarck's son-in-law Rantzau had hinted 'that the *Chancellor* was *in favour* of the recall', whereupon the Bavarian government had decided to allow the Redemptorists to return to Bavaria. This triumph for the ultramontane party was a serious blow to Lutz's position, Eulenburg told the Kaiser. But he also warned him of the next, inevitable step: soon Bavaria would propose a motion in the Bundesrat for the suspension of the relevant Bundesrat resolution, thereby enabling the return of the Redemptorists throughout Germany. 'And now that the Chancellor, in response to Rantzau's direct reports, has actually taken the step of moving closer to the Centre by agreeing to the recall, the more malleable elements in the Bundesrat could easily reach a decision which would greatly surprise Your Majesty. The timing of this action in Munich was cleverly planned to coincide with Your Majesty's absence. Everything therefore depends on a declaration of Your Majesty's will, to enlighten the Bundesrat on Your Majesty's views about the recall of the Redemptorists into the Reich.' Eulenburg strongly urged the Kaiser 'to destroy this letter at once – in strange hands it would look like an intrigue!' The Kaiser should not mention the letter to Herbert Bismarck – otherwise Eulenburg would be blacklisted – but he must refer to newspaper reports as his source. This was a matter of the greatest possible importance. 'I have often written to Your Majesty about Bavaria as "the Catholic hegemonial power in the Reich"; this is now a distinct possibility. If we support these tendencies we are doing ourselves damage, and *the day that we make concessions to Roman policy, Italy will cease to trust us.*' Eulenburg had already written to Lutz to inform him of the Kaiser's true opinion and to seek further information about other 'Catholic questions': the Kaiser had, after all, ordered him 'to keep in touch with matters in Bavaria'![36]

A few days later Eulenburg wrote a second letter to the Kaiser, who by now was on the way from Athens to Constantinople. In it he recounted that on a recent visit to Berlin he had found 'a state of immense alarm in the hallowed halls of the Wilhelmstrasse'. Holstein, Kiderlen-Wächter, Ludwig Raschdau, Dr Paul Kayser and the other counsellors in the Foreign Office were all 'at their wits' end' at the success of the 'coalition' between Rantzau in Munich and the 'ultramontane' Bavarian envoy in Berlin, Hugo Count von Lerchenfeld-Köfering. The Foreign Office men were 'of course powerless against the word put about by Friedrichsruh – and yet filled with the inner urge to bring Your Majesty's will to the fore in this important question'. All were of the opinion that it 'would be *of great importance* for the good of Your Majesty that, unless Your Majesty preferred to have the matter out with the Prince himself, or even *via Herbert*, Your Majesty should have the view which Your Majesty takes with regard to the recall of the Redemptorists clearly stated *before the representatives of the federal states, through Boetticher*'. The Reich Chancellor, Eulenburg asserted, would 'certainly not make this a resignation issue', for he could not face Germany with a threat to resign '*on account of the ultramontanes*'. Furthermore, 'the whole Ministry shares Your Majesty's point of view out of conviction'. Vice-President of the Ministry of State Heinrich von Boetticher and Minister for Ecclesiastical Affairs Gossler were particularly 'infuriated' by Bismarck's change of heart towards the Centre.[37] In support of his action Eulenburg also called on the Grand Duke of Baden to help by obtaining a 'formal declaration of the will of His Majesty The Kaiser to the federal governments (via Boetticher)', in order to put a stop to the 'intrigue' in which the Chancellor, Rantzau and Lerchenfeld were involved.[38] A head-on collision between the Kaiser and the Chancellor now seemed inevitable, for while there were increasing indications that Bismarck intended 'to go along with the ultramontanes', Holstein was able to observe from the Kaiser's marginal notes received at the Foreign Office on 4 November that the latter was 'still absolutely firm at the moment' on the church question.[39] He was nevertheless worried that as the Kaiser was 'unfortunately very impressionable', the Reich Chancellor might make him change his mind on his return. 'In a word', he warned Eulenburg on 5 November 1889, 'the Kaiser is not yet rated as a political factor. If he gives way this time too, his rating will sink down to near zero.'[40]

The very next day, as a direct result of Eulenburg's two letters, which the Kaiser had not received until after he had left Constantinople on 6 November,[41] two telegraphic ultimata arrived from Pera which put paid all to Holstein's doubts. The first telegram, addressed to Bismarck, was dictated by the Kaiser directly to the telegraph clerk. He did not mince his words. 'I see from articles in the *Kölnische* and the *National-Zeitung* that the recall of the Redemptorists has been proposed, not without prospects of success. I request you to declare that never in any circumstances can I or will I give my consent to this measure. Envoy in Munich notified direct, for information of government there.'[42] At the same time Boetticher received a telegram from Lucanus, just as Eulenburg had suggested, which read: 'His Majesty has telegraphed direct to the Reich Chancellor today concerning the Redemptorists. His Majesty asks Your Excellency to act in accordance with this telegram in the Bundesrat.'[43]

Holstein noted that, although the Kaiser's declaration had come too late to alter Lutz's position, which the latter had already made public under the impression that the Chancellor favoured the recall of the Redemptorists, it had nevertheless been in time to have a decisive influence on the Bundesrat and the federal governments, and was 'particularly well timed to influence our own ministers'. He hoped for a still more active participation by the Kaiser in the conduct of policy in the future, which would oblige the monarch to work more closely with individual ministers. 'Our All-Gracious Sovereign will *perhaps* learn from these events that he really must take some interest in work . . . It is not as if he will need to bury his imperial nose in papers for very long; it will be quite enough if he reads newspapers and asks *departmental ministers* to report to him from time to time.'[44] Eulenburg was confident that the Kaiser would not have his mind changed again on the Redemptorist question. 'The Redemptorists will *not* be forgotten by His Majesty – in any respect. I am certain of that', he assured Holstein. 'His Majesty has been warned and can no longer be easily trapped.'[45]

At the beginning of December 1889 Holstein recorded with satisfaction that the Kaiser, now back in Berlin, had spoken 'firmly' to the leader of the moderate, pro-Kartell wing of the Conservatives, Otto von Helldorff-Bedra, against the extreme right-wing tendency in the party, and had authorised him to make his views known.[46] He was even more pleased when Wilhelm praised the services of the Oberbürgermeister of Frankfurt am Main, the National Liberal Party leader Dr Johannes Miquel, in a speech on 9 December, for his toast was at once interpreted as a further demonstration of support for the Kartell. Jubilantly Holstein commented: 'It was the best speech that the Kaiser has made for a long time – the people can see from it that the sovereign is taking his work seriously.'[47]

Bismarck, who was 'rather taken aback' by the 'bluntness' of the imperial telegram,[48] tried to neutralise its effect as far as possible. He answered the Kaiser soothingly that he would act in accordance with his command as soon as the question came up in the Bundesrat or in the Reichstag; meanwhile he did not believe that the Bavarian government was likely to make a proposal respecting this matter in the Bundesrat.[49] He sent word to Lerchenfeld that he must forbid any declaration by the Bavarian government which would reveal the opposition of the other federal states to the return of the Redemptorists, for the confidentiality of the sittings of the Bundesrat must be preserved.[50] He published the imperial telegram to him, not least, Holstein suspected, in order to draw attention to the '*youthfulness*' of the Kaiser.[51]

Is it really true that Bismarck, shortly before his fall, pursued this plan to break up the Kartell coalition which suited the government so well, and replace it by a black–blue block in the Reichstag, as his secret enemies suspected? This thesis goes so much against the grain of orthodox research on Bismarck that it seems implausible even to Bismarck's great American biographer Otto Pflanze.[52] And yet there is no doubt of the authenticity of the sources cited above. The documents indisputably testify that men of such differing characters and political standpoints as Eulenburg and Holstein, Waldersee and Boetticher, Paul Kayser and the Grand Duke of Baden were convinced that this was Bismarck's aim, and were prepared to provoke direct conflict between the Kaiser and the Chancellor over it. Even if we accept that for a few of them – Holstein and Waldersee, for example – manipulative calculation may have played a part, and even if it cannot be denied that all the documentary evidence of Bismarck's sudden turning against the Kartell comes from the writings of the Kaiser's advisers, and none from the Bismarckians themselves, it is nevertheless scarcely conceivable that so many well-informed diplomats, civil servants, military officers and journalists should have fallen victim to a mere illusion. The likeliest explanation of the riddle seems to be that from the autumn of 1889 onwards Bismarck became increasingly aware of how much his position was threatened by the growing impatience of the young Hohenzollern ruler, and that, scenting danger, he began contemplating ways of frustrating the impending take-over of power by the Kaiser. The creation of a new majority in the Reichstag on a clerical–conservative basis was only one of the available solutions, and in itself not the most attractive. But the founder of the Reich must have foreseen – just as his secret opponents did – the internal and external tensions and crises that would be unleashed by such a change of coalition; and he would have calculated that the resultant chaos would make it impossible for the inexperienced monarch to dispense with his now almost 75-year-old Reich Chancellor. In the anxiety-ridden, paranoid world in which the conspirators supporting the Kaiser found themselves, these calculations on the part of Bismarck will certainly have been predicted and thought through to the end, even if Bismarck himself had not formulated them so fully. Hence also their subjective feeling – although objectively speaking they were the vanguard of a new regime under the personal leadership of the Kaiser – that they were mounting a heroic defence against the overly powerful Chancellor-Dictator Bismarck.

What is unmistakable is that within a short period there had been a massive shift of power. Six months earlier the Reich Chancellor had still held the monopoly of decision-making power in civil affairs; now, at the end of 1889, he could not even push through a minor religious measure, through which he might have hoped to engineer a new constellation of parties in the Reichstag, against the will of the Kaiser. From the autumn of 1889, instead of the dictatorship of the Chancellor which had existed hitherto, there was a dualism of power which could not last long. Nor was there much doubt as to which was the stronger element. 'Very few people suspect how weak the Chancellor is in relation to the Kaiser', Waldersee commented on 13 November. 'This is the clearest sign of his weakness. He is extremely worried about a conflict with the Kaiser because he fears that the Kaiser might dismiss him. The Kaiser knows this very well, and he uses it very skilfully to work on the Chancellor. He follows him in many things, but in others he remains absolutely independent.'[53] In so doing he received the advice of a heterogeneously assembled clique of whose existence the apparently all-powerful Reich Chancellor was completely ignorant.

THE CONSPIRATORS AGAINST BISMARCK

As we have seen, the numerous collisions between Kaiser and Chancellor from early in 1889, especially the violent disagreements over the coal miners' strike and the Russian conversion in the summer and over the Kartell in the autumn, had led to the formation of an influential group of advisers around the Kaiser, in addition to the civilian and military officials of his court. It remained an informal group: its members corresponded with each other but rarely met, and then almost always only in secret; and they had little in common either socially or ideologically. Their motives were varied and yet closely entwined. They combined personal ambition and objective conviction with a strong desire to increase the power of the Kaiser, of whom the members of the group had formed a totally idealised image, and to reduce the domination of the Bismarcks, which they saw increasingly as dictatorial. Although Waldersee and Eulenburg continued to be key figures in this group, the circle of Wilhelm II's advisers cannot in any way be described as orthodox-evangelical or reactionary-conservative, for even Waldersee recognised the tactical necessity of publicly aligning the Kaiser with the Kartell. Other members of the group of advisers, like the Grand Duke of Baden and his representative in Berlin, Adolf Freiherr Marschall von Bieberstein, were South-West German liberal–conservative in their thinking, while men such as the influential Friedrich von Holstein in the Foreign Office and the journalist Franz Fischer of the National Liberal *Kölnische Zeitung* represented an almost progressive tendency, or at any rate a constitutional, anti-clerical rationalist one. Indeed the group even included idealists such as Dr Georg Hinzpeter, the Kaiser's former tutor, and Dr Paul Kayser, the (Jewish) Colonial Director in the Foreign Office, who showed astonishing radicalism in putting forward a programme of state social reform for the benefit of the working classes. This heterogeneous group, and not the *Kreuzzeitung* faction, were the Kaiser's supporters in his historic conflict with the Bismarcks.[54]

After his return to Bielefeld Hinzpeter had not lost touch with Wilhelm, but continued to correspond with him and visit him occasionally in Berlin and Potsdam. At the beginning of 1889 Eulenburg, who since his posting to the North German courts often came into contact with the former tutor, warned Herbert Bismarck of the 'liberal' influence exercised by Hinzpeter, who had pushed himself forward 'in his discourteous and thick-skinned way' to speak to the Kaiser during a train journey from Bückeburg to Hanover. Commenting on the relationship between the 'Doctor' and his pupil, Eulenburg wrote: 'You know that he writes a great deal to the Kaiser, and that the Kaiser is sincerely attached to his old teacher . . . The one thing that strikes me as strange about all this is that we owe our glorious Kaiser to the education given him by this man!'[55]

In November 1889, however, it was precisely Hinzpeter's liberal views which Eulenburg and Holstein began to see as a useful weapon in the battle to defeat the Bismarcks and preserve the Kartell. Holstein urged Eulenburg to arrange for Hinzpeter to be in Berlin on the day the Kaiser returned from his oriental journey, 'because of the church question'.[56] Suddenly the tutor had become 'intelligent and sensible' in their eyes, 'an exceptionally clever man', if 'slightly inclined to strike a pose as an original'. On 9 November 1889 Eulenburg visited Hinzpeter in Bielefeld and subsequently reported to Holstein that the pedagogue shared 'our point of view and will put forward these opinions in Berlin'.[57] Not the least of their hopes was that Hinzpeter's influence would result in a less 'youthful' style in the Kaiser's telegrams.[58] In fact he had exactly the opposite effect.

Early in the new year a letter from Hinzpeter produced a violent reaction from the Kaiser in connection with the critical question of the Kartell. He wrote on 4 January 1890 from Bielefeld, commenting that the candidature there of Hammerstein, the editor of the *Kreuzzeitung*, in the forthcoming Reichstag elections was of serious concern, because 'it shows particularly clearly how the extreme Conservative Party is trying to destroy the Kartell's majority at all costs and in any way it can, and to force the Kaiser and his Chancellor to change the domestic policies it hates'. The Bielefeld Conservatives, he reported, had not only got in touch with the Centre Party leader, Ludwig Windthorst, to ensure his support for them, as half the voters in the constituency were Catholics; they were repeating 'with redoubled energy and clarity the insinuation . . . that the declaration in the *Reichsanzeiger* [of 2 October 1889] had been issued only under pressure from Prince Bismarck; the Kaiser was in complete agreement with the Kreuzzeitung party at heart; it would be rendering him a service to vote for the party'. Hinzpeter ensured that this information would make an even greater impact on the Kaiser by adding: 'That this insinuation constitutes gross *lèse-majesté*, for it makes the Kaiser look like a cowardly weakling who does not have the courage to think for himself, matters less to them in their passion for their party than a possible triumph of the party over their opponents.' It had even been claimed at an election meeting that 'the Kaiser read the *Kreuzzeitung* as he always did; it was even the only political newspaper that he read; the Kaiser was a Hammerstein man through and through!'[59]

Hinzpeter should have known what rage such language would unleash in his pupil. Wilhelm covered the letter with exclamation marks and comments. For days he continued to fume with anger. He telegraphed to Hinzpeter immediately, *en clair*, instructing him to 'let it be most distinctly understood' how 'outraged' he had been at the behaviour of the Conservative–clerical faction in Bielefeld.[60] He forwarded Hinzpeter's letter to the Reich Chancellor with orders 'to take the most appropriate and severe steps in my name to counteract immediately and vigorously the activities of Hammerstein which Hinzpeter has described so excellently'.[61] On 6 January he called on Herbert Bismarck and spoke 'very indignantly' about the 'outrageous public fraternisation between Windthorst and Hammerstein', of which he had heard through Hinzpeter. As Herbert reported to Friedrichsruh, the Kaiser expressed 'furious resentment over the †Zeitung and that creature of the Guelphs, Hammerstein, whom he described as a "traitor"'; he expected the Chancellor to put forward suggestions as to how to 'publicly brand this shameful alliance between the hypocritical †Zeitung and Windthorst, the enemy of the Reich, which is aimed purely against the Kartell, and to turn it to account for the elections'. As he put it: 'After all, I have repeatedly declared in public, and had it declared on my behalf, that I regard the †Zeitung gang as enemies, that I stand by the Kartell absolutely; and yet people still have the temerity to claim that I read the †Zeitung and do not take any other paper: it really is a scandal that I should be defenceless against such lies!'[62] The Chancellor tried to dissuade the Kaiser from issuing another declaration against the *Kreuzzeitung* party with the argument that it would not be feasible to take the initiative publicly since 'the grounds for it are not apparent in the form of a public statement of the untruths which are to be denied'.[63] But Wilhelm was not satisfied with this answer. On 8 January he again called on Herbert Bismarck, and in his presence dictated 'an order to the Flügeladjutant that the Kreuzzeitung should no longer be subscribed to or found in any palace or adjutant's room; the Adjutant was also to write to Her Majesty's Cabinet Counsellor to say that H.M. must get rid of the Kreuzzeitung. Finally he had it noted for Hahnke's benefit that subscribing to the Kreuzzeitung was to be forbidden for all officers.'[64] Not satisfied with that, the next day the Kaiser once more called on the Foreign Secretary and instructed him to telegraph to the Reich Chancellor to say that with the imperial ban on the *Kreuzzeitung* 'the starting-point described as necessary' in his submission had now been provided.[65] This left Bismarck with no alternative but to order a second imperial declaration in the press in support of the Kartell and against the *Kreuzzeitung* party. Unlike Herbert, who wanted to have the notice published in the semi-official *Norddeutsche Allgemeine Zeitung*, the Chancellor preferred the 'less official' *Berliner Politische Nachrichten*, and the declaration appeared there on 10 January.[66] Marshall von Bieberstein, who was spending the day with Holstein, noted in his diary: 'Imperial declaration about *Kreuzzeitung* imminent. Count Bismarck wanted it in *Nord. Allg.* Reich Chancellor ordered "*Pol. Nachrichten*". Incredible.'[67]

Only a few days later Hinzpeter sent a second letter to Wilhelm, in which he warned him against receiving a deputation of Conservative voters from Bielefeld, as such a distinction would only be seen as support for Hammerstein's candidature.[68] When Bismarck, to whom Hinzpeter had sent a copy of his letter, likewise advised against receiving the deputation, Wilhelm wired to him on 17 January: 'Quite agree with your opinion. Had already telegraphed to Hinzpeter refusing.' He added a menacingly autocratic declaration of principle: 'I am accustomed to being obeyed, but I will not enter into discussions with such people. A kaiser's word should not be twisted or quibbled with.'[69]

If the Bismarcks were less affected by the imperial anger with the *Kreuzzeitung* party than many people had expected, the reason was that Wilhelm's telegrams and orders were met with 'shouts of triumph over the defeat of Waldersee'.[70] Clearly Wilhelm's anger at the right-wing Conservatives had strengthened his resolve, apparent since December, not to appoint Waldersee as chancellor after all. Only a short time later, on 1 February 1890, the Kaiser summoned General von Caprivi from Hanover and ordered him to be prepared if necessary to become Bismarck's successor.[71] Understandably therefore the Chief of the General Staff was not exactly delighted by the 'quite useless and ill-considered interference' by Hinzpeter in a matter which was none of his business. On 11 January he commented that Wilhelm had wildly overreacted to Hinzpeter's report that the right-wing Conservatives in Bielefeld had claimed that 'the Kaiser was still on the side of the Kreuzzeitung and against the Chancellor in his heart', by wielding 'another blow, and again very overhastily, against the Kreuzzeitung'. The embargo on the *Kreuzzeitung* at court would 'create a colossal uproar' and was 'undoubtedly a mistaken measure'. It was altogether wrong of the Kaiser to have intervened in the election campaign. 'If he had used his influence to get rid of Hammerstein, I should have had no objection; but to take this action against the newspaper will do great damage.'[72] Not long afterwards Waldersee noted that Bismarck was trying 'to exploit the punishment of the Kreuzzeitung by the Kaiser for his own ends'.[73]

With the recruitment of the liberal commoner Hinzpeter to the ranks of the secret group of imperial advisers, tensions had inevitably arisen among them which could be masked only temporarily by tactical considerations. Waldersee in particular, intent as he was on combating socialism through a religious revival, disapproved of Hinzpeter's pro-worker and anti-clerical influence on the Kaiser.[74] Even so the two men, tutor and General, found they had much in common in their assessment of the situation when they met on 11 January 1890 after the funeral of Kaiserin Augusta. Like Waldersee, Hinzpeter thought that the relationship between the Kaiser and the Chancellor was becoming 'worse by degrees'. Waldersee was gratified to hear from Hinzpeter 'that the Kaiser has at last realised that much is kept from him, that the Chancellor therefore acts behind his back; he [Wilhelm II] also knows of the rumours that the Chancellor is involved in financial dealings in grand style – in short, light is beginning to dawn'. Hinzpeter and Waldersee were also at one in their criticism of Herbert

Bismarck. The General was delighted to hear from Hinzpeter 'that the Kaiser speaks of him [Herbert Bismarck] with little respect and actually values him only for his jokes'. Hinzpeter dropped meaningful hints to the Chief of the General Staff, that he should not waste his energy on little things, but should 'keep himself fresh and intact for great ones'.[75] For his part, Waldersee commented in his diary on the great influence which Hinzpeter exercised over the Kaiser. The tutor was now 'very much in the foreground', he noted in mid-February 1890, four weeks before Bismarck's fall; and Hinzpeter had 'probably been principally responsible for persuading the Kaiser to take action on the worker question and also corresponded on many other matters with the Kaiser'.[76]

Remarkably, in this meeting between the schoolmaster and the Chief of the General Staff, anxieties about Wilhelm II's mental state also came to the fore, based in both cases on many years of experience. Immediately on arriving Hinzpeter remarked: 'What do you think of the Kaiser's nervous state? He is increasingly overhasty. It is true I wrote and told him that his name was being misused in Bielefeld, and asked him whether he still stood by the declaration in the Reichsanzeiger – which by the way I never approved of and thought greatly exaggerated – and he telegraphed to me at once in the most vehement terms; as it was an open telegram of course it soon became known, and now he has taken the step of banning the newspaper at court.' After this explanation Waldersee was in no doubt that Hinzpeter had 'set the thing in train and therefore bears a considerable part of the blame'.[77] 'Herr Hinzpeter, the actual originator of the whole wretched affair, no doubt greatly regrets his action', he noted.[78]

No less important for the development of the Bismarck crisis was the contact established on 3 January 1890 between Waldersee and the Grand Duke of Baden. 'We had a very frank conversation and I spoke my mind freely, because I know that the Grand Duke truly means well by the Kaiser and because he is discreet', the General recorded after their first meeting.[79] When Kaiserin Augusta died Waldersee was at first afraid that this might weaken the bond between Wilhelm II and the Grand Duke and Grand Duchess Luise, who was Augusta's daughter. 'In both the Kaiser has true and devoted friends, who wish only the best for him', he wrote.[80] In the event Grand Duke Friedrich, who had already played a significant part in the Wohlgemuth affair, in the conflict over the Russian conversion and in bringing about the Kaiser's pro-Kartell declaration, now took centre stage in the role of fatherly adviser to Wilhelm at the start of the most critical phase of the crisis. Through the monarchical and militaristic instincts which set him against what was perceived as the dangerous 'omnipotence' of the Bismarcks, he gave a semblance of legitimacy to Wilhelm II's rapidly growing hunger for power.

14. Grand Duke Friedrich I of Baden

The significance of his alliance with Waldersee became clear when the Grand Duke showed the Kaiser an article in the official *Norddeutsche Allgemeine Zeitung* in which Waldersee was again attacked as anti-Russian. On 12 January 1890 the Grand Duke was able to assure the latter that 'the attack against you has been beaten off. I showed the Kaiser the article, which had of course been withheld from him, and he soon saw that it is a product of the House of Bismarck and is in actual fact directed against him. He has now given very strict orders that the newspaper retract it.'[81] And indeed Herbert Bismarck found himself obliged to publish an apologetic explanation on 12 January stating that Prince Bismarck's peace policy was naturally 'the Kaiser's policy' and that there was no question of contrary views in the political or military sphere within the All-Highest entourage.[82]

In a second meeting on 12 January, Waldersee and the Grand Duke exchanged frank views on the relationship between Bismarck and the Kaiser. Waldersee said he had the impression that 'the Kaiser is now beginning to see clearly how Bismarck is deceiving him, and I believe that Your Royal Highness has greatly contributed to this'. The Grand Duke replied that he had certainly talked things over 'thoroughly' with the Kaiser, and had observed that Wilhelm was 'absolutely determined to keep the Chancellor, even *against the latter's will*, but he wants to look more closely into affairs so as to remain in control of them'.[83] A week later the Grand Duke received Waldersee for another lengthy discussion during which he assured the General 'that the Kaiser now sees things more clearly and wants to free himself from the Chancellor's absolute power'.[84]

The Grand Duke had a permanent representative in Berlin in the person of Adolf Freiherr Marschall von Bieberstein, Baden's envoy and delegate to the Bundesrat, who soon developed an excellent rapport with the Grey Eminence of the Foreign Office, Friedrich von Holstein – and who was to become Herbert Bismarck's successor as foreign secretary. On 18 March 1890, the day on which the founder of the Reich was forced to tender his resignation, Herbert Bismarck confronted Marschall with the words: 'Your Grand Duke is the involuntary gravedigger of the Reich Chancellor.' The Chancellor's son later recalled that this remark threw Marschall into a state of 'very evident dismay' and he protested 'stammeringly' that 'his Grand Duke had always been a warm supporter of His Highness. I interrupted him with the words "I did say *involuntary*; if you prefer it, *unwitting*; the Grand Duke did not realise what he was bringing about and how he was made use of during the 3 quiet weeks after the death of Kaiserin Augusta. But I know exactly what happened, so you can spare yourself the trouble of an apology."' As Herbert did not learn of Marschall's own 'undermining work in league with Bötticher and the wives of both' until later, he was at first surprised by the Baden envoy's overreaction. It was only when Lerchenfeld remarked that he had hit the target more precisely than he realised, for it was Marschall himself who 'has influenced and led his Grand Duke all along', that Herbert saw why the envoy had been so strangely affected. And to understand Marschall's efforts to stir up trouble, Herbert believed, one needed to know about his wife's ambition.[85] Neither Herbert Bismarck nor his father, however, ever discovered how much Marschall came under the intellectual domination of Holstein during the crisis.

King Albert of Saxony, who had already come forward as a critic of Bismarck during the Russian conversion crisis, became another member of the group of imperial advisers in January 1890. In mid-January he had a long and 'very anxious' conversation with General von Verdy, the Minister of War, and he saw 'as clearly as the Grand Duke of Baden how things stand'.[86] The Kaiser's secret advisers recognised the significance – but also the danger – of this monarchical alliance against the Chancellor. On 17 January Eulenburg wrote to Waldersee: 'I hope that in the present very difficult situation the Kaiser will form a closer link with the King of Saxony and the Grand Duke of Baden, and that with Boetticher's support he will find in them a kind of substitute for the lack of support he is receiving from his Cabinet chiefs.'[87] 'The Kaiser's determination

15. King Albert of Saxony

to ally himself more closely with the princes is of such inestimable value', he commented in a letter to Holstein, 'that I shudder to think what would happen if he met with a setback and was thereby distracted from the path he has just begun to follow. I know the Kaiser and I am well aware how much the lively and joyful pursuit of an idea he has had captivates him and "makes him happy". Everything now depends on persuading the King of Saxony and the Grand Duke of Baden to make *written guarantees*, in order to consolidate the new alliance round the Kaiser against possible encroachments by the Chancellor.'[88] Before long, however, Eulenburg was dismayed to realise that the monarchical alliance which he had encouraged had led to a kind of 'Anti-Chancellor League'. 'Princes of the blood are prone to join forces and take extreme steps against major domos', he warned in some alarm.[89]

THE WITCH-HUNT AGAINST THE BISMARCKIANS AT COURT

The aim of the clique assembled by Holstein, Eulenburg and Waldersee in support of Wilhelm II was not, in fact, Bismarck's dismissal – the secret advisers were far too conscious of the consequences which a spectacular resignation by the founder of the Reich would entail – but the gradual build-up of the Kaiser's power behind the façade of Bismarck's chancellorship, which they hoped would continue for the time being. To give the Kaiser more freedom in making decisions there were two prerequisites: first, the Bismarckians among the court officials and Flügeladjutanten must be removed from the Kaiser's entourage; and, second, the Bismarcks' monopoly on information must be broken. As the Chancellor had an extraordinarily strong desire for power or, as he himself would put it, a highly developed sense of his responsibility, in addition to a deep conviction that his complicated constitutional creation must in no way be tampered with, these aims were of course far from easy to achieve.

Among the mainstays of Bismarck's power at court were General Wilhelm von Hahnke, the Chief of the Military Cabinet, Hermann von Lucanus, the Chief of the Civil Cabinet, and three of the Flügeladjutanten, Carl Count von Wedel, Gustav von Kessel and Adolf von Bülow. But the key figure was Eduard von Liebenau, the Oberhof- und Hausmarschall.[90] For Waldersee there was no doubt that Liebenau had 'sworn complete allegiance' and 'sold' himself 'to the House of Bismarck'.[91] He had 'curried favour with Count [Herbert] Bismarck from the beginning, he receives gross flattery from that quarter and naturally he always stands up for Count Bismarck with the Kaiser'.[92] For the Bismarck family it was of course 'very pleasant' to have a Hofmarschall who represented 'their interests to the Kaiser'.[93] Vice-Senior Master of Ceremonies August Count zu Eulenburg, a cousin of Philipp, likewise complained at the beginning of 1890 that Liebenau was behaving 'more and more as the agent of Casa Bismarck'.[94] From the autumn of 1889 onwards all possible means were tried to remove Liebenau from the vicinity of the Kaiser. The witch-hunt against this Bismarckian at the Wilhelmine court provides a perfect example of the conspiratorial methods which became common from now on under the new Kaiser's rule.

The lengthy campaign carried out against Liebenau seemed to have achieved its goal in November 1889. During his journey to Athens and Constantinople the Kaiser treated Liebenau badly and on one occasion even 'gave him such a dressing-down that the stupid man' – as Waldersee heard – 'went to the Kaiserin to ask for her protection! To the Kaiserin, towards whom he has behaved uncouthly, whom he has often offended and who loathes him. One can see what a truly wretched fellow he is.'[95] Finding the Kaiser in a 'very irritable mood' against Liebenau on 17 November 1889, Waldersee seized the opportunity of expressing his opinion that the Senior Marshal was 'quite unsuited' to his post. He told the Kaiser frankly: 'He makes trouble with everyone apart from Count [Carl von] Wedel who thinks it to his advantage to stand by Liebenau; there is not *one* person in the entire court who would not have complaints, and justifiable complaints, about him. In society there are protests from the highest to the lowest about his inconsiderate behaviour and his coarse manners. In the end Your Majesty is inevitably harmed by this, because the world believes the inconsiderate conduct in fact stems from Your Majesty. He suffers from delusions of grandeur and thinks he has to be the first person at court and in society. He has no political tact at all . . . I think Y.M. would spare yourself much annoyance if You made a change. The directorship of the Gardens would be quite a good way of getting rid of him.' The Kaiser said nothing in Liebenau's defence, but confirmed some of Waldersee's comments, leaving the latter with the impression 'that a change is not far off, but it might turn into a real rumpus, which I would be happy to see, in the Kaiser's interest. Liebenau does the Kaiser real harm; the sovereign is rather inclined to inconsiderate behaviour himself and he must at all costs have people around him who try to have a softening influence. What is more he [Liebenau] is nothing but a creature of Bismarck and is guilty of the greatest disloyalty towards his real master.'[96]

A week later the Kaiser discussed the matter with August Eulenburg, using Waldersee's arguments almost word for word. Eulenburg immediately hurried to report to Waldersee, who repeated his advice that Liebenau should be relieved of his post as Oberhof- und Hausmarschall and appointed director-general of the Royal Gardens; the 'best successor' to Liebenau would be August Eulenburg himself. 'I asked him just to speak to his cousin Philipp, whom the Kaiser is visiting at the moment and has doubtless told about this business.'[97]

The days of the hapless Oberhofmarschall now seemed to be numbered. 'The Kaiser was . . . very worked up yesterday and the day before about blunders made by Liebenau and is now in a mood in which the latter cannot do anything right in his eyes', Waldersee recorded at the end of November. Wilhelm discussed the situation with Philipp Eulenburg, who had the 'happy idea' of saying to the Kaiser: 'Do you know that Liebenau is in very good odour with the Bismarck family?', which provoked the defiant reply from Wilhelm: 'I care nothing about that, I choose my own Hofmarschall.'[98] Yet although the two Eulenburgs and Waldersee continued to take systematic action against Liebenau, and although Wilhelm repeatedly expressed himself highly dissatisfied with him, Liebenau's fall did not materialise.[99] When Waldersee arrived at the Schloss for lunch on 11 December the Kaiser summoned him in front of the other guests and immediately began discussing Liebenau, who he said had behaved very discourteously towards the ambassador in Paris, Count Münster. 'I reprimanded him severely for his impoliteness only a week ago and gave him a good dressing-down, but it does not seem to help', Wilhelm said. But there was still no sign of any firm intention to get rid of Liebenau.[100]

Over the Christmas period Philipp Eulenburg, on the prompting of his cousin August,[101] brought the Grand Duke and Hinzpeter into the cabal to give it further support. He told them of the Kaiser's complaints that Liebenau 'found a thousand excuses to keep away from him, systematically, all the people whom he, the Kaiser, would like to see at luncheon or dinner!' Yet as it was '*a matter of life and death* for us all that His Majesty should be in frequent contact with men of politics, high officials &c'. 'What a splendid effect contact with an influence of this kind would have on the development of our most beloved Kaiser's great intelligence and quick understanding, his great qualities of leadership!', sighed Wilhelm's bosom friend. Liebenau had to be removed at all costs, Eulenburg insisted. He begged the Grand Duke to provide the Kaiser with further material 'to document the impossibility of keeping Liebenau'; the Kaiser would be grateful to him for this, for he himself was trying to break free from this connection. Hinzpeter too had deplored the '*the bantering manners of an adjutant*' which were the hallmark of Liebenau's exercise of his office. Eulenburg warned the Grand Duke, however, that the Kaiser should not be allowed to detect any 'collaboration' between them.[102]

Even with this additional support, and in spite of several more clashes, Wilhelm still hesitated to dismiss his Oberhofmarschall. When Philipp Eulenburg brought up this vexatious subject again early in 1890 the Kaiser replied: 'The pot is gradually coming to the boil, and I shall find the right moment to get rid of him.'[103] A few days later the Kaiser, in conversation with August Eulenburg, gave vent to 'another very violent outburst against Liebenau'. This time the cause was truly ridiculous: the Kaiser had given Kamilla von Mirbach, the Belgian wife of the Kaiserin's Oberhofmeister, permission to spend the night in her husband's quarters in the Schloss after court parties. Liebenau had protested against this, but under pressure from the Kaiserin the Kaiser had upheld his 'command'. On 18 January the Kaiser received, as August Eulenburg recorded, 'a 20-page report from Liebenau in which he explained the impossibility of carrying out the All-Highest command, emphasising the wrongful use of bed-linen, crockery – His Majesty thought even chamber-pots had been mentioned'. The Kaiser covered Liebenau's memorandum 'with very strong comments' such as 'rubbish, nonsense, ridiculous', so that the Flügeladjutant on duty, Kessel, hesitated to send the original back to Liebenau. He drew up a reply which was dispatched to the Oberhofmarschall with the All-Highest signature. 'The report itself is to remain in the Schloss, in the keeping of the adjutants, for the amusement of its occupants', August Eulenburg mockingly recorded. 'The Kaiser described all this in the most forceful terms, using the strongest expressions, . . . in short he held forth . . . in a way that would have made anyone seeing such a scene for the first time swear that the final break was bound to happen that very day.' But the result had been as disappointing as on all the previous occasions, the Vice-Senior Marshal of Ceremonies complained, 'except that the impossibility of the situation and the damage to the dignity of the court is becoming more and more widely known and is talked about on the street, where people say that Liebenau must have some dangerous knowledge, on account of which His Majesty does not dare put an end to the business'.[104]

The longer the Liebenau crisis lasted, the more people racked their brains to find 'the mysterious reasons . . . which prevented action being taken against Liebenau'.[105] And from the evidence of the Kaiser's ambiguous attitude to him – violent animosity on one side, and indecisiveness and hesitancy on the other – it is indeed not easy to form a clear picture of what was going on. Did feelings of loyalty, gratitude, complacency or fear of awkward secrets coming to light outweigh the pressure from his advisers to sack Liebenau? Did Wilhelm want to maintain his independence from them too? The sources reveal nothing of his inner motives. What is beyond doubt, however, is that Liebenau's influence on Wilhelm in the critical weeks before Bismarck's dismissal was greatly weakened by the numerous attacks on him from all sides.[106] Only after the Chancellor's dismissal was Liebenau willing to recognise how untenable his position at court had become. But even then the Kaiser put off Liebenau's dismissal until May 1890.[107]

Almost as venomous, but in this case completely unsuccessful, was Waldersee's campaign against Cabinet Chiefs von Hahnke and von Lucanus, which also began in the autumn of 1889. The importance of the intermediary role played in the crisis by the Chief of the Military Cabinet is revealed by Herbert Bismarck's remark in December 1889 that Hahnke was 'the most reliable of all the generals', who saw the Kaiser most frequently and to whom he, Herbert, could speak daily.[108] This of course was the very reason why Waldersee's plans required him to be removed from the Kaiser's entourage. On 24 September Herbert Bismarck warned his father: 'As Liebenau told me in the strictest confidence, Waldersee is mounting an assault on Hahnke, who is to be given a Corps, to get him out of the Military Cabinet. Liebenau wants to appeal direct to H.M. not to go along with this.'[109] Six weeks later Waldersee did indeed try to convince the Kaiser that Hahnke was 'not a good chief of Cabinet', had no 'understanding of the significance of his position' and was doing the army 'much damage'. But Wilhelm was determined to keep him and remarked that 'he could not easily do without Hahnke, having just got used to working with him'.[110] Wilhelm's answer by no means satisfied the Chief of the General Staff, who continued his intrigues, convinced that the monarch had not yet seen through Hahnke's weaknesses. The Kaiser was 'definitely not well advised' by Hahnke, who was 'a very mediocre Cabinet Chief', Waldersee wrote. 'Unfortunately the Kaiser still does not see this at all and I am afraid that much damage will be done.'[111]

In mid-January 1890 Waldersee set out to win the support of the Grand Duke of Baden and Philipp Eulenburg in his campaign against Hahnke and Lucanus, as he had done in Liebenau's case. But although the Grand Duke professed himself convinced that 'Lucanus must go because he is the Chancellor's right-hand man, not the Kaiser's',[112] he refused to take any concrete steps against the Chief of the Civil Cabinet. In a letter to Eulenburg of 13 January Waldersee bemoaned the fact that the two officials who were supposed to be the Kaiser's 'right-hand men', namely the chiefs of the Military and Civil Cabinets, were failing to fulfil their duties adequately. Hahnke was certainly a 'thoroughly decent, honourable man', but he was incompetent. He did not have a good command of his responsibilities, '*he does not even fully comprehend them* and, what is now so important, he has no judgement and understanding of important questions and is quite unqualified to give His Majesty advice on serious matters'. Lucanus, on the other hand, was 'an unprincipled man and completely the Chancellor's creature at the moment'. Waldersee pressed for Eulenburg's help in this matter, with a view to enabling the Kaiser to become 'more independent of the Chancellor'.[113] Eulenburg agreed with Waldersee's views in principle but could see no prospect of bringing the two Cabinet Chiefs down, given the good relationship both enjoyed with Wilhelm.[114] He did not add that in these two cases, unlike that of Liebenau, there was no family advantage to be gained for him.

The second important prerequisite for enabling the Kaiser to play a more active role was the breaking of the monopoly on information exercised by the Chancellor. Wilhelm did not even read a daily newspaper, but only extracts which were selected for him in the Foreign Office under Herbert Bismarck's supervision, Holstein complained in December 1889; nor did he read his ambassadors' reports, but heard about their contents, with the appropriate anti-Austrian, anti-Italian and pro-Russian gloss, only through Herbert's oral reports. As a chancellor crisis must at all costs be avoided, one must proceed by small steps, the wily diplomat cautioned, suggesting that 'instead of newspaper extracts, or in addition to these, the Kaiser must *read a whole newspaper*'. Philipp Eulenburg or Hinzpeter would have to persuade him to subscribe either to the Free Conservative *Post* or – better still, because it was less subject to Bismarck's influence – the National Liberal *Kölnische Zeitung*. The *Nationalzeitung* was out of the question, because it was an organ of Bleichröder, Bismarck's banker. If it were too much to expect the Kaiser to read a daily newspaper, he must order the weekly edition of the *Kölnische*, in which at least the most important articles were reprinted. 'But this is the minimum that the sovereign can do. He *must* read *one* newspaper in order to make himself independent of the trends and tendencies which influence the choice of the extracts. This is *urgent*!', Holstein emphasised. It was a matter 'of the *most immense* importance for the Kaiser and for the Reich'.[115] The Grey Eminence of the Wilhelmstrasse failed to reveal that he had close links to the Berlin representative of the *Kölnische Zeitung*, Dr Franz Fischer.

In his subtle, manipulative way, Eulenburg used his New Year letter to the Kaiser to put forward a plea that in order to free himself from the possibility of being influenced by the selection of the newspaper extracts submitted to him, the monarch might perhaps '*read through* one of the more important newspapers each day'. Following Holstein's lead, he advised against the 'narrow-minded Kreuzzeitung', as also the *Post* and the *Nationalzeitung*, and suggested the *Kölnische Zeitung* as the best solution; if this were too long for the Kaiser, however, he could 'at least regularly read the whole weekly edition'.[116] After the Grand Duke of Baden had also hinted to the Kaiser that 'much was being withheld from him', Wilhelm did indeed order the weekly review of the National Liberal *Kölnische Zeitung* on 13 January 1890.[117]

Simultaneously Waldersee too took steps to break through the system by which Bismarck manipulated the Kaiser. In a furious diary entry of 14 January 1890 he recorded: 'A deception of the most flagrant kind is practised by the Chancellor in submitting newspaper extracts to the Kaiser. One would think that the Kaiser had such extracts laid before him because he did not have time to read whole newspapers but nevertheless wished to have an overall picture of the opinions expressed in the press. One would also think that the authorities responsible would have the moral and sacred duty to act with the greatest conscientiousness. But that is a mistake. The Chancellor demands that only those newspaper reports that meet his approval should be laid before the Kaiser. He quite deliberately carries out a shameless deception.' The extracts were selected, Waldersee continued, by Rudolf Lindau in the Foreign Office, 'a completely spineless fellow' (Waldersee's choice of words indicates that Lindau was of Jewish extraction), and by Konstantin Rössler in the Prussian Ministry of State, 'an elderly civil servant and a good man' who in spite of inner struggles was in the end obliged to submit to the Bismarcks.[118]

On 18 January 1890 Waldersee made use of his scheduled audience as chief of the General Staff to show the Kaiser a large number of extracts from Russian newspapers and to tell him that he was 'in a good position to do this more often', even if it aroused the Chancellor's hostility towards him. Wilhelm replied: 'Let that be my concern; from now on I wish you to bring me Russian newspaper extracts every week.' It was apparent from the tone of the whole conversation, the General reported, 'that there is great ill-feeling towards the Chancellor'.[119] At the end of January the Kaiser openly told Adjutant-General Max von Versen that he was having Russian newspapers submitted to him through the General Staff, since he received 'nothing or only mangled reports' from the Foreign Office.[120]

Bismarck's attempts to control the information on foreign affairs which reached Wilhelm II were under threat not only from Waldersee's extracts from Russian newspapers. The confidential reports sent by military attachés to the General Staff or directly to the Kaiser led to violent arguments with the Chancellor and the Foreign Secretary, especially as Waldersee was aiming to build up this military correspondence into a kind of rival diplomacy, and Wilhelm had no scruples in expressly allowing some military attachés to write to him personally. At the end of 1889 both Waldersee and Verdy predicted that Bismarck would soon try to undermine the position of the military attaché in Paris, Freiherr von Huene, who had been conducting a secret military-political correspondence with the Kaiser for the past year. As Waldersee explained, 'Count [Carl] Wedel found out that Huene has written to me . . . a few times and that I gave the letters to the Kaiser to read, and unfortunately he told the Chancellor; it is really a scandal and a betrayal of his master, whom he should serve exclusively. He [Wedel] even knows that the Kaiser has invited Huene to write confidentially to him directly. Whether he approves of this or not he ought not to tell anyone about it. In his suspicious way the Chancellor has now built himself up a great framework of fantasies; he thinks I run a regular political bureau in which correspondence with all the military attachés plays a central role.'[121]

However serious a threat Waldersee and his military attachés posed, however, it was not they but the clique led by Eulenburg, Holstein, Hinzpeter, Marschall von Bieberstein and the Grand Duke of Baden who represented the greatest danger for the Bismarcks in the final phase of the dismissal crisis, in which the principal point at issue was not foreign policy but the future course of domestic politics.

CHAPTER ELEVEN

Constitutional conflicts

BISMARCK'S PLANS FOR A *COUP D'ETAT*

IN May 1889, at the time of the serious disturbances among the Rhineland coal miners, the Kaiser had whispered to Philipp Eulenburg during his first official visit to Brunswick: 'I am having fearful difficulties with the Prince; constitutional change and other things.'[1] Subsequently, however, this problem, important though it was, had receded into the background again. During the winter months Wilhelm had made frequent and belligerent comments about the socialist threat, so that until December 1889 there was very little perceptible difference between his views on 'the social question' and those of the Chancellor. Towards the end of November Waldersee noted that on all sides the opinion was current 'that social democracy is making great strides and that, as massive strikes were to be expected next year, there might easily be bloodshed'.[2] On 24 November he recorded that Wilhelm II was 'not yet at all anxious' about the internal political crises which were clearly in the offing. He had recently remarked: 'My grandfather reigned for several years while there was constitutional conflict, and I shall be able to manage it too.' The General pointed out that circumstances had become infinitely more complicated since the Prussian conflict of the 1860s. 'A conflict in Prussia would not greatly alarm me either', he commented. 'But in the Reich it is highly dangerous.'[3] He admitted, however, that 'serious people' were 'very worried about the way we are developing and saying: "It cannot go on like this."'[4]

Only the recognition that a massive political clash would greatly strengthen Bismarck's position deterred Wilhelm, Waldersee and a few of the Kaiser's other advisers from pursuing the idea of a *coup d'état*, which otherwise seemed to them highly desirable. In the turmoil of a constitutional conflict or even a bloody *coup d'état* it would be impossible to manage without the founder of the Reich, who would then be able to impose harsh conditions on the Kaiser.[5] Waldersee suspected Bismarck of making secret plans along these very lines as early as mid-December 1889, when the Kaiser told him of the Chancellor's intention of strengthening the Berlin garrison 'in case of socialist uprisings'. The Chief of the General Staff thought this idea 'utterly foolish' and expressed the view to the Kaiser that 'we really have not come to that yet; there can be absolutely no question of uprisings yet; conditions would have to be very different for that to happen'. On the contrary, it was in fact the main aim of the Socialists 'to remain absolutely quiet'. In any case, Waldersee pointed out, there was no room for an increased military presence in the barracks of Berlin. He could see, however, that Bismarck's idea had 'made an impression' on the Kaiser.[6] On 26 December he remarked that in making such suggestions Bismarck was obviously trying 'to frighten the Kaiser and make him mistrust the workers, because in his view the Kaiser took the workers' side too much over the strike. At any rate the big industrialists, and in particular Herr Bleichröder, have been working on the Chancellor not to give way to the workers any more', the General claimed.[7] At the beginning of 1890 he noted that 'there is a difference of opinion just now between the Kaiser and the Chancellor over the handling of the strikes, and of the social question in general. The Kaiser favours the workers too much for the Chancellor. I have heard it said that the Chancellor would like the strikes to reach the point where the troops will intervene and shoot, so as to bring the Kaiser round. It is quite clear that he is trying to frighten him with socialist disturbances.'[8]

These differences of opinion had already become apparent in the previous weeks. In the negotiations with the Kartell parties over the renewal of the Anti-Socialist Bill, which Bismarck had made still more severe by adding an expulsion clause, the Chancellor gave vent to his deep displeasure with the National Liberals, who were prepared to support the extension only of a watered-down version of the bill beyond 1890. As the Conservatives, by contrast, welcomed the strengthening of the bill, the Kartell, with which the Kaiser had repeatedly and publicly identified himself, threatened to split apart over this question.[9] When Otto von Helldorff, the leader of the moderate Conservatives, went to Friedrichsruh for discussions at the end of November 1889, he found Bismarck 'in an angry mood over the attitude of the National Liberals to the Anti-Socialist Bill question'. Helldorff urged the Chancellor 'very strongly' 'to show forbearance to the National Liberals so as not to destroy the Kartell'.[10] Only when the three participating parties made a joint declaration, deferring the question of the renewal of the Anti-Socialist Bill until after the Reichstag elections, did the Kartell crisis seem to have been overcome.[11] But the conflicts within the coalition soon broke out again, fanned by the power struggle between the Kaiser and the Chancellor.

On 13 January 1890 the Kaiser took Philipp Eulenburg for a two-hour walk through the Tiergarten in the rain, and poured out his heart to him. 'I have scarcely ever discussed such important and far-reaching matters with the Kaiser before', Eulenburg commented afterwards. 'To have such wholehearted trust placed in me made me happy – but how painful it is to have to see him in difficulties that would dishearten the most experienced of men.' The Kaiser complained that 'the Chancellor wanted a *strong Anti-Socialist Bill*, which in itself was not a bad thing, but its severity would certainly lead to a conflict, which would – yes, is perhaps *intended to* – bring a constitutional change in its wake. He, the Kaiser, was in a really dreadful position, for he thought it dangerous to begin his reign with a sort of revolution, shooting and other measures of force.' For Eulenburg, who had conferred at length with Holstein, Hinzpeter and August Eulenburg in the preceding days, Bismarck's 'secret' purpose was no mystery. The Chancellor wanted to 'keep the reins of government and *the young Kaiser in his hand*', and had calculated 'that through a constitutional change with all the attendant furore, the consequences, the continual feuds – in a word: that through the *helplessness* to which this would reduce the Kaiser, the latter would certainly very soon find himself actually dependent on the Chancellor'. Eulenburg therefore advised the Kaiser that 'how he wanted to deal with the Social Democrats *eventually* was his affair; but I considered it extremely unwise for his *first* significant act of government to be a constitutional change – that is, in this case, a *coup d'état* – in the Reich. That would be to inflict upon the expectant German people a severe disappointment whose effects would be lasting, and to intensify the *opposition* which has always existed between Prussia and the whole of West and South Germany – to become a complete foreigner as Kaiser there.' The Kaiser, Eulenburg recalled, replied by saying that he wished 'to show the people, and especially the workmen, my good will, and to help them, but I have no intention of shooting at them!'[12] A year after Bismarck's dismissal Wilhelm II recalled the dilemma of those crisis-filled weeks in a marginal note. 'When the Prince demanded that I dissolve the Reichstag, let the masses go on strike and rise up in revolt, and then fire on them, I declared that there was time enough for that, but that some of their demands were justified, and these would have to be satisfied first. Then I should have "a clear conscience", and give the orders to shoot if necessary.'[13]

WILHELM'S SOCIAL WELFARE PLANS

These differences, at first sight more tactical than fundamental, over the handling of the social democratic threat became potentially explosive differences of principle from the beginning of 1890, when Kaiser Wilhelm II, under the influence of Hinzpeter, threw himself enthusiastically behind a comprehensive programme for social welfare. Greeted with universal astonishment, the programme promised – as was no doubt Wilhelm's main aim – to make him immensely popular with the people. But to combine this with the kind of *coup d'état* policy demanded by the Reich Chancellor would not only torpedo the programme but also expose the young Kaiser to universal public ridicule. In the wake of the crucial struggle for power which now began between Kaiser and Chancellor, Wilhelm's initially hesitant and unfocused ideas on social reform acquired a radicalism that went far beyond his original intentions.[14]

The former tutor's pro-labour influence on the Kaiser in this difficult area was deplored by almost everyone else in the group of advisers around Wilhelm, and criticised as one-sided and doctrinaire. Holstein, who had recruited Hinzpeter as an ally in the battle to save the Kartell, now described him as 'an undirected force of nature', 'a pike in the fishpond, nothing more, for he has no political judgement'. He reproached Hinzpeter with wanting 'to achieve fame at the Kaiser's side through the social question'.[15] The Empress Frederick took much the same view, lamenting that she had to stand silently by while her own son plunged rashly into things that he did not understand. 'He listens to Hintzpeter [sic] on these subjects, who is a very passionate "Christian Socialist". His ideas are very good to listen to, but alas too doctrinaire & theoretical to be the *only* ones to go by! After all Hintzpeter is *not* a political man by profession; he is kind & charitable to the poor, but he has a *one* sided view of the question.'[16] The tutor had won Wilhelm over to his social reform programme, she believed, with the argument that 'he would find it "a mine of popularity for himself", & it would make him a gt. man etc.'[17] Queen Victoria forwarded these comments to Lord Salisbury, whose reaction was similar: 'If the Empress is right in thinking that he [Wilhelm] has fallen under the guidance of two or three fanciful favourites, there are dangerous times in front of him. He is meddling with very sharp edge tools indeed.'[18] To his own surprise Waldersee, like the Empress Frederick, found himself on Bismarck's side in this question. The latter, as the General recorded on 20 January, was 'furious' that Hinzpeter 'is advising the Kaiser on the labour question. The Chancellor's view is that there should not be too great concessions to the workers, and I think he is right.'[19] On 8 February Waldersee had to admit that 'this time, to my sincere regret, I am more on the side of the Chancellor'.[20] Many years later he recalled the ambiguity of the Kaiser's attitude at this critical stage, recording that he had told Wilhelm at the beginning of 1890 that 'he would not have any success with his ideas of social reform, for in their greed people would ask for more and more and would never be satisfied; he replied that I might well be right, but he considered it his duty to try everything;

if people were still not satisfied, then – and he launched into threats of what he would do to punish them'.[21]

At the Kaiser's request Hinzpeter had a long conversation with the Chancellor at the end of January 1890 on Wilhelm's ideas for workers' welfare, in the course of which Bismarck contemptuously remarked, revealing his annoyance at the extent of Hinzpeter's influence, 'I shall propose you as Reich Chancellor.'[22] Hinzpeter certainly considered the social welfare question of vital importance to the whole nation,[23] but he justified his role in spurring the Kaiser on, as his adviser on the subject, with a strange mixture of pedagogical ambition and blindly monarchist zeal. As Eulenburg admitted to Holstein, the tutor was combining what was in itself the very laudable idea of 'making the Kaiser *work*' with the social welfare question.[24] Hinzpeter and he, Eulenburg, were striving to foster and develop the 'self-reliance' of the 'most beloved' Kaiser, and that could be achieved to some extent by his playing his part confidently in the social welfare question.[25] In his own letters Hinzpeter openly admitted that his first concern was for the 'spiritual salvation' of the Kaiser; the welfare question itself was of secondary importance to him.[26] After the first clash between Wilhelm and Bismarck over the social welfare question, Hinzpeter wrote to Eulenburg saying that he had 'pitied and at the same time admired the object of our mutual concern' – meaning of course the Kaiser. 'Bitter feelings were aroused in the principal person [Wilhelm II], not excluding that of humiliation, and the half victory for which he fought was dearly bought. But on the whole I believe that he has come out of this first severe contest stronger, more confident and also wiser. As this last was especially desirable, he has at least achieved some personal progress . . . Besides everything else it was also a spiritual crisis', the tutor claimed.[27] In his correspondence with Eulenburg, Hinzpeter referred to the Kaiser as the 'master' and to the Chancellor as the 'servant'.[28] The Chancellor's attitude, Hinzpeter discovered during his conversation with Bismarck, was that 'the master had no right to order important measures to be taken without consulting his servant and without having secured his consent: a surprise attack with the help of other authorities was inadmissible'.[29]

Prompted by Hinzpeter and with Eulenburg's support, Wilhelm II had persuaded the King of Saxony, who had come to Berlin for the funeral of Kaiserin Augusta, to put forward a proposal to the Bundesrat for a Social Welfare Bill. At the same time he ordered Boetticher to cast the Prussian vote in the Bundesrat for the Saxon proposal.[30] With that, open conflict with the Reich Chancellor was a foregone conclusion, since the Chancellor bore sole responsibility for Reich policy, while in his capacity as Prussian minister of foreign affairs he alone had the right to direct the Prussian vote in the Bundesrat. 'What do you think of my position?', Boetticher complained in despair. 'The Chancellor tells me to refuse, the Kaiser tells me to agree.'[31] The pro-Bismarck Bavarian envoy, Count Lerchenfeld, reported scathingly to Munich: 'It is not yet known how the Chancellor has reacted to the fact that his sovereign has ordered a draft bill which runs counter to his, the Chancellor's, convictions, from another federal prince.'[32] Eulenburg, however, worked himself up into a fever over the 'newly formed monarchical alliance' of the Kaiser with the King of Saxony and the Grand Duke of Baden, which he hoped would strengthen the Kaiser's power significantly.[33] In order to bring Bavaria into the new alliance as well, Eulenburg wrote on 14 January to the influential Bavarian Adjutant-General, Freiherr von Freyschlag, the confidant of Prince Regent Luitpold, assuring him that the Kaiser was as committed as ever to the Kartell and would therefore be 'painfully affected' if Bavaria were to agree in the Bundesrat to the strengthening of the Anti-Socialist Bill as desired by Bismarck, which the National Liberals could not accept. Both in this letter and in another to Minister-President Lutz he urged the Bavarian state government 'in terms which could not be misunderstood' to support the measures planned by the Kaiser in the sphere of social welfare legislation, including the forthcoming Saxon proposal.[34] 'The federal governments are aware of Your Majesty's views on the labour question', Eulenburg assured the Kaiser on 20 January.[35]

CONFLICT IN THE CROWN COUNCIL

As the Reichstag had to be dissolved on 25 January 1890 to enable the elections to take place, the crisis over the way forward – *coup d'état* or reform programme – came to a head rapidly. Only if the Chancellor stayed away until after the dissolution of the Reichstag, and only if the Anti-Socialist Bill were modified or postponed until after the elections, in the view of the Kaiser's advisers, could conflict between the Kaiser and the Chancellor be held off.[36] After a conversation with Holstein on 19 January Marschall noted in his diary: 'The Kaiser must not give way.' Discussing the situation with the Grand Duke of Baden he expressed the fear 'that the Reich Chancellor . . . wants to crush the Kaiser. The Kaiser must win the battle now.' The Grand Duke also became convinced, as a result of a meeting with Herbert Bismarck on 21 January, that the Reich Chancellor was planning 'to use force against the workers'. The Kaiser, however, was standing firm; it was his 'unalterable will' either to postpone the Anti-Socialist Bill until after the Reichstag elections, or to amend it in such a way as to make it acceptable to all three Kartell parties.[37] A collision between Kaiser and Chancellor, who were pursuing diametrically opposed programmes, seemed inevitable and indeed imminent, particularly as Bismarck was planning to return to Berlin after an absence of eight months.

On 22 January the Kaiser took Waldersee for a walk in the Tiergarten and again discussed the Anti-Socialist Bill with him. Wilhelm repeated his view that the bill should be brought before the next Reichstag, not the present one. The two men called on Herbert Bismarck and raised the question with him, only to receive the response: 'My father is coming tomorrow and he will deal with the Anti-Socialist Bill.' The Kaiser replied: 'I have heard nothing about that, and would consider it most deplorable; I thought that it was to be left to the new Reichstag when it is elected. Just now it can only cause harm and we shall ruin our chances in the elections.' Herbert hinted that his father was prepared to make this a resignation issue and said: 'My father is very experienced in these things and I think it will be as well to follow his advice.' As Waldersee noted, however, 'the Kaiser was of course not to be won over, and the majority of the Reichstag is probably on his side'.[38] Late on the morning of 23 January, a Thursday, Boetticher, as Vice-President of the Prussian State Ministry, was informed by one of the Flügeladjutanten that the Kaiser had ordered a council of ministers at which he would preside – in other words, a Crown Council – for the following evening at 6 o'clock in the royal palace. The agenda for the meeting was not divulged.[39] Herbert Bismarck and the other ministers could only suppose that the Kaiser wished to discuss social policy issues at this Crown Council.[40] Herbert telegraphed his father asking whether he would wish to request an audience with the Kaiser before the Crown Council, to which the Chancellor answered in the affirmative and prepared to leave for Berlin at once.[41] The Foreign Secretary thereupon decided to go to the Kaiser himself in order to clarify matters for his father, and to ask 'what the Council was to consider' and whether the Chancellor's presence was desired. The Kaiser replied, 'I have summoned the Council in order to put my ideas about the handling of the labour question to the ministers; if your father wishes to take part, I shall be very pleased.' Herbert then asked whether the Prince might be granted an audience a quarter of an hour before the Crown Council, to which the Kaiser agreed 'most readily'. In his telegram to Friedrichsruh recommending immediate departure for Berlin, Herbert expressed the optimistic belief that it would be 'not at all difficult' for the Chancellor to come to an agreement privately with the Kaiser on the social welfare question; moreover it was surely important that he should 'hear for himself what H.M. has to say to the ministers, and *how*'. If, on the other hand, he was not concerned with how the session went, the Chancellor could put off coming to Berlin until Saturday. But Herbert warned his father that, if experience were anything to go by, in the Kaiser's presence the other ministers would all 'remain silent, or agree with reservations to a greater or lesser degree'.[42]

Bismarck's arrival in Berlin on 24 January marked the beginning of the last phase of the crisis which was to end seven weeks later with the angry resignation of both Bismarcks. Everyone sensed the gathering storm. On the eve of his father's arrival Herbert Bismarck spent two hours in a state of great agitation at Holstein's bedside, the latter being ill with influenza at the time.[43] Eulenburg, smitten with the same illness in North Germany, commented that the thought of the crisis in Berlin gave him 'the oppressive feeling of an approaching thunderstorm'.[44] The very first act of the elderly Prince on reaching Berlin demonstrated that he was in no mood to shirk the struggle for power with the Kaiser. Before his audience he called the Ministry of State together for a 'confidential' consultation and made the ministers promise to support him against the monarch in the Crown Council. Waldersee recorded later that Bismarck had been 'very unfriendly towards his colleagues' during the meeting,[45] but from other sources we know that 'almost all the ministers' were deeply discontented with the Kaiser's treatment of them, and some of them were in any case thinking of resigning. Minister for Ecclesiastical Affairs Gossler had complained at length to Herbert Bismarck about the Kaiser a few days earlier, and had declared only at the last minute that he had decided to withdraw his threat to resign. Public Works Minister von Maybach also continued to feel 'piqued with H.M.', and in constant anxiety that he might 'receive a rude note' from the Kaiser. Both ministers supported Bismarck and thought that the Kaiser's attitude to the Ministry of State would in any case make it impossible for them to stay on much longer.[46] Nevertheless the ministers found themselves facing an acute dilemma. Although they were not at all accustomed to defying the all-powerful founder of the Reich, as non-political government officials they were deeply imbued with the monarchical principle, and to reject the express wish of the King of Prussia seemed to them simply inconceivable. How were they to vote in the forthcoming Crown Council?

Immediately before the Council met at 6 p.m. Bismarck had his audience with the Kaiser. The Prussian ministers – except Maybach, who did not attend – then assembled and Wilhelm II took the chair. He opened the session with a lengthy exposition contrasting the rapid industrialisation of Germany with the slower economic development of Britain. The rapid industrialisation of Germany had led to the exploitation and proletarianisation of the workforce in the major cities, which in turn had made them receptive to the doctrines of social democracy. Since almost all revolutions arose from the failure to make timely reforms, he wished to issue a proclamation to the Ministry of State demonstrating his 'heartfelt interest in the welfare of the workers' and announcing improvements in the sphere of Sunday and night work and of women's and children's work. He wished to introduce workers' committees in conjunction with government-run factory inspections, to set up arbitration bureaus for strike situations, and to build savings banks, churches, schools and hospitals for workers. He suggested that an international agreement on these questions should be sought and proposed that a congress be convened in Berlin with representatives from all industrialised countries. Finally he asked Boetticher to read out a programme written in his own hand, setting out all the points he had mentioned.[47]

How had Wilhelm II suddenly got hold of this astonishingly modern, indeed ground-breaking programme, that was criticised even by his liberal mother as 'state socialism'?[48] Immediately after Bismarck's dismissal the Kaiser called on the British ambassador, Malet, and claimed that he had written the document himself. He had certainly had 'the advantage of the advice of able and disinterested men', but the social policy programme was his own work. 'One night I sat up for two hours by myself and wrote down my views as to what ought to be done.' The Kaiser emphatically denied the widely believed rumour 'that Dr Hinzpeter was the author of that document'; Hinzpeter had seen the programme for the first time when Bismarck showed it to him at the end of January.[49] In the Crown Council itself, in addition to Hinzpeter Wilhelm named his advisers on this question as Hans Hermann Freiherr von Berlepsch, Oberpräsident of the Rhine Province, Count Hugo von Douglas and the painter and former director of mines August von Heyden, and these four names quickly found their way into the contemporary press and into Bismarck's memoirs, and thence into countless works of history until the present day.[50] In fact, however, no indications have been found that these alleged advisers influenced the Kaiser on social policy questions, with the sole exception of Hinzpeter. In all likelihood Wilhelm mentioned these names only to distract attention from those who were actually responsible. For the same reason he copied out in his own hand the memorandum which he brought to the Crown Council and had read out there. In reality the author of the famous imperial social welfare programme was not the Kaiser but a man named Kayser.

Dr Paul Kayser, one of the very few (baptised) Jews who had achieved acceptance in the highest ranks of the Reich administration, had been tutor in the 1870s to Philipp Eulenburg and Axel Varnbüler and also to Bismarck's two sons. Since then he had risen to become Director of the Colonial Department, at that time still a subdivision of the Foreign Office.[51] It was there that Eulenburg sought him out on 13 January, immediately after his memorable walk in the Tiergarten with the Kaiser, and asked him to draft 'an exposé' on the social question for the Kaiser, 'as quickly as possible'.[52] When Holstein heard of this he approached the journalist Dr Franz Fischer of the *Kölnische Zeitung* with the request that he too draw up a social policy programme and send it secretly to Eulenburg.[53]

Two days later both Kayser and Fischer had completed their reform plans. Both sent their work to Eulenburg on 15 January, asking not to be named as authors, and Eulenburg forwarded Kayser's exposé to Wilhelm II on 20 January.[54] (It is not clear whether he also sent Fischer's project.) It is an incontestable fact that the Kaiser copied out Kayser's exposé in full, for the Colonial Director's social programme is identical, word for word, with the 'Proposal by H.M. the Kaiser on the labour question' of 22 January which Wilhelm brought to the Crown Council with two other memoranda and had read out by Boetticher. The Bismarcks never found out that the programme had been worked out under their noses in the Foreign Office by one of their closest colleagues. It was not until the 1970s, when Eulenburg's correspondence was published, that the true identity of the author came to light.[55]

After the ministers had resolved (as arranged in their prior meeting with Bismarck) to examine the various social policy proposals in the memorandum and to draft the text of an imperial proclamation, the Kaiser moved on to the second part of the agenda, the Anti-Socialist Bill. 'In a thorough and well-reasoned manner', he expounded his view that the government ought to make a formal declaration that the Anti-Socialist Bill was acceptable to it even without the controversial expulsion clause, for only through such a declaration could the consent of the German Conservative Party be obtained and the Kartell prevented from disintegrating. Without the declaration not only would the Anti-Socialist Bill be lost, but the Kartell would be in complete disarray in the forthcoming elections. When the Kaiser had finished speaking, Bismarck rose to his feet and declared bluntly, without entering into the monarch's arguments, that in these circumstances he had no alternative but to ask to be relieved of all his offices, as he could not agree with the Kaiser's views. Boetticher later recalled that the Chancellor had spoken 'particularly vehemently', and had become increasingly heated in his exchanges with the Kaiser, finally remarking with great bitterness: 'I see more and more that this is no longer the place for me!' The Kaiser 'remained calm and measured', and asked each of the ministers in turn for his opinion. All supported the Reich Chancellor's view. The Kaiser was forced to give in, with the result that on the following day the Anti-Socialist Bill was rejected by 167 votes to 98; the German Conservatives voted with the Centre Party and the Social Democrats against the bill, and the Kartell entered the election campaign in a badly damaged state.[56] The Kaiser complained bitterly afterwards that 'the ministers are not my ministers, they are Prince Bismarck's ministers'.[57]

As Wilhelm insisted, against Bismarck's express wish, on dissolving the Reichstag with a speech from the throne in the White Hall of the Schloss, expressing his warm personal interest in the workers,[58] the conflict between Chancellor and Kaiser was plain for all to see. The mere fact that Bismarck stayed away from the ceremony was enough to demonstrate the rift.[59] 'Great excitement', Marschall noted in his diary that day. 'No one understands the

Reich Chancellor.'[60] Waldersee observed that 'the antagonism between Kaiser and Chancellor is growing. It is caused . . . by the fact that the Kaiser wishes to go on trying to calm the workers by making concessions and showing his interest in them, while the Chancellor thinks there should be no more giving way; instead, the workers must be shown the greatest severity, including firing on them.'[61] The Kaiser was indignant, Waldersee recorded, that Bismarck 'has no qualms about misusing the army to fire on the workers, and he thinks that in this matter too the Chancellor wants to rob him of the glory of having settled this very important question himself'.[62] With this remark the Chief of the General Staff had shrewdly hit upon the core of the conflict between the Kaiser and the Chancellor. Neither was concerned about the issue itself. It had long since become a pawn in their contest for decisive power in the Reich.

TOWARDS A NEW SYSTEM

Although the Chancellor's behaviour on 24 and 25 January caused passions to run high – for days Wilhelm 'repeatedly let fly with furious complaints about it'[63] – none of the Kaiser's advisers was in favour of dismissing Bismarck at this juncture. They were all still agreed that a grand exit by the founder of the Reich would cause great harm to the monarchy; they also realised that, although the personal prestige of the young Kaiser was growing, it had not yet reached the point at which personal rule by the monarch would be politically acceptable. Holstein summed up his views in a letter to Eulenburg on 27 January: 'The Kaiser has shown magnificent self-control. He was right not to let the Chancellor go yet. His Majesty must first create a personal position for himself; this is by no means identical with the inherited position of sovereign.'[64] The Kaiser had rightly perceived that time was on his side, not the Chancellor's; he would soon – in months, not years – have everything for himself.[65] The relationship between Wilhelm and Bismarck was therefore, according to Holstein, 'a question of *time*'. On 4 February he stressed the point again: 'There will always be opportunities for a break. *Today* it is too early. Today the Kaiser is not yet well enough known among the people. But he is rapidly gaining in stature and the Chancellor is losing.' Wilhelm had realised that 'it is not enough to be Kaiser in order to rule; one must also be a man of character . . . In the present difficult situation the Kaiser is learning more and more each day. He has already achieved considerable self-control in the past ten days.'[66]

All the Kaiser's advisers shared Holstein's opinion. The Baden envoy, Marschall, recognised the vital importance of avoiding a break with Bismarck before the Reichstag elections.[67] Holstein came to the conclusion – wrongly, as it turned out – that Hinzpeter had advised the Kaiser to break with Bismarck

before the elections so as to be able to win a triumphant victory by making a proclamation to the people, and was lost for words to describe the apparent stupidity of this 'schoolmaster who has gone mad'.[68] Hinzpeter, however, was also firmly convinced of the necessity of creating 'the possibility of further co-operation on a basis of trust, through mutual concessions' between Bismarck and the Kaiser.[69] As Eulenburg was able to assure Holstein, he was 'profoundly convinced that the Kaiser [was] *not capable of surviving without* the Chancellor'. 'He did not commit the *folly* of advising the Kaiser to dismiss the Chancellor before the elections', Eulenburg continued. 'Such thoughts may arise in the *unpolitical* minds of those who hate the Chancellor, military or princely minds.' Eulenburg himself was also working 'with all my strength to keep the two "great men" together', not least because he had realised that Bismarck's dismissal – for whatever reasons – would 'still be regarded by the people as a terrible thing to happen'.[70]

Even Paul Kayser, the author of the social welfare programme, could see that keeping Bismarck in office was more important than the Kaiser's social policy initiative. 'If the Chancellor goes before the elections the Kartell cannot be saved; its individual components will be sent spinning around like atoms in space. It would not matter in the least . . . for what reason he left; the fact itself would be enough to unsettle the whole population of the Reich.' It was significant, the Colonial Director warned, that 'even quite outspoken political opponents of the Chancellor' were now afraid 'that he might go very soon'.[71] He no doubt had Waldersee in mind, and it was indeed true that, although the Chief of the General Staff considered the dismissal of Bismarck inevitable in the long run, he did not think the moment yet ripe. On 12 January 1890, when for the first time in his diary he posed the question of how the dismissal of the Chancellor should be handled tactically, he commented: 'It is essential to give the Kaiser the right advice, so that he is not outmanoeuvred; however much he may have declined, the Chancellor is still a clever and very cunning man. His strength always was in his ability to find a way out of difficult situations. The break must at all costs be handled in such a way that it does not look to the world as if the Kaiser had wanted to get rid of the Chancellor.'[72] In a letter of 13 January the General commented that, if Bismarck noticed that 'a new era is about to begin, he will try to get out altogether', but he would try to ensure the succession for his son, 'and this, I think, will give His Majesty the chance to play for time . . . For as long as it can be managed the Chancellor must be kept in office; if he should in fact want to break away . . . he will choose the moment in such a way as to keep public opinion on his side. This must not be allowed; the goal must be to achieve the opposite effect.'[73] Now, after the clash of 24 January, the General expressed his admiration for the Kaiser in resisting the temptation

to turn the Reich Chancellor out on the spot. 'In fact it is almost incredible', he observed, 'that he should be willing to continue working with a man who he believes is capable of bringing about bloodbaths simply to safeguard his own reputation! . . . I do not think that this will go on for much longer, but I am in favour of it and I am trying . . . to make sure that the Kaiser does not allow the Chancellor to go or send him away, but waits for the elections to take place first!'[74]

In adopting this attitude the Kaiser's secret advisers were nevertheless assuming that Wilhelm would be able to determine many matters of policy even with Bismarck as chancellor. They privately hoped that Bismarck would withdraw into the domain of foreign affairs, which would at least give the Kaiser room to manoeuvre in questions of domestic policy.[75] Eulenburg accurately summarised their principal aim in stating that Bismarck must be kept on for the time being, but that '*at the same time* [we must] make the Kaiser *independent* . . . because we *need* him to be independent for what lies before us when the old Chancellor is living out his dotage'.[76] In this concept of a reduced degree of power for the Reich Chancellor the characteristics of the later Wilhelmine regime were already beginning to appear in the last stage of the Bismarck era. As we shall see, it was precisely Bismarck's refusal to accept such a system that finally provoked the Kaiser into dismissing him.

In their optimistic assessment of the situation Wilhelm's advisers were encouraged by the thought that the young monarch had after all managed to achieve a considerable amount, despite the opposition of the Iron Chancellor. He had closed the Reichstag with a speech expressing support for the workers and had set in train ministerial consultations on his social welfare programme. At a session of the Ministry of State on 26 January Bismarck gave orders for an imperial proclamation to be drawn up in accordance with the Kaiser's proposals, and he himself took on the responsibility of convening an international congress.[77] Though Wilhelm's wish to appoint Johannes von Miquel as Prussian minister of trade was thwarted, he did succeed in obtaining the appointment for Oberpräsident Hans Hermann Freiherr von Berlepsch, whom Bismarck had bitterly criticised for his pro-labour attitude in the crisis caused by the strikes of the previous year. The promotion of Berlepsch was the first ministerial appointment in the civilian domain that was entirely due to Wilhelm II's initiative; it was Berlepsch's express responsibility to put the Kaiser's social welfare programme into action. The Kaiser's success was all the more striking in that the Ministry of Trade had to this point been administered by Bismarck himself. What was more, the new minister's responsibilities were extended by the transfer of the Departments of Mining, Metallurgy and Saltworks from Maybach's Ministry of Public Works to the Ministry of Trade and Commerce.[78]

To develop this new system further the Kaiser now decided to receive the departmental ministers and other officials individually once a week or once a fortnight to hear their reports.[79] Like Holstein, Eulenburg thought this decision 'all the more admirable' because at the Crown Council on 24 January the ministers had presented 'a wretched display of lack of independence'; but he acknowledged that as a result 'the question of Chancellor or Kaiser' had become much more acute.[80] On 31 January Wilhelm arrived unannounced at a meeting of the Ministry of State and enquired about the progress of the consultations on his social welfare programme; he expressed satisfaction at Bismarck's response that the ministers had just resolved to draft two proclamations – one to the Reich Chancellor on the international congress to be convened, the other to the ministers of public works and of trade and commerce.[81] The 'little Kayser' too was able to note with satisfaction that through this resolution the proposals that he had submitted to the monarch in his secret memorandum would now be carried out. He wrote delightedly to Eulenburg, 'I think the success of the Kaiser's initiative is remarkable; in my opinion it has achieved the right thing in objective terms, and if the Proclamations prove to be noble in form as well as content, I am convinced that we shall soon see hymns of praise to the Kaiser in all the newspapers both here and abroad . . . I see the forthcoming proclamations of Kaiser Wilhelm II as the new leitmotiv which we need.'[82] The author of the second memorandum written for Wilhelm on the social welfare question, Franz Fischer, took a similar view: 'Our youthful Kaiser has indeed taken a great, bold step by his courageous, wise intervention in these questions, and he will surely succeed.'[83] When the two imperial proclamations were published – significantly without Bismarck's countersignature – on 4 February, Kayser's jubilation over the 'phenomenal success' of the young monarch knew no bounds, especially as he recognised in the proclamation to the Prussian ministers all the ideas he had included in his own programme. Bismarck would be forced to acknowledge, he thought, 'that *he* was blind and the Kaiser could see. All the successes and all the recognition which he could at least have shared with his sovereign will now come to the latter alone. Things will get better still. The press has not yet entered into it in enough detail. It is still gazing with amazement at the mere fact, like some traveller high in the Alps when the mist clears and he is blinded by the shining landscape in the valley below.'[84] Eulenburg was no less enthusiastic. He wrote to Paul Kayser: 'How wonderfully have we been guided into a way which leads to the noblest of goals: to the salvation of the poor and the wretched! The spark which *at best* glows in our hearts has become a burning flame in the heart of the beloved Kaiser – we should thank God for it from the depths of our souls.'[85] He sent fervent congratulations to Wilhelm II: 'This magnificent success will richly repay Your Majesty for much that has been difficult in recent days – but

Your Majesty's personal success has been so great that it gives me some concern to imagine the mood of the Chancellor and to ponder the consequences.' He sent the Kaiser extracts from the 'little' Kayser's letters and suggested that Kayser be appointed under-secretary of the Prussian Council of State which was to be convened to consider social welfare legislation, as announced in the proclamations. Kayser, he argued, was not only 'the best-qualified and most knowledgeable' man in this sphere, but was also 'congenial and useful' and above all 'not independent by nature', so he would 'simply implement Your Majesty's thoughts'.[86]

A few days later Bismarck and the Kaiser met to decide on the composition of the Council of State. Wilhelm agreed that Dr Robert Bosse, the under-secretary of state in the Reich Office of the Interior, who had been suggested by Bismarck and the ministers, should be appointed secretary of state of the Council of State. He insisted, however, on the appointment of Dr Kayser as acting secretary of state. In addition he demanded that the Prince of Pless, the industrial magnates Ritter, Krupp, Jencke and Stumm-Halberg, the conservative Centre Party member Freiherr von Huene and his own tutor Hinzpeter should be invited to join the Council, which Bismarck was obliged to concede.[87] Thus in the first half of February it actually looked as if Bismarck had reconciled himself to the new power structure, in which it was no longer he but Wilhelm II who would lay down the guidelines of domestic policy and take the most important decisions on government appointments. How was it that a crisis nevertheless arose and ended in Bismarck's abrupt dismissal on 18 March?

CHAPTER TWELVE

Bismarck's fall from power

THE REICH CHANCELLOR'S TACTICS

SINCE his violent clashes with the Kaiser on 24 January 1890 the Reich Chancellor's position had changed radically. On that day Wilhelm had taken over the chairmanship of the Prussian Ministry of State, laid a precisely worked-out programme of social welfare reform before the ministers and succeeded in appointing a new minister of trade with responsibility for implementing the programme. At the dissolution of the Reichstag he had made a speech favourable to workers' causes and had insisted on announcing his programme of social reform in the two imperial proclamations of 4 February. At his instigation an international conference to regulate social welfare legislation in all industrialised nations had been arranged, and the Prussian Council of State had been convened under his personal chairmanship. In addition, Wilhelm had decided to receive individual ministers as well as lesser officials on a regular basis, with the intention of circumventing Bismarck's control over them and of making them *his* ministers and *his* tools. But how would the 75-year-old founder of the Reich react to such a fundamental undermining of his power?

Several possible courses of action were open to Bismarck, who reflected on each in turn without ever quite making up his mind. First, he could simply have accepted the new power structure, withdrawn to Friedrichsruh or Varzin and lent his name to the Kaiser's initiatives, regardless of whether he thought them right or wrong. Second, he considered reducing the formal powers of the Reich chancellor, so as to retain the responsibility for making decisions without the monarch's interference at least in those areas still under his control – for instance in foreign policy. A third option consisted in resigning all his offices – Reich chancellor, Prussian minister-president and Prussian foreign minister – thus drawing a line under his political activity after twenty-eight eventful years at the top. Finally, however, it was also open to him to take up the fight for power with the young Kaiser. The fact that by following this course he might bring about the destruction of his life's work, the mighty Hohenzollern empire with its personal monarchy virtually independent of parliament, and might even provoke a civil war with foreign intervention, was plain for all to see.

For a short time after the Crown Council of 24 January Bismarck seemed to have reconciled himself to the new situation. At the next ministerial meeting he remarked resignedly that 'the moods of a monarch are like fair and foul weather, one cannot escape them, one takes an umbrella and still gets wet. I think we must make the best of it!'[1] When the Chancellor accompanied the other ministers to congratulate the Kaiser on his birthday on 27 January his behaviour was so friendly that everyone had the impression that the differences of opinion that had arisen at the Crown Council would not have any further repercussions. The Kaiser seemed to have come to an understanding with the Chancellor and even announced his intention of attending a parliamentary dinner to be given by Bismarck on 4 February.[2] The members of the Kaiser's entourage were well aware, however, that the truce was only outward and certainly would not last.[3]

The idea, widely aired at this time, that the excessively powerful position enjoyed by Bismarck as chancellor ought to be dismantled in formal terms, originated in the Kaiser's camp, where thoughts were already turning towards getting rid of the post of Reich chancellor altogether.[4] As the crisis became more acute, Wilhelm II expressed the wish, on 27 January, that Bismarck should restrict himself to the Foreign Office and hand over control of internal policy in the Reich and in Prussia to someone else.[5] Significantly, however, similar ideas were also being considered by the Bismarcks and their circle. On 30 January 1890 the Chancellor informed the Saxon envoy, Count Hohenthal, that he intended 'to "crumble away" his post bit by bit', and to limit himself to the 'retirement zone', by which he meant foreign affairs and his work in the Bundesrat.[6] He went even further in a conversation with the Bavarian envoy, Count Lerchenfeld, to whom he said that he needed to put the 'excessively puffed-up Reich chancellor' on a slimming diet, give up his offices gradually and retain only the 'rump', the control of foreign affairs. Once he had quitted his Prussian posts – including that of Prussian foreign minister – the Prussian vote in the Bundesrat could even be cast against his views, if the occasion arose.[7] On 8 February Bismarck told the Kaiser in the course of a 'violent argument' that he wished to withdraw from Prussia completely and to remain in office only as Reich chancellor. He suggested that General von Caprivi, of all people – whom the monarch had himself received a few days earlier – be appointed minister-president, and his son Herbert Prussian foreign minister.[8] 'The Chancellor put on a very good act and even wept', Waldersee commented in his diary. 'He was particularly eager to find out who the Kaiser might have in mind as his successor. The Kaiser was wise enough not to name anyone to him; the Chancellor would have set out at once to ruin the reputation of whoever it was . . . The Kaiser is in a very agitated state and complains bitterly about the Chancellor, whose lies he has now experienced for himself; he is also very annoyed with Lerchenfeld, who – undoubtedly at the Chancellor's instigation – has written to Munich about a chancellor crisis, as a result of which the importance of keeping the Chancellor is being emphasised there. Can this state of affairs last long without causing great damage?', the Chief of the General Staff asked himself.[9] The next day Bismarck declared in the Ministry of State that his departure from the two Prussian posts had become unavoidable, as the Kaiser had plans with regard to domestic policy which he, Bismarck, could not agree to implement. He had already agreed on this 'irrevocable' decision with the Kaiser; his withdrawal to the 'retirement zone of the Foreign Office' must be publicly announced on 20 February, the day of the forthcoming Reichstag elections. Thereafter he would have no more contact with the Prussian ministers than with the Bavarians. The ministers did not protest, because they too saw this withdrawal as the only way to 'prevent a total break'.[10] Waldersee claimed to have heard after this meeting, however, that 'the Chancellor criticised the ministers without sparing a single one of them, and told them that none of them could be his successor'.[11]

On 8 February Lerchenfeld heard further details of the Chancellor's intentions from Herbert Bismarck. It should not be forgotten, said Herbert, 'that the present Reich Chancellor would not live for ever, and since it was unlikely that a successor of his calibre would ever be available again, it was perhaps a good thing that the transition should be made during his lifetime'.[12] It was generally assumed that Boetticher would be appointed Prussian minister-president, but meanwhile nothing further could be done than 'to duly await the decision of His Majesty the King'.[13] On 10 February Prince Bismarck himself remarked to the British ambassador, Sir Edward Malet, 'I cannot approve or agree in what the Emperor is doing and it has come to this that I have made up my mind to resign all the offices I hold except those of Chancellor of the Empire and Minister for Foreign Affairs of the Empire . . . My position will be a difficult one.' He added, presciently, 'Herr von Boetticher as Prussian Prime Minister will be over me in the Bundesrath while as Chancellor I direct the policy of the Empire. I doubt its working and it will probably end in my complete retirement.'[14]

That very evening, however, Bismarck had a discussion with Lerchenfeld as a result of which he was to abandon all these considerations. The Chancellor began by stating that he was intending to give up both his Prussian offices 'so as to have nothing more to do with the instruction of the Prussian vote' in the

Bundesrat. He wanted to build up a new bastion of authority for himself, based on the Bundesrat and the Reich offices. But the conversation constantly came back to the central difficulty in this combination, namely 'what would happen if the Prussian vote [in the Bundesrat] were instructed against his opinion'. The non-Prussian states would certainly not be in favour of such a solution, Lerchenfeld declared, for 'the confidence of the states in the central executive rested on their knowledge that what Prussia wanted, the Chancellor wanted too, and what the Chancellor wanted, Prussia wanted too. This was the cement which held the German states together'. The Bavarian envoy subsequently reported to Munich 'that the Reich Chancellor is not yet clear in his own mind about his future position. But he is well aware of the difficulties the new arrangement would create everywhere'.[15] That same evening Bismarck gave up his plans. He summoned Boetticher and told him that he had no intention of allowing himself to be 'buried alive'; he could not accept a situation in which Boetticher would sit beside him as Prussian minster-president and tell him how to vote in the Bundesrat.[16] The next day he informed a visibly displeased Kaiser that on no account would he retire from his Prussian offices. When August Eulenburg heard that the planned separation of offices had been abandoned, he deplored Bismarck's decision, arguing that it was '*he* of all people [who] should have tried to draw the dividing line [between the Reich chancellorship and the Prussian Ministry] which will surely be needed in the future, and in the short term this has the evil effect that from now on with every storm that breaks we shall have to be prepared for a full-blown chancellor crisis'.[17]

In parallel with these complicated deliberations on reducing his sphere of authority, Bismarck was also seriously considering the possibility of simply retiring altogether. He had already threatened to retire from all his offices at the Crown Council of 24 January. He fired a second warning shot on 30 January, telling Hohenthal that the day on which the Saxon government brought the proposal on social welfare instigated by the Kaiser before the Bundesrat would be his last day in office.[18] Shortly afterwards, when he was due to meet the Kaiser on 3 February to discuss the two social welfare proclamations, he warned that his desire to hand in his resignation was growing 'as fast as a hothouse plant'.[19] When the differences between the Kaiser and the Chancellor on social policy became public knowledge many people wondered why Bismarck remained in office in such circumstances, and why the Kaiser did not simply dismiss him. The British ambassador, for instance, assumed that the Kaiser must have overriding reasons for wishing to keep Bismarck. 'He is the creator of the Empire: and it is everywhere felt he is a guarantee of peace. On the other hand it is thought even in Germany that if the Emperor's ambition does not find vent peacefully it will take a more dangerous form, and he himself knows that public confidence would be shaken if Prince Bismarck were to leave his side. The Chancellor probably feels it would be unchivalrous & also unsafe to do so.'[20] But these were only guesses. Two days later, on 10 February 1890, Malet was to hear from Bismarck himself how he really felt.

In his conversation with the ambassador that day the Reich Chancellor made no secret of his thoughts. With astonishing frankness, he condemned the system of personal rule which the Kaiser had introduced. As Malet stated in a confidential letter to Lord Salisbury, the gist of Bismarck's comments was that 'the Emperor deemed himself quite able to stand alone and to direct the internal and external affairs of the Empire without any anxiety as to his own capacity'. The Chancellor gave repeated indications that he considered his days were over, even if he remained in office for form's sake. He was in a nervous and exhausted state because of the untenable position in which he found himself. His decision to retire from his Prussian offices had been made inevitable by the Kaiser's desire to govern himself; but the division of offices could not be permanent. It was only out of consideration for the monarch's wishes and because of the forthcoming Reichstag elections that he remained in office at all. As he put it to Malet, 'I would retire altogether, but the Emperor wishes me to remain and I cannot refuse – for if I were to go now on the eve of the elections it might have an effect upon them which I desire as little now as before the present situation arose. I have had no hand in making it. The young Emperor has been launched in the path which he has taken by outsiders. Those chiefly responsible', Bismarck claimed, 'are Hinzpeter, his old tutor, Freiherr von Douglas, whom he has created Graf, the Grand Duke of Baden and the King of Saxony.' In their personal relations the Kaiser was extremely friendly, 'but he wishes to govern himself. He has not realised the utility of having a screen between himself and his subjects on which the blows of unpopularity may fall without injuring him', the Chancellor explained. 'He has no doubts. He thinks he can do all things and he wishes to have the entire credit all to himself. He was not in the least annoyed when I declined to countersign the rescript . . . He does not foresee any reaction in popularity when his scheme fails.' Even in foreign affairs the Kaiser believed he could overcome any problem himself. 'He is elated with what he considers to be the success of his visit to Russia and other countries. He only wants me to remain in order that I may make speeches in the Reichstag and induce it to vote money', Bismarck observed bitterly. He was glad, he said, 'that seventy five years are behind me and not before me – my work is done – but it is sad to see the edifice which I have raised brick by brick in danger of crumbling'. The new system would also cause difficulties in Anglo-German relations which Bismarck himself would no longer be able to prevent.[21]

Six days later Herbert confirmed in a letter to his brother Bill that their father had abandoned the idea of separating his offices as impracticable and was now thinking of 'complete retirement on 1 April', his seventy-fifth birthday. 'I do not know what will happen in the end, but more animosity is being stirred up than ever, and as this of course destroys confidence, the constant living together becomes more and more difficult.'[22] The Reich Chancellor's attitude was much the same when he and his wife visited the Empress Frederick a few days later. He spoke to her at length about Wilhelm's social policy 'coup' and about his own forthcoming retirement, and said 'he could not keep pace with innovations so suddenly resolved on and carried out in such a hurry and on the advice of people he thought in no way competent to give it'. The Dowager Empress found him 'looking remarkably strong & well & inclined to take things very philosophically. He is exceedingly fond of W. . . . but I fancy he is *uneasy* at the *very* great self-confidence and the naiveté with wh. he exercises his will & takes responsibilities, and also at the curious people who have access to him & are listened to.' She believed that Bismarck's intention to retire was entirely honest, but doubted whether the Kaiser would accept his resignation.[23]

THE DECISION APPROACHES

As we have seen, in the first weeks of 1890 Wilhelm II's advisers remained convinced of the necessity of keeping Bismarck in office. For all their criticism of his style of government and his policies, they were too afraid of the repercussions of a spectacular departure by the founder of the Reich to be willing to risk a violent break with him. So Bismarck might well have remained in office for a few months longer[24] had not the idea taken root in the Kaiser's camp during the month of February that the Chancellor was pursuing tactics which would make it acutely dangerous for the monarchy for him to stay on. Whether Bismarck had in fact decided to go on the offensive against Wilhelm and the crown, as the monarch's secret advisers feared, or whether their anxieties were mere fantasy, is still a matter of debate among historians today.[25] What seems quite clear from the wealth of sources available, however, is that Wilhelm's close advisers – *all* of them – were *sincerely* convinced of their interpretation of Bismarck's intentions, not least because after years of working with him they knew his mentality and methods, and could only too easily put themselves in his place.[26] This fact in itself is enough to explain the fatal process which led Kaiser Wilhelm II to decide, as early as mid-February, that Bismarck had to be dismissed in the near future.

Immediately after the Crown Council of 24 January and the subsequent fiasco in the Reichstag, when the Kartell broke apart over the strengthened anti-socialist law, Holstein became convinced that he had seen through Bismarck's tactics, which were highly dangerous for the Kaiser. He wrote to Eulenburg in some agitation: 'What just happened has the purpose of making the Centre into a government party. As a result of the disgrace which the Chancellor has brought on the Kartell parties in the socialist question, the elections cannot but turn out badly, i.e. the Kartell will lose its majority. The Chancellor will then say to the Kaiser: "We must move closer to the Centre, therefore it is necessary for Your Majesty to appoint a member of the Centre – for instance Huene – to the Ministry." The moment that the Centre becomes a government party in *Prussia*, it also becomes a government party in Bavaria. Then Lutz will be out within a few months . . . I do not think the consequences of that could ever be reversed. A good Reichstag can always follow on a bad one. But if the ultramontanes are once at the helm in Bavaria and are given the chance to alter the constituency boundaries to suit themselves, we shall never get them out – not without violence, anyway.' Holstein pointed to the similarity between the domestic and foreign policies of the Chancellor: 'In both cases he holds sway over a group of three. But he despises and mistreats them in order to run after a fourth which has its back turned to *him*. The fourth is the Centre in the one instance and Russia in the other.'[27]

With the publication of the two imperial proclamations on 4 February 1890 the ever-suspicious Holstein scented new dangers. Not only had Bismarck refused to countersign the rescripts, he had strengthened their wording in such a way as to cause disquiet among German industrialists and foreign governments. 'The longer of the two imperial proclamations has had the effect of alarming the propertied classes because of the general way in which it is expressed, which does not come from the Kaiser *but from the corrections made by the Chancellor*', Holstein warned on 7 February. 'It has led people to believe that a general redistribution of property is in the offing. It will not be a good thing for this uncertainty to continue for long, since the Kaiser needs the support of the propertied classes. The Kaiser, who is certainly an intelligent man, will know how to counteract these anxieties', he added. In this letter, parts of which Eulenburg forwarded to the Kaiser, Holstein commented that Bismarck had not made the corrections with the deliberate aim of putting the Kaiser in a difficult position, and added that 'much of what looks like deceitfulness in the Chancellor is simply the consequence of mental weakness, which is rapidly increasing'.[28] After discussions with Adolf Marschall von Bieberstein, Paul Kayser and Franz Fischer, however, he became convinced that Bismarck was consciously deceiving the Kaiser. 'The fact that the Chancellor undoubtedly feels deeply hurt by the Kaiser's independent conduct makes the situation very dangerous', Holstein warned on 10 February. 'He will certainly not rest until he has succeeded in making the Kaiser see it as a mistake to have acted alone and not followed his Chancellor. Even in the proclamations of the 4th of this month he made various additions which will make it more difficult for them to be carried out, so as to wreck the Kaiser's ideas, leaving the latter – as is the plan – entirely in the hands of the Chancellor.'[29] The Baden envoy, Marschall, was also of the opinion that Bismarck was trying to portray the Kaiser as 'a heaven-storming idealist', while he himself was 'the protector of the propertied classes'. The Chancellor was proclaiming his 'innocence of the Kaiser's action on all sides'.[30] Marschall indignantly recorded on 12 February that Bismarck had put the question of the eight-hour day on the agenda of the international social welfare conference, although he was very well aware that this point was anathema, particularly for the British government.[31]

The secret advisers were afraid, on the one hand, that Bismarck would exaggerate the Kaiser's ideas on social reform to the point of absurdity, while on the other hand they suspected him of intending to sabotage the work of both the Prussian Council of State and the international conference in Berlin, in order to wreck Wilhelm's initiative. 'There is no need to point out how dangerous it would be for the reputation of the Kaiser and of the monarchy if the proclamations were to remain nothing more than fine words', Holstein, Kayser and Fischer warned in a joint memorandum of 10 February. 'There is no doubt that the Kaiser is now in a very serious position . . . As things are now, a fiasco is very much to be feared. The Council of State could become a noose for the Kaiser.'[32] The longer this atmosphere of mistrust persisted, the more unbearable it became to both sides. In mid-February 1890 Wilhelm II and his group of advisers gradually came round to the idea that Bismarck would have to be dismissed in the interest of the monarchy. On 30 January Waldersee had noted contemptuously in his diary that the Chancellor no longer knew what he wanted: 'he wavers, tells lies and makes everyone who works with him ill-tempered'; but most significantly, 'he will not resign on any account and so . . . the great man will end his days as a small man'.[33] Ten days later the Chief of the General Staff was already pondering the question of when and under what circumstances Bismarck should be dismissed. Although on 9 February he still took the view that the inevitable 'great clash' should be delayed 'until the Reichstag has met and the Chancellor has reached a deadlock with it',[34] the following day he expressed the opinion that the present moment would be advantageous for the Kaiser. His thinking was almost schizophrenic at this point, for he sided entirely with Bismarck on the questions actually at issue, as he repeatedly admitted.[35] The social reform initiative had caused great disquiet among the middle parties, precisely the people whose support the Kaiser needed. 'Yesterday's stock exchange gave the Kaiser the answer to his proclamation by panicking over industrial securities!', Waldersee observed in alarm. It was a case of 'the big capitalists, including the Jews, against the Kaiser's ideas. The

Kaiser will realise that there are some very unreliable people among the National Liberals and Free Conservatives, to whom he recently gave much support. They love the Kaiser as long as he helps them do good business, and [he] is in for many disappointments.'[36] Like Holstein, Waldersee became convinced that Bismarck would exploit the agitated mood among the propertied classes for his own benefit, and would 'not hesitate to attack the Kaiser' in the process. This perception marked a decisive moment for the Chief of the General Staff.[37] On 10 February he wrote for the first time: 'I do not doubt for a moment that the Chancellor must fall, and the present moment is really not a bad one for the Kaiser.'[38] It was two days later that he first told others of his view that the Kaiser must dismiss Bismarck immediately after the opening of the new Reichstag. 'Believe me', he wrote to Philipp Eulenburg on 12 February, 'it cannot go on like *this* for long.' It was obvious to everyone that the Chancellor was opposed to Wilhelm II's social reform policy. 'How can our beloved sovereign possibly continue the great work which is engaging the whole world's attention now, if the Chancellor does not commit himself to supporting it with all the power and influence he has?', he asked. 'It cannot help but give rise to a completely untenable situation, and therefore in my opinion we are now facing a really serious chancellor crisis.' Bismarck, he had heard, was outwardly calm, but inwardly very agitated. 'I think that the Chancellor must at all costs remain until the Reichstag has assembled and he has shown his face; but I *doubt* that it would be in the Kaiser's interest for him to remain any longer than that.'[39] In a conversation with Hinzpeter on 15 February the Chief of the General Staff stated frankly that he thought Bismarck must be got rid of soon, for if he remained in office for long the Kaiser would not be able to carry out many of his ideas and would therefore suffer a defeat at the hands of the Chancellor. At this point Waldersee did not consider that there was much danger of a pro-Bismarck movement against the Kaiser after the dismissal. 'Once the Chancellor has gone', he observed contemptuously, 'he will collapse under the weight of his 75 years and his present friends will desert him; they will have no interest in seeing him return to power, and in any case the Kaiser would never lend his hand to that.'[40] Johannes von Miquel, to whom Waldersee spoke at length, also considered Bismarck's days to be numbered. In contrast to the Chief of the General Staff, however, he preferred 'a slow fading away'.[41]

On 18 February 1890, two days before the first round of the Reichstag elections, the Kaiser had in fact already made up his mind. 'The chancellor crisis is in full swing', Waldersee recorded in his diary. 'The Chancellor is wavering – another indication of how he has become an old man – at one moment he wants to go, the next he wants to stay. But he is indignant because he is convinced that the Kaiser would like to be rid of him.' The latter was certainly 'quite prepared in his mind' for the dismissal, Waldersee wrote, 'and my experience tells me that he will not go back on it now'.[42] 'The Kaiser has definitely had done with the Chancellor now', he recorded after further discussions with Wilhelm on 19 and 20 February.[43] Not only had the manner in which Bismarck 'ran down' his ministerial colleagues left the Kaiser 'outraged'; but 'the dishonesty of the Chancellor has at last become too much for the Kaiser, and he is beginning to realise that he has frequently been deceived by the great man'.[44]

On the day of the elections Wilhelm called on the Chief of the General Staff during his morning walk and discussed with him the issues which were still unresolved. Waldersee's account of this conversation shows once again how much he was in fact on his arch-enemy Bismarck's side on the social welfare question. His diary states: 'In the labour question he [the Kaiser] believes that the Council of State will put forward good proposals and that much can be achieved. This is quite true: some things will be achieved, such as workers' committees, limitation of Sunday working hours and women's working hours. Unfortunately, however, this will nowhere near satisfy the workers. He has stimulated their greed and brought unrest throughout the world of labour. The outcome can only be great disappointment for him. He has unleashed a current which will be hard to stem. If all workers were noble or even merely sensible people, all would be very well. But they are not, unfortunately, and, what is more, socialists will now stir them up into further madness. It will not be long before the Kaiser greatly regrets what he has done.'[45] In spite of this fatalistic prediction, which could have come from Bismarck himself, Waldersee continued to press for the Chancellor to be dismissed soon.

On 22 February the Kaiser again took Waldersee for a walk and almost immediately began to speak of the chancellor crisis, this time with 'a considerable degree of bitterness'. Wilhelm was 'completely convinced', according to the General, 'that the Chancellor is working against him with the greatest possible audacity in the labour question; it sounds incredible, but it is unfortunately true'. Bismarck, in the Kaiser's words, had 'wanted to go and then withdrawn his request the next day. I will not tolerate such games; from now on I shall decide on the date when he shall go and he must first wait a while. His misfortune is his immeasurable desire to rule. So he has gradually got everyone under his thumb and he is spoilt. But with me, for once, he is tackling the wrong person.'[46]

'COMPLETE VICTORY FOR THE KAISER'S CAUSE'

On 14 February 1890 Philipp Eulenburg arrived in Berlin, at almost the same moment as Hinzpeter. 'It is my impression that he [Eulenburg] is the only really close friend and I am sure that he gives wise advice', Waldersee commented.[47]

In the course of the following days and nights Eulenburg had numerous long conversations with Hinzpeter and Waldersee, Paul Kayser, Holstein, Marschall von Bieberstein, Rudolf Lindau, his two cousins August Eulenburg and Gustav Kessel, his intimate friends Axel Varnbüler and Kuno Moltke, and many of Wilhelm II's other advisers. Eulenburg also had several 'endless dialogues' with his old friend Herbert Bismarck, which 'unfortunately were not calculated to reassure me about the present state of affairs'.[48] Above all, however, he had daily 'discussions of all kinds of important matters' with Kaiser Wilhelm. They talked of the 'complicated state of social affairs', of the 'important questions of social welfare legislation and of foreign policy which are causing such a stir at the moment', as well as of colonial policy. Whereas Herbert gave the impression of being 'tense' and 'out of sorts', Wilhelm was in the best of spirits. 'The Kaiser was very merry and cheerful about several welcome developments in domestic politics', his friend recorded in his diary on 22 February 1890.[49]

The smouldering crisis now moved swiftly towards its climax. On the night of 22–3 February Eulenburg reported to the Kaiser: 'Today the whole Ministry was ransacked for an old regulation according to which ministers are supposed to have the right to speak to the King only in the presence of the minister-president. But all that was found was a ruling that in the case of a dispute between two ministers they were to appear before the king with the president.'[50] Bismarck was in fact looking for a Cabinet Order of 1 September 1852, and sent Counsellor Dr Erich von Schwarzkoppen to the Ministry of State's secretariat and then to the Finance Ministry to search for it. It was found the next day, 23 February. Barely three weeks later it became the direct cause of the final break between the Kaiser and the Chancellor.[51] Eulenburg's letter to the Kaiser also contained the news that Bismarck would ask for an audience in the next few days. He wanted to discuss the Reichstag elections, the programme for the international conference and 'general matters' with the Kaiser. Eulenburg urged the Kaiser to take care that the Chancellor did not introduce any discussion of the regulation of working hours into the conference programme, as it would alienate foreign powers. Following this warning, Wilhelm at once wrote to Queen Victoria deriding the 'nonsense' about fixing standard working hours as 'unadulterated blarney from beginning to end', which went some way towards reassuring the British government.[52] Eulenburg also reported to the Kaiser that Bismarck was aiming to cut down the social welfare bill and to delay it until after the international conference.[53]

Thus when a submission by Bismarck containing just such proposals arrived at the Schloss on 26 February the Kaiser had been amply forewarned.[54] He wrote furious marginal notes on the memorandum objecting to the postponement of the bill. He sent for Paul Kayser and complained that Bismarck was trying to put 'a spoke in his wheel', which he would not allow.[55] At Waldersee's audience on 1 March the Kaiser told him that Bismarck had suggested 'waiting until after the international conference had finished its work before bringing bills for the protection of labour before the Reichstag; the Kaiser at once realised that his intention was to put the whole thing off, and wrote to him to say that the bills should be put before the Reichstag as quickly as possible'.[56] The Kaiser's firm stand led some to hope that Bismarck would now give way. 'There is a growing hope that with Your Majesty using such tactics the Chancellor will be more circumspect in his wishes, and an understanding with him will be possible', Eulenburg assured the Kaiser.[57]

At the same time, however, there were increasing indications that Bismarck was aiming at a policy of force which would completely overshadow the Kaiser's initiative on social welfare, and with it his political power and independence. Waldersee was enraged when he heard on 16 February that Bismarck had attempted to convince the Kaiser that a revolution was imminent. The Chancellor, he noted, was constantly trying to frighten Wilhelm by predicting 'danger for the Kaiser himself and revolution in the spring or at the latest in the summer'. But this was 'all sheer trickery. The Socialists consider that the Kaiser is working in their interest and want to keep him alive at all costs. We shall have strikes, perhaps even bloody conflicts, but anyone who foresees a revolution now is either too inclined to see ghosts or is a swindler. A great deal more would have to happen before we reached that point.'[58] When Marschall met Eulenburg on 20 February both agreed that, although Bismarck was willing to concede the social reform proclaimed by the Kaiser, he was simultaneously aiming at a 'change of system': in other words, a reactionary constitutional change imposed forcibly from above.[59] Otto Prince von Stolberg-Wernigerode likewise expressed the view that, while the Kaiser was in favour of making a last attempt at satisfying public opinion before 'extreme measures' were employed, Bismarck on the other hand was convinced 'that we have already more or less reached the extreme point'.[60] On 26 February Waldersee and Minister of War General von Verdy du Vernois, after reporting to the Kaiser at a joint audience, heard from him what had happened at Bismarck's audience the previous day. Bismarck had declared that he was 'filled with anxiety that if there were serious disturbances and perhaps even riots the Kaiser would not take vigorous action and would not allow any shooting. When the Kaiser responded that he might set his mind at rest, for if it were necessary he would not shrink even from the most extreme measures, the Chancellor was gratified and said that he was now free of the nightmare which had oppressed him night after night.' Waldersee commented acidly: 'Both Verdy and I laughed and said that this was nothing but another piece of play-acting, in other words humbug, and I believe the Kaiser is of the same opinion.' Wilhelm went on to relate that Bismarck had spoken about the election result which had destroyed the Kartell[61] and had expressed the opinion that the National Liberals had lost votes because 'many of them were dissatisfied with the imperial proclamations and had voted with the Left-Liberals. This is an outrageous lie!' In the end, Wilhelm reported, Bismarck had said that as far as the social welfare question was concerned, from now on he was prepared 'to go along with the Kaiser, although he doubted that anything would be achieved. So the Chancellor gave way to the Kaiser, and the latter was very pleased', Waldersee commented.[62] Marschall reported to Karlsruhe on similar lines, stating that at his audience on 25 February the Chancellor had promised his support for the social welfare bill on condition that the Kaiser declared himself ready, 'in the event that the disastrous consequences feared by the Prince should materialise, to use force and to allow shooting'.[63]

During the last days of February Kaiser Wilhelm II demonstrated his powers of leadership to a broad and knowledgeable forum by presiding daily at the Prussian Council of State which he had convened, a task which he carried out 'with great confidence throughout'.[64] Even Waldersee, whose attitude to the Kaiser's initiative was sceptical in the extreme, saw great benefit in the fact 'that the Kaiser has the feeling of having promoted a good cause, and that many people who scarcely knew him before have come to know and appreciate him'.[65] Paul Kayser, the actual originator of the social welfare programme, described the result of the consultations as 'complete victory for the Kaiser's cause'. 'All the points in His Majesty's programme' had been accepted in principle by the Council of State, he commented enthusiastically. 'But the victory of the Kaiser's cause is a *personal* victory for His Majesty, and what Wilhelm I said of the organisation of the army, Wilhelm II will say of the social welfare legislation: "It is my own work."' Writing to Eulenburg, the 'little Kayser' observed that the previous few days must have been 'a time of severe trial' for the Kaiser, but it was also 'a time of apprenticeship culminating in mastery'. Eulenburg lost no time in forwarding this flattering letter to the monarch. During the sessions of the Council Wilhelm had 'shown qualities which it would truly be a sin to keep confined to the relatively small circle [of the Council of State]. The Kaiser maintained such a lively interest for seven hours each day that at the end of the day he seemed as fresh as he had been in the morning. He was always dignified, even in jest; he kept to the subject all the time with skill and flexibility, like any grey-haired parliamentary chairman, well-informed, and cutting short all unnecessary speeches. Finally, without any preparation . . . he set out his views on the latest major strike in a speech as perfect in form as in content, showing such profound awareness of his responsibility as head of state, such a warm, patriotic heart, such calm and objectivity, that the whole Council of State was spellbound. It would really be a very good thing if this were reported in the press . . . The nation has a right, after all, to know what kind of a ruler it has.'[66] After the last session of the Council of State '*the great Kaiser*' singled out '*the little Kayser*' for special attention; he '*was even permitted to kiss the Kaiserin's hand, and she thanked him for his support!*', Holstein reported to Eulenburg with amusement. This was an early and conspicuous example of the corrupting effect of the 'kingship mechanism' on Prussian–German officialdom. 'You should hear the little fellow now!', Holstein commented mockingly. He was 'more imperial than the Emperor'.[67]

Wilhelm's success in the Council of State was generally considered to have strengthened his position vis-à-vis Bismarck to a significant degree.[68] After the Council's sessions Waldersee observed that the Chancellor was 'unbearable' and 'in the blackest of moods'. While he continued to behave with the utmost affability towards outsiders, he treated his closest colleagues abominably, Waldersee reported on 1 March.[69] On the same day Paul Kayser commented that Bismarck must have realised that his days of glory were over; but he could not get over 'the defeat he had suffered at the hands of his Imperial Master' and was still laying mines 'to bring down the work which has begun and lead the Kaiser into absurd extremes'.[70] By way of a defensive strategy, Kayser and Holstein suggested that the Kaiser should refuse to sign any bills until the Council of State had discussed the measures in full. In this way Wilhelm could win in the Council of State the support which the Ministry of State ought to give him, but which, under Bismarck's domination, it 'cannot provide at present'.[71] Waldersee too was confident that with his success in the Council of State Wilhelm had outmanoeuvred Bismarck. 'I no longer feel anxious that he might be able to cause the Kaiser great problems in the labour question, now that I have seen how the Kaiser has remained master of the situation and has overpowered the Chancellor. The Kaiser's great success is also plain for all to see. No one now speaks of the Chancellor's social policy, but only of the Kaiser's. People can see who is master; this is vital, and an outstanding achievement for the Kaiser.' In well-informed circles it was now the universal opinion that Bismarck's days were numbered, for it was obvious that the rift between him and the Kaiser was too deep and the bitterness on both sides too great ever to be healed. Wilhelm had clearly emerged as the victor in the struggle over social welfare legislation. Confident that the end was not far off, Waldersee expressed the opinion that 'a clash in the near future should be avoided, and the Chancellor should be left to bring about his own ruin little by little'. Bismarck would have to appear before the newly elected Reichstag first, however, for 'its unsatisfactory composition is after all chiefly his fault', Waldersee opined.[72]

NEW PLANS FOR A *COUP D'ETAT*

In parallel with the Kaiser's victorious initiative on social policy, however, another development took place which convinced a growing number of the governing elite that Bismarck was right to advocate a reactionary revision of the constitution of the Reich. The Kaiser's policy was overshadowed by the results of the Reichstag elections, which were disastrous for the Prussian–German monarchy. The results came out shortly before the Council of State began its sittings, and were confirmed by the second round at the beginning of March. Of the 7,261,600 votes cast in all, the former opposition parties won no less than 4,658,900. The Kartell parties' mandate was reduced from 220 to 135 seats. The losses of the National Liberals and Free Conservatives, for whom the Kaiser had repeatedly demonstrated his support, were particularly painful. The left-liberal Progressive Party, however, climbed from 32 to 66 seats, while the Catholic Centre reached its peak in the history of the Reich with 106 seats. Without its support, or that of the Progressive Party, there could no longer be a government majority in the Reichstag. If that was not alarming enough, the electoral success of the Social Democrats was seen as little short of a catastrophe. With 1,427,000 votes (19.7 per cent) this republican and Marxist popular movement had become the strongest party in the German Reich in terms of numbers of votes.[73] All told, enormous numbers of people had turned against the Hohenzollern monarchy and called the existing political and social order into question.[74] How could the new situation be brought under control? Was this really the ideal moment to dismiss the founder of the Reich and to put the power of the state into the hands of an untried and impulsive young Kaiser?

When the results of the first round of the elections were announced on 23 February, Waldersee's furious reaction showed clearly, once again, that regardless of all personal rivalry he fully shared the fatalistic views of the old Reich Chancellor. After a meeting with the Kaiser at Potsdam on that day he noted: 'The election result turns out to be even worse than anyone thought. The Kartell has disintegrated, as there is no question of it having a majority any more. The true picture will not emerge until after the second-round elections, of which there will be more than a hundred, but it is certain that the Social Democrats have gained enormously. The Kaiser has not yet realised the full significance of all this; we are facing an important turning-point. This Reichstag will certainly not approve the demands of the military; its dissolution can hardly improve matters.'[75] After the results of the second round were declared the Chief of the General Staff summed up the new internal political situation thus: 'The elections . . . have resulted in the complete defeat of the National Liberals and probably of the Free Conservatives too, and thus in the annihilation of the Kartell's position: it is very far off having a majority now. The Conservatives have lost a few seats, but they still form a very strong group. The Progressives have increased greatly, likewise the Social Democrats. More alarming than the number of socialists in the Reichstag, however, is the number of socialist votes in the country. Never has an election aroused so much passion as this one, and it has shown very clearly the disturbing aspects of universal suffrage.' The way in which the Social Democratic movement was spreading into rural constituencies was particularly dangerous, in Waldersee's eyes. The reawakening of republican aspirations in South Germany was also alarming. 'It is a question that must be treated with the utmost seriousness and remedies must be considered', Waldersee commented anxiously.[76]

Waldersee laid the blame for the electoral catastrophe partly on Bismarck, who 'as the responsible, almost supreme, ruling Reich Chancellor' had suffered 'a terrible defeat'. He also blamed the Kaiser, who had 'given powerful stimulation to the greed of the working population' through his proclamations on social reform.[77] The political effect of the imperial proclamations would be very slow in materialising, if it happened at all. 'I am afraid that they do not help at all – and the Kaiser himself already shares this view.' Waldersee had 'not the slightest doubt that the general dissatisfaction will increase, and I see no reason to believe that future elections will turn out much better; the only thing that would help would be if the socialists rushed ahead too quickly and it came to bloodshed; then it is possible that property-owners and people living from the state would make common cause'.[78] For the Chief of the General Staff there could be no doubt about the conclusions which must be drawn from 'the bad elections'. As early as 23 February he commented in his diary: 'There is no alternative but to abolish universal suffrage. I shall be glad to help with that.'[79] A week later he went further: 'In the event that the Kaiser should wish to suspend universal suffrage and cannot do this through the Chancellor', he himself was ready to take Bismarck's place and would 'gladly risk my life for it'.[80] The leader of the Free Conservative Party, Wilhelm von Kardorff, was likewise 'very anxious about the future' and believed that 'the abolition of universal suffrage [was] necessary'; but in his opinion Bismarck's presence was 'absolutely essential' if a *coup d'état* of that nature were to be carried out.[81]

The dilemma that the election results represented for the Kaiser was stark: should he continue with the pro-worker policy of social reform which he had already proclaimed *urbi et orbi*? Could he risk Bismarck's resignation in the changed political situation? If he gave way – if he switched course towards a policy of force – would he not then be admitting his mistake and capitulating to the Chancellor? Wilhelm seems at first not to have understood the position of constraint in which he found himself. At the closure of the Council of State he announced emphatically: 'As far as the battle against social democracy is concerned, that is *my* own affair.' This he said in such a tone, according to Waldersee, 'that everyone felt that he was determined to take strong measures'.[82] The election results prompted thoughts of a *coup d'état* even in the otherwise soft-hearted Eulenburg, who did not at first give much thought to the political disadvantages of such a course for the Kaiser. In a letter to Holstein of 28 February he summed up the basic parliamentary dilemma for all future governments with the observation that the new Reichstag would not accept either a new Anti-Socialist Bill or an Army Bill unless, as a quid pro quo for the latter, the two-year military service demanded by the Progressive Party were granted, or political concessions were made to the Catholic Church. But '*the Chancellor is too old, the Kaiser too young and the Social Democrats too strong* for us to govern with a minority. To govern with the Centre means the *permanent* presence of Rome in the Bundesrat and the end of the Italian alliance . . . How about a demand by all the federal princes for a revision of the electoral law under the Chancellor's leadership? I admit that this would amount to a *coup d'état* and that shooting could hardly be avoided. But I almost think the Prince [Bismarck] could be persuaded to accept such a course, for *at bottom he is dissatisfied with his electoral law*. If the [German] princes supported it, he would become their standard-bearer "in the crisis of the Fatherland".'[83]

This was precisely what Bismarck proposed at the sitting of the Prussian Ministry of State on 2 March 1890. He told the ministers that he intended to bring into the new Reichstag an Anti-Socialist Bill which would be even more stringent than the one rejected by the Kartell-dominated Reichstag five weeks earlier. Social Democrat agitators would be disqualified from voting or standing for election and could be exiled. Bismarck's aim was plainly to dissolve the Reichstag several times in succession after it had rejected the bill, as it could be relied upon to do, and then finally to proclaim a new electoral law. He put forward the theory that the German Reich was founded on an alliance of *princes*, not of *states*. Therefore 'the princes . . . could decide, if need be, to withdraw from the joint treaty'. According to the minutes of the meeting he went on to say: 'In this way it would be possible to free oneself from the Reichstag if the results of the elections continued to be bad.'[84]

It was Holstein, Kayser and Marschall who drew attention to the dangerous consequences of such a policy for the Kaiser. 'We must save the Kaiser from that fate', Marschall noted in his diary on 2 March after a visit from Holstein.[85] The introduction of a more severe Anti-Socialist Bill would be 'a dark blot on the imperial chronicle, which has looked very good hitherto'. The Social Welfare Bill, on which the Council of State had just finished deliberating, would become 'an object of derision' if a more severe Anti-Socialist Bill including expulsion provisions were introduced at the same time.[86] For Paul Kayser, Bismarck's plan to persuade the Kaiser to introduce a more stringent Anti-Socialist Bill was 'the most masterful move in the whole game of chess; it means checkmate for the King', he warned. Writing to Eulenburg, he commented: 'It is unthinkable that the next Reichstag would accept a more severe Socialist Bill . . . Its reintroduction would therefore be a useless demonstration which would destroy the last remnant of the Kartell and throw down the apple of discord again. After the rejection of such a bill there would be no point in a dissolution. On the contrary, that would unleash the worst possible elements. For it would be death to the Kaiser's proclamations . . . The whole social welfare legislative programme would collapse and the blessing hoped for and rightly expected from it would become a curse.' A more severe Anti-Socialist Bill at this moment would be like a slap in the face for the Kaiser, and it was precisely that, and not the defeat of social democracy, which was quite obviously the Reich Chancellor's real aim. The next consequence of the introduction of such a bill would be the resignation of newly appointed Minister of Trade von Berlepsch, which would be closely followed by that of Heinrich von Boetticher. 'The Chancellor's haste is understandable', Kayser maintained. 'He is 75 years old and no longer has much to lose. But the Kaiser is the *future* of the country; a blow against *him strikes home*!' Since the ministers were still in awe of Bismarck, the Kaiser would have to convene the Council of State again. That would be the only way to prevent surprise attacks from the Chancellor. Then His Majesty would be on the crest of the hill at last. 'Post nubila Phoebus!'[87] Eulenburg sent extracts from this letter to the Kaiser on 4 March. He too warned that a more severe Anti-Socialist Bill would weaken 'the great impact of Your Majesty's proclamations' at home and abroad. The Chancellor's policy was aimed at 'destroying the existing state of affairs, and the sooner the better'.[88] The Kaiser's advisers were incensed to discover that Bismarck had told the ministers at the sitting of the Ministry of State on 2 March that the new, drastic Anti-Socialist Bill was 'particularly wished by the Kaiser'.[89] The crisis came rapidly to a head. In a conversation with Wilhelm II on 3 March, Marschall gained the impression that, although the monarch was thoroughly convinced of the dangers of Bismarck's policy, he was afraid to dismiss the Chancellor. Marschall therefore went with Holstein to visit the leader of the moderate wing of the Conservatives, Helldorff, and persuaded him to talk to the Kaiser the next day. After Helldorff's audience Marschall was able to note with satisfaction that the Kaiser was 'firm and ready for a break'. On 4 March he telegraphed to the Grand Duke of Baden:

'Helldorff saw the Kaiser today, pleaded with him not to concede the bill: this was a totally unnecessary provocation of the people, a policy aimed at creating scandal whereas calm was essential, and would destroy everything the Kaiser was trying to do. Helldorff has the impression that the Kaiser is firm.'[90]

His resolve stiffened by Eulenburg's and Kayser's letters and by Helldorff's visit, Wilhelm II took an uncompromising stand in his confrontation with Bismarck on 4 March. Immediately after the audience Marschall telegraphed to the Grand Duke of Baden reporting that the Kaiser had been determined 'not to give way, and to telegraph to Caprivi if the Chancellor threatened to resign'.[91] On 5 March Wilhelm wired to Eulenburg to say that Bismarck had 'finally given up the Anti-Socialist Bill yesterday upon my urging him to do so'.[92] 'So it would seem that we are over the worst', Eulenburg commented with relief to Holstein, praising the Kaiser's 'steadfastness'.[93] In the next few days, indeed, relations between the Kaiser and the Chancellor seemed to improve. Through Eulenburg's influence Paul Kayser was appointed secretary and personal observer for the Kaiser at the international social welfare conference.[94] At the same time Wilhelm showed that he recognised the difficult position in which the Prussian ministers found themselves by conferring the Order of the Black Eagle on Vice-President and Vice-Chancellor Heinrich von Boetticher, on the anniversary of the death of Kaiser Wilhelm I.[95] There was renewed hope that a *modus vivendi* might be reached, on the basis of mutual willingness to compromise and recognition of the Kaiser's need 'to rule according to his conscience'.[96] In conversation on 10 March 1890 with his banker, Gerson Bleichröder, as the latter reported to the Rothschilds in London, the Reich Chancellor indicated that he would not resign for the time being. 'Whether this will still be the case in a few months' time will depend on circumstances which cannot be foreseen today.'[97] Eulenburg's mood was optimistic, not least because he had just received news of his appointment as Prussian envoy at Stuttgart – a further sign of the Kaiser's increasing power and of the gradual decline of Bismarck's influence.[98]

THE LAST DAYS OF BISMARCK'S RULE

Soon, however, news of a very different order reached Eulenburg from Berlin, and it became only too clear that the Reich Chancellor was determined to go on fighting against the Kaiser. Bismarck was still refusing to introduce the Social Welfare Bill in the Reichstag before the end of the international conference; he had informed the participating nations that they could send an unlimited number of delegates; he had at first invited the larger German federal states, but had then withdrawn the invitations to certain states, including the highly industrialised kingdom of Saxony; he had invited countries such as Spain and Portugal which had no export industry and no labour problems; and he had deliberately selected nonentities to serve on the Prussian delegation and had left out the best candidates, for example Theodor Lohmann and Robert Bosse. It was perfectly clear that Bismarck was out to wreck the international conference, which was 'His Majesty's own work'. The group around Holstein became convinced 'that everything is coming to a head and *the Prince is pushing for conflict*'.[99] As one of them aptly remarked, it was 'as if there were two governments, and whatever the Kaiser does, others undo'.[100]

Suddenly a new controversy arose to add to their troubles: a 'massive Army Bill' drafted by Verdy, the War Minister, under which the Reichstag was expected to approve expenditure of 280 million marks with a yearly increase of 70 million marks. It was plain that the Reichstag would refuse unless major concessions were made in return.[101] Verdy, who was in any case reputed to sympathise with the democratic Progressive Party, seemed ready to concede two-year in place of three-year military service, but in this he was fiercely opposed by Holstein and Marschall, who again appealed for Eulenburg's help. The concession of two-year military service would permanently increase the power of the Reichstag and would certainly add to the social democratic threat within the army, they argued. But to introduce such a major Army Bill without being prepared to make concessions of this kind would be to invite conflict with the Reichstag. 'Today, at the age of 75, Bismarck would no longer fight to get it through; *on the contrary*, the Kaiser would have to fight for Bismarck's policy and in the process he would lose the trust that he has gradually earned through his calm and moderate conduct hitherto, and which must underpin the personal position of every true ruler.' The Kaiser's advisers in the Foreign Office warned strongly of 'the Chancellor's current liking for extreme situations in which he wishes to involve not only himself but the Kaiser'.[102] Just as they had brought first the federal princes and then the Council of State into the campaign against the centralised power of the state controlled by Bismarck, they now advised the Kaiser to summon the commanding generals to a council over which Wilhelm himself should preside. 'This would have a dual advantage', Holstein urged. 'First the Kaiser would hear the wisest views, for the quota of intelligence to be found among our commanders . . . does not exist in *any* other group of people in Germany. . . . Second, for their part the generals would have the chance to get to know their Kaiser. If His Majesty presides as he did in the Council of State he will make an impression on the leaders of his army which they are unlikely ever to forget . . . It is important the the army should acquire such an impression of its Supreme Commander, perhaps soon, for who knows what fate has in store for us!'[103] As was his wont, Eulenburg immediately forwarded this letter to the Kaiser, and when the Grand Duke of Baden returned to Berlin on

8 March Marshall succeeded in convincing him too that 'the Kaiser ought to listen to the commanding generals'. As we shall see, they were summoned to Berlin for 18 March 1890.[104]

The arrival of the Grand Duke in Berlin after an absence of two months led to a marked intensification of the long-drawn-out crisis in which the German government had found itself since the autumn of 1889. Waldersee noted in his diary that the Grand Duke was 'very concerned about the state of affairs' which he had found in the capital, and was pressing for decisive action. 'He sees the full extent of the rift between Kaiser and Chancellor; the former has also talked quite openly about it and said that the Chancellor is continuing to intrigue against him; but nevertheless he wants to keep him for the time being and behave outwardly as if he were on good terms with him. The Grand Duke now thinks, quite rightly, that this state of affairs ought not to go on for too long, and that it is not a good thing for it to become gradually obvious to the whole world that the Kaiser is play-acting. The Grand Duke is very worried about the elections; he says that in South Germany a very marked hostility to the Reich has emerged', partly due to 'antagonism against the Chancellor'.[105] Grand Duke Friedrich was also anxious, however, about the planned Army Bill. On 11 March he had a discussion with Waldersee and the War Minister which shocked him deeply and which was to trigger the last phase of the chancellor crisis. He discovered that these two very senior military figures, and other generals too, were by no means unwilling to make far-reaching concessions to the Reichstag in order to obtain the expansion of the army which they sought. 'What kind of men are these?', exclaimed Marschall with horror on hearing this from the Grand Duke. 'Two-year military service, the strength of the army establishment to be fixed annually – they are prepared to trade it all off!'[106] The Baden envoy hurried to the Foreign Office to convey his master's urgent plea to Holstein that he should get Eulenburg to leave Oldenburg for Berlin at once. The Grand Duke had 'learned from highly placed people that there is an immediate danger of a conflict either at home or abroad'. Marschall and Holstein, 'seriously perturbed and worried', joined in the plea to Eulenburg to come quickly; they were convinced that 'every hour is of value for the Kaiser'. Eulenburg should pretend that his sudden departure was a matter of private business; he should telegraph from Hanover, not from Oldenburg, to let them know the time of his arrival in Berlin, and even then he should not sign his name. Paul Kayser was to know nothing of the visit. Holstein gave no hint of the reason for this panic-stricken appeal; he merely told Eulenburg that the summons had nothing to do with the international social welfare conference. 'Much may go against our wishes at the conference, but the Kaiser's future does not depend upon it.'[107]

Eulenburg hesitated. He complained of nervous ailments in the head and stomach, said that he was expecting a visit from his mother and maintained that he could in any case do no more than mediate, which the Grand Duke of Baden, as the Kaiser's uncle, could do just as well. On the other hand, he cautioned, the Grand Duke was inclined 'to look on the dark side', and the 'feverish atmosphere' which pervaded Berlin had evidently affected and alarmed him, so that he was convinced that a great crisis was inevitable. But of course if the Kaiser had decided to dismiss Bismarck, Eulenburg observed, his own mediation would not be needed.[108]

Even before he had time to find out more about the concessions Verdy and Waldersee were willing to make on the Army Bill, news reached Eulenburg of another sensational development in Berlin. On 12 March 1890 Bismarck had a meeting with the leader of the Catholic Centre, Ludwig Windthorst, to discuss co-operation with the latter's party in the Reichstag. During the interview Bismarck made no secret of the fact that he needed the support of the Centre 'in order to hold his own'. It would be impossible for him to remain in office unless 'the Centre supported his policy in the Reichstag and in the [Prussian] Chamber of Deputies', he admitted. 'Would Windthorst be inclined to do this?' The two erstwhile arch-enemies then discussed the terms of a possible collaboration and seemed to agree in principle. Subsequently Bismarck brought up the demands of the Centre with regard to schools and the possible readmission of the Redemptorist order to Germany for discussion in the Prussian Ministry of State. After Bismarck's dismissal Windthorst remarked to a colleague that 'from our point of view B[ismarck] certainly left office too soon'; only he would have had the necessary authority to dismantle the *Kulturkampf* legislation.[109]

The news of this meeting, which Bismarck made no effort to keep secret, aroused 'anxious misgivings' in the 'patriotic souls' of the Holstein circle. 'Saving the Reich with the aid of the Jesuitical Guelphs is really the limit!', exclaimed Paul Kayser indignantly. The worst thing, however, was that Bismarck had set his 'machinations' in train '*without the Kaiser having any idea of it*'. The Chancellor was exploiting the fact that the Kaiser was in favour of the Army Bill in order to lure him on to thin ice. The Progressive Party was prepared to accept the Army Bill in return for the concession of two-year military service and annual army budgets; Windthorst would accept triennial budgets, but only in return for concessions in the sphere of ecclesiastical policy. 'At home he will perhaps be content with the Redemptorists and the fall of Lutz's government [in Bavaria] – the Jesuits will be next, and the schools will become the victims of Catholicism. Abroad some concession will have to be made to the Papacy ... That will jeopardise the alliance with Italy, and the way overtures are being made to

Russia shows that in any case it is considered unimportant if the Triple Alliance is destroyed.' The common denominator in all these intrigues was Bismarck's desire to deprive the Kaiser of his power. 'Every possible trick is being tried in the art of complicating matters', Kayser warned. 'It is a time-honoured practice, and it left Wilhelm I helpless, so that he had to give his Minister complete freedom of action. Will Wilhelm II be the same? A great turning-point seems to have been reached. For the Kaiser, the crossing of the Rubicon would in this case be a pilgrimage to Canossa.'[110]

In Marschall's view it was now beyond doubt that 'Prince Bismarck is continuing to use all possible methods not only to prevent a positive success for His Majesty in the field of social welfare, but to oppose His Majesty's intentions in everything, leading domestic policy not on to the path of reform but into the ways of scandal, provocation and confusion. As soon as his plan for a more severe Anti-Socialist Bill had failed, the Reich Chancellor threw himself with his characteristic energy into promoting the Army Bill . . . I am no army officer . . . but in *political* terms I am absolutely convinced that if this bomb is thrown into the new Reichstag, not only will His Majesty's initiative in social welfare policy be pushed completely into the background, but we shall be heading for chaos at home.' The concessions which were required in order to obtain a majority for the bill could not be made. But if an *Army* Bill were rejected, the Reichstag would have to be dissolved, Marschall observed. 'One does not need much political foresight to imagine the consequences which a dissolution of the Reichstag, say in June this year, would entail. To hold a new election which the opposition could turn into a plebiscite on the length of military service would in my view be no less than to destroy the already weakened government parties and to create a Reichstag which it would be impossible to live with.' The introduction of the huge Army Bill would amount to a direct negation of the calm, measured, reforming policy which the Kaiser had rightly initiated, Marschall argued. 'That the Reich Chancellor is now rushing ahead with the bill although until recently he opposed it, is symptomatic of the situation. The Social Welfare Bill is to be destroyed, and the Reichstag blown sky-high – then the time could come when the propertied classes regard the Reich Chancellor as the only saviour in their hour of need.'[111] A Prussian minister with whom Marschall discussed the situation agreed with him in thinking that Bismarck was trying to use the Army Bill, as he had done with the Anti-Socialist Bill earlier, to destroy the Kaiser's power. His 'transparent' aim was 'to blow up the Reichstag so as to wreck the Kaiser's social welfare programme and to create the measure of confusion in Germany which Bismarck needs in order to make himself indispensable'.[112] Eulenburg's decision to forward this letter from Marschall to the Kaiser on 14 March also contributed to the final break between Wilhelm and Bismarck, particularly as even Waldersee wrote to the Kaiser to say that in his opinion Bismarck's plans to dissolve the Reichstag again were 'an absolute disaster'.[113] So the moment of truth was now imminent.

'THE GREAT CLASH'

After a meeting with Wilhelm II on 9 March 1890 Waldersee recorded that the Kaiser, who had good sources of information, had heard that Bismarck had complained about him to the French ambassador. Wilhelm had nevertheless declared: 'I think it expedient to keep him on for the time being, and I am therefore acting as if I had not noticed how badly he is behaving towards me; and I shall dine with him again very soon so that people think that we are on good terms.'[114] Four days later the Chief of the General Staff was still able to observe that only a very small inner circle was aware of the 'breach between the Kaiser and the Chancellor'. Even in the Foreign Office there was 'a certain sense of oppression, without anyone believing that a catastrophe was in the offing'. True, a few people were expecting that the Chancellor would 'quit the helm'; but others believed that 'the Chancellor could still recover the upper hand'. On the same day, however, Waldersee noted that Windthorst's visit to the Chancellor had had a decisive effect 'in all right-thinking circles'.[115]

The news of Bismarck's interview with the leader of the Centre Party was like a red rag to a bull in its effect on Wilhelm II. But his anger grew beyond all bounds when he learned that the meeting had come about through the mediation of the Jewish banker Bleichröder. In the eyes of Wilhelm and Waldersee the Bismarck-Bleichröder-Windthorst combination confirmed their own lurid vision of Bismarck's rule as a thoroughly corrupt system under Jewish domination. At the end of January 1890, after the first bitter argument with the Chancellor, Waldersee had already noted that there were 'other reasons for ill-feeling' between Wilhelm II and Bismarck, and that these were 'the most important thing'. 'It has become clear to the Kaiser that the Chancellor has considerable dealings with the stock exchange, especially with Bleichröder, and furthermore he thinks it quite likely – as indeed everyone claims – that the reason why the Chancellor constantly postpones the long-promised tax reform is merely that he has discovered that he will have to pay far more, and in particular he will have to declare the extent of his fortune, which is probably colossal.'[116] A few observers who knew Bismarck well went further still, the Chief of the General Staff recorded in February. 'People are even wicked enough to believe that he makes Bleichröder work in his interest. Herr Schwabach has also been received by the Chancellor.'[117] In view of such suspicions one can hardly be surprised at the Kaiser's remark on 15 March 1890 to Generals Waldersee, Hahnke and

Wittich that he was convinced that Bleichröder's mediation between Bismarck and Windthorst meant that there was 'collusion between the Jesuits and the rich Jews'.[118]

The anti-Semitic Chief of the General Staff, as we know, had long harboured a deep personal antipathy towards Bismarck. Now, in the last phase of the chancellor crisis, he exercised a malign influence over Wilhelm through a constant and poisonous whispering campaign against his enemy. At the beginning of March 1890 he wrote contemptuously of Bismarck: 'I have not the least desire ever to have any dealings with him again; hc has acted despicably towards me too often and I want no more to do with him; the only question, now that he is in decline, is whether I should forgive him everything or call him to account for it.'[119] Waldersee found it 'incomprehensible' that Bismarck had not resigned long ago. The only reason he could think of for the Chancellor's remaining, he said, was 'that he cannot leave because he is afraid of his successor and of the anger which will be unleashed in many whom he has oppressed, lied to and deceived. One already hears it said that he has a very bad character; he has not hesitated to disclaim his friends and those who have helped him most; lying has become a habit with him; he has made use of his official position to enrich himself on a colossal scale and has had his sons promoted with unbelievable ruthlessness although no one thinks them competent!' In spite of Bismarck's outstanding achievements at times of conflict and in the wars of unification, which would ensure him 'an exalted place in history for all times', the Chancellor's 'great errors of judgement' had begun immediately after the establishment of the Reich. Since 1871 his government had been 'permeated with too much evil passion, which detracts greatly from its fine appearance'. In the end, history would judge Bismarck harshly, Waldersee predicted.[120] When the Berlin stock exchange suffered a sharp fall at the beginning of March in the wake of the Bismarck crisis, Waldersee saw that too as 'a trumped-up show by the big capitalists, above all Bleichröder', behind whom stood Bismarck. One could be sure, he claimed, that the Jews would use Bismarck's fall to bring about 'a massive slump', for 'no harm can come to these people as a result; on the contrary, fluctuations always bring them profit; the people who pay the bill are the small speculators and people of limited means, who are naturally very anxious'.[121] The Jews were 'mostly fellows with no homeland, who have no interest in anything but making money, and who – wonder of wonders – mostly support the Progressives, and often even vote for Socialists at elections'.[122] On this point too the Kaiser was of the same mind as Waldersee and saw 'the downward trend of the stock exchange . . . as a trumped-up show', although he was content to 'let it take its course'.[123]

It was yet another issue which proved to be the last straw. This was the Cabinet Order of 1852, mentioned earlier in this chapter, under which the minister-president had to be informed whenever another minister wished to make a report to the monarch. After many days of searching the order was eventually found, and Bismarck read it out to the Prussian ministers – with the exception of the minister for war, who had direct access to the Kaiser as supreme war lord – and asked them to act accordingly.[124] At the same time he pointed out that 'in the German Reich . . . according to the constitution, the Reich Chancellor is the only minister', which meant that direct contact between the secretaries of state of the Reich offices and the Kaiser was not permissible.[125] It was the Chancellor's last attempt to recover his monopoly of power over the central apparatus of government.[126]

On 15 March 1890 Waldersee recorded in his diary: 'The great clash has come!'[127] Bismarck's own description of the dramatic collision between him and the Kaiser in his son's official residence that morning is among the most striking and best-known passages in his memoirs. At 9 a.m. he had been awoken with the news that the Kaiser expected to receive him in audience at the Foreign Office in half an hour's time. The 75-year-old Chancellor did not believe Wilhelm's excuse that he had given the order the previous afternoon. Bismarck began the audience by stating that Windthorst had come to see him, whereupon the Kaiser exclaimed: 'Well, of course you had him thrown out?' The Reich Chancellor tried to explain that he had not only had the right, but the official duty, to find out the intentions of the leader of what was now the strongest party in the Reichstag. When the Kaiser insisted that the Chancellor ought to have 'asked him about it first', Bismarck that he could not allow 'my personal freedom of movement in my own house' to be subject to control. The Kaiser's characteristic retort was 'Not even when your sovereign commands it?' Wilhelm went on to complain that Windthorst's visit had been arranged through Bleichröder – 'Jews and Jesuits', he declared, always acted hand in glove. The Chancellor pointed out that it was not he but the Centre Party leader who had asked Bleichröder to mediate. The Kaiser then suddenly changed the subject to the Cabinet Order of 1852, which he described as 'a dusty old order . . . which had been long forgotten'. Bismarck stood his ground with the young monarch, maintaining that the order 'had been in force as long as our constitution had existed' and that a minister-president could not carry the entire responsibility for government policy without such a regulation. 'His Majesty's three predecessors' had all governed the country under this order. Wilhelm II, however, maintained that it restricted his royal prerogative and peremptorily demanded its withdrawal. Finally Bismarck seized the offensive: taking up a handful of secret reports on Tsar Alexander's current attitude, he strongly advised the Kaiser to cancel the visit to Russia which he had announced. Wilhelm took the papers from Bismarck's hand and was 'quite rightly hurt' and 'undoubtedly deeply insulted' by the reports they contained of remarks made by Alexander III about him and his last visit to St Petersburg. The whole audience, according to Bismarck's account, gave him the impression 'that the Kaiser wanted to be rid of me, that he had altered his intention of going through the initial negotiations with the new Reichstag with me, and of waiting until early summer, when it would be clear whether the dissolution of the new Reichstag would be necessary or not, before deciding on the question of parting with me'. Nevertheless the Chancellor refused to allow himself to be provoked into resigning by the 'ungracious behaviour' of his King. He was still of the opinion, he wrote, 'that it is not for me to take the initiative and thereby assume the responsibility for my own departure'.[128]

So much for Bismarck's own, highly compelling account of the clash on 15 March 1890, from which Wilhelm II emerges with little credit. But how did Wilhelm himself see this decisive quarrel? Only a few days after what had probably been one of the most powerful scenes ever played out in the Wilhelmstrasse, the Kaiser gave the British ambassador an emotional account of how Bismarck had become 'so violent . . . that I did not know whether he would not throw the inkstand at my head'.[129] Years later the Kaiser still recalled that the Chancellor had lost his self-control to such an extent that he might almost have struck him, his King, dead. 'The "Old Man" was raging with anger that day – whipped up beforehand by those villains Herbert and Lerchenfeld – so much so that I could not get a word in, and he seized the inkstand and threatened me with it, quite senseless with fury and forgetting himself completely. That was the first "outward" cause of his dismissal', he wrote in his own hand in the winter of 1903.[130]

Immediately after the violent set-to between Wilhelm and the Reich Chancellor, Waldersee arrived at the Schloss with Hahnke, the Chief of the Military Cabinet, and Adjutant-General von Wittich to report to the Kaiser. The Chief of the General Staff wrote a lengthy and hitherto largely unknown account of what had happened at the Wilhelmstrasse. According to this account, Wilhelm began the meeting with his top generals by announcing 'I have just come from seeing the Chancellor.' He went on: 'We had a serious dispute which will probably bring things to an end, and it is my wish that you should be aware of the whole episode. I told the Chancellor that it was not acceptable for him to enter into negotiations with Herr Windhorst [sic] at the present moment without telling me about it beforehand. He responded at once with great vehemence that as responsible minister-president he could receive whomever he wished; moreover he was cleverer than Windhorst and had no one else who could negotiate with him. On my replying that I must insist on my right to be informed in advance, and that at this moment negotiations with the Centre would make the worst possible impression in the country, and that I had heard only last night from Herr v.

Helldorff that the Conservatives would dissociate themselves from government policy at once, and that it gave me great cause for concern that Bleichröder had been the intermediary, he became absolutely enraged and talked nonsense, for instance saying that he was surrounded by spies, forgetting that his meeting with Windhorst had been reported in all the newspapers; sometimes he became subdued and even wept.' 'I am convinced', the Kaiser continued, 'that there is collusion between the Jesuits and the rich Jews. A second cause of conflict was his order to the ministers – on the basis of a Cabinet Order of 1852 – not to make any direct reports to me. After a violent argument I ordered him to submit a new order to me, suspending that of 1852.' Following this peroration from the Kaiser, Waldersee ventured the opinion that 'in spite of all this the Chancellor would not hand in his resignation, for he clung to office too much; he could not go because he would be leaving too much dirty linen behind, his son would not be his successor after all, and finally because by holding on he still hoped to regain the upper hand, and unfortunately also because he was too closely allied with the Jews and could not escape from them. But as the present situation was quite untenable it would be best, if he did not seek his own discharge, to *give* it to him, and as soon as possible.' 'Both Wittich and Hahnke agreed with that', Waldersee recorded. 'But the Kaiser replied that he would prefer the Chancellor to tender his own resignation, as it would look better in the eyes of the world.' At this, Waldersee made his final pitch against Bismarck by giving the Kaiser 'for the first time, a frank account of my views on the Chancellor, without sparing him anything. Hahnke and Wittich were astonished, but the Kaiser not at all; he was very well aware of the various accusations I had to make. Among other things I said: he has been very skilful in maintaining his reputation as a past master in handling foreign affairs, and as irreplaceable in that respect. But it seems to me that the situation in which we find ourselves is a not a good one; with all his skill he has not succeeded in preventing both France and Russia becoming our enemies, and growing stronger and bolder year by year. This is really a very serious state of affairs! Your Majesty has taken over the Reich at a very difficult time.' To this the Kaiser replied: 'It is strange, Herr v. Helldorff put exactly the same argument to me yesterday evening and thought it absolutely justified. Things are indeed going very badly with Russia at the moment; strong opposition is being whipped up against me and Tsar Alexander speaks of me in the most derogatory way; among other things he says I am mad. The ill-feeling towards us is constantly growing, and I shall not go to Krasnoe under any circumstances. As for our domestic affairs, everyone can see that things look very bad. What is the great Chancellor doing about that? What services has he rendered?' Waldersee derived some satisfaction from these remarks, since 'this was how I had seen the situation for years, but I was rarely believed'. He went on to record how the conversation ended. 'After much discussion of the Chancellor's outrageous behaviour in making derogatory remarks about the Kaiser to foreign diplomats I said finally: "As things stand, it is impossible for Your Majesty to do business with the Chancellor any longer. The present situation is intolerable for You and harmful for the Fatherland. I suggest that Your Majesty should come to a firm decision on the persons to appoint, and then act." In conclusion I added: "God grant that Your Majesty may hold Your head up high and direct Your choice to the right people." He then gave me his hand and said, "I think all will be well", adding "Good hunting!" in his usual lively manner. I replied in kind, "Good hunting."'[131]

Since Bismarck refused to resign of his own accord and Wilhelm was reluctant to take on himself the odium of dismissing the Chancellor, the next two days saw both attempting to seize the tactical advantage. On 16 March the Kaiser sent the Chief of his Military Cabinet to Bismarck with instructions to demand 'the order . . . or his resignation!'[132] Bismarck repeated his contention of the previous day that the minister-president could not function without the rights assured him by the Cabinet Order; if the Kaiser abolished the order, 'the title "president of the Ministry of State" would have to be abolished too'. Hahnke returned to the Schloss convinced that he would nevertheless succeed in accomplishing his mission. But when he called on Bismarck again the following morning, this time with the specific order – as the Chancellor recalled – 'that I was to hand in my resignation; I was to come to the Schloss in the afternoon to be released from my duties', Bismarck claimed to be too unwell to do so and said he would write instead.[133] Early that afternoon the Kaiser was able to inform the Chief of the General Staff that 'everything is in order. Hahnke has been to see the Chancellor; he is not sending an order, but his resignation.'[134] When the resignation letter failed to arrive Wilhelm sent Chief of the Civil Cabinet Hermann von Lucanus to the Reich Chancellor's palace to enquire 'why the letter of resignation demanded in the morning has not yet been delivered'. Bismarck responded that the Kaiser could dismiss him at any time without the need for him to tender his resignation. But he was intending, he said, to draw up the letter in such a way that he could publish it. 'I would not presume', he said, 'to take the responsibility for my resignation on myself, but would leave it to His Majesty.' Although Lucanus disputed Bismarck's justification for 'public explanation of the genesis' of his resignation, the Chancellor reiterated that he was tendering his resignation only on the orders of the Kaiser.[135]

One reason for Wilhelm's impatience to secure Bismarck's resignation was that a number of consular reports from Kiev, some of them long out of date, happened to be submitted to him at precisely this tense moment. Not only did these lead him to believe that a Russian attack on Austria was imminent; he was also convinced that Bismarck had tried to conceal the reports from him. On the morning of 17 March he told Waldersee that he had 'discovered that the Chancellor had held back many reports from Russia from him, and that things looked much worse there than he could have known; only recently the Chancellor had had an important report from Odessa about armaments filed away without submitting it to him first'. Instead of trying to calm the Kaiser down Waldersee fanned the flames of his anger, no doubt still hoping that his own appointment as Reich chancellor was not far off. 'A great deal of wrong has been done in this way and that is one of the reasons why the Chancellor cannot leave his post. He has too much dirty linen to hide', he told the Kaiser.[136] Bismarck was incredulous at these new accusations from the Kaiser, especially as they concerned his conduct of foreign policy, which he had always regarded as his forte. 'So, because a consul had reported on a few military events within his purview, some of them three months old . . . Austria was to be put on the alert, Russia threatened, preparations for war put in hand, and the visit which His Majesty himself had announced on his own initiative, cancelled; and because the consul's reports had arrived late, I was implicitly accused of high treason', he wrote in his memoirs. He went on to point to the 'caprices of fate' by which, the very day on which he was accused of suppressing the news of an imminent Russian attack, the Russian ambassador had called on him to announce that he had been authorised to negotiate with him the renewal of the Reinsurance Treaty, which was due to run out in June 1890![137] At the special sitting of the Ministry of State which Bismarck convened in his official residence on 17 March, the Reich Chancellor was thus able to make it appear that the fact 'that he could no longer support His Majesty's foreign policy' was the principal reason for his resignation.[138] He did it again with a vengeance in his celebrated resignation letter of 18 March 1890.[139]

That evening all the commanding generals and inspector generals of the army assembled in the Schloss. In an address lasting some twenty minutes, spoken according to Waldersee 'with great composure', the Kaiser set out his differences with the Reich Chancellor, and explained 'that in order to remain master of the situation he had had to issue an ultimatum to him insisting that he submit. The Chancellor had promised to tender his resignation; he would accept it and appoint Gen[eral] Caprivi Chancellor. Then the Kaiser spoke of our relations with Russia, which were bad, but about which the Chancellor had tried to deceive him.' He would never permit the Russians to invade Bulgaria, he added, for he had 'sworn to keep faith with the Emperor of Austria the previous year, and would hold to it'.[140] None of those present contradicted the Supreme War Lord. Only as they left was the old Field Marshal General von Moltke overheard to remark: 'Our young master will give us a lot to think about.'[141]

THE BISMARCK CRISIS: A POWER STRUGGLE BETWEEN KAISER AND CHANCELLOR

The confused story of Bismarck's dismissal is often, and quite rightly, presented as the end of a long and successful epoch, the story of the 'decline and fall from power' of the founder of the Reich.[142] But at the same time it was also a beginning, the story of a young ruler's assumption of power, the birth of a new, completely different system of government with new values and a new style. Indeed Bismarck's dismissal can only be understood at all as the conflict between two powerful men and the two systems of government which they embodied. For some of those involved – second-ranking figures like Paul Kayser, for instance – the conflict may have been about a particular cause; but for the main actors in the drama almost all differences of opinion on objective issues were secondary, as we have repeatedly seen. Throughout the months over which the dismissal crisis was played out the real issue was the struggle for power. On both sides the attitude taken to the ostensible cause of conflict usually depended on the advantage that could be derived from it and the damage that could be inflicted on the other camp. The true cause of the bitter feud can be detected in numerous remarks by those involved. At the end of January 1890, for instance, when the crisis had just entered its most acute phase, Wilhelm II's closest confidant, Waldersee, noted: 'The Kaiser is hurt that the Chancellor always makes his newspapers refer to "the Chancellor's policy"; he wants to be regarded as having the overall direction of policy himself.'[143] Prince Heinrich, the Kaiser's brother, rightly observed that Bismarck's resignation was 'the natural outcome of the growing independence of the Kaiser'.[144] Wilhelm himself confirmed this view of the crisis when he spoke in April 1890 of Bismarck's 'open disobedience and perfidy', and exclaimed: 'The Hohenzollern dynasty is surely worth enough not to have to give way to Bismarck.'[145] The Kaiser's tutor Hinzpeter was convinced that the only historically correct interpretation of Bismarck's dismissal was that it was 'an act of self-defence of the monarchy against the looming danger of strangulation by a bureaucracy which had become too powerful in the person of the Chancellor'.[146] Hinzpeter went further still. Perhaps the person who knew his pupil's character better than anyone else, he commented later that, for Wilhelm II, Bismarck's dismissal had been 'a natural desire to save his own personality'. This desire was 'particularly strong' in him, 'even stronger than all other tendencies and intentions, as was indeed seen in 1890'.[147] Similar psychological interpretations of Bismarck's attitude were also current. Paul Kayser, for instance, expressed the opinion in early March 1890, when the Chancellor's dismissal had already been decided upon and was only a matter of time, that his resistance to the social welfare programme could really only be explained as the 'psychological process in the soul of a powerful man . . . who after years of absolute monarchy [sic] feels threatened in his *monopoly* of power'.[148]

There were, admittedly, a few exceptional instances on both sides when inner conviction or personal passion played a greater role than tactical manoeuvring for political advantage. Wilhelm II's suspicions about the intentions of the Russian army or his constantly recurring feelings of hatred for 'Jesuits and Jews' are examples of this. It is more difficult to find instances when Bismarck was driven by conviction. His insistence, during the final clash with Wilhelm, on his right to receive any member of parliament whenever he wished in his own house, and on his right as minister-president to be informed of the intentions of any of his ministerial colleagues before they reported personally to the Kaiser, may seem to have been a matter of immutable conviction. Yet in both cases it was a question of repudiating imperial demands which would have strengthened the power of the monarch enormously while reducing that of the chancellor beyond recognition, had Bismarck given way. Moreover, in this final conflict between the Kaiser and the Chancellor on 15 March 1890 it was already so evident that Wilhelm intended to dismiss Bismarck that the latter would certainly have been on the look-out for tactically advantageous reasons for his retirement. The claim which was put about from Friedrichsruh very soon after the dismissal, and which proved so damaging to the Hohenzollern monarchy, namely that Bismarck had been anxious about Wilhelm's sanity and had wanted to stay in office only in order to rescue his life's work, was only the logical continuation of this struggle for power by other means.

But apart from these very few exceptions, was there a single difference of opinion between the young monarch and his 75-year-old opponent which was not determined principally by tactical considerations, in the long crisis which lasted from the summer of 1889 until March 1890? Ever since Wilhelm attended the Stoecker meeting of November 1887, his and his wife's intellectual affinity with the orthodox right wing of the German Conservative Party had been common knowledge. That he continued to hold these views inwardly is very clearly confirmed by Waldersee's diary. Yet at the instigation of Holstein, Eulenburg and Hinzpeter, who suspected that Bismarck might change tack and form a blue–black coalition dangerous to the monarchy, the Kaiser repeatedly and demonstratively spoke up for the Kartell parties, thus securing himself the support of the moderate Conservative and National Liberal bourgeoisie in the coming conflict with the Chancellor. Wilhelm II's initiative on social welfare policy may well have had its roots in earlier Christian charitable impulses implanted by Hinzpeter, but the actual motive behind it in January 1890 was his quest for popularity among the masses and his determination to take personal control of policy in this field at least. Wilhelm himself revealed how small a part inner conviction played in this dramatic intervention in social policy when he declared to Waldersee on 7 February 1890: 'I want to do all I can and I would not wish to be accused later of having failed to do anything. We shall soon see; if it does not work, at least I have done my duty.'[149] And yet from the moment he took action – from the Crown Council on 24 January and the issue of the uncountersigned imperial proclamations on 4 February 1890 – there was no going back for Wilhelm II. From then on Bismarck's every action was judged not for its intrinsic worth but only for its capacity to endanger, or even merely to overshadow, the Kaiser's social reform initiative and hence his reputation and future position. We have seen how several members of the Kaiser's coterie of advisers, Waldersee and Eulenburg in particular, were convinced of the need for a *coup d'état* to revise the constitution and do away with universal suffrage, especially after the catastrophic Reichstag election results of 20 and 27 February. But all agreed that this solution was impracticable *for the time being*, since a policy of force would simply have restored Bismarck's power and made the Kaiser's social reform initiative look like a ludicrous blunder. In the Kaiser's camp there was a strong feeling that a major Army Bill was needed – indeed, barely a year later the Kaiser was to give orders to the new Chancellor Caprivi to introduce just such a bill[150] – yet when on 12 March 1890 Bismarck put out feelers towards the Centre Party, whose support in the Reichstag was now indispensable for all legislation, his action was regarded by all the Kaiser's advisers as an attempt to destabilise the situation and permanently undermine the Kaiser's authority. As Marschall von Bieberstein put it: the Chancellor's goal was 'to blow the Reichstag sky-high, so as to frustrate the Kaiser's social reform plans and cause the degree of confusion in Germany which Prince Bismarck needs in order to prove himself indispensable'.[151]

Clearly, in this struggle for power the Kaiser had seized the initiative while the Chancellor had been forced on to the defensive from the start. Wilhelm's starting-point had been his inherited power, above all as king of Prussia and supreme war lord, and his extensive military and aristocratic court. Since mid-1889, however, an influential clique of advisers had formed around Friedrich von Holstein and Count Philipp zu Eulenburg, which had been joined by the envoy from Baden, Adolf Freiherr Marschall von Bieberstein, the Director of the Colonial Department Dr Paul Kayser and the journalist Dr Franz Fischer, and which also maintained close contact with Waldersee as Chief of the General Staff, Wilhelm II's former tutor Hinzpeter, and Grand Duke Friedrich of Baden. This secret and completely informal group was almost always extremely well informed about Bismarck's views and intentions – it was not for nothing that he complained of being surrounded by spies – while he had scarcely an inkling of the existence of the clique and its machinations. He blamed people who had no part whatever in the crisis, such as Count Douglas and the painter Heyden, and regarded his unfortunate deputy, Heinrich von Boetticher, as the worst of traitors. Even when writing his memoirs in exile at Friedrichsruh he had to admit that he had still not discovered, 'with any reliable certainty, the true reason for the break'.[152]

In the battle for power Bismarck's strength – apart, of course, from his incomparable political skill and his enormous prestige after twenty-eight years in power – lay in his rigorous control of the central apparatus of the state. We have seen how the Kaiser tried to circumvent this stronghold in the first weeks of the conflict by appealing for the support of the federal princes, in particular the Grand Duke of Baden and the King of Saxony. He pursued the same aim later by convening the Prussian Council of State and the international conference on social welfare, and finally by holding the meeting of commanding generals. But inevitably the real core of the conflict was the struggle for control of the power of the state. All Bismarck's attempts to reduce his own power to a degree acceptable to Wilhelm by giving up his Prussian offices were frustrated, both by the indissolubility of the symbiotic relationship between Prussia and the Reich and by the personal 'constitutional responsibility' which he held.[153] The stronghold of state power could not be bypassed. It was thus entirely within the logic of the situation that the final clash between Kaiser and Chancellor on 15 March was not a question of material differences of opinion, but of which of them had the right to negotiate with the Prussian ministers and the party leaders in the Reichstag.

With Bismarck's removal the way now lay open for Kaiser Wilhelm II to take over personal control of German policy. The dangers which this entailed for the monarchy were clearly recognised by none other than his tutor Hinzpeter. Writing to the future Minister for Ecclesiastical Affairs, Konrad Studt, five years after the event, he said of Bismarck's dismissal: 'That catastrophe was nothing other than an attempt by the monarchy to free itself from the suffocating hold of the bureaucracy. It seems that by an extraordinary effort of strength the attempt succeeded. The responsibility which the monarchy thereby took upon itself is very great. If it does not prove itself equal to this responsibility, it has incurred a danger which may bring about its destruction.'[154]

CHAPTER THIRTEEN

The haphazard transition: from the Bismarcks to the New Course

'WHAT A DAGGER IN MY HEART!'

IMMEDIATELY after the fall of Bismarck the battle to win the approval of governments abroad and of public opinion at home, which had been fundamental to the dismissal crisis from the outset, was resumed by both sides by other means. The founder of the Reich appeared in public more often in the eight days following his resignation than had been his custom in a whole year.[1] Deeply offended and angry, he took care to spread his own version of his downfall, according to which he had been 'thrown out' by a Kaiser who was perhaps not entirely of sound mind.[2] Herbert Bismarck had resigned with his father, and he and his mother talked of the Kaiser 'in the most disrespectful terms'.[3] Even before he left Berlin Prince Bismarck made it clear that he intended to create a 'Fronde' against the Kaiser. He had a 'stormy' interview with the Grand Duke of Baden during which he accused the Kaiser's uncle of interfering in matters which concerned only the Reich Chancellor. Tempers ran so high that the Grand Duke stormed out of the room.[4] After months of attempts by Bismarck and Wilhelm to take advantage of each other, Wilhelm suddenly could not do enough to soften the disastrous impression that the dismissal of the founder of the Reich was bound to make. He heaped Bismarck with titles and honours, offered him the Reich Chancellor's palace as his permanent home[5] and sent telegrams and lengthy self-justifying letters to Emperor Franz Joseph, Queen Victoria and others, in which he portrayed the resignation as an act of mercy which had become necessary for the sake of a highly meritorious but now regrettably dangerously unwell old servant. He went so far that even his Flügeladjutanten occasionally had the impression that he would have preferred to call Bismarck back.[6]

The Kaiser's tactic of killing his opponent with kindness was already apparent at the moment of Bismarck's dismissal. On 20 March 1890 Cabinet Chiefs von Hahnke and von Lucanus appeared at the Reich Chancellor's palace with two formal letters in which the Kaiser, while accepting the Prince's resignation 'with a heavy heart', conferred on him the Dukedom of Lauenburg (which Bismarck declined with the caustic remark that, at most, he would use the title if ever he wanted to travel incognito)[7] and announced that he had been appointed colonel-general with the rank of field marshal.[8] On 22 March (his grandfather's birthday) he sent Hinzpeter a hypocritical telegram undoubtedly intended for publication in which he lamented: 'My heart is as heavy as if I had lost my grandfather again! But it was destined by God, and so I must bear it, even if it should destroy me. The position of Officer of the Watch on the Ship of State has fallen to me. We must continue to steer the old course: and so full steam ahead!'[9]

If the reaction to Bismarck's fall was at first astonishingly calm in most parts of Germany, in the capitals of Europe there was considerable anxiety at the idea that the powerful German Reich was now to be governed by an 'immature youth' of thirty-one. The mood in Russia was carefully analysed by a young diplomat in St Petersburg who was destined to have the task, as ambassador in 1914, of transmitting the German declaration of war to Russia. In a private letter to Friedrich von Holstein of 20–1 March 1890, the First Secretary at the German embassy, Count Friedrich von Pourtalès, stated that there was great consternation in the Russian capital at events in Berlin, not only because Bismarck had pursued a distinctly pro-Russian policy and had served as the guarantor of peaceful existence between the two empires – the Reich Chancellor even told the Russian ambassador that he had been dismissed because of his pro-Russian policy[10] – but also because the relationship between Alexander III and Wilhelm II, which was in any case tense, had been further strained by Wilhelm's pro-labour initiative. There was 'the utmost astonishment' in St Petersburg, Pourtalès reported, 'that the Kaiser can have brought himself to accept his [Bismarck's] resignation'. The decision had given the Russian government all the more cause for concern because 'the Kaiser's recent activities in the field of social welfare' had seemed 'highly suspect to the Tsar' and had alienated him even more from the German monarch. Pourtalès also described the recent rapid change in Russian opinion of 'our young ruler'. 'Only a few months ago all that was known of him was that he liked travelling about and spreading alarm, and would occasionally deliver a fiery speech.' But since Bismarck was in charge, this behaviour had been regarded 'with a certain rather malicious pleasure'. 'Things are now substantially changed. Various events have taught the Russians that Kaiser Wilhelm II represents a force to be reckoned with. But his most recent conduct has at the same time provoked a change in anti-German circles here.' Among the Russian middle classes, in which 'democratic ideas are widely disseminated', the traditional hatred for Germany had given way to a more positive mood as a result of Willhelm II's initiative on social welfare. 'On the other hand recent events in Germany have caused some shaking of heads in conservative circles . . . where monarchical considerations still caused many people to desire co-operation with Germany; these events . . . have destroyed any remaining spark . . . of desire to be on good terms with us. These circles see our young Kaiser's behaviour as the action of a frivolous young man who is thereby undermining his own throne and the monarchical principle in general. In these circles I heard it said repeatedly yesterday that Prince Bismarck's resignation makes war seem more likely.'[11]

Another future ambassador, Count Anton Monts, who was at this time serving at the German embassy in Vienna, sent an equally pessimistic report on 24 March on the reaction in the Austrian capital to Bismarck's dismissal. Writing privately to the imperial Flügeladjutant Carl Count von Wedel, he commented that events in Berlin had made Kaiser Franz Joseph 'very worried; he is afraid of war and of unpleasant surprises, particularly if our sovereign's social policy should fail and he decided to try his hand at foreign politics'. According to Monts the Austrian people had at first been very much inclined to take Bismarck's side against Wilhelm II, but had since been reassured by the calm atmosphere in Berlin and the striking indifference towards the fall of Bismarck in South Germany. While there was much talk in the Danube monarchy of the ingratitude of the young Kaiser, 'the masses' were visibly impressed by the 'courage and independence of His Majesty'. Statesmen and politicians in Vienna, on the other hand, were 'all very discouraged', Monts continued. 'Uncertainty lay ahead, one was at the mercy of an immature youth who had shown the most unbelievable ingratitude towards his father and towards the benefactor of his dynasty.' The end result, Monts observed, was that 'the deplorable circumstances of the departure, the Fronde which it is unfortunately all too clear the Chancellor has already begun, the immaturity of the decisions taken, in short the last 8 days have used up the entire capital of authority which the great and good Kaiser Wilhelm I and his still greater Chancellor had built up over decades, the interest on which would have sustained the successors of both, and the organs dependent on it, for many years.' It was 'quite impossible to predict' what further consequences there might be, Monts warned, in view of Italy's doubtful fidelity as an ally. 'It is also an immense loss to Salisbury.' 'It is high time', his jeremiad concluded, 'that the sun shone on Germany once more. Since the passing of Kaiser Wilhelm we have had little to rejoice about.'[12] No less concerned than Monts was his superior, the ambassador Prince Heinrich VII von Reuss, who in a letter of 2 April deplored the 'alarming impression' created by Bismarck's dismissal 'in South Germany and abroad', and commented: 'It is impossible not to be anxious about it still, and it has been very severely criticised.'[13]

The Habsburg Emperor was still further unsettled by his correspondence with Bismarck, which left no room for doubt as to the true reasons for the Chancellor's dismissal. In a particularly friendly letter of 22 March Kaiser Franz Joseph (who, it must be remembered, knew nothing of the Russo-German Reinsurance Treaty) thanked Bismarck for his 'consistent and loyal co-operation' and expressed the hope that the 'firm bonds of friendship between Austria and Germany' which Bismarck had forged would be maintained 'in the difficult times in which we live . . . as a sure defence not only for the Allies, but also for the peace of Europe'.[14] The ex-Reich Chancellor's reply made no bones about revealing not only that he had been dismissed against his will, but also that he viewed the future with the greatest possible anxiety. He declared himself convinced that the continuation of the Austro-German alliance would be 'unaffected by any change of ministry' because it was based on 'the unalterable needs of both empires and of their peoples'; but he regretted being unable to carry on working to further 'strengthen and develop these relations and those of the Reich and its princes'. That, however, had not been 'the wish of my all-gracious sovereign', he wrote. 'I have always striven', Bismarck proudly proclaimed, 'to win for our personal monarchy, *qui règne* <u>*et*</u> *qui gouverne*, its constitutional right, which has been wrongly obscured, and when I think back to the day in September 1862 when I found my late lamented sovereign in front of the deed of abdication he had himself drawn up, and became his minister, I can claim that since then the authority of the monarchy in Prussia and in the rest of Germany has regained its strength. To strengthen it still further I would gladly have continued to serve my all-gracious sovereign, and my health is still good enough to do so. His Majesty did not allow it, and now I can support our great sovereign only with my prayers.' After which the ex-Chancellor stated, with unequivocal emphasis, 'that I am too dutiful an officer and vassal of my sovereign to desert my post *of my own free will*, in view of the crises which we seem to be facing in the internal affairs of our country'.[15]

It was now Wilhelm II's turn to justify his conduct to the Habsburg monarch.[16] On 3 April he sat down to compose a twenty-page letter, 'written in his own hand and composed absolutely without assistance',[17] which took him two days to finish. This letter, which was delivered to the Austrian Emperor by Carl Wedel, is not only the most complete account of the Bismarck crisis from the young Kaiser's point of view, but also the longest document he wrote during these critical years, and it provides several interesting insights into his mentality at the time. As Prince Bülow was to remark in his memoirs, the document contains numerous 'exaggerations and fantastic notions, not to speak of some palpable untruths'.[18]

Wilhelm began with the affirmation that he considered it his duty, 'given the deep and warm bonds of friendship which unite our countries and above all ourselves', to give his 'dear friend' Franz Joseph 'frankly and clearly a confidential survey of what lay behind, and eventually led to, the resignation of Prince Bismarck'. Contrary to the 'spate of suppositions and deductions in the press' and the 'official and semi-official hints' he wanted to give his imperial ally 'no more than a simple description of the course of events . . . without polemics or criticism'. He wished to observe at the outset, Wilhelm wrote, 'that it was not a matter of foreign policy which gave rise to differences of opinion between the Prince and myself, but purely internal and mostly tactical points of view'. Then he went back in time, tracing the beginning of the crisis to the spring of 1889. 'When the coal strike broke out in May last year and quickly grew to dimensions which threatened the whole nation in every aspect of its working life, once the normal security measures had been taken through troop deployments etc., the causes of the strike were investigated, as is natural. There were consultations in the Ministry of State, with which I did not concern myself at first; meanwhile, however, through my friends – particularly through my tutor, Privy Councillor Hinzpeter, who is Westphalian and lived on the spot – I instigated surveys and investigations concerning the relationship between employers and workmen, the state of the industry etc. . . . Soon, however, the ministers asked me to attend their consultations, as the Prince was quite intractable and the negotiations were not making the slightest progress', Wilhelm claimed, continuing: 'I appeared and assisted. It became clear at once that the Prince took a diametrically opposite view from myself and the ministers. He wanted to let the strike "rage and burn itself out completely" throughout the country. He rejected all notions of the state exercising its authority to intervene and expressed the opinion that that was the industry's own business; it must be allowed to fight out its private feuds. I, on the other hand, was of the opinion that this movement had already gone beyond the bounds of a private industrial dispute, and I found myself in agreement with the entire Ministry of State that if the King did not speedily take this matter in hand, no end of damage and mischief would be done to the country. Accordingly the old officials, who had lost their heads and had only made the confusion worse, were dismissed and replaced by excellent, qualified staff. As soon as that had happened I received deputations from the workers and the proprietors of the mines, which, as is well known, was successful. The Prince disapproved of this action too, for it was clear that he was increasingly inclined to take the side of the big industrialists, and regarded the workers' movement as partly revolutionary and totally unjustified, and thought that it should be curbed and cured with "Blood and Iron" alone, that is to say with grape-shot and repeating rifles.'

Once this crisis was over, Wilhelm continued, 'the Prince withdrew to the country, where he remained for 8–9 months until 25 [sic] January of this year'; what was more, 'during this time he had as good as cut himself off from the rest of the country, and the only person with whom he was in contact regarding the social welfare initiative was the old Commercial Councillor Baare – one of our biggest employers – who was the sworn enemy of this idea'. Wilhelm himself, however, had used this time 'to collect information on social welfare legislation; I sought guidance from all sides on the situation of the workers and on their possible and impossible wishes; I made contact with the Reichstag through its leaders etc. In the autumn I reached the clear conclusion and conviction that time was precious and that it was imperative to tackle the law on social welfare quickly, so that the Social Democrats could not steal a march on us and take up the same cause, which I had clear information that they intended to do. Therefore in the course of the autumn and continuing into January, on three different occasions, I first asked the Prince, then begged him, and finally informed him that it was my wish, that he should start work on amending social welfare legislation and submit an order to me on the subject for publication. He refused this three times very abruptly, saying that he did not wish it, that he was fundamentally opposed to it, and that was the end of the matter. Thereupon I sat down [the Kaiser continued] and worked through 2 nights on a memorandum giving a historical description of industrial relations in our country and indicating a series of principal points which in everyone's view constituted the worst evils, which must be tackled immediately through legislation. As soon as I had finished the memorandum I convened a Council of Ministers and summoned the Prince from Friedrichsruh. During this time the debates on the Anti-Socialist Bill were taking place in the Reichstag. They were very unpleasant, and in the course of them the Kartell parties went over to the opposition, driven to it by the Chancellor's inflexible self-will. They had undertaken to get the bill through for him if he would only declare that the expulsion clause would be formulated as "for consideration" not dropped. On 25 [sic] January I held the Council of Ministers of State and explained my views on the basis of my memorandum, and concluded with the request that the Ministry should discuss the points in it under the chairmanship of the Prince, and also that an international conference should be convened, and that they should submit two proclamations to me on the subject, for publication. This was followed by a discussion during which the Prince at once repeated, with emphasis, the hostile viewpoint he had taken in the spring [of 1889], and condemned the whole affair as impracticable. The ministers were so afraid of him that none dared venture an opinion on the matter. Finally I came to the expulsion clause in the Anti-Socialist Bill, which was to be accepted or rejected the next day, and I earnestly implored the Prince to help the government parties and save the affair from ending on a note of discord in the Reichstag in such a lamentable way, by holding out the prospect, at the final vote, that the clause would be treated as "to be considered". At the same time I mentioned that I had received the most earnest entreaties on the subject directly from men who were loyal to the King and to the government. By way of answer he flung his resignation at my feet – it hurts me to use the expression – most disrespectfully and curtly. The ministers remained silent and gave me no support. I of course did not accept his resignation, the Prince got his way, the bill was rejected and the Reichstag broke up in an atmosphere of universal rage and displeasure, about which I was regaled with a variety of comments about feebleness etc., so this was the mood that was spread around the country by way of preparation for the new elections. The direct results of this are now absolutely plain for us to see.'

With that, the Kaiser recalled, the Bismarck crisis had entered into an acute phase. 'You will surely be able to imagine the deep pain I felt from that moment on, when I was forced to recognise that the Prince *did not want* to work together with me', the Kaiser's letter to Franz Joseph continues. 'Now began a terrible time for me. While the proclamations were being discussed, he tried to bring in all kinds of other things, and constantly angered the ministers. When he finally brought me the 2 proclamations to sign he declared that he was absolutely opposed to them; they would bring ruin and disaster on to the Fatherland and he advised against them. If I nevertheless signed them, he would co-operate with this policy only *as long as* he found it compatible with his own views. If that were no longer the case he would go. The proclamations were published and the enormous success they had showed the Prince, who was utterly taken aback, that he had been altogether on the wrong track, that his opposition had been useless and that I had been right. Now began the preparations for the invitations to the conference, the convening of the Council of State under my chairmanship. He immediately started a little war against me, behind the scenes, not always by the most honourable means. It pained me most bitterly, but I accepted it calmly. On the one hand I was too proud to allow myself to be drawn into it; on the other I was still too fond of the man whom I worshipped! Soon, however, the conflicts multiplied on all sides. He suddenly prevented the ministers from reporting directly to me by digging out an unknown order which had been buried for 30 years. He took away all the work of the Reich secretaries of state and wanted to do everything and countersign everything himself. Meanwhile his health worsened from week to week, he could no longer sleep, his nerves gave way. He had weeping fits in the night and sometimes also during audiences. His doctor declared that if this state of affairs continued for 3 weeks longer the Prince would die of apoplexy! Eventually towards the end of February the Prince announced to me during an audience that he could not continue with his nerves and health in the state they were in, and asked to be released from some of his duties. I asked him to suggest to me what should be done, entirely in accordance with his own will and wishes, as I wished to avoid giving even the slightest impression that I was sending him away, or looking forward to his departure.

After lengthy negotiations he came to an agreement with the Chief of my Civil Cabinet, whom he had been to see for this purpose, that he would give up the presidency of the Ministry of State and wished only to retain the chancellorship and foreign affairs. After a few weeks he wanted to give that up too and retire completely on about 20 February or at the beginning of March. I agreed to his proposals with a heavy heart', declared Wilhelm, 'and accordingly an order was drawn up in accordance with his instructions and completed, except for the date, which he reserved the right to determine. He himself told me that he was quite satisfied with this solution and declared that he would now inform the Council of Ministers of what he had done. 2 days later he came to report to me and to my immense astonishment he abruptly informed me that he had absolutely no intention of going; he was staying! When I asked him the reason, in bewilderment, he said it was because when he had informed them of his departure the ministers of state had not immediately begged him to stay at all costs, and that the gentlemen's expressions had been "too pleased" about it. From this he had concluded that they wished to get rid of him, and that had aroused the old spirit of resistance in him, and he would now certainly stay on "just to annoy the ministers"! That was all he said. I could only reply that I was very glad to know that he would remain at my side, but hoped that the increasing burden of work and worry would not harm his health.'

That, according to the Kaiser's account, was the beginning of the last stage of the crisis. 'From that day onward war broke out', he wrote. 'In every report the Prince sought to discredit the Ministry; the gentlemen whom he had himself chosen 12 years ago and trained, he now vilified in the most abominable way, and tried to force me to dismiss them *en masse*, to which I did not agree. The time of the conference was approaching and he used all the resources of diplomacy to try to prevent it from happening. Then, when the sittings of the Council of State went so splendidly, and the results were such striking proof that I was on the right track with the above-mentioned memorandum and its proposals, the Prince was overcome with jealousy of his poor young Kaiser and he resolved to ruin his success! At first, behind my back, he tried to induce various diplomats to send reports home opposing the conference, and finally he tried to persuade the Swiss minister to ask the Bern government not to give up their conference for my sake, so that my conference would fall through. The Swiss minister, a good, honest fellow – whom I happen to know well – was incensed at such deceitful, unpatriotic conduct towards the German Kaiser, and telegraphed at once to the government in Bern to say that if the official cancellation of the Swiss conference was not in his hands within 12 hours he would hand in his resignation, but he would also say why. The next morning the requested announcement arrived and my conference was saved! When this plan failed the Prince tried another.

The new Reichstag had been elected; he was furious at the election results and wanted to break it up as soon as possible. The Anti-Socialist Bill would have to serve the purpose again. He told me he proposed to bring in a new, harsher bill. The Reichstag would reject it and he would dissolve the Reichstag. The people, he said, were already unsettled, the socialists would vent their anger by causing a putsch, there would be the beginnings of a revolution and then it would be up to me to let rifles and cannon-fire do their worst. In the process – and this was his secret intention – the conference and the Social Welfare Bill would of course be lost, and ruled out for a long time as an election manoeuvre or a Utopian idea. I refused to accept this but declared roundly that it was impossible to advise a young king at the beginning of his reign – who was suspected of all kinds of things – to answer the requests and wishes of his working-class subjects with rapid fire and grape-shot. At this he became very angry. He declared that it would have to come to shooting in the end, and so the sooner the better, and if I were unwilling, he would resign at once. So once again I was faced with a crisis!', bewailed the Kaiser. 'I summoned the leaders of the Kartell parties to see me and asked them whether or not I should introduce an Anti-Socialist Bill and break up the Reichstag. With one voice they declared themselves against this. They said that the proclamations and the Council of State were already having a calming effect, and the conference would do the same. There was no question of putsches or revolutionary movements and the social welfare legislation would pass the Reichstag with ease; if the bills put before it were not too severe it would behave quite reasonably. They authorised me to convey this to the Prince as the opinion of their voters, and to warn him against any attempt to force Anti-Socialist Bills upon them, as he would not receive a single vote for them. The Prince then came, and with some anxiety as to the outcome of the interview I explained to him that I could not consent to his bringing in the bill. Thereupon he declared that the whole affair did not matter to him in the least! and if I did not wish to introduce the bill, there was no more to be said! The stand which he had taken towards me only a few days earlier had completely vanished from his memory! And as for the business over which he had managed to keep the ministers, me and the government parties in a state of the greatest agitation for over 4 weeks, and for which he had been willing to bring down ministers and stir up conflicts, he dropped it as if it were a mere trifle! But as a result of all these machinations and intrigues, disagreements and quarrels on every possible subject, and also of the failure of his little "ambuscades" the Prince had got into the most extraordinary state of agitation. The ministers had to endure outbursts of fury and the rudest possible behaviour, until they refused to go on with their work. Business ground to a halt and piled up, nothing was settled; no project, however urgent, could be submitted to me since the ministers were forbidden – N.B.

behind my back – to report directly to me. Everything had to be submitted to him first, and what he did not want he simply rejected and did not allow it to reach me. There was general dissatisfaction among the civil servants, which also spread to parliamentary circles. In addition I was informed by my personal physician that the Prince's doctor was greatly concerned by his state of health, and that he was in danger of a total collapse which would end in nervous fever and apoplexy! All my attempts to bring the Prince some relief by taking a greater part in business, he took as attempts to push him out. Any gentlemen and councillors whom I summoned to discuss matters with him, he treated with disfavour and suspected them of intriguing with *me* against him! Eventually matters came to a climax and all the pent-up electricity discharged itself on to my "guilty head"! The Prince, eager for battle and inspired by the motives I mentioned above, to the horror of those in the know and in spite of my orders to the contrary, began secretly preparing a campaign against the new Reichstag. Everyone was to be roused to anger and given a beating. First the Kartell parties were to be over-trumped and then the Socialists provoked, until the whole Reichstag was blown sky-high and H.M. would after all be forced, willy-nilly, to shoot! Then came the interview with Windhorst [sic], engineered by the Jew Bleichröder, which let loose a storm of indignation in the country, and was officially shrouded in mystery, which gave rise to all kinds of speculations. Furthermore an attempt was made to give the impression that I had known about it and approved it. Whereas I heard of it only 3 days later from the newspapers and from shocked enquiries which I received from all sides. When I saw the Prince on the third day after this affair, which had ever-increasing repercussions and began to take on a very unpleasant aspect for him, he brought up the subject of Windhorst's visit and described it in such a way as to make it appear that the latter had appeared almost unannounced in his antechamber and taken him by surprise. I had however heard for certain that Bleichröeder [sic] had arranged this interview with his consent. When I told the Prince this and asked him to give me notice of such important matters at least by sending me a note or a verbal message through his secretary, the storm broke! Without the slightest courtesy or consideration he told me that he would not be kept in leading reins by me; he refused once and for all to accept such treatment from me; I understood nothing of parliamentary life; it was most certainly not for me to give him any orders in such matters etc. etc. When he eventually stopped raging at me, I tried to explain to him that it was not a question of orders, but that what mattered to me was that when he took such important steps, which could result in decisions that might be binding on me, and which I could not evade, I should hear about it from him, and not discover it from the press after the event, so that I should know what was going on. But that did not help at all. When I put it to him how much confusion and disturbance this visit had caused among the people, who were still in an excitable state because of the elections, and said that this surely cannot have been his intention, he let slip the fateful remark: "On the contrary, that is exactly my intention! There must be such utter confusion and chaos in the country that no one can tell what the aim of the Kaiser's policy is!!" When I declared that this was not at all my intention, but that my policy must be open and as clear as daylight to my subjects, he announced that he had nothing more to say, and angrily threw his resignation at my feet. I did not react to this *third* scene in the course of 6 weeks. Instead I changed the subject and spoke of the Council of Ministers, and the Cabinet Order through which he had prevented them from reporting directly to me. He declared that he did not trust "his" ministers; they went behind his back to bring things to me which "he" could not approve, and he had therefore cautioned them against this. When I pointed out that this was a grave insult to me, his sovereign, who was so loyally and sincerely devoted to him, and whom he was accusing of secret intrigues behind his back, he refused to admit it. But he said that if I demanded it, he would send me the order at once, in the course of the day, so that it could be revoked, for it did not really matter. When I again asked him – purely with the intention of relieving a man who was plainly very ill and under excessive nervous strain from some of his work and anxieties – to let me take a greater part in business, to take me into his confidence and consult me when there were important decisions to be made, he refused categorically, with the comment that it was necessary for him to take decisions in advance, before he came to me! I now realised, with great pain and a sore heart, that lust for power had taken a demonic hold on this noble, great man, and that he was using every opportunity, no matter what it was, to pursue his battle against the Kaiser. He wanted to do everything alone and rule alone and the Kaiser was not even to be allowed to work with him. In that moment it was clear to me that we must separate, or everything would be morally ruined and destroyed. God is my witness that for many a night in my prayers I strove and pleaded to be able to reach the heart of this man, and to be spared the fearful prospect of sending him away. But it was not to be! When the order that was to be revoked still had not arrived from the Prince 2 days later, I sent a message to him enquiring whether he would send it. He replied that he had no intention of doing so; he needed to use it against "his" ministers! At that I lost my patience and the old family pride of the Hohenzollerns was aroused in me. The stubborn old man now had to be forced to obey or else it must come to a parting of the ways, for now it was a question of who should prevail, the Kaiser or the Chancellor. I asked him once more to send the revocation of the order and to comply with the wishes and requests I had earlier expressed to him. He refused point-blank. With that the drama was over.'

Finally, in tones of deep self-pity, Wilhelm stressed how hard the decision had been for him. 'The man whom I had idolised all my life, for whom I had endured the torments of hell through moral persecution in my parents' home; the man for whom I *alone* had thrown myself into the breach to keep him on after my grandfather's death, which earned me my dying father's anger and my mother's ineradicable hatred: this man thought nothing of all that and pushed me out of his way because I would not do his bidding! What a dagger in my heart! The boundless contempt for humankind with which he regarded everyone, even those who would have gone to their deaths for him, did him an evil turn in making him despise even his sovereign, whom he wished to reduce to the role of his minion. When he announced to me that he was leaving and accused me of driving him away I remained silent and said nothing, but after he had gone I broke down – I am ashamed to say – and wept.'[19]

On the basis of this description of 'the barely credible events' which had taken place, Franz Joseph conceded that Wilhelm could not have acted otherwise than to dismiss Bismarck. After reading the long letter he could indeed imagine, he wrote, 'how difficult it must have been for you to take this decision and what painful times you have had to go through since the beginning of the crisis. Although I share your deep regret that it had to come to this, no less do I deplore the fact that such a great man, who has served Prussia, Germany and peace with such merit, could allow himself to be so carried away as to act towards his Kaiser and sovereign in a manner for which the ultimate reasons could be more easily explained than excused.' That Wilhelm, 'guided by wise impartiality and clear judgement', had chosen Caprivi as Bismarck's successor Franz Joseph saw as a particular guarantee that judicious policies would prevail, 'although on an earlier occasion you did have a difference of opinion with him'. With Caprivi's support Wilhelm would certainly 'keep a firm hand on the helm in both domestic and foreign affairs, and face the present immensely difficult period with quiet prudence and deliberation'.[20] In his reply on 14 April Wilhelm II, alluding to the secret Reinsurance Treaty with Russia, the decision not to renew which was being taken at precisely this time, commented that for him the struggle with Bismarck had been 'immensely difficult and bitter', 'but it is better thus; and better also for our relationship to one another, for the Prince acted with such independence and at the same time with such secrecy that unfortunately I would not have been able to discover for certain what steps he was taking in foreign policy without my knowledge, and how these could be justified to my allies'.[21]

In parallel with his attempts to justify himself in the eyes of the Habsburg monarch, Kaiser Wilhelm II also sought to put the best construction on the historic events which had led to Bismarck's dismissal for the benefit of his grandmother and the British government. With the obvious intention of representing the Chancellor's resignation as a regrettable necessity rendered inevitable by failing health and old age, Wilhelm telegraphed to Queen Victoria as early as 19 March 1890: 'I deeply regret to have to announce to you that Prince Bismarck has placed his resignation in my hands. His nerves and strength having given out and beginning to fail in the hope of preserving and refreshing his broken health I have accepted his resignation hoping to be able to consult him in any difficult question when he is better and to have him as Councillor as long as he lives. My policy will undergo no change whatever. Gen. v. Caprivi has been named Chancellor. William I. R.'[22] Three days later he called on the British ambassador, Sir Edward Malet, and gave him a lengthy explanation of the reasons for his decision. Malet's confidential report to the Queen – he sent a copy privately to Lord Salisbury at the same time – is eighteen pages long and quotes the Kaiser in direct speech throughout. According to the report, Wilhelm expounded the long history of his differences with the Reich Chancellor on the social welfare question and the Anti-Socialist Bill. He reproached Bismarck with deliberately seeking to drive the class conflict to the limit so as to be able to 'sweep the streets with grape shot'. His grandfather would have had sufficient respect in the country to be able to approve such a policy if necessary, Wilhelm declared; 'but for me, a young Monarch, just come to the throne, to have allowed my people to be shot down in the streets, without making an effort first of all to examine their grievances, would have been disastrous to me and my whole House. It would have been said that my only idea of governing was by bayonets.' But instead of listening to him Bismarck had tried to undermine his social reform initiative behind his back and treated him 'like a schoolboy'. 'The moment came when I was obliged to think of my own dignity.' As he had told Franz Joseph, the Kaiser repeated to Malet: 'I was assured by the doctors that his state of mental excitement was such that it might end in a crisis at any moment . . . I finally decided that, if I wished his life to be preserved, I must relieve him of his duties. He and all his family are at present intensely incensed against me, but I hope that in a few months they will see that they have reason to thank me.' He went on: 'I cannot tell you the pain and anxiety I have gone through this winter on his account. I have always had the greatest admiration for him, and when I was Prince I went through bitter moments from taking his side. I used to say to myself: "Ah! if, when I am Emperor, I could have such a Minister!" – for of course I never thought that I should come to the throne before I was sixty. When I became Emperor, I was overjoyed at having him as my Minister, and I looked forward to keeping [him] at my side until old age should force him to retire and now my real aim is to keep him alive for the sake of Germany and of Europe. I look upon him as a Capital of which Europe enjoys the interest. The universal confidence felt in him is so great, that a word from him in difficult moments of European Conflicts can arrest a crisis, and I still hope to make use of the enormous power which he thus wields for our benefit. One thing which has really cut me to the quick is that in all these painful moments he has never seemed to have the slightest personal regard for me or gratitude for the little efforts which I made to help him when I was Prince. He unfortunately regards everybody as a machine; sentiment in dealing with people is not in his character.'[23] The ambassador was rightly suspicious of Wilhelm's claim that he had parted with Bismarck in the latter's own interest and only on the grounds of urgent medical advice. In his covering letter to Queen Victoria he remarked sceptically: 'The point which remains at issue is whether the health of the Chancellor was really in so precarious a state as the Doctors asserted . . . Probably the vehemence of Prince Bismarck's manner to the Emperor led His Majesty to believe that a crisis was imminent when it was not. For the Prince has carried his point over and over again, especially with the late Emperor William, by violence of manner and he very likely, in his discussions with the young Emperor, assumed this manner in the hopes that it would have like effect.'[24] The ambassador did not venture to suggest to the Queen that the whole story of the doctors' urgent warning was purely an attempt by his imperial visitor to justify himself.

A few days after his conversation with Malet the Kaiser repeated his version of events in a personal letter to his grandmother. He claimed hypocritically that he had parted from the founder of the Reich in tears after a 'warm embrace'. In this letter Bismarck's apparently shattered health – his life had been in danger – stood as the only reason for the 'very difficult' decision which Wilhelm had had to take. 'I hope & trust that the woods of Friedrichsruh will do him good & help to recruit his forces, & strengthen his nerves; for he was very much shaken. I spoke to his doctor 2 days ago [sic], who assured me, that if the Chancellor had kept on a few weeks longer, he would infallibly have died of apoplexy. The nights he could not sleep, & in day time, as well as in bed, even sometimes when he worked with me, he suddenly would break down, with crying fits. After this had gone on for a month I became afraid of the consequences & after much discussion & with deep regret, I resolved to part from him, in order to keep him alive.' Here the Kaiser repeated the metaphor he had used to Malet: 'I look upon the Prince as an international European capital, which I must try to keep going as long as possible, & not use him up in guerilla warfare with the Reichstag. It was a very hard trial, but the Lord's will be done. I have been educated politically by the Prince, & now I must show what I can do.'[25] When the Queen sent Salisbury the Kaiser's letter he commented with perfect accuracy, as Holstein had done two years earlier: 'It is a curious Nemesis on Bismarck. The very qualities which he fostered in the Emperor in order to strengthen himself when the Emperor

Frederick should come to the throne, have been the qualities by which he has been overthrown.'[26]

Several years later, during his Scandinavian cruise in July 1896, while reminiscing with his best friend Philipp Eulenburg on Bismarck's dismissal, Wilhelm II still spoke with self-pity of the shattering events of March 1890. 'How I loved Prince Bismarck!', he exclaimed. 'What sacrifices I made for him! I sacrificed *my parental home* to him. For his sake I was *maltreated* for years of my life, and I endured it because I regarded him as the living expression of our Prussian Fatherland. And he repaid me for all that with *hatred*! *That* I cannot forget!' It was only during the preparations for the international conference on social welfare that he had recognised the necessity of dismissing Bismarck, he claimed. 'I had considered it the highest of my duties to take care of the welfare of the workers and of old people. The conference was convened in Berlin. There I discovered that the Prince had been to see the French, English and Italian ambassadors to persuade them to ensure that their countries *refused to attend* the conference to which I had invited them . . . This behaviour of the Prince towards me was *open rebellion, which was paraded before the whole of Europe*. I *owed* it to the crown to part with the man! As to the manner in which the Prince's resignation took place, I sent accounts in my own hand to my grandmother in England and to Kaiser Franz Joseph. They will be published under my will, for then I wish to be shown as *justified*. As long as live I shall bear the burden – I have no wish to spoil the German people's ideal for them.'[27]

THE NEW REICH CHANCELLOR

The appointment of the Commanding General of the X Army Corps in Hanover, Georg Leo von Caprivi, to be Bismarck's successor as Reich chancellor, Prussian minister-president and Prussian minister of foreign affairs, was one of the most genuinely personal decisions that Wilhelm II ever took in his thirty-year reign. Not even his brother knew of his intention. Indeed, when Prince Heinrich, who was with his ship *Irene* in the Mediterranean, read an announcement in a local Spanish newspaper that Caprivi had been appointed Reich chancellor and that Waldersee was to remain Chief of the General Staff, he exclaimed in astonishment to his officers: 'But how ridiculous – that is what it says . . . and yet we all know that it is the other way round!'[28] This anecdote, recorded by Baroness Spitzemberg in 1908, is borne out by the diary of one of the naval officers present. Vice-Admiral Paul Hoffmann noted several times that Prince Heinrich had commented that Caprivi was 'the last person my brother would take'. When the news of Caprivi's appointment was subsequently confirmed, there had been 'great surprise on the part of the Prince' and, 'at any rate, not a pleasant surprise'.[29]

When and why did the young Kaiser come to the surprising decision to appoint Caprivi, of all people, whom he thought 'stiff and rigid', and who had resigned as head of the Admiralty in 1888 in protest at Wilhelm's interference in naval affairs, to be Bismarck's successor? The question of the timing of the decision is more easily answered than that of the reasons behind it. In late January 1890 – that is, shortly after the relationship between Wilhelm and Waldersee had grown several degrees cooler[30] – the Baden envoy and intimate of Holstein, Baron Adolf Marschall von Bieberstein, was able to record in his diary: 'Today I tell the Grand Duke that Caprivi will be chancellor.'[31] It is known that Caprivi had been summoned to Berlin by the Kaiser at the beginning of February and had discussed the question of Bismarck's successor with him for an hour.[32] On 18 February 1890 Wilhelm called on Waldersee to take him for a walk and, although he made no direct reference to the question, the Chief of the General Staff already suspected at this time that the Kaiser's choice had fallen on Caprivi. Waldersee, who claimed not to wish to be Bismarck's immediate successor, seemed at first to welcome this decision: Caprivi was perhaps not particularly talented, but he was nevertheless a 'thoroughly honourable man of steadfast character' who was greatly respected in the Reichstag. The only question was 'whether he will be able to work well with the Kaiser', for Caprivi had 'a very strong will' and was the kind of person who could not take any opposition. Furthermore, 'he left the Admiralty as soon as the Kaiser, whose inclinations towards the navy he knew very well, came to the throne'.[33] It will surely have been no coincidence that Philipp Eulenburg called on the stubborn General in Hanover on 6 March on his way back to Oldenburg from Berlin. The Kaiser's friend was impressed by Caprivi's 'cleverness' and his 'calm, dignified, objective attitude', but could find no 'exceptionally intelligent insights in the political sphere'; above all, however, Eulenburg was astonished by Caprivi's radical views – he revealed himself as a firm opponent of the Anti-Socialist Bill and a proponent of the reduction of military service, at least for the infantry, from three to two years.[34] Eulenburg later remarked that he would have been in favour of the appointment of the ambassador Lothar von Schweinitz or of the Statthalter of Alsace-Lorraine Prince Chlodwig zu Hohenlohe-Schillingsfürst as Reich chancellor, but he would '*never* have thought' of Caprivi.[35]

Very little is recorded about Wilhelm's motives for this surprising choice. Eulenburg discovered that the Kaiser 'had retained a great admiration for Caprivi from the time when he was still in charge of the navy', although for this admiration to have survived Caprivi's angry resignation as head of the Admiralty is something of a rarity in the history of Wilhelm II's reign.[36] Another factor which should not be underestimated is the high regard in which Kaiserin Auguste Viktoria and her entire entourage held Caprivi, who – without means of his own and unmarried – had reached the rank of commanding general purely on the strength of his ability and his upright character.[37] And indifferent as Wilhelm was to his mother's opinion, she too thought well of Caprivi, although she considered him better fitted for the post of minister of war than for that of Reich chancellor. 'He is an honest straightforward respectable man, of *great* energy – a *very* stubborn and determined will – not given to any compromises – and rather violent. I should not think he understood *politics* in the least, but he is incapable of saying what he does not mean, or of an intrigue of any kind!'[38] Three days later, after the newly appointed Chancellor had called to introduce himself to her, the Empress Frederick wrote to Queen Victoria saying that Caprivi was 'extremely sensible', but she doubted whether he would get on with Wilhelm, for he was 'a *very consciencious* [sic] man, and thoroughly in earnest, – & if W. means (as he says sometimes) merely to have people who "obey him" and "carry out his orders" – I fear he will find it very difficult – almost impossible – to fulfill [sic] all the duties of his office'.[39] Such views of Caprivi's honesty and integrity were widely held.[40] But no one was more aware than Caprivi himself of how little these admirable qualities would help him to master the domestic and international political situation in which the German Reich found itself after Bismarck's departure. He described politics as a mysterious dark-room and said of Bismarck's sophisticated alliance system that he, unlike his predecessor, would never be able to juggle with several balls at once.[41] In December 1891 the Hereditary Prince of Saxe-Meiningen, the Kaiser's brother-in-law, although still acknowledging the Reich Chancellor to be 'a thoroughly distinguished, upright and noble-minded man', had to admit that there was a widespread view among high-ranking officers in the army that Caprivi did not possess the necessary abilities 'to direct the policy of a great power of the first rank'.[42] It is small wonder that some observers of the old school regarded the appointment of the unsophisticated General to the most responsible position in the entire Reich as a disastrous mistake. Kurd von Schlözer, the Prussian envoy to the Holy See, was quick to pass disparaging judgement on the new Chancellor, whom he described as a man 'of almost crass ignorance in non-military questions' who did not understand political affairs 'because he has practically never been out of Berlin circles and has no knowledge of human nature. One might just as well make any battalion commander chancellor. All that so-called uprightness of character is as good as useless!'[43] That Bismarck in his bitterness should also belittle his successor as 'a man with a small horizon' who had failed even as a soldier is no surprise.[44]

In the end the choice of the honest General can probably only be explained as a provisional solution which Wilhelm had intended as such from the outset. For the difficult transition period in the immediate aftermath of Bismarck's resignation he needed as Reich chancellor a strong-willed military man who in this case

16. General Georg Leo von Caprivi, Reich Chancellor 1890–1894

had the added benefit of a bald head and a white moustache which made him look not unlike his predecessor. Thereafter the excessive power of the chancellor would be gradually dismantled, perhaps even stripped away entirely; or a 'man of straw' – as Caprivi's successor Prince Hohenlohe was to call himself – would be installed. The Kaiser did not conceal from Caprivi that these were his intentions. As the newly appointed Chancellor confided to the Chief of the General Staff on 20 March 1890, the Kaiser had 'a new organisation in mind': he had appointed Caprivi only 'to put this into effect, and had told him . . . that he was not to remain for long'.[45] At any rate there is no doubt that Wilhelm now intended to rule himself, and to determine personally the guidelines of domestic and foreign policy.

The very fact that even today we know so little about the background to the appointment of Caprivi as chancellor is evidence that in taking this significant political step Wilhelm II was carrying out a totally personal decision. Even if Waldersee and other men around Holstein got wind of Caprivi's forthcoming appointment in late January or early February, it was only ever in the form of rumours about the Kaiser's intentions, and did not indicate that they had played any active part in the choice. Nearly everyone – whether for or against the appointment – was taken aback by the choice. Almost as if he were afraid that the influence of others would limit his freedom of action, Wilhelm kept his decision to offer the appointment to Caprivi to himself until the last minute.

Only his wife, Waldersee and his intimate – and in this case somewhat sceptical – friend Philipp Eulenburg seem to have been let in on the secret. Thus this most important of all decisions in the field of public appointments was made without consultation and without the agreement of anyone else. Nor was it, of course, ratified after the event by any committee or parliamentary body. In acting in this way, however, the Kaiser had infringed neither the Prussian constitution nor that of the Reich. Bismarck himself had after all seen to it that the 'Personal Monarchy' in Germany did not exist merely on paper but was the real and present form of government.

THE NEW SECRETARY OF STATE FOR FOREIGN AFFAIRS

During his walk with Waldersee on 18 February 1890 the Kaiser hinted that he intended not only to appoint Caprivi as Reich chancellor, but also, at least in the short term and for tactical reasons, to retain Herbert Bismarck as secretary of state at the Foreign Office. The Chief of the General Staff welcomed this decision and acknowledged that the retirement of the founder of the Reich would 'look better in the eyes of the world if the son does not follow until some time later'. That Herbert would also have to be dismissed sooner or later, however, Waldersee did not doubt, for 'every chancellor has to keep control of foreign policy and he will soon rid himself of this boorish, most unpleasant man who is so dangerous for our international position'. Waldersee even took a certain malicious pleasure in anticipating conflict between Caprivi and Herbert Bismarck. 'I should be happy to see Caprivi becoming his [Herbert's] Chancellor', he wrote. 'He knows him very well and considers him far from competent . . . – and he utterly detests his behaviour and activities.'[46]

Simultaneously with his ultimatum to Prince Bismarck demanding his letter of resignation, on 18 March, Wilhelm II therefore took action to keep Herbert in office. He offered him, in addition to the Foreign Office, the post of Prussian minister of foreign affairs, which brought with it responsibility for casting the Prussian vote in the Bundesrat.[47] He sent Philipp Eulenburg to Herbert several times to try to persuade him to remain, but in vain. 'He is in an alarming mood', Eulenburg observed in his diary.[48] There were long consultations in the Foreign Office itself, but nothing would deflect Herbert from his firm resolve to resign at the same time as his father. Holstein went so far as to call on Bleichröder in the hope that he would be able to persuade the departing Chancellor to put pressure on his son to remain. But the elder Bismarck told his banker that he could not advise Herbert to do so. In direct discussions with Herbert Bismarck himself, Holstein pointed out that as secretary of state he could play an important intermediary role between his father and the Kaiser, and that he would thereby enjoy 'considerable status with the latter'. Herbert continued to hold out, and gave as a reason for his departure, in addition to his health and the state of affairs in the Reichstag, 'His Majesty's methods of work'. On 21 March 1890 he handed in his resignation.[49]

The Kaiser had nevertheless not abandoned all hope of persuading the younger Bismarck to resume his office after a few months' leave. Paul Hatzfeldt could stand in for him in the meantime, Wilhelm told Waldersee on the evening of 22 March.[50] Again on 28 March, although Freiherr Marschall von Bieberstein's appointment as foreign secretary had already been confirmed, the Kaiser remarked that Herbert must 'come back as soon as he has recovered, and take up the place he held before! . . . He valued Herbert's efficiency, could work with him and they were friends.' As one of the Flügeladjutanten observed, 'our young sovereign' showed by such comments that he was not satisfied with the choice of Marschall and was already thinking of replacing him.[51] Wilhelm also informed the British ambassador on 28 March that Marschall's appointment was only 'for the present & that he hoped to have Count [Herbert] Bismarck back again in 6 or 8 months'.[52]

In contrast with his conduct over the choice of the new Reich chancellor, therefore, Wilhelm II had evidently given no thought to the possibility – likely as it had always been – that the younger Bismarck would also resign, and that it would therefore be necessary to appoint a new foreign secretary. Although the Bismarck crisis had lasted for several months and had been at an acute stage since at least 24 January, a successor to Herbert Bismarck had to be found in great haste. The choice of a new foreign secretary was all the more crucial because the new Chancellor had absolutely no experience in the realm of foreign policy. Of the seven German ambassadors, however, none seemed to fit the post. In the course of a detailed discussion on 17 March of 'what advice to give the Kaiser' Eulenburg and Waldersee came to the conclusion 'that here too we lack able people'. Count Hatzfeldt, in London, was impossible because of his personal circumstances. Count Münster, in Paris, was too old and 'his mental powers are declining'. Prince Reuss, in Vienna, had 'not the slightest desire' for the post; and Lothar von Schweinitz in St Petersburg was 'disliked by the Kaiser'.[53] Joseph Maria von Radowitz, the ambassador in Constantinople, was a very gifted man but because of his Russian wife he was 'impossible, since Austria, England and Italy consider him a Russian spy', Eulenburg informed the Kaiser.[54] Count Eberhard zu Solms-Sonnenwalde, the ambassador in Rome, was in Waldersee's and Eulenburg's eyes 'too insignificant' for this important appointment. Holstein pointed out that, although Solms might have had the right qualities, he was sixty-five years old and would not be willing to give up his comfortable post.[55]

Holstein too was forced to admit, in his own telling phrase, that 'new growth has not flourished beneath Bismarck's oak, and there is nobody one could recommend'.[56] Like many others, he thought at first of Count Friedrich Johann von Alvensleben, the German envoy in Brussels. But when Caprivi summoned him to Berlin Alvensleben proved to be 'in a state of nervous tension and generally so "broken down" in health . . . that there could be no question of appointing him'.[57] Alvensleben himself told Count Otto von Stolberg-Wernigerode that he had refused the post because he had been away from the world of high politics for too long to take over such a responsibility, 'which is even heavier now, since Caprivi himself is not yet familiar with the work'.[58]

Alvensleben having definitely turned down the appointment, numerous other candidates were considered. The Under-Secretary of State at the Foreign Office, the Bavarian Count Maximilian von Berchem, was turned down by Caprivi as 'too Catholic'.[59] Bernhard von Bülow 'or even Eulenburg (Ph.)' were suggested as possible secretaries of state.[60] However, both Berchem and Holstein spoke out strongly against the choice of Bülow. '*Bülow* is unacceptable to the Kaiserin because of his [divorced] wife, let alone his social qualities, that is to say, failings', Holstein asserted.[61] Prince Bismarck himself had described Bülow to Caprivi as a 'not very reliable [he had first written "completely unreliable"] man'.[62] Bismarck's recommendation that the position should be offered to the Conservative member of the Prussian Landtag, Count Friedrich Wilhelm zu Limburg-Stirum, who had been the temporary head of the Foreign Office briefly in 1880, Holstein considered 'pure mockery', which merely showed 'which way Bismarck would like to push things', and Eulenburg likewise found the suggestion unacceptable, as Limburg-Stirum was 'of Jewish extraction' on his mother's side, 'which permeates his being'. Indeed, Eulenburg went so far as to consider Bismarck's recommendation of Limburg as positively 'perfidious'.[63]

The fact that the choice eventually fell on Baden's envoy and representative in the Bundesrat, Adolf Freiherr Marschall von Bieberstein, completely inexperienced in foreign affairs as he was, was chiefly due to the widespread desire in diplomatic circles to find someone who would work successfully with Friedrich von Holstein. Why Holstein himself did not wish to become secretary of state remains one of the unsolved riddles of Germany history. Philipp Eulenburg, who must have known what he was talking about, wrote of this solitary bachelor that there were such 'marked peculiarities' about his private life that it would be impossible for him to accept the top office, and that the Kaiser also knew this.[64] After Bismarck's departure, however, Holstein's knowledge and ability were considered indispensable. As Count Monts wrote on 24 March 1890 from Vienna, 'Here we are afraid that it may not be possible to come to an arrangement which would allow Holstein to stay on as right-hand man to the chief, or perhaps that H. may not feel able to continue to serve after the events of recent days. And in my opinion this man is now the only one capable of keeping the diplomacy of the Reich on course.'[65] Others, however, viewed the idea of entrusting the Foreign Office to Marschall, as a South German and a non-diplomat, with consternation. Herbert Bismarck exclaimed in horror that it would now be necessary 'to hand over all secret papers to a fellow from Baden!' And Carl Wedel, equally appalled, commented: 'My God, I really do not like the arrangement at the Foreign Office! Marschall is a completely unknown quantity and not even of the Bismarckian school. He has no standing at all in Europe.'[66] Everyone predicted that under Marschall the Foreign Office would be completely under the influence of Holstein, which was indeed the aim of many of those involved.

Despite all misgivings, Marschall's candidature gained ground day by day. As early as 17 March 1890 Philipp Eulenburg had described the Baron as 'very capable, fresh and well-disposed' and had pondered his suitability to take charge of the Foreign Office. Subsequently Waldersee had recommended Marschall to the new Chancellor. Count Berchem also spoke up for Marschall's appointment. Two days later Eulenburg informed Marschall that the Kaiser wished to appoint him secretary of state if Herbert Bismarck did in fact step down. Marschall pointed out that Caprivi could not expect much support from him in foreign policy initially and also that the older officials at the Foreign Office might resent the appointment of 'a young non-Prussian'. But he did not refuse the post in principle.[67] Caprivi told the Chief of the General Staff on 22 March that he liked Marschall very much, but that in his view it was 'questionable whether it was wise to begin with a novice'.[68] A further difficulty was that the Grand Duke of Baden expressed decided opposition to Marschall's appointment. Holstein, on the other hand, made no bones about his preference for the Baden envoy. He dismissed the criticism made of Marschall that he did not know 'the courts', for 'to be well-informed, one does not need to have been on the spot oneself'. Marschall would 'soon, very soon, work his way into the great questions of foreign policy: he already knows their general outlines', Holstein stated confidently in a letter to Eulenburg.[69] As so often in the long crisis which had begun in the early summer of 1889, the influence of Eulenburg, who passed on Holstein's thoughts to the Kaiser, proved decisive. In a letter to Wilhelm of 26 March he ruled out both Radowitz and Limburg-Stirum; he praised Bernhard von Bülow's talents, but warned that Holstein, who was indispensable, would resign if Bülow were appointed. Alvensleben, he said, was universally respected and might perhaps overcome his doubts if the Kaiser expressed it as his personal wish. The best man for the job, however, would undoubtedly be the 'excellent' Marschall, who as a South German would make an attractive partnership with the Old Prussian Caprivi. He was a good speaker and practised in dealing with the Reichstag;

he had a very sound knowledge of German internal affairs and would also 'very soon master the great questions of foreign policy'. Eulenburg closed his letter by repeating his warning that it was particularly important in the Kaiser's own interest 'that Holstein, who is familiar with all the secret paths of Bismarckian policy, should stay. Odd creature that he is, he would certainly pack his bags if for instance Bülow, Radowitz or Stirum were appointed.'[70] The next morning Caprivi – 'after he had received the Kaiser's consent' – offered Marschall the Foreign Office, arguing that the great powers would see his appointment as proof 'that the Kaiser is not pursuing any warlike policy'.[71]

At the last moment an embarrassing incident occurred which was symptomatic of the confusion inherent in the new distribution of power between the Kaiser and the Chancellor. On 27 March, after Marschall had accepted the post offered to him by Caprivi on the Kaiser's instructions, Wilhelm II, with strong support from his Flügeladjutant Carl Wedel, began to negotiate with Alvensleben again, pressing him to accept the post 'at least on trial'. Alvensleben would not be persuaded, and even threatened to resign from the diplomatic service altogether 'when the Kaiser began to apply a kind of coercion'. But what would have happened if Alvensleben had given in? When Caprivi heard of the renewed negotiations with Alvensleben, he told Herbert Bismarck, General von Hahnke and Lothar von Schweinitz, who were all waiting to be received in audience in the Schloss, that if Alvensleben accepted he himself would have to resign, for he was now comitted to Marschall.[72] 'What a strange sign of the way things are done now, and what chaos this must lead to!', Wedel remarked.[73]

The decision to appoint a Badenese lawyer without any experience of foreign affairs as secretary of state at the Foreign Office may indeed have reassured the other powers, as Caprivi hoped, that the Kaiser harboured no warlike intentions. It undoubtedly had the effect not only of safeguarding Friedrich von Holstein's influence on the formulation of German foreign policy in the years that followed, but even of greatly increasing it, particularly as the new Chancellor likewise had no experience in this field. Stolberg showed a perspicacious grasp of the situation in a letter of 28 March 1890 to the German ambassador in Vienna, remarking that Marschall was 'a very intelligent, efficient, reliable and respectable man, and to have gained him for the service is in itself to be welcomed; but with the best will in the world he cannot yet know anything about the actual work'. In the Bundesrat and in the Reichstag he was greatly respected, but not in the Foreign Service. 'By and large everything will depend on Holstein at first; but how far he can be relied on remains to be seen.'[74] After only a few weeks the dominance of Holstein and his South German confidants Marschall and Kiderlen-Wächter (who was from Württemberg) in the Foreign Office was so great that the Under-Secretary of State Count Berchem resigned in protest.[75] One

17. Adolf Freiherr Marschall von Bieberstein, German Foreign Secretary 1890–1897

of the Flügeladjutanten made the far-sighted observation that 'with Berchem one of the strongest corner-stones has been torn out of the building, and we are moving more and more towards a downhill road! Servility and self-seeking are more and more in evidence! One cannot help wondering whether this is purely the product of our new circumstances or whether it is also partly a pernicious result of the Bismarckian system which suppressed all independence.'[76] The paucity of experience and political skill at this time, in comparison to the period of the Bismarck family's ascendancy, was certainly immense, and all the talent and secret knowledge of the cranky Privy Councillor Friedrich von Holstein, who shrank from the glare of publicity, could not make up for it. As a South German Prince wrote from Berlin to his father in April 1890: 'It is a pity that the two most important posts are now occupied by people who are not yet at all familiar with the responsibilities of the offices which they hold.'[77] Nature abhors a vacuum. The void which now opened up in the field of foreign policy acted on the eager young Kaiser like a lure.

THE NEW MINISTERS AND SECRETARIES OF STATE

A fundamental difference between the constitutional practice of Personal Monarchy as established by Bismarck in Prussia–Germany and the parliamentary system prevailing in northern, western and southern Europe was the absence in the German Reich of the principle of collective responsibility of ministers of the Crown. The Prussian Ministry of State, as Holstein complacently observed in February 1890, was 'a Ministry of Civil Servants and as such more comfortable . . . for the monarch than a parliamentary Cabinet'.[78] When Bismarck was dismissed none of the Prussian ministers or the secretaries of state of the Reich offices – apart from Herbert – declared their solidarity with him. Heinrich von Boetticher above all, Vice-President of the Ministry of State and Secretary of State at the Reich Office of the Interior, earned himself Bismarck's irreconcilable hatred by showing himself ready to fulfil Wilhelm's wishes, for which he was rewarded with the Order of the Black Eagle. In the Bismarck family Boetticher's wife was maliciously accused of having designs on the post of minister-president for her husband, as she would then have field marshal's rank and would enjoy the highest precedence at court.[79]

When Bismarck convened the Prussian Ministry of State for the last time on 17 March 1890 three of the ministers spoke out – albeit without great conviction – in favour of resignation *en bloc*. Finance Minister Adolf von Scholz stated his opinion that in view of the forthcoming dismissal of Prince Bismarck the Ministry of State should consider 'whether it ought not to join in this step' and even ventured to suggest that a collective resignation might yet avert the 'fateful event'. Scholz was seconded by Minister of Public Works Albert von Maybach (who was in any case tired of his job), who put forward the view that all the ministers should 'put their offices at His Majesty's disposal and he, at least, was determined to do so'. Minister of Agriculture Robert Freiherr Lucius von Ballhausen also declared that he would resign if Bismarck wished.[80] If it had come to collective resignation by these ministers, Minister of Ecclesiastical Affairs Gustav von Gossler would probably have joined them. But when the Prussian ministers assembled again that evening at Boetticher's house, this time without the two Bismarcks, none of them spoke out in favour of resigning *en bloc*. In a declaration characteristic of the non-political, civil-service mentality of the ministers, Boetticher stated 'that it was not in the Prussian tradition for ministers to put their portfolios at the king's disposal *in corpore*. In the present situation there was in any case no reason to do so, since the All-Highest wishes concerning the choice of the new minister-president were still quite unknown. There was no reason to believe that His Majesty wished to pursue policies at home which were opposed in principle to those pursued under Prince Bismarck. Besides, so far as one could see His Majesty would not accept the resignations of the ministers of state. The entire step would thus have a merely formal character which would not be in keeping with the position and prestige of the Ministry of State. If His Majesty desired changes in other offices, one could expect him to take the initiative himself.'[81] Nothing shows more clearly than the passive arguments and deferential turn of phrase adopted by Boetticher how much the Prussian ministers of state took it for granted that the young Kaiser would determine policy guidelines and appointments to the principal positions in Prussia and the Reich from now on. With the exception of Herbert Bismarck, therefore, all the Prussian ministers and the secretaries of state of the Reich offices at first remained at their posts in March 1890. (Five of them – Scholz, Lucius, Maybach, Gossler and Minister of War von Verdy – were to resign in the course of the next few months.) The unpolitical attitude demonstrated by the ministers, which Holstein at this time still welcomed but was later bitterly to regret, was an essential prerequisite to the growing personal power of the Kaiser.

Wilhelm's mother also recognised the lack of a strong, politically self-confident collective ministry both as the chief weakness of the Bismarckian system and as the basis of the increasingly 'despotic' power of Wilhelm II's position. On 22 March 1890 she wrote to Queen Victoria saying that Bismarck's resignation would have been no great misfortune if only his regime had been replaced by a collective ministry with Prince Chlodwig zu Hohenlohe as Reich chancellor, Count Paul von Hatzfeldt as foreign minister, Caprivi as minister of war and a liberal statesman as minister of the interior. Had that happened, 'wise and experienced & conciliatory men would have had the confidence of Germany & *of Europe*, & in time *I am sure* would have had the best influence on W.'. With a collective government of that kind an era of peace and stability would have begun, just as would have happened under her husband. But as things actually stood she could see 'nothing but confusion – sudden resolution not sufficiently considered – suddenly carried out, with truly Bismarkian [sic] contempt for people's feelings – but without the "coup d'oeil de maitre" wh. B. often had!'[82]

LIEBENAU'S DOWNFALL

If on the one hand the Prusso-German system of government lacked a responsible Reich Cabinet, on the other the men at court – the Oberhofmarschall, the military entourage and the three cabinets – played a political role which would not have been tolerated in a parliamentary system. Given that relations between the Wilhelmstrasse and the court were therefore closely intertwined, the departure of the two Bismarcks could not but affect the composition of the Kaiser's entourage. Above all, the long-heralded fall of the Oberhofmarschall, Eduard von Liebenau, became inevitable the moment that the protection of the Bismarck family vanished.

At the height of the Bismarck crisis Liebenau informed the Kaiser that the state of his nerves was so poor that he wished to ask for other employment. In response Wilhelm, following the advice of Liebenau's adversaries, Waldersee and

18. Count August zu Eulenburg, Senior Marshal of the Court and Household 1890–1918

Eulenburg, offered him the post of intendant of the Royal Gardens. Liebenau, however, requested an ambassadorial appointment; he had already come to an agreement with Herbert Bismarck over this, he said. 'The Kaiser laughed out loud' and asked what qualifications he had for such a post. Liebenau declined the Intendancy of the Royal Gardens, stating that he could not demean himself to take it. 'The Kaiser rubbed his hands and was glad that the initiative to leave had come from L. himself.'[83]

Aside from the witch-hunt conducted by his enemies, Liebenau's notorious incompetence also contributed to his downfall. He provided further proof of it when at the end of March 1890, to the Kaiser's utter bafflement, he failed to invite the new Reich Chancellor to a state banquet for the Prince of Wales, on the grounds that Caprivi had not yet officially informed him – Liebenau – of his return from a visit to Hanover.[84] Finally, on 23 May 1890, Wilhelm II took the decision to sack his longstanding companion.[85] He cited a series of other gaffes – Liebenau had for instance failed to inform him that during a night journey through Elbing a large group of shipyard workers were waiting to greet him as the train passed through the station[86] – as the reason for dismissing the Oberhofmarschall and appointing August Eulenburg in his place. Waldersee commented that there was 'unanimous' satisfaction at Liebenau's departure. He had congratulated the Kaiserin upon it, for she above all had 'had to suffer a great deal from the wretched man'.[87] A deeply embittered Liebenau left Potsdam the instant the question of his retirement pay had been settled.[88]

Liebenau's dismissal was regretted not only by the Bismarcks at Friedrichsruh but also by some at the Hohenzollern court. The Flügeladjutant Carl von Wedel commented bitterly on the Oberhofmarschall's fall: 'So that is what the Kaiser thinks of a man like Liebenau, who has stood by him loyally for fourteen years . . .! He drives him out of the house like a lackey, throws him away like a worn-out glove! – My God, the English blood has brought nothing good to the Hohenzollern dynasty, for our young sovereign has inherited this heartlessness from his mother! It will show itself in much worse outbreaks in the course of time!'[89] In actual fact, the Kaiser's English mother herself deeply deplored the change. She acknowledged that Liebenau was difficult and had offended many people; but he had kept the staff under control and had supervised court functions efficiently, which could not be expected from his successor.[90]

That Liebenau should be replaced as Oberhofmarschall by her old enemy August Eulenburg, who combined three other court posts (he became senior house marshal and senior master of ceremonies at the same time, and later also minister of the royal house), the Empress Frederick found extremely disturbing. Eulenburg was not to be trusted. 'He is very grasping & ambitious but he can make himself very agreeable & has the manners of the world . . . I do not think matters *can* go smoothly as long as many elements are still about William who have done such endless mischief!'[91] Waldersee, of course, saw the role of the new Oberhofmarschall in another light. August Eulenburg, he wrote, was extremely cautious and understood perfectly how to judge the Kaiser's mood and to act accordingly. Not until later did the General remark with some surprise that the Oberhofmarschall was trying to exert an independent influence on the monarch.[92] Though he bore no constitutional responsibility, he was soon to become one of the most influential of the figures at the Hohenzollern court, someone whose advice Wilhelm often chose to follow.

CHAPTER FOURTEEN

In Bismarck's footsteps: the conduct of foreign policy under the New Course

THE NON-RENEWAL OF THE RUSSIAN REINSURANCE TREATY

Wilhelm's repeated assurances, in the letters that he wrote to his fellow sovereigns justifying his conduct, that Bismarck's dismissal had been the result of differences of opinion of an exclusively domestic nature were not, as we have seen, the whole truth. At the very outset of the chancellor crisis a violent confrontation between Bismarck and the Kaiser over the Russian loan conversion had come about as a result of constant warnings from Waldersee, and indirectly also from Holstein, that the Bismarcks were relying to a dangerous degree on Russia while neglecting the two powers to which Germany was bound by the Triple Alliance, Austria-Hungary and Italy. And in the final phase of the crisis, in mid-March 1890, Waldersee succeeded in convincing the Kaiser, on the basis of the delayed consular reports from Kiev, that Bismarck had deliberately withheld important information from him about Russian military preparations for an attack on the Triple Alliance. In addition, Wilhelm's personal relationship with the Russian imperial family had continued to deteriorate during these crucial months. This put the relations between Kaiser and Chancellor under renewed strain, notably during the weeks before Alexander III's visit to Berlin in October 1889, and again in December that year during the hunt at Springe with the Tsarevich Nicholas. In the last violent quarrel with Wilhelm on 15 March Bismarck took his revenge by showing him the Tsar's deeply insulting remarks about him, from which the young Kaiser was forced to conclude that Alexander III thought him 'insane'. As luck would have it, it was precisely at this time, in the highly unsettled days immediately after Bismarck's fall and before the diplomatically inexperienced new men Caprivi and Marschall had had the chance to take stock of their responsibilities, that one of the most momentous decisions in the history of German foreign policy was made: the non-renewal of the secret Reinsurance Treaty with Russia. What part did Kaiser Wilhelm II play in this fateful step, which historians have almost unanimously seen as one of stupidity verging on the suicidal?[1] Although the records reveal that ultimately the young monarch followed the advice of the responsible statesmen of the Wilhelmstrasse, they also convey an alarming picture of the confusion and superficiality that prevailed behind the glittering façade of the German Reich during this transitional period.

The Reinsurance Treaty with Russia, entered into by Bismarck on 18 June 1887 for a three-year period, was intended to replace the Three Emperors' League between Germany, Russia and Austria-Hungary which was shortly to expire.[2] The treaty bound Germany to remain neutral in the event of an Austrian attack on Russia and Russia to do the same if France attacked Germany. On the instigation of the Russian ambassador, Count Shuvalov, Bismarck had also agreed to a 'very secret' additional clause in the treaty which was irreconcilable with the Dual Alliance between Germany and Austria-Hungary. According to this clause the German Reich was to provide diplomatic and moral support to Russia in the event that it took possession of the Straits or protected its interests by a military invasion of Bulgaria.[3] As the date for the renewal of the treaty drew near, at the end of 1889, neither the Tsar nor his Foreign Minister Giers had any doubt that the extension of so advantageous an agreement would be in Russia's interest.[4] Early the next year, on 10 February 1890, Shuvalov approached Bismarck about the renewal.[5]

As was recorded in an earlier chapter, Prince Bismarck had let Wilhelm II in on the secrets of his Russian policy very soon after his accession, and in so doing had informed him of the existence of the treaty.[6] Now – in mid-February 1890 – according to Herbert Bismarck the Kaiser gave his consent to the renewal of the treaty, whereupon Shuvalov left for St Petersburg in order to set the appropriate negotiations in train. When he returned to Berlin on 17 March with the official authorisation of the Tsar and was immediately invited to hold conversations with Bismarck, he was met with the disturbing news that the Reich Chancellor – not least because of differences with the Kaiser over policy towards Russia – was about to be dismissed. Bismarck evidently wished to exploit to his own tactical advantage the irate handwritten note about troop movements in Kiev which the Kaiser had sent to him on the morning of 17 March, by painting a graphic picture of the momentous international consequences of his fall for the benefit of the Russian ambassador. Shuvalov was duly appalled, and telegraphed the news immediately to St Petersburg.[7]

On 19 March the ambassador also expressed his anxiety to Herbert Bismarck, who was still in office as secretary of state, 'over a glass of beer'. Russia's disquiet was recognised by Herbert as an opportunity, if not of preventing his father's dismissal, then at least of showing it in an advantageous light. In a letter to Wilhelm he went so far as to claim that Tsar Alexander was keen to reach agreement on the renewal of the treaty with new conditions that would be extremely favourable to Germany, but that he would do so only with Prince Bismarck. The letter, dating from 20 March and steeped in cynical self-interest, stated: 'This is to inform Your Majesty, with my most humble duty, that the Russian ambassador told me last night in confidence that he had been authorised by the Tsar to extend for six years the secret treaty between Russia and Germany which guarantees us Russia's neutrality in the event of a French attack, and which expires in June of this year; indeed the intention is that the agreement should be regarded as permanent. Count Shuvalov therefore called on the Reich Chancellor immediately on his return from St Petersburg, on the 17th of the month, to give him the above-mentioned information. He then learned that Your Majesty had sent word to the Reich Chancellor that same morning to say that You were expecting the Reich Chancellor's letter of resignation. Count Shuvalov thereupon withdrew his proposals; when he heard last night that Your Majesty would not hesitate to proceed with the dismissal of Prince Bismarck, he said Tsar Alexander would give up the renewal of the secret treaty, as such a secret matter could not be negotiated with a new Reich chancellor.'[8] This was of course pure fabrication.

We know of the Kaiser's furious mood during these crucial days from the notes kept by his Flügeladjutant Carl von Wedel. He saw Wilhelm on the evening of 17 March and recorded that he had been greeted with the words 'that it was all over with the Prince!' The Kaiser had then told him that Bismarck had 'started up an intrigue with Shuvalov', the aim of which was 'to make Russia invade Bulgaria, thereby causing Austria to launch an attack, and then for us to dispute the *casus foederis*. Such an intrigue carried on behind his back "was the absolute limit". Austria would have fallen into a terrible trap, and poor Kaiser Franz Joseph, to whom he had pledged his word and intended to keep it, would have had to pay the price. Of course it would never have come to that, for on the day that Russia made as if to do anything of the sort, he would have given the order for mobilisation, together with Austria. He now understood why the Chancellor had for several days been waiting "so longingly" for Shuvalov's return. Luckily he had been informed about this whole intrigue – for Shuvalov had brought back approval for Bismarck's plan from St Petersburg – in good time: he had been warned this morning and the whole plan had been revealed to him.' Wedel assumed that this warning had been conveyed to the Kaiser by the Grand Duke of Baden, who had received the news from Holstein.[9]

When he received Herbert Bismarck's letter three days later, however, Wilhelm seems to have forgotten the devilish conspiracy between Bismarck and Shuvalov to start an Austro-Russian war, for he wrote at the top of the letter: 'I agree to the renewal of the treaty and authorise you to inform Shuvalov 20.III.90.' He did not understand, however, why the renewal could be negotiated only with the old Reich Chancellor, as his 'Why?' at the end of the letter indicates. He sent the letter back to the Foreign Secretary with these marginal notes. As the Kaiser clearly had not grasped the real purpose of his letter, Herbert wrote him a second one, in which he again drew attention, mendaciously, to Shuvalov's insistence that Prince Bismarck was indispensable to the renewal of the treaty. 'From Your Majesty's All-Highest marginal note on my respectfully submitted report of this morning, which I return herewith, I observe that it was not sufficiently clearly expressed, and I therefore take the liberty of submitting the following explanation to Your Majesty, with my humble duty. Already before Count Shuvalov's departure for St Petersburg Your Majesty had authorised Prince Bismarck to say to the Russian ambassador, as then confidentially suggested by him, that it was the All-Highest inclination to renew the secret treaty which was due to expire in three months, and this was conveyed to Count Shuvalov at that time. The latter had intended to enter into negotiations with Prince Bismarck at the present time, as the Tsar had authorised him to do. In the meantime, however, Prince Bismarck has been relieved of his offices by Your Majesty, and Count Shuvalov, as I respectfully stated in the last sentence of the enclosed report, has now informed me that on the Russian side the extension of the treaty would be given up. After this notification from Count Shuvalov I therefore cannot return to this matter, since he is aware that Your Majesty had previously authorised negotiations for the renewal of the secret agreement, and yet in spite of this he spoke to me last night in the negative sense which I respectfully reported to Your Majesty in the enclosed report.'[10]

The Kaiser probably did not receive this second letter from Herbert until late in the evening. For a second time, perhaps deliberately, he failed to see that this was a ploy by the son to force him to reinstate the father, and he still refused to understand why the negotiations could not be continued without Bismarck. Dramatically asserting his own claim to be in charge, he had Shuvalov woken that same night with a message 'summoning the Count to see the Kaiser the next morning at a quarter to nine in frock coat'.[11] During their early morning discussion Wilhelm assured Shuvalov, as the future Russian Foreign Minister Count Vladimir Nikolaievich Lamsdorff recorded in his diary, 'that apart from a few differences of opinion on domestic policy, which I ascribe to the very agitated state of mind of the Chancellor, nothing, really absolutely nothing at all, has changed with regard to our foreign policy, which I have directed until now and shall continue to direct . . . Herbert Bismarck told me that you have had a discussion with his father about the renewal of the secret treaty and that your Emperor, like myself, is favourably inclined towards the renewal of our agreement; Herbert added that you no longer wished, in view of recent events, to continue the discussions. I should deeply regret that and I ask you to tell your Emperor that I, for my part, stand by our obligations. I am ready to renew them in complete agreement with the wishes of His Majesty. Our policy was after all not his, that is to say Bismarck's, policy. It was the policy of my grandfather and it has remained mine.'[12]

An hour later thc Kaiser saw Herbert Bismarck at the station, where he had gone to meet the Prince of Wales, and told him that he had just spoken to Shuvalov. Herbert had misunderstood Shuvalov, Wilhelm said. He would call in the afternoon and 'set the matter straight'.[13] Herbert, however, refused to take any further part in these proceedings, as he intended to hand in his own resignation. The Kaiser was at first unwilling to take any notice of this. When the resignation letter arrived the same afternoon, however, Wilhelm instructed the new Chancellor, General von Caprivi, to arrange the renewal of the treaty. He did not of course inform Caprivi that he had already given Shuvalov his express approval.

In the turmoil of the next few days Kaiser Wilhelm seems temporarily to have put the affair out of his mind. When he found himself again confronted with the question of the renewal, several days later, the tide had turned completely. Caprivi had gone immediately to the Political Department of the Foreign Office on 22 March, to consult the relevant documents on the secret treaty and to seek advice. Holstein, who was the only person present on that day, a Saturday,[14] recorded that 'on the day of the investiture Caprivi asked to see a secret document which the Kaiser wished to discuss with him immediately afterwards. I let him have the document, to Herbert's unbounded fury.'[15] From Caprivi Holstein learned for the first time of the Kaiser's intention of renewing the treaty. He immediately took steps to prevent this, convincing the new Chancellor that he should not reach any decision until he had heard the views of the experts in the Foreign Office, and arranged a consultative meeting for the next day. He wrote at once to Philipp Eulenburg: 'Tomorrow at 10 Berchem, Raschdau and I will have a joint interview with Caprivi, at my suggestion. Then we shall have to see what Caprivi can do.'[16]

On the Sunday the consultation went according to Holstein's plan. The Privy Councillor convinced not only Caprivi, who immediately took a dislike to the double-dealing represented by the Reinsurance Treaty, but also Ludwig Raschdau, who at first argued for a more cautious approach, and the Under-Secretary Count Berchem, who was initially unsure of his opinion.[17] The latter drew up a joint report on the results of the consultation, in which he cited reasons for non-renewal which certainly could not be rejected out of hand. The treaty had 'the aim of provoking warlike incidents which it is highly unlikely to be possible to localise'. It was 'in direct contradiction, if not of the letter, then at least of the spirit of the Triple Alliance'. Moreover, it did not guarantee any 'reciprocity'; all the advantage was on the Russian side. The report gives a detailed picture of the diplomatic chaos which would ensue if Russia actually launched the 'oriental war' which was 'the object of the treaty'. From now on a 'calm, clear and loyal policy' must be pursued, for the 'lively enthusiasm' of the people, which was so important in modern times, would be entirely lacking for such a war. Therefore, the report concluded, Germany must withdraw 'in an amicable way' from the agreement with Russia.[18] It did not mention what was probably the most important reason why the leading advisers in the Foreign Office were convinced that the renewal of the secret treaty would be too risky. Simply by threatening to reveal its existence the vindictive former Chancellor would be able to blackmail the Kaiser and the new leadership of the Reich, perhaps even to bring about his own return to power. Worse still: by betraying the dark secret Bismarck would be able to wreck the Triple Alliance between Germany, Austria and Italy and thus prolong, after his dismissal, the very policy of chaos with which he had almost succeeded in checkmating the Kaiser.

Before the Kaiser and Caprivi were able to discuss the matter, Wilhelm II met Herbert Bismarck by chance on the evening of Monday 24 March. Herbert, who was enraged by Holstein's involvement in the affair,[19] submitted a new proposal to the monarch which was designed to cancel out the influence of Holstein and the other advisers: namely that the question of extending the Reinsurance Treaty could best be settled in St Petersburg between the Russian Foreign Minister, Giers, and the German ambassador, General Lothar von Schweinitz. The Kaiser also approved this suggestion, as Shuvalov telegraphed to Giers on 26 March. The ambassador went on to inform the astonished Russian Foreign Minister that Herbert Bismarck had persuaded the Kaiser to accept his new proposal with the argument that 'in view of the complete ignorance of this question on the part of those who would have to conduct the negotiations after him, it would be better to put the negotiations in the hands of persons who have already participated in them on both sides, and who know the situation well'.[20]

Schweinitz was in Germany at this time, but was soon to return to St Petersburg. As he was to take a message from Wilhelm to the Tsar, he was invited to tea with the Kaiser; but when the latter showed no signs of wishing to talk about the treaty, the ambassador was forced to bring up the subject himself. As he recalled: 'I was invited to take tea with Their Majesties at 8 o'clock in the evening with General Caprivi and the Statthalter Prince Hohenlohe. When we took our leave the Kaiser said goodbye to me, but I said that I must beg for a reply to the message which I had brought him at the request of Tsar Alexander.

At this he commanded the Chancellor and myself to come to see him the next day at a quarter to one.' So the Kaiser was prepared to lavish as much as a quarter of an hour (before lunch at 1 p.m.) on this vital matter![21]

Not until he called on Caprivi at 10.30 the next day did Schweinitz learn of the latter's view that the renewal should be rejected. The ambassador listened in alarm to the new Chancellor, whom he described as 'unassuming, honest and serious', explaining 'that the greatest problem facing him now was the question of the renewal of the Russian treaty, for unlike Prince Bismarck, in Kaiser Wilhelm I's famous metaphor, he could not be a juggler playing with five glass balls, he could only hold two glass balls at one and the same time'. Schweinitz, who knew from Shuvalov that the Kaiser had already given his consent to the renewal, was now confronted with a difficult decision. In the end the elderly ambassador – 'the model of Old Prussian loyalty and statesmanship'[22] – decided to stand by the Reich Chancellor. Indeed, he even allowed himself to be convinced, for the time being at least, by Caprivi's arguments. Describing this decisive moment, he wrote: 'He showed me the secret treaty with Romania, of whose existence I was aware without knowing anything of its contents. I could not but be convinced that it was impossible to reconcile the obligations which we had undertaken towards King Karl [of Romania] with the provisions of the Russian treaty. I fully recognised that under the changed circumstances it was dangerous to continue with such an ambiguous policy as Prince Bismarck had pursued in order to "dig out" Russia's mines on every side.' He went on: 'If Bismarck had remained at the helm, I would have been in favour of renewing the treaty without the [additional] clause. After a short and earnest discussion, in which Count Berchem was later also invited to take part, I declared to the Reich Chancellor that I should not wish to raise any objection if he put it to the Kaiser that the treaty should not be renewed.'[23]

Afterwards, when Caprivi and Schweinitz arrived for their joint audience with the Kaiser, the Chancellor delivered himself of a 'short, cogent report in which he maintained that he could not reconcile the various secret agreements with each other and therefore proposed that the Russian treaty should not be renewed, especially in order to avoid the danger to which we should be exposed if it became known; if it were brought to the knowledge of the Cabinet in Vienna, whether deliberately or by chance, it would alienate them from us. The Kaiser listened in silence and then asked: "Well now, what does the ambassador say to this?" I replied [Schweinitz wrote] that I accepted the arguments of the Chancellor . . . With this the matter was settled.'[24]

Wilhelm took the volte-face in his stride, commenting 'Well, then it won't do, although I am sorry about it.'[25] For his own part he explained his decision to Shuvalov a few years later by saying that Caprivi had threatened to resign if the treaty were renewed. He himself had 'only just come to the throne and had appointed a new Chancellor. The latter had confronted him with an ultimatum. There had been no question of provoking yet another ministerial crisis within 24 hours. He had given in to the stubborn obstinacy of Count [sic] Caprivi. Caprivi had been an honest, but at the same time stubborn person and he lacked insight.'[26] Leaving aside the misrepresentation and exaggeration in this piece of retrospective self-justification, it is certainly remarkable how quickly and uncomplainingly the Kaiser gave in to the Chancellor. That Wilhelm – who spent the afternoon drawing a design for a flower arrangement for Bismarck's retirement ceremony[27] – had not even begun to understand the consequences of this abrupt change of mind became clear when the question of the renewal of the treaty came up again in May 1890.

After his discussion with the Kaiser and Caprivi, Schweinitz had to muster all his diplomatic skill to explain the sudden about-turn in Germany's position, first to Shuvalov in Berlin and then to Giers and Lamsdorff in St Petersburg, who were understandably confused and annoyed, and to placate them. In so doing he laid particular stress on the fact that in all the confusion caused by the Bismarck crisis the new government was not yet in a position to take such major decisions, but that nothing had changed as far as relations between Germany and Russia as a whole were concerned. In May 1890 Giers therefore made a second attempt to obtain the renewal of the Reinsurance Treaty from the German government. On 16 May Schweinitz – who in the meantime had come to see the non-renewal as a serious error – sent a dispatch-rider to Berlin to take Caprivi an official report, the text of Shuvalov's telegram in French of 21 March, reproduced from memory by Schweinitz himself, and a private letter in his own hand, all pointing out valid arguments for renewal.[28]

The reaction from Berlin, however, was even now brusque and unfriendly. Holstein, Marschall von Bieberstein, Kiderlen-Wächter and Raschdau prepared formal opinions on 20 May, in which they set out their arguments of March even more forcibly, in support of Caprivi's decision to turn down the offer a second time.[29] Caprivi wrote a summary on 23 May explaining his decision as follows: 'In the reply to General von Schweinitz his attention should be drawn to the impossibility, with regard to public opinion in this country, of maintaining a complicated system of alliances after the departure of Prince Bismarck. To conclude secret alliances is all the more out of the question now because the conduct of the former Reich Chancellor in any case makes the danger of indiscretions more likely and fosters uncertainties and misunderstandings. But our policy can only be, and should be, a simple one.' The Kaiser, to whom Caprivi submitted the document, commented 'in a most resolute manner' that he 'would not commit himself to any verbal or written statement to the Tsar which did not conform with the sense of the summary'.[30] Wilhelm II therefore gave orders on 23 May 1890 that the Russian request for renewal of the treaty should again be refused. Like Caprivi, he considered that this could be done 'in a polite and friendly way', 'so that no ill-feeling remains on the Russian side'.[31]

In spite of the courteous tone of the refusal and the validity of some of the objections expressed by Holstein and Caprivi – above all the irreconcilability of the secret treaty with the Dual Alliance and the danger of betrayal by the resentful Bismarck – the non-renewal of the Reinsurance Treaty is still today considered by most historians to have been one of the most fatal decisions of the 'New Course'.[32] It is true that there is less inclination now than in the Bismarckian orthodoxy of earlier years to see in this abandonment of the 'direct line to Russia' the beginning of the catastrophic events leading up to the First World War. Nevertheless the 'enormous consequences', in the longer term, of the non-renewal of the treaty in the spring of 1890 are unmistakable, as Thomas Nipperdey has rightly perceived. 'It cleared the way to the anti-German alliance of 1893/4 with France for the Russian radical nationalists; . . . but even for the moderates the isolation of Russia vis-à-vis Germany and England was not acceptable. The decision put the Reich on a new footing with Austria, its only ally and a weak one at that: it made the Reich more susceptible to Austrian decisions and Austrian pressure because there was no Russian "counterweight", something which Bismarck had wanted to avoid at all costs. Finally, it gave European politics a stronger impetus towards the creation of power blocs and confrontation; it was more offensive than Bismarck's defensive tactics. There were good reasons for the decision . . . Nevertheless it was still a calamity.'[33]

Not only the decision itself, but also the rushed and clumsy way in which it was taken could not but cause offence to Russia. The solution repeatedly put forward by Raschdau, of entering into serious negotiations in the hope of achieving a different, less risky agreement with Russia, would undoubtedly have been more tactful, even if it had failed in the end. Whatever one thinks of the cynical conduct of the Bismarcks, the astute and pessimistic arguments of Holstein and the straightforward, unpolitical standpoint of the newly appointed Chancellor-General, the role of Kaiser Wilhelm II in the process which brought about a decision of such importance in international history is shocking in its superficiality and inconsistency. He stated his claim – not only verbally but also, and above all, by his dismissal of Bismarck – to decide policy guidelines himself, and yet lost track of this matter for days on end. He gave his consent several times for the renewal, apparently without having grasped the full significance of the secret alliance, and then changed his mind equally hastily, partly following the advice of his official advisers, but partly driven by feelings of wounded *amour propre* and especially by his desire to prove his personal authority. This combination of authoritarian conduct and outrageous frivolity did not bode well for the new era without Bismarck.

THE KAISER AND THE FOREIGN POLICY OF THE NEW COURSE

When he dismissed Bismarck, did the Kaiser have foreign policy goals or even firm plans which he could now pursue unhindered? His very first actions stood in such marked contrast to his attitude during the dismissal crisis that they aroused the suspicion that his differences with Bismarck had been mere pretexts in a tactical struggle for power. Wilhelm's acceptance of Tsar Alexander's invitation to take part in the Russian army manoeuvres only days after he had declared that he would not go to Russia under any circumstances; the appeal to the despised Bleichröder to act as intermediary in the attempt to give financial assistance to Germany's ally Italy; his order to the commanding generals in the Ruhr 'to let the repeating rifles do their work at the first opportunity' after his frequent solemn declarations that he could not besmirch the first years of his reign with the blood of his subjects – all this and more led to the conviction that the young Kaiser was vacillating and lacking in a sense of purpose.[34]

The Chief of the General Staff, Count Waldersee, above all, was struck by Wilhelm II's apparent aimlessness. 'Many people thought that the Kaiser would have ascended the throne with firm and great plans for the future', he wrote less than six months after Bismarck's dismissal. 'It is quite plain to me that this is not the case. His thoughts were certainly lofty, he hoped to win great esteem for the German Reich and no doubt also to do great deeds himself. But he had no firm plan as to how to achieve this.'[35] As early as April 1890 he commented after a discussion with Philipp Eulenburg: 'What makes me most anxious is to observe that the Kaiser still has no firm opinions. He wavers back and forth . . . How often I have seen him wavering between Austria and Russia! How many different opinions I have heard him express about certain people! A man can be excellent today but in a few days' time he will be quite worthless, and vice versa! . . . He is still too impressionable.'[36] A few months later the General again noted in his diary: 'The Kaiser still has no firm opinions on any subject and does not know what he is aiming at. Anyone reasonably clever can easily influence him and he makes the most surprising leaps in all directions.'[37] By the autumn of 1891 the tenor of Waldersee's comments on the Kaiser had not greatly changed: 'And he is supposed to be an energetic man who knows exactly what he wants! Unfortunately he is like a reed before the wind who is still unsure of where he wants to be, in whatever sphere.'[38] 'When I think of how the Kaiser assured me 2–3 years ago that he would never stray from the path he had chosen, after I had begged him to be consistent, as nothing was more harmful than constant changes of system!'[39]

Other observers in the immediate circle around the Kaiser, it must be said, took a much more favourable view of his activities in the field of foreign affairs at this time. His Flügeladjutant Count von Wedel, for instance, following a conversation with Wilhelm II not long after Bismarck's dismissal, commented: 'My God, what outstanding qualities of mind our young sovereign has! What great and original ideas he has, and yet how irritated one can be sometimes by the strange course he adopts! May Heaven grant its blessing!'[40] On several occasions Wedel praised the 'clear, calm and cool judgement of the Kaiser . . . in questions of foreign policy', and even after two and a half years' service in close attendance on the Kaiser, in May 1891, he still felt able to comment that 'our young sovereign . . . pursues his plans with rare thoughtfulness and consistency'.[41] Similarly, the Kaiser's friend Philipp Eulenburg, as Waldersee concluded after a meeting with him, was an 'idealist' who firmly believed 'that Providence has great things in store for the Kaiser'; he was 'full of hope and confidence'.[42] The clearest testimony to the Kaiser's strongly developed will-power was given by those who suffered most directly from it. Reich Chancellor von Caprivi, for one, complained on 6 June 1890 of 'the constant torrent of new plans from the Kaiser and the millions needed, but not available, to carry them out'.[43]

If the disappointed Waldersee complained of the Kaiser's lack of a clear foreign policy plan, he too had no illusions about Wilhelm's immense power, his steely determination to shine in the eyes of the world, to rule in person and especially to decide on the foreign policy pursued by the German Reich. His own diary bears witness to the energetic and self-confident political activity of the young monarch. As he commented not long after Bismarck's fall, '*One* thought runs through . . . all his actions – concern for his personal position. He wants to be popular!'[44] Even if Wilhelm II's aims were still vague and unclear, one thing is certain: the Chief of the General Staff was to have direct experience of the Kaiser's power from at least the autumn of 1890, when his own position began to be threatened. A year after Bismarck's departure Waldersee observed that the Reich Chancellor was having 'a dreadfully difficult time with the Kaiser' because 'the good man cannot resist governing alongside the government; he would like to control everything and above all to make the world believe that he is the man who has everything under his control. So he goes gaily on, negotiating directly with this person or that, on his own initiative; this must lead to confusion and I cannot conceive how much Caprivi has to put up with.'[45] Six months later the General noted anxiously that an increasing number of observers were coming to think that, however many mistakes Caprivi had made, the principal blame for policy failures lay not with the Chancellor but with the Kaiser himself. 'With a Kaiser of such limited experience, so few clear opinions, such a changeable nature and at the same time such a strong inclination do everything himself and such a firm conviction that he knows everything better – no chancellor can govern. None could help but fail.'[46] Above all, Wilhelm had a deplorable habit of corresponding with the military attachés at various German embassies abroad, and of using them to make contact with foreign monarchs while bypassing the Reich Chancellor and the Foreign Office. Even in the eyes of Waldersee, who as chief of the General Staff had tried to build up just such an information network via the military attachés, this could not but create further chaos. In January 1892 he recorded indignantly that the Chancellor, 'in a towering rage', had made a vain attempt to prevent the Kaiser from corresponding with King Umberto of Italy through the military attaché Engelbrecht.[47] Not long afterwards the Kaiser revealed to Waldersee that he had bypassed the Foreign Office and the ambassador in St Petersburg by asking the military attaché there, Villaume, to assure Tsar Alexander on his behalf that he had no thoughts of war. Waldersee was horrified when, after a discussion of the military repercussions of the Russian famine, the Kaiser remarked to him: 'Hinzpeter takes a similar view.' 'So this wretched, scheming schoolmaster discusses such matters too', the General commented in disgust.[48] Caprivi was only indirectly to blame for the unhappy mood and serious state of the country, Waldersee considered, for 'the Kaiser wants to rule in person and to put everything to rights by himself. In that case he must also bear the responsibility, and indeed he can be shown to have been the cause of all the discontent.' The Chancellor was responsible for the bungled state of affairs only insofar as 'he does not provide a strong counterweight'.[49]

Waldersee saw more clearly than anyone that the Kaiser's opportunities for making his mark in the field of foreign policy were greater, more direct and considerably more dangerous than in the internal affairs of Prussia and the Reich. Shortly before his dismissal from the post of chief of the General Staff in January 1891 Waldersee held discussions with the ambassadors Reuss (Vienna), Radowitz (Constantinople) and Münster (Paris) during which all four acquiesced in the view that Bismarck's retirement had left a vulnerable gap in foreign policy which Caprivi had been unable to fill. Previously, the ambassadors agreed, all foreign governments, whenever a crisis arose, had asked what Bismarck's view was. But that was no longer so, for no one asked the opinion of the new German Chancellor. In general the representatives of the Reich abroad had come to think that Caprivi himself 'does not know in which direction he should steer'. While the Chancellor held back too much, the Kaiser, for his part, showed 'a strong inclination to dabble in politics himself', the diplomats commented. Although he pointed out that Wilhelm was by no means pursuing 'a completely new aim', Waldersee conceded that the Kaiser loved 'sudden jumps' and acted 'quickly, often too quickly, on momentary impulses'. 'If such a thing is dangerous anywhere, it is in the field of high politics', he warned. The ambassadors' comments also gave Waldersee the distinct impression that the views expressed by the Kaiser often did not coincide with those of the Chancellor and his advisers in the Wilhelmstrasse.[50]

Nothing demonstrates more clearly the growing anxiety of the Chief of the General Staff over the Kaiser's intervention in foreign policy than his comments on relations with Russia. Waldersee himself, of course, deeply distrusted Russia, and the relationship between the two countries had grown increasingly tense since Bismarck's fall and the non-renewal of the Reinsurance Treaty. The General recognised, however, that on the German side a policy of steadiness, caution and courtesy was required if the gigantic Russian Empire was not to be driven entirely into the arms of France. Instead, Wilhelm allowed his own vanity and thoughtlessness to mislead him into making such critical comments about Russia both in private conversation and in public speeches that his advisers were often horrified and sometimes even doubted his sanity. Only a few days after Bismarck's dismissal Waldersee noted with disapproval that in his attitude towards Russia Wilhelm was indulging in 'frequent and rapid scene-changes'. He was referring to Wilhelm's wish to go to Russia after all, to attend the manoeuvres in the summer. '8 days ago the Kaiser was absolutely determined not to go to Russia . . . under any circumstances, although shortly before that the contrary was the case.' But then Schweinitz had come to Berlin with the Tsar's personal invitation, and the Kaiser instantly decided 'that Russia is not so bad after all and now he does want to go there!'[51] In spite of the fact that he was about to visit Russia, in May 1890 Wilhelm was provoked by interviews which Bismarck had given to Russian and French journalists into exclaiming that Austria could not be grateful enough to him for having 'turned out Bismarck', as the latter had been planning the 'destruction' of the Danube monarchy in collusion with Russia. As even the loyal Carl Wedel remarked: 'With such stories I often wonder how far it is the Kaiser's imagination which is playing tricks on him, or how far it is libellous busybodies who are secretly telling him untrue or distorted, exaggerated things, for such tales surely cannot be really true!!'[52]

Still in this overexcited state of mind, Wilhelm II made a speech during a visit to the province of East Prussia which gave cause for alarm in Russian government and army circles. He would do his best to safeguard peace, he declared in Königsberg; 'but if it should be God's will that I be required to defend myself and to protect my country's borders, East Prussia's sword will show no less keen an edge in fighting the enemy than it did in 1870'. Those who dared to overturn peace would not be spared a lesson 'which they will not forget in a hundred years', he added threateningly, and promised: 'I shall not let the province be touched, but if any one should attempt it, my sovereignty will stand against it like a rocher de bronce.'[53] These bombastic words, clearly destined for Russian ears, were received with anger and alarm. Waldersee noted on 21 May that people in Russia were 'highly incensed by the Kaiser's speech'. But then again the Kaiser had taken the Russian reaction 'very badly', so that the 'much-discussed journey' to Russia was once more hanging in the balance.[54]

Wilhelm was particularly enraged by the russification of Finland, which began with the Tsar's manifesto on the Finnish postal system in the summer of 1890. He considered this to be a 'constitutional violation'. In a marginal note on a report from St Petersburg he commented that he fully shared the pessimistic view that the Russian government's aim was to turn Finland gradually into a province of the Russian Empire. One could not blame the Finnish people, let alone the Swedes, for resenting the Russian measures.[55] Neither the German nor the Russian side, therefore, pinned any great hopes on the Kaiser's visit to the Russian capital in August 1890. After Wilhelm's return to Potsdam on 27 August Prince Ernst zu Hohenlohe-Langenburg, who was working in the Foreign Office, commented that 'everything went off satisfactorily in Russia and at least the personal relations between the two Emperors became warmer'. But there had been no political benefits worth recording.[56] In fact, however, what had once been a good relationship between the two Emperors could not be restored, and gave way to an ill-natured indifference on both sides.[57]

In the course of a conversation during which the Kaiser explained to the Tsar why Germany had not been able to renew the secret Reinsurance Treaty, the two had a curious argument about France, triggered by the Tsar's remark that in their common struggle against nihilism and socialism the central and eastern European monarchies would be greatly assisted if the monarchy could be restored in France; at any rate, he said, he would never ally himself with France as long as it remained a republic.[58] The acting Chief of the French General Staff, General Boisdeffre, who was in Russia at the time, also came out with the assertion that the senior officers in his country were planning to restore the monarchy.[59] When the Flügeladjutant Carl Wedel, in discussing the question with the Kaiser, adopted Bismarck's well-known view that the French Republic would always have more difficulty than a French monarchy in finding allies among the other great powers, and went on to express the opinion that a restored French monarchy would try to legitimise itself through a war against Germany to recover Alsace and Lorraine, Wilhelm strongly contradicted this view. 'Strange to say, the Kaiser did not share this opinion', Wedel recorded with surprise. 'He thinks that we can enter into peaceful discussions more easily with the French monarchy, and that the latter will see its primary aim as the inner rebirth of the nation, rather than risking an adventure abroad.'[60] To Waldersee too the Kaiser expressed his conviction 'that Boisdeffre is right and would be supported by the Russians'. The Chief of the German General Staff commented: 'I cannot share in this beautiful dream, although unrest and civil war in France would certainly be the best thing that could happen for us.'[61] In the next few months Wilhelm II continued his attempts 'to hook the French'. He treated Jules Simon, President Carnot, the French ambassador Herbette and General Gallifet with particular courtesy and corresponded openly with Boisdeffre and with the widow of the painter Jean-Louis-Ernest Meissonier, who died at the beginning of 1891.[62]

In February 1891 Count Münster, the German ambassador in Paris, put forward a proposal that a number of French artists should be persuaded to exhibit their pictures in Berlin. Wilhelm and Caprivi took up the suggestion and had the idea of sending the Empress Frederick incognito – she took the name Countess Lingen on such occasions – to Paris for this purpose. She should make contact with the French artists and select appropriate pictures.[63] Waldersee, who suspected that the influence of Hinzpeter's French wife was behind this swing towards France, had nothing but scorn for this 'childish idea' which the Kaiser and Caprivi were pursuing 'to conciliate the French! We are making ourselves a laughing-stock for the whole world!', he wrote in his diary. In fact the attempted reconciliation proved all too soon to be 'a fiasco on a grand scale'.[64] Above all, the idea that the presence of the Empress Frederick and her daughter Margarethe could be kept a secret in Paris was a complete illusion. In a letter to her son written on 21 February from her apartments in the German embassy in Paris, Vicky commented: 'The newspaper reporters are a perfect pest, they follow us about everywhere like our shadows but we succeed in dogging [sic] them now & then, & it is rather amusing to read in the newspapers of the places we have visited (many of which we have never gone near).'[65] They were less amused, however, when chauvinistic articles appeared in the press calling for mass demonstrations against the 'German' Empress, who was forced to leave hastily for England. The German ambassador expressed the opinion that the Empress's visit would have been very useful if it had lasted four days, as planned, instead of ten, and 'if the royal lady had brought a more tactful escort than Seckendorff'.[66] 'I only hope in Germany they will not attach too much importance to it, and this ridiculous excitement will in the end not prevent the French artists from sending their things to the Berlin Exhibition!', the Empress Frederick wrote to her daughter Moretta.[67] But the right-wing Berlin newspapers brought out a stream of articles, apparently originating in diplomatic circles, stirring up feeling against the liberal Empress. Among other things, she was falsely accused of refusing to receive the Russian ambassador Mohrenheim and his wife, of going to see mainly Jewish art collections in Paris, of buying nothing in the shops, and of failing to give a single *sou* to the poor of Paris although she had inherited millions from the Duchesse de Galliera.[68] In Conservative circles the joke went around that after this debacle the Empress Frederick would not call herself Countess Lingen any more, but only Miss Lingen – a pun on the German word for failure.[69]

In spite of being on such bad terms with his mother, the Kaiser threatened serious repercussions if there were any 'unseemly behaviour towards the All-Highest person of the Kaiserin' in Paris. In response to the demonstrations against her in Paris, he ordered the rule requiring passports to be carried in Alsace-Lorraine to be enforced with the utmost severity, which the '*very distressed*' Empress Frederick condemned as 'so unnecessary and as *wrong-headed* as could be'.[70] After the Paris incident Wilhelm resumed his hostile attitude towards France. He took a malicious delight in the news that the 'French luminaries' who wanted to exhibit their pictures in Moscow had been forced to flee from Russia.[71] He spoke contemptuously of the French republicans – calling them 'infamous curs!' – and completely failed to recognise the looming threat of an alliance between France and the Russian Empire. 'The saying "Do not come near your Prince unless you are called!" is also valid for the Frenchmen in Moscow', he wrote on a diplomatic report of May 1891. When the Austrian Foreign Minister, Count Kálnoky, pointed out that Russia was convinced that even without a treaty of alliance France would take the Russian side in a war between Russia and Germany, the Kaiser remarked scornfully: 'Like a whore, without having to marry her.'[72]

Wilhelm's relationship with Denmark, which would play a strategically important role in a European war, remained tense and full of mistrust. A 'diffident' request directed by King Christian IX to the Kaiser, asking for the return of a few towns and villages in North Schleswig, was politely but firmly turned down by Wilhelm after consultation with Caprivi.[73] Discussing the matter with Waldersee in November 1890, however, the Kaiser spoke 'quite sharply about the Queen of Denmark and Prince Waldemar and about the intrigues of all the female relations, that is to say the Empress of Russia, the Princess of Wales and the Duchess of Cumberland. I had gathered from earlier remarks', Waldersee wrote, 'that the mood was gradually improving in that regard, but certainly only on condition that the King completely accepted the present situation. As this does not now seem to be the case, we shall after all be well advised to reckon with Denmark in the next war. The Kaiser thinks it best to send a fleet to Copenhagen straight away with an ultimatum. If we have time, this is of course an appropriate action to take, but only if war should break out surprisingly just at a time when we had a suitable fleet to hand! The fact that the Norwegians have not been on very friendly terms with the Danes lately seemed to please the Kaiser particularly. But unfortunately it is of very little significance.'[74]

On 18 April 1891, at a large military banquet in the White Hall of the Berlin Schloss, Kaiser Wilhelm II made a speech which according to Wedel sounded 'very warlike'. In an allusion to Bismarck's famous saying that the German Reich had been established not by parliamentary majorities and resolutions but by blood and iron, Wilhelm proclaimed that 'the German Reich was forged not by parliamentary majorities and resolutions but by the soldier and the army. My trust lies in the army. We live in solemn times, and there may be bad times before us in the next few years . . . Whatever the future brings, we shall hold aloft our flags and our traditions, mindful of the words and deeds of [the Margrave of Brandenburg] Albrecht Achilles, who said: "I know of no nobler place to die than in the midst of one's enemies." This is also my own heartfelt belief, and on it rests my unshakeable faith in the loyalty, courage and dedication of My army, particularly of all the comrades who are posted at our borders.'[75] Although government and court officials tried to pass on a watered-down version of the speech to the press, the actual text reached the German newspapers via the *Pester Lloyd* and the Viennese *Fremdenblatt*. Both the Reich Chancellor and the Flügeladjutanten Wedel and Zitzewitz wondered what had caused the 'warlike mood of the Kaiser'. Wedel thought that it was just another insignificant gaffe, and that Wilhelm's 'inclination to make speeches in general and his lack of preparation' were alone to blame. But he was forced to retract this opinion when he discovered that Wilhelm had recently given orders for a mobile headquarters to be built and a mobilisation plan to be worked out for the royal stables.[76]

Waldersee also expressed scathing opinions about the Kaiser's weakness for handing out new colours and standards 'which are mostly far-fetched', and for accompanying these with fulminating speeches. Commenting on the belligerent speech of 18 April 1891 he said that the strong emphasis on Luther had offended the Catholics, and that Wilhelm had 'once again needlessly roused' enemies within. Such remarks, together with the Kaiser's pessimistic utterances about a future war, Waldersee noted, made 'all discerning listeners . . . go hot and cold'. 'But the Kaiser himself believes he made an outstanding speech and the next morning he was indignant not to find it reported word for word in the newspaper.'[77] Meanwhile the Russians were quietly continuing to pursue their goals without a care for anyone, the General observed. 'They are trying to suppress completely and neutralise all nationalities which are not pure Russian, and at the same time to increase their military power and to be prepared for a rapid transition to war. The Kaiser, who is of course very easily influenced, at one moment thinks that everything looks perfectly peaceful, and at the next – when he receives news such as the redeployment of a division or something of the sort – that war is imminent.' Caprivi wavered back and forth with the Kaiser and had neither the intelligence nor the energy to pursue a consistent policy, Waldersee complained. For him it was clear that there were only two courses open to Germany, and one must be chosen. 'Either we say to ourselves that war must certainly come, and in that case we must go on arming ourselves with all our might; it would be best for us to fix our eyes firmly on a date when we can bring about the decisive moment ourselves. Or we believe that there is a way round, and in that case we must take it, i.e. try to turn our policy in a different direction: get out of the Triple Alliance and have a rapprochement with Russia, with or without Austria. But what do we do? Militarily, nothing; politically, we ride around on the broken-backed Triple Alliance and flirt in the clumsiest possible way with the Poles. Meanwhile we are most certainly rushing towards our doom.'[78] In May 1891 the General confided to his diary that the Kaiser was 'in an anxious state of mind and he believes there will be war before this year is out . . . He is naturally not at all pleased that the French are sending a fleet to Kronstadt in the summer. He sees clearly that Russia is quite openly making a show of her friendship with France for our benefit.'[79]

FIRST STEPS IN COLONIAL AND NAVAL POLICY

Caprivi's foreign policy was directed principally towards ensuring the security of the German Reich on the continent of Europe.[80] To this end he wished first of all to strengthen the Triple Alliance between Germany, Austria-Hungary and Italy, prerequisites for which were the abandonment of the secret agreements with Russia and the establishment of relations of trust with Great Britain. This last, in turn, depended on a policy of restraint in German colonial affairs and particularly in naval armament. The advocates of this straightforward and moderate continental policy with its main focus on central Europe included not only Friedrich von Holstein, Adolf Marschall von Bieberstein and Alfred von Kiderlen-Wächter at the Foreign Office, but also – at first – the General Staff under Waldersee. The latter had a long conversation on 1 May 1890 with Caprivi about foreign policy, in the course of which he set out for the Chancellor the strategy and tactics to be adopted in any future war.[81] Caprivi expressed himself entirely satisfied with the Chief of the General Staff's analysis, and for his part Waldersee was glad to observe that Caprivi wanted 'to stick firmly to Austria'. 'I am sure that we shall get along well', noted Waldersee, who for the time being was still optimistic about his relationship with his fellow General at the Reich chancellor's palace.[82] The newly appointed Chancellor also told General Carl von Wedel that he wished to create both a 'Central European Customs Union' between Germany, Austria-Hungary and Italy and a maritime alliance with Denmark and the Netherlands, for 'if we can win over both [the latter] countries with their large sea-faring populations we shall be in a position to build up and maintain a large, powerful battlefleet, for which Germany does not have sufficient manpower on her own'.[83] As far as overseas policy was concerned, Caprivi professed himself an 'opponent' of colonies in principle, although he could not deny that to give up the colonies acquired so far by Germany would not be compatible with the national honour.[84] Like Caprivi, Waldersee rejected not only German colonialism, but also the 'whole colonial policy' of the European powers, and predicted, far-sightedly, that their imperialism would not last long. 'Africa belongs to the Africans! That is my view; and I consider it shameful on the part of the European powers simply to cut up the continent between them', he declared in the summer of 1890. 'The Mohammedans are working slowly but surely to convert the inhabitants to their creed and are far superior to us in their methods. I believe that Africa will belong to them and that after a few decades all Europeans will be driven out!'[85] In spite of the restraint in German colonial policy, however, Waldersee was concerned to note as early as June 1890 that the mood in Britain was becoming increasingly hostile towards Germany.[86]

The calm and restrained attitude required by Caprivi's cautious continental policy did not accord well either with Kaiser Wilhelm II's temperament or with his long-held, overweening ambitions. In the immediate aftermath of Bismarck's dismissal there was widespread 'fear of surprises, and unexpected escapades by our Kaiser', who was thought perfectly capable of 'going off at a tangent in the field of foreign policy'. In particular it was feared that Wilhelm might, 'if his social welfare policy were to run aground, perhaps start experimenting with diplomacy'.[87] It seemed at first that Wilhelm would favour an expansive colonial policy in Africa and thereby come into conflict with the Chancellor and the Foreign Office. In early April 1890 Caprivi confided to the Flügeladjutant Wedel that, as far as colonial policy was concerned, the Kaiser was pursuing 'far-reaching goals, with which he did not agree'.[88] From Waldersee's diaries we know that the Kaiser was 'determined' at this time 'not to give ground' to Britain in East Africa, but to extend and establish German influence 'as far as the lakes, to annex Zanzibar or at least to acquire a protectorate over it'. He had often expressed such sentiments recently, Waldersee recorded, as a result of which he had roused great hopes in colonial circles.[89] The Chief of the General Staff immediately recognised that this was an area 'in which differences of opinion between the Kaiser and the Chancellor' might arise.[90] When at the end of April 1890 Eulenburg drew the Kaiser's attention to the danger of an agreement between France and Britain which would cause 'great difficulties in Africa' for Germany, the impression made upon Wilhelm by this remark was unmistakable.[91]

In May 1890 Major Eduard von Liebert, the future governor of German East Africa, returned to Berlin from Africa in high spirits, after German troops had come to the aid of the German East Africa Company to put down the Bushiri rebellion in 1889. Liebert spoke so enthusiastically about the potential for development of this German protectorate (the present Tanzania) that the Foreign Office was seriously concerned that 'he might make the Kaiser even more enthusiastic and thereby cause us problems with England'.[92] The African explorer Hermann Wissmann also came to Berlin in the summer of 1890 – he received his patent of nobility at the Neues Palais on 24 June – and spoke 'very despondently of the bureaucratic attitude of the Chancellor and his people in dealing with African affairs'.[93] Seen from the point of view of the Wilhelmstrasse it was also alarming that Waldersee, who hitherto had argued for restraint in colonial policy, was now taking a vigorously defiant attitude towards Great Britain, whose expansionist aims in Africa certainly strike us as excessive today.[94] 'I do not wish to pass judgement on our relations with England', he wrote in his diary on 4 May 1890; 'I only know that we have been unbelievably submissive and timid hitherto, and also that the English have persisted in behaving in the most shameless and ruthless way in East Africa.'[95] In the conflict which was thus beginning to develop between the military and the Wilhelmstrasse it was of course the Kaiser's attitude which tipped the balance. Waldersee wrote on 24 May: 'In our East African policy we are coming up against many difficulties with England and would certainly give way if the Kaiser did not vigorously oppose it. All the officers who come here from Africa are full of hopes for the successful development of the colony so long as matters are handled with resolution here in Berlin.'[96] After Wilhelm had had a long conversation with Liebert during a military review at Spandau, Waldersee was able to note that the Kaiser had 'not the slightest inclination to give in to the excessive demands of the English, and there will be a difficult passage with Caprivi'.[97] The international repercussions of the Kaiser's attitude soon made themselves felt: on 23 May Lord Salisbury had to inform Queen Victoria that his negotiations with the German ambassador Count Hatzfeldt had failed, as the Kaiser wanted to cut Britain off from the great inland lake, 'which we cannot admit'.[98] In August 1890, after a conversation with the Prime Minister, the Queen complained of the great difficulties which her government was having with Germany over Africa.[99]

Wilhelm's confidence in Wissmann and Liebert was so great that he was also quite prepared to defend these officers against the colonial enthusiasts who advocated still more ambitious aims. An example of this occurred in December 1890 when Wissmann felt compelled to urge restraint on the celebrated African explorer and adventurer Eduard Schnitzer, who called himself Emin Pasha and who was leading a German expedition to Lake Nyanza in central Africa, now Lake Victoria.[100] The president of the German Colonial Society was none other that the Kaiserin's uncle, Prince Hermann zu Hohenlohe-Langenburg, who sent an urgent petition to the Kaiser on 25 December following newspaper reports of the recall of Emin Pasha. In it Hohenlohe gave a prophetic warning that the news of Emin's recall would spread like wildfire in central Africa and would put the man himself in grave danger. The chiefs of the warrior tribes and the Arabs who had entered into treaties with him as the Kaiser's representative and had ceded territory to Germany would 'look upon the disgraced Pasha as a deceiver and a traitor' and his authority would be severely damaged. 'If Emin should fall victim to the greed and vengeance of the natives . . . the imperial government would be accused of responsibility for the downfall of the man whom Europe admired for the selflessness with which he has devoted his life to winning Africa for the cause of civilisation.' In any case, Hohenlohe argued, the possibility should be borne in mind that on receiving the news of his recall Emin 'will feel compelled to throw himself into the arms of the English, which [would have] the worst possible consequences for the entire future development of the East African territories'. 'The high reputation and powerful position of the German Reich would be seriously impaired in the eyes of the natives' and grave problems would inevitably ensue. 'Everything which has been gained in East Africa over the past year, with so much effort and such heavy sacrifices, could hang in the balance. Filled with a passionate desire to see Germany's colonial possession achieving rapid and prosperous development', Hohenlohe appealed 'most humbly' to the Kaiser to cancel Emin Pasha's recall.[101]

Kaiser Wilhelm II replied to this petition with characteristic firmness on 27 December 1890: 'My dearest Uncle, I have received your letter concerning Emin Pasha and I hasten to answer it. First of all I must clear up a misunderstanding which is affecting you as well as many of our colonial friends. Emin *is not at all in disgrace* and has therefore not been recalled. Following orders from here Emin has by agreement placed himself under the Reich Commissar [Wissmann] and has pledged himself to obey his instructions. That he is in the service of the Reich at all is my own work for I instructed the former Reich Chancellor to take him on. Orders were given to Emin to advance to Nyanza equipped with a certain sum of money, gifts and bearers, to set up one or more bases – at any rate one fully equipped one – there, and then to return home. He was also ordered not to get involved in any fighting and only to enter into treaties, so as to have as reassuring an effect as possible. Only attacks on the caravan itself, of course, should be vigorously resisted. That is why the detachment which is accompanying him is so small, sufficient to protect the bearers but not for purposes of war. Emin has been at Nyanza for a long time now and has had plenty of time to look around, set up bases, explore the land around them and make it secure as agreed. As he was not making any preparations to come home, despite his orders, he received a second, repeated order to return after the completion of his task. Where is the disgrace or recall in that?! Reports and more detailed information cannot be expected for some time yet, to show what Emin has achieved. But if differences have arisen as a result of disobedience by Emin, giving the Reich Commissar cause to remind Emin of his orders again, that is simply Wissmann's duty. He bears the whole responsibility for military protection and for the measures that are required. The whole of Wissmann's past life provides a guarantee that he thoroughly understands what he is doing in Africa. But if there is to be any hope for the prosperous development of the colony, then the first requirement is that there should be discipline among those involved there. Without discipline no national institution can flourish and develop. So if anyone is sent out with instructions from the Commissar and then does not follow his instructions, he must learn to knuckle under, whether his name is Emin, Schulze or Lehmann. For if we cannot do without obedience in Europe, how much more necessary is it in Africa, where everything still has a half-military, warlike aspect, and the smallest mistake may be disastrous and could cost the lives of many. For the government here to interfere with Wissmann's orders, at a distance, ex officio, without more detailed reports, is absolutely out of the question. I have the fullest confidence in Wissmann and in his judgement. Emin must also learn that "Bravery even a Mameluke displays; obedience earns the Christian praise", which is quite good in relation to the Arabs. Your devoted nephew Wilhelm I. R. Happy New Year.'[102] Prince Hohenlohe declared himself reassured that all the rumours that Wilhelm was dissatisfied with Wissmann had been proved baseless by this confidential explanation from his 'All-Gracious Kaiser and Sovereign'.[103] As is well known, the ill-starred expedition led by Emin Pasha, who refused to acknowledge Wissmann's orders to return, ended with his murder by Arab slave-traders on 23 October 1892, just as Hohenlohe had predicted.[104]

If Wilhelm II had one aim that he pursued with even more enthusiasm than the expansion of the German colonial empire in Africa, it was to build up a large naval force. That this was his first priority became apparent in June 1890 when the astonishing agreement with Britain, under which Germany received the strategically important North Sea island of Heligoland in exchange for Zanzibar, became known.[105] The colonial enthusiasts and representatives of the export trade, who were chiefly to be found in the ranks of the Free Conservatives and National Liberals, were suddenly bitterly resentful of the Kaiser, by whom they felt they had been first 'urged on and then disowned and deserted'. A certain malicious hostility towards Caprivi also began to spread, and eventually led to the foundation of the Pan-German League.[106] The Kaiser, however, was enthusiastically in favour of the Heligoland–Zanzibar Treaty. At the beginning of his Scandinavian cruise in late June 1890, he discoursed on the strategic value of Heligoland in the presence of the newly appointed Secretary of State of the Reich Navy Office, Admiral Friedrich von Hollmann. He emphasised that if necessary he would have given up 'even more' than Zanzibar to obtain this little North Sea island, for he valued Heligoland 'very highly and he intends to set up cannon or howitzers on it. The English had no idea of Heligoland's strategic importance, and in order that they should remain in the dark he had also instructed that the German press should not mention it until the treaty had been ratified. Once that had been done he would soon show them how valuable the island was for us.'[107]

The Heligoland–Zanzibar Treaty and the Kaiser's brief visit to the South of England at the beginning of August 1890 led to a noticeable improvement in Anglo-German relations that summer. 'The English and the Belgians seem to be quite delighted with him [Wilhelm II]', as Ernst Hohenlohe at the Foreign Office reported to his father on 13 August 1890. 'The English newspapers are vying with each other in their panegyrics and assurances of friendship for Germany. The "Standard", the organ of the government, even says that England belongs in effect, if not formally, to the Triple Alliance and will stand shoulder to shoulder with it. The reception given to the Kaiser is said to have been very warm, as [August] Eulenburg told me. The notorious [Heligoland–Zanzibar] Treaty has had one good effect after all, in that the present mood towards us in England, even on the part of Gladstone and the opposition, has changed entirely for the better; and that makes up for a great deal. I thought the pompous handing-over of Heligoland [on 10 August 1890] rather ridiculous, and the comparison between this act and the Battle of Wörth in the imperial speech not very appropriate.'[108] This last remark referred to the Kaiser's address to his 'Comrades in the Navy!', in which he declaimed 'Four days have passed since we celebrated the memorable day of the Battle of Wörth, when under My late, revered grandfather the first hammer-blow in the construction of the new German Reich was struck by My father. Today, twenty years later, I reincorporate this island, as the last piece of German soil, into the German Fatherland, with neither battle nor blood. The island's destiny is to become a bulwark on the sea, a refuge for German fishermen, a base for my warships, a stronghold and a shield for Germany's seas against any enemy who should venture into them.'[109]

The ambivalence of Wilhelm II's attitude towards Great Britain, and the wishful thinking with which it was permeated, is exemplified by the fact that he chose the moment of taking possession of Heligoland to propose a toast in which he described Britain as a 'kindred land' and his grandmother as the 'noble lady whom we have to thank for the fact that the island is German once more. The Queen reigns over her country with far-seeing vision and lofty wisdom, and she attaches great importance to living in friendship with Me and My people. She values German officers, German melodies. Long live the Queen of England!'[110]

It was entirely in keeping with this attitude that the German Kaiser made use of the – supposed – permission which he had from the Queen to hand out advice to her on naval matters. As he wrote to her in February 1891, he had 'very grave doubts in the soundness & efficiency of the heaviest guns' in the battleships of the Royal Navy. The root of the evil lay in the English supply system, under which the fleet was obliged to accept the guns manufactured by Sir William Armstrong in Elswick. His advice was that the navy should manufacture its own guns in the Woolwich Arsenal, or at least have the right to inform the manufacturers of its requirements. In Germany the naval experts would say to Krupps: 'Such & such a gun is wanted in the fleet; it must have a certain number of qualities, which are named; then when the gun is made it is tested at the works. First by Krupp, then by a committee of officers, & if that is not enough it is sent to the Government Range.' He recommended that the German procurement system should be introduced in the Royal Navy.[111] Ten days later he sent his grandmother a warmly worded telegram of congratulations on the launch of two new Royal Navy ships, again emphasising the affinity between the two fleets. 'May the two fine new ships built by British hands prove a powerful addition to the Royal Navy and may they always . . . uphold the honour of the British flag. The whole of my Navy feels with me the honour done to our comrades in arms and begs to lay the most respectful congratulations at Your Majesty's feet.' He signed the telegram 'William German Emperor, King of Prussia, Admiral of the Fleet'.[112] Lord Salisbury suggested to the Queen that she confer a high-ranking British order on Admiral von der Goltz, who attended the launch as the Kaiser's representative, 'merely as a matter of policy'. It had the desired effect on Wilhelm.[113]

No observer could fail to see how much at home the Kaiser felt in the company of senior German naval officers. When he arrived at Wilhelmshaven on 22 April 1890 in the *Lahn* he was to have disembarked, according to the agreed programme. But instead he went on board the *Deutschland*, which had just returned from the Mediterranean. 'There were hearty greetings', Vice-Admiral Paul Hoffmann recorded in his diary. The Kaiser inspected the crew and 'then betook himself to the cabin, where he felt so much at home that he said to his suite: "Come in to my cabin, gentlemen."' There he presented Admiral Friedrich von Hollmann with his patent of appointment as secretary of state at the Reich Navy Office and talked to him at length about Bismarck's dismissal. The monarch's conspicuous camaraderie with Hollmann was demonstrated again that evening when the latter was on his way to the Casino club with Hoffmann. Just as the two admirals were about to enter the building the Kaiser's carriage drew up at the door. Hoffmann recorded: 'We hid behind a sentry box, but when we followed the Kaiser in he noticed us and was very pleased. Because of the rain the Admiral [Hollmann] was wearing large rubber galoshes, with his trousers rolled up, and a grey cloth draped round his neck over his raincoat. The Kaiser gave him no time to take off his coat. "I shall take my guest in", he said, taking the Admiral by the arm, and led him up the stairs into the brightly lit corridor where many officers were standing, and then said: "Hollmann, old fellow, take your overshoes off now, I will hold you steady", so that the Admiral was overcome with embarrassment at the way he was dressed.'[114]

This friendly relationship was bound to have its consequences. Naval officers, above all Chief of the Naval Cabinet Freiherr von Senden-Bibran and Navy Secretary von Hollmann, who was in the habit of lying in wait for the Kaiser during his morning walks in the Tiergarten, did their best to encourage Wilhelm's personal commitment to the enlargement of the fleet. Prince Heinrich, stationed in Kiel and enthusiastically dedicated to his career as a naval officer, also exercised an anti-British influence on his brother. When his mother praised British rule in Egypt in his presence in early 1891 Heinrich became 'quite savage'. 'He is alas *so* jealous of England as a nation & as a power!', she commented.[115] During the Scandinavian cruise in the summer of 1890 Alfred von Kiderlen-Wächter, the Foreign Office representative in attendance, became aware of the disproportionate influence of the navy on the Kaiser. He deplored the fact that Wilhelm was accommodated not in the *Hohenzollern* but in the bigger and more comfortable *Kaiser*, where he was completely under the thumb of the navy and was rarely available to receive Kiderlen-Wächter's reports on foreign affairs. Even Chief of the Military Cabinet General von Hahnke, who was travelling in the *Kaiser*, 'being of course a mere general' was 'treated badly by the sailors!'[116] Kiderlen urged the Reich Chancellor to come to Wilhelmshaven, where the Kaiser intended to spend four or five days after the Scandinavian cruise, in order to prevent Wilhelm from 'falling entirely into the navy's clutches'.[117] After the combined military and naval manoeuvres in Schleswig-Holstein in the late summer of 1890 Kiderlen reported that 'of course praise was again lavished on the navy' by the Kaiser, to the extent that 'even Hollmann felt quite embarrassed'.[118] In March 1891 Caprivi complained that it was 'altogether very difficult to get anywhere with the navy people, as he knew from his own experience how obsessed they were and how they suffered from delusions of grandeur'.[119] Only weeks later there were 'very serious arguments between the Kaiser and the senior naval officers' in Kiel, triggered by the officers' avowal 'that our navy is in fact incapable of anything and is by no means equal to its task if it came to the test'. The naval officers even hinted to the Kaiser 'that his endless travelling around with the fleet was seriously prejudicing the training of commanders, officers and crew', which made him 'furiously angry'. Carl Wedel, who described this clash in his diary, commented far-sightedly: 'This will not bring about any change, for unfortunately the navy is and will remain a malady of our All-Highest Master which can be cured only by a war and the insignificance of the navy which would no doubt be revealed in the course of it.'[120] Waldersee, who was also at the naval review in Kiel, described the same episode in equally critical terms: 'It is quite extraordinary how the Kaiser has developed a stronger passion for naval affairs than for anything else, and here one can see very clearly that he is much more interested in the navy than in the army. Unfortunately there are many people in the navy who exploit this and drive the Kaiser even further; I would not exclude Prince Heinrich from their number. But most of the older officers are sensible and are very well aware of how the Kaiser exaggerates and overestimates the importance of the navy. Admiral Hollmann, the new Secretary of State, is already seriously worried; the Kaiser has far-reaching ideas which naturally cost a great deal, while the Reichstag is not at all inclined to spend large sums of money on the navy.'[121]

Wilhelm II's enthusiasm for the navy represented a constant if as yet latent threat to the cautious continental policy which Caprivi was trying to pursue. Directly after the appointment of the new Chancellor Waldersee predicted 'an immediate explosion caused by differing opinions on the development of our navy', for 'Caprivi will probably not yield and the Kaiser will do so only with the utmost reluctance.'[122] And indeed in the autumn of 1890 he learned that Wilhelm II was pressing for an enlargement of the fleet at the expense of the army, noting almost disbelievingly in his diary that the Kaiser was making 'very substantially increased demands for the navy', and was inclined to give these naval requirements 'very decided priority' if approval could not be obtained in the Reichstag at the same time for army expenditure. He continued anxiously: 'Heligoland of course needs to be fortified and this will not be cheap. And now the Kaiser comes up with the bizarre idea of turning Memel [now Klaipeda, Lithuania] into a naval base, which he justifies on the grounds that Russia is building a naval base at Libau [now Liepaja, Latvia]. If he brings it up before the National Defence Committee he will find scarcely anyone willing to vote for it. Another outrageous plan is to give up Wilhelmshaven and to make Cuxhaven a naval base!!! I hope he does not put that forward seriously.'[123]

Wilhelm's idea of fortifying Memel as a naval base was by no means allowed to lapse, as Waldersee had hoped. While the Chief of the General Staff was making his report to the Kaiser on 20 December 1890 the latter remarked almost casually that he had given orders for Memel to be fortified and had instructed the Finance Minister, Miquel, to find the money for it. Waldersee – who once again assumed that naval officers had won the Kaiser over to the idea – still considered the plan senseless, for fortification would require a significant amount of artillery and a large garrison, and would thereby 'swallow up considerable millions'. And for all that it would be 'scarcely a cannon-shot' away from the Russian frontier. 'How are we to get a garrison there if the Russians do not want to let us do it?', he asked in despair. Despite his misgivings, however, the Chief of the General Staff decided to act as if he had not rightly understood the Kaiser – 'which was not difficult, as the Kaiser himself immediately embarked on another subject' – since he had already expressed contradictory views on several other questions.[124] On that winter morning at least, Waldersee is unlikely to have thought the young Kaiser 'too impressionable' or lacking opinions of his own. Only a few days later Wilhelm dismissed him as chief of the General Staff.

THE VISIT TO ENGLAND IN JULY 1891

At the end of June 1891 the Kaiser visited Heligoland. During the journey he jubilantly announced to Waldersee that the Triple Alliance had recently been renewed for six years.[125] The General did not share the monarch's satisfaction at this, for he had radically different ideas. 'I should consider it a much greater political success if we had freed ourselves from the Triple Alliance and had come to an understanding with Russia and Austria instead', he wrote. 'Shared conservative interests and solidarity against the republicanism threatening us from the Romance nations, probably combined with the partition of Turkey, could well provide the basis for it, in my view.' He was well aware, he commented, that the Kaiser would try to persuade the United Kingdom to join the Alliance during his forthcoming visit. But this was 'a vain effort', for 'England will never bind herself, and why should she? If she is attacked in Asia, she will be very glad to have the Triple Alliance as her ally; but until then she will keep a free hand, for which no one can blame her. She can still take sides at the last minute.'[126] In the General's view the British were businessmen who considered their options dispassionately, and 'they are very unlikely to do what the Kaiser wants – to join the Triple Alliance'.[127] Waldersee's sceptical opinion was to prove only too accurate, for Wilhelm's excitable and unpredictable attempts to press for an Anglo-German alliance aroused suspicion in London and, even in these early days, led the British government to wonder whether they might not do better to form an alliance with France and Russia.

At the beginning of July 1891 the Kaiser travelled via Holland, where the size of his entourage caused some astonishment, to Britain.[128] For the crossing from Flushing he wore his British admiral's uniform with the blue riband and collar of the Order of the Garter. He was accompanied not only by Kaiserin Auguste Viktoria but also by the latter's brother Duke Ernst Günther of Schleswig-Holstein.

The imperial suite included Secretary of State Adolf Marschall von Bieberstein, the three Chiefs of the Military, Naval and Civil Cabinets, General von Hahnke, Gustav Freiherr von Senden-Bibran and Dr Hermann von Lucanus, the Kaiser's physician Dr Rudolf Leuthold, the new Oberhofmarschall Count August zu Eulenburg, Adjutant-General Adolf von Wittich, the Flügeladjutanten von Kessel, von Scholl, von Brandis and von Hülsen. The Kaiserin's suite consisted as usual of Baron Mirbach, Bodo von dem Knesebeck, Countess Brockdorff and Fräulein von Gersdorff. When the *Hohenzollern* arrived at Sheerness (then called Port Victoria) on 4 July 1891, the German guests were met by the Prince of Wales, his elder son the Duke of Clarence, known as Eddy, and his brother Prince Arthur, Duke of Connaught. All three wore Prussian Hussar uniform with the yellow sash of the Order of the Black Eagle.[129] Also present at Sheerness were Count Hatzfeldt, Count Wolff-Metternich and Prince Pless, from the German embassy in London. The party travelled together by train to Windsor, where the ladies and members of the wider royal family were waiting at the station. In the course of the next few days the Kaiser received the diplomatic corps and a delegation from the extensive German colony in London.[130]

On 10 July Kaiser Wilhelm II and the Kaiserin were the guests of the Lord Mayor and the financial world of the City of London, and on the way from Buckingham Palace to the Guildhall and back, driving through streets decked with flags and lined with troops in colourful uniforms, they were cheered for the first time by considerable numbers of the ordinary people of London. Wilhelm wore the white uniform of the Cuirassier Guards, a golden helmet adorned with a silver eagle, and the blue Garter riband. He saluted repeatedly in response to the enthusiastic cheers of the crowd, which drowned out the protests of a few German socialists gathered at Ludgate Circus. He was accorded the respect due to 'the most powerful of Continental monarchs' and it was proudly reported that no foreign ruler – not even Napoleon III on his visit after the victorious war in the Crimea in 1855 – had ever been greeted with anything approaching the warmth shown to Kaiser Wilhelm II.[131]

During the visit crowds assembled daily in front of Buckingham Palace. Wherever the young Kaiser with the gleaming eagle helmet went – a military review in Wimbledon, a visit to the Crystal Palace, dinner with Lord Salisbury at Hatfield House – the yellow and black imperial standard with the device 'Gott mit uns, 1870' was hoisted, transparencies and triumphal arches of branches and flowers were set up, bearing slogans (of which many were in German) such as 'England and Germany; the peace of Europe', and everywhere the national anthem – the tune was of course the same in both countries – was enthusiastically sung and played.[132] The visit also gave rise to lengthy disquisitions in the London press on the successful co-operation between Germany and Great Britain, particularly in the Seven Years War and in the campaign against Napoleon, to maintain peace and stability in Europe.[133] This was precisely the theme of the important speech which Wilhelm made, looking 'pale and rather exhausted' as one observer commented, to the ruling elite of Britain assembled in the Guildhall. 'I have always felt at home in this lovely country, being the grandson of a Queen whose name will ever be remembered as the most noble character, and a lady great in the wisdom of her counsels, and whose reign has conferred lasting blessings on England', he declared. 'Moreover, the same blood runs in English and German veins. Following the examples of my grandfather and of my ever-lamented father, I shall always, as far as it is in my power, maintain the historical friendship between these two our nations, which . . . have so often been seen side by side in defence of liberty and justice . . . My aim is above all the maintenance of peace – (cheers) – for peace alone can give the confidence which is necessary to the healthy development of science, art and trade . . . You may rest assured, therefore, that I shall continue to do my best to maintain and constantly to increase the good relations between Germany and the other nations, and that I shall always be found ready to unite with you and them in a common labour for peaceful progress, friendly intercourse, and the advancement of civilisation. (Loud applause).'[134] The Berlin newspapers of every political hue judged the Kaiser's speech very favourably, and its conciliatory tone impressed even the Russian and French press, even if a few Moscow journalists voiced the suspicion that an Anglo-German agreement would only make more German troops available to attack Russia.[135]

As a souvenir of his state visit the Kaiser presented the Lord Mayor with a lifesize portrait of himself as British admiral of the fleet, with the Order of the Garter, painted by Rudolf Wimmer. He had given the Queen a similar portrait by the same painter a few weeks earlier.[136] On the night of 13–14 July he travelled by night train to Edinburgh, where he joined the *Hohenzollern* and departed for his annual Scandinavian cruise. Meanwhile the Kaiserin, who unlike Wilhelm had created 'a very disagreeable impression by her stiffness, rudeness and arrogance towards the royal family and even towards the Queen',[137] went to Felixstowe on the east coast of England, where the *Hohenzollern* had taken her five eldest sons for a seaside holiday on the 'German Ocean', as the North Sea was still called in Britain at that time.

Wilhelm's visit was undoubtedly a great success, and in his telegram of thanks to his grandmother from Scotland he had good reason to express the hope that he had helped 'to bring our two Nations a step nearer to each other'.[138] He followed this with a letter from Norwegian waters on 20 July. 'The stay in England was to me a great treat, & I am deeply touched by the friendly & warm reception I met at the hands of Your loyal subjects', he told the Queen. 'The old & strong feeling for their Dynasty & monarchical principles showed itself in all its vigour in the bearing of the people wherever one met them. It showed the loyal & devoted love the British cherish for their beloved Sovereign, as well as the wish to make me feel quite at home among them, beeing [sic], as I am a good deal of an Englishman myself. But what also accrued the feeling of contentment in me, was the open approval by the thinking people of the unswerving & honest labour I am given to, for the maintenance of peace & the development of good will amongst all nations – nota bene – as far as it is possible. This approval is a great recompense, & makes one forget all the trouble & disagreeable moments one has had to pull through. I hope & feel sure that my visit has been for the good of our subjects, & that it has tended to bring our two nations into nearer & warmer relations, which will be a benefit to them & the world at large. With Your benevolent encouragement & kindly interest . . . I hope to be able to persue [sic] my way, which Providence has marked down, & continue to strive for the fullfilment [sic] of those great problems which were so ably begun by dear Grandpapa Albert. With much love & respect I kiss Your hands & remain Ever Your most dutiful & aff^{ate} grandson Willy.'[139] When the completion of the first direct cable link between Britain and Germany was announced to him on his return to Kiel he sent a telegram to his grandmother expressing 'the confident hope that this successful result of the diligence and hard work of our officials and subjects on both sides may form a new link in the chain of friendly and sincere relations between our two realms'.[140] In another letter he described her as 'the "Nestor" or the "Sybilla" of Europe's sovereigns, venered [sic] & revered by all; feared only by the bad'.[141]

The burgeoning Anglo-German friendship gave rise to some embarrassment on the British side when it emerged that Prince Heinrich and his wife would be staying with Queen Victoria at Osborne at the time when a French naval squadron was due to pay an official visit to Cowes. It was therefore with some relief that an offer by the Kaiser's brother to go on an excursion in the Prince of Wales's yacht *Aline* from 19 to 21 August was accepted.[142] During his 'diplomatic' absence, however, Heinrich decided to take the opportunity of inspecting the 'arch-enemy squadron' from the outside, as he reported to his 'dear big brother'.[143] A few days later he was able to send the result of his spying activities, during which Princess Irène had 'diligently' photographed the French warships, to Berlin. 'The ships make no claim to any beauty, but are certainly well prepared for war. They are armour-plated throughout, and well equipped with large numbers of guns. The "Marceau" looks like a town shimmering with lights from a distance – and offers an excellent target with its high superstructure. Fore and aft and on both sides there are *long* heavy-calibre cannon (about 28 cm) in extended supports, but they are quite unprotected. In the battery there are 16 quick-loading cannon, apparently 15cm or 17cm, that is to say 8 on each side. Wherever else there is any space the ship is dotted with small quick-firing cannon, as are the other ships too, although there is a great variety of models and calibres. The "Furieux" has certain similarities with our "Siegfried" class . . . Neither "Requin" nor "Furieux" seems to have anti-torpedo cladding, or at least I have not been able to detect any! The torpedo boats are very inferior craft and have a certain similarity with the English ones.'[144] Wilhelm immediately passed on this secret information to his Naval Cabinet.

The press reports of the Kaiser's enthusiastic reception in Britain reminded Waldersee of past times when Wilhelm's thoughts about his second home had been of quite another order. 'If I think back to many a time in Prince Wilhelm's life, but also since he became Kaiser, when the English were treated as the most good-for-nothing people, as our enemies, as miserable shopkeepers, when English affairs were discussed in the most contemptuous way and the army was described as completely worthless, I have to keep reminding myself never to be surprised by anything.'[145] At the end of the visit the General was sceptical about its likely effect on British policy. As Konrad Canis has recently pointed out, the very fact that following the break with Russia Germany seemed to be more dependent on London than before convinced Lord Salisbury that Britain's independent stance towards the Triple Alliance could be maintained without compromise.[146] Waldersee was well aware, however, that the display of wealth and pomp which he had seen on his visit would impress the Kaiser, 'and his own love of pomp will have increased even more'. It was also to be feared, he thought, that the Kaiser's visit to Great Britain would arouse 'suspicion in France and Russia' which could be exploited against Germany.[147]

WILHELM AND THE FRANCO-RUSSIAN RAPPROCHEMENT

For all his personal success in London, however, on the international political scene Wilhelm II's assurances of friendship for Great Britain, together with the Heligoland–Zanzibar accord and the ostentatious renewal of the Triple Alliance, contributed to the emergence of a closer relationship between Russia and France.[148] It was no coincidence that the conspicuous public fraternisation between the absolutist empire of the tsars and the west European republic – during the visit of the French navy to Kronstadt the Tsar stood bare-headed while the revolutionary Marseillaise was played – took place immediately after the Kaiser's visit to London, which had seemed to herald a closer bond between Britain and the Triple Alliance. The Empress Frederick saw a clear causal link between these two historic events. 'William's reception in England certainly brought about the Cronstadt demonstration', she commented, and expressed horror at what she saw as the danger of a European war latent in this development. 'The more one thinks of a European war the more *terrible* & awful the thought is. – *No* one has anything to *gain*, & the destruction of our civilisation would be very near.'[149] The reaction of the Kaiserin Auguste Viktoria was quite different. In a confused letter to her husband from Felixstowe she asked: 'What do you say to the French Russian friendship, they are really making such a fuss, the Tsar stood up for the Marseillaise.'[150] After the Kronstadt demonstrations, which he likewise interpreted as a response to Wilhelm II's visit to Britain, Waldersee was in no further doubt as to Germany's endangered position. 'What a lot of nonsense has been babbled about the Triple Alliance', he exclaimed on 27 August 1891. 'How certain everyone was only a few weeks ago that the Kaiser's visit to England would be enough to make the Triple Alliance a Quadruple Alliance! . . . All reports indicate that the Franco-Russian friendship must be taken very seriously this time and that it would take only a trifle to make the French decide to go to war . . . The shamelessness of the Russians and their presumption are quite outrageous.'[151] The first years of Wilhelm II's reign had led to a foreign policy fiasco, Waldersee complained, 'for what we needed to avoid, what we have been trying to prevent for 15 years, namely a Franco-Russian alliance, has happened.'[152] In September 1891 he received a report from Paris that leading figures there were anxious 'at all costs to avoid war now', as time was clearly working in France's favour. 'The Kaiser is gradually taking the German Reich further and further backwards.'[153]

The Kaiser himself reacted to these signs of a Franco-Russian alliance nervously and with increased aggressiveness. During a long conversation with Waldersee on board the *Hohenzollern* on 12 August 1891 he declared that 'the Kronstadt celebrations had opened people's eyes, and it was not impossible, especially as the French are in a highly excited state, that a war could quickly develop out of some incident or another'.[154] Not long afterwards, during the autumn manoeuvres, Waldersee observed that the Kaiser was still 'extremely irritated with the Russians and the French'.[155] When Bernhard von Bülow reported from Romania in August 1891 on a conversation with a French observer who had said that after securing initial successes in a European war the French would come to an arrangement with Germany in order to avoid Russian hegemony, Wilhelm wrote scornfully on the report: 'Now hold on, old boy, I shall certainly come to an arrangement with you people! But I wonder if it will really suit you too!?' It was clear from the report, the Kaiser commented, 'how sure the two nations are of their position with regard to us. And how dangerous the resultant arrogance and presumption are.' They were behaving as if Germany were already 'beaten and divided up'.[156] In other marginal notes he reviled the French and the Russians, whom he referred to from now on as 'Gauls and Slavs!', calling them 'insolent dogs'.[157]

Wilhelm's anger also made itself felt in the aggressive anti-French toast which he proposed on 14 September 1891 in Erfurt at the gala dinner of the IV Army Corps. He proclaimed: 'In this place the Corsican *parvenu* utterly humiliated us and dishonoured us abominably, but it was from this place too that the beam of avenging light shone out in 1813 and struck him to the ground.'[158] The German ambassador in Paris reported that the entire French press was 'very indignant' and even the usually moderate *Temps* considered 'the language belligerent and the revival of memories on which the irreconcilable rancour of German chauvinism feeds, very dangerous . . . The yellow press, particularly the Bonapartist papers, use it as a pretext for outrageously violent attacks on the Kaiser, who dared to gravely abuse France.' When Caprivi forwarded this report to Wilhelm II in cipher and without comment, the latter wrote on the telegram: 'Thank God. How annoyed they must have been!' He followed this with a defiant telegram *en clair* to the Reich Chancellor: 'Many thanks for telegram . . . which gave me and all present boundless pleasure. Wilhelm.'[159] The Kaiser's mother was scandalised and wrote to Queen Victoria: 'W. has made one of his most unfortunate speeches at Erfurt in wh. he calls Napoleon the "Corsican parvenu". The Ministers thought this would not do, & had it changed into "Corsican conqueror" – but rather late as everyone knew the first! Erfurt of all places was the last to say such a thing because it was the place where almost all the German sovereigns cringed & grovelled in the dust before Napoleon. – Besides the year 1870 & 1871 settled accounts – with Napoleon's descendants & army, so it is not necessary or good taste or becoming especially in an Emperor to allude to him in these terms.'[160]

THE KAISER'S AIM: 'A SORT OF NAPOLEONIC SUPREMACY' IN EUROPE?

Is it possible to detect, in Wilhelm II's numerous and often seemingly contradictory utterances and actions in the field of foreign, colonial and armament policy during these first years after Bismarck's fall, any common denominators which might point to a fundamental idea behind them? Waldersee's comments, quoted earlier, to the effect that for all his inconsistency Wilhelm held fast to two goals – he wanted 'to win great esteem for the German Reich'[161] and to reinforce and win popular suppport for 'his personal position'[162] – contain the germ of two dominant ideas which we can perhaps consider as something of a key to his activity in foreign affairs. The two themes reappear as a leitmotiv in a number of Wilhelm's statements dating from the early part of his reign.

During a two-hour walk with Philipp Eulenburg near Tromsø in northern Norway in July 1892 Wilhelm II revealed the 'underlying thought' behind his policy, namely the establishment of German domination in Europe, although unlike Napoleon he intended to achieve this goal by peaceful means. 'I hope', he said according to Eulenburg's account, 'that Europe will gradually see the underlying purpose of my policy: leadership in the peaceful sense – a sort of Napoleonic supremacy – a policy which expressed its ideas by force of arms – in the peaceful sense.'[163] Six months later Kaiser Wilhelm felt moved to define the outlines of his foreign policy again, this time in several conversations with the heir to the Russian throne, the Tsarevich Nicholas, who had come to Berlin in January 1893 for the wedding of Wilhelm's youngest sister. In a letter to his 'dear friend' Kaiser Franz Joseph, Wilhelm set out the main points of his interviews with the future Tsar, which he hoped would help bring about a shift in Russian policy against France and in favour of the Triple Alliance. He had discussed the aims of the Alliance in several very detailed conversations with 'the young heir to the throne'. Germany, Austria-Hungary and Italy had been motivated only by their desire for self-preservation, he said. 'In concluding the alliance the powers had borne in mind the serious dangers which threatened the monarchies from the Republic of France, through the spread of republican propaganda. It was after all chiefly from Paris that all revolutionary teaching was spread, the aim of which was to undermine monarchical traditions in all states in a wide variety of [missing word]. These shared dangers were thus to be seen in the first place as the basis of the Alliance, and any power which wished to defend the interests of peace and of monarchy in the same way could join this Alliance at any time. Political matters were not, however, in any way the only area in which the three Allied powers shared common interests', the Kaiser continued. 'On the contrary, the Triple Alliance also wanted to bring the European powers closer to each other in the economic field as well, through trade agreements. In this way the Alliance wished both to reduce the causes of friction between European states and also, above all, to confront the dangers which threatened the entire trade of Europe through the fact that the Republic of North America was showing ever more inclination to seize all trade, including that of South America, for itself.'[164] In thus setting out his main lines of thought Kaiser Wilhelm revealed that his goal was to unite the European continent under German leadership against French republicanism and the growing economic power of the United States. Franz Joseph likewise expressed the hope 'that Russia will also feel at one with the other monarchical powers. May the necessity for this be realised in good time', he wrote in answer to Wilhelm's letter.[165]

In April 1893 the Kaiser went to Rome for the silver wedding celebrations of the King and Queen of Italy and took the opportunity of having a lengthy discussion with Pope Leo XIII, in the course of which he repeatedly stressed his ideology of monarchical leadership in Europe and his determined rejection of the democratic tendencies prevalent in France, in the Catholic Church and the Centre Party. Philipp Eulenburg had encouraged him to take up this theme, pointing out that 'the idea of the great monarchical bastion against revolution' had come from him, the Kaiser, and that he had 'voiced it many times' in their discussions.[166] In his interview with Leo XIII, Wilhelm emphasised 'that all the monarchies of Europe found themselves compelled by the advance of radical ideas to make a stand, in monarchical solidarity, against these ideas wherever they came from. Radicalism was the enemy of a well-ordered monarchy, and at the same time also the sworn enemy of the Church. Radicalism did not stand on its own, however, but was rooted in republicanism, for republican tendencies were its real basis. It was therefore equally dangerous to all monarchies, whether they were worldly or spiritual.' He, the Kaiser, therefore could not understand why the Pope supported the republican form of government in France. This attitude could not but seem strange to him, as to any observer. The policy of the Holy See towards France, Wilhelm told the Pope, had 'caused much anxiety to a great number of my fellow monarchs, as well as myself personally, since he was giving support to a republic, even if only in appearance. We monarchs represented the divine right of kings and conservative policy. The republic and radicalism, on the other hand, were based on regicide and the abolition of God and their purpose was to overturn the existing order . . . The [French] people could not find peace and stability because they had cut off the head of the King whom God had set over them, dishonoured the Church and scorned the Deity. The curse of the Lord lay on their land, and these afflictions were his punishment.' Moreover, by his attitude the Pope had 'caused uncertainty among a not insignificant proportion of my Catholic subjects, who saw him supporting an anti-monarchical type of state in another country', the Kaiser complained. Recently democratic leaders of the Centre Party like Lieber, Daller, Fusangel and Orterer had been elected to the Reichstag, he said, and they were causing problems to the government. 'They had used language towards My government which was scandalous and surely not in accordance with the spirit of the Pope, for they incited the masses to direct disobedience against the government.'[167]

In February 1894, while presiding over a meeting of the Prussian Ministry of State at which the commercial treaty with Russia was on the agenda, the Kaiser declared: 'Our supremacy must be demonstrated to Europe not only by our army, but also through commercial policy.' The policy pursued by the United States of America of cutting off Germany from its 'principal market, South America' must be counteracted by establishing a unified European tariff policy under German leadership. The Reich's aspirations towards supremacy must, however, be kept strictly secret from the Russians, for otherwise Russia's objections to the recent expansion of the German army would only be increased. Canis has rightly drawn attention to the similarity between these sentiments and the 'Napoleonic' ideas of domination which the Kaiser had expressed on his Scandinavian voyage in July 1892.[168]

The idea of an international monarchical order in Europe in which he as German kaiser would play the leading role was evidently a central aim of Wilhelm II. But this ambitious goal could not be achieved by peaceful means alone, any more than the Napoleonic system of a hundred years earlier. It was much more likely, as Wilhelm himself no doubt suspected, that military means would have to be used to create the new European order under German leadership. The military implication of his grand idea emerges in the final quotation from the Kaiser which we shall examine here. In the summer of 1895, by which time he had adopted a racist vocabulary and regularly referred to the Germans as Teutons, the Russians as Slavs and the French as Gauls or Romance peoples, the Kaiser defined the supreme aim of his foreign policy as the leadership of the entire Germanic population of Europe, including Scandinavia, against the Slav threat emanating from Russia. In a letter of 25 July 1895 he confided to the Crown Prince of Sweden: 'All my ideas and endeavours and all my policies are directed towards bringing the Germanic peoples in the world, especially in Europe, closer together and forging a stronger relationship between them, so as to guard ourselves more securely against the Slavic–Czech invasion which threatens us all to a most dangerous degree. Sweden–Norway is one of the principal factors in this league of Germanic peoples. What will become of us if this great northern union of states suddenly falls away and is perhaps absorbed by the Slavs (Russians)? In this respect the whole Germanic North of Europe constitutes the left flank of Germany, or rather of Europe, and is therefore of great importance for our security. Its disappearance would leave our flank exposed and would represent a grave danger for us all.'[169]

Given his notoriously erratic temperament, it would certainly be going too far to see such utterances as evidence that Wilhelm II had developed a coherent foreign policy strategy which he was now attempting to put into action. Nevertheless these declarations, scattered as they were over a number of years, reveal a recurring pattern of thought which suggests that his extremely active and strong-willed intervention in the field of diplomacy and arms policy was motivated by more than the vainglorious wish to be seen to be in control of Germany's foreign policy. The very fact that his remarks were tactically geared to suit each of his interlocutors gives particular significance to the consistency with which two basic elements – *German* supremacy in a *monarchically* structured Europe – come into play in each of the Kaiser's utterances. Wilhelm II's emotionally charged, fragmented thinking in the earliest phase of his reign, exemplified in the coarse marginal notes confiscated by Bismarck, had now resolved itself, after Bismarck's departure, if not by any means into a clear programme, then at least into an instinctive perception of his future role as German kaiser and king of Prussia. In later chapters we shall trace the way in which Wilhelmine policy on international affairs and naval armament evolved out of this initial stage. But first we shall examine the role which Kaiser Wilhelm II played in domestic politics during the first years of the New Course.

CHAPTER FIFTEEN

The dualism of power

THE KAISER AND THE 'RESPONSIBLE GOVERNMENT'

THE dominant role Wilhelm II was already able to play in foreign and military policy soon after Bismarck's fall should not, of course, mislead us into seeing him as a dictator or a tyrant. He was certainly determined to make full use of the enormous inherited power of the Prusso-German military monarchy, and he was imbued with strange anachronistic ideas of his divine right, which he proclaimed increasingly loudly and often, to the consternation of his advisers. But he was and remained a legitimate king and Kaiser, who was obliged to co-operate in matters of domestic policy with the government authorities and parliaments prescribed by the constitutions of Prussia and the Reich. As a result, with the monarch on the one hand and the 'responsible government' (that is to say the Reich chancellor, the Prussian departmental ministers and the secretaries of state of the Reich offices) on the other, a dualism of power emerged, which was inevitably characterised by conflicts, crises and growing frustration on both sides. In tracing the development of this phenomenon we shall see how the balance of power within the leadership of the country gradually shifted in favour of the Kaiser and his court. We shall begin by taking a closer look at the tensions between Wilhelm II and his 'responsible' advisers in the period immediately following Bismarck's fall.

In this post-Bismarck system of parallel power centres, the newly appointed Reich Chancellor and Minister-President, General von Caprivi, was at first at an advantage in that Wilhelm could not immediately dismiss him. As Waldersee observed in May 1890, Caprivi's strength lay in the fact the Kaiser could not afford 'to change chancellors rapidly' and was therefore obliged 'to treat him with great care' and even 'to give way to him sometimes'.[1] Indeed the controversial Cabinet order of 1852, which Bismarck had invoked in order to strengthen his authority over his ministerial colleagues vis-à-vis the crown and which had provided the final impetus for his fall, was replaced by a new decree giving the minister-president much the same rights.[2] And at first Wilhelm and the Kaiserin were in fact, as they frequently declared, 'quite delighted' with the 'new regime', which they compared favourably with the latter days of Bismarck's government.[3] In December 1890 the Kaiser wrote effusively to Queen Victoria about the new Reich Chancellor: 'Here we are getting on very well with Caprivi who is already adored by friends & revered by his opposition. I think he is one of the finest charakters [sic] Germany ever produced, & am sure You would immensely like him as soon as You saw him.'[4] Privy Councillor von Holstein also recorded with satisfaction at this time that Caprivi enjoyed the confidence of the Kaiser,[5] while even Waldersee was forced to admit that the Kaiser was 'still very pleased' with the Chancellor, even if this had been possible only because Caprivi had so far 'skilfully made concessions and given way'.[6]

This relationship was not destined to last long, however, for the further the Bismarck crisis receded into the past, the more self-confident the Kaiser became. Furthermore, despite his conciliatory attitude the new Chancellor was certainly not a man to 'cling' to his thankless task whatever the circumstances. Thus Waldersee predicted as early as 20 April 1890 that after the 'initial' harmony between the Kaiser and the Chancellor it was inevitable that they would come into serious conflict, for Caprivi was 'an earnest, conscientious man who holds fast to his opinion once he has decided that it is right, and who is not cut out for intrigues. It really will not be easy for him to get along with the Kaiser, who is lively, likes to be cheerful and is sometimes even childishly high-spirited, taking pleasure in all kinds of tomfoolery, but also inclined to obstinacy. Anyone who knows the Kaiser well realises that one can achieve much by temporarily giving in to him and waiting a little, and occasionally by a joke; but that is not Caprivi's style.'[7] Furthermore the new Chancellor's state of health gave cause for concern. 'He seemed to me to have aged, and in particular he had grown very thin', Waldersee noted soon after Caprivi's appointment, adding that he had heard confidentially that the Chancellor 'is said to be suffering from quite a severe case of diabetes. The doctor at Carlsbad is said to have stated that he could not last long unless he spared himself and avoided straining his nervous system too much; otherwise he would not survive beyond this autumn.'[8] From the outset Caprivi recognised that the greatest difficulty facing him in carrying out the functions of his office lay in the Kaiser's character and behaviour.[9] In June 1890 he was already complaining about the expensive, ceaseless 'torrent of new plans from the Kaiser'.[10] It was entirely characteristic of this conscientious and honourable man, however, that he shouldered all troublesome questions himself in order to relieve his colleagues. As Arthur von Brauer, Baden's new representative in the Bundesrat, perceptively reported to Karlsruhe in December 1890, Caprivi's position 'in relation to the All-Highest is so strong, and he feels so secure in it, that it has become positively the custom among the ministers to pass on any delicate matters, or anything that might displease His Majesty, to the Minister-President so that he can report on it, a task which he willingly takes upon himself. The General uses H.M.'s regard for him to protect his ministerial colleagues – conduct which certainly does great credit to the Chancellor's character, but which also carries the risk that he will wear himself out over trifles and sometimes arouse the imperial ire over things which, in terms of their importance and of the existing division of ministerial responsibilities, are not the concern of the minister-president or the chancellor.'[11] All too soon the honeymoon period with the Kaiser came to an end and gave way to the day-to-day realities of life.

The problems which Caprivi and the ministers and secretaries of state faced with the Kaiser in the initial period of the New Course were not merely a matter of ill-considered speeches and actions on his part, much as they deplored these. Wilhelm intervened increasingly in the conduct of affairs of state, or created difficulties by his refusal to accept compromises with political parties which the 'responsible' statesmen considered necessary. One of the first clashes with the Chancellor occurred in June 1890 over the Army Bill introduced by Minister of War General von Verdy du Vernois. Caprivi asked for the Kaiser's authority to offer concessions to the Catholic Centre Party or the two Left Liberal Progressive parties if this proved necessary (in spite of the reduction of the proposed increase from 30,000 to 10,000 men), in order to get the bill through the Reichstag.[12] In particular, Caprivi asked for the abolition of the septennate (under which the Reichstag was entitled to debate the army establishment only at seven-year intervals) and for the introduction of two-year (instead of three-year) military service for certain army units. The Chancellor was reported to have returned from Potsdam 'in a very depressed frame of mind', as the Kaiser had 'rejected everything out of hand and expressed himself very strongly', Waldersee learned. He commented that Caprivi 'is now very pessimistic; he is probably put out by his first difference of opinion with the Kaiser and does not know what to do next . . . I share Verdy's opinion that the Reichstag will give in if one shows that one is really in earnest and threatens it with dissolution; none of the parties wants new elections, and arrangements will be made so that the bills are adopted with very small majorities. Caprivi does not agree and wants to avoid a dissolution in all circumstances. If the Kaiser wants one, he will resign. I can already see that his nerves will not hold out, and the situation is therefore really serious. How the Bismarck family will rejoice to hear it.'[13]

It was not only the relations between Wilhelm II and his new Chancellor which were under strain. The first victim of the monarch's growing claims to power was the Minister of War. The conflict over the Army Bill in the Reichstag in the summer of 1890 led to the dismissal of Verdy, in which the Kaiser and Chief of the Military Cabinet von Hahnke were instrumental. Verdy had long since made himself unpopular with the Kaiserin by his tactlessness, and Bismarck's insinuations that the General had democratic inclinations had not failed to have the desired effect on the Kaiser. Only four weeks after Bismarck's fall Waldersee recognised the first signs that Wilhelm had taken against Verdy, which he attributed to Hahnke's influence. Verdy had not succeeded in 'curbing Hahnke's influence', he wrote. 'On the contrary, Hahnke has entrenched himself very firmly with the Kaiser.'[14] Then in June 1890, during the parliamentary negotiations over the Army Bill, the Minister of War was accused of 'the grossest errors and blunders'. It was said in government circles that by revealing his future plans, which would require even greater increases in the strength of the army, he had 'jeopardised the whole development' and 'welded the disintegrating Progressive Party together again'. In the Reichstag too the War Minister did not command 'the least respect', and when Caprivi disavowed him by declaring that he, the Chancellor, knew nothing of any further plans to enlarge the army, the General's fate was sealed.[15]

A violent disagreement between the Kaiser and the Minister of War as to whether heavy artillery guns should be cast in steel or bronze – Wilhelm considered himself qualified to pass judgement even on this – brought matters to a head. The Kaiser sided with the Essen firm of Krupp in favour of steel, while Verdy considered bronze to have an overwhelming advantage because it would not crack. When three bronze guns nevertheless did crack, it was generally held to be 'a great triumph for His Majesty' and a fiasco for Verdy.[16] After his return from Essen Wilhelm took action against Verdy 'with the greatest severity': on his instructions the Minister of War was given three months' leave with the proviso that he was then to return to his post *pro forma* for eight days, and resign on 1 October 1890, in order to avoid creating the undesirable impression that a Prussian minister of war had given way to the Reichstag.[17] The Chief of the General Staff acknowledged that his protégé had made mistakes and had often given offence, but he insisted that Verdy was a highly competent minister of war and that the army owed him a great deal. He had received no thanks for this. 'In fact the Kaiser has been extremely unfriendly and ungrateful towards him', he wrote, adding that Verdy might have done better to give way less and from the first to stand up to the Kaiser more.[18] Waldersee deplored the fall of his friend for several reasons, not least because his own position had been weakened by it.[19] He advised the outgoing Minister of War not to attend the autumn manoeuvres or the military review at Pasewalk, as he would only invite 'unfriendly attention from the Kaiser', which was indeed precisely what happened.[20] Waldersee predicted, however, that Wilhelm would have occasion to regret Verdy's departure, as he would never find 'a more compliant minister of war'.[21] When Verdy duly handed in his resignation on 1 October Waldersee commented that he was 'extremely embittered, with every right to be so', and saw trouble ahead. 'I am afraid that in his private capacity he will be unable to resist making skilful use of his pen. In him the Kaiser has made himself an enemy who should not be underestimated.'[22]

The Chief of the General Staff was hurt to realise that the Kaiser was avoiding discussing the subject of Verdy's successor with him. It was not until September that he finally heard from Caprivi what had been decided. He discovered that the Chancellor had first suggested the Kaiser's Adjutant-General, Adolf von Wittich, for the post and that the Kaiser had indeed offered it to Wittich, although 'by no means pressing him to accept'.[23] But when the Flügeladjutant Count Wedel also proposed Wittich for the War Ministry, Wilhelm told him that this was an impossible choice because Wilhelm himself and Wittich were 'both such strong characters that before three days had gone by they would be disagreeing violently with each other'.[24] He suggested to Wittich that he remain at court until the autumn, when he would appoint him commanding general of the Corps of Guards.[25] The Ministry of War was then offered to General Hans von Kaltenborn-Stachau, who held the post until 1893. Waldersee at first considered him 'a very competent and experienced general and a clever and high-minded man' who was too good for the post. But he too would soon be at loggerheads with the Kaiser, Waldersee predicted, for 'no minister of war who has any self-esteem can work with him for long'.[26] In the Chief of the General Staff's view the whole affair proved once again that the Kaiser had 'far too low an opinion of the importance of the Ministry of War'. What Wilhelm really wanted, Waldersee considered, was to run the army exactly as he wished; the war minister's role was simply to 'obey' and 'carry out everything' ordered by the monarch. The Kaiser had Hahnke's backing in taking this line, and Verdy had not had the energy to combat it.[27] Moreover Wilhelm had not had 'the slightest anxiety' over the change of minister, for it mattered very little to him who was minister of war. He had told Caprivi that 'the new minister will probably not last longer than nine months anyway!' 'Is this not appalling?', exclaimed Waldersee. 'Where are we heading? The whole world will lose all respect for us if the most important appointments change hands so often. Every such change harms the Kaiser.'[28] Shortly after taking over the Ministry of War Kaltenborn – just like Caprivi and Waldersee – came to recognise that 'the problems lie mainly with the Kaiser'.[29]

The extent to which the Kaiser took it for granted that the choice of civilian ministers was also the prerogative of the crown was demonstrated repeatedly in these first months after the fall of Bismarck. When Prussian Minister of Finance Adolf von Scholz handed in his resignation on 14 June 1890, Wilhelm II took advantage of his visit to Essen on 20 June to offer this key appointment to the director-general of the Krupp works, Johann Friedrich Jencke, without consulting Caprivi in advance. Jencke had earned his respect during the meetings of the Prussian Council of State, even though he had vigorously opposed Wilhelm's social welfare plans. The Prussian ministers, especially Boetticher, were incensed. They considered Jencke to be a tool of heavy industrial interests who would try to undermine the government's policies on social welfare and railways. Only one of them – it was of all people the reform-minded Minister of Trade Freiherr von Berlepsch – was confident that Jencke would 'completely dissolve his ties with Krupp and industry'. Caprivi also had serious reservations about the choice of Jencke, but he was convinced that the Kaiser would not rescind his decision and argued that it would therefore be wiser for the Ministry to accept it.[30] It was not on account of any ministerial protest, however, but rather because Jencke turned down the finance portfolio – he declared that he would prefer the Ministry of Public Works, which was responsible for Prussian railways – that the Kaiser gave up the idea of this appointment, thus avoiding a serious clash with the Chancellor. The issue was one of principle for, apart from objecting to Jencke's political views, Caprivi claimed the right at least 'to be consulted [and] if possible to make suggestions himself' where appointments of his close colleagues were concerned.[31] Instead of Jencke, Wilhelm offered the post to the National Liberal mayor of Frankfurt am Main, Johannes von Miquel, whom he had also met in the Council of State and who had asked the Flügeladjutant, Wedel, to present him to the Kaiser during a musical soirée at the Neues Palais on 12 June – just before Scholz's resignation.[32] Miquel's appointment was unanimously welcomed by the Prussian ministers on 22 June, and even Caprivi found it acceptable at first, until Miquel proved to be something of a cuckoo in the nest. The new Finance Minister, who was to become one of the most powerful figures in German politics in the next few years, rapidly embraced right-wing policies and made himself highly uncongenial to the Chancellor by becoming the champion of agrarian, heavy industrial and Bismarckian interests within the Ministry of State. This ministerial appointment thus did nothing to reinforce the cohesion of the government.[33] It was not long before the Kaiser had also changed his mind about Miquel. In December 1890 he complained that Miquel was 'a real know-all Hanoverian', prompting a comment by Waldersee that this was 'rather quick, considering the delight with which the Kaiser spoke [of him] only a few months ago, but a real sign of the times'.[34] Two and a half years later the General recorded that the Kaiser had 'alternated between praising [Miquel] to the skies and criticising him, trusting and then mistrusting him'.[35] Nonetheless many people predicted as early as 1891 that Miquel would soon take Caprivi's place as chancellor.[36]

Wilhelm II's influence was predominant in two further ministerial appointments in the winter of 1890. In November the Prussian Minister of Agriculture, Robert Freiherr Lucius von Ballhausen, proffered his resignation on the grounds that he could not accept 'the retreat of agricultural interests behind those of industry' and left office with Caprivi's agreement and, according to Lucanus, 'extremely amicably'.[37] But his successor Wilhelm von Heyden-Cadow, a major landowner and hitherto Oberpräsident in Frankfurt an der Oder, owed his appointment above all to the fact that he was 'a particular favourite of the Kaiser'.[38]

The departure of the Minister of Public Works, Albert von Maybach, at first took a similarly smooth course. Maybach handed in his resignation at the beginning of 1891 as a result of the growing criticism which his railway policy was attracting from the General Staff, industrialists and the general public.[39] The Kaiser immediately demanded that he be replaced by Jencke, who had staked his claim to this post in the summer, as we have seen. Since the Kaiser's undertaking to Jencke at that time, Caprivi had certainly always reckoned with the latter's eventual appointment,[40] but this time he had more opportunity to ensure that his own views were taken into account. He convened a secret meeting of the Ministry of State and with the help of his colleagues framed a list of conditions which Jencke would have to fulfil if he wished to become minister of public works. He would have to explain his views on social welfare, particularly in relation to railway workers, clarify his position vis-à-vis Prussia and the Reich (Jencke was a Saxon) and above all guarantee that he would not 'protect industry unfairly at the expense of agriculture'. Caprivi took this list to the Kaiser and asked him to keep Maybach in office until Jencke had given his answer on the points at issue.[41] In the end the director of the Krupp works was not appointed to the post; it is not clear why. Both Holstein and Eulenburg, who were anxious to 'rally the propertied classes, big business etc. as much as possible' and to 'reassure the big industrialists' through Jencke's appointment, were disappointed.[42] All that is known is that the Kaiser, the Chancellor and the ministers eventually agreed on the choice of a worthy but non-political railway official, Karl Thielen (*von* Thielen from 1900), who remained in office until 1902 but who very soon fell completely under the influence of Miquel.

Another 'very serious' conflict, again between the Kaiser and the entire Prussian Ministry, arose in September 1890 when the monarch refused to ratify the re-election of the Mayor of Berlin, the Radical Max von Forckenbeck. 'I'll never confirm that fellow in office', he told Conservatives in Breslau, and when the Chief of his Civil Cabinet submitted the matter to the Kaiser during manoeuvres at Rohnstock in Silesia he was met with a point-blank refusal.[43] Caprivi and the other ministers were unanimously in favour of ratification, since otherwise any bills laid before the Prussian Landtag, and in particular the new local government bill, would be jeopardised.[44] In Silesia the Reich Chancellor spelt out to the Kaiser the consequences of non-ratification: the people of Berlin would re-elect Forckenbeck or vote for someone 'even worse', in which case the only thing to do would be to govern the capital by special commission; this would lead to the resignation of the majority of the honorary city councillors and the situation would become completely untenable. Still the Kaiser refused to budge.[45] As Waldersee tersely commented, if Wilhelm persisted in his attitude, 'there will be a ministerial crisis, perhaps even a chancellor crisis'.[46] Caprivi spoke threateningly of a resignation of the whole Ministry, since it clearly no longer enjoyed the confidence of the monarch. With the support of Lucanus, who argued that Forckenbeck was not worth such a crisis, an arrangement was reached by which the question would be submitted once more to the decision of the Ministry of State, and the Chancellor became 'quite cheerful again', commenting to Kiderlen-Wächter: 'The Kaiser will give in this time.'[47] On 10 October Waldersee recorded in his diary: 'Today is an important day. Caprivi has gone to Potsdam to ask the Kaiser to ratify Forckenbeck's appointment. If he refuses the whole Ministry will resign.'[48] The next day he registered the first 'victory' of the Reich Chancellor over the monarch, commenting: 'Caprivi obtained Forckenbeck's ratification yesterday. I do not know whether it was a hard struggle. At any rate it is a great victory for the Chancellor. The only question is whether the Kaiser will take it very badly and whether it will mean that there is ill-feeling from now on.'[49]

The Chancellor's 'victory' in this case was the exception that proves the rule. A year later another incident showed all too clearly who even at this early stage was calling the tune as far as appointments were concerned. Deputy Reich Chancellor and Secretary of State of the Reich Office of the Interior Heinrich von Boetticher and Foreign Secretary Baron Marschall von Bieberstein sent a cipher telegram to the Flügeladjutant on duty, asking almost timidly for the 'All-Highest decision' on how the Prussian ministers of state should respond to an invitation to a banquet in honour of the Mayor of Berlin, Forckenbeck, and of the world-renowned physician and Progressive Party parliamentarian, Rudolf Virchow. In reply the Kaiser wired imperiously from his hunting box at Hubertusstock: 'Participation in banquet for Forckenbeck *and* Virchow *not appropriate*. I agree to personal *congratulations* to Forckenbeck. Virchow is to be ignored. He has completely forgotten and failed in his duty as a civil servant. W.'[50] Not until November 1891 did relations between the Kaiser and the Mayor of Berlin improve for a time. At the inauguration of the fountain designed by Begas, Wilhelm singled Forckenbeck out for particularly courteous treatment, a volte-face which prompted Waldersee to observe that 'two years ago he maltreated this same man when he offered the fountain as a gift from the city, and even described him on several occasions as one of the most wicked democrats. O quae mutatio rerum!'[51]

The position of the government in the Reichstag, where the Centre Party held a key position between right and left thanks to its strong mandate in the elections of February 1890, was even more critical than in the Prussian parliament. On 14 June 1890 Wilhelm told one of his Flügeladjutanten that he had 'had his first battle with the new Reich Chancellor today'. Caprivi had asked him to invite the leader of the Centre Party, Ludwig Windthorst, to the forthcoming soirée for members of parliament on the Pfaueninsel, the lake island in the Wannsee. He had 'emphatically refused, for Windthorst was a most dangerous man of hostile intentions, who had already done fearful harm to his [Wilhelm's] family'. Caprivi recorded later that he had had 'fought hard' with the Kaiser until Wilhelm had put an end to the argument by declaring that it was 'against his honour to invite Windthorst'. When Carl Wedel went to the Reichstag on 20 June he heard regrets expressed on all sides that 'the Kaiser allowed himself to be carried away by personal antipathies into politically unwise rebuffs'. Miquel, above all, pointed out that Wilhelm would have to reckon with the Centre Party 'if he wanted to avoid a *coup d'état*'.[52]

In the autumn of 1890 Caprivi made a renewed attempt to bring the Kaiser and Windthorst together at a parliamentary dinner, this time successfully, although, as Waldersee recorded, 'the Kaiser hesitated for a long time before he made up his mind to meet Windthorst at Caprivi's house'. In spite of his reservations, he talked at some length with the leader of the Centre on this occasion. The Chief of the General Staff was convinced, however, that Windthorst and his party would not co-operate with the government in the Reichstag unless they received a quid pro quo such as the recall of the Jesuits or the implementation of clerical school reform in Prussia, neither of which the government could concede.[53] Nevertheless his meeting with Windthorst led to a change in the Kaiser's tactical attitude to the Centre. When the party leader fell down stairs and injured himself in January 1891 the Kaiser wondered whether it would be overdoing it to send a Flügeladjutant to enquire after his health. And not long afterwards, while checking the invitation list for the next court ball, he asked, 'Why is Windthorst not on the list?'[54] Once again Waldersee was amazed by this rapid change of mind and complained that Germany was now 'dancing entirely to the Catholic tune' and would 'certainly come off worst' in the process.[55] At the other end of the political spectrum the Empress Frederick was no less struck by the change. When Windthorst died in March 1891 she wrote in astonishment to Frau von Stockmar that 'Flowers were sent to him from the Court etc . . . That would no doubt *not* happen if one of our German-Progressives fell ill, or for our very deserving Mayor etc . . . In spite of all his opposition, all his Guelph sympathies and Ultramontanism, the conservative Windhorst [sic] is nevertheless closer to those in power now than a free, independent, clear-thinking liberal man.'[56]

The tactical rapprochement of the Kaiser and the government in general with both the Centre and the Left Liberal Parties in the Reichstag, which will be discussed in more detail below, undermined the position of the Prussian Minister for Ecclesiastical Affairs, Education and Medicine, Gustav von Gossler, who had worked with the Conservative–National Liberal Kartell under Bismarck. In the summer of 1890 members of the Centre Party were already complaining vociferously and 'furiously' about the minister and demanding that he be replaced by an ultra-conservative more sympathetic to them, such as the Silesian Count Robert von Zedlitz-Trützschler.[57] It was not the wishes of the parliamentary parties, however, but those of the monarch and of his irresponsible advisers behind the scenes which determined the composition of the closely intertwined governments of Prussia and the Reich. Thus while on a train journey to Kiel with the Kaiser, Waldersee was able to bring the conversation round to the social question and to demand the dismissal of both Gossler and Minister of the Interior Ernst Ludwig Herrfurth. The evil of socialism must be rooted out, he urged the Kaiser, and to that end church and school must co-operate instead of fighting one another. Wilhelm listened calmly, but complained that he could not 'make the slightest progress on school reform', as Gossler wanted 'preliminary consultations with countless people, i.e. to put off the whole matter indefinitely'.[58]

The dominant role played by Wilhelm II in December 1890 in the school reform question is an example of the way in which, with the help of his unofficial advisers, he was able quite simply to bypass the state bureaucracy in matters which interested him personally.[59] It was he, and not the minister responsible, Gossler, who invited forty-five delegates to Berlin for a conference on education lasting several days. Instead of seeking Gossler's advice he listened to his former tutor Dr Hinzpeter, who – as the generals and the Prussian and Reich bureaucrats discovered to their consternation – was suddenly spending all his time at court.[60] What was more, for weeks on end Wilhelm granted regular audiences to Professor Konrad Schottmüller, who drafted the 'powerful' speech which the Kaiser made at the opening of the school conference on 4 December 1890 and which still attracts both guarded admiration and passionate criticism today.[61] In this startling speech Wilhelm inveighed against 'excessive education', with express reference to his own experience at Kassel, and coined the influential watchword: 'We ought to educate young Germans, not young Greeks and Romans.' The task of the secondary schools was to take up 'the fight against social democracy' and against the 'centrifugal tendencies' in the Reich, he declared.[62]

It is not difficult to imagine the enormous sensation which this speech caused. Arthur von Brauer reported that it was 'almost the only subject of conversation in corridors and salons, and one cannot be surprised that a speech from such an exalted source, which goes into so much detail and contains a quite specific, rather radical programme, will give rise to the most diverse opinions, and that in private and more restricted circles criticism will not always be kept within the bounds that should be observed at all costs when the speaker is so illustrious a personage. Firstly, many people think it highly undesirable that the monarch should have entered so much into details and that he should have publicly expressed such a firmly rooted preconceived opinion, so that in the course of further discussion, any frank expression of opinion which leads to different conclusions will seem almost like a deliberate rebellion against the All-Highest will . . . In two diametrically opposed camps there is much dissatisfaction with the speech even on purely formal grounds: among the strict Conservatives, because they fear that by descending into the arena where burning issues of the day are fought out the sovereign will damage the reputation of the monarchy; and among the strict constitutionalists, because as a matter of principle they reject personal intervention by the monarch, on doctrinaire grounds . . . No German who has any thought and feeling for the upbringing of children and for schools can fail to be powerfully affected by the Kaiser's words. He will not have changed the minds of any of the fanatical supporters of one side or another with his speech. On the other hand, among the wider public his positive suggestions have in fact met with enthusiastic support. It is well known that His Majesty prides himself on understanding the mood and wishes of the younger generation better than his ministers and advisers, and the success of his latest utterances seems to confirm this claim. At any rate I have frequently heard the opinion expressed, particularly among younger people, that the Kaiser has hit the nail on the head again this time, and they expect the abolition of the Realgymnasien [secondary schools for Latin, modern languages and sciences] and the elimination of Latin composition in the Gymnasien [grammar schools] to bring about a miraculously rapid end to all the shortcomings of our schools . . . On the other hand the resounding patriotic tone of the speech has awakened a joyous echo on many sides. The idea that in the interest of the schools' national duty, the German language and especially German composition and the history of the Fatherland should be moved to a dominant position at the heart of the curriculum, has really taken hold . . . It is natural and only to be expected that His Majesty's remarks about journalists, whom he described as "starving paupers" and "old Gymnasium boys down on their luck", should be unfavourably commented on in the press of all shades of opinion . . . What is incomprehensible is that apparently neither the Ministry of Ecclesiastical Affairs and Education nor the Civil Cabinet thought it prudent to examine H.M.'s speech before it was sent to be printed. At least I assume they did not, or it would be even more incomprehensible that the passage about the press and much else besides could be allowed to remain . . . Nothing is known for certain about the authorship of the imperial speech. What is clear is that no *official* adviser knew anything about it beforehand. Hinzpeter had nothing to do with it either. He spoke to me directly after it and commented on the content in terms that were anything but laudatory. Professor Schottmüller is said by many to have been the author . . . I have the impression that in essence the speech was the product of His Majesty's own direct inspiration, for his words clearly expressed the resentment which took root in him at the Gymnasium in Kassel against excessive classical formalism imposed by unskilful and pedantic teachers. In general one can say of the Kaiser's speech, as is true of many of the things his sense of power and his youthful enthusiasm have inspired him to say and do hitherto, that H.M. has earned more criticism and head-shaking than approval in official, cautious-minded circles, but that among the broad mass of the people his words have been greeted with jubilation and enthusiasm.'[63]

Waldersee was one of the many people who regretted that the Kaiser had put forward his own views so prominently, and thought he would have done better not to try to influence the deliberations of the conference.[64] 'The Kaiser's conduct at the opening of the school conference has displeased many people', he wrote. 'They think that he put himself and his own views forward altogether too much, and this is quite true.'[65] In the course of the conference the Kaiser said to Gossler, after a dinner at the Schloss, 'In the next few days I shall go back once more and take the chair; I am told on all sides that the gentlemen are very pleased when I speak.' Waldersee commented acidly: 'It is a mixed blessing for them, to say the least, and once again one can see how sycophantic individuals have managed to get at the Kaiser.'[66] In the speech which he made to the closing session of the school conference on 17 December 1890 Wilhelm spoke complacently of his 'full appreciation' of the fact that the learned delegates, after their free exchange of ideas, had finally 'arrived at the conclusions to which I pointed the way'.[67]

In these circumstances it is scarcely surprising that the Minister of Ecclesiastical Affairs and Education lost the desire to remain in office. During the school conference all the participants were 'absolutely horrified . . . by the wretched performance of the Minister, Gossler'.[68] When in January 1891 a bill rescinding discriminatory measures against Catholics, through which the government hoped to purchase the support of the Centre for the trade treaty with Austria, was introduced, Gossler again played 'a most wretched' role. 'It is my belief that he cannot possibly stay on as minister', Waldersee commented.[69] Wilhelm stood by his Minister at first. He sent him his portrait and publicly declared that he was the best minister of ecclesiastical affairs that Prussia had ever had. But then he suddenly dropped him when it was reported to him that Gossler's wife had strongly criticised the inscription 'sic volo sic jubeo' ('my will is my command') which the Kaiser had written on the portrait.[70] Finally, when Gossler was heavily defeated on his anti-clerical School Bill in the Prussian Landtag in March 1891 Wilhelm approved his dismissal. The Kaiser promised to appoint him Oberpräsident of the province of East Prussia by way of compensation, but dropped this idea too after members of the East Prussian aristocracy carried out a 'putsch' against Gossler, who was 'not grand enough' for them, while the Kaiser was hunting at Prökelwitz.[71] On 10 March Caprivi received the Kaiser's authority to offer the Prussian Ministry of Ecclesiastical Affairs to Count Robert von Zedlitz-Trützschler, who accepted.[72]

Looking back over all the decisions on appointments made during the first twelve months after Bismarck's fall, two characteristic features are immediately apparent: vacillation and uncertainty within the 'responsible' government, and overwhelming strength of will on the part of the young sovereign. When the Grand Duke of Baden came to Berlin in October 1890 for the ninetieth birthday of Field Marshal Count von Moltke he stressed the importance of 'more stability in high places', for 'the rapid changes and the persistent rumours of further changes' were having a highly unsettling effect.[73] Wilhelm von Rauchhaupt, the leader of the Conservatives, also thought 'the worst thing is the constant unease, the frequent changes and the resultant insecurity in many areas'.[74] No one doubted that the Kaiser was to blame for the unease. In June 1890 a member of the Reichstag observed that it was incredible 'how much the Kaiser is feared in the Reichstag',[75] and the same was even more true in the Reich and Prussian governments. With a self-assurance bordering on frivolity the monarch took it upon himself to offer the highest offices in the land to men whom he had happened to meet somewhere and who had won his personal favour. It is no wonder that the leadership of the country quickly lost any kind of unity through this behaviour and became increasingly psychologically dependent on the monarch. It was only very hesitantly – and usually only when there was a stalemate because the Kaiser's preferred candidate (Herbert Bismarck, Alvensleben, Jencke) turned down the post offered to him – that Caprivi, in conjunction with the Ministry of State, sometimes took the opportunity to come forward with counter-proposals. But without the right to decide on appointments and dismissals within the Prusso-German government, the authority of the Reich chancellor and minister-president inevitably lost ground in comparison with that of the kaiser and king.

THE FIRST CHANCELLOR CRISIS OF THE NEW COURSE

From early 1891 onwards, as if one year after the fall of Bismarck he now felt that he had held himself back long enough, Wilhelm II's high-handed behaviour and utterances increased alarmingly. He made a succession of speeches which left little room for doubt of his autocratic intentions and which displayed notions of ancestor-worship and divine right so out of tune with the times that they caused universal shock and dismay. The absolutist motto 'sic volo sic jubeo' which the Kaiser had written on the portrait he gave to Gustav von Gossler proved to be no momentary aberration but an authentic reflection of the young Hohenzollern ruler's innermost convictions. On 20 February 1891, in a speech to the provincial diet of Brandenburg, he railed against 'the spirit of disobedience' which was rife throughout the land and invoked the spirit of the Great Elector, calling on his audience to follow him on the path 'which I tread, and which I have marked out for myself, to lead you and all of us towards My goal and towards the salvation of the whole nation . . . You know that I regard My position and My task as one which Heaven has laid upon me and that I am called to serve a Higher One, by whom one day I shall be called to account', he declaimed.[76] A similar peroration followed in May 1891, on the occasion of a festival performance of *Barbarossa* in the concert hall at Düsseldorf. 'I for My part must not flinch from the paths marked out for me, for which I alone must answer to My conscience and to My God', he proclaimed, finishing with the declaration that there was only one ruler in the Reich and that was himself; he would tolerate no other.[77] On 24 February 1892 these absolutist pronouncements reached new heights in a speech (again at a banquet held by the provincial diet of Brandenburg) which aroused universal indignation and disapproval, especially as it was published in the official *Reichsanzeiger*. 'Unfortunately', the Kaiser declaimed, 'it has of late become customary to find fault with everything that the government does. On the slightest pretext the tranquillity of the people is disturbed and their pleasure in the existence, vigour, and prosperity of our great Fatherland is embittered. All this carping and fault-finding finally gives rise in the minds of many people to the idea that our country is the most unfortunate and the worst-governed in the world, and indeed that it is torture to live in it. That this is not the case we, of course, know perfectly well, but would it not be better if these dissatisfied grumblers shook the dust of Germany from their feet and withdrew as quickly as possible from these miserable and distressful surroundings? They would thus be put out of their misery, and they would at the same time do us a great favour. We live in a state of transition! Germany is gradually growing out of her childhood and will soon be entering upon her period of youth. It is, therefore, high time that we should throw off our childish ailments. We are passing through difficult and exciting times, in which, unfortunately, the judgement of the great majority of people is wanting in objectivity. They will be followed by more tranquil days if only our people will earnestly concentrate their energies, realise their duties, and, refusing to be led away by outside influences, will place their trust in God and in the unsparing diligence and paternal care of their hereditary

Ruler . . . Brandenburgers, a great future is still reserved for us, and I am still leading you on to a glorious destiny. Only do not let a gloom be cast on your outlook into the future, or the pleasure you take in united effort be lessened by mere carping criticism and discontented partisan talk. Catchwords alone are powerless to effect anything, and to the incessant, captious criticisms of the new course of our policy and those who are responsible for it I quietly but firmly reply, "My course is the right one, and in it I shall continue to steer."'[78]

Brauer was so dismayed by these utterances that he felt compelled to offer some sort of psychological explanation of them to his Grand Duke. 'It is not easy to detect the direct motivation behind decisions which are so much His Majesty's own', he commented. 'But perhaps it would not be too far off the mark to see the immediate reason not so much in a current political objective as in H.M.'s habit of surprising the world with an extraordinary oratorical performance at every banquet of the provincial diet of Brandenburg . . . The speech . . . was made on the initiative of the Kaiser himself, who said to a gentleman in his entourage the day before yesterday: "I shall make a thundering good speech today." The speech was sent from the Schloss to the *Reichsanzeiger* printers in the morning, before it had been delivered. This was fortunate, for if it had been transcribed after the event, even more extraordinary remarks would have come to public notice. For instance, His Majesty allowed himself to be carried away into saying "Everything that is finest and best in Germany we owe to the House of Hohenzollern which brought it all about, and so it shall be in my reign too"! . . . What concerns me more than anything else, for the future of our political life, is the fact that nothing will dissuade H.M. from believing that these speeches of his have the most excellent effect possible. When honest people have drawn his attention, or tried to do so, to the dangers of the Kaiser making such remarks, his response has been to say triumphantly: "My speech has hit home; they are making a fearful fuss about it in the newspapers." In fact His Majesty particularly enjoys reading the opposition newspapers after such speeches. When he was shown the perfidious article in the *Berliner Tageblatt*, which reproduced H.M.'s speech without any criticism, simply printing Article 27 of the Prussian Constitution (the right of every Prussian to free expression of opinion) in full beneath it, H.M. laughed and said: "They ought to have added that the *King* also has the right to express his opinion freely"! . . . The respectable newspapers, on the whole, show admirable restraint. But reading between the lines one can see deep resentment or a strong sense of discouragement, according to temperament or political standpoint. In particular one can detect the underlying thought that the days of enlightened despotism are over and that it is no longer possible for any prince nowadays to determine what course the state and the national cultural life should follow purely on the basis of his own ideas. In government circles and among the parliamentarians of the moderate tendency with whom I have occasion to be in contact, I have not found a single person prepared to defend the Kaiser's words on principle. But in these circles people are naturally careful about what they say. Unpleasant incidents have shown that even confidential private remarks are not immune from being repeated maliciously to the Kaiser, and everyone knows that while H.M. is indifferent to press comments, however disparaging, he is very sensitive indeed to the slightest criticism from his officials or other people in the higher ranks of society. Hence the great reserve, which of course contrasts strongly with the disrespectful remarks which are said to be doing the rounds in other circles.'[79]

The increasingly overbearing behaviour of the young monarch made itself felt even more strongly behind the scenes, in his handling of civilian statesmen. In March 1891 Friedrich von Holstein deplored the fact that 'the relations between Kaiser and Chancellor . . . have suddenly changed for the worse'. For example, he commented, 'His Majesty is now suddenly dissatisfied about Gossler's departure – for which the Reich Chancellor was not responsible – and today he said to the latter, in reference to Boetticher, "as if we had not had enough changes already", and then commanded the Reich Chancellor in so many words "to restore the Kartell".'[80] By treating the Chancellor in this way the Kaiser was only encouraging the rumours that Caprivi's position was 'shaky'; Philipp Eulenburg warned him that 'nothing is more dangerous than that for Your Majesty at the moment'.[81] 'The worst thing about the whole business', Holstein warned in his turn, 'is that in parliamentary circles the Kaiser is increasingly thought to be not in his right mind. An elderly member of parliament, Hobrecht, said to me "It is the Kaiser himself who is the cause of all the uncertainty." If His Majesty takes on another new chancellor now the mood in the country will be unbelievably bad. That may mean that His Majesty finds very strange things happening.'[82]

Alfred von Kiderlen-Wächter, who accompanied the Kaiser on his Scandinavian cruise in the summer of 1891 as representative of the Foreign Office,[83] reported anxiously to Berlin that 'since last year H.M.'s autocratic tendencies have decidedly increased. The *sic volo sic jubeo* makes itself felt in matters both great and small.' Kiderlen thought it a telling sign of the All-Highest attitude that the Kaiser had grown a full beard (as he had done two years earlier), and had remarked that 'with a beard like this you could thump on the table so hard that your ministers would fall down with fright and lie flat on their faces'. The worst aspect of this autocratic attitude was that it was 'not accompanied by any serious scrutiny or weighing of the facts; he just talks himself into an opinion. Anyone in favour of it is then quoted as an authority; anyone who differs from it "is being fooled"', Kiderlen complained.[84] Wilhelm's idea of the appropriate relationship between a Reich chancellor and his kaiser was further illustrated by a marginal note of September 1891 in which he criticised Bismarck for having been 'guilty of disobedience towards his sovereign'.[85] It was again apparent in the wording of a letter he wrote to the King of Saxony in February 1892: 'I have already fulfilled your wish regarding the lottery and have given the ministers instructions.'[86] Equally typical of Wilhelm II's attitude, in the eyes of his contemporaries, was the 'very stern' telegram which he sent from Rominten, his East Prussian hunting lodge, to the Minister of Justice, Hermann von Schelling, and the sensational imperial proclamation on prostitution issued in the autumn of 1891, which he had composed with the help of Lucanus, the Chief of his Civil Cabinet, and with which he set himself up as 'supreme guardian of the law'.[87]

In the military sphere, where as supreme war lord he could exercise unlimited, extra-constitutional authority, Wilhelm II's increasingly autocratic behaviour was particularly noticeable. In July 1891 it was the main topic of conversation between Waldersee and the former Adjutant-General, Max von Versen. Waldersee, by now also dismissed from his post as chief of the General Staff, was informed by Versen that at a sitting of the Commission on Home Defence chaired by the Kaiser himself Wilhelm had insisted on transferring the chairmanship of this important body to Prince Albrecht of Prussia in spite of the fact that the latter had firmly refused to accept the appointment, 'being quite rightly convinced that he was in no way equal to the post'. Wilhelm had dismissed his uncle's objections and declared that he thought it particularly important to have a member of his family as chairman. When the fortification of Heligoland subsequently came up for discussion by the committee Wilhelm allowed neither discussion nor a vote, but simply declared the draft proposal which he had earlier agreed with an admiral and a general to be the final decision. Moreover, Wilhelm decided, against the advice of the General Staff, to abandon the fortress at Dirschau (now Tczew) on the Vistula, on which work had already begun, and to fortify Marienburg instead, as it would not be right to expose 'the beautiful Schloss Marienburg to destruction by the Cossacks'. Versen assumed that while the Kaiser was hunting at Prökelwitz, 'where many a matter has been settled before', East Prussian influences had been brought to bear against the military arguments of the General Staff. 'Sic volo, sic jubeo!' Waldersee angrily exclaimed on hearing of the manner in which these decisions were taken.[88]

The General was rendered almost speechless by a remark made by the Kaiser in the autumn of 1891 to the leader of the Conservatives, Rauchhaupt. When the latter tried to thank him for an order conferred on him, Wilhelm replied 'My dear Rauchhaupt, voluntas regis suprema lex', and turned away from him, remarking to another member of parliament, 'I really let the old greybeard have it.' 'What sort of behaviour is that, and how should Conservatives, who thank God are all still loyal monarchists here, react?', Waldersee asked. 'It really will dishearten some people, but also push many others into the anti-monarchist camp.'[89] When it was discovered that Wilhelm had already written the same absolutist maxim in the Golden Book of the City of Munich (see illus. 25), Waldersee was appalled and commented that the inscription would 'naturally cause a great scandal and do *nothing* but harm. It is quite a good thing, although extremely sad, that for some time now the Kaiser's utterances have no longer been taken seriously.'[90] The blame for the utter confusion of both domestic and foreign policy lay 'with the Kaiser *alone*, for he thinks that he can reign alone, that he understands *everything* and knows *everything* better than anyone else, while in fact he has no clear idea about *any* goal, nor does he thoroughly understand *anything*. The fact that the Chancellor and the ministers remain in their posts only proves that they are all characterless people. Perhaps the good Lord has better things in mind for us than we deserve, and the "voluntas regis suprema lex" will help to clear the air. All parties seem to be agreed in condemning this pronouncement, and indeed it could not be otherwise. But can a Ministry exist if it condones such a remark? I think not; it cannot but fall victim to contempt.'[91]

Simultaneously with the growth of Wilhelm II's appetite for autocracy, the conviction was spreading that Caprivi was a spent force, despite his considerable success in parliament, and would soon resign. The dismissed Minister of War, General von Verdy du Vernois, wrote to Waldersee in May 1891: 'More and more people are beginning to see the light about Caprivi: in the commercial world people think that the sooner he goes the better it will be; the groups who are still most attached to him acknowledge his hard work, his eloquence etc., but claim that he has an irremediable fault: he has become dull.' The Chancellor's diabetes, although officially denied, was evident in his increasing irritability, Verdy reported. He was 'under extreme nervous strain because of pressure of work' and was evidently close to 'forgetting himself completely', which was noticeable in his catastrophic relationship with the War Office. 'Having recently treated the Ministry of War in the most ruthless way', Verdy wrote in May 1891, Caprivi had 'added insult to injury' by making a written statement on the Pension Bill which 'was more merciless and insulting than the most extreme of Bismarck's pronouncements. And this statement was not only sent to Kaltenborn, but circulated to all the other ministers too. As a result the Minister of War is said to be extraordinarily . . . upset and those in the know in the Ministry expect that he will ask to be relieved of his office.' 'It certainly would be madness for him to stay!', commented Kaltenborn's predecessor.[92]

It was no coincidence that the Reich Chancellor's first threat to resign was over a question of military policy, especially as Wilhelm II's identification of himself with his 'grandpapa', the 'powerful' Emperor-Hero Wilhelm I, made itself felt most strongly in military matters.[93] On 15 June 1891, the third anniversary of his accession, the Kaiser ordered the introduction of a huge new Army Bill, which was destined to be at the centre of political conflict for the next two years. The Kaiser's orders were contained in two lengthy memoranda, written in his own hand and addressed to the Minister of War and the Reich Chancellor respectively, which cast a highly revealing light on his way of thinking as a military monarch.[94] In the memorandum to Kaltenborn the Kaiser began with the question which had long been a matter of lively debate among the generals, in the Reich government and in the parties in the Reichstag: whether the enlargement of the army should be linked to a reduction in military service from three to two years, as demanded by the parties of the left and above all by the Centre Party. Modelling himself on the unyielding attitude of his grandfather in the Prussian constitutional conflict three decades earlier, the young Wilhelm held out resolutely for the retention of three-year service. His memorandum acknowledged that there were 'undoubtedly tempting advantages' in two-year service. 'But there is no denying that the reliability, efficiency and especially the discipline of the army will be put at considerable risk by this measure (two years instead of three) and may indeed actually be damaged. I have thought it over again, conscientiously examining the reasons for and against, and I have again come to the conclusion that all the arguments of my grandfather of most blessed memory are still absolutely valid today. Moreover, success in three victorious campaigns is a momentous and weighty argument in favour of the three-year service, to which it bears shining testimony. But there are also other significant factors, in my view', the Kaiser continued. 'The question of two-year service is no longer purely a matter of military expediency . . . No; from the time of the conflict onwards parliament has turned it into a trial of strength, to which it constantly reverts at any given moment in order to force the King eventually to accept it. But I, as my grandfather's grandson, as war lord of the Prussian Army, who strives to defend his prerogative as supreme commander of the army and to give what orders he sees fit – within the bounds of *tradition* – to promote the welfare of his army, I cannot act differently from Grandpapa. You will surely understand and share my feelings, my dear Kaltenborn. And I tell you frankly, on the very anniversary of the death of the victor of Wörth and Sedan [i.e. Kaiser Friedrich III], that I will never accept two-year military service; come what may, whatever the consequences, I cannot and will not do it.' After this solemn, emotional and very determined declaration Wilhelm got to the point. His aim, he explained, was on the one hand to increase the infantry establishment and on the other to train officers for reserve units. After giving a precise list of orders the Kaiser concluded: 'These, my dear General, are the guidelines which you are to follow. As soon as you have acquainted yourself with them and worked through them I shall expect your report . . . There will certainly be a fight with those rogues in parliament. But that will not do any harm; in 1860–6 it was just the same and Grandpapa found his Bismarck and Roon and won, and was then magnificently justified by success in war. In the same way I expect to gain my victory with the help of Caprivi and Kaltenborn, and then – with God's help – to prove the rightness of my principles in the next battle for Germany's existence, which will surely come. Knowing the man that you are, I have no doubt that you will stand fast by me and fight hard for our cause. I remain, my dear Kaltenborn, your affectionate King Wilhelm R.'[95]

On the same anniversary Wilhelm II also wrote a long letter to the Reich Chancellor, the immediate outcome of which was a letter of resignation from the latter. The imperial missive began with an expression of his 'most sincere royal thanks' for the passage of the three great reform bills in the latest round of legislation, namely the Social Welfare Bill, the Prussian Finance Bill and the Rural District Bill. Wilhelm wrote: 'You were as intrepid as you were politically skilful in supporting your colleagues when it was needed, and the selfless cool-headedness which you have shown in the face of great adversity and difficulties of all kinds and from all sides is mainly responsible for the fact that in one session three such important laws, the like of which no government of any state has succeeded in launching before, were successfully passed. You have made yourself an undying name for all time in the annals of our country's history, and may the grateful thanks of your King give you some consolation and satisfaction after all the unpleasantness which unfortunately you were often obliged to endure.' After these three bills it was necessary to have a pause in domestic legislation for the time being, the Kaiser decreed. 'Therefore my concern is now once again for the army, and I wish to return to the questions of reorganisation which I postponed last year at your particular request until you had fought out your winter campaign.' In similar terms to those of his letter to Kaltenborn, and again alluding to his late grandfather, Wilhelm explained that he was firmly opposed to 'the experiment with two-year military service'. 'Even if the army gained in numbers it would lose in inner worth and discipline. Moreover, as a result of the conflict of 1860–6 the question of two-year service has since then been turned into a trial of strength in which parliament – particularly the radical Democrats – is constantly trying to force its will upon the King and compel him to accept it. My grandfather was criticised and blamed for the conflict by these very same people, and they still consider that it has not been fought to the finish. I, as the grandson of this great man, who grew up in his traditions and under his teaching, cannot act in any other way than exactly as he did. I should be sinning against His Name, His Memory, if I were not to follow in his footsteps completely, and continue to build where he left off.' The Kaiser expressed his utmost determination 'never to accept two-year service'. He had 'personally worked out a reorganisation which – with the knowledge of no one else – I have sent to the War Minister today, with a letter in which I laid down my principles, according to which he is to proceed henceforth. I have instructed him to be in touch with you in order to prepare a bill for the autumn and next winter, postponing all other army bills except for Heligoland . . . The whole thing, with the greatest possible economy of means, amounts to a sum of 40 million. The War Minister will show you the scheme and my letter. We have reached the same point as in the year [18]60, when my grandfather was forced by the high age of the recruits to the reserve units and the territorial army to abolish both of these, so setting the ball rolling. I do not for one moment doubt that we shall have to fight for this bill as much as Grandpapa fought for his. But that does not matter. This is a good opportunity; a firm stand will clear the air both in parliament and among the people, like a thunderstorm. In his day Grandpapa found Bismarck, who helped him to fight and win. Today I have Caprivi, and I trust him absolutely to fight and win with me now. In addition we have the advantage of being able to point to the rightness of Grandpapa's principles as shown by the glorious wars. And if Bismarck really wishes for a rapprochement with the King, or to show that he sometimes takes the King's side, he may perhaps use his remaining strength to fight with us too. After all, he joined in the fight for the three-year service. It would have been different the other way round. So I look with confidence to you as my active collaborator in the absolutely essential and vital work to whose success you will contribute. Your truly devoted King Wilhelm R.'[96]

One can imagine Caprivi's feelings on receiving this directive, particularly when he learned that the Kaiser had worked out his plan with one of his Flügeladjutanten: according to information given to Waldersee by Adolf von Bülow the culprit was the 'stupid' Freiherr Gustav von Seckendorff, brother of Prince Heinrich's Hofmarschall.[97] The Chancellor was not only faced with the most crass example he had yet experienced of 'personal rule', with the monarch coolly handing down orders to the constitutionally 'responsible' Reich Chancellor and Minister of War instructing them to introduce and push through the Reichstag a legislative measure which he had worked out in detail and whose consequences were incalculable. Worse still, in direct emulation of the 'sacred' decisions of his grandfather in the Prussian constitutional conflict, Wilhelm II had ordered a fight with the Reichstag, which – as Caprivi was well aware – could easily lead to a *coup d'état* and even civil war, with the risk of foreign intervention. Holstein threw up his hands in despair when he heard of Wilhelm's peremptory demand. It was proof, he wrote to Eulenburg, 'that His Majesty has no idea of the difference between the Prussian constitution and the covenant between the German ruling princes on which the constitution of the Reich is based. Do *you* believe that Saxony, Bavaria and Württemberg would side *with* the Kaiser if he demanded an increase in the army when the Reichstag had voted *against* it? And if they do not, we shall have civil war with Russia and France looking on; the latter would certainly seize the opportunity of recapturing the Reich territories [Alsace and Lorraine].'[98]

Caprivi's immediate response was to submit his resignation, the first time he had taken this step since his appointment in March 1890. 'In the light of Your Majesty's firm resolve it is not for me to put forward the different opinion I have of the advisability of the measure', he wrote on 16 June 1891. 'On the other hand I have had to ask myself whether and to what extent the All-Highest intention could be carried out and whether it was in my power to do so. In my humble opinion there would be difficulties which I consider insuperable. In the course of last year's negotiations in the Reichstag I became convinced that to burden the Reich with considerable additional expenditure for military purposes would be very difficult in any event, and with the present Reichstag it is only possible if two-year military service for the infantry etc. is conceded. I have already taken the liberty, on a former occasion, of respectfully stating that I consider two-year service, with certain compensations, better than the present state of affairs with half-pay leave and reserves.' Even if the Reichstag were dissolved and new elections held, this would not change the fundamental situation, the Reich Chancellor asserted, for in order to get the Army Bill through the new Reichstag two-year military service would still have to be conceded. 'In pursuing Your chosen path Your Majesty would then be faced with the question of whether and how, by some means akin to a *coup d'état* – perhaps by altering the electoral law – it might be possible to achieve the aim in view. It would be irresponsible on my part to conceal from Your Majesty my conviction that such a *coup d'état* in the loosely knit German Reich and under the present circumstances would bring that Reich close to disintegration. The political situation is quite different from that which obtained in Prussia in the 1860s. I doubt that the federal governments would agree to all the increased expenditure planned by Your Majesty, but I have no doubt at all that they would not favour a *coup d'état*, whatever form it took, at present. Therefore not only am I unable to foresee a successful outcome to Your Majesty's planned course of action, but I am even more sure that it is well beyond my powers to carry out the All-Highest intentions. Taking a risk of this order can only succeed when the person who carries the official responsibility for the matter believes it can succeed. But I lack this belief, as well as other qualities essential to the fulfilment of so difficult a task. Nor can I conceal the fact that because it has pleased Your Majesty to take crucial decisions in a question of such great political importance without seeking my views beforehand, I have become convinced that I do not possess the All-Highest confidence to the degree without which I should be hindered at every step by the fear of making a false move.' Caprivi therefore asked to be relieved of all his offices.[99] Holstein, to whom the Chancellor showed both letters, urged Eulenburg to come to Berlin under whatever private pretext he could find, in order to avert a crisis which in his view could easily result in Waldersee's appointment as Reich chancellor and that of Herbert Bismarck as foreign secretary.[100]

The Kaiser sent a conciliatory reply to Caprivi's letter. 'When you accepted the position you now hold, you said "Whether one lays down one's life as a soldier in the face of the enemy or by exhausting one's strength in the interest of the Reich; both are of equal value in the end." At that time these words gave me the assurance that you would devote all your strength, trusting in God and your King, to the execution of the office which I entrusted to you. We should not and must not be separated by a misunderstanding – for such it undoubtedly is – of what I committed to paper yesterday in a solemn and hallowed moment. There can be no question of lack of confidence. You received yesterday, my dear Caprivi, certain indications of the way in which my army must be expanded in the future, which I consider essential in order to defend Prussia and the Reich as may be necessary in grave and critical situations. On the basis of these directives I expect a detailed report from the man whom I have chosen as my adviser, once he has studied the question thoroughly. I rely fully on you – as you must rely on me – and with God's help we cannot fail. The serious political objections which you have raised may, if necessary, result in the postponement of the decisions I have made after careful consideration, but never in their abandonment. I therefore expect you, my dear Caprivi, at 9.15 tomorrow morning to make your report. I gladly repeat today what I have already said, that you will always have my full confidence and my deepest admiration for the services you have rendered me as Reich Chancellor.'[101] In his audience of 17 June the Kaiser accused Caprivi of sending him 'a nasty letter' and complained: 'You made yesterday very unpleasant for me', to which the Chancellor showed both courage and dignity in replying: 'I shall always write to Your Majesty like that when I see that You wish to do something which will harm the Reich or Your Majesty Yourself.' After the Kaiser had declared himself ready to put off the large army increase until the following year – though still insisting on retaining three-year military service – Caprivi withdrew his resignation.[102] But this compromise did little to resolve the crisis.

When the Kaiser landed in Kiel on 10 August 1891 after his Scandinavian cruise, Caprivi was waiting for him in order to discuss the Army Bill. After the audience Kiderlen wrote to Holstein in some relief that he was now 'much easier in my mind', for 'with H.M.'s incessant talking himself into the need for an Army Bill, which he discussed constantly with his "restricted retinue" for no other reason than to win authoritative approval', the situation had seemed 'really serious'.[103] On 12 August Wilhelm received Waldersee on board the *Hohenzollern*. The General impressed upon him his strong objections to the abolition of three-year military service, maintaining that in view of the grave political situation Germany had no time 'for experimenting, but that the aim ought to be to strengthen the army if any means could be found of doing so'. Waldersee adroitly reminded the Kaiser of the Prussian constitutional conflict of the 1860s when his grandfather 'held fast to the three-year service and carried through the increase of the army against a hostile Landtag, against the advice of countless supposedly well-meaning people, against the opposition of the Crown Prince and against the opinion of many generals'. Retention of the three-year service and immediate enlargement of the army were indispensable, he insisted, since 'we are probably . . . facing a struggle for our existence'. If Germany lost this war the Reich would break apart, the monarchy would be jeopardised and Prussia would be destroyed, which would result in the most appalling state of affairs within the country. For him, the Kaiser, there was no better battlefield than the expansion of the army, for any party which persisted in voting against it would be digging its own grave. The Kaiser acknowledged the gravity of the international situation and promised Waldersee that 'under *no* circumstances would there be any question' of giving up three-year service.[104]

At the end of August 1891 the Reich Chancellor handed the Minister of War a long memorandum setting out his view of how Germany's military strength should be developed, a view diametrically opposed to that of the Kaiser. He agreed that since war was inevitable 'in the long or the short run' the German armed forces should be increased 'to the greatest extent permissible'. In order to achieve this increase, however, 'the support of the nation' must be won, and this could be achieved only through the timely concession of two-year service. Without this concession, conflict with the Reichstag leading to a *coup d'état* would be inevitable, but this must 'at all costs be avoided'. Caprivi argued that 'before one takes the first step towards a conflict one must be clear whether one is able and willing to take the eventual consequences of it'. One could argue about the advisability of a *coup d'état* for other reasons, for instance in order to achieve a change in electoral law after street battles. 'But if it were for the sake of three-year service it could lead to the disintegration of the German Reich and might be more catastrophic than losing a military campaign.' Like Friedrich von Holstein, Caprivi was convinced that the 'loosely knit Reich' would not tolerate extreme measures, and that 'any serious dispute between the federal governments or between them and the people . . . would be of benefit only to hostile foreign countries. To our enemies abroad a conflict at the present moment would look like a sign of weakness, an invitation to make war.' The Chancellor brusquely rejected Wilhelm's emotional evocation of his grandfather's steadfastness in the Prussian constitutional conflict with the comment: 'Just because this question became a trial of strength 30 years ago in Prussia in completely different circumstances it does not follow that it must do the same today.' This view of the importance of avoiding conflict with the Reichstag was, however, of fundamental significance for the proposed Army Bill, Caprivi continued, since 'from the answer to the question, whether one can or cannot risk a *coup d'état*, one must work backwards, for everything thereafter depends on it. If the answer is no, everything else is reduced to the question of whether to keep the *status quo* in military matters or to accept two-year service.' The question at issue was quite simple, Caprivi remarked. 'Is a significant increase of our military strength attainable without conceding two-year service?' This question too had to be answered with a firm negative, he said. By his reckoning, given the present composition of the Reichstag scarcely anyone other than the seventy-one Conservatives – and some of them only very unwillingly – would vote with the government if two-year military service were not conceded. Caprivi stressed, however, that he favoured the reduction of military service to two years for important reasons of military policy and not purely on grounds of parliamentary tactics. 'The closer one believes one is to war', he argued, the more vital it was 'to make the population feel that what is required of them is unavoidable'. The increase in the army must therefore take place in a way that was 'understood and approved at least by the majority of the people'. The implementation of compulsory military service for all was just such a 'popular, more general idea' and two-year service was a step on the way to the Scharnhorstian principle of the nation in arms. The Chancellor-General went even further and wondered whether the government should not also give up the septennate when it brought in the great Army Bill. He pointed out that in practice it was rare for seven years to go by without the government being obliged to break the septennate of its own accord. He thought it worth considering 'whether a peacetime military establishment calculated as a percentage of the population' could be achieved. Finally, Caprivi expressed his objections to the description of the Army Bill as a 'reorganisation'. That word, used by the Kaiser in an allusion to the reconstruction of the Prussian army in 1861, would create the dangerous impression abroad that Germany had been on the wrong track before and was therefore weak now, which was not the case. But even if these views of his were approved, the Chancellor concluded, 'very serious parliamentary battles' could still ensue. Even if the path of compromise were followed, the dissolution of the Reichstag might still be inevitable. 'But the prospects for new elections will be quite different if a longstanding wish of broad sections of the population is granted by the concession of two-year service. I have no doubt that this wish can be granted without military disadvantages. The fiction that the present so-called three-year service is better than a well-implemented two-year service cannot be maintained . . . But however that may be, it is my opinion that the decision lies in the answer to the question: Can we afford to risk conflict among the legislative bodies of the German Reich at this time for the sake of three-year military service? And I believe this question must be resolutely answered in the negative.'[105] As we shall see hereafter, the struggle over the form of the Army Bill, and especially the thorny question of the length of military service, would cast a dark shadow over the relationship between Wilhelm II and the Reich Chancellor until the summer of 1893.[106]

From the summer of 1891 the responsible government lived with the constant possibility of receiving some order from above which would put its entire legislative strategy at risk. Holstein, who had 'lost all sense of security', decided that one of the Flügeladjutanten – he thought first of Gustav von Kessel – would have to be recruited to tell the Kaiser unwelcome truths and draw his attention to the consequences of his orders. 'As the Kaiser generally lives only with his military entourage it is very important that there should be one among them who can occasionally teach him something', he argued.[107] When in December 1891 the Trade Treaties were passed by the Reichstag with a majority of 195 votes, to the delight of the Liberals and of the general population,[108] the Kaiser raised Caprivi to an earldom.[109] He conferred on Privy Councillor von Holstein – on his own birthday, in order to make the honour 'still more precious' – the cross of a commander of the Royal Order of the House of Hohenzollern which, as Caprivi explained, 'will by His Majesty's wish henceforth rank above the other orders'.[110] Philipp Eulenburg wrote to Holstein to congratulate him, emphasising that this distinction was 'a *quite exceptional* mark of His Majesty's confidence . . . If you but knew *what value* he attaches to his Order of the Household! It almost borders on eccentricity – and how the class of the decoration enhances the value!'[111] In spite of such outward signs of the Kaiser's confidence, however, there was still a general feeling of insecurity and dissatisfaction.[112] Not without reason Waldersee observed, after shooting with the Kaiser at Springe in November 1891, that the chancellor crisis had already begun, for Wilhelm had several times spoken of the choice of a successor and in the process had expressed the opinion, not exactly flattering to Caprivi, that anyone who had not yet become a 'nasty brute' in the post of chancellor would certainly become one in time.[113]

THE KAISER AND DOMESTIC POLICY

Is it possible to detect, in the Kaiser's numerous speeches, marginal comments, letters and actions during the months following Bismarck's dismissal, anything approaching a consistent line on domestic policy which would throw light on his fundamental ideological stance? As in foreign policy, here too there were endless contradictions, which led many an observer to see Wilhelm as vacillating helplessly over internal policy and having no clearly defined goals, except perhaps that of making himself popular. Waldersee, for one, commented irritably on the Kaiser shortly after Bismarck's fall: 'With regard to his attitude to the political parties I have already seen him showing ultra-Conservative, Free Conservative and National Liberal leanings; many people claim he is now working in favour of the Progressive Party, while others of course say that he is going to sacrifice all to the interests of the workers. It has almost reached the point that no party knows where it stands with him. If there were a definite plan behind this it would be an excellent thing, for he must not belong to any party and can deal with them all best if each believes he favours it. *But no such plan exists!* . . . His desire to make himself popular, which is becoming increasingly obvious, is most regrettable.'[114] In the ensuing months the Chief of the General Staff was to revise this early verdict on Wilhelm II's political inconsistency, as we shall see, but on one important point – the Kaiser's attitude towards the worker question – he was essentially right.

Once Bismarck was out of the way, little more was heard of the Kaiser's great social reform initiative. On the contrary, fear of a socialist revolution made him increasingly inclined to favour a *coup d'état*, which he had angrily rejected when Bismarck had suggested it. Within a few days of the latter's dismissal the first signs of a volte-face became apparent. When an article appeared in the *Hamburger Nachrichten* on 26 March 1890 calling for 'firm and resolute use of the power of the state in all its rigour' to counter the 'increasingly arrogant conduct' of the workers' movement, Wilhelm, furious at the non-renewal of the anti-socialist law which Bismarck had deliberately engineered, scribbled bitterly in the margin of the article: 'You rejected the Anti-Socialist Bill! Now you can draw your conclusions!'[115] The next day he told one of his Flügeladjutanten that another strike had begun at Gelsenkirchen, and that he had instructed the commanding generals 'to let loose with repeating rifles at the first opportunity!' The Flügeladjutant commented in astonishment: 'How strangely the Kaiser changes . . . What has happened to his high-sounding vow "that he would not besmirch the first years of his reign with the blood of his subjects"?'[116] Wilhelm's willingness to crush the Social Democratic workers' movement with military force was also apparent in his Proclamation to the Army of April 1890, in which in principle he granted the patriotic bourgeoisie the same opportunities as the aristocracy to take up the career of an officer. As he explained to his Flügeladjutant Carl Wedel, his intention was 'to win over the bourgeoisie, which is now more than ever looking to the crown for help, out of fear of social democracy, and to bind it to him. It was important to bring together as many elements as possible in the struggle against social democracy.' Wedel ridiculed the Kaiser's remark, commenting that 'the bourgeois elements . . . in the army are in any case far and away the majority' and that 'there has never been a glaring difference between aristocracy and bourgeoisie in the army anyway'. Nevertheless he was deeply impressed by the decisiveness which the young monarch displayed in this question.[117]

The change which was beginning to emerge in Wilhelm's attitude towards the workers was welcomed by Waldersee, who commented after his meeting with Philipp Eulenburg on 24 April 1890 that it had now 'gradually become plain to everyone that the Kaiser has not achieved his planned goal. I myself have been of this opinion from the beginning and have therefore been on Bismarck's side in this matter; on the other hand I have no objection to the Kaiser trying to solve the problem, but he has gone about it in the wrong way. There are bound to be many disappointments, for the workers can never be satisfied and have now evidently lost all sense of reality.'[118] Waldersee continued to be a determined advocate of the view that the social democratic threat could be combated only with military force, for which riots and violence in the streets would afford a welcome pretext. When the Socialists declared 1 May 1890 an international holiday the Chief of the General Staff commented revealingly: 'I am afraid that it will all go off peacefully, except perhaps for some brawling; at least here in Germany.'[119] The day having passed without any disturbance anywhere, he admitted: 'I do not like this; it shows . . . that the leaders of the movement are in control.'[120] The General lost no opportunity of trying to convince the young Kaiser of the soundness of his views on class conflict.

At the end of April 1890, while on a journey to Bremen, Wilhelmshaven and Oldenburg together, Wilhelm II and his Chief of General Staff discussed 'the possibility of disturbances'. The Kaiser expressed 'very determined views' on the subject. 'I pointed out', Waldersee wrote, 'that it was very difficult to recognise the right moment to intervene in earnest, so that as a rule it was missed; I also mentioned that people have scruples about giving orders to fire, for fear that there might be many innocent people, women and children, in the line of fire, and that experience showed that difficulties arose from this.'[121] At Potsdam in May 1890 Waldersee observed that Wilhelm was 'more serious than usual', apparently because 'many things are not going as smoothly as he thought. The worker question is causing him much anxiety; he sees that people are quite shameless and always want more.'[122] During the manoeuvres that autumn the Social Democrats were a frequent topic of conversation with the Kaiser, and Waldersee was gratified to note that Wilhelm was beginning 'to realise that these people are superbly organised and increasingly dangerous'.[123] As the National Liberal Johannes von Miquel also noticed shortly after his appointment as Prussian minister of finance in the summer of 1890, the Kaiser was 'really afraid only of the Social Democrats'.[124] Members of the Kaiser's closer entourage, Waldersee learned, were convinced that the monarch's recently adopted hard line against the workers arose particularly from 'concern for his personal safety'.[125]

In the late summer of 1890 the Kaiser repeatedly expressed the opinion that it was dangerous to remove troops from Berlin and other major cities, as always happened for the manoeuvres, and suggested that at such times reserve battalions should be sent to the cities.[126] When he returned to Berlin from the manoeuvres in East Prussia at the end of August, Lucanus told him of the violence which had broken out recently following a Social Democratic rally in the suburb of Friedrichshagen, but which had been easily brought under control by the local police. In disbelief Waldersee recorded: 'The Kaiser, without consulting anyone, immediately sent the Flügeladjutant to the 3rd Army Corps with orders to send a brigade to Berlin; even Hahnke was passed over, and Verdy and Caprivi by the same token; the Governor [of Berlin] was not even asked whether he needed troops!' Waldersee condemned Wilhelm's action as 'a very overhasty act' which would create a general sensation and would be derided by the Social Democrats. If a revolution had really been in the offing, he commented, the guard detachment of about 700 men who had remained behind after the Corps of Guards had left for the manoeuvres would in any case have been insufficient to control the situation, even with the assistance of the 3,000-strong Berlin police force. But there was no question of that, for the leaders of the Social Democrats had not the slightest intention of attempting to overthrow the government. So the Kaiser's order had been completely superfluous. 'It seems as if something has made the Kaiser nervous', Waldersee wrote.[127]

On 20 November 1890 Kaiser Wilhelm made one of the most notorious speeches of his long reign at the swearing-in of recruits to the Corps of Guards in Berlin. He told the young recruits that they were called to fight not only against foreign enemies, but also against enemies within. 'The spirit of opposition, rebellion and insurrection was stalking the land', he declared, and he warned the young men against 'listening to the deceivers and agitators. They belonged to him now and must be prepared perhaps even to fire on their fathers and brothers, if he commanded it.' Wedel, who heard the speech, commented: 'As almost always, our sovereign's speech was lively and gripping, and yet I think it a great pity that in such speeches he lowers himself too much from his exalted position, and puts himself too much on the level of a mere superior officer. On the other hand remarks like these are psychologically interesting because they show how concerned the Kaiser is about the destructive tendencies of the Social Democrats and how concerned he is personally to put a stop to their agitation.'[128] Even in military circles the Kaiser's conduct was 'very strongly condemned'. Nevertheless he repeated his admonitions the next day when he swore in more recruits at Potsdam.[129]

It is thus indisputable that a radical change took place in Wilhelm's attitude towards the workers' movement as soon as he had dismissed Bismarck. But this fact should not lead to a generalised assumption that he pursued no aims at all in domestic policy. His closest friend, Philipp Eulenburg, had the impression at the beginning of 1892 that 'His Majesty's ideal' was to form two major parties in the Landtag and the Reichstag. Eulenburg welcomed this, although he had to admit that 'in Germany more than in any other country such desires are quite illusory!'[130] At about the same time Waldersee recorded in his diary that the Kaiser had recently declared to the gentlemen of his suite that 'differences between parties did not matter to him at all; he saw only two differences: those who were loyal to the King or monarchically inclined, and those who were not!'[131] On the second anniversary of Bismarck's dismissal Waldersee, who in 1890 had described the young Kaiser as completely aimless and as 'a shapeless piece of elastic that could be stretched in any direction',[132] was forced to acknowledge that a succession of measures of considerable importance could now be attributed to Wilhelm II's personal initiative. He listed these as: 'eradication of social democracy through improvement in the welfare of the working classes, international action on the same lines, school reform, raising of moral standards, strengthening of the position of the German emperor, reconciliation with Russia, firm friendship with England, friendly relations with Denmark, improvement of the army'.[133]

What is surprising is that the internal policy measures cited by Waldersee were carried out not so much with the support of the right-wing parties as with that of the Catholic Centre and of the two Left–Liberal Progressive parties. In February 1891 the Badenese envoy reported in astonishment that public life in Berlin now presented 'one of the most curious phenomena that has perhaps ever been seen since we have lived under a constitution. We have a Government which faces a parliament elected under another Government by a very large oppositional vote. And yet the present Government . . . meets with curiously little difficulty in this oppositional parliament. Its budget is passed almost without cuts; its bills are accepted – but in the strangest manner.' The government 'naturally raises no objections to this assistance from the Left and the Centre', Brauer reported to Karlsruhe, even if accepting the 'assistance of the entire German democratic movement' seemed 'a dangerous policy in the eyes of many a friend of the Fatherland'.[134] The moderate policy of the New Course is undoubtedly to be attributed mostly to the constellation of parties in the Reichstag, for without the support of the Left Liberals and/or of the Catholic Centre no legislative measure could have been passed. It is nevertheless astonishing that not only General von

Caprivi but also Kaiser Wilhelm II was evidently much in favour of co-operation with these erstwhile opposition parties.

We have seen how the Kaiser overcame what were at first very strong prejudices against the Centre Party leader, Ludwig Windthorst, in the summer of 1890. As we have also seen, Wilhelm even agreed – although very reluctantly – to the re-election of the Left Liberal Max Forckenbeck as Mayor of Berlin. When Caprivi set to work on the renewal of the Trade Treaties in the late autumn of 1890 and in so doing moved increasingly towards the pro-free trade position of industry and away from the protectionism favoured by agriculture, the question inevitably arose of how the Kaiser would react to this change of direction. On 16 November, when it became known that the Prussian Minister of Agriculture, Robert Lucius von Ballhausen, had resigned in protest against precisely this development, Waldersee recorded in his diary: 'Unfortunately there are already people who believe that the Kaiser, who has expressed warm support for the agricultural interest in several speeches, for instance in Königsberg and Münster, might nevertheless become a free-trader, at least with regard to corn duties.'[135] Barely two weeks later one of the Flügeladjutanten remarked that the public had the impression that the Kaiser's relations with his mother had greatly improved and that her liberal influence over him was growing.[136]

The most striking evidence of a rapprochement between Wilhelm II and the Liberals, however, was his dismissal of the anti-Semitic Court Preacher Adolf Stoecker in November 1890, carried out despite the opposition of Kaiserin Auguste Viktoria and during the absence of her Oberhofmeister Baron Mirbach. In October the ultra-conservative Protestant Waldersee had learned that Stoecker had 'again spoken out vigorously' at the conference of provincial synods, but in so doing he had 'aroused the anger of the Kaiser, whipped up by Hinzpeter'.[137] Moreover Stoecker had made a number of speeches in South Germany which, according to Waldersee, had received 'enormous acclaim', but for which he had been officially reprimanded in Berlin.[138] After an audience on 1 November 1890 Waldersee and Wilhelm discussed Stoecker, whom the Kaiser criticised 'severely'. 'So there must have been another campaign against him recently', the Chief of the General Staff concluded in alarm, quick to realise the implications for his own position.[139] To cap it all, Wilhelm had nominated the preacher Ernst von Dryander to deputise for the ailing Kögel at the forthcoming marriage of his sister Viktoria (Moretta) with Prince Adolf of Schaumburg-Lippe, which Stoecker, who as court preacher had the right to conduct all services in the Court Chapel, took as an affront to himself. He and two other court preachers asked to be relieved of their offices, on the assumption that Wilhelm would turn down their request.[140] The Kaiser, however, told Wedel during a hunt on 4 November that Stoecker had promised to refrain from political agitation, but that he had not kept this promise and must now go. Everyone was aware that he, the Kaiser, did not approve of Stoecker's behaviour, but the Court Preacher 'clung to the Kaiserin and to Mirbach's coat-tails, and a stop must be put to that'.[141]

The anti-Semitic Court Preacher's spectacular fall from grace was widely greeted as a liberation, and not least by many people at the court. 'Bravo, bravo!', Wedel exclaimed in delight, a sentiment echoed by Adjutant-General von Wittich. 'It is a pleasure to see our young sovereign showing such deep, clear and thoughtful insight!'[142] The Empress Frederick reported joyfully to her mother: 'The pleasure in the town & the public is great at his [Stoecker's] dismissal fr. Court.' She thought it a 'pleasing accident' that it was the marriage of her daughter which had provided the occasion for the dreaded Court Preacher's dismissal. On hearing of Kögel's illness she had declared 'I would *not* go to Church if that evil man "Stöcker" officiated – & Vicky [Moretta] said she refused to be married by him. – However, our opinion was not needed, as the matter was already decided.' For fifty years the court preachers in Berlin had been 'a wretched set & clique' and had created much mischief under Friedrich Wilhelm IV and Wilhelm I, she claimed. 'They are a strong Party as all that is orthodox and reactionary, the whole of the "Kreuzzeitung" & aristocracy belong to them; *Dona* & her court especially.'[143]

Those in the know suspected that this sensational decision by the Kaiser was principally due to a revival of the influence of his former tutor Georg Ernst Hinzpeter. Wittich claimed to have 'fairly precise information' that Hinzpeter 'was behind this affair and had been the prime mover in it'.[144] Waldersee also gathered through the Oberhofmarschall, Count August zu Eulenburg, and other court officials 'that Hinzpeter had campaigned strongly' against Stoecker.[145] In addition to Hinzpeter the Chief of the Civil Cabinet, Hermann von Lucanus, was thought to have played a part in prompting the dismissal. Waldersee, who could see 'nothing but a victory for Jewry' in Stoecker's fall, laid 'most of the blame on that unprincipled fellow Lucanus'. The duty of the Chief of the Civil Cabinet ought to have been 'to give the Kaiser accurate information, but he gave way to pressure from Mirbach's enemies and even egged the Kaiser on . . . The Kaiser will take note of the wholehearted applause of the Jews, Social Democrats and Liberals of all shades which he will receive for this! Will it agree with him?'[146] 'The Kaiser's action is not at all laudable and cannot lead to anything good', the General angrily complained on 7 November. 'Unfortunately Minister Gossler is a very weak man and offered no resistance at all; on the contrary, he is said to be very glad to be rid of Stoecker. But the fault lies mainly with the wretched Lucanus, who I now know is utterly false.'[147] On 23 November Waldersee was still lamenting the episode. 'The saddest thing about it, as ever, is that the Kaiser allows himself to be so easily influenced; in this case – naturally without suspecting it – he gave in to Jewish pressure; he is in the process of undermining the foundations on which he stands, by paralysing one of his best resources at a time when as he himself has often said the forces of insurrection are at work. If Stöcker were not such a dangerous enemy of revolution he would certainly not be attracting so much hostility.'[148]

In the immediate aftermath of the fall of the Court Preacher it became positively dangerous for members of the imperial entourage to be seen at Stoecker's church services. Both Wilhelm von Wedell, the House Minister, and the Hofmarschall Maximilian Freiherr von Lyncker thought themselves lucky not to have been noticed in the cathedral when Stoecker preached a sermon condemned by the Kaiser as 'quite outrageous' in January 1891. Waldersee's presence, on the other hand, was at once relayed to the Kaiser. The latter sought him out directly after a dinner for the commanding generals and remarked that the sermon had been 'of such a kind that a few gentlemen would have liked to walk out'. 'It is really scandalous what a disgraceful campaign has been whipped up again here!', the Chief of the General Staff wrote indignantly afterwards. 'These worthless cowards who do not dare to come forward openly but worm their way to the Kaiser by flattery! Oh if I could but once catch someone *in flagrante*!'[149]

In the course of a deeply pessimistic survey of the eventful year of 1890 Waldersee observed, shortly before his own dismissal as chief of the General Staff, that the Kaiser was still in the process of 'learning the lesson that one can get nowhere with the Liberals; . . . I have no doubt that he will eventually return to the Conservatives, as many another king has done before him.'[150] But this prophecy was not to come true for a long time yet. In fact in the spring of 1891 violent disagreements arose between the Kaiser and the German Conservative Party. The Conservative leader, Rauchhaupt, commented despairingly that 'in the party and in the country there is real sadness about much that is happening, nor can it be denied that Germany is in truth divided into two camps, and that the Kaiser has many opponents, particularly among the great industrialists in the West and in South Germany'.[151] The immediate cause of the conflict with the monarch was the refusal of the Conservatives to approve the 22 million marks required for the construction of 'his long-cherished favourite wish', the cathedral in Berlin.[152] There were, however, several more profound issues over which the Kaiser took a surprisingly modern view, bringing him into conflict with the East Elbian landed nobility.

At the end of 1890 the Conservatives began a campaign designed to replace the Prussian Minister of the Interior, Ernst Ludwig Herrfurth, a middle-class career bureaucrat with liberal leanings, with one of their own. They accused him of appointing too many liberal commoners to the influential post of Landrat in the Prussian provincial administration. Waldersee complained that Herrfurth was 'a clumsy bureaucrat' who did not know 'the country and its needs' and was trying 'to bring his creatures into the administration', while persecuting Conservative officials.[153] Wilhelm not only rejected these accusations out of hand; he also criticised the Conservative provincial officials for their opposition to the rural district reform plans being championed by Herrfurth. During a hunt in the Grunewald in January 1891 the Kaiser expressed support for Caprivi's decision to take disciplinary measures against the rebellious Landräte led by Count Kanitz. In Waldersee's opinion 'Minister Herrfurth's utter worthlessness' was demonstrated by the fact that he had shown the Kaiser an article published by Kanitz in the *Kreuzzeitung*.[154] A week later he recorded that the Conservatives were being 'constantly insulted' by Wilhelm. 'He is happy to hob-nob with *feeble Liberals*, who pay court to him shamelessly.'[155] In June 1891 Wilhelm II again had cause to complain about the 'supercilious and brusque behaviour' of the Conservatives towards the bourgeois Minister of the Interior, who was 'a thorn in their side, just because he is plain Herr Herrfurth and not a count', as Carl Wedel observed. The Conservative leaders charged that Herrfurth was 'acting in the Liberal interest in his appointment of provincial officials, for he is an avowed enemy of the aristocracy. The Kaiser rejected this', the Flügeladjutant recorded, 'by saying that the appointment of provincial officials was carried out through him; their suitability was examined by the Civil Cabinet, and Lucanus made sure that Conservative principles were taken into account.' Under no circumstances would he be prepared to drop Herrfurth. Wedel could not help wondering, in his account of this episode, whether the Kaiser was unaware 'that Herrfurth and Lucanus are close friends and that the latter also has strong Liberal leanings?'[156]

Waldersee saw yet another sign of the Kaiser's inclination to act against the Conservative Party 'in the most passionate way' when he went to Berlin in June 1891 to attend a session of the Upper House of the Prussian Landtag. Shortly before the Landtag closed Wilhelm invited a large number of members of both houses to the Pfaueninsel on the Wannsee. He received them personally at the lakeside, where two steamers awaited them, and accompanied them back there afterwards. Throughout the excursion he treated them with the utmost affability, with the intention of obtaining a majority both for the Game Damage Bill and Herrfurth's Rural District Bill. He went from one parliamentarian to another stressing that he set great store by the passage of these bills, with the result that both were indeed passed. 'Without his intervention the opposite would have happened', Waldersee observed, although in his view neither law was important enough to justify the sovereign's use of 'the full weight of his personal influence' in support of them, for had he failed he would have caused immeasurable damage. 'If there is serious trouble over a matter of great moment I have no objection to the King intervening personally', he wrote; but such action was 'dangerous in any circumstances'.[157] In the months that followed Waldersee complained repeatedly of the Kaiser's tendency to rule with the liberal 'middle parties' against the Conservatives. 'It is quite plain to me that we are steering strongly towards the left, and under Caprivi's deliberate leadership', he noted angrily in November 1891.[158]

HINZPETER *REDIVIVUS*

If General Count Alfred von Waldersee blamed Wilhelm II's surprising and to him highly unwelcome swing towards the liberal middle parties on the Kaiser's need to co-operate with a Reich administration that in its turn had to answer to a Reichstag in which the balance of parties had shifted fundamentally since early 1890, in his eyes this was not the principal cause of the problem. Much more to blame for the monarch's left-wing tendencies, as Waldersee saw it, was the influence of two men who had already played a decisive role in the Bismarck crisis, namely Wilhelm's former tutor Dr Georg Ernst Hinzpeter and his ally Dr Paul Kayser, the (Jewish) Director of the Colonial Department at the Foreign Office. This astonishing conclusion cannot be seen merely as a delusion which had arisen in the mind of a Chief of General Staff whose position at court was becoming more and more insecure, for Waldersee was by no means alone in this opinion. In June 1890 the Flügeladjutant Count Carl von Wedel warned the Chancellor of the intrigues which Hinzpeter and Kayser were devising. Both men, he said, were 'unscrupulous intriguers' who enabled 'secret sources' to reach the Kaiser. Wedel pointed particularly to Hinzpeter as 'the mediator, who abuses the Kaiser's trust'.[159] On a train journey to Kiel an incensed Hermann von Lucanus told the Flügeladjutant that the Kaiser's recent speech at Essen had been corrected and sent back to Wilhelm by Hinzpeter. Lucanus, Wedel recorded, was 'furious and asked whether it was Hinzpeter or he, Lucanus, who was responsible to the Kaiser and the Ministry for his speeches . . . Since then Hinzpeter had of course been doubly hostile towards him and was seeking to overthrow him at the earliest opportunity.'[160] The commanding generals, who came to Berlin for Moltke's ninetieth birthday, likewise had the former tutor in mind when they expressed the unanimous opinion that 'the depressed and uncertain mood' in the country had come about 'principally . . . because of the Kaiser's inclination to consult private and unofficial individuals'.[161]

Hinzpeter's influence on Wilhelm, and through him on matters of domestic policy and public appointments, became apparent (as we have seen) when the Kaiser dismissed Court Preacher Adolf Stoecker.[162] It reached its zenith a few weeks later when the Kaiser intervened dramatically in the Prussian school reform question. Waldersee was horrified to find Hinzpeter back in Berlin in

19. Dr Georg Ernst Hinzpeter, the Kaiser's tutor from 1866 to 1877

the winter of 1890–1 and now described the schoolmaster from Bielefeld as the most dangerous of Wilhelm II's irresponsible advisers 'because he has contrived to keep himself in a strong position'.[163] 'The driving force is said to be the wife', he noted in his diary. 'She is ambitious and wants to get whatever she can for her husband. She is French, and although she has lived in Germany for 20 years she can still hardly speak German.'[164] The General predicted bitterly 'that people around the Kaiser will soon notice that this evil fellow has been here; in fact many people now see him for what he is. I am intrigued to see how Caprivi will get on with him.'[165] At present it was Minister of Ecclesiastical Affairs Gustav von Gossler who was suffering most from the misdeeds of this irresponsible adviser, Waldersee observed, for the 'Hinzpeter influence' was proving 'uncomfortably strong with regard to the question of school reform'.[166]

With Hinzpeter's arrival in Berlin at the beginning of December 1890 to attend the conference on school reform, his 'great influence on the Kaiser' became apparent to everyone.[167] Holstein and Eulenburg seriously wondered whether it would not be wise to appoint Hinzpeter to an under-secretaryship of state or a ministerial directorship in the Prussian Ministry of Ecclesiastical Affairs and Education. 'His relations with H.M. are the same – whether from Berlin or Bielefeld', they reasoned. 'The advantage of a man who has been using his influence anonymously having to answer for it publicly is obvious . . . Much gossip would be silenced . . . Anyone who disliked Hinzpeter would have the satisfaction of knowing that as under-secretary of state he will at least be given enough rope to hang himself.'[168]

However unmistakable the influence exercised by the former tutor, suddenly re-emerging from his Westphalian retreat into the limelight, on his now powerful pupil, the nature of the relationship between Hinzpeter and the Kaiser baffled contemporaries and is still hard to fathom today. Did Wilhelm II see the stern preceptor as a substitute for his dead father, and perhaps even for his deeply venerated grandfather, whose praise and encouragement he now needed more than ever? Had 'the Doctor' learned how to manipulate Wilhelm and direct his mood at will, during his long years in attendance on the young Prince?[169] Whether they liked or loathed the influence of this middle-class pedagogue over their high-handed monarch, contemporary observers agreed in sensing a strange mixture of domination and subjection on both sides of the relationship. Thus Waldersee remarked in December 1890: 'Hinzpeter's influence is undoubtedly in the ascendant; the Kaiser sees a great deal of this obsequious hypocrite. Several times when he has been with their Majesties he has not been deterred even by the presence of the entourage from his malicious intrigues. In this instance, too, the Kaiser is incredibly unaware of what is going on and, just as with Herbert Bismarck, he does not notice how much he allows Hinzpeter to influence and push him. The Kaiserin is wiser, as always, and hates the villain.'[170] When, at a dinner with the Chancellor and several ministers, the Chief of the General Staff deplored the tutor's hold over the Kaiser, Gossler and Boetticher, of all people, sprang to Hinzpeter's defence with the argument that 'anyone who disparages the Kaiser as strongly and criticises him as openly as Hinzpeter does can hardly have much influence' over him. Waldersee refused to be convinced, however, and retorted: 'This shows just how false Hinzpeter is; he is in fact very often with the Kaiser, who pays much attention to his views, and then immediately afterwards he criticises his former pupil!'[171] When in January 1891 Professor Schottmüller lamented the fact that the Kaiser began many things but never finished anything, Hinzpeter replied self-importantly: '*That does not matter in the least; the main thing is that I always keep him in suspense*; unless there is always something new he becomes apathetic.' Waldersee, who recorded this remark in his diary, commented: 'So the scoundrel gives himself credit for controlling the Kaiser completely, and does not care in the least what becomes of the Fatherland.'[172] Not long afterwards he wrote angrily, 'Wherever one goes one comes across people who think Hinzpeter dangerous; I have just been assured by eyewitnesses that on several occasions this wretched fellow has kissed the Kaiser's hand in front of many witnesses! It just shows what unscrupulous wretches these liberal gentlemen are. On the one hand outrageously presumptuous, [on the other] crawling to the Kaiser and prostrating themselves before him.'[173] In 1893 Poultney Bigelow, the American friend of the Kaiser's youth, also expressed his surprise at the tutor's attitude to his pupil after meeting Hinzpeter at the American Legation. 'I think he has gone mad', he reported to the Kaiser. 'He accused me of having compared you with *Alexander* and of attacking you in a periodical called Die Zukunft! . . . I think the dear old gentleman has "a screw loose" – he was so serious and critical!'[174]

Observers of all shades of opinion saw Hinzpeter's influence over appointments as particularly dangerous, especially as he openly boasted that the Kaiser even discussed changes in ministerial appointments with him. In January 1891, for instance, he told Kropatschek, the editor of the *Kreuzzeitung*, with whom he was working on the Commission for School Reform, that he had 'talked to the Kaiser about replacing Ministers Gossler and Herrfurth', and that the Kaiser had exclaimed: 'Finding replacements for ministers is difficult; when I get rid of generals I can find as many new ones as I want.'[175] Although Waldersee had nothing but contempt for Lucanus, whom he considered 'thoroughly unprincipled' and a 'very poor Chief of Cabinet' who gave the Kaiser bad advice,[176] he was outraged when it appeared that as a result of one of Hinzpeter's intrigues Lucanus might be replaced by Dr Paul Kayser. The three Generals, Caprivi, Waldersee and Wedel, otherwise at daggers drawn, agreed that this must be prevented at all costs. The Chancellor contrived to get 'the notorious Dr Kayser' appointed director of the Colonial Department at the Foreign Office, not least in the hope of 'pinning him down and rendering him harmless', as Carl Wedel put it. Wedel doubted, however, that this tactic would succeed, for 'the secret channels leading to the Kaiser via Hinzpeter will still be open to Kayser, and both have such a thirst for illicit intrigues and interference that they will certainly keep these channels accessible'.[177] The anti-Semitic Waldersee, above all, was adamant that Paul Kayser should not be given the influential Civil Cabinet post. 'He is an extraordinarily clever man but I would deeply deplore his appointment', he wrote, citing as his reason that Kayser was 'of Jewish extraction and has a Jewish wife'.[178] In October 1890 he commented: 'If Herr Hinzpeter protects such wretched fellows [as Kayser] he proves that he is himself good for nothing. For a time Herr Kayser's relationship with His Majesty went to his head and I think it quite credible that he planned – with Hinzpeter's support, of course – to get himself appointed Chief of the Civil Cabinet. I have reached the point where I think everything is possible, so what is to prevent a Jew obtaining even this position.'[179]

After the school conference Hinzpeter's influence seems to have receded a little. At the beginning of 1891 the 'Doctor' expressed the intention of staying on in Berlin for a while in order to bring about a reconciliation between the Kaiser and his mother, although Waldersee saw this as mere self-glorification, for Hinzpeter knew as well as he did that this was a hopeless task.[180] In the autumn of 1894, however, Waldersee heard from Count Botho zu Eulenburg, at that time Prussian Minister-President and Minister of the Interior, that Hinzpeter was 'unfortunately' still keeping in touch with the Kaiser 'in the same way as before', and in later years Philipp Eulenburg was likewise constantly discovering to his surprise that Wilhelm was maintaining close contact with his waspish tutor.[181] In the winter of 1890–1, at any rate, Hinzpeter was undoubtedly in great favour with the Kaiser, and under his influence Wilhelm II was able to nerve himself to take the momentous decision to part company with his longstanding, arch-conservative mentor Waldersee. As Flügeladjutant Carl von Wedel remarked shortly after Waldersee's dismissal, the fact that Hinzpeter had been involved in bringing about the fall of the Chief of the General Staff was 'beyond all doubt, just as I suspect that he has a hand in *all* these things'.[182]

CHAPTER SIXTEEN

The fall of the court generals

THE fundamental cause of the break with the two Bismarcks and the dismissal of Liebenau as Oberhofmarschall was Wilhelm II's ambition to take the most important decisions for himself. This conclusion is reinforced by the fact that only a year later Wilhelm consolidated the independence of his position by dismissing Waldersee and moving the two generals in his suite, Count Carl von Wedel and Adolf von Wittich, to other posts. All three officers had stood by him loyally in his conflict first with his parents and then with the Bismarcks. The first two, moreover, had handled delicate negotiations on his behalf in the murky affairs involving Anna Homolatsch and Elisabeth Countess Wedel, the sister-in-law of the Flügeladjutant. The knowledge they shared of very private political and personal events made the removal of the three generals from the court a highly dangerous manoeuvre, and Wilhelm felt compelled to avoid the slightest semblance of a clash in carrying it out. It was important that these influential officers should continue to feel beholden to him even when they were no longer in court service. That he nevertheless insisted on their removal from his immediate entourage is symptomatic of the growing self-confidence of the young Kaiser, who would not tolerate any strong-willed adviser at his side and was already bragging that he intended to be his own chief of General Staff. The story of Waldersee's dismissal, above all, demonstrates some of the least admirable aspects of Wilhelm II's personality: his inability to accept criticism, his craving for admiration and his sense of injured vanity at any unfavourable comment about him.

WALDERSEE'S 'DOWNFALL'

The dismissal of Count Alfred von Waldersee as chief of the General Staff was undoubtedly one of the most significant decisions taken by Kaiser Wilhelm II in the early part of his reign. The first volume of this biography described in detail the role of surrogate father played by Waldersee in the 1880s, when Wilhelm stood in venomous opposition to his liberal parents.[1] At the height of this conflict in 1887–8 Wilhelm entrusted certain private and secret papers to Waldersee for safekeeping. Precisely when and why the young Kaiser began to feel hampered by his close relationship with the powerful General cannot be clearly established, but it is not impossible that the first rift occurred when Waldersee tried to mediate in the blackmail episode involving Miss Love in April 1889. It will be remembered that Wilhelm had at first denied any intimate relationship with this woman but was later caught out when the Bismarcks felt compelled to buy back his letters to her.[2] At any rate, the first sign of trouble in the relationship between Wilhelm and Waldersee appeared in the spring of 1889. In March 1890 when Prince Heinrich heard what seemed to him the surprising news that Waldersee had not been appointed Reich chancellor, he remarked in front of his fellow naval officers that he had in fact heard of differences between his brother and Waldersee 'as much as a year ago'.[3] As the Flügeladjutant Colonel von Lippe also reported, the Kaiser complained several times in the course of 1889 about Waldersee 'interfering in everything' and 'stirring up trouble' everywhere.[4]

As we have already seen, the relationship came under further strain when the Bismarck crisis became acute towards the end of 1889 and Wilhelm was forced to consider whom he would appoint as Bismarck's successor.[5] After his dismissal as chief of the General Staff, Waldersee, looking back over the past year, noted in his diary that since January 1890 he had been 'pushed aside' by the Kaiser, which he attributed principally to the influence of Chief of the Military Cabinet General Wilhelm von Hahnke.[6] But the decisive break came – certainly not by coincidence – in mid-March 1890, at the height of the Bismarck crisis.

Sharp-eyed observers noticed that Waldersee's star was on the wane soon after Bismarck's dismissal. They dated the change in the Kaiser's mood from an incident in spring 1890 when the Kaiser criticised Waldersee in front of all his subordinates at the General Staff headquarters. Immediately beforehand Wilhelm had taken the Chief of the General Staff with him as usual for his morning walk, during which Waldersee must have made political comments which displeased the Kaiser, for on the way back Wilhelm remarked angrily to one of his adjutants: 'I don't understand W.; he's my Chief of General Staff, and ought not to worry his head about things which don't concern him.' After his harsh criticism of Waldersee at the General Staff headquarters, the Kaiser told the Chief of the Military Cabinet that he had wanted to show Waldersee that he could live without him.[7] The Flügeladjutant Wedel's view of this incident was that in accepting this insult from the Kaiser without protest, Waldersee had shown weakness, and that it had 'lowered him irretrievably in the eyes of the Kaiser'.[8] During his Scandinavian cruise in 1890 Wilhelm told his close friend Eulenburg that Waldersee was 'an intriguer pure and simple and wishes to become chancellor, although I have told him *I* shall *never* appoint him'. Politics were no business of the Chief of the General Staff; they were a matter for the Kaiser himself. Waldersee should 'kindly refrain from meddling in such matters'. Commenting on the Kaiser's changed attitude to Waldersee, Eulenburg observed: 'What H.M. resents is the way he [Waldersee] has exploited his close relationship with H.M. for his own purposes. H.M. holds the very correct view that people should work for *him* and not for themselves.'[9] Wilhelm also held Waldersee responsible for recommending Minister of War von Verdy, whom he had 'recognised too late' for what he was, and whom he had wanted to dismiss since the summer of 1890, as we have seen.[10] An increasing number of generals, including the Chancellor himself, expressed the opinion that Waldersee's 'worst crime' had been to advise the Kaiser to appoint Verdy minister of war.[11]

Waldersee naturally felt Wilhelm's growing coolness towards him in their almost daily meetings. After making his report to the Kaiser on 8 June 1890, in the course of which he obtained imperial approval for the deployment of troops against France and also for 'a few personal matters', Waldersee reflected that as far as the Kaiser was concerned he had never had any 'problems on official [!] business until now [!]'. The double qualification of this remark makes it clear that he had sensed a deterioration in his relationship with Wilhelm. He noted in his diary that 'in spite of his great amiability I thought I detected an underlying mood of gravity'.[12] The more coolly the Kaiser behaved towards him, the more critical became the General's opinion of the Kaiser. Thus at the manoeuvres at Memel Waldersee gained 'a very distressing impression' of Wilhelm's 'relationship with his entourages' and of the 'hasty and harsh judgements' which he passed on others in the presence of his Hofmarschall, his personal physician and the servants. Such frivolous conduct would never have been possible with the old Kaiser Wilhelm.[13] After the autumn manoeuvres in Silesia, which were particularly difficult to carry out because of floods, the Kaiser upbraided the participating generals 'very sharply' in the presence of young officers and foreign guests, which Waldersee condemned, since neither 'lack of judgement nor laziness' was to blame. 'Everyone has the best intentions, everyone is completely dedicated and does the best he can; people work for months at full stretch only to be rewarded with a harsh, overhasty verdict', he wrote, complaining that 'the Kaiser does not take the trouble to look at the circumstances . . . Those who are treated so mercilessly go away full of bitterness and lose all pleasure in their service. The subordinates lose all respect for their superiors, and so authority is undermined.'[14] Waldersee took it as a bad sign that commanding generals who visited the capital no longer reported to the Kaiser on arrival. 'Would it have been conceivable two years ago', he wondered in dismay, 'for a commanding general to come to Berlin and not to wish above all else to see the Kaiser?'[15]

Waldersee – and he was not alone – was appalled by Wilhelm II's practice of making sure that the side which he himself led in the manoeuvres was always victorious. When Philipp Eulenburg broached this sensitive subject with Wilhelm during the Scandinavian cruise of 1890 the Kaiser became '*very heated* and proved to me by an exposition of the military situation that he had won "according to the rules"'. Eulenburg argued that the army would always try to let the Kaiser win if he were participating, but Wilhelm refused to accept this, and claimed that it was 'a grave affront to all my commanding generals, who likewise regard me simply as a commanding general . . . Moreover if any commanding general opposing me in battle were to act dishonestly towards me, I should *dismiss him from the service at once.*' Eulenburg's suggestion that the Kaiser should in future abstain from active participation in the manoeuvres fell on deaf ears. Wilhelm considered it his right to 'undergo military training' and remarked 'that he hoped the army was satisfied he was *capable* of victory'. The Kaiser, Eulenburg reported, 'was so heated that my secretary . . . asked me whether H.M. had been reading me some document – the Kaiser's voice had sounded so consistently loud'. Wilhelm's friend was also amazed 'that no soldier has yet had the courage to discuss this question with H.M. The fact is they all go in holy terror of H.M.'[16] The Kaiser's personal intervention in manoeuvres, which was to cause trouble until the eve of the First World War, played an important part in Waldersee's dismissal, as we are about to see.

TRANSFER TO STUTTGART AS 'VICEROY OF SOUTH GERMANY'?

When Waldersee returned to Berlin on 9 August 1890 after a period of leave, it was at once clear to him that a battle with the Reich Chancellor and the Foreign Office was looming, for Caprivi, 'spurred on' by that 'scoundrel' Friedrich von Holstein, as the General maintained, had issued an order that military attachés would henceforth be answerable to their heads of mission and that they should not allude to political questions in their reports. As Waldersee had been trying for years to raise the status of the attachés and to make them independent of the ambassadors, the order signified a direct challenge to him as chief of the General Staff.[17] The attitude of the Kaiser, who had previously been 'in complete agreement with me', as Waldersee remarked, was naturally of crucial importance in this question. 'I must handle the whole military attaché affair with the greatest care, and first try to find out what the Kaiser's view is', he wrote.[18] 'I shall probably see the Kaiser in the next few days and then I shall be able to tell if there is any specific reason behind this business.'[19] He now became increasingly convinced that he would do well to distance himself from Wilhelm II and the current regime. He noted resignedly in his diary on 11 September 1890: 'To my way of thinking everything is going so badly and I see such a decline that it has long been my wish not to be regarded as one of the chief advisers, sharing the responsibility for it. I am also convinced that things will go further and further downhill; why should I allow myself to be buried by the avalanche which I cannot halt?'[20]

It was after the joint manoeuvres of the XIII Army Corps and the navy in Schleswig-Holstein at the beginning of September 1890 that the Kaiser mentioned for the first time the possibility that Waldersee might take over the command of the Württemberg Army Corps at Stuttgart. The Swabian diplomat, Alfred von Kiderlen-Wächter, who as always was accompanying the Kaiser as the Foreign Office's representative, was surprised to find himself suddenly being asked by the Kaiser, on the train, whether he should send Waldersee to Stuttgart as corps commander. 'As you can imagine I was completely dumbfounded', he told Friedrich von Holstein. Kiderlen had the impression that Wilhelm did not relish the idea of informing Waldersee of his transfer to Stuttgart. He was obviously afraid that the General would hand in his resignation and had hit upon the idea of offering Waldersee the inspectorship of the two Bavarian Army Corps at the same time, 'to soften the blow'. Wilhelm even asked Kiderlen to sound out Waldersee about this solution, which the young diplomat declined to do. Instead he reported the Kaiser's intention to the Chancellor. Through an oversight Caprivi's answer to Kiderlen, written in his own hand and addressed to the latter in person, fell into the Kaiser's hands. Wilhelm opened the Chancellor's letter with the words: 'Let's commit a breach of confidence for once!' Luckily Caprivi had declared himself in favour of Waldersee's transfer to Stuttgart. But Kiderlen wondered whether 'the secret urge to have a peep behind the scenes for a change' had played a part in the imperial 'breach of confidence'.[21]

On his arrival in Breslau for the Silesian manoeuvres, the Kaiser had an initial, hesitant discussion with Waldersee on 17 September 1890 as to whether the General might move to Stuttgart, where he would to a certain extent be 'viceroy of South Germany'.[22] According to Kiderlen 'the crafty W. at once pretended he was very glad to go but pointed out to H.M. that Prince Leopold of Bavaria was senior to him in the service and could therefore not be subordinated to him'. After this Wilhelm left the difficult task of resolving the matter to the Reich Chancellor, only remarking that he did not want to 'force' Waldersee to go.[23] In an hour-long interview with the Chief of the General Staff, Caprivi made it clear that he wished Waldersee to take on the Stuttgart post. He was the right man and the only man for the job, and his wife too was 'ideally suited to help him'. The Chancellor declared that the Kaiser also wished for this transfer, although only on condition that Waldersee was happy to go to Stuttgart, and that he should not 'in any circumstances' gain the impression, as Waldersee noted in his diary, that the Kaiser 'wanted to part company with me'. He 'was to receive the Order of the Black Eagle immediately, and the Kaiser would also agree to my other wishes'.[24] Kiderlen discovered that Caprivi had told the Chief of the General Staff 'pretty bluntly' that 'there was simply no room for him' in Berlin.[25]

Waldersee spent a restless night weighing up these proposals. Next morning the Kaiser burst into his bedroom, whereupon the Chief of the General Staff told him he could not accept the Stuttgart post. The people of Württemberg, he explained, hated the arrangement by which a Prussian general always held the command of the Württemberg Army Corps. The South German particularist movement, which had in any case grown stronger of late, would be given further momentum by Waldersee's appointment, for the entire liberal and Catholic press would denounce him as a 'Stoeckerite, a bigot and a *Kreuzzeitung* man'. As a result he would have to take severe military measures from the start, which would only create more dissatisfaction. Furthermore he had absolutely no desire to mix with the Württemberg court, which was 'contemptible in some respects'; in particular he could not tolerate the 'way of life' of the (homosexual) King Karl. Waldersee therefore suggested to the Kaiser that he should break with former practice and entrust a reliable Württemberger general with the Stuttgart command. The Kaiser reacted favourably to this proposal, but said that he also wished to discuss with the Chancellor a suggestion made by the King of Saxony that Prince Wilhelm, the heir to the throne of Württemberg, should take over the command.[26] Following this conversation Waldersee wrote to the Chancellor stating that he would not be able to bring the necessary enthusiasm to the post in Stuttgart. Both Caprivi and the Chief of the Military Cabinet thought Waldersee's decision a mistake, since he would not be able to hold out for much longer in Berlin, and 'H.M. would not make him such an offer a second time.'[27]

All in all Waldersee had the impression that Caprivi would not be displeased if he left Berlin. 'I do not believe that the Kaiser wanted to get rid of me', he reflected, 'but at the same time I realise that I cannot stay with him permanently. He is beginning to feel sure of himself in military matters, and does not want to appear dependent on me. In the same way that he thinks his own judgement is sound on every imaginable subject and has already caused much head-shaking, he fails to recognise that he is nothing but a dilettante in the military sphere. If he should take it into his head that he could command major troop formations himself it would really be very dangerous, even disastrous!'[28]

THE FATEFUL IMPERIAL MANOEUVRES ON THE NEISSE

Waldersee's bedroom conversation with the Kaiser about the Stuttgart command took place on 18 September 1890 at Rohnstock in Silesia. The next day a conflict arose between him and Wilhelm that was to seal the end of the General's career as chief of the Great General Staff and favourite of the monarch. The imperial manoeuvres, in Waldersee's disparaging words, had always gone well 'as long as the Kaiser was not in command . . . As soon as he entered the fray, however, everything became very artificial; indeed one would not hesitate to call it a childish game. Last year it was far better; but now he has gained confidence and really believes he knows something about commanding troops. In my opinion he has a certain grasp of parade-ground exercises, but not of the actual command of troops; of course he has no war experience at all and he refuses to believe that the cavalry has only limited usefulness in battle. He is extraordinarily restless and rushes about; he is usually right up at the front, interfering in the generals' command, issuing innumerable, often contradictory orders and paying scant attention to his advisers. Added to this he is extremely vain: he always wants to win and takes it very badly whenever the judges' decision goes against him.'[29] Waldersee particularly derided Wilhelm II's practice of sending his Flügeladjutanten 'quite brazenly over to the enemy' before the manoeuvre began. They then galloped back with secret reports to convey to the Kaiser, 'who later boasts in all seriousness of having been supplied with excellent information'.[30]

As ill luck would have it, on 19 September the Kaiser insisted on commanding the VI Army Corps, quite unaware that according to the plan for the manoeuvre drawn up by Waldersee this very corps would in all probability be thoroughly defeated. This was indeed what happened, particularly as the Kaiser's troop dispositions were 'extremely bad': he finished up with the river Neisse separating the two divisions he commanded.[31] The Chancellor-General also acknowledged that the Kaiser had made the unfavourable position of his side in the exercise even worse by 'one or two gross errors'.[32] The Chief of the General Staff's private verdict on Wilhelm's competence in command was terse. 'Extremely inadequate leadership, immature ideas, inexperience combined with a very self-confident manner. Playing to the gallery, in other words – a mere game, and no seriousness!'[33] Many of those taking part, among them the King of Saxony and the Hereditary Prince of Saxe-Meiningen (the Kaiser's brother-in-law), Waldersee recorded, had greeted the defeat of the monarch in the manoeuvres as a 'great stroke of luck' which might make him more circumspect in future.[34]

At the end of the exercise the hapless Waldersee was obliged to give a critical assessment of the manoeuvre in the presence of the Emperor of Austria, King

Albert and Prince Georg of Saxony, Prince Ludwig of Bavaria, the Austrian General Beck, Adjutant-General von Wittich and countless other officers and Flügeladjutanten. He did his best to mention all mistakes without giving offence. 'The Kaiser, whom people were of course watching closely, is said to have looked rather surprised and then very serious', Waldersee wrote. 'When I had finished he took over. He said first that he agreed with all I had said, but then began trying to make excuses for himself. Unfortunately his explanations were very weak, indeed lame. In the evening I could see that he was rather out of sorts and I also heard that he had been very angry and was in the mood to put the blame for his poor command on me.'[35] All those present noticed that Wilhelm did not shake hands with Waldersee as usual. A clumsy attempt by General von Versen to redeem the situation by flattery failed completely. Kiderlen reported that 'a high-ranking person' summed up the scene with the words: 'Versen was trying to lick H.M.'s boots but he wasn't having any!'[36] Wedel thought his rival Waldersee's assessment of the manoeuvre a masterpiece, as it 'dressed up strong criticism of the Kaiser in such a tactful form', but he suspected that the monarch had been deeply wounded by it nonetheless. 'If only Waldersee had spoken out as clearly and distinctly from the very beginning', the Flügeladjutant sighed, 'what a good effect he might then have had upon the Kaiser's military development! As it is, two years have gone by during which the Kaiser has constantly had his praises sung for his "talent as a military commander", while his most blatant mistakes have been passed over instead of being pointed out to him. Is it any wonder, particularly with his character and with the great gifts that he undoubtedly possesses, that he has gone astray and has finally come to believe in his own much-praised "talent"! Is it any wonder, and is not the psychological explanation only too easy to see, if he finds today's criticism disagreeable if not insulting, however tactfully expressed?' That evening Wedel heard that 'the Kaiser spoke very disparagingly of Waldersee's assessment'. Waldersee had 'risen somewhat' in the estimation of Wedel himself, 'although his star may have waned in the All-Highest circle because of his criticism'.[37]

When Wilhelm II returned to Berlin from Austria on 9 October he had reached 'a critical point', in the view of many observers, which Waldersee defined as follows: 'If he bears me a grudge because of his poor command of troops he shows that he lacks greatness of spirit and deserves to be treated as a little man. But if he has allowed himself to reflect calmly and to examine himself a little, he cannot but be grateful to me and tell himself that I acted without fear and did my duty. Then one could say: there is nobility in him after all.'[38] Although the newspapers were full of rumours of Waldersee's dismissal after this incident, Wilhelm II showed great friendliness at their next few meetings. The Chief of the General Staff nevertheless suspected that the Kaiser planned to move him to another post in the spring of 1891. 'Perhaps the Kaiser . . . will want to get rid of me then', he wrote, for Wilhelm was after all 'clever enough to think that it is not yet the right moment to make the break'.[39] Carl Wedel also realised that 'Waldersee's star is really sinking now.' 'Knowing the Kaiser's determined will, I am convinced that he has made up his mind to send him [Waldersee] away and that he will carry out this intention fairly soon, in one way or another. Who would have thought that the catastrophe would come so soon!?'[40]

WALDERSEE'S DISMISSAL

As several weeks passed without any change in the Kaiser's behaviour towards him, Waldersee's anxieties were gradually stilled. But in mid-December 1890, when Wilhelm gave his approval to Caprivi's directive that military attachés should abstain from political reporting, the Chief of the General Staff's feeling that he no longer possessed the Kaiser's confidence was reawakened.[41] Bitterly he complained: 'As usual Hahnke has not understood what is happening and has failed to warn the Kaiser of the consequences, so that the Kaiser has now decided to do exactly the opposite of what he considered to be right in the spring.' Waldersee was also disappointed with the attitude of Minister of War von Kaltenborn, who was likewise annoyed about the directive but claimed that, 'since the Kaiser has decided', he could do nothing.[42] On 16 December Waldersee had 'a conversation that at times became somewhat heated' with the Kaiser on the subject. Hahnke was also brought in on the discussion. The monarch, with Hahnke's backing, claimed that those attachés who were also Flügeladjutanten – that is to say the most important of them, in St Petersburg, Vienna, Paris and Rome – still had the right, and indeed the duty, to report directly to him, and he intended to 'impress this on them thoroughly now, as he would be seeing them again for the New Year'. Waldersee was perplexed by this and noted in his diary that it showed 'that the Kaiser does not know exactly what is in the directive, and that Hahnke has not understood it at all either', for it was precisely Caprivi's aim to cut off direct and indirect contact between the military attachés and the Kaiser. On Hahnke he commented: 'And this is supposed to be a Cabinet chief and adviser? It is truly pitiful!'[43] After this argument with Waldersee the Kaiser loudly declared: 'I've just repulsed an attack on Caprivi. There is no hope whatever of driving a wedge between C. and myself. I know perfectly well why Waldersee always tries to do so: he'd very much like to become chancellor himself.'[44]

Waldersee felt he understood the inner conflict which was now troubling Wilhelm. Writing at the beginning of 1891 he remarked: 'I can often tell that the Kaiser is struggling with himself inwardly. Partly he is his old self, and tells himself that I am too; partly his vanity makes him feel uncomfortable with me; partly he still feels stung by the business of the imperial manoeuvres.'[45] Others occasionally noticed 'an irritable mood' towards Waldersee on the Kaiser's part, for instance when he discovered from a newspaper that the General had been present at Stoecker's farewell sermon.[46] In mid-January 1891 Waldersee detected a further sign that the final crisis was near: he was no longer invited to dinners given by other people for the Kaiser. As the guest lists were always submitted to the monarch and Wilhelm made ample use of the opportunity to strike out the names of those he did not wish to see, the Chief of the General Staff had the distinct feeling that the Kaiser was behind his exclusion from these occasions.[47]

The precise moment at which the Kaiser decided to dismiss Waldersee as chief of the General Staff was recorded by the Flügeladjutant Carl von Wedel, to whom Wilhelm recounted in some agitation a conversation he had had with the Chancellor and the moderate leader of the Conservatives, Otto von Helldorff. The latter had complained that he no longer had any influence in the party, since he had been 'completely pushed out' by Count Waldersee. Asked by the Kaiser for his views, Caprivi then said that he 'considered it absolutely essential to send Waldersee away from Berlin at least for a time, because he would not stop his scheming'. The Kaiser had thereupon decided to move Waldersee to another post.[48] He nevertheless maintained a markedly friendly tone in his personal contacts with his Chief of Staff, so that the end came as an unpleasant shock for the latter.

During Waldersee's regular audience on 24 January 1891 the conversation came round to the imperial manoeuvres planned for the following autumn. They were to take place in Bavaria following the manoeuvres in Austria. It emerged that the Kaiser, without asking anyone, had fixed on 9, 10 and 11 September for the Bavarian manoeuvres, on the assumption that the Austrian manoeuvres would take place from 2 to 8 September, immediately *before* the German ones. In fact, however, the Austrian manoeuvres had been planned for 11 to 15 September. The Chief of the General Staff's attempts to bring forward the Bavarian manoeuvres were thwarted by the Bavarian Minister of War's urgent request that the appointed dates should not be changed. Prince Regent Luitpold, he explained, for whom it was 'not at all convenient that the Kaiser was coming', refused to hear 'so much as another word about the manoeuvres'. At his audience Waldersee therefore asked Wilhelm II to accept the Austrian invitation with a day's postponement. The Kaiser, however, stood firm and ordered Waldersee to write to the military attaché Adolf von Deines in Vienna saying that Wilhelm could come to Austria only from 2 to 8 September, and that he therefore requested that the manoeuvres there be brought forward. The reason for the Kaiser's uncompromising attitude was, as Waldersee despairingly recorded in his diary, that he wanted '*come what may* to go hunting on the Rominter Heide on 23 September!' The General was 'thoroughly depressed' as he left the audience. 'Stags, in other words pleasure, take first place!', he sighed. 'And such great interests are harmed by this desire to pursue pleasure!'[49]

That same evening the Kaiser asked Waldersee to 'return to him certain private papers concerning him which he had given me for safekeeping in 1888, and said the reason was that he had now obtained an iron safe. He seems to have had a dark suspicion that it might be awkward for him if these papers remained in my possession after our separation', Waldersee noted in a later account of his dismissal.[50] In spite of this omen the Chief of the General Staff was taken completely by surprise when the Kaiser, after conferring on him the Cross of Grand Commander of the Order of Hohenzollern on 27 January 1891, Wilhelm's thirty-second birthday, announced that he wanted to make Waldersee happy, and that he therefore intended to give him command of an army corps. During a meal with his officers – among them Count Alfred von Schlieffen and the future Minister of War, Heinrich von Gossler – Waldersee made up his mind to ask the Kaiser to be allowed to retire from the army. He was determined 'to make my view of the situation very plain . . . I believe I shall be doing a service to the army and to the Kaiser himself if I act firmly and show him that there are still people who do not simply submit to his will without further ado', he wrote.[51]

The Kaiser told Wedel during a drive in the Grunewald that in dismissing Waldersee he wished to show 'that the Chief of the General Staff was only his agent and as such was certainly not in command of the commanding generals'. In Wedel's view, however, Wilhelm was merely deceiving himself about his real motives in bringing out this argument. On the other hand he agreed with the opinion expressed by the Kaiser that Waldersee no longer enjoyed the confidence of the army as a whole. He also recorded that all the military officers in the inner circle at court had the impression that 'since the criticism [by the Kaiser] of the work of the General Staff in the previous spring [1890] relations [between Wilhelm and Waldersee] had changed, and also that there was a general feeling that the coldness between them had grown since Waldersee's criticism of last autumn'. Wedel gave no credence to Wilhelm's claim that he had not taken Waldersee's criticism amiss. 'So now Waldersee has also met his downfall', he reflected; 'and another star of the new era, which shone so brightly at first, has set!'[52]

In the course of a violent argument which took place on 28 January 1891 in the Berlin Schloss Wilhelm refused point-blank to accept Waldersee's resignation from the army. The Chief of the General Staff listened in astonishment as the Kaiser sought to justify his transfer to the army corps at Altona on the grounds that both Bismarck and the 'socialist conspirators' who had their headquarters in

Hamburg must be kept under surveillance, and that developments at the North German courts were beginning to take a 'very alarming' turn; moreover Waldersee's wife (the widow of Prince Friedrich of Schleswig-Holstein-Sonderburg-Augustenburg, Prince of Noer) would surely be glad to return to her old homeland. And finally, the Kaiser explained, it was his wish to lower the status of the chief of the General Staff in relation to the Prussian minister of war and the commanding generals. As he put it, 'the chief of the General Staff is to be no more than a kind of amanuensis for me, and for that reason I need a younger man'. Waldersee was undeterred, and told the Kaiser bluntly that he was convinced he would be doing him and the army a greater service by resigning than by staying on. 'Whether he grasped the meaning of these words I do not know, but in time he will understand.' After the audience Waldersee met his predecessor, Field Marshal Count von Moltke. He too found the proposed transfer of Waldersee to Altona 'quite incomprehensible, and expressed great concern about the Kaiser'. Waldersee reflected with distress that 'the saddest thing' about his dismissal as chief of the General Staff was the fact 'that everyone agrees with me – even the stupid Hahnke – that the Kaiser's misfortune at the manoeuvre is the real reason why he wants to get rid of me and that Caprivi has made skilful use of this. The whole world will see it as we do. It will be so damaging for the Kaiser!' He saw the future as 'very black'. 'The Kaiser wants to be his own chief of General Staff! God protect the Fatherland!'[53]

Waldersee's removal was regarded by many, not only in the army and in the Conservative Party, as a mistake of considerable political significance, although others naturally greeted it as a great relief. The old Field Marshal Count von Moltke, the King of Saxony and Kaiser Franz Joseph all expressed deep regret at his departure. For the first time for three years Waldersee was received kindly by the Empress Frederick. Wedel, however, saw the dismissal as a positive step, for Waldersee had always been 'a corrosive element' who interfered in everything 'using all kinds of underhand methods' and who was filled with 'immeasurable ambition'. Chief of the Military Cabinet General von Hahnke told Wedel that 'Waldersee had been in league with Verdy to bring down not only Bismarck but also him [Hahnke].'[54]

When he received Waldersee's resignation letter on 31 January 1891 the Kaiser became 'extremely agitated' and hurried to Caprivi, who gained the impression that Wilhelm was toying with the idea of keeping Waldersee on as chief of the General Staff after all. 'Caprivi sees it as a sign of the Kaiser's tender heart', Wedel wrote after a conversation with the Chancellor. Both generals – Caprivi and Wedel – nevertheless agreed that the Kaiser must stand firm. 'If Waldersee stayed in his present post', Wedel commented, 'it would be . . . no more than papering over the cracks. Before long the Kaiser would certainly feel that it was a humiliation to have given in, and in a few months an abrupt break would be more than likely – the Bismarck episode is a very instructive precedent for this.'[55]

On the same day Waldersee had a 'fierce battle' with Wilhelm II, which in many ways resembled the violent argument between the latter and Bismarck on 15 March 1890. Waldersee was summoned to make his report, and when this was over he asked for Hahnke and Wittich to be sent away as he wished to speak to the monarch alone. Face to face with the Kaiser, he told him that an independent chief of the General Staff was a source of strength for the king, for on the one hand both the minister of war and the Reich chancellor were dependent on parliamentary majorities and could therefore easily be brought down; but, on the other, both were also in a position to play off these majorities against the monarch. He declared bluntly that 'hardly anyone ever dares give a frank opinion to Your Majesty; I therefore think it my duty to do so now', whether or not it was hurtful to the Kaiser. First, to his regret, he had to say that since the accession the army had deteriorated. 'The ideal relationship between the war lord and the officer corps which he had inherited no longer existed; there was a prevailing sense of uncertainty, ill-feeling and lack of zest for service in the upper echelons, and this mood had spread quite a long way downwards, while at the bottom a youthful element was growing up which was anything but well-disposed', for respect for authority was disappearing because of the rapid changes in senior appointments. Yet the principle of authority was the foundation on which the army was built, and he, the Kaiser and King, 'depended entirely on the support of the army'. Wilhelm replied: 'No one has ever said anything like that to me', whereupon Waldersee retorted: 'General von Hahnke ought to have done so long ago and I blame the former Minister of War [Verdy] for this too. Neither of them had the courage to do it.' The Kaiser again tried to persuade Waldersee to accept the Altona command: he would show the world, he said, 'what it meant to be a friend of the German Emperor. Anyone who dared say a word against me would be destroyed; he would send the press packing and so on and so forth, in a way that would scarcely have impressed a lieutenant. Eventually he even begged me to agree, with the tenderest of gestures: he took my hand and said: "You will accept, won't you? Your Kaiser begs you." But I remained adamant and I thank God that he gave me the strength to do so', Waldersee boasted. The General went on to tell the Kaiser that the whole crisis had been instigated by the Reich Chancellor, who not only wanted him away from Berlin but was also trying to bring the General Staff under the control of the War Ministry. 'His aim was also to reduce the minister of war to a secretary of state', Waldersee continued, 'in the same way as he was trying to reduce the power of the High Command of the Navy in favour of the secretary of state, so as to have a free hand in the army and the navy.' The Kaiser strongly disputed this assertion, but Waldersee maintained that he was right, and named not only Caprivi but also Holstein, Kiderlen, Lindau and Raschdau 'and all the rest of them' as his enemies, who had been working for his downfall for weeks. At the close of the interview the Kaiser became 'elegiac' and said 'It is too sad to think of what I have already had to go through. My best friends are deserting me!' Waldersee returned home 'very depressed' by the interview, and consoled himself with the thought that it was the carnival season, so that it was no wonder 'if madness were in the air, for the whole business is crazy and senseless. I am to leave the job which I do well, so as to make way for someone more insignificant.'[56]

Waldersee's frankness had its effect – at least temporarily – on Wilhelm. When August Eulenburg called on the departing Chief of Staff on 1 February he too was 'very depressed about the Kaiser' and told Waldersee that after his audience the monarch 'had been in a more serious mood than he had ever seen him in before; he had cancelled all other audiences and had given everyone in his entourage the impression of being in a highly emotional state'.[57] Although Waldersee was increasingly inclined to blame his downfall on the hostile machinations of Caprivi, who had 'made skilful use of the Kaiser's sensitivity over his misfortune at the manoeuvres' in order to get rid of Waldersee, in his eyes the intrigue cast a harsh light on the character of the Kaiser himself, who was 'so gullible' as to have been convinced that he must send Waldersee away from Berlin 'in the interest of his reputation'.[58] The Kaiser had shown himself weak and indecisive. 'He could not summon up the courage to speak openly to me, so that we could part as good friends and in a way that would have done him no harm. The fact that in the end he did not even have the courage to tell me the real reason, and what I am supposed to have done – that is unbelievably cowardly; he was obviously too ashamed to make any accusation, because he knew how groundless it would have been. But it is profoundly sad to see what a weak man he is; he has allowed himself to be systematically manipulated against me without even noticing, much less having any sense of how unworthy this is.'[59]

Waldersee then accepted the post of commanding general of the IX Army Corps in Altona after all, which led Wilhelm to remark scornfully: 'I should jolly well think so, too – the man thinks himself as grand as a grand mogul.' When Wedel resumed his duties after a few days' absence and took a drive with the Kaiser, the latter gave him his version of the 'fierce battle' with Waldersee, exclaiming as he climbed into the carriage: 'Well, since I last saw you terrible things have happened!' The Kaiser talked for an hour about the events of the past few days. It was worst on 31 January, he said, when Waldersee had lectured him for an hour and a half. 'That was when he revealed himself in his true colours and showed what a great man he thinks he is', the Kaiser mocked. Waldersee had claimed that he could not accept the army corps 'because it meant a reduction in rank for him, for the chief of the General Staff has much greater weight than a commanding general', which he said Field Marshal von Moltke had confirmed. When Wilhelm retorted that this was not the case, and that the exceptionally high standing of the chief of the General Staff was entirely due to the personal prestige of Moltke, Waldersee replied 'that he had been the person who had reorganised the General Staff'; the high regard which the General Staff now enjoyed and its good relationship with the Ministry of War were entirely his own work. 'If he now took command of a corps the whole world would shout about it and he would be depicted as "fallen" and "removed", for he had held "a position in the world".' To this, as Wedel recorded, the Kaiser had replied 'that he had dismissed Bismarck without fear of the press' and that he gave Waldersee 'his word of honour that he would protect him as his friend against any attack, from the press or elsewhere'. Wilhelm categorically rejected Waldersee's 'vicious attack' on Caprivi, who Waldersee claimed had always hated him and had now brought about his downfall; he declared that 'the Chancellor, who was so calm and high-minded in both thought and action, and whom he, the Kaiser, had chosen for himself', had 'never said so much as a word against him, Waldersee'. He had objected equally vigorously to Waldersee's claim that there had been a conspiracy against him among the officials in the Foreign Office, who were now congratulating themselves on his fall. When Waldersee, 'having been evasive, was eventually pressed by the Kaiser into naming Holstein, Kiderlen and Lindau', Wilhelm declared that his suspicion was 'absolutely unfounded'. In the course of the conversation Waldersee had become 'more and more critical and impertinent towards him', the Kaiser complained. As a result Wilhelm had finally told him that the only reason why he had not been transferred to the Stuttgart command the previous autumn had been that the King of Württemberg had declared that 'they could send him any general so long as it was not Waldersee'. Although Waldersee had obviously been deeply hurt by this remark, he had become increasingly harsh in his criticisms and had finally accused the Kaiser of causing a feeling of 'extreme insecurity' in the army. 'At this the Kaiser indignantly interrupted the Count and asked him how he, who claimed to be his friend and who enjoyed his full confidence, could have concealed such a fact from him for so long, if it were true.' And when Waldersee had gone on to claim that the whole army would be 'very deeply affected' by his retirement, Wilhelm had replied that he was deceiving himself, for he knew from his 'very wide range of sources' that Waldersee did not enjoy the confidence of the army. The Kaiser had again listed the reasons why it was desirable for Waldersee to go to Altona: the particular honour of being in command of the Kaiserin's home corps; the need to keep the social democratic movement in Hamburg under surveillance; the fact that various contingents from small North German princely states belonged to this army corps; and finally the necessity of keeping an eye on Bismarck, 'the hermit of Friedrichsruh'. Waldersee had rejected all these arguments, again alluding to his 'position in the world', and on taking his leave had said: 'I am sure that Your Majesty will think the matter over again.'

Not long after this heated argument, as the Kaiser went on to relate, he had received from Waldersee a letter 'whose impertinence exceeded all bounds, the like of which had probably never before been written by a Prussian general to his king'. The letter ended with a request to retire from the army. Immediately afterwards Hahnke came to see the Kaiser with the news that he had just received a visit from an intermediary sent by Waldersee to make the necessary arrangements for him to take over the army corps in Altona. The Kaiser showed Hahnke Waldersee's letter, whereupon the 'thunderstruck' Chief of the Military Cabinet brought his fist down on the table and exclaimed: 'Your Majesty must dismiss him at once; he really is going too far!'[60]

The Kaiser, 'as his friend', nevertheless stood by his decision to send Waldersee to Altona. But he gave vent to 'violent outbursts' over press reports containing details of their private conversation, which he believed could only have come from Waldersee. The latter's farewell audience on 4 February lasted barely two minutes, and afterwards the monarch merely remarked: 'Well, he seems to have calmed down.' After a conversation with his old rival Wedel wrote: 'Every word betrayed how swollen with venom Waldersee was. He had probably felt too secure ever to dream of such a downfall.'[61] If the fall of his adversary gave the Flügeladjutant any malicious pleasure, for which he could hardly be blamed, the pleasure was short-lived, for only a few weeks later he too was forced to leave his post at court.

COUNT VON WEDEL'S TRANSFER TO THE FOREIGN OFFICE

As we have already seen, Count Carl von Wedel played an exceptional role among the Flügeladjutanten in the first years of Kaiser Wilhelm II's reign, possibly because he knew certain secrets of the Kaiser's private life from his time as military attaché in Vienna.[62] Philipp Eulenburg was quick to recognise Wedel's important position in the imperial entourage but was relieved to establish that Wilhelm had no 'inward interest' in the General.[63] His other rival, Waldersee, condemned Wedel as a 'blatant egoist' who had been 'foolish' enough to 'give himself great airs and push himself forward when he became a serving Flügeladjutant; he thought he could supplant General Wittich at that time'.[64] The mere fact that Wedel loyally supported Bismarck, then Caprivi and finally the Oberhofmarschall Liebenau too, was sufficient to arouse Waldersee's anger. Immediately after Bismarck's fall the Chief of the General Staff remarked maliciously that Wedel had 'become quite a changed man since he saw the Chancellor fall; he gives the impression of being completely defeated. The Kaiser has at last recognised his falseness and wants to be rid of him; unfortunately for him the crisis happened too quickly, so that he could not distance himself from the Chancellor; otherwise he would no doubt have disavowed him.'[65] Waldersee's gloating prophecy turned out at first to be a grave mistake. Wedel not only held out until the summer of 1891, but he had daily discussions with the Kaiser on vital questions of foreign and domestic policy and public appointments, was sent on confidential missions to Friedrichsruh, Vienna, Copenhagen and Stockholm, and accompanied the Kaiser on his trip to Russia in August 1890. All in all, he played a significant role in the first years of Caprivi's chancellorship, and not least in Waldersee's dismissal.[66] His position at the Hohenzollern court was even strengthened temporarily by the appointment of his cousin Ernst as Oberstallmeister (senior master of the stables) to the Kaiser in the autumn of 1890.[67] But then he too found his time in the imperial entourage running out. Wedel was reluctantly obliged to leave court service for the diplomatic service because he too had spoken his mind too often and too freely.

The strange story of how the transfer of a Hanoverian general and Flügeladjutant to a diplomatic career came about, and what role the personal wishes of the Kaiser played in it, emerges from Friedrich von Holstein's correspondence with Philipp Eulenburg and from a retrospective memorandum written by Wedel himself in the summer of 1894. According to Holstein, on 10 May 1891 the Chancellor received a telegram several pages long from the Kaiser, who was in Darmstadt. In it he claimed that the former military attaché in Paris, Ernst Freiherr von Hoiningen-Huene, with whom he had had a discussion in Karlsruhe about French intentions to begin a war, was better informed about France 'than the entire present embassy put together'. He therefore demanded that the ambassador Count Münster, whom he had long since condemned for 'hollow phrase-mongering with no serious background' and 'naivety and childish trust', be immediately replaced by his Flügeladjutant Carl von Wedel.[68] The Chancellor's first response was to ask for an audience, but he was determined to resign if the Kaiser insisted on sending a general to Paris, which in the eyes of the world would look like a threat of war. Holstein warned that throughout Germany people would say: 'Prince Bismarck is right when he says that His Majesty is mad.'[69] At his audience Caprivi succeeded in persuading the Kaiser to postpone Münster's recall until the autumn – in fact he remained ambassador in Paris until 1900 – but Wilhelm insisted that Wedel be appointed to the foreign service anyway. The Reich Chancellor sent for the Flügeladjutant and told him that for a long time the Kaiser had been considering appointing him to an embassy. Wilhelm's wish was also partly prompted, the Chancellor revealed, 'by his intention . . . to place another, younger general in command of His headquarters', and Wedel was an obstacle to this plan. There was, however, a problem with the choice of the embassy to be allocated to the General. Recently the Kaiser had been so strongly influenced against Münster in Paris that he had wanted to sack him and put Wedel in his place. But the Chancellor had political scruples about sending a general as ambassador to Paris, 'because an ulterior motive could easily be seen in it'. Caprivi had therefore advised the Kaiser to send Wedel to St Petersburg instead of Paris, for the recall of the ambassador there, Lothar von Schweinitz, would fall due soon. But the Kaiser had wanted to stick to his original plan, 'and so he, Caprivi, did not wish to raise any further difficulties; as far as he was concerned the French could think what they wanted'. Nevertheless the Chancellor had suggested to the Kaiser that Wedel should be seconded to the Foreign Office as a kind of transition, because this would somewhat reduce the sensitivity of his appointment as ambassador in Paris later. The Kaiser had eagerly seized on this suggestion and had ordered Wedel's secondment on 15 June 1891, the third anniversary of his accession, the Chancellor stated.

The unhappy Flügeladjutant then admitted that he too had sensed for some time that the Kaiser was 'not comfortable' with him. But he did not feel in any sense qualified to be ambassador in Paris and wanted to discuss his future with the monarch himself. That same evening Wedel went to the Neues Palais to take his turn in waiting as Flügeladjutant. The Kaiser 'soon appeared in our office, dictated a few telegrams to me and then asked me, as Major von Scholl was present, whether I could spare some time (!) as He would like to discuss something with me. We went into the next room, the one in which Kaiser Friedrich died, and He began the conversation by saying straight out that it was His aim to make use of me in *higher* diplomacy, for which not only He, but also others, thought me particularly suited. When I protested at this He remarked that I could not deny that I had a better understanding in this domain than the others, which was why He had often discussed politics with me. In the course of the conversation H.M. told me He had decided to recall Münster, and then He declared that he intended to send me to Paris in place of Münster, while reserving the right to send me to Petersburg later, when Schweinitz left, as He must be represented by an adjutant-general there.' After Wedel had asked for time to think it over, Wilhelm ended the interview with the remark that in the event of his acceptance he would second Wedel to the Foreign Office initially. Three days later, while driving with Wilhelm from the Neues Palais to the carnival of flowers in Westend, Wedel told the Kaiser that he did not feel equal to the post of ambassador in Paris. The Kaiser did not raise any objections, Wedel recalled, but asked him 'what I would think of St Petersburg, although he was not yet planning to recall Schweinitz from there', whereupon the General replied that he would 'at least venture an attempt' there, although he had doubts about his suitability. He was quite prepared to be seconded to the Foreign Office by way of transition – a step which, as he admitted in his memorandum of June 1894, he had since 'seen only too clearly as a grave mistake, and regretted'.[70] During a dinner to mark his departure from court service on 19 June 1891, at which the Kaiser appeared, the latter referred to Wedel several times as ambassador and remarked: 'Who knows, maybe he will be the ambassador who will spark a war for us one day.'[71]

Following the autumn manoeuvres in Austria in September 1891, Caprivi asked Wedel whether he would consider taking on a post as an envoy for a short time. To his 'utter astonishment' the Chancellor then told him on 18 October that 'H.M. has commanded that I should go to Belgrade as envoy.' The next morning Wedel turned down this 'second-class' post and asked the 'visibly embarrassed' Reich Chancellor 'in very definite terms' to be allowed to return to the army.[72] When he told the Kaiser that evening after a dinner at the Neues Palais that he had refused the Belgrade posting, Wilhelm replied: 'My dear friend, do not imagine that it was my idea. In your place I should also have refused, and there is an end to the matter.' Nevertheless the Kaiser rejected Wedel's urgent plea to be allowed to return to the army with the remark that something would certainly be found for him.

Wedel's transfer to the the diplomatic corps was not exactly welcomed at the Foreign Office. The influential Privy Councillor Holstein complained that 'he did not know how he was going to get rid of him [Wedel], and spoke very disparagingly about the generals who were being given diplomatic appointments on All-Highest orders, and were thereby ruining the service and putting off decent people from going into diplomacy'.[73] Wedel's secondment caused even more alarm among diplomats themselves, when Eulenburg discovered that he even had 'a sort of promise from H.M. of an embassy'.[74] Yet although Wedel's position in the Foreign Office became 'increasingly difficult and almost laughable', Chief of the Military Cabinet von Hahnke persisted in blocking his return to the army on the grounds that the Kaiser wished to obtain a suitable diplomatic post for him. It was not until 19 June 1892 that Wedel managed to get himself invited to luncheon at the Neues Palais through the good offices of his friend Friedrich von Scholl, who was on duty as Flügeladjutant at the time. While he was in the ante-room removing his sword, Wilhelm came in and asked 'Well, what is the matter?', to which the General replied that his position at the Foreign Office had become impossible and that he must therefore beg to be put in command of a division. By way of an answer the Kaiser asked whether Wedel would be prepared to accept the post of envoy at Stockholm as '*a short transition* to an embassy', adding that 'I would be doing him a favour if I accepted the offer, as he had very few capable people in the foreign service and I showed a particular gift for diplomacy.'[75] Much against his will Wedel agreed to take on the Stockholm post, and then only after the Kaiser had 'firmly and solemnly' repeated his promise that this post would be 'only a short transition' to an embassy. 'H.M. immediately afterwards spoke to the Secretary of State [Marschall von Bieberstein], who was also at the luncheon.'[76]

The next day, while riding in the Tiergarten, Wedel informed Caprivi of his agreement with the Kaiser. An uncomfortable conversation ensued. Caprivi made it plain that he still considered Wedel's refusal of the Belgrade post the previous year a mistake; since then he had abstained from any intervention, nor could he give Wedel any advice now. Wedel pointed out that his own hands were tied by the Kaiser's solemn promise, 'as I certainly could not have told him that I did not believe his promises'. The Chancellor agreed but observed that no embassy would be available for the foreseeable future. 'Vienna, where he would have preferred to send me and where I should be *persona gratissima*, was not free and one could not turn out an old and meritorious public servant like [Prince Heinrich VII] Reuss.' On 24 June 1892 Caprivi summoned the General and informed him that the Kaiser had appointed him envoy at Stockholm and had also directed that he 'should go to Vienna later in place of Reuss'. When and how Reuss would leave his post was still undecided, but the Prince might perhaps be offered the St Petersburg embassy. Wedel pointed out that this solution was impossible as Reuss was married to Princess Maria (Sitta) of Saxe-Weimar-Eisenach, whose grandmother was a Russian Grand Duchess. And it was unlikely that he, Wedel, could any longer be considered a candidate for St Petersburg either, as he had heard that in the previous year the Tsarina of Russia, whose sister was married to the Duke of Cumberland, had 'made extremely derogatory remarks about the Kaiser and had described it as a callous act on his part to entrust me, a Hanoverian, with a mission to the Danish court. In view of the influence the Tsarina had over her husband, her attitude needed to be taken very seriously, for in consequence I could hardly be considered persona grata in St Petersburg', Wedel declared.[77]

Wedel took up his post in Stockholm with bad grace and consoled himself with the thought of the 'Vienna post officially promised' to him.[78] According to Eulenburg, who was '*incensed* by such ingratitude towards the Kaiser', Wedel complained to a friend that 'the Kaiser mistreats me, the Foreign Office likewise. I have *never* aspired to anything but a division. (!!) I am thrown out and given a *paltry* legation like Stockholm.' Eulenburg's Swedish wife also heard that Wedel had tactlessly told people in Stockholm that he would be staying there only for a very short time.[79] He was so sure of his imminent nomination as ambassador in Vienna that he ordered a magnificent dinner service adorned with his monogram and a count's coronet from a porcelain shop in the Austrian capital. 'That is for Count Wedel, who is to be German ambassador here!', the shop's proprietor told the astonished Princess Reuss when she enquired after the owner of the service.[80]

Wedel was deeply dismayed when he heard soon afterwards of the appointment of the Hofmarschall Hugo Prince Radolin as ambassador at Constantinople. He rightly attributed this – to him – surprising development to Radolin's 'particular friendship [with] the camarilla in the Wilhelmstrasse', of which Holstein was the ringleader. Wedel was again passed over when Schweinitz was replaced at St Petersburg in January 1893 by General von Werder. When he spoke to the Kaiser in April that year in Berlin, Wilhelm was at pains to point out that it had not been his fault, remarking: 'Well yes, in fact I did want to send you there, but I had to accept the direct request of the Tsar [for Werder]. But in any case it is perhaps quite a good thing, for as a Hanoverian you would not have been persona grata to the Tsarina.' Neither the Kaiser nor the General suspected that the 'direct request' of Tsar Alexander for General von Werder had been engineered by the 'camarilla in the Wilhelmstrasse'.[81] But Wedel's greatest disappointment, of course, came when the Vienna post which he so ardently desired was allotted not to him but to Count Philipp zu Eulenburg, the Kaiser's best friend and at that time still in very close collaboration with Holstein.

Rumours began to reach Stockholm as early as the autumn of 1893 that Eulenburg was behaving quite openly as future ambassador in Vienna. His close acquaintances maintained that he had set his sights on this post because Vienna was the only place where he could have his children educated. When Wedel came to Berlin in November 1893 he called on the Reich Chancellor to challenge him on these rumours. Caprivi, visibly embarrassed by Wedel's complaints, responded that he knew nothing of the matter, although he had recently heard a remark by the Kaiser to the effect that Eulenburg would be 'very well received in Vienna'. Four weeks later, when he took leave of the Chancellor on his return to Stockholm, Caprivi offered words of reassurance to the disconsolate envoy: '*I give you my word* that you will become an ambassador. You are now first in line, as Count Berchem has refused the post in Washington, but there is nothing available and Vienna is out of the question for you.' With great bitterness Wedel reminded the Chancellor of his promise regarding the Vienna embassy when he had accepted the Stockholm legation in the summer of 1892. 'I had not doubted that the Kaiser would forget about it; his [Caprivi's] words, however, had given me a guarantee.' The Reich Chancellor was visibly discomposed at this, Wedel later recalled, and replied with a shrug of the shoulders 'that he was powerless where Count Eulenburg was concerned, and I must content myself with another embassy'. Rome was a possibility in the near future, for Count Solms had 'lost favour' – he had aroused the Kaiser's anger during the latter's visit to Rome by failing to tell him of a garden party given by an English nobleman[82] – and was likely to be recalled soon. No sooner had he returned to Stockholm in December 1893, however, than Wedel was astounded to hear that Solms had been replaced by the former envoy in Bucharest, Bernhard von Bülow. This new blow was the final proof for Wedel 'that the highest official in the Reich has no influence . . . at all' and that his promises were therefore worthless. Wedel bitterly told the Chief of the Military Cabinet that he had decided to resign the following summer. Hahnke broached the subject with the Kaiser, pointing out that Wedel had reason to be surprised that preference had been given to Bülow, who was several years younger, whereupon Wilhelm replied: 'Count Wedel will not be forgotten, but I have to be free to appoint ambassadors as the need dictates; moreover as a diplomat W. is younger than Bülow.'[83] As it turned out, the former Flügeladjutant remained in his post in Stockholm but eventually achieved promotion to the higher echelons of the diplomatic service in 1897, when he succeeded Bülow as ambassador in Rome. Thereafter he was appointed to Vienna in 1902 in succession to Philipp Eulenburg, finally becoming governor of Alsace-Lorraine in 1907. A few weeks before the outbreak of war in 1914 Kaiser Wilhelm II created Count Wedel a Prince (Fürst) with the title 'Highness'.

When Wedel looked back over the two and a half years he had spent as Flügeladjutant, in the course of a conversation with Hahnke in May 1891, he agreed with the latter that 'at the beginning everything had gone splendidly, until I opposed the Kaiser in the Bismarck and Liebenau affairs . . . which convinced him that I did not approve of everything he did. There is certainly something very peculiar about the favour of our All-Highest Master.'[84] Wedel affirmed that he had long felt 'that I was an embarrassment to the Kaiser and I fully accepted that he, as sovereign, had the absolute and exclusive right to choose his entourage entirely according to his own preference.' Hahnke disclosed to him that various people, especially Waldersee, had been intriguing against him. In his record of the conversation Wedel added that he had felt for a long time that even Hahnke no longer took his side. He went on to comment: 'I am glad that I am out of it all, for my character does not suit the Kaiser! He demands creatures for whom he has no respect but with whom he is at ease. At least I am leaving with a clear conscience in the knowledge that even if it was not always permissible or possible for me to say what I thought, I never consciously said anything that I did not think, nor ever agreed to any opinion or action that I did not approve. Furthermore in the 2 ½ years of my service it was my constant and to a certain degree successful endeavour, regardless of whom I had to deal with, to curb the uncouth tone and the indecent stories circulating in the Kaiser's entourage . . . My intentions were honest and true, but honourable people do not suit the young court.'[85]

THE DEPARTURE OF ADJUTANT-GENERAL ADOLF VON WITTICH

Shortly after Waldersee's dismissal as chief of the General Staff and the transfer of the reluctant Wedel to the diplomatic service, Kaiser Wilhelm II dismissed the second Adjutant-General whom he had appointed at his accession – the first, Max von Versen, having already been transferred to Metz in March 1890.[86] Unlike Wedel, Adolf von Wittich was sent back to active service as commander of the XI Army Corps in Kassel. In the first years of the Wilhelmine era both he and Versen had exercised a controversial influence both on the decisions of the Kaiser and on the prevailing tone of his immediate entourage. Wittich, it is true, had from the beginning found it difficult to adapt himself to life at court, and soon after the accession of the young Kaiser Carl Wedel had confided to the Austrian ambassador that the Adjutant-General had been quite unable to 'reconcile himself to the official duties of his present post'. Wittich constantly complained of the 'great, pointless waste of time that the numerous court festivities caused him' and consequently suffered from 'nervous irritability which became increasingly apparent every day'. It could not be denied, Wedel maintained, that 'General von Wittich is cut out for a purely military sphere of activity, and the sooner he gets it the better.' Not only Wittich himself and the Flügeladjutanten around him, but also Kaiser Wilhelm, felt that his 'personal qualities are not those which the duties of his present post require'.[87] Waldersee likewise recorded as early as August 1888 that Wittich was not happy at court. 'He is under immense strain, so that he will scarcely be able to endure it for long, and has little taste for court life', he wrote.[88] The root of the problem lay in Wittich's personal relationship with the Kaiser. As Waldersee remarked, the Adjutant-General was 'of a critical disposition, and is therefore inclined to raise objections; the Kaiser knows this very well and contradicts him frequently and apparently deliberately'. At any rate Wittich could not claim to 'lead [the Kaiser] in any way'.[89] 'That Wittich has no position at all, or rather has one which is unworthy of a man of any competence, is now clear to everyone', Waldersee noted in November 1888. 'Now he is to resume his lectures on military history for the Kaiser and he will do this very well, but to keep someone on as a serving adjutant-general just for this purpose really is an extraordinary luxury.'[90] The surprising thing about this state of affairs was that Wittich was kept on in his post at the imperial court until mid-1892. His own frustration showed itself meanwhile in a growing cynicism which had a very unfavourable effect on the character of the Kaiser and on the

20. General Adolf von Wittich, Adjutant-General and Commander of the All-Highest Headquarters 1888–1892

general tone of the court. In the summer of 1890 Waldersee commented that Wittich was responsible for many of the Kaiser's boorish remarks. It was 'really infuriating', he complained, how instead of moderating the Kaiser's disparaging utterances the Adjutant-General gave them 'even greater harshness'.[91] Wittich enjoyed 'biting criticism', Waldersee commented disapprovingly. 'He has not a good word to say for anyone and has a deplorable weakness for openly criticising and finding fault in the presence of younger officers... He has already done great harm by his way of behaving . . . for the Kaiser is very ready to accept criticisms of people and is himself very critical. Wittich is very largely responsible for the fact that the Kaiser has lost all respect for age and experience; he has greatly sinned!'[92] The Adjutant-General even committed the outrage, Waldersee wrote, of 'speaking in the most disparaging way about the Kaiser, and not in private among trustworthy people, but quite openly!'[93]

Rumours thus arose soon after his appointment as adjutant-general that Wittich would shortly be sent back to active service, whether as minister of war, chief of the General Staff or commander of an infantry division. When the fall of Minister of War von Verdy was in the offing in summer 1890 it looked for a time as if the Kaiser would indeed appoint his difficult Adjutant-General to the post. Waldersee reported that 'Wittich had prepared to make a lengthy speech as to why he did not wish to accept the post; as soon as he had begun, and the Kaiser realised what he was getting at, he said "I am truly glad that you do not want to leave me", clapped him on the shoulder and left him standing there with his speech.' Waldersee rightly suspected that Wilhelm was afraid of coming into conflict with Wittich as minister of war, and that he was relieved to be able to tell the Chancellor that the Adjutant-General had refused the post.[94] In May 1891 Wittich called on Verdy to discuss his future with him. As he had got wind of the conflict between Kaltenborn and Caprivi and had also noticed that the new Chief of the General Staff, Alfred von Schlieffen, 'is already beginning to bore the Kaiser',[95] Wittich had declared that he would never accept the post of chief of the General Staff, but that he would perhaps agree to take on that of minister of war, 'although only under certain conditions, such as private audiences with His Majesty, etc.!!' The experienced Verdy's comment on this was: 'I should be sorry if he were to take over this post; within a few weeks he would be in the most terrible conflict with the Kaiser and Caprivi!'[96]

In spite of these soundings Wittich's dismissal and transfer to Kassel in the summer of 1892 took everyone by surprise. Waldersee knew only that the Kaiser, who had suddenly lost his temper with the Adjutant-General, had declared: 'I have given that poltroon an A[rmy] Corps.'[97] The Empress Frederick regretted Wittich's departure from the court, for he was 'an honest, straightforward, trustworthy man – not afraid to speak his mind', although she doubted that he was particularly enlightened in his views or had much influence on Wilhelm.[98] The loss of Wittich also affected Chancellor von Caprivi, for whom he had been a source of strong support close to the Kaiser.

Wittich was replaced as adjutant-general by the elegant Hans von Plessen, who had served as Flügeladjutant to the old Kaiser Wilhelm from 1879 to 1888, and who was to remain 'a good comrade and a dear friend' to Wilhelm II from 1892 until the end of the Hohenzollern monarchy.[99] Even he, however, did not succeed in improving the general tone in the Kaiser's entourage. 'The Kaiser treats him with little consideration and has already attacked him rudely on several occasions', Waldersee noted in autumn 1894.[100] Plessen joined in the 'customary brash tone' at court and was 'not the man to influence the Kaiser in any direction'.[101]

The general opinion on the removal of Generals Versen, Waldersee, Wedel and Wittich from the entourage was that it must now be feared that, even more than hitherto, Wilhelm would decide what course to follow in all political and military matters on his own authority and without the advice of experienced older men. Immediately after Waldersee's fall the British military attaché, Colonel Russell, commented perceptively in a report to London that the Kaiser evidently wanted to be 'his own Chancellor, Chief of the Staff & his own Min[iste]r in every department – a task beyond any human Powers'. As far as public opinion

21. General Hans von Plessen, Adjutant-General and Commander of the All-Highest Headquarters 1892–1918

was concerned, people often expressed more regret over the downfall of the Chief of the General Staff than over that of Prince Bismarck eleven months earlier, Russell reported. The new man, Count von Schlieffen, certainly came from the same school as Waldersee, so that a change in military policy need not be expected; but Schlieffen was extremely reserved, did not have a strong personality and would therefore not acquire any great influence over Wilhelm II.[102] A few years after his fall Waldersee himself was distressed to have to record, after close observation of the imperial entourage, that all its members were afraid of the Kaiser and 'do not dare make the slightest protest or objection'. Wilhelm liked talking a great deal, and 'all agree with him in the most servile fashion'.[103] What was certainly true was that he had no experienced and courageous general at his side when in the spring of 1892 he was confronted with the biggest political crisis of the New Course. It was to lead to the separation of the office of Reich chancellor from that of Prussian minister-president and thereby to an aggravation of the crisis of government which had begun soon after Wilhelm II's accession.

CHAPTER SEVENTEEN

The School Bill crisis and the fragmentation of power

THE SCHOOL BILL CRISIS IN PRUSSIA

IT is not without irony that soon after Waldersee's departure from Berlin the otherwise liberal-minded Reich Chancellor veered towards a conservative–clerical course which most of his contemporaries found difficult to comprehend and which went much too far even for the fallen Chief of General Staff. This shift occurred exactly two years after Bismarck's dismissal and caused the most serious crisis of Caprivi's chancellorship. Two factors seem to have persuaded the Chancellor-General, who made no secret of his growing frustration, to make this dramatic change of course: the need to secure a majority in the Reichstag for the huge Army Bill ordered by the Kaiser in July 1891, and a defiant sense of solidarity with the newly appointed Conservative Prussian Minister of Ecclesiastical Affairs, Count Robert von Zedlitz-Trützschler, whom he had recommended to the Kaiser in the spring of 1891 as Gossler's successor. In shifting towards the right and against the liberal middle parties the Chancellor was unwittingly touching on one of the most sensitive points in the whole structure of the 'kleindeutsch' Prusso-German Reich, which had played an extraordinary role in the Kulturkampf and (as we have observed) also in the Bismarck crisis in 1889–90. This was the deep-rooted perception within the ruling elite, and not least at court, that both the Reich and the Triple Alliance could disintegrate if the Berlin government were to move closer to the Catholic Centre Party. In May 1891 Caprivi was reported to have stated to the 'very unpleasantly surprised' Kaiser that 'the only party from which we can seek support was the Centre, for it formed the only cohesive body!' Immediately Waldersee condemned this suggestion as 'a really hair-raising idea', for according to him the Centre consisted 'for the most part of notorious enemies of the Reich', on whom one could not possibly rely for support.[1] Caprivi's suggestion, the General wrote to his friend Verdy, 'certainly annoyed our royal master very much, but as you know that will not rule out his ordering a pact to be made with these people at a moment's notice'.[2]

As soon as he took over the Prussian Ministry of Ecclesiastical Affairs, Education and Medicine, Zedlitz-Trützschler personally set about devising a new Primary School Bill. It caused a furore when it was laid before the Prussian House of Representatives because it was of a strongly clerical nature and could have been passed only with the support of the German Conservatives and of the Catholic Centre Party against the liberals of all shades of opinion. It declared religion to be the highest goal of education and placed schools under the authority of the two main Christian churches, stipulating that the children of dissenters be compelled to attend religious education classes; the churches were given direct control over religious education in state schools and were also permitted to set up their own confessional schools.[3] The draft bill, which was seen as deeply reactionary, aroused unprecedented protests throughout Germany and above all among the nationalistically minded middle classes, who together with the East Elbian landed aristocracy formed the mainstay of the Reich. In liberal circles both right and left the bill was condemned as a 'disastrous mistake'.[4] Rudolf von Bennigsen made a sensational speech in the Reichstag on 22 January 1892 in which he urged all the liberal parties to unite against the new policy.[5] From Munich, where the Kaiser had recently appointed him Prussian envoy, Philipp Eulenburg expressed grave concern about the School Bill, which could become 'of decisive significance for the existence of the government'.[6] Ernst Prince zu Hohenlohe-Langenburg commented angrily that 'the concessions which have been made to ultramontanism and Polish interests for the sake of temporary advantages are surely very ill-advised, for history shows where such weakness towards the greatest enemies of our national existence has led and must lead, especially at a time when we should all be joining forces within the country in order to counter the dangers from abroad'.[7] Even Waldersee, who admired the ultra-conservative Zedlitz and was generally in favour of strengthening the Christian element in education, considered the bill a serious mistake which would lead to a dangerous increase in the influence of the Catholic Church and perhaps even to the return of the Jesuits to Germany. After all, he claimed, the Catholic Church was 'the fiercest opponent of the German Reich' and 'its destruction would be her greatest triumph'.[8] On the other hand there were misguided hotheads around the Kaiser who were hoping that the Zedlitz initiative would bring about a 'violent and brutal . . . war . . . on all fronts' between the government which was 'spoiling for a fight' and the entire German liberal movement, leading to a 'very healthy' solution to the 'mental apathy' of the 'stagnating masses'.[9]

So once again at the beginning of 1892 all eyes were on the Kaiser, to see whether he would withstand the 'liberal clamour' against the School Bill or disavow the Minister of Ecclesiastical Affairs and the Reich Chancellor who supported him. Philipp Eulenburg was convinced, just as he had been two and a half years earlier when Bismarck began to move closer to the Centre, that a deal between the Berlin government and the Catholic Church would bring about the fall of the liberal minority government in Munich, and with it the disintegration of the 'kleindeutsch' Reich.[10] He wrote urgently to the Kaiser on 21 January 1892 warning him of the dangers that would ensue for Crailsheim's government and the liberal parties in Bavaria, who provided 'the only support for the unity of the Reich' there, if the Centre were raised to the status of a government party in Berlin. He doubted whether Zedlitz had properly weighed up these dangers and advised Wilhelm to discuss the matter confidentially with him, so as to make clear to him that he, the Kaiser, took the moderate view on this question.[11] On 24 January Holstein was able to report to Munich that on 22 January in the afternoon, before he had received Eulenburg's letter, the Kaiser had gone straight from the station to see the Chancellor to find out what was happening, presumably because he had heard 'all kinds of strange things about the situation' from certain quarters. 'Then yesterday [23 January – that is, after receiving Eulenburg's letter] Zedlitz was suddenly summoned to luncheon. His Majesty told him that He would come and have a glass of beer with him that evening, and asked him to invite Helldorff, Manteuffel, Miquel, Benda, Tiedemann-Bomst and Douglas. Whether others were mentioned I do not know. In the evening the conversation was of almost nothing but the School Bill. His Majesty declared that "He would never accept a School Bill which was brought to him only by the Conservatives and the Centre. He demanded that the Free Conservatives and the majority of the National Liberals should also agree to it. If he made himself dependent on the Centre the next thing would be that the Redemptorists and then the Jesuits would come back. Moreover, in South Germany every minister who was not an ultramontane would fall, etc." His Majesty told Marschall all this at his audience this morning. Marschall replied that as a member of the Reichstag he had fought against the National Liberals, but that he could not deny that outside Prussia it was the National Liberals who were the supporters of the Reich idea. Marschall expressed to me his anxiety about the influence that Zedlitz – who *had been in complete agreement* with the Kaiser yesterday evening – would now exercise over Caprivi. Z[edlitz] went to see C[aprivi] this morning; then Marschall went over and found the R[eich] C[hancellor] in an agitated state and convinced that it was his duty as a decent man to stand by Zedlitz.'[12] Three days later Holstein's fears were confirmed. During a conversation with the Chancellor on 27 January, the Kaiser's birthday, the diplomat was 'very sad' to note that Caprivi 'wants to go along with the Centre; he thinks he will find stronger support there than with the National Liberals. The Conservatives are already firmly engaged as well . . . He, the R[eich] C[hancellor], therefore thinks that the committee will approve the bill without amendment. As I gather from [Franz] Fischer, in that case not only Miquel but also Bennigsen will go, and then the *reductio ad absurdum* of the conservative–clerical principle will come into play: that is to say, the Centre will bring forward its claims. How long do you think the German people will tolerate that, and how do you think the next elections will turn out?'[13]

A massive tug-of-war now began both within the government and in the public at large over Wilhelm II's decision. As Waldersee wrote at the end of January: 'On both sides people are trying to influence the Kaiser, and he has not yet reached a decision on which way to go.' 'As he . . . has no intention of giving up his unfortunate inclination to direct everything himself . . . he creates great uncertainty and unrest. He has expressed the most widely differing views to many people, and then when they compare notes it turns out that he has said very different things, which were mostly what those concerned wanted to hear!' It was a strange state of affairs and the outcome was not at all clear. 'The Liberals, who praised Caprivi to the skies only a few weeks ago, are now bitterly attacking him', Waldersee noted. 'In the Ministry there seem to be people who are ready to give way, and the Kaiser's favourite advisers like Hinzpeter and Helldorf[f] would have to deny their whole past if they were not trying to fish in murky waters again here.' In Waldersee's eyes, of course, the most desirable outcome of the School Bill crisis would have been the dismissal of Caprivi, with whom more and more people were becoming disillusioned. 'Lucanus called him two-faced not long ago; Boetticher has more or less fallen out with him; Miquel hates him' and was working to bring him down.[14] A 'wretched game of intrigue' was afoot, he noted on 21 February, 'with Caprivi, Boetticher, Miquel, Zedlitz and Lucanus each trying to go his own way; they do not trust each other and no one knows exactly what the Kaiser wants.' Clearly the latter had had no idea of how much unrest the bill would cause and now wanted to find a solution that would be acceptable to all parties.[15] 'Probably the only thing that is certain is that the Kaiser does not know which way to steer; he is beginning to be completely incapable of making a decision', Waldersee commented.[16]

Wilhelm's attitude towards religion in general and towards the Protestant zealots such as Zedlitz in particular kept everyone guessing. Early assumptions that he, like the Kaiserin, as a proponent of the Stoeckerite tendency, had been shaken by the dismissal of the anti-Semitic Court Preacher and the fall of Waldersee. But then came his solemn declaration, in his closing speech to the conference on school reform in December 1890: 'It goes without saying that I will, as king of Prussia, as well as *summus episcopus* of my Church, make it my most sacred duty to take every care that the religious sentiments and the Christian spirit are fostered and increased in the schools. May the school respect and revere the Church, and may the Church in its turn assist the school and facilitate the further accomplishment of its tasks. Then we shall be able to educate our young people up to the requirements of our modern national life.'[17] The speculation as to which side the monarch would take in the national crisis that had arisen over Zedlitz's bill began to take an even more serious form when 'the question of the Kaiser's madness', as Eulenburg noted anxiously in Munich, began to 'haunt us . . . again' in the winter of 1891–2.[18]

For those hoping for a clear lead from the monarch, the first signs were anything but reassuring. At a dinner given by Caprivi on 3 February, and again ten days later at another given by Boetticher, the Kaiser spoke out as he had originally done over the glass of beer at Zedlitz's house: he would approve the School Bill only if it received the support of the liberal middle parties. But then, angered by the attitude of the National Liberals and the two Progressive parties on military matters, he had spoken so disparagingly about them that his remarks were interpreted as support for Zedlitz's clerical School Bill.[19] The confusion reached its climax with Wilhelm II's alarming speech to the provincial diet of Brandenburg on 24 February 1892 in which he challenged 'the dissatisfied grumblers' to shake the dust of Germany off their feet and leave the country as soon as possible, which was generally taken to refer to the anti-clerical opponents of the School Bill.[20]

Simultaneously with the growing pressure on the Kaiser, who had the final power of decision in this momentous question – all were at least agreed on that – a small group close to the Chancellor was secretly exercising its influence to thwart the far-reaching concessions to political Catholicism Caprivi was planning to make, just as had happened two and a half years earlier in response to the Bismarcks' attempts at a rapprochement with the Centre Party.[21] This group, led like the earlier one by Holstein, was all the more dangerous to the Chancellor because it had direct access to Wilhelm II through Eulenburg. Writing to the latter on 27 January, Holstein admitted that he could see only two ways out of the present danger: 'First, for the Kaiser to ask the Reich Chancellor, as a favour to him, not to oppose an amendment of the bill at the committee stage but to leave the whole thing to Zedlitz . . . 2.: for His Majesty to get hold of a few Conservatives, members of the committee, and try to influence them quietly. His Majesty asked for members only of the Reichstag and of the Upper House [of the Prussian Landtag] (apart from Benda) to be invited for a glass of beer at Zedlitz's house, so members of [Prussian] House of Representatives were offended, on top of everything else. Helldorff and Manteuffel are both in the Upper House. As a last resort the Upper House could of course amend the bill to make it more moderate and conciliatory, in accordance with the Kaiser's wish, and then send the amended version back to the House of Representatives. The Upper House would gain a great deal of prestige in the country by doing so . . . I am afraid that the Reich Chancellor will speak [in the House of Representatives] tomorrow and commit himself completely. If he does not, it may perhaps still be possible to make a few suggestions to His Majesty.'[22]

With his long experience in Bavaria Eulenburg did not need to be told by Holstein of the dangers of a swing towards the Centre. In his view, such a shift could be justified only 'if the government is firmly resolved, directly they have attained their ends [in parliament], to begin a war and to alter the electoral system [i.e. undertake a *coup d'état*] etc., etc.' If this were not the intention, the reaction of the enraged liberal middle parties and of the social democratic workers' movement against black–blue clericalism would inevitably lead to parliamentary government and 'other embarrassments for the monarch'. 'The struggle of the united Liberals will be the more violent', Eulenburg shrewdly warned at the end of January 1892, 'in that their banner will read *suprema lex salus republicae* as against *suprema lex regis voluntas*. His Majesty's vigorous emphasis on the personality of the monarch will stamp the imprint of absolutism on the brow of any Conservative–Centre government' and thereby lead to the loss of the moderate parties 'which in my view form the natural basis nowadays for monarchical government'. The danger for Germany, which had been 'rocked to her foundation since the death of the old Kaiser', was plain for all to see, he warned. 'In Bavaria and Württemberg the Reich idea is enshrined only in Liberal hearts.' If the government in Berlin supported Zedlitz's School Bill it would 'lose the whole of Württemberg, the greater part of Bavaria and probably the greater part of the other federal states too'; they would all go over to the Bismarck camp 'with drums beating'. If the National Liberal Finance Minister Johannes von Miquel should resign in protest against the new policy he would 'enjoy *enormous prestige* throughout the Reich', which Bismarck could exploit for his purposes. 'We shall have some appalling experiences in the elections.' All these dangers were so great that the tactical gain would bear no relation to the loss. 'There are all kinds of surprises in store for Caprivi!'[23]

The dilemma for Holstein and Eulenburg in this rapidly developing crisis lay in the fact that any attempt to influence the Kaiser against the Zedlitz School Bill could easily provoke Caprivi's resignation. Although he considered the situation almost 'hopeless',[24] under pressure from Holstein Eulenburg agreed in principle to write a second letter to the Kaiser. He pointed out, however, that 'the more

H.M.'s attention is drawn to the dangerous consequences of the School Bill, the more easily may annoyance with Caprivi gain a hold. But that simply must be avoided. You know H.M. is inclined to criticise Caprivi – and I assume he is already silently reproaching him over this school question.' As a solution to the dilemma Eulenburg suggested the removal of the Prussian Interior Minister, Herrfurth; Zedlitz could then take his place and the School Bill could be withdrawn under the pretext of the appointment of a new minister of ecclesiastical affairs. 'I cannot imagine the Chancellor would stand by Herrfurth', Eulenburg argued, 'if H.M. urged this solution on him.'[25] But after a conversation with the Chief of the Reich Chancellery Karl Goering, Holstein, who was better able to gauge the prevailing mood in the Wilhelmstrasse, warned that Caprivi would certainly resign if the School Bill were not passed or if the Kaiser refused to approve it. The only way out, in his eyes, was therefore to make sure that the bill became 'bogged down' at the committee stage in the Landtag.[26] In his second letter of warning to the Kaiser on 10 March, therefore, Eulenburg expressed himself with the greatest possible caution, pointing out that the dilemma confronting Wilhelm was acute, for on the one hand Caprivi's resignation was '*impossible*' as far as the Reich was concerned, but on the other hand a School Bill which was accepted only by the Centre Party and the Conservatives was '*equally impossible*', for it would 'provoke conflict in the Reich with consequences which would be disastrous for Your Majesty's position'. Therefore the bill must at all costs be 'buried' in committee, if no compromise could be found, 'although boundless discretion will be required if we are not to cause Caprivi's fall!'[27] Although Eulenburg later expressed his alarm at the 'bomb-like' effect which his letter seemed to have, it was actually the intervention of another 'irresponsible adviser' which was to be the immediate cause of the furore that followed.

As in the Bismarck crisis two years earlier – and just as Waldersee had predicted – the influence of the moderate leader of the Conservatives, Otto von Helldorff-Bedra, on the Kaiser now made itself felt. Helldorff was still convinced that the right wing of his own party, led by Hammerstein, was working with the Catholic Centre to transform the German Reich into a clerical state. Determined to resist, he declared, 'I am fighting with all my strength for this one fundamental idea, that the Reich can only be governed and kept in existence if it is supported by the forces through which the Hohenzollerns have attained their position – the Protestant North – and if we keep the real and irreconcilable enemy, the policy of Rome, in our sights.'[28] On 7 March 1892 he wrote urgently to Eulenburg: 'How can we save the situation? We must keep Caprivi – and correct the Kaiser's position.' To achieve these goals a compromise acceptable both to the National Liberals and to the Centre must first be sought. 'That is Kartell policy on a grand scale – doing justice to the needs of the Catholics and allowing them to remain in the national arena. – That is in fact also the fundamental idea of the bill, which Caprivi supports, but which has been dangerously distorted by Zedlitz and his *Kreuzzeitung* advisers.' If no such consensus could be achieved there was only one other way out of the wilderness, and this was the 'adjournment' of the bill, in other words 'a decent burial'. 'This also corresponds to the Kaiser's original view. – But it is a question of how to bring it about, and I think that direct intervention by His Majesty is necessary. – I think he should speak out, not publicly – not in a speech (he must be strongly warned against that) – but to carefully selected individuals. He must make his real aims clear and contradict the misrepresentation of his [Brandenburg] speech. He does not need to tell everyone everything, only what each needs to know.' Helldorff insisted that it was 'absolutely imperative that I should explain the situation to His Majesty as I see it, for at moments like this there is no shortage of people around him who are very skilful at exploiting likes and dislikes or temporary bouts of rage, against Bennigsen for instance. I shall therefore ask for an audience today.'[29] Because of a mysterious illness from which the Kaiser had been suffering for weeks it was not until the evening of 16 March 1892 that Helldorff managed to speak to the monarch. But then his intervention did indeed work 'like a bomb'.

What was the matter with the Kaiser in this, the most serious domestic crisis since Bismarck's fall? Not least as a result of his disastrous Brandenburg speech on 24 February, in which he promised to lead his people towards 'a glorious destiny' and challenged his critics to leave Germany,[30] the warning uttered by Eulenburg in January about 'the question of the Kaiser's madness' seemed to be proving accurate at the very moment when the School Bill crisis was approaching its climax. For some time now observers had noticed a strange indisposition in the Kaiser which seemed to undermine his ability to take decisions. The official explanation given was that he had a cold. However, the Empress Frederick reported to London that the illness was not serious; Wilhelm looked well and had gone away to Hubertusstock, where he would be able to spend the whole day in the fresh air. 'His ear is in no way affected, he has not had an "Ohren Catarrh" for some time now', she wrote, although it was true that Wilhelm had been very upset at the dreadful press comments on his latest Brandenburg speech.[31] Many people suspected some kind of nervous breakdown.[32] The Oberhofmarschall, Count August zu Eulenburg, confided to his cousin Philipp that the heated arguments about his speech combined with the 'high degree of agitation' over the School Bill had affected the Kaiser 'so badly that His latest indisposition [had been] far more psychological than physical'.[33] Wilhelm's personal physician, Rudolf Leuthold, spoke of a 'marked inclination to fall asleep' combined with physical fatigue and 'slight feverishness with a mild cold'. He considered this to be 'simply the result of a certain nervous fatigue, which was to be expected given the many demands on our All-Highest Master . . . and his present psychic state, which is probably not quite balanced'. He forbade his imperial patient to go out or do any work and prescribed 'several days of absence from Berlin and the whole business of government', preferably at Hubertusstock. There was no cause for alarm over his condition and there was no question of his ear being involved, Leuthold confirmed.[34] The British ambassador reported to London that the illness was feverish and that the Kaiser was 'quite unnerved by worry',[35] while Wilhelm himself sent a telegram to his grandmother admitting that 'I was too much overworked and the doctor wishes some rest for me.'[36]

Whatever the cause of the Kaiser's indisposition, the 'long and detailed conversation' that Wilhelm had with Helldorff after dinner on 16 March unleashed a serious crisis the next day. The moderate leader of the Conservative Party, who was to lose his position not long afterwards, said of the evening's interview with the Kaiser that he had found him 'still looking rather unwell – and in a very depressed mood'. 'The state of affairs, the malicious criticism of his speech etc. . . . had deeply upset him – he is said not to have slept for several nights.' Helldorff 'explained his view of the situation and told His Majesty . . . that in my opinion the only sensible way of resolving this difficult situation was for an appropriate group of Conservatives in the House of Representatives to amend the excessively pro-Catholic points [in the bill], which would pave the way either for an understanding on a broader basis or for a decent burial. As others were working with all their might to use the Centre and the Conservatives to force the bill through if possible without amendment, and to obscure the Kaiser's true intentions, I considered it important that His Majesty should inform Miquel that he wished the bill to be handled in this way and ask him to bring the National Liberals round to an agreement or a conciliatory attitude – and then to speak confidentially to a few Conservatives, principally . . . His Majesty agreed to this . . . – I had no idea that there would be a Crown Council the next morning.'[37]

Although the School Bill was not on the agenda and the Kaiser had not intended to state his personal views on it until after the first reading in committee, he nevertheless brought up the subject of the situation created by the bill at the end of the Crown Council on 17 March. The manner in which he did so not only went down badly with the Minister of Ecclesiastical Affairs but also offended Caprivi as minister-president.[38] Wilhelm's remarks, which (as Helldorff was at pains to stress) 'came purely from his own initiative', and in which he repeated that he would approve the School Bill only if it were accepted in the Prussian House of Representatives with the support of the liberal middle parties, had worked 'like a bomb', for they had 'blown Zedlitz and his whole policy sky-high' – and 'unfortunately almost did the same to Caprivi'.[39] As Zedlitz asked for time to consider his position after the Crown Council, members of the Kaiser's entourage were at first confident that the minister would stay in office and that the threat of a chancellor crisis could therefore also be discounted.[40] These hopes were however dashed by the indiscretion of other ministers, who regaled their fellow guests at a dinner given by the Württemberg envoy with accounts of the Kaiser's dramatic intervention, as a result of which 'the whole affair was spread about town'. That evening Zedlitz handed in his resignation, and on the morning of 18 March 1892 (two years to the day after the dismissal of Bismarck) Caprivi handed in his own resignation as Reich chancellor, Prussian minister-president and Prussian minister of foreign affairs. The Kaiser tried in vain to persuade Zedlitz to stay by writing him 'an explanatory letter . . . about the meaning of the All-Highest remarks at the Crown Council'. Zedlitz asked for two days to think it over, but Countess Zedlitz declared openly that 'they were leaving in any case'. The crucial question now was whether Caprivi could be persuaded to stay. 'A chancellor crisis at this moment would really be more than disastrous', August Eulenburg commented.[41] On 19 March the Kaiser wrote on Caprivi's letter of resignation: 'No. I wouldn't dream of it. It is not nice to drive the cart into the mud and leave the Kaiser sitting in it. Caprivi has made a mistake; that can happen to anyone. His departure now would be a national disaster and is out of the question.'[42]

In Munich Philipp Eulenburg received the news of the dramatic developments in Berlin through a cipher telegram from Holstein on 18 March. 'At yesterday's Crown Council the Kaiser again expressed the wish that moderate elements should join in working on the School Bill, and that Kartell policy should be followed as before. The rest of the session passed uneventfully. But afterwards Zedlitz came to the Reich Chancellor and declared that he could not cut himself off from the Centre and had therefore sent in his resignation. Von Caprivi then did the same, with a long, not uncivil explanation. He told me "that he could not let Zedlitz drown at his side" . . . The Kaiser left for Hubertusstock this afternoon for eight days, having turned down Zedlitz's resignation in writing and that of the Reich Chancellor verbally through Lucanus. But neither has yet relented, and there is no one available who would be in a position to mediate. Best arrangement would be to make Count Zedlitz minister of the interior or postpone whole thing until after first reading. If Zedlitz stays the Chancellor will stay too.'[43] Eulenburg immediately forwarded these comments to the Kaiser and pointed out another argument against a change of chancellor. There was a real danger that at any moment a publication in Zürich would reveal the fact that Bismarck had purchased the agreement of King Ludwig II of Bavaria to the foundation of the Reich in 1870 with millions from the secret Guelph Fund. If this 'Ludwig II–Bismarck dirty trick', which could shake Bavarian national consciousness to its core and lead to very dangerous debates about the events surrounding the creation of the Reich, were to become public knowledge at a time when the Kaiser was about to dismiss the Chancellor, there would be 'a great upheaval in South Germany' which could put both Kaiser and Reich in grave danger. 'To allow *two such disturbing events to coincide is simply impossible*', the Kaiser's friend warned.[44] At the same time Eulenburg wrote to Caprivi drawing his attention to the inadvisability of a chancellor crisis at a moment when the Guelph Fund revelations were looming.[45] Others also tried to placate Caprivi, among them Prince Heinrich and the respected President of the Prussian House of Representatives, Georg von Köller, but without success.[46] All attempts to mediate were thwarted by the embittered resolve of the departing Minister of Ecclesiastical Affairs, who cast 'all the blame on the Kaiser'.[47] 'Majesty must have left Z[edlitz] badly in the lurch', Verdy wrote to Waldersee, adding with bitter irony: 'We know all about that, through thick and thin!'[48]

Holstein, whose principal fear was that his *bête noire* Waldersee might be appointed Reich chancellor, found it 'quite horrifying that at this terribly important moment His Majesty is sitting at Hubertusstock with a couple of adjutants. There is a real similarity with Louis XVI. And who knows what decisions are being made there, without a single sensible person to consult.' The military suite, with which the Kaiser once again found himself sitting 'for 8 days . . . alone in the forest', would think it quite natural if their master appointed another general as Reich chancellor, while all politicians considered such a decision 'impossible', the diplomat observed. Furthermore, to his alarm, the Kaiserin also expressed herself in favour of Waldersee. The only person who could break through the 'magic circle' surrounding Wilhelm was Philipp Eulenburg.[49]

Caprivi's solidarity with the former Minister of Ecclesiastical Affairs was admired by some as the 'upright conduct' and 'comradely kindness' of a 'man of unswerving honour'. At the same time it was criticised as evidence of his lack of political instinct. The Chancellor seemed to Holstein, Marschall, Helldorff and many others to have been 'hypnotised' and to have behaved 'as if shackled to Zedlitz'. The Chief of the Reich Chancellery, Goering, said of him that 'since the beginning of this year a complete change has taken place in him'.[50] Others discerned an element of calculation in the Chancellor's attitude: he was convinced that 'in a few weeks or at most months he would come to blows with the Kaiser over military demands and the two-year service which might be linked to them', and he thought he could make a better exit now.[51]

When Caprivi went to visit the Kaiser at Hubertusstock on 20 March he had already made up his mind that, if he were stay in office at all, he would remain only as Reich chancellor and give up both his Prussian offices, the minister-presidency and the position of Prussian foreign minister. Wilhelm seems to have accepted this suggestion, for after Caprivi's return from Hubertusstock Marschall von Bieberstein was obliged to point out to him that he would have to retain at least the post of Prussian foreign minister, since the latter was responsible for instructing the Prussian vote in the Bundesrat.[52] As Caprivi would therefore continue to be a member of the Prussian Ministry of State, and as the minister-president did not enjoy any special constitutional privileges in Prussia, his supporters hoped to have found a modus vivendi, especially if in addition Marschall could be appointed to the Prussian Ministry of State as minister without portfolio. But everything would depend on who was appointed as the new minister-president.[53] Upon Caprivi's return from Hubertusstock August Eulenburg reported to his cousin: 'C. suggested several names to the Kaiser today, none of which aroused particular enthusiasm at first. Boetticher is *on no account* to become minister-president, *nor* Miquel or any other National Liberal, but a Conservative who could pursue a Kartell policy.' Caprivi finally settled on Count Botho zu Eulenburg, the Oberhofmarschall's brother, and asked the latter '*without the Kaiser's consent*', to invite Botho to Berlin as unobtrusively as possible.[54] The latter accepted the minister-presidency after Caprivi had paid a second visit to Hubertusstock on 23 March, during which he succeeded in 'fulfilling Botho's chief condition', namely the appointment of the Secretary of State in the Reich Justice Office, Dr Robert Bosse, as the new Prussian minister of ecclesiastical affairs. Both Caprivi and Bosse himself expressed scruples about this change of office, 'but the Kaiser decided on him at once', August Eulenburg reported to Munich.[55] Bosse pointed out that he was a civil servant, not a politician, and that he could not speak in parliament without embarrassment. The Kaiser, however, declared bluntly 'Bosse and no other'. And with that and Caprivi's admonition that 'as a decent man' he must obey the Kaiser's command, Bosse had little alternative but to accept the thorny office proffered to him.[56]

On 24 March the envoys of Baden and Saxony, Brauer and Hohenthal, received orders to go to Hubertusstock next day. Wilhelm explained to them his position on the School Bill crisis, with a view to their informing the Grand Duke and the King respectively. Brauer wrote to his master: 'It was a great relief to me to discover, contrary to the many foolish rumours, how well H.M. looked and obviously felt. H.M. was extremely cheerful and his conversation at table was very lively and spiced with jokes . . . "I have sent for you", H.M. soon began in his unaffected, jovial way, "so that you can report to my uncles on the course of the crisis, and what *really* happened, and on my view of the situation. From the beginning I have consistently held the opinion that the draft of the School Bill was certainly a suitable basis on which to work out a successful result with the co-operation of all moderate parties. But I have never left anyone in any doubt that I would not ratify any bill which was acceptable only to the Centre and the extreme Conservatives. I often asked Zedlitz whether he was sure of *all* Conservatives, and he always assured me that he had them safely under control . . . I had no reason, after the Crown Council, to suppose that my words would lead to a crisis. I was very surprised when I received Zedlitz's letter of resignation that very evening and another from Caprivi next day. Particularly in the latter case there was no reason for it. Caprivi is a touchy, pigheaded fellow. He has often threatened to leave me in the lurch and tried to resign before now. I have given in countless times in order to keep him. I had warned him, indeed I had *ordered* him, not to commit himself to the School Bill. What was the use, he did it anyway! This time too I have implored him to stay. But I still could not prevent him giving up the presidency of the Ministry of State, out of utterly needless sensitivity. – It was just the same last autumn, when I asked the Reich Chancellor to demand the resignation of Herrfurth, who is now so unpopular with the Conservatives, from his Ministry after the Rural District Bill had been approved. The Chancellor absolutely refused to do it. But I was right! I am not as stupid as I look! Now we cannot get rid of him for the time being, because the liberal newspapers would immediately make a fuss" . . . The first task of the new minister-president, H.M. stated, would be "to break in the Conservative nags again: those fellows have become nothing but a bunch of snipers; he must turn them back into a regular battalion". Finally H.M. remarked: "Tell your gracious masters how difficult it is to manage with Caprivi. It could do no harm if your King or your Grand Duke gave him a good talking-to from time to time. I am too young to impress him"!'[57]

Wilhelm was also anxious to prove to the British government that he had not, as was universally claimed, 'behaved like a tyrant'. Summoning Sir Edward Malet, he gave him his version of the events leading up to the crisis. He began by complaining that Caprivi had not informed him of the new School Bill until the previous autumn; in his view he should have been consulted much earlier. He stated that he would have refused his consent to the bill there and then, had not Lucanus persuaded him to give in, since otherwise the Reich Chancellor and several of the ministers would have resigned. Public disquiet during the winter had only confirmed him in his view, Wilhelm told the British ambassador, that it would be a mistake to force the Zedlitz bill through in the Landtag. On 27 January, his birthday, he had gone personally to the Ministry of Ecclesiastical Affairs and had made it clear to Zedlitz that he did not wish the School Bill to be a party-political measure and that a compromise with the liberal parties must be found. Three days later he had begged Caprivi not to identify himself with the bill. His surprise and annoyance had therefore been all the greater, the Kaiser declared, when Caprivi had thrown down the gauntlet in favour of the bill in his speech in the House of Representatives. 'The feeling had now spread beyond Prussia. The National Liberal party in Germany, the backbone of the Empire outside Prussia, espoused the cause of the National Liberals in Prussia. General Caprivi & Count Zedlitz were throwing themselves blindly into the arms of the right fractions of Conservative and Clerical parties.' He, Wilhelm, had therefore called for reports from all parts of Germany, which proved to be unanimously against the bill. When Sir Edward Malet intervened to ask the Kaiser whether he had not in fact supported the School Bill in his Brandenburg speech, the latter replied that his intention had been exactly the opposite. He had meant that he would fight to the bitter end to prevent his ministers pushing through the School Bill in this form. According to Malet's report, the Kaiser described the end of the crisis as follows: 'Then came the famous Crown Council: at which the Emperor asked Count Zedlitz if he could not adopt some compromise. The latter looked annoyed and said nothing. The Emperor asked Count Caprivi his opinion, who said he believed the Bill met with the approval of the Majority of the House [of Representatives]: *only* the Minister of Finance thought it should be modified. The next day to his intense surprise the Emperor received the resignation first of Count Zedlitz, then of Count Caprivi.' Meanwhile the Kaiser had learned that Zedlitz drafted the bill himself; the officials in his ministry had refused to assist. 'If Count Zedlitz had had his way', the Kaiser told Malet, 'the Culturkampf would have been renewed in its most aggressive form & the National Liberal party alienated.' Malet's report concluded: 'The Emperor attributed Count Caprivi's action to lack of Parliamentary Education and to chivalry to his colleague. It was a fine sentiment but had led him into a mire from which attention to his Sovereign would have saved him.' Although the Kaiser told Malet he hoped that Caprivi would resume the Prussian minister-presidency in the autumn, in the ambassador's view it was obvious that the relationship of trust between Wilhelm II and his second chancellor, which was in any case fragile, had been irreparably damaged.[58] Philipp Eulenburg thought likewise. 'I am afraid that Caprivi's prestige has suffered greatly in the Kaiser's eyes', he wrote on 26 March 1892.[59]

As with Bismarck's dismissal, Wilhelm felt the need to show his own role in this chancellor crisis in the best possible light for the benefit of his grandmother. On 12 April he wrote to the Queen, who was staying in the South of France, saying: 'I had rather a bad winter this year caused by the trouble given me in the school law . . . The whole thing was very badly managed by the Ministry as a whole & by Count Zedlitz in person. He did the direct contrary to what I from the beginning told him to do; snubbed the liberals, instigated the Centre & Ultra-Conservatives († Zeitung) & finally behind my back & against my expressed wish entangled the Chancellor in the whole affair, a week after the latter had promised me not to make a speech alluding to the law or to take any active part in the debate. When I warned C[ount] Zedlitz for the last time before the dangers he was incurring he made no answer but simply resigned ab irato & without any positive reasons; pulling the Chancellor after him; who in, what the French call un exces de vertu, thought he was bound to keep the other Minister company & resigned too. It was rather hard lines upon me, considering that I was opposed to the law from the first moment I had cognizance of it; & considering that I had done my utmost to keep the Count from making a mess of it! The most extraordinary thing in this whole affair was, that the Ministry never even agreed when they sanctioned the law last autumn by signing it; & that now they were all dead against it excepting the Chancellor & Zedlitz who seem to have overruled their colleagues to such a degree, that they did not dare open their mouths or send word to me that my Ministry was hopelessly split, which they ought to have done at once.'[60]

THE SEPARATION OF OFFICES

With the ill-considered, almost panic-stricken resolution of the School Bill fiasco by means of the separation of the Reich chancellorship from the Prussian minister-presidency in March 1892, the Kaiser's ability to intervene in government business was greatly increased. Instead of having to contend with a head of government who periodically, with a soldierly integrity sometimes verging on misguided obstinacy, refused to obey 'commands' of his monarch which he considered disastrous, Wilhelm II was now faced with a state apparatus with two leaders, one of whom had only reluctantly remained in office.[61] The other, meanwhile, closely linked to the Kaiser, the court and the Conservative Party and initially free of the burdens of running a ministry, naturally hoped the two posts would eventually be reunited in his hands. 'I am *very* sorry that Ct. Botho Eulenburg (the Brother of the Hofmarschall) is to be Vice Chancellor [sic] & I am sure it will not be long before he takes Caprivi's place too', the Empress Frederick predicted on 24 March 1892.[62]

Five months after his appointment as Prussian minister-president, Botho Eulenburg's position was considerably strengthened when he was given the additional appointment of minister of the interior. Once again the Kaiser was responsible for this appointment: as Marschall recorded in his diary on 12 April, Wilhelm had spoken out 'very strongly against Herrfurth, whom he wishes to replace with B. Eulenburg'. The next day he noted: 'H.M. very angry with R[eich] C[hancellor], who has already twice threatened to resign. Has had a letter sent to B. Eulenburg about Herrfurth.' Caprivi, who had protected the liberal Minister of the Interior from the Kaiser the previous year, eventually abandoned his opposition to Herrfurth's dismissal and the appointment of Botho Eulenburg to this influential Prussian post, whereupon the Kaiser triumphantly announced to the Chief of the Civil Cabinet: 'My dear Lucanus! I went to see the Chancellor yesterday and sounded him out carefully in the course of the conversation about our further progress in ministerial appointments. I am happy to say he was quite resigned and prepared for Herrfurth's departure, for which he no longer wished to set a date; and when I indicated that Eulenburg would be the best successor to H. he at once gave his agreement and expressed the hope that he would accept the post. So all our anxieties on this score are settled and you may inform Eulenburg of it immediately. W. I. R.' The Kaiser agreed, however, to postpone the change of minister until after the closure of the Landtag.[63] The Empress Frederick thought the departure of Herrfurth from the government regrettable from every point of view and saw Botho Eulenburg's promotion as a further weakening of the Chancellor, for Eulenburg was 'a very oily slippery ambitious sly man'.[64]

Caprivi's position after the separation of offices was widely seen as untenable in the long run, a view he himself shared. In April 1892 Marschall noted in his diary that the Reich Chancellor was 'very depressed' and had 'nothing to do'.[65] In May Verdy reported to Waldersee from Berlin that the Chancellor was seen 'as a dead man, not only in the eyes of the many but also by the privileged few in the know. Lucanus, for instance, told me 14 days ago when I was dining privately with him and a few ministers at Douglas's house, that he gave him another 6 weeks and then things would get hot again.'[66] Waldersee himself at first thought that Caprivi would soon retire and be replaced as chancellor too, by Botho Eulenburg. 'As soon as the Kaiser has reason to be pleased with Eulenburg . . . he will dismiss Caprivi on the most trivial grounds; with the Kaiser the new has the advantage over the old', besides which the malleable Botho Eulenburg could count on 'very valuable support from his brother August, his cousin Philipp and the whole East Prussian clique'.[67] It was only after a visit to Berlin, when he found the Kaiser in a more serious mood than usual, that the General realised that in appointing Botho Eulenburg the Kaiser's intention had not in fact been to make him Reich chancellor later. He was still looking for a suitable candidate in the event of Caprivi's resignation and had entered into negotiations with the Governor of Alsace-Lorraine, Prince Chlodwig zu Hohenlohe-Schillingsfürst. In Waldersee's opinion Hohenlohe was already a 'worn-out man with no energy' at this time, although the fact that he had turned down the post of chancellor at least proved 'that he still [possesses] some self-knowledge'. (Two and a half years later Hohenlohe was to succeed Caprivi and Botho Eulenburg as Reich chancellor and Prussian minister-president respectively.) Wilhelm had also discussed the chancellorship with the ambassador at Vienna, Heinrich VII Prince Reuss. 'The worst thing, as always, is that the Kaiser has no definite system at all in mind when he makes new appointments', Waldersee very justifiably complained. Nor had he 'any understanding at all of the fact that a Ministry needs to be unified and harmonious. The School Bill catastrophe has made him completely unsure of himself; he does not know where he is going or with whom he should ally himself. The ship is adrift! He really is not the right captain to steer it through such tricky waters!'[68] He, Waldersee, who was often approached about it, had no intention of seeking the post of Reich chancellor. That would only be possible 'if the Kaiser recognised that his way of ruling will not work, and decided not to lend his ear to secret advisers and to pursue policies without the Chancellor and the ministers knowing anything about them'.[69]

Caprivi's position was further weakened by the fact that the two most influential military officers at court, Adjutant-General von Wittich and Chief of the Military Cabinet von Hahnke, who had hitherto admired and supported him, were 'finished' with him, as Waldersee recorded after a train journey with his two fellow generals in April 1892.[70] Wittich, who took a 'very pessimistic' view, reported that Caprivi was bitter and rarely appeared in public, while Marschall was no longer taken seriously anywhere, 'not even by the Kaiser'. The Foreign Office was ruled by Holstein, who had allied himself with Kiderlen-Wächter 'because of his relationship with the Kaiser'.[71] 'I found Hahnke also in the mood to make critical judgements', Waldersee observed.[72] Among the Flügeladjutanten, who had long since ceased to admire Caprivi, a positively hostile attitude towards the Chancellor was spreading. Helmuth von Moltke, the future Chief of the General Staff, confided to Waldersee: 'The more I get to know Caprivi, the more convinced I am that he is a man of very questionable character.' In support of this judgement Moltke 'indignantly' recounted that Caprivi had prejudiced the Kaiser against Zedlitz and had thrown all the blame for the crisis on him, with the result that the Kaiser now spoke of Zedlitz, whom he had recently regarded as the man of the future, 'in the most dismissive manner'. 'Really not nice, either of the Kaiser or of Caprivi!', Waldersee commented. 'It seems to me that it must gradually dawn on everyone who has ever had anything to do with the Kaiser how fickle he is. In fact I do not know anyone with whom he has been on intimate terms about whom he has not spoken disparagingly; Philipp Eulenburg is perhaps the exception; but there are some who deny even that.'[73]

Similar views on the confused state of affairs were now to be heard from all parts of the political spectrum. On a visit to Berlin in May 1892 Waldersee was shocked by how critically the Free Conservatives and the National Liberals in particular spoke of Caprivi and Marschall, who were generally considered 'completely finished'. The Finance Minister, Miquel, complained that there were ministers but no ministry; everything was going in different directions. Waldersee recorded that there was open talk of a possible reconciliation between the Kaiser and Bismarck, which would inevitably lead to Caprivi's resignation and

Botho Eulenburg's appointment as chancellor. It was presumed that Marschall's successor at the Foreign Office would be the ambassador in Madrid, Ferdinand Freiherr von Stumm.[74] In the ensuing months new names were constantly mentioned for both key posts; Waldersee was repeatedly named, as was the Catholic General Walther Freiherr von Loë.[75] Significantly, no one expected leadership qualities from the civil service types who headed the Prussian ministries. There was a total lack of 'independent, self-assured men' in the Ministry of State, Waldersee lamented in October 1892, for 'Thielen, Bosse, Schelling, Berlepsch, Kaltenborn, Heiden are all only second-rate people.'[76]

For the first few months after the separation of the top offices in the Reich and Prussia the Kaiser behaved with restraint, recognising that another chancellor crisis would seriously harm the prestige of the monarchy. He did his best to avoid conflicts and sent the new Minister-President instructions via Philipp Eulenburg 'to maintain the closest contact with Caprivi'.[77] Added to this was the fact that, as Wittich and Hahnke confided to Waldersee on their train journey in April, the Kaiser had to a certain extent lost his nerve as a result of the crisis. Wittich, Waldersee noted, judged the Kaiser 'increasingly severely, and says he is quite incapable of making decisions; his main aim at present is to be spared any serious business, nor does he want to be reminded of serious consequences; he does his best to divert himself with all kinds of distractions'.[78] Information of this kind from the immediate entourage of the Kaiser led Waldersee to the conclusion that there would after all be no change of chancellor for the time being. The Kaiser wanted 'to keep moving for several months', he wrote, and all the disruption associated with a change of chancellor was 'too troublesome' for him, besides which he would find it 'very difficult' to make any major decision.[79]

At this time of latent crisis and nervous indecision Kaiserin Auguste Viktoria proved herself a loyal ally to Caprivi. During Wilhelm's absence on his Scandinavian cruise in 1892 she invited the increasingly isolated Chancellor to dinner and discussed the Bismarck Fronde and the domestic and foreign situation with him in detail. 'I liked him so much for his decent and unselfish attitude to everything', she wrote to her husband.[80] Briefly, during that summer, the Kaiserin became an intermediary for letters and messages between the Kaiser and the Chancellor, particularly in the quarrel over whether an international exhibition should be staged in Berlin. Caprivi's dealings with industrialists had convinced him of the benefits of such an exhibition and he had already written both to the German federal governments and to the French government. The Kaiser, on the other hand, made his antagonism to the idea plain in a succession of 'very sharp remarks' and marginal comments. On the basis of a newspaper report Wilhelm had formed the opinion that an exhibition of this kind would result in great economic disadvantages for Germany. He was also afraid of the effect of 'workmen crowding to Berlin'. In July 1892 he said to Alfred von Kiderlen-Wächter, who was again serving as representative of the Foreign Office, 'Six months ago when there was first talk of an international exposition I urgently requested the Chancellor to put a stop to it. He then wanted to wait and see. I have however repeatedly told him that I was opposed to it; but he has always shut up like an oyster and now we have a pretty kettle of fish; if it is now stated that I do not want it, then people will again say I do not know what I want and had changed my mind; and yet from the outset I left no doubt that I was against it.' The Kaiser was 'more depressed than angry' as he spoke, the diplomat reported. In Kiderlen's opinion it would have been better for the Chancellor to give way and avoid a conflict, since he was well aware of the Kaiser's views and must have known he would not yield.[81] Immediately after this conversation on board the *Hohenzollern* Wilhelm wrote to his wife, who hastened to reply: 'I *wrote at once* to Caprivi with your message about the exhibition in Berlin as I do not know when he will have time to come here again, and it was urgent . . . I enclose Caprivi's answer . . . but he tells me that he had already written to you saying the same thing. If only things would calm down again; this agitation from all sides is unbearable!'[82]

The Kaiser replied to his Chancellor's letter on 20 July 1892 in tones of defiant, autocratic self-satisfaction which marked the end of the close season as far as relations between them were concerned. He abruptly dismissed the Chancellor's arguments in favour of an international exhibition. He was 'absolutely opposed to it'. Two and a half years earlier Bismarck had turned down a similar proposal after an audience with him, the Kaiser, and 'all outcry in favour of an exposition thereupon ceased'. Caprivi should have responded to the renewed demands for an exhibition with a 'decisive No. In this connection a reference to the decision taken by Prince [Bismarck] after an audience with the Kaiser would not have failed to be of effect, and would probably have nipped every movement in the bud.' One must not give in to the avarice of the Berlin *demi-monde* and underworld, Wilhelm declared. 'The fame of the Parisians gives the Berliners no peace', he commented mockingly. 'Berlin is a great city, cosmopolitan (perhaps?), therefore it must also have an exposition! It is easy to see that this line of thought is very understandable and acceptable for *Berlin* hotels, theatres, music-halls, etc. They will be the only ones to profit from it! Hence the propaganda. But the *proton pseudes* is to be found in the conclusion that it was the tourist traffic alone that brought Paris such good returns. This is completely false. The hundreds of real millionaires who settled there to *live*, and to amuse themselves for months, and who attract new acquaintances from all countries – it is all this that has fattened the calf. Paris is after all – what I hope Berlin will never be – the greatest whore-house in the world, hence another attraction quite apart from the exposition. There is nothing in Berlin to hold the foreigner with the exception of a few museums, palaces and the soldiers. He has seen everything with the red book in his hand in six days . . . The Berliner does not understand this and would be thoroughly offended if one told him so.' Almost sarcastically Wilhelm then turned to Caprivi's attempts to restore the Kaiser's popularity by encouraging nationalist feeling. 'I am deeply moved', he wrote, 'to see from your proposal – to place me at the head – how you are trying, in these times that are so difficult for us both, to find a good opportunity to increase my subjects' affection for me, and I warmly thank you. But my will is firm as a rock, and I will hold fast to what I hold right, and no devil – not even Prince Bismarck – can dissuade me. I will not have an exposition because it can injure my Fatherland and my capital! Moreover we Hohenzollerns are accustomed only to advance slowly and painfully amidst trouble, conflicts, party divisions, and lack of appreciation. How often have my ancestors, most recently my grandfather who rests with God, had to battle for measures in direct opposition to the will of the uncomprehending populace which first opposed, then criticised, but finally blessed them. *What do I care about popularity!* For as the guiding principles of my actions, I have only the dictates of my duty and the responsibility of my clear conscience towards God. Dear Caprivi, think of it, I was prepared for bullets and dynamite when I ascended the throne and I am still alive! Yes – even the Socialists say that one can talk to me; well! More cannot be expected in two years after Prince [Bismarck] and the disappearance of the Socialist law. Our time will come as for everyone. Let us be patient and persevering, let us do our duty whether people are annoyed or not, it is all the same. Respect will come, it is already "on the march" owing to your distinction and my trust in God. Only *trust in my leadership* and fight bravely where I point the way, and we will have no trouble in managing the *canaille* either in this affair or later on in *rebus militariis!* . . . Hence don't worry! As my Berliners say, Ausstellung is nich [exposition? No way!]. Farewell, best greetings to your colleagues and counsellors, as well as to Holstein. Your very affectionate King Wilhelm I. R.'[83] Philipp Eulenburg and Kiderlen-Wächter were well aware that the 'sometimes crude, sometimes arrogant' tone of this letter would be insulting to the Reich Chancellor and maintained that it was 'at all events *intended* to be friendly' and not intended to be 'ironical'. Even Wilhelm was at pains to avoid the dreaded possibility of the Chancellor handing in his resignation, and he explained that he had 'put this in writing because in that way I can more emphatically emphasise my position as king and father of his country; it might easily seem provocative if I as a young man were to say anything like this orally to someone much older'. Nonetheless Kiderlen thought it advisable to conceal from the Chancellor that he and Eulenburg had read the letter and that Holstein had also received a copy on the Kaiser's instructions.[84] The international exhibition did not take place.

When Robert Zelle of the Progressive Party was elected to succeed the late Forckenbeck as mayor of Berlin in September 1892 it was generally expected that the Kaiser, having only very reluctantly agreed to ratify Forckenbeck's election under pressure from the Ministry of State, would hold out against this appointment likewise. Instead, he ratified Zelle's election at once and even sent him a congratulatory telegram.[85] This surprising change of heart can be attributed to the influence of Eulenburg and Holstein, who had impressed on the Kaiser that to delay ratification would 'cause *very* bad blood and would be *thoroughly* exploited against His Majesty'. Holstein had written despairingly to Eulenburg: 'My God, if only our sovereign would use the gifts which nature has so richly bestowed on him to increase his power by winning popularity! It is not a question of vanity but of *power*! By flattering Zelle in showing how willing he is to ratify the appointment, H.M. will make him more or less *his* candidate.'[86] Waldersee was thunderstruck when the Kaiser's telegram appeared in the press. 'Once again one can only say that we should be prepared for anything at any time and that nothing should surprise us', he commented. 'How often and in what terms I myself have heard the Kaiser abusing the Progressive Party and its people! But he can be blown in any direction, from one day to the next; he who is in disgrace today can be praised to the skies tomorrow, and vice versa.'[87] The impression that the Kaiser was anxious to make a good showing with the Progressives was strengthened by his friendly reception of Rudolf Virchow when the latter was appointed rector of the University of Berlin. 'That would not have been possible a year ago', Waldersee exclaimed.[88] The reason for this 'flirtation' with the Progressive Party was, however, clear to all – it was intended to win their support for the Army Bill.[89]

THE GREAT ARMY BILL

The demand for a huge increase in the army put forward by the Kaiser in July 1891 continued to overshadow domestic politics in Germany more than any other question during these years.[90] As we have just seen, Caprivi's swing towards conservative–clerical school reform in Prussia had been prompted largely by his desire to win the votes of the Catholic Centre party for the Army Bill in the Reichstag. With the withdrawal of Zedlitz's School Bill in March 1892 the Chancellor was again confronted with the question of how to achieve a majority for the Army Bill. Caprivi himself was from the outset in favour of reducing the statutory period of military service from three to two years, at least for the infantry. He took this view not only for reasons of parliamentary tactics, although he certainly recognised that the Reichstag would never approve an increase in the army if three-year service, which was deeply unpopular in the country, were retained. The Chancellor-General also favoured the reduction of military service to two years on military grounds, seeing it as a welcome step on the path towards universal conscription and as a means of rejuvenating the infantry.[91] Other senior military figures – not least the former Minister of War General von Verdy and the future Field Marshal Colmar Freiherr von der Goltz – were equally keen on the idea of a nation in arms and therefore supported the introduction of two-year service as a long overdue measure to modernise and popularise the army.[92] The Kaiser, however, had shown almost religious zeal in categorically rejecting the concession of two-year service in his letters to Kaltenborn and Caprivi in July 1891,[93] and now agreed only with great reluctance to a compromise put forward by the Chancellor, under which two-year service would indeed be introduced but only for certain units. Everyone was aware of how unconvincing a solution this was, and of the precarious situation it had brought about. As the compromise by no means satisfied the Left Liberals – the parties of the left held the view that the increase in the army constituted an unacceptable burden for the population as a whole[94] – it was evident that the Chancellor would in any case be forced to make further concessions, either to the two Progressive parties or to the Catholic Centre, in order to achieve a majority.

Waldersee, naturally enough, was one of the most determined opponents of Caprivi's reform plans. From the beginning the influential former Chief of General Staff recorded in his diary that to his 'horror' Caprivi had declared himself in favour of two-year service and had won round Minister of War Hans von Kaltenborn to his point of view.[95] The Chancellor would now start 'working on the Kaiser in this sense', he predicted. 'Until recently the Kaiser very strongly supported 3-year military service; but regrettably I no longer have any confidence in his reliability and I do not think that there is any subject on which a change of heart can be ruled out.'[96] In early 1892 Waldersee welcomed a report that the Kaiser had 'recently spoken out decisively against two-year service', but he added: 'If only I could be more confident that he would keep to it!'[97] Caprivi's willingness to exploit his contacts with the press on this question reminded Waldersee of the last phase of Bismarck's domination and seemed to him 'one of the most brazen and vile acts' imaginable; if 'the Chancellor and General' had indeed 'started a press campaign against his Kaiser and War Lord', both civil and court martial proceedings ought to be instituted against him.[98] A new chancellor crisis seemed inevitable, despite the fact that the Kaiser told General von Versen at the end of 1891 that he really could not 'let [Caprivi] go now'.[99]

In the Kaiser's entourage Caprivi could at first count on the support of Adjutant-General von Wittich (although after the School Bill crisis even he turned his back on the Chancellor),[100] but otherwise his plans were regarded with hostility both at court and in army circles, and certain of the ruling federal princes also urged the Kaiser to retain three-year service for the entire army. They argued that one could not train young recruits to be reliable soldiers loyal to the monarchy in under three years. Some even predicted the downfall of the Prusso-German military monarchy and the advent of parliamentary government if the concessions proposed by Caprivi went through. Prince Bernhard of Saxe-Meiningen, who was at this time working in the General Staff under Schlieffen, expressed the opinion that the average German was 'a refractory type who [could] be turned into a competent soldier only by strict discipline maintained for a considerable time', hence the need to retain three-year service.[101] In early 1892 Prince Bernhard felt able to reassure Waldersee that the Kaiser had spoken to him 'in such strong terms' against two-year service that he was convinced that 'he would not let himself be won round by the Caprivi-Kaltenborn-Wittich triad'. He, Bernhard, respected all three Generals greatly, but he would 'never' understand how they could champion such an idea, for 'a measure so damaging to the efficiency of the army . . . could bring about the downfall of Germany's sovereign independence and might perhaps even cost the Kaiser his crown'.[102] In the months that followed, Wilhelm's brother-in-law repeatedly expressed his 'patriotic anxieties and apprehensions . . . with regard to the great Army Reorganisation Bill'.[103] The Grand Duke of Baden and the new King of Württemberg also proved to be determined opponents of two-year military service and had 'a violent argument with Caprivi' about it.[104] Both South German monarchs took it upon themselves in the summer of 1892 to implore the Kaiser never to give up three-year military service.[105] Waldersee told the Grand Duke that he found it 'incomprehensible that Caprivi could persist in working against the views of his Kaiser, although he was perfectly well aware of them'.[106] Foreign Secretary Marschall feared that it might come to 'a second edition of the School Bill', with the Chancellor and the Kaiser pursuing irreconcilable policies, 'for on the main issue, two-year service, H.M. is intransigent'.[107] After a conversation with the Chancellor on 16 September the Foreign Secretary noted in his diary: 'The Army Bill has been forced on us by H.M., as I expected. I am worried about the coming winter and fear there will be a Reich chancellor crisis.'[108]

The difficulty for Wilhelm lay in the fact that he could not risk Caprivi's resignation, for after the debacle over the School Bill a chancellor crisis over the Army Bill would have been nothing less than a catastrophe amounting to 'bankruptcy for the Kaiser'.[109] Waldersee summed up the monarch's dilemma thus: 'If it is true that the Kaiser will not hear of concessions, a row can scarcely be avoided; if he makes significant concessions – and without them nothing is likely to be achieved – the army will suffer damage and his own reputation will be diminished.'[110] All eyes were therefore on the Kaiser when on 18 August 1892 he made a passionate speech declaring his preference for three-year military service and going so far as to say that he would prefer a small, efficient army with three-year service to 'a great horde with two years'.[111] Although Waldersee was among the most prominent champions of three-year service, he roundly condemned this remark as 'totally useless'. As he pointed out, the Kaiser had 'declared himself – as he has already done at least a dozen times in the course of the last few years – against two-year service, while at the same time the Ministry of War, with his consent, has planned the reorganisation of the army on the basis of two-year service', and this plan was shortly to be laid before the Reichstag. 'Now he suddenly bursts in with his speech, so that no one knows what is actually happening.' After the speech, Caprivi had at first thought that he would have to resign, but then he had 'soon calmed down', Waldersee reported, adding contemptuously, 'treacherous scoundrel that he is!'[112]

In the following weeks, to the disgust of the opponents of army reform, Wilhelm again allowed Caprivi to talk him into a compromise solution by which two-year service would be introduced for infantry regiments but not for the cavalry.[113] As Prince Bernhard wrote indignantly to Waldersee: 'As recently as 18 August His Majesty made his aversion to the introduction of two-year service absolutely clear in his speech; very soon afterwards those who rule the roost at the moment must have succeeded in allaying our All-Highest sovereign's fears and converting him to the opposite view . . . I am astonished . . . at their boldness in wanting to introduce something completely untried all at once, while giving in to misguided public opinion, clearly to the disadvantage of the real efficiency of the principal arm of our defence forces, and all for the sake of achieving other gains. This yielding to the pressure of public opinion I find alarming, for there is nothing one should be more careful to avoid than giving in to this treacherous power when it is motivated not by the desire for improvement, but by the selfish wish to rid itself of a burden. The most popular solution would of course be zero-year military service or at any rate a militia system on the Swiss model. There is no doubt where that would lead us.' In the opinion of the Kaiser's brother-in-law Germany was approaching an exceptionally critical moment in its military history. It was even possible that 'the incomparable work of Kaiser Wilhelm I will be irreparably ruined and the country will be spending billions on an army which is incapable of success on the battlefield'.[114]

The credibility of the government now depended on its maintaining a consistent line and adhering to the compromise which had been achieved with such difficulty, even at the risk of conflict with the Reichstag. The Kaiser, however, continued to manoeuvre indecisively between the different camps. After a discussion at Rominten, his East Prussian hunting lodge, with Philipp Eulenburg in September 1892, he gave orders for the political parties to be informed that he would not hesitate to dissolve the Reichstag if the Army Bill were rejected.[115] Barely three weeks later he indicated that he was determined to stick to Caprivi 'through thick and thin', and that he had accepted the necessity of two-year service for the infantry. Waldersee noted incredulously in his diary: 'The Kaiser has given up the struggle completely and has suddenly become a decided supporter of two-year service! As he is a master of the art of deceiving himself and freeing himself from blame, he now says that he had never before known that our infantry served for only 2 ¼ years, not 3!!! . . . Did anyone ever pride himself more than he did on knowing everything about the army, he who of course knows everything better than experienced men? And now he confesses to complete ignorance on one of the most important questions in the army, on which he has for years, on countless occasions, stated his opinion! . . . What are we to expect next from a man of such a character!'[116]

The Kaiser's support for Caprivi's bill proved decisive, at least within the military and civilian elite. Many observers were troubled by the spineless attitude towards the monarch shown even by the highest-ranking army officers over this issue. The leading members of the army establishment now knew what was wanted 'on high' and adjusted their reports accordingly, Arthur von Brauer complained. 'On whom can the Kaiser still depend, if such highly placed officers simply echo his words?', this Bismarckian statesman commented. 'The number of men who still have the courage of their own convictions grow fewer and fewer.'[117] However, the parliamentary fate of the bill remained extremely uncertain even after the decision to concede two-year service for some army units. The Centre Party, on whom the adoption or rejection of the bill ultimately depended, concealed its hand in order to extract as many concessions as it could for itself and the Catholic Church in Germany. On the other hand no one except the Social Democrats and the Progressives wanted to face new elections. The leaders of the Centre Party hoped to do further business with Caprivi and recognised that rejection of the Army Bill would lead to his dismissal; the Conservatives and Free Conservatives were unwilling to come into conflict with the Kaiser over a military question; and the National Liberals feared that they would lose votes to the anti-militaristic Left Liberals if the Reichstag were dissolved.[118] Following a visit to the capital in October 1892 Waldersee recorded: 'The Kaiser of course takes the matter very lightly and goes on living without a care in his head.'[119]

When the much-debated Army Bill was brought before the Reichstag at the end of November 1892 few trusted the Kaiser's resolve.[120] Doubtless for that very reason Wilhelm at first identified himself fully with the bill as being 'his own work' and let it be known that he would regard any opposition to it 'with great displeasure'.[121] In a letter to Franz Joseph he expressed confidence that the 'opposition' to 'our bill' was in decline everywhere, while the people's appreciation of it and 'hence its chances of success' were growing daily.[122] This staunch support, however, also attracted bitter criticism, and there was renewed talk of the downfall of the monarchy and even of the German Reich. Waldersee condemned the Kaiser's 'unfortunate passion for making all such causes his own; instead of putting the ministers in the forefront he allows them to push him forwards, and indeed probably pushes himself forwards. If the bill is rejected', he warned, 'it will now be a defeat for him!' The Kaiser had no inkling 'of what a grave situation he is facing; it is by far the biggest step that he has taken since he came to the throne and a complete failure . . . would mean the beginning of the break-up of the Reich', he predicted. The Kaiser was reported to have said that if the bill did not go through the first time it would perhaps do so at the fourth or fifth attempt. 'What immature, childish ideas', Waldersee complained.[123]

There were thus conflicting reactions to the Kaiser's New Year address to the commanding generals in January 1893, which Eulenburg, urged on by Holstein and Kiderlen, had advised him to make.[124] Holstein thought the effect of Eulenburg's letter had been 'almost too powerful', for the Kaiser had gone so far as to speak of 'sweeping away the Reichstag',[125] which had made an undesirable impression particularly on the non-Prussian generals present – Prince Georg of Saxony, Prince Arnulf of Bavaria, the Bavarian General von Parsefal and General Wilhelm von Wölckern from Württemberg. Waldersee, who also heard the address, recorded in his diary that the Kaiser had claimed that the Army Bill had been 'carefully considered; all the princes had agreed to it; it was essential and he would see it through. He stated that it had come to his attention that in the army it was being adversely criticised by some individuals; he could not allow this; when he put forward a bill it was the duty of every officer not to find fault with it under any circumstances; he laid on us the duty of ensuring that the officers desisted from any criticism. Then he became livelier and turned his attention to the Reichstag, saying in effect: I shall get the bill through whatever it takes; what does this bunch of civilians know about military matters. I shall not lose a single man or a single mark, and I shall *send this half-crazy Reichstag to the devil if it opposes me!*' Commenting on the apprehensive mood among senior army officers after the speech Waldersee noted: 'When the Kaiser had gone groups immediately formed and the universal impression seems to have been that under *no circumstances* should a single word of the speech become public knowledge . . . But both the princes, and Generals Parceval and Wölckern too, are in duty bound to report the speech to their sovereigns; they in turn will speak to their ministers, and thus it will inevitably become public knowledge in the end. What might happen if the remarks about the Reichstag should become known! It would be an incalculable disaster!'[126] It was deplorable that the Kaiser had again committed himself so personally to the bill, for 'after what he has said it is impossible for him to retreat . . . unless, that is, he has fully realised the consequences of a dissolution of the Reichstag and has decided to fight the thing out. But he lacks the energy to do so.'[127] In fact the Kaiser's derogatory remarks did not remain a secret. The Hereditary Prince zu Hohenlohe-Langenburg, for example, wrote to his father about them, commenting: 'That the Kaiser expressed his firm determination not to give way over the army question pleases me very much. A surrender would damage his prestige at home and abroad. I think we need strong action by the government above all, even at the risk of a conflict. A repetition of the retreat over the School Bill would only help the opposition and would badly damage the standing of the Reich.'[128]

At the end of April 1893, after negotiations with Karl Freiherr von Hoyningen-Huene, the leader of the right wing of the Centre Party, Caprivi had hopes that a compromise would be reached, especially as the Kaiser declared himself willing to give up his visit to the Görtz family in Schlitz and shorten his stay in Karlsruhe in order to return to Berlin for the critical vote in the Reichstag. 'That will make a good impression', the Chancellor believed. 'God grant that we shall see Kaiser and Reich emerge unscathed from this difficult situation.'[129] Eulenburg urged the Kaiser 'to come roaring back to Berlin like Jupiter *tonans* from Karlsruhe or wherever else *within* Germany' he might be for the vote. Kiderlen likewise argued that '*for the sake of the Kaiser's prestige*' the dissolution of the Reichstag must not take place during his absence in Italy. 'If H.M. is in Karlsruhe it does not matter – on the contrary, it would make a colossal impression if he came storming back unexpectedly from there at the crucial moment.'[130] Wilhelm returned to Berlin on 3 May in time for the decisive parliamentary vote.

On 6 May 1893 the Army Bill was rejected by 210 votes to 162 and the Reichstag was dissolved. Germany now faced one of the most severe political crises since the fall of Bismarck.[131] The situation was drifting 'perilously close to a *coup d'état*', Waldersee commented.[132] Eulenburg, who had come from Munich to Berlin for the vote, noted in his diary that the Kaiser was 'very serious and violently attacked the Centre and the Bavarian wing, declaring that the Centre had burnt its bridges for at least 10 years. He intended to deliver to the Bavarian government, in the person of Crailsheim, who is to dine at the Neues Palais tomorrow with the other prime ministers who have come to Berlin for the dissolution, a stern reminder of its national obligations. I begged him to speak only in "Olympian" generalities, as kaiser, and to leave the details to me, which he agreed to do. I tried to calm the Kaiser by reminding him that the dissolution had taken place in much more favourable circumstances than could have been expected even eight days ago.'[133] When the ministers and envoys of the federal states arrived at the Neues Palais for a reception on 7 May the Kaiser made his views plain. To the Minister-President of Baden he said: 'When I returned to Germany [from Italy] the situation was clear. Acceptance of Huene's proposal or dissolution! I had already told the Grand Duke this when I was in Karlsruhe; in Berlin it was even plainer to me than from afar that this was inevitable . . . With a shortfall of 48 votes there was no alternative to dissolution.' To the Württemberg envoy Wilhelm remarked that 'these gentleman now had to be shown that we were in earnest. It would be no bad thing if a large number of Social Democrats were to get into the Reichstag, for it would give the philistines a fright and in the end they would themselves beg the government to free them from the impossible electoral system.' The Kaiser was in a good mood, Brauer reported to the Grand Duke of Baden, and had expressed 'his particular satisfaction that the situation was now "clear-cut" and the haggling was over'.[134]

Two days later Wilhelm made an angry speech to his generals on the Tempelhof Field which had the effect of intensifying the crisis to a dangerous degree. Referring to the Army Bill, he said: 'I could not have anticipated its rejection, and had hoped that it would have been passed unconditionally by the patriotic spirit of the Reichstag. I was, unfortunately, deceived in that expectation. A minority of patriotically minded men was unable to do anything against the majority. In the debates passionate words were spoken which are not heard with pleasure among educated men. I had to proceed to a dissolution, and I hope to obtain from a new Reichstag the approval of the Army Bill. Should, however, I be disappointed in this hope too, I am determined to do all I can to attain my object, for I am too strongly convinced that these proposals are necessary in order that I may be able to preserve the general peace. There has been some talk of inciting the masses. I do not believe that the German people will allow themselves to be stirred up by outsiders. On the contrary, I know that in these army proposals I am at one with the federal princes, with the people and with the army.'[135]

This speech, like most of Wilhelm's previous outbursts, did not escape Waldersee's censure. 'If only he could stop speaking without thinking! Such a speech cannot help in the slightest, but it can do great harm.'[136] Wilhelm's mother was outraged by this new proof of his anti-constitutional attitude and wrote: 'I was grieved and horrified to hear of the frightful speech! – I still *cannot* understand how a child of mine – a grandson of my father's – so utterly fails to understand the *importance*, the meaning and the value of a constitution, – and how there is such a world of difference between *all* his opinions and mine!'[137]

With the dissolution of the Reichstag in May 1893 the question arose of what would happen if the newly elected parliament again rejected the Army Bill. Caprivi threatened to seek another dissolution. The Prussian Minister of Finance,

Johannes von Miquel, who was becoming increasingly hostile to Caprivi,[138] warned that a constitutional conflict of this kind would only strengthen Caprivi's faltering relations with Wilhelm II. Herbert Bismarck, who had been elected to the Reichstag, was able to report to his father on a secret conversation between Miquel and the Free Conservative Party leader, Wilhelm von Kardorff, in which Miquel had declared 'that Caprivi's position with the Kaiser would only be strengthened if the bill were rejected and there were another dissolution; that was in the Kaiser's character. But if the bill were accepted by a few votes, H.M., who had already had enough of Caprivi, would grow so tired of him that Caprivi would be defeated in the regular autumn session by the determined onslaught which the Agrarians were planning to make on him then, and thus the country would be rid of the incompetent Caprivi within 5 months. The Kaiser had recently said [according to Miquel] that "he was beginning to realise that the commercial treaties recommended by Caprivi had been a stupid mistake". H.M. was beginning to doubt Caprivi's competence; this process would be interrupted if H.M.'s anger were aroused and he granted Caprivi a second dissolution. Miquel added [Herbert wrote] that he wanted to get rid of Caprivi because his narrow-mindedness was doing us harm both inside and outside the country; and he believed that this could most safely be achieved by leaving him in peace now, letting the Army Bill through by under 10 votes, and deferring the attack until the autumn.' Miquel claimed 'to know H.M. better now than anyone else', Herbert Bismarck added.[139]

The new elections resulted in heavy losses for the two Conservative parties (19 per cent down) but also for both Left Liberal parties (21 per cent down), while the Centre and the National Liberals maintained their share of the vote and the Social Democrats increased theirs by a further 18 per cent to 1.79 million (23.3 per cent of all votes). A worrying development was the increase in the number of Agrarians in the German Conservative Party and the election of sixteen overt anti-Semites to the Reichstag.[140] In spite of this polarisation the right wing of the Centre Party voted with the Conservatives, the anti-Semites and the Poles in favour of the Army Bill on 14 July 1893 to produce a small majority. The passage of the bill temporarily strengthened Caprivi's relations with the Kaiser.[141] The latter telegraphed his delight to his grandmother in Windsor: 'Army Bill passed easily. Great victory for my Government. Are all very happy.'[142] Philipp Eulenburg wrote enthusiastically from Munich to his imperial friend saying that he wished he could have left at once for Potsdam and the Neues Palais, for 'I am quite overcome with the desire to kiss my beloved Kaiser's hand after all the trouble we have had with this child of sorrows, the Army Bill! It is as though a heavy burden has been lifted from my soul after all the difficulty and anxieties which the child of sorrows caused even for me here in Bavaria! I see the adoption of the bill as a great political success. With this the first firm stone has been laid in the work of reconstruction after the unfortunate School Bill.'[143]

THE KAISER AND THE CONSERVATIVES

As has already been shown in an earlier chapter Kaiser Wilhelm, for all his inconsistency, tended towards the liberal rather than the conservative end of the party spectrum in domestic politics during the Caprivi years,[144] and the same can be said of the period after the School Bill crisis. Indeed it was most strikingly demonstrated in relation to the troublesome question of mass anti-Semitism, which came to a head once more around 1892. Although full of anti-Semitic prejudices to which he occasionally gave crude expression, the Kaiser was very far from approving the loutish anti-Semitism displayed by the likes of Hermann Ahlwardt, who was elected to the Reichstag in the winter of 1892 but was at the same time condemned to a harsh prison sentence for publishing the inflammatory pamphlet *Judenflinten*.[145] The trial of Ahlwardt coincided with the conference of the German Conservative Party at Berlin-Tivoli, at which the members adopted by a large majority an anti-Semitic clause in the official party manifesto. Waldersee, although himself a vehement anti-Semite, had for some time warned of the threat posed by the radical anti-Semitic movement, which he saw as a new form of socialism. Nonetheless he greeted this controversial decision by the Conservatives as the best tactic for taking the wind out of the sails of rabble-rousers such as Ahlwardt, Liebermann von Sonnenberg and Böckel. In his diary for 18 December 1892 the General reflected on this worrying development: 'A new element, which has long existed but has been disregarded by our national leaders, has suddenly appeared in the political affairs of this country and has rapidly acquired importance; it is anti-Semitism! I have been observing its development for years and in my opinion its leaders, at least in so far as they have appeared in public, are very dubious creatures, like Liebermann, Ahlwardt and their cronies. But I also have the impression that the movement is deeply anchored in the population, and I am convinced that it has much justification. Its aims are not clear, but if it is allowed to grow unchecked and unmodified it must inevitably lead to socialism and become like the social democratic movement. I have often warned Conservatives against throwing in their lot with anti-Semitism, because with such disreputable leaders alliances of that kind are far too easy to exploit for attacks. The Conservatives have recently held a conference in Berlin at which expressions of anti-Semitic beliefs received overwhelming support. People said that this is now the mood in broad sections of the population of this country; we must take account of it if we want to keep our hold over them.' Doing so, Waldersee considered, was 'fully justified'.[146]

The attitude adopted by the Reich Chancellor and the Foreign Office towards this thorny question was diametrically opposed to Waldersee's – Caprivi made a speech in the Reichstag against anti-Semitism for which he received much praise – and they strove, not without success, to persuade the Kaiser to publicly condemn the new development in the Conservative Party. Holstein warned that Waldersee's position had been greatly strengthened by the adoption of the anti-Semitic clause in the party manifesto. Until recently the General had had 'a negligibly small hold' on the Conservatives because they considered him an 'unprincipled self-seeker'; in fact he had had the backing of 'none but the anti-Semites'. But now, since the Tivoli conference, the Conservative Party was 'overrun by anti-Semites: Waldersee's position has thus been strengthened at a stroke; he sees the moment coming when he will be able to dictate laws and conditions to the Kaiser'.[147] The Kaiser must therefore be induced to make a public declaration against anti-Semitism, Holstein argued. He welcomed the fact that the Chancellor had 'asked His Majesty to express his disapproval to the Conservatives of the demagogic attitude of the party conference'. But instead of doing so in unmistakable terms, Holstein complained, the Kaiser had 'called out laughingly to Manteuffel when they next met: "You and your friends ought all to be hanged!" Naturally Manteuffel does not see that as a criticism', he reported to Eulenburg.[148] The latter therefore stressed, in a very persuasive letter he wrote to the Kaiser on 17 December 1892, his conviction that 'in the interests of Your Majesty it is *essential* to adopt a clear-cut attitude towards anti-Semitism – *in whatever form*, to demonstrate very seriously that You stand by the Reich Chancellor's attitude'. 'I am no lover of the Jews, nor is Your Majesty – we have exchanged views on that from time to time', Eulenburg recalled. But as a result of the Conservative Party joining forces with the demagogues of the anti-Semitic movement, the political situation in Germany had been given a 'crooked face' which needed a timely box on the ears to straighten it out. 'The ugly face I mean is the latest move of the Conservative ultras towards anti-Semitism', wrote the Count in his inimitable style. 'It sticks out its tongue at all of us – at Your Majesty too!' The Reich Chancellor had already ensured that anti-Semitism had been 'knocked over the head'; 'but everything depends on there being *not the slightest doubt* that Your Majesty is behind him – otherwise Your Majesty's sympathy for the anti-Semites will be used to drive a wedge between Your Majesty and the Reich Chancellor . . . There are serious reservations about anti-Semitism itself, in the form in which it appears. When Your Majesty has successfully weathered the political transition brought about by the departure of Prince Bismarck and his faction, *then* it may perhaps be possible to establish a means of attending to the justified complaints of anti-Semitism.' But on grounds of self-preservation alone the state must make a strong stand against mass anti-Semitism in its present form, 'for the slightest suspicion of silent complicity on Your Majesty's part would shake the monarchical principle *to its foundations*. Your Majesty would become a "*roi des gueux*" – and with that Hohenzollern is lost . . . If the word is spread everywhere . . . that everything has its price – military and civilian – if respect for authority is taken away from country people, clergy, minor officials – by the anti-Semites in competition with the Social Democrats – the seed will soon bear fruit which sees all *property-owners* as corrupt . . .'[149]

The determined and impressive action taken by Caprivi, Holstein and Eulenburg did not fail to have the desired effect on the Kaiser's attitude to the German Conservatives, as Waldersee was to be the first to discover. The General complained bitterly that the Chancellor had 'had the bad manners, or indeed the baseness, to speak about it in the Reichstag and he has of course worked on the Kaiser in the same sense. The latter is naturally furious with the Conservatives again and has willingly harnessed himself to Caprivi's carriage which is deep in the mud. A few days ago . . . he was very unfriendly to Manteuffel, the leader of the Conservatives on this question, and on the 16th he had a long, heated conversation with Count Schulenburg Beetzendorf on the subject. There is no question that he will personally try to work on other Conservatives as well.'[150] In the Kaiser's hostile attitude to the Conservatives Waldersee saw the confirmation of his earlier suspicion that Wilhelm in fact wanted to throw in his lot with the liberals.[151] 'What I tried to foretell six months ago now seems to me to be becoming more probable: the Kaiser is heading in the liberal direction, and if he does not get as far as Eugen Richter [on the radical left] immediately, he will at least catch up with [the more moderate leader Heinrich] Rickert. I think that would in fact be the best possible thing', he added bitterly. 'He must first find out what these people are like, and I shall not be surprised if he soon becomes a friend of the Jews and in the end even allows those Jewish scoundrels to become officers.'[152]

After some initial hesitation, Wilhelm took similar action on another question that aroused party political passions at this time: the commercial treaties with Austria-Hungary, Italy and Russia. Since the autumn of 1891 Caprivi, who had no experience of economic matters, had had to concern himself with the renewal of these trade agreements which Bismarck had made ten years earlier. He soon became convinced that in view of Germany's rapid industrialisation and urbanisation the high corn duties which his predecessor had conceded to the great landowners east of the Elbe could no longer be justified. In taking this view the Chancellor had the backing of his closest adviser, the Chief of the Reich Chancellery Karl Goering, who was credited with Left Liberal leanings.

Naturally he came into increasing conflict with the protectionist interests of the East Elbian landed aristocracy, who formed a powerful nucleus within the German Conservative Party, and likewise with Miquel and other Prussian ministers.[153] When the commercial treaties with the Danube monarchy, Italy and Belgium were ratified on 18 December 1891 the Kaiser expressed his warm thanks to Caprivi in a speech at Teltow, to the fury of the Conservatives, and raised 'this simple Prussian General', as he described him, to the rank of count.[154]

At the beginning of 1893, when it became known that Caprivi's government intended to conclude a commercial treaty with Russia as well, the Prussian landowners formed the Agrarian League to try to protect their interests. Waldersee noted with bitterness the Kaiser's hostile attitude to this 'powerful' popular movement and the landed classes it represented, which had hitherto been the mainstay of the Prussian monarchy. 'That *this* Kaiser would lend his hand to the ruin of the agricultural community, in which his most reliable people are still to be found, from which the state gets its strongest support, I would have thought impossible a few years ago. When I think of the promises he has made, how he has sworn never to allow any harm to come to agriculture! What will become of his reputation; no one can rely on him and on his word again.' The General gloomily predicted a speeding-up of the process by which the families of the old nobility were abandoning their estates and being replaced by people who had made their money in trade and industry. 'The army will take this hard too', he feared.[155]

It was perhaps characteristic of Kaiser Wilhelm II that he chose to see the protests of the Agrarian League against Caprivi's free trade policy as directed against himself.[156] Before long a wide gulf had opened up between him and the Conservative landed aristocracy of Prussia, a development that temporarily improved relations between him and Caprivi. As the Postmaster-General, Heinrich von Stephan, told Waldersee in October 1893, the Agrarians' attacks were less of a threat for Caprivi, 'because the Kaiser is in complete agreement with him on the agricultural question'.[157] Press reports of these attacks were immediately shown to the Kaiser who, as Miquel confirmed, was 'infuriated' by them.[158] Waldersee commented with dismay that it would be 'very sad' if the Kaiser, who had often declared that he would protect agriculture, 'let down the people who trusted his word on this question too'.[159] When in late November 1893 the Kaiser received the news that the vote for the return to Germany of the Jesuits had been carried in the Reichstag with the help of a few Conservatives, he remarked to Waldersee that the names of these deputies should be taken down, and 'none of them will ever be invited to the Schloss again. Then he expressed his displeasure with the Conservatives in general.' He went on to say that 'they did not know what they wanted and were really no longer a coherent party at all'.[160] During the parliamentary debates on the commercial treaties there were fierce clashes between the Conservatives and the Chancellor, in which the Kaiser did not hesitate to intervene. 'It would be right and proper for him not to interfere', Waldersee wrote on 15 December 1893. 'But he cannot resist doing so, and this is one of the causes of the confused state of affairs we are in.'[161]

Ultimately the ferocity of Wilhelm's antagonism towards the East Elbian landed nobility in the winter of 1893–4 can be explained only in terms of the foreign policy hopes he pinned on the trade treaty with Russia. 'The Kaiser himself was quite deliberately courting the Tsar now', Thomas Nipperdey writes, rightly pointing out that the trade treaty represented an attempt to renew the rapprochement with Russia, and was intended 'in effect to replace the Reinsurance Treaty' which had lapsed in 1890.[162] When the ratification of the Russian trade treaty by the Reichstag seemed in jeopardy, Wilhelm personally took up the cudgels for the treaty and informed the Conservative deputy and Speaker of the Reichstag, Albert von Levetzow, 'that anyone who wore military uniform and frequented the court must vote for the government, or at least abstain'.[163] At a dinner for members of parliament given by the Chancellor the Kaiser again expressed his support for the trade treaty, remarking that he had no desire 'to wage war with Russia for the sake of 100 stupid Junkers', whereupon Levetzow, otherwise a quiet, venerable figure, countered by declaring to the Kaiser's face 'with raised voice and great determination' that the loyalty of the Conservatives to the monarchy was beyond all doubt, even if after careful examination they should consider it their duty to vote against the treaty. He added 'with deep emotion' that this day was the saddest of his life. 'And truly the Kaiser has no more loyal subject!', Waldersee lamented, adding that the other Conservatives present also left the dinner 'deeply dismayed'.[164] The Kaiser had even 'uttered threats' against Count Kanitz, the Master of Ceremonies at court, Waldersee recorded.[165] A few weeks later Wilhelm said, again addressing Kanitz: 'The trade treaty must go through: I have given Tsar Alexander my word!!!!'[166] He informed Duke Johann Albrecht of Mecklenburg that he was having all known Agrarians who came to Berlin watched by the police in order to form an impression of their alleged state of distress. Waldersee doubted that the monarch would go quite so far, although he admitted 'that the Kaiser encourages informers to report to him on this subject, as he always enjoys doing, and that there are certainly plenty of scoundrels around who add fuel to the flames by such gossip'.[167] In spite of threats of this kind, when the trade treaty with Russia came to the vote on 16 March 1894, the Conservatives rejected it, with only a few exceptions – Count Dönhoff and Prince Karl zu Hohenlohe-Schillingsfürst, the son of the Statthalter of Alsace-Lorraine, voted in favour, according to Waldersee, purely 'out of fear of the Kaiser'.[168] When the treaty was nevertheless ratified with the support of the Centre, the Polish deputies, the two Progressive parties and the Social Democrats, Wilhelm again hastened to send the glad news by telegraph to his grandmother.[169] Philipp Eulenburg was not entirely wide of the mark when he wrote to the Kaiser on 11 March saying that the ratification of the Russian trade treaty, which was 'of epoch-making importance both politically and economically', was his '*great, personal* achievement'.[170]

After this success the Kaiser, while vacationing at Abbazia on the Adriatic, was so taken up with the pleasures of 'water sports, tennis, vulture-shooting and the fleet' that for the time being he took no further interest in matters of domestic legislation, as his companion Eulenburg, whom he had just promoted to the post of ambassador in Vienna, reported.[171] On his return, however, a fresh source of conflict between him and the East Elbian landowners arose in the form of the canal question. As Eulenburg recalled: 'When . . . the Kaiser came back from Abbazia to Berlin in mid-April his mood . . . was dominated particularly by the fact that the Conservatives, as a result of the undeniably serious plight of agriculture, had adopted a position diametrically opposed to the government, which at the instigation of the Kaiser had . . . put forward draft legislation for the construction of canals.' Eulenburg rightly stressed that this canal policy had been 'inspired by the Kaiser himself and pursued with his impulsive zeal'. The German Conservative Party, which had only reluctantly agreed to the construction of the Elbe–Trave canal, now refused to approve a further 55 million marks for the building of the Dortmund–Ems canal. But the Kaiser, as Eulenburg recorded, 'took this refusal as an insult aimed at him *personally* (!), ordered the names of all members of the Conservative Party who had voted against the canal to be struck off the court list (!) and wanted to "remove from office" the court chamberlains among them, which was only prevented thanks to the intervention of my cousin, the Oberhofmarschall Count August Eulenburg. What increased the Kaiser's rage (I cannot use any other word) even more was that with his impulsive instinct he sensed behind the attitude of the agrarian Conservatives – probably not without reason – machinations by Prince Bismarck. The blows which certain notorious hotheads aimed at the government during the negotiations in an extremely blunt (not to say *brutal Junkerish*) tone were really the last straw. For the Kaiser, who had unfortunately made an all too *personal* commitment to the construction of the canals, took this tone as a personal rebuff *to him* by the Conservatives and would not be placated. The cruel way in which the Great Elector had crushed the aristocracy and Frederick the Great's treatment of refractory nobles were frequently cited as the only proper way of teaching the insolent aristocracy a lesson.'[172] An eloquent example of this method of dealing with the aristocracy occurred in February 1894 when the Kaiser was informed that the Catholic Count Hompesch had said the Centre Party would not approve more than 2 million marks for the monument to Kaiser Wilhelm I. In the hearing of a large number of people Wilhelm II declared: 'I shall have the fellow thrown down the stairs if he comes into the Schloss.'[173] The domineering tone which Wilhelm II used towards the aristocracy was no different, as we shall see, from that which he used in his dealings with the Reich Chancellor.

A survey of the crisis-ridden relationship between the Kaiser and his Chancellor in the middle years of Caprivi's period of office thus continues to reveal a contradictory picture in which neither the monarch nor the leading statesman of the Reich appears unequivocally dominant. Certainly, Wilhelm II succeeded in halting Count Zedlitz's highly controversial Prussian School Bill in March 1892. But this was only after a long period of indecision and by unwittingly provoking a serious ministerial crisis which caused considerable damage to the reputation of the monarchy and to the cohesion of the Prussian and Reich government. The great army increase of the summer of 1893 was unquestionably the product of the Kaiser's personal initiative, but its passage through the Reichstag was only achieved by the abandonment of the three-year service which the Kaiser, remembering his 'grandpapa's' heroic stand in the Prussian constitutional conflict of the 1860s, had sworn never to give up. With the aid of contemporary correspondence and diaries we have been able to follow month after month of upheavals caused in both these instances by the dualistic division of power between Kaiser and Chancellor. As the sources also show, in a third extremely controversial instance, the trade treaty with Russia, Wilhelm II stood shoulder to shoulder with Caprivi against the East Elbian landed aristocracy which should have formed the natural mainstay of the Hohenzollern monarchy and had become the sworn enemy of the Chancellor they called 'the man without an inch of land or a blade of grass'.

It was a matter of universal speculation how long this crisis-ridden system of divided power could go on. No one seriously doubted, however, that it would be the Kaiser and not the Chancellor who emerged victorious, for while the position of the unpretentious Chancellor-General was becoming visibly weaker, that of the young Kaiser was growing stronger. Caprivi's surprising commitment to Zedlitz and his reactionary School Bill had shaken the confidence of the anticlerical Liberal bourgeoisie of all shades of opinion in his judgement, without winning him the support of the orthodox Protestants or of the Catholic population. Through his obstinate insistence on giving up the minister-presidency of Prussia to Botho Eulenburg he lost control over the Prussian Ministry of State, a fatal development which was aggravated by Wilhelm II's wilful manner of making appointments. In the endless battle over two-year military service

Caprivi proceeded to make bitter enemies of the influential military at court and at the 'front', who had the backing of some of the federal princes and members of the imperial family. Finally the new elections in the summer of 1893 brought Agrarians and anti-Semites into the Reichstag, who literally persecuted the increasingly isolated Chancellor with fanatical hatred. The future, on the other hand, belonged to the young monarch. Since he could not afford a second change of chancellor in the first few years following Bismarck's dismissal he had been obliged, for the time being, to hold on to the stubborn Caprivi and support the measures he demanded, even sometimes against his will. But at the latest by the spring of 1894, four years after Bismarck's departure, when the Kaiser was widely considered to have given sufficient proof of steadiness, a change of chancellor might again be contemplated. Even in the dualist phase which we examined in chapter 15 and in this chapter the enormous potential power which the crown possessed through the 'kingship mechanism' had already become apparent. Senior officers refused to speak of the Army Bill except in laudatory terms because they knew that the Kaiser saw 'every contradiction almost as an act of malice';[174] the officials of the Wilhelmstrasse became increasingly reserved even in their 'intimate private utterances' because they were aware 'that H.M. . . . is very sensitive to the slightest criticism';[175] and even experienced parliamentarians expressed surprise at 'the fear of the Kaiser that reigns in the Reichstag'.[176] If Wilhelm II thus had enormous potential to exercise both direct and indirect influence on the formulation of domestic policy in the Reich and in Prussia even under Caprivi, his freedom of action in the sphere of foreign policy, where he was not obliged to compromise with the Chancellor and was not subject to any sort of parliamentary control, was infinitely greater.

CHAPTER EIGHTEEN

Dynastic diplomacy

WILHELM II BETWEEN RUSSIA AND GREAT BRITAIN

IN chapter 14 we were able to trace the way in which the non-renewal of the secret treaty with Russia, the attempted rapprochement between Germany and Britain and the renegotiation of the Triple Alliance with Austria-Hungary and Italy in May 1891 helped bring about a strengthening of ties between the Tsarist empire and the French Republic, a menacing development to which Wilhelm II reacted with apprehension and anger. In the period following this Franco-Russian fraternisation, which found its most dramatic expression in the enthusiastic reception of the French fleet at Kronstadt in July–August 1891, the Kaiser tried desperately to restore the former friendly relationship between the two empires through his dynastic ties with the Russian imperial family – the sister of his grandfather Wilhelm I was the grandmother of Tsar Alexander III. His advances were repeatedly met with a contemptuous and wounding rebuff. He was particularly offended by the fact that in the autumn of 1891 the Tsar bypassed Berlin on his way from Copenhagen to St Petersburg without taking the slightest notice of him. He complained angrily to his grandmother Queen Victoria that the Tsar had thought fit 'to snub us in the most untoward manner, without any cause whatever, thereby putting aside the simplest rules of international courtesy en vogue between European courts'.[1] Following conversations with Wilhelm in Hanover and Springe, Waldersee noted: 'The Kaiser is very bitter against Tsar Alexander and quite rightly so, but also through his own fault. He had in fact hoped for a visit in Berlin and unfortunately went so far as to propose travelling to Danzig, but once again the Tsar evaded it under the pretext of being in deep mourning. This will naturally increase Russian arrogance still more, while we feel increasingly humiliated.'[2] There were alarming signs that internationally, and above all in Russia, Germany and its Kaiser were 'no longer as highly regarded as they were a few years ago'. Prince Ernst zu Hohenlohe, who had meanwhile been appointed to the German embassy in St Petersburg, reported in January 1892 that on Wilhelm II's birthday 'not a single grand duke [had] appeared at the embassy to offer congratulations, and very few other people called, whereas previously the house had been overrun on such days'. The ambassador, General von Schweinitz, had walked up and down his room in his uniform, 'smiling grimly', while the servants in full ceremonial dress stood expectantly around. Schweinitz commented: 'I feel like a lady who is receiving, and who is put to shame in front of her own servants because no one comes.'[3] In the autumn of 1892 Wilhelm himself was forced to admit that 'my relations with St Petersburg really no longer exist'.[4] The military representative in Russia, Lieutenant-Colonel Karl von Villaume, remarked in a conversation with Waldersee that it was only abroad that one could judge how much Germany's standing had fallen since Bismarck's dismissal. 'The only good thing was that people credited the Kaiser with great energy, and thought that he might suddenly let fly or bring off some unexpected coup', to which Waldersee retorted: 'One can only wonder what will happen when the world realises that he is basically a weak man from whom one can indeed expect strong words, but no deeds.'[5] Looking back on developments in foreign affairs since Bismarck's fall, Waldersee commented dejectedly at the end of 1893 that 'things can seldom have changed as rapidly and completely between 1890 and now as in this sphere'.[6]

If Wilhelm's Russian relations saw his attempts at rapprochement and his protestations of peaceful intentions[7] as dishonest and manipulative, it was not least because of the great military expansion which he had pursued since July 1891, as well as his own frequent belligerent utterances, which found their way to Russia through countless different channels. As Waldersee was dismayed to hear during a visit to Berlin in April 1892, Tsar Alexander himself was convinced that Germany was devising 'plans of attack' on Russia. 'Unfortunately careful investigations have shown that our Kaiser is probably to blame for the Russian attitude', he wrote, for Wilhelm had 'repeatedly and most imprudently made remarks against Russia and about how he intended to beat the Russians. I know of 3 or 4 such instances, and the last one I heard of was in Remplin, at the wedding of the Prince [Albert] of [Saxe-]Altenburg and the Princess [Helene] of [Mecklenburg-]Strelitz, at which numerous Russians were present. I do not doubt that such remarks are made even more frequently in the circle of his family, and they are of course passed around at once. The Empress Frederick and her daughters alone would ensure they were spread about . . . It is small wonder if such a dull-witted and timorous man as Tsar Alexander actually becomes

obsessed with the idea that we wanted to launch a war. But it is terrible that Kaiser Wilhelm can cause such damage by making rash speeches.'[8] Months later Prince Albert of Saxe-Altenburg confirmed to Waldersee that Alexander III had told him 'that Kaiser Wilhelm II had given grounds for suspicion [in Russia] by his remarks, to most of which he had probably given no thought whatsoever'. The General commented that the Tsar was absolutely right.[9]

The threat to Germany represented by the Franco-Russian alliance seemed all the more dangerous to Wilhelm II because the Vatican, and equally the Catholic Centre Party and the South German particularists, drew new hope from it for their own aspirations. The Kaiser saw in the increasingly self-assured attitude of the Vatican 'yet another proof that the ultramontanes are totally crazy and intoxicated with confidence and victorious frenzy. This has been brought about by their friendship with the French, and by the hope, nourished by this friendship, that the Triple Alliance will soon be destroyed by France *and Russia.* Exactly the same mood as was shown in the shameless and quite indescribable speeches and conduct [of the Catholic conference] in Mainz . . . We must take vigorous steps against the ultras and teach them a sharp lesson, firmly and coolly, with Protestant-German confidence, and this must also be aired *in the press*, clearly and unmistakably, the sooner the better.'[10] Waldersee likewise pointed to the significance of reports in circulation that 'the Holy Father [had] allied himself to our enemies'. 'If the Pope openly sides with our enemies in the event of war, our Catholics will be put in an extremely difficult position and one cannot expect a surge of patriotic feeling from the majority of them.' He wondered, moreover, 'how we are to incite the Poles to rebel if we do not have the Catholic clergy firmly on our side!' 'Our situation is really getting worse and worse!', he commented in despair.[11]

In the early summer of 1892 wounded dynastic pride threatened to take a heavy toll on relations between Germany and Russia when Tsar Alexander declared that during his forthcoming visit to Kiel he did not wish to come to Berlin or Potsdam, which so deeply offended Wilhelm II that he refused to countenance visiting the Tsar in Kiel. For days the Reich Chancellor and the Foreign Office strove in vain to convince the Kaiser that to refuse the Tsar's suggestion outright, as he proposed to do, would be '*too* rude'; a postponement of his visit to Kiel for a few days would be quite 'rude enough'. 'Can you imagine the effect of our Kaiser putting off the Tsar for 4 days because He has told a Silesian Count [Hans von Oppersdorff] that he is coming to stay with him!', Holstein asked Philipp Eulenburg, who was staying with the Kaiser in Prökelwitz in East Prussia. 'Do you think that such a thing has ever happened to the Tsar?'[12] Unmoved by the barrage of telegrams from Berlin, Wilhelm refused to alter his decision not to go to Kiel 'under any circumstances' and demanded that the

22. Three Emperors in a boat – Wilhelm II, Alexander III and the future Tsar Nicholas II in Kiel on 7 June 1892

Foreign Office draft his negative response to the Tsar. The Foreign Secretary telegraphed to Eulenburg to say that this was quite impossible, 'as the European situation and the position of Germany are at stake'. Finally the Kaiser agreed that Marschall could come and see him at Prökelwitz. The latter described his difficult negotiations with Wilhelm in his diary. 'I make my report to him, explain the historical relationship between Russia and Prussia and warn him against *cutting off the link completely*, for that would be playing into France's hands.' By way of a counter-proposal Marschall suggested a visit by the Kaiser to Kiel from 5 to 8 June, as the French were planning to hold a chauvinistic gymnastic festival in Nancy at that time. 'That strikes home. H.M. accepts. Immediate telegram to R[eich] C[hancellor]. Thank God!' It was 'a memorable day', Marshall later recorded.[13]

On 7 June 1892 Alexander III and Wilhelm II finally met on their yachts amidst a thunder of gun salutes and cheers in the flag-bedecked harbour at Kiel, accompanied by Prince Heinrich and the heir to the Russian throne, Tsarevich Nicholas. Wilhelm described the meeting to his mother: 'The Czar's visit went off very well. He was in the best of spirits & extreemly [sic] amiable. His son Niky has

greatly developed & is a charming well bred boy with agreeable manners. The weather was glorious & the old place looked its best. The Yacht is an enormous vessel, rather like a mailsteamer from the outside, but beautiful inside, all done up in Maple wood.'[14] After a conversation with Marschall the British ambassador reported that the meeting of the Emperors had passed off extremely satisfactorily for both sides and had made a considerable contribution to the safeguarding of peace in Europe.[15] The German Foreign Office informed the embassy in St Petersburg that the interview had 'gone off very well'. 'Both Emperors were in a very good mood and behaved much more warmly towards each other than previously.' The young Prince Hohenlohe reflected: 'Even if the meeting does not produce any fundamental change, the good impression that it made on the Tsar is of some value, as the tension which existed before was largely due to personal causes.'[16] Queen Victoria expressed her satisfaction to her grandson that the harmony between the two monarchs had made peace secure and dispelled the anxieties 'which an estrangement between you & him seemed to have caused'.[17]

In reality, as Waldersee heard, both monarchs were play-acting. The General doubted whether they had been alone even for five minutes; at any rate no political discussions had taken place. Not only had Tsar Alexander been determined to avoid them, but the Tsarevich Nicholas, who had not left his father's side for an instant, had also ensured that his father could not be inveigled into any serious conversation by Wilhelm II. Waldersee found it regrettable that Wilhelm had insisted on taking the *Hohenzollern* out to meet the Tsar, although the latter had expressly tried to prevent this by telegraphing that he wanted to visit Kiel and would not leave his ship until he arrived there. 'In spite of this our Kaiser could not resist sailing a little way out towards the Tsar and then leading him in', the General complained.[18] In the autumn of 1894, when Alexander lay dying, Waldersee was still bemoaning the fact that Wilhelm had 'often bored [the Tsar] dreadfully by his . . . importunity', adding that the heir to the Russian throne, Nicholas, had also 'remained completely indifferent to the many kindnesses and attentions and advances' towards him 'because I am sure he did not think they were sincere'.[19]

Whatever the meeting of the Emperors in Kiel achieved, it was overshadowed by the fact that at precisely the same time the Russian Grand Duke Constantine had attended the mass gathering of French chauvinists in Nancy to which Marschall had drawn the Kaiser's attention. The apparently pro-German attitude of the Tsar in Kiel took on a different aspect when it transpired that Alexander III had not only been informed of Constantine's intention of visiting Nancy, but had written to him to give his approval. Ernst Hohenlohe reported that it was 'widely assumed that the idea originated with the Russian ambassador in Paris, Mohrenheim (a Jew), and that he had intended it to reduce the impact of

the Kiel meeting, a cleverly thought out and successful chess move.'[20] Herbert Bismarck, disheartened by the turn German policy was taking, predicted in July 1893 that 'H.M. will find himself at war with Russia and the result will certainly be disastrous.'[21]

THE RUSSIAN FAMINE

With the Franco-Russian encirclement of Germany growing ever more constrictive, an unhoped-for way out seemed suddenly to be provided in the shape of a natural disaster on an enormous scale. The catastrophic food crisis in Russia caused by a succession of bad harvests had, since the summer of 1891, led to much conjecture in the courts, chancelleries and general staffs of Europe as to whether Russia was still in a position to wage war at all, or indeed whether the country was not heading inexorably towards its ruin. The Empress Frederick hoped that the Russian famine would bring a few years of peace during which the French would also be compelled to keep quiet, and that more trusting relationships in Europe would develop as a result of these conditions. 'If a *few years* of quiet development are passed, I always trust war will become more & more unlikely – only *alas* one must be prepared for it, & these preparations when on a scale as they are here are an *immense* evil & danger', she wrote to her mother at the end of 1891.[22] Friedrich von Holstein too thought the threat posed by Russia much reduced, as it 'will be destroyed in the foreseeable future by the internal damage it has suffered'. 'The only danger for us', he told Prince Hohenlohe, 'is French chauvinism', but with no hope of Russian support the French would have no desire for a war of revenge.[23] For his part Hohenlohe confirmed from the St Petersburg embassy that in view of the famine neither the Tsar nor Foreign Minister Giers had any wish to start a war. But he stressed that in the Russian capital 'there was widespread apprehension that, with our Kaiser's unpredictable nature, Russia's plight might be exploited for an aggressive move on the part of the Triple Alliance'. For that very reason the pacific comments made by Caprivi in his speech in Osnabrück, which had attracted much attention, had been received with relief.[24] Privately, however, even Caprivi expressed the opinion that if it should come to a revolution in Russia Germany should not stand idly by, but must strive 'to destroy the mighty empire, and therefore not to shrink from taking a risk'. The 'final goal' of his policy, the Chancellor indicated in early 1892, was the creation of the 'United States of Europe', in order to make Europe economically independent of America.[25]

From the outset the news of the famine in Russia had a peculiar fascination for Wilhelm II, for he too saw in the catastrophe the possibility of the total collapse of the mighty Russian Empire, the danger of a diversionary Russian

attack outwards and finally also a chance to eliminate the weakened Russian rival permanently by military means. During a hunting trip to Rominten in East Prussia in October 1891 he criticised the Chancellor's pacific speech in Osnabrück. 'Now, what is this speech aiming at? First, the Chancellor could have mentioned it to me beforehand, second in such grave circumstances he exposes himself to the danger that next time he will be driven *ad absurdum* by public opinion, and third we *need* a mood of unrest in Germany so as to push through the unavoidably necessary military increases. Such a mood suits me perfectly, for if the German lights his pipe and puts on his dressing gown he simply cannot be governed . . . The key to the whole situation is the urgent necessity for strengthening the army. That is the goal which *must* be reached. I have sought the opinions of the King of Saxony, the Regent of Bavaria, etc., and have received, particularly from the former, wholehearted agreement. The Prince Regent also agreed with me – but he is like an old peasant who finds everything too expensive. The situation in Russia is incalculable. No one can tell the outcome of this famine. According to a secret report the Warsaw hotels are already being used as winter quarters for the troops! That is not merely military activity, it is mobilisation! . . . Austria's inaction throws everything on to *our shoulders*. They do not stand by us, they simply leave us in the lurch in military matters! *We* are now more than ever obliged to go ahead – and as his contribution the Chancellor makes a peace speech in Osnabrück!'[26] The following month Waldersee lamented that 'in his present mood' the Kaiser was 'naturally' reacting to the 'very alarming' reports on the famine in Russia 'with delight'. 'He interprets them to suit himself, and exaggerates a great deal too; news of the effects of the failed [Russian] attempt to raise a loan in France also interested him greatly, but here too he interpreted everything in his own favour.'[27]

It is striking how often and at what length Wilhelm discussed the implications of the Russian crisis in his letters to his grandmother. In December 1891 he wrote saying: 'I think this fearful calamity will – with Gods help – for some time to come keep the Russians from making war upon their unsuspecting neighbours.'[28] In a long letter of 12 April 1892 he commented: 'The news that reached me from Russia are manifold; but all end with the same terrible gloomy outlook . . . Notwithstanding all what the Czar may say & fancy, the quite hopeless condition in which 17 of the Governments of the best "Black Earth" are, is perfectly appalling. A gentleman well versed in matters of national economy & finance, & who lived in Petersburg for over 20 years, told me that Russia was marching with giant strides towards its ruin *if* the Czar did not at last wake up. If the country is to be saved from a catastrophe, the only way for it is the immediate reducing of the army and navy on a large scale. This is the only manner to save

money, neither Germany or France will give Russia a rouble more. Should the present Minister of Finance die or be disabled, his successor will have to place this fact plainly before the Czar. It is more than probable, the Ruler of all the Russians will not accept the proposal of reducing his forces, as it would create a terrible uproar in the war party & the Slavophil camp, inwardly; & outwardly give the world at large the spectacle of the "Colossus on feet of Clay". Then says my informant the country is utterly lost; & in a shorter or longer lapse of time the whole Realm will come to grief with a fearful crash. As for making war on other people, he went on, that was quite out of the question, as the whole of the transport system on the lines have been utterly muddled [sic], so that a lot of engines & waggons have up [to] the number of 500 & 14000 respectively been broken down or spoiled by snowdrifts [sic], so that they are beyond hope of repair. – Government has done much – as much as money can do – to help the people; they have payed [sic] 190 millions of roubles to the districts 20 mill: to General Annekoff for labour schemes; & in private a like sum has been raised. But how about next winter nobody knows; as the conditions will be the same & the money will probably fail! Nearly the same report has been made to our Foreign Office by H. v. Bleichröder here, who seems to be backing out of Russia with his money.'[29]

Lord Salisbury, to whom the Queen showed the Kaiser's letter, agreed with his opinion of the situation in Russia but pointed out that the illness of the two strongest men in the Russian government, Vishnegradski and Giers, made the outcome difficult to foresee. 'It is impossible to say now who it is that is guiding that vast machine', he wrote, '& under such conditions, a reckless policy is quite a possibility.' If the Queen had an opportunity, on her forthcoming visit of condolence to Darmstadt, to speak to the Kaiser, she should urge greater 'calmness' on him, both in his actions and in his all too frequent speeches.[30]

Not much calmness was in evidence for the time being, however. In his birthday letter to his grandmother, written on 22 May 1892 while he was hunting at Schlobitten in East Prussia, the Kaiser again brought up the subject of the Russian threat. 'The way in which the Russians prepare for their future inroad into Germany is illustrated by the fact not withstanding the terrible calamity at home they are pushing up 5 new Infantry Divisions against the frontier of Eastern Prussia & Galizia, including the formation of the new 18th Army Corps at Dunaburg. The Divisions come from Finland, Saratow, Caucasus respectively. General Gowrko proudly remarked to a french [sic] officer, who passed Warsaw a few weeks ago: "Je suis le poing de la Russie sur la poitrine de l'Allemagne, un mot de Czar et je l'enfonce!!! Puisse't-il se passer que ce mot d'ordre soit bientôt donné"!! An agreeable neighbour! – Besides I am informed by a Russian friend of mine – a gentleman of the good old school & traditions – that in March,

when they thought Caprivi would leave & Bismarck come in, he heared [sic] two Generals in the Emperor's Palace say: "Si cela arrive on a ordonné en haut, d'immédiatement publier le texte du traité secret entre la France et nous"! My faith in the Peace is however *not* shaken but I trust in Prussia, my good conscience & my army.'[31]

HOPES OF AN ALLIANCE WITH BRITAIN

The strategic purpose that lay behind this correspondence with Queen Victoria is not difficult to discern. No one recognised more clearly than Wilhelm II that with the Franco-Russian pact it had become more desirable than ever to win the support – at least tacit – of Great Britain. After the Kaiser's visit to Britain in the summer of 1893 Waldersee quite rightly observed that 'he still wishes to persuade England to join the Triple Alliance',[32] and six months later Wilhelm himself declared: 'We cannot manage on the continent without England . . . England must play its part in European politics.'[33] Undaunted by the change of government in London, where Gladstone's Liberal ministry (with Lord Rosebery as foreign secretary) had replaced the supposedly more pro-German Conservative government led by Lord Salisbury,[34] the Kaiser continued his efforts to use his close dynastic ties with the British royal family to gain political advantages for his country. But it was precisely on the dynastic level that countless points of friction and rivalry stood in the way of achieving the goal of an Anglo-German alliance.[35] In the rejection of the Kaiser's attempts at rapprochement, which were seen as importunate in Windsor too, we may detect one possible germ of the Anglo-German naval rivalry through which, a mere ten years later, the British Empire would be driven into the arms of France and Russia.

Wilhelm II's tendency to impose himself where he was not welcome became all too apparent in the spring of 1891 when Queen Victoria asked him to delay his visit to England until after the wedding of Prince Aribert of Anhalt and Princess Louise of Schleswig-Holstein, the younger daughter of Prince and Princess Christian (Lenchen), on 6 July, as the presence of the Kaiser would put the bridegroom's father, Duke Friedrich of Anhalt, too much in the shade.[36] The Kaiser, however, was determined to attend the wedding, which he had been instrumental in bringing about, and wrote to his uncle, Arthur Duke of Connaught: 'Now this I must say is very extraordinary, & very surprising! With Aunt Lenchens wish & permission *I alone* brought the whole match together, I had the whole of the writing & arguing etc. to do, I coaxed the Duke into an answer affirmative & lastly I pay out of my own pocket just as much as Grandmama though the two do not belong to my house or family! I have allready

[sic] promised Aribert to assist at his wedding & also the Anhalts so that they would be very much mortified if we were not there for the ceremony. Now for the "etiquette" question. The fear of the Queen that she could not honour the Duke enough in my presence is ungrounded. 1). At my court – & so in all the courts of the continent – the parents of the bridal pair are placed on the same rank as ours for the festivities. 2). So it was for the old Anhalts at Uncle Frederic Charles wedding, at which the late king of Hanover was present they ranked with him under the form of "unbeschadet des Ranges" taken from the old French court regulations "sauf le rang". 3). The same was done for my mother in law at Yayas wedding, when the Kings of Saxony & Greece were present; also the Prince & Princess of Lippe at Vickys wedding wer[e] treated in the same manner. It is "ganz selbstverständlich" [quite taken for granted] with us that the "Schwiegereltern haben den pas vor allen anderen Herrschaften die eingeladen sind 'unbeschadet des Ranges'" [parents-in-law take precedence over all other ladies and gentlemen who are invited, "rank notwithstanding"]. So that I should think that there is not the slightest difficulty.' Wilhelm assured his uncle that the Senior Master of Ceremonies, Count August zu Eulenburg, who knew everything about such matters, had confirmed his opinion and could provide any further information required. 'By these lines', he concluded, 'you will see that after all I have gone through for Louise & Aribert for whose happiness I worked from sheer [sic] friendship & love to Aunt Helena & the dear girl, that I feel rather hurt at beeing [sic] "abgewinkt" [put off] only out of an etiquette question which in reality is none.'[37] Queen Victoria was compelled to give way and invite her grandson both to the wedding of the young couple and to the silver wedding celebrations, held at the same time, of the bride's parents.[38] Nevertheless, as we have already seen, the Kaiser's visit to Britain that summer was universally considered a great success.[39]

Building on this success, Wilhelm hoped to consolidate the improved relations between the two countries through regular visits to his grandmother. But in June 1892, when he announced his intention of visiting her again on the Isle of Wight, Queen Victoria made it unmistakably plain that such visits were not at all to her liking. She sent word to the British ambassador in Berlin that 'the Queen *never invited the Emperor*. On the contrary she merely said if he came in his yacht she would be glad to see him at luncheon or dinner but could not lodge him at Osborne this year. But she wd. be very thankful if he did *not* come.'[40] Nor had she any desire, so soon after the death of her grandson the Duke of Clarence, to hear the brass band which Wilhelm wished to bring to perform for her during his visit.[41] To Wilhelm himself she wrote on 29 June: 'You understand, dear Willy, that I *cannot* lodge you this year, *at Osborne*, – but shall be delighted to see you to luncheon or dinner. It is most kind of you to offer to bring over the fine Band, but I have not had any Military Band whatever this year on account of our deep mourning & wd. therefore advise their not coming this year.'[42]

Wilhelm did nevertheless visit his grandmother at Osborne for a week at the beginning of August 1892. He slept on board the *Kaiseradler*, as the old *Hohenzollern* was now called, but joined the Queen and the rest of the royal family for several dinner parties at Osborne, dressed in his British admiral's uniform. He took part in the Cowes regatta in his new yacht *Meteor*, which he proudly boasted was the finest in all England (and which was almost entirely manned by British sailors), and at first won almost every prize available.[43] He wrote delightedly to his mother: 'I am so glad to see dear Grandmama again & to be able to breathe Osborne's balmy air, together with having the pleasure of sailing a good yacht of my own in the Royal Yacht Squadron Fleet. Meteor has made a brave show of herself this summer & has won 7 first prizes & 3 second ones; . . . the yacht has been pronounced as the swiftest of the whole of England.'[44] He was bitterly disappointed when over the next few days his yacht was beaten by *Iverna*, *Queen Mab* and the smaller *Corsair*, which belonged to Admiral Victor Montagu. 'Fortune has not smiled upon the German Emperor during the week of yacht-racing', commented *The Times* on 6 August 1892.[45] As a result of these experiences at Cowes in the summer of 1892 the Kaiser put forward a proposal to his uncle the Prince of Wales, as commodore of the Royal Yacht Squadron, for a new race exclusively for larger yachts of over 100 tons – *Meteor* was 116 tons – for which he would give a Challenge Shield to the value of £90.[46] The offer was gratefully accepted.[47] The sailing rivalry between Wilhelm and his uncle which began in this way was, however, to have a fateful effect on their political relationship.

The marriage of the Duke of York (now heir presumptive and later King George V) and Princess Victoria Mary (May) of Teck, who had been engaged to his late brother Eddy, was to take place on 6 July 1893. As the day approached, Wilhelm was confident that he would be invited to the festivities at Windsor. But Queen Victoria again instructed her ambassador in Berlin to make it clear to the German government 'that it will be quite impossible . . . to invite the Emperor to the wedding of the Duke of York'; she would not be able to show the numerous guests sufficient honour if the German Emperor were present.[48] The decided tone of this instruction wounded Wilhelm all the more deeply because shortly afterwards his brother Heinrich received a letter from the Queen herself warmly inviting him and his wife Irène – another of the Queen's grandchildren – to the wedding. Both wished 'very much' to attend but they could not accept the invitation without the permission of the Kaiser. And Wilhelm refused, on the grounds that Heinrich's duties in the navy 'excluded the undertaking of any external official or family engagement'.[49] Not without justification Prince

Heinrich's Marshal of the Court, Albert Freiherr von Seckendorff, suspected that 'palace intrigues and difficulties on both sides of the Channel' were at the bottom of this extraordinary ruling.[50] He blamed the British side for 'uncle–nephew trouble', but also assumed, quite correctly, that the Kaiser was annoyed that not he but Heinrich had been invited to the wedding. The sudden emphasis on the Prince's naval duties was strange, he thought, considering the protracted Mediterranean cruise and 'many other pleasurable episodes' which Heinrich had been allowed to enjoy. During the past two summers Heinrich had made 'lengthy pleasure-trips to England' with the Kaiser's approval. As the Kaiser's brother Prince Heinrich was expected to go to Berlin for every court ball. Seckendorff could not think that a five-day trip to London for a family wedding could have serious military consequences.[51] Senden-Bibran, the Chief of the Naval Cabinet, confirmed Seckendorff's assumption that the reason for Wilhelm II's attitude lay in wounded vanity. 'The Kaiser is angry', he wrote, 'perhaps because He would prefer to go himself, and is not wanted.'[52] Seckendorff stressed that Heinrich and Irène had behaved quite correctly and had an absolutely clear conscience. To his direct question to Princess Irène, 'whether anything had been done by her or by the Prince [Heinrich] to bring about the invitation from England', she had responded with an unqualified 'No'.[53] This time the Kaiser gave in. On 14 June the Queen sent him a telegram from Balmoral thanking him for giving his brother and Irène permission to attend the wedding.[54] Wilhelm himself stayed away until the Cowes regatta at the end of July 1893.

Outwardly the Kaiser's eight-day visit, a sumptuous affair, passed off very satisfactorily. The new *Hohenzollern* – 'an enormous ship, painted white', as Queen Victoria observed after sailing around her – was universally admired; the German ship's band, this time consisting exclusively of string instruments, played 'beautifully' during the meal and the newly appointed Flügeladjutant Count Kuno von Moltke delighted his distinguished audience by playing an improvised piece on the piano, weaving together his own compositions, tunes from Wagner and 'God Save the Queen' (or 'Heil Dir im Siegerkranz').[55] Even on the water Anglo-German understanding seemed to be in the ascendant when the Prince of Wales's racing yacht *Britannia* won the Kaiser's Challenge Cup. Prince Hohenlohe-Langenburg, who as a relative and now secretary at the German embassy had ample opportunity to observe Wilhelm at close quarters,[56] wrote admiringly of him: 'The English were quite delighted with the Kaiser. He was in the best of moods the whole time and charming to everyone. He really has something uncommonly fascinating about him, and I was especially struck by his sharp, lively and intelligent expression.' During a visit to the Portsmouth docks the Kaiser's knowledgeable questions and comments astonished the British naval officers.[57]

And yet behind the glittering scenes personal animosity, threat of war and – for a brief moment – imperial panic prevailed. Count Philipp zu Eulenburg, visiting England for the first (and last) time on this occasion, had a conversation with the Prince of Wales on the latter's yacht, while the Kaiser was absorbed in captaining the vessel.[58] The heir to the British throne spoke sarcastically of his nephew's 'colonial sport', which he claimed to find as puzzling as Wilhelm's love for the navy. The Prince of Wales, Eulenburg reported, commented with a malicious glance that 'it was certainly very nice that his nephew was interested in ships'. 'But when one saw him rushing about so much with his paralysed arm, like up on deck, one could not help fearing that he would do himself harm.'[59]

The situation suddenly took a serious turn when Britain and France found themselves on the brink of war in Siam (Thailand) and the prospect of Great Britain joining the Triple Alliance, as Wilhelm had hoped, seemed to be within reach.[60] On the night of 30 July 1893 the Queen sent her grandson a telegram from the Foreign Secretary, Lord Rosebery, reporting that the French government had sent an ultimatum demanding the withdrawal of the British gunboats lying off Bangkok; Rosebery had rejected the French demand and asked for an immediate interview with the German ambassador, Hatzfeldt.[61] The attitude of the German embassy was reflected in a letter written by Ernst Hohenlohe to his father immediately after the crisis: 'France and England were a hair's breadth away from war with each other. As I learnt from official correspondence Italy too would probably not have remained inactive in the event of war. We should not have stood idly by either, and there would have been a European war.'[62]

Philipp Eulenburg's diary contains a striking description of the almost panic-stricken reaction of the Kaiser to the threat of war which had so suddenly flared up. 'Immediately on our return to the *Hohenzollern* the Kaiser took me into his saloon and he had completely lost his nerve. In fact I have never seen him so agitated and had to summon all my wits to calm him down with reasoned arguments. After the French fleet's visit to Kronstadt in 1891 this was the second great shock to befall him following the non-renewal of the secret treaty with Russia . . . The Kaiser declared "that England's navy was weaker than the combined Russian and French navies. Even with the help of our small navy England would still be weaker. The French were now trying to drive Russia into action, and given Tsar Alexander's hostile attitude towards Him [Wilhelm II] this could well succeed. Our army was not yet strong enough to fight against France and Russia at the same time. The French had chosen their moment skilfully. To do nothing, and wait until the waves closed over one's head, was impossible. All Germany's prestige would be lost if we did not take a leading role, and if we were not a world power we would cut a sorry figure."' Eulenburg tried to console the Kaiser with the (accurate) observation that Britain would not fight France but would try to reach a settlement. If it nevertheless came to war, the Germans could for the present be amused spectators. 'If we succeeded in playing such a role in a war between the other powers, we should not do badly out of it. Above all the Kaiser should not let himself be trapped into making any kind of binding declaration, but should listen with a benevolent smile to what the English say, and keep as quiet as possible.' The diplomats Kiderlen-Wächter and Wolff-Metternich, summoned to join in the discussion, supported Eulenburg's arguments. 'When they left', the Kaiser's friend recalled, 'the Kaiser seemed calmer, but looked miserable – pale and biting his lip nervously. I felt terribly sorry for him. Having arrived over here with all his great song and dance about ships he suddenly felt driven into a modest little corner, and politically shut out.'[63] Rosebery's 'hesitant conduct' in the crisis certainly restored peace, but the 'diplomatic defeat' which Great Britain had suffered, in the Kaiser's eyes, shook his confidence in the Liberal Foreign Secretary who was soon to become Prime Minister. 'The unpleasant result', Prince Hohenlohe commented, 'is that in India the English now have France nearby, as well as Russia as a neighbour.'[64] Throughout the world as well as at home the British Empire was in crisis, the young diplomat wrote. 'One cannot resist taking a certain malicious pleasure in this. But it would of course be fatal if this became the source of a world conflagration.'[65]

Partly because Hatzfeldt fell ill, and partly because of Rosebery's cool nature, the Siam crisis did not lead to the close Anglo-German co-operation for which the Kaiser had hoped, and he left the British Foreign Secretary in no doubt of his dissatisfaction over this. When Hatzfeldt reported a conversation with Rosebery in November 1893, in which the latter stated that he had urged on the Cabinet the importance of increasing the navy and that with the prospect of serious complications arising, he would seek the support of Germany and its allies, Wilhelm took the entire credit for this apparent change in British foreign policy, writing on the report from London: 'The sudden new about-turn by Rosebery, which astonishes Hatzfeldt, reverting to complete openness and confidential consultation with us, stems from my initiative. I found a safe opportunity of sending a message to Rosebery telling him bluntly that I would not tolerate this game of hide-and-seek with me any longer and that if he still valued My friendship and affection at all and hoped for support from us, he must return to the old relationship of complete honesty. Otherwise I would not be prepared to talk to him again. It seems to have worked.'[66]

The extent to which the Kaiser regarded himself as a silent ally of Great Britain against the Franco-Russian alliance becomes clear from his critical remarks about the weakness of the Royal Navy in the Mediterranean. When the newly appointed German ambassador in St Petersburg, General von Werder, reported that the

Russians had sent five warships to Toulon which would remain in the area for a year, the Kaiser saw this as a threat to British supremacy in the Mediterranean and not, as in fact was the intention of Russia and France, as an action directed against Germany.[67] Wilhelm expressed his concern in a comment on Werder's report: 'This reinforcement will be welcome to France and brings the number of anti-English battleships up to 19. A raid on the Dardanelles is only a matter of time now.'[68] Four months later he passed on to the British ambassador secret information that a Russian general, who had recently been with Tsar Alexander at Fredensborg, had indicated that the Russian ships now in the Mediterranean were heading for Alexandretta (the present Turkish port of Iskenderun). As Russia's access to Constantinople through the Balkans was blocked, it had to create a route southwards through Asia Minor for itself, the Kaiser declared. When Malet observed that Cyprus might therefore prove itself useful at last, the Kaiser responded: 'Yes . . . if you have got the ships. I am convinced France is working for a triple alliance in the Mediterranean and trying to induce Spain to join it by promise of help towards the recovery of Gibraltar. You must be very wary as to what is passing in Morocco.'[69] In communicating information of this kind Wilhelm cherished hopes of deepening the opposition between Britain and France and driving a wedge between France and Russia. Similarly, Waldersee calculated at this juncture that 'the stronger France feels in the Mediterranean, the more she will strive for sole supremacy there, which will bring her into conflict with England at first – which can only be to our advantage – and later also with Russia, as soon as she rules over Constantinople'.[70]

THE KAISER'S 'THIRST FOR UNIFORMS'

Kaiser Wilhelm II's inner attachment to Britain and his British relations often showed itself in ceremonial forms, as for instance when he wrote to his grandmother on 12 January 1894 asking permission to appoint his uncle Albert Edward, Prince of Wales, à la suite of the German Dragoon Regiment of which he had made the Queen colonel-in-chief. 'My passion since my early days for H.M. Redcoats & their history', he wrote, 'has always been a very warm one; the more so as they so often fought side by side with our soldiers! This puts into my mind wether [sic] You, as 'Chef' of our Dragoon Guards, would allow me to place Uncle Bertie à la suite of Your Regiment on my birthday. This day beeing [sic] the 25th anniversary of my entry into the I Regiment of the Guards & as such, a list of military honours is to be issued on the 27th, & it would give me great pleasure as well as be a great honour to the Regiment. Father & son would then both be in Your "Queens Own Regiment".'[71] The intention behind this imperial initiative was no secret to the British military attaché Colonel

Leopold Swaine. On the same day he proposed the appointment of the Kaiser as honorary colonel-in-chief of the 1st Royal Dragoons. He pointed out that this was a regiment that never left Britain so that His Majesty could always inspect it on his visits there. 'But above all its uniform is of the typical colour.'[72] That the military attaché had guessed the Kaiser's most secret wish became clear the very next day in a conversation in which Wilhelm II emphasised that he did not want to make the suggestion himself, but that he did indeed wish to be able to wear British army uniform as he regretted having to appear on horseback at military reviews in Britain wearing his admiral's uniform. Swaine was, however, astonished to learn that he actually wished to be given the honorary command of a Scottish Highland regiment. During his visit to Wimbledon in the summer of 1891 he had admired the Argyll & Sutherland Highlanders and wished to be able to wear their uniform. When Swaine expressed doubts about this and pointed out that such an appointment had never been given to a foreign ruler, Wilhelm retorted 'that all rules are broken sooner or later & that perhaps an exception could be made for him both on family & political grounds'. If it were not in fact possible to make him colonel-in-chief of a British regiment, Swaine suggested that he could perhaps be appointed a British field marshal, but added that the regimental uniform would be more highly appreciated in Germany than the appointment as field marshal.[73] The Prince of Wales, who was flattered by Wilhelm's wish to place him à la suite of the German Dragoon Regiment, strongly urged his appointment as a British general or field marshal. To give him the honorary command of a regiment, on the other hand, the Prince conceded, would be difficult.[74] The Queen's Private Secretary Sir Henry Ponsonby pointed out that on an earlier occasion when the Kaiser had asked for the honorary command of a British regiment it had been found, upon enquiry, that such a measure would be most unpopular in the army as well as among the general public.[75] Queen Victoria herself made no attempt to hide her distaste for this 'thirst for uniforms'. She said of Wilhelm's wish for a regiment: 'This wd. never do & he is an Admiral. – The Queen thinks he is far too much spoilt already.'[76] Minister of War Henry Campbell-Bannerman, whose advice was sought at the Queen's request, also rejected the idea of a regiment. Wilhelm's appointment as general or field marshal would certainly pose no problems from a purely military point of view, but there would be a risk that other foreign monarchs would expect a similar distinction. Heaping an excessive number of British honours on the German Emperor could cause ill-will in other countries, he warned. The Minister of War questioned Swaine's judgement and blamed his indiscretion for the dilemma now confronting his masters in London.[77] The negative opinion of the Foreign Secretary Lord Rosebery carried even more weight: he argued that Wilhelm was already an honorary admiral, which for a 'naval nation' like Great

Britain was a higher honour than appointment as a field marshal, and moreover that to confer a new distinction upon him only four years after his appointment as admiral would be seen as a declaration of policy, which at the moment was highly undesirable. 'It would tend to defeat the very policy we should desire to promote. The Germans hate the French, but they are intensely jealous of the English. Anything that appears to make their princes or Sovereigns appear under English influence is snarled at, and tends to prevent and not promote cooperation between the two nations . . . I am quite sure that it would do even more harm in Germany than in France or Russia.'[78] But when the elderly Duke of Cambridge, Queen Victoria's cousin and Commander-in-Chief of the British army, telegraphed from Cannes to say he was decidedly in favour of the appointment of Wilhelm as field marshal, the Queen made no secret of her anger. She was 'decidedly against' this appointment, not least because Wilhelm, who as an admiral already believed he had a right to intervene in British naval policy, would start meddling in army matters too. 'I am much annoyed. It is quite unprecedented & the admiral is quite enough', she emphatically declared.[79] In conversations with Sir Henry Ponsonby she pointed out that Wilhelm had accomplished nothing in military terms to make him deserve such high rank in the British army; the twenty-fifth anniversary of his army service was by no means a sufficient reason for it. She remembered, she said, that Wilhelm had been appointed an admiral precisely in order to avoid giving him the army rank he wished for.[80] Nonetheless the Prince of Wales again pleaded the cause of Wilhelm's appointment as field marshal. As grandson of the Queen and head of the best army in the world the Kaiser could not receive anything less than the highest rank, he insisted, and those who told the Queen the contrary obviously did not understand German sensibilities. If the Emperors of Russia or Austria should in fact claim the same rank, which he did not believe they would, the Queen could always get round this by evoking the close family relationship with Wilhelm. Sir Francis Knollys, Private Secretary to the heir to the throne, wrote despairingly to Ponsonby that it was incredible that mother and son should again be at loggerheads. As far as Wilhelm was concerned, the Prince's closest adviser commented scathingly: 'I can't abide the young man myself and think he is a vain, theatrical & ridiculous creature, but I do think that he ought to have everything or nothing.'[81]

The Queen won, at first. Seizing upon a suggestion made by Ponsonby, she acted as if the idea of giving Wilhelm a rank in the British army had come from Swaine alone, without the Kaiser's knowledge; the military attaché was accordingly coolly informed on 24 January that it would be 'safer' not to follow up his suggestion, as there was no precedent for such a distinction.[82] In her letter of congratulations on his birthday and on the twenty-fifth anniversary of his military service, the Queen made no mention of the uniform question and even pointed out that the first five or six years which Wilhelm had spent in the army in his teens could hardly count as 'service'.[83] At the same time Ponsonby wrote on the Queen's behalf to Knollys, the Duke of Cambridge and Lord Rosebery, explaining 'that it would be impossible at the present time when there is no special reason to confer the highest military rank upon the Emperor'.[84] The Prince of Wales was furious and ordered his luckless Private Secretary to telegraph to the Queen that he very deeply regretted her decision and only wished that he had not been drawn into this unfortunate affair.[85] In a letter he added that the ministers who had advised against granting the distinction were less competent in such questions than himself, the Prince of Wales. The opinion of the Commander-in-Chief ought to have been taken into account as well. In his view neither Rosebery nor Campbell-Bannerman was 'infallible'; both were talking 'rank nonsense when they say that because the Queen makes Her grandson, who is likewise half an Englishman, a Field Marshal She must give other Sovereigns who are no relations the same rank'. The decision was even more painful to him because Wilhelm had just sent him his new Prussian uniform by a special envoy.[86]

Victoria was shaken by the strength of her son's obstinate reaction to her decision. In stark contrast to Wilhelm's attitude to his constitutional advisers the Queen pointed out that she was obliged to accept the advice of her responsible ministers.[87] The Duke of Cambridge likewise expressed his regret at the Queen's decision. Swaine, he said, would not have made his suggestion if he had not received a hint from the highest quarters that the Kaiser wished to wear British uniform on his visits to Britain. As Wilhelm was already an admiral there would be no difficulty about making him a field marshal too. By conferring regiments and uniforms on members of the British royal family the Kaiser had demonstrated his attachment to his grandmother and his uncles; he, the Duke, could not understand why his courtesies had not been answered by the simple gesture of conferring the uniform upon him, when it obviously meant so much to him.[88] Even the former Prime Minister Lord Salisbury spoke out in favour of conferring the distinction on Wilhelm. 'If present Foreign Policy is to keep well with Germany the grant would be wise for personal considerations weigh heavily', he commented.[89]

Although Queen Victoria had got her way, her ill-feeling towards Wilhelm was still unmistakable weeks later. After a visit to court in March 1894 Prince Hohenlohe reported to his father: 'The Queen's expression seemed rather stern when I conveyed the Kaiser's greetings to her, and she did not say a word in reply. Has something happened there again?'[90] In fact neither the Prince of Wales nor the Duke of Cambridge had given up the campaign on the Kaiser's behalf and Rosebery, who had meanwhile succeeded Gladstone as prime minister, felt compelled to strengthen his case against the grant of the distinction. This time he pointed out that, if in addition to his admiral's rank the Kaiser were also made a field marshal, there would be nothing left to give him in the future. Rosebery feared 'that it will simply leave the German Emperor unsatiated while yet young', which could have an undesirable effect, for 'so marked an additional honour should only be given at a time when it is necessary to display either signal gratitude or signal friendship. Now there is at the present moment no necessity for the display of any extraordinary friendship, still less for the demonstration of any extraordinary gratitude.'[91]

In spite of all this, however, Queen Victoria eventually felt obliged to fulfil her grandson's original and dearest wish: on the occasion of the marriage of her grandson Ernst Ludwig, Grand Duke of Hesse-Darmstadt, at Coburg on 21 April 1894 she appointed Kaiser Wilhelm II colonel-in-chief of the 1st Royal Dragoons, a cavalry regiment.[92] He sent ecstatic thanks for this exceptional distinction – it was the first time in British military history that a foreign sovereign had appeared in the Army List – writing to her from the Wartburg: 'The kindness & love with which You have treated me have quite overwhelmed me! I am afraid that under the impression of the so unexpected & joyful surprize You prepared for me, I did not find the right words to express my feelings. Deeply & sincerely do I thank You for the great honour which You conferred upon me by making me hon. Colonel of the Royals. I am moved, deeply moved at the idea that I now too can wear beside the naval uniform the traditional British "Red Coat". How many brave & brillant [sic] soldiers have worn it & before all my beloved Grandpapa! The congratulations I receive from all parts show me, how much this token of Your kindness is valued here, & how glad they are here that the bonds of friendship between our countries & armies have received a new addition. I shall be very happy to wear the uniform of a so distinguished & brave Regiment, & am impatiently looking forward to the moment when I can meet my brother officers.'[93]

When a deputation from the regiment arrived in Berlin in June 1894 to present itself to the Kaiser he telegraphed enthusiastically to his grandmother that he had found the detachment quite exemplary. 'We are very glad to see the Officers of the Royals as our Brothers in Arms in our midst.'[94] In a speech which was reported with satisfaction in the British press Wilhelm declared how proud he was to be Colonel-in-Chief of the regiment. 'I have thus become a member of the body of British officers, and a tie, which has for a long series of years united the British and Prussian Armies is thus bound anew.'[95] At a musical evening in the Neues Palais on 10 June attended by the British deputation the Kaiser wore the new red Dragoons uniform and his composition, 'Song to Ägir', was given its first performance.[96] Swaine was delighted with the warmth of the reception given

23. Kaiser Wilhelm II in the uniform of the First Royal Dragoons, Aldershot 1894

the British officers.[97] Kiel Week then provided a further opportunity for demonstrations of Anglo-German friendship. The new relationship was overshadowed only a little by an initiative taken by Rosebery on the Congo question, which was condemned in Berlin as shameless; the Kaiser, it was reported, was 'particularly resentful' because he suspected 'that Lord Rosebery assumed that as the Queen's grandson He would not put up a fight'.[98]

When the great white and gold yacht *Hohenzollern*, bedecked with the German national flag, the Kaiser's personal standard and the white flag of the Royal Yacht Squadron, once again moored at Cowes in August of that year for the annual regatta, the Kaiser was wearing the uniform of the 1st Royal Dragoons.[99] He was at first disappointed that no detachment of the regiment had appeared to welcome him, but no one had thought to remind the Royals that the Kaiser was now their Colonel-in-Chief.[100] Nonetheless before returning to Germany Wilhelm, accompanied by his retinue – Adjutant-General von Plessen and the Flügeladjutanten Count Kuno von Moltke, Count Arnim and Major von Scholl – inspected his regiment at Aldershot and led it in a march-past before his uncle Prince Arthur, Duke of Connaught. 'The sight of these fine troops under his command is never to be forgotten', he telegraphed enthusiastically to the Queen.[101]

One cannot doubt the genuineness of his gratitude to his grandmother nor the honesty of his sense of belonging to the British royal family and the British army, expressed in his letter to the Queen of 24 August 1894. 'Most beloved Grandmama', he wrote from the Neues Palais, 'Your great kindness to me & all my gentlemen has left us all full of the most agreeable & indelible souvenirs. The stay at Cowes was enhanced by the dinners at dear old Osborne which I so adore, & which my entourage has come to love as much as I do; fascinated by the kind & benevolent "acceuil" [sic] of the much revered Queen. The [Cowes] "Week" is my real holiday & I must with all my heart thank You for the kind & warm reception which I met at Your hands. I hope & trust that next year – should it please Providence that we all be well – I will be able to bear home that only trophy I really covet, & I work for through the whole of the year "Her Majesty's Cup". Aldershot was a great success & all the arrangements for our stay at the pavilion were perfect. What a delightful place that is, how splendid the view & the air so bracing! Uncle Arthur was attention itself & managed the Military affairs in a first rate manner. The review was a lovely sight & my officers who never had seen a British parade before were deeply impressed by it. I am glad to say the Squadron of the Royals which You so kindly ordered over for me looked splendid & won an all round applause. The field day was very attractive & showed the brilliant marching qualities as also of endurance of the troops who were well handled. It gave me a great pleasure to be able to associate with so many nice comrades in arms & to think that I was looked upon as one of their own, belonging to "the thin Red Line" of England too!'[102]

THE POLISH QUESTION AND THE RUSSIAN TRADE TREATY

Farcical as the Kaiser's search for colourful uniforms may seem to us, Wilhelm II's personal initiatives in foreign policy undoubtedly had their serious side even in these early years. His hopes for Britain's support in a possible military conflict with Russia and France were fanned by his private correspondence with the American friend of his youth, Poultney Bigelow, whom Hinzpeter angrily accused in early 1893 of having talked the Kaiser into believing he was a second Alexander the Great, destined to conquer the East.[103] Bigelow was a strict Protestant, an anti-Semite and an enemy of Russia, who referred to the present as the 'ante bellum' – the pre-war time – and who used his widely read books and articles (he had just published his book *The German Emperor and His Eastern Neighbours*) to pursue the goal of inciting the Americans and the British to make common cause with Germany in a war against Russia, by appealing to their shared Protestantism.[104] On 18 August 1892 he wrote to the Kaiser: 'Your English visit I see added much to your popularity there, particularly the manly way in which you measured yourself with them [at sailing] . . . No Prime Minister can unmake alliances that are founded in mutual personal admiration; & whether Gladstone rules or Salisbury I fancy that no Government there can do much to stem the popular tide that runs against Russia . . . It is only necessary to make the British Public understand that Russia's movements are, not merely hostile to Austria & Germany, but to her own mercantile interests.'[105] Bigelow passed on alarming news to the Kaiser to the effect that Russia was keeping only one army division east of Moscow, while all the others were stationed between that city and Russia's border with Germany and Austria.[106] He also stated that he was in close contact with the Russian exiles Volkovsky and Stepniak, who published the periodical *Free Russia* in London and whose only fault in his eyes was that they considered the German Emperor as a new Bonaparte who was intent only on increasing his own power. He suggested to the Kaiser – and this was 1892, not 1917 – that Germany should secretly infiltrate this anarchist movement and place skilful agents in Königsberg, Thorn, Cracow and Brodnica with the aim of working for the liberation of Poland and the reannexation of Russia's Baltic provinces to the Reich.[107] It was possible that Bigelow himself would soon be travelling to the Baltic, but first he was awaiting the visit of a Polish friend who would bring him important information from Warsaw.[108]

The alarming effect which the American millionaire's exhortations had on Wilhelm II became apparent not only in the latter's correspondence with Queen Victoria, as we saw earlier,[109] but also in a conversation with his friend Philipp Eulenburg during the Scandinavian cruise of July 1892. At Tromsø the Kaiser, who prided himself on having acted in this delicate question without the knowledge of his Foreign Office, confided to Eulenburg his hopes of a forthcoming union of Russian Poland with Germany. He boasted that he had good reason to be pro-Polish. 'Having already secretly learnt that the mood in Posen and Russian Poland has swung round entirely in my favour, remarkable revelations reached me through the American Bigelow, my old friend from my youth and student days, from Warsaw where he was carrying out studies for a book commissioned by his government. When he was admitted to some of the Polish clubs which have sprung up under the agonizing pressure of Russian domination and which he was able to enter through confidential contacts, he discovered to his astonishment that all the interest, all the hopes of the Poles are directed towards me, that every sitting begins with a toast to me. They are full of hope for liberation from the Russian yoke, and in a war with Russia all Poland would range itself on my side with the express intention of being annexed by *me*.' The Kaiser denied that the motive behind the Polish freedom movement was the establishment of an independent Polish empire. The educated elements had given up this hope in view of their political weakness, he claimed. 'They want to come under *Prussia*. Bigelow writes confirming this in a way that leaves *no room for doubt*. If necessary Poland could of course be made into a province of the Reich. Alsace and Lorraine have been kept in good order as such. I have *not* passed Bigelow's interesting communications to the Foreign Office. The Polish question would be taken under their political wing, so to speak – and I am considering the matter for the time being from a purely *military* standpoint. For the General Staff and the deployment of the army it is a very great gain to have Poland on our side.' A General Staff officer had recently travelled to Poland, the Kaiser confided, and had found Bigelow's information confirmed to the letter. Even if a few Poles were still working for national independence that was 'cura posterior' (a concern for the future), Wilhelm opined. '*For the time being I* am their goal, and we must remember that.'[110]

The hopes that Wilhelm cherished of a Polish uprising on the outbreak of war between Russia and Germany had a direct effect on both domestic and foreign policy. In 1893 he conferred an order on the leader of the Polish party in the Reichstag, Joseph von Koscielski and held up the Poles as models of Prussian loyalty.[111] Waldersee watched with consternation as this policy, disastrous in his eyes, took its course. When he was told in confidence by the pro-government editor of the *Hamburger Korrespondent* 'that everything has been arranged by us in such a way that in the event of a war with Russia an insurrection would break out throughout Poland!!', the former Chief of the General Staff could not believe his ears and wrote scornfully: 'With such simpletons as we now have in charge of our diplomacy I am afraid it is not impossible that such nonsense is believed.'[112] He foresaw with deep concern that 'there is scarcely any sphere in which the Kaiser will have to face worse disappointments than in this'. It would be impossible to satisfy the Poles, who naturally aspired towards an independent state. Only if there were a firm intention of waging war against Russia could the former Chief of the General Staff see any benefit in the Kaiser's Polish policy. 'If it were the case that we definitely intended to wage war at a given time, and to make use of the Poles, I should certainly be in favour of making careful preparations for our strike', he wrote in October 1893, 'but not in such a misguided way as now.'[113] The present German policy towards Poland 'would make sense only if we wanted to launch an attack in the spring, which of course we have no intention of doing'.[114]

The secret contacts which the German General Staff made with the anti-Russian nationalists in Russian Poland on the orders of the Kaiser did not remain secret for long and proved very inconvenient when the Kaiser and the Wilhelmstrasse were forced to concede, in the autumn of 1893, that the hopes they had cherished since 1890 of an alliance with Great Britain had foundered on the island kingdom's 'Splendid Isolation', and that a new rapprochement with Russia would therefore be desirable. Waldersee, who as we have seen was well informed about the subversive activities of the General Staff in Russian Poland, was in no doubt about the damaging effect of this on German–Russian relations. 'Insurrection in the kingdom [sic] has been discussed in the most irresponsible way with influential Poles, and all kinds of promises have been made to them. All this has been closely watched in Russia and of course as a result their mistrust of us has increased and their garrisons have been strengthened.'[115] 'The final result can only be that we shall make a lasting enemy of Russia without satisfying the Poles', he predicted.[116] In the quest for improved relations with Russia Wilhelm II himself seemed to grasp that through his secret contacts with the Polish independence movement he had succeeded only in arousing suspicion in St Petersburg. At all events, in Thorn on 22 September 1894 he made a speech which struck a completely different note. He now declared unequivocally that the Polish fellow citizens of this West Prussian town 'may count on my favour and support . . . but only if they truly feel themselves to be Prussian subjects . . . For only if we stand together shoulder to shoulder like a phalanx is it possible to bring the struggle with subversion to a triumphant end.' In words not designed to encourage the Poles on the other side of the Russian border to throw in their lot with Prussia he threatened: 'I can also be very unpleasant, and if it is required, that is what I shall be.'[117] It was not until the First World War that plans for fomenting revolution in Russian Poland (and elsewhere) were revived.

Wilhelm II pinned his hopes for better relations with Russia chiefly on his personal relationship with the heir to the throne, Tsarevich Nicholas, who despite the coolness and reserve he had shown hitherto, accepted an invitation to the wedding of the Kaiser's youngest sister Margarethe (Mossy) and Prince Friedrich Karl of Hesse-Kassel (Fischy) in Berlin in January 1893.[118] On 26 January the

Kaiser visited the barracks of the Tsar Alexander Regiment of Grenadier Guards with the Tsarevich, and proposed a cordial toast in the officers' mess to the latter's father, which was intended both as a gesture of personal friendship and as 'a serious appeal for the union of monarchies against revolution'.[119] Raising his glass, he declared 'All of us see in Your Imperial Father not only the exalted Chief of the Regiment, not only our most distinguished comrade, but above all the upholder of long-lasting monarchical tradition, of tried and tested friendship and of the inner bonds forged by close links with my illustrious predecessors, sealed in former times with the blood of both Russian and Prussian regiments on the battlefield before the enemy.'[120] As the Kaiser reported to his grandmother, he had the opportunity to talk at length to the Tsarevich, who only a few months later was to assume absolute sovereignty over the gigantic Russian Empire. 'Niki' had shown 'sound judgement & a quiet clear mind', and had proved that he 'understands European questions much better than most of his countrymen & family; I hope that on the whole his visit will have a good effect on our relations in general.'[121] In a letter to Kaiser Franz Joseph, Wilhelm summed up the main points of his discussions with the future Tsar, whom he expected to bring about a shift in Russian policy away from France and towards the Triple Alliance. 'Far from having aggressive tendencies, the Triple Alliance had purely defensive aims in view', Wilhelm assured the Tsarevich. 'When the treaty was concluded the powers had in mind the grave dangers which threatened the monarchies from France, through the dissemination of republican propaganda. It was after all chiefly from Paris that all revolutionary doctrines were being spread, the aim of which was to undermine monarchical traditions in all countries in all manner of [missing word]. These common dangers were therefore to be seen primarily as the basis of the Alliance, and any power that wished to defend the interests of peace and of monarchy in the same way could join this Alliance at any time. The common interests of the three Allied powers were, however, by no means limited to the political sphere. The Triple Alliance sought to bring the European powers closer to each other in the economic sphere as well, through trade treaties. In so doing it aimed both to reduce the causes of friction between European states and also, above all, to counteract the danger which threatened the entire trade of Europe, in that the republic of North America was proving increasingly inclined to snatch for itself the entire trade of South America as well.' Nicholas had listened to these observations with interest, Wilhelm reported to Vienna, and had shown 'a pronounced aversion towards France'. At his request he had given him 'the above-mentioned "aphorisms" on politics in writing as an aide-mémoire, as he wanted to report the contents to his father in minute detail', the Kaiser wrote. 'If I should have succeeded, by what I said, in removing some of the Tsar's suspicion of the Triple Alliance, and in proving its usefulness for the monarchical principle, I should be very glad. I may add that both the King of Saxony and the Grand Duke of Baden have had long conversations of similar import with the Tsarevich, which will certainly not have failed to make an impression on him. Much will depend on whether he manages to use the information he has gathered in an appropriate way vis-à-vis his father. May he succeed, for the good of all concerned.'[122] These declarations once again show the hegemonial aims which lay at the heart of the foreign policy pursued by Kaiser Wilhelm, however inconsistent the individual initiatives might appear to be.[123]

The visit to Berlin by the heir to the Russian throne was in fact to have momentous consequences, for during the marriage festivities Nicholas fell in love with the Kaiser's Hessian cousin Alix, whose sister Ella had been the great love of Wilhelm's youth, and whose other sister Irène was married to his brother Heinrich. In April the following year, when the young Grand Duke Ernst Ludwig of Hesse and by Rhine, brother of the Darmstadt princesses, married Princess Viktoria Melitta of Saxe-Coburg and Gotha, it was Kaiser Wilhelm himself who succeeded in overcoming the religious scruples of his Hessian cousin, which she had hitherto seen as an obstacle to marriage with the Tsarevich. At Coburg she agreed to marry Nicholas. As the Kaiser's eldest sister Charlotte recorded in a letter, this success put her two brothers Wilhelm and Heinrich 'in bounding spirits: the Emperor specially, after he had brought the Russian engagement on of my cousin Alix: they look nice & happy & may her future remain so: certainly outwardly it is a grand partie & match, but upon close & nearby reflexion, not an enviable position'.[124] Queen Victoria too had mixed feelings at the thought that her favourite granddaughter would soon be living so far away from her. 'I grieve to see her also carried off so far away', she wrote to the Kaiser.[125] Wilhelm, however, pinned great political hopes on his new connection with Nicholas, who had 'Germany on his mind!' and with whom he had 'priority' because 'I made the match for him!'[126]

In his public speeches likewise Wilhelm II adopted a tone towards Russia that had rarely been heard from him since the fall of Bismarck. In February 1894 he declared in an address to the officers of the First Regiment of Guards that 'he saw himself as guardian of the peace of Europe and had from his youth learned to see this as his role'. Waldersee was surprised, recalling the 'completely different tone' of earlier utterances in which Wilhelm had announced that he wanted 'to smash the French and the Russians and parade into Berlin etc. . . . What kind of feelings must the Kaiser have if he ever allows himself an hour of calm reflection and remembers everything he has wanted and said! We must be prepared for many more sudden shifts and about-turns.'[127]

The corner-stone of the policy of rapprochement with Russia which the German leadership was pursuing, in ignorance of the fact that after lengthy negotiations in December 1893 Russia and France had undertaken to give each other military support in the event of a German attack, was the trade treaty, which held considerable advantages for Germany's eastern neighbour and which was eventually accepted after bitter political argument by the Reichstag in March 1894.[128] Wilhelm II had great expectations of the Russian trade treaty, which he described as 'one of the greatest achievements ever attained for Prussia and Germany'. He hoped it would both improve German–Russian relations and turn Russia against the French Republic. Echoing the 'fundamental idea' he had expressed in July 1892 and again in January 1893 of Germany's 'Napoleonic' domination in Europe, he announced at a Crown Council held in the Berlin Schloss on 18 February 1894: 'Our supremacy is to be demonstrated to Europe not only by our army but also through our trade policy, and a customs union of European nations is to be considered the goal of our future development, in order to be able to take up the economic struggle with North America.' The latter country was seeking 'to push us out of our principal market, South America, by establishing itself there through pacts with democratic movements. The rebellion in Chile was led by North American envoys, and now they are playing the same game in Brazil.' While the USA was playing the democratic card in this way, the conclusion of the trade treaty with Russia, so important for Germany and Europe, had been 'of decisive significance for the relationship between the two dynasties', the Kaiser declared. It was his belief that the moment at which Russia showed Germany its willingness to co-operate on trade policy had been 'the moment when His Majesty the Tsar of Russia, returning from Denmark, had realised the sad state of Russian agriculture and had given the order for the conclusion of the treaty to be proceeded with in earnest'. Wilhelm welcomed the conclusion of the treaty 'with all my heart' and emphasised that 'a good political relationship between nations whose economic relations are bad cannot last long'. He expressed his hope to the ministers of state that the trade treaty would bring about a 'strengthening of relations between Russia and Germany and [a] loosening of those between Russia and France'.[129] These ambitious international expectations explain the Kaiser's determination to push through the trade treaty despite the virulent opposition of the East Elbian agrarians and his angry outburst of February 1894, quoted earlier, that he had no desire 'to wage war with Russia for the sake of 100 stupid Junkers'.[130]

For an anti-Russian hothead like Waldersee, Wilhelm's swing towards Russia was incomprehensible. He reacted with scorn to the Kaiser's assertion that the rejection of the trade treaty by the Reichstag would lead to war with Russia 'in 3 months at the latest', and that in the course of it he would have to surrender the right bank of the Vistula. 'So the Kaiser has reached the point not only of being afraid of the Russians – I have long known that – but also of admitting it openly', he wrote mockingly in his diary. 'He thinks he can maintain peace by giving in, and he will achieve the opposite. How cocksure the Russians must feel when they hear that we are afraid of them, that we are prepared to give up provinces without a fight; I am convinced that it brings us closer to war. What are our people to think when they hear such ideas from the Kaiser, when they were assured only six months ago that once the Army Bill was passed we would be equal to any enemy.' According to what the Kaiser had said, which had been telegraphed all over the country, it must be clear to everyone 'that Russia has us entirely in her power', the General feared. 'As soon as the Tsar makes a threat, we give in! That must lead to a setback, such a state of affairs is demeaning. What jubilation there will be in France.'[131] 'This is the Kaiser who smashes everything!', Waldersee exclaimed in disgust some days later. 'The German Reich was so great and so much feared only a few years ago, and abroad too, especially in Russia, people felt that Kaiser Wilhelm II was an extraordinarily energetic man and that one must be prepared for him to be very quick to draw his sword; people were really afraid of him and that was a great guarantee of peace. That is all over now.'[132]

The news that Alexander III was seriously ill with Bright's disease, which reached the Kaiser on his return from England in mid-August 1894, revived his fear that the weakness of the Tsar might be exploited by the Slavophile military leadership of Russia to begin major redeployments of troops against Germany. As the support of Britain seemed more essential than ever in this situation, it was certainly not by chance that Wilhelm passed on details of the threatening development on his eastern frontier to his grandmother. In spite of the serious cholera epidemic in Western Russia and Poland the Russians were in the process of setting up three new army corps on the German–Russian border and were moving one or even two divisions from the Caucasus to Poland, he wrote to her on 24 August 1894. 'The rest of the Army is beeing [sic] moved up into closer formations & quarters, so that by next summer I shall have 10 Corps & 8 Cavalry divisions on my frontier. An outlook we think rather serious as we only have 5 Corps & 1 Cavalry Division to dispose of. We fancy that they want to strike an unexpected blow in the East, & to keep us from helping anybody they are going to attack, they have marschalled [sic] the enormous forces on our frontier. This state of affairs with an ill if not half dying Emperor unable to control the Slavophile party is rather ticklish & we must with great care & attention observe what is going on at Petersburg.'[133] It was not until his friend Nicky succeeded to the throne in October and immediately afterwards married Wilhelm's cousin Alix, who bore the name Alexandra following her reception into the Russian Orthodox church, that there seemed to be any prospect of a new era in Russo-German relations.

WILHELM II AND FRANCE

The relatively peace-loving phase which the Kaiser can be seen to have gone through in 1893–4 had its effect even on his attitude towards France. As the co-operation between Russia and France became increasingly visible and was publicly and enthusiastically celebrated in both countries (in Kronstadt, Nancy and Toulon), the danger of a French war of revenge with Russian support loomed distinctly larger. Wilhelm reacted to this threat by alternating between conciliatory signals on the one hand and aggressively menacing gestures on the other. Thus when the former President of the Third Republic, the Comte de MacMahon, died on 17 October 1893, he ordered a wreath to be laid on the coffin, which Waldersee predictably saw as another sign of weakness. The 'hatred of the French for us' was as strong as ever, he considered, and their 'intention of seeking revenge in the end' was 'unalterably fixed'. The Kaiser, on the other hand, was 'only too ready to believe what he hopes'; by having the wreath laid on MacMahon's coffin he thought he would be able to accomplish 'moral victory', but this was 'a complete misunderstanding of the French character'. In the opinion of the former Chief of the General Staff, 'with Russians as with the French one must show one's strength and let them see that one is not afraid of them'.[134] Only a few weeks later in a telegram to Philipp Eulenburg Wilhelm gave vent to his feelings about a failed French assassination attempt against him and the Reich Chancellor. The 'Gallic blackguards', he declared, had 'sent Caprivi and me a time bomb each. Toulon-Paris has made them cocky. Likewise the incredible utterances of kleindeutsch particularism which is showing its ugly face everywhere. This is the result! But God has prevented it and I thank Him for the preservation of our good Caprivi! But these days it is doubly necessary for all Germans to rally round the banner of the Kaiser and the Reich! He who fails to do so is an enemy, and woe to him who wishes to be my enemy! Wilhelm I. R.'[135]

When the French President Sadi Carnot was stabbed to death by the Italian anarchist Caserio in Lyon on 24 June 1894 Wilhelm sent a telegram of condolence to the widow of the murdered politician which caused universal astonishment, especially in France, where the Kaiser had long been seen as the arch-enemy. In a sycophantic letter which Wilhelm proudly forwarded to Caprivi, Hinzpeter congratulated him on this apparently wondrous change in German policy towards France. The former tutor wrote: 'How far Your Majesty's telegram to Mme Carnot was combined with a political purpose I cannot of course know. But my

French correspondence shows so clearly that it has achieved such a purpose that I am taking the liberty of drawing attention to it. This correspondence fairly well represents the average ideas and attitudes of the French and comes from the same people who told me as their firm opinion, not very long ago, that Kaiser Wilhelm II had only one purpose in life, which was to destroy France and that he spent his days searching for an opportunity to achieve this. This conviction may have been shaken by other things, but by nothing so much as this telegram. This has transformed the picture they have of the Kaiser.'[136]

The impression initially created that the Kaiser wished to bring about a shift towards France in German policy was, however, an illusion. Wilhelm did indeed pardon two French officers who had been condemned to long prison sentences for espionage, and personally ensured that the pardon was announced on the same day as Carnot's funeral. Nonetheless he did so only in response to a request from the French ambassador, Jules Herbette, who had gone to Kiel to thank the Kaiser for the message of condolence to Carnot's widow; he had no thought of changing the direction of German foreign policy.[137] This is borne out by a conversation between Waldersee and Verdy at the Swiss resort of Engelberg in the summer of 1894, in the course of which the former Minister of War, who had continued to maintain close contacts with Schlieffen, told the latter's predecessor of the new requirements that the Kaiser had laid down for waging war in the West and in the East. 'The Kaiser wants to take the offensive against France at once and has therefore weakened the Eastern army by 2–3 corps!', Waldersee indignantly recorded after hearing Verdy's information. 'He is doing exactly what the French hope and what they have prepared for. We shall be throwing ourselves at the positions they have reinforced with permanent fortifications and we all have every prospect of being sent back with bloody heads.' During his time as Chief of the General Staff Waldersee had planned to entice the French forward and then to counterattack them; by using such tactics, he believed, there was a good prospect of breaking through the French line of fortifications, but not by the Kaiser's offensive strategy. 'In my opinion Schlieffen ought to be prepared to sacrifice his post to prevent the Kaiser from carrying out his immature ideas', he wrote.[138] When Waldersee met the Grand Duke of Baden on the island of Mainau not long afterwards, the latter was also surprised to hear of the Kaiser's intention 'to go on the offensive from Lorraine at an early stage'. Grand Duke Friedrich had thought that 'my proposals, which also had Schlieffen's approval, still held good', the General noted. 'I could only tell him that Schlieffen's views had certainly remained the same, but that he has given way to the will of the Kaiser. The great danger that we incur in this way was perfectly plain to the Grand Duke.'[139]

Waldersee, Verdy and the Grand Duke of Baden were by no means the only ones who felt profoundly unsettled by the manner – mercurial, ill-considered and emotionally charged but always imperious and intolerant of any opposition – in which Wilhelm II intervened in every aspect of domestic, foreign and military policy. Shortly after his accession, and particularly after Bismarck's dismissal, the conviction was spreading that this irresponsible manner of ruling the country could easily lead to the collapse of the Prusso-German monarchy and the end of the German Reich.

CHAPTER NINETEEN

The rude awakening

At the time of the young Kaiser's accession in 1888, the prevailing mood almost everywhere was favourable, sometimes even enthusiastic. To compare this with the growing sense of foreboding which was perceptible only a few years later in leading political circles, at all levels of society and in every region of the Reich as well as abroad, is to gain some idea of the damage which Wilhelm II, over a very short period, had succeeded in inflicting on the standing of the Hohenzollern monarchy. The hectic round of state visits on which he had embarked immediately after the death of his father, the fall of the Bismarcks followed by the dismissal of several ministers of the old school, the removal from the imperial entourage of Oberhofmarschall von Liebenau and the older generals – Versen, Waldersee, Wittich and Wedel – all this had helped open the eyes of the world to the dangerously vulnerable position into which this inexperienced and unpredictable young monarch had manoeuvred himself. Even for those who knew nothing of the aggressive, often obscene tone which characterised the imperial entourage and who had not read the (hurriedly locked away) autocratic and insulting marginal comments written by the Kaiser on diplomatic reports, there could be no doubt that these changes in the leadership of both Prussia and the Reich, at the imperial court and in the army, were motivated by his desire to exercise a kind of personal autocracy. Indeed, Wilhelm himself gave notice of his wholly anachronistic ideas of divine right, personal rule and dynastic ancestor-worship in countless maxims, speeches and edicts which alarmed and horrified all who heard them. There was the motto, 'sic volo sic jubeo', which he inscribed on his portrait for the Minister of Ecclesiastical Affairs, Gustav von Gossler, the Golden Book of the city of Munich in which he wrote 'suprema lex regis voluntas', the speech of 20 February 1891 to the provincial diet of Brandenburg in which he railed against 'the spirit of disobedience' and called on the people to follow him blindly in the task 'which Heaven has laid upon me', the Düsseldorf speech of May 1891 in which he declared that there was only one ruler in the Reich, and that was himself, the imperial rescript of October 1891 on prostitution which none of the ministers had countersigned, the Brandenburg speech of February 1892 in which he challenged the 'dissatisfied grumblers' to shake 'the dust of Germany from their feet' and invited the people to follow him on the path traced out for them by God – who had 'taken endless trouble with . . . Our House' – to a great and glorious future, the address to the commanding generals in which he threatened to drive out the 'half-crazy Reichstag' if it continued to oppose him. All in all, there was certainly no lack of public indications of Wilhelm II's extraordinary quasi-absolutist conception of his role as kaiser and king. A hundred years after the French Revolution, in defiance of the democratic trends of his day and in an empire which was rapidly developing into the foremost industrial power in the world, he set out to exercise the kind of 'personal rule' to which Charles I, Louis XIV, Napoleon, the Russian tsars and the kings of Prussia in the pre-parliamentary era had aspired.

THE LAMENTATIONS OF THE EMPRESS FREDERICK

No one who has read the first volume of this biography will be surprised to learn that in the eyes of the Kaiser's mother his words and deeds only confirmed the fears she had long harboured. As early as 15 March 1890, anticipating the disaster which was taking place that very morning in the Wilhelmstrasse, the Dowager Empress bewailed the lack of a firmly anchored constitutional structure in the Reich, as a result of which the entire decision-making authority in this powerful nation now rested in the hands of her impulsive son. 'How one does bless a constitution like the British one, when one sees a young man totally without knowledge & experience – playing the Despot, without anything to prevent him fr. running into danger or mischief', she wrote in distress to Queen Victoria. '*Not one* older man or older relation is there to give a little timely advice, to warn & to give a gentle hint – both in political & important matters, – or in *family* & *court* matters.'[1] A week later, after Bismarck's departure, she reflected: 'The *System* Bismark [sic] was intensely *corrupt* and *bad*, – *this* is however not the reason that W. wanted the change, & this he does not even see through. The genius & prestige of Pce B. might *still* have been useful & valuable for Germany & for the cause of *peace*, – *especially* with so inexperienced & imprudent a sovereign, – & I fear that he will be missed in that respect, as I *also* fear that the combination wh. is to replace him will *not* be strong enough! – W. fancies he can do every thing himself – *you know* he *cannot*. A little modesty and "Selbst Erkenntnis" [self-knowledge] wd. show him that he is *not* the genius

24. The Empress Frederick in 1900

or the "Frederick the Great" he imagines.' 'I fear he will get into trouble', she added, especially because his 'love of playing the Despot & of showing off is very great'.[2] On 25 March, after Prince and Princess Bismarck had called on her to take their leave, the Empress wrote to her mother: 'I am afraid W. is a most thorough despot & has some very queer ideas on this subject in his head.'[3] Without Bismarck Wilhelm would behave like a tsar and cause a sensation by ruling Germany by ukase, his mother predicted apprehensively. 'Everything must be done in a hurry & be *startling*!! – & emanate or seem to emanate from one source!'[4] Wilhelm was 'so unpredictable that one must always be prepared for some completely unexpected coup!'[5] Like many other people she found Wilhelm's fondness for pompous ceremonies both ridiculous and alarming. In December 1890 she described the grand memorial parade with military bands and cannon-fire which her son had organised in honour of the Great Elector, and commented that she thought it very far-fetched, but 'W. does do the very oddest things, the element of show off, noise and "sensation", dramatic effects etc. is very preponderant, and in these serious times seems to me *very* youthful.'[6] The ministry that would suit Wilhelm best, she remarked sarcastically, would be composed of fantasists and eccentrics like Jules Verne, Richard Wagner, Lord Randolph Churchill, Lord Charles Beresford and a few bold African explorers. She had no doubt that her son would be reckless in whatever course he pursued, and the future filled her with anxiety.[7]

By the spring of 1891, a year after Bismarck's dismissal, the Kaiser had taken to proclaiming his ideas of absolute rule even more loudly than before. His mother predicted terrible catastrophe for the monarchy, Germany and the whole of Europe, writing to Freifrau von Stockmar: 'My son . . . *enjoys* his "power" to the full – the exercise of "personal rule", the satisfaction of every one of his caprices. – Everyone yields, everyone flatters and honours him and he is almost a megalomaniac. *Nowhere* can I see any *deeper, more serious* aspirations, nor any more spontaneous, nobler impulses – modesty, goodwill and piety.'[8] 'How frightful the speech in Düsseldorf was!', she exclaimed in another letter of May 1891.[9] And on the third anniversary of Wilhelm's accession she commented in a letter to Queen Victoria: 'It is indeed misfortune for us *all* that W. was called to the position he occupies so unprepared – (worse than unprepared) – imbued with prejudices, false notions & mistaken ideas; – that so *unripe* of character & judgement power was put into his hands, which alas he so often abuses! To me the grief is that he *might* have developed into something so much better if his dear Father could have been at the head of affairs for 20 or even 15 years.'[10] In December 1891 she reflected sadly: 'Oh *how* different all might be if that vile party who brought on 1848 & drove F[riedrich]W[ilhelm] IV off his head; – terrorized my Father in Law, & formed the body guard of Bismarkism [sic], broke Fritz's heart, and *destroyed* all the work of *our* Lives, took *entire* possession of our Son, *hunted* me & all *our* friends down, – did *not* exist!'[11]

In November 1891, after Wilhelm had promulgated his rescript on prostitution in the official *Reichsanzeiger* without ministerial countersignature, in Russian style, the Empress Frederick wrote perceptively to Queen Victoria: 'It is *so* unconstitutional and I do fear that if he thinks Germany can be governed by "Ukases" he will find out his mistake! – It is imprudent to *strain* the law and overstep the bounds of monarchical power & authority, in *our* days – in *Europe* – it will *not do*, – what it is doing in Russia where personal government is *daily* showing its worst side every one sees! – There cannot fail to be a Crash *there* some day.'[12] At the same time she lamented to Baroness Stockmar: 'I am quite beside myself over this "Declaration"! We *are* after all a constitutional country – if only very little, unfortunately – but *sadly* this seems not to have been taken

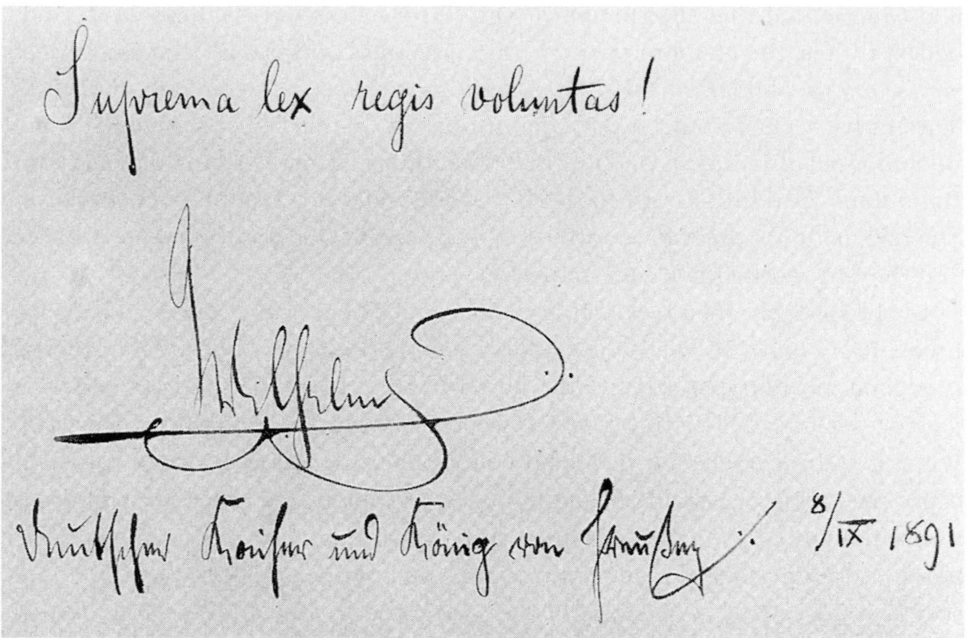

25. Suprema lex regis voluntas: Wilhelm II's entry in the Golden Book of the City of Munich, 1891

into account at all on high, and the idea seems to be that one can and must rule by "Ukases" – it is always the same deplorable trait that shows itself everywhere! "L'état c'est moi" – and it is – the *last* attempt at personal government in Europe! I am only afraid that this "last attempt" will be a miserable failure – for the time for such things is past, and I should be sorry if what could be useful in certain circumstances in having authority and power, – were brought to an end – because people refuse to accept the overstepping of that power.'[13] The Empress's worst fears were realised only days later when it became known that Wilhelm had written the absolutist dictum 'suprema lex regis voluntas' in the Golden Book of the city of Munich. As she wrote to the Queen, 'I think he *can* hardly understand *what* a bévue he is making when *he* writes such a thing!!! – A Czar, an infallible Pope, the Bourbons and our poor Charles I might have written such a sentence, – but a constitutional monarch in the 19th Century!! So young a man, – the son of *his* father – & *your grandson* – not to speak of a child of mine – should neither have nor express such a maxim!! It is like what he wrote under his Picture to Min: Gossler in Spring, wh. caused so much comment!'[14] When Wilhelm once again seized the opportunity to make a strong speech at the swearing-in of recruits at the end of the year, his mother lamented: 'W. has alas made a terrible new speech to the Recruits wh. is very freely criticised.'[15] A week later she wrote again to the Queen: 'If I *had* a *shadow* of influence I should implore W. to make no speeches in public for they are too terrible & not to write into books & under photographs – it makes ones hair stand on end! Here at Berlin people are becoming accustomed to the *very* strange utterances etc . . . & think it a peculiar style to wh. it is well not to attach too great importance, it is put down to ignorance & childish impetuosity.'[16]

After the *'most imprudent'* second Brandenburg speech of February 1892 in which he promised to lead his people towards 'a glorious destiny', the Empress Frederick, aghast, commented that Wilhelm was playing 'a most dangerous game' in exercising his rule in so irresponsible a manner. 'Here am I condemned to sit here and look on, in silence, without being able to say *one word* in warning, and knowing that the hideous mistakes made may lead to terrible consequences . . . I now watch as from *a grave* . . . the reckless course pursued by my own son . . . *no one* sensible has *any* influence – *no* one about him *warns* or *gives* advice. The worst of it is that *we* shall perhaps all have to pay for *his* ignorance & imprudence.'[17] Again referring to this speech, she wrote to her second daughter Moretta in Bonn: 'Willy has made a great speech again, wh. rather distresses me! . . . In *our* days a nation like Germany will *not* be led by the *will* of one person, especially that of a young & inexperienced man. – That perhaps did in *Prussia* in the days of Fred. Will. I & of Fred. the Great, but those days are *past*. *Modern* Life is too complicated for a sovereign to be like the Chieftain of a Clan in the olden time – personal govt. is no more possible or desirable! I sadly fear your Brothers do not quite see or understand this, and I am often *so* troubled and anxious! It is what I feared & foresaw when beloved Papa was taken who understood his times. Caprivi, whom I honour & respect & trust in every way, is too much of a *soldier* to be a *statesman* such as is needed in such ticklish times!'[18] In a letter to her mother she reverted to her earlier simile of a hen that has hatched a duck's egg. 'I wish I could put a padlock on his mouth for all occasions where speeches are made in public', she lamented. 'The Bismark [sic] education & the School of the Emp: William's entourage have made him what he is – & *their* teaching brings us these results.' She disclaimed all responsibility for the way her son had developed. 'His dear Father & *I* are in no ways responsible for his extraordinary ideas! We were for *constitutional* Liberty, for *quiet steady* progress, for *unobtrusive but unobstructed* evolution, – for *individualism* & the development of *culture*. *Not* for Imperialism, Caesarism, State Socialism etc . . .!!'[19] Two days later, in another lamentation addressed to the Queen, she described Wilhelm as 'a source of constant anxiety to me!', adding that 'the pernicious influence of the Bismarks [sic], of *certain* military circles & Junkers have so filled his head with ideas wh. I consider most false and dangerous, and which he takes up with the conviction & naiveté, of ignorance and inexperience. – There is *no* one there to advise or counteract the baneful turn given to his opinions! What *will* it come to!! – He was *snatched* out of our hands, all our wise friends were put to silence! . . . I assure you I tremble for him with all his rashness and obstinacy etc . . . He is a big Baby. Henry & Bernhard understand politics *no better* than he does! Some of his Aide de Camps [sic] were beside themselves with *enthusiasm* about this speech, wh. quite brought the perspiration to my forehead when I read it!! The speech was – alas – *not* an ebullition of the moment, he had written it all *down before* & took it with him & made Oberpräsident v. Achenbach prompt him! I should have *refused* & told him that such a speech was *impossible*. Afterwards the Ministers tried to weed out all the expressions wh. were a *great* deal stronger still – and therefore the "Staatsanzeiger" in which it was printed in its present form appeared 3 hours later than usual! . . . It is too despairing to see people rushing headlong into mistakes & on *quite* the wrong tack – & not to be able to stop them. *All* those who are blind enough to hate constitutional liberty admire and applaud him, and all the orthodox set! Why is it they do not see that they are playing the game of socialism (as Pce Bismark was!!).'[20]

At the end of March 1892, after the funeral of her equally liberal-minded brother-in-law, Grand Duke Ludwig IV of Hesse and by Rhine, the fourth anniversary of her husband's accession as Kaiser and the second anniversary of Bismarck's dismissal, the Empress Frederick looked back forlornly over political developments in Berlin. Of Wilhelm and his latest Brandenburg speech she wrote to her mother: 'He will *not* admit that the speech was a mistake in any way – & thinks the criticisms all pure spite and wickedness, but *some* that were shown him have annoyed him – wh. everybody is *thankful* for, as heretofore they have made *no* impression whatsoever – and it is hope[d] this may stagger him a little and make him a little more prudent & careful! I myself do *not* think so, – he is *so* imbued with false ideas that it would want a *constant* & *daily* & *powerful* influence to open his eyes, *explain* things in their true light. He does not understand *what* a Constitution *is*! He does not know a single member of the Liberal Party, he never reads *one* of the really good sensible newspapers!! If he only had the same political instinct that dear Louis [Ludwig IV of Hesse] had!!'[21]

In May 1892 Wilhelm II aroused widespread criticism when he commended a grenadier who had opened fire on some loutish passers-by, killing one and wounding another, in a speech to the regiment and rewarded him with a signed photograph of himself. The Empress Frederick was horrified by this fresh display of tactlessness on the part of her son. '*What* must people think when the sovereign *praises* the soldier openly for firing on another man, just as much his subject & entitled to as much protection. It is one more of the many bévues that startle &

alarm me! – He makes himself terribly unpopular.'[22] When Queen Victoria also expressed horror at the favour shown by Wilhelm to the soldier, Vicky commented that the incident had reminded her of some of her son's earlier actions: the award of the Order of the Black Eagle to Puttkamer, whom his dying father had dismissed; the telegram to Treitschke, who had so vilely attacked his, Wilhelm's, parents; the public praise for Gustav Freytag's inflammatory book. All these actions led in the same dangerous direction. 'Will there *ever* be an awakening from these mistaken ideas?', she asked. 'All the flattery etc . . . have made them take a firm root in W.'s mind!'[23]

With the crisis over the School Bill and the ensuing political turmoil, the worst fears which the widowed Empress, as a powerless critic of the Bismarck system, had harboured for decades, seemed to be realised. In a heartbreaking Cassandra-like lament addressed to Bogumilla von Stockmar at the end of April 1892 she wrote, virtually anticipating the entire catastrophe of German history yet to come: 'What an utterly *abominable* and *despicable* piece of anti-Semitism that miserable Norddeutsche Zeitung and the Conservative Party have achieved again. It is really shaming! – In Prussia generally things which a sensible, cultivated people can scarcely tolerate in modern times are gaining the upper hand! *Personal government* is exercised "ad absurdum" and the Conservative Party is behaving in such a grotesque and impossible way that no one can take them *au sérieux* any more! – But the *consolation is to be found in this* too – and my hopes are rooted in it. Perhaps the present government will have the effect that people will at last see and learn *what* must be done in order to guard against all possible extravagances – and how our constitution *must* develop and be extended – to protect freedom – and to protect the monarchy too! – Perhaps the Bismark [sic] system will be followed by a rebirth of the constitution of the German Reich on *solid* foundations – on firm principles of constitutional freedom . . . The government *will* gradually *be obliged*, if it does not wish to be still further endangered, to help in building up a strong, healthy structure on a *broad* base, if Germany is not to slide further and further down into evil ways and into a republic or even a socialist state. The latter state could never last; there would be chaos – and then reaction – dictatorship and God knows what other ravages!' Switching into English she expressed her feelings about Wilhelm's conduct: 'For *me* it is *most bitter hard* and disappointing to see all the aspirations and aims of our Lives *defeated & brought to nought* – and *that* by the very one who, it *might* have been *hoped*, would have carried them on, if he had been allowed to come under the influence of his Parents & their friends – instead of having his mind wharped & his opinions poisoned by others.' She begged her friend to burn her letter, 'for *even now* one cannot tell one's best friends in writing what one thinks, however good and loyal one's intentions are'.[24]

When the Army Bill was rejected by the Reichstag in May 1893 and Wilhelm gave vent to his anger in his speech at Tempelhof, his mother was incensed by this renewed demonstration of his anti-constitutional attitude: 'I was grieved and horrified to hear the frightful speech!', she wrote to Baroness Stockmar. 'I can *never* understand how a child of mine – a grandson of my father's – can understand nothing at all of the *meaning*, the sense and the value of a constitution, – and that *all* his views are so diametrically opposed to mine! . . . So much egoism, so much selfishness pains me deeply, – even if I am often disarmed by his truly childish naivety and ignorance! It is *not my* fault, and I am not responsible for this, it is the seed planted by Bismark [sic] which is coming up – and overgrowing everything! – The entire younger generation is in thrall to Bismark!'[25] She complained to her daughter Moretta on 18 May: 'I am miserable about things at Berlin. He [Wilhelm] is making one blunder after another! The speeches are quite dreadful, & the incessant whirl of amusement in wh. he lives in such serious times make a very bad impression, he neglects his duties and shows himself to be very selfish! For *me* his Mother it is very humiliating and distressing.'[26]

CRITICISM IN THE IMPERIAL FAMILY AND AT COURT

It would be a grave mistake to dismiss the Empress Frederick's harsh criticism of her son as little more than the expression of the bitter personal disappointment or of the 'un-Prussian' attitude of this liberal-minded 'Engländerin', for it is a demonstrable fact that other members of the imperial family, as well as many ruling princes at German and foreign courts, shared the same fears and voiced the same complaints, often with damaging effects on the reputation of the Hohenzollern monarchy. Thus for instance Wilhelm's sister Sophie, the Crown Princess of Greece, was reported by Waldersee to have 'spoken out in the most contemptuous and hostile way against her brother Wilhelm at St Petersburg, and said, among other things, that the whole family thought he was mad'.[27] During a visit to Berlin in the spring of 1892 the diplomat Bernhard von Bülow was shocked by the hypercritical mood which he found not only among foreign diplomats there, but also in the Kaiser's own family, among whom he named the Crown Princess of Greece and the Grand Duchess Ella of Russia in particular. He reported anxiously to his friend Philipp Eulenburg that the Empress Frederick was 'less unpopular than before in many circles. Prince Heinrich is also said to have joined in the general chorus of criticism against our most gracious sovereign in Darmstadt and in England. Et tu quique! His own relations – especially all the Russian, Danish, English and Greek elements and the whole Rumpenheim clique – are very hostile towards our most gracious sovereign.'[28]

Even Bernhard of Saxe-Meiningen, who was married to the Kaiser's elder sister Charlotte and who in fact shared Wilhelm's reactionary and militaristic views, was increasingly inclined to make openly critical remarks about his brother-in-law.[29] After Wilhelm's second speech to the provincial diet of Brandenburg the Kaiserin's brother, Duke Ernst Günther of Schleswig-Holstein, went so far as to criticise the 'not always very felicitous' speeches of the Kaiser.[30] Prince Ernst zu Hohenlohe-Langenburg, related both to the Kaiserin and to the British royal family, wrote to his father not long afterwards saying that things were 'getting more and more lively here [in Germany]'; Wilhelm's speech in Brandenburg resembled 'a declaration of infallibility and has been greeted everywhere with malicious pleasure, scorn or anger depending on which party people support'.[31] Prince Heinrich VII Reuss, who had been German ambassador in Vienna for many years and was married to a Princess of Saxe-Weimar-Eisenach, recognised Wilhelm II's good qualities even after he himself had been rudely dismissed from his post, but deeply deplored the fact that these qualities were 'so often cancelled out by inconsistency and excessive haste but above all by arrogance!' He had heard, he informed Carl Wedel, that 'Philli [Eulenburg] and his allies had succeeded in teaching their patron the Nietzschean theory of the Übermensch, and this had got somewhat out of control. This could explain many things. I try my best to see any sign of improvement, and if I find one, for instance that he talks less, something else immediately comes along to destroy my hopes.'[32] Duke Georg II of Saxe-Meiningen was unusually perceptive in commenting to a friend after the Kaiser's belligerent speech at Erfurt that it was 'very much in doubt whether things will remain as quiet as they have been since the [1870–1] war with France, or whether we are not in fact heading for colossal storms, perhaps even before my reign is over. It seems to me that there is a more and more chauvinist attitude in the highest place in this country – perhaps because of the realisation that armed peace is gradually ruining us, but perhaps also because the frequent sight of large numbers of German soldiers bolsters confidence in one's own strength. But what will we achieve, even at best, if we go on the offensive? The maintenance of the status quo. We already have that! But if it goes wrong, what then?'[33] The old Duke died a few weeks before war broke out in 1914.

It is particularly telling that the two German federal princes who had stood by the Kaiser in his bitter contest with the Bismarcks, Grand Duke Friedrich I of Baden and King Albert I of Saxony, also strongly criticised Wilhelm very soon after the dismissal of the founder of the Reich. In the summer of 1890, for instance, the Grand Duke and Grand Duchess expressed themselves in 'decidedly . . . anxious' terms about the Kaiser and bemoaned the fact that he very frequently took 'overhasty action' and was losing ground with the public. Although at that time they clung to the view that 'with all [Wilhelm's] good qualities one could nevertheless still hope for the best',[34] such hopes were all too quickly dashed. When the Grand Duke met Waldersee in October 1890 the latter recorded that the conversation immediately came round to the Kaiser, 'and the Grand Duke told me, after I had given him my impressions, that he regretted to have to say that he shared my views entirely'. Friedrich added that he had lately discussed the same subject with Caprivi in Baden and had expressed his fear 'that pushy people might easily acquire influence these days', whereupon the Reich Chancellor had answered that this was unfortunately already the case.[35] After spending an evening with the Grand Duke and his wife on the island of Mainau in the summer of 1894 Waldersee commented that as both of them had always set 'an outstanding example of devotion to duty and of regal dignity' it was 'really no wonder if they were deeply disturbed about the Kaiser and considered the present time a very serious one'.[36] The Grand Duchess Luise, Wilhelm's aunt, declared that what she felt most strongly with the Kaiser was 'the lack of imperial dignity which one cannot fail to notice if one has but seen the daily life of the Kaiser and his relations with his entourage'. She said that she fully recognised, as she had always done, 'the Kaiser's rare qualities, his quick understanding and his knowledge on all kinds of subjects . . . but she is shocked by his volatility and hastiness. His many ill-considered decisions are of course an inevitable result of this', General von Waldersee concluded. For his part the Grand Duke singled out for particular criticism the Kaiser's craving for pleasure and lack of seriousness. Wilhelm had been to visit the Grand Duke for three days (as Waldersee recounted) but had gone shooting every morning and had spent in all only half an hour in private conversation with his uncle. The Grand Duke described as particularly characteristic the manner in which Wilhelm had made up his mind on the important question of the return of the Redemptorist Order to Germany. His decision had been taken 'at Kiel, without interrupting the pleasures of the regatta week, during a 15-minute audience'.[37]

The King of Saxony was scarcely less critical in his comments on the new style of government. Six months after Bismarck's fall he told Caprivi that he was 'a little hurt' that 'I am almost never asked for my advice now, as I used to be in the days of the old Kaiser and under the present Kaiser too, in former days.'[38] While hunting at Königswusterhausen in December 1890 Waldersee had several conversations with the King, who fully shared the Chief of the General Staff's growing anxiety about the Kaiser.[39] The following autumn when King Albert met Waldersee at Wilhelmshöhe he remarked that although he thought highly of the Kaiser he knew 'many of his weaknesses and hoped he would gradually calm down'.[40] Towards the end of 1894 the King tried to persuade the Kaiser 'not to ask strangers for advice, but only the responsible ministers'. He had no success.[41]

It should perhaps be no surprise to find members of the Kaiser's closest entourage among his sternest critics, since they were the most likely to suffer from his changeable moods. After the Scandinavian cruise of 1890 a member of Wilhelm's accompanying suite commented to Waldersee: 'You should be glad that you were not with us; it was dreadful. Bad weather most of the time, and being crammed together in a small space soon became an ordeal. Moreover the Kaiser did not even have *one* serious conversation; apart for playing games with the navy, of which the latter is heartily sick, there was nothing but the most infantile fooling around. It is alarming to see how the Kaiser is deteriorating, and becoming increasingly nervy and irritable as well. Where will it lead?'[42] During the hunt at Grunewald in January 1891, to Waldersee's 'utter astonishment' the Chief of the Civil Cabinet, Hermann von Lucanus, engaged him in lengthy conversation 'about the Kaiser and above all about the general mood of dissatisfaction; this I could only confirm, unfortunately', the Chief of the General Staff recorded. 'Only I could not tell him that he himself was probably partly to blame because he gave the Kaiser such bad advice.'[43] The following year the General noted that he had again had a long conversation with Lucanus, who had 'complained bitterly about the Kaiser's dislike of work, which was growing stronger rather than decreasing'.[44] At the beginning of 1891 Waldersee recorded in his diary that the former tutor, Hinzpeter, 'in terms that one could scarcely credit', had strongly condemned the Kaiser's 'lack of respect for the truth'.[45] A senior official of the Kaiserin's household, the Oberhofmeister Freiherr von Mirbach, deeply deplored the Kaiser's 'erratic behaviour and lack of devotion to duty'.[46] At a service in Potsdam in December 1890 Wilhelm found the sermon tedious and demanded the sacking of the clergyman in question. The latter's superior conceded that he was not a gifted preacher, but he was still young and, moreover, had outstanding talent for pastoral care. 'And now he is to be got rid of without further ado, on the whim of our All-Highest Sovereign!', Waldersee commented angrily. 'Things are going from bad to worse!'[47]

For those in the know at the Hohenzollern court the appointment of Prince Chlodwig zu Hohenlohe-Schillingsfürst as Reich chancellor and Prussian minister-president in October 1894 represented a last chance to call a halt to the young Kaiser's autocratic behaviour, which was increasingly seen as disastrous. Thus for instance no less a person than Kaiserin Auguste Viktoria's chamberlain, Bodo von dem Knesebeck, courageously wrote to Prince Hohenlohe's son Alexander in the very early days of the new chancellorship, with the warning that 'the Kaiser does not see the consequences of the present situation because he does not yet understand how much he himself is to blame for it'. No one but Prince Hohenlohe was in a position to place conditions on the Kaiser which would protect the Chancellor from unpleasant surprises, the chamberlain argued.

He went on: 'The period during which the Prince is in control of affairs must become a period of training for the K[aiser], otherwise the situation will remain untenable. For although all other factors change the K[aiser] remains, and this constant factor must have its unpredictable element removed. I should like to think that there is still time to do this, and that it is not impossible, given the presence of such great gifts. But herein lies the kernel of all evil and part of the reason why this change is taking place under circumstances which have an element of tragedy. I cannot give any other name to the fact that to take on this office, which has no equal in Europe, is to make a sacrifice that is beyond measure!' Knesebeck added that he regarded the future 'with insurmountable anxiety'.[48]

In high-ranking army circles and above all among the military officers at court who were in constant attendance on the Kaiser, a particularly pessimistic mood prevailed. Adjutant-General von Wittich was among the most emphatic and outspoken critics of Wilhelm II. As early as the summer of 1890 Waldersee recorded that Wittich spoke 'in the most disparaging way' about the Kaiser, and indeed 'quite openly'.[49] Not long afterwards, following a long conversation with the Adjutant-General about the alarming development of the Kaiser's personality, the Chief of the General Staff noted that Wittich had 'expressed himself exactly in accordance with my own views on the Kaiser, his character and his deterioration and blamed it all on the Coburg blood. I believe he has hit the nail on the head.'[50] Waldersee also described in his diary how Wittich had criticised the Kaiser 'with a vengeance' after the New Year reception for the commanding generals in January 1892 and had complained that 'he was becoming more and more great and infallible'.[51] The views of Max von Versen, who as adjutant-general had been 'very close' to the Kaiser, were no less critical. When Waldersee met him on 19 December 1890 he found Versen 'very disappointed', and recorded that the General thought 'exactly as I do about the Kaiser's development'. Above all, Versen complained 'that the Kaiser had become far less dependable' – that no one ever knew 'how he stood with him'. 'Those immediately about him have reached the stage . . . where no one dares to say anything, neither the Cabinet chiefs nor Wittich, nor any of the Flügeladjutanten nor even a Hofmarschall. Everyone fears for his post.' Versen also agreed with Waldersee that Bismarck had 'suppressed people's personalities', but under Kaiser Wilhelm II 'we see the same thing happening, only in a stronger and more dangerous form'.[52] When the two Generals met again in the summer of 1891 Waldersee recorded that Versen was still 'very critical' of Wilhelm and entirely shared Waldersee's opinion 'that his development had taken quite a different course from what one might have expected'.[53] The new Adjutant-General, Hans von Plessen, whom many had expected to have a good influence on the Kaiser, was a disappointment, for he had 'as good as no influence at all' and was 'cowardly as well', Waldersee wrote in the summer of 1894. 'In short, wherever one looks there is dissatisfaction, ill-humour and discouragement, and not only in the higher ranks of the army, but among all independent, well-meaning people. Unfortunately I am not the only one who is convinced that we are going downhill all the time and that a catastrophe is inevitable.'[54] Even the once fiercely loyal Chief of the Military Cabinet, Wilhelm von Hahnke, eventually had no scruples in joining in the adverse criticism of the Kaiser expressed by his fellow generals.[55] Four years after Bismarck's dismissal Hahnke's enemy and rival Waldersee recorded that he was 'discontented' and 'anxious to get out'.[56] A year later Waldersee summed up the mood of the Chief of the Military Cabinet in a contemptuous note in his diary. 'Hahnke is very discontented but has not the energy to leave; he says quite openly that no one can tell today whether he will still be in favour tomorrow and that one has to be constantly prepared to hear that what is coal-black today is declared snow-white tomorrow.'[57] The former Minister of War General von Verdy was another who observed 'with great concern how everything is gradually deteriorating, and that it is the Kaiser himself who is responsible. In his view one of the worst things is that the Kaiser lacks nobility of mind. Unfortunately this is only too true, and the effects of this are truly disastrous', the former Chief of the General Staff commented.[58] Indeed, no less a man than the old Field Marshal Count Helmuth von Moltke 'expressed very great anxiety about the Kaiser' when Waldersee visited him in Kreisau in October 1890.[59] When Moltke died in April 1891 there were many who were deeply affected by his death, not only because they felt the loss of 'one of the last great men of the most glorious period of our history'. Prince Ernst zu Hohenlohe spoke for many when he expressed the view that 'under the present system of rushing ahead with everything' Moltke, 'as the guardian of the old traditions whom everyone respected, still seemed capable of preventing much of what the people of the new era come up with in their excess of enthusiasm'.[60]

COUNT VON WALDERSEE'S AWAKENING

The process of disenchantment was of course more painful for Waldersee himself than for anyone else. Once Wilhelm's closest confidant and the man who had done more than anyone to spur him on, first against his parents and then against the Bismarcks, it is a harrowing experience to read the unexpurgated version of his diaries and to watch the transformation which took place in his judgement of the young Kaiser. Like someone waking from a dream, Moltke's successor as chief of the General Staff was suddenly brought face to face with what he had failed to see before, and was able to recognise the disastrous mistake he had made.

In increasingly harsh terms he condemned the Kaiser's autocratic behaviour, his vanity and courting of popularity, his love of pomp, his extravagance, self-indulgence and superficiality, his constant interference in army business and in matters of civil administration, his inconsistency and thoughtlessness, his incessant and unfortunate speeches, his inconsiderate treatment of colleagues and subjects, the coarse and undignified tone of his entourage, his receptiveness to flattery and gossip relayed by irresponsible outsiders and not least his inability to tolerate well-meaning criticism.

The more negative his attitude to Wilhelm became and the more fatalistic a view he took of the future of Prussia–Germany, the more desperately he wondered whether he was not judging Wilhelm too harshly. In February 1894, for instance, he commented: 'I wish to God that I were able to write something good about the Kaiser. I sometimes wonder whether I am unjust and prejudiced in my judgement, but I cannot escape the conviction that our sovereign has no clear idea of his aims and is utterly unpredictable.'[61] On several occasions Waldersee asked himself whether he had any right to commit such critical comments on the Kaiser to paper at all, but again he concluded that, although he must hope that his remarks would 'remain undiscovered for a long time', it was nevertheless fitting 'that truth should be honoured and records such as this are necessary for the history which will have to be written one day of this very strange Kaiser and character'.[62] Demonstrably, there were certain experiences which the General shrank from recording in writing, although some of these were precisely the events which confirmed his general opinion of the Kaiser.[63] It goes without saying that Waldersee had no inkling that the editor of his diaries would later delete innumerable critical passages on Wilhelm in order to transmit to posterity an embellished picture of this sovereign and so distort the historical truth for decades to come.

It would certainly be a mistake for us to take every negative utterance of the fallen and increasingly bitter Chief of the General Staff at its face value. Not only because of the obvious frustration of his personal ambitions, but also because of very fundamental differences in attitude towards both domestic and foreign policy, we shall do well to ask ourselves from time to time whether we share the precepts on which Waldersee founded his criticisms of Wilhelm. On trade policy, for instance, the General took a view favouring the great landed proprietors and agriculture in general which, in the light of actual economic development, seems even more abstruse to us today than it did to his contemporaries. His only response to the steady growth of social democracy, as we shall see, was to advocate a highly dangerous policy of preventive war against the workers by which military force would be used to achieve the abolition of universal suffrage. The paranoid comments throughout his diary about the 'enormous power of the Jews' and his longing for 'a powerful personality . . . to make a clean sweep' occasionally make the General seem like a herald of a later nationalistic movement with which few would wish to be associated.[64] Nor should we forget, in assessing Waldersee's harsh criticism of the restrained foreign policy of the 'New Course', his own preferred solution to the strategic dilemma of the German Reich. The remedy which he repeatedly advocated was the formation of a Greater German Reich to which not only the German provinces of Austria but also a germanised Bohemia would be annexed. And that was not all: after the 'next' war against France he advocated that Germany should not claim any financial reparations but instead 'again take land'; in so doing 'we must remove the population from there to France and settle the land thus depopulated with people from Germany'.[65] Even taking his extreme views into account, however, we are left with such a formidable body of criticism of Wilhelm II's personality and style of rule that Waldersee's unexpurgated diaries must rank among the most revealing sources on the true events behind the scenes of the Wilhelmine Reich.

Waldersee's rude awakening can be dated to the summer of 1890, six months before his dismissal as chief of the General Staff, when he returned to Berlin after a long holiday and became aware of the criticism of the Kaiser which was being increasingly widely expressed. 'It is alarming what I have heard said about the Kaiser in 3 days in Berlin', he wrote in dismay on 12 August. 'There really is a universal sense of disappointment.'[66] Only one thought ran through all the Kaiser's actions, the Chief of the General Staff complained even at this early stage, and that was 'concern for his personal position. He wants to be popular!' The 'rapidly developed vanity' of the Kaiser was the basic weakness of his personality, Waldersee observed, adding: 'I thought his father a very vain man who enjoyed dressing up in fine clothes and posing; he set great store by outward ornament like orders and was delighted to be cheered by a few street boys. – But he [Wilhelm] outdoes all that by far. He fishes for ovations and – as of course follows from that – is very open to flattery. All this has developed so quickly that I am constantly surprised by what he does. As Prince Wilhelm he seemed to have many of the excellent qualities of his grandfather, he was modest and simple in his way of life, avoided attracting attention, did not like being escorted by adjutants and was very considerate towards old people. Now he loves splendour to an alarming degree and is beginning to behave like Louis XIV; he likes acting the patron, throws money away uselessly and does not worry himself at all about finances; he conducts himself with as much pomp as possible and likes nothing better than noisily cheering crowds; he is ruthless and odious towards the army; he has reached the point where he has a very high estimation of his abilities and thinks himself superior to everyone, but unfortunately this is a gross illusion and adds to the impression that there is very little behind it.

His great strength is a decided skill in dealing with people: he has a beguiling charm that few can resist. He often uses it with officers and thus wins hearts wherever he goes and does not stay long. But 2 years have been long enough to reveal the truth and the initial enthusiasm has cooled down now.'[67]

Not surprisingly, the clash with Wilhelm during the imperial manoeuvres in September 1890 sharpened the General's awareness of his weaknesses. After the manoeuvres Waldersee retreated to Russian Poland to stay with Count Guido Henckel von Donnersmark, and there reflected at length on what had become of the great hopes he had once harboured. Though it was scarcely six months since Bismarck's dismissal, for which he himself bore considerable responsibility, Waldersee too now recognised that Wilhelm II had 'come to the throne very much too early!' The realisation which was gradually dawning on him that Wilhelm 'was not, after all, the right man to lead the Fatherland out of the many dangers which threaten it' had now become a certainty. The Kaiser was certainly 'full of good intentions and idealistic ideas', but 'he did not have the clarity of vision to see that the approval of the masses, which he won surprisingly quickly, and the picture which those at home and abroad who did not know him formed of him and which the newspapers reflected back to him was not based on deeds, nor on the truth. His vanity having grown considerably, he very soon began to believe that he really was something quite special and clear signs of delusions of grandeur now began to appear.'[68] In his musings Waldersee constantly reverted to the 'extraordinarily highly developed . . . vanity' of the Kaiser, which he considered to be the 'mainspring of most of his behaviour'.[69]

For all his criticism General Waldersee at this time still emphasised that Wilhelm also had good qualities – a quick understanding, a good memory and a certain confidence in speaking – which enabled him to give an impression of being well informed in almost any domain. 'Furthermore, although he is outwardly hard, often very ruthless and indeed inclined to use ruthlessness to impress, he has a certain good nature and shows kindness towards those who have served him well.' The young Kaiser also had unusual aptitude for handling people, Waldersee acknowledged. 'He can win anyone over with beguiling charm and impress them with skilful, agreeable and often very amusing conversation. Anyone who does not know him very well falls under the spell of his personality, which of course is immensely enhanced by his position as Kaiser. Whether the Kaiser has real courage I do not yet know', the Chief of the General Staff commented in this character sketch of September 1890. 'He is a fearless rider and this is of course not to be dismissed, but whether he can show steadfastness in difficult times, take on great responsibilities and remain calm under enemy fire only experience will show. Hearing him speak one would be inclined to answer this question in the affirmative; but I have sufficient experience of life to have my doubts as yet. His concern for his personal safety is said to be great; his lack of consideration towards officers, civil servants and well-meaning people is matched by great consideration for the workers. He has already often threatened to smash the Centre, the Progressives, the feudal aristocracy and I do not know who else, but so far only people who are too weak or too monarchically inclined to resist have been given a taste of his power.'[70]

In time, however, the opinions which Waldersee expressed about his once so greatly admired Kaiser, King and Supreme War Lord became more and more pessimistic and scathing. Wilhelm was 'utterly unpredictable', he complained in an entry of 4 October 1890. 'Today he sings the praises of one party, tomorrow of another; he criticises the Jews and yet gets involved with people of Jewish origin; today he promises to support agriculture but tomorrow he does something likely to ruin it; he shows interest in industry and big business but is actually hostile and disparaging towards it; he dismisses Bismarck partly because he was supposed to have got involved with Windthorst and then stands calmly by while Caprivi enters into active dealings with him; he declares the aristocrats the noblest of his subjects and then threatens to shatter the factious aristocracy (which no one has heard of); today he is full of hatred and contempt for Russia and yet he pays court to the Russians in such a way that they themselves laugh at it; the Austrians are pitiful people on one occasion and his loyal and excellent allies on another. Is he a really pious man and a convinced Christian, or does he only pretend to be so; is he a trustworthy man or not – we do not know. These are all comments which one often hears now', the Chief of the General Staff concluded. He himself believed that the Kaiser 'actually has no firmly held convictions on any subject; he stumbles from one risky idea to another and there is no set plan behind his actions. There is *no subject* on which his opinions are clearly formed as yet.'[71] Another serious problem, as Waldersee observed, was that Wilhelm 'no longer [had] the slightest desire to do any work'. 'His amusements, whether playing about with the army or more especially the navy, travelling, shooting etc. come before everything else; and because these things come first he actually has scarcely any time for work. He reads very little, extracts from the newspaper being perhaps his most regular reading matter; he hardly writes any more himself, except for annotations on reports, and thinks the best audiences are those which are over soonest. Such a way of conducting business can last for quite a time before it becomes apparent to the outside world, but it cannot be kept up in the long run without causing very serious harm, and I believe it has reached that point already. It is truly scandalous how the court circular deceives the general public about the activities of the Kaiser. According to it he is at work from morning till night!'[72] Waldersee also had good grounds for observing that Wilhelm had 'an unusually lively imagination' and was thus 'very prone to exaggeration'; consequently his own duty as chief of the General Staff was 'to be doubly careful and to weigh things up calmly'.[73]

The Kaiser's weakness for flattery and for gossip peddled by irresponsible advisers made it infinitely more difficult for those who held political responsibility under the constitution to carry out their duties, Waldersee complained in November 1890. The old Kaiser Wilhelm had never allowed anyone to speak to him about third parties, 'unless it was within their competence. His entourage, with the exception of the Cabinet Chiefs, therefore had no influence. There was no one who could boast that he had the Kaiser's ear, or whose influence others feared.' Things were very different under Wilhelm II. The young Kaiser loved gossip 'and gives it free rein, allowing all those in his entourage to express opinions on others, and indeed encouraging them to do so. He thinks that by listening to people outside the official orbit he will be well informed about everything and know more than his ministers. He fails to see that 9/10 of his people simply tell him what he wants to hear! He does not notice how many people he offends, how many whose trust and therefore support he loses, how much he stirs people up against each other and all in all causes great damage.'[74] Waldersee accused not only the Kaiser's entourage but also outsiders of telling the Kaiser what he wanted to hear, which in turn he blamed on Wilhelm's vanity. 'I hear complaints from many sides of how much people flatter the Kaiser', he wrote in December 1890. But such eulogies were often 'provoked' by Wilhelm himself 'by his asking "What did you think of my speech?" or "Did I not do that well?", after which few people have the courage to tell him what they really think. But the good people often go straight up to him and come out with the most tedious flattery, sometimes about something he has done, sometimes in a more roundabout way by running down people against whom they know the Kaiser is prejudiced. All decent-thinking people are very concerned about this, if they know anything about what is going on, and many are deeply disheartened.'[75] On 10 January 1891, shortly before his dismissal as Chief of the General Staff, Waldersee reflected that 'these days . . . no one is sure of his life; that is what is so strange and disturbing about the state we are in; everyone feels unsafe because he does not know how far the Kaiser can be trusted; he who is in good odour today may be destroyed by gossip in a month! Gossip from the most insignificant people can be enough to bring discredit on respected people in high places, and the changes get faster and faster. It clearly shows a degree of nervous strain.'[76] 'Seldom has anyone disappointed expectations as much as he. Until he changes, until he realises that he is only a dilettante in all the countless areas into which he ventures with such confidence and makes up his mind to call on experienced people, until he decides not to tolerate gossip and to get rid of unqualified advisers, nothing can change and everything will get worse and worse.'[77] There was in fact no improvement and in later years the General again strongly condemned Wilhelm II's willingness to listen to slander, which in his eyes had become a cancer on the Prusso-German monarchy. 'This is the principal cause of our malady', he wrote in 1893. 'This is what produces the insecurity that weighs everything down; this is what breeds a false, bungling, contemptible, unprincipled spirit, and we may be destroyed by it.'[78] The mood in Berlin in general was worse than ever, Waldersee commented in January 1893, but above all at court. 'The closer one comes to the atmosphere of the court, the more one is repelled; there is an incredible stirring up of enmity and mistrust etc. Not a day goes by without someone being torn down and slandered.'[79]

For a short time in 1893 Waldersee allowed himself to hope that there had been a slight improvement in the 'mostly obscene' tone habitually used at the Hohenzollern court. In March that year he recorded that 'people who see the Kaiser often are reported to have remarked that he adopts a better tone in his conversations now. He had got into very bad habits through his contact with Herbert Bismarck and enjoyed hearing and telling dirty jokes, unfortunately also in the presence of servants. Among the people in his entourage Colonel Kessel also had a particularly bad effect on him. If there should really be a change for the better it would be an excellent thing', the General commented, clearly not quite convinced.[80] And indeed visitors to the imperial court continued to be struck by 'the bad tone among the members of the Kaiser's entourage'. During the visit paid by the Kaiser and Kaiserin to Rome in May 1893 the 'bad manners' and 'very arrogant, demanding and disagreeable' attitude of their suite were said to have made a particularly unpleasant impression. One could be sure, Waldersee warned on hearing various reports of the Kaiser's behaviour in Rome, that stories about it would do the rounds of all the courts of Europe, where they would be received with much malicious pleasure.[81]

Waldersee singled out the anniversaries of Kaiser Wilhelm I's death, Wilhelm II's accession and Bismarck's dismissal, as well as the young Kaiser's birthday, as opportunities to launch into brooding recollections or dark forebodings of the future of the Reich, repeatedly asking himself how he could have been so utterly deceived. In January 1891 he looked back with distress on the development of the political and religious views of the Kaiser, now aged thirty-two. In the spring of 1888, as crown prince, Wilhelm had maintained 'a firm, devout Lutheran attitude' and had had the courage to express this openly, Waldersee recollected. 'He had very decidedly conservative convictions of the kind that a sovereign who wishes to keep his throne today must have and cultivate, untroubled by the clamour of the unprincipled and unthinking masses. And what do we see barely 3 years later? A Kaiser who is led by fear of the Social Democrats, who believes he can create a position for himself by concessions to those who clamour for them, by courting popularity, by abandoning his firm Lutheran standpoint. He has been told that a king should not represent any strict confessional standpoint; there is some sense in that if one puts the emphasis on *represent*; he certainly should not persecute people with different beliefs, whether Catholics, Protestants, sectarians or Jews or anything else. But he should *have* a firm standpoint and not shrink from expressing it . . . He must show that he is the defender of the Lutheran Church and must hold unreservedly to the creed which is read out every Sunday in our Lutheran churches . . . Unfortunately Kaiser Wilhelm is now going the way of holding to no one, so he will eventually fall out with everyone.'[82] In Waldersee's eyes the Kaiser had quite the wrong conception of his role as supreme bishop of the Lutheran Church. 'How different I had imagined things would be when the Kaiser came to the throne', he sighed at the end of 1891. 'I dreamt of a Lutheran king in whom all fellow believers in Europe would see their powerful defender'; instead the Protestant camp was in conflict and the Catholics were being courted.[83] Waldersee was bewildered when he again received reports of the Kaiser's spiritualist leanings. Philipp Eulenburg, he wrote, had already been strongly urged, even before the first Scandinavian cruise, 'not to encourage the Kaiser in his spiritualist views, but to calm him and try to bring him round to a more level-headed way of thinking'. To begin with, Eulenburg had kept his word, but during the Kaiser's most recent visit to Munich Eulenburg had introduced the Kaiser to a spiritualist medium. When in a trance this woman had been asked by Wilhelm what attitude he should take to a friend in Russia, by which he of course meant Alexander III. 'Is it . . . not appalling that such a farce should be allowed?', Waldersee exclaimed. 'If our sovereign can be influenced in such a way, the welfare of the Fatherland is in the hands of swindlers and beyond hope. We may be facing a new edition of Friedrich Wilhelm II and Bischoffwerder!'[84]

Reminiscing at the beginning of 1891 on the past three years, the Chief of the General Staff commented bitterly on the disappointment of the hopes he had placed in the young Wilhelm. 'When I look back to the spring of 88, when the Kaiser was Crown Prince . . . and remember how he trusted me and how I hoped that he would become an excellent kaiser who would lead the country and show himself indifferent to the clamour of the masses and firm in his religious convictions, who would be devoted to the army and cherish all that is noble and good; when I think how it would have been the greatest possible satisfaction to me to help him in my field, and how well justified I was in having these hopes, I am bound to say that there has been a very great change indeed.' Looking back still further, Waldersee recalled: 'Prince Wilhelm was simple in his way of life, his speech; he did not like to be surrounded by a large suite, he was modest and considerate towards older people, had a certain respect for those who had served well in our wars; he was hard-working, liked writing and wrote a great deal – indeed sometimes too much – he had firmly held Lutheran beliefs and the courage to express them openly; he rejected excessive luxury and was very critical of gambling and the excesses of the racecourse.' 'And how do things look barely 3 years later?', Waldersee asked himself, gloomily answering his own question: 'He has become volatile, reluctant to work and unwilling to concentrate on one subject for long. He treats the army badly and is inconsiderate towards older people; his conviviality and bonhomie, which always made him popular at first, do not last, so that the good impression soon fades, especially among those who see him often. Quite unexpectedly, he has developed a very marked weakness for showy display, grand ceremonies and luxury.' The Kaiser undoubtedly had good sides – a quick understanding and 'a desire to attain what is good' – and Waldersee had not quite abandoned all hope that these aspects of his character would win through and overcome his undesirable tendencies.[85] Only a few days after this diary entry Waldersee was sacked as chief of the General Staff.

Even after his transfer to the command of the IX Army Corps in Altona, General Count Alfred von Waldersee remained on excellent terms with the military officers and senior officials at the Berlin court, with Prince Heinrich at Kiel, with his fellow generals and with the landed aristocracy. He continued to see and correspond with both current and former ministers as well as party leaders and journalists of various shades of opinion. He went to Berlin frequently and travelled the country on official tours of inspection or hunting trips. He must therefore continue to rank as one of the best-informed sources on the innermost workings of the court and of the world of high politics. It should therefore be no surprise that his open breach with Wilhelm II and his removal from the court in January 1891 did not result in any qualitative change in his attitude to his former hero. On the contrary, his judgement remained as sound – and as harsh – as ever.

Soon after his dismissal Waldersee commented with growing concern on Wilhelm II's despotic form of rule and frivolous lifestyle. 'Intelligent and very well-disposed people complain that he has less and less inclination to work or to listen to the opinion of others', he remarked. 'He does not give anyone the chance to speak, but expresses his own opinion with the greatest confidence and apparently will not allow any contradiction. With the life he leads there is indeed scarcely any time for work, especially as any free time is often lamentably wasted; for instance lately on the journey to Göhrde he played a very silly card game with Lucanus and Hahnke for almost 3 hours in the train. He could have got so much useful work done with those two . . . during that time.' For the first time Waldersee felt compelled to give serious consideration to the possibility that the Kaiser, of whom he had for years had such high hopes, might be mentally ill.[86]

'What is always the saddest thing for me', he wrote on Wilhelm II's thirty-third birthday, 'is to look into the Kaiser's heart and character.'[87]

Like everyone else the General was appalled when he read the newspaper report of the Kaiser's second speech to the Brandenburg provincial diet in February 1892. His remarks would 'quite justifiably [cause] bad blood' everywhere, Waldersee wrote in dismay, for they bore witness yet again 'to such boundless arrogance! He wants to lead Germany to glory and happiness and has done nothing so far but to drag her down. How great was the old Kaiser Wilhelm in his modesty, compared to this.'[88] It was 'deeply sad' to see 'how rapidly the standing of the Kaiser has diminished'. The foreign press was writing 'quite dreadful things' about him. 'In foreign newspapers one finds more and more comments about the Kaiser's mental state; many say quite openly that he has already gone mad!' As the speech coincided with the radical protest by all the liberal parties against the School Bill and with widespread street demonstrations by workers, it had given many people 'goose-flesh'. There was many an echo of the period leading up to the revolutionary events of 1848: 'The indecisiveness of those in charge, the confusion at the very top, where all kinds of unauthorised people have their say – all this is a repeat of the situation which brought about the events of '48', Waldersee remarked (just as the Empress Frederick had done) in his diary.[89]

On 1 March 1893 Wilhelm II gave his annual address at the gala dinner of the Brandenburg provincial diet. This time, in spite of the usual emphasis on his divine right, the speech had a rueful note. The Kaiser expressed his gratitude to his former mentor, Oberpräsident Heinrich von Achenbach, who had made a flattering introductory speech, for the 'firm confidence' placed in him as 'father of the people'. This was 'the finest reward which I and My trusted counsellors [can receive] in our work'. He acknowledged that the present time could not be compared with the 'glorious days gone by' of his grandfather. 'The noble form of our great departed Kaiser Wilhelm is ever present to our eyes with his mighty triumphs', he proclaimed; but those successes had been possible only because 'my grandfather had the most firm belief in his divinely appointed office combined with the most unwearied devotion to duty . . . Well, gentlemen, I grew up and was reared by him in these traditions', Wilhelm II declared. 'I too cherish the same belief. My highest reward is, therefore, to labour night and day for my people and their welfare. But I do not disguise from myself the fact that I can never make all the members of our nation equally happy and contented.'[90] 'Does he really believe that people trust him and his advisers?', Waldersee wondered when he read the speech in the newspaper. 'It is precisely because there is no trust, but general mistrust instead, that we are in such a bad way. The reference to his divine mission will hardly please the liberal gentlemen.

He claims to work day and night for the welfare of his people! Unfortunately all too many people know that he really does no work at all. In general there is rather a melancholy flavour to the speech', the General commented. 'Not very long ago he wanted to make the whole world happy; now the numbers are shrinking considerably.'[91] Waldersee found it 'comical' how every sentence of the Kaiser's speech was highlighted and analysed in the press, and wrote: 'If the people suspected how little thought is given to the Kaiser's individual words on such occasions! Often he actually does not know what he has said, and yet as is well known he has often said things that were highly questionable, but none were taken any further. He has often made threats, but he has never carried out a threat.' The elegiac tone which characterised this speech, the General claimed, was symptomatic. 'I have been aware of this mood for some time now', he commented. And he predicted, 'It only needs some great disappointment to come along, and with his character there is no question that despondency will follow, although at first he will not have the slightest hesitation in throwing the blame on others.'[92]

Looking back critically on the first five years of the young Kaiser's rule in June 1893 Waldersee wrote: 'He has often gone as far as to depict his grandfather as a tired old man who scarcely knew what was going on around him in his latter years. In fact the old gentleman was of incalculable value to us all, particularly as far as the socialist question was concerned. His venerable personality, known throughout the world for the conscientiousness and devotion to duty he showed until his dying day, with all his past, and the fact that he never gave any reason for malicious tongues to wag, was for us a bulwark against the tide of revolution. The present Kaiser, on the other hand, positively cultivates Social Democrats. Many people, particularly abroad, credit him with great energy, but the great majority in this country no longer do. People have realised that he is inclined to vacillate and therefore cannot to be relied on to stick to any course on which he has started; countless thoughtless remarks . . . have greatly encouraged the suspicion that he is a ruthless autocrat at heart; in the eyes of many, his versatility is merely a sign of superficiality; his private life is watched closely, and people conclude that he devotes most of his time to pleasure. His love of splendour is compared with the simplicity of his grandfather. No ruler in our times can be accused of such things and escape unscathed. They create a harmful atmosphere and this expresses itself by people becoming inflamed and voting for social democracy. It would not be fair to say that the Kaiser is chiefly to blame, but he is unquestionably very much to blame, as are all those who advise him. There was no need for us to have been brought so low as we have been!'[93] The wish, evident on all sides, for independent-minded men to come to the fore was quite clearly directed against the Kaiser, as it was believed that no member of the present regime would dare to tell him the truth, for fear of losing his position. 'I am sorry to say that there is undoubtedly much truth in this', Waldersee commented in the summer of 1894. 'The Kaiser thinks his opinion is the only right one, even if he often changes it. Apart from Bismarck and Minister Zedlitz I cannot think of anyone who faced up to him firmly and persisted in his own opinion.'[94] Waldersee reacted with disbelief to the passage in the Kaiser's speech at Königsberg on 6 September 1894, in which he claimed that his door was open to all. Such a remark might impress the unthinking masses, he said, but 'anyone in the know is all too well aware that for all his outward openness, the Kaiser is practically inaccessible. How many people have already tried to give him their views! They almost never manage it because the Kaiser will not listen to anyone who he thinks wants to instruct him or give him good advice. Even people in his closest entourage often do not succeed in saying a single word to him in private. And yet he boldly asserts that his door is open to everyone.'[95]

Among those closer to the Kaiser, Waldersee recorded in January 1895, the conviction was growing that it was 'impossible for any chancellor to work with the Kaiser so long as he conducts business as he has done so far'. Caprivi had swallowed far too much, and his successor Prince Chlodwig zu Hohenlohe-Schillingsfürst had already 'had many difficult moments because of the Kaiser's high-handed interference. The view is gradually gaining ground that the constitution must be developed further in order to keep greater control of the Kaiser's powers! This is how far we have come, 7 years after Wilhelm I's death!'[96] Similar despondent remarks occur in the General's diary on the fifth anniversary of Bismarck's dismissal. 'What twists and turns the Kaiser has already brought about!', he sighed in the spring of 1895. 'Neither Caprivi nor Hohenlohe is chiefly to blame for the lack of stability, but only the Kaiser. And it will go on until he realises that he must stop intervening personally in every matter of importance. It is quite remarkable how this can have gone on for 5 years already, and that both Caprivi and Hohenlohe, as well as the other ministers involved, have accepted it. They are none of them men of character; Zedlitz was the only one. How harshly history will judge this time in years to come! And at this grave moment when only strength of character can help us the Kaiser's chief occupation is breaking characters. One day he will come to suffer greatly for what he has done.'[97] When Waldersee was approached several times by conservative and even liberal politicians in Berlin who remarked, with a view to his becoming a candidate for the Reich chancellorship, that what was needed was 'above all an energetic man who will also be able to cope with the Kaiser', the General turned down this improbable request with the characteristic argument that 'the Kaiser has not yet reached the point of realising that he cannot continue to do business like this for much longer, and he has no intention of giving way to a chancellor or, to put it a better way, to let him work for him and stay in the background. That would require more experience.'[98]

Although his personal relationship to the Kaiser and Kaiserin slowly improved after Wilhelm conferred the Order of the Black Eagle on him in January 1895, to the point that even outsiders began to fear his influence again,[99] Waldersee did not alter his pessimistic view of Wilhelm's personality and style of rule. 'The most alarming thing, in my view, is that I still think that the Kaiser does not know what to do', he wrote at the beginning of 1896, after listing the major problems that confronted the Reich in both domestic and foreign policy. 'He has certain aims, but they are not clearly thought out, nor does he know how and by what means he wants to attain them. And as he still does not fully recognise the seriousness of the domestic situation, nor, in my opinion, properly understand the significance of the parties, and wants to reign Himself, intervening first in one place and then in another, often very rashly, and as there is no strong chancellor nor a united Ministry, it really is no wonder if we are regarded as being on a very uncertain course.'[100]

In the summer of 1896 Waldersee recorded that the Kaiser's mood was gradually becoming 'very embittered; he considers that most people are bad and complains of disloyalty and lack of gratitude . . . and refuses to see that the truth is the other way about. He himself is to blame: he has offended, deceived and let down so many loyal, able people that it is no wonder that many of them are disappointed and deeply hurt and leave him. Unlike his grandfather he does not have the gift of winning hearts and binding them to him; to begin with he captivates and enchants them with his charm and kindness, but it gradually fades and eventually he repels them. Years ago I put forward the theory', he wrote, 'that the Kaiser was ageing prematurely and would become an embittered man at odds with the world, and today I am convinced that he is well on the way to this.' Reflecting the views of someone 'very close to the Kaiser' Waldersee continued: 'in addition, he is as convinced as ever of his infallibility and superiority; he never makes a mistake; if anything goes wrong others are always to blame. Unfortunately he is not getting any more conscientious or hard-working; on the contrary, his desire to work is constantly dwindling; there is no sign of thoroughness and conscientiousness. With his exceptional gifts he can understand anything he wants to understand, but he never goes below the surface.'[101] In June 1896 Waldersee used the occasion of the eighth anniversary of Wilhelm's accession to take stock of the external and internal position of the Reich and of the Kaiser's achievements. The result was anything but encouraging. The world was like a great army camp living in a state of unnatural tension that would soon erupt, he observed. The danger of an international conflict was all the greater because the old European powers had now been joined by Japan, China and the

United States. Within Germany too the state of affairs was becoming increasingly untenable everywhere. 'Where the first clash will be God only knows', he wrote, adding prophetically that 'the next decades have every prospect of seeing great catastrophes'. Commenting on Kaiser Wilhelm himself, who ought to have taken control of this difficult situation, he wrote: 'Tomorrow is the anniversary of Kaiser Friedrich's death! The present Kaiser has been on the throne for 8 years now. He took up his duties with such fine and inspired ideas, and what has he achieved? Have we gone forwards or backwards? I have absolutely no hesitation in saying backwards. If the Kaiser were really the man he thought himself, if he had the gift of being able to use his exceptional abilities in the right way, if he were less vain, if he were more conscientious, if he had more knowledge of people and of the world and less inclination to surround himself with flatterers and yes-men, and if above all he had more courage, he could well have steered the world in a different direction, given the colossal power of the position he has inherited.' Instead, Waldersee had found it 'almost inconceivable how much he is impressed by outward splendour, great spectacles etc., and he commits the fatal error of believing others are equally impressed. A great parade, a ceremonial entry with a colour escort, 1000 bellowing louts and 10,000 foolish people waving handkerchiefs and shouting hurrah give him a feeling of well-being and confidence.' The magnificent reception which Wilhelm had put on for the Chinese Viceroy Li Hung Tschang had given him huge pleasure. The Kaiser was convinced 'that he had achieved something great, made an enormous impression, served the Fatherland well', Waldersee scoffed.[102]

After closely observing the Kaiser and his entourage 'in the broader sense' at Kiel and Travemünde that summer Waldersee wrote a scathing account of his impressions in his diary. In the imperial entourage no one trusted anyone else, he observed, and all were 'afraid of the Kaiser. He is quite well aware of this and encourages it, because in this way he hears a great deal of gossip and keeps control; only in one sphere does he fail to see how he is deceived and is up against a king rat, and that is foreign affairs, where the old sinners [by which he meant Friedrich von Holstein above all] continue to rule the roost and know how to influence him as they want . . . In Kiel, where Lucanus was also present, I was again much struck by the strange circumstances in which we find ourselves, and by the foolishness of the great mass of mankind . . . The Kaiser has always been of an autocratic nature, and although he has often let slip autocratic remarks he is regarded nevertheless as a sovereign who respects the constitutional procedures and on the whole is guided by them. But that is unfortunately untrue. He despises these procedures and actually practises absolute rule in any domain in which he cares to involve himself, which means that he does this regularly in foreign affairs, the army and the navy, and in other areas according to the circumstances, whenever he feels like seizing upon them. Uncle Chlodwig [the Reich Chancellor, Prince Hohenlohe] only puts up a mild resistance from time to time, but is in general absolutely compliant; the chiefs of Cabinet work on the other areas. Now this would be quite all right if the Kaiser were as industrious and clear-thinking a sovereign as Frederick the Great, whose Cabinet chiefs were only the obedient agents of his will . . . He could rule like that because he knew very few pleasures and was always at work. But now pleasure has become the most important thing, time for work is reduced to the minimum and at the same time the influence of the Cabinet chiefs is far-reaching and too great. We saw the same thing with Fried[rich] Wilhelm II and III and evil consequences inevitably followed.'[103] In mid-August 1896 Waldersee noted in his diary: 'The Kaiser intends to direct foreign policy and command the army as an autocrat. In the long run that is not feasible and it must lead to constant friction.'[104]

In October 1896 the General commented scathingly that it was now becoming clear 'what the Kaiser has achieved in his short reign by way of destroying characters, creating confusion in the country, stirring up the parties against each other and cultivating despicable people. The sycophancy and lack of character of the members of his entourage and all those with whom he surrounds himself is said to be on the increase. To complete this attractive picture of the court, no one trusts anyone else. If two people converse quietly together, the others immediately become suspicious and feel uncomfortable. No one feels really at ease! Philipp Eulenburg is still considered his best and most true friend. Once when someone urged him to fulfil his duty as a friend and speak frankly to the Kaiser, he burst into tears and said: Oh no, I cannot say unpleasant things to him! And this is the stuff of which our men of influence are made!'[105] In March 1897, during celebrations in Berlin for the centenary of the birth of Kaiser Wilhelm I, Waldersee spoke to Counts August and Botho Eulenburg, the Oberstkämmerer (Principal Chamberlain) Christian Kraft, Hereditary Prince of Hohenlohe-Oehringen (a nephew of the Reich Chancellor), the House Minister von Wedell-Piesdorf, Chief of the Naval Cabinet Admiral Freiherr von Senden-Bibran, Commanding General of the VIII Army Corps in Koblenz Vogel von Falckenstein, former Minister of War von Verdy du Vernois and several members of the Reichstag and other old acquaintances. He recorded that all of them felt 'great anxiety' and 'deep discontent' with regard to the Kaiser. Their chief grounds for complaint were Wilhelm's interference in foreign policy, the accelerated expansion of the fleet and his latest speech to the Brandenburg provincial diet, which had 'caused offence to all parties'. Summing up his impressions, the General commented: 'The feeling against the Kaiser is said to be very bad among broad sections of the people.' 'In fact the parties are all either hostile or deeply discontented and disturbed. It has now reached the point where everything that our sovereign does and says or might have said is bitterly criticised. His autocratic tendencies have gradually been recognised for what they are, and people know that he no longer pays heed to anyone else's opinion, knows *everything* better and is a wiser judge of *every* subject than anyone else. I should be less concerned if his nerves were in a perfectly healthy state; one could then be certain that he would learn from experience. But I am convinced that he is not capable of facing serious reversals; his nerves would give way and he would collapse, which would be the worst thing for us.'[106]

A review of the many highly critical comments about the Kaiser which General von Waldersee made in his diary over the years from 1890 to 1897 reveals that they constantly revolve around the same points. Only in one regard is there a distinct linear development, which in fact reflects the actual course of events. In the later years the diary accurately records the dominant role which the monarch was now in a position to play, above all in the formulation of foreign policy and in military and naval affairs. Otherwise Waldersee's constant criticism, despite all the allowances he makes for the Kaiser's positive qualities (his quick understanding, his good memory, his engaging friendliness in personal contacts, his amusing conversation and confidence in speaking), can be generally summed up as follows:

- The autocratic tendency verging on megalomania which Wilhelm II displayed in his speeches and sayings as the foremost champion of the principle of 'personal monarchy' led to a dangerous overestimation of his abilities. Deluding himself that he was infallible, he refused to tolerate any opposition.
- In spite of his exaggerated opinion of himself Wilhelm in Waldersee's eyes had no firm goals. The driving force behind his actions was vanity and desire for popularity, combined with an overactive, perhaps even diseased imagination. This gave rise to the Kaiser's notorious unpredictability.
- The Kaiser's autocratic pretensions were not based on the conscientious exercise of his role as monarch; on the contrary, they were accompanied by reluctance to work and a love of distraction and amusement. The result was superficiality and dilettantism.
- His autocratic attitude expressed itself in ruthlessness and severity towards his responsible advisers in the Reich and Prussian governments as well as in the army. Strong characters were broken; weaker ones became despondent and submissive. Wilhelm preferred to rule with mediocrities rather than with the ablest men in the land.
- In order to win personal prestige and to be able to assert himself over his ministers the Kaiser relied on information brought him by irresponsible outsiders and unauthorised members of his entourage, who of course generally told

him what he wanted to hear. The result of this gossip-mongering was general discord and a poisoned atmosphere, above all at court, and a pronounced feeling of insecurity among all those who held high office. Waldersee castigated Wilhelm II's inability to arouse a sense of loyalty in his advisers.

- As far as Wilhelm's character was concerned, the General deeply deplored his susceptibility to flattery, his love of pomp and luxury and his tendency towards obscenity.
- Like many of his contemporaries in Germany and abroad, Waldersee was increasingly inclined to wonder whether the Kaiser ought still to be regarded as mentally normal at all. I shall return to this point below.

For all his insight into the problematical nature of 'personal monarchy' under a ruler like Wilhelm II, Waldersee had no real solution of the dilemma to suggest, unlike Holstein and Marschall von Bieberstein, who (as we shall see) tried to force the Kaiser to abandon his disastrous style of rule by confronting him with repeated collective resignations by the Reich Chancellor and the entire Ministry of State. His only, paradoxical-sounding response was that Wilhelm should suffer a few serious setbacks, since only through such calamities could one hope for an improvement in his character. It had long been his hope, he stated in May 1890, 'that the year 1890 would bring sombre experiences, and with them, regeneration'; but it was already 'almost June, without anything of the kind having happened'.[107] Like Hinzpeter at the Gymnasium in Kassel more than fifteen years earlier, Waldersee now saw a salutary taste of humiliation as the only remedy for imperial arrogance. He even sent up a prayer: 'If God in His goodness would but . . . send him disappointments and anxieties! He would render great service to our Fatherland and spare us much evil.'[108] A year later, when the international situation was becoming acute and the sense of crisis was taking hold within the country too, Waldersee again drew hope from the thought that 'the Kaiser will for once be confronted by the seriousness of life; hitherto almost everything he has done has succeeded, which has spoilt him and given him an immense opinion of his own ability. He can only be helped by serious experiences and setbacks. May God grant that they come, without damaging our interests too much meanwhile.'[109] But the time never came for the 'self-examination and self-recognition' on which the General pinned his hopes. 'Unfortunately we are still a very long way from that', he admitted helplessly in December 1892.[110] It is perhaps not altogether fanciful, however, for us to imagine the ageing General sitting up in bed at dead of night to wonder whether the anachronistic 'Personal Monarchy' which Bismarck had bequeathed to the Germans was really the most appropriate system of government for a vibrant and rapidly industrialising nation of 60 million people at the threshold to the twentieth century.

CONSTERNATION IN THE WILHELMSTRASSE

After March 1890 Friedrich von Holstein, the Grey Eminence of the Foreign Office, who together with Waldersee and others had secretly advised the young monarch against Bismarck in the dismissal crisis, had rapidly become the most embittered enemy of the Chief of the General Staff. Yet he found himself learning the same painful lessons of disappointment as Waldersee. Holstein was particularly shocked by the Kaiser's undignified public appearances and his high-handed absolutist speeches, which led him to fear that Wilhelm might squander his inheritance of monarchical capital within a very short time. In early 1891 he warned that an unhappy and dangerous mood was beginning to spread throughout the Reich, and that, although it could be blamed partly on slander, the Kaiser himself bore much of the responsibility. As he wrote to Eulenburg, everyone was afraid that the Kaiser would go 'mad' like Ludwig II of Bavaria. It was said that he was suffering from 'delusions of grandeur, but had no moral courage'.[111] Unfortunately, Holstein stated, he could not but confirm 'that the position of His Majesty in the country is not improving; on the contrary. The serious-minded North German simply cannot understand why the Kaiser would order the troops in Potsdam and Berlin to the lake shore to watch him sailing in a torpedo boat in his admiral's uniform. There have been the most despicable jokes about it this last week – and that is the worst thing: people laugh.'[112] The Kaiser's speech to the Brandenburg provincial diet on 20 February 1891 completely shattered Holstein's confidence. He wrote in despair: 'It shows a misunderstanding of the times we live in. In 1891, when nothing is believed and everything is doubted, blind trust "through thick and thin" is not so easy to achieve.' Like many other observers Holstein thought the Kaiser's speech more suited to a historical melodrama than to the present political scene. 'This last speech is the most inopportune thing he has yet managed to do. The "Progressives" pretend to admire it in their newspapers. But when they talk, I mean the Left Liberals, it sounds quite different. Today one of them said: "If one listens to the speech with one's eyes shut, one thinks one is hearing [the melodramatic playwright Ernst von] Wildenbruch." . . . After the speech the talk about Ludwig II has started up again.' Holstein asked Eulenburg, as the Kaiser's friend, whether there was anyone in the imperial entourage, which 'criticises him so bravely behind his back', who 'does it to his face? Surely they cannot all approve of the speech.'[113] 'The worst thing . . . is that the Kaiser is increasingly regarded in parliamentary circles as mentally unwell', he warned.[114] At the end of 1891 Holstein again expressed his concern. 'It is sad that His Majesty still treats everything like a game. He will pay dearly for it before long, for things happen very quickly these days.'[115]

As head of the Political Department of the Foreign Office, Holstein was particularly alarmed by the Kaiser's increasing tendency to intervene personally in foreign affairs without consulting the appropriate authorities. A few days after the first Brandenburg speech he wrote to Eulenburg complaining that Wilhelm's action in sending the Empress Frederick to Paris had 'greatly increased mistrust of the Kaiser and of the way he rules'. The 'repeated conduct of His Majesty in foreign affairs' whereby he took political decisions 'as if he were dealing with court matters' was causing great disquiet. As examples of such intervention Holstein named not only the Paris incident but also Wilhelm's letter of condolence to the widow of the French painter Meissonier and the Flügeladjutant Wedel's mission to Copenhagen. 'All that is decreed without paying much attention to the Foreign [Office]', he objected. 'The public is beginning to realise that His Majesty is his own foreign minister, and this causes great uneasiness . . . In Germany this has greatly increased mistrust of the Kaiser and of the way he rules', he warned, and predicted that if Caprivi and Marschall did not stand up more firmly to the Kaiser 'we shall see all kinds of strange things happening'.[116] Within the diplomatic corps discontent arose from the fact that since Bismarck's dismissal almost every post as envoy had been given to someone from the army or the consular body.[117] Wilhelm's propensity to treat 'diplomats as a corps of irregulars which is to get its troops and its training through elements from God knows where' offended Holstein so much that in September 1892 he threatened to resign. The pleasure he took in his work was disappearing fast, he said.[118]

Like Waldersee, the Geheimrat saw Wilhelm's restlessness and lack of seriousness as diminishing the prestige of the monarchy, even as a threat to the survival of the Hohenzollern throne. He expressed his concern to Eulenburg in February 1891: 'On the 28th His Majesty is going to the Kaiserhof to attend the celebrations of the Bonn student fraternity Borussia. I have not met a single person . . . who does not think this a very regrettable idea. By going everywhere, eating everywhere, speaking everywhere His Majesty lowers himself too much. If he then suddenly decides to exert his authority, like the Great Elector or Frederick the Great, he is not taken seriously . . . You know that it goes against my innermost nature to criticise the Kaiser. But I am afraid I cannot do otherwise. Much of what is happening is the Kaiser's own fault.'[119] Holstein's comment can be compared with Waldersee's reaction to Wilhelm's behaviour a few weeks later in Bonn, 'where he goes out drinking with the student corps with a student's cap on his head and directs the whole nonsense himself'. 'It is simply unworthy of a kaiser!', the General raged. Such 'childish nonsense' would 'deeply offend pro-monarchical feeling and the Kaiser will inevitably lose all respect', he feared. 'Today in ermine robes, tomorrow dressed for the beer garden – it will not do, we cannot put up with it in the long run. It deeply offends monarchical feeling and gradually destroys it. Crown princes have been able to get away with such behaviour; they were forgiven for it as soon as they showed their seriousness on the throne. Sovereigns cannot do so with impunity.' What appalled Waldersee most was the fact that Wilhelm accompanied Grand Duke Adolf of Luxemburg to the station in Bonn wearing student fraternity costume. In his eyes, the Kaiser's 'undignified behaviour' had reached its peak. It was 'a blow in the face for all sovereigns. The German ruling princes, among whom there is already so much bad feeling, will be deeply offended and will tell themselves that such a thing could happen to them too.'[120] The anger aroused by Wilhelm's unseemly behaviour in Bonn had wide repercussions, as Holstein and Waldersee feared. Arthur von Brauer, the envoy from Baden, criticised the Kaiser's speech there, commenting that the public did not understand 'how H.M. could publicly praise institutions like student duelling which are now offences under the law of the land and which the Catholic Church punishes with lesser excommunication . . . The idleness and extravagance in the corps also give little cause for this institution, which the overwhelming majority of students detest, to be glorified in the highest quarters.'[121]

Just as the Empress Frederick had done, Holstein described the saying 'suprema lex regis voluntas' inscribed by the Kaiser in the Golden Book of Munich as a 'disaster for the nation'. 'He makes himself not so much hated as ridiculous', he wrote to Eulenburg. 'He is held in increasing contempt.'[122] Eulenburg was distressed to find himself compelled to write warning letters to his 'hotly beloved Kaiser' and to advise him to use the motto 'Ich Dien' at the first possible opportunity – forgetting that this was the motto of the Prince of Wales and that Wilhelm had recently used precisely this phrase to mock his uncle.[123]

At the Foreign Office no one doubted that socialists, democrats, Catholics and particularists were doing all they could to stir up ill-will towards the Kaiser, and that the Bismarck faction was doing the same, often to even greater effect. For that very reason, however, Wilhelm's frivolous and restless style of rule was strongly condemned as providing an easy target for his enemies. As Foreign Secretary Adolf Marschall von Bieberstein warned immediately after the Düsseldorf speech of 4 May 1891: 'The tendency to damage and discredit the present government, and more particularly His Majesty, by systematically spreading confusion and *anxiety* among the people, is becoming so evident everywhere, and is so clearly the weapon used by our enemies, that only *one* tactic is open to the government, and that is to act always and everywhere as a *calming* influence. Every word that is likely to create anxiety furthers our enemies' plans, and most of all when it comes from the highest quarters . . . In this connection I would mention as an example the fact that one of our foremost stockbrokers, when asked by an acquaintance of mine what effect the speech had had on the stock market, replied with a smile "None at all; we have long been accustomed to such things." What can we expect to happen at times when matters are serious and every word should be seen as an appeal to defend the most sacred possessions of the Fatherland?'[124] Not only at the Foreign Office but also among the Prussian ministers the uncomfortable feeling prevailed that the country was sailing helplessly towards a catastrophe. Thus, after a long conversation with Finance Minister Miquel Waldersee recorded that the latter was also 'very pessimistic and also has no illusions about the Kaiser'.[125]

Wilhelm's second Brandenburg speech of 24 February 1892 could not but have a disastrous effect on the mood among responsible statesmen. 'Oh these wretched speeches!', sighed the leader of the Conservative Party, Otto von Helldorff-Bedra, whose loyalties were normally with the Kaiser. 'When will he learn the caution that is so vital?'[126] The Reich Chancellor saw in the reception given to the speech the 'proof of how suspicious of autocratic tendencies people in all circles have become. They are drawing conclusions from all the Kaiser's speeches, taken together, which they find unacceptable. All parties agree on that.' The general critical mood had been in evidence again over Wilhelm's plans for the Berlin cathedral, Caprivi continued. 'It becomes very difficult to carry out business objectively when one constantly comes up against the suspicion that we are dealing with whims and caprices of the monarch, and when there is less and less inclination to co-operate with the monarch. I do not take a pessimistic view and I believe that this will be overcome, but we must not stretch our credit too much to begin with.'[127]

A few months later Wilhelm's plans to hold a lottery to raise funds to embellish the surroundings of the Berlin Schloss provoked such violent criticism that the idea had to be abandoned. Holstein warned despairingly that building plans of this kind inevitably revived the memory of the fairy-tale castles of the deranged King Ludwig II of Bavaria. He declared: 'The Kaiser is not in a position to be able to do such things at present.' If the Ministry of State gave way to the Kaiser on this question it would be 'the object of public contempt'.[128] Philipp Eulenburg likewise thought the planned lottery 'highly dangerous' and warned that with the present angry mood in Bavaria the Kaiser's building plans 'almost forcibly' reminded people of the castles built by the late King. Reporting in his official capacity to Marschall, he stated his opinion that 'in the interest of our All-gracious sovereign it is very much to be hoped . . . that the fantastic projects . . . will prove to be only fantasies, or that they will at least be reduced to a modest scale'. Wilhelm wrote bitterly in the margin of Eulenburg's report: 'Really, Philipp! et tu Brute!'[129]

If such anxieties were shared even by the Kaiser's closest friend, who according to Waldersee had a 'completely idealised picture of the Kaiser',[130] it is scarcely surprising that his enemies were delighted. In the Bismarck camp malicious jokes circulated about the latest symptoms of illness said to have been noticed in the Kaiser.[131] From the Wilhelmstrasse came a stream of increasingly urgent warnings that Wilhelm must be on his guard against poisonous attacks by the pro-Bismarck faction if he did not want to risk his throne. As Holstein wrote to Eulenburg in April 1892: 'His Majesty knows that his enemies are watching him. The April issue of the Contemporary Review contains an article, "William". Translated from German. The cleverest and bitterest thing which has yet been written attacking the Kaiser. His Majesty has read the article ... You will see from it that the Kaiser must weigh his words and deeds more carefully than hitherto if he does not want – metaphorically speaking – to be thrown to the dogs.'[132] The following month Alfred von Kiderlen-Wächter sent the same article, with the Kaiser's annotations, to Eulenburg, commenting that the latter would see from it 'what insidious attacks are still being made against His Majesty'. Certain turns of phrase such as 'burden of vanity', 'sword of the Prussian officer' and the like were, he said, unmistakably expressions used by the Bismarcks – 'they *could not* be invented by an *Englishman*'.[133] Eventually Eulenburg felt compelled to warn the Kaiser of the acute risk he was running by his autocratic behaviour. 'Our good German people will never favour a single large national party', he wrote to Wilhelm in August 1892. 'But Bismarck as the standard-bearer of parliament against the Kaiser with "absolutist tendencies" (which the people will see in every word that Your Majesty utters and in all Your many visits and journeys at home and abroad, because our philistine countrymen will never get it into their heads that Your Majesty can work during these changes of scenery!) – *that* is a watchword that has caught on.'[134]

Eulenburg's letter made little impression and changed nothing in the Kaiser's way of life or style of rule. Waldersee noted at the end of 1892 that he frequently heard the expression 'le roi s'amuse' in connection with the shooting expeditions on which Wilhelm had been engaged for weeks on end. Socialists and Progressives had long been running a book on how much time he would have left for audiences.[135] In the summer of 1894 calculations appeared in the press showing that in the course of the past twelve months Wilhelm had spent no less than 199 days on his travels, as a result of which the joke on the streets of Berlin was that the Kaiser had said 'I have no time to rule.'[136]

In the Wilhelmstrasse the premonition was gaining ground that Wilhelm II's reign might turn out to be the prelude to a republic and could even end with his execution. As early as September 1892 Holstein commented in dismay that 'our Master' had 'no instinct for the masses' and exclaimed: 'What humiliations await him!'[137] 'Would we have thought it possible 5 years ago for . . . the court to descend to the depths it has reached today?', he asked in a letter to Eulenburg, adding that the Kaiser 'even treats ruling as a sport. Is he likely to die on the throne?? He is not the man, nor is this the time, to play around with the people like a giant toy. I am much more inclined to believe in the coming of the German Republic – which Bismarck has already prepared the way for – than [as Eulenburg feared] in the disintegration of the Reich. The Liberals who are loyal to the Reich and who are being antagonised are much more likely to turn to the republic than to particularism.'[138]

A year later Holstein wrote to Eulenburg: 'You cannot imagine how much the feeling against H.M. in the country has intensified in the last 4 weeks. "Travel mania, avoidance of work, frivolity" stand out above all the other accusations against him. The R[eich] C[hancellor] is accused of not being firm enough with the Kaiser. The future belongs to a R[eich] c[hancellor] who knows how to "keep the whims of the Kaiser in check" . . . I am very depressed. We have worked all these years to create or maintain a position for H.M. in the eyes of the outside world. But it is all wasted effort if he ruins his own position within the country. Nor should we overlook the fact – and Marschall made the same observation in very bad humour yesterday – that the respect and recognition which H.H. [Bismarck] receives from the public is on the increase once more, because people are saying to themselves: "The man is right, one cannot do business with the Kaiser."'[139] In a further letter to Eulenburg at this time Holstein commented dejectedly: 'I am afraid that his constantly rising unpopularity means a more dismal future for H.M. than he suspects.'[140]

After Caprivi's dismissal in October 1894 the sense of crisis in the country reached a new peak. In the *Zukunft* Maximilian Harden published an article entitled 'The Giant Toy', in which he compared the Kaiser with King Charles I, who was beheaded in 1649. Holstein bewailed the fact that 'the feeling against the Kaiser among the public is quite appalling. People talk more and more of Ludwig II. *It cannot go on much longer like this* . . . The Kaiser suspects nothing of all this; he behaves with such carelessness that even I am sometimes worried about his mental equilibrium.' The latest chancellor crisis had done him great harm, Holstein concluded.[141] It was Wilhelm II's misfortune, he declared in another letter to Eulenburg at the beginning of 1895, 'that his glow-worm character reminds people of Friedrich Wilhelm IV and Ludwig II. Neither of us has any influence over the conclusions which the German people will draw from this similarity.'[142] Only those who 'long for a Bismarckian dictatorship or a republic' could draw any comfort from the present situation in Germany, he wrote to Eulenburg during the first weeks of Hohenlohe's chancellorship. 'That the regime of His Majesty Wilhelm the Second is the prelude to one of these two forms of government is a possibility which I am afraid I cannot exclude. His Majesty retained a certain respect for our stubborn Caprivi. He claps "Uncle" Hohenlohe on the back and says: "Well, you certainly have brought a long wish list with you" and leaves after two minutes, whereupon Hohenlohe returns from Wildpark [station in Potsdam] to Berlin with his fat portfolio full of things that have not been dealt with. That, dear friend, is an operetta regime, but not one that a European nation at the end of the 19th century will put up with.'[143] The young Kaiser was unconscious of any danger. Convinced of his own infallibility he continued, unperturbed, to build up his personal power. But how *did* German society at the end of the nineteenth century react to this anachronistic 'operetta regime'?

CHAPTER TWENTY

The predictable disaster: Wilhelm II and the 'public soul' of Germany

'THERE MAY BE EVIL TIMES TO COME'

No more than a few months after Bismarck's dismissal, well-placed observers like the Empress Frederick, Chief of the General Staff Count von Waldersee and Geheimrat Friedrich von Holstein – as should be more than evident from the preceding chapter – were already haunted by the fear that the superficial, impulsive and wholly anachronistic style of Wilhelm II's rule might lead to the downfall of the Hohenzollern monarchy, the proclamation of a republic or the imposition of a dictatorship. His grandiose aspirations to the role of autocratic sovereign by divine right in an increasingly critical, industrialised mass age made the danger of failure all too plain. When in January 1891 a portrait of Wilhelm II by Max Koner, which the Kaiser had commissioned for the German embassy in Paris, was exhibited in honour of the Kaiser's birthday, it made a deeply disquieting impression on many people. The Baden envoy Arthur von Brauer reported in surprise to Karlsruhe that the life-size oil painting was causing a stir in Berlin. 'The pose chosen by the artist is thought too theatrical, the stance is almost provocative and the field marshal's baton is unusually large and prominent. [The ambassador to France] Count Münster, who saw the picture here, gave a very disparaging opinion of it; in conception and manner it was very inappropriate for Paris. He even expressed his doubts in the highest quarters, but without success. The picture went off to Paris. Meanwhile Countess Asseburg had taken the opportunity to tell H.M. with the kind of ruthless frankness that only a lady can get away with that the picture was "very unattractive and absolutely impossible for Paris" because it would invite mockery from the Parisians. Her words made an impression; H.M. sent a telegram to Paris ordering the picture not to be unpacked and to be returned.'[1] On Waldersee the portrait made a

26. Max Koner's portrait of Kaiser Wilhelm II, January 1891 – a declaration of war?

'sad impression' because it showed the monarch 'as he really thinks and feels – immensely vain and self-confident!' The Kaiser was shown standing 'in an incredibly challenging attitude, in Garde du Corps uniform with a black cuirass and a purple mantle, leaning on a long field marshal's baton. It gives everyone the feeling that he is trying to impress the French and is saying to them: "beware of picking a fight with me".' One would not be able to give a true verdict on the portrait for another ten or twenty years, the Chief of the General Staff opined. 'If he has achieved great deeds by then, it will be an excellent picture; if things go differently, it will be simply ridiculous.'[2] Only three years later, when Wilhelm II's unpopularity at all levels of society throughout the Reich had reached critical proportions, Waldersee could no longer doubt that the throne of the Hohenzollerns, in which he had once had so much faith as a bastion against democratisation, had become very shaky. 'The Kaiser is now reaping what he sowed', he declared. 'In the first few years everything seemed to go very well and he was obviously very lucky too; regrettably this soon gave him arrogant and conceited feelings which made him blind to the real circumstances; he thought he could do everything himself, that he understood everything and could control the politics of the whole world, and what was the result? May the good Lord not turn aside from us completely, but protect us. There may be evil times to come.'[3] When the fifth volume of Treitschke's *German History* was published later that year many readers were struck by the resemblance between the first years of the reign of the Prussian King Friedrich Wilhelm IV and the present. Distraught, Waldersee asked himself: 'Are we to find ourselves in the same terrible plight as in 1848? God forbid, but it is very probable that we are heading in that direction.'[4]

Observers abroad often saw most clearly the immense risks inherent in the all-or-nothing game in which the Kaiser was engaged. As early as May 1890 the French historian Ernest Lavisse drew attention to the ambitious aims of the young Kaiser in an influential article in the *Figaro*, in which he emphasised the dangers that lay not only in the Kaiser's exceptionally powerful position but also in his character. 'In the person of Wilhelm II power and youth form a fascinating contrast', he wrote. 'Wilhelm II is not content with reigning; he wants to govern, like his forefathers who put Prussia together piece by piece with their own royal hands.' For this reason alone the dismissal of Bismarck had been a simple, quite natural necessity in Wilhelm's eyes: 'His property was in the hands of another: he took his property for himself.' The Germany whose government he had now personally taken over had 'immense reserves of loyalty, fidelity and patience' towards Wilhelm II. 'It is no longer at odds with itself, as it once was, for ruling princes both great and small, kings, grand dukes and dukes, are gradually disappearing in the majesty of empire . . . The parties are fragments which will not unite, but Germany wishes to be united and indivisible and to remain so. The only form which this unity can take at the present time is monarchy; the only monarch is the King of Prussia. Germany is personified in Kaiser Wilhelm II, and Germany gives him such power that he is able to take great risks.' And so Wilhelm, 'he alone and always he', would 'bear the burden of the day. In political battles he will always be personally engaged, exposed and visible to all. It is for him to find the solution to all questions . . . But not all questions are soluble . . . Should expectations remain unfulfilled, who will bear the responsibility? The Kaiser. This responsibility is not without danger', Lavisse warned. A further difficulty, in the historian's eyes, lay in the fact that, although Wilhelm had a modern air, he was 'a contemporary of the past'; he was indeed 'an anachronism' in his innermost convictions. 'Kaiser by the grace of God! I can assure you that this title is no empty formula for Wilhelm II', Lavisse stressed. 'His belief is entirely sincere.' Other characteristics of the monarch would have an equally marked impact on the style and content of German policy. 'Rapidity in action, a predilection for oratory, a lively mind and a thirst for greatness. The Kaiser seems even to have a taste for the impossible.' As to foreign policy, Wilhelm spoke of the Triple Alliance politely but without enthusiasm, as if he thought the alliance with Austria-Hungary and Italy a little *passé*. 'But to pursue colonial policy on a par with England and to come to agreements with her, to pay visits to the Queen of England and the Emperor of Russia, to repeat these visits, to review the English Navy in the uniform of an English admiral and to take the salute of the Russian Army in the uniform of the Russian Army . . . – that is fine, that is grand, that is impossible!' The future of Wilhelm II and of Germany was therefore still unclear and uncertain. 'The young Kaiser . . . attracts universal attention to a high degree; he is in favour with public opinion and deserves this distinction above all because of the grandeur of his vision and . . . also most especially because of the presentiment of certain dangers which he has expressed. Let the danger come: the Kaiser will not flinch. He is courageous to the point of audacity, he looks ahead with confidence, he strides onwards, the figure of an ideological soldier, into the unknown.'[5]

The uneasy feeling of witnessing a highly risky game of chance which Lavisse conveys here in somewhat romanticised and veiled terms was given much sharper expression by the Portuguese writer and diplomat José Maria Eça de Queirós shortly afterwards. In a most perceptive comment on the recklessly uncompromising attitude which flowed from the Wilhelmine concept of the role of the kaiser he wrote in 1891: 'Wilhelm II literally gambles with those terrible iron dice of which Bismarck once spoke.' He too predicted that Wilhelm would one day *either* 'calmly and majestically direct the destiny of Europe from his palace in Berlin', *or else* find himself sitting in a London hotel room rummaging 'for the battered double crown of Germany and Prussia in his little exile's travelling

27. 'The Kaiser's Dream'. Cartoon in the British journal *Truth*, December 1890, showing Wilhelm II's progress from birth (bottom right) to exile (top left) and the partition of Germany after a lost war

bag'.[6] As early as the end of 1890 the British republican journal *Truth* had published a caricature entitled 'The Kaiser's Dream' which predicted the future course of Wilhelm II's reign with breathtaking perspicacity (fig. 27). A series of images shows the political crisis after Bismarck's fall, the dangerously growing discontent of the German population, the attempt to avert internal crisis by an external war and the subsequent defeat and disintegration of the Reich into several republics. The Kaiser and the other deposed monarchs seek asylum in London with their exiles' travelling bags. In fact it was only in this last detail – the place of exile – that the caricaturist's uncannily accurate prediction failed to come true.

After Bismarck's departure – the celebrated 'Dropping the Pilot' caricature by Sir John Tenniel in *Punch* had already hinted at Wilhelm's foolhardy role in the dismissal crisis – it was the Kaiser's Brandenburg speech of February 1892 which did the most to set alarm bells ringing abroad. The Baden envoy in Berlin reported in disgust, but not without anxiety, that 'the imperial speech and the riots have of course provided the foreign press with a very welcome opportunity to paint our situation in the blackest colours and dish up the most incredible lies about both events for foreign readers . . . The most shameless of them is, as always, the Paris Figaro, which claims to have had telegraphic reports that as a result of the speech people in Berlin are saying que l'Empereur "ne dispose pas entièrement de ses facultés"! Unfortunately such idiocies are widely believed in foreign countries which are less well disposed towards us, and so they help make the general political situation worse, however filthy and nonsensical such products of the press are in themselves.'[7] In London, in an article entitled 'The Holy German Empire' published on 26 February 1892, the *St James's Gazette* drew attention to the alarming anachronism of the Kaiser's autocratic perorations, although at the same time putting forward the view that strong leadership would be welcomed by the German people as long as it brought unity and order. 'There is something very strange, and at first sight very much out of date, in this young man's enunciation, in words stronger than any English King has used since the days of Henry VIII, of the principle, *Quod principi placet, legis habet vigorem*; and no wonder that in Germany and elsewhere men are asking what the end of it will be. Is the Emperor a hot-headed young man inflated with an inordinate conception of his own power and capacity, and destined out of sheer obstinacy and vanity to jeopardize his throne and endanger his country? . . . The majority of the German people care little for Constitutional Government. They want a strong man at the head of affairs, an efficient administration, and a settled policy . . . If it is believed that the Emperor's policy can keep Germany strong and united, and prevent the forces of discord and disorder gaining ground, it will have the support of the people. But no one will deny that the oratorical recklessness which is one of the Kaiser's most marked characteristics is in itself a source of danger. As long as he confines his frankness to questions of internal policy no great harm is done; but let him make a speech in the same strain . . . on a subject of international difficulty, and the whole of Europe would be in a blaze. His glorification of Rossbach and Dennewitz was not very wise; and a large section of Frenchmen have not forgotten or forgiven the allusion in a recent speech to "the Corsican *parvenu*". It is true that the Emperor frequently insists on his desire for peace, but unless he can keep a watch on his lips it is quite possible that he may himself be the cause of war . . . A highly interesting figure is this young ruler, with his unbounded belief in himself and his firm conviction that Providence is his ally and will never desert him, or Prussia. He is all the more picturesque because he is an anachronism. He talks the language of the old Hohenzollern tradition – with modern variations. But the experiment of governing a great civilized country under a patriarchate is a curious one to be attempting in the year 1892.'[8]

THE GERMAN 'PUBLIC SOUL' AND THE KAISER

But how did the German people react to the imperial announcement of divine right and autocracy? Was the German 'public soul' (Heinrich Mann) really so unpolitical, as Lavisse and the author of the article in the *St James's Gazette* claimed, as to care little for constitutional forms and to accept Wilhelm II's autocracy as the price which had to be paid for national unity and internal order? Did the German nation even see its own image in Kaiser Wilhelm II, as Walther Rathenau maintained after the fall of the monarchy, and as is still willingly believed by (crypto-)monarchists today? In Rathenau's famous and often-quoted comment on Wilhelm in 1919, 'This people, at this time, consciously and unconsciously, wanted him thus and not otherwise. They wanted themselves and him to be thus and not otherwise.' 'Never before has a symbolic human being been so perfectly reflected in an epoch, nor an epoch in a human being. Not for a single day could Germany have been governed as it was governed, without the consent of the people.'[9] In the light of the contemporary sources, which bear ample witness to the growing unrest and anger among all social classes throughout the county, Rathenau's retrospective judgement must be firmly rejected, not from personal prejudice or ideological bias but out of respect for the truth and for the real feelings of that generation of the German people.

Waldersee's 'rude awakening' in the summer of 1890 was causally linked to his perception of the growing unpopularity of the Kaiser among the citizenry. There was 'absolutely no question but that the feeling against the Kaiser [was] growing fast', he commented in his diary only a few weeks after Bismarck's fall.[10] In May 1890 Waldersee noted that in the army, the political parties and the population as a whole 'a feeling of dissatisfaction with him [Wilhelm II] has become more widespread; I thought that he would have very fanatical admirers among the young, but they of all people . . . are said to complain bitterly about him and criticise his vanity and love of playing soldiers'. This mood was all the more dangerous, Waldersee observed, because the Bismarcks were clearly doing their best to spread the message from Friedrichsruh 'that the Kaiser is mad'.[11] Three months later the Chief of the General Staff's anxiety was increasing. To judge from German newspapers, he said, one would think that there was 'no more popular man and ruler . . . than Kaiser Wilhelm'; yet this impression was deceptive. He had found that 'the number of dissatisfied people is immense; and yet this is not even hinted at in the press. It worries me greatly, but mostly because the army is strongly affected by this mood.'[12] He was shocked by the 'growing discontent with the Kaiser among wide sections of the population' and attributed this alarming development to the fact that Wilhelm had no sense of purpose and acted only out of egoism and vanity.[13] After the 'absurd' speech at Düsseldorf in which Wilhelm had declared that he alone was master in the Reich and that he would tolerate no other, the General commented: 'Unfortunately we have already reached the point that the Kaiser is no longer taken seriously;

people have already heard so many wonderful speeches from him that they are only amused by them.'[14] The Catholic General Walther Freiherr von Loë, who had come to Berlin in October 1890 on the occasion of the visit of King Leopold II of the Belgians, was astonished to find that the mood in the capital had changed so much in a year.[15] Three years later Loë had become more deeply pessimistic still about the state of Germany, describing it as 'almost beyond hope'.[16]

Waldersee's forebodings about the disastrous future facing the monarchy under Wilhelm II were reinforced by what he himself observed while travelling around the country on tours of inspection or on leave. In September 1890 he commented that during recent months he had journeyed widely, whether in South Germany, East Prussia, Schleswig or Silesia or on the Rhine, had spoken to a great variety of people – the Grand Duke of Baden, King Albert and Prince Georg of Saxony, Prince Albrecht of Prussia, countless generals, big landowners and industrialists, ministers, provincial administrators and other senior officials – and had found the same feeling of insecurity and apprehension in all of them. 'Everywhere anxiety about the future and doubts about the Kaiser being the man to take us forward; most people even thought that he is having a destructive effect.' Waldersee added that he would be prepared to swear 'that lately I have found *not a single person* who had any confidence'.[17]

In October 1890 Waldersee recorded with concern that he was 'constantly receiving reports of the growing feeling against the Kaiser; it has now reached the best middle-class circles. In South Germany it is growing, and recently in Saxony too; in Bavaria I hear people are saying quite openly that he is mentally ill.'[18] In November, when members of parliament returned to Berlin for the new legislative term, Waldersee heard 'more opinions from the provinces, of course; unfortunately they bear witness to widespread discontent, especially with the Kaiser; he has in fact disappointed everyone – with the exception of the Social Democrats'.[19] Not long afterwards Waldersee recorded that he had spoken to Franz Fischer, the Berlin correspondent of the National Liberal *Kölnische Zeitung*, as well as to a banker (whom he did not further identify) and a general. 'With all 3 the conversation soon came round to the Kaiser and they were of one mind in thinking that the mood is getting persistently worse. They represent completely different milieux', the General commented.[20] When he met Theobald von Bethmann Hollweg early the next year while hunting in Buckow, the future Reich Chancellor assured him that 'in all sections of the population the mood is unfavourable'.[21]

In January 1891 Waldersee was horrified to observe that even among the ordinary people of Berlin the enthusiasm for the Kaiser which had been apparent until recently had completely vanished. At the consecration of the Friedenskirche he was struck by 'how coolly the people in the streets behaved towards the Kaiser; only 6 months ago all the streets leading to the church would have been decorated; this time there were only a few flags on the houses closest to the church and at most a few 100 people'. When the Chief of the General Staff commented on this to the Berlin Chief of Police the latter shrugged and said that this was simply 'an expression of the general mood'.[22] A year later Waldersee noticed that on the Kaiser's birthday the usual illuminations were limited to Unter den Linden and a very few other streets, 'as all Berliners are aware of a gradual waning of enthusiasm for the Kaiser'.[23] After talking the situation over in detail with civil servants and politicians in Berlin Waldersee was distressed to have to acknowledge that 'the increasing ill-feeling is unquestionably the main topic of all conversation'.[24]

Holstein, forced to watch helplessly while Wilhelm alienated his own people with one autocratic pronouncement after another and put the future of the monarchy at risk, took a no less critical view. He wrote despondently to Philipp Eulenburg in November 1891 reporting that in response to Wilhelm's disastrous inscription in the Golden Book of the City of Munich, 'suprema lex regis voluntas',[25] the anthem 'Heil Dir im Siegerkranz' had been shouted down in public. 'That is a sign of the times. Bismarck's position is much stronger than six months ago. People were also heard shouting "salus publica suprema lex" at the station', Holstein registered in dismay. 'More and more people are deserting the Kaiser. He has not yet noticed it, but come the moment when he needs to call on the devotion of the people and their trust in him – then he will notice what he has brought upon himself. Our sovereign seems predestined to suffer terrible defeats, for the ruler's success depends on the trust and love of those who do his bidding. No intelligent person can want an absolute state in Germany today. All the suspicions about the Kaiser's mental health are emerging again like a poisonous miasma . . . Things are not going badly in general – only His Majesty!'[26] Scarcely had Holstein uttered this lament when the Kaiser made a speech which again sparked rumours about his 'state of mental health', and this time – as Holstein pointed out – in all the parties.[27] When an anonymous satirical article appeared in the Berlin *Lokalanzeiger* at about this time Holstein saw it as a sad symptom of the general mood of the population that this edition of the newspaper, which normally had a circulation of 12,000, was so quickly sold out that it was changing hands for five marks. He sent a copy to Eulenburg with the remark that the malicious article owed its success to the fact that it 'corresponds to the picture that people have of H. Majesty'. It was significant that the paper had been sold principally to officers, for that showed 'how the Kaiser's position has been undermined, especially in the army', he wrote. 'His Majesty has seen the lampoon and is very angry about it', Holstein continued. 'To counteract this kind of thing His Majesty must show *seriousness of character* . . . I am beginning to get tired of these incidents.'[28]

The journalist Dr Franz Fischer, who as Berlin correspondent of the quasi-official *Kölnische Zeitung* had good contacts both at the Foreign Office and among the Rhenish-Westphalian industrialists, was elegiac in his comments. In a broad survey of public opinion a year after Bismarck's dismissal he declared: 'When our Kaiser came to the throne everyone breathed freely again, all hearts were drawn to him, a new confidence and a lively spirit of enterprise took hold everywhere; the upsurge in our industry, which had lain fallow for so long, was unmistakable. Today the exact opposite is true . . . And what is worst is that the reason for the general mistrust and uneasiness is precisely the picture which numerous people in all types of profession, the army, the civil service, the academic world, but above all the world of commerce, have formed of His Majesty the Kaiser, with or without good reason. Our government, especially the Reich Chancellor himself, may do their utmost to counteract the present dissatisfaction by proposing good laws and taking excellent administrative measures, but it will not achieve its goal. For the conviction is steadily growing that the entire body of ministers are not independent men who act in good faith, but more or less puppets, who blindly follow the whims and caprices of their imperial master. But such incredible legends are circulated about the Kaiser – and what is still more insane – they are so completely believed, even in royalist circles, that people think the most impossible things and the most extraordinary administrative measures are not only quite possible but very probable. At the very least the Kaiser is universally portrayed as a gentleman who is extremely erratic in his thinking, irresolute in his decisions, rash in his speeches, inconsistent in his preferences and unable to tolerate opposition. Private attempts to refute these false ideas are fruitless. On the contrary, the smallest trifles, such as the frequency with which the garrisons visited by the Kaiser are called out, or the recent transfer of the torpedo boat to Berlin, only provide fresh reasons for making a mountain out of a molehill. One is constantly hearing: the old Kaiser would never have done such-and-such!!'[29]

The well-informed Baden envoy described the mood of the people in equally sombre terms in December 1891, looking back over developments since Bismarck's departure. 'The theme of "suprema lex regis voluntas" was very far from played out', Brauer commented sarcastically, 'when news of the Kaiser's speech at the swearing-in of the recruits in Potsdam reached the wider public. At first people thought it was a hoax or a malicious invention. But soon it was confirmed by people who had heard it with their own ears that H.M. really did say the words attributed to him. The Commanding General of the Corps of Guards, General von Meerscheidt-Hüllessem, immediately realised what dangerous consequences the Kaiser's words could have and after H.M. had left he made a short speech to the assembled troops in which he said it was the duty of everyone present not to allow any of the all-highest remarks to become public knowledge. Of course this well-meaning command, addressed to such a large number of soldiers among whom there were probably Social Democrats, was doomed to failure. The all-highest words caused utter panic among the most timid spirits, and I have the impression that they were expressly exploited by the Kaiser's enemies to stir up fear and indignation – among the propertied classes, as if the Kaiser could make rash decisions dangerous to trade and commerce – and among the workers, as if the Kaiser might feel the inclination or the need to have them "shot down".' Commenting on the abrupt change in Wilhelm II's attitude to the workers Brauer wrote: 'At the beginning of his reign, imbued with ardent, optimistic dedication to his exalted calling, the Kaiser was full of joyous confidence that he had a mission to resolve the social question. The great popularity which he undoubtedly enjoyed at first deluded him into believing that it would be easy for him to outshine the Bebels, the Liebknechts etc. with the mass of workers, and to win their confidence. Hence the great energy which he put into organising the international social welfare conference, showing a personal interest in social welfare legislation, attending the Council of State and getting rid of the Anti-Socialist Bill. There have been more than enough utterances by H.M. to show that he expected his efforts to succeed very quickly and to bring about a sudden and complete transformation in the attitude of the misguided masses. He may therefore have felt bitter that in this respect he had nothing but disappointments: social democracy is becoming a more and more significant factor, and the early signs of the formation of an imperial workers' party, which the Kaiser hoped for, have had no lasting result. Given H.M.'s character it is very understandable, psychologically, if these disappointments have brought about a complete reversal of his attitude. I am afraid the Kaiser has now become convinced that nothing can be achieved amicably and that the worker question can be resolved only by force and on the principle of "the sooner the better". Thus for some time now he has repeatedly referred to "dangers from within", culminating in his recent remark about "shooting down brothers and friends".'[30]

Even before the Brandenburg speech of February 1892 a dangerously hostile mood, which came close to erupting in street protests, had built up towards the Kaiser. The Empress Frederick's inclination was to withdraw to the Taunus mountains or even to England, but when she reluctantly returned to Berlin in December 1891 for the season she observed with distress that her son was 'not at all popular!' and that his 'public utterances' were 'much criticized'.[31] 'I am miserable about affairs here – & W's growing great unpopularity, wh. *cannot astonish* me; he does not feel it or observe it & is as self confident as ever alas', she wrote.[32] The widowed Empress's assessment of the situation was amply borne out by the foreign diplomats stationed in the various capital cities within Germany. The British minister in Dresden, for instance, reported to London in December 1891 that 'uneasiness exists in Germany among patriotic & loyal people because their faith in the discretion & wisdom of the Crown has been so rudely & often shaken'.[33] A mocking character sketch of Kaiser Wilhelm in the London journal *Truth*, which was widely sold and discussed in Germany, attracted the comment from Waldersee (who had been given it by the Adjutant-General von Wittich, of all people) that it contained 'incredible things about the Kaiser and could only have been written by people with very close knowledge of him and the Prince of Wales, who comes out of it worst, and of the life and atmosphere in the English [royal] family'. The General had to admit that after reading it he was distressed 'that so much ridicule can be heaped on the Kaiser; of course it is immensely exaggerated, but unfortunately it identifies weaknesses in him with absolute accuracy'.[34]

Following this, Wilhelm II's speech to the Brandenburg provincial diet on 24 February 1892 provoked a disastrous public reaction both in Germany and abroad. Because the speech coincided, as already mentioned, with widespread agitation over the School Bill and major street protests by the unemployed in Berlin, the Kaiser's autocratic utterances seemed almost deliberately framed with class conflict in mind. The Reich Chancellor remarked philosophically that the street riots were 'of no importance', and that it had been an excellent thing that the Kaiser 'rode out once . . . in order to show courage. A repetition of the insolence shown on that occasion would be undesirable, however, since it would not be preventable without a police escort. So I was glad that the Kaiser gave up the idea of riding out again yesterday.'[35] According to Arthur von Brauer, a 'grumbling, pessimistic mood' had now 'taken hold of almost the entire nation . . . and poisoned our public life'.[36] The Kaiser's mother welcomed the news that Wilhelm had withdrawn to Hubertusstock, for otherwise there would probably have been more street demonstrations on the second anniversary of Bismarck's dismissal.[37] The British ambassador, reporting on the Kaiser's speech, commented that it had been strongly condemned everywhere, and deservedly so. Its critics maintained that Wilhelm had thrown down the gauntlet of imperial will to his people and had made it impossible for them to look up to him as an impartial sovereign. Even Wilhelm's supporters deplored the superficiality of his remarks; but, according to Sir Edward Malet, they warned against reading too much into them. The offensive phrase about the grumblers who were to shake the dust of Germany from their feet, for instance, was merely an allusion to an allegorical play by Ernst von Wildenbruch which the Kaiser had admired recently. Similarly the passage (particularly badly received in Britain) in which the Kaiser had spoken of the Almighty as 'our old ally' of Rossbach and Dennewitz had scarcely attracted any attention in Germany, as it was simply a well-known quotation from Field Marshal Blücher. In spite of all these explanations and excuses, however, the speech had been very strongly criticised, Malet reported, 'and His Majesty's Ministers, to whom it came as a complete surprise, will no doubt endeavour to guard against such an occurrence in the future. All the comments in the papers have been laid before His Majesty and it is hoped that the general condemnation of the Speech in the English papers of all shades will have a salutary effect.'[38] And indeed General von Verdy was able to report to Waldersee, from conversations with members of the inner circle of the imperial entourage, that 'His Majesty has lately read *everything* which has been published about his recent speech and the School Bill . . . and is said to be very depressed about it – at least for the moment.'[39] A few months later, when the British military attaché Leopold Swaine returned to Berlin after an absence of three years, he was shocked by the extent to which public opinion had changed. 'What strikes me more than anything, comparing the present time with that which I left behind me 3 years ago, is the way in which the Emperor is abused – openly abused – by all classes of society, even by the Army. The Country feels and knows that the Emperor lacks good advisers. The People are unsettled and no one knows what the next move is likely to bring.' It was very much to be feared, he warned, that 'matters will drift from bad to worse, and all confidence in the existing régime is fast leaking out'.[40]

The Kaiser was also attacked in parliament for the first time when the Left Liberal leader Eugen Richter made a critical speech, which won support from all the parties, in the Prussian House of Representatives in May 1892. 'The speech will certainly have a very profound effect on the Kaiser', Waldersee commented, for 'it is a harsh lesson which the country is teaching the Kaiser, and yet it is very sad that we have already reached this point'.[41] The following year it was the critical speeches of the Social Democrat August Bebel which created the most alarm in royalist circles. Waldersee found it 'positively outrageous . . . how people attack the Kaiser himself nowadays. Bebel was clever enough not to name him directly, but no one could possibly have misunderstood him.'[42] A shocked Brauer reported from Berlin at the end of 1892: 'There has probably never been such widespread and deep-rooted dissatisfaction with the existing state of affairs since the Reich was founded, and – what is worse – such despondency. Only the generations to come will be able to appreciate fully the extent of the monarchical feeling, national pride, calm and objective judgement and love of Kaiser and Reich which has been destroyed in the past two years. We who are living through it can gauge the extent of what we have lost from outward manifestations only; the inward transformation of the feelings and attitudes of the German people remains concealed from us as yet.'[43]

Not long after Bismarck's dismissal it became plain to observers in Berlin that despite their initial enthusiasm foreign diplomats and journalists were also taking an increasingly negative view.[44] After the second Brandenburg speech and the School Bill crisis of the spring of 1892 this foreign criticism reached new and disquieting heights. The future Foreign Secretary and Reich Chancellor Bernhard von Bülow, on a visit to Berlin, was horrified by the 'shameless' way in which foreign diplomats in the German capital found fault with the Kaiser and the Hohenzollern court. 'They criticise everything and spread the most incredible news about us', he wrote indignantly.[45] The reports, which reached Paris and St Petersburg from Berlin at this time, did indeed draw an alarming picture, not only of the political situation in Germany but also of the Kaiser himself. In March 1892 Holstein read a secret agent's report from Paris according to which the French Prime and Foreign Minister Alexandre Ribot considered the Kaiser 'mentally unsound, on the basis of ambassadorial and other reports' from Berlin. The agent concluded that 'the relevant ambassadorial report must have indicated that the "retraite" of the Kaiser to Hubertusstock was necessary on account of the mental state of the monarch; in addition the Quai d'Orsay is receiving reports from Berlin . . . that the view that the Kaiser is temporarily of unsound mind is gaining increasing acceptance in "exalted" circles in Berlin, especially in international diplomatic circles there. Some of the articles in the Parisian press, which for months has been maintaining at regular intervals that the German Kaiser is mad, are . . . inspired by the Foreign Ministry here [in Paris].'[46] Holstein commented tersely to Philipp Eulenburg that he knew 'not from Schweinitz but from private letters from St Petersburg' that Tsar Alexander III thought exactly the same.[47]

There was of course no need for secret reports from abroad to document the decline of confidence in the Prusso-German monarchy. The Reichstag elections, held by universal secret ballot for men, sent all too clear a message. The growth of the Marxist–republican Social Democratic Party seen in the Reichstag elections of February 1890, disastrous enough in itself as far as the monarchy was concerned, proved in the elections of the summer of 1893 to be only the beginning of an inexorable upward trend which continued until after the First World War. Since the Catholic Centre maintained its share of the vote (19 per cent) undiminished until 1912, the result was that the reservoir of parties loyal to the Kaiser continued to shrink – like the 'Peau de Chagrin' in Balzac's portrait of the Bourbon monarchy in post-revolutionary France – in spite of the losses suffered by the internally divided Left Liberals. In 1893 there were 4.35 million votes for the opposition parties which Bismarck had condemned as 'enemies of the Reich', as against 2.48 million for the former Kartell parties which had been regarded as 'staatserhaltend' or state-supporting. Five years later, in the Reichstag elections of 1898, this trend persisted: 4.43 million voted for the SPD, the Centre and the Left Liberal Parties while the former Kartell managed to attract no more than 2.17 million votes.[48]

Wilhelm's monarchist critics were of course aware that this decline could not be attributed solely to the Kaiser; they acknowledged that the growing criticism of the monarchy in public opinion was part of a general move towards democratisation which could be seen to be happening in other countries too. In their analyses, nevertheless, they laid much of the blame at his door. Contemplating the confused state of domestic politics in Germany, it did not take Waldersee long to realise that the hopes which he had once pinned on the Hohenzollern monarchy as a bulwark against the democratic tendencies of the day would not be fulfilled under Wilhelm II. When the Kaiser came to power, Waldersee observed, he had still hoped that the German Reich would become a 'stronghold' against the socialist movement, from which all the other European states could derive support. But already by September 1890 he had to admit that 'this hope has been bitterly disappointed'.[49] Two years later he wrote reflectively: 'I no longer have any doubt that we are facing great catastrophes which will shake the whole of Europe and may bring about a completely new era. I am far from throwing all the blame on the Kaiser now; but he did much to set in train the movement towards the abyss. While his grandfather was alive the whole world looked up to us, sensing that Germany was a pillar of firm, healthy strength, a rock on which the waves of revolution would break, a house beneath whose roof shelter might be found in troubled times. All this the grandson has destroyed in an astonishingly short time. He is continuing his work of destruction and still believes, in his immensely exaggerated estimation of his own capabilities, that he is the right man at the helm. On all sides one hears people expressing the wish for a real man, and nothing speaks more eloquently for the fact that no one believes in him any more.'[50] In the summer of 1894 the General again took stock of the international situation, observing no less despondently that things looked 'pretty bad in the world'. Many countries – Hungary, Italy, France, Britain, Serbia and Bulgaria – were in the grip of government crises. 'One would think that conditions here were ideal; but unfortunately this is not the case. Thank God we do not have parliamentary rule yet; the comings and goings of ministers still depend largely on the sovereign, or at least not on the Chambers. But it is true to say that the world is simmering and seething and I am inclined to think that great turmoil lies ahead. In the lifetime of Kaiser

Wilhelm I people had much confidence in Germany and saw her as a firmly grounded, robust power on which one could depend for support in bad times. Sadly that has now changed. In only a few days' time the present Kaiser's reign will have lasted for six years. What must he think of his achievements, if he takes the time to reflect calmly on the past? Many illusions of the early days have probably vanished. He wanted to make Germany greater and more feared and to make the country happy internally. Have we become greater? No, our reputation has declined and our enemies have become arrogant; Russia now believes she rules the world. And how do things stand within the country? No one is really content; countless people are embittered and offended; the Social Democrats scornfully proclaim that their cause is flourishing . . . It may be that the old ways in our country and throughout Europe have become outworn and must pass away. Nevertheless I think that men of action could still turn events in a different direction.'[51] Waldersee was repeatedly confronted with evidence that the growing ill-feeling was directed against the Kaiser himself. 'Within the country things look no better; indeed they look decidedly worse than a year ago, when it was clear that there had already been a deterioration', he noted in his diary at the end of 1894. 'The mood of dissatisfaction, ill-feeling and pessimism has undoubtedly spread and is more vociferously expressed, especially in South Germany; it is also directed much more pointedly against the Kaiser . . . Absolutely no one has any real confidence in him any more; this is inexpressibly sad, but unfortunately true.'[52] The ill-feeling was very widespread and the perception that the Kaiser was to blame was gaining ground everywhere. 'Seldom has anyone disappointed expectations as much as he.'[53]

The growing discontent of the population with Wilhelm II's autocratic conduct in the years following the fall of Bismarck was exemplified by the huge success of the satirical pamphlet *Caligula. A Study of Roman Caesaromania*, in which the liberal historian Ludwig Quidde drew an unmistakable and extremely critical portrait of Wilhelm II under the pretext of a scholarly analysis of classical sources. His aim, as he said himself, was to point out 'the dangers of the unrestrained exercise of a morbidly affected . . . autocratic mind'.[54] With thirty-four editions and hundreds of thousands of copies sold, Quidde's *Caligula* was by far the most successful political publication in the history of the Reich. It led to an intense debate – albeit conducted in coded terms – in newspapers of every party political slant on the personality and style of rule of the young Kaiser.[55] Quidde had insisted on his name appearing as the author. His calculation that no public prosecutor would be able to admit publicly to having spontaneously thought of Kaiser Wilhelm II in reading this analysis of a deranged Roman emperor did not quite work out as he hoped. Although he did not receive a jail sentence of four to five years, as his lawyer had feared, he was compelled to serve three months in Stadelheim prison and his career as a historian was ruined.[56]

Quidde's detention happened at a moment when convictions for contraventions of the laws of *lèse-majesté* (paragraphs 94–101 of the penal code) reached their peak. According to the lists of cases compiled annually by the Imperial Office of Statistics, in the thirty years of the last German Emperor's reign there were 12,196 prosecutions and no less than 9,212 convictions for *lèse-majesté* in all the German federal states, but the cases were by no means evenly distributed across the three decades (see table 1).[57] Although in the last years of Bismarck's rule and the first years of the New Course prosecutions remained below 800 annually and convictions as a rule below 600, both figures shot upwards from 1892 onwards: there were 922 prosecutions with 670 convictions in 1893, 952 prosecutions with 720 convictions in 1894, 858 prosecutions with 644 convictions in 1895 and 844 prosecutions with 623 convictions in 1896. Thereafter the number of trials for *lèse-majesté* declined rapidly, however: the number of guilty verdicts sank to below 300 around the turn of the century, and by the time of the outbreak of war only about twenty people a year were being convicted of this crime. This striking decline can be partly attributed to changed methods of collecting statistics.[58] Another reason for it, however, lay in the growing realisation of the Kaiser and the government authorities that these sensational trials could prove very harmful for the reputation of the monarchy, particularly as it was almost impossible to secure convictions in South Germany, while even in Prussia the judges were less and less inclined to impose the full sentence. In April 1893 Maximilian Harden was tried for *lèse-majesté* but acquitted by the presiding judge on the grounds that respect for a prince could also be shown by upholding the truth before him.[59] In the end the Kaiser announced through an All-Highest proclamation of 27 January 1907, his birthday, that instances of *lèse-majesté* committed through ignorance, thoughtlessness or otherwise without malice would not in future be punished.[60] That the decline in the numbers punished for infringements of the *lèse-majesté* laws cannot be seen as indicating a resurgence of Wilhelm II's popularity is, again, demonstrated by the sources. In 1897 Waldersee was still recording that 'the feeling against the Kaiser . . . is unfortunately widespread'; 'how much worse it has gradually become!'[61]

Waldersee and other commentators were repeatedly struck by the lenience shown towards the Kaiser by the Liberal and Social Democratic press; it was no wonder, they thought, that he had false ideas of his popularity. The explanation for this strange phenomenon, according to the General, was that the press and 'all the parties flirt with him' in the hope of winning his support. 'But if one speaks to party members of any persuasion' one only ever heard expressions of 'doubt and despair'. Waldersee added apprehensively: 'Where it will all lead is

Table 1 *Prosecutions and convictions for infringements of the laws relating to* lèse-majesté, *1882–1918*

Year	Prosecutions	Convictions
1882	587	487
1883	564	443
1884	538	438
1885	537	412
1886	558	456
1887	763	615
1888	820	654
1889	750	557
1890	794	581
1891	783	593
1892	788	581
1893	922	670
1894	952	720
1895	858	644
1896	844	623
1897	602	429
1898	632	486
1899	545	402
1900	385	300
1901	394	299
1902	348	284
1903	370	287
1904	338	258
1905	216	179
1906	235	193
1907	132	107
1908	50	34
1909	21	14
1910	21	17
1911	24	14
1912	30	23
1913	21	19
1914	81	64
1915	96	75
1916	71	55
1917	42	29
1918	31	21
Total	**15743**	**12063**

28. The Kaiser and his sons – an unconventional portrait

not yet quite clear to me, but that things are going badly I do not doubt for an instant. We are gradually going down in the estimation of other countries; all our enemies are beginning to feel pleased with themselves while at home pessimism is taking over.'[62] Four years later he commented: 'From the Liberals down to the Social Democrats, all are convinced that he [Wilhelm II] is working in their interest and they cover him with compliments, or at least do not attack him.

They are well aware of what he really is at heart, his arrogance, voluntas regis etc.'[63]

When the Reichstag elections of 1893 brought an increase of almost 25 per cent in Social Democrat votes Waldersee commented that the Kaiser ought really to see now 'that all his plans for reform and all his fine intentions have failed and that his advisers à la Douglas, Hinzpeter, Helldorf [sic] and his gang have advised him badly. He is hardly likely to acknowledge that he has contributed a great deal to the development of social democracy himself, but it is . . . undoubtedly true.'[64] After the speech of November 1893 in which the Kaiser had proclaimed to the newly sworn-in recruits of the Berlin garrison: 'You have *no will* of your own from now on, there is only *one will* for you and that is *my will*!', Waldersee predicted that the Social Democrats would be delighted with it, for 'no one brings them more recruits than the Kaiser'.[65]

THE 'STATE-SUPPORTING' PARTIES LOSE CONFIDENCE

Almost more alarming than the steady growth of the so-called anti-Reich parties during the first years of Wilhelm II's reign was the mounting discontent among those sections of the population on which the monarchy depended for its support. Only a matter of months after Bismarck's dismissal the pro-Bismarck Fronde had already begun to take root, especially in the southern and western regions of the Reich and among the students (the elite of the future), and to create enmity between influential bourgeois circles and the Kaiser. During these same years, as we have seen, the anti-Semitic campaign made great advances among the rural population and caused the venerable Conservative Party, the party of the landowning Prussian aristocracy, to embrace anti-Semitism in its official Tivoli programme in 1892. Meanwhile the Pan-German League, founded in 1890, was putting the cautious foreign and colonial policies of the New Course under increasing nationalistic pressure. And finally the Agrarian League, founded at much the same time, became the focus of all discontented elements in rural areas, uniting them in a dangerous mass agitation against the moderate policies of the government.

These menacing developments were of course regarded with the greatest anxiety by all observers with a political axe to grind. After a conversation with the right-wing editor of the *Deutsches Wochenblatt*, Arendt, Waldersee recorded in the winter of 1891 that the latter was 'very pessimistic', especially about 'the decline of the prestige of the monarchy, in particular in the circles hitherto considered well-disposed . . . even including civil servants and officers'.[66] Watching the growing dissatisfaction among the very parties which were formerly regarded as being loyally supportive of the state – the Conservatives, Free Conservatives and National Liberals – the General repeatedly commented that 'the Kaiser alone' was to blame for this development, for 'without realising what he was doing', he was working for 'the disintegration of the well-disposed parties'.[67] At the beginning of 1892 Waldersee was dismayed to observe that the mood throughout the country was 'very gloomy and uneasy among all who consider themselves the supporters of the state, in other words among the mass of the rural population, whether from the peasant or the great landowning classes, among all civil servants, in the army, among most industrialists, in academic circles, but also among countless ordinary people who are good patriots at heart. The fundamental cause is as regrettable as it is clear to me: it lies *entirely in the Kaiser himself*.'[68] Two years later this aristocratic, arch-conservative General noted in his diary: 'In all the circles in which I move, whose thinking I share and to which, through my personal circumstances, I belong, the prevalent mood is one of the most pronounced dissatisfaction and ill-will.' This 'great ill-will towards the Kaiser . . . however, has spread to much wider circles; I would even say that it is general, although this is not apparent outwardly; but that is not surprising, because the press is highly circumspect, which is also the reason why the Kaiser no doubt suspects very little of all this.'[69]

In South Germany, where Catholic particularism was continuing to gather momentum and veneration for Bismarck among the nationalist-minded bourgeoisie was growing to an alarming degree, the mood of antagonism towards the Kaiser was universal. As early as March 1891 a Bavarian Count, writing to Eulenburg, warned of the serious decline of belief in the idea of the Reich, which was being further accelerated by 'skilfully circulated mistrust of the Kaiser'. 'Last year the talk over beer, wine and tea was of the Kaiser's foolishness; this year only of his excitability, overhastiness etc.'[70] At a banquet in honour of Bismarck in Stuttgart in 1891, Eulenburg reported to Berlin, many people remained seated when the Kaiser's health was drunk.[71] The Baden envoy often found himself compelled to point out that pro-Bismarck rallies turned out to be 'not so much to honour Bismarck as to demonstrate against Wilhelm II'.[72] 'H.M. is unfortunately blamed more and more by the nation for everything that is wrong, and thus also for the Bismarck scandals!', he reported in July 1892, after Bismarck's triumphal progress to Vienna on the occasion of his son Herbert's wedding had led to a series of mass ovations for the former Chancellor. 'There is a desire to put the Kaiser in the wrong, and the criticism of the former Chancellor for his present conduct is nearly always accompanied by the corollary that "none of this would have happened if H.M. had kept the Prince in office!" . . . But in private conversation both among the upper classes and among lesser folk a deep grudge against H.M. comes to the fore everywhere, and it is our duty not to conceal the fact that H.M.'s popularity among the people has again suffered a heavy loss on account of recent events.'[73] Prince Bernhard of Saxe-Meiningen, who was married to Wilhelm II's sister, complained in 1894 that in South Germany even army officers made a show of supporting particularism. In the learned circles to which the Kaiser's brother-in-law also had access the feeling against the Kaiser was 'dreadful'. The general verdict was that this state of affairs could not go on much longer. 'He was sure that even now the Kaiser had not the slightest idea of how people really felt', Waldersee noted after a conversation with the Prince.[74]

The middle classes in the great cities of the west and north of Germany were no less angry and alienated. In 1892 Waldersee recorded that employers, especially industrialists, were 'in great anxiety'; there was insecurity throughout the commercial classes and even the teaching profession, writers and some lawyers felt 'deeply offended'.[75] In 1894, as commanding general in Altona, he had a confidential conversation with a 'well-intentioned and intelligent' citizen of Hamburg who told him 'how unpopular the Kaiser has become and how openly this is discussed at the higher levels of society; although I am well aware of this', Waldersee observed, 'I thought it expedient to show surprise'.[76] Shortly before this he had recorded in his diary, after a meeting with the military author Major Johannes Scheibert, that this 'absolutely reliable, unselfish, loyal man' was 'deeply shocked at the situation in Berlin' and had 'nothing but scorn and contempt . . . for the Kaiser'. 'He says that it is absolutely extraordinary how people talk about the Kaiser in the most disrespectful way in restaurants, even of the better class, in bars, in trams etc.'[77]

The bitter feud between Wilhelm II and Bismarck not only had disastrous consequences for the Kaiser in its effect on the National Liberal bourgeoisie; it also contributed to the growing alienation between Wilhelm and the traditional supporters of the Hohenzollern monarchy, the Conservative landed aristocracy of Prussia. Early in 1891 Holstein was dismayed to observe that the Conservatives under Limburg-Stirum were beginning to form a faction not only against the Chancellor but also against the Kaiser. In this 'open defiance' of their King, he said, 'the rumours of mental illness in the Kaiser play a part. It is scarcely ever mentioned, but many people, in all parties, think that the "carnival" cannot last much longer . . .'[78] In November 1891 Holstein reported to Eulenburg with horror that while conferring an order on Wilhelm von Rauchhaupt, the leader of the Conservatives, the Kaiser had said: 'But as you know, dear Rauchhaupt, regis voluntas suprema lex.' It was indicative of the mood among the Conservatives, the diplomat commented, that Rauchhaupt had expressed his intention of resigning his seat.[79] The Kaiser was 'very angry with the Conservatives', and there would be no lack of ill-considered speeches in the future, Waldersee predicted bitterly in 1892.[80] In 1894, as already mentioned, the normally placid Conservative

President of the Reichstag Albert von Levetzow was moved to raise his voice against the Kaiser and afterwards observed that this day was the saddest of his life; the other Conservatives present had also left the meeting with their King 'deeply upset' by it.[81] 'What great damage the Kaiser has done to himself and to the whole situation with his unfortunate passion for acting as spokesman on all matters of importance!', Waldersee exclaimed.[82] He reacted with bitter irony on hearing the following month that the Kaiser had sent the 'wretched' Count Dönhoff, who had voted for the Russian trade treaty 'purely out of fear of the Kaiser', a congratulatory telegram which read: 'Bravo! Done like a true nobleman.' 'That is the kind of person that H.M. likes', was Waldersee's scathing comment. 'Kaiser Friedrich and Kaiser Wilhelm I, who had a true sense of what is fitting in a nobleman, must be turning in their graves . . . If one could still take the Kaiser at all seriously in what he says, now would be the right moment to renounce noble rank.'[83] It was very much to be feared, Waldersee believed, 'that Germany has already passed its highest point', although, as he acknowledged, 'when Kaiser Wilhelm II came to the throne I certainly had a very different idea of how the country would develop'.[84]

The General became increasingly concerned about the steadily worsening mood in country areas, a mood which he compared with that of the years between 1808 and 1812, when Prussian history had reached what was then its lowest ebb. In the autumn of 1892 he recorded that he was receiving more and more comments from Silesia, Brandenburg and Saxony which testified to deep unhappiness, sometimes even despair, in rural circles. Even in monarchist families there the opinion had taken hold 'that the blame for the confused state of affairs in the country, for the general sense of insecurity, lay entirely with the Kaiser. Even in the circles that I have in mind royalist feeling is declining', he noted. 'People feel forgotten, and they see how the Kaiser is playing into the hands of the revolutionary parties.'[85] After a hunting trip to Silesia that winter he recorded that 'critical comments on the Kaiser are increasingly frequent and at the same time it is very apparent that people are remembering his grandfather more and more'.[86]

The crisis in the relationship between Wilhelm II and Prussia's Conservative rural population deepened dramatically in the winter of 1893–4 when the Kaiser gave his support to Caprivi's tariff policy favouring industry, commerce and labour. Waldersee watched in consternation as Wilhelm put himself at odds with the natural allies of the crown. Running down the Junkers had always been a favourite pastime on the Liberal side, the General commented scathingly, 'but now they are really flushed with victory, as the Kaiser has also spoken out against the Junkers . . . No one likes to mention the fact that the whole peasant class stands behind the Junkers.'[87] The longstanding mainstays of the government, namely the peasants and the small landed proprietors of the old Prussian provinces, were gradually being driven into the most extreme opposition. The Kaiser was now convinced that he was 'fighting a battle with the rebellious Junkers, like his forefathers. Catchwords like that always impress him', Waldersee mocked in early 1894.[88] When the Kaiser surprisingly proclaimed himself the protector of the aristocracy and of agriculture in his speech at Königsberg in September 1894 Waldersee could hardly believe his ears. 'What kind of nonsense is our gracious master babbling now', he exclaimed. 'His Liberal so-called admirers will pull very sour faces when they hear that he identifies himself with the aristocracy . . . But what are people to think, when he proudly claims to have kept his word that he would protect agriculture, whereas in fact he has dealt it the hardest of blows with the trade treaties.'[89] The fact was, Waldersee commented, that not only the landed nobility was disgruntled, 'but the entire section of the population, amounting to millions more, who live from agriculture'. Until a few years ago, in his view, the peasant class, or at least the Lutheran part of it, constituted 'the most dependable Conservative element that could be found'. In the old Prussian provinces this group were 'good Hohenzollernites', and in Mecklenburg, Holstein and Hanover they were still at least 'thoroughly conservative'. If its leaders were now intimidated by the government, he warned, the whole peasant class would switch over to the anti-government camp and would in future vote for the Progressive Democrats, the Anti-Semites or the Socialists.[90] Waldersee was repeatedly confronted with evidence that the ill-feeling in the country was directed against the Kaiser. This was 'really very sad, but no wonder . . . with his tendency to put himself forward everywhere'.[91]

The Berlin correspondent of the *Kölnische Zeitung* painted an alarming picture of the mood in the Reich in 1895, above all among the increasingly radical small farmers, writing: 'From ever more widespread circles we are receiving reliable information about a growing ill-feeling towards the person of His Majesty our Kaiser; in South Germany, where the upsurge of national feeling had been strongest in the last twenty years, radical elements are growing in the most alarming way; the numbers in Württemberg bear eloquent witness to this. In North Germany the Agrarian League has sown a seed which will produce the most dangerous of all fruit. For the time being the present Conservative leaders still believe they are in control of the movement, but plenty of indications that they are mistaken are already beginning to appear: in Bavaria with the peasant leaguers, on the Rhine in agricultural circles linked to the Centre Party. The Conservatives have thought it necessary to harness the small farmers to their cause . . . It cannot end well . . . The present movement can be held back for another year; the great landowners may also be able to rein in the Agrarian League for another year. But then our small farmers will notice that their leaders have been leading them around by the nose, and then they will become as bad and as dangerous as our present revolutionary parties. The game that the present Conservative Party is playing is equally dangerous for the state and the monarchy.'[92]

DISCONTENT IN THE ARMY

In its original text Waldersee's diary also provides us with an excellent source on the mood of the army, about which the Chief of the General Staff was particularly concerned, since he saw the armed forces as the natural and essential mainstay of the Hohenzollern throne. We have already seen the extent to which the other military officers closest to the Kaiser shared Waldersee's critical attitude.[93] The generals at court were not alone, however. Like them, the commanding generals of the seventeen army corps, the officers of the General Staff and of the Prussian Ministry of War and even younger officers who were to succeed to positions of leadership in the coming years were even now, in the aftermath of Bismarck's dismissal, expressing harsh views about their Supreme War Lord. Waldersee's diary records that at the 1892 New Year reception the Kaiser made a deplorable impression on the commanding generals by simply rejecting as 'nonsense' the changes to army uniform which many of them had requested. Afterwards, he noted, Adjutant-General von Wittich 'roundly' criticised the Kaiser and said 'he was getting more and more grand and infallible'.[94] After the Kaiser's New Year address to the commanding generals two years later Waldersee recounted that the generals had 'mostly looked embarrassed' and had 'smiled strangely'; nor had any of them had the slightest scruples afterwards about making derogatory remarks about the monarch. 'I was very distressed', the former Chief of the General Staff admitted, 'for never has the Kaiser expressed such feeble, I might even say childish, views in these speeches; he really lacks knowledge and experience and the worst thing is that he does not feel the least need to broaden them, for he thinks he knows everything very well and better than anyone else.'[95] On the same occasion in January 1895 the critical comments of the commanding generals were 'quite openly expressed in the most appalling way'. The Commanding General of the Guards made a 'particularly audacious' contribution, Waldersee commented.[96] Five years after Bismarck's fall he described the mood among senior officers as 'despondent', adding that 'Hahnke is highly discontented . . . General Winterfeld, the great Comm[anding] General of the Corps of Guards, is also very depressed about the Kaiser, and even people like Bissing are beginning to lose heart.'[97]

Even at this early stage Waldersee made repeated complaints about the sycophantic behaviour of the army officers attached to the court and the other favourites whom the Kaiser had collected around him. The behaviour of Freiherr Gustav von Meerscheidt-Hüllessem since his appointment as commanding general of the Corps of Guards was in his view thoroughly symptomatic of the corrupt atmosphere at the Hohenzollern court which Wilhelm had encouraged. Hüllessem accepted 'completely and unquestioningly everything the Kaiser did with the corps, which is often absolutely incredible', and he even encouraged the Kaiser in this habit. 'He is very often in the Kaiser's company, especially at the countless meals for the officers, and he tries to amuse the Kaiser with jokes which are always coarse and mostly obscene. Even young officers think it is unbecoming for him to lower himself to act as a jester', Waldersee noted disapprovingly. The award of the Order of the Black Eagle to Hüllessem on the occasion of the military review of 18 August had caused universal head-shaking, he wrote. 'We have not been accustomed to such things at reviews, and it will be seen more as a reward for unconditional obedience.' As Hüllessem's private life was 'very questionable' – he lived with a woman to whom he was not married – such an exalted order 'really did not look well' upon him.[98] Commenting on Winterfeld, Meerscheidt-Hüllessem's successor as commanding general of the Guards, Waldersee recorded that there was much indignation in well-informed circles, for 'as soon as the Kaiser [can] be seen on the horizon he [becomes] coarse, obviously in the hope of impressing our gracious master thereby . . . But he is just the sort of man to like an appointment close to the Kaiser', Waldersee added caustically.[99] In later years too Waldersee strongly criticised the attitude of the Kaiser, which led senior officers to curry favour with him by rudeness towards subordinates.[100]

A further reason for the growing unrest among the generals was Wilhelm II's personal interference in appointments and promotions in the officer corps. There was universal dismay, for instance, when Prince Friedrich of Hohenzollern-Sigmaringen, whom Wilhelm, under the influence of Countess Elisabeth Wedel, had formerly abhorred as a dangerous intriguer,[101] was promoted to commanding general of the III Army Corps in the autumn of 1893. Waldersee strongly condemned the preferential treatment shown to this 'indolent' Prince, for he was 'a very insignificant man, without a trace of passion for his career, without ambition – not even of smart appearance, and very poorly trained as a soldier'. For the proud former Chief of the General Staff it was 'truly not an uplifting feeling to hold the same post as such a lamentable General'.[102] When shortly after this the Kaiser gave Prince Hohenzollern yet another appointment, as a member of the National Defence Committee, Waldersee commented scathingly: 'If the Kaiser intended to diminish the significance and prestige of the Committee in the eyes of the army he could not have made a better choice than to nominate this General, who is as uneducated as he is insignificant and uninteresting, to the Committee. The sole reason is to have another more compliant man on it.'[103] Waldersee condemned out of hand the Kaiser's inclination to give preferential treatment in the army to 'minor princes'. 'I have nothing against the Hereditary Prince of Meiningen, although I am afraid that he loses his head easily', he wrote; 'but Pce Hohenzollern, Hereditary G[rand] D[uke Friedrich] of Baden and [Hereditary Duke Peter of] Oldenburg, Prince Leopold [of Bavaria] etc. are quite useless as senior troop commanders.' The Kaiser was altogether 'a very incompetent judge of men' and deluded himself greatly in selecting officers for the most senior positions. 'What mistakes he has already made!'[104]

Waldersee was particularly shocked to see how much his successor as chief of the General Staff, Count Alfred von Schlieffen, was dominated by the Kaiser and even allowed the latter to dictate the strategic guidelines for a future war to him.[105] In March 1892 the Kaiser made such 'derogatory' remarks about Schlieffen in front of numerous General Staff officers in the Tiergarten that in his predecessor's view Schlieffen ought to have resigned.[106] In 1892, in the hope of curing him of the offensive habit of personally evaluating military exercises in the General Staff offices, Wilhelm was 'very cunningly' informed that there was an outbreak of measles there![107] With a view to avoiding such difficulties in the future the officers developed a system by which a Flügeladjutant asked the General Staff confidentially for Schlieffen's solution and then, on the basis of this information, worked out with the Kaiser what opinion he should express. At the evaluation session the Kaiser could then demonstrate that he had found exactly the right solution![108] The 'lamentable' part of the Kaiser's evaluations, according to Waldersee, was that he was 'either guided by preconceived ideas or by momentary and always hastily formed impressions'; that was also the reason why most army commanders became 'uncertain and agitated' whenever the Kaiser was in the vicinity.[109] 'He obviously has not the least idea of the damage he does to himself and to everything', the General commented anxiously.[110] In later years Wilhelm II continued to do as he pleased in military matters, and the most senior officers followed his bidding. When in the spring of 1895, after weeks of preparation, Schlieffen made his report to the Kaiser on the forthcoming manoeuvres in the Uckermark and happened to mention Prenzlau, where the Prussian army had surrendered to the French in 1806, the Supreme War Lord declared 'that in view of the sad associations of this name he did not wish to have any manoeuvre there! Now poor Schlieffen has to abandon the very laborious preparations he has made and start all over again. The Kaiser also added', Waldersee recorded in his diary, 'that according to the map the terrain in the Uckermark did not seem suitable for cavalry, but he intended to take command himself, with large masses of cavalry, and anyway he would like to have the manoeuvres in Pomerania.' The former Chief of the General Staff commented bitterly: 'I hope he does not suspect what toil and trouble he causes with such caprices.'[111]

The growing discontent of the officer corps and the increasingly critical attitude of the younger officers in particular made Waldersee especially fatalistic. In the summer of 1890, when he was still Chief of the General Staff, he recorded that it was sad but 'unfortunately . . . already to a great extent the case', that the army was deeply dissatisfied with the Kaiser. 'Great mistrust has grown up, and there is no affection at all; there is a feeling of coldness and of great ill-will . . . Above all, the Kaiser should have bound the army to himself; such a thing cannot be bequeathed; it must be won. Kaiser Wilhelm I did this and what a fine relationship there was between him and the army, how profound was the feeling of solidarity and mutual security!'[112] Not long after Bismarck's dismissal Waldersee drew up a list in his diary of the manifold reasons for Wilhelm II's unpopularity in the army. His 'great preference' for the navy and the Guards and his consequent neglect of the line regiments, and especially of the infantry, were resented. There were complaints that the Kaiser showed 'considerably less courtesy' towards senior officers 'than they had been accustomed to with his grandfather', and that he passed harsh judgements 'most of which are probably expressed quite thoughtlessly and which verge on brutality'. In general the Kaiser's excessively strong expression of his will 'about things which the sovereign does not thoroughly understand', coupled with his arrogant attitude towards the judgement of experienced people, was responsible for frequent transfers and even reprimands for senior officers, which 'are regarded as being based entirely on his personal feelings'. Not least, Wilhelm's 'habit of speaking openly to junior officers about their superiors' caused deep dissatisfaction. Finally, the Chief of the General Staff recorded, the Kaiser had squandered the confidence which the officer corps had initially shown towards him by indulging his 'passion for playing soldiers', which expressed itself principally in frequent and utterly senseless orders for troops to be called out. A feeling of insecurity and discontent was apparent everywhere among senior officers, who spoke openly of the 'tactlessness' of the Kaiser and of the 'ruthlessness' of his Military Cabinet.[113]

A year after Bismarck's fall and shortly after his own transfer to Altona, Waldersee inveighed against the Kaiser's constant and harmful interference in army matters. Wilhelm had already ordered 'countless innovations', he complained, 'mostly little more than tomfoolery', among other things the 'unfortunate' arming of the cavalry regiments with lances. He had been 'very premature in trusting his own judgement', 'which only caused offence and did not impress anyone, because it was based on dilettantism'. His 'currying favour with the masses even at the expense of the army' had deeply offended the officer corps. 'Now, after 3 years, the inevitable result is that the army feels that its relationship to its War Lord has fundamentally changed and has of course become much cooler and more distant than before, and that confidence in the Kaiser's military talents has not grown at all; on the contrary, lack of confidence in his ability has spread', the General wrote.[114] He also strongly criticised the Kaiser's demand that new recruits should be paraded before him. If they had not practised this the parades went badly and the Kaiser was dissatisfied; if on the other hand the soldiers practised the parade drill so as to please the Kaiser, valuable weeks of basic training were wasted, which 'turned all the principles of our military training upside down. What utter confusion everyone is thrown into, and how it ruins people's characters', Waldersee commented despairingly.[115] He went on to condemn, as a well-meaning but 'very unfortunate . . . idea of the Kaiser', the encouragement of competition in the army through the award of medals for shooting and imperial prizes for riding, which had caused much 'envy, quarrelling, cheating, deception and discord'. The General emphasised that these 'evils, which are already very evident', were giving rise to widespread anxiety.[116] However much a militarist at heart, even Waldersee could not condone Wilhelm's praise for the grenadier who had shot and fatally wounded civilians. The Kaiser's action would 'cause great and very lasting harm', he predicted; this was 'yet another sign of utter immaturity of judgement'.[117] Wilhelm's behaviour in 1893, when numerous officers were caught gambling and he took a strongly censorious attitude, was considered 'incredible and injudicious' by all the army commanders, especially as he had recently not only allowed gambling during court hunting expeditions but had himself watched it.[118] In the summer of 1895 Waldersee again recorded with deep concern that 'the Kaiser could not be less popular in the army; he lacks his grandfather's great talent for winning hearts. He captivates people at first, but then repels them.'[119] Numerous observers, including the journalist Franz Fischer, the diplomat Philipp Eulenburg and the Empress Frederick, expressed similar views about the growing discontent within the army.[120]

Senior officers were particularly disgusted by Wilhelm II's liking for surrounding himself with cheering crowds when he went on manoeuvre. Thus for instance when the imperial manoeuvres of 1893 were held in the French-speaking part of Alsace-Lorraine the women, schoolchildren and servants at the German garrison had to be taken to the troop reviews to cheer, and this had to be done in such a way as to enable the Kaiser to see them several times on the same day.[121] This had been 'a complete and utter sham', wrote Waldersee,[122] who was very far from being the only one who regarded the Kaiser's taking personal command at manoeuvres as a dangerous farce. The saying 'the Kaiser always wins' was mockingly peddled around the entire army, and the arbiters at manoeuvres were indeed given direct instructions to ensure that he did.[123] The Kaiser's brother-in-law Prince Bernhard, who took part in the disastrous autumn manoeuvres in Silesia in 1890 in command of the Second Guards Infantry Division, deplored not only the harm which this undesirable practice did to the army, but also the effect on the Kaiser himself. It was spoiling the 'gracious sovereign' and doing him no favour, he declared, and 'nobody who truly wished him well should do such a thing'.[124] After the above-mentioned cavalry manoeuvre in Lorraine in the autumn of 1893 the Kaiser's personal direction of the attack was 'unanimously unfavourably criticised' by the officers – although not in such a way that Wilhelm heard about it.[125] In a conversation with Waldersee the Grand Duke of Baden remarked that his nephew had assembled the troops in too much of a hurry, ordered them to move before they had finished assembling and then moved them forwards without sufficient reconnaissance. The result had been that the mounted troops on one flank had ridden into marshy ground in which numerous horses had got stuck. To cap it all Wilhelm had attacked his own troops! At first the generals had hoped, according to the Grand Duke, that Wilhelm would conclude from this experience that to command cavalry in great numbers was no easy task. But Waldersee had to break the news to Wilhelm's uncle that although at first the Kaiser had been 'fully under the impression of having failed, as soon as he began the ride home he was told by some despicable flatterers that he had in fact led his troops very well, and his good mood returned immediately'.[126] General Viktor von Podbielski, who was taking a cure at Karlsbad at the same time as Reich Chancellor von Caprivi, informed Waldersee afterwards that during those imperial manoeuvres the Kaiser had 'concerned himself almost exclusively with the cavalry and put it to the most unbelievable use'. Podbielski had also heard complaints from several senior cavalry officers 'about constant orders and counter-orders' from the Kaiser; there had never been such great confusion. The attack which the Kaiser led with twelve regiments was, moreover, 'nothing but a copy of an identical attack which General Krosigk had performed before the Kaiser shortly beforehand near Salzwedel, the only difference being that the one in Lorraine had missed the enemy', the two Generals told each other in consternation.[127] How little Wilhelm had learned from this experience became evident a year later, when at the manoeuvres in East Prussia he again insisted on commanding a cavalry corps, with equally unfortunate results.[128] Again, before the imperial manoeuvres in the early autumn of 1895 in Western Pomerania – Wilhelm had personally chosen this region in the belief that it was particularly suitable for cavalry attacks – Waldersee wrote of the participating officers' fear of the Kaiser's 'passion' for commanding 'enormous contingents of cavalry', although he had always come to grief in doing so hitherto. 'For a cavalry commander he is far too reckless and inexperienced.'[129] 'Commanding large units of cavalry is an unfortunate obsession of the Kaiser's; he hopes to create an impressive effect by doing so, but he achieves the opposite.'[130] In addition the General made the very valid point that with the 'enormously increased firepower' of modern weapons it was 'absolute folly' to set so much store by the cavalry.[131]

If the generals were unanimous in considering the Kaiser's personal command at manoeuvres in peacetime highly damaging, the idea of his taking command himself in a war filled them with utter consternation. Waldersee, it is true, had his doubts as to whether at the decisive moment, when required 'to draw the sword and cut the knot', Wilhelm would really have the courage to start a war, in spite of all his bloodthirsty utterances.[132] Nevertheless for him, as for all the other army commanders, it was beyond doubt that Wilhelm II, as supreme war lord, would personally assume overall command over the nation's armed forces in the event of war breaking out. 'The possibility that the Kaiser will wish to direct operations himself in a war is a thought that fills one with grave concern', Waldersee confided to his diary in the spring of 1891.[133] A few months later the King of Saxony confirmed that it was of course true that Kaiser Wilhelm II 'must take overall command in war wherever he goes', 'but he must do it like his grandfather, that is, he must take advice'. The King agreed with Waldersee 'that in spite of all his passion for military matters and all his abilities the Kaiser does not have what it takes to command an army'.[134] Waldersee, of course, knew Wilhelm's excessive self-confidence too well to be able to believe that he would in fact listen to the generals whose advice he should follow. His constantly recurring nightmare was therefore that 'in his immense overestimation of his own competence' Wilhelm might take over command himself if a major war broke out.[135] 'I believe many people are now terrified at the thought that the Kaiser will want to take command himself in war!', Waldersee wrote after the imperial manoeuvres in Lorraine in the autumn of 1893.[136] The unanimous opinion among the generals was that 'the Kaiser himself is not the man for this task', he wrote. As the Austrian generals likewise had 'not the slightest confidence in the ability of the Kaiser to direct major operations', the Austrian army had already distanced itself to a regrettable degree from its German counterpart, and the realisation that Schlieffen had no influence on the monarch had also considerably increased the danger of a 'lukewarm war effort' on the part of Germany's ally.[137] The most that Waldersee dared hope for himself was that in the event of war he would be appointed to a command in the east against the Russians, where he would be able to use his own initiative in discharging his responsibilities. 'In the west my army is tied to other armies and all are under the direct control of the Kaiser, in whose leadership I, like many others, unfortunately have no confidence at all', he wrote in July 1894. 'With his rashness, his certainty of knowing better than anyone else and his vanity he is capable of creating the most appalling confusion and Schlieffen is not the man to guide him. If the Chancellor, the Minister of War and Hahnke too – as Verdy heard from Hahnke – were also to join the high command I foresee things getting very bad indeed.'[138] Waldersee's desire to have the eastern command was given short shrift by the Kaiser, however, with the remark: 'I do not like the idea of having the King of Saxony in the west. Just give him command in the east.' As Waldersee commented in despair, 'trivial personal interests thus become the deciding factor in a matter on which the good of the Fatherland might well depend'.[139]

It was not Count Schlieffen but Kaiser Wilhelm II who determined the deployment and strategy of the Prusso-German army. A year after his dismissal as chief of the General Staff Waldersee recorded in his diary: 'The Kaiser wants to send more troops to the west in case of war – which I told Schlieffen he would – and it seems the latter is willing to give in and even thinks in general that nothing needs to be done in the east.' Waldersee bitterly predicted that these plans would lead to the loss of the right bank of the Vistula and the province of Silesia.[140] Even the belligerent Waldersee was shocked by the Kaiser's aggressive predilection for an attacking strategy. Wilhelm's belief that one ought 'always to attack and then to follow up rapidly' could lead 'in a very short time to the destruction of whole divisions', the General warned.[141] In his address to the commanding generals in January 1895 Wilhelm demanded that the policy 'of attacking the enemy wherever we see him, regardless of whether he has even 4 times the strength!' must be maintained. The generals assembled in the Berlin Schloss found it particularly offensive that on this occasion the Kaiser held up as a model to them the victorious Japanese, 'who went on the attack everywhere with tremendous daring, and knocked the Chinese down, although they were superior in number'. The monarch had completely overlooked the fact, they objected, 'that there is nothing approaching an organised Chinese army, but only a mass of troops who are lamentably badly armed and even more badly led'. Indignantly Waldersee asked: 'Has any army ever understood offensive tactics better than us in the last 3 wars? Have troops ever gone on the attack more bravely than ours? Truly not! And now the miserable Japanese are held up as an example to us! That is indeed hard and we do not deserve it!'[142]

OF 'UNBOUNDED LOVE' FOR 'THE BEST OF ALL KINGS'

It was not of course Holstein, Marschall or Grand Duke Friedrich of Baden, nor the Empress Frederick or Waldersee, and certainly not the inexorably growing mass of 'grumblers' in the general population, who were to be the decisive influence on the course of the government crisis in the next few years. This role was to be played by two men of quite a different stamp who idolised the Kaiser. The ambassador Count Philipp zu Eulenburg and his close friend Bernhard von Bülow were very well aware of the extent to which they differed from most statesmen, officials, diplomats and officers in their estimation of their young ruler, but they revelled in their 'ideal' alternative view. After a secret meeting between the two men in the South Tyrol in January 1896 Eulenburg wrote revealingly to Wilhelm II: 'Our discussions and exchange of feelings had one sure foundation: unbounded love for our King . . . If the great mass of politicians had heard our conversations during these days they would not have believed their ears, it would have seemed *impossible* to them – for personal, human love for the best of all kings and natural warm friendship between ourselves – how is that to be rightly understood in our complicated world?'[143]

Neither man held back in expressing his love for Wilhelm, and at least as far as Eulenburg is concerned there can be no doubt that his assurances of love and gratitude were deeply felt and sincerely meant. It did not embarrass him to tell the Kaiser to his face that he was the 'kindest, best sovereign that Prussia has ever had'.[144] For him, Wilhelm II was 'the kindest, most gracious of Kaisers and the most sympathetic of friends!',[145] for whom he constantly 'yearned'. 'Who understands me better than Yr Majesty? – and how many things only come to fruition as a result of talking them over with so understanding and kind a Kaiser!'[146] Even after many years of their friendship the sight of the Kaiser's 'beloved' handwriting continued to throw Eulenburg into a state of ecstasy. 'Oh what joy!', he wrote in the summer of 1897 in answer to a letter from Wilhelm. 'Your Imperial and Royal Majesty cannot conceive how happy Your long, gracious, *immensely interesting* letter has made me! Yesterday evening after receiving it I was quite unable to sleep. Happiness, gratitude, the contents – my head and heart were spinning with it all!'[147] At Christmas 1895 after receiving a letter from the Kaiser he had written: 'I read the words of the first pages, which contained so much kindness and friendship, over and over again and thought how wonderfully God has ordered my life. But since He has directed it thus, I hope that He will also give me the mental and physical strength I need in order to serve Your Majesty in the way that I most fervently wish.'[148] When the Flügeladjutant and military attaché Georg von Hülsen brought another letter from Wilhelm to him to Vienna, the ambassador expressed his delight 'at Your Majesty's dear presence near me! . . . It was as if Your Majesty were sitting beside me and speaking to me – I heard Your voice – and especially the emphases, accompanied by lively gestures of the hand. In friendship, what pleasure one has in the *particular characteristics* of the person one loves.'[149] Philipp Eulenburg and his closest friends, the Flügeladjutant Count Kuno von Moltke and Axel Freiherr von Varnbüler, Württemberg's envoy and representative in the Bundesrat in Berlin, did indeed call Kaiser Wilhelm II 'das Liebchen' – the darling one – to each other, as their arch-enemies were later to claim.[150]

Eulenburg also took the greatest delight in the annual Scandinavian cruises and the other frequent excursions and hunting expeditions on which he accompanied the Kaiser, in spite of the considerable strain they entailed and the often tricky political problems which had to be solved in the course of them. 'I am still revelling in the memory of the *wonderful* trip and of all Your Majesty's kindness to me, and I never cease to think of my beloved Kaiser with the deepest gratitude', he wrote after the Scandinavian trip of the summer of 1893.[151] Only a few months later, after a journey to the Adriatic with Wilhelm, he wrote ecstatically: 'I must . . . once again tell Your Majesty how *indescribably grateful* I am for all the kindness which Your Majesty again showed me, and how the shining memory of our splendid Venetian journey will rank among the most wonderful experiences of my life!'[152] After the Scandinavian cruise of the summer of 1895 the ambassador and imperial favourite expressed his gratitude with the words: 'The sun of Your Majesty's *golden friendship* shines more and more brightly upon me. The best and highest ornament of a human being, the touchstone of his *true* worth. God bless my most beloved Kaiser for it.'[153]

Gratitude for his rapid promotion, first as envoy (Oldenburg, Stuttgart and Munich) and then to the post of ambassador at Vienna, naturally played an important part in Eulenburg's relationship with Wilhelm, but he repeatedly affirmed (and with some credibility) that the personal friendship of the Kaiser meant far more to him than the outward glamour or material advantages of his position. In his first letter after taking up his appointment as ambassador he wrote with evident pleasure: 'These are the first lines Your Majesty will receive from me from Vienna! . . . dictated by a sense of immeasurable gratitude towards my beloved Kaiser, who showers me with honours and distinctions and gives me even more than all this splendour: true friendship, the highest and most glorious blessing!'[154] Eulenburg considered himself and the Kaiser – as also Bülow – kindred souls, whose nature was different from that of their ordinary fellow humans. He wrote to Wilhelm bewailing the 'artificiality of city life' after a stay in a forest region with the Kaiser. 'Very few people understand these feelings as intensively as Yr Majesty feels them. For that one needs the *truly* dual nature of one who can both dream and be awake at the same time – like Yr Majesty, and perhaps myself too. We have a long way to go before it is common to find two souls in one breast!'[155]

Moved by what he felt sure was the Kaiser's reciprocal affection for him, in the winter of 1898 Eulenburg gave expression to this certainty in a letter to his imperial friend. In thanking him for a telegram from Damascus he declared: 'I know Yr Majesty's loyalty which, although it is becoming increasingly strange in these times characterised by egoism, gives such a noble stamp to Yr Majesty's personality. That is why I understand this sign of friendship and feel happy and grateful for it! But in this case I am almost inclined to believe in "transmitted effect", telepathy, for my thoughts have been so *constantly*, so *intensively* directed towards Yr Majesty at this time – with both anxiety and sympathy – that I believe Yr Majesty must have felt it!' Alluding to the fact that Wilhelm had said some comforting words to him about his brother Friedrich (Fredi), who had been forced to leave the army because he had been accused of homosexuality, the Kaiser's best friend continued: 'I should like to say something else to Yr Majesty – something quite personal which I did not dare say in Rominten because I was too deeply moved (which Yr Majesty does not like!). But my heart is so full of ardent gratitude that I cannot stay silent any longer. I know very well what Yr Majesty wished to convey to me by putting my picture – that old face – on your writing table (which I feel is almost too much!). I understood very well the *love* that out of deep human sympathy you wished to show to an old friend who was suffering inwardly from a mortal wound. The silent eloquence of this act touched me beyond words – did me an *infinite amount of good*! God bless Yr Majesty a thousand times for it!'[156]

The diplomat Alfred von Bülow had belonged to the 'Liebenberg Round Table' centred around Eulenburg since they had studied together in Leipzig and Strassburg; his brother Adolf, as we saw in the first volume of this biography, decisively influenced Prince Wilhelm's political education in an autocratic and militaristic sense as his personal adjutant.[157] Another brother of Alfred, the ambassador in Rome Bernhard von Bülow, was of course to play a more significant role still in the life of Wilhelm II as secretary of state at the Foreign Office from 1897 to 1900 and then from 1900 to 1909 as Reich chancellor and Prussian minister-president. It was chiefly thanks to his intimate friendship with Philipp Eulenburg, as we shall show, that Bülow rose to these heights.

The language which Bernhard von Bülow used in his correspondence with Eulenburg about the Kaiser was scarcely less effusive than that of the Kaiser's favourite, and yet one cannot escape the impression that the only purpose of Bülow's idolisation of the monarch was to manipulate Eulenburg to his own advantage. In his letters to 'dearest Philipp' Bülow constantly and cynically emphasised exactly what this 'sisterly soul' (Bülow's description of their relationship) wanted to hear, namely that Wilhelm II was a man 'of genius' and would accomplish great deeds in the future, always providing, of course, that he received the right advice. As early as 1892 he assured Eulenburg that he had 'confidence in the Kaiser's destiny. He will work his way through, not only to be a great ruler, but also to win recognition from many who misjudge him now.'[158] After a visit to Berlin he wrote enthusiastically: 'I found His Majesty kind, natural, looking well, fresh, full of life, intellect, ideas, and thoughtful and intelligent in everything he said. With the right people about him, serving and interpreting him, he will stand his ground, will be able to stand his ground.'[159] After another visit to court the following year Bülow wrote to his friend: 'Dearest Philipp, I have just come from the Neues Palais. His Majesty immensely kind – as brilliant, lively, interesting as possible. I was deeply moved when I was able to kiss his hand and thank him for so much graciousness. His face had such a kind expression, almost soft, in spite of all his energy. It pleases me to give him pleasure.'[160] In later years too the tone of Bülow's letters to Eulenburg remains precisely attuned to the latter's sensibilities. 'In my mind I have the dear Kaiser's fine, bright eyes constantly before me, with such a moving expression (because they are so trustful and fixed on the noblest of goals).'[161] 'It only makes me love our dear, beloved Kaiser all the more when I see him being attacked so much. I always think of his wide eyes, the eyes of both a child and a genius . . .'[162] Just like Eulenburg, Bülow wrote ecstatically of a letter from Wilhelm: 'Our beloved sovereign's letter to me has moved me profoundly. Even if he were not the Kaiser one could not but wish him well with all one's heart. What would I not do for such a sovereign! But for the very reason that I love him so much I do not want him *to do himself any harm.*'[163] 'I already belonged to our Kaiser body and soul', he wrote from Rome in 1896 after Wilhelm's visit to Italy; 'but this time he has utterly *and completely* conquered me.' 'It is impossible to be near him without being swept away by so much life, intellect, strength and goodness. The miasma of false perceptions and exaggerations, the fog of slander and malice fall away and his true image emerges in all its vigour and versatility. One cannot help being terribly fond of him *once one really knows him.*'[164]

Whether such declarations of love were sincerely meant or were merely a cynical pretence calculated to win, or at any rate to maintain, Eulenburg's confidence and support for Bülow himself, they led to a flight from reality into an idealised dream world from which neither man would awake until the damage they had done, in their blindness, was beyond repair. In February 1892, that is to say at a moment when the leading elite in Germany were throwing up their hands in horror at Wilhelm II's autocratic utterances and foreign newspapers were beginning to talk openly of his mental illness, Bülow was affirming his mystical belief in the Kaiser's spirit of genius and in Prussia's destiny to an infatuated Eulenburg. 'The doubts and complaints about our All-gracious sovereign are a sad – and alarming – phenomenon, and also politically harmful, both at home and abroad, but it has not made me lose my confidence in the future', he attested. The Kaiser, he maintained, possessed 'the qualities – such as steadfastness, courage, a quick understanding – which can neither be acquired nor instilled from outside;

His faults, which are mostly no more than the reverse side of these qualities, can be gradually reduced or completely cured by experience'.[165] Bülow dismissed Geheimrat von Holstein's 'prophecies of doom' as to the consequences of Wilhelm's absolutist speeches and style of government, merely remarking that the diplomat was too pessimistic. 'In people of great mind and temperament one must not take every word in such deadly earnest.'[166] 'I shall not allow all these long-drawn-out prophecies of doom to rob me of the complete confidence I have in our beloved sovereign. My overall impression is, as it has always been, that in general His Majesty not only has the right thing in mind, but also achieves it with clarity, freshness and brilliance of vision. It is only the form in which it is expressed which still occasionally lacks subtlety, and is sometimes a little careless too. Where strength and initiative abound, this usually happens. But it is people of precisely this nature who learn through their own experience.'[167] 'Do not let us be robbed either of our belief in the country's destiny, or of our confidence in the personality of the Kaiser', urged the future Reich Chancellor in a letter to Eulenburg in early 1895. 'Even if our dear Kaiser has occasionally slipped up over this or that on a matter of form, He has always reined in His strong will and lively temperament when reasons of state have been paramount, and he will continue to do so, which is the lodestone that has guided his house to greatness.'[168] In October 1895 Bülow gave further encouragement to Eulenburg's dangerous illusions with an effusive assurance of his fellow feeling: 'I understand so well that you are becoming increasingly attached to our All-Highest sovereign. I too, who do not know him nearly as well as you and am far less close to him, feel more and more at one with him inwardly. Not only because I am a royalist – not of the head but of the heart, the Prussian monarchy is in the end the most important thing for me – but also because of his individuality. His letter to the Tsar is a work of genius! How much he has to give! May God protect him!' Alluding to Holstein, Bülow added: 'malicious dwarves always abound where heroes hold sway'.[169] Even after taking up office as foreign secretary, apparently blind to reality, Bülow wrote his friend a letter of surpassing sycophancy which he no doubt hoped would be laid before the Kaiser. 'I grow fonder and fonder of the Kaiser. He is so important!! Together with the Great King and the Great Elector he is by far the most important Hohenzollern ever to have lived. In a way I have never seen before he combines genius – the most authentic and original genius – with the clearest *bon sens*. His vivid imagination lifts me like an eagle high above petty detail, yet he can judge soberly what is or is not possible and attainable. And what vitality! What a memory! How quick and sure his understanding! In the Crown Council this morning I was completely overwhelmed! He gave an exposé on the water questions, which are so complicated, and mentioned all the material and administrative aspects involved in as precise and detailed a way as any expert minister could have done, but with a vigour, vividness, broadness of overall vision, in short with a genius that no minister could even begin to achieve. May God preserve our great monarch and the noblest of men!'[170]

To counter the forcible return of Wilhelm II to the principle of reasons of state contemplated by Holstein, Marschall von Bieberstein, Bronsart von Schellendorf and Alexander Hohenlohe, Philipp Eulenburg and Bernhard Bülow together developed a strange, wholly irrational political philosophy centred on their romantic love for the Kaiser. Neither the state nor the welfare of the nation but their 'completely and utterly monarchist way of thinking' and above all their 'personal love and gratitude for our All-Gracious sovereign' formed the guiding principle of all political thought and action.[171] Eulenburg formulated the central tenet of this philosophy quite openly in a letter which he wrote to Kuno Moltke on 15 June 1895, the seventh anniversary of Wilhelm II's accession, declaring that as a good Prussian and a monarchist one could 'come to only one conclusion, to give unqualified support *to the Kaiser*. If we do not work towards considering *Him* as the personification of Germany – even if his qualities make it hard work for us! – we shall lose *everything*.' For him personally this 'idealistic' standpoint was the product not only of cerebral cogitation but also of his inmost feelings, Eulenburg confessed, since 'my heart and my emotions lead me to the very place where my logic as a supporter of the monarchical principle and of my King also leads me'.[172] The consequences of this attitude – as Eulenburg also recognised – were not particularly beneficial either to the idea of the state or to the welfare of the German people; indeed the contrary was true. He wrote to Bülow in the summer of 1896: 'The Kaiser has the kind of nature which wishes to experience things *for itself* and not to be guided by information about the experiences of others . . . Of course I shall never make this comment anywhere except here, as everyone would rebel at the idea that the Kaiser wants to find things out *for Himself*; i.e. in some cases even at the expense of the people. But I think that as long as we wish to be and are a monarchy the King's character must be taken into consideration. The republican trend of our times demands that the king conform to a pre-set pattern . . . The Kaiser combines in Himself, more than I have ever been able to observe in any other person, two completely different natures: the chivalrous – in the sense of the finest period of the Middle Ages, with its piety and mysticism – and the modern . . . The *chivalrous* side of the Kaiser is very dominant – perhaps more than H.M. Himself realises . . . If I now transpose the effect of this imperial personality into politics, I am obliged to conclude that H.M. must look for His *natural support* from the Conservatives, from the Conservatives in the good sense – and that is confirmed by looking at Prussian history. In the long run the king of Prussia cannot do without the Conservatives . . . I think therefore that the Kaiser's individuality will find its *equilibrium* in political life on a sensible Conservative basis. Everything about the Kaiser's strong, powerful nature stands in such a remarkable conflict with the domination of Liberal–Progressive or Liberal–Catholic (Centre) thinking that this in itself gives rise to the impression of indecision which all Germany is complaining about.'[173] Guided by these tenets, Eulenburg and Bülow piloted the Kaiser through the stormy waters of the endless government crises of the years 1894–7, thereby bringing him into still greater conflict with his own people. Not until many years later did they realise what a fatal mistake they had made.[174]

CHAPTER TWENTY-ONE

Caprivi's dismissal

THE KAISER AND HIS REICH CHANCELLOR

IF Leo von Caprivi had hoped that his withdrawal from the top post in Prussia would at least leave him freedom of action in the affairs of the Reich, this proved all too soon to be an illusion. In September 1893 the Chancellor made the painful discovery that the Kaiser would not even allow him to choose his closest colleagues in the Reich government. Having heard 'much good' of Count Arthur von Posadowsky-Wehner during a hunting expedition in Posen, Wilhelm insisted on appointing him successor to Freiherr von Maltzahn-Gültz as secretary of state at the Reich Treasury.[1] In political circles the choice of this outsider caused 'general astonishment';[2] it was presumed that one of the reasons why the Kaiser had hit upon Posadowsky was that he mistakenly believed that the Count was of Polish extraction. Wilhelm was still paying court to the Poles, Waldersee wrote, and thought that in choosing Posadowsky he had 'pulled off a skilful coup in this direction'; but the Poles were laughing heartily at this idea, for 'Posadowski [sic] comes from an old Silesian Lutheran family and had nothing Polish about him except the i.'[3] Whatever the truth of the matter, the Reich Chancellor found himself compelled to obey the Kaiser's command. The three candidates whom he had himself put forward for the Treasury after consulting Miquel were simply passed over. As he had never met Posadowsky, Caprivi made enquiries about him from the Oberpräsident of the province of Posen, Wilamowitz-Möllendorff. He could not of course have known that Lucanus, on the instructions of the Kaiser, had taken care that the Oberpräsident's response would be favourable.[4] After a personal meeting with Posadowsky in Berlin the Chancellor declared himself ready on 3 August to entrust him with the Treasury, but following a meeting with Miquel (who had wanted the appointment to go to Aschenborn) Posadowsky himself had second thoughts and asked Caprivi to choose someone else. Caprivi therefore telegraphed the Kaiser, who was taking part in the Cowes regatta with his yacht *Meteor*, asking 'May I tell the Count that Yr Majesty wishes to entrust the office to him and appeal to his patriotism?' 'Yes, most certainly', the Kaiser wired back, whereupon Posadowsky accepted the appointment.[5] But when the Chancellor sent the Kaiser a telegram two days later asking permission to announce Posadowsky's appointment, if necessary, he received the curt reply from England: 'Commission to be sent to me in Heligoland. Announcement without my signature is inadmissible.'[6]

With that the affair was by no means settled, for Posadowsky had not been able to familiarise himself with his tasks and declared that he was not in a position to chair the conference of all German finance ministers which was about to take place at Frankfurt am Main. The Reich Chancellor and Miquel therefore proposed that the handover be delayed until 1 September so that the outgoing Secretary of State, Maltzahn, could still preside at the Frankfurt session. This, however, again met with an insulting rejection from the Kaiser, who sent a telegram from Heligoland, dictated to Kiderlen, insisting that either Posadowsky or Miquel, but on no account Maltzahn, should take the chair at the ministerial conference in Frankfurt.[7] The Chancellor's patience finally snapped. In unusually resolute terms Caprivi pointed out that the conference in Frankfurt had already begun under Maltzahn's chairmanship. Posadowsky could not have presided at the conference, in the first place because he had not yet mastered the complicated business in hand, but in the second place also because he was not yet in the service of the Reich, since the Kaiser had not yet signed his commission of appointment. In Berlin all those concerned had decided on 1 September as the date on which Posadowsky would assume office, so that he had time to wind up his provincial duties and transfer to the service of the Reich. In the meantime Maltzahn would produce a written report on the deliberations in Frankfurt.[8]

On All-Highest orders Kiderlen informed the Reich Chancellor in a letter from Kiel on 12 August that His Majesty had been 'put under duress by the fact that the commission had been laid before him on 8th August, the day of the opening of the ministers' conference in Frankfurt am Main, and He had no longer been in a position, . . . confronted by a fait accompli, to come to a decision as to the chairmanship of the conference. H.M. expressed the view that a *proposal* submitted to Him was pointless if the proposed measure had already been taken . . . H.M. went on to express the hope that in future proposals to be submitted to Him by the Reich offices would be sent in sufficient time to make it unnecessary to start carrying out the measure submitted to the All-Highest decision until *after* the decision of His Majesty'[9] The Chancellor refused to accept this carefully worded reprimand, responding with a telegram which pointed out firmly that the start of the finance ministers' conference had been fixed 'weeks ago'. He, the Reich Chancellor, had 'considered himself justified in appointing the chairman of this purely preparatory meeting on his own initiative'. He had 'by no means mentioned the matter in order to seek [an] All-Highest decision, but only in answer to a question from His Majesty'.[10]

The Kaiser had nevertheless succeeded in getting his candidate appointed to the Reich Treasury and had thereby, for the first time, personally selected the head of one of the internal offices of the Reich. As a result Posadowsky enjoyed a degree of independence within the Reich leadership which further weakened the Chancellor's authority. This became apparent only a few months later when Caprivi reprimanded his new colleague for having 'obeyed a telegraphic order from His Majesty The Kaiser' without informing him, as his superior, in advance. Posadowsky emphatically rejected the charge of 'incorrect conduct in the exercise of his official duties' and asked the Reich Chancellor, 'in view of the . . . manner of the reproof issued to me . . . to have the goodness to recommend to His Majesty The Kaiser my immediate discharge from my office'. As Caprivi was unwilling to enter into any further argument with the Kaiser at this juncture, he had no alternative but to make his peace with Posadowsky,[11] who remained in government office until 1907 – at first at the Treasury and from 1897 at the Reich Office of the Interior – and became a significant driving force in domestic politics in Wilhelmine Germany. But Caprivi's days were numbered.

Posadowsky's appointment coincided with a violent conflict between Wilhelm and his Chancellor-General over a decision made by Caprivi to restrict the scope of the imperial manoeuvres in Baden and Württemberg, which were badly affected by drought. The monarch saw this as an inadmissible infringement by the civilian authorities of the Reich of his prerogative as supreme war lord. Waldersee, having discussed the matter with members of the imperial entourage – he expressly named Hahnke, August Eulenburg, Lucanus and Plessen as his informants – recorded in August 1893 that 'a break-up between the Kaiser and the Chancellor is imminent', as Caprivi's 'slackness' in restricting the manoeuvres in South-West Germany without obtaining the Kaiser's prior approval had gone down 'very badly' with the latter and had 'greatly embittered' him. From his own experience Waldersee knew 'very well that once the Kaiser is in the grip of a bad mood like this, it very easily spreads, especially when it is encouraged'.[12]

How bad relations between the Kaiser and his Chancellor had become after their arguments over Posadowsky's appointment and the Württemberg manoeuvres of the summer of 1893 is apparent from the long letter which Philipp Eulenburg wrote to Caprivi from the *Hohenzollern* after his visit to England with Wilhelm, in order to warn the Chancellor of the 'ill-feeling' towards him which the Kaiser had begun to display. Caprivi's proposal to appoint the moderate Centre leader Freiherr von Huene as secretary of state at the Reich Treasury had already greatly annoyed the Kaiser, Eulenburg reported. He went on: 'Although this matter was settled in accordance with His Majesty's wishes by the appointment of Count Posadowsky, it still irritates him now. The attitude adopted by the Centre in the Reichstag has put His Majesty in such a thoroughly bad mood that the Catholics will be "excommunicated" for a long time to come. His Majesty's dissatisfaction is related to this attitude.' But Wilhelm's annoyance at the abandonment of the Württemberg manoeuvres was more serious still, Eulenburg warned. This affair 'has made His Majesty . . . very angry indeed and he strongly resisted all attempts on my part to persuade him to take an objective view. His Majesty shares General von Hahnke's opinion in this matter and also has the support of the Grand Duke of Baden. The fact that the imperial authority in Württemberg was undermined by the changes made independently [by the civilian Reich authorities] to the arrangements for the manoeuvres, and the Foreign Office's handling of this matter, which His Majesty and his military suite consider as the inviolable domain of the army – these are two constantly recurring themes which arouse His Majesty's anger towards Your Excellency and which particularly affect Herr von Marschall too. I had hoped that the week at Cowes would mitigate these impressions. But this was an illusion. His Majesty returned to them with undiminished anger . . . His Majesty's anger . . . reached such a degree and his language was so forceful that I felt compelled to remind His Majesty very seriously of the dangers which could arise from such anger and from the possibility of a violent argument with Your Excellency.' Eulenburg strongly urged Caprivi not to think of resigning in his forthcoming meeting with the Kaiser. He had pointed out to the Kaiser, he wrote, 'that a sudden chancellor crisis just as he returned from a pleasure-trip would certainly cause an astonished Germany to put all the blame on him. I likewise drew His Majesty's attention to the fact that Your Excellency might say, by way of clearing up the ill-feeling prevailing between you, that government business had to go on when His Majesty was on pleasure-trips. Your Excellency does not need to be told that this unleashed a storm of indignation, but on this occasion too His Majesty kindly treated me with his habitual consideration. Now that I have given you this account of the state of affairs, which is strictly accurate, Your Excellency will be in a position to prepare yourself for "what is to come" and – I venture to express this fervent plea with the greatest respect! – I hope that Your Excellency, having all the facts at your disposal, will be able to steer clear of the rocks which threaten the Kaiser and us all if a storm should suddenly provoke the chancellor crisis which we dread . . . A conversation I have had with His Majesty today has left me the impression that my arguments have softened the intensity of

His Majesty's feelings. May God grant that it is so. I cannot bear to think what dangerous consequences for Germany and the Kaiser there would be if an acute crisis were suddenly to arise!'[13] Eulenburg was relieved to hear from his cousin the Oberhofmarschall on 17 August 1893 that the meeting between the Kaiser and Caprivi had passed off without a chancellor crisis. 'The interview with the Reich Chancellor went off well, fortunately, so that both parties are satisfied, although for very different reasons', August Eulenburg reported. 'H.M. says that the Chancellor said his "pater peccavi" [Father, I have sinned], while the latter maintains that H.M. did not have the courage even to mention "these trifles, which were really not worth a crisis", so that he, the Chancellor, had had to take the initiative himself.'[14]

The Kaiser's extreme anger over the affair of the manoeuvres had by no means subsided, however. On 16 September 1893, when the Chancellor was on the point of departing on leave, he was held back by a telegram from the Kaiser 'reproaching him in the most violent and impolite way for this alleged misdeed of his'.[15] After an encounter in October 1893 with the Finance Minister von Miquel and the Oberhofmarschall August Eulenburg at the unveiling ceremony of the memorial to Kaiser Wilhelm I in Bremen, Waldersee noted that the Kaiser had in fact 'inwardly' broken with Caprivi.[16] The ill-feeling was by no means one-sided, for the Chancellor was now also beginning to complain loudly about the difficulties he had to face because 'one could almost never be sure of the Kaiser; if he approved something, one could never be sure that a Flügeladjutant or a Cabinet chief would not arrive a few hours later with a different expression of his will'.[17] In the summer of 1892 Caprivi had still been able to say: 'Even if I am angry, when he (H.M.) looks at me kindly with his fine eyes, my anger disappears.' But this mood had long since been replaced by quite a different one.[18] From what he heard from Berlin, Waldersee accurately concluded that the Chancellor too was now 'in a highly irritable state'.[19]

It is difficult to imagine the cold arrogance with which the Kaiser had treated his Reich Chancellor since the School Bill crisis, and even more since the clash over the Württemberg manoeuvres. His hostile marginal comments on a report by Caprivi of 30 November 1893 give us an unusual insight into the relationship between the two men. Caprivi, who was after all constitutionally responsible for directing foreign policy, commented on a report by the military attaché in Rome, Colonel von Engelbrecht, on the Italian financial crisis. The monarch's scornful marginalia reached their peak in his response to Caprivi's offer to explain his own views in audience. 'No thank you!', Wilhelm wrote; he was 'perfectly well aware' of the Reich Chancellor's views.[20]

Waldersee saw a further sign of the growing estrangement between Wilhelm and Caprivi in the appointment of General Walter Bronsart von Schellendorf as Kaltenborn's successor as Prussian minister of war, for according to August Eulenburg the choice of this fiery General was not only made without the Chancellor's agreement, but 'was in fact even directed against him'. Waldersee found it incomprehensible that the Chief of the Military Cabinet should have recommended this choice, for Hahnke surely could not have wanted a strong minister of war. 'But the Kaiser will also realise that he will have to treat the Minister of War differently from now on', Waldersee predicted.[21] He was astonished to hear from Bronsart himself how much pressure the Kaiser had put on him to accept the War Ministry. Bronsart, 'after repeated attempts to resist, had had no alternative but to accept, after the way the Kaiser had appealed to him'.[22] The former Chief of the General Staff foresaw serious conflicts, since he could not imagine that Bronsart would accept 'the present state of affairs'.[23] And indeed it was not long before serious friction arose between Wilhelm and this General, who, unlike his two predecessors, was not prepared to suffer ill-treatment by the Kaiser in silence.

The Kaiser's arrogant lack of regard for the Reich Chancellor was conspicuous in other actions, as for instance in January 1894, when after discussing the situation in Cameroon in the officers' mess at Kiel he appointed a certain Captain von Natzmer of the Guards Rifles as governor of the colony on the recommendation of one of the officers. Out of the blue, on 2 January, the Chancellor received an All-Highest cipher telegram which left him and the Foreign Secretary speechless. It read: 'The recent events in Cameroon have once again proved the old, well-tried theory that civilian governors are not suitable for African conditions. The complete transformation that has taken place in East Africa since Herr [Colonel] von Schele arrived confirms this evidence. It is high time things changed in Cameroon too. Neither Governor Zimmerer, who is hated in all commercial circles as a loudmouth, nor a born subaltern like Herr Leist is capable of being of lasting use in Cameroon, with the breadth of vision required, as the population has virtually no respect at all for them. We also have to rely too much on force to be able to dispense with a military governor in Cameroon as well as elsewhere for the time being. In addition, I have heard that Leist has been guilty of a serious crime against morality by allowing gross immorality with Dahomeian women, which has completely destroyed his authority among the whites (above all the commandant of the garrison and his people) and the blacks. The new military governor must therefore go out immediately with the company of Marine Infantry which I am sending there. The most suitable candidate is Captain von Natzmer of the Guards Rifle Battalion, who has already been listed as a possible governor of East Africa because of his outstanding qualities. I have directed him to report to the Foreign Office immediately. Leist must be recalled at once. Wilhelm I. R.'[24] On investigation, Caprivi and Marschall discovered that the

Kaiser had heard both through the Chief of the Naval Cabinet and through his brother Prince Heinrich that the colonial troops from Dahomey, former slaves whose freedom had been purchased by Germany, were not paid any wages until they had 'served out' the equivalent of the purchase price, and were therefore compelled to make their wives earn money through prostitution.[25] On 2 January the Foreign Secretary noted in his diary 'The R[eich] C[hancellor] and I decided to go and see H.M. this afternoon and if necessary threaten to resign.'[26] In the end the two statesmen managed to convince the Kaiser that a detailed investigation into the accusations against Leist must be instituted first, and that to this end the Foreign Office should send a commission of inquiry out to Cameroon. They also persuaded him not to appoint Natzmer governor 'for the time being'.[27] Wilhelm was 'very gracious' on this occasion and cancelled Natzmer's appointment, Marschall recorded, but the confrontation had been 'necessary'.[28] The dispatch of a ship with a detachment of marine infantry, which the Kaiser had ordered at the same time, was not countermanded. 'The Kaiser has had his way after all', Waldersee noted tersely.[29] When Marschall remonstrated with Admiral von Senden for having 'gossiped to H.M. about affairs in Cameroon', he was rebuffed. 'He denies saying anything.'[30]

On the pretext that the Foreign Office was incompetent, Wilhelm II continued to intervene peremptorily in foreign policy and in diplomatic appointments. Philipp Eulenburg reported to his friend Bernhard von Bülow in the spring of 1894 that the Kaiser had 'thrown overboard *any* consideration' for the Foreign Office 'and only speaks of it with disdain'. He would not make any changes in its administration, however, because he 'derives increasing pleasure from deciding on policy matters *himself*, and sees justification for this in the "incompetence of the house"'.[31] Holstein too complained of the Kaiser's growing 'high-handedness' and observed that 'from his [Wilhelm's] point of view the best argument for keeping Caprivi must be that *no* successor to Caprivi would allow His Majesty so much freedom'.[32] The deep anxiety caused in the Foreign Office by an unsupervised exchange of telegrams between Wilhelm and the Tsar in late 1893 was exemplified by one diplomat's observation that it would be highly desirable 'if our All-Gracious sovereign were accompanied by a diplomatic adviser even on short journeys, or if he made it a strict rule never to send off politically important telegrams of that kind without previous checking by the Foreign Office'.[33] But as the disastrous 'Willy–Nicky' correspondence, which began only a few months later, was to prove, the Kaiser would not have dreamt of subjecting his correspondence with a foreign monarch to any form of official inspection.

The increasingly autocratic behaviour of the Kaiser in matters of foreign affairs was also repeatedly demonstrated in the appointments and dismissals of ambassadors and envoys, which as a rule went against the wishes of the Reich Chancellor. What Caprivi did not begin to suspect for some time (and even then only dimly) was the humiliating truth that in making these dispensations from above the Kaiser was often merely carrying out suggestions for appointments which had been secretly urged on him by Philipp Eulenburg, and which as often as not originated in the Political Department of the Foreign Office; the choices, in other words, of the Chancellor's direct subordinates. This strange mechanism can be seen operating in a whole succession of diplomatic appointments following the School Bill crisis. In one instance Marschall, Kiderlen-Wächter and Holstein wanted General Bernhard von Werder, who was on good terms with Tsar Alexander, appointed ambassador in St Petersburg in succession to Lothar von Schweinitz. Caprivi, however, was against the appointment of this elderly, unmarried General who had no particular diplomatic talent. In order to circumvent the Reich Chancellor's opposition the three top Foreign Office officials decided to persuade the Kaiser, through Eulenburg, to declare himself in favour of Werder. Eulenburg was to hint to Wilhelm that the Tsar had expressed a wish for Werder to be appointed.[34] The plot hit a snag, for the Kaiser was so determined to send Count Alvensleben to St Petersburg as ambassador that Eulenburg's question as to 'what the attitude of His Majesty might be if a particular wish of the Tsar had been received – concerning General von Werder, perhaps – provoked a very sharp rebuff'.[35] Only a few days later, however, the Kaiser telegraphed the Chancellor saying that the Russian ambassador Count Shuvalov had read him a telegram from the Tsar asking for Werder as ambassador. 'Our Kaiser replied that the most important thing for him was that his ambassador had the confidence of the Tsar', Holstein wrote triumphantly to Eulenburg. Werder was appointed.[36] The intrigue successfully over, a relieved Eulenburg was able to report to Bülow: 'After some anxious moments I heard that His Majesty had at once yielded to the wishes expressed and had now fully realised (almost *too* much – for we love superlatives!) the advantage we have gained from the present situation.'[37]

If the Reich Chancellor was duped and the Kaiser manipulated by the Holstein–Eulenburg team in this case, the type of intrigue they practised can be seen even more clearly in the dismissal of the ambassador in Rome, Count Solms, at the end of 1893, and his replacement by Bernhard von Bülow, who had manoeuvred himself into the good graces of the Kaiser's favourite. After Bülow had drawn attention to his own suitability for the post (he was married to an aristocratic Italian lady whose mother, he claimed, was King Umberto's best friend),[38] Eulenburg wrote to Wilhelm II suggesting that Solms be dismissed and Bülow appointed in his place. 'The critical remarks which Your

Majesty made to me in Berlin about Ambassador Solms', he wrote, 'have made me wonder whether Your Majesty is perhaps intending to prepare for a change in this post on the occasion of Your visit to Rome. All kinds of thoughts have occurred to me which I should like to lay before Your Majesty with my most humble duty for Your consideration: a kind of box of tools to be used in any way You wish. The question is, who is to be the successor. In my most humble opinion there is only one person, although Your Majesty has chosen him for Paris: Bernhard Bülow. For the time being old Münster with his long legs, long strides and booming voice can still play the part of the aurochs [extinct wild ox] in the monkey park so well that Bernhard could go to Rome *en attendant*.'[39] When, during the Kaiser's visit to Rome, Count Solms omitted to inform him of a garden party to be held by an English peer living there, the stage was set for a change of scene. Eulenburg wrote to the Kaiser again giving him advice for the audience which the Reich Chancellor was about to attend. 'May I most respectfully suggest to Your Majesty that You might defuse what would be a highly unpleasant crisis at this moment by *Yourself* bringing up the subject of Italy *first* and saying: "Solms will not do any longer – once he has gone matters will certainly improve." I have no scruples in giving Your Majesty this advice, because Solms has already sent his wine-cellar away – so he is beginning to settle his affairs in Rome. This turn of events puts the whole matter in quite a different light. Your Majesty knows whom *I* would recommend for Rome.'[40] The Kaiser answered at once, exactly as Eulenburg had hoped: 'All right! I have duly put paid to Solms, with the Chancellor's approval, and *Bülow's* appointment to Rome is to be announced!'[41] To take Bülow's place in Bucharest, at Holstein's urgent request Caprivi had given the Kaiser the choice of Count Casimir von Leyden or Count Anton Monts, which drew from Wilhelm II the response: 'I prefer Leyden; Monts is such a pessimist.'[42] Naturally Leyden was appointed. Malet reported perspicaciously to London that the ruthless dismissal of Count Solms and the surprising appointment of Bülow as ambassador in Rome had not come about through a decision by the Reich Chancellor or the Foreign Secretary, whose constitutional responsibility it was, but had been 'entirely due to the Emperor'.[43]

The Chancellor was understandably offended, not only by the manner of the ambassadors' appointments but also by the private correspondence which Wilhelm II conducted with some of them. But his protests had little effect.[44] Early in 1894 the Kaiser expressed his regret for having betrayed to the Chancellor a letter he had received from Bülow, a mistake he would never normally make. He resolved to be more careful in future, sending a message to Bülow through Philipp Eulenburg: 'But he *commands* you to write to Him whenever you consider it necessary. His Majesty has a similar arrangement with Hatzfeldt and He said that if He did not maintain direct contact with certain of His most distinguished diplomats in this way, He would soon be landed in trouble by the Reich Chancellor and the Secretary of State.' 'So you see', the Kaiser's best friend commented to the future Secretary of State and Reich Chancellor, 'I was right when I told you that your letter-*intermezzo* has had the desired effect.'[45]

Not only did the Kaiser continue to correspond with Bülow, but he gave him advance notice of his plan, which was to be kept 'absolutely secret from the Foreign Office', to meet King Umberto in Venice to discuss the question of a reconciliation between the Italian state and the Vatican, in which he wished to play the 'honest broker'. By taking this initiative, 'which could become a new milestone in the peace-making policy of the German Kaiser', Wilhelm hoped to strengthen monarchical feeling in Italy, reduce the French, American and Jesuit influence in the curia and – not least – improve his relationship with his own Catholic subjects. As he wrote to Bülow on 28 March 1894, shortly before his departure for Abbazia, 'It must not be forgotten that the idea of reconciliation could easily be seized upon by other people and either exploited for their own benefit or, as in the case of Gaul [i.e. France], undermined. Hence the wish that the German Kaiser might perhaps bring it about. He would also put himself in a very strong and impressive position vis-à-vis his Centre and fanatical priests in Germany if he were successful in this work of reconciliation. I must ask you to treat the Venice matter as *completely secret*, and also to make sure that the same is done on il Re's [the King's] side. I shall visit Pola and can dash over to Venice from there. I would inform il Re about the little trip in advance, and then he could also make a chance visit there without indicating any reason, and so we would meet! W.'[46] As Eulenburg, who accompanied the Kaiser to Abbazia, reported to his intimate friend Bülow, the Kaiser had become so excited in thinking about the idea of reconciliation 'with his characteristic enthusiasm' that the political advantage to be gained from it had become 'a shining vision' to him. 'In fact H.M. is like a child looking forward to the trip to Venice and to politicking with King Umberto and the two of us!' But Eulenburg was at a loss as to how the affair could be managed 'without creating an enormous uproar in the Foreign Office'. And besides: 'Is it even possible for H.M. to interfere in Italy's internal affairs in this way?'[47]

A more striking example still of the Kaiser's use of his personal power to make and unmake appointments was the replacement of Heinrich VII Prince Reuss, ambassador for many years in Vienna, by Philipp Eulenburg himself, whose appointment as Prussian envoy in Munich in 1891 had already been widely seen as an act of favouritism by Wilhelm II. With fine irony, but also hitting the nail on the head, Reuss wrote to Holstein on 3 January 1894: 'When one no longer has the confidence of one's sovereign, one can no longer render such good service as before. I wish my successor, who is armed with this invaluable possession, success in gaining the trust of the people here [in Vienna] so that the *cause* does not suffer.'[48] Eulenburg's influence on the Kaiser can be clearly seen in the fact that to all intents and purposes he picked his own successor as envoy in Munich, Max Freiherr von Thielmann. When he had second thoughts about his choice and wrote to the Kaiser, 'Your Majesty is quite right in describing Thielmann as unsuitable for Munich. His Jewish extraction and abrupt manners make him best fitted for a post where the nobility does not play such an important part as here [in Munich]', it was too late to rescind the appointment. The Kaiser telegraphed his friend to say that his letter had arrived half an hour too late. 'Everything has already been settled in accordance with your earlier suggestion! Alea jacta sunt!'[49]

The high degree of dependence of the leading German statesmen on the wishes and moods of the young Kaiser is further exemplified by the state of insecurity into which they were thrown when it was rumoured in January 1894 that Wilhelm II had had a conversation lasting an hour and a half with General Paul von Leszczynski. The rumour alone was enough to cause speculation that the General might be appointed Reich chancellor. In Waldersee's opinion, if the rumour of a meeting were confirmed 'it would certainly be significant and the Kaiser must have some plan for the General'. Leszczynski was undoubtedly a very competent and experienced general but as far as politics were concerned he was 'highly questionable'. Waldersee had 'often heard him talking complete nonsense; his Polish blood is all too obvious, he is very lively but always unsteady'. But he could talk 'like a book – and that is enough to impress the Kaiser'. 'I cannot think of him as a candidate for the chancellorship; in my view that should be an impossibility', but 'unfortunately these days one must always be prepared for such things'.[50] If a mere conversation between Wilhelm and a rather insignificant General could give rise to rumours of an imminent change of chancellor, anything that could be taken for a sign of a rapprochement between the Kaiser and the Bismarcks came close to provoking complete hysteria in the Wilhelmstrasse.

THE 'RECONCILIATION' WITH BISMARCK

The attacks made by the Bismarcks not only on the men of the New Course but also on himself were so insulting to the Kaiser that in the summer of 1891 he was still uttering furious threats that he would set 'his colleagues, the field marshals and commanding generals' on to the old 'Iron Chancellor' and have him stripped of his uniform. He was reported to have declared to the generals that 'if Prince Bismarck continued to behave as hitherto his patience would soon be at an end and he would not hesitate to send him straight to Spandau [prison] if necessary'.[51] At this time a reconciliation with the former Reich Chancellor still seemed absolutely unthinkable. But only six months later Friedrich von Holstein felt compelled to warn strongly against the Kaiser visiting Friedrichsruh. He suspected the Flügeladjutanten Gustav von Kessel and Count Carl von Wedel, the Adjutant-General von Wittich and the Kaiserin's Hofmeister Hugo Freiherr von Reischach of having persuaded the Kaiser that he was so strong and powerful that he could take the first step towards reconciliation with the Bismarcks without loss of face. But such a step, Holstein argued, would be seen by the nation as submission rather than reconciliation, and Wilhelm would thereby be subjecting himself to Prince Bismarck's authority. 'Can you see the Princess, Herbert, the Rantzaus and the Bills moving into 77 [Wilhelmstrasse] again?', Holstein asked Eulenburg. Kessel would then become 'senior medical attendant with a grand title' and Reischach would be made an ambassador.[52]

Another six months later the Wilhelmstrasse officials were panic-stricken when it became known that Prince Bismarck was planning to travel to Vienna for his son Herbert's marriage to Countess Marguerite Hoyos, taking advantage of this occasion to make a 'Greater Germany' tour by visiting Kaiser Franz Joseph, King Albert of Saxony and Prince Regent Luitpold of Bavaria. Under pressure from Caprivi and Marschall, who in turn had been urged to take a firm line by Holstein and Kiderlen-Wächter, Wilhelm wrote to Franz Joseph in mid-June 1892 asking his 'dear friend' not to receive the disgraced former Chancellor. At the same time Prince Reuss, the ambassador in Vienna, was instructed not to take any notice of the Bismarck wedding. Kiderlen described the Kaiser's letter to Franz Joseph as '*excellent*' and commented triumphantly that it was 'His Majesty's own work, for the pusillanimous Reich Chancellor would never have advised such a thing'.[53] In this letter Wilhelm reminded the Austrian Emperor that a 'major coup' achieved by Bismarck 'was the secret treaty – à double fond – with Russia, which was concluded behind Your back and which I dissolved'. 'Since his retirement', Wilhelm continued, 'the Prince has waged war on me, Caprivi, my ministers &c in the most perfidious manner . . . Incomprehensible though it is, he launches his fiercest attacks against the *Triple Alliance* (his own creation, of which he is so proud) and above all against our resolute solidarity with You and Your excellent people. His outrageous attitude towards Your country in the matter of the trade treaties is still too well known to waste words over it.' Bismarck's apparent 'desire for reconciliation' with Kaiser Franz Joseph, Wilhelm declared, was nothing but another 'humbug' on his part 'which is aimed only at arousing the love of sensation and the curiosity of the stupid masses'. Bismarck, the monarch complained, was 'as artful and cunning as can be in trying to turn things round so that it seems to the world that I am the one who is making the first approach'. With this purpose in mind Bismarck had devised the idea of an audience with Kaiser Franz Joseph as the 'climax of his programme', Wilhelm wrote. 'Ignoring my court and the Kaiserin in the rudest way he takes himself off to Dresden and Vienna to present himself there at once and play the part of the loyal old man.' Wilhelm therefore asked the Austrian Emperor, 'as a loyal friend', not to put him in a difficult position by receiving 'my disobedient subject' 'before he has made an approach to me and said his peccavi'. Wilhelm had made it clear to the people who had always been ready to mediate between them that he expected 'an unambiguous letter' from Bismarck, 'in which he asks to be received back into favour. I would not commit myself to anything before that.' As the Prince was still very far from taking this first step Wilhelm asked Franz Joseph not to receive him.[54] On 16 June 1892 the Austro-Hungarian Foreign Minister Count Kálnoky informed the German ambassador in Vienna that Kaiser Franz Joseph had refused to grant an audience to Bismarck and Herbert.[55]

The Bismarcks' campaign seemed to the statesmen in the Wilhelmstrasse all the more dangerous because even their former arch-enemy Waldersee, after a visit to Friedrichsruh, took up the cause of reconciliation between Wilhelm and Bismarck and, declaring that his 'sole endeavour' was to serve his All-Highest sovereign, proposed himself as Reich chancellor at this critical moment.[56] Caprivi hurried to Potsdam to persuade the Kaiser not to yield to this pressure for reconciliation. In his presence Wilhelm immediately wrote out a reply to Waldersee's initiative: 'Dear Count, The rumours which are sprouting like mushrooms in the press everywhere, alluding in various ways to "reconciliation" or "rapprochement" etc. between the Prince and Myself, make it seem advisable to come back to what you suggested to me in Kiel. You expressed the opinion that the Prince was tired of his campaign and yearned to be on pleasant terms with me again. But to "jump the ditch" – as the cavalry say – was very difficult for him. You said, however, that you were quite convinced that when the right time came for him you would be able to make it easier for him by offering to mediate. I should like to repeat, absolutely precisely, my answer at that time: that it was all very well, but the first step must be taken by the Prince, whatever happens. He must formulate his request or wish to be permitted to renew relations with me quite *unambiguously, in writing, directly to Me*. I shall not agree to anything else. You must keep this in mind as an unalterable principle. With many good wishes to your wife, Your sincerely devoted King Wilhelm R.'[57]

The Kaiser's displeasure at this escalation of the Bismarckian Fronde lingered on for some time yet. During the Scandinavian cruise that summer he complained to Eulenburg that it was shameful 'that the Kaiser is obliged to fight so hard against my former Chancellor and my Ambassador, who has been playing an infamous role'; Prince Reuss's days were therefore also numbered, he said.[58] In the Bismarck camp too hatred for the Kaiser and the men of the New Course reached new heights. In 1893 Philipp Eulenburg described the 'terrible fury of Herbert, Bill and Rantzau' as having given them a new family motto: 'May the blood of the Old Man be avenged on the Kaiser and his children.'[59] Faced with a growing pro-Bismarck cult, Wilhelm and his advisers were obliged to put on a semblance of goodwill in order to prevent the Bismarcks from getting the better of them. Thus there were earnest discussions behind the scenes as to how the Kaiser and the government should react to the death of the 'Iron Chancellor' in order to minimise the danger of the Bismarck family exploiting what would be an event of national importance to attack the crown. The idea of offering the dying Prince Bismarck an apartment in one of the Berlin palaces was rejected since it was clear, as Caprivi said, that 'the Prince will then become a political focus against which no power in the world can prevail'. Instead, on Caprivi's advice, the Kaiser sent a telegram from Hungary on 19 September 1893 offering Bismarck the imperial palace in Wiesbaden for the winter. The latter expressed his thanks but declined the offer.[60] The following month there was a further exchange of letters between Wilhelm II and Bismarck, which was seen as an attempt at rapprochement on the part of the Kaiser.[61]

The question of whether the Bismarcks possessed letters from Wilhelm 'which could be dangerous for His Majesty in the hands of [the journalist] Maximilian Harden' was particularly highly charged.[62] 'The most awkward and difficult question, which we must discuss, is the possible – perhaps even probable – *misuse of imperial correspondence* by the Bismarck family', Eulenburg wrote to Holstein on 7 October 1893. 'I spoke to His Majesty about it yesterday and asked him to tell me absolutely precisely what may have remained in the family's hands. The Kaiser spent a long time going over all the phases of his dealings with the Prince with me. There is apparently *nothing* for the time of Kaiser Friedrich (the most suspect to me!), as His Majesty firmly maintains. There were, however, dubious points concerning private letters from Prince Wilhelm from St Petersburg. These communications are of a *very* intimate nature – among other things they deal with the subject of the Prince *offering* Constantinople to the Russians. I doubt that the Prince has deposited the correspondence in the official files. But I do not doubt that the well-known, vigorous tone of the Kaiser's style will cause a quite extraordinary sensation in the world and will drive Austria, which at present is on better terms with Russia than we are, *even more* into Russia's arms out of retrospective shock.' The Kaiser's friend went on to consider how the letters could be recovered from the Bismarcks.[63]

In view of the vehemence with which the Kaiser rejected these attempts at rapprochement and mediation in the summer of 1892 and of the difficult discussions the following autumn about the dangers which the death of the founder of the Reich might bring for the monarch, the volte-face which Wilhelm performed in January 1894 when he ceremoniously received Prince Bismarck in the Berlin Schloss on the occasion of his thirty-fifth birthday was quite extraordinary. His action shows not only how successfully the Bismarckians in his entourage – his brother Prince Heinrich, his uncle the Prince Regent Albrecht of Brunswick, the Flügeladjutant Gustav von Kessel, the Oberhofmarschall August Count zu Eulenburg and his brother Botho, as well as the Finance Minister Johannes von Miquel – had worked to bring about the reconciliation, but also, and above all, how little the Kaiser cared by this time about keeping Caprivi's government in office. In April 1893, with the appointment as Flügeladjutant of Count Kuno (Tütü) von Moltke – who was an intimate friend of both 'Phili' Eulenburg and the new Württemberg envoy in Berlin, Axel Varnbüler, and the uncle of the painter Franz von Lenbach and thus on the best of terms with the Bismarckian clique in Munich[64] – yet another admirer of Bismarck entered the imperial entourage. He was to play a key role in the dramatic reconciliation of 26 January 1894.

As an unmistakable sign of his dissatisfaction with Caprivi and his advisers in the Wilhelmstrasse,[65] the Kaiser sent Kuno Moltke to Friedrichsruh on 22 January with a bottle of vintage Rhenish wine to invite the former Reich Chancellor to visit him in Berlin on his birthday. No sooner had Prince Bismarck accepted the invitation than Wilhelm II sent a telegram announcing the fact to his grandmother but completely distorting the truth of the matter: 'The congratulatory message I sent to Friedrichsruh through my Aide de Camp so touched the old Prince Bismarck that he immediately sent me an answer begging to be allowed to pay his respects to me tomorrow and to congratulate me on my jubilee. Am so thankful that this at last is possible.'[66] When the Foreign Secretary called on the Kaiser and Kaiserin on the morning of 23 January the Kaiser triumphantly showed him a letter and asked 'Whose handwriting is this?', whereupon Marschall replied 'Prince Bismarck'. The Secretary of State noted with dismay that 'H.M. has arranged this himself without any consultation.' At Marschall's request Wilhelm at first agreed that the news should be announced by the official government, but then sent Kuno Moltke to the Foreign Office 'with the Kaiser's orders that it should all be kept secret.' 'That will not do', the statesman commented, adding 'I go and see H.M. with Moltke and express my opinion strongly and frankly that the government itself must make the announcement so as not to appear to be put out and give the Bismarcks the chance to exploit the situation. Moltke contradicts. Rather a violent scene. H.M. gives way. 1 o'clock to the Reichstag. I tell everyone about it. Great joy.' Two days later, however, August Eulenburg came to the Foreign Office 'in rather an agitated state' with an article from the *Berliner Tageblatt* which alluded to collaboration between the Oberhofmarschall and the Foreign Secretary in the reconciliation with Bismarck. 'H.M. was furious about it', Eulenburg explained, 'and said it was to be denied in the *Reichsanzeiger*.' 'Done', Marschall noted tersely.[67] On the same day an announcement appeared in the official gazette indicating that Flügeladjutant von Moltke had been sent to Prince Bismarck on the Kaiser's own personal initiative and that even in government circles no one had been aware of the All-Highest 'magnanimous decision'. With that the disavowal of the Reich government was complete. In such circumstances, Prince Chlodwig zu Hohenlohe-Schillingsfürst commented in his journal, he would not wish to be the Reich chancellor. Barely ten months later, nevertheless, he was.[68]

It was characteristic of the lack of a sense of political responsibility in the Wilhelmstrasse that none of the government officials who had been exposed by the monarch in this way thought of resigning. Marschall was even relieved to be able to record, after conversations with Wilhelm and Philipp Eulenburg in the next few days, that 'H.M. is not thinking of making any changes in appointments.'[69] The Kaiser's favourite knew better, however, and recognised this clash between the Flügeladjutanten and the leading statesmen as the defining moment in which Marschall had irrevocably lost the monarch's confidence. The Kaiser told Eulenburg bluntly on 26 January 1894 that the Foreign Secretary had behaved so badly 'that my gentlemen thought he could not be kept on for much longer'.[70] Kuno Moltke wrote to Eulenburg describing Marschall as 'crazy and *insolent*'; he was no Prussian and therefore could not see German matters from a Prussian point of view. The Kaiser's standing among the people had changed overnight as a result of the reconciliation, Moltke enthused, going on to claim that 'the cheering of the people was tremendous yesterday [Wilhelm's birthday] again when the Kaiser went to give the password, and he realised that the crowds were not just a yelling *canaille*'. 'When I think of him tears still come to my eyes; in unselfish love for him let us rejoice as old friends and as loyal children of our Fatherland.'[71] Against such blind adulation the grey bureaucrats at their green baize desks in the Wilhelmstrasse stood little chance.

Waldersee was overwhelmed by this 'great event' and found it difficult to form a clear picture of what it meant. A reconciliation between the Kaiser and Bismarck must include Herbert Bismarck; but that was only possible, in his view, in conjunction with a 'complete change of system'. 'Might Caprivi really step down?', he wondered in some agitation. 'It may still be simply that the Kaiser hopes to win the Prince round by showing him great courtesy, and to persuade him to keep quiet.' Even then the ovations in the streets of Berlin would make a great impression on the Kaiser, the General wrote, for 'the old man is now

29. Prince Bismarck, supported by Prince Heinrich, in Berlin, January 1894

the popular one, while Caprivi is the dull philistine who does not know how to win people's hearts. How nervous they must be in the Wilhelmstrasse!' Abroad, where people had never understood why the Kaiser had got rid of such an asset as Bismarck, the reconciliation would be regarded as a perfectly natural step, Waldersee observed, while in Germany there were forced smiles to be seen among the Liberals, who were most anxious that Caprivi's position should not be undermined. Waldersee himself believed, however, that the Kaiser had 'finished' with Caprivi; all that remained was to arrange for the succession. Caprivi must get the Russian trade treaty through the Reichstag, but thereafter the Kaiser would certainly let him go. 'The Kaiser cannot be good friends with him and with

Bismarck and family at the same time', Waldersee commented, for 'the one excludes the other'. The time had clearly arrived for momentous decisions.[72] From a 'very subterranean source' Waldersee heard of a 'violent scene' which the Kaiser was said to have had with his mother in the Berlin Schloss. She had reproached her son 'for having sent a message to Bismarck without saying a word to her about it, and it was in any case much too soon for a reconciliation. The Kaiser is said to have replied that he needed Bismarck to get the Russian trade treaty through, and moreover circumstances were such that Caprivi would soon leave. The servants claim to have heard all this, in the Schloss itself; if it were really true, it would be a nice sign of the reliability and loyalty of these people', the General commented sarcastically.[73]

Several times in the course of these crucial days Wilhelm showed by deliberately slighting Herbert Bismarck that the reconciliation did not in fact include him, but only his father. At the investiture held at the Schloss on 21 January the Kaiser completely ignored his former Foreign Secretary, confidential adviser and drinking companion, which was all the more noticeable since numerous other royal personages, court officials and ministers treated Herbert with conspicuous friendliness. Observers were particularly struck by the repeated attempts which Prince Heinrich, Prince Regent Albrecht and the otherwise reserved August Eulenburg made to bring the younger Bismarck up to the Kaiser during the *cercle*. Although the Kaiser 'very obviously' avoided him, however, these attempts at a rapprochement at court were generally interpreted as the beginning of the end of the Caprivi era.[74] 'My appearance at the Order celebrations caused much excitement', Herbert reported to his father immediately afterwards. 'The Kaiserin spoke kindly to me, I had very long and friendly conversations with Prince Heinrich and Albrecht, both asked me to send you their greetings. During the *cercle* I stood in the last window embrasure, and was fetched from there on the instructions of S. Kanitz and A. Eulenburg; "I was to stay in the front row as H.M. wanted to see me." I followed and was asked twice more by both of them to stay at the front, "the Flügeladjutant knew about it" – but in the end H.M. was too embarrassed, he spoke to 4 or 5 indifferent people, 7 or 8 steps from me, talking to each for a conspicuously long time, and then turned to the opposite side from me on his way out of the hall. Botho Eulenburg greeted me with very noticeable warmth, but it was Miquel who compromised himself most with me. He talked very loudly, sent you greetings and wished you would come back, and showed contempt for the present leadership. With everyone so close together a great many people must have heard him.'[75]

If Herbert interpreted the Kaiser's conduct as embarrassment on this occasion, Wilhelm's insulting behaviour on his birthday – the day after he had received the ex-Chancellor at the Schloss – left no room for doubt that his disregard for the younger Bismarck was deliberate. On 28 January Herbert sent another report to Friedrichsruh: 'Yesterday after I had taken part in the *défilé* at court I waited for more than 20 minutes in the White Hall in case H.M. wanted to hear about your journey home. Everyone else left the Schloss immediately after their march-past. As the place was thinning out very much by now, I asked 2 friendly gentlemen of the court, of whom I sent one to A. Eulenburg, whether I should wait any longer. The answer came: "no, you are already attracting attention by staying, and if H.M. passes you later, when there are only royalties in the hall apart from you, you will be in trouble". A. Eulenburg undertook to tell H.M. that you arrived home safely, and that I would be at the gala opera performance. – During the very long interval at the opera I at first stayed behind, and various royalties spoke to me in a very friendly manner (the Albrechts, the Hermann Weimars, Hereditary Prince Meiningen etc.); the one who asked the most and was the most gracious was the King of Saxony; all sent you warm greetings. Afterwards I stood 3 or 4 steps away from the Kaiser, who spoke to Aug. Dönhoff for at least ¼ hour . . . then shook the hand of some nonentity standing in front of me and turned away. In the same way H.M. walked straight past me when he returned to his box . . . The impression it gave was that I was to be cut dead. Presumably H.M. thinks that *you* can be of use to him in his present difficulties, but that it makes no difference if I am badly treated.'[76]

THE *KLADDERADATSCH* ATTACKS

The hysterical reaction in the Wilhelmstrasse to the reconciliation between the Kaiser and Bismarck becomes more understandable still in the light of the fact that, not long beforehand, the satirical journal *Der Kladderadatsch* had published vicious attacks on three conspirators named Count Troubadour, Austernfreund (Oysterfriend) and Spätzle (Noodles), whom everyone immediately recognised as Philipp Eulenburg, Holstein and Kiderlen-Wächter.[77] As the articles were clearly based on inside information, a sense of almost panic-stricken insecurity and paranoia filled the corridors of the Foreign Office. 'Why are we *three* always named together?', Kiderlen asked Eulenburg. 'Who is it who knows so much about our business? Who is it who hates only us, and associates with the people Kladderadatsch praises, Schlözer, Reuss etc.? Who knows about such private matters as that Reuss proposed you as his successor? It seems to me that there is one natural suspicion, which I do not wish to express yet, but which I shall look into carefully.'[78] Holstein was firmly convinced that the attacks had been ordered by Herbert Bismarck and/or Waldersee, that the Silesian magnate Count Guido von Henckel-Donnersmarck was the paymaster and that sinister police agents were involved in the campaign. In mid-April the Grey

Eminence of the Wilhelmstrasse, who was blind in one eye, seconded by the former Flügeladjutant General Moritz von Bissing, challenged Count Henckel to a duel. He asked General Waldersee to be his second, but refused to take part in the duel with Holstein.[79] Although duelling was illegal, at the instigation of his military suite the Kaiser demanded that Kiderlen shoot it out with Henckel and the editor of *Kladderadatsch*, Wilhelm Polstorff, before there could be any question of his accompanying the Kaiser on his travels again. Henckel rejected Kiderlen's challenge, declaring that he had no connection with the newspaper attacks, but Polstorff demanded a postponement of the duel for four weeks. As a result Kiderlen was unable to accompany the Kaiser on his trip to the Adriatic. On 18 April 1894 the Swabian diplomat fought a duel with the editor in the Grunewald and seriously wounded him with the third bullet. Both men were condemned to four months' imprisonment, but were released after a few weeks.[80]

In the politically and socially exposed position in which he found himself, it was the lack of a single word of appreciation from the Kaiser which affected Friedrich von Holstein most deeply. Despondently he wrote to Eulenburg at the end of 1893: 'If, in my joyless existence, I now have to endure His Majesty standing by and smiling coldly while I am pelted with dirt, if he can tolerate the fact that the Berlin police are, and remain, notoriously hostile towards the New Course, he cannot be surprised if I hand over my cards to someone else . . . before long . . . It is not those wretched attacks against me which make me angry. It is His Majesty's attitude which offends me: he does not lift a finger, but seems to find it enjoyable when a lot of filth is let loose.'[81] In January 1894 Holstein lamented that he and his colleagues were 'quite helpless to deal with the present attacks, for one has no idea against whom one should take action'. 'And the longer the people behind the Kladderadatsch are convinced they cannot be found, the more insolent they will become in their libels . . . I am utterly outraged by the way the Kaiser lets his supporters down. Why does he not appoint a decent chief of police? We have been betrayed and sold.'[82] Two months later the Privy Councillor renewed his complaints: 'What is so discouraging about the situation is the Kaiser's coquetry and unreliability. There is no pleasure in serving a master who lets one down . . . His Majesty ruins everything . . . with his lust for political adultery.'[83] Holstein's grievances about lack of support from the Kaiser bear eloquent witness to the effectiveness of the 'kingship mechanism', by which the monarch could, with a mere word or a simple gesture, discriminate for or against the rival politico-social cliques.

Eulenburg tried to lure Wilhelm into action by arguing that the *Kladderadatsch* attacks were also directed against him, the Kaiser.[84] He pointed out that Holstein was suffering from sleeplessness and extreme nervous strain and urged Wilhelm to cheer him up by conferring an order on him, and likewise to bestow some mark of his favour on the Reich Chancellor, whose position had been undermined by the systematically circulated rumours of a crisis.[85] He also informed the Kaiser that the National Liberal Reichstag deputy, Heinrich Prince zu Schönaich-Carolath, was spreading the story among journalists that 'H.M. the Kaiser's penchant for spiritualism' had played a decisive role in the choice of Axel Varnbüler as envoy from Württemberg, and that Eulenburg too was supposedly a spiritualist and was 'in the highest favour with H.M. precisely because he shared this mentality'. Prince Carolath had named Count von Henckel-Donnersmarck as his source.[86] Replying to this letter on 11 January 1894 the Kaiser joked that Eulenburg's information had 'helped considerably to further my studies of so-called polite or court society'. He had acted rapidly upon it, putting Adjutant-General von Plessen in the picture without even hinting at Eulenburg's name. 'He wrote a very nice but firm letter, as if on his own accord, to Butter-Heinrich [Prince Carolath]. The enclosed reply from the Butterfürst [Anton, ruling Prince of Schönaich-Carolath] will show you the effect. Either – as the letter must have been written with the guidance of Isidor Lachmannski [this is an allusion to Henckel-Donnersmarck, who was married to the Jewish Frenchwoman Marquise Blanche de Païva, née Lachmann] – Butteranton is telling a pack of lies, or your informant has grossly deceived you. It would be interesting to find out which.'[87]

Since the Kaiser eventually sided with Holstein, Kiderlen and Eulenburg in this unpleasant affair, albeit without great enthusiasm,[88] Henckel and Waldersee found themselves in an increasingly difficult position. In May 1894 Henckel approached Eulenburg through the Bismarckian journalist Hugo Jacobi, asking him to intervene with the Kaiser on his behalf, which the newly appointed ambassador in Vienna declined to do, remarking that a man like Henckel, with his untold millions, must surely be in a position to prove himself innocent of the *Kladderadatsch* attacks by finding out the true identity of the perpetrator. The very next day Jacobi was authorised by Henckel to send Eulenburg a pneumatic dispatch informing him that the 'sources' were still in official employment and had even sat at the Reich Chancellor's table with Holstein and Kiderlen at the Kaiser's last birthday celebrations.[89] Armed with this information, Marschall was at once able to demand an explanation from two of the disaffected officials in the Foreign Office, Ernst Freiherr von Bothmer and Ludwig Raschdau, whereupon the attacks immediately ceased. But the fact that Marschall decided to take action as discreetly as possible against the culprits led Eulenburg to conclude, not without reason, that the attacks against his powerful subordinates, Holstein and Kiderlen, had not been entirely unwelcome to the Foreign Secretary.[90] As we have seen, it was not least as a result of the machinations of Holstein and

Kiderlen with Eulenburg and the Kaiser against Caprivi and Marschall that the internal cohesion between the men of the New Course had broken down.

An unbridgeable gulf now began to open not only between Marschall and Caprivi on the one hand and the top officials in the Political Department of the Foreign Office on the other, but also between Philipp Eulenburg and Holstein. 'In the past few days I have had an instinctive and very painful feeling that something has come between you and me', Holstein wrote to his erstwhile ally in April 1894.[91] The future course of the reign of Kaiser Wilhelm II was to be decisively influenced by the parting of the ways between Eulenburg – who in his blindness continued to support the Kaiser uncritically, with the backing of Bülow in Rome and of Kuno Moltke, Axel Varnbüler and August and Botho Eulenburg in Berlin – and Holstein, who was to make one last attempt to spur on the statesmen in the Wilhelmstrasse to resist the growing personal power of the young Kaiser.

Philipp Eulenburg's decision to abandon the alliance he had formed with Holstein back in 1886, electing instead to go his own way, can be dated almost precisely to 20 March 1894, the fourth anniversary of Bismarck's dismissal. It was on this day that he drew up a memorandum for Wilhelm II in which he argued that the Kaiser could now dispense with the services of Caprivi and the other men of the New Course if he wanted. 'To make the present situation clear', he stated, 'I must go back to the time when His Majesty parted company with Prince Bismarck in 1890. The *necessity* of dismissing Prince Bismarck with which His Majesty found himself confronted led to the appointment of civil servants who were *not congenial* to the Prince . . . As it was very widely believed that after the death of Prince Bismarck Germany would collapse, and as the complication now arose that the Prince not only left the political stage, but also threw his great weight behind an attack on the Kaiser's government, which his departure had put in a very difficult position, it was natural that his successors in the Foreign Office would be considered "inadequate".' He, Eulenburg, had at that time made it his mission in life to help 'guard the Kaiser and the Fatherland against the grave danger of his having to lay down his arms in the battle for public support waged by the Prince. His Majesty would *never* have been able to recover personally from the blow of a defeat; Prussia could not have made up for the setback to the monarchy except with difficulty, perhaps only after a successful war. It was therefore important not only to keep the "New Men" in the posts entrusted to them by His Majesty *much* longer than the entire opposition thought possible, but also to make their work as *successful* as possible. Both can be considered "accomplished" after an interval of 4 years. The "reconciliation" with Prince Bismarck in *Berlin* and the fact that he has actually left the ranks of those maliciously opposing His Majesty forms, as it were, the keystone of a building which reached its culmination in the Russian trade treaty . . . This success for the Prussian monarchy, this victory for His Majesty the Kaiser, in spite of the fact that it exhausted all the strength that was invested in it, is *no Pyrrhic victory*. The "King of Prussia" is now no longer "checkmated" if He sees fit to bring about a gradual change in the personalities who have been systematically fought against . . . The immense difficulty of keeping the "New Men" in place lay not only in the characters themselves, but also in their ignorance of public affairs. It also lay in the fact that His Majesty, with his profound sense of duty, was endeavouring to learn how to govern, and adequate experience can often be acquired only at one's own and other people's expense. It was natural that there should be friction, and there were countless things behind the scenes which had to be settled, prevented, got rid of or found – without coming to His Majesty's attention – in order to achieve our goal, which took quite a toll on my nerves . . . Only a year ago I should have been filled with the greatest anxiety by the realisation that His Majesty completely ignores the Reich Chancellor, that is, he regards the latter's ability to direct internal and external policy as non-existent, and by his complete loss of confidence in the Secretary of State. *Today* – after His Majesty's new government has stayed the course for four years – I do not see this fact as fatally damaging, because I no longer need fear for the Kaiser and the monarchy if it should come to a break. It is my opinion, however, that within a certain period either His Majesty must recover his confidence entirely or there *must indeed* be a change, because the *first requirement for fruitful co-operation* must be that the Kaiser has confidence in *the particular people* in government office who have to deal with the questions on which the weal and woe of Germany and Prussia depend.' Eulenburg fully recognised Holstein's merits: it was his experience and his unparalleled hard work alone which had saved Caprivi and Marschall in the first years of the New Course. But since 1892 the Chancellor and the Foreign Secretary had begun to act increasingly independently, 'like fledglings which have hopped out of the nest'. 'It was a period of many mistakes and blunders', the Kaiser's friend lamented. 'After this period mistakes were even made with a certain self-confidence.' The Kaiser's trust in the leadership of the Reich had been shaken on the one hand by these 'mistakes', but on the other hand also by the feeling that Holstein, 'who shows marked peculiarities in his private life (which have earned him the reputation of being "half-mad", and much hatred besides), actually held the reins in the sphere *most important* to His Majesty, rather than his Chief [Marschall]'. It was nevertheless absolutely essential that the Kaiser recover confidence in the Wilhelmstrasse. 'The Foreign Office must be the *natural, self-evident* tool, mouthpiece, machine with which the Kaiser works. From the moment the Kaiser has the feeling that the machine is not running smoothly, it *must* be oiled, for the Kaiser cannot find better machinery for His Own use. The impression which His Majesty has at present can be changed only if His Majesty has absolute trust in the Secretary of State, and – that is something which Marschall can no longer achieve. For the time being a change is no doubt impossible . . . But after a certain interval I would advise His Majesty to make a change for the reasons I have stated. In making a new appointment to this post His Majesty will no doubt, for reasons of gratitude as well, wish to treat Holstein considerately . . . To show gratitude to him it would be desirable to choose a secretary of state with whom he would be happy to work. But this secretary of state must nevertheless be regarded *publicly as an independent man.* The choice is narrow. I can suggest only Bülow . . . Bülow is held in the highest regard abroad, he is acceptable to nearly all parties and in any case he will suit the Reich Chancellor, *whoever he might be.* There is a certain feeling in the country that he would be the right man and, as I know him very well, I can also confirm that he would be the right man for His Majesty . . . Provided that the latter makes his complete peace with the House of Bismarck and finds an ambassadorial post for Herbert (perhaps in London). The Reich Chancellor is in a similar situation to Marschall. He has been considerably strengthened by the Russian trade treaty, and the gratitude of his Kaiser will secure his position for the time being. But whether the Chancellor will keep his nerve and stand fast in the face of a campaign waged more fiercely than ever before in the press and behind the scenes is another question. Even at the time of the worst conflict, when Prince Bismarck was in office, the press did not create such an uproar, and only the feeling that His Majesty the Kaiser will maintain his solidarity with his principal advisers can give him the courage to "stay the course".' The question was therefore becoming urgent, Eulenburg went on, as to 'whom His Majesty would chose as successor? A general would be faced with *very great difficulties*, since the best general had not proved satisfactory to His Majesty. Furthermore, if the idea should take hold in the army that the Reich chancellor *had* to be a general, politics would invade the highest ranks of the army and every officer would think less of carrying "the field marshal's baton" than "the inkwell of Wilhelmstrasse 77" in his kit bag.' The only solution which he could see, Eulenburg reasoned, was 'to put the political confusion within the country under the control of a man of *very great* experience both in parliamentary matters and in internal affairs, that is to say Minister-President Botho Eulenburg. On the one hand, he has a good relationship with Miquel as a colleague, and on the other he would have in B. Bülow, to whom he has been close for many years, a secretary of state at his side on whom he could rely in foreign policy matters. In making these observations I am guided by the fact that this combination would be accepted without misgivings by a surprisingly broad spectrum of opinion – and that this change would be the only one which could be carried out without great convulsions.'[92] Situating himself firmly in the power vacuum left since Bismarck's fall between the crown and the state, Philipp Eulenburg assumed the role of the omniscient arbitrator whose place it was to pronounce on the capacities and future of the constitutional 'responsible government'. In this document of March 1894 he sketched out for the first time the combination – Botho Eulenburg as Reich chancellor with Bernhard Bülow as foreign secretary – which he was to continue to advocate throughout the countless crises of the next three years. Rarely in modern history does one see as clearly as in this instance the destructive role which a favourite can play within a system of Personal Monarchy.

THE REVIVAL OF THE *COUP D'ETAT* PLANS

Immediately before the Scandinavian cruise of the summer of 1894 'considerable ill-feeling' prevailed between the Kaiser and Caprivi, which encouraged renewed speculation about a change of chancellor, particularly as there was growing concern over the general atmosphere of confusion and lack of leadership.[93] In accordance with the principle of the 'kingship mechanism', the monarch's loss of confidence in his principal advisers led to a further shrinking of Caprivi's power and at the same time to struggles and conflicts within the leadership of the Reich and the Prussian state. Arthur von Brauer, who had meanwhile returned to Karlsruhe as prime minister of Baden, gave a cogent analysis of the chaos behind the gleaming façade of Berlin in a letter to his Grand Duke in the autumn of 1894. 'To continue to speak of dualism in the government today would be to make oneself guilty of euphemism. It is not dualism, but at least "quadrilateralism": . . . Caprivi, Eulenburg, Miquel, Posadowsky, each with his supporters, face each other like four great enemy parties who agree only on one thing, that "it cannot go on like this and that H.M. must decide". "His Majesty's Government", in the sense of a unit, simply no longer exists . . . All those involved . . . cling so doggedly to their point of view that I believe that H.M. will *at last* have to come to a decision on the question of appointments. But it is characteristic that no party is certain of victory or even pretends to be so. No one knows which way H.M. will lean, and many people think a *tertius gaudens* [a third party] might well succeed both.'[94] After a visit to the capital Waldersee summed up the confusion behind the scenes in similar terms: 'The most profound mistrust and hatred between Caprivi and Miquel, Caprivi distrusts [Botho] Eulenburg and of course also the Kaiser, Bronsart on strained terms with Caprivi, but in serious conflict with the Kaiser, meanwhile all are the best of friends outwardly.'[95] In the Wilhelmstrasse Caprivi's position was regarded as in such danger that Philipp Eulenburg, at Holstein's instigation, was obliged to warn the Kaiser that if he conferred the Order of the Black Eagle on Miquel, as he apparently intended, the Chancellor would immediately resign. The Kaiser denied having any such intention and sent a telegram in answer to his friend's letter saying 'Had a good laugh over it! Who could think up such rubbish. And to think that you were taken in! Wilhelm I. R.'[96] Be that as it may, the sense of insecurity in the highest government circles was palpable.

Speculation about possible successors to Caprivi now grew increasingly lively. In March 1894 Minister of War Bronsart remarked during a visit to Waldersee in Altona that 'the separation of the posts of chancellor and minister-president causes continual problems; in fact the situation is untenable, but it nevertheless goes on'. Both Generals were of the opinion that Botho Eulenburg would be the best successor to Caprivi; if he were appointed Reich chancellor not only would the separation of the two offices be removed, but the lamentably bad relationship between the Conservatives and the Kaiser would also rapidly improve under him.[97] The Prussian Minister-President and Minister of the Interior, who like Caprivi complained that it was 'truly difficult to govern with this sovereign!', seemed, however, to lack the 'will to power'.[98] When in the following weeks it became increasingly obvious that Botho Eulenburg was a 'complete nonentity', other candidates for the chancellorship began to be discussed.[99] In the summer of 1894 rumours circulated that it was not Botho, but his cousin the Kaiser's favourite Philipp Eulenburg who was to be appointed to succeed Caprivi. The former Chief of the Admiralty, General von Stosch, told Rudolf von Bennigsen that the Kaiser had said to a close friend: 'I can get on with Caprivi, but I do not warm to him. The man lacks imagination, he does not understand when I tell him about further thoughts I have. I shall have a younger man as his successor, who is closer to me personally and to whom I owe no sort of debt as a result of his past. *He shall be my man only*.' Stosch presumed that the Kaiser had his friend Eulenburg in mind for the highest office, although he doubted that Eulenburg 'could be so reckless as to accept the chancellorship'.[100] Waldersee was among the many who regarded the favourite as the most unsuitable candidate imaginable for this responsible position. 'I will not presume to pass judgement on Eulenburg's diplomatic capabilities, but I am quite prepared to believe that he might be a passable foreign minister', he wrote. 'He is unsuitable to be chancellor because he has much too feminine a nature, without any vigour; and then there are his spiritualist leanings. I think he would lead us to rack and ruin. But he would suit the Kaiser very well, of course.'[101] The former Minister of Ecclesiastical Affairs, Count Robert von Zedlitz-Trützschler, who visited Waldersee at Altona in August that year, also saw an obstacle in Philipp Eulenburg's effete personality and in the spiritualist interests which he apparently shared with the Kaiser.[102]

'If Zedlitz were to become chancellor I would be happy and begin to hope that many things in this country might improve. The Kaiser could not choose anyone more appropriate', the arch-conservative Waldersee commented.[103]

At the end of June 1894 two events coincided that were to push the chancellor crisis, long since brewing, into an acute phase. At a by-election in the Hamburg constituency of Pinneberg a Social Democrat was elected, and in Lyon the French President, Sadi Carnot, was assassinated by an Italian anarchist. Waldersee commented on the first of these events that the Kaiser 'would not mind if quite a few more [Social Democrats] got into the Reichstag; for then the sensible people would pull themselves together', while he himself believed 'that we ought to make up our minds now to take vigorous action', as otherwise one might 'easily miss the right moment'.[104] Although the Kaiser was again talking about defeating the socialists and the anarchists the General was afraid that the monarch would lack the resolve to take such a step. 'The good Kaiser did indeed plan, under the influence of his immature youthful impulses, to make all mankind happy, and was even convinced that he could do it; what did he achieve? He made the masses greedier and class hatred has merely increased. Now he dreams of shooting down the Social Democrats, who have not the slightest intention of giving him the opportunity to do so.'[105]

The murder of Carnot led to deliberations in all European countries on how to combat anarchism most effectively. Waldersee had grave doubts as to whether the right means would be used. 'I do not believe that important decisions will be made, least of all here', he wrote on 6 July 1894. 'It is no use executing or locking up 100 scoundrels or so and keeping 1000 or so in fear of their lives; we must pursue the causes of the movement and put a stop to the way in which these people can openly and freely incite the lower classes against the upper classes, the unpropertied class against the property-owners, and bring religion into disrepute. But unfortunately through neglect we have already slipped a long way down, and if action is not taken soon we shall go rapidly downhill. The Kaiser, who surely wants what is best, does not begin to see the full picture; he likes to say that he will shoot down all ringleaders, but does not realise that it will never come to an uprising; the leaders of the movement see that they are making steady progress and they will be careful not to give any reason for shooting at them.' In Germany no one had had the courage to take action even against those Social Democrats who openly advocated the destruction of the existing order and left the room when toasts were drunk to the Kaiser. 'From my knowledge of the persons involved, fear of assassination attempts is already playing a part [in Wilhelm II's mind]', Waldersee complained. He would offer his services, he declared, if action was to be taken against the revolutionary parties, but 'the men of today in our government would of course have to disappear', and in general 'pretty well everything would have to be tackled differently from before'.[106]

The inevitable result of Wilhelm II's determination to use the general fear and anger aroused by the assassination in Lyon as an opportunity to take action against 'revolutionary tendencies' – by which he meant both anarchism and socialism – was to bring to a head the conflict between the Prussian Ministry of State under the leadership of Botho Eulenburg and Miquel on the one hand, and the Reich administration under Caprivi and Marschall on the other. The former advocated a large-scale operation in the Reich, which did not exclude the possibility of repeated dissolution of the Reichstag and, as a last resort, a *coup d'état*. The latter, however, considered it sufficient to introduce in the Prussian Landtag (not in the Reichstag) a law of association which would be acceptable to parliament. In mid-July 1894 Holstein and Caprivi explained to the Württemberg envoy, Varnbüler, the catastrophic consequences which would follow the introduction of a harsh anti-socialist bill into the democratically elected Reichstag. The Privy Councillor declared that 'His Majesty was in favour of making use of the present favourable moment and the fear of daggers and bombs in the middle classes; and since the Prussian Ministry of State, especially Botho Eulenburg, had spoken out against taking action through legislation in the [Prussian] Landtag, if their influence prevailed there was a danger that the Reich Chancellor would be forced to introduce into the Reichstag a bill which, in both his and Holstein's opinion, would certainly be rejected both by the present Reichstag and, after it had been dissolved, by a newly elected Reichstag.' The Reich Chancellor likewise stressed to Varnbüler that the Reichstag would not in any circumstances vote for a new anti-socialist bill; to take action in the Reich would therefore inevitably lead to the dissolution of parliament, to a *coup d'état* and to the imposition by force of a new electoral system. The consequences, it was feared in the Wilhelmstrasse, might be civil war and even the disintegration of the German Reich.[107]

The Kaiser thus found himself once again facing a choice between two rival groups representing opposing points of view. After lengthy discussions with Philipp Eulenburg and Kiderlen on the Scandinavian cruise he at first opted for the moderate course recommended by Caprivi and sent a telegram on 24 July to the Minister-President, ordering Prussian law to be tightened up. The chances of a bill against 'revolutionary tendencies' being accepted by the Reichstag were slight, he explained, echoing the views of the Reich government exactly. 'On the other hand it seems to me possible, for the time being, to achieve what Saxony already has by taking the path of legislation in *Prussia.* I wish appropriate proposals to be submitted to me through the Ministry of State. Absolute secrecy is essential. Audience on my return.'[108] On receiving this message Botho Eulenburg told his cousin Philipp that he would have a suitable bill drafted as instructed, but that he had asked the Kaiser to postpone 'the All-Highest decision as to whether to choose this or the more advisable, more comprehensive course of legislation for the Reich' until he was at Königsberg at the beginning of September, as until then almost all the ministers would be away on leave.[109]

After taking soundings in Berlin and discussing the subject at length with Botho in Munich on 29 August 1894, on his return to Vienna Philipp Eulenburg wrote the Kaiser a letter which gives us another clear glimpse of the 'kingship mechanism' and which was to be of decisive significance for the course of the crisis and the fate of Caprivi. As a result of his enquiries the Kaiser's close friend reported that 'a conflict has arisen between my cousin and the Reich Chancellor . . . over the question which Your Majesty commanded me to discuss with him. I must freely admit that after my conversation with him [Botho] I could not avoid the impression that his nerve is as strong as could possibly be desired, and even stronger than in another quarter . . . He told me that the danger from the anarchists and the socialists was much too serious to be counteracted otherwise than through the *Reichstag*. Only a rigorous anti-socialist law could provide the means of control which was so urgently needed. Of course this bill would *have* to be passed. It was not entirely out of the question that the present Reichstag, given the generally prevailing mood and the use of the whip, might accept the bill. But it was doubtful. At any rate a new Reichstag (in the event of resistance) would *also* have to be dissolved and if there were *another* rejection the government would have to bring in a *new electoral law*. It would be a matter of bending or breaking. The monarchy *needed* to act with rigour in such an important question . . . The idea of changing the law of association through the [Prussian] Landtag was wrong because it was a weak measure . . . A change in the law of association in Prussia would be nothing but a fiasco. This argument is *directly opposed* to that of the Reich Chancellor and is hard to refute. The Reich Chancellor will not contemplate another dissolution of the Reichstag *in any circumstances*, because he foresees difficulties with the federal states. So he would turn *this question* into *another* resigning issue. I do not doubt that the federal states would indeed make difficulties. Württemberg because it is democratic, Bavaria out of particularist aspirations. But on the other hand the question is too much of a *national* one for the federal governments not to co-operate if the Reich government shows it is in earnest and threatens reprisals. The Reich Chancellor would not "co-operate" in any case because he neither would nor could put the case for the bill in the Reichstag. Botho Eulenburg spoke *brilliantly* in favour of the anti-socialist laws in 1878. He would be able to do it. He would of course have to become an authorised member of the Bundesrat first. If we assume that Your Majesty wished to bring the question before the Reichstag, then a change of chancellor would come first. That would be a dramatic introduction to the proposed action. But the new Chancellor would reorganise the Ministry. (I doubt that Marschall would stay on as secretary of state in that case, – and I am not sure whether Botho Eulenburg would remain as minister-president. But only *he* could put the case for the bill.) – But now let us assume that the Reich Chancellor stays and Botho Eulenburg goes. In that case I can only repeat the *unanimous* opinion of statesmen and officials of all party groups, which is that dual government would then cease. It is only the remarkably conciliatory and at the same time loyal personality of my cousin that has enabled this form of government to survive . . . That is more or less what my enquiries have revealed. A conundrum, a vicious circle. I cannot give Your Majesty any advice – except perhaps to discuss the situation frankly with Botho Eulenburg. He is after all the most eminent *statesman* whom Your Majesty has, and he is also honourable and considerate. Marschall and Miquel are interested parties; the Reich Chancellor is obstinate. Lucanus (with whom I am on very close terms and whose diligence I admire) is not senior enough to be consulted on questions of *great* moment. Your Majesty will of course make the best of this complicated situation for yourself, but a discussion with an experienced statesman might reveal new points of view to Your Majesty which would be useful – and might perhaps help clarify the important choice to be made between the Reichstag and the Landtag in the anti-socialist question . . . It seems to me that the situation will not become acute until after the Chancellor's return from Karlsbad – that is, probably when Your Majesty is in Rominten [on 23 September].'[110]

On receiving this letter the Kaiser arranged to call on Botho Eulenburg on the afternoon of 2 September. They spent an hour discussing the situation. Afterwards the Minister-President reported to his cousin that the Kaiser had explained to him at great length 'that at first he had been in favour of taking the aforementioned action in Prussia, but that he had since become convinced – it was easy for me to see how – that this would not be adequate, but that the matter must be tackled in the Reich and that it was essential that it be carried through. He was well aware that it was a momentous matter, but he was determined to carry it through and he believed he could count on the co-operation of the federal allies, as he already knew he could on that of the Kings of Saxony and Württemberg . . . When I suggested that it would surely be necessary first to come to an understanding on the matter with the leading statesman, His Majesty agreed, but commented that if the latter would not see to it someone else would, and also expressed the opinion that only a general could be considered for that role. This gave me the chance to press the point that if this happened the opportunity should be taken of reversing the separation of the highest offices in the Reich and in Prussia, and that I myself would not stand in the way. His

Majesty protested very emphatically against the latter suggestion, however. The discussion came to an end on the point that all preparations should be made as agreed and that the decision should be taken after the return of the Chancellor and the Kaiser, that is in the second half of October.'[111] Botho Eulenburg's brother, the Oberhofmarschall August Eulenburg, confirmed in a letter to their cousin Philipp that the conversation which the latter had engineered between the Kaiser and the Minister-President had 'worked wonders'. After 'a very long and thorough discussion' the Kaiser had 'changed tack completely' and '*such* complete agreement . . . had been achieved that they even talked about the most far-reaching consequences of possible changes of appointments'.[112]

THE FALL OF CAPRIVI AND BOTHO EULENBURG

The day after his conversation with the Minister-President the Kaiser left Potsdam for East Prussia, where he was to direct the army manoeuvres. When Botho Eulenburg arrived in Königsberg, the East Prussian capital, the Kaiser invited him to luncheon with the King of Saxony, who was likewise in favour of large-scale action against 'revolutionary tendencies' in the Reich.[113] On 6 September 1894 Wilhelm made a speech at Königsberg which signalled the beginning of the end for Caprivi. 'With deep sadness in his heart', he said, he had been compelled to recognise 'that in the circles of the nobility close to me my best intentions have been misunderstood and sometimes challenged, and even the word opposition has reached my ears. Gentlemen! It is a nonsense for Prussian nobles to form an opposition against their king; it is only justified if the king is at their head, as the history of Our House teaches us. How often have my forefathers been obliged to confront misguided members of one class for the good of the whole! The successor of the man who became sovereign duke in Prussia in his own right will walk in the footsteps of his great ancestor; just as the first King [Friedrich I] said ex me mea nata corona [from me my crown is born] and his great son [Friedrich Wilhelm I] established his authority as firmly as a rocher de bronce [sic], I too, like my Imperial Grandfather, represent the Divine Right of Kings. Gentlemen! Whatever oppresses you, I feel too, for I am the greatest landowner in our nation and I know very well that we are going through difficult times. My thoughts are daily directed towards helping you; but you must support me in this, not by causing an uproar, not by using the methods of the professional parties of opposition that you have so often rightly fought against; no, but by addressing yourselves trustfully to your sovereign. My door is always open to each one of my subjects, to whom I gladly lend my ear. Follow that path from now on, and I shall consider everything that has happened as expunged!' Alluding to the memorial to his grandfather which he had unveiled that morning the Kaiser continued: 'Before us stands the statue of Kaiser Wilhelm I holding the sword of the Reich aloft in his right hand, a symbol of justice and order. It reminds us all of other duties, of the hard struggle against the efforts of those who attack the very foundations of our national and social life. Now, gentlemen, it is to you that My call is addressed. Go and fight for religion, for morality and order, against the parties seeking to overthrow them! Like the ivy which winds itself around the gnarled trunk of the oak tree, adorns it with its foliage and protects it when storms rage through its crown, the Prussian nobility encircles My House. May it, together with the entire nobility of the German nation, become a shining example to those of our people who are still hesitating. Come, then, let us go into this battle together! Onwards with God, and shame on him who lets his King down. In the hope that East Prussia will be the first province to join the front line of this war, I raise my glass!'[114] Two days later Wilhelm told Botho Eulenburg, when the latter took his leave in Marienburg, 'that He would inform the Chancellor of what He had decided by cipher telegram and that I myself might write to Karlsbad [to Caprivi] with further details, which I have already done'. 'I am very anxious to know what the answer will be', Botho wrote to his cousin, whom he asked to burn his letter at once.[115]

After the speech at Königsberg Waldersee had a sense that 'we are about to witness a change in the internal political scene and I think the Eulenburg family has plucked up courage to act against Caprivi'.[116] While the Kaiser was at Rominten for several days' shooting during the last week of September Philipp Eulenburg had a further opportunity to work on the monarch along the same lines as his cousin Botho.[117] Always inclined to regard the wishes and impulses of the Kaiser as a legitimate starting-point for any policy to be followed, the ambassador at Vienna, who had not been at his post for months, found himself increasingly at odds with his erstwhile ally Holstein.[118] He confided to his new confidant and prompter, Bülow, whom he looked upon as future Foreign Secretary and Reich Chancellor,[119] on 30 September 1894: 'I am . . . basically of the opinion that the Königsberg speech has had such a significant effect that it is impossible to take *half*-measures now. The Kaiser's appeal has electrified the people and if His government brings *weak* proposals before them or fails to bring any at all, the Kaiser's standing, which has just risen, will be *completely* and hopelessly ruined. Beside this consideration all Holstein's oversubtle, Bismarck-fearing calculations wither away to nothing.'[120]

At Rominten the Kaiser's friend drew up a detailed *aide-mémoire* in which he summed up the advantages and disadvantages of forcible action in the Reich, and read it out to the monarch on 27 September on the latter's return from a shooting expedition. In this highly revealing document he stated that 'the *advantages* of introducing a bill into the Reichstag with the object . . . of giving the Reich government a firm basis on which to combat socialism and the revolutionary parties are 1. that the external situation is eminently peaceful at the moment. The Tsar's serious illness guarantees peace for some time and at the same time puts a curb on the French, whose inclination is always to exploit any possibility of serious trouble in the German Reich to strike out at us . . . 2. The present combination of federal princes makes it advantageous to take action now: King Albert of Saxony can be counted on, King Wilhelm of Württemberg likewise, and the old Prince Regent of Bavaria is weighed down by great fear of the socialists. In a few years' time the picture may be quite different. The 3 Kingdoms will, in the foreseeable future, find themselves governed by princes who are less firm in their convictions, which may rule out fruitful action. 3. The moment is opportune in view of the confusion among the parties and the general discontent. It may not be possible to eliminate this discontent, but the slogan "War against revolution" has sufficient appeal to bring about a change for the better, rally those who have strayed and show them the path towards unity. The government is also in a position to show the firm hand and firm will which the misled masses have become accustomed to calling for. 4. The moment is opportune because the universal need for protection against revolution provides a useful electioneering slogan in the event of dissolution of the Reichstag.' Eulenburg set out equally clearly the disadvantages of a draconian measure in the Reichstag. 'The *dangers* of bringing in a strong bill in the Reichstag are undoubtedly serious, as one cannot count on its being accepted except in a more moderate form. As soon as it overstepped the mark, the majority of the Reichstag would oppose it and the Reich government would have to proceed to dissolution in order to maintain its own reputation. It is *possible*, but not certain, that the bill could be forced through the new Reichstag. It is much more likely that there would be conflict between the Reichstag and the Reich government, which could be brought to an end only by radical means (a change in the electoral system and perhaps force of arms). The difficulties which a deliberate alteration of the electoral system would entail in the Reich would be as follows: 1. As the constitution of the Reich, by its incorporation into the constitutions of the separate federal states, has become part of their constitutions, but there are *responsible ministers* in the federal states (as in Bavaria and Württemberg), these Ministries would be open to impeachment by their Landtage if they took sides with a Reich government that violated the constitution of the Reich. For this reason these federal governments will not give their support, and the Kings of Bavaria and Württemberg will have to take up the battle within their own domains if they dismiss their Ministries. Nor will they find any other Ministries to take on this battle; they will have to lead the fight *in person, against* the tide. Neither Prince Regent Luitpold nor King Wilhelm [of Württemberg] is the man to do this. Neither has the courage to fight his way through rebellions (these would be likely in Stuttgart). 2. The only way of coming to an agreement with the Reich government and *participating* in the conflict would be on the basis of a *revision* of the Reich constitution of 1871, whereby for instance the Bavarian government, in return for the alteration of the electoral system . . . might demand *advantages* which would inevitably change Bavaria's current position in the Reich to the detriment of Prussia's hegemony. That is the area in which the Bavarian Landtag, the Württemberg Landtag etc. would agree to negotiate. Prussia could not tolerate such a weakening of its position, and although the federal princes may honestly intend to stand by Prussia, out of this a conflict may arise which 3. requires force of arms. This would in fact lead to the spectacle of the Reich *in the process of disintegration.* The damage to our national and international standing would be immense. The danger of a French invasion would come to the fore at the same time. And there is another respect in which the conflict in the Reich would be dangerous. The concurrence of Prince Bismarck in the announcements of His Majesty in Königsberg, Thorn etc. may be sincere. In the event of a conflict in the Reich, with all its consequences, i.e. if the situation should become chaotic, the Kaiser would, *par la force des choses*, be *compelled* to call for Bismarck. That would be his Canossa, the effect of which would be to cause damage to the Prussian throne and to leave the Kaiser personally and politically played out.'[121]

As the Kaiser's stay at Rominten drew to a close, Eulenburg had the impression that Wilhelm had already decided on large-scale action in the Reich and hence on a change of chancellor. As he reported to Bülow, the dangers of a *coup d'état* in the Reich which he had listed in his *aide-mémoire* did not worry the Kaiser 'because he *seems* to want a *decision.* He does not say so openly – on the contrary, he would deny it.' The Cabinet Chiefs Lucanus and Hahnke, who had both come to Rominten, also seemed to Eulenburg to be trying to influence the Kaiser towards dismissing Caprivi. 'The ring around the Chancellor is tightening. Lucanus, the angry little terrier, has already got the big dog *firmly* by the hind leg. He was here yesterday. Hahnke is helping him. He was here today.'[122] On 28 September Eulenburg told his cousin the Minister-President that he could expect to take over as Reich chancellor in the near future. 'I have the feeling that His Majesty wants to "sort things out". Lucanus has been here, and he too seemed to me to be in a hurry. I have thought a great deal about the best moment for the change and am still not sure. It might be more effective for you to make your entrance after the dissolution of the Reichstag. On the other hand I should like to have avoided a chancellor departing as a result of a vote in the Reichstag. We have not yet reached that point.'[123] A conversation with the Kaiser on 29 September, in which the favourite attempted to mediate, nearly precipitated a crisis when it became apparent that important documents had been withheld from the monarch by the Chancellor and the Foreign Office, just as in the Bismarck crisis of 1890. Eulenburg recorded the conversation in dialogue form in a letter to Bülow:

Myself: In order to put Your Majesty fully in the picture about the situation I have brought with me the relevant reports received from the three kingdoms in the course of the summer. Although Y.M. is familiar with them, it would perhaps be a good thing to look through them again.

H.M.: What reports? I do not know of any.

Myself: (very shocked!): They will perhaps have been submitted to Y.M. in extenso . . .

H.M.: No, *most certainly* not. Well, let us see why these far from unimportant documents have been withheld from me!

Myself: (I read the reports, which are extremely important.)

H.M.: So it was because the federal states *agree* with taking action through the Reich! *That* was the reason! . . . I have been *systematically* kept in the dark in an irresponsible way, by Marschall *and* Caprivi. I am used to their doing that, of course; but this is the limit.

'So as you see, my old Bernhard', Eulenburg's letter continued, 'these people are digging their *own* graves! It used to make the Kaiser furious when Bismarck withheld important reports from him. Caprivi knows that as well as Marschall. With the sincere intention of enlightening H.M. on the unpredictable consequences of taking excessively strong action if there were conflict in the Reich I used the material officially at my disposal in order to impress the point on him!! Could anyone have imagined that reports from the three kingdoms would not be laid before the Kaiser when they concerned a question which has occupied H.M. almost exclusively for 4 months? Really, there is nothing one can do.'[124]

The Kaiser's irritation with Caprivi in these final days of the latter's chancellorship is evident from the audience he gave the Reich Chancellor on 5 October 1894 at his hunting lodge at Hubertusstock. Marschall recorded that 'His Majesty's antagonism towards the Reichstag . . . was immediately apparent when the ceremony to mark the completion of the new Reichstag building was mentioned. It was only with great difficulty that His Majesty could be persuaded to abide by his earlier decision to attend the ceremony. The discussion of the bill to be brought before the Reichstag after it had convened brought matters to a head. The Reich Chancellor declared himself ready in principle to introduce an anti-revolution bill, but considered it necessary that it should be announced in the speech from the throne. His Majesty categorically opposed this: the Reichstag must be taken completely *by surprise*, he said, as soon as the fiscal reform bill had been accepted, and the dissolution of the Reichstag must follow at once. The Reich Chancellor pointed out that this was not a practicable way of proceeding, as the revolution question had already been discussed in the press for weeks and questions would immediately be asked on the subject in the first days of the session . . . It would be impossible to hold back. At the same time C[aprivi] stated the view that the bill must be in a form that did not *a priori* exclude any possibility of acceptance by the Reichstag, for otherwise there would inevitably be convulsions and crises which would have the opposite effect to the political aim of the measure – the suppression of revolution. His Majesty now brought up the subject of a *coup d'état*, whereupon the Chancellor set out in detail the dangers of a *coup d'état* in the Reich, pointing out that it was impossible to destroy part of the constitution of the Reich without damaging the whole creation etc. His Majesty dismissed these arguments and launched into a long and enthusiastic exposition of the entire *coup d'état* programme which he had agreed with the King of Saxony. The Reich Chancellor tried to stress the merits of his point of view and finally remarked – as His Majesty remained obdurate – that His Majesty would have to choose another Reich Chancellor to carry out this policy, since his conscience and his convictions made it impossible for him to put forward a policy which he feared would "ruin the Kaiser and the Reich". His Majesty declared that he would not accept the Chancellor's resignation; it was for Him to decide what standpoint every official should adopt at a time of conflict – and the Kaiser, not the official, knew the soul of the German people and bore the responsibility before God; he appealed to the soldier who had once told Him that he would allow himself to be shot dead for His sake. Caprivi responded that he was still ready to do so, but that he could not act against his conscience and put forward a policy which he regarded as harmful to the Kaiser. There followed a long discussion. His Majesty expressed the view that Caprivi's resignation would allow his enemies to crow over him and would expose him – the Kaiser – to the suspicion that he had been "forced to get rid of a chancellor". Finally His Majesty alluded to the possibility of a deteriorating situation with regard to foreign relations – the death of the Tsar was imminent, complications were beginning to arise, C[aprivi] could not desert him at this time. C[aprivi] answered that the present state of affairs in the world of high politics in fact constituted a grave warning against following a domestic policy which would inevitably lead to crises and convulsions and might end in catastrophe. As C[aprivi] would not give way, His Majesty took a milder tone and asked for details of the draft anti-revolution bill. C[aprivi] said that it was still in preparation and that he would communicate with the Prussian Ministry of State about the details of it. His Majesty agreed, and with that the audience came to an end. The Reich Chancellor stayed for dinner. His Majesty led the conversation at table in a cheerful and relaxed way and took leave of C[aprivi] as kindly and graciously as always . . . The Reich Chancellor emphasised that the discussion was occasionally very heated, but there had been no hurtful words on either side. He is also convinced that His Majesty will not drop him now, although of course he is also *absolutely adamant* that he will refuse to co-operate in a *coup d'état*. The key to the situation now lies with the Prussian Ministry of State.'[125] On 12 October, at a session of the Crown Council, Caprivi spoke out against a harsh law and therefore against a *coup d'état* policy, while Botho Eulenburg advocated both. Astonishingly, Miquel took Caprivi's side, along with Bosse, the Minister of Ecclesiastical Affairs, and Berlepsch, the Trade Minister. The Kaiser, however, 'persisted in his advocacy of a rigorous measure'.[126]

The Chancellor was in a strange position, Waldersee noted at this time, for while he had publicly refused to take measures against 'revolutionary tendencies', the Kaiser had demanded just such measures. Conservatives, Free Conservatives and National Liberals were willing to co-operate but the parties which supported Caprivi – the Catholic Centre, the Left Liberals, Democrats and Social Democrats – could not support such a policy. 'The situation could not be more confused, and one would think that light must eventually dawn on the Kaiser', the former Chief of the General Staff commented. 'Leaving aside the Centre – although I think these people the most dangerous of all – if Eugen Richter, [Leopold] Sonnemann and [August] Bebel warmly support Caprivi in their newspapers and say that he is their man, it ought surely to open his eyes!'[127] The Kaiser had left Rominten for Hubertusstock, where both Caprivi and Botho Eulenburg had visited him, Waldersee noted, adding that negotiations were underway there, during which it was 'the specific aim' of the Minister-President to 'hoist the Chancellor out of the saddle so as to take his place'. 'The Eulenburg family' evidently thought the time had come to bring Caprivi down and replace him with Botho. 'They worked hard, and apparently successfully, during the manoeuvres in [East] Prussia and at Rominten.' The Chancellor's enemies had found an unexpected ally in the King of Saxony, who had previously supported Caprivi but who had now become afraid of the social democratic movement in his own country and who had convinced the Kaiser at Königsberg that things 'could not go on like this any longer'. 'Hence the constant emphasis on the fight against revolutionary tendencies in the imperial speeches.'[128] This last remark related to another speech of Wilhelm's calling for decisive action and undermining Caprivi's moderate policies. On 18 October 1894 he declared: 'Just as in 1861, when my grandfather undertook the reorganisation of His armed forces – misunderstood by many, contested by still more, the future proved him gloriously right – just as in those days, discord and mistrust prevails among the people today. The only pillar on which our Reich stood was the army. It is the same today!'[129] Even the militaristic Waldersee criticised Wilhelm for having emphasised 'too sharply' that 'the army also stands against internal enemies', by which he had 'again stirred up unrest quite unnecessarily'. The Kaiser's remark that the army was 'the only reliable support for the throne against revolution' had caused universal offence. 'It would be a sorry state of affairs for us if that were true', Waldersee commented, adding, 'unfortunately, however, he has a particular aptitude for offending and embittering the people who are his true and staunch supporters'.[130]

On 26 October 1894 both Caprivi and Botho Eulenburg were dismissed. It was no accident that this happened at Liebenberg, Philipp Eulenburg's country house. The direct cause was an article in the *Kölnische Zeitung*, obviously Wilhelmstrasse-inspired, in which the Chancellor made known the Kaiser's decision to support his moderate policy and not the provocative class-warfare policy of the Minister-President. The real reason, however, is to be found in Caprivi's long since untenable relationship with the Kaiser on the one hand, and with the Eulenburg clique on the other. 'The solution to the puzzle', Waldersee reflected when news of the double dismissal reached him, 'lies simply in the fact that the Kaiser had already had enough of his Chancellor a long time ago and was glad to be able to get rid of him now.'[131] The General rightly pointed out that 'relations between Caprivi and the Kaiser have been strained for almost a year and there has been much friction which has helped bring about the final clash'. The Chancellor had felt obliged 'to make representations to the Kaiser about the fact that he was negotiating with others behind his back', and this 'very much displeased' the Kaiser. After the dismissal, Waldersee recorded, Wilhelm declared that he had been compelled to part with Caprivi 'because he became increasingly uncongenial to me and wanted to have me under his tutelage'.[132] Later the Kaiser complained that Caprivi 'never did me a single favour'.[133] The dismissal showed, Waldersee commented, 'that it is difficult to work with the Kaiser, and for people who have some self-respect it is impossible in the long run. He is secretive; one soon has the impression that he is not being open, which naturally breeds mistrust. Caprivi allowed this to go on for far too long, chiefly out of vanity, because he felt powerful and great as chancellor, and now he has to bear the great disappointment of having the Kaiser declare his firm support for him, and then calmly dismiss him 5 days later, while saying very unkind things about him. Caprivi has also complained bitterly to various people and has said that it is really impossible to govern with our sovereign. He is very put out. And the fact that Caprivi's departure was clumsily handled, like almost everything that the Kaiser does – or thinks he does – on his own account, will damage the Kaiser's reputation even more. The whole world will say: what a way to behave, and what vacillation!'[134]

The simultaneous dismissal of Botho Eulenburg Waldersee attributed partly to the latter's mistaken tactics and partly to the Kaiser's fear of falling entirely into the hands of the Eulenburg family. In his opinion Botho, who had fully recognised 'the unhealthy aspect of the separation of offices', ought to have brought about the break with Caprivi far sooner. Instead he had held back cautiously 'until he went on the offensive a few months ago'. It had also been of decisive importance that the Minister-President had powerful opponents in his struggle with Caprivi. Above all, 'the entire entourage of the Kaiserin [was] on Caprivi's side . . . Might someone or other perhaps have told the Kaiser that he was completely in the hands of the Eulenburgs?', Waldersee wondered, answering his own question with the comment that 'it is possible that the Kaiserin might have done so'.[135] He himself could scarcely conceal his glee at Caprivi's fall. This 'great, and yet so small, man' had 'passed into oblivion full of bitterness' and had 'richly deserved this end', he wrote to Verdy, adding, 'May his ashes rest in peace! I hope never to hear or see anything of him again.' He reported that the Kaiser had exclaimed, in a similar mood: 'Thank God that I am rid of that tiresome fellow; I could not have borne him any longer.'[136] For his part, Caprivi showed his resentment at his treatment by failing to send his good wishes to the Kaiser either for the New Year or for his birthday.[137]

The Austro-Hungarian Foreign Minister Count Kálnoky had not the slightest doubt that Caprivi was the victim of an ingenious plot by the Eulenburgs which neither the Chancellor nor the Kaiser had seen through. When Philipp Eulenburg returned to his ambassadorial post in Vienna in mid-November 1894 after an absence of several months he gave a 'very extended' account, first to Kaiser Franz Joseph and then to him, Kálnoky, of the events leading up to the 'fall of Count Caprivi, staged at Liebenberg', which the Foreign Minister regarded as a 'catastrophe'. Kálnoky summarised the essential details of Eulenburg's account in a derisively mocking confidential letter to the Austrian ambassador in Berlin. Eulenburg, he said, had embroidered on the story of the dismissal 'at great length' and 'in a rather muddled way' with 'innumerable trivial details' and had dressed it up in 'particularly Eulenburgish wrappings'. The Kaiser had arrived at Liebenberg in the belief that his last attempt at reconciliation had succeeded and that Caprivi would therefore remain in office. 'So it was only at Liebenberg, where the Kaiser was surrounded exclusively by the influence of the Eulenburg clique, that the plot which had evidently been in preparation for a long time was carried out.' At Liebenberg Caprivi's comments in the press had been 'interpreted as unseemly crowing over Botho Eulenburg', and the Chancellor was depicted as the ally of the left-wing fractions and of the Centre Party. 'Whether the gentlemen intended Minister-President Count Eulenburg to fall likewise, I would doubt, and I am inclined to think that the speed with which Kaiser Wilhelm cut the knot and dismissed *both* rivals came as a surprise to the originators of the intrigue. In his impatience at being constantly burdened with these personal incompatibilities . . . Kaiser Wilhelm brought the whole situation to an end through what was perhaps too sudden a decision, but one whose logical justification cannot be denied. But various circumstances give me reason to believe that the clique was taken aback . . . The Kaiser is unwittingly caught in the web which has been spun around him – I hope the moment will nevertheless come when he will see through their activities and strike out at them. General Caprivi was no match for these people, he was too honest and too ponderous to be able to understand these cunning intrigues, nor, unfortunately, did he know how to behave with the young Kaiser. This business may cause much more mischief and I cannot see any improvement unless the Kaiser's good sense and good impulses find a way out.'[138] The faltering hopes for Wilhelm II's common sense which the Austrian Foreign Minister still clung to here were to vanish in the following years, under the weak chancellorship of Prince Chlodwig zu Hohenlohe-Schillingsfürst. But before we turn to the relationship between Kaiser and Chancellor under the so-called Newest Course, we shall investigate the largely unknown role of Wilhelm II as head of the House of Hohenzollern, a family beset with scandals.

CHAPTER TWENTY-TWO

Head of the family

WILHELM II was not only German Emperor and King of Prussia, Supreme War Lord and Principal Bishop of the Lutheran Church but also head of the Prussian royal family and all its branches. His power over the personal life of the numerous members of the House of Hohenzollern was great, whether exercised by direct command or indirectly through arranged marriages, promotions, allowances from his Privy Purse or the allocation of living quarters in royal palaces. Nor did he fail to take full advantage of it, unwelcome though this was to many of those affected. 'Of course W. *is* the head of the family both "de jure" and "de facto"', the Empress Frederick remarked in a letter to Queen Victoria. '*Alas* it is so, & I know it to my *cost*, & try to bear the fact as best I can! But it does not give him the *moral* right & authority to interfere, and therefore when he does so in family affairs it can only embitter.'[1] And indeed from the first years of his reign his attempts to maintain strict control over the private life of his close relatives led to bitter clashes and sensational scandals which not only damaged the reputation of the monarchy at home in Germany but were instantly relayed to his more distant relatives in England, Russia, Denmark, Greece and the other German states, arousing further hostility towards the young Kaiser and his wife.

Those most affected (apart from the reigning Kaiserin Auguste Viktoria and the imperial couple's seven children) after the death of Kaiserin Augusta in January 1890 were the Kaiser's widowed mother, who went to live as far as possible from Berlin, in the Taunus (at first at Bad Homburg and later, when Schloss Friedrichshof was completed, at Kronberg), his brother Prince Heinrich, who as a naval officer lived in Kiel with his young family, his eldest sister Charlotte, married to Hereditary Prince Bernhard of Saxe-Meiningen, his sister Sophie (Fozzie) who had married the Crown Prince of the Hellenes (Greece), Constantine (Tino),

The imperial family

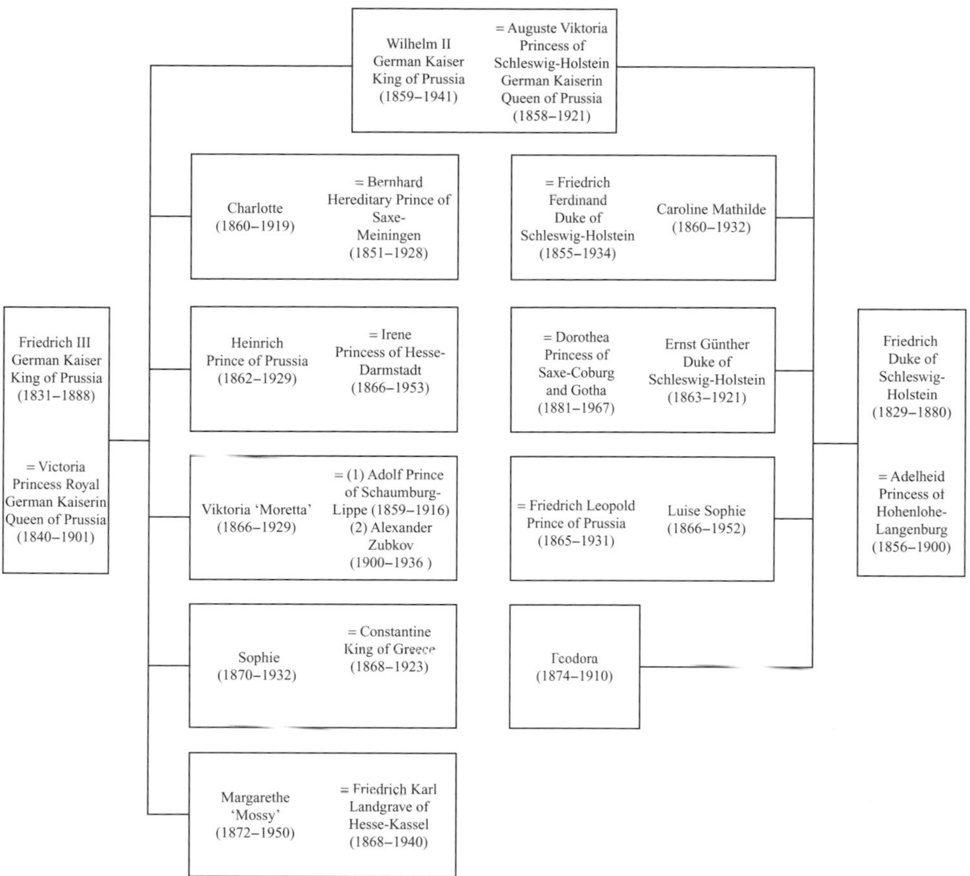

30. The imperial family tree

in October 1889 and lived in Athens, and his sisters Viktoria (Moretta) and Margarethe (Mossy), both still unmarried at the beginning of 1890 and living with their mother in Bad Homburg. The Kaiser's close family circle also included the Kaiserin's mentally disturbed mother, her brother and sisters and their spouses: Duchess Adelheid of Schleswig-Holstein, Duke Ernst Günther of Schleswig-Holstein-Sonderburg-Augustenburg, who was still single at this time; Princess Caroline Mathilde (Calma), who had married Friedrich Ferdinand, Duke of Schleswig-Holstein-Sonderburg-Glücksburg in 1885; Princess Luise Sophie (Yaya), who had married Prince Friedrich Leopold of Prussia in the summer of 1889; and the youngest sister Feodora (Baby), who lived only until 1910.

THE REIGNING KAISERIN

In the first years of the New Course the relationship between Wilhelm and Auguste Viktoria (Dona) was still close and affectionate.[2] On 22 October 1891, the Kaiserin's thirty-third birthday, Waldersee wrote: 'May God bestow rich blessings upon her and bring the Kaiser ever closer to her.' His wish seemed to be fulfilled.[3] The letters the Kaiser wrote to his wife have not survived, but her replies (although only a limited number remain) continually express great love and happiness, as well as sorrow at his frequent absences. 'Oh it is sometimes really like a dream!', she wrote in the summer of 1892. 'How many blessings, how much love I have from my darling. God has certainly been wonderfully gracious to us! And he will continue to help us, he will give me courage to stand by you in your difficult position! However much people attack you and try to make your life difficult, they will not come between us. – Love can do so much.'[4] 'Yes, my darling, my longing is as strong as ever', she wrote to her husband during his Scandinavian cruise in 1892. 'Poor darling', we read in a letter of 15 July 1892, 'I wish I could come and sooth[e] my darling now and then for love does sooth[e] the worries sometimes [original English]. Isn't that so, my darling?'[5] She heaped banal advice upon him, such as 'I hope you are taking good care of yourself, going to bed *early* and eating properly . . . I am glad my little husband goes swimming and doesn't read all the newspapers.'[6]

The few surviving letters from Auguste Viktoria to her husband give us unusual insights into the emotional life and the private domestic world of the imperial couple. They confirm Philipp Eulenburg's surprised observation around the turn of the century that the 'imperial marriage-bed' had been elevated into a true 'cult' by the Kaiserin.[7] When Wilhelm decided to spend a few days at Wilhelmshaven at the end of his 1890 Scandinavian cruise and before his journey to England, instead of hurrying back to his wife, she wrote to him reproachfully saying that it made her cry 'every time I think that I could have been with you for almost 3 days, oh my darling, it is fearfully hard when one loves one's little husband so much . . . God grant that our days together in Berlin will be all the nicer, as the children are all away and we can pretend that it is our first married year. What do you think, darling? . . . Your little wife, who loves you so passionately and longs for you very much, kisses you and strokes you quite softly as you like it.'[8] Alluding to his approaching visit to England she warned him: 'I hope you will not pay court to the beautiful daughters of Albion too much, or else I shall appear to you in your dreams again but not as nicely as before.'[9] A few weeks later, when the Kaiser was on his way to Vienna, she wrote to her 'dear, sweet darling': 'I kiss you all over & always pray for you, that you keep well and that nothing happens to you.'[10] Two years later she promised him, as in the first years of their marriage, 'loving kisses all over & on my place'.[11]

Even without the Kaiser's letters to her, it is possible to conclude from some of Dona's replies that the couple's erotic feelings were mutual. In the summer of 1892, for instance, they communicated their dreams to each other in the form of drawings. 'Your drawings make me feel quite excited', the Kaiserin wrote in one letter.[12] She willingly agreed to his particular wishes, which were evidently (as shown in his letters to his mother from Kassel and his correspondence with Miss Love)[13] inspired by a fetishist fascination with her hands and gloves. 'I shall let you have all your little pleasures', she promised him in expectation of his return. 'I always have gloves on at night now to take care of my hands! – How dear and nice of you to send me more drawings and I am looking forward to them so much . . . You naughty little husband for thinking that I might perhaps not give my little husband such a nice welcome as he imagines. You know how awfully much I love you, and if I am well, *how* willing I am to do everything. Come back as soon as you can my own darling and you will not be disappointed', she wrote in mid-July 1892, eight weeks before the birth of her daughter.[14]

In describing her dreams to Wilhelm, Dona was patently using feminine wiles to strengthen her hold on him. 'My dear, sweet darling', one of her letters of the summer of 1892 reads. 'Last night I dreamt of you so vividly, darling. At first you were quite out of my reach, then you came to me and I clung to you so tightly so as not to let you go again. You were wearing something very strange on your head, but in my dream I told you how well it suited you, and then of course you had to go away again.'[15] Some of the Kaiserin's dreams, however, reveal her deepest fears. Thus in September 1890 she told her husband: 'Last night I dreamt of the French Revolution, with hordes of women milling about, it was terribly upsetting. May God protect us from such things.'[16]

The influence which she exerted on Wilhelm II is rarely easily detected, but it is probably safe to assume that it was particularly active where social life and political appointments were concerned but also, principally, in questions of the children's upbringing and family matters in general.[17] It was demonstrated, for example, when in 1895 the Kaiser was urged both by the Chancellor and by Eulenburg to receive the second wife of Count Hatzfeldt, the German ambassador in London. In reply, Wilhelm told Eulenburg that he greatly regretted 'that nothing could be done in this matter. The Kaiserin had *refused in such decided terms* to receive the Countess that it was useless to make any further attempt. "We men take a less serious view", H.M. said, "and I sincerely wish that it could be done. But in these questions I cannot give orders to the Kaiserin. It is her domain. I cannot command her to receive someone who has made herself unacceptable to society." When H.M. appealed to me [Eulenburg wrote] and to my knowledge of the Kaiserin's character and conduct, I certainly could not deny that it was a difficult state of affairs.' It was a fact that the Kaiserin had a '*great* aversion to divorced women' and even the Kaiser could do nothing to change that.[18] As we shall see, Auguste Viktoria was equally intractable when Princess Sophie of Prussia and Princesses Ella and Alix of Hesse adopted the Greek and the Russian Orthodox faiths respectively.[19]

There are nevertheless indications that at this time, as in earlier years, Wilhelm found it tedious and constricting to be in constant proximity to his wife, heavily influenced as she was by her Oberhofmarschall von Mirbach and by the three pious '*Kreuzzeitung* ladies-in-waiting', Countess Therese von Brockdorff, Countess Mathilde von Keller and Claire von Gersdorff.[20] It is true that soon after his accession he had broken off the extramarital relationships in which he had indulged in the 1880s; but the fact that he still felt the need to get away from home is attested not only by Dona's constant complaints. Philipp Eulenburg, whose friendship with the Kaiser was now 'firmer than ever' and was observed jealously by the Kaiserin,[21] commented tellingly after the summer cruise of 1890: 'These Scandinavian voyages are a sort of fiord carnival. It is in fact H.M.'s "season" – minus feminine conversation, which is as he likes it.'[22] Early in 1891 the Kaiser brought together all the men who had accompanied him on the two previous Nordic cruises. A 'Nordland' Society was established, with the Kaiser as master and his travelling companions as members. On this occasion Waldersee met Count Emil Görtz, who 'entertained the company with humorous recitations'.[23] Such attempts to escape from the wife whom he sometimes found cloyingly sweet and clinging should not, however, mislead us into assuming – as many of his contemporaries and a few later commentators thought they could detect – that Wilhelm despised women in general and felt at ease only in the company of men.[24]

For Auguste Viktoria, more than for her husband, the building of churches was a major preoccupation. When the Church of the Ascension in Berlin was consecrated in June 1890 Waldersee noted approvingly that thanks to the 'energetic action of the Kaiser and Kaiserin . . . the appalling shortage of churches in Berlin [was] at last in the process of being remedied to some extent'. Since Wilhelm's accession, work had started, or was planned, on the construction of ten new churches, and the Kaiserin had single-handedly raised the money for the Church of the Ascension, Waldersee recorded.[25] Seven years later there were no less than twenty-two new churches in Berlin, which had been built with private funds collected by Mirbach – 'chiefly from Jews', it was said – in the name of the Kaiserin.[26] In the Foreign Office too it was well known that in such matters the Kaiserin was the driving force. Kiderlen complained in 1893 that 'all kinds of "Protestant" causes will approach Her Majesty the Kaiserin, and I am afraid of Her Majesty's inclinations and Mirbach's influence, which is very narrow-minded in these things'.[27] Auguste Viktoria's 'Protestant' attitude almost led to an international incident when she 'obstinately and for a considerable length of time' refused to call on the Pope during her visit to Rome in 1893. 'The gracious lady had religious scruples and thought that as the leading Protestant reigning lady she ought to ignore the Roman bishop. It took all H[is] M[ajesty]'s powers of persuasion and several reports from the Foreign Office to overcome H[er] M[ajesty]'s objections', Arthur von Brauer wrote confidentially to his Grand Duke.[28]

Contradictory as it may seem, the pietism which many derided (or simply regretted) in the Kaiserin did not by any means prevent her from displaying a lamentable arrogance and – with the passage of time – an excessive love of luxury. The Empress Frederick and her daughters, above all, were affronted by the Kaiserin's haughty behaviour.[29] 'Dona enjoys her position intensely & her whole face expresses the most intense satisfaction', Wilhelm's mother complained shortly before Bismarck's dismissal. 'Her pride is so great and she thinks she knows better than everyone because she is the Empress.' 'She meddles in *every* thing the family does, *every little* trifle is reported to her & she orders & directs in a way very galling for the others.'[30] It was simply 'offensive' for her daughters and her daughter-in-law Irène to be ordered about by Dona.[31] The young Kaiserin had 'something condescending & patronizing wh[ich] irritates me & rubs one up the wrong way; – & she orders her sisters in Law about as if she were the Empress Augusta, wh[ich] makes me *frantic* sometimes'.[32] In January 1891 the Empress Frederick had a further reproach for her daughter-in-law: 'Poor thing she seems to think that she has to exercise a sort of police supervision of the whole family in *all* things! This is not called for either by her position or her age or her experience. The Empress Augusta whose ideas were of another century also thought this, & Dona fancies she has stepped into *her* shoes (over my head) and must continue this! To me it is most galling!'[33] The blame for this presumptuous behaviour lay with the constant flattery heaped upon her by her entourage, against which no sensible influence could make itself felt at the Hohenzollern court. 'The flattery that is lavished on *both* of them is enough to turn anybody's head, & it is no wonder that theirs *is* turned!', Vicky wrote to Queen Victoria. 'Poor Dona' was therefore quite convinced 'that all William & she do & think & say is perfect'.[34] Dona's highly intelligent mother-in-law did not entertain the possibility that she herself gave the young Kaiserin a sense of inferiority which caused her to put on airs. But less enlightened members of the family – particularly the Kaiser's eldest sister Charlotte – also showed little understanding for Dona's arrogant behaviour.[35]

Soon after the fall of Bismarck complaints began to be heard, especially from Protestant critics like Waldersee, about the excessive love of luxury and pleasure which characterised the court of the young imperial couple. Family festivities became more and more grandiose and expensive and even the pious Kaiserin wore an extravagant quantity of jewellery.[36] It was often Wilhelm himself who encouraged his wife in her growing taste for splendour. In the winter 'season' of 1893–4 Philipp Eulenburg wrote to his mother describing a concert in the Marble Hall of the Neues Palais at which the Kaiserin had appeared 'in a gown of blue velvet combined with yellow muslin, large diamond and sapphire jewellery – not very well dressed'. By way of explanation he added: 'The Kaiser sometimes thinks up these things himself.'[37] Meanwhile the young sisters of the Kaiser and the Kaiserin went from one ball to another and caused much gossip by their behaviour.[38] Waldersee deplored the moral decline which he detected at the Berlin court and in the Kaiser's personal conduct too. 'There is no better illustration of how times have changed than the contrast between the Kaiser's prohibition of racing on Sunday in 1888 and his invitation, 2 years later, to a sleigh-ride and a ball on Sunday! Things are going backwards step by step', he wrote in early 1891.[39]

From 1894 onwards there are increasing indications that not only Wilhelm but also his wife, now thirty-five years old, suffered some kind of nervous crisis. She went on a starvation diet to lose weight, but although this was successful[40] she became increasingly irritable, particularly over her husband's 'incessant' and 'restless' activity.[41] 'Ostensibly because of her throat' but in fact 'on account of the state of her nerves', in 1894 she insisted on taking an Easter holiday with her children at Abbazia on the Adriatic. She was horrified, however, when the Kaiser, who had gladly agreed to her plan, wanted to give her a yacht, 'because Her Majesty thinks that this would increase the danger of this expedition becoming an annual event'.[42] The following year when Wilhelm announced that he intended to go on a Mediterranean cruise in an English yacht with the Kaiserin, she was '*in despair* and begged me to do what I could to prevent the journey', as Eulenburg recorded.[43] At the end of 1897 the Kaiserin's nervous exhaustion led to a complete breakdown. As August Eulenburg informed his cousin Philipp, she suffered 'such severe nervous shock that one cannot get any sense out of her. His Majesty, with his usual impetuosity, declared that a stay in the South (Riviera) was absolutely necessary, an idea which did not please the All-Highest patient at all, and now we are engaged in the cross-fire which we have seen so often, as you know.'[44] The marital crisis to which the Oberhofmarschall was alluding now became a persistent problem. During his Scandinavian cruise of 1898 the Kaiser discussed his relationship with the Kaiserin at length with his best friend, who immediately passed on what he had heard to the Foreign Secretary and future

Chancellor, Bernhard von Bülow. 'The poor woman's nerves have been so ruined by child-care and royal duties that H[is] M[ajesty] does not believe her present calm state can last . . . There was such a fight in Homburg that the Kaiserin almost lost all control of herself. Then there have been angry scenes between Therese Brockdorff and the Kaiser, which ended with the Kaiser dismissing the good Countess. H[is] M[ajesty] complained in the strongest possible terms about the intolerable disturbances which the Kaiserin inflicts on Him while he is working or resting by constantly bursting in and pestering him with nursery matters. I have the impression that peace prevails for the time being – and might also continue at Wilhelmshöhe, as they lead a country life when they stay there; but knowing the Kaiserin as I do, I fear the problems may increase in spite of all cures and precautionary measures.'[45]

The relationship between Wilhelm and Dona in the following years was shaken by so many crises that their doctors and members of their immediate entourage feared both might suffer a serious nervous breakdown.[46] In a letter to the Kaiserin's personal physician, the Kaiser complained that she had 'fired off unbelievable telegrams' to him because he had inspected a pigsty at Urville on his own, without her. 'The Kaiserin's nerves are in a condition which worries me very much', he wrote.[47] In the autumn of 1900 at Rominten there were such 'appalling scenes' between husband and wife over the education of their sons Auwi and Oskar that the Kaiser spoke to Eulenburg 'in great distress and anxiety' of the possibility that his wife might need to be sent to a clinic for nervous disorders. 'The poor dear Kaiserin really seems to be in a bad nervous condition', Eulenburg reported to Bülow after he had seen her rush after the Kaiser 'like a madwoman', when she had 'screamed and raved so much' that Wilhelm did not know what to do. On the train to Tilsit, where he was to unveil a memorial to Queen Louise of Prussia, the Kaiser launched into 'a very sad, painful outpouring of his feelings' which lasted for hours, and during which he told his friend that the Kaiserin had 'made scenes the *whole night*, crying and screaming'; it had been 'an absolute paroxysm'. He had 'begged her *on bended knee*' to be reasonable, but in vain. 'I am completely at loss', the Kaiser confessed. 'What am I to do? These crises and scenes are killing me. I cannot stand it. The Kaiserin is ill – ill because of a daily routine which is absolutely *impossible* for a Kaiserin. She cannot be a "bourgeois" mother, a loving wife and a ruling Kaiserin at the same time. She has reached the end of her tether and is becoming an "impossibility" to herself and to me. Advice must be sought . . . The thought of seeing the poor Kaiserin consigned to a cold-water treatment hospital is so frightful that I cannot take it in!' This reference to a mental hospital was evidence enough for Eulenburg of what the Kaiser really feared: that the Kaiserin might 'lapse into eccentricity in the same way as her mother'. But it was not only Auguste Viktoria's nerves which needed to be cured, the Kaiser said; in her outward appearance too it was already possible 'to see the effect of the shattered nerves in her lined, prematurely aged face and her grey hair'. Eulenburg tried to console his imperial friend with the thought that the Kaiserin was indeed suffering from a '*temporary* impairment of the nervous system', but that the doctors would undoubtedly find a way to cure it. Writing to Foreign Secretary von Bülow, however, he confessed his fear that these developments would 'unfortunately have a very distinct and severe effect on the private life of the Kaiser for the time being and may perhaps become significant in *political* matters by *affecting his nerves*'. Moreover, he warned, the imperial marriage itself was in danger if 'the Kaiserin *eventually* makes herself *unbearable* to the Kaiser by continuing to make intolerable scenes'. The 'comment about premature ageing' was 'a symptom which should not be overlooked'.[48] Eulenburg finally advised the Kaiser, who was 'so oppressed by fear of the dreadful scenes' made by the Kaiserin, to leave the 'imperial marriage-bed' as soon as a nocturnal scene began and to lock the doors to his own room.[49] Wilhelm accepted his friend's advice with 'childlike naivety', but nevertheless showed 'a certain masculine pride' when, surprisingly, the 43-year-old Kaiserin proved to be expecting a child a year later. The fact that she hid her condition from the Kaiser and defied her doctors' orders by accompanying her husband to the manoeuvres and afterwards to Cadinen and Rominten, where she suffered a miscarriage, earned the disparaging comment from the Kaiser's favourite that 'her love for H.M. is like the passion of a cook for her sweetheart who shows signs of cooling off. This method of forcing oneself upon him is certainly not the way to keep the beloved's affection.'[50]

PRINCE AND PRINCESS HEINRICH

Although he was stationed at Kiel and frequently stayed with his wife's family in Darmstadt, Wilhelm's 'devoted and obedient little brother' Heinrich, as he called himself,[51] was duty-bound to come to Berlin for all major court functions, all important state ceremonies, national festivals and ceremonies such as the opening and closing of the Reichstag and of the Landtag; he was also expected to attend major military reviews and the autumn manoeuvres. At important dinners and festivities his wife Irène was also obliged to be present. She was the sister of Princess Victoria of Battenberg, who lived in England, and also of the future Tsarina Alexandra and of Grand Duchess Sergei (Ella) of Russia. The frequent journeys to Berlin were troublesome enough, but Prince and Princess Heinrich had the additional problem of where to stay while they were there, since for many years they had no satisfactory accommodation in the capital. At first they stayed at the Empress Frederick's palace or in the Schloss itself, but this was only a temporary solution. After the death of Kaiserin Augusta, Wilhelm decided that his brother should take over the royal palace of Unter den Linden, where their grandfather had lived, as his permanent residence in Berlin. But the old Kaiser's daughter, Grand Duchess Luise of Baden, refused to remove her parents' furniture from the palace and to give up her own right of residence, with the result that Heinrich and his family never felt at home there. In 1893 the Kaiser decided to have the Niederländisches Palais fitted up for his brother, but the question of the old palace of Unter den Linden remained unresolved.[52] It was not until they acquired Schloss Hemmelmark in Schleswig in the summer of 1896 that Heinrich and Irène gained a greater degree of independence from the Berlin court, a source of much pride and pleasure to them.[53]

When the reigning Kaiserin asserted to Philipp Eulenburg in the autumn of 1895 that 'Prince Heinrich, who formerly loved his brother tenderly', had been 'turned against him by sister Charlotte',[54] the evidence suggests that this was at best only partly true, for signs of an estrangement between Wilhelm and Heinrich can be detected long before this.[55] We have already seen in an earlier chapter the ill-feeling which arose between the brothers when Heinrich and Irène, but not the Kaiser, were invited to England for the wedding of the Duke of York in 1893.[56] With characteristic modesty Heinrich strongly resisted Wilhelm's plan to promote him to the ranks of general and of admiral. As his Adjutant Freiherr von Seckendorff reported to the Chief of the Naval Cabinet, the Prince would be 'extremely upset' if this premature promotion were to take place. Since his appointment as naval captain and colonel in the army the Prince had been afraid that 'the pressure to go further would become increasingly strong and impossible to resist', Seckendorff wrote. Prince Heinrich, however, very much wished to make his own way in the navy, 'to which he is deeply attached by inclination and by his whole nature'. In the Prince's name his Adjutant pleaded 'that His Majesty might be prepared to postpone the promotion of His Royal Highness until the time when His Royal Highness is due, by *seniority*, to become a general', for it was not in fact possible 'to be a general in the army and a captain in the navy'. If the promotion should be imperative at the beginning of 1893, the Prince would have to 'accept with resignation'.[57]

Within only a few years of Wilhelm II's accession, Heinrich and his Hessian–English wife were regarded by insiders as belonging to the rebellious faction of the imperial family which had formed around the Empress Frederick. Although the widowed Empress had at first complained of Heinrich's habit of following Wilhelm everywhere and of visiting him almost every month while he never had time to visit his mother, by the summer of 1891 she was defending him against the Kaiser's attempts to control his life, protesting that 'nothing irritates me more than when Henry is pulled up, found fault with & interfered with

31. The Empress Frederick with her family at Friedrichshof. Standing, from left to right: Princess Irène of Prussia, Crown Princess Sophie of Greece, Princess Charlotte of Saxe-Meiningen, Kaiserin Auguste Viktoria, the Empress Frederick, Princess Viktoria of Schaumburg-Lippe, Princess Margarethe of Hesse. Sitting: Prince Adolf of Schaumburg-Lippe, Prince Friedrich Karl of Hesse, Prince Heinrich of Prussia, Kaiser Wilhelm II, Crown Prince Constantine of Greece, Prince Albert of Schleswig-Holstein, Grand Duke Ernst Ludwig of Hesse

by W. & his people, it makes me quite savage!'[58] The following spring Grand Duke Sergei returned from the funeral of Grand Duke Ludwig IV of Hesse at Darmstadt bringing 'very gloomy' accounts of what was happening in Germany: Kaiser Wilhelm was 'mentally ill' and had 'fallen out' with Princess Irène.[59] Bernhard Bülow was shocked to discover during a visit to Berlin that 'Prince Heinrich . . . [had] joined in the general chorus of criticism of our all-gracious sovereign in Darmstadt and in England'. 'His own relations – especially all the Russian, Danish, English and Greek ones, the whole Rumpenheim clique is very hostile towards our all-gracious sovereign.'[60]

At the beginning of 1893 open conflict finally broke out between the two brothers. Under the influence of his wife, and increasingly that of his mother too, Heinrich rebelled against the obligations which the Kaiser tried to impose on him. His request for an increase in his appanage was brusquely rejected by Wilhelm as 'quite out of the question'.[61] They also quarrelled violently 'about the royal family's religious affairs' – presumably in connection with the change of religion of Crown Princess Sophie of Greece (a sister of Wilhelm and Heinrich) and of Grand Duchess Ella (Princess Irène's sister).[62] The crisis finally reached its peak in early 1893 when instead of announcing his arrival in Berlin to the Kaiser, Heinrich committed the impropriety of sending him word through a servant that he would be lunching in the Querallee. The Kaiser insisted 'that the Prince should inform him of his arrival by telegram and seek his orders *or* have himself personally announced on arrival by the Flügeladjutant on duty; he had always followed this procedure with his grandfather'.[63]

The estrangement between the brothers prompted the Kaiser to ask the Chief of the Naval Cabinet in February 1893 to investigate the 'difficulties in the relationship' between him and Prince and Princess Heinrich and to suggest to him how 'these difficulties could be permanently put to rights'. After taking extensive soundings Admiral von Senden summarised his conclusions. 'In the first years of Y[our] M[ajesty]'s reign Prince H[einrich] was completely absorbed in his naval service. The Prince had only one ambition: to devote himself fully and completely to Y.M. through the All-H[ighest] service.' But this wish of Heinrich's had been seriously undermined by the influence of Princess Irène, who had grown up in completely un-Prussian surroundings in Darmstadt and England, the Admiral claimed. The Princess insisted on 'isolation' and 'completely free self-determination' and took 'a firm stand' against any influence from Berlin. 'Y.M. knows from experience', Senden wrote, 'that H[er] R[oyal] H[ighness] was brought up in circumstances in which the inner circle [of her family] had ideas which were quite at variance with all the traditions and time-honoured conventions of the [Prussian] court and of the [Hohenzollern] family . . . These ideas later gained the upper hand when H.R.H. married and sought to distance herself and her husband from Prussian ways, which she did with increasing success.' 'The former home of the Princess, where the simple, free habits of a happy childhood could be kept up, became the Prince's home too. Their rare visits to Berlin became merely an oppressive burden to them as they became increasingly estranged from the court there.' To solve the problem, Senden suggested that a three-man committee, consisting of the chiefs of the Military and Naval Cabinets and a third trustworthy person, should be entrusted with the task of conveying the commands and wishes of the Kaiser as head of the imperial family to the Prince and his wife, in order to achieve harmony at least in outward things.[64]

One important reason for Heinrich's reserve and Irène's wish to be allowed to lead as private a family life as possible was not mentioned in Senden's report:

even as a small child their son Waldemar, born in March 1889, was exhibiting the unmistakable signs of haemophilia. The Empress Frederick tried to console herself and her own mother with the thought that apart from this *one* defect in his constitution the child was strong and healthy. Furthermore, unlike her youngest brother Leopold, who had bled to death in Cannes in 1884, the little Prince Waldemar was of a quiet disposition, she claimed. 'I think it a very favorable sign that he has had no outward bleeding – & that he is of a quiet placid temperament, wh. poor dear Leo of course was not! This dear little man will not fret and worry about everything like our poor Leo, who was so sensitive & put himself so often needlessly into such a state.'[65]

In view of the strained relationship with his brother it is not surprising that the Kaiser very rarely entrusted him with important political tasks. Wilhelm II's proposal that Russia should join the Triple Alliance and seize the Straits, which Heinrich conveyed to his brother-in-law Nicholas II in November 1894, remained (at least until the first Russian revolution of 1905) an exception.[66] But the Kaiser also lacked confidence in Heinrich's capabilities, probably not without reason. Although Prince Heinrich would have become regent on behalf of the young Crown Prince in the event of the death or abdication (frequently contemplated in 1895–7) of Wilhelm II,[67] he was regarded as 'superficial, obstinate and insignificant' within the family and in political circles in general.[68] In the summer of 1896, after one of many clashes between the Prussian Prince and the heir to the Bavarian throne, Prince Ludwig, Philipp Eulenburg complained: 'Prince Heinrich's behaviour is terribly *immature*. Until all the courts in Europe have assessed him at his true, stupid worth and *properly* judged what he says and does, he can still do much harm. What kind of a regent would he make if – God forbid – something happened to our sovereign.'[69] Eulenburg was astounded, however, when shortly after this he heard from Plessen and Lyncker that Prince Heinrich had told them that Prince Ludwig was quite within his rights; the German ruling princes 'did not have to put up with anything from the Kaiser and had a perfect right to their own independence. The whole "German rubbish-heap" would not hold together for much longer anyway.' Wilhelm's best friend wondered whether he should tell the Kaiser about this remark, for it was 'certainly almost rebellion in his own house!'[70]

Only a year later the Prince proved to the world that such derogatory views of his intellectual capabilities were no exaggeration, when at the embarkation of German troops for China he proclaimed in the presence of his imperial brother: 'For me, fame and laurels have no allure; I am drawn by one thing alone: to bring the gospel of Your Majesty's sacred person to foreign lands, to preach it to everyone who wants to hear and also to those who do not want to hear.'[71] In July 1899 in conversation with Eulenburg Wilhelm expressed what the former considered 'an accurate opinion of Prince Heinrich', saying that he was 'a completely feminine, dependent character, and afflicted with anglomania' – a remark which perhaps says as much about the Kaiser's preoccupations as about the personality of his brother.[72]

THE KAISER'S SISTERS

In the early years of his reign Wilhelm II's relationship with his four sisters, Charlotte (Hereditary Princess of Saxe-Meiningen), Viktoria (Moretta), Sophie and Margarethe (Mossy), was characterised by violent upheavals which were caused partly by the continuing bitter conflict between the Kaiser and his mother, partly by the arrogant, status-conscious and at the same time petty-minded Protestant attitude of his wife, but also partly by the behaviour of some of his sisters, which did not always seem suitably royal. Soon after Wilhelm's accession Princess Sophie was married to Crown Prince Constantine (Tino) of Greece, Princess Viktoria to Prince Adolf of Schaumburg-Lippe and Princess Margarethe to Prince Friedrich Karl (Fischy) of Hesse-Kassel. Given the close relationship of the Hohenzollerns to the British, Russian, Danish and Greek royal families it was inevitable that family quarrels would also have serious consequences for international relationships between the powers.

While the three younger sisters, at first still unmarried, sided with their suffering mother, the eldest sister Charlotte and her husband Bernhard, Hereditary Prince of Saxe-Meiningen, took a decisive stand in favour of the young Kaiser. The Empress Frederick complained bitterly in a letter to Stockmar's widow in February 1891 that the hostility shown towards her by Wilhelm and Dona was shared by Charlotte and Bernhard (and at this time also by Prince Heinrich). 'My 3 eldest children *still* take the San Remo attitude – it makes such a *gulf* between me and them, and that hurts so much.'[73] 'There is *no* one to help me!', she wrote, in her isolation, to Queen Victoria on 5 December 1890. 'Henry *cannot* and also does *not* understand. Bernhard & Charlotte have worked against me for 3 years . . . W. & Dona have the power & I am nothing but a defenceless *widow*!'[74] She had received no sign of life 'from those whose mother I have the misfortune to be!', she wrote bitterly, on her return to the Rhine after a visit to England.[75] Although Prince Heinrich gradually came round to her side, the widowed Empress continued to find her eldest daughter's behaviour 'most odd'; she 'fights shy of me, hardly comes near me'.[76] Her son-in-law Bernhard was also 'impertinent' and 'ill-bred' towards her and took every opportunity of making 'a rude & sneering remark'.[77] Not long afterwards she again complained that 'the behaviour of W. & Dona – of Bernhard and Charlotte is a *constant* worry & trouble to me – but I have quite given *them* up, and trouble my head no more about them'.[78]

32. The Kaiser's four sisters. From top to bottom: Margarethe, Sophie, Viktoria and Charlotte

The close political and social contact which Bernhard and Charlotte maintained with the young Kaiser and Kaiserin, and which brought them certain career advantages (as we shall see), did not exclude conflict with Wilhelm and Auguste Viktoria, however. In 1896 Charlotte complained of the 'arrogant condescensions!!!!' of the Kaiserin, whom she ungraciously described as 'disgusting, old and ugly as if she were 50'.[79] The constant disputes over precedence between the Kaiser's eldest sister and his wife, the Hereditary Princess's erratic, pleasure-seeking lifestyle, together with her husband's often offensive manner, gave rise to gossip and criticism in the family (and well beyond it). The Empress Frederick often expressed sorrow at her daughter's superficial, restless life and was deeply concerned by her numerous extramarital relationships and her tendency towards malicious intrigues, through which Charlotte caused great mischief. 'Charlotte *must* go everywhere, and put herself "en evidence" – she *never* can keep quiet, and is always gadding about', she complained in a letter to her daughter Moretta.[80] When she heard that her eldest daughter had decided to learn Spanish, the Empress predicted that no more would come of it than of her earlier attempts to learn Italian and Russian, for Charlotte simply had no talent for languages. She welcomed her daughter's plan nevertheless, for it was better 'when she has a sensible occupation to keep her at home instead of constantly having visitors & paying visits and flying about from morning to night, and leading such a foolish, aimless life, wh[ich] ends in mischief-making'.[81] What she meant by this remark is evident from a letter in which she begged the newly married Moretta to be 'a tender loving good & unselfish little wife!' and 'never to be to him what Charlotte is to Bernhard!'[82] In a letter of 28 November 1890, which she asked Moretta to burn, she indicated which married princesses behaved dutifully despite marital problems and which could certainly not be regarded as good examples. 'I am so anxious for you to be a good wife as Dona, Calma, Yaya, Louisa Hohenzollern, Irene, Viktoria Battenberg, Ella & so many others are – and *not* like our poor Charlotte, or my poor sister Louise [Marchioness of Lorne] or Anastasia [Grand Duchess of Mecklenburg] etc . . . that give one such anxiety & do not make their Homes comfortable & happy – & are so much talked about in the world – & their names not mentioned with the respect one would wish. Aunt Minny [the Tsarina Marie Feodorovna], Aunt Olga [Queen of Greece], Elisabeth of Roumania, Pcss Wied are also such examples of good Hausfrauen – Aunt Beatrice [Battenberg], poor Aunt Helen [Duchess of Albany], the Queens of Holland & Spain. God knows *all* the Husbands of these Ladies were *not* patterns of perfection & excellence, therefore it must often have been a struggle to keep up to the mark & do their *duty*!'[83]

The Empress Frederick thought her eldest daughter's malicious tongue very dangerous, and not without reason. A compulsive gossip, she told the

Flügeladjutant Carl Wedel in early 1891 that while on a visit to Moretta in Bonn her mother had 'done all she could to stir up feelings against the Kaiser and told her son-in-law, Prince Adolf Schaumburg, dreadful stories about him'.[84] The Empress Frederick felt particularly threatened by the close relationship between Charlotte and the degenerate court of Duke Ernst II of Saxe-Coburg and Gotha, her late father's brother, whom the widowed Empress considered one of her worst enemies[85] and who was universally regarded as a 'morally very inferior man', 'thoroughly unprincipled, false, lying, a braggart and an intriguer'. In Waldersee's eyes, for instance, there was scarcely anyone 'who did less credit to his throne than this Coburger'.[86] Every time the Empress Frederick heard that her daughter was at Coburg she foresaw trouble. 'Charlotte is at Coburg again; I am always rather alarmed at all the nonsense, mischief & gossip she carries about, wh[ich] one believes, not knowing what confusion she always makes. There is a great deal of harm done in that way perhaps even unintentionally.'[87] That her anxiety was not unfounded we shall soon see from the events connected with her daughter Moretta's marriage plans and with the change of religion of Crown Princess Sophie of Greece.

In 1893, as a result of a violent quarrel between Bernhard of Saxe-Meiningen and his brother-in-law Wilhelm, the Hereditary Prince angrily resigned from the army and withdrew to Meiningen with his wife.[88] Not long afterwards, surprisingly, Bernhard accepted the Kaiser's offer of the command of the 22nd Division, hitherto held by Prince Friedrich of Hohenzollern, and moved to Kassel with Charlotte.[89] But the couple remained on bad terms with the Kaiser and Kaiserin. When Charlotte paid a visit to Prince Ernst zu Hohenlohe-Langenburg in London in January 1894 the latter reported to his father that the Kaiser's sister had 'poured out endless complaints. She and Bernhard seem to be very annoyed, and he wants to resign soon.'[90] The feud between Wilhelm and Bernhard was so serious that the Prince of Wales could not receive his niece Charlotte at Sandringham because of this 'unpleasantness'.[91] The couple were all the more delighted at the surprising appointment, on the occasion of the old Kaiser's birthday, of the Hereditary Prince as general in the infantry and commanding general of the VI Army Corps in Breslau. Charlotte wrote happily to a friend on 25 March 1895 in a mixture of English and German: 'These lines . . . come from a beaming Commandirende Generalin [commanding general's wife]. I wish you could have witnessed on the morning of 22nd the rushing screaming of Bernh. & his 2 aide de camps [sic], to my quiet writing table I was working at! Telephone from William Glückwunsch [congratulations] zum 6. Corps u. General d. Infanterie. I was stunned & tears rushed to my eyes, as the telegramm came: so fine in commemorating our beloved g[ran]d father etc.! I *never* before saw my husband *so* perfectly delighted & wild! A dream fulfilled & what a delightful future in every way! From all sides the gratulations pour in, really touching. And I!! well I feel proud of my husband, & only longed for *my* Father to witness our joy & his beloved cherished Silesia.'[92] In officer circles, not surprisingly, there was much annoyance at this preferential treatment of a brother-in-law, which gave the impression that the Kaiser could hand out senior army commands to members of his family like Easter presents. The promotion of the Hereditary Prince 'in one big leap' had 'caused much bad blood in the army, and quite rightly', Waldersee commented. 'We are quite happy to accept that Prussian princes should be advanced more rapidly, but if the minor princelings are given these advantages too it is a blow in the face for the army; . . . and what is more, these princes are almost without exception wretched creatures. The Meiningen fellow is perhaps the best of them, but he has considerable weaknesses; and I also think that as commanding general, and in Breslau of all places, he will soon cause much offence. Wherever he has served he has made himself universally disliked through his extreme rudeness; he has no manners and does not know how to behave in distinguished company. His tendency to make overhasty judgments and express them without due caution will cause many problems, particularly in Silesia; for instance he is very fond of indulging in brutal tirades against the Catholics, and that alone may soon create a rumpus.'[93]

Prince Bernhard repeatedly vented his feelings with great passion and violence not only against Catholics but also against Social Democrats, the liberal press and the parliamentary system in general. In his eyes, the Reichstag was a 'disgrace to German political life', 'the vilest representative body a great people has ever had in its history' and a 'collection of oxen and villains who take pleasure in keeping their nation small and powerless'.[94] The newly appointed Commanding General was to come into conflict above all with the 'swamp', the 'All-Highest, All-Gracious' Prussian Ministry of War.[95] When the Ministry refused him permission to build additional stabling behind the general headquarters at Breslau – a decision which put the Commanding General in a 'fuming rage' in which he wanted to 'throw everything over' – the Hereditary Princess defiantly approached a Jewish horse-dealer who had offered to build better stables for half the price. 'So the whole Province will see the Emperor's sister *helped by a kind Jew*', Charlotte wrote angrily to Baroness Heldburg.[96] Prince Bernhard took the view, which had more than a little bearing on Wilhelm II's future position of power, that the commanding generals should close ranks against 'interferences' by the War Ministry. Above all in questions of military training the commanding generals should take their orders 'only from His Majesty', he insisted.[97]

During their time at Breslau there was a noticeable cooling of relations between the Meiningens and the imperial court, above all where the Kaiserin, that 'conceited goose' as her sister-in-law called her, was concerned.[98] 'For various reasons' she did not 'fit in *there* any longer', Charlotte told her doctor.[99] In October 1896 Prince Bernhard and his wife received 'such an insulting and impudent' letter from the Kaiserin that they were tempted to withdraw from public life and retire to Meiningen. On the grounds that the (apparently distraught) wife of their Hofmarschall, Karl-August Freiherr Roeder von Diersburg, had left him, Auguste Viktoria had concluded that Charlotte was having an affair with Roeder. She issued an ultimatum that the Hofmarschall must resign from his position. According to Dona's letter 'the Kaiser did not yet know about it', 'but if there was no change she would have to tell him, and he would not stand for any nonsense'. Bernhard and Charlotte were determined, as the latter told her father-in-law, to show the world that they were 'above such *filth*' and 'would not drop a faithful servant on the basis of tittle-tattle'. 'This new impertinent interference by Her [the Kaiserin] is too bad and you can imagine how indignant B[ernhard] and I are. If she sends a rude answer and stirs up my brother against us, we shall come to Meiningen immediately, for She knows how to make W. flare up in no time at all! We have proofs enough!!! God knows who repeated this gossip to her, and it proves to us once again that one can *not* get on with her . . . Slander and gossip, envy and jealousy are now common currency in the "highest circle!" '[100] Prince Bernhard, equally enraged by the 'dreadfully rude, arrogant tone of the so-called gracious lady', protested that the Kaiserin had no right 'to watch over what Charlotte does and does not do as if she were "head of the family" and . . . to issue reprimands and orders to her'. That was the reason, he explained to his father, why he did not wish to remain in Prussia. 'In Berlin they maintain the fiction that every Prussian princess, even after her marriage, is still under the control of the king and queen as heads of the House [of Hohenzollern]. This fiction, which shows contempt for the honour and independence of the other German reigning houses, is what I have to stand up to.' Leaving the army and retiring to Meiningen was perhaps the only lasting protection against the demeaning attitude of the Kaiser and Kaiserin, the Hereditary Prince considered. 'It is after all not the first time that the Kaiserin has interfered in my domestic affairs in the most outrageous way', he complained. 'She does it to show that she has the upper hand even over princesses who have married out of the royal house.'[101] The incident threatened to have wide repercussions, especially as the King of Saxony offered to go and see the Kaiserin immediately and make it clear to her that Charlotte no longer belonged to the House of Hohenzollern. In the end matters were settled when Roeder brought his wife back to Breslau and made a point of appearing in society with her, but the anger aroused by Dona's presumptuous behaviour continued to rankle and was by no means confined to the

House of Meiningen.[102] When Charlotte learnt to ride a bicycle in the summer of 1898 she remarked that she was determined to resist any interference from the Kaiserin, who had already forbidden her sister Calma and her daughters to indulge in this 'indecent' sport.[103]

The life of Wilhelm II's eldest sister was increasingly overshadowed by porphyria, the rare metabolic disease which she had inherited from her mother, as has since been conclusively proved. The severe symptoms of this hereditary condition, which had been apparent since her twenty-fifth birthday – unbearable stomach pain, colic, paralysis in the arms and legs, sleeplessness, feverish sweats, constipation, palpitations, blistering skin rashes, dark red urine and periodic mental confusion – have been described in the first volume of this biography and more fully elsewhere.[104] The Empress Frederick was alarmed to hear of the signs of the illness which she had long predicted: Charlotte was suffering from 'fits of violent excitement'; she suddenly became 'so very thin' in 1893; her legs swelled up and were very painful; she had 'rheumatism' and daily headaches, was nervy and sleepless and had 'so little blood'. In August 1894 the Hereditary Princess sought the help of the celebrated doctor, Professor Ernst Schweninger, who was also Bismarck's personal physician. He diagnosed 'anaemia and nervous breakdown'. It was only many decades later that medical science identified porphyria as a distinguishable form of illness.[105] Until her death on 1 October 1919 Wilhelm II's eldest sister remained a patient of Schweninger and his colleague from Baden, the doctor and psychiatrist Georg Groddeck.

It cannot be ruled out that the second eldest sister Viktoria (called Moretta or Vicksey by her mother) also suffered from porphyria. At any rate she too made herself conspicuous by her behaviour throughout her life – she was widely regarded as a nymphomaniac – and had to be treated at a clinic in Bad Schwalbach in 1893 for 'persistent anaemia', disruption of haemoglobin production, stomach cramps, backache and loss of weight.[106] Soon after Wilhelm II's accession and after he had placed a definitive ban on Moretta's marriage to Prince Alexander of Battenberg,[107] the Kaiser began to look around for a suitable husband for his sister. He hit upon the youngest son of the ruling Prince of Schaumburg-Lippe, Prince Adolf, born in the same year as himself, good-looking and not without means, whom he had met in Bückeburg and who was at this time in military service as first lieutenant in the Bonn Hussars. As we have seen, at the beginning of 1889 Wilhelm commissioned Philipp Eulenburg to make secret investigations into the financial circumstances and private life of Prince Adolf. Eulenburg gave a positive report in February, although the question of the Lippe succession was still completely unresolved.[108] At the end of 1889 Eulenburg was sent on another secret mission to Bückeburg, where after protracted negotiations with Prince Adolf's father he succeeded in arranging for the young Prince to pay a discreet call on the Empress Frederick and her two unmarried daughters in Italy.[109]

At first the widowed Empress seemed to accept her son's decision as irrevocable. After a visit to Bad Homburg by Prince Adolf in June 1889 she wrote resignedly to Queen Victoria that he was certainly 'good looking & agreeable but of course not much of a parti!' The only other possible candidate for Moretta's hand was Prince Philipp of Württemberg, but as a Catholic he was less suitable, she commented at this time.[110] In reality, however, she continued working actively in secret to find what she considered a more suitable husband for her favourite daughter. As late as the summer of 1890, only a few months before the wedding which took place in November, she introduced Moretta to Hereditary Prince Ernst zu Hohenlohe-Langenburg, but this attempt was again frustrated by Wilhelm's determination to stick to his decision. 'My son has chosen *this* alliance', the widowed Empress wrote bitterly to Hermann Hohenlohe,[111] while in her letters to her own mother she complained furiously that Wilhelm was 'much pleased at having had *his* way! He has crushed & ruined without scruple *so much*, much – that he triumphs now of course.'[112] All three Victorias – the Queen, the Empress Frederick and her daughter Moretta – had the greatest hopes, however, for an English naval officer of noble birth named Captain Maurice Archibald Bourke.[113]

In a long letter of 12 June 1890 which the old Queen copied out, with numerous omissions, before destroying the middle pages of the original, the Empress Frederick had again seemed prepared to put a brave face on the Schaumburg-Lippe marriage. According to the Queen's copy she complained that she had 'cried so much I feel quite ill', but since 'William wishes this marriage particularly' she had accepted it as inevitable. She was also prepared to believe that Prince Adolf was 'thoroughly trustworthy & good & I am sure he will try to make her happy' and that Moretta too 'will try her utmost to do her duty'.[114] But what had she said in the many pages which the Queen did not copy out, but burnt? Strangely enough we know about their contents from a copy of another letter, which made the already disastrous relations between Kaiser Wilhelm II and his mother and sister even worse and could also have put his attachment to his grandmother the Queen under severe strain.

On 21 July 1890 Kaiserin Auguste Viktoria wrote a letter to Wilhelm II in which we find not only her own comments about the affair, but also those of the Duchess of Edinburgh and the Kaiser's sister Charlotte. From this remarkable and undoubtedly authentic document it becomes clear that Moretta was still in love with Captain Bourke after her engagement to Prince Adolf, that her mother favoured a match with this latest object of her daughter's affections, that the Empress Frederick had almost persuaded Queen Victoria to give her blessing to this step, and finally that Wilhelm and Dona had been fully informed about Moretta's new infatuation and the machinations of her mother and Queen Victoria since March 1890. In the letter the Kaiserin passed on to her husband an unwelcome piece of news which she had just heard from Charlotte. The latter, Dona explained to Wilhelm, had received a letter from her Aunt Marie of Edinburgh (a Russian Grand Duchess by birth, married to the heir to the Duchy of Coburg, Alfred, Duke of Edinburgh, the Empress Frederick's second brother). Charlotte gave the letter to Dona to copy – a striking example of the 'mischief-making' for which the Hereditary Princess had made herself feared. Duchess Marie's letter to Charlotte said: 'I must tell you that very disagreeable things are going on now, for poor Granny [Queen Victoria] has got once more completely bamboozled by your Mama [Empress Frederick], & the whole of the Bourke story was discussed with most disastrous results. First Granny talked [sic] it to Moretta, who took it quite lightly, said that she never took the episode au serieux ect. [sic] So Granny gave her a scolding and for once was quite sensible. But then she talked it over with your Mama, who told her a very different story: that Moretta was still madly in love with the gallant Captain, would have waited for him five or six years if necessary, followed him to the end of the world, and more of that sort of bosh and nonsense adding, that when Bourke broke it off, it nearly broke Moretta's heart. Whereupon dear . . . Granny shed romantic tears of real sorrow over the sad episode telling your Mama that if she had known it before this new engagement with Adolph . . . *she* (Granny) *would have certainly helped Moretta to accomplish this match!!!* Good heavens! The only thing that would rather have frightened her, she added, was *that dreadful tyrant William* who always takes things so badly & makes rows about anything. On his account also, Moretta was obliged to accept this new partie but would never forget her love for Bourke, the most charming individual in the world! Poor Granny wrote to me pages & pages about it, her mind is in a terrible state . . . She pretends now, that *you* poor Carolus [Charlotte] have created this mischief etc. She was evidently put up again by your Mama and begins again to hate William, but that will all disappear when he comes here [England] and again turns her round. I am going to write to him and prepare a little the ground for a careful & impressive talk with foolish Granny. – But what is downright wicked of your Mother, is so utterly to destroy Granny's present joy at the idea of Moretta's marrying at last und einen ächten Prinzen [and a real Prince], for she *was* very pleased, the old lady, & now her old mind is again tormented by romantic ideas of another thwarted attachment & eternal pinings for another man! But now only I *fully* see how right we were to stop the whole thing last March by telling William & how right our suppositions were that they both meant to settle the Bourke affair with

Granny when they came to England in June. Thank God a thousand times that we did it then, & prevented a most infernal row between your Mama Granny & William . . . *But* I should not be the least astonished if your Mama would try to bring it about again. God preserve us from further scandals, but somehow, I do not feel reassurance yet, that Moretta will really marry that very nice Adolph, so good looking, I think, & with such charming honest fine eyes. We are all quite taken with him & think her a lucky girl.' It was also a very bad thing, the Duchess wrote to Charlotte, that on their way to Princess Sophie's wedding in Athens the Empress Frederick and her daughters Moretta and Mossy had travelled in the *Surprise*, the very ship in which Captain Bourke was serving. 'What do you say to this story, darling?', Dona asked her husband indignantly after copying out the letter. 'But I told you recently that I did not think we could be sure of Vicky [Moretta] until the wedding had taken place. – I thought it my duty to write all this to you, darling, before you go to England, so that you can give Grandmama a good talking-to, for this is really going *too far*, when a girl, particularly a princess, *is engaged*, that her own mother is probably trying to break it off! You must also make it clear to Grandmama that Vicky [Moretta] must not travel back in the Surprise, for the constant contact with the man, when someone is like Vicky [Moretta] if she gets very worked up, she might well do something quite frightful and then tell you that there were reasons which compelled her to marry the fellow. With Victoria [Moretta] anything is possible, in my opinion! – Poor darling, I am so sorry, also because you had so much trouble with the whole Adolph business.'[115]

In the end all three Victorias yielded to the Kaiser's will. After Prince Adolf had travelled to Venice to meet the Empress Frederick and Moretta, the Empress reported to her mother fairly optimistically that 'Moretta seems to take great interest now in the preparation for the wedding & future Home, though she is very low sometimes and there is not much enthusiasm.'[116] And when Wilhelm arrived at Osborne on 4 August 1890 he was in fact successful, as the Duchess of Edinburgh had predicted, in convincing 'Granny' of the advantages of the match. Not least because he had brought about the marriage, he was in a particularly good mood and talked to the Queen at length about the reasons for Bismarck's dismissal, telling her 'how badly he had behaved, trying to intrigue against him with Russia'.[117]

In September 1890 the Empress Frederick reported that Moretta was again deeply depressed. 'The nearer the wedding approaches the more cast down she is', she wrote in one of her letters.[118] Once the wedding had taken place in the Berlin Schloss on 19 November 1890, however, the Empress Frederick urged her daughter to do her 'duty' and to be a faithful, 'good housewife', as we have seen.[119] She was relieved to be able to tell her mother at the end of 1890 that Moretta was now 'satisfied with her fate'.[120] Other members of court society who knew how much pressure Wilhelm had brought to bear in order to achieve this match were not optimistic about its future. 'I am quite sure that the couple do not suit each other and that the marriage can never be a happy one', Waldersee wrote on the day of the wedding.[121] Only a few years later Moretta's bizarre behaviour (and appearance) were causing indignation in the family. Her eccentricity reached its height in 1927 when as a widow aged sixty-one – Adolf had died in 1916 – she married a 27-year-old Russian gigolo, Alexander Zubkov, who moved into the Schaumburg Palace in Bonn with her and ruined her financially. It is not without irony that her second marriage ceremony was conducted according to the Russian Orthodox rite.[122]

Shortly after Moretta's first marriage there occurred one of the most unpleasant scandals in the history of the Hohenzollern family. It was triggered by the announcement made by the Crown Princess of Greece, Wilhelm's sister Sophie – during the wedding celebrations in Berlin in November 1890 – that she was to enter the Greek Orthodox Church.[123] The Kaiser's consternation at this is revealed by the savage marginal comments he wrote on the diplomatic dispatches from Athens, culminating in the exclamation: 'What a disgrace, what a humiliation for my whole house. What a sin against my father in his grave.'[124] Dona's outrage – which was further fuelled by the old rivalry between the Augustenburg and the Glücksburg lines, of which the latter had succeeded in winning the Greek throne[125] – at the Crown Princess's proposed change of religion was so great that the Kaiser told Waldersee that he blamed the premature birth of his son Joachim, born three weeks early, on the affair. He went on to complain that his mother had shown herself to be 'completely without religion', had 'nothing but scorn and contempt' for Christian belief and had even claimed that his sister's change of religion was no business of his, the Kaiser's, and that 'she could become a Jew if she felt like it'. Wilhelm had thereupon declared that he would forbid his sister to set foot on Prussian soil if she changed her religion, 'which naturally produced a further storm of indignation against the Kaiser' – his mother angrily declaring that in future she would neither visit nor receive him, if he carried out his threat to ban Sophie from Prussia. Wilhelm went on to speak 'very bitterly and despondently' about his mother and commented that she was 'only in Berlin to make mischief and to harm him'.[126] He told Hugo Freiherr von Reischach, his Hofmarschall, that his mother had treated him so badly that he had been reduced to tears.[127] The next day, when the Empress Frederick tried to visit the Kaiserin at the Schloss after her confinement, the Kaiser would not allow her to enter but promptly took her back to her carriage.[128] The gulf between Wilhelm's mother and his wife became even wider when Dona rejected the Empress Frederick's offer to hold the child during the baptism with

33. Princess Viktoria (Moretta), the Kaiser's sister, with her second husband Alexander Zubkov

the curt remark that 'William does not want it as you are not the Godmother.'[129] After this clash the Empress wrote to her daughter Moretta, who was on her honeymoon voyage to Malta and Egypt: 'I have been told *not* to go & see Dona at the Schloss so I leave for Kiel tomorrow with my Mossy! The parting from Sophie was terrible! *She* feels & I feel that she will never come back to Berlin, & it has nearly broken our hearts! There was *not* a *soul* of W's *or* Dona's court or a single official at the Railway Station, & W. was exceedingly rude to me again yesterday!' Only the thought that her youngest daughter was still unmarried and therefore needed her protection against the Kaiser as head of the family had prevented her from making a complete break with Wilhelm, she told Moretta.[130]

When Wilhelm nevertheless came to the station next day to say goodbye to his mother and Mossy, the Empress Frederick interpreted this gesture as a sign of remorse – 'perhaps that is his way of admitting that he has gone too far', she wrote.[131] When she arrived at Kiel she was relieved to find that her son Heinrich and his wife took her side in this family conflict.[132] 'I try & not think of the Schloss & of Berlin so much – that gives me such pain', she wrote to Moretta on Christmas Eve.[133] 'I was certain you would be as vexed & disgusted at the tiresome row & the utterly unnecessary quarrel Will: & Dona picked with poor dear Sophie & me! . . . I was *so* grieved & hurt for Tino, as it was so dreadfully rude to him.'[134] In order to counteract Wilhelm's 'contortion of *facts*', the widowed Empress commented to Queen Victoria that 'if he says that his sister is to be received *without Royal honours* in Prussia, because she has changed her religion, he simply shows he does not know the A.B.C. of his "métier" wh[ich] is to be civil to other royal families. It is rude & illbred & ridiculous, & makes me feel utterly ashamed of him; – & must disgust all sensible people.'[135] In her letters from Kiel she continued to complain of Wilhelm's hurtful behaviour towards the Greek Crown Prince and Princess, which had culminated in his failure to come to the station either to meet or to take leave of Sophie and Tino, or even to send a prince or a court official to represent him. At St Petersburg, on the other hand, the Tsar, the Tsarina, the whole imperial family and the entire court had gone to see the heir to the Greek throne and his wife on to their train; and the old Kaiser Wilhelm had always met his sisters, his daughter Luise, his aunts and his female cousins personally at the station.[136] When King George of Greece heard how the Kaiser had treated his heir and his wife, he would find it incredible 'that a man can be so foolish and childish when he is 32 years of age, and holds so high a position', she predicted.[137] She expressed the hope that Wilhelm would realise, with time, 'that he cannot carry out threats of the kind – if they are ever so solemnly made (as these were) – without consequences wh[ich] *must* destroy the peace of the family forever, – and show him in the light of a tyrant & bully – wh[ich] I think in spite of his love of showing his power & authority he would not like. That such heartlessness & "Rücksichtslosigkeit" [lack of consideration] has left a *deep* impression on me, you cannot wonder . . . But I feel his *rudeness* & undutiful conduct to *me* far less than I do his rudeness to his *sister* – who has gone away most deeply *disgusted* & hurt. – He has no heart, & she [Dona] has no tact – & they are both so convinced of their own perfection that they *will* run with their heads against the wall some day in all "naiveté".'[138] She passed on copies of the correspondence between Wilhelm and King George, which the latter had sent to her from Athens, to London, and did not mince her words over the Kaiser's 'preposterous' telegram to the King. 'He seems to be copying Peter the Great, Frederick William I, Napoleon or some such conspicuous tyrant', she wrote to Sophie. 'To a free-born Briton, as I thank God I am, such ideas, so little in harmony with the XIX Century and personal liberty and independence, are simply abhorrent; and this my own son!'[139] On the basis of this information Queen Victoria and the Prince of Wales naturally took sides strongly against Wilhelm and Dona.[140] The Queen asked her daughter to make known to the King of Greece 'that I deeply deplore what I think most *inconceivable* conduct on W's part', while the heir to the British throne condemned Wilhelm's attitude as 'pompous, ridiculous and unkind'.[141]

Wilhelm's and Dona's offensive behaviour only strengthened Crown Princess Sophie in her resolve to be received into the Greek Orthodox Church at Easter in 1891.[142] As her mother wrote to Bogumilla von Stockmar in March 1891, 'the scenes to which her brother and his wife subjected her have confirmed her in her decision and it will be carried out all the more quickly as a result!' Wilhelm, she commented, was so unpredictable that they must be prepared for another 'demonstration of fury and power', but on the other hand it was not impossible that he would at last realise 'how misguided it was to treat his poor sister so – and that by doing so he has achieved exactly what he wished to prevent'.[143] On 19 April 1891 Wilhelm read out to his Flügeladjutant Carl von Wedel a letter from the Crown Princess in which she confirmed her announced conversion to the Greek Orthodox Church. In his reply Wilhelm spoke of the 'disgrace' which his sister had brought upon the House of Hohenzollern and the country, of her 'violation of the oath she had taken at her confirmation' and of the 'insult to the memory of her father and grandfather'. He forbade her to cross the Prussian frontier for the next three years. Wedel's attempts to calm the Kaiser were rejected out of hand.[144]

While the Kaiser's eldest sister also expressed herself as 'very much enraged' by Sophie's conversion,[145] his younger sisters Moretta and Mossy were entirely on the Crown Princess's side and declared themselves 'absolutely *infuriated*' by Wilhelm's behaviour.[146] Moretta did her best, at first, to bridge the gulf between mother and son by pointing out that Wilhelm had 'a curious sort of character' and did not like to admit it 'when he has done a foolish act'.[147] But her attempts at mediation were frustrated by the hostile correspondence between her brother and Sophie. 'Dear me!', she sighed when her mother sent her copies of their letters. 'It's very sad – I hardly know what to say or think & am in a very difficult position wanting to help everybody . . . I *perfectly* understand her [Sophie's] feelings – & *he* [Wilhelm] may seem a tyrant etc. etc. but perhaps *his* answer was misunderstood & he may regret his hastyness & rough ways towards his

34. Crown Prince Constantine of Greece with his wife Sophie, the Kaiser's sister

little sister!' 'Really what "Zustände" [goings-on] there are in our family – will it never get any better?'[148]

The Kaiser's mother, above all, could not understand her son's intolerance. 'He has absolutely *no heart*', she commented in a letter to Sophie. 'It is perhaps *not* his fault . . . but it is a fact. There is no sentiment in his nature . . . He is, besides, not *learned* enough to understand that our Christian Religion has centuries ago divided into branches.' He was therefore convinced that 'the Prussian Church [is] the only right one'. But basically it was not the religious differences between the faiths but 'his silly vanity and pride of being "head of the family"' that was at the root of Wilhelm's conduct, she thought.[149] The 'foolish prohibition' banning his sister from crossing the Prussian frontier for the next three years infuriated her and robbed her of all sleep. 'It is really *too* unheard of', she wrote to Sophie. 'It is a piece of *tyranny* and *injustice* which fills me with *contempt*, indignation and disgust . . . It is a piece of despotism from which *every* right-minded person *must* turn away with pity and derision . . . It is really dreadful to think that he fancies he can do *all* and *everything* he likes!'[150] She complained bitterly in a letter to Baroness Stockmar: 'I have few joys to brighten my life, but that of having my Sophie with me ought not to be snatched from me by *him* who *should* be looking after me, and who has already brought so much suffering on me.'[151]

In spite of the prohibition Crown Prince Constantine visited his mother-in-law in the Taunus while the Crown Princess and their infant son at first stayed behind in Heidelberg, on the other side of the Prussian frontier. Only after Reich Chancellor von Caprivi interceded on her behalf did Wilhelm II allow his sister to visit her mother.[152] On 30 May 1891 the Empress Frederick was at last reunited with her daughter Sophie, her husband and their newborn son George at Bad Homburg.[153] They were visited there by the elderly King Christian IX of Denmark, father of the King of Greece. The widowed Empress reported proudly to Moretta in Bonn that she had written to Wilhelm to thank him, not for giving permission for her to see Sophie, since he had no right either to give or to refuse it, but because in the end he had not put any obstacles in the way of the visit.[154]

It was indicative of the power which Wilhelm could exercise as head of the family and also as supreme war lord that his brother-in-law Prince Adolf of Schaumburg-Lippe was fearful of being drawn into the family quarrel. When Moretta travelled to Bad Homburg to be with her mother and her two younger sisters for the third anniversary of her father's death, Prince Adolf excused himself on the pretext that he could not leave his squadron.[155] 'That is the price I have to pay for her [Moretta] having married a German prince who is only a Rittmeister [captain in the cavalry]. He has to do *all* W. wishes . . . & he *only* looks to the "Emperor", therefore he did *not* come here to see Sophie & Tino, though I invited him 3 times!!!', the Empress wrote bitterly to Queen Victoria.[156]

The following summer, when Sophie and Tino were staying in Copenhagen, Wilhelm sent his sister word that he hoped her forthcoming visit to Bad Homburg would do her good. Relieved, the Empress Frederick wrote to Moretta: 'Do not fear there will be disagreeables this time. W. has expressed his pleasure at hearing that Sophie is coming to Germany! If one only takes the matter simply & naturally *now*, it will be all right; & last year's mess quite forgotten.'[157] The Empress neverthless expressed complete agreement with the decision of the Greek Crown Prince and Princess not to attend her youngest daughter Mossy's wedding, which was to take place in Berlin in November 1892. 'Tino & Sophie are not thinking of coming to Mossy's wedding!', she told Moretta. 'I am sure *no* objection would be offered at Berlin; but . . . after the *insults* William has offered them and the affront his behaviour was to the King, [the] whole Greek family and to me; I should not think it becoming or dignified to go to the Court at Berlin.'[158] Not until the summer of 1894 was it possible for the Greek Crown Prince and Princess to visit the Kaiser in Berlin in circumstances that were at least outwardly normal.[159] Nonetheless a certain acrimony persisted in their relationship with the Kaiser and Kaiserin for years, which had a corrosive effect not only in Athens but also in St Petersburg, Copenhagen and London.[160]

Wilhelm's attitude to his sister's conversion naturally had wide repercussions and contributed to his already considerable unpopularity and isolation among the principal royal families of Europe. Queen Olga of Greece – a Russian Grand Duchess by birth – sent copies of Wilhelm's letter to Sophie to the Tsar and to Queen Victoria.[161] As we have seen, the Empress Frederick also sent her mother copies of the letters and telegrams which Wilhelm and King George had exchanged, together with a letter which she herself had received from the latter.[162] The incident had brought Prince Heinrich back on to his mother's side again, and Princess Charlotte expressed justifiable anxiety that 'the conduct of the Kaiser towards his sister' might be 'used in Homburg to create a rift between the two brothers'.[163]

The alienation of the German Kaiser and Kaiserin from their British, Russian, Danish, Greek and Hessian relations was intensified by the conversion of Grand Duchess Ella, Wilhelm II's Hessian cousin who was married to Grand Duke Sergei of Russia, to the Orthodox faith, which took place at the same time and aroused equally strong feelings in Berlin. Ella's father, the Kaiser claimed, had wept uncontrollably at the news of her conversion. For his part, Wilhelm said, he entirely shared the opinion of Grand Duchess Maria Pavlovna, the Mecklenburg born wife of Grand Duke Vladimir of Russia, that it was 'a disgrace for a German Protestant princess to go over to the Orthodox faith in Russia at a time when both the Lutheran Church and her compatriots there were being persecuted and oppressed in the most ruthless and unjust manner'. He wholeheartedly agreed with her comment that 'the great, powerful German Reich and its Kaiser' were now 'the stronghold and the refuge' of the Lutheran Church, so that Ella, by her conversion, would become 'an apostate and a traitor to her faith and her Fatherland'. Like the Grand Duchess, Wilhelm also believed that Ella's motives were 'an inordinate pursuit of popularity, a desire to improve her position at court, a great lack of intelligence and also a want of true religiousness and patriotic feeling'. She had never felt herself to be a German, he maintained, for as a child she had thought of herself as Hessian first and English second; and then after her marriage she had become Russian. The aim which Ella and her husband were pursuing through this dramatic step was to make themselves popular at the Russian court again, after they had 'completely lost their former favoured position through the thoroughly clumsy and tactless way in which they had set in train the project of marrying her sister [Alix] to the Grand Duke, the heir to the throne'. In spite of the Tsarina's refusal to allow Princess Alix ever to come to St Petersburg again, Grand Duke Sergei and his wife Ella had not given up hope of engineering the marriage between Alix and Nicholas, but they had realised that 'it would first be necessary to recover their former influential position of trust with Their All-Highest Majesties', and to accomplish this the Grand Duchess did not even shrink from changing her religion.[164] When Princess Alix likewise converted to the Russian Orthodox faith on her marriage to Tsar Nicholas II – Wilhelm prided himself on having brought about this match – the Kaiser and Kaiserin were shocked: the latter would '*never*' forgive this conversion, Eulenburg noted.[165]

By contrast, the search for a husband for the Kaiser's youngest sister Margarethe (Mossy) went reasonably smoothly. Princess Margarethe lived with her mother in Bad Homburg, but found the town 'like a tomb' and longed for the 'paradise' of social life in Berlin.[166] In German diplomatic circles it was rumoured for a time that the Empress Frederick was pursuing an alliance between her youngest daughter and the heir to the Russian throne, Nicholas, but it seems unlikely that this was true, given the Empress's strong anti-Russian feelings. Nevertheless when the ambassador Bernhard von Bülow got wind of this alleged plot he warned that if it happened Princess Margarethe would 'presumably soon turn into a full-blooded Russian and enemy of Germany, cf. Crown Princess Sophie of Greece and Grand Duchess Sergei [Ella]', and that she would 'stir up trouble for her Imperial brother from the very start'.[167] Be that as it may, for a long time Princess Mossy cherished ardent hopes of marrying Prince Max of Baden, the future Reich Chancellor, in which she was encouraged by both her mother and her sisters Moretta and Sophie, although they suspected that the attraction was not mutual and that the Prince 'would not dream' of marrying her.[168] After various other candidates had been considered, including Prince

Wilhelm Ernst of Saxe-Weimar-Eisenach and Prince Ernst Ludwig of Hesse-Darmstadt, the Princess's choice fell, surprisingly, on Max of Baden's best friend, Prince Friedrich Karl (Fischy) of Hesse, who proposed to her and was accepted on 19 June 1892 in the garden of Schloss Philippsruhe. There was now an anxious moment while the family awaited the reaction of the Kaiser, whose consent to the marriage had to be obtained. Prince Friedrich Karl set out at once for Berlin in some agitation; meanwhile the Empress Frederick wrote to her son explaining how the engagement had come about so suddenly and asking him to give it his blessing. Accompanied by Charlotte and Bernhard Meiningen, Louise and Aribert Anhalt and Albert Saxe-Altenburg, Friedrich Karl travelled from Berlin to Potsdam, where Wilhelm signified his approval of the engagement and announced it to the remainder of the family with evident pleasure. The Kaiserin's reaction on hearing the news, however, was 'cold as usual', which the Empress Frederick suspected was because 'it is not her brother Ernst Günther'.[169]

Wilhelm's agreement did not, however, signify the slightest willingness to consent to anything else his mother might wish. Shortly after her daughter's engagement the Empress asked Philipp Eulenburg to visit her secretly at Schloss Friedrichshof. She asked him to persuade her son during his Scandinavian cruise to allow her new son-in-law to leave the army. Wilhelm refused, arguing that with Prince Friedrich Karl's 'philosophising, melancholy' character, he needed 'contact with soldiers and military service' to provide 'the necessary counterbalance'.[170] The Kaiser's mother did nevertheless win one concession. When Wilhelm expressed his intention of inviting Duke Ernst II of Saxe-Coburg to his youngest sister's wedding, the bride's mother declared that in that case she would not attend the ceremony. No invitation was sent.[171]

THE 'RUTTING DUKE' ERNST GÜNTHER OF SCHLESWIG-HOLSTEIN

The relationship between the young imperial couple and Duke Ernst Günther, Kaiserin Auguste Viktoria's only brother, proved especially problematic. The Duke's secondment to the General Staff in the summer of 1890 had already attracted criticism;[172] but when a year later the Kaiser promoted him captain on the General Staff the officer corps was outraged. Waldersee, who had just been dismissed as chief of the General Staff, commented furiously: 'The Duke got himself seconded to the General Staff because he found winter service at Potsdam tedious and uncongenial.' It was no great misfortune, certainly, 'if someone of his type is nominally employed on the General Staff', for 'no one takes it seriously, and especially not in the present case, where the person in question is utterly lacking in both training and competence'. But that the Kaiser should make his brother-in-law a fully fledged General Staff officer was 'a slap in the face for the

General Staff, a humiliation of a corps which is highly respected throughout the world'.[173]

In spite of such privileged treatment the Duke always felt slighted, not only by Wilhelm II but also by the other members of the Hohenzollern family and by other royals of his own rank. In the spring of 1891 he complained that he had not been invited to the celebration of the engagement of his cousin Louise, Princess of Schleswig-Holstein, to Prince Aribert of Anhalt – 'and that the Kaiser responded by saying that I had not had any idea [about the forthcoming marriage]'. His irritation with Wilhelm's attitude towards him was also apparent from his remark that his uncle, Prince Christian of Schleswig-Holstein, who lived in England, was the only person of whom he could be sure that he was 'not constantly influenced by having to consider the wishes of the All-Highest personages'.[174] Ernst Günther was generally considered unusually immature and conceited. The Empress Frederick thought him 'a most foolish young man', although 'his sisters naturally think him perfection'.[175] His sister-in-law Moretta described him baldly as 'that idiot'.[176] Kiderlen-Wächter complained of the Duke's irresponsible gossiping, while Karl Samwer criticised his capriciousness and vanity. 'The only constant in him is egoism.'[177] Count von der Gröben, who was able to observe the young Duke closely during the imperial couple's visit to England in July 1891, gave a 'hair-raising' description of his behaviour there and deplored the fact that he had aroused 'universal indignation by his arrogance, rudeness and insensitivity'.[178] Prince Christian also complained bitterly about his nephew's outrageous behaviour. As Prince Hohenlohe reported in a letter to Langenburg, Ernst Günther had 'expressed himself in such a rude and stupid way in his correspondence with Uncle Christian' that 'the only possible explanation was that he was not quite in his right mind'. Christian had read out to him several passages from Ernst Günther's letters, 'which certainly defied description', Hohenlohe commented.[179]

On the marriage of his sister to the future Kaiser and King, Duke Ernst Günther had received a donation of a million marks, in addition to which he benefited from an annual appanage of 300,000 marks from the state as well as a private income of around 200,000 marks. In spite of this he was perpetually in trouble financially, partly as a result of ill-advised building projects. Even after his promotion to the rank of major in the summer of 1895, from which he had hoped he would achieve a certain 'liberation', it was said that he was still 'up to his ears in debt'.[180] He nevertheless aspired to a brilliant role at the centre of court society in Berlin, and to demonstrate this he organised a 'Corso-Fest' consisting of 'races, carnival of flowers etc.' on the Berlin trotting racecourse in the summer of 1891, which was attended by the Kaiser and Kaiserin.[181] As the only son of the late Duke Friedrich of Schleswig-Holstein-Sonderburg-Augustenburg he took it upon himself to try to settle his family's unresolved 'Hoheit question' (the title and precedence of 'Hoheit', 'Highness') but found it 'very difficult to get anywhere with the German ruling princes'. He complained that he had had to endure 'several very unkind remarks', especially from Duke Ernst II of Saxe-Coburg and Gotha. Like the Duke of Coburg, Duke Georg II of Saxe-Meiningen had opposed him on this question, whereas the Dukes of Anhalt and Saxe-Altenburg had expressed their agreement.[182] When the Kaiserin's brother took up the cudgels on the Schleswig-Holstein question in 1895 in a newspaper article appearing under a pseudonym, his initiative led to 'very strong attacks against Ernst Günther', whom Bismarck accused of 'aggressive insolence' in the *Hamburger Nachrichten*. Bismarck's criticisms in turn provoked 'great indignation in H.M.', as the Reich Chancellor Prince Hohenlohe noted.[183]

But what caused the greatest embarrassment of all to the Kaiser and Kaiserin was Ernst Günther's obsessive and tempestuous hunt for a bride. In the summer of 1890 he decided to go to England for the 'Season', 'not with the express intention of marrying *à tout prix*', he assured his uncle Christian, 'but to have an opportunity of looking around'. For obvious reasons he was not thinking of the daughters of the Prince of Wales, but he would 'very gladly . . . marry one of the daughters of the country, provided she were well brought up, pleasing etc. for with the lack of choice among Princesses this *still* seems to me to be the most favourable solution'. The upper echelons of the German aristocracy had nothing to offer a man in his position, he felt, since most girls from these circles had been very little out in the world, and moreover one would have 'a constant following of dependant [sic] relatives in the country'. Catholicism was the hindrance as far as Austrian young ladies were concerned. Ernst Günther therefore asked his uncle to give him 'certain hints' as to suitable marriageable Englishwomen.[184] His intention of marrying an Englishwoman was, however, thwarted by the Kaiser himself, for as the young Duke complained in October 1890, he 'would not hear of a *parti* with the English aristocracy'.[185]

When Princess Victoria Mary (May) of Teck, a princess of morganatic Württemberg descent on her father's side, whose mother was Queen Victoria's cousin Princess Mary Adelaide of Cambridge, was suggested as a possible wife for Ernst Günther, the Empress Frederick reported to Windsor that 'Dona was much offended & said to me that her Brother would not dream of making such a mésalliance!!!'[186] May Teck was destined for far greater things, however: she was betrothed in December 1891 to the (albeit rather strange) eldest son of the Prince of Wales, Albert Victor Duke of Clarence, known to his family as Eddy and next in line after his father to the throne of the United Kingdom of Great Britain and Ireland. When Wilhelm heard of the engagement he wrote hypocritically to his 'dearest Grandmother': 'What happy & interesting news! Eddy is engaged! I congratulate You with all my heart & am glad that the choice has met with your approval. He is indeed a lucky creature & may look forward to a happy life! For a handsomer & more accomplished young Princess is rarely to be found. I saw much of her last year, & I must say "Sie gefiel mir ausnehmend gut" ["I thought her exceptionally nice"]. I am sure the country at large will ring with joy & merry will this Xmas be for You & the whole of the United Kingdom. I am very glad for Uncle Bertie.'[187] After the Duke of Clarence's sudden death in January 1892 shortly before the wedding, in June 1893 Princess May of Teck married his brother the Duke of York, who succeeded to the British throne in 1910 as King George V and reigned until 1936 with Queen Mary at his side. Like the Battenberg marriage of Queen Victoria's youngest daughter Beatrice,[188] this marriage aroused disapproval in Germany, and not only in the imperial house. Cardinal Prince Gustav zu Hohenlohe-Schillingsfürst, the brother of the future Reich Chancellor, was astounded and wrote from the Vatican: 'I think the marriage of the Princess of Teck quite extraordinary . . . I cannot understand why the Queen has approved this marriage too.' He ironically advised his cousin Hermann to see that the Crown Prince of Italy was married off to the Kaiserin's youngest sister as soon as possible, 'otherwise yet another Principessa May will get in the way'.[189]

In the summer of 1893 Ernst Günther horrified Wilhelm II and the Kaiserin by proposing to the widowed Duchess of Aosta, Laetitia, a Princess Bonaparte by birth, in Paris, and it was only at the last moment that the German ambassador, Count Münster, succeeded in frustrating this plan.[190] A year later the Duke tried his luck with the beautiful and clever Princess Hélène of Orléans, whom he had met in Cairo. The late heir to the British throne, the Duke of Clarence, had been 'very much in love with her', it was said, and she would have had the prospect of becoming Queen of England had the Pope given his consent to this marriage.[191] Philipp Eulenburg, to whom the Kaiser and Kaiserin had entrusted the task of thwarting this latest venture by Ernst Günther, reported to his mother in April 1894 that 'the Kaiserin is in floods of tears. The Kaiser refuses to approve the match under any circumstances and sees it as an intrigue by the Orléans family, who "want to make themselves a good name through the House of Hohenzollern".'[192] Eulenburg commented despairingly to Bernhard Bülow that Ernst Günther was 'like a rutting stag in spring. Perhaps even stupider. The poor Kaiserin *wept* in front of me in her divine Silver Rococo Saloon! H.M. was only angry for political and personal reasons and has taken steps through the Empress Frederick to warn the Orléans off. I am afraid there will be a family row anyway, for the rutting Duke is giving himself airs as the head of the House of Holstein and showing off.'[193]

Like the Kaiser and Eulenburg, Bülow also saw 'enormous danger' for Germany in a marriage between the ambitious and 'perfidious' French princess

35. Duke Ernst Günther of Schleswig-Holstein, the Kaiserin's brother

and Duke Ernst Günther, who would be 'like wax in her hands'. It would lead to 'unpredictable consequences – and unpredictable trouble'. In the first place the Orléans family were zealous Catholics and would create a focus of anti-national ultramontane elements in Berlin; second they were 'passionately French' and would devote all their efforts towards furthering France's greatness, power and revenge. 'There is so much that she [Hélène] could find out, betray, prevent, stir up! What an opening she would give to our worst enemies! With her the Trojan horse would come within our walls', he warned. In the third place the family's position in France would be improved by the marriage, and for Germany there was no more dangerous form of rule in France than that of the Orléans. And finally, the scheming Princess Marie of Orléans, who had been married to Prince Waldemar of Denmark since 1885, would exploit her cousin Hélène's marriage 'to lay mines' in North-West Germany which could 'cause great destruction by exploding at critical moments'. For all these reasons it was essential that the marriage be prevented, Bülow considered, and this could best be done through Queen Victoria and Prince Christian of Schleswig-Holstein. Kaiser Wilhelm should declare, he advised, that he would never give his consent to the marriage and would never receive Ernst Günther and his wife; furthermore he would never allow the couple to live in Prussia, in particular in Schleswig-Holstein.

But although the prohibition of the marriage must be very firm, Bülow warned, it should be very carefully expressed so that it could not be exploited by the Danish royal family in order to turn Tsar Alexander III against Wilhelm II.[194]

Eulenburg set out the case to the German ambassador in Paris as follows: 'The Kaiser would see in such an alliance the desire of the Orléans to improve their position – he would also fear intrigues in Berlin . . . I should add that it would be difficult to persuade the Duke, whose heart is easily inflamed, to give up or desist from giving in to a *coup de tête*. It would therefore be a question of taking precautions on the other side and frightening off the Orléans from entering into an alliance with the Kaiser's brother-in-law . . . In my view much depends on the personality of the Princess. If she is of the same type as her sister [sic] Waldemar of Denmark one might have to be prepared for trouble, for the Duke is *not a very clever* man and is easily led. I imagine that you would probably know ways and means of conveying a placatory message to the Orléans. If they want the match, however, it will be impossible to do anything about it. In the end people always think that a *brouille* cannot last for ever and that it is only a question of time before the Duchess of Holstein-Orléans will be seen at the Berlin Schloss.'[195] From his experience of the womanising Duke in Paris the previous summer, in his reply to Eulenburg Münster could only confirm the worst fears of the Kaiser's friend. Unlike the Duchess of Aosta, Princess Hélène of Orléans would not be receptive to his arguments against the marriage, although Münster agreed that such an alliance would be 'extremely *harmful*'.[196] Eulenburg forwarded the ambassador's letter to the Kaiser with the comment that 'the business of Duke E[rnst] G[ünther]'s engagement is really *highly* dangerous . . . I think Prince Christian might after all be better as an intermediary with the Orléans than the Empress Frederick.'[197] For several more weeks the Kaiserin was tormented by the Duke's 'foolish rush to find a bride'.[198] Then on 25 June 1895 Princess Hélène married Emmanuel, Duke of Aosta, at Kingston-upon-Thames, leaving Ernst Günther to begin his search all over again.

The Duke's next project drove the Kaiser and his wife to distraction. In the spring of 1896 he announced his intention of marrying the eighteen-year-old Johanna von Spitzemberg, who was not of royal birth. She was the daughter of Baroness Hildegard von Spitzemberg and a niece of the Württemberg envoy Axel Freiherr von Varnbüler. In March 1896 Eulenburg found himself compelled, as he wrote, 'to torment [the Kaiser] yet again, alas, with a marriage project of Duke Ernst Günther's!' He sent Wilhelm a copy of a letter from Varnbüler but urged him not to react to Ernst Günther's plan until 'the news reaches Your Majesties *through* the Duke', for Varnbüler and Eulenburg would be put 'in an *impossible* position if the Duke found out that the matter had been brought to Your Majesty's attention directly'. Eulenburg also advised the Kaiser not to say anything to the Kaiserin about her brother's intention for the time being, for it would 'only fill Her Majesty with anxiety again and detract from the beneficial effect of the journey [to Italy]. It would also be very difficult to dissuade Her Majesty from writing – and then . . . Axel and I would be in a very unpleasant position!'[199] The Kaiser telegraphed in fury from Palermo: 'Am quite outraged at your news; all the more as the selfsame young man recently asked me to negotiate with another lady for him, which is already in progress. Find his behaviour indescribable. Please warn Axel at once. His niece is too good for such a wretch.'[200] Not until May did Ernst Günther admit to the Spitzemberg family, as Axel Varnbüler informed Eulenburg, that the Kaiser had 'suggested another girl [to him] – namely the daughter of our beloved lion, hippopotamus and dervish-howl-mimic' Count Emil Görtz. When it turned out that the eldest Görtz daughter was already spoken for, Countess Anna Görtz declared that she would invite the Duke to Schlitz soon, 'to give him the opportunity of meeting the second daughter'.[201] On 19 May Eulenburg wrote to Varnbüler from Prökelwitz on behalf of a grateful Kaiser to tell him that during a shooting trip to Schloss Primkenau, the Holstein family seat in Silesia, there had been 'serious arguments' with Ernst Günther. 'When the Duke tried to put off the match with the Görtz daughter for three more weeks "because his heart was engaged", the Kaiser answered that the Duke no doubt had the right to marry whom he wished, but that He and the Kaiserin were determined, if he entered into a marriage with a lady who was not of the appropriate rank, *never to receive her*. He must make his arrangements accordingly. Furthermore he [Ernst Günther] would prevent the Kaiserin from coming to Primkenau, because she would never set foot in the house if the marriage were not of appropriate rank. The Duke was quite flabbergasted.' According to the second sister of the Kaiserin and the Duke, Princess Calma of Schleswig-Holstein-Sonderburg-Glücksburg, Eulenburg continued, Ernst Günther had already lost interest in Johanna Spitzemberg; he had spoken definitely of marrying one of the Görtz daughters.[202] Varnbüler was not convinced, however, that 'the danger' was over, for Ernst Günther had renewed his wish to continue to pay court to his niece, and even if Johanna Spitzemberg had commented that she did not like Ernst Günther at all and could not imagine ever being able to love and marry him, Baroness Spitzemberg was nonetheless rather more taken with the ducal title than was desirable.[203] Eulenburg warned his friend that a marriage between Ernst Günther and Johanna would cost him his position as Württemberg's envoy and delegate to the Bundesrat. 'The Kaiserin, who is full of gratitude for your support for her views, would be overcome with thirst for revenge if she found herself in the position of perhaps having to call you and your wife Uncle and Aunt', he explained; and the Kaiser would also treat him with hostility if the marriage took place. 'I therefore beg you earnestly to go on lovingly fostering your niece's *anti*pathy. And that is not a difficult task given *this* personality. I have *never* met such a wayward fellow!'[204]

On 28 May 1896 Kaiser Wilhelm II finally contacted Varnbüler directly to put an end to the matter. 'My dear Axel', he wrote, 'I take up my pen with a heart full of distress and anxiety, to appeal to you in the matter of my brother in-law, about which you informed me through Phili. First of all, my dear Axel, accept my own and the poor, greatly suffering Kaiserin's warmest thanks for your loyal, honest attitude in this affair, which the Duke unfortunately kept carefully hidden from us. I expected nothing else from you, of course, but the Kaiserin's praise for your conduct must give you consolation and encouragement in a situation so painful for you too. You have behaved as an upright and honourable friend to me, which I shall not forget. But now to the matter itself. Phili has of course explained to you exactly what has passed between the Kaiserin and her brother as well as what I discussed with him in Primkenau . . . Now, my dear Axel, I have had enough of these constant charades and manoeuvres. The poor Kaiserin already has enough to worry and upset her, but this affair is breaking her heart. You know that the Kaiserin has told the Duke expressly that if he takes your niece she will have no more to do with him and she will be cut off once and for all from coming home to Primkenau. The thought of never returning to the home of her forefathers, never being able to pray beside the coffin of her dear, idolised father again, so overwhelmed the Kaiserin that she burst into floods of bitter tears. I cannot bear to see such despair and wretchedness any longer; it must be stopped. Only yesterday I again made clear to the Duke what a miserable situation would be created for him, for the family and above all for his intended bride. She would not be recognised, could never appear at court with him and never be seen in the family circle . . . I then told the Duke that as head of one of the oldest and most distinguished German princely houses, heaven had laid obligations upon him which compelled him to remain within certain limits, which I had already relaxed by allowing him to choose among the families of the old mediatised and ruling counts and princes, but to go further was quite out of the question. I ended the conversation by saying that I expected to hear from him at the beginning of next month that everything was settled. But you, dear Axel, must help with this. Advise your sister to go away at once and take your niece with her . . . So I ask you, dear Axel, to put a stop to the affair once and for all by a *fait accompli*. Your true friend and King, Wilhelm I. R.'[205]

At the same time Wilhelm revived his efforts to arrange a marriage between his brother-in-law and Countess Anna Görtz, the nineteen-year-old daughter of his friend the mediatised Reichsgraf (Imperial Count) 'Em' Görtz, her previous marriage plans having meanwhile come to nothing. On his return from

Prökelwitz the Kaiser held discussions with Countess Sophia Görtz, first at the Neues Palais and then at the Hotel Continental. His old friend was understandably agitated when news of this reached him at Schlitz, and on 23 May he wrote to the Kaiser: 'Your Majesty can well imagine how affected I was by the subject of your conversation, of which my wife informed me immediately – affected on the one hand by the joyful hope of seeing a fine and happy future ahead for my dear Anna, with God's help, but also by anxiety as to how the complication which *previously* arose, for reasons which my wife did not conceal from Your Majesty, would be settled. If I now take up my pen to write about this matter, it is in the conviction that no further impediment to Your Majesty's gracious intentions need be anticipated from another quarter. The conduct of the other party, at least, is such that we can scarcely believe that any further step will be taken from there. We are afraid that obstacles on grounds of health may be the reason. I can assure Your Majesty that we shall be happy, in these circumstances, to see the accomplishment of that other alliance indicated by Your Majesty. Naturally we are saying nothing to our daughter about it, nor shall we say anything until the gentleman himself wishes it. It seems to me that the first requirement is that it should happen as naturally as possible. In order to ensure this – that is to say, to ensure it for *both* sides – I am also particularly disinclined to try to bring about the rapprochement which we too now wish, by an invitation to Schlitz. As there has been so little contact hitherto, such an invitation would be difficult to explain in a non-committal manner, and it would certainly attract a great deal of attention! We have thought of another way of bringing about a closer acquaintance between them, as desired. There is no reason why my wife and I should not come to Berlin with Anna for a few days "to visit the two exhibitions there". That will give the gentleman the opportunity to call on us at our hotel, which in turn will lead naturally and quite inconspicuously to further contact in a way which is not at all binding on either party. If, as I may perhaps hope, this plan recommends itself to Your Majesty, it would only be a question of deciding on an appropriate moment and of letting the other side know.' Görtz asked the Kaiser to inform Ernst Günther of this plan, 'all the more because Your Majesty is the only person through whom this matter *can* be arranged!' In 'the interests of all sides!' he urged, Wilhelm should act, and 'the sooner the better'.[206]

The satyric saga of the 'Rutting Duke' did not come to an end, and that only temporarily, until he married Princess Dorothea (Dora) of Saxe-Coburg and Gotha in August 1898, the daughter of the Vienna-based Prince Philipp of Saxe-Coburg and Gotha. If a match between the German Kaiserin's brother and a sixteen-year-old Catholic Princess was problematic enough in itself – Prince Heinrich wrote from Vladivostok that he was glad his mother had not been present at the wedding of 'that d—d fool to a child!'[207] – the outrage was all the greater because the engagement coincided with a scandal of epic proportions in the bride's family.[208] Not only had Dora's mother, Princess Louise, a member of the Belgian branch of the Coburg family, forged the name of her sister Crown Princess Stephanie on countless bills of exchange, but in February 1898 the father of the young bride had fought a duel with his wife's Croatian lover, Lieutenant Mattachich-Keglevich. Shortly before her daughter's marriage to Duke Ernst Günther, Mattachich was arrested for fraud and Princess Louise of Saxe-Coburg and Gotha was shut away in a lunatic asylum and deprived of her rights. 'These are all most unpleasant matters!', Eulenburg exclaimed in despair and not without reason in a letter to the Kaiser.[209]

SCANDAL AT SCHLOSS GLIENICKE

Almost as scandalous as Ernst Günther's life was that of the Kaiserin's younger sister, Luise Sophie, known as Yaya, and her husband Prince Friedrich Leopold of Prussia, a cousin of the Kaiser and brother of Princess Louise Margaret, the wife of Queen Victoria's third son Prince Arthur, Duke of Connaught. Yaya and Fritz Leopold lived at Schloss Glienicke and had four children, although there was little sign of domestic bliss in the marriage. 'I fear there is much unhappiness', the Empress Frederick wrote regretfully about the relationship between Yaya and Fritz Leopold.[210] The Princess was distinctly plain – the widowed Empress commented with dismay in December 1891 that 'the poor thing is too distressingly ugly – really quite startling' – and although she dressed very well and had beautiful hair, she had 'such a temper that it never goes smooth there'.[211] Waldersee described Prince Fritz Leopold as a 'foolish, incompetent man' who treated his wife 'with appalling lack of consideration'.[212] Charlotte of Saxe-Meiningen spread the rumour that the Prince was 'not quite right in his high mind & intellect, or else he is a brute'.[213] Perhaps as a defiant reaction to her husband's maltreatment of her, the Kaiserin's younger sister became 'so fearfully arrogant . . . that even [Prince] Heinrich, who is certainly a good-natured fellow, doesn't speak to her any more'.[214]

The numerous critical remarks made by the couple about the Kaiser, which were deliberately reported to him, combined with Yaya's often improper behaviour, led in the winter of 1895–6 to one of the most sensational scandals in the history of the Prusso-German court. The indolent, difficult Fritz Leopold[215] committed the error of failing to send word immediately to the Kaiser, as head of the family, of a skating accident suffered by his wife, for which Wilhelm rewarded both his cousin and his sister-in-law Yaya with fourteen days of house arrest. Schloss Glienicke was surrounded by sentries from the 1st Guards Regiment, which led to wild speculation in the press that the Princess had attempted to commit suicide or that Fritz Leopold had been unmasked as a political conspirator. '*Who could* imagine the simple truth!', exclaimed the Empress Frederick, who saw Wilhelm's action as yet another expression of a reactionary spirit of the deepest dye. 'Despotic spirit and autocratic ways – & ideas of 200 years ago were not taught by Fritz & me – & not learnt from *us* – as *we* often suffered under a thralldom wh[ich] was very irksome, of wh[ich] [Wilhelm] *never knew anything*.'[216] Throughout the ruling houses of Germany the news that the Kaiserin's sister had been placed under house arrest was greeted with incredulity.[217]

The Senior Marshal of the Court and Palace, Count August zu Eulenburg, who shared the general view that the Kaiser's reaction had been excessive, gave his cousin Philipp an account of the affair. 'The outward cause of the conflict, which had been flickering in the embers for a long time because of constant disobedience and opposition from Glienicke, particularly on the female side, was the Princess's falling through the ice when she was yet again skating without an escort. This accident was then kept completely secret from the Neues Palais for the whole day, and when the Kaiserin happened to call at Glienicke that afternoon she was turned away on the pretext that the Princess had a bad cold and had gone to bed. A violent argument between His Majesty and the Prince then led to the imposition of house arrest . . . It was high time for them to be taught a proper lesson, even if it took an excessive form – as unfortunately frequently happens here.'[218] In a revealing comment on the background to the incident Philipp Eulenburg wrote: 'There is an open feud between Prince Friedrich – who is, to put it mildly, *quite* crazy – and his wife, the Kaiserin's sister, on the one hand, and the Kaiser and Kaiserin on the other. The Prince absolutely refuses to comply with the wishes and "commands" of his cousin, and his wife is equally unwilling to do what the Kaiserin wishes. A kind of minor rebellion in domestic affairs. The Friedrich Leopolds spent most of their time squabbling, threw every adjutant and Hofmarschall out of the house after a very short time, and behaved in an unsuitable manner. The Prince's open defiance finally led to the arrest – and the whole of Potsdam, or rather Berlin, is in a state of high excitement. Indeed this must surely be the first time since Frederick the Great that a member of the Prussian royal family has been placed under arrest with a military guard on his castle, and in this case one is as surprised at Prince Friedrich Leopold and his wife as at His Majesty himself.'[219] Eulenburg wrote to the Kaiser on 20 January 1896: 'The incredible affair of Prince Friedrich Leopold is upsetting people. I hope Their Royal Highnesses will go away for a long holiday, for *it is necessary for the peace of mind of Your Majesty and the Kaiserin* that you should be spared these tiresome rebellions at Glienicke, which are so trying on the nerves, for a while . . .'[220]

Through their close dynastic ties with the royal house of Prussia the British royal family was also drawn into the quarrel, which was not without danger in view of the fact that precisely at this time Anglo-German relations were under great strain because of the Transvaal crisis and the Kruger telegram.[221] Queen Victoria was kept up to date with events by almost daily letters from her daughter Vicky. On 1 January 1896 she wrote to her mother about this 'sad affair': 'I am much upset at William's having put Fritz Leopold & "Yaya" under arrest . . . This is a highhanded measure wh[ich] I am afraid will only exasperate Fritz L. and Louise-Sophie – & make relations very strained ever after.' Wilhelm's behaviour reminded her of painful times in the past – of his conduct in 1888 and 1889 and of his attitude to his sister Sophie's conversion. 'What a strange way of treating one's family!', she exclaimed in bewilderment. 'I cannot say *how* I regret this unseemly proceeding!!!' When she expressed her disapproval to the Kaiserin, Dona replied: 'Discipline must be enforced & kept up in the family.' But even leaving aside the inappropriateness of the punishment, the Empress Frederick pointed out, the airing of this family quarrel in the press was certainly very harmful for Wilhelm's reputation.[222]

Queen Victoria, who at first intended to stay out of the quarrel, nevertheless felt compelled to comment cautiously on it in response to a letter from Wilhelm justifying his conduct. 'It was very careless of her [Yaya] to attempt to walk on the ice, with no servant at hand, & it was very wrong of Fritz Leopold to speak disrespectfully of you both, but when people get into a passion they are often not aware of what they say, & I think those who repeat such expressions to you deserve great blame. It would surely have been better not to arouse any bitterness of feeling by such an arbitrary act as putting a Relation & a General Officer under arrest. Why not have sent for him or written to him to reprimand him? I fear the effect of the measures you have taken will not tend to the peace & harmony of the family & will cause a scandal.'[223] On 8 January 1896 Wilhelm was able to report to his grandmother that the measures against Fritz Leopold and Yaya had had a very beneficial effect. The Prince had shown remorse, so that the arrest could be rescinded.[224] But despite the ostensible reconciliation, inwardly Fritz Leopold continued to seethe with anger at his treatment. 'Unfortunately the storm continues to rage unpleasantly here, even if the waves have subsided outwardly', Captain von Heuduck wrote to August von Mackensen on 2 February 1896.[225] After a meeting with Fritz Leopold in October 1896 Waldersee too had the feeling that he would 'very soon have another clash with the Kaiser'. 'I could see again what an unhappy character the Prince has . . . He . . . allowed himself to make some very questionable remarks.'[226] Wilhelm himself expressed the view in 1899 that Friedrich Leopold was among those princes who 'might adopt a dangerous and hostile attitude under certain circumstances'. But his cousin would 'certainly be locked up in a madhouse soon', he commented.[227] Some years later he scribbled angrily on the margin of a petition from Friedrich Leopold: 'Unbelievable! I give the orders here! And that is that!'[228]

ARIBERT AND LOUISE OF ANHALT

Wilhelm II had played the matchmaker in another Berlin marriage in which the family relationships between the Houses of Hohenzollern, Schleswig-Holstein and Saxe-Coburg and Gotha were almost impossible to unravel. In December 1890 he personally announced the betrothal of Prince Aribert of Anhalt and Princess Louise Augusta of Schleswig-Holstein, born in 1872, daughter of Prince Christian and Princess Helena ('Lenchen').[229]

Wilhelm's role in bringing about this union, which was ill-starred from the beginning, had been far from merely ceremonial, however. A few days before the betrothal his Aunt Lenchen had begged him for financial support, as the marriage would be impossible without it. She had told him that 'Grandmama [Queen Victoria] is so keen about this *partie* [sic] coming off' that she had declared herself ready to give the bride 10,000 marks a year. As they, the bride's parents, could also give her 10,000 marks a year and Duke Friedrich of Anhalt had promised the Prince 30,000 marks, only 10,000 marks more were needed in order to reach the sum of 60,000 marks a year which would be required for the young couple to live on. Lenchen therefore asked Wilhelm for the remaining 10,000 marks and pointed out that the dowry of 20,000 marks that Louise would bring to the marriage from her grandmother and parents was 'far more than any P[rince]ss of Schleswig-Holstein has *ever* brought as a marriage portion, & I cannot help thinking, *more* than *very many German* princesses ever bring their husbands!'[230] Wilhelm agreed.[231] In spite of his generosity, as we have already seen, Queen Victoria asked the Kaiser not to come to England until after the wedding at the beginning of July 1891, ostensibly because his presence would put the bridegroom's father too much in the shade. Wilhelm, however, insisted on being present at the marriage which he had been chiefly responsible for bringing about. The Queen was obliged to give in and invite her grandson to the young couple's wedding.[232]

For a while Princess Louise of Anhalt became the 'direct channel from the Kaiser to the Queen [Victoria]'; it was, for example, through 'the Ariberts' that Wilhelm's enquiry about a possible visit to England in 1898, as well as the Queen's negative reply, were directed.[233] But all too soon this marriage ended in divorce. The Princess became dangerously anorexic[234] and separated from her husband around the turn of the century, when her husband's homosexuality could no longer be doubted. The separation became the subject of angry correspondence between the Kaiser and Kaiserin on the one hand and the British royal family on the other, in the course of which Christian of Schleswig-Holstein even threatened to appear before the Bundesrat to explain the grounds for the divorce, which the Kaiser would not permit 'under any circumstances'.[235] After the divorce in 1900 the Princess ostentatiously resumed her maiden name.

THE KOTZE AFFAIR

One Sunday in January 1891 Charlotte, Hereditary Princess of Saxe-Meiningen, and her brother-in-law Duke Ernst Günther organised a lively skating party and dance at a hunting lodge at Grunewald which was to have serious – and finally even fatal – consequences.[236] Among the fifteen participants, apart from the Kaiser's sister and her brother-in-law, were Prince Friedrich Karl (Fischy) of Hesse, Count Friedrich of Hohenau with his wife Charlotte (Lottka), Louis Ritter von Berger and his wife Sophie with her sister Eva, the officers Max and Paul Freiherren Schuler von Senden, Ludwig Freiherr von Knorring and the Masters of Ceremonies, Leberecht von Kotze and Karl Freiherr von Schrader with their respective wives, Elisabeth née von Treskow, and Alide née de Villier. The following morning some of the nocturnal revellers received disgusting anonymous letters, which not only contained obscene accusations of the worst kind but were also characterised by disdainful snobbery towards the guests who were not from the highest ranks of the aristocracy. But what was astounding about the letters, which were written in a disguised hand in block capitals and posted in Berlin, was that they described intimate events of which only eyewitnesses could have been aware. The assumption was that the writer had been at the skating party. But who could it have been?

In the course of the next four years several hundred such letters were sent. The recipients included Hereditary Princess Charlotte, Duke Ernst Günther, the Empress Frederick, Princess Margarethe (Mossy) of Prussia, Prince Friedrich Karl of Hesse, Prince and Princess Aribert of Anhalt, Prince Max of Baden, Prince Karl Egon zu Fürstenberg, Prince Hermann zu Stolberg-Wernigerode, Prince Friedrich Karl zu Hohenlohe-Oehringen, Count and Countess Eberhard von Dohna, Baron and Baroness Hugo von Reischach, the Chamberlain Edgard von Wedel, the Masters of Ceremonies von Kotze and von Schrader, and the brothers Friedrich and Wilhelm Hohenau, who were the sons of Prince Albrecht of Prussia (1809–72) by his morganatic marriage to Rosalie von Rauch. But the actual target of these truly repellent letters was Lottka, the strikingly beautiful, tall young wife of the homosexual Count Fritz Hohenau. She was Charlotte von der Decken by birth, and was contemptuously mocked as 'the stinking one', 'stinking Lotte' or 'Lottchen of Prussia' in the letters. She was accused of having pushed her way into the highest ranks of court society and of behaving in an unseemly way at court, but above all of leading a wild sexual life with high-born men and women: Prince 'Fichi' or 'Vischi' [sic] of Hesse, Prince Aribert of Anhalt, Prince Max of Baden, Prince Friedrich Karl zu Hohenlohe, Prince Hermann zu Stolberg, Prince Heinrich XIX Reuss of the junior branch, Prince Franz-Joseph of Battenberg, Eberhard Count von Dohna, the Freiherren Max and Paul Schuler von Senden, Hugo von Reischach and Ludwig von Knorring, as well as the English-born Princess Louise of Anhalt, were repeatedly named as her lovers and participants in her 'orgies' in the brothel of a certain Frau Hagenauer. Even the name of Herbert Bismarck crops up a few times in this connection. One typical letter of April 1892 to Princess Louise of Anhalt reads: 'The Countess's peculiar mania consists in a real lust to have the closest possible, that is to say sexual, relationship with all princes, and it gives her a special, irresistible thrill to steal a newly married husband from his young wife. As most men lack the energy to resist the captivating beauty of this woman, she has already caused untold mischief, especially as she suffers from a dreadfully infectious abdominal disease . . . which is almost invariably transmitted to the man through sexual relations, and on account of which Count Hohenau has long since been forbidden all sexual relations by his doctors.'[237] As the motive for these attacks the letters indicate expressly that what the writer (or writers) wanted was 'to destroy the reputation of Countess Hohenau, who has completely ruined people very close to us, in Berlin'.[238] If other people such as Hugo von Reischach and Louise von Anhalt were attacked, this was as a rule only in order to keep them away from Countess Hohenau. While Princess Louise was accused of having a lesbian relationship with the Countess, Reischach, the Empress Frederick's Hofmarschall, received a letter in February 1892 claiming that he was 'a Würt[t]emberger of the most suspect ancestry . . . whose father is said to have killed himself and whose mother, sister and brother have a highly dubious reputation'; he should therefore take care 'not to tread on the toes of Prussian people in high places' if he did not want to lose the 'brilliant position [he had] usurped here'.[239] One can imagine only too well the fury and frustration of the recipients of these letters that it proved impossible to find the source of the poison.

In the approximately thirty defamatory letters placed at the disposal of the military tribunal which later investigated the case – most had been destroyed at once by the recipients – the Kaiser himself was scarcely mentioned. There was only one letter, addressed to Ernst Günther on 23 January 1893 – just before the marriage of the Kaiser's youngest sister Mossy to Prince Friedrich Karl (Fischy) of Hesse – in which the libellous comments centred around the monarch. The letter claimed that Countess Hohenau, who was otherwise 'very extravagant', was holding back in the case of Ernst Günther, 'because she wishes Y[our]

H[ighness], as Her Majesty's brother, to continue to believe in her virtue. For the lovely Lottchen has . . . no less a goal than to be embraced by the strong arms of H.M. himself one day. She has been planning it for a long time, and whenever there is an opportunity she puts herself in H.M.'s way and pursues him with lascivious glances or shows her naked bosom as at the last court . . . But her masterstroke is to be launched tomorrow. Tomorrow afternoon at four o'clock L. will show herself off in the greatest splendour in front of H.M., appearing as a circus rider at a party which is to be staged with colossal pomp, and everyone agrees that she looks absolutely superb and plays her part to perfection. One cannot deny that she sets to work with wonderful energy and that her plan is excellently conceived, aiming first of all to captivate H.M. and to allow herself to be admired in front of all the princes of Europe, but also to awaken in the mind of the poor bridegroom old memories of earlier lovers' trysts, and to estrange his heart permanently from his wife.'[240] That the 'poor bridegroom' meant Prince Friedrich Karl of Hesse is clear from a postcard, sent a few days earlier to a Countess who is not further identified, which read: 'We hear that in order to keep her hold over Vischi [sic] and if possible to catch another prince as well, stinking Lotte has arranged a great party at which she will appear as a circus rider and only royalties will be invited. We shall not fail to elucidate those in high places further on the purpose of the exercise and the public danger involved in the enterprise, for which the impoverished Lieutenant is forced to sacrifice 1000 marks, so that Vischi will always have pleasant memories of those lovers' trysts in the Yorkstrasse [where the Hohenaus lived] and his heart will be permanently estranged from poor Mossi.'[241]

Who could have been the author of such letters? Whoever it was, he or she was familiar with the nicknames of princes and princesses and knew a great deal about the most private affairs of people of the highest social rank. Sometimes the letters were signed 'Bertha M.', and the writer is also identified elsewhere as a woman: it is indicated that the person in question was once connected with Princess Helena of Schleswig-Holstein, the mother of Princess Louise of Anhalt, in England.[242] Other letters suggest that several people are involved: they are signed 'The Committee' or 'One for many'. In March 1894 Countess Lottka Hohenau received a letter in which it was claimed that the writer of the anonymous letters was 'quite definitely the Duke of Holstein', the Kaiserin's brother, and Ernst Günther was also named as 'the anonymous letter writer' to House Minister Wilhelm von Wedell-Piesdorf. But as these denunciations were, paradoxically, also written in sinister block capitals, they were given little credence at first.[243]

On 16 June 1894 Master of Ceremonies Leberecht von Kotze was suddenly arrested on the orders of the Kaiser and personally conducted to the military prison in the Lindenstrasse by Chief of the Military Cabinet General Wilhelm von Hahnke, after Commandant Flügeladjutant von Scholl, who had been ordered to carry out the arrest, had failed to find the suspected culprit at his lodgings. The immediate reason for the arrest was the discovery of twelve pieces of blotting paper. They had been found by the group around Schrader and the Hohenaus on the desk of the Master of Ceremonies and in the Casino club and they showed several separate words, including the name 'Lottchen von Preussen', in capital letters. Kotze hotly protested his innocence, showed the Chief of the Military Cabinet two letters 'in the well-known hand' which he himself had received, and declared that he considered 'only women capable of the kind of thing being blamed on him'.[244]

The arrest of Kotze, which Hahnke had advised but which the Principal Chamberlain responsible for disciplinary matters at court, Prince Otto zu Stolberg-Wernigerode, had expressly warned against, caused a sensation and was, as Philipp Eulenburg recorded, 'deeply regretted in the interests of the monarchy' by all true royalists who had any understanding of the situation.[245] Only too soon it proved to be a grave mistake, not only politically and socially but also in legal terms, for since it was a case of libel, and thus of an offence requiring an offical complaint from the injured party, the arrest ordered by the Kaiser had been unlawful because no legal proceedings had been instituted. It was only several days later that Aribert Anhalt, Fritz Hohenau and Ernst Günther could be persuaded to bring an action for libel.[246] Since Kotze was a seconded captain of cavalry and was therefore subject to military jurisdiction, the military authorities were commissioned to investigate the case. This too turned out to have unforeseen political consequences, however, for the fact that the scandal now had to be dealt with in secret, in order to protect the Kaiser against accusations of perversion of justice, had more than a little to do with his and Hahnke's subsequent insistence on military prosecutions being held in camera, which in turn was to poison Wilhelm's relationship with the Reich Chancellor (as we shall see) for years to come.[247] The Bavarian envoy, Count Lerchenfeld, at once recognised the dangers for the monarchy which could arise from the overhasty and unlawful order by the Kaiser for the arrest. The press, he predicted, would use the case 'against the person of His Majesty the Kaiser . . . and find further material in it for attacks against the constitution of Prussian military courts'. He went on to bemoan the fact that the action taken by Wilhelm showed once again 'how ill-equipped His Majesty the Kaiser's closest entourage is for the task of drawing our sovereign's attention to the consequences of his decisions'.[248] Waldersee also regarded Kotze's arrest as a precipitate and ill-considered step. 'The whole affair is so entangled and some of it is of such a delicate nature that the best thing would be just to get rid of it.'[249] 'It is very regrettable', he wrote, 'that the affair is not being settled quietly; the arrest was certainly overhasty and probably quite illegal. But what is ever done with due consideration these days!'[250] Baroness Spitzemberg was no less outraged, and commented, with hindsight, that 'the whole horrible business, which did untold harm in our circles, could have been dealt with quietly if the Kaiser, instead of rushing in, had waited for Stolberg's memorandum and acted accordingly – frightful!'[251]

Kotze's arrest by order of the Kaiser was particularly surprising because his Master of Ceremonies had hitherto stood high in his favour. Described as good-looking, ridiculously dressed, hypercorrect and zealous in his duties and yet humorous and full of *joie de vivre*, he had been adept at entertaining the Kaiser, whom he 'idolised' – he is said once to have worn a startling green cravat for the sole reason that Wilhelm was going hunting – with piquant tittle-tattle from Berlin court society.[252] Although he had made many enemies at court through his affected behaviour it seemed doubtful from the outset that this naïve and 'fundamentally decent man' was at all capable of thinking up 'such sophisticated anonymous libels' as the letters contained.[253] At an officers' meeting at Breslau at the end of June 1894 opinion among army commanders was divided, but even those who thought Kotze guilty had to admit that he must have had 'other helpers'.[254]

The evidence of Kotze's guilt was always meagre and it soon collapsed altogether. The suspect sheets of blotting-paper with the disguised handwriting could have been used by someone else – by his adversary Baron Schrader, for instance, who used the same office. A graphologist was consulted and gave it as his expert opinion that the author of the libellous letters was not Kotze but must have been a woman – more precisely, 'a person of pathologically effeminate tendencies and of a vain and affected appearance'.[255] When the Chief of the Military Cabinet received another anonymous letter three days after the arrest – the Master of Ceremonies had no access to writing materials in prison! – even he realised that at worst Kotze could only have been guilty of aiding and abetting. 'We have not yet found the perpetrator', he admitted.[256] On 5 July 1894 the Governor of Berlin, General von Pape, as the responsible judicial authority, temporarily released the unfortunate courtier from captivity and reported to the Kaiser that the enquiries made indicated that there were 'insufficient grounds' for the suspicion that Kotze had written the anonymous letters.[257] Unwilling to accept that his own conduct should thus become subject to exposure, the Kaiser ordered that the investigation should continue under the supervision of the Commander of the III Army Corps, General Prince Friedrich von Hohenzollern-Sigmaringen. Both the Governor of Berlin and Principal Chamberlain Prince zu Stolberg-Wernigerode, who by reporting the blotting-paper evidence to the Kaiser had unintentionally provided the motive for Kotze's arrest, had no alternative but to resign.[258] After taking evidence in detail – the transcripts filled approximately a thousand pages – the court martial, which was composed of officers of the Railway Regiment, finally began proceedings against Kotze in March 1895. He was accused of having committed libel, 'or aided and abetted therein', 'in common with other persons'. After only a few days, however, the tribunal recommended that Kotze be acquitted on grounds of lack of evidence. After some hesitation the Kaiser, as supreme war lord, gave his consent.[259]

In striking contrast to his precipitate order for Kotze's arrest, the Kaiser now forced his brother-in-law Ernst Günther, Count Fritz Hohenau and Prince Aribert of Anhalt, who had been the principal plaintiffs, to make written apologies and then to call and leave their cards on Kotze. 'So the flow has been checked from above, a blessing', his sister Charlotte commented.[260] Wilhelm was supported in this conciliatory attitude by the Kaiserin, who now took a decided stand for Kotze. A private letter to a member of the British royal household from the military attaché, Colonel Swaine, indicates that Prince Christian of Schleswig-Holstein, Ernst Günther's uncle and father-in-law of Prince Aribert, also played a decisive role in putting an end to this sordid affair.[261] Waldersee noted that Count Schlieffen, as Duke Ernst Günther's superior officer, had also helped bring about the reconciliation – 'a sad mission indeed' for the Chief of the General Staff. He suspected, however, that there was no prospect of a satisfactory end to the whole unpleasant business yet, since it was still not known 'who the culprit is; many people still believe that it was Kotze . . . The affair has by no means been laid to rest.'[262] The Reich Chancellor, Prince Hohenlohe, also considered that the situation was still explosive, commenting to his son that battle continued to rage at court between 'the Wedell, Lucanus and Hahnke clique and the Schrader, Reischach, Christian Krafft [Hereditary Prince zu Hohenlohe-Oehringen] clique. – Hohenau, Aribert and Ernst Günther have made their peace, while the Schrader side is still unreconciled and bears a grudge against the aforementioned for having backed down. I am keeping to my resolve to do no more than observe.'[263]

With the Kaiser's confirmation of the acquittal the scandal entered its blood-stained final phase. As courts martial in Prussia took no oral evidence and conducted their proceedings in camera, Leberecht von Kotze now felt compelled to defend his honour publicly by challenging those of his accusers who had not apologised to fight it out with pistols.[264] It would be another case, Waldersee remarked bitterly, 'of an innocent man being put in the position of allowing himself to be shot dead in accordance with our so-called ideas of social standing; for if Kotze is to shoot it out with all those who maintained that he was the culprit, he will have so many duels on his hands that he will undoubtedly come to harm'. 'There would be far too many people who had sinned against Kotze . . . , including the Kaiser himself, since he ordered the arrest, which was utterly wrong.'[265] The

Chancellor urged the Kaiser to prevent the duels, as the middle classes would hold the Kaiser responsible for the original arrest and thus for the 'bad turn events had taken'. 'Since everything which affects the mood of the population towards Your Majesty is of great importance to me, and since I am always at pains to eliminate anything which could occasion ill-feeling towards Your M., the prospect of a duel with an unfortunate outcome for Herr von Kotze fills me with an anxiety which is all the greater because I am afraid that such an outcome could be exploited in a manner unfavourable to Y. M.', the elderly Prince wrote.[266] Even if the Kaiser had agreed to the Chancellor's request, the latter's well-meaning intervention came too late: on 13 April 1895, in a duel fought under very harsh conditions with Freiherr von Reischach, Kotze was shot in the thigh on the eighth exchange of bullets.[267] By way of a conciliatory gesture the Kaiser sent Kotze flowers and an Easter egg.

Full of sympathy for the wounded man and his wife Elisabeth, the Kaiser's sister Charlotte wrote: 'Kotze at last pronounced free, but since yesterday badly wounded, consequences of the first duel. I shudder when I think of the others that must come still: his wife is *so* courageous & behaves admirably; her letters all these days to me are "bewunderungswürdig" [admirable] from strength of mind & will: but the long 10 months strain must soon tell on her nerves: dear thing how I long to help & comfort her now!!'[268] Waldersee too was incensed at this new trial by fire which the unfortunate Kotze had to endure. 'The wretched Kotze affair has reached a temporary conclusion through the acquittal of this actually very foolish and harmless man, who was accused and arrested with such inordinately excessive haste', he wrote the day after the duel. 'What a quantity of dust and dirt has been stirred up to absolutely no purpose! There has already been a duel and Kotze has been shot in the leg by his chief accuser, Reischach. It makes me very angry that Kotze was obliged to fight a duel because of our completely misguided ideas . . . Are all the countless slanderers to go unpunished? That would be truly outrageous. In the decree introducing the disciplinary courts the Kaiser said: "I will not tolerate anyone in my army who impugns the honour of a comrade in a criminal manner." Why is no use made of this admirable sentiment?' Not without reason many newspapers expressed the opinion that, if laws were now to be passed to protect religion and morality, they would have to be applied first 'against the highest ranks of society'.[269]

As the next step towards his rehabilitation, on the advice of his Jewish defence lawyer, Fritz Friedmann, who was regarded as a brilliant man, Kotze lodged a formal complaint against Schrader for slander, but after his complaint had been repeatedly rejected by the public prosecutor's office, a duel between the two former colleagues seemed inevitable.[270] Waldersee had already predicted this outcome, which he strongly condemned, in August 1894 after a conversation with

Zedlitz-Trützschler. The former Minister of Ecclesiastical Affairs had expressed the opinion, the General noted, that 'nothing could be proved against Kotze, but suspicion continued to rest on him', so that it would probably be a matter of duels between him and his accusers in the end. The scandal was certainly 'deeply shaming and it throws a dark shadow over court society in Berlin', the former Chief of the General Staff wrote. But to settle it through a duel would be quite wrong, in his view. 'If Kotze is innocent his accusers must be treated as slanderers and hounded out; if he is guilty he must be got rid of. But what use is shooting?'[271]

In December 1895 Wilhelm II again intervened disastrously in the scandal when he angrily overturned the verdict of a disciplinary tribunal conducted by the Rathenow Hussars, which had found for Schrader and against Kotze, describing it as 'very partisan', and ordered a new investigation by a tribunal from the Hanoverian King's Uhlans. The officers whose verdict had been repudiated felt compelled to submit their resignations, and the Rathenow Hussar Regiment was further punished by receiving no All-Highest congratulatory telegram at its jubilee celebrations. General Prince Friedrich von Hohenzollern, whom the Kaiser blamed for the leaking of the verdict to the press, was dismissed in an insulting manner from his post as commander of the III Army Corps.[272]

Philipp Eulenburg congratulated the Kaiser on his rejection of the tribunal's verdict.[273] Other observers, however, were disturbed by the continuing escalation of the scandal and by the Kaiser's increasingly irritable mood. 'This affair really must be settled soon', Waldersee commented in March 1896.[274] The Austrian ambassador reported from Berlin on the widespread and growing nervousness caused by 'the lack of proper stability', while the Bavarian envoy, Lerchenfeld, pointed out that the precipitate dismissal of Prince Hohenzollern in direct connection with the Kotze affair involved a considerable risk for the crown. The Commander could have been unobtrusively retired three or four months later without any damage to the monarchy, but 'this affair, which has been dealt with in an unfortunate manner from the very start, has already caused a great deal of mischief. A number of people have lost their position in society. The court is divided and its reputation has suffered and the crown itself has not escaped unharmed, in consequence of the sudden change in the All-Highest attitude and the overhasty intervention of His Majesty the Kaiser.'[275] Count Monts reported to the German ambassador in Rome on the 'bad mood' of the Kaiser, commenting that it was directed principally against the Conservatives; but in addition to this 'the Jewish bankers and financiers are very offended about the cancelled procession at the subscription ball: there are no court balls, because H.M. is going to pay back Berlin society for the Kotze affair. In short, as in the days of Frederick William IV, the outlook's pretty gloomy.'[276]

Meanwhile the new disciplinary tribunal in Hanover reached the same conclusion as its predecessor at Rathenow. Schrader had not been guilty of defamation; Kotze, on the other hand, had dishonoured his profession and must resign his officer's commission. But in April 1896 the Kaiser rejected this verdict too, thereby confirming Kotze's right to demand satisfaction and clearing the way (albeit unintentionally) for a duel between his two Masters of Ceremonies. On Good Friday, 10 April 1896, Schrader was fatally wounded in an exchange of fire on the Ravensberg at Potsdam. He died the next day.[277] Eulenburg was with the Kaiser and Kaiserin on the *Hohenzollern* at Venice when Wilhelm read out the telegram announcing that Kotze had shot Schrader. The Kaiserin's face turned 'dark red and she immediately and vehemently exclaimed: "That is right!"'[278] The Empress Frederick reacted quite differently to the news of Schrader's death. 'We are horrified here at the Duel wh[ich] has taken place at Potsdam and that unfortunate H. v. Schrader has been killed by that horrid Kotze', she wrote to her mother.[279] Waldersee likewise deplored this fatal outcome, and particularly the lamentable fact 'that the Kaiser [had] actually encouraged duels'. 'Public opinion is considerably aroused and I think it very probable that the matter will be discussed in the Reichstag', he wrote.[280]

As a result of the sensational duels which Kotze fought with Reischach and Schrader the origin of the whole affair – the authorship of the obscene letters – had almost been forgotten. Indeed, among those in the know the suspicion was not infrequently voiced that this outcome was exactly what had been wished and intended from the outset – that Kotze had been induced to sacrifice himself to protect someone else, and had been 'rescued' by the Kaiser for this very reason.[281] In September 1894 Arthur von Brauer had written to the Grand Duke of Baden stating that the true culprit was 'a very highly placed person'; so much so, indeed, that he *ought* not to be found out. Brauer did not dare commit the name to paper.[282]

Although conclusive evidence is lacking and will probably never be found, in the light of research there are increasingly strong indications that the author of the anonymous letters was none other than Ernst Günther, the Kaiserin's brother. Philipp Eulenburg was in a particularly good position to discover the truth behind the scandal, not only because he was the Kaiser's closest confidant in this thorny affair, as in many others, and because he was also kept up to date with developments by his cousin August Eulenburg,[283] but also because he was a close friend of the brothers Dietrich and Georg von Hülsen, whose mother was the sister of Kotze's mother-in-law.[284] Furthermore he was repeatedly asked for advice in the affair by his old friend Christian Krafft, Hereditary Prince zu Hohenlohe-Oehringen, Stolberg's successor as principal chamberlain.[285] For Eulenburg, as for the Hülsens and also the Marshal of the Palace, Maximilian Freiherr von Lyncker, it was always clear that there was '*no question* of Kotze being guilty'. For a time the Kaiser's friend suspected the other Master of Ceremonies, Karl von Schrader, who had had numerous 'relationships' in exalted circles, of having written the obscene letters with the help of his '*very* pretty, *very* elegant' Dutch wife. During the Scandinavian cruise of 1894, however, Eulenburg heard that there was 'a trail leading to *Duke Ernst Günther*', who had connections with 'ladies of the *demi-monde*' with whom Schrader also consorted. Following this discovery Eulenburg discussed 'the unfortunate affair' with the Kaiser, whom he described as 'almost more agitated than I have ever seen him. The police at Berlin had told Him that they *had called a halt* to their investigation of the affair, because the trail led to the Duke.' The '*weak* and *extremely* confused' Ernst Günther had even written letters to the Kaiser, Eulenburg recorded, 'which *almost* described himself as guilty!!' Wilhelm had banned his brother-in-law from entering his house, but had used another scandalous story as the pretext, namely that the Duke had lost his badge of the Order of the Black Eagle, which was later found under the bed of a Berlin woman with whom he was infatuated. Eulenburg summed up the situation in a memorandum of April 1895 in which he recorded that the secret police officer Eugen von Tausch believed that Ernst Günther was guilty, while Kotze and his friends accused both the Duke and Schrader of being the perpetrators, claiming that 'actresses of *the most disreputable kind*' had been supplied 'by both of them (as has been established)' with information 'about all kinds of private matters at Fritz Hohenau's house' and also about 'all kinds of incidents in court society'.[286] In January 1895 Philipp Eulenburg reported to Friedrich von Holstein 'that the Kotze affair is said to have taken an unfavourable turn for Schrader lately. Police Lieutenant von Tausch has some incriminating evidence for which no one has yet asked him. I am afraid that out of consideration for *brother-in-law Duke E[rnst] G[ünther]* things will be kept quiet, and that consequently this evil affair will become completely obscured in the end, but it has dealt Berlin society a blow from which it will not recover.'[287]

Lieutenant von Tausch was also in close touch with General Waldersee, who by his own account had already heard the rumours of Ernst Günther's complicity many months before Kotze's arrest. He noted in mid-July 1894 that he had heard the suspicion voiced 'more than a year ago' that – as 'the whole world' was now whispering – the culprit was 'Duke Ernst Günther in collaboration with his mistress'. This being so, it was 'scarcely possible . . . to pursue the matter any further . . . No doubt there will be great efforts to kill off the affair.'[288] A month later he saw his prediction confirmed, noting that 'strenuous attempts are now being made' to prevent the Kotze affair from being solved, 'because the scandal affects the most exalted circles' and 'truly appalling facts' would come to light. 'Once again this seems to indicate that a great deal is rotten in the state.'[289]

Other thoroughly credible sources confirm the conclusions reached by Eulenburg and Waldersee. Soon after taking over the Reich chancellorship in the autumn of 1894 Prince Hohenlohe recorded that 'the Ernst Günther affair' was 'so complicated that I think it advisable to keep out of it'. The Kaiser had sent Hahnke to Ernst Günther to ask him whether he was the author of the anonymous letters. 'E.G. replied that if he had not been Chief of the Military Cabinet he would have answered him with a horsewhip, but in the circumstances he would merely show him the door.' According to the Chancellor the Kaiserin's brother was suspected not only by Hahnke but also by the Chief of Police in Berlin, Bernhard Freiherr von Richthofen, Adjutant-General von Plessen and the Kaiserin's Oberhofmeister von Mirbach. As if this coalition were not powerful enough, allegations against Ernst Günther were also made by Prince Aribert of Anhalt and his wife Louise, who accused the Kaiserin's brother of having taken part in the 'disgraceful deed'. Everything now depended on the Kaiser's attitude. But he hesitated. Hohenlohe noted in his journal: 'E.G. wanted to justify himself to the Kaiser, but the latter refuses to listen to what he has to say until Kotze's trial is over.'[290] It was perhaps after this confrontation that the 'confused' Duke began the correspondence with Wilhelm in which he 'almost' admitted his guilt. Nor do the self-pitying letters which Ernst Günther wrote to his uncle Christian demonstrate a convincing indignation at the widespread allegations against him.[291]

Further compromising details about the not exactly exemplary private life of the Kaiserin's brother – the lost Order of the Black Eagle, stories of fraudulent cheques, his demands for the return of presents given to his mistress and for her expulsion by the police, together with many other 'unpleasant matters' – threatened to come to light in the course of the trial of Eugen von Tausch in 1897. The newspapers took this opportunity of reporting openly that Tausch had named Ernst Günther to the Kaiser 'as the author of the anonymous letters, for which alleged evidence was produced', and had thereby caused 'great anger on the part of the Kaiser' against his brother-in-law.[292] The supposition widely expressed at this time that the obscene letters had been written by Duke Ernst Günther (and perhaps also by Karl Schrader) in collusion with 'actresses of *the most disreputable kind*' seems convincing. It explains much (although by no means everything) which has remained problematic about the affair: the uncanny inside knowledge, the snobbish attitude towards 'invaders' like Lottka Hohenau and Hugo Reischach, the allegedly 'feminine' handwriting and the 'pathologically effeminate' mentality of the letters, the horrified reaction of those in the know to the precipitate arrest ordered by the Kaiser and the conciliatory attitude suddenly adopted by Wilhelm towards Kotze on hearing of the strong suspicions of his own brother-in-law's guilt, and finally the way in which the explosive affair was 'obscured' in order to prevent the actual perpetrator being discovered.

Apart from Ernst Günther there was only one other 'very exalted personage' in the immediate proximity of the Kaiser who was suspected of having been involved in the sordid anonymous letter campaign: Wilhelm's eldest sister Charlotte, Hereditary Princess of Saxe-Meiningen, who was so fond of 'mischief-making'. In his memorandum on the Kotze case Philipp Eulenburg affirms that 'an important figure in Berlin society' is prepared to 'swear' that this 'fairly – depraved' Princess was the author of the letters, because she hated Lottka Hohenau for reasons unknown to him.[293] A recent study has revived this contemporary rumour with the suggestion – albeit without documentary evidence – that Charlotte might have collaborated with her brother's brother-in-law Ernst Günther.[294] This hypothesis is contradicted by the fact that on 26 June 1894 the Hereditary Princess wrote a revealing letter full of indignation about Kotze's arrest, in which she convincingly demonstrates not only his, but her own, innocence. As this private letter was certainly not destined for public consumption it can hardly be interpreted as an elaborate attempt to exculpate herself; we therefore have little alternative but to accept the Princess's protestations at face value. The letter, addressed to her father-in-law's third wife, Freifrau Ellen von Heldburg, reads: 'The Kotze scandal and his shocking arrest has affected me *very much*, especially because *she* [Elisabeth von Kotze] is my oldest and best friend. We know only what is in the newspapers and are until now firmly convinced of his innocence: we have known him for 16 years and until the court convicts him I consider [him] *not* guilty: he is such a loyal, slightly stupid man that *I* do not believe him capable of such nastiness, nor does Bernhard: the wretched pieces of blotting-paper do not prove anything; he would not commit such carefully worked out stuff to paper either in the Ceremonial Office or in the officers' mess. Nor could he have carried on with this *nastiness* systematically for 4 years, when we have travelled so much together and spent time together and often read the anonymous letters together. I have just received a pitiful letter from her [Elisabeth] K[otze]; she has seen her husband in prison and says he was admirably calm and firmly convinced that God would help him and show that he was innocent, and also that the truth would certainly come to light. Neither of them can yet grasp what is happening, or what will happen. There will be more than enough nastiness and filth, suspicions, slander etc. to come. I cannot understand my brother's *rash*!!! action in causing such a public scandal when he has always favoured him, and he an officer and holding an important post at court. Qui vivra verra! is all I can say. *I* had the first anonymous letter 4 years ago & up to last year they continued, some very vile, but never for one moment dreamed of accusing *him* poor man . . . But enough, the whole thing disgusts me too much.'[295]

Although many details of the Kotze affair remained concealed from the public at large, it created an appalling impression of moral decline at the court and even in the imperial family itself. With the fatal outcome of the affair, if not before, the Kaiser and his entourage were subjected to a barrage of criticism; there was 'deep indignation against the irresponsible advisers' who had counselled the Kaiser to interfere actively in the affair.[296] The conservative *Reichsbote* bewailed the fact that the Kotze case had 'destroyed more royalist feeling in the country than years of ideological work by true supporters of the monarchy could build up again'. The democratic *Berliner Tageblatt* pointed out that healthy popular sentiment could not but be harmed 'when men in the immediate entourage of the Kaiser flout all the laws of the land, all the prohibitions of the Church', and the liberal *Vossische Zeitung* was no less indignant that such things could take place at a time when 'the people are to be forced to pay more respect to the law by an anti-revolutionary measure; when scarcely a week goes by without the foundation stone of a new church being laid; and when social democracy is reproached for having a philosophy which offends against all morality and good sense'.[297] Maximilian Harden sharpened his pen to castigate the court as an obsolete establishment imposing itself 'like a mythological fossil . . . on to modern times'. The court scandal had shown up the 'whole danger' of an institution which 'keeps a modern ruler in the daintily tied bonds of a defunct rococo constraint', he wrote in *Die Zukunft*.[298] 'Berlin society', Philipp Eulenburg commented from Vienna in 1895, 'has probably never been as vulnerable to public criticism as now. Torn apart into factions which accuse each other of perpetrating the filthy anonymous attacks, it bears the very imprint of decay.'[299] It is not impossible that this scandal, which was discussed throughout the press and in countless sensational pamphlets,[300] even had an influence on the development of psychoanalysis. At any rate at the very time when Sigmund Freud recognised the power of the human sexual drive, numerous observers of the affair expressed very similar ideas. Thus Waldersee commented that the scandal had afforded 'sad insights into the highest circles in Berlin'. He was as convinced as ever, he wrote, 'that most of the numerous persons involved have a guilty conscience; each knows sordid stories about the other; they have made enemies of each other and now they are all as frightened as each other'.[301] And Charlotte Meiningen, Wilhelm II's sister, who had of course herself been suspected of involvement in the anonymous letters, expressed her sense of foreboding when she wrote: 'You cannot imagine the lies, the dirt, the nastiness of which there is now documentary evidence. And they came from the grandest people at court, and the leading figures of the *haute volée*! A sign of our times! Certainly sad and wretched enough. I shudder to think what harm humanity can do.'[302]

CHAPTER TWENTY-THREE

The Kaiser and the 'Newest Course'

THE APPOINTMENT OF 'UNCLE CHLODWIG' AS REICH CHANCELLOR AND MINISTER-PRESIDENT

On 26 October 1894 Kaiser Wilhelm II sent a telegram from Potsdam to Prince Chlodwig zu Hohenlohe Schillingsfürst, Statthalter of Alsace-Lorraine, reading: 'Request Your Highness to come here at once by the next express train and to bring Herr von Köller. Matters of importance to the Reich are concerned. Wilhelm I. R.'[1] Three days earlier, when he went to Liebenberg to shoot, Wilhelm had announced to his host Philipp Eulenburg the 'happy ending of the *ninth* crisis!', thus indicating that he still counted on both the Reich Chancellor, Count von Caprivi, and the Prussian Minister-President and Minister of the Interior, Count Botho Eulenburg, remaining in office. Not until the latter appeared at Liebenberg on 24 October to ask to be relieved of his duties did the Kaiser gradually realise that he would have to think of a successor for both posts. 'I have seldom seen him so helpless', commented Philipp Eulenburg in his diary.[2] How did Wilhelm II's choice come to fall on the 75-year-old Hohenlohe-Schillingsfürst, of all people? The Prince had been Bavarian Minister-President and Foreign Minister at the time of the foundation of the Reich (1866–70); but now, afflicted with infirmities of age and speech, a shrunken figure 'with his head bent to one side', he made a pitiful impression on all who saw him.[3] It is chiefly to Philipp Eulenburg's correspondence that we must look for further enlightenment.

The separation of the two highest offices, introduced in 1892 as an emergency solution, had been widely regarded from the outset as an arrangement that could not last. By the time open conflict broke out between Caprivi and Botho Eulenburg in the autumn of 1894 this view had been well and truly confirmed.

On 27 September the Baden envoy in Berlin, Eugen von Jagemann, reported that it was now clear that the divergence between the governmental hierarchies of the Reich and Prussia could not continue any longer; the press, he added, was talking of the forthcoming reunification of the two supreme offices and pointing to Philipp Eulenburg 'as the coming man, the only man to have any real influence on H.M. the Kaiser'.[4] At the same time the Grand Duke of Baden wrote from Lorraine to Philipp Eulenburg, who was with the Kaiser at Rominten at the time, urging that the 'functions of the Reich chancellor and the Prussian minister-president' should be reunited in one person; but that a new man would be required, for neither Caprivi nor Botho Eulenburg was fitted for both posts. In the whole of Germany there was only one statesman who stood above party and had 'a broad, clear vision, wise foresight and great firmness as a leader', and that was the Statthalter at Strassburg, Prince Chlodwig zu Hohenlohe-Schillingsfürst. He should become Reich Chancellor and Minister-President of Prussia. But since it was no longer possible for one person to manage the affairs of the Prussian state and those of the Reich on his own, Hohenlohe, who was old and must be spared, would have to have a 'vice-chancellor' and 'vice-minister-president' under him, who would later rise to become Reich chancellor and minister-president. For this post of vice-chancellor and vice-minister-president *cum jure succedendi* Philipp Eulenburg himself was the only possible candidate, the Grand Duke maintained.[5]

When news of this suggestion reached Bernhard von Bülow in Rome via a letter from Eulenburg, he immediately objected that Hohenlohe was too old to bear so great a responsibility; furthermore, as a Bavarian and former Minister-President of Bavaria he was not a suitable person to take over the Prussian minister-presidency. Bülow therefore adapted the Grand Duke of Baden's plan, suggesting that the supreme leadership of the country should be entrusted to *three* men: Prince Hohenlohe (either the Statthalter or his cousin Prince Hermann zu Hohenlohe-Langenburg) as Reich chancellor, Philipp Eulenburg as vice-chancellor and Botho Eulenburg as Prussian minister-president.[6] It was clear, however, that such a triumvirate would only increase the impression of the weakness of the Reich Chancellor.[7] With the fall of Botho and the (only too understandable) refusal of Philipp Eulenburg to take on high office in the harsh political limelight of Berlin, this proposed combination in any case came to nothing.

During the Kaiser's visit to Liebenberg Philipp Eulenburg was forced to recognise, after his cousin had announced his intention of resigning, that there was now no longer any prospect of the reconciliation for which the Kaiser still hoped between Botho and Caprivi. He therefore warned the Kaiser that it would be necessary to find a new man for both posts, which would not be easy, for this person must be 'neither Conservative nor Liberal . . . neither Ultramontane nor Radical . . . neither a churchman nor an atheist' – an 'impossible mythical creature', as the Kaiser remarked. Eulenburg went on to tell him that the Grand Duke of Baden had recently recommended Prince Chlodwig zu Hohenlohe-Schillingsfürst 'as a transition to someone else, who would have to be found'. The Kaiser seized on the idea at once and decided to write to Hohenlohe, as there could be no question of Caprivi staying on without Botho Eulenburg.[8] The very next day Philipp Eulenburg received from his cousin August a telegram which confirmed his supposition: 'Both going. The watchword is Chlodwig.'[9]

In telegraphing to Hohenlohe on 26 October Wilhelm not only asked the Statthalter to take the next express train to Berlin and to bring his Under-Secretary of State, Ernst Matthias von Köller, but also invited both men to stay at the Neues Palais. When Alexander, the Prince's younger son, who had also hurried to Potsdam, saw the magnificent apartments in the palace his first impression was of a beautiful cage in which his father was to be imprisoned.[10] At the same time Prince Hohenlohe received a letter from his close friend in the Foreign Office, Friedrich von Holstein, who strongly urged him not to accept the post of Reich chancellor definitely '*unless all your conditions are fulfilled with regard to government appointments*'.[11] In order to ensure that his advice was heard the Geheimrat had travelled out to Magdeburg to meet the Prince.[12] The actions of Wilhelm II and Holstein on 26 October 1894 mark the beginning of a conflict between the Kaiser and the Wilhelmstrasse which was to overshadow the next two and a half years, ending in the spring of 1897 with the victory of the Kaiser over the Reich Chancellor, the secretaries of state and the Prussian ministers. And it was indeed the control of public appointments, as will become evident, that proved the decisive weapon in the hands of the crown.

On his arrival at the Neues Palais Hohenlohe began by pointing out to the Kaiser the considerations which weighed against his taking over the Reich chancellorship and the Prussian minister-presidency respectively: '1. Age, poor memory, illness. 2. Poor public speaker. 3. Unfamiliarity with Prussian laws and politics. 4. Not a soldier. 5. Insufficient means. I could probably manage without the Statthalter's salary, but not in Berlin. I shall be ruined. My Russian connections [Princess Hohenlohe]. I have been in public life for almost thirty years, am seventy-five years old and do not wish to start something which I know will be too much for me.'[13] The Prince's wife Marie, who had inherited large estates in Western Russia, made a fruitless attempt to frustrate the appointment of her husband as Reich chancellor by sending a telegram to the Kaiserin, a niece of Chlodwig Hohenlohe. Auguste Viktoria merely wired back with a phrase that perhaps said more than she intended: 'The Prince is sacrificing himself for the Kaiser and the Reich . . .'[14] The Kaiser must nevertheless have promised the

Prince on this first occasion, in return for his acceptance of the chancellorship, to arrange for the 'removal of the financial difficulties' at the expense of the crown, by which means he was later able to secure Hohenlohe's dependence on him.[15] Soon after this he informed the new Reich Chancellor that he intended to address him as 'Du' and to treat him as his uncle, which 'of course' would be a one-sided arrangement only.[16]

In the confusion of the crisis Wilhelm seems at first to have intended to maintain the division between the supreme offices in the Reich and in Prussia and to appoint Heinrich von Boetticher minister-president of Prussia, for in his first interview with Hohenlohe he stipulated that Köller, who was designated for the post of Prussian interior minister, should become 'minister-president later . . . instead of Boetticher'. Boetticher's promotion, however, was opposed by Holstein, who wanted him 'out of the way' completely. After further negotiations with Holstein on the one hand and Wilhelm and Lucanus, the Chief of the Civil Cabinet, on the other, Hohenlohe recorded laconically: 'We'll leave Boetticher where he is, since Köller refuses and there is no one else.'[17] Wilhelm was able to telegraph to his grandmother in England in the late afternoon of 28 October the glad news that 'Uncle Chlodwig Hohenlohe has accepted the reunited post of Chancellor and President of Ministry. The outward policy remains exactly the same as it was before.'[18] There was no further reference to the idea of a vice-chancellor and a vice-minister-president.

Prince Hohenlohe also succeeded in appointing some of his younger relatives to provide necessary reinforcement for his team. Since Karl Goering, the former head of the Reich Chancellery, had resigned with Caprivi, Holstein suggested to Hohenlohe that the Vortragender Rat Heinrich von Günther be promoted to head of the Reich Chancellery and Hohenlohe's son Alexander, who had been a Free Conservative deputy in the Reichstag since 1893, appointed second-in-command of this small department.[19] Instead of Günther, Hohenlohe decided to appoint Freiherr Kurt von Wilmowski, who had hitherto worked in the Prussian Ministry of Agriculture, as head of the Reich Chancellery, while Alexander Hohenlohe was brought into the Foreign Office with the pro forma rank of counsellor; his task was to support his father in all his duties, 'like a kind of crown prince'.[20] Count Clemens von Schönborn-Wiesentheid, a cavalry captain whose brother had been married to Prince Hohenlohe's deceased daughter Stephanie, was also formally appointed to the Foreign Office, although he actually served as a personal adjutant to the new Reich Chancellor. Cardinal Gustav zu Hohenlohe, the Chancellor's brother – another brother was a member of the Prussian Upper House, a third was Senior Master of the Court at the Habsburg court – wrote to one of the federal princes giving equal weight to the prospects of success and failure facing the new regime: 'God grant Clodwig [sic] patience and good

36. Prince Chlodwig zu Hohenlohe-Schillingsfürst, Reich Chancellor 1894–1900

health. The honour is great but so is the burden, and I am worried about him, although I am also very glad that he has such an exalted post, and that he will be able to do much good and avert many an evil.'[21] For Waldersee it was more than doubtful that Hohenlohe was 'at all the right man' to overcome the difficult situation in which the Reich found itself. He was certainly 'wise, experienced, very cautious and perfectly suited to conduct a skilful defensive action. But that is not what is needed now', for it had become noticeable lately 'that from all sides dissatisfaction is focused on H.M.'.[22]

In spite of the assurances that he had received from Wilhelm the elderly Prince soon suspected that it had been 'perhaps a mistake, after all', to have yielded to the pressure put on him by the Kaiser. Alexander Hohenlohe, who knew the 'world of politics and the court at Berlin' well, openly admitted that his father lacked the necessary 'ruthless energy . . . without which a Reich chancellor cannot ward off his numerous political opponents and jealous enemies'. Chlodwig Hohenlohe did not have 'the initiative and the strength which the statesman there must possess in order to carry through his policy in parliament'. But the 75-year-old especially lacked the physical stamina 'to survive the gruelling life at the court of such a restless and self-willed monarch as the young Kaiser, who had ruled for only a few years'. Prince Alexander feared for his elderly father, who would now be 'exposed to the dangers and acts of malice' which were unavoidably bound up with the post of Reich chancellor. 'I sometimes felt as if I were sitting on the box next to an old coachman who could no longer get a proper grip on the reins in his hands, and yet I was not permitted to intervene and take the reins and whip out of his hands', he later recalled. It was abundantly clear to the Chancellor's son that those who had worked on the Kaiser behind the scenes to obtain his appointment wanted the post of chancellor to be held only by a man who would be a tool in their hands. 'I soon saw that his official life would from now on be a gruelling, day-to-day struggle against open attacks and intrigues, and even more from hidden sources; that in this milieu the vilest means and the most poisonous weapons were put to use and that a man like him, who . . . was not belligerent by nature . . . , must soon be defeated by the poisonous newts and toads among whom he found himself.'[23] That Hohenlohe saw himself as a 'transition to someone else' is clear from the mere fact that he extracted a promise from the Kaiser to guarantee his pension even if he were obliged to resign within two years.[24]

THE NEW MINISTERS

Just as they had done at the time of Bismarck's resignation in March 1890, after the fall of Caprivi and Botho Eulenburg, the Prussian ministers met on 27 October 1894 at Boetticher's house, and as before they discussed the crucial question of whether they could exert a collective influence on the formation of the new government, or would simply have to accept the will of the monarch. The Minister of Ecclesiastical Affairs, Dr Robert Bosse, expressed the opinion 'that we should all place our portfolios at His Majesty's disposal in the most loyal manner', while Boetticher (exactly as four and a half years earlier) pointed out that such a step 'was not the Prussian way and that we could not cause the Kaiser trouble at such a difficult time'. He informed his seven ministerial colleagues that the Kaiser had already summoned Hohenlohe and Köller to Potsdam and had decided to put an end to the division of offices – 'the root of all evil'. Bosse insisted that the ministers at least had the right to record their approval of the monarch's decisions regarding appointments, which Boetticher could not deny; but he insisted that the fact that the ministers had this right should not hinder the monarch from 'selecting and appointing His ministers personally'. The ministers were dismayed, however, to hear from the Minister of Agriculture, Wilhelm von Heyden-Cadow, that 'the not-yet-appointed future Minister of the Interior von Köller, on the instructions of the likewise not-yet-appointed future Minister-President Prince Hohenlohe', had put it to him that he should hand in his resignation. Bosse complained bitterly in early November of his colleagues' refusal to resign *en bloc* as he had suggested. 'Now the Ministry of State is acquiring an entirely different character from before, over our heads. But now it is too late . . .'[25]

In telegraphing his orders to Hohenlohe on 26 October to bring with him to Berlin the Under-Secretary of State responsible for the interior administration of the Reich province of Alsace-Lorraine, Ernst Matthias von Köller, the Kaiser had pre-empted the choice of the new Prussian minister of the interior. His intention from the outset was evidently to create a counterweight to the Catholic, liberal, South German Reich Chancellor in the person of the arch-conservative Köller. On 26 October Wilhelm told the Bavarian, Baden and Württemberg envoys that Köller was 'one of my staunchest Prussian Conservatives'.[26] With the appointment of this 'typical' Prussian Junker, as Alexander Hohenlohe rightly perceived, his father's government had acquired from the very outset 'an unpleasant appearance which did not accord with his political views nor with his entire political record'. 'My fears were later proved right to an even greater extent than I had anticipated', he wrote in his memoirs.[27] Even Waldersee, who in principle shared the Interior Minister's conservative standpoint, commented only a few weeks after his appointment that he thought Köller 'completely unsuited to his position'.[28] Köller's assumption of the role of the Kaiser's special representative in the Ministry of State was to lead to the first major government crisis of the 'Newest Course' before long.[29]

Although Prince Hohenlohe was unable to prevent the appointment of an arch-conservative Minister of the Interior, he was successful in resisting the equally right-wing candidates proposed by the Kaiser for the Ministries of Agriculture and Justice. Wilhelm was intent on appointing the Regierungspräsident of Breslau, Dr Georg von Heydebrand und der Lasa, 'personally known' to the Kaiser and brother of the future leader of the German Conservative Party, as the new minister of agriculture. 'He was active in Königsberg for a long time', the Kaiser wrote; he 'knows the agrarian campaign and its antics through and through, is himself a very great landowner and always behaved very correctly and calmly during the great agrarian agitation. He is fresh and vigorous, and a very suitable person.'[30] The previous day Hohenlohe had proposed the moderate Oberpräsident of Posen, Freiherr Hugo von Wilamowitz-Möllendorf, for the post and was most unwilling to accept the extreme right-wing Heydebrand as a member of his administration, particularly as the latter's superior, Oberpräsident Prince Hermann von Hatzfeldt-Trachenberg, described him as 'domineering, brusque and . . . very conceited' and commented ironically that he would be ideal as minister of the interior if one were aiming to stir up a constitutional conflict and as minister of agriculture if one wished to introduce a state monopoly on grain.[31] Hohenlohe therefore rejected Wilhelm's proposal with unusual firmness, on the grounds that Heydebrand was 'too much of a

Conservative Party man' who, combined with Köller, would create too strong a reactionary bias in the Ministry of State, thereby alienating all the other parties. 'If Heydebrand is taken on, I shall have to go', he resolutely declared. 'All my life I have been regarded, and have acted, as a moderate Liberal. I cannot become a Conservative Agrarian at the end of my life.'[32] On 6 November, when Hohenlohe came to see him at the Neues Palais, the Kaiser gave up his advocacy of Heydebrand after the Prince had argued 'with resounding success' that his appointment would be tantamount to a 'pater peccavi' on the part of the Kaiser vis-à-vis the agrarian movement.[33] Kaiser and Chancellor finally agreed on the Landesdirektor at Hanover, Ernst Freiherr von Hammerstein-Loxten, who had been suggested by the Oberpräsident of the province, the National Liberal Rudolf von Bennigsen. Hammerstein was also regarded, however, as 'such a determined agrarian' that even Germany's ally Austria-Hungary expected him to introduce an unacceptably protectionist tariff policy.[34]

Like Heyden, Prussian Minister of Justice Ludwig Hermann von Schelling was called on to hand in his resignation at this time of crisis. But while Hohenlohe had been responsible for the departure of the Minister of Agriculture, the dismissal of Schelling was entirely the work of the Kaiser, who sent the Chief of the Civil Cabinet to the Minister armed with the notorious blue letter. Hohenlohe begged the Kaiser to leave Schelling in office at least until the completion of his fifty years of service in December, but his request was curtly refused.[35] Wilhelm demanded the appointment of Hermann von Tessendorf, who had acted with extreme severity against the Centre Party and the Social Democrats during the Bismarck era. The Secretary of State of the Reich Justice Office, Arnold Nieberding, warned that the appointment of Tessendorf as Prussian minister of justice would be interpreted as 'signifying the introduction of harsh repression of the press, political associations and the right of public assembly, using all possible means'.[36] The newly appointed Reich Chancellor strongly urged the Kaiser to give up the idea of Tessendorf and instead to consider appointing the Director of the Reichsbank, Richard Koch, as minister of justice. Wilhelm abruptly turned down this request too. In his usual imperious style he wrote on Hohenlohe's submission: 'Koch is not the right man to reform the totally deplorable condition of the Prussian judiciary with the necessary ruthlessness and determination. His character is not suited for a ministerial post. Tessendorf must regain control over the law courts of Prussia and cleanse the Augean stables . . . I decided to appoint him justice minister fifteen months ago. The most reliable investigators have described him as the right man . . . Furthermore it is completely against my principles to appoint someone who is unknown to me to a Ministry, as I must always know exactly what I am letting myself in for.'[37] Hohenlohe therefore found himself compelled to offer the Ministry of Justice to Tessendorf. The latter was realistic enough, however, to see that as probably the most hated man in Germany he would face opposition from all parties. He refused the post. Hohenlohe now hesitantly suggested the President of the High Court at Celle, Karl Heinrich von Schönstedt, although he thought at first that Schönstedt 'might not satisfy His Majesty's requirements'. But Wilhelm declared himself willing to accept Schönstedt as minister of justice and worked harmoniously with him until 1905.[38]

Of crucial significance for the political conflicts between the Kaiser and the Reich administration over the next two and a half years was the decision in October 1894 to keep on Adolf Marschall von Bieberstein as secretary of state at the Foreign Office. Since the clash between Marschall and the Flügeladjutant Kuno von Moltke over Bismarck's visit to the Berlin Schloss in January 1894 vigorous attempts had been made to turn the Kaiser against the Foreign Secretary, as Holstein informed Prince Hohenlohe on 26 October. Holstein considered Marschall indispensable, however, and declared that he himself would resign if Marschall were forced to leave. Moreover in several letters to Philipp Eulenburg, as well as in his letter to Hohenlohe at this time, the Privy Councillor advocated the appointment of Marschall as Prussian minister without portfolio, arguing that it was 'a matter of political survival' for Hohenlohe himself to have him in the Ministry of State as a counterweight to Miquel.[39] Hohenlohe finally succeeded not only in keeping the Badenese statesman at the Foreign Office – 'Marschall is to stay, because otherwise the Foreign Office will fall apart', he noted after a conversation with the Kaiser on 28 October[40] – but also in making him a member of the Prussian Ministry of State. Because of the elderly Prince's lack of skill at public speaking, Marschall's talent as an orator in both the Reichstag and the Prussian Landtag, as well as his good relations with the Centre Party, also helped ensure that he remained in office.[41] Many people even assumed that he would soon be appointed Reich secretary for the interior instead of Boetticher.[42] But only days later the Kaiser's hostile behaviour towards the Foreign Secretary made it plain that Wilhelm 'obviously wants to pick a quarrel with Marschall'.[43] On 17 November 1894 Holstein had to advise the new Chancellor to take over the responsibility for reporting to the Kaiser on foreign policy 'until Marschall's standing with H.M. improves'.[44] It soon became apparent that Wilhelm was no longer willing to work with either Marschall or Boetticher.

Finally, a new appointment also had to be made to Chlodwig Hohenlohe's erstwhile post as Statthalter of Alsace-Lorraine. Originally the Kaiser had intended to nominate the departing Prussian Minister-President, Botho Eulenburg, to this position, which was much sought after because of the high salary and generous funds for official expenses attached to it, but Hohenlohe was determined not to accept the chancellorship if Eulenburg became Statthalter.[45] A letter from

Philipp Eulenburg convinced the Kaiser that it would be wiser to send a South German prince like Carl Egon zu Fürstenberg (although he was married to a French Princess) or Hermann zu Hohenlohe-Langenburg, the Kaiserin's uncle, to Strassburg as Statthalter. With a characteristic telegram from the Kaiser – 'Have appointed you my Statthalter in Alsace-Lorraine. Refusal not possible. Wilhelm I. R.' – Hohenlohe-Langenburg ('Uncle Hermann' as the Kaiser called him) was given the posting and Botho Eulenburg was left empty-handed.[46]

In contrast, another idea of the new Reich Chancellor which he hoped – perhaps erroneously – would strengthen his position, above all at the imperial court, failed to come to fruition. This was the appointment of the Kaiser's friend Philipp zu Eulenburg as minister of the royal house in place of Wedell-Piesdorf. It is probably correct to assume that this thought had its origin in the Grand Duke of Baden's suggestion that the elderly Chancellor should be given the support of a vice-chancellor who would be expected to succeed him. A first attempt by the new Chancellor to suggest Philipp Eulenburg to the Kaiser as house minister was frustrated when Wilhelm expressed doubts 'as to whether Eulenburg had sufficient experience in financial management' – a remark from which Holstein concluded that the Kaiser 'does not think Phil. Eul. suitable to be house minister. That is just going too far for H.M., as far as money matters are concerned.'[47] What was more, Wilhelm had 'a decidedly guilty conscience' with regard to Botho Eulenburg and was playing with the idea of appointing him, rather than Philipp Eulenburg, in place of Wedell-Piesdorf, 'whom His Majesty has in fact long wished . . . to be rid of'.[48] After further negotiations Hohenlohe was able to remind the ambassador in Vienna in January 1895 that he had shared his views 'on the importance of this post for H.M.' and had assured him of his readiness to 'make the patriotic sacrifice and take over the House Ministry'.[49] Eulenburg laid down both financial and institutional conditions, acting initially as if he intended to accept the post; but in the end he had to recognise that his special friendship with the Kaiser would not survive daily dealings with him on an official footing, and he saw to it that the negotiations came to nothing.[50]

In looking back over these momentous decisions on political appointments made in the late autumn of 1894, one is struck above all by the element of chance and the lack of forethought which governed the choice of the new men for the highest offices. The new Reich Chancellor and Minister-President Hohenlohe-Schillingsfürst did not come under consideration at all until a few days before his appointment, when the Kaiser suddenly saw himself compelled by Botho Eulenburg's unexpected resignation to think of a successor for both him and Caprivi. Hohenlohe, who from the beginning was regarded only as a 'transition to someone else who will have to be found', had been suggested by the Grand Duke of Baden, but even he only considered Hohenlohe's appointment feasible in conjunction with that of a younger statesman who would take over the two highest offices after a few months or years. When this solution proved impossible because of Philipp Eulenburg's refusal to take on such a responsible office Wilhelm II had recourse to the idea – we do not know who suggested it – of appointing the ultra-conservative Köller as minister of the interior in order to provide a Prussian counterbalance for the South German Reich Chancellor and Minister-President. Apart from Hohenlohe himself Köller remained the only minister whose selection in the autumn of 1894 can be attributed solely to the Kaiser. As we have seen, Wilhelm's attempt to appoint another two arch-conservatives, Heydebrand and Tessendorf, as ministers of agriculture and justice respectively, was unsuccessful, although it was indicative of the monarch's imperious and almost class-struggle-obsessed mood, with which the new Reich Chancellor most certainly had to reckon in making his counter-proposals (Hammerstein-Loxten and Schönstedt). In his negotiations, as a condition of his acceptance of the two supreme offices, Hohenlohe was not only able to extract several material promises from the Kaiser but also to secure the appointment of his son Alexander as his permanent adviser and of Count Schönborn as his personal adjutant. The choice of Prince Hermann zu Hohenlohe-Langenburg as Statthalter must also be seen as an impulsive choice by the Kaiser after his wish to reward Botho Eulenburg with this post had been frustrated by the new Chancellor's opposition. Lastly, Chlodwig Hohenlohe also succeeded, much against the will of Wilhelm II, in keeping on Heinrich von Boetticher as secretary of state of the interior and Adolf Freiherr Marschall von Bieberstein as secretary of state at the Foreign Office. Both were to play a central role in the forthcoming battle of the Wilhelmstrasse against the growing might of the Kaiser and his entourage.

THE KAISER AND THE HOHENLOHE GOVERNMENT

In what way had the relationship between the Kaiser and the executive shifted as a result of the change of chancellor and ministers in October 1894? How much power did Wilhelm II possess under the chancellorship of his wife's uncle; what influence did he exercise on the formulation of German domestic and foreign policy under the 'Newest Course'? How did the venerable Prince Hohenlohe regard his responsibilities as Reich chancellor and minister-president – did he see himself, in relation to the monarch, as supreme leader of the 'responsible government' or rather as the executive agent of the imperial will? And what role did the Prussian ministers and the secretaries of state of the Reich offices play in these decisive years? Were they – either singly or collectively – willing and able to take independent political decisions and to carry them through? The answers that one is compelled to give to these questions on the basis of the very explicit contemporary sources are shocking and depressing even today.

As soon as the new Reich Chancellor took office Friedrich von Holstein had urged upon him the importance of firmness with the Kaiser, in order to bring about, at least in foreign policy, 'a change for the better *from the start*', for 'the way in which H.M. has gradually become accustomed to discuss policy with every Tom, Dick or Harry cannot be compatible in the long run with the orderly pursuit of business'.[51] Despite isolated acts of protest, however, Hohenlohe hesitated to offer any resistance to the Kaiser's craving for power. At the close of 1894 he reflected that his relationship with the Kaiser was still too new 'for me to climb up on my high horse and reproach him about his conduct'.[52] After six months in office his hopes of an improvement in his relations with the Kaiser had become very modest. 'Monarchs are human like others and have an inclination towards a comfortable life, in which one should disturb them only in emergencies', he observed.[53] Not without irony he wrote on 7 January 1895, in a letter to Eulenburg: 'In my relationship with His Majesty all is well. I do not take it amiss if His Majesty gets one of his valets to telegraph to the cipher office and orders a dispatch-rider to be sent to Petersburg in the evening. What is bad is that the little, silent, cautious Tsar will soon be several lengths ahead of our Kaiser in public opinion. But nothing can be done about it.'[54] In January 1896 he is said to have remarked: 'I have taken a firm decision not to get angry about anything and I let everything take its course. If I wanted to do anything else I would have to resign at least once a week.'[55] When he came under increasing pressure from Holstein in the next few months to put a stop to the interferences of the Kaiser, his Cabinet chiefs and Flügeladjutanten, the Reich Chancellor rejected the diplomat's demand with the telling argument that in the end he would rather resign 'than govern with the Kaiser in an unfriendly relationship. The purpose of my existence in the Reich chancellor's palace is after all none other than to prevent overhasty decisions. An independent regime, in which the Kaiser walks alongside as a defeated man, so to speak, is unthinkable with this sovereign. It is also against my nature.'[56] Nor had he any desire, Hohenlohe added, 'to take on the role of victor over the defeated Kaiser, as it were. That is not for me.'[57]

From the outset there were rumours of a change of chancellor,[58] and as early as 24 November 1894 the Austrian ambassador reported from Berlin: 'As far as the stability of the new regime is concerned, there is much pessimism here in general; . . . many people say [Hohenlohe] is only a stand-in and they think that Kaiser Wilhelm intends Count Philipp Eulenburg to take over this post in the foreseeable future.'[59] In the spring of 1895 Waldersee expressed the opinion 'that Hohenlohe has long since had enough and has often regretted taking on the position'.[60] Although he 'very often' avowed that he was 'of a different opinion from the Kaiser on many questions', he had 'given in completely and renounced any idea of independence'.[61] In fact all observers already agreed at this time that the new Chancellor was 'not much more than a nullity'.[62] 'Hohenlohe scarcely even attracts mockery any more. To be precise, he does not come into the reckoning at all. The Kaiser could not have chosen a more pitiful Chancellor', Waldersee commented damningly in May 1895.[63] A week later he noted in his diary: 'If the question were asked in the Reichstag: what is your opinion of Hohenlohe's chancellorship or of the Newest Course? the *unanimous* verdict would be one of incompetence. Such a state of affairs has never existed before, and yet I can see no sign of a change.'[64]

Through his informants at court and in the officer corps Waldersee also became aware of the elderly Chancellor's increasing financial dependence on the Kaiser: Hohenlohe had abused his position in order to win far-reaching concessions for his wife's Russian property. 'Furthermore he has also managed to take very good care of the numerous branches of the Hohenlohe family', the General recorded.[65] Indeed it is one of the most scandalous events of the third Reich Chancellor's term of office that the Kaiser, at Hohenlohe's instigation, secretly used crown funds to triple his official annual salary of 54,000 marks. On 15 December 1894 an anonymous letter appeared in the press according to which Hohenlohe was to receive 100,000 marks from the Kaiser to top up his salary. Although a denial had to be issued, Hohenlohe and Philipp Eulenburg continued to negotiate over the secret subsidy. 'I have to make it clear to the Kaiser that the poor Reich Chancellor must not be allowed to suffer from yet another burden as well as that of his work', Eulenburg emphasised in January 1895.[66] At the beginning of February he sent a letter to the Kaiser pointing out that the Reich chancellor's salary was 'really *disgracefully* small'. The Prince's financial difficulties were made still more acute by the unpredictability of his 'miserly' wife Princess Marie, although she was obliged to support their sons Moritz and Alexander and their wives. 'So this puts the old Prince under pressure', Eulenburg wrote. 'Unfortunately since the news of Your Majesty's gracious intentions mysteriously leaked out . . . it has become impossible to help the Prince as planned. But Your Majesty will no doubt graciously keep this situation in mind. When he agreed to make the sacrifice of coming to Berlin the Prince was reassured by the certainty that he would be *provided for* in material terms. This turn of events has greatly affected him. It is *impossible* for Your Majesty to help from Your own pocket because the family is so large. But unforeseen advantages might *occasionally* arise which could be directed towards the Prince.'[67] It is safe to assume that the much-criticised political weakness of the Reich Chancellor was bound up with his secret financial dependence on the Kaiser. It was not long before Holstein was complaining that 'Hohenlohe [is being] worked upon by his wife and son to do nothing that might threaten the Kaiser's goodwill towards the Hohenlohe family.'[68]

Waldersee wrung his hands in despair over the way in which the German Reich was being governed. He commented acidly that Wilhelm II had realised after only a few weeks that with Hohenlohe's appointment he had 'dug up a semi-corpse', and that his behaviour had become correspondingly imperious.[69] In February 1895 he noted gloomily that Hohenlohe had now been Reich Chancellor for more than three months and that it should have become clear where he was heading. In reality, however, 'no firm programme [had been] settled with the Kaiser', who was continuing 'to make policy decisions on his own initiative'.[70] In a letter to Verdy in April 1895 Waldersee again complained of the confusion and indecision of the government, produced on the one hand by the constant interference of the Kaiser and on the other by the completely passive attitude of Hohenlohe. 'The wavering back and forth will go on until a mighty row breaks out', he prophesied. 'I had thought that Hohenlohe would quietly assess the situation first, but then come forward with some programme or some recognisable aim. But that was an illusion. Now I am convinced that he is glad just to get through without a row, and accepts everything.'[71] With every month of Hohenlohe's chancellorship, Waldersee wrote, the general conviction was growing stronger 'that the Kaiser's personal intervention is chiefly to blame for the sad state of affairs in which we find ourselves'. 'Previously it was only hinted at, but now people say it more clearly: that he takes no notice of the opinions of the ministers, but that they completely submit to his wishes and opinions; if these took a *consistent* course it would be all right; but as they change, and often very rapidly, one would think that the ministers must be in an impossible position. But this is not the case, for these gentlemen always knuckle under. They certainly complain that the Kaiser is impossible to get along with . . . but they do not leave. If there were a Ministry which was united under the leadership of the Chancellor, and in which they supported each other, the Kaiser would also be different; but he is clever enough to see that he would draw the short straw in that case, and he positively enjoys playing off one minister against another.'[72] Again in February 1896 Waldersee recalled Marschall's remark during the opening ceremony of the Kaiser Wilhelm Canal at Kiel the previous summer: 'This is too much for anyone to bear: one thing today and another tomorrow, and again a few days later.' Waldersee himself considered that under the present circumstances the Reich Chancellor and the Foreign Secretary were in a 'degrading' position which they ought not to accept, 'all the more because the whole world knows how things are. The Kaiser is his own chancellor and has only compliant ministers.' As a result it was not surprising that there was 'widespread mistrust', he said.

'No mortal being can tell us what the future will bring; but one thing I know for sure: we have no purposeful, clear and consistent leadership, either at home or in foreign affairs.'[73] During the laying of the foundation stone of the memorial for Kaiser Wilhelm I in Berlin on 18 August 1895 Waldersee had the opportunity of speaking to many acquaintances about the confused state of both domestic and foreign politics. All were agreed about the dire state of affairs and declared that it could not go on. 'Everyone is disgruntled and chasing in different directions; the ministers criticise each other and work only for themselves. Whether in domestic or foreign policy, there are orders today and counter-orders tomorrow and something else again the day after.' The General went on to express the opinion that there was absolutely no prospect of a change for the better in the near future. 'We shall go on wavering about in the same way for a long time and we shall gradually sink even lower in the process.'[74]

There is no doubt that under Prince Hohenlohe's chancellorship neither he nor the Prussian Ministry of State but Kaiser Wilhelm II – together with his advisers at court – was the decisive political factor.[75] Even if his personal power did not reach its peak until after the great reshuffle of secretaries of state and ministers in the summer of 1897, the sources show clearly the extent to which, during the first three years of the 'Newest Course', he was already not only setting the general political course of the country, but also intervening, sometimes down to the smallest detail, at least in the particular areas which interested him, which included both domestic and foreign policy, military and naval matters, art, science and, not least, the economy. At Prökelwitz in 1895 he gave Philipp Eulenburg a clear exposition of his conception of the proper relationship between himself, the Reich chancellor and the Prussian Ministry of State. The Kaiser had declared quite distinctly, Eulenburg reported to the astonished Reich Chancellor on 24 May, 'that he *quite categorically* refused to tolerate any quarrelling, discord and divergence within the Ministry. It was for Your Highness to ensure this homogeneity; decisions upon the principles of government were his own and Your Highness's business alone – he would "throw out" those who presumed to have *personal* views. Only Your Highness had the right to take a view and express it to His Majesty; the individual ministers must limit themselves to following the principles laid down . . . His Majesty stated further that he relied upon Your Highness always to inform him immediately of anything that happened in the government in any sphere.'[76] It was around this time that the Kaiser commented 'in very forceful tones that He would not tolerate individual political action by ministers'. 'It was for *Him* alone to govern with his minister-president . . .'[77]

This method of government did not of course produce the kind of objective decision-making which the Reich and its population deserved, but a regime that was characterised by personal passion and impulse, by court intrigue, backstairs influences and interdepartmental rivalries. Already in February 1895 the Foreign Secretary was complaining bitterly that it was 'a nice state of affairs that one constantly has to contend not only with foreign states but also with H.M. and his irresponsible advisers'.[78] At the same time Friedrich von Holstein recorded with consternation that 'the Kaiser is governing *with* the Cabinets *against* the constitutional organs of government. There is no one to act as intermediary between the latter and the Kaiser, as Wilmowski and Lehndorff did under Kaiser Wilhelm I. Without anyone to mediate there is bound to be a clash. I do not deny that I should like to be gone before the clash comes. *After* the clash, i.e. Hohenlohe's resignation, the Kaiser will go for a trial of strength. The more important federal princes, led by Bavaria, will oppose him and he will fail. That will destroy the Kaiser's prestige and he will be faced with a period of humiliation. *I* can do nothing to prevent it. For His Majesty now follows the advice of Hahnke against Bronsart (Minister of War), of Senden against [Admiral] Hollmann and of Lucanus against the whole Ministry of State.'[79]

Holstein could not believe his eyes or ears when he learned that on the occasion of his birthday in January 1895 the Kaiser intended to confer 'the rank of ministers of state' on his Oberhofmarschall, Count August zu Eulenburg, and the Chief of the Civil Cabinet, Hermann von Lucanus. The Ministry of State, without meeting for consultation, immediately and unanimously took the view that this was not permissible, as 'minister of state' was an *office* and not a rank or title, and Hohenlohe only just managed to put a stop to the Kaiser's plan. The Privy Councillor's verdict on the affair was that 'if the idea had been carried out the Hohenlohe regime would probably have succumbed from the sheer weight of ridicule. To have a fawning courtier as a minister – that has never happened in Prussian history, not even in the absolutist period.' He commented with deep concern that this affair was 'typical'. 'The Kaiser, who has no idea of constitutional law, can be persuaded to take the most utterly incredible decisions through the combined efforts of A[ugust] E[ulenburg], Hahnke and the apothecary [Lucanus].' He added sarcastically: 'Instead of minister of state August E. is being made lieutenant-general and Lucanus is getting the rank of "behind the ministers of state".'[80] A few weeks later Philipp Eulenburg surprised Holstein with a report on the relative power of the court officials, drawing particular attention to the rivalry between the increasingly influential Lucanus on the one hand and the Kaiser's military suite on the other. A strong movement against Lucanus had developed in the Kaiser's entourage, he claimed. '*He alone* is held responsible for *everything* the Kaiser says and all his political utterances . . . There is nothing that is not blamed on Lucanus. The bad mood among the maison militaire and other people in the imperial entourage has not sprung from any wish to put a particular successor in his place. There was *never* any question of that. I attribute it principally to the ever more powerful position of Lucanus.' Any attempt to get rid of the Chief of the Civil Cabinet would only have the opposite effect. 'His Majesty has grown *too* accustomed to him not to treat *any action* against him as a "nasty intrigue" and to strengthen his position still more.'[81]

By the early summer of 1896 the numerous ministerial crises and scandals were causing highly critical comments in the press and parliament about Kaiser Wilhelm II's style of government. People spoke 'quite openly about illegal influences on the Kaiser, in particular Hahnke and Lukanus [sic] and Flügel Adjutanten in general', noted Waldersee, who was forced to admit that there was 'much truth' in this criticism. 'The Kaiser wants to rule autocratically', he wrote on 3 May 1896, 'and in fact does so fully in military matters and foreign affairs; he only rarely intervenes in other areas, but there too his will is supposed to be paramount. Of course a great deal of hostility is building up against the Cabinet chiefs . . . The fact that other third persons also often join in the discussions with the Kaiser is beyond question. If Hohenlohe were an energetic man he would have gone long ago or else he would have insisted on a change.'[82] All such attacks on the irresponsible advisers were 'in fact directed against the Kaiser', Waldersee rightly observed.[83] Berlin was in a state of 'hopeless confusion', he commented not long afterwards. 'Hohenlohe is exhausted in every sense; he would like to get out as soon as possible and he lets everything take its own course. The ministers are each going their own ways, and are no doubt using the press in their own particular interests too. None of this seems to upset the Kaiser and he is only anxious to enable good old Uncle Chlodwig to stay on.'[84] A year later the mayor of Altona, Giese, reported after a visit to Berlin: 'No minister knows anything about the others; each claims that he does not know what the Kaiser actually wants; the Kaiser no longer communicates; Hohenlohe is completely indolent etc.'[85] This, then, is what the government of the Kaiserreich looked like from within, five years after Bismarck's dismissal. Only rarely were outsiders able to catch a glimpse of Wilhelm's autocratic style and the disturbing compliance of his ministers.

THE OPENING OF THE KAISER WILHELM CANAL IN JUNE 1895

The international celebrations to mark the opening of the Kiel Canal in June 1895, that is to say exactly seven years after his accession, provided a particularly clear illustration of the personal power of Wilhelm II in action, working in conjunction with the rapidly increasing commercial and industrial power of Germany. Not only at the ceremony itself, when he was the natural focus of attention during the laying of the foundation stone of the memorial at the eastern end of the canal, but also in the preparations for the gigantic and costly celebrations in Hamburg and Kiel, he played a dominant and decisive role. Even the technical problems which arose during the construction of the Kaiser Wilhelm Canal were overcome rather more rapidly than the experts thought advisable as a result of an imperial command. Waldersee, who as commanding general at Altona was responsible for security measures – the fear of assassination attempts being unusually great because so many prominent public figures would be present – commented in April 1895, 'the Kaiser has overcome the opposition of the technicians by giving definite orders that the canal absolutely must be ready at the beginning of June . . . But there is now great anxiety as to whether the procession through the canal on 20 June will be a success.' Countless pessimists claimed that the procession would not work, and after the event considerable damage was indeed discovered, caused by the premature opening procession through the canal.[86]

In March 1895 the Kaiser summoned a conference to discuss the opening ceremony without first approaching the long-serving Secretary of the Reich Office of the Interior, Heinrich von Boetticher. His action led to a letter of protest from Hohenlohe, who had to ask Wilhelm to receive Boetticher alone, before the conference, in recognition of his services in the construction of the canal. Boetticher complained bitterly of the influence of Plessen and the two Moltkes in the Kaiser's entourage, who drove the sovereign 'to more and more extreme ideas' and demanded enormous sums of money – there was talk of two million.[87] The Kaiser replied on 30 March to the Reich Chancellor's letter with the sarcastic mockery that characterised his manner with 'Uncle Chlodwig': 'Dear Uncle, I received your letter last night when my night-shirt was already flapping round me and I thought it meant we were about to mobilise! The contents were reassuring, however. It shall be as you wish. Instead of taking umbrage, by my reckoning the gentle Heinrich ought to be grateful that we want to help him carry the burden and share it out! Your letter quite rightly emphasises his great services for the canal, the project he lives for!, and far be it from me to belittle them. On the contrary! But in order that he can *devote himself entirely to his canal*, which he must, it is best that he should not concern himself with other things, otherwise the Privy Councillorish mind will be troubled with distractions which will have a harmful effect. The mere fact that in answer to almost all the questions of principle which were put to him from the naval side concerning the canal he could supply only inadequate information or none at all shows that he still needs to acquire a thorough knowledge of his "pet project". This is all the more evident because since we were both in Kiel in November 94 I have asked Boetticher to tell me in broad outline how the programme should be arranged, as he gave the end of June as the deadline! From November until now he has had 4 months! All the questions concerning the *canal* could have been committed to paper and submitted to me long since! From what you tell me about the two representatives of Lloyd's – the gentle Heinrich's doubts about their being called in on financial grounds – I gather that you support my summoning of a committee under you! This *doubtfulness* is precisely what is wrong with him and threatens to expose us to ridicule! He has so many doubts he never gets anywhere! This Lloyd's question could have been settled months ago by a short discussion with the gentlemen, so why was it not done? Now is the time to act, so I implore you to harden your heart against all objections! *It must be done!* If you can make the gentle Heinrich see it in the light of co-operation he will be content not to have to bear the enormous responsibility on his own! Let him worry about his canal and his ships and guests; we shall take care of the rest! Your devoted nephew Wilhelm I. R.'[88]

It was in the same autocratic spirit that Wilhelm conducted the meeting which he convened on 1 April 1895 to plan the opening ceremony. It was attended by civilian statesmen (Reich Chancellor Prince Hohenlohe, Secretary of State for the Interior von Boetticher, Prussian Minister of Public Works Karl von Thielen, Reich Postmaster-General Heinrich von Stephan and the Oberpräsident of Schleswig-Holstein, Georg Steinmann), admirals (Navy Secretary Friedrich von Hollmann, Commanding Admiral Wilhelm von Knorr, Chief of the Naval Cabinet Gustav Freiherr von Senden-Bibran), generals (Chief of the Military Cabinet Wilhelm von Hahnke, and Commanding General in Altona Alfred Count von Waldersee), Senior Marshal of the Court Count August zu Eulenburg, Albert Ballin, the owner of the Hamburg Packet line, a representative of the Bremen shipping company Lloyd and a landowner as a representative of agriculture. The Kaiser began by drawing attention to the difficulties which would be created in Kiel by the arrival of large crowds of people for the opening of the canal, and recommended the establishment of a committee with three subcommittees to take responsibility for accommodation, food and entertainment for the visiting masses. With the sole exception of the military precautions suggested by Waldersee, the Kaiser raised objections to the suggestions made by all the other participants, and his views were accepted. When Boetticher pointed out that two million marks would be required for the celebrations but that the Reichstag would not approve such a sum, the Kaiser ordered that it should be requested nonetheless. To Boetticher's suggestion that seventy-five members of the Reichstag should be invited the Kaiser responded that this was far too few, whereupon the figure was raised to 150. The elderly Reich Chancellor made 'a very pitiable impression' on Waldersee during the negotiations. As the General recorded, only Albert Ballin had the courage to stand up to the Kaiser in defence of his own point of view, although even he did not succeed in getting his way. In a particularly 'painful scene' Wilhelm II declared that the German princes must be accommodated on board the Lloyd's of Bremen ship during the celebrations, whereupon Ballin, as representative of the Hamburg Packet, objected that this would be a 'very painful' and 'very severe' blow to Hamburg and its company, which was providing two ships. The Kaiser 'flatly' refused Ballin's strong plea for a change of plan, remarking that he had given his word to Lloyd's of Bremen some time ago. 'The Kaiser then explained to Herr Ballin that he was quite wrong. Hamburg was extraordinarily favoured in having the honour of putting up the Kaiser himself (which is costing Hamburg over a million marks), in addition to which the *ambassadors* would be in a packet ship, and since they represented the sovereigns the Hamburg Packet was in fact receiving preferential treatment!!! When Ballin asked for the princes to be divided up, the Kaiser objected, very kindly for all those present, that such gentlemen could not be mixed with other people; they had to keep to their own society!' In describing the scene Waldersee commented that he thought 'everyone felt how unfair the Kaiser was being towards Hamburg, from which he was asking such large sacrifices; but unfortunately he has a talent for offending people in a quite extraordinary way'. Summing up his impressions of the day, the General wrote: 'the Kaiser's speech [was] very clear and his arguments, as usual, quite skilful, but the whole thing was characterised by a certain cursoriness which was probably very noticeable to all those present; there had no doubt been preliminary discussions with younger people who lack judgement . . . Some very extraordinary ideas were expressed.'[89]

Wilhelm presided over a further meeting to discuss the opening of the canal, this time in Kiel, which confirmed Waldersee's worst fears about the attitude of Ministers Boetticher, Köller and Hammerstein-Loxten to the Kaiser. He had expected nothing else, he said, than that Boetticher would say exactly what the Kaiser wanted to hear; but he had thought otherwise of Köller. And yet he too 'echoed the Kaiser's every word, nor did he have any scruples in talking of the agitation stirred up by the Conservatives. Hammerstein was perhaps the best of the three, but his desire to please the Kaiser was also obvious; none of the three let slip a single word, a single hint of an opinion differing from the Kaiser's; oh what a sorry trio!' It had been very depressing for him to observe yet again 'how people agree with everything the Kaiser says and no one makes even the most diffident attempt to put forward a different view'. Köller had particularly annoyed him with his way of 'persistently pushing himself forward' and submitting proposals to the Kaiser instead of taking decisions himself. After the meeting Waldersee recorded that the naval officers present complained that the Kaiser 'wants to decide everything himself and never alters his own opinions'.[90] In May 1895 he noted in his diary that he was 'genuinely sorry for the gentlemen from the navy, but especially the canal authorities. The problems mostly arise from the fact that the Kaiser wants to decide everything himself and gives hasty orders which soon turn out to be impracticable, and that too many authorities want to have a voice in the matter. Minister Boetticher, the Comm[anding] Admiral, the Chief of the Naval Cabinet, the Ob[er]Hofmarschall's office, Minister Thielen, the Minister of the Interior etc.: each wants to have his say and is partly under an obligation to do so anyway; sometimes agreement is reached with great difficulty, only to be overturned by the Kaiser.'[91]

Shortly before the opening celebrations Waldersee succeeded in preventing the Kaiser from committing a serious *faux pas*. The monarch had expressed the wish to be received on arrival in Hamburg by a guard of honour at the station and another at the town hall, and to be escorted by a squadron of Hussars. Waldersee had to point out to the Military Cabinet that Hamburg was not part of Prussia but a sovereign state which had invited the Kaiser and the other German ruling princes, and therefore had the right to decide for itself how it would receive them. Waldersee assumed, probably not without reason, that both Wilhelm and Hahnke would be annoyed by his intervention, but they conceded the point nonetheless and informed the city of Hamburg through its envoy in Berlin that the sovereign would agree to the arrangements made by the Senate. Even the official reception of the German ruling princes, 'which the Kaiser had been very much against', had now been approved by him, Waldersee recorded. The city fathers were relieved, for their part, having feared that a conflict with the Kaiser might develop out of this question. The German princes were also worried that 'the Kaiser might act inconsiderately towards the federal state of Hamburg'.[92] In constantly making unfriendly remarks about Hamburg and its mayors, Waldersee observed disapprovingly, Wilhelm was completely failing to recognise that the city was taking the greatest possible trouble to give him a magnificent and very expensive reception.[93] Oberpräsident of Schleswig-Holstein Georg Steinmann later felt compelled to resign because the Kaiser persisted in treating him badly.[94]

The opening celebrations, staged at a cost of several million marks, were among the most magnificent ceremonies of the entire Wilhelmine era. 'It will be a more splendid gathering than has perhaps ever been seen and it should put the opening of the Suez Canal in the shade', Waldersee predicted in the spring.[95] The innumerable sovereigns, royalties and diplomats – including the future King George V of England, an Austrian Archduke and an Italian Prince – were invited to a luncheon in the Hamburg Zoological Gardens on 19 June, for the Kaiser had decided from the outset that the ceremony would begin in Hamburg; the city was reluctantly obliged to accept 'the inevitable'.[96] That evening, with a storm brewing, a great gala dinner took place in the unfinished town hall, presided over by the Kaiser who replied 'in suitable words' to a speech by the Mayor of Hamburg, Lehmann. After dinner, in pouring rain, the 'somewhat tipsy' guests were offered coffee on an island constructed for the occasion in the Alster lake at a cost of 700,000 marks, before being taken by train to Brunsbüttel and delivered to the ships – the Kaiser to the new white-and-gold *Hohenzollern* – in which they were to stay. The same night the ships, accompanied by cheering and brass band music, sailed down the Elbe to the canal. But as the convoy ran aground several times it did not arrive at Holtenau or Kiel until the later afternoon of 20 June. The weather was still very wet, so that the Navy Ball was 'dreadfully spoilt by rain'. On the morning of 21 June the ceremony of the laying of the foundation stone of the memorial at the Baltic Sea end of the canal took place. The Reich Chancellor escorted the Kaiser and Kaiserin to the podium, near which stood the sovereigns, ruling princes and other high office-holders. Hohenlohe read out the document that was to be placed in the foundation stone; the Bavarian Foreign Minister Freiherr von Crailsheim gave an address and handed the Kaiser the trowel; 'then all the designated people struck the stone with a hammer, after which I gave three cheers for the Kaiser', Hohenlohe wrote.[97] That evening a banquet for 1,050 people took place at Kiel, during which the Kaiser read out a lengthy speech written by Hohenlohe, Boetticher and Lucanus. It was warmly applauded. After dinner the Kaiser conversed 'very enthusiastically' with the French naval officers present, who were 'very surprised by the Kaiser's technical knowledge'; one of them said 'Votre Majesté parle comme un mécanicien.' Next day a great naval exercise was held, during which the German battleships sailed out in single line ahead towards Denmark to meet four other ships coming in the other direction. 'There was a lot of firing. After that had gone on for some time everyone returned to their previous anchorage for lunch', the Reich Chancellor noted drily, plainly relieved that there had been no assassination attempts during the celebrations.[98] The French and Russian fleets which had met in the Baltic and sailed in line ahead into the harbour at Kiel, keeping up a constant flow of reciprocal greetings, made themselves disagreeably conspicuous.[99] The 'shameless' spying activities of two Russian officers, who went around photographing all forts and batteries, also made a bad impression in Kiel and Wilhelm complained personally about them in a letter to the Tsar.[100] Holstein deplored the fact that as a result of the lavish celebrations French chauvinism, which had 'nodded off for lack of stimulus', was once again 'wide awake'. If the joint arrival of the two fleets proved to be symptomatic of a Franco-Russian alliance, he warned, 'they will also be keen to go on the attack soon'.[101] Waldersee, moreover, was full of contempt for Hinzpeter, who had said of the Kaiser during the celebrations that he was 'very good at making public appearances, but he can do nothing else', from which the

General concluded that Hinzpeter 'does not seem to be very much in favour with the Kaiser at the moment, or else he was trying to provoke others into making comments'.[102]

The lavish provisions made for the international press by order of Prussian Interior Minister von Köller, who had 'previously taken care to obtain the Kaiser's blessing' for them, were an indication of things to come. 'No one has ever paid such court to these people as we are seeing here', Waldersee commented acidly. 'A ship was specially designated for them, an officer was allocated to the ship to provide them with information, they were given special places everywhere and paid compliments on all sides . . . We have totally capitulated to this great power! Eyewitnesses have told me that a good half of the group consisted of Jews', Waldersee complained, unpleasantly surprised by the success of this press policy – and of the free drinks. The 'wretched rascals from the press' had 'gone into ecstasies' and had had no qualms about 'rating Wilhelm II's success above those of Wilhelm I', he commented in disgust.[103]

Philipp Eulenburg was enraptured by the great success of the canal celebrations which the Kaiser had personally organised and wrote to Holstein commenting that Wilhelm was with good reason 'in very high spirits' about it. 'Everything was absolutely splendid and a great success. The vast array of ships a magnificent sight. The passage through the canal and the laying of the foundation stone extraordinarily impressive.'[104] Even the usually sceptical Waldersee had to admit that the celebrations had been a triumph and that the Kaiser had played his part admirably. That the canal voyage had been accomplished at all, the General observed, was 'a personal success for the Kaiser', since he had simply overridden the technicians' reservations. The Kaiser's speech, the review and the naval exercise had created a 'powerful' impression. 'There is not the slightest doubt that the Kaiser himself did very well, which is not to be underestimated, seeing that the whole world sent representatives . . . In summing up my impressions . . . I have to say that the opening of the canal signifies a great success for the Kaiser; most of those who attended were very impressed both by the festivities and by his personality. Even an ill-disposed person could not deny that he played his part very well. How sad it is that his outstanding gifts are so much reduced by his many failings – almost exclusively weaknesses of character – and that no one, in fact, is really willing to serve him and happy in his service.'[105]

CHAPTER TWENTY-FOUR

An enemy of the people

KAISER WILHELM II VERSUS PARLIAMENT AND NATION

THE brilliant success of the Kiel Canal opening celebrations should not blind us to the fact that parliamentary and public opinion continued to be anything but favourable to the Kaiser.[1] In the Wilhelmstrasse, and above all in the Political Department of the Foreign Office which, with the Reich Chancellor, effectively controlled the state apparatus, there was deep concern over the future of the monarchy under Wilhelm II. On 11 November 1894 Holstein warned the newly appointed Chancellor that the manner in which Caprivi had been ousted had dealt a further severe blow to the Kaiser's standing with the people. Moreover 'H.M.'s extremely brusque treatment of the architect [Paul] Wallot', who was responsible for the new Reichstag building, had created a very bad impression. The 'outrageous tone' lately adopted towards the Kaiser by the press undoubtedly found an echo in public opinion, he warned. 'I consider the position of the Kaiser to be increasingly *endangered*, unless he decides to act with more circumspection than hitherto. By *endangered* I mean that no parliamentary majority can be obtained for anything any more. The moment may come, if the Kaiser continues to treat parliament and people as *quantités négligeables*.'[2] A few months later Holstein told the German ambassador in Rome, Bülow, that the Kaiser had 'no notion what kind of future he is preparing for himself. He does not suspect that the most popular man in Germany would be a chancellor who was known to press his thumb in the Kaiser's eye and who obviously did so.'[3] The Privy Councillor went on to quote a remark confided to him by an unnamed man who saw the Kaiser frequently: 'If one hears what is being said, if one sees how His Majesty is steering *blindly* towards his doom – it makes one weep.'[4] Alexander Hohenlohe likewise warned against the 'policy of force, hitting out blindly', which Wilhelm

II and his irresponsible advisers evidently favoured, and drew attention to the Kaiser's unpopularity, which was being exploited by the pro-Bismarck faction to stir up discontent at every possible opportunity. 'The fact that a conflict in the Reich would entail inestimable risks for the survival of the Reich itself, seeing how unpopular, indeed one could almost say discredited, H.M. has undeniably become, needs no further demonstration', the Reich Chancellor's son declared in February 1895. The Kaiser had been discredited, in Prince Alexander's view, chiefly because of 'the inconsistency of H.M.'s decisions and ideas, which is obvious to the general public, and the fear, caused by one surprise after another, of what the next day might bring'.[5]

Wilhelm seems to have been fully aware of the hostile mood among the people, for during a visit to the Reich chancellor's palace on 14 December 1894 he remarked to Hohenlohe that 'he knew very well that he was unpopular and had often heard that people called out insults or shook their fists at him when he was out in his carriage'.[6] But it never occurred to him to tone down his reactionary anti-parliamentary views. On the contrary, the dismissal of the reform-minded Caprivi had a liberating effect on him, as if he no longer needed to go against his own convictions and take account of the state of the parties in the Reichstag or the mood of the people. The contempt which the Kaiser felt for the Reichstag emerges clearly from a malicious letter he wrote to Philipp Eulenburg on 9 December 1894 after the ceremony to mark the completion of the new Reichstag building. 'The dedication of the Reich monkey-house went ahead with all due ceremony and splendour and never a false note. Wallot was in raptures and stuck close by me when he noticed that I neither knocked his top hat in nor made any rude remarks. Especially as he has become a privy councillor. His rapture was so great that when I commanded him to escort the Kaiserin, as she was in front of us and stopped at every door because she didn't know which way to go, he rushed up to her with the words "By special command of His Majesty" and publicly offered her *his arm*! How about that!'[7]

The Kaiser's ill-tempered attitude towards the Reichstag was reciprocated in full measure, for neither the Social Democrats nor the Centre Party, nor indeed the Left-Liberal parties, had any intention of meekly accepting Wilhelm II's increasingly autocratic tendencies, together with the ever more apparent influence on him of the three Cabinet Chiefs and the Flügeladjutanten and the policy of conflict and incitement of class antagonism which had been pursued by him and the court since mid-1894. At the first session of the new Reichstag on 6 December 1894 the Social Democrat fraction, who had hitherto been in the habit of leaving the chamber during the ceremonial opening of parliament, demonstrated their displeasure at the harsher new course by remaining ostentatiously seated when the Kaiser was cheered. Wilhelm reacted to this protest by the workers' party with aggressive indignation. 'The first session of parliament proved to my great delight to be an emphatic demonstration in favour of the [anti-revolutionary] bill. The Reichstag was in *uproar* at the outrageous rudeness of those left-wing scoundrels and 10 minutes later accepted the latter's first motion on the bill with the help of the Centre! In a *good* Berlin *bar* the Socialists would have been given a thorough thrashing and thrown out of the door. The Reichstag is *far* too cowardly. Soon it will reach the point where someone will propose abandoning the cheers for the Kaiser so as to spare the feelings of the Left.'[8]

After the turbulent session of 6 December the Prussian Ministry of State, led by Interior Minister von Köller, decided to institute proceedings against the deputy Wilhelm Liebknecht for *lèse-majesté*, for which the consent of the Reichstag was required. The Reich Chancellor pointed out that it would probably be impossible to obtain such consent, since it would be voted down principally by the Centre Party and the deputies from the Rhineland, Bavaria and Württemberg. In the event of rejection, however, 'a dissolution in the name of H.M.' could not be risked. In any case, Hohenlohe emphasised, he refused to begin his chancellorship with a defeat. He wrote to the Kaiser stating his case with great firmness. 'Action against the Social Democrats is necessary, but only when they give cause. Laws against them do not help. They lead to conflict with the Reichstag, dissolution and partial *coup d'état* and increase the power and influence of the Social Democrats. If Y.M. wishes to pursue such a policy it is not for me to say anything against it. But *I shall not participate in it*, and in that case Y.M. would do better to appoint a general as Reich chancellor. I shall be glad to return to private life.'[9] On 14 December the Kaiser acknowledged that neither the prosecution of Liebknecht nor the anti-revolutionary bill could be used as the pretext for a dissolution. Instead he devised another plan which could lead to a change in the electoral system: this was 'to draw up an economic, but not merely agrarian, programme and to dissolve on the basis of that'. He also played with the idea of having a single election for both Reichstag and Landtag.[10] In the cautious style of the courtier – 'Y.M.'s idea . . . is certainly a very happy one, . . . but . . .' – the Reich Chancellor had to convince the Kaiser of the impracticability of his plan for a dissolution on such grounds.[11]

Waldersee regarded this development as highly dangerous, seeing the demonstration of 6 December as 'a rejection of the Kaiser by the Social Democrats'. Unlike Hohenlohe, however, he welcomed Köller's resolute action in deciding to prosecute the Social Democrat deputy, for if the Reichstag rejected the motion to penalise the Social Democrats it would have to be dissolved, with all the consequences that would arise from dissolution. For the reactionary General this episode proved that the state must now counterattack against the socialist movement before it was too late. 'What will happen if many more Social Democrats get into the Reichstag?', he asked. 'In the end that will mean that the Reichstag rules the Kaiser. I hope we are fast approaching the decisive moment; but is our good Hohenlohe the man to play for high stakes? I say no.'[12] When he learned from the newspapers that the Reichstag committee, as Hohenlohe had predicted, had voted against the prosecution of the Social Democrats by a large majority, Waldersee was disappointed, commenting that 'it really is not a good beginning for a new course'.[13] He was therefore heartened, on meeting Wilhelm in Hanover in mid-December, to hear how preoccupied the Kaiser was with the whole question of the Social Democrats. But Wilhelm had not 'made it clear what is to be done about it, and in the end he always takes comfort in imagining street battles during which we quickly manage to shoot down the rebels. The idea of taking a firm decision to face all the consequences, even the most extreme, whether on this occasion or with the anti-revolutionary bill, is never mentioned, and so we shall probably go on in the same aimless way as before.'[14] However much the Kaiser's anger may have been aroused, however often he may have been sorely tempted to abolish universal suffrage through a violent *coup d'état*, to Waldersee's regret he never came to a firm decision to order such a breach of the constitution on his own initiative.

The Kaiser saw a second opportunity to vent his displeasure with the Reichstag when in the spring of 1895 the Centre, the Left Liberals, the Poles, Guelphs, Alsatians and Social Democrats voted against sending a telegram congratulating Prince Bismarck on his eightieth birthday. 'With his usual haste' the Kaiser – who could barely be prevented from dissolving parliament over the affair – telegraphed to Friedrichsruh on 23 March to give forcible expression to his indignation.[15] The 'outrage telegram', as Marschall dubbed it, ran: 'I wish to express to Your Highness my deep sense of outrage at the resolution just passed by the Reichstag. It is utterly contrary to the feelings of all German princes and their peoples.' It led many to wonder – not without reason, Marschall noted – 'why H.M. dismissed the Prince in 1890'.[16] Once again Waldersee criticised the Kaiser's unthinking and overhasty action, which had aggravated the situation unnecessarily and lost him the advantage, for had it not been for his telegram the greater part of the nation and 'in particular, by far the majority of people abroad' would have condemned the Reichstag's resolution. Instead, Waldersee predicted, there would be howls of indignation against the Kaiser in Catholic and democratic circles and counter-demonstrations could certainly be expected.[17] After speaking to numerous gentlemen from Berlin at the birthday celebrations at Friedrichsruh, Waldersee recorded that a state of 'great agitation' prevailed among them. It had been established that 'the Kaiser sent off the telegram without speaking to Hohenlohe or any of the ministers, who were naturally extremely surprised. With the public in such a highly excitable state of mind anything is possible . . . If matters become even more acute, however, momentous events might follow and might even lead to a *coup d'état* and to the dissolution of the German Reich. This is not a new idea: for years people have been saying "things cannot go on like this; the Reich was founded on the voluntary union of the princes and therefore they can equally well dissolve the contract". One's next thought is naturally that the Reich will be re-established, but that universal suffrage will be abolished at the same time. In fact a great rift *has* occurred in the unity of Germany since the memorable 23 March [1895]. The Kaiser has shown the greatest possible severity towards a majority in the Reichstag which in all probability cannot be got rid of by dissolving the Reichstag and holding new elections. What is to be done? Either he must yield or he must take the consequences of a complete break; perhaps this is the greatest decision with which he has ever been confronted. Many people are already saying that Hohenlohe must go, because life is beginning in earnest now.'[18] The Kaiser had even intended to make 'a speech attacking the Reichstag' during the banquet in honour of Prince Bismarck on 1 April 1895, which Hohenlohe only just managed to prevent.[19]

Waldersee watched in bewilderment as political affairs nevertheless continued to follow their usual course, and the Kaiser even invited the new President of the Reichstag, Rudolf Freiherr von Buol-Berenberg, a member of the Centre Party, to a dinner in honour of Bismarck at the Schloss. 'How can the Kaiser get on with a Reichstag whose attitude has deeply offended him and which he has insulted in front of the whole world?' The inevitable consequence of such inconsistency, he commented, would be that 'what the Kaiser says is no longer taken seriously'.[20] Later Waldersee discovered the real story of how the telegram to Bismarck had come about. The Kaiser had sent one of his Flügeladjutanten to the Reichstag to find out the result of the vote straight away. When he was told of the rejection of the congratulatory telegram he immediately drew up his telegram and took it with him to see the Reich Chancellor, from whom he demanded the dissolution of the Reichstag. When Hohenlohe realised (as he put it) 'that I could not talk him out of both, I confined myself to deterring him from dissolving the Reichstag'. 'This demonstrates with absolute clarity how little influence Hohenlohe has on him', Waldersee aptly remarked, 'and also that he would not dream of risking his position. I have no doubt at all that if Hohenlohe had stood firm the Kaiser would have given way and the telegram would not have been sent.'[21]

In addition to the dinner which he gave at the Berlin Schloss for Bismarck's birthday, Wilhelm II made the journey to Friedrichsruh on 26 March, accompanied by the Crown Prince and numerous dignitaries, to offer his congratulations to the old man.[22] Both for the benefit of outside observers and within his own family Wilhelm sought to present a highly romanticised picture of the visit and to exploit its propaganda possibilities. As he wrote to his mother on 5 April 1895 from the *Hohenzollern*, 'the day at Friedrichsruh went off very well, the old Prince looked very well but his nerves were much shattered. He was extremely affected & nearly always crying, very often kissing my hand, & lamented the death of his wife, whom he sorely misses, & whose decease made him as he said "a broken down man". He was very happy to see our boy & as often as he could stroked his head & hair.'[23] In spite of the fine speeches made on the occasion, however,[24] in Waldersee's view it was evident that the relationship between the Kaiser and Bismarck continued to be cool on both sides. 'They are both play-acting', he noted afterwards. 'The Kaiser is outdoing himself in attentiveness . . . but it is all mere pretence. The Kaiser does not speak to him at all about affairs of state and still harbours the old grudge instilled into him by Caprivi; Bismarck still has the feeling that no real amends have been made to him. And that will not change!'[25] Even the appointment of Count Wilhelm von Bismarck as Oberpräsident of the province of East Prussia, which was 'the Kaiser's own idea' – 'yet another interference in other people's business', as Waldersee grumbled – and which was taken as a sign of improved relations between the Kaiser and the Bismarck family, turned out to be a *faute de mieux* solution, for Wilhelm II had offered the post successively to Botho Eulenburg, Count August von Dönhoff and the ex-Minister Gustav von Gossler, and had resorted to Bill Bismarck only when all the others had turned it down.[26] Bismarckians like Count Carl von Wedel and Prince 'Septi' Reuss were quick to detect the self-interested aim behind Wilhelm's ostentatious attentions to Bismarck, namely to demonstrate publicly that the veneration of the founder of the Reich was the particular preserve of the Kaiser and the monarchy. After the visit to Friedrichsruh, which Wedel described in detail in a letter to Reuss, the latter replied: 'I think your comments about it are absolutely right, and also as regards the [Kaiser's] almost childish anger with the old man . . . The fact that he refuses to understand what B[ismarck] is and always will be to the German people is quite enough to make me doubt that he has normal powers of comprehension. You were quite right to try to combat this delusion he has, as if he were trying to put the "one and only" in the shade; and what you say is so true, that B. deserves to be measured by a different scale from other men. The fact that he cannot grasp this shows his small-mindedness.'[27] Similar views could be heard at this time even within inner government circles. Minister of War Bronsart told Philipp Eulenburg – 'naively', as the latter thought – that it was important 'that His Majesty ask old Prince Bismarck for advice from time to time';[28] Bronsart was 'very worked up' and had spoken to him 'very *foolishly* about the relationship between the Kaiser and Bismarck', Eulenburg wrote to the Reich Chancellor.[29] In a letter to his friend and cousin Kuno Moltke he confirmed that the Kaiser's so-called reconciliation with Bismarck had been no more than a '*chess move against the enemy*'.[30] The Kaiser and Kaiserin were nevertheless deeply hurt when the old Prince made a '*nasty*' speech to the people of Schleswig-Holstein on 26 May 1895.[31] From then on 'H.M.'s irascibility towards Friedrichsruh' was once again '*very great*'.[32]

The Reichstag's rejection, on 11 May 1895, of the 'anti-revolutionary bill' which had been brought in at the Kaiser's instigation – and which the Empress Frederick had described as a 'terrible error of judgement', 'as *harmful* as it could possibly be!'[33] – provoked a fresh outburst of anger on Wilhelm's part. He telegraphed *en clair* to the Reich Chancellor from the Bethmann Hollweg family estate of Hohenfinow: 'Many thanks for the news. So we are left with fire-hoses for everyday use and grape-shot as a last resort!'[34] After the failure of this reactionary initiative in the Reich the Kaiser hit upon the idea of striking a blow against the growing social democratic movement at least in Prussia, where the parliament elected under the three-class franchise could produce a more comfortable majority for the government. He wrote to the Reich Chancellor on 23 August 1895 from Wilhelmshöhe, beginning by alluding to the laying of the foundation stone of the national memorial to Kaiser Wilhelm I in Berlin: 'Dear Uncle, I am pleased to see from all the newspapers – including the foreign ones – what a lasting effect the ceremony of the laying of the foundation stone has had, and how well the contents of the document [laid in the stone] were received. But I am annoyed and deeply hurt to see the socialist papers, whose scurrilous vilification of the person of the great Kaiser, so sacred to all of us, is beyond anything. They do this perfectly calmly, being well aware of the complete failure of our system of justice and of their absolute impunity under our liberal–jewified laws. At the same time it shows clever calculation, as the socialist . . . [sic] know very well that at a ceremony commemorating our Great Era it is precisely the *person* of the old Kaiser that *brings back the spirit of unity*, which infuriates the revolutionaries. Nothing can be done about this in the Reich. All this Reichstag will take is the milk-and-water of the budget etc. But in *Prussia* something must be done, and the Reich will soon follow. I have a feeling that a briefly worded, strong law on association: § 1 all Soc[ial] Dem[ocrat] associations and meetings are banned, would meet the case best. There is also another political reason. There is a mood of great rejoicing and festivity; the indignation caused by the infamous Soc. Dem. press is growing steadily. The [Bismarckian] "Hamb[urger] Nachrich[t]en" is campaigning strongly against the Soc. Dem[ocrats], and successfully. It would be very dangerous if the "Old Man" saw this as the moment to exploit the situation to his own advantage and to set the tone as defender of the memory of "his old master", almost as if he were becoming the voice of the people, because the grandson and his advisers were too cowardly to take on the Soc. Dem[ocrats]. He must not be given this opportunity. I think, therefore, that a quick, sharp attack by the government press against the outrageously provocative behaviour of the Soc[ialists] would be greeted with jubilation in the country and would be a good preparation for a short and successful campaign in the House of Representatives. "In sacred memory of the person of Wilhelm the Great" shall be our watchword, and with it we shall unite the people behind us *now, this summer*, as if by a magic spell. Once we have brought them together they will remain united; they will be "full of fire". So let "the protection of the Great Kaiser's memory" be our battle cry, and let us get to work and fire away at the Soc. Dem[ocrats]. We have the German people behind us, and last not least the "Old Man" too. He joins in for his own amusement, but this may be to our advantage. And then, having spurred on our countrymen, when they are confronted with the anti-association bill in the autumn we shall get it through with flying colours! We Prussians will take a bold stand and the Reich cannot but be ashamed. Lucanus fully supports this view and is going to discuss it with Köller. Wishing you good hunting, Your devoted nephew Wilhelm I. R.'[35]

At the same time the Kaiser sent a telegram directly to Interior Minister von Köller demanding a 'sharp' riposte to the attacks against Wilhelm I in the Social Democratic press and the introduction of a 'strong' anti-association bill. 'So the old anti-revolutionary game is beginning again', Marschall commented wearily.[36] Holstein spoke dispiritedly of the 'present Young Russian style of government by ukasc' which could only lead to disaster. It would mean the appointment of a 'guardian' for the Kaiser, he predicted, which 'a large majority of the German people' wished for.[37]

Once again Hohenlohe found himself compelled to warn against precipitate action, although he lacked the power and courage to reject the Kaiser's initiative outright.[38] Having obtained the Reich Chancellor's agreement, on 31 August Wilhelm issued fresh instructions for the monarchist press campaign that he planned to set in train: 'Please instruct Köller accordingly and see that [the] press campaign against socialists is pursued with undiminished vigour. The constantly recurring point, to be repeatedly emphasised in what is written, must be the personality of Kaiser Wilhelm the Great. Any diminution of His memory, any criticism of His person or of His deeds, must be presented as an insult to the entire nation and the latter must be persuaded by skilful handling to gird itself up in defence of the old gentleman, or to demand that the government do so.'[39] The Chancellor reluctantly passed on the Kaiser's orders to Köller with the comment that an attempt must at least be made 'to influence the non government press as

H.M. wishes, but in an inconspicuous manner'.[40] He subsequently made it plain that although he was prepared, 'at the special request of His Majesty', to take whatever action was within the bounds of possibility, he himself was convinced that it would be best not to introduce any bill at all.[41]

On 2 September, the twenty-fifth anniversary of the battle of Sedan, as Marschall recorded in his diary, Wilhelm made 'a strong speech against the Social Democrats and then raised his glass to me in a toast. After dinner I toned the speech down a little with Lucanus and discovered from him that an anti-socialist law is planned, and is to be laid before the Reichstag.'[42] Even in the modified version of his speech the Kaiser described the Social Democrats as 'a traitorous gang' and as 'a band of men unworthy to bear the name of Germans'.[43] The ensuing protests from the Social Democrats gave the Kaiser what he saw as an opportunity to take advantage of the alleged national 'storm of indignation' against the workers' movement in order to get an additional short and sharp legislative measure, under which anyone who caused a public nuisance by reviling the memory of any deceased German prince would be punished with imprisonment, passed quickly by the Reichstag. He told Minister of Justice Schönstedt, as Marschall recorded, that 'he wanted only a single clause, so as to protect the reputation of the old Kaiser, for he was not so stupid as to give the Socialists the pleasure of an exceptional law'.[44]

The leading role which both Hermann von Lucanus, the Chief of the Civil Cabinet, and the Kaiser's friend Philipp Eulenburg played in engineering these reactionary measures demonstrates the extent to which the political initiative in such questions had passed from the Wilhelmstrasse to the irresponsible advisers at court. Together they became convinced that because of the tense external situation the Kaiser could not risk a resignation by Hohenlohe. Eulenburg therefore set himself the task of placating the old Chancellor with the argument that the measure in question was only a limited one, both in Prussia and in the Reich, and that it had good prospects of success. 'As His Majesty's measure is limited to single §§ [clauses], and the Kaiser is *absolutely* determined to do *something* – (His latest observations about the socialists culminated in the view that we must proceed *step by step*, using every opportunity that arises) – I think it is a good thing if the Kaiser is not opposed when He *limits* Himself, as He evidently has done . . . I can confirm, moreover, that *Lucanus* really does not want a major measure, but is doing his best to keep the All-Highest wishes in check.' Lucanus and Eulenburg took the view that the Kaiser must be talked out of his plan to spend the winter on an English yacht, which he had rented at great expense, in the Mediterranean, since the presence of the monarch in Berlin would be absolutely indispensable during the negotiations over the two bills.[45]

While shooting at Rominten in East Prussia in September 1895 the Kaiser had a long conversation with Eulenburg during which he spoke of the behaviour of the Social Democratic Party at the Sedan anniversary celebrations. The ambassador sent the Chancellor a detailed account of the Kaiser's remarks. 'He began with a passionate attack on Kaiser Wilhelm I's former companions and adjutants, who in his opinion should have intervened personally to defend the honour of the old Kaiser against insult. If they had beaten Herr Bebel and his gang over the head at the editorial offices of the *Vorwärts* the mood of the whole of Berlin would have been on their side and the people, fired with patriotism, would have destroyed the printing house and given the Social Democrats a fright for the first time. The Kaiser demands that this insult to His grandfather, which has gone unpunished, should be redressed by the clauses to be laid before the Reichstag. The Kaiser's intention is only to call for atonement for the offence to the late German Kaiser and he seemed to have little inclination to demand the same for deceased German princes in general . . . He wished it to be left to each individual Landtag of the German Reich to safeguard the memory of its own prince.' Wilhelm had rejected each of the numerous arguments which Eulenburg put forward against the measures he had demanded. He claimed that Köller had 'the whole [Prussian] Landtag in his pocket' and would get the laws through with very little trouble. In general, Eulenburg reported, the Kaiser harboured 'great illusions' about Köller. The Kaiser's friend could tell from many of his remarks 'that the main impulse behind the Kaiser's wish to put the heaviest possible shackles on the social democratic movement is the fear that some misfortune might befall the Kaiserin or the children. He wants to protect His family, as we would wish to protect ourselves from a mad dog, for as a monarch faced by this mob He finds himself in a position in which we personally do not find ourselves, or only in exceptional cases. With reference to the appeal to the people that He had made in his speeches, He said that, having given it full consideration, He had suggested that He should be accorded the legal means to protect the peaceful elements in the state (NB His children) from assassination attempts; if that did not work, the time would have to come when He would demand, and also achieve, these means. If there were then an outcry he would be in a position to say that he had no other alternative, after his appeal to the people had gone unheard.' As the Kaiser was well aware that in acting this way he would be falling into Bismarck's trap, Eulenburg believed that 'these things will not be seriously considered until after the death of Prince Bismarck – unless some atrocity by the anarchists gave the Kaiser an earlier opportunity to carry out what he wishes'.[46]

While Wilhelm was hunting at Rominten Eulenburg received a twelve-page letter from Marschall arguing strongly against the Kaiser's plans to strengthen the Prussian anti-association law. In modern society, he maintained, the formation of associations had become an absolutely indispensable weapon for those with less economic power against the superior strength of capital.[47] Eulenburg's description of the Kaiser's reaction to this speaks volumes. In a secret letter of 1 October he wrote to the Chancellor: 'As I had touched upon the bill in a conversation shortly beforehand and had found His Majesty not exactly inclined to give up his plans, I asked for an audience today and tried to read out Herr von Marschall's letter to him. The Kaiser is no lover of lengthy analyses, and the only way I could make sure that he paid at least a certain amount of attention to the carefully worked out conclusions was by telling him that the only reason for the Secretary of State's detailed letter was the following: Herr von Marschall was firmly convinced that the personal safety of His Majesty and of the imperial family would be threatened from the moment that the Kaiser took severe measures against the socialists and aroused their resentment against him . . . Even this explanation did not prevent His Majesty from making critical comments about it. He even let fall the remark that Herr von Marschall should not interfere in this question, which was entirely an internal matter; his South German views were inappropriate to Prussian affairs, which cried out for a revision of the law of association. His – the Kaiser's – personal safety was in danger with or without a law, and if the punishment or restriction of the Social Democrats' organisational activities could be achieved, that would answer a need for which the remedy had to be found.'[48] When a few days later the Kaiser learned from the newspapers that a factory-owner in Mülhausen had been murdered, he telegraphed angrily to the Statthalter in Strassburg: 'Yet another victim of the revolutionary movement stirred up by the socialists. If only our people would pull themselves together!'[49] In his speech to the Cuirassier Guards Regiment at Breslau on 2 December the Kaiser again expressed his fury at what he saw as the unpatriotic behaviour of the Social Democrats at the twenty-fifth anniversary celebrations in the autumn. He reaffirmed his solidarity with his 'comrades' in the army and, as Waldersee commented, was 'carried away in intemperate expressions about the Social Democrats', to the extent that 'everyone was in a state of high anxiety because the Ob[er]b[ür]germeister and also reporters were present; the speech was later pruned for the benefit of the latter'.[50]

The Kaiser's attitude to his Catholic subjects – a third of the population of the Reich – and to their parliamentary representatives was scarcely less aggressive than to the rapidly growing Social Democratic workers' movement.[51] When Dr Clemens Freiherr von Heeremann, a Centre Party member, attacked Minister of Ecclesiastical Affairs, Education and Medicine Robert Bosse during the debate on the Teachers' Income Bill in the Prussian House of Representatives on 30 January 1896, and shortly afterwards Viktor Rintelen, a Centre Party representative in the Reichstag, announced his party's demands for changes to family and marriage law during the Reichstag's discussions of the Civil Code, the Kaiser wrote furiously to Hohenlohe: 'Greatly distressed as I was by Heeremann's disgraceful, Jameson-like attack on Bosse, I am absolutely outraged by Rintelen's speech on the Civil Code, which I have just read . . . A piece of legislation which has been completed with the approval of the entire German people, and which was praised as the crowning glory of the edifice when you introduced it at **my** *command* [sic], is threatened with destruction by the Centre; of course they do it in the hope of *bargaining*! Any minister of mine who dares to hint at the slightest accommodation or even concession to such an unpatriotic rabble ceases that very instant to be a minister. And although it is after the event, since unfortunately *the government side said nothing*, I wish you to condemn Rintelen's declaration very strongly at the next session and to emphasise the broad national point of view. Wilhelm I. R.'[52] Summoning all his patience, Prince Hohenlohe had to point out to the Kaiser that Rintelen was 'a foolish old chatterbox' whose remarks could not be regarded as expressions of his party's opinion. There was no question of concessions to the Centre Party, which would accept the Civil Code anyway, and if the Kaiser regretted the lack of response to Rintelen from the government benches, the Prince 'most humbly' begged him to read 'the masterly, patriotic speech' which Professor Gottlieb Planck had made as chairman of the committee on the Civil Code.[53]

Wilhelm's outbursts of rage against the Social Democratic Party and the Centre, however, should not mislead us into assuming that he was now in sympathy with the extreme-right agrarian Conservatives. On the contrary, considerable personal and political differences continued to divide him from them. Although the speech he made on 18 February 1895 at a reception for the leaders of the Agrarian League gave the impression that he was now 'seriously intending to go over to the agrarian camp',[54] even in this speech the Kaiser deplored the agitation stirred up by the agrarian movement, which 'could not but deeply wound my heart as Father of My People'.[55] Referring to the speech in a letter to Philipp Eulenburg Wilhelm commented that 'the Conservatives and Agrarian League are completely mad . . . I received the delegation from the League, although very much *contre coeur*, because *Hohenlohe personally and repeatedly asked me to do so.* In my speech you will have read between the lines the rude remarks I gave them to swallow.'[56]

A week later, in his annual speech to the provincial diet of Brandenburg, Wilhelm spoke of the 'thorny task' he faced as 'Margrave', and emphatically rejected the demands of the agrarians. Although he promised to support the agricultural classes as strongly as he could, he warned them not to 'harbour exaggerated hopes, let alone to demand that Utopia be brought about'. No class could expect to be given preference at the expense of the others, he declared. 'It is the duty of the sovereign to weigh the interests of all classes against each other . . . so that the general interests of the great Fatherland are safeguarded.'[57]

Waldersee expressed his dismay in his diary. 'The good mood and confidence that were gradually beginning to take hold among the Conservatives will be badly shaken by a speech made by the Kaiser at the dinner of the Brandenburg provincial diet. I am convinced that he expressed himself in a way that he did not intend, for he surely cannot contradict what he said to the leaders of the Agrarian League scarcely 8 days ago. It is a most unfortunate idea to make speeches all the time; he has already done himself so much harm in this way. The Liberals, who were deeply depressed 8 days ago, are of course jubilant . . . He loves to call himself 'Father of the People', but unfortunately this is greeted only with smiles and mockery, and when he says that people should follow him with confidence, they naturally say: very gladly, but we should like to have some idea of where he is leading us.'[58] There was 'great despondency and even indignation' about the 'complete about-turn in the Kaiser's mood' among the Conservatives, Waldersee observed. People rightly suspected that the influence of Hinzpeter and Helldorff was behind it, but the result was that 'confidence in the Kaiser is dwindling'. 'He rallies the aristocracy to the struggle against revolution, but now, by withdrawing his interest from the country people, he is driving them towards the revolutionary parties in droves.'[59]

Serious disappointment was in store for Waldersee and the right-wing Conservatives on another score. In May 1896 the Kaiser sent Hinzpeter a strangely worded telegram which was intended to be passed on to the Saarland heavy industrialist, Carl Freiherr von Stumm-Halberg, and in which he attacked former Court Preacher Adolf Stoecker. 'Stoecker has met the end which I predicted years ago', Wilhelm wrote, alluding to Stoecker's departure from the Conservative Party. 'Political pastors are an absurdity', he declared, concluding with words which, as the Liberal newspapers commented critically, sounded like a marginal note by Frederick the Great or even his father the Soldier King. 'The reverend gentlemen should look after the souls of their flocks and leave politics out of it, since that is none of their business.'[60] Waldersee thought it showed 'unforgivable tactlessness on the part of Messrs Hinzpeter and Stumm to publish such a telegram; they both know the Kaiser very well and are aware that . . . he frequently expresses himself crudely in conversation, although afterwards he often scarcely knows what he said, and in the same way he is prone to dash off a telegram and does not weigh his words; if he had thought calmly about it the Kaiser would never have written such muddled stuff, and Stumm and Hinzpeter have done him a very bad service. Of course the whole of the Liberal press is exulting in it.'[61] Two days later Waldersee was more critical in his comments. 'The Kaiser's telegram has quite rightly caused a great scandal, and I think it will have bad after-effects. There will soon come a time when our gracious sovereign will badly need the help of the clergy. I am truly sorry for the Lutheran clergymen; they will not know what to do, but many will go in the very dangerous direction taken by [Friedrich] Naumann.'[62]

MINISTERIAL AND CHANCELLOR CRISES

Clearly, Wilhelm's increasingly autocratic behaviour, his insistence on personally determining every measure of internal and external policy and his flaunting of repressive, anti-parliamentary and anti-Catholic attitudes put the statesmen of the Reich and Prussia under great strain. As Hohenlohe was timid, elderly and frail and, with very rare exceptions, never crossed swords with the monarch, the ministers had little alternative but to enter the lists against the Kaiser. For a while, under Hohenlohe, the Prussian Ministry of State acquired an importance that it had not possessed since the appointment of Bismarck more than thirty years earlier. Foreign and naval policy, which were matters for the Reich, remained outside the competence of the Ministry; nor did it have any say at all in 'purely military' affairs, being a civil authority. But as far as domestic policy was concerned, not only in Prussia but in the Reich as a whole, in the first two-and-a-half years of the new chancellorship the ministers of state, now a committee of eleven, formed a kind of government cabinet which met to discuss and to try to reach decisions on every question that arose. That it was unable, particularly in these crisis-ridden years, to assert itself successfully against the iron will of Wilhelm II, was one of the most fateful developments of the Kaiser's reign and is hence one of the most important themes of this book.

The underlying weakness of the Prussian Ministry of State as an organ of government was its heterogeneity, which itself was the natural consequence of the impulsive and inconsistent manner in which the Kaiser made ministerial appointments. The ministry which Hohenlohe took over in November 1894 as Prussian minister-president and minister of foreign affairs consisted of men whose political opinions were as varied as their characters. Vice-President von Boetticher was a survival (although somewhat weakened since the embarrassing revelations of financial impropriety in 1891) from the Bismarck era. Minister of Trade Hans Hermann Freiherr von Berlepsch had been appointed to the Ministry of State by Wilhelm II in early 1890 with the express task of carrying through the Kaiser's social reform programme against Bismarck. Johannes von Miquel, appointed finance minister a few months later, the former leader of the National Liberal Party and Oberbürgermeister of Frankfurt, had moved decidedly towards the right since taking office; now, as a confidant of Herbert Bismarck, Waldersee and Count Henckel von Donnersmarck, he constituted a link to the agrarian movement and the Bismarckian Fronde. Dominated by Miquel, Thielen, who had been appointed minister of railways in 1891, scarcely played a political role at all. Bosse, promoted minister of ecclesiastical affairs, education and medicine after the School Bill crisis of 1892, was even less inclined to venture on to the slippery political ice. Minister of War Walter Bronsart von Schellendorf, however, whose older brother had held the same post in the latter days of the Bismarck era, was a fiery General possessed of great self-confidence and extensive connections. Karl Heinrich von Schönstedt, the Minister of Justice appointed with Hohenlohe in November 1894, and Freiherr von Hammerstein-Loxten, appointed minister of agriculture at the same time, lacked experience at first and played no distinctive role outside their particular domains. Ernst Matthias von Köller, on the other hand, who had been appointed minister of the interior in November 1894 by telegraphic order from the Kaiser, saw himself from the outset as the Kaiser's right-hand man in the Ministry of State. Finally there was Adolf Freiherr Marschall von Bieberstein, the Reich Foreign Secretary from South Baden, who in the course of his parliamentary activity in the previous seven years had developed a relationship of trust with the leaders of the Centre group in particular, and who had been appointed to the Prussian Ministry of State as minister without portfolio in the autumn of 1894, on the insistence of Holstein and as a condition for Hohenlohe's acceptance of the chancellorship. This 'fellow from Baden', as Wilhelm called him, almost inevitably incurred the anger of the increasingly 'Prussian' monarch.

After only a few weeks there was a violent dispute between the Kaiser and the Foreign Secretary which almost cost the latter his position. Irritated beyond measure because the Reichstag, and first and foremost the Centre Party, was threatening to reject his demands for the navy, and spurred on by the former Jesuit Count Paul Hoensbroech, who had converted to Protestantism and whom the Kaiser had received against the Reich Chancellor's advice,[63] Wilhelm gave orders to his Oberhofmarschall, August Eulenburg, never to invite '*any* member of the Centre to a court function ever again'. He accused Marschall, who had arranged for the leader of the Centre Party, Dr Ernst Lieber, to be invited to the last court ball on 6 February, of having repeatedly tried to make him talk to Lieber on that occasion, against his will. After the ball the Kaiser let loose a torrent of 'very strong language' against Marschall and said that he was 'sick of this fellow from Baden, who had no Prussian feeling in him'. On 14 February 1895 Alexander Hohenlohe warned his father that he could expect an 'onslaught against Marschall' when the Kaiser came to see him that day.[64] When the monarch arrived at the Foreign Office in the morning he behaved 'very ungraciously' towards the Secretary of State, refused to shake hands with him, complained 'in the strongest possible terms' about the Reichstag, which evidently wished 'to destroy the navy', and went over to speak to the Reich Chancellor.[65] Holstein also reported that the Kaiser was 'in a rotten mood' and had been whipped up against Marschall principally by Hoensbroech and Lucanus.[66]

The letter that Wilhelm II wrote on 12 February 1895 from his hunting lodge at Hubertusstock to his friend 'Phili' Eulenburg left the ambassador at Vienna in no doubt 'that a separation from Herr von Marschall is to be expected in the near future'.[67] 'You will be surprised that I am writing from here', the Kaiser began. 'But when such grave times are approaching, and everyone in Berlin gets more and more insane, angry and impossible, one has to get out for a moment to keep a cool head and sound judgement. For I want to be able to judge everything that happens absolutely fairly. The incident which compels me to write to you is unfortunately another one of a personal nature. Our friend Marschall's behaviour towards me at the recent court ball surpassed everything that had happened before. After he had made himself unpleasantly conspicuous to me and to other onlookers for a long time by standing in my way and blocking doorways, he forced me to talk to him. He then told me that Herr Lieber was standing just there at the back, near me. An excellent man who was very well disposed towards the navy and was the right person to help the bill through, if I would speak to him! I replied that it was highly improper for him to confront me with such situations in my own house. Herr Lieber was a downright scoundrel who had been invited without my knowledge or permission. When he attacked the Army Bill last year he called the Hohenzollerns and the Protestants a band of murderous fire-raisers and the Germans in the South as "willy-nilly Prussians" from which he acquired the nickname of the willy-nilly Prussian throughout Bavaria. If he was prepared to agree to the ships, he should do it out of conviction, for the Fatherland, not as a present to me in return for talking to him. After that I left him standing there looking very crestfallen. He then went off and – without saying a word about what had happened – tried to get [Prince] Heinrich involved with Lieber. He [Heinrich] answered by telling him to give the swine a kick in the a . . . My Marshal of the Court [Egloffstein] told me later that Lieber had been invited just before the ball by Marschall's personal arrangement, without my permission, and was not even on the list!! That is going too far, and there will be consequences!'[68] The day after the court ball, during a visit to the Wilhelmstrasse, Wilhelm expressed his anger that Hoensbroech had not been accepted into the diplomatic service; 'he had given up his religion so that he could remain loyal to his King', and he, Wilhelm, would make sure that he was given an appointment in Prussia. In the course of a conversation with Marschall the Kaiser became 'very abusive towards Catholics in general' and said that it was quite useless showing any consideration towards the Centre. 'He would perhaps – since we had been trying in vain to do business with the Centre for six years – rally the Protestants around Him and start the Thirty Years War again.'[69]

The incident at the court ball when the Kaiser had flown into 'a furious temper' and turned his back on Lieber, and Marschall had received 'an even stronger snub' from Prince Heinrich, was confirmed by the accounts given by two of the Flügeladjutanten who were with Wilhelm at Hubertusstock, Gustav von Kessel and Hans von Arnim, whom Wilhelm ordered to write to Eulenburg as well. 'His Majesty dislikes being waylaid by anyone at balls, but most of all by his officials who are trying to bring people forward for him to speak to, without telling him anything about it beforehand, in order to advance parliamentary or other causes. As German kaiser he will not lend himself to this kind of thing. And absolutely not in the case of this Dr L.' As a result of this incident the Kaiser's attitude towards Marschall was highly critical, Kessel gleefully remarked, 'and in my opinion a change is in the offing, sooner or later . . . I consider M. to be easily replaceable, a man of narrow vision who goes around flattering everybody.' The Kaiser was right to be angry at Marschall's tactics, the Flügeladjutant added. 'He has given orders that L. is never to be invited again.' Arnim likewise declared that he could only agree with the Kaiser's view of Marschall's conduct and had come to the conclusion that the Foreign Secretary's demeanour was 'typical of the whole character of this man, from whom it was impossible to expect a policy in the grand style'.[70] Philipp Eulenburg rightly observed, on receiving these letters, that they spoke volumes about the blatant influence of the Flügeladjutanten on the Kaiser. '*At last* the Flügeladjutanten and the Supreme Headquarters have shot Marschall down! They are completely at their ease in writing to me. On the *orders* of the Kaiser!! My good Arnim's views on the *duties of a secretary of state*, or the qualities that such a person should *not* have – say more than an entire book about what I call "adjutant politics" . . . What would Hohenlohe, Holstein and Marschall say if I showed them *these* letters!!'[71]

But even Eulenburg, ever loyal to the Kaiser, considered Marschall doomed because of the purely personal dislike which the Kaiser and his entourage felt towards him.[72] As he commented to Holstein, Marschall had 'just as little ability as Caprivi to win friends among the entourage of H.M.'. The Kaiser's entourage had not turned against Hohenlohe, he assured Holstein, 'because he has the kindly manner of the grand seigneur and is polite by nature. Although the narrow-minded court ladies may talk against the *Catholic* Chancellor, this issue is taken up by the military and civilian entourage of H.M. only in order to bring up Marschall's "undignified" flirtation with the Centre Party. You know that H.M. has felt a pronounced antipathy to Marschall for almost two years.' Eulenburg went on to declare that it was only in order not to lose Holstein himself that he had worked to have Marschall kept on in the Hohenlohe government the previous autumn. 'The weakness of the Hohenlohe Cabinet lay and lies in Marschall, because he has no friends in Prussia and because the Kaiser hasn't liked him for two years.' After the incident at the court ball Eulenburg no longer had the slightest doubt that the Kaiser would get rid of Marschall. 'I tell you frankly', he wrote to Holstein, 'that in my opinion Hohenlohe would be stronger if Marschall went, because in the long run the antipathy of H.M. will become unbearable and because the King after all has the right to select his own ministers.'[73] Eulenburg also wrote to the Reich Chancellor with a view to distancing him from Marschall, so as to avoid the danger of a chancellor crisis arising from Marschall's impending dismissal. Meanwhile he wrote to the Kaiser strongly advising him to appoint Bernhard von Bülow as Marschall's successor.[74]

The Kaiser's reaction to his friend's letter was, to put it mildly, unexpected. He sent a telegram dismissing the whole 'stupid story' of Marschall, Lieber and the court ball as a carnival joke and telling Eulenburg to stop trying to oust the 'wounded M.' from his post.[75] This was followed up with a letter in characteristic style commenting on the letters from Marschall and Alexander Hohenlohe giving their version of the incident, which Eulenburg had sent him: 'My dear Philippus, The contents of this letter have left me dumbfounded with astonishment! Such a farrago of nonsense as Crown Prince Alexander [i.e. the Reich Chancellor's son] has dreamt up has seldom been written down in black and white. He is doing far more harm than 100 newspaper articles in [Maximilian Harden's] Die Zukunft. How can any sensible person who knows me and sees me every day, knows exactly whom I see and who is seen with me, which can be verified at any moment, imagine such rubbish! . . . I am so happy with old Hohenlohe, for everything is working so well and comfortably; whenever either of us wants something, we immediately exchange notes and discuss it, so that nothing can happen behind the scenes! Which is a paradise for me. But now I suddenly find that I have been quite mistaken! And that the nonsense is worse than before! And that it comes precisely from the place where it ought to be found least! Why didn't Alexander give the tale-bearers a direct slap in the face instead of just believing this outrageous rubbish! . . . I had settled the entire ball affair *with the Prince* [Hohenlohe] long ago, and had forgotten all about it. And then you come along with your grand plans for reshuffles! Wretched creature, leave me in peace with such stuff! Write and tell these people that they should be ashamed to believe such rubbish, and leave all those whom you want to make dance around like puppets sitting quietly at their posts! Thank God it is carnival time, otherwise I should have been really rude to you. I can *assure* you that I have spoken to *no one* at all about government business, or about Hohenlohe, out of sheer concern for nonsense of this kind, if for no other reason. Please make the widest possible use of this letter! As far as the Centre are concerned, you will have seen that I have dined with Arenberg and Kopp. With the former I had a detailed political conversation about the Anti-Revolution Bill, which, according to a later report he made to me, had a very good effect on his party!! Well then!? Where does that leave the bad treatment of the Centre, 3 gentlemen from which were promoted to major on 27th!? It is only natural that I should not tolerate tactlessness at parties where I am host, and if the Catholic side wanted to try to link that with something, it would be even more unjust! . . . I couldn't care less about Marschall: for all I care he can stay till his fat coagulates, and I am letting Schele go for Hohenlohe's sake, although he is being sacrificed to the villainous tricks of Kayser, who wants to be minister for the colonies! So tell me please, dearest Philippus, what reason is there for all these insinuations? Please make sure that the truth of the matter is made *quite clear*, straight away, and be very rude! to all and sundry! If Hohenlohe goes, I'll go too! Your loyal friend Wilhelm I. R.'[76]

But if the Kaiser had decided to keep on Marschall – who denied having asked the Kaiser more than once to talk to Lieber and claimed not to have spoken to Prince Heinrich at all[77] – for the time being, the crisis soon came to affect the entire government. In Holstein's view 'the whole affair is a sneak attack, taking advantage of the nervous irritability of the Kaiser'.[78] Only a few days after the court ball incident Minister of War Bronsart had 'a scene with the Kaiser because the latter doubted the loyalty of the Catholic officers'. 'The Minister of War was so angry about it', Holstein telegraphed to Bülow in Rome, 'that he told Herr von Lucanus yesterday that in his opinion the entire Ministry should resign, or impose conditions.'[79] The idea of collective resignation by the entire Ministry of State, in order to force the Kaiser to give up his methods of personal rule, was gaining ground all the time.

In the early summer of 1895 Prince Hohenlohe found himself face to face with a second, well-nigh intractable ministerial crisis which arose from what had become irreconcilable differences between the Kaiser and Minister of Trade Hans Hermann von Berlepsch, a convinced advocate of social reform. After a joint audience with the reactionary Minister of the Interior, Köller, on 12 May 1895, Hohenlohe noted that 'the Kaiser has reservations about setting a standard working day'.[80] Three days later Berlepsch informed the Reich Chancellor that he intended 'to ask the Kaiser and King to relieve me of my office' because he had reached the conclusion 'that it has become impossible to make any further progress in the foreseeable future with the social reforms which I think absolutely indispensable'.[81] Not for the first time the Chancellor found himself compelled to call on the services of Philipp Eulenburg as mediator.[82] When Eulenburg, who was at Prökelwitz in East Prussia with the Kaiser, brought up the subject of the opposition between the pro-reform Berlepsch and the ultra-conservative Köller, 'His Majesty took sides against Herr von Berlepsch with some vigour' – as Eulenburg reported to Hohenlohe – 'and referred to the fact that He and Your

Excellency, in the presence of Herr von Köller, had established the basis on which "the process of governing" was to be conducted.' Eulenburg strongly advised Hohenlohe to postpone dealing with matters of contention within the Ministry of State until the parliamentary recess, or if this were impossible, to inform the Kaiser of them without delay, to avoid the impression of 'secretiveness'.[83] But Berlepsch was unwilling to remain in office against his own convictions and handed in his resignation. On 8 June 1895 the Chancellor described in his journal the humiliating treatment he received from the monarch. 'At 6 o'clock I received an invitation to dine at the Marmorpalais. After supper the Kaiser, who was waiting for me with Lucanus, summoned me and showed me Berlepsch's letter. He refuses to let Berlepsch go. He said he had appointed him to carry out social legislation in accordance with the resolutions of 1890. He still wished this to be done; but as the originator of these measures he could also order a slowing-down. Berlepsch's resignation would signify a wish to change direction, and that was not his intention. But German industry must not be too heavily burdened and must not be made incapable of competing with foreign countries. H.M. had written comments on Berlepsch's letter which he read out to me . . . I am to be sent a written communication as well.'[84]

The Kaiser ultimately succeeded in delaying the Trade Minister's resignation for the duration of the summer. On 28 June Berlepsch went to see Hohenlohe with the news that the Chief of the Civil Cabinet had written to him to say that 'it would be some time yet' before the Kaiser would take a decision on his resignation. Berlepsch declared, however, that he could not remain in the Ministry any longer. 'If the Kaiser thought a reduction of the pace of social legislation were necessary, he respected this point of view but could not share it, in view of all his endeavours hitherto, without discrediting himself in the eyes of the Reichstag and of the world in general', as Hohenlohe recorded. Since the Kaiser was away, however, Berlepsch admitted that he had no alternative but to wait. The Reich Chancellor, equally resigned, commented: 'For the moment I can do nothing, as the Kaiser cannot be reached.'[85] It was not until he returned from his Scandinavian cruise and just before his departure for England that Wilhelm found an opportunity to explain to Berlepsch why he refused to accept his resignation. In a letter from the Neues Palais on 31 July 1895, evidently drafted by Lucanus, he wrote: 'I am unable to grant your request to be allowed to resign from the office of minister of trade and industry. You rightly emphasise that to abandon the social policy set out by Me in my order of 4 February 1890 and repeatedly referred to in speeches from the throne would result in severe damage to the interests of the Fatherland. Nothing is further from my mind, however, than to give up this course, and least of all would I approve a policy which sought to introduce repression by force, instead of humane legislation and administration, in relation to questions of social welfare. But this mistaken notion would inevitably be aroused by your resignation, because people would be inclined to see the change of minister as the expression of a systemic change. If therefore there is no difference of principle between Me and you with regard to the aims of our social policy, the only matter in question is the way in which we should continue on the path we are treading, as best befits our political and economic circumstances as a whole. It is true that in this regard I consider the maintenance of a slow tempo to be required at present . . . Nevertheless I do not wish all activity in the sphere of social welfare to cease, even at present; but I wish to prevent the great excitement which has unfortunately been caused by the agrarian agitation being further increased by parliamentary discussion of great social policy questions, such as the regulation of maximum working hours and the establishment of representation for the workers. There is, however, another sphere of social policy which still needs to be settled and which can be tackled without danger, together with further work to develop present arrangements for the amelioration of the position of the workers. This is the question of the organisation of craftsmen. I do not doubt that for the sake of accomplishing this task, with which all the state-supporting parties have a certain sympathy, you will be willing to continue to place your well-tried skill at My disposal and that you will add new services to those you have already rendered in such great measure. You will have My full confidence and My gratitude as your King in this endeavour. I therefore look forward to your suggestions for the improvement and protection of the craftsman's profession as soon as possible. Wilhelm R.'[86]

In the midst of this ministerial crisis Miquel launched an attempt to exploit the Kaiser's current mood, which had again become markedly 'Prussian', to increase his own influence, as Prussian finance minister, on the financial affairs of the Reich. But this was met with determined opposition on Wilhelm II's part. In a secret letter dated 24 May 1895 Philipp Eulenburg told the Chancellor that, although the Kaiser had expressed himself 'very strongly' to the effect that he 'no longer felt disposed to sacrifice Prussian financial interests to the Reich', on the other hand he had '*categorically*' rejected an extension of the powers of the Prussian finance minister at the expense of the Reich chancellor and the Reich Treasury. 'I have the impression', Eulenburg wrote, 'that H. Majesty considers this All-Highest decision as definitive, and that He particularly hopes that it will provide the *very keenly* desired settlement, and that no further reference to the matter will be forthcoming.'[87]

The double ministerial crisis of May 1895 led to a general discussion in Prökelwitz on the fundamental question of how the Kaiser, the Reich Chancellor and the Ministry of State should determine the guidelines of Prusso-German domestic policy. Philipp Eulenburg drew the Kaiser's attention to the various shades of opinion within the Ministry of State, which could lead to more serious conflicts and thereby endanger 'the homogeneity of His Majesty's government'. As Eulenburg reported to Hohenlohe, the Kaiser had '*quite categorically*' forbidden any controversial disputes within the Ministry and called upon the Chancellor to ensure the 'homogeneity' of the leadership of both the Reich and Prussia. Only the chancellor had the right 'to take a view and express it to His Majesty; the individual ministers must limit themselves to following the principles laid down'.[88]

There was one minister, however, who was not prepared to allow the guidelines for his sphere of responsibility to be determined by the Kaiser, and who had indeed expressed grave doubts from the outset about whether the monarch could still be regarded as mentally normal at all. This was the self-confident Minister of War, General Walter Bronsart von Schellendorf. Bronsart had already aroused Wilhelm's anger in the autumn of 1894, as Waldersee recorded in his diary, by paying a visit to the Conservative member of the Reichstag, Count Julius von Mirbach-Sorquitten, whom the Kaiser detested. 'The Kaiser . . . reprimanded the Minister of War to his face for it and received a rather sharp answer from Bronsart, who also wrote at once to Hahnke saying that he had been very unwilling to become a minister, thought himself very ill-suited to it and would be glad to resign.' Waldersee thought it 'sad that the Kaiser is already on such a bad footing with the really very effective Minister of War, after barely a year'.[89] Another violent clash between the monarch and the Prussian Minister of War is vividly depicted in a letter from Marschall. On 17 February 1895 he wrote to Philipp Eulenburg: 'After a dinner at Hahnke's house, which I attended with my wife, I went into the smoking room where a lively conversation was going on between Bronsart and Lucanus, into which the former drew me. Bronsart was recounting a scene which he had on Thursday morning – the day on which His Majesty treated me so ungraciously – with the Kaiser: "he had never thought he would have to speak to his King in that way". During his report to the monarch he had described the progress of the budget committee's discussions of the military budget and had expressed himself very satisfied with and appreciative of the Centre, which had approved all the items which he had considered particularly important. At this His Majesty had become very ungracious and had expressed the greatest offence not only at the Centre but in particular at Catholic officers, who he said would throw off their uniforms and desert their King at the decisive moment. His Majesty's words had been so insulting that he, Bronsart, had risen to his feet and protested loudly against these insults; he thought he had turned deathly pale with emotion and had found it very difficult to control himself. His Majesty had not replied and the audience came to an end. He would never, and would never wish to, experience such a scene again. The blame for it lay with *the unscrupulous people who spur the Kaiser on – who they were, he did not know.* Lucanus kept very quiet and did not contradict Bronsart when he bitterly criticised the fact that Count Hoensbroech had been led up to speak to the Kaiser, when it was probably he who had stirred him up so much. These things make the position of the ministers utterly impossible; one wears oneself out in parliament trying to achieve something, and then anonymous advisers come along and ruin everything. Things cannot go on like this.'[90] The Foreign Secretary noted in his diary at this time that Bronsart had said to the Chief of the Civil Cabinet: 'The irresponsible advisers should all be sent to the scaffold.'[91] During the canal-opening ceremony at Kiel, Bronsart spoke 'in a highly conspicuous way' and 'with great irritation' about the Kaiser, and also suddenly burst out with the remark that Lucanus was 'the greatest criminal of the century'.[92]

General von Bronsart's renewed threat to resign was all the more dangerous for the Kaiser because the Minister of War, since taking office, had been working on a highly popular reform of military courts – in harmony with the wishes of the majority of the Reichstag and of the people he wished to introduce public trials, conducted *viva voce* rather than solely in writing – to which the Kaiser vehemently objected. It was not just that Bronsart would have gone out in a blaze of glory over of this issue, however. Neither the Reich Chancellor, who as Bavarian minister-president had introduced a similar measure and felt he could not now be 'more Prussian than a Prussian general', nor the majority of the Prussian ministers were prepared to remain in office if the Minister of War resigned. Hence the Bronsart crisis threatened to become a general government crisis which might have catastrophic consequences for the Kaiser's standing.[93] This question more than any other overshadowed German domestic politics during the first two years of Hohenlohe's chancellorship.

A week after the discussion between Wilhelm and Eulenburg at Prökelwitz in May 1895 about the principles governing the relationship between the Kaiser, the Chancellor and the Ministry of State, Hohenlohe tried in vain to bring up the subject of the fundamental differences between the monarch and the Minister of War on the increasingly highly charged issue of court martial reform. 'I was unable to carry out my intention of speaking to H.M. about Bronsart except in a limited way', he recorded with weary resignation on 31 May 1895. 'The Kaiser was not available to speak to throughout the boat trip, nor on the Pfaueninsel.' When the Chancellor approached the Kaiser just before dinner and began to speak of Bronsart, the court martial question and the danger of the Minister's resignation, 'he did not allow me to finish, but said that in these difficult times a minister ought not to abandon his king. He claimed that the Minister of War had declared himself in agreement with him; they were of one mind, and during a journey from Magdeburg to Berlin they had come to an understanding . . . After these remarks he went in to dinner and the conversation was over. I then had to ask the Minister of War about it, and he told me that it was all untrue.'[94] Later the Kaiser remarked to Philipp Eulenburg 'that he had half ruined his army with the two years' service and *under no circumstances* could he now be induced to ruin it *completely*'. Eulenburg was in no doubt that the Kaiser would be prepared to lose Hohenlohe over this issue.[95]

In the following weeks the crisis intensified dramatically when Bronsart wished to put before the Reichstag the proposed new regulations for courts martial, introducing public hearings on all questions not of purely military content, while the Kaiser, backed by Plessen and Hahnke, persisted in refusing to allow the bill to go forward in this form. When Bronsart again threatened to resign in mid-August 1895 the Kaiser's response, as Waldersee recorded, was to say to him: 'You cannot desert me now. And, after all, I gave you the Black Eagle very early.' Later Wilhelm made the War Minister a present of a French cannon, which led Waldersee to comment that 'the Kaiser is extraordinarily skilful in managing these things and has often been successful before. After such acts of kindness, which, it must also be said, he performs with irresistible charm, the reaction is always to say: the Kaiser has been so infinitely gracious, I really cannot go now.'[96] Wilhelm was now convinced that the whole question had been definitely laid aside and that Bronsart had given his assurance that he would 'let [the matter] rest completely'. Eulenburg was therefore horrified to hear in the autumn of 1895 that Bronsart had not given up his plans for reform and was threatening to take the Chancellor and the Ministry of State with him.[97] In an 'alarming' debate carried out in the press by both camps, Hahnke and Plessen – rightly, in Waldesee's view – were identified as the people responsible for 'hindering Bronsart's efforts with the Kaiser'. 'It is deplorable that the only real problems faced by such a man should be with the Kaiser', Waldersee commented, 'when the sovereign ought after all to do everything he can to retain such a powerful asset. But unfortunately it has got to the point where the only people in favour are those who accept the All-Highest whims without question, as the Cabinet chiefs do. For the moment H.M. does not have the sense to see that this destroys characters and causes untold harm, because in the end only pitiful wretches are left in power, but he will suffer badly from it when it is too late.'[98]

While the Kaiser was shooting at Liebenberg in October 1895 news arrived that the Minister of War, in spite of the Kaiser's veto, but with the support of the entire Ministry of State except Köller,[99] had introduced the issue of the regulation of courts martial into the Reichstag. 'It was extremely painful and distressing for me', Eulenburg recalled, 'to have to explain to my beloved Kaiser, while he was doing me the honour of being my guest, the position in which he now found himself.' The War Minister's conduct had acted on him 'like a bomb'. The Kaiser's friend eventually succeeded in persuading Wilhelm that to sack Bronsart was out of the question, since Hohenlohe and the Ministry of State had to be kept in office. Together they decided that the government should be ordered to consult the commanding generals, which would delay a decision for at least nine months.[100] Eulenburg wrote at once to the Chancellor reporting that the Kaiser was '*very angry indeed* at Bronsart's behaviour'. He had been absolutely confident that the question had been postponed and regarded it as an act of '*perfidy*' engineered from Friedrichsruh that it had been brought up again. On the other hand he did not intend 'to allow Bronsart to go *under any circumstances* at present, so as not to give him the opportunity of making a grand exit'. The Kaiser had also shown very little understanding for Hohenlohe's view that he could not remain in office if Bronsart should fall over this issue, Eulenburg stated. The only solution therefore lay in the postponement of the matter. 'The poor Kaiser has been very affected by this business – I hope that it will be possible to arrange for the postponement.'[101]

On his return to Berlin the Kaiser immediately went to see the Chancellor to vent his feelings on the matter. Hohenlohe heard him out before cautiously explaining his own point of view. As he had introduced public hearings for military courts as minister-president of Bavaria he could not now, as Reich chancellor, bring in a bill which excluded public hearings without appearing ridiculous and earning the contempt of the Reichstag. If Bronsart resigned over this bill he too must take his leave. The Kaiser recognised the force of this argument, 'but let out a stream of accusations against the Minister of War, complaining that the latter had sought the support of foreign sovereigns for his views and had persuaded the King of Saxony to use his influence with him, the Kaiser, in favour of public trials. That was high treason and he would have liked to take the Black Eagle away from the Minister of War and throw him out of the army.' In this 'fit of bad temper' Wilhelm repeated his accusation that Bismarck was behind the crisis; he was 'very angry' about it and his patience was at an end. When Hohenlohe pointed out that respected generals such as Loë and Albedyll were also in favour of the reform of military courts, the Kaiser disputed this, insisting that 'all the commanding generals were against it'. 'He summoned the help of Plessen, who was waiting in the ante-room. Plessen defended the Kaiser's point of view, but weakly. Finally the Kaiser decided to send Plessen to the Minister of War again, to induce him not to resign now and to agree that the matter be postponed until another session.'[102] That same evening the Kaiser was able to telegraph to Hohenlohe that he had had a long and friendly conversation with Bronsart, who had agreed to the postponement of the tiresome affair until after the commanding generals had been consulted.[103] Eulenburg was even able to report to the Grand Duke of Baden that during this discussion Bronsart had 'assured His Majesty of his loyalty with tears in his eyes and declared that he was willing to stay in office and to agree to the consultation'.[104]

But the truce proved short-lived. On 15 November Captain Count Bogdan von Hutten-Czapski, an influential member of the Prussian Upper House, reporting to Holstein on the Kaiser's mood, commented that the 'defenceless ancestors' were 'still alive and the poor soldiers will continue in future to be sentenced behind closed doors'. The Minister of War was complaining 'bitterly about the daily interference of H.M. in official matters, especially in armament questions which involved the senseless expenditure of large sums; irresponsible advisers always had some new invention on tap that caught the fancy of the Kaiser and brought disorder into the administration; Hahnke's influence in military and Lucanus's influence in political affairs made the routine transaction of official business impossible, etc.'. Bronsart took particular offence at the way in which the Kaiser had reacted to an article of 4 November 1895 in the *Münchener Neueste Nachrichten*, which had reported on the secret decision of the Ministry of State to introduce public military trials even against the Kaiser's will if necessary. General von Hahnke had been sent by Wilhelm to put the Minister of War 'on the carpet', and had given him a 'fairly severe' dressing-down about the article. 'In the past two years I have witnessed many fights between the Kaiser and Bronsart', Hutten-Czapski remarked, 'but have never seen the latter so wrought up. As I heard from the Chancellor, the Kaiser too is very irritated. This mood, or rather this bad mood, the importance of the problems at issue, and the volatile character of the participants arouses fears about a catastrophe; this would be all the more unfortunate because Bronsart is very well liked in the Reichstag, and above all he would fall as the protagonist of a reform that is popular in the whole Reich, which in turn would significantly increase the difficulties of the government.' Bronsart was also angry with Köller, who had been the only member of the Ministry of State to vote against public trials, so as to put himself across to the Kaiser as the champion of the traditions of the army, Hutten-Czapski reported. Bronsart suspected, probably rightly, that Köller was behind the Munich newspaper article, the purpose of which had been to 'damage' him and the Chancellor in the eyes of the Kaiser. Köller was an 'enfant terrible' who had frequently been very tactless in the Reichstag and who was now trying to 'win recognition from the Kaiser at the expense of his colleagues', the Count commented.[105]

On 18 November 1895, at a meeting of the Ministry of State, Bronsart accused the Minister of the Interior to his face not only of being responsible for the Munich newspaper article but also of having told the Kaiser and Generals Hahnke and Plessen, while the court was hunting at Letzlingen on 15 and

16 November, of his vote against court martial reform.[106] It was now unthinkable for Bronsart and Köller to continue to work together. Moreover even the least politically minded ministers were indignant at the Interior Minister's disloyal behaviour. Thielen and Schönstedt spoke out 'very firmly', and Hammerstein-Loxten reproached Köller for having given all the ministers 'the unpleasant impression . . . that he wanted to "curry favour" with H.M. by mentioning his dissenting vote and by subsequently talking to Hahnke and Plessen'. By 27 November 1895 it was clear to Hohenlohe that he would have to tell the Kaiser that the entire Ministry of State was united against Köller and demanded his resignation.[107] When he went to Potsdam the next day at the request of the Ministry to explain to the Kaiser why Köller and the other ministers could no longer work satisfactorily together, he found Wilhelm disinclined to take any notice of the matter. The Kaiser declared 'quite definitely, and in not very obliging terms, that he would not dismiss Köller. He was the only man on whose energy he could count if force ever needed to be used.' 'I then withdrew', the Reich Chancellor recorded, 'but stayed to luncheon in the Neues Palais, where the Kaiser resumed his usual manner. But the refusal affected my standing with the ministers and officials. Even Wilmowski took the view that I could not accept this. I now drew up my resignation letter.' But deeply angered as he was, Prince Hohenlohe did not send the letter, for he was unwilling to inflict such an embarrassment on the Kaiser just before the opening of the Reichstag.[108]

Instead he summoned the ministers to a meeting to report on the insulting treatment he had received in the Neues Palais and at the same time to tell them that he nevertheless had doubts as to whether it would be loyal to confront the Kaiser with a ministerial crisis three days before the opening of the Reichstag. Some of the ministers (Marschall, Boetticher, Berlepsch and Thielen) spoke out in favour of decisive action, others took a softer line; but as neither Miquel nor Bronsart was present the ministers postponed their decision until the following day.[109] On 1 December 1895 the Chancellor had to write to the Kaiser informing him that in the unanimous opinion of his ministerial colleagues no further profitable co-operation with Köller could be ensured, and that in the prevailing circumstances the latter had declared himself ready to submit his resignation to the Kaiser. He, Hohenlohe, must 'humbly submit' to the Kaiser 'that He should graciously accept the resignation of Minister von Köller'.[110]

The news of the crisis reached Philipp Eulenburg on 29 November at Merano. He wrote to the Kaiser expressing his sympathy, but warning strongly against a change of chancellor. 'I cannot bear Your Majesty to be tormented by continual crises', he wrote. 'I can still clearly remember the expression on Your Majesty's face when I took my leave in the Neues Palais and You said: "I cannot go through a third chancellor crisis." Now it has reached that point again.' As a result of this crisis, Eulenburg warned, Wilhelm faced the alternative of dismissing either Hohenlohe and the entire Ministry, or Köller alone. He must let Köller go, for otherwise he would 'bring about a crisis which, given the extremely complex European situation, would alarm Europe and brand Your Majesty as an inconstant and arbitrary ruler'. 'No chancellor crisis – *no change of chancellor* – at this moment', Eulenburg urged. 'That would create the impression for which Friedrichsruh has been lying in wait with eager anticipation: now the Kaiser has gone far enough! Now is the time for the Chancellor whom I, Bismarck, have in my hand, whether he is called Waldersee, Schweinitz or whatever!' The Kaiser must not dismiss Hohenlohe; 'the next change of chancellor must be left to a greater power than Your Majesty – death, to which the old Prince will unfortunately fall victim in the foreseeable future'.[111] The letter had the desired effect. On the morning of 2 December, on the train from Wildpark to Berlin, Köller had his last audience with the Kaiser, at the end of which he handed in his letter of resignation. It was accepted at once.[112] On the same day the Kaiser sent a telegram summoning Eulenburg to Breslau. Matters were not at all as Eulenburg believed, Wilhelm claimed.[113]

The 'mental state' of the Kaiser during their meeting at Breslau made a lasting impression on Eulenburg. 'I was so deeply affected, indeed shaken, . . . to see my beloved Kaiser suffering so much at Breslau', he afterwards wrote to Wilhelm himself.[114] 'I have never seen our beloved sovereign so agitated and so depressed at the same time', he wrote to Lucanus, echoing these words almost exactly in a letter to the Chancellor: 'I have never seen the Kaiser more agitated – but in particular, never *more depressed* than in Breslau.'[115] He telegraphed to Holstein expressing concern because 'His Majesty is deeply hurt.'[116] 'I happen to love H.M. personally very much', he wrote later, 'and to see a person suffer deeply whom one has taken to one's heart – that is not my strong point.' The Kaiser had really 'suffered *deeply*', and Eulenburg's meeting with him in Breslau had taken on 'the character of the heartfelt outpourings of a tormented spirit'. Eulenburg had therefore begged 'the dear old gentleman', the Reich Chancellor, to 'take into account this feeling of our suffering King'.[117]

At the Breslau meeting the Kaiser declared to his friend that the thought of another chancellor crisis had 'thrown Him into utter despair'. 'In fact he had never been reduced to such a dreadful state, because he had felt so deeply and painfully hurt by the *insult* to him which the action of the Ministry of State represented; he had never suffered a personal affront like this before, and he could not understand how his Uncle Hohenlohe had been able to lend himself to it: no doubt it was because of his age . . . He had completely lost his confidence in the entire Ministry.' Eulenburg believed that the Kaiser's antipathy towards the Ministry had been 'worked up to the maximum' under the influence of his military suite – Wilhelm had been accompanied to Breslau by both Moltkes, Kuno and Helmuth. If he, the Kaiser, stated that he continued to have confidence in Köller, the Ministry could not simply respond that Köller no longer had the confidence of the Ministry, Wilhelm declared. He was firmly convinced that such a 'rebellion' could not have originated 'in the mind of a *Prussian* minister'; 'Marschall was behind it, for he was constantly putting the constitution forward. The Kaiser also left it open to conjecture that the old democrat Miquel might have taken some pleasure in outvoting the King of Prussia.'[118] Eulenburg was also of the opinion that in this case the Prussian Ministry of State had 'in fact behaved like a Cabinet in a parliamentary state'. Behind it all he detected Marschall's Badenese attitudes, Hohenlohe's South German feelings, Bronsart's 'thirst for revenge' and the inner convictions of 'the old democrat' Miquel.[119]

In any event Kaiser Wilhelm II demanded satisfaction for the 'gross insubordination' of the Ministry of State. His idea was, he said, that the entire Ministry should hand in its resignation by way of atonement, which he would then refuse to accept.[120] On hearing of this suggestion by the Kaiser, Marschall at once commented, 'That is what we aimed above all to avoid by acting in the way we did, since H.M. would then in fact have been in a position of constraint.'[121] Eulenburg also pointed out to the Kaiser that the collective resignation of the Ministry of State would have to remain a secret, for 'if it became known that the entire Ministry had tendered its resignation, and the Ministry then stayed on while Köller left, public opinion would *inevitably* draw the conclusion that handing in their resignations was their method of getting rid of Köller – and that this method *succeeded*'. In addition, if a collective resignation were publicly announced, a chancellor crisis might well follow, which must be avoided at all costs because it would seriously threaten the stability of the country. The formal resignation of the Ministry was necessary as an example for the future, but for the present it must be kept strictly confidential, Eulenburg argued. He was, he added, 'quite miserable' because 'I cannot bear to see Your Majesty in torment any longer!'[122]

On 5 December 1895 the Chief of the Civil Cabinet was able to report to Eulenburg that his arguments against the public announcement of the collective resignation of all the ministers had persuaded the Kaiser to abandon his idea altogether. Instead he had agreed with the Reich Chancellor that Köller would not be dismissed for the time being, but would merely be sent away on holiday, and that the Ministry of State would receive a 'firm order' in which its 'improper conduct' would be 'censured, and the rights of the crown emphatically upheld'. 'I thank God', Lucanus wrote, 'that there is some progress and that our Imperial Master will be given satisfaction without any more grave crises arising from it. His Majesty's mood has also improved as a result.'[123] Waldersee, however, recorded after the Kaiser had been shooting at Springe on 7 December, that Wilhelm was still 'in a decidedly serious mood' because of Köller's enforced resignation.[124] His anger with the Reich Chancellor and the ministers expressed itself in an 'unguarded remark' he made at this time and which was repeated by one of his Flügeladjutanten. At a function attended, among others, by Hohenlohe, Boetticher, Marschall and Berlepsch, the Kaiser pointed at them and called out mockingly: 'Aha, look at the Convention sitting over there!'[125]

The 'firm order', probably drafted by Lucanus, that Wilhelm II addressed to the Prussian Ministry of State on 9 December 1895 very clearly reflects the Prussian conception of 'personal monarchy', as opposed to parliamentary monarchy. In his capacity as king of Prussia Wilhelm rebuked the Ministry of State for its behaviour in the Köller crisis, which was 'inappropriate to the position in which the Ministry of State in Prussia stands in relation to the king. In the Prussian monarchy the appointment and dismissal of ministers is the constitutional right of the king, and the king's confidence alone is decisive in that regard. If therefore the Ministry of State, although aware that My confidence in Minister von Köller remained undiminished, induced the latter to hand in his resignation, the Ministry of State has thereby created a situation which may well be not unusual in states ruled by a parliamentary system, but which has no precedent in Prussia. It is my intention, however, to maintain in full the historical and constitutional rights and prerogatives of My crown and in due course to transmit them undiminished to my successors. I am compelled by the present case to announce this to the Ministry of State, so that its significance as a precedent for all time may be recognised.'[126]

Wilhelm II continued to urge his uncle the Reich Chancellor to stick to the 'old Prussian point of view' in dealing with the ministers, namely 'that a Prussian minister is in the agreeable position of not having to worry about the success or failure of any bill put forward by him in case he might have to resign on account of it. For he has *not* achieved office at the wish or under the mandate of one or other party in the Chamber. On the contrary, my ministers are appointed quite freely by me as a mark of All-Highest confidence; and as long as they possess this confidence they need not concern themselves with anything else. They are in a better position than in other constitutional states. I should be very grateful to you if you would occasionally remind the ministers very specifically about this principle, since as a result of their morally corrupting dealings with parliaments they are occasionally affected by attacks of constitutionalism, or, as with Miquel, infected by an old parliamentarian . . . Your faithful Nephew and King Wilhelm I. R.'[127]

More than ever, the Kaiser was determined to give an unmistakable demonstration of his right to decide on appointments in choosing a new minister of the interior. His first thought, characteristically, was to nominate the Chief of his Civil Cabinet, Hermann von Lucanus, as successor to Köller, which – since all the ministers declared this choice 'impossible'[128] – reawakened the danger of collective resignation by the Ministry of State. Philipp Eulenburg warned urgently of the consequences of another conflict. If the Kaiser gained the impression that '*the Ministry to which he has had to give way over Köller is exploiting its victory*, either there will be a storm which will change everything and alarm the world, or it will put him into a mental state which I fear *even more*: loss of the self-confidence that we *need*, that is *essential*. That would mean that the Kaiser is *completely* in the power of Lucanus – or of some other person.'[129] But although Wilhelm gave up the idea of appointing Lucanus, he also rejected the unanimous suggestions of the ministers, who recommended the Oberpräsident of Westphalia, Konrad Studt, the Regierungspräsident in Cologne, Oswald Freiherr von Richthofen, and the Secretary of State at the Reich Treasury, Count Posadowsky.[130] On 8 December the Kaiser sent Hohenlohe a telegram informing him that after detailed discussions he had appointed the Regierungspräsident of Düsseldorf, Eberhard Freiherr von der Recke von dem Horst, as minister of the interior.[131] As the Reich Chancellor rightly observed, the Kaiser had chosen the new minister entirely by himself.[132]

After the Köller crisis Eulenburg fell back on his old formula, according to which the Kaiser should take the most important political decisions in agreement with the Reich Chancellor, after which the ministers simply had to carry them out. He put it to Holstein on 7 December that Wilhelm was convinced that in the Köller crisis he had overstepped the mark of 'what a self-confident Prussian king may do'. Care must therefore be taken in future 'that H.M. should never feel he is being overridden by the entire Ministry of State, but only that he is facing *the definite firm will of the Chancellor, man to man*'.[133] 'The main focus must be placed on the understanding between H.M. and the Chancellor. At the expense of the *whole* Ministry, as far as I am concerned', he declared in a letter to Bülow.[134] 'He [Hohenlohe] will always have Y.M.'s confidence if he takes decisions alone with Y.M.', Eulenburg wrote to the Kaiser. 'If he obeys Y.M.'s will, Y.M. will also take account of the wishes which *he* may perhaps express at some time with regard to appointments. In my Prussian heart the only conceivable alliance is "the king with his minister-president and chancellor" against the ministers, not "the minister-president with all the ministers" against the king – if it is inevitable that controversial questions arise.'[135] According to the Kaiser's favourite, whose influence, with both the Kaiser and the Chancellor, was now at its height, everything depended, as he told the latter, on ensuring that 'what happens from now on should be decided *in dialogue between H. Maj. and Yr. Highness.* If the ministers do not accept what Yr. Highness has discussed and agreed with H. Maj., another resignation by this or that minister would *strengthen Yr. Highness's position vis-à-vis His Majesty immensely*, because H. Majesty would have to recognise it as the result of firm support for *His* wishes.' This form of decision-making had always been what the Kaiser wanted, Eulenburg emphasised.[136] '*Je ne demande pas mieux*', the 'dear old gentleman' replied, but added awkwardly that he was often afraid that, even when he wanted to speak to the Kaiser urgently, he was a nuisance to him. 'If I am wrong in thinking so, I shall be all the more pleased.'[137] The treatment that he was to receive over the next few months both in foreign affairs and at home could only confirm his worst fears.

CHAPTER TWENTY-FIVE

Wilhelm and world politics

THE KAISER AND THE FOREIGN POLICY OF THE NEWEST COURSE

DURING the chancellorship of his 'Uncle Chlodwig', Wilhelm II controlled the foreign policy of the German Reich to an even greater extent than he had been able to do in Caprivi's time.[1] His activity in this sphere was no longer limited to an occasional intervention in the Wilhelmstrasse's otherwise carefully calculated diplomatic games of finesse; he now determined the guidelines of foreign policy himself in characteristically impetuous and emotional fashion, often after discussing them with this or that favourite, Flügeladjutant, naval officer, ambassador or other visitor to the court, and expected his orders to be carried out by the Reich Chancellor and the Foreign Office. In the chancelleries of the European powers – including those of Germany's allies – there was concern at the harsher, more aggressive note to be detected in German foreign policy, which was generally attributed to Wilhelm II's personal influence. Thus for instance in November 1894 the usually circumspect Austrian Foreign Minister, Count Kálnoky, complained to the British ambassador, Sir Edmund Monson, in an unmistakable reference to the Kaiser, that German policy was now characterised by 'sudden impulse'.[2] A year later Martin Gosselin, then British chargé d'affaires in Berlin, went a step further and reported that the rumour was rife in Berlin that the Kaiser was suffering from hallucinations. He added, not without reason, 'It becomes, indeed, a serious matter if a Sovereign who possesses a dominant voice in the foreign policy of the Empire is subject to hallucinations & influences which must in the long run warp his judgement, & render Him liable at any moment to sudden changes of opinion which no one can anticipate or provide against.'[3] Waldersee thought much the same, commenting in his diary in 1896: 'If only our Kaiser would keep his hands off foreign policy; it is a field which he thinks he has mastered, but in fact he does everything wrong. It must be a sad task to be chancellor or secretary of state for foreign affairs! But these good people are willing to accept any kind of mistreatment; in my opinion they sin against both the Kaiser and the Fatherland by doing so.'[4]

Some of the new Reich Chancellor's closest colleagues, however, would not have disagreed with these criticisms. In December 1895 Friedrich von Holstein spoke of the 'unprecedented direct interference of the Kaiser in Foreign Office business', which was not only a vote of no confidence in Hohenlohe, Marschall and Holstein himself, but constituted 'an immense danger, because of the Kaiser's naivety and lack of experience'. 'The Kaiser his own Reich chancellor' was a perilous formula whatever the circumstances, the Privy Councillor continued, 'but most definitely now, with this impulsive and unfortunately very superficial sovereign, who knows nothing of constitutional law, of political events, of diplomatic history and – of how to handle people'.[5] Marschall too voiced his discontent at this time. 'Things are going badly with H.M. He constantly interferes in foreign policy. A monarch must have the last word, H.M. always wants to have the first word, and this is a cardinal error.'[6]

At the end of 1894 or early in 1895 Prince Alexander Hohenlohe, at the request of his father, drafted a revealing letter of protest to the Kaiser. The first version of the letter contained a passage reading: 'There are two ways in which foreign policy can be conducted in monarchical states: either as in Russia, where the Tsar is his own minister of foreign affairs and the minister merely carries out the orders, or as here, where the minister of foreign affairs is responsible not only to H.M. the Kaiser, but also to the country.' As this emphasis on the constitution would certainly have aroused the anger of the Kaiser, this passage was crossed out and replaced by a milder version: 'I have always held fast to the principle that foreign policy must be conducted in accordance with Your M's commands. The fulfilment of these commands is the duty of the Reich chancellor and the Foreign Office respectively. The Reich chancellor must take care that the instructions sent by him or the Foreign Office to the ambassadors do not conflict with any commands which might be sent directly from Y.M. to the ambassadors, or with communications which Y.M. thinks fit to send to foreign sovereigns. The interests of Y.M. require that measures be taken to avoid any possibility of contradiction or misunderstanding. This can only be the case, however, if Y.M. is graciously pleased to inform me of any direct communications to foreign sovereigns, in so far as they concern the progress of diplomatic negotiations.' The fact that even this watered-down plea was never sent speaks volumes.[7] Instead, where foreign affairs were concerned Prince Hohenlohe adopted the attitude disparaged by his son as the 'Russian' way: that of the minister simply following the orders of the Kaiser. An example of this is provided by his 'reverent' phraseology in the summer of 1895 when he sought the instructions of his All-Highest Sovereign on a question affecting an alliance, which was after all a matter of vital importance to the Reich. 'Y.M. would place me under a debt of reverent thanks', he wrote, 'if Y.M. would be pleased to inform me by telegraph of Y.M.'s All-[Highest] will, so that, according to the circumstances, I would be in a position to adopt the attitude corresponding to Y.M.'s will.'[8]

As we have already had occasion to see, the influence which Kaiser Wilhelm II exercised on the formulation of foreign policy manifested itself in the highly personal way in which he took decisions on appointments. He alone decided – often after discussions with his friend Philipp Eulenburg, perhaps during his Scandinavian cruises or on hunting expeditions – who should go where as ambassador or envoy,[9] but he was just as likely to make such decisions of his own accord. He regarded the filling of senior diplomatic posts as the crown's prerogative; as a rule he had not the slightest qualms in ignoring the advice of Hohenlohe, despite the decades of diplomatic experience the Reich Chancellor had to offer as former Bavarian Minister-President, German ambassador in Paris and Statthalter in Strassburg. When for example in 1895 Hohenlohe recommended the then Prussian envoy in Stuttgart, Theodor von Holleben, for the difficult post of Beijing, Wilhelm turned down the Reich Chancellor's suggestion on the grounds that Holleben's recall from Stuttgart would be 'unwelcome' to the King of Württemberg.[10] The Kaiser was also chiefly responsible for the sudden replacement of the German ambassador at St Petersburg, General von Werder, by the former Marshal of the Court Hugo Prince von Radolin in the spring of 1895. The advice which Holstein gave his friend Radolin on his appointment is evidence of the pernicious influence which the gossip-loving, indiscreet Kaiser exercised on the diplomatic profession: 'take pains to write interesting reports on the Kaiser's account. But then be careful in your criticisms about things which, if repeated in St Petersburg, might cause trouble for you. One never knows.'[11]

There were a few instances when the Kaiser allowed himself to be persuaded by the Reich Chancellor's representations to give way over an appointment. But such concessions were exceptions which merely confirmed the rule: foreign policy during Prince Hohenlohe's chancellorship, as Wolfgang J. Mommsen has rightly commented, was dominated by the 'uncontrolled personal rule' of Wilhelm II.[12] As he had done before, the Kaiser invited German ambassadors abroad to correspond directly with him, bypassing the Reich Chancellor and the Foreign Office, and took it amiss if they did not respond to their sovereign's invitation. Count Paul von Hatzfeldt, who had been the German Reich's representative in London for many years, attributed the Kaiser's noticeably unfriendly attitude towards him to the fact that he had 'never made use of his repeated permission to write to him directly'. The ambassador justified his refusal to do so on the grounds that 'in the first place it does not accord with my way of doing business to make policy behind the chancellor's back', and 'furthermore, although I can keep quiet and blindly carry out what I am told to do, I cannot express specific agreement when it would not correspond with my convictions to do so', which had frequently been the case recently.[13]

What was more, the Kaiser repeatedly and without the knowledge of the Reich Chancellor or the Foreign Office called on the ambassadors and military attachés of foreign powers in Berlin to discuss vitally important and secret foreign relations matters with them, or to make revelations of breathtaking significance. Hohenlohe had hardly moved into the Reich chancellor's palace when he found that the Kaiser, in direct negotiations with the Japanese envoy Aoki, had proposed a trade treaty between Japan and Germany.[14] In January 1895 he asked the Russian ambassador, Count Shuvalov, the real reasons for the Franco-Russian rapprochement – receiving the answer that it was the result of the non-renewal of the secret Reinsurance Treaty in the spring of 1890 – and told the astonished ambassador that the Triple Alliance could become a Quadruple Alliance if Russia joined. The Reich Chancellor and the Foreign Secretary did not hear of these ventures into high politics until the next morning.[15] Waldersee was shocked by the manner in which Wilhelm spoke of the heir to the Bavarian throne in the presence of the Russian ambassador. Although Shuvalov was seated at the same small table as the Kaiser, Wilhelm gave vent to 'extremely critical remarks about Prince Ludwig of Bavaria, in the presence of a foreign ambassador! It sent cold shivers down the spines of those listening, and some of them tried to divert the conversation on to something else, but in vain. The Kaiser does not seem to realise that it is Shuvalov's duty, even if the Kaiser regards him as a personal friend, to send home a report of the conversation.'[16]

Early in the morning of 18 January 1895 the Kaiser paid a surprise visit – 'without the slightest word to the Chancellor'[17] – to the French embassy and complained vehemently to the ambassador, Jules Herbette, about the fall of President Casimir-Périer. It was events of this kind, he explained to the bewildered representative of the neighbouring republic, that had made him an enemy of parliamentarism.[18] At the same time, in marginal notes on diplomatic reports, the Kaiser urged that strictly confidential negotiations be held between the monarchies of Europe in order to agree 'a common line' against revolution which could become the basis of operations if indeed, as he expected, 'something surprising' should happen in Paris.[19] Both Hohenlohe and Holstein were furious at this wilful interference by the Kaiser in diplomatic affairs,[20] and Waldersee was no less indignant at Wilhelm's initiative, which showed that the Kaiser had no real idea of the effect which his interference had on foreign policy. 'He fails to understand that this is an important political act possibly entailing considerable consequences', the General complained, adding 'Hohenlohe was very unhappy about it himself, but seems to have said nothing.'[21] On 25 January the Kaiser appeared at the Reich Chancellor's palace; as Hohenlohe told Holstein, he was seeking to justify his marginal comments. 'He said he would hear nothing about an action by the powers against France. Therein he agreed with me. What H.M. is considering is the following: he fears that in the event of the formation of a revolutionary socialistic government, one or another of the *monarchical* governments might betray the solidarity of the monarchical governments . . . and establish relations with the revolutionary government in France. H.M. would like to prevent this and make certain that the solidarity of the monarchical governments did not "crumble". "How this is to be arranged", so says H.M., "is a problem for cleverer people than I."'[22] The following year Herbette, who had been French ambassador for ten years, had to be recalled from Berlin after an incident during a ball at the Opera House, when the Kaiser asked him whether the recall of the French military attaché, the Comte de Foucauld, could be cancelled. When Herbette answered in the negative, the Kaiser abruptly and ostentatiously turned on his heel, after which the ambassador's position, in spite of attempts at reconciliation on all sides, was fatally undermined.[23] Wilhelm even went so far as to discuss high politics with the French ambassador in London, Baron de Courcel, during a visit to the Isle of Wight.[24]

In the spring of 1896 the Secretary of State at the Foreign Office joined forces with Holstein in urging the Reich Chancellor to threaten resignation in order to stop the Kaiser's direct discussions with foreign ambassadors. 'They say I should make the Kaiser refrain from all conversation with the ambassadors. If the Kaiser did not agree I should have no alternative but to go', the elderly Prince noted in his journal. But he went on to comment resignedly: 'The Kaiser's conversations with the ambassadors cannot be prevented. If I had regarded that as a reason for resigning I should have had occasion enough to go. To leave him in the lurch now, when the situation is becoming serious, would be simple dereliction of duty.'[25]

Wilhelm of course continued to annotate diplomatic reports with insulting remarks, as for instance in September 1895 when the German envoy in Lisbon reported that the Marquis de Soveral, hitherto Portuguese ambassador in London, had been appointed foreign minister. 'The blue monkey!', Wilhelm II scribbled on the report. 'It is amazing what heights a mandrill can reach.' In 1897 he wrote in the margin of further reports from Lisbon: 'The blue monkey is either a villain in the pay of England, or a great fathead! . . . I think he is a fathead.'[26] Similarly, the Kaiser made no secret of his annoyance at the appointment of Lobanov as Russian foreign minister in February 1895.[27] Not surprisingly, his telegrams and aggressive speeches constantly created a stir, especially when they concerned sensitive international questions. In the spring of 1895 Waldersee observed that the mood in France was becoming 'more hostile towards us as a result of the Kaiser's latest speeches'.[28] In September, while shooting at Rominten, he sent telegrams to the Reich Chancellor and the Chief of the Military Cabinet ordering them to submit proposals to him as to how Germany should respond to recent French armament plans.[29] Three weeks later the General welcomed the 'very strong' speech which the Kaiser had made at the dedication of the memorial to his father on the battlefield at Wörth, and in which he had emphasised that Germany would never surrender the Reich territory of Alsace-Lorraine; the speech had been just what was needed, in response not only to the vociferous demands for revenge coming from the French, but also to the 'many feeble spirits' in Germany itself, who saw the surrender of Lorraine as a way out of the international dilemma.[30]

But Wilhelm II's most dangerous activity, in terms of foreign policy, was the personal correspondence which he conducted with other monarchs and which the Reich Chancellor often did not see at all, or did not see until weeks after it had been written.[31] The notorious 'Willy–Nicky' correspondence, which began with the accession of the young Tsar Nicholas II in November 1894 and which aroused universal anger when it was published by the Bolsheviks in 1918, will be examined later in this chapter. Somewhat less explosive was the correspondence between Wilhelm II and King Umberto of Italy, and yet in the spring of 1896, when the Italian army had suffered a catastrophic defeat near Adua in the Tigre province of Ethiopia, this correspondence led to a fundamental disagreement between Hohenlohe and the Kaiser over the latter's personal diplomacy.

On 8 March 1896 Hohenlohe felt compelled to object to the dispatch of the former military attaché in Rome, now Flügeladjutant Colonel von Engelbrecht, to King Umberto with a letter written by Wilhelm himself.[32] Once again it was Holstein who had impressed upon the Chancellor the dangers of surrender. 'If Your Highness gives way, this kind of thing will be repeated *de plus fort en plus fort* . . . If it should come to the point – which I doubt – of Your Highness having to threaten resignation, please consider my own also at your disposal.'[33] Not least because he felt obliged to protect the ambassador at Rome, Bülow, from Engelbrecht's intrigues, the Prince nerved himself to write to the monarch stating that the Flügeladjutant's mission could not be permitted to go ahead. Such a step, he said, would inaugurate a policy for which he could not accept responsibility.[34] The Kaiser's reply is characteristic of his attitude to the conduct of foreign policy in general. 'Dear Uncle', he wrote, 'Your letter, which I have just received, and which prevents me from carrying out my intention of relieving poor King Umberto's situation, came as a most painful surprise. In it you say that he would be hurt by such a step! I see from this that you are not familiar with what happened concerning the question of army formations. In Bismarck's time, that is to say 7 years ago, the question arose and *enquiries were made in Berlin* by Italy as to what the reaction here would be if there were a reduction of cadres. The Prince [Bismarck] was very worried about it and moved heaven and earth *to prevent it*, as a result of which, through heavy pressure and Engelbrecht's single-minded skill this question was eventually postponed and the *Italians retained*, with a heavy heart but *by Germany's express official desire*, the very expensive organisation they had. Indeed even the King gave me *his word* during my visit to Rome, jamais je ne laisserai toucher à l'Armée. Now he is in a position of constraint: he cannot take a decision to do something until I release him from his word and he is given a free hand from here, which he does *not* have now. As he is very gentle and warm-hearted and has such close links with me, my letter would console him, give him courage and calm him. Furthermore, it in fact contains the same phrase that Bülow used in speaking to [Foreign Minister] Sermoneta yesterday on my behalf in accordance with your instructions, that whatever they did, Italy's value as our ally would not be affected. But as we *forbade* Italy 7 years ago to reduce its corps, we now bear *full responsibility* if the King consequently has to face difficult inner conflicts or other problems. I therefore urgently beg you to allow the letter to be sent so as to release the King from his word and give him back his freedom of action. He has already been advised by telegram that my letter is coming. Your devoted nephew Wilhelm I. R.'[35]

This time, surprisingly, Hohenlohe stood firm. He pointed out to the Kaiser that a word to the Italian ambassador in Berlin would suffice to allay the scruples of the King of Italy, whereas to send Engelbrecht to Rome would undermine Bülow's position. He therefore considered the Colonel's mission 'inadmissible for political reasons'.[36] And as Holstein had predicted, the Kaiser did in fact climb down, although with such bad grace that Hohenlohe's worst fears were realised. In reply to the Chancellor's objections he wrote: 'Dear Uncle, I believe I detect in your second letter of last night a certain sharpness which I cannot entirely reconcile with the subject it concerns. The very strong opposition which you express, in a way I do not fully understand, against Engelbrecht's so-called "mission", and which at the end of the letter culminates in a kind of threatening rejection, of course obliges me to give up the idea of carrying out my wish. So he will not go. Although I had sent a telegram to the King [Umberto] announcing that he would be coming with a letter. Now I shall have to find some excuse. I acted with the best intentions and hoped to give my ally some comfort and reassurance, which he urgently needed. I had loyally kept you informed of everything and shown you everything in advance, and now it has all come to nothing. You must of course bear the responsibility for it and for whatever consequences it may have.

But one expression in your letter pained me deeply and that is the sentence: "Engelbrecht's mission would weaken Bülow's position!" I can only conclude from this that you think me capable of personally going behind the backs of my own officials, whom I have myself appointed. I need hardly say that Engelbrecht had orders to call on Bülow at once and talk to him. It is equally clear that my Flügeladjutant is a welcome visitor to the ambassador, especially when – as in this case – he has spent nearly 10 years in the embassy in question, and that as the bearer of a letter from his sovereign it redounds to the credit of the ambassador to help him gain an audience. Particularly as the matter in question was of a purely military-technical nature and one with which the adjutant had been familiar for many years, while the ambassador, in spite of all his experience, is only a civilian and therefore cannot judge or understand purely military matters in the same way as a professional soldier who, furthermore – in this case – has himself conducted all the negotiations over these matters. It was also quite clear that Bülow is my personal, intimate friend, upon whom I have just conferred a Grand Cross for his outstanding services, and whom I would have to be mad, or a wicked intriguer, to want to harm or weaken in his position. Having examined both my own motives and the step I contemplated, it seems to me that it would in fact have been a help to you. But I have received a harsh rejection from you and it has hurt me all the more because I have given you so many proofs of my complete dependence on you and confidence in you in the course of this last winter. I have not entertained a single political idea or taken a single step without anxiously discussing it in detail with you, at the risk of pestering you with my frequent visits, or of looking almost incapable of independent action in the eyes of the world. So it weighs all the more heavily on me for being so unexpected, that you should accuse me of taking a step that would prevent you from carrying out your policy and weaken the ambassador's position; in other words, first you think I am capable of hatching intrigues behind your back, and second you think a Royal Prussian Colonel and Flügeladjutant capable of lending his hand to such an activity. I am sure that if you consider this affair calmly you will give up your completely groundless suspicions and will realise that my plan to send Engelbrecht would no more have harmed Bülow than the "mission" of Colonel and Flügeladjutant von Moltke to the Tsar harmed Radolin's position. On the contrary, it was of the greatest value to him. I hope your confidence in me will return, for one can often hold very different opinions about things and also argue about them sometimes, but that is no reason to suspect each other of evil motives, especially not in your truly devoted nephew Wilhelm I. R.'[37] This time it was the Chancellor's turn to apologise 'most humbly' for the expressions used in his 'hurriedly written' letter. He had 'hitherto always tried', he wrote to the Kaiser during a meeting of the Ministry of State, 'to express myself to

Y.M. only in a manner which expressed the deep reverence and loyal devotion which I have in my heart for Y.M.'.[38] He had 'absolute confidence in Y.M.'s wisdom and prudence' and hoped that he would be able to give proof of this in the future. Although his behaviour had displeased the Kaiser, he particularly begged him not to be deterred by this from 'honouring me with All-Highest visits. The discussions I have with Y.M. in person are the highlights of my arduous existence, and to receive Y.M.'s commands orally certainly contributes more to the satisfactory conduct of affairs than lengthy written reports.'[39]

Two days later the Kaiser sent his letter to King Umberto to Rome after all, but through Bülow, with orders to give the letter to the King unopened. The ambassador was obliged to guess from the King's remarks what the letter contained. Bülow felt compelled to urge the Chancellor 'not to mention *to anyone* the fact that this letter had been sent', although he was prepared to accept that it was 'doubtless a sign of conciliatory feeling' that he, and not the special messenger Engelbrecht, had been entrusted with the transmittal of the letter in the end.[40] Just as Holstein had done, Bülow warned of the 'incommensurable dangers of "rule by adjutant"'.[41] In similar vein Count Münster, the longstanding ambassador in Paris, asked despairingly: 'What is the point of the Wilhelmstrasse, when its business is divided among the Flügeladjutanten? One of old Kaiser Wilhelm's greatest qualities was that he never interfered in the business of the various departments and officials of the state!'[42]

During his Mediterranean cruise the following month Wilhelm II had an opportunity to comfort 'His Majesty's friend and ally' Umberto in person over the defeat in Tigre. When the King lamented the fact that 200 officers whom he knew personally had died in vain in Africa, the Kaiser responded that it was the lot of the soldier and an honour to die a soldier's death. 'The Italian officers who had fallen in Africa had died loyal to their King and for the glory of their flag. The honour of the Italian army had been upheld.' The example of Frederick the Great showed that defeats could be overcome; it was only a matter of knowing how to 'make up for them at the opportune moment and in the right place'. 'Revenge was a dish which had to be served cold. Instead of battling *contre vent et marée* in Eritrea, Italy ought to collect and strengthen itself militarily, financially and in its internal political situation.' Since the French had helped the Abyssinians by sending them money, arms and even a large number of officers, Italy must take revenge on *them*. 'The right moment to pay back the French', the Kaiser commented, 'must . . . be awaited with a cool head; the reckoning would come in the end not in the mountains and ravines of Tigre but on Middle European terrain.' As far as the Russian flirtation with Abyssinia was concerned, in Wilhelm's view it was 'a product of the increasingly evident hostility between Russia and England. This historic antagonism, which dominated the overall situation in Europe, was becoming noticeable outside our continent. It was not in the interest of Italy to put itself between the bear and the whale.' Umberto was particularly impressed by the Kaiser's remarks about developments in the Muslim world, in which an 'explosion of Islamic fanaticism' might be in the offing, which would be of great significance for Anglo-Russian rivalry.[43] These utterances in themselves give a clear indication of the militaristic and belligerent direction in which Wilhelm was now making his growing influence felt. They were far from the only pronouncements by the Kaiser which seemed to anticipate the future course of German foreign policy in a sometimes alarming manner.

WILHELM II AND THE SCANDINAVIAN CRISIS

The crisis-ridden developments in many parts of the world – in the Far East, where Japan and China were at war, in the collapsing Turkish empire, in the Horn of Africa, where Italy's colonial aspirations were causing dissension among the great powers, and in Madagascar, where the French were establishing a hold – presented Wilhelm with ample opportunities to bring his own, or rather Germany's, ambitions to the fore. But the issue which particularly concerned him at this time, because of his personal acquaintance with the Swedish royal family, into which his cousin Viktoria of Baden had married, was that of the threat posed by the Norwegian independence movement to the unity of the Kingdom of Sweden and Norway. As always, he saw a popular rebellion of this kind as a menace to the monarchical principle. In addition to this he feared that the independence of Norway might lead to that country being divided between Russia and Britain, which he was not prepared to countenance without German involvement in the south of the country. During the Scandinavian cruise of July 1893 he had already gone so far as to give orders, in a telegram to Caprivi, for the occupation of southern Norway in the event of a national uprising against the rule of the King of Sweden. His message ran: 'In case it should come to an open rebellion in Norway and the King were not able to suppress it, and if he should appeal to us and other friendly monarchies for aid, Count von Wedel [the German envoy] should be instructed that I am naturally ready to help and My navy is at his disposal. It is our duty, wherever monarchies are under threat, to spring to their aid and support them. Just as we did so morally in Portugal, we must be ready in case we have to do so physically in Norway.'[44] Three years later, when the situation in Scandinavia had become even more acute, the Kaiser subjected Prince Hohenlohe to a long lecture on the Norwegian question, which he believed could lead to the break-up of the union with Sweden at any moment. King Oscar was indecisive and could not cope with the increasingly radical Norwegian population, Wilhelm maintained. 'Generous supplies of Russian money were circulating in the country and, if the union were abandoned, it was to be feared that Russia would carry out its long-cherished plan and occupy part of Norway in the region of Tromsoe. The English would not be happy with that, and so they would also occupy part of the country, and in that case we should have to make haste to take the southern part of Norway, before the Danes did so. For we could not tolerate Denmark establishing a hold there . . . We would have to maintain a close watch and instruct the consul to keep us abreast of the situation by daily telegrams, for it might be necessary to send out a fleet, if Russia and England went ahead.'[45] During the Scandinavian cruise that year Kaiser Wilhelm had a serious discussion with the King of Sweden and Norway, which Philipp Eulenburg summed up in a memorandum as follows: 'Germany is the natural friend of Scandinavia, because it cannot be in favour of Russian expansion in the Baltic. If Scandinavia falls apart, however, and Russia takes over the north of Norway, while England perhaps takes Bergen, Germany may be compelled to occupy the southern areas of Norway. In order to avoid this eventuality, which would be extremely troublesome for Germany, Scandinavia must restore order *itself*, and in a forceful manner.' The possibility could not be ruled out, however, that 'even if Sweden intends to restore order in Norway by force, Russia will use the unrest as an excuse to take action for her own part, and perhaps occupy Tromsö and Bodö . . . It can be assumed that England will then occupy Bergen. A peaceful settlement between England and Russia could no doubt still be achieved by holding a conference, but *an agreement between these two powers* over Norway would be *at our expense*. For the existence of Sweden would be endangered by an agreement which was reached on the basis of occupation of Norwegian ports. The country would soon fall into Russia's lap like a ripe fruit. Germany could not tolerate such a shift in the balance of power in the Baltic. But then we should find ourselves faced with an unpopular war. Accordingly, German interests would be best served if the *integrity* of the present Scandinavia were maintained. It cannot be safely assumed that an Anglo-Russian war will develop from the Norwegian question. If England acquires a Norwegian port, she will probably be content with that. In the event that Russia occupies a Norwegian port, we should probably have to remain as spectators and wait to see how the conflict of interests between Russia and England developed. If it reaches the point that England occupies a Norwegian port, we should have to occupy Stavanger, and probably Christiania [now Oslo] too, to protect our own interests. That will put us on a better footing as the situation develops. We could later attach conditions to the surrender of what we had taken by way of security, or indeed we could protect Sweden better from the Russian threat by maintaining our occupation. His Majesty *is inclined to favour* such an occupation in the event that England and Russia proceed to occupy ports.'[46]

That summer, on board the *Hohenzollern*, Wilhelm II dictated to Philipp Eulenburg a letter with racist overtones addressed to Crown Prince Gustaf of Sweden. He did not send a copy to the Reich Chancellor until two months later. The developments in Norway, the Kaiser's letter explained, filled him 'with genuine anxiety'. Norway wanted at all costs to break away from the union with Sweden, and was thereby threatening 'the survival of both kingdoms and the stability of European peace. Given the gentle disposition and sensitive conscience of the King [of Sweden], it is a popular trick to draw his attention to the apparently sacred nature of the Norwegian constitution and to his oath upon it, so as to set him at war with his own conscience . . . But in the end it must come to conflict. And that is because the Norwegian constitution has proved to be completely inadequate, and has put the survival of the union under threat by placing the King in a powerless position. In other words: the King of Norway must now give way to the King of the union. The latter has also taken an oath: to maintain the union whatever happens.' 'I beseech you', Wilhelm exhorted the Crown Prince, 'to keep the King in constant remembrance of this point . . . If the King does not succeed *soon and once for all* in bringing about calm and lasting stability in Norway, it is inevitable that foreign states will intervene. Russia, as your neighbour, will be the first to put its finger into the pie. Under the appearance of friendly help on the grounds that the persistently disorderly state of affairs in the country was a danger to her own frontier population, she would march into Norway in a charmingly friendly way and help restore calm there as she did before in Hungary [in 1849]. Perhaps there would then be a telegram to the Tsar from the General entrusted with the task, sounding just like the one sent on that occasion: "All Norway is at Your Majesty's feet!" Russian money and Russian intrigues have prepared the ground well enough in Norway, especially among the radicals, for a party to be established at an opportune moment, which will call for Russian help. Russia would make sure that her friendly services were paid for with the Norwegian port she has long desired to have, together with the obligatory hinterland. This would be a *casus belli* for England. But if the latter should decide, in accordance with the Falstaffian rule, that discretion is the better part of valour, she would also take possession of a Norwegian port as a security, by arrangement with Russia. Under no circumstances could I accept that Slavs and Britons should share sovereignty over the Germanic North Sea without asking me or without my permission. I should therefore be likewise compelled to occupy the southern part of Norway in order to safeguard my trade and my coast. The union would thus be destroyed, and what would become of Sweden? Locked in by Russia on all sides as far as the North Sea, she would gradually succumb to the embrace of this colossus and be absorbed by it. Now that I have seen the magnificent Angermanälv, I do not doubt for a moments that it will be the first thing that Russia takes from Sweden. So the fate of your House would be sealed, along with that of your country. It might perhaps be possible to save Sweden from being absorbed directly into Russia if she were to join the German Reich's customs union. This could guarantee the King the survival of his kingdom and of his dynasty – after the loss of Norway. But what would this be in comparison to the glorious position which your father now holds and which, God willing, will be held by many more descendants of his House? There is another side of this affair which I should like you to consider: all my aims and endeavours, all my ideas in the political sphere, are directed towards bringing the Germanic races in the world, especially in Europe, closer together and forging strong links between them, in order to protect ourselves more surely against the Slav–Czech invasion which threatens us all to a most dangerous degree. Sweden–Norway is one of the principal factors in this alliance of Germanic peoples. What will happen if this great Nordic union of states suddenly falls apart and is perhaps absorbed by the Slavs (Russians)? In this context, the whole European Germanic North constitutes Germany's left flank, or indeed that of Europe itself, and is therefore of great importance for our security. Its disappearance would expose our flank and would represent a serious threat for us all. The monarchical principle as such would also be very severely affected. The Kings of Portugal, Serbia and Greece have already brought it into disrepute. May your father be preserved from being counted among such a crew. His duty as monarch and king is to set aside his personal feelings and obey the dictates of his duty, which demands that he establish respect and obedience towards kingly authority in his domains. At this moment the peace of Europe, and its survival, lie in the hands of your father. May this peace not be threatened or even destroyed by his giving in at the wrong moment! I thought it my duty to explain the circumstances to you once more, quite frankly and honestly, because so much depends on the decisions of your father. My warning is that of a true friend, whose intentions towards your country and your House are honourable. My plea is that of one colleague to another, to help him in his work. May it not go unheeded! You may use this letter as you think best. Wilhelm.'[47] One wonders how the Crown Prince of Sweden and Norway reacted to this 'warning' from his powerful neighbour.

Two months later, at Rominten, Wilhelm again sat down with Eulenburg to compose words of advice for Crown Prince Gustaf. The King of Sweden should work only with the really loyal, pro-union elements in Norway and should on no account make deals with 'revolution', Wilhelm warned. 'No more than water [deals] with fire.' For the King it was vital that he choose as his advisers only such men as 'watch over the crown like knights with the Holy Grail'. As for King Oscar's desire for a defensive alliance with Germany, the Kaiser felt compelled to decline. 'Russia knows that she would find us in her way if she wanted to lay her hand on Norway. What difference would it make to this relationship if . . . I *bound* myself by *treaty* to use force of arms in response to a Russian invasion of Norway? It would mean only that Russia would get the impression that Germany *wanted* to set itself up against Russia, whereas I am glad to be able to state that there are *no direct* points of conflict between Russia and Germany at present. If Russia *creates* such points – fair enough, then Germany will fight a just war if she is attacked. But I would be taking a heavy responsibility on myself if *I* tried to create points of conflict which were aimed at injuring Russia, and which would moreover be highly unpopular in Germany.' Unlike in the July letter, Wilhelm this time stressed the 'natural alliance between Germany and England', both of whom wished to maintain the status quo in Scandinavia. Sweden, he assured the Crown Prince, could always count on the support of both powers against a Russian attack even without contractual agreements. 'But if Sweden feels the *need* to make public connections with powerful states, nothing would prevent her identifying herself with the peaceful aims which the Triple Alliance set itself. That would issue a warning to Russia *without any hostile intent* being shown, for Europe at last believes in the peaceful aims of the Triple Alliance, and the opinion of chauvinistic Frenchmen and pan-Slavic Russians is of no consequence in this regard.' As with the earlier letter, the Reich Chancellor – and only he – was informed of this missive, with its highly risky assurances, only after it had been sent.[48] How the letter was received in the Swedish–Norwegian royal house is once again not known.

THE KAISER AND THE WAR IN EAST ASIA

Ever since the outbreak of the Sino-Japanese war in the summer of 1894 Wilhelm II had followed the progress of the campaign by land and by sea with unusually intense interest. At the beginning of 1895 he gave two lengthy lectures on the sea battles of the war which were to be of great significance for the development of the German warship building programme, as we shall see in a later chapter.[49] From the beginning he took the Japanese side and expressed his views so strongly at every possible opportunity that everyone in his entourage felt obliged to take a similar line. The Chancellor and the Foreign Office tried in vain to warn of the dangers of such partisanship not only for future relations with China but also for relations with the other great powers, who had their own interests in East Asia to defend. Waldersee too, who saw 'something almost childish' in Wilhelm's partisan attitude, warned anxiously that such bias was highly unwise in political terms. 'In Berlin everyone from the Kaiser down, and in his case particularly ostentatiously, was on the side of Japan and they celebrated Japanese victories almost as if they were German. Anyone who was not for Japan was considered stupid or malicious', he wrote retrospectively when the war ended in the spring of 1895 with total defeat for China.[50] In the Foreign Office the Kaiser's open bias in favour of Japan had been 'very unwelcome', Waldersee noted, 'because people in China are aware of it, take it very much amiss and draw the natural conclusion that Germany's policy is hostile towards China. This is in fact not the case and it is in our interest to remain neutral for the time being. But our good master cannot resist not only giving free rein to his personal feelings on every question, but also expressing them publicly.' 'Japan is rapidly making itself independent of Europe; it is active in all spheres; and if it gets to the point of no longer using imports from us and closing itself to them, then we derive no benefit from our friendship. On the other hand, if China opens itself up to more contact with Europe, it can be assumed that our industry will be able to make massive sales there', commented the General who, five years later, after the suppression of the Boxer Rebellion, was to make a triumphal entry into Beijing as 'Weltmarschall'.[51]

Wilhelm II hoped that a victory by Japan over China would lead to the establishment of a German empire in East Asia. At an audience on 2 November 1894 the Kaiser had already instructed the newly appointed Chancellor to make sure that when peace was concluded Germany was given 'compensation through the cession of Formosa' (now Taiwan). 'Since we protected Japan from intervention by England', he argued, 'we had a right to compensation, and Formosa was a good coaling station. We must likewise try to get Mozambique. We were dependent on England for our coaling stations, and this must stop.' The Kaiser had yielded only on one point, Prince Hohenlohe recorded. He had given up the idea of selling two old ships to Japan, which he had suggested in a letter on 30 October, in view of the reservations expressed by the Chancellor.[52] While still in Strassburg Hohenlohe had had to point out to the Kaiser in a cipher telegram that, of all the powers, Britain had the strongest interest in keeping China in its entirety as a buffer state; but a German demand for Formosa might give the signal for the Chinese empire to be carved up. Furthermore, it was possible that Japan itself had its eye on the island of Formosa as a reward for victory. 'Any intervention by Germany' in East Asia, he warned, would inevitably 'awaken the suspicion of *all* the powers at this moment'.[53]

The following spring, when the war ended in an overwhelming victory for Japan, Wilhelm reverted to his demand for Formosa. In a pencilled note to the Chancellor dated 19 March 1895 he wrote: 'I have made a marginal comment about Formosa because I recently spoke about it to Admiral Knorr, who knows the island. The Pescadores Islands, which are very close by, provide excellently protected anchorage and also a harbour defended by forts. The island of Formosa is largely still in its original state, inhabited by somewhat belligerent natives. Has great treasures to offer, however, especially coal, primarily in the north-east, and undoubtedly valuable. If Japan has no *special* interest in it, it would surely be important to secure this stock of coal and a coaling station by treaty, as a reward for good behaviour as a neutral.' He doubted that France, which was completely absorbed in the annexation of Madagascar, would want to 'embroil itself' in 'a second Madagascar' with Formosa.[54]

When Japan, in the peace of Shimonoseki, succeeded in obtaining not only the independence of Korea and a full indemnity but also the cession of Formosa, the Pescadores and the peninsula of Liaotung to itself, the German Kaiser's enthusiasm for the Empire of the Rising Sun suddenly disappeared. Instead he now ordered the Chancellor to take action with Russia and France against Japan and Great Britain in the hope that by this means it might still be possible to acquire a coaling station – perhaps Wei-hai-wei on the Shantung peninsula. On 11 April 1895 Hohenlohe recorded in his journal: 'At today's audience H.M. observed that he still stood by his plan of co-operating with Russia. H.M. assumes that England has come to an understanding with Japan through secret agreements. Japan will have made commercial concessions to England, and in return England has undertaken not to oppose the peace conditions. If it comes to peace, England will see to it that the Chinese receive a few advantages and will then obtain concessions from China. This can be counteracted by concerted action by Russia, Germany and France. H.M. believes that we could obtain territory in Shantung, in particular Wei-hai-wei. For that it would be necessary to have the iron-clad *Kaiser* brought into service, so that by dispatching it to join the Russian fleet we could ensure that England did not take Wei-hai-wei away from us. Upon my remarking that one must first be sure that Russia would not abandon its war plans again, H.M. expressed the opinion that it could not do so. For us, the Kaiser commented, co-operation with Russia was a great advantage, and if France and Italy joined in, this would constitute an impressive force opposing England, against which country H.M. seems to be very embittered at the moment.'[55] A week later the Kaiser gave the Chief of his Naval Cabinet instructions to discuss with the Foreign Secretary and the former German envoy in Beijing, Maximilian von Brandt, the question of whether it would be useful to inform the Russian government of Wilhelm's intention of acquiring Wei-hai-wei as a base. The reason for the Kaiser's instructions was that Germany, as Senden-Bibran recorded, 'as a result of the very limited naval forces available on the spot' would probably not be strong enough to enforce its claims without Russian support.[56] Thus the momentous decision to collaborate with Russia and France in forcing the Japanese to withdraw from Liaotung and then to seek to establish a coaling station on the Chinese coast can be attributed beyond all doubt to Wilhelm II himself.[57] On this question, in which Anglo-German antagonism and even the future isolation of Germany began to emerge, the Reich Chancellor simply accepted the 'All-Highest will' of the Kaiser.[58]

The monarch had no scruples in coming forward with detailed instructions for the further handling of the East Asian question and in particular the evacuation of the Liaotung peninsula. In the process he showed that he not only grasped the diplomatic and military implications but also fully understood the political advantages that could be gained on the home front from a conclusive success in the Far East. He ordered that it should first be established in Paris 'with absolute certainty and authenticity' what attitude the French Foreign Minister, Hanotaux, took towards the Russian demands on Japan. 'From a conversation which I had with Baron de Courcel [the French ambassador in London] at Cowes I gathered that France has hitherto disagreed with Russia on the question of evacuation . . . Until the reply comes from Paris', the Kaiser continued, 'I would suggest that the matter should be treated in a dilatory manner as far as Russia is concerned. If France abandons its reservations, then we can join in too, and we must try to influence Japan in a conciliatory and reassuring way. If Russia should decide to occupy Korean territory or a seaport, that would be the moment for us to occupy Wei-hai-wei without delay, to prevent the English or the French forestalling us as in Africa. A *fait accompli* always wins more respect from other states than recriminations. Our people would be happy and enthusiastic: it would make an excellent impression on them and boost the nation's self-confidence very much. You know of course that during the spring I secured the Tsar's written agreement in advance to the occupation of an area in China. It would therefore also be necessary for our cruiser division to be instructed to cruise in the vicinity of Wei-hai-wei or in the Gulf of Petshili at the appropriate moment, in an innocent way but with the necessary telegraph connections, so that on receipt of telegraphic instructions they could at once raise the flag there. For this to be carried out smoothly as far as China is concerned, I recommend the Brand[t]–Li-Hung-Tshang connection. Naturally all this must remain absolutely secret.'[59] While staying at Rominten with Eulenburg in September 1895 the Kaiser had a message sent to the Reich Chancellor asking what progress had been made with the occupation of Wei-hai-wei. The elderly Prince's reply was curt: 'As long as the Japanese are there we cannot land. War with China and Japan at the same time is too much.'[60] In Waldersee's eyes, the Kaiser's startling volte-face in turning against Japan after the peace settlement when he had taken the Japanese side so strongly during the war was characteristic of Wilhelm's attitude to politics in general. The 'sudden change' proved once again, he wrote, 'that we must be prepared for surprising moves at any time; hitherto they happened more often in domestic policy; but this is a pretty powerful one in foreign policy'.[61]

THE UNSUCCESSFUL COURTSHIP OF RUSSIA

The co-operation which Wilhelm II had ordered with Russia and France after the peace of Shimonoseki was not aimed solely at the pursuit of his ambitions in East Asia. It was part of a highly personal and at times even secret initiative, the purpose of which was to bring Russia into the Triple Alliance, thereby restoring the Three Emperors' League of the Bismarck era. This enterprise arose from the Kaiser's conviction that thanks to his personal friendship with Nicholas II, who had succeeded to the throne on the death of his father on 1 November 1894, and thanks to the numerous family ties which linked him to the Russian imperial house, he would be able to effect a complete transformation of Russia's international political orientation. He could not only pride himself on having brought about the marriage between his Darmstadt cousin Alix of Hesse and the Tsarevich; Wilhelm's brother Heinrich was married to the young Tsarina's sister, while another sister, Ella, had been the love of Wilhelm's youth and now lived at the Russian imperial court as the wife of Grand Duke Sergei. As we shall see, in his private correspondence with 'Nicky' and in numerous oral communications Wilhelm repeatedly used two arguments which were intended to make alliance with Germany seem an enticing prospect to the young Tsar. These were that Nicholas must protect 'Christian' Europe from the 'Yellow Peril' in the East while Wilhelm made sure that peace was preserved in Europe; and that Russia could take Constantinople and the Straits whenever it wanted, and need fear no opposition from Germany or Austria-Hungary. Even under the most favourable conditions a secret dynastic policy of such breathtaking arrogance would have had little chance of succeeding in the last years of the nineteenth century; but as Russia had long since concluded not only a military convention but also a formal alliance with the French Republic Wilhelm II's grandiose initiative was completely hopeless.

When the news of Tsar Alexander III's serious illness reached Rominten in September 1894, Wilhelm had expressed his confident expectation that 'a total change in political circumstances' would now come about 'through the accession of the heir to the throne', over whom he had 'influence'; he was therefore determined to travel to Russia for the funeral, he had declared.[62] When Philipp Eulenburg sent a diplomatic dispatch from Vienna commenting on the ill-humour of Count Kálnoky, the Austro-Hungarian Foreign Minister, Wilhelm's hopes of Russian friendship were highlighted by his marginal comments:

37. Wilhelm II with his British relations and the newly engaged couple, Nicholas of Russia and Alix of Hesse, Coburg 1894

Nicholas II was 'German-minded! That is why Kalnocky [sic] is annoyed.' 'Yes, because the heir to the throne is on good terms with us and I arranged his marriage for him, I have "priority" with him! And that annoys Kalnocky.'[63] On 18 October 1894 he told the sceptical Waldersee that 'the heir to the throne is very well disposed towards Germany, I know him very well and the Grand Duke of Hesse has confirmed it to me'.[64]

Like Hohenlohe and the Foreign Office, Waldersee recognised both the illusory basis and the colossal dangers of Wilhelm's courtship of Russian favour. A few weeks before the accession of Nicholas II the General had already summed up the situation. 'The way he forces his friendship upon them has been seen there as dishonest; as it has also been seen – unfortunately rightly so – as an indication of fear, it has done us a great deal of harm. The present Tsar [Nicholas] has been clever enough for years to behave in a very reserved, watchful way towards our Kaiser, nor has he responded in the slightest to the almost importunate advances made to him. Certain events during his visit to Springe in '89, which I would rather not commit to paper, made a lasting impression on the then heir to the throne.'[65] Waldersee wrote mockingly of Wilhelm II in November 1894: 'As he considers himself a powerful politician, now that a new Tsar has ascended the Russian throne he has plenty of opportunity to develop his talent. England is courting Russian friendship with all her might, and is trying to compete with the French . . . The best course would be for us not to behave as yet another suitor but to remain elegantly aloof, confident of our strength; that is the only way in which we can impress the Russians. The young Tsar is in the agreeable position of allowing himself to be courted by all the world and waiting calmly, without taking a single step to be accommodating. His advisers will certainly encourage him in that direction.'[66]

The disappointment of the hopes which the Kaiser had pinned on the young Russian sovereign was not long in coming. When Wilhelm visited Hohenlohe in the Reich chancellor's palace for the first time on 14 December 1894 – he remarked that the place was still as dark as it had been in his predecessors' time – it was plain that considerable disillusionment had set in. He agreed with the Chancellor that it was important to maintain a very reserved attitude towards the new Tsar; matters would become clearer and German–Russian relations would improve with time.[67] The coolness shown by the Russians towards Prince Heinrich and the Prussian officers (Plessen, Helmuth von Moltke, Villaume and Saussin) who accompanied him to St Petersburg at the time of the Tsar's accession, together with the rejection of Wilhelm's suggestion that their respective military aides-de-camp should be appointed, as they had been in the past, members of each other's *maison militaire*, showed all too clearly that there was no sign of an improved relationship as yet; on the contrary, some observers expressed the opinion that Germany's position in Russia was 'lost for ever'. 'The Kaiser is the poorer by one illusion', Waldersee commented in December 1894.[68] A few weeks later he registered his delight at hearing that 'the Kaiser has completely abandoned his illusions about Russia and Tsar Nicholas, who loves us so tenderly; that is great progress, even if it must also have been a grave disappointment'.[69] By April 1895 at the latest Wilhelm was forced to recognise 'that the mood in Russia is the worst it could possibly be for us', Waldersee noted. He commented contemptuously of Wilhelm that 'six months ago he still wanted to put the young Tsar in his pocket and control him!'[70]

News from Russia confirmed the fear harboured from the outset that the young Tsar would turn out to be a weak man who allowed himself to be governed by his mother, not only politically but even in the details of daily life. Wilhelm was indignant at a report from Prince Albert of Saxe-Altenburg that the Dowager Tsarina Maria Feodorovna, born a Danish princess, had said on returning to St Petersburg 'that much had not gone as it should have during her absence'. Nicholas had been unable 'to summon up a single word in response!!', the Kaiser complained.[71] Waldersee too recorded that the Dowager Tsarina had 'no greater wish than to humiliate us and the game of intrigue against us will be taken up with new vigour in Copenhagen'. The young Tsarina Alexandra, of whom the Kaiser had had such high expectations, was obviously a completely insignificant woman.[72]

The fact that Wilhelm was already on bad terms with the Darmstadt family, which had been very much under the influence of Queen Victoria since the premature deaths first of Grand Duchess Alice and then of Grand Duke Ludwig IV, also contributed to the Tsar's coolness and reserve towards him. Nor was the domineering and characteristically prudish way in which Wilhelm tried to impose his will during the preparations for the Russian wedding calculated to prejudice Nicholas and Alexandra, as Princess Alix was now called, or her brother Ernst Ludwig, in the Kaiser's favour. On 2 November 1894 in a conversation with Hohenlohe Wilhelm spoke 'rather heatedly' about the forthcoming marriage between Nicholas and Alix, and 'criticised the Grand Duke [Ernst Ludwig], who was a baby, very severely for not having travelled to Livadia with his sister in order to make arrangements for the marriage to be concluded there, as it is being delayed at the instigation of the Queen of Denmark'. The Kaiser had already made 'serious representations' to the Russian ambassador about it, Hohenlohe noted, and had asserted that 'the Princess could not come back to Germany unmarried, that was simply impossible and would be taken very much amiss throughout Germany. But to stay there with her fiancé, unmarried, would go against moral traditions here. He would give Prince Heinrich, who was to go to the funeral at St Petersburg, instructions in this sense.'[73]

In spite of the rejection of his attempts to curry favour, Wilhelm continued his efforts, through private letters and other communications to Nicholas II, to divert Russian foreign policy away from Europe and towards the east and south. Within a few days of the Tsar's accession he had already sent Nicholas the first of a total of seventy-five private letters. In the next twelve months six more letters were to follow, which he showed to his 'responsible' advisers only after having sent them, if at all. In vain Holstein urged the Reich Chancellor to make it plain to the Kaiser 'that there is a limit beyond which Your Highness would not allow even His Majesty to make a habit of going'.[74] Eulenburg's assurance that

Wilhelm's letters concerned trifles, and that on any serious question he would ask the Chancellor's advice first, was very far from the truth, as he himself was soon to realise with horror.[75]

At the end of April 1895 Wilhelm wrote a letter to the young Tsar in which he encouraged Russia to expand further to the east and at the same time announced that Germany wished for a base in East Asia. 'I shall certainly do all in my power to keep Europe quiet and also guard the rear of Russia so that nobody shall hamper your action towards the Far East', he assured Nicholas. 'For that is clearly the great task of the future for Russia to cultivate the Asian Continent and to defend Europe from the inroads of the Great Yellow race. In this you will always find me on your side ready to help you as best I can. You have well understood that call of Providence and have quickly grasped the moment; it is of immense political and historical value and much good will come of it. I shall with interest await the further development of our action and hope that, just as I will gladly help you settle the question of eventual annexations of portions of territory for Russia, you will kindly see that Germany may also be able to acquire a Port somewhere where it does not "gêne" you.'[76] The Tsar's reply, in which he stated that he had no objection if Germany wanted to acquire 'something' in the Far East, was shown to the Reich Chancellor by Wilhelm II under the seal of secrecy.[77]

The celebrations for the opening of the Kaiser Wilhelm Canal at Kiel in the summer of 1895 provided the monarch with a further opportunity to pursue his attempts at manipulation. Philipp Eulenburg recorded with dismay during the Scandinavian cruise that Wilhelm had told him that during the Kiel festivities he had held discussions with Grand Duke Alexis Alexandrovich, the Tsar's Francophile uncle, and the Russian Admiral Schillings on the mission of Russian policy in the Far East. The Kaiser had put it to the Grand Duke, Eulenburg, wrote, 'that He would protect Russia's back as long as she fulfilled her Christian duties towards Mongols and Buddhists in East Asia. His Majesty had an even more wide-ranging conversation with Admiral Schillings, to whom he described the dangers which threaten Europe from the continuing development of the yellow race. *Only Russia* could fulfil the great East Asian cultural mission, and *Germany* would regard it as her duty to protect Russia's borders in the west unless Germany were attacked by other states with Russia's knowledge. He told Schillings he should say this in St Peterburg, also to the Tsar. Schillings was extremely surprised and delighted. His Majesty hopes his words will have a good effect on the attitude of Russian society.' The Kaiser went on to tell his astonished friend that he had written to the Tsar not long ago, 'after I had received your letter about your conversation with [the Russian ambassador] Osten-Sacken in Vienna. The Tsar had told you through Osten-Sacken that Russia understood

38. The Kaiser's original sketch of 30 April 1895, warning of the Yellow Peril in the East

that Germany had to use all her power to overcome radical tendencies. Russia would *guarantee* Germany peace on her eastern frontier if he (Kaiser Wilhelm) would undertake to maintain peace in Europe. I (Kaiser Wilhelm) replied to the Tsar as follows: "As you have informed me of your decisions through Philipp Eulenburg, I *pledge* myself for my part to cover your back while you are engaged in East Asia. My ships in East Asia are also at your disposal." The Tsar answered me in very affectionate and grateful terms. I (Kaiser Wilhelm) therefore think that the situation is clear. If the Tsar breaks his word, well, he must answer before God for it. For my part I consider myself bound by my word and I shall keep it.'[78] Only a few days later, on board the *Hohenzollern*, to Eulenburg's consternation Wilhelm II read him another letter which he had sent to the Tsar 'through the *Naval Attaché* (!) at St Petersburg, Kalau vom Hofe', and which was 'written in English and about 7 octavo pages long'.[79] In this letter the Kaiser commented on his conversations with Alexis and Schillings, with whom he said he had had 'a serious word about East Asian affairs'. In the course of these talks, he wrote, he had emphasised 'how our interests were entwined in the Far East, that my ships had been ordered to second yours in case of need when things looked doubtful, that Europe had to be thankful to you that you so quickly had perceived the

39. The sketch as completed by Hermann Knackfuss: 'Nations of Europe, protect your holiest possessions! Wilhelm I. R.'

great future for Russia in the cultivation of Asia and in the Defence of the Cross and the old Christian European culture against the inroads of the Mongols and Buddhism, that it was natural that if Russia was engaged in this tremendous work you wished to have Europe quiet and your back free; and that it was natural and without doubt that this would be my task and that I would let nobody try to interfere with you and attack from behind in Europe during the time you were fulfilling the great mission which Heaven has shaped for you. That was as sure as Amen in Church!'[80]

Eulenburg, who had hitherto regarded the private correspondence of the two Emperors as harmless, was thus suddenly forced to recognise its highly dangerous nature. Shocked by this revelation, he noted on 5 July that 'this correspondence has been kept absolutely secret *until now*. I am unfortunately not yet in a position to be able to inform the Reich Chancellor about it because I am under an obligation to the Kaiser to keep quiet. His Majesty *must and shall* release me from this obligation.'[81] A week later, after hearing about the latest letter to 'Nicky', he wrote in alarm: 'His Majesty has thus *committed himself – without Hohenlohe*. This gives me yet more problems to solve, which fill me with *dread*! Nicky will of course *not* keep quiet and *Alexei the Frenchman* even less. If Hohenlohe hears of the letter from someone *other than me*, he will go at once. And yet he *must* know about it!'[82] But worse was to come.

On 30 April 1895, while staying with his old friend Count Emil von Görtz at Schlitz in the Vogelsberg, Wilhelm II sketched the picture entitled 'Against the Yellow Peril', on which he wrote the caption: 'Nations of Europe, protect your holiest possessions!'[83] Philipp Eulenburg was delighted when he saw Hermann Knackfuss's finished version of it at Rominten. He wrote to Kaiserin Auguste Viktoria on 29 September: 'I am quite under the spell of this truly magnificent work. The idea is uplifting and the execution masterly. When I immerse myself in the contemplation of the picture I am filled with the feeling that all Europe must indeed answer the call of the beloved Kaiser to unite in peaceful harmony for the sake of the cross and the holiest goods, but then the uneasy feeling steals over me that the evil in those men who oppose as an enemy power the good which manifests itself in the Kaiser's being will now cause them to tear down and attack with their ingenious criticism that which has sprung from the high-mindedness and noble heart of the Kaiser.'[84] To his mother Eulenburg wrote enthusiastically: 'The Kaiser has given me a magnificent engraving of the *wonderful* allegorical picture executed by Prof. Knackfuss from His Majesty's sketch: the peoples of Europe, represented as female figures, are called upon by St Michael to defend *the cross* against unbelief, heathenism etc. You will like it . . . It is a beautiful idea in a beautiful form.'[85] The Kaiser sent this drawing to the Tsar from Rominten at the end of September 1895. In his accompanying letter he explained that he had had the idea for the picture during the joint German–Russian action in the Far East in the spring, when he had realised the 'danger to Europe and our Christian Faith' emanating from that part of the world. It showed 'the powers of Europe represented by their respective Genii called together by the Arch-Angel Michael, – sent from Heaven, – to *unite* in resisting the inroad of Buddhism, heathenism and barbarism for the Defence of the Cross. Stress is especially laid on the *united* resistance of *all* European powers, which is just as necessary also against our common internal foes, anarchism, republicanism, nihilism.'[86]

While the Kaiser was hunting at Rominten news arrived of a plan to reinforce the French army with Algerian troops, who were to be stationed on the eastern border of France. This roused Wilhelm II to declare that such a measure 'was in fact nothing less than a threat of war'. He considered it his duty to take counter-measures, for one could not 'watch motionlessly as hostile neighbours tighten the noose more and more'.[87] His letter to Nicholas II also referred to this question. The decision of the French parliament to recall an army corps from Algeria and Tunisia and 'to form a *new* continental Corps on *my Western Frontier*' had 'fallen like a thunderbolt on Germany and has created a deep feeling of alarm', he claimed. Together with the existing French forces, which were already overwhelmingly strong, the new corps represented 'a serious danger to my country'. The honours heaped on the Russian Foreign Minister Lobanov and

General Dragomirov by France at the same time had, furthermore, awakened the 'ugly' impression in the German people that 'Russia would like France to be offensive against Germany' with the hope of Russian help. 'Such a serious danger will cause me to strongly increase my army, to be able to cope with such fearful odds', Wilhelm II wrote to the Tsar. 'I perfectly know that you personally do not dream of attacking us, but still you cannot be astonished that the European Powers get alarmed how the presence of your officers and high officials in *official way* [sic] in France fans the inflammable Frenchman into a white heated passion and strengthens the cause of Chauvinism and Revanche! God knows that I have done all in my power to preserve the European Peace, but if France goes on openly or secretly encouraged like this to violate all rules of international courtesy and Peace in peace times, one fine day, my dearest Nicky, you will find yourself nolens volens suddenly embroiled in the most horrible of wars Europe ever saw! Which will by the masses and by history perhaps be fixed on you as the cause of it.' The Tsar, Wilhelm continued, was young and inexperienced; moreover, because he was in mourning he was living in seclusion and therefore did not know what was going on behind the scenes. 'I have some experience of Politics, and see certain unmistakable symptoms, so I hasten to you, my friend, to plead in the name of the Peace of Europe; if you are allied for better, for worse with the French, well then, keep those damned rascals in order and make them sit still, if not, then don't let your Men who go to France make the French believe that you are allied and get reckless and turn their heads till they lose them, and we have to fight in Europe instead [of] for it against the East! Think of the awful responsibility for the shocking bloodshed!'[88] Philipp Eulenburg, who was at Rominten with the Kaiser, had great difficulty in copying this explosive letter for the Reich Chancellor. 'I am a poor Englishman; I had to copy the letter in a great hurry and could not read all of the Kaiser's difficult handwriting', he admitted; Hohenlohe would therefore have to 'work it out' if something were not clear. The copy was intended for Hohenlohe alone and should not be made known to the Foreign Office. 'The fact of the letter being sent . . . will of course displease Holstein very much', Eulenburg predicted, anxious to have time to consider the best way of explaining the Kaiser's action to the Privy Councillor. Not that the 'vigorous language' of the letter gave the Kaiser's friend and ambassador cause for concern. 'His Majesty was angered by the news of the reinforcement of the French army, and as the famous picture was about to be sent to the Tsar anyway, the Kaiser enclosed the political admonition with it.' Furthermore the reinforcement would probably not take place, and 'as a result the Kaiser's present anger will probably also disappear'.[89]

It is not without irony that the man whom the Kaiser chose for the task of delivering this letter and the drawing to St Petersburg was his Flügeladjutant

Helmuth von Moltke, who was later to play a decisive role, as Chief of the General Staff, in the preparation and provocation of the First World War.[90] In the first audience he granted Moltke, the Tsar showed surprise at the news of a forthcoming enlargement of the French army, but he expressed himself fully in agreement with the Flügeladjutant when the latter, on the instructions of 'the most distinguished representative of the monarchical principle' (as Moltke described Kaiser Wilhelm II), drew attention to the danger 'that a possibly victorious France would engender a flood of republican spirit which could be disastrous for the already socially churned-up ground in which the old monarchies were rooted'. Moltke also discussed with other members of the Tsar's family the danger of a Franco-German war and the possibility of collaboration between 'the two Emperors'.[91] Nicholas II received the Colonel again on 3 October at Tsarskoe Selo, where they had a wide-ranging conversation in French on the domestic and foreign political situation. He repeatedly emphasised his peaceful intentions, above all towards Germany. The two countries had not waged war against each other for 150 years, nor did they have any 'conflicting interests' now. Russia was principally an agricultural country and would have to concentrate on internal development for a long time to come; it was dependent on trade with industrialised Germany. He would do everything in his power to keep the French quiet. 'For the time being the French have Madagascar on their plate', the Tsar remarked; and they would be absorbed with that for at least a year. Even after that he would never allow a war. Kaiser Wilhelm had evidently allowed himself to be too provoked by newspaper extracts in the seclusion of his hunting lodge at Rominten, Nicholas observed. He himself read one German (the *Kölnische Zeitung*), one French (*Le Temps*), one English and one Russian newspaper every day, but he did not set much store by newspapers, for he knew 'how they are made. Some Jew or other sits there making it his business to stir up the passions of different peoples against each other and the people, who mostly have no political opinion of their own, are guided by what they read. That is why I shall never set the Russian press free as long as I live. The Russian press shall write only what I want . . . and my will alone shall prevail throughout the country', he said.[92]

When Moltke's telegram arrived at Rominten with the news that Nicholas knew nothing of the French military plans, 'His Majesty was transfixed with astonishment.' The Tsar obviously had '*no idea* of what is going on!', he exclaimed.[93] The Kaiser dismissed Eulenburg's suspicion that the Russian sovereign had been 'play-acting' with Moltke. 'No, . . . that is quite out of the question', he declared. 'The Tsar is absolutely *genuine* by nature. It can only be assumed that people are playing a terribly dangerous game with him.'[94] It was not until a six-page 'schoolboyish' letter from Nicholas arrived at Hubertusstock that the Kaiser discovered, with some relief, that the Tsar had not taken his letter amiss and was willing to continue the confidential correspondence with his fellow monarch.[95] All in all, however, the Tsar's reply amounted to a disappointing rejection of the 'well-meaning suggestions of our All-gracious sovereign', as Holstein sarcastically commented.[96] The Privy Councillor confirmed that 'the impression that H.M. had of Franco-Russian intentions, after Colonel Moltke's return, is a very disturbing one'. 'H.M. made remarks', he reported to the German ambassador at St Petersburg, Prince Radolin, 'to the effect that the Tsar did not seem to be informed, that Lobanov might be up to all sorts of tricks'.[97]

On 25 October 1895, again without consulting the Reich Chancellor and the Foreign Office, Wilhelm wrote another – the seventh – highly political letter to Nicholas II. The Tsar's autocratic views on the press were 'exactly the same as mine', he declared; but its influence, although always harmful, mendacious and nonsensical, varied from country to country according to the 'spirit in which the People of the different races are brought up'. 'Your subjects and mine are slower at thought, sober and quieter in their conclusions they draw as [sic] for instance Southerners or the French. The Roman and Gallic races are more easily roused, incensed and more ready to jump to conclusions, and once having flared up are more dangerous to peace than the Teutonic or Russian Race.' It was true that 'every Sovereign is sole master of his countrie's [sic] interests and he shapes his policy accordingly', yet the danger to 'our Principle of Monarchism' represented by the constant visits by Russian princes, grand dukes, statesmen and generals to the French Republic should not be underestimated. In Russia and Germany 'Republicans [were] Revolutionists de natura' and were consequently 'treated – rightly too – as people who must be shot or hung'. The much-fêted visits by the Russian royalties to France thus only caused confusion, the Kaiser argued. 'The R[épublique] F[rançaise] is from the source of the great Revolution and propagates and is bound to do so, the ideas of it. Don't forget that [President] Faure . . . sits on the throne of the King and Queen of France "by the Grace of God" whose heads Frenchmen Republicans cut off! The Blood of their Majesties is still on that country! Look at it, has it since then ever been happy or quiet again? Has it not staggered from bloodshed to bloodshed? And in its great moments did it not go from war to war? till it soused all Europe in streams of blood? Till at last it had the Commune over again? Nicky take my word on it the curse of God has stricken that People forever! We Christian Kings and Emperors have only one holy duty imposed on us by Heaven, that is to uphold the Principle "von Gottes Gnaden" ["by the grace of God"], we can have good relations with the R.F. but never be intime with her!' Once again the German Kaiser undertook to protect Russia's back 'from anybody in Europe', in the event of its being 'seriously engaged' in the Far East, and promised, if that were the case, to ensure 'that all be kept quiet, and that nothing would happen from me also to France, provided I was not attacked'.[98]

THE BALKANS AND A BLANK CHEQUE FOR AUSTRIA

At the same time as making these repeated attempts to divert Russia's attention to the east, Kaiser Wilhelm II – again on his own initiative and in direct imitation of the mission to Brest-Litovsk which he had carried out in September 1886 on Bismarck's instructions[99] – embarked on the risky undertaking of encouraging the Russians to annex Constantinople and capture the Dardanelles. In his first conversation with Hohenlohe on foreign policy, on 2 November 1894, he made it clear that in the event of a Russian advance he intended to remain strictly neutral on this tricky question and to make his position known both in St Petersburg and in Vienna. He told the newly appointed Reich Chancellor that in Austria the possibility had been mentioned to him that Russia intended to force a passage through the Dardanelles, and that a wish had been expressed that Germany should take a stand on this. 'H.M. refused to do so, as it was his opinion . . . that we had no interest in risking the life a single grenadier for that', the Chancellor recorded. 'Russia needed the key to the door for her navy and this need not matter to us.'[100] Without the prior knowledge of the Reich Chancellor or the Foreign Office the Kaiser solemnly assured the Russian leadership, both through the Grand Duke of Mecklenburg-Schwerin and through his own brother Prince Heinrich, who went to St Petersburg for the funeral of Alexander III, that he was ready '*to come to an understanding with Russia and attached no importance to the Straits from the German point of view*'. To Geheimrat von Holstein it was quite clear that the official German policy would now have to follow the same lines as this imperial initiative. 'The question of whether His Majesty would have done better to wait until he was approached on the matter is no longer relevant', he wrote in December, when he heard of the Kaiser's action. 'It has happened and is presumably being discussed at all the All-Highest tea tables in St Petersburg, Copenhagen, London etc. In order that Germany should not appear frivolous or deceitful we must adapt all our official comments so as not to contradict what the Kaiser said.' Even Philipp Eulenburg, who liked to console himself with the thought that his 'beloved Kaiser' might perhaps sometimes be 'too precipitate in political action, but never does anything really foolish – indeed usually does something of genius', thought it 'bad' that Russia 'received by word of mouth a secret assurance from His Majesty that he had no objection to the capture of C[onstantinople]'.[101]

Wilhelm took a similarly pro-Russian and anti-Austrian line towards another potential source of conflict in the Balkans which threatened to explode at any moment: Bulgaria. In defiance of the official policy pursued by the Wilhelmstrasse, in the autumn of 1894 he instructed the military attaché at Vienna, Colonel von Deines, to make it clear to the Austrian Chief of the General Staff, Feldzeugmeister Beck, 'that in the event of complications concerning Bulgaria we should not feel obliged to intervene actively on Austria's behalf'. According to Waldersee, Deines carried out this order, but when Foreign Minister Kálnoky checked up on the matter with the Foreign Office in Berlin he was told that the military attaché was interfering in politics in an unauthorised manner. 'I cannot wait to hear how the Kaiser will react to this', the General exclaimed. 'What is strange is that last year in Güns he assured Feldzeugm. Beck of exactly the opposite.'[102] When Wilhelm outlined his ideas on foreign policy to the new Chancellor three weeks later he commented, with regard to the growing tension between Bulgaria and Russia, that it was 'still very questionable whether Bulgaria will accept Russian sovereignty. Austria could take what she wanted for herself. He spoke disparagingly of Kálnoky.'[103] During the Scandinavian cruise in July 1895 Wilhelm cherished hopes of 'a completely passive attitude on the part of Austria' in the Bulgarian question; the assassination of the former Prime Minister Stefan Stambulov, in his opinion, was to be blamed on the behaviour of the government of Prince Ferdinand of Saxe-Coburg and Gotha and ought not to lead to Austrian intervention.[104]

In the autumn of 1895, however, Wilhelm II's attitude towards Russia and Austria-Hungary with regard to Balkan policy suddenly underwent a momentous change. In October the Kaiser was still urging a policy of expansion both in East Asia and in the Near East on the Russian leadership and assuring them of the benevolent neutrality of Germany.[105] And when Eulenburg reported from Vienna in early November 1895 that Kálnoky's successor Count Goluchowski had said that the Straits question concerned Germany 'just as much' as Austria, Wilhelm wrote dismissively on the report: 'ahem! in simple terms: we should help prevent the Russians getting into Stamboul. I shall not risk the life of a Pomeranian grenadier any more for Stamboul than for London.'[106] In a three-sided conversation with Philipp Eulenburg and the Austrian ambassador Count Ladislaus von Szögyény on 6 November 1895 the Kaiser discussed the handling of the crisis which would be caused by the apparently imminent break-up of Turkey. With a map of the Balkan peninsula spread out in front of him he strongly urged Szögyény to persuade Austria to make its claims for compensation plain. As the first step towards dividing up the Ottoman Empire would probably come from England, he said, it would 'perhaps be opportune for Austria, Germany and Russia – the three Emperors – to come to an understanding on what was to be done'. The Austrian ambassador expressed reservations about an expansion of Austria too far south because of the proximity of Greece, of which the Kaiser spoke 'contemptuously'. Szögyény reacted most touchily when Wilhelm floated the idea that in the division Italy could be given Albania. Eulenburg summed up the impressions that the Austrian ambassador must have gained from this meeting in these terms: 'Kaiser W. is still very irritated with England and thinks her capable of every possible political enormity, but he will not go against the Berlin Treaty. The Kaiser will give us (Austrians) strong and emphatic support if it should come to a division of Turkey. If we attack or allow ourselves to be *driven* into a hostile action by England we cannot count on his help. The Kaiser thinks the situation in the Balkans is very serious, almost hopeless, and he shows a tendency to support Russia – although only if Austria comes away with the equivalent of what Russia gains.'[107]

Friedrich von Holstein was horrified when he saw Eulenburg's report on the conversation, for the Kaiser's remarks to Szögyény showed that Wilhelm was no longer prepared to defend Austria-Hungary's status as a great power, and moreover that he assumed that the Franco-Russian alliance was a dead letter and had been, or could be, replaced by a Russian–German combination. 'If the Austrians take H.M.'s utterances *au pied de la lettre* it will mean the end of the Triple Alliance, because it will be *meaningless* for Austria', he warned. 'But I doubt that H.M.'s utterances will be taken literally. Their credibility is belied by the obstinacy with which H.M. treats the Franco-Russian alliance, which is very much in existence, as a *quantité négligeable* – a view probably shared by no one else today.'[108] The fact that the imperial endorsement 'Agreed' then appeared on a memorandum by Marschall, in which he had put forward exactly the opposite point of view on the Balkan question, added to the consternation in the Foreign Office. 'This is another surprise. If the Kaiser approves the views of the Foreign Office – *who* is then responsible for the confusion? Phil. Eulenburg himself?? Or both?' In despair, Holstein urged the elderly Chancellor to ask the Kaiser at his next audience for permission to write to Eulenburg saying that he, the Kaiser, considered it premature to discuss 'future eventualities in the Balkans' at the present time. In spite of the sleepless nights the Kaiser's interferences had already caused him, the Privy Councillor vowed to do his utmost to prevent 'poets [meaning Eulenburg] and dilettantes [Wilhelm II] from destroying the Triple Alliance while the Franco-Russian liaison remains in force'.[109]

When the Kaiser appeared at the Reich Chancellor's palace on 13 November 1895 for the customary audience, Hohenlohe and Marschall felt obliged to ask him the fundamental question of whether the government's conduct of German foreign policy was 'in harmony with the intentions of H.M.' or not. Using Holstein's exact words, the Reich Chancellor asserted that the preservation of Austria-Hungary's status as a great power had been 'one of the foremost, if not the foremost, aim of our foreign policy for decades'. If Germany were now to declare that it would not under any circumstances take part in a war over the Straits, it might be taken to mean that the policy it had hitherto pursued towards Austria was being abandoned. Such a shift would be all the more risky because in Austria too there were political and ethnic groups who wished for co-operation with Russia rather than with Germany and Italy and who saw 'an alliance between Russia, Austria and France as an opportunity to destroy Hungary and recover the position Austria had lost in Germany (the restoration of the status quo before 1866)'. These parties were not in the majority in Austria, but they would gain increased support if it were widely believed that the Triple Alliance were unwilling 'to take steps at the critical moment to support Austria's position as a great power'. The danger of such a shift in German policy was also great because Germany's relations with Russia still left much to be desired, as had been evinced not only by the 'rather cold' reserve of the Tsar but also by 'the inconsiderate behaviour of the Russian Master of Ceremonies with regard to the rank to be accorded to Prince Heinrich at the coronation'. From this it could be deduced that Russia 'values her French entente more highly than intimacy with us'. Germany would therefore be wrong 'to risk our other friendly relationships for the sake of an uncertain Russian friendship'. At the end of the audience the Kaiser declared himself in agreement with the arguments put forward by Hohenlohe and Marschall.[110]

Wilhelm now swung violently against Russia, not only for the objective reasons presented by his responsible advisers but also because his advances to St Petersburg received a snub. He was especially deeply wounded by the rejection of the suggestion he had made by telegram on 8 November that Russia and Germany should make common cause in the Turkish question.[111] 'H.M. is beginning to be quite angry with the Tsar because of the repeated cool rebuffs', Holstein reported to the German ambassador in Russia. 'H.M. would like to restore the Holy Alliance, but Lobanov, who runs the Tsar, won't desert France . . . Our Kaiser was very irritated for instance when in reply to his telegraphic proposal for an agreement *à deux* about the present Turkish difficulties, the Tsar referred him very briefly and coolly to the ambassadors of the powers in Constantinople.' Radolin should say nothing in St Petersburg about Wilhelm's annoyance, Holstein advised. 'H.M. will do enough of that himself. He treated Grand Duke Vladimir miserably, always answered in German to his French, etc.'[112]

Anger with Russia was not the only reason behind the belligerent mood of Wilhelm II and his closest advisers at this time. Another was the hope that complications abroad could help solve the deepening political crisis at home. Philipp Eulenburg expressed the view to Bülow that the serious international situation in the east had 'a refreshing, distracting character' which made him hope, 'for the sake of the dear Fatherland', 'that the situation there would not be cleared up too quickly'.[113] He admitted to Holstein that 'in general my wish is that things will not calm down in the Orient yet. Not that it should go as far as war, into which we should be drawn by Austria, but that there should be a certain unrest to distract people at home. Occupying minds with a little gentle excitement is better than the most splendid government bills and the finest leading articles in the *Nord[deutsche] Z[eitun]g*.'[114] Waldersee was ready with similar advice for the monarch: 'in my opinion it [is] in our interest to let others quarrel'.[115] After talking to members of the imperial entourage the General was hopeful at the close of the year of 1895 that plans were afoot 'to complicate the situation still further'. He reaffirmed his opinion that 'foreign policy can have a good influence on the situation at home'; it was unfortunately 'an unhealthy sign that we cannot help ourselves from within'.[116] After a conversation with Wilhelm II he recorded with satisfaction in his diary that the Kaiser had declared 'there were many indications that things would get serious in the spring'.[117]

It was in this warlike, anti-Russian frame of mind that Kaiser Wilhelm II summoned the Austrian ambassador to see him late in the evening of 13 November 1895 at the station, in order to explain 'the ultimate goals of German policy . . . on matters affecting the East' to him. According to Szögyény the Kaiser emphasised that he was in complete agreement with the Reich Chancellor and Baron Marschall in making the following 'highly important remarks': 'The survival of Turkish rule in Europe seems to me very doubtful, as you know; nevertheless, in the present circumstances I consider the preservation of the status quo in the European Orient as the most urgent task facing all the powers at this time.' Britain, Austria-Hungary and Italy were those primarily involved; Germany's role was only secondary, 'but she is very firmly involved. I am perfectly convinced that Austria-Hungary will not allow herself to be driven to any acts of provocation; but if complications should nevertheless arise, as a consequence of which Austria-Hungary saw her position as threatened – then it would be for Austria-Hungary, not for us, to decide whether this were actually the case – and in this case I declare quite plainly that I shall stand at Austria-Hungary's side with all the forces at my disposal, without any further enquiry as to whether the *casus foederis* existed in accordance with our treaty of alliance . . . Your All-Highest sovereign may be quite sure that if at any moment the position of the Austro-Hungarian monarchy is at issue, as to which the decisive judgement rests solely and entirely with Him and His All-Highest advisers, my entire fighting forces will be immediately and unconditionally at his disposal.'[118]

Would it be going too far to see this 'quite open' definition of the official German attitude on the Balkan question as anticipating the fatal 'blank cheque' which Kaiser Wilhelm offered the same ambassador, Szögyény, on 5 July 1914 in reply to a letter from Kaiser Franz Joseph, and which demonstrably strengthened the will of the government in Vienna to go to war? Did the Kaiser remember his declaration of 13 November 1895 in later years, and did the government of Austria-Hungary regard the Kaiser's promise as a considered expression of official German government policy or only as an eccentric initiative on the part of the monarch? Thirteen years after the late-night meeting at the station, at the height of the crisis over the annexation of Bosnia, Wilhelm II still remembered perfectly clearly the fateful, solemn assurance which he had given Szögyény to transmit to his emperor in November 1895. At the reception held at the Berlin opera house on 21 October 1908 the Kaiser reminded the Austrian ambassador that in 1895 he had declared 'in the most categorical way' 'that he did not regard loyalty to the alliance simply as a matter of keeping to the letter of our treaty, but that he would stand loyally at our [Austria-Hungary's] side on every issue, whether great or small'. He had even said, Wilhelm II went on, that 'Kaiser Franz Joseph was a Prussian field marshal, and consequently he had only to command and the whole Prussian army would obey his orders.'[119]

Count Szögyény did not have to take the Kaiser's word that he was speaking also in the name of the Reich Chancellor and of the Foreign Secretary. On 14 November 1895 Hohenlohe informed him that he was 'in complete agreement' with the Kaiser Wilhelm's declaration.[120] It was no wonder that Goluchowski regarded the declaration as the 'most important enunciation' on the part of Germany 'since the conclusion of the Austro-German alliance', and commented that 'its warmth and resolve put all previous statements in the shade'.[121] Unlike his later successor as Foreign Minister, Count Berchtold, Goluchowski will have had no doubts as to who ruled in Berlin.

CHAPTER TWENTY-SIX

Great Britain and the spectre of encirclement

THE KAISER AND THE BRITISH

THE harder line which German foreign policy was now following and which was generally attributed to the growing influence of Wilhelm II naturally caused disquiet in London, where the weak Liberal ministry under Lord Rosebery lost the general election and Lord Salisbury returned to Downing Street in July 1895 as prime minister and foreign secretary. The dangerous interplay between the authorities in Berlin and the increasingly nationalistic German press was viewed with particular apprehension by British observers. Martin Gosselin, the perceptive chargé d'affaires at the embassy in Berlin, commented in a dispatch of 5 November 1894 that the semi-official press, the Kaiser and the Reich government were now all noticeably more hostile in their attitude towards Britain. He reported that the *Kölnische Zeitung*, which was close to the Foreign Office, had threatened that Germany would create difficulties for England if the empire continued to impede German colonial expansion, while Marschall, the Foreign Secretary, had recently remarked 'that so long as England disputes every scrap of territory with Germany he does not see how the relations of the two countries can improve. The Emperor is keenly in favour of German Colonial development and the new Govt. is believed to be [in favour] of a more forward policy than the last.'[1]

It was not only Germany's new colonial appetite but also Wilhelm II's attempted courtship of Russia, German collaboration in the Far East with Russia and France against Japan, the sudden incursions of German foreign policy into the highly sensitive Dardanelles question and, not least, the talk in Berlin of a continental league against Britain, which for the first time brought the government in London face to face with the urgent problem of whether the security of the island realm and of its global interests could continue to be safeguarded by a policy of 'splendid isolation'. And, if not, the question arose of whether, in abandoning its non-aligned position, the country might not achieve better protection by joining with France and Russia than with the Triple Alliance. By 1895 – several years before the alliance between the United Kingdom and Japan (1902) and the ententes with France (1904) and Russia (1907) – the international isolation of the German Reich was already beginning to emerge with alarming clarity. Towards the end of 1894 the long-serving German ambassador in London, Paul Hatzfeldt, warned that a shift in British public feeling towards the Triple Alliance would have consequences which no ministry could prevent. Britain was already 'casting about elsewhere', blaming this 'on our alienation of herself', and the result might be 'the formation of a group that would be extraordinarily dangerous' both for Germany and for Austria-Hungary.[2] At the same time the French ambassador in Berlin, Jules Herbette, who could scarcely conceal his delight at the antagonism between Germany and Britain, commented in conversations with Marschall on the likelihood that a triple alliance between France, Russia and Great Britain would be stronger than that between Germany, Austria and Italy.[3] Early in 1895 Waldersee, filled with dark forebodings of disaster, voiced the fear that the German collaboration with France and Russia in East Asia which the Kaiser had ordered might anger Britain and might even result in the encirclement of the Reich by all three great powers. 'If we make a determined enemy of England', he predicted, 'the danger is that Russia and France will then disown us.'[4] Presciently, he expressed anxiety that Britain had 'taken our action very much amiss', for although this was scarcely perceptible 'under Rosebery's slack and apparently moribund Ministry', 'we should soon feel it if Salisbury took the helm'.[5]

The conflict of interest between Germany and Britain in the eastern Mediterranean, repeatedly exacerbated by Kaiser Wilhelm's personal diplomacy, had also put the previously harmonious relationship between the two countries under visible strain since the end of 1894. In a 'highly confidential' conversation with the military attaché Colonel Swaine in November the Kaiser brought up the subject of the Straits and tried to find out, by dropping what he thought was a 'harmless' question into the conversation, whether the understanding reached between the British government and Russia 'concerned the Dardanelles as well as India'. Swaine's reply clearly indicated, according to Wilhelm, 'that essentially that was the case, and that the question of opening up [the Straits] is already being seriously considered'. 'At the end of the conversation I remarked *en passant*', Wilhelm continued, 'that if England was really absolutely serious about wanting to ensure peace and be good neighbours with Russia in Asia by opening up the Dardanelles, I hoped she would inform the other powers of it in good time and not spring a surprise à la Congo Treaty on Europe; that would ill befit John Bull. The Colonel entirely shared this view and promised to take the hint, if he should hear any more about the question.' To this the Kaiser appended the instruction: 'Hatzfeldt to be informed of this at once. Petersburg, Rome, Stamboul, Vienna to be informed.'[6]

Rosebery's replacement by the Marquess of Salisbury in July 1895, which the Kaiser at first enthusiastically welcomed,[7] most certainly did not lead to a more trusting relationship between Wilhelm and the British government. After a dinner on 5 August 1895 'with Queenie', as Kiderlen disrespectfully called her, the Kaiser held a long conversation at Cowes with Salisbury about the partition of the Ottoman Empire as the most effective solution of the Near Eastern question.[8] Hatzfeldt was relieved to be able to assure Holstein that this time Wilhelm had not allowed himself to be 'carried away into anything dangerous'. 'The most important part seems to have been that H.M. defended the idea that Turkey wasn't going to collapse as quickly as Salisbury thought.'[9] Wilhelm told the Reich Chancellor, however, that when the Ottoman Empire came to be divided Salisbury evidently wanted to give the Italians Tripoli, Morocco 'and all kinds of things', and to let the Austrians have Salonika and the Russians Constantinople.[10] After the interview the Kaiser 'quite unexpectedly' gave Hatzfeldt what amounted to an order 'to write to him directly and personally, if anything important happened here [in England] with regard to the Near East and a possible partition'.[11]

The Kaiser remained deeply suspicious of Salisbury's Near Eastern policy over the next few months. When news of more Armenian massacres came through in October 1895 – Wilhelm indignantly complained that the role played by the European great powers vis-à-vis the Sultan was 'more than pathetic' and that they would do better to fire 'a cannon-ball into Yildiz'[12] – he suspected that Britain might use the shedding of 'Christian blood' in Turkey as a pretext to allow Russian ships to sail through the Dardanelles without consulting Germany and the other signatories to the Berlin Treaty.[13] When the Empress Frederick also urged her son to act against the 'Turkish business',[14] Wilhelm sent the Chancellor a telegram demanding vigorous German intervention in the Near Eastern question, contrary to the course being followed by the Foreign Office. 'If Russia and England come to an understanding over Egypt and the Dardanelles so that one gets the former and the other the latter, where does that leave us?', he demanded impatiently, urging that measures be taken to prevent an Anglo-Russian agreement.[15] Holstein, schooled in Bismarckian ways, pointed out how advantageous such an accord between Great Britain and Russia would be for Germany, for it would create a deep gulf between Russia and France. 'So why should *we* get involved in trying to prevent it?', he asked. 'Whether it becomes a reality or not, this plan holds no dangers for us. But there is *serious* danger for Germany in H.M.'s intended anti-English attitude and mistrustful opposition of England, which, in view of the apparent *Russian-French-German* agreement on Mediterranean questions, might so soften up England and Italy that the former would give up Egypt and the Mediterranean, and the latter give up the Triple Alliance.'[16] The Privy Councillor's arguments went unheeded. A few days later the Kaiser again called on Colonel Swaine and complained that Salisbury's strange attitude since his return to power was creating the suspicion everywhere that Britain intended to change its Mediterranean policy. The universal view, he now claimed, was that Britain wanted 'to give Constantinople to Russia, to win France round by concessions in Egypt and to take the Dardanelles herself'. 'I looked the Colonel straight in the eye as I said this', the Kaiser recorded; 'he was visibly startled.' Neither Germany nor the other signatory powers would tolerate such a breach of the Berlin Treaty, Wilhelm warned the dismayed military attaché.[17]

Wilhelm seems not to have shared the fears of Holstein, Waldersee and others of an imminent encirclement of Germany brought about by collaboration between Britain, France and Russia. One of the ideas which dominated his thinking at this time was that Britain was facing an international crisis and could be forced by German pressure to associate itself formally with the Triple Alliance. It was in this expectation that in the spring of 1895, when news arrived that the Russian Mediterranean squadron had suddenly departed for the Far East, he commented jubilantly: 'I hear England is alarmed at this. Which is quite good for us. Rosebery's famous entente cordiale with Russia does not seem very sound. And in the end England will return penitently into the arms of the Triple Alliance.'[18] A long private letter which he received in September 1895 from Sir Edward Sullivan, a rich Englishman with whom he had spoken on various occasions at Cowes, made a strong impression on Wilhelm. Following up on their discussions, Sullivan's letter contained observations on the state of the British army and navy and on British policy which the Kaiser found extremely interesting. Philipp Eulenburg, who was staying with him at Rominten when the letter arrived, recorded that the information supplied by the Englishman amounted to 'a damning criticism of the present situation and leads to the conclusion that England's total isolation will force her to seek support in the form of an alliance, for the days when England made others do her work for her are over. England's natural allies are the members of the Triple Alliance, especially Germany. Sullivan's portrayal of how England would react to a French invasion is also of great interest. He comes to the sad conclusion, for England, that it would be an easy matter for France to catch England off her guard. The French have a better fleet than the English, and with combined action by the Russians the English fleet could be defeated in the Mediterranean, or cornered so effectively that practically nothing would stand in the way of a French landing at the unfortified ports. Very shortly afterwards France could dictate peace terms to London.' Eulenburg went on to record that 'after reading this letter the Kaiser embarked on a long discussion of English policy and the state of affairs there. He told me that He had always made it very plain, not to the Queen and her ministers but to persons who He knew would pass on His remarks, that England can only expect support from the Triple Alliance, should she decide to ally herself formally with it. Without written, binding agreements she had no hope . . . The gloomy picture that Sullivan sketched of England's powerlessness brought the thought of England's downfall vividly before His Majesty's mind's eye, and in view of the common factors of race and religion, and of many and various other important links, His Majesty came to the conclusion that in every respect the consequences of such an eventuality could be incalculably great for us, and that it required the most serious consideration.'[19] To put the future Foreign Secretary and Reich Chancellor Bülow into the picture, Philipp Eulenburg gave him the comforting news that 'His Majesty would be quite happy for England to be "shot and wounded" – but he does *not* want England's downfall. That is absolutely certain.'[20]

The threatening tone of German policy – above all the German expectation that Great Britain would bind itself by treaty to the Triple Alliance against France and Russia, together with the Reich's excessive expansionist claims all over the world – was attributed by the British government principally to Wilhelm II himself, whom Lord Salisbury, on the basis of recent reports from Berlin according to which the Kaiser suffered from hallucinations, regarded as mentally unsound.[21] Indeed, recalling John Erichsen's strong warnings in the spring of 1888,[22] he even feared that Queen Victoria's grandson might go completely out of his mind. According to a memorandum by the new British ambassador in Berlin, Sir Frank Lascelles, the Prime Minister commented despondently to him on 4 December 1895 that 'The conduct of the German Emperor is very mysterious and difficult to explain. There is a danger of his going completely off his head . . . In commercial and colonial matters Germany was most disagreeable. Her demand for the left bank of the Volta was outrageous, so much so that Lord Salisbury thought it must have been the idea of the Emperor himself as no responsible statesman could have put it forward. The rudeness of German communications, much increased since Bismarck's time, was perhaps due to the wish of smaller men to keep up the traditions of the great Chancellor . . . In the Far East, the Germans are up to every sort of intrigue, asking for concessions & privileges of all sorts, with a view to cutting us out. The only way of meeting them is by countermining, & we are in a position to do so.'[23] The spectre of encirclement was slowly but surely taking shape.

BRITISH INSULTS

Wilhelm II's *Schadenfreude* at the possible 'downfall of England' was in part a reaction to various insulting rebuffs which he had received at the hands of his royal relations, the British government and the London press and which contributed in no small measure to a completely unnecessary exacerbation of the conflict of interests which did in fact exist between the two countries.[24] A typical example of the kind of wounding snub delivered to him by the British royal family occurred in connection with the inauguration of the Kiel Canal in June 1895. Wilhelm was almost childish in his enthusiasm to show off the great maritime achievement represented by the North Sea–Baltic canal for the benefit of his English relations. In a telegram to his grandmother he expressed confidence that the canal would further 'the union of nations' and contribute to the 'peaceful development of their wealth'. He thanked her for sending Prince George and the entire Channel Squadron to take part in the celebrations and informed her that on 20 June, the anniversary of her accession, the German fleet had joined with the British fleet in firing a salute.[25] Again on 12 July Wilhelm wrote enthusiastically to the Queen from Stockholm about the successful co-operation between the German and British fleets at Kiel. 'Let me thank You sincerely once more for Your having kindly sent Georgy to represent You at Kiel; & for the appearing of the fine Channel Squadron under the Flag of Lord W. Kerr. The sight of the whole assembled ships was really worth seeing & as the neucleus [sic] of the gathering the sight of the Channel Squadron in its uniformity of Type was most imposing . . . The relations between our officers were cordial & intimate & proved the ties of esteem & friendship excisting [sic] between our Navies.'[26] In reality such assurances of friendship bordered on hypocrisy, for the participation of the Duke of York, the future King George V, in the inauguration ceremony had been preceded by almost unbelievably insulting behaviour on the part of the Prince of Wales and his son. When Swaine suggested that in view of the fact that the Kaiser was an honorary admiral of the Royal Navy and that Austria and Italy were to be represented at Kiel by members of their royal families, a British prince should be sent to the ceremony,[27] the Prince of Wales asked in all seriousness for enquiries to be made as to whether the date of the inauguration of the canal could be postponed, as otherwise the ceremony would clash with Ascot Week.[28] Prince George's attitude to his German cousin's invitation was scarcely less insulting. Before his departure for Hamburg the young duke had informed the British ambassador in Berlin that he would be embarrassed if the Kaiser conferred an honorary rank in the Imperial Navy on him; he would be glad to accept any other German distinction but it was not possible for him to wear German naval uniform. When the Kaiser instructed the ambassador to be asked the reason for this astonishing statement the answer conveyed to Wilhelm by Hohenlohe was that 'the responsibility for this attitude undoubtedly lay with the Duke's mother, who as Your Majesty knows is very anti-German and obstinate, and to whom the Duke probably gave his promise on the matter'.[29] Even during Wilhelm's visit to Cowes in August 1895, to which he had looked forward like a child,[30] the Prince of Wales distinguished himself by his gross discourtesy, according to the imperial entourage. Kiderlen-Wächter reported to Berlin: 'Fat old Wales has again been inconceivably rude to H.M. On our arrival he let H.M. wait three-quarters of an hour before he came on board. When H.M. was talking with Salisb[ury] the day before yesterday, Wales came up twice to interrupt the conversation. The first time he completely failed, but the second time he succeeded in separating them.'[31] Waldersee, who was astonishingly accurately informed about events in Britain, heard not only that the Prince of Wales had made fun of the Kaiser 'in the most brazen way', but also that the Queen's third son Arthur, Duke of Connaught, had refused an invitation from the Kaiser to attend manoeuvres on the grounds that he had something else to do. 'They evidently think over there that they can cope with the Kaiser better if they treat him badly; they are well aware that he is impressed by English life, English wealth and luxury and can easily be fobbed off with the multitude of pleasures offered to him. I have been able to watch for years how he, who had nothing but mockery for England as Prince Wilhelm, has gradually become an anglomane', Waldersee noted regretfully. His own anglophobia, unlike that of the Kaiser, was not inclined to swing repeatedly over into anglomania.[32]

There were other reasons why Wilhelm's extended stay on the Isle of Wight in unremittingly rainy weather was not an unalloyed pleasure either for him or for his hosts.[33] As a result of new competition rules the Kaiser lost the races both for the Queen's Cup, which he coveted, and for the Meteor Challenge Shield, to his rival, his Uncle Bertie. The monarch had brought a thirty-strong military band with him, which played at every banquet, whether on the terrace in front of Osborne House or at the Royal Yacht Squadron in Cowes. Kiderlen, the Swabian diplomat known for his biting mockery, reported sardonically from Cowes that the Kaiser had given the British 'a special treat by bringing along a fleet of four battleships and a dispatch boat. They block the course of the racing vessels, every few moments they get an attack of *salutirium*, the sailors are flooding Cowes, the Queen has to invite the commanders, etc.!'[34]

The relationship between Kaiser Wilhelm and his grandmother, on the other hand, took a decidedly favourable turn when they agreed together that General Viscount Wolseley should be appointed to succeed the outgoing British ambassador in Berlin, Sir Edward Malet, who had asked to be relieved of his post after eleven stressful years.[35] It is astonishing to see the extent to which the Kaiser felt entitled to choose the new ambassador himself, in agreement with his grandmother. As we shall see, his restless activity on behalf of Wolseley cast a cloud over his relationship with Salisbury's government. The Kaiser turned down the Queen's first suggestion that Lord Cromer should go to Berlin, but he was positively enthusiastic when she brought up Wolseley's name. He said that he wanted a soldier and would welcome Wolseley's appointment with the greatest pleasure.[36] When Philip Eulenburg heard of the Kaiser's efforts to obtain Wolseley's appointment he groaned: 'A hero! A general! An Englishman! – who can withstand that!' The Russian ambassador in Berlin would not have a chance against such a rival.[37] The Kaiser's hidden motive, however, was probably quite different. Wolseley was also regarded at this time as the most likely candidate to take over the supreme command of the British army. His appointment as ambassador in Berlin would therefore leave the way open for the Kaiser's uncle, Arthur Duke of Connaught, to be appointed commander-in-chief, and this, or so Wilhelm believed, would bring potential advantages for Germany, since Connaught was married to a Prussian princess.

The trouble began when the Queen, who had meanwhile discovered that the new Conservative government did not after all want Wolseley as commander-in-chief,[38] sent word to Wilhelm that she would be grateful if he would grant Salisbury a second audience, whereupon at 3.30 p.m. the Kaiser summoned the Prime Minister by telephone to see him at 4 o'clock. Salisbury excused himself, saying that he had to see the Queen at Osborne House at that time. He did not say that he would come to see the Kaiser after his audience with the Queen. The Kaiser then sent him a message saying that he would wait for him on board the *Hohenzollern* until 6.30. This message, however, did not reach the Prime Minister, as he did not return to his rooms after seeing the Queen (or so he maintained later), but went into the park, where he was caught by a downpour until shortly before 8 p.m. The same evening Hatzfeldt wrote to Salisbury to say that he could come and see the Kaiser the next morning, 7 August, but the Prime Minister evaded this invitation too on the grounds that he unfortunately had to return to London very early in the morning.[39] 'William is a little sore at your not coming to see him', the Queen warned her Prime Minister in a cipher telegram of 8 August.[40]

Although Hatzfeldt was convinced that everything had happened as Salisbury claimed,[41] and Kiderlen was relieved that there had been no second conversation with the Kaiser – 'I am quite happy that the talk did not take place', he wrote to Holstein; 'I have more confidence in Hatzf[eldt]'s talks with Sal[isbury] than in those of H.M.!' – they nevertheless racked their brains over the reasons for Salisbury's behaviour. Kiderlen guessed that Queen Victoria's wish to appoint her son Arthur rather than Wolseley as commander-in-chief might well be the root cause. Writing to Holstein, this was his interpretation of the comedy of manners on the Isle of Wight: 'As you know, the Queen spoke with H.M. about Wolseley as the successor of Malet. The main purpose in sending Wolseley is to get rid of him here and to make Connaught commander-in-chief. H.M. of course knows that too and that is why he is so much for Wolseley for he thinks it is "most important" (!) that Connaught be made commander-in-chief . . . Now it is possible that the Queen worked out this little plan with her nephew without first asking Salisb[ury], in the certain belief that he would make no difficulties for her on this point as Rosebery would have done. Now perhaps Salisb. bucked after all – because it is certainly strange that he said nothing about it either to H.M. or to Hatzf[eldt]. In that case it is quite possible that it was for this reason that he did not want to see H.M. again, to avoid having to talk with him about Wols[eley].'[42]

The Queen's correspondence, published in 1931, shows that Kiderlen had hit the nail on the head. On 8 August 1895 she sent a message to Minister of War Lord Lansdowne, asking him to tell Wolseley 'that the Emperor is very anxious to have him at Berlin. H.M. fears Emperor will be greatly disappointed if he does not go.'[43] Wolseley, however, made no attempt to conceal the fact that he would prefer to be commander-in-chief.[44] The Queen was furious to hear that Lansdowne, without her permission, had offered the General both posts and had left the choice to him. 'The Emperor will be grievously disappointed.'[45] Salisbury and Lansdowne had to explain patiently to the Queen that the appointment of her son Arthur as commander-in-chief in succession to her cousin, the Duke of Cambridge, would have had serious political drawbacks.[46]

Such restriction of the sovereign's freedom of choice was exactly what Wilhelm II detested about the British constitutional system. He was mortified at the failure of his plans, telegraphing to Queen Victoria on 12 August: 'Very surprised and deeply grieved at the unexpected turn the matter we wished has taken.'[47] He informed Lord Salisbury through Hatzfeldt that he still wished to have 'a soldier or Grand Seigneur' as British ambassador at Berlin, causing the Prime Minister to protest that such interference from an outside source would rightly anger the diplomatic corps, added to which there were only two great landed proprietors who could even be considered, and in both cases the talent for diplomacy was 'more largely developed in the wife than in the husband'. He proposed at first to transfer the career diplomat Sir Edmund Monson from

Vienna to Berlin and to send Sir Nicholas O'Conor to Vienna as ambassador.[48] Queen Victoria countered by suggesting Lord Londonderry, Lord Jersey or Sir Francis Grenfell for the Berlin post.[49] Finally it was agreed that the ambassador at St Petersburg, Sir Frank Lascelles, should be transferred to Berlin. He seemed to satisfy the social demands of the Kaiser, even though his wife was not of exalted birth.[50] In a letter of 28 August to her grandson, after apologising at length for the confusion that had occurred, the Queen made great play of the excellent qualities possessed by Lascelles. 'Long & anxiously did I go over with Lord Salisbury various names we discussed', she assured Wilhelm; 'and at length we agreed that Sir F. Lascelles, now at St. Petersburg, wd. be the fittest. He is one of our best Diplomatists, a very able man of Lord Harewood's family; his Mother was a Sister of the D[uche]ss of Sutherland, whose Husband was a great friend of your Grandfather and Gt Uncle of Papa's, and my 1st Mistress of the Robes & great friend. He is also 1st Cousin to the Duke of Devonshire. – His wife is a clever agreeable person but she is not good looking, nor a *Grande Dame*. There might have been several *Grandes Dames*, but the husbands wd. not have done. – In this choice I have been most anxious to find someone who wd. be agreeable to you & wd. do all he cd. to maintain the best relations between the 2 Countries.'[51] The Kaiser accepted Lascelles but was nevertheless disappointed. 'H.M. had been looking forward to having a field-marshal with a suitable wife, and is peeved that this is not going to happen', Hatzfeldt reported to Berlin.[52]

The Kaiser suffered yet another painful setback during his visit to England in the summer of 1895, this time from the scathing personal attacks made on him by the press. Their effect on him was deeply and enduringly wounding, especially as the newspapers concerned, most of which had reported favourably on his earlier visits, were those that he regarded as close to the government. He did not understand the extent to which the British press had become a powerful and troublesome entity independent of the government. As Waldersee had predicted, neither Germany's collaboration with Russia and France against Japan nor the Wilhelmstrasse's newly declared interest in Southern Africa did Wilhelm II any favours in political and journalistic circles in London.[53] Most notably, an article in the *Standard* criticised German policy in the Far East and advised Wilhelm to take lessons in political wisdom from his grandmother during his visit to the Isle of Wight. 'The Kaiser has probably never been told the truth in such strong terms', Waldersee declared. He found it incomprehensible that Wilhelm continued to enjoy himself in England instead of leaving at once, for – as the General commented – he must have been very offended by the article, particularly by its insulting schoolmasterish tone, and could not have found it easy 'not to allow himself to be irritated by all these lectures and admonitions'.[54]

And indeed the barb implanted by the *Standard*'s malicious article left a deep wound, as we shall see.

On 10 August the Kaiser and his suite left Cowes amid cheering crowds and set out for Lowther Castle in Westmorland, the seat of Lord Lonsdale, a close friend since 1892, whom he had seen in September 1894 at the manoeuvres in East Prussia and more recently during the Kiel Week regatta, as well as at Cowes.[55] From Lowther, in spite of the bad weather, he went on expeditions to Ullswater and Windermere (clad, unusually for him, in a light-coloured civilian suit, a brown overcoat and a black bowler) with the other guests, who included Karl von Eisendecher and several members of the Churchill family, who were related to Lonsdale. He shot red grouse on Wemmersgill Moor and went out cycling.[56] Documents sent to Lord Lonsdale a year later and forwarded by him to the Chief Constable in Carlisle claimed that when the Kaiser arrived at Lowther a bomb attack on him planned from Switzerland had been foiled only just in time, but further police and diplomatic investigations revealed that it was more likely to have been an unsuccessful blackmail attempt.[57] More than a hundred years later no explanation can be found for this puzzling episode, of which the Kaiser was told nothing. For him, his stay in the rain-swept green hills of the Lake District remained an unforgettable experience. 'With a heavy heart I left Lowther and am unable to find words to thank you and Lonsdale for the kind hospitality shown to myself and my suite', he wrote to his hostess from the *Hohenzollern* on his way home. 'The memory of my stay will ever be most delightful . . . Compliments to all who made Lowther a second Paradise. William I. R.'[58] Staying with the Lonsdales was the only happy experience for Wilhelm during his visit to England in August 1895, from which he had hoped for so much.

This visit undoubtedly marked a turning-point in Anglo-German relations and hence in the international situation as a whole. In London, not only at court and in government circles but also in public opinion, impatience with Germany's conspicuously volatile and unreliable foreign policy, aggravated to no small degree by Kaiser Wilhelm's importunate behaviour, had given rise to a new, threatening tone which conveyed the unmistakable message: 'If you do not go along with us as we wish, we shall go against you.'[59] From this Waldersee concluded that it was more important than ever for Germany to maintain 'a reserved but confident stance and an independent policy'.[60] He strongly criticised the Kaiser's importunate efforts to win friends. 'The greatest failing of Kaiser Wilhelm's policy', he commented, 'is that he wants to be on good terms with the Russians, the French, the English and even the Sultan, and pays court to them in a dangerous way; he inevitably falls between two stools. I do not even wish to mention his attentions to Austria and Italy; but they too are quite excessive and no less unwise. If we stand proudly aloof, run after no one but give the clear impression that we are confident of our power, others will come and ask for our support. If we continue as we are, we cannot but lose public respect.'[61] The only pleasing aspect of the insulting attitude adopted in Great Britain, in the General's eyes, was the anger expressed about it 'more or less unanimously' throughout the German press. 'The whole nation' was positively 'outraged . . . by the impudence of the English', he commented in his diary of 13 August 1895, adding with satisfaction that at last 'some sort of national feeling' was beginning to stir in Germany.[62] From now on, indeed, reciprocal prejudices and feelings of hatred began to stir on both sides of the North Sea. Fanned by the press, they could no longer be held in check by the respective governments, especially as in the young German Reich the temptation to enter into a pact with the devil, in the shape of the dangerous new power of the media, not infrequently proved irresistible.

THE AIMS OF GERMAN COLONIAL POLICY

The German Empire had become a 'global empire', Kaiser Wilhelm II announced at the impressive state banquet held in the White Hall of the Berlin Schloss on 18 January 1896 to celebrate the twenty-fifth anniversary of the founding of the Reich.[63] Waldersee took a sceptical view of this claim, for it really could not be justified by 'our colonies, which are after all still very pitiful, . . . and although we keep up an extensive trade with all parts of the world, this surely does not qualify us as a global empire either. If the Kaiser aspired to such a thing, we should have to seize it; but where? and from whom? The millions of Germans who have emigrated are no longer German nationals and do not concern us, nor do most of them want anything from us.'[64] Still less of a colonial enthusiast in the accepted sense than Waldersee was the new Reich Chancellor Chlodwig Hohenlohe, for rather than seeking to acquire territories in the tropics, he called for the emigration of the surplus population of Germany to be encouraged, principally to Argentina and southern Brazil. In early 1895 he wrote to his cousin Hermann zu Hohenlohe-Langenburg condemning Bismarck's colonial policy: 'The Bismarckian conception was one of the great man's mistakes. When great men have stupid ideas, their stupidity is in proportion to their greatness. In heaven's name, what are we supposed to do with the 500,000–700,000 people added to the population of Germany every year? In the end the Germans will have no alternative but to eat each other up. I consider it one of the most urgent tasks of the Reich government to encourage emigration and to organise it as well as possible.'[65]

In October 1895 the Kaiser likewise declared the 'strengthening of the German presence' in southern Brazil to be an urgent objective of Germany's foreign policy.[66] His interest in expanding German colonial possessions in Africa had recently suffered a setback as a result of his discovery that neither he nor the government of the Reich but the German-African Company was the rightful owner of the colonies there. As Philipp Eulenburg told the Colonial Director in the Foreign Office, Paul Kayser, '*in complete confidence*' in September 1895, the Kaiser was very angry that 'according to the treaties concluded, the entire area of the colonies – and even any additional territory that is acquired – remains in the hands of the German-African Company and not in those of His Majesty's government. From the moment he learned this His Majesty's interest in the colonies was as good as extinguished. The Kaiser can neither direct the flood of German emigrants there and make plans for the colonies, nor take any pleasure in troops who fight only for a company of German capitalists.'[67]

Rather than the acquisition of territory overseas, Wilhelm II's attention was now directed principally towards finding suitable coaling stations and naval bases, which in turn was attributed to the influence exercised on him (and deplored by the civilian leadership of the Reich) by naval and army officers. During a visit to the Wilhelmstrasse in January 1895 he demanded the creation of an integrated organisation for colonial troops. The current system by which each colony organised its own troops must be replaced by a single colonial army of which the Marine Infantry would form the core.[68] The Secretary of State of the Reich Navy Office, Admiral Friedrich von Hollmann, found this demand unacceptable. He threatened to resign and was supported in his stance by the Reich Chancellor, who saw the Kaiser's demand as an infringement of his responsibility for the conduct of foreign policy.[69] Again in September 1895 Hohenlohe complained of the strong influence of the navy on the Kaiser in colonial matters when the latter blocked the creation of a separate Colonial Office. 'The navy should really not have anything to do with the administration of the colonies', he wrote to his cousin Hermann. 'But H.M. is still too much under the influence of the navy to accept a proposal aiming to bring that about.' Hohenlohe deplored the 'the military culture' in the colonies which led to a situation in which 'our colonies are being used to accommodate useless officers, when a governor with a few officials and a number of Sudanese would be quite adequate'. Against the Kaiser's will nothing could be done, however, and Hohenlohe also feared that Paul Kayser's 'personality' was 'not calculated to help change H.M.'s mind'.[70]

The influence of the navy and the army on the Kaiser's colonial policy showed itself principally in his repeated attempts to pressurise the Reich government into acquiring ports, coastal strips and islands which he considered would be useful to Germany's naval forces. At Hohenlohe's first audience on 2 November 1894 Wilhelm demanded, as we have seen, not only the annexation of the island of Formosa (Taiwan) but also the acquisition of Mozambique, as Germany was dependent on Britain for coaling stations and this must stop.[71] In July 1895 Waldersee noted in his diary that Germany now intended to take 'serious steps as regards Morocco'.[72] In the spring of 1896 Morocco again emerged as an objective of German colonial policy. After a conversation with Count Christian von Tattenbach, the German envoy in Tangier, the Kaiser called at the Reich chancellor's palace in a 'highly agitated' state, with the news that France wanted to buy the Canary Islands from Spain; Germany must therefore have something in Morocco by way of compensation. As the Chancellor recorded, Wilhelm began by discussing the advantages of a colony but in the end contented himself with the idea of acquiring a coaling station – 'and then we shall see what is to be done next'. At the Kaiser's suggestion Tattenbach was ordered 'to pursue the matter there'.[73] On All-Highest orders the warship *Moltke* was instructed to remain at Gibraltar 'in order to be ready if any action were to be undertaken on the Moroccan coast'.[74] A week later Hohenlohe had to urge the Kaiser to allow the *Moltke* to proceed on its way, as the rumour about the purchase of the Canary Islands by France had proved groundless. He went on to point out that the survival of the Triple Alliance might be threatened if Germany were to take action in Morocco immediately after the catastrophic defeat which the Italian army had suffered at Adua on 1 March.[75] The Kaiser, whom Hohenlohe found 'in a cheerful mood' this time, recognised the dangers to be feared from an action in Morocco and authorised the warship's departure from Gibraltar.[76] In July 1896 Wilhelm, anticipating not only a possible Anglo-French agreement on respective spheres of interest in North Africa, but also a Moroccan crisis unleashed by this, wrote on a report from Rome: 'The English squadron is *à portée* off Tangier! Might a little agreement with France at the expense of Italy and Spain be in the offing, with Morocco for England, Tunis and Tripoli for France? Please keep a lookout in Morocco! Because we shall have our say too!'[77] Only nine years later the German demand for compensation in Morocco for a very similar Anglo-French agreement in North Africa would unleash the first of the great pre-war crises.

In eastern and southern Africa, meanwhile, an increasingly critical situation was developing and the detached stance hitherto taken by Germany was fast becoming a thing of the past. In the summer of 1894 Sir Edward Malet drew attention to the fact that in the interest of Anglo-German understanding Wilhelm II's attitude over the demarcation of the two countries' possessions and spheres of influence in central Africa had been very moderate.[78] Only a few months later, however, it was only too plain to the statesmen of both countries that the Kaiser's attitude had changed. Remarks that he made to Swaine on 21 November 1894 led to a lively exchange between the British Foreign Secretary, Lord Kimberley, and the German ambassador in London, in the course of which Hatzfeldt deplored the growing antagonism between Germany and Britain in East Africa. 'Count Hatzfeldt said he had for 8 years tried to make a closer alliance between England and the triple alliance but that England refused and he was sorry to perceive there were signs of irritation in Germany against England's Colonial Policy.'[79] In his diary for 7 August 1895 Waldersee predicted that 'our Colonial people' would be 'very unhappy' at the new turn British policy was taking, for they would 'very soon realise in Africa . . . that England has no intention of giving up anything at all there; if *we* do not give up, there will be plenty of points of conflict there. Only if things go badly for England will she co-operate with us.'[80]

THE MALET INCIDENT

Far more dangerous than Anglo-German rivalry in East Africa, of course, was the tense relationship between the British Cape Colony under Cecil Rhodes and the Boer republics of Transvaal and the Orange Free State, in which not only the Kaiser but also the Reich government, backed by inflamed public opinion in Germany, interfered forcefully and with fatal results, as we shall soon see. Ever since the discovery of enormous gold and diamond deposits in the two Boer states in the 1860s the rural idyll was over for these descendants of Dutch immigrants. Thousands of 'Uitlanders', mainly British, had poured in to prospect for gold and precious stones. Among the most successful of them was Cecil Rhodes, who had come to South Africa at the age of seventeen and had since become one of the richest men in the world. In addition to his diamond business, which controlled 90 per cent of world production, Rhodes had risen to become Prime Minister of Cape Colony in 1890. As the first step on the way to bringing the whole of southern and eastern Africa under British rule Rhodes planned to gain control of the two Boer states with the help of an armed uprising by the Uitlanders. Four thousand rifles and three machine guns were smuggled into Johannesburg and plans were made to capture the arsenal in Pretoria. In October 1895 Leander Starr Jameson, a friend of Rhodes, led a private army of 500 men (including three serving British officers), with six machine guns and three field guns, to the western border of the Transvaal to await the moment for invasion. The rebellion which they were to support was planned for 28 December.[81]

The rapidly developing situation was watched with growing surprise and displeasure by both the Kaiser and the Foreign Office in Berlin. There were about 5,000 citizens of the German Reich living in the Transvaal at this time; German investments there amounted to nearly 500 million marks. In Germany public opinion identified itself to a large extent with the Boers, that small nation of 'Protestant farmers of Germanic origin' who were being threatened, as it seemed, by the 'plutocratic Anglo-Saxons'. The German Consul-General in Pretoria, Franz von Herff, had taken the opportunity of a banquet at the German Club in the town in honour of the Kaiser's birthday in January 1895, at which President of the Transvaal 'Ohm' Krüger was present, to assure the latter that the German Reich was not indifferent to fate of the Transvaal. The President had replied that, although his small country was like a child among the great powers, if the great power Britain tried to trample on his country, the great power Germany would be able to prevent this. These remarks had been enough to cause a serious diplomatic confrontation between London and Berlin: the British ambassador pointedly reminded the German Foreign Secretary of the status of the Transvaal under international law from the British point of view: Great Britain had annexed the country in 1877; four years later and again in 1884 it had recognised the sovereignty of the republic, but with the proviso that the Transvaal must not form any alliance with another state without British consent. Marschall contested this interpretation, made angry allusions to Cecil Rhodes's intentions and his anti-German utterances, and protested against the Cape Colony's plan to force the Transvaal into an economic union with it.[82] The repeated warnings of the government in London that as far as South Africa was concerned the most vital interests of the British Empire were at stake seemed not to have been understood in Berlin, probably because there was no desire to understand them.[83]

The following months brought further Anglo-German conflicts in this highly sensitive region. On the occasion of a visit by the German cruiser *Condor* Wilhelm II sent the President of the Transvaal the telegram which came to be known as 'the first Krüger telegram' and which, as Waldersee recorded, was 'particularly badly received' in London.[84] In October 1895 there was a second clash between Marschall and Malet when the latter, on the point of retiring as ambassador in Berlin, declared during his farewell visit to the Wilhelmstrasse that there was only one 'black spot' in the otherwise friendly relations between Germany and Britain, and that was the support which the Germans were giving the Boers in their hostile attitude towards Great Britain in the Transvaal. This was an intolerable state of affairs for Britain and if Germany persisted in its attitude serious complications could follow.[85] Marschall rejected Malet's critical remarks with such robustness that the Kaiser, for once, had nothing but praise for him for sending the ambassador away with a flea in his ear. Like the Foreign Secretary, Wilhelm poured scorn on the British threat, 'when they need us so much in Europe!' He too thought it outrageous that Malet should 'make us take the blame for Rhodes's arrogance!',[86] and considered the ambassador's warning an 'impertinence' which must be repudiated 'with the requisite energy'.[87] Writing to the Tsar, he commented that Malet had used 'very blustering words, about

Germany behaving badly to England in Africa', and had threatened 'that it [England] would not stand it any longer and that after buying off the French by concessions in Egypt they were at liberty to look after us. He was even so undiplomatic to utter the word "war"! saying that even England would not shrink from making war upon me if we did not knock down in Africa.'[88]

Infuriated beyond measure by the 'utterly incomprehensible' attitude of Lord Salisbury, who was unwilling to join the Triple Alliance, the Kaiser fiercely attacked British policy in a conversation with Leopold Swaine on 25 October 1895 that had a far-reaching effect on Anglo-German relations. He began by castigating Britain's Mediterranean policy and threatening to set up a continental league against it.[89] Then, according to his own account of the conversation, he told Swaine that Malet had made 'the most astonishing accusations of malicious conduct by us towards England' on his departure. 'Indeed, he had even gone so far as to utter the unbelievable word "war"; thus for the sake of a few square miles full of negroes and palm trees England had threatened her only real friend, the German Kaiser, grandson of Her Majesty the Queen of Great Britain and Ireland, with war.' Wilhelm rejected out of hand Swaine's attempt to explain away the words of the outgoing ambassador as a misunderstanding that ought to be cleared up as soon as possible. He retorted that 'this tone, even though it was surprising to hear it from so mild a man as Malet, matched the tone of the English press towards Germany', and went on to tell Swaine that 'the government press in particular behaved in the most unwarrantable way towards me. Germany and the Triple Alliance had been continually insulted and taunted, and a good part of my seven years of hard work to bring my Reich and England closer together on the basis of common interests and mutual respect in order to solve great cultural issues, had been destroyed . . . It was not compatible with the interests of my country to go along with all the whims of English policy and to react to vague hints and enigmatic utterances by English statesmen. Such conduct by England positively forced me to make common cause with France and Russia, both of which could have about a million men on my borders ready to invade, while England did not even have a kind word for me.' 'I ended the conversation', the Kaiser continued, 'with a strong warning that the only way England could escape from her present complete *isolement*, in which she found herself because of her "policy of selfishness and bullying", was to take an open and unqualified stand either on the side of the Triple Alliance or against it. The former course would have to be in the form customary among continental powers, i.e. under signed and sealed guarantee. The Colonel seemed profoundly shaken and affected.'[90] The Reich Chancellor, informing the German ambassadors in London and Rome of this conversation, softened the Kaiser's anti-British tone considerably by affirming that as soon as British policy adopted a more definite course 'His Majesty's remarks directed against England, in which He was expressing His thoughts about the current situation rather than voicing a fixed principle, will no longer have any practical application.'[91] By contrast, the impression made on the British government by Swaine's account of the Kaiser's comments was powerful in the extreme. The document was printed and circulated to all Cabinet ministers, one of whom wrote to the military attaché saying that it was 'the most important document you have ever sent to us from Berlin'.[92]

Although Hatzfeldt reported several times to Berlin that Malet had acted without instructions,[93] Philipp Eulenburg observed during the Kaiser's stay at Liebenberg at the end of October that Wilhelm was 'still very indignant' about Malet's behaviour.[94] The Kaiser expressed the opinion 'that he would not regard the Malet incident about the Transvaal as settled until he had been informed officially by some suitable person (unfortunately the new ambassador won't be here for another six weeks) that Lord Salisbury does *not* share Malet's views'.[95] The Kaiser's 'irritation' with Britain continued to manifest itself in the aggressive marginal comments which he wrote on diplomatic papers.[96] The Foreign Secretary had difficulty in preventing him from sending a telegram to Bülow in Rome with instructions to warn the Italians against Britain and 'cut them off' from Britain if possible. Holstein's comment was that in Marschall's place he would rather have walked out.[97] In mid-November 1895 Eulenburg reported to Bülow that although the Malet incident had been settled the Kaiser 'is very annoyed with England and will remain so'.[98] The Kaiser's irritation at 'England's brazenness' had wide repercussions and even led to a plea from Kaiser Franz Joseph that 'Kaiser Wilhelm should not be too hostile towards England, as that would put Austria into an extremely difficult position.'[99] It was not until 17 November that the Kaiser accepted that 'the British Premier . . . has formally said his *peccavi*', commenting with satisfaction that 'the rude answer . . . has fulfilled its purpose perfectly'.[100] In December 1895, however, Waldersee confided to his diary after long conversations with Wilhelm in Hanover and Springe that the Kaiser had 'criticised English policy . . . very sharply'.[101] Only two weeks later the most serious crisis of confidence in Anglo-German relations in the pre-war period broke out as a result of an act of 'unbelievable stupidity'.[102] How did it happen, and what role did the Kaiser play in it?

THE KRÜGER TELEGRAM

At the end of December 1895 the rapidly worsening situation in southern Africa had been causing 'great agitation' in the Wilhelmstrasse for days.[103] On 31 December at 1.30 p.m. the Foreign Office received news of the incursion into the Transvaal by Jameson and his irregular army. Consul-General von Herff telegraphed to say that President Krüger considered Jameson's action to be a breach of the London Convention of 1884 and was counting on the support of Germany and France. Herff asked for the protection of a landing party from the cruiser *Seeadler* to defend the Germans whose life and property seemed under serious threat. He described the Jameson Raid as 'unscrupulous land-grabbing'. On receiving this news the Foreign Secretary immediately sent a telegram to the Kaiser and then travelled to Potsdam with the Director of the Colonial Department, Paul Kayser, for consultations with the monarch. Wilhelm approved the decision to order the landing party of fifty marines on board the *Seeadler* to go to Pretoria to protect the Germans there. A second cruiser was dispatched to Delagoa Bay.[104] When a second telegram arrived from Herff on 1 January together with an open telegram from the Germans in Pretoria begging Wilhelm II 'to intervene at once to prevent untold misery and bloodshed', Marschall again went to the Schloss to inform the Kaiser.[105]

We know from the Reich Chancellor's journal that Wilhelm II was in a highly excitable state at this time, not least as a result of the Köller crisis, and that his behaviour towards the Minister of War was so violent that Bronsart had difficulty in restraining himself from 'drawing swords'; the General doubted that the Kaiser was 'entirely normal' mentally, and had 'grave concerns about the future'.[106] When the news from South Africa arrived, the Kaiser condemned Cecil Rhodes as a 'monstrous villain' and scrawled on a report from Cape Town: 'This is a gross stock exchange swindle cooked up by *German Jews*.'[107] Waldersee, who sat next to the Kaiser at the New Year dinner for the commanding generals, looked on as Wilhelm received 'a long telegram about the Transvaal, which caused him some agitation; he said that it had been established that uniformed English officers had taken part in Mr Jameson's expedition, and he foresaw serious complications'.[108] In a conversation with Lascelles on 1 January 1896 he expressed the hope that Jameson and all his followers would be captured and shot.[109] The *Times* correspondent Valentine Chirol informed the ambassador that 'the Emperor was so rabid when the news first reached him that he thought of sending back his English uniforms with the remark that they were only fit to be worn by South African bandits'. Wilhelm was also reported to have claimed that 'there was at all events one part of Africa over which England could not claim Suzerainty and that was Egypt'. Conveying these remarks in a private letter to Salisbury, Lascelles sighed: 'It will be a serious matter if the Emperor becomes actively hostile towards us, and he is so impulsive and impetuous that this is a contingency which must be considered.' The situation, Sir Frank added, was 'far more unsatisfactory than I could have believed possible'.[110]

On 2 January the Kaiser wrote to the Tsar taking up the idea of a continental league against Great Britain. 'The political horizon is peculiar just now. Armenia and Venezuela are open questions England brought up, and now suddenly the Transvaal Republic has been attacked in a most foul way as it seems not without England's knowledge. I have used very severe language in London and have opened communications with Paris for common defence of our endangered interests, as French and German colonists have immediately joined hands of their own accord to help the outraged boers. I hope you will also kindly consider the question, as it is one of principle of upholding treaties once concluded. I hope that all will come right, but come what may, I never shall allow the British to stamp out the Transvaal!'[111] In fact the British government had immediately distanced itself from the attack. On 30 December Colonial Secretary Joseph Chamberlain formally condemned Jameson's action and ordered the recall of his troops and the punishment of the British officers involved. On 1 January 1896 Hatzfeldt reported, after a conversation with Salisbury, that he was convinced that the government in London had not been behind the Jameson Raid and that it was being honest in distancing itself publicly from him.[112]

At 10 a.m. on 3 January 1896 the Kaiser, accompanied by Naval Secretary Admiral Friedrich von Hollmann, Commanding Admiral Eduard von Knorr and Chief of the Naval Cabinet Admiral Gustav Freiherr von Senden-Bibran, arrived at the Reich chancellor's palace, where Hohenlohe, Marschall and Paul Kayser were waiting for him. Geheimrat von Holstein was in a neighbouring room. 'The South African question was discussed', the Reich Chancellor noted in a brief account of the proceedings. 'Marschall proposed a telegram to Krüger, which was accepted. Then possible support for the Boers was discussed. Colonel Schele was summoned and ordered to travel to South Africa to investigate what kind of help might be given to the Boers. Holstein, Kayser and Marschall are against it.'[113] Marschall's diary throws rather more light on the strange process by which the decision was eventually reached to send the notorious Krüger telegram and above all indicates the pressure under which the 'responsible' statesmen, Hohenlohe, Marschall, Holstein and Kayser, were put by the monarch and the three Admirals. The Foreign Secretary writes: 'At 10 o'clock attended conference with H.M., together with Reich Chancellor, Hollmann, Knorr and Senden. H.M. put forward some weird and wonderful plans. Protectorate over the Transvaal, which I at once talked him out of. Mobilisation of the marines. Dispatch of troops to the Transvaal. When the Reich Chancellor objected: "That would mean war with England", H.M. replied: "Yes, but only on land." Then it was decided to send Schele on a reconnaissance mission to the Transvaal. Also an unfortunate idea. Finally, at my suggestion, H.M. sent a congratulatory telegram to President Krüger.'[114] The telegram, as is well known, was expressed in terms that were particularly offensive to Great Britain in that the Kaiser congratulated Krüger for having succeeded, 'without appealing for the help of friendly powers, in guarding the independence of your country against attacks from outside'.[115]

The impression given by these entries in the diaries of the statesmen involved is that the telegram, which was in fact suggested by Marschall and not by the Kaiser, was a hastily concocted compromise solution to prevent Wilhelm's more far-reaching and far more dangerous 'weird and wonderful' plans – the declaration of a German protectorate and the dispatch of troops, even at the risk of war with Britain – from being realised. Since the telegram was written in the first person and personally signed by Wilhelm II, in the eyes of the world he was seen from the outset as the person who actually initiated it. Historians too have frequently portrayed him, with his 'pathological urge to act', as solely responsible for it.[116] Perceptive contemporaries like the Empress Frederick, Sir Frank Lascelles, Valentine Chirol, Baroness Spitzemberg, Princess Daisy of Pless, the Bavarian envoy Count Hugo von Lerchenfeld and Hermann Freiherr von Eckardtstein, however, rightly emphasised the shared responsibility of the leaders of the Wilhelmstrasse for the telegram,[117] and more recent research has also demonstrated convincingly that well before 3 January Marschall and Holstein had turned strongly against Britain, that they expressed quite as much indignation as the Kaiser at the Jameson Raid and that initially they too were thoroughly satisfied with the effect of the congratulatory telegram to Krüger.[118]

It is indisputable that in the context of the Transvaal crisis the German Foreign Office, as much as the Kaiser himself, envisaged a kind of continental league between the Triple Alliance and the Franco-Russian alliance, with the aim of teaching Britain a lesson and demonstrating the value of German friendship. On 30 December 1895 Marschall had noted: 'Holstein has plans for a new anti-English policy.'[119] The very next day – but before news of the Jameson Raid arrived – the Foreign Secretary, adopting Holstein's line of thought, warned the new British ambassador, Lascelles, 'that if England overestimated the antagonism between the alliances of European states and thought she could get away with anything, this could prove to be a mistaken calculation, and in the end the idea that the continental powers could come to an agreement and enrich themselves at England's expense might find fertile ground in which to grow'.[120] On 1 January 1896 Marschall explained his 'train of thought' to the French ambassador, Herbette, as follows: 'England is counting on the antagonism between the continental powers and believes she can do anything she pleases. Could not the great continental groups agree together on specific purposes, and set aside all issues which might lead to a European war?'[121]

When the Jameson Raid brought the Transvaal crisis to a head, Holstein and Marschall, quite independently of the Kaiser, made it plain that they too considered the challenge to the independence of the Boer state as *casus foederis*, and therefore as activating treaty obligations. On 2 January the Foreign Secretary commented in his diary: 'It is a very frightening situation. If the Boers are beaten, we shall have to take action.'[122] He had already instructed Hatzfeldt, on receiving Herff's first telegram, to make official enquiries in London as to whether the Salisbury government approved of Jameson's action and warned that if the occasion arose he might have to break off diplomatic relations with Britain.[123] In spite of the ambassador's assurances that the British government had sincerely and publicly distanced itself from Jameson, Marschall raised the stakes with a note which Hatzfeldt was to hand over in London, and which culminated in the declaration 'that we protest against the invasion and cannot accept infringement of the independence of the Transvaal'. Not until the telegram announcing the defeat of Jameson arrived in Berlin that evening did the Foreign Secretary step back from the edge of the abyss, hastily instructing Hatzfeldt not to hand over the note with its ultimatum.[124] Fortunately, because Lord Salisbury happened to be out of London, the document was lying unread at the Foreign Office and the ambassador succeeded in recovering it before it was opened.[125]

Both Marschall and Holstein also agreed that the Krüger telegram should be published immediately – a logical step, since one of the main aims of official German policy was (and remained for the next few years) to teach Britain the 'lesson' that 'they need us'.[126] Indeed Marschall was more than satisfied with the initial public reaction to the telegram, as his diary entry for 3 January reveals. 'Our press is magnificent. All the parties are of one mind; even Auntie Voss [the liberal *Vossische Zeitung*] wants to fight.' He was also struck by the 'remarkably favourable' reaction of the French press.[127] But it was a different matter when the 'outrageous', 'insolent', 'foolish' and 'utterly insane' articles in the British press, even threatening 'a switch to the Franco-Russian side by England', reached Berlin. Marschall was obliged to write replies to the attacks for the *Kölnische Zeitung* and the *Norddeutsche Allgemeine*, and tried to prove first to the Kaiser and then to the British ambassador, on the basis of official documents, that the telegram had been factually correct: Britain's sovereignty over the Transvaal had indeed been stipulated in 1881, but in his view it had been revoked by the 1884 convention.[128]

Looking back over the workings of the decision-making process in Berlin in these first few days of 1896, it does in fact seem highly probable that the Krüger telegram came to be written only because the civilian statesmen in the Wilhelmstrasse wanted to prevent Wilhelm II from making infinitely worse demands and were put under enormous pressure by him in terms both of time and of his expectations. It is even possible that following Jameson's surrender and Hatzfeldt's retrieval of the German note from Downing Street the Anglo-German crisis might have been reasonably peacefully settled on 2 January, had not the Kaiser appeared at the Reich chancellor's palace the next morning with his 'extraordinary plans'. At the same time the evidence now available clearly shows that the responsible Reich authorities, above all Marschall and Holstein, must bear a much greater share of responsibility for the telegram than public opinion at the time, and subsequent historical writing, attributed to them. Their 'new anti-English policy', which aimed to force Britain to attach itself to the Triple Alliance against its will, would have driven it into the arms of France and Russia even without the Krüger telegram – just as was to happen when the German blackmail policy was pursued further through the intensified battleship building programme.[129] But of course the fact that Wilhelm II was nevertheless seen as the sole, or the main, culprit responsible for writing the fateful telegram had its roots in the whole egocentric style of personal rule which he exercised. The young Anglo-German Eyre Crowe, who was making a name as an expert on Germany in the British Foreign Office at the time, commented quite rightly at the height of the Transvaal crisis: 'When the Emperor complains, as he does to us, of being *personally* attacked in the press here, surely that is the natural and legitimate consequence of his very *personal* way of governing and mixing himself up so prominently in every act of govt. Why, for instance, that *personal* telegram? . . . If you want to govern with a big "I", you may have all the glory, but you must also take the defeat.'[130]

REACTIONS TO THE KAISER'S TELEGRAM

The Kaiser's congratulatory telegram to Krüger, and above all the reference to 'the independence of your country towards the outside world' and the insinuation that 'friendly powers' like Germany would have come to the Transvaal's aid if necessary, had the effect of 'a flash of lightning in a powder-keg' on public opinion in Britain.[131] The newspapers were full of vicious attacks on the Kaiser for days; he received numerous insulting anonymous letters and the officers of his own regiment, the 1st Royal Dragoons, turned his portrait to the wall. Only a few days after the telegram a new Royal Navy fighting unit of two battleships and four cruisers set sail for the Mediterranean.

The violence of the public reaction in Britain to the telegram came as an unpleasant surprise not only for the German Foreign Office but also for many other members of the German administration. The Reich Navy Secretary, Admiral von Hollmann, asked the British military attaché Swaine in consternation why people in Britain were so angry with Germany. The Kaiser's telegram had been 'perfectly harmless', he claimed, whereupon Swaine replied that the British reaction was not directed against the German nation but only against the Kaiser; against him, however, there was such strong feeling that his former popularity had completely disappeared, so much so 'that it would be impossible for His Majesty to visit Cowes this year'. The German Admiral was left to console himself with the thought that 'any idea of war between the two countries was out of the question, as "the Elephant could not fight the Whale"'. Hollmann had admitted that the Kaiser was still very prejudiced against Britain because of the article in the *Standard* the previous August, the military attaché added in his report on the conversation.[132]

As a result of the 'paroxysm of fury' which the telegram had unleashed in British public opinion, the mood in German diplomatic circles, where dreams of a continental league against Great Britain had so recently flourished, switched to an almost panic-stricken fear of a combination between Britain and France against Germany. From London, Hatzfeldt urged extreme caution, 'since if there were any further reason, however slight, England might come to an agreement with France and give up Egypt'. A visit to Cowes by Wilhelm would be extremely dangerous, he warned, for 'at the very least the Kaiser would be booed'.[133] Friedrich von Holstein heard that Courcel, the French ambassador in London, had thought there might be war between Britain and Germany and had assured Lord Salisbury that in that event Britain could be sure of the support of France, 'for France had only one enemy and that was Germany'.[134] Count Münster, the German ambassador in Paris, who had earlier been *en poste* in London for several years, condemned the telegram and warned that Britain might now turn away from Germany and towards France. The 'most tragic result' of the crisis was 'the mistrust and hatred that has grown up between the two great civilised states of Europe', he wrote to Holstein. 'I knew and felt for a long time that this existed latently . . . Our Kaiser saw only the surface of things, sand is always thrown in his eyes when he comes on a visit; this is how I explain how H.M. sent the telegram, without realising in advance that it was a match to set fire to an accumulation of inflammatory material; I do not believe that it will really come to war. I hope not. God preserve us from that. The Queen and the men in control are too level-headed for that.' But even without war the political and economic damage was very great. The British Admiralty had made use of the crisis to strengthen the Royal Navy enormously, and signs of an Anglo-French rapprochement could be seen everywhere.[135] When a report by Münster containing these warnings was submitted to the Kaiser he commented bitterly in the margin that the new British squadron would have created a sensation if it had appeared in the Mediterranean before. 'Now, under the pretext of "Transvaal–made in Germany" it is put into commission quite harmlessly and naturally and can reinforce the Mediterranean quite undisturbed. The English should be very grateful to me instead of criticising me.'[136]

Like the peace-loving Münster, the anglophobe General Count von Waldersee saw the outcry in Britain as the expression of a longstanding rivalry. It would certainly have been better, in his opinion, if Wilhelm had not sent the telegram, and yet it had not been 'a crime' and England could not have taken the Kaiser's action so badly unless 'she had a very guilty conscience'. The General took the view that the violent British reaction also reflected 'a long-held grudge' against Germany. He conceded that 'our colonial policy has made England uneasy and she persistently intrigues against us . . . The Kaiser has unquestionably acted overhastily in certain cases.' 'Why send ships to Delagoa Bay? They cannot be of any use, but they give England an excuse to be suspicious of us.'[137] Like Marschall, the General had at first believed that the opportunity of creating a continental league against Britain was within reach. 'So far Europe is still siding against England', he wrote on 7 January. The confused picture would soon clear, he predicted, since 'England in her madness' was driving Germany into getting closer to Russia and France. 'Will they respond?', he wondered anxiously. 'We may be facing very serious events and everything depends on skilful hands on our side. God grant that we have them.'[138] As the grave consequences of the imperial telegram began to emerge, however, Waldersee's thoughts became darker. He wondered how the German princes would react to the Kaiser's initiative. 'I think it not impossible that Bavaria might try to express reservations over the unilateral action of the Kaiser.'[139] The more the storm raged in the press the more Waldersee's anxiety grew, and he too began to see a war against Britain, France and Russia at the same time as within the realm of possibility. 'In England people seem to have gone quite mad', he wrote on 10 January. 'People rage against us in public, even in polite society, and speak and write about our Kaiser with outrageous insolence. The newspapers seem to have instructions to stir up and confuse public opinion and there are calls for naval armament as if a war were imminent. I hope we shall keep calm, and especially the Kaiser himself. But it is nevertheless possible that England, having brought so much disgrace on herself, is now planning a great coup. If we had skilful diplomats that would not be serious, but I am afraid that we do not. Hostile feelings towards us in France and Russia may gain the upper hand when people see that we are in an uncomfortable position. We cannot wage war against England; our strength rests entirely on the fact that we can either be a very valuable ally for England, or, if we can come to an understanding with her enemies, cause her great difficulties. But all this requires very careful work and a cool head.'[140]

Although he did not think the danger of war acute, the former Chief of the General Staff was compelled to acknowledge that Germany's international position had become significantly worse as a result of British hostility. In January 1896 he wrote: 'I had hoped that we would use the turmoil in the Far East to create divisions between the other powers, while we stood quietly by for the time being; but now in no time at all exactly the opposite has happened, and we are in the foreground! I am sure England is not seriously thinking of war, but the gravity of the situation is recognised there, and they are preparing themselves to face whatever arises.' Waldersee considered that a weakening of the Triple Alliance would be one of the direct results of British antagonism, for Italy would always have to co-operate with the United Kingdom, while Austria-Hungary would also want to remain on good terms with it. He personally would not be sorry if this happened, so long as it led to improved relations with Russia; but that was unlikely, for Russia was 'not at all inclined to come closer to us, quite apart from the fact that influential circles there would take the greatest possible pleasure in seeing us isolated'. So, following the Krüger telegram, the picture of an isolated Germany in an increasingly dangerous global situation was beginning to emerge. All countries were 'exerting their energies to the utmost' to arm themselves, and 'those on the continent have now been joined by England, Japan and China', Waldersee observed. Even the United States of America was making 'great efforts on the maritime front'. 'So there is a truly massive accumulation of explosive material and an apparently insignificant matter can easily set it ablaze.'[141]

When the storm in the British press began to die down on 12 January 1896 Waldersee likewise grew calmer in his assessment of the situation, although his comments showed the ambiguity of his attitude towards Britain. In his heart of hearts he welcomed the Anglo-German confrontation because it had at last proved to 'the many anglomanes . . . that English policy [is] two-faced, utterly false and exclusively egotistical'. It was unquestionably best for Germany 'to be permanently on a good footing with England, and if possible allied with her, but that is absolutely ruled out as long as England merely wants to exploit us and at the same time has the unbelievable temerity to make difficulties for us in our colonial policy *everywhere*. I hope that the outrageous spite and impertinence towards the Kaiser of which the English press has been guilty will never be forgotten.'[142] Waldersee, like Tirpitz and other enemies of Great Britain, often sought comfort in the thought that the vituperation from London would at least ensure 'that the Kaiser will be somewhat cured of his anglomania'.[143]

The Krüger telegram naturally led to bitter arguments not only in public but also within the royal family, putting Anglo-German relations under lasting strain. With the help of her daughter Beatrice, Queen Victoria gave her grandson a dressing-down, expressing her distress and surprise at his 'outrageous' and 'dreadful' telegram. 'Our great wish', she declared, 'has always been to keep on the best of terms with Germany, trying to act together, but I fear Your Agents in the Colonies do the very reverse which deeply grieves us.'[144] The Prince of

Wales spoke furiously of the hostile and completely unnecessary interference of the Kaiser in matters that did not concern him, which was all the more offensive because he was the Queen's grandson, constantly affirmed his love for England and held high rank in both the British army and the Royal Navy. It was out of the question for him to visit Cowes this year.[145] In February, in conversation with a high-ranking German delegation, the heir to the throne and his son George again expressed resentment at the Krüger telegram, which they saw as an insult, and furthermore one that had been delivered 'by a relation, an English admiral etc.'.[146] The Empress Frederick, who was in Berlin at this time, likewise criticised her son's telegram as 'a deplorable mistake'. She too emphasised how important it was 'for England & Germany to get on well together & be good friends!' and expressed the astonishingly progressive view, which would not be out of place in a historical conference today, that 'the whole foolish German Colonial Enterprize in Africa was encouraged by Pce Bismarck – as he quietly admitted – to be a thorn in England's side & an apple of discord – a question wh. he could put forward at any moment when he wanted to raise popular feeling here against England, to keep the nations asunder!' The German colonies were expensive and uneconomical, but they flattered the 'inordinate national vanity & chauvinism'.[147]

Wilhelm answered his grandmother's rebuke with a hypocritical assurance that his anger had been directed not against Britain but against the international 'mob of golddiggers' who had rebelled against the Queen. This letter, of which the Kaiser had discussed his own draft line for line with the Reich Chancellor,[148] is one of the strangest documents he ever penned in English and is given here in full. 'Most beloved Grandmama', he wrote from the Neues Palais on 8 January, 'Your kind letter just reached me in presence of Uncle Hohenlohe who had come for his "Vortrag" [audience], so I was able to immediatly [sic] speak to him about the affair. As you kindly express it so it is. Never was the telegram intended as a step against England or your government. By Sir Frank [Lascelles] as well as by the Embassy in London we knew that Government had done everything in its power to stop the freebooters, but that the latter had flatly refused to obey, & in a most unprecedented manner went & surprised a neighbouring country in deep peace! As Sir Frank himself told me when I asked him about the fact, on New Years day, these men were flibustiers [filibusters] or rebels, I hinted that we ought to join & cooperate in keeping them from doing mischief. But I never heard anything about the matter anymore. The reasons for the telegram were 3 fold. First in the name of "Peace" as such which had been suddenly violated, & which I always, following Your glorious example, try to maintain everywhere. This course of action has till now so often earned Your so valuable approval. Secondly for our Germans in Transvaal & our Bondholders at home with an invested capital of 250–300 millions & local commerce of the coast of 10–12 millions which were in danger in case fighting broke out in the towns. Thirdly. As Your Government & Ambassador had both made clear that the men were acting in open disobedience to Your orders they were rebels, I of cours[e] thought that they were a mixed mob of golddiggers quickly summoned together, who are generally known to be strongly mixed with the scum of all nations, never suspecting that there were real Englishmen or officers among them. Now to me rebels against the will of H. Most Gracious Majesty the Queen are to me the most execrable beeings [sic] in the world & I was so incensed at the idea of your orders having been disobeyed & thereby, Peace & the security also of my fellow countrymen endangered, that I thought it necessary to show that publicly! It has I am sorry to say been totally misunderstood by the British Press. I was standing up for law, order & obedience to a Sovereign whom I rever[e] & adore & whom to obey I thought paramount for her subjects. Those were my motives, & I challenge anybody who is a gentle man to point out where there is anything hostile to England in this!? The Secretary of Transvaal was even at his audience the day before yesterday cautioned by me to warn his Government on no account to do anything that could be interpreted as beeing [sic] hostile to England. The gunboat in Delagoa was only to land in case street fight & incendiarism broke out to protect the German consulate as they do in China or elsewhere but was forbidden to take any active part in the row; nothing more. As to the silly idea in the press that I was or wanted to behave hostilely to England I with a clear conscience refer to L. Salisbury, who has material enough in his hands from the last years to know my thoughts & what I do for England. But the English press has been rather rash in its conjectures & having since some months freely lavished its displeasure on our devoted heads, the home press still sore about certain Standard articles which appeared when I was at Cowes, & which were unkind to me personally – which wounded German amour propre more deeply than the authors may have thought – this made people rather hot & rash. But I hope & trust that this will soon pass away, as it is simply nonsense that two great nations nearly related in kinsmanship & religion, should stand aside & view each other askance, with the rest of Europe as lookers on, what would the Duke of Wellington & old Blücher say if they saw this? As to the wishes You expressed regarding the relations in Afrika [sic] in general & the question of the Agents I have immediatly [sic] communicated to Prince Hohenlohe who will give the matter his serious attention. So I hope that soon all will come right.'[149] The Queen found her grandson's attempts to exculpate himself 'lame and illogical', but after discussing it with Salisbury thought it advisable not to pursue the matter, as the 'real rebuke' he deserved would only annoy him still more.[150]

THE KAISER AND BRITAIN AFTER THE KRÜGER TELEGRAM

Wilhelm II's ingratiating response to his grandmother's reproaches suggests that he was trying to make good the massive damage caused to Germany's relations with Great Britain by his telegram to Krüger. There was further evidence of this in numerous other gestures he made, which seemed to destroy the hopes of anti-British statesmen in Germany that the Kaiser's 'anglomania' was at last over. When for instance the long-serving British military attaché, Leopold Swaine, took his leave at the end of February 1896 Wilhelm spoke effusively of his regret at the departure of this 'personal friend', and caused general astonishment by conferring the star of the Order of the Red Eagle Second Class on Swaine.[151] For the funeral of Prince Henry ('Liko') Battenberg, who had died while serving in the Ashanti war on the Gold Coast, Wilhelm sent such a large deputation to London that even the Empress Frederick was embarrassed.[152] Waldersee was deeply disappointed, and saw this action as a sign that 'our gracious sovereign [has] no idea how to exercise restraint'. The British would only see the gesture as 'a kind of apology', he feared, adding: 'the people in Berlin who thought the Kaiser had been cured of his love of England for a long time do not know him well. I truly do not want us to be on bad terms with England, but I do not want us to run after her.'[153] Count Anton Monts, Philipp Eulenburg's successor as Prussian envoy in Munich, who like many others at this time questioned the Kaiser's sanity, commented mockingly: 'H.M. is very upset about . . . the estrangement of the English, as he can hardly now go to Cowes!! That is why he is running after these people again, alas, which is exactly the opposite of what he should be doing! He sends an entirely superfluous deputation of the Gardes-du-Corps, a private letter to his tippling grandmother, and a message as unknown to the Foreign Ministry as the letter is, to be delivered personally to the old hucksteress by his Flügeladjutant Arnim . . . In short, the outlook's pretty gloomy, just as in the days of Friedrich Wilhelm IV.'[154] Even the elderly Reich Chancellor thought it risible that Flügeladjutant von Arnim, whom the Kaiser had sent to London for the Battenberg funeral, would come back with favourable but completely untrue reports on the mood there, 'because he has to do so in order to give H.M. courage and revive his hopes of Cowes'.[155]

Although the prospect of a visit to Cowes was ruled out for the time being, the Kaiser stuck to his plan to go on a Mediterranean cruise – albeit not as originally intended in an English yacht rented through Lord Lonsdale but in the *Hohenzollern*[156] – in the spring.[157] The Empress Frederick was astonished at the idea, but nevertheless suggested to Wilhelm that while in Genoa he might pay an hour's visit to Queen Victoria and his widowed aunt Beatrice in Cimiez.[158] The Kaiser replied that he had already discussed this idea with 'Uncle

Chancellor', who had advised against a visit to France. 'He quite understood the feelings prompting me, but positively . . . declared that such an interview was *impossible* if intended on *French territory*. The Southerner is most excitable, incognito quite out of the question & an orange peel or apple enough to create international difficulties. So that if it is impossible to meet *on the water* (Bay of Villefranche?) I am unable to see dear Grandmama as long as she is in *France* on terra firma.'[159] After further negotiations with 'Uncle Hohenlohe' Wilhelm expressed the hope that the Queen would agree to meet him on the Italian side of the border.[160] As the Kaiser did not wish to put the Queen under any obligation to come, he asked his mother to sound out the Queen, and although in the end the meeting between the two sovereigns did not take place, the Empress Frederick's mediation nevertheless helped improve relations between Wilhelm and his mother.[161] She was able to report to the Queen in the spring of 1896 that the personal relationship with her son was no longer strained. 'He is quite nice to me and I have forgiven him with all my heart the cruel wrong he did me; but of course we meet seldom and I am a complete outsider, powerless to prevent the countless blunders & mistakes that grieve me so much – unable to do any good, – in constant anxiety about him & the course of events and the future. I even perceive that William tries to give me pleasure here and there and I am always very grateful for it – and pleased when things go well with him and when he does the right thing no matter whether it be great or small.'[162]

It seems, therefore, that after the Krüger telegram the Kaiser was advocating a conciliatory policy towards Britain with which leading spirits in the German army and diplomatic service did not agree. But does this picture of a Kaiser who in his heart of hearts was in fact an 'anglomane' accord with reality? Were not his feelings for his mother's country altogether more complicated, contradictory and impulsive? In fact, comments made by Wilhelm II during these very weeks, while tension was running high, clearly indicate that his friendly gestures were purely manipulative and designed to deceive British policy-makers as to his real purpose: to force Britain into a formal association with the Triple Alliance, including obligations in the event of war.

When Waldersee saw the Kaiser again on 18 January 1896 at the annual celebration of the founding of the Reich he noted that Wilhelm was 'deeply disgruntled' with Great Britain, a feeling which his entourage thought likely to last. Although he had been reconciled with his 'quite sensible' grandmother to a certain extent, Waldersee commented, he had 'had to swallow really outrageous rudeness from England', which he would not forget. He had become convinced 'on the basis of good information', the General continued, 'that we shall never be able to count on England with any certainty, but that the English have the immense effrontery not only to demand that we help them, but also to do nothing at all in return. He has probably given up his favourite idea of making England join the Triple Alliance, in which intelligent people have probably never been able to believe. But what is to happen now is quite uncertain. Russia and France have been more on our side in general, but that by no means presupposes that they would also be prepared to act with us. In both countries there is very widespread antagonism towards us and it will depend which of us proceeds more skilfully, we or the English. If it is the latter, we may find ourselves very isolated.' Waldersee welcomed not only Wilhelm's critical view of Britain, but also the 'pleasing' fact that the Krüger telegram had made the Kaiser 'popular . . . throughout Germany'.[163]

On the evening of 3 March 1896 – the day on which news arrived of the devastating defeat of the Italian army at Adua in Abyssinia – Wilhelm II received Sir Frank Lascelles for a two-hour, wide-ranging conversation which the diplomat, as Hohenlohe noted, saw as 'initiating good relations with England'.[164] As the Kaiser told the Chancellor the next day, he had pointed out to the ambassador that it was impossible 'that England could come to an understanding with France, as France's links with Russia were far too strong', and Russia 'considered it her principal task to destroy England'. The great mistake of British policy, the Kaiser had said, was that instead of supporting Italy in Africa Britain made difficulties for it there. 'Franco-Russian policy was working against Italy's action in Abyssinia with a view to gaining a firm foothold or power there with which to block the way to India for the English.' After the Italian defeat at Adua Britain must give Italy either military assistance from Egypt or financial help. Hohenlohe's summary of the conversation continues: 'After His Majesty had drawn attention to the danger threatening England that her route to India might be cut off, he stressed that England would have only the route round the Cape. But here too dangers loomed, for the French intended to buy the Canary Islands from the Spaniards and to make difficulties for England on that route too.'[165] The statesmen officially in control of German foreign policy received further information about Wilhelm's conversation with Lascelles from the ambassador himself, who sent Marschall a copy of his telegram to Salisbury reporting on the interview. According to this, the Kaiser had told the ambassador that Russia's aim was 'to destroy' both Austria-Hungary and Great Britain and, although peaceful means were to be used initially, Russia was also fully prepared 'to wage war, even if it lasted ten years'. Russia had declared to France that the recovery of Alsace-Lorraine was out of the question. Instead the Russian plan was 'to annex Bulgaria and the Balkan states as well as Austria's Slavic territories, and to detach Germany from Austria by offering her [i.e. Germany] the German provinces'. Russia would also drive the Italians out of the Horn of Africa and establish itself there, in order to 'get control of the sea route to India'. At the same time it would raise the Egyptian question again and also compensate the French with the Canary Islands. 'This plan', the Kaiser had claimed, 'was not only approved by Russian statesmen but had been sanctioned by Tsar Nicholas. Although His Majesty had been badly treated by the English press and by certain English statesmen, he considered it his duty to draw England's attention to this danger; he was delighted that the English fleet had been expanded and expected that England would join the Triple Alliance or at any rate come to Italy's aid in her desperate situation.'[166]

The Kaiser's conversation with Lascelles once again threw the Wilhelmstrasse into a state of bad-tempered confusion. In his journal the Reich Chancellor noted that both Marschall and Holstein were 'full of resentment against H.M.', because he had advised the British ambassador that Britain should (perhaps by sending an expedition to Dongola in the Sudan) go to the help of the Italians after their defeat. Giving this advice after the Transvaal crisis was, in their view, 'an action contrary to the dignity of Germany', which might even lead to war between Germany and both France and Russia. The elderly Prince Hohenlohe saw the situation differently, although he too described it as 'serious'. 'The ill-humour of the English towards us is not the result of the conversation between H.M. and Sir Frank and of the call to England to help the Italians, but it is the consequence of our colonial policy and of the prosperity of our industry and the competition it has created in world trade. The activities we have pursued for years in South Africa, and which culminated in the Kaiser's telegram, have finally aroused English resentment. But to blame H.M. for it, when the Foreign Office is responsible . . . , would be unjust.'[167] Hohenlohe may have judged correctly to some extent, but one cannot help wondering whether he fully recognised the momentous nature of the Kaiser's initiative. Not only was it an attempt to help Italy and restore better relations with Britain, but it also had the dual aim of preventing an Anglo-French rapprochement, which would amount to a *de facto* entente between the three world powers, Britain, France and Russia, against Germany, and forcing Britain into a binding union with the Triple Alliance. The expedition to Dongola which Britain did in fact mount in March 1896 was regarded by Wilhelm II as entirely due to him – Salisbury had simply followed the advice he had given to Lascelles, he claimed; he had never thought 'that the English would fall into the trap so easily'.[168] The Kaiser himself put his machiavellian motives for suggesting the Dongola expedition into these words: 'The goal has been achieved. England has gone into action, she is compromised and the flirtation with Gallo-Russia has been interrupted. It is all I wanted. I am satisfied. W.'[169] He remarked gleefully to the Reich Chancellor that he thought the British would 'get a hiding in their Dongola campaign', and told Prince Alexander Hohenlohe that although Britain had undertaken the expedition to

Dongola 'for purely selfish reasons', they would soon 'come crawling to us on their knees, if we just let them wriggle'.[170] The goal (which was also to give rise to the future battleship building race) of making life difficult for the 'arrogant Britons'[171] in the expectation that they would join the Triple Alliance was a constant but extremely costly miscalculation at the heart of the new German foreign policy.[172]

After the conversation with Lascelles the Kaiser boasted that 'he had never made such rude remarks about an ambassador's country to his face as he had to Lascelles'.[173] Only later did Holstein hear privately from Hatzfeldt, who as he said would sooner have had his hand cut off than report officially on it, about the British Prime Minister's distressed reaction to Wilhelm II's conversation with the ambassador. Salisbury, 'who had just begun once again to become more candid and more confidential', was 'literally horrified . . . because he saw herein proof that he would again be confronted with all sorts of demands and that he could not hope for a calmer conduct of policy on our part'. Salisbury had ended by saying: '*Je vous avoue que cette agitation croissante m'inquiète vivement* [I confess that this growing agitation worries me greatly].' Since then Hatzfeldt had sensed a renewed reserve in Salisbury in discussing political questions, which made him very anxious with regard to Anglo-German relations.[174] Even in the restrained official language of diplomacy there was no mistaking the Prime Minister's despair when, in response to the Kaiser's remarks to Lascelles, he once again tried to make the principles of British policy in Europe clear to the Wilhelmstrasse. Great Britain wished to be on a friendly footing with Germany as in earlier days, and was indeed willing to come closer to the Triple Alliance, Salisbury explained. But it would 'never make a promise which entailed the obligation to go to war in any future eventuality'. Such a commitment, which in any case was unnecessary thanks to the country's position as an island, would never be accepted by public opinion. 'Whether this policy was sensible or not it was the only one possible in England; during his previous ministry from 1886 to 1892 . . . His Majesty the Emperor had been content with it; why was he not so now?'[175] One answer to this question must be that since that time the power of Kaiser Wilhelm II had grown enormously. In the first half of 1896, as the next chapter will show, he succeeded in making the breakthrough to Personal Monarchy in the final battle against the 'responsible government'.

CHAPTER TWENTY-SEVEN

Endgame: the breakthrough to decisive personal power

A HIGHLY CHARGED EMOTIONAL STATE

OBSERVERS both in Germany and abroad were at one in the opinion that Wilhelm II was particularly on edge and excitable in the first few months of 1896. His close family circle, court society (with the house arrest at Schloss Glienicke and the Kotze scandal) and even the cabinets of Europe (with the Dardanelles question, the Transvaal crisis, relations with Britain and naval expansion) witnessed outbursts of rage and impetuous actions which could not but be regarded as extremely worrying, indeed pathological, in their nature. But the event which angered him most was the ultimatum he was given by the Reich Chancellor and the entire Ministry of State in the Köller crisis of December 1895, which he regarded as an inadmissible infringement of his most sacred prerogatives as king of Prussia. In the months that followed Wilhelm II not only showed through his aggressive, autocratic attitude and vindictive comments how deeply offended he was by the collective action of the constitutional authorities; he also made it absolutely plain that he would not hesitate to seize supreme decision-making power in Prussia and in the Reich once and for all. On the advice of his best friend Philipp Eulenburg, his three Cabinet Chiefs Hahnke, Lucanus and Senden and the ubiquitous Flügeladjutanten, he sacked the supposed ringleaders of the 'ministerial revolt' – Bronsart von Schellendorf, Marschall von Bieberstein and Boetticher – and replaced them with men like Bülow who had promised to be nothing but 'executive tools of His Majesty'. With the dismissal of General Bronsart as minister of war in August 1896 the 'responsible government' finally lost all scope for independent action and the way was clear for Kaiser Wilhelm II to exercise 'real personal rule'.

When Hinzpeter visited Berlin in January 1896 he at once noticed that a fundamental shift of power had taken place since the Köller crisis. He told Boetticher that 'the position of the Reich Chancellor was no longer what it had been'; on the other hand that of the Chief of the Civil Cabinet, Lucanus, had 'grown much bigger'.[1] The Chancellor complained to Eulenburg that he had received no answer to a written request for an audience and had been so hurt by this that he felt he might have to distance himself more from the Kaiser and avoid private audiences with him altogether.[2] During a visit to Berlin in March 1896 Eulenburg noted that the Kaiser's antipathy towards Marschall was now '*insurmountable*'. When Eulenburg drew attention to the Foreign Secretary's parliamentary successes Wilhelm replied: 'That is just the trouble: his position gets better and better and I dislike him more and more. What can one do?'[3] The monarch's rancour towards the Chancellor and the ministers made itself powerfully felt at a dinner on 30 April 1896 when he rounded on all the ministers except Thielen and Hammerstein. He asked the new Minister of the Interior, von der Recke, for news of 'good, honest Köller', and commented: 'He was brought down by the same sort of disgraceful intrigues that are going on now.' Addressing the Foreign Secretary, he said: 'I tell you, Marschall, I shall not tolerate these press campaigns much longer. I demand that you put a stop to this business. If it goes on I shall move my headquarters to Berlin and arm my Flügeladjutanten with pistols to go and kill the villain who is behind it. The attacks are not against Hahnke but against me.' Marschall's reply was blunt: 'Yes, they are indeed against Your Majesty.' The Kaiser had a similar exchange with Bronsart. When Wilhelm concluded his remarks by saying that things could not go on as they were, the Minister of War answered frankly: 'I quite agree with Your Majesty, things cannot go on like this.' The Kaiser turned away in a fury, retorting: 'Yes, but I do not mean it in the same way as you.'[4] Wilhelm displayed petulant contempt for the responsible statesmen at every possible opportunity. In May, instead of enquiring after the health of the elderly Reich Chancellor, who was ill, he summoned him to Potsdam in bad weather, although he himself was in Berlin every day. His reaction to a perfectly reasonable suggestion from the Ministry of State that three noblemen and three wealthy commoners be appointed to the Prussian Upper House was to decide, after a report from Lucanus, to approve the appointment of the nobles and reject that of the three middle-class candidates. Holstein concluded disconsolately that 'Reich Chancellor and Ministry of State *ne comptent pour rien* these days. Nothing is done except by His Majesty and the Cabinets.'[5]

If Wilhelm yielded to representations from the statesmen of the Wilhelmstrasse it was only rarely and under enormous pressure. A typical example occurred in the early summer of 1896 when the situation in the German colony of South West Africa became so serious that the colonial officials there made an urgent request for a few hundred mounted infantrymen to be sent out. They also demanded the introduction of compulsory military service for citizens of the Reich living in the colony. Hohenlohe, Marschall, Bronsart, Admiral Hollmann and the Commanding Admiral of the Navy, Eduard von Knorr, were firmly in favour of sending the requested number of Volunteers to the colony. The Chief of the Naval Cabinet alone recommended instead that a naval battalion – 'notwithstanding the fact that no troops have had less training in riding', as Holstein acidly remarked – be sent out to this desert terrain instead, 'and – His Majesty supported Senden'. Furthermore, the Military Cabinet under Hahnke held out against the introduction of compulsory military service in South West Africa. Only after a joint audience with the Kaiser attended by the Reich Chancellor, the Foreign Secretary, the Director of the Colonial Department Dr Kayser and the Minister of War, together with Senden and Hahnke, was the 'responsible government' able to assert its authority in both questions.[6]

The vexed question of the succession to the minor princedom of Lippe gave rise to a dispute which provides a perfect example of Wilhelm's attitude towards his most senior officials and towards the federal constitution of the Reich in general. Since the death of the childless Prince Woldemar of Lippe-Detmold on 20 March 1895, Wilhelm II's brother-in-law, Prince Adolf of Schaumburg-Lippe, had assumed the regency on behalf of the mentally ill Prince Alexander, on the basis of a decree of 1890. This settlement, however, was stated by Count zu Lippe-Biesterfeld to be invalid, and the Landtag of Lippe also doubted its legality. The Landtag's decision to have the dispute over the succession settled by the Reich court in Leipzig was challenged by several of the German federal princes, who saw it as a curtailment in principle of their sovereignty. They demanded instead that the dispute be settled through a compromise reached between all parties involved. Hohenlohe informed the Kaiser in a submission of 20 April 1896, drafted by Marschall, that as Reich chancellor he had been charged by the Bundesrat with the task of bringing about such a compromise. He had ascertained that all the parties to the dispute were prepared to accept the decision of an arbitration tribunal presided over by the King of Saxony, who would have the right to select a number of judges from the Reich court for this purpose. Hohenlohe had taken the necessary steps to establish whether and under what conditions the King of Saxony would be prepared to preside over the tribunal.[7]

The Kaiser's furious reaction to this independent action by the Reich Chancellor shows how strongly he felt that he had been bypassed, especially in a matter of this kind. 'Do I still rule at all!?', he wrote in the margin of the submission. The King of Saxony certainly could not choose the judges 'without even asking or informing me!' The tribunal must be held 'under a chairman to be selected by *me* from 2 Princes to be chosen by each of the parties'. 'I am astonished that such a highly important question has been settled in this very surprising manner, without consulting me and obtaining my agreement. Only *one* man in the Reich can be the arbitrator in this question and that is *the German Kaiser*! Any other alternative is *unthinkable*.' All the other princes to whom he had spoken took 'absolutely the same view!', he claimed. Only Duke Georg II of Saxe-Meiningen, whose youngest son Friedrich was married to a Countess zu Lippe-Biesterfeld, had expressed support for the claims of the Biesterfeld line, the Kaiser declared, adding, 'The indignation here at the Duke of Meiningen is very great. Wilhelm I. R.'[8] The Kaiser's sisters added fuel to the fire by persuading him that a decision in favour of the Biesterfeld line would be 'a victory for democracy'. Both Charlotte and Moretta had no scruples in spreading the rumour that the mother of the reigning Countess zu Lippe-Biesterfeld had formerly worked as a shop assistant, while her father had been an American small-holder. The 'Biests' were '*disgusting, common*' people who had taken on princely airs. 'Can the German princes tolerate *such* a pack of worthless wretches?? . . . By this criterion any Herr Müller or Fräulein Schulz could become regent . . . What would Grandpapa or Papa have said?'[9] Wilhelm II undoubtedly thought of the succession dispute in similar terms.[10]

On 22 April 1896 Lucanus came to the Chancellor with the news that the Kaiser was 'very annoyed' and had declared in no uncertain terms that there was only one arbitrator in the Reich, and that was himself. The Chief of the Civil Cabinet suggested that the approach to the King of Saxony be withdrawn 'until the Kaiser had calmed down', which the Chancellor declined to do on the grounds that the Bundesrat, inclusive of the Prussian vote, had entrusted him with the task of bringing about an agreement between the parties involved by means of an arbitration tribunal, and that he had carried out this task. 'It is not clear to me what occasion there would have been for *the Kaiser's* intervention.'[11] In a further submission to the Kaiser he warned that it would be unwise to call into question the solution which had been reached. It would inevitably wreck the chances of any compromise and would thus perpetuate the present state of affairs and seriously damage the monarchical principle.[12] This missive was likewise returned to Hohenlohe with an angry comment scrawled on it in pencil. 'The request which I sanctioned related only to the discussion in the Bundesrat. The further steps which the Reich Chancellor wanted to take or took were not possible without my opinion being ascertained and needed to be discussed with me first. That was not done. I have once again been confronted with an extremely painful *fait accompli*. It is unacceptable that in a question with such serious consequences the Reich Chancellor should negotiate independently over the Kaiser's head with German princes, and even offer them the position of arbitrator, without my knowledge and permission. I stand by my opinion that in this matter the only right course is a princely court with judicial advisers under a chairman appointed by me. Moreover an opportunity could perfectly well have been found in Vienna [in mid-April 1896] during the audience at the German embassy to discuss this matter with me; why did this not happen? Quarrels between Reich princes cannot be settled by any court; it can only be done by the princes presided over by the Kaiser or by a representative, e.g. the Reich chancellor, designated chairman by All-Highest order. Wilhelm I. R.'[13]

The Chancellor reacted unusually firmly to this 'most ungracious marginal comment', declaring: 'With the best will in the world I cannot do the man *this* favour.'[14] He sent a cipher telegram to the Kaiser explaining the complicated legal situation again and persisting in the opinion he had previously expressed. There was no prospect of the parties in dispute accepting a princely court in the form demanded by the Kaiser. He 'reverently and urgently' begged the monarch not to do anything to disavow the negotiations which he, as Reich chancellor, had conducted on the instructions of the federal governments. Hohenlohe, who was himself a mediatised prince, noted privately at the time that this was not a quarrel between Reich princes but a dispute over a succession within one and the same princely house, whose members could come to whatever agreement they wished. But even in a quarrel between princes of the Reich the Kaiser did not hold the position of arbitrator. 'In the old days the Kaiser was the supreme feudal lord and the princes were his vassals. That has changed now. The Kaiser is *primus inter pares* and the princes are his allies.'[15] On 4 May 1896 Lucanus finally informed the Chancellor that the Kaiser realised that he would have to accept a tribunal, but to save face he insisted that some way should be found of 'acknowledging the Kaiser's authority'.[16]

Holstein and Marschall were brought to the brink of despair by this latest manifestation of the Kaiser's anachronistic views. What would have happened, they wondered, if the government had yielded to Wilhelm's desire to have the same power of disposal over Lippe and other federal lands of the German Reich 'as Charlemagne had over vacant fiefs?' Holstein doubted 'whether the Kaiser knows that the present German princes are his allies, while the German princes at the time of Charlemagne were his vassals'. Bismarck had designed the constitution of the Reich in such a way that the federal princes would rely on the support of the Reich government against the particularist elements in their states. 'But if the instincts of our present Kaiser were translated into action, the feeling of security within the Reich on the part of the individual federal princes would be wholly lost.'[17] Equally worrying for the Wilhelmstrasse was the

Kaiser's impulsive behaviour, of which this latest incident was yet another crass example. The Kaiser had the 'unfortunate habit', Holstein commented bitterly, 'of talking all the more rapidly and incautiously the more a matter interests him'. He went on to reflect that 'the chief danger in the life of Kaiser Wilhelm II is that he is and remains absolutely unconscious of the effect which his speech and actions have upon princes, public men, and the masses'. Consequently, Holstein observed, the most vital task for any government under Wilhelm's rule would be to counteract this danger and neutralise the effects of the Kaiser's character as far as possible.[18]

There was an unpleasant epilogue to this conflict during the Scandinavian cruise of July 1896. 'With deep sorrow and grief', Wilhelm wrote to his 'dear Lucanus' of the 'unbelievable stage' which the 'wretched Lippe affair' had reached. 'Yesterday I received by messenger the draft agreement for the arbitration tribunal in the Lippe affair, together with the final record of proceedings, which were sent for me to see and put into execution if I thought fit.' He had not signed the document, as he wanted it amended. 'In the passage referring to the possibility of the king not acting as arbitrator for any reason it is stated that the parties should be free to choose another prince of the Reich as they see fit! I crossed that out and with Ct. Eulenburg's agreement I wrote "the parties are to apply to the Kaiser, who will select another arbitrator". I had just finished reading it through when a telegram arrived from the Foreign Office to the effect that "through an indiscretion the agreement had already been published in the †Zeitung and I should therefore hurry up and sign it as soon as possible"! I am outraged at this unbelievable incident! The leaking *in politicis* at the higher levels is quite unprecedented. I ciphered the Reich Chancellor at once to say that I could not yet sign in such circumstances, because of an indiscretion, and that I was contemplating further amendments. He should at once order an investigation and establish who the culprit was and punish him as appropriate. Or if people in high places should be compromised, he should disgrace them publicly. I would not sign until the investigation, which should be announced, was over. Please see to anything further which might be necessary . . . Your affectionate Wilhelm I. R.'[19]

Although he was eventually forced to give way over the Lippe succession dispute,[20] the autocratic attitude which Wilhelm II displayed towards the federal princes in this affair as in many others – for instance his decree to the German princes of November 1887, which Bismarck ordered 'to be burnt without delay'[21] – manifested itself alarmingly in yet another case in the summer of 1896. During the festivities in Moscow for the coronation of Tsar Nicholas II, which Prince Ludwig of Bavaria was attending as representative of Prince Luitpold, the Prince Regent, Prince Ludwig protested vociferously against a speech in which he and the other German princes present had been described as the 'suite' of Prince Heinrich, the Kaiser's brother. 'We are not vassals but allies of the German Kaiser', he objected. Thinking to take the sting out of this embarrassing incident,[22] the Bavarian Count Konrad von Preysing-Lichtenegg-Moos suggested that Prince Ludwig should 'take the road to Canossa', as it were, by going straight from Moscow to see the Kaiser, accompanied by Preysing himself. Preysing's suggestion was strongly supported by the Foreign Office, Philipp Eulenburg and Monts, the Prussian envoy in Munich, since it was feared that otherwise Prince Ludwig would become 'the focus of inflammatory particularist demonstrations' on his return to Bavaria.[23] But in spite of the Reich Chancellor's carefully argued support for this course, Wilhelm II refused to receive Count Preysing, who had previously been a Centre Party delegate in the Reichstag.[24] When Eulenburg asked him about it, the Kaiser declared roundly: 'I do not receive a man who has openly lied to me' – an allusion to Preysing's vote against the Army Bill in the Reichstag in the summer of 1893.[25]

The Kaiser's meeting with the heir to the Bavarian throne, the future Ludwig III, on 29 June at Kiel, was not exactly a harmonious occasion, not least because, as the Bavarian envoy Count Lerchenfeld knew from experience, it was 'impossible to get anywhere' with Wilhelm when he was 'with his fleet among the sailors'.[26] Just before he left for his Scandinavian cruise the Kaiser telegraphed to the Chancellor from Wilhelmshaven: 'Prince Ludwig of Bavaria made his journey to Canossa yesterday. The course which the conversation took showed that far from regretting his unpatriotic and un-German conduct the Prince laboured under the impression that he had done a great deed for the German princes in general. He used the expression that it had made his blood boil to hear the word "suite", and he had therefore given vent to his oppressed feelings. He maintained that they were not vassals but were unfortunately often treated, or had been treated, as such. I at once dismissed this assertion very sharply and challenged him to name one instance which proved what he had said, but he could not do so and corrected himself, but claimed to know that there was a certain party whose aim was to have the princes of the Reich treated in that way (i.e. as vassals). My comment that I knew nothing of a princes' party put an end to all further comment. He then went on to refer to the fact that they had all been sent to Moscow as representatives of their independently ruling dynasties. But that did not seem to have been quite clear to the Russians, as they had been treated summarily, in spite of the fact that they had their own envoys, while my representative had been accorded the most honourable status. This prompted me to give the noble gentleman a lecture on the fact that the individual South German states were not – thank God – required to represent the Reich abroad nor to take responsibility for war and peace. Bavaria, Württemberg and

Saxony did not conduct great international policy; the fact that they had been allowed to keep their envoys was an act of courtesy and their main purpose was to maintain good family relations between the courts. The whole responsibility for the welfare of the Reich, its relations with foreign countries, were a matter for the Kaiser alone and I was therefore not prepared to allow any meddling, and for that reason I deplored the fact that through his remark, even if made *bona fide*, the impression had been created abroad that things were not going as well as they should in the German Reich. Wrong conclusions would be drawn, and wrong conclusions could easily lead to fatal decisions which might possibly lead to serious consequences.' After an interruption, the Kaiser continued this remarkable account of his conversation with Prince Ludwig. 'The Prince then mentioned that he had been pleased to see, from the great enthusiasm for his courageous support for the threatened right [of the princes] expressed in telegrams and letters sent to him from Bavaria and also from elsewhere in the Reich, that he had in fact done the right thing. There would also be great celebrations and ovations in Bavaria on his return, and he had therefore come here first, so that these ovations should not be misunderstood. The Centre had unfortunately treated the affair rashly and used it to its own advantage, which greatly displeased him. I responded that the ovations in Bavaria were a matter of complete indifference to me; as far as the Centre was concerned, it had at last shown how inimical it really was to the Reich in its thinking, and had opened the eyes of even the most stupid among the German people to its ultimate goals. In doing so it had forfeited all confidence and respect and had thoroughly confirmed its anti-Reich tendencies . . . I asked the Prince, in case he intended to make any speeches on the occasion of the expected ovations for his courageous support for the German princes whose freedom had been threatened, to give his audience the correct explanation of the expression "allies" or "allied governments". People seemed to have a completely wrong idea of it in his Fatherland. Bavaria was not an ally of Prussia or of another state in the sense of e.g. Austria or Italy with Germany, it did not have the power or freedom to loosen or dissolve its ties with the German Reich and to act as a free agent. It ought not to be forgotten that apart from the bonds created by the blood which the German tribes shed together 25 years ago to recover their unity, we were bound by the iron bond of a common Reich constitution, just as we shared a common system of justice and common finances. But above all we had a common Reich army under a war lord, and that was the Kaiser. Anyone who sought to detach himself from such ties, for whatever reason, was committing a breach of the constitution and would bring the full consequences of this on his own head. The Prince was silent, apparently very dismayed, and looked extremely embarrassed. Moreover, I went on to say, his colleagues were not at all edified by his behaviour, as far as I had heard. He maintained that he knew the opposite to be true, and even believed that they were very grateful to him. To cut off further long-winded repetitive arguments which threatened to go round in circles, I observed that I had just spoken to the German princes at Kyffhäuser and had received quite a different picture of the situation from their comments about him, but asked him to excuse me from passing them on to him. I was grateful to him, however, for coming and for discussing the matter with Me. The question did not concern My Person, but the German Reich, which I had the honour to represent and personify, and to which he had offered the requisite satisfaction by his visit. He took his leave, thanking me for receiving him and remarking that he considered that he had done his duty and regretted that it had been wrongly interpreted. Wilhelm I. R.'[27] Prince Ludwig, who had expected a friendlier reception after having done penance, understandably took lasting offence at this arrogant lecture and complained that the Kaiser had 'bawled him out'.[28] When the Kaiser repeated the 'very clear' words of his telegram to Eulenburg during the Scandinavian cruise, the latter concluded that this dressing-down would serve only to deepen the gulf which already existed between Wilhelm and the 'very touchy and resentful Wittelsbacher'.[29]

THE BRONSART CRISIS AND THE HOLSTEIN PLOT

The decisive battle for power which took place in the first half of 1896 was not, however, between Wilhelm and the federal princes but between him and the 'responsible government', that is to say the Reich Chancellor and the ministers of state. Among the key figures in the dispute was the argumentative War Minister General Bronsart von Schellendorf, whose reform plans had been at the heart of the Köller crisis and still threatened to trigger an explosion even after the forced dismissal of the Interior Minister.[30] At the beginning of the year, after a conversation with Wilhelm, Waldersee accurately summed up the situation in his diary, noting that 'the Kaiser was decidedly put out . . . by the fact that the entire Ministry forced him to choose between dismissing Köller or the Ministry. He gave in, but as Bronsart played the most important part in the affair, he is *particularly* put out with him.'[31] Bronsart was also 'very disgruntled', he added, and ready to resign if the Kaiser continued to oppose his policy for the reform of courts martial.[32] The conflict erupted stormily after the Kaiser's speech at the New Year reception of the commanding generals. 'Perhaps without fully realising the consequences of what he was saying', Waldersee commented, Wilhelm had spoken out about 'several serious issues', thereby provoking the Minister of War into submitting his resignation. Referring to the kernel of Bronsart's reform programme, the new regulations for courts martial under which public (instead of secret) and oral (instead of written) prosecutions would be introduced for the army, the Kaiser said 'he firmly maintained the stand adopted by his grandfather in rejecting public hearings, and was confirmed in this by many expert opinions'. He then declared that the fourth half-battalions introduced by Caprivi as a concession to the Reichstag in 1893 must be abolished 'in the very near future'; in 1899 increases in the army estimates would in any case have to come, and 'then we should be faced with the burning question of whether it would not be as well to *reintroduce 3-year military service*'. Finally the Supreme War Lord expressed his 'clear and well thought-out' (according to Waldersee) views on the question of armaments. At present, he said, the Italian army had a rifle that could be regarded as the best in the world; 'whether we can proceed to adopt it is a question that still needs to be investigated; it depends also on changing to a new gunpowder and the question of cost would be very momentous'. A new rapid-firing gun had been built for the artillery and had proved effective; he had ordered 200 to be made initially, which were destined for the frontier corps and would 'ensure great superiority for us'.[33]

The reception was scarcely over before the Minister of War told Waldersee angrily that the Kaiser's remarks about the courts martial, which showed that he had taken an 'absolutely definite' stand on the question, were unacceptable to him and 'he would be handing in his resignation today'.[34] The violent thunderstorm which broke that New Year's afternoon in 1896, and which brings to mind the inkstand scene of 15 March 1890, was so alarming that even the loyal Chief of the Military Cabinet had to admit that 'the Kaiser had gone too far and had lost all sense of moderation'. The Chancellor, who received various reports of the incident, commented: 'The scene . . . was such that the War Minister had difficulty in restraining himself.' Bronsart himself told Hohenlohe that 'he would have reached for his sword if it had been anyone else'. It caused consternation in the General, who expressed the opinion 'that H.M. did not seem quite normal and he was extremely worried for the future'.[35] During the dinner to which Walter von Loë had invited his fellow generals on the evening of 1 January Bronsart was still burning with anger.[36] It was not until the next day that the Kaiser saw the error of his ways and invited Bronsart to dinner, during which he told him in front of Hahnke and Plessen that he had been 'too hasty' and asked his forgiveness, whereupon the Minister of War withdrew his resignation.[37] In Waldersee's eyes the Kaiser had acted wisely in making a tactical surrender, for if Bronsart had resigned at that time over the issue of public courts martial it would have 'aroused widespread feeling against the Kaiser outside the army'. Nevertheless Waldersee had no doubt that the 'rift' between Wilhelm II and Bronsart was now irreparable, for 'mutual trust' had been 'considerably undermined by a whole series of differences of opinion'. The Kaiser would never forget 'a thing like that'

40. General Walter Bronsart von Schellendorf, Prussian War Minister 1893–1896

and would 'wait for a more favourable moment to rid himself of Bronsart', he predicted.[38]

The long-drawn-out conflict of the next few months was thus preordained, for while Wilhelm continued to refuse steadfastly to concede public courts martial, the Minister of War remained 'very much annoyed' by this All-Highest attitude, and determined to hand in his resignation 'certainly before the end of this year'.[39] But as Bronsart could count on the ardent support of the overwhelming majority of the German people and of the parties represented in the Reichstag for his modernising measures, neither the Reich Chancellor nor the other Prussian ministers felt able to take a less reformist line on this question than the fiery Prussian General himself. If he resigned because the Kaiser refused his assent to the reforms, Hohenlohe and all the ministers of state would have to resign too. The fatal deadlock which had forced the Kaiser to dismiss his favourite minister in the Köller crisis had arisen again. But this time it was to have a different outcome.

At first the Kaiser tried to avoid a decision on the explosive question of court martial reform on the pretext that he wanted to seek expert opinions on the subject from the commanding generals, but that these would not be ready until the end of the year. When the Reich Chancellor submitted a laboriously worked-out compromise solution to him in March 1896, Wilhelm was at first 'stunned and *extremely incensed* with Hohenlohe',[40] but under Philipp Eulenburg's soothing influence he turned down the suggestion on the grounds of the consultation with the generals. 'In accordance with what you had agreed I settled with the War Minister that in a question of such fundamental importance the army, i.e. the comm. generals, should be consulted first', he wrote to his 'dear Uncle' on 14 March. 'After he had thanked me, touched by my willingness to co-operate, and declared that whichever way the army voted he would accept it unconditionally, I sent for the documents from the War Min[istry] with a view to preparing a memorandum for the commanding generals. The documents were handed over hesitantly and late and were very voluminous, so that the work will require months. The date for the replies was then set at the winter of 96, for which your agreement was also obtained, so as to give those who were being consulted sufficient time and latitude to answer. Consequently nothing can be done in the matter until this year is out, I shall not comment until the army has spoken, until then I am not in a position to speak.'[41] In an audience on 16 March 1896 Wilhelm warned the elderly Chancellor against Bronsart and Marschall. The former was aiming to create a rift between Hohenlohe and the Kaiser while the latter wanted to become Reich chancellor.[42]

Friedrich von Holstein and Foreign Secretary Freiherr von Marschall were so enraged by the autocratic and militaristic behaviour of the Kaiser on the one hand and the spineless manoeuvrings of the Reich Chancellor on the other – Holstein complained that Wilhelm and his advisers at court had 'softened the old gentleman . . . so much' that 'he has lost all trace of resistance' – that they became the *de facto* allies of the War Minister against Wilhelm and his favourite Eulenburg.[43] Since 1895 Holstein had been conducting a fascinating debate with Eulenburg on constitutional developments since Bismarck's fall and in particular on the dangers of Wilhelm II's personal style of government, which as we shall shortly see was to have serious practical consequences in the spring and summer of 1896. Not long after Hohenlohe had taken office Holstein had written to Eulenburg setting out the fundamental difference between their ideas of the role of the monarchy in modern society. 'You support the principle . . . that the people must submit to the Kaiser. I consider this principle impracticable, as far as the *German* people and the *German* Kaiser are concerned . . . But the Kaiser, unfortunately, shares *your* views . . . in that at every possible opportunity he flies in the face of public opinion and behaves – much more than Nicholas II does – like an autocrat.'[44] Since then the disagreement between these former allies had intensified almost to the point of an open breach. At the end of 1895 Holstein commented in a letter to Eulenburg, looking back, that 'for the first time in the new year both of us, you and I, were *not* pulling in the same direction. I certainly consider myself a royalist, but I do not favour absolute submission to His Majesty's

41. The Grey Eminence of the German Foreign Office: Geheimrat Friedrich von Holstein

every line of thought in the same degree as you. As I know how honest you are I do not reproach you at all, and I honestly hope that you will not come to reproach yourself one day, when you see where all this leads.'[45] When the Bronsart crisis entered its most acute phase at the beginning of 1896, Holstein delivered a lecture to the Kaiser's favourite on his attitude. 'You instinctively incline to an autocratic regime no matter whether it be Russian patriarchal or *despotisme éclairé* on the French model. I am in favour of a moderate use of a practicable system of constitutional co-operative government which, with the exception of St Petersburg and Constantinople, is in operation in the rest of the European and civilised world. My opinion is, I know, unfashionable at the court here. "A strong government which can manage without the Reichstag" is Admiral von Senden's ideal, and not his alone. You also belong, perhaps without knowing it, to those who believe that every political, military and legal question is best decided directly by the Kaiser. The old English Cavaliers held similar beliefs . . . To be sure they first ruined the Stuarts, and then they died or ruined themselves in misery and want; but from an ethical standpoint they remain disinterested types of superb chivalry. I am not so chivalrous, I am for the possible, and since it seems to me that "governing without a Reichstag" is at present impossible in Germany, I would prefer that this Chancellor's political existence between this Kaiser and this Reichstag were not made impossible.'[46]

In his blinkered monarchism and personal devotion to the Kaiser Eulenburg failed to recognise the dangers to which his erstwhile ally was alluding in this letter. In his reply he strenuously denied having joined the 'autocratic tendency'; it was not he but Holstein, who did not know the Kaiser personally, whose political views had undergone a change since Wilhelm's accession, the Kaiser's friend claimed. 'The Holstein of 1888 with his Old Prussian loyalty to the King has certainly not become an anti-monarchist, but he has become a parliamentarian.' He and Holstein nevertheless continued to share the same political ground. 'We both serve . . . the Kaiser', Eulenburg maintained. 'One with love, the other without love.' Holstein should never forget the intimate personal bond which united them, he insisted. 'For that to be possible it is necessary to have a certain – I would almost say feminine – sensitiveness that is peculiar to us both.'[47] But in his letters to Bernhard von Bülow, whom he had recommended to the Kaiser as a future Reich chancellor, he adopted a harsher tone. He dismissed Holstein's growing disquiet with the all too simple psychological explanation that the Privy Councillor had become enraged by 'the progressively developing political sense of the Kaiser' because he was 'no longer in sole control'.[48]

This fundamental conflict led to an enthralling final battle for the future of the monarchy in the German Reich, fought out in the spring and summer of 1896. With the help of Marschall and later also of Alexander Hohenlohe, Holstein worked on a plan which he hoped would break the autonomy of the Prussian crown to the advantage of the 'responsible government' and by extension to that of parliament. In the event that the Kaiser refused to approve Bronsart's popular reform programme, Hohenlohe should threaten to resign along with the entire government, as in the Köller crisis. Philipp Eulenburg, on the other hand, saw it as his principal task to undermine this disgraceful plot, in which he was sure that Bronsart was also complicit, in order to rescue the unlimited decision-making power of his beloved Kaiser from the constitutional control exerted by the Wilhelmstrasse and the Reichstag.

As early as 1 February 1896 Holstein went to see the Reich Chancellor and urged the necessity of 'forcing the Kaiser to approve the court martial bill on the lines proposed by Bronsart'. He had no success at all. 'I shall take care not to provoke another conflict now', Prince Hohenlohe noted in his journal at the time.[49] A few weeks later the Chancellor commented that 'H[olstein]'s proposal amounts to using the present situation in order to put H.M. in a position of constraint by means of a demand by the entire Ministry of State that only a Court Martial Reform Bill conceding public trials be brought into the Reichstag. In Holstein's opinion, H.M. will not now be able to withstand the Ministry's pressure and will therefore give way, and my position will thereby become so strong that no one would dare simply to show me the door, as they now intend.'[50] Only days later the Reich Chancellor returned to the subject, noting that 'the proposal to exploit the present situation to push through the court martial question with

public trials which the Ministry of State has determined upon is to be carried out by first raising the matter in the press and the Reichstag and then sending a categorical demand from the Ministry of State to H.M. that he allow the draft bill . . . to be laid before the Bundesrat. Now, it is said, the Kaiser would be forced to give way. If the present opportunity is missed, the fear is that H.M. will dismiss the Ministry in the autumn without further ado, and will try to push through the court martial bill without public trials, as well as his fleet plans, with the help of a different Reich chancellor and a different Ministry. H.M. ought therefore to be put in a position of constraint, as in the Köller case.'[51] In June Holstein again implored the Chancellor to take advantage of this propitious moment to curb the personal power of the Kaiser. On 15 June 1896 – the eighth anniversary of Wilhelm II's accession – he came into Hohenlohe's office, as the Chancellor recorded, and insisted 'that things could not go on like this, that I must take a firmer stand with the Kaiser . . . The Kaiser – in his opinion – thinks he can do anything he likes and does not have the necessary respect for me.'[52]

Eulenburg got wind of this conspiracy on a visit to Berlin in March. He was outraged: Holstein, Marschall and Bronsart had a '*very evil attitude*' towards the Kaiser – the word 'very' was underlined three times. All three were working towards a decisive crisis 'which will destroy His Maj[esty]'s authority and deliver him bound hand and foot'. As in the Köller crisis 'the three conspirators' wanted to bring about a situation in which the Reich Chancellor would have to declare his '*solidarity* with the Ministry against H. Maj[esty]'. Hohenlohe, who wished to remain in office for the time being and depart on peaceful terms with the Kaiser later on, recognised the danger which lay in the court martial question above all. Like Eulenburg, he was afraid that Holstein, Marschall and Bronsart would use the press and the Reichstag to try to force him to take a stand against the Kaiser. In their present opposition to the Kaiser the conspirators were 'quite unrestrained, and shrinking from nothing, like certain people (including Holstein) before the dismissal of Prince Bismarck. I can only tell you that I am *horrified*!', Eulenburg wrote to Bülow.[53] Three days later he reported exhaustedly: 'These have been very difficult days, dearest Bernhard! . . . Only His Maj[esty]'s love and his gratitude for the patching up I have done can compensate for the real damage to my nerves . . . There is *no doubt whatsoever* about the existence of the Holstein-Marschall-Bronsart *League* . . . The exploitation of positions of constraint in order to pull the rope tight around the Kaiser's neck has become *systematic* and premeditated. Holstein goes so far as to use every petty detail for this practice, which *revolts* me *deeply* . . . It is *beyond all doubt* that this group is provoking rows *of all kinds* in order to bring Hohenlohe into conflict with the Kaiser in such a way as to demonstrate Hohenlohe's solidarity with the others and so *force* His Majesty to give way.' Eulenburg had nevertheless succeeded, he reported, in separating Hohenlohe from the conspirators and bringing him together with the Kaiser. He added that he had also achieved a partial reconciliation with Holstein on the basis of 'mutual *personal* inclination' and of foreign policy, although their differences on domestic policy were greater than ever. Holstein had even advised him 'to give up all mediation in the interest of H. Maj. now and altogether!!', the Count commented indignantly. 'So the system feels threatened by Hohenlohe's willingness to give in and my friendly influence!'[54]

While the Kaiser was visiting Abbazia, Eulenburg and Bülow met secretly in Venice to discuss their strategy against the 'conspirators' of the Wilhelmstrasse.[55] Meanwhile, Holstein, Marschall and Bronsart were using their contacts with the press to suggest that Eulenburg would succeed Hohenlohe on the basis of his opposition to court martial reform, while maintaining that Hohenlohe and the entire Ministry of State were totally united behind modernisation. Eulenburg had no doubt that the Kaiser would connect this press campaign with the Foreign Office. 'The "positions of constraint" which we have discussed and feared are coming to the fore, and this game can really only end with a violent solution', he wrote to Bülow on 29 April. At the very moment that he made this prediction he received 'very angry' and '*extremely* excited' telegrams from the Kaiser which took the words out of his mouth. 'Prince Hohenlohe very wretched and overwrought. M. and B. are playing their crazy game worse than ever in the most impudent way. All hell is let loose and there will have to be a thunderbolt. Wilhelm I. R.' For the first time Eulenburg wondered whether in these circumstances it would not be better for Bülow to come to Berlin as secretary of state at the Foreign Office initially, instead of becoming Reich chancellor straight away. If the Kaiser appointed Botho Eulenburg as Hohenlohe's immediate successor, the whole Foreign Office – including Holstein – would be 'blown sky high' and Bülow would therefore have an easier start on his ascent to the supreme position in the Reich, becoming chancellor in a few years' time. 'Farewell, my dearest Bernhard. How hard it is not to have you with me at such times as this!', lamented Eulenburg, who was about to leave for Budapest.[56] He begged the Kaiser to avoid a crisis at all costs until he returned. 'I do not want Your Majesty to face this infamous intrigue . . . without my help. After all I know the terrain in all its nuances better than anyone else close to Your Majesty.'[57]

Holstein, Marschall and others in Berlin tried in vain to bring the Kaiser's favourite back down to earth by emphasising the dangers for the survival of the monarchy which a ministerial and chancellor crisis would entail. If the Kaiser were to dismiss Bronsart or even allow him to resign without having first conceded public courts martial, 'a crisis such as the Kaiser had never even come near to experiencing' would arise, they warned. A partial ministerial crisis was out of the question because the departure of the Minister of War would make the departure of the whole Ministry of State inevitable and 'the situation of the whole Reich and of the Kaiser in particular' would then 'with great speed . . . become very serious'.[58] Throughout Germany 'the worst possible impression' had been created by the fact that the promised consultation of the commanding generals on the question of court martial reform had still not started after six months, while leading advocates of the reform among the generals had been compulsorily pensioned off, and the Kotze affair had not only drawn universal attention to the 'disastrous activities' of the imperial entourage but also given a practical demonstration of the iniquity of secret hearings. Yet in spite of the public indignation, Marschall pointed out to Eulenburg, the Kaiser's entourage was pushing him into 'such a highly charged emotional state that he is blaming his ministers for the press and demanding *the impossible* from them, namely to silence the press'. If this attitude on the Kaiser's part caused a ministerial crisis he would be in '*appalling danger*', for no politician who had a modicum of reputation to lose would be willing to enter a government that would inevitably be considered by public opinion as advocating '*the predominance of the Cabinets and the abandonment of court martial reform*'. 'And amid all the pandemonium that this would inevitably let loose in Germany the wearer of the crown would undoubtedly be the injured party.'[59] The Secretary for the Interior, Boetticher, sounded a similar warning. 'The German Kaiser has never experienced a crisis such as the one that is now on the way. He has no idea what it will be like!'[60] Holstein looked back philosophically at the Kaiser's development since the conflict with Bismarck over the Stoecker meeting in the winter of 1887, when Wilhelm, as he put it, had completely lost his 'stomach for politics' thanks to the '*toujours* Chancellor'. 'As a result of that he has taken to Cabinet pudding, but he will find that even more indigestible in the long run. He cannot govern the German Reich for long with third- and fourth-rate irresponsible advisers.'[61]

The Holstein–Marschall plot – there is no proof for the suspicion harboured by Eulenburg and the Kaiser that Bronsart was part of it – was thwarted by Chlodwig Hohenlohe's refusal to take action against the monarch. The 77-year-old Prince, who prided himself on being particularly well placed to deflect the Kaiser's worst decisions thanks to his relationship with the imperial family, his illustrious position in society and his long and distinguished political career,[62] was not the man to make a show of strength against the Kaiser. He admitted that Holstein's plan might possibly succeed, but he had 'serious reservations', above all in view of the consequences for himself, for 'the result of such a forcible action would ... be to make my position with the Kaiser intolerable. H.M. saw our action in the Köller affair as an infringement of his rights as king and still bears a grudge against the Ministry for it. A second *coup* of the same kind would aggravate this impression and create mistrust in H.M. which would make my task more difficult. It cannot be supposed that the Kaiser would accept the suppression of his role in the long term. The clash would not be prevented but only postponed.'[63] In the end he would rather go 'than govern with the Kaiser in an unfriendly relationship', the Reich Chancellor reflected. The previous eighteen months had surely shown that one could also govern peaceably with the Kaiser, he observed with short-sighted self-satisfaction. At any rate he had 'no desire to take on the role of the conqueror, as it were, ruling over the conquered Kaiser. That is not my style.' In a passage which betrays the wholly passive attitude which he took to his obligations as Reich chancellor and to his responsibilities towards the German nation, Prince Hohenlohe wrote in his diary: 'The purpose of my existence in the Reich chancellor's palace is in fact none other than to hold back precipitate decisions. An independent regime with the Kaiser at its side in the role of a conquered man, as it were, is unthinkable with this sovereign. It also goes against my nature.'[64] In June the elderly Prince was still refusing to 'stage a clash' on his own initiative. 'If the Kaiser wants to get rid of Bronsart I cannot change that', he wrote resignedly. 'The Kaiser will stay the way he is . . . So why make a great fuss now?'[65] The constitutional crisis over Bronsart and the new court martial regulations was nevertheless moving ineluctably towards its finale, with or without the 'conspiracy' in the Wilhelmstrasse.

CONFRONTATION IN PRÖKELWITZ

The situation became acute when the War Minister announced at the meeting of the Ministry of State on 3 May 1896 that he would hand in his resignation at the end of the parliamentary session. The whole Ministry thereupon decided 'to advise H.M. to approve the introduction of the court martial regulations' as otherwise the situation for 'those remaining' would be '*untenable*'.[66] During a 'satisfactory' audience the following morning, Marschall gained the impression that the Kaiser would yield, but this very soon proved to be an illusion.[67] On that same day Wilhelm telegraphed to Eulenburg at Budapest saying 'Shall never seek popularity with the street rabble at the expense of my army, stand by my often repeated view: it is the same as that of my late lamented grandfather.'[68] An audience of an hour and a half which the Reich Chancellor had on 8 May brought no agreement either. Instead of embarking on the 'burning' question of the reform of military courts the Kaiser made derogatory remarks about the War Minister, on whom he laid all the blame for the crisis, but who could not be dismissed now as it would make him more dangerous still. Alexander Hohenlohe deeply regretted that his father, in his 'gentle and considerate way', had not succeeded in having the main issue out with the Kaiser. The Prince

would have to make the government statement in the Reichstag himself, and in it he must make a firm announcement of the introduction of public courts martial, the Chancellor's son argued, for any weak or dilatory declaration would be received with scornful laughter in parliament, while public opinion would continue its assault on the 'irresponsible advisers' of the Kaiser. And that was not all. If the Reich Chancellor made a statement that did not satisfy the Reichstag he would inevitably expose himself to the danger that the War Minister, who by nature was certainly not likely to show the same consideration for the Kaiser, would unilaterally make such a positive declaration – 'with clattering spurs and rattling sabre . . . but to the delight of the Reichstag' – that Hohenlohe and the whole Ministry would be made fools of for all time. As for the Kaiser, 'however much H.M. rejects parliamentarism in general', Prince Alexander maintained, 'and however much he is inclined to belittle the power of public opinion, he still has too much English blood in his veins not to have an acute sense of such nuances'. But Prince Chlodwig zu Hohenlohe, who had 'sacrificed even the last years of his life in unparalleled patriotic devotion to his Kaiser – would be condemned to sit on the ruins of his well-earned reputation, which he was forced to allow to be destroyed out of consideration for the Kaiser's mood over a question which was not even important', and would have to console himself 'with some diamond-hung chain or other'. In Alexander's view his father did not deserve such a fate. The Kaiser would have to give in, if he wanted to avoid a crisis 'which could shake the whole state edifice as never before'.[69] Waldersee, who would have considered Bronsart's resignation an 'absolutely inconceivable' loss for the government, also took the view that the Kaiser would give in. 'He will do it as he did with the School Bill and the 2 year military service', he predicted.[70] He was mightily mistaken.

Philipp Eulenburg now employed truly machiavellian cunning to develop plans to distance the Reich Chancellor and the other ministers from Bronsart. He began by suggesting to the Kaiser that he should have the courts martial bill submitted to the Bundesrat not, as the normal procedure would be, as a Prussian proposal but as a presidial bill – that is to say, a measure put forward by the Reich government. The bill would of course have to accord with the decisions taken by the Ministry of State, as otherwise Bronsart would resign, and Hohenlohe and all the ministers would do likewise. But the moment that the reform bill was laid before the Bundesrat, the Kaiser could dismiss Bronsart without risking a collective resignation by the other ministers. The presidial bill could then be amended later by the Prussian government with reference to the expert opinions of the commanding generals. 'The authorisation for the introduction of this bill, which is granted by the *Kaiser*, does not bind the *King of Prussia*', Eulenburg argued.[71]

A few days later he thought up another even more sophisticated feint. The Reichstag was shortly to debate the fourth battalions, in the course of which an interpellation was to be expected on the military courts issue. The Reich Chancellor should therefore announce to the Reichstag that this military matter was being examined by all army and corps commanders, who would submit their findings to the Kaiser in the autumn. After this announcement, Eulenburg observed, Bronsart would be able to remain in office to begin with – 'especially if Yr. Majesty strokes him occasionally'. In the autumn, however, the Kaiser should suddenly convene an 'Areopagus of the most senior Prussian military', in which the royal princes, the Reich Chancellor and possibly a few federal princes would also take part. In order to be effective, however, this 'second act' must take the form of a 'surprise attack' on Bronsart. The 'Areopagus' could then declare itself against the introduction of public courts martial, for, as Eulenburg believed, 'there is still sufficient military feeling in Germany to vindicate the right of *such* an assembly, *presided over by the Kaiser*, to decide on the question of public trials'.[72]

Eulenburg's ploy was in danger of being thwarted by the firm conviction of the statesmen in the Wilhelmstrasse that, if the responsible government were not to be totally discredited, the Reich Chancellor would have to make a definite promise in his statement to the Reichstag that the bill to be introduced in the autumn would ensure public courts martial. If the statement did not say this, Marschall warned, it would amount to an admission that 'the views of the *responsible* government had not yet been accepted at the highest level, but rather that the irresponsible advisers had gained the upper hand'. This would become the signal for the Reichstag to debate the position of the Cabinets in the German system of government. If the Kaiser did not give his consent, therefore, 'a fatal position of constraint' would arise as a matter of course in a few days' time.[73]

For Eulenburg, who was passing through Berlin 'with a heavy heart' on his way to visit the Kaiser on the Dohna family's estate of Prökelwitz in East Prussia, as he did every year at this time,[74] the responsible government's proposed course was nothing but a South German liberal conspiracy against the Prussian militarist monarchy. The 'well-known' group (in which he now included Prince Alexander Hohenlohe) had driven the Reich Chancellor 'into a corner' and was presenting the Kaiser with 'what amounts to an ultimatum', he wrote to Bülow in alarm. 'There is something Bavarian-Badenese-Radical in the air which plans to do violence to the King of Prussia . . . The way out which I took so much trouble to find will not work, because Hohenlohe refuses.'[75] Secret consultations with his cousin Botho in Berlin only confirmed Eulenburg's impression that the proposed – '*completely unnecessary*' – statement by the Chancellor was both a 'measure of force against the Kaiser' and a symptom of the 'South

German–liberal desire to represent the Ministry as the stronghold of liberal opinion'. After further discussions with Holstein, Marschall and Prince Alexander he reported to Rome: 'In spite of all their fury with the War Minister the entire band is hypnotised by him and I see absolutely no way out now – unless the Kaiser gives way (which I hardly consider possible) or else manages by means of a very urgent expression of his will to change the Chancellor's decision concerning the declaration . . . That too seems improbable and I am expecting a showdown.'[76] On his arrival in East Prussia, on the Kaiser's instructions Eulenburg appealed for the help of the Chief of the Civil Cabinet. Lucanus was asked to get in touch with Hohenlohe in order to 'prevent the statement by the Reich Chancellor which the Ministry is demanding, or more precisely to strengthen Prince Hohenlohe's resistance to what is quite obviously an intrigue by the Ministry, which cannot but lead irretrievably to a break, for His Majesty has no intention of changing his views in the face of a *position of constraint* . . . I am afraid, however, that it will not succeed.'[77] Meanwhile, the prophecies of doom arriving from Holstein in Berlin provided a reminder that the future of Germany and of the Hohenzollern monarchy was at stake in this dangerous game of poker. In a cipher telegram of 16 May he warned that 'an acute constitutional conflict in the most unfavourable conditions imaginable is imminent at any moment'.[78]

The same day the Kaiser wrote to the Reich Chancellor from Prökelwitz warning him expressly of the possibility 'that one or other of my ministers who is frightened of parliament might press for a joint statement by the Ministry of State' on court martial reform. This, he insisted, was impossible for three reasons: first, the matter could be dealt with only by him and Hohenlohe as minister-president; in the second place it was not a political but a 'purely technical, military' question which could be decided 'only by the supreme war lord'; and finally the consultation of the generals was in process and must not be prejudiced by a statement in parliament. 'I cannot rid myself of the fear', the Kaiser continued, 'that the restless spirits in the Ministry are preparing some *coup* behind our backs and want to take you by surprise with a joint statement which you will be expected to make to parliament and which will probably go against our agreements of the 8th [of May]. But that would be a serious offence against the crown and its wearer. The Ministry cannot and may not make any statement on questions of principle without previously submitting the wording to me and obtaining my agreement after I have examined it. If individual ministers tried to do this nevertheless, they would be liable to prosecution.' Wilhelm went on to say that he considered 'the interpellation which has already been announced as a ploy ordered by one of the troublemakers in the Ministry so as to plunge you and all of us into conflicts which are quite unnecessary and in the course of which the person in question is aiming to take your place . . . So watch out for ambushes whether by individuals or by the entire Ministry and do not allow yourself to be led astray or forced into declarations which put me in a difficult position.' The previous autumn, the Kaiser added, the War Minister had given him his word of honour that he would on no account allow the question of court martial reform to be combined with that of the fourth battalions in the Reichstag. 'I shall hold him to that and you must play that card too if he gets up to any tricks', the Kaiser ordered, with a final exhortation to the Chancellor 'not to worry too much about all the rubbish other people and a few ill-mannered louts in the Ministry like to indulge in; good hunting for Monday, make sure you shoot the interpellation down like a good stag'.[79] The Kaiser's appeal was seconded by Eulenburg, who sent the Reich Chancellor a cipher telegram making it clear that Wilhelm would regard a statement to the Reichstag in the name of the whole Ministry, without his authorisation, 'as a constraint, as in the Köller crisis'. It was essential that Hohenlohe's statement be couched in evasive terms and it must not contain the 'extreme wishes' of the Minister of War.[80] Even the Kaiserin was mobilised to put pressure on her elderly uncle. In a note of 17 May she urged the Chancellor to stand up for the Kaiser when he spoke to the Reichstag, and to take control of the ministers who were trying to make trouble again. 'The Kaiser has often given way for your sake, even when it was difficult for him, but in military matters I know he takes a firm stand and will continue to do so', she assured him. 'For the good of all', and because it was clear 'how difficult the ministers often make things for the Kaiser', he, the Chancellor, should give way this time.[81]

Before the letters from East Prussia had arrived in Berlin, Hohenlohe had written to the Kaiser saying that he intended to reply to any interpellation in the Reichstag on the subject by stating that the new court martial bill, which would be 'based on the principles of modern legal opinion', was now in preparation and would be ready to be laid before parliament at its next session in the autumn.[82] The Kaiser peremptorily wired back via Eulenburg ordering that the phrase 'draft bill based on modern legal opinion' should be 'omitted, as it is superfluous and will lead to misunderstandings; above all it is an insult to the army as it exposes the previous procedure as not in accordance with modern legal opinion and would cause great anger'. Hohenlohe could state that the reform bill would represent a considerable step forward and would take account of modern legal opinion in so far as it was compatible with 'military organisation and the particular requirements of the army' and did not have any damaging effect on discipline but maintained it 'absolutely intact'. Furthermore the Chancellor should not commit himself to any date for the completion of the bill but at the very most say that it would 'probably' be laid before the Reichstag in its next session. Finally, the Kaiser 'ordered' the Reich Chancellor to inform the Chief of the Military Cabinet, General von Hahnke, 'at once' of the contents of his telegram.[83] Eulenburg sent Holstein an unequivocal warning that the Kaiser was 'absolutely intractable' on this question; it was entirely in Hohenlohe's hands what course the crisis would now take. 'According to the way he decides it the future will take shape.'[84]

It is not difficult to imagine the reaction of the leading statesmen in the Wilhelmstrasse to these All-Highest orders from the hunting lodge at Prökelwitz. Marschall described the Kaiser's letter of 17 May as 'incredible'. In it the Kaiser warned Prince Hohenlohe 'against resolutions by the Ministry as a whole which might impose declarations on him' and maintained 'that the court martial regulations are a purely military question which only he can decide upon, etc.'.[85] When the Kaiser's telegram arrived on 18 May Marschall noted furiously that the Kaiser was dictating 'word for word' what the Reich Chancellor should say in the Reichstag, including 'a phrase which, not in so many words but in its sense, is directed against public hearings. H[is] H[ighness] is determined not to have anything to do with it.'[86] Holstein telegraphed to Eulenburg that he had 'never seen the Prince so furious in the 22 years I have known him. Reason for anger: continuing bad treatment and unfair mistrust on the part of His Majesty.'[87] On 18 May, with the help of his 'General Staff', as Holstein called the small group of advisers around the Reich Chancellor, Hohenlohe drew up the statement which he would make that afternoon in answer to the interpellation in the Reichstag. The moment of truth had arrived.

In response to a question from the leader of the Centre Party, Dr Ernst Lieber, about the slow progress of the work on court martial reform, the Reich Chancellor, adopting some of the Kaiser's wording but nevertheless remaining faithful to his own views and those of the Ministry of State on the subject, stated that he 'definitely expected' to be able to lay the new regulations before the Reichstag in the coming autumn, and that the bill would be 'based on the principles of modern legal opinion, subject to the particular requirements of military organisation'.[88] That same evening, in response to Hohenlohe's telegram informing the Kaiser of the terms of his statement, a second cipher telegram arrived from Eulenburg conveying the monarch's bitterly disappointed reaction. The wording used by the Reich Chancellor had gone much further than the text which Wilhelm, 'after careful consideration, had Himself composed'. It would arouse expectations, 'particularly with regard to public hearings', 'which in the interests of the army His Majesty could not fulfil'. It had also been the Kaiser's deliberate intention to avoid specifying the deadline to which Hohenlohe had now committed himself.[89] Eulenburg followed this up with a letter to the Chancellor revealing that the Kaiser had been 'very upset' and 'quite beside himself' on receiving his telegram. As Eulenburg had feared 'an outburst of passion' if the Kaiser were to wire a response himself, he had persuaded Wilhelm to go out stalking and leave the telegram to him. It was to be hoped, he said, that 'the storm will have died down on his return from the forest'. But the Chancellor should be aware that the monarch was still 'absolutely opposed' to making any concession on the question of public hearings. 'As far as *public access* is concerned there is . . . *nothing* to be done', the Kaiser's friend warned. 'On that the Kaiser is absolutely *committed*.'[90] The elderly Prince's answer to Eulenburg's reproaches, which he had formulated himself before Marschall arrived to see him at 10 o'clock in the morning, was unusually acrimonious. It read: 'I am the Reich Chancellor and not a Chancellery official, and ought to know what to say.'[91]

Proud words indeed from the holder of the highest offices in Prussia and the Reich, and a refreshing contrast to the constant spectacle of a 'responsible government' giving way. If Prince Hohenlohe had acted in the spirit of this principle – if he had followed the advice of Holstein, Marschall, Bronsart and his own son and asserted himself against the Kaiser or resigned his office – would the Wilhelmstrasse have become the decision-making centre of German policy again, as it had been in Bismarck's day? The answer that probably has to be given to this question is doubly depressing. In the previous seven years Wilhelm II had acquired so much power that it would no longer have been possible to put the clock back. For the Reich Chancellor to have insisted on his primacy in decision-making, at a time when he had no reliable backing in the Reichstag or from public opinion, would only have led to his dismissal and that of the other ministers, and hence to a national crisis with unforeseeable consequences. But in any case Hohenlohe simply did not act in accordance with the admirable principle he had so proudly expressed. Once again, step by step, he capitulated. Within weeks the promising clarion call of 19 May 1896 was being drowned out by the roll of drums beating the retreat.

THE RECONSTITUTION OF THE GOVERNMENT BY THE CROWN

Eulenburg's first concern was to avert an immediate government crisis. The same day, 19 May 1896, he sent the Chancellor a placatory message from the imperial hunting lodge. 'The Kaiser was at first perplexed by your answer today. Then he was upset for a moment, but afterwards his attitude became calmer, and remains so . . . The anger has abated. He is in a good, cheerful mood again.'[92] He urged Prince Alexander to persuade his father to remain in office. 'The Kaiser's excitable state' must not give the Reich Chancellor reason to conclude that he should resign, although this would be only too understandable given the 'volatile nature' of the monarch. 'I am glad that it turned out in the way it did!', Eulenburg asserted. 'But I *really do not* think the time has come for far-reaching decisions by the Prince.'[93] In a secret note intended only for Chlodwig Hohenlohe, however,

Eulenburg informed him that, although the Kaiser wished to keep him, the Chancellor, in office, he did intend to dismiss Bronsart and Marschall. 'The resentment towards the Minister of War – but also towards Marschall – is *very* great and will probably lead to a clash in the not too distant future.'[94]

Although it suited Wilhelm to keep Hohenlohe in office – indeed he was to remain Reich chancellor and Prussian minister-president until October 1900 – the statement made by the Chancellor on his own initiative in the Reichstag on 18 May had nonetheless 'left a wound in the Kaiser's heart'[95] which had perceptibly diminished Wilhelm's willingness to co-operate even with him. To Philipp Eulenburg, the monarch's loss of trust seemed reason enough for a major government reshuffle. The aim which he and the Kaiser's other advisers at court now pursued was therefore to re-structure the Reich government from above, so that in future the will of the monarch, rather than that of the Reich chancellor, the secretaries of state and the ministers of state, would determine Germany's destiny. The role which Eulenburg played as Wilhelm II's confidant in this national crisis was so dominant that Holstein remarked that he sometimes could not tell 'what was H.M. and what was Phil. Eul.'.[96] Another diplomat, casting doubt on rumours that Eulenburg himself would shortly be appointed Reich chancellor, commented that he had always regarded the favourite as 'a charming dilettante' who would not be so 'foolish' as to 'step out of the shadows of the irresponsible adviser's role'. He could not imagine, either, that it would suit the Kaiser to let him out of 'the behind-the-scenes business . . . which he seems to manage to the All-Highest satisfaction'. Whomever the Kaiser appointed as Reich chancellor, however, the sad truth was that he would not follow the principle of 'who is the most effective, but who is the most obedient'.[97]

The chief difficulty which Wilhelm and Eulenburg faced in the summer of 1896 was that of Bronsart's proposed reform of the military judicial system, supported as it was both by the Chancellor and the Ministry of State and by public opinion, while the Kaiser had absolutely no intention of conceding it. On the Scandinavian cruise Wilhelm, 'with obvious irritation but also with very marked energy', declared that he would never be persuaded to change his mind. Although the danger of a change of chancellor filled him 'with extreme concern', he would take even this upon himself, since 'I would despise myself if I were to sacrifice my convictions on this question.'[98] Wilhelm remained obdurate in his antipathy for the Minister of War and the resignation of this key figure could not be far off. Hohenlohe complained that the Kaiser spoke 'in such derogatory terms' about Bronsart that the General felt he 'could not honourably continue to serve'. A few friendly words from the Kaiser would be enough to pacify him, the Chancellor commented; but he knew that 'H.M. will . . . not want to do that.'[99] On a visit to the court at the end of May Waldersee too observed that the Kaiser had 'completely broken' with the War Minister and 'often says so in the harshest possible terms'. Wilhelm accused Bronsart of 'inciting the other ministers against him and attacking the Military Cabinet and the influence of the Flügeladjutanten in the press, and he says he is absolutely certain of this... It is the same old story: the Kaiser listens to gossip and believes everything!'[100] As he had previously announced, Bronsart handed in his resignation on 17 June after the closure of the Reichstag session, but hinted that he was prepared to go on leave for the time being, in case the Kaiser should refuse to allow him to resign immediately.[101] A few days later Waldersee heard that Bronsart was to take three months' leave, but – like Verdy five years earlier[102] – would not return to his post thereafter. 'He has completely fallen out with the Kaiser', he commented. 'We shall soon see the Kaiser's 5th minister of war: a good turnover in 8 years!'[103] Wilhelm and Eulenburg had thus gained three months in which to drive a wedge between the Reich Chancellor and the outgoing War Minister.

Nothing shows more clearly the extent to which the most important decisions in the German Reich were now being taken by the Kaiser and his confidential advisers than the discussions which Eulenburg held in May 1896 with the Kaiser and Lucanus on the future composition of the Reich government and the course which domestic policy should follow in consequence. In two letters to Bülow of 24 May and 8 June 1896 the Kaiser's favourite reported, with astonishing condescension, on the progress of these discussions, which were to seal the fate of the Hohenlohe regime. After his conversations with Wilhelm in Prökelwitz Eulenburg was convinced that the relationship of trust between Kaiser and Chancellor could never be restored. 'But what can one do if the Kaiser no longer wants to ride this horse?', he asked. 'It is tired and will not take any more ditches. There is no point in my singing its praises. I know how gladly it carries its master, how well the old animal has ridden in the past, how exactly it understands every signal during exercises, how well known it is and how soft and good its name sounds! If H.M. has lost the desire to ride it, it will not be given the right aids and will not jump as it should.' Eulenburg expressed horror at the prospect of the crises which a third change of chancellor would bring, shaking Germany and its neighbours 'to the core', but he maintained that the sole decisive factor for him was the welfare of the 'beloved' Kaiser. 'I look at what it means for the intellectual development of our beloved Kaiser, and I try to decide on the basis of events and of His character how one should carry on life for the best, how can *the Kaiser* carry on his life for the best?' Eulenburg was convinced that 'the King's character must be taken into consideration as long as we are a monarchy and wish to remain one'. But Kaiser Wilhelm II, his friend declared, represented to an astonishing degree a combination of 'two totally different natures': 'the knightly – reminiscent of the finest days of the Middle Ages, with all their piety and mysticism – and the modern', although the knightly side was 'very predominant'. 'If I now transfer the effect of the Kaiser's personality on to politics', Eulenburg continued revealingly, 'I come to the inevitable conclusion that H.M. must find His *natural point of support* in the Conservatives, the Conservatives in the good sense – and that is confirmed by looking at Prussian history. The king of Prussia cannot long remain without the Conservatives . . . I think, therefore, that the Kaiser's individuality will find its equilibrium in politics on a sensible Conservative basis. The whole of the Kaiser's vital, dynamic being stands in such remarkable contradiction to the predominance of Liberal-Progressive or Liberal-Catholic (Centre) ideas that this has been one of the main reasons for the feeling of unsteadiness about which the whole of Germany is complaining.' The Conservative Party, for its part, would have no choice but to find its way back to the monarchy, Eulenburg predicted, for 'a king of Prussia who, like our dear King, is *la fine fleur* of kingship, and a Conservative Party that adopts democratic and demagogical airs, are mutually incompatible – but must also be brought together again *par la force des choses*'. The conclusion to be drawn from these reflections was obvious to Eulenburg. In the new government, above all, '*Prussians* must now dominate', and 'the single will of the Kaiser in His Ministry' must be made perfectly clear. Of course for the sake of the German Reich the wheel must not be turned too far to the right, he pointed out; on the contrary, extremely skilful men must be appointed who could master the art of ruling 'conservatively in Prussia and liberally in the Reich'. If this talented, predominantly Prussian, moderate Conservative, unified government were but in place, Eulenburg maintained, the 'myth of adjutant-politics', so damaging for the monarchy, could be scotched, for on the one hand the new government would be at peace with the adjutants, 'because there would be the greatest possible harmony of political views between it and H.M.'s entourage', and on the other the Kaiser would 'treat the Ministry with the same confidence as His adjutants'.[104]

On the practical question of choosing successors to the present government, Eulenburg and Wilhelm II at first had different ideas. Eulenburg's suggestion that his cousin Botho should be appointed Reich chancellor with Bernhard von Bülow as foreign secretary was initially turned down by Wilhelm because Lucanus had told him that as minister of the interior Botho Eulenburg had shown so little vigour in dealing with the Conservatives 'that it was inconceivable that he should hold the foremost position in the Reich in such difficult circumstances'. For this reason, Wilhelm declared, he had decided to appoint Bülow Reich chancellor immediately. 'He was tired of crises, He needed a chancellor who would stay for 20 years and more.' He dismissed Philipp Eulenburg's argument that he should not make use of Bülow in the present crisis especially if he wanted to retain him as chancellor for a long time; 'it seemed to be a firm decision',

Eulenburg recorded, and 'I did not insist.' But as he was still convinced of the advantages of a 'strongly Prussian' chancellorship under Botho Eulenburg, with Bülow as foreign secretary and Posadowsky as secretary of the interior, the Kaiser's friend decided to call on the Chief of the Civil Cabinet at Potsdam on his way back from East Prussia to Vienna on 24 May, in order to talk him out of his fear of the 'excessive power of the Eulenburgs'.[105]

Like Eulenburg, Bülow believed that in the final analysis the conflict between the crown and the Wilhelmstrasse turned on the question of who should have the final power of decision in the land. 'What is really at issue?', he asked. 'It is this: that His Majesty's prestige and plenitude of power, which was usurped [by Bismarck] in 1890 and which as a result of various events in the last year seems threatened, if not shaken, should be restored in the eyes of the army and the country.' The Kaiser should not, however, be forced into choosing between subordinating himself to the Ministry of State led by Bronsart and Marschall on the one hand, and provoking a '*casse-cou* policy' with 'dissolutions of the Reichstag, *coup d'état*, break-up of the Reich federation and general chaos' on the other. It was also open to him to dismiss the rebellious ministers and, having shown the army and the people that he would not allow his will to be broken, he could then come to a compromise solution with the new ministers on the question of court martial reform. The Kaiser's candidate for the chancellorship was well aware under what ordinance he would have to succeed Marschall or Hohenlohe. 'My standpoint will always be the same', he assured Eulenburg, who by now was cruising in the North Sea with the Kaiser. 'I am always at our dear sovereign's disposal – wherever, whenever and however he wishes . . . All I want is to be a card in the Kaiser's hand with which he wins as many tricks as possible.'[106] A week later, knowing full well that Eulenburg would read his letter out to the Kaiser, Bülow enlarged on his theme in terms that could not fail to be understood. 'I am filled solely and purely with burning love for our revered Kaiser and sovereign, profound devotion to Prussia and Germany . . . I would be a different kind of chancellor from my predecessors. Bismarck was a power in his own right, a Pepin, a Richelieu. Caprivi and Hohenlohe regarded or regard themselves as the representatives of the "government" and to a certain extent of the parliament in relation to His Majesty. I would regard myself as the executive tool of His Majesty, so to speak his political Chief of Staff. With me personal rule – in the good sense – would really begin.' But for precisely that reason he should be brought in as Reich chancellor at the right moment and not too early, he warned, for 'if this attempt at real personal rule were to fail, things would look black for our beloved Kaiser!'[107]

In the course of endless conversations during the Scandinavian cruise in July 1896 Wilhelm and Eulenburg devised a series of carefully thought-out plans addressing every eventuality – Hohenlohe's resignation *with* a subsequent *coup d'état*, Hohenlohe's resignation *without* a *coup d'état*, Hohenlohe's continuation in office with the proviso of sacrificing Bronsart, Marschall and Boetticher. There was only one possibility which they did not consider: a surrender by the Kaiser on the court martial reform question. To him this signified the preservation of 'the dignity of the crown' and the maintenance of his personal power as kaiser, king and supreme war lord. In numerous letters from the *Hohenzollern* the Kaiser's best friend set out to convince the two key figures in the 'responsible government', Hohenlohe and Holstein, of Wilhelm's absolute determination, and therefore of the futility of their resistance, on this question. And since Eulenburg's threats of the catastrophic consequences of a ministerial and chancellor crisis were far from being empty attempts at intimidation, the statesmen of the Wilhelmstrasse, conscious of their responsibility, eventually fell victim to their own prophecies of doom.

In their machinations Wilhelm and his friend began by recognising, on the one hand, that the 'clash' must not take place during the Scandinavian cruise, as this would put Eulenburg and the adjutants accompanying the Kaiser into a 'disagreeable position' and would awaken the impression that the Kaiser was allowing himself 'to be influenced in an outrageous manner'. On the other hand they realised that they must get their way by the end of August or the beginning of September at the latest, that is to say before the Reichstag reassembled and Bronsart finally departed.[108] At the very start of the cruise it became obvious to Eulenburg that Wilhelm had 'no clear sense' of the full implications of his attitude – he wanted to make immediate changes, replacing Hohenlohe by Bülow, Marschall by Brinken and Bronsart by Hähnisch.[109] It was therefore 'with some hesitation' that the Kaiser's friend wrote to the Chancellor on the Kaiser's instructions on 5 July informing him that Wilhelm would '*on no account*' give way on the court martial question. A catastrophe was therefore inevitable in the autumn unless Hohenlohe himself was prepared to give way and sanction the dismissal of Bronsart and Marschall. 'The King has the right to be influenced by personal feelings in forming an opinion of his ministers', the favourite again insisted. Marschall would be given an embassy (Constantinople) and Bernhard Bülow would be appointed secretary of state at the Foreign Office. Once a new minister of war had been appointed, Hohenlohe could modify his position on public courts martial for the Kaiser's sake. If he did so he would be able to look forward to a 'peaceful, dignified' retirement in the future. Otherwise there would be a serious constitutional conflict for which Hohenlohe would have to bear the responsibility. A passage dictated by Wilhelm, which Eulenburg later toned down to spare the elderly Chancellor's feelings, contained the threat that 'in the event of very serious attack' the Kaiser 'might even be *compelled* to vindicate his crown . . . and use forcible means'. 'Although it would make his heart bleed, he could announce to the world that the declaration of principle made by his Uncle the Chancellor in parliament accorded neither with his wishes nor with the orders which he had conveyed to Your Highness in the form of words which he himself had drawn up after he was consulted.'[110] Eulenburg also wrote to Holstein stating that after his conversation with Wilhelm in Norway it was clear to him that the Hohenlohe system in the form it had taken hitherto was 'done for', and that Marschall must be dismissed and Bülow brought into the government as future chancellor. 'The unremitting agitation against Marschall has aroused what appears to me to be such insurmountable dislike against him on the part of H.M.', he wrote. On the other hand Bülow must not be branded as an anti-public courts martial candidate for the chancellorship. If Bülow should be tarred with this brush and therefore refuse even an '*urgent request*' from the Kaiser to come to Berlin, there was a chance that 'H.M. out of ill-humour would suddenly appoint Wald[ersee] or would turn to Botho E[ulenburg].'[111] Eulenburg repeatedly warned the leading spirits in the Wilhelmstrasse that the Kaiser would not yield on the question of court martial reform.[112] Although he fully recognised the benefits to be gained from Bülow's appointment as foreign secretary under Hohenlohe – 'It would be a blessing for you to work with Hohenlohe for a little while longer', he wrote to his friend in Rome – Eulenburg was convinced that the Prince's response would be negative and that Bülow would therefore be appointed Reich chancellor immediately.[113]

Writing to his son, the Chancellor commented bitterly that the air of the North Sea had evidently stimulated 'the All-Highest vitality'. Eulenburg had informed him in a letter from Norway 'that H.M. has not changed his opinion with regard to public access [to courts martial], that he wants to get rid of Marschall and Bronsart and that he expects me to submit to all of this'.[114] In reply to Eulenburg's letter Hohenlohe stated that to give up the principle of open courts martial would be political suicide for him, for he would be heaped with 'scorn and ridicule' and 'a ridiculed Reich chancellor' would no longer be able to render any worthwhile service to the Kaiser. He would therefore continue to try to find a formula that would allow the Kaiser to give his consent to the public military trials which had been promised to the Reichstag. If such a compromise could be found, however, Bronsart would be 'the only man' who could make this formula palatable to the Reichstag. He therefore had serious reservations about the immediate departure of Bronsart and equally about the removal of Marschall at this juncture.[115] The 77-year-old Reich Chancellor also hinted indirectly to the Kaiser's friend that he cherished certain hopes which might induce him to leave office in a manner that would avoid conflict. 'If it could be arranged for me to be transferred back to Strassburg [as Statthalter] that would of course be the most pleasing solution and it would prevent any suggestion of a rift. I should be able to celebrate my golden wedding there next winter.'[116]

The only person who firmly resisted the far-reaching plans contemplated by Wilhelm and Eulenburg was Geheimrat Friedrich von Holstein. He continued to warn of the disastrous consequences of the national crisis, akin to civil war, which would ensue if the Kaiser did not give way on the court martial question and Bronsart, Hohenlohe, Marschall and the entire Ministry of State were to resign as a result.[117] The Kaiser must on no account bring about a constitutional conflict of this order with the Reichstag unless he had secured a written commitment from the principal German federal princes that they would support him 'through thick and thin'. If a number of ruling princes, headed by Bavaria, should suddenly refuse to co-operate with him in taking further action against parliament once the conflict had begun, 'the Kaiser will find himself in a serious position because the power of the princes would be augmented by that of all the opposition parties . . . This, my dear E[ulenburg], is the heart and soul of the whole matter, because it is here a question of the preservation or weakening of the imperial authority. All the other questions – for example ministerial appointments – are of secondary importance.'[118] If the Kaiser provoked the conflict without making quite sure of the support of the rulers of Bavaria, Saxony and Württemberg, he would be heading for 'a political defeat the consequences of which will be felt during his entire reign', Holstein warned.[119] The Kaiser himself was 'too eminent a person' to expose himself to the risk of a rejection from Munich, Dresden or Stuttgart, but Eulenburg could take soundings with the three rulers on his behalf.[120]

The persistent warnings and threats in Holstein's letters did not go unheeded in the imperial camp. Even Adjutant-General von Plessen spoke of the urgent need to avoid a chancellor crisis now; the 'growing possibility of a clash filled him with anxiety'.[121] In a memorandum of 21 July Eulenburg acknowledged that the resignation of Hohenlohe, Marschall and Bronsart would 'cause the greatest astonishment and agitation throughout Germany'. '*It will not be understood and it will show His Majesty in the wrong light.*' After such a sensational clash the new chancellor would be plunged into a situation of conflict in which he would inevitably be defeated. He would have no party in the Reichstag on his side, for the Centre and the Liberals would be fiercely opposed to him, while the Conservatives, who were still very much in conflict with the crown, would likewise withhold their support from Hohenlohe's successor. The Reichstag would therefore refuse to approve any more expenditure; dissolution would be inevitable, but new elections would only make the situation worse. With Holstein's warnings in mind, Eulenburg concluded: 'The conflict with the Reichstag can be carried through only with the full agreement of the federal princes, above all that of the [three] kings. They must make binding, written declarations, so that they do not try to extract advantages for themselves – (revision of the Treaty of Versailles, for instance) – out of Prussia's difficulties at the crucial moment. That would spread the conflict to the federal states and would entail the risk that foreign countries might exploit the strife to attack Germany.'[122] When Eulenburg put these concerns to the Kaiser he was unperturbed. The King of Saxony and the Grand Duke of Baden, he asserted, were indeed on Bronsart's side on the question of open courts martial, but they would not pose any danger in the event of a conflict, and even Bavaria, where the Wittelsbach princes might be tempted to make common cause with the Liberals, nationalists and particularists to 'trip up Prussia', did not greatly worry him. 'I have seventeen army corps and am therefore the master of the situation', he declared coolly.[123]

Although this train of thought by no means persuaded Wilhelm and Eulenburg to abandon their plans, they were nevertheless forced to consider whether it would not be preferable to postpone Bülow's candidacy for the chancellorship and appoint a transitional chancellor who would be prepared to see the constitutional conflict through to the end, if Hohenlohe remained intransigent.[124] In a decisive conversation in the *Hohenzollern* on 23 July 1896 Eulenburg told the Kaiser of the Reich Chancellor's negative response to his letter of 5 July and discussed the dangerous state of affairs which would arise if Hohenlohe and the entire government of the Reich and Prussia resigned in protest at the pressure on them. Wilhelm continued to maintain that surrender on public courts martial was 'completely out of the question' for him. 'I have scarcely ever seen the Kaiser so strongly affected', Eulenburg reported to Bülow. 'His eyes were brimming with tears when he said "I should think myself a miserable creature if I gave in on a question which my forefathers recognised as necessary and right! How could I show my face before them in Heaven? – although I am certainly not worthy to go there! And how would I stand *with the army*, which would see it as the destruction of all its protective walls, the army in whose eyes I already bear the heavy responsibility of having accepted the two-year service!"' For a whole hour the Kaiser listened 'with an attention that was really touching' while his friend set out the 'serious dangers' which would arise from Hohenlohe's resignation and the resultant conflict with the Reichstag. 'It made a very profound impression on His Majesty. The beloved Kaiser had not fully realised the possible consequences', Eulenburg commented.[125]

At a later stage in this memorable conversation, Eulenburg argued that in view of the foreseeable political complications Wilhelm II must find another successor for Hohenlohe, instead of putting Bülow, who was to be 'a Reich chancellor who would *last*', in an almost hopeless position. When Eulenburg went on to name his cousin Botho or, of all people, General von Bronsart to take on the role of transitional chancellor, the Kaiser exclaimed in astonishment: 'That reminds me of a suggestion by old Hahnke, who told me I should make the Minister of War my adjutant-general in waiting, to shut his mouth. I indignantly refused at that time – but today, the way you have described it makes me wonder whether this suggestion might perhaps be the solution in the present state of affairs? Bronsart is not dismissed, but *promoted* and receives an *exceptional mark* of favour from me. Then Hohenlohe has no reason to go. Bronsart is removed from office and replaced – but because I appoint him to *my* service, not because of the wretched court martial reform.' The 'German Reich Chancellor Hohenlohe' could then 'make a concession to the Prussian Prime Minister Hohenlohe' without burdening his conscience, the Kaiser observed happily. Eulenburg declared himself ready to pass on these suggestions at once to Hohenlohe and was further authorised by the Kaiser to offer him the postponement of Marschall's dismissal as well.[126]

In a letter to the Chancellor on 23 July, couched in terms as 'warm' as possible, Eulenburg emphasised that on the question of public hearings the monarch was still unable to give way on any account, for he must remain true to his 'most sacred conviction', to which he was bound by the well-tried tradition upheld by his forebears. Wilhelm had said, 'with tears in his eyes', 'I *cannot* sacrifice my conviction – I would think myself wicked! My uncle will sacrifice his Bavarian traditions when he sees my sincere goodwill to be as accommodating as possible with him!' But the Kaiser was prepared to make 'a great sacrifice' to the Chancellor on two other points. First, the replacement of Marschall by Bülow could be postponed and, second, on quitting the Ministry of War Bronsart could be promoted to adjutant-general in waiting to the Kaiser. Any idea that the Military Cabinet and the Kaiser's military suite had won a victory over Bronsart would be counteracted by his promotion, and the Reich Chancellor would thus be free to accept the opinion of the new minister of war and of the commanding generals of all seventeen Prussian army corps on the question of the court martial bill.[127]

The ensnaring of Prince Hohenlohe was, however, only one of the options that the Kaiser and his good friend 'Phili' were keeping open. At the same time they prepared themselves for conflict with the Reichstag and even for a possible *coup d'état* with all its consequences – civil war and military intervention by foreign powers. In his conversation of 23 July 1896 with Wilhelm, Eulenburg particularly emphasised that he had '*not the slightest scruples*' about entering into conflict with the parliament 'because Prussia is strong enough to fight it out'. If Hohenlohe refused to accept the Kaiser's offer a crisis would follow, Eulenburg noted calmly that evening. 'In that case Botho Eulenburg with Bernhard B. as secretary of state seems to me to have the best chance. But it is not inconceivable that Bronsart could emerge as "war chancellor" – and that might not be a bad thing, in a really hopeless situation.'[128]

THE BREAKTHROUGH

At the end of July 1896 two non-political developments compelled Wilhelm to break off his Scandinavian cruise early and return to Germany. First, he received news – prematurely, as it turned out – that Bismarck was dying, and he was determined to exploit the propaganda value of this historic event by making a dramatic appearance at his deathbed. The other reason was that his right ear was giving trouble and there were fears of inflammation of the brain. Accompanied by Eulenburg, Wilhelm hurried to Schloss Wilhelmshöhe overlooking Kassel, where Dr Moritz Trautmann was secretly waiting and was to carry out a major operation on his right middle ear.[129] With the Kaiser's return the decisive battle was imminent.

During the train journey from Kiel to Kassel, Eulenburg worked out the programme for the future drawing together the threads of the discussions of the past four weeks. On 1 August 1896, during a sunny walk in the park at Wilhelmshöhe, he read it out to the Kaiser, subsequently handing it to him as an *aide-mémoire*. This breathtaking exposé, which demonstrates with startling clarity the extent to which the imperial camp held the upper hand in its final battle with the 'responsible' government of the Reich, is undoubtedly one of the most impressive documents of the Wilhelmine era, showing as it does how carefully the Kaiser and his secret advisers had prepared themselves for every eventuality. If Hohenlohe were to give way and remain in office, Eulenburg's memorandum stated, Bronsart would be replaced by General von Hähnisch, Boetticher by Posadowsky and – after a few months – Marschall by Radolin or Brinken. But if Hohenlohe insisted on resigning, Bronsart, Marschall and Boetticher must be dismissed together by the beginning of September at the latest. Thereafter there were two possibilities for the Kaiser. In the event that 'His Majesty *does not want a policy of force*' he should appoint Botho Eulenburg as Reich chancellor and Bernhard Bülow as foreign secretary and vice-president of the Prussian Ministry of State. This combination would mean 'a Prussian, moderate course with an anti-revolutionary tendency – [but] not anti-constitutional or anti-Reichstag'. Botho would outlive his usefulness 'after a few years' and Bülow could then 'ease himself in' to the post of Reich chancellor. But the Kaiser and his friend also calmly faced the possibility of a breach of the constitution. 'In the event that His Majesty is not afraid of a policy of *forcible settlement* or cannot avoid it', the memorandum provided for Bülow to be kept in reserve until after the 'fierce battles' were over. In this case the Kaiser could take either Bronsart or

Waldersee as Reich chancellor instead of Botho Eulenburg. The appointment of a '*belligerent*' general of this kind as chancellor would in itself signify 'the government throwing down the gauntlet, a marked shift towards so-called reaction, which would arouse the *strongest* opposition throughout the Reich and might even lead to the dissolution of the Reichstag; it would at any rate exacerbate the difficulties with the Bavarians, Württembergers and Saxons enormously', as Eulenburg observed. He himself preferred the 'great combination (Eulenburg–Bülow)' as 'the more favourable solution', but acknowledged that the 'sudden cessation of the Hohenlohe era' would hold great dangers. The '*greatest possible* efforts' should therefore be made to keep the old Prince in office, as far as this was compatible with 'His Majesty's principles and dignity'.[130] After his walk with the Kaiser Eulenburg noted enthusiastically: 'For me this walk will always remain most memorable; it will be decisive in many ways. How kind, clever, simple and clear the Kaiser was. I had the feeling that we were closer to each other than before.' He had explained the content of his memorandum 'very thoroughly' to Wilhelm, 'entering into every detail. He agreed *unreservedly*, so that the exposé . . . must be considered as his programme.'[131]

The soothing telegrams and letters which he had received from the *Hohenzollern* led Holstein to conclude, rightly, that Wilhelm and Eulenburg intended to delay the decisive moment until they had spoken to the Reich Chancellor himself. With the modest means at his disposal – for apart from his cynical intellect and the power of his pen he had no way to exert influence – the Privy Councillor led the intellectual resistance of the 'responsible government' to the monarch's plans to seize power. He informed the Vortragender Rat, Karl von Lindenau, who was with Hohenlohe at Alt-Aussee in Austria, that he no longer believed there would be 'a row overnight, but deduce[d] from the various Imperial pronouncements the intention of saving up some crucial questions – business or personal – for the next *verbal* discussion with the Chancellor in the hope of being able to bowl him over with his innate *Suada* aided by some dramatic gestures. I imagine that any such discussion with the Kaiser must be very embarrassing for *anyone who wants to be polite* for the simple reason that H.M. can come closer to the limits of parliamentary freedom of speech than another mortal.' Holstein therefore advised that all outstanding questions should be settled in writing by the Reich Chancellor, as in that way the risky 'effects of the moment' would not play such an important role as in a discussion face to face. In order to avoid a meeting with the Kaiser Hohenlohe should therefore stay somewhere in South Germany while the crucial decisions were taken.[132]

On 26 July 1896 the Kaiser telegraphed to the Reich Chancellor to say that he had decided to return from his Scandinavian cruise a few days earlier than planned, and to go to Wilhelmshöhe. 'Philipp Eulenburg will accompany me and travel by way of Kassel directly to Aussee in order to discuss the situation with you.'[133] When the true reason for the sudden curtailment of the cruise – the resurgence of the Kaiser's ear trouble – became known in the Foreign Office, Holstein saw it as confirmation of a suspicion that he had harboured for weeks and had mentioned to those closest to him, namely that 'the events of the last few days were not free from pathological features'. How else could one describe the Kaiser's 'temperamental remarks' on 'the Lippe case, Bronsart, Marschall and the fleet programme' which had reached Berlin from the *Hohenzollern* over the past four weeks, he asked. But if the Kaiser were in any way not of sound mind there was an even more urgent need than before 'to uphold the personality and the *prestige* of the Chancellor as the *fixed point* around which the whole conception of the Reich and its supporters can come together'. The 'political personality' of Prince Hohenlohe was 'today even more than a month ago an indispensable, almost irreplaceable, political working capital for the German Reich'. Court martial reform and the 'enormous budget for the limitless fleet programme' – like the resistance against two-year military service under Caprivi – had been 'blown up artificially' by the Kaiser's entourage in order to reduce the Reich Chancellor's power, which they found uncomfortable, Holstein argued. In his forthcoming negotiations with Eulenburg the Reich Chancellor must therefore take care, for his own sake and that of the German people, that his position was not diminished, for the moment that he abandoned the standpoint he had taken in his statement to the Reichstag and capitulated to the imperial entourage, Hohenlohe would cease to 'represent anything firm and definite, and from that day onwards the Kaiser can get rid of him without fearing public indignation'. On the other hand, the more resolute the Chancellor was in what he said, the more careful the Kaiser would be. 'All unbalanced persons are like that!'[134] Holstein even advised the Reich Chancellor to allude to the danger of the Kaiser being declared incapable in his discussion with Eulenburg. Hohenlohe should tell him that 'a sick Kaiser must be more circumspect than a healthy one, since otherwise *the idea of placing him under guardianship* would be mooted in some corner of the Reich at the first opportunity, and the thought would never die out'. In mentioning Wilhelm's illness the Chancellor could 'point out frankly and unequivocally that the rule of the Kaiser, both with regard to the German princes and the German people, would be made much more difficult if the existence of this illness became known, and that this should be a warning to the Kaiser to be careful'.[135] The news that 'the illness has reappeared after lying quiescent for several years' would make a profound impression, the Privy Councillor urged in a last letter before Eulenburg's arrival at Alt-Aussee. 'Throughout the Reich there will be a general feeling of uncertainty, which may lead to disintegration and revolution if it is allowed to develop.' It was therefore Hohenlohe's duty to declare to the Kaiser 'with absolute firmness' that he must stay in office, 'but under conditions that correspond to the interests of the monarchy, namely that I [the Chancellor] remain at the head of a Cabinet capable of dealing with the problems brought before it in the interests of the crown and the country'.[136]

For his part, Hohenlohe assured Holstein that he would tell Eulenburg 'categorically' that the abandonment of public courts martial was an 'impossibility', and that he would consider the departure of Bronsart and Marschall 'disastrous'.[137] Privately, however, he noted that he wished to find 'a way out', 'out of consideration for the personality and position of the Kaiser'. He would gladly resign, but he could see that the Kaiser would be in a difficult position if he resigned over court martial reform.[138] When the Kaiser telegraphed from Wilhelmshöhe to tell him of his ear operation the Chancellor wired back with his 'humble' thanks for the 'gracious' telegram and asked 'most respectfully for news of Your Majesty's health soon'.[139] So even before Eulenburg's arrival the elderly Prince seems to have inwardly abandoned the idea of resistance on grounds of principle. In any case he had nothing with which to counter the programme so firmly decided upon by the Kaiser and his friend at Wilhelmshöhe.

In the meantime Eulenburg had travelled from Wilhelmshöhe via Hallstadt, where he met Bülow, and joined Prince Hohenlohe at Alt-Aussee, with the task of bringing about an ultimate solution to the long-running government crisis.[140] During their encounter of 3–5 August 1896, in which, as Hohenlohe himself acknowledged, it was a question of 'to be or not to be', Eulenburg behaved as if he were certain of victory and was not in the least embarrassed to be regarded 'as a torpedo in the imminent crisis'. As he had agreed in his conversations with the Kaiser on the *Hohenzollern* he explained to the Chancellor that the monarch could not concede open courts martial in any circumstances; he considered that his 'bounden duty'. But he was sincerely ready to consider a reform of the old court martial regulations taking in a few other features of 'modern' thinking and, if the reform bill were rejected by the Bundesrat or the Reichstag because it did not include public hearings, it could be introduced in Prussia. This reform was 'in accordance not only with the convictions of the Supreme War Lord but also of the Prussian army' and therefore could be supported by the civil government with a good conscience. As far as Bronsart was concerned, the Kaiser would relieve him of his functions as war minister and appoint him his adjutant-general in waiting, which would prevent the General's resignation from appearing to reflect sharp differences of opinion between the Kaiser and the government. 'I replied to Eulenburg', Hohenlohe recorded after the meeting, 'that the appointment of Bronsart as adjutant-general would not change the fact that he was no longer minister of war. Furthermore that these complicated suggestions would not cancel the fact that I had given up open courts martial. Bronsart's position would be enhanced, and I would be disgraced. I drew attention to the looming conflict with the Reichstag; for since 1871 no question had preoccupied public opinion as much as this one. If I gave up my standpoint, in political terms I would be a lost man.' At the same time the Prince recognised that this was 'a matter of complete indifference both to H.M. and Count Philipp Eulenburg'. 'What happens to me does not interest them, so long as the Kaiser's caprice and the Military Cabinet's will is carried out.' Hohenlohe even had the impression that Eulenburg wanted to remove him from his post, 'whether for his own motives or to please the Kaiser'. At any rate Eulenburg had made it clear to him that he could remain in office only provided that he 'submitted to the All-Highest will. That I shall not do.'[141]

For his part, Eulenburg gained the impression that Hohenlohe had taken many of his arguments on board. The Prince had become 'very pensive' and needed a few days to reflect, he reported to the Kaiser. If pressure were put on him to make up his mind immediately he would take it as 'a rude ultimatum from Your Majesty' and would throw all arguments overboard again. 'I beg Your Majesty's gracious forgiveness for this decision – but I could not thoughtlessly provoke a rejection.' But a 'very friendly letter from Yr. Majesty' could have a good effect on the Chancellor, Eulenburg advised. The news of the imminent death of Bismarck had already affected him deeply.[142] When he had taken his leave Hohenlohe had thanked Eulenburg 'with feeling' and had again expressed the hope that he would still be able to find 'some way out'.[143]

Eulenburg had barely left Alt-Aussee when a telegram arrived from Wilhelmshöhe with a pressing invitation from the Kaiser to the Reich Chancellor to go there. The invitation – 'obviously something arranged by E[ulenburg] with H.M.' – was expressed in such a way that Hohenlohe could not refuse, nor even return to Berlin first.[144] On 8 August the elderly Chancellor and his son Alexander therefore made the journey to Wilhelmshöhe via Munich. He found the Kaiser, who was clearly afraid that Hohenlohe would refuse to accept Bronsart's appointment as adjutant-general, 'very much on edge'. But on this point the Prince yielded at once without dissension. The Kaiser then set out 'the whole plan as drawn up by Eulenburg'. In the end the Chancellor agreed to wait for the expert opinion of the generals on the contentious question of public courts martial, although he knew very well that 'the commanding generals will say only what the Kaiser orders'. Meanwhile, the Kaiser said, Hohenlohe need have no scruples about going away to Werki, his wife's Russian estate, since a final answer on the court martial question could not be expected until the beginning of October.[145] In thus agreeing to postpone the critical moment until the autumn the Prince's desire to seek his own material advantage again played a discreditable role. In order to 'look after his Russian interests' he asked permission to be present at the Kaiser's forthcoming meeting with the Tsar and Tsarina at Breslau; he also expressed the wish that his son Alexander be appointed Bezirkspräsident (district president) of Colmar in Alsace.[146]

Wilhelm II himself left a typically heartless account of these decisive negotiations with the Reich Chancellor and his son. In a letter of 14 August to Eulenburg he reported that 'old Uncle' had been 'hale and hearty and very sensible'. 'Especially after his long conferences with Lucanus and Hahnke. In order to get him out of the state which you rightly describe as stagnation in his support for public hearings, it was made very clear to him that the main issue was not public access but the Central Military Court presided over by a general directly under the Kaiser. The South Germans were opposed to this. But as long as it was not conceded the entire reform bill would collapse and would not come before the Reichstag. This point was an absolutely indispensable condition for the Kaiser and he must get it accepted first. This was something quite new for the old man and he was grateful when he saw that the rejection which could be expected from the Bundesrat would give him a bridge over which he could retreat. Alexander was no use at all here. He talked all kinds of rubbish, stirred his father up the whole time and in the end had to be very roughly silenced by Lucanus that evening, in a violent dispute in which he got very worked up arguing for public access. This grass-green young clown, who has never served in the army and couldn't tell a Pioneer from a Garde du Corps, does a great deal of damage and compromises the old man by spreading a lot of irresponsible gossip which he invents. I therefore spoke very strongly and seriously to my uncle about his offspring, which shocked the old man very much, but he was very grateful. He went away with the impression that I was as steady as a rock and that public hearings were a nonsense, and with my permission to think about finding a way out in the above-mentioned manner. To bolster him up against his "colleagues" I gave him a written assurance that Bronsart's dismissal would not affect the declarations in the Reichstag. For the rest I insisted very strongly and warmly on his importance and on the need for him to stay for the sake of political affairs, and asked him to stick to what we had both agreed and not let himself be browbeaten by every unauthorised "adviser" who came along. He promised to do all this and as he left he said, "I shall let the matter rest entirely, speak to no one about it and keep very quiet in Berlin." He went on his way and we were very pleased.'[147]

When his three months of leave were over the normally so defiant General Bronsart von Schellendorf endured the undignified farce of 'promotion' to adjutant-general to the Kaiser, which no one could take seriously.[148] To succeed him as minister of war the Kaiser chose not General von Hähnisch but the Commander of the 25th Division, General Heinrich von Gossler, of whom he telegraphed delightedly to Eulenburg that he was 'in agreement with me on all questions' and wanted 'only to be his Kaiser's general'.[149] Gossler, he commented in a later message, was 'an old friend of mine, and we have already fought many a battle in the Ministry of War together'.[150] The public reaction to Bronsart's dismissal and Gossler's appointment was disastrous.[151] The Reich Chancellor noted in his journal that both decisions had caused 'an enormous uproar' and the Kaiser had done himself 'terrible harm' by his action. It was 'very sad', he commented, adding that the future looked 'black' to him. But instead of learning from this experience Wilhelm condemned the attacks in the press 'against government by the Cabinets, which of course is basically no more than personal rule', as 'expressions of high treason'.[152] Waldersee also expressed indignation at the decisions. 'There is universal regret at Bronsart's resignation', he commented, 'and dismay over the successor is almost as widespread. Indeed I believe the Kaiser could not have made an unluckier choice.'[153] Germany could never have a more effective war minister than Bronsart, in his view, and Gossler was 'a very dubious successor', who was perhaps hard-working and competent but was 'an extremely unprincipled and ambitious person'. 'He can be turned in any direction and certainly suits the Kaiser down to the ground', he wrote. 'I am very worried by this choice.'[154] The rift between Bronsart and the Kaiser could not of course be disguised by his appointment to the court. Only a few months later, in his next New Year address to the commanding generals, the Supreme War Lord declared with uncontrolled fury that he had 'felt compelled to dismiss the General from the army, relieve him of his post as adjutant-general and request him to absent himself from the celebrations of the Order of the Black Eagle. Thereupon he abruptly turned on his heel, said goodbye and vanished.' It was only with difficulty that the 'very agitated' Chief of the Military Cabinet succeeded in persuading the Kaiser that the order to Bronsart should be toned down.[155] Months afterwards it was still being said in political circles that the former War Minister had been forbidden by imperial order to go within a two-mile radius of Berlin.[156]

The violent public reaction to Bronsart's fall might still have enabled Hohenlohe, even at this late stage, to preserve the power of the chancellorship by taking a firm stand. Not only Holstein and other Foreign Office officials but also his own family and members of the South German professorate were urging him to stand by his principles, and their efforts were not entirely unsuccessful.[157] As Eulenburg had to explain with some anxiety to the Kaiser, their earlier hope that the Prince could retire without reference to court martial reform was now ruled out. Extreme caution was necessary in dealing with the Chancellor, for he had very strong feelings about the 'historical figure' he cut, and if he should gain the impression that he might be dismissed in the autumn he would 'much rather go *now* in a blaze of glory'. It was also most important, the Kaiser's friend advised, to delay the appointment of Alexander Hohenlohe as Bezirkspräsident until after the chancellor crisis was finally over. 'The prospect of relieving the burden on the Princess [Hohenlohe]'s purse is a strong reason to stay on for this rather cynical part mediatised, part Sarmatian princely family – a kind of patent axle on the Chancellor's carriage.'[158] But when the Reich Chancellor sent a telegram from Berlin seeking to make clear that he still stood by the 'principle of public access' Eulenburg did not hesitate to tell him plainly that this declaration, which had 'awakened memories of the Köller crisis', had 'annoyed' the Kaiser. The monarch was 'very certain, very *firm*' on this question and might easily be provoked to 'sudden, very extreme steps' by continued resistance from the Chancellor.[159] Hohenlohe again gave way, and Wilhelm and his best friend observed with relief that 'the Hohenlohe crisis . . . is over'.[160] And with the 'disappearance of the crisis' they were again confident that 'Hohenlohe will be able to spend the rest of his life in office. Then other arrangements will also be more peaceful and simple to make.'[161] And indeed, in the course of the next few months Wilhelm II and Philipp Eulenburg, together with the other advisers at court – August Eulenburg, the three Cabinet Chiefs and the adjutant-generals and Flügeladjutanten – were able to make almost all the changes in personnel which they had worked out together during the Scandinavian cruise and subsequently at Wilhelmshöhe.

THE REICH CHANCELLOR AS THE KAISER'S 'STRAW DOLL'

When he looked back a year later over the events of the summer of 1896 Prince Hohenlohe was only too painfully aware that in those crucial weeks he had lost the last vestiges of the Reich chancellor's independent power to the Kaiser. He wrote ruefully to his son Prince Alexander: 'If I had offered my resignation at that time, I would either have departed a popular man, or forced the Kaiser to give way . . . Now there is little to be done.'[162] He noted privately on 22 May 1897 that 'the government cannot go on as it is now. If H.M. wants to rule by himself, I cannot act as his straw man. If H.M. had approved the court martial bill years ago . . . the whole storm would not have arisen. If H.M. chooses all the ministers himself, the government will lose more and more consistency and respect . . . Without authority, government is impossible. If I cannot get the Kaiser's consent to measures I regard as necessary, then I have no authority . . . I cannot stay if H.M. removes Marschall against my will and if Holstein goes. Likewise I cannot stay if the Kaiser appoints ministers without consulting me . . . I cannot govern against public opinion as well as against the Kaiser. To govern against the Kaiser and the public is to hang in mid-air. That is impossible.'[163]

During his summer holiday on his estate in Bohemia he recorded that 'The entire attitude of H.M. in recent months . . . proves that the Kaiser is determined to choose another Reich chancellor, or at least to inaugurate a new system. He wants to support himself more on the right, to use Miquel to this end, and to tolerate me for the time being as a straw man until he has found the right chancellor. If I accept this plan I shall . . . be utterly discredited.'[164] 'The fact is', he recognised, 'that I have lost my authority over the Kaiser, and as a result I have no *raison d'être*.'[165] To remain in office as a 'half-dead Reich chancellor' was impossible for him, he declared with apparent resolve in mid-June 1897.[166] 'The role of the Reich chancellor who merely carries out the orders of H.M. does not appeal to me at all', he wrote to his son,[167] echoing this sentiment in a clear-sighted letter to a South German friend: 'If the Kaiser wants to be his own Reich chancellor he will have to appoint a straw doll. I have no desire to be one.'[168]

Seldom has the desolate situation of the Reich chancellor under Wilhelm II been as unsparingly laid bare as in these self-critical comments by Hohenlohe. But the elderly Prince did not act in accordance with the insight he showed. Admittedly, from time to time he noted down proud declarations which he intended to make to the Kaiser – in one such, written in March 1897, he expressed regret that Wilhelm was not the Tsar of Russia; the constitution was not of his making but he must uphold it – but it is more than doubtful that he ever made use of them.[169] He threatened to resign when Wilhelm II appointed Tirpitz in place of Admiral Friedrich Hollmann, the Secretary of State at the Reich Navy Office and as such responsible to the Reich Chancellor, but he did not succeed in reversing the Kaiser's decision.[170] And he went so far as to hand in his resignation on 31 May 1897 when the Kaiser peremptorily demanded the immediate resignation of Heinrich von Boetticher as Reich secretary for the interior and vice-president of the Prussian Ministry of State, but backed down at once after receiving a few flattering lines from the Kaiser brought to him by Lucanus.[171] The Reich Chancellor, now aged seventy-eight, constantly found reasons why it seemed 'easier' or more effective to postpone his retirement.[172]

Until his departure in October 1900 'Chlodwig, the old mummy', as his opponents unkindly nicknamed him,[173] endured the most outrageous rudeness from the Kaiser without reacting as he owed it to himself and to the German people to do. In the summer of 1897, as the so-called responsible Reich chancellor, he accepted a series of appointments and dismissals of momentous significance, which transformed the Reich and Prussian government into an administrative machine with the sole purpose of carrying out the Kaiser's will. Marschall von Bieberstein was replaced as foreign secretary by Bernhard von Bülow and shipped off to Constantinople as ambassador. Count Arthur von Posadowsky was appointed secretary at the Reich Office of the Interior in place of Boetticher; the former ambassador in Washington, Max Freiherr von Thielmann, was nominated to the Reich Treasury Office and General Viktor von Podbielski, a Conservative member of the Reichstag, to the Reich Post Office. Finance Minister Johannes von Miquel was promoted to vice-president of the Prussian Ministry of State, with instructions to steer a 'more Prussian' course in internal affairs; Admiral Alfred Tirpitz, with his plan to build a 'giant fleet', enthusiastically supported by Wilhelm II, took Hollmann's place as navy secretary. Bülow, meanwhile, prepared himself – at first still as secretary of state at the Foreign Office – to take over the Reich chancellorship, a post in which he himself had said that he would act as the 'executive tool' of the Kaiser and 'so to speak his political Chief of Staff'. With that, 'personal rule – in the good sense – would really begin'.[174] If, as was said of Hohenlohe, the 'historical figure' he cut in German history was his main motivation, it is certainly clear that in the summer of 1896, in his narrow-minded obsession with his personal reputation and his constant search for a 'good' way of quitting his post, he shamefully gambled away what little remained of the autonomy of the 'responsible government', which he had taken over from the upright General von Caprivi barely two years earlier. In the last four years of Prince Hohenlohe's period in office the German Reich chancellor was downgraded to the 'straw doll of the crown'.

Hohenlohe's pathetic spinelessness towards the Kaiser is all the more shocking because he was repeatedly reminded by old friends of the principles at stake in the conflict with the monarch. The Bavarian public servant, Hohenlohe's former colleague Otto Freiherr von Völderndorff-Waradein, writing to the Prince in June 1897, fully acknowledged that the position of Reich chancellor as defined by Bismarck's Reich constitution had been far beyond the powers of any man and that it needed to be reduced, perhaps by delegating Prussian internal affairs to a reliable deputy. But he pointed out that 'it is quite a different matter when the Kaiser wants to be his own Reich chancellor. That is of course even more beyond the powers of any man and would be sheer self-deception. For in practice decisions would be taken only by irresponsible secondary figures, who usually have only a one-sided knowledge and opinion of the real state of affairs, and who would make their views seem plausible to the monarch.' Völderndorff went on to say that he would deeply regret it if Prince Hohenlohe ended his 'glorious' life as a 'delivery boy', and advised him to resign from all his offices. 'The elections to the next Reichstag will most certainly turn out very badly', he added as a further reason for resignation.[175] Another old friend, Léon Dacheux, the Superior of the great seminary at Strassburg, wrote to Hohenlohe saying that in his eyes what was going on in Berlin was a constitutional conflict between the system of absolutism and a nineteenth-, almost twentieth-century system. 'But the system of Louis XIV, which has continued under the successors of Frederick the Great, no longer accords with the ideas of our own times.' A *coup d'état* such as that carried out by Louis Napoleon on 2 December 1851 could of course be attempted, but it would not provide a lasting solution; on the contrary, a violent measure of that kind would be the death-blow for the Hohenzollern monarchy. 'I like to believe that the German people will not put up barricades, but an abyss will open between the people and the throne, at least in the western provinces, and the bourgeoisie and the lower classes, especially in the Rhineland towns, will become increasingly republican, and the love which the nation has for the monarchy will be struck dead.'[176] A few weeks later Abbé Dacheux expressed regret that Hohenlohe had not resigned with Marschall and Boetticher. The honourable name of the Prince was too good to serve as camouflage for the introduction of 'more compliant' people 'who lend themselves to anything that the monarch demands, according to his own impulses or the influence of others'. With unerring accuracy the wise old priest warned that the Chancellor would achieve nothing by giving in. 'You are being demolished bit by bit.'[177]

CHAPTER TWENTY-EIGHT

Personal Monarchy: Wilhelm II at the summit of his power

THE FACE OF PERSONAL MONARCHY

HAVING got off relatively lightly with the dismissal of War Minister Bronsart von Schellendorf – the much-feared major government crisis had not after all materialised – Kaiser Wilhelm II had reached the zenith of his personal power, objectively speaking. And yet he was clearly far from satisfied. He showed less restraint than ever in giving vent to his autocratic, indeed aggressive claims to power in defiance of the Reich and Prussian government, the Reichstag and the overwhelming majority of the German people, whose attitude towards him was becoming increasingly critical. Every restriction placed on his power irritated him: he wanted to be in command and expected unconditional obedience. 'I know no constitution, I know only what I want', he exclaimed.[1] 'All of you know nothing', he asserted peremptorily to his admirals. 'I alone know something, I alone decide!'[2] Count August Eulenburg, the Oberhofmarschall, commented at the beginning of 1897 that the links between the court and the Wilhelmstrasse were 'in effect completely severed', and 'if any attempt is made to resume contact sparks fly every time, if there is not an actual explosion'.[3] It was becoming clear to those in high places who had dealings with him that 'the Kaiser could not get accustomed to the idea that his power was restricted by the Reichstag. The idea of ruling without the Reichstag never left his mind, and was encouraged by one sector of his entourage.'[4] The Reich Chancellor tried in vain to persuade the Kaiser that his power was limited by the constitution. The longstanding rights which he had inherited as king of Prussia still held good, Prince Hohenlohe acknowledged, 'in so far as the Prussian constitution had not placed limits on them. [But] in the Reich the Kaiser had only the rights which the Reich constitution accorded to him.' To this Wilhelm retorted that 'he did not care about public opinion. He knew that he was not loved and that people criticised him; but that did not deter him . . . The South German democratic states did not trouble him. He had 18 army corps and was quite able to deal with the South Germans.'[5] In May 1897 the British ambassador reported to London with growing anxiety on the unrest that was taking hold in Germany. He commented that the parliamentary debates on the laws on *lèse-majesté* and the reactionary law of association, the backstage intrigues at court seeking to influence the government's policies, and not least the manner in which the Kaiser repeatedly dealt out public insults to the Reichstag, the Social Democrats, the pro-Bismarck movement and even the Conservatives, had caused a gulf of mistrust to open up between Wilhelm II and the German people.[6]

Wilhelm's aggressiveness can be partly explained by the fact that he had been forced to put off until mid-1897 the dismissal of the 'loathed' and 'detested' Marschall, the Foreign Secretary, and likewise that of the Secretary for the Interior, Boetticher, of whom he had grown tired. Marschall's decision to 'go public' – in the winter of 1896–7 he appeared as chief prosecution witness in the case against Police Commissioner Eugen von Tausch, the police agent Lützow and the journalist Leckert, in the course of which highly compromising information about Ernst Günther and the Kotze affair, the Miss Love scandal, Mirbach's dishonest activities over church subscriptions and underhand machinations by Waldersee came to light – was the last straw.[7] In October 1896 Wilhelm exclaimed bitterly: 'I cannot stand Marschall any longer!'; his patience was at an end, regardless of what further parliamentary successes the Foreign Secretary might achieve.[8] While the court was hunting at Springe in December he telegraphed to the Reich Chancellor stating that he must get rid of Marschall and added, tellingly, that 'his entire entourage was of the same opinion'. At the beginning of 1897 Waldersee commented that 'everyone in the Kaiser's personal entourage including Hahnke, who has returned from leave', was against keeping Marschall and that they had 'repeatedly said so'. When an unsolicited report of some kind arrived from South Germany, sharply criticising 'Marschall's tactlessness and arrogance', Wilhelm declared: 'I see it is quite impossible for me to work with these South German ministers, they do not understand me, I shall find myself some North Germans or Old Prussians again; I shall get rid of the other lot by spring at the latest.'[9] He wrote angrily to his 'dearest Phili' complaining that in Berlin 'as usual people constantly invent things, slander, criticise etc. but above all refuse to obey orders. In the Ministry, as Boetticher recently put it, showing how well he knows his own faults, everyone "muddles along" calmly, except that rage and fury against Marschall has quite taken hold among his colleagues and has united them firmly against him. They will thank me if they live to see the day when that d . . . d South German vanishes from the scene.'[10]

The Kaiser reacted angrily to every flicker of independent spirit from the Reich Chancellor. It was only half in jest that he remarked reproachfully in October 1897, as Hohenlohe reported to his son in some surprise, that the old Chancellor 'had tyrannised him!'[11] Several times Wilhelm complained of the Chancellor's habit of accepting the 'orders' of his Kaiser only to question them again subsequently. 'My old uncle is dear and kind as always, but more and more reluctant to keep any promise he has made to me. In the last few months there have been several more flagrant instances when I had arranged what was to happen verbally and sometimes even in writing, only to be most kindly informed in those well-known little *billets doux* that to his very great regret what we had *agreed* was quite out of the question. The worthy gentleman calls the *orders* of his King and Master *agreements*! Well, I shall let things go on as they are until the spring, but then, dear Phili, even if you fling yourself on your knees before me, too bad! Then there must be a clean sweep, or we shall be ruined! Morally and physically! I shall not be able to bear it any longer by then!'[12]

The sharp orders which Wilhelm fired off at the Chancellor in tones of military command are probably without equal in the history of European monarchy. The flavour of these dispatches, letters, marginal notes and speeches is exemplified by an open telegram to Hohenlohe of 4 January 1897, in which the Kaiser rejected his advice not to respond to an item in an insignificant local newspaper concerning his attitude to duelling. It reads: 'What the *public* think or do not think of me is quite immaterial. The report . . . will *confuse the army* and other *decent people*. The *démenti* is therefore to be issued immediately. As we know from famous examples, publicity is of course most salutary, and I have as much right to it as my subjects. My answer. W.'[13] The Kaiser's attitude to duels had already given rise to a brusque telegram from the Neues Palais a few weeks earlier. 'I have seen in the press that the Reichstag intends to use an interpellation to criticise and challenge My right to grant pardon in cases of duels', Wilhelm telegraphed to the Chancellor. That was, he said, 'a piece of shamelessness which represents a direct attack on My personal rights and privileges. On no account will I allow this. In the event that they do have such an intention, therefore, there must be immediate talks with the members of the Reichstag to make it clear to them that this interpellation must be stopped. If the Reichstag should nevertheless insist on bringing in the interpellation, I would regard this as a direct attack on My Person and the crown, whose constitutional rights I have to defend, and I would respond to this attack immediately by dissolving the Reichstag. Wilhelm I. R.'[14] In similar tones Wilhelm denounced the *Kölnische Zeitung*, of which the Foreign Office often made use for its own diplomatic purposes and which, according to the Kaiser, 'in spite of all warnings, admonitions etc.', had not only 'had the effrontery' to publish articles 'insulting to my adjutant-generals and a

Commanding General of my army [i.e. Waldersee], but had furthermore instituted direct personal attacks on Adjutant-General von Hahnke again. I therefore command that the offices of this newspaper should immediately be barred by the Foreign Office until further orders from me, and that all authorities in the country should be instructed not to give the paper any further news or to receive any of its representatives. Anyone disobeying this command can regard himself as dismissed from my service.'[15] In August 1897 the Kaiser was still raging against the 'nonsense in the press', which was the work of 'the remnant of the men who used to cause mayhem in the Foreign Office' – in this case he meant Holstein – and demanded that this 'stop immediately and the press must be given orders to that effect!'[16] When President of the Reichstag Count Ballestrem presumed to reprimand Prussian Minister of Trade Ludwig Brefeld the Kaiser sent a thunderous telegram to the Chancellor declaring peremptorily that 'he has no right to reprimand one of My ministers and I will not have it, once and for all. He should be told this. Brefeld is to be praised for speaking so well in favour of the bill.'[17]

In the summer of 1898 Wilhelm caused much bad blood by sending an offensive telegram *en clair* in response to a petition from the Count Regent Ernst of Lippe-Biesterfeld that the commanding officer of the troops stationed in Detmold should accord him and his family the honours customarily paid to a ruling prince. 'Your letter received', the Kaiser wired back. 'Commanding General's orders are as approved by me after consultation. The regent gets what is due to a regent, nothing more. Furthermore I wish it to be understood once and for all that I will not tolerate the tone in which you thought fit to write to me. W. R.'[18] Hohenlohe complained that the Kaiser was still obsessed with the idea of a tribunal of princes to settle these succession disputes, consisting of three sovereigns on each side with him, the Kaiser, holding the casting vote as chairman. 'Naturally no German sovereign would agree to that', he commented.[19] The stir which this episode caused among the public, who did not always fully understand the esoteric subtleties of these princely squabbles, was great and did nothing to improve the standing of the monarchy.[20] The 'state-supporting' newspapers warned that the German Reich, united by blood, iron and tears, would fall apart if the dignity and rights of the federal princes were not upheld; the opposition press, on the other hand, made little attempt to hide its glee at the imperial gift to particularists, republicans and socialists.[21] Public opinion was scarcely less surprised and affronted by the Kaiser's telegram to Hinzpeter of 1899, in which he compared himself with the Great Elector and declared that 'just as with my ancestor, there is in me an indomitable will to follow the path which I have recognised as right, regardless of all opposition, without wavering. Wilhelm I. R.'[22]

In view of the increasing criticism directed at him from all sides it is understandable that Wilhelm II took refuge in the support of the Liebenberg circle around Philipp Eulenburg, in which he was idolised, as well as of the trusted members of his (predominantly military) entourage and of the army, supposedly still loyal to him. In a revealing speech to the regiments of the imperial bodyguard – the First Guards Regiment of Foot, the Gardes-du-Corps and the Hussar Body Guards – on the tenth anniversary of his accession in June 1898, he said in tones of self-pity: 'Seldom, indeed, has a time of such trouble passed over the head of a successor to a throne, who had to see his grandfather and his father die within so short a time. The crown was burdened with heavy anxieties. On all sides men doubted me; on all sides I encountered misconceptions. Only in *one* quarter was there confidence and belief in me, and that was in the army. And supported by it, and relying upon our God as of old, I undertook my heavy office, knowing well that the army was the main support of my country, the main pillar of the Prussian throne, to which God's decree had called me.'[23]

Like a feudal lord 'bound by common bonds of loyalty' to his army and navy, albeit on the threshold of the twentieth century, Kaiser Wilhelm tried to shield them from all political criticism. When the Chancellor warned him in November 1897 that the speech he was about to make at the swearing-in of naval recruits at Kiel would be read with close attention in St Petersburg and in Paris, where people would be on the look-out for grounds for suspicion of his foreign policy, the Supreme War Lord snarled back at him insultingly that 'speeches at swearings-in of recruits are *never* political in content, for politics have no place in either army or navy, and therefore do not concern Paris or Petersburg in the least. As I cannot but suppose that Y[our] H[ighness] was prompted by someone else to write the final passage, I would ask Y.H. to deliver the appropriate reprimand from Your sovereign and Kaiser in the clearest possible terms to the gentleman in question. Such promptings are highly improper and I refuse once and for all to tolerate them.'[24] Prince Hohenlohe courageously replied that as principal adviser to the Kaiser he had felt it his duty to draw his sovereign's attention to the fact that his speeches would attract malevolent attention abroad; he had required no prompting to recognise this danger.[25]

If the Kaiser's offensively peremptory tone in dealing with the constitutional organs of government was bad enough, his aggressive tendencies were even more noticeable in his comments on the elected representatives of the people. At the beginning of 1897 he learned that the seventy-year-old Social Democrat Wilhelm Liebknecht had made a 'fiery speech' in Holland, and saw it as a 'glaring confirmation' of the necessity of suppressing the workers' movement by military means. He ordered the Reich Chancellor to submit proposals as to 'what can be done against active members of the Reichstag who preach revolution abroad during the parliamentary session'.[26] A few weeks later, in a speech to the Brandenburg provincial diet which aroused indignation throughout the country, he again called for a '*fight against revolution*, with every means at our command'. The Social Democratic Party, which dared to 'attack the foundations of the state, which rebels against religion, and which does not even spare the person of the All-Highest sovereign, must be crushed. I shall rejoice to know that every man's hand is in mine, be he workman, prince or sovereign, if only I have his help in this conflict! . . . Then we shall work in the right spirit, and will not desist in the struggle to free our country from this malady [in fact he described the workers' movement as a "plague"[27]], which is invading not only our people but also our family life, and, above all, is striving to shake the position of woman, the most sacred thing that we Germans know. I hope, therefore, to see my Brandenburgers rally round me if the fiery signal should be raised.'[28]

In April 1897, after the Reichstag had rejected the greater part of the naval estimates, Kaiser Wilhelm sent an ill-advised letter to his brother Heinrich, which the latter read out to the crew of his cruiser. 'I deeply regret', the Kaiser wrote, 'that I cannot place a better ship at your disposal for the celebrations [of the Diamond Jubilee of Queen Victoria in London] than the *König Wilhelm*, while other nations will put on a splendid show with their proud battleships. This is the sad consequence of the behaviour of those unpatriotic people who succeeded in preventing the provision of the most badly needed ships. But I shall not rest until I have brought My navy up to the same level as the army.'[29] According to other accounts the Kaiser described the Reichstag deputies as 'scoundrels without a fatherland' or 'rogues'. At a crisis meeting of the Ministry of State Prince Hohenlohe was asked to send the Kaiser newspaper extracts showing the effect of his letter and to request permission to issue a *démenti*, since otherwise an interpellation in the Reichstag on the Kaiser's personal rule was to be feared. 'It is a very unpleasant business', the Chancellor commented. He was relieved to hear that the leader of the Centre Party was prepared to postpone the interpellation in view of the mood abroad.[30]

The Kaiser, who was staying with his friend Emil Görtz at Schlitz in the Vogelsberg, was of course adamant that there should be no *démenti*. Kiderlen-Wächter, who had gone with the Kaiser as representative of the Foreign Office, reported to the Reich Chancellor that at the first mention of a possible interpellation 'His Majesty flared up: what He and his brother wrote to each other was no one else's business, least of all the Reichstag's; that was the only answer the government should give. After the initial storm had died down I told His Majesty that I was sure Your Highness would give the Reichstag the answer His Majesty wished, but that a negative answer from the government would not prevent the Reichstag from indulging in remarks that would inevitably have a bad effect both at home and abroad.' The Kaiser had certainly welcomed the news given him by Kiderlen that Hohenlohe was negotiating with the Centre with a view to preventing the interpellation. But when Kiderlen 'went on to say that all kinds of different versions [of the Kaiser's letter] were circulating and that certain expressions in particular (I cited "scoundrels without a fatherland") had caused bad blood, and that it was therefore thought desirable at least to be able to deny that these expressions had been used, His Majesty reverted to the claim that the letter was a private matter between Him and his brother. When I interjected that the letter had acquired a different character because of the publicity given it by Prince Heinrich, His Majesty refused to accept this but simply went on repeating that He could write what He wished to his brother.' Summing up his discussion with the Kaiser Kiderlen concluded 'that His Majesty realises how inopportune the letter was, but will not admit it. Nor have I any doubt that the strong expressions are authentic; for when I cited them His Majesty was visibly very embarrassed.'[31]

Even before Kiderlen's report arrived in Berlin the Reich Chancellor received an open telegram from the Kaiser which provided clear confirmation of Kiderlen's account. 'Yr Highness's letter and enclosures received. I was already familiar with the latter. Making private expressions of opinion subjects for discussion in the press is somewhat reminiscent of the beginning of the Convention; possible interpellations and debates are even more so; it betrays a determination on the part of parliament to interfere in personal affairs, which has never arisen except in revolutionary times . . . I do not care in the slightest what the press writes or does not write about me. But I will not allow any interference in my personal affairs. It is a good thing that the proposed demand was turned down: if it had come to an interpellation I would have responded with dissolution. There must be no *démenti*. Wilhelm I. R.'[32] The unbridgeable gulf which had opened up between the monarch on the one side, and the constitution as understood by the Reich government and the parliament on the other, is vividly illustrated by the note which Hohenlohe made after receiving the Kaiser's telegram: 'This is not a question of private expressions of opinion but of an insult to the Reichstag. The private expression of opinion loses this character when it is read out *officially* to the crew of an imperial warship by an admiral. The dissolution of the Reichstag would not protect Prince Heinrich from a libel suit. If I were a member of the Reichstag I would start legal proceedings against H.R.H. [Prince Heinrich]. § 24 of the Reich constitution. For the dissolution of the Reichstag during the parliamentary term a resolution of the Bundesrat with the consent of the Kaiser is required.'[33]

The loudest outcry, however, was reserved for Wilhelm II's remark in his Brandenburg speech of 26 February 1897 that in comparison with his grandfather,

42. Wilhelm Imperator Rex: the Kaiser at the height of his personal power

who would have been canonised in the Middle Ages, Bismarck and Moltke had been nothing but 'lackeys and pygmies', who had the honour of carrying out the ideas of 'Kaiser Wilhelm the Great'.[34] As soon as the speech was over, worried Flügeladjutanten hurried from chair to chair, asking the guests not to mention the 'salient parts' of the speech.[35] Although this initially prevented the worst expressions from becoming public knowledge, supporters of the monarchy were deeply alarmed by them. Holstein quoted the remark of a prominent

Conservative to the effect that the Kaiser could have given Bismarck whatever names he liked, 'but to call him a pygmy – No! he *must* be wrong in the head.'[36] The Prussian envoy in Munich, Count Monts, reported with consternation on the mood in Bavaria: 'Our enemies here find it scarcely necessary . . . to hide their delight any longer over the exalted orator, who really is no longer responsible for his actions. The nationalist-minded people here are like a flock of frightened chickens. The average cultured South German politician, even the clerical, is indignant at the falsification of history so beloved by His Majesty, and at his description of Moltke and Bismarck as minions of the illustrious sovereign.' The monarchy itself was undermining 'all monarchical convictions . . . from above', the envoy warned, and 'the only patch of dry ground which the representative of the Reich still has to stand on in Bavaria . . . is shrinking fast. Another flood like the torrent of words from Brandenburg will wash it away completely.'[37] Utterly discouraged, Monts wrote to Philipp Eulenburg commenting that 'the mood in the west and south of Germany is deplorable. Our numerous enemies are jubilant, and are quietly preparing for the disintegration of the Reich. Our friends are disgusted with the Kaiser. The bitterness goes deeper than ever before . . . Many are saying secretly that H.M. is insane; already there are hints to this effect in the press . . . What I think of H.M. I dare not even say, but I fear that he is completely finished now, here in the south. Perhaps the masses will still shout hurrah during the manoeuvres, but the hearts of the patriotic middle classes are now surely lost to him for ever . . . Does His Majesty know all this, and that his vague romantic ideas are completely at variance with the mood of the nation? The great capital which he inherited from Wilhelm I is dwindling away all the time, even the most loyal, most monarchically minded men are losing faith in everything that was most sacred to them, and terrifying prospects are opening up.'[38]

In the Reichstag, as was to be expected, criticism of the Kaiser in the spring of 1897 was the strongest yet heard. In mid-March Julius Bachem, the editor of the Catholic *Kölnische Volkszeitung*, suggested to the leaders of the Centre Party that they should use the parliamentary debate on the naval estimates as an opportunity to protest against Wilhelm II's personal rule. 'Everyone now realises that the Kaiser's personality and intervention lie at the root of our present crisis', he wrote. 'Someone in the Reichstag should say very plainly indeed how perturbed the people are at this excess of imperial initiative; how urgently they desire that the monarch take more notice of the realities of political life . . . In a word: a most loyal but earnest declaration should be directed at the Kaiser . . . The speech would be greeted by the people as an act of deliverance . . . When similar situations arose in England and Holland, men raised their voices in determined and glorious protest. Why should not that happen here?'[39] As the Centre Party rejected this suggestion on the grounds that such an action would only strengthen the Kaiser's prejudice against the Catholics, it was left to the leader of the Left Liberal Party, Eugen Richter, to make an impressive speech pillorying the system of Personal Monarchy. On 18 May 1897 in the Reichstag this democrat of the old school put into words exactly what millions of people in the country, right up to the highest echelons of the civil and military administration, were feeling. 'Where today can we find a unified, purposeful will, which is not affected by sudden impulses, but is able to pursue its goal steadily, with prudence and good sense? Where are the ministers? Wherever you look there are only compliant courtiers who agree with every opinion from above, promoted bureaucrats, dashing Hussars turned politician, lackeys, but in the conventional sense of the word. The [Free Conservative] Deputy von Kardorff has spoken of the decline of monarchical feeling since the death of Kaiser Friedrich. In Germany the monarchical system can expect to last for a long time, because the monarchy is bound up with the evolution of the nation itself. It is all the more regrettable that the stock of monarchical feeling is being used up in a way which I should not have thought possible ten years ago, and not as a result of Social Democratic agitation, no – but as a result of events which are beyond the reach of parliamentary discussion, events which provoke criticism, not only among ordinary citizens but also deep within the civil service and the officer corps. Germany is a monarchical, constitutional country, but although it may still be possible to rule Russia according to the principles of *sic volo sic jubeo* or *regis voluntas suprema lex*, the German people will not allow themselves to be ruled like that for long.' Richter's remarks were greeted with a storm of applause both from the deputies and from the public galleries.[40] 'The speech created a huge sensation and unfortunately received widespread enthusiastic applause', Waldersee noted. Richter had 'torn the whole government apart in a way that had never been done before . . . Of course the main target of the attacks was the Kaiser.'[41] As the 'wretched' Boetticher had not been able to summon up the words to repudiate Richter's 'insults', the 'highly incensed' and 'embittered' Kaiser demanded his immediate resignation.[42]

THE KAISER AND WALDERSEE'S *COUP D'ETAT* PLANS

It is not surprising that, faced with this situation and this atmosphere, Wilhelm should again have toyed with the idea of freeing himself from the democratically elected Reichstag by means of forcible constitutional change. Already in November 1896 the newly appointed War Minister, General von Gossler, had gone to the Reichstag with the Kaiser's instructions, as he told Boetticher, 'to provoke a conflict'.[43] The Saarland industrialist and Free Conservative Reichstag deputy

Karl Ferdinand Freiherr von Stumm-Halberg was likewise given 'orders' 'in a very heated manner' by Wilhelm while walking with him in the Tiergarten, to go to the Reichstag and say that 'if the Navy Bill were not accepted, the Kaiser would cause a great rumpus [and] kick the ministers out'.[44] He had other representatives such as the Prince zu Wied, the Conservative Party leader Otto Freiherr von Manteuffel and the National Liberal deputy Friedrich Hammacher sounded out in the spring of 1897 as to what conditions the former Kartell parties would set in exchange for their support for a *coup d'état* policy and a naval expansion programme.[45] 'Never before have we been threatened so openly with a *coup d'état*', observed a member of the Centre Party.[46] At the end of March 1897 the Reich Chancellor recorded his impressions after a visit to the Schloss, noting that it was 'evident that the Kaiser is under the influence of men who tell him that he can inaugurate a great era of conflict, change the Reich constitution, get rid of universal suffrage and build countless cruisers'.[47] But who were these people who were pressing for a *coup d'état*, and how seriously should we take Wilhelm's threats to use force against the Reichstag?

In his immediate entourage influential people like Admiral von Senden and Adjutant-General von Plessen spoke out in favour of repression of the workers' movement and abolition of the Reichstag.[48] Wilhelm's brother-in-law Hereditary Prince Bernhard of Saxe-Meiningen, the Commanding General at Breslau, described the creation of 'a protective barrier against the socialist threat' as the most urgent task of domestic policy. The ordinary German was 'a narrow-minded brute totally lacking in political instinct' who in his 'pettifoggery and dullness' had committed the enormous mistake of putting 'parliamentary decisions into the horny hands of the stupidest, worst and vilest elements', that is to say 'those who are nothing, have nothing, know nothing and can do nothing.' 'Socialism is nothing more than the instinctive hatred of the poor and stupid for the rich and clever', he declared. 'Such crack-brained, grotesque ideas could only appear and gain influence among people as politically stupid as the Germans!' The chief failing of those in power, in the eyes of this royal Prince and General, was 'that they did not get on with it a long time ago, without worrying about existing laws on paper, and do what the public good requires, i.e. send the socialists about their business'. 'Why does intelligence and efficiency allow itself to be ruled by stupidity and madness?', he angrily demanded.[49]

Apart from these influences at court it was demonstrably Waldersee, above all, who advised the Kaiser to adopt a *coup d'état* policy – albeit in the belief that Wilhelm would never dare resort to a breach of the constitution. As early as 1895, as commanding general in what was then the Prussian port and industrial city of Altona, Waldersee had had occasion to discuss the 'totally disorganised state of affairs' with Conservative members of the Reichstag as well as with the former War Minister Verdy, who had contacts in liberal financial and academic circles. All were agreed that 'momentous decisions' would soon have to be faced and that Waldersee, and not Hohenlohe, was the right chancellor for this task. 'The gentlemen who talk of a *coup d'état* and say that things cannot go on as they are do not realise that the Kaiser is still very far from believing that momentous decisions need to be taken, and especially far from recognising that he himself is principally to blame for the situation in which we find ourselves.' Since the Kaiser 'would never stop dabbling in politics himself and becoming actively involved', Waldersee observed, he himself had no desire to become Reich chancellor; nor was the Kaiser intending to offer him the post.[50] One of Waldersee's sources who spent much time at court confirmed the General's view that Wilhelm II was not planning to carry out a *coup d'état*. 'Believe me, however resolutely he speaks about it, if things really become serious with the Social Democrats he will never allow shooting!'[51]

In the course of 1895 Waldersee's views on the Kaiser began to lose some of their earlier severity. The distinctions and attentions he received – promotion to colonel-general with the rank of field marshal, the Order of the Black Eagle and two ostentatious visits to him at Altona by the Kaiser – helped soften his criticism. He now laid much more blame on bad advisers around the monarch and recorded, after Wilhelm's visit on 16 December 1895, that he had the impression 'that the Kaiser has at last realised that he has been greatly deceived by swindlers and malicious busybodies'. Waldersee was now ready to put past grievances behind him and believed for a time that he was 'back on the old footing with the Kaiser again'.[52] The reasons for the change in the tone of the General's comments, however, are to be found in his own situation in life rather than in any change in the Kaiser's political views. Illness, together with social and financial problems, induced a certain melancholy in Waldersee. He was increasingly preoccupied with his imminent retirement from the army, with his withdrawal from politics and society and with thoughts of his death, which he supposed was now near. In the winter of 1896–7 he faced a serious crisis when Marschall revealed the contacts he had had with Police Commissioner von Tausch and other dubious individuals. In this painful situation the General was only too well aware of how dependent he was on the protection of the Kaiser. On the occasion of 27 January 1897, Wilhelm's thirty-eighth birthday, he wrote in his diary: 'When I look back over the past year in the Kaiser's life, I have reason only to be grateful; he has been kind and affectionate towards me every time we have met. God grant that he remain consistent now and protect me from the intrigues of Marschall and his people.'[53] Three weeks later he heard that 'darker and darker clouds' were gathering over his 'affair', and noted that 'in the end everything will depend on whether the Kaiser remains firm'.[54] Through his improved relationship with

Wilhelm, however, Waldersee regained his influence over him and thereby over the course of political events. And this, as we shall shortly see, very nearly led to violent consequences. Shooting down workmen, violation of the constitution, civil war, dissolution of the German Reich and foreign war – all came under discussion.

Waldersee thought he detected hopeful signs of a change in the Kaiser's attitude at the time of the unveiling of the memorial to Wilhelm I at Kiel on 24 November 1896. He found Wilhelm 'in good spirits', but 'very irritated with Bismarck', who had recently betrayed the secret of the Reinsurance Treaty with Russia in the *Hamburger Nachrichten.* In his annoyance, in the presence of hundreds of young officers the Kaiser declared that 'very highly placed people' had committed 'high treason against him and the nation'.[55] On this occasion Waldersee noted that the Kaiser considered universal suffrage a misfortune, and was 'preoccupied with the question of what should take its place. I made no bones about saying "Your Majesty is strong enough to undertake anything; only it must be tackled with vigour."'[56] The impressively well-organised strike of 18,000 dock workers in the Hamburg area, which had begun in the autumn of 1896 and was to last for several months, did not fail to have an effect on Wilhelm. During his second visit to Waldersee at Altona the Kaiser demanded 'energetic action' against the strikers, and his parting words to the General at the station were 'Just deal firmly with it, and there is no need to ask permission.' The Commanding General had to explain to the Kaiser that he could not lawfully intervene as long as the workers did not disturb the peace. 'We need laws against agitation etc.', which the present Reichstag would never approve, he complained.[57] Privately he noted that 'using armed force against hungry workers is really no pleasure. There is no question of really rebellious movements such as the Kaiser seems to think we are facing . . . If they should happen I would suppress them with the utmost severity.'[58]

Not only the Hamburg strike but also what seemed to be an increasingly hopeless parliamentary situation, which the Reichstag elections, due in the summer of 1898, would only make worse, strengthened Waldersee in his conviction that 'it is now high time for us to take big decisions'. He calculated that after the new elections the composition of the Reichstag would render the country virtually ungovernable. The Conservatives would lose a considerable number of votes to the Agrarian League and the Anti-Semites, the National Liberals would lose most of their seats to the progressive parties, the Centre, which was becoming increasingly democratic, would get rid of 'its few good elements', and the Social Democrats, who would win votes from all parties, could easily double their vote. 'How can anyone govern with that?', the General wondered despondently. 'Why shut one's eyes to it! If energetic action is not taken soon, the German Reich may be in a very parlous state in a few years. If anything needs to be done, then the sooner it is done, the better; any delay is harmful.'[59]

On 22 January 1897, bypassing the Reich and Prussian governments, Waldersee sent the Kaiser a memorandum which demanded a preventive war against the workers' movement as soon as possible.[60] He drew attention to the fact that the 'party of revolution' was steadily gaining ground and 'while carefully avoiding any disturbance or rebellion, eventually it will choose the moment to pit its strength against that of the state'.[61] 'But if the battle is inevitable', he went on to infer, with military logic, 'the state has nothing to gain by waiting', for 'the longer the revolutionary party is allowed to develop its organisation, the stronger it will get; the further it spreads, the more difficult the battle will be and the more harmful its consequences for the people'.[62] As always, it was for the Kaiser to decide on the future of the Reich. 'I cannot wait to hear what the Kaiser will say about it', Waldersee wrote after delivering the memorandum. 'I think he will do nothing, for if he wants to do anything he will have to appoint new ministers and embark on the great struggle with a clear vision.'[63]

When Waldersee went to Kiel on 30 January 1897 for the baptism of Prince Sigismund of Prussia (Prince Heinrich's son), Hahnke told him that the Kaiser had been 'very pleased' with the memorandum. 'As soon as the Kaiser saw me', Waldersee's diary continues, 'he shook my hand and said: I thank you very much for the memorandum; I am very glad that someone is at last telling the truth openly. I read it out at the recent meeting of the Ministry of State. You should have seen the gentlemen's faces: they looked as if they wished the earth would swallow them up.' This remark, loudly voiced over the lunch table, was heard by many of the guests and 'caused quite a sensation'. Even more significant was the 45-minute conversation between the two men which took place the same evening. In the course of it the Kaiser indicated to Waldersee that he did not want to take any action on the socialist question for the time being but that he was thinking of using it as an election slogan and was 'quite convinced that it would be necessary to tackle the matter firmly'. Wilhelm concluded by saying to Waldersee: 'I think the immediate future is very serious; if it becomes too hot for me, you must take your turn.' This, the General recorded, was the first time that the Kaiser had offered him (if only as a hint) the post of chancellor. Wilhelm then added: 'I know that you will do the job well if shooting becomes necessary.' On taking his leave Waldersee assured the Kaiser 'that if he wanted to make use of me for firm action, he would not be disappointed; but it must not be postponed for too long, or I would be too old', to which Wilhelm answered: 'Well, we shall see.' On the basis of this conversation and other impressions he received at Kiel, Waldersee became convinced 'that the Kaiser really intends to make me chancellor if certain circumstances arise. He is thoroughly dissatisfied with the attitude of Hohenlohe and the Ministry at the moment, and he lets them feel it too.'[64] The ministers would certainly complain that he was interfering in things which did not concern him; 'but they cannot very well do so knowing that the Kaiser was very pleased with my memorandum'.[65] A few days after the meeting at Kiel Waldersee stated with some confidence in a letter to Verdy that he thought 'the conviction has taken hold in the highest quarters that we cannot go on like this any longer; but I also know that from there to the point of resolving to take a decisive step a great deal of time may pass'.[66]

The Kaiser's Brandenburg speech on 26 February, in which he called for a 'fight against revolution, with all the means at our command', aroused universal fears that emergency laws against social democracy were about to be introduced. 'People in all the parties that make up the majority in the Reichstag, that is to say the anti-Reich parties, are terrified of the Kaiser taking energetic measures; there is a gloomy feeling that something bad is in the air', Waldersee recorded.[67] This was the moment when, as we have seen, the party leaders Stumm, Wied, Manteuffel and Hammacher were sent to the Reichstag by Wilhelm with threats of a *coup d'état*.[68] Nevertheless Waldersee was again doubtful of whether the Kaiser would find 'the courage to take a great decision'.[69] 'I am convinced', he wrote on 1 April 1897, 'that the Kaiser will do nothing.'[70] 'My general impression is that the Kaiser has not yet come to any firm decision, and as far as I am concerned, he is not thinking of appointing me [as Reich chancellor] at the moment. As he has become decidedly irresolute, I also think that if he closes the Reichstag soon and nothing very out of the ordinary happens, he will soon be distracted in his ideas by other impressions, journeys etc. and will simply leave everything as it is.'[71]

Only two weeks later Waldersee heard about the Kaiser's letter to Prince Heinrich describing the majority in the Reichstag as 'rogues' or 'scoundrels without a fatherland'. Although on the one hand the General was indignant at this latest gaffe – 'It really is a misfortune that the Kaiser lets himself be so carried away with fury', he wrote – on the other hand he could not help hoping that this 'powerful' aggravation of the situation might bring about the desired constitutional conflict. 'What will the Reichstag do?', he was eager to know. 'Possibly it will come to conflict very soon. Will not Hohenlohe go at once now? . . . But what an inheritance for his successor to take on!'[72] In the next few days 'powerful antagonism' towards the Kaiser made itself felt in the press, which in Waldersee's eyes indicated 'that the pace is quickening towards a crisis'.[73] The General was surprised, however, when the Centre Party decided to put up with the 'kick' from the Kaiser and take no action. 'The Kaiser must think . . . that he can trample on them at the moment', Waldersee commented.[74] Once again this gave him hope that Wilhelm would steel himself for the great coup against the Reichstag. He wondered 'whether the Kaiser will act soon or not until the autumn; at most he has until the autumn, but it would be better if he sent the Reichstag home soon . . . then appointed a new chancellor and gave him the time until November to prepare the really serious campaign'.[75]

How would Waldersee have acted if he had been appointed chancellor in the spring of 1897, which for a time seemed entirely possible? In his reply to the War Minister's request to the commanding generals to put forward their views on the handling of an internal crisis in case of need – a request which was 'obviously prompted by the memorandum I presented to the Kaiser', Waldersee believed – he demanded the abolition of universal suffrage for Reichstag elections, on the grounds that it had 'demonstrably' failed to prove its value. If the federal states did not *all* agree with this measure, Prussia would have to withdraw from the German Reich and re-create the Reich on a new basis.[76] (Bismarck had considered a similar *coup d'état* plan just before his dismissal, as we saw earlier.)[77] In conjunction with this internal policy of force, Waldersee was in favour of engineering a European war, in which Germany would not of course be involved at first. 'Such a war, which skilful diplomacy ought to be able to bring about, would be the best thing that could happen for us at this truly difficult time', he told Verdy, adding: 'I believe it would lead to better economic circumstances, a better internal situation and also better external affiliations . . . If matters should eventually reach a point where we had to intervene actively, we would bring such a powerful weight into the balance that it would be resolved in our favour.' 'The idea is morally questionable', the General admitted, 'but I think it is worth considering nevertheless. If we go on living as we have done hitherto, carelessly and without purpose, we shall slide, gradually at first, but soon much faster, down the slippery slope on which we find ourselves.'[78] Verdy agreed with his friend and expressed similar views. 'The ideal goal is still to strengthen the internal structure of our country on a healthy basis . . . and to reinforce the power of our position in external relations.' Domestic policy must be aimed at bringing about 'the collapse of the constitutional parliamentary institutions', 'which undermine all authority and reduce the state to the plaything of the parties; in the process both roots and branches must be cut away; both universal suffrage and freedom of the press etc. etc.' As far as foreign policy was concerned, the aim was 'to recover the prestige which we enjoyed after 1870'; this could be achieved only by the 'reunion' with Germany of the German provinces of the Habsburg empire, for which of course the support of Russia would be required. This would have to be purchased with the offer of Constantinople and of further Russian expansion in Asia.[79]

It was decisive for the future of Germany that the willingness to contemplate a *coup d'état* which Kaiser Wilhelm II had so clearly demonstrated in the first months of 1897 had meanwhile receded. Evidence of this is provided by the imperial marginalia on a letter from Eulenburg of 8 April, in which the Kaiser's friend referred to the soundings which Wilhelm had taken with Wied, Stumm and Manteuffel, news of which had reached his ears. 'Absolute lies!', the monarch scrawled in the margin in angry denial. 'I have never spoken to these men about it! Just as I have never mentioned the word *coup d'état* to the Prince [Hohenlohe]. The whole story is pure *fabrication*! I gave no such instructions and have received no answers.'[80] He telegraphed to Eulenburg *en clair* saying that 'not a syllable' of this information was true; but he was not surprised at it, 'since it is now considered good form, in order to cover up their own weakness and debility, for people to invent the most fabulous lies about me and spread them around. Wilhelm I. R.'[81] When he met Eulenburg in Vienna on 21 April the Kaiser gave his friend his word 'that He had *never* said a word either to Stumm or to Manteuffel or to Wied in the sense alleged, and had scarcely spoken to these gentlemen at all . . . If old Hohenlohe imagines that I am planning a *coup d'état*, he is playing a dishonest game too. He *knows* that I have no such intention . . . I perhaps spoke to him once in confidence about the possibility or impossibility of changing the electoral law – but I am sure you have done that too, we have *all* done that! To deduce a policy of force from that – that is bad faith! . . . I have only once spoken lately to the Grand Duke of Baden about the electoral law *in an academic sense*, without any ulterior motive, purely by chance.' The whole story of his *coup d'etat* plans had been invented by the men in the Wilhelmstrasse, who were operating 'a thoroughly *malicious* system' to further their own purposes 'at the *expense of their Kaiser*', Wilhelm claimed. Eulenburg's 'blood boiled' when his beloved Kaiser told him of this tissue of lies.[82]

It is by no means easy to unravel the tangle of conflicting sources on this critical episode. One is disinclined to take even Eulenburg's indignant protestations to Bülow and Holstein at face value, given the insouciance with which he had discussed the option of a *coup d'état* and all its consequences with the Kaiser in 1896, and had planned the appointment of either Waldersee or Bronsart as Reich chancellor in the event of just such a coup.[83] It seems probable, however, that the instructions given to Wied, Manteuffel and Stumm, of which Holstein had been told by Count Monts, who in turn had heard of them from the Conservative Party leader Heydebrand,[84] did not spring from any actual *coup d'état* plan but were intended chiefly as attempts to intimidate the Reichstag. Furthermore, in his purely 'academic' discussion of the question with the Reich Chancellor and the Grand Duke of Baden, the Kaiser's lively temper and the aggressive tone in which he customarily spoke of parliament and the opposition parties might easily have given rise to the false impression that he had serious intentions of embarking on a policy of force in violation of the constitution. Waldersee's often repeated view, shared by other members of the military, that the Kaiser would never give orders to shoot on his own initiative, rings all the more true because the General deeply deplored this attitude. It is at any rate clear that, although Wilhelm may from time to time have toyed with the idea of a *coup d'état*, under the influence of Waldersee, Senden and other officers, he had dropped the idea by mid-April 1897 at the latest. Instead he now pinned his hopes on the newly appointed statesmen Bülow and Tirpitz, with whose assistance he intended to pursue an exciting new foreign and naval policy. But at the same time, as far as domestic policy was concerned, he put his trust in the efficacy of a backward-looking monarchical ideology at the centre of which stood the grandfather whom he revered above everything, the heroic figure of the 'Heldenkaiser'.

THE CULT OF 'KAISER WILHELM THE GREAT'

What was going on in Wilhelm II's mind when he insulted his principal advisers, spoke disparagingly in public about Bismarck and Moltke, criticised the Reichstag and the nation in the most violent terms and praised his ancestors, and above all his grandfather, to the skies? In response to the steadily rising democratic tide and the increasingly vociferous criticism both of the Kaiser himself and of his quasi-absolutist rule, he propagated an amalgam of dynastic ancestor-worship, falsification of history along militaristic and chauvinistic Prusso-German lines and Christian manichaeism – a monarchical ideology which can only be described as utterly inappropriate for a modern, multi-cultural industrial society with strong federalist features. What were Bavarians, Saxons, Württembergers, Badeners, Hessians, Guelphs, Hanseatic citizens, Alsatians, Poles and Danes to make of the Kaiser's speech of 26 February 1897, in which he said of his 'Royal Grandfather of most blessed memory' that there must be a special significance in the fact that God had 'selected' a Brandenburger as his 'instrument' in the creation of the German Reich? What will their feelings have been on reading the passages in the speech in which he declared that 'Wilhelm the Great' had prepared himself for his calling for years, 'with the great thoughts already fully formed in his mind, which were to enable him to bring the Reich into being once more. We see how he first of all raised an army of the hired peasant sons of his provinces, and constituted them into a powerful, splendidly equipped force. We see how, by means of his army, he succeeded gradually in acquiring predominant power in Germany and in raising Brandenburg–Prussia to the leading position. This having been done, the moment arrived for him to call upon the entire Fatherland to rally round him, and on the field of battle he united those who had been adversaries.' What were Prussian Protestants, let alone Catholics, Jews and atheists, to make of the declaration by their sovereign and principal bishop that Kaiser Wilhelm I, 'if [he] had lived in the Middle Ages . . . would have been canonised, and bands of pilgrims would have travelled from all lands to offer up prayers at his tomb'? Fortunately this was still the case, the hero's grandson asserted, adding that 'the door of his sepulchre stands open' and his loyal subjects could daily delight in 'seeing this splendid old man and his statues'.[85]

That such utterances were not a momentary aberration but articles of faith in a monarchical ideology dressed up as a religion is demonstrated by numerous other speeches which the young Kaiser made at the dedication of memorial churches or monuments to commemorate the hundredth birthday of 'Wilhelm the Great' – for instance on Mount Kyffhäuser in the Harz or at the Deutsches Eck in Koblenz. Even the birthday of his father, whom he scarcely ever mentioned otherwise, provided him with the opportunity to preach his dynastic-Christian cult to the army in 1896 and 1897.[86] The Protestant element of the ideology was also much in evidence in the speech which the Kaiser made at the swearing-in of recruits in Berlin on 18 November 1897, giving great offence both in Germany and abroad. 'He who is not a good Christian is neither a good man nor a Prussian soldier, and can under no circumstances perform what is required of a soldier in the Prussian army', he proclaimed. 'Your duty is not light; it demands of you self-discipline and self-denial, the two highest qualities of the Christian; also absolute obedience and submission to the will of your superiors . . . On you my glorious ancestors look down from heaven; on you the statues of the kings and above all the monument to the great Kaiser look down . . . Your duty now is to show your loyalty to Me and to defend our highest goods, whether against an enemy abroad or at home, to obey when I command, and to stand by Me.'[87] We have had too many opportunities to observe the real depth of Wilhelm II's veneration for his grandfather to doubt that such declamations reflected genuine feeling on his part. But one does not need to be a Machiavelli to recognise that these speeches were also motivated by a desire to manipulate his subjects' Christian beliefs and their patriotic vision of their history for his own benefit and that of the Hohenzollern monarchy. Lavish and expensive publications glorifying the 'Heldenkaiser' also appeared, aimed at bolstering the unsteady throne.[88] The memory of the old Kaiser, Wilhelm II believed, was 'the best way of collecting the parties which support the state together in a patriotic decision'.[89]

This aim came to the fore particularly strongly when in the autumn of 1896 Wilhelm demanded that a commemorative medal should be issued, 'as a Reich matter', on the occasion of the unveiling of the memorial on 22 March 1897, the hundredth birthday of his grandfather, to all members of the army and the navy and to every Prussian and Reich government official, and not only to those who were currently in active service but also to all those who had served under the old Kaiser. To no avail, Prince Hohenlohe drew the Kaiser's attention to the constitutional and financial objections to this idea.[90] In the German Reich orders and decorations were regarded, 'in accordance with the constitutional structure of the Reich, as flowing from the territorial sovereignty of the rulers of the individual states', he pointed out. This was the reason why there had been no orders or decorations emanating from the Reich itself. The Chancellor went on to point out what an enormous quantity of medals the Kaiser was proposing to distribute. 'Even if only the total number of *active* members of the German army and the Imperial Navy (607,105 persons) and the total number of *active* Prussian officials and Reich officials living in Prussia (460,648 persons) were taken into account, the number of recipients would amount to more than a million. The cost of a copper medal with an eyelet, loop and ribbon is approximately 3 marks 10 pfennigs; the total cost of production for a million recipients would thus exceed three million marks.' If in addition to this it were wished to confer the medal on the civil servants and members of the armed forces who had served under Wilhelm I, as the Kaiser demanded, the number of recipients would rise to between three and four million. The requisite sums could be made available from Reich resources only by legislation, the Chancellor emphasised, which would require a bill to be submitted to the Bundesrat and then to the Reichstag. Even if no problems arose in the Bundesrat – although both Bavaria and Württemberg had already voiced reservations – it was out of the question that the Reichstag would approve several million marks for the purpose. On the contrary, 'the discussion would be highly unpleasant and would in all probability end in a defeat for the government, which would be all the more painful because the opposition would be directed at a personal wish of Your Majesty'. Finally the Chancellor referred to the 'regrettable, but indubitable fact' that 'numerous members of the army become members of the Social Democratic Party again after their discharge', and that in the lower ranks of the civil service there were also elements hostile to the state, so that the danger could not be ruled out that 'these people might get up to mischief with the medals conferred upon them, throw them away or sell them'. For all these reasons, as principal adviser to the crown, Hohenlohe argued that the commemorative medal should be instituted as a Prussian decoration, and that it should be issued only to the actual participants in the ceremony of the unveiling of the memorial on 22 March. If the number of recipients could be limited to about 50,000, the requisite sum of 155,000 marks could be taken out of the 'Kaiserlicher Dispositionsfonds' – a solution already suggested by Posadowsky in October – 'thereby avoiding parliamentary involvement altogether'.[91]

The Kaiser's response to this well-reasoned report by the Reich Chancellor was to write a curt note on it saying 'No. My order stands. W.'[92] A telegraph message from Bückeburg followed: 'The celebration of the hundredth anniversary of the birthday of the first *German* Kaiser of a Reich of the *German* nation, from the *German* House of Hohenzollern, has quite a different significance for Germany and the whole world from that of the coronation of the King of Prussia in 1861. This is a unique phenomenon such as has never occurred in history before, and very strong emphasis must therefore be placed on it, particularly for the benefit of the subversive elements. For the ordinary soldier who has lived through this year and who wears the portrait of the great Kaiser on his breast, it will be a talisman against unpatriotic temptations when he leaves the army.'[93] The Kaiser returned to the charge in a further letter, in which he declared himself 'most painfully surprised' by the contents of the Chancellor's memorandum. 'Having announced my will on this matter at the meeting of the Ministry of State at Hubertusstock [in October], I had a conversation 4 weeks later with Herr von Bötticher in which I discussed all eventualities and procedures and gave him all the necessary instructions. As a result of this a model of the medal was submitted to Me, which I accepted after making minor alterations, and which is therefore already in preparation. The question of obtaining funds was also considered at that time, and I had occasion then to point out to Herr von Bötticher the impropriety of using the Allerhöchster Dispositionsfond, which is chiefly intended for the support of the poor, widows and orphans and for artistic purposes, to this end. That is quite out of the question for me! The Reichstag must therefore be approached and asked for the money. In order to keep the amount down as much as possible, however, I am willing to allow a limitation. As I have already personally informed the army and the navy that they are to receive such a medal, none of the armed forces can be excluded. All the more so because it is a question of commemorating the Supreme War Lord, the Supreme Campaign Commander and the creator of the Reich, the victor of three wars! For the civil service, on the other hand, it is not necessary, as they did not have such direct contact with the great Kaiser as My Army. I will therefore strike the 460,648 civil servants off the list; this is the equivalent of a sum of 1,428,009 marks. Subtract this from 3,000,000 marks, which leaves 1,571,991. This sum is surely not too high for the commemoration of My grandfather, in view of the enormous surpluses this year and the increase of the national wealth by 3 billion marks! Asking for this small sum in the Reichstag should be a real pleasure! If the Reichstag should be so miserable and unpatriotic as to refuse it, they can be sure of a getting a beating in the streets and they will lose the last remnant of the people's respect. It would be an honourable defeat for the government, which would do us credit in the eyes of all. Besides, the memory of the Kaiser is the best way of collecting the parties which support the state together in a patriotic decision. The Centre will take care not to be left sitting alone on a bench with Bebel and company again, opposing their *Old Sovereign*. The matter only needs arranging with a bit of tact and care.' Wilhelm indignantly rejected Hohenlohe's argument that Social Democrats among the soldiers would not treat the commemorative medal with respect. 'As for the implied accusation against the army and the attitude of its members that it produces Social Democrats *en masse*', he angrily declared, 'as supreme war lord I categorically repudiate it as a serious insult!' 'It is certainly an undeniable fact that many are Social Democrats when they are taken on. But the percentage of those converted by its grand ideals, its educational influence and the moral and religious invigoration which it provides, is – praise God! – great. – And even if there are bad apples here and there, the great majority of those discharged join veterans' associations and continue to behave in a patriotic spirit. If it came to a comparison between the army and the 460,000 civil servants as to which was the more utterly reliable in their views and in their loyalty to Kaiser and Reich, the army would certainly not come off worst.' As supreme war lord he had the indisputable right to confer medals on the army and the navy. 'If the sum should be refused I shall take care of the rest. Wilhelm I. R.'[94] At the beginning of the New Year the Kaiser was able to report to his best friend in Vienna that 'the question of the medals for 22 March is settled, as Miquel has most kindly placed unexpected revenue surpluses at my disposal, so that this affair can be dealt with on a *Prussian* footing.'[95]

The cult of Kaiser Wilhelm I was of course not only an expression of his genuine veneration for his beloved grandfather, nor was it only calculated political propaganda for the benefit of the increasingly threatened Hohenzollern monarchy. It was also in good measure a projection of the vision he cherished of his own role as ruler, and of his supposed historical mission, on to the revered, transcendental figures of his Hohenzollern ancestors. This manifested itself almost literally in the grand historical costumed festivities which he held in the Berlin Schloss and at which he himself, as was avidly reported in all the newspapers, appeared as the Great Elector or as Frederick the Great. For the hundredth birthday of his venerated grandfather in March 1897 it was inevitable that there would be a court ball in the costumes of a hundred years earlier. To the roll of drums, Wilhelm II processed from the Picture Gallery to the White Hall at the head of the Company of Life Guards and the Bodyguard of Gendarmerie and seated himself on the throne. The adjutant-generals and Flügeladjutanten then marched in, halting at the throne on the imperial command and presenting arms. Very much against her will, at the Kaiser's command 'the poor Kaiserin, . . . red-faced and with quivering lips', had to inspect the line-up of adjutants. Minuets, gavottes, waltzes and Ländler were danced to music of the period taken from the Royal Library. The result was a glittering picture of the

43. The Kaiser and his entourage dressed in the style of 1797, the year of his grandfather's birth. The diminutive figure in a frock coat to the left of the monarch is the painter Adolf von Menzel

last years of the eighteenth century re-created at the threshold of the twentieth, a spectacle of magnificent uniforms, lavish costumes and rich colours exceeding anything the participants had ever seen. One of them, the Flügeladjutant Count Kuno von Moltke, commented to his friend Eulenburg that 'the darling [i.e. Kaiser Wilhelm II] had too low a forehead, but otherwise he looked very good, and happy', whereas Chief of the Naval Cabinet Freiherr von Senden-Bibran, who had dressed up as an English pirate admiral in a narrow-shouldered coat with his epaulettes askew and a piece of loose white cotton-wool on his head as a wig, merely looked ridiculous.[96]

If this historical masquerade was greeted with astonishment by public opinion, the response to Ernst von Wildenbruch's allegorical drama 'Willehalm', which received its first performance at a gala evening likewise in honour of the hundredth birthday of Kaiser Wilhelm I, was one of sheer embarrassment, especially since it was rumoured that whole sections of this piece of Germanic hagiography, in which the 'enslaved German soul' was set free by the 'young King's son Willehalm', had been partly written by Wilhelm II himself.[97] What is undoubtedly true is that the final scene planned by Wildenbruch, in which the heroic old Kaiser entered Valhalla, was cut out at the behest of Wilhelm II and his artistically inclined Flügeladjutanten, Kuno Moltke and Georg von

Hülsen. Instead, on their orders, the play ended with the death of the old hero, while the 'soul of Germany' bent over the dead man and placed the consecrating kiss of all Germany on his forehead, to the sound of a funeral march. Although Moltke found the play 'profoundly serious and affecting' and praised Wildenbruch for producing a work worthy of the 'grief, not yet dulled, for the noble, beloved departed one',[98] Prince Hohenlohe was certainly expressing the general opinion when he described the piece as 'deplorable'.[99] At any rate the ruling German princes, who had come to Berlin for the centenary, considered the performance 'the greatest humiliation inflicted on them since 1866'.[100] Abroad, parallels were drawn between Wilhelm II, Friedrich Nietzsche, who was now mentally deranged, and the destructive urge, with all its overtones of narcissistic resentment, which had emerged in the early compositions of Richard Strauss; heads were shaken, and it was said that in Germany there was 'a scent of Nero in the air'.[101]

THE 'GREAT FARCE' OVER BISMARCK'S DEATH

On 30 July 1898, ten years after the death of 'his' Kaiser, Prince Otto von Bismarck died at Friedrichsruh, aged eighty-three. Until the last moment it was Wilhelm II's 'dearest wish', as Bülow called it, to exploit the internationally historic importance of this event as propaganda for himself and his dynasty, by hurrying to the bedside of the national idol. The danger had long been recognised within the imperial entourage that otherwise 'the mighty man' could 'take millions of hearts with him to the grave', which 'would no longer beat for our Kaiser'.[102] Since 1892 at the latest Wilhelm had realised, as one historian has rightly observed, that he would have to give Bismarck 'a magnificent burial if he wanted to inherit from him'.[103] And to the end Bismarck and those around him denied the hated monarch – the 'stupid boy', as the founder of the Reich still called him on his deathbed – this satisfaction.[104]

Since Bismarck's dismissal, Wilhelm's attitude to his dangerously powerful, fractious subject had fluctuated wildly, as we have seen. The initial show of concern had soon given way to helpless rage, culminating in his publicly expressed threat in the summer of 1891 that he would lock Bismarck up in Spandau.[105] Again, during the 'grand tour of Greater Germany' made by the founder of the Reich in 1892 on the occasion of his son Herbert's marriage in Vienna, the Kaiser's anger against him knew no bounds.[106] Yet only a year later, while Bismarck lay ill at Kissingen, Wilhelm sent him a telegram from Hungary making a point of offering one of the royal residences for his convalescence.[107] This phase of ostentatious reconciliation – encompassing, as we have seen, Bismarck's visit to Berlin for the Kaiser's birthday in 1894, the Kaiser's indignant telegram and his emotional visit to Friedrichsruh in March 1895 after the Reichstag had refused to congratulate the former Reich Chancellor on his eightieth birthday[108] – came to an abrupt end in October 1896 when Bismarck revealed to the press the state secret of the Russian Reinsurance Treaty and its non-renewal. Wilhelm II telegraphed to Kaiser Franz Joseph saying that the newspaper articles contained 'what I told you of at our first meeting after his dismissal, and you and the world will now understand even better why I dismissed the Prince'.[109] Beside himself with anger – he regarded the disclosure of the secret as a betrayal of his grandfather[110] – he again spoke of imprisonment and confinement in a fortress.[111] He declared to his friend Eulenburg that 'in order to preserve his crown and his personal dignity' he would have to resort to '*forcible measures*' against the Prince if he betrayed any more secrets. The Kaiser was above all afraid of his letter of 1886 concerning the offer of Constantinople to Russia becoming known, and considered that its publication would 'ruin him in the eyes of Europe and also of Germany'. In any case the bond between him and Prince Bismarck was now severed and there could never be 'any sign of a connection ever again'.[112] Hohenlohe tried to calm the enraged Kaiser down by pointing out that, if he were incarcerated, the eighty-year-old founder of the Reich might die in prison. 'Then the question of the funeral would arise. The Kaiser would naturally want to make it a ceremonial occasion and would of course be present at it. Would it be worthy of so great a monarch to allow the funeral of his first and most famous chancellor to take place in a second-class fortress?' At this Wilhelm gradually regained his composure.[113] His fury nevertheless broke out again in the notorious speech of 26 February 1897 in which he called Bismarck the 'pygmy' and 'lackey' of Kaiser Wilhelm the Great. His refusal to send Bismarck a telegram of good wishes on his eighty-second birthday on 1 April 1897 was a further sign of his ill-will and was widely deplored, even in the Centre Party.[114]

Thus in the spring of 1897 Wilhelm's relations with the Bismarck family had sunk to a new low. When the diplomat Rear Admiral Karl von Eisendecher discussed 'the Friedrichsruh question' with the Kaiser, he found that the subject produced 'a very bad mood in our Royal master'. He warned Wilhelm that he would never succeed in 'estranging the old Chancellor from the heart of the people'. The Kaiser, Eisendecher recalled, 'violently resisted every attempt at a rapprochement and explained with some passion the reasons why an initiative on his part was completely out of the question now'. Eisendecher had therefore been delighted to hear that a new effort was to be made after all to bring about reconciliation. Although a 'really good, friendly relationship' would never be possible, given the lack of 'forbearance . . . in both gentlemen', who could not 'forgive and forget', it was nevertheless imperative that an 'outward bridge' should be built, for 'the old Prince will close his eyes for the last time in the foreseeable future, [and] the situation would be disastrous for H.M. if matters were still as they have been since the Russian treaty bombshell'.[115]

On 19 July 1897 Wilhelm II instructed the Chief of his Naval Cabinet to inform Prince Bismarck that the ship *Ersatz Leipzig* was to be named after him on 25 September, and to ask whether Bismarck's daughter Marie, Countess Rantzau, would perform the ceremony and the Prince himself attend the launch of the ship as the Kaiser's guest.[116] Bismarck turned down the imperial invitation with the harsh comment that he had no wish to serve as a table ornament.[117] The Kaiser, understandably, was deeply hurt by his 'curt refusal' on the grounds that his daughter was not well enough to take part in the launch, which was a full two months away. On board the *Hohenzollern* the chief fear was that Bismarck would make it known through the press that he had turned down the invitation, which would put 'our All-Highest sovereign in a painful situation'.[118] This did not happen, however, and Bülow advised that another attempt should be made to persuade the Bismarck family to attend the launch. At the beginning of August, therefore, Tirpitz, who was originally to have gone to Friedrichsruh in September to canvass Bismarck's support for his fleet building programme, was asked instead to visit the Prince as soon as possible, in the hope that he would be able to prevail upon him to allow one of his daughters-in-law to perform the naming ceremony.[119] 'This setting to rights', as Tirpitz put it, was 'a difficult and *thankless* task, but it had to be done for the sake of His Majesty and of the launch itself. I succeeded as far as it was at all possible to succeed given the circumstances and personalities involved, especially with people as obsessed as they are at Friedrichsruh.'[120] In his private record of the visit on 22 August, the newly appointed Navy Secretary noted that Bismarck had drunk one and a half bottles of champagne while he was there. He had remarked spitefully of Caprivi, whom he described as 'wooden', that as Reich chancellor it was only to be expected that he would carry out a campaign of revenge against landowners; after all he had 'spent 22 years as a lieutenant in Berlin without a penny by way of an allowance, and had always seen the wealthy cavalry officers whose fathers had landed property'. He spoke 'affectionately' of Kaiser Friedrich III, who had 'stuck up for him in spite of Vicky, even during his illness', but he refused to countenance any rapprochement with Wilhelm II. Speaking English because of the coachman's presence he said to Tirpitz: 'Tell your master I only want to be let alone ("let alone" was constantly repeated). I have no wish for myself any more. My task is done. I want to die in peace. There is no future left for me, no hope.'[121]

In spite of the Prince's request to be left in peace the Kaiser did not give up his ambition to turn Bismarck's funeral into a great state occasion for the benefit of the monarchy. By giving the first large iron-clad the name of *Bismarck* – in the end a few members of the family did in fact attend the ceremony at Kiel on 25 September 1897 – and by subsequently sending a flattering telegram to the Prince, he restored the link with Friedrichsruh.[122] In December 1897 he found an excuse to send his brother there to enquire after Bismarck's state of health and to extract a promise from the Prince's personal physician, Ernst Schweninger, that he would warn the Kaiser at once by telegraph of any deterioration in his elderly patient's condition.[123] Contrary to the doctors' optimistic assurances, Prince Heinrich gained the impression that 'the end was near', and advised the Kaiser to come to Friedrichsruh himself, whereupon Wilhelm sent a telegram from Rendsburg announcing his intention of coming to dinner on 16 December with fourteen people, among whom were Bülow, Tirpitz, Miquel, Lucanus, Helmuth von Moltke and the imperial physician Rudolf Leuthold.[124] When the conversation at table degenerated into the usual exchange of petty anecdotes, and the Kaiser even dragged up 'a few old barrack-room jokes', Moltke whispered in dismay to Tirpitz that it was 'frightful' to behave in such a superficial and undignified way in the presence of the great man. Suddenly, and quite out of context, Bismarck made a remark which imprinted itself on the memory of all those present. 'Your Majesty', he said, 'as long as You have this corps of officers You can of course do whatever You like; if that were no longer the case, it would be quite a different matter.'[125] By an enormous effort of will the dying Prince had succeeded in duping the Kaiser and his retinue one last time about his state of health. Reassured, August Eulenburg reported to his cousin Philipp that Bismarck was after all not in such a bad way; he had shown great 'mental alertness' at the dinner.[126]

On 28 July 1898, two days before the death of the founder of the Reich, Schweninger caused great surprise by leaving Friedrichsruh, thereby successfully creating the impression intended by the Bismarck family that his patient's condition was 'not quite as bad' as the newspapers had reported.[127] When he arrived in Berlin the doctor confided to his friend Maximilian Harden that Bismarck's death would mark the start of 'a great farce'; it was 'appalling to think of it'. The Kaiser was already sending constant telegrams from Norway.[128] Early in the morning of 31 July – Bismarck had died that night – Wilhelm telegraphed to Waldersee from Bergen. 'Prince Bismarck is said to be gravely ill', the message ran. 'As Schweninger told me two days ago that he was well I can only suppose that he lied to me and is deceiving me about [his patient's] condition. Go at once on *my orders* to Friedrichsruh and find out how His Highness is, and telegraph to me at Reval where I shall arrive on 2nd, or here if possible by this evening, or if it is urgent to the Consul at Elsinore.'[129] Only minutes later the news of Bismarck's death was received on the *Hohenzollern*. 'Deeply distressed', the Kaiser telegraphed to the Reich Chancellor: 'I had intended, when unfavourable news suddenly reached Us on 27th from newspaper dispatches, to sail directly to Kiel. Unfortunately on the same day I received a telegram from Schweninger telling me that the news was pure invention, that the Prince's condition had not changed! . . . Unfortunately that turns out to have been a lie!'[130] The Bismarck clique had succeeded in making the Kaiser appear indifferent to what was a deeply affecting event for the whole of Germany. Wilhelm nevertheless gave orders for an immediate return home. He intended to be at Kiel on 1 August and to stand by the Prince's coffin at Friedrichsruh a day later.[131]

With the obvious aim of placing himself at the head of the Bismarck movement, which had been so damaging for him hitherto, Wilhelm now set in train plans for a spectacular national act of mourning. In an effusive telegram to Herbert Bismarck he bewailed the loss 'of Germany's great son, whose loyal service in the work of reuniting the Fatherland won him the lifelong friendship of My grandfather, His Majesty the great Kaiser who rests in the Lord, and the undying gratitude of the entire German people for all time. I shall prepare the last resting place for his earthly remains in the cathedral in Berlin beside My ancestors. Wilhelm I. R.'[132] In an official proclamation to the Reich Chancellor he declared that he felt compelled to 'give expression before the world to the unanimous grief and the grateful admiration with which the entire nation is filled today, and to make a vow in the name of the nation to uphold and to build upon what he, the great Chancellor, had created under Kaiser Wilhelm the Great, and if need be to sacrifice life and property to defend it'.[133] While still in Norwegian waters he ordered the half-masting of flags and court and army mourning, sent a sixty-strong guard of honour to Friedrichsruh and ordered Senior Master of Ceremonies August Eulenburg to prepare the programme for a great funeral ceremony in Berlin.[134] He sent a telegram to the sculptor Reinhold Begas asking him to design a sarcophagus for the cathedral in Berlin. He personally gave directions to Begas, Ernst Ihne and Anton von Werner as to the artistic arrangements for the grandiose funeral service, to which all the federal princes and countless prominent personages were to be invited. His idea was that the catafalque should be erected on the terrace in front of the Reichstag, and that after a military ceremony in the Königsplatz the funeral procession, with Wilhelm at its head, should then proceed through the Brandenburg Gate along Unter den Linden to the cathedral. 'At least some of the mythical power of the dead man was to be transferred on to the person of the young Kaiser through this symbolic act', Lothar Machtan writes. The burial of the national hero in the Hohenzollern vault was to help raise the monarchy 'to a higher form of legitimacy'.[135]

But the entire plan in all its pomp and ceremony was thwarted by Bismarck's testamentary stipulation, dating from 1896, that he should be buried very simply at Friedrichsruh, beneath a gravestone with the inscription: *Bismarck, born 1.IV.1815, a loyal German servant of Kaiser Wilhelm*, a wish which his family was determined to respect.[136] 'Am I expected to hand over my father's body to be dragged off to Berlin so that the Kaiser can use it to fish for popularity?', Herbert Bismarck asked a former colleague of the Prince.[137] Nonetheless the Kaiser, arriving 'in a tense mood' at Kiel, decided to travel to Friedrichsruh the next day so that he could at least attend the funeral there. He forced the Kaiserin, who had just arrived from Wilhelmshöhe and was intending to travel to Coburg for her brother Ernst Günther's wedding,[138] to dress in mourning for Bismarck immediately, treating her to 'a long lecture' in front of Bülow, Philipp Eulenburg and others to the effect that the German people would never forgive her if she showed the slightest lack of admiration for the late Chancellor.[139]

At this juncture Wilhelm had not yet given up hope of being able to bury Bismarck in the cathedral in Berlin, in a sarcophagus donated by him and designed by Begas, and the preliminary works for a grand commemorative ceremony in front of the Reichstag building at first went ahead as planned. In the end, however, he had to be content with a modest memorial service in the Kaiser Wilhelm Gedächtniskirche.[140] His disappointment was even greater at Friedrichsruh itself. When he arrived there on the evening of 2 August with the Kaiserin, Bismarck's coffin had already been sealed, and everyone guessed the family's motives. Baroness Spitzemberg, who had revered the Prince, noted in her diary: 'The fact that the coffin was closed before the Kaiser arrived was undoubtedly prompted by the wish not to allow the man who had inflicted such a burning injury on the dead man to see him – and perhaps they did not believe they had the self-control to keep calm when they saw the living Kaiser looking upon their dead father!! . . . Blood is blood, and the Bismarcks are defiant, violent people, untamed by education or culture, and not of a noble disposition. If the Princess [Johanna von Bismarck] had been alive she would never have allowed him to look, but would have stood by the body of her Siegfried and cursed Hagen, who fatally insulted and harmed him!!'[141] Even from the grave Bismarck continued his fight for revenge against the Hohenzollerns with the publication of the first two volumes of his *Gedanken und Erinnerungen*, which were to be followed in 1918 by the third volume, directed particularly against Wilhelm II.

Nonetheless the 'Great Stealer of our People's hearts' was dead, and the young Kaiser felt as if he had awoken from a nightmare. After the memorial service at the Gedächtniskirche on 4 August he summoned the Reich Chancellor and all the ministers to the sacristy and declared: 'Today the curtain has come down on a long act in our history. Now a new act begins, in which the leading role falls to me.'[142] But no document from the Kaiser's pen testifies more eloquently to the relief he felt at his supposedly final victory in the long and bitter struggle against his overpowering rival in the Sachsenwald than the triumphant letter which he wrote, in English, to his mother from Rominten on 25 September 1898. In it he claimed that Bismarck's lifelong aim had been 'to be glorified and admired and worshipped by the People at the expense of our Dynasty and House, which he made the good People of Germany believe *he* was ever ready to die for, & which *he* had raised to the German Imperial Throne! . . . His ultimate ends are correctly stated by you, and the ways and means he adopted. But against one complaint you utter in your letter I must strongly protest! He is sayed [sic] to have turned away the hearts from their Parents of your 3 eldest children!! What the 2 others have to say for themselves, I do not know, but for my part I simply but firmly & with a clear consience [sic] am able to answer: "No!" He never dared, & I never should have allowed him, to talk about either you or dear Papa in my presence! But if you mean to allude to the possibility of my lending a hand to the overthrowal [sic] of the then allmighty [sic] Chancellor in the days of dear Papa's Reign, I quite openly confess that I was dead against it, & for a very good reason. The death of Grandpapa had so totally upset & even unnerved the country, that it was quite out of its mind; and in a state of hysterics. In this state it looked at *B[ismarck]* & *not* at *us* as the sole transmitter and *keeper* of the old tradition – it was wholly wrong, & was his own crafty doing – but it was *a fact*! Had Papa – & I with him – sent B. home, then such a storm would have broken lose [sic] against him & *you*, that we would have simply been powerless to stay it & you would have embitterd [sic] poor Papa's last days, spoilt the splendid, ineffaceable figure he had in his People's eye & fancy & endangered your *stay* with us, yes perhaps made it *impossible*. For the moment B. was the Master of the situation & the Empire! And the House of Hohenzollern was nowhere! Had we only even tried to touch him, the whole of the German Princes – I was secretly informed of this – would have arisen like one man & would have made us take back the Chancellor, to whom we – & especially later I – would have been delivered over bound hand & foot! The situation was simply impossible. I from that moment perfectly understood the terrible task, you then did not foresee, which Heaven had shaped for me; the task of rescuing the Crown from the overwhelming shadow of its Minister, to set the person of the Monarch in the first row at "his" place, to save the honour & the future of our House from the corrupting influence of the Great Stealer of our People's hearts & to make "him" atone for what he harmed Papa, you & even Grandpapa. Appaling [sic] enough for a young man of 30! to have to begin a reign with, after such a glorious one, having just passed! I however felt what was my duty & thank God He helped me! Without Him I was lost. When the strife waxed hot & B. began his most daring tricks against me, not recoiling before even High Treason, I sent a message to him saying: It seems to me as if he was riding into the lists *against* the House of

Hohenzollern *for* his *own family*; if it were so I warned him, that this was useless as in that he *must* be the loser. The reply was what I had expected; & I felled him, stretching him in the sand, for the sake of my Crown, & our House! Now since that terrible year, I had to bear up with the storm of Germany's feelings, & the vilest tricks of the enraged & passionate B! The same poor Papa & you would have had to bear! I bore it quietly, without flinching, the Royal standard firmly in my hand, the shield with the Black & White quarterings on my arm & God above, alone I bore it for 8 long years! Where is he now? The storm has calmed, the Standard waves high in the breeze, comforting every anxious look cast upwards; the Crown sends its rays "by the Grace of God" into Palace & hut, & – pardon me if I say so – Europe & the World listen to hear "what does the German Emperor say or think", & not what is the will of his Chancellor! To my notion in one point Papa's theory of the continuation of the Old Empire in the new one is right; he allways [sic] maintained & so do I! for ever & for ever, there is only *one real Emperor* in the world, & that is the *German*, regardless of his Person & qualities, but by *right* of a *thousand years tradition*. And his Chancellor has to *obey*!'[143] As Bülow very pertinently remarked, this letter, veering back and forth between self-pity and romantic monarchical arrogance, is perhaps more indicative of Wilhelm II's mentality than anything else he wrote.[144]

THE KAISER AS 'HIS OWN REICH CHANCELLOR'

Since Bronsart's dismissal, and even more since the forced resignations of Marschall and Boetticher, Wilhelm II had maintained such direct personal control over German domestic, foreign and arms policy that he was widely regarded as 'his own Reich chancellor'. As we have seen, as soon as the Reichstag had adjourned in the summer of 1897 he brought about the long-planned restructuring of the Reich and Prussian government entirely on his own initiative. Not only was he directly responsible for the appointment of Heinrich von Gossler as minister of war and of Alfred Tirpitz as secretary of state at the Reich Navy Office, but the most important posts in the civil administration also changed hands entirely in accordance with his wishes.[145] Waldersee, in Potsdam for the funeral of General von Albedyll on 16 June 1897, learned of the details of the great reshuffle. 'The strangest change of recent days' and of Wilhelm's personal style of government, he commented, was the appointment of General Viktor von Podbielski as successor to the late Heinrich von Stephan as secretary of state at the Reich Post Office. Podbielski, as Waldersee remarked, had long been 'decidedly out of favour and it is only as a result of his conduct as a [Conservative] member of the Reichstag that he has risen a little; but the Kaiser has nevertheless always treated him with marked coldness, until he called him over during the races at Hoppegarten on 14th [June 1897] and told him he wanted to make

44. Bernhard von Bülow, German Foreign Secretary 1897–1900 and Reich Chancellor 1900–1909

him a secretary of state.'[146] This became the pattern for Wilhelm II's selection of most ministers and secretaries of state from now onwards, and in 1909 and 1917 even for that of the Reich chancellor. On the whole, the arch-conservative Waldersee welcomed the changes among the leading figures of the government in the summer of 1897 as a sign 'that the Kaiser now intends to choose different, much better people than before; he has turned against the South Germans and wants Old Prussians; God grant that he continue in this direction: he will find plenty of able people.'[147] The South German Reich Chancellor was soon complaining that he could not hold his own against the Prussian excellencies and Junkers; they were 'too numerous, too powerful and have the kingdom and the army on their side'.[148]

The new system, in which the monarch and his entourage were at least outwardly in harmony with the ministers, seemed to be working smoothly at first. Bülow reported enthusiastically to his patron Philipp Eulenburg in February 1898 that everything was going 'wonderfully well' in terms of politics. 'The mood in the country has completely changed', which was not in any way due to his own talent, but to the Kaiser's genius. 'My only real merit is that I understand the aims and intentions of our dear master', the Foreign Secretary and future Reich Chancellor asserted hypocritically. 'Where are the eternal chancellor crises now?', he asked. 'How the once stubborn, refractory and conspiratorial ministers have changed into the meekest little lambs! What has happened to all the agitation against the "Cabinets", the alleged camarilla, the army, the fleet plans and the "limitless" aims! . . . And have not the other much more dangerous threats been averted with God's help? I cannot remember them without shuddering, for those threats were real.'[149] Only with time did Bülow and Eulenburg come to realise what a price they had paid for the apparent harmony between the crown and the government. Only with time did they recognise that the 'much more dangerous threats' of which Holstein, Marschall and Bronsart had warned had not disappeared.

In countless details of German policy the 'will' of the Kaiser, ten years after his accession, could now be seen to be the decisive factor, with the Reich and Prussian governments as mere administrative organs carrying out the imperial 'commands'.[150] Although the Ministry of State held discussions as to whether it might not be tactically advisable, in view of the forthcoming Reichstag elections, to make a concession to the Centre by allowing the return of a few religious orders, the question was decided by the Kaiser, who took the line of 'absolute rejection'.[151] Equally, despite the fact that the ministers and secretaries of state conducted lengthy negotiations with the party leaders in the Landtag or Reichstag committees on the Accountancy Bill, the 'Lex Heinze' against prostitution, the peacetime army establishment or the Meat Inspection Bill, many of their decisions were overturned 'in a mood of extreme indignation' by Wilhelm II 'on the All-Highest initiative and with great animation'.[152] Even with regard to the choice of the next archbishop of Cologne, which the Ministry of State discussed in July 1899, the Kaiser stipulated which two bishops were the only ones who could be considered for the post.[153] No question was too trivial for an expression of the imperial will. In a five-page letter Wilhelm personally explained to his aunt Luise why the Augusta Foundation for young girls would have to be moved to another site because of the rapid industrial expansion in Berlin. The safety, and above all 'the undisturbed blossoming of the tender young plants', was 'quite inconceivable' in the midst of 'thousands of factory workers and lightermen', he wrote.[154] Often, however, the decisions taken by the Kaiser concerned matters of the utmost gravity for German policy.

Wilhelm summoned Crown Councils or appeared unannounced at meetings of the Ministry of State far more frequently than before, in order to inform the ministers of his wishes.[155] Occasionally he overwhelmed the Ministry with detailed instructions. 'A strange sort of Crown Council', one experienced minister noted after one such meeting at the Neues Palais. 'Yesterday none of us was asked for our counsel on anything.'[156] The way in which the monarch now controlled the machinery of government is strikingly demonstrated by the Crown Council held on 15 February 1898 in the Schloss. The meeting was almost entirely taken up by an imperial speech. Wilhelm spoke first of the measures which were necessary to prevent flood damage, and which he ordered the government to put before the Prussian Landtag in the autumn as a draft bill. Then he ordered a redistribution of responsibilities between various Prussian ministries: the Waterways and Buildings Department was to be separated from the Ministry of Public Works and divided up between the other ministries, while the Ministry of Ecclesiastical Affairs, Education and Medicine should hand over its responsibility for medicine to the Interior Ministry. The only duty remaining for the Ministry of State was to 'work out the details' of these imperial ideas. Urgent matters such as the courts martial bill, which the Reich Chancellor wished to discuss, were postponed by Wilhelm until 'later'.[157]

Even if the elderly Prince Hohenlohe was surprised by this new style of government, his preordained successor, Bülow, who prefaced his letters to the Kaiser in bombastic style: 'Most Serene, most Mighty Kaiser and King! All-gracious Kaiser, King and Sovereign!',[158] appeared to be carried away by the 'genius' which Wilhelm II, this 'most noble and most important of all princes', had shown at the Crown Council. 'I grow fonder and fonder of the Kaiser', he wrote to Eulenburg after the meeting on 15 February. 'He is so important!! Together with the Great King and the Great Elector he is by far the most important Hohenzollern ever to have lived. In a way I have never seen before he combines genius – the most authentic and original genius – with the clearest *bon sens.* His vivid imagination lifts me like an eagle high above every petty detail, yet he can soberly judge what is or is not possible and attainable. And what vitality! What a memory! How quick and sure his understanding! In the Crown Council this morning I was completely overwhelmed! He gave an *exposé* of the terribly complicated waterways question – with all that that entailed in the way of material and departmental problems – which no departmental minister could have equalled for precision and accuracy. Yet it was done with a freshness, an attractiveness, a breadth of vision, in short with a brilliance far beyond the reach of any minister. God preserve this great monarch and noblest of men for us!'[159]

The Kaiser sought to control the Ministry of State, that is to say the only collective decision-making body in the Prusso-German government, in various ways. In the Crown Council of 7 October 1898 he impressed upon the ministers that they 'ought not to offer to resign over every difference of opinion', as one of them recorded in his diary. They had 'the honour to be his ministers; he had chosen us because we were the best he could find. We were not parliamentary ministers but ministers of the King of Prussia, and so long as he was satisfied with us, our task was simply to do our duty.'[160] Wilhelm naturally insisted more than ever on his right to dismiss ministers who no longer enjoyed the 'All-Highest confidence', and to replace them with men of his choice; in the ministerial reshuffle of September 1899 the 'responsible government' only just succeeded in persuading the monarch to concede that 'even in cases in which the Ministry of State [had not been] formally instructed to make suggestions for appointments to vacant ministerial posts', they would at least be informed *pro forma* of the names of their new colleagues.[161] In addition, Wilhelm brought in measures intended to strengthen his hold over the government machine. In September 1898 Miquel, in his capacity as vice-president, informed the other ministers of state that 'His Majesty the Kaiser and King has been pleased to stipulate . . . that from now on all decisions of the Royal Ministry of State concerning Reich affairs, and especially those relating to the instruction of the Prussian vote in the Bundesrat, must receive All-Highest approval before being put into effect.'[162] Although some of the ministers doubted whether it would always be possible to carry out this command, given the Kaiser's frequent absences from Berlin, the Ministry nevertheless decided 'to take cognizance of His Majesty's order and to act accordingly'.[163] Hohenlohe succeeded in thwarting a further controlling measure which Wilhelm thought up, but it was nevertheless indicative of Wilhelm's perception of what was within his power as king and kaiser (and also of the great confidence enjoyed by the 'rising sun', Bülow).[164] This was the wish expressed by the Kaiser in the autumn of 1898 to appoint an official of his Civil Cabinet to keep the minutes of the Ministry of State, 'in order to remain *au courant* about what is said' there. As the Reich Chancellor realised, he would have been 'compromised in the eyes of the entire senior bureaucracy' if he had given in to this ploy. He therefore 'urged Bülow . . . to dissuade the Kaiser from the idea. If he does not succeed, I shall go . . . If Bülow really wants me to stay, he must use his influence with the Kaiser, so that the latter does not make it impossible for me to stay.'[165] Hohenlohe's counter-proposal that Head of the Reich Chancellery Karl Freiherr von Wilmowski should be allotted the task of keeping the Ministry's minutes was rejected by the Kaiser for the reason – also typical of Wilhelm – that Wilmowski had encouraged the Chancellor to write 'rude letters to H.M.'.[166]

It was almost inevitable that Wilhelm II's authoritarian rule would undermine both the unity of the Reich and Prussian leadership and the authority of the Reich Chancellor. The Kaiser's habit of dealing directly with individual ministers or secretaries of state, who then carried out his wishes on his instructions and without further discussion, had a particularly detrimental effect. When, for instance, the Minister of War bypassed the Reich Chancellor to deliver 'what was unmistakably a reprimand from the Kaiser' to the Statthalter of Alsace-Lorraine, Prince Hermann zu Hohenlohe-Langenburg saw this insulting action as representing a danger that he, as Statthalter, might be reduced to the status of a mere Oberpräsident and the Reich territory of Alsace-Lorraine to that of a Prussian province.[167] In 1899 even the loyal General von Hahnke felt compelled to protest against the War Minister's habit of putting 'All-Highest decisions' into practice unilaterally and without previous consultation with the government.[168]

Despite the firm opposition of the ministers – who for the first time since Bronsart's dismissal discussed the possibility of a 'formal protest of the Ministry of State *in corpore*' – in March 1898 the Kaiser succeeded in appointing Reich Navy Secretary Admiral Alfred Tirpitz as a member of the Prussian Ministry of State with full voting rights. In vain Miquel, Posadowsky and most of the other ministers pointed out that the navy was a *Reich* organisation and nothing to do with Prussian affairs, and that the Kaiser's wish therefore amounted to 'an organisational change in the supreme departments of state which was contradictory both to the constitution and to the realities of political life', and that public opinion both in Prussia and in the other federal states would react with indignation to such a step. Both Hohenlohe and Bülow emphasised that the Kaiser would not back down and that any attempt to change his mind was 'hopeless'.[169]

Another piece of bureaucratic reorganisation, peremptorily demanded by Tirpitz and ordered by Wilhelm II after violent clashes over departmental responsibilities, proved positively disastrous. The tripartite administration of the navy introduced in 1889 was abolished in 1899 when the High Command was dissolved, to make it easier for the Kaiser to intervene directly in naval affairs.[170] But the break-up of the Naval High Command into several posts with direct access to the Kaiser, as we shall see, was only one aspect of a massive increase in the power of the Reich Navy Office, which was to give Tirpitz – for as long as he had the Kaiser's support – a tremendous amount of influence in shaping German foreign policy. The Foreign Office perceptively drew attention to the dangers inherent in Tirpitz's proposed reorganisation of the naval administration in a memorandum of 25 June 1898. 'If the Secretary of State at the Reich Navy Office extends his sphere of action, as he wishes, to all German overseas interests, and furthermore receives a power of command which is free of all parliamentary and other constitutional responsibility, he will acquire an authority superior to that of all other Reich Offices. Foreign policy, at least as far as it is overseas, will henceforth become a condominium between the Reich Navy Office and the Foreign Office.'[171] Wilhelm II nevertheless decided in favour of the reorganisation demanded by Tirpitz.

As far as the authority of the Reich Chancellor was concerned, Hohenlohe initially consoled himself with the thought that even if he could no longer take the initiative, he could still prevent decisions which he considered wrong or harmful. With time he lost even this justification for his existence. He complained increasingly loudly that the post of Reich chancellor had become 'nothing but a façade'; 'if H.M. wants to play Reich chancellor himself, there is nothing for me to do'.[172] As Hohenlohe's son Alexander clearly recognised, the chancellor's status had been reduced to that of an imperial 'scapegoat'. 'And if H.M. needs another scapegoat he will easily find one. The future Reich chancellor will not and cannot be anything else, unless circumstances change completely. And there is little prospect of that.'[173] He wrote to his father expressing his irritation that the newspapers never reported 'that the Kaiser has been to see you or that you have been to see the Kaiser, whereas I am constantly reading that Bülow has been to see him and he has been to see Bülow'.[174] 'A Reich chancellor whom the Kaiser avoids is in a ridiculous position', the old Prince himself admitted, complaining that he was obliged to spend more and more time simply to avoid 'putting H.M. into a bad mood with me'.[175] Hohenlohe's requests for audiences were often ignored – 'he finds Bülow more comfortable', the Chancellor commented. Often he had to make his report to the Kaiser during dinner at the Schloss or on a train journey to some hunting lodge or other, and to use this time to seek the Kaiser's 'commands' on the most important questions of domestic and foreign policy, such as when the Reichstag or Landtag should be convened, whether they should be dissolved or merely closed, and who should be given which appointment as Oberpräsident. Before the Kaiser departed on his journey to Constantinople, Palestine and Damascus in 1898 the Reich Chancellor was obliged to ask for an All-Highest decision on what was to be done in the event of war breaking out during the Kaiser's absence.[176] In domestic affairs too, the decisions of the Kaiser held a central place in the political life of the nation.

TWO DIFFERENT CHALLENGES: THE HARD LABOUR BILL AND THE CANAL REBELLION

On 6 September 1898, while he was on manoeuvres, without so much as a word to the Reich Chancellor in advance, Wilhelm made a speech at Oeynhausen in Westphalia in which he announced harsh legislation 'for the protection of those willing to work'. Under its provisions, he declared, 'anyone – whoever he is and whatever he is called – who seeks to prevent a German workman who is willing to carry out his work from doing so, or even to incite him to strike, will be punished with imprisonment'.[177] In his posthumously published memoirs Bülow explained the 'exaggerated harshness of the Kaiser's language' as a defiant reaction on Wilhelm's part to the support for the workers expressed by Hinzpeter, who had been sitting opposite him at table.[178] However that may be, this unexpected initiative on the part of the monarch plunged the 'responsible government', the Reichstag and German public opinion in general into a crisis that was to last for over a year.

The reaction of the government authorities to the Kaiser's intervention in the delicate question of workers' disputes in a modern industrial society speaks volumes. The luckless Chancellor commented sarcastically that it was certainly true that it would be useful for anti-strike legislation to be tightened up, 'but for that one needs the federal governments and the Reichstag, not an imperial speech at a manoeuvre. I have less and less desire to serve under this sovereign.' He could not take responsibility 'if the Kaiser insists that incitement to strike be punished by *hard labour*', he declared.[179] Posadowsky, who was responsible for social policy, was beside himself with anger at the Kaiser's interference. The head of the Reich Chancellery, Wilmowski, found the Secretary for the Interior 'in a deeply depressed and agitated state' because the Oeynhausen speech had 'ruined all his plans'. Posadowsky, like the Chancellor, was considering resignation, Wilmowski reported.[180] Nevertheless, after detailed discussions in the Ministry of State, he set about drafting a bill for the 'protection of free working conditions'. The Chancellor forwarded the draft to Bülow, who was accompanying the Kaiser on his Near Eastern voyage, on 10 November 1898, quite rightly doubting that it would 'completely satisfy H.M.'.[181] That his pessimism was thoroughly justified is evinced by the All-Highest marginal comments on the draft bill. Where it provided for a penalty of 'imprisonment for at least one month', the Kaiser demanded 'at least one year'; where 'imprisonment for at least six months' was proposed, he increased the punishment to 'hard labour for not less than two years', and he changed clause 8, which stipulated 'imprisonment for at least three months' for an offence to 'hard labour for up to three years'.[182] After Wilhelm's return from Palestine Hohenlohe commented ironically that 'the All-Highest proposals' still went 'far beyond the comprehension of serious legal experts'.[183] 'H.M. is still convinced that the workers must be protected by a harsh law'; it would therefore be necessary to 'proceed very gently and cautiously . . . to bring about a law which is halfway reasonable'.[184]

All the details of the handling of the controversial bill in the Reichstag were decided by the Kaiser himself. Even the Chief of the Civil Cabinet was reluctant to mention the so-called Hard Labour Bill while making his report to the Kaiser on 1 May 1899, but when the Kaiser asked a direct question about it he had no alternative but to explain the serious reservations about the bill. Wilhelm dismissed these out of hand and said that the government should continue to try to get the bill through the Reichstag, showing how completely he misjudged the mood in the country. In the Kaiser's opinion, if the bill came to grief in parliament, 'the blame would fall on the Reichstag', the Chancellor recorded. 'This would damage the Reichstag in the eyes of the many people, particularly in industrial circles, who supported measures against Social Democratic terrorism . . . H.M. is not afraid of any criticism of the bill which may arise during the summer; "people will of course criticise it". He does not care how much or how little. So he intends to let matters take their course, and then if the bill fails because there is no quorum, he will decide whether to adjourn or to close. If H.M. stands by his decision, we shall try to get the bill through the Bundesrat and then submit it to the Reichstag.'[185] This was done on 2 June. When the Reichstag adjourned four weeks later without having reached a decision on this hotly debated measure, the Kaiser telegraphed from Kiel expressing his warm appreciation to the Chancellor for the vigorous manner in which he had supported the Hard Labour Bill. 'The bourgeoisie has apparently capitulated for the moment, against its better judgement, to socialism, which *attacked* it *en rase campagne*. The government must fight all the harder, therefore the bill *must* be reintroduced in the autumn, and the House must accept it or else disappear.'[186] Obediently Hohenlohe promised to carry out the 'All-Highest commands'.[187] Neither the unanimous view of the 'responsible government' that there was no prospect of getting the bill through the Reichstag, nor a memorandum by the Grand Duke of Baden pointing out that the 'horse-trading' associated with it would divide the parties which supported the state and strengthen social democracy, had any effect in persuading the Kaiser to abandon the bill.[188]

Even Bülow was shocked to be told in July 1899 by Philipp Eulenburg, who was with the Kaiser on his Scandinavian cruise, of the open telegram that Wilhelm had sent to the Civil Cabinet. Accepting a request by the mayor of Dortmund that the Kaiser postpone his visit for the opening of the Dortmund–Ems Canal, Wilhelm telegraphed: 'Agree with proposal and very touched by feelings of citizens of Dortmund. Will visit, if at all this year, only after the bill for the protection of those willing to work has been accepted. I have promised the willing workers in Westphalia this and will not go there until I can bring them this gift.' Bülow tried desperately to suppress at least the second half of this telegram.[189] Although Wilhelm eventually allowed himself to be persuaded to take part in the festivities in Dortmund, no one could induce him to give up the Hard Labour Bill. At the end of October, after discussing the matter at length with the monarch, Bülow told the Chancellor that he regarded 'the withdrawal of the bill by His Majesty as completely out of the question'.[190] The result was another heavy defeat for the government and above all for the Kaiser. On 20 November 1899 the bill which he had personally announced fourteen months earlier in Oeynhausen was comprehensively – and even unanimously, as far as the hard labour clause was concerned – rejected by the Reichstag.

Not content with uniting the democratically elected German parliament against him over a bill which he had personally announced without the prior knowledge or approval of anyone in the government, the Kaiser managed simultaneously to provoke the 'Old Prussian' landed nobility, the officials of the Prussian provincial administration and even some of great peers of the realm into open rebellion against him. His passionate support for the plan to build a central canal linking the Rhine, the Ems, the Weser and the Elbe for the benefit of trade and industry brought him into conflict with the Conservative majority in the Prussian Lower House which had been elected under the three-class franchise and which under normal circumstances could be regarded as a pillar of loyal support for crown.[191] In November 1898, just a few weeks after announcing the Hard Labour Bill in Oeynhausen, the Kaiser left Hohenlohe in no doubt that he had 'set his mind' on getting the canal through in spite of the firm opposition of the East Elbian agrarians.[192] In April 1899 the Chancellor noted with alarm that Wilhelm was 'still enthusiastically in favour' of the Canal Bill and was determined to dissolve the Prussian Landtag if it rejected the measure.[193] The monarch even threatened those officials of the provincial adminsitration – the Regierungspräsidenten and Landräte – who had seats in the House with dishonourable discharge from their posts if they voted against the project,[194] thereby turning the Canal Bill into a deep crisis of confidence between the crown and the Conservatives, many of whom, it was claimed, opposed it chiefly 'out of hatred for H.M.'.[195]

The question of what should be done if the Landtag rejected the bill, so openly supported by the monarch, split the Prussian Ministry into two camps, as Hohenlohe and the Foreign Office considered the dissolution of the parliament essential, while Miquel and his followers were anxious at all costs to avoid an election campaign in which the state would be pitted against the Conservatives. Wilhelm was at first fully persuaded of the necessity of a dissolution in order to uphold the dignity of the crown. In May 1899, in marginal comments on a report from Munich, he expressly threatened the ministers Miquel, Recke, Hammerstein and Thielen with summary dismissal. Blind to his own lamentable lack of backbone, the Chancellor commented: 'If such a thing were said to me, my resignation letter would be in the All-Highest hands the next day. But these gentlemen think differently.'[196] The following month Bülow informed Hohenlohe that in the Kaiser's view dangerous consequences 'for the monarchical principle' would follow if the government accepted the defeat of a bill 'which had received the endorsement of the All-Highest person' without insisting on the dissolution of parliament. 'If the House of Deputies does not accept the Canal Bill, he must and will dissolve the House in order to uphold his personal dignity and in fulfilment of his duties as sovereign.' Bülow even had the impression, from the 'All-Highest ill-humour', that 'further opposition to the Canal Bill on the part of the Conservatives [would] influence the whole attitude of His Majesty towards the Conservatives for a long time to come'. This consideration would have to be borne in mind, the Foreign Secretary and future Reich Chancellor warned, in regard to 'all further legislative action in the next few years (in particular the Trade Treaties)'.[197]

In an audience at Wilhelmshöhe Miquel succeeded in persuading the Kaiser, in spite of his earlier telegram refusing to go Westphalia until the Hard Labour Bill had been accepted, to attend the festivities for the opening of the Dortmund–Ems Canal after all. On 10 August 1899 Wilhelm therefore set out for the Ruhr where, accompanied by the Reich Chancellor and the ministers, he spent the night at the Krupps' magnificent house, the Villa Hügel.[198] The next day he made a speech at the opening ceremony in Dortmund in which he committed himself 'with all possible vigour' personally to the construction of the central canal as part of an 'absolutely necessary . . . great work' which the Great Elector and Frederick the Great had begun and which, following 'the achievements of My grandfather' in re-establishing 'a strong, united Reich obeying a single will', must now be extended across the whole of Germany.[199]

Only five days after the Dortmund speech the Chancellor had to inform the Kaiser that the Canal Bill would almost certainly be defeated in the Landtag. Voting against it would be 128 Conservatives, 47 Free Conservatives and 26 Centre Party members, so that its passage would depend on the 13 Polish deputies whose attitude was not yet clear.[200] Wilhelm rapidly grasped the significance of these figures: many of the deputies who held posts in the Prussian provincial administration were prepared to vote against the government's bill. In a telegram from Wilhelmshöhe he stated that it was clear from Hohenlohe's figures 'that a number of My officials are among the opponents. This gives me occasion to point out once again that I am determined not to tolerate opposition on principle from political officials in a matter of such great and general importance and that the deputies in question must therefore understand that they will have to bear the consequences of their conduct . . . Furthermore I am sure that My Ministry of State, both as a whole and in all its constituent parts, will continue to devote all its strength to defending the bill.'[201] That very day the construction of the central canal, which Wilhelm had so strongly and publicly advocated, was rejected on the second reading, in a roll-call vote, by a majority of 228 to 126 with 65 abstentions.[202] The Kaiser immediately telegraphed from Metz ordering Interior Minister von der Recke to summon the Regierungspräsidenten and Landräte who had voted against the bill and to tell them in his name that at the third reading they must at least abstain from voting, or else they would lose their posts. He considered it incompatible with the duties of a royal official, he wrote, to vote against a government bill 'as long as they are in My service'.[203]

The opposition to the Canal Bill, which was led by the leader of the Conservatives, Count Limburg-Stirum, and the Centre Party deputy Count Ballestrem, was regarded by the Kaiser as an insult directed at himself. In his anger he fired off open dispatches viciously attacking Limburg-Stirum, who had a Jewish mother, to Bülow, although as foreign secretary he had no responsibility at all in the matter. 'The Conservative parties in their boundless narrowness and Junker arrogance have thrown down a gauntlet in challenge to me; I shall take it up', he threatened. 'This is rank stupidity, mixed with ill-will, and exploited by a Jew-boy. I am determined to make the party feel my anger through social punishments, and so to compel it to do its work after all.' The punishment, he declared, would be the 'exclusion of Limburg and his associates from society'.[204] In a further telegram to Bülow he gave orders that in all the newspapers to which the government had access 'the attention of the Conservative Party should be very emphatically drawn to the consequences which the rejection of the Canal Bill at the third reading would have for our internal political situation and especially for relations between the Conservatives and the crown'.[205] This was followed in September by yet another telegram: 'The traditional pillars of throne and altar, which have always been spoilt by the Royal House, have turned against their Lord, and that under the lead of that fellow Limburg of Jewish descent. Let loose all your press dogs and rain down cudgel blows on the party.'[206] The Kaiserin was so anxious about her husband's agitated state of mind that in her 'fear' she appealed to Bülow to write a calming letter to Wilhelm. 'It is really needed! . . . Ah, it has been a bad summer! May God continue to help us.'[207]

Suddenly, to the consternation of Hohenlohe and Holstein, the Kaiser gave up all talk of dissolving the Landtag because he believed he would need the votes of the Conservatives in the Reichstag for the Hard Labour Bill, despite the fact that it had no hope of success.[208] Both statesmen were firmly convinced that if the Canal Bill were definitively rejected the Landtag would have to be dissolved 'to strengthen the authority of the government and save the reputation of the crown'.[209] They were supported in this view by most of the ministers, who again discussed whether they should all resign, or at least place their portfolios at the Kaiser's disposal if he refused to grant a dissolution.[210] But such decisions had long since ceased to lie with the government; they depended entirely on the will of the monarch. Hohenlohe's diary note after the meeting of the Ministry of State – 'We shall see what Lucanus brings from the Kaiser and then meet again tomorrow' – was symptomatic of the balance of power which now prevailed.[211] When the Chief of the Civil Cabinet brought him the imperial decision, which was against dissolution of the Landtag, the old Prince nerved himself to send

Wilhelm a telegram warning urgently of the danger of 'serious damage to monarchical authority in Prussia and in the Reich' if 'a pronouncement like Count [Limburg-]Stirum's answer to Your Majesty's Dortmund speech' were allowed to go unpunished.[212] In all political circles the predominant opinion was that, after the Kaiser had spoken out so decidedly in favour of the bill, failure to dissolve parliament would be taken as a sure sign that imperial utterances need no longer be taken seriously.[213] Wilhelm II nevertheless decided against dissolution.

On 25 August 1899 the Kaiser summoned a Crown Council at which Hohenlohe and all the ministers with the exception of Miquel 'urgently' recommended dissolution.[214] Wilhelm responded with a long speech in which, as Bülow recalled, he treated the whole question from a military point of view. 'When a regiment mutinies . . . it is not disbanded on that account, for that would be an injury to the army . . . But the ringleaders are brought to the front and shot. On the analogy of this all officials, especially the Regierungspräsidenten and the Landräte, who had voted against the Canal Bill must now be dismissed.'[215] Instead of arranging for new elections, at which the Conservatives, as Hohenlohe hoped, would have suffered 'a healthy defeat', the Kaiser ordered the suspension of the so-called canal rebels from their posts, a ban on participation in the Agrarian League for all officials and the sacking of the ministers von der Recke and Bosse.[216] The two replacement ministers – Freiherr von Rheinbaben as minister of the interior and Dr Studt as minister of ecclesiastical affairs – were of course 'selected by the Kaiser'.[217] Once again the Reich Chancellor wondered whether the Kaiser had any intention of keeping him in office, if this was how he was to be treated. 'The difficulty is that I do not know what H.M. has in mind', he complained in October 1899, a year before he finally retired.[218]

The disciplinary action taken against the provincial officials caused much bad blood, and even within the imperial family it was condemned as 'simply outrageous and *fin de siècle*!'[219] When Philipp Eulenburg joined the Kaiser at Rominten in October 1899 he gained the impression that Wilhelm himself suspected that he had 'aggravated the situation most unfortunately by his personal intervention'. 'Not everyone learns the lessons of time', the Kaiser's confidant commented philosophically; but he had not yet given up hope 'that our Master and Kaiser will *not* be one of those who do not learn!'[220] The serious consequences of the disciplinary measures against the recalcitrant officials manifested themselves when two of the highest officers of the Prusso-German court, the Oberstjägermeister and Chancellor of the High Order of the Black Eagle, Duke Heinrich XI of Pless, and the Oberstkämmerer Prince Christian Krafft zu Hohenlohe-Oehringen (Ujest) resigned on the grounds that they too were opponents of the Canal Bill and did not wish to be treated any better than the officials who had been banned from court because of the way they had voted.[221] The deep impression that this affair made on the East Elbian aristocrats is clearly demonstrated by a bitter petition addressed to the Kaiser by Count Finck von Finckenstein auf Simnau, complaining 'that the blow which Your Majesty has dealt us and which found its expression in the banishment of the court officials has pierced us all to the heart. I can assure Your Majesty that I still feel the effects of that blow today, because there is something profoundly humiliating, almost degrading, in it.'[222] Wilhelm summoned Bülow to the Neues Palais, 'in visible agitation', and greeted him with the words: 'The great ones of my court are abandoning me.'[223] In his replies to Pless and Ujest, Wilhelm stressed that he did not wish to forbid his court officials on principle to vote against government bills; in the 'very special' case of the Canal Bill, however, what had been at stake was legislation of very great economic significance for the whole of Germany, for which he had pledged his personal support. 'In My speech at Dortmund I publicly stated in clear and unambiguous terms that I considered the Canal Bill as a subject in which I took the greatest possible personal interest and to which I gave my personal support; and I see its implementation as an essential requisite for the healthy, prosperous development of the entire economic life of our nation. In questions of such outstanding, crucial importance for our whole country, to which I think it my duty to give my personal support, I can and must expect that the officers of My court will not place themselves in direct and hostile opposition to Myself.' He went on to express the hope that Pless and Ujest would reconsider their decision.[224] In the end the Kaiser won over Pless by offering his son-in-law Prince Solms the post of principal chamberlain vacated by Ujest.[225]

Wilhelm II's constant demands on the two Berlin parliaments, unmatched by any corresponding willingness to entertain the wishes of the parties whose votes he needed, drove the elderly Chancellor to the brink of despair. The notes which he made for himself in November 1899 for an audience with the Kaiser sum up the government's dilemma: 'We demand the increase of the fleet, the central canal and the capitulation of the Reichstag, which is to agree to the Bill for the Protection of the Workers. But the demands which parliament makes are rejected. How then can a majority be obtained for what the government wishes.'[226] Although he constantly pondered whether he would not do better to retire, given these difficult and humiliating circumstances, Prince Hohenlohe remained in office until October 1900.

Ever since the appointment of Bülow as foreign secretary and Posadowsky as interior secretary in the summer of 1897, Hohenlohe had had the feeling that his position as Reich chancellor was under threat. Just before his resignation, looking back over his period of office, he recorded that 'when Bülow took Marschall's place, I had a rival beside me. Likewise with Posadowsky, although he was unable to usurp my position with the Kaiser. Bülow worked slowly and cautiously, but unremittingly, to achieve his goal of taking my place with the Kaiser. There was nothing I could do to prevent it. I could not unseat him from his position with the Kaiser, who preferred him. So fate had to take its course, until catastrophe came.'[227] What the old Prince could not have known was that Bülow's appointment as Reich chancellor and Prussian minister-president had been a foregone conclusion since the Scandinavian cruise of July 1896 at the latest, long before his arrival in the Wilhelmstrasse. Indeed Hohenlohe's role in the last four years of his term of office as chancellor, after he had given in to the forced dismissal of War Minister von Bronsart, had little but the name in common with Bismarck's tenure of that post. The most important questions of both domestic and foreign policy were no longer decided at the green baize tables of the Wilhelmstrasse but by the Kaiser in person; and he listened to Philipp Eulenburg and Bülow, August Eulenburg and Lucanus, Senden-Bibran, Plessen and the Flügeladjutanten, but only seldom, and usually reluctantly, accepted the advice of the chief 'responsible' statesman. One last time, in a memorandum of October 1900 giving the reasons for his resignation, Hohenlohe described the omnipotence of the Kaiser and the powerlessness of the Reich chancellor. The decision to send troops to China to suppress the Boxer Rebellion had been taken without his participation, he declared. He had not even been informed in advance of Waldersee's appointment as commander-in-chief of the international expedition against the rebels. 'Everything relating to foreign policy is discussed and decided by H.M. and Bülow. The departmental heads work out questions of domestic policy without my participation because they know H.M. does not listen to my advice . . . All personnel questions are decided without my advice and even without my knowledge.' In the press and in the Reichstag he was held responsible for policies about which he knew nothing.[228] On 17 October 1900 at an audience in Bad Homburg, Kaiser Wilhelm II gratefully accepted Hohenlohe's resignation from all his offices.[229] 'Uncle Chlodwig' had served his purpose, and Bülow was able to slip smoothly into his shoes, just as he and Eulenburg and the Kaiser had always planned.

CHAPTER TWENTY-NINE

The Kaiser, art and architecture

WILHELM II AND THE 'NATIONAL ROLE' OF THE ARTS

WILHELM II's rise to the summit of his personal power at the turn of the twentieth century cannot be adequately explained in terms of the structure of the Reich constitution established by Bismarck, in which the hereditary prerogatives of the Prussian crown, with its extra-constitutional power of military command, had remained intact, although this factor was undeniably central to the campaign waged by the Kaiser and his entourage against the organs of 'responsible' government. Nor is it enough to put the psychological and material venality, the weakness of character and the submissiveness of the civilian government officials under the microscope, highly significant as these considerations also were in the decisive power struggles of the 1890s, as the preceding chapters have shown. In order to comprehend Wilhelm's remarkable success against his adversaries in the Wilhelmstrasse it is also essential to take account of the almost superhuman energy and versatility of this extraordinarily forceful monarch, which, combined with his impressive memory and his articulacy, made it easy for him to get the better of almost everyone he spoke to – especially since most of those concerned were still dazzled by the lustre of the crown.

It was not only in politics in the narrower sense, however, that this ability to assert himself proved a sharp sword with which the Kaiser could cut through the Gordian knot of government bureaucracy. There was scarcely an aspect of public life on which he did not think himself entitled to pronounce, indeed to decide. Untroubled by self-doubt, he intervened powerfully in matters of science and technology, ship-building and canal construction, architecture, memorial design, sculpture and painting, costume and stage design, music, crafts and horticulture. He composed songs, drew pictures, designed new uniforms for the navy, for hunting, for overseas colonial troops. Archaeological digs, for instance at the Saalburg in the Taunus, at the Citadel in Metz,[1] at Troy and Baalbek or later on Corfu, particularly attracted his attention. Paul Seidel could not get over his astonishment at the variety of Wilhelm II's artistic activity. 'Whether it be the show-pieces of military spectacle, such as flags, banners and drums, whether it be costly gold or silver prizes for races, sailing or rowing regattas, or magnificent chains of office conferred by Him on rectors of Universities, mayors and conductors of victorious male-voice choirs in singing competitions, abbesses' staves for the heads of charitable institutions which have grown out of former convents, or whether it be merely simple photograph frames for His personal use to preserve memories of people and travels – even to the smallest such task the Kaiser devotes a keen interest, prefers to explain His wishes personally to the artists who are to carry them out, and indeed often produces a design in His own hand as a basis for the work.'[2] A British caricature published in February 1898 shows a day in the life of the restless German sovereign (see illustration 45). He rises at 4 a.m., holds a military review before breakfast, writes a play and composes an opera. After breakfast he paints a picture – the theme being 'The Subjugation of the World' – gives his children military drill, changes the map of Africa and China and sends a telegram to Krüger. Then he makes a fiery speech to the Reichstag, teaches a boot-maker his trade, meets 'a few Emperors', gives an interview to the editor of a satirical journal, dines with Lord Lonsdale and spends the rest of the evening smoking, drinking, singing and duelling until 2 a.m.[3] Critics accused most monarchs of this era of ignorance of the world, preoccupation with their own pleasure and indifference to the sufferings of their subjects. But not Wilhelm II. In his case more calm, reserve, modesty and dignified thoughtfulness would have been more appropriate and at any rate less dangerous for the survival of the monarchy.

It goes without saying that Wilhelm's interventions in the artistic world were hardly less controversial than his role in politics. Here too, by emphasising the themes of Christian chivalry, Germanic nationalism and Brandenburg–Prussian dynastic history, he strove to strengthen the Hohenzollern monarchy and to demonise the forces in the nation which he considered his enemies – Catholics, Jews, South Germans, democrats, socialists and other 'grumblers' and 'scoundrels without a fatherland'. His drawing of 1896, 'Niemand zu Liebe, Niemand zu Leide!' ('beholden to none, harming none') (illustration 46) shows a young knight in shining armour with the black-and-white cross of the Teutonic Order on his breast and his tunic adorned with the black eagle of Prussia, sword and shield in hand to protect the women and children who are playing violins and flutes in the idyllic setting of a Romanesque church behind him. In the foreground,

45. 'A Day with the German Emperor'. Alfred Bryan's cartoon in *Moonshine*, 26 February 1898

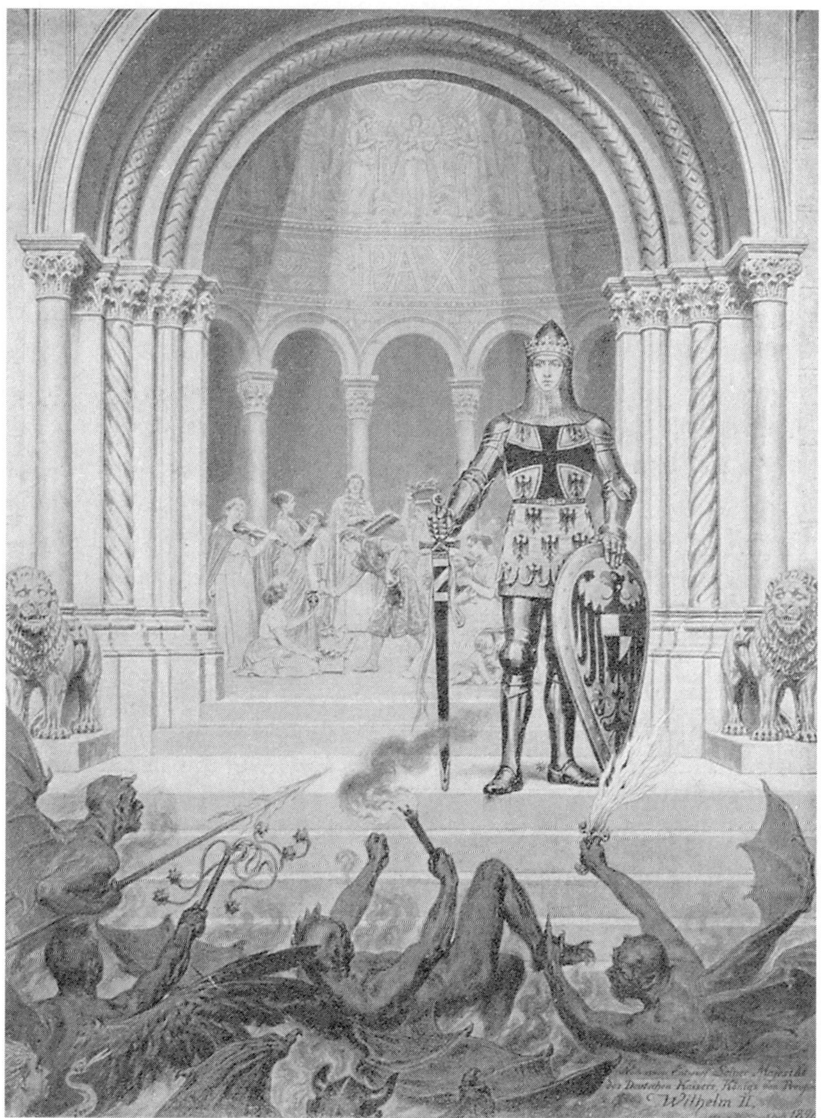

46. 'Beholden to none, harming none'. The Kaiser's manichaean sketch of 1896, as completed by Hermann Knackfuss

at the feet of the Knight of the Grail, dark devilish figures with horns and vampire wings writhe helplessly in hellfire. In November 1896 the German Kaiser and King of Prussia had this highly symbolic drawing published by the official Reich printers in a version completed by Hermann Knackfuss.[4] Any lingering doubts that the youthful knight was intended to represent the Kaiser himself should be dispelled by Wilhelm's Christmas present to Reich Chancellor Prince Hohenlohe in 1898: a bronze statuette of himself in the guise of a Teutonic knight leaning on his sword, with the inscription 'Credo'.[5] There was indisputable evidence of 'political trains of thought', running 'like red threads . . . through all the Kaiser's patronage of art', as the contemporary art historian Gerhard Malkowsky commented in his book *Art in the Service of the State*.[6]

Wilhelm's political aims emerge clearly in the restoration of the old German knights' castles – such as Marienburg in the east and Hohkönigsburg[7] in the west – for which he showed great enthusiasm. At the 'dedication ceremony in characteristic knightly style' in the restored West Prussian castle of Marienburg on 5 June 1902 Wilhelm II called on the German people 'to defend their national goods' against 'Polish presumption'.[8] But it was above all the restoration of the Saalburg, the Roman frontier fort near Bad Homburg, which gave him the opportunity, at the beginning of the new century, to voice the ambitious expectations which underpinned his patronage of this archaeological activity. On 11 October 1900 Kaiser and Kaiserin, 'greeted by fanfares of Roman tubas, entered the flower-bedecked gateway and made their way along the *Via triumphalis* to the square in front of the shrine'. At the laying of the foundation stone of what was to be the restored fort, Wilhelm addressed the thousands of guests and spectators. 'My first thought today is one of sorrowful gratitude to my unforgettable father, Kaiser Friedrich III. The Saalburg owes its resurrection to His energy, His enthusiastic purpose. Just as far away in the east of the Kingdom the mighty castle [the Marienburg] which once brought the seed of German culture to the East rose up anew at his bidding and is now close to its completion, so too, on the heights of the soaring Taunus, the old Roman fort has risen again like the phoenix from its ashes, bearing witness to Roman power, a link in the powerful iron chain which Rome's legions laid around the mighty empire, and which, at the bidding of the Roman Emperor, Caesar Augustus, imposed his will on the world and opened the whole world to Roman culture, which found fertile ground above all in Germania. So I dedicate this stone with a first stroke in memory of Kaiser Friedrich III; a second for German youth, the younger generation who I trust will learn here, in the museum that is coming into being, what a world empire means; and a third for the future of our German Fatherland, to which I trust it may be granted in future times, through the united efforts of princes and peoples, their sovereigns and their citizens, to become as powerful, as firmly united and as influential as the Roman world empire once was, so that in the future the saying which in the old days was: *civis Romanus sum*, may henceforward be: I am a German citizen.'[9]

Since in the artistic world he had no Bismarck, Caprivi, Holstein or Bronsart to contend with, nor any collective forum like the Prussian Ministry of State, and since the imperial initiatives in architecture, memorial design or painting needed no approval from the elected representatives of the people, Wilhelm II was comparatively free to bring his personal influence to bear on these matters in Prussia, and above all in Berlin itself. Outside Prussia too he could exert a direct influence as 'supreme head' of the Reich railways and postal service, monitoring and directing the architecture of stations and post offices as he thought fit.[10] Many of the artists, moreover, fought for the monarch's favour, and were even less inclined than the government officials to oppose the wishes of their Kaiser. They were hopelessly divided among themselves, which also made it easier for Wilhelm to intervene. He was able to look down on them all with lofty condescension when he reported to his mother in February 1893 on the progress of the numerous building projects under way in Berlin. 'At the wish of [Lord Mayor] Zelle in the name of Berlin we went & had a look at the models of Grandmama's statue. [The sculptor] Prof. [Friedrich] Schaper has made the best model so his was chosen, it was the same the town council had hit upon. Its place will be between Opera & Gr^d papa's Palace . . . The art circles are all more or less ablaze with rival strifes. [Reinhold] Begas is having a shindy with [Ernst] Ihne about Grandpapa's Monument. [The sculptor Alexander] Calandrelli is beeing [sic] pulled down because his model for Frederic the first Elector has been accepted. [The painter August] v. Heyden has fallen out with [Anton von] Werner. Zelle & the Council of Berlin are beeing [sic] pitched into by other wise ones, because they want to enlargen [sic] the Königstein Bridge & Schlossplatz, & because they chose Schaper & not the other ones. Alltogether [sic] a happy family!'[11]

A vivid example of the powerful influence exerted by the Kaiser on the artistic world is provided by the visit which he paid to the dress rehearsal of Ernst Wichert's play *Aus eignem Recht* (*In his own right*), accompanied by Lucanus and the Flügeladjutanten Dietrich von Hülsen and Count Kuno von Moltke, in late 1893. Wichert and the director, Ludwig Barnay, were delighted when the Kaiser not only praised the play but also suggested a few improvements in the script, which were eagerly accepted. Wichert noted enthusiastically afterwards: 'It was the most impressive theatrical performance I have ever seen. The hushed silence in the festively lit room, the ardent efforts of all the actors to give of their best, the concentration of attention on one single spectator whose verdict would be decisive, the natural inner agitation of being constantly aware of *this* spectator in *this* play, and finally the growing certainty of success in the expression of such a magnanimous judge – all combined to make an unforgettable impression, beside which the question of what fate lay in store for my play became quite secondary for the moment. That I did not allow my pleasure to be spoilt by foreseeing that now I could be really sure of the most violent attacks will, I think, be understood.'[12]

THE KAISER AND ARCHITECTURE

However unbridgeable the gulf between Wilhelm and his mother in their political views, in the artistic sphere they saw eye to eye, and above all in the Empress Frederick's last years, when she was suffering from cancer and in great pain, the Kaiser's correspondence with her about the renovation of the royal palaces and the progress of the many building projects in the centre of Berlin gave the Kaiser the opportunity to behave in a conciliatory and even solicitous way towards her. The letters which he wrote to her at the end of 1899 show the close co-operation between mother and son and the leading personal role which the Kaiser played in architectural questions. On 4 November he wrote from the Neues Palais: 'Most beloved Mama. The valuable hints which You kindly have given me in your letter about the library and the Monbijou Museum, were most interesting for me. They were all the more à propos as the matter is now under serious consideration of the Government, & as the Hausminister has now formally concluded an agreement with Minister Miquel as to the sum which is to be paid by the state to the Crown for the "Akademieviertel". After having studied the matter I have sent for the Cultusminister, who reported to me today. It seems according to his tale that Your proposals coincide with the views of the Ministers . . . I then told the Minister that Ihne would be the best man for the new library buildings & that it was my wish that he should be given the order to draw up the plans. With reference to the Alte Museum, the plaster casts can not yet be removed till the new house has been built, which has not yet been begun. The Ateliers & the School for Musik are in course of construction now first; so that for the present there will be no room for the removal of the Hohenzollern Museum to the Old Museum. But the Minister has promised to keep his eye on the question & to report to me as soon as the revirement will be possible. I am charmed at the opportunity of beeing [sic] able to turn Monbijou into a habitable residence & think that it will turn out to be an excellent Villino . . . As soon as the new "Dom" is finished – which by the bye [sic] has turned out a most magnificent building – the "Interimskirche" & a house next to it will vanish & allow of the enlarging of the grounds. The "Kaiser Friedrich Museum" is progressing most favourably & is all ready showing what a fine ornament it will be to that part of the town.'[13]

Only two weeks later he wrote to her full of gratitude for her suggestions for the renovation of the Neues Palais. 'All the interesting remarks You made & plans You developed in Your last letters have been put before the "High Court" of Ihnes eversmiling person & are beeing [sic] pondered upon by him. The Apollo-Saal plan is the first & most important. He . . . is in the act of preparing a sketch of the whole thing. The work of cleansing the gobelins has been most carefully done & everyone who has seen those finished is astonished at the brillancy [sic] of the colours & the fine drawing of the figures! The sums which will I am afraid cost something about a quarter of a million must be "floated" by the sale of the old mews & Academy to the State, & I hope to realise a good sum by it. Perhaps it will even be possible to polish up the stairs leading to our rooms & to simplify the too multifarious colours with which the taste of former Schloss- & Bau-Räthe have bedombed [sic] the ceilings & walls under the name of "Abtönung" [colour gradation] i.e. for instance certain light blue skies with "little stars" in gold twinkling down upon the astonished guests who must "wonder what they are", & how they came there! We have begun to hang some large pictures on the walls & find that they do very well!'[14]

Not only the interiors of the royal castles and palaces, but also the appearance of the city of Berlin, which was changing at lightning speed – between 1880 and 1900 the population of the capital doubled to two million – and of other German cities was personally influenced by Wilhelm II to a degree scarcely even suspected today. Although the Berlin Schloss itself was blown up in 1950 on the orders of the Politbüro, numerous other landmarks of the capital which survived the Allied bombing and the Red Army onslaught in some form or other, or were restored after 1945 – the Dom, the Kaiser Wilhelm Gedächtniskirche, several other churches and a number of government buildings – have features which can be traced directly to the last Kaiser.[15] To be precise, up until the end of 1904, through his personal involvement the Kaiser played a fundamental role in the appearance of a total of 163 new buildings for the government of the Reich or the Prussian state.[16] In and around Berlin alone, 58 new churches were built in this period, all of them owing their foundation and design to the Kaiser and Kaiserin. As the total cost of this church-building campaign ran to more than 34 million marks, not a few Berliners will occasionally have breathed the prayer: 'Oh Lord, hold back Thy blessings!' In the Prussian provinces the number of churches built in the first half of Wilhelm II's reign is estimated at several hundred.[17] Well over sixty Reich Post Office buildings (for instance in Strasburg, Ülzen, Cologne, Hanover-Linden, Berlin Tempelhofer Ufer, Karlsruhe, Memel and Königsberg) and numerous stations (including Hamburg Central Station, designed in 1903) owed their final form to instructions given by the Kaiser in his own hand.[18]

On the basis of the very full comments which he wrote in the margins of the designs and which were always meticulously carried out by the commissioning authorities, we can see how radically and in what detail Wilhelm intervened in the architects' plans, and also what guidelines he was following. For the Brandenburg provincial government building in Potsdam, for instance, he ordered that 'the gables over the windows of the 1st floor (1 flight up) should be made rounded at the top, not broken but full, as marked, with uninterrupted mouldings (see Neues Palais), leaving out the intermediary cartouches. The balusters at the windows on the same floor should not be rectangular but round. The small mouldings above the windows of the 1st floor should be uninterrupted, the gables over the windows should be kept flat, not projecting. Consoles should be omitted. On the upper floor the ornament between the windows should be removed. The mouldings should be uninterrupted. On the cupola the torches should be removed. The upper section should be a round lantern like on the wings of the Neues Palais (Frederick the Great's apartments and Ladies' Wing). Light arrangement of columns in the same material as shown in the drawing. The cartouches planned for the alcoves on the central building should be omitted. Over the entrance a projecting, broad balcony should be added, leaving out the heavy cartouches and figures, like in the central building of the Neues Palais (garden front). The Imperial monogram should be placed in the cartouches on the gables of both projecting wings. The projecting bay window in the right wing should be roofed with copper and kept simple without cartouches. The railing round the lantern on the cupola should be kept light and transparent. It is too heavy. Possibly wrought iron. Examples: balustrade at Wilhelmsthal near Kassel – Neues Palais, window railings. At the top beneath the two groups on the central building the projections should be omitted keeping an uninterrupted line. The sloping roofs over the windows in the projecting wings should be omitted. The cupola over the stairwell in the left lateral wing should be round, and the form of the upper section correspondingly round. It is noticeable that the upper part of the roof structure is too high in relation to the cupola. If possible, lower it in order to give more emphasis to the cupola. A coloured tracing containing the alterations should be prepared and glued on to the design for consideration, after which the design should be re-submitted.'[19] Otherwise, one hopes, the Kaiser was quite pleased with the architect's work! Wilhelm added similar instructions to countless designs by experienced architects.[20] One may argue with the Kaiser's taste, but there is no denying his direct, personal involvement in the architecture of numerous major buildings. Even in cases where the buildings no longer exist today, the architects' second plans for each show that the Kaiser's orders were always faithfully carried out.[21]

As a rule Wilhelm II's concern was above all to preserve the historical character of a city and to ensure, as he wrote, that all building projects took account of 'the character and peculiarities of the towns for which the buildings were proposed', and that at the same time 'all stereotyping . . . should be avoided'.[22] For major government buildings the Kaiser preferred the neo-Gothic style, as for instance for the Post Office building in Strassburg, and made sure that all the elements of a building were stylistically correct.[23] On one occasion he criticised the round

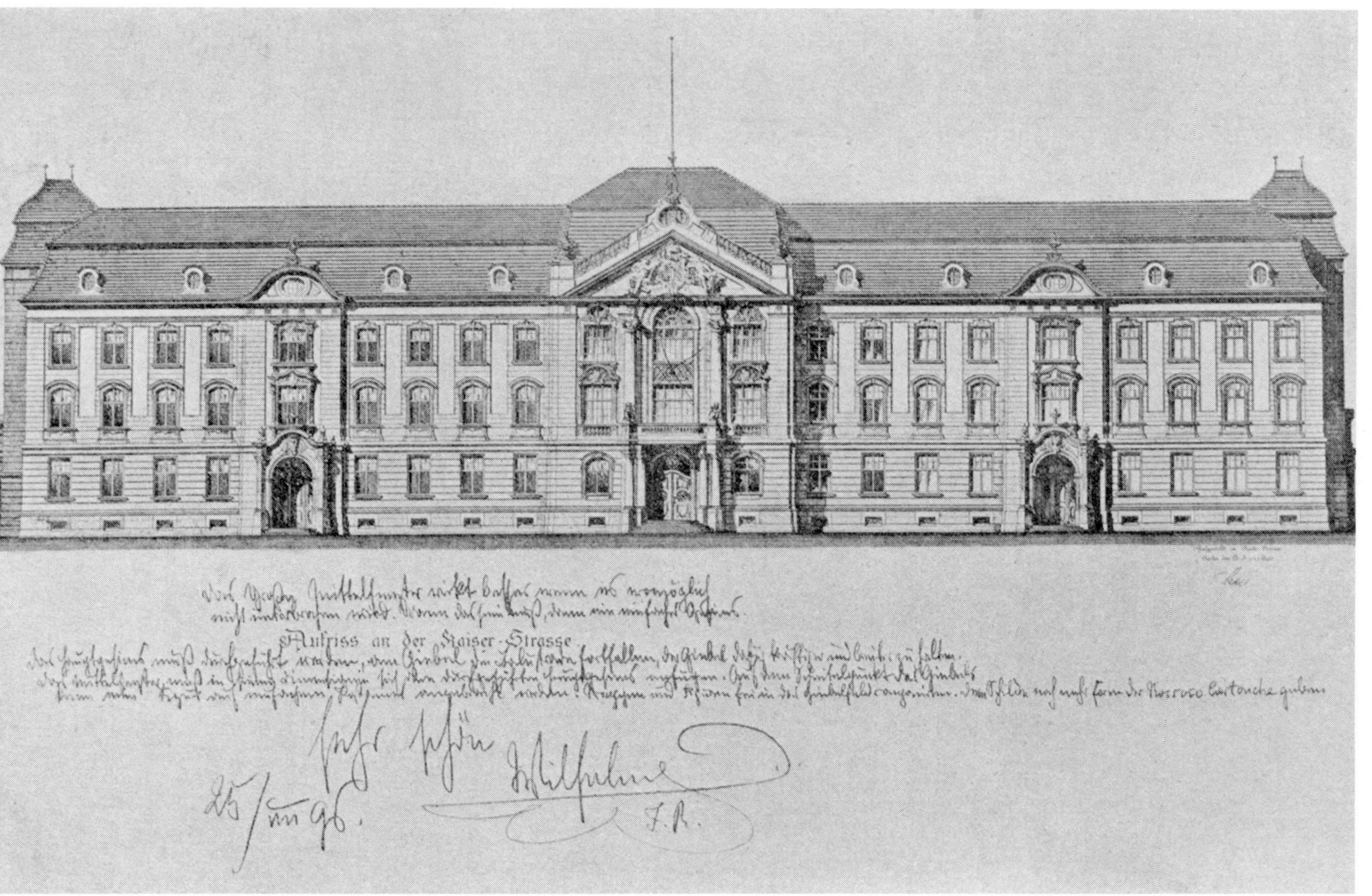

47. The architect's design for the main office of the postal service in Karlsruhe, with the Kaiser's commanded alterations

windows which the architects had included in buildings which were otherwise Gothic in style. 'Straight window heads are more correct in the Gothic, if pointed arches are not used', he noted reprovingly on one plan. 'Straight windows or arched windows are used in the Gothic style, but not rounded arches', he wrote on another, and 'round arches have no place in the Gothic' on a third.[24] For the new Post Office in Ülzen he specified that 'the rosettes would perhaps be best set right into the window frame as in Gothic models. The slanting window-sills would look good in glazed bricks, if not too expensive.'[25] The Kaiser rejected the first design for the Post Office in Hanover-Linden with the remark: 'The front of the house does not look Gothic enough and does not correspond with the fine, strictly Gothic gable.'[26]

Where the hundreds of churches that he had built in Berlin and the Prussian provinces were concerned, Wilhelm II exercised his personal influence 'to a far-reaching extent', as the loyal Seidel wrote admiringly in 1907. Here he drew 'almost without exception on the glorious works of the Middle Ages', taking as his principal model the 'Brandenburg–Gothic brick church of the Cistercians, which the Kaiser loved'.[27] Thus with the 'church-building movement inspired by the throne', according to Seidel, began 'the aspiration, both on the part of the authorities and among many private architects, with regard to the question of style and architectural design, no longer to try to create something modern and new with their plans, but to rely once more on medieval Gothic models'. Seidel goes on to inform us that at the turn of the year in 1889/90 Wilhelm took an intense interest in the Romanesque buildings on the Rhine. 'He had a large number of photographs of the famous Romanesque churches of the Rhineland sent to him, including reproductions of numerous details, especially the many different types of ornament.' He was particularly struck by the parish church and the imperial palace in Gelnhausen, the Dom at Limburg, the monastery of Maria-Laach in the Eifel, the Bonn Münster and the celebrated Romanesque churches at Andernach, Sinzig and of course Cologne. Later he brought back photographs from his journeys to Norway and Italy and put these too at the disposal of the architects Max Spitta and Franz Schwechten. In this way Wilhelm arrived at what seems to us the somewhat paradoxical conviction that the Romanesque style of architecture was 'particularly . . . suitable for Lutheran churches'.[28]

The influence of the medieval churches on the Rhine was already apparent in Spitta's design for the Gnadenkirche, the church that Wilhelm decided to build in 1890 in memory of the late Kaiserin Augusta, as also in Schwechten's Romanesque design for the Kaiser Wilhelm Gedächtniskirche. Every detail – 'both of the exterior of the building and of the internal decoration' – of these two churches, which Seidel described as 'favourite works of the Kaiser', was

planned by Wilhelm II himself. Both architects, Spitta and Schwechten, were expected to report to him at least once a month, 'when He discussed everything with them. He preferred the stricter early Romanesque character of Spitta's Gnadenkirche and consequently had changes made more often to the Kaiser Wilhelm Gedächtniskirche. The choir at Gelnhausen was used as the basis for the choirs of both churches. The Kaiser himself chose the site of the Gnadenkirche in the Invalidenpark, and likewise the much-disputed site of the Kaiser Wilhelm Gedächtniskirche, which the building authorities wished to place on an axis to the Hauptstrasse which passed by it. The Kaiser, on the other hand, decided on the . . . diagonal position, on the model of the Bonn Münster. While the churches were under construction He frequently visited the sites and the studios erected there for the sculptors, and held consultations with the architects, making use of the constantly increasing knowledge he had gained from his travels and his studies.'[29] In the next few years numerous other churches were built, again 'under the special personal care and concern of the Kaiser', in Berlin and the surrounding area, as also in Danzig (the Lutherkirche), Bad Homburg (the Church of the Redeemer), Hanover (the Garrison Church) and elsewhere.[30] To mention one last instance of the Kaiser's church-building activity, from 1892 he worked 'energetically' and 'with particular care and love', as Seidel states, to promote the construction of the Romanesque Church of the Redeemer in Jerusalem. Here too the Kaiser directed the project 'in every detail' in consultation with the architect, in this case Adler. He even drew the massive Romanesque tower himself, based on photographs he had taken in Tivoli during his visit to Rome in 1893.[31]

Wilhelm II devoted his remarkable energy not only to the building of new churches, but also to the restoration of old ones. Seidel lists the cathedrals at Havelberg, Wesel, Magdeburg, Brandenburg, Metz, Maria-Laach and Trier and states that the Kaiser always studied the restoration plans closely and took a strict line on 'stylistically correct works appropriate to each epoch'. He also intervened actively in the renovation of the garrison churches in Berlin and Potsdam, insisting on the removal of 'inartistic and tasteless elements' and the restoration of the 'perfectly formed art of their time'. The high altar at Maria-Laach was a gift from Wilhelm II, designed by Spitta on the basis of his instructions.[32] He was also personally responsible for the decoration of St Elizabeth's apartments in the Wartburg.[33]

Thanks to his interest in the decorative elements of church interiors, including stained glass, mural painting, mosaics and sculpture, these arts enjoyed a great revival during his reign, as the faithful Seidel records. New stained-glass works were founded in Freiburg, Frankfurt and Munich.[34] According to Seidel, 'the art of glass mosaics for church decoration owes its introduction and flourishing

development in Germany to the epoch-making church-building movement instigated by the Kaiser and Kaiserin'.[35] The small glass mosaic company of Puhl and Wagner was expanded with the Kaiser's support, thanks to which, according to Seidel, within ten years it had overtaken all its European rivals. The Kaiser scrutinised its designs for the Berlin churches, frequently visited its workshops and promoted its work by giving it major commissions.[36]

He was in fact tireless in his interest in all the details of church building. 'Always one of the first donors', he was responsible for the installation of church bells and organs in many churches, which gave 'an extraordinary stimulus' to organ-building and to church music in general. 'In the larger churches, above all in the Kaiser Wilhelm Gedächtniskirche and the [Berlin] Dom, the Kaiser ordered the erection of large organ galleries, so that the glorious old church music, much of which had been forgotten, could be revived for the Church and for church services.'[37] Indeed the Dom above all, the ground plans of which he had taken over from his parents and which, after long years of controversy over its construction, was finally completed in February 1905, must be regarded as largely Wilhelm II's own architectural creation.

THE 'DEAREST WISH OF THE ALL-HIGHEST': THE BUILDING OF THE DOM AT BERLIN

The construction of 'a cathedral worthy of the Protestant Church and of the Reich' was from the outset rightly regarded as 'a long-cherished and favourite wish of His Majesty the Kaiser'.[38] Wilhelm's parents had devoted much time to working out the details of the projected Dom in consultation with the architect Julius Raschdorff, but in the end it was left to their energetic son to bring their plans to fruition. As early as 26 June 1888, immediately after his accession, Wilhelm II asked for a report on the progress of the project. Although no decision had yet been taken on the shape of the Dom and its link with the Schloss, he issued a Cabinet Order of 9 July 1888 stating that 'It is My will that the project for the erection of a Dom in My capital city and residence of Berlin . . . should be vigorously pursued. The execution of this plan in accordance with the intentions of the late lamented Kaiser and King Friedrich is for me a sacred legacy. I wish the building to mark the climax of the work on the Dom project in which His Majesty the departed Kaiser and King was engaged for many years.'[39] Wilhelm began by rejecting out of hand the unanimous recommendation of the imperial commission on the project that Raschdorff's planned link between the Schloss and the Dom should not be accepted. It was no business of the commission, he commented, 'how I go to church!' He likewise rejected the idea of announcing a new architectural competition (the earlier one, in 1867, had not produced a satisfactory winning design), declaring that it was his wish 'that the design

produced by Geheim Rat Professor Raschdorff in accordance with the intentions of Their Majesties my father, Kaiser and King Friedrich, who rests in God, and my mother, Kaiserin and Queen Friedrich, should be the basis of future proceedings, without prejudice to modifications which may prove to be necessary on closer inspection.'[40] Seidel relates triumphantly that 'the opposition which Kaiser Wilhelm the Great was unable to overcome was thrust aside by our Kaiser's youthful energy. He no longer allowed the project to be buried in commissions and competition juries, but placed Himself . . . at the head of the authorities supervising the building, by reserving for Himself the right to determine and personally decide on every question of any significance.'[41]

Although Raschdorff was retained as architect, in spite of serious reservations on the part of the Conservatives,[42] his design was considerably revised. The original estimate for the cost of the project was 22 million marks in all, but it soon became apparent that the Landtag would never provide such a sum and would approve 10 million at the most. In an order to Raschdorff of 12 December 1890 the Kaiser therefore stipulated 'that the project for the Dom to be built here will be limited to the construction of a worthy House of God corresponding to the needs of the congregation, and a Royal Vault, and the total costs, including the cost of the interim church and of the demolition of the present Dom, must not exceed 10 million marks'.[43] Raschdorff stated that the Dom could be built to his design for this sum, provided that it was reduced in size. A building committee was set up to ensure that expenditure did not exceed 10 million marks. With Raschdorff as superintendent of works and Wilhelm II as commissioning client, contracts were signed allowing the monarch considerable freedom of action. 'The right is reserved to Your Majesty to order alterations to the plan, to suspend building works or to dismiss the architect of the Dom at any time.'[44]

At every stage of the project Kaiser Wilhelm, who often presided in person over the meetings of the building committee, put forward his ideas, and he also contributed directly to the decorative details by personally drawing revisions from time to time. 'You need only let Me know, and I will always be ready to come', he told the superintendents of works and artists.[45] After one visit to the studio of the sculptor Schott, who was making models for the angel figures which were to encircle the cupola, the Kaiser gave instructions 'that instead of the rococo form, the more severe Renaissance form should be chosen'.[46] A press report of this visit stated that 'the Kaiser examined every detail and with powerful strokes drew the cupola, and an angel figure in the Renaissance style, in the sculptor's sketchbook'.[47] For the painting of the cupola Wilhelm suggested landscapes with biblical features rather than figurative scenes. Even minor details like the weather vane did not escape the imperial attention.[48] Seidel, who was able to consult the minutes of the building committee and artists' diaries, comments admiringly on the 'burden of work' which 'His Majesty took upon himself and saw through to the end'. 'Step by step he took the final decision as to the execution of the details of the . . . project.'[49] The Kaiser took a particular interest in the vault, which he ordained should be as accessible as possible to the public and should 'help the visitor, by means of monuments . . . and inscriptions, to understand the development of the Prussian royal house, and with it that of the Brandenburg–Prussian state, in its historical sequence'.[50] His enthusiasm for building was only rarely disturbed by the idea that with every such intervention he would inevitably arouse controversy and criticism and thereby lay the monarchy open to attack.[51]

In April 1893 Wilhelm reported to his mother on the demolition of the old building. 'The old Dome is now attacked by Raschdorff & his Myrmidons & is coming down with a vengeance. The whole of the roof is gone, the two small towers too; the cupola has disappeared & they ar[e] taking off the cross now. In a month nothing more will have been left of it.'[52] But he had to turn down the Empress Frederick's attempts to make a few more alterations to the new Dom at the last minute. 'I am very sorry if it does not correspond to Your ideas!', he wrote in the summer of 1894; 'but all the wishes You expressed in former years were immediately transmitted to the Committee which superintends the construction, for the benefit of Raschdorff. And as far as was possible without fundamentally altering the plans he has acted on these suggestions. But on the other hand it was impossible to wait any longer for the beginning [of the building work] as Parliament would not have voted the money; & besides the plans had had time enough to be matured with regard to the "Grundriss" [ground plan] & the general dimensions. Considering that Papa himself told me that Raschdorff had been working on the plans by his commands since the last 10 or 15 years & that the plans were what he liked best. The size was given by the very large numbers of the "Domgemeinde" [congregation of the Dom] which has grown so very large. As for the details nothing has been settled at all & I would be thankful for any advice which You would give me, or to the Committee, or Raschdorff himself.'[53] On 17 July 1894 the laying of the foundation stone at last took place, in the presence of the Kaiser and Kaiserin; a document written by the Kaiser was sealed into the wall at the same time.[54]

It was originally hoped that the Dom would be completed by the turn of the century, but there were already delays in the summer of 1896, caused by strikes. Another factor was the increasing influence on the project exerted by Anton von Werner, thanks to the confidence placed in him by Wilhelm. His interventions led to alterations in the design which caused further delays.[55] In the end, the building took eleven years instead of six, as had been planned. While the final work was still in progress, preparations began for the consecration: as always, it was the Kaiser's chief concern that this should be a magnificent affair.

Since the Berlin Dom was intended to be seen as 'the central church for German Protestantism as a whole', all the Protestant federal princes and representatives of all Protestant churches in Germany and abroad were invited to the consecration. On 27 February 1905 the court, led by the 'Most Illustrious couple in Protestant Christendom' (whose twenty-fourth wedding anniversary it happened to be), entered the Dom for the consecration ceremony.[56]

Although the Kaiser was thoroughly satisfied with the building he had created, and received praise from many quarters for it, some naturally took a critical view. Hofmarschall Count Robert von Zedlitz-Trützschler noted in his diary: 'First, the site is badly chosen, for it is a large building standing in far too small a space. The cupola is too big for the lower section, and particularly inside the absence of the double nave which one is accustomed to seeing in a Dom of this kind is a great disadvantage. In general the whole thing is pretentious and invites comparison with other great cathedrals.' Of the 'outwardly splendid' consecration ceremony this court official, always of a critical turn of mind, wrote: 'Just as if the whole thing were a reflection of our time, I did not feel in an especially uplifted or spiritual mood during the service. Both clergymen gave magnificent addresses, but there was much sycophancy in their words which troubled me. I wonder if they themselves believed a quarter of the things they made such a fuss about? They particularly enjoyed praising the glorious work and its creator, by which of course they meant the Kaiser.'[57] Baroness Spitzemberg's mood on this 'divinely beautiful day' was not dissimilar. 'Princes and "notabilities" had been drummed up from all sides, as the usually so cautious [diplomat, Eduard von] Derenthall said, "as if it were for the consecration of a St Peter's, not a hideous Berlin church!" For a purely Prussian, at the most German Protestant occasion, why this exaggerated pomp, this cosmopolitan, pretentious behaviour which no one believes is genuine and which therefore makes one inwardly protest at it? I can only wonder that the population of a great city can have such leisure, curiosity and patience, which manifests itself afresh at every such spectacle; today they were there in their tens of thousands from the Schloss to the Stern, just to see a few court carriages rattling past, *voilà tout*.'[58]

'YOUR MAJESTY, THAT WILL NOT DO!': PAUL WALLOT AND THE REICHSTAG BUILDING

There were of course limits to the power of the Kaiser, and in architectural policy just as in domestic policy in general these limits were set in the first instance by the Reichstag, whose building, opened in December 1894, remains one of the most notable and historically significant buildings in Berlin today. Unlike the Dom and numerous churches and government buildings of the Wilhelmine era, however, the Reichstag building bears no signs of imperial interference, not least because of the timing of its construction, for the architect Paul Wallot had started work on the building on the eastern side of the Königsplatz back in 1884. When Wilhelm II succeeded to the throne the basic structure was already complete up to roof height, and the facing was already in a relatively advanced state, despite repeated strikes. Wallot was nevertheless well aware of the significance of the Kaiser's attitude and wrote to his friend and colleague Friedrich Bluntschli in January 1889: 'The Kaiser . . . took an evident and really knowledgeable interest in my poor efforts, and was moreover very gracious. It is of the greatest importance for the progress of the work that the Kaiser took this attitude.'[59]

Shortly thereafter, however, when Wallot was received in audience by the Kaiser at the Neues Palais to report on the building, the first disagreements arose. According to the record of this incident written by the architect Eugen Bracht, this project and that of the Dom were fundamentally different as far as the legal position was concerned, for the Reichstag itself, or rather the building committee elected by it, was the 'commissioning client', and neither the Kaiser nor the Reich government had any right to 'meddle' in it. 'For the Kaiser, however, "participation" always played a major role; he saw it as his mission to exert his influence on this most important building of his reign, and was obviously accustomed to official architects such as Schwechten etc. accepting the interference of the imperial blue pencil . . . After Wallot had shown the plans to H.M. and touched on the existing problems, the latter was at once ready with his opinion, and clapping Wallot on the shoulder he said, confident of victory: "My son – we shall do it like this", and was about to start drawing, or had already started, when Wallot, in his determined way, drawing himself up to his full height, replied: "Your Majesty, that will not do!" There must have been something, not only in the words, but also in the tone, that H.M. was not used to hearing – he saw himself being rejected as fellow builder and, with that, Wallot had acquired as implacable an enemy as can possibly be imagined.'[60] So much for Bracht's recollection. Probably more authentic is the description which Wallot himself gave of the successfully resisted attempt by the Kaiser to influence the Reichstag building, in a letter of 28 January 1889 to his colleague August Reichensperger. 'Then . . . the subject of the cupola came up. Here I was the victor . . . Quite seriously, in the whole affair and especially with the cupola a huge responsibility has been placed on me and I must fulfil it if I do not want to become a bedfellow of the late Herostratus . . . It took place at the Neues Palais in Potsdam; it was quite interesting to see the "Juvenis imperator" living in the rooms used by his great ancestors. H.M. was also quite gracious and he even called me "My son", but then he wrote an ordinance in which he declared that he was not satisfied. When this was read out to the Reichstag building committee some of these pillars immediately collapsed; although it was not in fact difficult to stand them up again afterwards, once the "experts" had spoken up for me.'[61]

This memorable clash at Potsdam in January 1889 soured the relationship between Wilhelm II and Wallot. While work on the building rapidly progressed – the cupola was completed at the end of 1891 and the glazing began the following spring – the Kaiser, characteristically, did not mince his words in criticising it. He wrote to his mother in February 1893: 'The Houses of Parliament outside the Brandenburg gate are growing more & more hideous, the scaffolding having mostly disappeared, the Reichstags-Bau-Commission, whose lack of taste vyes [sic] with its moneysquandering propensity, is fighting pitched battles among the members, because they cannot make up their minds whose statues are to be placed inside – what the Berliners call – the Reichs-Treibhaus [the Reich hothouse]. The Centre wont have Luther there & want a Pope, Conser: & Liberals declare they would see the Centre d—d first, before the statue of a Pope entered the precincts of the mansion. And all this at last comes before the patient ears of your poor firstborn offspring!'[62]

Although these private remarks by the Kaiser did not become public knowledge, when he condemned the Reichstag building as 'the height of bad taste' during his visit to Rome a few weeks later he aroused indignation throughout Europe. As Michael Cullen has shown, all artists and architects at once took Wallot's side, and not only in Germany, where the Kaiser's unpopularity was greater than ever before.[63] Understandably annoyed, Wallot wrote to Bluntschli on 6 May 1893: 'Only a few words today, so that you can see that I am still alive and that the blast from the imperial whipper snapper hasn't knocked me down. What can one do when one is attacked like that, and from such a quarter? Although I am glad to say that the imperial loudmouth – who is always trying to be astonishing and witty, without really being so at all – is pretty well alone in his view, it is not a trifling matter to be pilloried in front of all the world like that. The shabbiest thing about it is that the imperial prattler came out with his criticism in front of people of whom only the smallest proportion is in a position to check up on his judgement, i.e. to see the building in Berlin for themselves. What is more, people who are far away do not know how things really are here. Begas is the slanderer who has his ear and whose genius and judgement the Kaiser and his dear Mama . . . absolutely swear by . . . And all the charming things that Reinhold [Begas] has poured into Wilhelm's ears over the years, he now bawls out in his barrack-room style in Rome and at any rate thought – . . . enough said – he is a common, despicable brute, for whom Germany will have to pay the price in other ways. For one can probably assume that the Kaiser behaves in exactly the same way in other fields, for instance the military, as he does in the artistic sphere. It is merely easier for us to see what he gets up to in the latter sphere.'[64]

While Wallot was showered with honours by both German and foreign institutions and committees – not least as a defiant reaction to the Kaiser's hostile attitude – Wilhelm showed him his displeasure through petty acts of spite which often created a huge furore. In 1894, instead of awarding the great Gold Medal of the Berlin Exhibition to the architect, as the jury had unanimously decided and Minister of Ecclesiastical Affairs Bosse had recommended, the Kaiser gave it to the society painter Vilma Parlaghy. Instead of the Order of the Red Eagle, for which Wallot's name had been put forward by the responsible authorities, the Kaiser merely conferred the title of Geheimer Baurat on him when the massive Reichstag building was at last completed.[65] A contemporary art critic commented with cynical irony: 'I remember . . . that in 1894 the Kaiser struck off the great Gold Medal, for which Wallot's name had been unanimously put forward, from the list submitted to him, and conferred this same great Gold Medal of 1894, for which Wallot did not seem to him to be sufficiently well qualified, on Vilma Parlaghy . . . The Kaiser showed the Reichstag architect how benevolent and gracious a judge of art he was by awarding him . . . the small Gold Medal, and as appreciation must be given a scale, he awarded the great Gold Medal to Vilma Parlaghy, for which not a single one of the jurors had put her name forward. In so saying I mean only that in matters of art the Kaiser has always been absolutely constant in following his sacred conviction.'[66]

By virtue of the 'kingship mechanism', the monarch's displeasure proved a stumbling block for the architect in many ways. Thus in November 1892 Wallot complained in a letter to Reichensperger that 'the ultimate cause of all this bother is H. Majesty, who has withdrawn his favour both from this building and from myself. How have I deserved this? I don't know. But let's forget it.'[67] At the end of 1893 the architect commented perceptively on the difficulties with which he had found himself struggling since the quarrel with the Kaiser. 'I say that the man up there is to blame and that is true. This pressurising is quite pointless; it is only because he is full of his own importance; he does not wish the building well and he wants to make his power felt over it. So he attacks the Minister and the latter bears down on me . . . What is more, I am not alone in my anger. In a mere 3 to 4 years the Kaiser has managed to squander all the great wealth of devotion and monarchical feeling in the nation which he inherited.'[68] Six months later the irritated Wallot again wondered why 'His Majesty heaps all his kindness on men like Schwechten and Raschdorff. And why does he pour the whole torrent of his crass ignorance over my poor head? I don't know. But probably he was very hurt when I rejected an apotheosis of his grandfather and father

painted by Keller-Karlsruhe as inappropriate for the Reichstag building. As I discovered afterwards, the Grand Duke of Baden had used his influence very strongly on behalf of this picture and the National Gallery had to swallow it in the end . . . Bötticher will have said to His Majesty: "Well, Your Majesty, as far as I am concerned – I would of course have thought the picture very appropriate for the new Reichstag building – but one cannot do anything with this stubborn architect Wallot – the fellow is really obstinate." And soon afterwards His Majesty was making remarks – I was a wilful, pigheaded person.'[69]

We are already familiar with the aggressive, anti-parliamentary account, mockingly directed against the commoner Wallot, of the 'dedication of the Reich monkey-house' on 5 December 1894, which Wilhelm II sent to his friend 'Phili' Eulenburg.[70] Wallot gives us a rather different picture of the grandiose ceremony on the completion of the building. He writes: 'The dedication took place on the 5th – the building was ready for use. A "throne" was constructed in great haste in the rotunda in the lobby, opposite the entrance, and stands erected up above . . . At 12 o'clock H.M. appeared. He tried to shorten the time from the carriage into the lobby and make it amusing, by talking about all the money we had been rolling in. At the time, I could see, he was not quite sure how I would behave towards his exalted person after all that had gone before. When he did the rounds after the ceremony – which lasted over an hour – he tried to be kind. He avoided any direct praise – even of things which everyone likes. He got out of it by saying something like: "Yes of course, my mother told me about it." His mother, my old anti-patroness, being a clever woman, had taken care to visit the House at the eleventh hour, i.e. a few days before the dedication. When we were leaving the refreshment rooms and the Kaiserin [Auguste Viktoria], who was in front, hesitated for a moment, unsure which way she should go, H.M. said: "Wallot, give my wife your arm, otherwise we shan't get anywhere." Naturally I had to walk on ahead, and I told H[er] M[ajesty] of H.M.'s command, and equally naturally and fortunately H[er] M[ajesty] did not react to my kind offer. I begged H[er] M[ajesty] to be kind enough to excuse me – I had been obliged to take the Kaiser's wish as a command – if I had done something stupid, I hoped She would put it down to my inexperience in such matters. The situation was utterly ridiculous – what do you think of this "Majesty" – this revoker of medals? And when it was all over and the soldiers down below were marching past in goose-step, and the great Lord took his leave, he said to H[err] v. Bötticher: "Well, shall I give him the great Gold medal after all?" So as not to need to have heard this magnificent generosity on the part of the sovereign, I quickly slipped behind a curtain at the side. You ask whether H.M.'s attitude to the building would be more favourable from now on – it would of course be very nice for the

building . . . and for me too. But this will certainly not be the case. Of course – if I were a power to be reckoned with – but such a little mite of an architect, there is no reason for one to modify one's imperial opinion for him.'[71]

THE KAISER AND PAINTING

No one who is aware of the great interest which Wilhelm II's parents took in art, and above all of the Empress Frederick's undoubted talent for painting,[72] will be surprised that their son followed developments in painting with close attention. Here too he repeatedly, and in the end to the detriment of the monarchy, interfered in the factional struggles of artists. During his time at the gymnasium in Kassel he had attended drawing lessons at the academy of art, and only a few years later – modesty was never a hindrance with him – to his father's regret he publicly exhibited an oil painting depicting a German ship engaged in firing practice off the coast of Japan.[73] With his usual uncritical admiration for his Kaiser, Seidel tells us of the Monarch's 'favourite art'. 'Already early in his life a burning interest in marine painting was awakened in the soul of the young Prince Wilhelm.' Two teachers, the marine painters Carl Salzmann and Hans Bohrdt, fostered this early passion. Seidel goes so far as to claim that 'with the Kaiser's great gift . . . only the fulfilment of higher duties have prevented Him from acquiring the technical mastery necessary for an exclusively artistic activity in this field'.[74] Both Salzmann and Bohrdt have left striking descriptions of the Kaiser's activity as a painter. The former recalled how 'one autumn evening in 1893 I received a telegram summoning me to Berlin. I took the first and fastest train, but did not arrive at the Royal Schloss until a quarter past ten. I was at once taken to the Kaiser, who called out humorously: "So here he is at last! Where have you been hiding, I have had the whole of Berlin searched for you." The royal party had probably just returned from a theatre or a concert, for Her Majesty the Kaiserin and her ladies were in full evening dress, and likewise the gentlemen in attendance. The Kaiser, wearing the Litewka [an officer's off-duty jacket], was standing in front of an easel and was trying to draw a marine picture in charcoal and chalk on a large board. With the explanatory words "We are building a new iron-clad soon, 'Ersatz Preussen', and I would like to publish a pictorial representation of it, showing how the ship will look in the water; help me with it" and with short, energetic strokes the Kaiser drew a small profile view at the edge of the board, showing what he knew of the ship from the plan. In the middle of the beautiful, brightly lit room stood one of the adjutants, who was reading aloud from a large open newspaper a speech made by an opposition leader in the Reichstag, on which the Kaiser frequently made brief comments, half turning away from the easel.'[75] Hans Bohrdt describes a Scandinavian cruise

on which he accompanied the Kaiser, and during which Wilhelm was occupied with designing a commemorative certificate for the navy. 'He showed the same keen enthusiasm as when he painted at the Schloss, except that his work was frequently interrupted to look at the beauties of nature which the "Hohenzollern" was passing. Then work would stop for quite a long time, and after the joys of nature the Kaiser found it hard to get back to the design; sometimes he put the drawing aside for the whole day.'[76]

Wilhelm II's passion for marine painting was of course an expression of his obsession with the navy as a whole, but it also had direct consequences for painting and crafts. As supreme war lord, according to Salzmann he personally gave permission for 'artists who wish to perfect their skills specifically in naval painting to be taken on voyages of some length in training ships, and also to be taken to sea exercises in the large iron-clads. The big merchant navy companies also frequently and most obligingly grant artists absolutely free passage in their great steamers sailing both to the North and to southern countries. It is likewise due to the great interest which the Kaiser takes in this branch of art that in 1894 on the Kaiser's own initiative a course of instruction in sea- and ship-painting was established at the Royal Academic College for graphic arts in Berlin.'[77] And just as he scrutinised every detail and made corrections to architectural plans for government buildings and churches, Wilhelm decided on the interior arrangement of the new warships which were built at his instigation. According to Seidel, he strove to 'combine the stern necessities of military purposes with a simple but tasteful appearance. His Majesty personally decides upon and determines every detail of these living quarters, whether it is the composition of the walls, the materials for the furniture or the porcelain table service which is under consideration.'[78]

Wilhelm II's best-known and most reproduced drawing, however, was not a depiction of some sea-battle but the sketch which has already been mentioned and illustrated, 'Nations of Europe, protect your holiest possessions', which was completed by the painter Hermann Knackfuss and was intended to convince Tsar Nicholas II of his mission as protector of Christendom against the 'Yellow Peril' in the East.[79] In a letter to the Tsar which the future Chief of the General Staff Helmuth von Moltke delivered to St Petersburg, the Kaiser explained the deeper meaning of the drawing. 'Dearest Nicky', he wrote, in English as always, 'I worked it [the sketch] out with an artist – a first class draughtsman – and after it was finished had it engraved for public use. It shows the powers of Europe represented by their respective Genii called together by the Arch-Angel Michael, – sent from Heaven, – to *unite* in resisting the inroad of Buddhism, heathenism and barbarism for the Defence of the Cross. Stress is especially laid on the *united* resistance of *all* European powers, which is just as necessary also against our

common internal foes, anarchism, republicanism, nihilism. I venture to send you an engraving begging you to accept it as a token of my warm and sincere friendship to you and Russia.'[80] The next day the Tsar telegraphed his thanks for 'the charming picture' which Moltke had brought him. Recent research in the Moscow Archives has revealed the Tsar's laconic diary entry on his receipt of the drawing. It reads: 'Received the Kaiser's aide-de-camp, Moltke, with a letter and an engraving for me from the boring Mr Wilhelm.'[81] The Kaiser had copies of the lithograph sent to other crowned heads of Europe, and gave it to his wife for Christmas in 1896.[82] Knackfuss also 'completed' numerous other sketches by the Kaiser, as for instance the provocative 'Beholden to none, harming none', mentioned earlier in this chapter.[83] As Ferdinand Avenarius bitingly remarked in 1901, Knackfuss's drawings were so 'weak . . . that we could not understand why the opinion that they came from the Kaiser himself was not contradicted in the Kaiser's interest'.[84]

The instructions that the Kaiser gave between 1893 and 1899 for the painting of the throne room in the Palazzo Caffarelli, the German Embassy on the Capitol in Rome, showed little consideration for the feelings of Italians. The painter Hermann Prell was ordered to paint a cycle of seasons of the year simultaneously relating the Edda saga with the express intention of impressing the Italians.[85] The project not only testifies anew to Wilhelm II's dynastic and nationalistic understanding of art and his love of the world of the ancient Germanic sagas; it also demonstrates his ambivalent attitude towards Italian art, which fluctuated between admiration and arrogance.[86] The new masterpiece on the Capitol was also intended to outshine the French. As Prell told his wife, the Kaiser was obsessed with the wish that 'if the French allow their Pal[azzo] Farnese to fall into decay', 'an all the more beautiful, truly German palace' should be created on the Capitol, 'for if the Romans see a stylish palace up there, they will conclude that we are a stylish people'.[87]

Wilhelm of course followed the planning and execution of the project with great interest, and here too he intervened in the artistic formulation of the work.[88] He spoke with enthusiasm of the advantages of mural painting. The 'genre rubbish' depicting scenes from common life displayed in exhibitions nowadays was 'all stupid stuff', he commented. 'Only wall painting, even if it isn't fresco, can produce great art today.' After Prell had submitted preliminary sketches to him in June 1894 and had expressed doubts as to whether the Germanic sagas were really suitable for the throne room in Rome, Wilhelm admitted that he had himself been anxious 'about whether a Renaissance palace, where no one would understand them, was the right place for Odin [i.e. Wotan] and Thor; but the myths of the seasons, which all nations used allegorically and in which the names of the gods did not arise, would be easily understood by Italian viewers too'.[89]

When he visited Prell's Dresden studio in September 1896 to inspect the work in its early stages, the Kaiser expressed his satisfaction with it, only suggesting a few alterations to *Winter*, as Prell reported to his wife. 'What he said was so good and right, the concluding sequence was not pleasing, he would like to have the calm of the first one repeated, no more brown fellows, a blue-green mood – in short, all my own opinions.'[90] The opinions of the Kaiser and the artist were less harmonious, however, when it came to a nude of the Valkyrie seen from the back, which Wilhelm, no doubt under 'feminine influence', ordered to be removed in December 1898.[91] When Prell objected that very similar nudes could be found even in the Vatican, Wilhelm's answer was that, unlike the depictions in the Vatican, the naked Valkyrie was seated, 'on a fiery steed'. 'In this position the buttocks are spread apart and broadened, and therefore in her movement the naked rider, being so large and with the all too voluptuous lower part of her broad back, makes an unaesthetic, offensive impression on the viewer', and must therefore be removed.[92]

It was not until 1903, four years after the dedication of the *Gesamtkunstwerk* initiated by him on the Capitol, that Wilhelm was able to see the painted throne room for himself. Count Monts, by then the German ambassador in Rome, described his visit. 'H.M. was burning with curiosity to see the Throne Room which had been restored, if by no means beautified, under his aegis. He appeared at the Embassy unannounced and quite alone on the first morning of his stay in Rome. First he had all the windows in the room darkened, so as to be able to judge the effect of the frescoes by [artificial] light. Knowing how fond he was of expressing his feelings, I expect Wilhelm II would have declared himself satisfied if he had liked it . . . We were alone in the room; we calmly discussed the restoration work and, indirectly, its failure.'[93]

Although such attempts to exert political influence at home and abroad in the 1890s by means of pictures that were little better than caricatures tended to prove counterproductive, Wilhelm was nevertheless able, through the 'kingship mechanism', to control artistic policy very effectively, and to stem the tide of Impressionism and other modern movements in favour of historicism, which had been the only officially sanctioned style in Germany hitherto, at least for a few more years – although it may be wondered at what price. Another monarch might have thought it wise to remain aloof from the battle of the artists. Wilhelm intervened personally.

Hugo von Tschudi, one of the principal promoters of modern art in Germany, had become director of the National Gallery in Berlin in 1896. He was convinced that knowledge of recent French art was absolutely essential to the understanding of developments in contemporary German art. As director, in spite of the strictly nationalist and conservative policy of the National Art Committee,

which controlled the art collections and purchase funds, he managed, by means of donations, to build up a considerable collection of modern European art, which included among other things French Impressionists and modern German painters.[94] As the gallery lacked space, by 1898 a rearrangement of the paintings had become indispensable. But Tschudi's decision to give preferential treatment to the French works and to hang the older German pictures on the upper floors provoked strong opposition from official artistic circles. The traditionalists, drawing attention to the inscription 'To German Art' on the National Gallery, insisted that it should be dedicated exclusively to the art of the Fatherland.[95] The matter was also hotly and repeatedly debated in the Landtag. On 11 April 1899 the Kaiser paid a visit to the National Gallery to investigate the rumours circulating about the modern paintings. He was accompanied by the Minister of Ecclesiastical Affairs, Dr Bosse, the Chief of the Civil Cabinet, Lucanus, and two Flügeladjutanten, General Friedrich von Scholl and Captain Count Oskar von Platen zu Hallermund. Anton von Werner had also been ordered to attend, and as Wilhelm entered the building he said to the painter, 'It is time I inspected this place.' Werner, the Kaiser's favourite artist and chairman of the National Art Committee, has left us a striking account of the conversation between Wilhelm and Tschudi as they toured the Gallery. 'Herr von Tschudi explained the pictures. On seeing Böcklin's portrait with the skeleton . . . H.M. said he would rather see a genius or something of the kind instead of the skeleton. H.M. did not conceal his dislike of Böcklin. Nature did not look like that, He did not like the fantastic, the untrue . . . Coming to Ad[olf von] Menzel's "Iron Rolling Mill" H.M. to Tschudi: "You see, that is genuine and truthful, a celebration of work, and not all that filthy, degenerate, repulsive side which the modern painters always emphasise in their depictions of workers now" . . . H.M. wanted nothing to do with [Max] Liebermann and F[ritz] v. Uhde . . . With Millet, in response to H.M.'s challenge: "Now just tell me what you see in that", v. Tsch[udi] tried to make Millet out as a kind of Rembrandt. H.M.: "No no, in that case we would rather have the real Rembrandt." The others too, such as Fragiacomo and Segantini, were met with head-shaking and opposition from H.M.'[96] The departmental head in the Prussian Ministry of Ecclesiastical Affairs and Education, Ludwig Pallat, blamed Wilhelm's hostile reaction on Anton von Werner and Tschudi's other enemies, commenting: 'He did not conceal his disapproval of the modern art, especially the French, but seems not to have expressed himself particularly strongly about it. On the other hand, the artists hostile to Tschudi, led by Anton von Werner, spread stories of disapproving remarks by the Kaiser – the Frenchification must stop, the French must be removed from the gallery and suchlike. Even if the Kaiser did not say such things, he seems to have been influenced by those circles towards their views.'[97]

Whatever the truth of that, four months after his visit to the National Gallery, Wilhelm issued an order in which he took a decided stand against Tschudi and the Impressionists. 'In the course of my visit to the National Gallery in the spring of this year I observed that paintings which, by virtue of the subjects they depict, seem particularly likely to have an educational influence on the visitors, and which through their artistic merit represent the art of the nation in an outstanding fashion, have been removed from their advantageous positions and replaced by examples of the modern tendency in painting. I do not approve of these changes, and I wish the works in question to be put back in their old places and the more recent pictures to be moved to a less prominent place. At the same time I require that in future all acquisitions for the National Gallery, whether by purchase or donation, should first be submitted for my approval. Please make the necessary arrangements accordingly. Signed. Wilhelm R. Neues Palais, 29 August 1899.'[98] Tschudi had no choice but to obey, and before the end of the year Bosse's successor as minister for ecclesiastical affairs, Konrad Studt, was able to announce to the Kaiser that the re-hang of the Impressionists on the second floor had been completed. Wilhelm's ban on Tschudi's previous freedom to make purchases, however, proved much more draconian in its effect. In 1902 Alfred Lichtwark described the paradoxical situation in which the Director of the National Gallery now found himself. 'He has at his disposal the richest private means that a museum in Europe can attract, and he could bring together a first-rate collection in no time at all, except that his superior authorities don't dare accept it as a gift or recommend its acceptance to the Kaiser.'[99] By means of the 'kingship mechanism', Wilhelm had accorded the powerful protection of the state to the old salon art of the historical and battle painters, but at what cost! The glorious splendour of German Expressionism, which is still universally marvelled at as a brightly gleaming 'gash of fire' (Christos Joachimides) in the darkness, was shut out; it flourished, not in the capital of the Reich, but in Munich and Dresden, Darmstadt and Dachau, Worpswede, Weimar and Wuppertal, Düsseldorf and Hagen.[100]

Just as in politics Wilhelm had to rely on the advice of unofficial favourites such as Philipp Eulenburg in order to get his way against the constitutional authorities, in the world of artistic politics he had his 'artists' camarilla', which enabled him to exercise his personal rule. Apart from Ihne, Raschdorff and Schwechten, Salzmann and Bohrdt, Begas and Knackfuss, he listened above all to the advice of Anton von Werner, who had already played an important role as court painter to Wilhelm's grandfather and his parents, but who now, under Wilhelm II, rose to be unquestionably the most important influence on official artistic policy. The young Kaiser not only heaped honours upon him; he gave him the most important posts in the artistic world in Berlin. He became Director of the Academy, member of the Prussian National Art Committee, Chairman of the Union of Berlin Artists, Chairman of the Members' Association of the Academy, Chairman of the Berlin Local Association and Principal Chairman of the United German Art Association.[101] Werner's enemies accused him of being more imperial than the Kaiser.[102] As in the political sphere, the relationship between Wilhelm and his favourite in the artistic world was founded on mutual idolisation, as Werner's diaries testify. '[It is] amazing what expert knowledge, objectivity and total lack of preconceived ideas the Kaiser brings to the examination of such artistic questions, always with sound common sense', we read in one entry. And in another: 'I am always happy to be able to see and speak to the Kaiser on such occasions, and to take pleasure in his clear gaze and his fresh, lively nature. In his whole nature he is always so unaffected and natural, and even surprises like this one [Wilhelm had unexpectedly conferred an order on a young painter] spring from a right feeling or have a carefully considered purpose.'[103]

The Kaiser showed a keen interest not only in painting, but also in graphic art, photography and the new medium of film, and here too he was ready to intervene with imperial advice and instructions. Seidel writes admiringly of his detailed discussions of the various methods of graphic reproduction with the director of the chalcographic department of the Reich printing works, in which he showed a remarkable breadth of knowledge and gave expert instructions on the execution of commissions. In the sphere of photography the Kaiser took a lively interest in the problem of producing coloured photographs directly from nature. 'By hearing reports from the experts, particularly . . . from the Technical College in Charlottenburg, and by having the relevant experiments demonstrated to him, the Kaiser is fully informed about the various more or less successful work in this field, and if any progress is reported the Kaiser always demands a report on it', Seidel tells us, going on to describe approvingly how for one of the exterior shots the monarch wore 'his green hunting coat with black breeches, the orange sash of the Order of the Black Eagle over his white waistcoat and a red nosegay in his buttonhole'.[104]

Although Wilhelm often gave offence by the unusual intensity and ubiquitousness of his personal interventions in the sphere of art, he also earned himself great credit in the more traditional princely role of collector and patron.[105] It was thanks to him that the collections of the Gemäldegalerie, the Skulpturenabteilung and the Kupferstichkabinett in Berlin achieved international status during his reign. As early as April 1889 Wilhelm gave the Gemäldegalerie a special grant from the Allerhöchster Dispositionsfonds which enabled it to buy a large collection of the early works of Adolf von Menzel.[106] Collections of non-European art, above all the Islamic and East Asian Department, were also established at this time, and several very fruitful archaeological expeditions to the Aegean, the Near East and Egypt took place entirely as a result of his personal support.[107]

The fact that egotistical Realpolitik could be at work even in such apparently selfless patronage is shown by the case of the Schack Gallery, which the Kaiser magnanimously allowed the city of Munich to keep in 1894. He sent a telegram to the Mayor saying: 'I see from the telegrams that Count Schack has bequeathed his picture gallery to Me. This art treasure, which has become so dear to the artists and citizens of Munich as well as to all Germans, shall be preserved for Munich. May the people of Munich see in this a new proof of My Imperial favour and of My interest in their well-being, just as I take pleasure in owning a house in your beautiful city, as an Imperial symbol, in whose halls every devotee of art will be My welcome guest.'[108] It was Philipp Eulenburg who had drawn the attention of his imperial friend to the political advantages of leaving Count Schack's art collection in the Bavarian capital – albeit as imperial property and in a specially erected building, so that it was plain for all to see that it belonged to the Kaiser – instead of removing it to Berlin, as he was under strong pressure to do. Such an action, Eulenburg pointed out, would make Wilhelm popular in Bavaria and throughout South Germany. 'There would be a storm of gratitude and appreciation', he promised, and 'an imperial *property* in Munich, recognisable by emblems etc., would help to promote the idea of the Reich'. In addition to this, Eulenburg stressed, 'the withdrawal of the favour would always be a sword of Damocles hanging over the city', which would allow to Kaiser to exert a certain pressure on Bavaria. These political advantages, in his opinion, outweighed the purely artistic benefits of transferring the collection to Berlin.[109] Soon afterwards Eulenburg was able to report to his protégé Bülow in Rome that 'the affair of the Schack picture gallery went off splendidly. H.M. was really magnificent and bought the house for 400,000 without blinking. Nothing could have made him more popular in the South than this coup! The particularist–ultramontane party is beside itself over this *Kaiser's House* in Munich.'[110]

THE SIEGESALLEE AND 'GUTTER ART'

Not only because he saw himself as a sculptor *manqué* – according to Seidel he once said that 'if he had not been Kaiser, he would most like to have been a sculptor' – but also because he believed that the most immediate way of reaching ordinary people was through memorial art, Wilhelm II had always taken an intense interest in sculpture. He produced numerous designs of his own (for instance for the First Regiment of Foot Guards memorial on the battlefield at St Privat, and for the national memorial in Memel) and personally chose the sites for memorials (such as the equestrian statue of Kaiser Wilhelm I on the

Deutsches Eck at Koblenz).[111] As with all other forms of art, he intervened even in the smallest details of whatever project had taken his interest, especially where the supposed didactic effect of the work on his subjects was concerned.[112]

It was no coincidence that in memorial sculpture as in architecture, opposition to the Kaiser's ideas for putting dynastic history to pedagogical use should again come from the Reichstag. The parliamentary committee responsible for the memorial to Kaiser Wilhelm I that was to be erected in front of the Berlin Schloss had originally allowed Wilhelm II to choose both the site and the style of the equestrian statue of his grandfather. But in January 1894 the design which Wilhelm had approved came under harsh criticism from the committee, which, not surprisingly, he bitterly resented. 'The time when everything went well and people did everything to please him has gone', Waldersee commented on hearing of this, although he was confident that the Kaiser would get his way in the end.[113] His prediction was to be more than fulfilled with the row of statues of all thirty-two Hohenzollerns in front of the new Reichstag building.

Of all the artistic projects with which Kaiser Wilhelm II was associated, the Siegesallee, begun in 1895 and completed in 1901, was the most peculiarly his own. His idea of stimulating national pride and admiration for the Hohenzollern dynasty through a form of sculpture park had manifested itself early on. In Kiel 'a series of monuments commemorating the naval heroes who founded our navy in the most difficult of circumstances, or who won the first laurels in the wreath of its fame', commissioned by Wilhelm, was erected in front of the Naval Academy. The Kaiser himself produced most of the designs for these monuments.[114] He placed statues of his ancestors from the House of Orange, for which he likewise prepared the designs, on the terrace of the Berlin Schloss.[115] On 27 January 1895, his thirty-sixth birthday, he had an announcement published in a special edition of the *Reichsanzeiger* and *Königlich Preussischer Staatsanzeiger* which read: 'Almost a quarter of a century has passed since the German people, following the call of their princes, rose in unison to repel the foreign attacker, and through glorious victories, albeit fought with heavy sacrifices, achieved the unification of the Fatherland and the re-establishment of the Reich. My capital city and residence of Berlin played a full part in the development of German cities which ensued . . . As a mark of my appreciation of the city and in commemoration of the glorious past of our Fatherland I therefore wish to create a lasting adornment for My capital city and residence of Berlin, which will portray the development of the history of the Fatherland from the foundation of the Mark of Brandenburg to the restoration of the Reich. The object of my plan is to erect in the Siegesallee marble statues of the princes of Brandenburg and Prussia, beginning with Albrecht the Bear and ending with Kaiser and King Wilhelm I, and beside each of them the image of a man specially characteristic of his time, whether he be

48. The Siegesallee in Berlin

soldier, statesman or citizen, in a continuous series. I shall meet the costs of the entire work from My Privy Purse. Reserving for Myself the further decisions to be made, it is My pleasure to acquaint the Magistrate and the City Councillors with this plan on My birthday today.'[116] Wilhelm succeeded in taking the city fathers completely by surprise with this imperial 'gift'. As Seidel points out, his aim was to create a pictorial history lesson, which, like the Hohenzollern vault in Berlin Dom, would bring the history of Brandenburg–Prussia vividly to life for the people through its ruling princes. The Hohenzollern monarchy would gain stability, and the young Prusso-German empire historical legitimacy, through the depiction of a gradual evolution passing from generation to generation of rulers and culminating in the German Reich. The whole concept, as Uta Lehnert has convincingly argued, was a grandiloquent 'Réclame Royale'.[117]

The Kaiser was more directly involved with this building project than with any other. He entrusted the historical and artistic supervision of the scheme to the court historian Reinhold Koser and his old favourite Reinhold Begas respectively. He also had the last word in the choice of the twenty-five other sculptors employed, and commissioned 'Em' Görtz, the friend of his youth, to execute the statue of Margrave Ludwig II. Equally, it was entirely his decision to position the statues in two lines, which were to extend from the Königsplatz to the Kemperplatz and back.[118] He even issued detailed instructions for the execution of each monument. In consultation with Begas he prescribed that the statues, thirty-two in all, should be erected at equal intervals, in chronological order. Each statue should be framed by a semi-circular bench surmounted with busts of two characteristic contemporary figures. After inspecting an early model, the Kaiser ordered the statues in their surrounds to be moved further back into the Tiergarten and the steps to be raised.[119] He also decided on the dimensions of the figures and chose the material to be used, which was Carrara marble. He gave instructions that the statues of the rulers were to show them as young men; they were not to be open to mockery, and they must be historically accurate: every uniform button, it was said, must be correct.[120] Under Koser's supervision historians were commissioned to supply precise information on the appearance, dress, character and historical background of each figure.[121]

As soon as the preliminary works were completed a small model had to be produced first, followed by a large one; the final version was then carved in stone, usually not by the sculptors themselves but by Italian or German stonemasonry firms. Each of the models had to be approved by the Kaiser and even the fees required imperial authorisation for each instalment. This in itself was reason enough for the artists to be afraid to oppose the Kaiser's decisions.[122] He was always accompanied by a large entourage on his visits to their studios; his wife often came, and Begas almost always.[123] Here again Seidel emphasises the Kaiser's technical interest. 'At every stage of the work His Majesty took care to see for himself the progress made and never tired of encouraging and helping with word and deed. Many an afternoon the Kaiser would spend visiting one studio after another, and this work with the artists gave him some of his pleasantest hours of relaxation.'[124] Not all the artists regarded the imperial visits as an honour. When, for example, Rudolf Siemering announced to the Kaiser in the spring of 1899 that he had completed his clay model of Friedrich Wilhelm I, he asked to be given timely warning of Wilhelm's visit so as to be able to prepare himself; furthermore, he said, he did not wish to see Herr Begas in his studio ever again![125] Fritz Schaper, to whom the Kaiser had given the commission for the statue of the Great Elector, suffered a nervous breakdown in 1899 and had to spend several months in a sanatorium.[126]

The first three monuments were unveiled on 22 March 1898, the birthday of the old Kaiser Wilhelm. At the All-Highest request this ceremony was a simple affair without music or speeches. The expense of the unveiling ceremonies depended on the historical significance which Wilhelm II accorded to each ruler, but he was always at pains to create a period ambience. He selected the guests, instructing the College of Heralds to search for living descendants of the figures represented in the monuments to be invited to the ceremonies. He dressed in the clothes of each period, as if he were about to meet the prince immortalised in marble in person. At particularly special ceremonies he even gave a military salute to his stone ancestors; he is said to have stood for a full minute saluting Frederick the Great.[127] On 30 March 1901 the unveiling of the monuments to the Great Elector, King Friedrich Wilhelm III and Kaiser Wilhelm I took place. Lehnert aptly describes this spectacle as a 'showpiece for the imperial troops', through which the 'pre-eminent importance of the military as the mainstay of the monarchy' was to be demonstrated. 'The artistic message, by contrast, was completely drowned out.'[128] This spectacular event had been planned originally for 22 March, Wilhelm I's birthday, but it had to be postponed for eight days because Wilhelm II had been wounded in an assassination attempt at Bremen.

With the unveiling of the last of the thirty-two statues in December 1901 the realisation of Wilhelm's ambitious idea had been completed, and the didactic effect which he hoped it would have on the population did not entirely fail to materialise. According to some accounts it became a highly popular pastime to go for walks along the Siegesallee and play 'cheerful guessing games' as to who the figures were. But among the many million inhabitants of Berlin, not a few found the retrospective self-glorification of the Hohenzollern dynasty tasteless and ridiculous, especially as the ancestors represented included figures such as Heinrich the Child or Otto the Lazy, who positively invited mockery. It was not long before the lines of statues had acquired the nicknames of Corpse Avenue or Puppets' Parade among the people, and as a result of damage to the sculptures measures to protect the monuments at night were considered. In any case workers in Berlin, the great majority of whom voted Social Democrat, and the rural population in the surrounding provinces, never went into the Tiergarten, or only very rarely. Among educated people, and above all among artists and art critics, the predominant reaction was one of anger or mockery. Count Ernst zu Reventlow spoke of 'artistic sycophancy', Ferdinand Avenarius of 'pseudo-art with a political purpose, for the glorification of the dynasty', Karl Scheffler of trivial and meaningless 'court art'.[129] The poet Wilhelm Holzamer described the Siegesallee bitterly as the greatest obstacle that could have been placed in the way of art. 'It was an *imperial* commission', he wrote, but 'My God, what can it be, this avenue of white benches with little busts on them . . . A broad street, marked out on each side into 16 equal parts . . . and each of the 32 marked-out places filled with the same arrangement: bench, busts, pedestal, figure of a ruler . . . That was the directive – and the directive was that this history of Brandenburg should not be the people's history, nor intellectual or cultural history, nor a history of deeds, but the history of a dynasty . . . , an avenue to show off a family in marble, in which a portrait of a prince was itself the meaning – and meaning enough.'[130] 'What depths German art has reached', Fritz Schaper lamented after one of the unveiling ceremonies in the Siegesallee.[131] Max Liebermann went further still and called the Siegesallee a crime against good taste which he could not bear to look at except through dark glasses.[132] And after the First World War Heinrich Vogeler posed the question: 'Can we use the word culture at all, in describing the period in the life of our people before 1914, when the Siegesallee was created?'[133] Far from reconciling the German people with the late-absolutist Hohenzollern monarchy as represented by Wilhelm II, the Siegesallee has come to be regarded as a 'preposterous' and 'outlandish' symbol of the contrasts and illusions which were to be the downfall of Wilhelmine Germany.[134]

On 18 December 1901 the Kaiser invited all the artists who had rendered meritorious service by their work in the Siegesallee or in the Pergamon Museum to a gala dinner in the Schloss, at which Begas and Schaper were accorded the particular distinction of being seated next to the Kaiser. This was the occasion on which Wilhelm II made the half-hour speech which has gone down in history as the 'gutter speech', in which he explained the thinking behind the Siegesallee and expressed his abhorrence for modern art. He said:

> I joyfully take this opportunity to express to you in the first place my congratulations, and in the second place my thanks for the admirable way in which you have helped me to realise my original idea. The drawing up of the programme for the Siegesallee occupied a number of years, and it was Prof. Dr Koser, the esteemed historiographer of my House, who enabled me to set you gentlemen your task in a comprehensible form. When once the historical basis had been found, it was possible to take a further step, and directly the personalities of the princes had been settled, it was possible also, resting on historical research, to settle on the more important men who helped them in their work. In this manner originated the groups, and, to a certain extent influenced by history, the form which the group should take was arranged. When once this portion of the programme had been completed, naturally came the most difficult part of all, the question: will it be possible, as I hoped, to find so many artists in Berlin who will be able to work on a uniform scheme to carry out this programme? In approaching the solution of this problem, I had in my mind, in the event of its successful accomplishment, to show the world that the most favourable condition for the solution of an artistic problem was not to be found in the calling of committees, or in the appointment of all kinds of prize courts and competitions, but that the old way practised in ancient classical times, and also subsequently in the Middle Ages, that is to say, the direct intercourse of the person who gives the commission with the artist, affords the best guarantee for an artistic result and successful performance of the task. I am therefore specially obliged to Professor Reinhold Begas, because when I approached him with this thought in my mind, he declared to me without hesitation that there was absolutely no doubt that enough artists could always be found in Berlin to carry out such an idea without difficulty, and with his assistance, and thanks to the acquaintances which I had made among the sculptors of this

city by means of my visits to exhibitions and studios, I have indeed succeeded in gathering together a staff, the greater part of which I see assembled round me today, with the help of which I could undertake this task. I feel sure that you cannot refuse to bear me witness that with regard to the programme I have worked out I have made the treatment of it as easy as possible for you, that I have set and laid down the limits of your commission in general terms; but in every other respect I have given you the most absolute freedom, not only freedom in combination and composition, but even freedom to throw into the work so much of your own individuality, as every artist must do in order to impart to a work of art his own individual character; for every true work of art should contain in itself a grain of the artist's individual character. I believe, if I may use the term, that on the completion of the Siegesallee, we may record this experiment as successful. It only needed personal intercourse between him who gave the commission and the artist who executed it to banish every doubt, to answer every question, and no difficulties of a more serious kind presented themselves. I think, therefore, that from this point of view we may altogether look back on our Siegesallee with satisfaction. You have each in your own way performed the task to the best of your ability, and I am conscious that I have helped you by allowing the fullest measure of freedom and leisure, as indeed I consider necessary for the artist. I have never entered into details, but have contented myself with merely giving the initiative, the original impulse. With pride and joy the thought fills me today that Berlin can boast to the whole world of a body of artists who have been able to accomplish this magnificent work. It shows that the Berlin School of Sculpture stands upon a level which can scarcely have been surpassed, even in the time of the Renaissance, and I think that every one of you will admit without a touch of envy that the active example of Reinhold Begas, and his conception of the whole subject, based upon his knowledge of the antique, has been the guide of many of you in the execution of this great task. In this respect, too, we might draw a parallel with the great artistic performances of the Middle Ages and the Italians, inasmuch as the sovereign and art-loving Prince, who gave the commission to the artists, also found the masters, to whom a number of young men attached themselves, so that a definite school thereby grew up and enabled them to perform excellent work. Well, gentlemen, today, at the same hour, the Pergamon Museum has been opened at Berlin. I consider this, too, to be a very important episode in the history of our art, and a good omen and a fortunate coincidence. What will be presented in this building to the admiring visitor is a wealth of beauty, the most splendid that can be conceived collected in one place.

How is it with art in general throughout the world? It takes its models and draws from the springs of great mother Nature, and she, Nature, in spite of her great, apparently unrestricted, boundless freedom yet moves according to everlasting laws, which the Creator has set for Himself, and which can never be transgressed or broken without endangering the development of the world. Just so it is with art, and when we look at the splendid remains of classical antiquity there comes across us again the selfsame feeling: Here, too, prevails an eternal permanent law – the law of beauty, the law of harmony, the law of aesthetic. This law is expressed by ancients in such a marvellous, overpowering way and such perfect form, that we, with all our modern delicacy of feeling and all our practical skill, are proud if we are told, with regard to some specially excellent performance, 'this is almost as good as was done nineteen hundred years ago' – almost! With this thought in mind I should like to urgently impress upon you the idea that sculpture has still for the most part remained untouched by the so-called modern tendencies and movements. It still stands there noble and sublime; maintain it so, and do not allow yourselves to be induced by any judgement of men or any far-fetched doctrine to forsake those principles on which it is based. Art which exceeds the laws and limits which I have indicated is no longer art, but mere mechanical skill, mere craftsmen's work, and that art must never become. Under the much-used word freedom, and under its banner, artists often degenerate into monstrosity and exaggeration and conceit. The man who breaks away from the law of beauty, the feeling for aesthetic and harmony of which every human heart is sensible, even when it is unable to give it expression, and finds his main principle in the thought of some special tendency, some definite solution of what are rather technical problems, sins against the prime spring and origin of art. Yet again, art must be helpful, must influence our people in an educative way; it must also make it possible for the lower classes of society, when hard toil and labour are over, to rise again to ideals. The great ideals have become for us Germans a permanent possession, while other nations have more or less lost them. The German nation is now the only people left which is called upon in the first place to protect and cultivate and promote these great ideals, and one of these great ideals is that we should render it possible for our working and toiling classes to take pleasure in the beautiful, and to work up and out of their everyday range of thought. If art, as is frequently the case now, does nothing more than represent misery as still more hideous than it actually is, it thereby sins against the German people. The cultivation of ideals is at the same time the greatest work of civilisation; and if in this respect we wish to be and remain a model to other nations, the whole German people must assist in the work; and if civilisation is to fully perform its task, it must permeate to the lowest ranks of the people. That result can only be brought about if Art lends her hand to the task, if she elevates instead of sinking into the gutter. As the sovereign of this country I often feel with some vexation that art in the person of its masters does not combat such tendencies with sufficient energy. I do not for one moment fail to see that many a strenuous character is to be found among the adherents of these tendencies, whose intentions are perhaps of the best, but nonetheless, he is on a false track. The true artist needs no mountebank tricks, no puffs in the press, no connection. I do not believe that the great masters of art whom you regard as your models, either in ancient Greece or in Italy, or in the time of the Renaissance, ever resorted to advertising, as it is today frequently practised in the press, to give special prominence to their ideas. They worked

> according to the abilities that God gave them, and for the rest they let people talk. The honest and true artist must also act on like principles. Art which descends to advertising is no longer art, even if it be lauded a hundred or a thousand times over. Every man, however simple-minded he may be, has a feeling for what is beautiful or ugly, and to cultivate this feeling still further among our people, I need the help of all of you, and for producing in this Siegesallee a piece of such work I give you my special thanks. Gentlemen, even now I can inform you that the impression which the Siegesallee makes upon strangers is quite overpowering. All over the world an extreme respect for German art is observable. May it ever remain at this high level, and may my grand and great-grandchildren, if I should have any, ever see such masters standing at their side. Then, I am convinced, will our people be able to love what is beautiful and ever to prize their ideals.[135]

Had the Kaiser limited himself to the battle against modern tendencies in art, his backward-looking influence would certainly have been great, and regrettable from an aesthetic point of view. His activity in this sphere, however, would have been an internal matter with which his own people would have had to contend. But he was not content merely to throw down the gauntlet to French Impressionism, the Jugendstil, Expressionism and modern architecture, nor just to challenge the alleged 'enemies of the Reich' among the German political parties with a succession of provocative speeches and open telegrams. With his new *Weltpolitik* and the fleet building programme begun in 1897 he drove the oceanic world power of Great Britain into the arms of France and Russia and thereby set in train the fatal encirclement of the German Reich. Along with the dismissal of Bismarck in 1890 and the triggering of war in July 1914, this development must undoubtedly count among the most critical errors of judgement in the thirty years of Wilhelm's reign. We shall now look more closely into how it came about.

CHAPTER THIRTY

The challenge: from continental policy to *Weltpolitik*

KAISER Wilhelm II's success in securing supreme decision-making power for himself in the summer of 1896 coincided with Germany's transition to an ambitious foreign policy which left the careful, self-imposed limitations of the Bismarck and Caprivi eras far behind. Its goal was the elevation of the German Reich to the status of a world power on a level with the three established imperial powers, Great Britain, France and Russia, and the two new arrivals on the scene, the United States of America and Japan. The driving forces behind Germany's shift from a European-based policy to one which looked out to the world beyond have been the subject of close scholarly analysis in recent decades and have often given rise to controversy. But although there will continue to be differences in the evaluation of the individual causes of this development, historians are unanimous in agreeing that economic, ideological and domestic factors were at work that together made a change of direction towards an expansive foreign policy well-nigh inevitable.[1] When Wilhelm II proudly announced, at the spectacular state banquet in the White Hall of the Berlin Schloss in celebration of the twenty-fifth anniversary of the foundation of the Reich on 18 January 1896, that the German Empire had become a world empire,[2] he was alluding not only to Germany's enormous economic upsurge and the impressive increase in its population, through which the country had become the leading European industrial power. The Kaiser's words were also calculated to win approval among those who were untroubled by the Bismarckian nightmare of encirclement and impatient to throw off the cautious continental security policy pursued under the New Course, act with strength and determination throughout the world and ensure that at long last Germany too had a fitting 'place in the sun'. As we have already seen on frequent occasions in previous chapters, the expectation that spectacular foreign policy successes would restore the damaged image of the monarchy also played its part in this transition to *Weltpolitik*.[3]

For anyone who examines the rich documentary sources on this development, however, it will quickly become apparent that, although the urge for expansion was enthusiastically supported by part of the German population, the idiosyncratic character of the foreign policy that was actually pursued was not the result of irresistible pressure from below, but arose instead from government decisions in which the Kaiser and his entourage played a dominant role.[4] In other words, had Germany been a country governed by a parliamentary system, the impulse towards expansionism would not have been strong enough to overcome what was in fact considerable resistance among the population to a reckless *Weltpolitik*, as exemplified by the Cassandra call of Prince Hohenlohe's former Bavarian colleague, Otto Freiherr von Völderndorff, writing in November 1897. He warned that 'unless the colonial and world power policy [was] abandoned, the ruin of Germany and the most grave political complications must inevitably follow'.[5] But even if the constitutionally responsible authorities in the Wilhelmstrasse had merely retained their previous power, the result would certainly have been a more cautious German foreign policy than that which in fact came about under the pressure of Wilhelm II's constant restlessness and his irresistible urge to stir up trouble. As Paul Kennedy has very rightly observed, 'Even if one suspects that Germany without Wilhelm would have pursued an expansionist course in any case, the fact surely remains that he gave a "push" and a sense of urgency to the whole programme of *Weltpolitik*, especially in its distant aim of becoming the "successor to England" in the colonial field.'[6] Thomas Nipperdey also points out cogently the extent to which German *Weltpolitik*, characterised as it was by 'aimlessness and hectic restlessness, constantly changing plans and ideas', by 'hyperactivity' and 'the desire to be everywhere', was also imbued with 'a tendency imposed upon it by the Kaiser to show off and to seek prestige'.[7] In this chapter I shall attempt to show, on the basis of the documentary sources, the appalling degree to which Wilhelm II was responsible for the fact that in the eyes of the other great powers the German Reich rapidly became a malevolent rogue state, which was not prepared to abide by the recognised rules of the international community but which, on the contrary, was lying in wait for any opportunity to overturn the existing world order to its own advantage.

WILHELM AND WELTPOLITIK

It is scarcely surprising that the establishment of the system of 'personal monarchy' had an effect on foreign policy that was even more serious than its impact on domestic affairs. In the field of international relations Wilhelm had a free hand to do as he wished, without bureaucratic – let alone parliamentary – let or hindrance. In institutional terms alone, the right he claimed to decide everything personally prevented the formation of a collective forum in which the various foreign policy options open to the Reich could be objectively discussed. But in psychological terms too the effects of the Kaiser's domination over the 'responsible' authorities of the Wilhelmstrasse were immeasurable. The British military attaché, Lieutenant-Colonel Sir James Grierson, quite rightly observed in 1899 that 'His Majesty, what with the doctrine of divine right and the state of abject discipline to which he has reduced all around him, is getting into a state of "Grössenwahnsinn" [megalomania], and expects the policy of every other state to be changed to meet his views or something very like it, under penalty of his displeasure.'[8]

If Philipp Eulenburg's reaction to the dizzying speed with which the Kaiser's foreign policy developed was to retreat more and more into his shell, there was as always no lack of voices encouraging the Kaiser to believe that he was an extraordinary genius or even a saint, whose historic mission was to lead Germany to world domination. When the Kaiser went to Kiel in December 1897 for the embarkation of the German troops who were being sent to capture Kiaochow on the north-eastern coast of China, his brother Prince Heinrich, who was in command of the squadron conveying the troops, declared: 'Neither fame nor laurels hold any allure for me; I am drawn by only one thing: to announce the gospel of Your Majesty abroad, to preach it to everyone who wants to hear it, and also to those who do not want to hear it.'[9] 'It was as if we were embarking on a Crusade!', groaned Eulenburg,[10] while Baroness Spitzemberg commented that it was 'a dreadful thing, this passionate way of speaking and this bragging'.[11] Poultney Bigelow, the American writer and friend of Wilhelm's youth, congratulated him on the seizure of Kiaochow and spurred him to even greater deeds: he should reform and modernise the whole of China, and 'when you have cleaned up the Chinese Augean stable you will be a bigger man than Napoleon ever was – then come to America & be proclaimed Emperor of Yankeeland . . . – we shall soon have civil war again and every man with money in the bank will welcome a dictator – a strong reforming man', he added jokingly.[12]

Among those most deeply implicated in cultivating the Kaiser's dreams of world power was Dr Hinzpeter, former tutor to the two Hohenzollern brothers, now living in retirement in Bielefeld. Expressing his regret that ill-health had prevented him from attending Prince Heinrich's embarkation at Kiel, he wrote enthusiastically to Wilhelm II in December 1897: 'It would have been a sublime sight for me, to watch the Kaiser sending his fleet on a bold voyage and his brother on a magnificent adventure, Germany's first step on the path of truly global policy, now under the leadership of the Hohenzollerns in a double sense.'[13] A year later, when the Reich acquired the Caroline Islands in the Pacific, Hinzpeter felt compelled to inform his Kaiser that 'to calculate the value of the object by the number of square kilometres and inhabitants is really a very feeble attempt to denigrate it. Of course Germany is unfortunately in the first stage of its colonial policy, where it is only a matter of getting a firm footing in those places where the future of the world will be decided. And that, after all, is what the occupation of these groups of islands has achieved in such an outstanding way that it must be obvious even to the lay person. It truly warms my heart to see the figure of the Kaiser growing gradually but steadily in stature and brilliance.'[14] And for his imperial pupil's fortieth birthday Hinzpeter sent his congratulations with a remark pitched with unerring accuracy to match Wilhelm II's inner motivation: 'May it be granted to Your Majesty to create a place in the world for the German Reich like that which Your Majesty's grandfather created for it in Europe.'[15]

With his grandiose desire to make his mark, Wilhelm not only drove the new *Weltpolitik* forwards at a far greater speed that the experienced officials in the Reich Chancellor's palace and the Foreign Office considered wise. His sudden and unpredictable initiatives on the world stage also had the effect of greatly irritating and alarming the governments of the other powers. Repeatedly driven to the edge of desperation, the statesmen in London, Paris and St Petersburg saw themselves confronted not only with a dangerous international rival but also with an excessively powerful and hyperactive ruler who seemed to be not quite responsible for his actions. No less disturbed were the German diplomats, who had occasion enough to throw up their hands in horror at the Kaiser's inconsistencies. In November 1896 Friedrich von Holstein listed Wilhelm II's contradictory interventions in international politics over the previous few months and wondered despairingly what the outcome would be. 'On 30 August the Kaiser warns . . . the English against the Russians. On 25 October the Kaiser telegraphs the Chancellor saying it is necessary to ally ourselves with Russia and France, as a security for our colonies against the threat of English attack. On 12 November the Kaiser telegraphs the Reich Chancellor to say that he has warned Grand Duke Vladimir [of Russia] against England. On 21 November he tells the English ambassador that he will always stand by England and is also prepared, in order to bring about a better understanding, to exchange the greater part of the German colonies for a coaling station. Where will it all end?'[16] It was obvious 'that a great state cannot be ruled in this way, and that a great people will not be ruled in this way', Holstein observed.[17] When the German ambassador at St Petersburg, Hugo Prince Radolin, who had been Wilhelm's Oberhofmarschall ten years before as Count von Radolinski, complained at the beginning of 1897 of 'how dangerous this personal intervention' of the Kaiser in foreign policy was,[18] Holstein could only reply 'That H.M.'s interference makes our work infinitely more difficult is something to make one cry out to heaven. Though I am tough, I am gradually losing my nerve. The international situation in itself is not at all unfavourable for us . . . But to turn it to advantage one must know how to wait. This is something H.M. doesn't know how to do at all.'[19] In the spring of 1897 Holstein complained to Philipp Eulenburg that there was 'much that is sad in the Imperial marginalia to the report from London no. 38 which is being sent to you . . . Here H.M. emphasises that "it is our duty to establish closer relations with Gaul" and "to aid [Gaul] in its task as a main bulwark of European culture against the barbaric preponderance of the East". So here we have the third foreign policy programme within six months: first, closer relations with Russia and France to protect our colonies against England; then, the cession of our colonies to that same England with the sole exception of East Africa; now, after the Darmstadt fiasco and the refusal regarding the Jubilee, both Russia and England are over and done with and we are to seek our salvation with Gaul. We are simply dealing with a sensitive character who gives vent to *personal* displeasure in *practical* affairs. What material these three programmes in six months would have afforded a Bismarck for handing in his resignation!'[20] Chancellor Prince Hohenlohe urged Eulenburg to take a tolerant view of Holstein's 'disgruntled mood', for he was after all 'from the old Bismarckian school, where everything was weighed up and decided in the gloomy ground-floor room of the Reich chancellor's palace, and where the monarch took care not to disturb Bismarck's *circulos*. Then along comes a sovereign who is eager for action, who wants to have his say and who creates confusion by his contacts with ambassadors, for he creates contradictions between what he says and what the Foreign Office says, and sometimes even contradicts his own remarks to the various ambassadors. The diplomat of the old school is not used to that. Moreover H[olstein] does not know H.M. and does not feel the softening influence of that personal sympathy which our gracious sovereign's kindness calls forth.'[21] For Holstein it was not of course a matter of injured vanity, but of war and peace and the future of the German Reich. He complained again and again of the Kaiser's interventions, which were likely to cause trouble in European political affairs and to 'focus the mistrust of the other great powers on to Germany'. In his draft for a letter of protest from the Reich Chancellor to the Kaiser, which Hohenlohe did not in the end send, Holstein wrote: 'Apart from the fact that it is strictly incompatible with my position as responsible adviser for Y[our] M[ajesty] to decide on war and peace without my knowledge, Your Majesty . . . will recognise what harm and even danger is entailed in giving direct advice to foreign ambassadors.'[22]

THE WILHELMINE CONTINENTAL POLICY

Wilhelm II's oscillation between Russia, France and Great Britain, as reflected in Holstein's alarmed comments, shows with depressing clarity the dangers of war which the ill-considered European policy of the Reich already threatened to bring about, even before it ventured out on to the high seas and into the great wide world. What was the Kaiser aiming to achieve with his constantly changing attempts to curry favour with the other European powers? Were they motivated by any concept, any recognisable unifying idea? Let us begin by looking at the relationship with Russia and the hope for a union of continental powers under German leadership.

The idea of a continental league against Britain and/or the United States had, as we have seen, held a certain fascination for Wilhelm from the outset. As early as 1892 he had spoken of 'Napoleonic supremacy' as his actual goal in Europe,[23] and in 1899 he was still comparing himself unashamedly with Napoleon and Lord Nelson too, writing to his mother: 'Nelson is for me as far as Naval Policy & strategy are concerned "the Master" & I shape my naval ideas & plans from his! The same as I have learnt most of my military principles I adopt & follow from Napoleon the first. These two great rivals, yet masters in their own way are my task masters!'[24] At the Crown Council of 18 February 1894 he had expressed the idea, in regard to the Russo-German trade treaty, that German 'supremacy' in Europe should be demonstrated not only through military strength but also through a fiscal union of all European states against America.[25] In the summer of 1896 he reverted to this idea, not least as a result of the deterioration of Anglo-German relations following the Krüger telegram – he was looking forward to having his English uncles 'crawling around in front of me again soon', he wrote[26] – and also of the election of the Republican protectionist William McKinley as president of the United States. Tsar Nicholas II told the astonished Reich Chancellor Hohenlohe in early September 1896, after his meeting with Wilhelm II at Breslau, that Wilhelm had spoken to him 'of a customs union against the United States . . . and had appealed to him to canvass opinion in Paris in favour of it'.[27] In the confident belief that he could be 'sure' of the Tsar, Wilhelm triumphantly announced to his friend Eulenburg that on his visit to Paris Nicholas would declare himself ready 'to go hand in hand with France to the defence of the continent of Europe'. 'Our programme is: . . . to bring Europe together in the struggle against MacKinley [sic] and America in a common defensive customs union, with or without England, depending on the circumstances.'[28] And in a bombastic speech at Görlitz on 7 September he said of Nicholas II: 'We are all still under the spell of the fresh young figure of the knightly Emperor . . . He, the War Lord commanding the most powerful army, wishes only to see his troops used in the service of civilisation and for the protection of peace. In perfect accord with Me, his efforts are directed towards bringing together all the peoples of the European continent in order to unite them on the basis of common interests, for the protection of our holiest possessions.'[29] As this effusive toast had supposedly been approved in advance by the Tsar, Wilhelm and Eulenburg saw it as 'the point of departure [for] the unification of Europe' and 'the programme, as it were, for the immediate future of the European world'. The 'old friendship' between Wilhelm and Nicholas had 'come to life again', Eulenburg recorded after his arrival at the imperial hunting lodge of Hubertusstock on 4 October 1896. The Tsar was 'full of sincerity', the Kaiser had 'great trust in him and thinks the days at Breslau and Görlitz will bear good fruit'.[30] According to Eulenburg, who was in a position to know, the 'continental union against America, possibly also against England', was a favourite goal of Wilhelm II's foreign policy. 'How grateful the German people must be to such a Kaiser', he wrote enthusiastically, alluding to the friendship between Wilhelm and the Tsar.[31]

In reality, the meeting of the two Emperors aroused the impression in almost all commentators that Germany was running after the Tsar out of fear of Russian power. And that was not all: Wilhelm's attempts to play off the world empires, Russia and Britain, against each other only strengthened the determination of the Tsar, Queen Victoria and Lord Salisbury to bank on reaching mutual understanding in future crises.[32] During his meeting with the British Prime Minister at Balmoral on 27 September 1896 the Tsar, as Salisbury recorded in a 'very secret' memorandum for the Cabinet, was very critical of the German Emperor. 'He said that the Emperor was a very nervous man; he (the Emperor of Russia) was a quiet man, and he could not stand nervous men. He could not endure a long conversation with the Emperor William, as he never knew what he would do or say. I understood him to say that the Emperor William's manners were bad; that he would poke him in the ribs, and slap him on the back like a schoolboy.'[33] These remarks by the Tsar will have done little to dispel Salisbury's long-held concerns about the mental state of Queen Victoria's grandson. In Paris too Nicholas complained about Wilhelm II's nervous restlessness. 'The Emperor William . . . is a great talker', he told the Foreign Minister Gabriel Hanotaux. 'All one needs do is listen to him, which is what I did. He talked to me of everything, with a great desire to make himself agreeable to me and captivate me.' In complete contrast to the Kaiser's expectations, the Tsar declared his intention of standing by the alliance with France and not allowing himself to be put off by Wilhelm. He even spoke publicly of a 'deep feeling of being brothers in arms' between the two armies.[34] Although it was now scarcely possible to doubt the existence of a real Franco-Russian alliance, the Kaiser assured the Austrian ambassador Szögyény that there was 'at the most' a military convention, and certainly not a treaty of alliance, between the two countries.[35] The full significance of the Franco-Russian alliance as the basis of a Triple Entente to secure peace, directed against Germany and in particular against the Kaiser, is shown by Salisbury's satisfied comment that the Dual Alliance was 'a decided check to the Emperor William, who, if he had elbow-room, would certainly be nasty for us. Family quarrels have made him really bitter against this country, and I am always glad to see him "hobbled".'[36]

The widespread impression that Wilhelm was running after the young Tsar was made a great deal worse by Wilhelm himself when in October 1896 he decided to go to Wiesbaden for a few days while the Tsar and Tsarina were staying at Darmstadt, on the assumption that it would be easy to arrange another encounter with Nicholas there. Both Wilhelm and Hohenlohe were therefore deeply shocked when the Tsar sent a telegram turning down the suggestion of a meeting at Wiesbaden. Nicholas and Alexandra, who were in Bad Homburg on 17 October for the dedication of the Russian Chapel and visited the Empress Frederick in neighbouring Kronberg, were not even prepared to wait two or three hours the following day until Wilhelm arrived at Wiesbaden.[37] On the advice of the Chancellor, who was concerned about the negative impression that it would make abroad if the two Emperors did not meet at all, Wilhelm announced his intention of paying a short visit to Darmstadt on 19 October, which the Tsar and Tsarina returned the following day by calling on Wilhelm briefly at Wiesbaden.[38] But both meetings turned out to be diplomatic disasters. In Darmstadt, where in addition to Nicholas and Alexandra the latter's sister Ella, her brother Ernst Ludwig and Grand Duke Sergei (Ella's husband and Nicholas's uncle) were staying, the German Emperor, who still looked pale and depressed after the operation on his ear, was made to feel the cold contempt of his Russian and Hessian relatives. Eulenburg was indignant at the deliberate humiliation to which Wilhelm was subjected, and reported to Bülow: 'Those individuals the Grand Duke of Hesse and Grand Duke Sergei are *the absolute worst* in Darmstadt, and undoubtedly influence the Emperor Nicholas's view of things . . . Sergei hates our master, and the young Grand Duke would take the greatest of pleasure in accepting a Kingdom of Hesse, enlarged by Baden, from the Tsar and his allied Republic [France]. One should not forget that a German prince who is not a nationalist, like this one, always signifies a certain danger.'[39] The return visit to Wiesbaden the next day by the Tsar and his wife, accompanied by Ernst Ludwig, was conspicuously short and did nothing to improve relations. The memorandum which Wilhelm handed to the Tsar at Wiesbaden, 'On the need to form a politico-mercantile coalition of the European states against the USA', lay neglected by Nicholas for months before he decided to shelve it, on the advice of his ministers.[40]

As Eulenburg was able to observe, this experience with Nicholas had 'deeply affected' his imperial friend Wilhelm. 'I could see very clearly in the features of the beloved Kaiser the *degree* of the moral impression it made on him. H.M. looked *shocking* when I saw him on 22 October after his visit to Wiesbaden. The Tsar's refusal of the invitation to Wiesbaden was very brusque. The subsequent visit he paid to the Empress [Alexandra] in Darmstadt hurt H.M. because of the way in which he was received. The Tsar, the Empress [Alexandra], the Grand Duke [of Hesse] and Grand Duke Sergei and their wives were choosing precious stones which they had had sent from France. Our Kaiser stood by – a seventh person on the sideline. After that the Tsar's visit to Wiesbaden brought no warmth. It was timed to the minute. H.M. told me very little about it. I could feel how it tormented Him.'[41] Waldersee found Wilhelm's behaviour towards the Tsar undignified and incomprehensible, and expressed deep anxiety over it in his diary after a meeting with Wilhelm at Schwerin. 'The Kaiser insists on running after his cousin, one could even say throwing himself at him, and puts up with it when the latter coldly sends him packing. Even at the court at Darmstadt it was plain to see that people have nothing but scorn for our Kaiser's behaviour; they are under partly English, partly Russian influence there and have not the slightest trace of feeling for us [Prussia], not even for Germany. The Russians have nothing but scorn and derision; it really is profoundly sad and the attitude of our Kaiser is almost incomprehensible; in fact it can only be explained as fear of a war. But in this respect the worthy gentleman has followed the worst possible path: a firm, confident attitude will give us much more security than this miserable grovelling!'[42]

As if all that were not enough, the Tsar sent a message, delivered to the Kaiser by Grand Duke Vladimir at a family wedding, asking him not to write any more letters in future; Nicholas would not answer them. His father – Tsar Alexander III – would never have permitted such a correspondence, he declared, especially since Wilhelm's letters, as he had now discovered, had been written without the previous knowledge of the Reich Chancellor.[43] As is well known, Wilhelm took no notice of the Tsar's request and carried on regardless with the highly political 'Willy–Nicky' correspondence. He attributed Nicholas's hostile behaviour to the influence of his Danish mother. As he commented maliciously in a letter to Emperor Franz Joseph in early 1898, Russian foreign policy was obviously based on the principle '*fiat voluntas – Imperatricis Matris – pereat mundus*! [The Empress Mother's will be done – let the world perish!] The poor Emperor [Nicholas] is to be pitied and has no easy task, but I trust him more than his Minister!'[44] Others, like Baroness Spitzemberg, were more inclined to attribute the Tsar's unfriendliness to Wilhelm II's importunate behaviour, which discomfited those at whom it was directed, compelled them to respond with reserve or a direct rebuff, and put them into 'a complicated, false position, out of which a grudge develops against the person who caused it – very sad, because it has very serious consequences', she observed.[45] Unquestionably, the exclusion of Wilhelm from the Russian-Hessian-English family circle foreshadowed the later diplomatic encirclement of the Wilhelmine Reich.

Not for nothing was mistrust of Wilhelm II and his seemingly unfathomable foreign policy growing in St Petersburg, Paris and London, for at the same time as he was proposing a continental coalition against the Atlantic powers, America and Britain, Wilhelm was putting out feelers towards France, and not long after the Wiesbaden fiasco he went so far as to launch a 'new programme', 'that is to say, co-operation between Germany and France against Russia'.[46] In August 1896, in a marginal note on a report by Eulenburg, he directed that Germany must persuade the French Foreign Minister of the virtues of mutual co-operation, and 'give a warm welcome to any attempt at rapprochement with us which Hanotaux shows signs of making'.[47] These hopes of a Franco-German rapprochement were fostered by Eulenburg, who was on the best of terms with his 'old friend' Raymond Lecomte at the French embassy in Berlin, and introduced him to the Kaiser.[48] Eulenburg was evidently endeavouring to avert the threat of a world war through such contacts. Although he tried to soothe the anxieties of Franz Joseph over the Franco-Russian alliance by assuring him that 'in Europe *Germany and Austria* are the rulers, and our alliance paralyses the madness of the Franco-Russian extravagance', he was warned by Count Anton von Wolkenstein, who had been Austrian ambassador at St Petersburg for many years, that France would now have Russia behind it if it 'struck out'. Passing on the warning to Wilhelm, Eulenburg wrote: 'An untenable state of affairs in Paris could seek an outlet through war and rapidly set the world on fire; for it is no longer a question of a duel between Germany and France [alone].'[49]

Shortly after the meeting between the two Emperors at Breslau, in a conversation with the British military attaché Sir James Grierson, Kaiser Wilhelm expressed anxiety about the possibility of a French attack on Germany, in connection with the anticipated effect of the Tsar's visit to Paris on the monarchist and Bonapartist movements there. Supporters of these movements would greet the Tsar with shouts of 'Vive l'Empereur', which would make them nostalgic for the old 'idées napoléoniennes', Wilhelm remarked, 'and the reestablishment of an Empire or Monarchy in France would, he fears, mean war with Germany, as, to solidify the dynasty, probably the first thing would be an attempt to recover Alsace and Lorraine'.[50] Early in 1897, when Wilhelm again urged the Reich

Chancellor, in a series of letters and marginal comments, to attempt a rapprochement with France, Hohenlohe had to point out to him that there would be little prospect of success unless Germany were prepared to return Alsace-Lorraine to France.[51] Holstein commented sarcastically on this latest idea of the Kaiser's in a letter to Eulenburg: 'The moment H.M. already sees approaching when France will seek our alliance as a *substitute* for the alliance with Russia – that moment is still a long way off so far as human reckoning can judge. Before that time comes – if it ever does – we will have to get around a good many sharp corners. H.M. is only harming himself by speaking about this possibility with politicians at this early date.'[52] Eulenburg, to whom Hohenlohe appealed for support in this situation, assured him that he need not take the Kaiser's marginal notes seriously. 'An *entente* with France is something which His Majesty and I have frequently discussed', he admitted, but 'always as something for the future, very far off, and perhaps *never* to be a useful eventuality. But of course the *conversation itself* always had the character of the *activity*, as is characteristic of His Majesty. The marginal notes are also of *this character*. His Majesty will be easily persuaded. And what harm has His Majesty's restlessness done, in the end? We are stronger than ever at the moment as far as international affairs are concerned!'[53] In the next few years Wilhelm's thoughts repeatedly returned to the idea of a Franco-German agreement against Russia, especially on occasions when the young Tsar and Tsarina tried to avoid a meeting with him or Russian military activity on the frontier made him anxious.[54] In September 1899 he spoke to the Alsatian politician Hugo Freiherr Zorn von Bulach 'of co-operation with France'; that was 'the only salvation for France', he said. On this occasion he referred to the Russians as 'riff-raff'.[55]

A POLICY FOR THE FUTURE

How then did these continental aims – and it should be remembered that they were never abandoned – relate to the new overseas policy that was now coming increasingly to the fore? In this context the memorandum entitled 'Future policy', written in 1896 by Prince Heinrich's personal adjutant Georg Alexander Müller, who was later to become Chief of the Kaiser's Naval Cabinet, is a most revealing document. In it Müller boldly stated that world history was now dominated by the economic struggle. This struggle was at its most intense in Europe, where 'above all as a result of the world domination of England' the peoples living there had no room for free expansion. 'The war which could – and many say must – result from this situation of conflict would according to the generally accepted opinion in our country have the aim of breaking England's world domination in order to open up the necessary colonial possessions for the central

European states which need to expand', that is to say Germany, Austria-Hungary and Italy, and perhaps also the Scandinavian countries and Switzerland. 'But Germany stands far ahead in the need and indeed the right to expand', Müller claimed. And yet even in combination with the other Triple Alliance powers Germany would not have anything like the strength required 'in this great battle for economic survival . . . to break England's world domination'. It would therefore be necessary to join with France and Russia to form a continental league. 'In France we would gain a considerable increase in naval power; and in Russia an ally who would be able to attack the British Empire on land.' But in Müller's eyes the disadvantages of such a course were obvious. The 'destruction of the British world empire' would benefit not Germany, but the United States, France and above all Russia, which 'would naturally take India and therewith the lion's share of the spoils'. The Germans, at the very latest the next generation of them, would have to suffer 'because we permitted Russia's power to grow to the skies'. 'No', Prince Heinrich's Adjutant exclaimed, 'at that price we would rather not become a colonial power at all.' In an astonishing about-turn of the kind that was to become characteristic of the general ambiguity of Wilhelmine foreign policy until about 1912, Müller considered the possibility that instead of acting *against* Britain, Germany might act *with* it, in order to become a great colonial and world power. As Britain had the same interest in keeping down Russia as Germany itself, the latter could perhaps count on British support 'in the acquisition of territory outside Europe', he reasoned. By a happy chance Britain was after all 'our natural ally because we are both of the same race', which would add an idealistic element to the shared economic struggle, that is to say 'the preservation of the Germanic race against the Slav and Romance peoples'. Granted, if both Germanic races were to have a share in world domination Britain would probably expand more rapidly than Germany thanks to its much more powerful armaments, 'but even if Great Britain became stronger still, it could never become as dangerous for us as a stronger Russia would be', Müller declared. Reich Chancellor von Caprivi, he stated, had believed that 'Germany had no chance at all of becoming a world power, and consequently his policy was designed only to maintain [Germany's] position on the European continent.' This policy, 'now so widely ridiculed', would have been brilliantly vindicated 'if the German people were not coming to accept an entirely different opinion of their ability and duty to expand', the naval officer argued. In view of this development, however, the swing towards expansion either *with* Britain against the Franco-Russian alliance or as the leading power in a continental league *against* Britain was indispensable. Here too it was a case of 'all or nothing', Müller maintained. Either Germany must 'harness the total strength of the nation, ruthlessly, even if it means accepting the risk of a great war, or we limit ourselves to continental power alone. The middle way of contenting ourselves with a few leftover pieces of East Africa and the South Sea Islands without . . . suitability for settlement by Germans; of maintaining a fleet too strong for the mere defence of our coastline yet too weak for the pursuance of *Weltpolitik* – all this implies a dispersal of our strength . . . which Caprivi's policy logically wished to see diverted to the army.' Müller went on to ask whether Caprivi's policy would turn out to have been right. He hoped not. 'It would admittedly bring the present nation comfortable days without serious conflicts and excitements, but as soon as our exports began noticeably to decline the artificial economic edifice would start to crumble and existence therein would become very unpleasant indeed. Now, the Caprivi policy has been officially abandoned, and the new Reich government will hesitantly put to the nation the question – in the form of the new Navy Bill [planned by Wilhelm II and Tirpitz] – of whether the other policy, *Weltpolitik*, really can be adopted. Let us hope that this question receives an enthusiastic "Yes" for an answer, but also that then a change comes over our external relations in favour of an understanding with England, beside whom there is still a lot of room on this earth which is available or could be made available.'[56]

The memorandum by Georg Müller, who, though destined soon to become one of the most powerful men in German politics, was at this time a mere corvette-captain, was certainly not an officially binding document, but the ideas that are considered in it with such breathtaking insouciance are absolutely characteristic of the 'bid for world power' for which Wilhelmine Germany was preparing itself even before the turn of the century, and are therefore worth closer analysis here. The first combination discussed by Müller, that of a continental league of the Triple Alliance and the Dual Alliance against Great Britain, immediately raises the question: why should France and Russia accept German leadership? Had they not created the Dual Alliance only a few years earlier precisely in order to protect their security and sovereignty against the mightiest military power in the world? Both France and Russia had already become gigantic empires and their potential for expansion – in Asia, in the Near East and in Africa – was by no means exhausted. Why should they be tempted to participate in a break-up of the British world empire, the purpose of which was to raise the German Reich to the status of a hegemonial world power? After the 'great battle' both would be helplessly vulnerable to the superior power of Germany. The Kaiser's siren song of a European economic union against the United States of America could not prevail against the well-justified reservations of the French and the Russians. And if France and Russia were not willing to enter into the idea of a continental league under German leadership, would it not become necessary to use force to compel them to join the Triple Alliance? And where would that leave 'the problem of England'?

Even if Müller's first scenario of a continental league were not expressly designed to destroy British world domination, Britain would have been compelled to make use of all the power and influence at its disposal to prevent the European powers combining. The maintenance of the 'balance of power' on the continent had after all been the indispensable guiding principle of British foreign policy for centuries. Great Britain's interests in this respect undoubtedly coincided with the French and Russian desire for security, however great the differences between these three world powers might have been overseas. Thus even to get the continental league off the ground Germany would have had to overcome British supremacy, which was indeed the actual intention behind the battleship building programme, the early stages of which we shall examine shortly. And it was precisely this massive fleet building plan which – quite logically – was to lead to the increasingly close ties between Britain, France and Russia to protect the security of all three against Germany. None other than Wilhelm's uncle King Edward VII summed up the compelling logic of the British policy of *entente* when he remarked that if Britain dropped France and Russia out of fear of the German navy, it would give Germany the opportunity 'of demolishing her enemies, one by one, with us sitting by with folded arms, & she would then probably proceed to attack us'.[57] Müller's first suggestion thus led to an impasse, and not only for the reason he himself adduced, that Russia might become too strong by taking over India.

The prospects of success were no better, however, for the second possibility reviewed and preferred by him, that of an Anglo-German alliance to 'keep down' Russia and its western ally. Would not the establishment of a shared world domination by the two 'Germanic' world empires necessarily have entailed a war against the Dual Alliance which would naturally have given the German Reich supremacy on the continent of Europe? But even if Anglo-German collaboration had been limited to overseas acquisitions, as Müller suggested, the expansion of German power could not be expected to stop at that, for Berlin would have been exposed to the irresistible temptation, once it had acquired an enormous colonial empire, to broaden its narrow power base in Europe at the expense of France and Russia. And what form would the relationship between the two 'Germanic world empires' take? At one point in his memorandum Müller muses that it was perhaps not impossible 'that two so-to-speak satiated colonial powers . . . could coexist peaceably and use any surplus political power to further the struggle for the supremacy of the Germanic race', but this was not his considered view. Rather, he felt sure that an enlarged Britain and Germany would 'sooner or later but with absolute inevitability have to go to war to decide which of the two should dominate'. Moreover, he fantasised, it was by no means certain 'that the British world empire would be the stronger of the two Germanic states, for the lead which Great Britain now has over us means that she will be faced correspondingly sooner with the natural tendency for colonies to break away to form independent states'. One can all too readily understand, therefore, why the British government was unwilling to allow Germany 'the areas she needs for expansion whenever foreign empires collapse'.[58]

But if each of the routes to world power and 'world domination' indicated by Müller was as presumptuous and impracticable as the other, what is clear beyond any doubt is that the intensification of Germany's efforts in *both* directions – towards the creation of a continental league together with Russia and France against Great Britain and the forced imposition of an Anglo-German alliance against France and Russia – *simultaneously* was bound to lead to disaster. Even before the turn of the century, through the fatal zigzag course of expansionism it was pursuing all over the globe, the German Reich was coming into conflict with the world empires of Great Britain, France, Russia, Japan and the United States, all of which were to be among its enemies in the First World War.

ARMENIAN ATROCITIES AND THE CRETAN CRISIS

The kind of wilful, unpredictable, emotionally charged policy which both the Wilhelmstrasse and the governments of the established powers now had to face manifested itself as early as 1896 and 1897 in Wilhelm II's furious interventions in the critical situation in the eastern Mediterranean, where the German Reich had previously remained strictly aloof. The atrocities carried out in and around Constantinople in the summer of 1896, when up to 8,000 Christian Armenians were murdered, caused horror throughout Europe.[59] The regard in which the Kaiser had held Sultan Abdul Hamid II since his visit to Yildiz in the autumn of 1889 had completely vanished as a result of the massacres, at least for the time being.[60] He repeatedly demanded that the Turkish ruler, whom he now described as a 'wretched scoundrel', be deposed.[61] Wilhelm believed he could see through the Sultan's motives for the mass murder of the Armenians, and that they were part of a cynical diversionary strategy. In September 1896 he wrote on a report by the German ambassador, Baron Saurma, from Constantinople, that even more serious than the Armenian question in itself was 'the growing discontent of the Mohammedans in all circles, especially the army. Effendimis [the Sultan] knows it very well, and at first resorted to arrests and drownings. But as that did not help, he lets the Mohammedans loose on the Armenians as Christians, so as to curry favour with them.'[62] In December 1896 in a letter to Queen Victoria he wrote of the Sultan: 'May Allah soon take him to were [sic] it is very hot – [he] seems to be bent upon amusing himself aux depens of the Powers, "les 6 *im*=Puissances", as he calls them.'[63]

From the outset the Kaiser poured scorn on the inactivity of the great powers. At the end of July 1896, when the position of the Armenians in Turkey was becoming dangerous, Wilhelm had expressed the fear that 'all the Christians' there might be 'slaughtered', and criticised the lack of international intervention. 'And the Christian powers are to look calmly on, and if possible support it by a blockade!! Shame on us all!'[64] But when the crisis became acute he categorically rejected Marschall's urgent recommendation that German warships be sent to the eastern Mediterranean. 'No. They would in any case be too late and could not go through the Dardanelles', he wrote on the Foreign Secretary's cipher telegram. 'Anyway, England and France are there!' 'If the whole of the English and French, as well as the Russian, Italian etc., Mediterranean squadrons have been no use, a German ship will not be any use either.'[65] Only months later Wilhelm's attitude to the Sultan and the Ottoman Empire was to change completely again.

In early 1897 the rebellion against Turkish rule in Crete flared up anew. The leaders proclaimed union with Greece, and war between Greece and Turkey seemed inevitable. Under pressure from public opinion the Greek government sent a warship and a cargo ship to Crete on 6 February; a few days later the Greek fleet was mobilised, and Prince George of Greece, the second son of the King, crossed over to Crete with four torpedo boats.[66] During this crisis the Kaiser allowed himself to be led by strong anti-British and anti-Greek feelings. He distrusted London's intentions and was convinced that Britain was aiming to take advantage of the collapse of the Ottoman Empire.[67] When Hatzfeldt reported in January 1897 that Salisbury doubted that the ambassadors' conference meeting in Constantinople to solve the crisis could succeed, and that he thought it probable that in the end Austria would have to step in to prevent action by Russia, the Kaiser commented on this news with an explosive marginal note: 'I see, just as I thought, for want of an army Austria must act as slave labour in England's interests, and once that is under way, the Triple Alliance will have to follow, after which Salisbury will set Gaul on our heels! Then *vogue la galère* and *adieu* Africa for us?!'[68] On another report from Hatzfeldt, in which he described the proposal of the international conference to set up a supreme council of state in Turkey, the Kaiser expressed his suspicion that Britain was merely trying 'to compel Russia to intervene, so as to be able to play off Austria against her'.[69] Obsessed with his mistrust of Salisbury, he constantly tried to 'unmask' the real intentions of the British in the crisis.[70] He suspected, for instance, that Salisbury had made promises to Italy as to how it could profit from the collapse of Turkey at the expense of Austria.[71] Lascelles, no doubt recalling Lord Salisbury's warning of December 1895, reported that Wilhelm's behaviour appeared 'to verge on insanity'. He was 'violently excited against England' and 'in a wild state of excitement' over Crete, the ambassador informed the British Prime Minister, bemoaning the fact that 'His Majesty should have elected to be his own Minister for Foreign Affairs during this crisis.'[72]

Wilhelm favoured severe measures against Greece, which he likewise deeply distrusted. On 14 February 1897 he wrote to his mother: 'Everybody is very preoccupied here by the sudden & unexpected turn affairs have taken in Crete by the most untoward immixtion [sic] of the Greeks, who are in the act of setting fire to the whole of Europe if they persist in their rashness! For I just got the news that the King has given the order to mobilize his Army! If Heaven does not avert the catastrophy [sic], we are on the eve of very grave events & may before the year is many months older be plunged in the most terrific broils!'[73] Against the wishes of the Chancellor and the Foreign Office, who were in favour of an intervention by the powers on the spot, Wilhelm suggested a joint blockade of Piraeus, in direct conversations with the British, Austrian and Russian ambassadors.[74] He wanted to prove to the world, he wrote, that things could be managed 'even without England', for 'the continent must show the British for once that they are not the best at running it'.[75] Colonel Grierson described the Kaiser's hectic attempts to make the running in this affair in a letter to Sir Arthur Bigge of 20 February 1897. 'The Emperor is constantly visiting the ambassadors . . . and trying to win the Powers round to his proposal of a blockade of the Greek coast. The other day he went to the Austrian Embassy at 9 a.m. and found the Ambassador in bed. After rousing him, he came on to the British Embassy where the same thing was repeated. Apart from public opinion in England it seems to me all the same that the Emperor is right in his idea of nipping the danger to the peace of Europe in the bud, but I cannot see why Germany, whose interests in the Mediterranean are infinitesimal compared to those of other Powers, and who at this moment has not a single ship in Cretan or Greek waters (though one is *en route*), should take the lead . . . I very much fear that there will now be a new flood of vituperations against British policy in the German papers.'[76] When the German warship mentioned in Grierson's letter, the *Kaiserin Augusta*, arrived in the Eastern Mediterranean, the Kaiser demanded that its commander be instructed 'to act above all in unison with the *Russian* and *Austrian* admirals, and not to hesitate to take the severest measures – including shooting with live ammunition – if Greece pays no heed to warnings'.[77] Action at sea would make sense only if 'instead of a few cannon shots, a sharp naval engagement puts an abrupt end to the Greek fleet. For it has powerful torpedoes, and in return for "a few cannon shots" it could sink the ships with a few torpedoes during the night.'[78]

Just as after the Krüger telegram a year earlier, the Kaiser bewailed the fact that Germany's navy was not strong enough for international crises. 'This shows one once again how much Germany suffers from the *lack* of a *strong fleet*, as it cannot make itself felt effectively in the Concert [of Europe]. If, instead of one ship, we had had a strong cruiser division with armoured cruisers in Crete, Germany could immediately have blockaded Athens on her own initiative in February, *alone*, and thereby swept along the other powers with us and forced them willy-nilly to join in. As it is, in the end nothing happened, and the one who thwarts all plans, paralyses all energy and to whom after all that consideration is shown is England! And why? *Because she has the strongest fleet!* Our 1,000,000 Grenadiers are *no* help to us there!'[79]

In this crisis Kaiser Wilhelm found himself in conflict not only with his own Foreign Office and the agreed policy of the great powers, but also with the passionate feelings of his sister Crown Princess Sophie and his mother and grandmother. His contempt for the Greek royal family, with its close ties with Copenhagen, London and St Petersburg, played a decisive role in his hostile attitude towards Greek aspirations. He described the sons of King George I as 'louts without any education', and commenting on the influence of the Dowager Empress Marie of Russia, the Princess of Wales and the ladies of the Danish court on the Greek question he said: 'These petticoats should keep their fingers out of things.'[80] He scornfully rejected the humanitarian arguments of his sister, mother and grandmother, who were shocked by the atrocities committed by the Turks against the Cretan population.[81] Once again the Empress Frederick could only consider her son's heartlessness as an incomprehensible barrier to a better relationship between them. 'Of course I do not mention the subject to William, whose violence is unabated', she wrote to Queen Victoria.[82] 'What it is to me – as a Mother – to see my own son embarked on a course which threatens ruin & destruction to his own sister & brother in law & their family, I cannot tell you!!!'[83] After a visit which the Kaiser had paid to her at Kronberg she reported to Windsor: 'William . . . did not disguise that all his sympathies were with the Turks. – He believes everything that is in *disfavour* of the Greeks.'[84]

In spite of the blockade instigated by Wilhelm II, Greek troops crossed the Thessalian frontier on 10–11 April 1897, setting in motion the war with Turkey.[85] Fears were now beginning to be expressed in all the capitals of Europe about the escalation of the crisis, especially when the Greek army suffered a series of defeats. On 25 April Queen Victoria appealed to the Tsar – a further sign of British co-operation with the Franco-Russian alliance – 'to use all your powerful influence to bring about an armistice and terminate this disastrous war which, I am sure, must distress you as much as it does me. I earnestly trust that you will be able to agree to Lord Salisbury's proposal for joint action with you and France.'[86] At the beginning of May the representatives of the Russian, British, French and Italian governments in Constantinople suggested an armistice between Greece and Turkey, although without making it a condition that Athens should recognise the autonomy of Crete and the Greek troops should be withdrawn from the island, both points on which Wilhelm II had insisted as prerequisites. Instead, as Hohenlohe informed the Kaiser, the ambassadors 'merely expressed the hope that the Greek government, out of gratitude for the successful mediation, would agree to those demands later'. In view of this fact, the Chancellor advised the Kaiser 'that His Majesty should maintain the attitude He has held hitherto in the interest of European peace, and also impress this point of view upon the Austro-Hungarian ambassador. For it is beyond question that the Greek government will not agree to the demands of the powers except under the pressure of a direct threat of war.' The Kaiser unreservedly accepted Hohenlohe's suggestion. 'Have spoken to the ambassador in the sense indicated below and quite unambiguously, leaving him in no doubt that I shall not participate unless Greece has knuckled under and promised unconditional withdrawal of troops from and autonomy for Crete, and has submitted unconditionally to the verdict of the great powers.'[87]

In consequence, Queen Victoria's efforts to persuade Wilhelm that it was necessary for all the powers to negotiate an armistice as soon as possible[88] received short shrift from the Kaiser. In reply to a telegram from the Queen he wrote: 'The Powers have arranged that the Proposal of an Armistice must be begged for by Greece, with the promise to be added by the same Power that she is ready unconditionally to submit to the verdict of the Powers and immediately to recall her Troops from Creta, the Autonomy of which she has to accept. This is conditio sine qua non. Before Greece has not expressed her will in the above named way, intervention is out of the question. It is the Russian government who, I think, is most fit to take the lead in this matter. William I. R.'[89] Taken aback, the Queen wrote in her diary: 'Received a rude answer from William, sent en clair, whilst my telegram was in cypher.'[90] Even when he received a cry for help from his sister Sophie, begging him to prevent 'further bloodshed' and 'to hasten the mediation which has been proposed by the powers, for my sake',[91] the Kaiser answered that he could do nothing so long as Greece refused to submit to the powers, recognise the autonomy of Crete and 'restore the rights she has infringed, by withdrawing her troops from there'. Not until Greece showed itself willing to accept these conditions did the German ambassadors in Athens and Constantinople have the Kaiser's instructions 'to enter into discussions with the representatives of the other powers regarding mediation', as he replied to his sister on 12 May.[92]

A few days later Wilhelm was in a position to report triumphantly to his grandmother that Greece had agreed to the conditions he had specified. 'I am happy to be able to communicate to you that after the King and the Government had begged for my intervention through Sophy and after having officially notified to my minister and again through Sophy to me personally that they unconditionally accepted the conditions I had proposed I have ordered Baron v. Plessen to take the necessary steps to restore Peace in conjunction with the representatives of the other powers. William I. R.'[93] Once again the Queen noted with astonishment in her journal, 'Received another grandiloquent telegram, also en clair, from William saying that Willy of Greece [King George I of Greece] had begged through Sophy for his intervention.'[94] Wilhelm's self-satisfaction was even more evident in the telegram that he sent to his mother the same day: 'After Sophie had begged for mediation and had been acquainted with my conditions which she transmitted to the King – the Gov^{t} have unconditionally accepted them and I have unconditionally instructed my Minister at Athens to join the Powers in the work of restoring peace. – War is at an end now!' The Empress Frederick sent a copy of this 'curious telegram from him, in the usual strangely bombastic style', to her mother, and added anxiously: 'What a strange way of doing things – why all *himself* instead of his foreign office!!'[95]

When the peace negotiations came to an end in September 1897 the Chancellor congratulated the monarch on the success of his policy. In reply, the Kaiser wrote: 'If, in Your Highness's opinion, my firm stand helped to further the cause of peace, its success is essentially due to the prudence of Your Highness and of the agencies which succeeded in putting my ideas, as agreed with Your Highness, into practical effect.'[96] In his own estimation, of course, his role during the crisis had been so crucial that he was the only person who could, or should, make any statement to the Reichstag about it. In a telegram from Hubertusstock he complained about Marschall, who had promised the Reichstag to provide more information about Germany's Cretan policy in due course. 'There should be no question of this without my orders and without previous consultation with me.' The Kaiser was 'the only person whose place it is to inform the Reichstag about it', and he had already ordered Hohenlohe to convene a meeting of the Reichstag at the Schloss for this purpose. 'Your Highness thought this idea a good one in every respect and approved it, but thought that it was too soon to make a statement to the Reichstag. That has nevertheless happened, without my prior knowledge and consent, and I must express my astonishment at it.' From now on, Wilhelm insisted, he would himself keep parliament informed on the crisis. 'All interpellations or discussions concerning Crete in which the Reichstag requests further information must be notified to me, with an indication of the proposed reply . . . I have very consciously taken a step which is intended to preserve peace in Europe once more, and I am firmly resolved to continue to take charge of the matter myself . . . I have no doubt that I shall succeed with God's help in once again preventing a world conflagration, if it is possible.'[97]

The following year, when Wilhelm visited the Sultan in Yildiz during his spectacular Near Eastern tour, both the bitter rancour and the cunning political calculation which had motivated him in the Cretan crisis found expression in an emotionally charged letter to the Tsar. Full of hatred for Britain, the 'meddlesome Power' which he held responsible for the rebellion on the island, he told Nicholas II that he had involved his fleet in the crisis 'because I felt and saw that a certain Power was using us all others as catspaws to get us to help her to take Crete or Suda bay, and I would not be of the party who are expected to appear with bread and salt and on the top the keys of Crete praying the said Power to kindly look after the welfare of these poor darling "Cretans who may one and all roast in hell"!' If the Christians in Crete succeeded in driving the Muslims off the island the effect would be disastrous, the Kaiser declared. 'What a terrible blow to the prestige of the Christian in general in the eyes of the Mussulman and renewal of hatred you can hardly imagine! The Powers concerned in Crete have played a foolish and a most dangerous game.' The Tsar should never forget 'that the Mahometans [are] a tremendous card in our game in case you or I were suddenly confronted by a war with the certain meddlesome Power! You as the master of millions of Mahometans must be the best judge of this.' 'If you quietly go on following the lead of the other Power in Crete as has been done till now', Wilhelm warned the Tsar, 'the effect will be deplorable upon your own Mahometan subjects and on Turkey . . . Therefore I implore you to give this matter once more your most serious attention and if possible find means by which you can save the Sultan from a dangerous and compromising situation envers ses sujets and solve the Cretan question in a manner acceptable to him. Don't forget that his Army fought valiantly and victoriously *for Crete* . . . and reconquered the Province. It would never forget or forgive another Power the expulsion of their brothers in arms and their Master from a reconquered Province! What a splendid opportunity for you to step in and save the Sultan from disgrace, the world from bloody war and gain the gratitude of all Mahometans! Otherwise revolution may come, and the Sultan's blood may one day be at your door! . . . All hoping eyes are turned to the great Emperor of the East; will he bring the hoped for solution?'[98]

WILHELM'S PLANS FOR A GERMAN-JEWISH HOMELAND IN PALESTINE

The Kaiser's journey to Constantinople, Haifa, Jerusalem, Bethlehem, Beirut and Damascus in the autumn of 1898 is certainly one of the strangest episodes in his not exactly uneventful life, especially as he not only met the 'wretched scoundrel' Abdul Hamid, whom had recently wished to see deposed, and set himself up as protector of all the Muslims in the world, but also contemplated, for a short time but in all seriousness, proclaiming a Jewish homeland in Palestine under his patronage. How, and through whom, did this astonishing idea come to the German Kaiser, who in spite of his friendship with a few prominent German Jews was anything but pro-Jewish by inclination? What was his aim, and what prevented the plan from taking off? As this failed project affects one of the sorest points of German–Jewish relations in the past hundred years, it is not surprising that it has been the subject of much detailed research.[99]

With their enormous suite – the Kaiserin's three ladies in waiting, her Oberhofmeister Mirbach, the Oberhofmarschall August Eulenburg and the Oberstallmeister Ernst von Wedel, two personal physicians, Leuthold and Ilberg, the Adjutant-Generals and Flügeladjutanten Plessen, Kessel, Scholl, Mackensen and Pritzelwitz, the three Cabinet Chiefs Hahnke, Lucanus and Senden, the Foreign Secretary Bernhard Bülow and many others, including eighty maids, servants and bodyguards – the German Kaiser and Kaiserin left Berlin on 11 October 1898 in the imperial train. The ostensible reason for the journey to the Orient was the dedication of the Church of the Redeemer in Jerusalem on 31 October. More far-reaching hopes and plans, however, lay behind it. In memory of the journey through the Holy Land which his father had made in 1869 on the way to the opening of the Suez Canal, Wilhelm now wished to visit the holy places in his turn, in a kind of pilgrimage with the Kaiserin, while at the same time using the opportunity to make contact with the various German settlers in Palestine.[100] As always, the desire to promote the image of the monarchy played its part: through the dissemination in Germany of pictures and descriptions of the journey, useful propaganda was to be derived from the iconography of the imperial visit to the sacred sites of Christianity. On the orders of the Kaiser a tropical uniform was designed expressly for the tour, and the horses, specially broken in for him, which he would need for his theatrical entry into Jerusalem, were sent out in advance by sea. Poets (Frank Wedekind), caricaturists (Thomas Theodor Heine) and publishers (Albert Langen) who made fun of the imperial crusader in the satirical journal *Simplicissimus* were punished with long prison sentences or had to flee abroad.[101]

In view of the smouldering national and religious conflicts in the Near East the greatest caution was of course necessary if Wilhelm II's tour was not to prove the spark that set the powder keg alight.[102] The Kaiser's original plan to include a visit to Cairo had to be abandoned because of the serious danger of an assassination attempt – the Empress Elisabeth of Austria, whom Wilhelm had worshipped in his youth, was murdered by an Italian anarchist on 10 September 1898 on Lake Geneva – and because of the Anglo-French conflict in the upper Nile.[103] Nonetheless, after his help in the Cretan crisis Wilhelm expected his visit to Sultan Abdul Hamid II at Constantinople to bring about a decided improvement in relations between Germany and the Ottoman Empire, which, as everyone was well aware, might easily lead to complications with Russia, France and Britain. It is therefore all the more surprising that Wilhelm was willing, shortly before his journey, to consider the proclamation of a German protectorate over a Jewish homeland in Palestine, and to suggest this idea to the Sultan.

Documentary sources show that the initiative for this came not from Wilhelm but from Theodor Herzl, the founder of modern Zionism. In June 1895, shortly before he began drafting his influential book *Der Judenstaat*, he declared: 'I shall go to the German Kaiser, and he will understand me, as he has been brought up to appreciate great things. I shall say to the German Kaiser: Let our people go.'[104] Although he was well aware that a Jewish state called into being with Germany's help would have to pay 'the most usurious interest',[105] the German route to the realisation of his far-reaching plans was Herzl's preferred option. The German Reich would provide not only protection, but the model for the internal structure of the aristocratic Jewish republic which he wished to found. 'To live under the protection of this strong, great, moral, splendidly governed, tightly organised Germany can only have the most salutary effect on the Jewish national character', he wrote in his diary. 'At one stroke we would obtain a completely ordered internal and external legal status.' And the Germans would also profit from this alliance, for 'through Zionism it will again become possible for Jews to love this Germany, to which our hearts have been attached despite everything!'[106]

It was of course one thing to appeal to the Kaiser, but quite another to persuade the monarch to grant him an audience. In fact it was William H. Hechler, the eccentric Anglo-German chaplain at the British embassy in Vienna, who paved the way for Herzl's approach to the imperial court. He sent the Grand Duke of Baden three copies of *Der Judenstaat* and persuaded the Kaiser's uncle to receive Herzl at Karlsruhe on 22 April 1896.[107] In time Grand Duke Friedrich proved to be a benevolent, if hesitant, patron of the Zionists. For his part Wilhelm II had hitherto shown little interest in Zionism, although he was aware of the central aim of the movement and occasionally – albeit mockingly – expressed himself in favour of it. In May 1891, when he heard of the negotiations between America and Russia over the acquisition of land in Argentina for the persecuted Jews of Russia, he remarked 'Oh, if only we could send ours there too.'[108] In April 1896 the Kaiser visited Karlsruhe, and when the Grand Duke and Hechler tried to draw his attention to Herzl's ideas he called out jokingly to his uncle's guest: 'Hechler, I hear you want to become a minister of the Jewish State.'[109] A year later, on reading a report on the First Zionist Congress, Wilhelm wrote in the margin: 'I am very much in favour of the Mauschels going to Palestine, the sooner they clear off there the better. I shall put no obstacles in their way.'[110] The All-Highest sovereign's anti-Semitic scrawl does at least show that common ground existed between Herzl and the Kaiser. As the Zionist leader had noted in 1895, 'the anti-Semites are becoming our most reliable friends'.[111] When Herzl wrote to the Kaiser on the Grand Duke's advice, asking for an audience, he received the reply that the monarch could not receive him, but asked him to submit a written report on the Zionist Congress. On 1 December 1897 Herzl sent the Kaiser his pamphlet *The Basel Congress*.[112] The first contact had been made.

The fact that for a short while during the autumn of 1898 Wilhelm showed rather more sympathy for the Zionist dream was largely thanks to the efforts of Hechler, the Grand Duke of Baden and Philipp Eulenburg. Hechler was convinced that the Ark of the Covenant, containing the Ten Commandments inscribed by the hand of God and the original manuscript of the first part of the Old Testament written by Moses, was awaiting discovery on Mount Nebo to the east of the River Jordan. Kaiser Wilhelm, the Anglican chaplain urged, should persuade the Sultan to cede the Holy Land with Transjordania to him, so as to be able to prove to the world at last that God and Moses had personally written these Judaeo-Christian texts.[113] On 28 July 1898 Grand Duke Friedrich, after repeated requests from Hechler and Herzl, finally wrote to his imperial nephew. Although he enclosed material about the Zionist movement, his letter was mainly concerned with Hechler's 'discovery' of the whereabouts of the Ark of the Covenant. The Grand Duke urged the Kaiser to make use of his forthcoming tour of the Near East to persuade the Sultan to cede the territory in question to him, naturally without revealing the real reason for his request, the Ark.[114]

The Kaiser's reply was guarded. He would send the enclosed Zionist material to Eulenburg, who would make him an oral report on it.[115] Despite this initial caution, events now began to move rapidly. On 2 September the Grand Duke received Hechler and Herzl for a two-hour audience on the Isle of Mainau in Lake Constance. The Zionist leader was amazed by the Grand Duke's 'grandiose' willingness to discuss with him 'the most secret German political matters and . . . the intentions of the Kaiser'. The Grand Duke had learned from Adolf Marschall von Bieberstein, who had meanwhile been accredited as German ambassador in Constantinople, that the Sultan was favourably disposed towards the Zionist cause. As a result of Germany's support for Turkey in the crisis over Crete, Abdul Hamid had complete confidence in the Kaiser, Grand Duke Friedrich continued, and so German influence in Constantinople was now unlimited. 'If our Kaiser drops a word to the Sultan, it will certainly be heeded', he asserted. He went on to disclose that the Kaiser had shown a very marked interest in the

Ark of the Covenant, the discovery of which would be a world sensation. The Grand Duke asked Herzl whether he had the intention of founding a state in Palestine and advised him to recognise the Sultan's overlordship for the time being; a generation later one would be able to see what could be done. But he warned Herzl to be patient. The other powers had suddenly become suspicious of German intentions in Palestine, and Herzl would only add fuel to the fire if he pressed for an audience with the Kaiser.[116]

On 8 September Herzl appealed to the German ambassador in Vienna, Philipp Eulenburg. He offered to provide the Kaiser's friend with further information about the Zionist movement, and stressed that he needed to speak to the Kaiser before his visit to the Near East.[117] Since Wilhelm was to be in Vienna for the funeral of the murdered Empress Elisabeth on 17 September, Herzl hoped he might be granted an audience on that day.[118] As he described his plans to Eulenburg on 16 September, the ambassador became 'visibly fascinated'. What made the strongest impression was Herzl's hint that the Zionists would have to turn to Britain if German support were not forthcoming. He strongly urged that Herzl should meet his friend Bernhard von Bülow, who would be coming to Vienna with the Kaiser.[119]

Following an unproductive meeting with the new Foreign Secretary, Herzl left for Paris, Amsterdam and London without having met the Kaiser. From Paris he wrote to Eulenburg setting out his principal arguments again. He emphasised that with the establishment of a Jewish homeland in Palestine 'an element of German culture' would be brought to the eastern shore of the Mediterranean; that an 'ordered exodus' of the Jewish proletariat would take the wind out of the sails of the socialist movement in Europe; and that, with the decrease of anti-Semitic agitation that would follow, wealthy Jews would be happy to remain in their present countries. Moreover 'the influx of an intelligent, economically energetic people' would bring new strength to the Ottoman Empire. France was in turmoil over the Dreyfus affair and was in no position to oppose a German initiative. With even greater urgency Herzl renewed his plea for an audience with the Kaiser before his journey to Constantinople. 'One word from the Kaiser can have the greatest consequences for the shaping of things in the Orient. His journey to the Holy Land can attain the significance of a historic turning-point in the Orient, if the return of the Jews is initiated', he urged.[120]

Eulenburg, who was staying with the Kaiser at Rominten, evidently used Herzl's letter as an *aide-mémoire* in his conversations with Wilhelm. At any rate it is clear that the monarch adopted Herzl's arguments in the letter he wrote to the Grand Duke of Baden on 29 September 1898. He began by thanking his uncle for sending the material on Zionism, which he had worked through with Eulenburg. 'The result of my researches is now as follows: . . . The basic idea [of the movement] has always interested me and even aroused my sympathy. After studying the material you kindly sent me I have now come to the conclusion that we are dealing here with a question of the most far-reaching significance. I have therefore had careful approaches made to the promoters of this idea and have thus been able to find out that the transfer to the land of Palestine of those Israelites who are ready to go has been most excellently prepared for and is even perfectly financially sound in every respect. I have therefore said in response to a question from the Zionists whether I would be prepared to receive a delegation of them in audience, that I would be happy to receive a deputation in Jerusalem on the occasion of our visit there.' He was convinced, the Kaiser continued, 'that the settlement of the Holy Land by the wealthy and hard-working nation of Israel would soon bring to the former unsuspected prosperity' – and thus also lead to a significant economic revival for Turkey. 'The Turk would then get well again, i.e. he would receive money in the natural way, *without needing to borrow*, then he will no longer be sick, will be able to build his own roads and railways without having to rely on foreign companies and then it won't be so easy for him to be partitioned. Q.E.D! In addition, the energy, creativity and efficiency of the tribe of Sem would be diverted to worthier goals than the sucking dry of the Christians, and many an oppositional Semite now supporting the Social Democrats would clear off to the East, where there is more rewarding work to be done . . . Now I realise that nine-tenths of all Germans would recoil in horror if they were to discover that I sympathised with the Zionists or would even, as I intend to do if asked, place them under my protection!' But Wilhelm had his retort ready: 'Our dear God knows even better than we do that the Jews killed Our Saviour, and he has punished them accordingly. But neither the anti-Semites nor others, myself included, have been asked or empowered by Him to bully these people after our own fashion *in majorem Dei Gloriam*!' One must love one's enemies, the Kaiser piously reminded his uncle. And besides, 'from an earthly, realistic political standpoint it should not be forgotten that, considering the immense and extremely dangerous power which International Jewish capital represents, it would after all be of huge advantage to Germany if the world of the Hebrews looked up to it in gratitude?! Everywhere the hydra of the coarsest, ghastliest anti-Semitism is raising its dreadful head, and the terrified Jews – ready to leave those countries where danger threatens – are looking around for a protector. Well then, those who have returned to the Holy Land shall enjoy protection and security, and I shall intercede on their behalf with the Sultan.'[121]

As one can readily understand, Herzl was utterly overwhelmed when he learned from a letter from Eulenburg that the Kaiser was ready to take on the protectorate of the Jewish state. Wilhelm had shown a 'complete and profound understanding' of the movement and was ready to take up its cause '*urgently*' with the Sultan. The monarch did not wish to meet Herzl before his departure for Constantinople, as such a meeting could not be kept secret, but he was looking forward to receiving a Zionist deputation in Palestine. In a 'top secret' postscript Eulenburg suggested that Herzl should nevertheless come to Constantinople on 17 October in case the Kaiser needed a final briefing before he spoke to the Sultan. In addition, he offered to receive Herzl at Liebenberg, his estate in the Mark Brandenburg, before his departure for the Orient.[122]

At Liebenberg, Eulenburg repeated his view that Wilhelm had now become quite accustomed to the idea of a protectorate, and did not doubt that the Sultan would welcome his proposal. 'Wonderful, wonderful!', Herzl wrote delightedly. When Eulenburg warned that Germany would not go to war on behalf of the Zionists, a discussion followed on the probable reaction of the great powers to German sponsorship of a Jewish homeland in the Near East. Eulenburg's view was that 'since a protectorate was involved, the matter could not remain concealed for long. Therefore he thought it best to come right out with it, immediately and demonstratively.' As far as Russia was concerned, the ambassador was more sanguine than the Zionist. 'If the worst comes to the worst, our Kaiser could write [the Tsar] a letter and win him over to Zionism', Eulenburg assured Herzl. 'Since Russia has no objections to the departure of the Jews, no obstacles will be placed in the way of the cause.'[123]

Herzl spent the next day on tenterhooks at the Neues Palais, where Eulenburg and the Grand Duke were conferring with the Kaiser. The Grand Duke subsequently confirmed Eulenburg's impression that the Kaiser was 'full of enthusiasm' for the Zionist cause. On the grounds of a positive report received from Marschall, he was sure that the Sultan would react favourably. 'The Kaiser has now undertaken to mediate and he intends to go through with it', the Grand Duke assured Herzl. Wilhelm intended to receive Herzl both at Constantinople and at Jerusalem. Later, when Herzl met Bülow and Hohenlohe, the chill wind of political realism made itself felt for the first time. The elderly Chancellor wanted to know how much territory Herzl was claiming: whether as far north as Beirut or even beyond. He asked whether Herzl wanted to found a state, and how he thought Turkey would respond to such an initiative. The Foreign Secretary denied all knowledge of a favourable report from Marschall. Herzl had to console himself with the thought that in the end what really counted was the Kaiser himself.[124]

On 13 October 1898 Herzl and four other Zionist leaders left Vienna on the Orient Express. On the way they agreed to ask for the territory between Gaza and the Euphrates, for which they would seek autonomous status within the Ottoman Empire. On their arrival at Constantinople, however, things did not go well at first. Marschall brusquely declared that he did not know Herzl and had no time to receive him. Herzl then appealed to August Eulenburg, and sent him a letter for the Kaiser in which he begged for an audience.[125] The long-awaited meeting, at which Bülow was also present, finally took place on the evening of 18 October 1898.

Herzl was at first irritated by the anti-Semitic undertone of the hour-long conversation which ensued. The Kaiser declared that he would very much welcome the settlement of some German-Jewish 'elements' in Palestine. 'I am thinking of Hesse, for example, where there are usurers at work among the rural population. If these people took their possessions and went to settle in the colonies, they could be more useful.' Bülow added to the anti-Semitic tone by complaining that 'the Jews' had shown themselves ungrateful to the House of Hohenzollern and were now to be found among the oppositional parties. If the Kaiser proclaimed a protectorate in Palestine, he would expect the Jews to show their gratitude. Nevertheless, Herzl succeeded in getting his 'entire plan' across. 'Everything, everything', he later noted with relief. The Kaiser 'listened to me magnificently' and agreed that the plan for a Jewish homeland was 'completely natural'. And when Bülow expressed doubts about the Turkish reaction, Wilhelm exclaimed: 'But surely it will make an impression if the German Kaiser concerns himself with it and shows an interest in it . . . After all, I am the only one who still sticks by the Sultan. He puts stock in me.' He asked Herzl what he should demand of the Sultan, to which Herzl replied: 'A Chartered Company – under German protection'.[126] Sensing that he was 'approaching the climax of my tragic [sic!] enterprise', Herzl and his team left for Palestine.

We do not know in which of his conversations with the Sultan the Kaiser raised the subject of a German protectorate in Palestine. All that is known is that Abdul Hamid rejected the idea so brusquely that his imperial guest did not feel able to pursue the matter further.[127] But the Zionist plan was of course nothing new to the Turkish government, which had long since agreed upon its own policy on the matter. In spite of Marschall's (alleged) report to the contrary, the Sultan had never wavered in his opposition to the project. According to his daughter he declared, 'I cannot sell even a foot of land for it does not belong to me but my people . . . The Jews may spare their millions. When my empire is divided, perhaps they will get Palestine for nothing. But only our corpse can be divided. I will never consent to vivisection.'[128] As the records in the Yildiz Archives show, the Sultan's position was unanimously supported by his advisers. All were opposed to the idea of burdening the empire with yet another national and religious problem which the European powers would only exploit to their own advantage. The Ottoman ambassadors abroad warned that the real aim of the Zionists was the formation of an independent Jewish state which would not be limited to Palestine; such a state would become the focus of worldwide

Jewish activity. They also realised that the establishment of a Jewish state in the Holy Land would fuel the 'Arab Awakening' which would destroy the Ottoman Empire. On the orders of Abdul Hamid, the Council of Ministers had developed a programme to counter the Zionist danger at home and abroad.[129] Against this background, the Sultan's determined attitude should have come as no surprise to Wilhelm II.[130]

Although Herzl was not yet aware of it, Wilhelm's attitude to Zionism had in any case abruptly gone into reverse. Suddenly overwhelmed with admiration for the quasi-absolutist Sultan and the monotheistic rigour of Islam – a sentiment which found expression in the letter of 20 October to the Tsar quoted earlier in this chapter[131]– and out of reach of the influence of Eulenburg and the Grand Duke of Baden, but open to the dictates of *raison d'état* as presented to him by Tewfik Pasha, the Turkish ambassador in Berlin who was travelling with him, and Bülow, both of whom pointed out to him that Britain, France and Russia would never tolerate a German satellite state in the Near East, Wilhelm reverted to the indifferent, almost scoffing opposition to the Zionist idea which had characterised his attitude until his conversion by Eulenburg at Rominten three weeks earlier. His interest in Zionism proved fleeting and superficial. Motivated from the beginning by anti-Semitic prejudice and egoistical calculation, his intention of proclaiming a German protectorate over a Jewish homeland in Palestine rapidly faded away after Abdul Hamid's rebuff.

The Kaiser and his suite sailed into Haifa on 25 October 1898. When Wilhelm landed that afternoon, it was, as the official account of the voyage announced, the first time since the visit of Friedrich II of Hohenstaufen in 1228 that a German emperor had set foot on the soil of the Holy Land. By the time the imperial party set out on the dusty ride to Jaffa, they had been joined by numerous clergymen and over 500 others who had arrived in four steamers. The cortege needed no less than 230 tents, 120 carriages, 1,300 horses and mules, 100 coachmen and 600 drivers, 12 cooks and 60 waiters. This colossal caravan was guarded by a regiment of the Turkish army, and the German warships which had accompanied the travellers fired off thunderous salvoes every time the imperial standard appeared on the horizon.[132] On the way Wilhelm had a brief meeting by the roadside with Herzl before travelling on to Jerusalem for the principal event of the journey – the dedication of the Church of the Redeemer on 31 October.[133]

When he received Herzl and his four companions, who had all received a blessing from the Reverend William Hechler beforehand, in his encampment outside Jerusalem, Wilhelm addressed them only in non-committal platitudes. 'The land needs . . . water and shade', he told the Zionists. 'The settlements which I have seen, both the German ones and those of your countrymen, may serve as a model of what can be done with the land. There is room for all . . . The idea behind your movement . . . is a healthy one.' When Herzl remarked that the water supply could be secured by damming the river Jordan, although it would be very expensive, the Kaiser replied in his familiar tone: 'Well, you have plenty of money . . . More money than all of us.' 'He said neither Yes nor No', Herzl commented disconsolately after the meeting. He nevertheless took comfort in the thought that his efforts had not been in vain. 'This short audience will be preserved for time immemorial in the history of the Jewish people, and it is not impossible that it will also have historic consequences.'[134]

Although Wilhelm had lost interest in the Zionist cause, Herzl's idea of proclaiming a German protectorate in the Orient was too alluring to be abandoned altogether. From Jerusalem the Kaiser travelled via Beirut to Damascus, where on 8 November he astounded the world by announcing that he considered himself the protector of all the Muslims in the world. Intoxicated by his reception in the Syrian city, Wilhelm declared himself 'moved at the thought of standing in the place where one of the most chivalrous rulers of all time, the great Sultan Saladin, once stayed'. Today, he added, Sultan Abdul Hamid and 'the 300 million Mahomedans who live scattered throughout the world and who revere him as their Caliph, [may] rest assured that the German Kaiser will be their friend for all time!'[135] He telegraphed to the Grand Duke of Baden telling him that Damascus was 'a pearl among the cities of the world'; his entry into the city had been 'the most overwhelmingly enthusiastic which I have ever experienced, with the exception of [Buda]Pest'. He interpreted the enthusiasm as 'an expression of the gratitude of the Muslim world . . . to the German Reich for its loyalty and friendship towards the Sultan, a handsome reward for our often scorned policy'.[136] The All-Highest crusader made no secret of his disappointment with what he had seen in Palestine, however. The land was 'a dismal arid heap of stones', he wrote to his mother on his way home. 'The want of shade & water is appalling . . . Jerusalem is very much spoilt by the large quite modern suburbs, which are mostly formed by the numerous Jewish colonies . . . 60,000 of these people were there, greasy & squalid, cringing & abject, doing nothing but making themselves obnoxious equally to Christian & Mussulman by trying to fleece these neighbours from every farthing they manage to earn. Shylocks by the score!'[137]

If the project of an imperial protectorate over a Jewish homeland in Palestine was an illusion from the outset, for Wilhelm his Near Eastern tour was an unforgettable landmark in his life, which he was also determined to exploit publicly. His sister Charlotte could not believe her ears when she heard of his intention of making a ceremonial entry into Berlin, as if he had won a war.[138]

A sumptuously illustrated volume, together with cheaper popular publications and colourful children's books, made sure that the myth of the imperial pilgrimage was disseminated throughout the country.[139] And in his next speech to the Brandenburg provincial diet Wilhelm took care to make capital out of his supposed triumph in Palestine to attack the enemies of the Reich within its borders, by declaring: 'Of all the impressions I received, the most sublime and moving was . . . that of standing on the Mount of Olives and seeing the place, at its foot, where the mightiest battle which has even been fought on earth, the battle for the salvation of humanity, was fought by the One. This fact moved Me on that day to swear My oath of allegiance to heaven as it were anew, that I would leave nothing untried in order to unite My People and to remove whatever might divide them. But during my stay in that distant land and in all those different places, where the woods and the water which are so dear and so lovely to us Germans are absent, my thoughts turned to the Brandenburg lakes with their clear dark waters and the Brandenburg forest of oak and pine, and I thought to Myself that although we are sometimes looked down upon in Europe, we are much better off in the Mark than abroad . . . Yes, gentlemen, the tree which we see growing, and which we have to foster, is the German Imperial oak . . . My journey to the Holy Land and its sacred places will also be valuable to me in my task of protecting, tending and cultivating this tree, and, like a good gardener, of pruning the superfluous branches and attacking the vermin which gnaw at its roots in order to exterminate them. I hope, therefore, it will be vouchsafed to me to see this picture: the oak grown into a magnificent tree, and the German Michael standing before it, his hand on his sword-hilt, gazing across the frontier, ready to protect it. Secure is that peace which stands behind the shield and under the sword of the German Michael . . . Let it, therefore, be the aim of us Germans at least to stand together firm as a solid rock! Against this *rocher de bronze* of the German nation, both far beyond the seas and here at home in Europe, may every wave that threatens peace dash itself in vain!'[140]

THE ANNEXATION OF KIAOCHOW

In the interval between the conclusion of peace negotiations over Crete in September 1897 and Wilhelm II's voyage to the Orient in the autumn of 1898, another opportunity had arisen for him to indulge his compulsive urge for action, this time drawing him far out into the Pacific and the South Seas – and also into potential conflict with Russia, China, Japan, the United States and Britain. Since 1894 he had talked of the need for a German naval base on Formosa or on the Chinese coast, and since November 1896 he and the naval authorities had identified the bay of Kiaochow, on the Shantung peninsula in north-eastern

China, as the most suitable starting-point for a German empire in East Asia. Since then Berlin had in effect been waiting for China to provide Germany with a pretext – through some violation of German rights – for seizing Kiaochow with the little port of Tsingtau.[141] During his visit to Peterhof in August 1897 Wilhelm had expressed keen interest in acquiring this port as a coaling station, and had gained the impression that Russia would agree in principle to 'shared use' of Kiaochow with Germany for this purpose.[142] When on 6 November 1897 the news reached Berlin that two German missionaries had been murdered in Shantung, Wilhelm saw this as the long-awaited pretext to gain a foothold in China by the occupation of Kiaochow. He wrote excitedly to the Foreign Office: 'I have just read in the press the news of the attack on the German Catholic Mission in Shantung, which is under my protection. Full atonement for this must be exacted through vigorous intervention by the fleet . . . I am now quite determined to give up our excessively cautious policy, which is already regarded as weak throughout East Asia, and to use all severity and if necessary the most brutal ruthlessness towards the Chinese, to show at long last that the German Kaiser is not to be trifled with, and that it is a bad thing to have him as an enemy . . . Energetic action is all the more called for because it will enable me to prove once again to my Catholic subjects, including the ultramontanes, that their well-being is as close to my heart as that of, and that they can count on my protection as much as, my other subjects.'[143] He gave instructions for an order to be sent to Admiral Otto von Diederichs as follows: 'Proceed at once [to] Kiaochow with your whole squadron, occupy appropriate positions and places there and then exact full atonement in the way which seems the most appropriate to you. The greatest possible vigour is ordered. The goal of your voyage is to be kept secret.'[144]

The Kaiser's aggressive zeal and 'driving energy', now piling the pressure on the Wilhelmstrasse once again,[145] also manifested itself in the telegram which he sent to Bernhard von Bülow, who was still in Rome, on 7 November. 'Our conversation about Kiaochow, . . . at the end of which you emphasised that it was high time to give our wavering, lukewarm policy in East Asia more vigour, has had rapid results, more rapid than we expected . . . The Chinese have at last given us the reason and "incident" for which your predecessor Marschall waited so long. I decided to take action at once . . . The eyes of all, both Asians and Europeans living there, are upon us, and everyone is wondering whether we will put up with it or not.'[146] As Bülow's biographer Gerd Fesser points out, the newly appointed Foreign Secretary was not overjoyed at the prospect of beginning his term of office with a Russo-German conflict. 'At a safe distance in Rome . . . he went to ground, in effect, and remained stubbornly silent.'[147] It was therefore left to Hohenlohe to calm the excited Kaiser down and prevent him from taking any disastrous steps. In accordance with Wilhelm's orders the Chancellor instructed the German envoy in China to demand full satisfaction from the Chinese government; but he warned the Kaiser of the dangerous consequences of a quarrel with Russia over Kiaochow. 'If Your Majesty . . . wishes to give the Squadron Commander orders to take action at once, it might be necessary to choose somewhere other than Kiaochow, as in order to occupy Kiaochow in accordance with the agreement reached between Your Majesty and the Emperor of Russia at Peterhof, Russian consent would have to be sought.'[148]

As a result of Hohenlohe's warning, the Kaiser did at least take the precaution of personally consulting the Tsar again on his attitude to this delicate question. The ambiguous reply which he received to his enquiry – Nicholas telegraphed: 'Can neither give nor withhold consent to your sending a German squadron to Kiaochow as I have just heard that this port was only temporarily ours in 1895–6' – Wilhelm happily interpreted as indicating approval of a German occupation, and he again demanded swift and energetic action. 'We must use this excellent opportunity without delay, before another great power provokes China or comes to her aid! Now or never', he wrote to the Reich Chancellor.[149] For Bülow's benefit he set out in hypocritical terms the justification for his willingness to clarify Russia's position by means of a telegram to Nicholas. 'Deeply humiliating as it is that the German Reich should practically have to ask permission from St Petersburg to protect and avenge its Christian dependants in China, and to use for that purpose a place which out of excessive modesty it did not occupy three years ago, although this could easily have been done, I have not hesitated for a moment to take this step, for the good of my country.' Wilhelm went on to declare that he did not share the Tsar's concern 'that severe punishments in the east of China will perhaps cause agitation and insecurity and widen the gulf between the Chinese and the Christians'. On the contrary: 'thousands of German Christians will breathe a sigh of relief when they hear that the German Kaiser's ships are near by, hundreds of German traders will exult in the knowledge that the German Reich has at last won a firm foothold in Asia, hundreds of thousands of Chinese will tremble when they feel the iron fist of the German Reich bearing down on their necks, and the whole German Reich will rejoice that its government has done a manly deed . . . But let the world learn the lesson once and for all from this incident, that where I am concerned: "Nemo me impune lacessit." Wilhelm I. R.'[150] Like many of his utterances from this period, these words read like a draft of the notorious 'Hun speech' of 1900.

The concern felt by the Chancellor and the Foreign Office that with his precipitate order to Diederichs the Kaiser had disregarded Russian reservations to a dangerous degree soon proved justified when on 9 November news reached Berlin that Russia insisted at the very least on maintaining its prior right of anchorage ('priorité de mouiller') in Kiaochow.[151] As the deputy Secretary of State at the Foreign Office, Rotenhan, explained to the Kaiser on 10 November, this declaration by Russian Foreign Minister Muraviev meant 'that if a foreign power were allowed to take the port, Russia's priority must in all circumstances be secured. In order to safeguard this right the commander of the Russian squadron in the Far East had been given orders that as soon as German ships entered the port, Russian ships should be sent there likewise.'[152] Hohenlohe was dismayed, and remonstrated with the Russian ambassador. 'It has been our intention from the first to come to an agreement with you, and to do nothing that was unwelcome to you. If your Emperor had indicated to us from the first that our action was unwelcome to him, I [would have] very firmly opposed the navy's plans. The contents of the Emperor [Nicholas]'s telegram persuaded me otherwise. Now that our Kaiser is committed, I am bound to make the dignity of the Kaiser and of the Reich my first consideration.'[153] The danger of a Russo-German conflict was particularly acute because Diederichs, who had set sail on the Kaiser's orders, could no longer be reached by telegraph. 'Things really look very bad', Hohenlohe lamented on 10 November 1897. He tried to console himself with the thought that 'the Russians are [only] trying to frighten us. I cannot believe that the Emperor of Russia will declare war on us because of Kiaochow Bay. There is a danger that H.M. will send a telegram to the Tsar at once. And what will it say?'[154] Holstein cautioned Hohenlohe that 'the Russian declaration is so brutally explicit that it scarcely seems necessary to give the Kaiser any advice. He alone will know whether he wants war with Russia or not. We shall now have to be very careful with our action in China.'[155] Even Tirpitz believed that war with Russia was a possibility, and advised calling a halt to the 'dangerous' action.[156] Nevertheless the Chancellor avoided a confrontation with Wilhelm. 'I shall not write or telegraph to the Kaiser, but wait until he brings up the subject himself', he told Rotenhan.[157] He was to regret his passivity later.[158] But the Kaiser did not allow the Russian objections to trouble him greatly. He wrote to the Foreign Office: 'Count Muraviev's note corresponds perfectly to the character of this mendacious gentleman . . . We should attempt to come to an arrangement with Russia to acquire the rights to Kiaochow, if necessary by purchase. Even Russia will yield to a *fait accompli*, and will certainly not start a war on account of Kiaochow, as she needs us in the East.'[159]

The cruiser division under the command of Diederichs arrived in the Bay of Kiaochow on 14 November 1897 and landed a force of 30 officers, 77 petty officers and 610 sailors, although this was to have been done only 'if the Chinese reply is unsatisfactory.'[160] The Chancellor succeeded in persuading the Kaiser 'that the Squadron Commander should desist from making a proclamation or occupying Chinese territory until the Chinese reply arrives and is unsatisfactory'.[161]

Nevertheless, at a hastily convened meeting at the Reich chancellor's palace on 15 November, presided over by Wilhelm, it was decided that an unsatisfactory reply should be engineered by deliberately setting unacceptable conditions. The minutes of the meeting stated quite openly: 'Our demands to China are to be set at such a level that they cannot be fulfilled and therefore justify further seizure.'[162] During the discussion the Kaiser made his aims in Kiaochow clear: 'Permanent occupation of the bay is to be envisaged. His Majesty remarked that he stood by the fact that the Emperor Nicholas had given telegraphic approval. Two years ago the Tsar had already expressed his agreement to Germany taking a port in China, while thanking him for our support for Russian policy in the Far East . . . His Majesty therefore does not believe there will be war with Russia, and is convinced that the seizure will be welcomed with jubilation by public opinion in Germany, and that in the Reichstag too the Centre will appreciate and support this imperial intervention for the protection of the Catholic missions.'[163]

In spite of the Kaiser's confidence the threat of conflict with Russia continued to loom. On 18 November Hohenlohe had to inform Wilhelm that both the Russian government and the Tsar himself had spoken out against the German claims to a port in China. 'The tone and content of the Russian document leave no room for doubt that the Emperor Nicholas has been persuaded to take the view that Your Majesty intended to make improper use of his telegram, to the detriment of Russian rights, which it had never been the Emperor's intention to relinquish. The *amour propre* of the Emperor will not easily get over this impression', Hohenlohe warned. He therefore suggested that delaying tactics be used for the time being; the Tsar should be told 'that the Russian fleet can of course remain in Kiaochow Bay too, and that we would undoubtedly come to an agreement over future plans'.[164] The Kaiser annotated this document: 'Completely agree with Yr. H[ighness]'s suggestion. Their famous right of *premier mouillage* will in no way be infringed by our occupation and later seizure. Russians can stay anchored there until they are blue in the face. But that cannot prevent us building a coaling station and docks there.'[165]

It was still uncertain not only how Russia would react, but also what the response of the Chinese government would be to the annexation of a sizeable piece of territory and the dispatch of several warships under the command of the Kaiser's brother Prince Heinrich. But Wilhelm ignored the reservations of both the Wilhelmstrasse and the Minister of War. On 24 November 1897 he wrote to the Foreign Office: 'That the Chinese know exactly what we want is certain; that they will wage war is highly unlikely, as they have neither ships nor money and the number of troops in Shantung is not great. The fact that Heinrich is being sent and the second division formed must of course be mentioned [in a telegram to China], as everyone knows it and it shows that the Imperial House does not for a moment hesitate to risk the lives of its members for the honour of Germany. Wilhelm I. R.'[166] Two days later he wrote to Hohenlohe: 'In about three weeks the *Kaiserin Augusta* will arrive [in China]; Prince Heinrich and the rest of the ships not until February [1898]. So that the crews do not remain away from their ships any longer than absolutely necessary, the moment has now come to form the colonial force, charter the steamer and embark them as soon as possible. I expect an answer tomorrow morning so that I can give my orders. No one is in any more doubt about our intentions. A longer delay is impossible. That it is not for Russia to say anything in Kiaochow is as clear as daylight. The battle at Nikki [in West Africa] between France and England will bring about an interesting situation; at any rate it is favourable for us.'[167] On 29 November 1897 the Reich Chancellor noted with concern that War Minister General Heinrich von Gossler had told him 'that he had received an order to make available two heavy batteries for dispatch to China, together with about 1,000 men. The War Minister is very worried about it and says that this gives the Kiaochow affair a very serious character, and that H.M. cannot have fully realised its consequences. For if these troops were thrown out by the Chinese we should be at war with China. He has no doubt that the Chinese are strong enough to sweep our troops away, as there are troops trained by our officers in Tientsin. But the Prussian army could not tolerate a defeat. War with China would therefore be inevitable, and this would cost hundreds of millions. Even sending the batteries and troops will cost 10 million. The other ministers present shared the concerns of the War Minister.' 'I know nothing about the dispatch of the batteries', the Reich Chancellor admitted helplessly. 'It looks as if the whole War Ministry is in uproar. The ministers take the view that the Kiaochow question will have a very unfavourable effect on the Navy Bill. To judge from the War Minister's remarks, the mood in the army does not seem favourable to the Chinese venture. What will the "federal governments" say to it? If things continue at this pace, we shall have a war with China. The powers will see to that.'[168] Not until the occupation of Port Arthur (now Lushun) by Russia in December 1897 did Bülow succeed in defusing the Russo-German crisis.[169]

The 'driving' role which Kaiser Wilhelm II personally – and sometimes in uncanny anticipation of his fatal orders at the beginning of the crisis of July 1914 – played in November 1897 over the seizure of Kiaochow is unmistakable. Putting his trust in his friendly relationship with Nicholas II, he repeatedly disregarded the reservations expressed by those who foresaw disaster – not only the Reich Chancellor and the Foreign Office but also the army and the navy – and confronted his experienced advisers with one *fait accompli* after another through his inflammatory orders.[170] During the Boxer Rebellion of 1900 Hohenlohe admitted, in retrospect, that, although he had not supported the Kaiser's occupation policy in the Kiaochow episode, he had not opposed it resolutely enough either. 'You are quite right', he told his son, 'when you say that we would have done better to leave the Chinese alone. The seizure of Kiaochow at any rate added to the exasperation of the yellow beasts. That occupation, which the navy inspired the Kaiser to undertake, is of no use to us. But once Tirpitz and Bülow had excited H.M.'s interest (in 1897), there was nothing I could do, and I did not want to incur the usual accusation of "being in a blue funk". But I was wrong and should have protested against it with the Ministry.'[171] Wilhelm, on the other hand, notched up the capture of Kiaochow as a triumph for his personal diplomacy, and his friend Eulenburg congratulated him 'from the bottom of [his] heart' on his success, 'for which we have to thank Your Majesty's energetic initiative, *and nothing else*'.[172] On 8 January 1898, writing to the Austrian Emperor, Wilhelm commented: 'Our Chinese affair has gone off well, thanks to the previous agreement with the Emperor [Nicholas II] *in person* and has shown the pigtailed gentlemen of the East that the good missionaries are not to be regarded as mere wild game to be hunted down',[173] to which Franz Joseph replied by sending his congratulations 'on the new, promising world position in which Your Reich finds itself'.[174] Meanwhile Wilhelm congratulated Nicholas II on the occupation of Port Arthur, expressing his hope for 'Christian' co-operation with the words: 'Russia and Germany at the entrance of the Yellow Sea may be taken as represented by St George and St Michael shielding the Holy Cross in the Far East and guarding the Gates to the Continent of Asia. May you be able fully to realise the plans you often unrolled to me; my sympathy and help shall not fail in case of need.'[175]

In Kiel on 15 December 1897 to bid farewell to the troops departing for the occupation of Kiaochow under the command of Prince Heinrich, the Kaiser addressed his brother with the words: 'If anyone . . . should ever venture to wish to hurt or harm us in the due exercise of our rights, smash him with an iron fist! and, if it be God's will, weave a laurel wreath around your young brow which no one in the whole German Reich will begrudge you!'[176] Heinrich had to forgo his laurel wreath, for on 4 January 1898 China signed a lease by which the German Reich received the Bay of Kiaochow and the surrounding area, together with railway and mining concessions in the province of Shantung and tariff privileges, for ninety-nine years. A territory of 150,000 square kilometres with more than 33 million inhabitants had become a German zone of influence under the administration of the Reich Navy Office (not the Foreign Office).[177] With this step, however, Germany had entered into latent conflict not only with the other European great powers, as before, but also with the aspiring world powers in the Pacific: Japan and the United States.

49. The Kaiser and his three eldest sons at Kiel with Prince Heinrich on the latter's embarkation for China in December 1897

PRINCE HEINRICH OF PRUSSIA IN THE PACIFIC OCEAN

The letters of 'Henry the Navigator' (as the Kaiser's brother was jokingly called at the Foreign Office) from the Far East to Wilhelm and his mother in Kronberg give us a good picture of the geopolitical struggle in the Pacific Ocean, in which the German Reich was now keen to play a part. 'Once in the Far East', Prince Heinrich wrote from the *Deutschland*, which had arrived in Hong Kong, in April 1898, 'the world looks very much different from what it does in Berlin, London or Friedrichshof!'[178] As commandant of the East Asian squadron Heinrich had great hopes for Germany's future in the Far East and the South Seas. In a letter to his mother of February 1899 he observed: 'I doubt it not that Germany's future lies in the East.'[179] And at the same time he wrote to his brother from Amoy (now Xiamen): 'In itself China has few pleasures and only momentary satisfaction to offer the European, but our future no doubt lies here all the same.' The Manchu dynasty was obviously coming to an end, he commented, and it was not impossible that there would be a rebellion sooner or later in this 'gigantic empire', and that 'people would try to proclaim a Chinaman as ruler. – What then? How will the great powers react to this! I do not believe there will be a peaceful partition of China; jealousy between the individual European nations is too great for that! England would certainly claim the Yangtze region and thus acquire the lion's share right into the heart of China. Russia will certainly move towards Peking under some pretext or other. – I can see only one way out of this dilemma, consisting in an alliance between England, Germany and Russia. All the other nations, including Japan and America, are absolutely unreliable!' The prospects of acquiring the island of Formosa, which had been occupied by Japan since the Sino-Japanese war, seemed particularly promising to Prince Heinrich. He suggested entering into negotiations with Japan for the purchase of this 'enormously rich' island; the German government could raise the money for it by selling German firms concessions for the export of the island's produce. 'Formosa would not need to become a colony at first', the Prince explained to his brother, who in this case was sceptical. 'It could however be placed under German protection as a means of getting our hands on it for the time being.' Later, admittedly, the 'wild tribe . . . (of Malayan origin)', which would not allow any foreigner into the interior of the island, would have to be 'tamed' by Chinese troops trained by German officers. The chief object for the moment, however, was simply that 'the protection of our Reich would be made possible', Heinrich urged. 'The Japanese are said to have realised already how weak they are, and are bleeding to death financially. If you were willing to authorise me, I should gladly negotiate with the Emperor of Japan in due course, or put out a feeler.' Although Wilhelm II rejected this request in a typically biting marginal comment – 'Heaven preserve me from that! Heinrich's political ideas are still very Utopian!' – only a few years earlier, as we have seen, he had himself peremptorily demanded the annexation of Formosa.[180]

Not only was there no desire in the Wilhelmstrasse to negotiate with Japan for the acquisition of Formosa; it was even feared, at first, that the seizure of Kiaochow might lead to war with the Empire of the Rising Sun, a war which Germany could hardly win. Soon after taking office Bülow had to warn the Kaiser of the hopelessness of such a war and to suggest that it might be advisable to take over a less controversial port than Kiaochow. 'The probable consequences of a German–Japanese war are . . . so unfavourable that German policy must set itself the task of preventing, as far as possible, the disruption of German–Japanese relations, which for our enemies would be an unexpected stroke of luck', he cautioned. Germany could avoid this danger, however, if it made it plain to the Japanese 'that we are no longer opposed *in principle* to their establishing themselves *anywhere* on the East Asian *mainland*, as we were in 1895, but rather that we are inclined to recognise the legitimacy of "live and let live" for Japan too'.[181] The Kaiser went even further in a note he wrote on Bülow's letter: 'We can also give moral support to Japan's protest over [the American appropriation of] Hawaii and thereby draw it away from America.' But he rejected Bülow's suggestion that another East Asian port should be taken over as a base, pointing out that 'Amoy and Samsah . . . , like Korea, are closer to Tokyo or Formosa than Kiaochow.' Nor did he share the Foreign Secretary's anxiety about war with Japan, for the reason that – as he wrote – 'Our fleet is most definitely superior to the Japanese.'[182] On 12 December 1897, in consultation with the Naval High Command, he set the course 'for a war which is to be expected shortly with Japan'.[183]

Although this particular war was in fact avoided until 1914, armed conflict did break out in the Far East only a few weeks later, from which Germany expected to profit. On the eve of the Spanish–American war in April 1898 Prince Heinrich reported to his mother: 'Everything looks most warlike out here, & the American Commodore is constantly waiting for the declaration of war between Spain & the U.S. Japan is supposed to prepare for war to side with England against Russia, so they say here. God knows what the end will be.'[184] Germany's relations with the United States, in addition to those with Japan, were beginning to play a more important role in the thinking of Prince Heinrich and his fellow officers. He was gratified to hear of the first signs of a reawakening of patriotic feeling among the millions of ethnic Germans who had emigrated to America. The new German *Weltpolitik*, he commented to his mother in June 1898, had roused the Germans in the United States to their senses at last, 'hereby proving that the population of the U.S. does not *only* consist of english people, as has been lately professed by the two "english speaking races."' He had nothing but contempt, of course, for Germany's transatlantic rivals on the world stage. 'Have the Yanks . . . in any way shown their superiority in the way of colonisation to the Spaniards', he asked after the American capture of the Philippines. The 40,000-strong US army in Manila was nothing but 'a set of illdisciplined cowboys & bushrangers & a perfect set of ruffians . . . Their officers and generals have been described to me as unfit & untrained for their work!'[185] When Germany acquired the Caroline, Palau and Mariana Islands in the Pacific after the defeat of Spain in the war against the United States, Heinrich expressed his satisfaction 'that we have taken over a new colony in this part of the world', even if the direct economic advantages should prove minimal. 'We are badly in need of a coaling station & as such, they will be of great value', he declared. 'Furthermore this new possession, I hope, will help to increase our navy in general, as well as our squadron out here.'[186]

In his geopolitical aspirations, the brother whom Wilhelm II considered to be an anglophile was torn between his admiration for the British Empire and his goal of helping Germany to attain the status of a world power. 'I wish I could give many of our narrowminded countrymen at home an idea of Great Britain's position on this globe of ours! For it is a marvel to see how that British nation has settled all over the world', he wrote to his mother in February 1898. At the same time, he admitted, 'involuntarily one wishes the same for ones own country, the position of which alas is so very much different & so very much more difficult'.[187] But his admiration for the British Empire could rapidly swing over into rivalry and hostility, as his threatening comment only a few weeks later shows. 'Until now I have tried to make friends with the English, but should I find, that they intend harming us in any way out here, I shall stop that policy of mine . . . That England is to a degree jealous of "beastly Germany" having taken possession of the very best harbour in China is known to all of us! . . . I am trying to serve my country out here & I mean to do it thoroughly! I shall "consider every man my enemy who speaks ill of my sovereign", to use Nelson's words!'[188] Later that year Anglo-American co-operation in Samoa gave him further cause for complaint. 'The way certain other nations try to cut Germany out in every way does not seem over friendly . . . I am fully aware of the one fact though, that politics have got nothing to do with sympathies or antipathies, but that every nation has to seek its own interest where or when it is to be found!'[189]

In anticipation of a great struggle between Britain and Russia in the Far East Prince Heinrich instinctively sided with the former, but only as long as he saw advantages for his own country in so doing. He welcomed the news that Russia had seized Port Arthur and that Britain had taken Wei-hai-wei in retaliation as a symptom of increasing antagonism between the two countries. 'I am right glad that England has undertaken this task, thereby also ridding us of the presence of the Japanese in China, who have no business to be among us Europeans here', he wrote in April 1898. 'I surmise that Russia will take the whole of Korea & comfortably settle down there, thereby increasing her empire still more in size. Anyhow I hope our people at home will have the good sense to see things in this light & let both nations fight it out, if necessary! It would be such fun! The British fleet out here is strong enough to hold its own against the Russian fleet & they would be quite a match for each other.'[190] A few months later his hopes for a war in which Britain and Germany would combine against Russia seemed about to be realised. Writing to his mother in October 1898 from Kiaochow, he commented: 'I have seen them of late, those beloved Russians, & know pretty well, what they are aiming at! Peking on one side, Korea on the other, that's their "little plan". They are the greatest danger to us & Great Britain in the Far East & longing for a brush with the latter! So they openly told us! . . . I have a great mistrust in the Russians & am convinced they mean mischief! . . . If Russia moves towards Peking her influence over the Chinese Empire will be very great & Great Britain's and Germany's commerce endangered.'[191] And two weeks later he remarked: 'The only nation who is able to stop Russia by force is England! I wish to God, I was allowed to fight on the British side against those Russian barbarians!'[192] It was precisely the British Empire, however, that was to prove the greatest impediment to the realisation of Wilhelmine Germany's dreams of world power.

CHAPTER THIRTY-ONE

The Kaiser and England

THE 'WORLD-SAVING IDEA' OF AN ANGLO-GERMAN ALLIANCE

In spite of the dozens of intimate letters to his mother, grandmother and other English relations, Wilhelm's real feelings about Britain, and the inner motivation for his policy towards the country, sometimes seem well-nigh impossible to work out. His comments contradict each other so often and so completely that one is tempted – as has already been remarked – to regard any positive comment or gesture as mere camouflage to disguise his real intentions.[1] But is such an explanation, even if it contained a core of truth, sufficient to clarify the confusion, or were the Kaiser's true feelings towards his mother's homeland characterised by a muddled, profoundly ambivalent love–hatred, as is often claimed? Did his bombastic self-glorification really conceal a deeply felt longing to earn the respect of the British, and in particular of his revered grandmother? Was his aggressive behaviour only the result of his hurt feelings because the much-admired world empire allegedly treated him as a *quantité négligeable*? Could not the British royal family and the government in London have shown more consideration for Wilhelm II's peculiar sensitivity, in order to ease Anglo-German tensions? Was there, as Bernhard von Bülow claimed, a 'fundamentally fanatical' conspiracy against the Kaiser, characterised by 'very unfair misjudgement and latent hatred', throughout the courts of Europe, embracing almost the entire 'English-Battenberg-Hessian-Danish . . . *cousinage*', which sought to undermine Wilhelm's efforts to improve relations?[2] We shall continue to be confronted with these problems, both in connection with the battlefleet building programme and the complications which dogged Anglo-German relations right up until the outbreak of war in 1914. For Wilhelm's biographer it is small comfort that even those who were closest to the Kaiser were often unable to solve the riddle.

It is undoubtedly true that certain comments and actions on the British side hurt the Kaiser unnecessarily and rarely failed to evoke a defiant response from him. Queen Victoria, the Prince of Wales and, increasingly, Lord Salisbury too, who had previously been prepared to make concessions to Wilhelm's apparently insatiable conceit and love of pomp by conferring titles and other honours upon him, gave up such attempts at placating him after the Krüger telegram, and instead expressed their displeasure through insulting rebuffs. The Kaiser showed little understanding of his grandmother's decision not to invite him to Cowes in the summer of 1896 because of the anger aroused by the notorious telegram. His sense of injury was increased when he heard that the elderly King of Saxony had been invited to England, although he was not related to the royal family. He expressed his feelings in a note in the margin of the report from Dresden containing the news: 'I am not to go to Cowes but the King is even going to London etc.! It seems he is the First throughout the Reich now!'[3]

But what hurt Wilhelm particularly was the negative reply he received from Queen Victoria to his request in January 1897 to be allowed to bring some of his children to her Diamond Jubilee celebrations in June of that year.[4] He should not come over for her Jubilee at all, the old Queen wrote, but wait until the usual time for his visit in the late summer, and come not to London but to Osborne House on the Isle of Wight. And that was not all: since Wilhelm had not nominated his brother to be his official representative at the celebrations in London, she would invite Heinrich and Irène to the Jubilee 'as my grandchildren', for on this important occasion she wanted to have as many of her grandchildren around her as possible. Wilhelm noted bitterly on this letter: 'and I am the eldest grandchild'.[5] His distress at his grandmother's unjust behaviour made itself felt in many of his comments, and not least in his maltreatment of his brother.[6] Deeply disappointed, he wrote to the Queen on 10 June 1897: 'As . . . I shall not have the pleasure of offering my congratulations for Your Jubilee as I did 10 Years ago, I must with a heavy heart resort to pen & ink. To be the first & eldest of Your grandchildren & yet to be precluded from taking part in this unique fête, while cousins & far relations will have the privilege of surrounding You & cheering You during the coming happy days, whilst I may not be with them, nay the first of all is deeply mortifying; & I feel like a charger chained in the stables who hears the bugle sounding, & stamps & champs his bit, because he cannot follow his Regiment! I had hoped to lead the Royals as their Colonel past their Sovereign, if not as her Escort & to join their cheers when they salute their Queen in the exuberance of their loyal pride, & that only a few days after "our Waterloo" day, & in the great final charge I would have borne my sword proudly before the saluting point at the head of that magnificent Regiment side by side with Uncle George[,] Uncle Arthur, Uncle Bertie & so many others,

3 generations in arms! But it was all idle dreams! But such dreams are hard to give up for a passionate soldier!'[7]

No less significant for the relations between the two countries was the growing irritation of the Prime Minister, Lord Salisbury, with the Kaiser, whom he not only considered vain, impulsive and unreliable, but now also saw as a malevolent warmonger.[8] The Krüger telegram and the conflicts with the German Reich over Zanzibar in the late summer of 1896, then over the Cretan crisis and the Pacific, did little to improve Anglo-German relations, especially as the press on both sides did its best to inflame public opinion. Shortly after the seizure of Kiaochow Hatzfeldt reported from London that Germany could not at present count on much goodwill on the part of the British cabinet, and added: 'Certain remarks by Lord Salisbury . . . give me reason to believe that our opponents here have lately been trying to turn him against us again. That is the only explanation I can find for the fact that yesterday he reverted to what are supposedly old grievances, namely the Krüger telegram and our action against the pretendant in Zanzibar, with a certain bitterness, although he admitted that the press here has done us a great deal of wrong.'[9]

It is worth noting that Wilhelm made visible efforts, at least from time to time, to influence both German and British public opinion in a more favourable direction. During the conflict over the controversial succession in Zanzibar, for instance, there was malicious anti-British comment in the German press which led the Empress Frederick to remark despairingly: 'The German newspapers are not pleasant reading – this is a fresh opportunity for the most bitter unjust & violent attacks on England. It makes me so savage that sometimes I cannot finish reading the articles – they are so unfair & sneering & nasty.'[10] On this occasion the Kaiser not only shared his mother's anger but even took demonstrative action against the anglophobic press campaign. Clad in the red uniform of the Royal Dragoons, he received the British ambassador, his family and all the members of the embassy at the Schloss for a theatrical performance and a dinner and declared that the invitation was intended to be seen as a 'pro-English demonstration', in view of the anti-British attitude of the Foreign Office under Marschall and of the regrettable mood in the country, which was still being fuelled by Bismarck in the press. At this time the strange rumour was also doing the rounds that the Kaiser intended to give up all the colonies except German East Africa, because the cost of administering them was more than the value of all the trade with them, and because no more than about 3,000 Europeans in all, including the colonial officials and the security forces, had settled in the German colonies.[11] On 22 November 1896 Sir Frank Lascelles secretly informed Lord Salisbury that the Kaiser had proposed that, as Germany could not develop all the colonies it possessed, it might be wiser for it 'to concentrate her energies on a single one

50. Kaiser Wilhelm II in England after the death of his grandmother Queen Victoria in January 1901

and give up the rest'; he hoped that Britain might then compensate it with a coaling station.[12]

The message was unmistakable; and yet only a few weeks earlier Wilhelm had expressed quite different sentiments with regard to Britain. In a letter of 24 October 1896 to her mother, which she asked the Queen to burn immediately, the Empress Frederick gave a moving but fatalistic *tour d'horizon* of Anglo-German relations, which she had recently discussed with Wilhelm during a visit he had paid her from Wiesbaden. As daughter of the Prince Consort and widow of Kaiser Friedrich III she felt only love for Germany, but she was '*miserable*' over the latest development in Germany, and as she was now nothing but 'my poor

William's mother', she could only watch helplessly while he acted in a way which was 'so often deluded & mistaken!!' 'I can do *nothing, nothing*', she lamented. As ever, in her eyes the root of the evil lay in the warping of the German spirit caused by Bismarck's foundation of the Reich and his long years of authoritarian rule. 'The vanity of the Chauvinists etc. has been *so flattered & cultivated* by P^ce^ Bismarck that it has now overrun the country like a noxious weed.' As to the foreign policy of the Newest Course, she found it extremely distressing to see Germany 'pursue a policy of systematic unfriendliness to England, but so it *is* . . . I perceive a systematic *enmity* in Germany to the "Welt Stellung" [world status] of the British Empire, and an endeavour to take advantage of this commercial rivalry by making all sorts of *political waves* in order to hinder and checkmate England (vide Zanzibar) . . . Instead of the bands of *closest* friendship such as *our – now* fast disappearing – generation had *fondly dreamt of* & worked & lived for with untiring devotion and enthusiasm I see a system of *enmity* with *an aim* & a *purpose*.' The widowed Empress did not dispute that her son was playing the leading role in this system, but she persuaded herself that his anti-British policy was a kind of misunderstanding. 'William admires England very much and is very fond of you – & especially enjoys himself *in* England, – but he is not steady & coolheaded and farsighted enough to see that to strain every nerve for Germany to succeed in outdoing England, & *wrest* fr[om] her the position of supremacy she has in the world, is simply *nonsense*! It is a ridiculous, fantastic and wild idea; – but appeals to his imagination & love of the *marvellous*, the *sensational* & exaggerated!' She went on to warn the Queen of the unlimited fleet building plans which her son cherished and which were aimed against Britain. 'I hear that "Krupp" – the *greatest* manufacturer we have, who has a colossal fortune – has been ordered to buy some Docks near Wilhelmshafen, "Germania Docks", in order (it is supposed) to be able to have english shipbuilders & employ engineers etc. to design & build ships for the German Navy that shall beat the English.' She herself naturally wished Germany to be trusted and respected in the world, and hoped 'to see her people *free*, happy & prosperous, her liberty & culture progressing in every way, – but *not* embarking *now*, when there is still *so* much to be done at home on adventurous enterprizes, quarrel with her best friends & make herself universally disliked'.[13]

A year later the former British military attaché in Berlin, Colonel Sir Leopold Swaine, who had once been very close to the Kaiser, found himself in Berlin again and was struck by the anti-British feeling that had now become widespread in Germany, a development which he blamed in the first instance on Wilhelm. 'The feeling between the two countries has not been of the best for some time', he commented. 'Colonial antagonism as well as commercial rivalries have been partly to blame for this; but the principal culprit is the Emperor who cannot recover from his annoyance that the "famous Kruger" telegram had exactly the opposite result to what he anticipated.' What was more, in recent times the Kaiser had gone further than German public opinion and the Foreign Office wanted in his policy, which was directed principally against Britain, the Colonel observed. In view of German sensitivities and of the general reverence for Bismarck, who was probably dying, it would be advisable, he believed, for Queen Victoria to appoint someone soon to represent her at the funeral of the founder of the Reich. 'The veneration for Bismarck in Germany is as great as ever it was and has, if anything, been enhanced by the mess things have got into since he left office for which the Emperor, whether rightly or wrongly, is solely blamed. Any show, in spite of the angry abuse showered upon us by the German press, which we can make of honouring their great Statesman is certain of having a salutary effect on all the right-minded people in Germany and on those whose opinions we value. We are an old nation and can afford to be magnanimous vis-à-vis to an upstart one.'[14]

It is not without irony that at precisely the time when anglophobic sentiment was flaring up in Germany, influential circles in Britain were putting out feelers towards an alliance with Germany. The negotiations for an Anglo-German alliance which were conducted behind the scenes in the spring of 1898 and again in 1901 are the subject of a still unresolved controversy between historians, in which one side interprets the talks as the great 'missed opportunity' of pre-war diplomacy, while the other sees them as no more than a series of illusions and misunderstandings.[15] How did Kaiser Wilhelm behave in this potentially crucial moment in world history? How, and through what channels, did he learn of the British soundings, and what attitude did he take to the possibility of an alliance with his mother's homeland, before this opportunity proved to be an illusion? The correspondence of the royal family, which is generally neglected by diplomatic historians, can throw new light on this well-researched and yet still controversial question.

In mid-January 1898, while shooting at Buckow, the Kaiser took the British military attaché Grierson aside and discussed the threats of Anglo-French conflict on the upper Nile with him, as well as the German occupation of Kiaochow. He then switched to the subject of Anglo-German relations and said with obvious annoyance, as Grierson reported to London, 'that for eight years he had tried to be friends and allies with us but had failed and now had to go on alone and forward German interests. We would never again have such a chance of an ally on the continent, for never again would a grandson of the Queen of Great Britain be on the German throne, and more to the same effect. Then he said to me "Is there a British policy at all? What is your idea of the policy?" So I said that I was no politician, but that my idea was that our government would join neither the Triple nor the Dual alliance, as to join one would embroil us with the other and we did not want to embroil ourselves with anybody. We were strong enough to hold our own against either group, and it seemed unlikely that they would combine. Then he said: "There you are mistaken. They can combine and they *shall* combine. Socialism will compel the monarchs of Europe to combine for mutual assistance, and the great danger to be guarded against is the yellow race of the Far East. My allegorical picture ["Nations of Europe"] will yet be realized." Then he asked if I had seen his picture in "Punch" as Emperor of China, and said that the Empress was furious about it, but he thought it rather a good joke. "But", he said, "your people do not realize how monarchs are looked on on the continent, and so long as those journal attacks are made on me you cannot expect the German press to remain silent."'[16]

Before Wilhelm expressed this complaint, Queen Victoria had attempted on her own initiative to persuade the London press to take a more positive line. Early in 1898 she asked Sir Theodore Martin to contact the leading editors and urge them to be less hostile and more objective in their coverage of Germany and the Kaiser. On 13 January Martin told the Queen that all the editors to whom he had spoken, and above all the editor of *The Times*, had deplored the 'very bitter feeling against the Germans' which prevailed all over Britain and which could be attributed only in a small degree to trade rivalry. Even the humorous journal *Punch* would moderate its mockery and its caricatures, Martin assured the Queen.[17] The restraint of the British press soon had a noticeable effect on the German newspapers, and in Berlin court society the English Princess Aribert of Anhalt, who was credited with having great influence on Wilhelm II, contributed considerably to the improvement in Anglo-German relations. In addition, the unusually warm reception which Prince Heinrich was given in Gibraltar, Aden, India and Hong Kong during his voyage to the Far East was appreciated at the Berlin court as a welcome gesture of friendship.[18] Finally, the old Duke of Cambridge, Queen Victoria's cousin, expressed support for an Anglo-German rapprochement. At Cannes, where he had met the German ambassador in Paris, Count Münster, the Duke had pointed out that, although the Prince of Wales had for a time succeeded in prejudicing the Queen against Germany, which had prevented relations between the countries improving, 'that had not lasted, and at heart the old Queen loves and in some respects admires her grandson very much'. It was well known that the heir to the British throne did not harbour either anger or love for long, Münster told the Reich Chancellor in a secret letter, and therefore it was not impossible that even Albert Edward would now seek a better relationship with Germany.[19] Thus on a monarchical level friendly feelers had been put out on both sides of the North Sea before the statesmen and diplomats began their tentative political talks in March 1898.

In the imperial entourage, however, forces were at work to try to sabotage the improvement in Anglo-German relations which was gathering impetus. Grierson complained that there were those about the Kaiser who 'who stick at nothing to damage England and everything English in the Emperor's eyes'.[20] These irresponsible mischief-makers did great damage by spreading false and malicious rumours which the Kaiser only too willingly believed.[21] One of the chief culprits in this respect was the Chief of the Naval Cabinet, Gustav Freiherr von Senden-Bibran,[22] who told the Kaiser after a visit to England that, although he had been very kindly received by the Queen, the Prince of Wales had been deliberately discourteous to him. He had received him in the equerries' room, had been unfriendly and had not once enquired after the Kaiser and Kaiserin. As a result of this assertion by Senden, the Kaiser called on the British ambassador in February 1898 'in a most excited state', in order to express his fury at the poor treatment of the Admiral. 'His Majesty told Sir Frank Lascelles that this continued hostility evinced towards him by the Prince of Wales would prevent his coming to Cowes & would even possibly have serious results upon the relations between the two countries.'[23] The ambassador immediately reported the incident to Salisbury, who relayed the telegram to the Prince of Wales. The latter wrote to Lascelles rejecting Senden's assertions and instructing the luckless ambassador to tell the Kaiser that he, the heir to the British throne, refused to have anything further to do with Senden.[24] As had happened over the Vienna incident almost ten years earlier,[25] Prince Christian of Schleswig-Holstein, who was about to visit his daughter Princess Aribert in Berlin, was asked by the Queen to have a serious word with Wilhelm about his relationship with the Prince of Wales.[26]

When Lascelles carried out the Prince of Wales's instructions on 16 March 1898, the Kaiser expressed his regret, but said that he 'that he had great hopes that an improvement in the relations of the two Courts would soon take place, as important Personages were attempting to bring about an arrangement'. He named his mother as the driving force behind the initiative. Two days later, when the members of the British embassy were received by the Empress Frederick, Lascelles raised the subject of the Kaiser's remarks with her, whereupon the Empress expressed surprise that her name had been mentioned in this connection. She indicated, however, that she had had the idea 'that The Emperor should run over to Scotland during his annual cruise in the North Sea, and pay a visit to The Queen at Balmoral'.[27]

The Empress Frederick and other advocates of a rapprochement between Germany and Great Britain regretted that Wilhelm would not be going to Cowes in the summer of 1898 as usual, because of the Senden incident. But they had great hopes for the Kaiser's visit to Egypt, Malta and Gibraltar, which he had planned to make immediately following his journey to Constantinople,

Palestine and Damascus in the autumn of 1898.[28] Staff at the British embassy were astonished to discover that the Kaiser had not only heard that the improved tone of the London newspapers had been brought about by the personal orders of Queen Victoria, but even knew the name of her intermediary. 'He must have some very well informed correspondents at home', remarked Grierson, who was constantly surprised by the Kaiser's marked friendliness towards him.[29] There were reports of admiring comments made by Wilhelm II in May on the deeds of the British army on the upper Nile, and of his wish for a closer connection between the two countries. He spoke to Grierson of the successes of the 'Yankees' in the war against Spain, and commented 'that Europe will have to combine against them'. But in order for that to happen, England and Germany would have to co-operate, for 'we can do nothing against America unless England is with us'. As the military attaché noted with surprise, 'he never seems to be free from this dream of an European coalition'.[30]

It was only now that the politicians stepped in to the Anglo-German talks which had been going on, if only intermittently, at the monarchical level since the beginning of the year. The German seizure of Kiaochow and the Russian occupation of Port Arthur had roused fears in London that a general partition of China was in the offing, and this, together with the French advances into the Niger region and on the upper Nile, led the British government to question the wisdom of maintaining its policy of isolation. On 25 March 1898 an initial exploratory conversation took place between Arthur Balfour, who had taken charge at the Foreign Office during Salisbury's illness, and Hatzfeldt, the German ambassador.[31] Four days later the influential Colonial Secretary, Joseph Chamberlain, met the ambassador at Alfred de Rothschild's house, and to Hatzfeldt's astonishment proposed a full alliance between Great Britain and the Triple Alliance against France and Russia. When Salisbury was informed by his nephew Balfour of Chamberlain's unauthorised diplomatic initiative, the Prime Minister's response was scathing, not only on the disloyal conduct of his Cabinet colleague, but also, and principally, on the subject of Wilhelm II, whom he described as the greatest impediment to improved relations with Germany. 'The one object of the German Emperor since he has been on the throne has been to get us into a war with France', he declared bluntly. He regarded an alliance with Germany 'with some dismay, for Germany will blackmail us heavily'.[32] But in spite of his mistrust of Germany, and in particular of the Kaiser, after his return to Downing Street Salisbury nevertheless indicated in a conversation with the Austro-Hungarian ambassador in London, Count Deym, that the time might be ripe for Britain to enter into alliances to protect its interests. The British government, Hatzfeldt reported to Berlin, was however afraid that Germany would place unacceptable conditions on its friendship, 'particularly as regards colonies'. The Kaiser, as usual, covered Hatzfeldt's dispatch with marginal comments, which clearly reveal his own attitude to the idea of an Anglo-German agreement. Where reference was made to the fact that Salisbury was considering possible alliances, Wilhelm wrote: 'Here he is directly contradicting all the statements he made to us at the end of the eighties and beginning of the nineties, when we were trying to persuade him to join the Triple Alliance, or rather to draw closer to Italy! England must therefore be finding the situation sticky now.' Beside a reference to the likelihood that the German price for alliance with Britain would be too high, he commented: 'But we haven't been properly asked yet, or received a request.' Apart from the concession of British possessions in Borneo, Germany must also, 'if possible', get 'Samoa, [the] Carolines and one of the Philippine Islands'. He did not understand, Wilhelm commented finally, why Salisbury had dropped these hints to Deym, 'and not to Hatzfeldt or to me by Lascelles'.[33]

A few days later Wilhelm received a nineteen-page letter from his mother – now ill with cancer at Kronberg – which she asked him to burn at once, and in which she argued that an alliance between Germany and Britain might well be the last chance for world peace, and begged her son to accept the offer of alliance which she was sure London would make very soon. 'I *know* for a fact that in England part of the Govt. are *seriously inclined* to enter into a *real alliance* with Germany, a thing wh[ich] has never happened before, – and to my idea, and that of 100.000[s] of Germans – the *most blessed* that could happen *not only* for the 2 Countries but for the *world* and civilization!! – The moment *never* may come again – how grand it would be, & how *wise* – if you grasped the outstretched hand wh[ich] if I understand rightly Mr. Chamberlain is seriously & honestly offering!! – For *yourself*, your *own* position, your *own* future, for Germany, I could conceive no more magnificent opportunity. Misunderstandings would be swept away – and *peace* secured! We need not fear the Russians or the French, – nor both of them together and can *afford* to be on the *best* of terms with them. All the *great standing questions* of the day could be solved quietly & amicably & *without* war! – The *security* & prosperity of Europe increased *immensely*. – That Italy & Austria would only be too pleased to join is without doubt, and this combination would be so strong, that there would be no need to feel uncomfortable about Russia & France, who might be left to enjoy their alliance unharassed & undisturbed, & who then would *not* be likely to wish to attack *any* one, for with the German Army & english Fleet combined, *who* would take up the gauntlet?' Although she otherwise stayed out of politics, the Empress Frederick said, she had decided to take this step because neither Hohenlohe nor Bülow knew Great Britain, 'and therefore perhaps are not so keenly alive to the *immense* importance of an alliance between the 2 great Germanic & Protestant nations, wh[ich] for 50 years has been the dream of so many true patriots but prevented by the vicissitudes of political phases of a momentary kind! *All* quarrels all misunderstandings *disappear* & *melt* like snow – wh[ich] seems *so entirely natural* & right. It seems to me as if you can have this ripe fruit, of inestimable value, in the hollow of your hand if you will & can but seize it! Time presses and I am so afraid that if english statesmen see Germany does not care to respond to the idea they must look *else where* – that would be disastrous, *now* is the time & *now* the hour!' Obviously suspecting that Wilhelm would steer a narrow nationalistic course whatever happened, his mother repeatedly stressed the advantages which an alliance with the United Kingdom, which Japan and the United States could also join, would have not only for humanity in general but also for the German Reich. 'The moment has *come*!', she urged. It was a historic, deeply significant, richly promising opportunity which should be seized with both hands. The Anglo-German alliance embodied 'a world-saving idea', which Wilhelm must take hold of, not only for the benefit of mankind, Christianity and civilisation but also in the interests of Germany, for through it German influence in the world would become greater than ever before. The new century would have a more promising start than had ever been seen. '*My* father's dream & *your* father's dream, and what they worked & slaved for, would come true, though *they* are no more! For *you* it would be the most immense satisfaction & for dear Grandmama too! *No one* would exult more than Uncle Bertie – & as for me, I could sing "nunc dimittis"' and die at peace', Wilhelm's mother concluded.[34] No one in Britain knew that she was aware of these plans, which had originated with Chamberlain and which Salisbury might not support at all, the Empress Frederick confided to her son. He must therefore keep the matter absolutely secret.[35]

Wilhelm II's reaction to this 'world-saving' opportunity was both triumphant and fearful. Without consulting the Reich Chancellor or the Foreign Office, on 30 May 1898 he wrote a letter to Tsar Nicholas II in which he spoke of the 'enormous offers' which the British were ready to make to him, and then suggested that the Tsar might like to outbid the British offer! 'Dearest Nicky', he wrote, 'With a suddenness wholely [sic] unexpected to me I am placed before a grave decision which is of vital importance to my country, and which is so far reaching that I cannot foresee the ultimate consequences. The traditions in which I was reared by my beloved Grandfather of blessed memory as regards our two houses and countries, have as you will own, always been kept up by me as a holy bequest from him, and my loyalty to you and your family is, I flatter myself, above any suspicion. I therefore come to you as my friend and "confident" [sic] to lay the affairs before you as one who expects a frank and loyal answer to a frank and loyal question . . . About Easter a Celebrated Politician [Chamberlain] proprio motu suddenly sent for my Ambassador and à brûle pourpoint offered him a *treaty of* Alliance with England! Count Hatzfeldt, utterly astonished, said he could not quite make out how that could be after all that had passed between us since 1895. The answer was that the offer was made in real earnest and was sincerely meant. My Ambassador said he would report, but that he doubted very much whether Parliament would ever ratify such a treaty England till now always having made clear to anybody who wished to hear it, that it never by any means would make an Alliance with any Continental Power whoever it may be! Because it wished to keep its liberty of action . . . The answer was that the prospect had completely changed and that this offer was the consequence. After Easter the request was *urgently* renewed but by my commands cooly [sic] and dilatorily answered in a colourless manner. I thought the affair had ended. Now however the request has been renewed for the third time in such an unmistakable manner, putting a *certain short term* to my definite answer and accompanied by such enormous offers showing a wide and great future opening for my country, that I think it my duty to Germany duly to reflect before I answer. Now before I do it, I frankly and openly come to you, my esteemed friend and cousin, to inform you, as I feel that it is a question so to say of life and death. We two have the same opinions, we want peace, and we have sustained and upheld it till now! What the tendence [sic] of the Alliance is, you will well understand, as I am informed that the Alliance is to be with the Triple Alliance and with the addition of Japan and America, with whom pourparlers have already been opened! What the chances are for us in refusing or accepting you may calculate yourself! Now as my old and trusted friend I beg you to tell me what you can offer me and will do if I refuse. Before I take my final decision and send my answer in this difficult position, I must be able to see clearly, and clear and open without any backthoughts must your proposal be, so that I can judge and weigh in my mind before God, as I should, what is for the good of the Peace of my Fatherland and of the world. You need not fear for your Ally, in any proposal you make should he be placed in a combination wished by you. With this letter dearest Nicky I place my whole faith in your silence and discretion to *everybody* and write as in old times my Grandfather would have written to your Grandfather Nicholas I! May God help you to find the right solution and decision! It is for the next generation! But time is pressing so, please answer soon! Your devoted friend Willy. P. S. Should you like to meet me anywhere to arrange by mouth I am ready every moment at sea, or on land to meet!'[36] The Tsar, supremely indifferent to the Kaiser's dramatic letter, replied that the British had recently made similar approaches to him too.[37]

Although they were well aware that the contents of the Kaiser's letter, which they themselves had seen only after the event and in an inaccurate version, would be reported to London, the experienced and supposedly 'responsible' diplomats in the Wilhelmstrasse merely shrugged their shoulders over Wilhelm II's intervention. On Bülow's instructions the compromising document was quietly filed away in the confidential records of the Foreign Office. Friedrich von Holstein, who eighteen months earlier had seemed ready to go the barricades in protest against personal rule, now scoffed at the 'excited' Kaiser's exaggerated expectations, and at his evident fear of attack by the Royal Navy if Britain's offers were rejected. His dread of incurring British enmity, Holstein believed, had been apparent in a memorandum written by the Kaiser on 31 May, which had read: 'At the beginning of the next century we would have a battlefleet which, with *others* which will also have grown, could represent a real danger to England's fleet. Hence the intention either to force us into an alliance or to destroy us before we have become strong, like Holland in times past.'[38] Hatzfeldt too shook his head over the Kaiser's expectations and anxieties. He saw no sign of the 'enormous offers' of which the Kaiser dreamt. 'The feared naval attack is even more improbable. As far as can be humanly foreseen it could take place only if we ourselves first took up a hostile attitude', the ambassador commented – a remark which, no doubt unwittingly, expressed Wilhelm's guilt complex over his anti-British fleet building plans. From a conversation with Salisbury, Hatzfeldt gained the impression that, although the British government wished to be on good terms with Germany, it had no intention of suggesting a formal agreement.[39]

The reply, bristling with anti-British resentment, which Wilhelm wrote to his mother's letter on 1 June 1898, and of which he subsequently sent a copy to the Wilhelmstrasse, represents what is probably the most revealing assessment of Anglo-German relations, as he saw them, since his accession. In it he acknowledged that he had been fascinated by what his mother had said. 'The idea of an Alliance of the Anglo-Saxon race is not new, the accession of Germany to it however is so, as far at least as the English Government is concerned. Let me make a short sketch of our relations. In the first 6 years of my reign I tried to the very utmost of my powers . . . to elicit from L[ord] S[alisbury] a word implying approval of the idea of a[n] Anglo German cooperation & Convention. But it was utterly without any result, as he invariably allways [sic] ended in the same refrain: "An English Government cannot & never will form an Alliance with *any continental power* for the simple reason, that Parliament would hardly ever ratify such an instrument & because England prefers to keep its *liberty of action*, therefore I am unable to fullfill [sic] your wishes."!! So I let the matter drop & with a heavy heart gave up the task, which was a difficult one, though dear to me seeing that I worked on the same lines dear Papa, & Grandpapa (Consort) had shaped. In numerous phases of Foreign Affairs, notably in the Siamese Imbroglio (under the Liberal Government of L[ord] Rosebery) I staunchly stood by England & volunteered my help L[ord] R[osebery] so warmly begged for with Grandmama's consent at Osborne in 1894 [sic]! But instead of thanks or of help in our colonising enterprises, I got nothing whatever. I for the last 3 years have been abused, ill-treated & a butt to any bad joke any musikhall [sic] singer or fishmonger or pressman thought fit to let fly at me! – Notwithstanding all this two years ago I tried hard to have L[ord] S[alisbury] help to give us a coaling station in China, he flatly refused in a language that only Hatzfeld knew how to interpret so that no serious action came out of it! So pushed back, illtreated & riled by Grt Britain & her Prime Minister instead from her I got from Russia in a few conversations with the Emperor all I wanted & even more than I ever hoped for! This as "eine kleine Orientirung" [to set the record straight a little]. Now with respect to what you wrote about the Alliance of England-Amerika & Germany, this has very much interested me. The idea has been ventilated in the papers since 2 months & also sundry allusions & suggestions from Mr. Ch[amberlain] have been wafted over here by the same breezes. But as they were not officially transmitted as coming from the government or Prime Minister nobody very much heeded them as they seemed to be merely a repetition of the articles in the Press. By your letter I see for the first time that the thing is ment [sic] in earnest, & purports to be an overture, at least so it was ment [sic] by Mr. Ch[amberlain]. If that is the case & if as I gather there is a certain speed wished in the treatment of this question, then why in the name of all that is diplomatic use & sense, does not the *Prime Minister* make a *real* proposal? Why does Cabinet not make real propositions to serve as base for pourparlers? Why does the Cabinet not empower L[ord] S[alisbury] or L[ord] S[alisbury] Mr. Ch[amberlain] under his authority to expound the terms of a treaty to me? Private conversations & even statements before others are all very well, but do not represent the right way to a Treaty of Alliance! Besides if even Mr. Ch[amberlain] & as it seems *part* of the Cabinet are in earnest & begin to treat with me in the above informal manner, who will ever guard me against a sudden desaveu in the House of Lords by the Premier, or in Parliament by Balfour if they found popular feeling not in the lines they expected, as long as L[ord] S[alisbury] is in such a bad way with Mr. Ch[amberlain] & has not implicitly bound himself to this affair by officially authorising his Ministers to enter into negotiations?! England would not feel anything, but a *miscarried* try at an Alliance with her brings Russia & France down on my head & over my frontier on the same day!? – These are some of the difficulties which have cropped up in my mind since I have given your letters serious attention, & are the consequences of the treatment I have gone through at the hands of the British Government & notably L[ord] S[alisbury], & the result of the experience I had in the 10 years of my reign of

British Foreign Politics! Should Government wish to get out of the "splendid isolation", promote the idea of a "rapprochement" to me and the formation of an Alliance, then let the British Premier speak out openly & manly & officially as it is "d'usage" among Great Powers, & I will with pleasure listen & consider! But he can never expect me to "slip in by the back door" like a thief at night whom one does not like to own before ones richer friends. I will be most grateful if I can have any information how the things are going on.'[40]

In her reply of 3 June 1898 the Empress Frederick tried to reassure her son. No one would expect him to go in 'by the back door' in a matter of such importance. She could sympathise with his complaints about many of the things said against him by public opinion in Britain, but in Britain too certain comments by the German press had caused offence and it might in fact take some time for people there to recover their trust in German intentions. 'But these are, I hope, *passing* shadows, and I wish with all my heart that the idea floating in people's minds of an alliance may take *shape* & *form* and be brought forward in an official & decided way.' One could only wait and hope that nothing would happen in Berlin or London to ruin this opportunity.[41] When weeks again went by without further progress, however, the widowed Empress was forced to acknowledge that, while Wilhelm hoped, as she did, for an understanding in the near future, leading statesmen in both capitals were at best prepared to handle the matter only in a dilatory fashion. She was certain, she wrote to her mother, 'that William *is most* anxious for a rapprochement with England and *hopes* with all his heart that England *will come* forward in some sort of way & meet him halfway'. Prince Hohenlohe was not exactly anti-British, but if only on account of his Russian property, which he wished to keep, he was pro-Russian. Bülow was in favour of an agreement between Germany and Britain, but not for many years to come. In Britain, she believed, Chamberlain supported the idea of an understanding with Germany, while Salisbury was against it. If the Prime Minister would only come forward with a concrete proposal, she was sure Wilhelm would be ready to agree to it; Lascelles knew this too. 'Wm. thinks the moment propitious – & would be anxious for the idea to take shape & form! . . . I am sure W. wd. make every endeavour to "enfiler" a "rapprochement" with England, – but fears it wd. not be reciprocated by Ld. S[alisbury].'[42]

When Queen Victoria showed this letter to Salisbury, he strongly disputed the Empress Frederick's interpretation of events. 'It appears that she, and the German Emperor are of the opinion that in the communications which are going on about colonial territory between England and Germany, Mr. Chamberlain is anxious to meet the wishes of Germany while Lord Salisbury is opposing them. Lord Salisbury never wishes to discuss the actions of a colleague. But he feels bound to inform Your Majesty that this view of the comparative action of Lord Salisbury and Mr. Chamberlain is not correct – it is the very reverse of the truth.'[43] This distinctly unclear pronouncement by Salisbury touched on the Anglo-German negotiations over the future of the Portuguese colonies but left the central question of a full alliance between Germany and Great Britain, which was what mattered to the Empress Frederick, unanswered.

Wilhelm's hopes for 'enormous offers' from Britain, encouraged principally by his mother's well-meaning letters, were reflected in a memorandum written in the Wilhelmstrasse and approved by him on 15 June 1898. In it Germany's desiderata in its negotiations with Britain on the Portuguese colonies were listed.[44] These were, in West Africa, a naval base in the Canaries or the Cape Verde Islands; the island of Fernando Po in the Gulf of Guinea; the re-drawing of the frontier between Togo and the British colony of the Gold Coast; Angola, or at least the southern regions of Mossamedes and Benguela; and the British enclave of Walvis Bay on the Namibian coast. In East Africa Germany hoped to acquire Zanzibar and Pemba, together with the extension of the southern limits of German East Africa down to the Zambezi, which would have meant the assimilation of the Malawi of today, with its capital of Blantyre. In Asia the Kaiser and the Wilhelmstrasse expected to gain Portuguese Timor, the Sulu archipelago (with Jolo) in the South Philippines and at least one of the larger islands in the Philippines, preferably Mindanao. And, finally, they counted on acquiring exclusive possession of the Caroline Islands and the Samoan Islands in the South Pacific.[45] These far-reaching expectations – which in fact represented no more than a rough outline of the geopolitical goals of the new German leadership and did not begin to take into account Germany's possible ambitions in other parts of the world such as the Near East, the Caribbean and South America – were likewise thwarted by Salisbury's determined opposition. At the end of July 1898 he informed the Queen that in view of public opinion in Britain and the colonies he had rejected Hatzfeldt's demands for Blantyre and Walvis Bay, Samoa and Tonga in compensation for the British seizure of Delagoa Bay. First, he stated, Britain had already secured the right to Delagoa Bay in the event of Portugal giving it up, and second it could at most give up Walvis Bay in exchange for Togo. 'C[oun]t Hatzfeldt expressed great regret at Lord Salisbury's decision & hinted that if Germany could find no resource in friendship with England she must turn to Russia.'[46]

Wilhelm II's disappointment and fury with Salisbury knew no bounds. He scrawled marginal notes on Hatzfeldt's report calling the British Prime Minister a 'shameless scoundrel!' and describing his behaviour as 'positively Jesuitical, monstrous and insolent!' 'If he has to take the feeling in his colonies into consideration', the Kaiser raged, 'I have to take the feeling of the German people into consideration, and that is what counts for *me*. It shows once again that the noble

Lord is trifling with us and shifting around simply because he is not afraid of us, because *we have no fleet*, which those stupid donkeys in the Reichstag have constantly denied me throughout the 10 years of my reign. Well, I stay here on my bond and shall make *no more concessions*.'[47] During the Scandinavian cruise that summer Eulenburg recorded that 'His Majesty is furious with the English and to my regret particularly enjoys expressing himself very frankly in front of foolish naval officers.'[48] Wilhelm remarked bitterly that he expected that 'England (allied to America?) will take our colonies away one day, because she is bound to come to the conclusion that we cannot take on a war.'[49] He also gave vent to his feelings in an angry telegram to his mother which she forwarded to Windsor. The Empress Frederick was at first convinced that there must have been a misunderstanding and wrote to her son: 'I think I may say without indiscretion – *trusting* that you will keep it *to yourself*, – that in England there *has* been an impression that you had been particularly kind and friendly in your expressions to different people you had met, naval & others, while at the same time the ambassador maintained gt. stiffness & coldness. I dare say this puzzled very much . . . and most likely made people not quite know wh. way the wind blew.' She acknowledged that Great Britain had to remain on good terms with Russia, 'but on the other hand I cannot bring myself to believe that Russia's intentions *can* be very friendly to England though the Emperor (*Nicky*) is so devoted to Grandmama & likes England so much'.[50] She must have reported to her mother on Wilhelm's disappointment in similar terms, for in returning the Kaiser's telegram and the Empress Frederick's letters to the Queen Salisbury remarked incredulously: 'The Emperor William is wonderful.'[51] Hatzfeldt's conduct at the negotiations 'did not at all correspond with the Emperor's strange telegram', he commented; on the contrary, he had remained friendly regardless of the difference in the positions taken by the two countries.[52] But Salisbury was anxious that the Empress Frederick should not pass on his own comments to the Kaiser. The latter would certainly tell others about them and thereby create the impression that Salisbury was intriguing with the Kaiser against Chamberlain. 'The German Emperor takes offence very easily', he commented in exasperation, adding that he did not understand 'what it is he refers to when he speaks of his overtures having been received with "something between a joke and a snub" . . . The truth is that on questions of territorial cession the German Emperor and public opinion here take very opposite views. It would be impossible to do what the German Emperor desires without incurring the reproach of deserting British interests and making undue concessions . . . When the public is excited on the subject as they are now, it is too strong to be resisted.'[53] When Salisbury again fell ill, it was left to Balfour to bring the negotiations on the Portuguese colonies to a tolerably satisfactory conclusion on the basis that in return for British recognition of the German claim to Timor, Germany would abandon all claims to the Boer republics in South Africa and to the island of Zanzibar.[54]

In an attempt to re-launch the stalled talks on the idea of an alliance, the Empress Frederick invited the British ambassador to Schloss Friedrichshof at Kronberg while the Kaiser was visiting her there. Again on this occasion Wilhelm made no secret of his disappointment and anger 'at the manner in which He considered that He had been treated by Your Majesty's Government', as Lascelles reported to the Queen after his meeting with the Kaiser on 21 August 1898. 'He complained at the scant consideration shown in England for German interests, and the curt refusal which His demands usually met with.' The ambassador seized the opportunity to explain to the Kaiser that his demands for colonial expansion in Africa were generally considered 'exorbitant' in Britain. But he also indicated that there was a genuine desire in Britain for better relations with Germany, and that some influential circles were even prepared to envisage a formal, if strictly defensive, alliance under which treaty obligations would come into force only if one of the two countries were attacked by two powers simultaneously. In the following days and months the ambassador was more than once surprised by the exaggerated importance which the Kaiser attributed to this somewhat hypothetical statement which had not even been formally committed to paper. The day after the meeting, the Empress Frederick showed Lascelles a telegram she had just received from Wilhelm, in which he expressed gratitude for the ambassador's 'energetic intercession'. It had decidely altered the position, he said, and ensured that there would be a favourable conclusion for both sides. This was followed next day by a telegram direct to Lascelles from the Kaiser, in which Wilhelm described their discussion at Kronberg as highly satisfactory and stated that he had telegraphed appropriate instructions to Berlin and London. The attempts Lascelles made to correct the false impression the Kaiser had evidently gained from the conversation at Kronberg had no effect, as events were to show.[55]

Meanwhile the Kaiser's advisers in the Wilhelmstrasse, confident in the expectation that Britain would in time offer far more favourable conditions for an alliance with Germany, set out to persuade their master that Germany should maintain its neutrality between Britain and Russia, so that at the right moment – and with its new battlefleet – it could intervene for its own benefit in the war that would inevitably break out between these two rival world empires. German policy, as Bülow explained to the Kaiser, must maintain 'independence from both sides'; war between Russia and Britain would come 'all the sooner, the less both sides believe that we wished for any such thing!'[56] On 19 August 1898 Bülow described the aim of his foreign policy to Wilhelm as a 'firm and independent position between England and Russia, independent of both sides, but with the possibility of co-operating with one or the other, as soon as it suits Y.M.'.[57] By the end of the year Holstein was able to assure his confidant Hatzfeldt in London that they had steered the Kaiser back to the official line: the 'free hand' policy. 'H.M. realises the advantages there are for us in playing the part of spectators for as long as possible.'[58] Postponing the day of reckoning did not imply that Germany's designs had been abandoned, however, and by August 1900 Bülow was assuring his Kaiser of the victorious role he was destined to play. 'How true it is that in Your Majesty's reign the British are playing the same role as the French did under the Great Elector, and the Austrians under Frederick the Great. Dealing with the British is immensely laborious, immensely difficult and requires immense patience and skill. But just as the Hohenzollern eagle drove the two-headed Austrian eagle from the battlefield and clipped the wings of the Gallic cockerel, so will it vanquish the English leopard with God's help and Your Majesty's strength and wisdom.'[59]

THE IMPERIAL WARMONGER

Patience, wisdom and skill were not exactly the Kaiser's strong points. After Salisbury had put paid to his colonial aspirations, Wilhelm turned back to the continent, and in particular to Russia. On 18 August 1898 he wrote to Tsar Nicholas castigating the perfidious English, who had conducted negotiations with him but had 'never quite uncovered [their] hand; they are trying hard, as far as I can make out, to find a continental army to fight for their interests! But I fancy they won't easily find one, at least not *mine*! Their newest move is the wish to gain France over from you, and they in consequence have suddenly decided to send the Duke of Connaught to the French Army Manoeuvres.'[60] During his tour of the Orient the Kaiser at first tried to incite the Tsar against Britain, and then set his sights on the goal he had accused the British of pursuing, by attempting to drive a wedge between Russia and its ally France. On 28 September 1898 the French had given way to the British at Fashoda, as France was unwilling to risk taking on Britain's overwhelming naval power.[61] Wilhelm wrote maliciously to his 'dearest Nicky' from Damascus: 'The hatred of the English is strong and growing more and more intense – no wonder – whilst in the same time apace with it grows the open contempt of France, which has lost all the respect it once possessed of old! That is the unavoidable consequence of the terrible quagmire the French are now floundering about in their interior affairs, splashing the dirt right and left till the whole of Europe reeks with the stench! . . . Here [in the Near East] people look upon them as on a dying nation, especially since the last and most ignominious retreat of the French from Faschoda! What on earth has possessed them?! After such a first rate well arranged and plucky expedition of poor and brave [Major] Marchand? They were in a first rate position and able to help us others all in Africa who are so[r]ely in need of strong help! The news here have [sic] come as a thunderbolt on the Eastern people, nobody would believe them! At all events if it is true, what the papers say, that Count Mouravieff [the Russian Foreign Minister] councelled [sic] France to take this foolish step he was singularly and exceptionally ill advised, as it has given your *"friends and allies"* a mortal blow here and brought down their ancient prestige here never to rise again! The Moslems call it France's second Sedan.'[62]

At the same time Wilhelm pursued his efforts to incite Britain to go to war with France. On his way home from the Orient he wrote to his mother, who was visiting her native land (for the last time), from Messina on 20 November 1898, saying that the British fleet, which he had admired at Malta, was 'ready for imminent use should Britain want its ships for a quick blow. The speeches of the British ministers seem to show pretty clearly that Great Britain means buisiness [sic] this time & that the moment has come for her to settle the accounts with France on the whole globe. This *may* lead to war. But the moment – militarily spoken – is well chosen as nobody will dream of helping France – who has understood to make itself disagreeable to every body in the world –, by its perpetual intrigues & the self asserting vanity with which it upholds its pretentions. Should it come to war, I of course in private as Grandmama's grandson will pray for the success of her arms with all my heart – which is to me of no doubt whatever – as France is no equal to England on the Sea. Officially as head of the German Empire I would uphold a strict & benevolent neutrality. Should a second Power think fit to attack England from the rear, whilst it is fighting, I would act according to our arrangements made with Sir Frank Lascelles.'[63] When Queen Victoria had this letter sent to him, Lord Salisbury was understandably surprised to read of agreements of whose existence he knew nothing but which had apparently been made between the German sovereign and the British ambassador in Berlin.[64]

Lascelles again found himself compelled to make it plain to the Kaiser that the spontaneous remarks he had made during their meeting at Kronberg, from which Wilhelm had evidently drawn false conclusions, had been purely informative and in no way binding.[65] On 19 December he dined with the Kaiser at Potsdam and brought up the subject of the letter that Wilhelm had written to his mother from Messina. Once again Wilhelm persuaded himself that he had come to 'most satisfactory agreement' with Lascelles, writing to the Empress Frederick afterwards: 'I was able to tell him, that I had a conversation with the Russian Ambassador [Count Osten-Sacken], who seemed much alarmed at the situation in Europe & elsewhere, & that in the end I was able to make him understand, that in the case England was in the necessity of "settling accounts" with France, no general conflagration in Europe would be the result for the simple reason that the fight would be *purely on the water*. But we two beeing [sic] eminently land powers with fleets not to speak of, the very best thing would be to sit as quiet as possible & to look on; the French army not beeing [sic] engaged it would do no great harm to France if she lost a few ships & colonies & it would do none to Russia or us! "Ruhe ist die erste Bürgerpflicht." ["The main thing is to keep calm."] This greatly relieved the old gentleman who left me saying "Ah il me tombe une pierre du Coeur."'[66]

When a report arrived at Potsdam from Hatzfeldt two days later describing Salisbury's attitude to France as conciliatory and peaceable, Wilhelm dismissed the Prime Minister's words as pure sophistry in view of the British naval preparations which he knew to be under way, and wrote an angry marginal comment rejecting Salisbury's assertion that these measures had been planned the previous spring, long before the Fashoda crisis, and therefore had nothing to do with the present situation. The measures were 'being carried out only now! Furthermore Fashoda could have been foreseen a year ago! I told Grierson [in January 1898] during the hare shoot at Buckow that we had received news that Marchand had reached the upper Nile and was preparing to sail down it. Which gave him a colossal fright; then he pulled himself together quickly, laughed and said it was nonsense!' Against Salisbury's joking remark, reported by Hatzfeldt, that if the Admiralty were still making war preparations it must be doing so privately in its own room, as he, the Prime Minister, knew nothing about it, Wilhelm II noted: 'That I can quite believe! Since H. Excellency has not even the faintest glimmer of an idea about the navy and ships, as I discovered for myself, the last part is absolutely right!' As supreme war lord the German Kaiser knew much more about such things than the civilian Prime Minister of a parliamentary nation. 'Mobilisation is generally planned and prepared in a room and not on the street, nor in Tsung-li-Yamen!!' – by which he meant the parliament at Westminster. When Salisbury referred to Russia's peaceful intentions and expounded his favourite theory that a war between Britain and Russia was scarcely conceivable if only because there was no battlefield on which the two powers could fight each other, Wilhelm remarked spitefully: 'Well! When the Caucasian grenadiers come knocking on the door at Herat he will change his tune.' The Prime Minister's conviction that Britain need not fear a Russian attack on India received nothing but scorn from the Kaiser, who annotated this passage: 'Well! That is at the very least colossally irresponsible and shows he is badly informed about Russian troop movements in the East!' The only matter which gave the British leader cause for anxiety was the unpredictable situation in Macedonia, on which the Kaiser commented furiously: 'Quite right. And perhaps the English war preparations are also intended to encourage the Macedonians, and then, when there is a rebellion there they are hoping for a general mess so that they can get Crete, Suda [Bay], Dardanelles, French fleet etc. in one fell swoop; St Petersburg must therefore be urged to keep things *quiet* in the Balkans; and above all to hobble that scoundrel [Prince] Ferdinand [of Bulgaria] by his mutton legs. For he *is planning something*, and the House of Coburg reaches far up into the North! If Russia becomes embroiled in the Balkans – which Salisbury certainly hopes – she will be out of the way in Asia and he is rid of the problem, while the Austrians and Germans have it on their plate! – So *Effendimis* [Sultan Abdul Hamid II] must be encouraged at all costs to keep his European troops ready to fight on the Bulgarian frontier and to strengthen them so as to teach Ferdinand a lesson at once. To make Sir Fr. Lascelles aware of the real seriousness of the situation, since he treats the Macedonians more like a bad joke and refused to take the Bulgarian seriously, I told him that Ferdinand would undoubtedly try to create havoc down there; if he did so and provoked a European conflagration in the process, *I* would *have him murdered without further ado*, for "*fire raisers*" of that kind must be got rid of at once, and I did not go to the Orient for nothing!! The ambassador's face was a picture! But I think the "Coburger" may well be *warned*! – In his usual masterly way Salisbury has said nothing that he might have to *take back* later or that was *absolutely untrue*, but between the lines he says a great deal, and glosses over everything! – In the passage about India he is perhaps being dishonest in hiding his anxiety.'[67]

Bülow had scarcely had time to order these imperial outbursts to be locked away with the Foreign Office's secret records when he found himself compelled to consign another batch of correspondence, bedecked with revealing marginal notes in the All-Highest hand, to the Wilhelmstrasse's iron safe. In February 1899 Cecil Rhodes, who was to pass through Berlin on his return from Egypt, sent the Kaiser two books with a flattering accompanying letter in which he said that Wilhelm must feel very alone, since the Germans did not understand their Kaiser's 'big ideas'.[68] Although Rhodes was extremely unpopular in Germany, on Bülow's advice Wilhelm agreed to receive him, scrawling a characteristic note on the Foreign Secretary's submission: 'Well! It will create a splendid scandal among my dim-witted subjects, but I don't care. If I could I would hang Cecil Rh:, but as that isn't possible I shall make use of him. But it will cause a great sensation!'[69] In fact, as Rhodes had predicted, the two men got on extremely well at their meeting on 11 March. On the Kaiser's orders, less than three days later a treaty had been drawn up and signed, granting Rhodes the right to lay his Cape-to-Cairo telegraph and railway line through German East Africa. Rhodes asked the Kaiser 'why he did not go for Mesopotamia as a colonizing ground to which H.M. replied that this was a project he had had for years'. At dinner later at the British embassy the conversation came round to the Krüger telegram.

Lascelles's blood froze in his veins with terror, but Rhodes apologised for his role in the Jameson Raid and said to the Kaiser: 'I was a naughty boy, I did behave badly; you were quite right, but my people . . . thought that if I was to be punished, it was they who should do it and that no one else should interfere.' Finally the Kaiser gave him a large signed portrait of himself.[70] Writing to his mother about the meeting, Wilhelm described Rhodes as 'a most energetic man & marvellous organiser. I have of course promised to help him as far as is in my power so that he may be able to see the wish of his life fullfilled.'[71] To Lascelles he expressed the view that, 'with a man like Rhodes for a minister, he would do anything'.[72] Rhodes, for his part, also did his best to bring about warmer relations between Britain and Germany, and in particular between Wilhelm and his uncle, who it was safe to assume would succeed to the British throne before very long. In a letter to the Prince of Wales he expressed his confidence in the Kaiser's goodwill. 'I feel sure he is most anxious to work with England, and I think he is fond of the English; he must be so, for after all he is half an Englishman. I think he is very sensitive, for he spoke about the way the English papers had abused him. I heard in Berlin, on good authority, and I am sure, Sir, you will not mind my repeating it, that he thinks you do not like him, and that he is very anxious to gain your good opinion . . . I think . . . we ought to try and work with Germany, and the Emperor is really Germany, at least, it appeared to me to be so when I was in Berlin, Ministers doing just as he desires and the Reichstag most docile . . . It seems to me that in view of the complications in the world we must work with some nation, and Germany seems the best . . . I am sure of this that, if you showed him good feeling when he came to England, it would immensely influence his mind.'[73]

Although the Kaiser had been forced to give up his hopes for a war between Britain and France for the time being, he continued to make no secret of his view that the former had missed a historic opportunity. 'How funny it is to read about you all in the South of France when I remember our correspondence in October & November about the forthcoming probable war with France!', he wrote to his mother on 26 March 1899. 'In ways of course this state of things is to be preferred & I fully understand Grandmama not wanting to finish her Reign with fighting. But simply taken from a cool political point of view: England has missed a grand opportunity which will never come again. France unprepared, the Continent uncertain; Russia not ready, Germany a Friendly neutral, & England herself ready & stronger than ever! What a great pity! With every year the odds will be heavier! For France has awakened & is quietly but thoroughly preparing herself for the great struggle.'[74] The Empress Frederick, who had meanwhile arrived in Bordighera, replied that England would indeed have had a good chance of success if war had broken out with France the previous November or December, 'but the aim of politics is to avoid such struggles *if possible* and it *was* possible'. One must always hope for a friendly solution of outstanding questions, she commented, 'without plunging nations into the misery of war; no one can tell when and how they will end, and the responsability [sic] is too great to risk bringing on a struggle because one's chances are good, and those of one's adversary are not'.[75] The Empress's humane arguments made no impression whatever. Lieutenant-Colonel Grierson was astonished to hear from the Kaiser during a dinner in late March 1899 what a mistake Britain had made 'in not going to war with France over Fashoda. He said that we should never have such a chance again, and that the situation was just like that before Frederick the Great at the beginning of the 1st Silesian War. *He* had arranged everything to keep Russia out of the war', and now the Russians were cocking a snook at him. 'Every nation in Europe was building battleships and though we [British] might build we could not find men to put into our ships, and we should end by losing our supremacy. He said that England was no longer Nelson's England, and all this "because Grandmama wants to go to Cimiez"!' Commenting on Salisbury's agreements with France and Russia, the Kaiser asserted that England had made great concessions to the former and should not expect Russia to keep its promises.[76]

THE COBURG SUCCESSION

As if there were not already enough contentious issues between Britain and Germany in the sphere of *Weltpolitik* and between the two ruling houses, the sudden death of the young Prince Alfred of Saxe-Coburg and Gotha, only son of the reigning Duke Alfred (Affie), who was himself gravely ill, precipitated a conflict over the succession to the two Thuringian duchies of Coburg and Gotha which caused further strain between Wilhelm and his English relations. Duke Alfred's next successor would have been his younger brother Prince Arthur, Duke of Connaught, who had the advantage of being married to a Prussian princess, but who was unwilling to give up his career in the British army. Connaught's son Arthur, born in 1883, could have taken over the succession; after him, the next in line was Charles Edward (Charlie), Duke of Albany, the still underage son of Prince Leopold, Queen Victoria's youngest son, who had died of haemophilia in 1884.[77] From the outset the question of the succession threatened to become an affair of state, for while Queen Victoria was adamant that according to Coburg laws of succession it was absolutely nothing to do with the Kaiser, Wilhelm himself saw it from a German national perspective and insisted that the young successor – whether Arthur or Charles Edward – must at least be educated in Germany.[78] Wilhelm felt thoroughly snubbed when he heard in March 1899 that several members of the English Coburg family had had a meeting in the

South of France, where the Queen was on holiday, to discuss the succession. He pointed out to Grierson that he too was 'a member of the Royal Family and happened also to be German Emperor'; he had therefore expected to be asked for his opinion on a question of the succession to a German throne. He had not even been directly informed of the family's decision, he complained, but believed that the choice had fallen upon young Arthur of Connaught. He had no objection to this decision, but must insist, as German kaiser, that the Prince move to Germany if only to learn the language, which in spite of his Prussian mother he did not speak perfectly, and as a German prince he must join the German army. As Lascelles reported to Lord Salisbury, Grierson had received the impression from this conversation 'that the Emperor felt he had been slighted about the question of the Coburg Gotha succession and was sore in consequence'. The ambassador had also taken soundings at the German Foreign Office, which left no room for doubt that this was an affair 'upon which the Emperor felt very strongly'. He drew the Prime Minister's attention to the difficult position in which the young Prince Arthur would find himself if he moved to Germany as heir to the Duchy without the Kaiser's blessing.[79] The Queen, the Empress Frederick and the Prince of Wales were not at all surprised by Wilhelm's annoyance, but thought it would be wrong to make any apology. 'He seems greatly annoyed at not having been consulted about the Coburg-succession which "unter uns" [between ourselves] is no real concern of his, but as you know only *too* well [he] *must* meddle with everything', the heir to the British throne complained to his sister.[80]

During a meeting arranged at the suggestion of the Empress Frederick between the Duke of Coburg, his brother the Duke of Connaught and Wilhelm II at the Wartburg on 21 April 1899, the latter reiterated his conditions on the succession question and added threateningly that, if the family did not accept them, he would have a bill introduced into the Reichstag excluding foreign princes from succeeding to any German throne. Later he dropped this threat, presumably, as Grierson commented, because he realised that the bill would not receive sufficient support in the Reichstag.[81] But the Duke of Connaught was unwilling to give up his life in Britain and his position in the British army 'to begin life anew as a German and a German officer', as the Kaiser insisted. He was also uncomfortably aware that Wilhelm had the support of the German press and public opinion, which was hostile to a 'foreigner' succeeding in Coburg, and therefore declared himself ready to renounce the Coburg throne both for himself and for his son.[82] The other English members of the family feared the effect of Wilhelm II's emotional outbursts on the international political scene and agreed that it would be advisable for the Connaughts to renounce their rights in favour of the young Duke of Albany, especially in view of the danger that the family might lose its right to the Coburg succession altogether.[83] After a further meeting between Wilhelm and his uncle Arthur, this time in Kassel on 27 May, at least an outward semblance of harmony was achieved with the agreement that both Connaughts, father and son, would renounce the succession and that it would fall to 'little Albany'.[84] It was only a little over a year later, on the death of his uncle Alfred on 30 July 1900, that Charles Edward was to succeed to the Duchy of Saxe-Coburg and Gotha, although under the guardianship of Hereditary Prince Ernst zu Hohenlohe-Langenburg until 1905. Other problems which arose at the same time as the succession crisis in Coburg were not so easily overcome.

THE KAISER, LORD SALISBURY AND QUEEN VICTORIA

For some time now Wilhelm's contempt for the British Prime Minister and Foreign Secretary, which was entirely mutual, had assumed dangerous dimensions and put a strain on relations between the two countries. In November 1896 the Kaiser had accused the British government of giving financial support to the Hamburg dockers' strike for its own economic benefit. Salisbury had responded with a sharply worded dispatch to Lascelles which the Kaiser took very much amiss. As the Reich Chancellor noted with concern on 2 December 1896, the Kaiser was determined to go and see the British ambassador and make it plain to him 'that, given the feeling against England prevalent in Germany now, it would have the worst possible effect in Germany if it were discovered that England was encouraging a strike that deeply affected Germany's economic situation. Lascelles would then be exposed to the danger, according to Wilhelm, that the Berliners might break his windows.' The Kaiser had rejected the ambassador's claim that it was not London financiers, but perhaps British socialists, who were supporting the strike, on the grounds that the latter 'did not have . . . enough money and must have received funds from the English financial barons'. One can imagine the effect that these accusations had on Salisbury and Queen Victoria.[85] A few years later, not least because of the disappointment of Germany's hopes in the Samoan question, open conflict was to break out between Wilhelm II and the Prime Minister.[86]

Queen Victoria marked the fortieth birthday of her firstborn grandson on 27 January 1899 with an anxious comment in her journal: 'I wish he were more prudent and less impulsive at such an age!'[87] Wilhelm himself had a surprisingly accurate grasp of his grandmother's opinion of him and wrote to her on the occasion of his birthday saying: 'I fully understand how extraordinary the fact must seem to you that the tiny, weeny little brat you often had in your arms and dear Grandpapa swung about in a napkin has reached the forties! Just half of your prosperous and successful life! [My life] is full of heavy unceasing work, and animated with my untiring trials to emulate such grand example as you set us all. It is often full of moments when I fancy the strain is too strong and the burden too heavy to bear . . . I venture to believe that, where the Sovereign will sometimes shake her wisc head often over the tricks of her queer and impetuous colleague, the good and genial heart of my Grandmother will step in and show that, if he sometimes fails, it is never from want of goodwill, honesty, or truthfulness, and thus mitigate the shake of the head by a genial smile of warm sympathy and interest!' He went on to say how delighted he was by her invitation to him to visit her at Cowes or Balmoral.[88]

Like Salisbury, the old Queen did not take Wilhelm's expressions of friendship entirely at their face value. In a conversation with the Prime Minister on 17 February 1899 she agreed with him that, 'while William appeared to wish to be on good terms with us, he did not wish that we should be so with other countries and in particular Russia, whom he was always trying to set against us. Lord Salisbury does not at all believe the stories he tells.'[89] On 1 March 1899, shortly before her departure for the South of France, Queen Victoria wrote to the Tsar warning him, in words that presaged future developments, against the Kaiser's double-dealing. Wilhelm, she said, 'takes every opportunity of impressing upon Sir F. Lascelles that Russia is doing all in her power to work against us; that she offers alliances to other Powers and has made one with the Ameer of Afghanistan against us. I need not say that I do not believe a word of this, neither do Lord Salisbury nor Sir F. Lascelles. But I am afraid William may go and tell things against us to you, just as he does about you to us. If so, pray tell me openly and confidentially. It is so important that we should understand each other, and that such mischievous and unstraightforward proceedings should be put a stop to. You are so true yourself, that I am sure you will be shocked at this.'[90]

A few weeks later, in conversation with the British military attaché Grierson, the Kaiser delivered himself of an extraordinary tirade on the subject of Salisbury and the 'disgraceful' foreign policy of Great Britain. He was evidently feeling deeply hurt again, this time at not being invited to Queen Victoria's eightieth birthday on 24 May 1899 – for a time he even threatened to turn up uninvited[91] – but wounded feelings aside, the attack was of such virulence that at the British embassy, the Wilhelmstrasse, Downing Street and Windsor Castle there were serious concerns about his sanity. As Grierson reported to his ambassador, Lascelles, the Kaiser began by venting his feelings on the Coburg succession question, and then suddenly declared bluntly that he found British policy 'quite incomprehensible'. 'He said that for years he had been the one true friend to Great Britain on the continent of Europe, and had done everything to help her policy and assist her, and that he had received nothing in return but ingratitude, culminating in our behaviour in the Samoan affair, which, he said, had undone all the good he had effected in the way of cultivating friendly relations between the two countries. Some day, when it was too late, we should regret it. He had particularly desired to be in England on her Majesty's [80th] Birthday, and present all his children to the Queen, but that was now impossible. His consistent enemy throughout had been Lord Salisbury, and, while the latter remained Prime Minister, it would be impossible for him to come to England. Formerly it was the people of Great Britain who kept back the Government from war, now it was the Government which was holding back the people. The City and Mr. Chamberlain were all for war. They had failed to get war with France, and now they wanted war with Germany, which would be an easier thing for them as she had fewer ships than France.' The military attaché went on to suggest to his superior, with careful circumspection, that 'from the above Your Excellency will not fail to see that his Majesty was talking somewhat at random, for, by his own statement, his "enemy" Lord Salisbury was keeping back the people from making war on Germany!' Grierson felt compelled to add that, nevertheless, 'His Majesty's tone throughout this conversation was calm and dispassionate.'[92] During the conversation, which was in English, Wilhelm II had put on 'his gravest face and most impressive manner', but had otherwise been extremely friendly. But the military attaché seriously wondered whether the All-Highest sovereign might not be unhinged. He told the Queen's Private Secretary that he could not help fearing, given the doctrine of divine right and the state of subservient discipline to which the Kaiser had reduced all those around him, that he might be in a state of 'Grössenwahnsinn' (megalomania), in which he expected every other nation to change its policy in accordance with his views. Grierson did not think it wise to commit all his impressions to paper and asked Sir Arthur Bigge for a personal meeting during his forthcoming visit to London.[93] When Sir Frank Lascelles discussed the Kaiser's outburst with Bülow, the Foreign Secretary was visibly embarrassed. As the ambassador reported to Salisbury, Bülow declared 'that it was not for him to criticize in any way the language of his sovereign, but I, who knew the Emperor so well, must know that His Majesty's impetuosity sometimes led to exaggeration of expression . . . His Majesty was in fact more than half an Englishman, and was extraordinarily sensitive to anything which He could regard as a slight either from the Royal Family or from Her Majesty's Government.'[94] Salisbury himself was stunned by the Kaiser's completely unfounded accusations, which he regarded as the expression of a guilty conscience. 'So groundless is the charge', he wrote to Lascelles, 'that I cannot help fearing that it indicates a consciousness on the part of His Majesty that he cherishes some design which is bound to make me his enemy – and that he looks forward to the satisfaction of saying I told you so. It is a great nuisance that one of the main factors in the European calculation should be so ultra human. He is as jealous as a woman because he does not think the Queen pays him enough attention.'[95]

The Kaiser's angry assertion that he would not be able to come to Britain as long as Salisbury was still in occupation at Downing Street as prime minister provoked his grandmother to write a stern letter to her grandson on 18 May 1899, which served only to deepen the gulf between the two monarchs. She was hearing 'such very strange reports', she wrote, according to which he would not be welcome at Cowes this year. He should remember, however, that she had personally written to invite him and had also told his Flügeladjutant, Löwenfeld, when he was at Osborne, how much she was looking forward to Wilhelm's visit in August. 'I feel I must write to tell you how grieved I am at these false reports which have all emanated from Germany. As I am convinced that you cannot possibly be aware of this, I feel I must write this to you in order that you can make enquiries as to how such unfounded statements have got about, & put a stop to them. I am quite sure that you will be as much pained as I am, for I am convinced that you will be the first to assist me in clearing up a matter which may so easily make bad blood.'[96]

Queen Victoria's unusual tone was not without effect on Wilhelm. He telegraphed to Bülow on 20 May in dismay, stating that he had just 'received a rather angry letter' from the Queen, in which she complained of the false rumour that he had not been invited to Cowes this year and asked for the matter to be set straight, so that there should be no bad blood between the two countries. 'The letter gives me the impression', the Kaiser commented, 'that it was written in some agitation. Her Majesty's assertion, as far as I can remember, is completely invalid. Since My return from the Orient I have never seen any remark in any newspaper which even hinted at the absence of an invitation to Me from the Queen. I think it is more likely that some unpleasant intrigue is at work here. The absence of an invitation to her eightieth birthday (which even My Mother tried to obtain for Me in Cimiez) has been skilfully confused with the visit to Cowes. That may have been combined recently with the announcement conveyed by Lascelles on my instructions that I could not come to Cowes in view of the agitated state of public opinion in Germany because of Samoa, and that I therefore deeply regretted that in consequence of Her Majesty's refusal to allow Me and My family to visit her for her eightieth birthday, I had to forgo the pleasure of paying my respects to My grandmother in this important year. As Her Majesty can no longer read herself, I am afraid that the royal lady who acts as her secretary will have artfully confused these facts in reporting them to her, which has given rise to this strange accusation by the Queen. It is very striking that Her Majesty, who speaks to Me *for the first time in her life* of public opinion in

Germany, and apparently sets much store by it, makes not the slightest mention of the Samoa affair, although it has been so enormously upsetting and hurtful for our public.'[97] In view of Wilhelm's claim that the Queen's accusation must have been due to a misunderstanding, it may be worth pointing out that he too had misunderstood his grandmother's letter, for nowhere in it does she refer to newspaper reports or to German public opinion.

Wilhelm's pique over his family's exclusion from his grandmother's eightieth birthday celebrations manifested itself again at the banquet which he held in Berlin in honour of the occasion. He expressed his fury to the British ambassador over the bombardment of Samoa by British and American ships, which had caused considerable damage to German possessions, 'thereby causing the greatest indignation among all classes in Germany against England. Tho[ugh] England evidently treated Germany as a nonentity and ignored her Fleet, yet the time might come when England would have to respect it, and he only hoped that then it would not be too late, and that Germany might have formed combinations not agreeable to England. The feeling of irritation in Germany was so bitter, that he could not visit Cowes this year. All the good relations which existed between the two countries at the beginning of the year were now completely destroyed by British action in Samoa. His Majesty alluded to the large sums of money sent from England to bribe the American press to attack Germany & said Her Majesty's Government ought to have known of it, and they had taken no steps to counteract this evil influence. He praised Mr. Rhodes, and regretted more influential Englishmen did not visit Berlin. The Emperor's farewell words to the Ambassador were, "Tell your people to behave themselves properly", after which he quoted the speech of Dean Liddell to an undergraduate "You have not only imperilled your immortal soul, but you have incurred my serious displeasure" – the impression conveyed on the Ambassador being that the apologue conveyed a fairly acurate description of His Majesty's mind.'[98] Queen Victoria described Lascelles's report as 'an extraordinary account of William's impertinent & outrageous language'.[99]

On 27 May 1899 the Kaiser wrote a nine-page letter to his grandmother from Wilhelmshöhe which she described as 'a most violent attack on Ld. Salisbury & our supposed treatment of Germany' and which is indeed a document probably without parallel in modern European history.[100] 'I think it my duty to point out that public feeling over here has been very much agitated & stirred to its depths by the most unhappy way in which Lord Salisbury has treated Germany in the Samoan buisiness [sic]. After we had formed the South African agreement – which I settled very much against the will of our Colonial circles – the People in Germany thought that a new base had been laid for mutual understanding & goodwill in foreign & colonial questions. Then came our military demonstrations he Waterloo Place in Hannover for the Victors of Omdurman, which was gn of the warm interest taken by our Army in the deeds of their British others in Arms. My visit to Malta – as alas I could not extend my journey to gypt – was a sign of affectionate interest in your Fleet & your Flag, of which am so proud to be an Admiral. And last not least, with an utter disregard for public opinion – which was very sore about this – & in the teeth of a most violent opposition from all ranks of society in Germany I received Sir [sic] Cecil Rhodes. Only showing thereby that I thought it my duty to do all in my power to help your Government in the work of peace & goodwill, for the benefit of my country. As a "rendu" for all this Lord Salisbury has treated Germany in the Samoan Question in a way which was utterly at variance with the manners, which regulate the relations between great Powers according to European rules of civility. He not only left my Government for months without an answer to its proposals, dating from Autumn last year, but he even refrained from expressing his or the Governments regrets after the first acts of violence by Commander Sturdee & the other ships had occurred at Samoa. A fact the more unintelligible as the President of the U.S.A. immediately sent word to say how sorry he was such acts had happened on the part of American officers & men. On the contrary the British ships went on for days & weeks bombarding socalled "positions" of socalled "rebels" – though noone knows against whom they "rebelled" – & thereby burning & destroying plantations & houses belonging to my subjects with a loss of hundreds of thousands of marks without even so much as an excuse having been made, & that on an Island which by three fourths is in German hands. This way of treating Germany's feelings & interests has come upon the People like an electric shock & has evoked the impression that Lord Salisbury cares for us no more than for Portugal, Chili [sic], or the Patagonians, & out of this impression the feeling has arisen that Germany was beeing [sic] despised by his government, & this has stung my subjects to the quick. This fact is looked upon as a taint to the National honour & to their feelings of self respect. Therefore I am most sorry to have to state that popular feeling in Germany is just now very bitter on England, & as I found out during my spring visit to the South of Germany that feeling is the same with the simple labourer as with their Princes, it is unanimous. Most disheartening for my honest labours to bring the two countries to understand eachother [sic] better. If this sort of highhanded reatment of German affairs by Lord Salisbury is suffered to continue, I am afraid hat there will be a permanent source of misunderstandings & recriminations between the two Nations, which may in the end lead to bad blood. I of course have been silent as to what I have *personally* gone through these last six months, the shame & pain I have suffered, & how my heart has bled when to my despair I had to watch how the arduous work of years was destroyed – to make the two

Nations understand eachother & respect their aspirations & wishes – by one blow by the highhanded & disdainful treatment of Ministers who have never come over to stay here & to study our institutions, & People, & hardly ever have given themselves the trouble to understand them. Lord Salisbury's Government must learn to respect & treat us as equals, as long as he cannot be brought to do that, People over here will remain distrustful & a sort of coolness will be the unavoidable result. It is very probable that the news of the dissatisfaction over here with regard to the Samoan affairs may have reached your informant in a roundabout way & led him to believe it had to do with the visit to Cowes, but that is not the case. What a great pity it is that you could not pass by Strassburg or any other station where I could have met you & had a quiet talk with you in your carriage, about all this grievous mess. Now you will understand dear Grandmama why I so ardently hoped to be able to go over for your birthday. That visit would have been perfectly understood over here, as the duty of the grandson to his grandmother, putting "Emperor" etc. apart, & as according to the family ties, & nobody would have said a word against it. The more so as the children were to be shown greatgrandmama. But a pleasure trip to Cowes after all that has happened, & with respect to the temperature of our public opinion here is utterly impossible now. I had not the heart to write about all these disagreeable matters to you, as I did not want to worry you, & because I hoped that Lord Salisbury would change his mind again, & therefore gulped everything down & held my tongue. But as you have yourself kindly inquired about the state of public opinion in Germany, I thought it my duty to state the facts as they are. I can assure you there is no man more deeply grieved & unhappy than me! and all that on account of a stupid Island which is a hairpin to England compared to the thousands of square miles she is annexing right & left unopposed every year . . . Goodbye most beloved Grandmama. With much love & respect believe me ever your most dutiful & devoted grandson William I. R.'[101]

It is not particularly surprising that the Prime Minister categorically and comprehensively rejected the Kaiser's outrageous accusations. He assured the Queen that he had sent for the relevant records and had found not a single ground for Wilhelm's complaints. 'Either the German Emperor has been misled by certain correspondents he has in England – mostly intriguing men; or he is simply angry that England should have made agreements with Russia and France. His outspoken desire that there should be a war between England and France this spring probably indicates that his indignation has been created by recent agreements. He [Salisbury] entirely concurs with Your Majesty in thinking that it is quite new for a Sovereign to attack in a private letter the Minister of another Sovereign; especially one to whom he is so closely related. It is not a desirable innovation and might produce some confusion.'[102] The eighty-year-old Queen delivered an indignant reprimand to her grandson. She had been extremely surprised by his letter, she wrote on 12 June. 'The tone in which you write about Lord Salisbury I can only attribute to a temporary irritation on your part, as I do not think you would otherwise have written in such a manner, & I doubt whether any Sovereign ever wrote in such terms to another Sovereign, – & that Sovereign his own Grandmother, about their Prime Minister. I never should do such a thing, & I never personally attacked, or complained of Prince Bismarck, though I knew well what a bitter enemy he was to England & all the harm he did.' Salisbury had completely refuted the accusations that Wilhelm had made against him, in a memorandum which she was enclosing with her letter, she added.[103] 'Old Victoria's rude letter has hurt Him unutterably deeply!', Philipp Eulenburg reported to his friend Bülow.[104] In July 1899 the Kaiser was still speaking of his 'deeply wounded feelings over the way he has been treated by his grandmother and her family'.[105]

Anyone tempted to think that with this brusque reprimand of Kaiser Wilhelm II by Queen Victoria Anglo-German relations must have reached their lowest possible level should remember that the secret aims of the German battlefleet building programme, initiated in 1897 by the Kaiser and Admiral Alfred Tirpitz and directed against Great Britain, were still hidden from London. Once it dawned on the British what kind of menacing superweapon was being created on the other side of the North Sea for use against them, it would only be a matter of time before Britain joined the Franco-Russian Dual Alliance, thus completing the 'encirclement' of Germany.

CHAPTER THIRTY-TWO

Wilhelm and the birth of the German battlefleet

THE KAISER'S 'LIMITLESS FLEET PLANS'

THE resolve gradually formed by Kaiser Wilhelm II in the 1890s to build a 'giant fleet' of battleships with which to confront Great Britain was one of the most momentous and catastrophic decisions of his long reign, along with his dismissal of Bismarck in early 1890 and his role in the outbreak of the First World War in 1914. 'From the beginning to the end of the *Flottenpolitik*, the Kaiser played a critical and fatal role', writes Paul Kennedy, whose masterly analysis of the Anglo-German antagonism before 1914 is second to none.[1] Relations between Germany and Britain, already corroded by the mutual distrust which had characterised them since 1896, were irreparably damaged by this decision and, having made the painful discovery of its vulnerability on the world political stage during the Boer War,[2] Britain was rapidly driven into the arms of Japan, France and Russia. But how did Wilhelm's personal passion for the sea come to provide the impetus for a fatal armaments race between the two neighbours on the North Sea? What was the Kaiser's purpose in pursuing this 'limitless' naval policy, which he personally instigated and carried through against the will of parliament, the statesmen of the Wilhelmstrasse and even some leading members of the officer corps?

The young Kaiser's 'passion for the navy', which as we have seen can be traced back to his childhood, had cast its shadow over the policy of his government from the start.[3] Even his brother Heinrich, whose whole life revolved around his career as a naval officer, complained in a letter to Tirpitz in 1893 that it was 'impossible to avoid conversations about the navy . . . in one's day-to-day dealings with His Majesty!'[4] Wilhelm's frequent and vociferous demands for the expansion of the fleet, however, had proved fruitless – with the exception of four battleships accepted in the estimates for 1889 – for it was not long before the impression had arisen in the Reichstag that naval construction was more a question of the Kaiser's predilections than of political and strategic necessity. The Left Liberal Reichstag deputy Eugen Richter coined the expressive phrase 'limitless fleet plans' for these demands, arguing that they must be blocked on political and financial grounds. The Reichstag consistently rejected all Wilhelm II's expansion plans from 1890 onwards, especially after the naval authorities had provided the Kaiser's critics with additional ammunition by demanding a large sum of money for the refurbishment of the imperial yacht *Hohenzollern*.[5]

It was not only the Kaiser's personal enthusiasm which proved an impediment to the expansion of the German navy at first. His plans also lacked any systematic basis, and he annoyed the Reichstag by unnecessary confrontations. A further obstacle was the ineptitude of Secretary of State at the Reich Navy Office Admiral Friedrich Hollmann, who enjoyed the Kaiser's favour, indeed even his friendship, but completely failed to persuade the Reichstag of the necessity of expanding the fleet, possibly because he himself was not convinced of it.[6] Admittedly, naval planning was rendered particularly complicated at this period by the fact that tactical conceptions varied widely, against a background of extraordinarily rapid technological development in warship building. But that Hollmann could stand up in front of the Reichstag in early 1894 and declare that the funds requested for the building of new ships had been included in the estimates by the naval authorities 'without any plan' is rightly described by the military historian Wilhelm Deist as 'disarmingly naïve and helpless'.[7] In several respects Hollmann's attitude resembles that of the Kaiser: both wanted to enlarge the navy, both favoured a cruiser fleet and neither was able to put forward a convincing strategic scenario in which this fleet would serve a purpose commensurate with the expenditure involved in creating it. Consequently the German fleet stagnated in the first years of Wilhelm II's reign. It was old-fashioned and consisted of a motley collection of ships of different types. Its relative strength even decreased during this period, and it fell to fifth or sixth place in the world.

Wilhelm found this development extremely painful, especially since he saw it as the purpose of his reign and his historic task to raise Germany to the status of a naval power. In his memoirs Bülow relates a conversation in which he compared the Kaiser with the Soldier King. Just as the latter created the army which had been indispensable to the rise of Prussia, Wilhelm aimed to create an equivalent naval power to ensure the continued advancement of Prussia–Germany. 'You have often said yourself that your ideal would be, like that of Friedrich Wilhelm I, to forge the weapon which your son or, still better, your grandson will one day use', he told Wilhelm.[8] The comparison with the 'Heldenkaiser', Wilhelm I, was even more compelling, however. The young monarch constantly compared himself with his grandfather, who at the time of the constitutional conflict had strengthened the army in the face of what had at first seemed a hopeless struggle against the parliament. In fact it was precisely this vision of his own role that was largely to blame for Wilhelm II's problems, since he could only picture the expansion of the fleet as a victory wrung out of the 'stupid' Reichstag after a fierce battle, if necessary through dissolution or even a *coup d'état*. It occurred to him either not at all or only sporadically that he could have got his way by a systematic effort to win over the opposition, by co-operation rather than confrontation, and there was no one in his entourage who would have given him the necessary encouragement. On the contrary, Senden, Plessen and others constantly stoked up his prejudices against parliamentarism.

From the outset Wilhelm repeatedly gave notice that he would not rest until he had achieved his aim of enlarging the fleet. Soon after his accession he declared to a group of naval officers that, just as his grandfather had once said that his last thoughts would be with the army, he promised them that his last thoughts would be with the navy.[9] In 1897 he assured his brother Heinrich that he would not be content 'until I have brought My navy up to the same level as the army'.[10] The army officers in the Kaiser's entourage were very soon afraid that Wilhelm's passion for the navy would be to the detriment of the land forces. Chief of the Military Cabinet General von Hahnke remarked anxiously in 1892 'that the Kaiser would abandon the strengthening of the army for the sake of a few warships without a second thought'.[11] Waldersee too, as we have seen, was alarmed by the 'fanatical love for the navy', with which the Kaiser was obsessed.[12] Wilhelm 'no longer [has] the passion he should for the army . . . ; he has had the chance to look over the fleet and is full of it'.[13] Writing to his friend the former War Minister von Verdy, in April 1895, Waldersee asked sarcastically: 'Are you not sorry that you did not send your son into the navy? Anyone who is now not of the opinion that the main focus of all warfare is on the sea is an ignoramus and has not moved with the times.'[14] In all political circles the word was that 'the Kaiser's only real interest is in the navy'; everything else was 'pretty well indifferent' to him.[15]

If Wilhelm II was unable to make any progress with the plans he pursued with such ardour for the expansion of the fleet, it was not through any lack of ideas; his fault was rather that he had too many ideas and failed to bring them together into any systematic order. As early as the first year of his reign he had been prompt to suggest what the navy needed. He devoted a disproportionate amount of time to the theatrical aspect of the fleet, and to questions of prestige.[16] While receiving reports from the navy secretary or the chief of the Naval Cabinet Wilhelm liked to be shown plans and photographs of German or foreign warships, as the later chief of the Naval Cabinet, Georg Alexander von Müller, reported. Müller considered this a waste of time: 'The major part of the valuable time for reports [was] usually taken up with such things, which had little significance except to entertain the Kaiser. There were rarely decisions to be taken on them, and if there were, they related to quite unimportant trifles like the position of the smokestacks of a cruiser, from an aesthetic point of view, the form of a ship's bow or the decoration of the bow and stern of new ships that were to be built.'[17] His plans for the fleet were equally dilettantish and amounted to little more than demanding more money for ships from the Reichstag. 'Naves esse aedificandas', he said simply, and this was symptomatic of his attitude to the fleet in the Hollmann era.[18] In February 1895 he remarked that a general had advised him to take out a loan of a billion for the navy 'and then start building ships with it'. It did not occur to Wilhelm to work out an organic plan for the increases, develop a tactical and strategic concept to aim at, as happened with the later battlefleet programme, and then put forward systematic demands for the supply of the types of ship needed.[19]

By the autumn of 1893 there were increasing signs that the Kaiser at last meant business with the fleet plans he had so long cherished. 'I see . . . to my regret from His Majesty's marginal comments', Holstein wrote, 'that He is considering the idea [of an expansion of the fleet] – I hope only theoretically for the moment.'[20] Kiderlen had the same impression and wrote on 1 October 1893: 'From the All-Highest margin[alia] it almost seems that he is intending to make increased demands for the navy *this* year. That could cause new conflicts with the Reichstag.' But having recently approved the great Army Bill, he pointed out, the Reichstag ought 'to be allowed a little time for a breather'.[21] Political observers agreed that the Reichstag would reject these plans, but Waldersee already suspected that the Kaiser would get his way in the end. The imperial wishes, he wrote, had been 'indignantly' turned down by all and sundry, but 'to judge from experience, many of those who are now beating their breasts and declaring "not a single mark more for the navy" will be a lot tamer later, after they have been subjected to all kinds of influence in Berlin, and these are naturally mostly Conservatives who do not like denying the Kaiser anything'.[22] For the time being, however, the fleet plans came to nothing. After Miquel had submitted a memorandum on the financial position of Prussia and the Reich to him, Wilhelm declared himself ready 'to give up all increased demands for the navy for this year'. Waldersee welcomed this decision, commenting that the money would 'not have been granted in any circumstances', and that the battles it would have caused would have been 'extremely unpleasant'.[23]

FROM CRUISERS TO THE BATTLEFLEET

Wilhelm's enthusiasm for the navy was nourished by both political and literary events. In his eyes, his belief that the navy was indispensable to the future of Germany and to its continued rise in the world was confirmed by Alfred Thayer Mahan's work, *The Influence of Sea Power upon History*. Its thesis of the decisive significance of sea power was exactly to his taste. He wrote to an American friend: 'I am just now not reading but devouring Captain Mahan's book and am trying to learn it by heart. It is a first-class book and classical on all points.'[24] The Sino-Japanese war of 1894–5 seemed to him to provide compelling proof of the accuracy of Mahan's theories, and he followed its course with fascination.[25] The naval engagements were his principal interest, and in particular the battle of the Yalu estuary on 17 September 1894. The state of feverish excitement into which Wilhelm was projected caused a certain alarm. Marschall, the Foreign Secretary, recorded in his diary on 2 February 1895: 'H.M. is again very much preoccupied with the navy. I asked him to keep quiet and trust me, as I had hopes.' Three days later he noted: 'In the morning H.M. He has nothing but the navy in his head.'[26]

Having been provided with such a striking example of the significance of sea power by events in the Far East, Wilhelm was anxious to make positive progress with his naval construction plans at long last. The conclusions he was prepared to draw for German armaments policy from the superiority of the Japanese in their planning and conduct of the war can be seen in the two-hour extempore lecture which he gave to about 500 officers at the Royal War Academy in Berlin on 8 February 1895, using numerous drawings and tables. He began by expressing his admiration for the prudence with which the Japanese had prepared for the campaign over a ten-year period, 'in a completely modern way'. Of decisive importance, in his view, was the close co-operation between the warships, the merchant navy and the army over the landing in Korea, which had made the Japanese victories on the mainland possible. Using merchant ships, 12,000–13,000 men had been landed in $1\frac{1}{2}$ days. The Kaiser had high praise for the Japanese landing operation: 'Here, for the first time, very large numbers of troops were transported overseas; no nation has ever carried out a practice exercise for this before. It is astonishing how well Japan worked out and provided for everything in this respect.' But Wilhelm placed most weight on the decisive sea battle of the two fleets at the mouth of the Yalu. The Japanese fleet, he emphasised, had been numerically smaller but of better quality than the Chinese. It had 'consisted only of modern armoured cruisers', which had been built in Britain, France and, using these as models, partly also in Japan itself. The Chinese fleet, on the other hand, had consisted of two large and three smaller armoured warships, five cruisers and a few small ships, which were all obsolete and inferior to the Japanese ships, especially in their speed. A whole series of other shortcomings counted against the Chinese fleet: there were not enough exploding shells for the Chinese artillery, the experience and tactical training of the commanders left much to be desired, the communications system was inadequate, there was no integrated command of the fleet at all. It was quite a different matter with the Japanese, whose admiral, in Wilhelm II's opinion, 'acted in general in accordance with the same tactical principles which we have recognised as correct as a result of our exercises'. But, above all, the ships themselves were the decisive factor. The victory of the Japanese could be attributed to the fact that 'they were equipped with a series of cruisers in good order, which were capable of reaching a speed of 20 knots, while the Chinese could not achieve more than 10 knots'. Moreover 'the warlike spirit inherent in the Japanese, in contrast to their enemies . . . which makes their whole conduct in this war so appealing to us', had likewise contributed to their victory. 'Towards 5 o'clock the battle had reached its conclusion and the rays of the setting sun lit up the flag of the Rising Sun of the Mikado, fluttering victoriously in the wind . . . Through this sea-battle the Japanese succeeded in securing an immensely important victory for their country, i.e. through their supremacy at sea they paved the way for the further operations of the army, which now strides ahead from victory to victory. The Japanese capture of the enemy capital of Peking should be certain.'

From this analysis of the course of the war in the Far East Wilhelm drew two significant conclusions: first, that a good fleet 'can be of very critical importance in a campaign', and second 'that even if the fleet is smaller, it must above all consist of excellent *matériel* which fulfils all modern requirements'.[27] Both conclusions were calculated to lend weight to the Kaiser's wish to strengthen the German fleet. From various contemporary sources we know that on numerous occasions in the winter of 1894–5 Wilhelm brought out his shipping tables, drew attention to the shortfall in the German fleet by making comparisons with the numbers of ships possessed by different nations, and presented his analysis of the great sea-battle at the mouth of the Yalu. It is equally clear from his remarks that at this juncture he was seeking to build a fleet of modern armoured cruisers, which would destroy the enemy fleets in a war fought principally on land against Russia, France and possibly Denmark, and which would be in a position to keep the trade routes across the Atlantic open. It is important to note, however, that the idea of throwing down the gauntlet to the British Empire by building a large fleet of battleships was still very far from his mind.[28] True, he pressed for a rapid expansion of the navy, but the demand for four new cruisers which he put forward through the Navy Secretary in 1895 was comparatively modest.[29]

Another incident demonstrates how much the Kaiser's naval plans were dominated by the desire for impressive German warships to be seen in all four corners of the earth – a policy of prestige, in other words. On 31 January 1895 he appeared at the Reich chancellor's palace and demanded the withdrawal of the lone German warship stationed off Samoa on the grounds that the British ship that was also there was three times as big. 'If we wanted to increase the diminished German influence in Samoa we should have to make our presence felt with several German ships; this was not possible in the present state of the navy. H.M. does not want to give up Samoa', Hohenlohe remarked, 'but he thinks the presence of the ship there now is inadequate and therefore superfluous.'[30]

Wilhelm was now also making efforts to persuade the Reichstag to share his enthusiasm for the fleet. On 8 January 1895 he invited several of its members, mainly from the Conservative Party, together with the delegate of the Hanseatic cities in the Bundesrat, Krüger, to Potsdam. The Badenese envoy, Eugen von Jagemann, reported to Karlsruhe that the Kaiser had again given a 'two-hour extempore lecture' to this audience, in which he had shown 'great expertise and prudence'. He had 'described the German armed forces in relation to the Russian and French', and had then discussed 'the state of the navy, pointing out the more important German and foreign types of ship, often using his own drawings'. The obvious purpose of the lecture, Jagemann commented, was to influence opinion in the Reichstag in favour of approving the requested cruiser corvettes, for in the Kaiser's view the number of cruisers was insufficient for service abroad and for the protection of the merchant navy. In the last half hour of the lecture, however, the Kaiser had set out, 'as a future prospect, the need to acquire more iron-clads, which are decisive in sea-battles'.[31] In contrast to his previous tactics, therefore, the Kaiser was trying to win over the Reichstag by argument; but he made the mistake, according to Jagemann, of inviting only deputies who needed no persuading.

The Kaiser wanted more ships; but what exactly did he want? Was the reference to iron-clads in his lecture an early hint of what was to come? We know from Tirpitz's memoirs that this reference can be traced to a report that he himself had made to the Kaiser the previous day and which had left such a strong impression on the latter that he adopted elements of it for his lecture.[32] Nonetheless, at this time Wilhelm thought above all in terms of a cruiser fleet. This is abundantly clear from both of the lectures he gave, on 8 January and 8 February 1895, in which he had dealt not only with the sea-battle off the Yalu estuary but also with the types of ship in the German navy. He began by deploring the fact that, although Germany had the second-largest merchant fleet in the world, the country had no adequate cruiser fleet capable of protecting German interests worldwide. As he regarded his audience as landlubbers, he explained the differences between the various types of warship, first from a historical perspective and then progressing to current ship-building policy. Using drawings and tables he carefully explained the difference between battleships of the first class or ships of the line on the one hand, and armoured cruisers or armoured deck cruisers on the other. To give the officers of the War Academy 'at least some idea' of what a modern ship of the line looked like, the Kaiser showed them a picture which he had drawn with Salzmann's help showing the *Ersatz Preussen*, which was in the process of being built.[33] Such ships of the line fought in large tactical groups on the battle line, according to Wilhelm II, and could not be used separately. By way of contrast he showed them the armoured cruiser *Ersatz Leipzig*, which was also equipped with heavy guns and had belt armour, but unlike the ship of the line it was designed 'to engage with several opponents on its own, to chase them, and to do as much damage to them as it could with its big, long-range guns, and to keep the area allocated to it clear'. Of this type of ship the French had thirteen, the Russians ten, Italy five, the United States three and 'Germany none'. 'It is very difficult, strangely enough', the Kaiser remarked, 'to make any progress in this country on the question of armoured cruisers, as a secret terror fills the people in the Reichstag at the very mention of this name [*Panzerkreuzer*]. They are afraid that the wool is being pulled over their eyes, and that an absolutely monstrous weapon is to be built.' He went on to point out that the position was no better with more lightly protected cruisers. Here too the Franco-Russian Dual Alliance enjoyed a crushing superiority over the Triple Alliance. Taking everything into account, the Triple Alliance powers had five armoured cruisers, thirty-two armoured deck cruisers and thirty-one unprotected cruisers (which of course could play only a secondary role in war), while France, Russia and Denmark together possessed twenty-three armoured cruisers, forty-six armoured deck cruisers and forty-one unprotected cruisers. 'That means that there are 68 battle units of the Triple Alliance against 110 on the other side.' If one added the big armoured battleships or ships of the line, there would be a total of 133 'ships of all types usable in battle . . . on the side of the Triple Alliance against 201 of the other trio'. From these figures one could see 'how enormously the French and Russian fleets [had] developed'.[34]

These remarks again left no room for doubt that in the Kaiser's eyes the antagonist, as far as the German fleet was concerned, was the Franco-Russian Dual Alliance, and that he set most store by the cruiser fleet, ships of the line being mentioned only incidentally. He was particularly preoccupied with armoured cruisers. Deploring the hopeless state of the German fleet, he went on to comment that the German ships, with the exception of the *Irene*, were 'absolutely unfitted' to engage in 'any serious battle . . . and if any major conflicts should arise in the course of the Sino-Japanese war, or attempts at partition between the European states, the German Reich would be absolutely incapable of enforcing a single claim'. The three ships which were representing the German Reich abroad at the moment, he said, were still equipped with masts and rigging like the old frigates of the time of Nelson and had no rapid-fire guns or armoured deck, so that they would be defenceless against a modern cruiser. 'A single Japanese armoured cruiser would be enough to shoot our entire German cruiser squadron to pieces', Wilhelm declared. The situation was so bad that it could be assumed that 'in 1896 the German Reich would not have a single ship sailing out there in the world'.[35]

What conclusions did the Kaiser draw from this? He demanded that cruisers be built, namely the replacement of nine first-class armoured cruisers, fifteen second-class cruisers and twelve fourth-class cruisers 'so as to be able at least to provide moderately decent protection for our overseas trade and to get the necessary respect for our flag'. Together with the smaller ships this would mean 36 new ships with 756 guns, including 3 flagships. 'The German Reich could quite easily have had this number, if in the last 10 years 1 armoured cruiser and perhaps 2 second-class cruisers had been requested and gradually built each year. Instead, the Reichstag has . . . consistently struck out every single replacement ship, and so we have arrived at this deplorable state of affairs.' At present there was '*no more than one* ship being built' in Germany, the Kaiser complained. 'If things go on as they are at the moment, i.e. absolute refusal on the part of our parliament', the Chief of Staff of the Naval High Command had explained to him 'that by 1899 the German battlefleet will no longer be capable of leaving harbour at all. The German fleet will then be confronted with 10 brand new Russian first-class battleships and about 21 new French first-class battleships in the Channel, while Germany at the moment has only 5 armoured battleships. All the rest are second-class and according to modern principles no longer fit for a battle at sea.'[36]

Once again the question is: what did the Kaiser want? He spoke about decisive sea-battles and battleships, in other words a battlefleet. But he also spoke of armoured cruisers, lighter armed cruisers and small cruisers which were to enable German interests to be enforced worldwide. He demanded several types of ship for various uses on the high sea. So at this point it cannot yet be said that his ideas were fixed on a fleet of battleships; the armoured cruiser still had priority for him. How, and against whom, did he see these ships being used in a war?

Wilhelm's audience in the War Academy consisted of army officers who on the whole did not share the imperial enthusiasm for the navy. He tried to arouse their interest by making observations on 'the use of the fleet in relation to the army'. Following the logic of his remarks on the Sino-Japanese war, he made the axiomatic pronouncement that 'only he who controls the sea is in a position well and truly to get the better of his opponent and thus to maintain untrammelled freedom of action in military operations, which otherwise he could not afford'. If the French fleet, vastly superior to the German, should drive back or destroy the few German ships confronting it in a future war, it could blockade the German coast, stop all trade and even enable troops to be landed on the German coast. 'If we imagine the situation in reverse, if we were confronting France with our fleet in the same way as it confronts us, it would be the most natural thing for us to use our fleet to defeat the French fleet wherever it appeared, to blockade French ports, to transport a large part of the army and to land them rapidly so that they could quickly advance on Paris and be within the walls of the capital within a few days, while the enemy awaited attack from the east behind the triple reinforcements of their defensive forts. In spite of its inferior strength, in spite of its inadequate *matériel*, the German fleet has decided to pursue this idea nonetheless, in so far as it is at all feasible', the Kaiser affirmed, in an access of wishful thinking that was entirely typical of him and some of the leading naval officers. 'The intention is to bring all the ships together, hurl them against the enemy and if possible force him to fight, and, if God gives us victory, to gain victory too.' The Kaiser emphasised not only this strategically important use of the fleet in the event of a war on the continent, but also its value for 'providing food for our great armed forces'. 'The German army needs to eat', he pointed out. In 1870, on its advance into France, the German army had had no threat from the rear and could therefore obtain food supplies from Hungary, Russia and southern Austria. 'That will not be the case now. Not a single grain will come over the Russian border; Austria needs her own supplies herself. So we have no alternative but to bring in supplies by sea. But our supply route lies directly off the great French ports of Cherbourg, Brest, Rochefort, where the French have their armoured cruisers etc. etc. lying in wait to intercept any foreign ship. It therefore naturally follows that our fleet, instead of staying where it is for so-called coastal protection, as the Reichstag always envisaged, should aim to gain control of the Channel so as to repulse any interception by the French energetically from the outset. For that we need a whole series of fast armoured cruisers.' Its value in a war against France, Russia and probably Denmark was thus the principal reason for a strong German fleet. But the Kaiser also had visions of another possible use of the navy: war at sea against the United States. America, with its thirty-seven cruisers, was a country 'with which we could find ourselves in collision at any moment, partly because of what we might wish for abroad, e.g. because of Samoa or the like, partly because of wishes we might have at home, if for example our agriculture had a particular wish to export something, or to prevent some import which was unwelcome to it'. Without an effective battlefleet 'the German Reich would be absolutely incapable of getting its way at all against America, if America did not wish it'.[37]

One has the impression from this document that Wilhelm, presumably with the assistance of the Chief of his Naval Cabinet Senden, marshalled every conceivable argument to justify the naval expansion which he so ardently desired. But what is significant, and of particular importance for us here, is that in conjuring up this wide-ranging and often unrealistic strategic panorama in February 1895 the Kaiser made no mention at all of war with Great Britain. True, he commented that with its 205 warships Britain was only a little stronger than Russia, France and Denmark put together, and that 'former claims of English supremacy at sea' no longer had any validity. But the difference between the strength of the German and British navies was so great that Wilhelm simply left it out of account in his comparative tables. Indeed he expressly ruled out any parallels, saying that Britain 'cannot serve as a point of comparison for us, for as a marine nation it has completely abnormal criteria and is based entirely on the fleet'. Although he nevertheless mentioned the numbers of British ships being built it was only in order to underline the enormous disparity between the two countries.[38] But it was precisely in this context, as we shall see, that a disastrous transformation was to take place in his attitude over the next eighteen months.

The personal initiative which the Kaiser had taken in this question was in itself enough to arouse criticism from many of his compatriots, especially in the army. Waldersee noted sceptically in his diary for 10 February 1895: 'It is certainly a new idea for a Kaiser to speak on such a matter to such a large audience . . . That the Kaiser can speak well, that he has a great love for the navy and knows a great deal about its affairs, and that he considers a considerable increase necessary – all that is well known. That any member of the Reichstag, or at least an appreciable number of them, will become more amenable as a result of his lecture, is completely out of the question. But it may well provoke a great deal of criticism, and then I can see it having a bad effect in that people will become suspicious and say the Kaiser is gradually demanding more and more, and such a colossal increase that it will be beyond our financial means. I also believe that it is impossible for us to achieve the aim of creating a first-class fleet which would also be capable of protecting German commerce all over the world. For that we should first have to win great victories on land, which would enable us to maintain a smaller army. But is there any prospect of that?'[39] Waldersee considered the Kaiser's naval armament plans excessive, and did not think himself alone in taking this view. Two days later he noted: 'The Kaiser, who is always particularly obsessed by some question or other, is now completely absorbed with the expansion of the navy and as usual is inclined to go too far. In fact no one agrees with what he wants or thinks it is even possible, except a few young naval officers. The people in the Reichstag who go along with it probably do it only to keep on good terms with the Kaiser.'[40] The pro-government Berlin correspondent of the National Liberal *Kölnische Zeitung*, Franz Fischer, shared this view, and reported to Eulenburg on 17 February 1895 that the 'apparently enormous demands for funds for the navy on which His Majesty's two lectures are said to have been based', and which were estimated at a billion marks – to be raised by 1899 – were being discussed with alarm in the Reichstag.[41] It was perhaps precisely against this background of huge sums for the navy that the comparatively modest demand for the four new cruisers was approved, by the Reichstag with very little fuss.[42] Wilhelm had evidently doubted that even this demand would succeed, and had therefore sent his Flügeladjutant Kuno Moltke to Friedrichsruh again on 11 January 1895 to take Bismarck drawings he himself had made of American warships. He had 'only one thought', as Eulenburg assured Holstein, which was 'the approval of the cruisers. That was the actual motive for Moltke's journey and for all his friendly gestures.'[43] When they were approved, Wilhelm was delighted and telegraphed to Hohenlohe on 1 March 1895: 'Am overjoyed at the good news. What a splendid result in the vote. The Reichstag has the pleasant feeling of having done a good deed for the preservation of the greatness of the Fatherland and for our industry. It has given me great pleasure.'[44] He was, as Holstein recorded, 'as happy as a lark about his four cruisers'.[45]

And yet, not long ago, the Kaiser had declared that thirty-six cruisers were absolutely essential. He showed no more consistency in the key question of the battlefleet (which Senden demanded) as against the cruiser fleet (which Hollmann advocated).[46] But he remained obdurate on one point: the creation of a strong fleet in the longer term. And at the same time a highly significant change came about as to the purpose for which the fleet was to be built up. Hitherto the Kaiser's fleet plans had been formulated with an eye to co-operation between the Triple Alliance powers and Great Britain. The German fleet, as the Kaiser had often stressed, should work with the Royal Navy and Britain should remain 'Mistress of the Seas',[47] and his previous plans for the expansion of the fleet had not deviated from this basic idea. But now his thoughts began to turn in a new and fatal direction, for at the beginning of 1896 it occurred to the Kaiser that he could exploit events in South Africa, the Jameson Raid, the Krüger telegram and the strong British reaction to it, to pursue his naval construction plans.[48] The mechanism was exactly the same as he had used during the Sino-Japanese war: he tried to take advantage of actual crises to make it clear to public opinion how urgently Germany needed a strong fleet. But this time, unlike on the earlier occasion, his conduct had a strongly anti-British note. Twice during the Chancellor's audiences in early January 1896, when the battle between the British and the German press over the Krüger telegram was at its height, Kaiser Wilhelm – anxious to make the most of the 'heightened political mood of the moment' – peremptorily demanded that a loan of 'several hundred million' be taken out to build a battle fleet. At the same time he called for a committee to be established, consisting of the secretaries of state at the Foreign Office, the Reich Treasury and the Reich Navy Office, together with the Prussian Finance Minister, to discuss 'the enlargement of the fleet'.[49] He rejected Hohenlohe's objections to both demands in a revealing document of 8 January 1896, in which he commented that 'a distinction should be made between the firm plan for the expansion of the navy adopted as a result of the report of the High Command of 1 December, and my aims, which have arisen because of the Transvaal and about which I spoke to you. The two plans complement each other. But while the expansion plan must be worked out well in advance, in order to take the general position of the Reich in the world into account, the idea which the Transvaal has engendered, and which is based on urgent short-term needs, is intended to fill a momentary, but serious, gap. In many parts of the world, where the situation is getting increasingly critical, the Reich has inadequate or non-existent means of upholding its considerable interests. In the Mediterranean there is neither a squadron . . . nor even *one ship* fit to uphold our interests. We have had to resort to a training ship. In the whole of America, whether east, central or west, there is not a single gunboat to protect hundreds of thousands of Germans and millions of marks of our capital; and in Venezuela the situation could at any moment turn into one in which the fleet [would need] to show the flag to protect German property. In East Asia our only fighting division has been tied down for almost a year pursuing special local interests, with no prospect of being released. And now we have the Transvaal question too, which has shown very clearly indeed that the ever-shrinking navy does not have a sufficient number of ships to do justice to the world position of the Reich at all. We shall never again have such a favourable opportunity of proving to the country that the navy cannot continue in its present state. The movement which is sweeping through our people is a deep one, and the Reichstag, even if it hesitates, will be forced to take this mood into account! But with every day, every week that passes without any action, once the Reichstag has assembled, the optimistic mood will fade, the Reichstag will sink into the mud courtesy of Kanitz, Stoecker, Hammerstein etc., the parties which are uplifted and united now as a result of the 18th Jan. [an allusion to Wilhelm's speech on the 25th anniversary of the foundation of the Reich] will disintegrate, and if the poor navy's estimates are discussed and chewed over then, there will be such discord and lack of enthusiasm that even the better elements in the Reichstag will not be able to be persuaded of the necessity of enlarging the navy. For present needs it is not necessary to have a great plan; all that is needed is immediate replacement, not by building new ships at home but by *purchasing* armoured cruisers and cruisers as rapidly as possible wherever we can get them. Meanwhile cruisers would have to be laid down at home, the number of which depends on the efficiency of our shipyards. All the cruisers would then have to be taken into account in the programme which is still being worked out and which has to be submitted next year, to which reference would have to be made. These thoughts which I have briefly sketched out for you should give you the basis on which I wish the question to be considered and handled. Whether the Reichstag will have any inclination or enthusiasm for it is beside the point; our *duty* is to [make] the situation perfectly plain to them even at the risk of rejection. Wilhelm I. R.'[50]

One thing is more than clear from this document. The Kaiser's almost hysterical eagerness to do something for the navy outweighed any consideration of the political wisdom of what he was advocating. Without stopping to think, he leapt at the chance of exploiting the anglophobic current of German public opinion for the indiscriminate, unsystematic purchase of ships, his only concern being that this apparently favourable moment might slip through his fingers. Marschall von Bieberstein complained on 8 January 1896 that 'H.M. wants to get "a few hundred millions" from the Reichstag for ship-building. Hohenlohe is to make a big speech in the Reichstag to take advantage of the present mood.'[51] Again, when the Foreign Secretary had a long conversation with the Kaiser on 11 January about his correspondence with Queen Victoria on the Transvaal crisis, he noted afterwards that the constant 'refrain' had been 'enlargement of the fleet'.[52] Holstein's explanation for the Kaiser's 'pathological excitement' was that 'the illusion that he will get a few hundred millions – he ranges between 100 and 300 in what he says – at a stroke, has made him incapable of judging anything calmly . . . All this makes for an uncanny impression of excessive haste. The Kaiser's nerves have never been stimulated by anything so much as by the temptations of "the great fleet programme".'[53] Eulenburg did his best to calm the Geheimrat by arguing that 'despite all H.M.'s nervousness' the Kaiser was pushing ahead so far with his demands for the fleet merely 'because only by means of shock tactics applied to all branches of the administration is he able to achieve any results at all'. 'H.M. wants a lot, but not the impossible', he assured Holstein.[54] In the Wilhelmstrasse the imperial entourage was suspected of adding fuel to these demands for the fleet. Holstein also blamed 'the little Kayser', who wanted an independent Reich Colonial Office and was using his influence on the monarch through Hinzpeter, and Admiral Senden for fanning the flames of Wilhelm's enthusiasm for the navy. The Kaiser was 'in a state of pathological excitement' and was demanding 'at least 100 million immediately for the acquisition of ships', he wrote. As well as Kuno Moltke, Senden had now been sent to Friedrichsruh with naval plans, and Bismarck had promised his support.[55] The Kaiser's aversion, which was already considerable, to the idea of having to wring his fleet programme out of the Reichstag, was further inflamed by the anti-British and anti-parliamentarian promptings of his entourage.[56]

Given the situation in 1896, Wilhelm of course had no alternative at first but to trust to his Chancellor. The elderly Prince Hohenlohe was likewise obliged, against his own convictions, to undertake the task of sounding out opinion in the Reichstag on the Kaiser's idea of a loan for the fleet. The result, as was to be expected, was devastating. The leaders of the Centre, the National Liberals and the Conservatives in the Reichstag categorically rejected the Kaiser's wish for a loan. Hohenlohe wrote to him on 14 January 1896 to report on his findings. Fritzen, the leader of the Centre Party, had said that if his group agreed to the loan they would vanish from the scene at the next elections, for the population was already weighed down by the burden of taxation and it was impossible to impose further burdens on the country. The leader of the Conservatives, Levetzow, had replied that agriculture was in such a parlous state that an exceptional appropriation for the navy would not receive a single vote from his party. Both Fritzen and Levetzow had also pointed out that such a demand at the present moment would be seen by the people as a prelude to war with the British Empire, and it would have to be turned down for that reason alone. Only Oberpräsident von Bennigsen, the leader of the National Liberals, had been prepared to consult the members of his group, although he was not hopeful of a positive result. From these soundings Hohenlohe concluded 'that there [was] no trace of enthusiasm in the Reichstag for an enlargement of the fleet itself at this moment'. The introduction of a Loan Bill would inevitably lead to a massive defeat for the government, which would be greeted with *Schadenfreude* in other European countries and especially in Great Britain, 'as a personal defeat for Y[our M[ajesty]'.[57] Together with Hollmann and Marschall, Hohenlohe urged on the Kaiser the view that it was necessary to work out a long-term 'plan for the setting up of the fleet', which could be carried out only 'over the years'.[58] He was anxious to put the naval construction plans on to a systematic basis, which they had hitherto lacked, and thereby to counter the general impression that naval expansion was only an imperial hobby. 'I am eager to know how he will react', Marschall noted, having concluded from the Kaiser's good mood on 14 January that he evidently did not yet know anything 'about the result of the enquiries about the navy'.[59]

The perceptive Friedrich von Holstein, meanwhile, foresaw complications indicative of the political mood in Germany, where there was no shortage of advocates of naval expansion in nationalist circles. He viewed with alarm a scenario in which the Kaiser's initiative might be supported principally by chauvinist groups. Colonialist and pan-German elements led by Duke Johann Albrecht of Mecklenburg and the Colonial Director Paul Kayser had already convened a mass meeting at the Kroll opera house in Berlin, at which 'rousing speeches and resolutions in favour of [a] grand fleet expansion programme' were planned, Holstein wrote to Philipp Eulenburg on 14 January 1896.[60] Once again the ambassador tried to protect Wilhelm and pointed out that with his 'exaggerated, pushing demands' he was only trying to obtain 'the largest possible naval appropriation'. By strongly supporting the Kaiser's wishes Marschall would also be able to strengthen his own position, which was very necessary, Eulenburg added.[61]

When the Kaiser heard the next day how the parties in the Reichstag stood on the naval expansion question he reacted with fury and defiance. Oberhofmarschall Count August zu Eulenburg recorded that on the morning of 15 January Wilhelm had been 'very depressed about his disappointed hopes for the navy and of course he had little good to say about the Reichstag'.[62] If the members of parliament would not agree to his fleet plans, then he would show them what it was like to be without a fleet, he said. The following day, 16 January 1896, he gave orders for all German warships to be withdrawn from China and Delagoa Bay. The Kaiser's order, which the Chief of the Naval Cabinet conveyed to the Naval High Command, read: 'Have informed Freiherr von Marschall today that, as the Reichstag has so little sense of what the honour of our Fatherland means in relation to England abroad, and no one has any feeling for or interest in the navy, I can no longer allow My Home Fleet to remain in a weakened state in view of the uncertain prospects this spring. *Kaiser* should therefore return immediately, after repairs, and likewise the cruiser squadron, which should follow next month, with the exception of *Arcona*. The cruisers from Lourenço Marques should be sent to Zanzibar, as I do not wish them to stay there because of the concentration of English ships. The High Command is to give orders accordingly. Wilhelm I. R.'[63] It was only with difficulty that Hohenlohe and Marschall succeeded in having the imperial order countermanded.[64]

Wilhelm's anger at the rejection of his plans for naval construction rebounded on those who, in his opinion, had not given sufficient support to his ideas. Chief among them was the Reich Chancellor; and in the second place the Secretary of State at the Reich Navy Office, Hollmann. After a visit to the capital Waldersee noted that the Kaiser was 'enraged' with the Chancellor, adding: 'the reason for his anger with Hohenlohe is that the Kaiser wanted a big loan for naval purposes – 300 million – and had already begun negotiations with the "Vulkan" [shipyard] on that basis, and Hohenlohe explained to him that it would not work; he had consulted Fritzen, Bennigsen and Levetzow, that is, the leaders of the Centre, the National Liberals and the Conservatives, and they told him that there was no chance of getting such a bill through'. Waldersee, who was on the best of terms with the Conservatives, was however of the opinion that

Hohenlohe had not made a serious attempt to negotiate with the party leaders, as he 'personally had no desire, naturally enough, to tackle such a big question'. But now, Waldersee continued, there might be dangerous political consequences. The Kaiser had been 'beside himself', he wrote; his language suggested that 'the decision to change chancellors is definite'. In his displeasure the Kaiser had threatened 'to bring back all the cruisers, then there will soon be great uproar in the seaports, and all the Colonial people will join in'. In Waldersee's view Admiral Hollmann would also have to to resign as Reich navy secretary, as he had said 'that we did not have enough manpower to man a much larger fleet'.[65] Holstein shared the view that the Kaiser would try to get rid of all those who had let him down over the fleet plan, and thought that Wilhelm would use the question of the large demand for additional funds for the navy as a lever to bring about Hohenlohe's fall in the same way as he had used the Anti-Revolutionary Bill to engineer Caprivi's dismissal.[66]

The signs of the Kaiser's deep displeasure were not long in manifesting themselves. When the elderly Prince Hohenlohe asked for a joint audience for himself, Hollmann, Marschall and Admiral Knorr to discuss naval affairs, he received no answer for days. He then asked Hollmann 'if he knew anything'. As Hohenlohe recorded in his diary, the Admiral replied that he had spoken to the Kaiser at a beer party given by Senden, and that Wilhelm had said that Hohenlohe and Hollmann 'had let the Kaiser down and did not want to do anything. He would therefore have to wait . . . At the same time, the Kaiser expressed himself in such a way that Hollmann is convinced that the Kaiser hopes to find a Reich chancellor who will put forward large naval demands, dissolve the Reichstag if necessary and execute a *coup d'état*.' Hohenlohe took this news calmly. 'That is all right with me', he noted. 'In fact, though, I cannot see whom the Kaiser can find to carry out this experiment.'[67] Once again Holstein sent off an alarmed message to Eulenburg in Vienna. 'The Kaiser is extremely worked up. He wants to demand enormous funds for the Navy, *at least* 100 million, and thinks that by dissolving the Reichstag he will be able to get a more compliant parliament.' The Privy Councillor considered the Kaiser's plan quite impracticable and warned that 'the next Reichstag will be dreadful, it will not approve anything, and the German princes will *not* join in a Reich *coup d'état* campaign. Then the Kaiser will have his political Jena and will not recover from it any better than Friedrich Wilhelm IV from his ride [through the streets of Berlin] with the [revolutionary] flag on 20 March 1848.'[68]

Not until early February was a kind of armistice achieved between Kaiser and Chancellor on the fleet question, when Wilhelm visited the old Prince, who instead of being 'rude', as the 'embittered elements in the Ministry' demanded, only touched 'in passing and in courteous terms on the discourtesy of the Kaiser' in ignoring his request for an audience. In the course of their tête-à-tête the Kaiser expressed his willingness to leave the ship *Kaiser* in China. He also gave his consent for the Reichstag to be told that no further demands for funds for the navy would be made during the current session. Instead, as Wilhelm now agreed, a systematic procedure should be followed and new ships built on the basis of a carefully worked-out programme.[69] Waldersee suspected that the Kaiser had only seemingly abandoned his far-reaching ambitions for the fleet, noting in his diary on 9 February 1896 that Wilhelm had indeed temporarily postponed his naval expansion plans until the autumn, 'but that he should give up his plans I think is out of the question, nor will he forgive Hohenlohe, Marschall and Hollmann for having caused him problems in this'. The entire liberal camp, according to Waldersee, was terrified that the Kaiser might appoint a 'bold general' as chancellor, with the aim of bringing about a reactionary shift in internal policy and of carrying out his fleet programme. 'To judge from what they say the Kaiser must be surrounded with dangerous intriguers who constantly encourage him in his ideas, whereas the wretched hypocrites know perfectly well that the Kaiser himself is the instigator and leader of the movement.'[70]

Naturally these setbacks in no way dimmed the ardour of Wilhelm II's enthusiasm for the navy. On a train journey to Genoa on 23–4 May 1896 the Kaiser, who was in a good mood, talked 'chiefly about the navy'. Kiderlen-Wächter, who was accompanying the Kaiser on his Mediterranean journey as representative of the Foreign Office, reported to Berlin that Wilhelm had 'just received a plan from a certain Captain Diederichs for building an armoured vessel in a new style – obviously commissioned work! Entirely in accordance with H.M.'s ideas, with particularly powerful armament which is H.M.'s hobby-horse. How the Reichstag will rejoice, and Hollmann, who already had so much trouble in talking H.M. out of his last idea, will be happier still. That last idea of the Kaiser's was very lovely, only it could not float.'[71] The Kaiser as ship-builder might have been a butt for the diplomat's secret mockery, but no one laughed at his visions of naval power. His 'limitless' fleet plans were still creating a 'somewhat eerie feeling' in Berlin in March 1896, especially as there were still fears that the Reichstag might be dissolved on account of them. 'That the Kaiser will pursue his goal resolutely I do not doubt for a moment', Waldersee wrote on 3 March 1896. Even among naval officers there were 'divided opinions', as numerous 'well-respected officers [thought] such a great enlargement scarcely feasible'.[72]

TIRPITZ ANTE PORTAS

Thus in the spring of 1896 the battle lines were drawn. The Kaiser wanted to enlarge the fleet, but was confronted by the direct opposition of the Reichstag and the rather more passive resistance of his Chancellor, the Foreign Office and even the Secretary of State of the Reich Navy Office, who, as Holstein reported to Vienna, was openly saying 'the way the country has been governed hitherto cannot go on'.[73] In early July 1896, during an audience at Wilhelmshöhe lasting several hours and also attended by Senden and Plessen, Hollmann set out 'all his reservations about the submission to the Reichstag of a fleet plan spread out over years'. The parliament would never agree to such a plan, the Secretary of State stressed. It would lead to repeated dissolutions of the Reichstag, but a *coup d'état* was an impossibility in a federal Reich. Hollmann therefore proposed that 'the fleet plan should be completely abandoned and [we] should limit [ourselves] to demanding from the Reichstag only what was necessary each year to maintain and renew the fleet'. Asked by the Kaiser for their opinion, Plessen said that he agreed with Hollmann, while Senden vigorously opposed his suggestions. Surprisingly, the Kaiser eventually declared that he could see that Hollmann was right. 'It was too late now. Last year's enthusiasm and the impression made by his speech on 18 January [1896] should have been used . . . Now the great plan would have to be abandoned, considered as a study and laid aside, and there should be no more talk of it.'[74] During the Scandinavian cruise of 1896 too it was clear that the opposition from all sides had momentarily destroyed his faith in his plan. As he told Philipp Eulenburg, his demands now consisted only 'in the idea of "replacement" of the existing old ships with new, whereby the old ships [could] still [be] used in the second line in an auxiliary role'. If the Reichstag granted these 'itemised demands' each year the target would gradually be reached. Germans demanded 'strong action in overseas countries, with the appropriate ships', Wilhelm claimed. It would be illogical for them 'to demand an energetic German policy all over the world and then refuse to grant the navy what it needs'. 'England's uncouth behaviour in the Transvaal question, on which the whole of Germany was united, would not have happened if Germany had been in a position, with a reasonably good fleet, to have had the slightest say in the matter.' The Kaiser's ideas of expanding the fleet through 'acceptable' annual demands were in express contrast to the 'absolutely inordinate and very unclear' demands for the navy put forward by Senden, whom Wilhelm criticised 'very harshly', and whose 'boundless obstinacy' and complete 'obsession' he strongly condemned on this occasion.[75]

These remarks by Wilhelm were, however, prompted by the circumstances of the moment and by no means reflected his longer-term intentions. With his declaration Hollmann had after all gone back on the agreements reached in February and had pronounced any enlargement of the navy practically impossible. The Kaiser was now forming the resolve, as Waldersee had already guessed, to put the realisation of his fleet plans into other hands. The replacement of the responsible authorities, above all Hollmann, did not take place at once, but the search for suitable successors was in full swing, and Senden, as chief of the Naval Cabinet, played a major role in it. This fanatical navalist constantly encouraged Wilhelm in his fleet plans and tried to win him over to a change of personnel. In a submission to the Kaiser in the spring of 1896 he argued that Hollmann should be replaced as secretary at the Reich Navy Office by Rear-Admiral Alfred Tirpitz. Hollmann had shown no initiative in regard to either the Reichstag or the Reich Chancellor and had thereby proved that he was not equal to the far-reaching tasks which would fall to the holder of his office in the future. 'He lacks the qualities required to promote such an important question on his own, to take the lead in it, to weigh up the means to achieve the goal and to blaze the trail, supported by the commands of His Majesty', the Chief of the Naval Cabinet wrote. 'Given the opposition that a great naval programme will come up against in the Reichstag, the navy needs determined action on the part of the Reich Chancellor (R[eich N[avy O[ffice]), a man who will constantly stimulate interest, who will set the resources of the country in motion and who will be inventive in finding ways to promote such a goal.' Admittedly, Senden acknowledged, it would be sensible to wait until the navy estimates for 1896–7 had been accepted before replacing Hollmann. But his successor should be found and prepared in advance.[76]

Senden had already picked out Tirpitz for the post of navy secretary some time before; indeed he had systematically built him up for the role. In the abovementioned memorandum he drew attention to Tirpitz's professional qualifications for the post, which were undeniably outstanding, but at the same time he warned that his preferred candidate had a difficult character. 'If the choice should fall on Adm[iral] Tirpitz the unusual personality of this officer would have to be taken into account', he cautioned. Tirpitz had 'a very energetic, one might say ruthless nature, which needs curbing rather than driving forwards. He is ambitious, not particular about the means he uses, of a sanguine disposition: elated at one moment, depressed the next, but never letting up in his energetic activity, however despondent he might seem . . . He could not be treated like Adm[iral] H[ollmann], he is too self-confident and convinced of his own quality for that. He is not a compliant character who accepts an opinion at once at face value. On the contrary, he considers everything very carefully and then expresses his reservations very clearly and objectively. If his objections are not accepted then he accepts the decision and carries it out unflinchingly in accordance with his superior's wishes . . . He has been very spoilt in his naval career and has probably never had a superior officer who was his equal among the chiefs of the Admiralty . . . His Majesty will have to do full justice to his competence, follow his suggestions, give him room to manoeuvre, if they are to work profitably

51. Admiral Alfred von Tirpitz, State Secretary of the Reich Navy Office 1897–1916

together.' In spite of these difficulties the Chief of the Naval Cabinet considered Tirpitz 'the most suitable man . . . to bring the navy forward in this difficult situation'.[77]

Apart from Senden's support, the influence of Prince Heinrich on the appointment of Tirpitz should not be underestimated. He described the admiral as his 'master' and himself as his 'pupil' and 'truest and most gratefully devoted friend and comrade'. With the backing of his personal adjutant, Captain Georg Alexander Müller, who was to succeed Senden as chief of the Naval Cabinet, from 1896 the Kaiser's brother warmly advocated not only Tirpitz's appointment but also the creation of the battlefleet which Tirpitz demanded.[78] Like Tirpitz, Prince Heinrich strongly condemned Wilhelm II's Krüger telegram. He even dismissed the Kaiser's hopes of being able to exploit the excited state of public opinion in the aftermath of the telegram to expand the navy, since 'our failure . . . was not great enough to prove to the public at large how stupid and incredibly short-sighted it was'. 'If anyone at all shows any enthusiasm for the fleet question', the Prince wrote to Tirpitz in February 1896, 'he speaks of "cruisers" and his conscience is clear!' But it was obvious, he went on, 'that cruisers are a nonsense without a strong core provided by the battlefleet at home'. He expressed his hope that Tirpitz would soon take Hollmann's place and that the 'cruiser danger' would thereby be avoided. 'The Monarch's love for His Navy is unfortunately not always helpful to its sound development!', Wilhelm's brother wrote. But, as he recognised, certain '*naval officers* are *also* to blame for this state of affairs'.[79]

Senden had already asked Tirpitz, in December 1895, to set out his ideas on the future development of the battlefleet for submission to the Kaiser, as a counterpart to a memorandum produced by the Naval High Command. Tirpitz's views are well known. He was convinced that Germany's position in the world could be secured only by a large battlefleet, but not by cruisers; that a battlefleet would enhance the Reich's value as an ally, also in Britain's estimation; that a skilful propaganda campaign should be conducted to win over the German people, who were still sceptical, to a fleet programme of this magnitude; and that the Reich Navy Office must become the central authority in control of the navy in order to carry out this bold long-term project, which at this time Tirpitz saw as taking twelve years to complete. 'If we had an absolute monarchy the Kaiser would have the liberty to decide, on the basis of his advisers' proposals, exactly how strong the navy should be', he argued in February 1896 in a letter to Senden. In those circumstances the present division of responsibilities among the senior naval authorities might still make sense. But the fact was 'that the Kaiser cannot simply act according to his will in creating his navy, but has to reckon with a series of other factors and with the will . . . of the nation', and therefore all naval interests should be brought together under one authority, the Reich Navy Office, which must create a stronger position for itself in relation to the other government authorities, the Bundesrat, the Reichstag and the population in general.[80] This was all the more necessary, Tirpitz wrote to his mentor Admiral von Stosch, for '*our policy hitherto has been completely lacking in a sense of the political significance of sea power* . . . Our policy does not recognise that Germany's value as an ally even for European states often lies not in our army but in the navy . . . In my opinion Germany will rapidly fall from her position as a great power in the next century unless these general naval interests are energetically, promptly and systematically pursued now. Not least also because there is a strong palliative against educated and uneducated Social Democrats in the great new national task and the economic profit associated with it.'[81] In his memorandum for the Kaiser, Tirpitz put forward the convincing argument that with a fleet of seventeen battleships even the biggest naval power – Britain – would adopt an accommodating attitude towards Germany, which could never be achieved with ocean-going cruisers. But without a battlefleet the German people would be perpetually 'subject to the whims of the Anglo-Saxons'.[82] As Jonathan Steinberg rightly pointed out in 1965 and as Volker Berghahn, Wilhelm Deist, Paul Kennedy and Michael Epkenhans have subsequently proved beyond all doubt: 'This was the programme which Tirpitz pursued without rest from that day until the outbreak of war in 1914.'[83]

Alfred Tirpitz was no stranger to the Kaiser. Wilhelm had already pronounced him 'the future mainstay of the navy' in 1891.[84] As Prince Heinrich's remarks suggest, however, Wilhelm and Tirpitz were by no means of one mind on everything. The Kaiser wanted a fleet, but for a long time, as we have seen, he had no intention of directing it against Great Britain. Tirpitz, on the other hand, wanted to build his fleet expressly to confront Britain. Wilhelm wanted armoured ships for home defence, principally against the Franco-Russian Dual Alliance, and a cruiser fleet to show the flag and protect German interests overseas – in other words, different types of ships for different purposes. Tirpitz, however, was vehemently opposed to the cruiser fleet. It seemed to him unusable on principle, because it would always be dependent on British goodwill. On one thing, however, the Kaiser and the Admiral were agreed: Germany needed a large fleet if it did not wish to lose its status as a great power in the coming decades. Tirpitz had no choice but to work with the Kaiser, although he did not disguise his contempt for the monarch. And in 1896, after his numerous failed attempts to launch, in grand style, the fleet building programme he so passionately desired, Wilhelm could see no alternative but to dismiss the inept and recalcitrant Hollmann and entrust the task to the most energetic personality in the navy. And the best candidate by a long chalk, according to the Naval Cabinet and the Kaiser's own brother, was Tirpitz. At the end of January 1896 Senden arranged an audience for Tirpitz, in the course of which the Kaiser informed the Admiral that in the near future he would able to take over the Reich Navy Office as Hollmann's successor. At this meeting Tirpitz, showing great self-confidence, presented his plans and demands to the Kaiser.[85]

His appointment hung fire, however, because in March 1896 Marschall and Hollmann succeeded in obtaining the Reichstag's approval for a few cruisers, while for parliamentary reasons Hohenlohe was opposed to any change at the Reich Navy Office. Tirpitz saw this as a threat to the battlefleet plan which had already been approved in principle by the Kaiser, and wrote despondently to Senden on 20 March 1896: 'The way our interests are being represented in the Reichstag, indeed the whole situation, could hardly be less favourable. The Secretary of State at the F[oreign] O[ffice] supports the Navy Bill, blows a fanfare for cruisers and shoots at sparrows with a cannon. What kind of gunpowder is going to make any impression later when really serious demands are put forward? The Secretary of State at the Reich Navy Office beats his breast and says: as long as I stand here no limitless plan will be brought in, i.e. no iron-clads. Then the wicked uncle will turn up in April. The Reichstag sees what is happening and . . . at the very least undermines future demands by approving a few ships this time . . . According to people who know what goes on in the Reichstag, this manner of proceeding will have made it almost impossible for any iron-clads to be approved for years . . . That is the situation and you will of course know best whether the best solution might not be to leave the present secretary of state where he is for a few more years. The worst thing is that the whole nation has been poisoned with this nonsense about cruisers.' He went on to ask, in obvious irritation, whether 'His Majesty would adhere to his earlier decision nevertheless'.[86] The Kaiser set about rebuilding Tirpitz's morale. On 31 March Senden, writing from the *Hohenzollern*, informed the Admiral that 'His Majesty shares the view expressed in your letter of 20 March that a new secretary of state would arouse mistrust in the present situation in parliament . . . His Majesty will therefore defer making any change for the moment, but asks me to tell you expressly that postponement does not mean cancellation, and that His Majesty will make a point of using his influence in line with the ideas which you expressed both in your report of 3/1 96 and in the discussion at the end of January.'[87] When the Kaiser returned to Berlin on 29 April after his Mediterranean cruise, Tirpitz was on the point of leaving for China to take over command of the East Asian cruiser squadron.

Whether at this time Tirpitz really believed that he would still have a chance to realise his plans is doubtful. But in fact his appointment and his great fleet building programme had only been postponed, as the Kaiser had said, especially as naval officers and the imperial entourage, above all Senden, constantly kept up the pressure for expansion. Holstein wrote that Wilhelm was being encouraged by 'Senden and Co.' to believe that 'it is only due to the ineptitude, ill-will and lack of interest on the part of his government that the money for the gigantic fleet hasn't yet been granted, or that the first steps have not yet been taken to secure the appropriations – for instance, a *coup d'état*, change of constitution, etc. Hence the imperial bad mood against the government that is aired in nasty marginal comments.'[88] After meeting the Kaiser at Kiel and Travemünde in late June 1896 Waldersee recorded that Wilhelm 'evidently felt happiest at sea and among navy people . . . I now have no doubt that the Kaiser is sticking to his great navy plans, which he hinted at in January, and that he will come up with very big demands in the autumn. Of course he is constantly spurred on by the naval officers who flatter him appallingly and to whom he continues to show great friendliness. What concerns me about the whole thing is that the Chief of the Naval Cabinet, Senden, is a singularly insignificant man.'[89] The Empress Frederick was deeply shocked to realise, after a visit from Wilhelm in October 1896, that her son's intentions were now to seize for himself the hegemonial position hitherto held by Great Britain by building 'ships for the German Navy that shall beat the english', as she reported to Queen Victoria. 'William admires England very much and is very fond of you', she wrote; 'but he is not steady & cool headed and far sighted enough to see that to strain every nerve for Germany to succeed in outdoing England & *wresting* fr[om] her the position of supremacy she has in the world – is simply *nonsense*!'[90]

In the imperial entourage, meanwhile, efforts were being made not only to get rid of Hollmann, but also to undermine the position of the Chancellor, and here again the prime mover was the Chief of the Naval Cabinet. In the summer of 1896, during a visit to England, Senden declared that 'the great fleet programme must be pushed through and that the Kaiser intended to push it through. Hohenlohe, however, was too old for this sort of thing, and therefore the Kaiser was often talking about bringing in new blood.' Those who heard him assumed he meant Bernhard von Bülow.[91] The Kaiser expressed similar sentiments to Waldersee at Kiel in late January 1897 in the conversation described in chapter 28.[92] After his remarks about the possible need for a violent *coup d'état*, Wilhelm drew a graph in coloured pencils 'showing the growth of the French and German navies, and began talking about the need for the expansion of our fleet. Finally he gave me the drawing and recommended me to study it', Waldersee recorded.[93] He suspected that Wilhelm would have to appoint a new chancellor, as much to implement a domestic policy of force as to carry out the expansion of the fleet, and wrote: 'The fact that the Kaiser gave me the drawing and emphasised strongly that a great deal would have to be done for the navy soon, combined with Hohenlohe's opposition to the naval plans, suggests to me that a change of chancellor might arise out of this question.' Even if, 'with this very lively gentleman' one could never know 'what else might crop up to change things', one should never forget 'that with the Kaiser developments sometimes happen very quickly, and I know that in recent weeks he has often spoken of his intention of making a major change in the spring'.[94] These documents show clearly how determined the Kaiser already was, even *before* Tirpitz's appointment, which it was generally thought would happen in the autumn,[95] to implement a 'gigantic fleet plan' at all costs.

Meanwhile it was becoming increasingly clear that the Reichstag, not least out of opposition to Wilhelm II, would reject the comparatively moderate addition to the fleet which the government was seeking.[96] Hollmann had a last chance in March 1897, and presented the Reichstag with a new plan. He asked for 40 million marks for cruisers, torpedo boats and an iron-clad. The political omens were bad and there was little hope that the Secretary would win approval for the estimates. As early as 4 February Hohenlohe informed Eulenburg: 'The Kaiser is thinking of letting Hollmann go at the end of the year because he is not the right man to put the gigantic fleet plan into effect. I can already tell you today that that gigantic fleet plan is a practical impossibility. The present Reichstag will not accept it, at least not so soon before the elections; and if the voters even suspect that something like that is afoot, the next Reichstag will be even worse than the present one.'[97] Holstein too was again beset by anxiety that the Kaiser was heading towards a 'Reich *coup*', which the federal princes would never support. He regarded the 'gigantic fleet' as the 'real goal . . . to achieve which the Kaiser is prepared to gamble the peace, one might even say the existence, of the Reich'.[98] These were no imaginary fears, for Wilhelm did not hesitate to issue threats and attempt to intimidate his opponents. He clashed with the Reich Chancellor and Hollmann when he set up imperial committees on the naval question without their consent.[99] He threatened military intervention if the South Germans opposed his plans.[100] And on 5 March he told Hohenlohe that it was his duty, just as it had been his grandfather's with the army, to ensure that the navy was strong enough to fulfil its tasks.[101] If the Reichstag refused him the funds required, 'he would go ahead with construction and send the bill to the Reichstag later'.[102]

As we have already seen, Wilhelm also tried to intimidate the members of parliament, above all the Centre Party, whose support was crucial if the fleet plans were to receive parliamentary approval, by threatening dissolution and a *coup d'état*.[103] But this proved counterproductive and failed to achieve anything. Hollmann did not succeed in getting the estimates accepted by the Reichstag;[104] instead, they were cut by some 12 million marks. On the grounds of this drastic reduction Wilhelm demanded that the luckless Secretary of State tender his resignation – which, as Hohenlohe pointed out to the monarch, was inadmissible, since under the Reich constitution the secretary of state at the Reich Navy Office was answerable to the Reich chancellor, and therefore it was he, the Chancellor, and not Hollmann, who would have to resign.[105]

Wilhelm II was beside himself with rage. When Waldersee came to Berlin in mid-March 1897 for the confirmation of the youngest son of the Prince Regent of Brunswick he found a state of 'growing confusion'. The Kaiser came up to him at once and held an animated discussion about the naval estimates, which had just been rejected, and about foreign affairs in general. 'What do you say to the rejection of the ships in the Reichstag committee?', he asked angrily. As Waldersee recorded in his diary, 'a conversation now developed from which it was plain that the Kaiser is extremely embittered. He told me that both Hollmann and the Chancellor had tendered their resignations, that he would accept Hollmann's but that in Hohenlohe's case he wanted to think it over; he would close the Reichstag as soon as the budget debate was over. He expressed himself particularly strongly about the Centre and said "The whole thing is a fight by the Catholic Church against the Protestant Kaiserdom." Unfortunately this revelation has come to him a little late', wrote Waldersee. 'I then asked him how he stood with the German princes, in case relations with the Reichstag deteriorated still further, and he replied that the Prince Regent of Bavaria could not be trusted at all. I told him that the Centre was undoubtedly motivated by malice; it had taken offence against him for various reasons and was taking this opportunity to pick a quarrel, to show him its power, and moreover it knew very well that the rejection of 2 cruisers and a few torpedo boats was not sufficiently important to enable Him to use it as grounds for dissolution . . . From this conversation and from other information I received it was clear to me that the Kaiser does not know what to do; he is deeply hurt and also very dissatisfied with both Hohenlohe and Marschall, but does not yet know what further action to take. Evidently he has not yet decided on a great step, but is still considering that he might *perhaps* have a change of chancellor after the closure of the Reichstag.'[106]

No one was in any doubt that the Kaiser alone was the driving force behind the expansion of the fleet. Waldersee recorded on 19 March 1897 that Wilhelm had 'not accepted' the rejection of the two cruisers by the Reichstag committee and had sent Hohenlohe, Marschall and Hollmann 'into the fray', but again without success. 'What will the Kaiser do?', the General wondered, when the increased naval estimates were rejected in committee for the third time. 'After everything that has happened he ought to do something energetic and Hohenlohe really should not stay on. But I think nothing will happen, nor will Hohenlohe go, so that the dangerous situation will continue. It is very sad how freely the Kaiser is drawn into the Reichstag debates; people say quite calmly that he alone is pushing for the expansion of the navy, and unfortunately that is quite true.'[107] The Kaiser's defiant attitude towards the Reichstag was also apparent in the 'strange' proposal, which had been 'quite seriously discussed', that the naval estimates should be submitted to the Prussian Landtag. 'I have got used to thinking that anything is possible', Waldersee sighed. 'So why shouldn't such an idea find its supporters too?'[108]

THE FIRST NAVY BILL

Wilhelm did indeed seem to be ready to consider anything to prevent his fleet programme from foundering altogether. He contemplated dismissing Hohenlohe and talked of his grandfather and the situation in 1861; on the other hand, he was heartily tired of haggling with the Reichstag over every detail of the naval estimates.[109] The only way out seemed to be to draw up a new plan providing for a fleet of a fixed strength, and this was the course he now followed, even before Tirpitz took over the Reich Navy Office.[110] He gave instructions for a Navy Bill to be drafted, stipulating that the German fleet must be half as large as the combined Russian and French fleets.[111] On 7 April 1897 he consulted Commander-in-Chief of the Navy Admiral Eduard von Knorr and the interim head of the Reich Navy Office, Admiral Büchsel, as to how large such a fleet would need to be. Knorr, who had at his disposal the calculations made by his former Chief of Staff, Tirpitz, submitted a plan which specified two squadrons of ships of the line and twelve additional ships, that is to say a total of twenty-eight ships of the line, seven large and twenty-one small cruisers, as the necessary strength. In addition to this there were the twelve existing armoured coastal ships.[112] The Reich Navy Office, meanwhile, worked out the substantial sums that would be required for this construction plan.[113] At a conference held on 19 May 1897 Wilhelm had the plan significantly modified, in particular ordering a considerable increase in the cruiser component. In his view the composition of the fleet should be statutorily fixed at twenty-five ships of the line, eight first-class cruisers, thirty second-class cruisers, sixteen fourth-class cruisers, five gunboats, fourteen torpedo division boats and ninety-six torpedo boats. The whole programme was to be completed by 1 April 1910. At the same time a replacement programme for ships that were to be put out of service was to be implemented. The costs would amount to 54 million marks a year, reaching a total of 833 million by 1910. The crucial question, of course, was how the Reichstag could be induced to pass such a law.[114]

In fact the Kaiser himself was still the greatest obstacle to the acceptance of what were regarded as his limitless demands for the navy. In August 1897 Philipp Eulenburg felt compelled to inform his beloved friend that opinion throughout Germany was unfavourable and that the Reichstag was arming itself 'for a kind of policy of resistance' against his naval plans. 'The resistance is unfortunately directed very much against Your Majesty personally, because the German philistines see the approaching Navy Bill more as pandering to a hobby of Your Majesty's than fulfilling Germany's need. The opposition of all political shades is taking this line – a *lamentable* picture of German narrow-mindedness! In view of this undeniable fact it seems to me *tactically necessary* that the *professional* standpoint and that *alone* should be brought to the fore. And Tirpitz is the man for that. I think that it is important for *tactical* reasons that *Your Majesty* should now do as little as possible in the matter, so as to leave the emphasis on the *professionals* in public.' To spare his friend's feelings, Eulenburg added: 'Your Majesty knows very well that I – and all who have some insight into naval affairs – cannot but consider that it is precisely Your Majesty whose *professional* opinion carries the most weight, but the great masses *do not believe it*.'[115] On the very day that this letter was written Alfred Tirpitz, who had meanwhile returned from East Asia, came to see the Kaiser at Wilhelmshöhe for his first audience as secretary of state at the Reich Navy Office. A difficult and disastrous partnership had begun.

At first the Kaiser had hardly been able to wait for the Admiral to take office. Tirpitz, he wrote on 2 April 1897, was 'one of the most talented and notable personalities he had ever come across. He could become anything, even Reich chancellor.' He was a 'tough customer' and would not allow himself to be 'fobbed off' by either the Reich Treasury Office, the Reich Chancellor, the Reichstag or the federal governments as Hollmann had done.[116] The object of all this praise was less enthusiastic about his All-Highest master. After their crucial first meeting at Wilhelmshöhe on 18–19 August 1897 Tirpitz noted that the Kaiser 'made many remarks that showed that he does not live in the real world. The Germans only needed to be properly shaken up and to feel the reins. He would throw a firm anti-socialist law and an army cadre law at them for the next elections, and that would soon make them vote for the navy. He let fly with outrageous attacks on Bronsart, saying he had turned the whole country against him, and that was his thanks for the Order of the Black Eagle.' Wilhelm did not agree to the new Navy Secretary's request that he should give up his opposition to the new court martial regulations so as to encourage the Reichstag to be more accommodating on the fleet question, but he approved all the Admiral's proposals relating to the Navy Bill itself.[117] According to its provisions, the fleet was at first to comprise twenty-one ships of the line, eight (existing) armoured coastal ships, sixteen large and thirty small cruisers – considerably fewer than the Kaiser had stipulated only a few weeks earlier.[118] Prince Heinrich was surprised by the modesty of the naval estimates. They were 'cut down to the minimum', he wrote to Tirpitz after studying the bill, but were also 'based on an absolutely realistic and objective approach . . . Common sense tells one that any child must be able to see the necessity of it.' But Tirpitz's task would be no easy one, he warned, promising that 'you may be sure of my help whenever it is needed'.[119] In a memorandum which he drew up for the Kaiser after his audience Tirpitz made it plain that the fleet he built would be directed not only against Russia and France but at the same time against Great Britain.[120]

The support given by Wilhelm II and his brother to Tirpitz's plans, which in fact amounted to a fundamental shift in German foreign policy, was all the more critical because the new Navy Secretary, who in constitutional terms was under the direct control of the Reich chancellor, was being pressed by Prince Hohenlohe and other members of the Reich government to give up his dramatic fleet plans. Tirpitz told the Chief of the Naval Cabinet in confidence on 11 August 1897 'that attempts have been made to induce me to abandon the larger, longer-term demands for the navy and to limit my demands to replacing a few ships. I turned down this request very firmly and threatened to resign.'[121] Support for the gigantic fleet plan could certainly not be expected from Bismarck, whom Tirpitz visited on 22 August at Friedrichsruh, as we have seen. After his conversation with the 82-year-old former Chancellor the Admiral noted despondently that Bismarck had not grasped the fact that it was in Great Britain's power to 'wrestle Germany down' by imposing a strict blockade. 'He was thinking of the political situation of England in 1869 and does not understand how powerful her position is in 1897.'[122] The diplomat Rear-Admiral Karl von Eisendecher made a similar observation. 'From what I know of the Prince [Bismarck], I am not surprised by his somewhat cool attitude, prejudiced by ideas adopted from Roon, towards our efforts for the navy. He always treated the fleet in a rather offhand manner, and above all without any knowledge of the subject, and he was never enthusiastic about developing the fleet.'[123] Only the Grand Duke of Baden, whom Tirpitz met while taking a cure at Bad Ems, declared himself ready to give the Navy Bill his warm support.[124]

Wilhelm II was soon proved right in thinking that Tirpitz would be very different from Hollmann when it came to dealing with the political authorities. First, he succeeded in systematically building up enthusiasm for the navy, thereby putting the Reichstag under pressure 'from below', as it were. In June 1897 he unleashed a propaganda campaign, about which there had been previous deliberations, but which was nevertheless genuinely his own work.[125] The Kaiser reported enthusiastically to Eulenburg: 'Tirpitz has just organized a huge office which . . . will look after *maritima* in some 1,000–1,500 newspapers and magazines. In the great university towns all over the country the professor class has met us willingly and is going to co-operate by speaking, writing, and teaching about Germany's need to possess a strong fleet.'[126] Next, Tirpitz took the members of the Reichstag by surprise with the clarity of his language and the logic of his demands for two battle squadrons. The deputies expressed satisfaction with the new style in which they were being treated. As Volker Berghahn writes, Tirpitz succeeded in winning them over with tactics that combined concealment of his true plans, pressure, the element of surprise produced by his sensible and systematic approach, clarity of presentation and moderate initial demands. They were no longer confronted with demands for a hotchpotch of different types of ship, the purpose of which was never explained, as in the Hollmann era. Instead, they were presented with a clear concept, the aim of which was home defence with moderately sized battle squadrons, and the deputies had no desire to deny their consent to this well-integrated plan.[127] Wilhelm did what he could to smooth Tirpitz's path. He helped him both against the opposition within the navy and against Miquel.[128]

After the Navy Bill had been passed by the Reichstag by 272 votes to 139 on 28 March 1898, giving the navy a budget of 408 million marks, Tirpitz was fêted by the Kaiser, who invited him to dinner and, as we have seen, appointed him a Prussian minister of state with a seat and a vote in the Ministry of State, in spite of the strong resistance of the other ministers. This appointment was to help him strengthen his position at the expense of the civil government of the Reich.[129] Wilhelm swept aside the constitutional objections to his action, declaring in a letter to Hohenlohe: 'Certainly the navy is not under the control of the Prussian authorities, but it is under that of the King of Prussia, who is German Emperor. And like the army it is a component part of our national defences, with absolutely equal rights and equal value, and may therefore also be so bold as to crave a hearing at the illustrious gatherings of the Ministry of State. The magnificent development of our trade, its repercussions on Prussian affairs, the opening up of great areas in the market, in the national economy etc., will give Prussia an increasingly predominant position. Moreover the Admiral, who had only just come home from China, cheerfully took on the enormous task *alone*, in spite of being in poor health, of educating an *entire* population of 50 million unruly, uninformed, bad-tempered Germans and of converting them to a completely opposite point of view, and in eight months he achieved what seemed incredible to everyone. A powerful man indeed! Anyone who can carry out such a gigantic undertaking so splendidly, even contrary to the opinion of most of the ministers, must be in my Ministry, and with full rights! Nor can it be anything but an honour for this most learned body to be able to count such a man among its number and, I hope, to be inspired to energetic work by the fresh impetus and the broad outlook that he will bring to it.'[130]

Kaiser Wilhelm continued to give Tirpitz his support, as for instance over the elimination of the navy's other decision-making authority. In 1899, as we have seen, he dismantled the Naval High Command, replacing it with several separate posts answerable directly to him, on the grounds that this would make it easier for him to intervene in naval affairs. Wilhelm told his brother Heinrich that the 'principal motive' for the reform was 'so that the navy can feel that the monarch is actually its chief, as with the army'. Prince Heinrich replied from Shanghai expressing his bewilderment at this arrangement, commenting at the same time in a letter to the Chief of the Naval Cabinet that he had 'never yet seen anyone who had any doubts about the direct relationship between the monarch and the navy, but that now he might easily be held responsible for much that was not his fault at all! Somebody must tell the Kaiser the plain truth!' The new organisation was 'incomprehensible' to him; he and his officers had been 'speechless' when they heard the news, especially since before his departure from Germany he had written advocating the retention of the High Command. He did not understand, he wrote, how anyone could take on the responsibility of recommending a reorganisation of this kind.[131]

THE 'GIGANTIC FLEET PLAN' AND ITS AIMS

Tirpitz seems from the outset to have regarded the Navy Bill of 1898, which many considered astonishingly moderate, as only the first step in an ambitious plan that was intended to secure for Germany the status of a world power alongside Great Britain, the United States and Russia.[132] Before his meeting with Wilhelm II at Wilhelmshöhe in August 1897 he had explained to the Chief of the Naval Cabinet his reason for holding back initially: there was no need, he said, 'to argue now for funds which need not be requested until later'.[133] Soon after the Navy Bill was passed, therefore, he was planning the next stage. In April 1898, when he demanded the break-up of the Naval High Command and the transfer of the entire responsibility for the navy into his own hands, he argued that 'in the economic struggle in which the nations will inevitably be engaged in the next century . . . it will become increasingly necessary to defend Germany's maritime interests with military power'; this being the case, the Navy Bill certainly provided a 'secure basis for the Reich's naval power', but it must be brought up to 'the required strength'.[134] The ominous nature of the 'Tirpitz plan' is much more evident in these reflections on future developments than in the apparently modest Navy Bill of 1898. Nor was it long before the British recognised that the types of ship being built by Tirpitz represented a direct threat to their worldwide maritime power at its most sensitive point, that is to say in the North Sea, and began to take far-reaching strategic and diplomatic measures to counteract it. The fatal 'road into the abyss', as Prince Lichnowsky called it, lay open.

As Senden had warned, the new Secretary of State at the Reich Navy Office proved to be an extremely difficult character. Almost the complete antithesis of Bernhard von Bülow, 'slippery as an eel' by nature, who was appointed foreign secretary at the same time, Tirpitz was determined to get what he wanted or resign. Even in these early years he repeatedly held a pistol at the Kaiser's head with threats or actual requests to resign, in order to force Wilhelm to give in. When the Reich Chancellor told Wilhelm in October 1898 that Tirpitz was thinking of resigning, Wilhelm commented that the Admiral was a 'neurasthenic' who was unable to 'obey'.[135] He also repeatedly used these same Bismarckian methods to force his supporters to implore him to stay. 'You are *indispensable* and *irreplaceable*!', Prince Heinrich assured him after one such threat to resign. 'Try to speak privately to the Kaiser, I beg you. He is human as we all are, and will listen to calm, objective arguments from you! Objectively speaking you are right; but I beg you to make allowances for humans and their failings . . . I myself fully and completely support you, your work in the past and I hope in the future for the benefit of the navy, as I have hitherto.'[136]

On the other hand it is undeniable that in spite of his prickly character Tirpitz tried to make allowances for the Kaiser's vanity, the better to achieve his own ends. Before every audience he made detailed notes for himself, not only on the subject to be discussed but also on the form of words in which he could most effectively convey his wishes to the monarch. He made frequent use of the courtier's device of implying that the Kaiser himself was the initiator of a proposed measure. A typical note, in this instance for an audience on 28 November 1898, reads: 'If I have understood Y[our] M[ajesty] correctly, in our conduct over the next few years we should aim to create a situation which will facilitate the introduction of an amendment to the Navy Law.'[137] In the course of this audience Tirpitz persuaded Wilhelm to put the decisions which he took on the basis of their joint discussions into the form of All-Highest Cabinet Orders, 'so that they cannot be departed from without careful examination and a new decision by Your Majesty'.[138] In other words, Tirpitz institutionalised the Personal Monarchy in his own interest by establishing a system of imperial commands. The reason which he gave for this move was that the future expansion of the fleet must proceed along planned lines in order to ensure its success. The previous years had shown 'how much the progress of our fleet has suffered from the fact that no specific target for each stage and no corresponding systematic procedure' had existed.[139]

The frequent audiences and memoranda in which Tirpitz set out his far-reaching ideas to the Kaiser leave no room for doubt that Wilhelm II was well aware of the anti-British aims of the 'Tirpitz plan' and also that he was fully in sympathy with them. At the above-mentioned audience of 28 November 1898 he agreed to the Admiral's suggestion that the next stage should be to request a third squadron 'complete with cruiser groups and torpedo boat divisions, as a squadron to be at Your Majesty's disposal for use abroad'. In order to push ahead as fast as possible with the expansion of the fleet the Kaiser approved the postponement of coastal defence works, as Tirpitz had persuaded him that these could not be carried out at the same time as 'the rapid creation of a fleet'. 'For the sake of Germany's political position and to make it easier to preserve peace while avoiding the risk of serious political humiliation' as well as 'in case of war' it was 'incomparably more important to create a fleet quickly than to

52. Wilhelm II as admiral of the fleet in 1899

strengthen coastal defences', he urged.[140] In another report Tirpitz pointed out to the Kaiser how little the importance of the navy in warfare had been appreciated since 1870. No one had understood that 'supremacy at sea' must be 'the primary goal', therefore no one had accepted the obvious consequences of this in terms of the types of ship and the structure of the fleet; 'no one has grasped that coasts, colonies and commercial interests can only be protected by [a] battlefleet with a unified structure under a unified command'; no one had seen what history proved, that is to say 'that everything is decided by (concentration of forces) [and] battle'. As long as these old-fashioned ideas prevailed Britain would continue to enjoy supremacy at sea, he argued, referring to Mahan's theories. Germany must therefore concentrate its efforts entirely on the rapid creation of a battlefleet. The cruiser warfare which had hitherto been expected to happen would indeed play 'a certain role in the future too, but it will be at the great bases Suez, Gibraltar, Singapore, the Channel, that is to say at strategic points, that concentration of forces and therefore battle will take place'.[141]

Tirpitz's audience at the East Prussian hunting lodge of Rominten on 28 September 1899 marked another crucial stage. This time the Admiral proposed that the next phase of the battlefleet building programme be brought forward. He emphasised that it was his intention, as far as possible, to include 'all unpopular demands' in the next Navy Bill, and 'to state the purpose of the development from the outset'. For the home fleet he proposed to ask for statutory funding for forty ships of the line, eight large cruisers, twenty-four small cruisers and ninety-six large torpedo boats. For the overseas fleet he planned to demand five ships of the line, one large cruiser, three small cruisers, four to six gunboats and a flotilla of torpedo boats for Asia, and two to three large cruisers, three small cruisers and one to two gunboats for America. This target was to be reached in two phases. In the first phase the focus would be on the Third Squadron, the overseas ships and the modernisation of older ships, while in the second phase the ships of the Siegfried class would be replaced by ships of the line. The Secretary of State dangled an enticing, glorious and completely illusory future before his sovereign, whose obsession with the fleet and craving for recognition he knew well how to exploit. 'As soon as [the] goal has been achieved', he promised, 'Your Majesty will have an effective strength of 45 fully equipped ships of the line. A strength so considerable that it is inferior to none but England. But even in relation to England we undoubtedly have a good chance of success in terms of our geographical situation, defence system, mobilisation, torpedo boats, tactical training, planned organisational development, integrated command through the monarch. Apart from the fact that in a conflict our prospects would by no means be hopeless, England will have lost, both on general political grounds and simply from the sober point of view of the businessman, any inclination to attack us, and will consequently have conceded to Your Majesty such a degree of sea power that Your Majesty will be able to pursue a grandiose overseas policy.' According to his own notes, Tirpitz went on to say: 'If Your Majesty agrees and commands that I should proceed with this aim in mind, I promise Your Majesty to devote all my efforts to accomplishing it. [There will, however, be] no possibility of success unless all naval measures are in accordance with and subservient to this great aim . . . The creation of an effective fleet [is] such an indispensable necessity for Germany that without it the country would face ruin. 4 World Powers. Russia, England, America and Germany. Because 2 of these World Powers can be reached only by sea, national power at sea is of prime importance.' 'Salisbury's dictum – the great states are becoming greater and stronger, the small ones smaller and weaker – is my view too', Tirpitz affirmed, adding, 'As Germany has lagged behind particularly in relation to sea power, [it is] a question of life and death for Germany as a world power and a great civilised state to make up for what has been missed. Both so as to create and maintain sea power in the narrower sense (fleet) and because it signifies power in itself, Germany must keep her population German and develop further into an industrial and commercial nation on a worldwide scale . . . With commercial and industrial development on this scale the points of contact and conflict with other nations will grow, and therefore power, sea power, is indispensable if Germany is not to go rapidly downwards.' The Kaiser, the Secretary noted with satisfaction, expressed his agreement with this point of view and authorised him to proceed along these lines. He had already discussed the question with Senden, who was of the same mind but feared that it would not be easy to get the plan through parliament. 'His Majesty commented that these difficulties were not important', Tirpitz noted; 'if need be, the iron pot will then smash the earthenware pot.'[142]

THE NAVY BILL OF 1900

During his audience at Rominten Tirpitz had spoken of bringing forward the second phase of his great plan, but by this he had meant that the second Navy Bill should be introduced in 1901 or 1902. He had even asked the Kaiser not to speak publicly about it too soon, but had underestimated both Wilhelm's passion for the navy and his machiavellian tendency. As in 1896 after the Transvaal crisis, he saw Salisbury's intransigent stand on the question of the Portuguese colonies, which he described as a 'violation of German colonial interests', as an opportunity to exploit the disappointment of German expectations to drum up support for fleet building and to spike the guns of the 'stupid Reichstag'.[143] And as feelings were again running high against Britain after the outbreak of the Boer War on 11 October 1899, Wilhelm was sure, as before, that the ideal moment to strengthen the fleet had arrived. At the launch of the ship of the line *Karl der Grosse* in Hamburg on 18 October 1899, without the prior knowledge of the Navy Secretary he made a speech which included the celebrated remark: 'A strong German fleet is sorely needed.'[144] At his next audience on 23 October Tirpitz was obliged to express his gratitude for this mark of imperial support, but in reality he was indignant at the Kaiser's impulsive action. He needed more time for the careful preparation of the bill and also to build up popular support for it. He sought the Kaiser's permission to declare publicly 'that no new bill was being planned for 1900' and to hint at 1901 instead.[145] But the ball had been set rolling and could not be stopped, especially as questions were bound to be asked in the Reichstag about the Kaiser's remark. In the end Tirpitz decided not to apply the brakes but to go ahead, if public opinion were indeed thought to be favourable as a result of the events in South Africa.[146] He assured Hohenlohe that he had not known about the Kaiser's speech in advance, but that it must now be followed up. In his view a second Navy Bill would now have to be introduced in the autumn of 1900, and also it would be necessary to offer the Centre Party the repeal of the anti-Jesuit law in exchange for their support.[147] Hohenlohe, on the other hand, was sceptical and considered the moment inopportune, for numerous parliamentary reasons. In addition, he did not believe that German public opinion really shared the Kaiser's conviction that a strong fleet was necessary. 'I do not think the enthusiasm produced by His Majesty's speech in Hamburg was so great that public opinion and the Reichstag will happily agree to the idea behind the new bill', he wrote.[148] In the end Hohenlohe's desire to postpone the matter was overtaken by the publication in the *Norddeutsche Allgemeine Zeitung* of a remark by Tirpitz which gave the impression that 'the plan to increase the navy [had] already been *decided* by H.M.'.[149] In this way, as Berghahn comments, Tirpitz had thrust aside the Reich Chancellor's objections 'in a manner . . . that was characteristic of the constitutional practice [he] adopted of direct interaction with the monarch'.[150] Wilhelm was more convinced than anyone that the favourable moment must be seized. On 2 November he sent telegrams to the German federal princes stressing the 'necessity' of the enlargement of the fleet and making clear 'that He intended to have a new bill introduced in the Reichstag this winter'.[151] The Reich Chancellor had no alternative but to write to the governments of all the German states himself, as also to the Statthalter of Alsace-Lorraine, to explain the necessity of strengthening the fleet and inform them that the new bill would be introduced soon.[152]

But unlike Hohenlohe and Tirpitz, who intended to bring about a parliamentary majority for the bill by careful and conciliatory methods, the Kaiser continued on his usual collision course with the Reichstag. He had grave reservations about the Admiral's suggestion that the repeal of the anti-Jesuit law should be offered to the Centre as 'compensation', he declared.[153] It was at least his spoken intention to dissolve the Reichstag if necessary. Wilhelm's complete refusal to think in terms of parliamentary feasibility and to make concessions to the parties whose support was crucial drove the old Reich Chancellor to the verge of distraction.[154] Even Eulenburg, Bülow and Lucanus were now complaining about the Kaiser's habit of 'frequently [making] pronouncements without contacting the government at all, so that no one in official circles is prepared for them, and they then have to be modified or changed to restore harmony'.[155] But Wilhelm was untroubled by any scruples. On his return from his visit to England in November 1899 he telegraphed to the Chancellor from Flushing: 'I see from German press commentaries shown to me at Port Victoria that there are doubts as to whether the Navy Bill will be accepted by the Reichstag. Your Highness knows from My expressions of My will in the Crown Council and from what I recently told Your Highness at your house after dinner, that I am determined to dissolve the Reichstag if it should refuse to agree to the strengthening of our naval armed forces, which is absolutely essential for our security and our future. All other thoughts and considerations must take second place to this question, which is one of life or death for the Reich. In this question there is no going back for Me, just as for My revered late grandfather there was none in the question of army reorganisation. If we did not use the present moment to complete and expand the Navy Bill of 1898 we should be guilty of a political mistake which could never be rectified. From the standpoint of foreign policy too, the present moment is favourable and right. Not only did the Emperor of Russia recently congratulate Me on my speech at Hamburg and on the planned strengthening of the fleet, but during My visit to England, which has gone off in the most satisfactory manner in every respect, I have received assurances both from the court and the English ministers and from the officers of the Royal Navy with whom I have been in contact that the proposed strengthening of our sea power would be welcomed in England. I therefore fully expect that My government will be fully united and show absolute determination in putting forward the Navy Bill, so as to ensure that it is accepted for the good of the Fatherland. Wilhelm I. R.'[156]

At the turn of the century two German mail steamers were seized by the British on suspicion of carrying contraband for the Boers, a suspicion which turned out to be false.[157] A 'furious' wave of anti-British emotion swept through Germany and the Kaiser's first thought was once again to turn this opportunity to advantage.[158] Years later, in exile, he still took malicious pleasure in recalling the moment when Bülow received the telegram containing this very welcome piece of news, in his own and Tirpitz's presence. He had quoted the English proverb, 'It's an ill wind that blows nobody good', he wrote, and Tirpitz had exclaimed: 'Now we have the wind we need to bring our ship into harbour; the Navy Bill will go through. Your Majesty ought to confer an Order on the English commander as a reward for getting the Navy Bill through.' He then noted that the Reich Chancellor (sic!) ordered champagne, 'and so we three drank to the bill, its passage and the future German fleet, in gratitude to the English navy which had been so helpful'.[159]

The extent to which this triumphant retrospective description was coloured by wishful thinking is shown by the impatient telegram which Wilhelm actually sent on 10 January 1900 to Prince Hohenlohe, who was of course still Reich Chancellor at the time, and in which he showed how little comprehension he had of Tirpitz's careful parliamentary preparations. 'I have today personally given the Secretary of State at the Reich Navy Office, Admiral Tirpitz, the clearest instructions to introduce the Navy Bill within 8 days at the latest. The Admiral made some difficulties on the grounds that it had not yet been "polished" in every detail, and said that he wished to wait a little longer. I consider this wrong. Opinion throughout the population has become so aroused by the capture of the German Reich post steamers by England, and the anger over this insult is growing so much that sooner or later it will try to make itself felt with elemental violence. The first thing will be popular meetings, protest meetings and even more vicious attacks against England and the Queen herself than have already been made. That would considerably disrupt our negotiations with regard to the steamers and might drive the English to extremes against us if there are further setbacks. So the strong build-up of "explosive pressure" here must find an outlet in a way that puts patriotism to the best use for our country. And that is the immediate introduction of the Navy Bill. The Reichstag is in a position of constraint between the crown, the Bundesrat and the people, who are demanding the Navy Bill with increasing vehemence. The Admiral is rather conceited, and wants to "cut a good figure", as one says. So he is still tinkering with and working over his speeches and his bill, to prepare himself for all possible questions and objections. But all that is quite superfluous now. He has had plenty of time from 18 October (Hamburg) until the middle of January to get the bill ready in general outline. It is well known that under Prince Bismarck bills were sometimes drafted and settled in one night. With this [bill] it will not be a question of elaborate arguments or elegant verbal duels, but the bill will simply be introduced and the Reichstag will refer it to its committees, where it will in any case be favourably settled, with the help of pressure from the entire nation. The Admiral can show off all his flowery turns of phrase and all his skill in defence of the bill in the committees. The shorter and drier the speeches are kept at its introduction the better it will be. At such major political moments as the present one, the great stream of the German national idea, which is at last flowing strongly, must be used quickly. Count von Bülow and Herr von Lucanus are fully informed and share my view in every respect. I therefore beg Your Highness to pursue this matter in the way which I have indicated . . . Wilhelm I. R.'[160]

The bill was completed on 25 January 1900. The Kaiser again applied pressure for it to make faster progress, and arranged for it to go straight to the Bundesrat, 'bypassing the Prussian Ministry of State'.[161] In public too he continued to stir up support for the navy, declaring for instance, at the reception on 13 February 1900 for his brother on his return from East Asia, that the 'joyful and enthusiastic' welcome given to Prince Heinrich by 'all classes of people in My residence of Berlin' was an 'unambiguous sign of how great the support for the strengthening of our sea power [had] become'. 'The German people are at one with their princes and their Kaiser', he went on, 'in their wish to set up a new milestone in their mighty development by creating a great fleet to fulfil their needs. Just as Wilhelm the Great created the weapon with the help of which we have become Black, White and Red again [sic!], so the German people are setting out to forge the weapon through which, if God wills, they will remain Black, White and Red throughout all eternity, both at home and abroad.' Thanks to Tirpitz's skilful tactics in making the additional demands for funds look plausible to the Reichstag with reference to unforeseeable developments in arms technology, and above all in winning over the Centre, the government succeeded in getting the second Navy Bill through.[162] Carried along on a wave of anti-British emotion, the Reichstag voted in favour of the measure on 12 June 1900.[163] With this, the Anglo-German antagonism that had already begun to manifest itself to some extent in recent years became an inexorable, indeed merciless conflict which, as we know, was in the end to be fought out not on the high seas but in the trenches of the First World War.

The unequal and uncomfortable partnership between Wilhelm II and his 'evil spirit', Tirpitz, which began in 1895 and was to continue until it came to a stormy end in 1916, proved catastrophic for Germany and the world.[164] The forced battleship building programme robbed German foreign policy of its freedom of manoeuvre. Only a few years after the Navy Bill the diplomats of the Wilhelmstrasse were forced to accept the 'encirclement' of the German Reich by Great Britain, France and Russia, as helplessly as if they had been in chains. Yet the 'limitless' battlefleet programme, driven forward incessantly by Tirpitz with the support of Wilhelm II, continued unabated, and neither the increasingly vociferous warnings of the German ambassadors in London nor the efforts of the Reich chancellors, Bülow and Bethmann Hollweg, nor even the opponents of the Tirpitz plan within the navy and the army, were able to persuade the Kaiser to give way. Wilhelm's passion for the navy, fed by his urge to gain recognition and by dark feelings of hatred for his mother's homeland, had led him, like an addict, into the hands of another naval fanatic who alone seemed able to carry through the 'gigantic fleet plan' which he so fervently desired, in defiance of the Reich government and the Prussian Ministry of State, the Reichstag and the majority of the German people.[165] By withdrawing the trust he had placed in Tirpitz Wilhelm II had it in his power at any time to put a stop to the disastrous battleship building programme, directed as it was towards the overthrow of Great Britain's position as a world power and the supplanting of the balance of power in Europe by German domination. But he did so only after the Iron Dice of War had begun to roll.

CHAPTER THIRTY-THREE

'Young Germany, Your Kaiser!', or what was wrong with Wilhelm II?

THE NEW CENTURY

THE beginning of a new century, then as now, was an occasion for recollection of the past and conjecture about the future, and at the turn of the twentieth century Kaiser Wilhelm II, who now stood at the zenith of his power and yet seemed, so far, not to have himself under control, naturally played an important part in these reflections both in Germany and further afield. What did people think of him, and how did this affect their hopes and fears for the future? Contemporary comments vary widely in their assessment of him, and we shall examine just a few of them. In the admiring eyes of the Archduke Franz Ferdinand of Austria-Hungary, Wilhelm II was 'the grandest fellow in Europe',[1] while Tolstoy reviled him as 'one of the most abhorrent, not to say the most ridiculous representatives of imperial power'.[2] Among his most ardent admirers was an Englishman living in Vienna, who enthusiastically embraced the German language and German culture and was later to marry Wagner's daughter, but who in 1900 had not yet met the Kaiser: it was not until the winter of 1901 that Philipp Eulenburg was to introduce the racial theorist Houston Stewart Chamberlain to Wilhelm II at Liebenberg.[3] Writing in the weekly periodical *Jugend* in the summer of 1900, Chamberlain, the author of *The Foundations of the Nineteenth Century*, predicted a glorious future for the German Reich under its energetic, youthful Kaiser. 'The reign of Wilhelm II has the character of the dawning of a new day', was the jubilant comment of this 'evangelist of race', who twenty-three years later was to praise Adolf Hitler at Bayreuth as the saviour of Germany in its hour of greatest need.[4] Wilhelm II, according to Chamberlain, was 'in fact the first German Kaiser'. He had understood that it was his historic mission to 'ennoble' the world through 'German knowledge, German philosophy, German art and – if

God wills – German religion'. 'Only a Kaiser who undertakes this task is a true Kaiser of the German people', he declared. Wilhelm had recognised that the spread of the 'higher' German language and culture could be achieved only by power, and that power meant sea power above all. The son of an English admiral, Chamberlain asserted confidently that 'without a fleet nothing can be done'. But 'equipped with a great fleet, Germany is embarking on the course to which Cromwell showed England the way, and she can and must steer resolutely towards the goal of becoming the first power in the world. She has the moral justification for it and therefore also the duty.'[5] The picture of Wilhelm II which the militaristic Young Germany League drew in its propaganda piece, *Jung Deutschland, Dein Kaiser!* ('Young Germany, Your Kaiser!'), on the eve of the First World War was to be very similar.[6]

A considerably more sober, indeed more anxious, evaluation of the future prospects of the German Reich comes from the pen of the long-standing Austro-Hungarian ambassador in Berlin, Count Ladislaus von Szögyény. In a report of 5 February 1900 he too drew attention to the battlefleet plans and visions of world domination which now set the tone of imperial Germany's policy, but he did not fail to recognise the enormous dangers of such a challenge to the established world powers. 'The leading German statesmen, and above all Kaiser Wilhelm, have their eyes fixed on the distant future; their efforts are directed towards making Germany's position as a world power, which has taken great strides forward in recent times, a dominant one and in due course they are counting on becoming England's heirs in this respect. But in Berlin people are well aware that Germany is not in a position to come into this inheritance today, nor for a long time to come, and for this reason the collapse of English world power in the near future would not be at all welcome, as no one is in the slightest doubt that Germany's ambitious plans can at the moment only be a dream for the future. – Nevertheless Germany is already eagerly working with all her might to prepare herself for her self-appointed future mission.' In this connection Szögyény drew particular attention to the 'incessant preoccupation with the increase of German naval forces, and to the constantly repeated remarks referring to this aim which are made by H.M. the German Kaiser and his leading statesmen'. But this policy, he pointed out, was planned on a long-term basis. That both Wilhelm and Bülow were reluctant to give in to the present 'wave of popular feeling' against Britain on account of the Boer War and adopt a position that was 'too antagonistic towards England' could be attributed to their pragmatic recognition that they could not expect to enter into the 'possible inheritance which they so ardently desired', that of 'Great Britain's colonial power', except in 'the very distant future'. 'As to the period of time which will elapse before then, Kaiser Wilhelm has the least illusions of all, as he repeatedly indicated that he would certainly not live to see the realisation of these ambitious plans Himself, but He considered it His duty to prepare His country as well as possible for what could be expected to happen.' The ambassador went on to emphasise that German aspirations for world power had caused anxiety not only in Britain but also in Russia, and were indeed seen there as a mortal danger. He could well understand, he commented, why Russia 'would very much like to put a stop to the all too rapid realisation of Germany's aspirations before it is too late', for 'there can be no doubt of how dangerous it would be for Russia if Germany achieved an overwhelming position not only as the strongest land power in central Europe but also in the whole world'. Rightly predicting what was to happen, Szögyény pointed to the possibility that Britain and Russia would come to an agreement on the many unresolved questions that divided the two world empires. 'The spectre of a general coalition against Germany which always haunted Prince Bismarck still has a hold over people's minds here today', he reported. 'There is probably no need to spell out what role England would play in such an eventuality, and for this reason alone it seems quite natural that on the German side strenuous efforts will be made to avoid any serious damage to the relations between the cabinets of Berlin and London.'[7]

The Bismarckian Maximilian Harden was certainly among the most embittered critics of the Wilhelmine *Weltpolitik*. For him it was absolutely axiomatic that the full responsibility for that policy lay with Wilhelm II, together with the Liebenberg circle around Philipp Eulenburg, whom the Kaiser raised to the rank of prince on 1 January 1900. 'The Kaiser is his own Reich chancellor', he wrote in the periodical *Die Zukunft*, of which he was the editor, in 1902. 'All the important political decisions of the past twelve years have been made by him.'[8] For Harden too, the turn of the century and in particular the speech which 'the imperial War Lord' made on 27 July 1900 at Bremerhaven on the embarkation of the German troops who were to suppress the Boxer Rebellion under Waldersee's command, provided him with the occasion to express his views on the Kaiser. He launched his fiercest attack to date upon Wilhelm in his article 'The Fight with the Dragon', which was immediately confiscated, while Harden was sentenced to six months' imprisonment.[9] In his so-called Hun speech the Kaiser, alluding to his 'great immortal grandfather' and the Great Elector, had again used the sort of bloodthirsty militaristic turn of phrase which is familiar to us from his marginal comments and his earlier speeches, and which this time, in spite of the attempts of Bülow and the Flügeladjutanten to suppress the actual words, leaked out and shocked the world. The use of the word 'Huns' throughout the English-speaking world as a term of abuse for the Germans, particularly in both world wars and sometimes even today, can be traced back directly to the Kaiser's command to his troops in 1900 to conduct themselves in China 'like the Huns under their King Attila a thousand years ago', so that 'the name of Germany [shall] become known in China to such effect that no Chinaman will ever again

53. The Kaiser making his 'Hun Speech' in Bremerhaven on 27 July 1900

dare so much as to look askance at a German'. 'Pardon will not be given, prisoners will not be taken', he proclaimed. 'Whoever falls into your hands will fall to your sword.' The soldiers should 'set an example of manly self-control and discipline', and of 'traditional German excellence' as they had learned 'from our military history'; 'God's blessing' would ensure 'that Christianity becomes established in that country. That is what you stand for with your oath of allegiance!'[10]

54. A French comment on Wilhelm II's 'Hun Speech'

This speech, which aroused 'anxious unease' everywhere, convinced Harden that Germany had reached a 'monarchical crisis' as a result of Wilhelm II's Personal Monarchy. In this system there was no one who could force 'the widely roaming fantasy' of the Kaiser 'into the narrow bounds of common reality'. 'No one chases away sweet illusions and warns against setting too much store by the skilfully created but also artificial magnificence of the Reich. Everyone tries to gild the lily. Germany is immeasurably rich; Germany has a mission to become the first among all industrial and commercial nations, and in order to come closer to this goal she must carry her victorious standards overseas; and the Kaiser of the Germans, like the crowned heroes in the days of the crusades, must endeavour to conquer the world for the Gospel. The dreamers and the sly speculators whisper these things into the sovereign's ear . . . In times of peace and calm it is a harmless fallacy, or at least it causes no serious mischief; in times of chaos and confusion it may become a catastrophe.' With unerring accuracy Harden recognised the dynastic cult of the Hohenzollerns practised by Wilhelm, who had always been taught to see his ancestors 'only in legendary, transfigured form', as a major cause of his isolation from the real world and of his growing alienation from the German people, which – he expressly warned – could lead to assassination attempts. 'He who was raised to the throne so early, who takes pride in being the son of his forefathers, looks back and compares. What small means his ancestors had, and what mighty things they achieved nonetheless! Shall he alone, the richly endowed heir of their accumulated strength, be allotted none of the tasks without which the life of the monarch is not worth living and which transform the *roi fainéant* into the benefactor of the Reich?' 'It is perhaps this kind of mood', Harden suggested, 'that has given rise to the general tone of the speeches calling for revenge.' The Kaiser would probably be very surprised to discover how differently most Germans saw the world and their history, how little they had in common with the 'exuberance' of his 'romantic dreams of crusades'. To them 'the splendour of the Reich is in some danger. They see it in a difficult territorial position, surrounded by lurking suspicion and envy, unprepared at home, dependent on unreliable or weak allies abroad, with rapidly increasing prosperity, but lacking the riches which could allow the Reich to risk the gigantic battle with Great Britain, North America and Russia for world power and worldwide commercial supremacy.' The danger of assassination attempts by anarchists was rooted in the alienation between a monarch with autocratic inclinations and his people, for 'where the belief is allowed to grow that all political action springs from the mind of the monarch, the delusion will always take root in some sick or overheated brain that the violent elimination of a ruler who is hated by the impoverished masses is a heroic act from which good will come for the people.' And Harden quite rightly concluded from this observation that the monarchical form of government could be tolerated in the twentieth century only if the crown limited itself to its representational functions under the constitution. 'Kings and emperors can escape the flatterer's poison and the murderer's dagger only if they content themselves with the role that has been allotted to them since the constitutional struggles of our century: the role of the ceremonial representative of nationhood, removed from the squabbles of everyday life, safeguarded by special laws behind golden bars, whose carefully considered words are his deeds, who can do good and can never be held responsible for evil.'[11]

THE BROKEN SPELL: PHILIPP EULENBURG AND HIS 'LIEBCHEN'

The system of Personal Monarchy practised by Wilhelm II and seen by him as especially 'Prussian' represented the exact opposite of the constitutional form of monarchical government. By the turn of the century there was scarcely a single matter of personnel, domestic, foreign or military policy which was not decided by the Kaiser, scarcely ever a public occasion which he allowed to pass without making a speech, scarcely a field in which he did not feel called to make his strongly held opinions known, scarcely a 'colleague' among the monarchs of Europe whom he did not pester with correspondence and visits and offend with jocular digs in the ribs, scarcely a single diplomatic document which he did not bedeck with crude marginal comments. It is therefore hardly surprising that he himself, and still more those who had to live and work with him, began to complain of the great burden of work thus accumulated. His life was 'full of heavy and unceasing work', he wrote to his grandmother in 1899, and there were often 'moments when I fancy the strain is too strong and the burden too heavy to bear'.[12] That winter, he told his friend Philipp Eulenburg, he had had 'a difficult time politically', during which without Bülow he would have found himself 'in a terrible position'. 'What you are for him and for *everyone*, *he* knows best', Eulenburg told Bülow, reporting a remark of Wilhelm's.[13] On 22 June 1899 the Kaiser raised Bülow to the rank of count.

With Bülow's appointment as secretary of state at the Foreign Office in the summer of 1897 (and then as Hohenlohe's successor as Reich chancellor and Prussian minister-president in October 1900) Eulenburg had attained the goal for which he had worked tirelessly behind the scenes, against all the prophecies of doom emanating from the Wilhelmstrasse and throughout countless government crises: the establishment of the Personal Monarchy of Wilhelm II. Nor did the Kaiser fail to give generous recognition to his closest friend's loyal efforts. On the tenth anniversary of his accession he sent Eulenburg an affectionate telegram thanking him warmly for his friendship and support over the past years.[14] Eulenburg expressed his delight in a letter to Hermann von Lucanus, the Chief of the Civil Cabinet. 'The Kaiser has sent me such a moving telegram of thanks, in such – I can only say exuberant – terms, that I am deeply touched and quite overcome. It is quite true that I have *thoroughly* worn myself out and tormented myself for him for 10 years. But . . . he is a *unique, wonderful* Master – we both know that *very* well!'[15]

And indeed the new system, under Bülow's auspices, seemed at first to function very well. August Eulenburg reported from Berlin that since Bülow had taken office 'everything looks rosy', and following the successful passage of the first Navy Bill his cousin Philipp jubilantly exclaimed: 'Could anyone imagine a more magnificent demonstration of consistency, initiative and vision than the way the "limitless plans" of our poor dear Master have turned out, when he was called impulsive, inconsistent and foolish?'[16] Yet the triumphant feelings of the Kaiser's secret advisers were short-lived. For Eulenburg this was the beginning of a time of cruel suffering which was to lead to his enforced resignation from his post as ambassador in 1902 and to culminate four years later in the sensational trials for perjury and homosexual offences which were sparked by Harden's attacks (supported by Holstein) on him and Count Kuno von Moltke in *Die Zukunft*.[17] Blackmail by a Viennese bath attendant in the spring of 1896[18] was followed by the scandal-ridden divorces of his brother Friedrich and his intimate friend Kuno Moltke, whom the Kaiser had appointed military attaché at the Viennese embassy shortly beforehand. In both cases 'Phili' Eulenburg had himself been in danger of being unmasked, but in both cases Kaiser Wilhelm – unlike later, in the terrible days of the trials – showed his loyalty. The imperial favourite was relieved to observe that in spite of Friedrich Eulenburg's enforced dismissal from the army Wilhelm spoke to Philipp 'quite in the old way' while they were together on board the *Hohenzollern* in July 1898. 'In his manner and conduct His Majesty has kept up the old friendly ways with me and even when the full entourage is present he speaks mainly to me', Eulenburg wrote in relief to Bülow.[19] And even when Kuno Moltke's disastrous marriage failed because of the all too intimate relationship between the newly appointed military attaché and the ambassador, Eulenburg himself, the warm relations between the Kaiser and his best friend remained intact. Hearing from Eulenburg of Moltke's marriage problems, another friend, the Württemberg envoy Axel Freiherr von Varnbüler, assured Moltke that the Kaiser would understand everything and protect him. 'The Liebchen', Varnbüler wrote, 'is man enough to put a stop to nasty gossip – and he knows and loves you too well in your peculiarity to allow even the shadow of blame to fall upon you.'[20] Varnbüler's prediction was to prove correct.[21] In fact it was not the Kaiser who lost faith in Eulenburg but Eulenburg who lost faith in the Kaiser.

In the summer of 1897 the tone of Eulenburg's references to his 'dearly beloved' Kaiser in his letters to Bülow was still that of earlier days: 'I felt that my love for him . . . was bursting into flame.'[22] He was 'deeply moved' to hear Wilhelm's self-critical lament that he did not feel anything when he dismissed officers or officials, and that he reproached himself bitterly for this. 'Something is missing in me that others have. All poetic feeling in me is dead – has been killed', the Kaiser had confessed. 'Experiences and experiments in his youth have artificially dammed the balancing qualities in his nature', Eulenburg observed sympathetically. But he still believed that the Kaiser's nature would prove stronger than these constraints.[23] Soon, however, Eulenburg saw his 'Liebchen' in quite another light, as if scales had fallen from his eyes. From the spring of 1898 he began to show concern over the effect of the Kaiser's personality abroad. Over the next months and years this concern was to grow into a panic-stricken terror. He warned of the increasing dislike of Wilhelm felt not only by the Tsar and Tsarina but also at the German courts, like Coburg and Darmstadt, which had close family ties with Britain, and of the threat which this represented.[24] 'We must be very careful', he wrote to Bülow. 'And especially our beloved sovereign. His manner annoys insignificant and weak characters even more than it irritates those who are naturally opposed to him. Even his strong handshake makes people uncomfortable, and that Darmstadt alliance which he himself brought about [Nicholas and Alexandra] is turning into a dangerous breeding ground for hysterical modern bacilli which find nourishment wherever Siegfried's fist is feared or is considered "vulgar".'[25] The new Foreign Secretary agreed that the greatest care was needed, for 'there is much jealousy and antipathy towards the German people abroad, among the nations from Paris to Moscow and Prague and from London to New York, and at the courts there is much very unfair misjudgement and lurking hatred againt His Majesty. Not only Princess Beatrice but almost the entire English-Battenbergian-Hessian-Danish etc. *cousinage* is quietly plotting against His Majesty . . . We shall gain the day only if we work fearlessly but also wisely, with clear vision but also with a skilful hand. Now is not the time for rash experiments; too violent swings of the pendulum could have disastrous results; we need steadiness, calm and shrewdness.'[26] Although Bülow had recently compared Wilhelm II with Frederick the Great he now had premonitions of trouble. 'It is a misfortune', he lamented, 'that our beloved, highly talented Kaiser so easily goes too far and gives too much rein to his temperament and sometimes to his imagination';[27] and Eulenburg, no less anxious, complained of 'that energetic mood' which his imperial friend so often displayed and which made it so hard to maintain a calm, confidence-building foreign policy.[28] In the following months Eulenburg and Bülow exchanged increasingly worried comments about the trouble Wilhelm was causing – his 'frequent visits to ambassadors', his demand for 'a "German" Asia Minor', the 'most extraordinary comments' he had made on the Russian succession on hearing of the sudden death of the Tsar's brother in 1899 – all of which put Bülow's efforts at risk. 'I follow it all with deep feelings of friendship with you; I *feel* what you are suffering', Eulenburg wrote to the Foreign Secretary in February 1899.[29]

The two former admirers of the Kaiser also discussed his 'irritation . . . with Austria and the Austrian court' with concern as early as 1898.[30] In particular, there was deep animosity between Wilhelm and the heir to the Austrian throne, Franz Ferdinand, as a result of the Kaiser's jokes, which had wounded the Archduke '*terribly* deeply', the ambassador reported. At the station in Berlin the Kaiser had greeted the haughty Habsburger with the words: 'Don't imagine that I have come to receive *you*. I am expecting the Crown Prince of Italy.' Worse, in Budapest after a conversation with the Archduke, Wilhelm had remarked that he had not thought that he would be so clever, whereupon Franz Ferdinand, 'white with rage', had exclaimed: 'Did he take me for a cretin?' But in Eulenburg's opinion the chief reason for their mutual antipathy lay in the Archduke's jealousy of Germany's growing power. 'A strong Germany with a ruler of genius is too favourable a soil for the evil character bacilli that dominate the heir to the Habsburg throne', he wrote to Bülow in May 1899.[31] A more serious issue than the relationship with the Danube monarchy, however, was that of Germany's relations with Russia and Britain. During the Scandinavian cruise of 1899 Eulenburg set out to persuade the Kaiser, who was enraged with both world empires, 'that in the end it is preferable to *run after* Russia *and* England than to *anger* them both'.[32]

This voyage was also one of painful disillusionment for the Kaiser's friend. While going for a walk along the seashore on 14 July 1899 Wilhelm and Eulenburg had a memorable conversation about the danger of an enforced abdication by the Kaiser, which not unnaturally made a deep impression on the latter. Wilhelm's friend strongly urged him to show more caution 'at home and abroad'. 'In a dangerous situation of the kind which could perhaps be caused or aggravated by an incautious act on His Majesty's part, the government [could be] *forced out*. That could be followed by a campaign in the Reich to make the Kaiser abdicate or to deprive him of his rights. A structure like the German state was a delicate, subtle creation – a work of art in a glass case . . . If the work of art were not treated with sufficient care the anger of the people would be aroused. The Kaiser grew very grave at this remark and asked again *who* could have such ideas?' Eulenburg refused to name names but quoted a comment by the late Cardinal Prince Gustav zu Hohenlohe-Schillingsfürst, the Reich Chancellor's brother, who had warned that 'the Kaiser should be *very much* on his guard, *very* careful. I know for a fact that the idea of declaring him of unsound mind is in many minds, and that very many people, including highly placed personalities, would be glad to help institute proceedings to do so.' Eulenburg should warn the Kaiser, the Cardinal had said. 'Very much against his usual practice', the ambassador recorded, 'the Kaiser did not end this conversation with a joke or an energetic verbal jab of the broadsword à la 1st Guards Regiment but remained lost in thought.' Later Wilhelm reverted to this conversation and admitted that 'even close to him there were elements who might possibly adopt a dangerous and hostile attitude, e.g. the bigoted, Hanoverian-minded Prince Albrecht [of Prussia, the Regent of Brunswick], also Prince Friedrich Leopold [of Prussia] – who incidentally was bound to be locked up in a madhouse soon'.[33]

But Eulenburg's exhortations to calm and caution were of no avail. 'The life of a kaiser like our master functions like a natural element. It is a cloud that is sometimes white, sometimes grey, sometimes black, and contains rain, hail, storms – and in particular a great deal of electricity', he sighed.[34] A few days after the conversation on the seashore the Kaiser's favourite heard of Wilhelm's telegrams to Hinzpeter at Bielefeld and to the city of Dortmund, which horrified him.[35] He wrote despairingly to Bülow that it was 'really a mystery to me why the Kaiser has to adopt that loud, resounding, energetic tone just because of a memorial to the Great Elector, and why Hinzpeter publishes the telegram!'[36] Shortly afterwards he reported that 'the bomb of the Bielefeld telegram' had evidently not been enough; the Kaiser had now brought the telegram to the city of Dortmund to show him in his cabin, 'with the satisfied smile of a wise statesman', Eulenburg wrote to the Foreign Secretary. 'I must admit that it filled me with consternation . . . The consequences are obvious – and the Kaiser does not see them! . . . An action like that makes the prospects for the future look so hopeless that it makes me very sad indeed!'[37] Not only he but the entire suite on board the *Hohenzollern* were perplexed and appalled by the telegrams, he added. 'I see a kind of bitterness emerging everywhere. In the past I used to argue with two or at most three grumblers; now everyone is grumbling, without exception, in an exhausted, hopeless way that gives the whole suite an oriental character of fatalism – and of bad-tempered fear of the Sultan.' 'Seeing this makes me very melancholy', Eulenburg confessed. 'There is so much I should like to say to him – then his Caliphate chokes me.'[38]

It was only now that Eulenburg recognised the brutal aggressiveness and also the shameful political ignorance of the Kaiser he had idolised for so long. When news arrived on 21 July 1899 of unrest among workmen at Augsburg, Wilhelm reacted with 'great satisfaction' and loudly declared at table: 'Good! . . . Carry on! The moment is coming where we must *act*. And I shall let *nothing* hold me back – not even the Ministry, which will simply be "out on its ear" if it does not co-operate. Be so good as to read my speeches which have been printed and published since my accession. You will clearly see that I have warned the German people, at first in a friendly way and then in earnest, of the dangers which threaten them within the country. The German middle class has completely *failed*! The government must *act*, otherwise all will be lost. If a serious conflict has arisen abroad and there is a possibility that half of the army is tied down within the country by a general strike, we are *lost*. And at the last Hamburg strike England already had its finger in the pie, and did quite well out of it. So the time has come to take action. I have already made enquiries as to how far my military powers go in relation to the constitution. *The Minster of War has told me* that I can declare a state of emergency throughout *the entire Reich at any time*. (!!!) Matters will not improve until the troops drag the Social Democratic leaders out of the Reichstag and gun them down. We need a law under which to be a Social Democrat is enough to get oneself exiled to the *Carolines*.' The Kaiser '*hoped*' there would be 'looting' during the unrest among workers in Berlin and other major cities, he said, because then 'but only after a few hundred middle-class shops have been destroyed' it would be possible to carry out 'a *very* severe blood-letting'.[39]

On some evenings the conversation at dinner in the *Hohenzollern* switched to foreign affairs. The Kaiser spoke at length about his meeting with Cecil Rhodes, who he said had advised him to consider the 'acquisition and opening up of *Mesopotamia*' as his allotted 'task'. The Englishman had told him he should 'build the railway through Asia Minor to the Euphrates – the land route to India', and had remarked that the Kaiser had surely not gone to Jerusalem because of the holy places but had pursued 'other aims' there. The Kaiser had been delighted at this, he said, and had answered: 'You have guessed right! I shall build this railway and reopen these lands of ancient civilisation to the world!'[40]

Hearing these effusions from the monarch whom he had blindly supported through endless struggles and crises since his accession made Philipp Eulenburg feel 'thoroughly unwell', he admitted. The realisation was at last beginning to dawn on him that the Bismarcks, Holstein, Caprivi, Marschall, Waldersee, Wittich, Bronsart and all the other warning voices with their predictions of disaster had not been entirely wrong in trying to put an end to the personal rule of *this* Kaiser. He wrote to his political ally Bülow, whom he had worked for years to raise to the highest offices in the Reich, on 21 July 1899, feeling 'out of sorts and sad': 'In the 11 years of his reign the Kaiser has become very much quieter as far as his outer being is concerned', he commented. 'Spiritually, however, there has not been the slightest development. He is unchanged in his explosive manner. Indeed even harsher and more sudden, as his self-esteem has grown with experience – which is no experience. For his "individuality" is stronger than the effect of experience. That could be a way of describing something else – but it is *not* anything else. He does not belong in our era.' 'Such strong natures cannot help but cause convulsions when they are at the head of a nation', the Kaiser's friend continued. 'We are approaching a moment when a decision will be reached as to whether the epoch or the Kaiser will be the stronger. I am afraid that he will be the loser.'[41]

Shortly before the end of the cruise Eulenburg found another opportunity to warn the Kaiser of the dangers that lay in his autocratic tendencies. The widespread and growing dissatisfaction was beginning to give him 'an uncanny feeling', he told Wilhelm, 'because the parties which are usually so divided are coming together in their common resentment towards Your Majesty'. The old battle with the Bismarcks was still playing a part in the present situation, he went on. 'It culminates in a dangerous conflict between Your Majesty's *personality* and the people as a whole. The undoubtedly *modern* side of Your Majesty . . . is almost progressive in character, but it is paralysed by an energy which makes itself felt too *harshly* in public. By your speeches and telegrams Your Majesty gives the impression of wanting to resurrect the *absolute* monarchy. But that is no longer understood by *any* party throughout the Reich . . . Parliamentarism is deeply rooted in all Germans.' The Kaiser retorted 'rather sharply' that he claimed for himself 'the same freedom of speech as every German man! I must *say* what I want, so that the sensible elements know what to do and whom to *obey*. If I were *silent*, the completely "finished" middle classes would have no idea what to do! . . . You are only afraid that I might use force against parliament.' But 'somehow or other it must come to a clash one day. Everything is going that way, and so one has to *accept* that there will be a fight.' Eulenburg warned strongly against this attitude, which would only rouse the people even more against the Kaiser. 'The anger will be directed against the absolute Kaiser', and he must do everything to avoid 'arousing disastrous and dangerous conflicts!'[42]

The next few months brought no improvement in the sombre mood of the Kaiser's friend, now Prince zu Eulenburg-Hertefeld. He wrote anxiously to the Chief of the Civil Cabinet: 'As long as you, Your Excellency, work closely with Bernhard Bülow and watch over the interests of our beloved sovereign, I sleep soundly. But I tell you frankly that my blood sometimes runs cold when I think of the possibility of a change in this regard!'[43] Shortly afterwards he sent Lucanus another 'heartfelt sigh over the internal situation in Germany'. He was 'not without concern for our beloved sovereign. His position is becoming more and more precarious at home – and more and more splendid abroad. But the latter can balance out the former only *for a time*.'[44] 'I feel how much our *genuine* love for our beloved sovereign binds us together – our *caring* love for Him! . . . I *tremble* at the thought that the Kaiser might lose you! . . . I sometimes wonder whether he realises what we have to bear for Him? I am almost inclined to say no.'[45]

At the end of 1899 Eulenburg called on the Grand Duke of Baden and told him of his serious concern 'about the increasing irritation of public opinion against the Kaiser'. 'He had had frequent opportunities to discuss this state of affairs with the Kaiser and to draw his attention to the dangers for the future', the Grand Duke afterwards reported to the Reich Chancellor. 'Eulenburg went on to tell me that he had had a long talk with Lucanus about this whole business and had urged him to make it increasingly clear to the Kaiser how difficult his position was in relation to public opinion. Lucanus then said that he had done so . . . Count Bülow has also spoken to Eulenburg in the same sense, and has complained very much about the difficulties he faces.'[46] How must these men have felt when they looked back at the endless ministerial and chancellor crises of the past ten years, in which they had consistently taken sides with the crown against the responsible statesmen!

The new century brought no relief, but rather the reverse. Wilhelm took the Boxer Rebellion and the murder of the German envoy in China, as Eulenburg remarked, 'as a *personal* insult', for which he was determined to seek '*revenge*'.[47] He telegraphed to a horrified Bernhard Bülow saying: 'The German envoy must be avenged by my troops. Peking must be razed to the ground.'[48] When he realised that the troops under Waldersee's command would arrive too late to capture Beijing, according to Bülow he was 'so beside himself . . . that he completely lost control' and instructed Eulenburg at Kiel to convey his orders to the Foreign Office to conclude a defensive and offensive alliance with Japan, which Eulenburg only managed with great difficulty – he called it his 'great battle' – to prevent.[49] The Scandinavian cruise of July 1900 which followed this episode was, if possible, even more worrying than that of the previous year. In an attempt to avoid rousing the Kaiser's feelings, Eulenburg begged all his companions on board ship to 'regale H.M. with the most harmless possible stories and to avoid politics', and he himself, 'with the greatest impudence', changed the subject from foreign politics to 'the most banal and trivial matters'. In vain! Already on 15 July Eulenburg had to report to Bülow that there had been 'a violent outbreak' that morning, which 'worries me very much'. He had gone for a walk on deck with the Kaiser and Georg Hülsen and had been telling harmless theatre stories, when the Kaiser suddenly started talking furiously about Berlin society and especially about the Conservatives. 'His violence in conversation was terrifying', Eulenburg commented. 'I must admit that I saw an *abyss of hate and bitterness* within him, which nothing can change . . . H.M. *cannot control himself any more* when he is filled with anger. Yesterday he did not even notice that there were sailors near by, he simply raved on, so that they could hear every word he said. Hülsen was so terrified that he was taken ill . . . I consider our present situation a very dangerous one; I see no way out . . . I feel as though I were sitting on a barrel of gunpowder.' Eulenburg could think of no other alternative 'but to await events and to pray to God that His Majesty will not be obliged to confront any complicated problems, for repeated scenes, like the one I witnessed in Kiel, would surely lead to a nervous crisis, the character of which cannot be foretold . . . These things affect me very much. I had such faith in the Kaiser's ability – and in the good that time would do, but now I am losing faith in both and I see a human being whom I love deeply suffering so much without being able to help.'[50] 'I am sometimes very frightened, and I am seriously worried. I am worried about you too, my dearest Bernhard', he wrote to Bülow. 'How are you going to be able to steer the ship with a Captain who is not well??'[51] 'I am very, very anxious about the future.' It was clear to him, Eulenburg added, that 'His Majesty must not remain a single day without reliable diplomatic advice.'[52] Just one week later Wilhelm made his rabid 'Hun speech' at Bremerhaven.

55. Philipp Eulenburg's last Scandinavian cruise with the Kaiser, July 1903. Eulenburg is on the extreme left of the main group, holding a champagne glass; Wilhelm is in the centre, wearing a white cap and white shoes.

During the last Scandinavian cruise in which he was to take part, that of the summer of 1903,[53] Eulenburg's experiences were so shattering that he had the nightmarish feeling of having put the fate of his countrymen into the hands of a madman. The Kaiser's 'nervousness', he told Bülow, now Reich chancellor, had changed considerably in the past two or three years. 'Although in the past (I am thinking of the war in China) he gave me much cause for concern with his fearful passion, his speeches etc., at least his behaviour was caused by his being "completely obsessed" with *one* idea . . . Today things are different: he is no longer painfully obsessed with any *one* thing which weighs down on him and will not let him rest; he is difficult to handle and complicated in *all* things no matter how trivial. *No one* can make even the most harmless remark about the weather, travelling plans or other completely indifferent things without provoking a violent objection, an insulting response or even an outburst of rage.' 'The Kaiser's *earnestness*' had become 'a *real* torment' to him, Eulenburg wrote, 'when he builds up his houses of cards and violently abuses people and things, constantly contradicting himself. How long can it be before his outbursts of hatred against "the dirty, obstinate, stupid, poor, badly dressed Germans, who should follow the example of Englishmen and Americans", will leak out into the world through naval officers, paymasters, engineers, sailors and messenger boys?' He had talked 'at great length' with the imperial physician, Rudolf Leuthold, about 'our poor dear Master', 'who wanders around as if in a dream world and boosts his ego into an ever greater phantom. I asked him [Leuthold] whether I had been *right* in observing that his nervousness is expressing itself increasingly strongly while his self-control is becoming weaker and weaker? Leuthold shares this view, and his anxiety has increased.' Bülow should therefore be prepared for 'a slow transformation of the mental and emotional condition of our dear Master'. 'Ask God for *strength*', he advised the Reich Chancellor, 'for without His help you will be almost in despair.' He himself, the Kaiser's friend wrote, often felt 'the tears welling up in me, when I hear our dear, kind Master (whom I still thank for *so much*, especially his *loyalty*, in spite of all the suffering he has caused me) – making wild attacks against all kinds of windmills, and see his face quite distorted with anger. Almost every day there are conversations like that, remarks like that . . . There is no question of self-control any more. Sometimes he seems to have *completely* lost all self-discipline.'[54] Frequently, as Eulenburg continued in his report to Bülow in the next few days, the Kaiser would be roused to 'explosions of high spirits' by jokes and funny stories told by his travelling companions on board the *Hohenzollern*, but lighter moments like these could not disguise the fact that the Kaiser was seriously unwell and was not in control of himself. With a 'completely distorted . . . pale, nervous face' and 'gesticulating excitedly', he would speak of war and revenge. 'The poor, poor Kaiser', sighed his closest friend. 'How he destroys everything around him that should be his mainstay and his pride!' All those on board were 'shocked by the fact, which becomes more apparent all the time, that H.M. sees and judges *all* things and *all* men purely from his personal standpoint. Objectivity is lost completely; subjectivity rides on a biting and stamping stallion – contradiction celebrates daily triumphs in his own utterances.' 'Pale, ranting wildly, looking restlessly about him and piling lie on lie', Wilhelm made 'such a terrible impression' on Eulenburg and all those on the Scandinavian cruise that, as Eulenburg exclaimed, 'I still cannot get over it! I could hardly sleep a wink at night. "Not in good health" is perhaps the gentlest way one can put it.' 'I am filled with deep sadness', Eulenburg confessed. 'I can hardly wait for the hour of my release from this royal cage . . . Leuthold is at his wits' end. But he considers all his medical plans more or less impracticable. So *nothing* happens. It is an appalling wait for *some* kind of crisis. And everyone here on board waits with him.'[55]

NEURASTHENIA, MENTAL ILLNESS, BAD BLOOD – WHAT WAS WRONG WITH WILHELM II?

Philipp Eulenburg's increasingly anguished expressions of alarm from 1899 onwards are striking, but in fact only reflect, if rather more dramatically than usual, what other well-informed observers had been lamenting for some time. Blinded by his love for the Kaiser, it took him much longer to recognise the central problem of the Personal Monarchy, which lay in the character of Wilhelm II. It will not have escaped the reader's notice how regularly the suspicion had arisen since the beginning of his reign that Wilhelm II could not be entirely sound in body and mind, and that such suspicions were expressed in all circles around the Kaiser, whether by his own family, his more distant royal relations, his entourage of courtiers, court society in Berlin, the officer corps, higher civil servants, political parties of all shades of opinion or foreign diplomats and journalists. Anxious comments had been made by relatives and other observers about his outward appearance even in the very early days following his accession. Queen Victoria's cousin Princess Augusta Caroline of Cambridge, the Hereditary Grand Duchess of Mecklenburg-Strelitz, wrote in 1888 after a visit from the young Kaiser: 'I don't think he looks well, very brown-yellow, pulled out yet in some way, puffed.'[56] Similarly when he travelled through Munich a few weeks later it was remarked that he did not look well.[57] A year later Dr Leuthold was expressing anxiety about the detrimental effect on Wilhelm's health of his hectic lifestyle. Although conceding that the Kaiser had got over a bout of influenza without the accompanying nervous symptoms from which he usually suffered, Leuthold ascribed his illness to exhaustion caused by his ceaseless round of shooting expeditions. There was no denying, he declared, that 'the travelling, the official functions, the shooting [have] gradually become too much for the Kaiser'. Already in these early days Leuthold urged Eulenburg to persuade Wilhelm to lead a quieter life.[58] After Bismarck's dismissal concerns of this kind began to be much more frequently expressed in the imperial entourage. 'The Kaiser looked very seedy', a senior naval officer recorded after seeing him on a visit to Wilhelmshaven in the spring of 1890.[59] After the Scandinavian cruise in July that year Eulenburg reported that the sea voyage had not done the Kaiser any good. 'His restlessness has increased rather than diminished and he takes too little sleep.' Leuthold had stated bluntly that 'things could not go on like that', and Eulenburg too had the feeling that 'H.M. is doing himself harm and simply must have eight hours' sleep if he wants to keep going any longer.'[60] During the following year's cruise the Kaiser slipped on a wet deck and fell on his back, dislocating his right kneecap. After the kneecap had been put back in place a plaster cast had to be applied, although the knee was swollen and bleeding.[61] For his mother, of course, this accident was nothing new. 'William has put out his knee pan several times!', she wrote to the Queen. 'He is very knock kneed and as a child he had not strong "Knie Bänder" . . . He has *not* his balance as one side is lighter than the other in consequence of the arm – wh[ich] makes him awkward on his legs . . . He sprained one leg fencing at Bonn (when I went to look after him) – *once* dancing & once swimming.'[62] In the winter of 1896–7, when the Kaiser's nervous irritability gave particular cause for concern, diplomats reported that he was suffering badly from a carbuncle on the knee and looked 'very careworn and thin'.[63] Philipp Eulenburg was shocked by Wilhelm's appearance in October 1896, after his ear operation and the humiliating encounter with the Russian imperial family at Darmstadt.[64]

It was not only the Kaiser's physical health that gave rise to anxiety from early on, but also, and principally, his mental condition. Very much in the spirit of the times, those about him often explained the monarch's alarming state of mind in terms such as 'nervous tension' or 'neurasthenia'.[65] When Waldersee saw Wilhelm at Kiel in 1891 and thought him 'rather worn out', he commented that it seemed to him 'as others had already claimed – that a certain nervous depression had set in'.[66] A little later, after meeting the Kaiser and Kaiserin again he wrote: 'This time I noticed that the Kaiser was not as fresh as usual; he complained of being overtired, and the Kaiserin said to me twice: "Wilhelm must have more sleep, he is completely exhausted." But there is in fact no real reason for the Kaiser to be overtired; I am afraid it is his nerves which are beginning to give way.'[67] In November 1892 Waldersee again noted in his diary: 'The Kaiser has a cold and has cried off two shoots; it is said that he is also suffering from a certain moral depression, as happened last year.'[68]

Gradually the opinion took hold that the Kaiser's restless and peculiar behaviour might be not the cause but the symptom of a serious emotional disorder. Hinzpeter, who had earlier seen the 'salutary humiliation' of his arrogant pupil as the only way of improving his character, had adopted a completely different approach since Wilhelm's accession. It was not repression but constant encouragement that was now required, in order to ward off the danger of a nervous breakdown. In a conversation with Schottmüller in January 1891, in which the latter complained that the Kaiser began many things but finished none, Hinzpeter is reported to have said: '*That does not matter at all; the main thing is that I keep him busy all the time*; if he does not have something new all the time he falls into apathy.' Waldersee, who recorded this remark, continued: 'He [Hinzpeter] then spoke utterly cynically about the Kaiser's ear complaint, saying that it could easily lead to death or mental disorder, so Prince Heinrich ought really to live in Berlin to prepare himself for a possible regency.'[69]

In March 1892 – this was the time of the School Bill crisis, during which Caprivi resigned as Prussian minister-president[70] – when Waldersee heard that the Kaiser had been unwell for several days, he saw this as the beginning of the nervous breakdown which he had for a long time expected to happen. Wilhelm was not actually ill, he said, 'but only very pulled down, fatigued, disgruntled'; he complained of too much work. 'If his nerves were better, I would hope that he would begin to improve now, but as they are notoriously bad I am afraid it will soon get to the point where he loses heart completely.' Wilhelm had come to the throne with the ambitious aim of being recognised as an 'acclaimed ruler, esteemed and feared by the whole world'. 'He himself, who became quite giddy with the apparent successes of the early days, who believes he knows everything better than anyone else, who thinks his opinion on every subject is the right one, now sees that many things are going badly in Germany, that the mood is turning against him. Now he finds himself involved in serious internal conflicts; at the moment it is true that he is still inclined to blame everything on others, and is still as self-confident and arrogant as before in what he says, but I think that this is nothing but bragging; his inner mood is beginning to change.' It was to be feared, Waldersee commented, that the Kaiser's 'so easily overtaxed nerves' might fail him completely. 'What will happen if he begins to waver even more, if he becomes really discouraged, I do not know', he wrote. 'I only know that we shall go down even more.'[71] 'To my very great regret I have to say that he is an unreliable, weak character.'[72] In the summer of 1892 Waldersee went so far as to express the apparently paradoxical hope that a nervous breakdown on the part of the monarch might save the Reich from the ruin that otherwise lay in store for it. Deeply anxious, he wrote of Wilhelm II that 'he may well destroy all his grandfather's work; if he goes on for a few more years as he has done up to now, it will be inevitable. Perhaps our salvation lies in his nerves breaking down so that he will let others take over.'[73] After the chancellor crisis of October 1894, during which the bad state of the Kaiser's nerves was 'often' mentioned, Waldersee commented: 'I said long ago that there was absolutely no doubt he would soon come to the end of his tether.'[74]

When Waldersee went to Berlin in January 1896 for the twenty-fifth anniversary of the founding of the Reich and for the chapter meeting of the Order of the Black Eagle, he recorded after several conversations with the Kaiser that Wilhelm was 'deeply out of sorts in himself and his nerves are again beginning to make themselves felt'. The reason for his irritability, according to the General, was 'the affair of Prince [Friedrich] Leopold, in which he has again been much too hasty', and which had 'very much upset' both the Kaiser and the Kaiserin. In addition to this there were his 'anger at the Köller affair, which he has not yet quite got over, [and] his quarrel with Bronsart at the New Year'; finally, the Transvaal crisis and 'the behaviour of the English' had left him 'deeply disgruntled'.[75] At this time, as we have seen, the signs of an 'unhealthy excitability' in the Kaiser were increasing, and were becoming apparent in an extremely aggressive attitude towards both his own people and foreign powers. Holstein commented that it filled him with alarm to hear how the Kaiser had been speaking recently. 'Everything makes an uncanny, rushed impression', the Privy Councillor wrote. 'The Kaiser's nerves' seemed more tense than ever before.[76] A few weeks later Wilhelm felt compelled to withdraw to his hunting lodge at Hubertusstock because of a nervous attack. Waldersee recorded in his diary on 16 February 1896: 'The Kaiser is very annoyed and his nerves are in a bad way, as I was already aware from what people close to him had told me in January. He has gone to Hubertusstock to recover; it is no doubt quite a good thing that it has become necessary, but sad. How will our sovereign be able to go on, if he already has nervous disorders at the age of 36. He should simply hold back more, and not try to control everything. That is just what he cannot manage and it has already done us much harm. But I scarcely think that he has reached the point of realising that *he alone* is to blame if the cart has got stuck in the mud.'[77]

On 23 October 1896 the wedding of the Hereditary Grand Duke of Oldenburg to Duchess Elisabeth of Mecklenburg took place at Schwerin, an occasion on which Waldersee saw the Kaiser and many old acquaintances from the imperial entourage. 'I found the Kaiser not looking well', he wrote afterwards. 'Years ago I predicted that his nerves would soon let him down; I think that moment has almost come. First of all the condition of his ear has deteriorated again since the Scandinavian cruise, and the reappearance of this complaint has depressed him very much. In addition to that all kinds of annoyances, partly political, partly domestic, have arisen, and his nervous state has *often* been apparent in a melancholy mood, lack of energy, emotional outbursts etc. To combat this there is a constant search for distraction and restlessness, but it is clear that this is not a good idea and that it is more likely to have the opposite effect. If anything serious should arise now, e.g. major political disappointments, which can very easily happen, there will be a breakdown! And then what? We could be in a really serious situation with this worn-out old Chancellor, these ministers, most of whom are pathetic, the lamentable Cabinet chiefs and finally the effeminate and distinctly second-rate Prince Heinrich, who is the only stand-in available!'[78] Similar sentiments were expressed a few days later by Finance Minister Johannes von Miquel, of whom Waldersee wrote: 'For the first time I found him in an anxious mood about the Kaiser, both as to his influence on the machinery of state and as to his health.'[79] 'I should be less worried if his nerves were quite healthy', the General wrote. 'Then one could say that experience would have a purging effect on him. But I am convinced that he is not capable of enduring serious setbacks; his nerves would give way and he would collapse, which would be the most terrible thing for us.'[80]

Naturally it seemed no more satisfactory to his contemporaries than it does to us to explain Wilhelm II's unstable state of health merely in terms of his 'nerves'. Many went further and expressed the suspicion that he was 'mentally ill' or would become so in time. The uncomfortable feeling that Wilhelm II was in some way not normal, a feeling which every speech, telegram, marginal note or other conspicuously odd action helped to nurture, spread throughout Europe with alarming speed after his accession. Already in 1890 highly placed people who had been in his grandfather's entourage were speaking of the young Wilhelm's 'hereditary taint'.[81] Immediately following Bismarck's dismissal Holstein warned that the Kaiser was considered to be 'frankly speaking not quite right in the head', and predicted that 'this spectre of mental illness will haunt the sovereign all his life'.[82] Hinzpeter and others soon began to speak of 'Caesaromania', a diagnosis that was on everyone's lips after the publication of Ludwig Quidde's *Caligula* pamphlet.[83] Wilhelm's own sister Sophie, the Crown Princess of Greece, as we have seen, expressed herself 'in the most contemptuous and hostile way' about him at St Petersburg and added that 'the whole family thought he was mad'. In Berlin the Empress Frederick and her daughters were suspected of spreading such rumours.[84] Grand Duke Sergei of Russia, who was married to Princess Ella of Hesse-Darmstadt, stated bluntly that he thought the Kaiser was 'mentally ill',[85] while at the same time the French Foreign Ministry came to the conclusion that Wilhelm was 'emotionally disturbed' and 'temporarily of unsound mind'.[86]

Early in 1891 Waldersee felt compelled to register the 'extremely sad but very significant fact' that 'serious, patriotically minded men are really considering the idea that the Kaiser is very gradually succumbing to a mental disorder. This would certainly be the greatest misfortune that could befall the Fatherland, especially if the development were slow. May we be spared this by God's grace!'[87] In December that year the General wrote again in his diary: 'It is said that the question is being openly and widely discussed, especially among doctors, of whether a mental illness is slowly developing, perhaps in connection with the ear complaint. This would be the most terrible thing that could happen, first for the Kaiser himself . . . but then for the Fatherland. What misfortunes may occur until then and what else may arise from it is completely unforeseeable.'[88]

After the dismissal of Caprivi and Botho Eulenburg, which was decided upon at the Eulenburg estate of Liebenberg, such fears intensified. The chargé d'affaires at the British embassy, Martin Gosselin, reported in November 1895, shortly before the Köller crisis came to a head, that there were 'curious rumours going about Berlin as to the Emperor's health. Count Phi[lip]p Eulenburg . . . is known to have a wonderful influence over His Majesty; & not infrequently this is attributed . . . to the mesmeric power which it is asserted His Excellency is able to exert over his Sovereign.' Gosselin had also heard that the Kaiser was suffering from hallucinations, which found their expression in such things as the famous 'yellow peril' picture. The diplomat continued: 'If there should be any truth in these rumours, it would account for much that is otherwise inexplicable: it becomes, indeed, a serious matter if a Sovereign who possesses a dominant voice in the foreign policy of the Empire is subject to hallucinations & influences which must in the long run warp his judgement, & render Him liable at any moment to sudden changes of opinion which no one can anticipate or provide against.'[89]

In Germany comparisons were frequently made between Wilhelm II's 'glow-worm character' and the mentally deranged Kings Friedrich Wilhelm IV of Prussia and Ludwig II of Bavaria.[90] Prussian War Minister General Bronsart von Schellendorf commented after his alarming clash with the Kaiser in January 1896, as we have seen, 'that H.M. did not seem quite normal' and that it was very worrying for the future.[91] After the Brandenburg speech of February 1897, in which he had described Bismarck and Moltke as pygmies and lackeys of his sainted grandfather, and the historical fancy-dress party for the hundredth birthday of Wilhelm I on 22 March, the rumours that Wilhelm II was mentally deranged became common currency. The East Elbian leaders of the German Conservative Party declared that they considered the Kaiser 'not quite normal'; the King of Saxony commented that he was obviously 'unstable', and his uncle, Grand Duke Friedrich of Baden, spoke 'in a very worrying way about the psychological side of the matter, about the loss of touch with reality'.[92] Count Monts reported with horror from Munich that in South Germany the Kaiser was considered 'clearly no longer a responsible person'.[93] The resentment against him among patriotic people was deeper than ever before, and people were saying secretly that 'H.M. is insane', he wrote.[94] 'What is the point of all this work and effort. Nobody believes one when one describes the Kaiser as he was, for example on his visit here, so simple, understanding, clear, moderate and quiet. – It is as if from time to time our master were seized by an evil spirit, benighting his mind and compelling him to make speeches which deeply insult the nation.'[95] It was out of the question, Monts continued, 'that the disorder could be anything but psychological'; he could see Wilhelm II 'going the same way . . . as King Friedrich Wilhelm IV, who like H.M. had become excited and intoxicated by his own words, after which a state of despondency usually followed, which made it necessary for him to shut himself away . . . for days on end'.[96] Bismarck told a Bavarian writer that he had recognised 'the Kaiser's abnormal mental condition' as early as 1888, and had resisted his dismissal as chancellor only because he had been afraid of a national catastrophe.[97] After meeting Philipp Eulenburg in April 1897 Bülow noted that Eulenburg had spoken of the 'inclination' which existed even in the Kaiser's family 'to put H.M. into care', and had mentioned the Empress Frederick and Prince Heinrich. The view 'that H.M. is mentally ill' was being spread both from the Bismarckian side and from the Foreign Office. Eulenburg had discussed this danger with Leuthold, who (according to hasty notes made by Bülow) had expressed the opinion 'that H.M. did not have any maniacal characteristics; on the contrary, it was more that he was too fragmented in his ideas' and 'changeable'; in the view of the Kaiser's physician the 'only danger was that H.M. would put too much strain on his nerves [and] then break down'.[98] During the Scandinavian cruise in the summer of 1900 Leuthold still believed that the Kaiser's alarming outbursts of fury could be attributed to 'a certain weakness of the nervous system', but 'he emphatically denies any possibility of mental disorder'.[99] On the basis of such assurances Bülow felt able to calm the Reich Chancellor's fears when, upon Bülow's taking office as foreign secretary in October 1897, Prince Hohenlohe asked him if he considered Wilhelm II to be 'mentally quite normal'.[100] When he took Hohenlohe's place three years later, however, Bülow had to warn the Kaiser through Eulenburg that there was a danger that the next incautious move on the part of the Kaiser might lead to a coalition between the German federal princes and the Reichstag with the object of forcing him to abdicate.[101]

Members of Wilhelm's own family felt that the 'lackey' and 'Hun' speeches, along with numerous other instances of aggressive and peculiar behaviour, confirmed their earlier forebodings about him.[102] His eldest sister Charlotte spoke in March 1897 of her '*serious* fears', and later repeatedly expressed the opinion that her brother really ought to be in Professor Schweninger's clinic, where she herself was receiving treatment for her painful hereditary illness.[103] In 1908 she decided that 'as a true Prussian' she should set about having Wilhelm placed under a kind of collective regency of all the federal princes. She wrote to her doctor: 'I want to persuade the German princes to go to the Kaiser in a body . . . and to offer him their help, in the interest of the Reich and in the name of their peoples, under clearly stated conditions. I think it is imperative that they should work together and keep together, and it is the only thing which could still make an impression. [Prince] Ludwig [of Bavaria] would have to be the spokesman, in the name of his father and of the grand dukes, [and of] Saxony and Württemberg.'[104] The lamentations of Wilhelm's mother, his other sisters and his own wife quoted in the chapter on the imperial family show that they were no less anxious and despondent.[105] In later years even Wilhelm II's eldest son admitted to doubts about his father's mental state. 'I know my father very well . . . It is claimed that genius and madness are very close to each other. There have been moments when, seeing the strange expression in his eyes, I have doubted his sanity.'[106]

Naturally enough the question of the mental condition of the eccentric German Kaiser was the subject of lively and often wildly speculative discussion in the foreign press. After the Brandenburg speech of February 1897 Waldersee commented that the British and American newspapers were spreading the story 'with the greatest impudence' that the Kaiser was 'of unsound mind'.[107] When Wilhelm hurt his eye during the Scandinavian cruise that summer, and at the same time the 26-year-old son of General von Hahnke, who was travelling with the Kaiser's party, was killed in an accident during a bicycling expedition on shore, an article appeared in the Canadian press – and was widely reprinted elsewhere – linking these two accidents, with appalling insinuations against the Kaiser. The young Gustav von Hahnke, the writer of the article deduced, had hit the Kaiser in the face after being insulted by him, and had been made to pay for it with his life.[108] The *New York Times* expressed the opinion at about this time that the German monarch, already a 'surprising' man, would 'sooner or later "go queer" altogether'.[109] Newspaper articles and pamphlets with titles such as 'Is Kaiser Wilhelm of Normal Mind?'[110] frequently appeared. After the outbreak of war in 1914 such publications became a veritable flood.

Conjectures of this kind by irresponsible journalists were by no means insignificant in an age of unrestricted media attention, but naturally the opinions of the crowned heads and the leading statesmen of Europe had a more direct effect on relations between the great powers in the period leading up to the First World War. And these opinions, as this biography has shown, were scarcely more favourable to Wilhelm II. Russian General A. A. Mossolov, who was for many years the head of the Court Chancellery under Nicholas II, tellingly records in his memoirs that both the imperial couple and their entourage considered the Kaiser, whom they found importunate and very often quite insufferable, 'raving mad'.[111] As we have seen in many instances above, Wilhelm's personality, which was regarded as disturbed, had a direct effect on British foreign policy.[112] As early as the autumn of 1888 Lord Salisbury had remarked, à propos of the Vienna incident, that he thought 'the Emperor William must be a little off his head'.[113] The British Prime Minister and Foreign Secretary not only developed a strong personal antipathy towards the Kaiser but also became convinced that Wilhelm could be neither trusted nor believed.[114] He frequently wondered whether the Kaiser was perhaps 'not "all there"'; he considered him a danger to peace and 'the most dangerous enemy we had in Europe', because he was 'mad enough for anything'.[115] In December 1895, after reading Martin Gosselin's report on the Kaiser's apparent hallucinations and the hypnotic influence of Philipp Eulenburg over him, Salisbury warned the new British ambassador in Berlin, Sir Frank Lascelles, of the danger that Wilhelm might in time go 'completely off his head'.[116] It did not take long for the new ambassador to report that the German Emperor's behaviour appeared to 'verge on insanity'.[117] Upon what were these forebodings based?

Apart from the 'neurasthenia' and 'mental disorder' to which Kaiser Wilhelm II's strangeness was attributed, contemporary observers also ascribed it to a hereditary or organic condition. Besides his ear complaint, which at a very early stage had given rise to fears of an infection penetrating the wall of the skull, with resultant brain damage,[118] a second theory as to the possible cause of his odd behaviour was in circulation at the Hohenzollern court, namely that it was due to his Coburg or Guelphic 'blood'. In October 1890 Waldersee recorded that Adjutant-General Adolf von Wittich had expressed 'exactly the same opinion as I have about the Kaiser, his character and his decline, and blamed everything on the Coburg blood. I think he has hit the nail on the head.'[119] Again in early 1892 he wrote: 'Oh, this infamous Coburg blood, mixed with the Guelphic, which makes itself so sadly felt in reckless behaviour!'[120] Two years later Waldersee was even more convinced that the 'Coburg blood' was responsible for the bizarre personality of the Kaiser, whom he compared with his great-uncle, Duke Ernst II of Saxe-Coburg and Gotha. He commented: 'For me, having had the advantage of knowing Duke Ernst of Coburg and having lived through both his rise and his decline, the similarity of the two characters becomes increasingly apparent; but at the same time I cannot but recognise that at heart the Kaiser is much nobler, and that is the Hohenzollern in him. At the moment, however, the Coburg blood is causing very dangerous impulses to emerge.'[121] When Quidde's *Caligula* pamphlet with its unmistakable portrayal of Wilhelm II's megalomaniac, power-loving personality appeared shortly afterwards, Waldersee recalled similar comments which members of the imperial entourage had made in the past. 'On looking through it I was reminded', he wrote, 'that 4 years ago General

Wittich referred to tendencies of Caesaromania, and that he also put forward the assertion that the Kaiser visibly derives an agreeable nervous stimulus from the mention of massive losses when he makes speeches about military history. Nevertheless, in spite of the numerous really striking analogies, I should like to think that the way in which our Kaiser is developing need not lead to a sad end. It is true that I very rarely see him now, but when I do I observe him very closely, and I cannot say that I can detect any real disorder. If he did have any really alarming tendencies they would have developed much more in the past 6 years. What has worried me for a long time is the evident predominance of the Guelphic–Coburg blood, which is certainly disturbing enough in itself.'[122]

The news that the Kaiser had to have an operation to remove a small cyst on his cheek on 1 June 1894 led to further pessimistic forecasts. Waldersee noted anxiously in his diary: 'A rumour had been going round for 3 months that he had a swelling under one ear, and all sorts of other rumours were immediately linked with this. According to what I recently heard from people who had seen the Kaiser it is true that a clearly visible swelling had appeared. It is perhaps not surprising that this gives rise to serious thoughts. What has God in store for us?'[123] At the height of the power struggle between the Kaiser and the Chancellor in the summer of 1896, when Trautmann had to carry out a radical operation on the Kaiser's right ear at Wilhelmshöhe,[124] Friedrich von Holstein recalled the comments of Sir Felix Semon, the German-born throat specialist who practised at the British court and was also close to Bismarck. Four or five years earlier Semon had claimed, according to Holstein, that 'the restlessness of the present Kaiser [was] the precisely definable first stage of a psychiatric condition, but one which in the beginning should be considered and treated from the physiological rather than from the psychological standpoint'.[125] What half-physiological, half-psychiatric illness can Semon have meant by this?

A document quoted in the first volume of this biography is of particular relevance in this connection. It refers to a conversation which Sir Schomberg McDonnell, private secretary to the Prime Minister Lord Salisbury at the time, had with Surgeon-General John Erichsen in March 1888, just after the death of Kaiser Wilhelm I and the accession of the fatally ill Kaiser Friedrich III. McDonnell immediately reported the conversation verbally to Salisbury. As he later recalled, Erichsen came to see him to bring some extremely important and alarming news for the Prime Minister. He told McDonnell that 'when Prince William of Prussia was 14 or 16 . . . his condition gave rise to some anxiety', and that German doctors had sent him detailed notes on the case at the time. These had convinced him that Wilhelm 'was not, and never would be, a normal man'; that 'he would always be subject to sudden accesses of anger; and that when angry he would be quite incapable of forming a reasonable or temperate judgement on the subject under consideration'; that 'while it was not probable that he would actually become insane, some of his actions would probably be those of a man not wholly sane'. From this diagnosis Erichsen drew the worrying conclusion that the future Kaiser Wilhelm II could 'possibly be a danger for Europe' if he came to the throne. When McDonnell passed on Erichsen's comments to Salisbury, the Prime Minister 'was of course immensely interested', the former private secretary recounted, adding that on several occasions in later years 'when the Emperor committed some indiscretion he used to say privately the single word "Erichsen"'.[126]

The information brought by Erichsen explains the importance which Lord Salisbury accorded to Gosselin's report on the 'hallucinations' of the Kaiser and the apparently hypnotic effect of Eulenburg upon him, and also his warning to Lascelles in December 1895 that Wilhelm might eventually become 'completely mad'. But what was the illness which Erichsen believed he had diagnosed in the young Wilhelm on the basis of the medical notes sent him by his German colleagues? In the first volume of this biography attention was drawn to the fact that the British royal family suffered from two hereditary diseases: haemophilia, which manifested itself for the first time in some of Queen Victoria's children and was passed by them to the ruling families of Hesse-Darmstadt, Prussia (Prince Heinrich's sons), Russia (Tsarevich Alexei) and Spain,[127] and porphyria, a hereditary metabolic disorder which causes attacks of mental confusion and painful physical symptoms. The best-known case of porphyria is that of King George III, Queen Victoria's grandfather and great-great-grandfather of Wilhelm II. Unlike haemophilia, which is sex-specifically transmitted, porphyria is passed on by a dominant gene: about half the children of an affected parent will inherit the mutation, whether they are male or female. Recent DNA research has shown beyond doubt, as has already been mentioned, that Kaiser Wilhelm II's eldest sister Charlotte, the Hereditary Princess of Saxe-Meiningen, and her daughter Feodora, Princess Reuss of the younger line, suffered from porphyria variegata.[128] Could it be that Charlotte's 'big brother' also inherited this mutation from his mother?

Unlike in his sister's case, there seems to be virtually no evidence that Wilhelm II suffered from the physical symptoms of the illness – unbearable abdominal pain and headaches, sleeplessness, vomiting, colic, feverish sweating, constipation, rapid heart beat, paralysis in the arms and legs, blister-like skin rashes and dark red urine. On the other hand the attacks of mental and emotional disorder so frequently complained of would accord with such a hypothesis. Perhaps one day DNA research will bring the certainty, one way or the other, which presently eludes the historian. But the possibility that a documentary-based scholarly approach might shed light even on this complex medical question should not be underestimated. On 31 May 1918, when all Europe lay in ruins and the worst fears of the Empress Frederick, the Bismarcks, Holstein, Bronsart and in the end even Eulenburg were beginning to come true, Wilhelm II's youngest son was examined by the distinguished psychiatrist Professor Robert Gaupp. Prince Joachim, Gaupp concluded, was incurably ill, both mentally and physically. Two years later, in a state of frenzy, he was to take his own life. The Kaiser's son, the psychiatrist wrote, spoke too fast; he had sudden facial tics; he was extremely easily emotionally and sexually aroused; he was inclined to 'violent, uncontrollably exploding outbursts of anger in which all self-control [was] lost'; his thinking, in spite of a quick understanding, was superficial; his behaviour was 'determined by momentary impulses'; his 'highly developed ego' meant that he felt 'unjustly injured' by any resistance he met to the pursuit of his passionate wishes, and reacted 'with outbursts of rage'. Also symptomatic of Joachim's disorder was his inability to keep to decisions he had taken, and 'a certain weakness in giving way to the temptations of the outside world and to impulsive urges and moods'. In addition, he had 'vaso-motor disorders' and 'a rapid fluctuation in the blood supply to the face'. The overall condition of 'physical and psychological irregularity', Gaupp concluded in his report, 'points to an *innate abnormal tendency*'.[129] Had both father and son, even if not to the same degree, and of course over and above all their other psychological and physical problems, inherited the royal hereditary malady of porphyria from their Guelph ancestors? In the present state of knowledge this possibility, which would explain much, cannot be excluded. At the British court, at any rate, there was support for this opinion. In November 1908 Lord Esher, a confidant of King Edward VII, wrote in his diary of Kaiser Wilhelm II: 'I am sure that the taint of George III is in his blood.'[130]

Notes

PREFACE TO THE ENGLISH EDITION

1 See the contributions in Annika Mombauer and Wilhelm Deist, eds., *The Kaiser. New Research on Wilhelm II's Role in Imperial Germany*, Cambridge 2003.

2 Christian Baechler, *Guillaume II le Kaiser*, Paris 2003. Other recent studies include Jost Rebentisch, *Die vielen Gesichter des Kaisers. Wilhelm II. in der deutschen und britischen Karikatur (1888–1918)*, Berlin 2000; Lothar Reinermann, *Der Kaiser in England. Wilhelm II. und sein Bild in der britischen Öffentlichkeit*, Paderborn, Munich, Vienna, Zürich 2001; Stefan Samerski, ed., *Wilhelm II. und die Religion. Facetten einer Persönlichkeit und ihres Umfelds*, Berlin 2001; Thomas Hartmut Benner, *Die Strahlen der Krone. Die religiöse Dimension des Kaisertums unter Wilhelm II. vor dem Hintergrund der Orientreise 1898*, Marburg 2001; Douglas Mark Klahr, 'The Kaiser Builds in Berlin. Expressing National and Dynastic Identity in the Early Building Projects of Wilhelm II', dissertation, Providence, RI 2002; Holger Afflerbach, ed., *Wilhelm II. als Oberster Kriegsherr im Ersten Weltkrieg. Quellen aus der militärischen Umgebung des Kaisers 1914–1918*, Munich 2004.

3 Quoted in John C. G. Röhl, 'The Kaiser's Germany as Seen from Beachy Head on a Clear Day. Autobiographical Reminiscences of an Anglo-German Historian', in Stefan Berger, Peter Lambert and Peter Schumann, eds., *Historikerdialoge. Geschichte, Mythos und Gedächtnis im deutsch-britischen kulturellen Austausch 1750–2000*, Göttingen 2003, p. 128.

4 Max Weber to Friedrich Naumann, 14 December 1906, quoted in Wolfgang J. Mommsen, *War der Kaiser an allem schuld? Wilhelm II. und die preußisch-deutschen Machteliten*, Munich 2002, p. 7.

PREFACE TO THE GERMAN EDITION

1 John C. G. Röhl, *Young Wilhelm. The Kaiser's Early Life 1859–1888*, translated from the German by Jeremy Gaines and Rebecca Wallach, Cambridge 1998.

2 John C. G. Röhl, *Germany without Bismarck. The Crisis of Government in the Second Reich, 1890–1900*, London 1967.

3 John C. G. Röhl, ed., *Philipp Eulenburgs politische Korrespondenz*, 3 vols., Boppard-am-Rhein 1976–83.

4 See Hans-Ulrich Wehler, *Deutsche Gesellschaftsgeschichte*, vol. III, *Von der 'Deutschen Doppelrevolution' bis zum Beginn des Ersten Weltkrieges, 1849–1914*, Munich 1995, pp. 1016–20.

5 See for example Nicolaus Sombart, *Wilhelm II. Sündenbock und Herr der Mitte*, Berlin 1996, and Giles MacDonogh, *The Last Kaiser. William the Impetuous*, London 2000.
6 Geoff Eley, 'The View from the Throne: The Personal Rule of Kaiser Wilhelm II', *Historical Journal*, 28, 2 (1985), pp. 469–85.
7 See e.g. Norman Rich, *Friedrich von Holstein. Politics and Diplomacy in the Era of Bismarck and Wilhelm II*, 2 vols., Cambridge 1965; Isabel V. Hull, *The Entourage of Kaiser Wilhelm II, 1888–1918*, Cambridge 1982; Hull, 'Persönliches Regiment', in John C. G. Röhl, ed., *Der Ort Kaiser Wilhelms II. in der deutschen Geschichte*, Munich 1991; Paul Kennedy, 'The Kaiser and German Weltpolitik. Reflexions on Wilhelm II's Place in the Making of German Foreign Policy', in John C. G. Röhl and Nicolaus Sombart, eds., *Kaiser Wilhelm II – New Interpretations*, Cambridge 1982; Katharine A. Lerman, *The Chancellor as Courtier. Bernhard von Bülow and the Governance of Germany 1900–1909*, Cambridge 1990; Thomas A. Kohut, *Wilhelm II and the Germans. A Study in Leadership*, Oxford, New York 1991; Lamar Cecil, *Wilhelm II*, 2 vols., Chapel Hill, London 1989–96; Christopher Clark, *Kaiser Wilhelm II*, London 2000; Roderick R. McLean, *Royalty and Diplomacy in Europe, 1890–1914*, Cambridge 2001; Holger Afflerbach, 'Wilhelm II as Supreme Warlord in the First World War', *War in History*, 5 (1998), pp. 427–49.
8 But see Thomas Nipperdey, *Deutsche Geschichte 1866–1918*, 2 vols., Munich 1990–2, vol. II, *Machtstaat vor der Demokratie*, pp. 475–85.
9 Heinrich Otto Meisner, *Denkwürdigkeiten des General-Feldmarschalls Alfred Grafen von Waldersee*, 3 vols., Stuttgart, Berlin 1922–3.

1 THE ACCESSION

1 Kaiser Wilhelm II, Armee-Befehl, 15 June 1888, Geheimes Staatsarchiv (GStA) Berlin, Rep. 92 Scholz Papers, No. 11.
2 Kaiser Wilhelm II, An die Marine!, proclamation of 15 June 1888, ibid.
3 Admiral Gustav Freiherr von Senden-Bibran, Aufzeichnungen über das Jahr 1888, Bundesarchiv-Militärarchiv (BA-MA) Freiburg, Senden-Bibran Papers N160/11, p. 21.
4 Kaiser Wilhelm II, An Mein Volk!, proclamation of 18 June 1888, GStA Berlin, Rep. 92, Scholz Papers, No. 11.
5 Széchényi to Kálnoky, 19 June 1888, Haus-, Hof- und Staatsarchiv (HHStA) Vienna, PA III 134. Cf. *Truth*, 23, 599, 21 June 1888, p. 1065. Reactions to Wilhelm's proclamations are collected together in Politisches Archiv des Auswärtigen Amtes (PA AA), R 3446.
6 Helmuth von Moltke to Eliza von Moltke, 17 and 19 June 1888, Eliza von Moltke, ed., *Generaloberst Helmuth von Moltke. Erinnerungen, Briefe, Dokumente 1877–1916. Ein Bild vom Kriegsausbruch, erster Kriegsführung und Persönlichkeit des ersten militärischen Führers des Krieges*, Stuttgart 1922, pp. 141–2.
7 See Crown Prince Wilhelm to Waldersee, 14 June 1888, GStA Berlin, Waldersee Papers, B I No. 42.
8 Count Alfred von Waldersee, diary entry for 19 June 1888, ibid.; printed in Heinrich Otto Meisner, ed., *Denkwürdigkeiten des General-Feldmarschalls Alfred Grafen von Waldersee*, 3 vols., Stuttgart, Berlin 1922–3, I, p. 405 (cited below as Meisner).
9 Rudolf Vierhaus, ed., *Das Tagebuch der Baronin Spitzemberg. Aufzeichnungen aus der Hofgesellschaft des Hohenzollernreiches*, Göttingen 1960, diary entry for 28 June 1888, p. 253.
10 Spitzemberg, diary entry for 28 June 1888, ibid., p. 252.
11 Széchényi to Kálnoky, 24 June 1888, HHStA Vienna, PA III 134.
12 Ibid.; Spitzemberg, diary entry for 28 June 1888, *Tagebuch*, p. 252; Prince Wilhelm of Württemberg to Prince Adolphus von Teck, 2 July 1888, Royal Archives (RA) Geo V. CC50/176. On the situation at the Stuttgart court, see Philipp Eulenburg to Kaiser Wilhelm II,

5 July 1888, printed in John C. G. Röhl, ed., *Philipp Eulenburgs politische Korrespondenz*, 3 vols., Boppard am Rhein 1976–83, I, No. 185.

13 Széchényi to Kálnoky, 24 June 1888, HHStA Vienna, PA III 134.

14 Helmuth von Moltke to Eliza von Moltke, 25–7 June 1888, Moltke, *Erinnerungen, Briefe, Dokumente*, pp. 142–7.

15 See e.g. *The Standard*, 26 June 1888.

16 Kaiser Wilhelm II to Kögel, 22 June 1888, GStA Berlin, Kögel Papers No. 5.

17 Helmuth von Moltke to Eliza von Moltke, 26 June 1888, Moltke, *Erinnerungen, Briefe, Dokumente*, pp. 143ff.

18 Ibid.

19 Széchényi to Kálnoky, 24 June 1888, HHStA Vienna, PA III 134; Kaiser Wilhelm II, Thronrede bei der Eröffnung des Deutschen Reichstages, 25 June 1888, *Centralblatt Deutschlands*, 25 June 1888.

20 Helmuth von Moltke to Eliza von Moltke, 26 June 1888, Moltke, *Erinnerungen, Briefe, Dokumente*, pp. 143ff.

21 Die Thronrede Kaiser Wilhelms II., *Centralblatt Deutschlands*, 25 June 1888. The speech from the throne is printed in Johannes Penzler, ed., *Die Reden Kaiser Wilhelms II. in den Jahren 1888–1895*, Leipzig (n.d.), pp. 11ff.

22 *The Standard*, 26 June 1888.

23 Ibid., 26 and 27 June 1888.

24 Kaiser Wilhelm II to Kaiserin Augusta, 11 July 1888, GStA Berlin, BPHA Rep. 53J Lit. P No. 14a.

25 Spitzemberg, diary entry for 28 June 1888, *Tagebuch*, pp. 252–3.

26 Empress Frederick to Queen Victoria, 29 June 1888, RA Z42/5, printed in Sir Frederick Ponsonby, ed., *Letters of the Empress Frederick*, London 1928, pp. 322–3.

27 Empress Frederick, diary entry for 27 June 1888, GStA Berlin BPH Rep. 52 No. 3.

28 Kaiser Wilhelm II, Thronrede bei Eröffnung des Landtages, 27 June 1888, Penzler, *Reden Kaiser Wilhelms II. in den Jahren 1888–1895*, pp. 15ff.

29 Spitzemberg, diary entry for 28 June 1888, *Tagebuch*, pp. 252–3.

30 Herbert Bismarck's notes, 5 October 1888, in J. Lepsius, A. Mendelssohn-Bartholdy and F. Thimme, eds., *Die Große Politik der europäischen Kabinette 1871–1914*, 40 vols., Berlin 1922–7, VI, No. 1352; Prince Bismarck to Herbert Bismarck, 5 October 1888, cited ibid., p. 346, footnote.

31 Philipp Eulenburg to Kaiser Wilhelm II, 15 October 1888, *Eulenburgs Korrespondenz*, I, No. 199. See Eulenburg to his mother, 4 October 1888, ibid., No. 197.

32 Empress Frederick, diary entry for 25 December 1888, GStA Berlin, BPH Rep. 52 No. 3.

33 Waldersee, diary entry for 14 October 1888, GStA Berlin, Waldersee Papers; cf. Meisner, II, p. 7.

34 Douglas, speech of 4 October 1888, printed in *Norddeutsche Allgemeine Zeitung*, No. 474, 7 October 1888.

35 See the fully documented account of Wilhelm's education in Kassel and Bonn in John C. G. Röhl, *Young Wilhelm. The Kaiser's Early Life, 1859–1888*, Cambridge 1998, pp. 201–28 and 274–305.

36 Douglas, speech of 4 October 1888, *Norddeutsche Allgemeine Zeitung*, No. 474, 7 October 1888.

37 Cited in F. Hellwig, *Carl Freiherr von Stumm-Halberg*, Heidelberg, Saarbrücken 1936, p. 516.

38 Cf. Nicolaus Sombart, '"Ich sage, untergehen." Zum zweiten Band von Philipp Eulenburgs politischer Korrespondenz', *Merkur*, 385 (June 1980), pp. 542–54.

39 Eulenburg to Count Kuno von Moltke, 15 June 1895, *Eulenburgs Korrespondenz*, III, No. 1112.

40 Eissenstein to Kálnoky, 29 October 1888, HHStA Vienna, PA III 134.
41 Empress Frederick to Queen Victoria, 9 April 1889, RA Z44/26.
42 Waldersee, diary entry for 13 November 1889, GStA Berlin, Waldersee Papers; cf. Meisner, II, pp. 76–7.
43 Wedell to Kaiser Wilhelm II, 19 November 1890, cited in John C. G. Röhl, *The Kaiser and His Court. Wilhelm II and the Government of Germany*, Cambridge 3rd edn 1996, pp. 86 and 237.
44 Waldersee, diary entry for 2 December 1888, GStA Berlin, Waldersee Papers; omitted from Meisner, II, pp. 24–5.
45 Waldersee, diary entry for 2 December 1888, GStA Berlin, Waldersee Papers; cf. Meisner, II, pp. 24–5.
46 Waldersee, diary entry for 12 December 1888, GStA Berlin, Waldersee Papers; omitted from Meisner, II, pp. 25–6.
47 *Stenographische Berichte des Preußischen Abgeordnetenhauses*, 13th session, 9 and 12 February 1889, pp. 319ff., 351–2.
48 Waldersee, diary entry for 2 December 1888, GStA Berlin, Waldersee Papers; Meisner, II, pp. 24–5.
49 See Peter Domann, *Sozialdemokratie und Kaisertum unter Wilhelm II. Die Auseinandersetzung der Partei mit dem monarchischen System, seinen gesellschafts- und verfassungspolitischen Voraussetzungen*, Wiesbaden 1974; Marina Cataruzza, 'Das Kaiserbild in der Arbeiterschaft am Beispiel der Werftarbeiter in Hamburg und Stettin', in John C. G. Röhl, ed., *Der Ort Kaiser Wilhelms II. in der deutschen Geschichte*, Munich 1991, pp. 131–44.
50 See the article in favour of an imperial civil list in *Der Schwabische Merkur*, 7 February 1889, PA AA, Preußen 1 No. 1d, Bd 1 secr., together with Bismarck's sceptical marginal comment.
51 Note of 26 January 1889, GStA Berlin, 2.2.1. No. 3085. See Waldersee, diary entry for 13 November 1889, GStA Berlin, Waldersee Papers; cf. Meisner, II, pp. 76–7.
52 Lerchenfeld to Berchem, 16 August 1888, PA AA, R 3474.
53 Rantzau to Berchem, 18 August 1888, PA AA, R 3474.
54 Eissenstein to Kálnoky, 3 November 1888, HHStA Vienna, PA III 134.
55 See Röhl, *Kaiser and His Court*, pp. 73–4.
56 The loyal address of the Catholic archbishops and bishops of Prussia of 29 August 1888 together with Kaiser Wilhelm II's reply is to be found in PA AA, R 3448.
57 It is a measure of the worldwide significance attached to the accession of Wilhelm II that the imperial house of Japan requested an account of the educational principles which Hinzpeter had used in preparing him for the throne: Count Robert von Zedlitz-Trützschler to Kaiser Wilhelm II, 17 October 1891, Lucanus to Zedlitz, 19 October 1891, Bosse to Kaiser Wilhelm II, 13 September 1892, Prince T. Arisugawa of Japan to Kaiser Wilhelm II, 11 October 1892, GStA Berlin, 2.2.1. No. 3085.
58 See the article 'Kaiser Wilhelm II. und die Süddeutschen', *Der Schwäbische Merkur*, 5 September 1888, PA AA, R 3447.
59 Ibid.
60 Dr Paul Kayser, memorandum of 7 September 1889, PA AA, R 3448.
61 *Badische Presse*, 4 October 1889, PA AA, R 3448.
62 Edouard Simon, *L'Empereur Guillaume II et la première année de son règne*, Paris 1889. See Edouard Simon to Kaiser Wilhelm II, 27 July 1889, Lucanus to Bismarck, 27 August 1889, Berchem to Bismarck, 31 August 1889, PA AA, R 3448.
63 Ernest Lavisse, 'L'Empereur d'Allemagne', *Figaro*, 21 May 1890, PA AA, R 3448.
64 *Perseveranza*, 27 May 1890, PA AA, R 3448.
65 Poultney Bigelow, translation of an article in *Kurjer Codzienny*, 11 May 1890, PA AA, Nordamerika 6 No. 4.

66 Harold Frederic, *The Young Kaiser. William II of Germany. A Study in Character Development on a Throne*, London 1891. See Empress Frederick to Queen Victoria, 10 July 1891, RA Z50/57.
67 Sir Edward Malet to Lord Salisbury, 14 June 1888, RA I56/86.
68 Empress Frederick to Queen Victoria, 15 April 1889, RA Z44/30. See also Lord Salisbury to Queen Victoria, 9 March 1889, RA A67/51. See below, p. 91.
69 Empress Frederick to Queen Victoria, 27 April 1889, RA Z44/33.
70 Empress Frederick to Queen Victoria, 30 October 1889, RA Z46/31.
71 Széchényi to Kálnoky, 4 July 1888, HHStA Vienna, PA III 134.
72 Spitzemberg, diary entry for 19 August 1888, *Tagebuch*, pp. 253–4.
73 Waldersee, diary entry for 7 July 1888, GStA Berlin, Waldersee Papers; Meisner, I, pp. 411–12.
74 Waldersee, diary entry for 26 August 1888, GStA Berlin, Waldersee Papers; cf. Meisner, II, p. 1.
75 Széchényi to Kálnoky, 26 January 1889, HHStA Vienna, PA III 136.
76 Kaiser Wilhelm II, order to the Reich Chancellor of 28 January 1889, Széchényi to Kálnoky, 30 January 1889, HHStA Vienna, PA III 136.
77 Széchényi to Kálnoky, 9 February 1889, ibid. See below, p. 181.
78 Széchényi to Kálnoky, 26 March 1889, HHStA Vienna, PA III 136.
79 Count August zu Eulenburg to Philipp Eulenburg, 27 March 1889, *Eulenburgs Korrespondenz*, I, No. 216.
80 Malet, report of 27 March 1889, RA I57/19.
81 Széchényi to Kálnoky, 28 March 1889, HHStA Vienna, PA III 136.
82 Széchényi to Kálnoky, 9 January 1889, ibid. Ordered by Kálnoky, the report was sent to the embassies in St Petersburg, London, Paris, Rome and Constantinople.

2 THE FIRST STATE VISITS

1 Kaiser Wilhelm II, marginal comment on Bülow's report of 26 December 1904, PA AA, R 18858.
2 Széchényi to Kálnoky, 24 June 1888, HHStA Vienna, PA III 134.
3 Kaiser Franz Joseph to Kaiser Wilhelm II, draft telegram, June 1888, HHStA Vienna, Kabinettsarchiv, Geheimakten 2.
4 Kaiser Franz Joseph to Kaiser Wilhelm II, 17 June 1888, ibid. The exchange of letters between Franz Joseph and Wilhelm was published in *Neue Freie Presse* on 31 July and 7 August 1921.
5 Kaiser Wilhelm II to Kaiser Franz Joseph, 23 June 1888, HHStA Vienna, Kabinettsarchiv, Geheimakten 2. See *Große Politik*, VI, No. 1342. A draft of Franz Joseph's reply of 10 July 1888 is located in HHStA Vienna, Kabinettsarchiv, Geheimakten 2.
6 Waldersee, diary entry for 23 June 1888, Meisner, I, pp. 406ff. See Széchényi to Kálnoky, 21 June 1888, HHStA Vienna, PA III 134.
7 Kaiser Wilhelm II to Kaiserin Augusta, 11 July 1888, GStA Berlin, BPHA Rep. 53J Lit. P No. 14a.
8 Kaiser Wilhelm II to Crown Prince Rudolf, 12 July 1888, newspaper clipping from *L'écho de Paris*, 3 September 1895, PA AA, Österreich 86 No. 1. See Brigitte Hamann, 'Das Leben des Kronprinzen Rudolf von Österreich-Ungarn nach neuen Quellen', diss., Vienna 1977. A copy of Wilhelm's letter was sent by Rudolf to the Quai d'Orsay in July 1888. In the autumn of 1895 a French translation was then published in a Paris newspaper: Brigitte Hamann, *Rudolf, Kronprinz und Rebell*, Vienna, Munich 1978, pp. 361–2.
9 Queen Victoria, diary entry for 27 June 1888, RA QVJ, in George Earle Buckle, ed., *The Letters of Queen Victoria*, Third Series, 3 vols., London 1930, I, p. 421. See Empress Frederick to Queen Victoria, 8 July 1888, RA Z42/9. Cf. Albert Edward Prince of Wales to Empress Frederick, 11 July 1888, RA Add A5.

10 Queen Victoria to Kaiser Wilhelm II, 3 July 1888, RA I56/82, in Buckle, *Letters of Queen Victoria*, I, pp. 423–4.
11 Kaiser Wilhelm II to Queen Victoria, 6 July 1888, RA I56/84, in Buckle, *Letters of Queen Victoria*, I, pp. 424–5.
12 See *Young Wilhelm*, pp. 805–8.
13 Count Herbert Bismarck, Aufzeichnungen aus dem Herbst 1891, BA Koblenz, Bismarck Papers FC 3018 N; Tsar Alexander III to Kaiser Wilhelm II, 24 June 1888, PA AA, R 3613.
14 Kaiser Wilhelm II to Crown Prince Rudolf, 12 July 1888, newspaper clipping from *L'écho de Paris*, 3 September 1895, PA AA, Österreich 86 No. 1. Cf. Hamann, *Rudolf, Kronprinz und Rebell*, p. 361.
15 Senden-Bibran, Aufzeichnungen über das Jahr 1888, BA-MA Freiburg, Senden-Bibran Papers N160/11, p. 21.
16 Empress Frederick, diary entry for 13 July 1888, GStA Berlin, BPH Rep. 52 No. 3.
17 Széchényi to Kálnoky, 4 July 1888, HHStA Vienna, PA III 134.
18 Senden-Bibran, Aufzeichnungen über das Jahr 1888, BA-MA Freiburg, Senden-Bibran Papers N160/11, p. 22.
19 Alfred von Kiderlen-Wächter to Holstein, 16 July 1888, in Norman Rich and M. H. Fisher, eds., *The Holstein Papers. The Memoirs, Diaries and Correspondence of Friedrich von Holstein 1837–1909*, 4 vols., Cambridge 1956–63, III, No. 271.
20 Waldersee, diary entry for 1 July 1888, Meisner, I, p. 410.
21 Herbert Bismarck to his father, 14 July 1888, BA Koblenz, Bismarck Papers FC 3005 N.
22 Herbert Bismarck to his father, 16 July 1888, ibid.; Kiderlen-Wächter to Holstein, 16 July 1888, *Holstein Papers*, III, No. 271.
23 Count Herbert Bismarck, Aufzeichnungen aus dem Herbst 1891, BA Koblenz, Bismarck Papers FC 3018 N.
24 Kiderlen-Wächter to Holstein, 19 July 1888, *Holstein Papers*, III, No. 272; Senden-Bibran, Aufzeichnungen über das Jahr 1888, BA-MA Freiburg, Senden-Bibran Papers N160/11, p. 24. See also Kiderlen-Wächter to Holstein, 25 July 1888, *Holstein Papers*, III, No. 273.
25 Senden-Bibran, Aufzeichnungen über das Jahr 1888, BA-MA Freiburg, Senden-Bibran Papers N160/11, p. 24.
26 Herbert Bismarck, memorandum, 25 July 1888, *Große Politik*, VI, No. 1346; Herbert Bismarck, Aufzeichnungen aus dem Herbst 1891, BA Koblenz, Bismarck Papers FC 3018 N. See also Kiderlen-Wächter to Holstein, 25 July 1888, *Holstein Papers*, III, No. 273.
27 Senden-Bibran, Aufzeichnungen über das Jahr 1888, BA-MA Freiburg, Senden-Bibran Papers N160/11, p. 25.
28 Herbert Bismarck to Rantzau, 22 July 1888, BA Koblenz, Bismarck Papers FC 3014 N.
29 Kiderlen-Wächter to Holstein, 25 July 1888, *Holstein Papers*, III, No. 273. Cf. the description in Senden-Bibran, Aufzeichnungen über das Jahr 1888, BA-MA Freiburg, Senden-Bibran Papers N160/11, pp. 25–6.
30 Kiderlen-Wächter to Holstein, 19 and 25 July 1888, *Holstein Papers*, III, No. 272 and 273.
31 Kiderlen-Wächter to Holstein, 25 July 1888, ibid., No. 273.
32 Senden-Bibran, Aufzeichnungen über das Jahr 1888, BA-MA Freiburg, Senden-Bibran Papers N160/11, p. 26. For the relationship between Wilhelm and Ella, see *Young Wilhelm*, pp. 325–37.
33 Kiderlen-Wächter to Holstein, 25 July 1888, *Holstein Papers*, III, No. 273; Herbert Bismarck to Rantzau, 26 July 1888, BA Koblenz, Bismarck Papers FC 3014 N.
34 Herbert Bismarck to Rantzau, 26 July 1888, BA Koblenz, Bismarck Papers FC 3014 N.
35 Kiderlen-Wächter to Holstein, 25 July 1888, *Holstein Papers*, III, No. 273.
36 Waldersee, diary entry for 10 July 1888, Meisner, I, p. 412; Waldersee to Holstein, 26 July 1888, *Holstein Papers*, III, No. 275.

37 Pourtalès to Holstein, 25 July 1888, *Holstein Papers*, III, No. 274.
38 Schweinitz to Bismarck, 25 July 1888, *Große Politik*, VI, No. 1347.
39 Count W. N. Lamsdorff, diary entries for 30 January/11 February 1889, cited in George F. Kennan, *The Decline of Bismarck's European Order. Franco-Russian Relations, 1879–1890*, Princeton 1979, p. 398.
40 Sir Robert Morier to Sir Henry Ponsonby, 22 January 1889, RA I57/11; Ponsonby to Queen Victoria, 31 January 1889, RA I57/12.
41 Széchényi to Kálnoky, 2 August 1888, HHStA Vienna, PA III 134.
42 Senden-Bibran, Aufzeichnungen über das Jahr 1888, BA-MA Freiburg, Senden-Bibran Papers N160/11, pp. 24–5.
43 Waldersee to Verdy, 13 July 1888, GStA Berlin, Waldersee Papers B1 No. 53.
44 Waldersee to Holstein, 26 July 1888, *Holstein Papers*, III, No. 275.
45 Empress Frederick to Queen Victoria, 16 August 1888, RA Z42/31; Senden-Bibran, Aufzeichnungen über das Jahr 1888, BA-MA Freiburg, Senden-Bibran Papers N160/11, p. 28.
46 Empress Frederick to Queen Victoria, 16 August 1888, RA Z42/31.
47 Senden-Bibran, Aufzeichnungen über das Jahr 1888, BA-MA Freiburg, Senden-Bibran Papers N160/11, p. 27.
48 Herbert Bismarck to Rantzau, 29 July 1888, BA Koblenz, Bismarck Papers FC 3014 N.
49 Kaiser Wilhelm II to Eulenburg, 28 August–4 September 1888, *Eulenburgs Korrespondenz*, I, No. 194.
50 See Birgit Marschall, *Reisen and Regieren. Die Nordlandfahrten Kaiser Wilhelms II.*, Heidelberg 1991, pp. 80–101; also Klaus von See, *Barbar Germane Arier. Die Suche nach der Identität der Deutschen*, Heidelberg 1994, p. 16.
51 Kiderlen-Wächter to Holstein, 25 July 1888, *Holstein Papers*, III, No. 273.
52 Senden-Bibran, Aufzeichnungen über das Jahr 1888, BA-MA Freiburg, Senden-Bibran Papers N160/11, p. 28.
53 Kiderlen-Wächter to Holstein, 25 July 1888, *Holstein Papers*, III, No. 273.
54 See Waldersee to Holstein, 26 July 1888, ibid., No. 275.
55 Senden-Bibran, Aufzeichnungen über das Jahr 1888, BA-MA Freiburg, Senden-Bibran Papers N160/11, p. 29.
56 Széchényi to Kálnoky, 2 August 1888, HHStA Vienna, PA III 134.
57 See below, pp. 67–72.
58 Kaiser Wilhelm II to Kaiserin Augusta, 9 August 1888, GStA Berlin, Brand.-Preuß. Hausarchiv, Rep. 53J Lit. P. No. 14a.
59 Herbert Bismarck to Rantzau, 29 July 1888, BA Koblenz, Bismarck Papers FC 3014 N.
60 Kaiser Wilhelm II to Kaiserin Augusta, 9 August 1888, GStA Berlin, Brand.-Preuß. Hausarchiv, Rep. 53J Lit. P. No. 14a.
61 Rantzau to Berchem, 16 August 1994, PA AA, R 3474.
62 Eissenstein to Kálnoky, 26 September 1888, HHStA Vienna, PA III 134.
63 Kaiser Wilhelm II to Kaiserin Augusta, 10 October 1888, GStA Berlin, Brand.-Preuß. Hausarchiv Rep. 53J Lit. P. No. 14a.
64 Raschdau to Holstein, 13 October 1888, *Holstein Papers*, III, No. 279.
65 Prince Ernst zu Hohenlohe-Langenburg to his father, 6 January 1889, Hohenlohe-Zentralarchiv Neuenstein, Hermann Hohenlohe-Langenburg Papers, Bü. 56.
66 Waldersee, diary entry for 21 October 1888, GStA Berlin, Waldersee Papers; Meisner, II, p. 10.
67 Kaiserin Auguste Viktoria to Kaiser Wilhelm II, 13 October 1888, GStA Berlin, BPHA Rep. 53T Preußen: An Kaiser Wilhelm II., Bd III; Herbert Bismarck to Wilhelm Bismarck, 19 October 1888, in Walter Bußmann, ed., *Staatssekretär Graf Herbert von Bismarck. Aus seiner politischen Privatkorrespondenz*, Göttingen 1964, No. 369.

68 Christopher Duggan, *Creare la nazione. Vita di Francesco Crispi*, Rome, Bari 2000, pp. 664–7.
69 Cardinal Gustav von Hohenlohe-Schillingsfürst to Duke Georg II of Saxe-Meiningen, 27 December 1888, Thüringisches Staatsarchiv Meiningen, HA 361.
70 Raschdau to Holstein, 13 October 1888, *Holstein Papers*, III, No. 279.
71 Herbert Bismarck to Wilhelm Bismarck, 19 October 1888, in Bußmann, *Staatssekretär*, No. 369.
72 Herbert Bismarck to Holstein, 15 and 17 October 1888, *Holstein Papers*, III, Nos. 280–1.
73 Empress Frederick to Queen Victoria, 27 April 1889, RA Z44/33.
74 Holstein, diary entry for 24 October 1888, *Holstein Papers*, II, pp. 379–80.
75 Salisbury to Queen Victoria, 23 October 1888, RA A67/14.
76 Holstein, diary entry for 24 October 1888, *Holstein Papers*, II, pp. 379–80.
77 Cited in Duggan, *Creare la nazione*, p. 667.
78 Cf. Heinz-Joachim Fischer, 'Kaiserliche Eskapaden in der Ewigen Stadt. Was das Vatikanische Geheimarchiv über den Besuch Wilhelms II. in Rom enthüllt', *Frankfurter Allgemeine Zeitung*, 7 August 1996, p. 6.
79 Holstein, diary entry for 22 October 1888, *Holstein Papers*, II, pp. 378–9.
80 Cited in Christoph Weber, *Quellen und Studien zur Kurie und zur vatikanischen Politik unter Leo XIII.*, Tübingen 1973, pp. 40–1. See also Waldersee, diary entries for 16 and 21 October and 1, 2, 4, 7 and 26 November 1888, GStA Berlin, Waldersee Papers; also Raschdau to Holstein, 13 October 1888, *Holstein Papers*, III, No. 279. See Herbert Bismarck to Kaiser Wilhelm II, 21 October 1888, BA Koblenz, Bismarck Papers FC 2986 N; partially printed in Bußmann, *Staatssekretär*, No. 370. Cf. Count zu Solms to Holstein, 2 November 1888, *Holstein Papers*, III, No. 282. See also Kurd von Schlözer, *Letzte römische Briefe 1882–1894*, Stuttgart, Berlin, Leipzig 1924, pp. 119–37; Friedrich Noack, *Das Deutschtum in Rom seit dem Ausgang des Mittelalters*, 2 vols., Berlin, Leipzig 1927, I, pp. 665–6, 689–90, II, p. 646; James E. Ward, 'Leo XIII and Bismarck. The Kaiser's Vatican Visit of 1888', *Review of Politics*, 24 (1962), pp. 392–414.
81 Herbert Bismarck to Holstein, 15 October 1888, *Holstein Papers*, III, No. 280.
82 Waldersee, diary entry for 21 October 1888, GStA Berlin, Waldersee Papers; cf. the abridged and distorted version in Meisner, II, p. 10.
83 Waldersee, diary entries for 1 4, and 7 November 1888, GStA Berlin, Waldersee Papers; omitted from Meisner, II, pp. 13, 15–16.
84 Cardinal Gustav von Hohenlohe-Schillingsfürst to Duke Georg II of Saxe-Meiningen, 27 December 1888, Thüringisches Staatsarchiv Meiningen, HA 361.
85 Holstein, diary entry for 24 October 1888, *Holstein Papers*, II, pp. 379–80.
86 Queen Victoria to Sir Theodore Martin, 4 November 1888, RA Y172/82.
87 Queen Victoria to Empress Frederick, 6 November 1888, RA Add U32/612.
88 Grand Duke Ludwig IV of Hesse-Darmstadt to Empress Frederick, 13 October 1888, GStA Berlin, BPH Rep. 52 No. 3, pp. 338–44.
89 Herbert Bismarck to his father, 9 October 1888, in Bußmann, *Staatssekretär*, No. 368.
90 Ibid.
91 Eissenstein to Kálnoky, 20 October 1888, HHStA Vienna, PA III 134.

3 THE KAISER AND HIS MOTHER

1 Empress Frederick to Queen Victoria, 18, 20, 23 and 24 June, 5 and 21 July 1888, RA Z41/65–6, Z42/2–3, 7, 19; Ponsonby, *Letters of the Empress Frederick*, pp. 319–27.
2 Empress Frederick to Queen Victoria, 27 and 30 July, 13, 29 and 30 August 1888, RA Z42/23–4, 29, 40–1. See Rainer von Hessen, ed., *Victoria Kaiserin Friedrich. Mission und Schicksal einer englischen Prinzessin in Deutschland*, Frankfurt, New York 2002.

3 Empress Frederick to Queen Victoria, 18 June 1888, in Ponsonby, *Letters of the Empress Frederick*, pp. 319–21; see also 30 August and 1 September 1888, RA Z42/41–2.
4 Empress Frederick to Queen Victoria, 29 September 1888, RA Z43/11.
5 Empress Frederick, diary entry for 4 July 1888, GStA Berlin BPH Rep. 52 No. 3; Empress Frederick to Queen Victoria, 5 July 1888, in Ponsonby, *Letters of the Empress Frederick*, pp. 325–6.
6 Empress Frederick to Queen Victoria, 28 September 1888, RA Z43/10.
7 Empress Frederick to Queen Victoria, 22 August 1888, RA Z42/36, in Ponsonby, *Letters of the Empress Frederick*, pp. 329–31.
8 Empress Frederick to Queen Victoria, 21 August 1888, RA Z42/34.
9 Empress Frederick to Queen Victoria, 30 August 1888, RA Z42/41.
10 Empress Frederick to Queen Victoria, 27 July 1888, RA Z42/23.
11 Empress Frederick to Queen Victoria, 16 August 1888, RA Z42/31. See above, ch. 2.
12 Empress Frederick to her daughter Princess Viktoria of Schaumburg-Lippe, 28 January 1891, Archiv der Hessischen Hausstiftung (AdHH) Schloss Fasanerie. The Empress commented along similar lines to her daughter Sophie. See Arthur Gould Lee, *The Empress Frederick Writes to Sophie, Her Daughter, Crown Princess and Later Queen of the Hellenes*, London 1955, p. 77. See also Queen Victoria to Kaiser Wilhelm II, 24 January and 4 March 1891, GStA Berlin, BPHA Rep. 52 W3 No. 11.
13 See Empress Frederick to Queen Victoria, 13 December 1890, RA Z49/43.
14 Empress Frederick to Queen Victoria, 22 August 1888, RA Z42/36, in Ponsonby, *Letters of the Empress Frederick*, pp. 329–30.
15 Empress Frederick to Queen Victoria, 7 July 1888, RA Z42/8; Empress Frederick to Queen Victoria, 13 July 1888, AdHH Schloss Fasanerie; Empress Frederick, diary entry for 30 July 1888, GStA Berlin BPH Rep. 52 No. 3.
16 Empress Frederick to Queen Victoria, 23 August 1888, RA Z42/37.
17 Waldersee, diary entry for 18 December 1890, GStA Berlin, Waldersee Papers; cf. Meisner, II, p. 167.
18 Empress Frederick to Queen Victoria, 16 August 1888, RA Z42/31.
19 Empress Frederick to Queen Victoria, 28 September 1888, RA Z43/10.
20 Empress Frederick to Queen Victoria, 12 September 1888, RA Z42/46.
21 Empress Frederick to Queen Victoria, 10 December 1890, RA Z49/39. See below, pp. 505–12.
22 Empress Frederick to Queen Victoria, 2 November 1888, RA Z43/32.
23 Empress Frederick to Queen Victoria, 5 July 1888, in Ponsonby, *Letters of the Empress Frederick*, pp. 325–6.
24 Empress Frederick, diary entry for 20 June 1888, GStA Berlin BPH Rep. 52 No. 3; cited in Egon Caesar Conte Corti, *The English Empress. A Study in the Relations between Queen Victoria and Her Eldest Daughter, Empress Frederick of Germany*, London 1957, p. 306. See also Empress Frederick to Queen Victoria, 5 July 1888, RA Z42/7.
25 Empress Frederick to Queen Victoria, 30 July 1888, RA Z42/24.
26 See Empress Frederick to Queen Victoria, 5 July 1888, RA Z42/7.
27 Empress Frederick, diary entry for 17 June 1888, GStA Berlin BPH Rep. 52 No. 3, quoted in part in Corti, *English Empress*, p. 304.
28 Empress Frederick, diary entry for 25 June 1888, GStA Berlin BPH Rep. 52 No. 3, quoted in Corti, *English Empress*, p. 306; Empress Frederick to Queen Victoria, 5, 7, 12 and 16 July 1888, RA Z42/7–8, 11 and 13. See also Empress Frederick to Queen Victoria, 23, 25 and 29 August and 14 September 1888, RA Z42/37–8, 40, 43.
29 Empress Frederick to Queen Victoria, 29 August 1888, RA Z42/40.
30 Kaiser Wilhelm II to Minister of Justice von Friedberg and House Minister von Wedell, 3 December 1888, GStA Berlin, 2.2.1. No. 3115; Prince Ernst zu Hohenlohe-Langenburg

to his father, 6 January 1889, Hohenlohe-Zentralarchiv Neuenstein, Hermann Hohenlohe-Langenburg Papers, Bü. 56; Empress Frederick to Queen Victoria, 21 January 1889, AdHH Schloss Fasanerie (copy in RA Z60A/10); Széchényi to Kálnoky, 30 January 1889, HHStA Vienna, PA III 136. See Ponsonby, *Letters of the Empress Frederick*, pp. 339–65.

31 Empress Frederick to Queen Victoria, 25 October 1888, AdHH Schloss Fasanerie; Empress Frederick to Queen Victoria, 27 April 1889, RA Z44/33.

32 Holstein, diary entry for 24 October 1888, *Holstein Papers*, II, pp. 379–80.

33 Friedberg to Lucanus, 21 October 1888, Kaiser Wilhelm II to Friedberg, 27 October 1888, Lucanus to Friedberg, 27 October 1888, GStA Berlin, 2.2.1. No. 3115; Empress Frederick to Queen Victoria, 25 October 1888, AdHH Schloss Fasanerie; Eissenstein to Kálnoky, 29 October 1888, HHStA Vienna, PA III 134; Empress Frederick, diary entries for 17 and 28 October and 2 November 1888, GStA Berlin, BPH Rep. 52 No. 3; Bogumilla Freifrau von Stockmar to Empress Frederick, 2 November 1888, ibid., pp. 392–4; Lucanus to Tessendorff, 8 November 1888, GStA Berlin, 2.2.1. No. 3115. Many documents on the 'Morier affair' are to be found at Windsor Castle in RA I/57.

34 Empress Frederick to Queen Victoria, 20 and 22 March 1889, RA Z44/15 and 17.

35 Empress Frederick to Queen Victoria, 30 October 1888, RA Z43/31.

36 Empress Frederick to Queen Victoria, 29 March 1889, RA Z44/20, cited in Ponsonby, *Letters of the Empress Frederick*, pp. 369–70.

37 Empress Frederick, diary entries for 17 and 28 October 1888, GStA Berlin, BPH Rep. 52 No. 3; Empress Frederick to Queen Victoria, 16, 17 and 20 October 1888, RA Z43/24, 25 and 27; Grand Duke Ludwig IV of Hesse-Darmstadt to Empress Frederick, 17 October and 3 November 1888, GStA Berlin, BPH Rep. 52 No. 3, pp. 349–52 and 396–9. Empress Frederick to Queen Victoria, 27 April 1889, RA Z44/33.

38 Empress Frederick to Queen Victoria, 30 October 1888, RA Z43/31.

39 Empress Frederick to Queen Victoria, 2 November 1888, RA Z43/32, in Ponsonby, *Letters of the Empress Frederick*, pp. 357–61.

40 Empress Frederick, diary entries for 5–7 and 12–13 November and 8 December 1888, GStA Berlin BPH Rep. 52. No. 3; Empress Frederick to Kaiser Wilhelm II, 7 November 1888, GStA Berlin, BPHA Rep. 52T No. 13.

41 Empress Frederick to Queen Victoria, 2 November 1888, RA Z43/32, in Ponsonby, *Letters of the Empress Frederick*, pp. 357–61.

42 Ibid.

43 Empress Frederick to Queen Victoria, 10 November 1888, RA Z43/36, in Ponsonby, *Letters of the Empress Frederick*, pp. 361–4.

44 Empress Frederick to Queen Victoria, 23 June 1888, RA Z42/2.

45 Empress Frederick to Queen Victoria, 13 August 1888, RA Z42/29.

46 Empress Frederick to Queen Victoria, 2 November 1888, RA Z43/32, in Ponsonby, *Letters of the Empress Frederick*, pp. 357–61.

47 Empress Frederick to Queen Victoria, 26 September 1888, RA Z43/8.

48 Empress Frederick to Queen Victoria, 29 June 1888, RA Z42/5, in Ponsonby, *Letters of the Empress Frederick*, pp. 322–3.

49 Empress Frederick to Queen Victoria, 16 August 1888, RA Z42/31.

50 Empress Frederick, diary entries for 23 and 29 December 1888, GStA Berlin, BPH Rep. 52 No. 3.

51 Empress Frederick to Queen Victoria, 2 November 1888, RA Z43/32, in Ponsonby, *Letters of the Empress Frederick*, pp. 357–61.

52 Ibid.

53 Queen Victoria to Kaiser Wilhelm II, 3 July 1888, RA I56/82; Empress Frederick to Queen Victoria, 7 July 1888, RA Z42/8.

54 Empress Frederick to Queen Victoria, 13 August 1888, RA Z42/29.
55 Empress Frederick to Queen Victoria, 5 July 1888, RA Z42/7, in Ponsonby, *Letters of the Empress Frederick*, pp. 325–6. See also Empress Frederick to Queen Victoria, 27 July 1888, RA Z42/23.
56 Empress Frederick to Queen Victoria, 26 September 1888, RA Z43/8.
57 Empress Frederick to Queen Victoria, 3 September 1888, RA Z42/43.
58 Empress Frederick to Queen Victoria, 23 June 1888, RA Z42/2.
59 Albert Edward Prince of Wales to Empress Frederick, 29 June 1888, RA Add A5/479/20.
60 Albert Edward Prince of Wales to Empress Frederick, 4 July 1888, RA Add A5/479/21.
61 Albert Edward Prince of Wales to Empress Frederick, 29 June 1888, RA Add A5/479/20.
62 Albert Edward Prince of Wales to Empress Frederick, 4 July 1888, RA Add A5/479/21.
63 Albert Edward Prince of Wales to Empress Frederick, 27 June 1888, RA Add A5.
64 Empress Frederick, diary entry for 22 June 1888, GStA Berlin BPH Rep. 52 No. 3.
65 Ibid.
66 Empress Frederick, diary entry for 28 June 1888, ibid.
67 Empress Frederick to Queen Victoria, 7 July 1888, RA Z42/8; Empress Frederick to Queen Victoria, 12 October 1888, RA Z43/21.
68 Empress Frederick, diary entry for 28 June 1888, GStA Berlin BPH Rep. 52 No. 3, quoted in Corti, *English Empress*, p. 309.
69 Empress Frederick, diary entry for 2 July 1888, GStA Berlin BPH Rep. 52 No. 3.
70 Empress Frederick to Queen Victoria, 27 August 1890, RA Z49/10.
71 See *Young Wilhelm*, pp. 666–8.
72 Empress Frederick, diary entry for 2 July 1888, GStA Berlin BPH Rep. 52 No. 3; Empress Frederick to Queen Victoria, 29 June and 5 July 1888, in Ponsonby, *Letters of the Empress Frederick*, pp. 322–3, 325.
73 Empress Frederick to Queen Victoria, 13 and 16 August, 1 and 24 September 1888, RA Z42/29, 31, 42 and Z43/7; Empress Frederick, diary entry for 1 September 1888, GStA Berlin BPH Rep. 52 No. 3; Empress Frederick to Prince Hermann zu Hohenlohe-Langenburg, 22 September 1888, Hohenlohe-Zentralarchiv Schloss Neuenstein, Hermann Hohenlohe-Langenburg Papers, Bü. 105.
74 Empress Frederick to Queen Victoria, 7 July and 29 September 1888, RA Z42/8 and Z43/11.
75 Empress Frederick to Queen Victoria, 15, 20 and 27 April 1889, RA Z44/30, 31 and 33.
76 Empress Frederick, diary entries for 30 September and 1 October 1888, GStA Berlin BPH Rep. 52 No. 3; Empress Frederick to Queen Victoria, 28 September 1888, RA Z43/10.
77 *Young Wilhelm*, p. 679.
78 Empress Frederick, diary entry for 20 June 1888, GStA Berlin, BPH Rep. 52 No. 3; Empress Frederick to Queen Victoria, 20 June 1888, RA Z41/66; Empress Frederick to Queen Victoria, 12 July and 24 October 1888, AdHH Schloss Fasanerie.
79 Empress Frederick to Queen Victoria, 20 and 24 June 1888, RA Z41/66, Z42/3.
80 Empress Frederick, diary entry for 10 July 1888, GStA Berlin, BPH Rep. 52 No. 3; Empress Frederick to Queen Victoria, 16 July 1888, RA Z42/13.
81 Empress Frederick to Queen Victoria, 18 July 1888, RA Z42/14; Empress Frederick, diary entry for 18 July 1888, GStA Berlin, BPH Rep. 52 No. 3.
82 Empress Frederick to Queen Victoria, 12 July 1888, AdHH Schloss Fasanerie; Empress Frederick to Queen Victoria, 16 and 18 July 1888, RA Z42/13, 15.
83 Empress Frederick, diary entry for 19 July 1888, GStA Berlin, BPH Rep. 52 No. 3.
84 Empress Frederick to Queen Victoria, 20 July 1888, RA Z42/18.
85 Empress Frederick, diary entry for 21 July 1888, GStA Berlin, BPH Rep. 52 No. 3; Empress Frederick to Queen Victoria, 21 July 1888, RA Z42/19. See also the Empress Frederick's notes of 14 August 1888, GStA Berlin, BPHA Rep. 52J, general, No. 5; Wedell to Kaiser Wilhelm II, 30 August 1888, GStA Berlin, 2.2.1. No. 3085.

86 Herbert Bismarck to his father, 4 August 1888, BA Koblenz, Bismarck Papers FC 3005 N, printed in part in Bußmann, *Staatssekretär*, No. 363.
87 Empress Frederick to Queen Victoria, 20 July 1888, RA Z42/18.
88 Empress Frederick to Queen Victoria, 18 July 1888, RA Z42/15.
89 The Empress handed over another collection of her letters to Windsor in the spring of 1891: Empress Frederick to Bogumilla Freifrau von Stockmar, 17 February 1891, AdHH Schloss Fasanerie.
90 Empress Frederick, diary entries for 22, 23 and 26 July 1888, GStA Berlin, BPH Rep. 52 No. 3. Cf. Queen Victoria to Empress Frederick, 25 July 1888, RA Add U32/570.
91 Empress Frederick to Queen Victoria, 13 September [?] 1888, RA Z43/2. In January 1891 the Empress Frederick received back the letters which she had written to the Kaiserin Augusta: Empress Frederick to Queen Victoria, 13 January 1891, RA Z50/5.
92 'Secret. Memo'd concerning the collecting of material for a Life of Fritz', Empress Frederick to Queen Victoria, 13 September [?] 1888, RA Z43/2; Empress Frederick to Queen Victoria, 14 September 1888, RA Z43/3, in Ponsonby, *Letters of the Empress Frederick*, pp. 341–3. See Empress Frederick to Queen Victoria, 17 and 22 September 1888, RA Z43/4 and Z43/6.
93 Empress Frederick, diary entry for 20 September 1888, GStA Berlin, BPH Rep. 52 No. 3.
94 Empress Frederick, diary entry for 23 September 1888, GStA Berlin, BPH Rep. 52 No. 3.
95 Empress Frederick to Queen Victoria, 24 September 1888, RA Z43/7, in Ponsonby, *Letters of the Empress Frederick*, pp. 344–5. See also Empress Frederick, diary entry for 25 September 1888, GStA Berlin, BPH Rep. 52 No. 3.
96 See Ponsonby, *Letters of the Empress Frederick*, pp. 359–60.
97 Empress Frederick, diary entry for 26 September 1888, GStA Berlin, BPH Rep. 52 No. 3; Wedell to Kaiser Wilhelm II, 25 September 1888, GStA Berlin, 2.2.1. No. 3115; Empress Frederick to Queen Victoria, 27 September 1888, RA Z43/9.
98 Eissenstein to Kálnoky, 23 September 1888, HHStA Vienna, PA III 134.
99 Empress Frederick to Queen Victoria, 24 September 1888, RA Z43/7, in Ponsonby, *Letters of the Empress Frederick*, pp. 364–5; also Empress Frederick, diary entry for 25 September 1888, GStA Berlin, BPH Rep. 52 No. 3.
100 Empress Frederick to Queen Victoria, 26 September 1888, RA Z43/8, in Ponsonby, *Letters of the Empress Frederick*, pp. 345–6.
101 Empress Frederick to Queen Victoria, 29 September 1888, RA Z43/11, in Ponsonby, *Letters of the Empress Frederick*, pp. 348–50.
102 Herbert Bismarck to Holstein, 30 September 1888, *Holstein Papers*, III, No. 276.
103 Empress Frederick, diary entry for 1 October 1888, GStA Berlin, BPH Rep. 52 No. 3.
104 Empress Frederick to Queen Victoria, 2 October 1888, RA Z43/13, in Ponsonby, *Letters of the Empress Frederick*, pp. 350–1.
105 Empress Frederick to Queen Victoria, 11 October 1888, RA Z43/20, in Ponsonby, *Letters of the Empress Frederick*, pp. 351–2.
106 Empress Frederick to Queen Victoria, 12 October 1888, RA Z43/21, in Ponsonby, *Letters of the Empress Frederick*, p. 352.
107 Empress Frederick to Queen Victoria, 20 October 1888, RA Z43/27, in Ponsonby, *Letters of the Empress Frederick*, pp. 353–4.
108 Empress Frederick to Queen Victoria, 30 October 1888, RA Z43/31, in Ponsonby, *Letters of the Empress Frederick*, pp. 355–6.
109 Empress Frederick to Kaiser Wilhelm II, 29 October 1888, sent on 8 November, GStA Berlin, BPHA Rep. 52T No. 13; copy ibid., BPH Rep. 52 No. 3, pp. 384–7.
110 Empress Frederick, diary entries for 5, 6 and 12 November 1888, GStA Berlin, BPH Rep. 52 No. 3.
111 Empress Frederick to Kaiser Wilhelm II, 7 November 1888, GStA Berlin, BPHA Rep. 52T No. 13. See also Empress Frederick, diary entry for 13 November 1888, GStA Berlin, BPH

Rep. 52 No. 3. Cf. Holstein, diary entry for 11 November 1888, *Holstein Papers*, II, pp. 382–3. See also Kaiserin Auguste Viktoria to Kaiserin Augusta, 16 November 1888, GStA Berlin, BPHA Rep. 52 W3 No. 4 Victoria, Dok. No. 45.

112 Empress Frederick to Queen Victoria, 2 November 1888, RA Z43/32, in Ponsonby, *Letters of the Empress Frederick*, pp. 357–61.

113 See e.g. Princess Victoria of Battenberg to Queen Victoria, 16 and 29 June 1888, RA Add U166/50–1; Princess Viktoria (Moretta) of Prussia to Queen Victoria, 27 June 1888, RA Z83/27; Princess Margarethe of Prussia to Queen Victoria, 28 June 1888, RA Z83/28. Cf. also Grand Duchess Augusta Caroline of Mecklenburg-Strelitz to Eulenburg, 7 August 1888, in *Eulenburgs Korrespondenz*, I, No. 189.

114 Queen Victoria's Journal (QVJ), 27 June 1888, printed in Buckle, *Letters of Queen Victoria*, I, p. 421. See also Queen Victoria to Lord Salisbury, 28 June 1888, RA I56/79.

115 Queen Victoria to Princess Victoria of Battenberg, RA Add U173/143.

116 Queen Victoria to Princess Maria Anna of Prussia, 30 July 1888, RA Vic Addl Mss A7/259.

117 The exchange of letters is printed in Buckle, *Letters of Queen Victoria*, I, pp. 423–5. See above, ch. 2.

118 Queen Victoria to Albert Edward Prince of Wales, 17 July 1888, RA I56/86a.

119 Queen Victoria to Albert Edward Prince of Wales, 24 July 1888, RA I56/86b.

120 Queen Victoria to Empress Frederick, 25 July 1888, RA Add U32/570.

121 Queen Victoria to Kaiser Wilhelm II, 3 July 1888, RA I56/82. See Albert Edward Prince of Wales to Empress Frederick, 4 July 1888, RA Add A5/479/21.

122 Kaiser Wilhelm II to Queen Victoria, 6 July 1888, RA I56/84.

123 See above, pp. 33–4.

124 See above, p. 29; Kaiser Wilhelm II, marginal comments, 4 July 1888, PA AA, Asservat No. 4.

125 Empress Frederick to Queen Victoria, 20 June 1888, RA Z41/67, quoted in Corti, *English Empress*, p. 308. See also Empress Frederick, diary entry for 20 June 1888, GStA Berlin BPH Rep. 52 No. 3.

126 See *Truth*, 16 August 1888, p. 277.

127 Winterfeldt to Herbert Bismarck, 30 June 1888, PA AA, R 5668.

128 Kaiser Wilhelm II, marginal comment, 5 July 1888, PA AA, Asservat No. 4.1.

129 Leopold Swaine to Sir Henry Ponsonby, 4 July 1888, in Ponsonby, *Letters of the Empress Frederick*, p. 324.

130 Queen Victoria to Sir Henry Ponsonby, [7 July 1888], RA I56/77. Cf. Ponsonby, *Letters of the Empress Frederick*, p. 324.

131 Sir Edward Malet to Lord Salisbury, 14 July 1888, RA I56/86; also Count Herbert Bismarck, diary, 3 and 7 July 1888, BA Koblenz, Bismarck Papers FC 3018 N.

4 AN OMINOUS FAMILY QUARREL

1 Kaiser Wilhelm II, speech of 16 August 1888, Penzler, *Reden Kaiser Wilhelms II. in den Jahren 1888–1895*, pp. 19–21. Cf. Count Georg von Werthern to Eulenburg, 9 September 1888, *Eulenburgs Korrespondenz*, I, No. 195.

2 Empress Frederick to Queen Victoria, 25 August 1888, RA Z42/38, in Ponsonby, *Letters of the Empress Frederick*, pp. 333–4.

3 Empress Frederick to Queen Victoria, 25 August 1888, RA Z42/38. See Ponsonby, *Letters of the Empress Frederick*, pp. 333–4.

4 Empress Frederick to Queen Victoria, 26 August 1888, RA Z42/39, in Ponsonby, *Letters of the Empress Frederick*, pp. 336–7.

5 Albert Edward Prince of Wales to Prince Christian of Schleswig-Holstein, 3 April 1889, in Buckle, *Letters of Queen Victoria*, I, p. 489. See Ponsonby, *Letters of the Empress Frederick*, p. 338; Corti, *English Empress*, p. 305.
6 Rantzau to Berchem, 28 August 1888, PA AA, R 3447.
7 Count Herbert Bismarck, Aufzeichnungen aus dem Jahre 1891, BA Koblenz, Bismarck Papers FC 3018 N.
8 Herbert Bismarck, memorandum, 25 July 1888, *Große Politik*, VI, No. 1346.
9 Pourtalès to Herbert Bismarck, 21 September 1888, PA AA, R 3447.
10 Swaine to Albert Edward Prince of Wales, 3 September 1888, RA Z281/1.
11 Empress Frederick, diary entries for 3 and 5 September 1888, GStA Berlin BPH Rep. 52 No. 3; Albert Edward Prince of Wales to Empress Frederick, 6 September 1888, RA Add A5. See above, p. 41.
12 Swaine to Albert Edward Prince of Wales, 3 September 1888, RA Z281/1.
13 Ellis to Swaine, 12 September 1888, RA Z281/2.
14 Swaine to Albert Edward Prince of Wales, 20 September 1888, RA Z281/3.
15 Swaine to Albert Edward Prince of Wales, 5 October 1888, RA Z281/12; Count Herbert Bismarck, diary entry for 19 February 1889, BA Koblenz, Bismarck Papers FC 3018 N.
16 Ellis to Swaine, 25 September 1888, RA Z281/4.
17 Queen Victoria to Arthur Duke of Connaught, 27 September 1888, RA Vic Addl Mss A15/5166. See also Queen Victoria to Princess Victoria of Battenberg, 2 October 1888, RA Add U173/145.
18 See *Truth*, 27 September 1888, p. 529.
19 Crown Prince Rudolf to Crown Princess Stephanie, 12 September 1888, cited in Hamann, *Rudolf, Kronprinz and Rebell*, p. 363.
20 Paget to Albert Edward Prince of Wales, 25 September 1888, RA Z281/5.
21 Salisbury to Paget, cipher telegram, 30 September 1888, RA Z281/7.
22 Paget to Salisbury, cipher telegram, 1 October 1888, RA Z281/8.
23 Paget to Albert Edward Prince of Wales, 1 and 5 October 1888, RA Z281/9, 11.
24 Paget to Sir Frank Lascelles, cipher telegram, 4 October 1888, RA Z281/10; Albert Edward Prince of Wales to Queen Victoria, 5 October 1888, RA Z280/65; Paget to Albert Edward Prince of Wales, 5 October 1888, RA Z281/11.
25 Queen Victoria to Salisbury, 15 October 1888, RA T9/111; Albert Edward Prince of Wales to Empress Frederick, 31 October 1888, RA Add A5/479/34.
26 Swaine to Ellis, 27 September 1888, RA Z281/6.
27 Salisbury, memorandum, 13 October 1888, RA T9/110, in Buckle, *Letters of Queen Victoria*, I, pp. 438–40.
28 Salisbury to Queen Victoria, 15 October 1888, RA T9/112, in Buckle, *Letters of Queen Victoria*, I, pp. 441–2.
29 Queen Victoria to Salisbury, 15 October 1888, RA T9/111, in Buckle, *Letters of Queen Victoria*, I, pp. 440–1. See also Queen Victoria to Salisbury, 13 November 1888, RA Z280/67.
30 Paget to Ponsonby, 13 November 1888, RA Z280/68; Salisbury to Queen Victoria, 18 November 1888, RA Z280/69.
31 Queen Victoria to Kaiserin Augusta, 14 November 1888, GStA BPHA Rep. 52 W3 No. 4 Victoria, Dok. No. 47.
32 Kaiserin Augusta to Queen Victoria, 17 November 1888, ibid., Dok. No. 48. On Wilhelm's treatment during his stay in London in the summer of 1887, see *Young Wilhelm*, pp. 682–9.
33 Albert Edward Prince of Wales to Queen Victoria, 1 November 1888, RA Z280/66.
34 Albert Edward Prince of Wales to Empress Frederick, 31 October 1888, RA Add A5/479/34.
35 Alexandra Princess of Wales to Prince George, 17 October 1888, RA Geo V, AA31/1.

36 Albert Edward Prince of Wales to Empress Frederick, 31 October 1888, RA Add A5/479/34. See also Albert Edward Prince of Wales to Empress Frederick, 23 November 1888, ibid.
37 Arthur Duke of Connaught to Queen Victoria, 25 October 1888, RA Z185/25.
38 Count Herbert Bismarck, Aufzeichnungen aus dem Jahre 1891, BA Koblenz, Bismarck Papers FC 3018 N.
39 Count Herbert Bismarck, diary entries for 31 August–16 September 1888, ibid.
40 Count Herbert Bismarck, diary entries for August–October 1888, ibid.
41 Adolf von Deines, report of August 1888, PA AA, R 8593.
42 Kaiser Wilhelm II, marginal notes on Deines's report of 30 August 1888, PA AA, Asservat No. 4.
43 Rantzau to Auswärtiges Amt, 4 September 1888, PA AA, R 5900, Geheime Acten betreffend den Prinzen von Wales.
44 Berchem, draft of 6 September 1888, ibid. See above, pp. 73–7.
45 Reuss, report of 13 September 1888, PA AA, R 5900.
46 See also Bismarck to Kaiser Wilhelm II, 20 September 1888, *Große Politik*, VI, No. 1351.
47 Herbert Bismarck to his father, 16 September 1888, BA Koblenz, Bismarck Papers FC 3005 N.
48 Count Herbert Bismarck, diary, August–October 1888, BA Koblenz, Bismarck Papers FC 3018 N. See also the retrospective note in Herbert Bismarck, Aufzeichnung, pp. 56–7, ibid.
49 Empress Frederick to Queen Victoria, 20 April 1889, RA Z44/31, in Ponsonby, *Letters of the Empress Frederick*, pp. 370–3. See Prince Christian of Schleswig-Holstein to Sir Henry Ponsonby, 13 May 1889, RA Z281/46; also Prince Ernst zu Hohenlohe-Langenburg to his father, 9 June 1889, Hohenlohe Zentralarchiv Neuenstein, Hermann Hohenlohe-Langenburg Papers, Bü. 56.
50 Queen Victoria to Albert Edward Prince of Wales, 7 February 1889, RA I57/17, in Buckle, *Letters of Queen Victoria*, I, pp. 467–8. See also Ponsonby to Knollys, 12 and 15 February 1889, RA Z281/13 and 14.
51 Albert Edward Prince of Wales to Queen Victoria, 8 February 1889, RA Z498/49.
52 Queen Victoria to Salisbury, 27 February 1889, RA A67/46, in Buckle, *Letters of Queen Victoria*, I, p. 473.
53 Salisbury to Queen Victoria, 9 March 1889, RA A67/51. See Prince Ernst zu Hohenlohe-Langenburg to his father, 4 March 1889, Hohenlohe Zentralarchiv Neuenstein, Hermann Hohenlohe-Langenburg Papers, Bü. 56.
54 Count Herbert Bismarck, Aufzeichnung, pp. 64ff., BA Koblenz, Bismarck Papers FC 3018 N; diary entry for 6 March 1889, ibid.
55 Herbert Bismarck to Rantzau, 30 July 1889, ibid., FC 3014 N, in Bußmann, *Staatssekretär*, No. 386. See also Count Herbert Bismarck, Aufzeichnung, pp. 66ff., BA Koblenz, Bismarck Papers, FC 3018 N.
56 Queen Victoria to Empress Frederick, 12 March 1889, RA Vic Add Mss U32/619.
57 Prince Ernst zu Hohenlohe-Langenburg to his father, 19 May 1889, Hohenlohe Zentralarchiv Neuenstein, Hermann Hohenlohe-Langenburg Papers, Bü. 56.
58 Knollys to Ponsonby, 13 March 1889, RA Addl Mss A12/1641. See also George Duke of Cambridge to Ponsonby, 16 March 1889, RA W11/36.
59 Herbert Bismarck to Kaiser Wilhelm II with marginal comments by the Kaiser, 17 March 1889, BA Koblenz, Bismarck Papers FC 2986 N; Salisbury to Queen Victoria, 20 March 1889, RA A67/59; Waldersee, diary entry for 25 March 1889, GStA Berlin, Waldersee Papers; omitted from Meisner, II, p. 47.
60 Count Herbert Bismarck, Aufzeichnung, pp. 55ff., BA Koblenz, Bismarck Papers FC 3018 N.
61 Salisbury to Queen Victoria, 29 March 1889, RA A67/48, in Buckle, *Letters of Queen Victoria*, I, pp. 483–5; Prince Ernst zu Hohenlohe-Langenburg to his father, 19 May 1889, Hohenlohe

Zentralarchiv Neuenstein, Hermann Hohenlohe-Langenburg Papers, Bü. 56; Count Herbert Bismarck, diary entries for 21–31 March 1889, BA Koblenz, Bismarck Papers FC 3018 N.

62 Albert Edward Prince of Wales to Prince Christian of Schleswig-Holstein, 3 April 1889, RA T9/119, in Buckle, *Letters of Queen Victoria*, I, pp. 487–9.

63 Count Herbert Bismarck, diary entry for 8 April 1889, BA Koblenz, Bismarck Papers FC 3018 N.

64 Prince Christian of Schleswig-Holstein, Report on my conversations with the Emperor, 16 April 1889, RA Z281/34. See Buckle, *Letters of Queen Victoria*, I, pp. 491–3.

65 Count Herbert Bismarck, diary entry for 9 April 1889, BA Koblenz, Bismarck Papers FC 3018 N.

66 Malet to Albert Edward Prince of Wales, cipher telegram, 9 April 1889, RA Z281/17.

67 Albert Edward Prince of Wales to Prince Christian of Schleswig-Holstein, 10 April 1889, RA Z281/18 and 19, in Buckle, *Letters of Queen Victoria*, I, pp. 492–3.

68 Salisbury to Knollys, 10 April 1889, RA Z281/20.

69 Salisbury to Albert Edward Prince of Wales, 11 April 1889, RA Z281/22.

70 Prince Christian of Schleswig-Holstein, Report on my conversation with the Emperor, 16 April 1889, RA Z281/34; Malet to Albert Edward Prince of Wales, cipher telegram, 11 April 1889, RA Z281/21.

71 Malet to Albert Edward Prince of Wales, cipher telegram, 12 April 1889, RA Z281/26.

72 Albert Edward Prince of Wales to Malet, cipher telegram, 13 April 1889, RA Z281/27.

73 Albert Edward Prince of Wales to Prince Christian of Schleswig-Holstein, 16 April 1889, RA Z281/33.

74 Prince Christian of Schleswig-Holstein, Report on my conversations with the Emperor, 16 April 1889, RA Z281/34.

75 Ibid.

76 Malet to Salisbury, 13 April 1889, RA Z281/28.

77 Ponsonby to Queen Victoria, 20 April 1889, RA I57/24, in Buckle, *Letters of Queen Victoria*, I, p. 494.

78 Albert Edward Prince of Wales to Salisbury, 16 April 1889, RA Z281/32.

79 Albert Edward Prince of Wales to Prince Christian of Schleswig-Holstein, 19 April 1889, RA Z281/37.

80 Albert Edward Prince of Wales to Prince Christian of Schleswig-Holstein, 19 April 1889, RA Z281/36.

81 Queen Victoria to Albert Edward Prince of Wales, 21 April 1889, RA Z281/38.

82 Empress Frederick to Queen Victoria, 20 and 27 April 1889, RA Z44/31 and 33. See Ponsonby, *Letters of the Empress Frederick*, p. 370.

83 Prince Ernst zu Hohenlohe-Langenburg to his father, 28 April 1889, Hohenlohe Zentralarchiv Neuenstein, Hermann Hohenlohe-Langenburg Papers, Bü. 56.

84 Queen Victoria to Albert Edward Prince of Wales, 21 April 1889, RA Z281/38.

85 Prince Christian of Schleswig-Holstein to Albert Edward Prince of Wales, 28 April 1889, RA Z281/39.

86 Prince Ernst zu Hohenlohe-Langenburg to his father, 19 May 1889, Hohenlohe Zentralarchiv Neuenstein, Hermann Hohenlohe-Langenburg Papers, Bü. 56.

87 Summary of 'Incident at Vienna', May 1889, RA Z281/40.

88 Prince Ernst zu Hohenlohe-Langenburg to his father, 11 and 19 May 1889, Hohenlohe Zentralarchiv Neuenstein, Hermann Hohenlohe-Langenburg Papers, Bü. 56. Cf. Knollys to Ponsonby, 23 May 1889, RA Z281/66.

89 Lord Salisbury to Queen Victoria, 11 May 1889, RA Z280/71. Cf. Count Herbert Bismarck, Aufzeichnung, p. 59, BA Koblenz, Bismarck Papers FC 3018 N.

90 Ponsonby to Queen Victoria, 13 May 1889, RA Z280/72.

91 Knollys to Ponsonby, 12 May 1889, RA Z281/45.
92 Ponsonby to Queen Victoria, 13 May 1889, RA Z280/72; Prince Christian of Schleswig-Holstein to Ponsonby, 13 May 1889, RA Z281/46.
93 Drafts by Salisbury and Ponsonby, amended by Queen Victoria, May 1889, RA Z281/55; Ponsonby to Knollys, 22 May 1889, RA Z281/60; Knollys to Ponsonby, 22 May 1889, RA Z281/61; Ponsonby to Queen Victoria, Queen Victoria to Ponsonby, 23 May 1889, RA Z281/67.
94 Knollys to Ponsonby, 23 May 1889, RA Z281/66; Ponsonby to Albert Edward Prince of Wales, 23 May 1889, RA E63/42.
95 Ponsonby to Knollys, 23 May 1889, RA Z281/64. Cf. Ponsonby to Prince Christian of Schleswig-Holstein, 25 May 1889, in Buckle, *Letters of Queen Victoria*, I, p. 500.
96 Queen Victoria to Kaiser Wilhelm II, telegram, 24 May 1889, GStA Berlin, BPHA Rep. 52 W3 No. 11.
97 Queen Victoria to Kaiser Wilhelm II, 25 May 1889, GStA Berlin, BPH Rep. 53 No. 156; copy by the Prince of Wales in RA Addl Mss A/4/11.
98 Albert Edward Prince of Wales to Empress Frederick, 29 May 1889, RA Addl Mss A/4/12.
99 Herbert Bismarck to Kaiser Wilhelm II, 28 May 1889, GStA Berlin, BPH Rep. 53 No. 156; diary entry for 28 May 1889, BA Koblenz, Bismarck Papers FC 3018 N.
100 Ponsonby to Prince Christian of Schleswig-Holstein, 1 June 1889, in Buckle, *Letters of Queen Victoria*, I, p. 501.
101 Kaiser Wilhelm II to Queen Victoria, 28 May 1889, copy, RA T9/161.
102 Queen Victoria to Albert Edward Prince of Wales, copy, 1 June 1889, RA Z280/74; Arthur Duke of Connaught to Queen Victoria, 3 June 1889, RA Z186/32.
103 Knollys to Ponsonby, 2 and 4 June 1889, RA Z281/86 and 89; Ponsonby to Queen Victoria, 5 June 1889, RA Z280/75.
104 Empress Frederick to Queen Victoria, 3 June 1889, RA Z45/10.
105 Prince Ernst zu Hohenlohe-Langenburg to his father, 9 June 1889, Hohenlohe Zentralarchiv Neuenstein, Hermann Hohenlohe-Langenburg Papers, Bü. 56.
106 Salisbury's draft of a letter from Queen Victoria to Kaiser Wilhelm II, 4 June 1889, copy in Ponsonby's hand with amendments by the Queen, RA Z281/88.
107 Ibid.
108 Marlborough House draft, 7 June 1889, RA Z281/99; Ponsonby to Salisbury, 7 June 1889, RA Z281/95.
109 Salisbury to Ponsonby, 7 June 1889, RA Z281/98.
110 Ponsonby to Prince Christian of Schleswig-Holstein, 8 June 1889, in Buckle, *Letters of Queen Victoria*, I, p. 501.
111 Knollys to Prince Christian of Schleswig-Holstein, 8 June 1889, in Buckle, *Letters of Queen Victoria*, I, pp. 501–2; Prince Christian of Schleswig-Holstein to Ponsonby and to Knollys, 11 June 1889, RA Z281/103 and 104.
112 Ponsonby to Prince Christian of Schleswig-Holstein, 8 June 1889, in Buckle, *Letters of Queen Victoria*, I, p. 501.
113 Count Herbert Bismarck, Aufzeichnung, pp. 60–1, BA Koblenz, Bismarck Papers FC 3018 N.
114 Malet to Queen Victoria, 15 June 1889, in Buckle, *Letters of Queen Victoria*, I, pp. 503–4; RA QVJ, 23 July 1889; Ponsonby to Queen Victoria, 23 July 1889, RA A67/82; Queen Victoria to Princess Victoria of Hesse, 7 August 1889, RA Add U173/154.
115 Salisbury to Queen Victoria, 27 July 1889, RA I57/44, in Buckle, *Letters of Queen Victoria*, I, p. 518. Bismarck's letter of 12 October 1889 expressing his gratitude to Queen Victoria is in RA I57/64.

116 Kaiser Wilhelm II to Malet, 14 June 1889, RA I57/33, in Buckle, *Letters of Queen Victoria*, I, p. 504.
117 Count Herbert Bismarck, Aufzeichnung, pp. 67–8, BA Koblenz, Bismarck Papers FC 3018 N.
118 Eulenburg to Herbert Bismarck, 17 July 1889, *Eulenburgs Korrespondenz*, I, No. 228.
119 Arthur Duke of Connaught to Sir Howard Elphinstone, 5 August 1889, RA Vic Addl Mss A25/795.
120 Count Herbert Bismarck, Aufzeichnung, pp. 70ff., BA Koblenz, Bismarck Papers FC 3018 N.
121 Casimir Count von Leyden to Ponsonby, 15 June 1889, RA Z281/107.
122 Kaiser Wilhelm II to Queen Victoria, 23 June 1889, RA I57/36, in Buckle, *Letters of Queen Victoria*, I, p. 505.
123 Queen Victoria to Princess Victoria of Battenberg, 7 August 1889, RA Add U173/154.
124 Count Herbert Bismarck, Aufzeichnung, p. 70, BA Koblenz, Bismarck Papers FC 3018 N.
125 RA QVJ, 2–8 August 1889, in Buckle, *Letters of Queen Victoria*, I, pp. 520–2.
126 RA QVJ, 2 August 1889.
127 Queen Victoria to Princess Victoria of Battenberg, 7 August 1889, RA Add U173/154.
128 E. Phipps, report from Vienna, 20 August 1889, RA I57/55.
129 RA QVJ, 5 and 6 August 1889.
130 RA QVJ, 7 August 1889.
131 Count Herbert Bismarck, Aufzeichnung, pp. 72–3, BA Koblenz, Bismarck Papers FC 3018 N.
132 RA QVJ, 2–8 August 1889, in Buckle, *Letters of Queen Victoria*, I, pp. 520–2; Queen Victoria to Arthur Duke of Connaught, 9 August 1889, RA Vic Addl Mss A15/5343.
133 Count Herbert Bismarck, Aufzeichnung, pp. 75–6, BA Koblenz, Bismarck Papers FC 3018 N.
134 See Lothar Reinermann, *Der Kaiser in England. Wilhelm II. und sein Bild in der britischen Öffentlichkeit*, Paderborn, Munich, Vienna, Zürich 2001, pp. 89ff.
135 RA QVJ, 2–8 August 1889, in Buckle, *Letters of Queen Victoria*, I, pp. 520–2; Queen Victoria to Arthur Duke of Connaught, 9 August 1889, RA Vic Addl Mss A15/5343.
136 Salisbury to Queen Victoria, 8 August 1889, in Buckle, *Letters of Queen Victoria*, I, p. 523.
137 Kaiser Wilhelm II to Queen Victoria, 10 August 1889, RA I57/49, in Buckle, *Letters of Queen Victoria*, I, pp. 523–4; Queen Victoria's reply, ibid., p. 524.
138 W. Beauclerk to Lord Salisbury, 17 August 1889, RA I57/54.
139 Kaiser Wilhelm II to Queen Victoria, 17 August 1889, RA I57/53, in Buckle, *Letters of Queen Victoria*, I, pp. 526–7. See below, pp. 156–8.
140 Herbert Bismarck to his father, 5 October 1889, in Bußmann, *Staatssekretär*, No. 390; Kaiser Wilhelm II to Queen Victoria, telegram, 9 October 1889; Queen Victoria to Kaiser Wilhelm II, 9 October 1889, RA I57/60–1.
141 Vice Admiral Paul Hoffmann, diary entries for 26–8 October 1889, Hoffmann Papers, in the possession of Dr Margot Leo-Hoffmann, Freiburg.
142 Kaiser Wilhelm II to Queen Victoria, telegram, 30 October 1889; Queen Victoria to Kaiser Wilhelm II, 31 October 1889, RA I57/67–8.
143 Salisbury to Queen Victoria, 31 October 1889; Sir Edmund Monson to Salisbury, 31 October 1889, RA I57/69–70.
144 On Erichsen's report to Salisbury on Wilhelm's mental health, see *Young Wilhelm*, pp. 322–3 and below, pp. 1065–6.
145 Salisbury to Queen Victoria, 24 October 1889, RA I57/66.
146 Salisbury to Queen Victoria, 2 November 1889, RA I57/71.

147 Prince Ernst zu Hohenlohe-Langenburg to his father, 16 October 1889, Hohenlohe-Zentralarchiv Neuenstein, Hermann Hohenlohe-Langenburg Papers, Bü. 56.
148 Phipps, report from Vienna, 20 August 1889, RA I57/55.
149 Sir Arthur Nicolson, report from Budapest, 27 August 1889, RA I57/55.
150 Herbert Bismarck to Rantzau, 30 July 1889, BA Koblenz, Bismarck Papers, FC 3014 N, in Bußmann, *Staatssekretär*, No. 386.
151 Ponsonby to Queen Victoria, 8 August 1889, in Buckle, *Letters of Queen Victoria*, I, p. 523.
152 Count Herbert Bismarck, Aufzeichnung, pp. 76ff., BA Koblenz, Bismarck Papers FC 3018 N.
153 Empress Frederick to Queen Victoria, 17 August 1889, RA Z45/41.

5 THE YOUNG KAISER: A SKETCH DRAWN FROM LIFE

1 See above, pp. 12–16.
2 Dr G. Hinzpeter, *Kaiser Wilhelm II. Eine Skizze nach der Natur gezeichnet*, Bielefeld 1888, p. 3. Hinzpeter's role and influence on the future Kaiser are dealt with in detail in *Young Wilhelm*.
3 Hinzpeter, *Kaiser Wilhelm II.*, pp. 3–15.
4 Kaiser Wilhelm II to Kaiserin Augusta, 11 July 1888, GStA Berlin, BPHA Rep. 53J Lit. P No. 14a.
5 Kaiser Wilhelm II to Queen Victoria, 6 July 1888, RA I56/84.
6 Kiderlen-Wächter to Holstein, 16 July 1888, *Holstein Papers*, III, No. 271.
7 Kiderlen-Wächter to Holstein, 19 July 1888, ibid., No. 272. Bismarck's promemoria for Wilhelm II is printed in *Große Politik*, VI, No. 1343, pp. 311–14.
8 Count Herbert Bismarck, Aufzeichnungen aus dem Herbst 1891, BA Koblenz, Bismarck Papers FC 3018 N. The memorandum on German–Russian relations the Chancellor had drafted for Wilhelm is located in PA AA, Deutschland 131 secr.
9 Kiderlen-Wächter to Holstein, 25 July 1888, *Holstein Papers*, III, No. 273.
10 Ibid.
11 Ibid.
12 Kiderlen-Wächter to Holstein, 16 July 1888, ibid., No. 271.
13 Széchényi to Kálnoky, 2 August 1888, HHStA Vienna, PA III 134.
14 Waldersee to Holstein, 26 July 1888, *Holstein Papers*, III, No. 275.
15 Waldersee, diary entry for 26 August 1888, GStA Berlin, Waldersee Papers; omitted from Meisner, II, p. 1.
16 Waldersee, diary entry for 11 October 1888, GStA Berlin, Waldersee Papers; cf. Meisner, II, pp. 6–7.
17 Waldersee, diary entry for 13 October 1888, GStA Berlin, Waldersee Papers; cf. Meisner, II, p. 7.
18 Széchényi to Kálnoky, 26 January 1889, HHStA Vienna, PA III 136.
19 Empress Frederick to Queen Victoria, 2 November 1888, RA Z43/32.
20 Herbert Bismarck to his father, 5 October 1888, BA Koblenz, Bismarck Papers FC 3005 N, in Bußmann, *Staatssekretär*, No. 366.
21 Herbert Bismarck to Rantzau, 4 July 1889, BA Koblenz, Bismarck Papers FC 3014 N, in Bußmann, *Staatssekretär*, No. 382.
22 Holstein to Radolin, 28 November 1889, *Holstein Papers*, III, No. 300.
23 Holstein, diary entry for 24 October 1888, ibid., II, pp. 427–8.
24 Prince Ernst zu Hohenlohe-Langenburg to his father, 1 November 1888, Hohenlohe-Zentralarchiv Neuenstein, Hermann Hohenlohe-Langenburg Papers, Bü. 55.
25 Eulenburg to Holstein, 6 August 1889, *Eulenburgs Korrespondenz*, I, No. 230.

26 Vice-Admiral Paul Hoffmann, diary entry for 23 October 1889, Hoffmann Papers, Freiburg.
27 Ibid.
28 Hoffmann, diary entries for 24–5 October and 9 and 11 November 1889, ibid.
29 Hoffmann, diary entry for 25 October 1889, ibid.
30 Hoffmann, diary entry for 7 November 1889, ibid.
31 Waldersee, diary entry for 13 November 1889, GStA Berlin, Waldersee Papers; cf. Meisner, II, pp. 76–7.
32 Waldersee, diary entry for 13 November 1889, GStA Berlin, Waldersee Papers; cf. Meisner, II, pp. 76–7. See also Waldersee, diary entry for 3 January 1890, GStA Berlin, Waldersee Papers; omitted from Meisner, II, p. 87.
33 Waldersee, diary entry for 5 January 1890, GStA Berlin, Waldersee Papers; cf. Meisner, II, p. 87.
34 Kaiser Wilhelm II to Bismarck, telegram, 17 January 1890, cited in *Eulenburgs Korrespondenz*, I, p. 420.
35 Kaiser Wilhelm II to Empress Frederick, 25 September 1898, AdHH Schloss Fasanerie. A copy is printed in *The Memoirs of Prince von Bülow*, 4 vols., London and New York 1931–2, I, pp. 230–2. See below, pp. 871–3.
36 Kaiser Wilhelm II, dictated notes on a conversation with Pope Leo XIII, 23 April 1893, in Chlodwig Fürst zu Hohenlohe-Schillingsfürst, Karl Alexander von Müller, ed., *Denkwürdigkeiten der Reichskanzlerzeit*, Stuttgart, Berlin 1931, pp. 608–11.
37 Charles Seymour, ed., *The Intimate Papers of Colonel House*, 4 vols., Boston, New York 1926–8, II, p. 139.
38 Kaiser Wilhelm II to Crown Prince Gustaf of Sweden and Norway, 25 July 1895, printed in Hohenlohe, *Denkwürdigkeiten der Reichskanzlerzeit*, pp. 102–5.
39 Bismarck noted on Prince Reuss's report of 28 April 1888: 'To be locked away on account of the marginalia of H[is] I[mperial] H[ighness]' (*Große Politik*, VI, pp. 301–2).
40 See PA AA, Asservat No. 4.1, 2–5.
41 On Bülow's report of 1 March 1889, Bismarck noted: 'This document is to be locked away, a copy is to be kept with the non-secret files under A 3728 without the margin[alia]' (PA AA, Asservat No. 4).
42 Kaiser Wilhelm II, marginal note on a report of 9 January 1892 from Petropolis, ibid.
43 Kaiser Wilhelm II, marginal note on a report of 24 June 1893 from Luxemburg, ibid.
44 Kaiser Wilhelm II, marginal notes on Schlözer's report of 4 October 1890 and Bülow's report of 22 October 1890, ibid.
45 Kaiser Wilhelm II, marginal notes on Bülow's reports of 14, 16, 21, 24 February, 7 March, 13 April, 15 May, 6 and 12 November 1889, 22 November 1891, as well as on Herbert Bismarck's report of 6 April 1889, ibid.
46 Kaiser Wilhelm II, marginal note on Alvensleben's report from Brussels of 8 March 1889, ibid.
47 Kaiser Wilhelm II, marginal note on Bülow's report of 1 June 1889, ibid.
48 Kaiser Wilhelm II, marginal note on Lichnowsky's report of 15 July 1891, ibid.
49 Kaiser Wilhelm II, marginal note on Bülow's report of 17 March 1889, ibid. See also Kaiser Wilhelm II, marginal note on Reuss's report of 23 May 1891, and his marginal note on Eulenburg's report of 7 May 1896, ibid.
50 Kaiser Wilhelm II, marginal note on Bülow's report of 14 April 1889, PA AA, R 9845.
51 Kaiser Wilhelm II, marginal note on Bülow's report of 11 March 1889, PA AA, Asservat No. 4.1.
52 Kaiser Wilhelm II, marginal note on Bülow's report of 20 December 1888, ibid. See Waldersee, diary entry for 20 April 1889, GStA Berlin, Waldersee Papers; cf. Meisner, II, p. 49.
53 Wilhelm II, marginal note on report A8118, BA Koblenz, Bismarck Papers FC 2986 N.

54 Kaiser Wilhelm II, marginal note on Reuss's report of 6 November 1888, PA AA, Asservat No. 4.1.
55 Kaiser Wilhelm II, marginal note on Reuss's report of 22 December 1888 with a comment by Bismarck, ibid.
56 Kaiser Wilhelm II, marginal note on a report from Belgrade of 2 January 1889, ibid.
57 Kaiser Wilhelm II, marginal note on Reuss's and Bray's reports of 23 January, 8 and 23 February and 3 March 1889, PA AA, R 11618.
58 Kaiser Wilhelm II, marginal note on Consul Schroeder's report from Beirut of 28 May 1889, PA AA, Asservat No. 4. See also Kaiser Wilhelm II, marginal note on Reuss's report from Vienna of 22 March 1889, ibid.
59 Kaiser Wilhelm II, marginal note on Radowitz's report from Therapia of 23 October 1891, ibid.
60 Kaiser Wilhelm II, marginal note on Alvensleben's report from Brussels of 8 March 1889, ibid. On Ferdinand's marital plans, see Wilhelm's marginal note on Bülow's report of 5 May 1889, ibid.
61 Kaiser Wilhelm II, marginal note on Lichnowsky's report of 15 July 1891, ibid.
62 Kaiser Wilhelm II, marginal note on articles in the *Berliner Tageblatt* of 28 February and 20 March 1889, ibid.
63 Kaiser Wilhelm II, marginal note on Schweinitz's report of 3 July 1889, ibid.
64 Kaiser Wilhelm II, marginal note on Schweinitz's report of 24 May 1891, ibid.
65 Kaiser Wilhelm II, marginal note on Leyden's report of 8 December 1890, ibid.
66 Kaiser Wilhelm II, marginal note on Schweinitz's report of 30 December 1891, ibid.
67 Kaiser Wilhelm II, marginal note on Alfred von Bülow's report from St Petersburg of 20 July 1891, ibid.
68 Kaiser Wilhelm II, marginal note on Count Rex's report from St Petersburg of 14 September 1892, ibid.
69 Kaiser Wilhelm II, marginal note on de Maistre's report from Athens of 25 September 1889, ibid.
70 Kaiser Wilhelm II, marginal note on an article in the *Standard* of 25 December 1889, ibid.
71 Kaiser Wilhelm II, marginal note on Brincken's report of 23 October 1888, PA AA, Asservat No. 4.1.
72 Empress Frederick to Queen Victoria, 1 August 1890, RA Z48/44.
73 For a more detailed assessment, see below, pp. 643–50.
74 Kaiser Wilhelm II, marginal note on Wesdehlen's report of 10 November 1890, PA AA, Asservat No. 4.
75 Kaiser Wilhelm II, marginal note on Kiderlen-Wächter's report from Copenhagen of 9 March 1897, ibid.
76 Kaiser Wilhelm II, marginal note on Wesdehlen's report of 3 February 1894, ibid.
77 Kaiser Wilhelm II, marginal note on Wesdehlen's report of 9 November 1892, ibid.
78 Kaiser Wilhelm II, marginal note on Kiderlen-Wächter's from Copenhagen of 9 March 1897, ibid.
79 Kaiser Wilhelm II, marginal note on Wesdehlen's report of 17 February and 2 March 1892, ibid.
80 Kaiser Wilhelm II, marginal note on Wesdehlen's report of 28 January 1891, ibid.
81 Kaiser Wilhelm II, marginal note on Waecker-Gotter's report of 31 January 1891, ibid.
82 Hoffmann, diary entry for 26 October 1889, Hoffmann Papers, Freiburg. See also Kaiser Wilhelm II, marginal note on Wesdehlen's report of 28 January 1891, PA AA, Asservat No. 4.1.
83 Kaiser Wilhelm II, marginal note on Wesdehlen's report of 10 November 1890, ibid.
84 Hoffmann, diary entries for 28–30 October 1889, Hoffmann Papers, Freiburg.

85 Hoffmann, diary entry for 31 October–3 November 1889, ibid.
86 Hoffmann, diary entry for 3 November 1889, ibid.
87 Hoffmann, diary entry for 5 November 1889, ibid.
88 Hoffmann, diary entry for 4 November 1889, ibid.
89 See Margarete Jarchow, *Hofgeschenke. Wilhelm II. zwischen Diplomatie und Dynastie 1888–1914*, Hamburg 1998, pp. 35 and 183.
90 Hoffmann, diary entry for 6 November 1889, Hoffmann Papers, Freiburg.
91 Hoffmann, diary entry for 4 November 1889, ibid.
92 Hoffmann, diary entry for 6 November 1889, ibid.
93 Ibid.
94 Empress Frederick to Queen Victoria, 4 November 1889, RA Z46/32.
95 Hoffmann, diary entry for 1 November 1889, Hoffmann Papers, Freiburg.
96 Kaiser Wilhelm II to Prince Otto von Bismarck, 12 June 1889, BA Koblenz, Bismarck Papers FC 2986 N.
97 Malet to Salisbury, 11 June 1889, RA I57/29.
98 Kaiser Wilhelm II, marginal notes on newspaper clippings of 24 May 1890, PA AA, Asservat No. 4.1.
99 Waldersee, diary entry for 15 January 1892, GStA Berlin, Waldersee Papers; omitted from Meisner, II, p. 229.
100 Swaine to Ponsonby, 1 July 1892, RA I59/89.
101 Kaiser Wilhelm II, marginal note on Eulenburg's report from Munich of 18 January 1894, PA AA, Asservat No. 4.
102 Kaiser Wilhelm II, marginal note on Schweinitz's report of 3 March 1889, PA AA, Asservat No. 4.1.
103 Waldersee, diary entry for 12 October 1893, GStA Berlin, Waldersee Papers; cf. Meisner, II, p. 295.
104 Kaiser Wilhelm II, marginal note on Wesdehlen's report of 25 August 1888, PA AA, Asservat No. 4.1.
105 Waldersee, diary entry for 10 October 1891, GStA Berlin Waldersee Papers; omitted from Meisner II, p. 218.
106 Herbert Bismarck to Rantzau, 4 January 1889, BA Koblenz, Bismarck Papers FC 3014 N.
107 Eulenburg to Herbert Bismarck, 3 January 1889, *Eulenburgs Korrespondenz*, I, No. 209. See also Herbert Bismarck to his father, 30 December 1888, BA Koblenz, Bismarck Papers FC 3005 N.
108 Kaiser Wilhelm II to Eulenburg, 27 February 1889, *Eulenburgs Korrespondenz*, I, No. 213. See Eulenburg's reply of 1 March 1889, ibid., No. 214.
109 Herbert Bismarck to Rantzau, 3 February 1889, BA Koblenz, Bismarck Papers FC 3014 N.
110 Széchényi to Kálnoky, 31 January and mid-February 1889, HHStA Vienna, PA III 136.
111 Kaiser Wilhelm II to Queen Victoria, 14 February 1889, RA Z500/3.
112 Kaiserin Auguste Viktoria to Kaiser Wilhelm II, 23 September 1890, GStA Berlin, BPHA Rep. 53T Preußen: An Kaiser Wilhelm II., Bd IV.
113 Hoffmann, diary entry for 8 November 1889, Hoffmann Papers, Freiburg.
114 Morley, quoted in Jonathan Steinberg, 'Kaiser Wilhelm and the British', in Röhl and Sombart, *Kaiser Wilhelm II – New Interpretations*, p. 127.
115 Max Freiherr von Holzing-Berstett to his wife, 21 May 1910, Generallandesarchiv Karlsruhe, Holzing Papers, No. 116/21.
116 Kaiser Wilhelm II, speech of 23 August 1888, Penzler, *Reden Kaiser Wilhelms II. in den Jahren 1888–1895*, pp. 21–3.
117 Werthern to Eulenburg, 9 September 1888, *Eulenburgs Korrespondenz*, I, No. 195.
118 Holstein, diary entry for 11 November 1888, *Holstein Papers*, II, pp. 382–3.

119 Kaiser Wilhelm II, marginal note on Hohenlohe's report of 16 April 1898, PA AA, Asservat No. 4.
120 Kaiser Wilhelm II, marginal note on von der Goltz's report of 12 April 1891, ibid.
121 Kaiser Wilhelm II, marginal note on Hohenlohe's report of 13 March 1897, ibid.
122 Holstein, diary entry for 11 November 1888, *Holstein Papers*, II, pp. 382–3.
123 Kaiser Wilhelm II, marginal note on Radowitz's report from Madrid of 14 May 1893, PA AA, Asservat No. 4.
124 Kaiser Wilhelm II, marginal note on Count Rex's report of 14 September 1892, ibid.
125 Kaiser Wilhelm II, marginal note on an article in *Le Temps* of 29 March 1889, ibid.
126 Kaiser Wilhelm II, marginal note on an article in *Hamburger Korrespondent* of 25 August 1889, ibid.
127 Bismarck's comment on the article 'Zum 18. October' in *Berliner Zeitung*, PA AA, R 3447. See Empress Frederick to Queen Victoria, 21 October 1888, RA Z43/28. See also Grand Duke Ludwig IV of Hesse-Darmstadt to Empress Frederick, 17 October 1888, GStA Berlin, BPH Rep. 52 No. 3, pp. 349–52.
128 Kaiser Wilhelm II, speech of 27 October 1888, Penzler, *Reden Kaiser Wilhelms II. in den Jahren 1888–1895*, pp. 27–9.
129 Press reactions to the Kaiser's speech are collected in PA AA, R 3447.
130 Eissenstein to Kálnoky, 29 October 1888, HHStA Vienna, PA III 134.
131 Beauclerk to Salisbury, 28 October 1888, RA I56/97.
132 Holstein, diary entry for 29 October 1888, *Holstein Papers*, II, pp. 380–1.
133 Herbert Bismarck to his father, 31 October 1888, PA AA, R 3447.
134 Rottenburg to Herbert Bismarck, 1 November 1888, ibid.
135 Kaiser Wilhelm II, marginal note on Dönhoff's report of 2 January 1889, PA AA, Asservat No. 4.
136 Kaiser Wilhelm II, marginal note on Rantzau's report from Munich of 24 September 1889, ibid.
137 Kaiser Wilhelm II, marginal note on Rantzau's report from Munich of 3 June 1889, ibid.
138 Kaiser Wilhelm II, marginal note on Eisendecher's report from Karlsruhe of 13 July 1889 as well as on Münster's report from Paris of 21 October 1892, ibid.
139 Kaiser Wilhelm II, marginal note on Otto Bülow's and Below's reports of 24 November and 20 December 1898 from Rome, ibid.
140 Kaiser Wilhelm II, marginal note on Reuss's report from Vienna of 10 June 1889, ibid.
141 Waldersee, diary entry for 13 November 1888, GStA Berlin, Waldersee Papers; omitted from Meisner, II, pp. 18–19.
142 Waldersee, diary entry for 7 September 1888, GStA Berlin, Waldersee Papers. This passage was heavily crossed out by Meisner in the original diary and omitted from the published version. See Meisner, II, p. 3.
143 Waldersee, diary entry for 11 October 1888, GStA Berlin, Waldersee Papers. The original diary has at this point been so badly mutilated by Meisner that the text cannot be deciphered with any certainty. Cf. Meisner, II, p. 6.
144 Waldersee, diary entry for 29 October 1891, GStA Berlin, Waldersee Papers; omitted from Meisner, II, pp. 219–20.
145 Kaiser Wilhelm II, marginal note on a report in the *Daily News* of 29 March 1889, PA AA, Asservat No. 4.
146 Kaiser Wilhelm II, marginal note on a Vienna newspaper article of 12 May 1889, ibid.
147 Kaiser Wilhelm II, marginal note on Schlözer's report from Rome of 23 November 1889, ibid.

148 Prince Ernst zu Hohenlohe-Langenburg to his father, 24 May 1889, Hohenlohe-Zentralarchiv Neuenstein, Hermann Hohenlohe-Langenburg Papers, Bü. 56.
149 Kaiser Wilhelm II, marginal note on an article in the *Daily News* of 15 June 1889, PA AA, Asservat No. 4.
150 Kaiser Wilhelm II, marginal note on Schlözer's report of 4 October 1890, ibid.
151 Kaiser Wilhelm II, marginal note on Schweinitz's report of 24 May 1891, ibid.
152 Waldersee, diary entry for 29 October 1888, GStA Berlin, Waldersee Papers; Meisner, II, pp. 12–13.
153 Cosima Wagner to Kaiser Wilhelm II, 23 August 1888, PA AA, R 3474.
154 Lucanus to Prince Otto von Bismarck, 18 September 1888, PA AA, R 3474.
155 See *Eulenburgs Korrespondenz*, I, pp. 191–4.
156 *Hamburger Nachrichten*, quoted in *Münchener Neueste Nachrichten*, 12 October 1888, PA AA, Preußen 1 No. 1d, Bd I.
157 Prince Otto von Bismarck to Kaiser Wilhelm II, 20 September 1888, PA AA, R 3474.
158 Note by Holstein, 15 and 21 October 1888, ibid.
159 Herbert Bismarck to Lucanus, 31 October 1888, ibid.
160 Herbert Bismarck to Rantzau, 5 August 1888, BA Koblenz, Bismarck Papers FC 3014 N.
161 Rantzau to Herbert Bismarck, 6 August 1888, BA Koblenz, Bismarck Papers FC 3028 N; Herbert Bismarck to Rantzau, 12 August 1888, ibid., FC 3014 N. Cf. Hans Rall, *Wilhelm II. Eine Biographie*, Graz, Vienna, Cologne 1995, pp. 84ff.
162 Waldersee, diary entry for 27 February 1894, GStA Berlin, Waldersee Papers; omitted from Meisner, II, p. 309.

6 FIRST STEPS IN FOREIGN AFFAIRS

1 PA AA, Asservat No. 4.1.
2 Prince Wilhelm of Prussia, marginal notes on Deines's report from Vienna of 5 February 1888, PA AA, Asservat No. 4.1.
3 Herbert Bismarck to Kaiser Wilhelm II, 11 November 1888, BA Koblenz, Bismarck Papers FC 2986 N.
4 Waldersee, diary entries for 19–22 November 1888, GStA Berlin, Waldersee Papers; omitted from Meisner, II, p. 21.
5 Waldersee, diary entry for 27 January 1889, GStA Berlin, Waldersee Papers; cf. Meisner, II, p. 34.
6 Waldersee, diary entry for 25 March 1889, GStA Berlin, Waldersee Papers; cf. Meisner, II, p. 47.
7 Prince Wilhelm of Prussia, marginal note on Deines's report from Vienna of 5 February 1888, PA AA, Asservat No. 4.1.
8 Kaiser Wilhelm II, marginal note on Yorck's report of 31 July 1888, ibid.; cited in *Große Politik*, VI, p. 341.
9 Kaiser Wilhelm II, marginal note on Yorck's report of 14 October 1888, PA AA, Asservat No. 4.1.
10 Kaiser Wilhelm II, marginal note on the military attaché's report from Paris of 14 November 1888, ibid.
11 Kaiser Wilhelm II, marginal note on a report from St Petersburg of 5 November 1888, ibid.
12 Kaiser Wilhelm II, marginal note on Rechenberg's report from Warsaw of 10 November 1888, ibid.; Waldersee, diary entry for 15 November 1888, GStA Berlin, Waldersee Papers; Meisner, II, p. 19. Cf. the Kaiser's marginal notes on a report from St Petersburg of 27 June 1888, PA AA, Asservat No. 4.1.

13 Prince Otto von Bismarck to Kaiser Wilhelm II, 19 August 1888, *Große Politik*, VI, No. 1350.
14 Waldersee, diary entry for 1 January 1889, GStA Berlin, Waldersee Papers; omitted from Meisner, II, p. 27. See below, pp. 210–16 and 218–25.
15 Kaiser Wilhelm II, speech of 16 August 1888, Penzler, *Reden Kaiser Wilhelms II. in den Jahren 1888–1895*, pp. 19–21. See above, p. 73.
16 See Kaiser Wilhelm II, marginal note on a report from Monts of 28 August 1888, PA AA, Asservat No. 4.1.
17 Kaiser Wilhelm II, note of 3 November 1888, BA Koblenz, Bismarck Papers FC 2986 N.
18 Kaiser Wilhelm II, marginal note dated 21 December 1889 on Schweinitz's report of 11 December 1889, PA AA, Asservat No. 4.
19 Waldersee, diary entry for 8 September 1888, GStA Berlin, Waldersee Papers; omitted from Meisner, II, p. 4.
20 Waldersee, diary entry for 2 March 1889, GStA Berlin, Waldersee Papers; largely omitted from Meisner, II, p. 42.
21 Waldersee, diary entry for 15 April 1889, GStA Berlin, Waldersee Papers; omitted from Meisner, II, p. 48.
22 Waldersee, diary entry for 20 April 1889, GStA Berlin, Waldersee Papers; omitted from Meisner, II, p. 49.
23 Waldersee, diary entry for 28 October 1888, GStA Berlin, Waldersee Papers; cf. Meisner, II, p. 12.
24 Waldersee, diary entry for 27 October 1888, GStA Berlin, Waldersee Papers; Meisner, II, p. 12.
25 Waldersee, diary entry for 28 October 1888, GStA Berlin, Waldersee Papers; cf. Meisner, II, p. 12.
26 Waldersee, diary entry for 1 November 1888, GStA Berlin, Waldersee Papers; cf. Meisner, II, p. 13, where key passages have been omitted.
27 Waldersee, diary entry for 3 November 1888, GStA Berlin, Waldersee Papers; numerous passages omitted from Meisner, II, p. 14.
28 Waldersee, diary entry for 6 November 1888, GStA Berlin, Waldersee Papers; abridged in Meisner, II, p. 15.
29 Waldersee, diary entry for 11 November 1888, GStA Berlin, Waldersee Papers; abridged and distorted in Meisner, II, pp. 17–18.
30 Waldersee, diary entry for 13 November 1888, GStA Berlin, Waldersee Papers; Meisner, II, pp. 18–19; Waldersee to Verdy du Vernois, 26 November 1888, GStA Berlin, Waldersee Papers, B I No. 53. See also Holstein, diary entry for 11 November 1888, *Holstein Papers*, II, pp. 382–3; Herbert Bismarck to his father, 30 December 1888, BA Koblenz, Bismarck Papers FC 3005 N.
31 Prince Otto von Bismarck to Waldersee, 24 November 1888 and Waldersee's coded reply of the same day, PA AA, R 1009.
32 Waldersee, diary entry for 27 November 1888, GStA Berlin, Waldersee Papers; omitted from Meisner, II, p. 23.
33 Waldersee, diary entry for 12 December 1888, GStA Berlin, Waldersee Papers; omitted from Meisner, II, pp. 25–6.
34 Waldersee, diary entry for 1 January 1889, GStA Berlin, Waldersee Papers; omitted from Meisner, II, p. 27.
35 Waldersee, diary entries for 21 and 26 January 1889, GStA Berlin, Waldersee Papers; cf. Meisner, II, pp. 32–3.
36 Eulenburg's note of 17 February 1889, *Eulenburgs Korrespondenz*, I, pp. 328–9.
37 Holstein to Eulenburg, 6 February 1889, ibid., No. 211.

38 Eulenburg's note of 17 February 1889, ibid., pp. 328–9. See Herbert Bismarck to Bülow, 28 October 1888, in Bußmann, *Staatssekretär*, No. 371. For Herbert Bismarck's view, see Salisbury to Queen Victoria, 29 March 1889, RA A67/48.
39 See Waldersee, diary entries for 21 October, 9 and 25 November and 2 December 1888, GStA Berlin, Waldersee Papers; Meisner, II, pp. 10, 16–17 and 22ff.
40 Waldersee, diary entry for 21 January 1889, GStA Berlin, Waldersee Papers; omitted from Meisner, II, p. 32.
41 Waldersee, diary entries for 1 and 3 November 1888, GStA Berlin, Waldersee Papers; cf. Meisner, II, pp. 13–14.
42 Waldersee, diary entry for 9 November 1888, GStA Berlin, Waldersee Papers; printed only in part in Meisner, II, pp. 16–17.
43 Waldersee, diary entry for 12 November 1888, GStA Berlin, Waldersee Papers; cf. Meisner, II, p. 18.
44 Kaiser Wilhelm II, marginal note on Deines's report of 20 December 1888, PA AA, R 8594.
45 Kaiser Wilhelm II, marginal note on Deines's report of 20 December 1888, PA AA, Asservat No. 4.1.
46 Kaiser Wilhelm II, marginal note on a report from Vienna of 19 October 1888, ibid.
47 Kaiser Wilhelm II, marginal note on Count Monts's report of 18 January 1889, ibid.
48 Kaiser Wilhelm II, marginal note on Reuss's report of 22 December 1888, ibid. See above, pp. 121–2.
49 Paul Hoffmann, diary entry for 10 November 1889, Hoffmann Papers, Freiburg.
50 Waldersee, diary entry for 22 April 1889, GStA Berlin, Waldersee Papers; cf. Meisner, II, p. 49.
51 Waldersee, diary entries for 1 and 21 January 1889, GStA Berlin, Waldersee Papers; cf. Meisner, II, pp. 27, 32.
52 Waldersee, diary entry for 25 January 1889, GStA Berlin, Waldersee Papers; omitted from Meisner, II, p. 33.
53 Waldersee, diary entry for 1 January 1889, GStA Berlin, Waldersee Papers; omitted from Meisner, II, p. 27.
54 Waldersee, diary entry for 21 January 1889, GStA Berlin, Waldersee Papers; cf. Meisner, II, p. 32.
55 Ibid.
56 Waldersee, diary entry for 25 March 1889, GStA Berlin, Waldersee Papers; omitted from Meisner, II, p. 47.
57 Waldersee, diary entry for 5 April 1889, GStA Berlin, Waldersee Papers; largely omitted from Meisner, II, p. 48.
58 Colonel Euan Smith to Salisbury, 20 November 1888, RA I56/99–100.
59 Prince Ernst zu Hohenlohe-Langenburg to his father, 10 September 1889, Hohenlohe-Zentralarchiv, Schloss Neuenstein, Hermann Hohenlohe-Langenburg Papers, Bü. 56. See also the exchange between the Sultan of Zanzibar and Kaiser Wilhelm II of 2 June and 15 August 1889, in PA AA, R 3579.
60 Herbert Bismarck to his father, 16 September 1888, BA Koblenz, Bismarck Papers FC 3005 N. Cf. Herbert Bismarck to Holstein, 30 September 1888, *Holstein Papers*, III, No. 276.
61 Herbert Bismarck to Rantzau, 30 July 1889, BA Koblenz, Bismarck Papers FC 3014 N, in Bußmann, *Staatssekretär*, No. 386.
62 Kaiser Wilhelm II, marginal note on his conversation with Admiral von Knorr of 11 September 1888, PA AA, Asservat No. 4.1.
63 Kaiser Wilhelm II, marginal note on Hatzfeldt's report of 30 November 1888, ibid.
64 Kaiser Wilhelm II, marginal note on Hatzfeldt's report of 12 December 1888, ibid.

65 Kaiser Wilhelm II, Gedankensplitter über den Krieg in Transvaal, 21 December 1899, RA W60/28, and Weitere Gedankensplitter über den Transvaalkrieg, 4 February 1900, RA W60/67.

66 Count Herbert Bismarck, Aufzeichnungen, pp. 62–70, BA Koblenz, Bismarck Papers FC 3018 N.

67 Kaiser Wilhelm II, marginal note on report No. A 17256/89 from East Africa, PA AA, Asservat No. 4.1.

68 Herbert Bismarck, diary entries for 15 and 17 February 1890, BA Koblenz, Bismarck Papers FC 3018 N.

69 Empress Frederick to Queen Victoria, 20 April 1889, RA Z44/31, in Ponsonby, *Letters of the Empress Frederick*, pp. 370–3.

70 Herbert Bismarck to Rantzau, 30 July 1889, BA Koblenz, Bismarck Papers FC 3014 N, in Bußmann, *Staatssekretär*, No. 386. See Kaiserin Auguste Viktoria to Kaiser Wilhelm II, 1 July 1889, GStA BPHA Rep. 53T Preußen: An Kaiser Wilhelm II., Bd III.

71 Kiderlen-Wächter to Holstein, 19 July 1888, *Holstein Papers*, III, No. 272.

72 Senden-Bibran, Aufzeichnungen über das Jahr 1888, BA-MA Freiburg, Senden-Bibran Papers N160/11, p. 27.

73 Hoffmann, diary entries for 21 October and 11 November 1889, Hoffmann Papers, Freiburg.

74 Hoffmann, diary entry for 9 November 1889, ibid.

75 Hoffmann, diary entries for 22 October and 11 November 1889, ibid.

76 Hoffmann, diary entry for 8 November 1889, also 25 October 1889, ibid.

77 Kaiser Wilhelm II, Jugenderinnerungen, dictated on 2 April 1926, Mewes Papers, in the possession of Oberst Joachim von Natzmer, Munich. Cf. Kaiser Wilhelm II, *Aus meinem Leben 1859–1888*, Berlin, Leipzig 1927, pp. 274–5.

78 Kaiser Wilhelm II, Jugenderinnerungen, dictated on 2 April 1926, Mewes Papers, in the possession of Oberst Joachim von Natzmer, Munich.

79 See Hoffmann, diary entry for 22 October 1889, Hoffmann Papers, Freiburg. For the relationship between Wilhelm and his brother Heinrich, see below, pp. 628–33.

80 Prince Heinrich of Prussia to Kaiser Wilhelm II, telegram, 31 March 1889, GStA Berlin, BPHA Rep. 52 V1 No. 13.

81 Waldersee to Holstein, 26 July 1888, *Holstein Papers*, III, No. 275.

82 Waldersee, diary entry for 9 October 1888, GStA Berlin, Waldersee Papers; cf. Meisner, II, p. 4.

83 Waldersee, diary entry for 6 November 1888, GStA Berlin, Waldersee Papers; omitted from Meisner, II, p. 15.

84 Kaiser Wilhelm II, Jugenderinnerungen, dictated on 2 April 1926, Mewes Papers, Munich. Cf. Kaiser Wilhelm II, *Aus meinem Leben*, p. 263. 'Not one word' of the ex-Kaiser's dictated memoirs were used by the ghost writers of the book. See Kurt Jagow to Friedrich Mewes, 9 April 1926, Mewes Papers, Munich.

85 Bülow, *Memoirs*, II, p. 30.

86 Waldersee, diary entry for 11 November 1888, GStA Berlin, Waldersee Papers; abridged and distorted in Meisner, II, pp. 17–18.

87 Prince Christian of Schleswig-Holstein, Report on my conversations with the Emperor, 16 April 1889, RA Z281/34; Malet to Salisbury, 3 February 1889, RA I57/14; Herbert Bismarck to Kaiser Wilhelm II, 17 March 1889, with marginal notes of the Kaiser, BA Koblenz, Bismarck Papers FC 2986 N.

88 Kaiser Wilhelm II to Queen Victoria, 17 August 1889, RA I57/53, in Buckle, *Letters of Queen Victoria*, I, pp. 526–7.

89 Kaiser Wilhelm II, marginal note on an article in the *Manchester Guardian*, 3 August 1889, PA AA, Asservat No. 4.

90 Kaiser Wilhelm II to Queen Victoria, 22 December 1889, RA I57/75.

91 Grand Duchess Augusta Caroline of Mecklenburg-Strelitz to Count Adolphus von Teck, 7 September 1888, RA Geo V CC50/184.

7 THE PILLARS OF IMPERIAL POWER

1 Holstein to Eulenburg, 5 May 1896, *Eulenburgs Korrespondenz*, III, No. 546.

2 See Röhl, *Kaiser and His Court*, pp. 70–106.

3 See Konrad Breitenborn, *Im Dienste Bismarcks. Die politische Karriere des Grafen Otto zu Stolberg-Wernigerode*, Leipzig 3rd edn 1986, pp. 331–2; Breitenborn, ed., *Die Lebenserinnerungen des Fürsten Otto zu Stolberg-Wernigerode*, Wernigerode 1996, pp. xvi–xviii.

4 See Bismarck to Crown Prince Wilhelm, 3 May 1888, Albedyll to Kaiser Wilhelm II, 23 June 1888, PA AA, R 3433. See also Breitenborn, *Im Dienste Bismarcks*, pp. 341–2; Eissenstein to Kálnoky, 3 November 1888, HHStA Vienna, PA III 134; Holstein, diary entry for 22 October 1888, *Holstein Papers*, II, pp. 380–1.

5 Széchényi to Kálnoky, 30 June and 4 July 1888, HHStA Vienna, PA III 134. Cf. Holstein, diary entry for 22 October 1888, *Holstein Papers*, II, pp. 380–1.

6 Waldersee, diary entry for 7 July 1888, GStA Berlin, Waldersee Papers; Meisner, I, p. 411.

7 Kaiser Wilhelm II to Kaiserin Augusta, 11 July 1888, GStA Berlin, BPHA Rep. 53J Lit. P No. 14a.

8 Waldersee, diary entry for 20 November 1896, GStA Berlin, Waldersee Papers; omitted from Meisner, II, p. 377.

9 Széchényi to Kálnoky, 2 August 1888, HHStA Vienna, PA III 134.

10 Empress Frederick to Queen Victoria, 10 July 1891, RA Z50/57.

11 Széchényi to Kálnoky, 30 June 1888, HHStA Vienna, PA III 134.

12 See Holstein, diary entry for 22 October 1888, *Holstein Papers*, II, pp. 380–1.

13 Széchényi to Kálnoky, 4 July 1888, HHStA Vienna, PA III 134.

14 For the career of Versen, who died in early 1893, see Alfred Freiherr von Werthern, *General von Versen. Ein militärisches Zeit- und Lebensbild. Aus hinterlassenen Briefen und Aufzeichnungen*, Berlin 1898.

15 Waldersee, diary entries for 20 July 1891 and 8 October 1893, GStA Berlin, Waldersee Papers; cf. Meisner, II, pp. 212–13 and 294–5.

16 Empress Frederick to Queen Victoria, 9 January 1892, RA Z52/14.

17 Swaine to Ponsonby, 1 July 1892, RA I59/89.

18 Waldersee, diary entry for 29 March 1889, GStA Berlin, Waldersee Papers; omitted from Meisner, II, p. 48. See also ibid., pp. 294–5.

19 For details of Wittich's dismissal and his role at the court, see below, ch. 16.

20 Széchényi to Kálnoky, 21 June 1888, HHStA Vienna, PA III 134.

21 Swaine to Ponsonby, 1 July 1892, RA 59/89.

22 Holstein to Eulenburg, 7/8 April 1895, *Eulenburgs Korrespondenz*, III, No. 1101; Eulenburg to Holstein, 16 April 1895, ibid., No. 1104.

23 Empress Frederick to Queen Victoria, 13 August 1890, RA Z49/5. On Zitzewitz's sudden death, see Empress Frederick to her daughter Princess Viktoria of Schaumburg-Lippe, 27 February 1892, AdHH Schloss Fasanerie.

24 Empress Frederick to Queen Victoria, 4 July 1891, RA Z50/55.

25 Empress Frederick to Queen Victoria, 9 November 1891, RA Z51/43. See Holstein to Eulenburg, 7/8 April 1895, *Eulenburgs Korrespondenz*, III, No. 1101; Eulenburg to Holstein, 16 April 1895, ibid., No. 1104.

26 Waldersee, diary entry for 8 December 1888, GStA Berlin, Waldersee Papers; Meisner, II, p. 25.

27 Rudolf Schmidt-Bückeburg, *Das Militärkabinett der preußischen Könige und deutschen Kaiser. Seine geschichtliche Entwicklung und staatsrechtliche Stellung 1787–1918*, Berlin 1933, pp. 177ff.; Waldersee, diary entry for 23 June 1888, GStA Berlin, Waldersee Papers; Meisner, I, p. 406.
28 Rudolf Graf von Stillfried-Alcántara, *Ceremonial-Buch für den Königlich-Preußischen Hof*, Berlin 1871–8, p. iv.
29 Graf Carl von Wedel, *Zwischen Kaiser und Kanzler*, Leipzig 1943, p. 187; also Wedel, promemoria of 13 June 1894, in the possession of Graf Gustav von Wedel, Frankfurt a.M.
30 Waldersee, diary entry for 22 October 1894, GStA Berlin, Waldersee Papers; omitted from Meisner, II, p. 327.
31 Senden-Bibran, Aufzeichnungen über das Jahr 1888, BA-MA Freiburg, Senden-Bibran Papers N160/11, pp. 38–40.
32 Waldersee, diary entry for 26 August 1888, GStA Berlin, Waldersee Papers; cf. Meisner, II, p. 1.
33 Princess Viktoria of Schaumburg-Lippe to Empress Frederick, 9 May 1891, AdHH Schloss Fasanerie.
34 Waldersee, diary entry for 17 September 1891, GStA Berlin, Waldersee Papers; omitted from Meisner, II, pp. 216–17.
35 Bodman to Brauer, 7 March 1897, in Walther Peter Fuchs, ed., *Großherzog Friedrich I. von Baden und die Reichspolitik 1871–1907*, 4 vols., Stuttgart 1968–80, III, No. 1657.
36 Eulenburg to Kaiser Wilhelm II, 15 April 1893, *Eulenburgs Korrespondenz*, II, No. 791.
37 Holzing-Berstett to his mother, 12 August 1905, Generallandesarchiv Karlsruhe, Holzing Papers, No. 116/11.
38 Holzing-Berstett to his mother, 3 September 1905, ibid.
39 Heinrich Prinz von Schönburg-Waldenburg, *Erinnerungen aus kaiserlicher Zeit*, Leipzig 1929, pp. 132–3.
40 Walter Görlitz, ed., *Der Kaiser . . . Aufzeichnungen des Chefs des Marinekabinetts Admiral Georg Alexander v. Müller über die Ära Wilhelms II.*, Göttingen 1965, pp. 188–9.
41 Waldersee, diary entry for 22 October 1888, GStA Berlin, Waldersee Papers; cf. Meisner, II, p. 10.
42 Hoffmann, diary entries for 4 and 12 November 1889, Hoffmann Papers, Freiburg.
43 Waldersee, diary entry for 22 October 1888, GStA Berlin, Waldersee Papers; cf. Meisner, II, p. 10.
44 Waldersee, diary entry for 19 November 1888, GStA Berlin, Waldersee Papers; cf. Meisner, II, p. 21.
45 Waldersee, diary entry for 29 August 1888, GStA Berlin, Waldersee Papers; cf. Meisner, II, p. 2.
46 Waldersee, diary entry for 5 November 1888, GStA Berlin, Waldersee Papers; cf. Meisner, II, p. 15.
47 Waldersee, diary entry for 26 August 1888, GStA Berlin, Waldersee Papers; cf. Meisner, II, p. 1; Széchényi to Kálnoky, 26 January 1889, HHStA Vienna, PA III 136.
48 Waldersee, diary entry for 26 November 1888, GStA Berlin, Waldersee Papers; cf. Meisner, II, pp. 22–3.
49 Waldersee, diary entry for 29 August 1888, GStA Berlin, Waldersee Papers; cf. Meisner, II, p. 2.
50 See Mathilde Gräfin von Keller, *Vierzig Jahre im Dienst der Kaiserin. Ein Kulturbild aus den Jahren 1881–1921*, Leipzig 1935, pp. 131–2; also *Young Wilhelm*, pp. 257ff.
51 Waldersee, diary entry for 26 November 1888, GStA Berlin, Waldersee Papers; cf. Meisner, II, pp. 22–3. See also the entries for 22 October and 19 November 1888, GStA Berlin, Waldersee Papers; cf. Meisner, II, pp. 10 and 21.

52 August Eulenburg to Philipp Eulenburg, 27 March 1889, *Eulenburgs Korrespondenz*, I, No. 216; Waldersee, diary entry for 29 March 1889, GStA Berlin, Waldersee Papers; omitted from Meisner, II, p. 48.
53 Waldersee, diary entry for 14 April 1889, GStA Berlin, Waldersee Papers; omitted from Meisner, II, p. 48.
54 Széchényi to Kálnoky, 16 June 1888, HHStA Vienna, PA III 134.
55 Kaiser Wilhelm II to Kaiserin Augusta, 11 July 1888, GStA Berlin, BPHA Rep. 53J Lit. P No. 14a.
56 Waldersee, diary entry for 11 October 1888, GStA Berlin, Waldersee Papers; Meisner, II, p. 6.
57 Holstein to Eulenburg, 23 September 1895, cited in *Eulenburgs Korrespondenz*, III, p. 1579.
58 Waldersee, diary entry for 26 October 1894, GStA Berlin, Waldersee Papers; omitted from Meisner, II, pp. 327–8.
59 See the brief character sketch in Röhl, *Kaiser and His Court*, pp. 142–3.
60 Kaiser Wilhelm II to Miessner, 13 December 1888, GStA Berlin, BPH Rep. 53 No. 400.
61 See *Young Wilhelm*, pp. 456 and 503–4.
62 See Holstein, diary entry for 22 October 1888, *Holstein Papers*, II, pp. 380–1. For Wilhelm's campaign against the Union Club, see *Young Wilhelm*, pp. 507–15.
63 Waldersee, diary entry for 16 June 1888, GStA Berlin, Waldersee Papers; cf. Meisner, I, p. 405.
64 Waldersee, diary entry for 20 June 1888, GStA Berlin, Waldersee Papers. The text printed in Meisner, I, p. 405, is incomplete. Kaiser Wilhelm II, *Ereignisse und Gestalten aus den Jahren 1878–1918*, Berlin, Leipzig 1922, p. 20. Wilhelm von Hahnke was married to Josefine von Bülow (1842–1911).
65 RA QVJ, 5 August 1889.
66 Waldersee, diary entry for 2 February 1891, GStA Berlin, Waldersee Papers; omitted from Meisner, II, p. 184.
67 Waldersee, diary entry for 10 October 1890, GStA Berlin, Waldersee Papers; cf. Meisner, II, p. 153.
68 Waldersee, diary entry for 19 January 1891, GStA Berlin, Waldersee Papers; omitted from Meisner, II, p. 176.
69 Waldersee, diary entry for 17 May 1891, GStA Berlin, Waldersee Papers; omitted from Meisner, II, p. 207.
70 See Walther Hubatsch, *Der Admiralstab und die obersten Marinebehörden in Deutschland 1884–1945*, Frankfurt a.M. 1958; Jonathan Steinberg, *Yesterday's Deterrent. Tirpitz and the Birth of the German Battle Fleet*, London 1966; Volker R. Berghahn, *Der Tirpitz-Plan. Genesis und Verfall einer innenpolitischen Krisenstrategie unter Wilhelm II.*, Düsseldorf 1971.
71 Jörg-Uwe Fischer, *Admiral des Kaisers. Georg Alexander von Müller als Chef des Marinekabinetts Wilhelms II.*, Frankfurt a.M. 1992, pp. 29–30.
72 See Alfred von Tirpitz, *Erinnerungen*, Leipzig 1919, p. 36.
73 Müller to Tirpitz, 8 December 1889, cited in Fischer, *Admiral des Kaisers*, p. 34.
74 Széchényi to Kálnoky, 4 July 1888, HHStA Vienna, PA III 134.
75 Senden-Bibran, Aufzeichnungen über das Jahr 1888, BA-MA Freiburg, Senden-Bibran Papers N160/11, pp. 13ff.
76 Kaiser Wilhelm II, *Aus meinem Leben*, p. 273.
77 Senden-Bibran, Aufzeichnungen über das Jahr 1888, BA-MA Freiburg, Senden-Bibran Papers N160/11, pp. 33ff.
78 Ibid., p. 36.
79 Ibid.
80 Ibid., p. 38.
81 Bülow, *Memoirs*, II, p. 65.
82 Eulenburg to Holstein, 16 July 1896, cited in Röhl, *Kaiser and His Court*, p. 236.

83 Berghahn, *Tirpitz-Plan*, p. 189 n. 79.
84 Holstein to Bülow, 6 June 1896, cited in Isabel V. Hull, *The Entourage of Kaiser Wilhelm II, 1888–1918*, Cambridge 1982, p. 179.
85 See *Truth*, 22 November 1888, p. 898.
86 Waldersee, diary entry for 5 July 1888, GStA Berlin, Waldersee Papers; cf. Meisner, I, pp. 410–11.
87 Waldersee, diary entry for early August 1888, GStA Berlin, Waldersee Papers; cf. Meisner, I, p. 414.
88 See above, pp. 154–5.
89 Herbert Bismarck to his father, BA Koblenz, Bismarck Papers FC 3005 N.
90 Waldersee to Verdy, 26 November 1888, GStA Berlin, Waldersee Papers, B I No. 53.
91 See Schmidt-Bückeburg, *Militärkabinett*, pp. 175–6.
92 Waldersee to Verdy, 2 December 1888, GStA Berlin, Waldersee Papers, B I No. 53. See also Herbert Bismarck to his father, 30 December 1888, BA Koblenz, Bismarck Papers FC 3005 N.
93 Waldersee, diary entries for 8 and 26 April 1889, GStA Berlin, Waldersee Papers; cf. Meisner, II, pp. 48ff.
94 Bismarck's marginal comments on Herbert Bismarck's notes of April 1890, BA Koblenz, Bismarck Papers.
95 Hereditary Prince Bernhard of Saxe-Meiningen to Colmar Freiherr von der Goltz, 20 October 1898 and 9 April 1899, BA-MA Freiburg, v. d. Goltz Papers N737, Zug. 161/95.
96 Holzing-Berstett to his father, 4 January 1904, Generallandesarchiv Karlsruhe, Holzing Papers, No. 116/11.
97 Hereditary Prince Bernhard of Saxe-Meiningen to von der Goltz, 20 October 1898, BA-MA Freiburg, v. d. Goltz Papers N737, Zug. 161/95.
98 See below, p. 301.
99 Waldersee, diary entry for 23 June 1888, GStA Berlin, Waldersee Papers; cf. Meisner, I, p. 406.
100 Waldersee, diary entry for 9 October 1888, GStA Berlin, Waldersee Papers; omitted from Meisner, II, pp. 4–5.
101 Wedel, *Zwischen Kaiser und Kanzler*, p. 137.
102 Swaine to Ponsonby, 1 July 1892, RA I59/89.
103 Széchényi to Kálnoky, 29 December 1888, HHStA Vienna, PA III 134.
104 Széchényi to Kálnoky, 9 January 1889, ibid., PA III 136.

8 THE DOMINATION OF THE BISMARCKS

1 Herbert Bismarck to his father, 9 October 1888, BA Koblenz, Bismarck Papers FC 3005 N; in Bußmann, *Staatssekretär*, No. 368.
2 See *Young Wilhelm*, pp. 720–37.
3 Count Herbert Bismarck, 'Notizen V', BA Koblenz, Bismarck Papers FC 3018 N.
4 Herbert Bismarck to Rantzau, 17 June 1888, ibid., FC 3014 N.
5 Herbert Bismarck to his brother, 23 June 1888, in Bußmann, *Staatssekretär*, No. 362.
6 Herbert Bismarck to Holstein, 15 October 1888, *Holstein Papers*, III, No. 280.
7 Herbert Bismarck to Holstein, 17 October 1888, ibid., No. 281.
8 Rantzau to Herbert Bismarck, 29 August 1888, BA Koblenz, Bismarck Papers FC 3028 N.
9 Herbert Bismarck to Rantzau, 30 August 1888, ibid., FC 3014 N.
10 Prince Otto von Bismarck to Herbert Bismarck, 7 December 1888, PA AA, R 1009.
11 Herbert Bismarck to his father, 30 December 1888, BA Koblenz, Bismarck Papers FC 3005 N.

12 Bismarck left for Varzin on 8 June 1889 and did not return to Berlin until 24 January 1890: Count Herbert Bismarck, diary 1888–90, BA Koblenz, Bismarck Papers FC 3018 N.
13 Herbert Bismarck to his father, 9 June 1889, ibid., FC 3005 N; in Bußmann, *Staatssekretär*, No. 376.
14 Salisbury to Queen Victoria, 15 October 1888, RA T9/112, in Buckle, *Letters of Queen Victoria*, I, pp. 441–2.
15 Széchényi to Kálnoky, 9 February 1889, HHStA Vienna, PA III 136. See above, p. 24.
16 Széchényi to Kálnoky, 23 February 1889, HHStA Vienna, PA III 136.
17 Empress Frederick to Queen Victoria, 20 April 1889, RA Z44/31.
18 In the entire period from 12 July 1888 to 24 January 1890 the Chancellor was in Berlin only for the first six months of 1889. See Waldersee, diary entry for 27 March 1889, GStA Berlin, Waldersee Papers; cf. Meisner, II, p. 47.
19 Heinrich von Eckardt, Aufzeichnung vom 8. November 1889, GStA Berlin, BPH Rep. 53 No. 29.
20 Count Herbert Bismarck, diary 1888–90, BA Koblenz, Bismarck Papers FC 3018 N.
21 Széchényi to Kálnoky, 26 January 1889, HHStA Vienna, PA III 136.
22 Empress Frederick to Queen Victoria, 20 April 1889, RA Z44/31.
23 See e.g. Herbert Bismarck to Kaiser Wilhelm II, 11 November 1888, BA Koblenz, Bismarck Papers FC 2986 N. For earlier examples, see *Young Wilhelm*, pp. 417–18.
24 Herbert Bismarck to Rantzau, 17 June 1888, BA Koblenz, Bismarck Papers FC 3014 N.
25 Waldersee, diary entry for 13 November 1889, GStA Berlin, Waldersee Papers; cf. Meisner, II, pp. 76–7. See also Waldersee, diary entry for 3 January 1890, GStA Berlin, Waldersee Papers; omitted from Meisner, II, p. 87.
26 Széchényi to Kálnoky, 26 January 1889, HHStA Vienna, PA III 136. For the Geffcken case, see above, pp. 62–6.
27 Széchényi to Kálnoky, 26 January 1889, HHStA Vienna, PA III 136.
28 See Philipp Graf zu Eulenburg-Hertefeld, *Aus 50 Jahren. Erinnerungen, Tagebücher und Briefe aus dem Nachlaß des Fürsten*, Berlin 2nd edn 1925, pp. 81–107; Louis L. Snyder, 'Political Implications of Herbert von Bismarck's Marital Affairs, 1881, 1892', *Journal of Modern History*, 36 (1964), pp. 155–69; Otto Pflanze, *Bismarck and the Development of Germany*, 3 vols., Princeton 1963–90, III, pp. 55–6.
29 Count Herbert Bismarck, diary 1888–90, BA Koblenz, Bismarck Papers FC 3018 N.
30 Waldersee, diary entry for 28 February 1889, GStA Berlin, Waldersee Papers; largely omitted from Meisner, II, p. 41. On 27 February 1889 Herbert Bismarck wrote in his diary: 'In the evening ball at my place. 67 people. To bed at 6 a.m.': Count Herbert Bismarck, diary 1888–90, BA Koblenz, Bismarck Papers FC 3018 N.
31 Waldersee, diary entry for 28 February 1889, GStA Berlin, Waldersee Papers; largely omitted from Meisner, II, p. 41.
32 Waldersee, diary entry for 9 June 1889, GStA Berlin, Waldersee Papers; cf. Meisner, II, p. 53.
33 Waldersee, diary entry for 12 October 1889, GStA Berlin, Waldersee Papers; omitted from Meisner, II, p. 70.
34 See Herbert Bismarck to Rantzau, 24 July 1889, BA Koblenz, Bismarck Papers FC 3014 N.
35 Waldersee, diary entry for 12 October 1889, GStA Berlin, Waldersee Papers; omitted from Meisner, II, p. 70.
36 Waldersee, diary entry for 28 February 1889, GStA Berlin, Waldersee Papers; abridged and distorted in Meisner, II, pp. 41–2.
37 Waldersee, retrospective diary entry, GStA Berlin, Waldersee Papers; cf. Meisner, II, pp. 63–4.
38 Waldersee, diary entry for 12 October 1889, GStA Berlin, Waldersee Papers; omitted from Meisner, II, p. 70.

39 Waldersee, diary entry for 13 November 1889, GStA Berlin, Waldersee Papers; Meisner, II, p. 77.
40 Waldersee, diary entry for 15 April 1889, GStA Berlin, Waldersee Papers; cf. Meisner, II, p. 48.
41 Waldersee, diary entry for 16 October 1889, GStA Berlin, Waldersee Papers; omitted from Meisner, II, p. 73. See Waldersee, diary entry for 10 November 1889, GStA Berlin, Waldersee Papers; printed under an incorrect date in Meisner, II, p. 76.
42 Count Herbert Bismarck, diary entry for 7 June 1889, BA Koblenz, Bismarck Papers FC 3018 N.
43 Waldersee, diary entry for 2 January 1890, GStA Berlin, Waldersee Papers; largely omitted from Meisner, II, p. 86.
44 Waldersee, diary entry for 18 January 1890, GStA Berlin, Waldersee Papers; cf. Meisner, II, p. 95.
45 See *Young Wilhelm*, pp. 713–15.
46 Herbert Bismarck to his father, 5 October 1888, BA Koblenz, Bismarck Papers FC 3005 N; in Bußmann, *Staatssekretär*, No. 366.
47 Eulenburg to Kaiser Wilhelm II, 5 July 1888, *Eulenburgs Korrespondenz*, I, No. 185.
48 Kaiser Wilhelm II to Eulenburg, 28 July 1888, Eulenburg to Kaiser Wilhelm II, 1 August 1888, ibid., I, No. 187. A facsimile of this Skaldengesang, published in 1892, is printed in See, *Barbar Germane Arier*, pp. 18–19.
49 Kaiser Wilhelm II to Eulenburg, 28 August–4 September 1888, *Eulenburgs Korrespondenz*, I, No. 194.
50 Eulenburg to Kaiser Wilhelm II, 11 September 1888, ibid., No. 196.
51 Eulenburg to his mother, 4 October 1888, ibid., No. 197.
52 Herbert Bismarck to his father, 5 October 1888, in Bußmann, *Staatssekretär*, No. 366.
53 Eulenburg to his mother, 4 October 1888, *Eulenburgs Korrespondenz*, I, No. 197.
54 Eulenburg to Kaiser Wilhelm II, 15 October 1888, ibid., No. 199.
55 Eulenburg to his mother, 8 July 1888, ibid., No. 186. See also Herbert Bismarck to Rantzau, 12 September 1888, BA Koblenz, Bismarck Papers FC 3014 N.
56 Herbert Bismarck to his father, 5 October 1888, in Bußmann, *Staatssekretär*, No. 366.
57 Eulenburg to his mother, 4 October 1888, *Eulenburgs Korrespondenz*, I, No. 197; Herbert Bismarck to his father, 5 October 1888, in Bußmann, *Staatssekretär*, No. 366.
58 Herbert Bismarck to his father, 5 October 1888, in Bußmann, *Staatssekretär*, No. 366.
59 Prince Otto von Bismarck to Herbert Bismarck, 7 October 1888, ibid., No. 367. See also the Chancellor's marginal notes on his son's letter, ibid., pp. 524–5.
60 Prince Otto von Bismarck to Herbert Bismarck, 7 October 1888, ibid., No. 367; the Chancellor's marginal notes, ibid., pp. 524–5. Herbert Bismarck's reply of 8 October 1888 is located in BA Koblenz, Bismarck Papers FC 3005 N. Cf. Herbert Bismarck to Holstein, 15 October 1888, *Holstein Papers*, III, No. 280.
61 Eulenburg to Herbert Bismarck, 15 October 1888, *Eulenburgs Korrespondenz*, I, No. 198.
62 Eulenburg to Kaiser Wilhelm II, 15 October 1888, ibid., I, No. 199.
63 Kaiser Wilhelm II to Eulenburg, 31 October 1888, in Eulenburg, *Aus 50 Jahren*, p. 197.
64 Eulenburg to Kaiser Wilhelm II, 1 November 1888, *Eulenburgs Korrespondenz*, I, No. 201.
65 Eulenburg to Herbert Bismarck, 1 November 1888, ibid., No. 202.
66 Rantzau to Herbert Bismarck, 5 [?] November 1888, cited ibid., p. 320.
67 Eulenburg to his mother, 17 December 1888, ibid., No. 207.
68 Ibid.
69 Eulenburg to Grand Duchess Augusta Caroline of Mecklenburg-Strelitz, 8 April 1889, ibid., No. 217.
70 Eulenburg to Kaiser Wilhelm II, 22 December 1888, ibid., No. 208.

71 Eulenburg to Kaiser Wilhelm II, 15 October 1888, ibid., No. 199. See *Young Wilhelm*, pp. 713–15.
72 Eulenburg to Holstein, 4 February 1889, *Holstein Papers*, III, No. 287.
73 Eulenburg to Kaiser Wilhelm II, 15 October 1888, *Eulenburgs Korrespondenz*, I, No. 199.
74 Eulenburg to Kaiser Wilhelm II, 19 November 1888, ibid., No. 205.
75 Eulenburg to his father, 21 November 1888, ibid., No. 206.
76 Széchényi to Kálnoky, 9 January 1889, HHStA Vienna, PA III 136.
77 Eulenburg, diary entry for 17 January 1889, *Eulenburgs Korrespondenz*, I, p. 327.
78 Empress Frederick to Queen Victoria, 1 and 6 April 1889, RA Z44/21, 24 and 25. See below, pp. 639–43.
79 Kaiser Wilhelm II to Eulenburg, 27 February 1889, *Eulenburgs Korrespondenz*, I, No. 213.
80 Eulenburg to Kaiser Wilhelm II, 1 March 1889, ibid., No. 214.
81 See *Young Wilhelm*, pp. 485–8.
82 Ibid., pp. 502–7.
83 See ibid., pp. 455–63.
84 Count Herbert Bismarck, diary entry for 9 May 1889, BA Koblenz, Bismarck Papers FC 3018 N.
85 Eulenburg to Holstein, 18 February 1894, *Eulenburgs Korrespondenz*, II, No. 911. See also Waldersee, diary entry for 26 November 1888, GStA Berlin, Waldersee Papers; cf. Meisner, II, pp. 22–3.
86 In the six months from December 1888 to May 1889, the Kaiser received Herbert Bismarck in audience fifty-seven times; the corresponding figure for the six months from June to November 1889 is twenty-two: Count Herbert Bismarck, diary 1888–90, BA Koblenz, Bismarck Papers FC 3018 N; also Herbert Bismarck, 'Notizen V', ibid.

9 THE BISMARCK CRISIS BEGINS

1 Waldersee, diary entry for 3 June 1889, GStA Berlin, Waldersee Papers; cf. Meisner, II, p. 53.
2 See Kaiser Wilhelm II to Kaiserin Augusta, 28 September 1889, GStA Berlin, Rep. 5J Lit. P No. 14a.
3 *Badische Presse*, Karlsruhe, 4 October 1889. See above, ch. 1.
4 Waldersee, diary entry for 4 May 1889, GStA Berlin, Waldersee Papers; omitted from Meisner, II, p. 50.
5 See Waldersee, diary entries for 24 April and 4 May 1889, GStA Berlin, Waldersee Papers; omitted from Meisner, II, pp. 49–50.
6 Széchényi to Kálnoky, 26 January 1889, HHStA Vienna, PA III 136.
7 Waldersee, diary entry for 13 November 1889, GStA Berlin, Waldersee Papers; cf. Meisner, II, pp. 76–7.
8 See above, ch. 7.
9 Wilhelm's decision against the widening of the remit of the Reich Railway Office serves as an example. See below, pp. 225–6.
10 Waldersee, diary entry for 25 February 1889, GStA Berlin, Waldersee Papers; cf. Meisner, II, p. 40.
11 For example on 21 and 26 February and 13 and 23 March 1889: Waldersee, diary entries for those dates, GStA Berlin, Waldersee Papers.
12 Waldersee, diary entry for 18 March 1889, ibid.; Meisner, II, pp. 45–6.
13 For example on 4, 6 and 21 February and 13 March 1889: Count Herbert Bismarck, diary, BA Koblenz, Bismarck Papers FC 3018 N.
14 Waldersee, diary entry for 1 April 1889, GStA Berlin, Waldersee Papers; Meisner, II, p. 48.

15 Waldersee, diary entry for 2 March 1889, GStA Berlin, Waldersee Papers; cf. Meisner, II, p. 42. See also the diary entry for 12 March 1889, ibid., pp. 44–5. See Wilhelm's marginal notes on Deines's (5 February and 20 December 1888), Yorck's (14 October 1888) and Huene's reports (14 November 1888), PA AA, Asservat 4.1.
16 Count Herbert Bismarck, 'Notizen V', BA Koblenz, Bismarck Papers FC 3018 N.
17 Waldersee, diary entry for 3 February 1890, GStA Berlin, Waldersee Papers; omitted from Meisner, II, p. 99.
18 Waldersee, diary entry for 1 May 1889, GStA Berlin, Waldersee Papers; omitted from Meisner, II, p. 50.
19 Waldersee, diary entry for 15 April 1889, GStA Berlin, Waldersee Papers; cf. Meisner, II, p. 48.
20 See Herbert Bismarck to Rantzau, 27 June 1889, BA Koblenz, Bismarck Papers FC 3014 N; in Bußmann, *Staatssekretär*, No. 380.
21 Waldersee, diary entry for 26 March 1889, GStA Berlin, Waldersee Papers; cf. Meisner, II, p. 47.
22 Waldersee, diary entry for 19 March 1889, GStA Berlin, Waldersee Papers; omitted from Meisner, II, p. 46.
23 See above, pp. 191–7.
24 See above, pp. 173–6.
25 Count Herbert Bismarck, diary, 29 January and 19 February 1889, BA Koblenz, Bismarck Papers FC 3018 N.
26 Waldersee, diary entry for 19 and 25 February and 18 March 1889, GStA Berlin, Waldersee Papers; cf. Meisner, II, pp. 39–40, p. 46.
27 Waldersee, diary entry for 25 April and 1 May 1889, GStA Berlin, Waldersee Papers; omitted from Meisner, II, pp. 49–50.
28 Waldersee, diary entry for 6 April 1889, GStA Berlin, Waldersee Papers; largely omitted from Meisner, II, p. 48.
29 Wilhelm II, marginal note of April 1889, PA AA, Asservat 4.1.
30 Wilhelm II, marginal note on Brandt's report from Beijing of 15 July 1890, ibid.
31 Waldersee, diary entry for 6 April 1889, GStA Berlin, Waldersee Papers; omitted from Meisner, II, p. 48.
32 Waldersee, diary entry for 12 May 1889, GStA Berlin, Waldersee Papers; omitted from Meisner, II, p. 50.
33 Count Herbert Bismarck, diary, 22 July 1889, BA Koblenz, Bismarck Papers FC 3018 N.
34 Count Herbert Bismarck, diary, 20 May and 15 August 1889, ibid.
35 Count Herbert Bismarck, diary, 14 and 16 May 1889, ibid.
36 Waldersee, diary entry for 19 May 1889, GStA Berlin, Waldersee Papers; cf. Meisner, II, pp. 50–1.
37 Waldersee, diary entry for 20 May 1889, GStA Berlin, Waldersee Papers; cf. Meisner, II, pp. 51–2.
38 Waldersee, diary entry for 2 June 1889, GStA Berlin, Waldersee Papers; omitted from Meisner, II, p. 53.
39 Holstein to Eulenburg, 28 September 1889, *Eulenburgs Korrespondenz*, I, No. 235.
40 Waldersee, diary entry for 4 May 1889, GStA Berlin, Waldersee Papers; Meisner, II, p. 50.
41 See below, pp. 225–7.
42 Count Herbert Bismarck, 'Notizen V', BA Koblenz, Bismarck Papers FC 3018 N; diary, 12 and 15 May 1889, ibid.
43 See the Kaiser's speeches of 14 and 16 May 1889, Penzler, *Reden Kaiser Wilhelms II. in den Jahren 1888–1895*, pp. 53–7. See Christopher Clark, *Kaiser Wilhelm II*, London 2000, pp. 37ff., who rightly points out the unprecedented nature of the young monarch's intervention.

44 Waldersee, diary entry for 19 May 1889, GStA Berlin, Waldersee Papers; cf. Meisner, II, p. 50.
45 Bismarck to Kaiser Wilhelm II, 25 May 1889, in Meisner, II, p. 451.
46 Robert Freiherr Lucius von Ballhausen, *Bismarck-Erinnerungen*, Stuttgart, Berlin 1920, p. 497. For the machiavellian motives behind Bismarck's tactics during the miners' strike, see Lothar Gall, *Bismarck. Der weisse Revolutionär*, Frankfurt a.M., Berlin, Vienna 1980, pp. 690–1.
47 Eulenburg to Herbert Bismarck, 19 May 1889, *Eulenburgs Korrespondenz*, I, No. 221; Count Herbert Bismarck, diary, 20 May 1889, BA Koblenz, Bismarck Papers FC 3018 N.
48 Holstein to Eulenburg, 23 May 1889, *Eulenburgs Korrespondenz*, I, No. 222.
49 Count Herbert Bismarck, 'Notizen V', BA Koblenz, Bismarck Papers FC 3018 N.
50 Ibid.
51 Waldersee, diary entry for 19 May 1889, GStA Berlin, Waldersee Papers; cf. Meisner, II, p. 51.
52 Waldersee, diary entry for 2 and 3 June 1889, GStA Berlin, Waldersee Papers; cf. Meisner, II, p. 53; Herbert Bismarck to Rantzau, 4 July 1889, BA Koblenz, Bismarck Papers FC 3014 N, in Bußmann, *Staatssekretär*, No. 382; Count Herbert Bismarck, diary, 5 June 1889, BA Koblenz, Bismarck Papers FC 3018. See also Waldersee, diary entry for 9 June 1889, GStA Berlin, Waldersee Papers; omitted from Meisner, II, p. 53, but cf. Meisner, II, pp. 63–4.
53 Count Herbert Bismarck, diary, 8 May and 4 June 1889, BA Koblenz, Bismarck Papers FC 3018.
54 Waldersee, restrospective diary entry written in later years, GStA Berlin, Waldersee Papers; cf. Meisner, II, pp. 63–4. Cf. Herbert Bismarck to Rantzau, 4 July 1889, BA Koblenz, Bismarck Papers FC 3014 N; in Bußmann, *Staatssekretär*, No. 382.
55 Holstein to Eisendecher, 5 July 1889, *Holstein Papers*, III, No. 292.
56 Prince Otto von Bismarck to Eisendecher, 4 July 1889, cited ibid., p. 313, n. 1.
57 Holstein to Eisendecher, 11 July 1889, ibid., No. 293.
58 Waldersee, retrospective diary entry written in later years, GStA Berlin, Waldersee Papers; cf. Meisner, II, pp. 63–4. Cf. Herbert Bismarck to Rantzau, 4 July 1889, BA Koblenz, Bismarck Papers FC 3014 N; in Bußmann, *Staatssekretär*, No. 382.
59 Count Herbert Bismarck, 'Notizen V', BA Koblenz, Bismarck Papers FC 3018 N.
60 Waldersee, retrospective diary entry written in 1892, GStA Berlin, Waldersee Papers; cf. Meisner, II, pp. 55–6.
61 Cited above, p. 204.
62 Waldersee, diary entry for 25 March 1889, GStA Berlin, Waldersee Papers; omitted from Meisner, II, p. 47.
63 Waldersee, diary entry for 15 April 1889, GStA Berlin, Waldersee Papers; omitted from Meisner, II, p. 48.
64 Waldersee, diary entry for 7 July 1889, GStA Berlin, Waldersee Papers; cf. Meisner, II, pp. 54–9.
65 Count Herbert Bismarck, 'Notizen V', BA Koblenz, Bismarck Papers FC 3018 N; Count Herbert Bismarck, diary, 11 June 1889, ibid.
66 Kaiser Wilhelm II to Prince Otto von Bismarck, 12 June 1889, ibid., FC 2986 N.
67 Waldersee, diary entry for 7 July 1889, GStA Berlin, Waldersee Papers; cf. Meisner, II, pp. 54–9.
68 Waldersee, retrospective diary entry written in 1892, GStA Berlin, Waldersee Papers; cf. Meisner, II, pp. 55–6. See also Herbert Bismarck to his father, 14 June 1889, BA Koblenz, Bismarck Papers FC 3005 N; in Bußmann, *Staatssekretär*, No. 377.
69 Rantzau to Herbert Bismarck, 1 July 1889, cited in Bußmann, *Staatssekretär*, pp. 541–2, n. 5.
70 Herbert Bismarck to Rantzau, 4 July 1889, BA Koblenz, Bismarck Papers FC 3014 N; in Bußmann, *Staatssekretär*, No. 382.

71 Herbert Bismarck to his father, 14 June 1889, BA Koblenz, Bismarck Papers FC 3005 N; in Bußmann, *Staatssekretär*, No. 377.
72 Herbert Bismarck to Rantzau, 4 July 1889, BA Koblenz, Bismarck Papers FC 3014 N; in Bußmann, *Staatssekretär*, No. 382.
73 Eulenburg to Holstein, 6 August 1889, *Eulenburgs Korrespondenz*, I, No. 230.
74 Count Herbert Bismarck, 'Notizen V', BA Koblenz, Bismarck Papers FC 3018 N.
75 Waldersee, diary entry for 7 July 1889, GStA Berlin, Waldersee Papers; cf. Meisner, II, pp. 54–9.
76 Holstein to Eulenburg, 3 July 1889, *Eulenburgs Korrespondenz*, I, No. 225.
77 Holstein to Herbert Bismarck, late June 1889, in Bußmann, *Staatssekretär*, No. 379.
78 Herbert Bismarck to Rantzau, 27 June 1889, BA Koblenz, Bismarck Papers FC 3014 N; in Bußmann, *Staatssekretär*, No. 380. See also Herbert Bismarck to Eulenburg, 24 June 1889, *Eulenburgs Korrespondenz*, I, No. 224.
79 Herbert Bismarck to Rantzau, 4 July 1889, BA Koblenz, Bismarck Papers FC 3014 N; in Bußmann, *Staatssekretär*, No. 382.
80 Ibid.
81 Ibid.
82 Prince Otto von Bismarck to Eisendecher, 1 July 1889, cited from the Foreign Office files in *Holstein Papers*, III, p. 313, n. 3.
83 Rantzau's marginal notes on Herbert Bismarck to Rantzau, 6 July 1889, BA Koblenz, Bismarck Papers FC 3014 N; in Bußmann, *Staatssekretär*, No. 383.
84 Herbert Bismarck to Rantzau, 8 July 1889, BA Koblenz, Bismarck Papers FC 3014 N; in Bußmann, *Staatssekretär*, No. 385.
85 Holstein to Eisendecher, 5 July 1889, *Holstein Papers*, III, No. 292.
86 Herbert Bismarck to Rantzau, 6 July 1889, BA Koblenz, Bismarck Papers FC 3014 N; in Bußmann, *Staatssekretär*, No. 383.
87 Eulenburg to Herbert Bismarck, 17 July 1889, Holstein to Eulenburg, 2 August 1889, Eulenburg to Holstein, 6 August 1889, *Eulenburgs Korrespondenz*, I, Nos. 228–30; Kaiserin Auguste Viktoria to Kaiser Wilhelm II, 12, 13, 15, 16, 17 and 18 July 1889, GStA Berlin, BPHA Rep. 53T Preußen: An Kaiser Wilhelm II., Bd III.
88 Waldersee, retrospective addendum to 9 July 1889, GStA Berlin, Waldersee Papers; cf. Meisner, II, pp. 58–9.
89 Waldersee, restrospective note written in 1892, GStA Berlin, Waldersee Papers; cf. Meisner, II, pp. 55–6.
90 *Hamburger Nachrichten*, 19 June 1889. See *Berliner Tageblatt* of 20 June 1889.
91 Waldersee, diary entry for 7 July 1889, GStA Berlin, Waldersee Papers; omitted from Meisner, II, p. 54.
92 Ibid.; cf. Meisner, II, pp. 54–9.
93 Wilhelm II, marginal note, 20 June 1889, PA AA, R 1009.
94 See Meisner, II, pp. 60ff.
95 Eulenburg to Holstein, 6 August 1889, *Eulenburgs Korrespondenz*, I, No. 230.
96 Waldersee, diary entry for 7 July 1889, GStA Berlin, Waldersee Papers; cf. Meisner, II, pp. 54–9.
97 Waldersee, diary entry for 26 December 1889, GStA Berlin, Waldersee Papers; largely omitted from Meisner, II, p. 85.
98 Waldersee, diary entry for 7 December 1889, GStA Berlin, Waldersee Papers; omitted from Meisner, II, p. 81.
99 Waldersee, diary entry for 25 March 1889, GStA Berlin, Waldersee Papers; cf. Meisner, II, p. 47.
100 Herbert Bismarck to his father, 14 June 1889, BA Koblenz, Bismarck Papers FC 3005 N.

101 Waldersee, diary entries for 2 and 3 June 1889, GStA Berlin, Waldersee Papers; cf. Meisner, II, p. 53.
102 Waldersee, diary entry for 6 October 1889, GStA Berlin, Waldersee Papers; cf. Meisner, II, p. 69.
103 Count Herbert Bismarck, Aufzeichnung, pp. 76ff., BA Koblenz, Bismarck Papers FC 3018. See above, pp. 107–8.
104 Prince Ernst zu Hohenlohe-Langenburg to his father, 13 November 1889, Hohenlohe-Zentralarchiv Neuenstein, Hermann Hohenlohe-Langenburg Papers, Bü. 56.
105 Count Herbert Bismarck, Aufzeichnung, pp. 84–5, BA Koblenz, Bismarck Papers FC 3018. Cf. Salisbury to Queen Victoria, 24 October 1889, RA I57/66.
106 Waldersee, diary entry for 16 October 1889, GStA Berlin, Waldersee Papers; omitted from Meisner, II, p. 73.
107 Waldersee, diary entry for 21 October 1889, GStA Berlin, Waldersee Papers; omitted from Meisner, II, p. 73.
108 Waldersee, diary entry for 10 October 1889, GStA Berlin, Waldersee Papers; cf. Meisner, II, p. 69.
109 Count Herbert Bismarck, Aufzeichnung, pp. 84ff., BA Koblenz, Bismarck Papers FC 3018 N. See Herbert Bismarck to his father, 5 October 1889, in Bußmann, *Staatssekretär*, No. 390.
110 Herbert Bismarck's note on the imperial hunt at Hubertusstock, October 1889, BA Koblenz, Bismarck Papers FC 3005.
111 Holstein to Radolin, 15 October 1889, *Holstein Papers*, VI, No. 294.
112 Waldersee, diary entry for 12 and 16 October 1889, GStA Berlin, Waldersee Papers; Meisner, II, pp. 70 and 73.
113 Waldersee, diary entry for 12 October 1889, GStA Berlin, Waldersee Papers; Meisner, II, p. 70. Holstein to Radolin, 15 October 1889, *Holstein Papers*, III, No. 294.
114 Count Herbert Bismarck, Aufzeichnung, p. 86, BA Koblenz, Bismarck Papers FC 3018.
115 Waldersee, diary entry for 13 October 1889, GStA Berlin, Waldersee Papers; omitted from Meisner, II, p. 71.
116 Waldersee, retrospective note written in later years, GStA Berlin, Waldersee Papers; cf. Meisner, II, p. 70. Cf. Holstein to Radolin, 15 October 1889, Reuss to Holstein, 5 November 1889, *Holstein Papers*, III, Nos. 294 and 299.
117 Grand Duchess Olga, the daughter of Alexander III, cited in Ian Vorres, *The Last Grand Duchess*, London 1964, p. 66. The episode is recounted almost verbatim in Alexander Grand Duke of Russia, *Once a Grand Duke*, New York 1932, p. 174. For the growing concern of the Tsar over the mental health of Wilhelm II, see below, pp. 297–301.
118 Count Herbert Bismarck, Aufzeichnung, pp. 88–9, BA Koblenz, Bismarck Papers FC 3018 N.
119 Waldersee, diary entry for 15 October 1889, GStA Berlin, Waldersee Papers; Meisner, II, p. 70.
120 Waldersee, retrospective note written in later years, GStA Berlin, Waldersee Papers; Meisner, II, p. 72.
121 Count Herbert Bismarck, Aufzeichnung, p. 89, BA Koblenz, Bismarck Papers FC 3018 N.
122 Count Herbert Bismarck, Aufzeichnung, pp. 89–90, ibid. See Tsar Alexander III to Kaiser Wilhelm II, 18 November 1889, Kaiser Wilhelm II to Tsar Alexander III, 5 December 1889, PA AA, R 3571.
123 Waldersee, diary entry for 14 December 1894, GStA Berlin, Waldersee Papers; cf. Meisner, II, pp. 332–3.
124 Count Herbert Bismarck, Aufzeichnung, pp. 90–1, BA Koblenz, Bismarck Papers FC 3018 N.

125 Waldersee, diary entries for 3 and 5 November 1889, GStA Berlin, Waldersee Papers; cf. Meisner, II, p. 75.
126 See Kaiser Wilhelm II to Queen Victoria, telegram, 17 October 1889, RA I57/65.
127 Kaiser Wilhelm II to Kaiser Franz Joseph, 20 October 1889, HHStA Vienna, Kabinettsarchiv, Geheimakten 2.
128 Waldersee, diary entry for 23 October 1889, GStA Berlin, Waldersee Papers; Meisner, II, p. 74.
129 Kaiser Franz Joseph to Kaiser Wilhelm II, 24 October 1889, HHStA Vienna, Kabinettsarchiv, Geheimakten 2.
130 Waldersee, diary entry for 5 November 1889, GStA Berlin, Waldersee Papers; omitted from Meisner, II, p. 75.
131 Waldersee, diary entry for 17 November 1889, GStA Berlin, Waldersee Papers; Meisner, II, pp. 77–8.
132 Herbert Bismarck to Holstein, 21 October 1889, Solms to Holstein, 26 October 1889, *Holstein Papers*, III, Nos. 295–6.
133 Waldersee, diary entry for 17 November 1889, GStA Berlin, Waldersee Papers; omitted from Meisner, II, p. 78. See above, pp. 125–6.
134 Waldersee, diary entry for 3 November 1889, GStA Berlin, Waldersee Papers; omitted from Meisner, II, p. 75.
135 Waldersee, diary entry for 8 December 1889, GStA Berlin, Waldersee Papers; Meisner, II, pp. 81ff.
136 Waldersee, diary entry for 27 October 1889, GStA Berlin, Waldersee Papers; largely omitted from Meisner, II, p. 75.
137 Waldersee, diary entry for 5 November 1889, GStA Berlin, Waldersee Papers; cf. Meisner, II, pp. 75–6.
138 Waldersee, diary entry for 31 October 1889, GStA Berlin, Waldersee Papers; omitted from Meisner, II, p. 75.
139 Waldersee, diary entry for 7 December 1889, GStA Berlin, Waldersee Papers; printed in Meisner, II, p. 81, under an incorrect date.
140 Waldersee, diary entry for 2 January 1890, GStA Berlin, Waldersee Papers; cf. Meisner, II, pp. 85–6.
141 Ibid.
142 Waldersee, diary entry for 23 December 1889, GStA Berlin, Waldersee Papers; omitted from Meisner, II, pp. 83ff.
143 Waldersee, diary entry for 8 December 1889, GStA Berlin, Waldersee Papers; Meisner, II, p. 81.
144 Holstein to Radolin, 5 December 1889, *Holstein Papers*, III, No. 301.
145 Herbert Bismarck to his father, 9 December 1889, BA Koblenz, Bismarck Papers FC 3005 N. See also Bußmann, *Staatssekretär*, Nos. 551–3.
146 Count Herbert Bismarck, 'Notizen V', BA Koblenz, Bismarck Papers FC 3018 N; also Prince Otto von Bismarck to Herbert Bismarck, 10 December 1889, telegram, PA AA, R 1009; Herbert Bismarck to his father, 9 and 10 December 1889, with the Chancellor's marginal notes, BA Koblenz, Bismarck Papers FC 3005 N.
147 Herbert Bismarck to his father, 10 December 1889, BA Koblenz, Bismarck Papers FC 3005 N.
148 Herbert Bismarck to his father, 9 December 1889, ibid.
149 Verdy to Waldersee, 10 December 1889, GStA Berlin, Waldersee Papers, No. 53; also Waldersee, diary entry for 10 December 1889, GStA Berlin, Waldersee Papers; omitted from Meisner, II, p. 82.
150 Herbert Bismarck to his father, 9 and 10 December 1889, with marginal comments by the Chancellor, BA Koblenz, Bismarck Papers FC 3005 N.

151 Waldersee, diary entry for 10 December 1889, GStA Berlin, Waldersee Papers; omitted from Meisner, II, p. 82.
152 Count Herbert Bismarck, 'Notizen V', BA Koblenz, Bismarck Papers FC 3018 N.
153 Waldersee, diary entry for 11 December 1889, GStA Berlin, Waldersee Papers; omitted from Meisner, II, p. 83.
154 Count Herbert Bismarck, 'Notizen V', BA Koblenz, Bismarck Papers FC 3018 N; diary, 13 December 1889, ibid.
155 Herbert Bismarck to his father, 13 December 1889, ibid., FC 3005 N; in Bußmann, *Staatssekretär*, No. 394.
156 Herbert Bismarck to his father, 31 December 1889, BA Koblenz, Bismarck Papers FC 3018 N; in Bußmann, *Staatssekretär*, No. 395.
157 Herbert Bismarck to his father, 7 January 1890, BA Koblenz, Bismarck Papers FC 3018 N.
158 Holstein to Eulenburg, 26 December 1889, *Eulenburgs Korrespondenz*, I, No. 270.
159 Eulenburg to Kaiserin Auguste Viktoria, 30 December 1889, ibid., No. 274.
160 Eulenburg to Holstein, 28 December 1889, ibid., No. 272.

10 THE KAISER, THE CHANCELLOR AND THE KARTELL

1 Bunsen to Morier, 14 September 1889, Balliol College Oxford, Morier Papers, Box 42.
2 Herbert Bismarck to his father, 4 August 1888, BA Koblenz, Bismarck Papers FC 3005 N; in Bußmann, *Staatssekretär*, No. 363. See Hermann Oncken, *Rudolf von Bennigsen. Ein deutscher liberaler Politiker. Nach seinen Briefen und hinterlassenen Papieren*, 2 vols., Stuttgart, Leipzig 1910, II, pp. 543–7.
3 Herbert Bismarck to Rantzau, 5 August 1888, BA Koblenz, Bismarck Papers FC 3014 N.
4 Herbert Bismarck to Rantzau, 12 August 1888, ibid.; Rantzau to Herbert Bismarck, 29 August 1888, ibid., FC 3028 N; Werthern to Eulenburg, 9 September 1888, *Eulenburgs Korrespondenz*, I, No. 195; Lucius von Ballhausen, *Bismarck-Erinnerungen*, pp. 583–4.
5 See above, p. 12.
6 Douglas, Speech of 4 October 1888, *Norddeutsche Allgemeine Zeitung*, No. 474, 7 October 1888.
7 See Eissenstein to Kálnoky, 13 October 1888, HHStA Vienna, PA III 134.
8 Douglas, speech of 4 October 1888, *Norddeutsche Allgemeine Zeitung*, No. 474, 7 October 1888.
9 Herbert Bismarck to Wilhelm Bismarck, 19 October 1888, in Bußmann, *Staatssekretär*, No. 369.
10 Stoecker to Waldersee, 22 March 1889, GStA Berlin, Waldersee Papers. See above, p. 204.
11 Eissenstein to Kálnoky, 13 October 1888, HHStA Vienna, PA III 134.
12 Kaiserin Auguste Viktoria to Kaiser Wilhelm II, 11 October 1888, GStA Berlin, BPHA Rep. 53T Preußen: An Kaiser Wilhelm II., Bd III.
13 Waldersee, diary entry for 7 September 1888, GStA Berlin, Waldersee Papers; cf. Meisner, II, p. 3.
14 Waldersee, diary entry for 7 November 1888, GStA Berlin, Waldersee Papers; Meisner, II, p. 16.
15 Waldersee, diary entry for 23 October 1888, GStA Berlin, Waldersee Papers; cf. Meisner, II, p. 11.
16 Waldersee, diary entries for 11, 14, 17, and 23 October 1888, GStA Berlin, Waldersee Papers; cf. Meisner, II, pp. 5–11.
17 Freiherr von Hammerstein, 'Das monarchische Gefühl', *Kreuzzeitung*, late January 1889.
18 Széchényi to Kálnoky, 30 January 1889, HHStA Vienna, PA III 136.
19 Waldersee, diary entry for 19 March 1889, GStA Berlin, Waldersee Papers; omitted from Meisner, II, p. 46.

20 Waldersee, diary entry for 9 October 1889, GStA Berlin, Waldersee Papers; omitted from Meisner, II, p. 69.
21 Waldersee, diary entry for 9 July 1889, GStA Berlin, Waldersee Papers; Meisner, II, p. 60.
22 Herbert Bismarck to Rantzau, 27 June 1889, BA Koblenz, Bismarck Papers FC 3014; in Bußmann, *Staatssekretär*, No. 380.
23 Prince Otto von Bismarck to Lutz, 6 August 1889, Bismarck, *Die gesammelten Werke*, 15 vols., Berlin 1923–35, VIc, p. 416.
24 Waldersee, diary entry for 9 July 1889, GStA Berlin, Waldersee Papers; cf. Meisner, II, p. 60.
25 Holstein to Eulenburg, 28 September 1889, *Eulenburgs Korrespondenz*, I, No. 235.
26 Holstein to Eulenburg, 28 and 30 September 1889, ibid., Nos. 235–6.
27 Holstein to Eulenburg, 30 September 1889, ibid., No. 236.
28 See Otto Gradenwitz, *Bismarcks letzter Kampf, 1888–1898*, Berlin 1924, p. 76; Lucius von Ballhausen, *Bismarck-Erinnerungen*, pp. 503–4; Waldersee, diary entry for 6 October 1889, GStA Berlin, Waldersee Papers; cf. Meisner, II, p. 69.
29 Herbert Bismarck to Prince Otto von Bismarck, 5 October 1889, BA Koblenz, Bismarck Papers FC 3005 N; in Bußmann, *Staatssekretär*, No. 390.
30 Holstein to Eulenburg, 8 October 1889, *Eulenburgs Korrespondenz*, I, No. 237.
31 Waldersee, diary entry for 27 October and 1 and 25 November 1889, GStA Berlin, Waldersee Papers; cf. Meisner, II, pp. 75, 80.
32 Waldersee, diary entry for 4 November 1889, GStA Berlin, Waldersee Papers; omitted from Meisner, II, p. 75.
33 Waldersee, diary entry for 21 October 1889, GStA Berlin, Waldersee Papers; Meisner, II, pp. 73–4.
34 See Meisner, II, p. 73, n. 2.
35 See for example Holstein to Eulenburg, 6 and 12 November 1889, *Eulenburgs Korrespondenz*, I, Nos. 246 and 250.
36 Eulenburg to Kaiser Wilhelm II, 25 October 1889, ibid, No. 241.
37 Eulenburg to Kaiser Wilhelm II, 29 October 1889, ibid., No. 242.
38 Eulenburg to Grand Duke Friedrich I of Baden, 30 October 1889, ibid., No. 243.
39 Holstein to Eulenburg, 4 November 1889, ibid., No. 244.
40 Holstein to Eulenburg, 5 November 1889, ibid., No. 245.
41 Eulenburg to Holstein, 10 November 1889, ibid., No. 248.
42 Kaiser Wilhelm II to Prince Otto von Bismarck, telegram, 6 November 1889, cited in Holstein to Eulenburg, 9 November 1889, ibid., No. 247.
43 Lucanus to Boetticher, telegram, 6 November 1889, cited in Holstein to Eulenburg, 6 November 1889, ibid., No. 246.
44 Holstein to Eulenburg, 9 November 1889, ibid., No. 247. See also Eulenburg to Kaiser Wilhelm II, 10 November 1889, ibid., No. 249.
45 Eulenburg to Holstein, 19 November 1889, ibid., No. 254.
46 Holstein to Radolin, 5 December 1889, *Holstein Papers*, III, No. 301.
47 Holstein to Eulenburg, 12 December 1889, *Eulenburgs Korrespondenz*, I, No. 265. The Frankfurt speech is printed in Penzler, *Reden Kaiser Wilhelms II. in den Jahren 1888–1895*, pp. 82–4.
48 Waldersee, diary entry for 17 November 1889, GStA Berlin, Waldersee Papers; cf. Meisner, II, p. 78.
49 Prince Otto von Bismarck to Kaiser Wilhelm II, 9 November 1889, BA Berlin, Reichskanzlei, No. 863, cited in *Eulenburgs Korrespondenz*, I, No. 247, n. 1.
50 Ibid.; also Brauer to Lerchenfeld, 27 October 1889, BA Berlin, Reichskanzlei, No. 863, cited in *Eulenburgs Korrespondenz*, I, No. 243, n. 3.

51 Holstein to Eulenburg, 12 November 1889, *Eulenburgs Korrespondenz*, I, No. 250. See also Holstein to Ida von Stülpnagel, 13 November 1889, Helmuth Rogge, *Friedrich von Holstein. Lebensbekenntnis in Briefen an eine Frau*, Berlin 1932, p. 152.

52 Pflanze, *Bismarck and the Development of Germany*, III, pp. 368–9. See also below, pp. 436–50.

53 Waldersee, diary entry for 13 November 1889, GStA Berlin, Waldersee Papers; cf. Meisner, II, pp. 76–7.

54 Count Herbert Bismarck, 'Notizen II', April 1890, pp. 48–9, BA Koblenz, Bismarck Papers FC 3018 N.

55 Eulenburg to Herbert Bismarck, 25 January 1889, *Eulenburgs Korrespondenz*, I, No. 210.

56 Holstein to Eulenburg, 4 November 1889, ibid., No. 244.

57 Eulenburg to Holstein, 10 November 1889, ibid., No. 248.

58 Eulenburg to Holstein, 19 November 1889, Holstein to Eulenburg, 12 December 1889, ibid., Nos. 254 and 265.

59 Hinzpeter to Kaiser Wilhelm II, 4 January 1890, BA Berlin, Reichskanzlei, No. 1816, cited in *Eulenburgs Korrespondenz*, I, p. 403. See also Hinzpeter to Bismarck, 7 January 1890, BA Berlin, Reichskanzlei, No. 1816.

60 Hinzpeter to Eulenburg, 8 January 1890, *Eulenburgs Korrespondenz*, I, No. 277.

61 Kaiser Wilhelm II, marginal notes on Hinzpeter's letter of 4 January 1890, BA Berlin, Reichskanzlei, No. 1816, cited in *Eulenburgs Korrespondenz*, I, p. 403.

62 Herbert Bismarck to his father, 6 January 1890, BA Koblenz, Bismarck Papers FC 3005 N; in Bußmann, *Staatssekretär*, No. 396.

63 Bismarck to Hinzpeter, 5 January 1890, Bismarck to Kaiser Wilhelm II, 7 January 1890, BA Berlin, Reichskanzlei, No. 1816, cited in *Eulenburgs Korrespondenz*, I, p. 403.

64 Herbert Bismarck to his father, 8 January 1890, BA Berlin, Reichskanzlei, No. 1816, cited in *Eulenburgs Korrespondenz*, I, p. 403.

65 Herbert Bismarck to his father, 9 January 1890, BA Berlin, Reichskanzlei, No. 1816, cited in *Eulenburgs Korrespondenz*, I, p. 403.

66 Schwartzkoppen to Herbert Bismarck, 9 January 1890, BA Berlin, Reichskanzlei, No. 1816, cited in *Eulenburgs Korrespondenz*, I, p. 403.

67 Marschall, diary entry for 10 January 1890, cited in *Eulenburgs Korrespondenz*, I, p. 403.

68 Hinzpeter to Kaiser Wilhelm II, 15 January 1890, BA Berlin, Reichskanzlei, No. 1860.

69 Kaiser Wilhelm II to Bismarck, telegram, 17 January 1890, cited in *Eulenburgs Korrespondenz*, I, p. 420.

70 Eulenburg to Hinzpeter, 9 January 1890, ibid., No. 278.

71 Lerchenfeld, report of 2 April 1890, in Karl Alexander von Müller, 'Die Entlassung. Nach den bayerischen Gesandtschaftsberichten', *Süddeutsche Monatshefte*, 19/1 (December 1921), p. 140. See below, p. 321.

72 Waldersee, diary entries for 11 and 12 January 1890, GStA Berlin, Waldersee Papers; cf. Meisner, II, pp. 88–9.

73 Waldersee, diary entries for 20 and 21 January 1890, GStA Berlin, Waldersee Papers; omitted from Meisner, II, p. 95.

74 See Holstein to Eulenburg, 26 December 1889, Eulenburg to Holstein, 28 December 1889, *Eulenburgs Korrespondenz*, I, Nos. 270 and 272.

75 Waldersee, diary entry for 11 January 1890, GStA Berlin, Waldersee Papers; cf. Meisner, II, pp. 88–9.

76 Waldersee, diary entry for 15 February 1890, GStA Berlin, Waldersee Papers; omitted from Meisner, II, p. 102.

77 Waldersee, diary entry for 11 January 1890, GStA Berlin, Waldersee Papers; cf. Meisner, II, pp. 88–9.

78 Waldersee, diary entry for 20 and 21 January 1890, GStA Berlin, Waldersee Papers; omitted from Meisner, II, p. 95.
79 Waldersee, diary entry for 3 January 1890, GStA Berlin, Waldersee Papers; cf. Meisner, II, p. 86.
80 Waldersee, diary entry for 8 January 1890, GStA Berlin, Waldersee Papers; cf. Meisner, II, p. 87.
81 Waldersee, diary entry for 12 January 1890, GStA Berlin, Waldersee Papers; cf. Meisner, II, p. 89.
82 *Norddeutsche Allgemeine Zeitung*, 12 January 1890, cited in Meisner, II, p. 89, n. 3. See alsoWaldersee, diary entry for 13 January 1890, GStA Berlin, Waldersee Papers; Meisner, II, p. 90.
83 Waldersee, diary entry for 12 January 1890, GStA Berlin, Waldersee Papers; cf. Meisner, II, p. 89.
84 Waldersee, diary entry for 18 January 1890, GStA Berlin, Waldersee Papers; Meisner, II, p. 94.
85 Herbert Bismarck, notes of April 1890, pp. 48–50, with the Reich Chancellor's marginal comment 'cherchez la femme', BA Koblenz, Bismarck Papers.
86 Waldersee, diary entry for 12 January 1890, GStA Berlin, Waldersee Papers; cf. Meisner, II, p. 90.
87 Eulenburg to Waldersee, 17 January 1890, in Meisner, II, p. 92.
88 Eulenburg to Holstein, 17 January 1890, *Eulenburgs Korrespondenz*, I, No. 284. See Röhl, *Germany without Bismarck. The Crisis of Government in the Second Reich, 1890–1900*, London, Berkeley and Los Angeles 1967, p. 35.
89 Eulenburg to Holstein, 29 January 1890, Eulenburg to Hinzpeter, 3 February 1890, *Eulenburgs Korrespondenz*, I, Nos. 294 and 304.
90 The correspondence between Herbert Bismarck and his father reveals the close co-operation of both with Liebenau at the court. See Bußmann, *Staatssekretär*, Nos. 366, 368 and 390.
91 Waldersee, diary entry for 29 August and 22 October 1888, GStA Berlin, Waldersee Papers; cf. Meisner, II, pp. 2 and 10.
92 Waldersee, diary entry for 5 November 1888, GStA Berlin, Waldersee Papers; cf. Meisner, II, p. 15.
93 Waldersee, diary entry for 22 October 1888, GStA Berlin, Waldersee Papers; cf. Meisner, II, p. 10.
94 August Eulenburg to Philipp Eulenburg, 1 February 1890, *Eulenburgs Korrespondenz*, I, No. 298.
95 Waldersee, diary entry for 17 November 1889, GStA Berlin, Waldersee Papers; omitted from Meisner, II, p. 77.
96 Ibid. Cf. Meisner II, p. 77.
97 Waldersee, diary entry for 25 November 1889, GStA Berlin, Waldersee Papers; cf. Meisner, II, p. 80.
98 Waldersee, diary entry for 27 November 1889, GStA Berlin, Waldersee Papers; largely omitted from Meisner, II, p. 80.
99 Eulenburg to Kaiser Wilhelm II, 3 December 1889, *Eulenburgs Korrespondenz*, I, No. 262; Waldersee, diary entries for 2 and 7 December 1889, GStA Berlin, Waldersee Papers; omitted from Meisner, II, p. 81; Waldersee to Eulenburg, 8 December 1889, *Eulenburgs Korrespondenz*, I, No. 263; August Eulenburg to Philipp Eulenburg, 12 December 1889, ibid., No. 264.
100 Waldersee, diary entries for 11 and 23 December 1889, GStA Berlin, Waldersee Papers; omitted from Meisner, II, pp. 82–3. See Waldersee to Eulenburg, 8 December 1889, *Eulenburgs Korrespondenz*, I, No. 263. See also August Eulenburg to Philipp Eulenburg, 12 December 1889, ibid., No. 264.

101 August Eulenburg to Philipp Eulenburg, 23 December 1889, cited in *Eulenburgs Korrespondenz*, I, p. 393.
102 Eulenburg to Grand Duke of Baden, 25 December 1889, ibid., No. 269.
103 Waldersee, diary entry for 9 January 1890, GStA Berlin, Waldersee Papers; cf. Meisner, II, p. 87.
104 August Eulenburg to Philipp Eulenburg, 21 January 1890, *Eulenburgs Korrespondenz*, I, No. 289.
105 August Eulenburg to Philipp Eulenburg, 15 February 1890, ibid., No. 323.
106 See Stolberg-Wernigerode to Reuss, 5 June 1890, cited ibid., p. 543.
107 On Liebenau's dismissal, see below, pp. 331–3.
108 Herbert Bismarck to his father, 9 and 10 December 1889, BA Koblenz, Bismarck Papers, FC 3005 N.
109 Herbert Bismarck to his father, 24 September 1889, ibid.
110 Waldersee, diary entry for 11 December 1889, GStA Berlin, Waldersee Papers; cf. Meisner, II, p. 82.
111 Waldersee, diary entry for 31 January 1890, GStA Berlin, Waldersee Papers; omitted from Meisner, II, p. 99.
112 Waldersee, diary entry for 18 January 1890, GStA Berlin, Waldersee Papers; Meisner, II, p. 94.
113 Waldersee to Eulenburg, 13 January 1890, printed with omissions in Meisner, II, pp. 90–2. See *Eulenburgs Korrespondenz*, I, p. 453; also Waldersee to Eulenburg, 12 February 1890, ibid., No. 321.
114 Eulenburg to Waldersee, 17 January 1890, printed with omissions in Meisner, II, p. 92.
115 Holstein to Eulenburg, 12 December 1889, *Eulenburgs Korrespondenz*, I, No. 265.
116 Eulenburg to Kaiser Wilhelm II, 1 January 1890, ibid., No. 276.
117 Marschall, diary entries for 5 and 13 January 1890, cited ibid., p. 402; Eulenburg to Kaiser Wilhelm II, 15 January 1890, ibid., No. 280; but cf. Lindau to Eulenburg, 17 January 1890, ibid., No. 285.
118 Waldersee, diary entry for 14 January 1890, GStA Berlin, Waldersee Papers; largely omitted from Meisner, II, p. 93.
119 Waldersee, diary entry for 18 January 1890, GStA Berlin, Waldersee Papers; cf. Meisner, II, p. 94.
120 Waldersee, diary entry for 30 January 1890, GStA Berlin, Waldersee Papers; Meisner, II, p. 98.
121 Waldersee, diary entries for 26 and 28 December 1889, GStA Berlin, Waldersee Papers; cf. Meisner, II, pp. 83–4.

11 CONSTITUTIONAL CONFLICTS

1 Eulenburg, note of January 1914, cited in *Eulenburgs Korrespondenz*, I, p. 406.
2 Waldersee, diary entries for 27 November and 7 December 1889, GStA Berlin, Waldersee Papers; omitted from Meisner, II, pp. 80–1.
3 Waldersee, diary entry for 24 November 1889, GStA Berlin, Waldersee Papers; Meisner, II, pp. 79–80.
4 Waldersee, diary entry for 25 November 1889, GStA Berlin, Waldersee Papers; omitted from Meisner, II, p. 80.
5 For details of Bismarck's *coup d'état* plans in the dismissal crisis, see Hans Delbrück, 'Staatsstreichpläne als Ursachen von Bismarcks Rücktritt. Sensationelle Enthüllungen', *Neues Wiener Journal*, 11 December 1913; Egmont Zechlin, *Staatsstreichpläne Bismarcks und Wilhelms II., 1890–1894*, Stuttgart, Berlin 1929; Werner Pöls, *Sozialistenfrage und Revolutionsfurcht in ihrem Zusammenhang mit den angeblichen Staatsstreichplänen*

Bismarcks, Lübeck, Hamburg 1960; John C. G. Röhl, 'Staatsstreichplan oder Staatsstreichbereitschaft? Bismarcks Politik in der Entlassungskrise', *Historische Zeitschrift*, 203 (1966), pp. 610–24; also Clark, *Kaiser Wilhelm II*, pp. 39ff.

6 Waldersee, diary entry for 11 December 1889, GStA Berlin, Waldersee Papers; cf. Meisner, II, p. 83.

7 Waldersee, diary entry for 26 December 1889, GStA Berlin, Waldersee Papers; largely omitted from Meisner, II, p. 85.

8 Waldersee, diary entry for 2 January 1890, GStA Berlin, Waldersee Papers; cf. Meisner, II, p. 86.

9 Waldersee, diary entry for 17 November 1889, GStA Berlin, Waldersee Papers; Meisner, II, p. 78.

10 Waldersee, diary entry for 27 November 1889, GStA Berlin, Waldersee Papers; omitted from Meisner, II, p. 80.

11 See Waldersee, diary entries for 4 and 8 December 1889, GStA Berlin, Waldersee Papers; omitted from Meisner, II, p. 81.

12 Eulenburg, note of January 1914, cited in *Eulenburgs Korrespondenz*, I, pp. 406–7. See Eulenburg to Kaiser Wilhelm II, 15 January 1890, ibid., No. 280; Eulenburg to Holstein, 17 January 1890, ibid., No. 284.

13 Kaiser Wilhelm II, marginal notes on *Münchener Neuesten Nachrichten*, 5 March 1891, PA AA, Asservat No. 4.

14 For the historiography of Wilhelm II's social welfare initiative, see Lamar Cecil, *Wilhelm II, vol. I, Prince and Emperor, 1859–1900*, Chapel Hill, London 1989, p. 133, and Clark, *Kaiser Wilhelm II*, pp. 38–9, who portrays Wilhelm's motives in a rather idealistic light.

15 Holstein to Eulenburg, 1 and 4 February 1890, *Eulenburgs Korrespondenz*, I, Nos. 299 and 306.

16 Empress Frederick to Queen Victoria, 15 February 1890, RA Z47/32.

17 Empress Frederick to Queen Victoria, 19 February 1890, RA Z47/34.

18 Salisbury to Queen Victoria, 24 February 1890, RA L16/36.

19 Waldersee, diary entries for 20 and 21 January 1890, GStA Berlin, Waldersee Papers; omitted from Meisner, II, p. 95. See Holstein to Eulenburg, 26 December 1889, *Eulenburgs Korrespondenz*, I, No. 270.

20 Waldersee, diary entries for 3 and 8 February 1890, GStA Berlin, Waldersee Papers; largely omitted from Meisner, II, p. 100.

21 Waldersee, diary entry for 22 December 1894, GStA Berlin, Waldersee Papers; omitted from Meisner, II, p. 334.

22 Marschall, diary entry for 29 January 1890, cited in *Eulenburgs Korrespondenz*, I, p. 428. See Holstein to Eulenburg, 4 February 1890, ibid., No. 306; Lucius von Ballhausen, *Bismarck-Erinnerungen*, p. 514.

23 Hinzpeter to Eulenburg, 23 January 1890, *Eulenburgs Korrespondenz*, I, No. 291.

24 Eulenburg to Holstein, 3 February 1890, ibid., No. 303.

25 Eulenburg to Hinzpeter, 3 February 1890, ibid., No. 304.

26 Hinzpeter to Eulenburg, 8 January 1890, ibid., No. 277.

27 Hinzpeter to Eulenburg, 2 February 1890, ibid., No. 302.

28 Hinzpeter to Eulenburg, 23 January and 6 February 1890, ibid., Nos. 291 and 307.

29 Hinzpeter to Eulenburg, 6 February 1890, ibid., No. 307.

30 Marschall, diary entry for 12 January 1890, cited ibid., p. 411; Marschall's report of 15 January 1890, in Gradenwitz, *Bismarcks letzter Kampf*, pp. 120–1.

31 Holstein to Eulenburg, 15 January 1890, *Eulenburgs Korrespondenz*, I, No. 283.

32 In Hugo Graf Lerchenfeld-Koefering, *Erinnerungen and Denkwürdigkeiten, 1843–1925*, Berlin 1934, pp. 355ff.

33 Eulenburg to Holstein, 17 January 1890, *Eulenburgs Korrespondenz*, I, No. 284. See above, pp. 248–9.
34 Eulenburg to Freyschlag, 14 January 1890, *Eulenburgs Korrespondenz*, I, No. 279; Eulenburg to Lutz, 18 January 1890, ibid., No. 286.
35 Eulenburg to Kaiser Wilhelm II, 20 January 1890, ibid., No. 288.
36 Holstein to Eulenburg, 15 January 1890, ibid., No. 283.
37 Marschall, diary entries for 19–23 January 1890, cited ibid., p. 412.
38 Waldersee, diary entry for 22 January 1890, GStA Berlin, Waldersee Papers; cf. Meisner, II, pp. 95–6.
39 Herbert Bismarck to his father, 23 January 1890, BA Koblenz, Bismarck Papers FC 3005 N.
40 Boetticher to Prince Otto von Bismarck, cipher telegram, 23 January 1890, ibid. See Georg Freiherr von Eppstein, ed., *Fürst Bismarcks Entlassung. Nach den hinterlassenen, bisher unveröffentlichten Aufzeichnungen des Staatssekretärs des Innern, Staatsministers Dr. Karl Heinrich von Boetticher und des Chefs der Reichskanzlei unter dem Fürsten Bismarck Dr. Franz Johannes von Rottenburg*, Berlin 1920, pp. 143–5.
41 Herbert Bismarck to his father, cipher telegram, 23 January 1890, BA Koblenz, Bismarck Papers FC 3005 N.
42 Herbert Bismarck to his father, letter and telegrams, 23 January 1890, ibid.
43 Holstein to Eulenburg, 27 January 1890, *Eulenburgs Korrespondenz*, I, No. 293; Holstein to Herbert Bismarck, 24 January 1890, *Holstein Papers*, III, No. 302.
44 Eulenburg to Holstein, 25 January 1890, *Eulenburgs Korrespondenz*, I, No. 292.
45 Waldersee, diary entry for 25 January 1890, GStA Berlin, Waldersee Papers; Meisner, II, p. 96.
46 Herbert Bismarck to his father, 19 and 20 January 1890, BA Koblenz, Bismarck Papers FC 3005 N.
47 Minutes of the Crown Council meeting of 24 January 1890, in Eppstein, *Bismarcks Entlassung*, pp. 157–65.
48 Empress Frederick to Queen Victoria, 15 February 1890, RA Z47/32; also Ponsonby to Queen Victoria, 15 February 1890, RA I58/18, and Salisbury to Queen Victoria, 15 February 1890, RA L16/33.
49 Malet to Queen Victoria, 22 March 1890, RA I58/33–4.
50 Minutes of the Crown Council meeting of 24 January 1890, in Eppstein, *Bismarcks Entlassung*, p. 157; Bismarck, *Die gesammelten Werke*, XV, pp. 491ff. See Ernst Engelberg, *Bismarck. Das Reich in der Mitte Europas*, Berlin 1990, p. 561; Rall, *Wilhelm II*, pp. 90–1.; Clark, *Kaiser Wilhelm II*, pp. 41–2.
51 On Kayser's relationship to the German-Jewish community on the one hand, and the Bismarcks on the other, see Spitzemberg, diary entry for 14 February 1898, *Tagebuch*, p. 365.
52 Kayser to Eulenburg, 15 January 1890, *Eulenburgs Korrespondenz*, I, No. 281.
53 Franz Fischer, note of 15 January 1890, ibid., No. 282; Holstein to Eulenburg, 15 January 1890, ibid., No. 283.
54 Kayser to Eulenburg, 15 January 1890, ibid., No. 281. The original memorandum was located in the files of the Hausarchiv, now in GStA Berlin, BPHA Rep. 53 EIII No. 3. In Merseburg it was subsequently incorporated into the files of the Civil Cabinet as 2.2.1. No. 29960/1.
55 The text of Kayser's exposé is printed in Eppstein, *Bismarcks Entlassung*, pp. 146–51. See Bismarck, *Die gesammelten Werke*, XV, pp. 491ff. See also Eulenburg to Holstein, 9 March 1890, *Eulenburgs Korrespondenz*, I, No. 341.
56 Holstein to Eulenburg, 27 January 1890, *Eulenburgs Korrespondenz*, I, No. 293; Boetticher, Zur Geschichte der Entlassung des Fürsten Bismarck am 20. March 1890, in Eppstein,

Bismarcks Entlassung, pp. 33–78. In the official transcipt there is no mention of Bismarck's threat to resign: ibid., pp. 162–5.

57 Eppstein, *Bismarcks Entlassung*, p. 47.

58 Kaiser Wilhelms II, speech of 25 January 1890, Penzler, *Reden Kaiser Wilhelms II. in den Jahren 1888–1895*, pp. 87–9. The speech had been written by Boetticher and Bosse. See Eppstein, *Bismarcks Entlassung*, pp. 47–8; Waldersee, diary entry for 25 January 1890, GStA Berlin, Waldersee Papers; cf. Meisner, II, p. 96.

59 Waldersee, diary entry for 25 January 1890, GStA Berlin, Waldersee Papers; Meisner, II, p. 97; Malet, report of 8 February 1890, RA I58/15.

60 Marschall, diary entry for 25 January 1890, cited in *Eulenburgs Korrespondenz*, I, p. 423.

61 Waldersee, diary entry for 25 January 1890, GStA Berlin, Waldersee Papers; cf. Meisner, II, p. 96.

62 Waldersee, diary entry for 25 January 1890, GStA Berlin, Waldersee Papers; cf. Meisner, II, p. 97.

63 Holstein to Eulenburg, 1 February 1890, *Eulenburgs Korrespondenz*, I, No. 299.

64 Holstein to Eulenburg, 27 January 1890, ibid., No. 293; Eulenburg to Holstein, 29 January 1890, ibid., No. 294; Eulenburg to Kaiser Wilhelm II, 30 January 1890, ibid., No. 296.

65 Holstein to Eulenburg, 1 and 4 February 1890, ibid., Nos. 299 and 306.

66 Holstein to Eulenburg, 4 February 1890, ibid., No. 306.

67 Marschall, diary entry for 30 January 1890, cited ibid., p. 426.

68 Holstein to Eulenburg, 1 and 4 February 1890, ibid., Nos. 299 and 306.

69 Hinzpeter to Eulenburg, 6 February 1890, ibid., No. 307.

70 Eulenburg to Holstein, 3 February 1890, ibid., No. 303; Eulenburg to Hinzpeter, 3 February 1890, ibid., No. 304.

71 Kayser to Eulenburg, 1 February 1890, ibid., No. 300.

72 Waldersee, diary entry for 12 January 1890, GStA Berlin, Waldersee Papers; cf. Meisner, II, p. 90.

73 Waldersee to Eulenburg, 13 January 1890, printed with omissions in Meisner, II, pp. 90–2.

74 Waldersee, diary entry for 25 January 1890, GStA Berlin, Waldersee Papers; cf. Meisner, II, p. 97.

75 On 27 January 1890 Marschall noted in his diary: 'The Kaiser wants him [Bismarck] to confine himself to the Foreign Office and give up domestic affairs:' cited in *Eulenburgs Korrespondenz*, I, p. 427. See below, pp. 273–5.

76 Eulenburg to Holstein, 3 February 1890; Eulenburg to Hinzpeter, 3 February 1890, *Eulenburgs Korrespondenz*, I, Nos. 303 and 304.

77 Minutes of the Crown Council meeting of 26 January 1890, in Eppstein, *Bismarcks Entlassung*, pp. 166–7.

78 Holstein to Eulenburg, 31 January 1890, August Eulenburg to Philipp Eulenburg, 1 February 1890, Franz Fischer to Eulenburg, 2 February 1890, *Eulenburgs Korrespondenz*, I, Nos. 297, 298 and 301; cf. *Holstein Papers*, III, pp. 324–5; Waldersee, diary entry for 27 January 1890, GStA Berlin, Waldersee Papers; partly omitted from Meisner, II, p. 98.

79 Holstein to Eulenburg, 27 January and 12 February 1890, *Eulenburgs Korrespondenz*, I, Nos. 293 and 320; Rudolf Lindau to Eulenburg, 6 February 1890, ibid., No. 308.

80 Eulenburg to Holstein, 29 January 1890, ibid., No. 294.

81 Boetticher, Geschichte der Entlassung, in Eppstein, *Bismarcks Entlassung*, pp. 51–2.

82 Kayser to Eulenburg, 1 February 1890, *Eulenburgs Korrespondenz*, I, No. 300.

83 Fischer to Eulenburg, 2 February 1890, ibid., No. 301.

84 Kayser to Eulenburg, 6 and 7 February 1890, ibid., Nos. 310 and 314. Cf. Holstein to Eulenburg, 6 and 7 February 1890, ibid., Nos. 309 and 313.

85 Eulenburg to Kayser, 7 February 1890, ibid., I, No. 312.

86 Eulenburg to Kaiser Wilhelm II, 7 February 1890, Eulenburg to Kayser, 7 February 1890, Eulenburg to Holstein, 9 February 1890, ibid., Nos. 311, 312 and 316.
87 Minutes of the Ministry of State, 7 and 9 February 1890, quoted ibid., p. 443. A list of the members of the Prussian Council of State can be found in Eppstein, *Bismarcks Entlassung*, pp. 211–17.

12 BISMARCK'S FALL FROM POWER

1 Boetticher, Geschichte der Entlassung, in Eppstein, *Bismarcks Entlassung*, p. 49.
2 Ibid.; Holstein to Eulenburg, 4 February 1890, *Eulenburgs Korrespondenz*, I, No. 306.
3 August Eulenburg to Philipp Eulenburg, 1 February 1890, ibid., No. 298.
4 See Röhl, *Germany without Bismarck*, pp. 37–8.
5 Marschall, diary entry for 27 January 1890, cited in *Eulenburgs Korrespondenz*, I, p. 427.
6 Ernst Gagliardi, *Bismarcks Entlassung*, 2 vols., Tübingen 1927–41, pp. 69–70.
7 Lerchenfeld, report of 30 January 1890, Lerchenfeld, *Erinnerungen*, pp. 357–8.
8 Lucius von Ballhausen, *Bismarck-Erinnerungen*, p. 516; Bismarck, *Die gesammelten Werke*, XV, pp. 505–6; Holstein to Eulenburg, 9 and 10 February 1890, *Eulenburgs Korrespondenz*, I, Nos. 317–18.
9 Waldersee, diary entry for 9 February 1890, GStA Berlin, Waldersee Papers; largely omitted from Meisner, II, pp. 101–2.
10 Lucius von Ballhausen, *Bismarck-Erinnerungen*, p. 515; Boetticher, Geschichte der Entlassung, in Eppstein, *Bismarcks Entlassung*, pp. 60ff.
11 Waldersee, diary entry for 9 February 1890, GStA Berlin, Waldersee Papers; largely omitted from Meisner, II, pp. 101–2.
12 Lerchenfeld, report of 8 February 1890, in Müller, 'Die Entlassung', pp. 146ff.
13 Boetticher, Geschichte der Entlassung, in Eppstein, *Bismarcks Entlassung*, pp. 54 and 62.
14 Malet to Salisbury, 12 February 1890, RA I58/16.
15 Lerchenfeld, report of 10 February 1890, Lerchenfeld, *Erinnerungen*, pp. 359ff.; Boetticher, Geschichte der Entlassung, in Eppstein, *Bismarcks Entlassung*, pp. 62–3.
16 Marschall, report of 11 February 1890 in Gradenwitz, *Bismarcks letzter Kampf*, pp. 127ff.; Holstein to Eulenburg, 12 February 1890, *Eulenburgs Korrespondenz*, I, No. 320.
17 August Eulenburg to Philipp Eulenburg, 9 March 1890, *Eulenburgs Korrespondenz*, I, No. 342.
18 Boetticher, Geschichte der Entlassung, in Eppstein, *Bismarcks Entlassung*, p. 50.
19 Holstein to Eulenburg, 4 February 1890, *Eulenburgs Korrespondenz*, I, No. 306; Herbert Bismarck to Eulenburg, 3 February 1890, Eulenburg, *Aus 50 Jahren*, p. 291.
20 Malet, report of 8 February 1890, RA I58/15.
21 Malet to Salisbury, 12 February 1890, RA I58/16. The Kaiser and his advisers were incensed by Bismarck's habit of discussing his differences with Wilhelm with foreign ambassadors. See Holstein to Reuss, 5 March 1890, *Berliner Monatshefte*, No. 15, 1937, pp. 327–8.
22 Herbert Bismarck to his brother, 16 February 1890, in Bußmann, *Staatssekretär*, No. 402.
23 Empress Frederick to Queen Victoria, 19–21 February 1890, RA Z47/34.
24 See Miquel's assumption in Waldersee, diary entry for 8 February 1890, GStA Berlin, Waldersee Papers; Meisner, II, p. 100.
25 See Pflanze, *Bismarck and the Development of Germany*, III, pp. 368–9; cf. Gall, *Bismarck*, pp. 689ff.
26 See above, pp. 233–4.
27 Holstein to Eulenburg, 27 January 1890, *Eulenburgs Korrespondenz*, I, No. 293.
28 Holstein to Eulenburg, 7 February 1890, ibid., No. 313; Eulenburg to Kaiser Wilhelm II, 8 February 1890, ibid., No. 315.

29 Holstein to Eulenburg, 10 and 26 February 1890, ibid., Nos. 318 and 325.
30 Marschall, diary entry for 12 February 1890, ibid., p. 450; Waldersee, diary entries for 3 and 5 February 1890, GStA Berlin, Waldersee Papers; cf. Meisner, II, pp. 99–100.
31 Marschall, report of 12 February 1890, in Gradenwitz, *Bismarcks letzter Kampf*, pp. 131–2; Salisbury to Queen Victoria, 15 February 1890, RA L16/33.
32 Holstein to Eulenburg, 7 and 10 February 1890, *Eulenburgs Korrespondenz*, I, Nos. 313 and 318. See Kayser to Eulenburg, 7 February 1890, Eulenburg to Kaiser Wilhelm II, 8 February 1890, ibid., Nos. 314–15. After a meeting with Holstein and Paul Kayser on 7 February, Marschall noted: 'I am convinced the Reich Chancellor is aiming for a fiasco' (cited ibid., p. 444).
33 Waldersee, diary entry for 30 January 1890, GStA Berlin, Waldersee Papers; partly omitted from Meisner, II, p. 99.
34 Waldersee, diary entry for 9 February 1890, GStA Berlin, Waldersee Papers; cf. Meisner, II, p. 101.
35 Waldersee, diary entries for 3, 8, 9 and 10 February 1890, GStA Berlin, Waldersee Papers; largely omitted from Meisner, II, p. 100.
36 Waldersee, diary entries for 3, 8, 9 and 19 February 1890, GStA Berlin, Waldersee Papers; largely omitted from Meisner, II, pp. 100, and 104.
37 Waldersee, diary entry for 8 February 1890, GStA Berlin, Waldersee Papers; omitted from Meisner, II, p. 100.
38 Waldersee, diary entry for 10 February 1890, GStA Berlin, Waldersee Papers; several important passages omitted from Meisner, II, p. 101.
39 Waldersee to Eulenburg, 12 February 1890, *Eulenburgs Korrespondenz*, I, No. 321. See Waldersee, diary entry for 15 February 1890, GStA Berlin, Waldersee Papers; omitted from Meisner, II, p. 102.
40 Waldersee, diary entry for 15 February 1890, GStA Berlin, Waldersee Papers; cf. Meisner, II, p. 102.
41 Waldersee, diary entry for 16 February 1890, GStA Berlin, Waldersee Papers; distorted in Meisner, II, p. 102.
42 Waldersee, diary entry for 18 February 1890, GStA Berlin, Waldersee Papers; largely omitted from Meisner, II, p. 103.
43 Waldersee, diary entry for 20 February 1890, GStA Berlin, Waldersee Papers; Meisner, II, pp. 104–5.
44 Waldersee, diary entry for 20 February 1890, GStA Berlin, Waldersee Papers; omitted from Meisner, II, p. 105.
45 Waldersee, diary entry for 20 February 1890, GStA Berlin, Waldersee Papers; omitted almost entirely from Meisner, II, p. 104.
46 Waldersee, diary entry for 23 February 1890, GStA Berlin, Waldersee Papers; cf. Meisner, II, pp. 105–6.
47 Ibid.
48 For Eulenburg's activities in Berlin in the period 14–23 February 1890, see *Eulenburgs Korrespondenz*, I, pp. 454–6.
49 Eulenburg, diary entries for 14–22 February 1890, ibid.
50 Eulenburg to Kaiser Wilhelm II, 22 February 1890, ibid., No. 324.
51 See Gradenwitz, *Bismarcks letzter Kampf*, p. 110; Röhl, *Germany without Bismarck*, pp. 42–3.
52 Kaiser Wilhelm II to Queen Victoria, 24 February 1890, RA I58/21.
53 Eulenburg to Kaiser Wilhelm II, 22 February 1890, *Eulenburgs Korrespondenz*, I, No. 324.
54 Bismarck's report to Kaiser Wilhelm II of 26 February 1890, in Bismarck, *Die gesammelten Werke*, VIc, pp. 432–3.
55 Kayser to Holstein, 28 February 1890, *Holstein Papers*, III, No. 306.

56 Waldersee, diary entry for 1 March 1890, GStA Berlin, Waldersee Papers; cf. Meisner, II, p. 108.
57 Eulenburg to Kaiser Wilhelm II, 22 February 1890, *Eulenburgs Korrespondenz*, I, No. 324.
58 Waldersee, diary entry for 16 February 1890, GStA Berlin, Waldersee Papers; omitted from Meisner, II, p. 103.
59 Marschall, diary entry for 20 February 1890, cited in *Eulenburgs Korrespondenz*, I, p. 455.
60 Stolberg-Wernigerode to Reuss, 22 February 1890, cited ibid.
61 See below, pp. 286–7.
62 Waldersee, diary entry for 26 February 1890, GStA Berlin, Waldersee Papers; cf. Meisner, II, pp. 106–7.
63 Marschall, report of 27 February 1890, Gradenwitz, *Bismarcks letzter Kampf*, p. 140; also Boetticher, Geschichte der Entlassung, in Eppstein, *Bismarcks Entlassung*, pp. 55–8.
64 Waldersee, diary entry for 1 March 1890, GStA Berlin, Waldersee Papers; cf. Meisner, II, p. 107. The Kaiser's speech is printed in Eppstein, *Bismarcks Entlassung*, pp. 173–6. See Boetticher's account of the debates of the Council of State, Geschichte der Entlassung, ibid., pp. 58ff.
65 Waldersee, diary entry for 1 March 1890, GStA Berlin, Waldersee Papers; cf. Meisner, II, p. 107.
66 Kayser to Eulenburg, 1 March 1890, *Eulenburgs Korrespondenz*, I, No. 328. Eulenburg sent this letter to the Kaiser on 3 March. See ibid., No. 332; GStA Berlin, BPHA Rep. 53 E III No. 3.
67 Holstein to Eulenburg, 2 March 1890, *Eulenburgs Korrespondenz*, I, No. 329. See Kayser to Eulenburg, 2 March 1890, ibid., No. 330.
68 Eulenburg to Holstein, 28 February 1890, ibid., No. 326.
69 Waldersee, diary entry for 1 March 1890, GStA Berlin, Waldersee Papers; omitted from Meisner, II, p. 108.
70 Kayser to Eulenburg, 1 March 1890, *Eulenburgs Korrespondenz*, I, No. 328.
71 Holstein to Eulenburg, 2 March 1890, ibid., No. 329.
72 Waldersee, diary entry for 1 March 1890, GStA Berlin, Waldersee Papers; cf. Meisner, II, p. 108.
73 Bernhard Vogel, Dieter Nohlen and Rainer-Olaf Schultze, *Wahlen in Deutschland. Theorie-Geschichte-Dokumente, 1848–1970*, Berlin 1971, pp. 290–1.
74 See Rauchhaupt to Hammerstein, 20 February 1890, Hans Leuß, *Wilhelm Freiherr von Hammerstein*, Berlin 1905, pp. 83–4.
75 Waldersee, diary entry for 23 February 1890, GStA Berlin, Waldersee Papers; cf. Meisner, II, p. 106.
76 Waldersee, diary entry for 5 March 1890, GStA Berlin, Waldersee Papers; omitted from Meisner, II, pp. 111–12.
77 Waldersee, diary entry for 2 March 1890, GStA Berlin, Waldersee Papers; cf. Meisner, II, p. 109.
78 Waldersee, diary entry for 5 March 1890, GStA Berlin, Waldersee Papers; omitted from Meisner, II, pp. 111–12.
79 Waldersee, diary entry for 23 February 1890, GStA Berlin, Waldersee Papers; cf. Meisner, II, p. 106.
80 Waldersee, diary entry for 1 March 1890, GStA Berlin, Waldersee Papers; important passages omitted from Meisner, II, pp. 108–9.
81 Waldersee, diary entry for 4 March 1890, GStA Berlin, Waldersee Papers; cf. Meisner, II, p. 110.
82 Waldersee, diary entry for 1 March 1890, GStA Berlin, Waldersee Papers; cf. Meisner, II, pp. 107–8.

83 Eulenburg to Holstein, 28 February 1890, *Eulenburgs Korrespondenz*, I, No. 326.
84 Minutes of the meeting of the Prussian Ministry of State of 2 March 1890, printed in Zechlin, *Staatsstreichpläne Bismarcks und Wilhelms II.*, pp. 178ff. The last sentence was crossed out by Bismarck, evidently because it revealed too much of his intentions.
85 Marschall, diary entry for 2 March 1890, cited in *Eulenburgs Korrespondenz*, I, p. 463.
86 Holstein to Eulenburg, 2 March 1890, ibid., No. 329.
87 Kayser to Eulenburg, 2 and 4 March 1890, ibid., Nos. 330 and 333.
88 Eulenburg to Kaiser Wilhelm II, 3 March 1890, ibid., No. 332.
89 Marschall, diary entry for 3 March 1890, cited ibid., p. 476.
90 Marschall, diary entries for 3–4 March 1890, cited in Röhl, *Germany without Bismarck*, p. 52.
91 Marschall, diary entry for 4 March 1890, cited ibid., p. 53.
92 Kaiser Wilhelm II to Eulenburg, 5 March 1890, *Eulenburgs Korrespondenz*, I, p. 477.
93 Eulenburg to Holstein, 5 March 1890, ibid., No. 334.
94 Eulenburg to Kaiser Wilhelm II, 10 March 1890, ibid., No. 344.
95 The Kaiser's note and Boetticher's reply are printed in Eppstein, *Bismarcks Entlassung*, pp. 177–8. See also August Eulenburg to Philipp Eulenburg, 9 March 1890, *Eulenburgs Korrespondenz*, I, No. 342. For Bismarck's furious reaction to the decoration of Boetticher, see Bismarck, *Die gesammelten Werke*, XV, pp. 509–10.
96 Eulenburg to Herbert Bismarck, 8 March 1890, *Eulenburgs Korrespondenz*, I, No. 340.
97 Bleichröder to Lord Rothschild, 10 March 1890, Rothschild Archives London, RAL XI/64/1. See also the similar letter of 7 March 1890, ibid.
98 Eulenburg to Kaiser Wilhelm II, 7 March 1890, Eulenburg to Herbert Bismarck, 8 March 1890, *Eulenburgs Korrespondenz*, I, Nos. 336 and 340. For Eulenburg's appointment, see above, pp. 191–7.
99 Kayser to Eulenburg, 5 and 7 March 1890, Eulenburg to Kaiser Wilhelm II, 7 March 1890, Marschall to Eulenburg, 12 March 1890, ibid., Nos. 335–7 and 349.
100 Kayser to Eulenburg, 12 March 1890, ibid., No. 347.
101 See Bleichröder to Lord Rothschild, 7 and 10 March 1890, Rothschild Archives London, RAL XI/64/1.
102 Holstein to Eulenburg, 7 March 1890, *Eulenburgs Korrespondenz*, I, No. 338. See Stig Förster, *Der doppelte Militarismus. Die deutsche Heeresrüstungspolitik zwischen Status-quo-Sicherung und Aggression 1890–1913*, Stuttgart 1985, p. 31.
103 Holstein to Eulenburg, 7 March 1890, *Eulenburgs Korrespondenz*, I, No. 338.
104 Eulenburg to Kaiser Wilhelm II, 8 March 1890, ibid., No. 339; Marschall, diary entry for 8 March 1890, cited ibid., p. 482.
105 Waldersee, diary entries for 8 and 10 March 1890, GStA Berlin, Waldersee Papers; cf. Meisner, II, p. 113.
106 Marschall, diary entry for 11 March 1890, cited in *Eulenburgs Korrespondenz*, I, p. 489.
107 Holstein to Eulenburg, 11 March 1890, ibid., No. 345.
108 Eulenburg to Holstein, 12 March 1890, Eulenburg to Grand Duke of Baden, 13 March 1890, ibid., Nos. 334 and 350. Cf. Eulenburg's note of 12 March, ibid., p. 489.
109 Karl Bachem, *Vorgeschichte, Geschichte und Politik der Deutschen Zentrumspartei*, 9 vols., Cologne 1927–32, V, pp. 116ff.; IX, pp. 93ff.; Marschall, diary entry for 13 March 1890, cited in *Eulenburgs Korrespondenz*, I, p. 492.
110 Kayser to Eulenburg, 12 March 1890, *Eulenburgs Korrespondenz*, I, No. 348.
111 Marschall to Eulenburg, 12 March 1890, ibid., No. 349.
112 Marschall, report of 15 March 1890, in Gradenwitz, *Bismarcks letzter Kampf*, pp. 147–8.
113 Waldersee, diary entry for 12 March 1890, GStA Berlin, Waldersee Papers; Meisner, II, p. 114.

114 Waldersee, diary entry for 9 March 1890, GStA Berlin, Waldersee Papers; Meisner, II, p. 113.
115 Waldersee, diary entry for 13 March 1890, GStA Berlin, Waldersee Papers; partly omitted from Meisner, II, p. 114.
116 Waldersee, diary entry for 25 January 1890, GStA Berlin, Waldersee Papers; largely omitted from Meisner, II, pp. 96–7.
117 Waldersee, diary entry for 15 February 1890, GStA Berlin, Waldersee Papers; omitted from Meisner, II, p. 102.
118 Waldersee, diary entry for 15 March 1890, GStA Berlin, Waldersee Papers; cf. Meisner, II, pp. 114–16.
119 Waldersee, diary entry for 2 March 1890, GStA Berlin, Waldersee Papers; cf. Meisner, II, p. 109.
120 Waldersee, diary entry for 5 March 1890, GStA Berlin, Waldersee Papers; largely omitted from Meisner, II, pp. 111–12.
121 Waldersee, diary entry for 8 March 1890, GStA Berlin, Waldersee Papers; omitted from Meisner, II, p. 112.
122 Waldersee, diary entry for 5 March 1890, GStA Berlin, Waldersee Papers; omitted from Meisner, II, pp. 111–12.
123 Waldersee, diary entry for 8 March 1890, GStA Berlin, Waldersee Papers; omitted from Meisner, II, p. 113. Cf. Bleichröder to Rothschild, 19 and 20 March 1890, Rothschild Archives London, RAL XI/64/1.
124 Kayser to Eulenburg, 5 and 7 March 1890, Eulenburg to Kaiser Wilhelm II, 7 March 1890, *Eulenburgs Korrespondenz*, I, Nos. 335–7.
125 Bismarck to August Eulenburg, 14 March 1890, cited ibid., p. 499.
126 Waldersee, diary entry for 4 March 1890, GStA Berlin, Waldersee Papers; cf. Meisner, II, p. 110.
127 Waldersee, diary entry for 15 March 1890, GStA Berlin, Waldersee Papers; Meisner, II, p. 114.
128 Bismarck, *Gedanken und Erinnerungen*, III, pp. 81–7. For the following, see Bernard Miall, transl., *New Chapters of Bismarck's Autobiography*, London, 1921, pp. 166–76.
129 Malet to Queen Victoria, 22 March 1890, RA I58/33–4.
130 Kaiser Wilhelm II, marginal notes on articles in *Vossische Zeitung* of 30 November 1903 and in *Berliner Tageblatt* of 2 December 1903, PA AA, R 1009.
131 Waldersee, diary entry for 15 March 1890, GStA Berlin, Waldersee Papers; cf. Meisner, II, pp. 114–16.
132 Waldersee, diary entry for 17 March 1890, GStA Berlin, Waldersee Papers; cf. Meisner, II, pp. 117–18.
133 Bismarck, *Gedanken und Erinnerungen*, III, pp. 87–8.
134 Waldersee, diary entry for 17 March 1890, GStA Berlin, Waldersee Papers; cf. Meisner, II, pp. 117–18.
135 Bismarck, *Gedanken und Erinnerungen*, III, p. 94; Bleichröder to Rothschild, 17, 18 and 20 March 1890, Rothschild Archives London, RAL XI/64/1.
136 Waldersee, diary entry for 17 March 1890, GStA Berlin, Waldersee Papers; cf. Meisner, II, pp. 117–18.
137 Bismarck, *Gedanken und Erinnerungen*, III, pp. 89–90. The decision not to renew the Reinsurance Treaty with Russia is discussed below, pp. 334–43.
138 The minutes are printed in Bismarck, *Gedanken und Erinnerungen*, III, pp. 163–70.
139 Printed in Bismarck, *Gedanken und Erinnerungen*, III, pp. 95–100.
140 Waldersee, diary entry for 19 March 1890, GStA Berlin, Waldersee Papers; cf. Meisner, II, pp. 118–19. On the choice of Caprivi as Bismarck's successor, see below, pp. 320–4.

141 Bismarck, *Gedanken und Erinnerungen*, III, p. 100. See also the comments of General Paul Bronsart von Schellendorf in Arnold Oskar Meyer, *Bismarck. Der Mensch und der Staatsmann*, Stuttgart 2nd edn 1949, p. 654.
142 Engelberg, *Bismarck. Das Reich in der Mitte Europas*, pp. 557ff.
143 Waldersee, diary entry for 25 January 1890, GStA Berlin, Waldersee Papers; largely omitted from Meisner, II, pp. 96–7.
144 Hoffmann, diary, 22 March 1890, Hoffmann Papers, Freiburg.
145 Hoffmann, diary, 23 April 1890, ibid.
146 Hinzpeter to Kaiser Wilhelm II, 22 May 1899, GStA Berlin, BPHA Rep. 53J Lit. H No. 1.
147 Hinzpeter to Studt, 9 March 1895, Stadt- und Landesbibliothek Dortmund, Studt Papers 7824.
148 Kayser to Eulenburg, 2 March 1890, *Eulenburgs Korrespondenz*, I, No. 330.
149 Waldersee, diary entry for 8 February 1890, GStA Berlin, Waldersee Papers; largely omitted from Meisner, II, p. 100.
150 See below, pp. 388–96.
151 Marschall, report of 15 March 1890, in Gradenwitz, *Bismarcks letzter Kampf*, pp. 147–8.
152 Bismarck, *Gedanken und Erinnerungen*, III, p. 80.
153 Leonhard von Muralt, *Bismarcks Verantwortlichkeit*, Göttingen 1955.
154 Hinzpeter to Studt, 9 March 1895, Stadt- und Landesbibliothek Dortmund, Studt Papers 7824.

13 THE HAPHAZARD TRANSITION: FROM THE BISMARCKS TO THE NEW COURSE

1 Waldersee claimed that Bismarck's behaviour had 'clearly displayed the hallmark of the most disgusting *Jewish tricks of the trade* [den Stempel widerwärtigster *jüdischer Mache*]': Waldersee, diary 1890, fol. 48, GStA Berlin, Waldersee Papers; cf. Meisner, II, p. 123.
2 Wedel, *Zwischen Kaiser und Kanzler*, pp. 72ff.
3 Waldersee, diary 1890, fol. 48, GStA Berlin, Waldersee Papers; omitted from Meisner, II, p. 122; Wedel, *Zwischen Kaiser und Kanzler*, pp. 72ff.
4 Malet to Queen Victoria, 29 March 1890, RA I58/36.
5 Wedel, *Zwischen Kaiser und Kanzler*, p. 71.
6 Ibid., p. 68.
7 Malet to Queen Victoria, 29 March 1890, RA I58/36.
8 Printed in Bismarck, *Gedanken und Erinnerungen*, III, pp. 101–5.
9 Kaiser Wilhelm II to Hinzpeter, 22 March 1890, printed in Wedel, *Zwischen Kaiser und Kanzler*, pp. 62–3. On Hinzpeter's advice the telegram was published as if it had been addressed to Emil Görtz; see ibid., pp. 99–100 and 115–16.
10 Ibid., pp. 91ff. See below, pp. 335–8.
11 Pourtalès to Holstein, 20 or 21 March 1890, *Holstein Papers*, III, No. 308.
12 Count Anton Monts to Wedel, 24 March 1890, Wedel Papers, Frankfurt a.M. The letter is printed in Wedel, *Zwischen Kaiser und Kanzler*, pp. 56–9.
13 Reuss to Wedel, 2 April 1890, Wedel Papers, Frankfurt a.M.; also Eulenburg's letters to Holstein and to Kaiser Wilhelm II of 8 April 1890, *Holstein Papers*, III, No. 313 with enclosure.
14 Kaiser Franz Joseph to Prince Otto von Bismarck, 22 March 1890, HHStA Vienna, Kabinettsarchiv Geheimakten 2.
15 Prince Otto von Bismarck to Kaiser Franz Joseph, 26 March 1890, ibid.
16 For the following, see Wilhelm von Schweinitz, ed., *Denkwürdigkeiten des Botschafters General Hans Lothar von Schweinitz*, 2 vols., Berlin 1927, II, p. 265.

17 Eulenburg to Holstein, 25 April 1890, *Holstein Papers*, III, No. 317.
18 Bülow, *Memoirs*, I, p. 234.
19 Kaiser Wilhelm II to Kaiser Franz Joseph, 3–5 April 1890, HHStA Vienna, Kabinettsarchiv Geheimakten 2. The letter is printed in Hanns Schlitter, 'Briefe Kaiser Franz Josephs I. und Kaiser Wilhelms II. über Bismarcks Rücktritt' *Österreichische Rundschau*, 58 (1919), pp. 100ff.
20 Kaiser Franz Joseph to Kaiser Wilhelm II, 12 April 1890, Entwurf, HHStA Vienna, Kabinettsarchiv Geheimakten 2; printed in Schlitter, 'Briefe', pp. 100ff.
21 Kaiser Wilhelm II to Kaiser Franz Joseph, 14 April 1890, HHStA Vienna, Kabinettsarchiv Geheimakten 2.
22 Kaiser Wilhelm II to Queen Victoria, telegram, 19 March 1890, RA I58/27. See Salisbury to Queen Victoria, cipher telegram, 20 March 1890, RA I58/30.
23 Malet to Queen Victoria, memorandum, 22 March 1890, RA I58/34.
24 Malet to Queen Victoria, 22 March 1890, RA I58/33. See the Queen's diary entry for 29 March 1890, RA QVJ.
25 Kaiser Wilhelm II to Queen Victoria, 27 March 1890, RA 158/32. See the Queen's reply of 31 March 1890, GStA Berlin, BPHA Rep. 52 W3 No. 11.
26 Salisbury to Queen Victoria, 7 April 1890, RA I58/39.
27 Eulenburg, notes of 12 July 1896, *Eulenburgs Korrespondenz*, III, No. 1239.
28 Spitzemberg, entry of 10 September 1908, *Tagebuch*, p. 487.
29 Hoffmann, diary entries for 22 and 24 March and 23 April 1890, Hoffmann Papers, Freiburg. Hoffmann was present on 22 April 1890 when Prince Heinrich told the Kaiser at Wilhelmshaven: 'At the time I said that Caprivi was the last person you would take.'
30 See above, pp. 225–7.
31 Marschall, diary entry for 29 January 1890, cited in Röhl, *Germany without Bismarck*, p. 59.
32 See Bismarck's later comments on these events, *Norddeutsche Allgemeine Zeitung*, 28 June 1892.
33 Waldersee, diary entry for 18 February 1890, GStA Berlin, Waldersee Papers; in Meisner this passage is omitted, II, p. 103.
34 Eulenburg to Kaiser Wilhelm II, 7 and 8 March 1890, *Eulenburgs Korrespondenz*, I, Nos. 336 and 339; Eulenburg to Holstein, 9 March 1890, ibid., No. 341.
35 See ibid., p. 503.
36 Ibid.
37 Waldersee, diary entry for 3 October 1894, GStA Berlin, Waldersee Papers; cf. Meisner, II, p. 330.
38 Empress Frederick to Queen Victoria, 22 March 1890, RA Z48/5.
39 Empress Frederick to Queen Victoria, 25 March 1890, RA Z48/6; also Empress Frederick to Queen Victoria, 29 March 1890, RA Z48/8.
40 See e.g. Wedel, *Zwischen Kaiser und Kanzler*, pp. 66, 78ff., 111–12, 125.
41 See Engelberg, *Bismarck. Das Reich in der Mitte Europas*, pp. 587–8.
42 Hereditary Prince Bernhard of Saxe-Meiningen to Waldersee, 29 November 1891, GStA Berlin, Waldersee Papers No. 36.
43 Schlözer, *Letzte römische Briefe, 1882–1894*, p. 158.
44 Cited in Waldersee, diary entry for 31 August 1890, GStA Berlin, Waldersee Papers; omitted from Meisner, II, p. 142.
45 Waldersee, diary entry for 20 March 1890, GStA Berlin, Waldersee Papers; cf. Meisner, II, p. 119.
46 Waldersee, diary entry for 18 February 1890, GStA Berlin, Waldersee Papers; largely omitted from Meisner, II, p. 103; also the diary entries for 19 and 20 March 1890, GStA Berlin, Waldersee Papers; omitted from Meisner, II, pp. 118ff.

47 See Röhl, *Germany without Bismarck*, p. 44.
48 *Eulenburgs Korrepondenz*, I, p. 506; also Waldersee, diary entry for 20 March 1890, GStA Berlin, Waldersee Papers; omitted from Meisner, II, p. 119.
49 Holstein to Eisendecher, 26 March 1890, *Holstein Papers*, III, No. 309. Herbert Bismarck's letter of resignation is printed in Bußmann, *Staatssekretär*, No. 405.
50 Waldersee, diary entry for 23 March 1890, GStA Berlin, Waldersee Papers; omitted from Meisner, II, p. 122.
51 Wedel, *Zwischen Kaiser und Kanzler*, pp. 68–71.
52 Malet to Queen Victoria, 29 March 1890, RA I58/36.
53 Waldersee, diary entry for 17 March 1890, GStA Berlin, Waldersee Papers; omitted from Meisner, II, p. 118.
54 Eulenburg to Kaiser Wilhelm II, 26 March 1890, *Eulenburgs Korrepondenz*, I, No. 369. See Waldersee, diary entry for 17 March 1890, GStA Berlin, Waldersee Papers; omitted from Meisner, II, p. 118.
55 Waldersee, diary entry for 17 March 1890, GStA Berlin, Waldersee Papers; omitted from Meisner, II, p. 118; Holstein to Eulenburg, 26 March 1890, *Eulenburgs Korrespondenz*, I, No. 368.
56 Holstein to Eulenburg, 26 March 1890, *Eulenburgs Korrespondenz*, I, No. 368.
57 August Eulenburg to Philipp Eulenburg, 22 March 1890, ibid., No. 363; see *Holstein Papers*, I, p. 148; Holstein to Eulenburg, 26 March 1890, *Eulenburgs Korrespondenz*, I, No. 368; Marschall, report of 27 March 1890, ibid., p. 512.
58 Stolberg-Wernigerode to Reuss, 28 March 1890, ibid., pp. 512–13.
59 Wedel, *Zwischen Kaiser und Kanzler*, p. 65.
60 Monts to Wedel, 24 March 1890, Wedel Papers, Frankfurt a.M.
61 Holstein to Eulenburg, 26 March 1890, *Eulenburgs Korrespondenz*, I, No. 368. In 1885 Bernhard von Bülow had married the divorced Countess Marie Dönhoff, an Italian friend of the then Crown Princess Victoria. See *Young Wilhelm*, p. 549.
62 Marschall, report of 27 March 1890, *Eulenburgs Korrespondenz*, I, p. 512.
63 Holstein to Eulenburg, 26 March 1890, ibid., No. 368; Eulenburg to Kaiser Wilhelm II, 26 March 1890, ibid., No. 369. According to Wedel, *Zwischen Kaiser und Kanzler*, p. 66, Limburg-Stirum was also recommended to the Kaiser by Alvensleben.
64 Eulenburg's memorandum for Kaiser Wilhelm II of 20 March 1894, *Eulenburgs Korrespondenz*, II, No. 933.
65 Monts to Wedel, 24 March 1890, Wedel Papers, Frankfurt a.M., printed in Wedel, *Zwischen Kaiser und Kanzler*, pp. 56–9.
66 Wedel, *Zwischen Kaiser und Kanzler*, pp. 64–7.
67 Marschall, report of 27 March 1890, *Eulenburgs Korrespondenz*, I, p. 512.
68 Waldersee, diary entry for 22 March 1890, GStA Berlin, Waldersee Papers; omitted from Meisner, II, p. 121.
69 Holstein to Eulenburg, 26 March 1890, *Eulenburgs Korrespondenz*, I, No. 368.
70 Eulenburg to Kaiser Wilhelm II, 26 March 1890, ibid., No. 369.
71 Marschall, report of 27 March 1890, ibid., p. 512; Waldersee, diary 1890, fol. 44, GStA Berlin, Waldersee Papers; cf. Meisner, II, p. 122.
72 Wedel, *Zwischen Kaiser und Kanzler*, pp. 62ff.; Waldersee, diary 1890, fol. 44, GStA Berlin, Waldersee Papers; cf. Meisner, II, p. 122.
73 Wedel, *Zwischen Kaiser und Kanzler*, pp. 64–5.
74 Stolberg-Wernigerode to Reuss, 28 March 1890, *Eulenburgs Korrespondenz*, I, pp. 512–13.
75 Waldersee, diary entries for 27 and 30 May 1890, GStA Berlin, Waldersee Papers; Meisner, II, p. 129.
76 For details, see Wedel, *Zwischen Kaiser und Kanzler*, pp. 102–8.

77 Prince Ernst zu Hohenlohe-Langenburg to his father, 2 April 1890, Hohenlohe-Zentralarchiv Neuenstein, Hermann Hohenlohe-Langenburg Papers, Bü. 57.
78 Holstein to Eulenburg, 10 February 1890, *Eulenburgs Korrespondenz*, I, No. 318.
79 Wedel, *Zwischen Kaiser und Kanzler*, p. 74; also above, p. 248.
80 Minutes of the meeting of the Prussian Ministry of State of 17 March 1890, printed in Bismarck, *Die gesammelten Werke*, XV, pp. 570ff. and Eppstein, *Bismarcks Entlassung*, pp. 179ff.
81 Eppstein, *Bismarcks Entlassung*, pp. 68–9.
82 Empress Frederick to Queen Victoria, 22 March 1890, RA Z48/5.
83 Waldersee, diary entry for 20 April 1890, GStA Berlin, Waldersee Papers; omitted from Meisner, II, p. 124; August Eulenburg to Philipp Eulenburg, 27 March 1890, *Eulenburgs Korrespondenz*, I, No. 371.
84 Kiderlen-Wächter to Holstein, 1 July 1890, *Holstein Papers*, III, No. 324.
85 A copy of the All-Highest Order of 23 May 1890 is to be found in the files of the Civil Cabinet, GStA Berlin, 2.2.1 No. 3307/1. See Waldersee, diary entries for 23–4 May 1890, GStA Berlin, Waldersee Papers; cf. Meisner, II, pp. 128–9; also Stolberg-Wernigerode to Reuss, 5 June 1890, cited in *Eulenburgs Korrespondenz*, I, p. 543.
86 Wedel, *Zwischen Kaiser und Kanzler*, pp. 108–9.
87 Waldersee, diary entries for 23 and 24 May 1890, GStA Berlin, Waldersee Papers; cf. Meisner, II, pp. 128–9.
88 Liebenau to House Minister Wilhelm von Wedell, 4 June 1890, Liebenau to Flügeladjutant Count Carl von Wedel, 4 June 1890, Wedel Papers, Frankfurt a.M.
89 Wedel, *Zwischen Kaiser und Kanzler*, p. 109.
90 Empress Frederick to Queen Victoria, 27 May and 4 June 1890, RA Z48/24 and 27.
91 Empress Frederick to Queen Victoria, 27 May, 4 and 19 June 1890, RA Z48/24, 27 and 34–5.
92 Waldersee, diary entry for 22 January 1894, GStA Berlin, Waldersee Papers; cf. Meisner, II, p. 304.

14 IN BISMARCK'S FOOTSTEPS: THE CONDUCT OF FOREIGN POLICY UNDER THE NEW COURSE

1 See Hans Hallmann, ed., *Zur Geschichte und Problematik des Deutsch-Russischen Rückversicherungsvertrages von 1887*, Darmstadt 1968; Norman Rich, *Friedrich von Holstein. Politics and Diplomacy in the Era of Bismarck and Wilhelm II*, 2 vols., Cambridge 1965; George F. Kennan, *The Decline of Bismarck's European Order. Franco-Russian Relations, 1879–1890*, Princeton 1979; Klaus Hildebrand, *Das vergangene Reich. Deutsche Außenpolitik von Bismarck bis Hitler, 1871–1945*, Stuttgart 1995, pp. 118ff. and 155ff.; Konrad Canis, *Von Bismarck zur Weltpolitik. Deutsche Außenpolitik 1890 bis 1902*, Berlin 1997; Volker Ullrich, *Die nervöse Großmacht 1871–1918. Aufstieg und Untergang des deutschen Kaiserreichs*, Frankfurt a.M. 1997.
2 See Hildebrand, *Das vergangene Reich*, pp. 118–22.
3 Canis, *Von Bismarck zur Weltpolitik*, p. 26.
4 Diary of the Russian diplomat Count Lamsdorff, 19 December 1889, *Berliner Monatshefte* 9, 1931, pp. 158–77; Hallmann, *Rückversicherungsvertrag*, p. 167.
5 See *Große Politik*, VII, No. 1924.
6 See Bismarck's memorandum for Kaiser Wilhelm II of 19 August 1888, *Große Politik*, VI, No. 1350. See above, pp. 141–2. Kaiser Friedrich III on the other hand was not informed of the existence of the secret treaty with Russia. See Lamsdorff, diary entry for 4 March 1889, in Hallmann, *Rückversicherungsvertrag*, p. 165.
7 *Große Politik*, VII, No. 1367.
8 Ibid., No. 1366.

9 Wedel, *Zwischen Kaiser und Kanzler*, pp. 36–7.
10 *Große Politik*, VII, No. 1367.
11 Schweinitz, *Denkwürdigkeiten des Botschafters*, II, pp. 396–406.
12 Lamsdorff, diary entries for 9/21 March 1890, in Hallmann, *Rückversicherungsvertrag*, pp. 167–9.
13 Ibid., p. xxxv.
14 Rich, *Friedrich von Holstein*, I, p. 310; Hallmann, *Rückversicherungsvertrag*, p. xxxv.
15 Holstein to Eisendecher, 16 April 1890, *Holstein Papers*, III, No. 315.
16 Holstein to Eulenburg, 22 March 1890, *Eulenburgs Korrespondenz*, I, No. 364.
17 Hallmann, *Rückversicherungsvertrag*, p. xxxviii.
18 *Große Politik*, VII, No. 1368.
19 See Holstein to Eisendecher, 26 March 1890, Holstein to Herbert Bismarck, 5 April 1890, and Herbert Bismarck to Holstein, 5 April 1890, in *Holstein Papers*, III, Nos. 309, 311–12.
20 Lamsdorff, 14/26 March 1890, in Hallmann, *Rückversicherungsvertrag*, p. 172.
21 Schweinitz, *Denkwürdigkeiten*, ibid., p. 197.
22 Ibid., p. xlii.
23 Ibid., pp. 197–8.
24 Ibid., pp. 198–9; also Schweinitz, *Briefwechsel*, pp. 264–6.
25 *Große Politik*, VII, No. 1392.
26 Lamsdorff, diary for 7 January 1895, in Hallmann, *Rückversicherungsvertrag*, p. 186.
27 Wedel, *Zwischen Kaiser und Kanzler*, pp. 68–9.
28 See *Große Politik*, VII, Nos. 1372–3.
29 Ibid., Nos. 1374–7.
30 Ibid., No. 1378.
31 Ibid.
32 See e.g. Rich, *Friedrich von Holstein*, I, pp. 322ff.
33 Thomas Nipperdey, *Deutsche Geschichte 1866–1918*, Vol. II, *Machtstaat vor der Demokratie*, Munich 1992, p. 622.
34 Waldersee, diary entry for 2 February 1891, GStA Berlin, Waldersee Papers; cf. Meisner, II, p. 184.
35 Waldersee, diary entry for 25 September 1890, GStA Berlin, Waldersee Papers; cf. Meisner, II, pp. 149–50.
36 Waldersee, diary entry for 24 April 1890, GStA Berlin, Waldersee Papers; cf. Meisner, II, p. 124.
37 Waldersee, diary entry for 11 August 1890, GStA Berlin, Waldersee Papers; cf. Meisner, II, pp. 137–8.
38 Waldersee, diary entry for 13 September 1891, GStA Berlin, Waldersee Papers; omitted from Meisner, II, p. 216.
39 Waldersee, diary entry for 5 November 1891, GStA Berlin, Waldersee Papers; cf. Meisner, II, p. 221.
40 Wedel, *Zwischen Kaiser und Kanzler*, pp. 95–6.
41 Ibid., pp. 155 and 185.
42 Waldersee, diary entry for 24 April 1890, GStA Berlin, Waldersee Papers; cf. Meisner, II, p. 124. See Eulenburg to Holstein, 25 April 1890, *Holstein Papers*, III, No. 317.
43 Wedel, *Zwischen Kaiser und Kanzler*, p. 109. Cf. Solms to Holstein, 13 June 1890, *Holstein Papers*, III, No. 320.
44 Waldersee, diary entry for 11 August 1890, GStA Berlin, Waldersee Papers; cf. Meisner, II, pp. 137–8.
45 Waldersee, diary entry for 6 March 1891, GStA Berlin, Waldersee Papers; cf. Meisner, II, p. 198.

46 Waldersee, diary entry for 27 November 1891, GStA Berlin, Waldersee Papers; omitted from Meisner, II, p. 223.
47 Waldersee, diary entry for 4 January 1892, GStA Berlin, Waldersee Papers; omitted from Meisner, II, pp. 228–9.
48 Waldersee, diary entry for 30 January 1892, GStA Berlin, Waldersee Papers; cf. Meisner, II, p. 232.
49 Waldersee, diary entry for 30 November 1891, GStA Berlin, Waldersee Papers; omitted from Meisner, II, p. 224.
50 Waldersee, diary entry for 19 January 1891, GStA Berlin, Waldersee Papers; omitted from Meisner, II, p. 176. See Canis, *Von Bismarck zur Weltpolitik*, p. 68.
51 Waldersee, diary entry for 23 March 1890, GStA Berlin, Waldersee Papers; cf. Meisner, II, p. 122.
52 Wedel, *Zwischen Kaiser und Kanzler*, p. 101.
53 Kaiser Wilhelm II, speeches of 14 and 15 May 1890, Penzler, *Reden Kaiser Wilhelms II. in den Jahren 1888–1895*, pp. 112–16.
54 Waldersee, diary entry for 21 May 1890, GStA Berlin, Waldersee Papers; omitted from Meisner, II, p. 128.
55 Kaiser Wilhelm II, marginal notes on Pourtalès's report of 8 July 1890, PA AA, Asservat No. 4.
56 Prince Ernst zu Hohenlohe-Langenburg to his father, 29 August 1890, Hohenlohe-Zentralarchiv Neuenstein, Prince Hermann zu Hohenlohe-Langenburg Papers, Bü. 57.
57 See Kaiser Wilhelm II, marginal note on consular telegram of 31 October 1891 on Alexander III's travel plans: 'Couldn't care less' (PA AA, Asservat No. 4).
58 Wedel, *Zwischen Kaiser und Kanzler*, pp. 121–2; Waldersee, diary entry for 22 December 1890, GStA Berlin, Waldersee Papers; Meisner, II, p. 170.
59 Waldersee, diary entry for 1 September 1890, GStA Berlin, Waldersee Papers; Meisner, II, p. 142.
60 Wedel, *Zwischen Kaiser und Kanzler*, pp. 121–2.
61 Waldersee, diary entry for 1 September 1890, GStA Berlin, Waldersee Papers; cf. Meisner, II, p. 142.
62 See Waldersee's diary entry for 6 March 1891, in Meisner, II, pp. 197–8.
63 Empress Frederick to Queen Victoria, 3 February 1891, RA Z50/12. For the background to these attempts to reach an understanding with France, see Canis, *Von Bismarck zur Weltpolitik*, p. 90. But cf. Brauer to Turban, 21 February 1891, Fuchs, *Großherzog von Baden*, III, No. 1104.
64 Waldersee, diary entry for 26 February 1891, GStA Berlin, Waldersee Papers; omitted from Meisner, II, pp. 196–7. But see Waldersee's diary entry for 6 March 1891, Meisner, II, pp. 197–8.
65 Empress Frederick to Kaiser Wilhelm II, 21 February 1891, GStA Berlin, Brand.-Preuß. Hausarchiv, Rep. 52T No. 13.
66 Münster to Holstein, 5 March 1891, *Holstein Papers*, III, No. 335; also Empress Frederick to Queen Victoria, 8 April 1891, RA Z50/37.
67 Empress Frederick to her daughter Princess Viktoria of Schaumburg-Lippe, 28 February 1891, AdHH Schloss Fasanerie.
68 Empress Frederick to Queen Victoria, 29/30 March and 2 April 1891, RA Z50/29 and 32.
69 Verdy to Waldersee, 24 May 1891, GStA Berlin, Waldersee Papers, No. 53.
70 Empress Frederick to Bogumilla Freifrau von Stockmar, 18 March 1891, AdHH Schloss Fasanerie; also Brauer to Turban, 1 March 1891, Fuchs, *Großherzog von Baden*, III, No. 1107; Empress Frederick to Queen Victoria, 24 March 1891, RA Z50/24.

71 Kaiser Wilhelm II, marginal notes on the report from Vienna of 23 May 1891, PA AA, Asservat No. 4.
72 Ibid.
73 Wedel, *Zwischen Kaiser und Kanzler*, pp. 156ff.
74 Waldersee, diary entry for 16 November 1890, GStA Berlin, Waldersee Papers; cf. Meisner, II, p. 160.
75 In Penzler, *Reden Kaiser Wilhelms II. in den Jahren 1888–1895*, pp. 174–5.
76 Wedel, *Zwischen Kaiser und Kanzler*, pp. 174–9; also Brauer to Turban, 23 April 1891, Fuchs, *Großherzog von Baden*, III, No. 1110.
77 Waldersee, diary entry for 21 April 1891, GStA Berlin, Waldersee Papers; Meisner, II, p. 204.
78 Waldersee, diary entry for 21 April 1891, GStA Berlin, Waldersee Papers; cf. Meisner, II, pp. 204–5.
79 Waldersee, diary entry for 17 May 1891, GStA Berlin, Waldersee Papers; omitted from Meisner, II, p. 207.
80 On the foreign policy of the New Course, see Rainer Lahme, *Deutsche Außenpolitik 1890–1894. Von der Gleichgewichtspolitik Bismarcks zur Allianzstrategie Caprivis*, Göttingen 1990; also Rich, *Friedrich von Holstein*, I, pp. 287ff.; Ralf Forsbach, *Alfred von Kiderlen-Wächter (1852–1912). Ein Diplomatenleben im Kaiserreich*, 2 vols., Göttingen 1997; Canis, *Von Bismarck zur Weltpolitik*, pp. 17–137; Ullrich, *Nervöse Großmacht*, pp. 182–8.
81 On 31 May 1890 Waldersee and War Minister von Verdy had a joint audience with the Kaiser at which the monarch approved the 'completely altered' and 'very bold' operations plans for the eastern front presented to him by the General Staff: Waldersee, diary entry for 31 May 1890, GStA Berlin, Waldersee Papers; cf. Meisner, II, p. 129.
82 Waldersee, diary entry for 1 May 1890, GStA Berlin, Waldersee Papers; omitted from Meisner, II, p. 126.
83 Wedel, *Zwischen Kaiser und Kanzler*, p. 83.
84 Ibid., p. 80.
85 Waldersee, diary entry for 25 July 1890, GStA Berlin, Waldersee Papers; omitted from Meisner, II, p. 132. Cf. Waldersee's similar comments of January 1889, above, p. 149.
86 Waldersee, diary entry for 8 June 1890, GStA Berlin, Waldersee Papers; omitted from Meisner, II, p. 131.
87 Wedel, *Zwischen Kaiser und Kanzler*, pp. 85 and 88. See Kaiser Franz Joseph's similar concerns, cited above, p. 308.
88 Wedel, *Zwischen Kaiser und Kanzler*, p. 80.
89 Waldersee, diary entry for 25 July 1890, GStA Berlin, Waldersee Papers; Meisner, II, p. 131.
90 Waldersee, diary entry for 4 May 1890, GStA Berlin, Waldersee Papers; cf. Meisner, II, p. 126.
91 Eulenburg to Holstein, 25 April 1890, *Holstein Papers*, III, No. 317.
92 Waldersee, diary entry for 4 May 1890, GStA Berlin, Waldersee Papers; cf. Meisner, II, p. 126.
93 Waldersee, diary entry for 28 October 1890, GStA Berlin, Waldersee Papers; omitted from Meisner, II, p. 156.
94 See Canis, *Von Bismarck zur Weltpolitik*, pp. 58–60.
95 Waldersee, diary entry for 4 May 1890, GStA Berlin, Waldersee Papers; cf. Meisner, II, p. 126.
96 Waldersee, diary entry for 24 May 1890, GStA Berlin, Waldersee Papers; omitted from Meisner, II, p. 129.
97 Waldersee, diary entry for 8 May 1890, GStA Berlin, Waldersee Papers; omitted from Meisner, II, p. 127.
98 Salisbury to Queen Victoria, 23 May 1890, RA I58/49. See Buckle, *Letters of Queen Victoria*, I, p. 606.
99 RA QVJ, 7 August 1890.

100 Emin Pasha's original orders were to extend the border between the German and the British spheres of influence in East Africa as far as possible in Germany's favour. See Wissmann's order to Emin of 30 March 1890 in Georg Schweitzer, *Emin Pasha. His Life and Work*, 2 vols., London 1898, II, pp. 41–2. The Anglo-German colonial treaty of 1 July 1890 committed the Reich government to accepting an agreed boundary. See Schmidt to Emin, 30 August and 7 September 1890, Schweitzer, *Emin Pasha*, II, pp. 97ff. For Emin Pasha's adventurous life and his death, see Gaetano Casati, *Ten Years in Equatoria*, London 1891; Carl Peters, *New Light on Dark Africa; Being the Narrative of the German Emin Pasha Expedition*, London, New York, Melbourne 1891; Vita Hassan, *Die Wahrheit über Emin Pascha*, 1893; F. Stuhlmann, *Mit Emin Pascha ins Herz von Afrika*, 1894; A. J. A. Symons, *Emin, Governor of Equatoria*, London 1928.

101 Prince Hermann zu Hohenlohe-Langenburg to Kaiser Wilhelm II, 25 December 1890, Hohenlohe-Zentralarchiv Neuenstein, Hermann Hohenlohe-Langenburg Papers, Bü. 109.

102 Kaiser Wilhelm II to Prince Hermann zu Hohenlohe-Langenburg, 27 December 1890, ibid.

103 Prince Hermann zu Hohenlohe-Langenburg to Kaiser Wilhelm II, December 1890, ibid.

104 Schweitzer, *Emin Pasha*, II, pp. 294–5.

105 See Canis, *Von Bismarck zur Weltpolitik*, pp. 57–63.

106 Waldersee, diary entry for 25 July 1890, GStA Berlin, Waldersee Papers; Meisner, II, pp. 131–2.

107 Wedel, *Zwischen Kaiser und Kanzler*, pp. 116–17. Cf. Queen Victoria to Salisbury, 11 and 12 June 1890, RA I58/50–1.

108 Prince Ernst zu Hohenlohe-Langenburg to his father, 13 August 1890, Hohenlohe-Zentralarchiv Neuenstein, Hermann Hohenlohe-Langenburg Papers, Bü. 57.

109 Kaiser Wilhelm II, speech of 10 August 1890, Penzler, *Reden Kaiser Wilhelms II. in den Jahren 1888–1895*, pp. 121–4.

110 Ibid., pp. 123–4.

111 Kaiser Wilhelm II to Queen Victoria, 16 February 1891, RA E56/40 and 41; Queen Victoria to Kaiser Wilhelm II, 4 March 1891, GStA Berlin, BPHA Rep. 52 W3 No. 11.

112 Kaiser Wilhelm II to Queen Victoria, telegram, 26 February 1891, RA I59/4.

113 Salisbury to Queen Victoria, March 1891, RA I59/9 and 14.

114 Hoffmann, diaries for 22–3 April 1890, Hoffmann Papers, Freiburg.

115 Empress Frederick to her daughter Princess Viktoria of Schaumburg-Lippe, 19 January 1891, AdHH Schloss Fasanerie. For Wilhelm's relationship with his brother, see below, pp. 628–33.

116 Kiderlen-Wächter to Holstein, 7 July 1890, *Holstein Papers*, III, No. 321.

117 Kiderlen-Wächter to Holstein, 19 July 1890, ibid., No. 324.

118 Kiderlen-Wächter to Holstein, 21 September 1890, ibid., No. 328.

119 Wedel, *Zwischen Kaiser und Kanzler*, p. 157.

120 Ibid., p. 173.

121 Waldersee, diary entry for 6 September 1890, GStA Berlin, Waldersee Papers; omitted from Meisner, II, p. 142.

122 Waldersee, diary entry for 18 February 1890, GStA Berlin, Waldersee Papers; in Meisner this last passage is omitted, II, p. 103.

123 Waldersee, diary entry for 11 October 1890, GStA Berlin, Waldersee Papers; omitted from Meisner, II, p. 154.

124 Waldersee, diary entries for 20 December 1890 and 17 January 1891, GStA Berlin, Waldersee Papers; cf. Meisner, II, pp. 168 and 176.

125 See Canis, *Von Bismarck zur Weltpolitik*, pp. 71–87.

126 Waldersee, diary entry for 3 July 1891, GStA Berlin, Waldersee Papers; cf. Meisner, II, pp. 211–12.

127 Waldersee, diary entry for14 July 1891, GStA Berlin, Waldersee Papers; omitted from Meisner, II, p. 212.
128 Rumbold to Ponsonby, 24 June 1891, RA I59/29.
129 See Roderick R. McLean, *Royalty and Diplomacy in Europe, 1890–1914*, Cambridge 2001, p. 87.
130 *Daily Telegraph*, 6 July 1891, *Times*, 10 July 1891, RA I59/31–3.
131 *Daily Telegraph*, 11 July 1891, RA I59/34–5.
132 *Morning Post*, 13 July 1891, RA I59/37.
133 See e.g. *Standard*, 4 July 1891, RA I59/30.
134 Kaiser Wilhelm II, speech of 10 July 1891, cited from *Daily Telegraph*, 11 July 1891, RA I59/34.
135 *Morning Post*, 13 July 1891, *Standard*, 14 July 1891, RA I59/37–8.
136 This portrait now hangs in Osborne House. See Jarchow, *Hofgeschenke*, pp. 52–3.
137 Prince Ernst zu Hohenlohe-Langenburg to his father, 16 October 1891, Hohenlohe-Zentralarchiv Schloss Neuenstein, Hermann Hohenlohe-Langenburg Papers, Bü. 58.
138 Kaiser Wilhelm II to Queen Victoria, 14 July 1891, RA I59/39.
139 Kaiser Wilhelm II to Queen Victoria, 20 July 1891, RA I59/42.
140 Kaiser Wilhelm II to Queen Victoria, 12 August 1891, RA I59/45.
141 Kaiser Wilhelm II to Queen Victoria, 22 May 1892, RA I59/80.
142 Empress Frederick to Queen Victoria, 26 August 1891, RA Z51/15.
143 Prince Heinrich of Prussia to Kaiser Wilhelm II, 16 August 1891, BA-MA Freiburg, Senden-Bibran Papers, N160/10.
144 Prince Heinrich of Prussia to Kaiser Wilhelm II, 21 August 1891, ibid.
145 Waldersee, diary entry for 15 July 1891, GStA Berlin, Waldersee Papers; omitted from Meisner, II, p. 212.
146 Canis, *Von Bismarck zur Weltpolitik*, pp. 67–71.
147 Waldersee, diary entry for 15 July 1891, GStA Berlin, Waldersee Papers; omitted from Meisner, II, p. 212.
148 Canis, *Von Bismarck zur Weltpolitik*, p. 63.
149 Empress Frederick to Queen Victoria, 8 September 1891, RA Z51/21.
150 Kaiserin Auguste Viktoria to Kaiser Wilhelm II, 26/28 July 1891, GStA Berlin, BPHA Rep. 53T Preußen: An Kaiser Wilhelm II., Bd IV.
151 Waldersee, diary entry for 27 August 1891, GStA Berlin, Waldersee Papers; omitted from Meisner, II, p. 215.
152 Ibid.
153 Waldersee, diary entry for 13 and 25 September 1891, GStA Berlin, Waldersee Papers; omitted from Meisner, II, pp. 216 and 218.
154 Waldersee, diary entry for 13 August 1891, GStA Berlin, Waldersee Papers; omitted from Meisner, II, pp. 214–15.
155 Waldersee, diary entry for 17 September 1891, GStA Berlin, Waldersee Papers; cf. Meisner, II, p. 217.
156 Kaiser Wilhelm II, marginal notes on Bülow's report of 11 August 1891 and on the report from Egypt of 6 April 1892, PA AA, Asservat No. 4.
157 Kaiser Wilhelm II, marginal notes on Bülow's report of 22 November 1891, ibid.
158 Kaiser Wilhelm II, speech at Erfurt, 14 September 1891, Penzler, *Reden Kaiser Wilhelms II. in den Jahren 1888–1895*, pp. 192–3.
159 Caprivi to Kaiser Wilhelm II, cipher telegram, 17 September 1891, Kaiser Wilhelm II to Caprivi, telegram *en clair*, 18 September 1891, PA AA, Asservat No. 4.
160 Empress Frederick to Queen Victoria, 19 September 1891, RA Z51/26.
161 Waldersee, diary entry for 25 September 1890, GStA Berlin, Waldersee Papers; cf. Meisner, II, pp. 149–50.

162 Waldersee, diary entries for 24 April and 11 August 1890, GStA Berlin, Waldersee Papers; cf. Meisner, II, pp. 124 and 137–8.
163 Eulenburg's notes of 11 July 1892, *Eulenburgs Korrespondenz*, II, No. 688.
164 Kaiser Wilhelm II to Kaiser Franz Joseph, 31 January 1893, HHStA Vienna, Kabinettsarchiv Geheimakten 2.
165 Kaiser Franz Joseph to Kaiser Wilhelm II, 5 February 1893, ibid.
166 Eulenburg to Kaiser Wilhelm II, 16 April 1893, *Eulenburgs Korrespondenz*, II, No. 793.
167 Kaiser Wilhelms II, dictated notes on a conversation with Pope Leo XIII, 23 April 1893, in Hohenlohe, *Denkwürdigkeiten der Reichskanzlerzeit*, pp. 608–11. See above, p. 118.
168 Minutes of the Crown Council of 18 February 1894, GStA Berlin; cited more fully below, p. 499. See Canis, *Von Bismarck zur Weltpolitik*, pp. 121–2.
169 Kaiser Wilhelm II to Crown Prince Gustaf of Sweden and Norway, 25 July 1895, printed in Hohenlohe, *Denkwürdigkeiten der Reichskanzlerzeit*, pp. 102–5.

15 THE DUALISM OF POWER

1 Waldersee, diary entry for 4 May 1890, GStA Berlin, Waldersee Papers; omitted from Meisner, II, p. 126.
2 Cabinet order of 14 April 1890, in Gradenwitz, *Bismarcks letzter Kampf*, p. 114.
3 Kiderlen-Wächter to Holstein, 7 and 15 July 1890, *Holstein Papers*, III, Nos. 321–2. See Forsbach, *Kiderlen-Wächter*, I, p. 98.
4 Kaiser Wilhelm II to Queen Victoria, 25 December 1890, RA I58/63.
5 Holstein to Brandt, 26 December 1890, *Holstein Papers*, III, No. 332.
6 Waldersee, diary entry for 26 December 1890, GStA Berlin, Waldersee Papers; omitted from Meisner, II, p. 171.
7 Waldersee, diary entry for 20 April 1890, GStA Berlin, Waldersee Papers; omitted from Meisner, II, p. 124.
8 Ibid.
9 Waldersee, diary entries for 20–2 March and 20 April 1890, GStA Berlin, Waldersee Papers; cf. Meisner, II, pp. 119ff. and 124.
10 Wedel, *Zwischen Kaiser und Kanzler*, p. 109. Cf. Solms to Holstein, 13 June 1890, *Holstein Papers*, III, No. 320.
11 Brauer to Turban, 17 December 1890, Fuchs, *Großherzog von Baden*, III, No. 1089.
12 For the genesis of Verdy's Army Bill, see above, pp. 291–2.
13 Waldersee, diary entry for 8 June 1890, GStA Berlin, Waldersee Papers; cf. Meisner, II, pp. 130–1.
14 Waldersee, diary entry for 20 April 1890, GStA Berlin, Waldersee Papers; omitted from Meisner, II, p. 124. Cf. Meisner, II, pp. 132ff.
15 Wedel, *Zwischen Kaiser und Kanzler*, p. 115. Cf. Waldersee, diary entry for 7 June 1890, GStA Berlin, Waldersee Papers; Meisner, II, p. 130. See Förster, *Der doppelte Militarismus*, pp. 31ff.
16 Kiderlen-Wächter to Holstein, 21 September 1890, *Holstein Papers*, III. No. 328; also Brauer to Turban, 11 October 1890, Fuchs, *Großherzog von Baden*, III, No. 1076.
17 Waldersee, diary entry for 25 July 1890, GStA Berlin, Waldersee Papers; Meisner, II, p. 133.
18 Waldersee, diary entry for 7 October 1890, GStA Berlin, Waldersee Papers; omitted from Meisner, II, p. 153.
19 But cf. Waldersee, diary entry for 2 October 1890, GStA Berlin, Waldersee Papers; omitted for the most part from Meisner, II, pp. 151–2. See below, p. 412.
20 Waldersee, diary entry for 3 September 1890, GStA Berlin, Waldersee Papers; omitted from Meisner, II, p. 142.
21 Waldersee, diary entry for 25 July 1890, GStA Berlin, Waldersee Papers; Meisner, II, p. 134.

22 Waldersee, diary entry for 2 October 1890, GStA Berlin, Waldersee Papers; omitted from Meisner, II, pp. 151–2.
23 See below, pp. 433–4.
24 Wedel, *Zwischen Kaiser und Kanzler*, p. 133.
25 Ibid., pp. 117–18.
26 Waldersee, diary entry for 2 October 1890, GStA Berlin, Waldersee Papers; omitted from Meisner, II, pp. 151–2.
27 Waldersee, diary entry for 19 August 1890, GStA Berlin, Waldersee Papers; cf. Meisner, II, p. 140.
28 Waldersee, diary entry for 2 October 1890, GStA Berlin, Waldersee Papers; cf. Meisner, II, pp. 151–2.
29 Waldersee, diary entry for 21 October 1890, GStA Berlin, Waldersee Papers; Meisner, II, p. 155.
30 Minutes of the Prussian Ministry of State, 16 June 1890, in Röhl, *Germany without Bismarck*, pp. 60–1.
31 Waldersee, diary entry for 25 July 1890, GStA Berlin, Waldersee Papers; Meisner, II, p. 134; Wedel, *Zwischen Kaiser und Kanzler*, p. 116.
32 Wedel, *Zwischen Kaiser und Kanzler*, pp. 109–10.
33 See Hans Herzfeld, *Johannes von Miquel. Sein Anteil am Ausbau des Deutschen Reiches bis zur Jahrhundertwende*, 2 vols., Detmold 1938; Röhl, *Germany without Bismarck*, pp. 61–2; Wedel, *Zwischen Kaiser und Kanzler*, pp. 130–1.
34 Waldersee, diary entry for 10 December 1890, GStA Berlin, Waldersee Papers; cf. Meisner, II, p. 165.
35 Waldersee, diary entry for 5 August 1893, GStA Berlin, Waldersee Papers; omitted from Meisner, II, pp. 291–2.
36 Empress Frederick to Bogumilla Freifrau von Stockmar, 11 and 18 April 1891, AdHH Schloss Fasanerie.
37 Minutes of the Prussian Ministry of State, 30 September 1890, in Röhl, *Germany without Bismarck*, p. 62; Wedel, *Zwischen Kaiser und Kanzler*, p. 130.
38 Minutes of the Prussian Ministry of State, 30 September 1890, in Röhl, *Germany without Bismarck*, p. 62; Wedel, *Zwischen Kaiser und Kanzler*, p. 130; also Waldersee, diary entry for 16 November 1890, GStA Berlin, Waldersee Papers; omitted from Meisner, II, p. 161.
39 Waldersee, diary entry for 19 January 1891, GStA Berlin, Waldersee Papers; omitted from Meisner, II, p. 176.
40 Wedel, *Zwischen Kaiser und Kanzler*, p. 125.
41 Röhl, *Germany without Bismarck*, p. 63.
42 Ibid., p. 77.
43 Kiderlen-Wächter to Holstein, 21 September 1890, *Holstein Papers*, III, No. 328.
44 Wedel, *Zwischen Kaiser und Kanzler*, p. 128.
45 Kiderlen-Wächter to Holstein, 21 September 1890, *Holstein Papers*, III, No. 328.
46 Waldersee, diary entry for 3 October 1890, GStA Berlin, Waldersee Papers; omitted from Meisner, II, p. 152.
47 Kiderlen-Wächter to Holstein, 21 September 1890, *Holstein Papers*, III, No. 328.
48 Waldersee, diary entry for 10 October 1890, GStA Berlin, Waldersee Papers; omitted from Meisner, II, p. 154.
49 Waldersee, diary entry for 11 October 1890, GStA Berlin, Waldersee Papers; omitted from Meisner, II, p. 154.
50 Boetticher and Marschall to Flügeladjutant on duty, 14 October 1891, with the Kaiser's marginal notes, PA AA, Preußen 1 No. 1d, Bd I; Kaiser Wilhelm II to Boetticher, 14 October 1891, PA AA, Asservat No. 4.

51 Waldersee, diary entry for 2 November 1891, GStA Berlin, Waldersee Papers; cf. Meisner, II, pp. 220–1.
52 Wedel, *Zwischen Kaiser und Kanzler*, pp. 110–15.
53 Waldersee, diary entry for 18 November 1890, GStA Berlin, Waldersee Papers; Meisner, II, p. 162.
54 Waldersee, diary entry for 1 February 1891, GStA Berlin, Waldersee Papers; Meisner, II, p. 184; also Wedel, *Zwischen Kaiser und Kanzler*, p. 152.
55 Waldersee, diary entry for 1 February 1891, GStA Berlin, Waldersee Papers; omitted from Meisner, II, p. 184.
56 Empress Frederick to Bogumilla Freifrau von Stockmar, 18 March 1891, AdHH Schloss Fasanerie.
57 Wedel, *Zwischen Kaiser und Kanzler*, p. 115.
58 Waldersee, diary entry for 17 August 1890, GStA Berlin, Waldersee Papers; cf. Meisner, II, p. 139.
59 Brauer, report of 7 December 1890, in Fuchs, *Großherzog von Baden*, III, No. 1088.
60 See below, pp. 405–9.
61 See e.g. Clark, *Kaiser Wilhelm* II, pp. 60–1. Cf. Nipperdey, *Deutsche Geschichte 1866–1918*, Vol. I, *Arbeitswelt und Bürgergeist*, Munich 1990, p. 535. For the effect of the Kaiser's speech on Prussian secondary schools, See, *Barbar Germane Arier*, p. 148.
62 Kaiser Wilhelm II, speech of 4 December 1890, Penzler, *Reden Kaiser Wilhelms II. in den Jahren 1888–1895*, pp. 152–62. Cf. Louis Elkind, transl., *The German Emperor's Speeches: Being a Selection from the Speeches, Edicts, Letters and Telegrams of the Emperor William II*, New York, London, Bombay 1904, pp. 160–1. See Waldersee, diary entry for 5 December 1890, GStA Berlin, Waldersee Papers; Meisner, II, p. 164.
63 Brauer to Turban, 7 December 1890, Fuchs, *Großherzog von Baden*, III, No. 1088.
64 Waldersee, diary entry for 6 December 1890, GStA Berlin, Waldersee Papers; omitted from Meisner, II, p. 164.
65 Waldersee, diary entry for 10 December 1890, GStA Berlin, Waldersee Papers; cf. Meisner, II, p. 165.
66 Waldersee, diary entry for 14 December 1890, GStA Berlin, Waldersee Papers; cf. Meisner, II, p. 166.
67 Kaiser Wilhelm II, speech of 17 December 1890, Penzler, *Reden Kaiser Wilhelms II. in den Jahren 1888–1895*, pp. 163–7; Elkind, *German Emperor's Speeches*, pp. 166–7.
68 Waldersee, diary entry for 14 January 1891, GStA Berlin, Waldersee Papers; Meisner, II, p. 175.
69 Waldersee, diary entry for 25 January 1891, GStA Berlin, Waldersee Papers; omitted from Meisner, II, p. 177.
70 Empress Frederick to Queen Victoria, 24 March 1891, RA Z50/24.
71 Wedel, *Zwischen Kaiser und Kanzler*, p. 187.
72 Röhl, *Germany without Bismarck*, p. 78, n. 2.
73 Waldersee, diary entry for 27 October 1890, GStA Berlin, Waldersee Papers; omitted from Meisner, II, p. 156.
74 Wedel, *Zwischen Kaiser und Kanzler*, p. 172.
75 Count Udo Stolberg, ibid., p. 115.
76 Kaiser Wilhelm II, speech at the Brandenburg provincial diet, 20 February 1891, Penzler, *Reden Kaiser Wilhelms II. in den Jahren 1888–1895*, pp. 168–71.
77 Kaiser Wilhelm II, speech at the Rhenish provincial diet, 4 May 1891, ibid., pp. 176–8; Wedel, *Zwischen Kaiser und Kanzler*, p. 179.
78 Kaiser Wilhelm II, speech of 24 February 1892, Penzler, *Reden Kaiser Wilhelms II. in den Jahren 1888–1895*, pp. 207–10; Elkind, *German Emperor's Speeches*, pp. 292–4. See Marschall, diary entry for 24 February 1892, cited in *Eulenburgs Korrespondenz*, II, p. 780.

79 Brauer to Grand Duke Friedrich I of Baden, 26 February 1892, Fuchs, *Großherzog von Baden*, III, No. 1159; also Brauer to Turban, 1 March 1892, ibid., No. 1162.
80 Holstein to Eulenburg, 16 March 1891, *Eulenburgs Korrespondenz*, I, No. 485.
81 Eulenburg to Kaiser Wilhelm II, 23 March 1891, ibid., No. 493.
82 Holstein to Eulenburg, 16 March 1891, ibid., No. 485.
83 On Kiderlen's role, see Forsbach, *Kiderlen-Wächter*, I, pp. 120ff.
84 Kiderlen-Wächter to Holstein, 3 August 1891, *Holstein Papers*, III, No. 345.
85 Kaiser Wilhelm II, marginal note on an article in the *Hamburger Nachrichten* of 21 September 1891, PA AA, Asservat No. 4.
86 Kaiser Wilhelm II to King Albert of Saxony, 1 February 1892, GStA Berlin, BPH Rep. 53 No. 377.
87 Waldersee, diary entry for 31 October 1891, GStA Berlin, Waldersee Papers; omitted from Meisner, II, p. 220; Brauer, report of 29 October 1891, Fuchs, *Großherzog von Baden*, III, No. 1137; also below, pp. 507–8.
88 Waldersee, diary entry for 20 July 1891, GStA Berlin, Waldersee Papers; Meisner, II, pp. 212–13.
89 Waldersee, diary entry for 31 October 1891, GStA Berlin, Waldersee Papers; cf. Meisner, II, p. 220.
90 Waldersee, diary entry for 17 November 1891, GStA Berlin, Waldersee Papers; cf. Meisner, II, p. 222.
91 Waldersee, diary entry for 18 November 1891, GStA Berlin, Waldersee Papers; printed in part in Meisner, II, p. 222.
92 Verdy to Waldersee, 24 May 1891, GStA Berlin, Waldersee Papers No. 53.
93 See Förster, *Der doppelte Militarismus*, pp. 36ff.
94 See Holstein to Eulenburg, 17 June 1891, *Eulenburgs Korrespondenz*, I, No. 525.
95 Kaiser Wilhelm II to Kaltenborn, 15 June 1891, BA-MA Freiburg, KGFA W10/50266.
96 Kaiser Wilhelm II to Caprivi, 15 June 1891, printed from the original in the files of the Reichskanzlei (Militärsachen 2b vol. I 1255 Mappe 13) in Heinrich Otto Meisner, 'Der Reichskanzler Caprivi', *Zeitschrift für die gesamte Staatswissenschaft*, 3 (1955), pp. 742–3.
97 Waldersee, diary entry for 6 December 1891, GStA Berlin, Waldersee Papers; omitted from Meisner, II, pp. 224–5.
98 Holstein to Eulenburg, 16 and 17 June 1891, *Eulenburgs Korrespondenz*, I, Nos. 523 and 525.
99 Caprivi to Kaiser Wilhelm II, 16 June 1891, in Meisner, 'Reichskanzler Caprivi', p. 744.
100 Holstein to Eulenburg, 16 June 1891, Eulenburg to Holstein, 17 June 1891, *Eulenburgs Korrespondenz*, I, Nos. 523 and 524.
101 Kaiser Wilhelm II to Caprivi, 16 June 1891, in Meisner, 'Reichskanzler Caprivi', pp. 744–5.
102 Holstein to Eulenburg, 17 June 1891, *Eulenburgs Korrespondenz*, I, No. 525.
103 Kiderlen-Wächter to Holstein, 10 August 1891, *Holstein Papers*, III, No. 346; also Eulenburg to Caprivi, 31 July 1891, *Eulenburgs Korrespondenz*, I, No. 530.
104 Waldersee, diary entry for 13 August 1891, GStA Berlin, Waldersee Papers; cf. Meisner, II, pp. 214–15.
105 Caprivi, Denkschrift betreffend die weitere Entwicklung der deutschen Wehrkraft, Caprivi to Kaltenborn, 27 August 1891, BA-MA Freiburg, W10/50266.
106 See below, pp. 456–65; also Förster, *Der doppelte Militarismus*, pp. 40–1.
107 Holstein to Eulenburg, 17 June 1891, *Eulenburgs Korrespondenz*, I, No. 525.
108 See Empress Frederick to Queen Victoria, 19 December 1891, RA Z52/7.
109 Kaiser Wilhelm II to Queen Victoria, 19 December 1891, RA I59/55. See Waldersee, diary entry for 19 December 1891, GStA Berlin, Waldersee Papers; omitted from Meisner, II, p. 227.

110 Caprivi to Holstein, 27 January 1892, *Holstein Papers*, III, No. 355. See Holstein to Kaiser Wilhelm II, 30 January 1892, ibid., No. 358.

111 Eulenburg to Holstein, 4 February 1892, ibid., No. 359.

112 Prince Ernst zu Hohenlohe-Langenburg to his father, 28 January 1892, Hohenlohe-Zentralarchiv Neuenstein, Hermann Hohenlohe-Langenburg Papers, Bü. 59.

113 Waldersee, diary entry for 21 November 1891, GStA Berlin, Waldersee Papers; Meisner, II, p. 223.

114 Waldersee, diary entry for 24 April 1890, GStA Berlin, Waldersee Papers; cf. Meisner, II, p. 124.

115 Kaiser Wilhelm II, marginal notes on the article 'Warnungszeichen' in *Hamburger Nachrichten*, 26 March 1890, Wedel Papers, Frankfurt a.M.; cited in Wedel, *Zwischen Kaiser und Kanzler*, pp. 67–8.

116 Wedel, *Zwischen Kaiser und Kanzler*, pp. 67–8.

117 Ibid., pp. 95–6.

118 Waldersee, diary entry for 24 April 1890, GStA Berlin, Waldersee Papers; cf. Meisner, II, p. 124.

119 Waldersee, diary entry for 1 May 1890, GStA Berlin, Waldersee Papers; omitted from Meisner, II, p. 126.

120 Waldersee, diary entry for 4 May 1890, GStA Berlin, Waldersee Papers; omitted from Meisner, II, p. 126.

121 Waldersee, diary entry for 24 April 1890, GStA Berlin, Waldersee Papers; omitted for the most part from Meisner, II, p. 124.

122 Waldersee, diary entry for 4 May 1890, GStA Berlin, Waldersee Papers; cf. Meisner, II, p. 126.

123 Waldersee, diary entry for 3 September 1890, GStA Berlin, Waldersee Papers; omitted from Meisner, II, p. 142.

124 Wedel, *Zwischen Kaiser und Kanzler*, pp. 110–15. Cf. ibid., p. 152.

125 Waldersee, diary entry for 25 September 1890, GStA Berlin, Waldersee Papers; cf. Meisner, II, p. 149.

126 Waldersee, diary entry for 1 September 1890, GStA Berlin, Waldersee Papers; cf. Meisner, II, p. 142.

127 Waldersee, diary entries for 30 August and 3 September 1890, GStA Berlin, Waldersee Papers; cf. Meisner, II, pp. 141–2.

128 Wedel, *Zwischen Kaiser und Kanzler*, p. 131.

129 Waldersee, diary entry for 22 November 1890, GStA Berlin, Waldersee Papers; omitted from Meisner, II, p. 163.

130 Eulenburg to Holstein, 30 January 1892, *Holstein Papers*, III, No. 357.

131 Waldersee, diary entry for 28 November 1891, GStA Berlin, Waldersee Papers; cf. Meisner, II, pp. 223–4.

132 Waldersee, diary entry for 24 April 1890, GStA Berlin, Waldersee Papers; cf. Meisner, II, p. 124.

133 Waldersee, diary entry for 16 March 1892, GStA Berlin, Waldersee Papers; Meisner, II, pp. 234–5.

134 Brauer, reports of 11 and 17 February and 19 December 1891, cited in Röhl, *Germany without Bismarck*, pp. 75–6.

135 Waldersee, diary entry for 16 November 1890, GStA Berlin, Waldersee Papers; omitted from Meisner, II, p. 161.

136 Wedel, *Zwischen Kaiser und Kanzler*, p. 132.

137 Waldersee, diary entry for 29 October 1890, GStA Berlin, Waldersee Papers; Meisner, II, p. 156.

138 Waldersee, diary entry for 4 November 1890, GStA Berlin, Waldersee Papers; omitted from Meisner, II, p. 157.
139 Waldersee, diary entry for 1 November 1890, GStA Berlin, Waldersee Papers; omitted from Meisner, II, p. 157.
140 Waldersee, diary entry for 4 November 1890, GStA Berlin, Waldersee Papers; omitted from Meisner, II, p. 157; Empress Frederick to Queen Victoria, 11 November 1890, RA Z49/33.
141 Wedel, *Zwischen Kaiser und Kanzler*, pp. 129–30 and 151.
142 Ibid.; also Brauer to Grand Duke Friedrich, 9 November 1890, Fuchs, *Großherzog von Baden*, III, No. 1081.
143 Empress Frederick to Queen Victoria, 11 November 1890, RA Z49/33.
144 Wedel, *Zwischen Kaiser und Kanzler*, p. 130.
145 Waldersee, diary entries for 16 and 18 November 1890, GStA Berlin, Waldersee Papers; cf. Meisner, II, pp. 161–2.
146 Waldersee, diary entry for 4 November 1890, GStA Berlin, Waldersee Papers; omitted from Meisner, II, p. 157.
147 Waldersee, diary entry for 7 November 1890, GStA Berlin, Waldersee Papers; omitted from Meisner, II, p. 158.
148 Waldersee, diary entry for 23 November 1890, GStA Berlin, Waldersee Papers; omitted from Meisner, II, p. 163. Cf. Waldersee's retrospective note of 21 November 1894 in Meisner, II, p. 332.
149 Waldersee, diary entry for 3 January 1891, GStA Berlin, Waldersee Papers; cf. Meisner, II, p. 174.
150 Waldersee, diary entry for 2 January 1891, GStA Berlin, Waldersee Papers; omitted from Meisner, II, p. 172.
151 Wedel, *Zwischen Kaiser und Kanzler*, p. 172.
152 Count Karl von Kalnein to Eulenburg, 8 March 1892, *Eulenburgs Korrespondenz*, II, No. 602. See below, ch. 29.
153 Waldersee, diary entry for 9 January 1891, GStA Berlin, Waldersee Papers; omitted from Meisner, II, p. 174.
154 Waldersee, diary entry for 10 January 1891, GStA Berlin, Waldersee Papers; cf. Meisner, II, p. 175.
155 Waldersee, diary entry for 18 January 1891, GStA Berlin, Waldersee Papers; omitted from Meisner, II, p. 176.
156 Wedel, *Zwischen Kaiser und Kanzler*, p. 192.
157 Waldersee, diary entry for 22 June 1891, GStA Berlin, Waldersee Papers; cf. Meisner, II, pp. 209–10.
158 Waldersee, diary entry for 2 November 1891, GStA Berlin, Waldersee Papers; cf. Meisner, II, pp. 220–1.
159 Wedel, *Zwischen Kaiser und Kanzler*, p. 108.
160 Ibid., p. 116.
161 Waldersee, diary entry for 27 October 1890, GStA Berlin, Waldersee Papers; omitted from Meisner, II, p. 156.
162 Above, pp. 401–3.
163 Waldersee, diary entry for 21 November 1890, GStA Berlin, Waldersee Papers; abridged in Meisner, II, p. 163.
164 Waldersee, diary entry for 4 January 1891, GStA Berlin, Waldersee Papers; cf. Meisner, II, p. 174.
165 Waldersee, diary entries for 19 and 24 November 1890, GStA Berlin, Waldersee Papers; cf. Meisner, II, pp. 162ff.

166 Waldersee, diary entry for 27 October 1890, GStA Berlin, Waldersee Papers; omitted from Meisner, II, p. 156.
167 Waldersee, diary entry for 14 January 1891, GStA Berlin, Waldersee Papers; cf. Meisner, II, p. 175.
168 Holstein to Eulenburg, 17 March 1891, Eulenburg to Holstein, 19 March 1891, *Eulenburgs Korrespondenz*, I, Nos. 489–90.
169 See Thomas A. Kohut, *Wilhelm II and the Germans. A Study in Leadership*, New York, Oxford 1991, p. 286.
170 Waldersee, diary entry for 7 December 1890, GStA Berlin, Waldersee Papers; cf. Meisner, II, p. 164.
171 Waldersee, diary entry for 10 December 1890, GStA Berlin, Waldersee Papers; cf. Meisner, II, pp. 164–5.
172 Waldersee, diary entry for 9 January 1891, GStA Berlin, Waldersee Papers; cf. Meisner, II, p. 174.
173 Waldersee, diary entry for 19 January 1891, GStA Berlin, Waldersee Papers; cf. Meisner, II, p. 176.
174 Bigelow to Kaiser Wilhelm II, 26 January 1893, GStA Berlin, Brand.-Preuß. Hausarchiv Rep. 53J Lit. B. No. 10 Bigelow 1.
175 Waldersee, diary entry for 14 January 1891, GStA Berlin, Waldersee Papers; cf. Meisner, II, p. 175.
176 Waldersee, diary entries for 11 and 27 April 1890, GStA Berlin, Waldersee Papers; omitted from Meisner, II, pp. 124 and 138; also above, pp. 168–9.
177 Wedel, *Zwischen Kaiser und Kanzler*, pp. 108, 116 and 125; Waldersee, diary entry for 1 June 1890, GStA Berlin, Waldersee Papers; Meisner, II, p. 129.
178 Waldersee, diary entry for 11 August 1890, GStA Berlin, Waldersee Papers; omitted from Meisner, II, p. 138.
179 Waldersee, diary entry for 4 October 1890, GStA Berlin, Waldersee Papers; cf. Meisner, II, pp. 151–2.
180 Waldersee, diary entry for 20 December 1890, GStA Berlin, Waldersee Papers; omitted from Meisner, II, p. 169.
181 Waldersee, diary entry for 21 October 1894, GStA Berlin, Waldersee Papers; cf. Meisner, II, p. 327. See *Eulenburgs Korrespondenz*, III, pp. 1535, 1639–40, 1848–9 and 1951–2.
182 Wedel, *Zwischen Kaiser und Kanzler*, p. 142.

16 THE FALL OF THE COURT GENERALS

1 See *Young Wilhelm*, pp. 424–30, 490–515, 599–629 *et passim*.
2 Ibid., pp. 455–62, as well as above, pp. 197 9.
3 Hoffmann, diary entries for 22 and 24 March and 23 April 1890, Hoffmann Papers, Freiburg. See above, p. 320.
4 Wedel, *Zwischen Kaiser und Kanzler*, p. 128.
5 See above, pp. 225–7; also Cecil, *Wilhelm II*, I, pp. 182–4.
6 Waldersee, diary entry for 29 January 1891, GStA Berlin, Waldersee Papers; Meisner, II, p. 180.
7 Kiderlen-Wächter to Holstein, 19 July and 21 September 1890, *Holstein Papers*, III, Nos. 324 and 328.
8 Wedel, *Zwischen Kaiser und Kanzler*, p. 129.
9 Eulenburg to Holstein, 1 August 1890, *Holstein Papers*, III, No. 327.
10 Wedel, *Zwischen Kaiser und Kanzler*, p. 129. See above, pp. 373–5.
11 Waldersee to Verdy, 30 October 1891, GStA Berlin, Waldersee Papers No. 53.

12 Waldersee, diary entry for 7 June 1890, GStA Berlin, Waldersee Papers; omitted from Meisner, II, p. 130.
13 Waldersee, diary entry for 30 August 1890, GStA Berlin, Waldersee Papers; cf. Meisner, II, pp. 141–2.
14 Waldersee, diary entry for 13 September 1890, GStA Berlin, Waldersee Papers; cf. Meisner, II, p. 144.
15 Waldersee, diary entry for 10 October 1890, GStA Berlin, Waldersee Papers; cf. Meisner, II, p. 153.
16 Eulenburg to Holstein, 1 August 1890, *Holstein Papers*, III, No. 327.
17 See Heinrich Otto Meisner, *Militärattachés und Militärbevollmächtigte in Preußen und im Deutschen Reich. Ein Beitrag zur Geschichte der Militärdiplomatie*, Berlin 1957, pp. 56–7 and 73ff.; Gerhard Ritter, *Die deutschen Militär-Attachés und das Auswärtige Amt. Aus den verbrannten Akten des Großen Generalstabes*, Heidelberg 1959, pp. 21ff. and 33ff. The letters from Holstein to Eulenburg of 6 and 27 November 1890 and 10 January 1891 are informative on this issue: *Eulenburgs Korrespondenz*, I, Nos. 433, 444 and 464; also Forsbach, *Kiderlen-Wächter*, I, pp. 104–5.
18 Waldersee, diary entry for 10 August 1890, GStA Berlin, Waldersee Papers; cf. Meisner, II, pp. 135–6.
19 Waldersee, diary entry for 10 August 1890, GStA Berlin, Waldersee Papers; omitted from Meisner, II, p. 137.
20 Waldersee, diary entry for 11 September 1890, GStA Berlin, Waldersee Papers; cf. Meisner, II, p. 144.
21 Kiderlen-Wächter to Holstein, 21 September 1890, *Holstein Papers*, III, No. 328.
22 Waldersee, diary entry for 11 September 1890, GStA Berlin, Waldersee Papers; cf. Meisner, II, p. 144.
23 Kiderlen-Wächter to Holstein, 21 September 1890, *Holstein Papers*, III, No. 328.
24 Waldersee, diary entries for 21 and 24 September 1890, GStA Berlin, Waldersee Papers; cf. Meisner, II, pp. 145–9; also Wedel, *Zwischen Kaiser und Kanzler*, pp. 126ff.
25 Kiderlen-Wächter to Holstein, 21 September 1890, *Holstein Papers*, III, No. 328.
26 Waldersee, diary entries for 21 and 24 September 1890, GStA Berlin, Waldersee Papers; cf. Meisner, II, pp. 145–9; also Wedel, *Zwischen Kaiser und Kanzler*, pp. 126ff.
27 Kiderlen-Wächter to Holstein, 21 September 1890, *Holstein Papers*, III, No. 328.
28 Waldersee, diary entry for 24 September 1890, GStA Berlin, Waldersee Papers; cf. Meisner, II, p. 148.
29 Waldersee, diary entry for 21 September 1890, GStA Berlin, Waldersee Papers; cf. Meisner, II, pp. 145ff.
30 Waldersee, diary entry for 25 September 1890, GStA Berlin, Waldersee Papers; omitted from Meisner, II, p. 151.
31 Waldersee, diary entry for 21 September 1890, GStA Berlin, Waldersee Papers; cf. Meisner, II, pp. 145ff. See Wedel, *Zwischen Kaiser und Kanzler*, pp. 125–6.
32 Kiderlen-Wächter to Holstein, 21 September 1890, *Holstein Papers*, III, No. 328.
33 Waldersee, diary entry for 11 September 1890, GStA Berlin, Waldersee Papers; omitted from Meisner, II, p. 145.
34 Waldersee, diary entry for 3 October 1890, GStA Berlin, Waldersee Papers; Meisner, II, p. 152.
35 Waldersee, diary entry for 21 September 1890, GStA Berlin, Waldersee Papers; cf. Meisner, II, pp. 145–6; also the diary entry of 9 October 1890, GStA Berlin, Waldersee Papers; omitted from Meisner II, p. 154.
36 Kiderlen-Wächter to Holstein, 21 September 1890, *Holstein Papers*, III, No. 328.
37 Wedel, *Zwischen Kaiser und Kanzler*, p. 126.

38 Waldersee, diary entry for 9 October 1890, GStA Berlin, Waldersee Papers; omitted from Meisner, II, p. 153.
39 Waldersee, diary entry for 16 October 1890, GStA Berlin, Waldersee Papers; omitted from Meisner, II, p. 155.
40 Wedel, *Zwischen Kaiser und Kanzler*, p. 128.
41 Caprivi's instruction of 11 December 1890 is printed in Meisner, *Militärattachés*, pp. 73ff.; see Waldersee's marginal notes in Ritter, *Militär-Attachés*, pp. 33ff.
42 Waldersee, diary entry for 13 December 1890, GStA Berlin, Waldersee Papers; Meisner, II, p. 165.
43 Waldersee, diary entry for 17 December 1890, GStA Berlin, Waldersee Papers; cf. Meisner, II, pp. 166–7.
44 Holstein to Brandt, 26 December 1890, *Holstein Papers*, III, No. 332.
45 Waldersee, diary entry for 2 January 1891, GStA Berlin, Waldersee Papers; omitted from Meisner, II, p. 173.
46 Wedel, *Zwischen Kaiser und Kanzler*, pp. 134 and 148. See above, p. 403.
47 Waldersee, diary entries for 15 and 16 January 1891, GStA Berlin, Waldersee Papers; cf. Meisner, II, pp. 175–6.
48 Wedel, *Zwischen Kaiser und Kanzler*, pp. 147–8.
49 Waldersee, diary entry for 25 January 1891, GStA Berlin, Waldersee Papers; omitted from Meisner, II, p. 177.
50 Waldersee, diary note of late January 1891, GStA Berlin, Waldersee Papers; omitted from Meisner, II, p. 177.
51 Waldersee, diary entry for 28 January 1891, GStA Berlin, Waldersee Papers; cf. Meisner, II, pp. 177–8.
52 Wedel, *Zwischen Kaiser und Kanzler*, p. 137.
53 Waldersee, diary entry for 28 January 1891, GStA Berlin, Waldersee Papers; cf. Meisner, II, p. 179; also Waldersee to Grand Duke of Baden, 28 January 1891, Fuchs, *Großherzog von Baden*, III, No. 1094.
54 Wedel, *Zwischen Kaiser und Kanzler*, pp. 139–40.
55 Ibid., p. 142.
56 Waldersee, diary entry for 31 January 1891, GStA Berlin, Waldersee Papers; cf. Meisner, II, pp. 180ff. On Holstein's and Philipp Eulenburg's role, and that played by the other Foreign Office officials mentioned in connection with Waldersee's downfall, see Holstein to Eulenburg, 10 January, 3 February and 27 March 1891, *Eulenburgs Korrespondenz*, I, Nos. 464, 470 and 495; August Eulenburg to Philipp Eulenburg, 1 February 1891, ibid., No. 469; Philipp Eulenburg to August Eulenburg, 14 February 1891, ibid., No. 472; and Waldersee to Eulenburg, 18 February 1891, ibid., No. 474. See also Waldersee to Verdy, 18 February 1891, GStA Berlin, Waldersee Papers.
57 Waldersee, diary entry for 1 February 1891, GStA Berlin, Waldersee Papers; cf. Meisner, II, p. 182.
58 Waldersee, diary entry for 2 February 1891, GStA Berlin, Waldersee Papers; omitted from Meisner, II, p. 184.
59 Waldersee, diary entry for 6 February 1891, GStA Berlin, Waldersee Papers; omitted from Meisner, II, pp. 185–6.
60 Wedel, *Zwischen Kaiser und Kanzler*, pp. 142–8.
61 Ibid., p. 150.
62 See Röhl, *Kaiser and His Court*, p. 128.
63 Eulenburg to Holstein, 6 August 1889, *Eulenburgs Korrespondenz*, I, No. 230.
64 Waldersee, diary entry for 17 November 1889, GStA Berlin, Waldersee Papers; omitted from Meisner, II, p. 78.

65 Waldersee, diary entry for 23 March 1890, GStA Berlin, Waldersee Papers; omitted from Meisner, II, p. 122.
66 See Wedel, *Zwischen Kaiser und Kanzler*, pp. 71–8, 81–93, 119–25, 157–62 *et passim*.
67 Ibid., p. 149. On the transfer of Count Ernst von Wedel from the service of Grand Duke Carl Alexander of Saxe-Weimar-Eisenach to the Kaiser's court as Oberstallmeister, see the unpublished memoirs of Wedel's daughter Countess Alice zu Lynar, in the possession of Count Peter von Wedel, Bad Driburg.
68 Kaiser Wilhelm II to Caprivi, 9 May 1891, *Große Politik*, VII, pp. 295–6. See Helmuth Rogge, *Holstein und Hohenlohe. Neue Beiträge zu Friedrich von Holsteins Tätigkeit als Mitarbeiter Bismarcks und als Ratgeber Hohenlohes*, Stuttgart 1957, pp. 358ff.; also Holstein to Eulenburg, 10 May 1891, *Eulenburgs Korrespondenz*, I, No. 514.
69 Holstein to Eulenburg, 10 May 1891, *Eulenburgs Korrespondenz*, I, No. 514.
70 Count Carl von Wedel, promemoria, 13 June 1894, Wedel Papers, Frankfurt a.M.
71 Ibid.
72 See Eulenburg to Holstein, 6 November 1891 and 27 April 1892, *Eulenburgs Korrespondenz*, I, No. 552, and II, No. 647.
73 Prince Ernst zu Hohenlohe-Langenburg to his father, 13 June 1892, Hohenlohe-Zentralarchiv Neuenstein, Hermann Hohenlohe-Langenburg Papers, Bü. 59.
74 Eulenburg to Holstein, 20 July 1892, *Holstein Papers*, III, No. 374.
75 Wedel, promemoria, 13 June 1894, Wedel Papers, Frankfurt a.M.
76 Ibid.
77 Ibid.
78 Ibid.
79 Eulenburg to Holstein, 2 September 1892, *Eulenburgs Korrespondenz*, II, No. 696.
80 Kiderlen-Wächter to Eulenburg, 3 March 1893, ibid., No. 778.
81 On this intrigue, see Holstein to Eulenburg, 7, 18 and 21 November 1892, *Eulenburgs Korrespondenz*, II, Nos. 727, 732 and 736; Eulenburg to Caprivi, 14 November 1892, ibid., No. 730; Eulenburg to Holstein, 19 November 1892, ibid., No. 733.
82 Malet to Lord Rosebery, 20 December 1893, RA I59/115.
83 Wedel, promemoria, 13 June 1894, Wedel Papers, Frankfurt a.M.
84 Ibid.
85 Wedel, *Zwischen Kaiser und Kanzler*, pp. 181–93.
86 See above, pp. 162–3.
87 Széchényi to Kálnoky, 23 February 1889, HHStA Vienna, PA III 136.
88 Waldersee, diary entry for 26 August 1888, GStA Berlin, Waldersee Papers; cf. Meisner, II, p. 1.
89 Waldersee, diary entry for 9 October 1888, GStA Berlin, Waldersee Papers; omitted from Meisner, II, pp. 4–5.
90 Waldersee, diary entry for 26 November 1888, GStA Berlin, Waldersee Papers; cf. Meisner, II, pp. 22–3.
91 Waldersee, diary entry for 30 August 1890, GStA Berlin, Waldersee Papers; cf. Meisner, II, pp. 141–2.
92 Waldersee, diary entry for 19 August 1890, GStA Berlin, Waldersee Papers; omitted from Meisner, II, p. 140; similarly, Waldersee to Verdy, 28 May 1891, GStA Berlin, Waldersee Papers, No. 53.
93 Waldersee, diary entry for 19 August 1890, GStA Berlin, Waldersee Papers; omitted from Meisner, II, p. 140.
94 Waldersee, diary entry for 19 August 1890, GStA Berlin, Waldersee Papers; omitted from Meisner, II, p. 140.
95 Waldersee to Verdy, 28 May 1891, GStA Berlin, Waldersee Papers, No. 53.

96 Verdy to Waldersee, 24 May 1892, ibid.
97 Waldersee, diary entry for 13 October 1892, ibid.; omitted from Meisner, II, pp. 265–6.
98 Empress Frederick to Queen Victoria, 13 August 1892, RA Z53/24. See Waldersee's diary entry for 29 August 1892, Meisner, II, p. 264.
99 Kaiser Wilhelm II to Hans von Plessen, telegram, 29 December 1892, BA-MA Freiburg, Plessen Papers, Msg 1/3117.
100 Waldersee, diary entry for 22 October 1894, GStA Berlin, Waldersee Papers; omitted from Meisner, II, p. 327.
101 Waldersee, diary entry for 14 December 1894, GStA Berlin, Waldersee Papers; omitted from Meisner, II, p. 333.
102 Colonel Russell, report of 13 February 1891, RA I59/2.
103 Waldersee, diary entry for 17 December 1894, GStA Berlin, Waldersee Papers; cf. Meisner, II, p. 334.

17 THE SCHOOL BILL CRISIS AND THE FRAGMENTATION OF POWER

1 Waldersee, diary entry for 2 May 1891, GStA Berlin, Waldersee Papers; omitted from Meisner, II, pp. 207–8.
2 Waldersee to Verdy, 28 May 1891, GStA Berlin, Waldersee Papers, No. 53.
3 See Nipperdey, *Deutsche Geschichte 1866–1918*, I, pp. 535–6.
4 Empress Frederick to her daughter Princess Viktoria of Schaumburg-Lippe, 1 March 1892 *et passim*, AdHH Schloss Fasanerie.
5 Oncken, *Bennigsen*, II, pp. 557–63.
6 Eulenburg to Holstein, 22 January 1892, *Eulenburgs Korrespondenz*, II, No. 574.
7 Prince Ernst zu Hohenlohe-Langenburg to his father, 7 March 1892, Hohenlohe-Zentralarchiv Neuenstein, Hermann Hohenlohe-Langenburg Papers, Bü. 59.
8 Waldersee, diary entry for 20 February 1892, GStA Berlin, Waldersee Papers; omitted from Meisner, II, p. 232.
9 Hereditary Prince Bernhard of Saxe-Meiningen to Waldersee, 30 January 1892, GStA Berlin, Waldersee Papers No. 36.
10 See above, pp. 233–4.
11 Eulenburg to Kaiser Wilhelm II, 21 [sic] January 1892, Johannes Haller, *Aus dem Leben des Fürsten Philipp zu Eulenburg-Hertefeld*, Berlin, 1924, p. 66.
12 Holstein to Eulenburg, 24 January 1892, *Eulenburgs Korrespondenz*, II, No. 575.
13 Holstein to Eulenburg, 27 January 1892, ibid., No. 578. See Oncken, *Bennigsen*, II, pp. 557–63.
14 Waldersee, diary entries for 30 January and 6 February 1892, GStA Berlin, Waldersee Papers; these passages are omitted from Meisner, II, p. 232; also Brauer to Turban, 1 February 1892, Fuchs, *Großherzog von Baden*, III, No. 1150.
15 Waldersee, diary entry for 21 February 1892, GStA Berlin, Waldersee Papers; cf. Meisner, II, pp. 232–3.
16 Waldersee, diary entry for 12 March 1892, GStA Berlin, Waldersee Papers; cf. Meisner, II, p. 234; also Brauer to Turban, 6 February 1892, Fuchs, *Großherzog von Baden*, III, No. 1151.
17 Kaiser Wilhelm II, speech of 17 December 1890, Penzler, *Reden Kaiser Wilhelms II. in den Jahren 1888–1895*, pp. 163–7; Elkind, *German Emperor's Speeches*, p. 167.
18 Eulenburg to Holstein, 22 January 1892, *Eulenburgs Korrespondenz*, II, No. 574.
19 Helldorff to Eulenburg, 7 March 1892, ibid., No. 600; also Brauer to Turban, 16 February 1892, Fuchs, *Großherzog von Baden*, III, No. 1154.
20 On Kaiser Wilhelm II's speech of 24 February 1892, see above, pp. 384–5.
21 Above, pp. 233–4.

22 Holstein to Eulenburg, 27 January 1892, *Eulenburgs Korrespondenz*, II, No. 578.
23 Eulenburg to Holstein, 28 and 29 January and 14 February 1892, *Holstein Papers*, III, Nos. 356, 357 and 361.
24 Eulenburg to Holstein, 28 and 29 January 1892, ibid., Nos. 356 and 357; also August Eulenburg to Philipp Eulenburg, 14 February 1892, *Eulenburgs Korrespondenz*, II, No. 587; Holstein to Eulenburg, 17, 18 and 27 February 1892, ibid., Nos. 588, 590 and 593.
25 Eulenburg to Holstein, 22 February 1892, *Holstein Papers*, III, No. 362.
26 Holstein to Eulenburg, 1 March 1892, *Eulenburgs Korrespondenz*, II, No. 596; Eulenburg to Holstein, 5 March 1892, *Holstein Papers*, III, No. 364.
27 Eulenburg to Kaiser Wilhelm II, 10 March 1892, *Eulenburgs Korrespondenz*, II, No. 604.
28 Helldorff to Eulenburg, 24 October 1892, ibid., No. 723.
29 Helldorff to Eulenburg, 7 March 1892, ibid., No. 600.
30 See above, pp. 384–5.
31 Empress Frederick to Queen Victoria, 20 March 1892, RA Z52/34. Wilhelm's inner-ear illness is described at some length in ch. 13 of *Young Wilhelm*, pp. 306–24.
32 Cf. Brauer to Turban, 18 March 1892, Fuchs, *Großherzog von Baden*, III, No. 1166.
33 August Eulenburg to Philipp Eulenburg, 19 March 1892, *Eulenburgs Korrespondenz*, II, No. 617.
34 Leuthold to Eulenburg, 15 March 1892, ibid., No. 607.
35 Malet to Ponsonby, 22 and 25 March 1892, RA I59/63 and 66; Malet to Salisbury, 2 April 1892, RA I59/68.
36 Kaiser Wilhelm II to Queen Victoria, 15 March 1892, RA S25/53.
37 Helldorff to Eulenburg, 24 March 1892, *Eulenburgs Korrespondenz*, II, No. 628.
38 August Eulenburg to Philipp Eulenburg, 19 March 1892, ibid., No. 617.
39 Helldorff to Eulenburg, 24 March 1892, ibid., No. 628.
40 Brauer to Turban, 19 March 1892, Fuchs, *Großherzog von Baden*, III, No. 1167.
41 August Eulenburg to Philipp Eulenburg, 19 March 1892, *Eulenburgs Korrespondenz*, II, No. 617; Verdy to Waldersee, Sunday, ? March 1892, GStA Berlin, Waldersee Papers No. 53.
42 Kaiser Wilhelm II, marginal note on Caprivi's letter of resignation of 19 March 1892 cited in Herzfeld, *Johannes von Miquel*, II, p. 306. See *Eulenburgs Korrespondenz*, II, p. 808.
43 Holstein to Eulenburg, 18 March 1892, *Eulenburgs Korrespondenz*, II, No. 613.
44 Eulenburg to Kaiser Wilhelm II, 19 March 1892, ibid., No. 615. On the Guelph Fund and the bribing of King Ludwig II of Bavaria in 1870, see ibid., p. 801; R. Nöll von der Nahmer, *Bismarcks Reptilienfonds*, Mainz 1968, pp. 142–61; H. Maatz, *Bismarck und Hannover 1866–1898*, Hildesheim 1971; S. A. Stehlin, *Bismarck and the Guelph Problem 1866–1890*, The Hague 1973; Hans Rall, *König Ludwig II. und Bismarcks Ringen um Bayern*, Munich 1973.
45 Eulenburg to Caprivi, 19 March 1892, *Eulenburgs Korrespondenz*, II, No. 616.
46 Verdy to Waldersee, March 1892, GStA Berlin, Waldersee Papers No. 53; Marschall's diary entry for 20 March 1892, *Eulenburgs Korrespondenz*, II, p. 813.
47 Holstein to Eulenburg, 31 March 1892, *Eulenburgs Korrespondenz*, II, No. 634.
48 Verdy to Waldersee, March 1892, GStA Berlin, Waldersee Papers No. 53.
49 Holstein to Eulenburg, 18 and 19 March 1892, *Eulenburgs Korrespondenz*, II, Nos. 613 and 618.
50 Holstein to Eulenburg, 19 March 1892, ibid., No. 618; Helldorff to Eulenburg, 24 March 1892, ibid., No. 628.
51 August Eulenburg to Philipp Eulenburg, 19 March 1892, ibid., No. 617.
52 August Eulenburg to Philipp Eulenburg, 20 March 1892, cited ibid., No. 619; Marschall's diary entry for 20 March 1892, ibid., p. 814.
53 Holstein to Eulenburg, 20 March 1892, ibid., No. 620; August Eulenburg to Philipp Eulenburg, 20 March 1892, ibid., No. 621.

54 August Eulenburg to Philipp Eulenburg, 20 March 1892, ibid., No. 621; Holstein to Eulenburg, 21 and 23 March 1892, ibid., Nos. 623 and 625; Marschall's diary entry for 21 March 1892, cited ibid., p. 819; Brauer to Turban, 20–2 March 1892, Fuchs, *Großherzog von Baden*, III, Nos. 1168–71.
55 August Eulenburg to Philipp Eulenburg, 23 March 1892, *Eulenburgs Korrespondenz*, II, No. 626.
56 On Bosse's appointment as minister for ecclesiastical affairs, see the extracts from his diary cited ibid., pp. 822–3. See also Helldorff to Eulenburg, 24 March 1892, ibid., No. 628.
57 Brauer to Grand Duke of Baden, 25 March 1892, Fuchs, *Großherzog von Baden*, III, No. 1174.
58 Malet to Salisbury, 2 April 1892, RA I59/68.
59 Eulenburg to Holstein, 26 March 1892, *Holstein Papers*, III, No. 366.
60 Kaiser Wilhelm II to Queen Victoria, 12 April 1892, RA I59/72.
61 See Eulenburg to Holstein, 26 March 1892, *Holstein Papers*, III, No. 366.
62 Empress Frederick to Queen Victoria, 24 March 1892, RA Z52/35.
63 Kaiser Wilhelm II to Lucanus, 18 April 1892, cited in *Eulenburgs Korrespondenz*, II, p. 855; Marschall, diary entries, 12–21 April, cited ibid.
64 Empress Frederick to Queen Victoria, 13 August 1892, RA Z53/24. See also Spitzemberg, *Tagebuch*, p. 301.
65 Marschall, diary entry for 12 April 1892, cited in *Eulenburgs Korrespondenz*, II, p. 844.
66 Verdy to Waldersee, 11 May 1892, GStA Berlin, Waldersee Papers No. 53.
67 Waldersee, diary entry for 28 March 1892, ibid.; Meisner, II, pp. 236–7.
68 Waldersee, diary entry for 14 April 1892, GStA Berlin, Waldersee Papers; cf. Meisner, II, pp. 238–9.
69 Waldersee, diary entry for 16 April 1892, GStA Berlin, Waldersee Papers; omitted from Meisner, II, pp. 239–40.
70 Waldersee, diary entry for 28 April 1892, GStA Berlin, Waldersee Papers; omitted from Meisner, II, p. 240.
71 Waldersee, diary entry for 12 June 1892, GStA Berlin, Waldersee Papers; cf. Meisner, II, pp. 244–5.
72 Waldersee, diary entry for 28 April 1892, GStA Berlin, Waldersee Papers; omitted from Meisner, II, p. 240.
73 Waldersee, diary entry for 29 April 1892, GStA Berlin, Waldersee Papers; omitted from Meisner, II, p. 240.
74 Waldersee, diary entry for 8 May 1892, GStA Berlin, Waldersee Papers; Meisner, II, pp. 240–1.
75 See e.g. Waldersee, diary entries for 7 October, 13 November and 1 December 1892, GStA Berlin, Waldersee Papers; Meisner, II, pp. 265 and 269.
76 Waldersee, diary entry for 7 October 1892, GStA Berlin, Waldersee Papers; omitted from Meisner, II, p. 265.
77 Eulenburg to Holstein, 20 July 1892, *Holstein Papers*, III, No. 374.
78 Waldersee, diary entry for 28 April 1892, GStA Berlin, Waldersee Papers; omitted from Meisner, II, p. 240.
79 Waldersee, diary entry for 29 April 1892, GStA Berlin, Waldersee Papers; omitted from Meisner, II, p. 240.
80 Kaiserin Auguste Viktoria to Kaiser Wilhelm II, 3–4 July 1892, GStA Berlin, BPHA Rep. 53T Preußen: An Kaiser Wilhelm II., Bd IV.
81 Kiderlen-Wächter to Holstein, 13 July 1892, *Holstein Papers*, III, No. 372.
82 Kaiserin Auguste Viktoria to Kaiser Wilhelm II, 14–15 July 1892, GStA Berlin, BPHA Rep. 53T Preußen: An Kaiser Wilhelm II., Bd IV.
83 Kaiser Wilhelm II to Caprivi, 20 July 1892, printed in *Holstein Papers*, III, No. 373, enclosure.

84 Kiderlen-Wächter to Holstein, 20 July 1892, ibid., No. 373; Eulenburg to Holstein, 20 July 1892, ibid., No. 374.
85 Kaiser Wilhelm II to Zelle, 7 October 1892, printed in *Eulenburgs Korrespondenz*, II, p. 945.
86 Holstein to Eulenburg, 27 September 1892, ibid., No. 707.
87 Waldersee, diary entries for 7, 9 and 13 October 1892, GStA Berlin, Waldersee Papers; omitted from Meisner, II, p. 265.
88 Waldersee, diary entry for 28 October 1892, GStA Berlin, Waldersee Papers; omitted from Meisner, II, p. 267. See also Empress Frederick to Queen Victoria, 24 October 1891, RA Z51/40.
89 Waldersee, diary entries for 7, 9 and 13 October 1892, GStA Berlin, Waldersee Papers; omitted from Meisner, II, p. 265.
90 See above, pp. 388–96; also Förster, *Der doppelte Militarismus*, pp. 36ff.
91 See e.g. Caprivi to Eulenburg, 28 February 1892, Eulenburg to Kaiser Wilhelm II, 14 August 1892, *Eulenburgs Korrespondenz*, II, Nos. 595 and 693; Caprivi to Bennigsen, 29 September 1892, Oncken, *Bennigsen*, II, pp. 577–9.
92 Waldersee, diary entry for 28 September 1892, GStA Berlin, Waldersee Papers; omitted from Meisner, II, p. 265; Kiderlen-Wächter to Colmar von der Goltz, 21 December 1892, BA-MA Freiburg, von der Goltz Papers, N737/21.
93 Both letters are cited above, pp. 389–91.
94 See Empress Frederick to her daughter Princess Viktoria of Schaumburg-Lippe, 28 October 1892, AdHH Schloss Fasanerie.
95 Waldersee, diary entry for 29 October 1891, GStA Berlin, Waldersee Papers; cf. the totally distorted version in Meisner, II, pp. 219–20. See also Waldersee, diary entries for 1 and 5 November 1892, GStA Berlin, Waldersee Papers; omitted from Meisner, II, pp. 267–8.
96 Waldersee, diary entry for 2 May 1891, GStA Berlin, Waldersee Papers; omitted from Meisner, II, p. 206.
97 Waldersee, diary entry for 20 January 1892, GStA Berlin, Waldersee Papers; Meisner, II, p. 231.
98 Waldersee, diary entries for 16 and 25 October 1891, GStA Berlin, Waldersee Papers; omitted from Meisner, II, pp. 218–19.
99 Waldersee, diary entry for 29 October 1891, GStA Berlin, Waldersee Papers; omitted from Meisner, II, pp. 218–19.
100 See Brauer to Turban, 29 June 1892, Fuchs, *Großherzog von Baden*, III, No. 1191.
101 Hereditary Prince Bernhard of Saxe-Meiningen to Waldersee, 9 October 1891, GStA Berlin, Waldersee Papers No. 36.
102 Hereditary Prince Bernhard of Saxe-Meiningen to Waldersee, 30 January 1892, ibid.
103 Hereditary Prince Bernhard of Saxe-Meiningen to Waldersee, 12 October 1892, ibid.
104 Waldersee, diary entry for 30 January 1892, GStA Berlin, Waldersee Papers; omitted from Meisner, II, p. 232.
105 Waldersee, diary entry for 10 August 1892, in Meisner, II, p. 259. See Fuchs, *Großherzog von Baden*, III, pp. 166–78, especially Kaiser Wilhelm II to Grand Duke of Baden, 6 November 1892, ibid., No. 1206, and Grand Duke of Baden to Kaiser Wilhelm II, 14 November 1892, ibid., No. 1209.
106 Waldersee, diary entry for 30 January 1892, GStA Berlin, Waldersee Papers; omitted from Meisner, II, p. 232.
107 Marschall, diary entry for 29 August 1892, cited in *Eulenburgs Korrespondenz*, II, p. 925.
108 Marschall, diary entry for 16 September 1892, cited ibid., p. 940.
109 See Eulenburg's notes of 24 September 1892, in Haller, *Eulenburg*, pp. 89–91; *Holstein Papers*, III, No. 376.
110 Waldersee and Verdy, 3 February 1893, GStA Berlin, Waldersee Papers No. 53.

111 Eulenburg's notes of 24 September 1892, in Haller, *Eulenburg*, p. 90.
112 Waldersee, diary entry for 31 August 1892, GStA Berlin, Waldersee Papers; omitted from Meisner, II, p. 264.
113 Brauer to Grand Duke of Baden, 22 October 1892, Fuchs, *Großherzog von Baden*, III, No. 1201.
114 Hereditary Prince Bernhard of Saxe-Meiningen to Waldersee, 12 October 1892, GStA Berlin, Waldersee Papers No. 36.
115 Eulenburg to Holstein, 30 September 1892, *Holstein Papers*, III, No. 376; also Eulenburg to Kaiser Wilhelm II, 10 September 1892, *Eulenburgs Korrespondenz*, II, No. 702.
116 Waldersee, diary entry for 19 October 1892, GStA Berlin, Waldersee Papers; cf. Meisner, II, p. 266.
117 Brauer, reports of 6, 17 and 20 November 1892, Fuchs, *Großherzog von Baden*, III, Nos. 1207, 1210–11.
118 Waldersee, diary entry for 1 November 1892, GStA Berlin, Waldersee Papers; omitted from Meisner, II, p. 267.
119 Ibid.
120 Waldersee, diary entry for 1 December 1892, GStA Berlin, Waldersee Papers; omitted from Meisner, II, p. 270.
121 Waldersee, diary entries for 20 November, 3 and 15 December 1892, GStA Berlin, Waldersee Papers; omitted from Meisner, II, p. 270; cf. p. 272.
122 Kaiser Wilhelm II to Kaiser Franz Joseph, 31 January 1893, HHStA Vienna, Kabinettsarchiv Geheimakten 2.
123 Waldersee, diary entry for 20 November 1892, GStA Berlin, Waldersee Papers, omitted from Meisner, II, p. 270; also Waldersee's entry of 3 December 1892, GStA Berlin, Waldersee Papers.
124 Kiderlen-Wächter to Eulenburg, 18 December 1892, Holstein to Eulenburg, 23 December 1892, Eulenburg to Kaiser Wilhelm II, 29 December 1892, *Eulenburgs Korrespondenz*, II, Nos. 752–4.
125 Holstein to Eulenburg, 1 January 1893, ibid., No. 755.
126 Waldersee, diary entry for 3 January 1893, GStA Berlin, Waldersee Papers; abridged in Meisner, II, p. 274.
127 Waldersee, diary entry for 6 January 1893, GStA Berlin, Waldersee Papers; omitted from Meisner, II, p. 275. See also Waldersee, diary entry for 19 March 1893, GStA Berlin, Waldersee Papers; cf. Meisner, II, pp. 287–8.
128 Prince Ernst zu Hohenlohe-Langenburg to his father, 26 January 1893, Hohenlohe-Zentralarchiv Neuenstein, Hermann Hohenlohe-Langenburg Papers, Bü. 60. Cf. Caprivi to Eulenburg, 17 January 1893, *Eulenburgs Korrespondenz*, II, No. 762; Brauer to Turban, 5 January 1893, Fuchs, *Großherzog von Baden*, III, No. 1218.
129 Caprivi to Eulenburg, 30 April 1893, *Eulenburgs Korrespondenz*, II, No. 799. See Brauer to Grand Duke of Baden, 1 May 1893, Fuchs, *Großherzog von Baden*, III, No. 1256.
130 Eulenburg to Kaiser Wilhelm II, 16 April 1893, *Eulenburgs Korrespondenz*, II, No. 793; Kiderlen-Wächter to Eulenburg, 15 April 1893, ibid., p. 1065.
131 See Förster, *Der doppelte Militarismus*, pp. 63ff.
132 Waldersee, diary entry for 11 May 1893, GStA Berlin, Waldersee Papers; omitted from Meisner, II, p. 289.
133 Eulenburg, diary entry for 6 May 1893, *Eulenburgs Korrespondenz*, II, p. 1074.
134 Brauer to Grand Duke of Baden, 7 May 1893, Fuchs, *Großherzog von Baden*, III, No. 1267.
135 Kaiser Wilhelm II, 9 May 1893, Penzler, *Reden Kaiser Wilhelms II. in den Jahren 1888–1895*, p. 230; Elkind, *German Emperor's Speeches*, pp. 216–17.

136 Waldersee, diary entry for 11 May 1893, GStA Berlin, Waldersee Papers; omitted from Meisner, II, p. 289.
137 Empress Frederick to Bogumilla Freifrau von Stockmar, 17 May 1893, AdHH Schloss Fasanerie.
138 See Marschall's diary entries for autumn 1892, cited in *Eulenburgs Korrespondenz*, II, p. 977.
139 Herbert Bismarck to his father, 8 July 1893, BA Koblenz, Bismarck Papers FC 3005 N.
140 See Gerhard A. Ritter and M. Niehuss, *Wahlgeschichtliches Arbeitsbuch. Materialien zur Statistik des Kaiserreichs, 1871–1918*, Munich 1980, pp. 40ff.
141 But cf. Förster, *Der doppelte Militarismus*, p. 74.
142 Kaiser Wilhelm II to Queen Victoria, telegram, 14 July 1893, RA I59/101.
143 Eulenburg to Kaiser Wilhelm II, 14 July 1893, *Eulenburgs Korrespondenz*, II, No. 812.
144 See above, pp. 400–9.
145 On Wilhelms II's anti-Semitic beliefs, see Röhl, 'Kaiser Wilhelm II and German Anti-Semitism' in Röhl, *Kaiser and His Court*, pp. 190–212. On anti-Semitism in Wilhelmine Germany in general, see Nipperdey, *Deutsche Geschichte 1866–1918*, II, pp. 289–311.
146 Waldersee, diary entry for 18 December 1892, GStA Berlin, Waldersee Papers; omitted from Meisner, II, p. 272.
147 Holstein to Eulenburg, 13 December 1892, *Eulenburgs Korrespondenz*, II, No. 749.
148 Holstein to Eulenburg, 16 December 1892, ibid., No. 750.
149 Eulenburg to Kaiser Wilhelm II, 17 December 1892, ibid., No. 751.
150 Waldersee, diary entry for 18 December 1892, GStA Berlin, Waldersee Papers; omitted from Meisner, II, p. 272.
151 See above, pp. 401–9.
152 Waldersee, diary entry for 18 December 1892, GStA Berlin, Waldersee Papers; omitted from Meisner, II, p. 272; also Waldersee, diary entry for 5 March 1893, GStA Berlin, Waldersee Papers; omitted from Meisner, II, p. 286.
153 See Empress Frederick to Queen Victoria, 27 October 1891, RA Z51/41.
154 Kaiser Wilhelm II, speech of 18 December 1891, Penzler, *Reden Kaiser Wilhelms II. in den Jahren 1888–1895*, pp. 202–4; Waldersee, diary entry for 19 December 1891, GStA Berlin, Waldersee Papers; omitted from Meisner, II, p. 227. See above, p. 396.
155 Waldersee, diary entry for 12 February 1893, GStA Berlin, Waldersee Papers; omitted from Meisner, II, p. 285.
156 See Waldersee, diary entry for 10 November 1893, GStA Berlin, Waldersee Papers; omitted from Meisner, II, p. 299.
157 Waldersee, diary entry for 12 October 1893, GStA Berlin, Waldersee Papers; omitted from Meisner, II, p. 295.
158 Waldersee, diary entry for 19 October 1893, GStA Berlin, Waldersee Papers; omitted from Meisner, II, p. 297.
159 Waldersee, diary entry for 12 October 1893, GStA Berlin, Waldersee Papers; omitted from Meisner, II, p. 295.
160 Waldersee, diary entry for 3 December 1893, GStA Berlin, Waldersee Papers; omitted from Meisner, II, p. 299.
161 Waldersee, diary entry for 15 December 1893, GStA Berlin, Waldersee Papers; omitted from Meisner, II, p. 300.
162 Nipperdey, *Deutsche Geschichte 1866–1918*, II, p. 627. See Klaus Hildebrand, *Deutsche Außenpolitik 1871–1918*, Munich 1989, p. 30.
163 Cited in Herbert Bismarck to his father, 10 February 1894, BA Koblenz, Bismarck Papers FC 3005 N.
164 Waldersee, diary entry for 20 February 1894, GStA Berlin, Waldersee Papers; omitted from Meisner, II, p. 308.

165 Waldersee, diary entry for 15 December 1893, GStA Berlin, Waldersee Papers; omitted from Meisner, II, p. 300.
166 Waldersee, diary entry for 12 March 1894, GStA Berlin, Waldersee Papers; cf. Meisner, II, p. 310.
167 Waldersee, diary entry for 13 March 1894, GStA Berlin, Waldersee Papers; Meisner, II, p. 310.
168 Waldersee, diary entries for 24 February and 11 March 1894, GStA Berlin, Waldersee Papers; cf. Meisner, II, pp. 309–10.
169 Kaiser Wilhelm II to Queen Victoria, telegram, 16 March 1894, RA I60/63.
170 Eulenburg to Kaiser Wilhelm II, 11 March 1894, *Eulenburgs Korrespondenz*, II, No. 925.
171 See the essay 'Zwei Kaiser und ein König auf dem Wasser', in Philipp Fürst zu Eulenburg-Hertefeld, *Erlebnisse an deutschen und fremden Höfen*, Leipzig 1934, pp. 9–35.
172 Philipp Fürst zu Eulenburg, Eine preußische Familiengeschichte, pp. 461ff., BA Koblenz, Eulenburg Papers.
173 Waldersee, diary entry for 10 February 1894, GStA Berlin, Waldersee Papers; cf. Meisner, II, pp. 307–8.
174 Brauer, reports of 17 and 20 November 1892, Fuchs, *Großherzog von Baden*, III, Nos. 1210–11.
175 Brauer, report of 26 February 1892, ibid., No. 1159.
176 Count Udo Stolberg in Wedel, *Zwischen Kaiser und Kanzler*, p. 115.

18 DYNASTIC DIPLOMACY

1 Kaiser Wilhelm II to Queen Victoria, 8 December 1891, RA I59/53. On the relationship between Alexander III and Wilhelm II, see McLean, *Royalty and Diplomacy*, pp. 15ff.
2 Waldersee, diary entry for 21 November 1891, GStA Berlin, Waldersee Papers; printed only in part in Meisner, II, p. 223; also Waldersee's note of 30 August 1893, cited in Meisner, II, p. 293.
3 Prince Ernst zu Hohenlohe-Langenburg to his father, 28 January 1892, Hohenlohe-Zentralarchiv Neuenstein, Hermann Hohenlohe-Langenburg Papers, Bü. 59.
4 Eulenburg to Holstein, 2 October 1892, *Holstein Papers*, III, No. 377.
5 Waldersee, diary entry for 12 June 1892, GStA Berlin, Waldersee Papers; omitted from Meisner, II, pp. 244–5.
6 Waldersee, diary entry for 30/31 December 1893, GStA Berlin, Waldersee Papers; omitted from Meisner, II, p. 301.
7 See Waldersee, diary entry for 30 January 1892, GStA Berlin, Waldersee Papers; cf. Meisner, II, p. 232.
8 Waldersee, diary entry for 16 April 1892, GStA Berlin, Waldersee Papers; cf. Meisner, II, pp. 239–40.
9 Waldersee, diary entry for 1 December 1892, GStA Berlin, Waldersee Papers; omitted from Meisner, II, p. 270.
10 Kaiser Wilhelm II, marginal notes on a report from Vienna of 31 August 1892, PA AA, Asservat No. 4.
11 Waldersee, diary entries for 21 and 29 November 1893, GStA Berlin, Waldersee Papers; omitted from Meisner, II, p. 299.
12 Holstein to Eulenburg, 24 May 1892, *Eulenburgs Korrespondenz*, II, No. 664.
13 Marschall, diary entries for 24 and 25 May, cited ibid., p. 878.
14 Kaiser Wilhelm II to Empress Frederick, 10 June 1892, AdHH Schloss Fasanerie.
15 Malet, report of 10 June 1892, RA I59/83.

16 Prince Ernst zu Hohenlohe-Langenburg to his father, 13 June 1892, Hohenlohe-Zentralarchiv Neuenstein, Hermann Hohelohe-Langenburg, Bü. 59. Cf. Canis, *Von Bismarck zur Weltpolitik*, p. 105.
17 Queen Victoria to Kaiser Wilhelm II, 14 and 29 June 1892, GStA Berlin, BPHA Rep. 52 W3 No. 11.
18 Waldersee, diary entry for 11 June 1892, GStA Berlin, Waldersee Papers; cf. Meisner, II, pp. 241–2.
19 Waldersee, diary entry for 9 October 1894, GStA Berlin, Waldersee Papers; cf. Meisner, II, pp. 323–4.
20 Prince Ernst zu Hohenlohe-Langenburg to his father, 13 June 1892, Hohenlohe-Zentralarchiv Neuenstein, Hermann Hohelohe-Langenburg, Bü. 59.
21 Herbert Bismarck to his father, 11 July 1893, BA Koblenz, Bismarck Papers FC 3005 N.
22 Empress Frederick to Queen Victoria, 19 December 1891, RA Z52/7.
23 Prince Ernst zu Hohenlohe-Langenburg to his father, 13 June 1892, Hohenlohe-Zentralarchiv Neuenstein, Hermann Hohelohe-Langenburg, Bü. 59.
24 Prince Ernst zu Hohenlohe-Langenburg to his father, 13 December 1891, ibid., Bü. 58.
25 Waldersee, diary entry for 19 January 1892, GStA Berlin, Waldersee Papers; cf. Meisner, II, pp. 229–30.
26 As reported in Eulenburg to Holstein, 8 October 1891, *Holstein Papers*, III, No. 349.
27 Waldersee, diary entry for 21 November 1891, GStA Berlin, Waldersee Papers; printed only in part in Meisner, II, p. 223.
28 Kaiser Wilhelm II to Queen Victoria, 8 December 1891, RA I59/53.
29 Kaiser Wilhelm II to Queen Victoria, 12 April 1892, RA I59/72.
30 Salisbury to Queen Victoria, 22 April 1892, RA I59/78.
31 Kaiser Wilhelm II to Queen Victoria, 22 May 1892, RA I59/80.
32 Waldersee, diary entry for 7 August 1893, GStA Berlin, Waldersee Papers; omitted from Meisner, II, p. 292.
33 Cited in *Holstein Papers*, III, p. 411.
34 See Paul Kennedy, *The Rise of the Anglo-German Antagonism 1860–1914*, London 1980, pp. 212ff.
35 See McLean, *Royalty and Diplomacy*, pp. 78ff.
36 The marriage of Aribert Anhalt and Louise Holstein is described below, pp. 663–4. See also Marie Louise Princess of Schleswig-Holstein, *My Memories of Six Reigns*, London 1956, pp. 66ff.
37 Kaiser Wilhelm II to Arthur Duke of Connaught, 11 March 1891, RA Addl Mss A15/5680.
38 Malet to Ponsonby, 19 March 1891, RA I59/10; Malet to August Eulenburg, 24 March 1891, RA I59/11.
39 See above, pp. 360–4.
40 Queen Victoria to Ponsonby, 15 June 1892, RA I59/86.
41 Ponsonby to Malet, 16 June 1892, RA I59/87.
42 Queen Victoria to Kaiser Wilhelm II, 29 June 1892, GStA Berlin, BPHA Rep. 52 W3 No. 11.
43 See Kristin Lammerting, *Meteor. Die kaiserlichen Segelyachten*, Cologne 1999, pp. 30–47.
44 Kaiser Wilhelm II to his mother, 28 July 1892, AdHH Schloss Fasanerie.
45 *Times*, 2–6 August 1892; *Standard*, 3–8 August 1892; *Daily Telegraph*, 4 August 1892, RA I59/91.
46 Kaiser Wilhelm II to Albert Edward Prince of Wales, 23 October 1892, RA L21/70.
47 Albert Edward Prince of Wales to Kaiser Wilhelm II, 31 October 1892, RA L21/73.
48 Malet to Queen Victoria, [late] May 1893, RA I59/100.
49 Seckendorff to Senden-Bibran, 14 June 1893, BA-MA Freiburg, Senden-Bibran Papers N160/2.

50 Ibid.
51 Seckendorff to Senden-Bibran, 16 June 1893, ibid.
52 Cited ibid.
53 Seckendorff to Senden-Bibran, 15 and 16 June 1893, ibid.
54 Queen Victoria to Kaiser Wilhelm II, telegram, 14 June 1893, GStA Berlin, BPHA Rep 52 W3 No. 11. See Heinrich's report to the Kaiser of 9 July 1893 on his experiences in London, BA-MA Freiburg, Senden-Bibran Papers N160/2.
55 RA QVJ, 29 July–6 August 1893.
56 See Prince Ernst zu Hohenlohe-Langenburg to his father, 16 July 1893, Hohenlohe-Zentralarchiv Neuenstein, Hermann Hohenlohe-Langenburg Papers, Bü. 60.
57 Prince Ernst zu Hohenlohe-Langenburg to his father, 10 August 1893, ibid.
58 See Eulenburg's notes, Am Hofe von England, in Philipp Fürst zu Eulenburg-Hertefeld, *Das Ende König Ludwigs II. und andere Erlebnisse*, Leipzig 1934, pp. 212–45.
59 See *Große Politik*, VIII, No. 1752.
60 On the Siam crisis, see Rich, *Friedrich von Holstein*, I, pp. 350ff.; Lahme, *Deutsche Außenpolitik*, pp. 395ff.; Canis, *Von Bismarck zur Weltmacht*, pp. 116–18.
61 See *Große Politik*, VIII, No. 1752.
62 Prince Ernst zu Hohenlohe-Langenburg to his father, 10 August 1893, Hohenlohe-Zentralarchiv Neuenstein, Hermann Hohenlohe-Langenburg Papers, Bü. 60. See also Prince Ernst's letter of 16 July 1893, ibid.
63 Cited in Haller, *Eulenburg*, pp. 84–5.
64 Prince Ernst zu Hohenlohe-Langenburg to his father, 10 August 1893, Hohenlohe-Zentralarchiv Neuenstein, Hermann Hohenlohe-Langenburg Papers, Bü. 60.
65 Prince Ernst zu Hohenlohe-Langenburg to his father, 28 August 1893, ibid.
66 Kaiser Wilhelm II, marginal notes on Hatzfeldt's report of 18 November 1893, cited in *Holstein Papers*, III, p. 446.
67 See Canis, *Von Bismarck zur Weltpolitik*, pp. 119–20.
68 Kaiser Wilhelm II, marginal notes on Werder's report from St Petersburg of 8 July 1893, PA AA, Asservat No. 4.
69 Malet to Rosebery, 7 November 1893, RA I59/113.
70 Waldersee, diary entry for 29 November 1893, GStA Berlin, Waldersee Papers; omitted from Meisner, II, p. 299.
71 Kaiser Wilhelm II to Queen Victoria, 12 January 1894, RA I60/1.
72 Swaine to Ponsonby, 12 January 1894, RA I60/2.
73 Swaine to Ponsonby, 13 January 1894, RA I60/3.
74 Albert Edward Prince of Wales to Queen Victoria, 16 and 19 January 1894, RA I60/5 and 14; Knollys to Ponsonby, 20 January RA I60/17.
75 Ponsonby to Queen Victoria, 16 January 1894, RA I60/6.
76 Queen Victoria, marginal notes on Ponsonby's letters of 15, 17 and 18 January 1894, RA I60/4, 8 and 10; Queen Victoria to Malet and to Prince of Wales, secret telegrams, 19 January 1894, RA I60/11; Malet's reply of 20 January 1894, RA I60/18.
77 Henry Campbell-Bannerman to Ponsonby, 18, 19 and 24 January 1894, RA I60/9, 12 and 37; Queen Victoria, marginal note of 27 January 1894, RA I60/46.
78 Rosebery to Ponsonby, 27 January 1894, RA I60/55.
79 George Duke of Cambridge to Ponsonby with Queen Victoria's marginal notes, 20 January 1894, RA I60/24.
80 Ponsonby, notes, 23 January 1894, RA I60/36; QVJ, 23 January 1894.
81 Knollys to Ponsonby, 22 and 24 January RA I60/31 and 39.
82 Ponsonby to Swaine, 24 January 1894, RA I60/41; Swaine's reply of 27 January 1894, RA I60/56.

83 Queen Victoria to Kaiser Wilhelm II, 25 January 1894, GStA Berlin, BPHA Rep. 52 W3 No. 11.
84 Ponsonby to Queen Victoria, 24 January 1894, RA I60/42.
85 Knollys to Ponsonby, 26 January 1894, RA I60/46.
86 Knollys to Ponsonby, 27 January 1894, RA I60/57.
87 Knollys to Ponsonby, 26 January 1894, RA I60/51; Ponsonby to Queen Victoria, 26 January 1894, RA I60/52 and 53.
88 George Duke of Cambridge to Ponsonby, 26 January 1894, RA I60/48.
89 Salisbury to Ponsonby, 26 January 1894, RA I60/47.
90 Prince Ernst zu Hohenlohe-Langenburg to his father, 8 March 1894, Hohenlohe-Zentralarchiv Neuenstein, Hermann Hohenlohe-Langenburg Papers, Bü. 61.
91 Rosebery to Queen Victoria, 11 April 1894, RA I60/62.
92 Queen Victoria to Kaiser Wilhelm II, 28 April 1894, GStA Berlin, BPHA Rep. 52 W3 No. 11.
93 Kaiser Wilhelm II to Queen Victoria, 24 April 1894, RA I60/64.
94 Kaiser Wilhelm II to Queen Victoria, telegram, 8 June 1894, RA I60/72.
95 Kaiser Wilhelm II, toast to Royal Dragoons, 7 June 1894, Penzler, *Reden Kaiser Wilhelms II. in den Jahren 1888–1895*, pp. 269–70; Engish version from *Standard*, 11 June 1894, RA I60/73.
96 See the article 'The Emperor William and the Dragoons', ibid.
97 Swaine to Queen Victoria, 15 June 1894, RA I60/75; Malet to Queen Victoria, 16 June 1894, RA I60/77.
98 Holstein to Eulenburg, 12 and 16 June 1894, Eulenburg to Holstein, 16 June 1894, Kálnoky to Eulenburg, 17 June 1894, Eulenburg to Kaiser Wilhelm II, 18 June 1894, *Eulenburgs Korrespondenz*, II, Nos. 975–80; also Kaiser Wilhelm II to his mother, 21 June 1894, AdHH Schloss Fasanerie.
99 RA QVJ, 6–13 August 1894; *Daily Telegraph*, 7 August 1894; *Times*, 7 August 1894, RA I60/81.
100 Major-General Sir Leopold Swaine, *Camp and Chancery in a Soldier's Life*, London 1926, pp. 215–16; McLean, *Royalty and Diplomacy*, pp. 78–9.
101 Kaiser Wilhelm II to Queen Victoria, 13 and 15 August 1894, RA I60/84 and 88; Kaiser Wilhelm II to Arthur Duke of Connaught, 15 August 1894, RA L4/26a; Kaiser Wilhelm II to Prince Albert Edward of Wales, 24 August 1894, RA L4/26b.
102 Kaiser Wilhelm II to Queen Victoria, 24 August 1894, RA I60/91.
103 Bigelow to Kaiser Wilhelm II, 26 January 1893, GStA Berlin, BPHA Rep. 53J Lit. B No. 10 Bigelow 1. See above, pp. 407–8. See Poultney Bigelow, *Prussian Memories 1864–1914*, New York 1916.
104 Bigelow to Kaiser Wilhelm II, 20 July 1892, GStA Berlin, BPHA Rep. 53J Lit. B No. 10 Bigelow 1. Bigelow enclosed his article 'The Czar's Western Frontier', which had just been published in *Harper's Magazine*.
105 Bigelow to Kaiser Wilhelm II, 18 August 1892, GStA Berlin, BPHA Rep. 53J Lit. B No. 10 Bigelow 1.
106 Bigelow to Kaiser Wilhelm II, 26 January 1893, ibid.
107 Bigelow to Kaiser Wilhelm II, 20 July 1892, ibid. Stepniak (in reality Sergei Kravchinski) had stabbed to death the chief of the Russian secret police in 1878.
108 Bigelow to Kaiser Wilhelm II, 18 August 1892, ibid.
109 See above, pp. 479–80.
110 Eulenburg's note of 11 July 1892, in *Eulenburgs Korrespondenz*, II, No. 688.
111 Waldersee, diary entry for 6 August 1893, GStA Berlin, Waldersee Papers; omitted from Meisner, II, p. 292.

112 Waldersee, diary entry for 3 November 1893, GStA Berlin, Waldersee Papers; omitted from Meisner, II, p. 299; but cf. ibid., p. 298.
113 Waldersee, diary entry for 16 October 1893, GStA Berlin, Waldersee Papers; omitted from Meisner, II, p. 296.
114 Waldersee, diary entry for 15 November 1893, GStA Berlin, Waldersee Papers; omitted from Meisner, II, p. 299.
115 Waldersee, diary entry for 30 September 1894, GStA Berlin, Waldersee Papers; cf. Meisner, II, p. 323.
116 Waldersee, diary entry for 16 October 1893, GStA Berlin, Waldersee Papers; omitted from Meisner, II, p. 296.
117 Kaiser Wilhelm II, speech of 22 September 1894, Penzler, *Reden Kaiser Wilhelms II. in den Jahren 1888–1895*, pp. 278–9.
118 On this marriage, see below, pp. 650–1.
119 Eulenburg to Kaiser Wilhelm II, 5 February 1893, *Eulenburgs Korrespondenz*, II, No. 767.
120 Kaiser Wilhelm II, toast of 26 January 1893, Penzler, *Reden Kaiser Wilhelms II. in den Jahren 1888–1895*, pp. 222–3.
121 Kaiser Wilhelm II to Queen Victoria, 28 January 1893, RA I59/98. See Waldersee, diary entry for 20 October 1894, GStA Berlin, Waldersee Papers; Meisner, II, p. 326.
122 Kaiser Wilhelm II to Kaiser Franz Joseph, 31 January 1893, HHStA Vienna, Kabinettsarchiv Geheimakten 2.
123 See above, pp. 366–70.
124 Hereditary Princess Charlotte of Saxe-Meiningen to Helene Freifrau von Heldburg, 26 April 1894, Thüringisches Staatsarchiv Meiningen, HA 342.
125 Queen Victoria to Kaiser Wilhelm II, 28 April 1894, GStA Berlin, BPHA Rep. 52 W3 No. 11.
126 Kaiser Wilhelm II, marginal notes on Eulenburg's report from Vienna, PA AA, Asservat No. 4.
127 Waldersee, diary entry for 24 February 1894, GStA Berlin, Waldersee Papers; printed only in part in Meisner, II, p. 309.
128 See Canis, *Von Bismarck zur Weltpolitik*, pp. 120–7.
129 Minutes of the Crown Council of 18 February 1894, GStA Berlin, cited in part in Canis, *Von Bismarck zur Weltpolitik*, pp. 121–2. Canis points out that the most telling sections of the minutes were omitted by the editors of *Große Politik*.
130 Waldersee, diary entry for 9 February 1894, GStA Berlin, Waldersee Papers; printed only in part in Meisner, II, pp. 306–7. See above, p. 469. See also Philipp Eulenburg to Botho Eulenburg, 9 February 1894, *Eulenburgs Korrespondenz*, II, No. 900.
131 Waldersee, diary entry for 9 February 1894, GStA Berlin, Waldersee Papers; printed only in part in Meisner, II, pp. 306–7.
132 Waldersee, diary entry for 15 February 1894, GStA Berlin, Waldersee Papers; heavily abridged and distorted in Meisner, II, p. 308.
133 Kaiser Wilhelm II to Queen Victoria, 24 August 1894, RA I60/91.
134 Waldersee, diary entry for 29 October 1893, GStA Berlin, Waldersee Papers; omitted from Meisner, II, p. 298.
135 Kaiser Wilhelm II to Eulenburg, 29 November 1893, telegram, *Eulenburgs Korrespondenz*, II, No. 852.
136 Hinzpeter to Kaiser Wilhelm II, 30 June 1894, PA AA, Asservat No. 4.
137 Malet to Queen Victoria, 14 July 1894, RA I60/80.
138 Waldersee, diary entry for 1 August 1894, GStA Berlin, Waldersee Papers; Meisner, II, pp. 318–19.
139 Waldersee, diary entry for 6 August 1894, GStA Berlin, Waldersee Papers; omitted from Meisner, II, p. 320.

19 THE RUDE AWAKENING

1 Empress Frederick to Queen Victoria, 15 March 1890, RA Z48/2.
2 Empress Frederick to Queen Victoria, 22 March 1890, RA Z48/5.
3 Empress Frederick to Queen Victoria, 25 March 1890, RA Z48/6.
4 Empress Frederick to Queen Victoria, 8 April 1890, RA Z48/11.
5 Empress Frederick to Bogumilla Freifrau von Stockmar, 29 March 1891, AdHH Schloss Fasanerie.
6 Empress Frederick to Queen Victoria, 1 December 1890, RA Z49/39; also Empress Frederick to her daughter Princess Viktoria of Schaumburg-Lippe, 1 March 1892, AdHH Schloss Fasanerie.
7 Empress Frederick to Queen Victoria, 8 April 1890, RA Z48/11.
8 Empress Frederick to Bogumilla Freifrau von Stockmar, 7 April 1891, AdHH Schloss Fasanerie.
9 Empress Frederick to Bogumilla Freifrau von Stockmar, 9 May 1891, ibid.
10 Empress Frederick to Queen Victoria, 12 June 1891, RA Z50/45.
11 Empress Frederick to Queen Victoria, 12 December 1891, RA Z52/5.
12 Empress Frederick to Queen Victoria, 4 November 1891, AdHH Schloss Fasanerie.
13 Empress Frederick to Bogumilla Freifrau von Stockmar, 4 November 1891, ibid.
14 Empress Frederick to Queen Victoria, 16 November 1891, RA Z51/46.
15 Empress Frederick to Queen Victoria, 5 December 1891, RA 52/2.
16 Empress Frederick to Queen Victoria, 12 December 1891, RA 52/5.
17 Empress Frederick to Queen Victoria, 16 February 1892, RA 52/20. See also Empress Frederick to Queen Victoria, 20 May 1892, RA Z52/51.
18 Empress Frederick to her daughter Princess Viktoria of Schaumburg-Lippe, 25 February 1892, AdHH Schloss Fasanerie.
19 Empress Frederick to Queen Victoria, 27 February 1892, RA 52/25.
20 Empress Frederick to Queen Victoria, 29 February 1892, RA 52/29; also Empress Frederick to her daughter Princess Viktoria of Schaumburg-Lippe, 8 March 1892, AdHH Schloss Fasanerie.
21 Empress Frederick to Queen Victoria, 20 March 1892, RA Z52/34.
22 Empress Frederick to Queen Victoria, 14 May 1892, RA Z52/48.
23 Empress Frederick to Queen Victoria, 20 May 1892, RA Z52/51. See also Empress Frederick to Queen Victoria, 24 October 1891, RA Z51/40.
24 Empress Frederick to Bogumilla Freifrau von Stockmar, 26 April 1892, AdHH Schloss Fasanerie.
25 Empress Frederick to Bogumilla Freifrau von Stockmar, 17 May 1893, ibid.
26 Empress Frederick to her daughter Princess Viktoria of Schaumburg-Lippe, 18 May 1893, AdHH Schloss Fasanerie.
27 Waldersee, diary entry for 16 April 1892, GStA Berlin, Waldersee Papers; cf. Meisner, II, pp. 239–40.
28 Bülow to Eulenburg, 6 April 1892, *Eulenburgs Korrespondenz*, II, No. 641.
29 Waldersee, diary entry for 30 December 1894, GStA Berlin, Waldersee Papers; omitted from Meisner, II, p. 334.
30 Duke Ernst Günther of Schleswig-Holstein to Prince Christian of Schleswig-Holstein, 16 December 1891, RA Add A18/G1.
31 Prince Ernst zu Hohenlohe-Langenburg to his father, 7 March 1892, Hohenlohe-Zentralarchiv Neuenstein, Hermann Hohenlohe-Langenburg Papers, Bü. 59.
32 Reuss to Wedel, 2 September 1895, Wedel Papers, Frankfurt a.M.
33 Georg II Duke of Saxe-Meiningen to Carl Werder, 21 September 1891, Thüringisches Staatsarchiv Meiningen, HA 395/II.

34 Waldersee, diary entry for 7 August 1890, GStA Berlin, Waldersee Papers; cf. Meisner, II, p. 135.
35 Waldersee, diary entry for 21 October 1890, GStA Berlin, Waldersee Papers; cf. Meisner, II, pp. 155–6.
36 Waldersee, diary entry for 3 August 1894, GStA Berlin, Waldersee Papers; omitted from Meisner, II, p. 319.
37 Waldersee, diary entry for 6 August 1894, GStA Berlin, Waldersee Papers; printed only in part in Meisner, II, pp. 319–20.
38 Wedel, *Zwischen Kaiser und Kanzler*, p. 127.
39 Waldersee, diary entry for 17 December 1890, GStA Berlin, Waldersee Papers; omitted from Meisner, II, p. 167.
40 Waldersee, diary entry for 13 September 1891, GStA Berlin, Waldersee Papers; omitted from Meisner, II, p. 216.
41 Hohenlohe, undated note, printed in Hohenlohe, *Denkwürdigkeiten der Reichskanzlerzeit*, p. 32.
42 Waldersee, diary entry for 12 August 1890, GStA Berlin, Waldersee Papers; omitted from Meisner, II, p. 138.
43 Waldersee, diary entry for 10 January 1891, GStA Berlin, Waldersee Papers; printed only in part in Meisner, II, pp. 174–5.
44 Waldersee, diary entry for 1 November 1892, GStA Berlin, Waldersee Papers; cf. Meisner, II, p. 267.
45 Waldersee, diary entry for 9 January 1891, GStA Berlin, Waldersee Papers; cf. Meisner, II, p. 174.
46 Waldersee, diary entry for 26 October 1894, GStA Berlin, Waldersee Papers; cf. Meisner, II, pp. 327–8.
47 Waldersee, diary entry for 7 December 1890, GStA Berlin, Waldersee Papers; omitted from Meisner, II, p. 164.
48 Bodo von dem Knesebeck to Alexander Hohenlohe, 29 October 1894, Hohenlohe, *Denkwürdigkeiten der Reichskanzlerzeit*, p. 4.
49 Waldersee, diary entry for 19 August 1890, GStA Berlin, Waldersee Papers; omitted from Meisner, II, p. 140.
50 Waldersee, diary entry for 11 October 1890, GStA Berlin, Waldersee Papers; omitted from Meisner, II, p. 154.
51 Waldersee, diary entry for 4 January 1892, GStA Berlin, Waldersee Papers; omitted from Meisner, II, pp. 228–9.
52 Waldersee, diary entry for 19 December 1890, GStA Berlin, Waldersee Papers; cf. Meisner, II, p. 168.
53 Waldersee, diary entry for 20 July 1891, GStA Berlin, Waldersee Papers; omitted from Meisner, II, pp. 212–13.
54 Waldersee, diary entry for 27 June 1894, GStA Berlin, Waldersee Papers; omitted from Meisner, II, p. 315.
55 Waldersee, diary entry for 5 January 1894, GStA Berlin, Waldersee Papers; Meisner, II, pp. 301ff.
56 Waldersee, diary entry for 27 June 1894, GStA Berlin, Waldersee Papers; omitted from Meisner, II, p. 315.
57 Waldersee, diary entry for 21 August 1895, GStA Berlin, Waldersee Papers; cf. Meisner, II, pp. 356–7.
58 Waldersee, diary entry for 18 July 1894, GStA Berlin, Waldersee Papers; omitted from Meisner, II, p. 317.
59 Waldersee, diary entry for 20 October 1890, GStA Berlin, Waldersee Papers; cf. Meisner, II, p. 155.

60 Prince Ernst zu Hohenlohe-Langenburg to his father, 1 May 1891, Hohenlohe-Zentralarchiv Neuenstein, Hermann Hohenlohe-Langenburg Papers, Bü. 58.
61 Waldersee, diary entry for 27 February 1894, GStA Berlin, Waldersee Papers; printed only in part in Meisner, II, p. 309.
62 Waldersee, diary entry for 5 January 1894, GStA Berlin, Waldersee Papers; Meisner, II, pp. 301ff.
63 See e.g. Waldersee, diary entry for 14 December 1894, GStA Berlin, Waldersee Papers; Meisner, II, pp. 332ff.
64 Waldersee, diary entry for 12 October 1895, GStA Berlin, Waldersee Papers; omitted from Meisner, II, p. 361.
65 Waldersee, diary entry for 10 August 1895, GStA Berlin, Waldersee Papers; omitted entirely from Meisner, II, p. 356.
66 Waldersee, diary entry for 12 August 1890, GStA Berlin, Waldersee Papers; omitted from Meisner, II, p. 138.
67 Waldersee, diary entry for 11 August 1890, GStA Berlin, Waldersee Papers; cf. Meisner, II, pp. 137–8.
68 Waldersee, diary entry for 25 September 1890, GStA Berlin, Waldersee Papers; cf. Meisner, II, p. 149.
69 Waldersee, diary entry for 25 September 1890, GStA Berlin, Waldersee Papers; omitted from Meisner, II, p. 149.
70 Waldersee, diary entry for 25 September 1890, GStA Berlin, Waldersee Papers; cf. Meisner, II, pp. 149–51.
71 Waldersee, diary entry for 4 October 1890, GStA Berlin, Waldersee Papers; printed only in part in Meisner, II, pp. 152–3.
72 Waldersee, diary entry for 4 October 1890, GStA Berlin, Waldersee Papers; cf. Meisner, II, p. 153.
73 Waldersee, diary entry for 16 November 1890, GStA Berlin, Waldersee Papers; cf. Meisner, II, p. 161.
74 Waldersee, diary entry for 24 November 1890, GStA Berlin, Waldersee Papers; cf. Meisner, II, pp. 163–4.
75 Waldersee, diary entry for 20 December 1890, GStA Berlin, Waldersee Papers; cf. Meisner, II, p. 169.
76 Waldersee, diary entry for 10 January 1891, GStA Berlin, Waldersee Papers; omitted from Meisner, II, p. 175.
77 Waldersee, diary entry for 3 January 1892, GStA Berlin, Waldersee Papers; omitted from Meisner, II, p. 228.
78 Waldersee, diary entry for 3 January 1893, GStA Berlin, Waldersee Papers; cf. Meisner, II, p. 274.
79 Waldersee, diary entry for 6 January 1893, GStA Berlin, Waldersee Papers; omitted from Meisner, II, p. 275.
80 Waldersee, diary entry for 17 March 1893, GStA Berlin, Waldersee Papers; omitted from Meisner, II, p. 286.
81 Waldersee, diary entries for 11, 12 and 16 June 1893, GStA Berlin, Waldersee Papers; cf. Meisner, II, p. 290.
82 Waldersee, diary entry for 18 January 1891, GStA Berlin, Waldersee Papers; omitted from Meisner, II, p. 176.
83 Waldersee, diary entry for 1 November 1891, GStA Berlin, Waldersee Papers; omitted from Meisner, II, p. 220.
84 Waldersee, diary entry for 18 November 1891, GStA Berlin, Waldersee Papers; cf. Meisner, II, p. 222.

85 Waldersee, diary entry for 18 January 1891, GStA Berlin, Waldersee Papers; omitted from Meisner, II, p. 176. See also the very negative judgement of 6 February 1891, omitted from Meisner, II, pp. 185–6.
86 Waldersee, diary entry for 22 December 1891, GStA Berlin, Waldersee Papers; cf. Meisner, II, p. 228.
87 Waldersee, diary entry for 27 January 1892, GStA Berlin, Waldersee Papers; omitted from Meisner, II, pp. 231–2.
88 Waldersee, diary entry for 26 February 1892, GStA Berlin, Waldersee Papers; Meisner, II, pp. 233–4.
89 Waldersee, diary entries for 3 and 7 March 1892, GStA Berlin, Waldersee Papers; both entries are omitted from Meisner, II, p. 234.
90 Kaiser Wilhelm II, speech of 1 March 1893, Penzler, *Reden Kaiser Wilhelms II. in den Jahren 1888–1895*, pp. 226–7; Elkind, *German Emperor's Speeches*, pp. 294–5.
91 Waldersee, diary entry for 3 March 1893, GStA Berlin, Waldersee Papers; omitted from Meisner, II, p. 286.
92 Waldersee, diary entry for 5 March 1893, GStA Berlin, Waldersee Papers; omitted from Meisner, II, p. 286.
93 Waldersee, diary entry for 23 June 1893, GStA Berlin, Waldersee Papers; cf. Meisner, II, pp. 290–1.
94 Waldersee, diary entry for 7 June 1894, GStA Berlin, Waldersee Papers; cf. Meisner, II, p. 314.
95 Waldersee, diary entry for 12 September 1894, GStA Berlin, Waldersee Papers, cf. Meisner, II, p. 322.
96 Waldersee, diary entry for 15 January 1895, GStA Berlin, Waldersee Papers; cf. Meisner, II, pp. 334–6.
97 Waldersee, diary entries for 14 and 17 March 1895, GStA Berlin, Waldersee Papers; cf. Meisner, II, pp. 339–40.
98 Waldersee, diary entry for 25 January 1895, GStA Berlin, Waldersee Papers; cf. Meisner, II, pp. 336–7.
99 Eulenburg to Holstein, 21 January 1895, *Eulenburgs Korrespondenz*, II, No. 1075; Holstein to Eulenburg, 7 April 1895, ibid., No. 1101; Waldersee, diary entry for 28 January 1895, GStA Berlin, Waldersee Papers; cf. Meisner, II, pp. 336–7.
100 Waldersee, diary entry for 21 January 1896, GStA Berlin, Waldersee Papers; omitted from Meisner, II, pp. 365–6.
101 Waldersee, diary entry for 8 June 1896, GStA Berlin, Waldersee Papers; cf. Meisner, II, pp. 370–1.
102 Waldersee, diary entry for 14 June 1896, GStA Berlin, Waldersee Papers; omitted from Meisner, II, p. 371.
103 Waldersee, diary entry for 28 June 1896, GStA Berlin, Waldersee Papers; cf. Meisner, II, pp. 371–2.
104 Waldersee, diary entry for 15 August 1896, GStA Berlin, Waldersee Papers; omitted from Meisner, II, p. 372.
105 Waldersee, diary entry for 25 October 1896, GStA Berlin, Waldersee Papers; cf. the abridged version in Meisner, II, pp. 374–5.
106 Waldersee, diary entry for 16 March 1897, GStA Berlin, Waldersee Papers; cf. Meisner, II, pp. 393ff.
107 Waldersee, diary entry for 20 May 1890, GStA Berlin, Waldersee Papers; omitted from Meisner, II, p. 128.
108 Waldersee, diary entry for 25 September 1890, GStA Berlin, Waldersee Papers; omitted from Meisner, II, pp. 149ff.

109 Waldersee, diary entry for 20 August 1891, GStA Berlin, Waldersee Papers; omitted from Meisner, II, p. 215.
110 Waldersee, diary entry for 19 December 1892, GStA Berlin, Waldersee Papers; omitted from Meisner, II, p. 272.
111 Holstein to Eulenburg, 10 January 1891, *Eulenburgs Korrespondenz*, I, No. 464.
112 Holstein's note of March 1891, BA Koblenz, Eulenburg Papers, 1891, p. 81.
113 Holstein to Eulenburg, 22 February 1891, *Eulenburgs Korrespondenz*, I, No. 477; cf. ibid., No. 479. But see Bülow to Eulenburg, 13 March 1893, ibid., II, No. 785, where Bülow describes the Kaiser's Brandenburg speech as 'excellent'.
114 Holstein to Eulenburg, 16 March 1891, ibid., I, No. 485.
115 Holstein to Eulenburg, 1 December 1891, ibid., No. 559.
116 Holstein to Eulenburg, 28 February 1891, ibid., No. 481.
117 See Kiderlen-Wächter to Eulenburg, 3 March 1893, ibid., II, No. 778.
118 Holstein to Eulenburg, 4 September 1892, ibid., No. 697.
119 Holstein to Eulenburg, 22 February 1891, ibid., I, No. 477; cf. ibid., No. 479.
120 Waldersee, diary entries for 11 and 17 May 1891, GStA Berlin, Waldersee Papers; heavily abridged in Meisner, II, p. 207.
121 Brauer to Turban, 9 May 1891, Fuchs, *Großherzog von Baden*, III, No. 1116.
122 Holstein to Eulenburg, 16 November 1891, *Eulenburgs Korrespondenz*, I, No. 555.
123 Eulenburg to Kaiser Wilhelm II, 28 November and 19 December 1891, ibid., Nos. 556 and 566.
124 Marschall to Eulenburg, 6 May 1891, ibid., No. 513.
125 Waldersee, diary entry for 4 January 1892, GStA Berlin, Waldersee Papers; omitted from Meisner, II, pp. 228–9.
126 Helldorff to Eulenburg, 7 March 1892, *Eulenburgs Korrespondenz*, II, No. 600.
127 Caprivi to Eulenburg, 28 February 1892, ibid., No. 595. See Count Karl von Kalnein to Eulenburg, 8 March 1892, ibid., No. 602.
128 Holstein to Eulenburg, 27 April and 7 May 1892, ibid., Nos. 646 and 650.
129 Eulenburg to Holstein, 27 April 1892, ibid., No. 647; Eulenburg to Marschall, 2 and 5 May 1892, PA AA, Preußen 1 No. 1d, Bd I; Holstein to Eulenburg, 3 and 7 May 1892, *Eulenburgs Korrespondenz*, II, Nos. 648 and 650.
130 Waldersee, diary entry for 13 January 1891, GStA Berlin, Waldersee Papers; cf. Meisner, II, p. 175.
131 Herbert Bismarck to his father, 23 August 1892, BA Koblenz, Bismarck Papers FC 3005N.
132 Holstein to Eulenburg, 27 April 1892, *Eulenburgs Korrespondenz*, II, No. 646.
133 Kiderlen-Wächter to Eulenburg, 10 May 1892, ibid., No. 651.
134 Eulenburg to Kaiser Wilhelm II, 12 August 1892, ibid., No. 691.
135 Waldersee, diary entry for 13 November 1892, GStA Berlin, Waldersee Papers; Meisner, II, p. 269.
136 Waldersee, diary entry for 18 August 1894, GStA Berlin, Waldersee Papers; Meisner, II, p. 320. See Richard J. Evans, ed., *Kneipengespräche im Kaiserreich. Stimmungsberichte der Hamburger Politischen Polizei 1892–1914*, Reinbek bei Hamburg 1989, pp. 328ff.
137 Holstein to Eulenburg, 4 September 1892, *Eulenburgs Korrespondenz*, II, No. 697.
138 Holstein to Eulenburg, 4 September 1892, ibid., No. 698.
139 Holstein to Eulenburg, 7 August 1893, ibid., No. 814.
140 Holstein to Eulenburg, 11 August 1893, ibid., No. 815.
141 Holstein to Eulenburg, 9 and 11 November 1894, ibid., Nos. 1045 and 1047.
142 Holstein to Eulenburg, 17 February 1895, ibid., III, No. 1089.
143 Holstein to Eulenburg, 27 November 1894, ibid., II, No. 1052.

20 THE PREDICTABLE DISASTER: WILHELM II AND THE 'PUBLIC SOUL' OF GERMANY

1 Brauer to Turban, 28 January 1891, Fuchs, *Großherzog von Baden*, III, No. 1096.
2 Waldersee, diary entry for 18 January 1891, GStA Berlin, Waldersee Papers; omitted from Meisner, II, p. 176.
3 Waldersee, diary entry for 20 October 1894, GStA Berlin, Waldersee Papers; omitted from Meisner, II, pp. 324ff.
4 Waldersee, diary entry for 17 December 1894, GStA Berlin, Waldersee Papers; cf. Meisner, II, pp. 333–4.
5 Ernest Lavisse, *Figaro*, May 1890; German translation in *Börsen-Courier*, 23 May 1890.
6 José Maria Eça de Queirós, 'O imperador Guilherme' (1891), in *Echos de Pariz*, Porto 4th edn 1920, translated by Professor Dr Erwin Koller.
7 Brauer to Turban, 1 March 1892, Fuchs, *Großherzog von Baden*, III, No. 1162.
8 'The Holy German Empire', *St James's Gazette*, 26 February 1892, RA 59/59.
9 Walther Rathenau, *Der Kaiser*, Berlin 1919, pp. 24–5. See Nicolaus Sombart, *Wilhelm II. Sündenbock und Herr der Mitte*, Berlin 1996, p. 11 *et passim*. Cf. Thomas Kohut, *Wilhelm II and the Germans*, pp. 125ff.
10 Waldersee, diary entry for 18 August 1890, GStA Berlin, Waldersee Papers; omitted from Meisner, II, p. 140.
11 Waldersee, diary entry for 18 May 1890, GStA Berlin, Waldersee Papers; omitted from Meisner, II, p. 128.
12 Waldersee, diary entry for 10 August 1890, GStA Berlin, Waldersee Papers; omitted from Meisner, II, pp. 136–7.
13 Waldersee, diary entry for 11 August 1890, GStA Berlin, Waldersee Papers; cf. Meisner, II, pp. 137–8.
14 Waldersee, diary entry for 11 May 1891, GStA Berlin, Waldersee Papers; omitted in large part from Meisner, II, pp. 206–7.
15 Waldersee, diary entry for 31 October 1890, GStA Berlin, Waldersee Papers; omitted from Meisner, II, p. 157.
16 Loë to Waldersee, 16 November 1893, extracts printed in Meisner, II, pp. 454–6. See Waldersee, diary entry for 18 November 1893, GStA Berlin, Waldersee Papers; the key passages are omitted from Meisner, II, p. 299.
17 Waldersee, diary entry for 24 September 1890, GStA Berlin, Waldersee Papers; cf. Meisner, II, p. 149.
18 Waldersee, diary entry for 3 October 1890, GStA Berlin, Waldersee Papers; omitted from Meisner, II, p. 152.
19 Waldersee, diary entry for 24 November 1890, GStA Berlin, Waldersee Papers; omitted from Meisner, II, p. 163.
20 Waldersee, diary entry for 12 December 1890, GStA Berlin, Waldersee Papers; omitted from Meisner, II, p. 165.
21 Waldersee, diary entry for 3 January 1891, GStA Berlin, Waldersee Papers; omitted from Meisner, II, p. 174.
22 Waldersee, diary entry for 19 January 1891, GStA Berlin, Waldersee Papers; omitted from Meisner, II, p. 176.
23 Waldersee, diary entry for 31 January 1892, GStA Berlin, Waldersee Papers; omitted from Meisner, II, p. 232.
24 Waldersee, diary entry for 26 February 1891, GStA Berlin, Waldersee Papers; omitted from Meisner, II, pp. 196–7.
25 See illus. 25, p. 508.

26 Holstein to Eulenburg, 16 November 1891, *Eulenburgs Korrespondenz*, I, No. 555.
27 Holstein to Eulenburg, 17 March and 19 April 1891, ibid., Nos. 489 and 503.
28 Holstein to Eulenburg, 27 March 1891, ibid., No. 495.
29 Franz Fischer to Eulenburg, 19 April 1891, ibid., No. 503.
30 Brauer to Turban, 6 December 1891, Fuchs, *Großherzog von Baden*, III, No. 1142.
31 Empress Frederick to Queen Victoria, 5 December 1891, RA Z52/2.
32 Empress Frederick to Queen Victoria, 8 March 1892, RA Z52/30.
33 Strachey, report from Dresden, 4 December 1891, RA I59/52.
34 Waldersee, diary entry for 22 January 1892, GStA Berlin, Waldersee Papers; omitted from Meisner, II, p. 231.
35 Caprivi to Eulenburg, 28 February 1892, *Eulenburgs Korrespondenz*, II, No. 595. See Kalnein to Eulenburg, 8 March 1892, ibid., No. 602.
36 Brauer to Turban, 6 March 1892, Fuchs, *Großherzog von Baden*, III, No. 1163.
37 Empress Frederick to Queen Victoria, 20 March 1892, RA Z52/34.
38 Malet to Salisbury, 5 March 1892, RA Z59/61.
39 Verdy to Waldersee, March 1892, GStA Berlin, Waldersee Papers No. 53.
40 Swaine to Ponsonby, 1 July 1892, RA I59/89.
41 Waldersee, diary entry for 11 May 1892, GStA Berlin, Waldersee Papers; omitted from Meisner, II, p. 241.
42 Waldersee, diary entry for 29 November 1893, GStA Berlin, Waldersee Papers; omitted from Meisner, II, p. 299.
43 Brauer to Grand Duke of Baden, 31 December 1892, Fuchs, *Großherzog von Baden*, III, No. 1217.
44 Waldersee, diary entry for 8 March 1891, GStA Berlin, Waldersee Papers; omitted from Meisner, II, p. 198.
45 Bülow to Eulenburg, 6 April 1892, *Eulenburgs Korrespondenz*, II, No. 641.
46 Holstein to Eulenburg, 1 April 1892, ibid., No. 638.
47 Holstein to Eulenburg, 31 March 1892, ibid., No. 634.
48 See Ritter and Niehuss, *Wahlgeschichtliches Arbeitsbuch*, pp. 38–42.
49 Waldersee, diary entry for 24 September 1890, GStA Berlin, Waldersee Papers; cf. Meisner, II, p. 149.
50 Waldersee, diary entry for 18 December 1892, GStA Berlin, Waldersee Papers; omitted from Meisner, II, p. 272.
51 Waldersee, diary entry for 7 June 1894, GStA Berlin, Waldersee Papers; cf. Meisner, II, p. 314.
52 Waldersee, diary entry for 30 December 1894, GStA Berlin, Waldersee Papers; omitted from Meisner, II, p. 334.
53 Waldersee, diary entry for 3 January 1892, GStA Berlin, Waldersee Papers; omitted from Meisner, II, p. 228.
54 Ludwig Quidde, *Caligula. Eine Studie über römischen Cäsarenwahnsinn*, Leipzig 12th edn 1894; Quidde, *Erinnerungen*, cited in Hans-Ulrich Wehler, ed., *Caligula. Schriften über Militarismus und Pazifismus*, Frankfurt a.M. 1977, p. 24.
55 The reaction to Quidde's pamphlet is analysed extensively in Martin Kohlrausch, 'Monarchie und Massenöffentlichkeit. Veränderungen in der Rezeption des wilhelminischen Kaisertums, 1890–1925', diss., Florence 2002. See Joachim Radkau, *Das Zeitalter der Nervosität. Deutschland zwischen Bismarck und Hitler*, Munich, Vienna 1998, pp. 275–6; Gisela Brude-Firnau, *Die literarische Deutung Kaiser Wilhelms II. zwischen 1889 und 1989*, Heidelberg 1997, pp. 32–9.
56 See Hans Wehberg, *Ludwig Quidde. Ein deutscher Demokrat und Vorkämpfer der Völkerverständigung*, Offenbach 1948; Utz-Friedbert Taube, *Ludwig Quidde. Ein Beitrag zur Geschichte des demokratischen Gedankens in Deutschland*, Munich 1963; Wehler, *Caligula*; John C. G. Röhl, *Kaiser Wilhelm II. Eine Studie über Cäsarenwahnsinn*, Munich 1989.

57 I am grateful to Dr Jost Rebentisch of Cologne for drawing my attention to these interesting official statistics. See Jost Rebentisch, *Die vielen Gesichter des Kaisers. Wilhelm II. in der deutschen und britischen Karikatur (1888–1918)*, Berlin 2000, pp. 58 and 60.

58 From 1897 onwards the official figures include only cases involving insults to the Kaiser or other ruling princes in Germany, not to lesser members of the royal families.

59 On Harden's article 'Monarchen-Erziehung' in the *Zukunft* of 31 December 1892 and the subsequent trial, see Bernd-Uwe Weller, *Maximilian Harden und die 'Zukunft'*, Bremen 1970, p. 109. Harden was sentenced to six months in prison for *lèse-majesté* on the basis of his articles 'Pudel Majestät', 'An den Kaiser' and 'Großvaters Uhr', published in the *Zukunft* in 1898: Hans Dieter Hellige and Ernst Schulin, eds., *Walther Rathenau Gesamtausgabe*, vol. VI, *Briefwechsel Walther Rathenau–Maximilian Harden*, Munich, Heidelberg 1983, pp. 314–21.

60 Kaiser Wilhelm II, All-Highest Proclamation of 27 January 1907. See 'Der Kaiser und die Majestätsbeleidigung' in *Vossische Zeitung* of 28 January 1907; also the minutes of the Prussian Ministry of State meeting of 23 March 1907, GStA Berlin.

61 Waldersee, diary entry for 19 March 1897, GStA Berlin, Waldersee Papers; omitted from Meisner, II, p. 395.

62 Waldersee, diary entry for 24 September 1890, GStA Berlin, Waldersee Papers; cf. Meisner, II, p. 149.

63 Waldersee, diary entry for 27 February 1894, GStA Berlin, Waldersee Papers; printed only in part in Meisner, II, p. 309; also the diary entry for 18 August 1894, GStA Berlin, Waldersee Papers; omitted from Meisner, II, pp. 320–1.

64 Waldersee, diary entry for 23 June 1893, GStA Berlin, Waldersee Papers; cf. Meisner II, pp. 290–1.

65 Waldersee, diary entries for 18 and 21 November 1893, GStA Berlin, Waldersee Papers; cf. Meisner, II, p. 299. See the watered-down version of the speech in Penzler, *Reden Kaiser Wilhelms II. in den Jahren 1888–1895*, p. 255.

66 Waldersee, diary entry for 22 December 1891, GStA Berlin, Waldersee Papers; cf. Meisner, II, p. 228.

67 Waldersee, diary entry for 29 October 1891, GStA Berlin, Waldersee Papers; omitted from Meisner, II, pp. 219–20.

68 Waldersee, diary entry for 3 January 1892, GStA Berlin, Waldersee Papers; omitted from Meisner, II, p. 228. See Brauer to Turban, 20 December 1891 and 22 May 1892, Fuchs, *Großherzog von Baden*, III, Nos. 1143 and 1184.

69 Waldersee, diary entry for 17 December 1893, GStA Berlin, Waldersee Papers; completely omitted from Meisner, II, p. 300.

70 Count Friedrich Yrsch to Eulenburg, March and 7 April 1891, *Eulenburgs Korrespondenz*, I, Nos. 487 and 497.

71 Eulenburg to Holstein, 12 April 1891, ibid., No. 498.

72 Brauer to Turban, 19 June 1892, Fuchs, *Großherzog von Baden*, III, No. 1190.

73 Brauer to Turban, 12 July 1892, ibid., No. 1193.

74 Waldersee, diary entry for 30 December 1894, GStA Berlin, Waldersee Papers; omitted from Meisner, II, p. 334. See Brauer to Turban, 8 January 1892, Fuchs, *Großherzog von Baden*, III, No. 1145.

75 Waldersee, diary entry for 9 March 1891, GStA Berlin, Waldersee Papers; omitted from Meisner, II, p. 198.

76 Waldersee, diary entry for 17 December 1894, GStA Berlin, Waldersee Papers; omitted from Meisner, II, pp. 333–4. On the popular mood in Hamburg, see Evans, *Kneipengespräche*, pp. 322ff. *et passim*.

77 Waldersee, diary entry for 23 August 1894, GStA Berlin, Waldersee Papers; omitted from Meisner, II, p. 322.

78 Holstein to Eulenburg, 17 March 1891, *Eulenburgs Korrespondenz*, I, No. 489.
79 Holstein to Eulenburg, 16 November 1891, ibid., No. 555.
80 Waldersee, diary entry for 3 January 1892, GStA Berlin, Waldersee Papers; omitted from Meisner, II, p. 228.
81 Waldersee, diary entry for 20 February 1894, GStA Berlin, Waldersee Papers; omitted from Meisner, II, p. 308. See above, p. 469.
82 Waldersee, diary entry for 14 February 1894, GStA Berlin, Waldersee Papers; omitted from Meisner, II, p. 308.
83 Waldersee, diary entry for 11 March 1894, GStA Berlin, Waldersee Papers; cf. Meisner, II, p. 310.
84 Waldersee, diary entry for 20 February 1894, GStA Berlin, Waldersee Papers; omitted from Meisner, II, p. 308.
85 Waldersee, diary entry for 13 November 1892, GStA Berlin, Waldersee Papers; omitted from Meisner, II, p. 269.
86 Waldersee, diary entry for 1 December 1892, GStA Berlin, Waldersee Papers; omitted from Meisner, II, p. 270.
87 Waldersee, diary entry for 14 February 1894, GStA Berlin, Waldersee Papers; omitted from Meisner, II, p. 308.
88 Waldersee, diary entry for 5 January 1894, GStA Berlin, Waldersee Papers; omitted from Meisner, II, p. 303.
89 Waldersee, diary entry for 9 September 1894, GStA Berlin, Waldersee Papers; omitted from Meisner, II, p. 322.
90 Waldersee, diary entry for 5 January 1894, GStA Berlin, Waldersee Papers; omitted from Meisner, II, p. 303.
91 Waldersee, diary entry for 18 December 1894, GStA Berlin, Waldersee Papers; omitted from Meisner, II, p. 334.
92 Fischer to Eulenburg, 17 February 1895, *Eulenburgs Korrespondenz*, III, No. 1090.
93 See above, chs. 16 and 19.
94 Waldersee, diary entry for 4 January 1892, GStA Berlin, Waldersee Papers; omitted from Meisner, II, pp. 228–9.
95 Waldersee, diary entry for 5 January 1894, GStA Berlin, Waldersee Papers; Meisner, II, pp. 301ff.
96 Waldersee, diary entry for 15 January 1895, GStA Berlin, Waldersee Papers; omitted from Meisner, II, pp. 334ff.
97 Waldersee, diary entry for 21 August 1895, GStA Berlin, Waldersee Papers; cf. Meisner, II, pp. 356–7.
98 Waldersee, diary entry for 6 September 1891, GStA Berlin, Waldersee Papers; cf. Meisner, II, pp. 215–16.
99 Waldersee, diary entry for 27 June 1894, GStA Berlin, Waldersee Papers; omitted from Meisner, II, p. 315.
100 Waldersee, diary entry for 21 July 1895, GStA Berlin, Waldersee Papers; cf. Meisner, II, pp. 353ff.
101 See *Young Wilhelm*, pp. 490ff.
102 Waldersee, diary entry for 15 October 1893, GStA Berlin, Waldersee Papers; cf. Meisner, II, pp. 295–6.
103 Waldersee, diary entry for 10 November 1893, GStA Berlin, Waldersee Papers; omitted from Meisner, II, p. 299.
104 Waldersee, diary entry for 21 July 1895, GStA Berlin, Waldersee Papers; omitted from Meisner, II, pp. 353–5.
105 Waldersee, diary entry for 4 January 1892, GStA Berlin, Waldersee Papers; omitted from Meisner, II, pp. 228–9. See below, p. 572.

106 Waldersee, diary entry for 13 March 1892, GStA Berlin, Waldersee Papers; cf. Meisner, II, p. 234.
107 Waldersee, diary entry for 13 March 1892, GStA Berlin, Waldersee Papers; cf. Meisner, II, p. 234.
108 Waldersee, diary entry for 6 March 1893, Meisner, II, p. 286.
109 Waldersee, diary entry for 21 August 1895, GStA Berlin, Waldersee Papers; omitted from Meisner, II, pp. 356–7.
110 Waldersee, diary entry for 13 March 1892, GStA Berlin, Waldersee Papers; cf. Meisner, II, p. 234.
111 Waldersee, diary entries for 6 April and 4 May 1895, GStA Berlin, Waldersee Papers; cf. Meisner, II, pp. 344–5 and 347.
112 Waldersee, diary entry for 10 August 1890, GStA Berlin, Waldersee Papers; omitted from Meisner, II, pp. 136–7.
113 Waldersee, diary entries for 6 May and 11 September 1890, GStA Berlin, Waldersee Papers; cf. Meisner, II, pp. 126–7 and 144.
114 Waldersee, diary entry for 9 March 1891, GStA Berlin, Waldersee Papers; omitted from Meisner, II, p. 198.
115 Waldersee, diary entry for 14 December 1894, GStA Berlin, Waldersee Papers; cf. Meisner, II, p. 333.
116 Waldersee, diary entries for 21 July and 8 August 1895, GStA Berlin, Waldersee Papers; cf. Meisner, II, pp. 353–5.
117 Waldersee, diary entry for 14 May 1892, GStA Berlin, Waldersee Papers; omitted from Meisner, II, p. 241.
118 Waldersee, diary entry for 3 December 1893, GStA Berlin, Waldersee Papers; omitted from Meisner, II, p. 299.
119 Waldersee, diary entry for 21 July 1895, GStA Berlin, Waldersee Papers; omitted from Meisner, II, pp. 353–5.
120 Waldersee, diary entries for 10 and 13 January 1891, GStA Berlin, Waldersee Papers; cf. Meisner, II, pp. 174–5; Empress Frederick to Queen Victoria, 29 August 1891, RA Z51/16.
121 Waldersee, diary entry for 10 November 1893, GStA Berlin, Waldersee Papers; distorted and printed under an incorrect date in Meisner, II, p. 298.
122 Waldersee, diary entry for 16 October 1893, GStA Berlin, Waldersee Papers; omitted from Meisner, II, p. 296. See also Loë to Waldersee, 16 November 1893, partially printed in Meisner, II, pp. 454–6.
123 Waldersee, diary entry for 9 October 1895, GStA Berlin, Waldersee Papers; omitted from Meisner, II, p. 360.
124 Hereditary Prince Bernhard of Saxe-Meiningen to Waldersee, 9 October 1891, GStA Berlin, Waldersee Papers No. 36.
125 Waldersee, diary entry for 4 October 1893, GStA Berlin, Waldersee Papers; omitted from Meisner, II, p. 294.
126 Waldersee, diary entry for 6 August 1894, GStA Berlin, Waldersee Papers; omitted from Meisner, II, p. 320.
127 Waldersee, diary entry for 15 October 1893, GStA Berlin, Waldersee Papers; cf. Meisner, II, pp. 295–6.
128 Waldersee, diary entry for 22 October 1894, GStA Berlin, Waldersee Papers; omitted from Meisner, II, p. 327.
129 Waldersee, diary entry for 12 July 1895, GStA Berlin, Waldersee Papers; Meisner, II, p. 352.
130 Waldersee, diary entries for 6 April and 4 May 1895, GStA Berlin, Waldersee Papers; cf. Meisner, II, pp. 344–5 and 347.
131 Waldersee, diary entry for 29 October 1895, GStA Berlin, Waldersee Papers; omitted from Meisner, II, p. 361.

132 Waldersee, diary entry for 27 August 1891, GStA Berlin, Waldersee Papers; omitted from Meisner, II, p. 215.
133 Waldersee, diary entry for 9 March 1891, GStA Berlin, Waldersee Papers; omitted from Meisner, II, p. 198.
134 Waldersee, diary entry for 17 September 1891, GStA Berlin, Waldersee Papers; cf. Meisner, II, p. 216.
135 Waldersee, diary entry for 23 May 1891, GStA Berlin, Waldersee Papers; omitted from Meisner, II, p. 209.
136 Waldersee, diary entry for 15 October 1893, GStA Berlin, Waldersee Papers; omitted from Meisner, II, pp. 295–6.
137 Waldersee, diary entry for 3 July 1891, GStA Berlin, Waldersee Papers; omitted from Meisner, II, pp. 211–12.
138 Waldersee, diary entry for 20 July 1894, GStA Berlin, Waldersee Papers; omitted from Meisner, II, p. 318.
139 Waldersee, diary entry for 1 August 1894, GStA Berlin, Waldersee Papers; printed only in part in Meisner, II, p. 319.
140 Waldersee, diary entry for 4 January 1892, GStA Berlin, Waldersee Papers; omitted from Meisner, II, pp. 228–9.
141 Waldersee, diary entry for 17 September 1891, GStA Berlin, Waldersee Papers; omitted from Meisner, II, pp. 216–17.
142 Waldersee, diary entry for 15 January 1895, GStA Berlin, Waldersee Papers; omitted from Meisner, II, pp. 334ff.
143 Eulenburg to Kaiser Wilhelm II, 20 January 1896, *Eulenburgs Korrespondenz*, III, No. 1190. Cf. Bülow's description of his meeting with Eulenburg in Merano: *Memoirs*, I, pp. 679ff.
144 Eulenburg to Kaiser Wilhelm II, 27 February 1896, *Eulenburgs Korrespondenz*, III, No. 1192.
145 Eulenburg to Kaiser Wilhelm II, 20 July 1892, ibid., II, No. 683.
146 Eulenburg to Kaiser Wilhelm II, 5 February 1893, ibid., No. 767.
147 Eulenburg to Kaiser Wilhelm II, 22 August 1897, ibid., III, No. 1340.
148 Eulenburg to Kaiser Wilhelm II, 31 December 1895, ibid., No. 1183.
149 Eulenburg to Kaiser Wilhelm II, 7 January 1897, ibid., No. 1288.
150 Varnbüler to Kuno Moltke, 15 April, 7 May and 4 June 1898, ibid., Nos. 1366 and 1373 with note 2; also Röhl, *Kaiser and* His *Court*, pp. 58–9.
151 Eulenburg to Kaiser Wilhelm II, 27 August 1893, *Eulenburgs Korrespondenz*, II, No. 819.
152 Eulenburg to Kaiser Wilhelm II, 16 April 1894, ibid., No. 953.
153 Eulenburg to Kaiser Wilhelm II, 2 August 1895, ibid., III, No. 1120.
154 Eulenburg to Kaiser Wilhelm II, 19 April 1894, ibid., II, No. 968.
155 Eulenburg to Kaiser Wilhelm II, 27 February 1896, ibid., III, No. 1192. See below, pp. 1045–56, and Eulenburg to his mother and to Kaiserin Auguste Viktoria, 29 September 1895, *Eulenburgs Korrespondenz*, III, p. 1549.
156 Eulenburg to Kaiser Wilhelm II, 15 November 1898, *Eulenburgs Korrespondenz*, III, No. 1389.
157 See *Young Wilhelm*, pp. 416ff.
158 Bülow to Eulenburg, 15 June 1892, *Eulenburgs Korrespondenz*, II, No. 681.
159 Bülow to Eulenburg, 6 April 1892, ibid., No. 641.
160 Bülow to Eulenburg, 19 December 1893, ibid., No. 866.
161 Bülow to Eulenburg, 30 April 1896, ibid., III, No. 1210.
162 Bülow to Eulenburg, 6 November 1896, ibid., No. 1271.
163 Bülow to Eulenburg, 20 March 1896, ibid., No. 1201.
164 Bülow to Eulenburg, 1 April 1896, ibid., No. 1205.
165 Bülow to Eulenburg, 8 February 1892, ibid., II, No. 583.

166 Bülow to Eulenburg, 9 February 1895, ibid., No. 1081.
167 Bülow to Eulenburg, 4 December 1896, ibid., III, No. 1281.
168 Bülow to Eulenburg, 6 February 1895, ibid., II, No. 1080.
169 Bülow to Eulenburg, 30 October 1895, ibid., III, No. 1151.
170 Bülow to Eulenburg, 15 February 1898, ibid., No. 1362.
171 Bülow to Eulenburg, 27 December 1895, ibid., No. 1181.
172 Eulenburg to Kuno Moltke, 15 June 1895, ibid., No. 1112.
173 Eulenburg to Bülow, 8 June 1896, ibid., No. 1233.
174 See e.g. Eulenburg to Bülow, 23 September 1900, ibid., No. 1427.

21 CAPRIVI'S DISMISSAL

1 See Maltzahn to Caprivi, 9, 12, 16 and 17 July 1893, Caprivi to Maltzahn, 16 July 1893, Caprivi to Marschall, 26 July 1893, BA Berlin, Reichskanzlei, No. 1621.
2 Waldersee, diary entry for 30 August 1893, GStA Berlin, Waldersee Papers; cf. Meisner, II, p. 293.
3 Waldersee, diary entry for 21 January 1894, GStA Berlin, Waldersee Papers; omitted from Meisner, II, p. 304.
4 Caprivi to Wilamowitz, 25 July 1893, Wilamowitz to Caprivi, 27 July 1893, BA Berlin, Reichskanzlei, No. 1621. See Röhl, *Germany without Bismarck*, p. 105. Cf. Hohenlohe, *Denkwürdigkeiten*, II, pp. 503–4.
5 Posadowsky to Caprivi, 3 August 1893, Caprivi to Kaiser Wilhelm II, 3 August 1893, the Kaiser's reply, 3 August 1893, BA Berlin, Reichskanzlei, No. 1621.
6 Caprivi to Kaiser Wilhelm II, 5 August 1893, Kaiser Wilhelm II to Caprivi, 6 August 1893, cited in Röhl, *Germany without Bismarck*, p. 105.
7 Kaiser Wilhelm II to Caprivi, 8 August 1893, BA Berlin, Reichskanzlei, No. 1621.
8 Caprivi to Kaiser Wilhelm II, 8 August 1893, ibid.
9 Kiderlen-Wächter to Caprivi, 12 August 1893, ibid.
10 Caprivi to Kiderlen-Wächter, 12 August 1893, ibid.
11 Posadowsky to Caprivi, 16 February 1894, ibid.
12 Waldersee, diary entries for 24 August and 23 September 1893, GStA Berlin, Waldersee Papers; omitted from Meisner, II, pp. 292ff.
13 Eulenburg to Caprivi, 12 August 1893, *Eulenburgs Korrespondenz*, II, No. 817.
14 August Eulenburg to Philipp Eulenburg, 17 August 1893, ibid., No. 818.
15 Spitzemberg, *Tagebuch*, p. 315. See Eulenburg to Caprivi, 17 September 1893, Caprivi to Eulenburg, 18 September 1893, *Eulenburgs Korrespondenz*, II, Nos. 824–5.
16 Waldersee, diary entry for 19 October 1893, GStA Berlin, Waldersee Papers; Meisner, II, p. 297.
17 Waldersee, diary entry for 15 October 1893, GStA Berlin, Waldersee Papers; cf. Meisner, II, pp. 295–6.
18 Holstein to Eulenburg, 11 August 1893, *Eulenburgs Korrespondenz*, II, No. 815.
19 Waldersee, diary entry for 25 October 1893, GStA Berlin, Waldersee Papers; omitted from Meisner, II, p. 298.
20 Kaiser Wilhelm II, marginal notes of 1 December 1893 on Caprivi's report of 30 November 1893, PA AA, Asservat No. 4. See Marschall, diary entry for 2 December 1893, cited in *Eulenburgs Korrespondenz*, II, p. 1154. Cf. the earlier clash between Wilhelm II and Caprivi over Engelbrecht's reports, above, p. 345.
21 Waldersee, diary entries for 19 and 25 October 1893, GStA Berlin, Waldersee Papers; cf. Meisner, II, pp. 297–8.
22 Waldersee, diary entry for 25 October 1893, GStA Berlin, Waldersee Papers; omitted from Meisner, II, p. 298.

23 Waldersee, diary entry for 19 October 1893, GStA Berlin, Waldersee Papers; Meisner, II, p. 297.
24 Kaiser Wilhelm II to Caprivi, cipher telegram, 2 January 1894, PA AA, Preußen 1 No. 1d, Bd I. See also Jagemann to Brauer, 14 January 1894, Fuchs, *Großherzog von Baden*, III, No. 1309.
25 Marschall to Caprivi, 4 January 189[4], PA AA, Preußen 1 Nr 1d, Bd I.
26 Marschall, diary entry for 2 January 1894, cited in *Eulenburgs Korrespondenz*, II. p. 1180. See also Spitzemberg, diary, p. 318.
27 Caprivi, draft telegram of 4 January 1894, PA AA, Preußen 1 No. 1d, Bd I.
28 Marschall, diary entry for 2 January 1894, cited in *Eulenburgs Korrespondenz*, II. p. 1180. See also Spitzemberg, diary, p. 318.
29 Waldersee, diary entry for 9 January 1894, GStA Berlin, Waldersee Papers; Meisner, II, p. 303.
30 Marschall, diary entry for 4 January 1894, cited in *Eulenburgs Korrespondenz*, II. p. 1180.
31 Eulenburg to Bülow, 2 April 1894, ibid., No. 948.
32 Holstein to Eulenburg, 7 December 1893, ibid., No. 859.
33 Pourtalès to Eulenburg, 2 December 1893, ibid., p. 1156.
34 Holstein to Eulenburg, 7 November 1892, ibid., No. 727.
35 Eulenburg to Caprivi, 14 November 1892, ibid., No. 730.
36 Holstein to Eulenburg, 18 November 1892, Eulenburg to Holstein, 19 November 1892, ibid., Nos. 732 and 733.
37 Eulenburg to Bülow, 28 February 1893, ibid., No. 776.
38 Bülow to Eulenburg, 27 March 1893, ibid., No. 789.
39 Eulenburg to Kaiser Wilhelm II, 15 April 1893, ibid., No. 790.
40 Eulenburg to Kaiser Wilhelm II, 4 December 1893, ibid., No. 857.
41 Kaiser Wilhelm II to Eulenburg, 6 December 1893, ibid., No. 858.
42 Holstein to Eulenburg, December 1893, ibid., No. 859.
43 Malet to Rosebery, abstract, 20 December 1893, RA I59/115; Reuss to Wedel, 9 January 1894, Wedel Papers, Frankfurt a.M.
44 See Caprivi's letter of protest to Bülow of 8 March 1894, *Eulenburgs Korrespondenz*, II, p. 1239.
45 Eulenburg to Bülow, 27 March 1894, ibid., No. 941. See also ibid., p. 1239, as well as Nos. 924, 929 and 930.
46 Kaiser Wilhelm II to Bülow, 28 March 1894, ibid., No. 944.
47 Eulenburg to Bülow, 28 March 1894, ibid., No. 943.
48 Reuss to Holstein, 3 January 1894, *Holstein Papers*, III, No. 397.
49 Eulenburg to Kaiser Wilhelm II, 7 March 1894, Kaiser Wilhelm II to Eulenburg, 8 March 1894, *Eulenburgs Korrespondenz*, II, Nos. 922–3.
50 Waldersee, diary entry for 21 January 1894, GStA Berlin, Waldersee Papers; omitted from Meisner, II, p. 304.
51 Brauer to Turban, 6 and 8 June 1891, Fuchs, *Großherzog von Baden*, III, Nos. 1121–2.
52 Holstein to Eulenburg, 7 and 10 December 1891, *Eulenburgs Korrespondenz*, I, Nos. 561–2; Empress Frederick to Queen Victoria, 5 and 12 December 1891, in Ponsonby, *Letters of the Empress Frederick*, pp. 430–2.
53 Kiderlen-Wächter to Eulenburg, 15 June 1892, *Eulenburgs Korrespondenz*, II, No. 680. See the correspondence on this crisis ibid., pp. 879–902.
54 Kaiser Wilhelm II to Kaiser Franz Joseph, 13 June 1892, in Gradenwitz, *Bismarcks letzter Kampf*, pp. 240–2.
55 Kálnoky to Reuss, 16 June 1892, in Otto Gradenwitz, ed., *Akten über Bismarcks großdeutsche Rundfahrt vom Jahre 1892*, Heidelberg 1922, No. 12.
56 Waldersee to Kaiser Wilhelm II, 13 June 1892, in Meisner, II, pp. 245–6; Gradenwitz, *Akten*, pp. 3–4. See Kiderlen-Wächter to Eulenburg, 15 June 1892, *Eulenburgs Korrespondenz*, II,

No. 680; also Eulenburg's notes on a conversation with Waldersee on 29 June 1892, ibid., p. 918.

57 Kaiser Wilhelm II to Waldersee, 10 June 1892, GStA Berlin, Waldersee Papers No. 42. See Holstein to Eulenburg, 9 June 1892, *Eulenburgs Korrespondenz*, II, No. 671.

58 Eulenburg's notes of 26 June 1892, *Eulenburgs Korrespondenz*, II, No. 685.

59 Eulenburg to Holstein, 2 October 1893, ibid., No. 828.

60 Caprivi to Eulenburg, 18 September 1893, ibid., No. 825.

61 Prince Otto von Bismarck to Kaiser Wilhelm II, 21 October 1893, ibid., p. 1118; Eulenburg to Kaiser Wilhelm II, 24 October 1893, ibid., No. 831.

62 Caprivi to Eulenburg, 18 September 1893, ibid., No. 825.

63 Eulenburg to Holstein, 7 October 1893, ibid., No. 830.

64 Eulenburg to Varnbüler, 19 April 1893, ibid., No. 796. See also Eulenburg to Kaiser Wilhelm II, 12 August 1892, ibid., No. 691.

65 See Eulenburg, *Aus 50 Jahren*, pp. 257–68. See also Hohenlohe, *Denkwürdigkeiten*, II, pp. 508ff.; Spitzemberg, *Tagebuch*, pp. 319ff.

66 Kaiser Wilhelm II to Queen Victoria, telegram, 25 January 1894, RA I60/45. See also Brauer to Jagemann, 26 January 1894, Fuchs, *Großherzog von Baden*, III, p. 288, footnote.

67 Marschall, diary entries for 22–5 January 1894, cited in Eulenburgs Korrespondenz, II, pp. 1197–8.

68 Hohenlohe, *Denkwürdigkeiten*, II, p. 510; Bülow, *Memoirs*, IV, pp. 656–6.

69 Marschall, diary entry for 26 January 1894, cited in *Eulenburgs Korrespondenz*, II, p. 1198.

70 Eulenburg, *Aus 50 Jahren*, p. 267.

71 Kuno Moltke to Eulenburg, 25 and 28 January 1894, *Eulenburgs Korrespondenz*, II, Nos. 889 and 890. Cf. Eulenburg's reply of 1 February 1894, ibid., No. 891. See Fuchs, *Großherzog von Baden*, III, No. 1322.

72 Waldersee, diary entry for 25, 26 and 28 January 1894, GStA Berlin, Waldersee Papers; printed only in part in Meisner, II, p. 305.

73 Waldersee, diary entry for 29 January 1894, GStA Berlin, Waldersee Papers; cf. Meisner, II, p. 305.

74 Waldersee, diary entries for 22 and 26 January 1894, GStA Berlin, Waldersee Papers; cf. Meisner, II, p. 304.

75 Count Herbert von Bismarck to his father, 21 January 1894, BA Koblenz, Bismarck Papers FC 3005 N. See also Jagemann to Brauer, 25 January 1894, Fuchs, *Großherzog von Baden*, III, No. 1316.

76 Herbert Bismarck to his father, 28 January 1894, BA Koblenz, Bismarck Papers FC 3005 N.

77 The first article, entitled 'Der vierte Mann im Skat', appeared on 24 December 1893. On this scandal, see above all Helmuth Rogge, 'Die Kladderadatsch-Affäre. Ein Beitrag zur inneren Geschichte des Wilhelminischen Reichs', *Historische Zeitschrift*, vol. 195/1 (August 1962), pp. 90–130.

78 Kiderlen-Wächter to Eulenburg, 14 January 1894, *Eulenburgs Korrespondenz*, II, No. 885.

79 Pourtalès to Eulenburg, 18 April 1894, ibid., No. 957.

80 See ibid., p. 1255; also Forsbach, *Kiderlen-Wächter*, I, pp. 129, 131, 137ff. On the question of duelling in imperial Germany, see Ute Frevert, *Ehrenmänner. Das Duell in der bürgerlichen Gesellschaft*, Munich 1991; Tobias C. Bringmann, *Reichstag und Zweikampf. Die Duellfrage als innenpolitischer Konflikt des deutschen Kaiserreichs 1871–1918*, Freiburg 1997.

81 Holstein to Eulenburg, 28 December 1893, *Eulenburgs Korrespondenz*, II, No. 872.

82 Holstein to Eulenburg, 10 January 1894, ibid., No. 881.

83 Holstein to Eulenburg, 22 March 1894, ibid., No. 936.

84 Eulenburg to Kaiser Wilhelm II, 9 January 1894, ibid., No. 880.

85 Eulenburg to Kaiser Wilhelm II, 23 January 1894, ibid., No. 887.

86 See Eulenburg to Varnbüler, 18 December 1893, ibid., No. 865; Eulenburg to Kaiser Wilhelm II, 13 January 1894, ibid., No. 884.
87 Kaiser Wilhelm II to Eulenburg, 11 January 1894, ibid., No. 882.
88 See the extensive correspondence on this issue ibid., particularly Nos. 949, 950, 960, 964, 972 and 974.
89 Eulenburg to Kaiser Wilhelm II, 7 May 1894, ibid., No. 962. See Eulenburg's account of these murky dealings in Eulenburg to Kuno Moltke, 15 June 1895, ibid., No. 1112.
90 See Marschall's diary entries for May 1894, cited ibid., p. 1300.
91 Holstein to Eulenburg, 18 April 1894, ibid., No. 956.
92 Eulenburg's note for Kaiser Wilhelm II of 20 March 1894, ibid., No. 933.
93 Waldersee, diary entry for 20 August 1894, GStA Berlin, Waldersee Papers; cf. Meisner, II, p. 321. See Prince Ernst zu Hohenlohe-Langenburg to his father, 2 April and 11 June 1893, Hohenlohe-Zentralarchiv Neuenstein, Hermann Hohenlohe-Langenburg Papers, Bü. 60.
94 Brauer to Grand Duke Friedrich I of Baden, 12 and 14 September 1894, Fuchs, *Großherzog von Baden*, III, No. 1341 and 1343.
95 Waldersee, diary entry for 20 October 1894, GStA Berlin, Waldersee Papers; cf. Meisner, II, p. 326.
96 Kaiser Wilhelm II to Eulenburg, telegram, 13 March 1894, *Eulenburgs Korrespondenz*, II, p. 1249.
97 Waldersee, diary entry for 30 March 1894, GStA Berlin, Waldersee Papers; omitted from Meisner, II, p. 312.
98 Waldersee, diary entry for 21 October 1894, GStA Berlin, Waldersee Papers; omitted from Meisner, II, p. 327.
99 Waldersee, diary entries for 8 and 29 August 1894, GStA Berlin, Waldersee Papers; Meisner, II, p. 322.
100 Stosch to Bennigsen, 3 July 1894, printed in Oncken, *Bennigsen*, II, p. 591.
101 Waldersee, diary entry for 16 July 1894, GStA Berlin, Waldersee Papers; omitted from Meisner, II, p. 317.
102 Waldersee, diary entry for 20 August 1894, GStA Berlin, Waldersee Papers; largely omitted from Meisner, II, p. 321.
103 Waldersee, diary entry for 20 August 1894, GStA Berlin, Waldersee Papers; cf. Meisner, II, p. 321.
104 Waldersee, diary entry for 25 June 1894, GStA Berlin, Waldersee Papers; omitted from Meisner, II, p. 315.
105 Waldersee, diary entry for 28 June 1894, GStA Berlin, Waldersee Papers; omitted from Meisner, II, p. 315.
106 Waldersee, diary entry for 6 July 1894, GStA Berlin, Waldersee Papers; cf. Meisner, II, pp. 315–16.
107 Varnbüler to Eulenburg, 16 July 1894, *Eulenburgs Korrespondenz*, II, No. 984; Fischer to Eulenburg, 16 July 1894, ibid., No. 985.
108 Kaiser Wilhelm II to Botho Eulenburg, 24 July 1894, ibid., p. 1333.
109 Botho Eulenburg to Philipp Eulenburg, 26 July 1894, ibid., No. 987.
110 Eulenburg to Kaiser Wilhelm II, 30 August 1894, ibid., No. 989. See also Eulenburg to Grand Duke of Baden, 5 September 1894, ibid., No. 990.
111 Botho Eulenburg to Philipp Eulenburg, 9 September 1894, ibid., No. 991.
112 August Eulenburg to Philipp Eulenburg, 9 September 1894, ibid., No. 992.
113 Botho Eulenburg to Philipp Eulenburg, 9 September 1894, ibid., No. 991.
114 Kaiser Wilhelm II, speech of 6 September 1894, Penzler, *Reden Kaiser Wilhelms II. in den Jahren 1888–1895*, pp. 274–7.
115 Botho Eulenburg to Philipp Eulenburg, 9 September 1894, *Eulenburgs Korrespondenz*, II, No. 991.

116 Waldersee, diary entry for 20 September 1894, GStA Berlin, Waldersee Papers; cf. Meisner, II, p. 323. See Major Ebmeyer, 'Caprivis Entlassung', *Deutsche Revue*, 47/4 (1922), pp. 193ff.; also Jagemann to Brauer, 20 and 28 September and 4 October 1894, Fuchs, *Großherzog von Baden*, III, Nos. 1347, 1351 and 1354.
117 See Eulenburg to Holstein, 25 September 1894, *Eulenburgs Korrespondenz*, II, No. 997.
118 For the growing conflict between Eulenburg and Holstein, especially in regard to their opinion of the Kaiser, see Eulenburg to Bülow, 12 October and 25 December 1894, *Eulenburgs Korrespondenz*, II, Nos. 1023–4 and 1069.
119 Eulenburg had been praising the extraordinary political talents of Bülow in his letters to the Kaiser since early 1892. See Eulenburg to Kaiser Wilhelm II, 12 March 1892, ibid., No. 605. See also above, pp. 575–9.
120 Eulenburg to Bülow, 30 September 1894, *Eulenburgs Korrespondenz*, II, No. 1006.
121 Eulenburg's note of 27 September 1894, ibid., No. 1002.
122 Eulenburg to Bülow, 30 September 1894, ibid., No. 1006.
123 Philipp Eulenburg to Botho Eulenburg, 28 September 1894, ibid., No. 1003.
124 Eulenburg to Bülow, 30 September 1894, ibid., No. 1006.
125 Marschall to Eulenburg, 6 October 1894, ibid., No. 1013. See also Eulenburg to Bülow, 6 October 1894, ibid., No. 1015.
126 Eulenburg's note of 13 October 1894, Haller, *Eulenburg*, p. 150.
127 Waldersee, diary entry for 10 October 1894, GStA Berlin, Waldersee Papers; omitted from Meisner, II, p. 324.
128 Waldersee, diary entry for 20 October 1894, GStA Berlin, Waldersee Papers; cf. Meisner, II, pp. 324ff.
129 Kaiser Wilhelm II, speech of 18 October 1894, cited from E. Schröder, *Zwanzig Jahre Regierungszeit. Ein Tagebuch Kaiser Wilhelms II. Vom Antritt der Regierung, 15. Juni 1888 bis zum 15. Juni 1908 nach Hof- und anderen Berichten*, Berlin 1909, pp. 183ff.
130 Waldersee, diary entries for 21 and 22 October 1894, GStA Berlin, Waldersee Papers; cf. Meisner, II, p. 327.
131 Waldersee, diary entry for 30 October 1894, GStA Berlin, Waldersee Papers; Meisner, II, pp. 329–30.
132 Waldersee, diary entry for 29 October 1894, GStA Berlin, Waldersee Papers; largely omitted from Meisner, II, pp. 328–9.
133 Cited in Hatzfeldt to Holstein, 6 March 1895, *Holstein Papers*, III, No. 453.
134 Waldersee, diary entry for 29 October 1894, GStA Berlin, Waldersee Papers; mostly omitted from Meisner, II, pp. 328–9.
135 Waldersee, diary entry for 31 October 1894, GStA Berlin, Waldersee Papers; cf. Meisner, II, p. 330.
136 Waldersee to Verdy, 9 January 1895, GStA Berlin, Waldersee Papers No. 53.
137 Waldersee, diary entry for 9 April 1895, ibid.; cf. Meisner, II, p. 345.
138 Kálnoky to Szögyény, 14 November 1894, copy, BA Berlin, Nowak Papers No. 47.

22 HEAD OF THE FAMILY

1 Empress Frederick to Queen Victoria, 27 August 1890, RA Z49/10.
2 The early years of Wilhelm's and Dona's marriage are recounted in the first volume of this biography. See *Young Wilhelm*, pp. 325–66 and 452ff.
3 Waldersee, diary entry for 22 October 1891, GStA Berlin, Waldersee Papers; omitted from Meisner, II, p. 218. See Princess Viktoria of Schaumburg-Lippe to Empress Frederick, 26 and 30 October 1891, AdHH Schloss Fasanerie.
4 Kaiserin Auguste Viktoria to Kaiser Wilhelm II, 31 July 1892, GStA Berlin, BPHA Rep. 53T Preußen: An Kaiser Wilhelm II., Bd IV.

5 Kaiserin Auguste Viktoria to Kaiser Wilhelm II, 14–15 July 1892, ibid.
6 Kaiserin Auguste Viktoria to Kaiser Wilhelm II, 3–5 July 1892, ibid.
7 Eulenburg to Bülow, 1 October 1900, *Eulenburgs Korrespondenz*, III, No. 1434.
8 Kaiserin Auguste Viktoria to Kaiser Wilhelm II, 21 June 1890, GStA Berlin, BPHA Rep. 53T Preußen: An Kaiser Wilhelm II., Bd IV.
9 Kaiserin Auguste Viktoria to Kaiser Wilhelm II, 24 July 1890, ibid.
10 Kaiserin Auguste Viktoria to Kaiser Wilhelm II, 23 September 1890, ibid.
11 Kaiserin Auguste Viktoria to Kaiser Wilhelm II, 3–5 July 1892, ibid.
12 Kaiserin Auguste Viktoria to Kaiser Wilhelm II, 14 July 1892, ibid.
13 See *Young Wilhelm*, pp. 235–42 and 460; also above, pp. 197–9.
14 Kaiserin Auguste Viktoria to Kaiser Wilhelm II, 19/20 July 1892, GStA Berlin, BPHA Rep. 53T Preußen: An Kaiser Wilhelm II., Bd IV.
15 Kaiserin Auguste Viktoria to Kaiser Wilhelm II, 5 July 1892, ibid.
16 Kaiserin Auguste Viktoria to Kaiser Wilhelm II, 21 September 1890, ibid.
17 See Eulenburg to Bülow, 23–5 September 1900 and 26 September 1901, *Eulenburgs Korrespondenz*, III, Nos. 1427–9 and 1454.
18 Eulenburg to Holstein, 22 June and 21 July 1895, ibid., Nos. 1114 and 1118.
19 See below, pp. 643–50.
20 See e.g. Eulenburg to Bülow, 29 September 1899, *Eulenburgs Korrespondenz*, III, No. 1403.
21 Kiderlen-Wächter to Holstein, 19 July 1890, *Holstein Papers*, III, No. 324; Eulenburg to Bülow, 1 October 1900, *Eulenburgs Korrespondenz*, III, No. 1434.
22 Eulenburg to Holstein, 1 August 1890, *Holstein Papers*, III, No. 327.
23 Waldersee, diary entry for 12 January 1891, GStA Berlin, Waldersee Papers; omitted from Meisner, II, p. 175. On Emil Görtz's efforts to amuse the Kaiser, see Röhl, *Kaiser and his Court*, p. 16.
24 Empress Frederick to Queen Victoria, 27 December 1890, RA Z49/48; Kaiser Wilhelm II to his mother, 14 September 1892, AdHH Schloss Fasanerie; Queen Victoria to Kaiser Wilhelm II, 24 January and 4 March 1891, GStA Berlin, BPHA Rep. 52 W3 No. 11; Empress Frederick to her daughter Princess Viktoria of Schaumburg-Lippe, 28 January 1891, AdHH Schloss Fasanerie; also Lee, *Empress Frederick Writes to Sophie*, p. 77; Nicolaus Sombart, 'The Kaiser in His Epoch. Some Reflexions on Wilhelmine Society, Sexuality and Culture', in Röhl and Sombart, *Kaiser Wilhelm II. New Interpretations*, pp. 287–311; Sombart, *Wilhelm II.*, pp. 66ff., 159ff. *et passim*.
25 Waldersee, diary entry for 2 June 1890, GStA Berlin, Waldersee Papers; omitted from Meisner, II, p. 130.
26 Empress Frederick to Queen Victoria, 7 January 1897, AdHH Schloss Fasanerie. The important part played by the Kaiser and Kaiserin in the building of new churches in Berlin and the provinces is discussed more fully in chapter 29 below.
27 Kiderlen-Wächter to Eulenburg, 16 April 1893, *Eulenburgs Korrespondenz*, II, No. 795.
28 Brauer to Grand Duke of Baden, 9 April 1893, Fuchs, *Großherzog von Baden*, III, No. 1238.
29 Empress Frederick to Queen Victoria, 20 August 1890, RA Z49/7.
30 Empress Frederick to Queen Victoria, 15 March 1890, RA Z48/2.
31 Empress Frederick to Queen Victoria, 27 August 1890, RA Z49/10.
32 Empress Frederick to Queen Victoria, 16 December 1890, RA Z49/44.
33 Empress Frederick to Queen Victoria, 20 January 1891, RA Z50/6.
34 Empress Frederick to Queen Victoria, 15 March 1890, RA Z48/2; also Empress Frederick to her daughter Princess Viktoria of Schaumburg-Lippe, 13 December 1890, AdHH Schloss Fasanerie.
35 Prince Ernst zu Hohenlohe-Langenburg to his father, 24 October 1890, Hohenlohe-Zentralarchiv Neuenstein, Hermann Hohenlohe-Langenburg Papers, Bü. 57. See below, pp. 633–9.

36 Empress Frederick to her daughter Princess Viktoria of Schaumburg-Lippe, 28 January 1891, AdHH Schloss Fasanerie.
37 Eulenburg to his mother, 13 December 1893, *Eulenburgs Korrespondenz*, II, No. 863.
38 See e.g. Prince Ernst zu Hohenlohe-Langenburg to his father, 8 February 1891, Hohenlohe-Zentralarchiv Neuenstein, Hermann Hohenlohe-Langenburg Papers, Bü. 58.
39 Waldersee, diary entries for 10 and 12 January 1891, GStA Berlin, Waldersee Papers; omitted from Meisner, II, p. 175; also the entry for 23 November 1890, Meisner, II, p. 162.
40 See Eulenburg's notes, 12/13 October 1895, *Eulenburgs Korrespondenz*, III, No. 1145.
41 Kaiserin Auguste Viktoria to Eulenburg, 22 December 1893, ibid., II, No. 868.
42 Eulenburg to Bülow, 9 March 1894, ibid., No. 924; also Holstein to Eulenburg, 16 February 1894, ibid., No. 910; Eulenburg to his mother, 6 March 1894, ibid., p. 1245.
43 Eulenburg's notes, 12/13 October 1895, ibid., III, No. 1145.
44 August Eulenburg to Philipp Eulenburg, 27 December 1897, ibid., No. 1352.
45 Eulenburg to Bülow, 20 July 1898, ibid., No. 1380.
46 Eulenburg to Bülow, 14 July 1900, ibid., No. 1419.
47 Kaiser Wilhelm II to Dr Zunker, 23 September 1900, cited in Eulenburg to Bülow, 25 September 1900, ibid., No. 1429.
48 Eulenburg to Bülow, 23, 24 and 25 September 1900, ibid., Nos. 1427–9.
49 Eulenburg to Bülow, 1 October 1900, ibid., No. 1434.
50 Eulenburg to Bülow, 29 September and 1 October 1901, ibid., Nos. 1455–6; Bülow, *Memoirs*, I, p. 607.
51 Prince Heinrich of Prussia to Kaiser Wilhelm II, 23 January 1893, BA-MA Freiburg, Senden-Bibran Papers N160/10.
52 Seckendorff to Senden-Bibran, 24 November 1890, 8 and 10 January 1891, ibid., N160/2; Empress Frederick to her daughter Princess Viktoria of Schaumburg-Lippe, 3 and 13 February 1891, AdHH Schloss Fasanerie; Empress Frederick to Bogumilla Freifrau von Stockmar, 29 March 1891, ibid.; Senden-Bibran to Seckendorff, 21 February 1893, Seckendorff to Senden-Bibran, 24 February and 25 March 1893, BA-MA Freiburg, Senden Papers N160/2.
53 Empress Frederick to Queen Victoria, 22 July 1896, AdHH Schloss Fasanerie.
54 Eulenburg's notes, 12/13 October 1895, *Eulenburgs Korrespondenz*, III, No. 1145.
55 See for example Seckendorff to Senden-Bibran, 26 April 1891, BA-MA Freiburg, Senden Papers N160/2.
56 Above, pp. 483–4. See Empress Frederick to Queen Victoria, 25 March 1897, AdHH Schloss Fasanerie, and below, pp. 967–8.
57 Seckendorff to Senden-Bibran, 12 January 1891, BA-MA Freiburg, Senden Papers N160/2; also Seckendorff's letter of 5 June 1893, ibid.
58 Empress Frederick to Queen Victoria, 22 August 1891, postscript, RA Z51/14; 26 August 1891, RA Z51/15.
59 Holstein to Eulenburg, 1 April 1892, *Eulenburgs Korrespondenz*, II, No. 638.
60 Bülow to Eulenburg, 6 April 1892, ibid., No. 641.
61 Senden-Bibran to Seckendorff, 21 February 1893, BA-MA Freiburg, Senden-Bibran Papers N160/2.
62 Senden-Bibran to Kaiser Wilhelm II, 11 February 1893, ibid., N160/1.
63 Senden-Bibran to Seckendorff, 21 February 1893, ibid., N160/2.
64 Senden-Bibran to Kaiser Wilhelm II, 11 February 1893, ibid., N160/1.
65 Empress Frederick to Queen Victoria, 17 and 22 August 1891, RA Z51/12 and 14. See Charlotte Zeepvat, *Prince Leopold. The Untold Story of Queen Victoria's Youngest Son*, London 1998.
66 For Wilhelm's offer to the Tsar, see below, p. 760.
67 See e.g. Holstein to Eulenburg, 27 September 1895, *Eulenburgs Korrespondenz*, III, No. 1135.
68 Eulenburg's memorandum, 12/13 October 1895, ibid., No. 1145.

69 Eulenburg to Holstein, 12 June 1896, ibid., p. 1698. See also Eulenburg to Kuno Moltke, 1 February 1894, ibid., II, No. 891.
70 Eulenburg to Holstein, 5 July 1896, appendix I, *Holstein Papers*, III, No. 554.
71 See Eulenburg to Bülow, 18 December 1897, *Eulenburgs Korrespondenz*, III, No. 1354, and August Eulenburg to Philipp Eulenburg, 27 December 1897, ibid., No. 1352.
72 Eulenburg to Bülow, 12 July 1899, ibid., No. 1397.
73 Empress Frederick to Bogumilla Freifrau von Stockmar, 8 February 1891, AdHH Schloss Fasanerie; also Empress Frederick to her daughter Princess Viktoria of Schaumburg-Lippe, 9 February 1891, AdHH Schloss Fasanerie.
74 Empress Frederick to Queen Victoria, 5 December 1890, RA Z49/40.
75 Empress Frederick to Bogumilla Freifrau von Stockmar, 11 April 1891, AdHH Schloss Fasanerie.
76 Empress Frederick to her daughter Princess Viktoria of Schaumburg-Lippe, 5 December 1890, ibid.
77 Empress Frederick to Queen Victoria, 5 April 1891, RA Z50/34; Empress Frederick to Bogumilla Freifrau von Stockmar, 7 April 1891, AdHH Schloss Fasanerie.
78 Empress Frederick to her daughter Princess Viktoria of Schaumburg-Lippe, 5 and 19 June 1891, ibid. Cf. Princess Viktoria of Schaumburg-Lippe to Empress Frederick, 4 June 1891, ibid.; also Lee, *Empress Frederick Writes to Sophie*, p. 89.
79 Hereditary Princess Charlotte of Saxe-Meiningen to Professor Ernst Schweninger, 7 September 1896, Schweninger Papers, BA Berlin; Hereditary Princess Charlotte of Saxe-Meiningen to Ellen Freifrau von Heldburg, 9 July 1896, Thüringisches Staatsarchiv Meiningen, HA 342.
80 Empress Frederick to her daughter Princess Viktoria of Schaumburg-Lippe, 31 October 1892 and 8 January 1893, AdHH Schloss Fasanerie.
81 Empress Frederick to her daughter Princess Viktoria of Schaumburg-Lippe, 15 November 1891, cited from the extract in the catalogue of the auctioneers Butterfield & Butterfield, Los Angeles, 17 June 1998, No. 2396.
82 Empress Frederick to her daughter Princess Viktoria of Schaumburg-Lippe, 24 February 1891, AdHH Schloss Fasanerie.
83 Empress Frederick to her daughter Princess Viktoria of Schaumburg-Lippe, 28 November 1890, ibid.
84 Wedel, *Zwischen Kaiser und Kanzler*, pp. 180–1.
85 See Lee, *Empress Frederick Writes to Sophie*, pp. 56–7.
86 Waldersee, diary entry for 12 October 1893, GStA Berlin, Waldersee Papers; cf. Meisner, II, p. 295.
87 Empress Frederick to Queen Victoria, 20 August 1890, RA Z49/8; also Empress Frederick to Queen Victoria, 31 March 1891, RA Z50/30.
88 Empress Frederick to her daughter Princess Viktoria of Schaumburg-Lippe, 9–12 July 1893, AdHH Schloss Fasanerie. See also Kiderlen to Eulenburg, 28 September 1892, *Eulenburgs Korrespondenz*, II, No. 709.
89 Empress Frederick to her daughter Princess Viktoria of Schaumburg-Lippe, 27 October 1893, AdHH Schloss Fasanerie.
90 Prince Ernst zu Hohenlohe-Langenburg to his father, 15 January 1894, Hohenlohe-Zentralarchiv Neuenstein, Hermann Hohenlohe-Langenburg Papers, Bü. 61.
91 Albert Edward Prince of Wales to Queen Victoria, 16 January 1894, RA I60/5.
92 Hereditary Princess Charlotte of Saxe-Meiningen to Ellen Freifrau von Heldburg, 25 March 1895, Thüringisches Staatsarchiv Meiningen, HA 342.
93 Waldersee, diary entry for 24 March 1895, GStA Berlin, Waldersee Papers; cf. Meisner, II, pp. 340–1.

94 Hereditary Prince Bernhard of Saxe-Meiningen to Colmar Freiherr von der Goltz, 28 September, 3 November and 23 December 1899, BA-MA Freiburg, von der Goltz Papers, N737 Zug. 161/95.
95 Hereditary Prince Bernhard of Saxe-Meiningen to Colmar Freiherr von der Goltz, 20 October 1898 and 9 April 1899, ibid.
96 Hereditary Princess Charlotte of Saxe-Meiningen to Ellen Freifrau von Heldburg, 14 June 1895, Thüringisches Staatsarchiv Meiningen, HA 342.
97 Hereditary Prince Bernhard of Saxe-Meiningen to Waldersee, 30 March 1895, GStA Berlin, Waldersee Papers.
98 Hereditary Princess Charlotte of Saxe-Meiningen to Ellen Freifrau von Heldburg, 7 October 1897, Thüringisches Staatsarchiv Meiningen HA 342.
99 Hereditary Princess Charlotte of Saxe-Meiningen to Schweninger, BA Berlin, Schweninger Papers No. 130.
100 Hereditary Princess Charlotte of Saxe-Meiningen to Duke Georg II of Saxe-Meiningen, 1 November 1896, Thüringisches Staatsarchiv Meiningen, HA 342; Hereditary Princess Charlotte of Saxe-Meiningen to Ellen Freifrau von Heldburg, 5, 7 and 11 November 1896, ibid.; also Ellen Freifrau von Heldburg to Hereditary Princess Charlotte of Saxe-Meiningen, 4 November 1896, ibid.
101 Hereditary Prince Bernhard of Saxe-Meiningen to his father, 6 November 1896, ibid., HA 341; Kaiserin Auguste Viktoria to Hereditary Prince Bernhard of Saxe-Meiningen, 5 November 1896, copy, ibid.
102 Hereditary Princess Charlotte of Saxe-Meiningen to Ellen Freifrau von Heldburg, 7 and 27 December 1896, 11, 15 and 27 January and 7 October 1897, ibid., HA 342.
103 Hereditary Princess Charlotte of Saxe-Meiningen to her mother, 1 and 5 June 1898, AdHH Schloss Fasanerie.
104 See *Young Wilhelm*, pp. 105ff., and John C. G. Röhl, Martin Warren and David Hunt, *Purple Secret. Genes, 'Madness' and the Royal Houses of Europe*, London 1998, especially ch. 7.
105 Röhl, Warren and Hunt, *Purple Secret*, pp. 139–40.
106 Empress Frederick to her daughter Princess Viktoria of Schaumburg-Lippe, 8 March, 3 April, 1 and 19 June 1892 and 19 October 1893, AdHH Schloss Fasanerie.
107 See *Young Wilhelm*, pp. 516–46 and 800–5.
108 Kaiser Wilhelm II to Eulenburg, 27 February 1889, *Eulenburgs Korrespondenz*, I, No. 213. See above, p. 197.
109 Eulenburg to Kaiser Wilhelm II, 17 December 1889, *Eulenburgs Korrespondenz*, I, No. 266.
110 Empress Frederick to Queen Victoria, 19 June 1889, RA Z45/17; also Lee, *Empress Frederick Writes to Sophie*, pp. 66–7.
111 Empress Frederick to Prince Hermann zu Hohenlohe-Langenburg, 12 June 1890, Hohenlohe-Zentralarchiv Neuenstein, Hermann Hohenlohe-Langenburg Papers, Bü. 105.
112 Empress Frederick to Queen Victoria, 19 June 1890, RA Z48/34–5; also Lee, *Empress Frederick Writes to Sophie*, pp. 66–7.
113 Captain Bourke (1853–1900), second son of the 6th Earl of Mayo, was much liked in the British royal family and was on good terms not only with the Empress Frederick but also with the Prince of Wales and the Duke of York, later King George V, whom he accompanied to the opening ceremony of the Kiel Canal in 1895. See Empress Frederick to Queen Victoria, 3 February 1892, RA Z52/17; Albert Edward Prince of Wales to Empress Frederick, 18 September 1900, RA Add A 4/178. For the anxieties of the Empress Frederick, see Hannah Pakula, *An Uncommon Woman. The Empress Frederick, Daughter of Queen Victoria, Wife of the Crown Prince of Prussia, Mother of Kaiser Wilhelm*, New York 1995, p. 537.
114 Empress Frederick to Queen Victoria, 12 June 1890, RA Z48/30.

115 Kaiserin Auguste Viktoria to Kaiser Wilhelm II, 21 July 1890, GStA Berlin, BPHA Rep. 53T Preußen: An Kaiser Wilhelm II., Bd IV.
116 Empress Frederick to Queen Victoria, 7 August 1890, RA Z49/3.
117 RA QVJ, 4–6 August 1890.
118 Empress Frederick to Queen Victoria, 10 September 1890, RA Z49/14.
119 Empress Frederick to her daughter Princess Viktoria of Schaumburg-Lippe, 24 February 1891, AdHH Schloss Fasanerie.
120 Empress Frederick to Queen Victoria, 5 December 1890, RA Z49/40.
121 Waldersee, diary entry for 18 November 1890, GStA Berlin, Waldersee Papers; Meisner, II, p. 162.
122 For this embarrassing episode, see J. J. Lynx, *The Great Hohenzollern Scandal*, London 1965.
123 The Baden political documents edited by Walther Peter Fuchs contain numerous references to this issue. See in particular Brauer to Turban, 23 April and 2 May 1891, Fuchs, *Großherzog von Baden*, III, Nos. 1110–11; Empress Frederick to Grand Duke of Baden, 6 and 15 May 1891, ibid., Nos. 1113 and 1117; Grand Duke of Baden to Empress Frederick, 16 May 1891, ibid., No. 1118.
124 Brauer to Hardeck, 27 June 1891, ibid., No. 1125.
125 Empress Frederick to Queen Victoria, 5 September 1891, RA Z51/20.
126 Waldersee, diary entry for 18 December 1890, GStA Berlin, Waldersee Papers; cf. Meisner, II, p. 167; also Lee, *Empress Frederick Writes to Sophie*, p. 73.
127 Lee, *Empress Frederick Writes to Sophie*, p. 73.
128 Waldersee, diary entries for 19 and 20 December 1890, GStA Berlin, Waldersee Papers; Meisner, II, pp. 167–8.
129 Empress Frederick to Queen Victoria, 25 and 28 January 1891, RA 50/8 and 10; also Lee, *Empress Frederick Writes to Sophie*, p. 77.
130 Empress Frederick to her daughter Princess Viktoria of Schaumburg-Lippe, 19 December 1890, AdHH Schloss Fasanerie.
131 Empress Frederick to her daughter Princess Viktoria of Schaumburg-Lippe, 21 December 1890, ibid.
132 Empress Frederick to her daughter Princess Viktoria of Schaumburg-Lippe, 22 December 1890, ibid., also Lee, *Empress Frederick Writes to Sophie*, p. 73.
133 Empress Frederick to her daughter Princess Viktoria of Schaumburg-Lippe, 24 December 1890, AdHH Schloss Fasanerie.
134 Empress Frederick to her daughter Princess Viktoria of Schaumburg-Lippe, 1 January 1891, ibid.
135 Empress Frederick to Queen Victoria, 12 June 1891, RA Z50/45; also Lee, *Empress Frederick Writes to Sophie*, p. 73.
136 Empress Frederick to Queen Victoria, 22 December 1890, RA Z49/45.
137 Lee, *Empress Frederick Writes to Sophie*, p. 74.
138 Empress Frederick to Queen Victoria, 27 December 1890, RA Z49/48; also Empress Frederick to Queen Victoria, 9 and 20 January 1891, RA Z50/3 and Z50/6; Empress Frederick to her daughter Princess Viktoria of Schaumburg-Lippe, 24 February 1891, AdHH Schloss Fasanerie.
139 Empress Frederick to Crown Princess Sophie, January 1891, cited in Lee, *Empress Frederick Writes to Sophie*, p. 76.
140 Ibid., p. 74.
141 Queen Victoria to Empress Frederick, January 1891, cited in Lee, *Empress Frederick Writes to Sophie*, p. 77; also ibid., pp. 88–9.
142 Empress Frederick to Queen Victoria, 29 March 1891, RA Z50/29.
143 Empress Frederick to Bogumilla Freifrau von Stockmar, 29 March 1891, AdHH Schloss Fasanerie.

144 Wedel, *Zwischen Kaiser und Kanzler*, pp. 175–6.
145 Ibid., p. 180.
146 Empress Frederick to Bogumilla Freifrau von Stockmar, 18 and 19 May 1891, AdHH Schloss Fasanerie.
147 Princess Viktoria of Schaumburg-Lippe to Empress Frederick, 2 June 1891, ibid.
148 Princess Viktoria of Schaumburg-Lippe to Empress Frederick, 6, 8, 9 and 18–22 May 1891, ibid.
149 Cited in Lee, *Empress Frederick Writes to Sophie*, pp. 85–6.
150 Cited ibid., pp. 86ff.
151 Empress Frederick to Bogumilla Freifrau von Stockmar, 18 and 19 May 1891, AdHH Schloss Fasanerie.
152 Wedel, *Zwischen Kaiser und Kanzler*, p. 190. See Princess Viktoria of Schaumburg-Lippe to Empress Frederick, 29 May 1891, AdHH Schloss Fasanerie.
153 Empress Frederick to Bogumilla Freifrau von Stockmar, 1 June 1891, Empress Frederick to Princess Viktoria of Schaumburg-Lippe, 1 June 1891, Princess Viktoria of Schaumburg-Lippe to Empress Frederick, 31 May 1891, ibid.; also Lee, *Empress Frederick Writes to Sophie*, p. 89.
154 Empress Frederick to her daughter Princess Viktoria of Schaumburg-Lippe, 2 June 1891, AdHH Schloss Fasanerie.
155 Empress Frederick to Queen Victoria, 15 June 1891, RA Z50/46.
156 Empress Frederick to Queen Victoria, 26 August 1891, RA Z51/15.
157 Empress Frederick to her daughter Princess Viktoria of Schaumburg-Lippe, 1 June 1892, AdHH Schloss Fasanerie.
158 Empress Frederick to her daughter Princess Viktoria of Schaumburg-Lippe, 28 June 1892, ibid.
159 Kaiser Wilhelm II to his mother, 21 June 1894, ibid.
160 See Empress Frederick to Queen Victoria, 6 August 1896, ibid.; Kaiser Wilhelm II to his mother, 13 September 1896, ibid.
161 Wedel, *Zwischen Kaiser und Kanzler*, p. 190. See Princess Viktoria of Schaumburg-Lippe to Empress Frederick, 29 May 1891, ibid.
162 Empress Frederick to Queen Victoria, 16 January 1891, RA Z50/4.
163 Wedel, *Zwischen Kaiser und Kanzler*, p. 190. See Princess Viktoria of Schaumburg-Lippe to Empress Frederick, 29 May 1891, AdHH Schloss Fasanerie.
164 Military attaché in St Petersburg to Kaiser Wilhelm II, 1 March 1891, with numerous marginal comments by the Kaiser, PA AA, Rußland 82 No. 1 secr; also Empress Frederick to her daughter Princess Viktoria of Schaumburg-Lippe, 9 March 1892, AdHH Schloss Fasanerie.
165 Eulenburg's notes, 12/13 October 1895, *Eulenburgs Korrespondenz*, III, No. 1145.
166 Empress Frederick to her daughter Princess Viktoria of Schaumburg-Lippe, 6 April 1892, AdHH Schloss Fasanerie; also Lee, *Empress Frederick Writes to Sophie*, p. 71.
167 Bülow to Eulenburg, 6 April 1892, *Eulenburgs Korrespondenz*, II, No. 641.
168 See Lee, *Empress Frederick Writes to Sophie*, p. 71.
169 Empress Frederick to Kaiser Wilhelm II, 20 June 1892, GStA Berlin, BPHA Rep. 52T No. 13; Empress Frederick to her daughter Princess Viktoria of Schaumburg-Lippe, 3, 20 and 22 June 1892, AdHH Schloss Fasanerie; Empress Frederick to Queen Victoria, 20 and 23 June 1892, RA Z53/7 and 8.
170 See *Eulenburgs Korrespondenz*, II, p. 903.
171 Waldersee, diary entry for 13 December 1890, GStA Berlin, Waldersee Papers; Meisner, II, p. 166.
172 Duke Ernst Günther of Schleswig-Holstein to Prince Christian of Schleswig-Holstein, 2 August 1890, RA Add A18/G1.

173 Waldersee, diary entry for 7 June 1891, GStA Berlin, Waldersee Papers; Meisner, II, p. 209.
174 Duke Ernst Günther of Schleswig-Holstein to Prince Christian of Schleswig-Holstein, 10 March 1891, RA Add A18/G1.
175 Empress Frederick to Queen Victoria, 16 December 1890, RA Z49/44.
176 Princess Viktoria of Schaumburg-Lippe to Empress Frederick, 18 May 1891, AdHH Schloss Fasanerie.
177 Kiderlen-Wächter to Eulenburg, 12 December 1892, *Eulenburgs Korrespondenz*, II, No. 748; Karl Samwer to Prince Christian of Schleswig-Holstein, 7 October 1891, RA Add A18/L2.
178 Prince Ernst zu Hohenlohe-Langenburg to his father, 16 October 1891, Hohenlohe-Zentralarchiv Schloss Neuenstein, Hermann Hohenlohe-Langenburg Papers, Bü. 58.
179 Prince Ernst zu Hohenlohe-Langenburg to his father, 27 July 1893, Hohenlohe-Zentralarchiv Neuenstein, Hermann Hohenlohe-Langenburg Papers, Bü. 60. See also Duke Ernst Günther of Schleswig-Holstein to Prince Christian of Schleswig-Holstein, 12 October 1893, RA Add A18/G1. See *Holstein Papers*, III, No. 385.
180 Eulenburg to Varnbüler, 10 and 19 May 1896, *Eulenburgs Korrespondenz*, III, No. 1227; Duke Ernst Günther of Schleswig-Holstein to Prince Christian of Schleswig-Holstein, 29 June 1895, RA Add A18/G1.
181 Prince Ernst zu Hohenlohe-Langenburg to his father, 9 May 1891, Hohenlohe-Zentralarchiv Schloss Neuenstein, Hermann Hohenlohe-Langenburg Papers, Bü. 58.
182 Duke Ernst Günther of Schleswig-Holstein to Prince Christian of Schleswig-Holstein, 10 March 1891, RA Add A18/G1.
183 Prince Chlodwig zu Hohenlohe to Prince Alexander Hohenlohe, 27 May 1895, Hohenlohe, *Denkwürdigkeiten der Reichskanzlerzeit*, pp. 72–3. See Bismarck's article 'Zur schleswig-holsteinschen Frage', *Hamburger Nachrichten*, 23 May 1895, reprinted in Hermann Hofmann, *Fürst Bismarck 1890–1898*, 2 vols., Stuttgart, Berlin, Leipzig 1913, II, pp. 300–2.
184 Duke Ernst Günther of Schleswig-Holstein to Prince Christian of Schleswig-Holstein, 2 August 1890, RA Add A18/G1.
185 Duke Ernst Günther of Schleswig-Holstein to Prince Christian of Schleswig-Holstein, 15 October 1890, RA Add A18/G1.
186 Empress Frederick to Queen Victoria, 26 November 1891, RA Z51/49.
187 Kaiser Wilhelm II to Queen Victoria, 8 December 1891, RA I59/53.
188 For the marriage of Princess Beatrice to Prince Heinrich (Liko) of Battenberg, see *Young Wilhelm*, pp. 520–3.
189 Cardinal Prince Gustav zu Hohenlohe-Schillingsfürst to Prince Hermann zu Hohenlohe-Langenburg, 25 July 1893, Hohenlohe-Zentralarchiv Schloss Neuenstein, Hermann Hohenlohe-Langenburg Papers, Bü. 85.
190 Münster to Eulenburg, 22 April 1894, *Eulenburgs Korrespondenz*, II, No. 959.
191 Ibid.
192 Eulenburg's note of 16 April 1894, ibid., p. 1287.
193 Eulenburg to Bülow, 7 May 1894, ibid., No. 963.
194 Bülow to Eulenburg, 21 April 1894, ibid., No. 958.
195 Eulenburg to Münster, 16 April 1894, ibid., No. 954.
196 Münster to Eulenburg, 22 April 1894, ibid., No. 959.
197 Eulenburg to Kaiser Wilhelm II, 26 April 1894, ibid., No. 960.
198 Eulenburg to Bülow, 7 May 1894, ibid., No. 963.
199 Eulenburg to Kaiser Wilhelm II, 29 March 1896, ibid., III, No. 1204; also Eulenburg to Varnbüler, 5 April 1896, ibid., No. 1207.
200 Kaiser Wilhelm II to Eulenburg, 4 April 1896, ibid., No. 1206.
201 Varnbüler to Eulenburg, 1 and 6 May 1896, cited ibid., p. 1661.

202 Eulenburg to Varnbüler, 19 May 1896, ibid., No. 1227.
203 Varnbüler to Eulenburg, [May 1896], ibid., No. 1229.
204 Eulenburg to Varnbüler, 25 May 1896, ibid., No. 1230.
205 Kaiser Wilhelm II to Varnbüler, 28 May 1896, ibid., No. 1231.
206 Görtz to Kaiser Wilhelm II, 23 May 1896, GStA Berlin, BPHA Rep. 53J Lit. G. No. 5.
207 Prince Heinrich of Prussia to his mother, 11 September 1898, AdHH Schloss Fasanerie. See Empress Frederick to Queen Victoria, 17 April 1897, ibid.
208 Duke Ernst Günther of Schleswig-Holstein to Prince Christian of Schleswig-Holstein, 27 April 1897 and 15 June 1898, RA Add A18/G1.
209 Eulenburg to Kaiser Wilhelm II, 4 April and 18 May 1898, *Eulenburgs Korrespondenz*, III, Nos. 1365 and 1371.
210 Empress Frederick to her daughter Princess Viktoria of Schaumburg-Lippe, 2 December 1891, AdHH Schloss Fasanerie; also her letter of 9 December 1892, ibid.
211 Empress Frederick to her daughter Princess Viktoria of Schaumburg-Lippe, 7 December 1891, ibid.; Empress Frederick to Queen Victoria, 12 December 1891, RA Z52/5.
212 Waldersee, diary entry for 23 November 1890, GStA Berlin, Waldersee Papers; cf. Meisner, II, p. 162.
213 Cited in Prince Ernst zu Hohenlohe-Langenburg to his father, 10 January 189[2], Hohenlohe-Zentralarchiv Neuenstein, Hermann Hohenlohe-Langenburg Papers, Bü. 59.
214 Prince Ernst zu Hohenlohe-Langenburg to his father, 14 June 1890, ibid., Bü. 57.
215 See the correspondence of Major von Krosigk with August von Mackensen, BA-MA Freiburg N39/45; also Heuduck to Mackensen, 9 June 1896, ibid.
216 Empress Frederick to Queen Victoria, 4 and 6 January 1896, AdHH Schloss Fasanerie.
217 Hereditary Princess Charlotte of Saxe-Meiningen to Ellen Freifrau von Heldburg, 11 November 1896, Thüringisches Staatsarchiv Meiningen, HA 342.
218 August Eulenburg to Philipp Eulenburg, 6 January 1896, *Eulenburgs Korrespondenz*, III, No. 1186.
219 See ibid., p. 1634.
220 Eulenburg to Kaiser Wilhelm II, 20 January 1896, ibid., No. 1190.
221 See below, ch. 26.
222 Empress Frederick to Queen Victoria, 1 January 1896, AdHH Schloss Fasanerie.
223 Queen Victoria to Kaiser Wilhelm II, 5 January 1896, RA O45/169.
224 Kaiser Wilhelm II to Queen Victoria, 8 January 1896, RA Z500/5; draft in the files of the Auswärtiges Amt, PA AA; printed in Hohenlohe, *Denkwürdigkeiten der Reichskanzlerzeit*, pp. 154–6.
225 Heuduck to Mackensen, 4 February 1896, Mackensen Papers, BA-MA N39/45.
226 Waldersee, diary entry for 31 October 1896, GStA Berlin, Waldersee Papers; cf. Meisner, II, pp. 375–6.
227 Eulenburg to Bülow, 12 July 1899, *Eulenburgs Korrespondenz*, III, No. 1397.
228 Kaiser Wilhelm II, marginal notes on Prince Friedrich Leopold of Prussia to Kaiser Wilhelm II, 5 July 1904, Burg Hohenzollern.
229 Duke Ernst Günther of Schleswig-Holstein to Prince Christian of Schleswig-Holstein, 10 March 1891, RA Add A18/G1; Empress Frederick to her daughter Princess Viktoria of Schaumburg-Lippe, December 1890, AdHH Schloss Fasanerie; also Empress Frederick to Queen Victoria, 5 December 1890, RA Z49/40. See Princess Marie Louise, *My Memories of Six Reigns*, pp. 66–109.
230 Princess Helena of Schleswig-Holstein to Kaiser Wilhelm II, 29 November 1890, GStA Berlin, BPHA Rep. 53J Lit. S No. 25; also Empress Frederick to her daughter Princess Viktoria of Schaumburg-Lippe, 10 December 1890, AdHH Schloss Fasanerie.
231 See Eulenburg to Bülow, 1 October 1900, *Eulenburgs Korrespondenz*, III, No. 1434.

232 See Kaiser Wilhelm II to Duke Arthur of Connaught, 11 March 1891, RA Addl Mss A15/5680, cited above, pp. 481–2; also Malet to Ponsonby, 19 March 1891, RA I59/10; Malet to August Eulenburg, 24 March 1891, RA I59/11.
233 Eulenburg to Bülow, 11 July 1898, *Eulenburgs Korrespondenz*, III, No. 1378; also Grierson to Bigge, 5 February 1898, RA I61/34.
234 Empress Frederick to her daughter Princess Viktoria of Schaumburg-Lippe, 19 April, 19 and 27 October 1893, AdHH Schloss Fasanerie.
235 Prince Christian of Schleswig-Holstein to Eulenburg, 1 October 1900, Eulenburg to Bülow, 1 October 1900, *Eulenburgs Korrespondenz*, III, Nos. 1433–4.
236 Waldersee, diary entries for 10 and 12 January 1891, GStA Berlin, Waldersee Papers; omitted from Meisner, II, p. 175. On the Kotze scandal, see Bringmann, *Reichstag und Zweikampf*, pp. 152–224. See the earlier accounts of the affair published by Caesar Schmidt in Zürich: Anon., *Das Geheimnis des Ceremoniemeisters. Hofroman aus der jüngsten Vergangenheit. Von Carl Fürst von ... (in Preußen verboten)*; Dr Fritz Friedmann, *Der deutsche Kaiser und die Hofkamarilla. I. Der Fall Kotze. II. Wilhelm II. und die Revolution von oben*, Zürich 1896; Anon. [H. von Langen-Allenstein], *Herr von Tausch und die Verfasser der anonymen Briefe der Hofgesellschaft*, Zürich 1897.
237 Anonymous to Princess Louise of Anhalt, 7 April 1892, GStA Berlin, PK 1, HA Rep. 89, No. 3307/10, fol. 40ff.; also the letter dated 15 June 1892 to Count Wilhelm Hohenau, cited in Bringmann, *Reichstag und Zweikampf*, p. 157.
238 Anonymous to Prince Aribert of Anhalt, 10 March 1893, GStA Berlin, PK 1, HA Rep. 89, No. 3307/10, fol. 83ff.
239 Ibid.; to Hugo Freiherr von Reischach, 5 February 1892, ibid., pp. 26–7. In another of the letters addressed to Reischach his mother was described as a 'Jewess and a professional'. See Bringmann, *Reichstag und Zweikampf*, p. 157.
240 Anonymous to Duke Ernst Günther of Schleswig-Holstein, 23 January 1893, GStA Berlin, PK 1, HA Rep. 89, No. 3307/10, fol. 78ff.
241 Anonymous to 'Frau Gräfin', 10 January 1893, ibid., fol. 76.
242 Anonymous to Princess Louise of Anhalt, 7 April 1892, ibid., fol. 40ff.; to Prince Aribert of Anhalt, 10 March 1893, ibid., fol. 83ff.
243 Anonymous to Countess Hohenau, 1 March 1894, ibid., letter No. 40; Anonymus to Wedell-Piesdorff, 17 May 1894, ibid., fol. 120.
244 Hahnke to Kaiser Wilhelm II, 16 June 1894, GStA Berlin, PK 1, HA Rep. 89, No. 3307/2.
245 Philipp Eulenburg, memorandum Der Fall Kotze, BA Koblenz, Eulenburg Papers, typescript for April 1895, pp. 332–354b.
246 See Brauer to Grand Duke of Baden, 13 September 1894, Fuchs, *Großherzog von Baden*, III, No. 1342.
247 See Marschall to Eulenburg, 30 April 1896, *Eulenburgs Korrespondenz*, III, No. 1211; Monts to Eulenburg, 7 August 1896, ibid., No. 1254.
248 Lerchenfeld's report 28 June 1894, cited in Bringmann, *Reichstag und Zweikampf*, p. 172.
249 Waldersee, diary entry for 24 June 1894, GStA Berlin, Waldersee Papers; omitted from Meisner, II, p. 315.
250 Waldersee, diary entry for 27 June 1894, GStA Berlin, Waldersee Papers; omitted from Meisner, II, p. 315.
251 Spitzemberg, diary, 12 April 1896, *Tagebuch*, p. 343.
252 For Kotze's character, see Bringmann, *Reichstag und Zweikampf*, pp. 166–7; Bülow, *Memoirs*, IV, p. 213; Philipp Eulenburg, Der Fall Kotze, in BA Koblenz, Eulenburg Papers, typescript for April 1895, pp. 332–354b.
253 See the documentation in Bringmann, *Reichstag und Zweikampf*, p. 173.
254 Waldersee, diary entry for 27 June 1894, GStA Berlin, Waldersee Papers; omitted from Meisner, II, p. 315.

255 Cited in Bringmann, *Reichstag und Zweikampf*, p. 173.
256 Hahnke to O. Meding, 4 August 1894, GStA Berlin, PK 1, HA Rep. 89, No. 3307/2.
257 Pape to Kaiser Wilhelm II, 5 July 1894, Bringmann, *Reichstag und Zweikampf*, p. 174. See Waldersee, diary entry for 5 July 1894, GStA Berlin, Waldersee Papers; omitted from Meisner, II, p. 315.
258 Bringmann, *Reichstag und Zweikampf*, pp. 174–5.
259 See ibid., pp. 180–3.
260 Hereditary Princess Charlotte of Saxe-Meiningen to Ellen Freifrau von Heldburg, 20 May 1895, Thüringisches Staatsarchiv Meiningen, HA 342.
261 Swaine to Sir Arthur Bigge, 9 June 1895, RA I60/109.
262 Waldersee, diary entry for 23 May 1895, GStA Berlin, Waldersee Papers; cf. Meisner, II, p. 349.
263 Chlodwig Hohenlohe to Alexander Hohenlohe, 27 May 1895, Hohenlohe, *Denkwürdigkeiten der Reichskanzlerzeit*, p. 72.
264 See Marschall to Eulenburg, 30 April 1896, *Eulenburgs Korrespondenz*, III, No. 1211.
265 Waldersee, diary entry for 20 August 1894, GStA Berlin, Waldersee Papers; cf. Meisner, II, p. 321.
266 Hohenlohe to Kaiser Wilhelm II, 12 April 1895, Hohenlohe, *Denkwürdigkeiten der Reichskanzlerzeit*, p. 59; also Holstein to Eulenburg, 26 January 1895, *Eulenburgs Korrespondenz*, II, No. 1078.
267 Chlodwig Hohenlohe to Prince Hermann zu Hohenlohe-Langenburg, 13 April [erroneously dated October] 1895, Hohenlohe-Zentralarchiv Neuenstein, Hermann Hohenlohe-Langenburg Papers, Bü. 86.
268 Hereditary Princess Charlotte of Saxe-Meiningen to Ellen Freifrau von Heldburg, 14 April 1895, Thüringisches Staatsarchiv Meiningen, HA 342.
269 Waldersee, diary entries for 14 and 18 April 1895, GStA Berlin, Waldersee Papers; cf. Meisner, II, pp. 345–6.
270 See August Eulenburg to Philipp Eulenburg, 17 December 1894, *Eulenburgs Korrespondenz*, II, No. 1066. See Bringmann, *Reichstag und Zweikampf*, pp. 178 and 189ff.
271 Waldersee, diary entry for 20 August 1894, GStA Berlin, Waldersee Papers; cf. Meisner, II, p. 321.
272 Kaiser Wilhelm II to Eulenburg, 25 December 1895, *Eulenburgs Korrespondenz*, III, No. 1178; Bringmann, *Reichstag und Zweikampf*, pp. 195ff. See Prince Friedrich of Hohenzollern to Colmar Freiherr von der Goltz, 24 February 1896, von der Goltz Papers, BA-MA N737/25.
273 Eulenburg to Kaiser Wilhelm II, 2 January 1896, *Eulenburgs Korrespondenz*, III, No. 1184.
274 Waldersee, diary entry for 3 March 1896, GStA Berlin, Waldersee Papers; omitted from Meisner, II, pp. 368–9.
275 Szögyeny's report of 6 February 1896, Lerchenfeld's report of 6 February 1896, cited in Bringmann, *Reichstag und Zweikampf*, pp. 196 and 198.
276 Monts to Bülow, 24 February 189[6], cited in Bülow, *Memoirs*, I, pp. 32ff.
277 Details in Bringmann, *Reichstag und Zweikampf*, pp. 197–202.
278 Philipp Eulenburg, Der Fall Kotze, in BA Koblenz, Eulenburg Papers, typescript for April 1895, pp. 351–2. On the occasion of the Kaiser's birthday in January 1902, Leberecht von Kotze was awarded the Order of the Red Eagle third class.
279 Empress Frederick to Queen Victoria, 12 April1896, AdHH Schloss Fasanerie.
280 Waldersee, diary entry for 15 April 1896, GStA Berlin, Waldersee Papers; omitted from Meisner, II, p. 369.
281 Count Bogdan von Hutten-Czapski, *Sechzig Jahre Politik und Gesellschaft*, 2 vols., Berlin 1936, I, p. 263; Lerchenfeld's report of 6 February 1896, cited in Bringmann, *Reichstag und Zweikampf*, p. 208.

282 Brauer to Grand Duke of Baden, 13 September 1894, Fuchs, *Großherzog von Baden*, III, p. 311.
283 August Eulenburg to Philipp Eulenburg, 17 December 1894, *Eulenburgs Korrespondenz*, II, No. 1066.
284 See Eulenburg to Holstein, 16 April 1895, ibid., III, No. 1104.
285 See Eulenburg to Kaiser Wilhelm II, 16 April 1896, BA Koblenz, Eulenburg Papers, typescript for April 1895, pp. 347–51.
286 Philipp Eulenburg, Der Fall Kotze, BA Koblenz, Eulenburg Papers, typescript for April 1895, pp. 345–6.
287 Eulenburg to Holstein, 21 January 1895, *Eulenburgs Korrespondenz*, II, No. 1075.
288 Waldersee, diary entry for 16 July 1894, GStA Berlin, Waldersee Papers; cf. Meisner, II, p. 317.
289 Waldersee, diary entries for 12, 18 and 20 August 1894, GStA Berlin, Waldersee Papers; not in Meisner, II, pp. 320–1.
290 Hohenlohe, undated note, Hohenlohe, *Denkwürdigkeiten der Reichskanzlerzeit*, pp. 35–6.
291 Duke Ernst Günther of Schleswig-Holstein to Prince Christian of Schleswig-Holstein, 29 June 1895 and 26 April 1896, RA Add A18/G1.
292 Karl Samwer to Prince Christian of Schleswig-Holstein, 10 February 1897, RA Add A18/L2.
293 Philipp Eulenburg, Der Fall Kotze, in BA Koblenz, Eulenburg Papers, typescript for April 1895, p. 353.
294 Bringmann, *Reichstag und Zweikampf*, pp. 208ff.
295 Hereditary Princess Charlotte of Saxe-Meiningen to Ellen Freifrau von Heldburg, 26 June 1894, Thüringisches Staatsarchiv Meiningen, HA 342; also Hereditary Princess Charlotte of Saxe-Meiningen to Ellen Freifrau von Heldburg, 14 April 1895, cited above, p. 670, and 1 November 1896, Thüringisches Staatsarchiv Meiningen, HA 342.
296 Marschall to Eulenburg, 30 April 1896, *Eulenburgs Korrespondenz*, III, No. 1211.
297 *Berliner Tageblatt*, 14 April 1895, *Vossische Zeitung*, 14 April 1895 and 11 April 1896, cited in Bringmann, *Reichstag und Zweikampf*, pp. 185 and 214.
298 Harden, *Die Zukunft*, 7 July 1894, cited in Bringmann, *Reichstag und Zweikampf*, p. 215.
299 Philipp Eulenburg, Der Fall Kotze, BA Koblenz, Eulenburg Papers, typescript for April 1895, p. 334.
300 See Bringmann, *Reichstag und Zweikampf*, pp. 210ff.
301 Waldersee, diary entry for 23 May 1895, GStA Berlin, Waldersee Papers; cf. Meisner, II, p. 349.
302 Hereditary Princess Charlotte of Saxe-Meiningen to Ellen Freifrau von Heldburg, 20 May 1895, Thüringisches Staatsarchiv Meiningen, HA 342.

23 THE KAISER AND THE 'NEWEST COURSE'

1 Kaiser Wilhelm II to Hohenlohe, telegram, 26 October 1894, in Hohenlohe, *Denkwürdigkeiten der Reichskanzlerzeit*, p. 1.
2 Haller, *Eulenburg*, pp. 153ff.
3 Otto Hammann, *Der neue Kurs*, Berlin 1918, p. 137.
4 Jagemann, report, 27 September 1894, cited in Röhl, *Germany without Bismarck*, p. 119.
5 Grand Duke of Baden to Eulenburg, 25 September 1894, *Eulenburgs Korrespondenz*, II, No. 998. See Eulenburg to Bülow, 30 September 1894, ibid., No. 1006.
6 Bülow to Eulenburg, 6 October 1894, ibid., No. 1016.
7 Eulenburg to Bülow, 27 October 1894, ibid., No. 1038.
8 Haller, *Eulenburg*, pp. 154, 157–8.

9 Ibid., p. 155. See also Nipperdey, *Deutsche Geschichte*, II, pp. 709–10; Ullrich, *Nervöse Großmacht*, pp. 155ff.
10 Alexander von Hohenlohe, *Aus meinem Leben*, Frankfurt a.M. 1925, p. 226.
11 Holstein to Hohenlohe, 26 October 1894, Hohenlohe, *Denkwürdigkeiten der Reichskanzlerzeit*, pp. 1–2; Grand Duke of Baden to Hohenlohe, 26 October 1894, Fuchs, *Großherzog von Baden*, III, No. 1371.
12 Alexander Hohenlohe, *Aus meinem Leben*, p. 226.
13 Hohenlohe, note of October 1894, Hohenlohe, *Denkwürdigkeiten der Reichskanzlerzeit*, p. 4.
14 Alexander Hohenlohe, *Aus meinem Leben*, p. 225.
15 Hohenlohe's notes of 27 October 1894, Hohenlohe, *Denkwürdigkeiten der Reichskanzlerzeit*, p. 3.
16 Hohenlohe's notes of 2 November 1894, ibid., p. 7.
17 Hohenlohe's notes of 27–8 October 1894, ibid., p. 3.
18 Kaiser Wilhelm II to Queen Victoria, telegram, 28 October 1894, RA I60/97.
19 Holstein to Hohenlohe, 26 October 1894, Hohenlohe, *Denkwürdigkeiten der Reichskanzlerzeit*, pp. 1–2.
20 Eulenburg to Kaiser Wilhelm II, 3 November 1894, *Eulenburgs Korrespondenz*, II, No. 1042.
21 Cardinal Prince Gustav zu Hohenlohe to Duke Georg II of Saxe-Meiningen, 7 November 1894, Thüringisches Staatsarchiv Meiningen, HA 361.
22 Waldersee to Verdy, 9 January 1895, GStA Berlin, Waldersee Papers No. 53.
23 Alexander Hohenlohe, *Aus meinem Leben*, pp. 235–40.
24 John C. G. Röhl, *Deutschland ohne Bismarck. Die Regierungskrise im Zweiten Kaiserreich 1890–1900*, Tübingen 1969, p. 270, n. 16.
25 Röhl, *Germany without Bismarck*, p. 123.
26 Jagemann, report of 26 October 1894, cited ibid.
27 Alexander Hohenlohe, *Aus meinem Leben*, pp. 224–5.
28 Waldersee to Verdy, 9 January 1895, GStA Berlin, Waldersee Papers No. 53.
29 See below, pp. 721–31.
30 Kaiser Wilhelm II to Hohenlohe, 5 November 1894, cited in Röhl, *Germany without Bismarck*, p. 124.
31 Hohenlohe, *Denkwürdigkeiten der Reichskanzlerzeit*, p. 134.
32 Hohenlohe, undated note, ibid., p. 10.
33 Hohenlohe's notes, 6 November 1894, ibid., pp. 10–11.
34 Kálnoky to Szögyény, 14 November 1894, copy, BA Berlin, Nowak Papers No. 47.
35 Schelling, letter of resignation, 2 November 1894. See Röhl, *Germany without Bismarck*, pp. 124–5.
36 Nieberding to Hohenlohe, 4 November 1894, Hohenlohe to Kaiser Wilhelm II, 4 November 1894, cited ibid., p. 125.
37 Kaiser Wilhelm II, marginal notes on Hohenlohe's report of 4 November 1894, Kaiser Wilhelm II to Hohenlohe, 5 November 1894, cited ibid.
38 On these negotiations, see ibid., pp. 125–6.
39 Holstein to Hohenlohe, 26 October 1894, Hohenlohe, *Denkwürdigkeiten der Reichskanzlerzeit*, pp. 1–2. See Eulenburg to Bülow, 6 and 12 October 1894, *Eulenburgs Korrespondenz*, II, Nos. 1015 and 1024.
40 Hohenlohe's note of 28 October 1894, Hohenlohe, *Denkwürdigkeiten der Reichskanzlerzeit*, p. 3.
41 See Count Anton Monts's remarks in Bülow, *Memoirs*, I, pp. 28ff.; also Gosselin's report of 3 November 1894, RA I60/98.
42 Kálnoky to Szögyény, 14 November 1894, copy, BA Berlin, Nowak Papers No. 47.

43 Holstein to Hohenlohe, 11 November 1894, Hohenlohe, *Denkwürdigkeiten der Reichskanzlerzeit*, p. 11.
44 Holstein to Hohenlohe, 17 November 1894, ibid., p. 15.
45 Alexander Hohenlohe, *Aus meinem Leben*, pp. 226–7.
46 Ibid., pp. 228; Hohenlohe's notes of 27 October 1894, Hohenlohe, *Denkwürdigkeiten der Reichskanzlerzeit*, p. 3.
47 Hohenlohe's notes of 6 November 1894, ibid., pp. 10–11.
48 August Eulenburg to Philipp Eulenburg, 17 December 1894, *Eulenburgs Korrespondenz*, II, No. 1066.
49 Hohenlohe to Eulenburg, 20 January 1895, Hohenlohe, *Denkwürdigkeiten der Reichskanzlerzeit*, p. 30.
50 Eulenburg to Bülow, 7 November 1894, *Eulenburgs Korrespondenz*, II, No. 1043; Eulenburg to Holstein, 7 November 1894, ibid., No. 1044; Bülow to Eulenburg, 10 November 1894, ibid., No. 1046; Eulenburg to Kaiser Wilhelm II, 14 November 1894 and 23 January 1895, ibid., Nos. 1049 and 1076; Eulenburg to Hohenlohe, 22 January 1895, Hohenlohe, *Denkwürdigkeiten der Reichskanzlerzeit*, p. 30; Eulenburg to Holstein, 22 January 1895, *Holstein Papers*, III, No. 436.
51 Holstein to Hohenlohe, 17 November 1894, Hohenlohe, *Denkwürdigkeiten der Reichskanzlerzeit*, p. 15.
52 Hohenlohe, diary entry for 31 December 1894, ibid., p. 27.
53 Chlodwig Hohenlohe to Prince Hermann zu Hohenlohe-Langenburg, 21 April 1895, Hohenlohe-Zentralarchiv Neuenstein, Hermann Hohenlohe-Langenburg Papers, Bü. 86.
54 Hohenlohe to Eulenburg, 7 January 1895, *Eulenburgs Korrespondenz*, II, p. 1441.
55 Waldersee, diary entry for 21 January 1896, GStA Berlin, Waldersee Papers; cf. Meisner, II, pp. 365–6.
56 Hohenlohe, note of 28 February 1896, Hohenlohe, *Denkwürdigkeiten der Reichskanzlerzeit*, pp. 181–2.
57 Hohenlohe, diary entry for 2 March 1896, ibid., p. 186.
58 See Bülow to Eulenburg, 5 January 1895, *Eulenburgs Korrespondenz*, II, No. 1073.
59 Szögyény to Kálnoky, 24 November 1894, cited ibid., p. 1401.
60 Waldersee, diary entry for 28 April 1895, GStA Berlin, Waldersee Papers; omitted from Meisner, II, p. 346.
61 Waldersee, diary entry for 8 May 1895, GStA Berlin, Waldersee Papers; omitted from Meisner, II, p. 348.
62 Waldersee, diary entry for 9 April 1895, GStA Berlin, Waldersee Papers; cf. Meisner, II, p. 345.
63 Waldersee, diary entry for 12 May 1895, GStA Berlin, Waldersee Papers; omitted from Meisner, II, p. 349.
64 Waldersee, diary entry for 19 May 1895, GStA Berlin, Waldersee Papers; omitted from Meisner, II, p. 349.
65 Waldersee, diary entry for 10 October 1895, GStA Berlin, Waldersee Papers; omitted from Meisner, II, p. 360.
66 Eulenburg to Holstein, 11 and 21 January 1895, *Eulenburgs Korrespondenz*, II, No. 1075; Alexander Hohenlohe to Eulenburg, 14 January 1895, ibid., p. 1447; also Jagemann to Brauer, 19 December 1894 and 5 January 1895, Fuchs, *Großherzog von Baden*, III, Nos. 1400 and 1405.
67 Eulenburg to Kaiser Wilhelm II, 5 February 1895, *Eulenburgs Korrespondenz*, II, No. 1079.
68 Holstein to Eulenburg, 22 December 1895, ibid., III, No. 1175.
69 Waldersee, diary entry for 11 June 1895, GStA Berlin, Waldersee Papers; omitted from Meisner, II, p. 350.

70 Waldersee, diary entry for 17 February 1895, GStA Berlin, Waldersee Papers; omitted from Meisner, II, p. 339.
71 Waldersee to Verdy, 12 April 1895, GStA Berlin, Waldersee Papers No. 53.
72 Waldersee, diary entry for 22 April 1895, GStA Berlin, Waldersee Papers; omitted from Meisner, II, p. 346.
73 Waldersee, diary entry for 4 February 1896, GStA Berlin, Waldersee Papers; printed only in part in Meisner, II, p. 367.
74 Waldersee, diary entry for 21 August 1895, GStA Berlin, Waldersee Papers; cf. Meisner, II, p. 356.
75 See Harden's article 'Hohenzollern oder Hohenlohe', *Die Zukunft*, February 1895.
76 Eulenburg to Hohenlohe, 24 May 1895, Hohenlohe, *Denkwürdigkeiten der Reichskanzlerzeit*, pp. 71–2.
77 Eulenburg to Holstein, 22 June 1895, *Eulenburgs Korrespondenz*, III, No. 1114.
78 Marschall, diary entry for 17 February 1895, cited ibid., p. 1469.
79 Holstein to Bülow, 7 February 1895, ibid., No. 1082.
80 Holstein to Eulenburg, 26 January 1895, ibid., No. 1078.
81 Eulenburg to Holstein, 16 April 1895, ibid., No. 1104.
82 Waldersee, diary entry for 3 May 1896, GStA Berlin, Waldersee Papers; omitted from Meisner, II, p. 369.
83 Waldersee, diary entry for 10 May 1896, GStA Berlin, Waldersee Papers; omitted from Meisner, II, p. 369.
84 Waldersee, diary entry for 14 May 1896, GStA Berlin, Waldersee Papers; printed only in part in Meisner, II, pp. 369–70.
85 Waldersee, diary entry for 31 May 1897, GStA Berlin, Waldersee Papers; omitted from Meisner, II, pp. 397–8.
86 Waldersee, diary entries for 28 April, 18 May, 28 June and 10 July 1895, GStA Berlin, Waldersee Papers; omitted from Meisner, II, pp. 346, 349 and 351–2.
87 Holstein to Eulenburg, 7 April 1895, *Eulenburgs Korrespondenz*, III, No. 1101. See Jagemann to Brauer, 8 April 1895, Fuchs, *Großherzog von Baden*, III, p. 422 note.
88 Kaiser Wilhelm II to Hohenlohe, 30 March 1895, Hohenlohe, *Denkwürdigkeiten der Reichskanzlerzeit*, pp. 57–8.
89 Waldersee, diary entry for 2 April 1895, GStA Berlin, Waldersee Papers; cf. Meisner, II, pp. 343–4.
90 Waldersee, diary entry for 6 April 1895, GStA Berlin, Waldersee Papers; cf. Meisner, II, pp. 344–5.
91 Waldersee, diary entry for 18 May 1895, GStA Berlin, Waldersee Papers; omitted from Meisner, II, p. 349.
92 Waldersee, diary entry for 12 June 1895, GStA Berlin, Waldersee Papers; cf. Meisner, II, p. 350.
93 Waldersee, diary entry for 6 April 1895, GStA Berlin, Waldersee Papers; cf. Meisner, II, pp. 344–5.
94 Waldersee, diary entry for 15 October 1896, GStA Berlin, Waldersee Papers; cf. Meisner, II, pp. 372–3.
95 Waldersee, diary entry for 23 February 1895, GStA Berlin, Waldersee Papers; omitted from Meisner, II, p. 339.
96 Waldersee, diary entry for 23 February 1895, GStA Berlin, Waldersee Papers; omitted from Meisner, II, p. 339. See Evans, *Kneipengespräche*, p. 138.
97 Hohenlohe, diary entry for 19 June 1895, Hohenlohe, *Denkwürdigkeiten der Reichskanzlerzeit*, pp. 79–80.
98 Hohenlohe, diary entry for 23 June 1895, ibid., p. 80.

99 Waldersee, diary entry for 23 June 1895, GStA Berlin, Waldersee Papers; Meisner, II, pp. 350–1.

100 Eulenburg, notes of 13 and 28 July 1895, *Eulenburgs Korrespondenz*, III, Nos. 1117 and 1119; Kaiser Wilhelm II to Tsar Nicholas II, 10 July 1895, Walter Goetz, ed., *Briefe Wilhelms II. an den Zaren 1894–1914*, Berlin 1920, pp. 292ff.

101 Holstein to Eulenburg, 17 June 1895, *Eulenburgs Korrespondenz*, III, No. 1113.

102 Waldersee, diary entry for 23 June 1895, GStA Berlin, Waldersee Papers; Meisner, II, pp. 350–1.

103 Waldersee, diary entries for 27 and 28 June 1895, GStA Berlin, Waldersee Papers; cf. Meisner, II, pp. 351–2.

104 Eulenburg to Holstein, 22 June 1895, *Eulenburgs Korrespondenz*, III, No. 1114.

105 Waldersee, diary entries for 23 and 28 June 1895, GStA Berlin, Waldersee Papers; cf. Meisner, II, pp. 350ff.

24 AN ENEMY OF THE PEOPLE

1 See Harden's attacks on Hohenlohe and the Kaiser in *Die Zukunft* of 14 December 1894. See *Eulenburgs Korrespondenz*, II, p. 1436.

2 Holstein to Hohenlohe, 11 November 1894, Hohenlohe, *Denkwürdigkeiten der Reichskanzlerzeit*, p. 11.

3 Holstein to Bülow, 15 February 1895, *Holstein Papers*, III, No. 443.

4 Holstein to Bülow, 21 February 1895, ibid., No. 447.

5 Alexander Hohenlohe to Eulenburg, 17 February 1895, Hohenlohe, *Denkwürdigkeiten der Reichskanzlerzeit*, pp. 42–4.

6 Hohenlohe, diary entry for 14 December 1894, ibid., p. 23.

7 Kaiser Wilhelm II to Eulenburg, 9 December 1894, *Eulenburgs Korrespondenz*, II, No. 1060. See Eulenburg's telling reply of 14 December 1894, ibid., No. 1063; also Jagemann to Brauer, 2 October 1894, Fuchs, *Großherzog von Baden*, III, No. 1353.

8 Kaiser Wilhelm II to Eulenburg, 9 December 1894, *Eulenburgs Korrespondenz*, II, No. 1060.

9 Hohenlohe, undated notes, Hohenlohe, *Denkwürdigkeiten der Reichskanzlerzeit*, pp. 20–1.

10 Hohenlohe, diary entry for 14 December 1894, ibid., p. 23.

11 Hohenlohe, undated notes, ibid., pp. 31–2.

12 Waldersee, diary entries for 7 and 9 December 1894, GStA Berlin, Waldersee Papers; omitted from Meisner, II, p. 332.

13 Waldersee, diary entry for 14 December 1894, GStA Berlin, Waldersee Papers; cf. Meisner, II, pp. 332–3.

14 Waldersee, diary entry for 14 December 1894, GStA Berlin, Waldersee Papers; omitted for the most part from Meisner, II, pp. 332–3.

15 Waldersee, diary entry for 24 March 1895, Meisner, II, p. 340. See Marschall's diary entry for 23 March 1895, Fuchs, *Großherzog von Baden*, III, p. 410, note.

16 Kaiser Wilhelm II to Bismarck, telegram, 23 March 1895, in Hohenlohe, *Denkwürdigkeiten der Reichskanzlerzeit*, p. 53. Bismarck's reply is to be found in GStA Berlin, BPH Rep. 53 No. 140; see also Marschall to Hohenlohe, 31 March 1895, Hohenlohe, *Denkwürdigkeiten der Reichskanzlerzeit*, p. 58.

17 Waldersee, diary entry for 24 March 1895, GStA Berlin, Waldersee Papers; cf. Meisner, II, pp. 340–1. See Jagemann to Brauer, 24 and 25 March 1895, Fuchs, *Großherzog von Baden*, III, Nos. 1429 and 1431.

18 Waldersee, diary entry for 25 March 1895, GStA Berlin, Waldersee Papers; omitted from Meisner, II, p. 341.

19 Holstein to Eulenburg, 7 April 1895, *Eulenburgs Korrespondenz*, III, No. 1101.

20 Waldersee, diary entry for 28 March 1895, GStA Berlin, Waldersee Papers; cf. Meisner, II, pp. 341–2.
21 Waldersee, diary entry for 2 April 1895, GStA Berlin, Waldersee Papers; cf. Meisner, II, p. 342. See Hohenlohe's notes on his conversation with the Kaiser of 24 March 1895, Hohenlohe, *Denkwürdigkeiten der Reichskanzlerzeit*, pp. 53–4.
22 Waldersee, diary entry for 28 March 1895, Meisner, II, pp. 341–2.
23 Kaiser Wilhelm II to his mother, 5 April 1895, AdHH Schloss Fasanerie.
24 Kaiser Wilhelm II's two speeches of 26 March 1895 are printed in Penzler, *Reden Kaiser Wilhelms II. in den Jahren 1888–1895*, pp. 301–3.
25 Waldersee, diary entry for 28 March 1895, GStA Berlin, Waldersee Papers; cf. Meisner, II, pp. 341–2.
26 Waldersee, diary entries for 17 March and 2 April 1895, GStA Berlin, Waldersee Papers; cf. Meisner, II, pp. 339–40 and 342.
27 Reuss to Wedel, 2 September 1895, Wedel Papers, Frankfurt a.M.
28 Eulenburg to Holstein, 22 June 1895, *Eulenburgs Korrespondenz*, III, No. 1114.
29 Eulenburg to Hohenlohe, 12 September 1895, Hohenlohe, *Denkwürdigkeiten der Reichskanzlerzeit*, pp. 97–8.
30 Eulenburg to Kuno Moltke, 15 June 1895, *Eulenburgs Korrespondenz*, III, No. 1112.
31 Eulenburg to Holstein, 22 June 1895, ibid., No. 1114. Bismarck's speech is printed in Bismarck, *Die gesammelten Werke*, XIII, pp. 605–8.
32 Eulenburg to Holstein, 1 October 1895, *Eulenburgs Korrespondenz*, III, No. 1141.
33 Empress Frederick to Bogumilla Freifrau von Stockmar, 14 April 1895, AdHH Schloss Fasanerie.
34 Kaiser Wilhelm II to Hohenlohe, 11 May 1895, Hohenlohe, *Denkwürdigkeiten der Reichskanzlerzeit*, p. 63. A facsimile of the telegram is printed ibid., pp. 64–5.
35 Kaiser Wilhelm II to Hohenlohe, 23 August 1895, ibid., pp. 92–3.
36 Marschall, diary entry for 25 August 1895, cited in *Eulenburgs Korrespondenz*, III, p. 1526.
37 Holstein to Eulenburg, 28 August 1895, ibid., No. 1124.
38 Hohenlohe to Kaiser Wilhelm II, undated, Hohenlohe, *Denkwürdigkeiten der Reichskanzlerzeit*, p. 93.
39 Kaiser Wilhelm II to Hohenlohe, 31 August 1895, ibid., p. 94.
40 Hohenlohe to Köller, 31 August 1895, ibid., p. 94.
41 Eulenburg to Hohenlohe, 21 September 1895, ibid., pp. 99–100; Waldersee, diary entry for 12 October 1895, GStA Berlin, Waldersee Papers; omitted from Meisner, II, pp. 361–2.
42 Marschall, diary entry for 2 September 1895, cited in *Eulenburgs Korrespondenz*, III, p. 1526.
43 Kaiser Wilhelm II, speech of 2 September 1895, in Schröder, *Tagebuch Kaiser Wilhelms II.*, pp. 216–17.
44 Marschall, diary entry for 6 September 1895, cited in *Eulenburgs Korrespondenz*, III, p. 1527.
45 Eulenburg to Hohenlohe, 12 September 1895, Hohenlohe, *Denkwürdigkeiten der Reichskanzlerzeit*, pp. 97–8.
46 Eulenburg to Hohenlohe, 21 September 1895, ibid., pp. 99–100.
47 Marschall to Eulenburg, 29 September 1895, *Eulenburgs Korrespondenz*, III, No. 1140.
48 Eulenburg to Hohenlohe, 1 October 1895, Hohenlohe, *Denkwürdigkeiten der Reichskanzlerzeit*, pp. 111–12. See also Eulenburg to Marschall, 2 October 1895, *Eulenburgs Korrespondenz*, III, No. 1142.
49 Kaiser Wilhelm II to Hohenlohe-Langenburg, 8 October 1895, Hohenlohe-Zentralarchiv Neuenstein, Hermann Hohenlohe-Langenburg Papers, Bü. 308.
50 Waldersee, diary entry for 9 December 1895, GStA Berlin, Waldersee Papers; omitted from Meisner, II, p. 362. Cf. the watered-down version of the speech in Penzler, *Reden Kaiser Wilhelms II. in den Jahren 1888–1895*, pp. 321–3.

51 See Holstein to Eulenburg, 21 February 1895, *Eulenburgs Korrespondenz*, III, No. 1093.
52 Kaiser Wilhelm II to Hohenlohe, 4 February 1896, Hohenlohe, *Denkwürdigkeiten der Reichskanzlerzeit*, pp. 164–5.
53 Hohenlohe to Kaiser Wilhelm II, 5 February 1896, ibid., p. 165.
54 Waldersee, diary entry for 12 February 1895, GStA Berlin, Waldersee Papers; omitted from Meisner, II, p. 338.
55 Kaiser Wilhelm II, speech to the leaders of the Agrarian League, 18 February 1895, Penzler, *Reden Kaiser Wilhelms II. in den Jahren 1888–1895*, pp. 294–5.
56 Kaiser Wilhelm II to Eulenburg, 21 February 1895, *Eulenburgs Korrespondenz*, III, No. 1094.
57 Kaiser Wilhelm II, speech of 24 February 1895 to the Brandenburg provincial diet, Penzler, *Reden Kaiser Wilhelms II. in den Jahren 1888–1895*, pp. 295–6.
58 Waldersee, diary entry for 25 January 1895, GStA Berlin, Waldersee Papers; cf. Meisner, II, pp. 336–7.
59 Waldersee, diary entry for 10 March 1895, GStA Berlin, Waldersee Papers; cf. Meisner, II, p. 339.
60 Erich Eyck, *Das Persönliche Regiment Wilhelms II. Politische Geschichte des Deutschen Kaiserreiches von 1890 bis 1914*, Zürich 1948, p. 157.
61 Waldersee, diary entry for 12 May 1896, GStA Berlin, Waldersee Papers; omitted from Meisner, II, p. 369.
62 Waldersee, diary entry for 14 May 1896, GStA Berlin, Waldersee Papers; omitted from Meisner, II, p. 369; Waldersee, diary entry for 17 May 1896, GStA Berlin, Waldersee Papers; omitted from Meisner, II, p. 370.
63 Cf. Chlodwig Hohenlohe to Prince Hermann zu Hohenlohe-Langenburg, 8 March 1895, Hohenlohe-Zentralarchiv Neuenstein, Hermann Hohenlohe-Langenburg Papers, Bü. 86.
64 Alexander Hohenlohe to his father, 14 February 1895, Hohenlohe, *Denkwürdigkeiten der Reichskanzlerzeit*, p. 39. See Hohenlohe's draft of a letter to the Kaiser, not sent, 12 February 1895, ibid., pp. 38–9.
65 Marschall to Eulenburg, 17 February 1895, *Eulenburgs Korrespondenz*, III, No. 1088.
66 Holstein to Bülow, 15 February 1895, *Holstein Papers*, III, No. 443.
67 Eulenburg to Hohenlohe, 16 February 1895, Hohenlohe, *Denkwürdigkeiten der Reichskanzlerzeit*, pp. 39–40.
68 Kaiser Wilhelm II to Eulenburg, 12 February 1895, *Eulenburgs Korrespondenz*, II, No. 1083.
69 Holstein to Bülow, 7 February 1895, ibid., No. 1082; Marschall to Eulenburg, 17 February 1895, ibid., No. 1088.
70 Gustav von Kessel to Eulenburg, 13 February 1895, Hans von Arnim to Eulenburg, 14 February 1895, ibid., Nos. 1084 and 1085.
71 Eulenburg, diary entry, cited ibid., II, p. 1462. See Eulenburg's reply to the Kaiser of 14 February 1895, ibid., No. 1086; Eulenburg to Hohenlohe, 16 February 1895, Hohenlohe, *Denkwürdigkeiten der Reichskanzlerzeit*, pp. 39–42.
72 Eulenburg to Kaiser Wilhelm II, 22 February 1895, *Eulenburgs Korrespondenz*, III, No. 1095.
73 Eulenburg to Holstein, 19 February 1895, *Holstein Papers*, III, No. 446.
74 Eulenburg to Hohenlohe, 16 February 1895, Hohenlohe, *Denkwürdigkeiten der Reichskanzlerzeit*, pp. 39–42; Eulenburg to Kaiser Wilhelm II, 18 February 1895, *Eulenburgs Korrespondenz*, III, No. 1091. See Hohenlohe's reply to Eulenburg of 21 February 1895, ibid., No. 1092.
75 Kaiser Wilhelm II to Eulenburg, telegram, 21 February 1895, ibid., III, p. 1481.
76 Kaiser Wilhelm II to Eulenburg, 21 February 1895, ibid., No. 1094.
77 Marschall to Eulenburg, 17 February 1895, ibid., No. 1088.
78 Holstein to Bülow, 21 February 1895, *Holstein Papers*, III, No. 447; Similar: Holstein to Eulenburg, 21 February 1895, *Eulenburgs Korrespondenz*, III, No. 1093.

79 Holstein to Bülow, 17 February 1895, *Holstein Papers*, III, No. 444; Holstein to Eulenburg, 17 February 1895, *Eulenburgs Korrespondenz*, III, No. 1089.
80 Hohenlohe's notes of 12 May 1895, Hohenlohe, *Denkwürdigkeiten der Reichskanzlerzeit*, pp. 63–4.
81 Berlepsch to Hohenlohe, 15 May 1895, ibid., p. 65.
82 Hohenlohe to Eulenburg, 22 May 1895, ibid., pp. 68–70.
83 Eulenburg to Hohenlohe, 24 May 1895, ibid., pp. 71–2.
84 Hohenlohe, diary entry for 8 June 1895, ibid., p. 75.
85 Hohenlohe, diary entry for 28 June 1895, ibid., pp. 82–3.
86 Kaiser Wilhelm II to Berlepsch, 31 July 1895, ibid., p. 86.
87 Eulenburg to Hohenlohe, 24 May 1895, ibid., pp. 70–1.
88 Eulenburg to Hohenlohe, 24 May 1895, ibid., pp. 71–2. See above, p. 691.
89 Waldersee, diary entry for 20 October 1894, GStA Berlin, Waldersee Papers; cf. Meisner, II, p. 326.
90 Marschall to Eulenburg, 17 February 1895, *Eulenburgs Korrespondenz*, III, No. 1088.
91 Marschall, diary entry for 16 February 1895, cited ibid., p. 1469; also *Holstein Papers*, III, No. 444.
92 Eulenburg to Holstein, 22 June 1895, *Eulenburgs Korrespondenz*, III, No. 1114.
93 Eulenburg to Kaiser Wilhelm II, 4 August 1895, ibid., No. 1121.
94 Hohenlohe, notes of 31 May 1895, Hohenlohe, *Denkwürdigkeiten der Reichskanzlerzeit*, p. 74. See the discrepant account in Eulenburg to Grand Duke of Baden, 6 November 1895, *Eulenburgs Korrespondenz*, III, No. 1153.
95 Eulenburg to Holstein, 29 February 1896, *Holstein Papers*, III, No. 530.
96 Waldersee, diary entry for 10 October 1895, GStA Berlin, Waldersee Papers; cf. Meisner, II, p. 360.
97 Grand Duke of Baden to Eulenburg, 26 October 1895, Eulenburg to Grand Duke of Baden, 6 November 1895, *Eulenburgs Korrespondenz*, III, Nos. 1149 and 1153.
98 Waldersee, diary entry for 25 November 1895, GStA Berlin, Waldersee Papers; omitted from Meisner, II, p. 361. See Marschall's diary entry for 17 November 1895, cited in *Eulenburgs Korrespondenz*, III, p. 1593.
99 See Grand Duke of Baden to Eulenburg, 26 October 1895, *Eulenburgs Korrespondenz*, III, No. 1149.
100 Eulenburg to Grand Duke of Baden, 6 November 1895, ibid., No. 1153.
101 Eulenburg to Hohenlohe, 29 October 1895, Hohenlohe, *Denkwürdigkeiten der Reichskanzlerzeit*, p. 114.
102 Hohenlohe, diary entry for 31 October 1895, ibid., pp. 114ff.
103 Kaiser Wilhelm II to Hohenlohe, 31 October 1895, ibid., p. 116.
104 Eulenburg to Grand Duke of Baden, 6 November 1895, *Eulenburgs Korrespondenz*, III, No. 1153.
105 Hutten-Czapski to Holstein, 15 November 1895, *Holstein Papers*, III, No. 500.
106 See Hohenlohe, *Denkwürdigkeiten der Reichskanzlerzeit*, pp. 124–5.
107 Hohenlohe, diary entry for 27 November 1895, ibid., p. 125; Marschall, diary entries for 26 and 29 November 1895, cited in *Eulenburgs Korrespondenz*, III, p. 1593; Holstein to Eulenburg, 29 November 1895, ibid., Nos. 1159–60. On the Köller crisis, see also Nipperdey, *Deutsche Geschichte*, II, pp. 710–11; Clark, *Kaiser Wilhelm II*, pp. 78–9.
108 Hohenlohe to Kaiser Wilhelm II, 29 November 1895, Hohenlohe, *Denkwürdigkeiten der Reichskanzlerzeit*, pp. 126–7. See Marschall to Eulenburg, 20 December 1895, *Eulenburgs Korrespondenz*, III, No. 1174.
109 Hohenlohe, diary entry for 29 November 1895, Hohenlohe, *Denkwürdigkeiten der Reichskanzlerzeit*, p. 127.

110 Hohenlohe to Kaiser Wilhelm II, 1 December 1895, ibid., pp. 129–30.
111 Eulenburg to Kaiser Wilhelm II, 29 November 1895, *Eulenburgs Korrespondenz*, III, No. 1161; Eulenburg to Hohenlohe, 29 November 1895, Hohenlohe, *Denkwürdigkeiten der Reichskanzlerzeit*, pp. 127–8. See Holstein to Bülow, 29 November 1895, *Holstein Papers*, III, No. 505; Eulenburg to Holstein, 29 November 1895, ibid., No. 506.
112 Holstein to Eulenburg, 2 December 1895, *Eulenburgs Korrespondenz*, III, No. 1163.
113 See Haller, *Eulenburg*, p. 160.
114 Eulenburg to Kaiser Wilhelm II, 6 December 1895, *Eulenburgs Korrespondenz*, III, No. 1168.
115 Eulenburg to Lucanus, 7 December 1895, ibid., No. 1170; Eulenburg to Hohenlohe, 6 December 1895, Hohenlohe, *Denkwürdigkeiten der Reichskanzlerzeit*, pp. 137–8.
116 Eulenburg to Holstein, 4 December 1895, *Eulenburgs Korrespondenz*, III, p. 1603.
117 Eulenburg to Holstein, 7 December 1895, *Holstein Papers*, III, No. 508.
118 Eulenburg's notes of 3 December 1895, Haller, *Eulenburg*, pp. 160–1.
119 Eulenburg to Bülow, 6 December 1895, *Eulenburgs Korrespondenz*, III, No. 1169.
120 Eulenburg's notes of 3 December 1895, Haller, *Eulenburg*, pp. 160–1.
121 Marschall, diary entry for 4 December 1895, cited in *Eulenburgs Korrespondenz*, III, p. 1602.
122 Eulenburg to Kaiser Wilhelm II, 3 December 1895, ibid., No. 1164. See Eulenburg to Holstein, 7 December 1895, *Holstein Papers*, III, No. 508.
123 Lucanus to Eulenburg, 5 December 1895, *Eulenburgs Korrespondenz*, III, No. 1166.
124 Waldersee, diary entry for 9 December 1895, GStA Berlin, Waldersee Papers; omitted from Meisner, II, p. 362.
125 Holstein to Eulenburg, 22 December 1895, *Eulenburgs Korrespondenz*, III, No. 1175.
126 Kaiser Wilhelm II, order to the Prussian Ministry of State, 9 December 1895, Hohenlohe, *Denkwürdigkeiten der Reichskanzlerzeit*, p. 139.
127 Kaiser Wilhelm II to Hohenlohe, 6 May 1896, ibid., p. 218.
128 Holstein to Eulenburg, 5 December 1895, *Eulenburgs Korrespondenz*, III, No. 1165. Cf. Hohenlohe to Eulenburg, 9 December 1895, Hohenlohe, *Denkwürdigkeiten der Reichskanzlerzeit*, pp. 138–9; also Eulenburg to Holstein, 7 December 1895, *Holstein Papers*, III, No. 508.
129 Eulenburg to Hohenlohe, 6 December 1895, Hohenlohe, *Denkwürdigkeiten der Reichskanzlerzeit*, pp. 137–8.
130 See ibid., p. 138. Cf. Eulenburg to Holstein, 7 December 1895, *Holstein Papers*, III, No. 508; Lucanus to Hohenlohe, 7 December 1895, Hohenlohe, *Denkwürdigkeiten der Reichskanzlerzeit*, p. 138.
131 Kaiser Wilhelm II to Hohenlohe, 8 December 1895, *Denkwürdigkeiten der Reichskanzlerzeit*, p. 138. See Hohenlohe to Eulenburg, 9 December 1895, ibid., pp. 138–9; also Marschall, diary entries for 5 and 8 December 1895, cited in *Eulenburgs Korrespondenz*, III, p. 1606.
132 Hohenlohe to Eulenburg, 9 December 1895, Hohenlohe, *Denkwürdigkeiten der Reichskanzlerzeit*, pp. 138–9.
133 Eulenburg to Holstein, 7 December 1895, *Holstein Papers*, III, No. 508.
134 Eulenburg to Bülow, 6 December 1895, *Eulenburgs Korrespondenz*, III, No. 1169.
135 Eulenburg to Kaiser Wilhelm II, 6 December 1895, ibid., No. 1168.
136 Eulenburg to Hohenlohe, 6 December 1895, Hohenlohe, *Denkwürdigkeiten der Reichskanzlerzeit*, pp. 137–8.
137 Hohenlohe to Eulenburg, 9 December 1895, ibid., pp. 138–9; also Chlodwig Hohenlohe to Prince Hermann zu Hohenlohe-Langenburg, 27 December 1895, Hohenlohe-Zentralarchiv Neuenstein, Hermann Hohenlohe-Langenburg Papers, Bü. 86.

25 WILHELM AND WORLD POLITICS

1 See Canis, *Von Bismarck zur Weltpolitik.*
2 Sir Edmund Monson, report of 8 November 1894, RA I60/101.
3 Gosselin to Salisbury, 29 November 1895, cited in *Eulenburgs Korrespondenz*, III, pp. 1484–5.
4 Waldersee, diary entry for 22 October 1896, GStA Berlin, Waldersee Papers; printed only in part in Meisner, II, pp. 373–4.
5 Holstein to Eulenburg, 26 December 1895, *Eulenburgs Korrespondenz*, III, No. 1180.
6 Marschall, diary entry for 25 December 1895, Marschall Papers, Schloss Neuershausen.
7 Alexander Hohenlohe, draft letter from his father to the Kaiser, 1894/5, Hohenlohe, *Denkwürdigkeiten der Reichskanzlerzeit*, p. 27.
8 Hohenlohe to Kaiser Wilhelm II, undated [August 1895], ibid., pp. 93–4.
9 See e.g. Eulenburg to Marschall, 2 October 1895, *Eulenburgs Korrespondenz*, III, No. 1142; also Eulenburg's notes of 11 July 1892, ibid., II, No. 688; Eulenburg to Holstein, 2 October 1892, *Holstein Papers*, III, No. 377.
10 Chlodwig Hohenlohe to Prince Hermann zu Hohenlohe-Langenburg, 29 September 1895, Hohenlohe-Zentralarchiv Neuenstein, Hermann Hohenlohe-Langenburg Papers, Bü. 86.
11 Holstein to Radolin, 4 June 1895, *Holstein Papers*, III, No. 464.
12 Wolfgang J. Mommsen, *Großmachtstellung und Weltpolitik, 1870–1914. Die Außenpolitik des Deutschen Reiches*, Frankfurt a.M. 1993, p. 123. See Canis, *Von Bismarck zur Weltpolitik*, p. 14.
13 Hatzfeldt to Holstein, 15 March 1896, *Holstein Papers*, III, No. 532.
14 Hohenlohe's notes of 2 November 1894, Hohenlohe, *Denkwürdigkeiten der Reichskanzlerzeit*, p. 8, n. 3.
15 Hohenlohe's notes of 12 January 1895, ibid., p. 29. See Kaiser Wilhelm II to Hohenlohe, 16 January 1895, Hohenlohe's undated notes, ibid., pp. 29–31.
16 Waldersee, diary entry for 25 January 1895, GStA Berlin, Waldersee Papers; cf. Meisner, II, p. 336.
17 Ibid.
18 Herbette's report of 18 January 1895, *Documents Diplomatiques Français 1871–1914*, Ministère des Affaires Etrangères, ed., 41 vols., Paris 1929–36, XI, pp. 541ff.
19 Kaiser Wilhelm II, marginal notes on Eulenburg's report from Vienna of 17 January 1895, cited in *Holstein Papers*, III, p. 491.
20 See Holstein to Bülow, 23 January 1895, ibid., III, No. 437.
21 Waldersee, diary entry for 25 January 1895, GStA Berlin, Waldersee Papers; cf. Meisner, II, p. 336.
22 Hohenlohe to Holstein, 25 January 1895, *Holstein Papers*, III, No. 438.
23 Waldersee, diary entry for 3 March 1896, GStA Berlin, Waldersee Papers; cf. Meisner, II, p. 368; Hohenlohe, diary entry for 5 March 1896, Hohenlohe, *Denkwürdigkeiten der Reichskanzlerzeit*, p. 187.
24 Kaiser Wilhelm II to Hohenlohe, 31 August 1895, *Denkwürdigkeiten der Reichskanzlerzeit*, pp. 94–5. See below, p. 748.
25 Hohenlohe's report of 7 March 1896, Hohenlohe, *Denkwürdigkeiten der Reichskanzlerzeit*, pp. 191–2.
26 Kaiser Wilhelm II, marginal notes on the reports of 21 September 1895 and 18 and 19 April 1897, PA AA, Asservat No. 4.
27 Monson, telegram from Vienna, 28 February 1895, RA I60/105. See Eulenburg to Hohenlohe, 5 March 1895, Hohenlohe, *Denkwürdigkeiten der Reichskanzlerzeit*, p. 50.
28 Waldersee, diary entry for 2 April 1895, GStA Berlin, Waldersee Papers; omitted from Meisner, II, pp. 342ff.

29 Eulenburg's notes, 27 September 1895, *Eulenburgs Korrespondenz*, III, No. 1136.
30 Waldersee, diary entry for 20 October 1895, GStA Berlin, Waldersee Papers; omitted from Meisner, II, p. 361.
31 See Hohenlohe, *Denkwürdigkeiten der Reichskanzlerzeit*, p. 26. See Goetz, *Briefe Wilhems II. an den Zaren*, p. 288.
32 Hohenlohe's notes of 8 March 1896, Hohenlohe, *Denkwürdigkeiten der Reichskanzlerzeit*, p. 193.
33 Holstein to Hohenlohe, 8 March 1896, ibid., pp. 193–4.
34 See Marschall to Eulenburg, two telegrams, 8 March 1896, *Eulenburgs Korrespondenz*, III, Nos. 1193 and 1194; Eulenburg to Kaiser Wilhelm II, 9 March 1896, ibid., No. 1196.
35 Kaiser Wilhelm II to Hohenlohe, 8 March 1896, Hohenlohe, *Denkwürdigkeiten der Reichskanzlerzeit*, p. 194.
36 Hohenlohe to Kaiser Wilhelm II, 8 March 1896, ibid., pp. 194–5.
37 Kaiser Wilhelm II to Hohenlohe, 9 March 1896, ibid., pp. 195–6. Cf. Eulenburg to Bülow, 13 March 1896, *Eulenburgs Korrespondenz*, III, No. 1198.
38 Hohenlohe to Kaiser Wilhelm II, 9 March 1896, Hohenlohe, *Denkwürdigkeiten der Reichskanzlerzeit*, p. 196.
39 Hohenlohe to Kaiser Wilhelm II, 10 March 1896, ibid., pp. 197–8.
40 Bülow to Hohenlohe, 19 March 1896, ibid., p. 201.
41 Bülow to Holstein, 19 March 1896, *Holstein Papers*, III, No. 533.
42 Münster to Holstein, 19 March 1896, ibid., No. 534.
43 Bülow to Hohenlohe, 13 April 1896, Hohenlohe, *Denkwürdigkeiten der Reichskanzlerzeit*, pp. 208–11.
44 Kaiser Wilhelm II to Caprivi, telegram, 12 July 1893, cited in Folke Lindberg, *Kunglig utrikespolitik. Studier i svensk utrikespolitik under Oscar II och fram till borggårdskrisen*, Stockholm 1966, p. 110. See Eulenburg's notes of 11 July 1892, in *Eulenburgs Korrespondenz*, II, No. 688.
45 Hohenlohe, diary entry for 22 February 1895, Hohenlohe, *Denkwürdigkeiten der Reichskanzlerzeit*, p. 45.
46 Eulenburg's notes of 22 July 1895, *Eulenburgs Korrespondenz*, III, pp. 1516–17. See also ibid., No. 1119.
47 Kaiser Wilhelm II to Crown Prince Gustav of Sweden and Norway, 25 July 1895, printed in Hohenlohe, *Denkwürdigkeiten der Reichskanzlerzeit*, pp. 102–5.
48 Kaiser Wilhelm II to Crown Prince Gustav of Sweden and Norway, 27 September 1895, printed ibid., pp. 108–10.
49 Below, ch. 32.
50 Waldersee, diary entries for 4 and 23 May 1895, GStA Berlin, Waldersee Papers; cf. Meisner, II, pp. 347 and 349.
51 Waldersee, diary entry for 15 February 1895, GStA Berlin, Waldersee Papers; largely omitted from Meisner, II, pp. 338–9.
52 Hohenlohe's notes of 2 November 1894, Hohenlohe, *Denkwürdigkeiten der Reichskanzlerzeit*, pp. 7–8. See Marschall's diary entry for 31 October 1894, cited in *Eulenburgs Korrespondenz*, II, p. 1406.
53 Hohenlohe to Kaiser Wilhelm II, 18 November 1894, Hohenlohe, *Denkwürdigkeiten der Reichskanzlerzeit*, pp. 15–16.
54 Kaiser Wilhelm II, undated notes, printed ibid., pp. 52–3. See Holstein's notes of 27 March 1895, ibid., pp. 55–6.
55 Hohenlohe, diary entry for 11 April 1895, ibid., p. 58.
56 Senden-Bibran to Kaiser Wilhelm II, 22 April 1895, BA-MA Freiburg, Senden-Bibran Papers.

57 Holstein to Radolin, 18 June 1895, *Holstein Papers*, III, No. 470. See Radolin to Holstein, 10 August 1895, ibid., No. 483. See also Hohenlohe to Holstein, 6 August 1895, ibid., No. 481; also *Große Politik*, IX, Nos. 2285–90. Cf. Hohenlohe, diary entry for 10 September 1895, Hohenlohe, *Denkwürdigkeiten der Reichskanzlerzeit*, p. 96.

58 Hohenlohe to Kaiser Wilhelm II, undated [August 1895], Hohenlohe, *Denkwürdigkeiten der Reichskanzlerzeit*, pp. 93–4. See above, pp. 733–4.

59 Kaiser Wilhelm II to Hohenlohe, 31 August 1895, Hohenlohe, *Denkwürdigkeiten der Reichskanzlerzeit*, pp. 94–5. Cf. Holstein to Eulenburg, 2 and 7/8 April 1895, *Eulenburgs Korrespondenz*, III, No. 1101.

60 Hohenlohe to Eulenburg, 24 September 1895, *Eulenburgs Korrespondenz*, III, No. 1131.

61 Waldersee, diary entries for 4 and 23 May 1895, GStA Berlin, Waldersee Papers; cf. Meisner, II, pp. 347 and 349.

62 Eulenburg to Marschall, 30 September 1894, *Eulenburgs Korrespondenz*, III, No. 1008.

63 Kaiser Wilhelm II, marginal notes on Eulenburg's report from Vienna, PA AA, Asservat No. 4.

64 Waldersee, diary entry for 20 October 1894, GStA Berlin, Waldersee Papers; cf. Meisner, II, p. 326.

65 Waldersee, diary entries for 6 and 14 December 1894, GStA Berlin, Waldersee Papers; cf. Meisner, II, pp. 332–3. See above, p. 221. Cf. Kaiser Wilhelm II to Tsar Nicholas II, 8 November 1894, in Goetz, *Briefe Wilhelms II. an den Zaren*, pp. 287–8; Tsar Nicholas II to Kaiser Wilhelm II, 8 November 1894, Hohenlohe, *Denkwürdigkeiten der Reichskanzlerzeit*, p. 8.

66 Waldersee, diary entry for 21 November 1894, GStA Berlin, Waldersee Papers; omitted from Meisner, II, p. 332.

67 Hohenlohe, diary entry for 14 December 1894, Hohenlohe, *Denkwürdigkeiten der Reichskanzlerzeit*, p. 23.

68 Waldersee, diary entries for 6 and 14 December 1894, GStA Berlin, Waldersee Papers; cf. Meisner, II, pp. 332–3. Cf. Kaiser Wilhelm II to Tsar Nicholas II, 8 November 1894, Goetz, *Briefe Wilhelms II. an den Zaren*, pp. 287–8; Tsar Nicholas II to Kaiser Wilhelm II, 8 November 1894, Hohenlohe, *Denkwürdigkeiten der Reichskanzlerzeit*, p. 8.

69 Waldersee, diary entry for 25 January 1895, GStA Berlin, Waldersee Papers; this passage is omitted from Meisner, II, pp. 336–7.

70 Waldersee, diary entry for 2 April 1895, GStA Berlin, Waldersee Papers; omitted from Meisner, II, pp. 342ff.

71 Eulenburg's notes of 28 July 1895, *Eulenburgs Korrespondenz*, III, No. 1119.

72 Waldersee, diary entry for 21 August 1895, GStA Berlin, Waldersee Papers; this passage is omitted from Meisner, II, pp. 356–7.

73 Hohenlohe's notes of 2 November 1894, Hohenlohe, *Denkwürdigkeiten der Reichskanzlerzeit*, p. 7.

74 Holstein to Hohenlohe, 25 December 1894, ibid., p. 26.

75 Hohenlohe, diary entry for 26 December 1894, ibid.

76 Kaiser Wilhelm II to Tsar Nicholas II, 26 April 1895, Goetz, *Briefe Wilhelms II. an den Zaren*, pp. 290–2.

77 Friedrich Curtius, ed., *Denkwürdigkeiten des Fürsten Chlodwig zu Hohenlohe-Schillingsfürst*, 2 vols., Stuttgart 1906, II, p. 521.

78 Eulenburg's notes, 5 July 1895, *Eulenburgs Korrespondenz*, III, No. 1116.

79 Eulenburg's notes, 13 July 1895, ibid., No. 1117.

80 Kaiser Wilhelm II to Tsar Nicholas II, 10 July 1895, Goetz, *Briefe Wilhelms II. an den Zaren*, pp. 292–4.

81 Eulenburg's notes, 5 July 1895, *Eulenburgs Korrespondenz*, III, No. 1116.

82 Eulenburg's notes, 13 July 1895, ibid., No. 1117; also Eulenburg's notes of 28 July 1895, ibid., No. 1119.

83 The original sketch of Wilhelm II's world-famous drawing sports the signature 'Em u W 30/IV 95 Schlitz'. It is reproduced in Hans Wilderotter and Klaus-D. Pohl, eds., *Der letzte Kaiser. Wilhelm II. im Exil*, Gütersloh, Munich 1991, p. 321.

84 Eulenburg to Kaiserin Auguste Viktoria, 29 September 1895, *Eulenburgs Korrespondenz*, III, p. 1549.

85 Eulenburg to his mother, late September 1895, ibid., p. 1549.

86 Kaiser Wilhelm II to Tsar Nicholas II, 26 September 1895, Goetz, *Briefe Wilhelms II. an den Zaren*, pp. 294–6.

87 Eulenburg's notes, 27 September 1895, *Eulenburgs Korrespondenz*, III, No. 1136.

88 Kaiser Wilhelm II to Tsar Nicholas II, 26 September 1895, Goetz, *Briefe Wilhelms II. an den Zaren*, pp. 294–6.

89 Eulenburg to Hohenlohe, 28 and 30 September 1895, Hohenlohe, *Denkwürdigkeiten der Reichskanzlerzeit*, p. 102; also Radolin to Holstein, 28 September 1895, *Holstein Papers*, III, No. 487.

90 For Moltke the Younger, see Annika Mombauer, *Helmuth von Moltke and the Origins of the First World War*, Cambridge 2001.

91 Helmuth Moltke to Kaiser Wilhelm II, telegram of 30 September 1895, printed in Hohenlohe, *Denkwürdigkeiten der Reichskanzlerzeit*, p. 110. See Moltke's report to the Kaiser in *Große Politik*, IX, pp. 365–6.

92 Helmuth Moltke to Kaiser Wilhelm II, report on his farewell audience with Nicholas II, 3 October 1895, GStA Berlin, BPH Rep. 53 No. 116.

93 Eulenburg to Holstein, 1 October 1895, *Eulenburgs Korrespondenz*, III, No. 1141. See Radolin to Eulenburg, 2 October 1895, ibid., No. 1143.

94 Eulenburg to Hohenlohe, 1 October 1895, Hohenlohe, *Denkwürdigkeiten der Reichskanzlerzeit*, pp. 111–12.

95 Eulenburg's notes, 12/13 October 1895, *Eulenburgs Korrespondenz*, III, No. 1145.

96 Holstein to Eulenburg, 19 November 1895, ibid., No. 1156.

97 Holstein to Radolin, 30 October 1895, *Holstein Papers*, III, No. 496.

98 Kaiser Wilhelm II to Tsar Nicholas II, 25 October 1895, Goetz, *Briefe Wilhelms II. an den Zaren*, pp. 297–300. The Reich Chancellor was not shown this letter until 30 October. See Hohenlohe, *Denkwürdigkeiten der Reichskanzlerzeit*, p. 116.

99 Eulenburg's notes of 28 July 1895, *Eulenburgs Korrespondenz*, III, No. 1119. On the meeting of the then Prince Wilhelm with Tsar Alexander III at Brest-Litovsk, see *Young Wilhelm*, pp. 570–84.

100 Hohenlohe's notes of 2 November 1894, Hohenlohe, *Denkwürdigkeiten der Reichskanzlerzeit*, pp. 7–8. See Marschall's diary entry for 31 October 1894, cited in *Eulenburgs Korrespondenz*, II, p. 1406.

101 Eulenburg to Bülow, 12 November 1895, *Eulenburgs Korrespondenz*, III, No. 1154.

102 Waldersee, diary entry for 9 October 1894, GStA Berlin, Waldersee Papers; omitted from Meisner, II, p. 324.

103 Hohenlohe's notes of 2 November 1894, Hohenlohe, *Denkwürdigkeiten der Reichskanzlerzeit*, pp. 7–8. See Marschall's diary entry for 31 October 1894, cited in *Eulenburgs Korrespondenz*, II, p. 1406.

104 Eulenburg's notes, 28 July 1895, *Eulenburgs Korrespondenz*, III, No. 1119.

105 See Lobanov's report to the French Foreign Minister Hanotaux in *Documents Diplomatiques Français*, XII, p. 264.

106 Kaiser Wilhelm II, marginal notes, *Große Politik*, X, p. 147.

107 Eulenburg to Marschall, 6 November 1895, Hohenlohe, *Denkwürdigkeiten der Reichskanzlerzeit*, pp. 117–18.

108 Holstein to Bülow, 8 November 1895, ibid., p. 119. Cf. Hatzfeldt to Holstein, 10 November 1895, *Holstein Papers*, III, No. 497.
109 Holstein to Hohenlohe, 12 November 1895, Hohenlohe, *Denkwürdigkeiten der Reichskanzlerzeit*, p. 120. See Eulenburg to Kaiser Wilhelm II, 31 December 1895, *Eulenburgs Korrespondenz*, III, No. 1183.
110 Hohenlohe's notes of 13 November 1895, Hohenlohe, *Denkwürdigkeiten der Reichskanzlerzeit*, pp. 120–1. See Hutten-Czapski to Holstein, 15 November 1895, *Holstein Papers*, III, No. 500.
111 See the exchange of telegrams of 8 and 9 November 1895, *Große Politik*, X, Nos. 2452–3.
112 Holstein to Radolin, 16 November 1895, *Holstein Papers*, III, No. 501.
113 Eulenburg to Bülow, 12 November 1895, *Eulenburgs Korrespondenz*, III, No. 1154.
114 Eulenburg to Holstein, 16 November 1895, ibid., No. 1155.
115 Waldersee, diary entry for 9 December 1895, GStA Berlin, Waldersee Papers; omitted from Meisner, II, p. 362.
116 Waldersee, diary entry for 31 December 1895, GStA Berlin, Waldersee Papers; these passages are omitted from Meisner, II, p. 363.
117 Waldersee, diary entry for 9 December 1895, GStA Berlin, Waldersee Papers; omitted from Meisner, II, p. 362.
118 Szögyény, telegram, 14 November 1895, cited in Helmut Krausnick, 'Holstein, Österreich-Ungarn und die Meerengenfrage im Herbst 1895. Persönliches Regiment oder Regierungspolitik?', in *Forschungen zu Staat und Verfassung. Festgabe für Fritz Hartung*, Berlin 1958, pp. 519–20. See *Große Politik*, X, pp. 203–7.
119 Szögyény, report of 21 October 1908, in *Österreich-Ungarns Außenpolitik von der Bosnischen Krise 1908 bis zum Kriegsausbruch 1914*, Vienna, Leipzig 1930, p. 278. See Krausnick, 'Holstein, Österreich-Ungarn und die Meerengenfrage', pp. 485ff.
120 Szögyény, telegram, 14 November 1895, Krausnick, 'Holstein, Österreich-Ungarn und die Meerengenfrage', p. 520.
121 *Große Politik*, X, pp. 205ff.

26 GREAT BRITAIN AND THE SPECTRE OF ENCIRCLEMENT

1 Gosselin, report of 5 November 1894, RA I60/100.
2 Hatzfeldt to Holstein, 19 December 1894, *Holstein Papers*, III, No. 430.
3 Malet, report of December 1894, RA I60/104.
4 Waldersee, diary entry for 25 April 1895, GStA Berlin, Waldersee Papers; omitted from Meisner, II, p. 346.
5 Waldersee, diary entry for 23 May 1895, GStA Berlin, Waldersee Papers; omitted from Meisner, II, p. 349.
6 Kaiser Wilhelm II, marginal comment on Hatzfeldt's report of 11 November 1894, *Große Politik*, IX, No. 2161.
7 See Kaiser Wilhelm II to Queen Victoria, 12 July 1895, RA I60/119, in Buckle, *Letters of Queen Victoria*, II, pp. 535–6.
8 *Große Politik*, X, No. 2385.
9 Kiderlen-Wächter to Holstein, 7 August 1895, Hatzfeldt to Holstein, 14 August 1895, *Holstein Papers*, III, Nos. 482 and 486.
10 Hohenlohe's notes of 18 August 1895, Hohenlohe, *Denkwürdigkeiten der Reichskanzlerzeit*, p. 88.
11 Hatzfeldt to Holstein, 14 August 1895, *Holstein Papers*, III, No. 486.
12 Kaiser Wilhelm II, marginal notes, *Große Politik*, X, pp. 85, 102, 119, 134 *et passim*.
13 Kaiser Wilhelm II to Hohenlohe, 9 October 1895, Hohenlohe, *Denkwürdigkeiten der Reichskanzlerzeit*, pp. 112–13.

14 See the Kaiser's notes on the conversation with his mother, ibid., p. 123; also Empress Frederick to Queen Victoria, 1 January 1896, AdHH Schloss Fasanerie.
15 Kaiser Wilhelm II to Hohenlohe, 20 October 1895, *Große Politik*, X, No. 2437.
16 Holstein to Bülow, 23 October 1895, *Holstein Papers*, III, No. 490.
17 Kaiser Wilhelm II to Marschall, 25 October 1895, *Große Politik*, XI, No. 2579.
18 Kaiser Wilhelm II, undated notes, printed in Hohenlohe, *Denkwürdigkeiten der Reichskanzlerzeit*, pp. 52–3.
19 Eulenburg's notes, 24 September 1895, *Eulenburgs Korrespondenz*, III, No. 1129. See Eulenburg's notes of 12/13 October 1895, ibid., No. 1145.
20 Eulenburg to Bülow, 12 November 1895, ibid., No. 1154.
21 See Gosselin to Salisbury, 29 November 1895, ibid., II, pp. 1484–5.
22 See *Young Wilhelm*, pp. 318–19, and below, pp. 1065–6.
23 Lascelles, notes on a conversation with Salisbury of 4 December 1895, cited in Kennedy, *Anglo-German Antagonism*, pp. 219–20.
24 For the hostility of the British press at this time, see Reinermann, *Der Kaiser in England*, pp. 145ff.
25 Kaiser Wilhelm II to Queen Victoria, telegram, 20 June 1895, RA I60/117.
26 Kaiser Wilhelm II to Queen Victoria, 12 July 1895, RA I60/119, in Buckle, *Letters of Queen Victoria*, II, pp. 535–6.
27 Swaine to Bigge, 24 March 1895, RA I60/106.
28 Knollys to Bigge, 18 April 1895, RA I/108.
29 Hohenlohe to Kaiser Wilhelm II, 10 June 1895, Hohenlohe, *Denkwürdigkeiten der Reichskanzlerzeit*, p. 76.
30 See Kaiser Wilhelm II to Queen Victoria, 12 July 1895, RA I60/119, in Buckle, *Letters of Queen Victoria*, II, pp. 535–6.
31 Kiderlen-Wächter to Holstein, 7 August 1895, *Holstein Papers*, III, No. 482.
32 Waldersee, diary entry for 21 August 1895, GStA Berlin, Waldersee Papers; cf. Meisner, II, pp. 356–7.
33 See Hatzfeldt to Holstein, 14 August 1895, *Holstein Papers*, III, No. 486.
34 Kiderlen-Wächter to Holstein, 7 August 1895, ibid., No. 482.
35 Malet to Bigge, 1 August 1895, in Buckle, *Letters of Queen Victoria*, II, pp. 542–3; also Swaine to Bigge, 16 August 1895, RA I60/129.
36 RA QVJ, 6 August 1895, in Buckle, *Letters of Queen Victoria*, II, pp. 544–5; Queen Victoria to Salisbury, 8 August 1895, RA I60/121, printed ibid., p. 547; Salisbury to Bigge, 9 August 1895, RA I60/125, ibid., p. 548.
37 Eulenburg to Holstein, 11 August 1895, *Holstein Papers*, III, No. 484.
38 Lansdowne to Queen Victoria, 7 August 1895, in Buckle, *Letters of Queen Victoria*, II, pp. 545–6.
39 Kiderlen-Wächter to Holstein, 7 August 1895, *Holstein Papers*, III, No. 482.
40 Queen Victoria to Salisbury, 8 August 1895, RA I60/121, Salisbury to Queen Victoria, 8 August 1895, RA I60/123, both in Buckle, *Letters of Queen Victoria*, II, pp. 547–8.
41 Hatzfeldt to Holstein, 14 August 1895, *Holstein Papers*, III, No. 486; Waldersee, diary entry for 21 August 1895, GStA Berlin, Waldersee Papers; cf. Meisner, II, p. 356.
42 Kiderlen-Wächter to Holstein, 7 August 1895, *Holstein Papers*, III, No. 482.
43 Bigge to Lansdowne, 8 August 1895, in Buckle, *Letters of Queen Victoria*, II, p. 547.
44 Lansdowne to Bigge, 10 August 1895, ibid., p. 548.
45 Queen Victoria to Salisbury, 11 and 12 August 1895, ibid., pp. 549–50.
46 Salisbury to Queen Victoria, 11 August 1895, Lansdowne to Queen Victoria, 14 August 1895, Salisbury to Bigge, 16 August 1895, ibid., pp. 549–54.
47 Kaiser Wilhelm II to Queen Victoria, 12 August 1895, RA I60/126.

48 Salisbury to Bigge, 16 and 19 August 1895, in Buckle, *Letters of Queen Victoria*, II, pp. 553–5.
49 Bigge to Queen Victoria, 20 August 1895, RA I60/130.
50 For Wilhelm II and the appointment of Lascelles as ambassador, see Willem-Alexander van't Padje, 'At the Heart of the Growing Anglo-German Imperialist Rivalry: Two British Ambassadors in Berlin, 1884–1908', DPhil. diss., Oxford 2001, pp. 29ff.
51 Queen Victoria to Kaiser Wilhelm II, 28 August 1895, GStA Berlin, Brand.-Preuß. Hausarchiv, Rep. 52 W3 No. 11; copy in RA I60/132. Cf. Buckle, *Letters of Queen Victoria*, II, pp. 560–1.
52 Hatzfeldt to Holstein, 14 August 1895, *Holstein Papers*, III, No. 486. Malet's farewell address was printed in *The Times*, 14 October 1895, RA I60/134.
53 See below, pp. 779–83.
54 Waldersee, diary entry for 7 August 1895, GStA Berlin, Waldersee Papers; omitted from Meisner, II, p. 356.
55 Kaiser Wilhelm II to Queen Victoria, 12 August 1895, RA I60/126.
56 Lonsdale Papers, Lowther.
57 J. Meyer to Lord Lonsdale, 10 June and 29 July 1896, Chief Constable P. Clarke to Lord Lonsdale, 15 August 1896, Clarke to Meyer, 15 August 1896, Frederick Adams to Clarke, 26 August 1896, Lonsdale Papers, Cumbria Record Office, Carlisle.
58 Kaiser Wilhelm II to Countess of Lonsdale, 15 August 1895, Lonsdale Papers, Cumbria Record Office, Carlisle. Cf. Waldersee, diary entry for 21 August 1895, GStA Berlin, Waldersee Papers; omitted from Meisner, II, pp. 356–7.
59 Waldersee, diary entry for 7 August 1895, GStA Berlin, Waldersee Papers; omitted from Meisner, II, p. 356.
60 Ibid.
61 Waldersee, diary entry for 8 August 1895, GStA Berlin, Waldersee Papers; cf. Meisner, II, p. 356.
62 Waldersee, diary entries for 10 and 13 August 1895, GStA Berlin, Waldersee Papers; omitted from Meisner, II, p. 356.
63 Kaiser Wilhelm II, speech of 18 January 1896, Penzler, *Reden Kaiser Wilhelms II. in den Jahren 1896–1900*, Leipzig 1904, pp. 9–10.
64 Waldersee, diary entry for 31 January 1896, GStA Berlin, Waldersee Papers; omitted from Meisner, II, pp. 366–7.
65 Chlodwig Hohenlohe to Prince Hermann zu Hohenlohe-Langenburg, 8 March 1895, Hohenlohe-Zentralarchiv Neuenstein, Hermann Hohenlohe-Langenburg Papers, Bü. 86.
66 Kaiser Wilhelm II, marginal notes of 6 October 1895, cited in Nancy Mitchell, *The Danger of Dreams. German and American Imperialism in Latin America*, Chapel Hill, London 1999, pp. 114–15.
67 Eulenburg to Kayser, 24 September 1895, *Eulenburgs Korrespondenz*, III, No. 1130.
68 Hohenlohe, note of 31 January 1895, Hohenlohe, *Denkwürdigkeiten der Reichskanzlerzeit*, p. 32.
69 Alexander Hohenlohe to Eulenburg, 17 February 1895, ibid., pp. 42–44.
70 Chlodwig Hohenlohe to Prince Hermann zu Hohenlohe-Langenburg, 29 September 1895, Hohenlohe-Zentralarchiv Neuenstein, Hermann Hohenlohe-Langenburg Papers, Bü. 86.
71 Hohenlohe, diary entry for 2 November 1894, Hohenlohe, *Denkwürdigkeiten der Reichskanzlerzeit*, pp. 7–8. See above, pp. 745–6.
72 Waldersee, diary entry for 10 July 1895, GStA Berlin, Waldersee Papers; omitted from Meisner, II, p. 352.
73 Hohenlohe, diary entry for 1 March 1896, Hohenlohe, *Denkwürdigkeiten der Reichskanzlerzeit*, p. 185.
74 Hohenlohe to Kaiser Wilhelm II, 7 March 1896, ibid., p. 188.
75 Ibid.

76 Ibid., pp. 190–1.
77 Kaiser Wilhelm II, marginal notes on Bülow's report from Rome of 21 July 1896, *Große Politik*, XI, p. 295.
78 Malet to Queen Victoria, 16 June 1894, RA I60/77.
79 Lord Kimberley, report of 21 November 1894, RA I60/102.
80 Waldersee, diary entry for 7 August 1895, GStA Berlin, Waldersee Papers; omitted from Meisner, II, p. 356.
81 Elizabeth Longford, *Jameson's Raid: The Prelude to the Boer War*, London 1984.
82 See Matthew S. Seligmann, *Rivalry in Southern Africa, 1893–1899. The Transformation of German Colonial Policy*, London 1998, pp. 70–4.
83 See Kennedy, *Anglo-German Antagonism*, p. 220.
84 Waldersee, diary entry for 21 August 1895, GStA Berlin, Waldersee Papers; omitted from Meisner, II, pp. 356–7. See G. W. F. Hallgarten, *Imperialismus vor 1914. Die soziologischen Grundlagen der Außenpolitik europäischer Großmächte vor dem Ersten Weltkrieg*, 2 vols., Munich 1963, I, pp. 372–3; Seligmann, *Rivalry*, pp. 74–5; Canis, *Von Bismarck zur Weltpolitik*, pp. 166–7.
85 Marschall's notes of 15 October 1895, *Große Politik*, XI, No. 2578. For the ensuing crisis, see van't Padje, 'Anglo-German Imperialist Rivalry', pp. 185–93.
86 Kaiser Wilhelm II, marginal notes, *Große Politik*, XI, No. 2578, p. 7.
87 Holstein to Eulenburg, 28 October 1895, *Holstein Papers*, III, No. 492.
88 Kaiser Wilhelm II to Tsar Nicholas II, 25 October 1895, Goetz, *Briefe Wilhelms II. an den Zaren*, pp. 297–300.
89 See above, pp. 759–60.
90 Kaiser Wilhelm II to Marschall, 25 October 1895, *Große Politik*, XI, No. 2579.
91 Proclamation of 27 October 1895, *Große Politik*, XI, No. 2579, pp. 10–11, note.
92 See *Große Politik*, XI, No. 2579, p. 11, note.
93 See the exchange of telegrams between 29 October and 17 November 1895 with the Kaiser's marginal notes, *Große Politik*, XI, Nos. 2581–4.
94 Eulenburg to Holstein, 28 October 1895, *Holstein Papers*, III, No. 492; also Eulenburg to Bülow, 12 November 1895, *Eulenburgs Korrespondenz*, III, No. 1154.
95 Eulenburg to Holstein, 29 October 1895, *Holstein Papers*, III. No. 493.
96 Holstein to Eulenburg, 2 November 1895, *Eulenburgs Korrespondenz*, III, No. 1152.
97 Ibid.
98 Eulenburg to Bülow, 12 November 1895, ibid., No. 1154.
99 Eulenburg to Marschall, 6 November 1895, in Hohenlohe, *Denkwürdigkeiten der Reichskanzlerzeit*, pp. 117ff. See Eulenburg to Holstein, 31 January 1896, *Holstein Papers*, III. No. 525.
100 Kaiser Wilhelm II, marginal notes on Marschall's report of 17 November 1895, *Große Politik*, XI, No. 2584.
101 Waldersee, diary entry for 9 December 1895, GStA Berlin, Waldersee Papers; omitted from Meisner, II, p. 362.
102 Gerd Fesser, 'Ohrfeige für England. Die brisante Depesche Wilhelms II. an Burenpräsidenten "Ohm" Krüger', *Die Zeit*, 5 January 1996, p. 28. See Rich, *Friedrich von Holstein*, II, p. 469; Kennedy, *Anglo-German Antagonism*, pp. 220ff.
103 Marschall, diary entries for 28 and 31 December 1895, Marschall Papers, Schloss Neuershausen. See below, pp. 786–8.
104 Marschall, diary entry for 31 December 1895, Marschall Papers, Schloss Neuershausen.
105 Marschall, diary entry for 1 January 1896, ibid.
106 Hohenlohe, diary entry for 3 January 1896, Hohenlohe, *Denkwürdigkeiten der Reichskanzlerzeit*, p. 151; also above, p. 727, and below, pp. 799–807.

107 Kaiser Wilhelm II, marginal notes on reports from Cape Town, cited in Jochen Laufer, 'Die deutsche Südafrikapolitik 1890–1898 im Spannungsfeld zwischen deutsch–englischen Beziehungen, Wirtschaftsinteressen und Expansionsforderungen in der bürgerlichen Öffentlichkeit', diss., Humboldt-Universität Berlin 1986, p. 213.

108 Waldersee, diary entry for 5 January 1896, GStA Berlin, Waldersee Papers; cf. Meisner, II, p. 363.

109 Eyre Crowe to his mother, 9 January 1896, Crowe Papers, Bodleian Library, Oxford.

110 Lascelles to Salisbury, 4 January 1896, quoted in van't Padje, 'Anglo-German Imperialist Rivalry', pp. 89–90.

111 Kaiser Wilhelm II to Tsar Nicholas II, 2 January 1896, Goetz, *Briefe Wilhelm II. an den Zaren*, pp. 300–1.

112 Canis, *Von Bismarck zur Weltpolitik*, p. 179.

113 Hohenlohe, diary entry for 3 January 1896, Hohenlohe, *Denkwürdigkeiten der Reichskanzlerzeit*, p. 151.

114 Marschall, diary entry for 3 January 1896, Marschall Papers, Schloss Neuershausen. Cf. Alexander Hohenlohe's retrospective account of February 1914 in Hohenlohe, *Denkwürdigkeiten der Reichskanzlerzeit*, pp. 612–13.

115 Kaiser Wilhelm II to President Krüger, telegram, 3 January 1896, *Große Politik*, XI, pp. 31–2.

116 See above all *Große Politik*, I, pp. 287–9; Eyck, *Das Persönliche Regiment*, pp. 131ff.; Michael Balfour, *The Kaiser and His Times*, London 1975, pp. 193–4; Robert K. Massie, *Dreadnought. Britain, Germany and the Coming of the Great War*, New York 1991, pp. 222–5; Cecil, *Wilhelm II*, I, pp. 285–8.

117 Empress Frederick to Queen Victoria, 4 January 1896, cited in Sir Sidney Lee, *King Edward VII: A Biography*, 2 vols., London 1925–7, I, p. 727; Lascelles to Salisbury, 7 January 1896, van't Padje, 'Anglo-German Imperialist Rivalry', pp. 94ff.; Spitzemberg, diary entry for 5 January 1896, *Tagebuch*, pp. 340–1; D. Chapman-Huston, ed., *The Private Diaries of Princess Daisy of Pless, 1873–1914*, London 1950, p. 50; Lerchenfeld to Crailsheim, 7 January 1896, cited in Ekkehard-Teja P. W. Wilke, *Political Decadence in Imperial Germany. Personnel–Political Aspects of the German Government Crisis 1894–1897*, Urbana, Chicago, London 1976, p. 168; Hugo Graf Lerchenfeld-Koefering, *Erinnerungen und Denkwürdigkeiten*, Berlin 1935, pp. 373 and 385; Baron von Eckardstein, *Ten Years at the Court of St James, 1895–1905*, London 1921, pp. 84–6.

118 See Laufer, 'Deutsche Südafrikapolitik', pp. 98ff.; Seligmann, *Rivalry*, pp. 92ff.; Canis, *Von Bismarck zur Weltpolitik*, pp. 178ff.; Clark, *Kaiser Wilhelm II*, p. 133.

119 Marschall, diary entry for 30 December 1895, Marschall Papers, Schloss Neuershausen. See Holstein to Hatzfeldt, 1 January 1896, in G. Ebel, *Botschafter Paul Graf von Hatzfeldt. Nachgelassene Papiere 1838–1901*, 2 vols., Boppard am Rhein 1976, pp. 1065–6.

120 Marschall, diary entry for 31 December 1895, Marschall Papers, Schloss Neuershausen.

121 Marschall, diary entry for 1 January 1896, ibid.

122 Marschall, diary entry for 2 January 1896, ibid.

123 Fesser, 'Ohrfeige für England', p. 28.

124 Marschall, diary entry for 2 January 1896, Marschall Papers, Schloss Neuershausen.

125 Ibid.

126 Canis, *Von Bismarck zur Weltpolitik*, p. 181.

127 Marschall, diary entries for 3 and 4 January 1896, Marschall Papers, Schloss Neuershausen. See Dönhoff's report of 6 January 1896 on the unanimous praise for the Krüger telegram in the Saxon press, PA AA. See Evans, *Kneipengespräche*, pp. 346–7.

128 Marschall, diary entries for 4–11 January 1896, Marschall Papers, Schloss Neuershausen.

129 See below, ch. 28.

130 Eyre Crowe to his mother, 9 January 1896, Crowe Papers, Bodleian Library, Oxford.
131 Fesser, 'Ohrfeige für England', p. 28.
132 Swaine, report of January 1896, RA I60/149.
133 Hohenlohe, diary entry for 9 February 1896, Hohenlohe, *Denkwürdigkeiten der Reichskanzlerzeit*, p. 169. See Monts to Bülow, 24 February 189[6], in Bülow, *Memoirs*, I, pp. 32ff.; Hatzfeldt to Holstein, 28 April 1896, *Holstein Papers*, III, No. 543.
134 Holstein to Bülow, 22 April 1896, *Holstein Papers*, III, No. 541. See Seligmann, *Rivalry*, pp. 128–9.
135 Münster to Holstein, 13 January 1896, *Holstein Papers*, III, No. 521.
136 Kaiser Wilhelm II, marginal comments on Münster's report from Paris of 16 January 1896, PA AA, Asservat No. 4, printed in *Große Politik*, XI, p. 82. See the Kaiser's angry comment 'all bosh' on Hatzfeldt's report from London of 17 January 1896, cited in *Holstein Papers*, III, p. 586.
137 Waldersee, diary entry for 7 January 1896, GStA Berlin, Waldersee Papers; omitted from Meisner, II, p. 364.
138 Ibid.
139 Waldersee, diary entry for 10 January 1896, GStA Berlin, Waldersee Papers; cf. Meisner, II, p. 364.
140 Ibid.
141 Waldersee, diary entry for 11 January 1896, GStA Berlin, Waldersee Papers; omitted for the most part from Meisner, II, p. 364.
142 Waldersee, diary entry for 12 January 1896, GStA Berlin, Waldersee Papers; cf. Meisner, II, pp. 364–5.
143 Waldersee, diary entry for 10 January 1896, GStA Berlin, Waldersee Papers; cf. Meisner, II, p. 364. Cf. Tirpitz to Stosch, 13 February 1896, printed in Tirpitz, *Erinnerungen*, pp. 54–6.
144 RA QVJ, 3 and 5 January 1896; Queen Victoria to Kaiser Wilhelm II, 5 January 1896, RA O45/169, in Buckle, *Letters of Queen Victoria*, III, pp. 7ff. Cf. the slightly different version in RA O45/55.
145 Knollys to Bigge, 4 January 1896, in Buckle, *Letters of Queen Victoria*, III, pp. 7–8.
146 Hohenlohe, diary entry for 9 February 1896, Hohenlohe, *Denkwürdigkeiten der Reichskanzlerzeit*, p. 169.
147 Empress Frederick to Queen Victoria, 4 and 11 January 1896, AdHH Schloss Fasanerie. See also Münster to Holstein, 13 January 1896, *Holstein Papers*, III, No. 521.
148 According to Anton Monts, the Kaiser's 'private' letter was not shown to the officials in the Wilhelmstrasse. See Bülow, *Memoirs*, I, p. 32, and below, p. 794. In actual fact the files of the Auswärtiges Amt contain Wilhelm II's original draft with numerous alterations in his own hand. The version printed in Hohenlohe's *Denkwürdigkeiten der Reichskanzlerzeit* (pp. 154–6) is based on a copy made by the Chancellor of this improved text. The letter eventually sent to Windsor differs again in some of its wording from that version. Cf. Cecil, *Wilhelm II*, I, p. 288.
149 Kaiser Wilhelm II to Queen Victoria, 8 January 1896, RA Z500/5. The Queen sent copies of Wilhelm's letter and of her reply to her eldest daughter: Empress Frederick to Queen Victoria, 18 January 1896, AdHH Schloss Fasanerie.
150 See Cecil, *Wilhelm II*, I, p. 288.
151 Kaiser Wilhelm II to Queen Victoria, Queen Victoria to Kaiser Wilhelm II, 23 February 1896, RA I60/146–7; Swaine to Bigge, 2 March 1896, RA I60/151.
152 Empress Frederick to Queen Victoria, 4 February 1896, AdHH Schloss Fasanerie.
153 Waldersee, diary entry for 4 February 1896, GStA Berlin, Waldersee Papers; cf. Meisner, II, p. 367. Cf. Holstein to Hohenlohe, 8 March 1896, Hohenlohe, *Denkwürdigkeiten der Reichskanzlerzeit*, p. 192.

154 Monts to Bülow, 24 February 189[6], in Bülow, *Denkwürdigkeiten*, I, pp. 34. Cf. Bülow, *Memoirs*, I, p. 32.

155 Hohenlohe, diary for 9 February 1896, Hohenlohe, *Denkwürdigkeiten der Reichskanzlerzeit*, p. 169. See Monts to Bülow, 24 February 189[6], in Bülow, *Memoirs*, I, pp. 32ff.; Hatzfeldt to Holstein, 28 April 1896, *Holstein Papers*, III, No. 543.

156 Eulenburg's notes of 12/13 October 1895, *Eulenburgs Korrespondenz*, III, No. 1145. Both the Kaiserin and Senden-Bibran were outraged at the Kaiser's decision to rent a yacht in England.

157 Waldersee, diary entry for 3 March 1896, GStA Berlin, Waldersee Papers; omitted from Meisner, II, pp. 368–9. In January 1896, Waldersee recorded that the Kaiser had now decided neither to visit Cowes in the summer nor to rent an English yacht for his Mediterranean cruise in the spring, saying: 'With that I am giving up the one thing that I really enjoy' (Waldersee, diary entry for 21 January 1896, GStA Berlin, Waldersee Papers; cf. Meisner, II, pp. 365–6). The Kaiser and Kaiserin began their Mediterranean cruise – in a German yacht – on 28 March 1896.

158 Empress Frederick to Queen Victoria, 29 February 1896, AdHH Schloss Fasanerie, Empress Frederick to Kaiser Wilhelm II, 13 March 1896, GStA Berlin, BPHA Rep. 52T No. 13.

159 Kaiser Wilhelm II to Empress Frederick, 14 March 1896, AdHH Schloss Fasanerie.

160 Kaiser Wilhelm II to his mother, undated [March 1896] and 18 March 1896, ibid.; Hohenlohe, diary entry for 16 March 1896, Hohenlohe, *Denkwürdigkeiten der Reichskanzlerzeit*, p. 199.

161 Empress Frederick to Queen Victoria, 19, 24, 26 and 28 March 1896, AdHH Schloss Fasanerie.

162 Empress Frederick to Queen Victoria, 12 April 1896, ibid.

163 Waldersee, diary entry for 21 January 1896, GStA Berlin, Waldersee Papers; cf. Meisner, II, pp. 365–6.

164 Hohenlohe, diary entry for 5 March 1896, Hohenlohe, *Denkwürdigkeiten der Reichskanzlerzeit*, p. 187.

165 Hohenlohe to Hatzfeldt, 4 March 1896, *Große Politik*, XI, No. 2770.

166 Marschall, note of 4 March 1896, ibid., No. 2771.

167 Hohenlohe, diary entry for 7 March 1896, Hohenlohe, *Denkwürdigkeiten der Reichskanzlerzeit*, pp. 191–2; also Hohenlohe to Holstein, 8 March 1896, ibid., pp. 192–3.

168 Kaiser Wilhelm II, marginal notes on Hatzfeldt's report from London of 12 March 1896, *Große Politik*, XI, p. 241; Kiderlen-Wächter to Holstein, 25 March 1896, *Holstein Papers*, III, No. 537. The editors of the *Große Politik der Europäischen Kabinette* were also convinced that the British Dongola expedition could be 'traced back directly to Wilhelm II': *Große Politik*, XI, p. 235.

169 Kaiser Wilhelm II, marginal notes on Radolin's report from St Petersburg of 21 March 1896, *Große Politik*, XI, p. 168.

170 Hohenlohe, diary entry for 16 March 1896 with note, Hohenlohe, *Denkwürdigkeiten der Reichskanzlerzeit*, p. 199. For Salisbury's motives, see *Große Politik*, XI, No. 2698ff., and Holstein to Radolin, 22 March 1896, *Holstein Papers*, III, No. 536.

171 Kaiser Wilhelm II, marginal notes on Bülow's report from Rome of 13 May 1896, *Große Politik*, XI, p. 253.

172 The part played by the Kaiser in Germany's decision to build a fleet of battleships is fully explored in chapter 32.

173 Kiderlen-Wächter to Holstein, 25 March 1896, *Holstein Papers*, III, No. 537.

174 Hatzfeldt to Holstein, 15 March 1896, ibid., No. 532.

175 Marschall's notes of 13 March 1896, *Große Politik*, XI, No. 2779.

27 ENDGAME: THE BREAKTHROUGH TO DECISIVE PERSONAL POWER

1 Marschall, diary entry for 25 January 1896, cited in *Eulenburgs Korrespondenz*, III, pp. 1639–40.
2 Eulenburg to Kaiser Wilhelm II, 27 February 1896, ibid., No. 1192.
3 Eulenburg to Bülow, 13 March 1896, ibid., No. 1198. See Bülow's reply of 20 March 1896, ibid., No. 1201.
4 Holstein to Eulenburg, 1 May 1896, ibid., No. 1213.
5 Holstein to Eulenburg, 6 May 1896, ibid., No. 1219.
6 Holstein to Eulenburg, 5 May 1896, *Holstein Papers*, III, No. 546.
7 Hohenlohe to Kaiser Wilhelm II, 20 April 1896, Hohenlohe, *Denkwürdigkeiten der Reichskanzlerzeit*, pp. 213–14.
8 Kaiser Wilhelm II, marginal notes of 21 April 1896, GStA Berlin, Zivilkabinett 2.2.1. No. 13162/1.
9 Princess Viktoria of Schaumburg-Lippe to Kaiser Wilhelm II, 21 July 1897, GStA Berlin, 2.2.1. No. 13162/1; Hereditary Princess Charlotte of Saxe-Meiningen to Ellen Freifrau von Heldburg, 14 April 1895, Thüringisches Staatsarchiv Meiningen, IIA 342. For Charlotte's intrigues in the Lippe succession dispute, see Röhl, *Kaiser and His Court*, p. 90.
10 Kaiser Wilhelm II to his mother, 5 April 1895, AdHH Schloss Fasanerie.
11 Hohenlohe, diary entry for 22 April 1896, Hohenlohe, *Denkwürdigkeiten der Reichskanzlerzeit*, p. 214; Hohenlohe to Lucanus, 22 April 1896, GStA Berlin, Zivilkabinett 2.2.1. No. 13162/1.
12 Hohenlohe to Kaiser Wilhelm II, 22 April 1896, Hohenlohe, *Denkwürdigkeiten der Reichskanzlerzeit*, p. 214.
13 Kaiser Wilhelm II to Hohenlohe, 23 April 1896, ibid., p. 215.
14 Holstein to Eulenburg, 5 May 1896, *Holstein Papers*, III, No. 546.
15 Hohenlohe to Kaiser Wilhelm II, 23 April 1896, GStA Berlin, Zivilkabinett 2.2.1. No. 13162/1; printed in part in Hohenlohe, *Denkwürdigkeiten der Reichskanzlerzeit*, pp. 215–16.
16 Hohenlohe, diary entry for 4 May 1896, Hohenlohe, *Denkwürdigkeiten der Reichskanzlerzeit*, pp. 215–16.
17 Holstein to Eulenburg, 5 May 1896, *Holstein Papers*, III, No. 546.
18 Ibid.
19 Kaiser Wilhelm II to Lucanus, 16 July 1896, GStA Berlin, Rep. 89, No. 13162/1.
20 The Lippe succession crisis would preoccupy the Kaiser for many months to come and would lead to further disputes with the Chancellor and the King of Saxony. See the documents and letters in GStA Berlin, 2.2.10. No. 2796; GStA Berlin, 2.2.1. No. 13162/1; Hohenlohe, *Denkwürdigkeiten der Reichskanzlerzeit*, pp. 390 and 407–8.
21 See *Young Wilhelm*, pp. 737–40.
22 See Karl Möckl, *Die Prinzregentenzeit. Gesellschaft und Politik während der Ära des Prinzregenten Luitpold in Bayern*, Munich 1972, pp. 393ff. See Prince Ernst zu Hohenlohe-Langenburg to his father, 13 June 1896, Hohenlohe-Zentralarchiv Neuenstein, Hermann Hohenlohe-Langenburg Papers, Bü. 62.
23 Marschall to Hohenlohe, 18 June 1896, Hohenlohe, *Denkwürdigkeiten der Reichskanzlerzeit*, pp. 236–7.
24 Hohenlohe to Kaiser Wilhelm II, 23 June 1896, ibid., pp. 237–8.
25 Eulenburg to Holstein, 5 July 1896, Enclosure I, *Holstein Papers*, III, No. 554.
26 Dieter Albrecht, ed., Hugo Graf Lerchenfeld-Kœfering, *Kaiser Wilhelm II. als Persönlichkeit und Herrscher*, Regensburger Historische Forschungen, Vol. 11, Kallmünz1985, pp. 29ff. For the Kaiser's mood in Kiel in June 1896 see below, p. 828.
27 Kaiser Wilhelm II to Hohenlohe, telegram, 30 June 1896, Hohenlohe, *Denkwürdigkeiten der Reichskanzlerzeit*, pp. 238–40.

28 Lerchenfeld-Koefering, *Kaiser Wilhelm II. als Persönlichkeit und Herrscher*, p. 31.
29 Eulenburg to Holstein, 5 July 1896, Enclosure I, *Holstein Papers*, III, No. 554.
30 See Hull, *Entourage*, pp. 216ff.
31 Waldersee, diary entry for 5 January 1896, GStA Berlin, Waldersee Papers; cf. Meisner, II, p. 363.
32 Waldersee, diary entry for 9 December 1895, GStA Berlin, Waldersee Papers; omitted from Meisner, II, p. 362.
33 Waldersee, diary entry for 5 January 1896, GStA Berlin, Waldersee Papers; cf. Meisner, II, p. 363.
34 Ibid.
35 Hohenlohe, diary entry for 3 January 1896, Hohenlohe, *Denkwürdigkeiten der Reichskanzlerzeit*, p. 151; also Holstein to Eulenburg, 2 January 1896, *Holstein Papers*, III, No. 1185.
36 Waldersee, diary entry for 5 January 1896, GStA Berlin, Waldersee Papers; cf. Meisner, II, p. 363.
37 Hohenlohe, diary entry for 3 January 1896, Hohenlohe, *Denkwürdigkeiten der Reichskanzlerzeit*, p. 151; also Holstein to Eulenburg, 2 January 1896, *Eulenburgs Korrespondenz*, III, No. 1185.
38 Waldersee, diary entry for 5 January 1896, GStA Berlin, Waldersee Papers; cf. Meisner, II, p. 363.
39 Waldersee, diary entry for 3 March 1896, GStA Berlin, Waldersee Papers; omitted from Meisner, II, pp. 368–9.
40 Eulenburg to Bülow, 13 March 1896, *Eulenburgs Korrespondenz*, III, No. 1198.
41 Kaiser Wilhelm II to Hohenlohe, 14 March 1896, Hohenlohe, *Denkwürdigkeiten der Reichskanzlerzeit*, pp. 198–9.
42 Hohenlohe, diary entry for 16 March 1896, ibid., p. 199.
43 Holstein to Eulenburg, 19 June 1896, *Eulenburgs Korrespondenz*, III, No. 1234.
44 Holstein to Eulenburg, 17 February 1895, ibid., No. 1089.
45 Holstein to Eulenburg, 17 December 1895, ibid., No. 1171. Several passages of this letter are printed in Haller, *Eulenburg*, pp. 181–2; also Holstein to Bülow, 18 December 1895, *Holstein Papers*, III, No. 511. See Eulenburg to Holstein, 19 December 1895, ibid., No. 513; Eulenburg to Hohenlohe, 19 December 1895, Hohenlohe, *Denkwürdigkeiten der Reichskanzlerzeit*, pp. 143–4.
46 Holstein to Eulenburg, 9 February 1896, *Holstein Papers*, III, No. 528.
47 Eulenburg to Holstein, 14 February 1896, Haller, *Eulenburg*, pp. 194–5; Eulenburg to Holstein, 19 February 1896, *Holstein Papers*, III, No. 529. See Otto Hammann, *Bilder aus der letzten Kaiserzeit*, Berlin 1922, pp. 14–15.
48 Eulenburg to Bülow, 12 November 1895, *Eulenburgs Korrespondenz*, III, No. 1154.
49 Hohenlohe, diary entry for 1 February 1896, Hohenlohe, *Denkwürdigkeiten der Reichskanzlerzeit*, p. 164.
50 Hohenlohe, note of 28 February 1896, ibid., pp. 181–2; also Hohenlohe's comments of 3 May 1896, ibid., pp. 216–17.
51 Hohenlohe, diary entry for 2 March 1896, ibid., p. 186.
52 Hohenlohe, diary entry for 15 June 1896, ibid., p. 235.
53 Eulenburg to Bülow, 13 March 1896, *Eulenburgs Korrespondenz*, III, No. 1198. See Bülow's reply of 20 March 1896, ibid., No. 1201.
54 Eulenburg to Bülow, 16 March 1896, ibid., No. 1199.
55 Eulenburg to Bülow, 29 March 1896, ibid., No. 1203; Bülow to Eulenburg, 1 April 1896, ibid., No. 1205.
56 Eulenburg to Bülow, 29 April 1896, ibid., No. 1208. A facsimile of Wilhelm's telegram to Eulenburg of 29 April 1896 is to be found in the Conrad Haussmann Papers, Stuttgart.

See also Eulenburg to Holstein, 30 April 1896, *Holstein Papers*, III, No. 544; Eulenburg to Hohenlohe, 1 May 1896, Hohenlohe, *Denkwürdigkeiten der Reichskanzlerzeit*, p. 216.

57 Eulenburg to Kaiser Wilhelm II, 29 April 1896, *Eulenburgs Korrespondenz*, III, No. 1209.

58 Holstein to Eulenburg, 1 May 1896, ibid., No. 1213 (Eulenburg sent this letter to the Kaiser from Pest); Eulenburg to Kaiser Wilhelm II, 3 May 1896, ibid., No. 1215.

59 Marschall to Eulenburg, 30 April 1896, ibid., No. 1211.

60 Cited in Holstein to Eulenburg, 1 May 1896, ibid., No. 1212.

61 Ibid.

62 See Alexander Hohenlohe to Völderndorff, 20 February 1896, Hohenlohe, *Denkwürdigkeiten der Reichskanzlerzeit*, pp. 175–6.

63 Hohenlohe, diary entry for 2 March 1896, ibid., p. 186.

64 Hohenlohe, note of 28 February 1896, ibid., pp. 181–2; Hohenlohe, diary entry for 2 March 1896, ibid., p. 186.

65 Hohenlohe, diary entry for 15 June 1896, ibid., p. 235.

66 Marschall, diary entry for 3 May 1896, cited in *Eulenburgs Korrespondenz*, III, p. 1673; Holstein to Eulenburg, telegram, 3 May 1896, cited ibid., p. 1677.

67 Marschall, diary entry for 4 May 1896, cited ibid., p. 1674.

68 Kaiser Wilhelm II to Eulenburg, telegram, 4 May 1896, ibid., No. 1217.

69 Alexander Hohenlohe to Eulenburg, 11 May 1896, Hohenlohe, *Denkwürdigkeiten der Reichskanzlerzeit*, pp. 220–3.

70 Waldersee, diary entry for 3 May 1896, GStA Berlin, Waldersee Papers; omitted from Meisner, II, p. 369.

71 Eulenburg to Kaiser Wilhelm II, 5 May 1896, *Eulenburgs Korrespondenz*, III, No. 1218.

72 Eulenburg to Kaiser Wilhelm II, 7 May 1896, ibid., No. 1220.

73 Marschall to Eulenburg, telegram, 9 May 1896, ibid., No. 1221; also Alexander Hohenlohe to Eulenburg, 11 May 1896, Hohenlohe, *Denkwürdigkeiten der Reichskanzlerzeit*, pp. 220–3.

74 For the Kaiser's shooting trips to Prökelwitz and Rominten, see Andreas Gautschi, *Wilhelm II. und das Waidwerk. Jagen und Jagden des letzten Deutschen Kaisers. Eine Bilanz*, Hanstedt 2000, pp. 30ff. and 44ff.

75 Eulenburg to Bülow, 13 May 1896, *Eulenburgs Korrespondenz*, III, No. 1222.

76 Eulenburg to Bülow, 16 May 1896, ibid., No. 1223.

77 Eulenburg to Lucanus, 16 May 1896, ibid., No. 1224.

78 Holstein to Eulenburg, telegram, 16 May 1896, ibid., No. 1225.

79 Kaiser Wilhelm II to Hohenlohe, 16 May 1896, Hohenlohe, *Denkwürdigkeiten der Reichskanzlerzeit*, pp. 225–6.

80 Eulenburg to Hohenlohe, 17 May 1896, ibid., p. 227.

81 Kaiserin Auguste Viktoria to Hohenlohe, 17 May 1896, ibid., pp. 226–7.

82 Hohenlohe to Kaiser Wilhelm II, 16 May 1896, ibid., pp. 224–5. See also the telegram of 17 May 1896 drafted for the Chancellor by Marschall, Holstein, Wilmowski and Alexander Hohenlohe, ibid., p. 227. See Marschall, diary entry for 16 May 1896, *Eulenburgs Korrespondenz*, III, p. 1684.

83 Eulenburg and Kaiser Wilhelm II to Hohenlohe, telegram, 17 May 1896, Hohenlohe, *Denkwürdigkeiten der Reichskanzlerzeit*, p. 228.

84 Eulenburg to Holstein, 17 May 1896, *Holstein Papers*, III, No. 548.

85 Marschall, diary entry for 17 May 1896, cited in *Eulenburgs Korrespondenz*, III, p. 1684.

86 Marschall, diary entry for 18 May 1896, cited ibid.

87 Holstein to Eulenburg, 19 May 1896, cited ibid.

88 Hohenlohe to Kaiser Wilhelm II, telegram, 18 May 1896, Hohenlohe, *Denkwürdigkeiten der Reichskanzlerzeit*, pp. 228–9.

89 Eulenburg to Hohenlohe, telegram, 17 May 1896, ibid., p. 229.

90 Eulenburg to Hohenlohe, 18 May 1896, ibid., pp. 229–31.
91 Hohenlohe to Eulenburg, telegram, 19 May 1896, ibid., p. 231; Marschall, diary entry for 19 May 1896, cited in *Eulenburgs Korrespondenz*, III, p. 1684.
92 Eulenburg to Hohenlohe, 19 May 1896, Hohenlohe, *Denkwürdigkeiten der Reichskanzlerzeit*, pp. 231–2.
93 Eulenburg to Alexander Hohenlohe, 19 May 1896, ibid., p. 232; also Eulenburg to Holstein, 19 May 1896, *Holstein Papers*, III, No. 549.
94 Eulenburg to Hohenlohe, 18 May 1896, Hohenlohe, *Denkwürdigkeiten der Reichskanzlerzeit*, pp. 229–31.
95 Eulenburg to Hohenlohe, 5 July 1896, *Eulenburgs Korrespondenz*, III, No. 1235.
96 Lindenau to Holstein, 5 August 1896, *Holstein Papers*, III, No. 580.
97 Heinrich VII Prince Reuss to Count Carl von Wedel, 18 July 1896, Wedel Papers, Frankfurt a.M.
98 Eulenburg to Holstein, 5 July 1896, Enclosure II, *Holstein Papers*, III, No. 554.
99 Hohenlohe to Eulenburg, 20 May 1896, Hohenlohe, *Denkwürdigkeiten der Reichskanzlerzeit*, p. 233.
100 Waldersee, diary entry for 8 June 1896, GStA Berlin, Waldersee Papers; cf. Meisner, II, pp. 370–1.
101 Alexander Hohenlohe to his father, 17 June 1896, Hohenlohe, *Denkwürdigkeiten der Reichskanzlerzeit*, pp. 235–6.
102 Above, p. 374.
103 Waldersee, diary entry for 21 June 1896, GStA Berlin, Waldersee Papers; omitted from Meisner, II, p. 371.
104 Eulenburg to Bülow, 8 June 1896, *Eulenburgs Korrespondenz*, III, No. 1233. Cf. Eulenburg to Kaiser Wilhelm II, 20 January 1896, ibid., No. 1190.
105 Eulenburg to Bülow, 24 May 1896, ibid., No. 1228.
106 Bülow to Eulenburg, 16 July 1896, ibid., No. 1242.
107 Bülow to Eulenburg, 23 July 1896, ibid., No. 1245. Eulenburg read Bülow's letter to the Kaiser on 1 August: ibid., p. 1713.
108 Eulenburg to Bülow, 7 and 23 July 1896, ibid., Nos. 1237 and 1246.
109 Eulenburg to Bülow, 23 July 1896, ibid., No. 1246.
110 Eulenburg to Hohenlohe, 5 and 6 July 1896, ibid., Nos. 1235–6.
111 Eulenburg to Holstein, 12 July 1896, *Holstein Papers*, III, No. 558; also Eulenburg to Holstein, 3 August 1896, *Eulenburgs Korrespondenz*, III, No. 1248; Eulenburg to Holstein, telegram, 4 August 1896, *Holstein Papers*, III, No. 576. See Hohenlohe to Holstein, 5 August 1896, ibid., No. 579; printed in part in Hohenlohe, *Denkwürdigkeiten der Reichskanzlerzeit*, pp. 250–1.
112 Eulenburg to Holstein, 26 July 1896, *Holstein Papers*, III, No. 570.
113 Eulenburg to Bülow, 7 July 1896, *Eulenburgs Korrespondenz*, III, No. 1237.
114 Hohenlohe to Alexander Hohenlohe, 18 July 1896, Hohenlohe, *Denkwürdigkeiten der Reichskanzlerzeit*, p. 242. See Hohenlohe, diary entry for 1 July 1896, ibid., p. 240.
115 Hohenlohe to Eulenburg, 16 July 1896, *Eulenburgs Korrespondenz*, III, No. 1240.
116 Hohenlohe to Eulenburg, 16 July 1896, ibid., No. 1241.
117 See Holstein to Eulenburg, 22 December 1895, ibid., No. 1175; Holstein to Eulenburg, 9 February 1896, *Holstein Papers*, III, No. 528.
118 Holstein to Eulenburg, 10 July 1896, *Holstein Papers*, III, p. 622.
119 Holstein to Eulenburg, 18 July 1896, ibid., No. 564 Enclosure.
120 Holstein to Eulenburg, 14 July 1896, ibid., No. 559; also Holstein to Kiderlen-Wächter, 10 July 1896, ibid., No. 557.
121 Quoted in Bülow to Eulenburg, 23 July 1896, *Eulenburgs Korrespondenz*, III, No. 1245.

122 Eulenburg, note of 21 July 1896, ibid., No. 1243; also Bülow to Eulenburg, 23 July 1896, ibid., No. 1245.
123 Eulenburg, note of 15 July 1896, *Holstein Papers*, III, No. 561.
124 Eulenburg, note of 21 July 1896, *Eulenburgs Korrespondenz*, III, No. 1243; also Bülow to Eulenburg, 23 July 1896, ibid., No. 1245.
125 Eulenburg to Bülow, 23 July 1896, ibid., No. 1246.
126 Ibid.
127 Eulenburg to Hohenlohe, 23 July 1896, ibid., No. 1244.
128 Eulenburg to Bülow, 23 July 1896, ibid., No. 1246.
129 For further details, see *Young Wilhelm*, pp. 319–20.
130 Eulenburg, note of 1 August 1896, *Eulenburgs Korrespondenz*, III, No. 1247.
131 Eulenburg, diary entry for 1 August 1896, ibid., III, p. 1720.
132 Holstein to Lindenau, 24 July 1896, *Holstein Papers*, III, No. 567.
133 Kaiser Wilhelm II to Hohenlohe, telegram, 26 July 1896, cited ibid., p. 633.
134 Holstein to Lindenau, 24 and 27 July 1896, ibid., Nos. 567 and 572; also Holstein to Eulenburg, telegram, 4 August 1896, ibid., No. 577; Monts to Eulenburg, 7 August 1896, *Eulenburgs Korrespondenz*, III, No. 1254.
135 Holstein to Lindenau, 29 July 1896, *Holstein Papers*, III, No. 573.
136 Holstein to Lindenau, 1 August 1896, ibid., No. 575.
137 Hohenlohe to Holstein, 27 July 1896, ibid., No. 571.
138 Hohenlohe, undated note, Hohenlohe, *Denkwürdigkeiten der Reichskanzlerzeit*, p. 248.
139 Hohenlohe to Kaiser Wilhelm II, cipher telegram, 4 August 1896, PA AA, Preußen 1 No. 1d, Bd I.
140 Eulenburg, diary entries for 1–3 August 1896, *Eulenburgs Korrespondenz*, III, pp. 1720–1.
141 Eulenburg, note of 4 August 1896, ibid., No. 1249; Hohenlohe, diary entry for 4 August 1896, Hohenlohe, *Denkwürdigkeiten der Reichskanzlerzeit*, pp. 249–50; also Hohenlohe to Holstein, 5 August 1896, *Holstein Papers*, III, No. 579. Cf. Hohenlohe, *Denkwürdigkeiten der Reichskanzlerzeit*, pp. 253–4.
142 Eulenburg to Kaiser Wilhelm II, telegram and letter of 4 August 1896, *Eulenburgs Korrespondenz*, III, Nos. 1251–2.
143 Eulenburg to Kaiser Wilhelm II, 5–6 August 1896, ibid., No. 1253.
144 Lindenau to Holstein, 5 August 1896, *Holstein Papers*, III, No. 580.
145 Hohenlohe, diary entry for 8 August 1896, Hohenlohe, *Denkwürdigkeiten der Reichskanzlerzeit*, pp. 251–2; also the undated note, ibid., p. 253. See Kiderlen-Wächter to Eulenburg, 10 August 1896, *Holstein Papers*, III, No. 581.
146 August Eulenburg to Philipp Eulenburg, 11 August 1896, *Eulenburgs Korrespondenz*, III, No. 1255.
147 Kaiser Wilhelm II to Eulenburg, 14 August 1896, ibid., No. 1259.
148 Eulenburg to Kaiser Wilhelm II, 12 and 18 August 1896, ibid., Nos. 1256 and 1261.
149 Kaiser Wilhelm II to Eulenburg, telegram, 13 August 1896, ibid., No. 1257.
150 Kaiser Wilhelm II to Eulenburg, 14 August 1896, ibid., No. 1259.
151 Empress Frederick to Queen Victoria, 17 August and 12 September 1896, AdHH Schloss Fasanerie. Marschall described Gossler as 'wholly incapable, especially in parliament': diary entry for 14 August 1896, *Eulenburgs Korrespondenz*, III, p. 1732.
152 Hohenlohe's diary for 24 August 1896, Hohenlohe, *Denkwürdigkeiten der Reichskanzlerzeit*, pp. 256–7.
153 Waldersee, diary entry for 20 August 1896, GStA Berlin, Waldersee Papers; cf. Meisner, II, p. 372.
154 Waldersee, diary entry for 16 August 1896, GStA Berlin, Waldersee Papers; omitted from Meisner, II, p. 372.

155 Waldersee, diary entries for 6 and 20 January 1897, GStA Berlin, Waldersee Papers; cf. Meisner, II, pp. 382ff.
156 Jagemann to Brauer, 8 April 1897, Fuchs, *Großherzog von Baden*, III, No. 1679.
157 Eulenburg to Kaiser Wilhelm II, 12 August 1896, *Eulenburgs Korrespondenz*, III, No. 1256. See the bitterly angry letter that Hohenlohe now wrote to Eulenburg but did not send: Hohenlohe, *Denkwürdigkeiten der Reichskanzlerzeit*, pp. 253–4; also Holstein to Eulenburg, 13 August 1896, *Holstein Papers*, III, No. 583. Marschall wrote in his diary on 14 August: 'Holstein is furious with Philipp Eulenburg' (*Eulenburgs Korrespondenz*, III, p. 1732.).
158 Eulenburg to Kaiser Wilhelm II, 12 and 24 August 1896, *Eulenburgs Korrespondenz*, III, Nos. 1256 and 1262.
159 Eulenburg to Hohenlohe, 24 August 1896, Hohenlohe, *Denkwürdigkeiten der Reichskanzlerzeit*, pp. 254–5.
160 Eulenburg, note of 4 October 1896, *Eulenburgs Korrespondenz*, III, No. 1265.
161 Eulenburg to Kaiser Wilhelm II, 14 and 24 August 1896, ibid., Nos. 1258 and 1262.
162 Hohenlohe to his son Alexander, 7 June 1897, Hohenlohe, *Denkwürdigkeiten der Reichskanzlerzeit*, p. 352.
163 Hohenlohe's diary for 22 May 1897, ibid., pp. 342–3.
164 Hohenlohe's diary for 5 June 1897, ibid., pp. 351–2.
165 Hohenlohe's diary for 3 June 1897, ibid., p. 350.
166 Hohenlohe's diary for 19 June 1897, ibid., pp. 358–9.
167 Hohenlohe to his son Alexander, 29 August 1897, ibid., pp. 379–80.
168 Hohenlohe to Otto Freiherr von Völderndorff-Waradein, undated [summer 1897], ibid., p. 344.
169 Hohenlohe, undated note [March 1897], ibid., p. 311.
170 Hohenlohe, diary entry for 8 March 1897, ibid., p. 312; Hohenlohe to Kaiser Wilhelm II, 9 March 1897, ibid., pp. 312–13.
171 Hohenlohe to Kaiser Wilhelm II, 31 May 1897, Lucanus to Hohenlohe, 1 June 1897, Kaiser Wilhelm II to Hohenlohe, 1 June 1897, ibid., pp. 346–7.
172 See for example Hohenlohe, diary entries for 1 and 22 June and 17 October 1897, ibid., pp. 347, 359 and 392–3.
173 Herbert Bismarck to Bülow, 18 October 1900, printed in facsimile in Bülow, *Memoirs*, I, p. 389.
174 See above, p. 826.
175 Völderndorff to Hohenlohe, 22 June 1897, Hohenlohe, *Denkwürdigkeiten der Reichskanzlerzeit*, pp. 359–60.
176 Dacheux to Hohenlohe, 19 May 1897, ibid., pp. 340–1.
177 Dacheux to Hohenlohe, 4 August 1897, ibid., pp. 374–5.

28 PERSONAL MONARCHY: WILHELM II AT THE SUMMIT OF HIS POWER

1 Holstein to Bülow, 5 March 1897, cited in Röhl, *Germany without Bismarck*, p. 212.
2 Holstein to Eulenburg, 3 March 1897, *Eulenburgs Korrespondenz*, III, No. 1300.
3 August Eulenburg to Philipp Eulenburg, 9 February 1897, ibid., p. 1773.
4 Holstein to Eulenburg, 30 November 1896, *Holstein Papers*, III, No. 587.
5 Hohenlohe, notes of 7 March 1897, Hohenlohe, *Denkwürdigeiten der Reichskanzlerzeit*, pp. 311–12.
6 Lascelles, report of 14 May 1897, RA I61/19.
7 See Dieter Fricke, 'Die Affäre Leckert-Lützow-Tausch und die Regierungskrise von 1897 in Deutschland', *Zeitschrift für Geschichtswissenschaft*, 7 (1960).

8 Eulenburg to Bülow, 26 October 1896, *Eulenburgs Korrespondenz*, III, No. 1268; also Eulenburg, notes of 4 October 1896, ibid., No. 1265; Hohenlohe, diary entries for 25 November 1896 and 22 May 1897, Hohenlohe, *Denkwürdigkeiten der Reichskanzlerzeit*, pp. 279 and 343; Eulenburg to Kaiser Wilhelm II, 8 December 1896, *Eulenburgs Korrespondenz*, III, No. 1283.

9 Waldersee, diary entry for 6 January 1897, GStA Berlin, Waldersee Papers; cf. Meisner, II, pp. 382–3.

10 Kaiser Wilhelm II to Eulenburg, 5 January 1897, *Eulenburgs Korrespondenz*, III, No. 1287.

11 Hohenlohe to his son Alexander, 31 October 1897, Hohenlohe, *Denkwürdigkeiten der Reichskanzlerzeit*, p. 398.

12 Kaiser Wilhelm II to Eulenburg, 5 January 1897, *Eulenburgs Korrespondenz*, III, No. 1287. See Hohenlohe, diary entries for 6 September and 17 October 1897, Hohenlohe, *Denkwürdigkeiten der Reichskanzlerzeit*, pp. 381–2 and 392.

13 Kaiser Wilhelm II to Hohenlohe, 4 January 1897, *Denkwürdigkeiten der Reichskanzlerzeit*, pp. 288–91.

14 Kaiser Wilhelm II to Hohenlohe, 15 November 1896, ibid., pp. 278–9.

15 Kaiser Wilhelm II to Hohenlohe, 7 January 1897, ibid., p. 291. See the Chancellor's reply, ibid., pp. 291–2.

16 Kaiser Wilhelm II, marginal comment on Eulenburg's report from Vienna of 16 August 1897, PA AA, Asservat No. 4.

17 Kaiser Wilhelm II to Hohenlohe, 26 June 1899, Hohenlohe, *Denkwürdigkeiten der Reichskanzlerzeit*, pp. 508–9.

18 Kaiser Wilhelm II to Count Ernst zu Lippe-Biesterfeld, 17 June 1898, ibid., pp. 454–5. See Hohenlohe to Völderndorff, 24 June 1898, as well as his diary entry for 2 August 1898, ibid., pp. 455–6.

19 Hohenlohe to Völderndorff, 26 November 1898, ibid., pp. 471–2.

20 Empress Frederick to Kaiser Wilhelm II, 25 November 1898 and 24 January 1899, GStA Berlin, BPHA Rep. 52T No. 13.

21 See Bülow, *Memoirs*, I, pp. 217–18.

22 Kaiser Wilhelm II to Hinzpeter, July 1899, cited in Schröder, *Tagebuch Kaiser Wilhelms II.*, pp. 328–9; also Prince Heinrich of Prussia to Empress Frederick, 29 August 1899, AdHH Schloss Fasanerie.

23 Kaiser Wilhelm II, speech of 16 June 1898, Penzler, *Die Reden Kaiser Wilhelms II. in den Jahren 1896–1900*, pp. 96–8; Elkind, *German Emperor's Speeches*, p. 226.

24 Kaiser Wilhelm II to Hohenlohe, 21 November 1897, Hohenlohe, *Denkwürdigkeiten der Reichskanzlerzeit*, p. 417. Cf. Kaiser Wilhelm II's marginal notes on Hohenlohe's report of 21 November 1897 in PA AA, Preußen 1 No. 1d, Bd I.

25 Hohenlohe to Kaiser Wilhelm II, 22 November 1897, Hohenlohe, *Denkwürdigkeiten der Reichskanzlerzeit*, p. 417.

26 Kaiser Wilhelm II to Hohenlohe, undated [January 1897], ibid., p. 296; printed in facsimile ibid., pp. 304–5.

27 See below, pp. 849–51. For the real text of the Kaiser's speech, see Kuno Moltke to Eulenburg, 1 March 1897, *Eulenburgs Korrespondenz*, III, No. 1299.

28 Kaiser Wilhelm II, speech of 26 February 1897, Penzler, *Reden Kaiser Wilhelms II. in den Jahren 1896–1900*, pp. 38–41; Elkind, *German Emperor's Speeches*, p. 301.

29 Kaiser Wilhelm II to Prince Heinrich of Prussia, April 1897, from the version in *Berliner Lokal-Anzeiger*, 25 April 1897, PA AA, Preußen 1 No. 1d, Bd I.

30 Hohenlohe, diary entry for 29 April 1897, Hohenlohe, *Denkwürdigkeiten der Reichskanzlerzeit*, pp. 332–3. See Hohenlohe to Marschall, 27 April 1897, PA AA, Preußen 1 No. 1d, Bd I.

31 Kiderlen-Wächter to Hohenlohe, 30 April 1897, Hohenlohe, *Denkwürdigkeiten der Reichskanzlerzeit*, p. 334.
32 Kaiser Wilhelm II to Hohenlohe, 30 April 1897, ibid., p. 333.
33 Hohenlohe, notes, undated [30 April 1897], ibid., p. 335.
34 Kaiser Wilhelm II, speech of 26 February 1897, Penzler, *Reden Kaiser Wilhelms II in den Jahren 1896–1900*, pp. 38–41. See Kuno Moltke to Eulenburg, 1 March 1897, *Eulenburgs Korrespondenz*, III, No. 1299.
35 Bülow, *Memoirs*, I, pp. 38–9. Cf. Bülow to Eulenburg, 17 March 1897, *Eulenburgs Korrespondenz*, III, No. 1307; also Kuno Moltke to Eulenburg, 1 March 1897, ibid., No. 1299.
36 Holstein to Bülow, 5 April 1897, cited in Röhl, *Germany without Bismarck*, p. 212; Spitzemberg, *Tagebuch*, pp. 352–3; Theodor Fontane, *Briefe an Georg Friedländer*, Heidelberg 1954, p. 311.
37 Monts to Holstein, 2 March 1897, *Eulenburgs Korrespondenz*, III, No. 1301.
38 Monts to Eulenburg, 20/21 March 1897, ibid., No. 1309.
39 Julius Bachem to Karl Bachem, 15 March 1897, cited in Röhl, *Germany without Bismarck*, p. 216.
40 Eugen Richter, speech of 18 May 1897, in Eyck, *Das Persönliche Regiment*, pp. 171–2.
41 Waldersee, diary entry for 24 May 1897, GStA Berlin, Waldersee Papers; omitted from Meisner, II, pp. 396–7.
42 Hohenlohe, diary entry for 31 May 1897, Hohenlohe, *Denkwürdigkeiten der Reichskanzlerzeit*, pp. 345–6.
43 Holstein to Bülow, 23 November 1896, cited in Röhl, *Germany without Bismarck*, p. 207. See also Holstein to Eulenburg, 24 November 1896, *Holstein Papers*, III, No. 586; Haller, *Eulenburg*, p. 204.
44 Hohenlohe to Eulenburg, 25 March 1897, Hohenlohe, *Denkwürdigkeiten der Reichskanzlerzeit*, pp. 322–3.
45 Eulenburg to Kaiser Wilhelm II, 8 April 1897, *Eulenburgs Korrespondenz*, III, No. 1313, Enclosure I.
46 Karl Bachem, cited in Röhl, *Germany without Bismarck*, p. 215.
47 Hohenlohe, diary entry for 30 March 1897, Hohenlohe, *Denkwürdigkeiten der Reichskanzlerzeit*, pp. 326–7.
48 See Holstein to Eulenburg, 9 February 1896, *Holstein Papers*, III, No. 528; Holstein to Bülow, 17 February 1897, ibid., IV, No. 605.
49 Hereditary Prince Bernhard of Saxe-Meiningen to Colmar Freiherr von der Goltz, 10 October 1899, BA-MA Freiburg, von der Goltz Papers N737 Zug. 161/95. See also Prince Bernhard's letter to his father, 16 August 1900, Thüringisches Staaatsarchiv Meiningen, HA 341.
50 Waldersee, diary entry for 7 February 1895, GStA Berlin, Waldersee Papers; cf. Meisner, II, p. 338.
51 Waldersee, diary entry for 31 January 1895, GStA Berlin, Waldersee Papers; cf. Meisner, II, p. 338, where the date given is incorrect.
52 Waldersee, diary entry for 17 December 1895, GStA Berlin, Waldersee Papers; cf. Meisner, II, p. 362.
53 Waldersee, diary entry for 28 January 1897, GStA Berlin, Waldersee Papers; omitted from Meisner, II, pp. 385ff.
54 Waldersee, diary entry for 16 February 1897, GStA Berlin, Waldersee Papers; cf. Meisner, II, p. 392.
55 Waldersee, diary entry for 25 November 1896, GStA Berlin, Waldersee Papers; cf. Meisner, II, p. 377.
56 Ibid.

57 Waldersee, diary entry for 29 November 1896, GStA Berlin, Waldersee Papers; printed only in part in Meisner, II, p. 377.
58 Waldersee, diary entry for 2 December 1896, GStA Berlin, Waldersee Papers; printed only in part in Meisner, II, p. 377.
59 Waldersee, diary entry for 28 January 1897, GStA Berlin, Waldersee Papers; omitted from Meisner, II, pp. 385ff.
60 Waldersee's memorandum of 22 January 1897 is printed in Meisner, II, pp. 386–9.
61 Waldersee, diary entry for 28 January 1897, GStA Berlin, Waldersee Papers; omitted from Meisner, II, p. 385.
62 Waldersee's memorandum of 22 January 1897, Meisner, II, pp. 386–9.
63 Waldersee, diary entry for 28 January 1897, GStA Berlin, Waldersee Papers; omitted from Meisner, II, pp. 385ff.
64 Waldersee, diary entry for 31 January 1897, GStA Berlin, Waldersee Papers; cf. Meisner, II, pp. 389ff.
65 Waldersee, diary entry for 19 February 1897, GStA Berlin, Waldersee Papers; omitted from Meisner, II, p. 392. See Waldersee to Verdy, 5 February 1897, GStA Berlin, Waldersee Papers No. 53.
66 Ibid. General Adolf von Deines also approved of Waldersee's memorandum: Waldersee, diary entry for 4 March 1897, GStA Berlin, Waldersee Papers; omitted from Meisner, II, p. 393.
67 Waldersee, diary entry for 1 April 1897, GStA Berlin, Waldersee Papers; omitted from Meisner, II, p. 396.
68 Above, pp. 852–60.
69 Waldersee, diary entry for 26 March 1897, GStA Berlin, Waldersee Papers; Meisner, II, p. 395.
70 Waldersee, diary entry for 1 April 1897, GStA Berlin, Waldersee Papers; omitted from Meisner, II, p. 396.
71 Waldersee, diary entry for 16 March 1897, GStA Berlin, Waldersee Papers; cf. Meisner, II, pp. 393ff.
72 Waldersee, diary entry for 26 April 1897, GStA Berlin, Waldersee Papers; omitted from Meisner, II, p. 396.
73 Waldersee, diary entry for 2 May 1897, GStA Berlin, Waldersee Papers; omitted from Meisner, II, p. 396.
74 Waldersee, diary entry for 16 May 1897, GStA Berlin, Waldersee Papers; omitted from Meisner, II, p. 396.
75 Waldersee, diary entry for 26 April 1897, GStA Berlin, Waldersee Papers; omitted from Meisner, II, p. 396.
76 Waldersee to Gossler, 20 February 1897, Meisner, II, pp. 388–9, note. See Waldersee, diary entry for 19 February 1897, GStA Berlin, Waldersee Papers; omitted from Meisner, II, p. 392.
77 See above, p. 288.
78 Waldersee to Verdy, 5 February 1897, GStA Berlin, Waldersee Papers No. 53.
79 Verdy to Waldersee, 25 February 1897, ibid.
80 Kaiser Wilhelm II, marginal comments on Eulenburg to Kaiser Wilhelm II, 8 April 1897, *Eulenburgs Korrespondenz*, III, No. 1313, Enclosure I.
81 Kaiser Wilhelm II to Eulenburg, 20 April 1897, quoted in Eulenburg to Bülow, 24 April 1897, ibid., No. 1317.
82 Eulenburg to Bülow, 24 April 1897, ibid.; also Eulenburg to Holstein, 11 May 1897, ibid., No. 1318.
83 See above, pp. 832–3.
84 See Marschall's diary entry for 2 April 1897, quoted in *Eulenburgs Korrespondenz*, III, p. 1813.
85 Kaiser Wilhelm II, speech of 26 February 1897, Penzler, *Reden Kaiser Wilhelms II. in den Jahren 1896–1900*, pp. 38–41; Elkind, *German Emperor's Speeches*, p. 300.

86 Kaiser Wilhelm II, speeches in Berlin on 12 November and in Kiel on 24 November 1896, speech in Berlin, 18 October 1897, Penzler, *Reden Kaiser Wilhelms II. in den Jahren 1896–1900*, pp. 35–6. and 67–8.
87 Kaiser Wilhelm II, speech in Berlin, 18 November 1897, ibid., pp. 70–2; Elkind, *German Emperor's Speeches*, p. 226.
88 See e.g. Wilhelm Oncken, ed., *Unser Heldenkaiser. Festschrift zum hundertjährigen Geburtstage Kaiser Wilhelms des Großen*, Berlin 1897; Bernhard von Kugler, *Deutschlands größter Held. Jubel-Ausgabe zur hundertjährigen Gedächtnisfeier des Geburtstags weiland Sr. Majestät Kaiser Wilhelm I.*, Dresden 1893.
89 Kaiser Wilhelm II to Hohenlohe, 6 December 1896, copy in Bundesarchiv Koblenz, Boetticher Papers, 23. See below, pp. 863–4.
90 Eulenburg, notes of 27 October 1896, *Eulenburgs Korrespondenz*, III, No. 1269.
91 Hohenlohe to Kaiser Wilhelm II, 5 December 1896, GStA Berlin, BPHA Rep. 53 Lit. H. No. 2. Cf. Hohenlohe, undated note [early December 1896], Hohenlohe, *Denkwürdigkeiten der Reichskanzlerzeit*, pp. 285–6; also Posadowsky to Boetticher, 20 October 1896, BA Koblenz, Boetticher Papers, 23.
92 Kaiser Wilhelm II, marginal notes on Hohenlohe's report of 5 December 1896, GStA Berlin, BPHA Rep. 53 Lit. H. No. 2.
93 Kaiser Wilhelm II to Hohenlohe, 6 December 1896, telegram, Hohenlohe, *Denkwürdigkeiten der Reichskanzlerzeit*, pp. 285–6.
94 Kaiser Wilhelm II to Hohenlohe, 6 December 1896, letter from Springe, copy in BA Koblenz, Boetticher Papers, 23. See Röhl, *Germany without Bismarck*, pp. 206–7.
95 Kaiser Wilhelm II to Eulenburg, 5 January 1897, *Eulenburgs Korrespondenz*, III, No. 1287.
96 Kuno Moltke to Eulenburg, 1 March 1897, ibid., No. 1299.
97 Hohenlohe, diary entry for 24 March 1897, Hohenlohe, *Denkwürdigkeiten der Reichskanzlerzeit*, pp. 321–2; Jagemann to Brauer, 8 April 1897, Fuchs, *Großherzog von Baden*, III, No. 1679.
98 Kuno Moltke to Eulenburg, 1 March 1897, *Eulenburgs Korrespondenz*, III, No. 1299.
99 Hohenlohe, diary entry for 24 March 1897, Hohenlohe, *Denkwürdigkeiten der Reichskanzlerzeit*, pp. 321–2.
100 Hohenlohe, diary entry for 27 March 1897, ibid., pp. 323–4.
101 Romain Rolland, 'Journal Intime, 22. Januar 1898', in *Cahiers Romain Rolland*, III, *Richard Strauß et Romain Rolland*, Paris 1951, p. 118.
102 Bodo von dem Knesebeck to Eulenburg, 9 December 1893, cited in Lothar Machtan, *Bismarcks Tod und Deutschlands Tränen. Reportage einer Tragödie*, Munich 1998, p. 89. See also ibid., p. 100.
103 Ibid., p. 99.
104 See ibid., pp. 75ff.
105 Wedel, *Zwischen Kaiser und Kanzler*, p. 192. See above, pp. 590–1.
106 See above, pp. 591–2.
107 Kaiser Wilhelm II to Bismarck, 19 September 1893, in Schröder, *Tagebuch Kaiser Wilhelms II.*, pp. 151–2.
108 Above, pp. 594–7 and 703–6.
109 Kaiser Wilhelm II to Kaiser Franz Joseph, 29 October 1896, HHStA Vienna, Kabinettsarchiv Geheimakten 2.
110 Eulenburg, note of 8 November 1896, Bülow, note of 7 April 1897, *Eulenburgs Korrespondenz*, III, Nos. 1272 and 1312.
111 Kaiser Wilhelm II to Hohenlohe, 28 October 1896, Hohenlohe to Kaiser Wilhelm II, 29 October 1896, Hohenlohe, *Denkwürdigkeiten der Reichskanzlerzeit*, pp. 270–1.

112 Eulenburg, notes of November 1896, *Eulenburgs Korrespondenz*, III, No. 1272. Wilhelm's letter to Bismarck of 11 September 1886 on his negotiations with Alexander III is printed in *Young Wilhelm*, pp. 580–3.
113 Bülow, *Memoirs*, IV, pp. 677–8. See also Eulenburg to Kaiser Wilhelm II, 26 November 1896, *Eulenburgs Korrespondenz*, III, No. 1280.
114 Hohenlohe, diary entry for 27 March 1897, Hohenlohe, *Denkwürdigkeiten der Reichskanzlerzeit*, pp. 323–4.; Jagemann to Brauer, 8 April 1897, Fuchs, *Großherzog von Baden*, III, No. 1679.
115 Eisendecher to Tirpitz, 1 September 1897, BA-MA Freiburg, Tirpitz Papers N253/4.
116 Senden to Tirpitz, 19 July 1897, ibid.
117 Tirpitz to Senden, 11 August 1897, ibid.
118 Tirpitz to Prince Heinrich of Prussia, 29 October 1897, ibid.
119 Senden to Tirpitz, 4 August 1897, ibid. See Wilmowski to Hohenlohe, 24 August 1897, Hohenlohe, *Denkwürdigkeiten der Reichskanzlerzeit*, p. 379.
120 Tirpitz to Prince Heinrich of Prussia, 29 October 1897, BA-MA Freiburg, Tirpitz Papers N253/4.
121 Tirpitz, notes on a visit to Friedrichsruh, 22 August 1897, ibid.
122 Senden to Tirpitz, 3 September 1897, ibid.; Hans Blum, *Fürst Bismarck und seine Zeit. Eine Biographie für das deutsche Volk*, Munich 1899, p. 61.
123 Machtan, *Bismarcks Tod*, p. 35.
124 August Eulenburg to Philipp Eulenburg, 27 December 1897, *Eulenburgs Korrespondenz*, III, No. 1352.
125 Tirpitz, *Erinnerungen*, pp. 93–4; Bülow, *Memoirs*, I, p. 204.
126 August Eulenburg to Philipp Eulenburg, 27 December 1897, *Eulenburgs Korrespondenz*, III, No. 1352; also Jagemann's report of 20 December 1897, Fuchs, *Großherzog von Baden*, III, No. 1793.
127 See Philipp Eulenburg's essay 'Bismarck stirbt', in Eulenburg, *Aus 50 Jahren*, pp. 270–80.
128 Maximilian Harden to K. Harden, 29 July 1898, cited in Machtan, *Bismarcks Tod*, p. 62.
129 Kaiser Wilhelm II to Waldersee, 31 July 1898, GStA Berlin, Waldersee Papers, BI No. 42. See Meisner, II, p. 417.
130 Kaiser Wilhelm II to Hohenlohe, 31 July 1898, quoted in Machtan, *Bismarcks Tod*, p. 70.
131 Ibid., p. 132.
132 Blum, *Fürst Bismarck und seine Zeit*, p. 139; Eulenburg, *Aus 50 Jahren*, p. 273.
133 Johannes Penzler, *Fürst Bismarck nach seiner Entlassung. Leben und Politik des Fürsten seit seinem Scheiden aus dem Amte auf Grund aller authentischen Kundgebungen*, 7 vols., Leipzig 1897–8, VII, p. 492.
134 Machtan, *Bismarcks Tod*, p. 133.
135 Ibid., pp. 133–4.
136 Posadowsky to Hohenlohe, 31 July 1898, cited ibid., pp. 134–5; Eulenburg, *Aus 50 Jahren*, p. 273.
137 Arthur von Brauer, *Im Dienste Bismarcks. Persönliche Erinnerungen*, Berlin 1936, pp. 403 and 407.
138 See Duke Ernst Günther of Schleswig-Holstein to Prince Christian of Schleswig-Holstein, 14 August 1898, RA Add A18/G1.
139 Bülow, *Memoirs*, I, pp. 224–5.
140 Hohenlohe, diary entry for 2 August 1898, Hohenlohe, *Denkwürdigkeiten der Reichskanzlerzeit*, pp. 456–8; Machtan, *Bismarcks Tod*, pp. 135–6.
141 Spitzemberg, diary entry for 2 August 1898, cited in Machtan, *Bismarcks Tod*, p. 138; Cf. Spitzemberg, *Tagebuch*, p. 373. See Bülow, *Memoirs*, I, pp. 224ff.

142 Kaiser Wilhelm II, speech of 4 August 1898, cited in Bülow, *Memoirs*, I, pp. 228–9.
143 Kaiser Wilhelm II to his mother, 25 September 1898, AdHH Schloss Fasanerie. Cf. Bülow, *Memoirs*, I, pp. 230–2. See Empress Frederick to Kaiser Wilhelm II, 22 and 27 September, 24 October and 25 November 1898, GStA Berlin, BPHA Rep. 52T No. 13 and 13a.
144 Bülow, *Memoirs*, I, p. 229.
145 See above, pp. 840–1.
146 Waldersee, diary entries for 16 and 17 June 1897, GStA Berlin, Waldersee Papers; cf. Meisner, II, p. 399.
147 Waldersee, diary entries for 16 and 17 June 1897, GStA Berlin, Waldersee Papers; cf. Meisner, II, p. 399.
148 Hohenlohe, diary entry for 15 December 1898, Hohenlohe, *Denkwürdigkeiten der Reichskanzlerzeit*, p. 473. But cf. Alexander Hohenlohe's comment, ibid., p. 474.
149 Bülow to Eulenburg, 15 February 1898, *Eulenburgs Korrespondenz*, III, No. 1362.
150 Hohenlohe to his son Alexander, 29 August 1897, Hohenlohe, *Denkwürdigkeiten der Reichskanzlerzeit*, pp. 379–80.
151 Hohenlohe, diary entry for 8 May 1898, ibid., p. 444.
152 Bülow to Hohenlohe, 12 October 1898, BA Berlin, Reichskanzlei No. 237; Hohenlohe, diary entry for 20 November 1898, Hohenlohe, *Denkwürdigkeiten der Reichskanzlerzeit*, pp. 469–70; Hohenlohe, diary entry for 27 February 1899, ibid., p. 486; Hohenlohe to Völderndorff, 11 March 1899, ibid., p. 489; also Grierson to Bigge, 18 March 1899, RA I62/6; Bülow to Hohenlohe, 15 March 1900, Hohenlohe, *Denkwürdigkeiten der Reichskanzlerzeit*, p. 568; Hohenlohe, diary entry for 15 March 1900, ibid.
153 See Hohenlohe, *Denkwürdigkeiten der Reichskanzlerzeit*, pp. 511–12.
154 Kaiser Wilhelm II to Grand Duchess Luise of Baden, 30 August 1899, GStA Berlin, BPH Rep. 53 No. 55.
155 Such meetings took place on 16 March, 7 October and 30 December 1896, 26 January, 13 February and 14 October 1897, 1 January, 15 February and 7 October 1898 and 13 June and 23 August 1899.
156 Bosse, diary entry for 14 October 1897, cited in Röhl, *Germany without Bismarck*, p. 259.
157 Hohenlohe, diary entry for 15 February 1898, Hohenlohe, *Denkwürdigkeiten der Reichskanzlerzeit*, p. 428. See BA Koblenz, files of the Prussian Ministry of Justice, P135/2050.
158 Bülow to Kaiser Wilhelm II, 10 March 1897, GStA Berlin, BPHA Rep. 53J Lit. B. No. 16a.
159 Bülow to Eulenburg, 15 February 1898, *Eulenburgs Korrespondenz*, III, No. 1362.
160 Bosse, diary entry for 7 October 1898, cited in Röhl, *Germany without Bismarck*, p. 259.
161 Miquel to all Prussian ministers, 2 September 1899, BA Berlin, Reichskanzlei No. 1461. See Thielen to Boetticher, 7 September 1899, BA Koblenz, Boetticher Papers.
162 Miquel to all Prussian ministers, 22 September 1898, BA Koblenz, files of the Prussian Ministry of Justice, P135/4385.
163 Minutes of the Prussian Ministry of State, 24 September 1898, ibid.
164 Alexander Hohenlohe to his father, 13 July 1897, Hohenlohe, *Denkwürdigkeiten der Reichskanzlerzeit*, p. 370.
165 Hohenlohe to Völderndorff, 25 October 1898, ibid., pp. 464–5.
166 Hohenlohe, diary entry for 20 November 1898, ibid., pp. 469–70.
167 Prince Hermann zu Hohenlohe-Langenburg to Prince Chlodwig zu Hohenlohe-Schillingsfürst, 7 March 1898, ibid., p. 433.
168 Hahnke, memorandum of 2 October 1899, BA Berlin, Reichskanzlei, Geschäftsgang 1 Vol. 6 No. 360.
169 Minutes of the Prussian Ministry of State, 22 and 28 March 1898, GStA Berlin.
170 See Tirpitz to Kaiser Wilhelm II, 24 April 1898, Knorr, memorandum of 21 May 1898 with Tirpitz's marginal notes, Tirpitz to Kaiser Wilhelm II, 28 May 1898, Tirpitz to Hohenlohe,

28 May 1898, Kaiser Wilhelm II to Tirpitz, 1 and 14 June 1898, Tirpitz to Hohenlohe, 14 June 1898, Tirpitz to Wilmowski, 16 June 1898, BA Berlin, Reichskanzlei No. 1612.

171 Memorandum entitled 'Bemerkungen zu der Differenz zwischen Oberkommando der Marine und Reichsmarineamt', 25 June 1898, BA Berlin, Reichskanzlei No. 1612.

172 Hohenlohe to Völderndorff, 5 October and 28 December 1898, Hohenlohe, *Denkwürdigkeiten der Reichskanzlerzeit*, pp. 462 and 475. See Völderndorff to Hohenlohe, 22 June 1897, ibid., pp. 359–60.

173 Alexander Hohenlohe to his father, 20 September 1900, ibid., p. 583.

174 Alexander Hohenlohe to his father, 10 January 1899, ibid., pp. 478–9.

175 Hohenlohe to his son Alexander, 7 January 1900, ibid., p. 554; Hohenlohe to Völderndorff, 18 March 1899, ibid., p. 490.

176 Hohenlohe to Völderndorff, 5 October 1898, ibid., p. 462; Hohenlohe, diary entries for 30 April, 6 October and 15 December 1898 and 27 February 1899, ibid., pp. 443–4, 462–3, 473–4 and 486; also Alexander Hohenlohe to Prince Hermann zu Hohenlohe-Langenburg, 19 June 1899, Hohenlohe-Zentralarchiv Schloss Neuenstein, Hermann Hohenlohe-Langenburg Papers, Bü. 294.

177 Kaiser Wilhelm II, speech of 6 September 1898 in Oeynhausen, Penzler, *Reden Kaiser Wilhelms II. in den Jahren 1896–1900*, pp. 111–13.

178 Bülow, *Memoirs*, I, p. 234.

179 Hohenlohe to Völderndorff, 25 October 1898, Hohenlohe, *Denkwürdigkeiten der Reichskanzlerzeit*, pp. 464–5; Hohenlohe to his son Alexander, 9 September 1898, ibid., p. 458; Hohenlohe to Grand Duke of Baden, [October 1899], ibid., p. 532.

180 Wilmowski to Hohenlohe, 12 September 1898, ibid., p. 459.

181 Hohenlohe to Bülow, 10 November 1898, ibid., pp. 468–9.

182 Röhl, *Germany without Bismarck*, p. 262.

183 Hohenlohe to his son Alexander, 26 November 1898, Hohenlohe, *Denkwürdigkeiten der Reichskanzlerzeit*, p. 471.

184 Hohenlohe to Völderndorff, 26 November 1898, ibid., pp. 471–2.

185 Hohenlohe to [Posadowsky?], 1 May 1899, ibid., p. 498. See Hohenlohe, diary entry for 11 May 1899 and Hohenlohe to his son Alexander, 17 May 1899, ibid., pp. 501–2.

186 Kaiser Wilhelm II to Hohenlohe, telegram, 26 June 1899, ibid., pp. 508–9.

187 Hohenlohe to Kaiser Wilhelm II, 26 June 1899, ibid., p. 509.

188 Hohenlohe to Grand Duke of Baden, undated, ibid., p. 532.

189 Kaiser Wilhelm II to Lucanus, telegram, quoted in Bülow to Auswärtiges Amt, 22 July 1899, ibid., pp. 512–13.

190 Wilmowski to Hohenlohe, 23 October 1899, ibid., pp. 532–3.

191 See Hannelore Horn, *Der Kampf um den Bau des Mittellandkanals. Staat und Politik*, Cologne, Opladen 1964.

192 Hohenlohe, diary entry for 20 November 1898, Hohenlohe, *Denkwürdigkeiten der Reichskanzlerzeit*, pp. 469–70.

193 Hohenlohe to his son Alexander, 13 April 1899, ibid., p. 495.

194 Bülow, *Memoirs*, I, p. 293.

195 Hohenlohe, diary entry, undated [May 1899], Hohenlohe, *Denkwürdigkeiten der Reichskanzlerzeit*, p. 501.

196 Hohenlohe to his son Alexander, 17 May 1899, ibid., pp. 501–2.

197 Bülow to Hohenlohe, 26 June 1899, ibid., p. 508.

198 Miquel to Hohenlohe, 8 August 1899, ibid., p. 515; Hohenlohe to his son Alexander, 11 August 1899, ibid., pp. 515–16.

199 Kaiser Wilhelm II, speech at Dortmund, 11 August 1899, Penzler, *Reden Kaiser Wilhelms II. in den Jahren 1896–1900*, pp. 159ff.; Elkind, *German Emperor's Speeches*, pp. 309–10.

200 Hohenlohe to Kaiser Wilhelm II, 16 August 1899, Hohenlohe, *Denkwürdigkeiten der Reichskanzlerzeit*, pp. 516–17.
201 Kaiser Wilhelm II to Hohenlohe, 17 August 1899, ibid., p. 517; Hohenlohe's reply, ibid.
202 Hohenlohe to Kaiser Wilhelm II, 17 August 1899, ibid., p. 518.
203 Kaiser Wilhelm II to von der Recke, 18 August 1899, ibid., p. 518.
204 Kaiser Wilhelm II to Bülow, August 1899, cited in Bülow, *Denkwürdigkeiten*, I, p. 295. Cf. the translation in Bülow, *Memoirs*, I, p. 293.
205 Bülow to Auswärtiges Amt, 18 August 1899, Hohenlohe, *Denkwürdigkeiten der Reichskanzlerzeit*, p. 518.
206 Kaiser Wilhelm II to Bülow, September 1899, cited in Bülow, *Denkwürdigkeiten*, I, p. 296; cf. *Memoirs*, I, p. 294.
207 Kaiserin Auguste Viktoria to Bülow, 18 August 1899, cited in Bülow, *Memoirs*, I, p. 293.
208 Hohenlohe, diary entry for 11 August 1899, Hohenlohe, *Denkwürdigkeiten der Reichskanzlerzeit*, p. 516; Holstein to Hohenlohe, 19 August 1899, ibid., p. 519.
209 Hohenlohe, diary entry for 20 August 1899, ibid., pp. 519–20.
210 Hohenlohe, diary entry for 22 August 1899, ibid., p. 522.
211 Hohenlohe, diary entry for 20 August 1899, ibid., pp. 519–20.
212 Hohenlohe to Kaiser Wilhelm II, 21 August 1899, ibid., pp. 521–2. See Alexander Hohenlohe to his father, 22 August 1899, ibid., p. 522.
213 Hutten-Czapski to Hohenlohe, 23 August 1899, ibid., pp. 469–70; Hohenlohe, diary entry for 27 August 1899, ibid., pp. 523–4.
214 Hohenlohe to his son Alexander, 25 August 1899, ibid., p. 523.
215 Bülow, *Memoirs*, I, p. 295.
216 Hohenlohe to his son Alexander, 25 August and 13 September 1899, Hohenlohe, *Denkwürdigkeiten der Reichskanzlerzeit*, pp. 523 and 526; Hohenlohe, diary entry for 7 March 1900, ibid., p. 567; Bülow, *Memoirs*, I, pp. 295–6; Lucanus to Kaiser Wilhelm II, 27 August 1899, GStA Berlin, BrPrHA Rep. 53J Lit. L No. 12.
217 Miquel to Hohenlohe, 4 September 1899, Hohenlohe, *Denkwürdigkeiten der Reichskanzlerzeit*, pp. 525–6.
218 Hohenlohe to his son Alexander, 6 October 1899, ibid., p. 531.
219 Hereditary Princess Charlotte of Saxe-Meiningen to Schweninger, BA Berlin, Schweninger Papers 90 Schw. 4 No. 130; Hereditary Prince Bernhard of Saxe-Meiningen to Colmar Freiherr von der Goltz, 10 October 1899, BA-MA Freiburg, von der Goltz Papers N737 Zug. 161/95.
220 Eulenburg to Hohenlohe, 2 October 1899, Hohenlohe, *Denkwürdigkeiten der Reichskanzlerzeit*, p. 531.
221 Duke Heinrich XI von Pless to Kaiser Wilhelm II, 21 October 1899, GStA Berlin, BPHA Rep. 53J Lit. P No. 1; Prince Christian Krafft zu Hohenlohe-Oehringen to August Eulenburg, 21 October 1899, GStA Berlin, BPHA Rep. 53 E. III No. 4.
222 Count Hans Finck von Finckenstein-Simnau to Kaiser Wilhelm II, 23 May 1900, Finck to August Eulenburg, 22 May 1900, August Eulenburg to Finck, 28 May 1900, GStA Berlin, BPHA Rep. 53 E. III No. 4.
223 Bülow, *Memoirs*, I, p. 295.
224 Kaiser Wilhelm II to Duke Heinrich XI von Pless, 24 October 1899, GStA Berlin, BPHA Rep. 53J Lit. P No. 3; Kaiser Wilhelm II to Prince Christian Krafft zu Hohenlohe-Oehringen, 25 October 1899, GStA Berlin, BPHA Rep. 53 E. III No. 4.
225 Bülow, *Memoirs*, I, p. 296.
226 Hohenlohe, notes of 6 November 1899, Hohenlohe, *Denkwürdigkeiten der Reichskanzlerzeit*, p. 538.
227 Hohenlohe, diary entry for 17 October 1900, ibid., p. 592.

228 Hohenlohe's undated notes, [October 1900], ibid., p. 582. Cf. the draft resignation letter ibid., pp. 581–2; also Hohenlohe to his son Alexander, 8 and 10 October 1900, ibid., p. 589.

229 Kaiser Wilhelm II to Hohenlohe, 17 October 1900, ibid., pp. 592–3; Hohenlohe, diary entry for 16 October 1900, ibid., pp. 591–2.

29 THE KAISER, ART AND ARCHITECTURE

1 Hammerstein to Kaiser Wilhelm II, 10 April 1899, Hohenlohe-Zentralarchiv Schloss Neuenstein, Hermann Hohenlohe-Langenburg Papers, Bü. 312.

2 Paul Seidel, *Der Kaiser und die Kunst*, Berlin 1907, p. 255.

3 Cartoon by Alfred Bryan in *Moonshine*, 26 February 1898, reprinted in Jost Rebentisch, *Die vielen Gesichter des Kaisers. Wilhelm II. in der deutschen und britischen Karikatur (1888–1918)*, Berlin 2000, p. 381.

4 Kaiser Wilhelm II, 'Niemand zu Liebe, Niemand zu Leibe!', signed 'Berlin, 28.XI.96, W.', GStA Berlin, BPHA Rep. 53 No. 270.

5 See Hohenlohe, *Denkwürdigkeiten der Reichskanzlerzeit*, p. 474.

6 Gerhard Malkowsky, *Die Kunst im Dienste der Staatsidee. Hohenzollernsche Kunstpolitik vom Großen Kurfürsten bis auf Wilhelm II*, Berlin 1912, p. 237.

7 See Alexander Hohenlohe to his father, 5 May 1899, Hohenlohe, *Denkwürdigkeiten der Reichskanzlerzeit*, p. 499.

8 See Hartmut Boockmann, *Die Marienburg im 19. Jahrhundert*, Frankfurt, Berlin, Vienna 1982, pp. 38–9 and 167ff. Cf. Seidel, *Kaiser und die Kunst*, pp. 60ff. and 66–70.

9 Kaiser Wilhelm II, speech at the Saalburg, 11 October 1900, Seidel, *Kaiser und die Kunst*, pp. 53–4.

10 Jarchow, *Hofgeschenke*, p. 86. See *Der Kunstwart*, 11th year, vol. 13 (April 1898), p. 36.

11 Kaiser Wilhelm II to his mother, 20 February 1893, AdHH Schloss Fasanerie. The Empress Frederick's reply of 25 February 1893 is located in GStA Berlin, BPHA Rep. 52T No. 13.

12 Ernst Wichert, 'Eine Generalprobe vor dem Kaiser', December 1893, GStA Berlin, BPHA Rep. 53 No. 179; also Poultney Bigelow to Kaiser Wilhelm II, December 1893, ibid., BPHA Rep. 53J Lit. B No. 10 Bigelow 1.

13 Kaiser Wilhelm II to Empress Frederick, 4 November 1899, AdHH Schloss Fasanerie.

14 Kaiser Wilhelm II to Empress Frederick, 18 November 1899, ibid.

15 See Wolf Jobst Siedler, *Abschied von Preußen*, Berlin 1991.

16 Seidel, *Kaiser und die Kunst*, p. 38.

17 Ibid., pp. 74–6.

18 Ibid., pp. 40ff.; Jürgen Julier, ed., *Kaiserlicher Kunstbesitz. Aus dem holländischen Exil Haus Doorn*, Berlin 1991, p. 259.

19 Seidel, *Kaiser und die Kunst*, p. 35.

20 See the illustrations ibid., pp. 32–3, 36–7, 40, 42–3 and 48.

21 Ibid., p. 38.

22 Seidel, *Kaiser und die Kunst*, p. 40. For further examples see ibid., pp. 31, 41 and 43.

23 Kaiser Wilhelm II, marginal notes on the design for the Post Office in Strassburg, 18 June 1895, ibid., pp. 40–1.

24 Kaiser Wilhelm II, marginal notes on the designs for the Post Offices in Güsten (Anhalt), Geestemünde and Königsberg, ibid., pp. 40–1 and 43.

25 Kaiser Wilhelm II, marginal notes on the design for the Post Office in Ülzen, ibid., p. 43.

26 Kaiser Wilhelm II, marginal notes on the design for the Post Office in Hanover-Linden, ibid., p. 43.

27 Ibid., p. 76.

28 Ibid., pp. 77–8.
29 Ibid., p. 80. Wilhelm also chose Schwechten as the architect for the Residenzschloss in Posen: ibid., p. 38.
30 See the list ibid., p. 82.
31 Ibid., p. 83.
32 Ibid., p. 98.
33 See ibid., pp. 102–4.
34 Ibid., pp. 98–9.
35 Ibid., p. 100.
36 Ibid., pp. 102–4.
37 Ibid., pp. 108–10.
38 Kalnein to Eulenburg, 8 March 1892, *Eulenburgs Korrespondenz*, II, No. 602.
39 Kaiser Wilhelm II, All-Highest Cabinet Order of 9 July 1888, Seidel, *Kaiser und die Kunst*, p. 92.
40 Carl-Wolfgang Schümann, *Der Berliner Dom im 19. Jahrhundert*, Berlin 1980, pp. 245–6.
41 Seidel, *Kaiser und die Kunst*, p. 92.
42 See Kalnein to Eulenburg, 8 March 1892, *Eulenburgs Korrespondenz*, II, No. 602.
43 Schümann, *Berliner Dom*, p. 248; also Julius Schneider, *Die Geschichte des Berliner Doms*, Berlin 1993, p. 67.
44 Schümann, *Berliner Dom*, p. 251.
45 Seidel, *Kaiser und die Kunst*, p. 93.
46 Schümann, *Berliner Dom*, p. 252.
47 Ibid., p. 252.
48 Ibid., p. 253.
49 Seidel, *Kaiser und die Kunst*, p. 93.
50 Schümann, *Berliner Dom*, p. 247.
51 Ibid., p. 252.
52 Kaiser Wilhelm II to his mother, 20 February 1893, AdHH Schloss Fasanerie. The Empress Frederick's reply of 25 February 1893 is located in GStA Berlin, BPHA Rep. 52T No. 13.
53 Kaiser Wilhelm II to his mother, 21 June 1894, AdHH Schloss Fasanerie.
54 The document is printed in Seidel, *Kaiser und die Kunst*, pp. 92–3.
55 Schümann, *Berliner Dom*, p. 253; also Schneider, *Geschichte des Berliner Doms*, pp. 68–9.
56 The sermon of Oberhof- and Domprediger Dryander is printed in Seidel, *Kaiser und die Kunst*, pp. 94–7.
57 Robert Graf von Zedlitz-Trützschler, *Zwölf Jahre am deutschen Kaiserhof*, Berlin, Leipzig 1923, pp. 114–15.
58 Spitzemberg, diary entry for 27 February 1905, *Tagebuch*, pp. 445–6.
59 Wallot to Bluntschli, 3 January 1889, quoted in Michael S. Cullen, *Der Reichstag. Die Geschichte eines Monumentes*, Berlin 1983, p. 201.
60 Cited ibid., pp. 202–3.
61 Wallot to Reichensperger, 28 January 1889, ibid., pp. 204–5.
62 Kaiser Wilhelm II to his mother, 20 February 1893, AdHH Schloss Fasanerie. The Empress Frederick's reply of 25 February 1893 is located in GStA Berlin, BPHA Rep. 52T No. 13.
63 See Hohenlohe, diary entry for 14 December 1894, Hohenlohe, *Denkwürdigkeiten der Reichskanzlerzeit*, p. 23.
64 Wallot to Bluntschli, 6 April 1893, cited in Cullen, *Reichstag*, pp. 219–20.
65 See ibid., pp. 226ff. and 232–3.
66 Maximilian Rapsilber, 'Der Kaiser als Kunstrichter', in L. Leipziger, ed., *Der Roland von Berlin. Eine Wochenschrift für das Berliner Leben*, vol. 2, part 25, 23 June 1904, p. 59; copy in GStA Berlin, 2.2.12., Oberhofmarschallamt, fasc. 135.

67 Wallot to Reichensperger, 27 November 1892, cited in Cullen, *Reichstag*, p. 215; Friedrich von Thiersch to Bluntschli, 6 November 1892, ibid., p. 216.
68 Wallot to Bluntschli, 28 December 1893, ibid., p. 220.
69 Wallot to Bluntschli, 27 June 1894, ibid., pp. 220–1.
70 Above, p. 701.
71 Wallot to Bluntschli, 16 January 1895, cited in Cullen, *Reichstag*, pp. 242–6.
72 See Inge Eichler, 'Victoria als Malerin, Sammlerin und Mäzenin', in Rainer von Hessen, ed., *Victoria Kaiserin Friedrich (1840–1901). Mission und Schicksal einer englischen Prinzessin in Deutschland*, Frankfurt a.M. 2002, pp. 134–50; Karoline Müller and Friedrich Rothe, eds., *Victoria von Preußen 1840–1901*, Berlin 2001, *passim*.
73 See *Young Wilhelm*, pp. 212–13 and 559–62. The oil painting exhibited in 1886 is reproduced in Seidel, *Kaiser und die Kunst*, p. 237. See also Wilhelm's work of 1876, ibid., p. 221.
74 Ibid., pp. 219–22.
75 Cited ibid., pp. 240–2.
76 Ibid., p. 250.
77 Ibid., pp. 242–3.
78 Ibid., p. 225.
79 See above, pp. 754–6.
80 Goetz, *Briefe Wilhelms II. an den Zaren*, pp. 294–6.
81 Cited in Elisabeth Heresch, *Nikolaus II. 'Feigheit, Lüge und Verrat'. Leben und Ende des letzten russischen Zaren*, Munich 1992, p. 101.
82 Jarchow, *Hofgeschenke*, pp. 69–70 and 141.
83 The finished work is reproduced above, p. 891.
84 Ferdinand Avenarius, 'Hofkunst und andere Kunst', *Der Kunstwart*, vol. 15, part 3 (November 1901), p. 87.
85 See Hartwig Fischer, *Ein Wilhelminisches Gesamtkunstwerk auf dem Kapitol. Hermann Prell und die Einrichtung des Thronsaals in der Deutschen Botschaft zu Rom 1894–1899*, Basel 1998.
86 See ibid., pp. 67–8 and 127.
87 Prell to his wife, 9 June 1896, ibid., pp. 52–3.
88 See Eulenburg to Bülow, 4 July 1898, *Eulenburgs Korrespondenz*, III, No. 1377.
89 Fischer, *Wilhelminisches Gesamtkunstwerk*, pp. 52–3.
90 Prell to his wife, 3 September 1896, cited ibid., pp. 91–2.
91 Ibid., p. 130.
92 The Kaiser's Privy Purse to Prell, 4 January 1899, cited ibid., p. 130.
93 Karl-Friedrich Nowak and Friedrich Thimme, eds., *Erinnerungen und Gedanken des Botschafters Anton Graf Monts*, Berlin 1932, p. 99. But cf. Wilhelm's positive comments in Fischer, *Wilhelminisches Gesamtkunstwerk*, p. 135.
94 Nicolaas Teeuwisse, *Vom Salon zur Sezession. Berliner Kunstleben zwischen Tradition und Aufbruch zur Moderne 1871–1900*, Berlin 1986, pp. 197–207.
95 Ibid., p. 207.
96 Excerpts from Anton von Werner's lost diaries, in Dominik Bartmann, *Anton von Werner. Zur Kunst und Kunstpolitik im Deutschen Kaiserreich*, Berlin 1985, p. 217.
97 Ludwig Pallat, *Richard Schöne, Generaldirektor der Königlichen Museen zu Berlin. Ein Beitrag zur Geschichte der preußischen Kunstverwaltung 1872–1905*, 1959, p. 327, cited in Bartmann, *Anton von Werner*, pp. 217–18.
98 Cited in Teeuwisse, *Vom Salon zur Sezession*, pp. 213–14.
99 Alfred Lichtwark, *Briefe an die Kommission für die Verwaltung der Kunsthalle*, vol. XI, 1903, p. 193, cited ibid., p. 214.
100 See Peter Paret, *Die Berliner Secession. Moderne Kunst und ihre Feinde im Kaiserlichen Deutschland*, Frankfurt a.M., Berlin, Vienna, 1983.

101 Bartmann, *Anton von Werner*, pp. 32–3.
102 Ibid., pp. 178 and 267.
103 Seidel, *Kaiser und die Kunst*, p. 198.
104 Ibid., pp. 212–14.
105 See Bartmann, *Anton von Werner*, p. 177.
106 Teeuwisse, *Vom Salon zur Sezession*, p. 95; Seidel, *Kaiser und die Kunst*, p. 194.
107 Teeuwisse, *Vom Salon zur Sezession*, p. 161.
108 Seidel, *Kaiser und die Kunst*, p. 144.
109 Eulenburg to Kaiser Wilhelm II, 16 April 1894, *Eulenburgs Korrespondenz*, II, No. 953.
110 Eulenburg to Bülow, 7 May 1894, ibid., No. 963. See also No. 966.
111 Seidel, *Kaiser und die Kunst*, pp. 161 and 176–7; Lutz Tittel, 'Monumentaldenkmäler von 1871 bis 1918', in Ekkehard Mai and Stephan Waetzoldt, eds., *Kunstverwaltung, Bau- und Denkmalkunst im Kaiserreich*, Berlin, 1981.
112 See for example Seidel, *Kaiser und die Kunst*, p. 181.
113 Waldersee, diary entry for 19 January 1894, GStA Berlin, Waldersee Papers; omitted from Meisner, II, p. 304.
114 Seidel, *Kaiser und die Kunst*, p. 228.
115 Ibid., p. 174.
116 *Reichs-Anzeiger*, special edition, Sunday, 27 January 1895, in Uta Lehnert, *Der Kaiser und die Siegesallee. Réclame Royale*, Berlin 1998, p. 22.
117 See ibid., pp. 15–22.
118 Ibid., p. 39.
119 Ibid., p. 67.
120 Ibid., p. 70.
121 Ibid., pp. 52 and 78.
122 Ibid., p. 79.
123 *Berliner Lokal-Anzeiger*, No. 571, 6 December 1899, cited ibid., p. 80.
124 Seidel, *Kaiser und die Kunst*, p. 168.
125 Lehnert, *Siegesallee*, p. 79.
126 Ibid., pp. 262 and 292–3.
127 Ibid., p. 245.
128 See the detailed account of the ceremony ibid., pp. 244–52.
129 Ibid., pp. 288–9.
130 Wilhelm Holzamer, *Die Siegesallee. Kunstbriefe an den deutschen Michel*, Leipzig 1902, cited ibid., p. 288.
131 Schaper to his wife, 3 May 1900, ibid., p. 292.
132 Barbara Tuchman, *Proud Tower*, New York 1966, p. 303.
133 Cited in Lehnert, *Siegesallee*, p. 293.
134 Gerhard Masur, *Imperial Berlin*, London 1971, p. 212; Ronald Taylor, *Berlin and Its Culture. A Historical Portrait*, New Haven, London 1997, p. 169; Alexandra Richie, *Faust's Metropolis. A History of Berlin*, London 1998, p. 231.
135 Kaiser Wilhelm II, speech of 18 December 1901, Johannes Penzler, ed., *Die Reden Kaiser Wilhelms II. in den Jahren 1901–Ende 1905*, Leipzig n.d., pp. 57–63; Elkind, *German Emperor's Speeches*, pp. 185–9.

30 THE CHALLENGE: FROM CONTINENTAL POLICY TO *WELTPOLITIK*

1 See Hildebrand, *Deutsche Außenpolitik 1871–1918*, pp. 32ff.; Hildebrand, *Das vergangene Reich*, pp. 190ff.; Canis, *Von Bismarck zur Weltpolitik*, pp. 138ff.
2 Kaiser Wilhelm II, speech of 18 January 1896, Penzler, *Reden Kaiser Wilhelms II. in den Jahren 1896–1900*, pp. 9–10. See also Elkind, *German Emperor's Speeches*, pp. 141–2.

3 See above, pp. 763–4 and 858 also Canis, *Von Bismarck zu Weltpolitik*, pp. 223ff.
4 Szögyény to Goluchowski, 5 February 1900, cited in Paul Kennedy, 'The Kaiser and German Weltpolitik. Reflexions on Wilhelm II's Place in the Making of German Foreign Policy', in Röhl and Sombart, *Kaiser Wilhelm II. New Interpretations*, p. 158.
5 Völderndorff to Hohenlohe, 9 November 1897, Hohenlohe, *Denkwürdigkeiten der Reichskanzlerzeit*, pp. 401–2.
6 Kennedy, 'Kaiser and German Weltpolitik', pp. 158–9.
7 Nipperdey, *Deutsche Geschichte*, II, p. 632.
8 Grierson to Bigge, 4 May 1899, RA I62/10a.
9 Prince Heinrich of Prussia, speech of 15 December 1897, Penzler, *Reden Kaiser Wilhelms II. in den Jahren 1896–1900*, pp. 80–1.
10 Eulenburg to Bülow, 18 December 1897, *Eulenburgs Korrespondenz*, III, No. 1354.
11 Spitzemberg, diary entry for 19 December 1897, *Tagebuch*, p. 362.
12 Bigelow to Kaiser Wilhelm II, January 1898, GStA Berlin, BPHA Rep. 53J Lit. B No. 10.
13 Hinzpeter to Kaiser Wilhelm II, 9 December 1897, ibid., BPHA Rep. 53J Lit. H. No. 1.
14 Hinzpeter to Kaiser Wilhelm II, 12 June 1899, ibid.
15 Hinzpeter to Kaiser Wilhelm II, 4 February 1899, ibid.
16 Holstein to Eulenburg, 24 November 1896, *Eulenburgs Korrespondenz*, III, No. 1279.
17 Holstein to Eulenburg, 24 November 1896, *Holstein Papers*, III, No. 586.
18 Radolin to Holstein, 18 January 1897, ibid., IV, No. 596.
19 Holstein to Radolin, 19 January 1897, ibid., No. 597.
20 Holstein to Eulenburg, 3 February 1897, ibid., No. 599. See *Große Politik*, XII, No. 3104. For Wilhelm's inclination towards France at this time, see also Holstein to Hatzfeldt, 14 April 1897, *Holstein Papers*, IV, No. 609.
21 Hohenlohe to Eulenburg, 3 March 1897, Hohenlohe, *Denkwürdigkeiten der Reichskanzlerzeit*, p. 309; Eulenburg to Hohenlohe, 1 March 1897, ibid., p. 308.
22 Holstein's undated draft, ibid., p. 309, n. 1.
23 See above, p. 367.
24 Kaiser Wilhelm II to Empress Frederick, 20 February 1899, AdHH Schloss Fasanerie.
25 See above, p. 499.
26 Kaiser Wilhelm II, marginal notes on Münster's report from Paris of 26 May 1896, *Große Politik*, XI, No. 2853.
27 Hohenlohe, diary entry for 6 September 1896, Hohenlohe, *Denkwürdigkeiten der Reichskanzlerzeit*, pp. 260–1. For the meeting of the two emperors in Breslau, see Canis, *Von Bismarck zur Weltpolitik*, pp. 205–6; McLean, *Royalty and Diplomacy*, pp. 31–2.
28 Kaiser Wilhelm II to Eulenburg, 9 September 1896, *Große Politik*, XI, No. 2861.
29 Kaiser Wilhelm II, speech in Görlitz, 7 September 1896, Penzler, *Reden Kaiser Wilhelms II. in den Jahren 1896–1900*, pp. 32–3.
30 Eulenburg, note of 4 October 1896, *Eulenburgs Korrespondenz*, III, No. 1265.
31 Eulenburg to Kaiser Wilhelm II, 22 August 1897, ibid., No. 1340.
32 Margaret M. Jefferson, 'Lord Salisbury's Conversations with the Tsar in Balmoral, 27 and 29 September 1896', *Slavonic and East European Review*, 39 (1960–1), pp. 216–22; Canis, *Von Bismarck zur Weltpolitik*, p. 207.
33 Jefferson, 'Salisbury's Conversations with the Tsar', p. 220; Andrew Roberts, *Salisbury. Victorian Titan*, London 2000, pp. 643–4.
34 Hanotaux, note of 12 October 1896, *Documents Diplomatiques Français*, XII, p. 781; McLean, *Royalty and Diplomacy*, p. 32; *Große Politik*, XI, p. 369. See also Nicholas II to his mother, 2 October 1896, in E. J. Bing, ed., *The Letters of the Tsar Nicholas and the Empress Marie*, London 1937, pp. 119–25.
35 Canis, *Von Bismarck zur Weltpolitik*, pp. 207–8.

36 Salisbury to Goschen, 6 September 1897, cited in Kennedy, *Anglo-German Antagonism*, p. 233.
37 Empress Frederick to Queen Victoria, 17 October 1896, AdHH Schloss Fasanerie; Alexander Hohenlohe to his father, 8 October 1896, Hohenlohe, *Denkwürdigkeiten der Reichskanzlerzeit*, pp. 388–9.
38 Hohenlohe to Kaiser Wilhelm II, 16 October 1896, Hohenlohe to his son Alexander, 17 October 1896, ibid., pp. 268–9; Kaiser Wilhelm II to Hohenlohe, 20 October 1896, *Große Politik*, XI, No. 2868.
39 Eulenburg to Bülow, 26 October 1896, *Eulenburgs Korrespondenz*, III, No. 1268.
40 Kaiser Wilhelm II, memorandum 'On the need to form a politico-mercantile coalition of the European states against the USA', printed in Russian translation in A. A. Fursenko, *Bor'ba za razdel Kitaya i amerikanskaya doktrina otkritich dverei 1895–1900*, Moscow, Leningrad 1956, pp. 209–12.
41 Eulenburg, note of 8 November 1896, *Eulenburgs Korrespondenz*, III, No. 1272.
42 Waldersee, diary entry for 25 October 1896, GStA Berlin, Waldersee Papers; badly distorted in Meisner, II, pp. 374–5.
43 Holstein to Hatzfeldt, 27/28 November 1896, Hatzfeldt, *Nachgelassene Papiere*, II, No. 692.
44 Kaiser Wilhelm II to Kaiser Franz Joseph, 8 January 1898, HHStA Vienna, Kabinettsarchiv Geheimakten 2. See Holstein to Hatzfeldt, 27/28 November 1896, Hatzfeldt, *Nachgelassene Papiere*, II, No. 692; McLean, *Royalty and Diplomacy*, p. 32.
45 Spitzemberg, diary entry for 18 November 1896, *Tagebuch*, p. 348.
46 Hohenlohe to Eulenburg, 4 February 1897, Hohenlohe, *Denkwürdigkeiten der Reichskanzlerzeit*, p. 297. See also Hohenlohe to Kaiser Wilhelm II, draft of January 1897, ibid., p. 296.
47 Kaiser Wilhelm II, marginal notes on Eulenburg's report of 19 August 1896, *Große Politik*, XI, p. 328.
48 Eulenburg to Kaiser Wilhelm II, 7 June 1896, *Eulenburgs Korrespondenz*, III, p. 1741; Eulenburg to Kaiserin Auguste Viktoria, 15 October 1896, ibid., No. 1266.
49 Eulenburg to Kaiser Wilhelm II, 14 November 1896, ibid., No. 1275.
50 Grierson to Bigge, 14 September 1896, RA I60/156a. But cf. above, pp. 347–8.
51 Hohenlohe to Kaiser Wilhelm II, January 1897, Hohenlohe, *Denkwürdigkeiten der Reichskanzlerzeit*, pp. 296–7.
52 Holstein to Eulenburg, 3 February 1897, *Holstein Papers*, IV, No. 599.
53 Hohenlohe to Eulenburg, 4 February 1897, Eulenburg to Hohenlohe, 8 February 1897, Hohenlohe, *Denkwürdigkeiten der Reichskanzlerzeit*, pp. 297–8.
54 Hohenlohe, diary entry for 27 February 1899, Hohenlohe to his son Alexander, 24 September 1899, ibid., pp. 486 and 527.
55 Alexander Hohenlohe to his father, 18 September 1899, ibid., pp. 526–7.
56 Georg Alexander (from 1900 von) Müller, memorandum of 1896 entitled 'Zukunftspolitik', in Görlitz, *Der Kaiser* pp. 36–41. See John C. G. Röhl, ed., *From Bismarck to Hitler. The Problem of Continuity in German History*, London 6th edn 1984, pp. 56–60.
57 Knollys to Hardinge, 13 November 1909, Cambridge University Library, Hardinge Papers, 18; cited in John C. G. Röhl, 'Der Kaiser und England', in Wilfrid Rogasch, ed., *Victoria & Albert, Vicky & The Kaiser. Ein Kapitel deutsch-englischer Familiengeschichte*, Berlin 1997, p. 175. See McLean, *Royalty and Diplomacy*, p. 155.
58 Müller, 'Zukunftspolitik', pp. 39–40.
59 Empress Frederick to Queen Victoria, 25 September 1896, AdHH Schloss Fasanerie. On the Armenian question, see Jost Dülffer, 'Die Kreta-Krise und der griechisch-türkische Krieg 1890–1898', in Jost Dülffer, Hans-Otto Mühleisen and Vera Torunsky, eds., *Inseln als Brennpunkte internationaler Politik*, Cologne 1986, pp. 30 and 36–39; William L. Langer, *The Diplomacy of Imperialism 1890–1902*, New York 2nd edn 1965, pp. 321–50.

60 Gerd Fesser, *Der Traum vom Platz an der Sonne. Deutsche 'Weltpolitik' 1897–1914*, Bremen 1996, p. 19; Lascelles to Salisbury, 28 August 1896, RA I60/154.

61 Kaiser Wilhelm II, marginal notes on Saurma's report from Constantinople, 22 August 1896, PA AA, Asservat No. 4; marginal notes on Marschall's telegrams of 28 and 29 August 1898, *Große Politik*, XII, 1, No. 2898 and 2901. See Dülffer, 'Kreta', p. 36; Langer, *Diplomacy*, p. 326.

62 Kaiser Wilhelm II, marginal notes on Saurma's report from Constantinople, 19 September 1896, *Große Politik*, XII, 1, No. 2904.

63 Kaiser Wilhelm II to Queen Victoria, 18 December 1896, RA I60/161. This passage has been omitted from Buckle, *Letters of Queen Victoria*, III, p. 109.

64 Kaiser Wilhelm II, marginal notes on Saurma's report from Constantinople, 29 July 1896, *Große Politik*, XII, 1, No. 2893.

65 Kaiser Wilhelm II, marginal notes on Marschall's telegrams of 29 August 1896, ibid., Nos. 2899 and 2900.

66 See *Große Politik*, XII, 2, especially Nos. 3137, 3150 and 3151. For the Crete crisis and its background, see Langer, *Diplomacy*, pp. 315ff.; Rich, *Friedrich von Holstein*, II, pp. 477ff., Dülffer, 'Kreta', pp. 13–59.

67 Kaiser Wilhelm II, marginal notes on the Empress Frederick's letter, 17 February 1897, printed in Hohenlohe, *Denkwürdigkeiten der Reichskanzlerzeit*, p. 301.

68 Hatzfeldt to Hohenlohe, 7 January 1897, *Große Politik*, XII, 1, No. 3100.

69 Hatzfeldt and Hohenlohe, 15 January 1897, ibid., No. 3102. See also Kaiser Wilhelm II to Hohenlohe, 15 January 1897, ibid., No. 2932.

70 See Marschall's notes of 21 February 1897 with Kaiser Wilhelm II's marginal notes, ibid., 2, No. 3166.

71 Ibid. See also Kaiser Wilhelm II, marginal notes on Hatzfeldt's report from London, 14 April 1897, ibid., No. 3222.

72 Lascelles to Salisbury, 20 February 1897, quoted in van't Padje, 'Anglo-German Imperialist Rivalry', p. 109.

73 Kaiser Wilhelm II to Empress Frederick, 14 February 1897, AdHH Schloss Fasanerie.

74 See *Große Politik*, XII, 2, No. 3152, especially the notes on p. 327. Cf. Kaiser Wilhelm II's marginal notes on Marschall's report of 21 February 1897, ibid., No. 3166; marginal notes on the report from the envoy in Athens, Freiherr von Scheel-Plessen to Hohenlohe, 28 March 1897, ibid., No. 3215.

75 Hohenlohe to Kaiser Wilhelm II, 22 February 1897, ibid., No. 3168, with marginal notes, p. 347.

76 Grierson to Bigge, 20 February 1897, RA I61/2a.

77 Kaiser Wilhelm II, marginal comments on Marschall's notes of 21 February 1897, *Große Politik*, XII, 2, No. 3166.

78 Hohenlohe to Kaiser Wilhelm II, 15 February 1897, ibid., No. 3152, with the Kaiser's marginal notes, p. 328.

79 Kaiser Wilhelm II, marginal notes on Scheel-Plessen's report from Athens, 28 March 1897, ibid., No. 3215.

80 Kaiser Wilhelm II, marginal notes on Kiderlen-Wächter's report from Copenhagen of 9 March 1897, PA AA, Asservat No. 4.

81 Crown Princess Sophie of Greece to Empress Frederick, 12 February 1897, enclosure to Empress Frederick to Kaiser Wilhelm II, 17 February 1897, printed in Hohenlohe, *Denkwürdigkeiten der Reichskanzlerzeit*, p. 303. See also Crown Princess Sophie of Greece to Empress Frederick, 23 April 1897, copy in Empress Frederick to Queen Victoria, 27 April 1897, AdHH Schloss Fasanerie; Queen Victoria to Wilhelm II, 17 February 1897, Hohenlohe, *Denkwürdigkeiten der Reichskanzlerzeit*, p. 303. See Buckle, *Letters of Queen Victoria*, III, pp. 135–6.

82 Empress Frederick to Queen Victoria, 23 March 1897, AdHH Schloss Fasanerie.
83 Empress Frederick to Queen Victoria, 4 April 1897, ibid.
84 Empress Frederick to Queen Victoria, 28 April 1897, ibid.
85 See *Große Politik*, XII, 2, Nos. 3219 and 3220.
86 Queen Victoria to Tsar Nicholas II, 25 April 1897, in Buckle, *Letters of Queen Victoria*, III, p. 154.
87 Hohenlohe to Kaiser Wilhelm II, 10 May 1897, *Große Politik*, XII, 2, No. 3231, with marginal notes, pp. 417–18.
88 Queen Victoria to Kaiser Wilhelm II, 7 May 1897, transmitted by Gough to August Eulenburg, GStA Berlin, BPHA Rep. 53 J. Lit. G. No. 13.
89 Kaiser Wilhelm II to Queen Victoria, draft, ibid.; text of the telegram in RA H40/22.
90 RA QVJ, 7/8 May 1897.
91 Crown Princess Sophie of Greece to Kaiser Wilhelm II, 9 May 1897, *Große Politik*, XII, 2, No. 3232; Wilhelm's reply, ibid., No. 3233.
92 Kaiser Wilhelm II to his sister Sophie, 12 May 1897, ibid., No. 3238.
93 Kaiser Wilhelm II to Queen Victoria, 13 May 1897, RA H40/56.
94 RA QVJ, 13 May 1897.
95 Empress Frederick to Queen Victoria, 14 May 1897, AdHH Schloss Fasanerie. For the peace negotations, see *Große Politik*, XII, 2, Nos. 3248, 3249 and 3250.
96 Kaiser Wilhelm II to Hohenlohe, 24 September 1897, Hohenlohe, *Denkwürdigkeiten der Reichskanzlerzeit*, p. 386.
97 Kaiser Wilhelm II to Hohenlohe, 23 February 1897, *Große Politik*, XII, 2, No. 3169. See also Hohenlohe, *Denkwürdigkeiten der Reichskanzlerzeit*, p. 305.
98 Kaiser Wilhelm II to Tsar Nicholas II, 20 October 1898, Goetz, *Briefe Wilhelms II. an den Zaren*, pp. 313–15. A somewhat inaccurate copy is to be found in PA AA, Preußen 1 No. 1d, Bd I.
99 On the following see Ernst Freiherr von Mirbach, ed., *Das deutsche Kaiserpaar im Heiligen Lande im Herbst 1898. Mit Allerhöchster Ermächtigung Seiner Majestät des Kaisers und Königs bearbeitet nach authentischen Berichten und Akten*, Berlin 1899; Theodor Herzl, *Gesammelte zionistische Werke*, 5 vols., Tel Aviv 1934; Hermann Ellern and Bessi Ellern, *Herzl, Hechler, the Grand Duke of Baden and the German Emperor 1896–1904*, Tel Aviv 1961; Alexander Bein, *Erinnerungen und Dokumente über Herzls Begegnung mit Wilhelm II. Zeitschrift für die Geschichte der Juden*, 1964; Julius H. Schoeps, *Theodor Herzl 1860–1904. Wenn ihr wollt, ist es kein Märchen. Eine Text-Bild-Monographie*, Vienna 1995; Jan Stefan Richter, *Die Orientreise Kaiser Wilhelms II. 1898*, Hamburg 1997; John C. G. Röhl, 'Herzl and Kaiser Wilhelm II. A German Protectorate in Palestine?', in Ritchie Robinson and Edward Timms, eds., *Theodor Herzl and the Origins of Zionism*, Edinburgh 1997, pp. 27–38; Alex Carmel and Ejal Jakob Eisler, *Der Kaiser reist ins Heilige Land. Die Palästinareise Wilhelms II. 1898. Eine illustrierte Dokumentation*, Stuttgart 1999.
100 See Kaiser Wilhelm II to Tsar Nicholas II, 18 August 1898, Goetz, *Briefe Wilhelms II. an den Zaren*, pp. 311–12. Cf. Grand Duke Carl Alexander of Saxe-Weimar-Eisenach to Kaiser Wilhelm II, 11 October 1898, GStA Berlin, BPHA Rep. 53J Lit. S No. 2.
101 See Wedekind's satirical poem 'Im heiligen Land' and Heine's cartoon 'Palästina' in *Simplicissimus*, 3, No. 31, September 1898. See Ernestine Koch, *Albert Langen. Ein Verleger in München*, Munich, Berlin 1969, pp. 8 and 92.
102 See Bismarck's warning, cited in Hellige and Schulin, *Rathenau und Harden*, pp. 322–3.
103 See Posadowsky to Hohenlohe, 25 September 1898, Hohenlohe to Posadowsky, 28 September 1898, Hohenlohe, *Denkwürdigkeiten der Reichskanzlerzeit*, pp. 460–1; see Hellige and Schulin, *Rathenau und Harden*, Nos. 14 and 15.
104 Cited in Schoeps, *Herzl*, p. 87.

105 Herzl, diary entry for 15 November 1898. Translations are taken from Raphael Patal, ed., *The Complete Diaries of Theodor Herzl*, translated by Harry Zohn, 5 vols., New York, London 1960.
106 Herzl, diary entry for 8 October 1898.
107 William H. Hechler to Grand Duke Friedrich I of Baden, 26 March and 18 April 1898, Ellern and Ellern, *Herzl*, Nos. 1 and 2; Fuchs, *Großherzog von Baden*, III, Nos. 1532–3; Herzl to Grand Duke of Baden, 26 April 1898, Ellern and Ellern, *Herzl*, No. 3; Fuchs, *Großherzog von Baden*, III, No. 1537.
108 Kaiser Wilhelm II, marginal notes on Schweinitz's report from St Petersburg, 24 May 1891, PA AA, Asservat No. 4.
109 Herzl, diary entry for 23 April 1896; Isaiah Friedman, *Germany, Turkey and Zionism 1897–1918*, Oxford 1977, p. 57.
110 Kaiser Wilhelm II, marginal notes of 1 October 1897, PA AA, No. R 529.
111 Herzl, diary entry for 12 June 1895.
112 Herzl to Kaiser Wilhelm II, 22 October and 1 December 1897, Ellern and Ellern, *Herzl*, Nos. 10 and 11; Herzl to Grand Duke of Baden, 22 October 1897, Fuchs, *Großherzog von Baden*, III, No. 1759.
113 Desmond Stewart, *Theodor Herzl, Artist and Politician*, London 1974, p. 260.
114 Grand Duke of Baden to Kaiser Wilhelm II, 28 July 1898, Ellern and Ellern, *Herzl*, No. 12; Fuchs, *Großherzog von Baden*, IV, No. 1879.
115 Kaiser Wilhelm II to Grand Duke of Baden, 29 August 1898, Fuchs, *Großherzog von Baden*, IV, No. 1884.
116 Herzl, diary entry for 3 September 1898. See Herzl to Grand Duke of Baden, 8 September 1898, Ellern and Ellern, *Herzl*, No. 14; Fuchs, *Großherzog von Baden*, IV, No. 1888.
117 Herzl, diary entry for 9 September 1898.
118 Herzl, diary entry for 15 September 1898.
119 Herzl, diary entry for 16 September 1898.
120 Herzl to Eulenburg, 24 September 1898, *Eulenburgs Korrespondenz*, III, No. 1386.
121 Kaiser Wilhelm II to Grand Duke of Baden, 29 September 1898, Ellern and Ellern, *Herzl*, No. 16; Bein, *Herzls Begegnung mit Wilhelm II.*, pp. 44ff.; Fuchs, *Großherzog von Baden*, IV, No. 1892. See *Eulenburgs Korrespondenz*, III, pp. 1920ff.
122 Eulenburg to Herzl, 27 September 1898, Bein, *Herzls Begegnung mit Wilhelm II.*, pp. 44ff.; Fuchs, *Großherzog von Baden*, IV, No. 1891. See *Eulenburgs Korrespondenz*, III, No. 1387. See Herzl, diary entry for 2 October 1898.
123 Herzl, diary entry for 6 October 1898.
124 Herzl, diary entry for 9 October 1898.
125 Herzl, diary entries for 15–17 October 1898; Herzl to Kaiser Wilhelm II, 18 October 1898, draft, ibid.
126 Herzl, diary entries for 19–21 October 1898.
127 Herzl, diary entry for 3 January 1901.
128 Ayse Osmanoglu, *Avec mon Père le Sultan Abdulhamid de son palais à sa prison*, Paris 1991, p. 54. See Philip Mansel, *Constantinople, City of the World's Desire, 1453–1924*, London 1995, pp. 321–2.
129 Bulent Mim Kemal Öke, 'Ottoman Policies towards Zionism', MPhil. diss., Cambridge 1979, pp. 61–92.
130 Bülow, *Memoirs*, I, p. 250. Cf. Friedman, *Germany, Turkey and Zionism*, p. 79; Öke, 'Ottoman Policies towards Zionism', p. 91.
131 See above, p. 944; also Bülow, *Memoirs*, I, pp. 253–4.
132 Mirbach, *Das deutsche Kaiserpaar im Heiligen Lande*, pp. 86 and 99–100.
133 Herzl, diary entry for 29 October 1898.

134 Herzl, diary entry for 2 November 1898.
135 Kaiser Wilhelm II, speech of 8 November 1898 in Damascus, Penzler, *Reden Kaiser Wilhelms II. in den Jahren 1896–1900*, pp. 126–7.
136 Kaiser Wilhelm II to Grand Duke of Baden, 7 November 1898, PA AA, Preußen 1 No. 1d, Bd I. Cf. Hereditary Princess Charlotte of Saxe-Meiningen to Empress Frederick, 13 November 1898, AdHH Schloss Fasanerie.
137 Kaiser Wilhelm II to Empress Frederick, 20 November 1898, AdHH Schloss Fasanerie, printed in facsimile in Carmel and Eisler, *Heilige Land*, pp. 173–4.
138 Hereditary Princess Charlotte of Saxe-Meiningen to Empress Frederick, 1 December 1898, AdHH Schloss Fasanerie. See also Hohenlohe, diary entry for 20 November 1898, Hohenlohe, *Denkwürdigkeiten der Reichskanzlerzeit*, pp. 469–70; Kaiser Wilhelm II, speech of 1 December 1898 on his return to Berlin, Penzler, *Reden Kaiser Wilhelms II. in den Jahren 1896–1900*, pp. 127–8.
139 See Mirbach, *Das deutsche Kaiserpaar im Heiligen Lande*; Hans Forsten, *Unser Kaiser in Palästina. Reise Kaiser Wilhelms II. und der Kaiserin Auguste Victoria nach dem gelobten Lande*, Berlin 1898; Adolf Meyer, *Ins Heilige Land. Reisebilder von der großen Festfahrt nach Jerusalem im Oktober und November 1898*, Berlin 1899; Pastor Ludwig Schneller, *Die Kaiserfahrt durch's heilige Land*, Leipzig 1899; Richard Schott, *Eine Fahrt nach dem Orient. Zur Erinnerung an den Einzug des deutschen Kaisers und der Kaiserin in Jerusalem, Herbst 1898*, Berlin, Eisenach, Leipzig 1898; Friedrich Zange, *Die Jerusalemfahrt Kaiser Wilhelms II. im Lichte der Geschichte*, Berlin 1899; Marie von Bodelschwingh, *Aus Heiligem Land. Den Kindern erzählt*, Bielefeld 1899; Paul von Frankenberg, *Kaiser Wilhelms II. Reise nach Jerusalem*, Berlin 1899. See further Thomas Hartmut Benner, *Die Strahlen der Krone. Die religiöse Dimension des Kaiserstums unter Wilhelm II. vor dem Hintergrund der Orientreise 1898*, Marburg 2001.
140 Kaiser Wilhelm II, speech of 3 February 1899, Penzler, *Reden Kaiser Wilhelms II. in den Jahren 1896–1900*, pp. 144–8; cf. Elkind, *German Emperor's Speeches*, p. 303. See *Young Wilhelm*, pp. 304–5.
141 Fesser, *Platz an der Sonne*, pp. 11–12; Jost Dülffer, Martin Kröger and Rolf-Harald Wippich, *Vermiedene Kriege. Deeskalation von Konflikten der Großmächte zwischen Krimkrieg und Erstem Weltkrieg 1865–1914*, Munich 1997, pp. 475–6. See also *Große Politik*, XIV, No. 3669, note.
142 *Große Politik*, XIV, Nos. 3679 and 3680. See Dülffer et al., *Vermiedene Kriege*, p. 477.
143 Kaiser Wilhelm II to Auswärtiges Amt, 6 November 1897, *Große Politik*, XIV, No. 3686.
144 Ibid., No. 3687.
145 See Dülffer et al., *Vermiedene Kriege*, p. 477.
146 Kaiser Wilhelm II to Bülow, 7 November 1897, *Große Politik*, XIV, No. 3690.
147 Fesser, *Platz an der Sonne*, p. 13.
148 Hohenlohe to Kaiser Wilhelm II, 6 November 1897, *Große Politik*, XIV, No. 3688; Holstein to Eulenburg, 23 November 1897, *Eulenburgs Korrespondenz*, III, No. 1353.
149 Kaiser Wilhelm II to Hohenlohe, 7 November 1897, *Große Politik*, XIV, No. 3689.
150 Kaiser Wilhelm II to Bülow, 7 November 1897, ibid., No. 3690.
151 Dülffer et al., *Vermiedene Kriege*, p. 478.
152 Rotenhan to Kaiser Wilhelm II, 10 November 1897, *Große Politik*, XIV, No. 3693.
153 Hohenlohe, *Denkwürdigkeiten der Reichskanzlerzeit*, p. 413.
154 Hohenlohe to Rotenhan, 10 November 1897, ibid., p. 412.
155 Holstein to Hohenlohe, 9 November 1897, ibid., p. 411. See Fesser, *Platz an der Sonne*, p. 13.
156 Tirpitz to Hohenlohe, 10 November 1897, Hohenlohe, *Denkwürdigkeiten der Reichskanzlerzeit*, p. 412.
157 Hohenlohe to Rotenhan, 10 November 1897, ibid.

158 See below, p. 960.
159 Cited in Rotenhan to Bülow, 11 November 1897, *Große Politik*, XIV, No. 3695; Dülffer et al., *Vermiedene Kriege*, p. 478.
160 Fesser, *Platz an der Sonne*, p. 13.
161 Hohenlohe to Kaiser Wilhelm II, 11 November 1897, *Große Politik*, XIV, No. 3696.
162 Memorandum of 15 November 1897, ibid., No. 3701. See Dülffer et al., *Vermiedene Kriege*, pp. 479–80.
163 Memorandum of 15 November 1897, *Große Politik*, XIV, No. 3701.
164 Hohenlohe to Kaiser Wilhelm II, 18 November 1897, ibid., No. 3707.
165 Kaiser Wilhelm II, marginal notes, Hohenlohe to Kaiser Wilhelm II, 18 November 1897, ibid. See Dülffer et al., *Vermiedene Kriege*, p. 481.
166 Kaiser Wilhelm II to Auswärtiges Amt, 24 November 1897, Hohenlohe, *Denkwürdigkeiten der Reichskanzlerzeit*, p. 418.
167 Kaiser Wilhelm II to Hohenlohe, 26 November 1897, ibid., p. 419.
168 Hohenlohe, diary entry for 29 November 1897, ibid.
169 Dülffer et al., *Vermiedene Kriege*, p. 481; Fesser, *Platz an der Sonne*, p. 14.
170 See e.g. Bülow to Hatzfeldt, 2 December 1897, *Holstein Papers*, IV, No. 636; also Hohenlohe to Holstein, 31 December 1897, ibid., No. 640.
171 Hohenlohe to his son Alexander, 8 July 1900, Hohenlohe, *Denkwürdigkeiten der Reichskanzlerzeit*, p. 578. Cf., however, ibid., p. 424.
172 Eulenburg to Kaiser Wilhelm II, 23 December 1897, *Eulenburgs Korrespondenz*, III, No. 1355.
173 Kaiser Wilhelm II to Kaiser Franz Joseph, 8 January 1898, HHStA Vienna, Kab. Archiv, Geheimakten 1.
174 Kaiser Franz Joseph to Kaiser Wilhelm II, 11 January 1898, ibid.
175 Kaiser Wilhelm II to Tsar Nicholas II, 19 December 1897, *Große Politik*, XIV, No. 3739.
176 Kaiser Wilhelm II, speech of 15 December 1897 in Kiel, Penzler, *Reden Kaiser Wilhelms II. in den Jahren 1896–1900*, pp. 78–80.
177 Fesser, *Platz an der Sonne*, pp. 14–15.
178 Prince Heinrich of Prussia to Empress Frederick, 10 April 1898, AdHH Schloss Fasanerie.
179 Prince Heinrich of Prussia to Empress Frederick, 20 February 1899, ibid.
180 Prince Heinrich of Prussia to Kaiser Wilhelm II, 4 February 1899, GStA Berlin, BPHA Rep. 52 V1 No. 13a. See above, pp. 745–9.
181 Bülow to Kaiser Wilhelm II, 13 December 1897, *Große Politik*, XIV, No. 3732.
182 Kaiser Wilhelm II, marginal notes on Bülow's report of 13 December 1897, ibid.
183 Cited in Dülffer et al., *Vermiedene Kriege*, p. 485.
184 Prince Heinrich of Prussia to Empress Frederick, 10 April 1898, AdHH Schloss Fasanerie.
185 Prince Heinrich of Prussia to Empress Frederick, 1 June 1898, ibid.
186 Prince Heinrich of Prussia to Empress Frederick, 6 May and 9 August 1899, ibid.
187 Prince Heinrich of Prussia to Empress Frederick, 8 February 1898, ibid.
188 Prince Heinrich of Prussia to Empress Frederick, 10 April 1898, ibid.
189 Prince Heinrich of Prussia to Empress Frederick, 1 June 1898, ibid.
190 Prince Heinrich of Prussia to Empress Frederick, 8 April 1898, ibid.
191 Prince Heinrich of Prussia to Empress Frederick, 24 October 1898, ibid.
192 Prince Heinrich of Prussia to Empress Frederick, 6 November 1898, ibid. See also Heinrich's letter of 27 December 1898, ibid.

31 THE KAISER AND ENGLAND

1 See above, ch. 26.
2 Bülow to Eulenburg, 20 July 1898, *Eulenburgs Korrespondenz*, III, No. 1381.

3 Kaiser Wilhelm II, marginal notes on Dönhoff's report of 5 May 1896, PA AA, Asservat No. 4.
4 Kaiser Wilhelm II to Queen Victoria, 2 January 1897, RA I61/1.
5 Queen Victoria to Kaiser Wilhelm II, 17 February 1897, in Hohenlohe, *Denkwürdigkeiten der Reichskanzlerzeit*, p. 303.
6 Grierson to Bigge, 20 February 1897, RA I61/2a. See above, pp. 628–33.
7 Kaiser Wilhelm II to Queen Victoria, 10 June 1897, RA Z500/7.
8 See above, p. 798, and below, pp. 991–8.
9 Hatzfeldt to Hohenlohe, 11 December 1897, *Große Politik*, XIV, No. 3730.
10 Empress Frederick to Queen Victoria, 29 August 1896, AdHH Schloss Fasanerie.
11 Grierson to Bigge, 24 November 1896, RA I60/160.
12 Lascelles to Salisbury, 22 November 1896, quoted in van't Padje, 'Anglo-German Imperialist Rivalry', pp. 102–3.
13 Empress Frederick to Queen Victoria, 24 October 1896, AdHH Schloss Fasanerie.
14 Swaine to Bigge, 16 November 1898, RA I61/27.
15 See Kennedy, *Anglo-German Antagonism*, pp. 231ff.; J. A. S. Grenville, *Lord Salisbury and Foreign Policy*, London 1964, pp. 148–76; J. L. Garvin and J. Amery, *Life of Joseph Chamberlain*, 6 vols., London 1932–69, III, pp. 251ff.; Friedrich Meinecke, *Geschichte des deutsch-englischen Bündnisproblems*, Munich, Berlin 1927, pp. 85–114; Gerhard Ritter, *Die Legende von der verschmähten englischen Freundschaft 1898/1901*, Leipzig 1929; Roberts, *Salisbury*, pp. 689ff.; Peter Winzen, *Bülows Weltmachtkonzept. Untersuchungen zur Frühphase seiner Außenpolitik 1897–1901*, Boppard am Rhein 1977, pp. 156ff.; Gregor Schöllgen, *Imperialismus und Gleichgewicht. Deutschland, England und die orientalische Frage 1871–1914*, Munich 1992, pp. 86ff.; Hildebrand, *Das vergangene Reich*, pp. 213ff.
16 Grierson to Bigge, 21 January 1898, RA I61/32a. The cartoon of Wilhelm II as Emperor of China appeared in *Punch* on 15 January 1898. See Röhl, 'Der Kaiser und England', p. 178.
17 Sir Theodore Martin to Queen Victoria, 13, 14 and 16 January 1898, RA I61/30–32; Bigge to Queen Victoria, 26 January 1898, RA I61/33.
18 Grierson to Bigge, 5 February 1898, RA I61/34.
19 Münster to Hohenlohe, 23 February 1898, Hohenlohe, *Denkwürdigleiten der Reichskanzlerzeit*, p. 431.
20 Grierson to Bigge, 26 February 1898, RA I61/35a.
21 See George Duke of York to Bigge, 9 March 1898, RA I61/36.
22 See Holstein to Eulenburg, 10 November 1897, *Eulenburgs Korrespondenz*, III, No. 1352.
23 Bigge to Queen Victoria, 14 March 1898, RA I61/37.
24 Grierson to Bigge, 26 February and 6 March 1898, RA I61/35a–b; Bigge to Queen Victoria, 14 March 1898, RA I61/37 and Add A18/Y10; Eulenburg to Bülow, 11 July 1898, *Eulenburgs Korrespondenz*, III, No. 1378.
25 See above, ch. 4.
26 Prince Christian of Schleswig-Holstein to Albert Edward Prince of Wales, undated [March 1898] RA Add A18/Y10.
27 Lascelles to Albert Edward Prince of Wales, 25 March 1898, RA I61/39.
28 Ibid.; Grierson to Bigge, 7 May 1898, RA I61/46a.
29 Grierson to Bigge, 28 May 1898, RA I61/48a. See Kaiser Wilhelm II to Tsar Nicholas II, 30 May 1898, in Goetz, *Briefe Wilhelms II. an den Zaren*, pp. 309–11. The Kaiser was in close touch in England with, among others, the Earl of Lonsdale und Sir Edward Sullivan. See Lonsdale to Kaiser Wilhelm II, 4 August and 28 September 1898, GStA Berlin, BPHA Rep. 53J Lit. L. No. 9 and Lit. V. No. 1-W No. 1; Sullivan to Kaiser Wilhelm II, 7 January 1899, ibid., Lit. S. Nos. 15–21.
30 Grierson to Bigge, 7 May 1898, RA I61/46a.
31 *Große Politik*, XIV, 1, No. 3781.

32 Cited in Roberts, *Salisbury*, p. 690. See Paul Kennedy, *The Samoan Tangle. A Study in Anglo-German-American Relations 1878–1900*, Dublin 1974, p. 157.
33 Hatzfeldt to Auswärtiges Amt, 22 May 1898, with Kaiser Wilhelm II's marginal notes, *Holstein Papers*, IV, No. 654; also Lonsdale to Kaiser Wilhelm II, 4 August 1898, GStA Berlin, BPHA Rep. 53J Lit. L. No. 9; Kennedy, *Samoan Tangle*, pp. 138–9 and 155ff.
34 Empress Frederick to Kaiser Wilhelm II, 29 May 1898, GStA Berlin, BPHA Rep. 52T No. 13. Cf. Empress Frederick to Hatzfeldt, 30 May 1898, Hatzfeldt, *Nachgelassene Papiere*, II, No. 725; Holstein to Hatzfeldt, 8 June 1898, ibid., No. 727.
35 Empress Frederick to Kaiser Wilhelm II, 31 May 1898, GStA Berlin, BPHA Rep. 52T No. 13.
36 Kaiser Wilhelm II to Tsar Nicholas II, 30 May 1898, Goetz, *Briefe Wilhelms II. an den Zaren*, pp. 309–11. A copy of a draft of this letter is to be found in the files of the Auswärtiges Amt, PA AA, Preußen 1 No. 1d Bd 1.
37 See *Große Politik*, XIV, 1, Nos. 3803–4; Rich, *Friedrich von Holstein*, II, pp. 582–5; Canis, *Von Bismarck zur Weltpolitik*, pp. 282 and 286.
38 Cited in Holstein to Hatzfeldt, 31 May 1898, *Holstein Papers*, IV, No. 656. See Kaiser Wilhelm II's note, *Große Politik*, XIV, 1, No. 3799.
39 Hatzfeldt to Holstein, 2 June 1898, *Holstein Papers*, IV, No. 658. See Hatzfeldt to Auswärtiges Amt, 2 June 1898, *Große Politik*, XIV, 1 No. 3800; Roberts, *Salisbury*, p. 691.
40 Kaiser Wilhelm II to Empress Frederick, 1 June 1898, draft, PA AA, Preußen 1 No. 1d, Bd I, printed in *Holstein Papers*, IV, No. 657. This document was placed in the secret files of the Auswärtiges Amt on Bülow's instructions.
41 Empress Frederick to Kaiser Wilhelm II, 3 June 1898, PA AA, Preußen 1 No. 1d, Bd I.
42 Empress Frederick to Queen Victoria, 15 July [June?] 1898, RA I61/52, in Buckle, *Letters of Queen Victoria*, III, pp. 258–9.
43 Salisbury to Queen Victoria, 21 June [July?] 1898, RA I61/50, in Buckle, *Letters of Queen Victoria*, III, pp. 259–60.
44 On the negotiations, see Rich, *Friedrich von Holstein*, II, pp. 586–7; Grenville, *Salisbury and Foreign Policy*, pp. 177–98; Kennedy, *Samoan Tangle*, pp. 130ff.
45 Bülow to Hatzfeldt, 8 June 1898, *Große Politik*, XIV, 1, Nos. 3804–6; also ibid., Nos. 3835–6. See Eugen Fischer, *Holsteins großes Nein. Die deutsch-englischen Bündnisverhandlungen von 1898–1901*, Berlin 1925, p. 65; A. S. Jerussalimski, *Die Außenpolitik und die Diplomatie des deutschen Imperialismus Ende des 19.* Jahrhunderts, Berlin 1954, pp. 614–15; Kennedy, *Samoan Tangle*, pp. 138–9 and 155ff.
46 Salisbury to Queen Victoria, 27 July 1898, RA I61/53.
47 Kaiser Wilhelm II, marginal notes on Richthofen to Eulenburg, 20 July 1898, PA AA, IA England 78 No. 1 secr., Bd. IV, R 5802; cited in Kennedy, *Samoan Tangle*, p. 130. The Kaiser's marginal comments were omitted by the editors of *Große Politik*.
48 Eulenburg to Bülow, 11 July 1898, *Eulenburgs Korrespondenz*, III, No. 1378.
49 Eulenburg to Bülow, 23 July 1898, ibid., No. 1383.
50 Empress Frederick to Kaiser Wilhelm II, 29 June 1898, GStA Berlin, BPHA Rep. 52T No. 13. See Lonsdale to Kaiser Wilhelm II, 4 August 1898, ibid., Rep. 53J Lit. L. No. 9.
51 Salisbury to Queen Victoria, 2 August 1898, RA I61/62.
52 Salisbury to Queen Victoria, 10 August 1898, RA I61/63, in Buckle, *Letters of Queen Victoria*, III, pp. 263–4.
53 Salisbury to Queen Victoria, 4 August 1898, in Buckle, *Letters of Queen Victoria*, III, pp. 262–3. Cf. the version of this letter, dated 1 August 1898, in RA I61/57.
54 Balfour to Queen Victoria, 26 August 1898, RA I61/65, 65a and b.
55 Lascelles to Queen Victoria, 9 December 1898, RA I61/78. See below, pp. 985–7.
56 Bülow to Kaiser Wilhelm II, 24 August 1898, *Große Politik*, XIV, 1, No. 3867.

57 Bülow to Kaiser Wilhelm II, 19 August 1898, GStA Berlin, BPHA Rep. 53J Lit. B. No. 16a, cited in Canis, *Von Bismarck zur Weltpolitik*, p. 288.

58 Holstein to Hatzfeldt, 22 December 1898, Canis, *Von Bismarck zur Weltpolitik*, p. 293.

59 Bülow to Kaiser Wilhelm II, 6 August 1900, GStA Berlin, BPHA Rep. 53J Lit. B. No. 16a, cited in Röhl, 'Kaiser und England', p. 174.

60 Kaiser Wilhelm II to Tsar Nicholas II, 18 August 1898, copy in PA AA, Preußen 1 No. 1d, Bd I, in Goetz, *Briefe Wilhelms II. an den Zaren*, pp. 311–12.

61 Kaiser Wilhelm II to Tsar Nicholas II, 20 October 1898, copy in PA AA, Preußen 1 No. 1d, Bd I, in Goetz, *Briefe Wilhelms II. an den Zaren*, pp. 313–15. See Ludwig Bittner, 'Neue Beiträge zur Haltung Kaiser Wilhelms II. in der Faschoda-Frage', *Historische Zeitschrift*, 162 (1940); Langer, *Diplomacy*, pp. 566–70; Kennedy, *Samoan Tangle*, p. 157.

62 Kaiser Wilhelm II to Tsar Nicholas II, 9 November 1898, in Goetz, *Briefe Wilhelms II. an den Zaren*, pp. 315–18. The copy of a draft of this letter, dated 8 November 1898, differs in several respects from the version actually sent. See PA AA, Preußen 1 No. 1d, Bd I.

63 Kaiser Wilhelm II to Empress Frederick, 20 November 1898, AdHH Schloss Fasanerie, printed in facsimile in Carmel and Eisler, *Der Kaiser reist ins Heilige Land*, pp. 173–4. See also Kaiser Wilhelm II to Queen Victoria, 16 November 1898, in Buckle, *Letters of Queen Victoria*, III, p. 311.

64 Salisbury to Queen Victoria, 26 November 1898, RA I61/77.

65 Lascelles to Queen Victoria, 9 December 1898, RA I61/78. See above, p. 983.

66 Kaiser Wilhelm II to Empress Frederick, 20 December 1898, AdHH Schloss Fasanerie. Cf. the copy in RA I61/80. See also Lascelles to Queen Victoria, 24 December 1898, RA I61/81, in Buckle, *Letters of Queen Victoria*, III, pp. 321–2; Empress Frederick to Kaiser Wilhelm II, 25 December 1898, GStA Berlin, BPHA Rep. 52T No. 13.

67 Kaiser Wilhelm II, marginal notes on Hatzfeldt's report from London of 22 December 1898, PA AA, Asservat No. 4. Hatzfeldt's report is printed without the Kaiser's marginal notes in *Große Politik*, XIV, 2, pp. 405ff. Cf. Bigge to Queen Victoria, 25 December 1898, RA I61/82, in Buckle, *Letters of Queen Victoria*, III, pp. 322–3; also Kaiser Wilhelm II to Queen Victoria, 29 December 1898, RA I61/83, ibid., pp. 323–4.

68 Rhodes to Kaiser Wilhelm II, 26 February 1899, GStA Berlin, BPHA Rep. 53J Lit R. No. 12.

69 Kaiser Wilhelm II, marginal notes on Bülow's report of 23 February 1899, PA AA, Asservat No. 4.

70 Grierson to Bigge, 18 March 1899, RA I62/6. On Wilhelm's wish for a 'German Asia Minor', see Eulenburg to Bülow, 10–11 July 1899, *Eulenburgs Korrespondenz*, III, No. 1396.

71 Kaiser Wilhelm II to Empress Frederick, 26 March 1899, AdHH Schloss Fasanerie.

72 Grierson to Bigge, 18 March 1899, RA I62/6; Kennedy, *Samoan Tangle*, p. 162.

73 Rhodes to Albert Edward Prince of Wales, undated [March 1899], in Buckle, *Letters of Queen Victoria*, III, pp. 349–51.

74 Kaiser Wilhelm II to Empress Frederick, 26 March 1899, AdHH Schloss Fasanerie.

75 Empress Frederick to Kaiser Wilhelm II, 30 March 1899, PA AA, Preußen 1 No. 1d, Bd I.

76 Grierson to Bigge, 1 April 1899, RA I62/9a.

77 See RA QVJ, 6 February 1899, in Buckle, *Letters of Queen Victoria*, III, p. 337. See Hartmut Pogge von Strandmann, 'Nationalisierungsdruck und königliche Namensänderung in England. Das Ende der Großfamilie europäischer Dynastien', in Gerhard A. Ritter and Peter Wende, eds., *Rivalität und Partnerschaft. Studien zu den deutsch-britischen Beziehugen im 19. und 20. Jahrhundert. Festschrift für Anthony J. Nicholls*, Paderborn 1999, pp. 69–91; also Kennedy, *Samoan Tangle*, pp. 180–1.

78 RA QVJ, 15 March and 9 April 1899, in Buckle, *Letters of Queen Victoria*, III, pp. 347–8 and 356; Pogge von Strandmann, 'Nationalisierungsdruck', pp. 74–5 and 77.

79 Lascelles to Salisbury, 31 March 1899, RA I62/9; Grierson to Bigge, 1 April 1899, RA I62/9a; Grierson to Lascelles, 3 May 1899, in Buckle, *Letters of Queen Victoria*, III, pp. 357ff. See Pogge von Strandmann, 'Nationalisierungsdruck', pp. 75–6.
80 Albert Edward Prince of Wales to Empress Frederick, 25 April 1899, cited in Pogge von Strandmann, 'Nationalisierungsdruck', p. 77; also Buckle, *Letters of Queen Victoria*, III, p. 356.
81 Grierson to Bigge, 4 May 1899, RA I62/10a.
82 Arthur Duke of Connaught to Queen Victoria, 22 April 1899, RA Vic/Add A 35/484, cited in Pogge von Strandmann, 'Nationalisierungsdruck', p. 77.
83 Queen Victoria to Empress Frederick, 3 May 1899, Albert Edward Prince of Wales to Empress Frederick, 2 May 1899, Alfred Duke of Coburg to Queen Victoria, 8 May 1899, cited in Pogge von Strandmann, 'Nationalisierungsdruck', pp. 78–9.
84 Kaiser Wilhelm II to Queen Victoria, 27 May 1899, cited in Pogge von Strandmann, 'Nationalisierungsdruck', p. 80. See Agatha Ramm, ed., *Beloved and Darling Child. Last Letters between Queen Victoria and Her Eldest Daughter 1886–1901*, Stroud 1990, pp. 231–2.
85 Hohenlohe, note of 2 December 1896, Hohenlohe, *Denkwürdigkeiten der Reichskanzlerzeit*, pp. 284–5. Wilhelm's accusations were discussed by Salisbury and the Queen on 2 December 1896: RA QVJ.
86 See above all Kennedy, *Samoan Tangle*, pp. 178ff.; Canis, *Von Bismarck zur Weltpolitik*, pp. 313ff.
87 RA QVJ, 27 January 1899, in Buckle, *Letters of Queen Victoria*, III, p. 336.
88 Kaiser Wilhelm II to Queen Victoria, 2 February 1899, in Buckle, *Letters of Queen Victoria*, III, pp. 336–7.
89 RA QVJ, 17 February 1899, in Buckle, *Letters of Queen Victoria*, III, pp. 340–1.
90 Queen Victoria to Tsar Nicholas II, [March?] 1899, RA I62/22, in Sir Sidney Lee, *King Edward VII. A Biography*, 2 vols., London 1925–7, I, p. 741, and in Buckle, *Letters of Queen Victoria*, III, pp. 343–4.
91 Lascelles to Salisbury, 11 March 1899, in Kennedy, *Samoan Tangle*, p. 180.
92 Grierson to Lascelles, 3 May 1899, RA I62/10, in Buckle, *Letters of Queen Victoria*, III, pp. 357–9.
93 Grierson to Bigge, 4 May 1899, RA I62/10a.
94 Lascelles to Salisbury, 5 May 1899, RA I62/11, printed in part in Buckle, *Letters of Queen Victoria*, III, pp. 359–60.
95 Salisbury to Lascelles, 10 May 1899, cited in Kennedy, *Samoan Tangle*, p. 182.
96 Queen Victoria to Kaiser Wilhelm II, 18 May 1899, GStA Berlin, BPHA Rep. 52 W3 No. 11.
97 Kaiser Wilhelm II to Bülow, 20 May 1899, quoted in Bülow to Hohenlohe, 21 May 1899, Hohenlohe, *Denkwürdigkeiten der Reichskanzlerzeit*, pp. 503–4.
98 Lascelles, report from Berlin, 26 May 1899, RA I62/12. The Kaiser was particularly hurt when he heard that his cousin Grand Duke Ernst Ludwig of Hesse-Darmstadt had been invited to attend the Queen's birthday celebrations: Holstein to Hatzfeldt, 13 May 1899, Hatzfeldt, *Nachgelassene Papiere*, II, No. 757; McLean, *Royalty and Diplomacy*, p. 92.
99 RA QVJ, 5 June 1899, in Buckle, *Letters of Queen Victoria*, III, p. 381.
100 RA QVJ, 1 June 1899.
101 Kaiser Wilhelm II to Queen Victoria, 27 May 1899, RA I62/14, in Buckle, *Letters of Queen Victoria*, III, pp. 375–9. See Eyck, *Das Persönliche Regiment*, p. 234.
102 Salisbury to Queen Victoria, 3 June 1899, RA I62/15, in Buckle, *Letters of Queen Victoria*, III, p. 379. See Salisbury's memorandum of the same day, ibid., pp. 379–81.
103 Queen Victoria to Kaiser Wilhelm II, 12 June 1899, RA I62/19, ibid., pp. 381–2. See Salisbury to Queen Victoria, 14 June 1899, RA I62/21.

104 Eulenburg to Bülow, 10–11 July 1899, *Eulenburgs Korrespondenz*, III, No. 1396.
105 Eulenburg to Bülow, 4 July 1899, ibid., No. 1395.

32 WILHELM AND THE BIRTH OF THE GERMAN BATTLEFLEET

1 Kennedy, 'The Kaiser and German Weltpolitik', p. 162.
2 On the eve of the Boer War, one German general wrote to another: 'If only our fleet were at the stage [. . .] where we could use the war in South Africa [. . .] for a surprise attack [on England]': Mudra to Colmar Freiherr von der Goltz, 2 October 1899, BA-MA Freiburg, von der Goltz Papers, N737 Zug. 228/95.
3 See above, pp. 153–8 and 355–60.
4 Prince Heinrich of Prussia to Tirpitz, 24 October 1893, BA-MA Freiburg, Tirpitz Papers N253/183.
5 Wilhelm Deist, *Flottenpolitik und Flottenpropaganda. Das Nachrichtenbureau des Reichsmarineamtes 1897–1914*, Stuttgart 1976, pp. 20–1.
6 See Waldersee, diary entry for 9 March 1897, GStA Berlin, Waldersee Papers; omitted from Meisner, II, p. 393.
7 Deist, *Flottenpolitik und Flottenpropaganda*, p. 23; Cecil, *Wilhelm II*, I, p. 303.
8 Bülow, *Memoirs*, II, p. 62.
9 Cecil, *Wilhelm II*, I, p. 298.
10 Kaiser Wilhelm II to Prince Heinrich of Prussia, April 1897, *Berliner Lokal-Anzeiger*, 25 April 1897, PA AA, Preußen 1 No. 1d, Bd I. See above, p. 848.
11 Waldersee, diary entry for 28 April 1892, GStA Berlin, Waldersee Papers; omitted from Meisner, II, p. 240.
12 Waldersee, diary entry for 29 April 1892, GStA Berlin, Waldersee Papers; omitted from Meisner, II, p. 240. See above, pp. 155 and 359.
13 Waldersee, diary entry for 10 September 1892, GStA Berlin, Waldersee Papers; omitted from Meisner, II, p. 264.
14 Waldersee to Verdy, 12 April 1895, GStA Berlin, Waldersee Papers.
15 Waldersee, diary entry for 10 March 1895, Meisner, II, p. 339.
16 Cecil, *Wilhelm II*, I, p. 313. See Massie, *Dreadnought*, pp. 150ff; Thomas Kohut, *Wilhelm II and the Germans*, pp. 177ff.; also above, pp. 33ff.
17 Görlitz, *Der Kaiser*, p. 29. Cf. Paul Simsa, *Marine intern. Entwicklung und Fehlentwicklung der deutschen Marine 1888–1939*, Stuttgart 1972, p. 23.
18 Hildebrand, *Das vergangene Reich*, p. 202.
19 Holstein to Bülow, 7 February 1895, *Eulenburgs Korrespondenz*, II, No. 1082.
20 See Holstein to Eulenburg, 5 October 1893, ibid., p. 1114.
21 Kiderlen-Wächter to Eulenburg, 1 October 1893, ibid., No. 827.
22 Waldersee, diary entry for 16 October 1893, GStA Berlin, Waldersee Papers; omitted from Meisner, II, p. 296.
23 Waldersee, diary entry for 19 October 1893, GStA Berlin, Waldersee Papers; cf. Meisner, II, p. 297.
24 Cecil, *Wilhelm II*, I, p. 299.
25 Waldersee, diary entry for 14 December 1894, GStA Berlin, Waldersee Papers; omitted from Meisner, II, pp. 332–3. See also the diary entry for 15 February 1895, Meisner, II, pp. 338–9.
26 Marschall, diary entries for 2–9 February 1895, cited in *Eulenburgs Korrespondenz*, II, p. 1459.
27 Kaiser Wilhelm II, lecture of 8 February 1895 in the Königliche Kriegs-Akademie, BA-MA Freiburg, N160/13.

28 See Cecil, *Wilhelm II*, I, p. 300.
29 Ibid., p. 303.
30 Hohenlohe, diary entry for 31 January 1895, Hohenlohe, *Denkwürdigkeiten der Reichskanzlerzeit*, p. 32.
31 Jagemann, report of 9 January 1895, Fuchs, *Großherzog von Baden*, III, No. 1407.
32 Tirpitz, *Erinnerungen*, p. 49; Steinberg, *Yesterday's Deterrent*, p. 72; Franz Uhle-Wettler, *Alfred von Tirpitz in seiner Zeit*, Hamburg, Berlin, Bonn 1998, p. 68.
33 See above, p. 908.
34 Kaiser Wilhelm II, lecture of 8 February 1895, BA-MA Freiburg, N160/13.
35 Ibid.
36 Ibid.
37 Ibid.
38 Ibid.
39 Waldersee, diary entry for 10 February 1895, GStA Berlin, Waldersee Papers; omitted from Meisner, II, p. 338.
40 Waldersee, diary entry for 12 February 1895, GStA Berlin, Waldersee Papers; omitted from Meisner, II, p. 338.
41 Fischer to Eulenburg, 17 February 1895, *Eulenburgs Korrespondenz*, III, No. 1090.
42 Cecil, *Wilhelm II*, I, p. 303.
43 Eulenburg to Holstein, 21 January 1895, *Eulenburgs Korrespondenz*, II, No. 1075.
44 Kaiser Wilhelm II to Hohenlohe, 1 March 1895, Hohenlohe, *Denkwürdigkeiten der Reichskanzlerzeit*, p. 49. See Waldersee, diary entry for 24 February 1895, GStA Berlin, Waldersee Papers; omitted from Meisner, II, p. 339.
45 Holstein to Radolin, 2 March 1895, *Holstein Papers*, III, No. 452.
46 Uhle-Wettler, *Tirpitz*, p. 87.
47 Cecil, *Wilhelm II*, I, p. 294.
48 See above, pp. 783ff.
49 Hohenlohe to Kaiser Wilhelm II, 7 January 1896, in Hohenlohe, *Denkwürdigkeiten der Reichskanzlerzeit*, pp. 152–3.
50 Kaiser Wilhelm II to Hohenlohe, 8 January 1896, ibid., pp. 153–4.
51 Marschall, diary entry for 8 January 1896, Marschall Papers, Schloss Neuershausen.
52 Marschall, diary entries for 8, 11 and 13 January 1896, ibid.
53 Holstein to Eulenburg, 25 January 1896, *Eulenburgs Korrespondenz*, III, No. 1191; *Holstein Papers*, III, p. 589, n. 2.
54 Eulenburg to Holstein, 31 January 1896, *Holstein Papers*, III, No. 525.
55 Holstein to Eulenburg, 13 and 25 January 1896, *Eulenburgs Korrespondenz*, III, Nos. 1188 and 1191.
56 Cf. Steinberg, *Yesterday's Deterrent*, pp. 90–1.
57 Hohenlohe to Kaiser Wilhelm II, 14 January 1896, Hohenlohe, *Denkwürdigkeiten der Reichskanzlerzeit*, pp. 156ff. .
58 Hohenlohe, undated note, Hohenlohe to Kaiser Wilhelm II, 7 January 1896 and diary entry for 22 January 1896, ibid., pp. 151–61. See Holstein to Eulenburg, 14 January 1896, *Eulenburgs Korrespondenz*, III, No. 1189.
59 Marschall, diary entries for 12–16 January 1896, Marschall Papers, Schloss Neuershausen. See *Eulenburgs Korrespondenz*, III, p. 1637.
60 Holstein to Eulenburg, 14 January 1896, *Eulenburgs Korrespondenz*, III, No. 1189.
61 Eulenburg to Holstein, 14 January 1896, cited ibid., p. 1637.
62 August Eulenburg to Hohenlohe, 15 January 1896, Hohenlohe, *Denkwürdigkeiten der Reichskanzlerzeit*, pp. 158–9.
63 Kaiser Wilhelm II to Senden, 16 January 1896, ibid., p. 159.

64 Marschall, diary entries for 13–16 January 1896, Marschall Papers, Schloss Neuershausen; Hohenlohe to Kaiser Wilhelm II, 17 January 1896, Kaiser Wilhelm II to Hohenlohe, 17 January 1896, Holstein to Hohenlohe, undated, and Hohenlohe, note for his son Alexander of 18 January 1896, Hohenlohe, *Denkwürdigkeiten der Reichskanzlerzeit*, pp. 160–1.
65 Waldersee, diary entry for 21 January 1896, GStA Berlin, Waldersee Papers; cf. Meisner, II, pp. 365–6.
66 Holstein to Hatzfeldt, 16 January 1896, *Holstein Papers*, III, No. 522.
67 Hohenlohe, diary for 25 January 1896, Hohenlohe, *Denkwürdigkeiten der Reichskanzlerzeit*, p. 162.
68 Holstein to Eulenburg, 25 January 1896, *Eulenburgs Korrespondenz*, III, No. 1191.
69 Hohenlohe, diary entry for 1 February 1896, Hohenlohe, *Denkwürdigkeiten der Reichskanzlerzeit*, p. 164.
70 Waldersee, diary entry for 9 February 1896, GStA Berlin, Waldersee Papers; printed only in part in Meisner, II, p. 367.
71 Kiderlen-Wächter to Holstein, 25 March 1896, *Holstein Papers*, III, No. 537.
72 Waldersee, diary entry for 3 March 1896, GStA Berlin, Waldersee Papers; printed only in part in Meisner, II, pp. 368–9.
73 Holstein to Eulenburg, 13 and 25 January 1896, *Eulenburgs Korrespondenz*, III, Nos. 1188 and 1191.
74 Hohenlohe, diary entry for 2 July 1896, Hohenlohe, *Denkwürdigkeiten der Reichskanzlerzeit*, pp. 240–1.
75 Eulenburg, note of 9 July 1896, *Eulenburgs Korrespondenz*, III, No. 1238.
76 Senden, undated memorandum [spring 1896], BA-MA Freiburg Senden Papers N160/11 fol. 45ff. See Steinberg, *Yesterday's Deterrent*, p. 94.
77 Senden, undated memorandum [spring 1896], BA-MA Freiburg, Senden Papers N160/11 fol. 45ff.
78 Cf. Müller's memorandum of 1896, above, pp. 934–8.
79 Prince Heinrich of Prussia to Tirpitz, 21 February 1896, BA-MA Freiburg, Tirpitz Papers N253/183.
80 Tirpitz to Senden, 15 February 1896, ibid., Senden Papers N160/5.
81 Tirpitz to Stosch, 21 December 1895 and 13 February 1896, Tirpitz, *Erinnerungen*, pp. 52–6.
82 Deist, *Flottenpolitik und Flottenpropaganda*, pp. 52–61.
83 Steinberg, *Yesterday's Deterrent*, p. 84. For Tirpitz's intentions, see above all Berghahn, *Der Tirpitz-Plan*; Paul M. Kennedy, 'Tirpitz, England and the Second Navy Law of 1900. A Strategical Critique', *Militärgeschichtliche Mitteilungen*, 2 (1970); Michael Epkenhans, *Die wilhelminische Flottenrüstung 1908–1914. Weltmachtstreben, industrieller Fortschritt, soziale Integration*, Munich 1991; Massie, *Dreadnought*, pp. 160ff.
84 Uhle-Wettler, *Tirpitz*, p. 53.
85 Steinberg, *Yesterday's Deterrent*, pp. 81ff.; Uhle-Wettler, *Tirpitz*, pp. 90–3.
86 Tirpitz to Senden, 20 March 1896, BA-MA Freiburg, Senden Papers N160/5. See Steinberg, *Yesterday's Deterrent*, p. 95.
87 Senden to Tirpitz, 31 March 1896, BA-MA Freiburg, Senden Papers N160/5. See Steinberg, *Yesterday's Deterrent*, p. 96.
88 Holstein to Bülow, 17 February 1897, *Holstein Papers*, IV, No. 605.
89 Waldersee, diary entry for 28 June 1896, GStA Berlin, Waldersee Papers; cf. Meisner, II, pp. 371–2.
90 Empress Frederick to Queen Victoria, 24 October 1896, AdHH Schloss Fasanerie; cited in Röhl, 'Der Kaiser und England', p. 169.
91 Holstein to Kiderlen-Wächter, 13 August 1896, *Holstein Papers*, III, No. 582.

92 See above, pp. 856–7.
93 Waldersee, diary entry for 31 January 1897, GStA Berlin, Waldersee Papers; Meisner, II, p. 390.
94 Waldersee, diary entry for 31 January 1897, GStA Berlin, Waldersee Papers; cf. Meisner, II, p. 391.
95 Berghahn, *Tirpitz-Plan*, p. 97.
96 Ibid., p. 95.
97 Ibid., pp. 97–8.
98 Ibid., p. 98.
99 Röhl, *Germany without Bismarck*, pp. 212ff.
100 Berghahn, *Tirpitz-Plan*, p. 100.
101 Ibid., p. 99.
102 Ibid.
103 See above, pp. 852–3.
104 Berghahn, *Tirpitz-Plan*, p. 100.
105 Hohenlohe, note of 8 March 1897, Hohenlohe to Kaiser Wilhelm II, 9 and 15 March 1897, Kaiser Wilhelm II to Hohenlohe, 15 and 16 March 1897, Hohenlohe, *Denkwürdigkeiten der Reichskanzlerzeit*, pp. 312–20.
106 Waldersee, diary entry for 16 March 1897, GStA Berlin, Waldersee Papers; cf. Meisner, II, pp. 393ff.
107 Waldersee, diary entry for 20 March 1897, GStA Berlin, Waldersee Papers; printed only in part in Meisner, II, p. 395.
108 Waldersee, diary entry for 30 May 1897, GStA Berlin, Waldersee Papers; cf. Meisner, II, p. 397.
109 Berghahn, *Tirpitz-Plan*, pp. 101–2.
110 Ibid., p. 102.
111 Ibid., p. 103.
112 Ibid., p. 105.
113 Ibid.
114 Ibid., p. 106.
115 Eulenburg to Kaiser Wilhelm II, 18 August 1897, *Eulenburgs Korrespondenz*, III, No. 1339.
116 Berghahn, *Tirpitz-Plan*, p. 106.
117 Tirpitz, notes on his visit to the Kaiser, 18–20 August 1897, BA-MA Freiburg, Tirpitz Papers N253/4.
118 Berghahn, *Tirpitz-Plan*, pp. 108–9.
119 Prince Heinrich of Prussia to Tirpitz, 17 October 1897, Tirpitz to Prince Heinrich of Prussia, 29 October 1897, BA-MA Freiburg, Tirpitz Papers N253/4.
120 Berghahn, *Tirpitz-Plan*, pp. 108–9.
121 Tirpitz to Senden, 11 August 1897, BA-MA Freiburg, Tirpitz Papers N253/4.
122 Tirpitz, notes on his visit to Bismarck, 22 August 1897, ibid.
123 Eisendecher to Tirpitz, 1 September 1897, ibid.
124 Tirpitz to Senden, 11 August 1897, ibid. See Kaiser Wilhelm II to Hohenlohe, 23 August 1897, Hohenlohe, *Denkwürdigkeiten der Reichskanzlerzeit*, pp. 377–8.
125 Berghahn, *Tirpitz-Plan*, p. 119.
126 Bülow, *Memoirs*, I, p. 133. See W. Marienfeld, *Wissenschaft und Schlachtflottenbau in Deutschland 1897–1906*, Frankfurt a.M. 1957; Deist, *Flottenpolitik und Flottenpropaganda*, pp. 71–145 *et passim*.
127 Bülow, *Memoirs*, I, pp. 111–12; Berghahn, *Tirpitz-Plan*, p. 123.
128 Berghahn, *Tirpitz-Plan*, p. 125.
129 Kaiser Wilhelm II to Tirpitz, 24 March 1898, BA-MA Freiburg, Tirpitz Papers N253/4; Cecil, *Wilhelm II*, I, p. 317. See above, p. 878.

130 Kaiser Wilhelm II to Hohenlohe, 27 March 1898, Hohenlohe, *Denkwürdigkeiten der Reichskanzlerzeit*, p. 436.
131 Prince Heinrich of Prussia to Senden, 19 April 1899, BA-MA Freiburg, Senden-Bibran Papers N160/4.
132 See Volker R. Berghahn, 'Zu den Zielen des deutschen Flottenbaus unter Wilhelm II.', *Historische Zeitschrift*, 210/1 (February 1970), pp. 46, 53 *et passim*.
133 Tirpitz to Senden, 11 August 1897, BA-MA Freiburg, Tirpitz Papers N253/4.
134 Tirpitz to Kaiser Wilhelm II, 24 April 1898, Hohenlohe, *Denkwürdigkeiten der Reichskanzlerzeit*, pp. 441–2. See above, p. 878.
135 Hohenlohe, diary entry for 6 October 1898, *Denkwürdigkeiten der Reichskanzlerzeit*, pp. 462–3.
136 Prince Heinrich of Prussia to Tirpitz, 8 June 1901, BA-MA Freiburg, Tirpitz Papers N253/183.
137 Tirpitz, notes for an audience with the Kaiser on 28 November 1898, ibid. N253/4.
138 Tirpitz, secret notes, presented to the Kaiser on 28 November 1898, ibid.
139 Ibid.
140 Ibid.
141 Tirpitz, notes for an audience with the Kaiser, undated, BA-MA Freiburg, Tirpitz Papers N253/4, fol. 227.
142 Tirpitz, notes of 29 September 1899 on an audience with the Kaiser in Rominten on 28 September 1899, ibid. N253/5.
143 Kaiser Wilhelm II, marginal notes on Richthofen to Eulenburg, 20 July 1898, Eulenburg to Richthofen, 22 July 1898, PA AA, IA England 78 No. 1 secr., Bd. IV, R 5802.
144 Wilmowski to Hohenlohe, 24 October 1899, Hohenlohe, *Denkwürdigkeiten der Reichskanzlerzeit*, p. 533; Hohenlohe, note of 26 October 1899, ibid., pp. 534–5; Hohenlohe, diary entry for 5 November 1899, ibid., pp. 537–8. See Tirpitz, *Erinnerungen*, pp. 104–5; also Kaiser Wilhelm II to Hohenlohe, 10 January 1900, Hohenlohe, *Denkwürdigkeiten der Reichskanzlerzeit*, pp. 555–6; Berghahn, *Tirpitz-Plan*, p. 211.
145 Tirpitz, notes on the events between 28 September and 17 November 1899, BA-MA Freiburg, Tirpitz Papers N253/5.
146 Berghahn, *Tirpitz-Plan*, p. 213.
147 Wilmowski to Hohenlohe, 25 October 1899, Hohenlohe, note of 26 October 1899, Hohenlohe, *Denkwürdigkeiten der Reichskanzlerzeit*, pp. 533–4.
148 Hohenlohe to Tirpitz, 26 October 1899, BA-MA Freiburg, Tirpitz Papers N253/6, printed in Hohenlohe, *Denkwürdigkeiten der Reichskanzlerzeit*, pp. 533–4.
149 Hohenlohe to Tirpitz, 29 October 1899, BA-MA Freiburg, Tirpitz Papers N253/6.
150 Berghahn, *Tirpitz-Plan*, p. 214.
151 Tirpitz, notes on the events between 28 September and 17 November 1899, BA-MA Freiburg, Tirpitz Papers N253/5.
152 Hohenlohe to the prime ministers of the individual German states, 6 November 1899, ibid. N253/6.
153 Bülow to Hohenlohe, 25 October 1899, Hohenlohe, *Denkwürdigkeiten der Reichskanzlerzeit*, p. 533.
154 Hohenlohe, note of 6 November 1899, ibid., p. 538. See above, p. 886.
155 Grand Duke of Baden to Hohenlohe, 10 and 15 November 1899, Hohenlohe, *Denkwürdigkeiten der Reichskanzlerzeit*, pp. 539 and 545.
156 Kaiser Wilhelm II to Hohenlohe, 29 November 1899, BA-MA Freiburg, Tirpitz Papers N253/5, in Hohenlohe, *Denkwürdigkeiten der Reichskanzlerzeit*, p. 547.
157 See *Große Politik*, XV, pp. 441ff.; Cecil, *Wilhelm II*, I, p. 334.
158 Grierson to Bigge, 6 January 1900, RA I62/76b. See Kennedy, *Anglo-German Antagonism*, pp. 239–40.

159 Kaiser Wilhelm II, *Ereignisse und Gestalten*, pp. 196–7.

160 Kaiser Wilhelm II to Hohenlohe, 10 January 1900, Hohenlohe, *Denkwürdigkeiten der Reichskanzlerzeit*, p. 547. See Tirpitz to Hohenlohe, 16 January 1900, BA-MA Freiburg, Tirpitz Papers N253/5.

161 Kaiser Wilhelm II to Hohenlohe, undated, Hohenlohe, *Denkwürdigkeiten der Reichskanzlerzeit*, p. 557.

162 Tirpitz to Hohenlohe, 27 April 1900, ibid., p. 571.

163 Cecil, *Wilhelm II*, I, p. 335.

164 See Spitzemberg, diary entry for 29 September 1912, *Tagebuch*, pp. 548–9.

165 See Holger Afflerbach, *Der Dreibund. Europäische Großmacht- und Allianzpolitik vor dem Ersten Weltkrieg*, Vienna, Cologne, Weimar 2002, pp. 363ff.

33 'YOUNG GERMANY, YOUR KAISER!', OR WHAT WAS WRONG WITH WILHELM II?

1 Archduke Franz Ferdinand of Austria-Hungary to Baron Beck, 28 August 1900, cited in Robert A. Kann, 'Kaiser Wilhelm II. und Thronfolger Franz Ferdinand in ihrer Korrespondenz', in Kann, *Erzherzog Franz Ferdinand Studien*, Vienna 1976, p. 50. But cf. Eulenburg to Bülow, 26 May 1899, *Eulenburgs Korrespondenz*, III, No. 1394 and below, p. 1049.

2 Tolstoy to Prince Grigory Wolkonsky, 4/16 December 1899, quoted in R. F. Christian, ed., *Tolstoy's Letters*, 2 vols., London 1978, II, No. 458.

3 See Eulenburg, *Erlebnisse an deutschen und fremden Höfen*, pp. 321–58; Houston Stewart Chamberlain, *Briefe 1882–1924 und Briefwechsel mit Kaiser Wilhelm II.*, 2 vols., Munich 1926, II, pp. 131–275; Geoffrey G. Field, *Evangelist of Race. The Germanic Vision of Houston Stewart Chamberlain*, New York 1981.

4 Houston Stewart Chamberlain, 'Kaiser Wilhelm II', *Jugend*, I, No. 22 (1900), pp. 370ff.; Chamberlain to Adolf Hitler, 7 October 1923, Chamberlain, *Briefe*, II, pp. 124–6.

5 Chamberlain, 'Kaiser Wilhelm II.', pp. 370ff.

6 Ernst Heinrich Bethge, *Jung Deutschland, Dein Kaiser!*, Langensalza 1913.

7 Szögyény, report of 5 February 1900, HHStA Vienna, PA III/153, cited in part in Kennedy, *Anglo-German Antagonism*, p. 241.

8 Maximilian Harden, *Die Zukunft*, No. 40, 1902, p. 340.

9 Weller, *Maximilian Harden*, pp. 113–14; Hellige, *Rathenau und Harden*, pp. 160ff. and 332.

10 See Bernd Sösemann, 'Die sog. Hunnenrede Wilhelms II. Textkritische und interpretatorische Bemerkungen zur Ansprache des Kaisers vom 27. Juli 1900 in Bremerhaven', *Historische Zeitschrift*, vol. 222 (1976), pp. 349–50; Sösemann, 'Wir sollen sein ein einig Volk von Schlächtern', *Frankfurter Allgemeine Zeitung*, 27 July 2000.

11 Maximilian Harden, 'Der Kampf mit dem Drachen', *Die Zukunft*, No. 32, 11 August 1900, pp. 225–36. For the generational shift more generally, see Martin Doerry, *Übergangsmenschen. Die Mentalität der Wilhelminer und die Krise des Kaiserreichs*, Weinheim, Munich 1986; Thomas Kohut, *Wilhelm II and the Germans*; Ullrich, *Nervöse Grossmacht*; Radkau, *Das Zeitalter der Nervosität.*

12 Kaiser Wilhelm II to Queen Victoria, 2 February 1899, in Buckle, *Letters of Queen Victoria*, III, pp. 336–7. See above, pp. 991–2.

13 Eulenburg to Bülow, 4–7 July 1899, *Eulenburgs Korrespondenz*, III, No. 1395. See Kaiser Wilhelm II to Eulenburg, 15 June 1899, ibid., p. 1941.

14 Kaiser Wilhelm II to Eulenburg, 15 June 1898, cited in Bülow, *Memoirs*, I, p. 218; cf. Eulenburg to Bülow, 15 June 1898, *Eulenburgs Korrespondenz*, III, No. 1375; also Eulenburg to Bülow, 12–19 July 1899, ibid., No. 1397.

15 Eulenburg to Lucanus, 16 June 1898, *Eulenburgs Korrespondenz*, III, No. 1376.

16 August Eulenburg to Philipp Eulenburg, 23 March 1898, ibid., p. 1887; Eulenburg to Bülow, 22 March 1898, ibid., No. 1364.

17 See Hull, *Entourage*, pp. 109–45, and Hull, 'Kaiser Wilhelm II and the "Liebenberg Circle"', in Röhl and Sombart, *Kaiser Wilhelm II. New Interpretations*, pp. 193–220; also Cecil, *Wilhelm II*, pp. 98–122. For the trials of Eulenburg and Kuno Moltke, see Karsten Hecht, 'Die Harden-Prozesse. Strafverfahren, Öffentlichkeit und Politik im Kaiserreich', diss., Munich 1997.

18 See *Eulenburgs Korrespondenz*, III, Nos. 1294 and 1422 and p. 1977.

19 Eulenburg to Bülow, 4 and 20 July 1898, ibid., Nos. 1377 and 1380. See Bülow to Eulenburg, 27 May 1898, ibid., No. 1372.

20 Varnbüler to Kuno Moltke, 15 April 1898, ibid., No. 1366.

21 See Varnbüler to Kuno Moltke, 4 June 1898, ibid., No. 1373; Eulenburg to Kaiser Wilhelm II, 16 August and 15 November 1898 and 9 May 1899, ibid., Nos. 1385, 1389 and 1393.

22 See e.g. Eulenburg to Bülow, 26 July 1897, ibid., No. 1335.

23 Eulenburg's note of 26 July 1897, *Eulenburgs Korrespondenz*, III, No. 1335.

24 See Roderick R. McLean, 'Kaiser Wilhelm II and His Hessian Cousins. Intra-state Relations in the German Empire and International Dynastic Politics, 1890–1918', *German History*, 19/1 (2001), pp. 28–53.

25 Eulenburg to Bülow, 22 March 1898, *Eulenburgs Korrespondenz*, III, No. 1364.

26 Bülow to Eulenburg, 20 July 1898, ibid., No. 1381.

27 Ibid.

28 Eulenburg to Bülow, 10–11 July 1899, ibid., No. 1396.

29 Eulenburg to Bülow, 11 February and 10–11 July 1899, ibid., Nos. 1390 and 1396.

30 Eulenburg to Bülow, 20 July 1898, ibid., No. 1380.

31 Eulenburg to Bülow, 26 May 1899, ibid., No. 1394. Cf. Bülow, *Memoirs*, I, pp. 398–9.

32 Eulenburg to Bülow, 10–11 July 1899, *Eulenburgs Korrespondenz*, III, No. 1396.

33 Eulenburg to Bülow, 12–19 July 1899, ibid., No. 1397.

34 Eulenburg to Bülow, 20–22 July 1899, ibid., No. 1399.

35 See above, p. 663.

36 Eulenburg to Bülow, 12–19 July 1899, *Eulenburgs Korrespondenz*, III, No. 1397.

37 Eulenburg to Bülow, 20–2 July 1899, ibid., No. 1399.

38 Eulenburg to Bülow, 26–7 July 1899, ibid., No. 1400.

39 Eulenburg to Bülow, 20–2 July 1899, ibid., No. 1399.

40 Ibid.

41 Ibid.

42 Eulenburg to Bülow, 26–27 July 1899, ibid., No. 1400.

43 Eulenburg to Lucanus, 5 December 1899, ibid., No. 1405.

44 Eulenburg to Lucanus, 10 December 1899, ibid., No. 1408.

45 Eulenburg to Lucanus, 2 January 1900, ibid., No. 1412.

46 Grand Duke of Baden to Hohenlohe, 10 November 1899, Hohenlohe, *Denkwürdigkeiten der Reichskanzlerzeit*, pp. 539–40.

47 Eulenburg, note of 27 July 1900, cited in Haller, *Eulenburg*, p. 257.

48 Kaiser Wilhelm II to Bülow, 19 June 1900, *Große Politik*, XVI, No. 4527.

49 See Bülow, *Memoirs*, I, p. 452; Eulenburg to Bülow, 14 July 1900, *Eulenburgs Korrespondenz*, III, No. 1419.

50 Eulenburg to Bülow, 14–15 July 1900, *Eulenburgs Korrespondenz*, III, No. 1419. Cf. Bülow, *Memoirs*, I, pp. 452–3.

51 Eulenburg to Bülow, 18 July 1900, *Eulenburgs Korrespondenz*, III, No. 1420.

52 Eulenburg to Bülow, 21 July 1900, ibid., No. 1421.

53 See Eulenburg, *Mit dem Kaiser als Staatsmann und Freund auf Nordlandreisen*, 2 vols., Dresden 1931, II, pp. 315–71.

54 Eulenburg to Bülow, 21 July 1903, *Eulenburgs Korrespondenz*, III, No. 1497.
55 Eulenburg to Bülow, 26–8 July and 9 August 1903, ibid., Nos. 1498 and 1499.
56 Grand Duchess Augusta Caroline of Mecklenburg-Strelitz to George Duke of Cambridge, 30 August 1888, RA Vic Addl Mss A8/2689.
57 Count Georg von Werthern to Eulenburg, 22 October 1888, *Eulenburgs Korrespondenz*, I, No. 200.
58 Dr Rudolf Leuthold to Eulenburg, 25 December 1889, ibid., No. 268.
59 Vice Admiral Paul Hoffmann, diary for 22–3 April 1890, Hoffmann Papers, Freiburg.
60 Eulenburg to Holstein, 1 August 1890, *Holstein Papers*, III, No. 327.
61 Kaiser Wilhelm II to Queen Victoria, telegram, 29 July 1891, RA I59/43.
62 Empress Frederick to Queen Victoria, 6 August 1891, RA Z51/7.
63 Grierson to Bigge, 2 April 1897, RA I61/4a.
64 See above, p. 932.
65 See Radkau, *Zeitalter der Nervosität*, particularly pp. 275ff.; also Ullrich, *Nervöse Grossmacht*.
66 Waldersee, diary entry for 20 August 1891, GStA Berlin, Waldersee Papers; omitted from Meisner, II, p. 215.
67 Waldersee, diary entry for 17 September 1891, GStA Berlin, Waldersee Papers; cf. Meisner, II, p. 217.
68 Waldersee, diary entry for 25 November 1892, GStA Berlin, Waldersee Papers; omitted from Meisner, II, p. 270.
69 Waldersee, diary entry for 9 January 1891, GStA Berlin, Waldersee Papers; cf. Meisner, II, p. 174.
70 See above, ch. 17.
71 Waldersee, diary entry for 16 March 1892, GStA Berlin, Waldersee Papers; printed in part in Meisner, II, pp. 234–5.
72 Waldersee, diary entry for 25 September 1892, GStA Berlin, Waldersee Papers; omitted from Meisner, II, p. 265.
73 Waldersee, diary entry for 31 August 1892, GStA Berlin, Waldersee Papers; omitted from Meisner, II, p. 264.
74 Waldersee, diary entry for 21 November 1894, GStA Berlin, Waldersee Papers; omitted from Meisner, II, p. 332.
75 Waldersee, diary entry for 21 January 1896, GStA Berlin, Waldersee Papers; cf. Meisner, II, pp. 365–6.
76 Holstein to Eulenburg, 25 January 1896, *Eulenburgs Korrespondenz*, III, No. 1191; Eulenburg to Holstein, 31 January 1896, *Holstein Papers*, III, No. 525.
77 Waldersee, diary entry for 16 February 1896, GStA Berlin, Waldersee Papers; omitted from Meisner, II, p. 367.
78 Waldersee, diary entry for 25 October 1896, GStA Berlin, Waldersee Papers; cf. the distorted version in Meisner, II, pp. 374–5.
79 Waldersee, diary entry for 31 October 1896, GStA Berlin, Waldersee Papers; cf. Meisner, II, pp. 375–6.
80 Waldersee, diary entry for 16 March 1897, GStA Berlin, Waldersee Papers; cf. Meisner, II, pp. 393ff.
81 Bodman, report of 7 March 1897, Fuchs, *Großherzog von Baden*, No. 1657.
82 See Holstein to Eulenburg, 4 and 11 April 1890, *Eulenburgs Korrespondenz*, I, Nos. 379 and 384. Cf. above, p. 534.
83 Holstein to Eulenburg, 6 December 1890, *Eulenburgs Korrespondenz*, I, No. 452. See above, pp. 556–7.
84 Waldersee, diary entry for 16 April 1892, GStA Berlin, Waldersee Papers; cf. Meisner, II, pp. 239–40. See above, p. 512.

85 Holstein to Eulenburg, 1 April 1892, *Eulenburgs Korrespondenz*, II, No. 638.
86 Report from Paris, 22 March 1892, ibid., pp. 839–40; also Hull, *Entourage*, p. 16.
87 Waldersee, diary entry for 23 February 1891, GStA Berlin, Waldersee Papers; omitted from Meisner, II, pp. 189ff.
88 Waldersee, diary entry for 22 December 1891, GStA Berlin, Waldersee Papers; cf. Meisner, II, p. 228.
89 Gosselin to Salisbury, 29 November 1895, cited in *Eulenburgs Korrespondenz*, III, pp. 1484–5.
90 Holstein to Eulenburg, 17 February 1895, ibid., No. 1089. See Bülow, *Memoirs*, I, pp. 135–6 and 174–5.
91 Hohenlohe, *Denkwürdigkeiten der Reichskanzlerzeit*, p. 151. See above, p. 808.
92 Holstein to Bülow, 24 March and 2 April 1897, cited in Röhl, *Kaiser and His Court*, p. 22.
93 Monts to Holstein, 2 March 1897, in *Eulenburgs Korrespondenz*, III, No. 1301; cf. Bülow, *Memoirs*, I, p. 39.
94 Monts to Eulenburg, 20/21 March 1897, *Eulenburgs Korrespondenz*, III, No. 1309.
95 Ibid.
96 Bodman, report of 4 March 1897, Fuchs, *Großherzog von Baden*, III, No. 1654.
97 Bodman, report of 7 March 1897, ibid., No. 1657.
98 Bülow, note of 7 April 1897, *Eulenburgs Korrespondenz*, III, No. 1312.
99 Eulenburg to Bülow, 14–15 July 1900, ibid., No. 1419. Cf. Bülow, *Memoirs*, I, p. 453.
100 Bülow, *Memoirs*, I, pp. 135 and 175.
101 Bülow to Eulenburg, 22 November 1900, Eulenburg to Kaiser Wilhelm II, 22 November 1900, *Eulenburgs Korrespondenz*, III, Nos. 1439 and 1440.
102 Hereditary Princess Charlotte of Saxe-Meiningen to Ellen Freifrau von Heldburg, 15 March 1897, Thüringisches Staatsarchiv Meiningen, HA 342.
103 Hereditary Princess Charlotte of Saxe-Meiningen to Schweninger, 26 March 1897, 4 April 1901 and 4 June 1903, cited in Röhl, *Cäsarenwahnsinn*, p. 28.
104 Hereditary Princess Charlotte of Saxe-Meiningen to Schweninger, 5 December 1908, cited ibid., pp. 28–9.
105 See above, ch. 22.
106 Hopman, report of 3 February 1915, cited in Holger Afflerbach, ed., *Wilhelm II. als Oberster Kriegsherr im Ersten Weltkrieg. Quellen aus der militärischen Umgebung des Kaisers 1914–1918*, Munich 2004.
107 Waldersee, diary entry for 6 April 1897, GStA Berlin, Waldersee Papers; cf. Meisner, II, p. 396.
108 Alvensleben, report from Brussels of 24 October 1897, PA AA, IA Preußen 1 No. 1d, Bd VIII. This version of events on board the *Hohenzollern* in July 1897 also gained currency in Germany. See Friedrich Percyval Reck-Malleczewen, *Tagebuch eines Verzweifelten*, Berlin, Bonn 1981, pp. 109–10. See Marschall, *Reisen und Regieren*, pp. 198ff.
109 *New York Times*, 28 November 1897, cited in Mitchell, *Danger of Dreams*, p. 22.
110 Wolf von Schierbrand, 'Is Kaiser Wilhelm II of Normal Mind?', *Lippincott*'s, 78 (November 1906), pp. 619–25.
111 A. A. Mossolov, *At the Court of the Last Tsar*, London 1935, pp. 202–10.
112 T. G. Otte, '"The Winston of Germany": The British Foreign Policy Élite and the Last German Emperor', *Canadian Journal of History*, December 2001, pp. 471–504.
113 Salisbury to Paget, 16 October 1888, cited in Otte, '"Winston"', p. 478.
114 Lord George Hamilton, *Parliamentary Reminiscences and Reflections, 1886–1906*, London 1922, p. 137.
115 Cited in Roberts, *Salisbury*, p. 555. See Otte, '"Winston"', pp. 479–80.
116 Lascelles, note of 4 December 1895, cited above, p. 770.

117 Lascelles to Salisbury, 16 February 1897, quoted in van't Padje, 'Anglo-German Imperialist Rivalry', p. 109. See above, pp. 992–4.

118 For the medical condition of Prince Wilhelm's ear, see ch. 13 of *Young Wilhelm.*

119 Waldersee, diary entry for 11 October 1890, GStA Berlin, Waldersee Papers; omitted from Meisner, II, p. 154.

120 Waldersee, diary entry for 21 March 1892, GStA Berlin, Waldersee Papers; omitted from Meisner, II, p. 236.

121 Waldersee, diary entry for 15 February 1894, GStA Berlin, Waldersee Papers; omitted from Meisner, II, p. 308.

122 Waldersee, diary entry for 29 April 1894, GStA Berlin, Waldersee Papers; cf. Meisner, II, p. 313.

123 Waldersee, diary entry for 2 June 1894, GStA Berlin, Waldersee Papers; omitted from Meisner, II, p. 314.

124 See above, p. 832.

125 Holstein to Lindenau, 29 July 1896, *Holstein Papers*, III, No. 573.

126 Sir Schomberg McDonnell to King George V, 26 October 1914, cited in *Young Wilhelm*, pp. 318–19.

127 See D. M. Potts and W. T. W. Potts, *Queen Victoria's Gene. Haemophilia and the Royal Family*, Stroud 1995, and Zeepvat, *Prince Leopold.*

128 See Röhl, Warren and Hunt, *Purple Secret*, and above, p. 639.

129 Professor Robert Gaupp, medical report of 31 May 1918, Archiv des vormals regierenden preußischen Königshauses, Burg Hohenzollern; my italics.

130 Lord Esher, diary entry for 21 November 1908, Esher Papers, Churchill Archives Centre, Cambridge, cited in Röhl, Warren and Hunt, *Purple Secret*, p. 311.

Archival sources

1. Geheimes Staatsarchiv (GStA) Berlin and Merseburg:
   Brandenburg-Preußisches Hausarchiv
   Files of the Civil Cabinet
   Minutes of the Prussian Crown Council
   Minutes of the Prussian Ministry of State
   Files of the Kotze Case
   Papers of Rudolf Kögel
   Papers of Karl Friedrich Nowak
   Papers of Adolf von Scholz
   Papers of Count Alfred von Waldersee
2. Bundesarchiv (BA) Berlin (formerly Zentrales Staatsarchiv Potsdam):
   Files of the Reichskanzlei
   Papers of Ernst Schweninger
3. Politisches Archiv des Auswärtigen Amtes (PA AA) Berlin (formerly in Bonn):
   Asservat No. 4
   Files of the German Foreign Office
4. Cambridge University Library:
   Papers of Sir Charles Hardinge
5. Churchill Archives Centre, Cambridge:
   Papers of Viscount Esher
6. Cumbria Record Office, Carlisle:
   Papers of Lord Lonsdale
7. Stadt- und Landesbibliothek Dortmund:
   Papers of Konrad Studt
8. Archiv der Hessischen Hausstiftung (AdHH) Schloss Fasanerie:
   Papers of Kaiserin Victoria (the Empress Frederick) and Kaiser Friedrich III
9. Bundesarchiv-Militärarchiv (BA-MA) Freiburg:
   Papers of Colmar Freiherr von der Goltz
   Papers of Hans von Plessen
   Papers of Gustav Freiherr von Senden-Bibran
   Papers of Alfred von Tirpitz

10. Bismarck-Archiv Schloss Friedrichsruh:
    Papers of Prince Otto von Bismarck
    Papers of Count Herbert von Bismarck
11. Archiv des vormals regierenden preußischen Königshauses, Burg Hohenzollern:
    Papers of Kaiser Wilhelm II
12. Generallandesarchiv (GLA) Karlsruhe:
    Papers of Max Freiherr von Holzing-Berstett
13. Bundesarchiv (BA) Koblenz:
    Files of the Prussian Ministry of Justice
    Papers of Prince Otto von Bismarck
    Papers of Count Herbert von Bismarck
    Papers of Heinrich von Boetticher
    Papers of Robert Bosse
    Papers of Bernhard von Bülow
    Papers of Count Philipp zu Eulenburg-Hertefeld
14. Rothschild Archives London (RAL):
    Letters of Gerson Bleichröder
15. Thüringisches Staatsarchiv Meiningen (ThStaMgn):
    House Archive of the ducal family of Saxe-Meiningen
16. Hohenlohe-Zentralarchiv Schloss Neuenstein:
    Papers of Prince Hermann zu Hohenlohe-Langenburg
    Papers of Prince Ernst zu Hohenlohe-Langenburg
17. Bodleian Library, Oxford:
    Papers of Sir Eyre Crowe
18. Haus-, Hof- und Staatsarchiv (HHStA) Vienna:
    Kabinettsarchiv Geheimakten
    Ambassadors' Reports from Berlin
19. Royal Archives (RA) Windsor:
    Papers of Queen Victoria
    Papers of Albert Edward Prince of Wales
    Papers of King George V
    Papers of Prince Christian of Schleswig-Holstein
20. Privately held papers:
    Papers of Vice-Admiral Paul Hoffmann, Freiburg im Breisgau
    Papers of Lord Lonsdale, Lowther, Cumbria
    Papers of Countess Alice zu Lynar, Bad Driburg
    Papers of Adolf Freiherr Marschall von Bieberstein, Neuershausen
    Papers of Friedrich Mewes, Munich
    Papers of Count Carl von Wedel, Frankfurt a.M.
    Papers of Count Ernst von Wedel, Bad Driburg

Select bibliography

Abret, Helga, *Die Majestätsbeleidigungsäffare des 'Simplicissimus'-Verlegers Albert Langen. Briefe und Dokumente zu Exil und Begnadigung 1898–1903*, Frankfurt, New York 1985

Adam, Juliette, *Guillaume II 1890–1899*, Paris 1917

The Schemes of the Kaiser, London 1917

Afflerbach, Holger, *Der Dreibund. Europäische Großmacht- und Allianzpolitik vor dem Ersten Weltkrieg*, Vienna 2002

'Wilhelm II as Supreme Warlord in the First World War', in Mombauer and Deist, *The Kaiser*

ed., *Wilhelm II. als Oberster Kriegsherr im Ersten Weltkrieg. Quellen aus der militärischen Umgebung des Kaisers 1914–1918*, Munich 2004

Albrecht, Dieter, ed., Hugo Graf Lerchenfeld-Köfering, *Kaiser Wilhelm II. als Persönlichkeit und Herrscher*, Regensburger Historische Forschungen, Vol. 11, 1985

Alexander Grand Duke of Russia, *Once a Grand Duke*, New York 1932

Anon. [H. von Langen-Allenstein], *Herr von Tausch und die Verfasser der anonymen Briefe der Hofgesellschaft*, Zürich 1897

Anon., *Das Geheimnis des Ceremoniemeisters. Hofroman aus der jüngsten Vergangenheit. Von Carl Fürst von . . . (in Preußen verboten)*

Kaiserreden. Reden und Erlasse, Briefe und Telegramme Kaiser Wilhelms des Zweiten, ein Charakterbild des deutschen Kaisers, Leipzig 1902

The Private Life of Two Emperors. William II of Germany and Francis Joseph of Austria, 2 vols., London 1905

Unser Kaiser und sein Volk! Deutsche Sorgen. Von einem Schwarzseher, Freiburg, Leipzig 7th edn 1906

Unser Kaiser, 25 Jahre der Regierung Kaiser Wilhelms II., 1888–1913, Berlin, Leipzig 1913

The Real Kaiser. An Illuminating Study, New York 1914

The Last of the War Lords. New Lights on the Life and Personality of Kaiser Wilhelm II, His Relations with Britons and Americans, with Artists and Writers, and Some Account of His Love-Affairs, London 1918

Semi-Imperator, 1888–1919, Munich 1919

Wilhelms II. Abschiedsbrief an das deutsche Volk. Den Deutschen ein Spiegel, Berlin 2nd edn 1922

Drei deutsche Kaiser. Wilhelm I., Friedrich III., Wilhelm II. Ihr Leben und ihre Zeit, 1858–1918, Freiburg, Würzburg 1987

Arren, Jules, *Wilhelm II. Was er sagt, was er denkt*, Leipzig 1911

Avenarius, Ferdinand, 'Hofkunst und andere Kunst', *Der Kunstwart*, vol. 15, part 3, November 1901

Bachem, Karl, *Vorgeschichte, Geschichte und Politik der Deutschen Zentrumspartei*, 9 vols., Cologne 1927–32

Baechler, Christian, *Guillaume II le Kaiser*, Paris 2003

Balfour, Michael, *The Kaiser and His Times*, London 1975

Bartmann, Dominik, *Anton von Werner. Zur Kunst und Kunstpolitik im Deutschen Kaiserreich*, Berlin 1985

Bein, Alexander, *Erinnerungen und Dokumente über Herzls Begegnung mit Wilhelm II., Zeitschrift für die Geschichte der Juden*, 1964

Benner, Thomas Hartmut, *Die Strahlen der Krone. Die religiöse Dimension des Kaisertums unter Wilhelm II. Vor dem Hintergrunde der Orientreise 1898*, Marburg 2001

Benson, E. F., *The Kaiser and His English Relations*, London, New York 1936

Bérard, Victor, *La France et Guillaume II*, Paris 1907

Berghahn, Volker R., 'Zu den Zielen des deutschen Flottenbaus unter Wilhelm II.', *Historische Zeitschrift*, 210/1, February 1970, pp. 34–100

Der Tirpitz-Plan. Genesis und Verfall einer innenpolitischen Krisenstrategie unter Wilhelm II., Düsseldorf 1971

'Des Kaisers Flotte und die Revolutionierung des Mächtesystems vor 1914', in Röhl, *Der Ort Kaiser Wilhelms II.*

Bernstein, Hermann, *The Willy–Nicky Correspondence, Being the Secret and Intimate Telegrams Exchanged between the Kaiser and the Tsar*, New York 1918

Beseler, Dora von, *Der Kaiser im englischen Urteil*, Stuttgart 1932

Der Kaiser in Vergangenheit und Gegenwart. Gedanken zum 75. Geburtstag Kaiser Wilhelm II., Leipzig 1934

Bethge, Ernst Heinrich, *Jung Deutschland, Dein Kaiser!*, Langensalza 1913

Beumelburg, Werner, *Wilhelm II. und Bülow*, Oldenburg 1932

Beynes, N., *L'Empereur Guillaume*, Paris 1915

Bigelow, Poultney, *The German Emperor and His Eastern Neighbours*, New York 1892

Prussian Memories 1864–1914, New York 1916

Prussianism and Pacifism. The Two Wilhelms between 1848 and 1918, New York, London, 1919

Bing, E. J., ed., *The Letters of the Tsar Nicholas and the Empress Marie*, London 1937

Bismarck, Otto Fürst von, *Die gesammelten Werke*, 15 vols., Berlin 1923–35

Bittner, Ludwig, 'Neue Beiträge zur Haltung Kaiser Wilhelms II. in der Faschoda-Frage', *Historische Zeitschrift*, 162 (1940), pp. 540–50

Blum, Hans, *Fürst Bismarck und seine Zeit. Eine Biographie für das deutsche Volk*, Munich 1899

Bismarck. Ein Buch für Deutschlands Jugend und Volk, Heidelberg 1903

Bodelschwingh, Marie von, *Aus Heiligem Land. Den Kindern erzählt*, Bielefeld 1899

Bodenheimer, M., *Die Zionisten im kaiserlichen Deutschland*, Bensberg 1972

Boelcke, Willi A., *Krupp und die Hohenzollern in Dokumenten. Krupp-Korrespondenz mit Kaisern, Kabinettschefs und Ministern 1850–1918*, Frankfurt a.M. 1970

Boockmann, Hartmut, *Die Marienburg im 19. Jahrhundert*, Frankfurt, Berlin, Vienna 1982

Borkenhagen, Hermann, 'Kaiser Wilhelms II. Schuld', *Flugschrift der DNVP*, Berlin 1920

Bornhak, Conrad, *Deutsche Geschichte unter Kaiser Wilhelm II.*, Leipzig, Erlangen 2nd edn 1921

Brauer, Arthur von, *Im Dienste Bismarcks. Persönliche Erinnerungen*, Berlin 1936

Breitenborn, Konrad, *Im Dienste Bismarcks. Die politische Karriere des Grafen Otto zu Stolberg-Wernigerode*, Leipzig 3rd edn 1986 ed., *Die Lebenserinnerungen des Fürsten Otto zu Stolberg-Wernigerode*, Wernigerode 1996

Bringmann, Tobias C., *Reichstag und Zweikampf. Die Duellfrage als innenpolitischer Konflikt des deutschen Kaiserreichs 1871–1918*, Freiburg 1997

Brude-Firnau, Gisela, *Die literarische Deutung Kaiser Wilhelms II. zwischen 1889 und 1989*, Heidelberg 1997

Buchner, Max, *Kaiser Wilhelm II., seine Weltanschauung und die Deutschen Katholiken*, Leipzig 1929

Buckle, George Earle, ed., *The Letters of Queen Victoria*, Third Series, 3 vols., London 1930

Bülow, Bernhard Fürst von, *Denkwürdigkeiten*, 4 vols., Berlin 1930–1

The Memoirs of Prince von Bülow, 4 vols., London and New York 1931–2

Burmeister, Hans W., *Prince Philipp zu Eulenburg-Hertefeld, 1847–1921. His Influence on Kaiser Wilhelm II and His Role in the German Government, 1888–1902*, Wiesbaden 1981

Bußmann, Walter, ed., *Staatssekretär Graf Herbert von Bismarck. Aus seiner politischen Privatkorrespondenz*, Göttingen 1964

Canis, Konrad, *Bismarck und Waldersee*, Berlin 1980

Von Bismarck zur Weltpolitik. Deutsche Außenpolitik 1890 bis 1902, Berlin 1997

Carmel, Alex and Eisler, Ejal Jakob, *Der Kaiser reist ins Heilige Land. Die Palästinareise Wilhelms II. 1898. Eine illustrierte Dokumentation*, Stuttgart 1999

Carnegie, Andrew, *William II, German Emperor & King of Prussia*, Dunfermline 1913

Casati, Gaetano, *Ten Years in Equatoria*, 2 vols., London 1891

Cataruzza, Marina, 'Das Kaiserbild in der Arbeiterschaft am Beispiel der Werftarbeiter in Hamburg und Stettin', in Röhl, *Der Ort Kaiser Wilhelms II.*

Catling, A. H., *The Kaiser under the Searchlight*, London 1914

Cecil, Lamar, *The German Diplomatic Service, 1871–1914*, Princeton 1976

Wilhelm II. volume I. *Prince and Emperor, 1859–1900*, Chapel Hill, London 1989

Wilhelm II. volume II. *Emperor and Exile, 1900–1941*, Chapel Hill, London 1996

Chamberlain, Houston Stewart, 'Kaiser Wilhelm II.' *Jugend*, 1900

Briefe 1882–1924 und Briefwechsel mit Kaiser Wilhelm II., 2 vols., Munich 1926

Chamier, Jacques D., *Ein Fabeltier unserer Zeit*, Zürich 1938

Chickering, Roger, *We Men Who Feel Most German. A Cultural Study of the Pan-German League, 1886–1914*, Boston, London, Sydney 1984

Christian, R. F., ed., *Tolstoy's Letters*, 2 vols., London 1978

Clark, Christopher, *Kaiser Wilhelm II*, London 2000

Conradi, Hermann, *Wilhelm II. und die junge Generation. Eine zeitpsychologische Betrachtung*, Leipzig 1889

Corti, Egon Caesar Conte, *Wenn ... Sendung und Schicksal einer Kaiserin*, Graz, Vienna, Cologne 1954

The English Empress. A Study in the Relations between Queen Victoria and Her Eldest Daughter, Empress Frederick of Germany, London 1957

Cowles, Virginia, *The Kaiser*, New York 1963

Craig, Gordon A., *The Politics of the Prussian Army, 1640–1945*, New York 1955

Cullen, Michael S., *Der Reichstag. Die Geschichte eines Monumentes*, Berlin 1983

Cunliffe-Owen, Marguerite, *Imperator et Rex, William II of Germany*, New York 1904

Curtius, Friedrich, ed., *Denkwürdigkeiten des Fürsten Chlodwig zu Hohenlohe-Schillingsfürst*, 2 vols., Stuttgart 1906

Daudet, Ernest, *Guillaume II et François Joseph*, Paris 1916

Davis, Arthur N., *The Kaiser I Knew. My Fourteen Years with the Kaiser*, London 1918

Deist, Wilhelm, *Flottenpolitik und Flottenpropaganda. Das Nachrichtenbureau des Reichsmarineamtes 1897–1914*, Stuttgart 1976

'Kaiser Wilhelm II. als Oberster Kriegsherr', in Röhl, *Der Ort Kaiser Wilhelms II.*

Militär, Staat und Gesellschaft. Studien zur preußisch-deutschen Militärgeschichte, Munich 1991

Delbrück, Hans, 'Staatsstreichpläne als Ursachen von Bismarcks Rücktritt. Sensationelle Enthüllungen', *Neues Wiener Journal*, 11 December 1913

Dickenson, Asa Don, *The Kaiser. A Book about the Most Interesting Man in Europe*, New York 1914

Documents Diplomatiques Français 1871–1914, Ministère des Affaires Etrangères, ed., 41 vols., Paris 1929–36

Dod, Karl C., *The Friendship of Wilhelm II and Nicholas II and Its International Implications, 1894–1905*, Urbana, IL, 1934

Doerry, Martin, *Übergangsmenschen. Die Mentalität der Wilhelminer und die Krise des Kaiserreichs*, Weinheim, Munich 1986

Domann, Peter, *Sozialdemokratie und Kaisertum unter Wilhelm II. Die Auseinandersetzung der Partei mit dem monarchischen System, seinen gesellschafts- und verfassungspolitischen Voraussetzungen*, Wiesbaden 1974

Dryander, Ernst von, *Wollte der Kaiser Krieg?*, Berlin 1919

Erinnerungen aus meinem Leben, Bielefeld, Leipzig 1922

Dubber, Ursula, *Wilhelm II und England, 1898–1914*, Heidelberg 1944

Duchhardt, Heinz, 'Der 18. Januar 1701 und die europäische Monarchie', *Majestas*, 10, 2002, pp. 151–66

Duggan, Christopher, *Creare la nazione. Vita di Francesco Crispi*, Rome, Bari 2000

Dülffer, Jost, 'Die Kreta-Krise und der griechisch-türkische Krieg 1890–1898', in Dülffer, Jost, Mühleisen, Hans-Otto and Torunsky, Vera, eds., *Inseln als Brennpunkte internationaler Politik*, Cologne 1986

Dülffer, Jost, Kröger, Martin and Wippich, Rolf-Harald, *Vermiedene Kriege. Deeskalation von Konflikten der Großmächte zwischen Krimkrieg und Erstem Weltkrieg 1865–1914*, Munich 1997

Ebel, G., *Botschafter Paul Graf von Hatzfeldt. Nachgelassene Papiere 1838–1901*, 2 vols., Boppard am Rhein 1976

Ebmeyer, Major, 'Caprivis Entlassung', *Deutsche Revue*, 47/4, 1922

Eça de Queirós, José Maria, 'O imperador Guilherme' (1891), in *Echos de Pariz*, Porto 4th edn 1920

Eckardstein, Baron von, *Ten Years at the Court of St James, 1895–1905*, London 1921

Eckardt, Julius von, *Aus den Tagen von Bismarcks Kampf gegen Caprivi*, Leipzig 1920

Elkind, Louis, ed., *The German Emperor's Speeches: Being a Selection from the Speeches, Edicts, Letters and Telegrams of the Emperor William II*, New York, London, Bombay 1904

Ellern, Hermann and Ellern, Bessi, *Herzl, Hechler, the Grand Duke of Baden and the German Emperor 1896–1904*, Tel Aviv 1961

Endres, Franz Carl, *Die Tragödie Deutschlands. Im Banne des Machtgedankens bis zum Zusammenbruch des Reiches. Von einem Deutschen*, Stuttgart 3rd edn 1924

Engelberg, Ernst, *Bismarck. Das Reich in der Mitte Europas*, Berlin 1990

Epkenhans, Michael, *Die wilhelminische Flottenrüstung 1908–1914. Weltmachtstreben, industrieller Fortschritt, soziale Integration*, Munich 1991

'Wilhelm II and "His" Navy, 1888–1918', in Mombauer and Deist, *The Kaiser*

Eppstein, Georg Freiherr von, ed., *Fürst Bismarcks Entlassung. Nach den hinterlassenen, bisher unveröffentlichten Aufzeichnungen des Staatssekretärs des Innern, Staatsministers Dr. Karl Heinrich von Boetticher und des Chefs der Reichskanzlei unter dem Fürsten Bismarck Dr. Franz Johannes von Rottenburg*, Berlin 1920

Erck, Alfred and Schneider, Hannelore, *Georg II. von Sachsen-Meiningen. Ein Leben zwischen ererbter Macht und künstlerischer Freiheit*, Zella-Mehlis, Meiningen 1997

Erfurt, Erich, *Bismarcks Sturz und die Änderung der deutschen Politik*, Berlin 1940
Eulenburg-Hertefeld, Philipp Fürst zu, *Aus 50 Jahren. Erinnerungen, Tagebücher und Briefe aus dem Nachlaß des Fürsten*, Berlin 2nd edn 1925
Mit dem Kaiser als Staatsman und Freund auf Nordlandreisen, 2 vols., Dresden 1931
Das Ende König Ludwigs II. und andere Erlebnisse, Leipzig 1934
Erlebnisse an deutschen und fremden Höfen, Leipzig 1934
Evans, Richard J., ed., *Kneipengespräche im Kaiserreich. Stimmungsberichte der Hamburger Politischen Polizei 1892–1914*, Reinbek bei Hamburg 1989
Everling, Friedrich, *Der Kaiser. Wie er war – wie er ist*, Berlin 1934
Eyck, Erich, *Die Monarchie Wilhelms II nach seinen Briefen, seinen Randbemerkungen und Zeugnissen seiner Freunde*, Berlin 1924
Das persönliche Regiment Wilhelms II. Politische Geschichte des Deutschen Kaiserreiches von 1890 bis 1914, Zürich 1948
Farrar, Lancelot L., *Divide and Conquer*, New York 1978
Fellner, Fritz, 'Wilhelm II. und das wilhelminische Deutschland im Urteil österreichischer Zeitgenossen', in Röhl, *Der Ort Kaiser Wilhelms II.*
Fesser, Gerd, *Reichskanzler Bernhard Fürst von Bülow. Eine Biographie*, Berlin 1991
'Ohrfeige für England. Die brisante Depesche Wilhelms II. an Burenpräsidenten "Ohm" Krüger', *Die Zeit*, 5 January 1996
Der Traum vom Platz an der Sonne. Deutsche 'Weltpolitik' 1897–1914, Bremen 1996
Die Kaiserzeit. Deutschland 1871–1918, Erfurt 2000
Fiebig-von Hase, Ragnhild, *Lateinamerika als Konfliktherd der deutsch-amerikanischen Beziehungen 1890–1903*, 2 vols., Göttingen 1986
'Die Rolle Kaiser Wilhelms II. in den deutsch-amerikanischen Beziehungen, 1890–1914', in Röhl, *Der Ort Kaiser Wilhelms II.*
'The Uses of "Friendship". The "Personal Regime" of Wilhelm II and Theodore Roosevelt, 1901–1909', in Mombauer and Deist, *The Kaiser*
Field, Geoffrey G., *Evangelist of Race. The Germanic Vision of Houston Stewart Chamberlain*, New York 1981
Fink, Carole, Hull, Isabel V. and Knox, MacGregor, eds., *German Nationalism and the European Response, 1890–1945*, Norman, OK, 1985
Fischer, Eugen, *Holsteins großes Nein. Die deutsch-englischen Bündnisverhandlungen von 1898–1901*, Berlin 1925
Fischer, Franz, 'Zur Entlassung Bismarcks. Unveröffentlichte Briefe aus dem Archiv der Kölnischen Zeitung', *Kölnische Zeitung*, 14/15 January 1921
Fischer, H. W. H., *Private Lives of William II and His Consort and the Secret History of the Court of Berlin 1888–1898 from the Papers and Diaries Extending over a Period Beginning June 1888 to the Spring of 1898, of Ursula, Countess of Eppinghoven*, New York 1898–9
Secret Life of the Kaiser from Birth to Exile, from the Private Papers and Diaries of Baroness von Larisch-Reddern, New York 1919
Behind the Scenes with the Kaiser 1888–1922. The True Story of the Kaiser as He Lived, Loved, Played and Warred, by the Baroness von Larisch, New York 1922
Fischer, Hartwig, *Ein Wilhelminisches Gesamtkunstwerk auf dem Kapitol. Hermann Prell und die Einrichtung des Thronsaals in der Deutschen Botschaft zu Rom 1894–1899*, Basel 1998
Fischer, Heinz-Joachim, 'Kaiserliche Eskapaden in der Ewigen Stadt. Was das Vatikanische Geheimarchiv über den Besuch Wilhelms II. in Rom enthüllt', *Frankfurter Allgemeine Zeitung*, 7 August 1996
Fischer, Jörg-Uwe, *Admiral des Kaisers. Georg Alexander von Müller als Chef des Marinekabinetts Wilhelms II.*, Frankfurt a.M. 1992

Fischer-Sallstein, Conrad, *Prinz Heinrich in Kiautschau. Reisen zu Wasser und zu Lande des Prinz-Admirals in Indien, China, Japan (1898–1900)*, Berlin 1900

Fontane, Theodor, *Briefe an Georg Friedländer*, Heidelberg 1954

Fontenoy, Marquise de, *William II, Germany: Francis Joseph, Austria-Hungary*, Philadelphia 1900

Forsbach, Ralf, *Alfred von Kiderlen-Wächter (1852–1912). Ein Diplomatenleben im Kaiserreich*, 2 vols., Göttingen 1997

Forsten, Hans, *Unser Kaiser in Palästina. Reise Kaiser Wilhelms II. und der Kaiserin Auguste Victoria nach dem gelobten Lande*, Berlin 1898

Förster, Stig, *Der doppelte Militarismus. Die deutsche Heeresrüstungspolitik zwischen Status-quo-Sicherung und Aggression 1890–1913*, Stuttgart 1985

Fox, Edward Lyell, *Wilhelm Hohenzollern & Co.*, New York 1917

Franke, Lydia, 'Die Randbemerkungen Wilhelms II. in den Akten der auswärtigen Politik als historische und psychologische Quelle', diss., Berlin 1933

Frankenberg, Paul von, *Kaiser Wilhelms II. Reise nach Jerusalem*, Berlin 1899

Franz, Eckhardt G., ed., *Erinnertes. Aufzeichnungen des letzten Großherzogs Ernst Ludwig von Hessen und bei Rhein*, Darmstadt 1983

Frederic, Harold, *The Young Emperor. William II of Germany. A Study in Character Development on a Throne*, London 1891

Freksa, Friedrich, *Menschliche Rechtfertigung Wilhelms II. Nach seinen Randbemerkungen in den Akten des Auswärtigen Amtes*, Munich 1920

Frevert, Ute, *Ehrenmänner. Das Duell in der bürgerlichen Gesellschaft*, Munich 1991

Fricke, Dieter, 'Die Affäre Leckert-Lützow-Tausch und die Regierungskrise von 1897 in Deutschland', *Zeitschrift für Geschichtswissenschaft*, 7, 1960

Fried, Alfred H., *The German Emperor and the Peace of the World*, London, New York 1912

Friedländer, Prof. Dr Adolf Albrecht, *Wilhelm II. Versuch einer psychologischen Analyse*, Halle 1919

Friedman, Isaiah, *Germany, Turkey and Zionism 1897–1918*, Oxford 1977

Friedmann, Fritz, *Der deutsche Kaiser und die Hofkamarilla. I. Der Fall Kotze, II. Wilhelm II. und die Revolution von oben*, Zürich 1896

Frobenius, Herman, *Die Hohenzollern. Geschichte Brandenburg-Preußens und des Deutschen Reiches unter den Hohenzollern*, Berlin n.d.

Fuchs, Georg, *Der Kaiser, die Kultur und die Kunst. Betrachtungen über die Zukunft des deutschen Volkes aus den Papieren eines Unverantwortlichen*, Munich, Leipzig 1904

Fuchs, Walther Peter, ed., *Großherzog Friedrich I. von Baden und die Reichspolitik 1871–1907*, 4 vols., Stuttgart 1968–80

Fursenko, A. A., *Bor'ba za razdel Kitaya i amerikanskaya doktrina otkritich dverei 1895–1900*, Moscow, Leningrad 1956

Gagliardi, Ernst, *Bismarcks Entlassung*, 2 vols., Tübingen 1927–41

Gall, Lothar, *Bismarck. Der weisse Revolutionär*, Frankfurt a.M., Berlin, Vienna 1980

Garvin, J. L. and Amery, J., *Life of Joseph Chamberlain*, 6 vols., London 1932–69

Gauss, Christian, *The German Emperor as Shown in His Public Utterances*, New York 1922

Gautschi, Andreas, *Wilhelm II. und das Waidwerk. Jagen und Jagden des letzten Deutschen Kaisers. Eine Bilanz*, Hanstedt 2000

Gerard, James W., *My Four Years in Germany*, London, New York, Toronto 1917

Face to Face with Kaiserism, London, New York, Toronto 1918

Gisevius, Hans B., *Der Anfang vom Ende. Wie es mit Wilhelm II. begann*, Zürich 1971

Goetz, Walter, ed., *Briefe Wilhelms II. an den Zaren 1894–1914*, Berlin 1920

Gollwitzer, Heinz, *Die Gelbe Gefahr. Geschichte eines Schlagworts. Studien zum imperialistischen Denken*, Göttingen 1962

Geschichte des weltpolitischen Denkens, 2 vols., Göttingen 1972–82
Görlitz, Walter, ed., *Regierte der Kaiser? Kriegstagebücher, Aufzeichnungen und Briefe des Chefs des Marine-Kabinetts Admiral G. A. von Müller, 1914–1918*, Göttingen 1959
ed., *Der Kaiser . . . Aufzeichnungen des Chefs des Marinekabinetts Admiral Georg Alexander v. Müller über die Ära Wilhelms II.*, Göttingen 1965
Gradenwitz, Otto, ed., *Akten über Bismarcks großdeutsche Rundfahrt vom Jahre 1892*, Heidelberg 1922
Bismarcks letzter Kampf, 1888–1898, Berlin 1924
Grand-Carteret, John, *Les célébrités vues par l'image 'Lui' devant l'objectif caricatural*, Paris 1906
Grenville, J. A. S., *Lord Salisbury and Foreign Policy*, London 1964
Grunder, Horst, *Christliche Mission und deutscher Imperialismus*, Paderborn 1982
Grunow, Alfred, *Der Kaiser und die Kaiserstadt*, Berlin 1970
Gutsche, Willibald, *Wilhelm II. Der letzte Kaiser des Deutschen Reiches. Eine Biographie*, Berlin 1991
Guttmann, Bernhard, *Schattenriß einer Generation 1888–1919*, Stuttgart 1950
Haller, Johannes, *Aus dem Leben des Fürsten Philipp zu Eulenburg-Hertefeld*, Berlin 1924
Hallgarten, G. W. F., *Imperialismus vor 1914. Die soziologischen Grundlagen der Außenpolitik europäischer Großmächte vor dem Ersten Weltkrieg*, 2 vols., Munich 1963
Hallmann, Hans, ed., *Zur Geschichte und Problematik des deutsch-russischen Rückversicherungsvertrages von 1887*, Darmstadt 1968
Hamann, Brigitte, 'Das Leben des Kronprinzen Rudolf von Österreich-Ungarn nach neuen Quellen', diss., Vienna 1977
Rudolf, Kronprinz und Rebell, Vienna, Munich 1978
Hammann, Otto, *Der neue Kurs*, Berlin 1918
Um den Kaiser, Erinnerungen aus den Jahren 1906–1909, Berlin 1919
Bilder aus der letzten Kaiserzeit, Berlin 1922
Hammer, S. C., *Wilhelm II, et blad au Tysklands nyeste historie*, Kristiania 1915
Hank, Manfred, *Kanzler ohne Amt. Fürst Bismarck nach seiner Entlassung 1890–1898*, Munich 1977
Hansen, Heinrich E., 'Die "Hunnenrede" Kaiser Wilhelms II. in Bremerhaven', *Jahrbuch der Männer vom Morgenstern*, 50, 1969, pp. 207–31
Harden, Maximilian, "Der Kampf mit dem Drachen", *Die Zukunft*, No. 32, 11 August 1900
Hartau, Friedrich, *Wilhelm II in Selbstzeugnissen und Bilddokumenten*, Reinbek bei Hamburg 1978
Hartung, Fritz, *Das persönliche Regiment Kaiser Wilhelms II.*, Berlin 1952
Hassan, Vita, *Die Wahrheit über Emin Pascha*, 1893
Hecht, Karsten, 'Die Harden-Prozesse. Strafverfahren, Öffentlichkeit und Politik im Kaiserreich', diss., Munich 1997
Heinsick, Paul, *Wilhelm II., Friedenskaiser oder nicht?*, Leipzig 1915
Helfritz, Hans, *Wilhelm II. als Kaiser und König. Eine historische Studie*, Zürich 1954
Hellige, Hans Dieter and Schulin, Ernst, eds., *Walther Rathenau Gesamtausgabe*, vol. VI, *Briefwechsel Walther Rathenau–Maximilian Harden*, Munich, Heidelberg 1983
Hellwig, F., *Carl Freiherr von Stumm-Halberg*, Heidelberg, Saarbrücken 1936
Hemmi, Beate, *Wilhelm II. und die Reichsregierung im Urteil schweizerischer diplomatischer Berichte*, Zürich 1964
Heresch, Elisabeth, *Nikolaus II. 'Feigheit, Lüge und Verrat'. Leben und Ende des letzten russischen Zaren*, Munich 1992
Herre, Franz, *Wilhelm II. Monarch zwischen den Zeiten*, Cologne 1993
Hervier, Paul-Louis, *The Two Williams. Studies of the Kaiser and the Crown Prince*, London 1916

Herwig, Holger H., *The German Naval Officer Corps. A Social and Political History, 1890–1918*, Oxford 1973

'Luxury' Fleet. The Imperial German Navy 1888–1918, London 1987

Herzfeld, Hans, *Johannes von Miquel. Sein Anteil am Ausbau des Deutschen Reiches bis zur Jahrhundertwende*, 2 vols., Detmold 1938

Herzl, Theodor, *Gesammelte zionistische Werke*, 5 vols., Tel Aviv 1934

Heyking, Elisabeth von, *Tagebücher aus vier Weltteilen, 1886–1904*, Leipzig 1926

Hilarus, *Jung Wilhelm. Heitere Bilder aus der Jugendzeit Kaiser Wilhelms II.*, Berlin 1910

Hildebrand, Klaus, *Deutsche Außenpolitik 1871–1918*, Munich 1989

Das vergangene Reich. Deutsche Außenpolitik von Bismarck bis Hitler, 1871–1945, Stuttgart 1995

Hill, David J., *Impressions of the Kaiser*, New York, London 1918

Hintze, Otto, *Die Hohenzollern und ihr Werk. Fünfhundert Jahre vaterländischer Geschichte*, Berlin 2nd edn 1915

Hinzpeter, Georg Ernst, *Kaiser Wilhelm II. Eine Skizze nach der Natur gezeichnet*, Bielefeld 1888

Hoche, Jules, *L'Empereur Guillaume II intim*, Paris 1913

Hofmann, Hermann, *Fürst Bismarck 1890–1898*, 2 vols., Stuttgart, Berlin, Leipzig 1913

Hohenlohe, Alexander von, *Aus meinem Leben*, Frankfurt a.M. 1925

Hohenlohe-Schillingsfürst, Chlodwig Fürst zu, *Denkwürdigkeiten*, 2 vols., Stuttgart, Leipzig 1907

Müller, Karl Alexander von, ed., *Denkwürdigkeiten der Reichskanzlerzeit*, Stuttgart, Berlin 1931

Horn, Hannelore, *Der Kampf um den Bau des Mittellandkanals. Staat und Politik*, Cologne, Opladen 1964

Hubatsch, Walther, *Der Admiralstab und die obersten Marinebehörden in Deutschland 1884–1945*, Frankfurt a.M. 1958

Hull, Isabel V., *The Entourage of Kaiser Wilhelm II, 1888–1918*, Cambridge 1982

'Kaiser Wilhelm II and the "Liebenberg Circle"', in Röhl and Sombart, *Kaiser Wilhelm II. New Interpretations*

'"Persönliches Regiment"', in Röhl, *Der Ort Kaiser Wilhelms II.*

'Military Culture, Wilhelm II and the End of the Monarchy in the First World War', in Mombauer and Deist, *The Kaiser*

Hutten-Czapski, Bogdan Graf von, *Sechzig Jahre Politik und Gesellschaft*, 2 vols., Berlin 1936

Jacks, William, *The Life of His Majesty William the Second, German Emperor, with a Sketch of His Hohenzollern Ancestors*, Glasgow 1904

Jarchow, Margarete, *Hofgeschenke. Wilhelm II. zwischen Diplomatie und Dynastie 1888–1914*, Hamburg 1998

Jefferson, Margaret M., 'Lord Salisbury's Conversations with the Tsar in Balmoral, 27 and 29 September 1896', *Slavonic and East European Review*, 39, 1960–1

Jerussalimski, A. S., *Die Außenpolitik und die Diplomatie des deutschen Imperialismus Ende des 19. Jahrhunderts*, Berlin 1954

Jonge, J. A., *Wilhelm II.*, Amsterdam 1986

Julier, Jürgen, ed., *Kaiserlicher Kunstbesitz. Aus dem holländischen Exil Haus Doorn*, Berlin 1991

Kanghi-Tschu, *Deutschland, der Kaiser und Simplizissimus. Aus den Berichten eines chinesischen Diplomaten*, Munich 1911

Kann, Robert A., 'Kaiser Wilhelm II. und Thronfolger Franz Ferdinand in ihrer Korrespondenz', in Kann, Robert A., *Erzherzog Franz Ferdinand Studien*, Vienna 1976, pp. 46–85

Kautsky, Karl, *The Guilt of Wilhelm Hohenzollern*, London 1920

Kautzsch, W., *Intimes des Berliner Hoflebens unter Wilhelm II.*, Berlin 1922

Keller, Mathilde Gräfin von, *Vierzig Jahre im Dienst der Kaiserin. Ein Kulturbild aus den Jahren 1881–1921*, Leipzig 1935

Kennan, George F., *The Decline of Bismarck's European Order. Franco-Russian Relations, 1879–1890*, Princeton 1979

The Fateful Alliance. France, Russia and the Coming of the First World War, New York 1984

Kennedy, J. M., *The War Lord: A Character Study of Kaiser Wilhelm II by Means of His Speeches, Letters and Telegrams*, London 1914

Kennedy, Paul, 'Tirpitz, England and the Second Navy Law of 1900. A Strategical Critique', *Militärgeschichtliche Mitteilungen*, 1970, 2

The Samoan Tangle. A Study in Anglo-German-American Relations 1878–1900, Dublin 1974

The Rise of the Anglo-German Antagonism 1860–1914, London 1980

'The Kaiser and German Weltpolitik. Reflexions on Wilhelm II's Place in the Making of German Foreign Policy', in Röhl and Sombart, *Kaiser Wilhelm II. New Interpretations*

Klahr, Douglas Mark, 'The Kaiser Builds in Berlin. Expressing National and Dynastic Identity in the Early Building Projects of Wilhelm II', diss., Providence, RI, 2002

Klaussmann, A. O., ed., *The Kaiser's Speeches, Forming a Character Portrait of Wilhelm II*, New York, London 1903

Koch, Ernestine, *Albert Langen. Ein Verleger in München*, Munich, Berlin 1969

Koffler, Dosio, *Wilhelm II.*, Berlin 1931

Kohlrausch, Martin, 'Monarchie und Massenöffentlichkeit. Veränderungen in der Rezeption des wilhelminischen Kaisertums, 1890–1925', diss., Florence 2002

Kohut, A., *Kaiser Wilhelm II als Denker. Goldene Worte und Aussprüche aus seinen Reden, Erlässen, Gesprächen, Briefen und Telegrammen*, Lüneburg 1913

Kohut, Thomas A., *Wilhelm II and the Germans. A Study in Leadership*, Oxford, New York, 1991

Kospoth, K. A., *Wie ich zu meinem Kaiser stand. Persönliche Erinnerungen an Kaiser Wilhelm II.*, Breslau 1924

Kracke, Friedrich, *Prinz und Kaiser. Wilhelm II. im Urteil seiner Zeit*, Munich 1960

Krausnick, Helmut, *Neue Bismarck-Gespräche. Vier unveröffentlichte politische Gespräche des Kanzlers mit österreich-ungarischen Staatsmännern sowie ein Gespräch Kaiser Wilhelms II.*, Hamburg 1940

Holsteins Geheimpolitik in der Ära Bismarck, 1886–1890, Hamburg 1942

'Holstein, Österreich-Ungarn und die Meerengenfrage im Herbst 1895. Persönliches Regiment oder Regierungspolitik?', in *Forschungen zu Staat und Verfassung. Festgabe für Fritz Hartung*, Berlin 1958

Krockow, Christian Graf von, *'Unser Kaiser'. Glanz und Sturz der Monarchie*, Munich 1996

Kaiser Wilhelm II. und seine Zeit. Biographie einer Epoche, Berlin 1999

Kroll, Frank-Lothar, ed., *Preussens Herrscher. Von den ersten Hohenzollern bis Wilhelm II.*, Munich 2000

Kugler, Bernhard von, *Deutschlands größter Held. Jubel-Ausgabe zur hundertjährigen Gedächtnisfeier des Geburtstags weiland Sr. Majestät Kaiser Wilhelm I.*, Dresden 1893

Kürenberg, Joachim von, *War alles falsch? Das Leben Kaiser Wilhelms II.*, Bonn 1951

Kurtz, Harold, *The Second Reich. Kaiser Wilhelm II and His Germany*, London, New York 1970

Lahme, Rainer, *Deutsche Außenpolitik 1890–1894. Von der Gleichgewichtspolitik Bismarcks zur Allianzstrategie Caprivis*, Göttingen 1990

Lamber, Juliette, *Guillaume II (1890–1899)*, Paris 1917

Lambi, Ivo Nikolai, *The Navy and German Power Politics, 1862–1914*, Boston, London, Sydney 1984

Lammerting, Kristin, *Meteor. Die kaiserlichen Segelyachten*, Cologne 1999

Lamprecht, Karl, *Der Kaiser. Versuch einer Charakteristik*, Berlin 1916

Lamsdorff, Wladimir N., *Dnevnik 1886–1892*, 2 vols., Moscow 1926–34

'Die Nichterneuerung des Rückversicherungsvertrages. Auszüge aus dem Tagebuch des russischen Außenministers Grafen Wladimir Nikolajewitsch Lamsdorff', *Berliner Monatshefte*, 9, 1931, pp. 158–77

Lanessan, J., *L'Empire germanique sous la direction de Bismarck et de Guillaume II*, Paris 1915

Langer, William L., *The Franco-Russian Alliance, 1890–1894*, Cambridge 1929

European Alliances and Alignments, 1871–1890, New York 2nd edn 1956

The Diplomacy of Imperialism 1890–1902, New York 2nd edn 1965

Laufer, Jochen, 'Die deutsche Südafrikapolitik 1890–1898 im Spannungsfeld zwischen deutsch-englischen Beziehungen, Wirtschaftsinteressen und Expansionsforderungen in der bürgerlichen Öffentlichkeit', diss., Humboldt-Universität Berlin 1986

Le Mang, Erich, *Die persönliche Schuld Wilhelms II. Ein zeitgemäßer Rückblick*, Dresden 1919

Lee, Arthur Gould, *The Empress Frederick Writes to Sophie, Her Daughter, Crown Princess and Later Queen of the Hellenes*, London 1955

Lee, Sir Sidney, *King Edward VII. A Biography*, 2 vols., London 1925–7

Legge, Edward, *The Public and Private Life of Kaiser Wilhelm II*, London 1915

King Edward, the Kaiser and the War, London 1917

Lehnert, Uta, *Der Kaiser und die Siegesallee. Réclame Royale*, Berlin 1998

Lepsius, J., Mendelssohn-Bartholdy, A. and Thimme, F., eds., *Die Große Politik der europäischen Kabinette, 1871–1914*, 40 vols., Berlin 1922–7

Lerchenfeld-Koefering, Hugo Graf, *Erinnerungen und Denkwürdigkeiten, 1843–1925*, Berlin 1934

Kaiser Wilhelm II. als Persönlichkeit und Herrscher, Kallmünz 1985

Lerman, Katharine Anne, *The Chancellor as Courtier. Bernhard von Bülow and the Governance of Germany 1900–1909*, Cambridge 1990

'The Chancellor as Courtier. The Position of the Responsible Government under Kaiser Wilhelm II, 1900–1909', in Röhl, *Der Ort Kaiser Wilhelms II.*

'The Kaiser's Elite? Wilhelm II and the Berlin Administration 1890–1914', in Mombauer and Deist, *The Kaiser*

Leudet, Maurice, *Guillaume II intim*, Paris 1897

Leuß, Hans, *Wilhelm Freiherr von Hammerstein*, Berlin 1905

Liman, Paul, *Der Kaiser 1888–1909. Ein Charakterbild Kaiser Wilhelms II.*, Leipzig 1909

Lindberg, Folke, *Kunglig utrikespolitik. Studier i svensk utrikespolitik under Oscar II och fram till borggårdskrisen*, Stockholm 1966

Longford, Elizabeth, *Jameson's Raid. The Prelude to the Boer War*, London 1984

Lowe, Charles, *The German Emperor, William II*, London 1895

Lucius von Ballhausen, Robert Freiherr, *Bismarck-Erinnerungen*, Stuttgart, Berlin 1920

Ludwig, Emil, *Wilhelm der Zweite*, Berlin 1926

Lugaro, Ernesto, *An Emperor's Madness or National Aberration?*, London, New York 1916

Lynx, J. J., *The Great Hohenzollern Scandal*, London 1965

Maatz, H., *Bismarck und Hannover 1866–1898*, Hildesheim 1971

Maccotta, Giuseppe, *Guillermo II, la Germania e l'Europa 1888–1914*, Rome 1934

MacDonogh, Giles, *The Last Kaiser. William the Impetuous*, London 2000

Machtan, Lothar, *Bismarcks Tod und Deutschlands Tränen. Reportage einer Tragödie*, Munich 1998

Malkowsky, Gerhard, *Die Kunst im Dienste der Staatsidee. Hohenzollernsche Kunstpolitik vom Großen Kurfürsten bis auf Wilhelm II.*, Berlin 1912

Mann, Golo, *Wilhelm II.*, Munich 1964

Mansel, Philip, *Constantinople, City of the World's Desire, 1453–1924*, London 1995

Marie Louise Princess of Schleswig-Holstein, *My Memories of Six Reigns*, London 1956

Marienfeld, W., *Wissenschaft und Schlachtflottenbau in Deutschland 1897–1906*, Frankfurt a.M. 1957

Marschall, Birgit, *Reisen und Regieren. Die Nordlandfahrten Kaiser Wilhelms II.*, Heidelberg 1991

Martin, Rudolf, *Deutsche Machthaber*, Berlin, Leipzig 1910

Massie, Robert K., *Dreadnought. Britain, Germany and the Coming of the Great War*, New York 1991

Masur, Gerhard, *Imperial Berlin*, London 1971

Maute, Hans-Ernst, *Die Februarerlasse Kaiser Wilhelms II. und ihre gesetzliche Ausführung. Unter besonderer Berücksichtigung der Berliner Internationalen Arbeiterschutzkonferenz von 1890*, Bielefeld 1984

Mazel, Henri, *La Psychologie du Kaiser*, Paris 1919

McCabe, Joseph, *The Kaiser. His Personality and Career*, London 1915

McLean, Roderick R., 'Kaiser Wilhelm II and His Hessian Cousins. Intra-state Relations in the German Empire and International Dynastic Politics, 1890–1918', *German History*, 19/1, 2001

Royalty and Diplomacy in Europe, 1890–1914, Cambridge 2001

'Dreams of a German Europe. Wilhelm II and the Treaty of Björkö of 1905', in Mombauer and Deist, *The Kaiser*

Mehnert, Ute, *Deutschland, Amerika und die 'Gelbe Gefahr'. Zur Karriere eines Schlagworts in der Großen Politik 1905–1917*, Stuttgart 1995

Meinecke, Friedrich, *Geschichte des deutsch-englischen Bündnisproblems*, Munich, Berlin 1927

Meinhold, Paul, *Wilhelm II., 25 Jahre Kaiser und König*, Berlin 1912

Meisner, Heinrich Otto, ed., *Denkwürdigkeiten des General-Feldmarschalls Alfred Grafen von Waldersee*, 3 vols., Stuttgart, Berlin 1922–3

Militärattachés und Militärbevollmächtigte in Preußen und im Deutschen Reich. Ein Beitrag zur Geschichte der Militärdiplomatie, Berlin 1957

Meyer, Adolf, *Ins Heilige Land. Reisebilder von der großen Festfahrt nach Jerusalem im Oktober und November 1898*, Berlin 1899

Meyer, Arnold Oskar, *Bismarck. Der Mensch und der Staatsmann*, Stuttgart 2nd edn 1949

Mirbach, Ernst Freiherr von, ed., *Das deutsche Kaiserpaar im Heiligen Lande im Herbst 1898. Mit Allerhöchster Ermächtigung Seiner Majestät des Kaisers und Königs bearbeitet nach authentischen Berichten und Akten*, Berlin 1899

Mitchell, Nancy, *The Danger of Dreams. German and American Imperialism in Latin America*, Chapel Hill, London 1999

Möckl, Karl, *Die Prinzregentenzeit. Gesellschaft und Politik während der Ära des Prinzregenten Luitpold in Bayern*, Munich 1972

Moltke, Eliza von, ed., *Generaloberst Helmuth von Moltke. Erinnerungen, Briefe, Dokumente 1877–1916. Ein Bild vom Kriegsausbruch, erster Kriegsführung und Persönlichkeit des ersten militärischen Führers des Krieges*, Stuttgart 1922

Mombauer, Annika, *Helmuth von Moltke and the Origins of the First World War*, Cambridge 2001

'Wilhelm, Waldersee and the Boxer Rebellion', in Mombauer, and Deist, *The Kaiser*

Mombauer, Annika and Deist, Wilhelm, eds., *The Kaiser. New Research on Wilhelm II's Role in Imperial Germany*, Cambridge 2003

Mommsen, Wilhelm, *Bismarcks Sturz und die Parteien*, Stuttgart 1924

Mommsen, Wolfgang J., *Der Autoritäre Nationalstaat. Verfassung, Gesellschaft und Kultur des deutschen Kaiserreiches*, Frankfurt a.M. 1990

'Kaiser Wilhelm II and German Politics', *Journal of Contemporary History*, 25/2–3, May–June 1990, pp. 289–316

Großmachtstellung und Weltpolitik, 1870–1914. Die Außenpolitik des Deutschen Reiches, Frankfurt a.M. 1993

War der Kaiser an allem schuld? Wilhelm II. und die preußisch-deutschen Machteliten, Munich 2002

Morré, Harold, *20 Jahre S. M. Heitere Bilder zu ernsten Ereignissen*, Berlin 1909

Mosse, Werner E., ed., *Juden im Wilhelminischen Deutschland 1890–1914*, Tübingen 1976

Mossolov, A. A., *At the Court of the Last Tsar*, London 1935

Müller, Karl Alexander von, 'Die Entlassung. Nach den bayerischen Gesandtschaftsberichten', *Süddeutsche Monatshefte*, 19/1, December 1921

Muralt, Leonhard von, *Bismarcks Verantwortlichkeit*, Göttingen 1955

Muret, Maurice, *Guillaume II*, Paris 1940

Naumann, Friedrich, *Demokratie und Kaisertum*, Berlin 3rd edn 1904

Neyen, Emil, *Wilhelm II. Sein Werk. 450 Millarden neuer Schulden und 21 fliehende Könige*, Berlin 1919

Nichols, J. Alden, *Germany after Bismarck. The Caprivi Era, 1890–1894*, Cambridge 1958

Niemann, Alfred, *Kaiser und Heer. Das Wesen der Kommandogewalt und ihre Ausübung durch Kaiser Wilhelm II.*, Berlin 1921

Wanderungen mit Kaiser Wilhelm II., Leipzig 1924

Nipperdey, Thomas, *Deutsche Geschichte 1866–1918*, Vol. I, *Arbeitswelt und Bürgergeist*, Munich 1990

Deutsche Geschichte 1866–1918, Vol. II, *Machtstaat vor der Demokratie*, Munich 1992

Noack, Friedrich, *Das Deutschtum in Rom seit dem Ausgang des Mittelalters*, 2 vols., Berlin, Leipzig 1927

Nöll von der Nahmer, R., *Bismarcks Reptilienfonds*, Mainz 1968

Nostiz, Herbert von, *Bismarcks unbotmäßiger Botschafter, Fürst Münster von Derneburg (1820–1902)*, Göttingen 1968

Noussanne, Henri de, *The Kaiser as He Is: or, the Real William II (Le Veritable Guillaume II)*, New York, London 1905

Nowak, Karl-Friedrich and Thimme, Friedrich, eds., *Erinnerungen und Gedanken des Botschafters Anton Graf Monts*, Berlin 1932

Öke, Bulent Mim Kemal, 'Ottoman Policies towards Zionism', MPhil. diss., Cambridge 1979

Oncken, Hermann, *Rudolf von Bennigsen. Ein deutscher liberaler Politiker. Nach seinen Briefen und hinterlassenen Papieren*, 2 vols., Stuttgart, Leipzig 1910

Oncken, Wilhelm, ed., *Unser Heldenkaiser. Festschrift zum hundertjährigen Geburtstage Kaiser Wilhelms des Großen*, Berlin 1897

Osmanoglu, Ayse, *Avec mon Père le Sultan Abdulhamid de son palais à sa prison*, Paris 1991

Österreich-Ungarns Außenpolitik von der Bosnischen Krise 1908 bis zum Kriegsausbruch 1914. Diplomatische Aktenstücke des österreichisch-ungarischen Ministerium des Äußern, selected by Ludwig Bittner, Alfred F. Pribram, Heinrich Srbik and Hans Uebersberger, edited by Ludwig Bittner and Hans Uebersberger, 9 vols., Vienna, Leipzig 1930

Otte, Thomas, '"The Winston of Germany." The British Foreign Policy Élite and the Last German Emperor', *Canadian Journal of History*, December 2001, pp. 471–504

Otto, Berthold, *Wilhelm II. und wir! Die Kaiserartikel des 'Deutschen Volksgeistes' aus den Jahren 1919–1925*, Berlin 1925

Padje, Willem-Alexander van't, 'At the Heart of the Growing Anglo-German Imperialist Rivalry. Two British Ambassadors in Berlin, 1884–1908', DPhil. diss., Oxford 2001

Pakula, Hannah, *An Uncommon Woman. The Empress Frederick, Daughter of Queen Victoria, Wife of the Crown Prince of Prussia, Mother of Kaiser Wilhelm*, New York 1995

Paléologue, Maurice, *Guillaume II et Nicolas II*, Paris 1934

Palmer, Alan, *The Kaiser. Warlord of the Second Reich*, New York 1916

Paulmann, Johannes, *Pomp und Politik. Monarchenbegegnungen in Europa zwischen Ancien Régime und Erstem Weltkrieg*, Paderborn, Munich, Vienna, Zürich 2000

Penzler, Johannes, ed., *Fürst Bismarck nach seiner Entlassung. Leben und Politik des Fürsten seit seinem Scheiden aus dem Amte auf Grund aller authentischen Kundgebungen*, 7 vols., Leipzig 1897–8

Die Reden Kaiser Wilhelms II. in den Jahren 1888–1895, Leipzig, n.d.

ed., *Die Reden Kaiser Wilhelms II. in den Jahren 1896–1900*, Leipzig 1904

Die Reden Kaiser Wilhelms II. in den Jahren 1901–Ende 1905, Leipzig n.d.

Perris, George, *Germany and the German Emperor*, London 1912

Peters, Carl, *New Light on Dark Africa; Being the Narrative of the German Emin Pasha Expedition*, London, New York, Melbourne 1891

Peters, Evelene, *Roosevelt und der Kaiser, ein Beitrag zur Geschichte der deutsch-amerikanischen Beziehungen, 1895–1906*, Leipzig 1936

Pezold, Dirk von, *Cäsaromanie und Byzantinismus bei Wilhelm II.*, Cologne 1971

Pfeil und Klein-Ellguth, Hans Graf von, *Mein Kaiser! Der Fall Zedlitz-Trützschler und Wilhelms II. wahres Gesicht*, Leipzig 1924

Pflanze, Otto, *Bismarck and the Development of Germany*, 3 vols., Princeton 1963–90

Philipp, Ferdinand, *Bismarcks vertrauliche Gespräche u. a. über Wilhelm II.*, Dresden 1927

Pless, Daisy Fürstin von, *Tanz auf dem Vulkan. Erinnerungen an Deutschlands und Englands Schicksalswende*, 2 vols., Dresden 2nd edn 1930

Pogge von Strandmann, Hartmut, 'Der Kaiser und die Industriellen', in Röhl, *Der Ort Kaiser Wilhelms II.*

'Nationalisierungsdruck und königliche Namensänderung in England. Das Ende der Großfamilie europäischer Dynastien', in Ritter, Gerhard A. and Wende, Peter, eds., *Rivalität und Partnerschaft. Studien zu den deutsch-britischen Beziehungen im 19. und 20. Jahrhundert. Festschrift für Anthony J. Nicholls*, Paderborn 1999

'Rathenau, Wilhelm II and the Perception of Wilhelminismus', in Mombauer and Deist, *The Kaiser*

Pöls, Werner, *Sozialistenfrage und Revolutionsfurcht in ihrem Zusammenhang mit den angeblichen Staatsstreichplänen Bismarcks*, Lübeck, Hamburg 1960

Pommerin, Reiner, *Der Kaiser und Amerika. Die USA in der Politik der Reichsleitung, 1890–1917*, Cologne 1986

Ponsonby, Sir Frederick, ed., *Letters of the Empress Frederick*, London 1928

Briefe der Kaiserin Friedrich, Berlin 1929

Posadowsky, Arthur Graf, *Weltwende. Gesammelte politische Aufsätze*, Stuttgart 2nd edn 1920

Potts, D. M. and Potts, W. T. W., *Queen Victoria's Gene. Haemophilia and the Royal Family*, Stroud 1995

Prince, Morton, *The Psychology of the Kaiser. A Study of His Sentiments and Obsessions*, London 1915

Pudor, Heinrich, *Kaiser Wilhelm II. und Rembrandt als Erzieher*, Dresden 2nd edn 1891

Quidde, Ludwig, *Caligula. Eine Studie über römischen Cäsarenwahnsinn*, Leipzig 12th edn 1894

Der deutsche Pazifismus während des Weltkrieges, 1914–1918, Boppard am Rhein 1979

Rachfahl, Felix, *Kaiser und Reich 1888–1913. 25 Jahre preußisch-deutsche Geschichte. Festschrift zum 25-jährigen Regierungsjubiläum Wilhelms II., deutschen Kaisers, Königs von Preussen*, Berlin 1913

Radkau, Joachim, *Das Zeitalter der Nervosität. Deutschland zwischen Bismarck und Hitler*, Munich, Vienna 1998

Radziwill, Princess Catherine, *Sovereigns and Statesmen of Europe*, London, New York, Toronto, Melbourne 1915

Radziwill, Princess Marie, *This Was Germany. An Observer at the Court of Berlin*, London 1937

Rall, Hans, *König Ludwig II. und Bismarcks Ringen um Bayern*, Munich 1973
Wilhelm II. Eine Biographie, Graz, Vienna, Cologne 1995
Ramm, Agatha, ed., *Beloved and Darling Child. Last Letters Between Queen Victoria and Her Eldest Daughter 1886–1901*, Stroud 1990
Rapsilber, Maximilian, 'Der Kaiser als Kunstrichter', in Leipziger, L., ed., *Der Roland von Berlin. Eine Wochenschrift für das Berliner Leben*, vol. 2, part 25, 3 June 1904
Raschdau, Ludwig, *Unter Bismarck und Caprivi. Erinnerungen eines deutschen Diplomaten aus den Jahren 1885–1894*, Berlin 1939
In Weimar als preußischer Gesandter. Ein Buch der Erinnerung an deutsche Fürstenhöfe, 1894–1897, Berlin 1939
Rathenau, Walther, *Der Kaiser*, Berlin 1919
Rebentisch, Jost, *Die vielen Gesichter des Kaisers. Wilhelm II. in der deutschen und britischen Karikatur (1888–1918)*, Berlin 2000
Reck-Malleczewen, Friedrich Percyval, *Tagebuch eines Verzweifelten*, Berlin, Bonn 1981
Reif, Heinz, *Adel im 19. und 20. Jahrhundert*, Munich 1999
ed., *Adel und Bürgertum in Deutschland*, 2 vols., Berlin 2000
Reifland, Ursula, 'Das persönliche Regiment Wilhelms II. in der Zeit der Kanzlerschaft Bülows', unpublished manuscript, Berlin 1951
Reinermann, Lothar, *Der Kaiser in England. Wilhelm II. und sein Bild in der britischen Öffentlichkeit*, Paderborn, Munich, Vienna, Zürich 2001
Reischach, Hugo Freiherr von, *Unter drei Kaisern*, Berlin 2nd edn 1925
Reventlow, Ernst Graf von, *Kaiser Wilhelm II. und die Byzantiner*, Munich 1906
Der Kaiser und die Monarchisten, Berlin 1913
Glanz und Tragödie Wilhelms II., Stuttgart 1938
Von Potsdam nach Doorn, Berlin 1940
Rich, Norman, *Friedrich von Holstein, Politics and Diplomacy in the Era of Bismarck and Wilhelm II*, 2 vols., Cambridge 1965
Rich, Norman and Fisher, M. H., eds., *Die Geheimen Papiere Friedrich von Holsteins*, 4 vols., Göttingen 1956–63
The Holstein Papers. The Memoirs, Diaries and Correspondence of Friedrich von Holstein 1837–1909, 4 vols., Cambridge 1956–63
Richie, Alexandra, *Faust's Metropolis. A History of Berlin*, London 1998
Richter, Jan Stefan, *Die Orientreise Kaiser Wilhelms II. 1898*, Hamburg 1997
Ritter, Gerhard, *Die Legende von der verschmähten englischen Freundschaft 1898/1901*, Leipzig 1929
Die deutschen Militär-Attachés und das Auswärtige Amt. Aus den verbrannten Akten des Großen Generalstabes, Heidelberg 1959
Ritter, Gerhard A., *Die Arbeiterbewegung im Wilhelminischen Reich*, Berlin 1959
Ritter, Gerhard A. and Niehuss, M., *Wahlgeschichtliches Arbeitsbuch. Materialien zur Statistik des Kaiserreichs, 1871–1918*, Munich 1980
Roberts, Andrew, *Salisbury. Victorian Titan*, London 2000
Robolsky, Hermann, *Kaiser Wilhelm II. und seine Leute*, Berlin 1891
Rogge, Helmuth, *Friedrich von Holstein. Lebensbekenntnis in Briefen an eine Frau*, Berlin 1932
Holstein und Hohenlohe. Neue Beiträge zu Friedrich von Holsteins Tätigkeit als Mitarbeiter Bismarcks und als Ratgeber Hohenlohes, Stuttgart 1957
'Die Kladderadatsch-Affäre. Ein Beitrag zur inneren Geschichte des Wilhelminischen Reichs', *Historische Zeitschrift*, 195/1, August 1962, pp. 90–130
Roggenbach, Franz von, *Im Ring der Gegner Bismarcks. Denkschriften und politischer Briefwechsel Franz von Roggenbachs mit Kaiserin Augusta und Albrecht von Stosch*, Stuttgart 1943

Röhl, John C. G., 'Staatsstreichplan oder Staatsstreichbereitschaft? Bismarcks Politik in der Entlassungskrise', *Historische Zeitschrift*, 203, 1966, pp. 610–24

Germany without Bismarct. The Crisis of Government in the Second Reich, 1890–1900, London, Berkeley, Los Angeles 1967

ed., *Philipp Eulenburgs politische Korrespondenz*, 3 vols., Boppard am Rhein 1976–83

Kaiser Wilhelms II. Eine Studie über Cäsarenwahnsinn, Munich 1989

ed., *Der Ort Kaiser Wilhelms II. in der deutschen Geschichte*, Munich 1991

The Kaiser and his Court. Wilhelm II and the Government of Germany, 3rd edn, Cambridge 1996

Young Wilhelm. The Kaiser's is Early Life, 1859–1888, Cambridge 1998

'Der Kaiser und England', in Rogasch, Wilfrid, ed., *Victoria & Albert, Vicky & The Kaiser. Ein Kapitel deutsch-englischer Familiengeschichte*, Berlin 1997

'Herzl and Kaiser Wilhelm II. A German Protectorate in Palestine?', in Robinson, Ritchie and Timms, Edward, eds., *Theodor Herzl and the Origins of Zionism*, Edinburgh 1997

'The Kaiser's Germany as Seen from Beachy Head on a Clear Day. Autobiographical Reminiscences of an Anglo-German Historian', in Berger, Stefan, Lambert, Peter and Schumann, Peter, eds., *Historikerdialoge. Geschichte, Mythos und Gedächtnis im deutsch-britischen kulturellen Austausch 1750–2000*, Göttingen 2003

Röhl, John C. G. and Sombart, Nicolaus, eds., *Kaiser Wilhelm II – New Interpretations*, Cambridge 1982

Röhl, John C. G., Warren, Martin and Hunt, David, *Purple Secret. Genes, 'Madness' and the Royal Houses of Europe*, London 1998

Rolland, Romain, 'Journal Intime, 22. Januar 1898', in *Cahiers Romain Rolland*, III, *Richard Strauß et Romain Rolland*, Paris 1951

Roth, Erwin, *Preußens Gloria in Heiligen Land: Die Deutschen und Jerusalem*, Munich 1973

Routier, Gaston, *Guillaume II à Londres et l'union franco-russe*, Paris 1894

Samerski, Stefan, ed., *Wilhelm II. und die Religion. Facetten einer Persönlichkeit und ihres Umfelds*, Berlin 2001

Saunders, George, *Builder and Blunderer. A Study of Emperor William's Character and Foreign Policy*, New York 1914

The Last of the Huns, London, New York 1914

Scheuner, Ulrich, 'Die Kunst als Staatsaufgabe im 19. Jahrhundert', in Mai, Ekkehard and Waetzoldt, Stephan, eds., *Kunstverwaltung, Bau und Denkmalkunst im Kaiserreich*, Berlin 1981

Schierbrand, Wolf von, *The Kaiser's Speeches*, New York 1903

'Is Kaiser Wilhelm II of Normal Mind?', *Lippincott's*, 78, November 1906

Schlitter, Hanns, 'Briefe Kaiser Franz Josephs I. und Kaiser Wilhelms II. über Bismarcks Rücktritt', *Österreichische Rundschau*, 58, 1919

Schlözer, Kurd von, *Letzte römische Briefe 1882–1894*, Stuttgart, Berlin, Leipzig 1924

Schmidt-Bückeburg, Rudolf, *Das Militärkabinett der preußischen Könige und deutschen Kaiser. Seine geschichtliche Entwicklung und staatsrechtliche Stellung 1787–1918*, Berlin 1933

Schmidt-Pauli, Edgar von, *Der Kaiser. Das wahre Gesicht Wilhelms II.*, Berlin 1928

Schneider, Julius, *Die Geschichte des Berliner Doms*, Berlin 1993

Schneller, Pastor Ludwig, *Die Kaiserfahrt durch's heilige Land*, Leipzig 1899

Schoeps, Julius H., *Theodor Herzl 1860–1904. Wenn ihr wollt, ist es kein Märchen. Eine Text-Bild-Monographie*, Vienna 1995

Schöllgen, Gregor, *Imperialismus und Gleichgewicht. Deutschland, England und die orientalische Frage 1871–1914*, Munich 1992

Schönburg-Waldenburg, Heinrich Prinz von, *Erinnerungen aus kaiserlicher Zeit*, Leipzig 1929

Schott, Richard, *Eine Fahrt nach dem Orient. Zur Erinnerung an den Einzug des deutschen Kaisers und der Kaiserin in Jerusalem, Herbst 1898*, Berlin, Eisenach, Leipzig 1898

Schottelius, Herbert and Deist, Wilhelm, eds., *Marine und Marinepolitik im kaiserlichen Deutschland 1871–1914*, Düsseldorf 1972

Schröder, E., *Zwanzig Jahre Regierungszeit. Ein Tagebuch Kaiser Wilhelms II. Vom Antritt der Regierung, 15. Juni 1888 bis zum 15. Juni 1908 nach Hof- und anderen Berichten*, Berlin 1909

Schröder, Wilhelm, *Das persönliche Regiment, Reden und sonstige öffentliche Äusserungen Wilhelms II.*, Munich 1907

Schümann, Carl-Wolfgang, *Der Berliner Dom im 19. Jahrhundert*, Berlin 1980

Schüssler, Wilhelm, *Bismarcks Sturz*, Leipzig 3rd edn 1922

Deutschland zwischen Rußland und England. Studien zur Außenpolitik des Bismarckischen Reiches, 1879–1914, Leipzig 1940

Kaiser Wilhem II., Schicksal und Schuld, Göttingen 1962

Schwabe, Klaus, ed., *Das diplomatische Korps, 1871–1945*, Boppard am Rhein 1985

Schweinitz, Wilhelm von, ed., *Denkwürdigkeiten des Botschafters General Hans Lothar von Schweinitz*, 2 vols., Berlin 1927

ed., *Briefwechsel des Botschafters General v. Schweinitz*, Berlin 1928

Schweitzer, Georg, *Emin Pasha. His Life and Work*, 2 vols., London 1898

Schwering, Axel von, *The Berlin Court under Wilhelm II*, London, New York 1915

See, Klaus von, *Barbar Germane Arier. Die Suche nach der Identität der Deutschen*, Heidelberg 1994

Seidel, Paul, *Der Kaiser und die Kunst*, Berlin 1907

Seligmann, Matthew S., *Rivalry in Southern Africa, 1893–1899. The Transformation of German Colonial Policy*, London 1998

'Military Diplomacy in a Military Monarchy? Kaiser Wilhelm II and His Relations with British Service Attachés', in Mombauer and Deist, *The Kaiser*

Sexau, Richard, *Kaiser oder Kanzler. Der Kampf um das Schicksal des Bismarck-Reiches*, Berlin 1936

Seymour, Charles, ed., *The Intimate Papers of Colonel House*, 4 vols., Boston, New York 1926–8

Shaw, Stanley, *William of Germany*, New York, London 1913

The Kaiser 1859–1914, London 1914

Shel'king, Eugeni, *The Game of Diplomacy*, London 1900

Siedler, Wolf Jobst, *Abschied von Preußen*, Berlin 1991

Simon, Christian, 'Kaiser Wilhelm II. und die deutsche Wissenschaft', in Röhl, *Der Ort Kaiser Wilhelms II.*

Simon, Edouard, *L'Empereur Guillaume II et la première année de son règne*, Paris 1889

Simsa, Paul, *Marine intern. Entwicklung und Fehlentwicklung der deutschen Marine 1888–1939*, Stuttgart 1972

Smith, Alson J., *In Preußen keine Pompadour. Wilhelm II. und die Gräfin Waldersee*, Stuttgart 1965

Snyder, Louis L., 'Political Implications of Herbert von Bismarck's Marital Affairs, 1881, 1892', *Journal of Modern History*, 36, 1964, pp. 155–69

Diplomacy in Iron. The Life of Herbert von Bismarck, Malabar, FL, 1985

Sombart, Nicolaus, '"Ich sage, untergehen." Zum zweiten Band von Philipp Eulenburgs politischer Korrespondenz', *Merkur*, 385, June 1980

'The Kaiser in His Epoch. Some Reflexions on Wilhelmine Society, Sexuality and Culture', in Röhl and Sombart, *Kaiser Wilhelm II – New Interpretations*

'Der Kaiser und seine Kritiker. Gedanken zur Problematik der Beurteilung Wilhelms II.', *Wissenschaftskolleg Jahrbuch*, 1982/3, pp. 310–34

Wilhelm II. Sündenbock und Herr der Mitte, Berlin 1996

Sonntag, Josef, *Schuld und Schicksal. Die Tragödie Wilhelms II.*, Leipzig 1927

Sösemann, Bernd, 'Die sog. Hunnenrede Wilhelms II. Textkritische und interpretatorische Bemerkungen zur Ansprache des Kaisers vom 27. Juli 1900 in Bremerhaven', *Historische Zeitschrift*, 222, 1976, pp. 342–58

ed., *Theodor Wolff. Die Wilhelminische Epoche. Fürst Bülow am Fenster und andere Begegnungen*, Frankfurt a.M. 1989

Theodor Wolff. Ein Leben mit der Zeitung, Munich 2000

'Hollow-Sounding Jubilees. Forms and Effects of Public Self-Display in Wilhelmine Germany', in Mombauer and Deist, *The Kaiser*

Stehlin, S. A., *Bismarck and the Guelph Problem 1866–1890*, The Hague 1973

Stein, Adolf, *Wilhelm II.*, Leipzig 1909

Bülow und der Kaiser, Berlin 1931

Steinberg, Jonathan, *Yesterday's Deterrent. Tirpitz and the Birth of the German Battle Fleet*, London 1966

'Kaiser Wilhelm and the British', in Röhl and Sombart, *Kaiser Wilhelm II – New Interpretations*

Stern, Fritz, *Gold and Iron. Bismarck, Bleichroeder and the Building of the German Empire*, New York 1977

Stewart, Desmond, *Theodor Herzl, Artist and Politician*, London 1974

Stibbe, Matthew, 'Kaiser Wilhelm II. The Hohenzollerns at War', in Hughes, Matthew and Seligmann, Matthew, eds., *Leadership in Conflict 1914–1918*, Barnsley 2000, pp. 265–83

'Germany's "Last Card". Wilhelm II and the Decision in Favour of Unrestricted Submarine Warfare in January 1917', in Mombauer and Deist, *The Kaiser*

Stillfried-Alcántara, Rudolf Graf von, *Ceremonial-Buch für den Königlich-Preußischen Hof*, Berlin 1871–8

Stöber, Gunda, *Pressepolitik als Notwendigkeit. Zum Verhältnis von Staat und Öffentlichkeit im Wilhelminischen Deutschland 1890–1914*, Stuttgart 2000

Stolberg-Wernigerode, Otto Graf von, *Wilhelm II.*, Lübeck 1933

Stöwer, Willy, *Zur See mit Pinsel und Palette, Erinnerungen von Prof. Willy Stöwer*, Braunschweig 1929

Straub, Eberhard, *Albert Ballin. Der Reeder des Kaisers*, Berlin 2001

Stribrny, Wolfgang, *Bismarck und die deutsche Politik nach seiner Entlassung (1890–1898)*, Paderborn 1977

Stuhlmann, F., *Mit Emin Pascha ins Herz von Afrika*, 1894

Svanström, Ragnar, *Kejsaren. En bok om Wilhelm II*, Stockholm 1978

Swaine, Sir Leopold, *Camp and Chancery in a Soldier's Life*, London 1926

Syberkrop, Louis, *From the Cradle to the Madhouse, a Biography of Wilhelm II and a Prophecy of His Demise*, Creston 1917

Symons, A. J. A., *Emin, Governor of Equatoria*, London 1928

Taube, Utz-Friedbert, *Ludwig Quidde. Ein Beitrag zur Geschichte des demokratischen Gedankens in Deutschland*, Munich 1963

Taylor, Ronald, *Berlin and Its Culture. A Historical Portrait*, New Haven, London 1997

Teeuwisse, Nicolaas, *Vom Salon zur Sezession. Berliner Kunstleben zwischen Tradition und Aufbruch zur Moderne 1871–1900*, Berlin 1986

Tesdorpf, Paul, *Die Krankheit Wilhelms II.*, Munich 1919

Tirpitz, Alfred von, *Erinnerungen*, Leipzig 1919

Der Aufbau der deutschen Weltmacht, Stuttgart 1924

Tittel, Lutz, 'Monumentaldenkmäler von 1871 bis 1918', in Mai, Ekkehard and Waetzoldt, Stephan, eds., *Kunstverwaltung, Bau- und Denkmalkunst im Kaiserreich*, Berlin 1981

Topham, Anne, *Memories of the Kaiser's Court*, New York 1914

Tuchman, Barbara, *Proud Tower*, New York 1966

Uhle-Wettler, Franz, *Alfred von Tirpitz in seiner Zeit*, Hamburg, Berlin, Bonn 1998

Ullrich, Volker, *Die nervöse Großmacht 1871–1918. Aufstieg und Untergang des deutschen Kaiserreichs*, Frankfurt a.M. 1997

Viereck, George Sylvester, *The Kaiser on Trial*, Richmond, VA, 1937

Vierhaus, Rudolf, ed., *Das Tagebuch der Baronin Spitzemberg. Aufzeichnungen aus der Hofgesellschaft des Hohenzollernreiches*, Göttingen 1960

Viktoria Luise, Herzogin von Braunschweig, *Deutschlands letzte Kaiserin*, Göttingen 1971

Vogel, Bernhard, Nohlen Dieter, and Schultze, Rainer-Olaf, *Wahlen in Deutschland. Theorie-Geschichte-Dokumente, 1848–1970*, Berlin 1971

Vollmar, Georg, *Socialism and the German Kaiser*, London 1903

Vorres, Ian, *The Last Grand Duchess*, London 1964

Ward, James E., 'Leo XIII and Bismarck. The Kaiser's Vatican Visit of 1888', *Review of Politics*, 24, 1962

Weber, Christoph, *Quellen und Studien zur Kurie und zur vatikanischen Politik unter Leo XIII.*, Tübingen 1973

Wedel, Graf Carl von, *Zwischen Kaiser und Kanzler*, Leipzig 1943

Wehberg, Hans, *Ludwig Quidde. Ein deutscher Demokrat und Vorkämpfer der Völkerverständigung*, Offenbach 1948

Wehler, Hans-Ulrich, ed., *Caligula. Schriften über Militarismus und Pazifismus*, Frankfurt a.M. 1977

Das Deutsche Kaiserreich 1871–1918, Göttingen 5th edn 1983

Deutsche Gesellschaftsgeschichte, vol. III, *Von der 'Deutschen Doppelrevolution' bis zum Beginn des Ersten Weltkrieges, 1849–1914*, Munich 1995

Weitowitz, Rolf, *Deutsche Politik und Handelspolitik unter Reichskanzler Leo von Caprivi, 1890–1894*, Düsseldorf 1978

Weller, Bernd-Uwe, *Maximilian Harden und die 'Zukunft'*, Bremen 1970

Wendel, Friedrich, *Wilhelm II. in der Karikatur*, Dresden 1926

Werthern, Alfred Freiherr von, *General von Versen. Ein militärisches Zeit- und Lebensbild. Aus hinterlassenen Briefen und Aufzeichnungen*, Berlin 1898

White, Arnold, *Is the Kaiser Insane? A Study of the Great Outlaw*, London 1915

Whittle, Tyler, *The Last Kaiser. A Biography of William II, German Emperor and King of Prussia*, London 1977

Wilderotter, Hans and Pohl, Klaus-D., eds., *Der letzte Kaiser. Wilhelm II. im Exil*, Gütersloh, Munich 1991

Wile, Frederic, *Men around the Kaiser. The Makers of Modern Germany*, Indianapolis 1914

Wilhelm II, *Vergleichende Geschichtstabellen von 1878 bis zum Kriegsausbruch, 1914*, Leipzig 1921

Ereignisse und Gestalten aus den Jahren 1878–1918, Berlin, Leipzig 1922

Aus meinem Leben 1859–1888, Berlin, Leipzig 1927

Wilke, Ekkehard-Teja P. W., *Political Decadence in Imperial Germany. Personnel–Political Aspects of the German Government Crisis 1894–1897*, Urbana, IL, London 1976

Wilson, Lawrence, *The Incredible Kaiser. A Portrait of William II*, New York 1965

Windelband, Wolfgang, *Herbert Bismarck als Mitarbeiter seines Vaters*, Stuttgart 1921

Winzen, Peter, *Bülows Weltmachtkonzept. Untersuchungen zur Frühphase seiner Außenpolitik 1897–1901*, Boppard am Rhein 1977

'Zur Genesis von Weltmachtkonzept und Weltpolitik', in Röhl, *Der Ort Kaiser Wilhelms II.*

Bernhard Fürst von Bülow. Weltmachtstratege ohne Fortune – Wegbereiter der großen Katastrophe, Göttingen, Zürich 2003

Ydewalle, Charles d', *Guillaume II*, Brussels 1972

Zange, Friedrich, *Die Jerusalemfahrt Kaiser Wilhelms II. im Lichte der Geschichte*, Berlin 1899

Zechlin, Egmont, *Staatsstreichpläne Bismarcks und Wilhelms II., 1890–1894*, Stuttgart, Berlin 1929

Zedlitz-Trützschler, Robert Graf von, *Zwölf Jahre am deutschen Kaiserhof*, Berlin, Leipzig 1923

Zeepvat, Charlotte, *Prince Leopold. The Untold Story of Queen Victoria's Youngest Son*, London 1998

Zelinsky, Hartmut, *Richard Wagner. Ein deutsches Thema. Eine Dokumentation zur Wirkungsgeschichte Richard Wagners 1876–1976*, Berlin, Vienna 1983

'Kaiser Wilhelm II., die Werk-Idee Richard Wagners und der "Weltkampf"', in Röhl, *Der Ort Kaiser Wilhelms II.*

Zentner, Kurt, *Kaiserliche Zeiten. Wilhelm II. und seine Ära in Bildern und Dokumenten*, Munich 1964

Index